P9-DNN-470
HERKIMER COUNTY COMMUNITY COLLEGE

AY 67 .N5 W7 2002

The World Almanac and book of facts 2002

Herkimer County
Community College Library
Herkimer, New York
13350

1. Books may be kept for three weeks and may be renewed once, except when otherwise noted.
2. Reference books, such as dictionaries and encyclopedias are to be used only in the Library.
3. A fine is charged for each day a book is not returned according to the above rule.
4. All injuries to books beyond reasonable wear and all losses shall be made good to the satisfaction of the Librarian.
5. Each borrower is held responsible for all books drawn on his card and for all fines accruing on the same.

THE AUTHORITY SINCE 1868

THE WORLD ALMANAC®

AND BOOK OF FACTS

2002

WORLD ALMANAC BOOKS

HERKIMER COUNTY COMMUNITY COLLEGE
LIBRARY
HERKIMER, NY 13350-9987

THE WORLD ALMANAC® AND BOOK OF FACTS 2002

Editorial Director: William A. McGeveran Jr.
Managing Editor: Lori P. Wiesenfeld
Desktop Production Manager: Elizabeth J. Lazzara
Senior Editor: Kevin Seabrooke
Associate Editors: Mette A. Bahde, Russell S. Cobb
Desktop Publishing Associate: Lloyd Sabin
Contributing Editors: Elizabeth Barden, David M. Faris, Richard Hantula, Kate Hill, Jane Hogan, Geoffrey M. Horn, Joshua Lustig, Andrea J. Pitluk, Jane Reynolds, Dr. Lee T. Shapiro, Elbert Ventura, Donald Young, Karen Young
Research: Meghan Berger, Rachael Mason, Catherine McHugh
Cover: Jay Jaffe/Bill SMITH STUDIO

WORLD ALMANAC EDUCATION GROUP
Chief Executive Officer, WRC Media Inc.: Martin E. Kenney Jr.
President, World Almanac Education Group: Alfred De Seta
Publisher: Ken Park
Director–Purchasing and Production: Edward A. Thomas
Associate Editor: Ileana Parvulescu; **Desktop Publishing Assistant:** Michael D. Jeter
Director of Indexing Services: Marjorie B. Bank; **Index Editor:** Walter Kronenberg
Facts On File World News Digest: Marion Farrier, Editor in Chief; Jonathan Taylor, Managing Editor
World Almanac Reference Database@*FACTS.com*: Louise Bloomfield, Dennis La Beau

WORLD ALMANAC BOOKS
Vice President–Sales and Marketing: James R. Keenley
Marketing Coordinator: Sarah De Vos

We acknowledge with thanks the many helpful letters and e-mails from readers of THE WORLD ALMANAC. Because of the volume of mail, it is not possible to reply to each one. However, every communication is read by the editors, and all suggestions receive careful attention. THE WORLD ALMANAC's e-mail address is Walmanac@waegroup.com.

The first edition of THE WORLD ALMANAC, a 120-page volume with 12 pages of advertising, was published by the New York World in 1868. Annual publication was suspended in 1876. Joseph Pulitzer, publisher of the *New York World*, revived THE WORLD ALMANAC in 1886 with the goal of making it a "compendium of universal knowledge." It has been published annually since then. THE WORLD ALMANAC does not decide wagers.

COVER PHOTOS: Globe, Corbis Stock Market; Lance Armstrong, AllSport; Euro, European Commission; Firefighters at WTC site, Thomas E. Franklin/Bergen Record/Corbis SABA; John Adams (also on page 813), *John Adams,* Alisa Mellon Bruce Fund, Photograph © 2001 Board ofTrustees, National Gallery of Art, Washington, c. 1821.

THE WORLD ALMANAC and BOOK OF FACTS 2002
Copyright © 2002 by World Almanac Education Group, Inc.
The World Almanac and The World Almanac and Book of Facts are registered trademarks of World Almanac Education Group
Library of Congress Catalog Card Number 4-3781
International Standard Serial Number (ISSN) 0084-1382
ISBN (softcover) 0-88687-872-1
ISBN (hardcover) 0-88687-874-8
Printed in the United States of America

The softcover and hardcover editions are distributed to the book trade by St. Martin's Press; the paperback edition is distributed to the magazine trade by TV Guide Distribution Inc.

WORLD ALMANAC BOOKS
A Division of World Almanac Education Group, Inc.
A WRC Media Company
512 Seventh Avenue
New York, NY 10018

CONTENTS

GENERAL INDEX

Note: Page numbers in **boldface** indicate key reference. Page numbers in *italics* indicate photos.

–D–

—I—

The World Almanac

and Book of Facts

2002

THE TOP TEN NEWS STORIES OF 2001

In the worst-ever terrorist attack on the U.S., hijackers Sept. 11 commandeered 4 U.S. commercial jetliners, crashing 2 of them into the twin towers of New York City's World Trade Center and 1 into the Pentagon outside Washington, DC. Both towers collapsed, with thousands of people still inside. A 4th jet crashed into a field in Pennsylvania. More than 5,000 people were believed to have died in the assaults, which U.S. Pres. George W. Bush called "acts of war." The prime suspect behind the attack was Osama bin Laden, a wealthy Saudi Arabian exile and Islamic extremist harbored in Afghanistan, alleged to be leader of the wide-ranging Al Qaeda terrorist network. Bush demanded that Afghanistan's ruling Taliban militia surrender Bin Laden or face attack. On Oct. 7, the U.S. and Great Britain, backed by an international coalition against terrorism, launched missile and air strikes against Afghanistan, the first offensive actions in what was expected to be a long war.

Right-wing Israeli Likud party leader Ariel Sharon was elected prime minister Feb. 6, defeating Labor Party incumbent Ehud Barak in the most lopsided election in Israeli history. Sharon formed a unity government with the Labor Party. Violent **clashes between Israeli and Palestinian forces** continued, despite periodic cease-fire agreements. By late September, after one year, the death toll had risen to nearly 800 people, of whom about 600 were Palestinians. Israeli and Palestinian leaders ordered another cease-fire Sept. 18, following U.S. pressure in the wake of the Sept. 11 terrorist attacks.

In inaugural ceremonies Jan. 20, **George W. Bush (R) was sworn in as 43d U.S. president.** His new cabinet included former Missouri Sen. John Ashcroft as attorney general, Colin L. Powell, the first African-American secretary of state, and Defense Sec. Donald Rumsfeld. In its first months, the Bush administration won congressional approval of a 10-year, $1.35 trillion tax cut, fulfilling one of Bush's major campaign promises. The new administration also pushed for a missile defense system aimed at intercepting intercontinental ballistic missile attacks.

The 50 Democrats in the Senate gained a majority after Vermont Sen. James Jeffords announced on May 24 that he was leaving the Republican Party to become an independent and would vote with the Democrats on organizational matters.

The U.S. experienced an **economic slowdown,** marked by stock market declines, particularly in the high-tech sector, widespread job losses, and fears of recession. In the 2d quarter, gross domestic product grew at a seasonally adjusted rate of 0.2%, the lowest since 1993. By late August, the Federal Reserve had lowered interest rates 7 times, bringing the federal funds rate to its lowest level since 1994. The Fed lowered interest rates 2 more times, following concerns about the economic impact of the Sept. 11 attack. The September unemployment rate was 4.9%, up from 4.0% at the start of the year

Former Yugoslav Pres. Slobodan Milosevic was arrested by Serbian authorities Apr. 1, turning himself in after a tense standoff in Belgrade. On June 28 he was extradited to the custody of the United Nations International Criminal Tribunal for the Former Yugoslavia in The Hague, the Netherlands. He was accused of war crimes and crimes against humanity during the 1999 Kosovo conflict; the indictment was expanded in October to include alleged war crimes in Croatia.

Convicted Oklahoma City bomber **Timothy McVeigh was executed** by lethal injection June 11 in Terre Haute, IN. In 1997, McVeigh had been convicted on federal charges for the 1995 bombing of the Alfred P. Murrah Federal Building in Oklahoma City, which killed 168 people.

A **U.S. spy plane made an emergency landing on Hainan Island in China** Apr. 1 after colliding with a Chinese fighter jet, which the U.S. contended had caused the collision by flying too close. The fighter jet and its pilot were lost. Chinese authorities detained the 24-member U.S. crew for 11 days, while U.S. and Chinese authorities negotiated China's demand for a full apology. The crew was released after U.S. Amb. Joseph Prueher signed a letter saying the U.S. was "very sorry" for the loss of the pilot and for the airplane's landing in China without "verbal permission."

U.S. Pres. George W. Bush announced Aug. 9 that he would allow federal funding for **embryonic stem-cell research** when based on existing stem cell lines, but not when based on creating new cell lines. The president had sought to strike a balance between many scientists who believed stem cell research could lead to treatments for diseases such as Alzheimer's and Parkinson's and conservative supporters who opposed any procedure involving the killing of embryos.

An outbreak of **foot-and-mouth disease,** a viral infection found in livestock, began in Great Britain in February. By March, the disease had spread to France, Ireland, and the Netherlands. British Prime Min. Tony Blair in March authorized a mass slaughter of healthy livestock. By the end of August more than 3.8 million animals had been slaughtered, and the epidemic's cost was estimated in billions of dollars.

America at War

By Geoffrey M. Horn

Geoffrey M. Horn is a freelance writer who frequently writes on current events.

The terrorist attack of Sept. 11, 2001, killed more than 5,000 people, over twice as many as died at Pearl Harbor, and accomplished what no foreign army had been able to do for nearly 2 centuries. Not since the War of 1812 had an overseas enemy managed to mount a direct assault on the American mainland. With oceans to the east and west and friendly borders to the north and south, many Americans had taken homeland security for granted. Not anymore.

In the aftermath of the attack, U.S. government suspicions quickly turned to the Al Qaeda terrorist network said to be headed by Osama bin Laden. (See "Osama Bin Laden—A Profile," on page 61.) On Oct. 7 the U.S. and Great Britain launched a series of retaliatory strikes against both Al Qaeda and Afghanistan's Taliban regime, which had given Bin Laden safe haven. In a letter to Congress 2 days later, Pres. George W. Bush wrote that it was "not possible to know at this time either the duration of combat operations or the scope and duration of the deployment of U.S. armed forces." Bush acknowledged, however, that the American campaign against terrorism would likely be "lengthy." Administration officials repeatedly cautioned that overt military operations were only one phase in a "new kind of war"—a war that would also require covert activities, diplomatic and financial pressure, and security measures that would directly impact Americans' lives.

See also "U.S. Struck by Terrorist Attack" in *Chronology of the Year's Events*, pages 59-60.

"Evil, Despicable Acts of Terror"

The Sept. 11 attack was simply conceived and methodically executed. The 19 hijackers—5 each on American Airlines Flights 11 and 77 and United Airlines Flight 175, and 4 on United Flight 93—chose aircraft scheduled to depart from east coast airports within a few minutes of each other, around 8 A.M. The aircraft were operationally similar, either

Boeing 757s or 767s; all carried relatively light loads of passengers and heavy loads of jet fuel, enough for a transcontinental journey. (See *Chronology of the Year's Events*, "150 Minutes That Changed the World," page 58.)

According to the FBI and airline records, the hijackers crisscrossed the country during the summer, testing cockpit security. Several made their flight reservations with credit cards over the Internet, dutifully logging their frequent-flier numbers, and on Sept. 11 some held top-dollar seats near the cockpit, in first or business class. To pass through airport screening, the terrorists flashed drivers' licenses with photo IDs they had obtained in Florida and Virginia. The box cutters, razor blades, and small knives that they carried were probably legal under Federal Aviation Administration guidelines—subsequently revoked—that allowed passengers to carry knives with blades up to 4 inches long.

Federal authorities knew early on that something had gone wrong with American Flight 11. But when the jetliner ploughed into the North Tower of the World Trade Center at 8:48 A.M. EDT, many onlookers attributed the crash to some terrible accident, not deliberate sabotage. The collision of United Flight 175 with the South Tower 15 minutes later—seen live worldwide through TV cameras already trained on the North Tower fire—left no room for doubt. Notified that the nation was under attack, Pres. Bush, in Sarasota, FL, made a brief statement and was later flown by Air Force One to secure locations in Louisiana and Nebraska. In Washington, Secret Service agents whisked Vice Pres. Richard Cheney from his office at the White House to a secure underground bunker, where, in communication with Bush and other top administration officials, he and National Security Adviser Condoleezza Rice began to coordinate the immediate U.S. response.

The crisis mood deepened when, at 9:37 A.M. (revised time issued by the Defense Dept.), a third hijacked aircraft, American Flight 77, slammed into the Pentagon, in Arlington, VA, and a fourth, United Flight 93, was reported to have made a U-turn over northern Ohio and to be headed straight for Washington, DC. While the Capitol and the White House were evacuated, passengers on board Flight 93, who were in telephone contact with the ground and knew what had happened at the World Trade Center, hatched plans to overcome the hijackers and take back the plane. Their Boeing 757 went down over western Pennsylvania, after what law enforcement officials characterized (based on evidence recovered from the cockpit voice recorder) as a desperate and wild struggle.

Back in New York City, at what soon would become known as Ground Zero, the overwhelming majority of employees and visitors below the 90th floor of the North Tower and the 80th floor of the South Tower were safely evacuated. On the uppermost floors of the 110-story buildings, however, the situation was hopeless. Eyewitnesses at street level reacted with horror as some people jumped to their deaths, including a man and woman holding hands; thousands of other victims were trapped by a fire so intense it weakened the twin towers' steel support structures. When the South Tower collapsed around 10 A.M. and the North Tower went down about half an hour later, the office workers—and hundreds of rescue workers as well—were crushed in a cascade of more than 1 million tons of steel and concrete. A 47-story office tower nearby, at 7 World Trade Center, collapsed before evening.

When Pres. Bush returned to Washington that night, smoke was still rising from the Pentagon, and 16 acres of lower Manhattan had been reduced to rubble. In his first formal address to an anguished, angry nation, he called the attacks "evil, despicable acts of terror" and pledged that the perpetrators and those who harbored them would be brought to justice.

Assessing the Damage

Almost immediately, public officials and business leaders began the grim task of calculating the damage. No hope remained for all 232 passengers (including the 19 hijackers) and 33 crew members on board the four hijacked aircraft. Among the dead were television producer David Angell, Akamai Technologies co-founder Daniel Lewin, and actress-photographer Berry Berenson Perkins, all on American Flight 11; Los Angeles Kings hockey scouts Mark Bavis and Garnet (Ace) Bailey, on United Flight 175; 3 schoolchildren and their teachers participating in a National Geographic field trip, on American Flight 77, which also carried conservative television commentator Barbara Olson, wife of U.S. Solicitor General Theodore Olson; and Todd Beamer, Mark Bingham, and Jeremy Glick, 3 of the men believed to have led the passenger revolt that brought down United Flight 93. In addition to those on board Flight 77, the assault on the Pentagon killed 125 people.

Much more difficult to assess was the death toll at Ground Zero, still smoldering a month after the attack. By then, more than 400 bodies had been recovered; in addition to those on board Flights 11 and 175, over 4,600 people, including citizens of more than 60 countries, were still missing and presumed dead. Some companies that had offices in the World Trade Center were devastated, including Cantor Fitzgerald, a bond brokerage firm that lost more than 600 employees from its offices in the North Tower, between the 101st and 105th floors. New York City lost a reported 343 firefighters, whose courage and professionalism in the moments after the planes hit had made them national heroes; many were lost in the collapsing buildings as they labored to carry 75 to 100 pounds of equipment up 80 flights of stairs.

Recovery and reconstruction costs were expected to exceed $100 billion. With near-unanimity, Congress on Sept. 14 approved $40 billion in emergency spending, of which half was allocated for disaster recovery operations in New York, Virginia, and Pennsylvania. On Oct. 9, New York City Mayor Rudolph Giuliani—whose energetic and inspirational leadership had buoyed the city's spirits—forecast that the attack would cost the city 100,000 jobs and billions in lost revenue. That same day, Giuliani and New York Gov. George Pataki, a longtime rival who had been working closely with the mayor since Sept. 11, asked the federal government to provide another $54 billion in incentives and subsidies to compensate the city for its losses.

The nation's ailing airlines were directly affected by the catastrophe. When commercial air traffic resumed on Sept. 13 after a 2-day shutdown, carriers faced combined losses of at least $100 million per day. The airlines estimated that losses from the shutdown and subsequent slump were at least $18 billion. Major operators such as Continental, American, United, US Airways, and Delta announced schedule cutbacks and layoffs of 20% or more. Ripple effects were felt among foreign airlines and in the aircraft manufacturing industry. On Sept. 21, Congress approved a $15 billion bailout bill to keep domestic carriers aloft.

Stock trading, halted on the morning of Sept. 11, resumed the following Monday after an unprecedented 6-day delay. Economic uncertainties prompted the worst one-week sell-off since the Great Depression, with a loss of $1.4 trillion in market value. Investors recouped some of their losses the following week, as the markets stabilized and recovered, but many economists believed the U.S. had already been plunged into recession. During the following weeks, the Federal Reserve Board moved aggressively to cut interest rates, as Pres. Bush and congressional leaders began shaping an economic stimulus package valued at up to $75 billion.

Appropriations of federal funds were matched by an outpouring of private support (an estimated $850 million by mid-October) to aid relief organizations and victims' families.

Response and Retaliation

The crisis response brought heightened vigilance at air terminals and other public facilities, including the presence of uniformed National Guard troops with M-16 assault rifles at airport checkpoints. In a nationally televised address to a joint session of Congress on Sept. 20, Pres. Bush named Pennsylvania Gov. Tom Ridge to head a newly established cabinet-level Office of Homeland Security; 19 days later, Ridge and Condoleezza Rice announced the creation of 2 additional positions, a special adviser for cyberspace security and a national director for combating terrorism. Immigration officials stepped up their scrutiny of foreigners entering the U.S.; new security measures were implemented at water supply systems, railroads, trucking companies, and nuclear power plants; and the U.S. Coast Guard mounted what it described as the largest port defense operation since World War II. Meanwhile, analysts began the quiet but essential task of determining why the Sept. 11 assault had caught U.S. intelligence agencies by surprise.

In the 4 weeks following the attack, federal authorities issued thousands of subpoenas and detained more than 600 people, of whom 220 were still being held, many for immigration violations. Related probes were carried out in Europe, notably in Hamburg, Germany, where Mohamed Atta, a 33-year-old Egyptian who was on board Flight 11 and is believed to have been a ringleader in the conspiracy, lived as a student in the 1990s. U.S. investigators uncovered evidence that, in addition to taking lessons at U.S. flight schools, the terrorists had explored the feasibility of using crop-dusting planes, perhaps as a delivery system for chemical and biological weapons. This threat was underlined on Sept. 25 when federal investigators revealed that they had charged 20 people with fraudulently obtaining commercial licenses to haul hazardous materials by truck. Several reported cases of anthrax in early October also heightened fears of bioterrorism (see *Chronology*, page 60, and "Biological and Chemical Weapons," pages 875-76).

Meanwhile, on Oct. 11, the FBI publicly indicated that there could be further terrorist attacks within the U.S. or against U.S. interests abroad, possibly within days.

To unravel international terrorism's financial support network, Pres. Bush on Sept. 24 signed an executive order targeting the assets of 27 organizations and individuals, including Al Qaeda, Bin Laden, and Bin Laden's aide Ayman al-Zawahiri, founder of an Egyptian terrorist group, Al-Jihad, that also appeared on the list. Bin Laden and Zawahiri topped the FBI's list of 22 most-wanted terrorist suspects, released by Pres. Bush on Oct. 10, and the assets of additional individuals and organizations were targeted Oct. 12.

U.S. government officials and other civic and religious leaders made clear that the U.S. had no quarrel with mainstream Islam or with the vast majority of the world's more than 1 billion Muslims. They were responding, in part, to a wave of harassment against Muslim Americans, including the desecration of mosques in at least 6 states in the first week after the attack. Some Sikh men, required by their religion to wear beards and turbans (though not actually Muslims), were also the targets of apparent hate crimes. Pres. Bush took the lead in reassuring the nation's nearly 6 million Muslims during a Sept. 17 visit to a mosque at Washington's Islamic Center.

Sec. of State Colin Powell and Defense Sec. Donald Rumsfeld sought to assemble a broad international backing for America's initial objective, the destruction of Al Qaeda and the Taliban regime. British Prime Min. Tony Blair was a strong early supporter. Also supportive was Russian Pres. Vladimir Putin, whose nation had been bloodied in wars against Muslim rebels in Afghanistan in the 1980s and Chechnya in the 1990s. Pres. Pervez Musharraf of Pakistan allowed use of the country's air space and apparently allowed deployment of military personnel and equipment; but his government denied that Pakistan was providing a base for offensive operations. Pakistan had been a principal supporter of the Taliban and had reportedly enlisted Bin Laden's aid in training guerrilla fighters for the war against India in Kashmir; an ending of U.S. sanctions, imposed in 1998 when Pakistan began testing nuclear weapons, helped provide an incentive for cooperation. To win support from Arab states such as Egypt, Jordan, and Saudi Arabia—and to defuse an inflammatory issue Bin Laden had used to gain followers among the Muslim masses—U.S. officials let it be known that Powell was preparing an Arab-Israeli peace initiative that included support for a Palestinian state.

For the first time in its 52-year history, NATO on Sept. 12 invoked the Collective Defense Clause of its founding treaty, pledging the 19 member states to assist the U.S.; on Oct. 8, NATO announced that it was sending 5 AWACS planes to patrol American skies, while U.S. early warning and surveillance aircraft were deployed in the Afghan combat. The UN lent support on Sept. 28, when the Security Council unanimously approved a U.S.-sponsored resolution requiring all member states to cooperate in the antiterrorist campaign. Still unresolved, as bombs and cruise missiles began to rain down over Kabul and other Afghan cities, were questions about the role the U.S. and UN might play in brokering a peace agreement among Afghanistan's hostile factions, and how these groups might establish an effective government if the Taliban were removed.

Facts About the September 11 Attack

- The exact number of fatalities may never be known. As of Oct. 12, the toll of dead and missing in the attacks was reported as 265 on the 4 hijacked planes (including the 19 hijackers), 125 at the Pentagon, and 5,080 at the World Trade Centers, for a total of 5,470.
- That ranked as either the worst single-day death toll in American history, or the 2d worst after the Civil War battle of Antietam on Sept. 17, 1862. According to the National Park Service, at least 3,650 soldiers on both sides were killed that day, and an unknown number of the estimated 19,000 wounded, missing, and captured also died. The death toll from the 1941 Japanese attack on Pearl Harbor is commonly reported as 2,300 to 2,400.
- Among the fatalities in New York were 343 city firefighters, 23 city police officers, and 37 officers of the Port Authority of New York and New Jersey.
- Among the 125 workers dead or missing at the Pentagon were 21 uniformed Army and 33 uniformed Navy personnel. About 1 out of 3 of the victims were women. The highest ranking officer killed was Army Lt. Gen. Timothy J. Maude, a 53-year-old Vietnam veteran, who had been heading up recruitment efforts.
- The bond-trading firm of Cantor Fitzgerald, headquartered on the top floors of the World Trade Center's North Tower, lost more than 600 of its 1,000 employees. Some 1,500 children of Cantor Fitzgerald employees lost parents.
- 79 workers at the famous Windows on the World restaurant on the 107th floor of the North Tower were among the dead and missing.
- At least 50 Arabs, including Arab Americans, were believed dead or missing in the WTC attack.
- Citizens of at least 62 nations perished in the attack.
- Five people were rescued from the WTC rubble Sept. 12, but none thereafter.
- Because most local TV broadcast antennas were atop the World Trade Center, New Yorkers without cable could receive only one or two channels in the weeks after the attack.
- The tower collapses destroyed New York's Emergency Command Center, on the 23d floor of Building 7 of the World Trade Center. Mayor Giuliani had been on his way there when the emergency "bunker" was evacuated after the second plane struck.
- The Pentagon area hit had been under construction, with some offices not yet re-occupied, meaning fewer people than normal were there when the plane struck.
- After the crash into the North Tower, some security officials in the South Tower reportedly informed workers that the South Tower was safe and that they should return to their offices.
- The World Trade Center towers were designed to collapse one floor on the top of the next, to avoid the additional devastation that could result if the skyscrapers toppled onto other buildings.
- Rubble from the collapse of the World Trade Center covered 16 acres and was estimated to weigh 1.2 million tons.
- A month after the attack, fire continued to smolder beneath the rubble, producing acrid smoke that wafted over much of lower Manhattan.
- Pres. Bush's approval rating rose to 90% in a CNN/USA Today Gallup poll conducted Sept. 21-22. That was up from 51% in a poll just before the terror attacks.
- Defense Sec. Donald Rumsfeld Sept. 27 confirmed that two Air Force generals would now have the authority to order the shooting down of any commercial airliners that threatened targets in the US.
- Estimates of the costs of plotting the attack ranged from $200,000 to $500,000. City Comptroller Alan Hevesi estimated that the cost to New York City's economy alone could exceed $100 billion.

CHRONOLOGY OF THE YEAR'S EVENTS

Reported Month by Month, Nov. 1, 2000, to Oct. 10, 2001

NOVEMBER 2000

National

Bush DWI Arrest Revealed Near Campaign's End— On **Nov. 2**, 5 days before Americans were to vote, Gov. George Bush (TX), the Republican presidential nominee, acknowledged that he had been arrested for driving under the influence of alcohol near Kennebunkport, ME, in 1976, near the Bush family's summer home. Bush had pleaded guilty, paid a $150 fine, and temporarily lost his right to drive in Maine. He was 30 at the time. The report of the 24-year-old arrest broke on Maine TV stations **Nov. 2**. The source for the story was Tom Connolly, a former Democrat candidate for governor in Maine.

Presidential Winner Unknown on Election Night— More than 100 million Americans went to the polls **Nov. 7**, but their choice for president remained in doubt. Neither Vice Pres. Al Gore, the Democratic candidate, nor Gov. George W. Bush (TX), the Republican nominee, had won the needed 270 electoral votes. Though the margins of victory were very narrow in a number of states, Gore appeared to emerge from the evening with 255 electoral votes to 246 for Bush, with the outcome in Florida, New Mexico, and Oregon uncertain. Florida's 25 electoral votes would determine the winner. In the initial tally, subject to a machine recount mandated by state law because of the narrow margin, Bush led in Florida by about 1,800 votes out of 6 million cast.

The television networks declared early in the evening that Gore had won Florida, then withdrew that projection. Early on the morning of **Nov. 8** they declared that Bush had carried Florida, then again withdrew their projections. Gore had phoned Bush to congratulate him, and was on his way to make a concession speech when informed that the vote in Florida was closer than had been thought; he phoned Bush a 2d time to withdraw his concession.

Late on election night, Gore and Bush each had about 48% of the nationwide popular vote, with Gore leading by about 200,000. Consumer advocate Ralph Nader, the Green Party nominee, had the largest share of the rest. He attracted mostly liberal supporters, apparently drawing far more votes from Gore than from Bush. In Florida, Nader had about 100,000 votes. Gore won a number of big states—New York, Pennsylvania, New Jersey, Massachusetts, Illinois, Michigan, and California—while Bush swept the South outside Florida (including Tennessee and Arkansas, home states of Gore and Clinton), and most of the Plains and Mountain states. Strong support from blacks and union members gave Michigan to Gore, and blacks (who supported Gore nationwide by 90%) also helped put Pennsylvania in Gore's column. Nationwide, exit polls showed that men supported Bush by a margin of about 11 percentage points, and women backed Gore by the same spread. White voters favored Bush by a margin of 10 percentage points, while Hispanics favored Gore 2 to 1.

For further presidential election developments, see separate boxes. Presidential Election – Key Events, for *November* and *December*.

Republicans emerged from election night assured of 50 seats in the U.S. Senate to 49 for the Democrats, with the result in Washington state in doubt. The GOP had gone into the election holding a 54–46 margin. Hillary Rodham Clinton, who had recently established a residence in New York, became the first wife of a president to be elected to political office when she defeated U.S. Rep. Rick Lazio (R) for the Senate seat of retiring Sen. Daniel P. Moynihan (D, NY). In Missouri, voters gave a majority to the late Gov. Mel Carnahan (D), who had died in a plane crash in October but remained on the ballot; acting Gov. Roger Wilson (D) had said he would appoint Carnahan's widow, Jean, to the Senate if Carnahan finished ahead of the GOP incumbent, John Ashcroft. Four incumbent senators lost. They were 79-year-old William Roth (R, DE), perhaps best known for establishing

Presidential Election

Nov. 7	Election Day. TV networks project Vice Pres. Al Gore, the Democratic nominee, as winner of Florida's electoral votes, then retract the projection.
Nov. 8	Networks project Gov. George Bush (TX), the Republican candidate, as the winner in Florida. That projection is retracted when Florida's race tightens further. Gore, following the Bush victory projection, telephones Bush to concede. Later, advised that the count is tightening, he calls Bush again to retract his concession. Results in 2 other states remain undecided, but Florida's 25 electoral votes will be decisive; Gore has won 255, and Bush 246.
Nov. 8-9	Voters in Florida's Palm Beach County file lawsuits challenging the county's results, contending that the design of the county's so-called butterfly ballot misled many Gore voters into voting for Reform Party candidate Pat Buchanan.
Nov. 9	An incomplete count gives Bush a Florida lead of 1,784 votes. The narrow margin prompts a mandatory statewide mechanical recount. Gore campaign requests manual recounts in 4 Florida counties: Palm Beach, Broward, Miami-Dade, and Volusia. Democrats maintain the machines likely failed to count many imperfectly punched ballots that a manual inspection would evaluate; Republicans argue that such a process is too subjective and vulnerable to bias. Palm Beach agrees to begin a manual recount of sample precincts.
Nov. 10	The machine recount, completed in all but one county, cuts Bush's Florida lead to 327.
Nov. 11	Republicans file a lawsuit in U.S. District Court seeking an injunction against manual recounts.
Nov. 12	The Palm Beach County canvassing board votes, 2-1, to launch a complete manual recount, after the sample recount shows a gain for Gore. Volusia County begins its recount.
Nov. 13	Florida Sec. of State Katherine Harris sets a deadline of 5 P.M. the next day for counties to submit recount results, citing a state law that she must certify results on the 7th day after Election Day. Gore sues in Leon County Circuit Court, Tallahassee, for a deadline extension. U.S. District Court denies the GOP request for an injunction against manual recounts. Broward County begins its recount.
Nov. 14	Volusia County completes its recount in time. Leon County Circuit Court, while upholding the deadline, directs that Harris exercise "discretion" in enforcing it. Harris gives 3 counties still counting until 2 P.M. Nov. 15 to justify an extension. Broward County officials vote to proceed with a countywide manual recount, reversing a decision not to do so. Miami-Dade County decides not to conduct a complete manual recount.
Nov. 15	Harris rejects counties' requests for extensions, saying she will exclude from official results tallies submitted after Nov. 14. Florida Supreme Court rejects Harris's request to enjoin Broward and Palm Beach from continuing their manual recounts. Broward County begins its recount. Miami-Dade County asks for more time to consider a complete manual recount. Gore proposes to drop all legal challenges if Bush accepts the results of recounts underway, or a statewide manual recount. Bush declines.
Nov. 16	Democrats sue in Leon County Circuit Court, contending Harris's decision was "arbitrary and unreasonable."
Nov. 17	Leon County Circuit Court upholds Harris's decision to exclude recount results beyond her deadline. Florida Supreme Court blocks Harris from certifying results, pending appeal. Miami-Dade officials announce they will conduct a countywide recount after all. U.S. 11th Circuit Court of Appeals denies GOP appeal of U.S. District Court's Nov. 13 decision. Overseas absentee ballots are due by midnight.

the Roth IRA accounts, to Gov. Thomas Carper; Rod Grams (R, MN), to former state auditor Mark Dayton; Spencer Abraham (R, MI), to U.S. Rep. Debbie Stabenow; and Charles Robb (D, VA), to former Gov. George Allen.

The Republicans went into the election with a 223–210 margin in the U.S. House, with 2 seats held by independents. On election night, GOP candidates won or were leading in 221 contests, and Democrats won or were ahead in 212. The 2 independents were reelected. Democrats won 8 of the 11 gubernatorial seats contested, a net gain of one seat (in West Virginia).

3 Convicted in LA Police Scandal—Three Los Angeles policemen were convicted **Nov. 15** by a California State Superior Court jury of obstructing justice and falsifying police reports. However, a 4th officer was found not guilty, and on **Dec. 23**, Los Angeles Superior Court Judge Jacqueline Connor overturned the convictions of the 3 policemen, saying that some jurors had followed instructions incorrectly and that evidence had been insufficient. In all, about 70 police officers were under investigation for possible wrongdoing in connection with the unfolding corruption scandal in the LAPD's anti-gang Rampart Division. Amid widespread evidence that the police had violated the rights of arrested persons, more than 100 convictions had been overturned. On **Nov. 21**, the City Council approved a $15 million settlement for a man paralyzed after being shot in the back by a police officer. The 2 police officers present at the time claimed that the wounded man, Javier Ovando, had tried to shoot them. Ovando spent 3 years in prison for assault. One of the officers later said that Ovando had been framed.

Coca Cola Settles Bias Suit for $192.5 Million—A class action racial discrimination suit against the Coca Cola Co. ended **Nov. 16** when the company agreed to a $192.5 million settlement with the plaintiffs, about 2,000 current and former black employees. A group of employees had alleged that they suffered discrimination in pay, promotions, and performance evaluations. The award was the largest ever in a racial discrimination suit in the U.S., topping the $176.1 million settlement reached with Texaco in 1996. The settlement awarded $113 million to the plaintiffs, set aside $43.5 million for salary adjustments, and pledged the remaining $36 million to monitor future employment practices.

Cheney Rallies Quickly From Heart Attack—Dick Cheney, the Republican nominee for vice president, suffered a mild heart attack **Nov. 22** and entered George Washington University Hospital; he was released **Nov. 24**. Cheney had suffered from heart trouble in the past, including 3 attacks since his late 30s, and had undergone quadruple bypass surgery in 1988, after his third attack.

U.S. Economy at a Glance: November 2000	
Unemployment rate	4.0%
Consumer prices (change over Oct.)	+0.2%
Producer prices (change over Oct.)	+0.1 %
Trade deficit	$33.13 bil
Dow Jones high (Nov. 6)	10,977.21
Dow Jones low (Nov. 22)	10,399.32
Index of leading economic indicators (change over Oct.)	–0.2%

International

Mideast Situation Deteriorates Amid Violence—Former Prime Min. Shimon Peres and the Palestinian leader Yasir Arafat met **Nov. 1** and agreed on steps to end the month-long eruptions of violence in their territories. Then, a bomb exploded in an outdoor market in West Jerusalem **Nov. 2,** killing 2 people, and plans to announce a truce were delayed. Pres. Bill Clinton met separately in Washington, DC, with Arafat **Nov. 9**, and Prime Min. Ehud Barak of Israel **Nov. 12**, with no result other than an agreement by the antagonists to resume talks. Arafat failed to persuade Clinton to support deployment of 2,000 UN peacekeepers between the Palestinians and Israeli troops. Arafat spoke to the UN Security Council **Nov. 10**, and made the same request. Through the month, attacks by Palestinians were followed by Israeli retaliations. A bomb killed 2 adults and injured several children on a bus in the Gaza Strip transporting the children of Israeli settlers, **Nov. 20**, and the Israelis then launched a missile attack on Palestinian military and police facilities in Gaza City. After a bomb killed an Israeli soldier

Key Events—November 2000

Nov. 17	A Democratic lawyer sues in Seminole County Circuit Court challenging inclusion of any absentee ballots from that county, where election officials let Republican workers correct errors in GOP absentee ballot application forms.
Nov. 18	Harris announces updated statewide results, including overseas absentee ballots, bringing Bush's lead to 930. Democrats accuse Harris of bending election laws to accommodate more absentee votes from members of the military; Republicans accuse Democrats of trying to throw out military votes.
Nov. 21	Florida Supreme Court unanimously rules that the state's official tally should include manual recount results from Broward, Miami-Dade, and Palm Beach. Orders Harris to accept recount results submitted by 5 P.M. Nov. 26, a Sunday, or, if her office is closed that day, 9 A.M. Nov. 27.
Nov. 22	Bush campaign files 2 appeals with the U.S. Supreme Court. One challenges the Florida Supreme Court's Nov. 21 decision requiring Harris to delay certifying the results in order to include the 3 counties' manual recounts. The other challenges the Nov. 17 rebuff by the U.S. 11th Circuit Court of Appeals.
	The Miami-Dade canvassing board halts its recount, saying it could not finish by the deadline. Democrats suggest the board was intimidated by GOP demonstrators.
	GOP state legislators meet with Florida Gov. Jeb Bush (R) to discuss calling a special session to appoint the Bush electors.
	Bush sues election officials in more than a dozen Florida counties, asking that they be forced to count 646 overseas military ballots rejected for technical reasons. The suit is withdrawn Nov. 25, after many counties began tallying the ballots. However, the Republicans Nov. 24 file 4 new suits against counties that did not.
	Florida's 3rd District Court of Appeal denies a Gore campaign's request for an order to resume the Miami-Dade recount.
	Palm Beach County Circuit Court refuses a Gore campaign request to order county election officials to consider all "dimpled" ballots, those on which voters had only indented the small squares of paper, called chads, that they were supposed to punch out completely.
Nov. 23	Florida Supreme Court upholds the 3rd District Court's Nov. 22 decision.
Nov. 24	U.S. Supreme Court agrees to consider whether the Florida Supreme Court, in its decision to include the manual recounts, had improperly rewritten state law or violated the U.S. Constitution.
	U.S. Supreme Court declines to consider Bush's appeal of the federal appeals court's Nov. 17 decision, because that case is still pending.
	Nassau County's canvassing board votes to submit to Harris its original Election Day machine tally rather that the results of its mechanical recount—which had given Gore a gain of 51 votes—saying the recount was less accurate.
Nov. 25	Broward County completes its recount, giving Gore a net gain of 567 votes.
Nov. 26	Harris certifies Bush as winner by 537 votes, including over 100 from military absentee ballots previously rejected.
	Palm Beach County fails to complete its manual recount by the 5 P.M. deadline.
Nov. 27	Lawyers for Gore file a formal "contest" of certified results in Leon County Circuit Court contesting results in Miami-Dade, Palm Beach and Nassau counties.
	Gore team asks Judge N. Sanders Sauls to restore 51 Gore votes in Nassau County and to appoint a special master to count 10,000 Miami-Dade and 3,000 Palm Beach ballots.
	Venue of the Seminole County absentee ballot case is moved to Leon County.
Nov. 28	Judge Sauls rejects request for an immediate court count of the 13,000 ballots, saying he wishes to hear arguments.
Nov. 29	Sauls grants a Republican motion to have all disputed ballots from Miami-Dade and Palm Beach transported to Tallahassee for safekeeping.
	Palm Beach County officials say their final count, not included in the certified tally, show net Gore gain of 188 votes.
Nov. 30	A select committee of the Florida legislature votes along party lines to call a session to ratify the slate of Bush electors.

near the Lebanese border, Israel attacked Hezbollah guerrilla targets in southern Lebanon, **Nov. 26**. Mary Robinson, UN high commissioner for human rights, urged Israel, **Nov. 27**, to accept international monitors and to remove some Jewish settlements in heavily populated areas.

First Crew Begins Work at Space Station—Three astronauts—2 Russians and 1 American—aboard a capsule launched by a Russian Soyuz rocket docked with the International Space Station **Nov. 2**. Less than an hour later they entered the station and began their duties as the station's first crew. U.S. Navy Capt. William Shepherd commanded the crew, which also included Yuri Gidzenko and Sergei Krikalev. The rocket had been launched from Kazakhstan **Oct. 31**. The crew was to perform maintenance work and remain aboard until February 2001. Assembly of the station was expected to continue for 5 more years.

Salvadorans Cleared in 1980 Deaths of 4 U.S. Women—Two former defense officials in El Salvador were found not guilty **Nov. 3** by a U.S. jury of complicity in the deaths of 4 American church workers in 1980. Salvadoran soldiers had raped and murdered the 4 women—3 nuns and a lay worker. The federal jury in West Palm Beach, FL, concluded that then-Defense Min. Jose Guillermo Garcia and the chief of the national guard at the time, Carlos Eugenio Vides Casanova, did not have "command responsibility"—knowledge that the crime was being committed coupled with failure to act. Relatives of the victims said **Nov. 7** that they would seek a retrial.

Adversaries in Sierra Leone Accept Cease-Fire—Sierra Leone's on-again, off-again civil war paused again **Nov. 10** when the government and the Revolutionary United Front agreed to a 30-day cease-fire. The 2 sides had stopped negotiations after the conflict resumed in May. Although India and Jordan had decided that they would withdraw their peacekeeping contingents by the end of the year, Britain increased its presence to 2,300 for the month of November.

Pinochet Linked to Killing of 2 in U.S.—Newly declassified and released U.S. documents **Nov. 13** implicated former Chilean dictator Gen. Augusto Pinochet in the murder of 2 people in Washington, DC, in 1976. A car-bomb had killed Orlando Letelier, a Chilean socialist who opposed the junta led by Pinochet, and his American secretary, Ronni Moffitt. The new documents included a cable from Pinochet to Pres. Alfredo Stroessner of Paraguay asking him to provide fake Paraguayan passports for 2 agents of the Chilean secret police in order to get them into the United States. Stroessner complied.

The United States barred the 2 men, but they did later enter successfully using Chilean passports with fake names. They eventually were convicted of involvement in the murders and were imprisoned in the United States.

Asian Nations Support New Global Trade Talks—The 21 member nations of the Asia-Pacific Economic Cooperation held its annual summit in Brunei, **Nov. 15–16**. Pres. Clinton got the nations, in their **Nov. 16** communiqué, to endorse a resumption of global trade discussions that would seek to end trade barriers. Plans for such talks had been abandoned after a tumultuous 1999 meeting of the World Trade Organization in Seattle. The summit also focused on such problems as the AIDS epidemic and high oil prices.

Clinton Visits Vietnam—Arriving in Hanoi **Nov. 16**, Bill Clinton became the first U.S. president to visit Vietnam while in office since Pres. Richard Nixon toured South Vietnam in 1969. He met separately **Nov. 17** with Pres. Tran Duc Luong and Prem. Phan Van Khai. Luong urged Clinton to provide help on a legacy of the war, 3.5 million mines and 300,000 tons of unexploded bombs that killed or injured 2,000 people each year. He said Vietnam believed Agent Orange, a defoliant sprayed by Americans to expose North Vietnamese positions, had caused thousands of birth defects. Speaking **Nov. 17** at Vietnam National University, Clinton noted the immense sacrifices by both sides in the war. Although he opposed the war at the time, he did not offer an apology for the U.S. role. He urged Vietnam to embrace free trade.

On **Nov. 18**, Pres. Clinton met with Le Kha Phieu, general secretary of the Communist Party. The same day, Clinton visited a site where U.S. forensic experts, working with Vietnamese laborers, were seeking remains of a U.S. pilot whose plane had crashed. Nearly 2,000 U.S. military personnel remained unaccounted for. Clinton provided documents that might help the Vietnamese find remains of some of their own 300,000 missing soldiers. He concluded his visit in Ho Chi Minh City (formerly Saigon) **Nov. 19**; there and elsewhere, he was greeted enthusiastically by crowds of Vietnamese.

Peru Leader's Stormy Rule Ends in Resignation—Pres. Alberto Fujimori of Peru resigned **Nov. 20**, 5 months before the scheduled election in which he said he would not be a candidate. Though he has been applauded for reviving the Peruvian economy and suppressing guerrilla movements, Fujimori was also accused of human rights abuses that had drawn international criticism. Peru's Congress voted, 62-9, **Nov. 21**, to reject his resignation and instead approved a motion to remove him from office. He faced a possible investigation for corruption and rights abuses. Since both vice presidents had already resigned, the president of Congress, Valentin Paniagua Corazao, who had the backing of Fujimori's foes, was sworn in as interim president **Nov. 22**.

Ex-President Aristide Regains Power in Haiti—Jean-Bertrand Aristide, a former Roman Catholic priest and former president of Haiti, reclaimed the presidency in a **Nov. 26** election. Aristide had become Haiti's first democratically elected president in 1990, but he was ousted in a military coup the following year. A U.S.-led military intervention restored him to power in 1994. Ineligible in 1995 for immediate reelection, Aristide supported Rene Prevel, who won. In the 2000 election, major opposition parties, charging the system was rigged, boycotted the contest. In results announced **Nov. 29**, Aristide won 92% of the votes. whose independence the party favored, but even in Quebec the Liberals won more seats. The Alliance won all but 2 of its seats in 4 conservative western provinces. The Liberals took 100 of 103 seats in Ontario.

Canadian Prime Minister Wins 3d Term—Prime Min. Jean Chrétien of Canada won a 3d term **Nov. 27** when his Liberal Party captured a lopsided majority in parliamentary elections. On **Oct. 22**, Chrétien called for an election, although his term had a year and a half to run. The Liberal platform, released **Nov. 1**, pledged deep tax cuts but also called for new spending on education and the environment. It advocated making the Internet available to all Canadians. Leaders of the parties debated **Nov. 8** and **9**. On **Nov. 27**, the Liberals won 172 of 301 seats in Parliament, an increase of 11 from their current standing. The new conservative party, the Canadian Alliance, won 66 seats, the Bloc Québecois 38, the New Democrats 13, and the Progressive Conservatives 12. The Liberals got 41% of the national popular vote. All the Bloc Québecois vote came from the province of Quebec,

Barak Calls for Election in Israel—Prime Min. Ehud Barak of Israel announced **Nov. 28** that he would dissolve the Knesset (parliament) and call for new elections. Although only 17 months into a 4-year term, Barak had seen his efforts for a peace settlement with the Palestinians fade in the face of renewed violence. His popularity had plummeted. Former Prime Min. Benjamin Netanyahu was expected to run against him. However the current leader of Netanyahu's Likud Party, Ariel Sharon, a strongly vocal opponent of concessions to the Palestinians, said **Nov. 29** that he would challenge Netanyahu for the right to run.

Netherlands Legalizes Voluntary Euthanasia—The lower house of the Dutch parliament voted, 104-40, **Nov. 28**, to legalize physician-assisted suicide, and the bill was approved by the upper house **Apr. 10, 2001,** making the Netherlands the first country in the world to officially sanction the practice. Under the controversial new law, doctors operate under fixed guidelines, including a stipulation that any euthanasia procedure has to be approved by a commission including a medical expert and a lawyer.

General

Kasparov Dethroned as Chess Champion—Gary Kasparov, who had held the world chess championship for 15 years and who was acclaimed by many as the best player ever, lost his title **Nov. 2**. In the 15th game of a match with

Presidential Election Key Events—December 2000

Dec. 1	U.S. Supreme Court hears arguments on appeal by GOP presidential candidate George W. Bush. Florida Supreme Court denies Democratic presidential candidate Al Gore's appeal of Judge N. Sanders Sauls's decision not to immediately begin a recount of 13,000 Palm Beach and Miami-Dade ballots. Florida Supreme Court rejects a Democratic appeal of a lower court decision denying a revote in Palm Beach County because of an allegedly confusing ballot. A Martin County voter sues to throw out all the county's absentee ballots, in a case similar to one filed in Seminole County.
Dec. 2-3	Trial of Gore's contest suit held. Gore's lawyers argue that problems in voting machines caused votes to be missed that could be discerned by manual inspection.
Dec. 4	Sauls dismisses Gore's contest lawsuit. U.S. Supreme Court vacates Florida Supreme Court decision extending the certification deadline and remands the case to the state court for clarification.
Dec. 6	Leaders of Florida's House and Senate sign a proclamation to convene a session to consider appointing the Bush electors. U.S. 11th Circuit Court of Appeals rejects Republican appeals to throw out the certified hand-recount results.
Dec. 8	Two Leon County judges dismiss the Seminole and Martin County absentee ballot suits. Florida Supreme Court reverses Sauls's decision and orders a manual recount of all ballots in which a vote for president was not recorded by machine. The recount begins. The court also restores 383 votes from partially completed recounts in Palm Beach and Miami-Dade, reducing Bush's lead to 154
Dec. 8	Bush team appeals to the U.S. Supreme Court.
Dec. 9	U.S. Supreme Court, in a 5-4 decision, halts the manual recounts, pending a hearing of Bush's appeal.
Dec. 10	Bush and Gore teams file briefs with U.S. Supreme Court. Bush's lawyers claim that since no standard was set for counting undervotes, the recount violated the Constitution's equal protection guarantee.
Dec. 11	U.S. Supreme Court hears oral arguments. Florida Supreme Court issues a rewritten version of the decision remanded to it, claiming it had merely tried to reconcile conflicts within state law.
Dec. 12	Deadline under federal law for states to certify electors. U.S. Supreme Court, 5-4, reverses Florida Supreme Court's order to proceed with the recount, with 7 justices agreeing that it violates the equal protection guarantee. The Court technically remands the decision to the Florida Supreme Court for review, while noting that there is no time for a recount. Florida's House votes to appoint the Bush electors.
Dec. 13	Gore, in a televised address, concedes to Bush. Bush addresses the nation as president-elect. The final official results showed that Gore remained slightly ahead of Bush with the total nationwide popular vote, with 51,003, 894 as against Bush's 50,495,211.

Vladimir Kramnik, Kasparov offered a draw, which gave the challenger the half-point he needed to ensure victory. Kasparov, in 1985, had become, at 22, the youngest champion ever. His only previous notable defeat had been by a computer in 1997.

One Siamese Twin Survives 20-Hour Operation—After a legal battle, Siamese twin girls underwent a 20-hour operation in England **Nov. 6-7**, and one of them survived. The girls, known as Jodie and Mary, had been born **Aug. 8** to a couple in Malta who then sought medical help in Britain. Their doctors believed that both twins would die within 6 months of birth unless separated, in which case one could live and the other would die. The parents, both Roman Catholics, objected to an operation, but doctors sought court permission to operate, and prevailed. The operation, in Manchester's St. Mary's Hospital, resulted soon in the death of the weaker twin, Mary. The other twin, Jodie, showed steady progress, and a hospital spokeswoman said **Nov. 9** that she was making "a rapid recovery." Years of further surgery and skin grafts would be required to assure her a normal life.

DECEMBER 2000

National

Senate to Be Split 50-50 Between Parties—The last U.S. Senate contest was decided **Dec. 1**, when a statewide recount gave the victory to Maria Cantwell (D) in Washington. She defeated incumbent Slade Gorton (R) by 2,229 votes. The Democrats and the Republicans now would have 50 seats each in the new Senate.

U.S. Shuttle Crew Works at Space Station—The U.S. space shuttle Endeavor docked with the International Space Station **Dec. 2**, and the crew subsequently attached large electricity-generating solar panels to the structure. The shuttle had lifted off from Cape Canaveral, FL, **Nov. 30**. The mission commander was Navy Cmdr. Brent Jett Jr. On **Dec. 8** the 5-man shuttle crew boarded the space station, becoming the station crew's first visitors. Heading home, the Endeavor touched down at Cape Canaveral **Dec. 11**.

Congress Adjourns After "Lame Duck" Session—Congress reconvened **Dec. 5** to complete its year's business in a "lame duck" session. Ten days later, on **Dec. 15**, the 106th Congress adjourned, having completed work on an omnibus spending bill covering all federal departments whose appropriations bills for the 2001 fiscal year had not already been cleared. The Senate, by voice vote, and the House, 292-60, approved the $450 billion bill. Pres. Bill Clinton failed to secure inclusion of a provision granting amnesty to illegal immigrants, though Congress did agree to allow some illegal immigrants to remain in the United States while they sought permanent residency.

Senators Choose Leaders—Republicans and Democrats elected their leaders for the new Senate **Dec. 5**. Trent Lott (MS) and Don Nickles (OK) were reelected without opposition as GOP majority leader and majority whip; prevailing in contested votes, Larry Craig (ID) and Rick Santorum (PA) were elected chairs of the Republican Policy Committee and Republican Conference, respectively. Bill Frist (TN) was named head of the National Republican Senatorial Committee. Without challenge, Democrats reelected Tom Daschle (SD) and Harry Reid (NV) as minority leader and minority whip. Patty Murray (WA) and Barbara Mikulski (MD) were reelected chair of the Democratic Senatorial Committee and secretary of the Democratic Caucus, respectively. Technically, the Democrats would be the majority party in the new Senate until **Jan. 20**, when Dick Cheney would replace Al Gore as vice president, and could vote in the event of ties.

Osprey Crash Casts Doubt—Four U.S. Marines were killed **Dec. 12** when a V-22 Osprey aircraft crashed during a training mission near Jacksonville, NC. The Osprey, considered an engineering marvel when first unveiled in 1991, was built with tilt-rotor engines that allow it to take off like a helicopter and fly like an airplane. However, the crash was the Osprey's 4th and its 2d in 2000. The Marine Corps, which had planned to buy 360 Osprey aircraft to form the backbone of its aviation operations, asked the Department of Defense the same day to postpone production of its order, which had been scheduled to begin the following week.

On **Dec. 18**, the presidential electors meeting in state capitals and in Washington, DC, cast 271 votes for Bush and 266 for Gore. One Gore elector in the District of Columbia had cast a blank ballot in protest against the fact that the District had no voting representative in Congress. On **Dec. 21**, Bush resigned as governor of Texas and was succeeded by Lt. Gov. Rick Perry (R).

In his concession, Gore said, "While I strongly disagree with the court's decision, I accept it." He urged all Americans to rally behind the new president. A short while later, Bush spoke in the chamber of the Texas House of Representatives. He said he had been elected not to serve one party, but to serve one nation. He reaffirmed his support for tax reduction, reform of Social Security and Medicare, and an improved educational system.

Gore Concedes to Bush, 36 Days After Election—On **Dec. 13**, in a televised address, Vice Pres. Al Gore, the

Democratic candidate, conceded the presidential election to Texas Gov. George W. Bush (R.). Not since 1876 had an election dragged on so long without a winner being determined. Gore ended his challenge after the U.S. Supreme Court, **Dec. 12**, stopped for good any manual recount of votes in Florida, leaving Bush's small edge in the certified popular vote in Florida unchanged. With Florida's 25 electoral votes in Bush's column, it appeared that the final tally in the Electoral College would be Bush 271, Gore 267.

Hillary Clinton Makes $8 Million Book Deal—First Lady and Senator-elect Hillary Rodham Clinton **Dec. 15** received an $8 million advance after reaching an agreement with Simon & Schuster to write a memoir of her years in the White House. The Congressional Accountability Project, a private watchdog group, expressed concern Dec. 18 about a possible conflict of interest because the publisher's parent firm, Viacom, had business before Congress and federal agencies. The $8 million was close to the non-fiction record $8.5 million that Pope John Paul II once received.

Bush Announces Choices for Cabinet—On **Dec. 16**, as expected, Pres.-elect George W. Bush named Gen. Colin Powell (ret.) to be secretary of state, taking the first step in the formation of his cabinet, which had been delayed by the election deadlock. The son of Jamaican parents, Powell had grown up in the New York City borough of the Bronx. He was the first black to head the National Security Council (under Pres. Ronald Reagan) and serve as chairman of the Joint Chiefs of Staff (under the elder Pres. Bush).

Prof. Condoleezza Rice of Stanford University, who had also been the school's provost, was named **Dec. 17** to be national security adviser. The same day, Bush said that Karen Hughes, his press spokesperson, would be counselor to the president and that Alberto Gonzales would be White House general counsel. On **Dec. 19**, Bush met separately with Pres. Clinton and with Vice. Pres. Gore. On **Dec. 20**, Bush announced he had chosen Paul O'Neill, chairman of Alcoa, the world's largest aluminum manufacturer, to be treasury secretary. Also on **Dec. 20**, Bush named 3 more cabinet secretaries: Melquiades Martinez, chairman of Orange County, FL (housing and urban development); Ann Veneman, former California food and agriculture secretary (agriculture); and Donald Evans, his campaign chairman (commerce).

On **Dec. 22**, Bush named Sen. John Ashcroft (R, MO), who had just been defeated for reelection, as attorney general. Ashcroft, a controversial figure, had the strong support of Republican conservatives. Bush also chose New Jersey Gov. Christie Whitman (R) of New Jersey to head the Environmental Protection Agency. **On Dec. 28** he named Donald Rumsfeld to be secretary of defense. Rumsfeld, who had held the same position under Pres. Gerald Ford, strongly supported construction of a shield to defend against nuclear missiles. On **Dec. 29**, he named 4 more cabinet members: Wisconsin Gov. Tommy Thompson (health and human services); Gale Norton, former attorney general of Colorado (interior); Rod Paige, superintendent of schools in Houston, TX (education); and Anthony Principi (veterans' affairs). All cabinet appointees were subject to approval by the Senate.

Office Worker Kills 7 in Rampage—Michael McDermott, a 42-year-old employee of a Malden, MA, Internet consulting firm allegedly went on a rampage **Dec. 26,** killing 7 coworkers with a semiautomatic rifle, a semiautomatic pistol, and a shotgun before sitting in the lobby to await arrest. He was arraigned **Dec. 27** and held without bail.

U.S. Population Rose by 13.2% During the 1990s—The U.S. Census Bureau **Dec. 28** put the resident population of the United States on Apr. 1, 2000, at 281,421,906, a 13.2% increase from 1990. On the basis of unadjusted state figures, as required by law in the U.S. House (starting in 2002), states gaining seats would be Arizona (2), California (1), Colorado (1), Florida (2), Georgia (2), Nevada (1), North Carolina (1), and Texas (2). States losing seats would be Connecticut (1), Illinois (1), Indiana (1), Michigan (1), Mississippi (1), New York (2), Ohio (1), Oklahoma (1), Pennsylvania (2), and Wisconsin (1). Based on these figures, California, already the most populous state, added 4,111,627 residents, more than any other state. Nevada had the highest rate of gain, 66.3%. North Dakota had the smallest total increase, 3,400, and the lowest rate of increase, 0.5%. The District of Columbia's population declined by 5.7%.

Stock Values Decline During 2000—When the markets closed on **Dec. 29**, the technology heavy NASDAQ composite index stood at 2,470.52, a fall of 39.3% over 12 months. The Dow Jones Industrial Average declined 6.2% in 2000, closing at 10,786.85. The Standard & Poor's 500-stock index declined 10.1% to 1,320.28.

U.S. Economy at a Glance: December 2000	
Unemployment rate	4.0%
Consumer prices (change over Nov.)	+0.2%
Producer prices (change over Nov.)	Unchanged
Trade deficit	$32.99 bill
Dow Jones high (Dec. 5)	10898.72
Dow Jones low (Dec. 20)	10,318.93
Index of leading economic indicators (change over Nov.)	-0.6%
4th Quarter GDP (at annual rate)	+1.0%

U.S. Economy at a Glance: Calendar Year 2000	
Unemployment rate	4.0%
Consumer prices (change over 1999)	+3.4%
Producer prices (change over 1999)	+3.6%
Trade deficit	$369.7 bill
Dow Jones high (Jan. 14)	11,722.98
Dow Jones low (Mar. 7)	9,796.03
GDP (change over 1999)	+5.0%

International

Vicente Fox Inaugurated as President of Mexico—In an unprecedented peaceful transfer of power from one party to another, Vicente Fox, of the opposition National Action Party, was sworn in as president of Mexico **Dec. 1**. He succeeded Pres. Ernesto Zedillo, latest in a long line of presidents from the Institutional Revolutionary Party (PRI), which had ruled the country since 1929. On **Dec. 1**, Fox swore in his cabinet members and made them pledge to adhere to a set of ethical rules and to reveal their assets. Fox, in his inaugural address, promised to end the corruption and human rights abuses and to improve education and the environment. He also pledged to end the rebellion in Chiapas province, and on **Dec. 1** ordered federal Army troops to begin pulling out. On **Dec. 2** the Zapatista National Liberation Army said it was ready to resume peace negotiations with the government.

Pinochet Charged and Put Under House Arrest—On **Dec. 1**, a judge in Chile formally charged the country's former dictator, Gen. Augusto Pinochet, in the kidnapping and murder of political opponents. Pinochet, who had just turned 85, was reportedly in poor health. An appeals court reversed the arrest order **Dec. 11**, saying that the judge should have interviewed Pinochet to determine fitness for trial. The Supreme Court, in appeal, dismissed the charges, **Dec. 20**, but ordered that Pinochet be interrogated, leaving the door open for another indictment.

UN Finds Palestinians in Dire Economic Straits—A UN report made available **Dec. 5** said that poverty and unemployment in Palestinian-controlled areas in the West Bank and the Gaza Strip were up sharply. Because of Israeli restrictions during the current strife, about 20% of the Palestinian workers had been unable to get to their jobs in Israel, causing much economic hardship. Pounding by Israeli weapons had caused substantial damage to buildings. Meanwhile, violent clashes continued. Ten people were killed on **Dec. 8**, the 13th anniversary of the first intifada, or Palestinian uprising. In a gun battle near a refugee camp in the Gaza Strip, **Dec. 13**, Israeli soldiers killed 4 Palestinians and wounded more than 25.

European Union Plans to Add New Members—Leaders of the 15 European Union members met in Nice, France, **Dec. 7-11**, to plan for the expansion of the EU in the coming years to 27 members. The 12 new members would include many former Communist countries. The EU agreed **Dec. 11** to restructure the commission to consist of 27 members, 1 from each country, regardless of size. Under a revised formula, the Council of Ministers, the EU's legislature, would continue to be structured on the basis of proportional representation.

Israel's Election for Prime Minister Underway—Ehud Barak of Israel resigned **Dec. 9**, opening the way for a new election. However, the voters would choose only a prime minister; membership in the Knesset (parliament) would not be contested in the election. Furthermore, only members of the present Knesset were to be eligible to run for prime minister; that provision excluded former Prime Min. Benjamin Netanyahu, who had resigned from Parliament after losing the election for prime minister in 1999. The Knesset, **Dec. 13**, gave preliminary approval, 67-35, to a bill that would allow Netanyahu to run. But Netanyahu said he would run only if the Knesset was also up for election. And the Knesset, **Dec. 19**, voted, 69-49, not to hold a general election. This cleared the way for Ariel Sharon, leader of the opposition Likud Party, to become Barak's challenger.

Convicted Pakistani Leader Goes Into Exile—Former Prime Min. Nawaz Sharif of Pakistan was released from prison **Dec. 9** and allowed to go into exile. He had been sentenced to life in prison principally for his effort to prevent Gen. Pervez Musharraf, the country's present military ruler, from taking power. Sharif arrived in exile in Saudi Arabia, **Dec. 10**, along with nearly 20 relatives, under an agreement to stay in exile for 10 years, not run for office for 21 years, and give up $10 million in property and other assets.

Ex-President Regains Power in Romania—After being out of power for 4 years, Ion Iliescu recaptured the presidency of Romania in a runoff election **Dec. 10**. He defeated an ultranationalist, Corneliu Vadim Tudor, by a 2-1 margin.

Ethiopia, Eritrea End Border War—Two warring African neighbors, Ethiopia and Eritrea, signed a peace treaty **Dec. 12** that formally ended their bloody border conflict. Both countries agreed to accept a border determined by a neutral commission in Geneva, Switzerland. Some 4,200 UN peacekeepers would be deployed along the border.

Russian President Meets With Castro in Cuba—Pres. Vladimir Putin of Russia visited Cuba, **Dec. 14-17**, the first time a Russian leader had come to the Communist-ruled island since the collapse of the Soviet Union in 1991. On **Dec. 14** Putin and Pres. Fidel Castro denounced the U.S. trade embargo imposed on Cuba. Putin was unsuccessful in persuading Castro to repay Cuba's Cold War debt owed to Russia, even though the Russian president had reportedly offered to forgive 70% of the approximately $20 billion debt. Putin visited Canada **Dec. 17–19** and met with Prime Min. Jean Chretien.

Chernobyl Closed—The Chernobyl nuclear power plant in Ukraine, site in 1986 of the world's most serious nuclear accident, was officially decommissioned **Dec. 15** by Pres. Leonid Kuchma. The European Bank for Reconstruction and Development had pledged $215 million, **Dec. 7**, for the completion of 2 more safely designed plants that would replace power lost from Chernobyl.

UN Imposes Sanctions on Afghanistan—The UN Security Council voted 13–0, **Dec. 19**, to impose sanctions on the Taliban regime that controlled most of Afghanistan. China and Malaysia abstained. The UN resolution demanded that the Taliban, a militant Islamic group, give up the terrorist Osama bin Laden, charged in the bombing of 2 U.S. embassies in Africa in 1998. The sanctions forced the closure of Taliban diplomatic offices around the world and barred aircraft from other countries from landing in Afghanistan. The UN also imposed an arms embargo on the Taliban regime.

Milosevic's Foes Win Control of Serb Parliament—The transfer of political power in Serbia was completed **Dec. 23**, when the 18-party coalition called the Democratic Opposition of Serbia won more than two-thirds of the seats in parliamentary elections. The same coalition had backed the successful candidate for president of Yugoslavia, Vojislav Kostunica, against the incumbent, Slobodan Milosevic, in September.

General

Kuerten Clinches Top Ranking With Victory—Brazilian Gustavo Kuerten, 24, defeated Andre Agassi, 6-4, 6-4, 6-4, **Dec. 3**, to capture the Tennis Masters Cup. The triumph also gave him the top spot in the Association of Tennis Professionals (ATP) year-end men's ranking for 2000.

Weinke Becomes Oldest Heisman Winner—Florida State quarterback Chris Weinke was awarded the Heisman Trophy **Dec. 9**, becoming the oldest player ever to win the award in its 66-year history. Weinke, 28, threw for 4,167 yards and 33 touchdowns, against just 11 interceptions. He received 369 first-place votes and 1,628 in all, while his closest competitor, Oklahoma quarterback Josh Heupel, received 286 first-place votes and 1,552 total, making it one of the closest races ever.

GM to Phase Out Oldsmobile—General Motors announced **Dec. 12** that the oldest U.S. auto brand, Oldsmobile, would soon reach the end of the road. Citing sales for the brand that had declined 18.5% from 1999, GM also announced it would also close 2 plants, one in Lansing, MI, and the other in Luton, England, and lay off 15,000 workers over the next year as part of a worldwide restructuring plan. GM, the world's largest carmaker, said that the division would be phased out over the next several years, and that current Oldsmobile models would continue to be built as long as they were economically viable.

JANUARY 2001

National

Bush Names Last Cabinet Nominees; Chavez Withdraws—Pres.-elect George W. Bush announced final nominees to fill out his cabinet, **Jan. 2**, although within a week one of them, Linda Chavez, conservative commentator and former civil rights commissioner, withdrew from consideration. Bush nominated Chavez, a Hispanic, for secretary of labor, as well as ex-Sen. Spencer Abraham (R, MI), a Lebanese-American, for secretary of energy and ex-Rep. Norman Mineta (D, CA), a Japanese–American, for secretary of transportation.

Chavez drew criticism from some quarters for having opposed affirmative action, the minimum wage, and bilingual education. Her withdrawal, **Jan. 9**, was prompted by the revelation that she had sheltered an illegal alien from Guatemala, to whom she had given about $1,500. Chavez said the woman had performed chores but was not an employee. In her place, Bush **Jan. 11** nominated Elaine Chao, a Taiwanese immigrant and former deputy transportation secretary; she is married to Sen. Mitch McConnell (R, KY).

107th Congress Convenes—The 107th Congress met for the first time **Jan. 3**. The Senate was divided equally, 50–50, between the parties. The Republicans held a majority in the House, 221–211, with 2 independents and 1 vacancy. Because Vice Pres. Al Gore (D) would preside over the Senate until Inauguration Day, **Jan. 20**, and could break any tie vote, the Democrats were able to name the committee chairs for 17 days. Republicans and Democrats, **Jan. 5**, agreed that Senate committees would contain an equal representation from both parties.

Fed Cuts Interest Rates—Responding to indications of a faltering economy, the Federal Reserve Board **Jan. 3** cut the federal funds rate, the overnight rate banks charge each other, from 6.5% to 6%. The cut followed a series of rate increases; the last, in May 2000, had sought to keep inflation in check. On **Jan. 4**, the Fed reduced the discount rate, charged by the Federal Reserve Bank to member banks, to 5.5%. On **Jan. 31**, the Fed cut the federal funds rate again, to 5.5%.

Clinton Acts to Protect Natural Areas—On **Jan. 5** Pres. Bill Clinton issued an executive order barring road-building and timber-cutting in 58.5 million acres of federally owned forest land, nearly one third of the total. Large areas in the Western states and Alaska were protected. Three days later, Mike Dombeck, director of the Forest Service, banned the logging of old-growth timber in U.S. national forests. On **Jan. 17**, Clinton created 6 new national monuments and expanded 2 others. In all, Clinton as president had established 17 national monuments and expanded 4. The new ones included Minidoka International NM (ID), commemorating the internment of Japanese–Americans during World War II; Pompeys Pillar NM (MT), a tall sandstone formation sighted by the Lewis and Clark Expedition; and the Upper Missouri River Breaks NM (MT), the only lengthy portion of the river still in a natural state.

Bush Certified Winner of Presidency—On **Jan. 6**, when Congress, in joint session, received the count of the electoral votes from the 50 states and the District of Columbia. The tally was 271 for former Texas Gov. George W.

Bush and 266 for Vice Pres. Al Gore, with 1 Gore elector abstaining to protest Washington DC's lack of congressional representation. Gore presided as Congress certified these results. House Democrats, led by members of the Congressional Black Caucus, sought to block the 25 votes for Bush from Florida, where, they claimed, many black voters had been disenfranchised. However, their challenge could not go forward because it was not backed by any senator, as required by law.

Energy Shortage Forces Blackouts in California—Parts of California were briefly plunged into darkness beginning **Jan. 17** after utility companies failed to deliver a sufficient amount of energy. Pacific Gas and Electric Co. and Southern California Edison Co— $12 billion in debt— had been forced to buy power at ever-increasing rates while being barred by law from passing higher costs on to their customers. The California Public Utilities Commission temporarily raised the price caps **Jan. 4**. Gov. Gray Davis (D) **Jan. 8** accused out-of-state generators of withholding supplies to force up prices; he proposed a $1 billion rescue plan that included construction of new power generating plants. TV sets and traffic lights went blank **Jan. 17-18,** as PGE ordered rolling blackouts that impacted 2 million people in northern California. Davis **Jan. 17** declared a state of emergency and directed the Dept. of Water Resources to buy energy and sell it cheaply to utilities.

Ex-Louisiana Governor Gets 10 Years in Prison—Edwin Edwards, a former governor of Louisiana, was sentenced **Jan. 8** to 10 years in prison and fined $250,000. He had been convicted in May 2000 of extorting money from applicants seeking riverboat casino licenses. Four other convicted defendants, including Edwards's son Stephen, received lesser prison terms. Edwards said he would appeal.

Liberals Oppose Nominee for Attorney General—A number of liberal organizations, including Planned Parenthood and the National Organization for Women, said **Jan. 9** that they would urge the Senate to reject the nomination of former Sen. John Ashcroft (R, MO) as attorney general. Ashcroft, who had just been defeated for reelection, strongly opposed abortion, gun control laws, and extension of hate-crime laws to gays. Testifying at his Senate confirmation hearings **Jan. 16**, he stressed that he would enforce the law, whether he agreed with it or not. Ashcroft noted that as senator he had voted for 26 of 27 minority nominees for federal judgeships; the exception, Judge Ronnie White, testifying **Jan. 18**, charged that Ashcroft, who had led the successful effort to defeat his nomination, had falsely painted him as soft on crime. The Senate, **Feb. 1**, approved Ashcroft's nomination, 58–42, with all opposing votes coming from Democrats.

AOL Time Warner Gets Final OK—The proposed creation of the world's largest media conglomerate cleared a final hurdle, **Jan. 11**, when the Federal Communications Commission, 5–0, approved the merger of America Online and Time Warner Inc. The European Union and the Federal Trade Commission had given their approvals. The FCC required that AOL Time Warner's instant messaging system be made compatible with that of at least one rival and that compatibility be provided to 2 other competitors within 6 months.

Previously, on **Dec. 14, 2000,** the merger was approved by the Federal Trade Commission. The FTC had stipulated that the combined company must open its high-speed cable lines to other Internet service providers. Also, the company would have to open its cable system to other interactive TV companies and could not interfere with their transmissions. Finally, the new company would have to offer Internet access over digital subscriber line wires where it would also offer access over its cable lines.

Riady Admits Role in Illegal Gifts—James Riady, a wealthy Indonesian businessman, agreed to plead guilty **Jan. 11** to involvement in a scheme to funnel illegal donations to U.S. politicians and influence U.S. international trade policies. He had reimbursed bank employees for their donations, most of which went to Democrats. The bank pleaded guilty to 86 counts of transmitting illegal contributions from outside the United States. Riady agreed to pay an $8.6 million fine and to cooperate in an ongoing investigation into contributions to the Democratic Party.

Jesse Jackson Admits to Illegitimate Child—The Rev. Jesse Jackson, the civil rights leader, acknowledged **Jan. 18** that he had fathered an illegitimate daughter born in 1999 to an employee of his Rainbow/PUSH Coalition, whom he was paying $3,000 a month for child support. His affair had occurred while Jackson was counseling Pres. Clinton after the latter's relationship with a White House intern became public. The Coalition disclosed **Jan. 19** that it had given a $35,000 "severance package" to the mother, Karin Stanford.

Clinton Admits Testifying Falsely—On **Jan. 19**, his last full day as president, Clinton announced an agreement with independent counsel Robert Ray that would spare him the risk of being indicted after he left office. In his agreement, Clinton stated for the first time that "certain of my responses to questions about Ms. Lewinsky were false," referring to a 1998 deposition he gave in the Paula Jones sexual misconduct suit. Clinton agreed to pay a $25,000 fine to the Arkansas Bar Assn. and to forfeit his law license for 5 years.

Clinton Announces Controversial Last-Minute Pardons—On the morning of **Jan. 20**, hours before his term ended, Pres. Clinton issued 176 pardons and commutations, exercising a presidential power under the U.S. Constitution.

Those pardoned included former CIA Director John Deutch, who faced prosecution for storing government secrets on his home computer; Susan McDougal, convicted in connection with the Whitewater real estate venture in which Pres. Clinton and his wife had become ensnarled, and later jailed for refusing to testify concerning the affair; Patricia Hearst Shaw, a member of the Hearst newspaper family who had long since completed a prison term for bank robbery while she was involved with the radical Symbionese Liberation Army; and Roger Clinton, the president's half-brother, who had pleaded guilty to a charge involving cocaine distribution. Clinton also commuted the sentences of 4 New York Hasidic Jews who had been convicted of embezzlement and who may have lobbied one of the state's Hasidic communities to vote for Hillary Clinton in her Senate bid.

The most controversial pardon was that of Marc Rich, a commodities trader who had fled to Switzerland in 1983. He had been charged with conspiracy, evading more than $48 million in taxes, racketeering, and trading with Iran while that country held U.S. hostages. His ex-wife, Denise, a significant contributor to the Democratic Party and to the Clinton Library, had appealed to Clinton to pardon Rich. So had Jack Quinn, Rich's attorney and a former White House counsel in the Clinton administration.

George W. Bush Inaugurated—George Walker Bush, former governor of Texas and son of the 41st president, was sworn in as the 43rd president at noon **Jan. 20,** on the steps of the Capitol in Washington, DC. U.S. Chief Justice William Rehnquist administered the oath. Former Defense Sec. Dick Cheney took the oath for vice president.

In a brief inaugural address, Pres. Bush, mindful of the divisive 2000 election and the partisan acrimony in the capital in recent years, appealed for civility among the nation's leaders and called on all Americans to be good citizens. Recalling his self-characterization as a compassionate conservative, Bush said that "compassion is the work of a nation, not just a government." America's grandest ideal, he added, was that "everyone belongs, that everyone deserves a chance, that no insignificant person was ever born."

Bush Unveils Education Plan—Pres. Bush **Jan. 23** proposed an education plan including an increased role for the federal government. As expected, the proposal included vouchers for parents to help them remove their children from failing schools and pay tuition for private or religious schools. A key objective of Bush's plan was to demand accountability from schools getting federal money, with standardized tests for students in grades 3 through 8.

The day before, in a controversial move, Bush issued an executive order barring federal finds for international agencies that perform abortions or provide abortion counseling.

Cabinet Members Ok'd; Interior Nominee Wins Approval—The U.S. Senate gave its approval to a number of Bush's cabinet picks beginning **Jan. 20**, when an unusual session was convened to approve Sec. of State Colin Powell, Defense Sec. Donald Rumsfeld, Treasury Sec. Paul O'Neill, Education Sec. Roderick Paige, Energy Sec. Spencer Abraham, Commerce Sec. Donald Evans, and Agriculture Sec. Ann Veneman. Housing Sec. Mel Martinez and Sec. of Veteran Affairs Anthony Principi were approved **Jan. 23**. Sec. of Health and Human Services Tommy Thompson was approved on **Jan. 24**, along with Transportation Sec. Norman Mineta. Labor Sec. Elaine Chao was approved **Jan. 29**.

The Senate **Jan. 30** approved, 75–24, the nomination of former Colorado Attorney Gen. Gale Norton as secretary of the interior. A coalition of environmental groups opposed her nomination, saying that she had favored allowing corporate polluters to regulate themselves and had opposed as unconstitutional laws protecting endangered species and restricting mining. Testifying **Jan. 18** at her confirmation hearings, Norton had said she no longer viewed these laws as unconstitutional. On **Feb. 1** Ashcroft was approved as attorney general, 58-42, completing Bush's cabinet.

U.S. Economy at a Glance: January 2001	
Unemployment rate	4.2%
Consumer prices (change over Dec.)	+0.6%
Producer prices (change over Dec.)	+1.1%
Trade deficit	$33.3 bill
Dow Jones high (Jan. 3)	10,945.75
Dow Jones low (Jan. 9)	10,572.55
Index of leading economic indicators (change over Dec.)	+0.8%

International

Palestinians Accept Peace Plan With Reservations—The Palestinian National Authority **Jan. 1** expressed concerns about Middle East peace plan principles set forth by Pres. Bill Clinton on **Dec. 23** and already accepted by Prime Min. Ehud Barak of Israel. The plan provided for Palestinian control over the Gaza strip, Arab neighborhoods in East Jerusalem, and 95% of the West Bank, with partial control of the Temple Mount. The PNA feared these principles would not produce a viable state, objected to provisions allowing for annexation of Israeli settlements in the West Bank, and protested that the plan would forgo the right of return for Palestinians who had become refugees following the formation of Israel in 1948 and subsequent wars. However, on **Jan. 2**, Palestinian leader Yasir Arafat met with Clinton in the White House and gave qualified acceptance to the plan.

'Mad Cow Disease' Causes Widespread Alarm—Australia, **Jan. 5**, banned the import of beef products from 30 European countries, adding to growing anxiety over the spread of Mad Cow Disease, formally called bovine spongiform encephalopathy (BSE) in cows. The disease is thought to be transmitted when cows consume feed that includes ground-up animal parts. Humans can acquire a variant of a closely-related ailment, Creutzfeldt–Jakob disease (CJD), by eating contaminated beef. Victims suffered from a deterioration of the brain, in which spongelike holes appear. By January, 80 people in Britain had died of CJD, and 2 had died in France. Sales of beef products fell sharply across Europe. The European Union had taken measures to contain BSE—as a precaution, cows more than 30 months old had been destroyed.

Investigations Into Ship Bombing Continue—It was reported **Jan. 7** that an investigation by the U.S. Navy into the October 2000 terrorist bombing of the destroyer *Cole* in Yemen concluded that neither the captain of the ship nor the crew should be punished for security lapses. A Defense Dept. commission **Jan. 9** recommended enhanced security measures for U.S. military forces overseas. On **Jan. 19**, Defense Sec. William Cohen said that responsibility in the bombing, which killed 17 sailors, extended throughout the Pentagon leadership, ranging from those on board to himself. He recommended that no one be disciplined.

U.S. Army Admits Soldiers Killed Korean Civilians—After a year-long inquiry, the U.S. Army found, **Jan. 11**, that American soldiers had shot and killed unarmed South Korean civilians in July 1950, during the Korean War. The incident, which had recently come to general attention, occurred near the village of No Gun Ri. South Korea claimed that 400 civilians had been killed or wounded, or were missing as a result of the shootings; the U.S. Army disputed whether any reliable estimate was possible. The Army said that there had been no order to kill civilians, and that the incident was the result of confusion and was not deliberate.

President of Congo Shot Dead by Bodyguard—Pres. Laurent Kabila of the Democratic Republic of the Congo was shot and killed **Jan. 16**. It was reported **Jan. 19** that he had been shot in the presidential palace by a bodyguard, who in turn was shot to death by other guards. In a unanimous vote of parliament, **Jan. 24**, Joseph Kabila, the president's 29-year-old son, was elected president; he was sworn in **Jan. 26**. In 1997, the elder Kabila had led the revolt that overthrew longtime ruler Mobutu Sese Seko. Under Kabila, poverty appeared to worsen, the mining industry was fading, and neighbors Rwanda and Uganda were backing a new rebel force inside Congo. In all, troops of 6 nations, some allied with Kabila, were on Congolese soil.

Public Outcry Forces Philippines President to Resign—Pres. Joseph Estrada of the Philippines resigned **Jan. 20**. Estrada had been accused of accepting more than $11 million in kickbacks from tobacco taxes and illegal gambling, and an impeachment trial was held in which senators sat as judges. On **Jan. 16**, the senators voted narrowly not to open bank records that could have proven Estrada's guilt, and the prosecutors resigned in protest. The trial was halted, but public opinion had turned against Estrada, and huge throngs massed at his residence. On **Jan. 19**, key military and political leaders abandoned him, leading to his resignation the following day. He was immediately succeeded by Vice Pres. Gloria Macapagal Arroyo.

Libyan Convicted in 1988 Bombing of Plane—A 3-judge Scottish court, meeting in the Netherlands, found a Libyan guilty **Jan. 31** in the 1988 bombing of Pan Am Flight 103, which claimed 270 lives. The plane exploded over Lockerbie, Scotland, killing all aboard, along with 11 people on the ground. International pressure had induced Libya to turn over 2 suspects to the court. One, Abdel Baset Ali Mohmed al-Megrahi, a Libyan intelligence officer, was found guilty and sentenced to life in prison. The other, Al Amin Khalifa Fhimah, a former airline manager, was acquitted.

General

Bowl Win Gives NCAA Football Title to Oklahoma—The University of Oklahoma Sooners, who had barely made the Top 25 in the preseason rankings, captured the NCAA Division I-A football title by defeating Florida State, 13–2, in the Orange Bowl, **Jan. 3**. Oklahoma had been the only Division 1–A team to go unbeaten during the regular season. The Sooners were led by quarterback Josh Heupel and the Seminoles by quarterback Chris Weinke, winner of the Heisman Trophy. Oklahoma was coached by Bob Stoops.

Monkey Is First Genetically Engineered Primate—The **Jan. 12** issue of *Science* magazine reported the birth of the first genetically engineered primate. The birth of the rhesus monkey was accomplished by scientists at Oregon Health Sciences University. They had inserted a jellyfish gene in 224 monkey ova. Six of these developed into fetuses, 3 of which were born alive. However, only one live monkey had the gene. The gene caused jellyfish to glow; the live monkey did not glow, though one that was stillborn showed evidence of fluorescent green fingernails and hair. Scientists hoped to use the breakthrough to find cures by creating monkeys with human diseases.

Earthquake Rocks El Salvador—Early in the morning of **Jan. 13**, an earthquake shook the coast of El Salvador, killing more than 800 people. By **Jan. 15** a relief effort was operating smoothly. On **Jan 18**, it was reported that the quake caused about $1 billion in damage, nearly half of the country's yearly budget.

Pope Names New Cardinals—Pope John Paul II, **Jan. 21**, named 37 new cardinals, a record number at one time. Three Americans were named: Archbishop Edward Michael Egan of New York; Archbishop Theodore McCarrick of Washington, DC.; and the Rev. Avery Dulles, a Jesuit theologian at Fordham University. Dulles, 82, was a convert to Catholicism and the son of former U.S. Sec.of State John Foster Dulles. The new cardinals included 10 from Latin

America. On **Jan. 28**, the pope named 7 more cardinals, bringing the total eligible to vote (under age 80) to 135, a record. The new cardinals were elevated on **Feb. 21**.

Earthquake Takes Heavy Human Toll in India—A 7.9-magnitude earthquake **Jan. 26** devastated the western Indian state of Gujarat, killing more than 20,000 people. The epicenter was near Bhuj, a city of 150,000, where thousands of buildings were destroyed.

Baltimore Wins Super Bowl With Powerful Defense—The Baltimore Ravens won Super Bowl XXXV **Jan. 28**, defeating the New York Giants, 34–7. The Ravens had not been a preseason favorite, but their strong defense had held regular-season opponents to just 165 points, a league record for 16 games. Their key defensive player was linebacker Ray Lewis, who had been jailed a year earlier in connection with a double murder and who had pleaded guilty to obstructing justice. He was named the game's most valuable player. Baltimore's quarterback was Trent Dilfer. In the championship game, the Ravens intercepted the Giants quarterback, Kerry Collins, 4 times. Baltimore was coached by Brian Billick.

FEBRUARY 2001

National

California Spends Big to End Energy Crisis—The California legislature **Feb. 1** completed action on a $10 billion plan to ease the state's energy crisis. Gov. Gray Davis (D) signed the bill the same day. It would permit the state to issue bonds to fund the purchase of power through long-term contracts. As more regional blackouts remained a threat, Davis also proposed to increase funding for energy conservation programs.

Clinton's Pardons Bring Blizzard of Questions—Pardons and commutations issued by Pres. Bill Clinton just before he left office continued to prompt questions. Marc Rich, a commodities trader living in Switzerland, had left the United States in 1983 with his business partner Pincus Green to avoid trial on charges of racketeering, violating the U.S. embargo on Iran, and evading $48 million in taxes. Both fugitives renounced their U.S. citizenship. Both received a pardon from Clinton. Rich's ex-wife, Denise, refused **Feb. 8** to testify before the House Committee on Government Reform on grounds of possible self-incrimination. She had made substantial contributions and pledges to the Democrats and to Clinton's presidential library, and had appealed directly to Clinton to pardon her ex-husband. Mary Jo White, the U.S. attorney in New York, said **Feb. 15** that her office and the FBI had begun investigating the pardon of Rich. It was reported the next day that Prime Min. Ehud Barak of Israel had asked Clinton by phone to pardon Rich. In an article in the *New York Times*, **Feb. 18**, Clinton sought to defend the pardon, denying there had been a quid pro quo. He wrote that 3 prominent Republican attorneys who had represented Rich had advocated the pardon, though all 3 quickly denied involvement. He also claimed that prosecutors should have brought civil, not criminal, charges against Rich.

Bill Clinton issued a statement **Feb. 21** saying that he and his wife, Sen. Hillary Rodham Clinton, had not been aware that her brother, Hugh Rodham, had received nearly $400,000 in payments in return for lobbying for a pardon and a prison commutation, respectively, for 2 convicted felons. The 2, Almon Braswell, a businessman convicted of mail fraud and perjury, and Carlos Vignali, a cocaine trafficker, had both received what they had sought. After the story broke, Rodham said he would return the money. Investigations into the pardons expanded **Feb. 21** after the revelation that Bill Clinton's half-brother Roger—himself the beneficiary of a pardon—had sought clemency for 6 people.

Other Controversies Follow Clintons From White House—Former Pres. Clinton said **Feb. 2** that he and Sen. Hillary Rodham Clinton would return $86,000 worth of gifts they received in 2000. However, Clinton said they would keep $104,000 in gifts received prior to 2000. Another controversy involved Clinton's choice of a penthouse in midtown Manhattan for his office, which would cost $650,000 or more a year, a record for a former president; he announced **Feb. 2** that his Presidential Library Foundation would pay about half of the rent. Next, a dispute flared over furnishings taken from the White House by the Clintons. Records of the National Park Service showed that at least $28,000 worth of furniture and accessories were donations to the permanent collection of the White House. On **Feb. 7** the Clintons returned the gifts shown in the NPS records, and some other gifts. Bill Clinton resolved the controversy over the cost of his office **Feb. 16**, by choosing instead to rent space in Harlem, a predominantly black neighborhood. The annual cost of the Harlem office space was put at $210,000.

Bush Sends His Tax Cut Proposal to Congress—As promised during his campaign, Pres. George W. Bush sent to Congress **Feb. 8** a proposal to lower federal taxes. The administration projected that the package would reduce revenues by $1.6 trillion over the next 10 years. Under existing law, the 5 tax brackets ranged from 15% to 39.6%. The new brackets would range from 10% to 33%—the highest for single taxpayers earning more than $136,000 per year and married taxpayers earning more than $166,500. If enacted into law, the proposal would double the child tax credit to $1,000 and eliminate the estate tax. The so-called marriage penalty would be reduced. Posing with a $46,000 Lexus and a $227 muffler, Senate Minority Leader Tom Daschle (D, SD) said that under the plan a millionaire could buy the car while the "typical working person" could afford only the muffler. Republicans argued that the tax cut benefitted all Americans, but with those who paid the most taxes receiving the most tax relief.

The tax cut was reflected in the $1.96 trillion budget for the 2002 fiscal year that the administration submitted to Congress **Feb. 28**. Surpluses for the next 10 years were projected at $5.6 trillion. Bush's plan included about $150 billion for prescription drugs for the elderly and increased defense spending by $14.2 billion to $310.5 billion. On the other hand, some executive departments, including Transportation, Agriculture, Interior, and Energy, would see their budgets cut. The overall budget contained a 4% increase in spending.

U.S. Spacecraft Lands on an Asteroid—A NASA spacecraft landed on the asteroid Eros, 196 million miles from Earth, **Feb. 12**. The 1,100-pound craft, the NEAR Shoemaker, had orbited Eros for a year, and sent back 160,000 photographs and other data. Eros is 21 miles long, 8 miles wide, and 8 miles thick. One purpose of the Eros project was to gain information on how earthlings might protect themselves from a collision with an asteroid, such as those that had created havoc when they impacted the planet in distant times.

FBI Agent Arrested As Spy For Russians—A senior FBI agent was arrested **Feb. 18** in a park near his Vienna, VA, home and charged 2 days later with espionage. A court affidavit filed **Feb. 20** asserted that the agent, Robert Hanssen, a 25-year veteran of the bureau, had begun giving information to the Soviet Union in 1985 and had continued to spy for Russia after the Soviet regime collapsed. Hanssen was arrested, officials said, after dropping off some classified documents in a park. At the time, they said, a bag with $50,000 was waiting for him nearby. The affidavit asserted that Hanssen had written a letter to an official in the Soviet Embassy in Washington, DC, in 1985 volunteering to spy. He provided the names of 3 Soviet agents who had been spying for the United States. The Soviet Union later executed 2 of the agents and imprisoned the 3d. FBI Director Louis Freeh said **Feb. 20** that he believed the damage caused by Hanssen "was exceptionally grave." FBI officials said Hanssen had been paid $1.4 million for his services.

Bush Speaks to Nation From House Chamber—Pres. Bush spoke to a joint session of Congress in the House chamber, **Feb. 27**, and to a national audience on television. In his first major address as president, Bush strongly advocated his 10-year $1.6 trillion tax cut. Bush contended that budget surpluses would permit this cut plus additional funding of his education reforms and a prescription drug benefit for elderly low-income Americans. Bush cited debt reduction and strengthening of Social Security and Medicare as other major priorities. Sen. Tom Daschle (SD) and Rep. Dick Gephardt (MO.), in the Democratic response, contended that the nation could not afford Bush's tax cut; they advocated a cut of about half as much, and weighted less toward the wealthy.

Quake in Seattle Area Injures 250—A 6.8 magnitude earthquake struck the Pacific Northwest, **Feb. 28**, injuring 250 people and causing at least $1 billion in damage. Since it occurred 30 miles underground, it was less destructive than recent heavy shocks in California. In Seattle, buildings swayed and some collapsed, and pavement buckled. The tremor occurred just as Mayor Paul Schell was about to begin a news conference to discuss damage caused by vandals during a Mardi Gras celebration the previous evening. Police had used tear gas and rubber bullets to disperse a crowd; 70 people were injured.

U.S. Economy at a Glance: February 2001	
Unemployment rate	4.2%
Consumer prices (change over Jan.)	+0.3%
Producer prices (change over Jan.)	+0.1%
Trade deficit	$26.99 bill
Dow Jones high (Feb. 1)	10,983.63
Dow Jones low (Feb. 23)	10,441.90
Index of leading economic indicators (change over Jan.)	-0.2%

International

Indonesian Parliament Rebukes President—The Parliament of Indonesia **Feb. 1** censured Pres. Abdurrahman Wahid for allegedly lying and abusing his powers. By a 393–4 vote, it accepted a commission report that concluded Wahid had probably been involved in embezzling $3.7 million from a food distribution agency and had testified falsely about his personal misuse of a $2 million gift from the Sultan of Brunei.

New Congo President Visits Washington—Pres. Joseph Kabila of the Democratic Republic of the Congo (formerly Zaire), who had succeeded his assassinated father in January, met **Feb. 1** in Washington, DC, with Sec. of State Colin Powell. Kabila said he wanted to reinstitute the cease-fire agreement, the Lusaka accords, that would permit foreign troops to leave Congolese soil. Kabila also met the same day with Pres. Paul Kagame of Rwanda, who had troops in Congo that opposed his father's government. On **Feb. 2**, Kabila told the UN Security Council that he would seek to end Congo's civil war, but that first the "armies of aggression" from Burundi, Rwanda, and Uganda must leave his country. At a meeting with rebel leaders in neighboring Zambia, **Feb. 15**, Kabila agreed to open talks with them. The Security Council, **Feb. 22**, adopted a resolution establishing a pullback and withdrawal of all foreign and rebel troops. On **Feb. 28**, soldiers from Rwanda and Uganda began withdrawing from front-line positions.

Ariel Sharon Elected Prime Minister of Israel—Ariel Sharon, a controversial public figure for decades, won a landslide victory **Feb. 6** in Israel's election for prime minister. The leader of the Likud party, Sharon defeated the Labor party's incumbent prime minister, Ehud Barak, 63% to 37%. Sharon had won an intense following during his successful military career, during which he had repeatedly helped Israel prevail in confrontations with the country's Arab neighbors. A commission of inquiry found him indirectly responsible for the 1982 massacre by the Lebanese Christian Phalangist militia of hundreds of Palestinians in 2 refugee camps. Sharon went on to serve in 6 cabinet positions, including foreign minister under Prime Min. Benjamin Netanyahu. When the latter lost to Barak in 1999 and resigned as Likud's leader, Sharon succeeded him. Barak, in turn, failed to reach a peace settlement with the Palestinians, leading to the worst violence in years. Some charged that Sharon had set off the lethal clashes, now known as the Al-Aqsa Intifada, when he visited the Temple Mount (known as the Al-Aqsa Mosque to Palestinians), in Jerusalem in October 2000.

On **Feb. 9**, Sharon, seeking to form a coalition government, asked Barak to become defense minister, and asked former Prime Min. Shimon Peres to serve as foreign minister. Barak **Feb. 20** rejected the invitation and resigned as Labor Party leader and from his Knesset seat. On **Feb. 26**, the Labor Party's governing body voted to join a unity government.

Colombian President Meets With Rebel Leader—The president of Colombia met with the leader of the nation's largest rebel group **Feb. 8 - 9** and agreed to resume peace talks. Manuel Marulanda, leader of the leftist Revolutionary Armed Forces of Colombia (FARC), the largest rebel group, had been fighting the government since the 1960s. The prospect of all-out war within Colombia had been growing, and the violent activities of right-wing paramilitary groups and the drug cartels had added to the chaos. After the meeting, Pres. Andres Pastrana declared that the peace process had been revived.

9 Die as Surfacing U.S. Sub Hits Fishing Boat—The U.S. Navy submarine *USS Greenville* collided with a Japanese fishing boat **Feb. 9** as the sub surfaced 9 miles from Pearl Harbor in Hawaii. The *Ehime Maru*, which carried 20 crew members, 13 high school students, and 2 teachers, sank. Although 26 people were rescued by the Coast Guard, 4 students, the 2 teachers, and 3 crew members were killed. The Navy said **Feb. 10** that the sub had surfaced quickly as part of a drill, and that 16 civilians were on board. On **Feb. 11**, Prime Min. Yoshiri Mori officially protested the accident to U.S. Ambassador Thomas Foley. The Navy said **Feb. 13** that civilians, under supervision of sailors, had been at 2 of the controls as the sub surfaced. Officials said **Feb. 14** that the civilians on the sub consisted mostly of donors to the *USS Missouri* Memorial Assn. The Japanese had surrendered on the battleship *Missouri* in 1945.

Israeli–Palestinian Violence Flares Again—The Israeli election did little to change the climate of violence in the Middle East. On **Feb. 13**, 2 Israeli helicopter gunships fired 4 missiles at the car of a Palestinian official, Massoud Ayyad, whom the Israelis regarded as a terrorist, killing him. On **Feb. 14**, a Palestinian bus driver drove into a crowd of soldiers and commuters at a bus stop in Azur, killing 7 soldiers and a civilian and injuring 20 others.

Pres. Bush Visits Mexico—Pres. Bush left the United States **Feb. 16** for the first time since he was inaugurated, visiting Mexico. He met with newly-inaugurated Pres. Vicente Fox in San Cristobal de los Ranchos; the 2 leaders dealt with differences relating to trade, immigration, and drug trafficking. Bush indicated that he was sympathetic to Mexico's displeasure at having to annually demonstrate its committment to anti-drug efforts in order to avoid economic sanctions.

U.S., British Planes Attack Iraqi Radar Sites—More than 3 dozen U.S. and British aircraft struck at Iraqi radar installations and antiaircraft sites near Baghdad, **Feb. 16**. The objective was to protect allied planes patrolling the southern no-fly zone. In January, the Iraqis had fired more surface-to-air missiles at allied patrols than during all of 2000. Pres. Bush, on a visit to Mexico, said **Feb. 16** that the air strikes were routine. At his first press conference as president, Bush said **Feb. 22** that he was concerned about evidence of Chinese assistance in the building of the Iraqi radar systems.

Hague Tribunal Convicts 3 of Rape—Three former Bosnian Serb soldiers were found guilty **Feb. 22** of raping and torturing Muslim women and girls during the Balkan conflicts of the 1990s. The decision was handed down by a 3-judge panel of the international tribunal at The Hague, The Netherlands, that was trying accused war criminals. The verdict, for the first time, defined rape as a crime against humanity. Two of the 3 men were also convicted of sexual slavery—they had kept women and girls, some as young as 12, for up to 8 months—the first time an international court had condemned this particular crime. The crimes occurred after Serbs seized the Bosnian town of Foca in 1992. Hundreds of women in the vicinity were thought to have been raped; 16 of them testified against the accused men at their trial. The convicted men received sentences of 12 to 28 years in prison. On **Feb. 26**, the tribunal sentenced Dario Kordic, an ally of the late Pres. Franjo Tudjman of Croatia, and Mario Cerkez, a former brigade commander of Croatians in Bosnia, to 25 and 15 years in prison, respectively, for crimes against humanity. The charges related to the killings of hundreds of Muslims in central Bosnia in 1992 and 1993.

Sec. of State Powell on 1st Overseas Trip—Colin Powell arrived in the Middle East **Feb. 24** on his first trip overseas as secretary of state. He met separately in Cairo with Pres. Hosni Mubarak of Egypt and Russian Foreign Min.

Igor Ivanov. Both agreed with him that Iraqi ruler Saddam Hussein must not be allowed to build weapons of mass destruction, but Mubarak contended that allied sanctions were harming only the Iraqi people, not Hussein. Powell and Ivanov explored differences on U.S. plans to build a missile shield and on Russia's conduct of the war against Chechen rebels. Powell met with King Abdullah of Jordan in Amman, **Feb. 25**, then separately with Prime Min.-elect Ariel Sharon of Israel and Yasir Arafat, the Palestinian leader. Sharon and Arafat were adamant that the violence must end before peace talks could take place. On **Feb. 26**, Powell met with Pres. Bashar al-Assad of Syria and with the Saudi foreign minister, Prince Saud al-Faisal. In Kuwait, Powell, along with ex-Pres. George H. W. Bush and Gen. H. Norman Schwarzkopf, participated in a ceremony observing the 10th anniversary of the Gulf War that had liberated Kuwait from an Iraqi occupation.

General

Stock Car Racing's Top Driver Dies in Crash—Dale Earnhardt Sr., the most successful and popular driver in stock-car racing, died in a crash **Feb. 18** on the final turn of the final lap of the sport's most important race, the Daytona 500. Ahead of him, finishing 1st and 2d, were Michael Waltrip, a member of Earnhardt's racing stable, and Earnhardt's son, Dale Jr. More than 200,000 fans at the Daytona Beach (FL) track, and millions on TV, watched the tragic climax to the race. Earnhardt, 49, had won NASCAR's annual Winston Cup championship 7 times and earned $41 million. An aggressive driver with an intense personal following, he had been nicknamed the Intimidator.

MARCH 2001

National

Congress Overturns Clinton's Work Safety Rules—Workplace safety regulations issued by Pres. Bill Clinton before he left office were overturned by the Senate, 56-44, **Mar. 6**, and by the House, 223-206, **Mar. 7**. Pres. George W. Bush had indicated he would sign the bill. The regulations would have required employers to provide a work environment that would protect them from repetitive stress injuries. Opponents complained that the regulations were unworkable and too costly to business.

Bush Meets with World Leaders—Pres. Bush met with South Korean Pres. Kim Dae Jung Mar. 7 in Washington, DC. He told Kim that he would not soon press forward with efforts by former Pres. Bill Clinton to reach an agreement with North Korea on missiles and to normalize relations with the Communist state. Kim was eager for the talks to continue. Bush met in Washington **Mar. 20** with Prime Min. Ariel Sharon of Israel. He assured Sharon that the United States remained solidly supportive of Israel. The 2 leaders agreed that peace talks could not be pressed during the current climate of violence.

U.S. Stocks Have a Bad Week—One of the worst weeks in the history of Wall Street began **Mar. 12** with a 436.37-point decline in the Dow Jones industrial average, a 4.1% loss. The NASDAQ index, loaded with high-tech companies, fell 6.3% **Mar. 12**, for a cumulative decline of 61% from its March 2000 peak. By Friday, **Mar. 16**, all the major indexes were down more than 6% for the week, with the Dow recording its worst week since 1989. The Federal Reserve Board, **Mar. 20**, cut interest rates by 0.5%, stating that there was "substantial risk" that the economy would be weak for some time. The Dow fell 238.35 points **Mar. 20**.

Vice President Hospitalized with Chest Pains—Vice Pres. Dick Cheney was hospitalized **Mar. 5** with chest pains. Doctors at George Washington University Hospital opened a partly blocked artery that had also been opened in November. Cheney had had 4 heart attacks, the most recent one just after the 2000 election, but doctors reported that the latest incident did not actually involve a heart attack. Cheney left the hospital **Mar. 6**, and resumed his normal work schedule **Mar. 7**.

Bush Takes Positions on Environmental Issues—In a letter to 4 Republican senators **Mar. 13**, Pres. Bush said he would not act to regulate carbon dioxide emissions from power plants, since this might harm consumers with an energy crisis looming. The decision, hailed by industry, reversed a campaign pledge by Bush and appeared to undercut his new Environmental Protection Agency chief, former NJ Gov. Christie Todd Whitman.

It was the first in a series of Bush decisions criticized by environmental groups. On **Mar. 20**, Bush repealed a last-minute Clinton administration decision to reduce the amount of arsenic traces allowed in drinking water. Christie Whitman defended the move, arguing that more information was needed to determine the costs faced by small communities in implementing the new standard. On **Mar. 27**, Bush removed the U.S. from the Kyoto protocol, the 1997 treaty on global warming that called for the U.S. to cut emissions by one-third by 2012. European leaders criticized Bush's decision. On **Mar. 29**, Bush declared "I will explain as clearly as I can, today and every other chance I get, that we will not do anything that harms our economy."

Senate Votes Bankruptcy Law Overhaul—A sweeping overhaul of federal bankruptcy laws was passed by the Senate 83-15 **Mar. 15**. The bill enacted more stringent guidelines for consumer debt, that tightened provisions for canceling debts. Rules for small business debtors were also tightened. The Senate defeated Democrat-sponsored amendments to soften the legislation, including one that would have exempted anyone driven into bankruptcy by medical bills. Competing Senate and House versions would go before a conference committee.

U.S. Expels Russian Diplomats—On **Mar. 23**, Sec. of State Colin Powell announced that he had ordered the expulsion of more than 40 Russian diplomats, saying that they were intelligence officials working as diplomats. The move broadened the fallout from the recent arrest of FBI agent Robert Hanssen, charged with spying for the Soviet Union and Russia. Russia retaliated by expelling 4 U.S. diplomats from Moscow Mar. 23 and asking 46 more to leave by July.

Campaign Finance Reformers Gain Momentum—On **Mar. 27** the Senate defeated, 60-40, an amendment to the controversial McCain-Feingold campaign finance reform bill that would have diluted it to allow individuals, corporations, and labor unions to make $60,000 in donations to national political parties per year. The bill, sponsored by Sen. John McCain (R, AZ) and Russell Feingold (D, WI), sought to completely ban "soft money" contributions to parties, calling such contributions a corrupting influence on the political process. Opponents of the bill argued that banning contributions would violate the First Amendment's free speech protections. They also objected to a provision that would restrict corporations (including most advocacy groups) from running television ads within 60 days of an election. On **Mar. 28**, the Senate did approve a provision, 57-43, raising the "hard money" contribution that an individual can donate to a specific candidate for federal office from $1,000 to $2,000. The McCain-Feingold bill, as amended, passed the Senate, 59-41, **Apr. 2**. House action remained pending.

U.S. Economy at a Glance: Mar. 2001	
Unemployment rate	4.3%
Consumer prices (change over Feb.)	+0.1%
Producer prices (change over Feb.)	–0.1%
Trade deficit	$31.17 bill
Dow Jones high (Mar. 8)	10,858.25
Dow Jones low (Mar. 22)	9,389.48
1st Quarter GDP (at annual rate)	+1.2%
Index of leading economic indicators (change over Feb.)	-0.3%

International

Sharon Completes His Cabinet—Prime Min.-elect Ariel Sharon completed construction of his cabinet in Israel. The rival Labor Party, now a partner in the prospective coalition government, **Mar. 2** chose Benjamin Ben-Eliezer, regarded as a hawk on Middle East issues, to be the defense minister. By **Mar. 5**, Sharon had filled all the slots in his government. As he took office, **Mar. 7**, Sharon pledged to bring security to Israel. The Israeli Knesset (parliament) voted that day to abandon the direct election of prime ministers, after only a 5-year experiment, and return to the parliamentary system in which Knesset members would choose the prime minister.

Foot-and-Mouth Disease Alarms Europe—Two farms in France and Belgium were quarantined **Mar. 3** because it was feared that they harbored foot-and-mouth disease. The disease, dangerous to animals while only rarely affecting humans, had already flared in Britain, where animals were slaughtered, parks and zoos closed, and many public events canceled to prevent spread of the disease. The virus is easily transported on clothing, or even on the wind. The French government **Mar. 5** stopped meat exports and moved forward with the slaughter of 50,000 animals that had come from Great Britain or been in contact with animals from Britain.

On **Mar. 13**, the United States banned imports of animals and animal products from all 15 countries in the European Union. Canada also did so, and added Argentina, which reported the disease, as well. Several European countries banned French meat. Prime Min. Tony Blair of Britain **Mar. 14** ordered a step-up in the slaughter of animals; some 180,000 pigs, sheep, and cows had already been killed during the 3 weeks since the outbreak had been discovered in Britain. On **Mar. 21**, after confirming a case of the disease, the Dutch government banned all livestock exports. After a case was confirmed in Ireland, **Mar. 22**, Irish authorities began slaughtering animals in the vicinity.

Albanians Are Center of New Balkan Tensions—Three Macedonian soldiers were killed **Mar. 4** during fights with ethnic Albanian rebels, the latest violent flare-up in the Balkans. Emboldened by the fall of the Milosevic regime in Serbia and the 1999 NATO intervention on behalf of the Kosovar Albanians, rebels sought to carve territory out of Macedonia for inclusion in a Greater Albania, which would include Kosovo and Albania. On **Mar. 9**, the Albanians killed a policeman and trapped a convoy that included 2 senior government leaders. That incident occurred a few miles from where U.S. peacekeeping troops had just forced some of the rebels from a town on the Macedonia-Kosovo border. NATO and Yugoslavia agreed **Mar. 12** to allow Yugoslav troops to return to a small part of the area along that border.

On **Mar. 22**, the rebels declared a cease-fire and reportedly began to withdraw from villages held in Macedonia. They asked for negotiations with the Macedonian government, which had rejected such overtures. On **Mar. 23**, Macedonia used attack helicopters against the rebels for the first time. Macedonians launched a ground offensive **Mar. 25**.

Navy Conducts Inquiry on Submarine Accident—The U.S. Navy **Mar. 5** opened an investigation into the collision between one of its submarines, the U.S.S. *Greeneville*, and a Japanese fishing trawler that killed 9 people on the trawler. Petty Officer First Class Patrick Seacrest testified, **Mar. 19**, that the sub's sonars had showed the trawler, the *Ehime Maru*, was nearby, but that he did not notify the ship's captain of a potential problem. Testifying **Mar. 20**, Cmdr. Scott Waddle, the captain, disregarding advice from his lawyers, accepted full responsibility for the accident, though adding that some of his subordinates had contributed to it by failing to follow standard procedures. Waddle was reprimanded and ordered to resign **Apr. 23**.

Bombing Accident Kills 6 in Kuwait—An errant bomb from a U.S. Navy fighter plane over Kuwait exploded **Mar. 12** at an observation post, killing 5 Americans—4 soldiers and an airman—and a New Zealand army major. Three Americans were seriously wounded and several Kuwaiti military personnel were also hurt.

Ugandan President Reelected—Yoweri Museveni was reelected president of Uganda, **Mar. 14**, with 69% of the vote. Coming to power in 1986, he had led Uganda away from civil war and toward peace and economic growth. His leading opponent, Kizza Besigte, charged vote fraud.

Oil Rig Collapses Into Atlantic—After workers fought a fierce 5-day battle to keep it afloat, the largest oil rig in the world collapsed off the coast of Brazil **Mar. 20**, sinking into the Atlantic Ocean. Three explosions **Mar. 15** had destroyed one of the structural support beams, causing the rig to list severely. Up to 395,000 gallons of diesel oil were on board the rig when it went down. Several workers died.

Chechens Hijack Plane—A Russian airplane en route from Istanbul to Moscow was hijacked by 3 Chechen terrorists **Mar. 15** and rerouted the plane to Medina, Saudi Arabia. The hijackers sought to have all Russian troops removed from the breakaway Russian province of Chechnya. After nearly 24 hours of fruitless negotiations, a Saudi security team successfully stormed the plane **Mar. 16**. Three people—a hijacker, a Russian flight attendant, and one Turkish passenger—were killed in the fighting.

Israeli–Palestinian Violence Continues—A suicide bomber killed 2 Israeli teenagers **Mar. 28** as they waited for a school bus in Jerusalem, continuing the cycle of violence that had plagued Israel since the collapse of peace talks in Sept. 2000. In immediate retaliation, Israeli helicopter gunships attacked the bases and camps of the personal security forces of Palestinian leader Yasir Arafat. On **Mar. 30**, Palestinians engaged in mass demonstrations against the Israeli air attacks, while 6 Palestinians were killed and 100 wounded in new clashes.

Serbian Forces Attempt to Arrest Milosevic—Before dawn on **Mar. 31**, Serbian police and security forces attempted to arrest former Pres. Slobodan Milosevic at his home in Belgrade. The arrest, for allegations of corruption while in office, was unrelated to his indictment for war crimes by an international tribunal. His supporters, fearing an attempt to arrest him, had massed at his compound. At 2 A.M., 100 or more police officers appeared at the compound, joined at 2:30 A.M. by special forces units. For a while, a Yugoslav army unit also barred access to the compound. Shots were exchanged by the authorities and by bodyguards and armed supporters of Milosevic in the compound. Two police officers were wounded. The standoff continued through the day, but Milosevic was peacefully taken into custody on **Apr. 1**.

General

School Shootings Shock Communities—15-year-old Charles Andrew Williams was arrested **Mar. 5** after having allegedly shot 15 people, 2 fatally, at a high school in Santee, CA. Both of those killed were students, and the wounded included 11 students, a security guard, and a student teacher. In Williamsport, PA, **Mar. 7**, a girl was arrested after allegedly shooting and wounding a classmate. Six miles from Santee, in El Cajon, CA, **Mar. 22**, a teen shot and wounded 5 people. The alleged assailant, Jason Hoffman, 18, was shot by police and seriously wounded.

Kwan Wins 4th World Championship—American Michelle Kwan **Mar. 24** captured her 4th career gold medal at the World Figure Skating Championships in Vancouver, Canada. Kwan won the silver. Sarah Hughes of the U.S. won the bronze.

***Gladiator* Wins Oscar as Best Picture**—*Gladiator*, a high-tech action film, was named 2000's best picture **Mar. 25** at the annual awards ceremony of the Academy of Motion Picture Arts and Sciences. Russell Crowe, who portrayed a Roman general who became a slave and then a gladiator, was named best actor. Julia Roberts was voted best actress for her portrayal, in *Erin Brockovich,* of a legal aide who takes on a utility company. *Traffic*, a story of the war on drugs, won Oscars for best director (Steven Soderbergh) and best supporting actor (Benicio Del Toro). Marcia Gay Harding was named best supporting actress for her portrayal of the artist Lee Krasner in *Pollack. Crouching Tiger, Hidden Dragon*, a magical martial-arts adventure from Taiwan, won 4 Oscars, including best foreign-language film. The ceremony was hosted by comedian, actor, and novelist Steve Martin.

APRIL 2001

National

Senate Approves Campaign Finance Bill—The Senate **Apr. 2** approved, 59-41, a bill that would ban the large unregulated political contributions known as soft money. Passage had seemed certain after supporters beat back several amendments in March. During the last election cycle nearly $500 million in soft money had poured into the political parties. All but 3 Democrats supported the bill, which also drew the support of 12 Republicans. The victory capped a long effort by Sen. John McCain (R, AZ), who had made campaign reform the centerpiece of his bid for the Republican presidential nomination in 2000. House action was pending.

Senate Trims Bush Tax Cut—The Senate **Apr. 6** approved a $1.2 trillion tax cut over 10 years, cut down somewhat from the $1.6 trillion that the House had approved and Pres. George W. Bush had advocated. The 65-35 vote reflected support from all Republicans and from 15 Democrats. The Democratic leadership had favored a cut in the $800 billion range, and had argued that too much of the money was going to the wealthiest taxpayers. Only one Democrat in the evenly divided Senate had supported the proposed $1.6 trillion figure, while 2 Republicans had opposed it. On **Apr. 9**, Bush unveiled his 2002 fiscal year budget, the outlines of which had been announced in February. He stuck with his commitment to a $1.6 trillion tax cut.

Algerian Convicted of Plot to Bomb U.S. Sites—A federal jury in Los Angeles **Apr. 6** found an Algerian guilty of 9 charges that included transporting explosives. Authorities said the explosives were to have been used during the millennium celebration, which occurred soon after the defendant, Ahmed Ressam, was arrested at the Canada-U.S. border in Washington on Dec. 14, 1999. Authorities found materials in his car that could have been used to construct bombs.

Protests Jolt Cincinnati After Police Shoot Teen—Violent demonstrations flared in Cincinnati after a white policeman shot and killed an unarmed black youth, Timothy Thomas, 19, on **Apr. 7**. Thomas had fled from police on foot after they tried to arrest him. There had been 14 outstanding warrants for Thomas's arrest, mostly for minor misdemeanor charges. The police said that the officer, Stephen Roach, thought Thomas was about to draw a weapon; however, none was found on the youth. Thomas was the 4th black suspect shot and killed in the city since November. As protests and looting erupted, more than 40 people sought treatment at hospitals and more than 100 arrests were made. Mayor Charles Luken, **Apr. 12**, imposed an 8 P.M.– 6 A.M. curfew. Luken lifted the curfew **Apr. 16**. On **May 7**, Roach was charged with 2 misdemeanors—negligent homicide and obstructing official business. On **Sept. 26**, a municipal court judge acquitted him, calling his actions "reasonable" in a "very dangerous situation."

TV Role in Oklahoma Bomber's Execution Debated—On **Apr. 12** Atty. Gen. John Ashcroft responded to requests by relatives of the 1995 Oklahoma City bombing victims to witness the execution of the convicted mastermind, Timothy McVeigh, set for **May 16.** Under Ashcroft's plan, 10 relatives and survivors would be able to attend the execution at the prison in Terre Haute, IN; other relatives of the bombing victims would be allowed to watch it on closed-circuit television. No recording of the TV images would be allowed. McVeigh, who had waived all appeals of his conviction and death sentence, had requested that his execution be shown on television. Federal regulations forbid any recording of federal executions.

Mississippi Keeps Confederate Symbol in Flag—Voters in Mississippi, by a 2-1 margin, decided **Apr. 17** to stick with their state flag which includes the Confederate battle cross in the upper left-hand corner. The alternative design that was rejected featured 20 white stars in a circle, symbolizing Mississippi's entrance into the union as the 20th state.

Interest Rate Cuts Boost Stocks—The Federal Reserve Board cut key short-term interest rates **Apr. 18**, and gave an immediate lift to stock values. The cuts were the 4th during 2001 and, unlike the previous ones, had not been anticipated. In its actions, the Fed dropped its federal funds rate on overnight loans between banks from 5% to 4.5%, and reduced the discount rate on loans to banks from the Federal Reserve system from 4.5% to 4%. The Fed acted on evidence that companies were scaling back their operations and laying off workers at an accelerated pace as the economy weakened. The Dow Jones industrial average closed up 399.10 on **Apr. 17**, finishing at 10,615.83, and the tech-heavy Nasdaq index closed up 156.22 to finish at 2,079.44.

Ex-Sen. Kerrey Discloses Involvement in Civilian Deaths—In a speech **Apr. 18** and in subsequent interviews, ex-Sen. Bob Kerrey (D, NE) disclosed that at least 13 civilians were killed by members of his U.S. Navy unit in a Vietnamese village in 1969. The former Nebraska senator and presidential contender, who was currently president of the New School University in New York City, revealed the Vietnam War incident as the *New York Times* prepared to publish an account of what reportedly happened.

In 1969, Lt. Kerrey was in command of a 6-man SEAL unit seeking a Viet Cong leader in the village of Thanh Phong. Thinking they were being fired upon, Kerrey said he ordered his men to fire on the village huts. But in the huts, the Americans subsequently found only the unarmed bodies of old men, women, and children. In the *Times* Sunday Magazine article, published **Apr. 29**, Gerhard Klann, a member of Kerrey's unit, claimed that the villagers had been rounded up and that Kerrey had ordered them shot so that they would not summon Viet Cong soldiers. The other 4 members of Kerrey's unit rejected Klann's account.

Kerrey received the bronze star for the Thanh Phong raid. He later received the medal of honor after he lost part of a leg in another battle.

U.S. Economy at a Glance: April 2001	
Unemployment rate	4.5%
Consumer prices (change over Mar.)	+0.3%
Producer prices (change over Mar.)	+0.3
Trade deficit	$32.0 bill
Dow Jones high (Apr. 27)	10,810.05
Dow Jones low (Apr. 3)	9,485.71
Index of leading economic indicators (change over Mar.)	+0.2%

International

China Holds U.S. Crew 11 Days After Midair Collision—A U.S. Navy spy plane collided with a Chinese fighter plane **Apr. 1**. The Chinese jet plunged into the South China Sea, and the U.S. plane, with a wing and an engine damaged, landed on the Chinese island of Hainan, where the 24 crew members were detained for 11 days. The fighter was one of 2 that had been flying close to the slow-moving U.S. propeller plane before the collision. U.S. officials said the incident had occurred 50 miles southeast of the island over international waters, and requested the return of the crew and the plane, which had advanced intelligence-gathering equipment. The Chinese claimed the collision had occurred in their air space. The Chinese pilot, Wang Wei, reportedly parachuted from his falling plane; he remained missing and presumed dead.

Pres. George W. Bush, **Apr. 2**, demanded the return of the crew and plane, saying that relations between the 2 countries were at stake. On **Apr. 3**, China blamed the United States for the collision, and called for an apology as an apparent precondition for the release of the crew. U.S. Sec. of State Colin Powell **Apr. 4** issued a statement expressing regret over the loss of the Chinese pilot, and he wrote Chinese officials proposing a settlement. However, Pres. Jiang Zemin **Apr. 4** insisted that China receive an official apology. Bush **Apr. 5** said he regretted the loss of the pilot and plane, and the next day, U.S. officials met with the crew. China said **Apr. 7** that the U.S. response so far was unacceptable.

The American ambassador to China, Joseph Prueher, presented a letter to Foreign Minister Tang Jiaxuan **Apr. 11** stating that the United States was "very sorry" that the Chinese pilot had died and that the U.S. plane had landed in Chinese territory without permission. That language was sufficient to end the impasse. On **Apr. 12**, the crew members returned to U.S. soil, landing at Hickam Air Force Base in Hawaii, for debriefing and medical checkups. Bush said **Apr. 12** that the U.S. plane did not cause the collision, and administration officials said that reconnaissance flights near Chinese territory would soon resume, Chinese protests notwithstanding. Defense Sec. Donald Rumsfeld said, **Apr. 13**, that the crew had destroyed much secret data in the 15 minutes between the plane's landing in China and the crew's surrender to the Chinese.The same day, the Pentagon released a tape made in January showing a Chinese plane, possibly the same one involved in the collision, flying very close to a U.S. spy plane. The crew—21 men and 3 women—received a warm welcome **Apr. 14** when they returned to their home base at Whidby Island Naval Air Station, WA.

The incident had triggered speculation that the United States would firm up its support for Taiwan. Bush unveiled a list of weapons, **Apr. 23,** that would be sold to Taiwan, including submarines and destroyers but not the sophisticated Aegis destroyers sought by the Taiwanese. Bush said, **Apr. 25**, that if Taiwan was attacked by China the United States would do "whatever it takes" to aid its defense. The statement appeared to depart from the customary ambiguity regarding the U.S. military commitment. However, State Dept. officials insisted there was no change in policy .

Ex-Yugoslav President Imprisoned in Belgrade—Ex-Pres. Slobodan Milosevic of Yugoslavia was arrested **Apr. 1** and imprisoned in Belgrade. His 10-year rule had ended with defeat at the polls in September and a popular uprising in October. Since then he had continued to live at the official presidential residence. He had been indicted for crimes against humanity by a war crimes tribunal in The Hague, the Netherlands, and Yugoslav officials had prepared unrelated charges of corruption. A 2-day standoff between Milosevic, with about 20 armed bodyguards and supporters, and police and security forces ended before dawn, after Milosevic received assurances that his arrest was not a prelude to his transfer to The Hague. Nonetheless, he became distraught, waving a pistol and threatening suicide, before surrendering. After appearing before the investigating judge, **Apr. 1**, Milosevic was ordered detained for 30 days.

The Bush administration had been urging Milosevic's arrest, and on **Apr. 2** Sec. of State Colin Powell, certifying that Yugoslavia was cooperating with The Hague tribunal, released economic aid to the battered Yugoslavs.

Israeli-Palestinian Talks Reopen—On **Apr. 1**, high-level talks between Israelis and Palestinians took place for the first time since Ariel Sharon became prime minister, as Sharon's son Omri met secretly with Palestinian leader, Yasir Arafat. Foreign Min. Shimon Peres began discussions **Apr. 4** in Athens with senior Palestinian officials. Israel, responding, it said, to mortar fire on a Jewish settlement, attacked a Palestinian refugee camp in the Gaza Strip, **Apr. 11**. Tanks and helicopter gunships led the assault on the Khan Yunis camp, and an ensuing gun battle left 2 Palestinians dead and 27 injured. Jordan's foreign minister, Abdallah al-Khatib, came to Jerusalem **Apr. 16** with a peace proposal, but his visit was overshadowed by an Israeli air strike **Apr. 15-16** on a Syrian radar installation in Lebanon that killed 3 soldiers. Israel said it was responding to recent incidents, the killing of 3 Israeli soldiers and the kidnapping of 3 others by Syrian-backed Hezbollah guerrillas in Lebanon.

In turmoil in the West Bank and Gaza, **Apr. 30**, 8 Palestinians died, several by bombs that they may have been constructing.

Former President of Philippines Indicted, Jailed—Joseph Estrada, former president of the Philippines, who had been ousted from office in January as a result of a popular uprising, was indicted **Apr. 4**. The government ombudsman charged that he had accepted $82 million in bribes and kickbacks. His wife and a son were also indicted. Although angry supporters demonstrated in his favor, Estrada was arrested Apr. 25, specifically on a charge of plunder, and imprisoned in the National Police compound in Manila. Estrada's supporters continued their protests, and on the night of **Apr. 30-May 1** they stormed the presidential palace, seeking to oust Pres. Gloria Macapagal Arroyo. Four people, including 2 policemen, were killed and scores were injured.

Protesters Sound Off at Summit of the Americas—The opening session of the Summit of the Americas was delayed **Apr. 20** as protesters massed in the streets of Quebec city. Canadian police had battled thousands of people for hours, trying to clear the way for the 34 presidents and prime ministers of Western Hemisphere nations who were attending. (Cuban Pres. Fidel Castro was not invited.) The demonstrators, including labor union members, environmentalists, and human rights advocates, were protesting the proposed Free Trade Area of the Americas, prime agenda item for the summit. Bush, speaking at the summit **Apr. 21** while 20,000 demonstrators filled the streets, said that after nearly a decade of delay it was time to create a free trade area for the entire hemisphere, while "protecting our environment and improving labor standards." The leaders **Apr. 22** reaffirmed their commitment to a hemispheric free-trade area.

U.S. Missionary, Daughter Die in Attack on Plane—A Peruvian air force fighter jet crew, believing it was intercepting contraband drugs, fired at a plane carrying a missionary family, **Apr. 20**, killing Veronica Bowers and her infant daughter Charity. Jim Bowers and his son Cory, 6, were unhurt. The pilot, Kevin Donaldson, though wounded, crash-landed the plane in the Amazon River. Villagers in dugout canoes rescued the survivors. The Bowers were missionaries for the Assn. of Baptists for World Evangelism.

A U.S. State Dept. official said **Apr. 21** that an unarmed U.S. government tracking plane had been nearby and had provided informationa for the Peruvians. The United States had been utilizing spotter planes to alert Peru interceptors to suspected drug runners. U.S. officials said **Apr. 22** that a Peruvian officer on the U.S. plane had called in the Peruvian fighter to attack the Cessna, but that the Americans on the plane, who were Central Intelligence Agency contract employees, had sought to stop it.

Underdog Candidate Becomes Japan's Premier—The Japanese Diet, **Apr. 25,** elected Junichiro Koizumi, a former health and welfare minister, as the country's new premier, even though he was not the choice of many leaders of the dominant Liberal Democratic Party. Premier Yoshiro Mori, who was widely viewed as unsuccessful in dealing with Japan's growing economic and social problems, and whose public statements included embarrassing gaffes, said **Apr. 5** that he would step down. On **Apr. 22-23**, for the first time, the LDP held a national primary. Koizumi won overwhelmingly with 123 chapter votes. On **Apr. 24**, LDP members of the Diet added their votes to those totals, and Koizumi emerged with 298 of 487. He had vowed to shake up the party—"Change the LDP to Change Japan" was his slogan. Koizumi had long disparaged the way that party elders had traded favors and cut deals behind the scenes.

Maneuvers on Puerto Rican Island Stir Protests—The Puerto Rican island of Vieques once again became a site of violent protests as the U.S. Navy resumed maneuvers that included firing shells from warships. The bombing exercises, which dated from 1941, had provoked protests in 1999 after a shell killed a civilian guard and wounded 4 people. On **Apr. 26** a federal judge in Washington declined to issue a temporary restraining order against the new maneuvers. Residents contended that the maneuvers were endangering their health and the island's environment. Since then 600 people had been arrested. On **Apr. 27**, 65 of several hundred demonstrators were arrested, and the bombing was suspended until the target area was cleared. Bombing exercises resumed **May 1**.

U.S. Businessman Becomes First Tourist in Space—Dennis Tito, wealthy chief executive of Wilshire Associates, a financial consulting company based in California, became the first tourist in space **Apr. 28**. A former NASA engineer, Tito had reportedly offered Russia's financially strapped Aviation and Space Agency up to $20 million to take a ride to the International Space Station. NASA opposed the idea, but joined all of its 17 space-station partners **Apr. 24** in agreeing to let Tito fulfill his lifelong dream. Tito had agreed to pay for any damage he might cause and to commit himself and his heirs to refrain from claiming damages for injuries or death. In Kazakhstan, **Apr. 28**, a Russian booster rocket sent Tito and 2 Russian astronauts on their way to the space station, which they reached **Apr. 30**.

General

Duke Wins 3d College Basketball Title—Duke won its 3d men's college basketball title **Apr. 2** by defeating Arizona, 82-72, in Minneapolis. Ahead by only 2 points at the half, Duke was led by forward Mike Dunleavy, who scored 18 of his 21 points after the intermission. He hit 5 3-point shots. Duke's Shane Battier, who scored 18 in the championship game, was named the most valuable player of the final 4. In semifinal games **Mar. 31**, Duke had overcome a 22-point deficit to defeat Maryland, 95-84, and Arizona had defeated defending champion Michigan State, 80-61.

In a contest between 2 teams from Indiana for the women's NCAA title, Notre Dame defeated Purdue, 68-66, in St. Louis **Apr. 1**.

Tiger Woods Wins 4th Straight Major Golf Title—Tiger Woods chalked up another unique achievement in his golf career **Apr. 9**, when he won the Masters at Augusta, GA, his 4th consecutive victory in a major professional tournament. In 2000 he had won the U.S. Open, British Open, and PGA. In the Masters, Woods's score was 272, 16 under par, for a margin of 2 strokes over David Duval. Although his victories were not in the same calendar year, Woods was the first to hold all 4 major pro titles simultaneously.

MAY 2001

National

Ex-Klansman Convicted in Deaths of 4 Girls—A former member of the Ku Klux Klan was convicted in a state court in Birmingham, AL, **May 1** in the 1963 bombing of the 16th Street Baptist Church that killed 4 black girls ages 11 to 14. The defendant, Thomas Blanton, was one of 4 men whom the FBI had believed were implicated in the crime. One, Robert Chambliss, had been convicted in 1977 and died in prison. A 2d, Herman Cash, died without ever being charged. A 3d, Bobby Frank Cherry, had been indicted with Blanton in 2000, but his trial was postponed after Judge James Garrett ruled that he was mentally incompetent. A jury of 8 whites and 4 blacks returned a verdict in 2 hours. Judge Garrett sentenced Blanton to 4 life terms in prison.

Bush Proposes Missile Shield—Pres. George W. Bush **May 1** proposed that the United States build a shield against intercontinental nuclear missiles and at the same time sharply reduce its own nuclear arsenal. The plan would violate the 1972 U.S.-Soviet Antiballistic Missile Treaty, but Bush said it might be possible to negotiate changes with Russia without amending the pact. Defense Sec. Donald Rumsfeld said **May 1** that the United States would go ahead with the plan despite some uncertainty as to the technology's effectiveness. The defense system proposed by Pres. Bush would be less comprehensive than Pres. Ronald Reagan's Strategic Defense Initiative, better known as Star Wars.

FBI Finds Documents, McVeigh Execution Delayed—The execution of Timothy McVeigh, who had been convicted in the 1995 Oklahoma City bombing that killed 168 people, was delayed after the FBI reported it had found 3,135 pages of documents that had not been shown to McVeigh's lawyers. On **May 10**, 6 days before McVeigh's scheduled execution in Terre Haute, IN, the Justice Dept. reported that FBI archivists in Oklahoma City had found the documents, which included reports, correspondence, tapes, and photographs. McVeigh, convicted in 1997, had waived any further appeals.

On **May 11**, Atty. Gen. John Ashcroft reset the execution for **June 11**, noting it was "clear that the FBI had failed to comply fully" with its obligation to provide relevant documents to the defense. He ordered an internal investigation but said he doubted any of the documents would cast reasonable doubt on McVeigh's guilt. Lawyers for Terry Nichols, also convicted in the bombing, said **May 12** that they had filed an appeal with the U.S. Supreme Court. FBI Director Louis Freeh, testifying before a House committee **May 16**, acknowledged that the FBI had made a "serious error." Freeh, also under fire for other FBI problems, had announced his forthcoming retirement **May 1**. Ashcroft said **May 24** that 900 more relevant documents had been found but that he would not postpone the execution again. On **May 31**, McVeigh's lawyers asked Judge Richard Matsch, who had sentenced him, for a stay of execution.

New Interest-Rate Cuts Lift Stocks Again—The Federal Reserve Board cut key short-term interest rates on **May 15** for the 5th time since January. The Fed dropped its federal funds rate on overnight loans between banks from 4.5% to 4%. It reduced the discount rate on loans to banks from the Federal Reserve system from 4% to 3.5%. The move reflected concern about slight increases in unemployment and the fact that many companies were cutting investments in new plants and equipment as profits slipped. On **May 16**, in apparent response to the rate cuts and a less-than-expected rise in consumer prices, stock values soared. The Dow Jones industrial average rose 342.95 points to 11,215.92.

Teen Convicted of a Murder—A 14-year-old Florida boy was convicted of 2d-degree murder in Palm Beach, **May 16**, a year after he shot a teacher to death in his classroom. In May 2000 the teacher, Barry Grunow, had expelled the boy, Nathaniel Brazill, from school on the last day of classes for throwing water balloons. Brazill returned with a gun, asked Grunow if he could see 2 classmates and, when that request was denied, shot him. Brazill claimed at his trial that he had intended only to threaten the teacher. He was sentenced **July 27** to 20 years in prison without parole. In January, another 14-year-old Florida boy, Lionel Tate, was convicted of 1st-degree murder for beating a 6-year-old playmate to death.

Blame Assessed for Colorado School Massacre—A report by the 20-member Columbine Review Commission assessed blame, **May 17**, for the 1999 high school bloodbath that left 14 students and a teacher dead. The commission, appointed by Gov. Bill Owens (R), strongly criticized Jefferson County Sheriff John Stone, saying his office had failed to respond to evidence of the "suicidal and violent tendencies" of the 2 gunmen, Eric Harris and Dylan Klebold. The evidence cited included the discovery of a pipe bomb behind Harris's home (after which no search was made of the house), an essay Klebold had written describing such an assault, and a videotape in which the 2 held weapons and talked of killing. The report noted that gunmen's parents apparently did not notice that bombs were being stockpiled in their homes. It also criticized school officials for failing to act on evidence of antisocial behavior and students for maintaining a code of silence in the face of suspicious activity. The sheriff's department was rebuked because "no efforts were made to engage, contain or capture the perpetrators" inside the school prior to their suicides.

Bush Outlines Broad Effort to Meet Energy Needs—Pres. George W. Bush unveiled proposals to confront an array of energy-related problems. The price of gas for automotive vehicles had been surging, and the costs for providing heat, air conditioning, and other energy needs to homes and offices had been climbing as well. California was experiencing state-controlled blackouts because companies that supplied energy could not meet demand.

The overview of Bush's national energy policy, released **May 16**, said that "America . . . faces the most serious energy shortage since the oil embargoes of the 1970's." Noting that the United States produced far less oil than it did in 1970, Bush's report forecast that in 20 years the country would be importing almost two-thirds of its oil. The report said that no major oil refineries had been built in nearly a generation, and called for building up to 1,900 new electric plants. On the plus side, the report noted that appliances today were more energy efficient than in the past.

Bush recommended a sharp increase in domestic production of coal as well as oil and natural gas. His report said nuclear energy could become a more important part of the mix—but that a national depository for nuclear waste would need to be established. He reaffirmed his call for drilling in the Arctic National Wildlife Refuge. His proposals, some of which could be accomplished by executive order, also included income tax credits for buyers of cars using alternative fuels and a loosening of federal regulatory controls.

Critics argued that little money was proposed for research into clean-energy technology, and that no short-term relief was offered for high energy bills.

Firestone, in Tire Dispute, Stops Selling to Ford—The longtime business relationship between Firestone and Ford ended **May 21** when Bridgestone/Firestone Inc. announced that it would no longer sell tires to Ford. In August 2000, at Ford's request, Firestone had recalled 6.5 million tires used on Ford Explorer sport utility vehicles. Since then the 2 companies had squabbled over whether the tire-related fatalities (more than 100) and injuries in Explorer accidents were the fault of the vehicle's design or the tires supplied by Firestone. Bridgestone/Firestone's **May 21** decision came after

Ford prepared to release a report stating that remaining Firestone tires were presenting problems. General Motors and Nissan Motor said **May 21** that they would continue to buy tires from Firestone. Citing safety concerns, Ford further announced **May 22** that it would replace 13 million Firestone tires on Explorers and other vehicles.

Senator Leaves GOP, Puts Democrats in Power—Citing "fundamental differences" with his party, Sen. James Jeffords (VT) announced **May 24** that he was leaving the Republican party, under whose banner he had served in Congress for 26 years, to become an independent. The Senate had been divided 50-50 between the 2 major parties, but the Republicans held all the committee chairmanships because Vice Pres. Dick Cheney could cast a tie-breaking vote on Senate organization. Jeffords said he would vote with the Democrats to organize the Senate. In his statement to cheering supporters in Burlington, VT, Jeffords, whose voting record put him in the moderate-to-liberal range, cited differences with Bush on the budget and taxation, abortion, judicial appointments, missile defense, energy, the environment, and education.

Congress Gives Final OK to $1.35 Trillion Tax Cut—On **May 26** Congress approved a $1.35 trillion tax cut spread over 10 years. Early in May, Congress had approved in broad outline a budget that included a tax cut of this magnitude, slightly less that the president's $1.6 trillion proposal.

The final approval of the tax plan came just before a Memorial Day weekend break. On **May 24**, the Senate approved its version of the tax bill, 62-38, with 12 Democrats joining all 50 Republicans in support. A conference committee reached the final compromise **May 25**

The final bill provided that those filing individual tax returns for 2000 would get $300 rebates, while those filing joint returns would get $600 (provided that they earned at least $6,000 and $12,000, respectively, in taxable income). Every tax bracket except the 15% bracket would be lowered in stages until 2006—with the top bracket dropping from 39.6% to 35%. A new 10% rate was established for income up to $6,000 for individuals and $12,000 for couples. The size of estates subject to an estate tax would gradually rise, and the estate tax would be repealed altogether by 2010. The child credit would rise in stages from $500 to $1,000 by 2010. Contributions allowed to IRAs and 401(k)-type plans would rise. Taxes for couples would be adjusted downward to eliminate the so-called marriage penalty.

This version received final approval in the House, 240-154, and in the Senate, 58-33. Only 2 Republicans (both in the Senate) voted against it, while 31 Democrats voted in favor of the plan. Bush signed the bill into law **June 7**.

U.S. Economy at a Glance: May 2001	
Unemployment rate	4.4%
Consumer prices (change over Apr.)	+0.4%
Producer prices (change over Apr.)	+0.1
Trade deficit	$28.3 bill
Dow Jones high (May 21)	11,337.92
Dow Jones low (May 3)	10,796.65
Index of leading economic indicators (change over Apr.)	+0.4%

International

U.S., China Argue Over Damaged Spy Plane—The United States and China continued to negotiate over the fate of the damaged U.S. Navy spy plane that had made a forced landing on China's Hainan Island after colliding with a Chinese fighter plane over the South China Sea. On **May 7**, the United States resumed unarmed reconnaissance flights off the coast of China. China said, **May 8**, that it would not let the plane be flown home, but did not rule out allowing it to be dismantled and taken away in pieces. Pentagon officials said **May 15** that the crew had been unable to destroy all classified documents before the Chinese boarded the craft. The U.S. Defense Dept. announced **May 29** that China had agreed to allow the plane to be disassembled and flown back to the U.S. in pieces.

U.S. Voted Off UN Human Rights Commission—The United States **May 3** failed to win reelection to the UN Human Rights Commission. This meant the United States would be missing from the commission for the first time since the UN was founded under U.S. leadership in 1947. In a secret vote by the UN's Economic and Social Council to fill 3 seats for vacancies among Western nations, the tally was France 52, Austria 41, Sweden 32, and the United States 29. On **May 7**, the State Dept. reported that the United States had also lost its seat on another UN panel, the International Narcotics Control Board. The U.S. House voted, 252-165 on **May 10**, to withhold $244 million in dues until the United States was restored to the Human Rights Commission.

Killings Continue in Mideast—A 4-month-old girl became the youngest known victim of violence in the Middle East, **May 7**, when she was killed by Israeli troops firing into a refugee camp in Gaza. The Israelis were retaliating for a mortar attack on a Jewish settlement. On **May 9**, 2 Israeli boys, aged 13 and 14, were stoned to death by Palestinians while on a hike near their home in Tekoa, another settlement. During a month of sporadic killing on both sides, the Israelis **May 14** shot and killed 5 Palestinian guards at a roadblock near Beituniya in the West Bank. On **May 18**, 5 Israelis were killed and more than 100 were injured by a suicide bomber at a shopping mall in Netanya. Israeli air strikes on Nablus and Ramallah, later that day, killed 9 Palestinians. On **May 21**, Sec. of State Colin Powell announced that he was sending an envoy, William Burns, to the Middle East to try to re-establish peace talks.

Wealthy Conservative Regains Power in Italy—Silvio Berlusconi, who had held the premiership of Italy for 7 months in 1994, regained that job in parliamentary elections **May 13**. His center-right coalition prevailed over the center-left coalition led by Francisco Rutelli, a former mayor of Rome. Berlusconi had promised to cut back Italy's huge government bureaucracy and aid the average citizen through lower taxes, increased pensions, and public works projects. His own substantial fortune—he was the richest man in Italy—included television networks and banking, insurance, and real estate interests.

Dalai Lama Meets With Bush—The Dalai Lama, the exiled spiritual leader of Tibet, met with Pres. George W. Bush in the White House **May 23**. The meeting, on the 50th anniversary of China's annexation of Tibet, brought a protest from the Communist regime, which was also displeased by the visit of Taiwan's Pres. Chen Shui-bian to New York on **May 22**. The Dalai Lama told reporters **May 23** that he was not seeking independence for Tibet.

Secretary of State Visits Africa—Colin Powell, the first African-American secretary of state, began an African tour **May 23**. In Mali, he called Africa "a priority for the Bush administration." He met with Pres. Thabo Mbeki of South Africa **May 24**, and later said Mbeki was committed to reducing the rate of the viral infection that caused AIDS. Speaking in Johannesburg, South Africa, **May 25**, Powell urged Pres. Robert Mugabe of Zimbabwe to allow a free presidential election to be held. In Kenya **May 26**, Powell met with Pres. Daniel arap Moi, 76, in power since 1973, and asked him to step down. In both Kenya and Uganda, **May 27**, Powell urged neighboring Sudan to relax its civil war against Christian and other non-Islamic elements in the south. Sudan was promised delivery of 40,000 tons of food to fend off famine caused by a prolonged drought.

4 Are Guilty in African Bombings That Killed 224—Four men were found guilty in federal district court in New York **May 29** in connection with the 1998 bombings of the U.S. embassies in Kenya and Tanzania that killed 224 people. All were found guilty of conspiracy. Two, Mohamed Rashed Daoud al-'Owhali and Khalfan Khamis Mohamed, were also found guilty of murder. Both were later sentenced to life in prison. A 3d was found guilty of aiding and abetting murder, and a 4th of perjury. In presenting their case, prosecutors sought to link the men and their deeds to the international terrorist organization run by Osama bin Laden.

General

Pope Follows Route of the Apostle Paul—Pope John Paul II completed a journey retracing the footsteps of St. Paul through historic lands. He began in Greece, **May 4,** and in a major move toward reconciliation with the Greek Orthodox Church, he apologized for attacks on Constantinople by Crusader knights beginning in 1204. John Paul was making the first papal visit to Greece since the churches broke apart, and many demonstrated against his presence. Archbishop Christodoulos, head of the Greek church, welcomed his statement that Roman Catholics had "sinned . . . against their Orthodox brothers and sisters."

The pope flew to Damascus, Syria, **May 5**. At the airport, Pres. Bashar al-Assad, in a hostile reference to Israelis, urged Christians and Muslims to unite against those who he said had betrayed Christ and were mistreating the Palestinians. On **May 6** John Paul became the first pope ever to enter a mosque—the Umayyad Mosque in Damascus. The pope concluded his trip by visiting Malta **May 8-9**.

Kentucky Derby Winner Has 2d-Best Time Ever— Monarchos won the Kentucky Derby in Louisville, **May 5**, and posted the 2d-best winning time ever, 1:59.97. The winner's jockey was Jorge Chavez and its trainer John Ward. Coming from behind, Monarchos prevailed by 4 3/4 lengths over a 55-1 long shot, Invisible Ink. Secretariat, which set the Derby record of 1:59.4 in 1973, went on in that year to win the Triple Crown. Monarchos's bid to emulate that feat was dashed, **May 19**, when Point Given won the 2d race in the series, the Preakness Stakes, in Baltimore; Monarchos finished 6th.

Soccer Stampedes Kill Scores—At least 123 people died after a stampede at a soccer match in Accra, Ghana, **May 9**. Near the end of the match some fans had torn up chairs and started throwing the pieces onto the field. The police fired tear gas, triggering the stampede. On **Apr. 11**, 43 people had died during an effort to push their way into a soccer stadium for a match in Johannesburg, South Africa. On **Apr. 29** in Lubumbashi, Congo, 10 were killed and 51 badly hurt when police sought to break up a riot at a soccer match.

14 Illegal Immigrants Die in Desert—The U.S. Border Patrol reported **May 23** that the bodies of 11 illegal immigrants had been found in the southern Arizona desert, along with 13 survivors, one of whom later died. The Mexicans, all men ages 16 to 35, had perished in temperatures that ran as high as 115 degrees. The survivors, who may have paid $1,000 apiece for the journey, said they had been smuggled across the border in a vehicle and told to walk to a highway, which proved to be 70 miles away. On **May 28**, as he left the hospital in Yuma, AZ, one of the survivors was arrested and charged with being an organizer of the border crossing.

Collapse of Dance Floor Kills 25, Injures 300—A wedding celebration in Jerusalem became a tragedy **May 24-25**, as a crowded 3d-floor dance hall collapsed and crashed onto the floors below. The death toll was put at 23, with 300 injured, including the bride and bridegroom. Miki Levi, Jerusalem's police chief, blamed a "serious engineering failure" for the collapse of the dance floor.

A Rookie Wins 500-Mile Race Again—For the 2d year in a row and only the 8th time ever, a rookie driver—Helio Castroneves—won the Indianapolis 500 race **May 27**. He edged his teammate, Gil de Ferran. Castroneves, who was from Brazil, lived in Miami.

JUNE 2001

National

Democrats Take Power in Senate—The Democrats gained control of the U.S. Senate **June 5**, when Sen. Jim Jeffords (VT) formally withdrew from the Republican Party and became an independent. With the switch, the Senate contained 50 Democrats, 49 Republicans, and 1 independent. In a parting shot as majority leader, Sen. Trent Lott (R, MS) said **June 2** that his party "must begin to wage the war" against the Democrats.

With the shift, Tom Daschle (D, SD) became majority leader, and Robert Byrd (D, WV) replaced Strom Thurmond (R, SC) as president pro tempore. Harry Reid (D, NV) became majority whip. Changes in chairs of major committees included: Judiciary, Patrick Leahy (D, VT) replaced Orrin Hatch (R, UT); Foreign Relations, Joseph Biden (D, DE) replaced Jesse Helms (R, NC); and Armed Services, Carl Levin (D, MI) replaced John Warner (R, VA). Jeffords became chair of the Environment and Public Works Committee.

Jury Awards $3 Billion to Ill Cigarette Smoker—A plaintiff who had smoked Marlboro cigarettes for 40 years and had lung cancer won $3 billion in punitive damages from the Philip Morris Companies **June 5**. The Los Angeles jury also awarded the smoker, Richard Boeken, $5.5 million in general damages. Boeken's lawyer argued that Philip Morris had glamorized smoking while knowing of its harmful effects; a lawyer for Philip Morris argued that Boeken had been aware of the risks. The punitive award, the highest ever for an individual, was being appealed.

Timothy McVeigh Is Executed—Timothy McVeigh, who built and delivered the bomb that killed 168 people in Oklahoma City in 1995, was executed **June 11** by lethal injection at the Federal Penitentiary in Terre Haute, IN. Witnesses included representatives of the media, survivors of the bombing, and relatives of the victims. In Oklahoma City, where McVeigh had parked a rented truck with a ticking 4,800-pound bomb in front of the targeted building, 232 witnesses watched on closed-circuit television.

On **June 6**, U.S. District Court Judge Richard Matsch had rejected an appeal by McVeigh's lawyers to grant a stay of execution. On **June 7**, in Denver, a 3-judge panel of the U.S. Court of Appeals for the 10th Circuit unanimously upheld Matsch's ruling; McVeigh then asked his lawyers to abandon legal efforts to postpone the execution. McVeigh, who wrote many letters in prison, had likened the bombing to a military mission and offered no apology.

New Mayor Elected in Los Angeles—James Hahn, the city attorney, was elected mayor of Los Angeles in a runoff **June 5**. Campaigning as a law and order candidate, he defeated Antonio Villaraigosa, the son of Mexican immigrants and president of the local American Civil Liberties Union chapter, 54%-46%. Both are Democrats. Villaraigoso had sought to become L.A.'s 1st Hispanic mayor.

Bush to End Vieques Bombing Exercises in 2003—The White House **June 14** announced that the Navy's practice bombing of the Puerto Rican island of Vieques would end in 2003. The Navy had maintained that the use of the site for the bombing was essential. In April about 180 people had been arrested during protests on the island. Puerto Ricans, led by Gov. Sila Maria Calderon, and others who opposed the bombing continued to demand that it stop immediately; the same position was taken by a majority of Vieques voters in a non-binding referendum **July 29**.

Appeals Court Halts Breakup of Microsoft—On **June 28**, a U.S. appeals court for the District of Columbia unanimously threw out a District Court ruling that the Microsoft Corp. be broken up. The appeals court upheld the lower court ruling that Microsoft had violated Section 2 of the Sherman Antitrust Act—that it had illegally exploited its monopoly position in personal computer operating systems. However, it dismissed the lower court determination that Microsoft had sought to monopolize the market for Internet browser software.

Only 7 judges participated; 3 others on the court had recused themselves. The signatories sharply rebuked District Judge Thomas Penfield Jackson, declaring that he had violated 3 judicial canons by talking about the case with journalists. In remanding the case in order to deal with unresolved issues, the appeals judges ordered that a district judge other than Jackson assume jurisdiction.

Vice President Receives Implant to Aid Heart—In a surgical procedure **June 30**, Vice Pres. Dick Cheney received an implanted cardioverter defibrillator to control an irregular heart beat. Cheney, 60, had long suffered from heart trouble, dating from the first of 4 heart attacks at age 37. Cheney left the hospital for his official residence by mid-afternoon; his cardiologist, Dr. Jonathan Reiner, said Cheney's ability to perform his official duties would not be impaired.

U.S. Economy at a Glance: June 2001

Unemployment rate	4.5%
Consumer prices (change over May)	+0.2%
Producer prices (change over May)	+0.1%
Trade deficit	$ 29.07 bil.
Dow Jones high (June 5)	11,175.84
Dow Jones low (June 27)	10,343.84
Index of leading economic indicators (change over May)	+0.7%
2d Quarter GDP (annual rate)	+0.2%

International

King and Queen of Nepal Shot to Death by Son—King Birendra and Queen Aiswarya of Nepal and 7 other members of the royal family were shot dead or mortally wounded with an automatic weapon in their palace in the capital, Kathmandu, **June 1**, while eating dinner. The gunman, according to authorities, was their son, Dipendra, 29, the crown prince, who then shot and fatally wounded himself. Dipendra reportedly had argued with his mother over the choice of a bride. The killings occurred at a time when Maoist guerrillas were mounting an offensive in the countryside. The state council announced **June 2** that Dipendra was the new king, but placed authority in the hands of the slain king's brother Gyanendra, who had been out of the city during the slaughter. Gyanendra **June 3** called the deaths an "accident" caused by an exploding weapon. On **June 4,** Dipendra died of his wounds and Gyanendra was proclaimed king. Riots erupted as many people disputed the attribution of the shootings to Dipendra, and police killed 3 who violated a curfew. An official report, issued **June 14**, placed sole responsibility for the shootings on Dipendra.

Bomb at Tel Aviv Nightclub Kills 22—21 people, 15 to 32 years old, were killed or mortally wounded **June 1** in Tel Aviv, Israel, when a bomber, who also died, set off an explosion at the entrance of a nightclub. More than 100 were wounded. The disaster was the worst act of terrorism since the latest round of violent conflict between Palestinians and Israelis had begun in fall 2000. On **June 2**, Yasir Arafat, the Palestinian leader, condemned the bombing and called for an immediate cease-fire. American officials said **June 13** that the Israelis and Palestinians had accepted a U.S. cease-fire proposal. Pres. Bush and Prime Min. Ariel Sharon of Israel met in Washington, DC, **June 26**. Replying to Bush's appeal that he take new steps for peace, Sharon reiterated the position that the Palestinians must first forgo acts of violence for at least 10 consecutive days. Sec. of State Colin Powell said, **June 28**, that Arafat and Sharon had agreed to a new timetable, requiring 7 days of nonviolence followed by a 30-day cooling off period. During the following weeks, however, sporadic violence continued.

Candidate Born Poor Elected President of Peru—Alejandro Toledo, the 8th of 16 children, a one-time shoeshine boy and son of a sheep-herder, was elected president of Peru **June 3**. He prevailed, 52%-48%, over former Pres. Alan Garcia. The runoff was held after no candidate won a majority in the first round in April. In 2000, Toledo had pulled out of a runoff for president with then-Pres. Alberto Fujimori, charging fraud. Fujimori later went into exile, amid widespread allegations of corruption. Garcia, too, had ended his presidency in exile in 1992, after charges of corruption. In the current campaign, both candidates pledged economic reform.

Scientists Call Global Warming Serious—11 atmospheric scientists in the National Academy of Sciences warned, **June 6**, that global temperatures were rising, and said human-induced warming and associated sea level rises were expected to continue through the 21st century. The White House had asked for the report. In March, Pres. Bush had rejected the Kyoto Protocol, a 1997 treaty negotiated in Japan that committed signatories to reduce emissions of greenhouse gases, believed responsible for rising temperatures.

Blair and Labour Party Retain Power in Britain—The British Labour Party retained its huge majority in parliamentary elections on **June 7**, thus assuring another term as prime minister for Tony Blair. In the 659-seat House of Commons, Labour emerged with a 167-seat majority, down slightly from 179. The Conservative candidate, William Hague, had promised a tax cut and warned that a Labour victory would bring an end to the venerable British currency, the pound, in favor of the euro. Neither issue caught fire. Blair pledged to restore public services that many voters believed needed major financial attention. Hague announced **June 8** that he would step down as Conservative Party leader.

Reform-Minded President Reelected in Iran—Pres. Mohammad Khatami, considered a reform-minded moderate, won a landslide reelection victory in Iran **June 8**, earning 76% of the vote in a field of 10 candidates. However, the conservative Islamic clergy, led by Ayatollah Ali Khamenei and backed by the armed forces and judiciary, continued to wield the strongest power.

Bush Meets With European Leaders—Pres. Bush traveled to Europe in June for the first time as president, and met with a number of national leaders, including Pres. Vladimir Putin of Russia. During his 1st meeting, in Spain **June 12**, he conferred with Prime Min. Jose Maria Aznar, and stated themes that he would reiterate during his trip: He would press for the construction of a missile defense shield, while treating the ABM treaty as "a relic of the past," and he would oppose the Kyoto Protocol on global warming, which he said was "not based on sound science" and posed a threat to the U.S. economy. At a NATO meeting in Brussels, **June 13**, Bush's stand on the missile shield drew criticism from some leaders.

In Goteborg, Sweden, **June 14**, at a meeting of the leaders of the European Union, others present declared support for the Kyoto Protocol. Demonstrations in Goteborg turned violent the next day, with hundreds arrested, dozens injured, and several shot. He met Putin for the 1st time ever, in Slovenia **June 16**. They appeared to get along surprisingly well on the personal level, but Putin said that Bush risked straining relations between the 2 countries if he built a missile shield without Russian consent.

Milosevic Faces War-Crimes Trial—Slobodan Milosevic, former president of Yugoslavia, was delivered to The Hague **June 29**, where he faced an indictment for war crimes. Milosevic had fallen from power in October after an election defeat and a popular uprising, and had been imprisoned on corruption charges in Belgrade since April. On **June 23**, the Yugoslav cabinet had adopted a decree committing itself to extraditing him to The Netherlands, where he had been indicted in 1999 for war crimes allegedly committed in the province of Kosovo. Many Serbs, including Yugoslav Pres. Vojislav Kostunica, opposed the extradition, claiming that the Hague tribunal was biased against Serbia. However, the United States and other nations, threatening to withhold economic assistance to ravaged Yugoslavia, had pressed for it. Yugoslav Prime Min. Zoran Zizic resigned in protest on **June 29**, and the Hague tribunal issued an expanded list of charges against Milosevic. A number of other prominent figures indicted for war crimes remained in Yugoslavia.

Annan Reelected UN Secretary General—Kofi Annan was reelected unanimously **June 29** as secretary general of the United Nations. The 189-member General Assembly acted with unusual dispatch, considering that his first term did not expire until the end of the year. Annan, who is from Ghana, declared in his acceptance speech that he would continue his emphasis on human rights.

General

***The Producers* Wins Top Theater Awards**—A musical comedy that holds Germany's Nazi regime up to zany ridicule captured most of the glory **June 3** at the 2001 Tony awards in New York. The production received a record 12 Tonys, including one for Best Musical. The musical is based on the 1968 movie of the same name; both were produced by Mel Brooks. They tell the story of 2 producers, played on Broadway by Nathan Lane and Matthew Broderick, who raise money from donors for a musical, *Springtime for Hitler,* that they expect will be a flop, allowing them to make a fortune through a devious plan. Instead, their play succeeds, and they are ruined. Lane was named best actor in a musical. *Proof* was named the best dramatic play.

Japanese School Children Stabbed—On **June 8** in a quiet suburb of Osaka, Japan, 8 children, 6 to 8 years old, were stabbed to death at an elementary school by a man wielding a kitchen knife. The man, Mamoru Takuma, who had a history of psychological problems, was subdued by school officials and arrested.

Mother Drowns Her 5 Children—In Houston, TX, **June 20**, Andrea Yates, was arrested in the drowning deaths of her 5 children, who ranged in ages from 6 months to 7 years. Yates was taken into custody and charged with capital murder after she called police to her home. Her husband,

Russell Yates, said that she suffered from severe depression and had attempted suicide 2 years prior to the drowning.

Colorado Wins NHL Title—The Colorado Avalanche, playing at home, won the 7th and decisive game in the National Hockey League title series, **June 9**, defeating the New Jersey Devils, 3–1. The winners received the Stanley Cup. Colorado goalie Patrick Roy was named most valuable player for the title series, a distinction he had already won twice while playing for Montreal. After having played 20 years for Boston, 40-year-old Ray Bourque, one of the NHL's all-time great defensemen, was in his 2d season with Colorado, and played on a championship team for the first time. The Devils were the defending champions; Colorado's only previous title was in 1996

LA Lakers Repeat as NBA Champs—The Los Angeles Lakers won their 2d straight National Basketball Assn. title **June 15**, defeating the Philadelphia 76ers, 108–96, and taking the championship series, 4 games to 1. The Lakers had gotten off to a slow start during the 2000–01 season, hampered by a feud between the team's 2 superstars, Shaquille O'Neal and Kobe Bryant. In the playoffs, however, the Lakers were in top form, sweeping through the first 3 rounds without a loss. After losing their first game against the 76ers, 107–101, in overtime on **June 6**, the Lakers won 4 in a row for the title. Their post-season winning percentage of .938 (15-1) was a playoff record. In the finale, O'Neal and Bryant had 29 and 26 points, respectively. O'Neal was named most valuable player of the series.

Goosen Wins U.S. Open Golf Title—Retief Goosen of South Africa won the U.S. Open golf championship **June 18** in a playoff against Mark Brooks. Tiger Woods, who had won 4 straight major tournaments, saw his string broken, as he finished with 283.

Pope Visits Ukraine Amid Opposition—Pope John Paul II arrived in Kiev, the capital of Ukraine, to begin a visit **June 23**. For the pope, reconciliation of the Eastern Orthodox and Roman Catholic churches—separated by a schism in 1054—was a high priority, although leaders of the Orthodox church had opposed the visit and refused to appear with him. Pres. Leonid Kuchma met him at the airport, where John Paul appealed to both churches to "ask forgiveness" of each other. On **June 24**, the pope visited the site where the Soviet secret police had executed more than 100,000 people between 1929 and 1941. The next day, at Babi Yar, he prayed at the site where the Nazis had killed 200,000 Jews and others during World War II. The pope, **June 27**, conducted a mass for 600,000 people in Lviv.

JULY 2001

National

Police Ask Congressman About Missing Intern—The case of a missing government intern became headline news in July, as police investigated her relationship with a 53-year-old married congressman, U.S. Rep. Gary Condit (D, CA). Chandra Levy, 24, a former intern with the Bureau of Prisons, had last been seen **Apr. 30**. Police determined she had left her Washington, DC, apartment, apparently the next day, carrying her keys but not her handbag or other possessions. They found her bags packed in her apartment, in preparation for a return to California for college graduation.

Linda Zamsky, an aunt of Chandra Levy, said **July 6** that Chandra had told her of an affair with Condit. After Condit's 3d interview by District of Columbia police, **July 6**, Executive Asst. Chief Terrance Gainer said **July 7** that he had answered their questions satisfactorily and that he was not a suspect in Levy's disappearance.

New FBI Chief Appointed—Pres. George W. Bush announced **July 5** that he was nominating Robert S. Mueller III to succeed Louis Freeh as director of the Federal Bureau of Investigation. Mueller, the U.S. attorney for San Francisco, had promised, in hearings before Congress **July 30-31**, to take steps to restore the agency's credibility and increase its efficiency, in the wake of various public embarrassments, including misplacing of evidence in the trial of Timothy McVeigh and disclosures of apparent espionage activities by longtime FBI agent Robert Hanssen. Mueller was confirmed by the Senate, **Aug. 2**, in a 98-0 vote.

Ex-FBI Agent, Spy for Russia, to Get Life Term—Robert Hanssen, a 25-year veteran of the FBI who had been arrested in February, pleaded guilty **July 6** to espionage, attempted espionage, and conspiracy. Sentencing was postponed until **Jan. 2002**; a plea-bargain agreement accepted by a U.S. District Court in Alexandria, VA, provided that he would be spared the death penalty if he cooperated with authorities and revealed details of his espionage activities. Hanssen had admitted to spying for the Soviet Union and Russia; his attorney, Plato Cacheris, said Hanssen's activities dated back to 1979 and continued, with interruptions, until his arrest. Under the agreement, he would be required to forfeit about $1 million in cash and diamonds that he allegedly received from the Russians, to the extent it was recoverable.

Campaign Finance Bill Stalled—The push in Congress for a campaign finance reform bill that would ban so-called soft money—unregulated donations to political parties—suffered a setback in the House **July 12**. GOP leaders, though opposed to the bill, had told its leading sponsors, Martin Meehan (D, MA) and Christopher Shays (R, CT), they would allow a House vote on it. Meehan and Shays had a package of 14 amendments designed to make the House bill conform closely to the Senate version, passed in April. But in ground rules laid down **July 11**, Speaker Dennis Hastert (R, IL) required that each amendment be voted on separately—which would make it more difficult for supporters to hold their shaky coalition together. After supporters gathered enough votes—208 Democrats, 19 Republicans, and 1 independent—to defeat the rules, Hastert declared he had no plans to bring the bill up again.

Interceptor Downs Missile in a U.S Test—In a successful test of a planned U.S. missile defense system, a prototype interceptor struck and destroyed a mock warhead above the Pacific Ocean **July 14**. An intercontinental missile carrying the warhead and the interceptor had been launched from Vandenberg Air Force Base in California and the Marshall Islands, respectively. Two similar tests had failed in 2000, and one had succeeded in 1999. The Bush administration was advocating the construction of a missile-defense system, based on the use of interceptors, to guard against attack from so-called rogue nations such as Iraq or North Korea.

U.S. Economy at a Glance: July 2001	
Unemployment rate	4.5%
Consumer prices (change over June)	-0.3%
Producer prices (change over June)	-0.9%
Trade deficit	$28.83 bill
Dow Jones high (July 19)	10,610.00
Dow Jones low (July 10)	10,175.64
Index of leading economic indicators (change over June)	+0.3%

International

Milosevic Arraigned at The Hague—Slobodan Milosevic, the first former head of state to face a trial on war crimes charges for acts committed while in office, rejected the authority of the international war-crimes tribunal at his arraignment **July 3** in The Hague for alleged war crimes against ethnic Albanians in the Serbian province of Kosovo. Judge Richard May of Britain presided. Milosevic, charged with murder and other crimes, refused to accept legal counsel. A plea of not guilty was entered.

European Union Rejects GE Purchase of Honeywell—The European Union's executive body, the European Commission, **July 3** rejected the proposed acquisition by General Electric Co. of Honeywell International Inc. Never before had the EU rejected a merger of 2 U.S. corporations that had been approved by U.S. regulators. The commission concluded that the $45 billion merger would stifle competition in both jet engines and avionics.

Macedonia and Albanian Rebels Sign Cease-Fire—The government of Macedonia and leaders of ethnic Albanian rebels signed a cease-fire agreement **July 5**. About one-third of Macedonia's 2 million people were Albanians, many of whom were demanding more representation in government as well as official recognition in the Constitution and approval of Albanian as an official language. The conflict had been raging for months, despite intense pressure from NATO and the European Union for a cease-fire.

Chile's Pinochet Spared Trial—An appeals court in Chile held **July 9**, 2–1, that former Chilean leader Gen. Augusto Pinochet was too ill to stand trial on charges he had covered up killings, abductions, and torture while he was in power. Pinochet, who ruled Chile for 17 years, was accused of a wide range of human rights abuses in more than 250 criminal complaints. The court found that he had a case of mild dementia and was unable to defend himself.

Beijing Awarded 2008 Summer Olympic Games—The International Olympic Committee **July 13** voted to hold the 2008 summer games in Beijing, China. The world's most populous nation had long coveted the games, and had narrowly lost out to Sydney, Australia, as the site of the 2000 summer games. This time, in voting in Moscow, Beijing won 56 of 105 votes on the deciding 2d ballot. Toronto placed 2d in the final ballot with 22, and Paris had 18. The United States had declared public neutrality on the choice of a city. Speculation was widespread as to whether awarding China the games would help improve the Communist country's poor human rights record. The committee's vote set off jubilant celebrations in China.

On **July 14**, a day after the Olympic committee decision in Beijing's favor, Li Shaomin, a naturalized U.S. citizen held on spy charges since February, was convicted and expelled from China. On **July 24**, a court in Beijing convicted 2 Chinese citizens who were permanent U.S. residents of spying for Taiwan. They were sentenced to 10 years in prison, but 2 days later China announced that they had been released for medical reasons and deported.

India–Pakistan Summit Ends in Stalemate—Prime Min. Atal Bashari Vajpayee of India and Gen. Pervez Musharraf, the military ruler of Pakistan, met in a summit in New Delhi and Agraa, India, **July 14-16**, but failed to reach an agreement in their ongoing dispute over Kashmir, mountainous region populated by Muslims, made a part of mostly Hindu India in 1947. Recently, Kashmiri militants had been crossing the border into India.

Russia and China Sign Friendship Treaty—Russia and China signed a treaty of "friendship and cooperation" in Moscow **July 16**. Pres. Jiang Zemin of China and Vladimir Putin of Russia embraced warmly at the signing ceremony. The treaty put the 2 countries on record against the Bush administration plan for a missile-defense system. It committed Russia to support the incorporation of Taiwan into China and obliged the signatories to work together closely if either was menaced by a third country.

Riots Mar Summit—Thousands of demonstrators, many of them violent, stole the spotlight from the summit in Genoa, Italy, of the world's 7 major industrial powers and Russia. In a now-familiar replay of the scene at other such events, protestors representing a variety of causes massed in the streets and sought to disrupt the summit. They were met by 16,000 police. On the opening day of the summit, **July 20**, one demonstrator was shot dead by police. Scores were arrested and over 200 people, including some police officers, were injured.

On **July 20**, the leaders announced that $1.2 billion had been pledged to a UN "Global Health Fund" that would primarily target the AIDS epidemic, a crisis in many poor countries. The division between the United States and other major nations on global warming continued. Prime Min. Jean Chretien of Canada said **July 21** that his government supported the Kyoto Protocol, which the Bush administration opposed. Meanwhile, the crowds of street protestors swelled by **July 21** to an estimated 50,000.

Indonesian President Ousted—The Indonesian legislature in July ended the rule of Pres. Abdurrahman Wahid. In 1999, in the first democratic transfer of power in the nation's history, the legislature had elected Wahid over Megawati Sukarnoputri, daughter of the nation's founder, Sukarno. She then became vice president, and it was she to whom the legislature now gave its support. Anticipating the action by the People's Consultative Assembly, the president, early on **July 23**, declared he was "freezing" the legislature. Nonetheless, the assembly promptly initiated proceedings against him, and military and police leaders said they would not support his decree. Indeed, tanks massed outside the presidential palace with their cannons aimed at the building.

The assembly, **July 23**, voted unanimously (with Wahid's party boycotting the vote) to oust the president, elevating Megawati in his stead. Wahid at first balked, but on **July 26** he left the palace and the country.

Bush, Putin Link Missile Tests to Cut in Weapons—Pres. Bush and Pres. Putin of Russia met privately for the 2d time in 2 months **July 22** in Genoa, and agreed to discuss further the issue of antimissile defense and the 1972 ABM treaty. Bush wanted to push ahead with an antimissile defense system that would violate the treaty, which Putin insisted remain inviolate. Putin seemed open to replacing the treaty if it could be linked to more cuts in nuclear weaponry. The leaders agreed only on future "consultations"; they were not called negotiations.

Pope Asks Bush to Reject Use of Human Embryos—Pres. Bush met with Pope John Paul II at his summer residence at Castel Gondolfo **July 23**. The pontiff appealed to Bush not to support the use of human embryos for research, which the church opposed as a violation of human life.

178 Nations (but not U.S.) in Climate Agreement—In Bonn, Germany, **July 23**, a new compromise agreement was reached on combating global warming. In all, 178 nations supported the accord, but the United States was not among them. The agreement modified the 1997 Kyoto Protocol, with the newer terms allowing industrial nations with high emissions more flexibility in making their reductions. Under the binding agreement, 38 industrial nations must cut their emissions by 2012. Developing nations still were not required to reduce emissions, however. Pres. Bush had declared that opposition to the original Kyoto protocol and was looking for alternative approaches.

Bush Visits Troops in Kosovo—Pres. Bush concluded a week in Europe **July 24** by visiting U.S. military forces stationed in Kosovo. Though he said at Camp Bondsteel that he wanted to "hasten the day" when the troops could all come home, he said he would abide by the U.S. commitment to participate in keeping the peace.

Bulgaria's Ex-King Becomes Its Premier—The parliament of Bulgaria **July 24** formally approved the appointment of its former king, Simeon II, as the country's new premier. Simeon Saxe-Coburg had become king as a child in 1943 after the death of his father. Bulgaria was then allied with Nazi Germany. In 1945, the Communists took over and executed his uncle, the regent. In 1946, Simeon and his mother went into exile. Returning to Bulgaria, Simeon in April 2001 founded a new political movement that gained power in June, winning half the seats in a parliamentary election.

Secretary of State Meets With Chinese Leaders—In Beijing, **July 28**, U.S. Sec. of State Colin Powell met with Pres. Jiang Zemin and Prime Min. Zhu Rongji. Powell sought to explain the Bush administration's position in favor of building a missile shield, which the Chinese opposed. He touched on human right and agreed to resume the formal dialogue on human rights with Chinese officials, broken off after a NATO bomb fell on the Chinese embassy in Yugoslavia in 1999.

Japanese Premier's Party Wins Election—The Liberal Democratic Party of Prime Min. Junichiro Koizumi won a landslide victory **July 29** in elections to the upper house of parliament. The popular new leader appeared to have given a lift to a party that had been losing support. Koizumi had vowed to achieve political reform by shaking up his own party, whose close relations with big business had led to a number of major scandals.

Month of Killings Follows Mideast Cease-Fire—A month after an Israeli-Palestinian agreement leading to new peace framework talks, the negotiations appeared no closer at hand, and many more were dead. The worst incident occurred **July 31**, when missiles from Israeli helicopters hit offices of the militant Hamas group in Nablus on the West Bank, killing 8 people, including 2 small boys. Two local Hamas leaders were also among those killed.

General

Patient Gets 1st Fully Implanted Artificial Heart—On **July 2**, Robert Tools, 59, a former telephone company worker and teacher, received the first fully implantable artificial heart, which operates on batteries. The 7-hour procedure was performed at the University of Louisville's Jewish Hospital in Louisville, KY. Surgeons announced **July 3** that the recipient was alert and resting comfortably. Prior to the operation, his life expectancy had been put at 30 days. On **July 3**, Dr. David Lederman, president of Abiomed, the Danvars, MA, company that manufactured the device, called an Abiocor, said that the initial goal was to double the life expectancy to at least 60 days.

Williams—Again—and 'Wild Card' Win Wimbledon—Venus Williams of the United States successfully defended her women's tennis singles title at Wimbledon, July 8. In a big surprise, by contrast, a so-called wild card, Goran Ivanisevic of Croatia, took the men's singles title July 9. Williams defeated Justine Henin of Belgium, 6–1, 3–6, 6–0.

Ivanisevic, ranked 125th in the world, defeated Australia's Patrick Rafter, 6–3, 3–6, 6–3, 2–6, 9–7. He was the first wild card ever to win at Wimbledon.

Armstrong Again Wins Tour de France—Lance Armstrong won the Tour de France for the 3d consecutive year, **July 29**. His total time of 86 hours, 17 minutes, and 28 seconds was the 3d fastest ever.

AUGUST 2001

National

House Passes 'Patients' Rights' Bill—The House **Aug. 2** approved, 226-203, the so-called Patients' Bill of Rights. The bill would go to a conference committee with the Senate, which had already approved a bill. The 2 versions differed sharply as to what a litigant could collect from a health-care provider. The Senate bill allowed a patient to collect $5 million in punitive damages and unlimited noneconomic damages, while the House capped punitive and noneconomic damages at $1.5 million. The cap had prompted all but 5 House Democrats to vote against the bill. Under both bills, patients could gain access to the nearest emergency room, and choose any pediatrician, gynecologist, or obstetrician without first seeking a referral. Health plans would have to allow patients to use out-of-network physicians, and would have to pay for any medicine that a doctor prescribed, even if it was not on an HMO's roster of preferred drugs.

Bush OKs Funding of Embryonic Stem-Cell Research—Pres. George W. Bush announced **Aug. 9** that he would allow federal funding of limited stem-cell research utilizing human embryos. In a televised address to the nation, Bush described his difficulty in reaching a decision. Researchers had predicted that the use of cells from embryos could create transplant tissues that could save the lives of those suffering from Alzheimer's disease, Parkinson's disease, juvenile diabetes, and other diseases. Transplanting the tissues would, however, kill the embryos, and many conservatives opposed the use of embryos as constituting a destruction of life. The president's compromise was to authorize government funding for research only on those embryonic stem-cell lines that had already been produced, but not on embryos stored in fertility clinics or created for experimentation. While many Americans approved Bush's compromise, many scientists feared that the existing cell lines would prove insufficient. While many conservative religious leaders, including the Revs. Jerry Falwell and Pat Robertson, applauded the decision, a spokesman for the U.S. Conference of Catholic Bishops denounced it.

Slowing Economy Becomes a Focus of Attention—Evidence that the long economic boom was fading became more apparent **Aug. 17** when the Ford Motor Co. said it would dismiss up to 5,000 of its salaried employees, or 10% of its managers and engineers—one of the biggest of a recent series of corporate payroll cutbacks. For the 7th time in 2001, the Federal Reserve Board **Aug. 21** cut a key interest rate in an ongoing effort to head off an economic recession. The board lowered the rate on overnight loans among banks by another quarter of a percentage point, to 3.5%. It also lowered the discount rate on loans to banks from the Federal Reserve system by one-quarter of a percentage to 3.0%. In making the announcement, the Fed noted a continuing weak business environment in the United States and slowing economic growth elsewhere. On **Aug. 30**, for the first time since April, the Dow Jones industrial average closed on Wall Street at below 10,000—9,919.58, to be exact. It had lost 10 percent of its value since June.

The Bush administration, **Aug. 22**, projected that the federal surplus for the 2001 fiscal year (not including the Social Security system) would be a meager $600 million, a considerable change from the $122 billion surplus projected in April. Since then, the economy had been slowing, and a major tax cut had been approved and partially implemented. Pres. Bush said, **Aug. 24**, that the disappearing surplus would force Congress to stop increasing spending.

Senator Helms Announces Retirement—Sen. Jesse Helms (NC), a leading Republican conservative in the Senate, announced his retirement **Aug. 22**, saying that he would not seek reelection in 2002. As chairman of the Foreign Relations Committee, he had been a frequent critic of the United Nations and of U.S. foreign policy in general. Helms, 79, had been in poor health.

Congressman Denies Knowing What Happened to Intern—Rep. Gary Condit (D, CA) denied in interviews that he had had anything to do with the disappearance of Chandra Levy, a 24-year-old intern who had not been seen since the end of April. In his interview with Connie Chung of ABC News, **Aug. 23**, Condit said he had a "very close" 5-month relationship with Levy, but declined to characterize it further. Billy Martin, the lawyer for Levy's parents, denied, **Aug. 23**, Condit's assertion to Chung that the parents had asked him not to discuss the details of the relationship. On **Aug. 24**, Rep. Dick Gephardt (D, MO), the House minority leader, said he found Condit's responses in the interview "disturbing and wrong . . . I didn't hear an apology."

U.S. Economy at a Glance: August 2001	
Unemployment rate	4.9%
Consumer prices (change over July)	+0.1%
Producer prices (change over July)	+0.4%
Trade deficit	$24.1 bill
Dow Jones high (Aug. 2)	10,551.18
Dow Jones low (Aug. 30)	9,919.58
Index of leading economic indicators (change over July)	+0.3%

International

Bosnian Serb Found Guilty in Massacre of 7,000—Radislav Krstic, a former Bosnian Serb general, was found guilty **Aug. 2** of genocide in connection with the massacre of 7,000 Bosnian Muslim men and boys. The massacre had occurred in the eastern Bosnian town of Srebrenica in **July 1995**. The UN had declared the town a safe haven for Bosnian Muslims, and it was lightly guarded by Dutch UN peacekeepers. The latter did not attempt to stop Bosnian Serbs who seized the town and assembled all the remaining men and boys who were between the ages of 13 and 70, and shot them. Krstic, who became the local Serb commander while the 10-day atrocity was underway, claimed that he did not learn about the killings until later.

In its judgment, the 3-member UN International Criminal Tribunal for the Former Yugoslavia also found Krstic, 53, guilty of other crimes against humanity. The tribunal, which could not impose the death penalty, sentenced him to 46 years in prison.

Thai Premier Acquitted, Remains in Office—Thailand's highest judicial body **Aug. 3** found Thai Premier Thaksin Shinawatra not guilty of charges that he had concealed millions of dollars in assets in 1997 while serving as deputy premier. The 8–7 decision came from the Constitutional Court, which had been established under the 1997 constitution to deal with political reform. Indicted in December 2000, Thaksin would have been forced from office had he been found guilty. The justices chose not to issue a written opinion.

Macedonian Agreement Seeks to End Insurgency — After making concessions, 2 Macedonian Slav political parties reached an agreement **Aug. 8** with 2 ethnic Albanian parties who had been demanding more civil rights for the minority Albanian population. During the past 6 months, Albanian rebels and the Macedonian military had clashed repeatedly, with more than 130 killed. To reach an accord, the Macedonian parties agreed that Albanian would be the official language in any area where ethnic Albanians constituted at least 20% of the population. Also, up to an additional 1,000 ethnic Albanians could join the 6,000-member Macedonian police force within 2 years. NATO would deploy 3,500 troops whose sole responsibility would be to disarm the rebels.

The agreement came amid more violence. Five ethnic Albanians were killed **Aug. 7**, and 10 Macedonian Slav soldiers were ambushed and killed Aug. 8. Eight more soldiers were killed **Aug. 10** in a land-mine explosion, and 5 ethnic Albanians were killed in one village **Aug. 12**. The 4 parties signed the peace accord Aug. 13, and the rebels Aug. 14 agreed to turn in their weapons. On **Aug. 17**, an advance contingent of 400 NATO troops from Britain began arriving in Macedonia. Eventually, troops from 11 of the 19 NATO countries would be represented in the force. The 300 U.S. soldiers would provide logistical support. A British soldier died **Aug. 27** after Macedonians hurled a chunk of concrete through the window of his jeep the day before. The collection of weapons began **Aug. 27**.

Palestinian Suicide Bomber Kills 16 in Jerusalem Restaurant—A Palestinian set off a bomb in a crowded restaurant in Jerusalem, **Aug. 9**, killing 16 people, including himself, and wounding more than 130. Islamic Jihad, a militant Palestinian group, claimed responsibility, saying it had done so in response to Israel's policy of assassinating Palestinians thought to be involved in attacks on Israelis. In its own retaliation, Israel, **Aug. 10**, seized Orient House, an unofficial political headquarters for the Palestinians in Jerusalem. Police arrested at least 7 Palestinian security guards and carried off computers, files, and maps. Also on **Aug. 10**, Israeli fighter jets destroyed a Palestinian police headquarters in Ramallah in the West Bank.

Another suicide bomber, **Aug. 12**, killed himself and wounded 15 other people in a café near Haifa. In a gun battle in Hebron, **Aug. 12,** Israelis shot and killed an 8-year-old girl and wounded 12 other people. In 3 separate attacks, **Aug. 25**, Palestinians killed 7 Israelis, including 3 soldiers.

The U.S. State Dept., **Aug. 27**, admonished Israel that the targeted killing of Palestinians was not ending the violence and was inflaming the conflict. Just hours earlier, the Israelis had killed Mustafa Zubari, former chief of the Popular Front for the Liberation of Palestine, in an airborne attack on his office in the West Bank. Israeli forces, **Aug. 28**, seized parts of the Palestinian town of Beit Jala; they pulled out of the town **Aug. 30**.

Northern Ireland Government Suspended Briefly—The failure of the Roman Catholic Provisional Irish Republican Army to turn in its weapons kept the future of self-governance in Northern Ireland in doubt. In July David Trimble, leader of the Ulster Union Party, the province's largest Protestant party, resigned from the fledgling government of Northern Ireland because the IRA had not begun to decommission, or give up, their arms, as they had agreed to do. To prevent a complete collapse of peace talks, John Reid, the British secretary for Northern Ireland, **Aug. 10** ordered a temporary suspension of the government—which lasted for 24 hours on **Aug. 12**—which triggered an automatic 6-week extension of peace negotiations. Angered, the IRA **Aug. 13** rescinded its latest offer to begin decommissioning. Trimble had not regarded the offer as serious.

Angolan Rebels Kill 250 Passengers on Train—About 250 people on a train were killed **Aug. 10** when a land mine placed on the track by UNITA rebels exploded about 90 miles from Luanda, the capital. Rebels, who also attacked the bombed train, said they would strike again until the government resumed peace talks. The government denied a rebel claim that the train had been carrying military equipment.

General

Over-age Player Costs Little League Team Its Glory—A Little League baseball team that had become the pride of the Bronx saw its season ruined **Aug. 31**, when the league's charter committee ordered that all its season's victories be forfeited. The team, led by pitcher Danny Almonte, had finished 3d in the Little League world series. But officials in the Dominican Republic confirmed that Almonte had been born on April 7, 1987, making him 14 years old; too old under league rules.

The Championship was won **Aug. 26** by Japan's Tokyo Kitasuna over the Apopka, FL, team, 2-1. It was Japan's 2d win in 3 years.

SEPTEMBER 2001

National

For events relating to the Sept. 11 terrorist attack; see pages 58-61 and 33-35.

Sen. Gramm Joins List of Retirees—Sen. Phil Gramm (R, TX) announced **Sept. 4** that he would not seek reelection in 2002. Elected to the House in 1978 as a Democrat, he had switched to the Republicans in 1983, and in 1984 had been elected to the first of 3 terms in the Senate. A fiscal conservative, Gramm was the 3d southern Republican senator, after Jesse Helms (NC) and Strom Thurmond (SC), who had chosen not to seek another term in 2002.

U.S. Drops Move to Break Up Microsoft—Reversing the approach of the Clinton administration, the Justice Dept. **Sept. 6** said it would no longer seek to split Microsoft Corp. into more than one company, or pursue the legal claim that the company had illegally tied its network browser to its operating system that permitted computers to run. In June, a U.S. court of appeals had sent these legal issues back to district court. State prosecutors who were plaintiffs along with the Justice Dept. said they supported the latter's decision. The department said it would seek to curb anti-competitive actions by Microsoft.

After Attacks, Fears for Economy Grow—The terrorist attacks accelerated fears for the economy, which already appeared headed toward recession. On **Sept. 17**, the Federal Reserve Bank cut key interest rates again; the federal funds rate for overnight loans between banks was cut from 3.5% to 3.0%. The stock market reopened on Monday, **Sept. 17**, and experienced a sharp decline, with values falling 14.3% by **Sept. 21**, as measured by the Dow Jones industrial average, its worst weekly percentage decline since the Great Depression. During the week of **Sept. 24-28**, the Dow Jones average recovered about half of the losses of the previous week.

The airlines, which were losing $100 million or more per day in the days after the attack because of disruption in service and a decline in the number of travelers choosing to fly, began laying off employees. Midway Airlines went out of business altogether **Sept. 12**, laying off 1,700. Continental announced **Sept. 15** it would dismiss 12,000. It was followed by US Airways (**Sept. 17**, 11,000), America West (**Sept. 18**, 2,000), American and United (**Sept. 19**, 20,000 each), Northwest (**Sept. 21**, 10,000), and Delta (**Sept. 26**, up to 13,000). The Boeing Co. said **Sept. 18** that it would lay off 20,000 to 30,000 by the end of 2002 because of fewer orders for aircraft. Congress, **Sept. 21**, approved a $15 billion bailout package for the airlines.

Retiring New York Mayor Wants to Stay On—Mayor Rudolph Giuliani (R) won widespread praise for his efforts to rally New York City after the terrorist attack, and some wanted him to continue as mayor beyond the expiration of his term at the end of 2001. However, Giuliani was not eligible for a 3d term. The **Sept. 11** primaries for mayor, postponed by the attack, were held **Sept. 25**. Michael Bloomberg, a wealthy businessman, won the GOP nomination. Among the Democrats, Bronx Borough Pres. Fernando Ferrer and city Public Advocate Mark Green qualified for a runoff, held **Oct. 11**, which Green won. Giuliani explored possibilities for attempting to extend his term through legal means.

150 Minutes That Changed the World

(All times EDT, Sept. 11, 2001; A.M. unless otherwise indicated.)

7:58 United Airlines Flight 175 (UAL 175), a Boeing 767 bound for Los Angeles, leaves the gate at Boston's Logan Intl. Airport; 56 passengers (including 5 hijackers), 9 crew.

7:59 American Airlines Flight 11 (AA 11), another Los Angeles-bound Boeing 767, takes off from Boston; 81 passengers (including 5 hijackers), 11 crew.

8:01 United Airlines Flight 93 (UAL 93), a Boeing 757, takes off from Newark, NJ, bound for San Francisco; 37 passengers (including 4 hijackers), 7 crew.

8:10 American Airlines Flight 77 (AA 77), a Los Angeles-bound Boeing 757, takes off from Dulles Intl. Airport, near Washington, DC; 58 passengers (including 5 hijackers), 6 crew.

8:15 AA 11 veers off course, heading south toward New York City.

8:47 UAL 175 deviates from its flight path, turning left over New Jersey; 12 minutes later, it makes another sharp left, turning north toward New York City.

8:48 AA 11 slams into the North Tower (Tower 1) of the World Trade Center, between the 96th and 103d floors, causing a huge explosion and fire. Rescue workers rush to the scene.

9:00 Air controllers lose radar contact with AA 77 over southern Ohio; at some point after that, the plane reverses course and heads back toward Washington.

9:03 UAL 175 smashes into the South Tower (Tower 2) of the World Trade Center, between floors 87 and 93, causing a 2d huge explosion and fire.

9:05 Pres. Bush, in Sarasota, FL, for a school event to promote his education program, is quietly told by his chief of staff, Andrew Card Jr., "America's under attack."

9:30 In Sarasota, Bush makes a brief public statement, saying the nation has experienced an "apparent terrorist attack" and vowing "to hunt down and to find those folks who committed this act."

9:37 UAL 93, over northern Ohio, diverges from its planned route and heads back east.

9:37 AA 77 ploughs into the west side of the Pentagon, causing a fire and major structural damage.

9:45 From UAL 93, passenger Todd Beamer contacts a GTE supervisor. During a 13-minute phone call, she tells him about the other hijackings; he tells her some passengers have decided to fight back. His last words heard on the call: "Are you guys ready? Let's roll!"

9:48 In Washington, orders are given to evacuate the Capitol and the White House.

9:49 The Federal Aviation Administration bans aircraft takeoffs throughout the U.S. and orders most international flights to land in Canada or return to their departure point.

9:55 Bush leaves Florida on board Air Force One, from which he communicates with Vice Pres. Cheney and other top U.S. officials in Washington. He stops at Air Force bases in Louisiana and Nebraska, returning to Washington about 7 P.M.

9:59 The South Tower of the World Trade Center collapses.

10:10 UAL 93 crashes near Shanksville, PA, about 80 mi SE of Pittsburgh.

10:28 The North Tower of the World Trade Center collapses.

International

U.S. Walks Out of Racism Conference—The United States and Israel, **Sept. 3**, walked out of the UN conference on racism in Durban, South Africa. The departures were prompted by the unwillingness of many delegates to agree to compromise language in a draft resolution that equated Zionism with racism and that denounced Israeli policy toward the Palestinians. Anticipating the confrontation, the U.S. State Dept. had announced **Aug. 29** that it would send only a small midlevel delegation to the conference. Ultimately, on **Sept. 8**, the delegates from some 160 countries accepted a resolution that recognized the inalienable right of the Palestinian people to . . . an independent state and "the right to security for all states in the region, including Israel." The characterization of Israel as a racist state was dropped. The final document deplored discrimination against women as well as indigenous people, gypsies, and other minorities. The resolution condemned slavery and the slave trade, but omitted a controversial call for reparations and apologies from former slave-holding nations.

Israeli-Palestinian Death Toll Continues to Rise—A Palestinian suicide bomber died **Sept.** 4 when his bomb went off in Jerusalem. The militant Palestinian group Hamas claimed responsibility for a **Sept. 9** bomb explosion in the Israeli town of Nahariya that killed the bomber and 3 others. In response, Israeli tanks **Sept. 11** rolled into Jericho and Jenin in the West Bank, and within 3 days at least 13 Palestinians had been killed.

On **Sept. 18**, the Palestinian leader Yasir Arafat and Prime Min. Ariel Sharon of Israel both ordered a halt in offensive actions. Israeli tanks and troops began to pull back in the vicinity of Jericho and Jenin. On **Sept. 26**, Arafat met with Foreign Min. Shimon Peres of Israel; they agreed on several small steps to ease the crisis.

Taliban Tries Christians; Anti-Taliban Leader Killed—Afghanistan's ruling Taliban militia **Sept. 4** put on trial 8 foreign aid workers on charges that they had sought to convert Afghan Muslims to Christianity. The accused—4 Germans, 2 Americans, and 2 Australians—worked for Shelter Now International. The aid workers denied the charges. On **Sept. 15**, the leader of the armed opposition to the Taliban died from a bomb attack at his field headquarters. The victim, Ahmed Shah Massoud, led the Northern Alliance, which continued to fight the Taliban even though it controlled about 10% of the country. It was believed that the bomb had been brought to the headquarters by 2 men posing as journalists who sought to interview Massoud. They and an aide to Massoud were also killed in the explosion.

President Fox of Mexico Visits U.S.—Pres. Vicente Fox of Mexico began a visit to the United States **Sept. 5** with an appeal to Pres. Bush to reach an agreement to legalize the status of 3.5 million Mexicans who had entered the country illegally. On **Sept. 6**, in an address to Congress, Fox renewed his appeal. While supporting Fox, Bush did not set a time frame for changing the status of the Mexicans, and many in Congress were expected to oppose granting them either permanent residence or status as guest workers. On **Sept. 7**, Fox urged the scrapping of a 1947 defense treaty among the United States and Latin nations that he said was obsolete following the end of the cold war. He favored a new treaty focusing on social problems.

Macedonia Parliament Supports Peace Agreement—The parliament of Macedonia **Sept. 6** voted 91-19 to draft constitutional changes in support of the August peace agreement worked out between the government and ethnic Albanian rebels. The changes would grant more rights to the Albanians. On **Sept. 3**, Prem. Ljubco Georgievski complained that the peace agreement rewarded military aggression, but that Macedonia must not go against the Western nations. The government said **Sept. 17** that it would agree to let NATO troops stay after they had disarmed the rebels in order to oversee the implementation of other aspects of the peace plan. On **Sept. 26**, NATO and Macedonia agreed on a 3-month stay by 700 troops. The rebels announced **Sept. 27** that they had disbanded. Some NATO troops began leaving Macedonia **Sept. 27**.

Britain's Tories Pick New Party Leader—The Conservative Party in Britain announced **Sept. 13** that party members had chosen Iain Duncan Smith to lead the Tories, who had lost badly in the June election. The previous leader, William Hague, had then resigned. Five men sought to head the party, and 3 were eliminated in ballots among Tory members of Parliament. In the final round, in a national vote prescribed by Hague, Duncan Smith defeated Kenneth Clarke 61% to 39%. Smith, a strong social conservative, opposed further British integration with the European Union.

Heirs to Communist Party Win Polish Election—The Democratic Left Alliance, successor to the Communist Party, won a plurality in Poland's parliamentary election **Sept. 23**. The governing Solidarity virtually disappeared, getting under 5% of the vote. A coalition government was expected.

General

Venus Williams Defeats Sister to Keep U.S. Tennis Title—Venus Williams retained her U.S. Open women's singles tennis title in New York City, **Sept. 9**, by defeating her sister, Serena Williams, 6-2, 6-4. Venus had been seeded 4th, Serena 10th. On **Sept. 7**, they had advanced to the final by defeating, respectively, Jennifer Capriati (the 2d seed) and Martina Hingis, the top seed. No sisters had met in a Grand Slam finale since 1884, and never before had 2 African-Americans

U.S. Struck by Terrorist Attack

A page turned in American history at 8:48 AM, **Sept. 11,** when American Airlines Flight 11, out of Boston and scheduled for Los Angeles, struck Tower 1 of the World Trade Center in the financial district of Lower Manhattan. After impact the plane remained imbedded in the building, and a long plume of black smoke rose from the tower. Fifteen minutes later, as onlookers in New York and TV viewers around the world stared at Tower 1 (North Tower), United Airlines Flight 175, also out of Boston and originally bound for LA, swooped low over Manhattan and struck Tower 2 (South Tower), creating an orange fireball as it exploded inside the building. Some 35 minutes later, American Flight 77, which had taken off from Dulles International Airport in Chantilly, VA, on a flight to LA, struck the western side of the Pentagon in a recently renovated area that was lightly occupied. About 25 minutes later, United Flight 93 from Newark and scheduled for San Francisco, crashed in a rural area of Somerset County, PA, while apparently headed toward Washington, DC.

More than 5,000 people—mostly Americans, but including people from more than 60 other nations—perished in the attacks, more than twice the toll at Pearl Harbor, making certain that **Sept. 11, 2001,** would join **Dec. 7, 1941,** as a date that will live in infamy. Pres. George W. Bush pledged to destroy the responsible international terrorist organizations and punish as well the regimes that supported and sheltered them. Military retaliation began the night of **Oct. 7,** when the United States and Britain struck from the air at Afghanistan, where the terrorist leaders and their training camps were harbored.

See also the separate box, "150 Minutes That Changed the World," and *America at War,* pages 33-35.

Pres. Bush was notified of the initial strikes while speaking to elementary-school children in Sarasota, FL. He left Florida on Air Force One which, for security reasons, pursued a zigzag course that took it over the Atlantic Ocean, then to stops at Barksdale Air Force Base in Shreveport, LA, and Offutt AFB in Omaha, NE, before it flew to Washington, DC. He arrived in Washington around 7 P.M., and later on **Sept. 11**, spoke to the nation from the White House, asserting that "our way of life, our very freedom came under attack . . . ." He said that "our nation saw evil" but had responded with "the best of America" in the daring of rescue workers and the caring for suffering neighbors and strangers. He vowed to bring the surviving perpetrators to justice, making "no distinction between the terrorists . . . and those who harbored them."

Towers 1 and 2 had collapsed 100 and 56 minutes, respectively, after impact, allowing sufficient time for thousands to flee, most of them down many flights of stairs. A number of people, trapped above the level of impact, and doomed by the flames, leaped to their deaths. Some 343 NYC firefighters and 60 police officers who had rushed into the towers to save others were killed when the buildings fell. The collapses were attributed to the intense heat from fires that weakened steel columns and trusses. The towers came straight down, each floor being flattened in turn by the combined weight of the floors above. On the ground, flying debris and billows of ash spread rapidly, overtaking hundreds fleeing down the streets. Several adjacent buildings, including a Marriott hotel, also collapsed, and others were shaky.

A massive around-the-clock rescue effort, supported by emergency units from as far away as California, yielded meager results—only 5 people were pulled live from the wreckage, none after **Sept. 12**. All hospitals within miles and several other buildings were converted into emergency medical centers, and some 600 survivors received treatment. Family members and friends of the missing waited for days and weeks in vain for news; many posted and passed out fliers with photographs of the missing. Pres. Bush visited "Ground Zero" in Lower Manhattan, **Sept. 14**, thanked the rescuers, and received a strong welcome in return.

Minutes after the 2d plane hit the World Trade Center, the Federal Aviation Administration grounded all commercial flights, stranding thousands of passengers. Most arriving international flights were diverted to Canada or sent back to their cities of origin. U.S. military aircraft patrolled the skies above New York and Washington. Navy ships able to mount air defenses moved toward West Coast cities. The FAA **Sept. 12** eliminated curbside check-ins and barred all kinds of cutting instruments, not previously forbidden, from carry-on luggage. The government said the same day that armed sky marshals would henceforth travel on some commercial flights. Long after other airports had been allowed to reopen on **Sept. 13,** Logan International Airport in Boston and Ronald Reagan National Airport near Washington, DC, were kept closed. (With the reopening of Reagan Airport on **Oct. 4,** all airports were providing service.) Vice Pres. Dick Cheney revealed that on **Sept. 11** Bush had ordered the military to shoot down any other airliner believed in the hands of hijackers and bound for a target.

Wall Street's financial markets closed quickly on **Sept. 11** and would not reopen until **Sept. 17**. Voting in primary elections for local offices throughout New York State, in progress at the time of the attacks, was put off until **Sept. 25**.

The assault prompted numberless declarations of patriotism and unity by Americans, and political leaders took a nonpartisan tack. Congress **Sept. 11** adopted a resolution giving the administration broad power to act against terror groups and nations protecting them. Bush **Sept. 12** called the attacks "acts of war," and Congress granted him $20 billion in emergency funding that day—doubling it to $40 billion **Sept. 14**. Both houses (the Senate 98–0 and the House 420–1) authorized Bush to use all "necessary and appropriate force" needed against those responsible. That same day, Bush authorized the call-up of 50,000 military reservists.

On **Sept. 13**, Sec. of State Colin Powell named Osama bin Laden, a wealthy, anti-American Saudi exile operating out of Afghanistan, as the prime suspect. Bin Laden, thought to run Al Qaeda, a shadowy terrorist organization with operatives in many countries, had been linked to the bombing of U.S. embassies in Kenya and Tanzania in 1998 and of the USS *Cole* in Yemen in 2000. He had left Saudi Arabia with a $250 million inheritance in 1991, and was estranged from his family.

Mullah Mohammed Omar, the leader of the Taliban, a Muslim fundamentalist group that ruled most of Afghanistan, criticized the attacks **Sept. 11** and claimed Bin Laden was not responsible. Most Middle East rulers joined national leaders throughout the world in deploring the attacks; Iraq's Saddam Hussein, however, called them the result of America's "evil policy." By **Sept. 12**, foreign aid workers and thousands of citizens were fleeing impoverished and war-ravaged Afghanistan. A huge refugee population massed across the border with Pakistan. The Taliban said **Sept. 13** they would turn Bin Laden over to an Islamic court if the United States could prove he was to blame. On **Sept. 14**, Omar called on Muslims to prepare for a jihad, or holy war.

Meanwhile, the Bush administration sought to organize an international response. On **Sept. 12**, for the first time ever, NATO invoked its mutual defense provision stating that an armed attack against any member would be treated as an attack against all, and, if the United States moved against the terrorists, would require a unified military response. Russia **Sept. 13** said it would work with the alliance. Pres. Jiang Zemin told Bush **Sept. 12** that China would join in anti-terrorist efforts. On **Sept. 14**, pressured by the United States, yet fearful of unrest among fundamentalist Muslims at home, Pakistan agreed to allow American access to its air space and military facilities.

By that day the FBI had the names (all of apparent Middle East origin) of all 19 hijackers, although later reports cast doubt some of the purported identities. It was soon determined that several of the hijackers had received instruction at flight-training schools in Florida. A roundup across the United States came rapidly, with the number of persons being detained rising to 75 by **Sept. 18**.

Pres. Bush declared **Sept. 14** a day of national mourning and remembrance; he and 4 former presidents were among those attending a service in Washington's National Cathedral. New York's Cardinal Edward Egan **Sept. 16** said a mass at St. Patrick's Cathedral for the victims. First Lady Laura Bush **Sept. 17** attended a memorial service near the Pennsylvania crash site. The same day, Pres. Bush visited a mosque at the Islamic Center in Washington and decried violent acts reported against Arab-Americans and those thought to be Arab-Americans. A Sikh man had been shot to death in Arizona a day before.

Bush, **Sept. 17**, said of Bin Laden that he was "Wanted: Dead or Alive." Meeting with Bush in Washington the next French Pres. Jacques Chirac said France stood in solidarity with the United States. Indonesia's new president, Megawati Sukarnoputri, with her own problems of Muslim unrest, met with Bush **Sept. 19** and denounced the attacks, but declined to go further. Japanese Prime Min. Junichiro Koizumi said that Japan would give what military help it could.

The U.S. Defense Dept. **Sept. 19** had ordered deployment of combat aircraft to the Persian Gulf, Indian Ocean, Uzbekistan, and Tajikistan; the next day the Army said ground troops were being sent to the region. Afghanistan's senior clerics issued an edict saying that Bin Laden should be asked to leave the country voluntarily. In a speech to a joint session of Congress, **Sept. 20**, Pres. Bush demanded that Afghanistan hand over Bin Laden; otherwise, he said, the Taliban would share his fate. He announced that Gov. Tom Ridge (R) of Pennsylvania would head a new Office of Homeland Security that would coordinate efforts by 40 U.S. agencies to thwart terror attacks within the U.S. British Prime Min. Tony Blair attended the speech and pledged solid support to the overall anti-terrorist effort. (Ridge was sworn in as homeland security chief, **Oct. 8.**)

A telecast fund-raiser featuring top movie stars and musicians, **Sept. 21**, was watched at least in part by an estimated 89 million people. By the end of the month, $500 million had been raised from all sources to help survivors of the victims.

As an apparent thank-you for its support, the Bush administration **Sept. 21** lifted economic sanctions against Pakistan—and against Pakistan's rival India as well. The sanctions had been imposed in 1998 after both had tested nuclear weapons. The same day, the United Arab Emirates broke relations with the Taliban. A 2d country, Saudi Arabia, broke relations **Sept. 25**, leaving Pakistan as the only country to recognize the Taliban. During a trip through Central Asia, Pope John Paul II, speaking in Kazakhstan **Sept. 23**, urged the United States not to make an overzealous military response. On **Sept. 24**, Pres. Vladimir Putin said Russia would provide intelligence data and military weapons as a part of any international move against the Taliban.

The FAA, **Sept. 23**, had grounded all crop-dusting planes, fearing they might be used to disperse toxic substances; one of the hijackers had reportedly been compiling information on crop-dusters. Bush, **Sept. 24**, ordered all U.S. assets of certain terrorist groups frozen. He later said foreign banks that did not cooperate with U.S. investigators would not be able to do business in the United States.

In Chicago **Sept. 27**, Pres. Bush asked governors for National Guard troops to begin protecting the nation's airports and said that armed sky marshals in plainclothes would ride on many flights. Doors would be fortified to make access to the cockpit more difficult. An administration official said that a new federal agency would supervise the 28,000 workers who screen passengers and baggage.

The UN Security Council, **Sept. 28**, unanimously passed a resolution requiring all 189 members to put a stop to the financing and training of terrorists within their borders and to cooperate in any military effort against them. Pakistani clerics of the same militant school of Islam as the Taliban met with the Taliban leader, Mullah Omar, on **Sept. 29** in Kandahar, Afghanistan, but did not persuade him to give up Bin Laden. The same day, before meeting with Jordan's King Abdullah, Bush signed a free-trade agreement Jordan had sought for a year. Abdullah pledged Jordan's support to the antiterrorist effort.

In a speech to Democrats in Des Moines, IA, **Sept. 29**, former Vice Pres. Al Gore declared, "George W. Bush is my commander in chief," and said everyone was united behind him. Polls showed that a large majority of Americans supported Bush's response to the attacks.

By month's end, about 500 people in the U.S. and elsewhere had been arrested or detained in connection with the attacks. It was reported **Sept. 30** that Bush had approved a secret plan to support efforts throughout Afghanistan by anti-Taliban forces to overthrow the regime. Bush also authorized $100 million for relief aid for Afghan refugees.

Addressing the UN General Assembly **Oct. 1,** New York City Mayor Rudolph Giuliani appealed to delegates not to remain neutral in the current conflict, to choose civilization over terrorism. The same day, a 3d U.S. aircraft carrier, the *Kitty Hawk*, left Japan for the Middle East; the *Enterprise* and *Carl Vinson* were already in the region.

NATO said **Oct. 2** that the United States had presented clear evidence, sufficient to justify NATO military action, that Osama bin Laden and his organization were responsible for the terror attacks. A document released by the British government **Oct. 4** asserted that one of Bin Laden's top associates had orchestrated the terror attacks and that at least 3 hijackers had been identified as associates of the Al Qaeda organization. Prime Min. Blair said no one could doubt Bin Laden's responsibility.

About 40 people were killed **Oct. 1** when a militant Muslim group attacked the Legislative Assembly building in Srinagar in the Indian province of Jammu and Kashmir.

Amid rising concerns about the use of lethal substances by terrorists, Sec. of Health and Human Services Tommy Thompson told a Senate committee **Oct. 3** that the government was planning to stockpile 40 million doses of smallpox vaccine. On **Oct. 5,** Robert Stevens, a photo editor for American Media Inc. in Boca Raton, FL, publisher of *The National Enquirer* and other tabloids, died after being infected with anthrax, a deadly bacterium. By Oct. 10, 2 co-workers were found to have been exposed to anthrax; a criminal investigation was launched, but there was no initial evidence of a direct connection with the Sept. 11 terrorist attacks. A worker at NBC News in New York was confirmed with a manageable case of anthrax **Oct. 12** that was believed to have been contracted from powder in a letter she opened.

Bush said **Oct. 4** that the United States would try to get $320 million in food and medical supplies to the Afghan people. Large numbers of refugees were now filling up refugee camps across the Pakistan border. About 1,000 U.S. troops in the 10th Mountain Division were sent to Uzbekistan **Oct. 5**—the first U.S. deployment in a former part of the Soviet Union.

After dark on **Oct. 7** and through the night, the United States and Britain struck at targets in Afghanistan, using cruise missiles and long-range bombers. The assault was directed at airports, air defenses, and communication and command centers. B-2 stealth bombers flew nonstop from Missouri to their objectives. In Kandahar, in S. Afghanistan, targets included the Taliban's headquarters and the compound of Mullah Omar, as well as Al Qaeda housing units. The Taliban defense headquarters near Kabul was also attacked. Targets were hit beginning about 12:30 PM EDT, or 9:00 PM in Afghanistan. Pres. Bush announced the attack 30 minutes after it began, warning that though the immediate focus was on Afghanistan, "the battle is broader." U.S. bombing continued nightly after **Oct. 7**.

In a prerecorded tape provided to the Al Jazeera television network in Qatar and played **Oct. 7,** Osama bin Laden warned, "America will not live in peace" until peace came to "Palestine" and until "the army of infidels depart the land of Muhammad." He charged that the United States bore responsibility for 80 years of "humiliation and disgrace" brought upon the Islamic world.

The rise in tensions and the bombing had disrupted the distribution of food in Afghanistan by relief agencies. On **Oct. 8,** U.S. transport planes dropped 37,000 meals into areas where mass starvation was feared imminent. On **Oct. 9,** the Pentagon reported the destruction of 7 terrorist training camps in Afghanistan, though acknowledging that they were probably vacant of personnel. The same day it was reported that 4 guards at a UN-sponsored land-mine-clearing operation near Kabul had been killed, apparently in the bombing raids. By **Oct. 9,** the Pentagon was claiming effective control of the skies over Afghanistan, and heavy air strikes were launched **Oct. 10** against Taliban garrisons and troop encampments. On **Oct. 10**, at an emergency meeting in Qatar of the Islamic Conference, 56 Islamic nations briefly condemned the Sept 11 attacks and avoided direct condemnations of the military action in Afghanistan.

OCTOBER 1-10, 2001

National

Key Interest Rate Cut to a 39-Year Low—For the 9th time in 2001, and the 2d time since the terror attacks, the Federal Reserve Board **Oct. 2** cut its rate on overnight loans among banks, this time from 3.0% to 2.5%. The rate had not been so low since 1962. The latest cut continued the Fed's efforts to breathe life into an economy apparently slipping into recession. The Fed also cut its rate on loans to banks from the Federal Reserve system from 2.5% to 2%. Pres. George W. Bush **Oct. 3** proposed a $75 billion economic stimulus package that included tax relief for businesses, help for the unemployed, and an accelerated timetable for the payout of tax cuts approved earlier in 2001. The Labor Dept. reported **Oct. 4** that 528,000 new claims for unemployment, a 9-year high, had been filed in the previous week.

International

Khaleda Zia Wins Power in Bangladesh; Elections Called in Sri Lanka—The Bangladesh Nationalist Party, led by Khaleda Zia, won 36% of the votes in general elections **Oct. 1**, while the rival Awami League, led by the incumbent Hasina Wazed (known as Sheikh Hasina), won 41%. Zia formed an alliance with 2 Islamic parties and another party to gather a plurality of 46% and form a government; she was sworn in as prime minister **Oct. 10**. Sheikh Hasina claimed the election had been unfair and said her party would boycott Parliament. In Sri Lanka, Pres. Chandrika Kumaratunga dissolved Parliament **Oct. 10** and called new elections for December, in response to an expected vote of no confidence, after a year of political instability.

Despite Truce, Israeli-Palestinian Strife Continues—The shaky cease-fire was tested **Oct. 2** when 2 Palestinian gunmen attacked the Israeli settlement of Alei Sinai, killing a couple and wounding 15. The gunmen were killed. In retaliation, Israeli tanks **Oct. 3** attacked the town of Beit Lahia; 6 Palestinians were killed. Killings on both sides continued in subsequent days. Meanwhile, on Oct. 5, 2 protesters in a demonstration in Gaza against U.S. airstrikes in Afghanistan were fatally shot by Palestinian police.

Explosion Kills Israelis on Russian Airliner—A Siberian Airlines jetliner exploded **Oct. 4** and plunged into the Black Sea, killing all 64 passengers and 12 crew members. The passengers included 51 Israelis taking a holiday trip to Siberia. U.S. military officials reported evidence **Oct. 5** that a missile from a Ukrainian military training exercise had hit the plane.

General

Baseball Season Ends With Memorable Moments—Major League Baseball's regular season concluded with the spotlight on several extraordinary players. On **Oct. 5,** in his 160th (of 162) games, Barry Bonds, left fielder for the San Francisco Giants, hit his 71st home run, the most ever for any player in one season. His blow, breaking the 1998 total of 70 by St. Louis's Mark McGwire, came in the first inning at home against Los Angeles. For good measure, Bonds added another homer in his next at-bat, in the 3d inning. He added number 73 on **Oct. 7,** the last day of the season.

On **Oct. 4,** Rickey Henderson of the San Diego Padres scored his 2,246th run—a home run, no less—to break Ty Cobb's major-league record. On **Oct. 7,** Henderson doubled for his 3,000th hit—the 25th player to attain that plateau. That same game—a 14-5 loss to Colorado—was the last in a 20-year career (all with San Diego) for outfielder Tony Gwynn, who finished with 3,141 hits. He had led the National League in hitting 8 times, and his career average was .338.

On **Oct. 5,** Cal Ripken Jr. played in his 3,000th game—all for Baltimore—becoming the 7th player to reach that milestone. Ripken retired on **Oct. 6,** to an ecstatic salute from his fans. His phenomenal achievements included playing in 2,632 consecutive games (1982-98), playing 8,243 consecutive innings over 904 games, playing in 19 All-Star games as a shortstop and 3d baseman, 3,184 hits, 431 home runs, and a career batting average of .276.

Osama bin Laden—A Profile

On Feb. 23, 1998, Osama bin Laden and several other Islamic extremist leaders, acting in the name of the World Islamic Front, issued a *fatwa*, or religious edict, holding that "to kill the Americans and their allies—civilians and military—is an individual duty for every Muslim who can do it in any country in which it is possible to do it." Explicitly declaring a *jihad*, or holy war, against "Jews and Crusaders," the document calls upon "every Muslim who believes in God and wishes to be rewarded to comply with God's order to kill the Americans and plunder their money."

The U.S. government contends that Bin Laden, in line with this agenda, has used his terrorist network, Al Qaeda ("The Base"), and allied groups to carry out a series of operations against the U.S. These include the near-simultaneous bombings of U.S. embassies in Kenya and Tanzania on Aug. 7, 1998, which claimed over 200 lives; the suicide bombing of the U.S.S. *Cole* at Aden, Yemen, on Oct. 12, 2000, killing 17 Americans; and the catastrophic aerial assault on the World Trade Center and the Pentagon on Sept. 11, 2001.

Bin Laden's Background. Born in 1957, Osama bin Muhammad bin Awad bin Laden was the 17th of more than 50 children of Muhammad bin Laden, a Yemeni laborer who became Saudi Arabia's wealthiest construction contractor. When his father died in a plane crash in 1968, he inherited between $20 million and $300 million.

After Soviet troops moved into Afghanistan in Dec. 1979, Bin Laden joined the Mujahedin, the Afghan resistance movement. During the 1980s, the final decade of the cold war, the U.S. and Bin Laden fought on the same side against the Soviet Union. The CIA funneled billions of dollars in funds and military hardware to the Afghan resistance, and Bin Laden supplied money, men, and equipment to the anti-Soviet cause. In the mid-1980s he co-founded Makhtab al-Khidamat (MAK), the "Services Office," which recruited, sheltered, and trained thousands of fighters from dozens of countries for the Afghan war. MAK formed the model for Al Qaeda, which he founded in the late 1980s.

Hatred of the U.S. After Soviet troops pulled out of Afghanistan in 1989, Bin Laden returned to Saudi Arabia and took an active role in opposing the monarchy. He was outraged when the kingdom joined the Persian Gulf War coalition against Iraq and allowed U.S. troops on the Arabian Peninsula.

When the Gulf War ended, Bin Laden fled to Sudan. He reportedly established at least 3 terrorist camps in N Sudan. Suspecting his involvement in terrorist incidents—including a 1993 bomb blast in a garage beneath the World Trade Center and the killing of 18 U.S. troops in Somalia—the U.S. put pressure on both the Saudi and Sudanese governments. In 1994 his Saudi passport was revoked, his Saudi assets were frozen, and the other branches of the Bin Laden family disowned him. Sudan expelled him 2 years later.

Al Qaeda and the Taliban. Bin Laden found refuge in Afghanistan, where he allied himself with the Taliban, an Islamic extremist faction that captured Kabul, the Afghan capital, in Sept. 1996. Bin Laden provided money and fighters for the Taliban cause; in return, the Taliban, whose leader, Mullah Mohammed Omar, was said to be a friend, allowed him to establish terrorist training camps on Afghan soil. Some of those trained fought against India in Kashmir.

From his haven in Afghanistan, Bin Laden in Aug. 1996 issued a "Declaration of Jihad," calling upon the Islamic world to expel the U.S. from the Arabian Peninsula.

By then the CIA had received presidential authorization to use any means available to destroy Al Qaeda. The U.S. made its first overt military strike against Bin Laden on Aug. 20, 1998, 13 days after the embassy bombings. U.S. cruise missiles hit Al Qaeda and allied camps in Afghanistan and a pharmaceutical plant in Khartoum, Sudan; independent inquiries later cast doubt on the U.S. claim that the factory was producing chemical weapons for Bin Laden. On Nov. 4, 1998, he was indicted in federal court in New York City for his role in the embassy bombings. A year after the Taliban refused to hand him over for trial, the UN imposed sanctions on Afghanistan.

Notable Supreme Court Decisions, 2000-2001

Following were some of the major cases of the Supreme Court's 2000–2001 term, which ended June 28:

Civil Rights: The justices Feb. 21 ruled, 5–4, that the 11th Amendment prevented private citizens from suing a state for damages in federal court for violating the Americans With Disabilities Act, unless the state consented *[Board of Trustees of the University of Alabama v. Garrett]*.

The court Apr. 24 ruled, 5–4, that under the 1964 Civil Rights Act, private citizens could sue federally funded state agencies only for acts of intentional discrimination, not for actions that merely had a disparate impact on persons of a particular race or ethnicity *[Alexander v. Sandoval]*.

The court May 29 ruled, 7–2, that under the Americans With Disabilities Act, the Professional Golfers' Association (PGA) Tour Inc. had to allow a disabled player to use a golf cart. The golfer, Casey Martin, had a circulatory disease that made it painful and dangerous for him to walk long distances between holes *[PGA Tour Inc. v. Martin]*.

Criminal Law: On June 4 the court, in a 6–3 decision, for the 2d time overturned the death sentence of Johnny Paul Penry, a retarded Texas inmate, on the grounds that the jury that sentenced him to death had been given wrong instructions *[Penry v. Johnson]*.

The justices May 14 ruled, 8–0, that there was no "medical necessity" exception to federal laws prohibiting the manufacture and distribution of marijuana *[U.S. v. Oakland Cannabis Buyers' Cooperative]*.

Election Law: The court, in a 5–4 decision, Dec. 12, 2000, ruled that manual recounts of the ballots cast in the 2000 presidential race in Florida could not proceed because the inconsistent standards used to evaluate ambiguously marked ballots in different counties violated the equal-protection clause of the U.S. Constitution. The recounts had been sought by Vice Pres. Al Gore (D); the ruling effectively decided the presidential race in favor of his opponent, Texas Gov. George W. Bush (R) *[Bush v. Gore]*.

The court Apr. 18 ruled, 5–4, that North Carolina's 12th Congressional District, whose irregular shape had been challenged as an unconstitutional racial gerrymander, was the permissible result of attempts to create a majority-Democrat district *[Easley v. Cromartie]*.

The court, in a 5–4 decision June 25, upheld federal election laws limiting the amount of money that political parties could spend in coordination with federal candidates *[Federal Election Commission v. Colorado Republican Federal Campaign Committee]*.

Environment: The court Feb. 27 unanimously upheld the power of the Environmental Protection Agency to establish national air quality standards under the 1990 Clean Air Act. The justices said the EPA could consider only public health, not cost-benefit considerations, in setting air quality rules *[Whitman v. American Trucking Association]*.

The justices Jan. 9 ruled, 5–4, that the 1972 Clean Water Act did not give the federal government jurisdiction to regulate the dredging and filling of isolated ponds and wetlands within a state's borders *[Solid Waste Agency v. U.S. Army Corps of Engineers]*.

Immigration: The court June 25 ruled, 5–4, that 2 1996 immigration laws had not deprived the courts of power to hear challenges to deportation orders *[Calcano-Martinez v. Immigration and Naturalization Service (INS); INS v. St. Cyr]*.

The court, June 11 rejected, 5–4, a challenge to an immigration law that made it easier for an out-of-wedlock child born overseas to become a U.S. citizen if the mother rather than the father was American *[Nguyen v. INS]*. The justices June 28 ruled, 5–4, that the U.S. government could not keep a deportable alien in indefinite detention if no country was willing to accept him *[Zadvydas v. Davis]*.

Labor: The court Mar. 21 ruled, 5–4, that most employers could require their workers to submit all work-related disputes to arbitration, rather than a lawsuit *[Circuit City Stores v. Adams]*.

Search and Seizure: The court, in a 5–4 decision Apr. 24, ruled that the police could take a person into custody for a minor infraction whose maximum penalty was a fine. The case involved a woman arrested and jailed for driving without a seat belt *[Atwater v. City of Lago Vista]*.

The justices Nov. 30, 2000, ruled, 6–3, that police roadblocks used to conduct random searches for illegal drugs were unconstitutional *[Indianapolis v. Edmond]*.

The court, in a 6–3 decision Mar. 21, ruled that public hospitals could not test pregnant women for illegal drug use without their consent if the purpose of the tests was to uncover violations of the drug laws, rather than to protect the fetus *[Ferguson v. City of Charleston*].

The court June 11 ruled, 5–4, that police needed a warrant to scan the outside of a home with a thermal imaging device to detect heat lamps used to grow marijuana. The justices held that the thermal imaging was equivalent to a physical intrusion into the home *[Kyllo v. U.S.*].

Speech and Press: The court June 11 ruled, 6–3, that religious and secular organizations were entitled to equal access to public elementary school grounds for after-school meetings. The justices held that a school district's rule banning religious groups violated the First Amendment *[Good News Club v. Milford Central School]*.

The court June 25 ruled, 7–2, that publishers had infringed the copyrights of freelance contributors by making their articles available on electronic databases without permission, after the articles had been published in newspapers and magazines *[New York Times Co. v. Tasini]*.

The court May 21 ruled, 6–3, that journalists could not be held liable for broadcasting an illegally intercepted cellular telephone conversation on a matter of public importance, if they did not play a role in illegal acquisition of the information *[Bartnicki v. Vopper]*.

The court, in a 5–4 decision, June 28 struck down a Massachusetts law that placed strict limits on advertisements for tobacco products, on the grounds that it violated the advertisers' free speech rights and was also partially preempted by federal law *[Lorillard Tobacco v. Reilly]*.

The 2001 Nobel Prizes

The 2001 Nobel Prize winners were announced Oct. 8-12. Each prize consisted of a large solid gold medal and a cash award worth 10 million Swedish kronor (about $950,000).

Chemistry: William S. Knowles of the U.S. and Ryoji Noyori of Japan shared one-half of the award for creating hydrogen-based catalysts that can distinguish between two forms of chiral (mirrored) molecules, a process used in making antibiotics, anti-inflammatory drugs, and heart medicines; K. Barry Sharpless of the U.S. received the other half for analogous work on oxygen-based catalysts.

Memorial Prize in Economic Science: Three Americans—George A. Akerlof, A. Michael Spence, and Joseph E. Stiglitz—were honored for their analyses of "markets with asymmetric information," in which one participant in a transaction has much better information than another.

Literature: V. S. Naipaul, a British writer born in Trinidad, received the award for uniting "perceptive narrative and incorruptible scrutiny" in works such as the novels *The Enigma of Arrival* and *A Way in the World*.

Peace: The prize went, in equal portions, to the UN and to its secretary general, Kofi Annan of Ghana.

Physics: Two Americans, Eric A. Cornell and Carl E. Wieman, and a German, Wolfgang Ketterle, shared the award for their research on Bose-Einstein condensate, a supercold state of matter with important applications for computing, nanotechnology, and precision instruments.

Physiology or Medicine: Two Britons, R. Timothy (Tim) Hunt and Paul M. Nurse, and one American, Leland H. Hartwell, won for describing the molecular mechanisms that regulate the cycle of cell development; the discoveries may lead to new cancer treatments.

Notable Quotes in 2001

"This is not only an attack on the United States, but an attack on the civilized world."

German *Chancellor Gerhard Schroeder,* responding to the Sept. 11 terrorist attacks on the U.S.

"We will make no distinction between the terrorists who committed these acts and those who harbored them."

Pres. George W. Bush, Sept. 11, after the terrorist attacks of the World Trade Center and the Pentagon.

"When I take action, I'm not going to fire a $2 million missile at a $10 empty tent and hit a camel in the butt. It's going to be decisive."

Pres. Bush, talking to a group of senators, Sept. 11, as reported by *Newsweek* magazine.

"Our responsibility to history is already clear: to answer these attacks and rid the world of evil. War has been waged against us by stealth and deceit and murder. This nation is peaceful, but fierce when stirred to anger. This conflict was begun on the timing and terms of others. It will end in a way and at an hour of our choosing. Our purpose as a nation is firm, yet our wounds as a people are recent and unhealed and lead us to pray."

Pres. Bush at a prayer service in Washington, DC, Sept. 14, honoring the victims of the terrorist attacks.

"I can hear *you!* The rest of the world can hear you, and the people who knocked these buildings down will hear from all of us soon."

Pres. Bush during a speech Sept. 14, at the site of the World Trade Center, after a rescue worker there shouted, "We can't hear you."

"Every little thing makes you paranoid. My girlfriends and I went to the movies. When the theater started to shake, I lost it until I realized that it was the subway going by."

Samia Watt, of Queens, NY, talking to a *New York Times* reporter, Sept. 23.

"Bush is my commander in chief."

Former Vice Pres. and presidential candidate *Al Gore Jr.*, speaking at the annual Jefferson Jackson dinner held by the Iowa Democratic Party in Des Moines, Sept. 29.

"The cold war, it took 50 years, plus or minus. It did not involve major battles. It involved continuous pressure. It involved cooperation by a host of nations. And when it ended, it ended not with a bang, but through internal collapse. It strikes me that might be a more appropriate way to think about what we are up against here, than would be any major conflict."

Sec. of State Donald Rumsfeld, in Cairo prior to the Oct. 7 military action against terrorism.

"Neither America nor the people who live in it will dream of security before we live it in Palestine, and not before all the infidel armies leave the land of Muhammad, peace be upon him."

Osama bin Laden (in translation) on a videotape released after the start of the U.S. assault on Afghanistan.

"More than two weeks ago, I gave Taliban leaders a series of clear and specific demands: Close terrorist training camps, hand over leaders of the al Qaeda network, and return all foreign nationals, including American citizens unjustly detained in your country. None of these demands were met. And now, the Taliban will pay a price... To all the men and women in our military... I say this: Your mission is defined. The objectives are clear. Your goal is just."

Pres. Bush, in an address to the nation following launching of military action in Afghanistan, Oct. 7.

"By the time it was over, our candidate had won the popular vote, and the only way they could win the election was to stop the voting in Florida."

Pres. Bill Clinton in a speech in Chicago during a Jan. 2001 farewell tour of the Midwest.

"He can say what he wants to say, but January the 20th I'll be honored to be sworn in as the President."

George W. Bush, in his response to Clinton's comment.

"The amazing thing about this job is that the job seems to follow you around."

Pres. Bush on vacation in Maine on his 55th birthday, July 6.

"The people of America have been overcharged, and on their behalf, I'm here asking for a refund."

Pres. Bush, campaigning for his tax cut plan.

"This budget places short-term, political partisan gratification ahead of the nation's needs."

Sen. Robert Byrd (D, WV) commenting on the budget resolution passed by the Senate, opposed by all but 5 Democrats.

"I tried to walk a line between acting lawfully and testifying falsely, but I now know that I did not fully accomplish that goal."

Former *Pres. Bill Clinton,* after his immunity deal with Independent Counsel Robert Ray.

"I have changed my party label but I have not changed my beliefs."

Vermont *Sen. James Jeffords* announcing that he would leave the Republican Party.

"I've got to leave a little bit early. I've got to mow Jim Jeffords's lawn this afternoon."

Senate Majority leader Tom Daschle at a meeting of the Democratic Leadership Council.

"If statistics are any indication, the system may well be allowing some innocent defendants to be executed."

Supreme Court Justice Sandra Day O'Connor, a longtime supporter of the death penalty.

"In the 90s it was irrational exuberance. Now it may be irrational doom and gloom."

Former *Labor Sec. Robert Reich* on the economy in mid-2001.

"It feels real heavy. The biggest thing is not having a heartbeat. As long as I hear the sound, I know I am still here."

The first recipient of a self-contained artificial heart, *Robert Tools*, describing life after the operation.

"Don't do it; please, kill me if you want."

Prince Nirajan, reportedly trying to stop his brother *Crown Prince Dipendra* from killing their mother during the June 1 massacre in Nepal; Dipendra shot and killed them both.

"It's good for the people to see the king and queen having a hamburger at Planet Hollywood."

King Abdullah of Jordan, commenting on his unceremonious eating habits.

"Lance Armstrong once said that Jan Ullrich is the greatest talent in cycling. This doesn't seem to be the case."

German cyclist *Jan Ullrich* after losing stage 11 of the Tour de France to Armstrong, who went on to win his 3d straight title.

"People will be able to ride a streetcar named Desire. One of the worst things we ever did, as a city, was get rid of the rail line."

New Orleans Mayor *Marc Morial* on plans to bring back the famed streetcar to the French Quarter.

"We expected this. You can't have a trade summit these days without tear gas; it would be like having a cheeseburger without the cheese."

A *senior Bush administration official* discussing protests in Quebec City against the Summit of the Americas, Apr. 20-22.

"I have just made a colossal mistake...Instead of losing 25 pounds, I gained them."

Opera singer *Luciano Pavarotti*, at a Feb. 14 news conference in London.

"We got a little closer than we wanted to."

Tim Nelson, a Penn State sophomore, on the group of 27 students who set an unofficial world record by cramming into a Volkswagen Beetle.

"When you get over 95, every day is your day."

Bob Hope, on celebrating his 98th birthday, May 29, which Los Angeles proclaimed as "Bob Hope Day."

OBITUARIES

Deaths, Oct. 31, 2000 — Oct. 10, 2001

A

Aaliyah, 22, promising hip-hop and rhythm and blues singer and actress: killed in a plane crash in the Bahamas, Aug. 25, 2001.

Adams, Douglas, 49, author of the popular comedic science fiction novel *The Hitchhiker's Guide to the Galaxy*; Santa Barbara, CA, May 11, 2001.

Adler, Larry, 87, U.S.-born harmonica virtuoso who elevated his instrument to concert status; London, England, Aug. 6, 2001.

Adler, Mortimer, 98, American scholar and philosopher who pioneered the "Great Books" program of learning; San Mateo, CA, June 28, 2001.

Agee, Tommie, 58, center fielder whose spectacular catches helped the "Miracle Mets" win the 1969 World Series; New York, NY, Jan. 22, 2001.

Amado, Jorge, 88, Brazil's best-known writer, whose novels captured the vitality of Afro-Brazilian culture; Salvador, Brazil, Aug. 6, 2001.

Ammons, A(rchie) R(andolph), 75, prize-winning poet best known for the long poem *Garbage*; Ithaca, NY, Feb. 25, 2001.

Anthony, Earl, 63, professional bowler who in the early 1980s became bowling's first million-dollar winner; New Berlin, WI., Aug. 14, 2001.

Atkins, Chet, 77, guitarist and producer who helped create the "Nashville sound" in country music; Nashville, TN, July 2, 2001.

Aumont, Jean-Pierre, 90, renowned French actor known for such films as Truffaut's *Day for Night* (1973); St. Tropez, France, Jan. 30, 2001.

B

Balthus (Balthasar Klossowski), 92, reclusive French artist, best known for his paintings of young girls; Le Rossiniere, Switzerland; Feb. 18, 2001.

Barnard, Christiaan, 78, South African surgeon who in 1967 performed the first human-to-human heart transplant; Paphos, Cyprus, Sept. 2, 2001.

Barty, Billy, 76, diminutive actor whose 7-decade career included *Willow* (1988) and *Day of the Locust* (1975); Glendale, CA, Dec. 23, 2000.

Birenda, King Bikram, 55, king of Nepal, shot in his palace along with other members of the royal family, by his son; guided Nepal to democracy; Kathmandu, Nepal, June 1, 2001.

Borge, Victor, 91, Danish-born pianist who blended comedy and classical music, to resounding success; Greenwich, CT, Dec. 23, 2000.

Boudreau, Lou, 84, baseball Hall of Famer; led the Cleveland Indians to their last Series title to date, 1948; Olympia Fields, IL, Aug. 10, 2001.

Brooks, Gwendolyn, 83, poet who was the first black American writer to win a Pulitzer; Chicago, IL, Dec. 3, 2000.

Brown, Les, 88, big "Band of Renown" leader, long associated with Bob Hope; Los Angeles, CA, Jan. 4, 2001.

C

Callaway, Ely, 82, founder of the company that became the world's largest golf club maker; Rancho Santa Fe, CA, July 5, 2001.

Caras, Roger, 72, author, broadcast personality, animal welfare advocate; Towson, MD, Feb. 18, 2001.

Chaban-Delmas, Jacques, 85, prime minister of France (1969-72); hero in the French Resistance during World War II; Paris, France, Nov. 10, 2000.

Coca, Imogene, 92, Sid Caesar's first comedy partner on NBC's classic *Your Show of Shows* in the 1950s; Westport, CT, June 2, 2001.

Como, Perry, 88, pop singer whose easy-going manner and breezy tunes made him a star beginning in the mid-1940s; Jupiter, FL, May 12, 2001.

Cormier, Robert E., 75, author of acclaimed books for young adults, including *The Chocolate War* (1974); Leominster, MA, Nov. 2, 2000.

Cranston, Alan, 86, longtime U.S. senator from California; implicated in savings & loan scandal; Los Altos, CA, Dec. 31, 2000.

D

Davis, James (Jimmie), 101(?), country music singer, twice Lousiana governor (1944-48, 1960-64); composed "You Are My Sunshine" (1940); Baton Rouge, LA, Nov. 5, 2000.

de Camp, L. Sprague, 92, prolific novelist; early recipient of the Grand Master Award for science fiction and fantasy; Plano, TX, Nov. 6, 2000.

De la Beckwith, Byron, 80, white supremacist convicted (1994) in 1963 killing of civil rights leader Medgar Evers, Jackson, MS, Jan. 21, 2001.

De Valois, Dame Ninette, 102, dancer, choreographer, and founding director of London's Royal Ballet; London, England, Mar. 8, 2001.

Dionne, Yvonne, 67, one of the quintuplets whose 1934 birth was hailed as a medical miracle; Montreal, Canada, June 23, 2001.

Donahue, Troy, 65, actor whose appeal to teenagers made him a superstar in the early 1960s; Santa Monica, CA, Sept. 2, 2001.

Downey Jr., Morton, 67, confrontational 80s talk show host; Los Angeles, CA, Mar. 12, 2001.

E

Earnhardt, Dale, 49, stock car racing great, killed in crash at Daytona Internat. Speedway, Daytona Beach, FL, Feb. 18, 2001.

Evans, Dale, 88, star of western movies and TV with her husband, Roy Rogers; Apple Valley, CA, Feb. 7, 2001.

Evans, Rowland, Jr., 79, syndicated conservative Washington columnist and commentator; Washington, DC, Mar. 23, 2001.

F

Francis, Arlene, 93, actress known for her bubbly wit, long featured on the TV game show *What's My Line?*; San Francisco, CA, May 31, 2001.

G

Gebel-Williams, Gunther, 66, German-born animal trainer whose death-defying acts thrilled circus-goers; Venice, FL, July 19, 2001.

Gierek, Edward, 88, Polish Communist leader (1970-1980), toppled by food protests and labor unrest; Cieszyn, Poland, July 29, 2001.

Gilliam, Joe, 49, one of the first black starting quarterbacks in the NFL, with the Pittsburgh Steelers in 1974; Nashville, TN, Dec. 25, 2000.

Goetz, Robert, 90, surgeon who in 1960 performed the first successful coronary bypass operation; Scarsdale, NY, Dec. 15, 2000.

Graham, Katharine, 84, pioneering publisher; made the *Washington Post* into one of the most important U.S. newspapers; Boise, ID, July 17, 2001.

Greco, Jose, 82, dancer and choreographer who popularized flamenco dance forms in the U.S.; Lancaster, PA, Dec. 31, 2000.

Greer, Jane, 76, film noir actress of the 1940s; Los Angeles, CA, Aug. 24, 2001.

Groza, Lou, 76, star place-kicker for the NFL Cleveland Browns during the 40s, 50s, and 60s; elected to the Pro Football Hall of Fame in 1974; Middleburgh Heights, OH, Dec. 6, 2000.

H

Hanna, William, 90, half of the Hanna-Barbera team that created *The Flintstones*, *The Jetsons*, and *Tom and Jerry;* North Hollywood, CA, Mar. 22, 2001.

Henderson, Joe, 64, jazz saxophonist and composer; San Francisco, CA, June 30, 2001.

Herblock (Herbert Lawerence Block), 91, *Washington Post* editorial cartoonist; his satirical targets included Sen. Joseph McCarthy and Pres. Richard Nixon; Washington, DC, Oct. 7, 2001.

Hewlett, William, 87, co-founder of global computer and electronics powerhouse Hewlett-Packard, Palo Alto, CA, Jan. 12, 2001.

Hooker, John Lee, 83, legendary bluesman whose roots were in the Mississippi Delta but whose influence became worldwide; Los Altos, CA, June 21, 2001.

Hoyle, Sir Fred, 86, British astronomer who explained how elements form in stars and coined the term "big bang"; Bournemouth, Eng, Aug. 20, 2001.

I

Ingrid, Queen Mother, 90, mother of Queen Margrethe II of Denmark, sat on Danish throne as wife of King Frederik IX, 1947-72; near Copenhagen, Denmark, Nov. 7, 2000.

J

Johnson, J. J., (James Louis), 77, one of the world's greatest jazz trombonists; Indianapolis, IN; Feb. 4, 2001.

K

Kabila, Laurent, 61, Congolese president who toppled dictator Mobutu Sese Seko in 1997; assassinated in Kinshasa, Congo, Jan. 16, 2001.

Kael, Pauline, 82, provocative film critic, long associated with the *New Yorker* magazine; Great Barrington, MA, Sept. 3, 2001.

Karnilova, Maria, 80, actress who earned a Tony for her portrayal of Golde in *Fiddler on the Roof* (1964); New York, NY, Apr. 20, 2001.

Ketcham, Hank, 81, cartoonist who created freckle-faced trouble-maker Dennis the Menace; Carmel, CA, June 1, 2001.

Kiam, Victor, 2d, 74, Remington CEO who promoted his shavers on TV; Stamford, CT, May 27, 2001.

Klemperer, Werner, 80, Emmy-winning actor; played Nazi Col. Wilhelm Klink on *Hogan's Heroes*; New York, NY, Dec. 6, 2000.

Kramer, Stanley, 87, Oscar-winning director whose films include *Guess Who's Coming to Dinner* (1967) and *Inherit the Wind* (1960); Woodland Hills, CA, Feb. 19, 2001.

L

Lardner, Ring Jr. 85, Oscar-winning screenwriter; blacklisted and jailed for refusing to testify before Congress as to alleged Communist affiliations; New York, NY, Oct. 31, 2000.

Lemmon, Jack, 76, screen legend and two-time Oscar winner; his films include *Some Like It Hot* (1959), *The Days of Wine and Roses* (1962), and *The Odd Couple* (1968); Los Angeles, CA, June 27, 2001.

Lewis, John, 80, pianist and founder of the Modern Jazz Quartet; New York, NY; Apr. 8, 2001.

Lindbergh, Anne Morrow, 94, aviation pioneer, best-selling author, and wife of aviator Charles Lindbergh; Passumpsic, VT, Feb. 7, 2001.

Lindsay, John, 79, mayor of New York City during the turbulent late 60s and early 70s; Hilton Head Island, SC, Dec. 19, 2000.

Ludlum, Robert, 73, author of best-selling suspense novels, starting with *The Scarlatti Inheritance* (1971); Naples, FL, Mar. 12, 2001.

M

Maas, Peter, 72, author; wrote about crime and corruption in *The Valachi Papers* (1969) and *Serpico* (1972); New York, NY, Aug. 23, 2001.

Mansfied Mike, 98, former U.S. senator (D, MT); was majority leader (1961-77) Washington, DC, Oct. 5, 2001.

Masters, William H., 85, scholar and author who teamed with colleague Virginia E. Johnson to revolutionize American attitudes about sex; Tucson, AZ, Feb. 16, 2001.

Mathews, Eddie, 69, Hall of Fame third baseman for the Braves who hit 512 home runs in the 50s and 60s; San Diego, CA, Feb. 18, 2001.

Maxim, Joey, 79, light heavyweight champ (1950-52); outlasted Sugar Ray Robinson in a famous 1952 fight; West Palm Beach, FL, June 2, 2001.

McGuire, Al, 72, Hall of Fame college basketball coach and broadcaster; Milwaukee, WI, Jan. 26, 2001.

McGuire, Dorothy, 83, stage and screen actress of the 1940s known for playing kind, soft-spoken women; Santa Monica, CA, Sept. 13, 2001.

Moakley, Joe (John Joseph), 74, popular Democratic congressman from Boston (1973-2001); Bethesda, MD, May 28, 2001.

Montgomery, George, 84, Hollywood Western actor; long married to singer Dinah Shore; Rancho Mirage, CA, Dec. 12, 2000.

Muhammad, Khalid Abdul, 53, controversial Nation of Islam leader dismissed in 1994 after an inflammatory speech; Marietta, GA, Feb. 17, 2001.

N

Narayan, R. K., 94, world-renowned Indian novelist, who set his stories in the fictional town of Malgudi; Madras, India, May 13, 2001.

O

O'Connor, Carroll, 76, classically trained actor who played working class Archie Bunker in the groundbreaking TV series *All in the Family*; Culver City, CA, June 21, 2001.

O'Farrill, Chico, 79, Cuban-born big band arranger, composer, and one of the creators of Afro-Cuban jazz; New York, NY, June 28, 2001.

Olson, Barbara, 45, conservative commentator; killed in crash of hijacked plane; Somerset Co., PA, Sept. 11, 2001.

P

Paz Estenssoro, Victor, 93, four-time Bolivian president who, in the 1980s, undid features of a revolution he crafted in the 1950s; Tarija, Bolivia, June 7, 2001.

Phillips, John, 65, singer-songwriter and founding member of the folk-pop group the Mamas and the Papas; Los Angeles, CA, Mar. 18, 2001.

Potamkin, Meyer, 91, banker, philanthropist, and art collector; Philadelphia, PA, July 8, 2001.

Q

Quine, W(illard) V(an) O(rman), 92, renowned, empirically oriented Harvard University logician and philosopher; Boston, MA, Dec. 25, 2000.

Quinn, Anthony, 86, Oscar-winning actor who played the title role in *Zorba the Greek*, among other big parts during a 6-decade career; Boston, MA, June 3, 2001.

R

Rabin, Leah, 72, widow of assassinated Israeli Prime Min. Yitzhak Rabin; a forceful campaigner for Middle East peace; Jerusalem, Nov. 12, 2000.

Ramone, Joey, 49, punk rock pioneer whose band, the Ramones, sparked a rock revolution; New York, NY, Apr. 15, 2001.

Reagan, Maureen, 60, politically active daughter of former U.S. Pres. Ronald Reagan; recently an advocate for victims of Alzheimer's disease; Granite Bay, CA, Aug. 8, 2001.

Rhodes, James, 91, Republican governor of Ohio who sent the National Guard to Kent State University in 1970, resulting in the deaths of 4 students; Columbus, OH, Mar. 4, 2001.

Richler, Mordecai, 70, satirical Jewish Canadian writer best known for his novel *The Apprenticeship of Duddy Kravitz* (1959); Montreal, Canada, July 2, 2001.

Robards, Jason, 78, renowned stage and film actor who won a Tony for *Disenchanted* (1959) and Academy Awards for *All the President's Men* (1977) and *Julia* (1978); Bridgeport, CT, Dec. 26, 2000.

Rogers, William P., 87, secretary of state under Pres. Richard Nixon (1969-73); Bethesda, MD, Jan. 2, 2001.

S

Schindler, Rabbi Alexander, 75, influential Reform Judaism leader (1973-96); Westport, CT, Nov. 15, 2000.

Sithole, Rev. Ndabaninge, African nationalist leader in the former Rhodesia; a rival to Zimbabwean ruler Robert Mugabe; near Philadelphia, PA, Dec. 12, 2000.

Son Sann, 89, former Cambodian prime minister, opponent of the Khmer Rouge; Paris, France, Dec. 19, 2000.

Sothern, Ann, 92, actress best known for her starring role as a brassy "private secretary" in a 1950s sitcom; Ketchum, ID, Mar. 15, 2001.

Stanley, Kim, 76, actress acclaimed for her Broadway roles in William Inge's *Bus Stop* (1953) and *Picnic* (1955); Santa Fe, NM, Aug. 20, 2001.

Stargell, Willie, 61, Hall of Fame slugger who played 21 seasons for the Pittsburgh Pirates; Wilmington, NC, Apr. 9, 2001.

Stassen, Harold, 93, Republican governor of Minnesota who ran at least 10 times for president of the U.S.; Bloomington, MN, Mar. 4, 2001.

Stern, Isaac, 81, world-renowned violinist known for his wide repertoire and efforts to save Carnegie Hall; New York, NY, Sept. 22, 2001

Sullivan, Rev. Leon, 78, civil rights leader whose *Principles* guided U.S. businesses in apartheid-era South Africa; Scottsdale, AZ, Apr. 24, 2001.

T

Thieu, Nguyen Van, 78, president of Vietnam (1967-75) during most of the war; Boston, MA, Sept. 30, 2001.

Trout, Robert, 91, legendary member of Edward R. Murrow's famed CBS news team; remembered for historic D-Day broadcast; New York, NY, Nov. 14, 2000.

Tupolev, Alexei Andreyevich, 75, Russian aircraft designer and industry magnate; designed the first supersonic passenger jet (the TU-144); Moscow, Russia, May 12, 2001.

W

Walston, Ray, 86, Tony and Emmy winning actor who played the alien on TV's *My Favorite Martian;* Beverly Hills, CA, Jan. 1, 2001.

Waugh, Auberon, 61, biting British journalist and essayist, son of Evelyn; Taunton, England, Jan. 16, 2001.

Welty, Eudora, 92, Pulitzer Prize-winning fiction writer who often focused on life in her native Mississippi; Jackson, MS, July 23, 2001.

Williams, Hosea, 74, civil rights activist and Martin Luther King lieutenant; Atlanta, GA, Nov. 16, 2000.

Windsor, Marie, 80, actress who played femme fatales in *Narrow Margin* (1952) and *The Killing* (1956); Beverly Hills, CA, Dec. 10, 2000.

X

Xenakis, Iannis, 78, composer of complex, largely electronic music; Paris France, Feb. 4, 2001.

Z

Zatopek, Emil, 78, Czech runner; set Olympic records in the 5,000m, 10,000m, and marathon, 1952; Prague, Czech Rep., Nov. 22, 2000.

Offbeat News Stories, 2001

Give Me an "S," Give Me a "P." In the highly competitive world of civic bragging rights, Springfield still reigns supreme—Springfield, Massachusetts, that is. According to the U.S. Board of Geographic names, there are more than 60 cities and neighborhoods across the country that share the name Springfield, and the latest figures from Census 2000 confirm that Springfield, MA, with a population of 152,082, outstrips all other Springfields, retaining its "census title" for at least another decade. Runner-up Springfield, MO, is gaining ground though, and trails by only 502 citizens, closing the gap 8% since 1990. Other top Springfields include Springfield, IL, with about 111,000 people; Springfield, OH, with about 65,000; and Springfield, OR, with about 53,000.

Mr. Showbiz. For an entire year, the man legally known as DotComGuy participated in his own brand of "reality" programming. DCG, as he is known to his friends, entered a house in Dallas, TX, with nothing but a computer and an Internet connection. For the next 12 months he documented his entire life via webcams (except in the bathroom). He furnished the house, shopped and surfed online, ate, slept, and gave hundreds of interviews, including one to TV personality Ed McMahon, without leaving the house. Though he had sponsors to help with the costs, and an estimated 300,000 to 400,000 Internet visitors a month, the DotComGuy project only managed to break even. "I'm not the most entertaining guy," he admitted. However, DCG's efforts were not wholly without reward. When he emerged from his Internet exile at 12:01 A.M. Central Time, Jan. 1, 2001, he had something he didn't have when he started: a fiancée.

Shoe City. The world's most infamous shoe collector may have finally run out of closet space. Former Philippines first Lady Imelda Marcos—who reportedly left behind about 1,200 pairs of shoes when she and her late husband ex-president Ferdinand Marcos fled the country in disgrace in 1986—has opened a museum dedicated to footwear, and boosting tourism. Located in Manila, the Marikina City Footwear Museum houses hundreds of pairs of shoes donated by Imelda, as well as others from local politicians and film stars. The former first lady has said on many occasions that the real reason she bought so many pairs of shoes was to help support the Marikina shoe industry.

Lizard Rock. *Masiakasaurus knopfleri* is one of the newest, and by far the most culturally hip, dinosaurs to be discovered. A team of scientists, led by Dr. Scott Sampson of the University of Utah, decided to name the 6-foot-long, meat-eating dinosaur after Mark Knopfler, singer/songwriter for the rock band Dire Straits, because the team listened to his music as they dug in 100-degree heat on the island of Madagascar and felt that it brought them luck in discovering the creature. For the record, the name translates roughly as "vicious lizard knopfler."

And the Loser Is . . . There were no limos, no expensive gowns, and no long acceptance speeches at this award ceremony. In fact, the "winners" never even showed up at the 21st annual Golden Raspberry Awards, traditionally held the day before the Academy Awards. The "Razzies" (Super-8 film reels spray-painted gold) are "awarded" to the worst that Hollywood has to offer. In 2001, the 535 members of the Golden Raspberry Award Foundation voted *Battlefield Earth*, an adaptation of an L. Ron Hubbard science fiction novel, as the worst film in a record-tying 7 categories, including picture, actor (John Travolta), supporting actress (Kelly Preston), and supporting actor (Barry Pepper). For the record, the leader in career Razzie awards is Sylvester Stallone, with 8.

A Pop Is Born. Just in time for the summer blockbuster movie season, Dum Dum Pops came up with a fresh new flavor of lollipop—Buttered Popcorn. (Care for a Root Beer or Cream Soda-flavored lollipop with that?) The 95-year-old Spangler Candy Company last introduced a new flavor (blue raspberry) more than five years ago.

Faucets of Water. It's doubtful an inspirational film like *Chariots of Fire* will ever be made about them, but the "Hand Washing Olympics" do promote good hygiene. Held annually at the Food Safety Summit and Expo in Washington, DC, the competition uses a plastic fluorescent particle solution instead of real germs and highlights a "dirty" secret: most people don't wash their hands as well or as often as they should, spending an average of 5-7 seconds on a task that should take 20 seconds. The most commonly missed areas are the cuticles, between the fingers, and the back of the hand. (Reportedly, about one out of every three people don't wash their hands at all after using a public restroom.) For the record, Michelle Samariya-Timm, a graduate student in public health at Montclair State University (NJ), came away with the gold medal, scoring a perfect "germ-free" 100.

Everybody's Got a Cell Phone These Days. The staff of Scotland's Blair Drummond Safari Park was plagued by a heavy-breathing prankster who phoned them dozens of times, at all hours, for 3 straight nights in July. The park manager was set to call the police when the mystery was solved-with a loud shriek. A search of the chimpanzee enclosure turned up a cell phone in the straw bedding of 11-year-old "Chippy." His keeper, Gary Gilmour, did notice that his phone was missing but never guessed it was Chippy who had picked his pocket. Chippy somehow accessed numbers stored in the phone and used the redial button. "Chimps are very intelligent and Chippy is certainly one of the smartest ones I've worked with," Gilmour said. Chippy seemed a little depressed when they took the phone away but appeared to be enjoying his newfound celebrity.

Who Needs Lemonade? During the long, hot days of summer enterprising kids have been known to make a few bucks peddling lemonade at a corner stand—but how about setting up a toll road and charging cars 50 cents to go through? That's what Sean Jordan, his siblings, Brandon and Emily, and their cousin Devon Johnson decided to do when road construction caused a huge traffic jam near their house near Corvalis, OR. The young entrepreneurs cleared an old railroad line that runs parallel to the main road and straight through the Jordan's property, made detour signs for the drivers, and started collecting. Though their money was stolen the first night (and their signs trashed), and some cars did speed by without paying, they made $270 in an 11-day period.

Only $25,000? The world's first marathon at the South Pole, scheduled for Jan. 2002, was accepting applications in the fall of 2001, but with a limit of 75 runners it might have been expected to fill up fast—or not. The fee was $25,000 and entrants had to meet strict requirements, including a medical evaluation before facing the ice, drifting snow, and 9,000-ft. course elevation. Event organizers likened running in polar snow to running in mud, and said athletes should expect to double the amount of time it would normally take them to finish. The southern summer does provide 24-hour daylight, but temperatures can still fall to –13 degrees. No mention was made of a psychological evaluation for entry.

Cheese Loophole. Though the Bush White House imposed a moratorium on implementing last-minute rules issued by the Clinton administration, the USDA was able to put through its new cheese standard. Calling it a "notice" instead of a regulation (the standards are technically voluntary), officials announced that beginning Feb. 22, 2001, the holes, or "eyes," in Grade A Swiss cheese could be as small as 3/8 inch in diameter. The smaller eyes will help keep the cheese from getting tangled in high-speed slicing machines. (Bacteria that emit carbon dioxide are responsible for the bubbles that form the holes in Swiss cheese.)

The Dog Ate My Tax Return. It was reported in August that at least 40,000 tax returns and payments totaling $810 million had disappeared from an IRS processing center run by Mellon Bank in Pittsburgh. The bank's chairman, Martin McGuinn, said employees, apparently feeling they were behind in their work, had hidden, and in some cases, shredded returns. The IRS terminated its contract with Mellon which fired several employees allegedly involved. A federal investigation was in progress.

Historical Anniversaries

1902 — 100 Years Ago

Pasadena hosts the first Rose Bowl game, and first post-season football game, **Jan. 1**.

Pres. Theodore Roosevelt announces, **Mar. 10**, that J.P. Morgan's Northern Security Company will be prosecuted by the federal government under the Sherman Antitrust Act.

Cecil Rhodes dies, **Mar. 26**, in Africa and leaves most of his fortune for Oxford University scholarships.

United Mine Workers president John Mitchell leads anthracite coal miners in a crippling 5-month strike that ends when workers agree to accept arbitration **Oct. 21**.

The U.S. withdraws troops, **May 20,** and Cuba becomes independent 4 years after the Spanish-American War.

Alphonso XIII, age 16, assumes full authority as king of Spain **May 16.**

Boer leaders sign the Treaty of Vereeniging **May 31,** ending the Boer War with Britain.

Congress passes a measure, **June 28,** providing for the construction of a canal across the isthmus of Panama.

Pres. Roosevelt suffers minor injuries, and a Secret Service agent is killed, when Roosevelt's coach collides with a trolley car in Pittsfield, MA, **Sept. 3.**

Art. *Horseman on the Beach* by Paul Gauguin; *Waterloo Bridge* by Claude Monet; *The Misses Hunter* and *Lord Ribblesdale,* by John Singer Sargent; bronze sculpture *Comin' Through the Rye* by Frederic Remington.

Literature. "Heart of Darkness" by Joseph Conrad; *The Hound of the Baskervilles* by Sir Arthur Conan Doyle; *The Wings of the Dove* by Henry James; *The Tale of Peter Rabbit* by Beatrix Potter; *The Virginian* by Owen Wister; *The Pothunters* by P.G. Wodehouse.

Nonfiction. *Varieties of Religious Experience* by William James; *The Story of My Life* by Helen Keller.

Theater. *The Admirable Crichton* by James M. Barrie; *The Girl With the Green Eyes* and *The Stubbornness of Geraldine* by Clyde Fitch; *The Lower Depths* by Maxim Gorky; *Old Heidelberg* by Wilhem Meyer-Forster; *Cathleen ni Houlihan* by William Butler Yeats; the musicals *Floradora*, *The Chinese Honeymoon,* and *Twirly Whirly.*

Music. *Prélude à après-midi d'un faune* by Claude Debussy; *Symphony No. 3 in D minor* by Gustav Mahler; *Pavane pour une infante defunte* by Maurice Ravel; *Symphony No. 2* by Jean Sibelius; the operas *Pelléas et Mélisande* by Claude Debussy, *The Girls of Vienna* by Franz Lehar, and *Le jongleur de Notre-Dame* by Jules Massenet.

Popular Songs. "Bill Bailey, Won't You Please Come Home" by Hughie Cannon; "In the Good Old Summertime" by George Evans.

Science and Technology. Hookworm disease is discovered by Dr. Charles W. Stiles; the Aswan Dam across the Nile River in Egypt is completed; the brassiere is invented by Charles R. Debevoise; Andrew Carnegie establishes the Carnegie Institution of Washington for the furthering of scientific research; the American Automobile Association is founded in Chicago; a new process of producing rayon is patented by chemist A.D. Little; the New York Central's *Twentieth Century Limited* and Pennsylvania Railroad's *Broadway Limited* go into service.

Sports. The U.S. tennis team defeats Britain to win the Davis Cup title; the National League baseball championship is won by Pittsburgh and the American League championship by Philadelphia.

Miscellaneous. Edwin Binney develops Crayola-brand crayons; the U.S. Army uniform changes from blue to olive drab; Enrico Caruso makes the first commercial phonograph recording; the Algonquin Hotel opens on 44th Street in New York City; Horn & Hardart opens its first "Automat," in Philadelphia; the National Biscuit Company introduces Barnum's Animal Crackers; fish-and-chip shops open in London; the Flatiron Building in New York City is completed.

1952 — 50 Years Ago

George VI of England dies **Feb. 6,** and his daughter Elizabeth becomes queen.

The first atomic submarine, *Nautilus,* is dedicated **June 14** at Groton, CT.

Egypt's King Farouk is forced to abdicate **July 26** after a coup led by Gen. Mohammed Naguib.

Pres. Harry S. Truman signs, **July 16,** a G.I. Bill of Rights for Korean War veterans.

The world's first hydrogen bomb is detonated, **Nov. 1,** by the United States on the Marshall Islands.

On **Nov. 4**, Dwight D. Eisenhower is elected U.S. president, with Sen. Richard M. Nixon as vice president.

Art. *The Green Night* by Marc Chagall; *The Pink Violin* by Raoul Dufy; *Mountains and Sea* by Helen Frankenthaler; *Woman I* by Willem de Kooning; *End of Autumn* by Georges Rouault; sculpture *Madonna and Child* by Jacob Epstein.

Literature. *Invisible Man* by Ralph Ellison; *Giant* by Edna Ferber; *Doting* by Henry Green; *The Old Man and the Sea* by Ernest Hemingway; *The Natural* by Bernard Malamud; *The Groves of Academe* by Mary McCarthy; *Wise Blood* by Flannery O'Connor; *East of Eden* by John Steinbeck; *The Long March* by William Styron; *Player Piano* by Kurt Vonnegut Jr.; *Charlotte's Web* by E. B. White; (play) *Waiting for Godot* by Samuel Beckett; (poetry) *Collected Poems 1917-1952* by Archibald MacLeish; "Do not go gentle into that good night" by Dylan Thomas.

Nonfiction. *Witness* by Whittaker Chambers; *The Power of Positive Thinking* by Norman Vincent Peale; *Amy Vanderbilt's Complete Book of Etiquette.*

Theater. *The Waltz of the Toreadors* by Jean Anouilh; *The Seven-Year Itch* by George Axelrod; *Jane* by S. N. Behrman; *The Grass Harp* by Truman Capote; *The Mousetrap* by Agatha Christie; *Quadrille* by Noel Coward; *Venus Observed* by Christopher Fry; *The Chairs* by Eugene Ionesco; *Dial 'M' for Murder* by Frederick Knott; *The Shrike* by Joseph Kramm; *The Deep Blue Sea* by Terence Rattigan; musicals: *New Faces*; *Wish You Were Here.*

Music. Boston Symphony makes its first European tour; *Trouble in Tahiti* by Leonard Bernstein; *Water Music* by John Cage; symphony *Die Harmonie der Welt* by Paul Hindemith; *West Point Suite* by Darius Milhaud; *Symphonie Concertante in E minor* by Sergei Prokofiev; *Sea Piece With Birds* by Virgil Thomson; George Gershwin's *Porgy and Bess* tours in Europe.

Popular Songs. "Blue Tango" by Leroy Anderson; "Count Your Blessings (Instead of Sheep)" by Irving Berlin; "I Saw Mommy Kissing Santa Claus" by Tommy Conner; "Lullabye of Birdland" by George Shearing; "Do Not Forsake Me" by Dmitri Tiomkin; "Your Cheatin' Heart" by Hank Williams; "Don't Let the Stars Get in Your Eyes" by Slim Willet.

Movies. *High Noon; Ikuru; Limelight; Pat and Mike; The Quiet Man; The Red Badge of Courage; The Seven Deadly Sins; Umberto D; Viva Zapata!;* musicals: *Hans Christian Andersen* with Danny Kaye, *Singin' in the Rain* with Gene Kelly and Debbie Reynolds. Academy Awards given out in 1952: best picture *An American in Paris;* best actor Humphrey Bogart in *The African Queen;* best actress Vivien Leigh in *A Streetcar Named Desire.*

Television. *The Today Show* debuts on NBC, starring Dave Garroway; *Dragnet* and *The Adventures of Ozzie and Harriet* also debut.

Science and Technology. The passenger liner *S.S. United States* sets a new record crossing the Atlantic Ocean; archaeologist Michael Ventris deciphers Mycenaean texts that date from 1450 BC; Jonas Salk tests his vaccine against polio; Chicago's G. D. Searle laboratories develop a contraceptive tablet.

Sports. U.S. Summer Olympic team wins 40 gold medals and finishes first in team rankings in Helsinki, Finland; the New York Yankees defeat the Brooklyn Dodgers in the World Series; Troy Ruttman, 22, becomes the youngest driver to win the Indianapolis 500; Sam Snead wins the Masters golf tournament; Rocky Marciano defeats "Jersey Joe" Walcott for the world heavyweight boxing championship; Jackie Robinson and Hank Sauer of the National League hit home runs to win the All-Star Game.

Miscellaneous. Kellogg's Sugar Frosted Flakes are introduced; the first Holiday Inn opens, on U.S. Highway 70 in Memphis; many "unidentified flying objects" are reported.

Helping Children Cope in Today's World

By Fred Rogers

Mr. Fred Rogers served as producer and host of the children's TV program *Mister Rogers' Neighborhood* for more than 30 years. While the last new episode was aired in 2001, the series continues to be broadcast to the latest generation of eager young viewers. An ordained Presbyterian minister, Fred Rogers has written widely about children and remains active in a variety of projects as head of Family Communications, Inc., in Pittsburgh.

Right beside my chair at Family Communications, Inc. is a piece of calligraphy which a good friend gave to me years ago. It's a quotation from *The Little Prince* by Antoine de St. Exupery: "L'essentiel est invisible pour les yeux." ("What is essential is invisible to the eyes.") Those may seem like unexpected words to highlight in *The World Almanac,* where trends and changes are made very visible; nevertheless, I find those words, and what they stand for, to be more valuable every day. I thought of them especially in the wake of the "attacks" of Sept. 11, 2001.

Looking Behind the Figures

It takes time to go beyond what our eyes are seeing and to begin to listen with our hearts; nevertheless, that's the only way that we'll come to know truly what's happening in our society—by looking and listening behind and beyond the statistics, attempting to understand what helps families cope with the world of today.

Our *Neighborhood* television program has been broadcast on national public television for 33 seasons. People often ask what we've changed in our broadcasts over the years. Our answer: "Not much at all." To be sure, the outsides of many children's lives have changed—the figures in this almanac attest to that—divorce, number of children living with grandparents or in foster care, homelessness. But the insides of children's lives have not changed much at all! We human beings evolve very slowly, and, of course, that which remains the same is that which is truly essential (and invisible)!! Deep inside us, we want...and need...to know that we are loved and lovable, that we are capable of giving and receiving love.

It is only through our relationships that we come to know that we are loved. Children inevitably become aware of events in the outside world that may be upsetting, even frightening. Every child needs a place—a home, a room, a lap—where, in the company of a caring adult, he or she can feel safe. In this sanctuary, a child's sense of being a valuable person, and the feeling that life is worth the effort to live, can begin to take shape. Hope and trust that the world is, on the whole, a good place starts here. That's the foundation for all learning and healthy growing.

Dealing with Feelings

In the safe harbor of a caring relationship with an adult, children learn healthy ways to deal with their feelings about the outside changes in their lives. Saying goodbye to a parent at the beginning of a workday, dealing with divorce or domestic violence, or moving to another city—all that evokes strong feelings, and those feelings can be overwhelming and frightening. In *Mister Rogers' Neighborhood,* we have a saying: "Feelings are mentionable, and whatever is mentionable can be more manageable." What a gift we all give to children when we encourage them to talk about what makes them happy, jealous, angry, shy, afraid, or proud.

There's a *Neighborhood* song that we sing many times on our programs. It too seems to be more important than ever in today's world: "What do you do with the mad that you feel?" If we understand that anger is a reaction to feeling powerless or helpless we can better understand children's anger and help them know that it's okay to be angry, but we must help them also know that it's not okay to hurt anyone. We need to encourage them to find constructive ways of dealing with their so-called negative feelings...ways like pounding clay, playing instruments, running fast, stomping, building with blocks, drawing angry pictures. We also need to make limits and rules in order to help children develop self-control so they can stop themselves when they're about to do something that may hurt themselves or others. What a perfect time to say, "I'm proud of you"—when a child is about to hit someone but hesitates, finds control, and holds back.

Of course, most very young children don't yet have the ability to talk at length about what they're feeling. It's mostly through their play that they are able to deal with the stresses in their lives. Children need open-ended playthings, so they can make their fantasy world be what they need. They need to be able to control what happens to their toys, cars, or stuffed animals, even though they aren't able to control many of the other things in their lives. Obviously, they can't "control" their parents' divorce or a move to another home or a new baby's arrival into the family. Children need to have a chance to re-work their feelings about what has happened in their lives, and they do this through their play. They can also "rehearse" something that's expected to take place in the future, so that they can manage better when it actually happens (like saying "goodbye" or having to get an immunization!!).

Keeping Some Things the Same

Another way to deal with the changes in today's world is to ensure that some things don't change. The *Neighborhood* programs produced in 2001 look much the same as the ones we made in the 1960s. I always put on a sweater and sneakers at the beginning of the television visit, and put them away at the end. I feed the fish. I encourage pretending through the Neighborhood of Make-Believe. The Trolley brings us back to reality. I close with the same song each day.

Children feel far more comfortable and secure when things happen predictably, with routines, rituals, and traditions. Those traditions, big or small, create anchors of stability, especially in rough seas. The wise psychologist Erik Erikson has said that "tradition is to humans what instinct is to animals." Traditions enable us to survive.

In family life, small everyday traditions can be just as important as the ones we typically remember from holiday times. I know of one family that has a "silly supper" every Wednesday evening: the children dress in pajamas, they eat bowls of cereal and fruit, and spend time talking about what happened to them that day—the silly things and also the tough things. The whole family looks forward to their "silly suppers."

At home, at child care, or in school, anywhere caring adults are with children, the possibilities for creating traditions are endless. What matters is the reassurance that a child can count on some things staying the same in a world where many things change quickly and unexpectedly.

The Role of Parents and Caregivers

We are asking so much of parents and caregivers in today's world: "Listen," "Encourage your child's play," "Develop routines," "Set limits," "Help your child feel secure and loved." And all that added to the other things parents must do in their lives! Look between the lines of the *Almanac* charts, and you'll find many parents and grandparents who feel severely pressured juggling work and family needs, also teenage parents, parents struggling with poverty or addictions, and families moving away from their loved ones. Many adults feel that they are falling short in one—if not all—parts of their lives; they often feel that they are failures.

Well, people are not failures when they're doing the best they can. If parents are managing to cover most of the important bases most of the time, then I think that they have every reason to feel good about who they are and what they're doing. Sometimes all it takes is one kind word to nourish another person. There's a wonderful "ripple effect" that is created when we nourish a child...or a parent or grandparent, a teacher or a child-care provider. One kind empathetic word has a way of turning into many!

Yes, it is our "relationships" that are primary in all of living. When the gusty winds blow and shake our lives, if we know that there are those who care about us, we may bend with the wind...but we won't break. That which has real value in life—in any millennium—is very simple...very deep and very simple!! It happens "inside" of us, in the "essential, invisible" part of us, and that is what allows everyone to be a potential "neighbor," no matter where that person happens to be listed on *World Almanac* charts.

CHILDREN: A STATISTICAL PORTRAIT

Population

Source: U.S. Bureau of the Census, International Data Base

Of the world's estimated population of 6.157 billion at midyear 2001, 38.9%, or some 2.393 billion, were 19 years old or younger. More than 87% of those youth were living in less developed nations. In the U.S., 28.3 % of the population was 19 or under in 2001.

2001 World Population by Age for Developed and Less Developed Nations[1]

Source: U.S. Bureau of the Census, International Data Base; numbers in millions

Population	Total		0-19		20-39		40-59		60-79		80+	
Less developed nations	4,968	100%	2,098	42.2%	1,605	32.3%	882	17.7%	349	7.0%	35	0.7%
Developed nations	1,189	100	295	24.8	341	28.7	318	26.7	196	26.7	38	3.2
TOTAL	**6,157**	**100**	**2,393**	**38.9**	**1,947**	**31.6**	**1,200**	**19.5**	**544**	**8.8**	**74**	**1.2**

(1) Excludes missing data in some categories for: Cook Islands, Micronesia, Federated States of Wallis and Futuna, and Western Sahara.

Percent of Population 19 Years or Younger, Selected Countries and Territories, 2001

Source: U.S. Bureau of the Census, International Data Base; estimated

	% 19 or younger
WORLD	**38.9**
Less developed countries	42.2
More developed countries	24.8
Africa	53.4
Sub-Saharan Africa	55.2
Northern Africa	45.2
Near East	46.4
Asia (whole continent)	39.2
North America	33.8
Latin America and the Caribbean	41.2
South America	39.8
Europe	24.1
Western Europe	22.7
Eastern Europe	26.0
Baltics	25.5
Commonwealth of Independent States	30.1
Oceania	32.9
20 most populous nations	
China	33.1
India	43.3
United States	28.3
Indonesia	40.0
Brazil	38.6
Russia	25.7
Pakistan	51.4
Bangladesh	48.6
Japan	20.4
Nigeria	54.6
Mexico	43.8
Germany	21.2
Philippines	47.2
Vietnam	43.4
Egypt	45.0
Turkey	38.5
Iran	46.0
Ethiopia	57.9
Thailand	32.2
United Kingdom	25.1

Population in the U.S., by Age, 2001

Source: U.S. Bureau of the Census

Age	Number	Percent
Total, all ages	**278,058,881**	**100.0**
0-4	18,899,320	6.8
5-9	19,546,220	7.0
10-14	20,270,387	7.3
15-19	20,076,841	7.2
20-24	19,094,725	6.9
25-29	17,482,479	6.3
30-34	19,683,222	7.1
35-39	21,836,820	7.9
40-44	22,909,306	8.2
45-49	20,461,593	7.4
50-54	18,185,074	6.5
55-59	13,596,270	4.9
60-64	10,952,622	3.9
65-69	9,411,443	3.4
70-74	8,744,053	3.1
75-79	7,401,414	2.7
80+	9,507,092	3.4

Children Ages 0-17 as a Proportion of the U.S. Population, 1950-2020

Source: *America's Children: Key National Indicators of Well-Being, 2001,* Federal Interagency Forum on Child and Family Statistics

1950	1960	1970	1980	1990	2000	2010	2020
31%	36%	34%	28%	26%	26%	24%	24%

Note: Figures for the years 1950 to 2000 are based on census data; later figures are projections.

Families

Average U.S. family size was 3.14 in 2000, down from 3.16 in 1990. According to the Children's Defense Fund, half of all American children will live in a single-parent family at some point in their childhood. One in 3 is born to unmarried parents, and one in four lives in a single-parent home. One in 24 lives with neither parent, as increasing numbers live with grandparents or in foster care.

Living Arrangements of Children Under 18 Years in the U.S., 2000

Source: U.S. Bureau of the Census
(in thousands)

In household, living with:	Total under 18 years	Under 1 year	1-2 years	3-5 years	6-8 years	9-11 years	12-14 years	15-17 years	Total under 6 years
Both parents	49,795	2,660	5,539	8,392	8,321	8,703	8,245	7,936	16,590
Mother only	16,162	869	1,671	2,578	2,829	2,945	2,705	2,565	5,118
Father only	3,058	218	352	450	499	537	509	493	1,020
Neither parent	2,981	121	279	446	414	508	512	702	846
TOTAL	**72,012**	**3,868**	**7,845**	**11,867**	**12,067**	**12,694**	**11,973**	**11,698**	**23,580**

Annual Expenditures on a Child by Husband-Wife Families in the U.S., 2000[1]

Source: *Expenditures on Children by Families, 2000 Annual Report,* USDA

Age of child	Total	Housing	Food	Transportation	Clothing	Health care	Child care and education	Misc.[2]
0-2	$8,740	$3,250	$1,060	$1,150	$440	$580	$1,310	$950
3-5	8,980	3,220	1,220	1,130	430	560	1,450	970
6-8	8,990	3,140	1,550	1,250	480	630	930	1,010
9-11	8,950	2,920	1,830	1,330	530	690	610	1,040
12-14	9,690	3,150	1,840	1,450	890	690	450	1,220
15-17	9,860	2,710	2,050	1,830	790	730	770	980
Total	**$165,630**	**$55,170**	**$28,650**	**$24,420**	**$10,680**	**$11,640**	**$16,560**	**$18,510**

Note: Before-tax income: $38,000 to $64,000 (Average = $50,600). (1) Figures are estimated expenses for a child in a 2-child family. To estimate expenses for an only child, multiply the figure by 1.24. To estimate expenses for each child in a family with 3 or more children, multiply the total expense for each appropriate age category by 0.77. (2) Miscellaneous expenses include personal care items, entertainment, and reading materials.

Percent of U.S. Children Under 18 Living Below Poverty, 1980-99

Source: *America's Children: Key National Indicators of Well-Being, 2001,* Federal Interagency Forum on Child and Family Statistics

	1980	1985	1990	1995	1996	1997	1998	1999
Children in all families								
Related children[1]	18	20	20	20	20	19	18	16
White, non-Hispanic	—	—	12	11	10	11	10	9
Black	42	43	44	42	40	37	36	33
Hispanic[2]	33	40	38	39	40	36	34	30
Children in married-couple families								
Related children	—	—	10	10	10	10	9	8
White, non-Hispanic	—	—	7	6	5	5	5	5
Black	—	—	18	13	14	13	12	11
Hispanic[2]	—	—	27	28	29	26	23	22
Children in female-householder families, no husband present								
Related children	51	54	53	50	49	49	46	42
White, non-Hispanic	—	—	40	34	35	37	33	29
Black	65	67	65	62	58	55	55	52
Hispanic[2]	65	72	68	66	67	63	60	52

— Indicates data unavailable. (1) Related children include all children in the household related to the householder by blood, adoption, or marriage. (2) Persons of Hispanic origin may be of any race. **Note:** Determinations of the poverty level do not count noncash benefits, such as food stamps. Poverty thresholds reflect family size and composition and are adjusted for inflation. The average poverty threshold for a family of 4 was $17,029 in 1999.

Distribution of Family Income for U.S. Children Under 18, 1980-99

Source: *America's Children: Key National Indicators of Well-Being, 2001,* Federal Interagency Forum on Child and Family Statistics

Poverty level	1980	1985	1990	1991	1992	1993	1994	1995	1996	1997	1998	1999
Extreme poverty	6.6	8.1	8.3	9.3	9.9	9.6	9.4	7.9	8.4	8.5	7.6	6.4
Below poverty, but above extreme poverty	11.3	12.0	11.6	11.8	11.7	12.4	11.9	12.2	11.4	10.8	10.7	10.0
Low income	24.0	22.8	21.8	22.2	22.0	22.2	22.0	22.5	22.7	21.4	21.2	21.7
Medium income	41.4	37.7	37.0	35.7	34.9	33.4	33.7	34.5	34.0	34.4	33.5	33.0
High income	16.8	19.4	21.3	21.0	21.5	22.3	23.1	22.8	23.5	25.0	27.0	29.0
Very high income	4.3	6.1	7.4	7.0	7.3	8.4	9.1	8.9	9.2	10.1	11.2	12.4

Note: Estimates included only children related to the householder. Extreme poverty is less than 50 percent of the poverty threshold (i.e., $8,515 for a family of 4 in 1999). Poverty is between 50 and 99 percent of the poverty threshold (i.e., between $8,515 and $17,028 for a family of 4 in 1999). Low income is between 100 and 199 percent of the poverty threshold (i.e., between $17,029 and $34,057 for a family of 4 in 1999). Medium income is between 200 and 399 percent of the poverty threshold (i.e., between $34,058 and $68,115 for a family of 4 in 1999). High income is 400 percent of the poverty threshold or more (i.e., $68,116 or more for a family of 4 in 1999). Very high income is 600 percent of the poverty threshold and over (i.e., $102,174 or more for a family of 4 in 1999).

Education

More than one-fourth of the U.S. population were enrolled in school in 1999: 8 million in nursery school and kindergarten, 33 million in elementary school, 16 million in high school, and 15 million in college. By 1999, the baby "boomlet," the influx of children of baby boomers, combined with increased immigration sent school enrollment to an all-time high. About two-thirds of elementary and high school students that year had at least one parent who was a baby boomer. One in five had at least one parent who was foreign-born.

School enrollments were more diverse, too. In 1999, only 63% of all schoolchildren were non-Hispanic whites, compared to 79% in 1972. The end of the 20th century also saw a great rise in nursery school enrollment. In 1964, only 5% of 3- and 4-year-olds went to nursery school; in 1999, 50% did.

In 1998, about 9% of 16 to 19-year-olds had dropped out of high school, according to the KIDS COUNT program of the Annie E. Casey Foundation. A survey by the U.S. Education Dept. estimated that about 1.7% of children in the U.S., or about 850,000, were being schooled at home in 1999.

Children's School Readiness Skills in the U.S., 1993, 1999

Source: National Center for Education Statistics

(in percent; for children 3 to 5 years old not yet enrolled in kindergarten)

	Recognizes all letters		Counts to 20 or higher		Writes name		Reads or pretends to read storybooks		Has 3-4 skills	
	1993	1999	1993	1999	1993	1999	1993	1999	1993	1999
TOTAL	**21**	**24**	**52**	**57**	**50**	**51**	**72**	**74**	**35**	**39**
Age										
3 years old	11	15	37	41	22	24	66	70	15	20
4 years old	28	28	62	67	70	70	75	76	49	50
5 years old	36	44	78	81	84	87	81	77	65	69
Sex										
Male	19	21	49	54	47	47	68	70	32	35
Female	23	27	56	60	53	56	76	77	39	43
Race/ethnicity										
White, non-Hispanic	23	25	56	60	52	54	76	79	39	42
Black, non-Hispanic	18	25	53	60	45	49	63	66	31	35
Hispanic	10	14	32	41	42	43	59	57	22	25
Mother's highest education										
Less than high school	8	7	30	36	40	32	55	53	19	15
High school	17	17	48	48	48	49	70	69	30	31
Vocational education or some college	23	25	59	60	51	52	79	79	39	42
College degree	31	35	68	73	58	61	84	84	52	54
Graduate/professional training or degree	39	40	68	73	59	64	83	83	55	57
Family type										
Two parents	22	26	54	58	51	53	74	75	37	41
None or one parent	18	19	49	54	47	48	65	69	31	33
Poverty status										
Above threshold	24	28	57	62	53	56	74	77	40	45
Below threshold	12	10	41	39	41	37	64	63	23	19

International Math and Science Achievement of 8th Graders, 1999

Source: National Center for Education Statistics

(The 38 countries listed are those that participated in the assessment program.)

Country	AVG. SCORE Math	AVG. SCORE Science	Country	AVG. SCORE Math	AVG. SCORE Science	Country	AVG. SCORE Math	AVG. SCORE Science
Australia	525	540	Israel	466	468	Philippines	345	345
Belgium (Flemish)	558	535	Italy	479	493	Romania	472	472
Bulgaria	511	518	Japan	579	550	Russia	526	529
Canada	531	533	Jordan	428	450	Singapore	604	568
Chile	392	420	Korea, South	587	549	Slovakia	534	535
Cyprus	476	460	Latvia[1]	505	503	Slovenia	530	533
Czech Republic	520	539	Lithuania	482	488	South Africa	275	243
England	496	538	Macedonia, Rep. of	447	458	Taiwan	585	569
Finland	520	535	Malaysia	519	492	Thailand	467	482
Hong Kong, SAR	582	530	Moldova	469	459	Tunisia	448	430
Hungary	532	552	Morocco	337	323	Turkey	429	433
Indonesia	403	435	Netherlands	540	545	**United States**	**502**	**515**
Iran	422	448	New Zealand	491	510			

(1) Latvian-speaking schools only.

Access to a Computer and Internet Use by U.S. Children, 3 to 17 Years, 2000

Source: U.S. Bureau of the Census

(in thousands)

	All children aged 3 to 17	Home computer access		Use Internet at home	
	NUMBER	NUMBER	PERCENT	NUMBER	PERCENT
TOTAL	**60,635**	**39,430**	**65.0**	**18,437**	**30.4**
Age					
3 to 5 years	11,915	6,905	58.0	864	7.3
6 to 11 years	24,837	15,924	64.1	6,135	24.7
12 to 17 years	23,884	16,600	69.5	11,439	47.9
Race and Hispanic origin					
White	47,433	33,062	69.7	15,940	33.6
White non-Hispanic	38,438	29,731	77.3	14,773	38.4
Black	9,779	4,161	42.5	1,441	14.7
Asian and Pacific Islander	2,581	1,855	71.9	909	35.2
Hispanic (of any race)	9,568	3,546	37.1	1,229	12.8

Community Service Participation of U.S. Students in Grades 6-12, 1999

Source: National Center for Education Statistics

Race/Ethnicity	% participating	Grades	% participating	Sex	% participating
White, non-Hispanic	56	6-8	48	Male	47
Black, non-Hispanic	47	9-10	50	Female	57
Hispanic	39	11-12	61		
Other race/ethnicity	53			**TOTAL**	**52**

Child Care/Day Care

Three out of five children in the U.S. under six are in child care. Half of those are in pre-schools or child care centers, one-third stay with relatives, and the remainder with other care givers. A national study conducted under the auspices of the Children's Defense Fund found that one-third of child care programs were rated inadequate.

While millions of school-age children are enrolled in after-school programs, some 7 million are home alone after school during the afternoon, hours when juvenile crime peaks.

Child Care Arrangements in the U.S., 1995, 1999

Source: *America's Children: Key National Indicators of Well-Being, 2001,* Federal Interagency Forum on Child and Family Statistics

	Parental care only		Type of nonparental care arrangement: Total in nonparental care[2]		Care in a home[1]: By a relative		Care in a home[1]: By a nonrelative		Center-based program[3]	
	1995	1999	1995	1999	1995	1999	1995	1999	1995	1999
TOTAL	49	46	51	54	20	22	15	14	23	27
Age/grade in school										
Ages 0-2	51	49	50	51	23	24	19	17	12	16
Ages 3-6, not yet in kindergarten	26	23	74	77	19	23	17	16	55	60
Kindergarten	56	52	44	48	18	20	14	13	16	22
1st–3d grade	62	57	38	43	18	21	10	9	13	18
Race and Hispanic origin										
White, non-Hispanic	49	48	51	53	17	19	17	16	24	28
Black, non-Hispanic	40	34	60	66	31	33	10	11	27	35
Hispanic[4]	58	52	42	48	23	24	10	11	13	19
Other	49	43	51	57	22	29	11	11	25	28
Mother's highest level of education[5]										
Less than high school graduate	67	59	33	41	18	21	6	9	13	16
High school graduate/GED	51	49	49	51	22	27	13	11	19	23
Vocational/technical or some college	44	43	56	57	22	23	17	16	25	29
College graduate	40	43	60	57	14	15	22	17	34	35
Mother's employment status[5]										
35 hours or more per week	22	22	78	78	32	33	25	22	33	37
Less than 35 hours per week	42	45	58	55	25	25	19	17	24	26

(1) May be the child's own home or another home. (2) Since some children have more than one type of nonparental care arrangement, details do not sum to the total percentage of children in nonparental care. (3) Center-based programs include day care centers, prekindergartens, nursery schools, Head Start programs, and other early childhood education programs. (4) Persons of Hispanic origin may be of any race. (5) Children without a mother in the home are not counted.

Health

Sources: UNICEF; CDC; Federal Interagency Forum on Child and Family Statistics

As immunization levels in the youngest U.S. children have risen at all economic levels, attention has turned recently to another health issue, their fitness. The percentage of children 6-11 who are overweight has increased 75% since 1980, and the percent of those 12-19 who are overweight has more than doubled. A chief factor is lack of physical activity. Walking and biking by children ages 5-15 dropped 40% between 1977 and 1995.

10.8 million American children 18 and under—14%, or 1 in 7—lacked health insurance coverage at some point during 1999. This figure constitutes a drop from the previous year, with 90% of the improvement among lower income children.

Overweight Children Ages 6-19 in the U.S.[1]

Source: CDC, National Center for Health Statistics

6-11 years of age	1963-1965[2]	1971-1974	1976-1980[3]	1988-1994
	Percent of population			
Both sexes	4.2	4.0	6.5	11.4
Boys	4.0	4.3	6.6	11.8
White	4.4	4.1	6.7	11.6
Black	1.6	5.3	6.7	12.3
Girls	4.5	3.6	6.4	11.0
White	4.5	3.7	5.7	9.8
Black	4.5	3.3	11.1	16.9

12-19 years of age	1966-1970[2]	1971-1974	1976-1980[3]	1988-1994
	Percent of population			
Both sexes	4.6	6.1	5.0	10.5
Boys	4.5	5.4	4.5	11.3
White	4.7	5.5	4.6	12.1
Black	3.1	5.0	4.8	10.4
Girls[3]	4.7	6.7	5.4	9.7
White	4.5	6.1	4.7	9.0
Black	6.4	10.1	10.0	16.3

NOTE: Overweight is defined as body mass index (BMI) at or above the sex- and age-specific 95th percentile. (1) The race groups, white and black, include persons both of Hispanic and of non-Hispanic origin. (2) Data for 1963-65 are for children 6-11 years of age; data for 1966-70 are for adolescents 12-17 years of age, not 12-19 years. (3) Excludes pregnant girls starting with 1971-74. Pregnancy status not available for 1963-65/1966-70. **Note:** Overweight is defined as body mass index (BMI) at or above the sex- and age-specific 95th percentile.

Percent of U.S. Students in Grades 9-12 Who Participated in Moderate to Vigorous Activity[1], 1999

Source: CDC, National Center for Chronic Disease Prevention and Health Promotion, Youth Risk Behavior Survey

Grade	% all students	% Males	% Females
9th	77.3	81.6	72.9
10th	68.9	76.3	61.6
11th	62.6	70.5	54.4
12th	67.6	75.6	59.7

Race	% all students	% Males	% Females
White, non-Hispanic	71.9	78.1	65.3
Black, non-Hispanic	60.0	68.1	52.5
Hispanic	64.9	75.7	54.3

(1) Moderate to vigorous activity is defined as activity which caused the person to sweat or breathe hard for at least 20 minutes on 3 or more of the previous 7 days or walking or biking for at least 30 minutes on 5 or more of the previous 7 days.

Risky Behavior

Alcohol is the most commonly used psychoactive substance during adolescence. In 1999 half of high school students reported drinking in the previous 30 days. Long-term statistics indicate that heavy drinking among high school seniors peaked in 1981, with 41% reporting having 5 or more drinks in a row during the previous 2 weeks—compared with 30% in 1999.

In 1999 more than one-third of high school students reported smoking on one or more of the previous 30 days. Smoking among adolescents increased 27% from 1991 to 1999.

Nearly 47% of high school students had used marijuana in 1999, a 50% increase from 1990. Use of any illicit drugs, including marijuana, increased some 12% from 1992 to 1997, and has been relatively stable (between 25 and 30%) since. (See also the Vital Statistics chapter, page 883.)

While the juvenile population grew throughout the 1990s, juvenile crime rates dropped 23% from 1995 to 1999, when juveniles accounted for 16% of arrests for serious violent crime. Violent youth crime peaks in the afternoon, when many children are home from school and unsupervised.

Percentage of U.S. High School Students Who Felt Sad or Hopeless or Attempted Suicide, 1999

Source: CDC, *Youth Risk Behavior Surveillance—United States, 1999*

	Felt sad or hopeless[1]			Attempted suicide[2]		
Race/Ethnicity	Male	Female	Total	Male	Female	Total
Non-Hispanic White	19.0	31.3	24.9	4.5	9.0	6.7
Non-Hispanic Black	19.6	37.7	28.9	7.1	7.5	7.3
Hispanic	27.7	46.1	37.0	6.6	18.9	12.8
TOTAL	**21.0**	**35.7**	**28.3**	**5.7**	**10.9**	**8.3**

(1) Almost every day for at least 2 weeks in a row during the 12 months preceding the survey. (2) One or more times.

Percentage of U.S. High School Students Who Engaged in Sexual Behavior, 1999

Source: CDC, *Youth Risk Behavior Surveillance—United States, 1999*

	Ever had sexual intercourse			1st sexual intercourse before age 13		
	Male	Female	Total	Male	Female	Total
Race/Ethnicity						
Non Hispanic White	45.4	44.8	45.1	7.5	3.5	5.5
Non Hispanic Black	75.7	66.9	71.2	29.9	11.4	20.5
Hispanic	62.9	45.5	54.1	14.2	4.4	9.2
Grade						
9	44.5	32.5	38.6	17.7	5.5	11.7
10	51.1	42.6	46.8	13.9	5.1	9.4
11	51.4	53.8	52.5	7.8	4.5	6.2
12	63.9	65.8	64.9	7.6	2.1	4.8
TOTAL	**52.2**	**47.7**	**49.9**	**12.2**	**4.4**	**8.3**

UNITED STATES GOVERNMENT

EXECUTIVE BRANCH	LEGISLATIVE BRANCH	JUDICIAL BRANCH
PRESIDENT	**CONGRESS**	**Supreme Court of the United States**
Vice President	**Senate House**	Courts of Appeals
Executive Office of the President	Architect of the Capitol	District Courts
White House Office	U.S. Botanic Garden	Territorial Courts
Office of the Vice President	General Accounting Office	Court of International Trade
Council of Economic Advisers	Government Printing Office	Court of Federal Claims
Council on Environmental Quality	Library of Congress	Court of Appeals for the Armed Forces
National Security Council	Congressional Budget Office	Court of Veterans Appeals
Office of Administration	Tax Court	Administrative Office of the Courts
Office of Management and Budget		Federal Judicial Center
Office of National Drug Control Policy		Sentencing Commission
Office of Policy Development		
Office of Science and Technology Policy		
Office of the U.S. Trade Representative		

The Bush Administration

As of Oct. 2001; mailing addresses are for Washington, DC.

Terms of office of the president and vice president: Jan. 20, 2001, to Jan. 20, 2005.

President — George W. Bush receives an annual salary of $400,000 (taxable), and an annual expense allowance of $50,000 (nontaxable) for costs resulting from official duties. In addition, up to $100,000 a year may be spent on travel expenses and $19,000 on official entertainment (both nontaxable), available for allocation within the Executive Office of the President.
Website: http://www.whitehouse.gov/president
E-mail: president@whitehouse.gov

Vice President — Dick Cheney receives an annual salary of $186,300 (taxable), plus $10,000 for expenses, (nontaxable).
Website: http://www.whitehouse.gov/vicepresident
E-mail: vice.president@whitehouse.gov

The Cabinet Department Heads

(Salary: $161,200 per year)

Secretary of State — Colin L. Powell
Secretary of the Treasury — Paul H. O'Neill
Secretary of Defense — Donald H. Rumsfeld
Attorney General — John Ashcroft
Secretary of the Interior — Gale Norton
Secretary of Agriculture — Ann M. Veneman
Secretary of Commerce — Donald L. Evans
Secretary of Labor — Elaine L. Chao
Secretary of Health and Human Services — Tommy Thompson
Secretary of Housing and Urban Development — Mel Martinez
Secretary of Transportation — Norman Y. Mineta
Secretary of Energy — Spencer Abraham
Secretary of Education — Roderick R. Paige
Secretary of Veterans Affairs — Anthony Principi

The White House Staff

1600 Pennsylvania Ave. NW 20500
Website: http://www.whitehouse.gov

Chief of Staff to the President — Andrew H. Card Jr.
Asst. to the President & Deputy Chief of Staff — Joseph W. Hagin II
Asst. to the President & Deputy Chief of Staff — Joshua Bolten
Office of Homeland Security — Tom Ridge, dir.
Assistants to the President:
Counsel to the President — Alberto R. Gonzalez
Deputy Counsel to the President — Timothy Flanigan
Domestic Policy Council — Margaret LaMontagne
Presidential Personnel — Clay S. Johnson
Press Secretary — L. Ari Fleischer
Legislative Affairs — Nicholas Calio
Counselor to the President/Communications — Karen Hughes
National Economic Policy — Lawrence Lindsay
Intergovernmental Affairs — Ruben S. Barrales
National Security — Condoleezza Rice
Staff Secretary — Harriet Miers
Political Affairs — Kenneth B. Mehlman
Public Liaison — Lezlee Westine
Management & Administration — Hector F. Irastorza Jr.
Cabinet Secretary — Albert Hawkins
Director of Presidential Scheduling — Bradley Blakeman
Director of Speechwriting — Michael Gerson
Chief of Staff to the First Lady — Andrea Ball
E-mail: first.lady@whitehouse.gov
Senior Advisor to the President — Karl Rove
Director of Advance — Brian D. Montgomery
Oval Office Operations — Linda Gambatesa
Faith-Based and Community Initiatives — John DiIulio
Office of National AIDS Policy — Scott Evertz

Executive Agencies

Council of Economic Advisers — Glenn Hubbard, chair
Website: http://www.whitehouse.gov/cea/index.html
Office of Administration — Philip Larsen, dir.
Website: http://www.whitehouse.gov/oa/index.html
Office of Science & Technology Policy — Rosina Bierbaum
Website: http://www.ostp.gov
Office of Nat. Drug Control Policy — Edward H. Jurith, act. dir.
Website: http://www.whitehousedrugpolicy.gov
Office of Management and Budget — Mitchell E. Daniels Jr., dir.
Website: http://www.whitehouse.gov/OMB/index.html
U.S. Trade Representative — Robert B. Zoellick
Website: http://www.ustr.gov
Council on Environ. Quality — James L. Connaughton, chair
Website: http://www.whitehouse.gov/ceq/index.html

Department of State

2201 C St. NW 20520
Website: http://www.state.gov

Secretary of State — Colin L. Powell
Deputy Secretary — Richard L. Armitage
Chief of Staff — Bill Smullen
U.S. Ambassador to the United Nations — John D. Negroponte
Under Sec. for Political Affairs — Marc Grossman
Under Sec. for Management — Grant S. Green
Under Sec. for Global Affairs — Paula J. Dobriansky
Under Sec. for Economic, Business, & Agricultural Affairs — Alan P. Larson
Under Sec. for Arms Control & International Security Affairs — John R. Bolton
Policy Planning Director — vacant
Chief of Protocol — Donald B. Ensenat
Inspector General — Clark Kent Ervin
Legal Adviser — William H. Taft IV
Director General of the Foreign Service & Director of Personnel — Edward W. Grehm Jr.
Assistant Secretaries for:
Administration — Willliam A. Eaton
African Affairs — Walter H. Kansteiner
Consular Affairs — Mary A. Ryan
Democracy, Human Rights, & Labor — Lorne W. Craner
Diplomatic Security — David Carpenter

East Asian & Pacific Affairs — James A. Kelly
Economic & Business Affairs — Earl Anthony Wayne
European & Canadian Affairs — A. Elizabeth Jones
Intelligence & Research — Carl W. Ford Jr.
International Narcotics & Law — Rand Beers
International Organization Affairs — David Welch
Legislative Affairs — Paul V. Kelly
Near Eastern Affairs — William Joseph Burns
Oceans, International Environmental, & Scientific Affairs — Kenneth S. Bill, act.
Politico-Military Affairs — Lincoln P. Bloomfield
Population, Refugees, & Migration — Alan J. Kreczko
Public Affairs — Richard A. Boucher
South Asian Affairs — Christina B. Rocca

Department of the Treasury

1500 Pennsylvania Ave. NW 20220
Website: http://www.ustreas.gov

Secretary of the Treasury — Paul H. O'Neill
Deputy Sec. of the Treasury — Kenneth Dam
Under Sec. for Domestic Finance — Peter R. Fischer
Under Sec. for International Affairs — John Taylor
Under Sec. for Enforcement — Jimmy Gurule
General Counsel — David Aufhauser
Inspector General — Jeffrey Rush
Inspector General for Tax Administration — David Williams
Assistant Secretaries for:
Economic Policy — vacant
Enforcement — vacant
Financial Institutions — Sheila Bair
Fiscal Affairs — Donald Hammond
International Affairs — vacant
Legislative Affairs — John Duncan
Management — Jim Flyzik, act.
Public Affairs — Michele Davis
Tax Policy — Mark Weinberger
Treasurer of the U.S. — Rosario Marin
Bureaus:
Alcohol, Tobacco, & Firearms — Bradley A. Buckles, dir.
Comptroller of the Currency — John Hawke, comm.
Customs — vacant
Engraving & Printing — Tom Ferguson, dir.
Federal Law Enforcement Training Center — W. Ralph Basham, dir.
Financial Management Service — Richard Gregg, comm.
Internal Revenue Service — Charles Rossotti, comm.
Mint — Henrietta Fore
Office of Thrift Supervision — Ellen S. Seidman
Public Debt — Van Zeck, comm.
U.S. Secret Service — Brian L. Stafford, dir.

Department of Defense

The Pentagon 20301
Website: http://www.defenselink.mil

Secretary of Defense — Donald H. Rumsfeld
Deputy Secretary — Paul D. Wolfowitz
Under Sec. for Acquis. and Technol. — E. C. "Pete" Aldridge
Under Sec. for Personnel & Readiness — David S. C. Chu
Under Sec. for Policy — Douglas J. Feith
Assistant Secretaries for:
Command, Control, Communications, & Intelligence — John P. Stenbit
Force Management — Charles S. Abell
Health Affairs — Dr. J. Jarrett Clinton
International Security Affairs — Peter W. Rodman
International Security Policy — Dr. J. D. Crouch II
Legislative Affairs — Powell A. Moore
Public Affairs — Victoria Clarke
Reserve Affairs — Craig W. Duehring, act.
Spec. Operations & Low-Intensity Conflict — Daniel J. Gallington, act.
Program Analysis & Evaluation — Barry D. Watts, dir.
Inspector General — Robert J. Lieberman
Comptroller — Dov S. Zakheim
General Counsel — William J. Haynes II
Intelligence Oversight — George B. Lotz II
Operational Test & Evaluation — Thomas P. Christie, dir.
Chairman, Joint Chiefs of Staff — Gen. Richard B. Myers
Secretary of the Army — Thomas E. White
Secretary of the Navy — Gordon R. England
Commandant of the Marine Corps — James Jones
Secretary of the Air Force — James G. Roche

Department of Justice

Constitution Ave. & 10th St. NW 20530
Website: http://www.usdoj.gov

Attorney General — John Ashcroft
Deputy Attorney General — Larry Thompson
Associate Attorney General — Jay Stephens
Office of Dispute Resolution — Peter R. Steenland Jr.
Solicitor General — Theodore B. Olson
Office of Inspector General — Glenn Fine
Assistants:
Administration — Janis Sposato
Antitrust Division — Charles James
Civil Division — Robert McCallum
Civil Rights Division — Ralph Boyd
Criminal Division — Michael Chertoff
Environ. & Nat. Resources Division — Thomas Sansonetti
Justice Programs — Deborah Daniels
Legal Counsel — Daniel Koffsky, act.
Legislative Affairs — Daniel J. Bryant
Policy Development — Viet Dinh
Tax Division — Loretta C. Argrett
Executive Secretariat — Kathie Harting
Office of Investigative Agency Policies — vacant
Office of Public Affairs — Mindy Tucker
Office of Information & Privacy — Richard L. Huff/Daniel J. Metcalfe
Community Oriented Policing Services — Ralph Justus, act.
Federal Bureau of Investigation — Thomas J. Pickard, act.
Exec. Off. for Immigration Review — Kevin D. Rooney, dir.
Bureau of Prisons — Kathleen Hawk Sawyer, dir.
Community Relations Service — Rose M. Ochi, dir.
Drug Enforcement Admin. — Asa Hutchinson
Office of Intelligence Policy & Review — James Baker, act.
Office of Professional Responsibility — H. Marshall Jarrett, counsel
Exec. Off. for U.S. Trustees — Martha Davis
Foreign Claims Settlement Comm. — John Lacy
Exec. Office for U.S. Attorneys — Ken Wainstein
Immigration & Naturalization Service — James W. Ziglar
Pardon Attorney — Roger C. Adams
U.S. Parole Commission — Edward F. Reilly
U.S. Marshals Service — Louis McKinney, act.
U.S. Natl. Cen. Bureau of INTERPOL — Edgar A. Adamson
Office of Intergovernmental Affairs — Lori Sharpe Day
Office of Tribal Justice — Todd Araujo
Violence Against Women Act — Catherine Pierce, act.
National Drug Intelligence Center — Michael T. Horn, dir.

Department of the Interior

1849 C St. NW 20240
Website: http://www.doi.gov

Secretary of the Interior — Gale Norton
Deputy Secretary — J. Steven Griles
Assistant Secretaries for:
Fish, Wildlife, & Parks — Craig Manson, nom.
Indian Affairs — Neal A. McCaleb
Land & Minerals — J. Steven Griles, act.
Policy, Management, & Budget — P. Lynn Scarlett
Water & Science — Bennett Raley
Bureau of Land Management — Kathleen Clarke, nom.
Bureau of Reclamation — John W. Keys III, comm.
Fish & Wildlife Service — Marshall P. Jones Jr., act.
Geological Survey — Charles Groat
Mineral Management Service — vacant
National Park Service — Fran P. Mainella, dir.
Surf. Mining Reclam. & Enforcement — Jeffrey Jarrett
Communications — Eric Ruff, dir.
Congressional & Legislative Affairs — David L. Bernhardt
Solicitor — William G. Myers
External Affairs — Kit Kimball
Exec. Secretariat & Regulatory Affairs — Fay Iudicello

Department of Agriculture

1400 Independence Ave. SW 20250
Website: http://www.usda.gov

Secretary of Agriculture — Ann M. Veneman
Deputy Secretary — James R. "Jim" Moseley
Under Secretaries for:
Farm & Foreign Agric. Services — J. B. Penn
Food, Nutrition, & Consumer Services — Eric M. Bost
Food Safety — Elsa A. Murano
Marketing & Regulatory Programs — William T. "Bill" Hawk
Natural Resources & Environment — Mark E. Rey
Research, Education, & Economics — Joseph Jen
Rural Development — Michael Neruda, act.
Assistant Secretaries for:
Administration — Lou Gallegos
Congressional Relations — Mary Waters
General Counsel — J. Michael Kelly, act.
Inspector General — Joyce N. Fleischman, act.
Chief Financial Officer — Edward R. McPherson
Chief Information Officer — Ira L. Hobbs, act.
Chief Economist — Keith Collins
Communications — Kevin Herglotz

Department of Commerce

14th St. between Constitution & Pennsylvania Ave. NW 20230
Website: http://www.doc.gov

Secretary of Commerce — Donald L. Evans
Deputy Secretary — Samuel Bodman
Chief of Staff — Laurie Fenton
General Counsel — Ted Kassinger
Assistant Secretaries:
Chief Financial Officer & Asst. Secretary for Admin. — Otto Wolff
Economic Development Admin. — David Sampson
Export Admin. — Jim Jochun
Export Enforcement — Michael Garcia
Import Administration — Faryar Shirzad
Legislative Affairs — Brenda Becker
Market Access & Compliance — William Lash
National Telecomm. Information Administration — Nancy Victory
Oceans & Atmosphere — vacant
Patent & Trademark Office — vacant
Trade Development — Linda Conlin
U.S. & Foreign Commercial Service — Maria Cino
Bureau of the Census — William G. Barron Jr., act. dir.
Under Sec. for Oceans & Atmosphere — vacant
Under Sec. for Export Admin. — Kenneth Juster
Under Sec. for International Trade — Grant Aldonas
Under Sec. for Econ. Affairs — Kathleen Cooper
Under Sec. for Technology — vacant
Natl. Institute of Standards & Tech. — vacant
Minority Business Dev. Agency — Ronald Langston
Public Affairs — Mary Crawford

Department of Labor

200 Constitution Ave. NW 20210
Website: http://www.dol.gov

Secretary of Labor — Elaine L. Chao
Deputy Secretary — D. Cameron Findlay
Chief of Staff — Steven J. Law
Assistant Secretaries for:
Admin. & Management — Patrick Pizzella
Congressional & Intergov. Affairs — Kristine Iverson
Employment & Training — Emily Stover DeRocco
Employment Standards — Joe Kennedy, act.
Occupational Safety & Health — John Henshaw
Mine Safety & Health — David Lauriski
Pension & Welfare Benefits — Ann Combs
Policy — Christopher Spear
Public Affairs — Stuart Roy, deputy asst. sec.
Veterans Employment & Training — Charles Ciccolella, deputy asst. sec.
Solicitor of Labor — Howard Radzley, act.
Bureau of International Affairs — Jorge Perez-Lopez, act.
Women's Bureau — Shinae Chun
Inspector General — Gordon S. Heddell
Bureau of Labor Statistics — Katharine G. Abraham

Department of Health and Human Services

200 Independence Ave. SW 20201
Website: http://www.os.dhhs.gov

Secretary of Health & Human Services — Tommy Thompson
Deputy Secretary — Claude A. Allen
Chief of Staff — Robert Wood
Assistant Secretaries for:
Aging — Josefina Carbonell
Children & Families — Wade F. Horn
Health — vacant
Legislation — Scott Whitaker
Management & Budget — Dennis Williams, act.
Planning & Evaluation — Bobby Piyush Jindal
Public Affairs — Kevin Keane
General Counsel — Alex Azar
Inspector General — Janet Rehnquist
Office for Civil Rights — Robinsue Frohboese, act. dir.
Surgeon General — David Satcher
Centers for Medicare and Medicaid Services — Tom Scully

Department of Housing and Urban Development

451 7th St. SW 20410
Website: http://www.hud.gov

Secretary of Housing & Urban Development — Mel Martinez
Deputy Secretary — Alphonso R. Jackson
Chief of Staff — Daniel R. Murphy
Assistant Secretaries for:
Community Planning & Development — Roy A. Bernardi
Cong. & Intergov. Relations — Melody H. Fennel
Fair Housing & Equal Opportunity — vacant
Housing & Federal Housing Comm. — John C. Weicher
Policy Development & Research — vacant
Public & Indian Housing — vacant
General Counsel — Richard A. Hauser
Chief Information Officer — Gloria R. Parker
Inspector General — David C. Williams, act.
Chief Financial Officer — Angela Antonelli
Government National Mortgage Assn. — Ronald Rosenfeld
Off. of Federal Housing Enterprise Oversight — Armando Falcon Jr.

Department of Transportation

400 7th St. SW 20590
Website: http://www.dot.gov

Secretary of Transportation — Norman Y. Mineta
Deputy Secretary — Michael P. Jackson
Assistant Secretaries for:
Administration — Melissa Allen
Aviation & International Affairs — Read Van de Water
Budget & Programs — Donna McLean
Governmental Affairs — Sean B. O'Hollaren
Public Affairs — Tim Arnade, act.
Transportation Policy — vacant
U.S. Coast Guard Commandant — Adm. James M. Loy
Federal Aviation Admin. — Jane Garvey
Federal Highway Admin. — Mary Peters
Federal Railroad Admin. — Allan Rutter
Maritime Admin. — vacant
Natl. Highway Traffic Safety Admin. — Dr. Jeffrey W. Runge
Federal Transit Admin. — Jennifer L. Dorn
Research & Special Programs Admin. — vacant
St. Lawrence Seaway Devel. Corp. — Albert Jacquez

Department of Energy

1000 Independence Ave. SW 20585
Website: http://www.energy.gov

Secretary of Energy — Spencer Abraham
Deputy Secretary — Frank Blake
Under Secretary — Robert Card
Chief of Staff — Kyle McStarrow
Deputy Chief of Staff for Intl. Policy — Joe McMonigie
General Counsel — Lee Otis
Inspector General — Gregory Friedman

Assistant Secretaries for:
Administration & Human Resource Management — Richard Farrell
Congressional & Intergov. Affairs — Dan Brouillette
Defense Programs — Kathleen Carlson, act.
Energy Efficiency & Renewable Energy — David Garman
Environment, Safety, & Health — Steven Carey, act.
Environmental Restoration & Waste Management — Jessie Roberson
Fossil Energy — Robert Kripowicz, act.
International Affairs — Vicky A. Bailey
Policy — vacant
Nuclear Energy — Bill Magwood, dir.
Energy Information Admin. — Mary Hutzler
Economic Impact & Diversity — Theresa Speake
Hearings & Appeals — George Breznay, dir.
Energy Research — vacant
Civilian Radioactive Waste Management — Lake H. Barrett, act. dir.
Nonproliferation & National Security — Ken Baker, act.
Chief Financial Officer — Bruce Carnes
Energy Advisory Board — vacant
Office of Public Affairs — Jeanne Lopatto

Department of Education

400 Maryland Ave., SW 20202
Website: http://www.ed.gov
Secretary of Education — Roderick R. Paige
Deputy Secretary — William D. Hansen
Chief of Staff — Terry R. Abbott
Inspector General — Lorraine P. Lewis
General Counsel — Brian W. Jones
Assistant Secretaries for:
Adult & Vocational Education — Carol D'Amico
Civil Rights — Gerald Reynolds
Educational Research & Improvement — Grover J. "Russ" Whitehurst
Elementary & Secondary Educ. — Susan B. Neuman
Intergov. & Interagency Affairs — Laurie M. Rich
Legislative & Congressional Affairs — Rebecca O. Campoverde
Postsecondary Education — Maureen McLaughlin
Special Educ. & Rehab. Services — Robert H. Pasternack
Bilingual Education & Minority Language Affairs — Arthur Love
Rehab. Services Admin. — Joanne M. Wilson, comm.
Education Statistics — Gary Phillips, act. comm.

Department of Veterans Affairs

810 Vermont Ave. NW 20420
Website: http://www.va.gov
Secretary of Veterans Affairs — Anthony Principi
Deputy Secretary — Leo Mackay, Jr.
Assistant Secretaries for:
Congressional Affairs — Gordon Mansfield
Management — D. Mark Catlett, act.
Human Resources & Admin. — Jacob Lozada, PhD.
Policy & Planning — Claude Kicklighter
Public & Intergovernmental Affairs — Maureen Cragin
Inspector General — Richard J. Griffin
Under Sec. for Benefits — Joseph Thompson
Under Sec. for Health — Thomas L. Garthwaite, M.D.
Under Sec. for Memorial Affairs — Robin Higgins
General Counsel — Tim McClain
Board of Veterans Appeals — Eligah Dane Clark, chair
Board of Contract Appeals — Guy H. McMichael III, chairman
Small & Disadvantaged Business Utilization — Scott S. Denniston, dir.
Veterans Service Organization Liaison — Allen F. Kent

Notable U.S. Government Agencies

Source: *The U.S. Government Manual*; National Archives and Records Administration; World Almanac research

All addresses are Washington, DC, unless otherwise noted; as of Oct. 2001
* = independent agency

Bureau of Alcohol, Tobacco, and Firearms — Bradley A. Buckles, dir. (Dept. of Treas., 650 Mass. Ave NW, 20226).
Website: http://www.atf.treas.gov

Bureau of the Census — William G. Barron Jr., act. dir. (Dept. of Commerce, 4700 Silver Hill Rd., Suitland, MD 20746).
Website: http://www.census.gov

Bureau of Economic Analysis — J. Steven Landerfeld, dir. (Dept. of Commerce, 1441 L St. NW, 20230).
Website: http://www.bea.doc.gov

Bureau of Indian Affairs — Neal A. McCaleb, asst. sec. (Dept. of the Interior, 1849 C St. NW, 20240).
Website: http://www.doi.gov/bureau-indian-affairs.html

Bureau of Prisons — Kathleen Hawk Sawyer, dir. (Dept. of Justice, 320 First St. NW, 20534).
Website: http://www.bop.gov

Centers for Disease Control & Prevention — Jeffrey P. Koplan, dir. (Dept. of HHS, 1600 Clifton Rd. NE, Atlanta, GA 30333).
Website: http://www.cdc.gov

*__Central Intelligence Agency__ — George J. Tenet, dir. (Wash., DC 20505).
Website: http://www.odci.gov

*__Commission on Civil Rights__ — Mary Frances Berry, chair (624 9th St. NW, 20425).
Website: http://www.usccr.gov

*__Commodity Futures Trading Commission__ — James E. Newsome, act. chair (3 Lafayette Centre, 1155 21st St. NW, 20581).
Website: http://www.cftc.gov

*__Consumer Product Safety Commission__ — Ann Brown, chair (East-West Towers, 4330 East-West Hwy., Bethesda, MD 20814).
Website: http://www.cpsc.gov

*__Environmental Protection Agency__ — Christine Todd Whitman, adm. (Ariel Rios Bldg., 1200 Pennsylvania Ave. NW, 20460).
Website: http://www.epa.gov

*__Equal Employment Opportunity Commission__ — Cari M. Dominguez, chair (1801 L St. NW, 20507).
Website: http://www.eeoc.gov

*__Export-Import Bank of the United States__ — John E. Robson, pres. and chair (811 Vermont Avenue NW, 20571).
Website: http://www.exim.gov

*__Farm Credit Administration__ — Michael M. Reyna, chair, Farm Credit Administration Board (1501 Farm Credit Drive, McLean, VA 22102).
Website: http://www.fca.gov

Federal Aviation Administration — Jane F. Garvey, adm. (Dept. of Trans., 800 Independence Ave. SW, 20591).
Website: http://www.faa.gov

Federal Bureau of Investigation — Thomas J. Pickard, act. dir. (Dept. of Justice, 935 Pennsylvania Ave. NW, 20535).
Website: http://www.fbi.gov

*__Federal Communications Commission__ — Michael K. Powell, chair (445 12th St. SW, 20554).
Website: http://www.fcc.gov

*__Federal Deposit Insurance Corporation__ — Donald E. Powell, chair (550 17th St. NW, 20429).
Website: http://www.fdic.gov

*__Federal Election Commission__ — Danny L. McDonald, chair (999 E St. NW, 20463).
Website: http://www.fec.gov

*__Federal Emergency Management Agency__ — Joe M. Allbaugh, dir. (500 C St. SW, 20472).
Website: http://www.fema.gov

*__Federal Energy Regulatory Commission__ — Curt Hébert Jr., chair (888 1st St. NE, 20426).
Website: http://www.ferc.fed.us

Federal Highway Administration — Mary Peters, adm. (Dept. of Trans., 400 7th St. SW, 20590).
Website: http://www.fhwa.dot.gov

*Federal Maritime Commission — Harold J. Creel Jr., chair (800 N. Capitol St. NW, 20573).
Website: http://www.fmc.gov

*Federal Mine Safety & Health Review Commission — Theodore F. Verheggen, chair (1730 K St. NW, 20006).
Website: http://www.fmshrc.gov

*Federal Reserve System — Alan Greenspan, chair, Board of Governors (20th St. & Constitution Ave. NW, 20551).
Website: http://www.federalreserve.gov

*Federal Trade Commission — Timothy J. Muris, chair (600 Pennsylvania Ave. NW, 20580).
Website: http://www.ftc.gov

Fish & Wildlife Service — Marshall P. Jones Jr., act. dir. (Dept. of the Interior, 1849 C St. NW, 20240).
Website: http://www.fws.gov

Food and Drug Administration — Bernard A. Schwetz, Ph.D., act. dep. comm. (5600 Fishers Lane, Rockville, MD 20857).
Website: http://www.fda.gov

Forest Service — Dale Bosworth, chief (Dept. of Agriculture, 201 14th St. SW, 20250).
Website: http://www.fs.fed.us

General Accounting Office — (cong. agency) David Michael Walker, comptroller gen. (441 G St. NW, 20548).
Website: http://www.gao.gov

*General Services Administration — Stephen A. Perry, adm. (1800 F St. NW, 20405).
Website: http://www.gsa.gov

Government Printing Office — (cong. agency) Michael F. DiMario, public printer (732 N. Capitol St. NW, 20401).
Website: http://www.gpo.gov

Immigration & Naturalization Service — James W. Ziglar, comm. (Dept. of Justice, 425 I St. NW, 20536).
Website: http://www.ins.usdoj.gov

*Inter-American Foundation — Frank Yturria, chair (901 N Stuart St., 10th floor, Arlington, VA 22203).
Website: http://www.iaf.gov

Internal Revenue Service — Charles Rossotti, comm. (Dept. of Treas., 1111 Constitution Ave. NW, 20224).
Website: http://www.irs.gov

Library of Congress — (cong. agency) Dr. James H. Billington, Librarian of Congress (101 Indep. Ave. SE, 20540).
Website: http://www.loc.gov

*National Aeronautics and Space Administration — Daniel S. Goldin, adm. (300 E St. SW, 20546).
Website: http://www.nasa.gov

*National Archives & Records Administration — John W. Carlin, archivist (700 Pennsylvania Ave. NW, 20408).
Website: http://www.nara.gov

*National Endowment for the Arts — vacant (1100 Pennsylvania Ave. NW, 20506).
Website: http://www.arts.gov

*National Endowment for the Humanities — William Ferris, chair (1100 Pennsylvania Ave. NW, 20506).
Website: http://www.neh.fed.us

National Institutes of Health — Dr. Ruth Kirchstein, act. dir. (9000 Rockville Pike, Bethesda, MD 20892).
Website: http://www.nih.gov

*National Labor Relations Board — vacant, chair (1099 14th St. NW, 20570).
Website: http://www.nlrb.gov

National Oceanic and Atmospheric Administration — Scott B. Gudes, act. undersec. (Dept. of Commerce, 14th & Constitution Ave. NW, 20230).
Website: http://www.noaa.gov

National Park Service — Fran B. Mainella, dir. (Dept. of the Interior, 1849 C St. NW, 20240).
Website: http://www.nps.gov

*National Railroad Passenger Corp. (Amtrak) — George Warrington, Pres. & CEO (60 Mass. Ave. NE, 20002).
Website: http://www.amtrak.com

*National Science Foundation — Dr. Rita Colwell, dir., National Science Foundation; Eamon Kelly, chair, National Science Board (4201 Wilson Blvd., Arlington, VA 22230).
Website: http://www.nsf.gov

*National Transportation Safety Board — Carol J. Carmody, chair (490 L'Enfant Plaza SW, 20594).
Website: http://www.ntsb.gov

*Nuclear Regulatory Commission — Richard A. Meserve, chair (11555 Rockville Pike, Rockville, MD 20852).
Website: http://www.nrc.gov

Occupational Safety & Health Administration — John Henshaw, asst. sec. (Dept. of Labor, 200 Constitution Ave. NW, 20210).
Website: http://www.osha.gov

*Occupational Safety & Health Review Commission — Thomasina V. Rogers, chair (1120 20th St. NW, 9th Floor, 20036).
Website: http://www.oshrc.gov

*Office of Government Ethics — Amy Comstock, dir. (1201 New York Ave. NW, Suite 500, 20005).
Website: http://www.usoge.gov

*Office of Personnel Management — Kay Coles James, dir. (1900 E St. NW, 20415-0001).
Website: http://www.opm.gov

*Office of Special Counsel — Elaine D. Kaplan, special counsel (1730 M St. NW, Suite 300, 20036).
Website: http://www.osc.gov

*Peace Corps — Charles R. Baquet III, act. dir. (1111 20th St., NW, 20526).
Website: http://www.peacecorps.gov/home.html

*Postal Rate Commission — George A. Omas, vice-chair (1333 H St. NW, Suite 300, 20268).
Website: http://www.prc.gov

*Securities and Exchange Commission — Harvey L. Pitt, chair (450 5th St. NW, 20549).
Website: http://www.sec.gov

*Selective Service System — Alfredo V. Rascon, dir. (National Headquarters, 1515 Wilson Blvd., Arlington, VA 22209-2425).
Website: http://www.sss.gov

*Small Business Administration — Hector V. Barrett, adm. (409 Third St. SW, 20416).
Website: http://www.sba.gov

Smithsonian Institution — (quasi-official agency) Lawrence Small, sec. (1000 Jefferson Dr. SW, Rm. 354, 20560-0033).
Website: http://www.si.edu

*Social Security Administration — Larry G. Massanari, act. comm. (6401 Security Blvd., Baltimore, MD 21235).
Website: http://www.ssa.gov

Surgeon General — Dr. David Satcher (Dept. of HHS, 200 Independence Ave. SW, 20201).
Website: http://www.surgeongeneral.gov

*Tennessee Valley Authority — Glenn L. McCollough Jr., chair, Board of Directors (400 W. Summit Hill Dr., Knoxville, TN 37902, and One Mass. Ave. NW, Suite 300, 20444). *Website:* http://www.tva.gov

*Trade and Development Agency — Thelma J. Askey, dir. (1621 N. Kent St., Suite 200, Arlington, VA 22209).
Website: http://www.tda.gov

United States Coast Guard — Adm. James M. Loy, commandant (Dept. of Trans., 2100 2d St. SW, 20593).
Website: http:/www.uscg.mil

United States Customs Service — Charles Winwood, comm. (1300 Pennsylvania Ave. NW, 20229).
Website: http://www.customs.treas.gov

*United States International Trade Commission — Stephen Koplan, chair (500 E St. SW, 20436).
Website: http://www.usitc.gov

United States Mint — Hennetta Holsman Fore, dir. (U.S. Mint Headquarters, 801 9th St., NW, 20002).
Website: http://www.usmint.gov

*United States Postal Service — John E. Potter, Postmaster General (475 L'Enfant Plaza SW, 20260).
Website: http://www.usps.gov

United States Secret Service — Brian L. Stafford, dir. (Dept. of Treas., 950 H St. NW, Ste. 8000, 20001).
Website: http://www.ustreas.gov/usss

CABINETS OF THE U.S.

The U.S. Cabinet and Its Role

The heads of major executive departments of government constitute the Cabinet. This institution, not provided for in the U.S. Constitution, developed as an advisory body out of the desire of presidents to consult on policy matters. Aside from its advisory role, the Cabinet as a body has no function and wields no executive authority. Individual members exercise authority as heads of their departments, reporting to the president.

In addition to the heads of federal departments as listed below, the Cabinet commonly includes other officials designated by the president as of Cabinet rank.

The officials so designated by Pres. George W. Bush include: Vice Pres. Dick Cheney, Chief of Staff to the President Andrew H. Card Jr., Environmental Protection Agency Administrator Christine Todd Whitman, Office of Management and Budget Director Mitchell E. Daniels Jr., Office of National Drug Control Policy Acting Director Edward H. Jurith, United States Trade Representative Robert B. Zoellick, and Office of Homeland Security Director Tom Ridge.

The Cabinet meets at times set by the president. Members of Pres. Bush's Cabinet listed in this chapter are as of Oct. 1, 2001.

Secretaries of State

The Department of Foreign Affairs was created by act of Congress on July 27, 1789, and the name changed to Department of State on Sept. 15, 1789.

President	Secretary	Home	Apptd.
Washington	Thomas Jefferson	VA	1789
"	Edmund Randolph	VA	1794
"	Timothy Pickering	PA	1795
Adams, J.	Timothy Pickering	PA	1797
"	John Marshall	VA	1800
Jefferson	James Madison	VA	1801
Madison	Robert Smith	MD	1809
"	James Monroe	VA	1811
Monroe	John Quincy Adams	MA	1817
Adams, J.Q.	Henry Clay	KY	1825
Jackson	Martin Van Buren	NY	1829
"	Edward Livingston	LA	1831
"	Louis McLane	DE	1833
"	John Forsyth	GA	1834
Van Buren	John Forsyth	GA	1837
Harrison, W.H.	Daniel Webster	MA	1841
Tyler	Daniel Webster	MA	1841
"	Abel P. Upshur	VA	1843
"	John C. Calhoun	SC	1844
Polk	John C. Calhoun	SC	1845
"	James Buchanan	PA	1845
Taylor	James Buchanan	PA	1849
"	John M. Clayton	DE	1849
Fillmore	John M. Clayton	DE	1850
"	Daniel Webster	MA	1850
"	Edward Everett	MA	1852
Pierce	William L. Marcy	NY	1853
Buchanan	William L. Marcy	NY	1857
"	Lewis Cass	MI	1857
"	Jeremiah S. Black	PA	1860
Lincoln	Jeremiah S. Black	PA	1861
"	William H. Seward	NY	1861
Johnson, A.	William H. Seward	NY	1865
Grant	Elihu B. Washburne	IL	1869
"	Hamilton Fish	NY	1869
Hayes	Hamilton Fish	NY	1877
"	William M. Evarts	NY	1877
Garfield	William M. Evarts	NY	1881
"	James G. Blaine	ME	1881
Arthur	James G. Blaine	ME	1881
"	F.T. Frelinghuysen	NJ	1881
Cleveland	F.T. Frelinghuysen	NJ	1885
"	Thomas F. Bayard	DE	1885
Harrison, B.	Thomas F. Bayard	DE	1889
Harrison, B.	James G. Blaine	ME	1889
"	John W. Foster	IN	1892
Cleveland	Walter Q. Gresham	IN	1893
"	Richard Olney	MA	1895
McKinley	Richard Olney	MA	1897
"	John Sherman	OH	1897
"	William R. Day	OH	1898
"	John Hay	DC	1898
Roosevelt, T.	John Hay	DC	1901
"	Elihu Root	NY	1905
"	Robert Bacon	NY	1909
Taft	Robert Bacon	NY	1909
"	Philander C. Knox	PA	1909
Wilson	Philander C. Knox	PA	1913
"	William J. Bryan	NE	1913
"	Robert Lansing	NY	1915
"	Bainbridge Colby	NY	1920
Harding	Charles E. Hughes	NY	1921
Coolidge	Charles E. Hughes	NY	1923
"	Frank B. Kellogg	MN	1925
Hoover	Frank B. Kellogg	MN	1929
"	Henry L. Stimson	NY	1929
Roosevelt, F.D.	Cordell Hull	TN	1933
"	E.R. Stettinius Jr.	VA	1944
Truman	E.R. Stettinius Jr.	VA	1945
"	James F. Byrnes	SC	1945
"	George C. Marshall	PA	1947
"	Dean G. Acheson	CT	1949
Eisenhower	John Foster Dulles	NY	1953
"	Christian A. Herter	MA	1959
Kennedy	Dean Rusk	NY	1961
Johnson, L.B.	Dean Rusk	NY	1963
Nixon	William P. Rogers	NY	1969
"	Henry A. Kissinger	DC	1973
Ford	Henry A. Kissinger	DC	1974
Carter	Cyrus R. Vance	NY	1977
"	Edmund S. Muskie	ME	1980
Reagan	Alexander M. Haig Jr.	CT	1981
"	George P. Shultz	CA	1982
Bush, G.H.W.	James A. Baker 3d	TX	1989
"	Lawrence S. Eagleburger	MI	1992
Clinton	Warren M. Christopher	CA	1993
"	Madeleine K. Albright	DC	1997
Bush, G.W.	Colin L. Powell	NY	2001

Secretaries of the Treasury

The Treasury Department was organized by act of Congress on Sept. 2, 1789.

President	Secretary	Home	Apptd.
Washington	Alexander Hamilton	NY	1789
"	Oliver Wolcott	CT	1795
Adams, J.	Oliver Wolcott	CT	1797
"	Samuel Dexter	MA	1801
Jefferson	Samuel Dexter	MA	1801
"	Albert Gallatin	PA	1801
Madison	Albert Gallatin	PA	1809
"	George W. Campbell	TN	1814
"	Alexander J. Dallas	PA	1814
"	William H. Crawford	GA	1816
Monroe	William H. Crawford	GA	1817
Adams, J.Q.	Richard Rush	PA	1825
Jackson	Samuel D. Ingham	PA	1829
"	Louis McLane	DE	1831
"	William J. Duane	PA	1833
"	Roger B. Taney	MD	1833
"	Levi Woodbury	NH	1834
Van Buren	Levi Woodbury	NH	1837
Harrison, W.H.	Thomas Ewing	OH	1841
Tyler	Thomas Ewing	OH	1841
"	Walter Forward	PA	1841
"	John C. Spencer	NY	1843
"	George M. Bibb	KY	1844
Polk	Robert J. Walker	MS	1845
Taylor	William M. Meredith	PA	1849
Fillmore	Thomas Corwin	OH	1850
Pierce	James Guthrie	KY	1853
Buchanan	Howell Cobb	GA	1857
"	Phillip F. Thomas	MD	1860
"	John A. Dix	NY	1861
Lincoln	Salmon P. Chase	OH	1861
"	William P. Fessenden	ME	1864
"	Hugh McCulloch	IN	1865
Johnson, A.	Hugh McCulloch	IN	1865

President	Secretary	Home	Apptd.
Grant	George S. Boutwell	MA	1869
"	William A. Richardson	MA	1873
"	Benjamin H. Bristow	KY	1874
"	Lot M. Morrill	ME	1876
Hayes	John Sherman	OH	1877
Garfield	William Windom	MN	1881
Arthur	Charles J. Folger	NY	1881
"	Walter Q. Gresham	IN	1884
"	Hugh McCulloch	IN	1884
Cleveland	Daniel Manning	NY	1885
"	Charles S. Fairchild	NY	1887
Harrison, B.	William Windom	MN	1889
"	Charles Foster	OH	1891
Cleveland	John G. Carlisle	KY	1893
McKinley	Lyman J. Gage	IL	1897
Roosevelt, T.	Lyman J. Gage	IL	1901
"	Leslie M. Shaw	IA	1902
"	George B. Cortelyou	NY	1907
Taft	Franklin MacVeagh	IL	1909
Wilson	William G. McAdoo	NY	1913
"	Carter Glass	VA	1918
"	David F. Houston	MO	1920
Harding	Andrew W. Mellon	PA	1921
Coolidge	Andrew W. Mellon	PA	1923
Hoover	Andrew W. Mellon	PA	1929
"	Ogden L. Mills	NY	1932
Roosevelt, F.D.	William H. Woodin	NY	1933
"	Henry Morgenthau, Jr.	NY	1934
Truman	Fred M. Vinson	KY	1945
"	John W. Snyder	MO	1946
Eisenhower	George M. Humphrey	OH	1953
"	Robert B. Anderson	CT	1957
Kennedy	C. Douglas Dillon	NJ	1961
Johnson, L.B.	C. Douglas Dillon	NJ	1963
"	Henry H. Fowler	VA	1965
"	Joseph W. Barr	IN	1968
Nixon	David M. Kennedy	IL	1969
"	John B. Connally	TX	1971
"	George P. Shultz	IL	1972
"	William E. Simon	NJ	1974
Ford	William E. Simon	NJ	1974
Carter	W. Michael Blumenthal	MI	1977
"	G. William Miller	RI	1979
Reagan	Donald T. Regan	NY	1981
"	James A. Baker 3d	TX	1985
"	Nicholas F. Brady	NJ	1988
Bush, G.H.W.	Nicholas F. Brady	NJ	1989
Clinton	Lloyd Bentsen	TX	1993
"	Robert E. Rubin	NY	1995
"	Lawrence H. Summers	CT	1999
Bush, G.W.	Paul H. O'Neill	PA	2001

Secretaries of Defense

The Department of Defense, originally designated the National Military Establishment, was created on Sept. 18, 1947. It is headed by the secretary of defense, who is a member of the president's Cabinet. The departments of the army, of the navy, and of the air force function within the Defense Department, and since 1947 the secretaries of these departments have not been members of the president's Cabinet.

President	Secretary	Home	Apptd.
Truman	James V. Forrestal	NY	1947
"	Louis A. Johnson	WV	1949
"	George C. Marshall	PA	1950
"	Robert A. Lovett	NY	1951
Eisenhower	Charles E. Wilson	MI	1953
"	Neil H. McElroy	OH	1957
"	Thomas S. Gates Jr.	PA	1959
Kennedy	Robert S. McNamara	MI	1961
Johnson, L.B.	Robert S. McNamara	MI	1963
"	Clark M. Clifford	MD	1968
Nixon	Melvin R. Laird	WI	1969
"	Elliot L. Richardson	MA	1973
"	James R. Schlesinger	VA	1973
Ford	James R. Schlesinger	VA	1974
"	Donald H. Rumsfeld	IL	1975
Carter	Harold Brown	CA	1977
Reagan	Caspar W. Weinberger	CA	1981
"	Frank C. Carlucci	PA	1987
Bush, G.H.W.	Richard B. Cheney	WY	1989
Clinton	Les Aspin	WI	1993
"	William J. Perry	CA	1994
"	William S. Cohen	ME	1997
Bush, G.W.	Donald H. Rumsfeld	IL	2001

Secretaries of War

The War Department (which included jurisdiction over the navy until 1798) was created by act of Congress on Aug. 7, 1789, and Gen. Henry Knox was commissioned secretary of war under that act on Sept. 12, 1789.

President	Secretary	Home	Apptd.
Washington	Henry Knox	MA	1789
"	Timothy Pickering	PA	1795
"	James McHenry	MD	1796
Adams, J.	James McHenry	MD	1797
"	Samuel Dexter	MA	1800
Jefferson	Henry Dearborn	MA	1801
Madison	William Eustis	MA	1809
"	John Armstrong	NY	1813
"	James Monroe	VA	1814
"	William H. Crawford	GA	1815
Monroe	John C. Calhoun	SC	1817
Adams, J.Q.	James Barbour	VA	1825
"	Peter B. Porter	NY	1828
Jackson	John H. Eaton	TN	1829
"	Lewis Cass	MI	1831
"	Benjamin F. Butler	NY	1837
Van Buren	Joel R. Poinsett	SC	1837
Harrison, W.H.	John Bell	TN	1841
Tyler	John Bell	TN	1841
"	John C. Spencer	NY	1841
"	James M. Porter	PA	1843
"	William Wilkins	PA	1844
Polk	William L. Marcy	NY	1845
Taylor	George W. Crawford	GA	1849
Fillmore	Charles M. Conrad	LA	1850
Pierce	Jefferson Davis	MS	1853
Buchanan	John B. Floyd	VA	1857
"	Joseph Holt	KY	1861
Lincoln	Simon Cameron	PA	1861
"	Edwin M. Stanton	PA	1862
Johnson, A.	Edwin M. Stanton	PA	1865
"	John M. Schofield	IL	1868
Grant	John A. Rawlins	IL	1869
"	William T. Sherman	OH	1869
"	William W. Belknap	IA	1869
"	Alphonso Taft	OH	1876
"	James D. Cameron	PA	1876
Hayes	George W. McCrary	IA	1877
"	Alexander Ramsey	MN	1879
Garfield	Robert T. Lincoln	IL	1881
Arthur	Robert T. Lincoln	IL	1881
Cleveland	William C. Endicott	MA	1885
Harrison, B.	Redfield Proctor	VT	1889
"	Stephen B. Elkins	WV	1891
Cleveland	Daniel S. Lamont	NY	1893
McKinley	Russel A. Alger	MI	1897
"	Elihu Root	NY	1899
Roosevelt, T.	Elihu Root	NY	1901
"	William H. Taft	OH	1904
"	Luke E. Wright	TN	1908
Taft	Jacob M. Dickinson	TN	1909
"	Henry L. Stimson	NY	1911
Wilson	Lindley M. Garrison	NJ	1913
"	Newton D. Baker	OH	1916
Harding	John W. Weeks	MA	1921
Coolidge	John W. Weeks	MA	1923
"	Dwight F. Davis	MO	1925
Hoover	James W. Good	IL	1929
"	Patrick J. Hurley	OK	1929
Roosevelt, F.D.	George H. Dern	UT	1933
"	Harry H. Woodring	KS	1937
"	Henry L. Stimson	NY	1940
Truman	Robert P. Patterson	NY	1945
"	Kenneth C. Royall[1]	NC	1947

(1) Last member of the Cabinet with this title. The War Department became the Department of the Army and became a branch of the Department of Defense in 1947.

Secretaries of the Navy

The Navy Department was created by act of Congress on Apr. 30, 1798.

President	Secretary	Home	Apptd.
Adams, J.	Benjamin Stoddert	MD	1798
Jefferson	Benjamin Stoddert	MD	1801
"	Robert Smith	MD	1801
Madison	Paul Hamilton	SC	1809
"	William Jones	PA	1813
"	Benjamin W. Crowninshield	MA	1814
Monroe	Benjamin W. Crowninshield	MA	1817
"	Smith Thompson	NY	1818
"	Samuel L. Southard	NJ	1823
Adams, J.Q.	Samuel L. Southard	NJ	1825
Jackson	John Branch	NC	1829
"	Levi Woodbury	NH	1831
"	Mahlon Dickerson	NJ	1834
Van Buren	Mahlon Dickerson	NJ	1837
"	James K. Paulding	NY	1838
Harrison, W.H.	George E. Badger	NC	1841
Tyler	George E. Badger	NC	1841
"	Abel P. Upshur	VA	1841
"	David Henshaw	MA	1843
"	Thomas W. Gilmer	VA	1844
"	John Y. Mason	VA	1844
Polk	George Bancroft	MA	1845
"	John Y. Mason	VA	1846
Taylor	William B. Preston	VA	1849
Fillmore	William A. Graham	NC	1850
"	John P. Kennedy	MD	1852
Pierce	James C. Dobbin	NC	1853
Buchanan	Isaac Toucey	CT	1857
Lincoln	Gideon Welles	CT	1861
Johnson, A.	Gideon Welles	CT	1865
Grant	Adolph E. Borie	PA	1869
"	George M. Robeson	NJ	1869
Hayes	Richard W. Thompson	IN	1877
"	Nathan Goff Jr.	WV	1881
Garfield	William H. Hunt	LA	1881
Arthur	William E. Chandler	NH	1882
Cleveland	William C. Whitney	NY	1885
Harrison, B.	Benjamin F. Tracy	NY	1889
Cleveland	Hilary A. Herbert	AL	1893
McKinley	John D. Long	MA	1897
Roosevelt, T.	John D. Long	MA	1901
"	William H. Moody	MA	1902
"	Paul Morton	IL	1904
"	Charles J. Bonaparte	MD	1905
"	Victor H. Metcalf	CA	1906
"	Truman H. Newberry	MI	1908
Taft	George von L. Meyer	MA	1909
Wilson	Josephus Daniels	NC	1913
Harding	Edwin Denby	MI	1921
Coolidge	Edwin Denby	MI	1923
"	Curtis D. Wilbur	CA	1924
Hoover	Charles Francis Adams	MA	1929
Roosevelt, F.D.	Claude A. Swanson	VA	1933
"	Charles Edison	NJ	1940
"	Frank Knox	IL	1940
"	James V. Forrestal	NY	1944
Truman	James V. Forrestal[1]	NY	1945

(1) Last member of Cabinet with this title. The Navy Department became a branch of the Department of Defense when the latter was created on Sept. 18, 1947.

Attorneys General

The Office of Attorney General was established by act of Congress on Sept. 24, 1789. It officially reached Cabinet rank in Mar. 1792, when the first attorney general, Edmund Randolph, attended his initial Cabinet meeting. The Department of Justice, headed by the attorney general, was created June 22, 1870.

President	Attorney General	Home	Apptd.
Washington	Edmund Randolph	VA	1789
"	William Bradford	PA	1794
"	Charles Lee	VA	1795
Adams, J.	Charles Lee	VA	1797
Jefferson	Levi Lincoln	MA	1801
"	John Breckenridge	KY	1805
"	Caesar A. Rodney	DE	1807
Madison	Caesar A. Rodney	DE	1807
"	William Pinkney	MD	1811
"	Richard Rush	PA	1814
Monroe	Richard Rush	PA	1817
"	William Wirt	VA	1817
Adams, J.Q.	William Wirt	VA	1825
Jackson	John M. Berrien	GA	1829
"	Roger B. Taney	MD	1831
"	Benjamin F. Butler	NY	1833
Van Buren	Benjamin F. Butler	NY	1837
"	Felix Grundy	TN	1838
"	Henry D. Gilpin	PA	1840
Harrison, W.H.	John J. Crittenden	KY	1841
Tyler	John J. Crittenden	KY	1841
"	Hugh S. Legare	SC	1841
"	John Nelson	MD	1843
Polk	John Y. Mason	VA	1845
"	Nathan Clifford	ME	1846
"	Isaac Toucey	CT	1848
Taylor	Reverdy Johnson	MD	1849
Fillmore	John J. Crittenden	KY	1850
Pierce	Caleb Cushing	MA	1853
Buchanan	Jeremiah S. Black	PA	1857
"	Edwin M. Stanton	PA	1860
Lincoln	Edward Bates	MO	1861
"	James Speed	KY	1864
Johnson, A.	James Speed	KY	1865
"	Henry Stanbery	OH	1866
"	William M. Evarts	NY	1868
Grant	Ebenezer R. Hoar	MA	1869
"	Amos T. Akerman	GA	1870
"	George H. Williams	OR	1871
"	Edwards Pierrepont	NY	1875
"	Alphonso Taft	OH	1876
Hayes	Charles Devens	MA	1877
Garfield	Wayne MacVeagh	PA	1881
Arthur	Benjamin H. Brewster	PA	1882
Cleveland	Augustus Garland	AR	1885
Harrison, B.	William H. H. Miller	IN	1889
Cleveland	Richard Olney	MA	1893
"	Judson Harmon	OH	1895
McKinley	Joseph McKenna	CA	1897
"	John W. Griggs	NJ	1898
"	Philander C. Knox	PA	1901
Roosevelt, T.	Philander C. Knox	PA	1901
"	William H. Moody	MA	1904
"	Charles J. Bonaparte	MD	1906
Taft	George W. Wickersham	NY	1909
Wilson	J.C. McReynolds	TN	1913
"	Thomas W. Gregory	TX	1914
"	A. Mitchell Palmer	PA	1919
Harding	Harry M. Daugherty	OH	1921
Coolidge	Harry M. Daugherty	OH	1923
"	Harlan F. Stone	NY	1924
"	John G. Sargent	VT	1925
Hoover	William D. Mitchell	MN	1929
Roosevelt, F.D.	Homer S. Cummings	CT	1933
"	Frank Murphy	MI	1939
"	Robert H. Jackson	NY	1940
"	Francis Biddle	PA	1941
Truman	Thomas C. Clark	TX	1945
"	J. Howard McGrath	RI	1949
"	J.P. McGranery	PA	1952
Eisenhower	Herbert Brownell Jr.	NY	1953
"	William P. Rogers	MD	1957
Kennedy	Robert F. Kennedy	MA	1961
Johnson, L.B.	Robert F. Kennedy	MA	1963
"	N. de B. Katzenbach	IL	1964
"	Ramsey Clark	TX	1967
Nixon	John N. Mitchell	NY	1969
"	Richard G. Kleindienst	AZ	1972
"	Elliot L. Richardson	MA	1973
"	William B. Saxbe	OH	1974
Ford	William B. Saxbe	OH	1974
"	Edward H. Levi	IL	1975
Carter	Griffin B. Bell	GA	1977
"	Benjamin R. Civiletti	MD	1979
Reagan	William French Smith	CA	1981
"	Edwin Meese 3d	CA	1985
"	Richard Thornburgh	PA	1988
Bush, G.H.W.	Richard Thornburgh	PA	1989
"	William P. Barr	NY	1991
Clinton	Janet Reno	FL	1993
Bush, G.W.	John Ashcroft	MO	2001

Secretaries of the Interior

The Department of the Interior was created by act of Congress on Mar. 3, 1849.

President	Secretary	Home	Apptd.
Taylor	Thomas Ewing	OH	1849
Fillmore	Thomas M. T. McKennan	PA	1850
"	Alex H. H. Stuart	VA	1850
Pierce	Robert McClelland	MI	1853
Buchanan	Jacob Thompson	MS	1857
Lincoln	Caleb B. Smith	IN	1861
"	John P. Usher	IN	1863
Johnson, A.	John P. Usher	IN	1865
"	James Harlan	IA	1865
"	Orville H. Browning	IL	1866
Grant	Jacob D. Cox	OH	1869
"	Columbus Delano	OH	1870
"	Zachariah Chandler	MI	1875
Hayes	Carl Schurz	MO	1877
Garfield	Samuel J. Kirkwood	IA	1881
Arthur	Henry M. Teller	CO	1882
Cleveland	Lucius Q.C. Lamar	MS	1885
"	William F. Vilas	WI	1888
Harrison, B.	John W. Noble	MO	1889
Cleveland	Hoke Smith	GA	1893
"	David R. Francis	MO	1896
McKinley	Cornelius N. Bliss	NY	1897
"	Ethan A. Hitchcock	MO	1898
Roosevelt, T.	Ethan A. Hitchcock	MO	1901
"	James R. Garfield	OH	1907
Taft	Richard A. Ballinger	WA	1909
"	Walter L. Fisher	IL	1911
Wilson	Franklin K. Lane	CA	1913
"	John B. Payne	IL	1920
Harding	Albert B. Fall	NM	1921
"	Hubert Work	CO	1923
Coolidge	Hubert Work	CO	1923
"	Roy O. West	IL	1929
Hoover	Ray Lyman Wilbur	CA	1929
Roosevelt, F.D.	Harold L. Ickes	IL	1933
Truman	Harold L. Ickes	IL	1945
"	Julius A. Krug	WI	1946
"	Oscar L. Chapman	CO	1949
Eisenhower	Douglas McKay	OR	1953
"	Fred A. Seaton	NE	1956
Kennedy	Stewart L. Udall	AZ	1961
Johnson, L.B.	Stewart L. Udall	AZ	1963
Nixon	Walter J. Hickel	AK	1969
"	Rogers C.B. Morton	MD	1971
Ford	Rogers C.B. Morton	MD	1971
"	Stanley K. Hathaway	WY	1975
"	Thomas S. Kleppe	ND	1975
Carter	Cecil D. Andrus	ID	1977
Reagan	James G. Watt	CO	1981
"	William P. Clark	CA	1983
"	Donald P. Hodel	OR	1985
Bush, G.H.W.	Manuel Lujan	NM	1989
Clinton	Bruce Babbitt	AZ	1993
Bush, G.W.	Gale Norton	CO	2001

Secretaries of Agriculture

The Department of Agriculture was created by act of Congress on May 15, 1862. On Feb. 8, 1889, its commissioner was renamed secretary of agriculture and became a member of the Cabinet.

President	Secretary	Home	Apptd.
Cleveland	Norman J. Colman	MO	1889
Harrison, B.	Jeremiah M. Rusk	WI	1889
Cleveland	J. Sterling Morton	NE	1893
McKinley	James Wilson	IA	1897
Roosevelt, T.	James Wilson	IA	1901
Taft	James Wilson	IA	1909
Wilson	David F. Houston	MO	1913
"	Edwin T. Meredith	IA	1920
Harding	Henry C. Wallace	IA	1921
Coolidge	Henry C. Wallace	IA	1923
"	Howard M. Gore	WV	1924
"	William M. Jardine	KS	1925
Hoover	Arthur M. Hyde	MO	1929
Roosevelt, F.D.	Henry A. Wallace	IA	1933
"	Claude R. Wickard	IN	1940
Truman	Clinton P. Anderson	NM	1945
Truman	Charles F. Brannan	CO	1948
Eisenhower	Ezra Taft Benson	UT	1953
Kennedy	Orville L. Freeman	MN	1961
Johnson, L.B.	Orville L. Freeman	MN	1963
Nixon	Clifford M. Hardin	IN	1969
"	Earl L. Butz	IN	1971
Ford	Earl L. Butz	IN	1974
"	John A. Knebel	VA	1976
Carter	Bob Bergland	MN	1977
Reagan	John R. Block	IL	1981
"	Richard E. Lyng	CA	1986
Bush, G.H.W.	Clayton K. Yeutter	NE	1989
"	Edward Madigan	IL	1991
Clinton	Mike Espy	MS	1993
"	Dan Glickman	KS	1995
Bush, G.W.	Ann M. Veneman	CA	2001

Secretaries of Commerce and Labor

The Department of Commerce and Labor, created by Congress on Feb. 14, 1903, was divided by Congress Mar. 4, 1913, into separate departments of Commerce and Labor. The secretary of each was made a Cabinet member.

Secretaries of Commerce and Labor

President	Secretary	Home	Apptd.
Roosevelt, T.	George B. Cortelyou	NY	1903
"	Victor H. Metcalf	CA	1904
"	Oscar S. Straus	NY	1906
Taft	Charles Nagel	MO	1909

Secretaries of Labor

President	Secretary	Home	Apptd.
Wilson	William B. Wilson	PA	1913
Harding	James J. Davis	PA	1921
Coolidge	James J. Davis	PA	1923
Hoover	James J. Davis	PA	1929
"	William N. Doak	VA	1930
Roosevelt, F.D.	Frances Perkins	NY	1933
Truman	L.B. Schwellenbach	WA	1945
"	Maurice J. Tobin	MA	1949
Eisenhower	Martin P. Durkin	IL	1953
"	James P. Mitchell	NJ	1953
Kennedy	Arthur J. Goldberg	IL	1961
"	W. Willard Wirtz	IL	1962
Johnson, L.B.	W. Willard Wirtz	IL	1963
Nixon	George P. Shultz	IL	1969
"	James D. Hodgson	CA	1970
"	Peter J. Brennan	NY	1973
Ford	Peter J. Brennan	NY	1974
"	John T. Dunlop	CA	1975
"	W.J. Usery Jr.	GA	1976
Carter	F. Ray Marshall	TX	1977
Reagan	Raymond J. Donovan	NJ	1981
"	William E. Brock	TN	1985
"	Ann D. McLaughlin	DC	1987
Bush, G.H.W.	Elizabeth Hanford Dole	NC	1989
"	Lynn Martin	IL	1991
Clinton	Robert B. Reich	MA	1993
"	Alexis M. Herman	AL	1997
Bush, G.W.	Elaine L. Chao	KY	2001

Secretaries of Commerce

President	Secretary	Home	Apptd.
Wilson	William C. Redfield	NY	1913
"	Joshua W. Alexander	MO	1919
Harding	Herbert C. Hoover	CA	1921
Coolidge	Herbert C. Hoover	CA	1923
"	William F. Whiting	MA	1928
Hoover	Robert P. Lamont	IL	1929
"	Roy D. Chapin	MI	1932
Roosevelt, F.D.	Daniel C. Roper	SC	1933
"	Harry L. Hopkins	NY	1939
"	Jesse Jones	TX	1940
"	Henry A. Wallace	IA	1945
Truman	Henry A. Wallace	IA	1945
"	W. Averell Harriman	NY	1947
"	Charles Sawyer	OH	1948
Eisenhower	Sinclair Weeks	MA	1953
"	Lewis L. Strauss	NY	1958
"	Frederick H. Mueller	MI	1959

President	Secretary	Home	Apptd.
Kennedy	Luther H. Hodges	NC	1961
Johnson, L.B.	Luther H. Hodges	NC	1963
"	John T. Connor	NJ	1965
"	Alex B. Trowbridge	NJ	1967
"	Cyrus R. Smith	NY	1968
Nixon	Maurice H. Stans	MN	1969
"	Peter G. Peterson	IL	1972
"	Frederick B. Dent	SC	1973
Ford	Frederick B. Dent	SC	1974
"	Rogers C.B. Morton	MD	1975
"	Elliot L. Richardson	MA	1975
Carter	Juanita M. Kreps	NC	1977
"	Philip M. Klutznick	IL	1979
Reagan	Malcolm Baldrige	CT	1981
"	C. William Verity Jr.	OH	1987
Bush, G.H.W.	Robert A. Mosbacher	TX	1989
"	Barbara H. Franklin	PA	1992
Clinton	Ronald H. Brown	DC	1993
"	Mickey Kantor	CA	1996
"	William M. Daley	IL	1997
"	Norman Y. Mineta	CA	2000
Bush, G.W.	Donald L. Evans	TX	2001

Secretaries of Housing and Urban Development

The Department of Housing and Urban Development was created by act of Congress on Sept. 9, 1965.

President	Secretary	Home	Apptd.
Johnson, L.B.	Robert C. Weaver	WA	1966
"	Robert C. Wood	MA	1969
Nixon	George W. Romney	MI	1969
"	James T. Lynn	OH	1973
Ford	James T. Lynn	OH	1974
"	Carla Anderson Hills	CA	1975
Carter	Patricia Roberts Harris	DC	1977
"	Moon Landrieu	LA	1979
Reagan	Samuel R. Pierce Jr.	NY	1981
Bush, G.H.W.	Jack F. Kemp	NY	1989
Clinton	Henry G. Cisneros	TX	1993
"	Andrew M. Cuomo	NY	1997
Bush, G.W.	Mel Martinez	FL	2001

Secretaries of Transportation

The Department of Transportation was created by act of Congress on Oct. 15, 1966.

President	Secretary	Home	Apptd.
Johnson, L.B.	Alan S. Boyd	FL	1966
Nixon	John A. Volpe	MA	1969
"	Claude S. Brinegar	CA	1973
Ford	Claude S. Brinegar	CA	1974
"	William T. Coleman Jr.	PA	1975
Carter	Brock Adams	WA	1977
"	Neil E. Goldschmidt	OR	1979
Reagan	Andrew L. Lewis Jr.	PA	1981
"	Elizabeth Hanford Dole	NC	1983
"	James H. Burnley	NC	1987
Bush, G.H.W.	Samuel K. Skinner	IL	1989
"	Andrew H. Card Jr.	MA	1992
Clinton	Federico F. Peña	CO	1993
"	Rodney E. Slater	AR	1997
Bush, G.W.	Norman Y. Mineta	CA	2001

Secretaries of Energy

The Department of Energy was created by federal law on Aug. 4, 1977.

President	Secretary	Home	Apptd.
Carter	James R. Schlesinger	VA	1977
"	Charles Duncan Jr.	WY	1979
Reagan	James B. Edwards	SC	1981
"	Donald P. Hodel	OR	1982
"	John S. Herrington	CA	1985
Bush, G.H.W.	James D. Watkins	CA	1989
Clinton	Hazel R. O'Leary	MN	1993
"	Federico F. Peña	CO	1997
"	Bill Richardson	NM	1998
Bush, G.W.	Spencer Abraham	MI	2001

Secretaries of Health, Education, and Welfare

The Department of Health, Education, and Welfare was created by Congress on Apr. 11, 1953. On Sept. 27, 1979, it was divided by Congress into the departments of Education and of Health and Human Services, with the secretary of each being a Cabinet member.

President	Secretary	Home	Apptd.
Eisenhower	Oveta Culp Hobby	TX	1953
"	Marion B. Folsom	NY	1955
"	Arthur S. Flemming	OH	1958
Kennedy	Abraham A. Ribicoff	CT	1961
"	Anthony J. Celebrezze	OH	1962
Johnson, L.B.	Anthony J. Celebrezze	OH	1963
"	John W. Gardner	NY	1965
"	Wilbur J. Cohen	MI	1968
Nixon	Robert H. Finch	CA	1969
"	Elliot L. Richardson	MA	1970
"	Caspar W. Weinberger	CA	1973
Ford	Caspar W. Weinberger	CA	1974
"	Forrest D. Mathews	AL	1975
Carter	Joseph A. Califano Jr.	DC	1977
"	Patricia Roberts Harris	DC	1979

Secretaries of Health and Human Services

President	Secretary	Home	Apptd.
Carter	Patricia Roberts Harris	DC	1979
Reagan	Richard S. Schweiker	PA	1981
"	Margaret M. Heckler	MA	1983
Reagan	Otis R. Bowen	IN	1985
Bush, G.H.W.	Louis W. Sullivan	GA	1989
Clinton	Donna E. Shalala	WI	1993
Bush, G.W.	Tommy Thompson	WI	2001

Secretaries of Education

President	Secretary	Home	Apptd.
Carter	Shirley Hufstedler	CA	1979
Reagan	Terrel Bell	UT	1981
"	William J. Bennett	NY	1985
"	Lauro F. Cavazos	TX	1988
Bush, G.H.W.	Lauro F. Cavazos	TX	1989
"	Lamar Alexander	TN	1991
Clinton	Richard W. Riley	SC	1993
Bush, G.W.	Roderick R. Paige	TX	2001

Secretaries of Veterans Affairs

The Department of Veterans Affairs was created on Oct. 25, 1988, when Pres. Ronald Reagan signed a bill that made the Veterans Administration into a Cabinet department, effective Mar. 15, 1989.

President	Secretary	Home	Apptd.
Bush, G.H.W.	Edward J. Derwinski	IL	1989
Clinton	Jesse Brown	IL	1993
Clinton	Togo D. West Jr.	NC	1998
"	Hershel W. Gober (acting)	AR	2000
Bush, G.W.	Anthony Principi	CA	2001

CONGRESS

The One Hundred and Seventh Congress With Official 2000 Election Results

Source: Voter News Service; World Almanac research

The 107th Congress convened on Jan. 3, 2001.

The Senate

Dem., 50; Rep., 49; Ind., 1; Total, 100. Boldface denotes the 2000 election winner. *Incumbent.

Terms are for 6 years and end Jan. 3 of the year preceding the senator's name in the following table. Annual salary, $145,100; President Pro Tempore, Majority Leader, and Minority Leader, $161,200. To be eligible for the Senate, one must be at least 30 years old, a U.S. citizen for at least 9 years, and a resident of the state from which chosen. Congress must meet annually on Jan. 3, unless it has, by law, appointed a different day.

The ZIP code of the Senate is 20510; the telephone number is 202-224-3121; the website is http://www.senate.gov

Senate officials in 2001 (after Democrats won control following the departure of Sen. James M. Jeffords (VT) from the Republican Party) were: President Pro Tempore, Robert C. Byrd (WV); Majority Leader, Tom Daschle (SD); Majority Whip, Harry Reid (NV); Minority Leader, Trent Lott (MS); Minority Whip, Don Nickles (OK).

D-Democrat; R-Republican; ACP-A Connecticut Party; C-Conservative; I-Independent; IN-Independence; L-Liberal; RL-Right to Life; WF-Working Families

Term ends	Senator (Party); Service from[1]	2000 Election
	Alabama	
2003	Jeff Sessions (R); 1/7/97	
2005	Richard Shelby (R); 1/6/87	
	Alaska	
2003	Ted Stevens (R); 12/24/68	
2005	Frank H. Murkowski (R); 1981	
	Arizona	
2005	John McCain (R); 1/6/87	
2007	**Jon Kyl*** (R); 1/4/95	**1,108,196**
	William Toel (I)	109,230
	Arkansas	
2003	Tim Hutchinson (R); 1/7/97	
2005	Blanche Lambert Lincoln (D); 1/6/99	
	California	
2005	Barbara Boxer (D); 1993	
2007	**Dianne Feinstein*** (D); 11/10/92	**5,932,522**
	Tom Campbell (R)	3,886,853
	Colorado	
2003	Wayne Allard (R); 1/7/97	
2005	Ben Nighthorse Campbell (R); 1993	
	Connecticut	
2005	Christopher J. Dodd (D); 1981	
2007	**Joe Lieberman*** (D); 1989	**828,902**
	Phil Giordano (R)	448,077
	Delaware	
2003	Joseph R. Biden, Jr. (D); 1973	
2007	**Thomas R. Carper** (D); 2001	**181,566**
	William V. Roth, Jr.* (R)	142,891
	Florida	
2005	Bob Graham (D); 1/6/87	
2007	**Bill Nelson** (D); 2001	**2,989,487**
	Bill McCollum (R)	2,705,348
	Georgia	
2003	Max Cleland (D); 1/7/97	
2005	**Zell Miller*** (D)[2]; 7/24/00	**1,413,224**
	Mack F. Mattingly (R)[2]	920,478
	Hawaii	
2005	Daniel K. Inouye (D); 1963	
2007	**Daniel K. Akaka*** (D); 4/28/90	**251,215**
	John Carroll (R)	84,701
	Idaho	
2003	Larry E. Craig (R); 1991	
2005	Mike Crapo (R); 1/6/99	
	Illinois	
2003	Richard J. Durbin (D); 1/7/97	
2005	Peter G. Fitzgerald (R); 1/6/99	
	Indiana	
2005	Evan Bayh (D); 1/6/99	
2007	**Richard G. Lugar*** (R); 1977	**1,427,944**
	David L. Johnson (D)	683,273
	Iowa	
2003	Tom Harkin (D); 1985	
2005	Chuck Grassley (R); 1981	
	Kansas	
2003	Pat Roberts (R); 1/7/97	
2005	Sam Brownback (R); 1/7/97	
	Kentucky	
2003	Mitch McConnell (R); 1985	
2005	Jim Bunning (R); 1/6/99	
	Louisiana	
2003	Mary L. Landrieu (D); 1/7/97	
2005	John B. Breaux (D); 1/6/87	
	Maine	
2003	Susan M. Collins (R); 1/7/97	
2007	**Olympia J. Snowe*** (R); 1/4/95	**437,689**
	Mark W. Lawrence (D)	197,183
	Maryland	
2005	Barbara Ann Mikulski (D); 1/6/87	
2007	**Paul S. Sarbanes*** (D); 1977	**1,230,013**
	Paul H. Rappaport (R)	715,178
	Massachusetts	
2003	John F. Kerry (D); 1/2/85	
2007	**Edward M. Kennedy*** (D); 11/7/62	**1,889,496**
	Jack E. Robinson, III (R)	334,341
	Michigan	
2003	Carl Levin (D); 1979	
2007	**Debbie Stabenow** (D); 2001	**2,061,952**
	Spence Abraham* (R)	1,994,693
	Minnesota	
2003	Paul David Wellstone (D); 1991	
2007	**Mark Dayton** (D); 2001	**1,181,553**
	Rod Grams* (R)	1,047,474
	Mississippi	
2003	Thad Cochran (R); 12/27/78	
2007	**Trent Lott*** (R); 1989	**654,941**
	Troy D. Brown, Sr. (D)	314,090
	Missouri	
2005	Christopher (Kit) Bond (R); 1/6/87	
2003	Jean Carnahan (D)[3]; 2001	
	Montana	
2003	Max Baucus (D); 12/15/78	
2007	**Conrad Burns*** (R); 1989	**208,082**
	Brian Schweitzer (D)	194,430
	Nebraska	
2003	Chuck Hagel (R); 1/7/97	
2007	**Ben Nelson** (D); 2001	**353,097**
	Don Stenberg (R)	337,967
	Nevada	
2005	Harry Reid (D); 1/6/87	
2007	**John Ensign** (R); 2001	**330,687**
	Ed Bernstein (D)	238,260
	New Hampshire	
2003	Robert Smith (R); 12/7/90	
2005	Judd Gregg (R); 1993	
	New Jersey	
2003	Robert G. Torricelli (D); 1/7/97	
2007	**Jon S. Corzine** (D); 2001	**1,511,237**
	Bob Franks (R)	1,420,267
	New Mexico	
2003	Pete V. Domenici (R); 1973	
2007	**Jeff Bingaman*** (D); 1983	**363,744**
	Bill Redmond (R)	225,517

Term ends	Senator (Party); Service from[1]	2000 Election
	New York	
2005	Charles E. Schumer (D,IN,L); 1/6/99	
2007	**Hillary Rodham Clinton** (D,L,WF); 2001	**3,751,716**
	Rick Lazio (R,C)	2,918,841
	North Carolina	
2003	Jesse Helms (R); 1973	
2005	John Edwards (D); 1/6/99	
	North Dakota	
2005	Byron L. Dorgan (D); 12/14/92	
2007	**Kent Conrad*** (D); 1/6/87	**176,470**
	Duane Sand (R)	110,420
	Ohio	
2005	George V. Voinovich (R); 1/6/99	
2007	**Mike DeWine*** (R); 1/4/95	**2,666,736**
	Ted Celeste (D)	1,597,122
	Oklahoma	
2003	James M. Inhofe (R); 11/21/94	
2005	Don Nickles (R); 1981	
	Oregon	
2003	Gordon Smith (R); 1/7/97	
2005	Ron Wyden (D); 2/6/96	
	Pennsylvania	
2005	Arlen Specter (R); 1981	
2007	**Rick Santorum*** (R); 1/4/95	**2,481,962**
	Ron Klink (D)	2,154,908
	Rhode Island	
2003	John F. Reed (D); 1/7/97	
2007	**Lincoln D. Chafee*** (R); 11/2/99	**222,588**
	Robert A. Weygand (D)	161,023
	South Carolina	
2003	Strom Thurmond (R); 11/7/56	
2005	Ernest Hollings (D); 11/9/66	
	South Dakota	
2003	Tim Johnson (D); 1/7/97	
2005	Tom Daschle (D); 1/6/87	
	Tennessee	
2003	Fred Thompson (R); 12/9/94	
2007	**Bill Frist*** (R); 1/4/95	**1,255,444**
	Jeff Clark (D)	621,152
	Texas	
2003	Phil Gramm (R); 1985	
2007	**Kay Bailey Hutchison*** (R); 6/5/93	**4,082,091**
	Gene Kelly (D)	2,030,315
	Utah	
2005	Robert F. Bennett (R); 1993	
2007	**Orrin G. Hatch*** (R); 1977	**504,803**
	Scott N. Howell (D)	242,569
	Vermont	
2005	Patrick Leahy (D); 1975	
2007	**James M. Jeffords*** (I)[4]; 1989	**189,133**
	Ed Flanagan (D)	73,352
	Virginia	
2003	John W. Warner (R); 1/2/79	
2007	**George F. Allen** (R); 2001	**1,420,460**
	Charles S. Robb* (D)	1,296,093
	Washington	
2005	Patty Murray (D); 1993	
2007	**Maria Cantwell** (D); 2001	**1,199,437**
	Slade Gorton* (R)	1,197,208
	West Virginia	
2003	John D. Rockefeller IV (D); 1/15/85	
2007	**Robert C. Byrd*** (D); 1959	**469,215**
	David T. Gallaher (R)	121,635
	Wisconsin	
2005	Russ Feingold (D); 1993	
2007	**Herbert H. Kohl*** (D); 1989	**1,563,238**
	John Gillespie (R)	940,744
	Wyoming	
2003	Michael B. Enzi (R); 1/7/97	
2007	**Craig Thomas*** (R); 1/4/95	**157,622**
	Mel Logan (D)	47,087

(1) Jan. 3, unless otherwise noted. (2) Zell Miller was appointed July 24, 2000, to fill the vacancy caused by the death of Paul Coverdell on July 18, 2000. The election on Nov. 7, 2000, was a special election to fill Sen. Coverdell's unexpired term. Although the ele ction was non-partisan, the presumed party labels are listed. (3) Candidate Mel Carnahan died Oct. 16, 2000. Under Missouri statutes, the deadline to replace a party nominee was Oct. 13, 2000, so Carnahan's name remained on the ballot. He received 1,191,812 votes and his nearest opponent, John Ashcroft, received 1,142,852. The governor appointed Jean Carnahan, Mel Carnahan's widow, to fill the vacancy through the November 2002 general election, when the seat would be up for election. (4) Republican Sen. James M. Jeffords changed his party designation to Independent on June 5, 2001.

The House of Representatives

Rep., 219; Dem., 210; Ind., 2; Vacant, 4; Total, 435. Boldface denotes the 2000 election winner. *Incumbent.

Terms are for 2 years ending Jan. 3, 2003. Annual salary, $145,100; Speaker of the House, $186,300; Majority Leader and Minority Leader, $161,200. To be eligible for membership, a person must be at least 25 years of age, a U.S. citizen for at least 7 years, and a resident of the state from which he or she is chosen. The ZIP code of the House is 20515; the telephone number is 202-225-3121. The website is http://www.house.gov

House officials in 2001 were: Speaker, J. Dennis Hastert; Majority Leader, Dick Armey; Majority Whip, Tom DeLay; Minority Leader, Richard A. Gephardt; Minority Whip, David E. Bonior.

D-Democrat; R-Republican; AI-American Independent; C-Conservative; CC-Conscience for Congress; CN-Constitution; GR-Green; I-Independent; IA-Independent American; IN-Independence; L-Liberal; LB-Libertarian; NL-Natural Law; RF-Reform; RL-Right to Life; SC-Social Choice; TX-Taxpayers; UC-United Citizens; WF-Working Families

Dist.	Representative (Party)	2000 Election
	Alabama	
1.	**H. L. "Sonny" Callahan*** (R)	**151,188**
	Dick Coffee (LB)	14,031
2.	**Terry Everett*** (R)	**151,830**
	Charles Woods (D)	64,958
3.	**Bob Riley*** (R)	**147,317**
	John Sophocleus (LB)	21,119
4.	**Robert Aderholt*** (R)	**140,009**
	Marsha Folsom (D)	86,400
5.	**Bud Cramer*** (D)	**186,059**
	Alan Barksdale (LB)	22,110
6.	**Spencer Bachus*** (R)	**212,751**
	Terry Reagin (LB)	28,189
7.	**Earl F. Hilliard*** (D)	**148,243**
	Ed Martin (R)	46,134
	Alaska	
	Don E. Young* (R)	**190,862**
	Clifford Mark Greene (D)	45,372
	Arizona	
1.	**Jeff Flake** (R)	**123,289**
	David Mendoza (D)	97,455
2.	**Ed Pastor*** (D)	**84,034**
	Bill Barenholtz (R)	32,990
3.	**Bob Stump*** (R)	**198,367**
	Gene Scharer (D)	94,676
4.	**John Shadegg*** (R)	**140,396**
	Ben Jankowski (D)	71,803
5.	**Jim Kolbe*** (R)	**172,986**
	George Cunningham (D)	101,564
6.	**J.D. Hayworth*** (R)	**186,687**
	Larry Nelson (D)	108,317

Dist.	Representative (Party)	2000 Election
	Arkansas	
1.	**Marion Berry*** (D)	**120,266**
	Susan Myshka (R)	79,437
2.	**Vic Snyder*** (D)	**126,957**
	Bob Thomas (R)	93,692
3.	**Vacant**[1]	
4.	**Mike Ross** (D)	**108,143**
	Jay Dickey* (R)	104,017
	California	
1.	**Mike Thompson*** (D)	**155,638**
	Russel J. "Jim" Chase (R)	66,987
2.	**Wally Herger*** (R)	**168,172**
	Stan Morgan (D)	72,075
3.	**Doug Ose*** (R)	**129,254**
	Bob Kent (D)	93,067
4.	**John T. Doolittle*** (R)	**197,503**
	Mark A. Norberg (D)	97,974
5.	**Robert T. Matsui*** (D)	**147,025**
	Ken Payne (R)	55,945
6.	**Lynn Woolsey*** (D)	**182,116**
	Ken McAuliffe (R)	80,169
7.	**George Miller*** (D)	**159,692**
	Christopher A. Hoffman (R)	44,154
8.	**Nancy Pelosi*** (D)	**181,847**
	Adam Sparks (R)	25,298
9.	**Barbara Lee*** (D)	**182,352**
	Arneze Washington (R)	21,033
10.	**Ellen O. Tauscher*** (D)	**160,429**
	Claude B. Hutchison, Jr. (R)	134,863
11.	**Richard W. Pombo*** (R)	**120,635**
	Tom Y. Santos (D)	79,539
12.	**Tom Lantos*** (D)	**158,404**
	Mike Garza (R)	44,162
13.	**Fortney Pete Stark*** (D)	**129,012**
	James R. (Jim) Goetz (R)	44,499
14.	**Anna G. Eshoo*** (D)	**161,720**
	Bill Quraishi (R)	59,338
15.	**Mike Honda** (D)	**128,545**
	Jim Cunneen (R)	99,866
16.	**Zoe Lofgren*** (D)	**115,118**
	Horace "Gene" Thayn (R)	37,213
17.	**Sam Farr*** (D)	**143,219**
	Clint Engler (R)	51,557
18.	**Gary A. Condit*** (D)	**121,003**
	Steve R. Wilson (R)	56,465
19.	**George Randanovich*** (R)	**144,517**
	Dan Rosenberg (D)	70,578
20.	**Cal Dooley*** (D)	**66,235**
	Rich Rodriguez (R)	57,563
21.	**Bill Thomas*** (R)	**142,539**
	Pedro "Pete" Martinez, Jr. (D)	49,318
22.	**Lois Capps*** (D)	**135,538**
	Mike Stoker (R)	113,094
23.	**Elton W. Gallegly*** (R)	**119,479**
	Michael Case (D)	89,918
24.	**Brad Sherman*** (D)	**155,398**
	Jerry Doyle (R)	70,169
25.	**Howard P. "Buck" McKeon*** (R)	**138,628**
	Sid Gold (D)	73,921
26.	**Howard L. Berman*** (D)	**96,500**
	Bill Farley (LB)	13,052
27.	**Adam Schiff** (D)	**113,708**
	James E. Rogan* (R)	94,518
28.	**David Dreier*** (R)	**116,557**
	Janice M. Nelson (D)	81,804
29.	**Henry A. Waxman*** (D)	**180,295**
	Jim Scileppi (R)	45,784
30.	**Xavier Becerra*** (D)	**83,223**
	Tony Goss (R)	11,788
31.	**Hilda L. Solis** (D)	**89,600**
	Krista Lieberg-Wong (GR)	10,294
32.	**Diane E. Watson** (D)[2]	**72,955**
	Noel Irwin Hentschel (R)	19,403
33.	**Lucille Roybal-Allard*** (D)	**60,510**
	Wayne Miller (R)	8,260
34.	**Grace Flores Napolitano*** (D)	**105,980**
	Robert Arthur Canales (R)	33,445
35.	**Maxine Waters*** (D)	**100,569**
	Carl McGill (R)	12,582
36.	**Jane Harman** (D)	**115,651**
	Steven T. Kuykendall* (R)	111,199
37.	**Juanita Millender-McDonald*** (D)	**93,269**
	Vernon Van (R)	12,762
38.	**Steve Horn*** (R)	**87,266**
	Gerrie Schipske (D)	85,498
39.	**Ed Royce*** (R)	**129,294**
	Gill G. Kanel (D)	64,938
40.	**Jerry Lewis*** (R)	**151,069**
	Frank N. Schmit (NL)	19,029
41.	**Gary G. Miller*** (R)	**104,695**
	Rodolfo G. Favila (D)	66,361
42.	**Joe Baca*** (D)	**90,585**
	Eli Pirozzi (R)	53,239
43.	**Ken Calvert*** (R)	**140,201**
	Bill Reed (LB)	29,755
44.	**Mary Bono*** (R)	**123,738**
	Ron Oden (D)	79,302
45.	**Dana Rohrabacher*** (R)	**136,275**
	Ted Crisell (D)	71,066
46.	**Loretta Sanchez*** (D)	**70,381**
	Gloria Matta Tuchman (R)	40,928
47.	**Christopher Cox*** (R)	**181,365**
	John Graham (D)	83,186
48.	**Darrell Issa** (R)	**160,627**
	Peter Kouvelis (D)	74,073
49.	**Susan A. Davis** (D)	**113,400**
	Brian P. Bilbray* (R)	105,515
50.	**Bob Filner*** (D)	**95,191**
	Bob Divine (R)	38,526
51.	**Randy "Duke" Cunningham*** (R)	**172,291**
	George "Jorge" Barraza (D)	81,408
52.	**Duncan Hunter*** (R)	**131,345**
	Craig Barkacs (D)	63,537
	Colorado	
1.	**Diana DeGette*** (D)	**141,831**
	Jesse L. Thomas (R)	56,291
2.	**Mark Udall*** (D)	**155,725**
	Carolyn Cox (R)	109,338
3.	**Scott McInnis*** (R)	**199,204**
	Curtis Imrie (D)	87,921
4.	**Bob Schaffer*** (R)	**209,078**
	Dan Sewell Ward (NL)	19,721
5.	**Joel Hefley*** (R)	**253,330**
	Kerry Kantor (LB)	37,719
6.	**Tom Tancredo*** (R)	**141,410**
	Kenneth A. Toltz (D)	110,568
	Connecticut	
1.	**John B. Larson*** (D)	**151,932**
	Bob Backlund (R)	59,331
2.	**Rob Simmons** (R)	**141,410**
	Sam Gejdenson* (D)	111,520
3.	**Rosa L. DeLauro*** (D)	**156,910**
	June M. Gold (R)	60,037
4.	**Christopher Shays*** (R)	**119,155**
	Stephanie Sanchez (D)	84,472
5.	**Jim Maloney*** (D)	**118,932**
	Mark Nielsen (R)	98,229
6.	**Nancy L. Johnson*** (R)	**143,698**
	Paul Valenti (D)	75,471
	Delaware	
	Michael N. Castle* (R)	**211,797**
	Michael C. Miller (D)	96,488
	Florida	
1.	**Vacant**[3]	
2.	**Allen Boyd*** (D)	**185,579**
	Doug Dodd (R)	71,754
3.	**Corrine Brown*** (D)	**102,143**
	Jennifer S. Carroll (R)	75,228

Dist.	Representative (Party)	2000 Election
4.	**Ander Crenshaw** (R)	**203,090**
	Tom Sullivan (D)	94,587
5.	**Karen L. Thurman*** (D)	**180,338**
	Pete Enwall (R)	100,244
6.	**Clifford (Cliff) B. Stearns*** (R)	**Unopposed**
7.	**John L. Mica*** (R)	**171,018**
	Dan Vaughen (D)	99,531
8.	**Ric Keller** (R)	**125,253**
	Linda W. Chapin (D)	121,295
9.	**Michael Bilirakis*** (R)	**210,318**
	Jon Duffey (RF)	46,474
10.	**C. W. Bill Young*** (R)	**146,799**
	Josette Green (NL)	26,908
11.	**Jim Davis*** (D)	**149,465**
	Charlie Westlake (LB)	27,197
12.	**Adam H. Putnam** (R)	**125,224**
	Mike Stedem (D)	94,395
13.	**Dan Miller*** (R)	**175,918**
	Daniel E. Dunn (D)	99,568
14.	**Porter Goss*** (R)	**242,614**
	Sam Farling (NL)	41,988
15.	**Dave Weldon*** (R)	**176,189**
	Patsy Ann Kurth (D)	117,511
16.	**Mark Foley*** (R)	**176,153**
	Jean Elliott Brown (D)	108,782
17.	**Carrie P. Meek*** (D)	**Unopposed**
18.	**Ileana Ros-Lehtinen*** (R)	**Unopposed**
19.	**Robert Wexler*** (D)	**171,080**
	Morris Kent Thompson (R)	67,789
20.	**Peter Deutsch*** (D)	**Unopposed**
21.	**Lincoln Diaz-Balart*** (R)	**Unopposed**
22.	**Clay Shaw***(R)	**105,855**
	Elaine Bloom (D)	105,256
23.	**Alcee L. Hastings*** (D)	**89,179**
	Bill Lambert (R)	27,630
	Georgia	
1.	**Jack Kingston*** (R)	**131,684**
	Joyce Marie Griggs (D)	58,776
2.	**Sanford Dixon Bishop, Jr.*** (D)	**96,430**
	Dylan Glenn (R)	83,870
3.	**Michael A. (Mac) Collins*** (R)	**150,200**
	Gail Notti (D)	86,309
4.	**Cynthia McKinney*** (D)	**139,579**
	Sunny Warren (R)	90,277
5.	**John Lewis*** (D)	**137,333**
	Hank Schwab (R)	40,606
6.	**Johnny Isakson*** (R)	**256,595**
	Brett DeHart (D)	86,666
7.	**Bob Barr*** (R)	**126,312**
	Roger Kahn (D)	102,272
8.	**Saxby Chambliss*** (R)	**113,380**
	Jim Marshall (D)	79,051
9.	**Nathan Deal*** (R)	**183,171**
	James Harrington (D)	60,360
10.	**Charlie Norwood*** (R)	**122,590**
	Denise Freeman (D)	71,309
11.	**John Linder*** (R)	**Unopposed**
	Hawaii	
1.	**Neil Abercrombie*** (D)	**108,517**
	Phil Meyers (R)	44,989
2.	**Patsy Takemoto Mink*** (D)	**112,856**
	Russ Francis (R)	65,906
	Idaho	
1.	**C. L. "Butch" Otter** (R)	**173,743**
	Linda Pall (D)	84,080
2.	**Michael Simpson*** (R)	**158,912**
	Craig Williams (D)	58,265
	Illinois	
1.	**Bobby L. Rush*** (D)	**172,271**
	Raymond G. Wardingley (R)	23,915
2.	**Jesse L. Jackson, Jr.*** (D)	**175,995**
	Robert Gordon, III (R)	19,906
3.	**William O. Lipinski*** (D)	**145,498**
	Karl Groth (R)	47,005
4.	**Luis V. Gutierrez*** (D)	**89,487**
	Stephanie Sailor (LB)	11,476
5.	**Rod R. Blagojevich*** (D)	**142,161**
	Matt Beauchamp (LB)	20,728
6.	**Henry J. Hyde*** (R)	**133,327**
	Brent Christensen (D)	92,880
7.	**Danny K. Davis*** (D)	**164,155**
	Robert Dallas (R)	26,872
8.	**Philip M. Crane*** (R)	**141,918**
	Lance Pressl (D)	90,777
9.	**Jan Schakowsky*** (D)	**147,002**
	Dennis J. Driscoll (R)	45,344
10.	**Mark Steven Kirk** (R)	**121,582**
	Lauren Beth Gash (D)	115,924
11.	**Gerald C. "Jerry" Weller*** (R)	**132,384**
	James P. Stevenson (D)	102,485
12.	**Jerry F. Costello*** (D)	**Unopposed**
13.	**Judy Biggert*** (R)	**193,250**
	Thomas Mason (D)	98,768
14.	**J. Dennis Hastert*** (R)	**188,597**
	Vern Deljonson (D)	66,309
15.	**Tim Johnson** (R)	**125,943**
	F. Michael "Mike" Kelleher, Jr. (D)	110,679
16.	**Donald A. Manzullo*** (R)	**178,174**
	Charles W. Hendrickson (D)	88,781
17.	**Lane A. Evans*** (D)	**132,494**
	Mark Baker (R)	108,853
18.	**Ray LaHood*** (R)	**173,706**
	Joyce Harant (D)	85,317
19.	**Dave Phelps*** (D)	**155,101**
	James "Jim" Eatherly (R)	85,137
20.	**John M. Shimkus*** (R)	**161,393**
	Jeffrey S. Cooper (D)	94,382
	Indiana	
1.	**Peter J. Visclosky*** (D)	**148,683**
	Jack Reynolds (R)	56,200
2.	**Mike Pence** (R)	**106,023**
	Robert W. Rock (D)	80,885
3.	**Tim Roemer*** (D)	**107,438**
	Chris Chocola (R)	98,822
4.	**Mark E. Souder*** (R)	**131,051**
	Michael "Mike" Dewayne Foster (D)	74,492
5.	**Steve Buyer*** (R)	**132,051**
	Greg Goodnight (D)	81,427
6.	**Dan Burton*** (R)	**199,207**
	Darin Patrick Griesey (D)	74,881
7.	**Brian D. Kerns** (R)	**135,869**
	Michael Douglas Graf (D)	66,764
8.	**John N. Hostettler*** (R)	**116,879**
	Paul E. Perry (D)	100,488
9.	**Baron Hill*** (D)	**126,420**
	Michael E. Bailey (R)	102,219
10.	**Julia M. Carson*** (D)	**91,689**
	Marvin B. Scott (R)	62,233
	Iowa	
1.	**Jim Leach*** (R)	**164,972**
	Bob Simpson (D)	96,283
2.	**Jim Nussle*** (R)	**139,906**
	Donna L. Smith (D)	110,327
3.	**Leonard L. Boswell*** (D)	**156,327**
	Jay Marcus (R)	83,810
4.	**Greg Ganske*** (R)	**169,267**
	Michael L. Huston (D)	101,112
5.	**Tom Latham*** (R)	**159,367**
	Mike Palecek (D)	67,593
	Kansas	
1.	**Jerry Moran*** (R)	**214,328**
	Jack Warner (LB)	25,581
2.	**Jim Ryun*** (R)	**164,951**
	Stanley Wiles (D)	71,709

Dist.	Representative (Party)	2000 Election
3.	**Dennis Moore*** (D)	**154,505**
	Phill Kline (R)	144,672
4.	**Todd Tiahrt*** (R)	**131,871**
	Carlos Nolla (D)	101,980

Kentucky

Dist.	Representative (Party)	2000 Election
1.	**Edward Whitfield*** (R)	**132,115**
	Brian S. Roy (D)	95,806
2.	**Ron Lewis*** (R)	**160,800**
	Brian Pedigo (D)	74,537
3.	**Anne Meagher Northup*** (R)	**142,106**
	Eleanor Jordan (D)	118,785
4.	**Ken Lucas*** (D)	**125,872**
	Don Bell (R)	100,943
5.	**Harold "Hal" Rogers*** (R)	**145,980**
	Sidney Jane Bailey (D)	52,495
6.	**Ernest Fletcher*** (R)	**142,971**
	Scotty Baesler (D)	94,167

Louisiana

Dist.	Representative (Party)	2000 Election
1.	**David Vitter*** (R)	**191,379**
	Michael A. Armato (D)	29,935
2.	**William J. Jefferson*** (D)	**Unopposed**
3.	**W.J. "Billy" Tauzin*** (R)	**143,446**
	Edwin J. "Eddie" Albares (I)	16,908
4.	**"Jim" McCrery*** (R)	**122,678**
	Phillip R. Green (D)	43,600
5.	**John C. Cooksey*** (R)	**123,975**
	Roger Beall (D)	42,977
6.	**Richard H. Baker*** (R)	**165,637**
	Kathy J. Rogillio (D)	72,192
7.	**Chris John*** (D)	**152,796**
	Michael P. Harris (I)	30,687

In Louisiana, all candidates of all parties ran against each other on Nov. 7, 2000, in a non-partisan primary, unless they were unopposed incumbents, in which case they were declared elected. In each district having an election Nov. 7, the winning candidate received more than the 50% of the vote needed to avoid a runoff.

Maine

Dist.	Representative (Party)	2000 Election
1.	**Thomas H. Allen*** (D)	**202,823**
	Jane A. Amero (R)	123,915
2.	**John E. Baldacci*** (D)	**219,783**
	Richard H. Campbell (R)	79,522

Maryland

Dist.	Representative (Party)	2000 Election
1.	**Wayne T. Gilchrest*** (R)	**165,293**
	Bennett Bozman (D)	91,022
2.	**Robert L. Ehrlich Jr.*** (R)	**178,556**
	Kenneth T. Bosley (D)	81,591
3.	**Benjamin L. Cardin*** (D)	**169,347**
	Colin Harby (R)	53,827
4.	**Albert R. Wynn*** (D)	**172,624**
	John B. Kimble (R)	24,973
5.	**Steny H. Hoyer*** (D)	**166,231**
	Thomas E. "Tim" Hutchins (R)	89,019
6.	**Roscoe G. Bartlett*** (R)	**168,624**
	Donald M. DeArmon (D)	109,136
7.	**Elijah E. Cummings*** (D)	**134,066**
	Kenneth Kondner (R)	19,773
8.	**Constance A. Morella*** (R)	**156,241**
	Terry Lierman (D)	136,840

Massachusetts

Dist.	Representative (Party)	2000 Election
1.	**John W. Olver*** (D)	**169,375**
	Peter J. Abair (R)	73,580
2.	**Richard E. Neal*** (D)	**Unopposed**
3.	**James P. McGovern*** (D)	**Unopposed**
4.	**Barney Frank*** (D)	**200,638**
	Martin D. Travis (R)	56,553
5.	**Martin T. Meehan*** (D)	**Unopposed**
6.	**John F. Tierney*** (D)	**205,324**
	Paul McCarthy (R)	83,501
7.	**Edward J. Markey*** (D)	**Unopposed**
8.	**Michael E. Capuano*** (D)	**Unopposed**
9.	**Vacant**[4]	
10.	**William D. Delahunt*** (D)	**234,675**
	Eric V. Bleicken (R)	75,536

Michigan

Dist.	Representative (Party)	2000 Election
1.	**Bart Stupak*** (D)	**169,649**
	Chuck Yob (R)	117,300
2.	**Peter Hoekstra*** (R)	**186,762**
	Bob Shrauger (D)	96,370
3.	**Vernon Ehlers*** (R)	**179,539**
	Timothy W. Steele (D)	91,309
4.	**Dave Camp*** (R)	**182,128**
	Lawrence D. Hollenbeck (D)	78,019
5.	**James A. Barcia*** (D)	**184,048**
	Ronald G. Actis (R)	59,274
6.	**Fred Upton*** (R)	**159,373**
	James Bupp (D)	68,532
7.	**Nick Smith*** (R)	**147,369**
	Jennie Crittendon (D)	86,080
8.	**Mike Rogers** (R)	**145,179**
	Dianne Byrum (D)	145,019
9.	**Dale E. Kildee*** (D)	**158,184**
	Grant Garrett (R)	92,926
10.	**David E. Bonior*** (D)	**181,818**
	Tom Turner (R)	93,713
11.	**Joe Knollenberg*** (R)	**170,790**
	Matthew Frumin (D)	124,053
12.	**Sander Levin*** (D)	**157,720**
	Bart Baron (R)	78,795
13.	**Lynn Nancy Rivers*** (D)	**160,084**
	Carl F. Berry (R)	79,445
14.	**John Conyers, Jr.*** (D)	**168,982**
	William A. Ashe (R)	17,582
15.	**Carolyn Cheeks Kilpatrick*** (D)	**140,609**
	Chrysanthea D. Boyd-Fields (R)	14,336
16.	**John D. Dingell*** (D)	**167,142**
	William Morse (R)	62,469

Minnesota

Dist.	Representative (Party)	2000 Election
1.	**Gil Gutknecht*** (R)	**159,835**
	Mary Rieder (D)	117,946
2.	**Mark Kennedy** (R)	**138,957**
	David Minge* (D)	138,802
3.	**Jim Ramstad*** (R)	**222,571**
	Sue Shuff (D)	98,219
4.	**Betty McCollum** (D)	**130,403**
	Linda Runbeck (R)	83,852
5.	**Martin Olav Sabo*** (D)	**176,629**
	Frank Taylor (R)	58,191
6.	**Bill Luther*** (D)	**176,340**
	John Kline (R)	170,900
7.	**Collin C. Peterson*** (D)	**185,771**
	Glen Menze (R)	79,175
8.	**James L. Oberstar*** (D)	**210,094**
	Bob Lemen (R)	79,890

Mississippi

Dist.	Representative (Party)	2000 Election
1.	**Roger F. Wicker*** (R)	**145,967**
	Joe T. "Joey" Grist, Jr. (D)	59,763
2.	**Bennie G. Thompson*** (D)	**112,777**
	Hardy Caraway (R)	54,090
3.	**Charles W. "Chip" Pickering, Jr.*** (R)	**153,899**
	William Clay Thrash (D)	54,151
4.	**Ronnie Shows*** (D)	**115,732**
	Dunn Lampton (R)	79,218
5.	**Gene Taylor*** (D)	**153,264**
	Randy McDonnell (R)	35,309

Missouri

Dist.	Representative (Party)	2000 Election
1.	**William Lacy Clay, Jr.** (D)	**149,173**
	Z. Dwight Billingsly (R)	42,730
2.	**Todd Akin** (R)	**164,926**
	Ted House (D)	126,441
3.	**Richard A. Gephardt*** (D)	**147,222**
	Bill Federer (R)	100,967
4.	**Ike Skelton*** (D)	**180,634**
	Jim Noland (R)	84,406
5.	**Karen McCarthy*** (D)	**159,826**
	Steve Gordon (R)	66,439
6.	**Samuel B. (Sam) Graves, Jr.** (R)	**138,925**
	Steve Danner (D)	127,792
7.	**Roy Blunt*** (R)	**202,305**
	Charles Christrup (D)	65,510
8.	**Jo Ann Emerson*** (R)	**162,239**
	Bob Camp (D)	67,760

Dist.	Representative (Party)	2000 Election
9.	**Kenny Hulshof*** (R)	**172,787**
	Steven R. Carroll (D)	111,662
	Montana	
	Dennis Rehberg (R)	**211,418**
	Nancy Keenan (D)	189,971
	Nebraska	
1.	**Doug Bereuter*** (R)	**155,485**
	Alan Jacobsen (D)	72,859
2.	**Lee Terry*** (R)	**148,911**
	Shelley Kiel (D)	70,268
3.	**Tom Osborne** (R)	**182,117**
	Roland E. Reynolds (D)	34,944
	Nevada	
1.	**Shelley Berkley*** (D)	**118,469**
	Jon Porter (R)	101,276
2.	**Jim Gibbons*** (R)	**229,608**
	Tierney Cahill (D)	106,379
	New Hampshire	
1.	**John E. Sununu*** (R)	**150,609**
	Martha Fuller Clark (D)	128,387
2.	**Charles Bass*** (R)	**152,581**
	Barney Brannen (D)	110,367
	New Jersey	
1.	**Robert E. Andrews*** (D)	**167,327**
	Charlene Cathcart (R)	46,455
2.	**Frank A. LoBiondo*** (R)	**155,187**
	Edward G. Janosik (D)	74,632
3.	**Jim Saxton*** (R)	**157,053**
	Susan Bass Levin (D)	112,848
4.	**Christopher H. Smith*** (R)	**158,515**
	Reed Gusciora (D)	87,956
5.	**Marge Roukema*** (R)	**175,546**
	Linda A. Mercurio (D)	81,715
6.	**Frank Pallone, Jr.*** (D)	**141,698**
	Brian T. Kennedy (R)	62,454
7.	**Mike Ferguson** (R)	**128,434**
	Maryanne Connelly (D)	113,479
8.	**Bill J. Pascrell, Jr.*** (D)	**134,074**
	Anthony Fusco, Jr. (R)	60,606
9.	**Steven R. Rothman*** (D)	**140,462**
	Joseph Tedeschi (R)	61,984
10.	**Donald M. Payne*** (D)	**133,073**
	Dirk B. Weber (R)	18,436
11.	**Rodney P. Frelinghuysen*** (R)	**186,140**
	John P. Scollo (D)	80,958
12.	**Rush Holt *** (D)	**146,162**
	Dick Zimmer (R)	145,511
13.	**Robert Menendez*** (D)	**117,856**
	Theresa de Leon (R)	27,849
	New Mexico	
1.	**Heather A. Wilson*** (R)	**107,296**
	John J. Kelly (D)	92,187
2.	**Joe R. Skeen*** (R)	**100,742**
	Michael A. Montoya (D)	72,614
3.	**Tom Udall*** (D)	**135,040**
	Lisa L. Lutz (R)	65,979
	New York	
1.	**Felix J. Grucci, Jr.** (R,IN,C,RL)	**133,020**
	Regina Seltzer (D)	97,299
2.	**Steve J. Israel** (D)	**90,438**
	Joan B. Johnson (R)	65,880
3.	**Peter T. King*** (R,IN,C,RL)	**143,126**
	Dal Lamagna (D,GR,WF)	95,787
4.	**Carolyn McCarthy*** (D,IN,WF)	**136,703**
	Gregory R. Becker (R,C,RL)	87,830
5.	**Gary L. Ackerman*** (D,IN,L,WF)	**137,684**
	Edward Elkowitz (R,C)	61,084
6.	**Gregory W. Meeks*** (D,WF)	**Unopposed**
7.	**Joseph Crowley*** (D)	**78,207**
	Rosa Robles Birtley (R)	24,592
8.	**Jerrold L. Nadler*** (D,L,WF)	**150,514**
	Marian S. Henry (R)	27,098
9.	**Anthony D. Weiner*** (D,L)	**98,983**
	Noach Dear (R,C)	45,649

Dist.	Representative (Party)	2000 Election
10.	**Edolphus Towns*** (D,L)	**120,700**
	Ernestine M. Brown (R)	6,852
11.	**Major R. Owens*** (D,WF)	**112,050**
	Susan Cleary (R,SC)	8,406
12.	**Nydia M. Velazquez*** (D,WF)	**86,290**
	Rosemary Markgraf (R)	10,052
13.	**Vito J. Fossella*** (R,C,RL)	**109,806**
	Katina M. Johnstone (D,WF)	57,603
14.	**Carolyn B. Maloney*** (D,L)	**152,167**
	C. Adrienne Rhodes (R)	47,254
15.	**Charles B. Rangel*** (D,L,WF)	**130,166**
	Jose Augustin Suero (R,RF)	7,346
16.	**Jose E. Serrano*** (D,L)	**103,041**
	Aaron Justice (R)	3,934
17.	**Eliot L. Engel*** (D,L)	**115,093**
	Patrick McManus (R,C)	13,201
18.	**Nita M. Lowey*** (D)	**126,878**
	John G. Vonglis (R,C)	58,022
19.	**Sue W. Kelly*** (R,C)	**145,532**
	Larry Otis Graham (D,L,WF)	85,871
20.	**Benjamin A. Gilman*** (R)	**136,016**
	Paul J. Feiner (D,L,GR,WF)	94,646
21.	**Michael R. McNulty*** (D,IN,C)	**175,339**
	Thomas G. Pillsworth (R)	60,333
22.	**John E. Sweeney*** (R,C)	**167,368**
	Kenneth F. McCallion (D,GR,WF)	79,111
23.	**Sherwood L. Boehlert*** (R,IN)	**124,132**
	David B. Vickers (C,RL)	42,854
24.	**John M. McHugh*** (R,C)	**138,322**
	Neil P. Tallon (D,WF)	42,698
25.	**James T. Walsh*** (R,IN,C)	**151,880**
	Francis J. Gavin (D)	64,533
26.	**Maurice D. Hinchey*** (D,IN,L,WF)	**140,395**
	Bob Moppert (R,C)	83,856
27.	**Thomas M. Reynolds*** (R,C)	**157,694**
	Thomas W. Pecoraro (D)	69,870
28.	**Louise M. Slaughter*** (D)	**151,688**
	Mark C. Johns (R,C)	75,348
29.	**John J. La Falce*** (D,IN,L)	**128,328**
	Brett M. Sommer (R,C,RL)	81,159
30.	**Jack Quinn*** (R,IN,C)	**138,452**
	John Fee (D,L,GR,WF)	67,819
31.	**Amo Houghton*** (R,C)	**154,238**
	Kisun J. Peters (D)	45,193
	North Carolina	
1.	**Eva M. Clayton*** (D)	**124,171**
	Duane E. Kratzer, Jr. (R)	62,198
2.	**Bob Etheridge*** (D)	**146,733**
	Doug Haynes (R)	103,011
3.	**Walter B. Jones*** (R)	**121,940**
	Leigh Harvey NcNairy (D)	74,058
4.	**David Price*** (D)	**200,885**
	Jess Ward (R)	119,412
5.	**Richard M. Burr*** (R)	**172,489**
	Steven Francis LeBoeuf (LB)	13,366
6.	**Howard Coble*** (R)	**195,727**
	Jeffrey Dean Bentley (LB)	18,726
7.	**Mike McIntyre*** (D)	**160,185**
	James R. Adams (R)	66,463
8.	**Robert C. (Robin) Hayes*** (R)	**111,950**
	Mike Taylor (D)	89,505
9.	**Sue Myrick*** (R)	**181,161**
	Ed McGuire (D)	79,382
10.	**T. Cass Ballenger*** (R)	**164,182**
	Delmas Parker (D)	70,877
11.	**Charles H. Taylor*** (R)	**146,677**
	Sam Neill (D)	112,234
12.	**Mel Watt*** (D)	**135,570**
	Chad Mitchell (R)	69,596
	North Dakota	
	Earl Pomeroy* (D)	**151,173**
	John Dorso (R)	127,251

Dist.	Representative (Party)	2000 Election
	Ohio	
1.	**Steve Chabot*** (R)	**116,768**
	John Cranley (D)	98,328
2.	**Rob Portman*** (R)	**204,184**
	Charles W. Sanders (D)	64,091
3.	**Tony P. Hall*** (D)	**177,731**
	Regina Burch (NL)	36,516
4.	**Michael G. Oxley*** (R)	**156,510**
	Daniel L. Dickman (D)	67,330
5.	**Paul E. Gillmor*** (R)	**169,857**
	Dannie Edmon (D)	62,138
6.	**Ted Strickland*** (D)	**138,849**
	Mike Azinger (R)	96,966
7.	**Dave Hobson*** (R)	**163,646**
	Donald E. Minor (D)	60,755
8.	**John A. Boehner*** (R)	**179,756**
	John G. Parks (D)	66,293
9.	**Marcy Kaptur*** (D)	**168,547**
	Dwight E. Bryan (R)	49,446
10.	**Dennis J. Kucinich*** (D)	**167,093**
	Bill Smith (R)	48,940
11.	**Stephanie Tubbs-Jones*** (D)	**166,691**
	James J. Sykora (R)	22,324
12.	**Pat Tiberi** (R)	**139,242**
	Maryellen O'Shaughnessy (D)	115,432
13.	**Sherrod Brown*** (D)	**170,059**
	Rick H. Jeric (R)	84,295
14.	**Thomas C. Sawyer*** (D)	**149,184**
	Rick Wood (R)	71,432
15.	**Deborah Pryce*** (R)	**156,792**
	Bill Buckel (D)	64,805
16.	**Ralph Regula*** (R)	**162,294**
	William Smith (D)	62,709
17.	**James A. Traficant, Jr.*** (D)	**120,333**
	Paul H. Alberty (R)	54,751
18.	**Bob Ney*** (R)	**152,325**
	Marc D. Guthrie (D)	79,232
19.	**Steven C. LaTourette*** (R)	**174,262**
	Dale Virgil Blanchard (D)	70,429
	Oklahoma	
1.	**Steve Largent*** (R)	**138,528**
	Dan Lowe (D)	58,493
2.	**Brad Carson** (D)	**107,273**
	Andy Ewing (R)	81,672
3.	**Wes Watkins*** (R)	**137,826**
	Angus W. Yandell, Jr. (I)	14,660
4.	**J.C. Watts, Jr.*** (R)	**114,000**
	Larry Weatherford (D)	54,808
5.	**Ernest Istook*** (R)	**134,159**
	Garland McWatters (D)	53,275
6.	**Frank D. Lucas*** (R)	**95,635**
	Randy Beutler (D)	63,106
	Oregon	
1.	**David Wu*** (D)	**176,902**
	Charles Starr (R)	115,303
2.	**Greg Walden*** (R)	**220,086**
	Walter Ponsford (D)	78,101
3.	**Earl Blumenauer*** (D)	**181,049**
	Jeffery L. Pollock (R)	64,128
4.	**Peter A. DeFazio*** (D)	**197,998**
	John Lindsey (R)	88,950
5.	**Darlene Hooley*** (D)	**156,315**
	Brian J. Boquist (R)	118,631
	Pennsylvania	
1.	**Robert A. Brady*** (D)	**149,621**
	Steven N. Kush (R)	19,920
2.	**Chaka Fattah*** (D)	**180,021**
	Kenneth V. Krawchuk (LB)	3,673
3.	**Robert A. Borski*** (D)	**130,528**
	Charles F. Dougherty (R)	59,343
4.	**Melissa Hart** (R)	**145,390**
	Terry E. Van Horne (D)	100,995
5.	**John E. Peterson*** (R)	**147,570**
	Thomas A. Martin (LB)	17,020
6.	**Tim Holden*** (D)	**140,084**
	Thomas G. Kopel (R)	71,227
7.	**Curt Weldon*** (R)	**172,569**
	Peter A. Lennon (D)	93,687

Dist.	Representative (Party)	2000 Election
8.	**Jim Greenwood*** (R)	**154,090**
	Ronald L. Strouse (D)	100,617
9.	**Bill Shuster** (R)[5]	**55,670**
	H. Scott Conklin (D)	47,220
10.	**Don Sherwood*** (R)	**124,830**
	Patrick Casey (D)	112,580
11.	**Paul E. Kanjorski*** (D)	**131,948**
	Stephen A. Urban (R)	66,699
12.	**John P. Murtha*** (D)	**145,538**
	Bill Choby (R)	56,575
13.	**Joseph M. Hoeffel*** (D)	**146,026**
	Stewart J. Greenleaf (R)	126,501
14.	**William J. Coyne*** (D)	**Unopposed**
15.	**Pat Toomey*** (R)	**118,307**
	Ed O'Brien (D)	103,864
16.	**Joseph R. Pitts*** (R)	**162,403**
	Bob Yorczyk (D)	80,177
17.	**George W. Gekas*** (R)	**166,236**
	Leslye Hess Herrmann (D)	66,190
18.	**Mike Doyle*** (D)	**156,131**
	Craig C. Stephens (R)	68,798
19.	**Todd Platts** (R)	**168,722**
	Jeff Sanders (D)	61,538
20.	**Frank Mascara*** (D)	**145,131**
	Ronald J. Davis (R)	80,312
21.	**Phil English*** (R)	**135,164**
	Marc A. Flitter (D)	87,018
	Rhode Island	
1.	**Patrick J. Kennedy*** (D)	**123,442**
	Stephen Cabral (R)	61,522
2.	**James R. Langevin** (D)	**123,805**
	Rodney D. Driver (CC)	42,625
	South Carolina	
1.	**Henry Brown** (R)	**139,597**
	Andy Brack (D)	82,622
2.	**Vacant**[6]	
3.	**Lindsey Graham*** (R)	**150,176**
	George Brightharp (D,UC)	67,174
4.	**Jim DeMint*** (R)	**150,436**
	Ted Adams (CN)	16,532
5.	**John Spratt*** (D)	**126,877**
	Carl L. Gullick (R)	85,247
6.	**James E. "Jim" Clyburn*** (D)	**138,053**
	Vince Ellison (R)	50,005
	South Dakota	
	John R. Thune* (R)	**231,083**
	Curt Hohn (D)	78,321
	Tennessee	
1.	**William L. "Bill" Jenkins*** (R)	**Unopposed**
2.	**John J. Duncan, Jr.*** (R)	**187,154**
	Kevin J. Rowland (LB)	22,304
3.	**Zach Wamp*** (R)	**139,840**
	William L. Callaway (D)	75,785
4.	**Van Hilleary*** (R)	**133,622**
	David Dunaway (D)	67,165
5.	**Bob Clement*** (D)	**149,277**
	Stan Scott (R)	50,386
6.	**Bart Gordon*** (D)	**168,861**
	David Charles (R)	97,169
7.	**Ed Bryant*** (R)	**171,056**
	Richard P. Sims (D)	71,587
8.	**John S. Tanner*** (D)	**143,127**
	Billy Yancy (R)	54,929
9.	**Harold E. Ford, Jr.*** (D)	**Unopposed**
	Texas	
1.	**Max Sandlin*** (D)	**118,157**
	Noble Willingham (R)	91,912
2.	**Jim Turner*** (D)	**162,891**
	Gary Lyndon Dye (LB)	15,939
3.	**Sam Johnson*** (R)	**187,486**
	Billy Wayne Zachary (D)	67,233
4.	**Ralph M. Hall*** (D)	**145,887**
	Jon Newton (R)	91,574
5.	**Pete Sessions*** (R)	**100,487**
	Regina Montoya Coggins (D)	82,629
6.	**Joe Barton*** (R)	**222,685**
	Frank Brady (LB)	30,056

Dist.	Representative (Party)	2000 Election
7.	**John Culberson** (R)	**183,712**
	Jeff Sell (D)	60,694
8.	**Kevin Brady*** (R)	**233,848**
	Gil Guillory (LB)	21,368
9.	**Nick Lampson*** (D)	**130,143**
	Paul Williams (R)	87,165
10.	**Lloyd Doggett*** (D)	**203,628**
	Michael Davis (LB)	37,203
11.	**Chet Edwards*** (D)	**105,782**
	Ramsey Farley (R)	85,546
12.	**Kay Granger*** (R)	**117,739**
	Mark Greene (D)	67,612
13.	**William "Mac" Thornberry*** (R)	**117,995**
	Curtis Clinesmith (D)	54,343
14.	**Ron Paul*** (R)	**137,370**
	Loy Sneary (D)	92,689
15.	**Ruben Hinojosa*** (D)	**106,570**
	Frank Jones (LB)	13,167
16.	**Silvestre Reyes*** (D)	**92,649**
	Daniel Power (R)	40,921
17.	**Charles Stenholm*** (D)	**120,670**
	Darrell Clements (R)	72,535
18.	**Sheila Jackson Lee*** (D)	**131,857**
	Bob Levy (R)	38,191
19.	**Larry Combest*** (R)	**170,319**
	John Turnbow (LB)	15,579
20.	**Charles A. Gonzalez*** (D)	**107,487**
	Alejandro (Alex) De Pena (LB)	15,087
21.	**Lamar Smith*** (R)	**251,049**
	Jim Green (D)	73,326
22.	**Tom DeLay*** (R)	**154,662**
	Jo Ann Matranga (D)	92,645
23.	**Henry Bonilla*** (R)	**119,679**
	Isidro Garza, Jr. (D)	78,274
24.	**Martin Frost*** (D)	**103,152**
	James "Bryndan" Wright (R)	61,235
25.	**Ken Bentsen*** (D)	**106,112**
	Phil Sudan (R)	68,010
26.	**Dick Armey*** (R)	**214,025**
	Steve Love (D)	75,601
27.	**Solomon P. Ortiz*** (D)	**102,088**
	Pat Ahumada (R)	54,660
28.	**Ciro D. Rodriguez*** (D)	**123,104**
	William A. (Bill) Stallknecht (LB)	15,156
29.	**Gene Green*** (D)	**84,665**
	Joe Vu (R)	29,606
30.	**Eddie Bernice Johnson*** (D)	**109,163**
	Kelly Rush (LB)	9,798
	Utah	
1.	**James V. Hansen*** (R)	**180,591**
	Kathleen McConkie Collinwood (D)	71,229
2.	**Jim Matheson** (D)	**145,021**
	Derek W. Smith (R)	107,114
3.	**Chris Cannon*** (R)	**138,943**
	Donald Dunn (D)	88,547
	Vermont	
	Bernard Sanders* (I)	**196,118**
	Karen Ann Kerin (R)	51,977
	Virginia	
1.	**Jo Ann S. Davis** (R)	**151,344**
	Lawrence A. Davies (D)	97,399
2.	**Edward L. "Ed" Schrock** (R)	**97,856**
	Jody M. Wagner (D)	90,328
3.	**Robert C. "Bobby" Scott*** (D)	**Unopposed**
4.	**J. Randy Forbes** (R)[7]	**70,917**
	Louise Lucas (D)	65,190
5.	**Virgil H. Goode, Jr.*** (I)	**143,312**
	John W. Boyd, Jr. (D)	65,387
6.	**Robert W. "Bob" Goodlatte*** (R)	**Unopposed**
7.	**Eric I. Cantor** (R)	**192,652**
	Warren A. Stewart (D)	94,935
8.	**James P. Moran, Jr.*** (D)	**164,178**
	Demaris H. Miller (R)	88,262
9.	**Frederick C. "Rick" Boucher*** (D)	**137,488**
	Michael C. "Oz" Osborne (R)	59,335
10.	**Frank R. Wolf*** (R)	**238,817**
	Brian M. Brown (I)	28,107
11.	**Thomas M. Davis, III*** (R)	**150,395**
	M. L. "Mike" Corrigan (D)	83,455
	Washington	
1.	**Jay Inslee*** (D)	**155,820**
	Dan McDonald (R)	121,823
2.	**Rick Larsen** (D)	**146,617**
	John Koster (R)	134,660
3.	**Brian Baird*** (D)	**159,428**
	Trent R. Matson (R)	114,861
4.	**Doc Hastings*** (R)	**143,259**
	Jim Davis (D)	87,585
5.	**George R. Nethercutt, Jr.*** (R)	**144,038**
	Tom Keefe (D)	97,703
6.	**Norm Dicks*** (D)	**164,853**
	Bob Lawrence (R)	79,215
7.	**Jim McDermott*** (D)	**193,470**
	Joe Szwaja (GR)	52,142
8.	**Jennifer Dunn*** (R)	**183,255**
	Heidi Behrens-Benedict (D)	104,944
9.	**Adam Smith*** (D)	**135,452**
	Chris Vance (R)	76,766
	West Virginia	
1.	**Alan B. Mollohan*** (D)	**170,974**
	Richard Kerr (LB)	23,797
2.	**Shelley Moore Capito** (R)	**108,769**
	Jim Humphreys (D)	103,003
3.	**Nick Joe Rahall, II*** (D)	**146,807**
	Jeff Robinson (LB)	13,979
	Wisconsin	
1.	**Paul D. Ryan*** (R)	**177,612**
	Jeffrey C. Thomas (D)	88,885
2.	**Tammy Baldwin*** (D)	**163,534**
	John Sharpless (R)	154,632
3.	**Ron Kind*** (D)	**173,505**
	Susan Tully (R)	97,741
4.	**Jerry Kleczka*** (D)	**163,622**
	Tim Riener (R)	101,811
5.	**Tom Barrett*** (D)	**173,893**
	Jonathan Smith (R)	49,296
6.	**Tom Petri*** (R)	**179,205**
	Dan Flaherty (D)	96,125
7.	**David R. Obey*** (D)	**173,007**
	Sean Cronin (R)	100,264
8.	**Mark Green*** (R)	**211,388**
	Dean Reich (D)	71,575
9.	**F. James Sensenbrenner, Jr.*** (R)	**239,498**
	Mike Clawson (D)	83,720
	Wyoming	
	Barbara Cubin* (R)	**141,848**
	Michael Allen Green (D)	60,638

The following members of Congress are nonvoting: Aníbal Acevedo Vilá (Popular Democratic Party), resident commissioner, Puerto Rico; Eleanor Holmes Norton (D), District of Columbia; Robert A. Underwood (D), Guam; Eni F. H. Faleomavaega (D), American Samoa; Donna M. Christian-Christensen (D), Virgin Islands. (1) Asa Hutchinson resigned on Aug. 6, 2001, to become administrator of the Drug Enforcement Administration. A special election to fill the seat was to be held Nov. 20, 2001. (2) Diane E. Watson won special election June 5, 2001, to fill seat left vacant after the death of Julian C. Dixon on Dec. 8, 2000. (3) Seat fell vacant after the resignation of Joe Scarborough. A special election to fill the seat was to be held Oct. 16, 2001. (4) Seat fell vacant after the death of John Joseph Moakley on May 28, 2001. A special election to fill the seat was to be held Oct. 16, 2001. (5) Bill Shuster won special election May 15, 2001, to fill the seat left vacant by the resignation of his father, Bud Shuster. (6) Seat fell vacant after the death of Floyd D. Spence on Aug. 16, 2001. A special election to fill the seat was to be held on Dec. 18, 2001. (7) J. Randy Forbes won special election June 19, 2001, to fill the seat left vacant after the death of Norman Sisisky on Mar. 29, 2001.

Congressional Committees

(as of Sept. 2001)

Rep. = Republican; Dem. = Democrat

Senate Standing Committees

Agriculture, Nutrition, and Forestry
Chairman: Tom Harkin, IA
Ranking Rep.: Richard G. Lugar, IN

Appropriations
Chairman: Robert C. Byrd, WV
Ranking Rep.: Ted Stevens, AK

Armed Services
Chairman: Carl Levin, MI
Ranking Rep.: John W. Warner, VA

Banking, Housing, and Urban Affairs
Chairman: Paul S. Sarbanes, MD
Ranking Rep.: Phil Gramm, TX

Budget
Chairman: Kent Conrad, ND
Ranking Rep.: Pete V. Domenici, NM

Commerce, Science, and Transportation
Chairman: Ernest Hollings, SC
Ranking Rep.: John McCain, AZ

Energy and Natural Resources
Chairman: Jeff Bingaman, NM
Ranking Rep.: Frank H. Murkowski, AK

Environment and Public Works
Chairman: James M. Jeffords, VT
Ranking Rep.: Robert Smith, NH

Finance
Chairman: Max Baucus, MT
Ranking Rep.: Chuck Grassley, IA

Foreign Relations
Chairman: Joseph R. Biden, Jr., DE
Ranking Rep.: Jesse Helms, NC

Governmental Affairs
Chairman: Joe Lieberman, CT
Ranking Rep.: Fred Thompson, TN

Health, Education, Labor, and Pensions
Chairman: Edward M. Kennedy, MA
Ranking Rep.: Judd Gregg, NH

Indian Affairs
Chairman: Daniel K. Inouye, HI
Ranking Rep.: Ben Nighthorse Campbell, CO

Judiciary
Chairman: Patrick Leahy, VT
Ranking Rep.: Orrin G. Hatch, UT

Rules and Administration
Chairman: Christopher J. Dodd, CT
Ranking Rep.: Mitch McConnell, KY

Small Business and Entrepreneurship
Chairman: John F. Kerry, MA
Ranking Rep.: Christopher (Kit) Bond, MO

Veterans' Affairs
Chairman: John D. Rockefeller IV, WV
Ranking Rep.: Arlen Specter, PA

Senate Special Committee

Aging
Chairman: John B. Breaux, LA
Ranking Rep.: Larry E. Craig, ID

Senate Select Committees

Ethics
Chairman: Harry Reid, NV
Ranking Representative.: Pat Roberts, KS

Intelligence
Chairman: Bob Graham, FL
Ranking Rep.: Richard Shelby, AL

Joint Committees of Congress

Economic
Chairman: Representative Jim Saxton (Rep., NJ)
Vice Chairman: Senator John F. Reed (Dem., RI)

Library
Chairman: Representative Vernon Ehlers (Rep., MI)
Vice Chairman: Senator Christopher J. Dodd (Dem., CT)

Printing
Chairman: Senator Dianne Feinstein (Dem., CA)
Vice Chairman: Representative Bob Ney (Rep., OH)

Taxation
Chairman: Representative Bill Thomas (Rep., CA)
Vice Chairman: Vacant

House Standing Committees

Agriculture
Chairman: Larry Combest, TX
Ranking Dem.: Charlie Stenholm, TX

Appropriations
Chairman: C. W. Bill Young, FL
Ranking Dem.: David R. Obey, WI

Armed Services
Chairman: Bob Stump, AZ
Ranking Dem.: Ike Skelton, MO

Budget
Chairman: Jim Nussle, IA
Ranking Dem.: John Spratt, SC

Education and the Workforce
Chairman: John A. Boehner, OH
Ranking Dem.: George Miller, CA

Energy and Commerce
Chairman: W.J. "Billy" Tauzin, LA
Ranking Dem.: John D. Dingell, MI

Financial Services
Chairman: Michael G. Oxley, OH
Ranking Dem.: John J. La Falce, NY

Government Reform
Chairman: Dan Burton, IN
Ranking Dem.: Henry A. Waxman, CA

House Administration
Chairman: Bob Ney, OH
Ranking Dem.: Steny H. Hoyer, MD

International Relations
Chairman: Henry J. Hyde, IL
Ranking Dem.: Tom Lantos, CA

Judiciary
Chairman: F. James Sensenbrenner, Jr., WI
Ranking Dem.: John Conyers, Jr., MI

Resources
Chairman: James V. Hansen, UT
Ranking Dem.: Nick Joe Rahall, II, WV

Rules
Chairman: David Dreier, CA
Ranking Dem.: Martin Frost, TX

Science
Chairman: Sherwood L. Boehlert, NY
Ranking Dem.: Ralph M. Hall, TX

Small Business
Chairman: Donald A. Manzullo, IL
Ranking Dem.: Nydia M. Velazquez, NY

Standards of Official Conduct
Chairman: Joel Hefley, CO
Ranking Dem.: Howard L. Berman, CA

Transportation and Infrastructure
Chairman: Don E. Young, AK
Ranking Dem.: James L. Oberstar, MN

Veterans' Affairs
Chairman: Christopher H. Smith, NJ
Ranking Dem.: Lane A. Evans, IL

Ways and Means
Chairman: Bill Thomas, CA
Ranking Dem.: Charles B. Rangel, NY

House Select Committee

Intelligence
Chairman: Porter Goss, FL
Ranking Dem.: Nancy Pelosi, CA

Congress divides its tasks among some 250 committees and subcommittees. Standing committees generally have legislative jurisdiction and operate with subcommittees that handle work in specific areas. Select and joint committees are chiefly for oversight or housekeeping. The chair of each House or Senate committee and a majority of its members come from the majority party, which, as of Oct. 2001, was the Republican Party in the House and Democratic Party in the Senate.

Political Divisions of the U.S. Senate and House of Representatives, 1901-2001

Source: *Congressional Directory*; Senate Library

(All figures reflect immediate post-election party breakdown; **boldface** denotes party in majority immediately after the election.)

		SENATE					HOUSE OF REPRESENTATIVES				
Congress	Years	Total Sens.	Demo-crats	Repub-licans	Other parties	Vacant	Total Members	Demo-crats	Repub-licans	Other parties	Vacant
57th	1901-03	90	29	**56**	3	2	357	153	**198**	5	1
58th	1903-05	90	32	**58**			386	178	**207**		1
59th	1905-07	90	32	**58**			386	136	**250**		
60th	1907-09	92	29	**61**		2	386	164	**222**		
61st	1909-11	92	32	**59**		1	391	172	**219**		
62d	1911-13	92	42	**49**		1	391	**228**	162	1	
63d	1913-15	96	**51**	44	1		435	**290**	127	18	
64th	1915-17	96	**56**	39	1		435	**231**	193	8	3
65th	1917-19	96	**53**	42	1		435	210[1]	**216**	9	
66th	1919-21	96	47	**48**	1		435	191	**237**	7	
67th	1921-23	96	37	**59**			435	132	**300**	1	2
68th	1923-25	96	43	**51**	2		435	207	**225**	3	
69th	1925-27	96	40	**54**	1	1	435	183	**247**	5	
70th	1927-29	96	47	**48**	1		435	195	**237**	3	
71st	1929-31	96	39	**56**	1		435	163	**267**	1	4
72d	1931-33	96	47	**48**	1		435	216[2]	**218**	1	
73d	1933-35	96	**59**	36	1		435	**313**	117	5	
74th	1935-37	96	**69**	25	2		435	**322**	103	10	
75th	1937-39	96	**75**	17	4		435	**333**	89	13	
76th	1939-41	96	**69**	23	4		435	**262**	169	4	
77th	1941-43	96	**66**	28	2		435	**267**	162	6	
78th	1943-45	96	**57**	38	1		435	**222**	209	4	
79th	1945-47	96	**57**	38	1		435	**243**	190	2	
80th	1947-49	96	45	**51**			435	188	**246**	1	
81st	1949-51	96	**54**	42			435	**263**	171	1	
82d	1951-53	96	**48**	47	1		435	**234**	199	2	
83d	1953-55	96	46	**48**	2		435	213	**221**	1	
84th	1955-57	96	**48**	47	1		435	**232**	203		
85th	1957-59	96	**49**	47			435	**234**	201		
86th	1959-61	98	**64**	34			436[3]	**283**	153		
87th	1961-63	100	**64**	36			437[4]	**262**	175		
88th	1963-65	100	**67**	33			435	**258**	176		1
89th	1965-67	100	**68**	32			435	**295**	140		
90th	1967-69	100	**64**	36			435	**248**	187		
91st	1969-71	100	**58**	42			435	**243**	192		
92d	1971-73	100	**54**	44	2		435	**255**	180		
93d	1973-75	100	**56**	42	2		435	**242**	192	1	
94th	1975-77	100	**60**	37	2		435	**291**	144	1	
95th	1977-79	100	**61**	38	1		435	**292**	143		
96th	1979-81	100	**58**	41	1		435	**277**	158		
97th	1981-83	100	46	**53**	1		435	**242**	192	1	
98th	1983-85	100	46	**54**			435	**269**	166		
99th	1985-87	100	47	**53**			435	**253**	182		
100th	1987-89	100	**55**	45			435	**258**	177		
101st	1989-91	100	**55**	45			435	**260**	175		
102d	1991-93	100	**56**	44			435	**267**	167	1	
103d	1993-95	100	**57**	43			435	**258**	176	1	
104th	1995-97	100	48	**52**			435	204	**230**	1	
105th	1997-99	100	45	**55**			435	207	**227**	1	
106th	1999-2001	100	45	**55**			435	211	**223**	1	
107th	2001-03	100	50	**50**[5]			435	212	**221**	2	

(1) Democrats organized House with help of other parties. (2) Democrats organized House because of Republican deaths. (3) Proclamation declaring Alaska a state issued Jan. 3, 1959. (4) Proclamation declaring Hawaii a state issued Aug. 21, 1959. (5) Republican Sen. James M. Jeffords changed his party designation to Independent on June 5, 2001, switching control of the Senate to Democrats from Republicans.

Floor Leaders in the U.S. Senate Since the 1920s

Majority Leaders

Name	Party	State	Tenure
Charles Curtis[1]	R	KS	1925-1929
James E. Watson	R	IN	1929-1933
Joseph T. Robinson	D	AR	1933-1937
Alben W. Barkley	D	KY	1937-1947
Wallace H. White	R	ME	1947-1949
Scott W. Lucas	D	IL	1949-1951
Ernest W. McFarland	D	AZ	1951-1953
Robert A. Taft	R	OH	1953
William F. Knowland	R	CA	1953-1955
Lyndon B. Johnson	D	TX	1955-1961
Mike Mansfield	D	MT	1961-1977
Robert C. Byrd	D	WV	1977-1981
Howard H. Baker Jr.	R	TN	1981-1985
Robert J. Dole	R	KS	1985-1987
Robert C. Byrd	D	WV	1987-1989
George J. Mitchell	D	ME	1989-1995
Robert J. Dole	R	KS	1995-1996
Trent Lott	R	MS	1996-2001
Thomas A. Daschle	D	SD	2001-

Minority Leaders

Name	Party	State	Tenure
Oscar W. Underwood[2]	D	AL	1920-1923
Joseph T. Robinson	D	AR	1923-1933
Charles L. McNary	R	OR	1933-1944
Wallace H. White	R	ME	1944-1947
Alben W. Barkley	D	KY	1947-1949
Kenneth S. Wherry	R	NE	1949-1951
Henry Styles Bridges	R	NH	1951-1953
Lyndon B. Johnson	D	TX	1953-1955
William F. Knowland	R	CA	1955-1959
Everett M. Dirksen	R	IL	1959-1969
Hugh D. Scott	R	PA	1969-1977
Howard H. Baker Jr.	R	TN	1977-1981
Robert C. Byrd	D	WV	1981-1987
Robert J. Dole	R	KS	1987-1995
Thomas A. Daschle	D	SD	1995-2001
Trent Lott	R	MS	2001-

Note: Majority and Minority Leaders through Oct. 2001. (1) First Republican to be designated floor leader. (2) First Democrat to be designated floor leader.

Speakers of the House of Representatives

(through Oct. 2001)

Party designations: A, American; D, Democratic; DR, Democratic-Republican; F, Federalist; R, Republican; W, Whig

Name	Party	State	Tenure
Frederick Muhlenberg	F	PA	1789-1791
Jonathan Trumbull	F	CT	1791-1793
Frederick Muhlenberg	F	PA	1793-1795
Jonathan Dayton	F	NJ	1795-1799
Theodore Sedgwick	F	MA	1799-1801
Nathaniel Macon	DR	NC	1801-1807
Joseph B. Varnum	DR	MA	1807-1811
Henry Clay	DR	KY	1811-1814
Langdon Cheves	DR	SC	1814-1815
Henry Clay	DR	KY	1815-1820
John W. Taylor	DR	NY	1820-1821
Philip P. Barbour	DR	VA	1821-1823
Henry Clay	DR	KY	1823-1825
John W. Taylor	D	NY	1825-1827
Andrew Stevenson	D	VA	1827-1834
John Bell	D	TN	1834-1835
James K. Polk	D	TN	1835-1839
Robert M. T. Hunter	D	VA	1839-1841
John White	W	KY	1841-1843
John W. Jones	D	VA	1843-1845
John W. Davis	D	IN	1845-1847
Robert C. Winthrop	W	MA	1847-1849
Howell Cobb	D	GA	1849-1851
Linn Boyd	D	KY	1851-1855
Nathaniel P. Banks	A	MA	1856-1857
James L. Orr	D	SC	1857-1859
William Pennington	R	NJ	1860-1861
Galusha A. Grow	R	PA	1861-1863
Schuyler Colfax	R	IN	1863-1869
Theodore M. Pomeroy	R	NY	1869
James G. Blaine	R	ME	1869-1875
Michael C. Kerr	D	IN	1875-1876
Samuel J. Randall	D	PA	1876-1881
Joseph W. Keifer	R	OH	1881-1883
John G. Carlisle	D	KY	1883-1889
Thomas B. Reed	R	ME	1889-1891
Charles F. Crisp	D	GA	1891-1895
Thomas B. Reed	R	ME	1895-1899
David B. Henderson	R	IA	1899-1903
Joseph G. Cannon	R	IL	1903-1911
Champ Clark	D	MO	1911-1919
Frederick H. Gillett	R	MA	1919-1925
Nicholas Longworth	R	OH	1925-1931
John N. Garner	D	TX	1931-1933
Henry T. Rainey	D	IL	1933-1935
Joseph W. Byrns	D	TN	1935-1936
William B. Bankhead	D	AL	1936-1940
Sam Rayburn	D	TX	1940-1947
Joseph W. Martin Jr.	R	MA	1947-1949
Sam Rayburn	D	TX	1949-1953
Joseph W. Martin Jr.	R	MA	1953-1955
Sam Rayburn	D	TX	1955-1961
John W. McCormack	D	MA	1962-1971
Carl Albert	D	OK	1971-1977
Thomas P. O'Neill Jr.	D	MA	1977-1987
James Wright	D	TX	1987-1989
Thomas S. Foley	D	WA	1989-1995
Newt Gingrich	R	GA	1995-1999
J. Dennis Hastert	R	IL	1999-

Congressional Bills Vetoed, 1789-2001

Source: Senate Library

President	Regular vetoes	Pocket vetoes	Total vetoes	Vetoes overridden
Washington	2	—	2	—
John Adams	—	—	—	—
Jefferson	—	—	—	—
Madison	5	2	7	—
Monroe	1	—	1	—
John Q. Adams	—	—	—	—
Jackson	5	7	12	—
Van Buren	—	1	1	—
William Harrison	—	—	—	—
Tyler	6	4	10	1
Polk	2	1	3	—
Taylor	—	—	—	—
Fillmore	—	—	—	—
Pierce	9	—	9	5
Buchanan	4	3	7	—
Lincoln	2	4	6	—
Andrew Johnson	21	8	29	15
Grant	45	48	93	4
Hayes	12	1	13	1
Garfield	—	—	—	—
Arthur	4	8	12	1
Cleveland[1]	304	110	414	2
Benjamin Harrison	19	25	44	1
Cleveland[2]	42	128	170	5
McKinley	6	36	42	—
Theodore Roosevelt	42	40	82	1
Taft	30	9	39	1
Wilson	33	11	44	6
Harding	5	1	6	—
Coolidge	20	30	50	4
Hoover	21	16	37	3
Franklin Roosevelt	372	263	635	9
Truman	180	70	250	12
Eisenhower	73	108	181	2
Kennedy	12	9	21	—
Lyndon Johnson	16	14	30	—
Nixon	26	17	43	7
Ford	48	18	66	12
Carter	13	18	31	2
Reagan	39	39	78	9
George H. W. Bush[3]	29	15	44	1
Clinton[4]	36	1	37	2
George W. Bush[5]	—	—	—	—
Total[3,4]	**1,484**	**1,065**	**2,549**	**106**

— = 0. (1) First term only. (2) Second term only. (3) Excluded from the figures are 2 additional bills, which Pres. George H. W. Bush claimed to be vetoed but Congress considered enacted into law because the president failed to return them to Congress during a recess period. (4) Does not include line-item veto, which was ruled unconstitutional by the Supreme Court on June 25, 1998. (5) As of Oct. 1, 2001.

Librarians of Congress

Librarian	Served	Appointed by President
John J. Beckley	1802-1807	Jefferson
Patrick Magruder	1807-1815	Jefferson
George Watterston	1815-1829	Madison
John Silva Meehan	1829-1861	Jackson
John G. Stephenson	1861-1864	Lincoln
Ainsworth Rand Spofford	1864-1897	Lincoln
John Russell Young	1897-1899	McKinley
Herbert Putnam	1899-1939	McKinley
Archibald MacLeish	1939-1944	F. D. Roosevelt
Luther H. Evans	1945-1953	Truman
L. Quincy Mumford	1954-1974	Eisenhower
Daniel J. Boorstin	1975-1987	Ford
James H. Billington	1987-	Reagan

U.S. SUPREME COURT

(data as of Oct. 2001)

Justices of the United States Supreme Court

The Supreme Court comprises the chief justice of the U.S. and 8 associate justices, all appointed by the president with advice and consent of the Senate. Salaries: chief justice, $186,300 annually; associate justice, $178,300 annually. The Supreme Court is at the U.S. Supreme Court Bldg., 1 First St. NE, Washington, DC 20543. The website for the Supreme Court is http://www.supremecourtus.gov

Members of the Supreme Court at the start of the 2001-2002 term (Oct. 1, 2001): Chief justice: William H. Rehnquist; associate justices: Stephen G. Breyer, Ruth Bader Ginsburg, Anthony M. Kennedy, Sandra Day O'Connor, Antonin Scalia, David H. Souter, John Paul Stevens, Clarence Thomas.

Name,[1] apptd. from	Service Term	Service Yrs	Born	Died
John Jay, NY	1789-1795	5	1745	1829
John Rutledge, SC	1789-1791	1	1739	1800
William Cushing, MA	1789-1810	20	1732	1810
James Wilson, PA	1789-1798	8	1742	1798
John Blair, VA	1789-1796	6	1732	1800
James Iredell, NC	1790-1799	9	1751	1799
Thomas Johnson, MD	1791-1793	1	1732	1819
William Paterson, NJ	1793-1806	13	1745	1806
John Rutledge[2], SC	1795	—	1739	1800
Samuel Chase, MD	1796-1811	15	1741	1811
Oliver Ellsworth, CT	1796-1800	4	1745	1807
Bushrod Washington, VA	1798-1829	31	1762	1829
Alfred Moore, NC	1799-1804	4	1755	1810
John Marshall, VA	1801-1835	34	1755	1835
William Johnson, SC	1804-1834	30	1771	1834
Henry B. Livingston, NY	1806-1823	16	1757	1823
Thomas Todd, KY	1807-1826	18	1765	1826
Joseph Story, MA	1811-1845	33	1779	1845
Gabriel Duval, MD	1811-1835	22	1752	1844
Smith Thompson, NY	1823-1843	20	1768	1843
Robert Trimble, KY	1826-1828	2	1777	1828
John McLean, OH	1829-1861	32	1785	1861
Henry Baldwin, PA	1830-1844	14	1780	1844
James M. Wayne, GA	1835-1867	32	1790	1867
Roger B. Taney, MD	1836-1864	28	1777	1864
Philip P. Barbour, VA	1836-1841	4	1783	1841
John Catron, TN	1837-1865	28	1786	1865
John McKinley, AL	1837-1852	15	1780	1852
Peter V. Daniel, VA	1841-1860	19	1784	1860
Samuel Nelson, NY	1845-1872	27	1792	1873
Levi Woodbury, NH	1845-1851	5	1789	1851
Robert C. Grier, PA	1846-1870	23	1794	1870
Benjamin R. Curtis, MA	1851-1857	6	1809	1874
John A. Campbell, AL	1853-1861	8	1811	1889
Nathan Clifford, ME	1858-1881	23	1803	1881
Noah H. Swayne, OH	1862-1881	18	1804	1884
Samuel F. Miller, IA	1862-1890	28	1816	1890
David Davis, IL	1862-1877	14	1815	1886
Stephen J. Field, CA	1863-1897	34	1816	1899
Salmon P. Chase, OH	1864-1873	8	1808	1873
William Strong, PA	1870-1880	10	1808	1895
Joseph P. Bradley, NJ	1870-1892	21	1813	1892
Ward Hunt, NY	1872-1882	9	1810	1886
Morrison R. Waite, OH	1874-1888	14	1816	1888
John M. Harlan, KY	1877-1911	34	1833	1911
William B. Woods, GA	1880-1887	6	1824	1887
Stanley Matthews, OH	1881-1889	7	1824	1889
Horace Gray, MA	1881-1902	20	1828	1902
Samuel Blatchford, NY	1882-1893	11	1820	1893
Lucius Q.C. Lamar, MS	1888-1893	5	1825	1893
Melville W. Fuller, IL	1888-1910	21	1833	1910
David J. Brewer, KS	1889-1910	20	1837	1910
Henry B. Brown, MI	1890-1906	15	1836	1913
George Shiras Jr., PA	1892-1903	10	1832	1924
Howell E. Jackson, TN	1893-1895	2	1832	1895
Edward D. White, LA	1894-1910	16	1845	1921
Rufus W. Peckham, NY	1895-1909	13	1838	1909
Joseph McKenna, CA	1898-1925	26	1843	1926
Oliver W. Holmes, MA	1902-1932	29	1841	1935
William R. Day, OH	1903-1922	19	1849	1923
William H. Moody, MA	1906-1910	3	1853	1917
Horace H. Lurton, TN	1909-1914	4	1844	1914
Charles E. Hughes, NY	1910-1916	5	1862	1948
Willis Van Devanter, WY	1910-1937	26	1859	1941
Joseph R. Lamar, GA	1910-1916	5	1857	1916
Edward D. White, LA	1910-1921	10	1845	1921
Mahlon Pitney, NJ	1912-1922	10	1858	1924
James C. McReynolds, TN	1914-1941	26	1862	1946
Louis D. Brandeis, MA	1916-1939	22	1856	1941
John H. Clarke, OH	1916-1922	5	1857	1945
William H. Taft, CT	1921-1930	8	1857	1930
George Sutherland, UT	1922-1938	15	1862	1942
Pierce Butler, MN	1922-1939	16	1866	1939
Edward T. Sanford, TN	1923-1930	7	1865	1930
Harlan F. Stone, NY	1925-1941	16	1872	1946
Charles E. Hughes, NY	1930-1941	11	1862	1948
Owen J. Roberts, PA	1930-1945	15	1875	1955
Benjamin N. Cardozo, NY	1932-1938	6	1870	1938
Hugo L. Black, AL	1937-1971	34	1886	1971
Stanley F. Reed, KY	1938-1957	19	1884	1980
Felix Frankfurter, MA	1939-1962	23	1882	1965
William O. Douglas, CT	1939-1975	36[3]	1898	1980
Frank Murphy, MI	1940-1949	9	1890	1949
Harlan F. Stone, NY	1941-1946	5	1872	1946
James F. Byrnes, SC	1941-1942	1	1879	1972
Robert H. Jackson, NY	1941-1954	12	1892	1954
Wiley B. Rutledge, IA	1943-1949	6	1894	1949
Harold H. Burton, OH	1945-1958	13	1888	1964
Fred M. Vinson, KY	1946-1953	7	1890	1953
Tom C. Clark, TX	1949-1967	18	1899	1977
Sherman Minton, IN	1949-1956	7	1890	1965
Earl Warren, CA	1953-1969	16	1891	1974
John Marshall Harlan, NY	1955-1971	16	1899	1971
William J. Brennan Jr., NJ	1956-1990	33	1906	1997
Charles E. Whittaker, MO	1957-1962	5	1901	1973
Potter Stewart, OH	1958-1981	23	1915	1985
Byron R. White, CO	1962-1993	31	1917	
Arthur J. Goldberg, IL	1962-1965	3	1908	1990
Abe Fortas, TN	1965-1969	4	1910	1982
Thurgood Marshall, NY	1967-1991	24	1908	1993
Warren E. Burger, VA	1969-1986	17	1907	1995
Harry A. Blackmun, MN	1970-1994	24	1908	1999
Lewis F. Powell Jr., VA	1971-1987	16	1907	1998
William H. Rehnquist, AZ	1971-1986	15	1924	
John Paul Stevens, IL	1975-		1920	
Sandra Day O'Connor, AZ	1981-		1930	
William H. Rehnquist, AZ	1986-		1924	
Antonin Scalia, VA	1986-		1936	
Anthony M. Kennedy, CA	1988-		1936	
David H. Souter, NH	1990-		1939	
Clarence Thomas, VA	1991-		1948	
Ruth Bader Ginsburg, DC	1993-		1933	
Stephen G. Breyer, MA	1994-		1938	

(1) Chief justices in italics. (2) Named as acting chief justice; confirmation rejected by the Senate, Dec. 15, 1795. (3) Longest term of service.

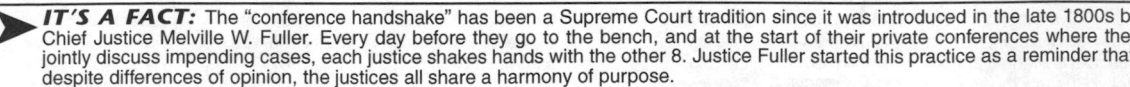

IT'S A FACT: The "conference handshake" has been a Supreme Court tradition since it was introduced in the late 1800s by Chief Justice Melville W. Fuller. Every day before they go to the bench, and at the start of their private conferences where they jointly discuss impending cases, each justice shakes hands with the other 8. Justice Fuller started this practice as a reminder that, despite differences of opinion, the justices all share a harmony of purpose.

STATE AND LOCAL GOVERNMENT

Mayors of Selected U.S. Cities

As of Oct. 10, 2001

D, Democrat; R, Republican; N-P, Non-Partisan; I, Independent; Prog. Coal., Progressive Coalition

City	Name	Next Election
Abilene, TX	Grady Barr, N-P	2002, May
Akron, OH	Donald L. Plusquellic, D	2003, Nov.
Alameda, CA	Ralph J. Appezzato, N-P	2002, Nov.
Albany, GA	Thomas Coleman, D	2001, Nov.
Albany, NY	Gerald D. Jennings, D	2001, Nov.
Albuquerque, NM	Jim Baca, D	2001, Nov.
Alexandria, LA	Edward G. Randolph Jr., D	2002, Nov.
Alexandria, VA	Kerry J. Donley, D	2003, May
Alhambra, CA	Paul T. Talbot, N-P	(1)
Allentown, PA	William Heydt, R	2001, Nov.
Amarillo, TX	Trent Sisemore, N-P	2003, May
Ames, IA	Ted Tedesco, N-P	2001, Nov.
Anaheim, CA	Tom Daly, N-P	2002, Nov.
Anchorage, AK	George Wuerch, R	2003, Apr.
Anderson, IN	J. Mark Lawler, D	2003, Nov.
Anderson, SC	Richard A. Shirley, N-P	2002, June
Ann Arbor, MI	John Hieftie, D	2002, Nov.
Annapolis, MD	Dean L. Johnson, R	2001, Nov.
Appleton, WI	Timothy M. Hanna, N-P	2004, Apr.
Arcadia, CA	Mickey Segal, N-P	2002, Apr.
Arlington, MA	John W. Hurd, N-P	2002, Apr.
Arlington, TX	Elzie Odom, N-P	2002, May
Arlington Hts., IL	Arlene J. Mulder, N-P	2005, Apr.
Arvada, CO	Ken Fellman, N-P	2003, Nov.
Asheville, NC	Leni Sitnick, N-P	2001, Nov.
Athens, GA	Doc Eldridge, D	2002, Nov.
Atlanta, GA	Bill Campbell, D	2001, Nov.
Atlantic City, NJ	James Whelan, N-P	2002, May
Augusta, GA	Bob Young, N-P	2002, Nov.
Augusta, ME	William E. Dowling, N-P	2002, Nov.
Aurora, CO	Paul E. Tauer, N-P	2003, Nov.
Aurora, IL	David L. Stover, N-P	2005, Apr.
Austin, TX	Kirk P. Watson, N-P	2003, May
Bakersfield, CA	Bob Price, N-P	2002, Mar.
Baldwin Park, CA	Manuel Lozano, N-P	2003, Mar.
Baltimore, MD	Martin O'Malley, D	2004, Nov.
Baton Rouge, LA	Bobby Simpson, R	2004, Nov.
Battle Creek, MI	Ted Dearing, N-P	(2)
Bayonne, NJ	Joseph V. Doria Jr., N-P	2002, May
Baytown, TX	Pete C. Alfaro, N-P	2002, May
Beaumont, TX	David W. Moore, N-P	2002, May
Belleville, IL	Mark Kern, N-P	2005, Apr.
Bellevue, WA	Chuck Mosher, N-P	2001, Nov.
Bellflower, CA	Ray Smith, N-P	2003, Mar.
Bellingham, WA	Mark Asmundson, N-P	2003, Nov.
Berkeley, CA	Shirley Dean, N-P	2002, Nov.
Bethlehem, PA	Donald T. Cunningham Jr., D	2001, Nov.
Beverly Hills, CA	Mark Egerman, N-P	2003, Mar.
Billings, MT	Charles F. Tooley, N-P	2001, Nov.
Biloxi, MS	A. J. Holloway, R	2005, June
Binghamton, NY	Richard A. Bucci, R	2001, Nov.
Birmingham, AL	Bernard Kincaid, D	2003, Nov.
Bismarck, ND	Bill Sorensen, R	2002, June
Bloomfield, NJ	John Bukowski Jr., R	2001, Nov.
Bloomington, IL	Judy Markowitz, N-P	2005, Apr.
Bloomington, IN	John Fernandez, D	2003, Nov.
Bloomington, MN	Gene Winstead, N-P	2003, Nov.
Boca Raton, FL	Steven L. Abrams, N-P	2003, Mar.
Boise, ID	H. Brent Coles, N-P	2001, Nov.
Bossier City, LA	George Dement, N-P	2005, Apr.
Boston, MA	Thomas M. Menino, D	2001, Nov.
Boulder, CO	Will Toor, N-P	2001, Nov.
Bridgeport, CT	Joseph P. Ganim, D	2003, Nov.
Bristol, CT	Frank N. Nicastro Sr., D	2001, Nov.
Brockton, MA	John T. Yunits Jr., D	2001, Nov.
Broken Arrow, OK	Jim Reynolds, N-P	2003, Apr.
Brooklyn Park, MN	Grace Arbogast, N-P	2002, Nov.
Brownsville, TX	Blanca S. Vela, N-P	2003, May
Bryan, TX	Jay Don Watson, N-P	2004, May
Buena Park, CA	Steve Berry, N-P	2004, Nov.
Buffalo, NY	Anthony M. Masiello, D	2001, Nov.
Burbank, CA	Bob Kramer, R	2003, May
Burlington, VT	Peter A. Clavelle, Prog. Coal.	2003, Mar.
Calumet City, IL	Gerome P. Genova, I	2005, Apr.
Camarillo, CA	William Q. Liebmann, N-P	(2)
Cambridge, MA	Anthony D. Galluccio, D	2002, Jan.
Camden, NJ	Gwendolyn A. Faison, D	2005, May
Canton, OH	Richard D. Watkins, R	2003, Nov.
Cape Coral, FL	Arnold Kempe, N-P	2004, Nov.
Carlsbad, CA	Claude A. Lewis, N-P	2002, Nov.
Carson, CA	Daryl Sweeney, N-P	2005, Mar.
Carson City, NV	Ray Masayko, N-P	2004, Nov.
Casper, WY	Dr. Thomas Walsh, N-P	(2)
Cedar Rapids, IA	Lee R. Clancey, N-P	2001, Nov.
Champaign, IL	Gerald Schweighart, N-P	2003, Apr.
Chandler, AZ	Jay Tibshraeny, N-P	2002, Mar.
Charleston, SC	Joseph P. Riley Jr., D	2003, Nov.
Charleston, WV	Jay Goldman, D	2003, May
Charlotte, NC	Patrick McCrory, R	2001, Nov.
Charlottesville, VA	James B. Caravati, D	2002, May
Chattanooga, TN	Bob Corker, N-P	2005, Mar.
Chesapeake, VA	William E. Ward, N-P	2004, May
Chester, PA	Dominic F. Pileggi, R	2003, Nov.
Cheyenne, WY	Jack Spiker, N-P	2004, Nov.
Chicago, IL	Richard M. Daley, D	2003, Feb.
Chicopee, MA	Richard J. Kos, N-P	2001, Nov.
Chino, CA	Eunice Ulloa, R	2004, Nov.
Chula Vista, CA	Shirley A. Horton, N-P	2002, June
Cicero, IL	Betty Loren-Maltese, R	2005, Apr.
Cincinnati, OH	Charlie Luken, D	2001, Nov.
Clarksville, TN	Johnny Piper, N-P	2002, Nov.
Clearwater, FL	Brian Aungst, N-P	2002, Mar.
Cleveland, OH	Michael R. White, D	2001, Nov.
Cleveland Hts., OH	Edward J. Kelley, N-P	2001, Nov.
Clinton, IA	La Metta Wynn, N-P	2003, Nov.
Clifton, NJ	James A. Anzaldi, R	2002, May
Colorado Spgs., CO	Mary Lou Makepeace, R	2003, Apr.
Columbia, MO	Darwin Hindman, N-P	2004, Apr.
Columbia, SC	Robert D. Coble, N-P	2002, Apr.
Columbus, GA	Bobby G. Peters, D	2002, Nov.
Columbus, OH	Michael B. Coleman, D	2003, Nov.
Compton, CA	Eric Perrodin, N-P	2005, Apr.
Concord, CA	Helen M. Allen, N-P	(2)
Concord, NH	William J. Veroneau, N-P	2001, Nov.
Coon Rapids, MN	Lonni McCauley, N-P	2002, Nov.
Coral Gables, FL	Don Slesnick, N-P	2003, Apr.
Coral Springs, FL	John Sommerer, N-P	2002, Mar.
Corona, CA	Jeffrey P. Bennett, N-P	(2)
Corpus Christi, TX	Samuel Loyd Neal, N-P	2003, Apr.
Costa Mesa, CA	Gary Monahan, N-P	2002, Nov.
Council Bluffs, IA	Tom Hanafan, N-P	2001, Nov.
Covington, KY	Irvin T. "Butch" Callery, N-P	2004, Nov.
Cranston, RI	John O'Leary, D	2002, Nov.
Cuyahoga Falls, OH	Donald L. Robart, R	2001, Nov.
Dallas, TX	Ronald Kirk, N-P	2003, May
Daly City, CA	Gonzalo "Sal" Torres, N-P	(2)
Danbury, CT	Gene F. Eriquez, D	2001, Nov.
Danville, VA	John C. Hamlin, N-P	2002, May
Davenport, IA	Phillip Yerington, N-P	2001, Nov.
Davis, CA	Ken Wagstaff, N-P	2002, Mar.
Dayton, OH	Michael R. Turner, N-P	2001, Nov.
Daytona Beach, FL	Baron H. Asher, N-P	2001, Nov.
Dearborn, MI	Michael A. Guido, N-P	2001, Nov.
Dearborn Hts., MI	Ruth A. Canfield, N-P	2001, Nov.
Decatur, IL	Terry M. Howley, N-P	2003, Apr.
Delray Beach, FL	David Schmidt, N-P	2002, Mar.
Denton, TX	Euline Brock, N-P	2002, May
Denver, CO	Wellington E. Webb, N-P	2003, May
Des Moines, IA	Preston A. Daniels, N-P	2001, Nov.
Des Plaines, IL	Tony Arredia, N-P	2005, Apr.
Detroit, MI	Dennis W. Archer, D	2001, Nov.
Dothan, AL	Chester L. Sowell, N-P	2005, July
Dover, DE	James L. Hutchinson, N-P	2002, Apr.
Downey, CA	Keith McCarthy, N-P	(2)
Dubuque, IA	Terrance M. Duggan, N-P	2001, Nov.
Duluth, MN	Gary L. Doty, N-P	2003, Nov.
Durham, NC	Nicholas J. Tennyson, N-P	2001, Nov.
East Hartford, CT	Timothy D. Larson, D	2001, Nov.

City	Name	Next Election
East Lansing, MI	Mark S. Meadows, N-P	2001, Nov.
East Orange, NJ	Robert L. Bowser, D	2001, Nov.
Edison, NJ	George Spadoro, D	2001, Nov.
Edmond, OK	Saundra Naifeh, N-P	2003, May
El Cajon, CA	Mark Lewis, N-P	2002, Nov.
Elgin, IL	Ed Schock, N-P	2003, Apr.
Elizabeth, NJ	J. Christian Bollwage, D	2004, Nov.
Elkhart, IN	David L. Miller, R	2003, Nov.
El Monte, CA	L. Rachel Montes, N-P	2003, Mar.
El Paso, TX	Raymond C. Caballero, N-P	2003, May
Elyria, OH	Bill Grace, D	2003, Nov.
Enfield, CT	Mary Lou Strom, R	2001, Nov.
Enid, OK	J. Doug Frantz, N-P	2003, Apr.
Erie, PA	Joyce Savocchio, D	2001, Nov.
Escondido, CA	Lori Holt Pfieler, N-P	2002, Nov.
Euclid, OH	Paul Oyaski, D	2003, Nov.
Eugene, OR	James D. Torrey, N-P	2004, Nov.
Evanston, IL	Lorraine Morton, N-P	2005, Apr.
Evansville, IN	Russell G. Lloyd, Jr., D	2003, Nov.
Everett, WA	Edward D. Hansen, N-P	2001, Nov.
Fairbanks, AK	Steve Thompson, N-P	2004, Oct.
Fairfield, CA	George Pettygrove, N-P	2001, Nov.
Fairfield, CT	John Metsopoulous, R	2001, Nov.
Fall River, MA	Edward M. Lambert Jr., D	2001, Nov.
Fargo, ND	Bruce W. Furness, N-P	2002, Apr.
Farmington Hills, MI	Nancy Bates, N-P	2003, Nov.
Fayetteville, NC	J. L. Dawkins, N-P	2001, Nov.
Fitchburg, MA	Mary Whitney, N-P	2001, Nov.
Flagstaff, AZ	Joe Donaldson, N-P	2002, May
Flint, MI	Woodrow Stanley, D	2003, Nov.
Florissant, MO	James J. Eagan, N-P	2003, Apr.
Fontana, CA	David Eshleman, D	2002, Nov.
Ft. Collins, CO	Ray Martinez, N-P	2003, Apr.
Ft. Lauderdale, FL	Jim Naugle, N-P	2003, Mar.
Ft. Smith, AR	C. Raymond Baker, N-P	2002, Nov.
Ft. Wayne, IN	Graham A. Richard, D	2003, Nov.
Ft. Worth, TX	Kenneth L. Barr, N-P	2003, May
Fountain Valley, CA	John Collins, N-P	(2)
Frankfort, KY	William I. May Jr., N-P	2004, Nov.
Fremont, CA	Gus Morrison, N-P	2004, Nov.
Fresno, CA	Alan Autoy, N-P	2005, Jan.
Fullerton, CA	F. Richard Jones, M.D., N-P	(2)
Gadsden, AL	Steven A. Means, N-P	2002, Nov.
Gainesville, FL	Paula M. DeLaney, N-P	2001, Nov
Galveston, TX	Roger R. "Bo" Quiorga, N-P	2002, May
Gardena, CA	Terrance Terauchi, N-P	2005, Mar.
Garden Grove, CA	Bruce A. Broadwater, N-P	2002, Nov.
Garland, TX	Jim Spence, N-P	2002, May
Gary, IN	Scott L. King, D	2003, Nov.
Gastonia, NC	Jennifer T. Stultz, N-P	2001, Nov.
Glendale, AZ	Elaine M. Scruggs, N-P	2002, May
Glendale, CA	Gus Gomez, N-P	2002, Apr.
Grand Forks, ND	Michael R. Brown, N-P	2004, June
Grand Prairie, TX	Charles V. England, N-P	2002, May
Grand Rapids, MI	John H. Logie, N-P	2003, Nov.
Greeley, CO	Jerry Wones, N-P	2001, Nov.
Green Bay, WI	Paul F. Jadin, N-P	2003, Apr.
Greensboro, NC	Keith Holliday, N-P	2001, Nov.
Greenville, SC	Knox H. White, R	2003, Nov.
Greenwich, CT	Lolly H. Prince, R	2001, Nov.
Groton, CT	Deloris Hauber, N-P	2001, Nov.
Gulfport, MS	Ken Combs, R	2005, June
Hamden, CT	Carl Amento, D	2001, Nov.
Hamilton, OH	Adolf Olivas, N-P	2001, Nov.
Hammond, IN	Duane W. Dedelow Jr., R	2003, Nov.
Hampton, VA	Mamie E. Locke, N-P	2004, May
Harrisburg, PA	Stephen R. Reed, D	2001, Nov.
Hartford, CT	Michael P. Peters, N-P	2001, Nov.
Haverhill, MA	James A. Rurak, D	2001, Nov.
Hawthorne, CA	Larry Guidi, N-P	2001, Nov.
Hayward, CA	Roberta Cooper, N-P	2002, Apr.
Helena, MT	Colleen McCarthy, N-P	2001, Nov.
Henderson, NV	James B. Gibson, N-P	2003, June
Hialeah, FL	Raul L. Martinez, R	2001, Nov.
High Point, NC	Arnold J. Koonce, Jr., N-P	2002, Nov.
Hoboken, NJ	David Roberts, N-P	2005, May
Hollywood, FL	Mara Giulianti, N-P	2004, Mar.
Holyoke, MA	Michael Sullivan, D	2001, Nov.
Honolulu, HI	Jeremy Harris, N-P	2004, Sept.

City	Name	Next Election
Houston, TX	Lee P. Brown, N-P	2001, Nov.
Huntington, WV	David Felinton, D	2004, Nov.
Huntington Beach, CA	Dave Garafalo, N-P	(2)
Huntington Park, CA	Richard Loya, N-P	2002, Mar.
Huntsville, AL	Loretta Spencer, N-P	2004, Aug.
Idaho Falls, ID	Linda Milam, N-P	2001, Nov.
Independence, MO	Ron Stewart, N-P	2002, Apr.
Indianapolis, IN	Bart Peterson, D	2003, Nov.
Inglewood, CA	Roosevelt F. Dorn, N-P	2002, Nov.
Iowa City, IA	Ernest W. Lehman, N-P	2001, Nov.
Irvine, CA	Larry Agran, N-P	2002, Nov.
Irving, TX	Joe Putnam, N-P	2002, May
Irvington, NJ	Sara B. Bost, D	2002, May
Jackson, MS	Harvey Johnson, D	2005, June
Jacksonville, FL	John A. Delaney, R	2003, May
Janesville, WI	Thomas J. Stehura, N-P	2002, Apr.
Jefferson City, MO	Thomas P. Rackers, R	2003, Apr.
Jersey City, NJ	Glenn Cunningham, D	2005, May
Johnson City, TN	Duffuie Jones, N-P	2003, May
Joliet, IL	Arthur Schultz, N-P	2003, Apr.
Juneau, AK	Sally Smith, N-P	2003, Oct.
Kalamazoo, MI	Robert B. Jones, N-P	2001, Nov.
Kansas City, KS	Carol S. Marinovich, N-P	2005, Apr.
Kansas City, MO	Kay Barnes, N-P	2003, Mar.
Kenner, LA	Louis J. Congemi, R	2002, Apr.
Kenosha, WI	John M. Antaramian, D	2002, Apr.
Kettering, OH	Marilou W. Smith, N-P	2001, Nov.
Killeen, TX	Fred L. Latham, N-P	2002, May
Knoxville, TN	Victor H. Ashe, R	2003, Nov.
Kokomo, IN	James E. Trobaugh, R	2003, Nov.
LaCrosse, WI	John D. Medinger, N-P	2005, Apr.
Lafayette, IN	Dave Heath, R	2003, Nov.
La Habra, CA	Steve Anderson, N-P	(2)
Lake Charles, LA	Randy Roach, D	2005, May
Lakeland, FL	Ralph L. Fletcher, N-P	2004, Nov.
Lakewood, CA	Robert Wagner, N-P	2003, Mar.
Lakewood, CO	Steve Burkholder, N-P	2003, Nov.
Lakewood, OH	Madeline Cain, D	2003, Nov.
La Mesa, CA	Arthur Madrid, N-P	2002, Nov.
La Mirada, CA	Pete Dames, N-P	2002, Mar.
Lancaster, CA	Frank C. Roberts, N-P	2002, Apr.
Lancaster, PA	Charlie Smithgall, R	2002, Nov.
Lansing, MI	David C. Hollister, N-P	2001, Nov.
Laredo, TX	Elizabeth G. "Betty" Flores, N-P	2002, May
Largo, FL	Robert E. Jackson, N-P	2003, Mar
Las Cruces, NM	Ruben A. Smith, D	2003, Nov.
Las Vegas, NV	Oscar B. Goodman, D	2003, June
Lawrence, KS	Mike Rundle, N-P	2002, Apr.
Lawrence, MA	Patricia Dowling, N-P	2001, Nov.
Lawton, OK	Cecil E. Powell, D	2003, Mar.
Lexington, KY	Pam Miller, N-P	2002, Nov.
Lima, OH	David J. Berger, N-P	2002, Nov.
Lincoln, NE	Don Wesely, N-P	2003, May
Little Rock, AR	Jim Dailey, N-P	2002, Nov.
Livermore, CA	Cathie Brown, N-P	2001, Nov.
Livonia, MI	Jack E. Kirksey, N-P	2003, Nov.
Lodi, CA	Stephen J. Mann, N-P	(2)
Long Beach, CA	Beverly O'Neill, N-P	2002, Apr.
Longmont, CO	Leona Stoecker, N-P	2001, Nov.
Longview, TX	Earl Roberts, N-P	2003, May
Lorain, OH	Craig Foltin, D	2003, Nov.
Los Angeles, CA	James Hahn, N-P	2005, June
Louisville, KY	David Armstrong, D	2002, Nov.
Lowell, MA	Eileen M. Donoghue, D	2001, Nov.
Lubbock, TX	Windy Sitton, R	2002, May
Lynchburg, VA	Carl B. Hutcherson, Jr., N-P	2002, May
Lynn, MA	Patrick J. McManus, D	2001, Nov.
Lynwood, CA	Louis Byrd, N-P	(2)
Macon, GA	C. Jack Ellis, D	2003, Nov.
Madison, WI	Susan J.M. Bauman, N-P	2003, Apr.
Malden, MA	Richard C. Howard, D	2001, Nov.
Manchester, CT	Stephen T. Cassano, N-P	2001, Nov.
Manchester, NH	Robert A. Baines, N-P	2001, Nov.
Mansfield, OH	Lydia J. Reid, D	2003, Nov.
Marietta, GA	Ansley L. Meaders, D	2003, Nov.
McAllen, TX	Leo Montalvo, R	2005, May
Medford, MA	Michael J. McGlynn, D	2001, Nov.
Medford, OR	Lindsay D. Berryman, N-P	2004, Nov.

City	Name	Next Election
Melbourne, FL	John Buckley, N-P	2004, Nov.
Memphis, TN	Willie W. Herenton, D	2003, Oct.
Mentor, OH	Richard Hennig, N-P	2002, Nov.
Merced, CA	MaryJo Knudsen, N-P	2001, Nov.
Meriden, CT	Joseph J. Marinan Jr., N-P	2001, Nov.
Meridian, MS	John Robert Smith, R	2005, June
Mesa, AZ	Keno Hawker, N-P	2004, Mar.
Mesquite, TX	Mike Anderson, N-P	2003, May
Miami, FL	Joe Carollo, N-P	2001, Nov.
Miami Beach, FL	Neisen O. Kasdin, N-P	2004, Nov.
Midland, TX	Michael J. Canon, N-P	2004, May
Midwest City, OK	Eddie O. Reed, N-P	2002, Apr.
Milford, CT	Frederick L. Lisman, R	2001, Nov.
Milpitas, CA	Henry C. Manayan, N-P	2004, Nov.
Milwaukee, WI	John O. Norquist, D	2004, Apr.
Minneapolis, MN	Sharon Sayles Belton, D	2001, Nov.
Minnetonka, MN	Karen J. Anderson, N-P	2001, Nov.
Mobile, AL	Michael C. Dow, R	2005, Aug.
Modesto, CA	Carmen Sabatino, N-P	2003, Nov.
Monroe, LA	Melvin L. Rambin, R	2004, Mar.
Montclair, NJ	Robert J. Russo, N-P	2004, May
Montebello, CA	Maryanne Saucedo, N-P	2001, Nov
Monterey Park, CA	Francisco Alonso, N-P	2002, Apr.
Montgomery, AL	Bobby N. Bright, D	2003, Nov.
Montpelier, VT	William J. Fraser, N-P	(3)
Moreno Valley, CA	Richard A. Stewart, N-P	(2)
Mt. Prospect, IL	Gerald L. "Skip" Farley, N-P	2005, Apr.
Mt. Vernon, NY	Ernest D. Davis, D	2003, Nov.
Mountain View, CA	Rosemary Stasek, N-P	(2)
Muncie, IN	Dan Cannan, R	2003, Nov.
Muskogee, OK	Hershel McBride, N-P	2002, Apr.
Napa, CA	Ed Henderson, N-P	2002, Apr..
Naperville, IL	George A. Pradel, N-P	2003, Apr.
Nashua, NH	Bernard A. Streeter, N-P	2003, Nov.
Nashville, TN	Bill Purcell, N-P	2003, Aug.
National City, CA	George H. Waters, R	2002, Nov.
Newark, NJ	Sharpe James, D	2002, May
New Bedford, MA	Frederick M. Kalisz Jr., N-P	2001, Nov.
New Britain, CT	Lucian J. Pawlak, D	2001, Nov.
New Haven, CT	John DeStefano Jr., D	2001, Nov.
New Orleans, LA	Marc H. Morial, D	2002, Feb.
Newport Beach, CA	John E. Noyes, N-P	(2)
Newport News, VA	Joe S. Frank, N-P	2002, May
New Rochelle, NY	Timothy Idoni, D	2003, Nov.
Newton, MA	David B. Cohen, N-P	2001, Nov.
New York, NY	Rudolph W. Giuliani, R	2001, Nov.
Niagara Falls, NY	Irene J. Elia, Ph. D., R	2003, Nov.
Norfolk, VA	Paul D. Fraim, N-P	2002, June
Norman, OK	Ron Henderson, N-P	2004, Apr.
N. Charleston, SC	R. Keith Summey, R	2003, June
N. Little Rock, AR	Patrick Henry Hays, N-P	2001, Nov.
Norwalk, CA	Gordon Stefenhaten, N-P	2002, Mar.
Norwalk, CT	Frank J. Esposito, R	2001, Nov.
Novato, CA	Pat Eklund, N-P	(2)
Oakland, CA	Jerry Brown, N-P	2002, Nov.
Oak Park, IL	Joanne Trapani, N-P	2005, Apr.
Oceanside, CA	Terry Johnson, N-P	2004, Nov.
Odessa, TX	Bill R. Hext, N-P	2002, May
Ogden, UT	Matthew Godfrey, N-P	2003, Nov.
Oklahoma City, OK	Kirk Humphreys, N-P	2002, Apr.
Olympia, WA	Stan Biles, N-P	2003, Nov.
Omaha, NE	Mike Fahey, D	2005, May
Ontario, CA	Gary C. Ovitt, N-P	2002, Nov.
Orange, CA	Mark Murphy, N-P	2002, Nov.
Orlando, FL	Glenda E. Hood, N-P	2002, Sept.
Oshkosh, WI	Jon Dell' Antonia, N-P	2002, Apr.
Overland Park, KS	Ed Eilert, R	2003, Apr.
Owensboro, KY	Waymond O. Morris, N-P	2003, Nov.
Oxnard, CA	Manuel M. Lopez, N-P	2002, Nov.
Palm Springs, CA	William G. Kleindienst, N-P	2003, Nov.
Palo Alto, CA	Liz Kniss, N-P	2001, Nov.
Parma, OH	Gerald M. Boldt, D	2003, Nov.
Pasadena, CA	Bill Bogaard, N-P	2003, May
Pasadena, TX	John Manlove, N-P	2005, May
Passaic, NJ	Samul (Sammy) Rivera, N-P	2005, May
Paterson, NJ	Martin G. Barnes, R	2002, May
Pawtucket, RI	James E. Doyle, D	2001, Nov.
Peabody, MA	Peter Torigian, D	2001, Nov.
Pembroke Pines, FL	Alex G. Fekete, N-P	2004, Mar.
Pensacola, FL	John R. Fogg, N-P	2003, June
Peoria, IL	David P. Ransburg, N-P	2005, Apr.
Philadelphia, PA	John F. Street, D	2003, Nov.
Phoenix, AZ	Skip Rimsza, R	2004, Sept.
Pico Rivera, CA	Pete Ramirez, N-P	2003, Mar.
Pierre, SD	Gary Drewes, N-P	2002, Apr.
Pine Bluff, AR	Dutch King, N-P	2005, Jan.
Pittsburgh, PA	Tom J. Murphy, D	2001, Nov.
Pittsfield, MA	Gerald S. Doyle Jr., N-P	2001, Nov.
Plainfield, NJ	Albert McWilliams, N-P	2002, Nov.
Plano, TX	Jeran Akers, N-P	2002, May
Plantation, FL	Rae Carole Armstrong, D	2003, Mar.
Pocatello, ID	Gregory R. Anderson, N-P	2001, Nov.
Pomona, CA	Edward S. Cortez, N-P	2004, Nov.
Pompano Beach, FL	William F. Griffin, N-P	2002, Mar.
Pontiac, MI	Walter Moore, N-P	2001, Nov.
Port Arthur, TX	Oscar Ortiz, D	2002, May
Portland, ME	Cheryl A. Leeman, N-P	2001, Nov.
Portland, OR	Vera Katz, N-P	2003, Nov.
Portsmouth, VA	James W. Holley III, N-P	2004, May
Providence, RI	Vincent A. Cianci Jr., I	2002, Nov.
Provo, UT	Lewis K. Billings, N-P	2001, Nov.
Quincy, IL	Charles W. Scholz, D	2005, Apr.
Quincy, MA	James A. Sheets, D	2001, Nov.
Racine, WI	James M. Smith, N-P	2003, Apr.
Raleigh, NC	Paul Y. Coble, N-P	2001, Nov.
Rancho Cucamonga, CA	William Alexander, N-P	2002, Nov.
Rapid City, SD	Jerry Munson, N-P	2003, May
Reading, PA	Joseph D. Eppihimer, D	2003, Nov.
Redding, CA	Bob Anderson, N-P	(2)
Redondo Beach, CA	Gregory C. Hill, N-P	2005, Mar.
Redwood City, CA	Ira Ruskin, N-P	2001, Nov.
Reno, NV	Jeff Griffin, N-P	2002, Nov.
Rialto, CA	Grace Vargas, D	2004, Nov.
Richardson, TX	Gary Slagel, N-P	2003, May
Richmond, CA	Rosemary M. Corbin, D	2001, Nov.
Richmond, VA	Timothy M. Kaine, N-P	2002, July
Riverside, CA	Ronald O. Loveridge, N-P	2001, Nov.
Roanoke, VA	Ralph K. Smith, R	2004, May
Rochester, MN	Charles J. Canfield, N-P	2002, Nov.
Rochester, NY	William A. Johnson Jr., D	2001, Nov.
Rochester Hills, MI	Patricia Somerville, N-P	2003, Nov.
Rock Hill, SC	Doug Echols, N-P	2001, Oct.
Rock Island, IL	Mark W. Schwiebert, N-P	2005, Apr.
Rockford, IL	Doug Scott, D	2005, Apr.
Rockville, MD	Rose G. Krasnow, N-P	2001, Nov.
Rome, NY	Joseph A. Griffo, R	2003, Nov.
Rosemead, CA	Jay Imperial, N-P	2002, Mar.
Roseville, MI	Gerald K. Alsip, N-P	2001, Nov.
Roswell, NM	Bill B. Owen, N-P	2002, Mar.
Royal Oak, MI	Dennis G. Cowan, N-P	2001, Nov.
Sacramento, CA	Heather Fargo, N-P	2004, Nov.
Saginaw, MI	Gary L. Loster, N-P	2001, Nov.
St. Charles, MO	Patricia M. York, N-P	2003, Apr.
St. Clair Shores, MI	Curtis L. Dumas, N-P	2003, Nov.
St. Cloud, MN	Larry Meyer, N-P	2001, Nov.
St. Joseph, MO	Larry R. Stobbs, N-P	2002, Apr.
St. Louis, MO	Francis Slay, D	2005, Mar.
St. Louis Park, MN	Jeff Jacobs, N-P	2003, Nov.
St. Paul, MN	Norm Coleman, N-P	2001, Nov.
St. Petersburg, FL	Rick Baker, N-P	2005, Mar.
Salem, OR	Michael Swaim, N-P	2002, Nov.
Salinas, CA	Anna M. Caballero, N-P	2002, Nov.
Salt Lake City, UT	Ross "Rocky" Anderson, D	2003, Nov.
San Angelo, TX	Rudy Izzard, N-P	2003, May
San Antonio, TX	Ed Garza, N-P	2003, May
San Bernardino, CA	Judith Valles, D	2001, Nov.
San Diego, CA	Dick Murphy, R	2004, Nov.
Sandy City, UT	Thomas M. Dolan, N-P	2001, Nov
San Francisco, CA	Willie L. Brown Jr., N-P	2003, Nov.
San Jose, CA	Ron Gonzales, D	2002, Nov.
San Leandro, CA	Shelia Young, N-P	2002, June
San Mateo, CA	Jan Epstein, N-P	(2)
San Rafael, CA	Albert J. Boro, N-P	2003, Nov.
Santa Ana, CA	Miguel Pulido, N-P	2002, Nov.
Santa Barbara, CA	Harriet Miller, N-P	2001, Nov.
Santa Clara, CA	Judy Nadler, N-P	2002, Nov.
Santa Clarita, CA	JoAnne Darcy, N-P	2002, Apr.

City	Name	Next Election
Santa Cruz, CA	Keith A. Sugar, N-P	(2)
Santa Fe, NM	Larry Delgado, N-P	2002, Mar.
Santa Maria, CA	Joe Centeno, N-P	2004, Nov.
Santa Monica, CA	Pam O'Connor, N-P	(2)
Santa Rosa, CA	Janet Condvon, N-P	(2)
Sarasota, FL	Mollie C. Cardamone, N-P	2003, Apr.
Savannah, GA	Floyd Adams Jr., N-P	2003, Nov.
Schaumburg, IL	Al Larson, N-P	2003, Apr.
Schenectady, NY	Albert P. Jurczynski, R	2003, Nov.
Scottsdale, AZ	Mary Manross, D	2004, May
Scranton, PA	James P. Connors, R	2001, Nov.
Seattle, WA	Paul Schell, D	2001, Nov.
Sheboygan, WI	James R. Schramm, N-P	2005, Apr.
Shreveport, LA	Keith Hightower, D	2002, Nov.
Simi Valley, CA	Bill Davis, N-P	2002, Nov.
Sioux City, IA	Martin Dougherty, N-P	2001, Nov.
Sioux Falls, SD	Gary Hanson, N-P	2002, Apr.
Skokie, IL	George Van Dusen, N-P	2005, Apr.
Somerville, MA	Dorothy A. KellyGay, N-P	2001, Nov.
South Bend, IN	Stephen J. Luecke, D	2003, Nov.
South Gate, CA	Raul Moriel, N-P	2003, Jan.
Southfield, MI	Donald F. Fracassi, R	2001, Nov.
Sparks, NV	Tony Armstrong, N-P	2003, June
Spartanburg, SC	James E. Talley, N-P	2001, Nov.
Spokane, WA	John Talbott, N-P	2001, Nov.
Springfield, IL	Karen Hasara, N-P	2003, May
Springfield, MA	Michael J. Albano, D	2001, Nov.
Springfield, MO	Thomas J. Carlson, N-P	2003, Apr.
Springfield, OH	Warren R. Copeland, N-P	2001, Nov.
Stamford, CT	Dannel P. Malloy, D	2001, Nov.
Sterling Hts., MI	Richard J. Notte, N-P	2001, Nov.
Stockton, CA	Gary Podesto, N-P	2004, Mar.
Stratford, CT	William O. Cabral, N-P	2001, Nov.
Suffolk, VA	E. Dana Dickens III, N-P	2002, July
Sunnyvale, CA	Patricia Vorreiter, N-P	(2)
Sunrise, FL	Steven B. Feren, N-P	2005, Mar.
Syracuse, NY	Roy A. Bernardi, R	2001, Nov.
Tacoma, WA	Brian Ebersole, N-P	2003, Nov.
Tallahassee, FL	Scott Maddox, N-P	2002, Feb.
Tampa, FL	Dick A. Greco, N-P	2002, Mar.
Taunton, MA	Thaddeus Strojny, N-P	2001, Nov.
Taylor, MI	Gregory E. Pitoniak, D	2001, Nov.
Tempe, AZ	Neil G. Giuliano, N-P	2004, May
Temple, TX	Keifer Marshall Jr., N-P	2002, May
Terre Haute, IN	Judy Anderson, D	2003, Nov.
Thornton, CO	Noel I. Busck, N-P	2003, Nov.
Thousand Oaks, CA	Dennis C. Gillette, N-P	(2)
Titusville, FL	Larry D. Bartley, N-P	2004, Nov.
Toledo, OH	Carty Finkbeiner, N-P	2001, Nov.
Topeka, KS	Harry Butch Feiker, N-P	2005, Apr.
Torrance, CA	Dee Hardison, N-P	2002, Mar.
Trenton, NJ	Douglas H. Palmer, N-P	2002, May
Troy, MI	Matt Pryor, N-P	2004, Apr.
Troy, NY	Mark Pattison, D	2001, Nov.
Tucson, AZ	Robert E. Walkup, R	2003, Nov.
Tulsa, OK	M. Susan Savage, D	2002, Mar.
Tuscaloosa, AL	Alvin DuPont, D	2005, Sept.
Tyler, TX	Kevin P. Eltife, N-P	2002, May
Union City, NJ	Raul "Rudy" Garcia, D	2002, May
Upland, CA	John Pomierski, N-P	2002, Nov.
Utica, NY	Edward A. Hanna, I	2003, Nov.
Vacaville, CA	David A. Fleming, N-P	2002, Nov.
Vallejo, CA	Anthony J. Intintoli Jr., N-P	2003, Nov.
Vancouver, WA	Royce E. Pollard, N-P	2001, Nov.
Vineland, NJ	Perry Barse, R	2004, June
Virginia Beach, VA	Meyera E. Oberndorf, I	2004, May
Visalia, CA	Don Landers, N-P	2001, Nov.
Vista, CA	Gloria E. McClellan, R	2002, Nov.
Waco, TX	Linda Ethridge, N-P	2002, May
Walnut Creek, CA	Kathy Hicks, N-P	2002, Nov.
Waltham, MA	David Gately, D	2003, Nov.
Warren, MI	Mark A. Steenbergh, N-P	2003, Nov.
Warren, OH	Henry Angelo, D	2003, Nov.
Warwick, RI	Scott Avedisian, R	2002, Nov.
Washington, DC	Anthony A. Williams, D	2002, Nov.
Waterbury, CT	Philip A. Giordano, R	2001, Nov.
Waterloo, IA	John R. Rooff III, R	2001, Nov.
Waukegan, IL	Daniel T. Drew, D	2005, Apr.
Waukesha, WI	Carol Lombardi, N-P	2002, Apr.
Wauwatosa, WI	Theresa M. Estness, N-P	2004, Apr.
W. Allis, WI	Jeannette Bell, N-P	2004, Mar.
W. Covina, CA	David Truax, N-P	2003, Mar.
W. Hartford, CT	Robert R. Bouvier, R	2001, Nov.
W. Haven, CT	H. Richard Borer Jr., D	2001, Nov.
W. Palm Beach, FL	Joel T. Daves, N-P	2003, Mar.
Westland, MI	Robert J. Thomas, D	2001, Nov.
Westminster, CA	Margis L. Rice, N-P	2002, Nov.
Westminster, CO	Nancy Heil, N-P	2003, Nov.
Wheaton, IL	C. James Carr, N-P	2003, Apr.
White Plains, NY	Joseph Delfino, R	2001, Nov.
Whittier, CA	Greg Nordbak, N-P	2003, Apr.
Wichita, KS	Bob Knight, N-P	2003, Apr.
Wichita Falls, TX	Jerry Lueck, N-P	2002, May
Wilkes-Barre, PA	Thomas McGroarty, D	2003, Nov.
Wilmington, DE	James M. Baker, D	2004, Nov.
Wilmington, NC	David L. Jones, N-P	2001, Nov.
Winston-Salem, NC	Jack Cavanagh Jr., N-P	2001, Nov.
Woodbridge, NJ	James E. McGreevey, D	2003, Nov.
Woonsocket, RI	Susan D. Menard, N-P	2001, Nov.
Worcester, MA	Raymond V. Mariano, N-P	2001, Nov.
Wyandotte, MI	Leonard T. Sabuda, N-P	2005, Apr.
Wyoming, MI	Douglas L. Hoekstra Jr., N-P	2001, Nov.
Yakima, WA	Mary Place, N-P	2002, Jan.
Yonkers, NY	John Spencer, R	2003, Nov.
York, PA	Charles Robertson, D	2001, Nov.
Youngstown, OH	George M. McKelvey, D	2001, Nov.
Yuma, AZ	Marilyn R. Young, N-P	2001, Nov.

(1) Mayoralty rotated among city council members every 9 mos. (2) Mayor elected by town council. (3) City manager; hired, not elected.

Races for Governor, 2000

Source: Voter News Service; Puerto Rico State Elections Commission; official results

State	Democrat	Vote	Republican	Vote	Other	Vote
DE	**Ruth Ann Minner**	**191,695**	John M. Burris	128,603		
IN	**Frank O'Bannon***	**1,232,525**	David McIntosh	908,285		
MO	**Bob Holden**	**1,152,752**	Jim Talent	1,131,307		
MT	Mark O'Keefe	193,131	**Judy Martz**	**209,135**		
NH	**Jeanne Shaheen***	**275,038**	Gordon Humphrey	246,952	Mary Brown (I)	35,904
NC	**Mike Easley**	**1,530,324**	Richard Vinroot	1,360,960		
ND	Heidi Heitkamp	130,144	**John Hoeven**	**159,255**		
UT	Bill Orton	321,979	**Michael O. Leavitt***	**424,837**		
VT	**Howard Dean***	**148,059**	Ruth Dwyer	111,359	Anthony Pollina (PG)	28,116
WA	**Gary Locke***	**1,441,973**	John Carlson	980,060		
WV	**Bob Wise**	**324,822**	Cecil H. Underwood*	305,926		
	New Progressive Party	Vote	Popular Democratic Party	Vote	Puerto Rican Independence Party	Vote
Puerto Rico	Carlos Pesquera	919,194	**Sila Calderón**	**978,860**	Ruben Berrios	104,705

*= Incumbent. **Boldface** denotes winner. (I) = Independent. (PG) = Progressive.

Governors of States and Puerto Rico

As of Oct. 2001.

State	Capital, ZIP Code	Governor	Party	Term years	Term expires	Annual salary[1]
Alabama	Montgomery 36130	Don Siegelman	Dem.	4	Jan. 2003	$94,655
Alaska	Juneau 99811	Tony Knowles	Dem.	4	Dec. 2002	83,280
Arizona	Phoenix 85007	Jane Dee Hull	Rep.	4	Jan. 2003	95,000
Arkansas	Little Rock 72201	Mike Huckabee	Rep.	4	Jan. 2003	71,738
California	Sacramento 95814	Gray Davis	Dem.	4	Jan. 2003	175,000
Colorado	Denver 80203	Bill Owens	Rep.	4	Jan. 2003	90,000
Connecticut	Hartford 06106	John G. Rowland	Rep.	4	Jan. 2003	78,000
Delaware	Dover 19901	Ruth Ann Minner	Dem.	4	Jan. 2005	107,000
Florida	Tallahassee 32399	Jeb Bush	Rep.	4	Jan. 2003	123,175
Georgia	Atlanta 30334	Roy E. Barnes	Dem.	4	Jan. 2003	122,998
Hawaii	Honolulu 96813	Ben Cayetano	Dem.	4	Dec. 2002	94,780
Idaho	Boise 83720	Dirk Kempthorne	Rep.	4	Jan. 2003	95,500
Illinois	Springfield 62706	George H. Ryan	Rep.	4	Jan. 2003	150,691
Indiana	Indianapolis 46204	Frank O'Bannon	Dem.	4	Jan. 2005	77,200
Iowa	Des Moines 50319	Tom Vilsack	Dem.	4	Jan. 2003	107,482
Kansas	Topeka 66612	Bill Graves	Rep.	4	Jan. 2003	94,446
Kentucky	Frankfort 40601	Paul Patton	Dem.	4	Dec. 2003	103,018
Louisiana	Baton Rouge 70804	M. J. "Mike" Foster Jr.	Rep.	4	Jan. 2004	95,000
Maine	Augusta 04333	Angus S. King Jr.	Ind.	4	Jan. 2003	70,000
Maryland	Annapolis 21401	Parris N. Glendening	Dem.	4	Jan. 2003	120,000
Massachusetts	Boston 02133	Jane Swift	Rep.	4	Jan. 2003	135,000
Michigan	Lansing 48909	John Engler	Rep.	4	Jan. 2003	177,000
Minnesota	St. Paul 55155	Jesse Ventura	IPM[2]	4	Jan. 2003	120,303
Mississippi	Jackson 39205	Ronnie Musgrove	Dem.	4	Jan. 2004	101,800
Missouri	Jefferson City 65102	Bob Holden	Dem.	4	Jan. 2005	119,982
Montana	Helena 59620	Judy Martz	Rep.	4	Jan. 2005	83,672
Nebraska	Lincoln 68509	Mike Johanns	Rep.	4	Jan. 2003	65,000
Nevada	Carson City 89710	Kenny C. Guinn	Rep.	4	Jan. 2003	117,000
New Hampshire	Concord 03301	Jeanne Shaheen	Dem.	2	Jan. 2003	96,060
New Jersey	Trenton 08625	Donald T. DiFrancesco, act.	Rep.	4	Jan. 2002	83,333
New Mexico	Santa Fe 87503	Gary E. Johnson	Rep.	4	Jan. 2003	90,000
New York	Albany 12224	George E. Pataki	Rep.	4	Jan. 2003	179,000
North Carolina	Raleigh 27603	Mike Easley	Dem.	4	Jan. 2005	118,430
North Dakota	Bismarck 58505	John Hoeven	Rep.	4	Jan. 2005	85,506
Ohio	Columbus 43266	Bob Taft	Rep.	4	Jan. 2003	126,496
Oklahoma	Oklahoma City 73105	Frank Keating	Rep.	4	Jan. 2003	101,140
Oregon	Salem 97310	John Kitzhaber	Dem.	4	Jan. 2003	88,300
Pennsylvania	Harrisburg 17120	Mark Schweiker	Rep.	4	Jan. 2003	105,035
Rhode Island	Providence 02903	Lincoln C. Almond	Rep.	4	Jan. 2003	95,000
South Carolina	Columbia 29211	Jim Hodges	Dem.	4	Jan. 2003	106,078
South Dakota	Pierre 57501	William J. Janklow	Rep.	4	Jan. 2003	95,389
Tennessee	Nashville 37243	Don Sundquist	Rep.	4	Jan. 2003	85,000
Texas	Austin 78711	Rick Perry	Rep.	4	Jan. 2003	115,345
Utah	Salt Lake City 84114	Michael O. Leavitt	Rep.	4	Jan. 2005	100,600
Vermont	Montpelier 05609	Howard Dean	Dem.	2	Jan. 2003	115,763
Virginia	Richmond 23219	James S. Gilmore III	Rep.	4	Jan. 2002	124,855
Washington	Olympia 98504	Gary Locke	Dem.	4	Jan. 2005	135,960
West Virginia	Charleston 25305	Bob Wise	Dem.	4	Jan. 2005	95,000
Wisconsin	Madison 53707	Scott McCallum	Rep.	4	Jan. 2003	122,406
Wyoming	Cheyenne 82002	Jim Geringer	Rep.	4	Jan. 2003	95,000
Puerto Rico	San Juan 00936	Sila Calderón	PDP[3]	4	Jan. 2005	70,000

(1) Salary in effect in 2001. (2) Independence Party of Minnesota. (3) Popular Democratic Party.

State Officials, Salaries, Party Membership

As of Oct. 2001; I=independent

Alabama

Governor — Don Siegelman, D, $94,655
Lt. Gov. — Steve Windom, R, $12 per day, plus $50 per day expenses, plus $3,780 per mo expenses
Sec. of State — Jim Bennett, R, $66,722
Atty. Gen. — William Pryor, R, $124,951
Treasurer — Lucy Baxley, D, $66,722
Legislature: meets annually at Montgomery the 1st Tues. in Mar., 1st year of term of office; 1st Tues. in Feb., 2d and 3d yr; 2d Tues. in Jan., 4th yr. Members receive $10 per day salary, plus $50 per day expenses, plus $2,280 per mo expenses.
Senate — Dem., 24; Rep., 11. Total, 35
House — Dem., 68; Rep., 37. Total, 105

Alaska

Governor — Tony Knowles, D, $83,280
Lt. Gov — Fran Ulmer, D, $77,712
Atty. General — Bruce Botelho, D, $88,548
Legislature: meets annually in Jan. at Juneau for 120 days with a 10-day extension possible upon 2/3 vote. First session in odd years. Members receive $24,012 annually, plus $173 per diem.
Senate — Dem., 6; Rep., 14. Total, 20
House — Dem., 10; Rep., 30. Total, 40

Arizona

Governor — Jane Dee Hull, R, $95,000
Sec. of State — Betsey Bayless, R, $70,000
Atty. Gen. — Janet Napolitano, D, $90,000
Treasurer — Carol Springer, R, $70,000
Legislature: meets annually in Jan. at Phoenix. Each member receives an annual salary of $24,000.
Senate — Dem., 15; Rep., 15. Total, 30
House — Dem., 24; Rep., 36. Total, 60

Arkansas

Governor — Mike Huckabee, R, $71,738
Lt. Gov. — Winthrop P. Rockefeller, R, $34,673
Sec. of State — Sharon Priest, D, $44,836
Atty. Gen. — Mark Pryor, D, $59,781
Treasurer — Jimmie Lou Fisher, D, $44,836
Auditor — Gus Wingfield, D, $44,836
General Assembly: meets odd years in Jan. at Little Rock. Members receive $13,100 annually.
Senate — Dem., 27; Rep., 8. Total, 35
House — Dem., 70; Rep., 30. Total, 100

California

Governor — Gray Davis, D, $175,000
Lt. Gov. — Cruz Bustamante, D, $131,250
Sec. of State — Bill Jones, R, $131,250
Controller — Kathleen Connell, D, $140,000
Treasurer — Phil Angelides, D, $140,000
Atty. Gen. — Bill Lockyer, D, $148,750
Legislature: meets at Sacramento on the 1st Mon. in Dec. of even-numbered years; each session lasts 2 years. Members receive $99,000 annually, plus $121 per diem.
Senate — Dem., 26; Rep., 14. Total, 40
Assembly — Dem., 50; Rep., 30. Total, 80

Colorado

Governor — Bill Owens, R, $90,000
Lt. Gov. — Joe Rogers, R, $68,500
Sec. of State — Donetta Davidson, R, $68,500
Atty. Gen. — Ken Salazar, D, $80,000
Treasurer — Mike Coffman, R, $68,500
General Assembly: meets annually in Jan. at Denver. Members receive $30,000 annually plus $99 per diem for attendance at interim committee meetings.
Senate — Dem., 18; Rep., 17. Total, 35
House — Dem., 30; Rep., 35. Total, 65

Connecticut

Governor — John G. Rowland, R, $78,000
Lt. Gov. — M. Jodi Rell, R, $71,500
Sec. of State — Susan Bysiewicz, D, $65,000
Treasurer — Denise Nappier, D, $70,000
Comptroller — Nancy S. Wyman, D, $65,000
Atty. Gen. — Richard Blumenthal, D, $75,000
General Assembly: meets annually odd years in Jan. and even years in Feb., at Hartford. Members receive $28,000 annually, plus $5,500 (senator), $4,500 (representative) per year for expenses.
Senate — Dem., 21; Rep., 15. Total, 36
House — Dem., 100; Rep., 51. Total, 151

Delaware

Governor — Ruth Ann Minner, D, $107,000
Lt. Gov. — John C. Carney Jr., D, $49,400
Sec. of State — Harriet Smith Windsor, D, $98,400
Atty. Gen. — M. Jane Brady, R, $108,400
Treasurer — Jack Markell, D, $87,300
General Assembly: meets annually the 2d Tues. in Jan. and continues until June 30, at Dover. Members receive $32,700 annually.
Senate — Dem., 13; Rep., 8. Total, 21
House — Dem., 15; Rep., 26. Total, 41

Florida

Governor — Jeb Bush, R, $123,175
Lt. Gov. —Frank T. Brogan, R, $117,990
Sec. of State — Katherine Harris, R, $121,931
Comptroller — Robert R. Milligan, R, $121,931
Atty. Gen. — Robert Butterworth, D, $121,931
Treasurer — Tom Gallagher, R, $121,931
Legislature: meets annually at Tallahassee. Members receive $27,900 annually, plus expense allowance.
Senate — Dem., 15; Rep., 25. Total, 40
House — Dem., 43; Rep., 77. Total, 120

Georgia

Governor — Roy E. Barnes, D, $122,998
Lt. Gov. — Mark Taylor, D, $77,996
Sec. of State — Cathy Cox, D, $93,120
Atty. Gen. — Thurbert Baker, D, $121,632
General Assembly: meets annually at Atlanta on 2d Mon. in Jan. Members receive $16,200 annually ($128 per diem and $7,000 annual expense reimbursement).
Senate — Dem., 32; Rep., 24. Total, 56
House — Dem., 105; Rep., 75. Total, 180

Hawaii

Governor — Ben Cayetano, D, $94,780
Lt. Gov. — Mazie K. Hirono, D, $90,041
Atty. Gen. — Earl I. Anzai, $85,302
Comptroller — Wayne Kimura, $85,302
Dir. of Budget & Finance — Neal H. Miyahira, $85,302
Legislature: meets annually on 3d Wed. in Jan. at Honolulu. Members receive $32,000 annually; presiding officers receive $37,000.
Senate — Dem., 22; Rep., 3. Total, 25
House — Dem., 32; Rep., 19. Total, 51

Idaho

Governor — Dirk Kempthorne, R, $95,500
Lt. Gov. — Jack Riggs, R, $25,250
Sec. of State — Pete T. Cenarrusa, R, $77,500
Treasurer — Ron Crane, R, $77,500
Atty. Gen. — Alan Lance, R, $85,500
Legislature: meets annually the Mon. on or nearest Jan. 9 at Boise. Members receive $14,760 annually, plus $75 per day during session if required to maintain a 2d residence, $40 if no 2d residence; plus $50 per day when engaged in legislative business when legislature is not in session.
Senate — Dem., 4; Rep., 31. Total, 35
House — Dem., 12; Rep., 58. Total, 70

Illinois

Governor — George H. Ryan, R, $150,691
Lt. Gov. — Corinne Wood, R, $115,235
Sec. of State — Jesse White, D, $132,963
Comptroller — Daniel Hynes, D, $115,235
Atty. Gen. — James Ryan, R, $132,963
Treasurer — Judy Baar Topinka, R, $115,235
General Assembly: meets annually in Nov. and Jan. at Springfield. Members receive $57,619 annually.
Senate — Dem., 27; Rep., 32. Total, 59
House — Dem., 62; Rep., 56. Total, 118

Indiana

Governor — Frank O'Bannon, D, $77,200
Lt. Gov. — Joseph E. Kernan, D, $64,000
Sec. of State — Sue Anne Gilroy, R, $66,000
Atty. Gen. — Steve Carter, R, $79,400
Treasurer — Tim Berry, R, $66,000
Auditor — Connie Kay Nass, R, $66,000
General Assembly: meets annually on the Tues. after the 2d Mon. in Jan. at Indianapolis. Members receive $11,600 annually, plus $112 per day while in session, $25 per day while not in session.
Senate — Dem., 19; Rep., 31. Total, 50
House — Dem., 53; Rep., 47. Total, 100

Iowa

Governor — Tom Vilsack, D, $107,482
Lt. Gov. — Sally Pederson, D, $76,698
Sec. of State — Chester J. Culver D, $87,990
Atty. Gen. — Tom Miller, D, $105,430
Treasurer — Michael L. Fitzgerald, D, $87,990
Auditor — Richard D. Johnson, R, $87,990
Sec. of Agriculture — Patty Judge, D, $87,990
General Assembly: meets annually in Jan. at Des Moines. Members receive $21,385 annually, plus expense allowance.
Senate — Dem., 20; Rep., 30. Total, 50
House — Dem., 44; Rep., 56 Total, 100

Kansas

Governor — Bill Graves, R, $94,446
Lt. Gov. — Gary Sherrer, R, $108,246
Sec. of State — Ron Thornburgh, R, $74,147
Atty. Gen. — Carla Stovall, R, $85,267
Treasurer — Tim Shallenburger, R, $74,147
Insurance Commissioner — Kathleen Sebelius, D, $74,147
Legislature: meets annually on the 2d Mon. of Jan. at Topeka. Members receive $78.75 per day salary, plus $85 per day expenses while in session, $5,400 total allowance while not in session.
Senate — Dem., 13; Rep., 27. Total, 40
House — Dem., 48; Rep., 77. Total, 125

Kentucky

Governor — Paul Patton, D, $103,018
Lt. Gov. — Steve Henry, D, $87,580
Sec. of State — John Y. Brown III, D, $87,580
Atty. Gen. — A. B. Chandler III, D, $87,580
Treasurer — Jonathan Miller, D, $87,580
Auditor — Ed Hatchett, D, $87,580
Sec. of Economic Dev. — Gene Strong, $162,750
General Assembly: meets even years in Jan. at Frankfort. Members receive $154 per day, plus $94 per day expenses during session and $1,459 per month for expenses for interim.
Senate — Dem., 17; Rep., 20, 1 vacancy. Total, 38
House — Dem., 66; Rep., 34. Total, 100

Louisiana

Governor — M. J. "Mike" Foster Jr., R, $95,000
Lt. Gov. — Kathleen Babineaux Blanco, D, $85,000
Sec. of State — W. Fox McKeithen, R, $85,000
Atty. Gen. — Richard Ieyoub, D, $85,000
Treasurer — John Kennedy, D, $85,000
Legislature: meets in odd-numbered years at Baton Rouge starting last Mon. in Mar., for 60 legislative days of 85 calendar days; meets in even-numbered years on last Mon. in Apr. for 30 days of 45 calendar days. Members receive $16,800 annually, plus $97 per day expenses while in session and $500 per month as an unvouchered expense allowance.
Senate — Dem., 27; Rep., 12. Total, 39
House — Dem., 75; Rep., 30. Total, 105

Maine

Governor — Angus S. King Jr., I, $70,000
Sec. of State — Dan A. Gwadosky, D, $65,874
Atty. Gen. — G. Steven Rowe, D, $92,082
Treasurer — Dale McCormick, D, $65,874
State Auditor — Gail M. Chase, D, $77,438
Legislature: meets in odd-numbered years at Augusta on first Wed. in Dec.; meets in even-numbered years on Wed. after first Tues. in Jan. Members receive $10,815 for first regular session, $7,725 for 2d, plus a daily expense allowance.
Senate — Dem., 17; Rep., 17; 1 ind. Total, 35
House — Dem., 89; Rep., 61; 1 ind. Total, 151

Maryland

Governor — Parris N. Glendening, D, $120,000
Lt. Gov. — Kathleen Kennedy Townsend, D, $100,000
Comptroller — William Donald Schaefer, D, $100,000
Atty. Gen. — J. Joseph Curran Jr., D, $100,000
Sec. of State — John Willis, D, $70,000
Treasurer — Richard N. Dixon, D, $100,000

General Assembly: meets 90 consecutive days annually beginning on 2d Wed. in Jan. at Annapolis. Members receive $31,509 annually, plus expenses.
Senate — Dem., 33; Rep., 14. Total, 47
House — Dem., 106; Rep., 35. Total, 141

Massachusetts

Governor — Jane Swift, R, $135,000
Lt. Gov. — vacant
Sec. of the Commonwealth — William F. Galvin, D, $120,000
Atty. Gen. — Thomas F. Reilly, D, $122,500
Treasurer and Receiver General — Shannon P. O'Brien, D, $120,000
State Auditor — A. Joseph DeNucci, D, $120,000
General Court (legislature): meets Jan. annually in Boston. Members receive $49,710 annually.
Senate — Dem., 34; Rep., 6. Total, 40
House — Dem., 136; Rep., 24. Total, 160

Michigan

Governor — John Engler, R, $177,000
Lt. Gov. — Dick Posthumus, R, $123,900
Sec. of State — Candice S. Miller, R, $124,900
Atty. Gen. — Jennifer M. Granholm, D, $124,900
Treasurer — Doug Roberts (appointed), $153,000
Legislature: meets annually in Jan. at Lansing. Members receive $79,650 annually.
Senate — Dem., 15; Rep., 23. Total, 38
House — Dem., 52; Rep., 58. Total, 110

Minnesota

(IPM=Independence Party of Minnesota; DFL=Democratic-Farmer-Labor Party)
Governor — Jesse Ventura, IPM, $120,303
Lt. Gov. — Mae Schunk, IPM, $66,168
Sec. of State — Mary Kiffmeyer, R, $66,168
Atty. Gen. — Michael Hatch, DFL, $93,983
Treasurer — Carol C. Johnson, DFL, $66,168
Auditor — Judith H. Dutcher, DFL, $72,187
Legislature: meets for a total of 120 days within every 2 years, at St. Paul. Members receive $31,140 annually, plus expense allowance during session.
Senate — DFL, 39; R, 27; 1 ind. Total, 67
House — DFL, 65; R, 69. Total, 134

Mississippi

Governor — Ronnie Musgrove, D, $101,800
Lt. Gov. — Amy Tuck, D, $60,000
Sec. of State — Eric Clark, D, $75,000
Atty. Gen. — Mike Moore, D, $90,800
Treasurer — Marshall Bennett, D, $75,000
Auditor — Phil Bryant, R, $75,000
Legislature: meets annually in Jan. at Jackson. Members receive $10,000 per regular session, plus travel allowance, and $1,500 per month when not in session.
Senate — Dem., 34; Rep., 18. Total, 52
House — Dem., 86; Rep., 33; 3 ind. Total, 122

Missouri

Governor — Bob Holden, D, $119,982
Lt. Gov. — Joe Maxwell, D, $77,079
Sec. of State — Matt Blunt, R, $96,350
Atty. Gen. — Jeremiah W. Nixon, D, $96,350
Treasurer — Nancy Farmer, D, $96,350
State Auditor — Claire McCaskill, D, $96,350
General Assembly: meets annually at Jefferson City beginning 1st Wed. after 1st Mon. in Jan. Members receive $31,246 annually.
Senate — Dem., 18; Rep., 16. Total, 34
House — Dem., 88; Rep., 75. Total, 163

Montana

Governor — Judy Martz, R, $83,672
Lt. Gov. — Karl Ohs, R, $58,961
Sec. of State — Bob Brown, R, $63,571
Atty. Gen. — Mike McGrath, D, $71,638
Legislative Assembly: meets odd years in Jan. at Helena. Members receive $71.83 per legislative day, plus $87.25 per day for expenses while in session.
Senate — Dem., 19; Rep., 31. Total, 50
House — Dem., 42; Rep., 58. Total, 100

Nebraska

Governor — Mike Johanns, R, $65,000
Lt. Gov. — Dave Maurstad, R, $47,000
Sec. of State — John A. Gale, R, $52,000
Atty. Gen. — Don Stenberg, R, $64,500
Treasurer — David Heineman, R, $49,500
State Auditor — Kate Witek, R, $49,500
Legislature: Unicameral body composed of 49 members who are elected on a nonpartisan ballot and are called senators; meets annually in Jan. at Lincoln. Members receive $12,000 annually, plus expenses.

Nevada

Governor — Kenny C. Guinn, R, $117,000
Lt. Gov. — Lorraine Hunt, R, $50,000
Sec. of State — Dean Heller, R, $80,000
Controller — Kathy Augustine, R, $80,000
Atty. Gen. — Frankie Sue Del Papa, D, $110,000
Treasurer — Brian Krolicki, R, $80,000
Legislature: meets at Carson City odd years starting on 1st Mon. in Feb. for 120 days. Members receive $130 per day salary, plus $85 per day expenses, while in session.
Senate — Dem., 9; Rep., 12. Total, 21
Assembly — Dem., 27; Rep., 15. Total, 42

New Hampshire

Governor — Jeanne Shaheen, D, $96,060
Sec. of State — William M. Gardner, D, $76,603
Atty. Gen. — Philip T. McLaughlin, D, $85,753
Treasurer — Georgie A. Thomas, R, $76,603
General Court (Legislature): meets every year in Jan. at Concord. Members receive $200, presiding officers $250, biannually.
Senate — Dem., 11; Rep., 13. Total, 24
House — Rep., 256; Dem., 138; 1 ind.; 5 vacancies. Total, 400

New Jersey

Governor (acting) — Donald T. DiFrancesco, R, $83,333
Sec. of State — DeForest B. Soaries, R, $133,000
Atty. Gen. — John J. Farmer Jr, R, $133,000
Treasurer — Peter R. Lawrance, act., N-P, $133,000
Legislature: meets throughout the year at Trenton. Members receive $35,000 annually, except president of Senate and speaker of Assembly, who receive 1/3 more.
Senate — Dem., 16; Rep., 24. Total, 40
Assembly — Dem., 35; Rep., 45. Total, 80

New Mexico

Governor — Gary E. Johnson, R, $90,000
Lt. Gov. — Walter Bradley, R, $65,000
Sec. of State — Rebecca Vigil-Giron, D, $65,000
Atty. Gen. — Patricia Madrid, D, $72,500
Treasurer — Michael A. Montoya, D, $65,000
Legislature: meets starting on the 3d Tues. in Jan. at Santa Fe; odd years for 60 days, even years for 30 days. Members receive $136 per day while in session.
Senate — Dem., 25; Rep., 17. Total, 42
House — Dem., 40; Rep., 30. Total, 70

New York

Governor — George E. Pataki, R, $179,000
Lt. Gov. — Mary O. Donohue, R, $151,500
Sec. of State — Randy A. Daniels, D, $120,800
Comptroller — H. Carl McCall, D, $151,500
Atty. Gen. — Eliot Spitzer, D, $151,500
Legislature: meets annually on the 1st Wed. after the 1st Mon. in Jan. at Albany. Members receive $79,500 annually, plus $138 per day expenses.
Senate — Dem., 25; Rep., 36. Total, 61
Assembly — Dem., 99; Rep., 51. Total, 150

North Carolina

Governor — Mike Easley, D, $118,430
Lt. Gov. — Beverly Perdue, D, $104,523
Sec. of State — Elaine F. Marshall, D, $104,523
Atty. Gen. — Roy Cooper, D, $104,523
Treasurer — Richard H. Moore, D, $104,523
General Assembly: meets odd years starting on the 3d Wed. following the 2d Mon. in Jan. at Raleigh. Members receive $13,951 annually and an expense allowance of $559 per month, plus subsistence and travel allowance while in session. Also meets in even years for a short session (about 6-8 weeks), usually in May.
Senate — Dem., 35; Rep., 15. Total, 50
House — Dem., 62; Rep., 58. Total, 120

North Dakota

Governor — John Hoeven, R, $85,506
Lt. Gov. — Jack Dalrymple, R, $66,380
Sec. of State — Alvin A. Jaeger, R, $64,742
Atty. Gen. — Wayne Stenehjem, R, $73,204
Treasurer — Kathi Gilmore, D, $62,974
Legislative Assembly: meets odd years in Jan. at Bismarck. Members receive $250 per month salary, plus $125 per calendar day salary during session and $45 per day expenses plus any additional state or local taxes on lodging, with a limit of $650 per month.
Senate — Dem., 17; Rep., 32. Total, 49
House — Dem., 29; Rep., 69. Total, 98

Ohio

Governor — Bob Taft, R, $126,496
Lt. Gov. — Maureen O'Connor, R, $66,306
Sec. of State — J. Kenneth Blackwell, R, $93,446
Atty. Gen. — Betty D. Montgomery, R, $93,446
Treasurer — Joseph T. Deters, R, $93,466
Auditor — Jim Petro, R, $93,466
General Assembly: begins odd years at Columbus starting on 1st Mon. in Jan. Members receive $51,674 annually.
Senate — Dem., 12; Rep., 21. Total, 33
House — Dem., 40; Rep., 59. Total, 99

Oklahoma

Governor — Frank Keating, R, $101,140
Lt. Gov. — Mary Fallin, R, $75,530
Sec. of State — Mike Hunter, R, $65,000
Atty. Gen. — Drew Edmondson, D, $94,349
Treasurer — Robert Butkin, D, $82,004
Auditor — Clifton Scott, D, $82,004
Legislature: meets annually at noon the first Mon. in Feb. at Oklahoma City. In odd-numbered years, the session includes one day (1st Tuesday after 1st Monday) in Jan. Members receive $38,400 annually.
Senate — Dem., 30; Rep., 18. Total, 48
House — Dem., 52; Rep., 49. Total, 101

Oregon

Governor — John Kitzhaber, D, $88,300
Sec. of State — Bill Bradbury, D, $67,900
Atty. Gen. — Hardy Myers, D, $72,800
Treasurer — Randall Edwards, D, $67,900
Legislative Assembly: meets odd years in Jan. at Salem. Members receive $1,283 monthly, $85 expenses per day during session and when attending meetings during the interim, plus between $400 and $550 expense account during interim.
Senate — Dem., 14; Rep., 16. Total, 30
House — Dem., 28; Rep., 32. Total, 60

Pennsylvania

Governor — Mark Schweiker, R, $105,035
Lt. Gov. — Robert C. Jubelirer, R, $83,027
Sec. of the Commonwealth — Kim Pizzingrilli, R, $95,346
Atty. Gen. — Mike Fisher, R, $112,785
Treasurer — Barbara Hafer, R, $112,785
General Assembly: convenes annually on the 1st Tues. in Jan. at Harrisburg. Members receive $61,889 annually, plus expenses.
Senate — Dem., 20; Rep., 29; 1 vacancy. Total, 50
House — Dem., 99; Rep., 104. Total, 203

Rhode Island

Governor — Lincoln C. Almond, R, $95,000
Lt. Gov. — Charles J. Fogarty, D, $80,000
Sec. of State — Edward S. Inman III, D, $80,000
Atty. Gen. — Sheldon Whitehouse, D, $85,000
Treasurer — Paul J. Tavares, D, $80,000
General Assembly: meets annually in Jan. at Providence. Members receive $10,000 annually.
Senate — Dem., 44; Rep., 6. Total, 50
House — Dem., 85; Rep., 15. Total, 100

South Carolina

Governor — Jim Hodges, D, $106,078
Lt. Gov. — Robert L. Peeler, R, $46,545
Sec. of State — Jim Miles, R, $92,007
Comptroller Gen. — James A. Lander, D, $92,007
Atty. Gen. — Charles M. Condon, R, $92,007
Treasurer — Grady L. Patterson Jr., $92,007
General Assembly: meets annually on the 2d Tues. in Jan. at Columbia. Members receive $10,400 annually, plus $130 per day for expenses.
Senate — Dem., 21; Rep., 24; 1 vacancy. Total, 46
House — Dem., 54; Rep., 70. Total, 124

South Dakota

Governor — William J. Janklow, R, $95,389
Lt. Gov. — Carole Hillard, R, $69,243
Sec. of State — Joyce Hazeltine, R, $64,813
Treasurer — Dick Butler, D, $64,813
Atty. Gen. — Mark Barnett, R, $80,995
Auditor — Vernon Larson, R, $64,813
Legislature: meets annually beginning the 2d Tues. in Jan. at Pierre, for 40-day session in odd-numbered years, and 35-day session in even-numbered years. Members receive $12,000 per 2-year term plus $110 per legislative day or $75 for a statute committee.
Senate — Dem., 11; Rep., 24. Total, 35
House — Dem., 20; Rep., 50. Total, 70

Tennessee

Governor — Don Sundquist, R, $85,000
Lt. Gov. — John S. Wilder, D, $49,500
Sec. of State — Riley C. Darnell, D, $124,200
Comptroller — John Morgan, D, $124,200
Atty. Gen. — Paul Summers, D, $114,528
General Assembly: meets annually on the 2d Tues. in Jan. at Nashville. Members receive $16,500 annual salary, plus $114 per day expenses while in session.
Senate — Dem., 18; Rep., 15. Total, 33
House — Dem., 58; Rep., 41. Total, 99

Texas

Governor — Rick Perry, R, $115,345
Lt. Gov. — Bill Ratliff, R, $7,200
Sec. of State — Henry Cuellar, D, $112,352
Comptroller — Carole Keeton Rylander, R, $92,217
Atty. Gen. — John Cornyn, R, $92,217
Railroad Commissioners — Michael L. Williams, R, Chair; Tony Garza, R; Charles R. Matthews, R; $92,217
Legislature: meets odd years in Jan. at Austin. Members receive $7,200 annually, plus $95 per day expenses while in session.
Senate — Dem., 15; Rep., 16. Total, 31
House — Dem., 78; Rep., 72. Total, 150

Utah

Governor — Michael O. Leavitt, R, $100,600
Lt. Gov. — Olene S. Walker, R, $78,200
Atty. Gen. — Mark Shurtleff, R, $84,600
Auditor — Auston G. Johnson, R, $80,700
Treasurer — Edward T. Alter, R, $72,200
Legislature: convenes for 45 days on 3d Mon. in Jan. each year at Salt Lake City. Members receive $120 per day, plus $38 a day expenses.
Senate — Dem., 9; Rep., 20. Total, 29
House — Dem., 24; Rep., 51. Total, 75

Vermont

Governor — Howard Dean, D, $115,763
Lt. Gov. — Douglas A. Racine, D, $48,258
Sec. of State — Deborah L. Markowitz, D, $72,845
Atty. Gen. — William H. Sorrell, D, $87,507
Treasurer — James H. Douglas, R, $72,845
Auditor — Elizabeth M. Ready, D, $72,845
General Assembly: meets in Jan. at Montpelier (annual and biennial session). Members receive $536 per week while in session plus $105 per day for special session, plus expenses.
Senate — Dem., 16; Rep., 14. Total, 30
House — Dem., 62; Rep., 83; Prog. Coalition, 4; 1 ind. Total, 150

Virginia

Governor — James S. Gilmore III, R, $124,855
Lt. Gov. — John H. Hager, R, $36,321
Atty. Gen. — Randolph A. Beales, R, $110,667
Sec. of the Commonwealth — Anne P. Petera, R, $124,435
Treasurer — Mary G. Morris, R, $112,653
General Assembly: meets annually in Jan. at Richmond. Members receive $18,000 (senate), $17,640 (assembly) annually, plus expense and mileage allowances.
Senate — Dem., 19; Rep., 21. Total, 40
House — Dem., 50; Rep., 49; 1 ind. Total, 100

Washington

Governor — Gary Locke, D, $135,960
Lt. Gov. — Brad Owen, D, $71,070
Sec. of State — Sam Reed, R, $78,177
Atty. Gen. — Christine Gregoire, D, $123,600
Treasurer — Mike Murphy, D, $95,275
Legislature: meets annually in Jan. at Olympia. Members receive $32,064 annually, plus $82 per diem while in session, and $82 per diem for attending meetings during interim.
Senate — Dem., 25; Rep., 24. Total, 49
House — Dem., 49; Rep., 49. Total, 98

West Virginia

Governor — Bob Wise, D, $95,000
Sec. of State — Joe Manchin III, D, $70,000
Atty. Gen. — Darrell McGraw, D, $80,000
Treasurer — John D. Perdue, D, $75,000
Comm. of Agric. — Gus Douglass, D, $75,000
Auditor — Glen B. Gainer III, D, $75,000
Legislature: meets annually in Jan. at Charleston, except after gubernatorial elections, when the legislature meets in Feb. Members receive $15,000 annually.
Senate — Dem., 28; Rep., 6. Total, 34
House — Dem., 75; Rep., 25. Total, 100

Wisconsin

Governor — Scott McCallum, R, $122,406
Lt. Gov. — Margaret A. Farrow, R, $63,672
Sec. of State — Douglas La Follette, D, $54,610
Treasurer — Jack Voight, R, $54,610
Atty. Gen. — James E. Doyle, D, $112,274
Legislature: meets in Jan. at Madison. Members receive $44,233 annually, plus $75 per day expenses.
Senate — Dem., 18; Rep., 15. Total, 33
Assembly — Dem., 43; Rep., 55; 1 vacancy. Total, 99

Wyoming

Governor — Jim Geringer, R, $95,000
Sec. of State — Joseph B. Meyer, R, $77,500
Atty. Gen. — Horace MacMillan, R, $89,067
Treasurer — Cynthia Lummis, R, $77,500
State Auditor — Max Maxfield, R, $77,500
Legislature: meets odd years in Jan., even years in Feb., at Cheyenne. Members receive $125 per day while in session, plus $80 per day for expenses.
Senate — Dem., 10; Rep., 20. Total, 30
House — Dem., 17; Rep., 43. Total, 60

ECONOMICS

Consumer Price Index

The Consumer Price Index (CPI) is a measure of the average change in prices over time of one or more kinds of basic consumer goods and services.

From Jan. 1978, the Bureau of Labor Statistics began publishing CPIs for 2 population groups: (1) a CPI for all urban consumers (CPI-U), which covers about 87% of the total population; and (2) a CPI for urban wage earners and clerical workers (CPI-W), which covers about 32% of the total population. The CPI-U includes, in addition to wage earners and clerical workers, groups such as professional, managerial, and technical workers, the self-employed, short-term workers, the unemployed, retirees, and others not in the labor force.

The CPI is based on prices of food, clothing, shelter, and fuels; transportation fares; charges for doctors' and dentists' services; drug prices; and prices of other goods and services bought for day-to-day living. The index currently measures price changes from a designated reference period, 1982-84, which equals 100.0. Use of this reference period began in Jan. 1988.

U.S. Consumer Price Indexes, 2000-2001

Source: Bureau of Labor Statistics, U.S. Dept. of Labor

(Data are semiannual averages of monthly figures. For all urban consumers.)

(1982–84=100)	1st half 2000	% change 2d half 1999 to 1st half 2000	2d half 2000	% change 1st half 2000 to 2d half 2000	1st half 2001	% change 2d half 2000 to 1st half 2001
ALL ITEMS	**170.7**	**1.7**	**173.6**	**1.6**	**176.6**	**1.7**
Food, beverages	167.2	1.2	169.5	1.4	172.4	1.7
Housing	167.5	1.6	171.4	2.1	175.5	2.4
Apparel	130.4	–0.3	128.8	–1.2	129.0	0.2
Transportation	152.2	3.7	154.5	1.5	156.1	1.0
Medical care	258.2	2.2	263.3	2.0	270.1	2.6
Recreation	102.9	0.9	103.8	0.9	104.6	0.8
Other goods, services	268.5	2.8	273.6	1.9	278.9	1.9
Services	193.1	1.6	197.3	2.1	201.9	2.3
SPECIAL INDEXES						
All items less food	171.5	1.9	174.4	1.6	177.5	1.8
Commodities less food	138.7	2.5	139.7	0.7	140.5	0.6
Nondurables	157.1	2.8	159.2	1.3	161.3	1.3
Energy	120.5	8.2	128.8	6.9	134.6	4.5
All items less energy	177.5	1.3	179.7	1.2	182.4	1.5

U.S. Consumer Price Indexes (CPI-U),[1] Annual Percent Change, 1989-2000

Source: Bureau of Labor Statistics, U.S. Dept. of Labor

	1989	1990	1991	1992	1993	1994	1995	1996	1997	1998	1999	2000
ALL ITEMS	**4.8**	**5.4**	**4.2**	**3.0**	**3.0**	**2.6**	**2.8**	**3.0**	**2.3**	**1.6**	**2.2**	**3.4**
Food	5.8	5.8	2.9	1.2	2.2	2.4	2.8	3.3	2.6	2.2	2.1	2.3
Shelter	4.5	5.4	4.5	3.3	3.0	3.1	3.2	3.2	3.1	3.3	2.9	3.3
Rent, residential	3.9	5.6	6.1	2.5	2.3	2.5	2.5	2.7	2.9	3.2	3.1	3.6
Fuel and other utilities	3.3	3.5	3.3	2.2	3.0	1.0	0.7	3.1	2.6	–1.8	0.2	7.1
Apparel and upkeep	2.8	4.6	3.7	2.5	1.4	–0.2	–1.0	–0.2	0.9	0.1	–1.3	–1.3
Private transportation	4.9	5.2	2.6	2.2	2.3	3.1	3.7	2.7	0.7	–2.2	1.9	6.1
New cars	2.0	1.8	3.8	2.5	2.4	3.4	2.2	1.7	0.2	–0.6	–0.3	–0.1
Gasoline	9.5	14.1	–1.8	–0.2	–1.3	0.5	1.6	6.1	–0.1	–13.4	9.3	28.5
Public transportation	5.0	10.1	4.4	1.7	10.3	3.0	2.3	3.4	2.6	1.9	3.9	6.0
Medical care	7.7	9.0	8.7	7.4	5.9	4.8	4.5	3.5	2.8	3.2	3.5	4.1
Entertainment	5.2	4.7	4.5	2.8	2.5	2.9	2.5	3.4	2.1	1.5	0.9	1.3
Commodities	4.7	5.2	4.2	2.0	1.9	1.7	1.9	2.6	1.4	0.1	1.8	3.3

(1) The Consumer Price Index CPI-U measures the average change in prices of goods and services purchased by all urban consumers.

Consumer Price Index, 1915-2001

Source: Bureau of Labor Statistics, U.S. Dept. of Labor

(1967 = 100. Annual averages of monthly figures, specified for all urban consumers.)

Prices as measured by the U.S. Consumer Price Index have risen steadily since World War II. What cost $1.00 in 1967 (the reference year) cost about 30 cents in 1915, 54 cents in 1945, and $5.29 by 2001.

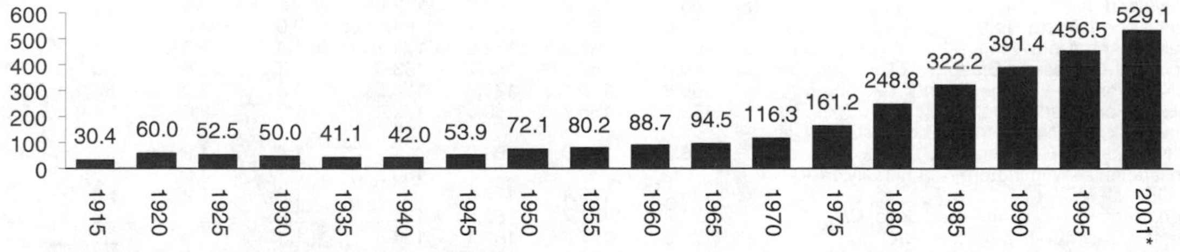

*Average for 1st half 2001.

U.S. Consumer Price Indexes for Selected Items and Groups, 1970-2000

Source: Bureau of Labor Statistics, U.S. Dept. of Labor

(1982-84 = 100, unless otherwise noted. Annual averages of monthly figures. For all urban consumers.)

	1970	1975	1980	1985	1990	1995	1998	1999	2000
ALL ITEMS	**38.8**	**53.8**	**82.4**	**107.6**	**130.7**	**152.4**	**163.0**	**166.6**	**172.2**
Food and beverages	**40.1**	**60.2**	**86.7**	**105.6**	**132.1**	**148.9**	**161.1**	**164.6**	**168.4**
Food	39.2	59.8	86.8	105.6	132.4	148.4	160.7	164.1	167.8
Food at home	39.9	61.8	88.4	104.3	132.3	148.8	161.1	164.2	167.9
Cereals and bakery products	37.1	62.9	83.9	107.9	140.0	167.5	181.1	185.0	188.3
Meats, poultry, fish, and eggs	44.6	67.0	92.0	100.1	130.0	138.8	147.3	147.9	154.5
Dairy products	44.7	62.6	90.9	103.2	126.5	132.8	150.8	159.6	160.7
Fruits and vegetables	37.8	56.9	82.1	106.4	149.0	177.7	198.2	203.1	204.6
Sugar and sweets	30.5	65.3	90.5	105.8	124.7	137.5	150.2	152.3	154.0
Fats and oils	39.2	73.5	89.3	106.9	126.3	137.3	146.9	148.3	147.4
Nonalcoholic beverages	27.1	41.3	91.4	104.3	113.5	131.7	133.0	134.3	137.8
Other foods	39.6	58.9	83.6	106.4	131.2	151.1	165.5	168.9	172.2
Food away from home	37.5	54.5	83.4	108.3	133.4	149.0	161.1	165.1	169.0
Alcoholic beverages	52.1	65.9	86.4	106.4	129.3	153.9	165.7	169.7	174.7
Housing	**36.4**	**50.7**	**81.1**	**107.7**	**128.5**	**148.5**	**160.4**	**163.9**	**169.6**
Shelter	35.5	48.8	81.0	109.8	140.0	165.7	182.1	187.3	193.4
Rent of primary residence[1]	46.5	58.0	80.9	111.8	138.4	157.8	172.1	177.5	183.9
Fuel and other utilities[1]	29.1	45.4	75.4	106.5	111.6	123.7	128.5	128.8	137.9
Gas (piped) and electricity	25.4	40.1	71.4	107.1	109.3	119.2	121.2	120.9	128.0
Household furnishings and operation	46.8	63.4	86.3	103.8	113.3	123.0	126.6	126.7	128.2
Apparel	**59.2**	**72.5**	**90.9**	**105.0**	**124.1**	**132.0**	**133.0**	**131.3**	**129.6**
Men's and boys'	62.2	75.5	89.4	105.0	120.4	126.2	131.8	131.1	129.7
Women's and girls'	71.8	85.5	96.0	104.9	122.6	126.9	126.0	123.3	121.5
Footwear	56.8	69.6	91.8	102.3	117.4	125.4	128.0	125.7	123.8
Transportation	**37.5**	**50.1**	**83.1**	**106.4**	**120.5**	**139.1**	**141.6**	**144.4**	**153.3**
Private	37.5	50.6	84.2	106.2	118.8	136.3	137.9	140.5	149.1
New vehicles	53.0	62.9	88.4	106.1	121.4	139.0	143.4	142.9	142.8
Used cars and trucks	31.2	43.8	62.3	113.7	117.6	156.5	150.6	152.0	155.8
Gasoline	27.9	45.1	97.5	98.6	101.0	99.8	91.6	100.1	128.6
Public	35.2	43.5	69.0	110.5	142.6	175.9	190.3	197.7	209.6
Medical care	**34.0**	**47.5**	**74.9**	**113.5**	**162.8**	**220.5**	**242.1**	**250.6**	**260.8**
Entertainment	**47.5**	**62.0**	**83.6**	**107.9**	**132.4**	**153.9**	**—[2]**	**—[2]**	**—[2]**
Other goods and services	**40.9**	**53.9**	**75.2**	**114.5**	**159.0**	**206.9**	**237.7**	**258.3**	**271.1**
Tobacco products	43.1	54.7	72.0	116.7	181.5	225.7	274.8	355.8	394.9
Personal care	43.5	57.9	81.9	106.3	130.4	147.1	156.7	161.1	165.6
Personal care products	42.7	58.0	79.6	107.6	128.2	143.1	148.3	151.8	153.7
Personal care services	44.2	57.7	83.7	108.9	132.8	151.5	166.0	171.4	178.1

(1) Dec. 1982 = 100. (2) The BLS stopped tracking this category after 1997, and began tracking a category classified as Recreation. The Recreation index for 2000 is 103.3; for 1999 is 102.0; for 1998, 99.6; for 1997, 97.4; for 1995, 94.5.

Consumer Price Indexes by Region and Selected Cities, 1999-2001[1]

Source: Bureau of Labor Statistics, U.S. Dept. of Labor

	Semiannual averages				Percent change from preceding semiannual average			
(1982-84 = 100)	2d half 1999	1st half 2000[R]	2d half 2000	1st half 2001	2d half 1999	1st half 2000[R]	2d half 2000	1st half 2001
U.S. CITY AVERAGE	**167.8**	**170.8**	**173.6**	**176.6**	**1.5**	**1.8**	**1.6**	**1.7**
Northeast urban	**174.8**	**178.0**	**180.7**	**183.8**	**1.5**	**1.8**	**1.5**	**1.7**
Size A—More than 1,500,000	175.8	178.8	181.6	184.7	1.5	1.7	1.6	1.7
Size B/C—50,000 to 1,500,000	104.9	107.1	108.5	110.4	1.5	2.1	1.3	1.8
Midwest urban	**164.0**	**167.0**	**169.6**	**172.8**	**1.5**	**1.8**	**1.6**	**1.9**
Size A—More than 1,500,000	165.3	168.5	171.1	174.3	1.5	1.9	1.5	1.9
Size B/C—50,000 to 1,500,000	104.9	106.8	108.3	110.4	1.6	1.8	1.4	1.9
Size D—Nonmetro. (less than 50,000)	158.4	161.1	164.2	166.9	1.5	1.7	1.9	1.6
South urban	**163.1**	**166.1**	**168.3**	**170.9**	**1.4**	**1.8**	**1.3**	**1.5**
Size A—More than 1,500,000	162.5	165.5	168.3	171.1	1.6	1.8	1.7	1.7
Size B/C—50,000 to 1,500,000	104.8	106.7	108.0	109.6	1.2	1.8	1.2	1.5
Size D—Nonmetro. (less than 50,000)	163.7	166.3	167.6	170.0	1.2	1.6	.8	1.4
West urban	**170.0**	**173.1**	**176.5**	**180.2**	**1.3**	**1.8**	**2.0**	**2.1**
Size A—More than 1,500,000	171.1	174.5	178.3	182.3	1.4	2.0	2.2	2.2
Size B/C—50,000 to 1,500,000	105.3	106.9	108.7	110.6	1.0	1.5	1.7	1.7
SELECTED AREAS								
Atlanta, GA	166.3	169.2	171.9	176.1	1.7	1.7	1.6	2.4
Boston–Brockton–Nashua, MA–NH–ME–CT	177.5	181.8	185.4	190.5	1.7	2.4	2.0	2.8
Chicago–Gary–Kenosha, IL–IN–WI	169.4	172.6	175.1	178.5	1.2	1.9	1.4	1.9
Cleveland–Akron, OH	163.7	166.3	169.6	172.6	1.5	1.6	2.0	1.8
Dallas–Fort Worth, TX	159.6	162.7	166.7	168.9	2.0	1.9	2.5	1.3
Detroit–Ann Arbor–Flint, MI	165.1	168.4	171.2	174.1	1.4	2.0	1.7	1.7
Houston–Galveston–Brazoria, TX	150.0	152.7	155.7	158.9	1.7	1.8	2.0	2.1
L.A.–Riverside–Orange County, CA	166.8	170.1	173.0	176.5	0.9	2.0	1.7	2.0
Miami–Fort Lauderdale, FL	163.4	166.7	169.0	172.4	1.2	2.0	1.4	2.0
New York, NY–Northern NJ–Long Island, NY–NJ–CT–PA	178.2	181.0	184.0	186.5	1.4	1.6	1.7	1.4
Philadelphia–Wilmington–Atlantic City, PA–DE–NJ–MD	173.4	175.4	177.6	180.5	1.8	1.2	1.3	1.6
San Francisco–Oakland–San Jose, CA	174.2	177.7	182.6	188.7	2.0	2.0	2.8	3.3
Seattle–Tacoma–Bremerton, WA	174.0	177.3	181.1	184.4	1.4	1.9	2.1	1.8
Washington–Baltimore, DC–MD–VA–WV	105.1	106.6	108.6	109.7	1.6	1.4	1.9	1.0

R = Revised. (1) For all urban consumers.

Percentage Change in Consumer Prices in Selected Countries

Source: International Monetary Fund

(annual averages)

COUNTRY	1975-1980	1980-1985	1991-1992	1992-1993	1993-1994	1994-1995	1995-1996	1996-1997	1997-1998	1998-1999	1999-2000
Canada	8.7	7.4	1.5	1.8	0.2	2.2	1.6	1.6	1.0	1.7	2.7
France	10.5	9.6	2.4	2.1	1.7	1.8	2.0	1.2	0.7	0.5	1.7
Germany	4.1	3.9	4.0	4.1	3.0	1.8	1.5	1.8	1.0	0.6	1.9
Italy	16.3	13.7	5.1	4.5	4.0	5.2	4.0	2.0	2.0	1.7	2.5
Japan	6.5	2.7	1.7	1.3	0.7	–0.1	0.1	1.7	0.6	–0.3	–0.6
Spain	18.6	12.2	5.9	4.6	4.7	4.7	3.6	2.0	1.8	2.3	3.4
Sweden	10.5	9.0	2.3	4.6	2.2	2.5	0.5	0.5	–0.1	0.5	1.0
Switzerland	2.3	4.3	4.1	3.3	0.8	1.8	0.8	0.5	0.1	0.7	1.6
United Kingdom	14.4	7.2	3.7	1.6	2.5	3.4	2.4	3.1	3.4	1.6	2.9
United States	8.9	5.5	3.0	3.0	2.6	2.8	3.0	2.3	1.6	2.2	3.4

Index of Leading Economic Indicators

Source: The Conference Board

The index of leading economic indicators is used to project the U.S. economy's performance. The index is made up of 10 measurements of economic activity that tend to change direction in advance of the overall economy. The index has predicted economic downturns from 8 to 20 months in advance and recoveries from 1 to 10 months in advance; however, it can be inconsistent, and has occasionally shown "false signals" of recessions.

Components

- Average weekly hours of production workers in manufacturing
- Average weekly initial claims for unemployment insurance, state programs
- Manufacturers' new orders for consumer goods and materials, adjusted for inflation
- Vendor performance (slower deliveries diffusion index)
- Manufacturers' new orders, nondefense capital goods industries, adjusted for inflation
- New private housing units authorized by local building permits
- Stock prices, 500 common stocks
- Money supply: M-2, adjusted for inflation
- Interest rate spread, 10-yr Treasury bonds less federal funds
- Consumer expectations (researched by Univ. of Michigan)

U.S. Gross Domestic Product, Gross National Product, Net National Product, National Income, and Personal Income

Source: Bureau of Economic Analysis, U.S. Dept. of Commerce

(billions of current dollars)

	1960	1970	1980	1990	1999R	2000
GROSS DOMESTIC PRODUCT	**—**	**—**	**—**	**$5,546.1**	**$9,268.6**	**$9,872.9**
GROSS NATIONAL PRODUCT	**$515.3**	**$1,015.5**	**$2,732.0**	**5,567.8**	**9,261.8**	**9,860.8**
Less: Consumption of fixed capital	46.4	88.8	303.8	602.7	1,151.4	1,241.3
Equals: Net national product	**468.9**	**926.6**	**2,428.1**	**4,965.1**	**8,110.4**	**8,619.5**
Less: Indirect business tax and nontax liability	45.3	94.0	213.3	444.0	713.1	762.7
Business transfer payments	2.0	4.1	12.1	26.8	41.3	43.9
Statistical discrepancy	–2.8	–1.1	4.9	7.8	–72.7	–130.4
Plus: Subsidies less current surplus of government enterprises	0.4	2.9	5.7	4.5	33.3	37.6
Equals: National income	**424.9**	**832.6**	**2,203.5**	**4,491.0**	**7,462.1**	**7,980.9**
Less: Corporate profits with inventory valuation and capital consumption adjustments	49.5	74.7	177.2	380.6	825.2	876.4
Net interest	11.3	41.2	200.9	463.7	506.5	532.7
Contributions for social insurance	21.9	62.2	216.5	503.1	660.7	701.5
Wage accruals less disbursements	0.0	0.0	0.0	0.1	5.2	.0
Plus: Personal interest income	27.5	81.8	312.6	666.3	950.0	1,000.6
Personal dividend income	24.9	69.3	271.9	698.2	343.1	379.2
Government transfer payments to persons	12.9	22.2	52.9	144.4	988.4	1,036.0
Business transfer payments	2.0	4.1	12.1	21.3	31.1	33.1
Equals: PERSONAL INCOME	**409.4**	**831.8**	**2,258.5**	**4,673.8**	**7,777.3**	**8,319.2**

Note: R = Revised.

U.S. Gross Domestic Product

Source: Bureau of Economic Analysis, U.S. Dept. of Commerce

(billions of current dollars)

	1990	2000[1]	Second Quarter 2001[1]
Gross domestic product	**$5,513.8**	**$9,872.9**	**$10,202.6**
Personal consumption expenditures	**3,742.6**	**6,728.4**	**7,044.6**
Durable goods	465.9	819.6	844.7
Nondurable goods	1,217.7	1,989.6	2,062.3
Services	2,059.0	3,919.2	4,137.6
Gross private domestic investment	**802.6**	**1,767.5**	**1,669.9**
Fixed investment	802.7	1,718.1	1,706.5
Nonresidential	587.0	1,293.1	1,260.2
Structures	198.7	313.6	338.6
Equipment and software	388.3	979.5	921.7
Residential	215.7	425.1	446.2
Change in private inventories	0	49.4	–36.6

	1990	2000[1]	Second Quarter 2001[1]
Net exports of goods and services	**$–74.4**	**$–364.0**	**$–347.4**
Exports	550.4	1,102.9	1,079.6
Goods	NA	785.6	754.4
Services	NA	317.3	325.2
Imports	624.8	1,466.9	1,427.0
Goods	NA	1,244.9	1,197.8
Services	NA	221.9	229.2
Government consumption expenditures and gross investment	**1,042.9**	**1,741.0**	**1,835.4**
Federal	424.9	590.2	609.9
National defense	313.4	375.4	396.1
Nondefense	111.5	214.8	213.8
State and local	618.0	1,150.8	1,225.5

(1) Seasonally adjusted at annual rates. NA = Not available.

Countries With Highest Gross Domestic Product and Per Capita GDP[1]

Source: Central Intelligence Agency, *The World Factbook 2001*

Gross Domestic Product

(billions of dollars; 1999 estimates)

1. United States	$9,255.0	21. Taiwan	$357.0
2. China[2]	4,800.0	22. Iran	347.6
3. Japan	2,955.0	23. South Africa	296.1
4. Germany	1,864.0	24. Philippines	282.0
5. India	1,805.0	24. Pakistan	282.0
6. France	1,373.0	26. Poland	276.5
7. United Kingdom	1,290.0	27. Colombia	245.1
8. Italy	1,212.0	28. Belgium	243.4
9. Brazil	1,057.0	29. Malaysia	229.1
10. Mexico	865.5	30. Egypt	200.0
11. Canada	722.3	31. Switzerland	197.0
12. Spain	677.5	32. Saudi Arabia	191.0
13. South Korea	625.7	33. Austria	190.6
14. Russia	620.3	34. Bangladesh	187.0
15. Indonesia	610.0	35. Chile	185.1
16. Australia	416.2	36. Sweden	184.0
17. Turkey	409.4	37. Venezuela	182.8
18. Thailand	388.7	38. Portugal	151.4
19. Argentina	367.0	39. Greece	149.2
20. Netherlands	365.1	40. Algeria	147.6

Per Capita Gross Domestic Product[3]

(dollars; 1999 estimates unless otherwise noted)

1. Luxembourg	$34,200	21. Finland	$21,000
2. United States	33,900	22. Sweden	20,700
3. Singapore	27,800	23. Ireland	20,300
4. Switzerland	27,100	24. Bahamas	20,000[4]
5. Monaco	27,000	25. San Marino	20,000[5]
6. Norway	25,100	26. Israel	18,300
7. Belgium	23,900	27. Andorra	18,000[6]
8. Denmark	23,800	28. U. Arab Emirates	17,700
9. Iceland	23,500	29. Brunei	17,400
10. Japan	23,400	New Zealand	17,400
Austria	23,400	31. Spain	17,300
12. France	23,300	32. Qatar	17,000
Canada	23,300	33. Taiwan	16,100
14. Netherlands	23,100	34. Cyprus	15,400[4,7]
15. Liechtenstein	23,000[4]	35. Portugal	15,300
16. Germany	22,700	36. New Caledonia[4]	15,000
17. Kuwait	22,500	Virgin Islands	15,000
18. Australia	22,200	38. Greece	13,900
19. United Kingdom	21,800	39. Bahrain	13,700
20. Italy	21,400	40. Malta	13,800

(1) U.S. data from *The World Factbook* may differ from data from the U.S. Bureau of Economic Analysis. International GDP estimates derive from purchasing power parity calculations, which involve the use of intl. dollar price weights applied to quantities of goods and services produced in a given economy. (2) Chinese government figures may substantially overstate the GDP. Hong Kong, a special administrative region of China since July 1, 1997, had a GDP of $158.2 billion and a per capita GDP of $23,100 in 1999. (3) These territories or former territories had large per capita GDPs: Bermuda (UK, 1999) $31,500, Cayman Islands (UK, 1997) $24,500, Aruba (Neth., 1998) $22,800, Gibraltar (UK, 1997) $17,500, Guam (U.S., 1996) $19,000, Greenland (Den., 1997) $16,100, Faroe Islands (Den., 1996) $16,000, Macau (Port., 1998) $16,000. (4) 1998 est. (5) 1997 est. (6) 1996 est. (7) Excludes Turkish-held area.

U.S. National Income by Industry[1]

Source: Bureau of Economic Analysis, U.S. Dept. of Commerce

(billions of current dollars)

	1960	1970	1980	1990	1998	1999	2000
National income without capital consumption adjustment	**$428.6**	**$835.1**	**$2,263.9**	**$4,513.6**	**$6,928.6**	**$7,416.5**	**$7,946.6**
Domestic industries	**425.1**	**827.8**	**2,216.3**	**4,492.0**	**6,949.3**	**7,423.3**	**7,958.7**
Private industries	**371.6**	**695.4**	**1,894.5**	**3,830.2**	**6,043.0**	**6,469.1**	**6,949.7**
Agriculture, forestry, fisheries	17.8	25.9	61.4	98.0	104.2	110.7	117.9
Mining	5.6	8.4	43.8	36.8	50.6	48.2	57.1
Construction	22.5	47.4	126.6	222.0	331.1	389.9	425.0
Manufacturing	125.3	215.6	532.1	859.5	1,168.7	1,189.8	1,237.5
Durable goods	73.4	127.7	313.7	483.1	684.2	697.7	723.2
Nondurable goods	52.0	87.9	218.4	376.3	484.4	492.1	514.3
Transportation, public utilities	35.8	64.4	177.3	326.3	500.8	515.4	555.4
Transportation	18.5	31.5	85.8	139.2	216.2	234.9	245.2
Communications	8.2	17.6	48.1	91.6	149.3	144.4	163.4
Electric, gas, sanitary services	9.1	86.8	43.4	95.5	135.3	136.1	146.7
Wholesale trade	25.0	47.5	143.3	261.7	409.2	444.8	479.7
Retail trade	41.3	79.9	189.4	392.3	580.0	621.8	663.5
Finance, insurance, real estate	51.3	96.4	279.5	684.2	1,273.5	1,379.9	1,476.6
Services	46.9	109.8	341.0	949.4	1,624.9	1,768.7	1,937.0
Government	**53.5**	**132.4**	**321.8**	**661.1**	**906.3**	**954.1**	**1,009.0**

NOTE: R = revised. (1) Figures may not add because of rounding. Total national income also includes income from outside the U.S.

U.S. National Income by Type of Income[1]

Source: Bureau of Economic Analysis, U.S. Dept. of Commerce

(billions of current dollars)

	1960	1970	1980	1990	1998	1999	2000
NATIONAL INCOME[2]	**$424.9**	**$832.6**	**$2,203.5**	**$4,491.0**	**$6,994.7**	**$7,462.1**	**$7,980.9**
Compensation of employees	**296.7**	**618.3**	**1,638.2**	**3,297.6**	**4,981.0**	**5,310.7**	**5,715.2**
Wages and salaries	272.8	551.5	1,372.0	2,745.0	4,153.9	4,477.4	4,837.2
Government	49.2	117.1	260.1	516.0	689.3	724.3	768.4
Other	223.7	434.3	1,111.8	2,229.0	3,464.6	3,753.1	4,068.8
Supplements to wages and salaries	23.8	66.8	266.3	552.5	827.1	833.4	878.0
Employer contrib. for social ins.	12.6	34.3	127.9	278.3	420.1	323.6	343.8
Other labor income	11.2	32.5	138.4	274.3	406.9	509.7	534.2
Proprietors' income with adjustments	**52.1**	**80.2**	**180.7**	**363.3**	**577.2**	**672.0**	**715.0**
Farm	11.6	14.7	20.5	41.9	28.7	26.6	30.6
Nonfarm	40.5	65.4	160.1	321.4	548.5	645.4	684.4
Rental income of persons, with capital consumption adjustment	**15.3**	**18.2**	**6.6**	**–14.2**	**162.6**	**147.7**	**141.6**
Corp. profits with inventory adjustment	**49.8**	**69.5**	**194.0**	**354.7**	**732.3**	**825.2**	**876.4**
Corp. profits before tax	49.9	76.0	237.1	365.7	717.8	776.3	845.4
Corp. profits tax liability	22.7	34.4	84.8	138.7	240.1	253.0	271.5
Corp. profits after tax	27.2	41.7	152.3	227.1	477.7	523.3	573.9
Dividends	12.9	22.5	54.7	153.5	279.2	343.5	379.6
Undistributed profits	14.3	19.2	97.6	73.6	198.5	179.8	194.3
Inventory valuation adjustment	–0.2	–6.6	–43.1	–11.0	14.5	–2.9	–12.4
Net interest	**11.3**	**41.2**	**200.9**	**463.7**	**449.3**	**506.5**	**532.7**

NOTE: R = revised. (1) Figures do not add, because of rounding and incomplete enumeration. (2) National income is the aggregate of labor and property earnings that arises in the production of goods and services. It is the sum of employee compensation, proprietors' income, rental income, adjusted corporate profits, and net interest. It measures the total factor costs of goods and services produced by the economy. Income is measured before deduction of taxes. Total national income figures include adjustments not itemized.

Selected Personal Consumption Expenditures in the U.S., 1993-2000

Source: Bureau of Economic Analysis, U.S. Dept. of Commerce

(billions of dollars)

	1993	1994	1995	1996	1997	1998	1999	2000
Personal Comsumption expenditures	**$4,454.7**	**$4,716.4**	**$4,969.0**	**$5,237.5**	**$5,529.3**	**$5,856.0**	**$6,250.2**	**$6,728.4**
Durable goods	**513.4**	**560.8**	**589.7**	**616.5**	**642.5**	**693.2**	**760.9**	**819.6**
Motor vehicles and parts	222.1	242.3	249.3	256.3	264.2	288.8	324.7	346.8
New autos	82.1	86.5	82.2	81.9	82.5	87.9	98.0	105.0
Net purchases of used autos	38.8	43.0	50.0	51.4	53.1	54.9	57.6	59.1
Other motor vehicles	68.9	77.7	80.2	84.3	89.0	104.5	124.7	136.5
Tires, tubes, accessories, and other parts	32.4	35.2	36.9	38.7	39.6	41.5	44.4	46.3
Furniture and household equipment	192.4	211.2	225.0	236.9	248.9	265.2	285.2	307.3
Furniture, including mattresses and bedsprings	41.8	45.1	47.5	50.9	53.8	56.7	60.0	64.1
Kitchen and other household appliances	25.9	27.5	29.1	30.0	30.8	32.1	34.1	36.3
China, glassware, tableware, and utensils	20.8	22.5	23.8	25.4	27.2	29.1	31.4	33.8
Video and audio goods, including musical instruments, and computer goods	62.6	71.0	77.0	80.0	83.7	90.3	98.0	106.9
Other durable house furnishings	41.3	45.1	47.7	50.5	53.5	57.1	61.7	66.1
Other (jewelry, books, sports equipment, and misc.)	98.9	107.2	115.4	123.3	129.4	139.3	151.0	165.5
Nondurable goods	**1,375.2**	**1,438.0**	**1,497.3**	**1,574.1**	**1,641.6**	**1,708.5**	**1,831.3**	**1,989.6**
Food	697.9	728.2	755.8	786.0	812.2	852.6	899.8	957.5
Food purchased for off-premise consumption	428.3	445.5	459.8	476.7	486.5	507.9	536.7	569.6
Purchased meals and beverages	261.7	274.5	287.5	300.5	316.6	335.4	353.4	378.0
Food furnished to employees (including military) and food produced and consumed on farms	7.9	8.2	8.4	8.7	9.0	9.3	9.6	9.9
Clothing and shoes	231.1	240.7	247.8	258.6	271.7	284.8	300.9	319.1
Shoes	34.0	35.8	37.1	38.8	40.1	42.4	44.8	46.8
Women's and children's apparel except shoes	128.4	132.3	135.5	140.8	148.0	154.6	164.0	175.1
Men's and boys' apparel except shoes	68.6	72.5	75.2	78.9	83.6	87.7	92.1	97.2
Gasoline, fuel oil, and other energy goods	119.4	122.5	127.4	139.7	143.2	127.9	143.1	183.2
Tobacco products	44.9	45.4	46.7	48.2	49.8	54.4	65.7	72.1
Toilet articles, drug preparations, and sundries	120.5	128.3	137.1	148.3	161.2	174.8	194.6	214.0
Other (household supplies, nondurable toys, and misc.)	161.5	172.9	182.6	193.3	203.4	214.2	227.2	243.7
Services	**2,566.1**	**2,717.6**	**2,882.0**	**3,047.0**	**3,245.2**	**3,454.3**	**3,658.0**	**3,919.2**
Housing	666.5	704.7	740.8	772.5	810.5	859.7	909.0	958.8
Owner-occupied nonfarm dwellings—space rent	475.2	502.6	529.3	555.4	585.5	625.0	664.6	702.7
Tenant-occupied nonfarm dwellings—rent	160.3	169.3	177.0	180.6	186.1	194.0	201.3	209.3
Rental value of farm dwellings	5.6	5.9	6.0	6.2	6.4	6.7	7.2	7.7
Other	25.5	27.0	28.5	30.2	32.5	34.0	35.9	39.1
Household operation	268.9	284.0	298.1	317.3	333.0	345.6	359.7	385.7
Gas, electricity, water, and other sanitary services	151.9	156.0	160.9	169.5	173.0	173.0	175.8	189.7
Telephone and telegraph	74.6	82.7	87.8	97.1	105.0	112.9	122.3	131.3
Other	42.4	45.3	49.4	50.7	55.1	59.7	61.5	64.7
Transportation	166.2	180.9	197.7	214.2	234.4	246.3	257.4	272.8
User-operated transportation	127.7	141.0	155.3	169.7	186.6	195.6	204.7	215.8
Repair, greasing, washing, parking, storage,	97.6	110.0	122.2	134.2	146.3	153.6	163.6	173.4
Other user-operated transportation	30.1	31.0	33.1	35.5	40.3	42.0	41.1	42.3
Purchased local transportation	9.5	10.0	10.4	11.2	11.6	12.3	12.4	13.0
Mass transit systems	6.7	7.1	7.1	7.7	7.8	8.3	8.6	9.0
Taxicab	2.8	3.0	3.2	3.5	3.7	4.1	3.8	3.9
Purchased intercity transportation	29.1	29.8	32.1	33.3	36.2	38.4	40.3	44.0
Railway	0.6	0.6	0.6	0.6	0.7	0.7	0.7	0.8
Bus	1.6	1.5	1.6	1.8	1.8	1.9	2.0	2.2
Airline	23.3	23.7	25.5	26.2	29.0	30.8	32.3	35.8
Other	3.6	4.0	4.3	4.7	4.7	4.9	5.3	5.1
Medical care	700.6	737.3	780.7	814.4	854.6	899.0	939.9	996.5
Physicians	173.0	181.0	192.4	199.1	208.8	220.5	231.2	245.6
Dentists	39.9	42.9	46.5	48.4	51.9	55.1	58.3	62.1
Other professional services	95.1	103.6	112.9	119.7	125.9	132.1	138.4	146.4
Hospitals and nursing homes	339.2	353.9	370.9	390.8	408.9	427.8	446.6	472.4
Health insurance	53.5	55.8	58.0	56.6	59.3	63.6	65.3	70.0
Recreation	151.2	160.0	176.0	191.1	206.2	221.0	238.9	256.2
Other	612.6	650.7	688.7	737.5	806.5	882.6	953.1	1,049.3
Personal care	48.9	51.0	53.8	58.0	60.6	65.4	70.6	77.5
Personal business	365.9	381.6	406.8	435.1	489.0	529.8	577.3	638.9
Education and research	101.2	107.2	114.5	122.3	130.5	140.2	149.5	159.9
Higher education	55.7	59.2	62.9	66.1	69.4	74.0	77.4	80.6
Nursery, elementary, and secondary schools	23.8	25.0	26.4	27.4	29.0	29.9	31.4	32.5
Other	21.8	23.0	25.2	28.8	32.1	36.3	40.7	46.8
Religious and welfare activities	116.6	127.9	134.9	146.8	149.5	163.9	173.0	190.3
Net foreign travel	–20.1	–17.0	–21.4	–24.8	–23.1	–16.6	–17.2	–17.3
Foreign travel by U.S. residents	48.7	53.0	54.1	57.6	63.6	68.8	72.3	80.7
Less: Expenditures in the United States by nonresidents	68.7	69.9	75.4	82.4	86.7	85.4	89.6	97.9

Note: Some subtotals may not add to totals because of rounding

Distribution of U.S. Total Personal Income[1]

Source: Bureau of Economic Analysis, U.S. Dept. of Commerce
(billions of current dollars)

Year	Personal income	Personal taxes	Disposable personal income	Personal outlays	Personal Savings: Amount	Personal Savings: As pct. of disposable income
1960	$411.7	$48.7	$362.9	$339.6	$23.3	6.4%
1965	555.8	61.9	493.9	456.2	37.8	7.6
1970	836.1	109.0	727.1	666.1	61.0	8.4
1975	1,315.6	156.4	1,159.2	1,054.8	104.4	9.0
1980	2,285.7	312.4	1,973.3	1,811.5	161.8	8.2
1985	3,439.6	437.7	3,002.0	2,795.8	206.2	6.9
1990	4,791.6	624.8	4,166.8	3,958.1	208.7	5.0
1991	4,968.5	624.8	4,343.7	4,097.4	246.4	5.7
1992	5,264.2	650.5	4,613.7	4,341.0	272.6	5.9
1993	5,480.1	689.9	4,790.2	4,575.8	214.4	4.5
1994	5,757.9	739.1	5,018.9	4,842.1	176.8	3.5
1995	6,072.1	795.0	5,277.0	5,097.2	179.8	3.4
1996	6,425.2	890.5	5,534.7	5,376.2	158.5	2.9
1997	6,784.0	989.0	5,795.1	5,674.1	121.0	2.1
1998	7,391.0	1,070.9	6,320.0	6,054.7	265.4	4.2
1999	7,777.3	1,159.2	6,618.0	6,457.2	160.9	2.4
2000	8,319.2	1,288.2	7,031.0	6,963.3	67.7	1.0

(1) Personal income minus taxes=disposable income; disposable income minus outlays=savings. Figures may not add because of rounding.

Banks in the U.S.—Number, Deposits

Source: Federal Deposit Insurance Corp. (as of Dec. 2000)

Comprises all FDIC-insured commercial and savings banks, including savings and loan institutions (S&Ls).

	TOTAL NUMBER OF BANKS					TOTAL DEPOSITS (millions of dollars)				
		Commercial banks[1]					Commercial banks[1]			
Year	ALL BANKS	Natl.	State	Non-members	All savings	ALL DEPOSITS	Natl.	State	Non-members	All savings
1935	**15,295**	5,386	1,001	7,735	1,173	**$45,102**[2]	$24,802	$13,653	$5,669	$978[2]
1940	**15,772**	5,144	1,342	6,956	2,330	**67,494**	35,787	20,642	7,040	4,025
1945	**15,969**	5,017	1,864	6,421	2,667	**151,524**	77,778	41,865	16,307	15,574
1950	**16,500**	4,958	1,912	6,576	3,054	**171,963**	84,941	41,602	19,726	25,694
1955	**17,001**	4,692	1,847	6,698	3,764	**235,211**	102,796	55,739	26,198	50,478
1960	**17,549**	4,530	1,641	6,955	4,423	**310,262**	120,242	65,487	34,369	90,164
1965	**18,384**	4,815	1,405	7,327	4,837	**467,633**	185,334	78,327	51,982	151,990
1970	**18,205**	4,621	1,147	7,743	4,694	**686,901**	285,436	101,512	95,566	204,367
1975	**18,792**	4,744	1,046	8,595	4,407	**1,157,648**	450,308	143,409	187,031	376,900
1980	**18,763**	4,425	997	9,013	4,328	**1,832,716**	656,752	191,183	344,311	640,470
1985	**18,033**	4,959	1,070	8,378	3,626	**3,140,827**	1,241,875	354,585	521,628	1,022,739
1990	**15,158**	3,979	1,009	7,355	2,815	**3,637,292**	1,558,915	397,797	693,438	987,142
1993	**13,220**	3,304	969	6,685	2,262	**3,528,487**	1,576,725	476,093	701,512	774,157
1994	**12,603**	3,075	976	6,400	2,152	**3,611,618**	1,630,171	533,261	711,006	737,180
1995	**11,970**	2,858	1,042	6,040	2,030	**3,769,477**	1,695,817	614,924	716,829	741,907
1996	**11,670**	2,763	1,024	5,902	1,981	**3,788,905**	1,795,110	567,809	698,497	727,489
1997	**10,922**	2,597	992	5,554	1,779	**4,125,811**	2,004,855	729,009	687,832	704,115
1998	**10,463**	2,456	994	5,324	1,689	**4,386,298**	2,137,946	810,471	733,027	704,855
1999	**10,221**	2,363	1,010	5,207	1,641	**4,538,036**	2,154,259	899,252	777,264	707,261
2000	**9,905**	2,230	991	5,094	1,590	**4,914,808**	2,250,464	1,032,110	894,000	738,234

(1) "Nonmembers" are banks that are not members of the Federal Reserve System; "National" and "State" institutions are members.
(2) Figures for 1935 do not include data for S&Ls (not available).

50 Largest U.S. Bank Holding Companies

Source: *American Banker* (as of Dec. 31, 2000)

Company Name	Total Assets ($ in thousands)
Citigroup Inc., New York, NY	$902,210,000
J.P. Morgan Chase & Co., New York, NY	715,348,000
Bank Of America Corp., Charlotte, NC	642,191,000
Wells Fargo & Co., San Francisco, CA	272,426,000
Bank One Corp., Chicago, IL	269,300,000
First Union Corp., Charlotte, NC	254,170,000
Taunus Corp., New York, NY	197,815,000
Fleet Boston Financial Corp., Boston, MA	179,519,000
Suntrust Banks, Inc., Atlanta, GA	103,496,380
National City Corp., Cleveland, OH	88,534,609
U.S. Bancorp, Minneapolis, MN	87,336,000
Keycorp, Cleveland, OH	87,164,924
Firstar Corp., Milwaukee, WI	77,584,892
Bank Of New York Co., New York, NY	77,113,797
Wachovia Corp., Winston-Salem, NC	74,031,652
PNC Financial Services Group, Pittsburgh, PA	69,915,985
State Street Corp., Boston, MA	69,298,347
BB&T Corp., Winston-Salem, NC	59,340,228
Mellon Financial Corp., Pittsburgh, PA	50,563,617
Fifth Third Bancorp, Cincinnati, OH	45,856,906
Southtrust Corp., Birmingham, AL	45,146,531
Regions Financial Corp., Birmingham, AL	43,909,839
Comerica Inc., Detroit, MI	42,032,150
Summit Bancorp., Princeton, NJ	39,668,367
Amsouth Bancorp., Birmingham, AL	38,968,133
MBNA Corp., Wilmington, DE	$38,666,212
Charles Schwab Corp., San Francisco, CA	38,154,765
Northern Trust Corp., Chicago, IL	36,022,222
Union Planters Corp., Memphis, TN	34,720,718
Charter One Financial, Inc., Cleveland, OH	33,033,878
M&T Bank Corp., Buffalo, NY	28,949,456
Huntington Bancshares, Columbus, OH	28,599,377
Popular Inc., San Juan, PR	28,058,000
Marshall & Ilsley Corp., Milwaukee, WI	26,077,739
Old Kent Financial Corp., Grand Rapids, MI	23,866,576
Zions Bancorp., Salt Lake City, UT	21,938,263
Compass Bancshares, Birmingham, AL	20,023,496
First Tennessee Natl Corp., Memphis, TN	18,556,099
Bancwest Corp., Honolulu, HI	18,457,066
Banknorth Group, Inc., Portland, ME	18,233,761
Hibernia Corp., New Orleans, LA	16,698,046
National Commerce Bancorp., Memphis, TN	16,636,536
Greenpoint Financial Corp., New York, NY	15,764,810
TD Waterhouse Holdings, Inc., New York, NY	15,330,045
North Fork Bancorp., Melville, NY	14,840,962
Pacific Century Financial Corp., Honolulu, HI	14,026,624
Provident Financial Group, Cincinnati, OH	13,978,252
Associated Banc-Corp, Green Bay, WI	13,128,394
Colonial BancGroup, Inc., Montgomery, AL	11,741,202
People's Mutual Holdings, Bridgeport, CT	11,608,026

U.S. Bank Failures[1]

Source: Federal Deposit Insurance Corp.

Comprises all FDIC-insured commercial and savings banks, including savings and loan institutions (S&Ls) 1980 and after.

Year	Closed or assisted	Year	Closed or assisted	Year	Closed or assisted	Year	Closed or assisted	Year	Closed or assisted
1934	9	1961	5	1972	2	1983	99	1992	181
1935	26	1963	2	1973	6	1984	106	1993	50
1936	69	1964	7	1975	13	1985	180	1994	15
1937	77	1965	5	1976	17	1986	204	1995	8
1938	74	1966	7	1978	7	1987	262	1996	6
1939	60	1967	4	1979	10	1988	465	1997	1
1940	43	1969	9	1980	22	1989	534	1998	3
1955	5	1970	7	1981	40	1990	382	1999	8
1959	3	1971	7	1982	119	1991	271	2000	7
1960	1								

(1) Does not include S&L failures prior to 1980.

World's 50 Largest Banking Companies[1]

Source: *American Banker* (Dec. 31, 2000)

Banks	Asssets (millions)
Citigroup Inc., USA	$902,210
Deutsche Bank AG, Germany	872,627
Bank of Tokyo-Mitsubishi Ltd., Japan	720,809
J.P. Morgan Chase & Co., USA	715,348
UBS, Switzerland	673,706
HSBC Holdings, United Kingdom	673,475
Bayerische Hypo-und Vereinsbank AG, Germany	653,862
BNP Paribas, France	652,411
Bank of America Corp., USA	642,191
ING NV, Netherlands	612,833
Credit Suisse Group, Switzerland	612,098
Fuji Bank Ltd., Japan	548,720
Sumitomo Banking Corp., Japan	515,238
ABN Amro Holding NV, Netherlands	509,733
Caisse Nationale de Credit Agricole, France	503,814
Dai-Ichi Kangyo Bank Ltd., Japan	499,408
Norinchukin Bank, Japan	498,438
Royal Bank of Scotland Group, United Kingdom	477,619
Barclays, United Kingdom	471,919
Sakura Bank Ltd., Japan	464,953
Dresdner Bank, Germany	454,803
Sanwa Bank Ltd., Japan	450,034
Axa SA, France	447,253
Societe Generale, France	429,042
Commerzbank, Germany	428,751
Fortis Bank, Belgium	412,897
Industrial Bank of Japan Ltd., Japan	$408,787
Westdeutsche Landesbank Girozentrale, Germany	377,042
Lloyds Tsb Group, United Kingdom	325,305
Rabobank, Netherlands	323,226
Banco Santander Central Hispanoamerican, Spain	321,204
Intesabci Spa, Italy	313,195
Abbey National, United Kingdom	305,018
Tokai Bank Ltd., Japan	293,689
Bayerische Landesbank Girozentrale, Germany	284,130
Asahi Bank Ltd., Japan	276,584
Banco Bilbao Vizcaya Argentaria, Spain	275,921
Wells Fargo & Co., USA	272,426
Halifax Group, United Kingdom	272,418
Bank One Corp., USA	269,300
First Union Corp., USA	254,170
Dexia, Belgium	242,458
Nordea AB, Sweden	210,719
Zenshinren Bank, Japan	208,948
Almanij NV, Belgium	205,029
Washington Mutual Inc., USA	194,716
Unicredito Italiano, Italy	190,973
Royal Bank of Canada, Canada	189,677
National Australia Bank Ltd., Australia	186,476
FleetBoston Financial Corp., USA	179,519

(1) Includes bank holding companies and commercial and savings banks. **NOTE:** Data for U.S. companies listed include assets not included in "50 Largest U.S. Bank Holding Companies" table.

Federal Deposit Insurance Corporation (FDIC)

The Federal Deposit Insurance Corporation (FDIC) is the independent deposit insurance agency created by Congress to maintain stability and public confidence in the nation's banking system. In its unique role as deposit insurer of banks and savings associations, and in cooperation with other federal and state regulatory agencies, the FDIC seeks to promote the safety and soundness of insured depository institutions in the U.S. financial system by identifying, monitoring, and addressing risks to the deposit insurance funds. The FDIC aims at promoting public understanding and sound public policies by providing financial and economic information and analyses. It seeks to minimize disruptive effects from the failure of banks and savings associations. It seeks to ensure fairness in the sale of financial products and the provision of financial services.

The FDIC's income consists of assessments on insured banks and income from investments. The Corporation may borrow from the U.S. Treasury, not to exceed $30 billion outstanding, but the agency has made no such borrowings since it was organized in 1933. The FDIC's Bank Insurance Fund was $31.7 billion (unaudited) and the Savings Association Insurance Fund stood at $10.8 billion (unaudited), as of June 30, 2001.

Federal Reserve Board Discount Rate

The discount rate is the rate of interest set by the Federal Reserve that member banks are charged when borrowing money through the Federal Reserve System. Includes any changes through Oct. 10, 2001.

Effective date	Rate
1980:	
Feb. 15	13
May 30	12
June 13	11
July 28	10
Sept. 26	11
Nov. 17	12
Dec. 5	13
1981:	
May 5	14
Nov. 2	13
Dec. 4	12
1982:	
July 20	11½
Aug. 2	11
Aug. 16	10
Aug. 27	10
Oct. 12	9½
Nov. 22	9
Dec. 15	8½
1984:	
April 9	9
Nov. 21	8½
Dec. 24	8
1985:	
May 20	7½
1986:	
March 7	7
April 21	6½
July 11	6
Aug. 21	5½
1987:	
Sept. 4	6
1988:	
Aug. 9	6½
1989:	
Feb. 24	7
1990:	
Dec. 18	6½
1991:	
Apr. 30	5½
Sept. 13	5
Nov. 6	4½
Dec. 20	3½
1992:	
July 2	3
1994:	
May 17	3½
Aug. 16	4
Nov. 15	4¾
1995:	
Feb. 1	5¼
1996:	
Jan. 31	5
1998:	
Oct. 15	4¾
Nov. 17	4½
1999:	
Aug. 24	4¾
Nov. 16	5
2000:	
Feb. 2	5¼
Mar. 21	5½
May 16	6
2001:	
Jan. 3	5¾
Jan. 31	5
Mar. 20	4½
Apr. 18	4
May 15	3½
June 27	3¼
Aug. 21	3
Sept. 17	2½
Oct. 2	2

Federal Reserve System

The Federal Reserve System is the central bank for the U.S. The system was established on Dec. 23, 1913, originally to give the country an elastic currency, to provide facilities for discounting commercial paper, and to improve the supervision of banking. Since then, the system's responsibilities have been broadened. Over the years, stability and growth of the economy, a high level of employment, stability in the purchasing power of the dollar, and reasonable balance in transactions with other countries have come to be recognized as primary objectives of governmental economic policy.

The Federal Reserve System consists of the Board of Governors, the 12 District Reserve Banks and their branch offices, and the Federal Open Market Committee. Several advisory councils help the board meet its varied responsibilities.

The hub of the system is the 7-member Board of Governors in Washington. The members of the board are appointed by the president and confirmed by the Senate, to serve 14-year terms. The president also appoints the chairman and vice chairman of the board from among the board members for 4-year terms that may be renewed. As of June 2001 the board members were: Alan Greenspan, Chair; Roger W. Ferguson Jr., Vice Chair; Edward W. Kelley Jr.; Laurence H. Meyer; and Edward M. Gramlich; there were two vacancies.

The board is the policy-making body. In addition to those responsibilities, it supervises the budget and operations of the Reserve Banks, approves the appointments of their presidents, and appoints 3 of each District Bank's directors, including the chairman and vice chairman of each Reserve Bank's board.

The 12 Reserve Banks and their branch offices serve as the decentralized portion of the system, carrying out day-to-day operations such as circulating currency and coin and providing fiscal agency functions and payments mechanism services. The District Banks are in Boston, New York, Philadelphia, Cleveland, Richmond, Atlanta, Chicago, St. Louis, Minneapolis, Kansas City, Dallas, and San Francisco.

The system's principal function is monetary policy, which it controls using 3 tools: reserve requirements, the discount rate, and open market operations. Uniform reserve requirements, set by the board, are applied to the transaction accounts and nonpersonal time deposits of all depository institutions.

Responsibility for setting the discount rate (the interest rate at which depository institutions can borrow money from the Reserve Banks) is shared by the Board of Governors and the Reserve Banks. Changes in the discount rate are recommended by the individual boards of directors of the Reserve Banks and are subject to approval by the Board of Governors.

The most important tool of monetary policy is open market operations (the purchase and sale of government securities). Responsibility for influencing the cost and availability of money and credit through the purchase and sale of government securities lies with the Federal Open Market Committee (FOMC), which is composed of the 7 members of the Board of Governors, the president of the Federal Reserve Bank of New York, and 4 other Federal Reserve Bank presidents, who each serve one-year terms on a rotating basis. The committee bases its decisions on economic and financial developments and outlook, setting yearly growth objectives for key measures of money supply and credit. The decisions of the committee are carried out by the Domestic Trading Desk of the Federal Reserve Bank of New York.

The Federal Reserve Act prescribes a Federal Advisory Council, consisting of 1 member from each Federal Reserve District, who is elected annually by the Board of Directors of each of the 12 Federal Reserve Banks. The council meets with the Federal Reserve Board 4 times a year to discuss business and financial conditions, as well as to make advisory recommendations.

The Consumer Advisory Council is a statutory body, including both consumer and creditor representatives, which advises the Board of Governors on its implementation of consumer regulations and other consumer-related matters.

Following the congressional passage of the Monetary Control Act of 1980, the Federal Reserve System's Board of Governors established the Thrift Institutions Advisory Council to provide information and perspectives on the special needs and problems of thrift institutions. This group is composed of representatives of mutual savings banks, savings and loan associations, and credit unions.

United States Mint

Source: United States Mint, U.S. Dept. of the Treasury

The United States Mint was created on Apr. 2, 1792, by an act of Congress, which established the U.S. national coinage system. Supervision of the mint was a function of the secretary of state, but in 1799 the mint became an independent agency reporting directly to the president. The mint was made a statutory bureau of the Treasury Department in 1873, with a director appointed by the president to oversee its operations.

The mint manufactures and ships all U.S. coins for circulation to Federal Reserve banks and branches, which in turn issue coins to the public and business community through depository institutions. The mint also safeguards the Treasury Department's stored gold and silver, as well as other monetary assets.

The composition of dimes, quarters, and half dollars, traditionally produced from silver, was changed by the Coinage Act of 1965, which mandated that these coins from here on in be minted from a cupronickel-clad alloy and reduced the silver content of the half dollar to 40%. In 1970, legislative action mandated that the half dollar and a dollar coin be minted from the same alloy.

The Eisenhower dollar was minted from 1971 through 1978, when legislation called for the minting of the smaller Susan B. Anthony dollar coin. The Anthony dollar, which was minted from 1979 through 1981, marked the first time that a woman, other than a mythical figure, appeared on a U.S. coin produced for general circulation. Authorized by the U.S. Dollar Coin Act of 1997 to replace the Susan B. Anthony dollar in 2000, is the Golden Dollar Coin. Golden in color, with a smooth edge and wide border, the obverse side depicts Sacagawea (a Shoshone woman who helped guide Lewis and Clark) and her infant son. The reverse shows an American eagle and 17 stars, one for each of the states at the time of the Lewis and Clark expedition.

Mint headquarters are in Washington, DC. Mint production facilities are in Philadelphia, Denver, San Francisco, and West Point, NY. In addition, the mint is responsible for the U.S. Bullion Depository at Fort Knox, KY.

Proof coin sets, silver proof coin sets, and uncirculated coin sets are available from the mint. The mint also produces ongoing series of national and historic medals in honor of significant persons, events, and sites.

Since 1982, the mint has produced the following congressionally authorized commemorative coins: 1982 George Washington half dollar; 1984 U.S. Olympic coins; 1986 U.S. Statue of Liberty coins; 1987 Bicentennial of the U.S. Constitution coins; 1989 U.S. Congressional coins; 1990 Eisenhower Centennial coin; 1991 United Services Organization 59th Anniversary coin; 1991 Korean War Memorial coin; 1991 Mount Rushmore Anniversary coins; 1992 U.S. Olympic coins; 1992 White House 200th Anniversary coin; 1992 Christopher Columbus Quincentenary coins; 1993 Bill of Rights coins; 1993 World War II 50th Anniversary coins; 1994 World Cup USA coins; Thomas Jefferson 250th Anniversary coin; U.S. Veterans coins (featuring the Prisoner of War coin, Vietnam Veterans Memorial coin, and Women in Military Service for America coin); Bicentennial of the U.S. Capitol Commemorative Silver Dollar; 1995 Civil War Battlefield coins; 1995/1996 U.S. Olympic Games of the Atlanta Centennial Games; 1997 U.S. Botanic Garden Silver

Dollar; 1997 Franklin Delano Roosevelt Gold coin; 1997 Gold and Silver Jackie Robinson Commemorative coins; 1997 National Law Enforcement Memorial Silver Dollar; Black Revolutionary War Patriots Silver Dollar; Robert F. Kennedy Silver Dollar; National Law Enforcement Officers Memorial Silver Dollar; 1999 Yellowstone National Park Silver Dollar; 1999 George Washington five-dollar gold coin; the Dolley Madison Silver Dollar; 2000 U.S. Leif Ericson Proof Silver Dollar; 2000 Icelandic Leif Ericson Proof Silver Krønur; the 2000 Library of Congress Commemorative Coin Program featuring the Proof Silver Dollar and the Proof Bi-metallic Gold and Platinum $10 coin; the 2001 American Buffalo Proof Siver Dollar; and the 2001 U.S. Capitol Visitor Center Commemorative Coin Program, featuring the Half Dollar Clad Proof coin, the Proof Silver Dollar, and the Proof Gold $5 coin.

The congressionally authorized American Eagle gold, platinum, and silver bullion coins are available through dealers worldwide. The gold and platinum eagles are sold in one-ounce, half-ounce, quarter-ounce, and one-tenth-ounce sizes. The American eagle silver bullion coin contains one troy ounce of .999 fine silver and is priced according to the daily market value of silver. These coins also are available directly from the mint in proof condition, separately priced.

The mint offers free public tours and operates sales centers at the U.S. mints in Denver and Philadelphia; it also operates a sales center at Union Station, in Washington, DC.

Further information is available from the U.S. Mint, Customer Service Center, 10003 Derekwood Ln., Lanham, MD 20706.

Telephone number: (800) USA-MINT.

Website: http://www. usmint.gov

Portraits on U.S. Treasury Bills, Bonds, Notes, and Savings Bonds

Denomination	Savings bonds	Treasury bills*	Treasury bonds*	Treasury notes*
$50	Washington		Jefferson	
75	Adams			
100	Jefferson		Jackson	
200	Madison			
500	Hamilton		Washington	
1,000	Franklin	H. McCulloch	Lincoln	Lincoln
5,000	Revere	J. G. Carlisle	Monroe	Monroe
10,000	J. Wilson	J. Sherman	Cleveland	Cleveland
50,000	C. Glass			
100,000		A. Gallatin	Grant	Grant
1,000,000		O. Wolcott	T. Roosevelt	T. Roosevelt
100,000,000				Madison
500,000,000				McKinley

*The U.S. Treasury discontinued issuing treasury bill, bond, and note certificates in 1986. Since then, all issues of marketable treasury securities have been available only in book-entry form, although some certificates remain in circulation.

New Commemorative State Quarters, 1999-2008

Source: United States Mint, U.S. Dept. of the Treasury

Beginning in Jan. 1999, a series of five quarter dollars with new reverses are being issued each year through 2008, celebrating each of the 50 states. To make room on the reverse of the commemorative quarters for each state's design, certain design elements have been moved, thereby creating a new obverse design as well. The coins are being issued in the sequence the states became part of the Union (date each state entered the union is shown below).

1999	**2000**	**2001**	**2002**	**2003**
Delaware Dec. 7, 1787	Massachusetts Feb. 6, 1788	New York July 26, 1788	Tennessee June 1, 1796	Illinois Dec. 3, 1818
Pennsylvania Dec. 12, 1787	Maryland Apr. 28, 1788	North Carolina Nov. 21, 1789	Ohio Mar. 1, 1803	Alabama Dec. 14, 1819
New Jersey Dec. 18, 1787	South Carolina May 23, 1788	Rhode Island May 29, 1790	Louisiana Apr. 30, 1812	Maine Mar. 15, 1820
Georgia Jan. 2, 1788	New Hampshire June 21, 1788	Vermont Mar. 4, 1791	Indiana Dec. 11, 1816	Missouri Aug. 10, 1821
Connecticut Jan. 9, 1788	Virginia June 25, 1788	Kentucky June 1, 1792	Mississippi Dec. 10, 1817	Arkansas June 15, 1836

2004	**2005**	**2006**	**2007**	**2008**
Michigan Jan. 26, 1837	California Sept. 9, 1850	Nevada Oct. 31, 1864	Montana Nov. 8, 1889	Oklahoma Nov. 16, 1907
Florida Mar. 3, 1845	Minnesota May 11, 1858	Nebraska Mar. 1, 1867	Washington Nov. 11, 1889	New Mexico Jan. 6, 1912
Texas Dec. 29, 1845	Oregon Feb. 14, 1859	Colorado Aug. 1, 1876	Idaho July 3, 1890	Arizona Feb. 14, 1912
Iowa Dec. 28, 1846	Kansas Jan. 29, 1861	North Dakota Nov. 2, 1889	Wyoming July 10, 1890	Alaska Jan. 3, 1959
Wisconsin May 29, 1848	West Virginia June 20, 1863	South Dakota Nov. 2, 1889	Utah Jan. 4, 1896	Hawaii Aug. 21, 1959

Denominations of U.S. Currency

Since 1969 the largest denomination of U.S. currency that has been issued is the $100 bill. As larger-denomination bills reach the Federal Reserve Bank, they are removed from circulation. Because some discontinued currency is expected to be in the hands of holders for many years, the description of the various denominations below is continued.

Amt.	Portrait	Embellishment on back	Amt.	Portrait	Embellishment on back
$1	Washington	Great Seal of U.S.	$100	Franklin	Independence Hall
2	Jefferson	Signers of Declaration	500	McKinley	Ornate denominational marking
5	Lincoln	Lincoln Memorial	1,000	Cleveland	Ornate denominational marking
10	Hamilton	U.S. Treasury	5,000	Madison	Ornate denominational marking
20	Jackson	White House	10,000	Chase	Ornate denominational marking
50	Grant	U.S. Capitol	100,000*	W. Wilson	Ornate denominational marking

*For use only in transactions between Federal Reserve System and Treasury Department.

U.S. Currency and Coin

Source: Financial Management Service, U.S. Dept. of the Treasury (June 29, 2001)

Amounts Outstanding and in Circulation, 2001

Currency	Total currency and coin	Total currency	Federal Reserve notes[1]	U.S. notes	Currency no longer issued
Amounts outstanding	$768,474,579,815	$736,322,681,117	$735,804,815,091	$264,592,916	$253,273,110
Less amounts held by:					
Treasury	70,916,613	16,668,613	16,458,890	20,739	188,984
Federal Reserve banks	171,356,926,707	170,231,225,063	170,231,222,038	—	3,025
Amounts in circulation	$596,746,736,495	$566,074,787,441	$565,557,134,163	$264,572,177	$253,081,101

Coins[2]	Total	Dollars[3]	Fractional coins
Amounts outstanding	$32,151,898,698	$3,462,931,398	$28,688,967,300
Less amounts held by:			
Treasury	354,248,000	330,472,000	23,776,000
Federal Reserve banks	1,125,701,644	288,507,443	837,194,201
Amounts in circulation	$30,671,949,054	$2,843,951,955	$27,827,997,099

(1) Issued on or after July 1, 1929. (2) Excludes coins sold to collectors at premium prices. (3) Includes $481,781,898 in standard silver dollars.

Currency in Circulation by Denominations

Denomination	Total currency in circulation	Federal Reserve notes[1]	U.S. notes	Currency no longer issued
$1	$7,464,822,256	$7,318,136,850	$143,481	$146,541,925
$2	1,248,351,650	1,115,953,108	132,385,966	12,576
$5	8,544,953,565	8,404,734,575	109,876,210	30,342,780
$10	13,801,299,850	13,779,246,280	5,950	22,047,620
$20	93,799,237,320	93,779,132,000	3,380	20,101,940
$50	53,524,106,800	53,512,612,850	—	11,493,950
$100	387,377,350,400	387,333,201,500	22,157,100	21,991,800
$500	143,070,000	142,882,000	—	188,000
$1,000	166,400,000	166,195,000	—	205,000
$5,000	1,755,000	1,700,000	—	55,000
$10,000	3,440,000	3,340,000	—	100,000
Fractional parts	485	—	—	485
Partial notes[2]	115	—	90	25
TOTAL CURRENCY	**$566,074,787,441**	**$565,557,134,163**	**$264,572,177**	**$253,081,101**

(1) Issued on or after July 1, 1929. (2) Represents the value of certain partial denominations not presented for redemption.

Comparative Totals of Money in Circulation — Selected Dates

Date	Dollars (in millions)	Per capita[1]	Date	Dollars (in millions)	Per capita[1]
Mar. 30, 2001	$585,916.0	$2,121.82	June 30, 1965	$39,719.8	$204.14
Mar. 31, 2000	562,949.0	2,050.00	June 30, 1960	32,064.6	177.47
Mar. 31, 1999	517,829.0	1,902.21	June 30, 1955	30,229.3	182.90
Mar. 31, 1998	474,979.0	1,762.42	June 30, 1950	27,156.3	179.03
Mar. 31, 1997	444,534.0	1,664.58	June 30, 1945	26,746.4	191.14
Mar. 31, 1996	416,280.0	1,573.15	June 30, 1940	7,847.5	59.40
Mar. 31, 1995	401,610.0	1,531.39	June 30, 1935	5,567.1	43.75
Mar. 31, 1990	257,664.4	1,028.71	June 30, 1930	4,522.0	36.74
June 30, 1985	185,890.7	778.58	June 30, 1925	4,815.2	41.56
June 30, 1980	127,097.2	558.28	June 30, 1920	5,467.6	51.36
June 30, 1975	81,196.4	380.08	June 30, 1915	3,319.6	33.01
June 30, 1970	54,351.0	265.39	June 30, 1910	3,148.7	34.07

(1) Based on Bureau of the Census estimates of population. The requirement for a gold reserve against U.S. notes was repealed by Public Law 90-269, approved Mar. 18, 1968. Silver certificates issued on and after July 1, 1929, became redeemable from the general fund on June 24, 1968. The amount of security after those dates has been reduced accordingly.

IT'S A FACT: The 1792 law that established the Mint made coin defacement, counterfeiting, and embezzlement by Mint employees crimes punishable by death.

New U.S. Currency Designs

On Mar. 25, 1996, the U.S. Treasury issued a redesigned $100 note incorporating many new and modified anticounterfeiting features. It was the first of the U.S. currency series to be redesigned. A new $50 note was issued Oct. 27, 1997, a new $20 bill was released into circulation Sept. 24, 1998, and new $10 and $5 notes were issued May 24, 2000; a new $1 note with a more modest redesign was to come next. Old notes are being removed from circulation as they are returned to the Federal Reserve.

The new $100 bill has a larger portrait, moved off-center; a watermark (seen only when held up to the light) to the right of the portrait, depicting the same person (Benjamin Franklin); a security thread that glows red when exposed to ultraviolet light in a dark environment; color-shifting ink that changes from green to black when viewed at different angles, to appear in the numeral on the lower, front right-hand corner of the bill; microprinting in the numeral in the note's lower, front left-hand corner and on the portrait; and other features for security, machine authentication, and processing of the currency. The redesigned $5, $10, $20, and $50 bills incorporate the same features as the $100 bill, with the notable addition of a low-vision feature, a large (14-mm high, as compared to 7.8-mm on the old design), dark numeral on a light background on the back of the note. (The security thread glows yellow in the $50, green in the $20, orange in the $10, and blue in the $5. There is no color-shifting ink on the $5 note.) More new currency information is available on the U.S. Treasury's website: http://www.ustreas.gov

Summary of Receipts, Outlays, and Surpluses or Deficits, 1936-2001

Source: Financial Management Service, U.S. Dept. of the Treasury
(millions of current dollars)

Fiscal Year[1]	Receipts	Outlays	Surplus or Deficit (–)[2]	Fiscal Year[1]	Receipts	Outlays	Surplus or Deficit (–)[2]
1936	$3,923	$8,228	$–4,304	1970	$192,807	$195,649	$–2,842
1937	5,387	7,580	–2,193	1971	187,139	210,172	–23,033
1938	6,751	6,840	–89	1972	207,309	230,681	–23,373
1939	6,295	9,141	–2,846	1973	230,799	245,707	–14,908
1940	6,548	9,468	–2,920	1974	263,224	269,359	–6,135
1941	8,712	13,653	–4,941	1975	279,090	332,332	–53,242
1942	14,634	35,137	–20,503	1976	298,060	371,779	–73,719
1943	24,001	78,555	–54,554	Transition quarter[3]	81,232	95,973	–14,741
1944	43,747	91,304	–47,557	1977	355,559	409,203	–53,644
1945	45,159	92,712	–47,553	1978	399,561	458,729	–59,168
1946	39,296	55,232	–15,936	1979	463,302	503,464	–40,162
1947	38,514	34,496	4,018	1980	517,112	590,920	–73,808
1948	41,560	29,764	11,796	1981	599,272	678,209	–78,936
1949	39,415	38,835	580	1982	617,766	745,706	–127,940
1950	39,443	42,562	–3,119	1983	600,562	808,327	–207,764
1951	51,616	45,514	6,102	1984	666,457	851,781	–185,324
1952	66,167	67,686	–1,519	1985	734,057	946,316	–212,260
1953	69,608	76,101	–6,493	1986	769,091	990,231	–221,140
1954	69,701	70,855	–1,154	1987	854,143	1,003,804	–149,661
1955	65,451	68,444	–2,993	1988	908,166	1,063,318	–155,151
1956	74,587	70,640	3,947	1989	990,701	1,144,020	–153,319
1957	79,990	76,578	3,412	1990	1,031,308	1,251,776	–220,469
1958	79,636	82,405	–2,769	1991	1,054,265	1,323,757	–269,492
1959	79,249	92,098	–12,849	1992	1,090,453	1,380,794	–290,340
1960	92,492	92,191	301	1993	1,153,226	1,408,532	–255,306
1961	94,388	97,723	–3,335	1994	1,257,451	1,460,553	–203,102
1962	99,676	106,821	–7,146	1995	1,351,495	1,515,412	–163,917
1963	106,560	111,316	–4,756	1996	1,452,763	1,560,094	–107,331
1964	112,613	118,528	–5,915	1997	1,578,955	1,600,911	–21,957
1965	116,817	118,228	–1,411	1998	1,721,421	1,652,224	+70,039
1966	130,835	134,532	–3,698	1999	1,827,302	1,704,942	+124,360
1967	148,822	157,464	–8,643	2000	2,025,038	1,788,045	+236,993
1968	152,973	178,134	–25,161	2001[4]	1,989,000	1,868,000	+121,000
1969	186,882	183,640	3,242				

(1) Fiscal years 1936 to 1976 end June 30; after 1976, fiscal years end Sept. 30. (2) May not equal difference between figures shown, because of rounding. (3) Transition quarter covers July 1, 1976-Sept. 30, 1976. (4) Congressional Budget Office estimate as of Sept. 26, 2001.

Budget Receipts and Outlays, 1789-1935

Source: U.S. Dept. of the Treasury; annual statements for years ending June 30 unless otherwise noted
(thousands of dollars)

Yearly Average	Receipts	Outlays	Yearly Average	Receipts	Outlays	Yearly Average	Receipts	Outlays
1789-1800[1]	$5,717	$5,776	1866-1870	$447,301	$377,642	1901-1905	$559,481	$535,559
1801-1810[2]	13,056	9,086	1871-1875	336,830	287,460	1906-1910	628,507	639,178
1811-1820[2]	21,032	23,943	1876-1880	288,124	255,598	1911-1915	710,227	720,252
1821-1830[2]	21,928	16,162	1881-1885	366,961	257,691	1916-1920	3,483,652	8,065,333
1831-1840[2]	30,461	24,495	1886-1890	375,448	279,134	1921-1925	4,306,673	3,578,989
1841-1850[2]	28,545	34,097	1891-1895	352,891	363,599	1926-1930	4,069,138	3,182,807
1851-1860	60,237	60,163	1896-1900	434,877	457,451	1931-1935	2,770,973	5,214,874
1861-1865	160,907	683,785						

(1) Average for period March 4, 1789, to Dec. 31, 1800. (2) Years from 1801 to 1842 end Dec. 31; average for 1841-1850 is for the period Jan. 1, 1841, to June 30, 1850.

Public Debt of the U.S.

Source: Bureau of Public Debt, U.S. Dept. of the Treasury; *World Almanac* research

Fiscal year	Debt (billions)	Debt per cap. (dollars)	Interest paid (billions)	% of federal outlays	Fiscal year	Debt (billions)	Debt per cap. (dollars)	Interest paid (billions)	% of federal outlays
1870	$2.4	$61.06	—	—	1982	$1,142.0	$4,913	$117.4	15.7
1880	2.0	41.60	—	—	1983	1,377.2	5,870	128.8	15.9
1890	1.1	17.80	—	—	1984	1,572.3	6,640	153.8	18.1
1900	1.2	16.60	—	—	1985	1,823.1	7,598	178.9	18.9
1910	1.1	12.41	—	—	1986	2,125.3	8,774	190.2	19.2
1920	24.2	228	—	—	1987	2,350.3	9,615	195.4	19.5
1930	16.1	131	—	—	1988	2,602.3	10,534	214.1	20.1
1940	43.0	325	$1.0	10.5	1989	2,857.4	11,545	240.9	21.0
1950	256.1	1,688	5.7	13.4	1990	3,233.3	13,000	264.8	21.1
1955	272.8	1,651	6.4	9.4	1991	3,665.3	14,436	285.5	21.6
1960	284.1	1,572	9.2	10.0	1992	4,064.6	15,846	292.3	21.2
1965	313.8	1,613	11.3	9.6	1993	4,411.5	17,105	292.5	20.8
1970	370.1	1,814	19.3	9.9	1994	4,692.8	18,025	296.3	20.3
1975	533.2	2,475	32.7	9.8	1995	4,974.0	18,930	332.4	22.0
1976	620.4	2,852	37.1	10.0	1996	5,224.8	19,805	344.0	22.0
1977	698.8	3,170	41.9	10.2	1997	5,413.1	20,026	355.8	22.2
1978	771.5	3,463	48.7	10.6	1998	5,526.2	20,443	363.8	22.0
1979	826.5	3,669	59.8	11.9	1999	5,656.3	20,746	353.5	20.7
1980	907.7	3,985	74.9	12.7	2000	5,674.2	20,591	362.0	20.3
1981	997.9	4,338	95.6	14.1	2001	5,807.5	20,353	359.5	20.8

Note: Through 1976 the fiscal year ended June 30. From 1977 on, the fiscal year ends Sept. 30.

U.S. Budget Receipts and Outlays, 1997-2001

Source: Financial Management Service, U.S. Dept. of the Treasury

As of Sept. 26, 2001, the estimate from the Congressional Budget Office of the total U.S. budget surplus for the fiscal year 2001was $121 billion, or 1.2% of GDP. The estimate was about half the size of the previous year's surplus, and was markedly lower than the August estimate of $153 billion. 2001 was the fourth consecutive year of a surplus, after 28 consecutive years of federal budget deficits.

(in millions of current dollars; many figures do not add to totals because of independent rounding or omitted subcategories, including some subcategories with negative values.)

	Fiscal 1997[1]	Fiscal 1998[1]	Fiscal 1999[1]	Fiscal 2000[1]
NET RECEIPTS				
Individual income taxes	$737,466	$828,597	$879,480	$1,004,461
Corporation income taxes	182,294	188,677	184,680	207,288
Social insurance taxes and contributions:				
Federal old-age and survivors insurance	336,728	358,784	383,559	411,676
Federal disability insurance	55,261	57,016	60,910	68,907
Federal hospital insurance	110,710	119,863	132,268	135,528
Railroad retirement fund	4,051	4,353	4,143	4,336
Total employment taxes and contributions	506,750	540,015	580,880	620,447
Other insurance and retirement:				
Unemployment	28,202	27,484	26,480	27,641
Federal employees retirement	4,344	4,261	4,399	4,693
Non-federal employees	74	74	73	70
Total social insurance taxes and contributions	**539,371**	**571,835**	**611,832**	**652,851**
Excise taxes	56,926	57,669	70,412	68,866
Estate and gift taxes	19,845	24,076	27,782	29,010
Customs duties	17,927	18,297	18,336	19,913
Deposits of earnings by Federal Reserve Banks	19,636	24,540	25,917	32,293
All other miscellaneous receipts	5,491	5,027	5,112	5,807
Net Budget Receipts	**1,578,955**	**1,721,421**	**1,827,302**	**2,025,038**
NET OUTLAYS				
Legislative Branch	2,362	2,600	2,612	2,913
The Judiciary	3,259	3,463	3,793	4,087
Executive Office of the President:				
The White House Office	39	46	51	53
Office of Management and Budget	56	56	59	64
Total Executive Office	**219**	**236**	**416**	**284**
International Assistance Program:				
International security assistance	4,403	4,950	5,405	6,534
Multilateral assistance	2,141	1,850	1,857	1,759
Agency for International Development	2,814	2,435	2,337	2,622
International Development Assistance	2,902	2,494	2,410	2,953
Total International Assistance Program	**10,128**	**8,980**	**10,061**	**12,083**
Agriculture Department:				
Food stamp program	22,857	20,141	19,005	18,295
Farm Service Agency	7,417	10,421	19,508	33,353
Forest Service	3,209	3,399	3,423	3,978
Total Agriculture Department	**52,549**	**53,950**	**62,839**	**75,728**
Commerce Department:				
Bureau of the Census	282	542	1,131	4,214
Total Commerce Department	**3,780**	**4,047**	**5,036**	**7,931**
Defense Department—Military:				
Military personnel	69,722	68,976	69,503	75,950
Operation and maintenance	92,465	93,473	96,420	105,871
Procurement	47,691	48,207	48,824	51,616
Research, development, test, evaluation	37,026	37,421	37,362	37,608
Military construction	6,188	6,046	5,519	5,111
Total Defense Department—Military	**258,330**	**256,124**	**261,379**	**281,233**
Defense Department—Civil	30,282	31,216	32,008	32,019
Education Department	30,014	31,498	32,435	33,308
Energy Department	14,470	14,444	16,054	15,010
Health and Human Services Department:				
Public Health Service	21,755	23,680	25,554	28,281
Health Care Financing Adm.	369,714	379,950	390,181	413,124
Food and Drug Administration	873	838	951	1,023
National Institutes of Health	11,199	12,501	13,815	15,415
Total Health and Human Services Dept.	**339,541**	**350,571**	**359,700**	**382,627**
Housing and Urban Development Department	27,525	30,224	32,736	30,830
Interior Department	6,722	7,232	7,814	8,036
Justice Department:				
Federal Bureau of Investigation	2,700	2,949	3,040	3,088
Drug Enforcement Administration	969	1,099	1,203	1,339
Immigration and Naturalization Service	2,770	3,593	3,775	4,163
Federal Prison System	2,939	2,682	3,204	3,708
Total Justice Department	**14,315**	**16,169**	**18,318**	**19,561**
Labor Department:				
Unemployment Trust Fund	24,299	23,408	24,870	24,149
Total Labor Department	**30,461**	**30,002**	**32,459**	**31,354**
State Department	5,245	5,373	6,463	6,849
Transportation Department:				
Federal Aviation Administration	8,815	9,242	9,507	9,561
Total Transportation Department	**39,835**	**39,467**	**41,836**	**46,030**
Treasury Department:				
Internal Revenue Service	31,386	33,153	37,087	37,986
Interest on the public debt	355,796	363,824	353,511	362,118
Total Treasury Department	**379,345**	**390,094**	**386,703**	**390,813**
Veterans Affairs Department	39,277	41,776	43,169	47,087
Environmental Protection Agency	6,167	6,288	6,752	7,236
General Services Administration	1,083	1,095	–46	25

	Fiscal 1997[1]	Fiscal 1998[1]	Fiscal 1999[1]	Fiscal 2000[1]
National Aeronautics and Space Administration	$14,358	$14,206	$13,665	$13,442
Office of Personnel Management	45,404	46,307	47,515	48,660
Small Business Administration	334	–78	58	–422
Social Security Administration	393,309	408,202	419,790	441,810
Other independent agencies:				
Corporation for Natl. and Community Service	564	591	609	684
Corporation for Public Broadcasting	260	250	281	316
District of Columbia	717	818	–2,910	312
Equal Employment Opportunity Commission	231	244	255	290
Export-Import Bank of the U.S.	–114	–208	–159	–743
Federal Communications Commission	1,001	1,769	3,293	4,073
Federal Deposit Insurance Corporation	–14,181	–4,122	–5,025	–2,837
Legal Services Corporation	282	285	298	301
National Archives & Records Adm.	198	210	225	201
National Foundation on the Arts and Humanities	230	207	217	218
National Labor Relations Board	175	177	182	198
National Science Foundation	3,131	3,188	3,285	3,487
Nuclear Regulatory Commission	51	38	37	33
Railroad Retirement Board	4,870	4,837	4,830	4,992
Securities and Exchange Commission	–20	–231	–255	–506
Smithsonian Institution	491	488	486	517
Tennessee Valley Authority	–337	–784	2	–307
Total other independent agencies	–2,489	10,653	6,943	10,526
Undistributed offsetting receipts	–154,970	–161,036	–159,080	–172,844
NET BUDGET OUTLAYS	**$1,600,911**	**$1,652,224**	**$1,704,942**	**$1,788,045**
Less net receipts	1,578,955	1,721,421	1,827,302	2,025,038
DEFICIT (-) OR SURPLUS (+)	**$–21,957**	**$+70,039**	**$+124,360**	**$+236,993**

(1) Fiscal year ends Sept. 30.

State Finances: Revenue, Expenditures, Debt, and Taxes

Source: Census Bureau, U.S. Dept. of Commerce

(fiscal year 1998)

STATE	Revenue (millions)	Expenditures (millions)	Debt (millions)	Per capita[1] debt	Per capita[1] taxes	Per capita[1] expenditures
Alabama	$15,501	$14,702	$4,467	$1,022	$1,380	$3,364
Alaska	7,313	6,141	3,911	6,307	1,460	9,904
Arizona	15,122	14,278	2,725	570	1,579	2,988
Arkansas	10,361	8,943	2,488	975	1,807	3,506
California	154,017	133,673	53,975	1,628	2,184	4,033
Colorado	14,158	13,148	4,059	1,001	1,619	3,242
Connecticut	16,437	15,213	17,505	5,334	2,932	4,635
Delaware	4,540	3,941	3,713	4,924	2,693	5,227
Florida	49,209	42,459	17,825	1,180	1,575	2,810
Georgia	27,639	23,203	6,269	805	1,600	2,979
Hawaii	6,646	6,266	5,421	4,575	2,672	5,288
Idaho	4,870	4,230	2,031	1,622	1,734	3,379
Illinois	43,294	38,314	26,582	2,192	1,749	3,159
Indiana	19,149	18,614	7,056	1,187	1,638	3,132
Iowa	11,629	10,321	2,472	862	1,697	3,597
Kansas	8,687	8,371	1,482	559	1,729	3,154
Kentucky	16,853	14,778	7,422	1,874	1,857	3,731
Louisiana	17,786	15,706	7,152	1,636	1,485	3,592
Maine	5,888	4,865	3,875	3,093	2,028	3,883
Maryland	19,613	17,593	11,201	2,166	1,837	3,402
Massachusetts	28,120	28,030	35,798	5,797	2,386	4,539
Michigan	46,724	38,798	16,189	1,641	2,216	3,933
Minnesota	25,089	20,388	5,485	1,148	2,613	4,269
Mississippi	10,701	9,927	3,224	1,164	1,652	3,585
Missouri	19,505	16,525	15,770	8,903	1,628	3,022
Montana	3,725	3,512	3,121	2,355	2,667	3,977
Nebraska	5,576	5,184	5,089	1,820	1,093	3,111
Nevada	7,573	6,103	5,013	2,997	1,657	3,374
New Hampshire	4,024	3,594	3,108	5,427	4,518	2,993
New Jersey	37,007	31,702	27,214	3,354	1,923	3,907
New Mexico	8,757	8,089	7,161	3,158	1,815	4,649
New York	102,242	92,584	80,042	76,562	4,207	5,088
North Carolina	34,064	26,830	25,312	8,227	1,075	3,507
North Dakota	2,936	2,675	2,552	1,329	2,097	4,218
Ohio	51,273	41,113	33,207	14,963	1,329	3,652
Oklahoma	11,935	10,655	9,684	5,564	1,657	3,173
Oregon	15,666	13,965	11,889	5,738	1,730	4,211
Pennsylvania	49,482	44,237	38,859	17,658	1,472	3,688
Rhode Island	5,478	4,378	3,994	5,464	5,514	4,418
South Carolina	14,566	14,483	12,378	5,082	1,308	3,727
South Dakota	2,886	2,272	2,179	2,105	2,872	3,099
Tennessee	16,904	15,890	14,635	3,321	606	2,898
Texas	71,649	54,761	52,545	14,736	735	2,732
Utah	8,742	7,810	7,043	3,781	1,775	3,666
Vermont	3,055	2,603	2,730	2,116	3,563	4,381
Virginia	26,138	22,739	20,867	11,877	1,728	3,308
Washington	28,737	24,230	20,242	11,080	1,925	4,209
West Virginia	8,034	7,336	6,584	3,689	2,042	4,060
Wisconsin	28,334	20,466	19,262	11,225	2,138	3,898
Wyoming	3,092	2,373	2,215	1,046	2,179	4,944
ALL STATES[2]	**$1,152,870**	**$998,365**	**$510,486**	**$1,876**	**$1,837**	**$3,668**

(1) Per capita amounts are based on population figures of the resident U.S. population (excluding the District of Columbia) as of July 1, 1996. (2) Totals in this line may not add because of rounding.

IT'S A FACT: Every Treasurer of the U.S. since the Truman administration has been a woman. The 41st treasurer, Rosario Marin, was sworn in on Aug. 16, 2001. She is the highest ranking Latina in the Bush administration.

State and Local Government Receipts and Current Expenditures

Source: Bureau of Economic Analysis, U.S. Dept. of Commerce

(billions of current dollars)

	1998R	1999	2000
Receipts	**$1,072.3**	**$1,142.7**	**$1,230.1**
Personal tax and nontax receipts	234.9	249.7	274.1
Income taxes	182.8	194.8	216.3
Nontaxes	33.1	35.1	37.3
Other	19.1	19.8	20.5
Corporate profits tax accruals	35.1	36.6	40.2
Indirect business tax and nontax accruals	583.1	617.5	661.2
Sales taxes	284.9	307.1	331.7
Property taxes	229.2	238.5	248.5
Other	69.0	71.9	81.0
Contributions for social insurance	10.0	9.6	9.9
Federal grants-in-aid	209.1	229.3	244.6
Current expenditures	**1,030.6**	**1,092.7**	**1,170.5**
Consumption expenditures	808.4	855.0	918.0
Transfer payments to persons	234.1	252.0	267.8
Net interest paid	–0.6	–3.0	–4.5
Interest paid	73.9	75.1	76.6
Less: Interest received by government	74.5	78.1	81.1
Less: Dividends received by government	0.4	0.4	0.4
Subsidies less current surplus of government enterprises	–10.9	–11.0	–10.5
Subsidies	0.4	0.5	0.5
Less: Current surplus of government enterprises	11.3	11.4	11.0
Less: Wage accruals less disbursements	0.0	0.0	0.0
Surplus or deficit (–), national income and product accounts	41.7	50.0	59.6

(R) Revised figures. (1) Seasonally adjusted at annual rates.

State and Local Government Current Expenditures and Gross Investment, by Function

Source: Bureau of Economic Analysis, U.S. Dept. of Commerce

(millions of dollars)

	1997			1998		
	Total[1]	Current Expends	Gross Investment[2]	Total[1]	Current Expends.	Gross Investment[2]
TOTAL	**$1,135,758**	**$960,147**	**$175,611**	**$1,028,681**	**$1,212,200**	**$183,519**
Central executive, legislative, and judicial activities	**71,725**	**68,150**	**3,575**	**71,502**	**76,521**	**5,019**
Administrative, legislative, and judicial activities	38,921	36,593	2,328	40,258	43,286	3,028
Tax collection and financial management	32,804	31,557	1,247	31,244	33,235	1,991
Civilian safety	**118,689**	**110,537**	**8,152**	**119,773**	**128,429**	**8,656**
Police	53,218	50,390	2,828	55,977	59,129	3,152
Fire	20,297	18,613	1,684	20,145	21,682	1,537
Correction	45,174	41,534	3,640	43,651	47,618	3,967
Education	**407,721**	**367,955**	**39,766**	**396,141**	**442,174**	**46,033**
Elementary and secondary	312,962	284,993	27,969	306,202	338,991	32,789
Higher	70,195	59,693	10,502	64,606	76,530	11,924
Libraries	5,747	5,104	643	5,906	6,704	798
Other	18,817	18,165	652	19,427	19,949	522
Health and hospitals	**27,450**	**21,827**	**5,623**	**23,936**	**32,575**	**8,639**
Health	26,408	24,406	2,002	25,578	27,990	2,412
Hospitals	1,042	–2,579	3,621	–1,642	4,585	6,227
Income support, social security, and welfare	**255,974**	**255,216**	**758**	**267,027**	**268,290**	**1,263**
Govt. employees retirement and disability	1,768	1,768	—	—	—	—
Workers' compensation and temporary disability insurance	10,021	10,021	—	13,138	13,138	0
Medical care	169,123	169,123	—	173,951	173,951	0
Welfare and social services	75,062	74,304	758	79,938	81,201	1,263
Veterans' benefits and services	**277**	**260**	**17**	**309**	**356**	**47**
Housing and community services	**30,877**	**5,525**	**25,352**	**6,588**	**30,451**	**23,863**
Housing, comm. dev., urban renewal	6,852	2,899	3,953	5,978	9,800	3,822
Water	6,308	–3,428	9,736	–4,445	4,758	9,203
Sewerage	10,612	597	10,015	–1,332	7,776	9,108
Sanitation	7,105	5,457	1,648	6,387	8,117	1,730
Recreational and cultural activities	**17,142**	**12,388**	**4,754**	**14,163**	**19,539**	**5,376**
Energy	**–3,250**	**–7,688**	**4,438**	**–7,658**	**–4,336**	**3,322**
Gas utilities	–1,139	–1,404	265	–679	–262	417
Electric utilities	–2,111	–6,284	4,173	–6,979	–4,074	2,905
Agriculture	**4,643**	**4,379**	**264**	**4,857**	**5,218**	**361**
Natural resources	**11,897**	**9,331**	**2,566**	**9,005**	**11,552**	**2,547**
Transportation	**134,408**	**67,866**	**66,542**	**76,950**	**139,958**	**63,008**
Highways	106,923	54,610	52,313	64,149	113,466	49,317
Water	1,705	71	1,634	–256	868	1,124
Air	2,741	–1,268	4,009	–2,062	2,073	4,135
Transit and railroad	23,039	14,453	8,586	15,119	23,551	8,432
Economic development, regulation, and services	**8,488**	**8,103**	**385**	**7,443**	**7,952**	**509**
Labor training and services	**5,474**	**5,345**	**129**	**5,744**	**5,990**	**246**
Commercial activities	**–13,919**	**–14,224**	**305**	**–14,641**	**–14,290**	**351**
Publicly owned liquor store systems	–648	–658	10	–726	–712	14
Govt.-administered lotteries, parimutuels	–13,527	–13,527	—	–13,744	–13,744	—
Other	256	–39	295	171	166	337
Net interest paid[2]	**–6,452**	**–6,452**	**—**	**–2,345**	**–2,345**	**—**
Other and unallocable	**64,614**	**51,629**	**12,985**	**49,887**	**64,166**	**14,279**

(1) Sum of current expenditures and gross investment. (2) Excludes interest received by social insurance funds, which is netted against expenditures for the appropriate functions.

Top U.S. Charities by Donation, 2000

Source: The Chronicle of Philanthropy

(in millions of dollars)

Rank	Organization	Private Support[1]	Total Income
1	Salvation Army (Alexandria, VA)	$1,396.9	$2,717.8
2	YMCA of the USA (Chicago)	693.3	3,606.0
3	American Red Cross (Falls Church, VA)	678.3	2,405.4
4	American Cancer Society (Atlanta)	620.0	672.0
5	Fidelity Investments Charitable Gift Fund (Boston)	573.4	670.2
6	Lutheran Services in America (St. Paul)	559.0	3,684.7
7	United Jewish Communities (New York)	524.2	533.5
8	America's Second Harvest (Chicago)	470.9	473.2
9	Habitat for Humanity International (Americus, GA)	466.7	466.7
10	Harvard University (Cambridge, MA)	451.7	2,862.1
11	Catholic Charities USA (Alexandria, VA)	446.3	2,309.0
12	Nature Conservancy (Arlington, VA)	403.4	704.0
13	Boys and Girls Clubs of America (Atlanta)	362.3	800.0
14	American Heart Association (Dallas)	$357.8	$436.4
15	Gifts In Kind International (Alexandria, VA)	346.5	349.7
16	Cornell University (Ithaca, NY)	341.4	1,785.2
17	World Vision (Federal Way, WA	331.4	408.1
18	Duke University (Durham, NC)	331.0	2,281.2
19	Stanford University (Palo Alto, CA)	319.6	2,587.0
20	Campus Crusade for Christ International (Orlando, FL)	315.0	360.1
21	Larry Jones International Ministries/ Feed the Children (Oklahoma City)	312.1	314.4
22	Columbia University (New York)	284.4	1,794.4
23	Boy Scouts of America (Irving, TX)	273.8	646.6
24	University of Pennsylvania (Philadelphia)	270.1	2,518.4
25	AmeriCares Foundation (New Canaan, CT)	267.0	268.1

(1) Private support consists of donations from individuals, foundations and corporations. Total income also included government funding and fees charged.

Consumer Credit Outstanding, 1998-2000

Source: Federal Reserve System

(billions of dollars)

Estimated amounts of credit outstanding as of end of year. Not seasonally adjusted.

	1998	1999	2000
TOTAL	**$1,331.7**	**$1,426.2**	**$1,569.0**
Major Holders			
Commercial banks	508.9	499.8	543.7
Finance companies	168.5	181.6	193.2
Credit unions	155.4	167.9	185.3
Savings institutions	51.8	61.5	64.0
Nonfinancial business	74.9	80.3	82.7
Pools of securitized assets[1]	372.4	435.1	500.1
Major Types of Credit[2]			
Revolving	586.5	623.2	693.2
Commercial banks	210.3	189.4	218.1
Finance companies	32.3	33.8	38.3
Credit unions	19.9	20.6	21.8
Savings institutions	12.5	15.8	16.6
Nonfinancial business	39.2	42.8	42.4
Pools of securitized assets[1]	272.3	320.8	356.1
Nonrevolving	745.2	802.9	875.8
Commercial banks	298.6	310.4	325.6
Finance companies	136.2	147.8	154.9
Credit unions	135.5	147.3	163.5
Savings institutions	39.2	45.7	47.5
Nonfinancial business	35.7	37.5	40.2
Pools of securitized assets[1]	100.1	114.2	144.0

(1) Outstanding balances of pools upon which securities have been issued; these balances are no longer carried on the balance sheets of the loan originators. (2) Includes estimates for holders that do not separately report consumer credit holding by type.

Leading U.S. Businesses in 2000

Source: FORTUNE Magazine

(millions of dollars in revenues)

Advertising, Marketing	
Omnicom Group	$6,154
Interpublic Group	5,626
Aerospace	
Boeing	$51,321
United Technologies	26,583
Lockheed Martin	25,329
Honeywell Intl.	25,023
Raytheon	18,321
Textron	13,090
General Dynamics	10,359
Northrop Grumman	8,287
Airlines	
AMR	$20,245
UAL	19,352
Delta Air Lines	15,888
NWA	11,415
Continental Airlines	9,899
US Airways Group	9,269
Southwest Airlines	5,650
Trans World Airlines	3,538
America West Holdings	2,344
Alaska Air Group	2,177
Apparel	
Nike	$8,995
VF	5,748
Jones Apparel Group	4,143
Liz Claiborne	3,104
Reebok International	2,865
Kellwood	2,362
Warnaco Group	2,309
Polo Ralph Lauren	1,956
Phillips-Van Heusen	1,456
Automotive Retailing, Services	
AutoNation	$22,331
Beverages	
Coca-Cola	$20,458
Pepsico	20,438
Coca-Cola Enterprises	14,750
Anheuser-Busch	12,262
The Pepsi Bottling Group	7,982
Pepsi Americas	2,528
Adolph Coors	2,414
Building Materials, Glass	
Owens-Illinois	$5,815
Owens-Corning	4,940
USG	3,781
Armstrong Holdings	3,272
Vulcan Materials	2,492
Texas Industries	1,306
Chemicals	
E. I. du Pont de Nemours	$29,202
Dow Chemical	23,008
PPG Industries	8,629
Ashland	8,190
Rohm & Haas	6,879
Union Carbide	6,526
Commercial Banks	
J.P. MorganChase	$60,065
Bank of America Corp.	57,747
Wells Fargo	27,568
Bank One Corp.	25,168
First Union Corp.	24,246
Fleet Boston	22,608
U.S. Bancorp	9,966
Computer and Data Services	
Electronic Data Systems	$19,227
Computer Sciences	9,371
America Online	6,886
Unisys	6,885
Automatic Data Proc.	6,288
Science Applications Intl.	6,092
Computer Peripherals	
EMC	$8,873
Seagate Technology	6,448
Quantum	4,727
Lexmark International	3,807
Computers, Office Equipment	
IBM	$88,396
Hewlett-Packard	48,782
Compaq Computer	42,383
Dell Computer	31,888
Xerox	18,632
Sun Microsystems	15,721
Gateway	9,601
Apple Computer	7,983
Computer Software	
Microsoft	$22,956
Oracle	10,130
Computer Assoc. Intl.	6,103
Diversified Financials	
General Electric	$129,853
Citigroup	111,826
Fannie Mae	44,089
Freddie Mac	30,000
American Express	23,675

Electronics, Electrical Equip.	
Emerson Electric	$15,545
Whirlpool	10,325
Eaton	8,988
Rockwell International	7,220
Energy	
Duke Energy	$49.318
Reliant Energy	29,339
Utillicorp	28,975
Engineering, Construction	
Haliburton	$13,344
Fluor	11,056
Entertainment	
Walt Disney	$25,402
Viacom	20,044
Food	
ConAgra	$25,386
Sara Lee	20,414
H. J. Heinz	9,408
Kellogg	6,955
General Mills	6,700
Campbell Soups	6,267
Smithfield Foods	5,151
Quaker Oats	5,041
Food and Drug Stores	
Kroger	$49,000
Albertson's	36,762
Safeway	31,997
Walgreen	21,207
CVS	20,088
Rite Aid	14,681
Publix	14,575
Winn-Dixie Stores	13,698
Food Production	
IBP	$16,950
Archer Daniels Midland	12,877
Farmland Industries	12,239
Tyson Foods	7,158
Food Services	
McDonald's	$14,243
Tricon Global Restaurants	7,093
Darden Restaurants	3,701
Wendy's International	2,237
Forest and Paper Products	
International Paper	$28,180
Georgia-Pacific	22,218
Weyerhaeuser	15,980
Kimberly-Clark	13,982
Smurfit-Stone Container	8,839
Boise Cascade	7,807
Furniture	
Leggett & Platt	$4,276
Steelcase	3,316
Furniture Brands Intl.	2,116
Hon Industries	2,046
General Merchandisers	
Wal-Mart Stores	$193,295
Sears Roebuck	40,937
Kmart	37,028
Target	36,903
J. C. Penney	32,965
Federated Dept. Stores	18,407
May Department Stores	14,511
Health Care	
Aetna	$26,819
UnitedHealth Group	21,122
Cigna	19,994
HCA Healthcare	16,670
Hotels, Casinos, Resorts	
Marriott International	$10,017
Park Place Entertainment	4,896
Crestline Capital	4,817
Starwood Hotels & Resorts	4,402
Industrial and Farm Equip.	
Caterpillar	$20,175
Deere	13,137
Ingersoll-Rand	9,220
American Standard	7,598
Cummins Engine	6,597
Insurance (Life and Health)	
TIAA-CREF[2]	$38,064
Metropolitan Life[1]	31,947
Prudential of America[1]	22,760
New York Life (Mutual)	21,450
Northwestern Mut. Life (Mut.)	16,974
Insurance (Property and Casualty)	
State Farm Ins. (Mutual)	$47,863
American Intl. Group (Stock)	45,972
Berkshire Hathaway (Stock)	33,976
Allstate (Stock)	29,134
Loews (Stock)	20,670
Liberty Mutual Group (Mutual)	16,438
Nationwide Ins. Enterprise1	14,762
Hartford Fin'l. Svces. (Stock)	14,703
Mail, Pkg., Freight Delivery	
United Parcel Svce.	$29,771
Fdx	18,257
Metal Products	
Gillette	$9,986
ITW	9,984
Crown Cork & Seal	7,289
Masco	7,243
Newell Rubbermaid	6,935
Fortune Brands	5,492
Metals	
Alcoa	$23,090
LTV	4,934
AK Steel Holding	4,612
Nucor	4,586
Phelps Dodge	4,525
Mining, Crude -Oil Production	
Occidental Petroleum	$14,543
Unocal	9,202
Motor Vehicles and Parts	
General Motors	$184,632
Ford Motor	180,598
Delphi Automotive Systems	29,139
TRW	17,231
Johnson Controls	17,155
Network Communications	
Lucent Technologies	$41,420
Motorola	37,580
Cisco Systems	18,928
Petroleum Refining	
Exxon Mobil	$210,392
Texaco	51,130
Chevron	48,069
USX	35,570
Conoco	32,513
Pharmaceuticals	
Merck	$40,363
Pfizer	29,574
Johnson & Johnson	29,139
Bristol-Myers Squibb	21,331
Pharmacia & Upjohn	18,150
American Home Products	13,810
Abbott Laboratories	13,746
Eli Lilly	10,862
Schering-Plough	9,815
Amgen	3,629
Pipelines	
Enron	$100,789
Dynergy	29,445
El Paso Corp.	21,950
Publishing & Printing	
Gannett	$6,244
R.R. Donnelley & Sons	5,764
Tribune	5,080
McGraw-Hill	4,281
New York Times	3,490
Knight-Ridder	3,212
Railroads	
Union Pacific	$11,878
Burlington Northern Santa Fe	9,205
CSX	8,526
Norfolk Southern	6,159
Rubber and Plastic Prods	
Goodyear Tire	$14,417
Cooper Tire & Rubber	3,472
Pactiv	3,134
Sealed Air	$3,068
Mark IV Industries	2,071
PolyOne	1,888
Scientific, Photo., and Control Equip.	
Minnesota Mining & Mfg.	$16,724
Eastman Kodak	13,994
Applied Materials	9,564
Securities	
Morgan Stanley/Dean Witter	$45,413
Merrill Lynch	44,872
Goldman Sachs Group	33,000
Lehman Bros. Holdings	26,447
Semiconductors	
Intel	$33,726
Solectron	14,138
Texas Instruments	11,875
Soaps, Cosmetics	
Procter & Gamble	$39,951
Colgate-Palmolive	9,358
Avon Products	5,715
Estee Lauder	4,367
Clorox	4,083
Specialty Retailers	
Home Depot	$45,738
Costco Wholesale	32,164
Lowe's	18,779
Gap	13,674
Circuit City Group	12,614
Best Buy	12,494
Office Depot	11,570
Toys "R" Us	11,332
Staples	10,674
Limited	10,105
Telecommunications	
AT&T	$65,981
Verizon	64,707
SBC Communications	51,476
WorldCom	39,090
BellSouth	26,151
Sprint	23,613
Qwest Communications	16,610
Temporary Help	
Manpower	$10,843
Kelly Services	4,487
Spherion	3,741
Textiles	
Mohawk Industries	$3,256
Springs Industries	2,275
WestPoint Stevens	1,816
Burlington Industries	1,620
Pillowtex	1,451
Interface	1,284
Tobacco	
Philip Morris	$63,276
R.J. Reynolds Tobacco	8,167
Universal	3,402
Toys, Sporting Goods	
Mattel	$5,590
Hasbro	3,787
Transportation Equipment	
Brunswick	$4,507
Harley-Davidson	2,906
Trinity Industries	2,741
Utilities, Gas and Electric	
Southern	$23,381
PG&E Corp.	22,483
TXU	22,009
American Electric Power	13,694
Edison International	11,635
Xcel Energy	11,592
Waste Management	
Waste Management	$12,492
Allied Waste Industries	5,708
Republic Services	2,103
Wholesalers	
McKesson HBOC	$37,101
Ingram Micro	30,715
Cardinal Health	29,871
Tech Data	20,428
Supervalu	20,339
Sysco	19,303

(1) Not a stock company, but reported financial data according to Generally Accepted Accounting Principle. (2) Not a mutual company, but reported financial data based on statutory accounting.

U.S. Corporations With Largest Revenues in 2000

Source: FORTUNE Magazine

(millions of dollars)

Company, headquarters	Revenues
Exxon Mobil, Irving, TX	$210,392
Wal-Mart Stores, Bentonville, AR	193,295
General Motors, Detroit, MI	184,632
Ford Motor, Dearborn, MI	180,598
General Electric, Fairfield, CT	129,853
Citigroup, New York, NY	111,826
Enron, Houston, TX	100,789
IBM, Armonk, NY	88,396
AT&T, New York, NY	65,981
Verizion Communications, New York, NY	64,707
Philip Morris, New York, NY	63,276
J.P. Morgan Chase, New York, NY	60.065
Bank of America Corp., Charlotte, NC	57,747
SBC Communications, San Antonio, TX	$51,476
Boeing, Seattle, WA	51,321
Texaco, White Plains, NY	51,130
Duke Energy, Charlotte, NC	49,318
Kroger, Cincinnati, OH	49,000
Hewlett-Packard, Palo Alto, CA	48,782
Chevron, San Francisco, CA	48,069
State Farm Insurance Cos., Bloomington, IL	47,863
American International Group, New York, NY	45,972
Home Depot, Atlanta, GA	45,738
Morgan Stanley Dean Witter, New York, NY	45,413
Merrill Lynch, New York, NY	44,872

Fastest-Growing U.S. Franchises in 2000[1]

Source: *Entrepreneur* Magazine, Jan. 2001

Company	Business	Minimum start-up cost[2]
7-Eleven Inc.	convenience stores	$12,500
McDonald's	hamburgers, chicken	489,900
Coverall North America Inc.	commercial cleaning	6,300
Taco Bell Corp.	Mexican fast food	236,400
Subway	submarine sandwiches	63,400
Jani-King	commercial cleaning	8,200
Mail Boxes Etc.	postal & business services	125,900
Quizno's Franchise Company	submarine sandwiches	170,000
Jiffy Lube Int'l. Inc.	oil-change services	174,000
Curves for Women	fitness businesses	20,600
WSI Internet	Internet services	34,000
Century 21 Real Estate Corp.	real estate	35,900
RE/MAX Int'l. Inc.	real estate services	20,000
Auntie Anne's Inc.	hand-rolled soft pretzels	179,800
Sonic Drive In Restaurants	drive-in restaurant	621,300
GNC Franchising Inc.	vitamin stores	132,700
Quik Internet	Internet services	63,700
Popeyes Chicken & Biscuits	fried chicken & biscuits	700,000
Great Clips Inc.	hair care	88,400
Church's Chicken	southern fried chicken and biscuits	$194.8
Liberty Tax Service	income-tax preparation services	33,100
Adventures in Advertising Inc.	promotional products/ advertising specialties	11,900
Papa Murphy's	take-&-bake pizza	148,800
PostNet Postal & Business Services	postal, buisiness & communications centers	96,700
Jan-Pro Franchising Int'l. Inc.	commercial cleaning	1,000
Denny's Inc.	full-service family restaurant	858,000
RadioShack	electronics stores	60,000
Christmas Decor Inc.	holiday & event decorating services	176,000
Sign-A-Rama Inc.	signs	103,000
Sylvan Learning Centers	supplemental education	121,100

(1) Based on the number of new franchise units added. (2) Not including franchise fee, which varies.

Largest Corporate Mergers or Acquisitions in U.S.

Source: Securities Data Co.

(as of Oct. 2001; an * denotes an announced merger or acquisition not yet complete; year = year effective or announced)

Company	Acquirer	Dollars	Year
Time Warner	America Online, Inc.	$181.6 bil	2001
Warner-Lambert	Pfizer, Inc.	89.7 bil	2000
Mobil Corp.	Exxon Corp.	86.4 bil	1999
Citicorp	Travelers Group Inc.	72.6 bil	1998
Ameritech Corp.	SBC Communications Inc.	72.4 bil	1999
GTE Corp.	Bell Atlantic Corp.	71.3 bil	2000
Tele-Communications	AT&T	69.9 bil	1999
AirTouch Communications	Vodafone Group PLC	65.8 bil	1999
BankAmerica Corp.	NationsBank Corp.	61.6 bil	1998
AT&T Broadband & Internet Services	Comcast Corp.	57.5 bil	2001
US WEST	Qwest Communication	56.3 bil	2000
Amoco Corp.	British Petroleum Co. PLC	55.0 bil	1998
MediaOne Group	AT&T	51.9 bil	2000
Liberty Media Group	shareholders	46.0 bil	2001
Texaco*	Chevron	43.0 bil	2000
MCI Communications	WorldCom Inc.	41.4 bil	1998
SDL Inc.	JDS Uniphase Corp.	41.0 bil	2001
CBS Corp.	Viacom	40.9 bil	2000
Chrysler Corp.	Daimler-Benz AG	40.5 bil	1998
Wells Fargo & Co.	Norwest Corp.	34.4 bil	1998
ARCO	BP Amoco PLC	33.7 bil	2000
J.P. Morgan & Co.	Chase Manhattan	33.6 bil	2000
US West Media Group	shareholders	31.7 bil	1998
Hughes Electronics*	Echo Star Communications	31.5 bil	2001
Agilent Technologies	shareholders	31.2 bil	2000
Associates First Capital	Citigroup	$31.0 bil	2000
VoiceStream Wireless Corp.	Deutsche Telekom AG	30.8 bil	2001
NYNEX	Bell Atlantic	30.8 bil	1997
Electronic Data Sys.	shareholders	29.7 bil	1996
First Chicago NBD	BANC ONE Corp.	29.6 bil	1998
RJR Nabisco	Kohlberg Kravis Roberts	29.4 bil	1989
Pharmacia & Upjohn	Monsanto Co.	26.9 bil	2000
Associates First Capital	shareholders	26.6 bil	1998
Lucent Technologies	shareholders	24.1 bil	1996
Bestfoods	Unilever PLC	23.7 bil	2000
Compaq Computer*	Hewlett-Packard	23.5 bil	2001
Amer. General Corp.	American Int'l Group	23.4 bil	2001
AMFM, Inc.	Clear Channel Communications	22.7 bil	2000
Pacific Telesis Group	SBC Communications	22.4 bil	1997
General Re Corp.	Berkshire Hathaway Inc.	22.3 bil	1998
US Bancorp, MN	Firstar Corp.	21.1 bil	2001
Ascend Communications	Lucent Technologies	21.1 bil	1999
Network Solutions, Inc.	VeriSign, Inc.	20.8 bil	2000
Waste Management	USA Waste Services	20.0 bil	1998
Nabisco Holdings*	Philip Morris	19.4 bil	2000
AT&T Wireless	shareholders	18.8 bil	2001
Capital Cities/ABC	Walt Disney	18.3 bil	1996
SunAmerica Inc.	American Int'l. Group	18.1 bil	1999
Palm Inc. (3 Com Corp)	shareholders	17.9 bil	2000
Seagate Technology	Veritas Software	17.7 bil	2000

2001 Federal Corporate Tax Rates

Taxable Income Amount	Tax Rate	Taxable Income Amount	Tax Rate	Taxable Income Amount	Tax Rate
Not more than $50,000 . . .	15%	$100,001 to $335,000	39%	$15,000,001 to $18,333,333	38%
$50,001 to $75,000	25%	$335,001 to $10,000,000. .	34%	More than $18,333,333. . . .	35%
$75,001 to $100,000	34%	$10,000,001 to $15,000,000	35%		

Personal service corporations (used by incorporated professionals such as attorneys and doctors) pay a flat rate of 35%.

U.S. Capital Gains Tax

Source: George W. Smith IV, CPA, Partner, George W. Smith & Company. P.C.; as of Oct. 2001

The following shows how the maximum tax rate on net long-term capital gains for individuals has changed since 1960.

Year	Max %	Year	Max %	Year	Max %
1960	25.0	1978	28.0	1990	28.0[3]
1970	29.5	1981	20.0	1997	20.0[4]
1971	32.5	1987	28.0	1999	20.0[5]
1972	35.0[1]	1988	33.0[2]	2001	20/18[6]

(1) From 1972 to 1976, the interplay of minimum tax and maximum tax resulted in a marginal rate of 49.125%. (2) Statutory maximum of 28%, but "phase-out" notch increased marginal rate to 33%; interplay of all "phase-outs" could have increased the effective marginal rate to 49.5%. (3) The Budget Act of 1990 increased the statutory rate to 31% and capped the marginal rate at 28%; however, some taxpayers faced effective marginal rates of more than 34% because of the phase-out of personal exemptions and itemized deductions. (4) New rate is for those who, after July 28, 1997, sell capital assets held for more than 18 mos (12 mos for sales after Dec. 31, 1997). A 10% capital gains rate applies to individuals in the 15% income tax bracket. (Those who, after July 28, 1997, but before Jan. 1,1998, sell capital assets held between 12 and 18 mos will be taxed at the old top rate of 28%. Those who sold capital assets after May 6, 1997, but before July 29, 1997, will be taxed at the 20% rate, so long as such assets were held for at least a year.) (5) The IRS Restructuring and Reform Act of 1998 repealed the more-than-18-month holding period for sales after Dec. 31, 1997. Beginning Jan. 1, 1998, capital assets need only be held 12 months to have the 20%/10% capital gains rates apply. (6) For capital assets bought after Dec. 31, 2000, and held for more than 5 years, the 20% minimum capital gains rate will be lowered to 18%. The 10% rate will be lowered to 8%, regardless of when the assets were bought. The capital gains rate for the sale of collectibles such as antiques remains 28%. Capital gains on the sale of certain depreciable real estate will be taxed at 25%.

Global Stock Markets

Source: The Conference Board; not seasonally adjusted

Stock price indexes (1990=100):	June 1, 1960	June 1, 1970	June 1, 1980	June 1, 1990	2000		2001	
					Jan. 1	June 1	Jan. 1	June 1
United States	17.1	21.9	34.3	107.6	419.2	430.1	410.6	368.0
Japan	4.4	7.3	23.8	110.8	67.8	54.6	48.0	45.0
Germany	36.1	27.5	30.5	111.1	404.2	425.1	401.8	358.2
France	16.3	15.6	23.8	112.0	311.4	360.0	330.0	287.5
United Kingdom	8.2	11.6	24.9	108.2	274.9	282.9	279.9	252.0
Italy	28.9	20.6	15.9	117.3	276.8	308.5	303.3	254.1
Canada	14.8	25.0	60.3	103.6	247.9	304.2	272.5	226.1

U.S. Holdings of Foreign Stocks

Source: Bureau of Economic Analysis, U.S. Dept. of Commerce

(billions of dollars)

	1997[R]	1998[R]	1999
Western Europe	**$721.1**	**$960.5**	**$1,167.8**
Of which: United Kingdom	217.5	295.6	374.6
Finland	14.8	45.6	160.2
France	85.0	130.4	183.2
Germany	65.0	104.4	117.6
Ireland	14.1	19.5	18.2
Italy	41.5	59.1	53.5
Netherlands	107.0	115.4	141.9
Spain	25.2	37.7	35.7
Sweden	38.8	43.7	74.8
Switzerland	61.9	73.6	64.3
Canada	**70.8**	**62.0**	**100.7**
Japan	**136.4**	**145.9**	**273.7**
Latin America	$92.5	$54.0	$89.1
Of which: Argentina	12.9	8.9	11.3
Brazil	31.3	17.4	28.9
Mexico	35.0	27.8	30.2
Other W. Hemisphere	45.8	77.8	129.0
Of which: Bermuda	22.6	37.2	45.9
Netherlands Antilles	15.8	24.8	26.7
Other countries and territories	**141.2**	**176.0**	**266.3**
Of which: Australia	31.1	34.3	39.2
Hong Kong	28.1	27.0	38.7
Singapore	10.2	10.3	16.3
TOTAL HOLDINGS	**$1,207.8**	**$1,476.2**	**$2,026.6**

(R) Revised figures.

Gold Reserves of Central Banks and Governments

Source: International Financial Statistics, IMF; million fine troy ounces

Year end	All countries[1]	United States	Belgium	Canada	France	Germany[2]	Italy	Japan	Nether-lands	Switzer-land	United Kingdom
1975	1,018.71	274.71	42.17	21.95	100.93	117.61	82.48	21.11	54.33	83.20	21.03
1980	952.99	264.32	34.18	20.98	81.85	95.18	66.67	24.23	43.94	83.28	18.84
1985	949.39	262.65	34.18	20.11	81.85	95.18	66.67	24.33	43.94	83.28	19.03
1990	939.01	261.91	30.23	14.76	81.85	95.18	66.67	24.23	43.94	83.28	18.94
1995	908.79	261.70	20.54	3.41	81.85	95.18	66.67	24.23	34.77	83.28	18.43
1996	906.10	261.66	15.32	3.09	81.85	95.18	66.67	24.23	34.77	83.28	18.43
1997	890.57	261.64	15.32	3.09	81.89	95.18	66.67	24.23	27.07	83.28	18.42
1998	966.15	261.61	9.52	2.49	102.37	118.98	83.36	24.23	33.83	83.28	23.00
1999	940.51	261.67	8.30	1.81	97.24	111.52	78.83	24.23	31.57	83.28	20.55
2000	950.63	261.61	8.30	1.18	97.25	111.52	78.83	24.55	29.32	77.79	15.67

(1) Covers IMF members with reported gold holdings. For countries not listed above, see International Monetary Fund's *International Financial Statistics Report.* (2) West Germany prior to 1991.

Record One-Day Gains and Losses on the Dow Jones Industrial Average

Source: Dow Jones & Co., Inc.; as of Sept. 30, 2001

GREATEST POINT GAINS

Rank	Date	Close	Net Chg	% Chg
1.	3/16/2000	10630.60	499.19	4.93
2.	4/5/2001	9918.05	402.63	4.23
3.	4/18/2001	10615.83	399.10	3.91
4.	9/8/1998	8020.78	380.53	4.98
5.	9/24/2001	8603.86	368.05	4.47
6.	5/16/2001	11215.92	342.95	3.15
7.	12/5/2000	10898.72	338.62	3.21
8.	10/28/1997	7498.32	337.17	4.71
9.	10/15/1998	8299.36	330.58	4.15
10.	3/15/2000	10131.41	320.17	3.26

GREATEST POINT LOSSES

Rank	Date	Close	Net Chg	% Chg
1.	9/17/2001	8920.70	–684.81	–7.13
2.	4/14/2000	10305.77	–617.78	–5.66
3.	10/27/1997	7161.15	–554.26	–7.19
4.	8/31/1998	7539.07	–512.61	–6.37
5.	10/19/1987	1738.74	–508.00	–22.61
6.	3/13/2001	10208.25	–436.37	–4.1
7.	9/20/2001	8376.21	–382.92	–4.37
8.	10/12/2000	10034.58	–379.21	–3.64
9.	3/7/2000	9796.03	–374.47	–3.68
10.	1/4/2000	10997.93	–359.58	–3.17

GREATEST % GAINS

Rank	Date	Close	Net Chg	% Chg
1.	10/6/1931	99.34	12.86	14.87
2.	10/30/1929	258.47	28.40	12.34
3.	9/21/1932	75.16	7.67	11.36
4.	10/21/1987	2,027.85	186.84	10.15
5.	8/3/1932	58.22	5.06	9.52
6.	2/11/3192	78.60	6.80	9.47
7.	11/14/1929	217.28	18.59	9.36
8.	12/18/1931	80.69	6.90	9.35
9.	2/13/1932	85.82	7.22	9.19
10.	5/6/1932	59.01	4.91	9.08

GREATEST % LOSSES

Rank	Date	Close	Net Chg	% Chg
1.	10/19/1987	1,738.74	–508.00	–22.61
2.	10/28/1929	260.54	–38.33	–12.82
3.	10/29/1929	230.07	–30.57	–11.73
4.	11/6/1929	232.13	–25.55	–9.92
5.	12/18/1899	58.27	–5.57	–8.72
6.	8/12/1932	63.11	–5.79	–8.40
7.	3/14/1907	76.23	–6.89	–8.29
8.	10/25/1987	1,793.93	–156.83	–8.04
9.	7/21/1933	88.71	–7.55	–7.84
10.	10/18/1937	125.73	–10.57	–7.75

Dow Jones Industrial Average Since 1963

High		YEAR	Low	
Dec. 18	767.21	**1963**	Jan. 2	646.79
Nov. 18	891.71	**1964**	Jan. 2	766.08
Dec. 31	969.26	**1965**	June 28	840.59
Feb. 9	995.15	**1966**	Oct. 7	744.32
Sept. 25	943.08	**1967**	Jan. 3	786.41
Dec. 3	985.21	**1968**	Mar. 21	825.13
May 14	968.85	**1969**	Dec. 17	769.93
Dec. 29	842.00	**1970**	May 6	631.16
Apr. 28	950.82	**1971**	Nov. 23	797.97
Dec. 11	1036.27	**1972**	Jan. 26	889.15
Jan. 11	1051.70	**1973**	Dec. 5	788.31
Mar. 13	891.66	**1974**	Dec. 6	577.60
July 15	881.81	**1975**	Jan. 2	632.04
Sept. 21	1014.79	**1976**	Jan. 2	858.71
Jan. 3	999.75	**1977**	Nov. 2	800.85
Sept. 8	907.74	**1978**	Feb. 28	742.12
Oct. 5	897.61	**1979**	Nov. 7	796.67
Nov. 20	1000.17	**1980**	Apr. 21	759.13
Apr. 27	1024.05	**1981**	Sept. 25	824.01

High		YEAR	Low	
Dec. 27	1070.55	**1982**	Aug. 12	776.92
Nov. 29	1287.20	**1983**	Jan. 3	1027.04
Jan. 6	1286.64	**1984**	July 24	1086.57
Dec. 16	1553.10	**1985**	Jan. 4	1184.96
Dec. 2	1955.57	**1986**	Jan. 22	1502.29
Aug. 25	2722.42	**1987**	Oct. 19	1738.74
Oct. 21	2183.50	**1988**	Jan. 20	1879.14
Oct. 9	2791.41	**1989**	Jan. 3	2144.64
July 16	2999.75	**1990**	Oct. 11	2365.10
Dec. 31	3168.83	**1991**	Jan. 9	2470.30
June 1	3413.21	**1992**	Oct. 9	3136.58
Dec. 29	3794.33	**1993**	Jan. 20	3241.95
Jan. 31	3978.36	**1994**	Apr. 4	3593.35
Dec. 13	5216.47	**1995**	Jan. 30	3832.08
Dec. 27	6560.91	**1996**	Jan. 10	5032.94
Aug. 6	8259.31	**1997**	Apr. 11	6391.69
Nov. 23	9374.27	**1998**	Aug. 31	7539.07
Dec. 31	11497.12	**1999**	Jan. 22	9120.67
Jan. 14	11722.98	**2000**	Mar. 7	9796.03

Milestones of the Dow Jones Industrial Average

(as of Sept. 30, 2001)

First close over...

100 .... Jan. 12, 1906
500 .... Mar. 12, 1956
1000 .... Nov. 14, 1972
1500 .... Dec. 11, 1985
2000 .... Jan. 8, 1987
2500 .... July 17, 1987
3000 .... April 17, 1991
3500 .... May 19, 1993
4000 .... Feb. 23, 1995
4500 .... June 16, 1995
5000 .... Nov. 21, 1995
5500 .... Feb. 8, 1996
6000 .... Oct. 14, 1996
6500 .... Nov. 25, 1996

First close over...

7000 .... Feb. 13, 1997
7500 .... June 10, 1997
8000 .... July 16, 1997
8100 .... July 24, 1997
8200 .... July 30, 1997
8100 .... July 24, 1997
8200 .... July 30, 1997
8300 .... Feb. 12, 1998
8400 .... Feb. 18, 1998
8300 .... Feb. 12, 1998
8400 .... Feb. 18, 1998
8500 .... Feb. 27, 1998
8600 .... Mar. 10, 1998
8700 .... Mar. 16, 1998

First close over...

8800 ... Mar. 19, 1998
8900 ... Mar. 20, 1998
9000 ... Apr. 6, 1998
9100 ... Apr. 14, 1998
9200 ... May 13, 1998
9300 ... July 16, 1998
9500 ... Jan. 6, 1999*
9600 ... Jan. 8, 1999
9700 ... Mar. 5, 1999
9800 ... Mar. 11, 1999
9900 ... Mar. 15, 1999
10000 ... Mar. 29, 1999
10100 ... Apr. 8, 1999
10300 ... Apr. 12, 1999*

First close over...

10400 ....Apr. 14, 1999
10500 ....Apr. 21, 1999
10700 ....Apr. 22, 1999*
10800 ....Apr. 27, 1999
11000 ....May 3, 1999*
11100 ....May 13, 1999
11200 ....July 12, 1999
11300 ....Aug. 25, 1999
11400 ....Dec. 23, 1999
11500 ....Jan. 7, 2000
11700 ....Jan. 14, 2000*

*9400, 10200, 10600, 10900, and 11600 are not listed because the Dow had risen another 100 points or more by the time the market closed for the day.

Components of the Dow Jones Averages

(as of Sept. 30, 2001)

Dow Jones Industrial Average

Aluminum Co. of America (Alcoa)
American Express
AT&T
Boeing
Caterpillar
Citigroup
Coca-Cola
DuPont
Eastman Kodak
Exxon Mobil
General Electric
General Motors
Hewlett-Packard
Home Depot*
Honeywell International
IBM
Intel*
International Paper
J.P. Morgan
Johnson & Johnson
McDonald's
Merck
Microsoft*
Minnesota Mining & Manufacturing
Philip Morris
Procter & Gamble
SBC Communications*
United Technologies
Wal-Mart
Walt Disney

*These companies became component stocks of the DJIA Nov. 1, 1999, replacing Chevron; Goodyear Tire & Rubber; Sears, Roebuck; and Union Carbide. The inclusion of Intel and Microsoft, both traded on the Nasdaq stock market, marked the first time a DJIA component has not been listed on the NYSE since the Dow's inception in 1896.

Dow Jones Utility Average

AES Corp.
American Electric Power
Columbia Energy Group
Consolidated Edison
Dominion Resources
Duke Energy
Edison International
Enron
PG&E
Public Service Enterprise Group
Reliant Energy
Southern Co.
TXU
Unicom
Williams Cos.

Dow Jones Transportation Average

Airborne Freight
Alexander & Baldwin
AMR (American Airlines)
Burlington Northern Santa Fe
CNF Transportation
CSX
Delta Air Lines
FDX
GATX
J.B. Hunt Transportation
Norfolk Southern
Northwest Airlines
Roadway Express
Ryder System
Southwest Air Lines
UAL (United Air Lines)
Union Pacific
US Airways
USFreightways
Yellow Corp.

Record One-Day Gains and Losses on the Nasdaq Stock Market

Source: Nasdaq Stock Market; as of Oct. 3, 2001

GREATEST POINT GAINS

Rank	Date	Point Change
1.	1/3/01	324.83
2.	12/5/00	274.05
3.	4/18/00	254.41
4.	5/30/00	254.37
5.	10/19/00	247.04
6.	10/13/00	242.09
7.	6/2/00	230.88
8.	4/25/00	228.75
9.	4/17/00	217.87
10.	6/1/00	181.59

GREATEST POINT LOSSES

Rank	Date	Point Change
1.	4/14/00	–355.49
2.	4/3/00	–349.15
3.	4/12/00	–28.27
4.	4/10/00	–258.25
5.	1/4/00	–229.46
6.	3/14/00	–200.61
7.	5/10/00	–200.28
8.	5/23/00	–199.66
9.	10/25/00	–190.22
10.	3/29/00	–189.22

GREATEST % GAINS

Rank	Date	% Change
1.	1/3/01	14.17
2.	12/5/00	10.48
3.	4/5/01	8.92
4.	4/18/01	8.12
5.	5/30/00	7.94
6.	10/13/00	7.87
7.	10/19/00	7.79
8.	12/22/00	7.56
9.	10/21/87	7.34
10.	4/18/00	7.19

GREATEST % LOSSES

Rank	Date	% Change
1.	10/19/87	–11.35
2.	4/14/00	–9.67
3.	10/20/87	–9.00
4.	10/26/87	–9.00
5.	8/31/98	–8.56
6.	4/3/00	–7.64
7.	1/2/01	–7.23
8.	12/20/00	–7.12
9.	4/12/00	–7.06
10.	4/10/00	–7.06

IT'S A FACT: Between Mar. 2000 and Aug. 2001, an estimated $5 trillion in stock market wealth vanished. That loss was likely quite concentrated: the top 1% of equity owners hold about 50% of all corporate stock, and the top 5% of owners hold about 80%, according to the American Enterprise Institute.

Nasdaq Stock Market Since 1971

High	YEAR	Low	High	YEAR	Low	High	YEAR	Low
114.12	**1971**	99.68	223.96	**1981**	170.80	586.35	**1991**	352.85
135.15	**1972**	113.65	241.63	**1982**	158.92	676.95	**1992**	545.85
136.84	**1973**	88.67	329.11	**1983**	229.88	790.56	**1993**	645.02
96.53	**1974**	54.87	288.41	**1984**	223.91	803.93	**1994**	691.23
88.00	**1975**	60.70	325.53	**1985**	245.82	1072.82	**1995**	740.53
97.88	**1976**	78.06	411.21	**1986**	322.14	1328.45	**1996**	978.17
105.05	**1977**	93.66	456.27	**1987**	288.49	1748.62	**1997**	1194.39
139.25	**1978**	99.09	397.54	**1988**	329.00	2200.63	**1998**	1357.09
152.29	**1979**	117.84	487.60	**1989**	376.87	4090.61	**1999**	2193.13
208.29	**1980**	124.09	470.30	**1990**	322.93	5048.62	**2000**	2332.78

Milestones of the Nasdaq Stock Market

Source: Nasdaq Stock Market; as of Oct. 15, 2001

First close over...		First close over...		First close over...	
100	Feb. 8, 1971	1,000	July 17, 1995	3,500	Dec. 3, 1999
200	Nov. 13, 1980	1,500	July 11, 1997	4,000	Dec. 29, 1999
300	May 6, 1986	2,000	July 16, 1998	4,500	Feb. 17, 2000
400	May 30, 1986	2,500	Jan. 29, 1999	5,000	Mar. 9, 2000
500	Apr. 12, 1991	3,000	Nov. 3, 1999		

Most Active Common Stocks in 2000

New York Exchange Volume (millions of shares)	
Lucent Technologies Inc.	3,972.4
Compaq	3,610.3
AT&T Corp.	3,417.6
America Online	3,106.0
Nortel Networks	2,863.0
Pfizer Inc.	2,457.7
Citigroup Inc.	2,366.6
General Electric	2,363.0
Nokia Corp.	2,191.8
Motorola, Inc.	1,178.5

American Exchange Volume (millions of shares)	
NASDAQ 100 Shares	4,929.9
Standard & Poor's SPDR Trust I	1,461.8
Diamonds Trust	308.9
Nabors Industries	296.7
IVAX Corp.	203.5
Merrill Lynch Internet HOLDRS	183.5
Standard & Poor's MidCap Trust I	176.4
Grey Wolf Inc.	174.7
Biotech Holdings	157.7
Devon Energy Corp.	150.0

NASDAQ Volume (millions of shares)	
Cisco Systems, Inc.	14,122.8
Intel Corp.	13,669.2
Oracle Corp.	13,565.6
Sun Microsystems, Inc.	10,404.0
Microsoft Corp.	10,324.6
WorldCom, Inc.	8,634.3
3Com Corp.	8,562.8
Dell Computer Corp.	8,317.6
JDS Uniphase Corp.	6,320.5
LM Ericsson Telephone Company	5,359.2

Average Yields of Long-Term Treasury, Corporate, and Municipal Bonds

Source: Office of Market Finance, U.S. Dept. of the Treasury

Period	Treasury 30-year bonds	New Aa corporate bonds[1]	New Aa municipal bonds[2]
1986			
June	7.57	9.39	7.75
Dec.	7.37	8.87	6.70
1987			
June	8.57	9.64	7.69
Dec.	9.12	10.22	7.83
1988			
June	9.00	10.08	7.67
Dec.	9.01	10.05	7.40
1989			
June	8.27	9.24	6.94
Dec.	7.90	9.23	6.76
1990			
June	8.46	9.69	6.98
Dec.	8.24	9.55	6.85
1991			
June	8.47	9.37	6.90
Dec.	7.70	8.55	6.43
1992			
June	7.84	8.45	6.32
Dec.	7.44	8.12	6.02
1993			
June	6.81	7.48	5.54
Dec.	6.25	7.22	5.27
1994			
June	7.40	8.16	5.96
Dec.	7.87	8.66	6.63
1995			
June	6.57	7.42	5.61
Dec.	6.06	7.02	5.46
1996			
June	7.06	8.00	5.82
Dec.	6.55	7.45	5.47
1997			
June	6.77	7.71	5.39
Dec.	5.99	6.68	5.07
1998			
Jun	5.70	6.43	5.01
Dec	5.06	6.13	4.90
1999			
Jun.	6.04	7.21	5.31
Dec.	6.35	7.55	5.91
2000			
June	5.93	7.75	5.74
Dec.	5.49	7.21	5.23
2001			
Jan.	5.54	7.15	5.07
Mar.	5.34	6.87	5.08
June	5.67	7.11	5.18

(1) Treasury series based on 3-week moving average of reoffering yields of new corporate bonds rated Aa by Moody's Investors Service with an original maturity of at least 20 years. (2) Index of new reoffering yields on 20-year general obligations rated Aa by Moody's Investors Service.

Performance of Mutual Funds by Type, 2000

Source: CDA/Wiesenberger, Rockville, MD, 800-232-2285

(data for period ending Sept. 30, 2000)

Fund Type/Fund Objective	AVERAGE RETURN 1-year	3-year	5-year
Diversified Stock			
Aggressive Growth	45.80%	21.33%	19.87%
Equity Income	8.67	7.21	13.79
Growth-Domestic	25.78	16.54	19.31
Growth & Income	13.57	10.61	16.43
Mid Cap	49.86	21.60	20.59
S&P 500 Index	12.69	15.86	21.17
Small Cap	39.39	12.14	16.89
Specialty Stock			
Sector-Energy/Natural Res	23.28	–0.09	10.82
Sector-Financial Services	23.62	10.49	19.29
Sector-Precious Metals	–28.60	–19.95	–15.19
Sector-Health/Biotechnology	91.10	25.13	24.21
Sector-Other	12.14	6.88	11.30
Sector-Real Estate	23.38	–0.54	10.36
Sector-Tech/Communications	61.66	42.82	31.64
Sector-Utilities	21.91	18.28	17.11
World Stock			
Emerging Market Equity	8.36	–7.41	0.22
Global Equity	20.59	11.53	14.09
Non-US Equity	12.45	7.06	8.36
Emerging Market Income	22.16	1.29	13.45
Hybrid			
Asset Allocation-Domestic	11.81%	9.62%	12.70%
Asset Allocation-Global	8.95	4.68	6.58
Balanced-Domestic	12.02	9.01	12.63
Balanced-Global	10.40	7.67	10.47
Bond			
Corporate-High Yield	–0.39	0.33	5.37
Corporate-Investment Grade	4.64	4.26	5.41
Convertible	30.92	12.78	15.75
General Bd-Investment Grade	5.52	4.58	5.46
General Bd-Long	5.81	4.80	6.05
General Bd-Short & Interm	5.56	4.97	5.57
General Mortgage	5.95	4.92	5.57
Global Income	1.38	1.28	4.48
Loan Participation	5.80	5.99	6.45
Multi-Sector Bond	4.24	2.87	5.91
US Government/Agency	5.74	4.74	5.25
US Government-Long	6.32	4.92	5.61
US Government-Short & Interm	5.53	4.81	5.24
US Treasury	7.56	5.83	6.09
Municipal Bond			
Municipal-High Yield	1.72	2.60	4.72
Municipal-Insured	5.43	3.48	4.77
Municipal-National	4.46	3.28	4.54
Municipal-Single State	4.71	3.27	4.64

Chicago Board of Trade, Contracts Traded 1991, 2000

Source: Chicago Board of Trade

	1991	2000	% change 1991-2000
FUTURES GROUP			
Agricultural	31,896,207	48,480,178	52.0
Financial	78,578,593	137,590,254	75.1
Stock index	702,927	3,572,461	408.2
Metals	133,606	19,514	–85.4
Total futures	**111,311,333**	**189,662,407**	**70.4**
OPTIONS GROUP			
Agricultural	5,105,826	11,823,282	131.6
Financial	23,014,276	31,842,462	38.4
Stock index	3,059	200,379	6440.5
Metals	2,804	22	–99.2
PCS insurance	—	6	100.0
Total options	**28,125,965**	**43,866,151**	**56.0**
COMBINED FUTURES AND OPTIONS			
Agricultural	37,002,033	60,303,460	62.9
Financial	101,529,869	169,432,716	66.8
Stock index	705,986	3,772,840	434.4
Metals	136,410	19,536	–85.7
PCS insurance	—	6	100.0
GRAND TOTAL	**139,437,298**	**233,528,558**	**67.5**

U.S. Mutual Fund Shareholders[1]

Source: The Investment Company Institute

Shareholder Characteristics	
Median age[2]	44
Median annual household Income	$55,000
Median household financial assets	$80,000
Median household mutual fund assets	$25,000
Median number of funds owned	4
Employed[2]	82%
Married or living with a partner	74%
Spouse or partner employed	75%
Four-year college degree or more[2]	50%
Owning:	
Equity funds	88%
Bond funds	42%
Hybrid funds	35%
Money market funds	48%
Variable annuity invested in mutual funds	23%

Year	Households owning mutual funds (in millions)
1980	4.6
1984	10.2
1988	22.2
1992	25.8
1994	30.2
1996	36.8
1998	44.4
1999	48.4
2000	50.6

(1) Data include households owning mutual funds inside and outside employer-sponsored retirement plans. (2) Refers to the household's responding financial decision maker for mutual fund investments. (3) Data from 1980-1988 exclude households owning mutual funds solely through employer-sponsored retirement pllans.

Distribution of Financial Assets of U.S. Families

Source: Federal Reserve System (by type of asset, in percent)

	1989	1992	1995	1998
TOTAL	100	100	100	100
Certificates of deposit	10.2	8.1	5.7	4.3
U.S. savings bonds	1.5	1.1	1.3	0.7
Bonds	10.2	8.4	6.3	4.3
Stocks	15.0	16.5	15.7	22.7
Mutual funds (excl. money market)	5.3	7.7	12.7	12.5
Retirement accounts	21.5	25.5	27.9	27.5
Cash value of life insurance	6.0	6.0	7.2	6.4
Other managed assets	6.6	5.4	5.9	8.6
Other financial	4.8	3.8	3.4	1.7
Financial assets as a percentage of total assets	30.4	31.5	36.6	40.6

Stock Ownership of U.S. Families, by Income and Age, 1989, 1992, 1995, and 1998

Source: Federal Reserve System

(in percent, except as noted)

		Families having direct or indirect stock holdings[1]				Median value of portfolios for families with stock holdings (thousands of 1998 dollars)				Stock holdings as share of fnancial assets[2]			
		1989	1992	1995	1998	1989	1992	1995	1998	1989	1992	1995	1998
All families		**31.6**	**36.7**	**40.4**	**48.8**	**$10.8**	**$12.0**	**$15.4**	**$25.0**	**27.8%**	**33.7%**	**40.0%**	**53.9%**
Annual Income (in thousands of dollars):	Under $10	*	6.8	5.4	7.7	*	6.2	3.2	4.0	*	15.9	12.9	24.8
	$10-25	12.7	17.8	22.2	24.7	6.4	4.6	6.4	9.0	11.7	15.3	26.7	27.5
	$25-50	31.5	40.2	45.4	52.7	6.0	7.2	8.5	11.5	16.9	23.7	30.3	39.1
	$50-100	51.5	62.5	65.4	74.3	10.2	15.4	23.6	35.7	23.2	33.5	39.9	48.8
	Over $100	81.8	78.3	81.6	91.0	53.5	71.9	85.5	150.0	35.3	40.2	46.4	63.0
By age of family head (years):	Under 35	22.4	28.3	36.6	40.7	3.8	4.0	5.4	7.0	20.2	24.8	27.2	44.8
	35-44	38.9	42.4	46.4	56.5	6.6	8.6	10.6	20.0	29.2	31.0	39.5	54.7
	45-54	41.8	46.4	48.9	58.6	16.7	17.1	27.6	38.0	33.5	40.6	42.9	55.7
	55-64	36.2	45.3	40.0	55.9	23.4	28.5	32.9	47.0	27.6	37.3	44.4	58.3
	65-74	26.7	30.2	34.4	42.6	25.8	18.3	36.1	56.0	26.0	31.6	35.8	51.3
	75 +	25.9	25.7	27.9	29.4	31.8	28.5	21.2	60.0	25.0	25.4	39.8	48.7

*Denotes insufficient data. (1) Indirect holdings are those in mutual funds, retirement accounts, and other managed assets. (2) Among stock holding families

Minerals

Source: U.S. Geological Survey, U.S. Dept. of the Interior; as of mid-2001

Aluminum: the second most abundant metallic element in the earth's crust. Bauxite is the main source of aluminum; convert to aluminum equivalent by multiplying by 0.232. Guinea, Brazil, and Australia have 58% of the world's reserves. Aluminum is used in the U.S. principally in transportation (37%), packaging (23%), and building (15%).

Chromium: about 3/4 of the world's production of chromite, the chief source of chromium, is in India, Kazakhstan, Turkey, and South Africa. The chemical and metallurgical industries use about 90% of all chromite consumed in the world.

Cobalt: used in superalloys for jet engines, chemicals (paint driers, glass and ceramics, catalysts, magnetic coatings, and rechargeable batteries), permanent magnets, and cemented carbides for cutting tools. Australia, Belgium, Canada, Congo (formerly Zaire), Finland, Morocco, Norway, Russia, and Zambia account for most of the world cobalt refinery production.

Columbium (niobium): used mostly as an additive in steelmaking and in superalloys. Brazil and Canada are the world's leading columbium raw materials (feedstock) producers. There is no U.S. columbium mining industry.

Copper: main uses of copper in the U.S. are in building construction (39%), electrical and electronic products (28%), transportation (11%), industrial machinery and equipment (11%), and consumer and general products (11%). The leading producer is Chile, followed by the U.S., Indonesia, Australia, Canada, China, Russia, Peru, Poland, Kazakhstan, Mexico, and Zambia. Principal mining states are Arizona, Utah, and New Mexico.

Gold: used in the U.S. in jewelry and the arts (85%), electrical and electronics (7%), other industries (5%), and dentistry (3%). South Africa has about half of the world's resources; significant quantities also are present in the U.S., Australia, Russia, Uzbekistan, Canada, and Brazil. Gold is mined in nearly all the Western U.S. states and in Alaska.

Iron ore: the source of primary iron for the world's iron and steel industries. Major iron ore producers include Australia, Brazil, China, India, Russia, the U.S., and Ukraine.

Lead: Australia, China, the U.S., Peru, Mexico, and Canada are the world's largest producers of lead. Transportation accounts for the major end use in the U.S., with 75% used in batteries, bearings, casting metals, and solders. Other uses include emergency power supply batteries, construction sheeting, sporting ammunition, and power cable coverings. The U.S. produces and consumes about 25% of the world's lead metal, including primary and recycled material.

Manganese: essential to iron and steel production. The U.S., Japan, and Western Europe have exhausted nearly all of their economically minable manganese. South Africa and the former Soviet Union have over 80% of the world's identified resources.

Nickel: vital to the stainless steel industry; used to make superalloys for the chemical and aerospace industries. Leading producers include Russia, Canada, Australia, New Caledonia, and Indonesia.

Platinum-Group Metals: the platinum group consists of 6 related metals: platinum, palladium, rhodium, ruthenium, iridium, and osmium. They commonly occur together in nature and are among the scarcest of the metallic elements. They are consumed in the U.S. by the following industries: automotive, electrical and electronic, chemical, and dental and medical. The automotive, chemical, and petroleum-refining industries use platinum-group metals mainly as catalysts. Russia and South Africa have most of the world's reserves.

Silver: used in the following U.S. industries: photography, electrical and electronic products, sterlingware, electroplated ware, and jewelry. Silver is mined in more than 60 countries. Nevada produces more than 40% of U.S. silver, Idaho 16%.

Tantalum: a refractory metal with unique electrical, chemical, and physical properties; used in the U.S. mostly to produce electronic components, mainly tantalum capacitors. Africa, Australia, Brazil, and Canada are the world's leading tantalum raw materials (feedstock) producers. There is no U.S. tantalum mining industry.

Titanium: Ilmenite and rutile are the major sources of titanium. Approximately 95% of titanium minerals is used to produce titanium dioxide pigments for paint, paper, and plastics. The remainder is used to produce metal, carbides and chemicals. Major mining operations are in Australia, Canada, Norway, and South Africa. U.S. mine production is in Florida and Virginia.

Vanadium: used as an alloying element in steel and aerospace titanium alloys, as a catalyst in the production of maleic and phthalic anhydride, and in the production of sulfuric acid. South Africa, Russia, and China are the world's largest producers of vanadium-bearing ores and concentrates.

Zinc: used as a protective coating on steel, as diecastings, as an alloying metal with copper to make brass, and as a component of chemical compounds in rubber and paints. It is mined in 41 countries. China is the leading producer, followed by, Australia, Canada, Peru, the U.S., and Mexico. In the U.S., mine production comes mostly from Alaska, Tennessee, New York, and Missouri.

World Mineral Reserve Base, 2001

Source: U.S. Geological Survey, U.S. Dept. of the Interior; as of mid-2001

Mineral	Reserve Base[1]	Mineral	Reserve Base[1]	Mineral	Reserve Base[1]
Aluminum	35,000 mil metric tons[2]	Iron ore	300,000 mil metric tons	Silver	420,000 metric tons
Chromium	7,600 mil metric tons	Lead	140 mil metric tons	Tantalum	73,000 metric tons
Cobalt	9.9 mil metric tons	Manganese	5,000 mil metric tons	Titanium	660 mil metric tons[4]
Columbium	5.5 mil metric tons	Nickel	150 mil metric tons	Vanadium	27 mil metric tons
Copper	650 mil metric tons	Platinum-Group Metals	80,000 metric tons	Zinc	430 mil metric tons
Gold	77,000 metric tons[3]				

(1) Includes demonstrated reserves that are currently economic or marginally economic, plus some that are currently subeconomic. (2) Bauxite. (3) Excludes China and some other countries for which reliable data were not available. (4) Titanium dioxide (TiO_2) content of ilmenite and rutile.

U.S. Nonfuel Mineral Production—10 Leading States in 2000

Source: U.S. Geological Survey, U.S. Dept. of the Interior

Rank/State	Value (mil of $)	Percent of U.S. total	Principal minerals, in order of value
1. California	3,350	8.34	Sand & gravel (construction), cement, boron, stone (crushed), gold
2. Nevada	2,800	6.96	Gold, sand & gravel (construction), silver, lime, diatomite
3. Arizona	2,550	6.35	Copper, sand & gravel (construction), cement, molybdenum concentrates, stone (crushed)
4. Texas	2,050	5.09	Cement, stone (crushed), sand & gravel (construction), lime, salt
5. Florida	1,920	4.78	Phosphate rock, stone (crushed), cement, sand & gravel (construction)
6. Michigan	1,670	4.17	Cement, iron ore, sand & gravel (construction), stone (crushed), magnesium compounds
7. Georgia	1,660	4.13	Clays, stone (crushed), cement, sand & gravel (construction)
8. Minnesota	1,570	3.90	Iron ore, sand & gravel (construction), stone (crushed), sand & gravel (industrial), stone (dimension)
9. Utah	1,420	3.53	Copper, gold, cement, sand & gravel (construction), magnesium metal
10. Missouri	1,320	3.28	Stone (crushed), lead, lime, zinc

U.S. Nonfuel Minerals Production

Source: U.S. Geological Survey, U.S. Dept. of the Interior

Production as measured by mine shipments, sales, or marketable production (including consumption by producers).

Mineral	Unit	1995	1996	1997	1998	1999	2000
Beryllium (metal equivalent)	metric tons	202	211	231	243	200	180
Copper (recoverable content of ores, etc.)	thousand metric tons	1,850	1,920	1,940	1,860	1,600	1,440
Gold (recoverable content of ores, etc.)	metric tons	317.0	326.0	362.0	366.0	341.0	353.0
Iron ore, usable (includes byproduct material)	million metric tons	62.5	62.1	63.0	62.9	57.7	63.1
Lead (in concentrate)	thousand metric tons	386	426	448	481	503	457
Magnesium metal (primary)	thousand metric tons	142	133	125	106	W	W
Molybdenum (content of ore and concentrate)	metric tons	58,000	56,000	58,900	53,300	43,000	41,000
Nickel (content of ore and concentrate)	metric tons	1,557	1,333	—	—	—	—
Silver (recoverable content of ores, etc.)	metric tons	1,560	1,570	2,180	2,060	1,950	1,860
Zinc (recoverable content of ores, etc.)	thousand metric tons	603	586	592	709	808	786
Asbestos	thousand metric tons	9	10	7	6	7	5
Barite	thousand metric tons	543	662	692	476	434	392
Boron minerals	thousand metric tons	728	581	604	587	618	564
Bromine	million kilograms	218	227	247	230	239	228
Cement (portland, masonry, etc.)	thousand metric tons	76,906	79,266	82,582	83,931	86,600[E]	87,846
Clays	thousand metric tons	43,000	43,100	41,800	41,900	42,200	40,800
Diatomite	thousand metric tons	722	729	773	725	747	677
Feldspar	thousand metric tons	880	890	900[E]	820[E]	875[E]	790[E]
Fluorspar	thousand metric tons	51	8	—	—	—	—
Garnet (industrial)	metric tons	46,300	60,900	64,900	74,000	60,700	60,200
Gemstones	million dollars	48.7	43.6	25.0	14.3	16.1	17.2
Gypsum	thousand metric tons	16,600	17,500	18,600	19,000	22,400	19,500
Helium (extracted from natural gas)	million cubic meters	101	103	116	112.0	118[E]	117[E]
Helium (Grade A sold)	million cubic meters	96	95	107	112.0	108[E]	125[E]
Iodine	thousand kilograms	1,220	1,270	1,320	1,490	1,620	1,470
Lime	thousand metric tons	18,530	19,225	19,678	20,132	19,565	19,555
Mica (scrap & flake)	thousand metric tons	108	97	114	87	104	101
Peat	thousand metric tons	648	549	661	685	731	755
Perlite (sold and used by producers)	thousand metric tons	700	684	706	685	711	672
Phosphate rock (marketable product)	thousand metric tons	43,500	45,400	45,900	44,200	40,600	38,600
Potash (K_2O equivalent)	thousand metric tons	1,480	1,390	1,400	1,300	1,200	1,300
Pumice and pumicite	thousand metric tons	529	612	577	583	643	697
Salt	thousand metric tons	40,800	42,900	40,600	40,800	41,000	43,300
Sand and gravel (construction)	thousand metric tons	907,000	914,000	961,000	1,080,000	1,080,000[E]	1,120
Sand and gravel (industrial)	thousand metric tons	28,200	27,800	28,500	28,200	28,900	28,400
Soda ash (sodium carbonate)	thousand metric tons	10,100	10,200	10,700	10,100	10,200	10,200
Sodium sulfate (natural)	thousand metric tons	327	306	318	290	NA	NA
Stone (crushed)	million metric tons	1,260	1,330	1,410	1,510	1,560[E]	1,560
Stone (dimension)	thousand metric tons	1,160	1,150	1,180	1,140	1,250[E]	1,254
Sulfur (in all forms)	thousand metric tons	11,800	12,000	12,000	11,600[E]	11,300	10,300
Talc	thousand metric tons	1,060	994	1,050	971	925	851

(W) Withheld to avoid disclosing company proprietary data. (—) No production. (E) Estimated. (NA) Not available.

U.S. Reliance on Foreign Supplies of Minerals

Source: U.S. Geological Survey, U.S. Dept. of the Interior

Mineral	% Imported in 2000	Major sources (1996-1999)	Major uses
Arsenic trioxide	100%	China, Chile, Mexico	Wood preservatives, herbicides, nonferrous alloys
Asbestos	100	Canada	Roofing products, gaskets, friction products
Bauxite & alumina	100	Australia, Guinea, Jamaica, Brazil	Aluminum production, refractories, abrasives, chemicals
Columbium (niobium)	100	Brazil, Canada, Germany, Russia	Steelmaking, superalloys
Fluorspar	100	China, S. Africa, Mexico	Hydrofluoric acid, aluminum fluoride, steelmaking
Graphite (natural)	100	China, Mexico, Canada	Refractories, brake linings, pencils
Manganese	100	S. Africa, Gabon, Australia, France	Steelmaking, batteries, agricultural chemicals
Mica, sheet (natural)	100	India, Belgium, Germany, China	Electronic & electrical equipment
Quartz Crystal	100	Brazil, Germany, Madagascar	Electronics, optical applications
Strontium	100	Mexico, Germany	Television picture tubes, ferrite magnets, pyrotechnics
Thallium	100	Belgium, Canada, Germany, UK	Superconductor materials, electronics, alloys, glass
Thorium	100	France	Ceramics, welding electrodes, catalysts
Yttrium	100	China, Hong Kong, France, UK	Television phosphors, fluorescent lights, oxygen sensors
Gemstones	99	Israel, India, Belgium	Jewelry, carvings, gem & mineral collections
Bismuth	95	Belgium, Mexico, UK, China	Pharmaceuticals, chemicals, alloys, metallurgical additives
Antimony	94	China, Mexico, S. Africa, Bolivia	Flame retardants, batteries, chemicals, ceramics & glass
Palladium	89	Russia, S. Africa, UK, Belgium	Catalysts, dental, electronics, electrical
Tin	86	China, Brazil, Peru, Bolivia	Solder, tinplate, chemicals, alloys
Platinum	83	S. Africa, UK, Russia, Germany	Catalysts, jewelry, dental & medical alloys
Stone (dimension)	80	Italy, Canada, Spain, India	Construction, monuments
Tantalum	80	Australia, China, Thailand, Japan	Capacitors, superalloys, cemented carbide tools
Chromium	78	S. Africa, Kazakhstan, Russia, Zimbabwe	Steel, chemicals, refractories
Titanium concentrates	76	S. Africa, Australia, Canada, India	Pigment, welding rod coatings, metal, carbides, chemicals
Cobalt	74	Norway, Finland, Zambia, Canada	Superalloys, cemented carbides, magnetic alloys, chemicals
Rare earths	72	China, France, Japan, UK	Catalysts, glass polishing, ceramics, magnets, metallurgy, phosphors
Barite	71	China, India, Mexico, Morocco	Oil & gas well drilling fluids, chemicals
Potash	70	Canada, Russia, Belarus	Fertilizers, chemicals
Iodine	69	Chile, Japan,Russia	Sanitation, pharmaceuticals, heat stabilizers, catalysts
Tungsten	68	China, Russia, Bolivia	Cemented carbides, electrical & electronic components, tool steels
Titanium (sponge)	62	Russia, Japan, Kazakhstan, China	Aerospace, armor, chemical processing, power generation
Zinc	60	Canada, Mexico, Peru	Galvanizing, zinc-base alloys, brass & bronze
Nickel	58	Canada, Norway, Russia, Australia	Stainless steel, low alloy steel, high-nickel alloys, plating
Peat	52	Canada	Horticulture, agriculture
Silver	52	Canada, Mexico, Peru	Photographic materials, electrical products, catalysts

U.S. Copper, Lead, and Zinc Production, 1950-2000

Source: U.S. Geological Survey, U.S. Dept. of the Interior

	Copper		Lead		Zinc			Copper		Lead		Zinc	
Year	Quantity (metric tons) (1,000)	Value ($1,000)	Quantity (metric tons)	Value ($1,000)	Quantity (metric tons)	Value ($1,000)	Year	Quantity (metric tons) (1,000)	Value ($1,000)	Quantity (metric tons)	Value ($1,000)	Quantity (metric tons)	Value ($1,000)
1950	827	379,122	390,839	113,078	565,516	167,000	1993	1,800	3,635,000	355,185	248,540	488,283	496,795
1960	1,037	733,706	223,774	57,722	395,013	112,365	1994	1,850	4,430,000	363,000	298,000	570,000	619,000
1970	1,560	1,984,484	518,698	178,609	484,560	163,650	1995	1,850	5,640,000	386,000	359,000	603,000	756,000
1975	1,282	1,814,763	563,783	267,230	425,792	366,097	1996	1,920	4,610,000	426,000	459,000	586,000	615,000
1980	1,181	2,666,931	550,366	515,189	317,103	261,671	1997	1,940	4,570,000	448,000	460,000	592,000	860,000
1985	1,105	1,631,000	413,955	174,008	226,545	201,607	1998	1,860	3,235,000	481,000	480,000	709,000	819,000
1990	1,586	431,000	483,704	490,750	515,355	847,485	1999	1,600	2,680,000	503,000	485,000	808,000	895,000
1991	1,630	3,931,000	465,931	343,907	517,804	602,426	2000	1,440	2,000,000	457,000	440,000	786,000	1,020,000
1992	1,760	4,179,000	397,076	307,337	523,430	673,800							

U.S. Pig Iron and Raw Steel Output, 1940-2000

Source: American Iron and Steel Institute
(net tons)

Year	Total pig iron	Raw steel[1]	Year	Total pig iron	Raw steel[1]	Year	Total pig iron	Raw steel[1]
1940	46,071,666	66,982,686	**1975**	79,923,000	116,642,000	**1994**	54,426,000	100,579,000
1945	53,223,169	79,701,648	**1980**	68,721,000	111,835,000	**1995**	56,097,000	104,930,000
1950	64,586,907	96,836,075	**1985**	50,446,000	88,259,000	**1996**	54,485,000	105,309,478
1955	76,857,417	117,036,085	**1990**	54,750,000	98,906,000	**1997**	54,679,000	108,561,182
1960	66,480,648	99,281,601	**1991**	48,637,000	87,896,000	**1998**	53,164,000	108,752,334
1965	88,184,901	131,461,601	**1992**	52,224,000	92,949,000	**1999**	51,002,000	107,395,010
1970	91,435,000	131,514,000	**1993**	53,082,000	97,877,000	**2000**	52,787,000	112,242,000

(1) Steel figures include only that portion of the capacity and production of steel for castings used by foundries operated by companies producing steel ingots.

World Gold Production, 1975-2000

Source: U.S. Geological Survey, U.S. Dept. of the Interior
(thousands of troy ounces)

		Africa			North and South America				Other			
Year	World prod.	South Africa	Ghana	Congo Dem. Rep	United States	Canada	Mexico	Colombia	Australia	China	Philip-pines	USSR/ Russia[1]
1975	38,476.4	22,937.8	523.9	115.7	1,052.3	1,653.6	144.7	308.8	526.8	NA	502.6	NA
1980	39,197.3	21,669.5	353.0	96.5	969.8	1,627.5	196.0	510.4	547.6	NA	753.5	8,425.0
1985	49,283.7	21,565.2	299.4	257.2	2,427.2	2,815.1	265.7	1,142.4	1,881.5	1,950.0	1,063.0	8,700.0
1986	51,534.1	20,513.7	287.1	257.2	3,739.0	3,364.7	250.6	1,285.9	2,413.8	2,100.0	1,296.4	8,850.0
1987	53,033.6	19,176.5	327.6	385.8	4,947.0	3,724.0	256.8	853.6	3,559.0	2,300.0	1,048.1	8,850.0
1988	60,309.0	19,965.6	355.6	401.9	6,459.5	4,334.3	292.5	932.8	5,046.1	2,507.8	980.0	8,925.0
1989	65,336.0	19,530.3	429.5	340.8	8,543.4	5,127.9	276.9	948.6	6,544.7	2,893.6	964.3	9,773.8
1990	70,206.9	19,454.4	541.4	299.0	9,458.4	5,446.7	311.3	943.7	7,849.2	3,215.1	790.6	9,709.5
1991	70,422.6	19,326.1	845.9	282.9	9,454.3	5,676.3	326.1	1,120.3	7,530.3	3,858.1	833.2	8,359.2
1992	73,529.6	19,742.8	997.7	225.1	10,616.6	5,189.2	318.0	1,032.6	7,825.5	4,501.1	729.9	8,231.6
1993	73,300.0	19,907.8	1,250.0	280.0	10,642.3	4,916.8	356.9	883.1	7,947.5	5,144.1	508.9	8,228.2
1994	72,500.0	16,650.0	1,400.0	357.0	10,500.0	4,710.0	446.9	668.0	8,236.6	4,240.0	870.0	8,172.7
1995	71,800.0	16,800.0	1,710.0	322.0	10,200.0	4,890.0	652.0	680.0	8,150.0	4,500.0	873.0	4,250.0
1996	74,000.0	16,000.0	1,580.0	264.0	10,500.0	5,350.0	787.0	710.0	9,310.0	4,660.0	1,020.0	3,950.0
1997	78,500.0	15,800.0	1,760.0	309.0	11,600.0	5,510.0	836.0	605.0	10,000.0	5,630.0	1,090.0	3,700.0
1998	80,400.0	14,900.0	2,330.0	154.0	11,800.0	5,320.0	817.0	605.0	10,000.0	5,720.0	1,100.0	3,670.0
1999	82,000.0	14,500.0	2,510.0	129.0	11,000.0	5,090.0	723.0	611.0	9,730.0	5,470.0	1,000.0	4,050.0
2000	81,900.0	13,800.0	2,320.0	129.0	11,300.0	4,940.0	848.0	611.0	9,530.0	5,790.0	965.0	4,500.0

(1) Figures for 1975-94 are for USSR as constituted prior to Dec. 1991; after 1994, Russia only. NA = Not available.

U.S. and World Silver Production, 1930-2000

Source: U.S. Geological Survey, U.S. Dept. of the Interior
(metric tons)

Year[1]	United States	World	Year[1]	United States	World	Year[1]	United States	World
1930	1,578	7,736	1970	1,400	9,670	1994	1,490	14,000
1935	1,428	6,865	1975	1,087	9,428	1995	1,560	15,100
1940	2,164	8,565	1980	1,006	10,556	1996	1,570	15,200
1945	904	5,039	1985	1,227	13,051	1997	2,180	16,400
1950	1,347	6,323	1990	2,120	16,600	1998	2,060	16,140
1955	1,134	9,967	1991	1,860	15,600	1999	1,950	16,100
1960	1,120	7,505	1992	1,800	14,600	2000	1,860	18,300
1965	1,238	8,007	1993	1,640	14,300			

(1) Largest production of silver in the United States was in 1915—2,332 metric tons.

Aluminum Summary, 1980-2000

Source: U.S. Geological Survey, U.S. Dept. of the Interior

Item	Unit	1980	1985	1990	1994	1995	1996[4]	1997[4]	1998[4]	1999[4]	2000[4]
U.S. production	1,000 metric tons	6,231	5,262	6,441	6,385	6,563	6,860	7,150	7,150	7,530	7,120
Primary aluminum	1,000 metric tons	4,654	3,500	4,048	3,299	3,375	3,577	3,603	3,713	3,779	3,668
Secondary aluminum[1]	1,000 metric tons	1,577	1,762	2,393	3,086	3,188	3,310	3,550	3,440	3,750	3,450
Primary aluminum value	Billion dollars	7.8	3.8	6.6	5.2	6.4	5.6	6.1	5.4	5.5	6.0
Price (Primary aluminum)[2]	Cents/pound	76.1	48.8	74.0	71.2	85.9	71.3	77.1	65.5	65.7	74.6
Imports for consumption[3]	1,000 metric tons	647	1,420	1,514	3,382	2,975	2,810	3,080	3,550	4,000	3,910
Exports[3]	1,000 metric tons	1,346	908	1,659	1,365	1,610	1,500	1,570	1,590	1,650	1,760
World production	1,000 metric tons	15,383	15,398	19,299	19,200	19,700	20,700	21,600	22,500	23,100	24,000

(1) Recoverable metal content from purchased scrap, old and new. (2) Average prices for primary aluminum, quoted by *Metals Week.* (3) Crude and semicrude (incl. metal and alloys, plates, bars, etc., and scrap). (4) All tonnage data, except primary production, have been rounded to 3 significant figures.

Economic and Financial Glossary

Source: Reviewed by William M. Gentry, Graduate School of Business, Columbia University

Annuity contract: An investment vehicle sold by insurance companies. Annuity buyers can elect to receive periodic payments for the rest of their lives. Annuities provide insurance against outliving one's wealth.

Arbitrage: A form of hedged investment meant to capture slight differences in the prices of 2 related securities—for example, buying gold in London and selling it at a higher price in New York.

Balanced budget: A budget is balanced when receipts equal expenditures. When receipts exceed expenditures, there is a **surplus;** when they fall short of expenditures, there is a **deficit.**

Balance of payments: The difference between all payments, for some categories of transactions, made to and from foreign countries over a set period of time. A *favorable* balance of payments exists when more payments are coming in than going out; an *unfavorable* balance of payments obtains when the reverse is true. Payments may include gold, the cost of merchandise and services, interest and dividend payments, money spent by travelers, and repayment of principal on loans.

Balance of trade (trade gap): The difference between exports and imports, in both actual funds and credit. A nation's balance of trade is *favorable* when exports exceed imports and *unfavorable* when the reverse is true.

Bear market: A market in which prices are falling.

Bearer bond: A bond issued in bearer form rather than being registered in a specific owner's name. Ownership is determined by possession.

Bond: A written promise, or IOU, by the issuer to repay a fixed amount of borrowed money on a specified date and generally to pay interest at regular intervals in the interim.

Bull market: A market in which prices are on the rise.

Capital gain (loss): An increase (decrease) in the market value of an asset over some period of time. For tax purposes, capital gains are typically calculated from when an asset is bought to when it is sold.

Commercial paper: An extremely short-term corporate IOU, generally due in 270 days or less.

Convertible bond: A corporate bond (see below) that may be converted into a stated number of shares of common stock. Its price tends to fluctuate along with fluctuations in the price of the stock and with changes in interest rates.

Consumer price index (CPI): A statistical measure of the change in the price of consumer goods.

Corporate bond: A bond issued by a corporation. The bond normally has a stated life and pays a fixed rate of interest. Considered safer than the common or preferred stock of the same company.

Cost of living: The cost of maintaining a standard of living measured in terms of purchased goods and services. Inflation typically measures changes in the cost of living.

Cost-of-living adjustments: Changes in promised payments, such as retirement benefits, to account for changes in the cost of living.

Credit crunch (liquidity crisis): A situation in which cash for lending is in short supply.

Debenture: An unsecured bond backed only by the general credit of the issuing corporation.

Deficit spending: Government spending in excess of revenues, generally financed with the sale of bonds. A deficit increases the government debt.

Deflation: A decrease in the level of prices.

Depression: A long period of economic decline when prices are low, unemployment is high, and there are many business failures.

Derivatives: Financial contracts, such as options, whose values are based on, or *derived* from, the price of an underlying financial asset or indicator such as a stock or an interest rate.

Devaluation: The official lowering of a nation's currency, decreasing its value in relation to foreign currencies.

Discount rate: The rate of interest set by the Federal Reserve that member banks are charged when borrowing money through the Federal Reserve System.

Disposable income: Income after taxes that is available to persons for spending and saving.

Diversification: Investing in more than one asset in order to reduce the riskiness of the overall asset portfolio. By holding more than one asset, losses on some assets may be offset by gains realized on other assets.

Dividend: Discretionary payment by a corporation to its shareholders, usually in the form of cash or stock shares.

Dow Jones Industrial Average: An index of stock market prices, based on the prices of 30 companies, 28 of which are on the New York Stock Exchange.

Econometrics: The use of statistical methods to study economic and financial data.

Federal Deposit Insurance Corporation (FDIC): A U.S. government-sponsored corporation that insures accounts in national banks and other qualified institutions against bank failures.

Federal Reserve System: The entire banking system of the U.S., incorporating 12 Federal Reserve banks (one in each of 12 Federal Reserve districts), 24 Federal Reserve branch banks, all national banks, and state-chartered commercial banks and trust companies that have been admitted to its membership. The governors of the system greatly influence the nation's monetary and credit policies.

Full employment: The economy is said to be at full employment when everyone who wishes to work at the going wage-rate for his or her type of labor is employed, save only for the small amount of unemployment due to the time it takes to switch from one job to another.

Futures: A futures contract is an agreement to buy or sell a specific amount of a commodity or financial instrument at a particular price at a set date in the future. For example, futures based on a stock index (such as the Dow Jones Industrial Average) are bets on the future price of that group of stocks.

Golden parachute: Provisions in contracts of some high-level executives guaranteeing substantial severance benefits if they lose their position in a corporate takeover.

Government bond: A bond issued by the U.S. Treasury, considered a safe investment. Government bonds are divided into 2 categories—those that are not marketable and those that are. *Savings bonds* cannot be bought and sold once the original purchase is made. Marketable bonds fall into several categories. *Treasury bills* are short-term U.S. obligations, maturing in 3, 6, or 12 months. *Treasury notes* mature in up to 10 years. *Treasury bonds* mature in 10 to 30 years. *Indexed bonds* are adjusted for inflation.

Greenmail: A company buying back its own shares for more than the going market price to avoid a threatened hostile takeover.

Gross domestic product (GDP): The market value of all goods and services that have been bought for final use during a period of time. It became the official measure of the size of the U.S. economy in 1991, replacing *gross national product (GNP),* in use since 1941. GDP covers workers and capital employed within the nation's borders. GNP covers production by U.S. residents regardless of where it takes place. The switch aligned U.S. terminology with that of most other industrialized countries.

Hedge fund: A flexible investment fund for a limited number of large investors (the minimum investment is typically $1 million). Hedge funds use a variety of investment techniques, including those forbidden to mutual funds, such as short-selling and heavy leveraging.

Hedging: Taking 2 positions whose gains and losses will offset each other if prices change, in order to limit risk.

Individual retirement account (IRA): A self-funded tax-advantaged retirement plan that allows employed individuals to contribute up to a maximum yearly sum. With a *traditional* IRA, individuals contribute pre-tax earnings and defer income taxes until retirement. With a *Roth* IRA, individuals contribute after-tax earnings but do not pay taxes on future withdrawals (the interest is never taxed). *401(k) plans* are employer-sponsored plans similar to traditional IRAs, but having higher contribution limits.

Inflation: An increase in the level of prices.

Insider information: Important facts about the condition or plans of a corporation that have not been released to the general public.

Interest: The cost of borrowing money.

Investment bank: A financial institution that arranges the initial issuance of stocks and bonds and offers companies advice about acquisitions and divestitures.

Junk bonds: Bonds issued by companies with low credit ratings. They typically pay relatively high interest rates because of the fear of default.

Leading indicators: A series of 11 indicators from different segments of the economy used by the U.S. Commerce Department to predict when changes in the level of economic activity will occur.

Leverage: The extent to which a purchase was paid for with borrowed money. Amplifies the potential gain or loss for the purchaser.

Leveraged buyout (LBO): An acquisition of a company in which much of the purchase price is borrowed, with the debt to be repaid from future profits or by subsequently selling off company assets. A leveraged buyout is typically carried out by a small group of investors, often including incumbent management.

Liquid assets: Assets consisting of cash and/or items that are easily converted into cash.

Margin account: A brokerage account that allows a person to trade securities on credit. A **margin call** is a demand for more collateral on the account.

Money supply: The currency held by the public, plus checking accounts in commercial banks and savings institutions.

Mortgage-backed securities: Created when a bank, builder, or government agency gathers together a group of mortgages and then sells bonds to other institutions and the public. The investors receive their proportionate share of the interest payments on the loans as well as the principal payments. Usually, the mortgages in question are guaranteed by the government.

Municipal bond: Issued by governmental units such as states, cities, local taxing authorities, and other agencies. Interest is exempt from U.S.—and sometimes state and local—income tax. *Municipal bond unit investment trusts* offer a portfolio of many different municipal bonds chosen by professionals. The income is exempt from federal income taxes.

Mutual fund: A portfolio of professionally bought and managed financial assets in which you pool your money along with that of many other people. A share price is based on net asset value, or the value of all the investments owned by the funds, less any debt, and divided by the total number of shares. The major advantage, relative to investing individually in only a small number of stocks, is less risk—the holdings are spread out over many assets and if one or two do badly the remainder may shield you from the losses. *Bond funds* are mutual funds that deal in the bond market exclusively. *Money market mutual funds* buy in the so-called money market—institutions that need to borrow large sums of money for short terms. These funds often offer special checking account advantages.

National debt: The debt of the national government, as distinguished from the debts of political subdivisions of the nation and of private business and individuals.

National debt ceiling: Total borrowing limit set by Congress beyond which the U.S. national debt cannot rise. This limit is periodically raised by congressional vote.

Option: A type of contractual agreement between a buyer and a seller to buy or sell shares of a security. A **call** option contract gives the right to purchase shares of a specific stock at a stated price within a given period of time. A **put** option contract gives the buyer the right to sell shares of a specific stock at a stated price within a given period of time.

Per capita income: The total income of a group divided by the number of people in the group.

Prime interest rate: The rate charged by banks on short-term loans to large commercial customers with the highest credit rating.

Producer price index: A statistical measure of the change in the price of wholesale goods. It is reported for 3 different stages of the production chain: crude, intermediate, and finished goods.

Program trading: Trading techniques involving large numbers and large blocks of stocks, usually used in conjunction with computer programs. Techniques include *index arbitrage,* in which traders profit from price differences between stocks and futures contracts on stock indexes, and *portfolio insurance,* which is the use of stock-index futures to protect stock investors from potentially large losses when the market drops.

Public debt: The total of a nation's debts owed by state, local, and national government. Increases in this sum, reflected in public-sector deficits, indicate how much of the nation's spending is being financed by borrowing rather than by taxation.

Recession: A mild decrease in economic activity marked by a decline in real (inflation-adjusted) GDP, employment, and trade, usually lasting from 6 months to a year, and marked by widespread decline in many sectors of the economy.

Savings Association Insurance Fund (SAIF): Created in 1989 to insure accounts in savings and loan associations up to $100,000.

Seasonal adjustment: Statistical changes made to compensate for regular fluctuations in data that are so great they tend to distort the statistics and make comparisons meaningless. For instance, seasonal adjustments are made for a slowdown in housing construction in midwinter and for the rise in farm income in the fall after summer crops are harvested.

Short-selling: Borrowing shares of stock from a brokerage firm and selling them, hoping to buy the shares back at a lower price, return them, and realize a profit from the decline in prices.

Stagnation: Economic slowdown in which there is little growth in the GDP, capital investment, and real income.

Stock: *Common stocks* are shares of ownership in a corporation. For publicly held firms, the stock typically trades on an exchange, such as the New York Stock Exchange; for closely held firms, the founders and managers own most of the stock. There can be wide swings in the prices of this kind of stock. *Preferred stock* is a type of stock on which a fixed dividend must be paid before holders of common stock are issued their share of the issuing corporation's earnings. Preferred stock is less risky than common stock. *Convertible preferred stock* can be converted into the common stock of the company that issued the preferred. *Over-the-counter stock* is not traded on the major or regional exchanges, but rather through dealers from whom you buy directly. *Blue chip* stocks are so called because they have been leading stocks for a long time. *Growth* stocks are from companies that reinvest their earnings, rather than pay dividends, with the expectation of future stock price appreciation.

Supply-side economics: A school of thinking about economic policy holding that lowering income tax rates will inevitably lead to enhanced economic growth and general revitalization of the economy.

Takeover: Acquisition of one company by another company or group by sale or merger. A *friendly takeover* occurs when the acquired company's management is agreeable to the merger; when management is opposed to the merger, it is a *hostile* takeover.

Tender offer: A public offer to buy a company's stock; usually priced at a premium above the market.

Zero coupon bond: A corporate or government bond that is issued at a deep discount from the maturity value and pays no interest during the life of the bond. It is redeemable at face value.

AGRICULTURE

U.S. Farms—Number and Acreage by State, 1999-2000

Source: National Agricultural Statistics Service, U.S. Dept. of Agriculture

STATE	Farms (1,000) 1999	Farms (1,000) 2000	Acreage (mil) 1999	Acreage (mil) 2000	Acreage per farm 1999	Acreage per farm 2000
Alabama	48.0	47.0	9.2	9.0	192	192
Alaska	0.6	0.58	0.91	0.92	1,596	1,586
Arizona	7.7	7.5	27.5	26.7	3,571	3,560
Arkansas	48.5	48.0	14.7	14.6	302	304
California	89.0	87.5	27.8	27.8	312	318
Colorado	29.0	29.0	31.8	31.6	1,097	1,090
Connecticut	4.0	3.9	0.4	0.36	93	92
Delaware	2.6	2.6	0.6	0.58	223	223
Florida	45.0	44.0	10.4	10.3	231	234
Georgia	50.0	50.0	11.2	11.1	224	222
Hawaii	5.5	5.7	1.4	1.44	262	253
Idaho	24.5	24.5	11.9	11.9	486	486
Illinois	79.0	78.0	27.7	27.7	351	355
Indiana	65.0	64.0	15.5	15.5	238	242
Iowa	96.0	95.0	33.0	32.8	344	345
Kansas	65.0	64.0	47.5	47.5	731	742
Kentucky	91.0	90.0	13.6	13.6	149	151
Louisiana	30.0	29.5	8.2	8.1	272	275
Maine	6.9	6.8	1.3	1.27	184	187
Maryland	12.4	12.4	2.1	2.1	169	169
Massachusetts	6.1	6.1	0.6	0.57	93	93
Michigan	53.0	52	10.4	10.4	196	200
Minnesota	81.0	79.0	28.8	28.6	356	362
Mississippi	43.0	43.0	11.4	11.1	265	258
Missouri	110.0	109.0	30.1	30.0	274	275
Montana	28.0	27.6	57.0	56.7	2,036	2,054
Nebraska	55.0	54.0	46.4	46.4	844	859
Nevada	3.0	3.0	6.8	6.8	2,267	2,267
New Hampshire	3.1	3.1	0.4	0.42	135	136
New Jersey	9.6	9.6	0.8	0.83	86	87
New Mexico	15.5	15.2	44.7	44.0	2,884	2,895
New York	39.0	38.0	7.8	7.7	200	203
N. Carolina	58.0	57.0	9.3	9.2	160	161
N. Dakota	30.5	30.3	39.4	39.4	1,292	1,300
Ohio	80.0	80.0	14.9	14.9	186	186
Oklahoma	84.0	85.0	34.0	34.0	405	400
Oregon	40.5	40.0	17.2	17.2	425	430
Pennsylvania	59.0	59.0	7.7	7.7	131	131
Rhode Island	0.7	0.7	0.1	0.06	86	86
S. Carolina	25.0	24.0	4.9	4.7	194	196
S. Dakota	32.5	32.5	44.0	44.0	1,354	1,354
Tennessee	91.0	90.0	11.9	11.7	131	130
Texas	227.0	226.0	130.5	130	575	575
Utah	15.5	15.5	11.6	11.6	748	748
Vermont	6.7	6.8	1.3	1.34	200	197
Virginia	50.0	49.0	8.6	8.7	172	178
Washington	40.0	40.0	15.7	15.7	393	393
W. Virginia	20.5	20.5	3.6	3.6	176	176
Wisconsin	78.0	77.0	16.3	16.2	209	210
Wyoming	9.2	9.2	34.6	34.6	3,761	3,761
UNITED STATES	**2,194**	**2,172**	**947**	**943**	**432**	**434**

U.S. Farms, 1940-2000

Source: National Agricultural Statistics Service, U.S. Dept. of Agriculture

The number of farms declined in 2000 (by about 1%), while the size of the average farm increased very slightly (by about 0.5%). These changes continued a decades-long trend toward fewer, but larger farming operations.

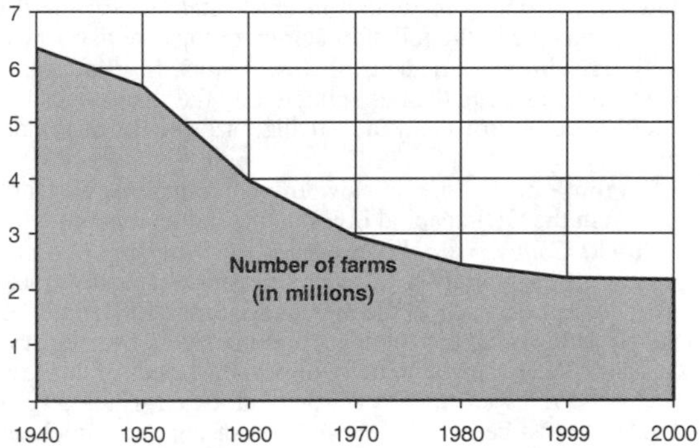

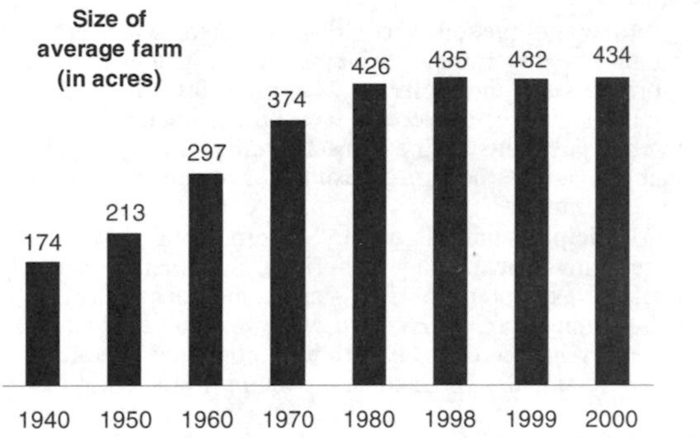

Decline in U.S. Farm Workers, 1820-1994*

Source: U.S. Dept. of Agriculture, Economic Research Service

Of the approximately 2.9 mil workers in the U.S. in 1820, 71.8%, or about 2.1 mil, were employed in farm occupations. The percentage of U.S. workers in farm occupations had declined drastically by the turn of the century, and by 1994 only 2.5% of all U.S. workers were employed in farm occupations.

(percent of total U.S. workers in farm occupations)

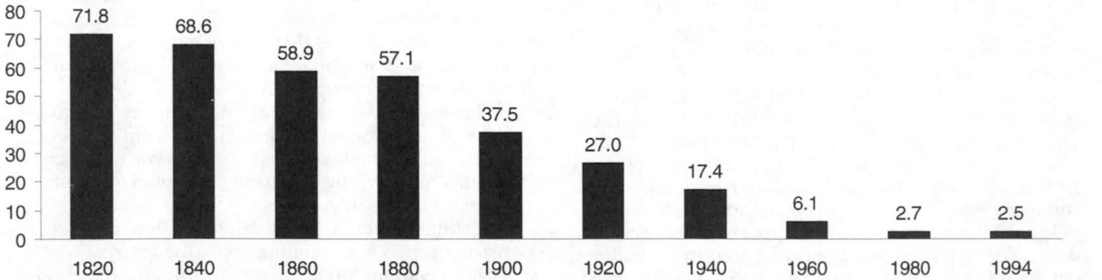

* Figures not compiled for years after 1994. Total workers for 1994 are employed workers age 15 and older; total workers for 1980 are members of the experienced civilian labor force ages 16 and older; total workers for 1900 to 1960 are members of the experienced civilian labor force 14 and older; total workers for 1820 to 1880 are gainfully employed workers 10 and older.

Eggs: U.S. Production, Price, and Value, 1999-2000[1]

Source: National Agricultural Statistics Service, U.S. Dept. of Agriculture

STATE	Eggs produced 1999 (mil)	Eggs produced 2000 (mil)	Price per dozen[2] 1999 (dollars)	Price per dozen[2] 2000 (dollars)	Value of Production 1999 (1,000 dollars)	Value of Production 2000 (1,000 dollars)
AL	2,450	2,378	1.230	1.310	251,125	259,598
AR	3,458	3,559	1.110	1.060	319,865	314,378
CA	6,606	6,293	0.479	0.453	263,690	237,561
CO	921	988	0.636	0.697	48,813	57,386
CT	828	866	0.578	0.555	39,882	40,053
DE	257	386	0.713	0.670	15,270	21,552
FL	2,776	2,723	0.465	0.478	107,570	108,466
GA	5,176	5,114	0.879	0.868	379,142	369,913
HI	149	143	0.870	0.890	10,803	10,636
ID	255	249	0.620	0.611	13,175	12,678
IL	876	944	0.472	0.472	34,456	37,131
IN	5,838	6,098	0.517	0.516	251,521	262,214
IA	6,754	7,554	0.380	0.383	213,877	241,099
KS	387	416	0.418	0.390	13,470	13,523
KY	922	940	0.785	0.904	60,314	70,813
LA	481	493	0.811	0.810	32,508	33,278
ME	1,356	1,135	0.622	0.596	70,286	56,377
MD	894	845	0.599	0.598	44,625	42,109
MA	109	93	0.654	0.627	5,946	4,875
MI	1,533	1,621	0.420	0.420	53,655	56,739
MN	3,138	3,271	0.438	0.428	114,537	116,666
MS	1,569	1,581	1.210	1.180	158,207	155,465
MO	1,690	1,614	0.516	0.520	72,670	69,940
MT	94	84	0.430	0.460	3,368	3,220
NE	2,837	2,999	0.395	0.375	93,385	93,719
NH	34	39	0.902	0.859	2,526	2,813
NJ	547	574	0.556	0.527	25,344	25,208
NY	1,017	1,113	0.556	0.564	47,121	52,311
NC	2,587	2,490	1.070	1.070	230,674	222,025
OH	8,193	8,163	0.517	0.503	352,982	342,166
OK	945	931	0.876	0.838	68,985	65,015
OR	774	805	0.499	0.482	32,198	32,334
PA	6,135	6,313	0.541	0.546	276,586	287,242
RI	15	12	0.661	0.625	826	625
SC	1,215	1,245	0.644	0.642	65,205	66,608
SD	592	585	0.373	0.354	18,401	17,258
TN	274	278	1.170	1.240	26,715	28,727
TX	4,413	4,423	0.654	0.697	240,509	256,903
UT	521	712	0.443	0.434	19,234	25,751
VT	65	68	0.619	0.598	3,327	3,389
VA	845	824	0.885	0.963	62,319	66,126
WA	1,312	1,310	0.540	0.549	59,031	59,942
WV	237	249	1.270	1.460	25,093	30,307
WI	1,031	1,225	0.416	0.479	35,741	48,898
WY	3.6	3.6	0.470	0.550	141	165
Other[3]	603	656	0.438	0.438	22,046	23,988
U.S.[4]	**82,715**	**84,412**	**0.622**	**0.618**	**4,287,164**	**4,347,190**

(1) Estimates cover the 12-month period from Dec. 1 of the previous year through Nov. 30. (2) Average of all eggs sold by producers, including hatching eggs. (3) AK, AZ, NM, and for 1999 ND and NV combined to avoid disclosure of individual operations; totals listed under "other." (4) Total states may not equal U.S. total because of rounding.

Livestock on Farms in the U.S., 1900-2001

Source: National Agricultural Statistics Service, U.S. Dept. of Agriculture

(in thousands)

Year (On Jan. 1)	All cattle[1]	Milk cows	Sheep and lambs	Hogs and pigs[2]
1900	59,739	16,544	48,105	51,055
1910	58,993	19,450	50,239	48,072
1920	70,400	21,455	40,743	60,159
1930	61,003	23,032	51,565	55,705
1940	68,309	24,940	52,107	61,165
1950	77,963	23,853	29,826	58,937
1955	96,592	23,462	31,582	50,474
1960	96,236	19,527	33,170	59,026
1965	109,000	16,981	25,127	56,106
1970	112,369	12,091	20,423	57,046
1975	132,028	11,220	14,515	54,693
1980	111,242	10,758	12,699	67,318
1985	109,582	10,777	10,716	54,073
1990	95,816	10,015	11,358	53,788
1991	96,393	9,966	11,174	54,416
1992	97,556	9,688	10,797	57,649
1993	99,176	9,581	10,906	58,795
1994	100,974	9,494	9,836	60,847
1995	102,755	9,487	8,886	57,150
1996	103,487	9,416	8,461	56,124
1997	101,656	9,318	8,024	57,366
1998	99,744	9,199	7,825	62,213
1999	99,115	9,133	7,215	60,896
2000	98,198	9,190	7,032	59,117
2001	97,309	9,203	6,915	59,081

(1) From 1966, includes milk cows and heifers that have calved. (2) 1900-95, as of Dec. 1 of preceding year; 1996-2000 as of June 1 of same year.

U.S. Meat Production and Consumption, 1940-2000

Source: Economic Research Service, U.S. Dept. of Agriculture

(in millions of pounds)

	Beef		Veal		Lamb and mutton		Pork		All red meats[1]		All Poultry	
Year	Production	Consumption[2]	Production	Consumption[2]	Production	Consumption[2]	Production	Consumption[2]	Production	Consumption[2]	Production	Consumption[2]
1940	7,175	7,257	981	981	876	873	10,044	9,701	19,076	18,812	NA	NA
1950	9,534	9,529	1,230	1,206	597	596	10,714	10,390	22,075	21,721	3,174	3,097
1960	14,728	15,465	1,109	1,118	769	857	13,905	14,057	30,511	31,497	6,310	6,168
1970	21,684	23,451	588	613	551	669	14,699	14,957	37,522	39,689	10,193	9,981
1980	21,643	23,560	400	420	318	351	16,617	16,838	38,978	41,170	14,173	13,525
1990	22,743	24,030	327	325	363	397	15,354	16,025	38,787	40,778	23,468	22,152
1991	22,917	24,115	306	305	363	397	15,999	16,392	39,585	41,209	24,701	23,272
1992	23,086	24,262	310	311	348	388	17,233	17,462	40,977	42,423	26,201	24,394
1993	23,049	24,006	285	286	337	381	17,088	17,408	40,759	42,081	27,328	25,097
1994	24,386	25,128	293	291	308	346	17,696	17,812	42,683	43,577	29,113	25,754
1995	25,222	25,534	319	319	285	346	17,849	17,768	43,675	43,967	30,393	25,944
1996	25,525	25,861	378	378	268	333	17,117	16,797	43,288	43,369	32,015	26,760
1997	25,490	25,611	334	333	260	332	17,274	16,823	43,358	43,099	32,964	27,261
1998	25,760	26,305	262	265	251	360	19,010	18,308	45,283	45,237	33,352	27,821
1999	26,493	26,937	235	235	248	358	19,308	18,946	46,284	46,476	35,252	29,584
2000	26,888	27,211	225	225	234	354	18,952	18,571	46,299	46,361	36,087	29,752

(1) Meats may not add to total because of rounding. (2) Consumption (also called total disappearance) is estimated as: production plus beginning stocks, plus imports, minus exports, minus ending stocks. NA = not available.

Total U.S. Government Agricultural Payments, by State, 1990-2000

Source: Economic Research Service, U.S. Dept. of Agriculture

(in thousands of dollars)

STATE	1990	1995	1996	1997	1998	1999	2000
Alabama	$82,226	$54,140	$75,663	$65,784	$96,924	$179,505	$170,852
Alaska	1,117	1,735	1,258	1,490	1,404	1,766	1,672
Arizona	43,349	9,456	57,283	46,575	78,784	107,899	107,066
Arkansas	312,696	383,783	362,026	274,938	472,333	815,267	900,648
California	252,333	239,809	300,748	220,475	357,340	668,501	667,466
Colorado	236,723	167,661	176,240	175,637	260,325	374,202	351,116
Connecticut	2,123	2,382	1,792	1,383	2,433	8,708	18,143
Delaware	3,213	3,150	4,900	5,725	10,848	19,850	25,028
Florida	37,155	55,778	22,842	19,044	24,882	76,859	56,741
Georgia	130,593	67,332	114,700	109,156	178,490	361,827	380,057
Hawaii	519	947	580	554	231	820	11,927
Idaho	133,431	89,482	115,781	110,401	198,933	210,657	261,297
Illinois	506,603	543,753	386,706	552,452	944,879	1,798,822	1,943,916
Indiana	244,170	246,026	213,649	265,132	468,917	852,051	938,464
Iowa	753,733	786,652	508,278	712,839	1,168,724	2,061,881	2,302,094
Kansas	834,746	422,226	554,988	529,786	879,853	1,401,286	1,231,923
Kentucky	81,610	67,382	74,673	83,056	140,474	232,109	448,473
Louisiana	154,631	164,251	199,023	157,343	258,518	420,630	451,831
Maine	6,982	14,114	4,647	4,191	6,496	11,650	13,851
Maryland	17,386	15,241	17,664	19,489	38,267	68,265	88,470
Massachusetts	3,023	2,490	1,556	1,196	1,733	10,186	10,973
Michigan	168,831	151,055	109,586	121,289	210,620	401,436	381,056
Minnesota	511,759	467,807	349,272	417,041	794,547	1,409,859	1,502,230
Mississippi	185,969	133,544	197,665	169,868	283,008	440,837	463,901
Missouri	299,065	256,629	291,395	278,025	427,995	717,096	869,390
Montana	299,599	189,809	240,720	230,857	360,673	492,057	490,002
Nebraska	624,646	507,302	388,738	454,549	814,690	1,411,884	1,406,971
Nevada	5,347	4,264	2,605	2,133	2,675	2,676	3,918
New Hampshire	1,856	1,216	1,102	890	1,858	3,944	4,768
New Jersey	15,744	5,491	3,250	3,629	5,564	10,258	22,481
New Mexico	63,840	55,134	58,989	38,995	60,809	92,378	79,495
New York	59,304	43,563	43,401	39,623	60,799	120,397	159,876
North Carolina	73,255	41,476	78,432	87,743	129,399	290,453	447,096
North Dakota	545,378	296,215	353,417	361,478	608,766	975,583	1,170,234
Ohio	197,006	167,351	163,109	186,431	315,578	650,237	678,104
Oklahoma	319,040	164,662	236,704	205,603	304,393	532,263	439,851
Oregon	89,137	52,145	73,299	63,461	100,732	105,641	137,401
Pennsylvania	41,414	41,096	37,111	35,470	45,630	95,717	147,848
Rhode Island	191	317	156	122	167	877	1,218
South Carolina	62,637	34,586	43,080	43,050	62,608	127,788	144,499
South Dakota	332,851	245,016	229,547	268,087	437,451	791,124	789,895
Tennessee	91,029	47,405	79,813	76,201	132,890	227,205	298,873
Texas	974,702	643,119	764,572	648,444	1,001,887	1,961,835	1,647,066
Utah	34,897	25,045	21,478	20,094	25,149	30,521	36,181
Vermont	5,793	4,334	4,015	3,093	4,466	12,221	26,093
Virginia	32,378	25,967	30,529	30,589	46,195	100,980	152,452
Washington	205,425	116,062	155,364	147,263	260,524	270,594	352,503
West Virginia	6,049	5,268	4,545	5,675	5,363	11,269	23,509
Wisconsin	181,243	184,350	158,153	176,552	256,082	503,046	603,213
Wyoming	31,283	31,432	24,524	22,390	28,710	40,203	34,302
UNITED STATES	**$9,298,030**	**$7,279,451**	**$7,339,570**	**$7,495,294**	**$12,380,016**	**$21,513,119**	**$22,896,433**

U.S. Federal Food Assistance Programs, 1990-2000[1]

Source: Food and Nutrition Service, U.S. Dept. of Agriculture

(in millions of dollars)

	1990	1991	1992	1993	1994	1995	1996	1997	1998	1999	2000
Food stamps[2]	$15,491	$18,769	$22,462	$23,653	$24,493	$24,620	$24,327	$21,487	$18,893	$17,698	$17,056
Puerto Rico nutrition asst.[3]	937	963	1,002	1,040	1,079	1,131	1,143	1,174	1,204	1,236	1,268
Natl. school lunch[4]	3,834	4,224	4,564	4,750	5,016	5,160	5,355	5,554	5,830	6,019	6,147
School breakfast[5]	596	685	787	869	959	1,048	1,119	1,214	1,272	1,345	1,393
WIC[6]	2,122	2,301	2,597	2,825	3,169	3,440	3,695	3,844	3,890	3,940	3,971
Summer food service[7]	164	182	204	220	230	237	250	244	263	268	268
Child/adult care[7]	813	945	1,094	1,225	1,354	1,464	1,534	1,572	1,553	1,621	1,684
Special milk	19	20	20	19	18	17	17	17	17	16	15
Nutrition for the elderly[4]	142	144	151	153	152	148	145	145	141	140	137
Food distrib. to Indian reserv.[7]	66	65	62	63	65	65	70	71	72	76	72
Commodity supp. food prog.[7]	85	93	105	113	107	99	100	99	94	98	95
Food dist.—charitable inst.[8]	104	93	116	91	105	64	11	6	9	3	2
Emergency food assistance[9]	334	301	272	271	264	135	44	192	234	270	225
TOTAL[10]	**$24,707**	**$27,985**	**$33,436**	**$35,292**	**$37,011**	**$37,628**	**$37,810**	**$35,619**	**$33,472**	**$32,730**	**$32,333**

(1) All data are for fiscal (not calendar) years. (2) Includes federal share of state administrative expenses and other federal costs. (3) Puerto Rico participated in the Food Stamp Program from FY 1975 until July 1982, when it initiated a separate grant program. (4) Includes cash payments and commodity costs (entitlement, bonus, and cash in lieu). (5) Excludes startup costs. (6) Includes the WIC Farmers Market Nutrition Program, program studies and special grants. (7) Includes commodity costs and administrative expenditures. (8) Includes summer camps. (9) Includes the Emergency Food Assistance Program (TEFAP) for all years, and the Soup Kitchens/Food Banks Program (1989-96). (10) Excludes Food Program Administration (federal) costs. Totals may not add because of rounding.

U.S. Farm Marketings by State, 1999-2000

Source: Economic Research Service, U.S. Dept. of Agriculture
(in thousands of dollars)

STATE	RANK, 2000	1999 FARM MARKETINGS Total	Crops	Livestock and products	2000 FARM MARKETINGS Total	Crops	Livestock and products
Alabama	(24)	$3,438,287	$661,564	$2,776,723	$3,272,295	$588,446	$2,683,849
Alaska	(49)	47,544	18,894	28,650	51,933	20,226	31,707
Arizona	(29)	2,178,036	1,190,779	987,257	2,289,754	1,226,432	1,063,322
Arkansas	(13)	5,259,413	1,862,815	3,396,598	4,887,463	1,639,079	3,248,384
California	(01)	24,800,669	18,087,031	6,713,638	25,509,829	19,240,740	6,269,089
Colorado	(16)	4,353,604	1,337,821	3,015,783	4,561,322	1,228,871	3,332,451
Connecticut	(44)	482,466	302,204	180,262	502,521	337,056	165,465
Delaware	(40)	718,258	152,609	565,649	741,179	183,801	557,378
Florida	(09)	7,065,634	5,702,203	1,363,431	6,951,096	5,573,093	1,378,003
Georgia	(12)	5,240,968	1,906,822	3,334,146	5,049,552	1,944,550	3,105,002
Hawaii	(41)	533,333	446,845	86,488	530,183	443,543	86,640
Idaho	(23)	3,347,324	1,744,409	1,602,915	3,389,246	1,761,120	1,628,126
Illinois	(08)	6,757,488	5,233,166	1,524,322	7,022,330	5,312,484	1,709,846
Indiana	(14)	4,373,126	2,792,334	1,580,792	4,580,756	2,886,234	1,694,522
Iowa	(03)	9,716,453	5,004,190	4,712,263	10,774,252	5,027,118	5,747,134
Kansas	(05)	7,616,027	2,607,252	5,008,775	7,905,407	2,417,203	5,488,204
Kentucky	(21)	3,456,149	1,297,699	2,158,450	3,605,477	1,270,606	2,334,871
Louisiana	(33)	1,847,599	1,227,563	620,036	1,819,807	1,166,533	653,274
Maine	(43)	515,207	229,331	285,876	503,594	241,811	261,783
Maryland	(36)	1,480,998	543,638	937,360	1,472,742	624,878	847,864
Massachusetts	(45)	396,130	295,378	100,752	391,875	300,964	90,911
Michigan	(22)	3,470,098	2,139,060	1,331,038	3,474,924	2,139,628	1,335,296
Minnesota	(06)	7,060,774	3,513,061	3,547,713	7,522,018	3,647,249	3,874,769
Mississippi	(27)	3,173,759	1,031,013	2,142,746	2,922,459	885,835	2,036,624
Missouri	(15)	4,255,850	1,779,318	2,476,532	4,566,967	1,890,453	2,676,514
Montana	(34)	1,716,225	788,506	927,719	1,806,371	704,315	1,102,056
Nebraska	(04)	8,555,037	3,130,167	5,424,870	8,951,881	3,029,152	5,922,729
Nevada	(47)	334,272	117,989	216,283	386,462	149,250	237,212
New Hampshire	(48)	153,135	90,083	63,052	154,371	94,216	60,155
New Jersey	(39)	740,337	553,597	186,739	812,247	619,146	193,101
New Mexico	(31)	1,953,423	512,634	1,440,789	2,086,411	473,304	1,613,107
New York	(25)	3,097,417	1,054,211	2,043,206	3,122,868	1,188,575	1,934,293
North Carolina	(07)	6,687,856	2,837,753	3,850,103	7,409,676	3,134,768	4,274,908
North Dakota	(28)	2,758,886	2,111,684	647,202	2,689,343	2,049,942	639,401
Ohio	(17)	4,428,837	2,642,582	1,786,255	4,404,604	2,653,890	1,750,714
Oklahoma	(18)	3,990,508	855,083	3,135,425	4,219,858	779,193	3,440,665
Oregon	(26)	3,052,453	2,262,383	790,070	3,049,277	2,222,981	826,296
Pennsylvania	(19)	4,070,341	1,193,080	2,877,261	4,033,373	1,252,000	2,781,373
Rhode Island	(50)	47,606	39,147	8,459	47,976	40,296	7,680
South Carolina	(35)	1,406,077	632,792	773,285	1,544,226	752,342	791,884
South Dakota	(20)	3,539,069	1,708,809	1,830,260	3,790,061	1,755,116	2,034,945
Tennessee	(32)	1,974,368	963,096	1,011,272	2,019,679	1,029,880	989,799
Texas	(02)	13,051,582	4,571,831	8,479,751	13,343,556	4,181,401	9,162,155
Utah	(37)	966,584	242,905	723,679	1,010,202	240,004	770,198
Vermont	(42)	540,699	68,062	472,637	507,903	66,759	441,144
Virginia	(30)	2,283,039	703,535	1,579,504	2,281,203	732,331	1,548,872
Washington	(11)	4,933,296	3,274,860	1,658,436	5,049,735	3,339,415	1,710,320
West Virginia	(46)	386,598	53,035	333,563	390,704	51,321	339,383
Wisconsin	(10)	5,596,072	1,446,753	4,149,319	5,220,527	1,416,149	3,804,378
Wyoming	(38)	851,672	172,062	679,610	954,360	159,649	794,711
UNITED STATES		**$188,609,610**	**$93,146,365**	**$95,463,245**	**$193,585,849**	**$94,113,346**	**$99,472,503**

Value of U.S. Agricultural Exports and Imports, 1977-2000

Source: Economic Research Service, U.S. Dept. of Agriculture
(in billions of dollars, except percent)

Year[1]	Trade surplus	Agric. exports	% of all exports	Agric. imports	% of all imports
1977	10.6	24.0	20	13.4	9
1978	13.4	27.3	21	13.9	8
1979	15.8	32.0	19	16.2	8
1980	23.2	40.5	19	17.3	7
1981	26.4	43.8	19	17.3	7
1982	23.6	39.1	18	15.5	6
1983	18.5	34.8	18	16.3	7
1984	19.1	38.0	18	18.9	6
1985	11.5	31.2	15	19.7	6
1986	5.4	26.3	13	20.9	6
1987	7.2	27.9	12	20.7	5
1988	14.3	35.3	12	21.0	5
1989	18.1	39.7	12	21.6	5
1990	17.7	40.4	11	22.7	5
1991	15.1	37.8	10	22.7	5
1992	18.2	42.6	10	24.5	5
1993	18.3	42.9	10	24.6	4
1994	17.4	44.0	9	26.6	4
1995	24.9	54.7	10	29.9	4
1996	27.3	59.9	10	32.6	4
1997	21.6	57.4	9	35.8	4
1998	16.6	53.6	8	37.0	4
1999[2]	11.8	49.1	8	37.2	4
2000[3]	11.9	50.9	7	38.9	3

(1) Fiscal year (Oct.-Sept.). (2) Revised. (3) Preliminary.

Farm Business Real Estate Debt Outstanding, by Lender Groups,[1] 1960-2000

Source: Economic Research Service, U.S. Dept. of Agriculture
(in thousands of dollars)

		AMOUNTS HELD BY PRINCIPAL LENDER GROUPS				
Dec. 31	Total farm real estate debt[2]	Farm Credit System[2]	Farm Services Agency[3]	Life insurance companies[4]	All operating banks	Other[5]
1960	$11,309,593	$2,222,301	$623,895	$2,651,587	$1,355,733	$4,456,068
1970	27,505,932	6,420,357	2,179,873	5,122,291	3,328,876	10,454,540
1980	89,692,429	33,224,684	7,435,059	11,997,922	7,765,058	29,269,705
1985	100,076,120	42,168,554	9,820,913	11,272,689	10,731,881	26,082,096
1988	77,832,498	28,445,452	8,979,749	9,039,395	14,433,688	16,934,218
1989	75,978,245	26,895,927	8,203,215	9,113,109	15,685,485	16,080,503
1990	74,731,876	25,924,490	7,639,490	9,703,958	16,288,128	15,169,299
1991	74,943,893	25,305,300	7,040,851	9,545,804	17,416,527	15,631,629
1992	75,421,255	25,407,547	6,394,446	8,765,021	18,756,851	16,095,415
1993	76,036,358	24,899,573	5,837,377	8,985,489	19,594,554	16,719,356
1994	77,679,838	24,596,715	5,465,063	9,025,132	21,079,145	17,513,779
1995	79,286,920	24,851,298	5,055,018	9,091,957	22,276,503	18,012,138
1996	81,657,044	25,729,867	4,701,970	9,468,069	23,275,938	18,481,196
1997	85,359,385	27,097,928	4,372,663	9,698,796	25,239,726	18,950,271
1998	89,615,293	28,887,735	4,073,399	10,723,206	27,168,314	18,762,640
1999	94,225,918	30,302,269	3,872,170	11,489,982	29,798,857	18,762,640
2000[6]	97,335,000	31,483,000	3,557,000	11,806,000	31,821,000	18,668,000

(1) Exclude operator households. (2) Includes data for joint stock land banks and real estate loans by Agricultural Credit Assn. (3) Includes loans made directly by Farm Services Agency for farm ownership, soil and water loans to individuals, Native American tribe land acquisition, grazing associations, and half of economic emergency loans. Also includes loans for rural housing on farm tracts and labor housing. (4) American Council of Life Insurance members. (5) Estimated by ERS, USDA. Includes Commodity Credit Corporation storage and drying facility loans. (6) Preliminary

Grain, Hay, Potato, Cotton, Soybean, Tobacco Production, by State, 2000

Source: National Agricultural Statistics Service, U.S. Dept. of Agriculture

STATE	Barley (1,000 bu)	Corn, grain (1,000 bu)	Cotton (Upland) (1,000 b)	All hay (1,000 t)	Oats (1,000 bu)	Potatoes (1,000 cwt)	Soybeans (1,000 bu)	Tobacco (1,000 lb)	All wheat (1,000 bu)
Alabama	—	10,725	540.0	1,296	—	697	2,880	—	4,860
Alaska	102,000	—	—	17,000	7,000	129,000	—	—	—
Arizona	—	6,468	760.0	1,870	—	2,520	—	—	8,775
Arkansas	4,104	22,750	1,450.0	2,879	—	—	83,200	—	59,400
California	5,780	39,950	2,200.0	8,568	1,875	16,355	—	—	34,200
Colorado	12,075	149,860	—	4,080	2,205	30,658	—	—	71,370
Connecticut	—	—	—	137	—	—	—	2,009	—
Delaware	2,268	25,272	—	63	—	1,128	9,159	—	4,158
Florida	—	2,100	100.0	675	—	8,423	285	11,475	441
Georgia	—	32,100	1,640.0	1,560	2,520	—	3,840	69,130	10,800
Hawaii	—	—	—	—	—	—	—	—	—
Idaho	55,480	9,120	—	5,292	1,050	152,320	—	—	108,450
Illinois	—	1,668,550	—	2,670	4,015	1,855	459,800	—	52,440
Indiana	—	815,850	—	2,627	1,950	784	258,980	7,980	35,190
Iowa	—	1,740,000	—	6,000	12,060	NE	459,240	—	846
Kansas	245	416,000	23.0	6,540	2,200	986	50,000	—	347,800
Kentucky	600	159,900	—	6,255	—	—	46,020	299,530	23,940
Louisiana	—	42,920	910.0	665	—	—	22,620	—	9,805
Maine	1,470	—	—	242	2,100	17,920	—	—	—
Maryland	4,100	62,775	—	711	NE	1,222	22,145	9,000	12,600
Massachusetts	—	—	—	197	—	638	—	400	—
Michigan	1,140	244,280	—	4,330	4,800	14,963	74,880	—	36,000
Minnesota	15,360	957,000	—	6,840	22,320	21,240	293,150	—	96,526
Mississippi	—	38,500	1,730.0	1,280		—	34,760		12,925
Missouri	—	396,110	540.0	6,657	1,590	1,678	175,000	2,912	49,400
Montana	38,000	2,520	—	3,560	2,600	3,503	—	—	135,210
Nebraska	162	1,014,300	—	6,055	1,890	10,127	173,850	—	59,400
Nevada	225	—	—	1,602	—	3,150	—	—	1,470
New Hampshire	—	—	—	101	—	—	—	—	—
New Jersey	312	10,050	—	260	—	713	3,920	—	1,995
New Mexico	—	11,680	130.0	1,670	—	3,770	—	—	4,200
New York	580	47,040	—	3,098	3,900	5,964	4,356	—	7,420
North Carolina	1,440	75,400	1,440.0	1,848	2,100	3,400	44,880	419,710	27,500
North Dakota	97,350	104,160	—	5,110	19,845	26,950	61,050	—	313,785
Ohio	1,014	485,100	—	4,521	6,840	1,134	186,480	13,200	79,920
Oklahoma	NE	37,800	155.0	4,869	660	—	4,650	—	142,800
Oregon	8,400	5,220	—	3,018	2,450	30,683	—	—	51,010
Pennsylvania	5,325	137,160	—	4,430	8,265	3,510	16,985	10,170	10,335
Rhode Island	—	—	—	20	—	138	—	—	—
South Carolina	NE	18,200	380.0	720	2,100	—	11,000	81,260	9,065
South Dakota	5,775	431,200	—	7,393	13,420	812	152,950		114,268
Tennessee	—	67,260	715.0	4,730		—	28,750	105,398	20,900
Texas	NE	235,600	3,950.0	8,880	4,300	5,196	7,020	—	66,000
Utah	5,460	3,024	—	2,500	490	435	—	—	6,850
Vermont	—	—	—	406	—	—	—	—	—
Virginia	5,785	48,180	—	3,240	—	1,292	19,110	64,130	12,915
Washington	34,300	18,500	159.0	3,249	1,125	108,000	—	—	164,880
West Virginia	—	4,550	—	1,315	NE	—	705	1,560	549
Wisconsin	3,200	363,000	—	6,000	19,040	33,800	58,000	2,020	8,730
Wyoming	7,885	8,184	—	2,154	1,485	NE	—	—	4,312
UNITED STATES	**317,865**	**9,968,358**	**3,950.0**	**152,183**	**149,195**	**515,964**	**2,769,665**	**1,099,884**	**2,223,440**

NE = Not estimated, bu = bushels, b = bales (480-lbs), t = tons, cwt = hundredweight.

Production of Principal U.S. Crops, 1989-2000

Source: National Agricultural Statistics Service, U.S. Dept. of Agriculture

Year	Corn for grain (1,000 bu)	Oats (1,000 bu)	Barley (1,000 bu)	Sorghum for grain (1,000 bu)	All wheat (1,000 bu)	Rye (1,000 bu)	Flaxseed (1,000 bu)	Upland Cotton (1,000 b)	Cottonseed (1,000 t)
1989......	7,531,953	373,587	404,203	615,420	2,036,618	13,647	1,215	12,196.6	4,677.4
1990......	7,934,028	357,654	422,196	573,303	2,729,778	10,176	3,812	15,505.4	5,968.5
1991......	7,474,765	243,851	464,326	584,860	1,980,139	9,734	6,200	17,614.3	6,925.5
1992......	9,476,698	294,229	455,090	875,022	2,466,798	11,440	3,288	16,219.5	6,230.1
1993......	6,336,470	206,770	398,041	534,172	2,396,440	10,340	3,480	16,134.6	6,343.2
1994......	10,102,735	229,008	374,862	649,206	2,320,981	11,341	2,922	19,662.0	7,603.9
1995......	7,373,876	162,027	359,562	460,373	2,182,591	10,064	2,211	17,532.2	6,848.7
1996......	9,293,435	155,273	395,751	802,974	2,285,133	9,016	1,602	18,413.5	7,143.5
1997......	9,206,832	167,246	359,878	633,545	2,481,466	8,132	2,420	18,245.0	6,934.6
1998......	9,758,685	165,981	352,125	519,933	2,547,321	12,161	6,708	13,475.9	5,365.4
1999[1].....	9,430,612	146,193	280,292	595,166	2,299,010	11,038	7,864	16,293.7	6,354.0
2000......	9,968,358	149,195	317,865	470,070	2,223,440	8,619	10,730	16,822.0	6,439.0

Year	Tobacco (1,000 lb)	All hay (1,000 t)	Beans, dry edible (1,000 cwt)	Peas, dry edible (1,000 cwt)	Peanuts[2] (1,000 lb)	Soybeans[3] (1,000 bu)	Potatoes (1,000 cwt)	Sweet potatoes (1,000 cwt)
1989......	1,367,188	144,706	23,729	3,883	3,989,995	1,923,666	370,444	11,358
1990......	1,626,380	146,212	32,379	2,372	3,602,770	1,925,947	402,110	12,594
1991......	1,664,372	152,073	33,765	3,715	4,926,570	1,986,539	417,622	11,203
1992......	1,721,671	146,903	22,615	2,535	4,284,416	2,190,354	425,367	12,005
1993......	1,613,319	146,799	21,913	3,292	3,392,415	1,870,958	428,693	11,053
1994......	1,582,896	150,060	29,028	2,255	4,247,455	2,516,694	467,054	13,395
1995......	1,268,538	154,166	30,812	4,765	4,247,455	2,176,814	443,606	12,906
1996......	1,517,334	149,457	27,960	2,671	3,661,205	2,382,364	498,633	13,456
1997......	1,787,399	152,536	29,370	5,752	3,539,380	2,688,750	467,091	13,327
1998......	1,479,867	151,780	30,418	5,934	3,963,440	2,741,014	475,771	12,382
1999[1].....	1,292,692	159,707	33,085	4,773	3,829,490	2,653,758	478,216	12,234
2000......	1,099,884	152,183	26,440	3,499	3,287,600	2,769,665	515,964	13,613

Year	Rice (1,000 cwt)	Sugarcane (1,000 t)	Sugar beets (1,000 t)	Pecans[4] (1,000 lb)	Apples (1,000 t)	Grapes (1,000 t)	Peaches (1,000 t)	Oranges[5] (1,000 bx)	Grapefruit[5] (1,000 bx)
1989......	154,487	29,426	25,131	250,500	4,958.4	5,930.9	1,181.5	209,050	69,500
1990......	156,088	28,136	27,513	205,000	4,828.4	5,659.9	1,121.1	184,415	49,300
1991......	159,367	30,252	28,203	299,000	4,853.4	5,555.9	1,347.8	178,950	55,500
1992......	179,658	30,363	29,143	166,000	5,284.3	6,052.1	1,336.0	209,610	55,265
1993......	156,110	31,101	26,249	365,000	5,342.4	6,023.2	1,330.1	255,760	68,375
1994......	197,779	30,929	31,853	199,000	5,667.8	5,870.6	1,253.3	240,450	65,100
1995......	173,871	30,944	27,954	268,000	5,292.5	5,922.3	1,150.8	263,605	71,050
1996......	171,321	29,462	26,680	221,500	5,196.0	5,554.3	1,058.2	263,890	66,200
1997......	182,992	31,709	29,886	335,000	5,161.9	7,290.9	1,312.3	292,620	70,200
1998......	184,443	32,743	32,499	73,200	5,381.3	5,816.4	1,162.8	315,525	63,150
1999[1].....	206,027	35,299	33,420	203,100	5,223.3	6,234.8	1,216.7	224,580	61,200
2000......	191,113	36,346	32,521	105,000	5,167.4	7,314.6	1,259.9	299,840	66,780

(1) Revised. (2) Harvested for nuts. (3) Harvested for beans. (4) Utilized production only. (5) Crop year ending in year cited.

Principal U.S. Crops: Area Planted and Harvested, 1998-2000

Source: National Agricultural Statistics Service, U.S. Dept. of Agriculture

(in thousand acres)

STATE	Area Planted[1] 1998	1999	2000	Area Harvested[1] 1998	1999	2000
AL.....	2,253	2,228	2,085	2,093	2,105	1,895
AZ.....	775	728	746	769	724	739
AR	8,550	8,458	8,490	8,263	8,289	8,234
CA	4,983	4,758	4,738	4,459	4,312	4,345
CO	6,291	6,638	6,418	5,942	6,316	5,996
CT	101	102	103	96	95	100
DE	519	498	500	505	480	493
FL.....	1,125	1,099	1,102	1,029	1,072	1,051
GA	4,041	3,859	3,908	3,408	3,357	3,348
HI	33	37	35	33	37	35
ID	4,504	4,516	4,502	4,356	4,362	4,323
IL	23,651	23,520	23,671	23,452	23,356	23,533
IN	12,929	12,722	12,697	12,596	12,578	12,602
IA	24,791	24,891	24,990	24,588	24,727	24,828
KS	23,065	22,911	22,899	22,144	21,759	21,642
KY	5,864	5,811	5,808	5,632	5,524	5,531
LA.....	4,055	3,790	3,775	3,752	3,740	3,673
ME	283	290	278	278	282	273
MD	1,470	1,489	1,531	1,415	1,421	1,496
MA	132	137	124	129	132	119
MI.....	6,776	6,880	6,768	6,653	6,730	6,653
MN	20,310	20,175	20,293	19,990	19,778	19,790
MS	4,810	4,905	4,770	4,717	4,812	4,607
MO	13,629	13,611	13,683	13,330	13,446	13,373
MT	9,791	9,794	8,883	9,188	9,301	8,078

STATE	Area Planted[1] 1998	1999	2000	Area Harvested[1] 1998	1999	2000
NE	18,955	19,325	19,199	18,570	18,789	18,637
NV	513	509	523	510	506	518
NH	71	77	73	70	77	72
NJ	450	416	368	408	357	359
NM....	1,232	1,250	1,294	946	1,073	896
NY	2,994	3,112	2,924	2,934	3,044	2,888
NC....	5,016	4,945	4,909	4,785	4,582	4,645
ND	20,751	20,058	21,722	20,081	18,701	20,281
OH....	10,651	10,571	10,657	10,520	10,320	10,546
OK....	10,607	11,013	10,467	8,592	8,254	7,934
OR....	2,236	2,288	2,300	2,158	2,168	2,236
PA	4,347	4,296	4,237	4,247	4,160	4,179
RI.....	14	12	12	14	12	12
SC	1,902	1,787	1,675	1,757	1,690	1,600
SD	16,495	16,523	17,290	16,093	16,179	16,870
TN	4,834	4,913	5,062	4,572	4,692	4,851
TX	23,785	25,033	23,309	16,804	20,189	16,124
UT	1,105	1,081	1,089	1,047	1,031	1,019
VT	357	351	320	352	338	315
VA	2,930	2,912	2,843	2,767	2,726	2,769
WA....	4,382	4,184	4,185	4,251	3,923	4,099
WV....	659	660	685	652	646	679
WI	8,082	8,368	7,809	7,792	8,078	7,587
WY....	1,779	1,834	1,703	1,692	1,775	1,623
U.S.[2] ..	**330,043**	**329,556**	**328,449**	**311,545**	**312,222**	**307,839**

(1) Crops included in area planted are corn, sorghum, oats, barley, winter wheat, rye, durum wheat, other spring wheat, rice, soybeans, peanuts, sunflower, cotton, dry edible beans, potatoes, canola, millet, and sugarbeets. Harvested acreage is used for all hay, tobacco, and sugarcane in computing total area planted. Includes double-cropped acres and unharvested small grains planted as cover crops. (2) State figures do not add to U.S. totals because of sunflower and canola unallocated acreage.

Average Prices Received by U.S. Farmers, 1940-2000

Source: National Agricultural Statistics Service, U.S. Dept. of Agriculture

Figures below represent dollars per 100 lb for hogs, beef cattle, veal calves, sheep, lamb, and milk (wholesale); dollars per head for milk cows; cents per lb for chickens, broilers, turkeys, and wool; cents per dozen for eggs; weighted calendar year prices for livestock and livestock products other than wool. For 1943-63, wool prices are weighted on marketing year basis. The marketing year was changed in 1964 from a calendar year to a Dec.-Nov. basis for hogs, chickens, broilers, and eggs.

Year	Hogs	Cattle (beef)	Calves (veal)	Sheep	Lambs	Milk cows	Milk	Chickens (excl. broilers)	Broilers	Turkeys	Eggs	Wool
1940...	5.39	7.56	8.83	3.95	8.10	61	1.82	13.0	17.3	15.2	18.0	28.4
1950...	18.00	23.30	26.30	11.60	25.10	198	3.89	22.2	27.4	32.8	36.3	62.1
1960...	15.30	20.40	22.90	5.61	17.90	223	4.21	12.2	16.9	25.4	36.1	42.0
1970...	22.70	27.10	34.50	7.51	26.40	332	5.71	9.1	13.6	22.6	39.1	35.4
1975...	46.10	32.20	27.20	11.30	42.10	412	8.75	9.9	26.3	34.8	54.5	44.8
1980...	38.00	62.40	76.80	21.30	63.60	1,190	13.05	11.0	27.7	41.3	56.3	88.1
1985...	44.00	53.70	62.10	23.90	67.70	860	12.76	14.8	30.1	49.1	57.1	63.3
1986...	49.30	52.60	61.10	25.60	69.00	820	12.51	12.5	34.5	47.1	61.6	66.8
1987...	51.20	61.10	78.50	29.50	77.60	920	12.54	11.0	28.7	34.8	54.9	91.7
1988...	42.30	66.60	89.20	25.60	69.10	990	12.26	9.2	33.1	38.6	52.8	138.0
1989...	42.50	69.50	90.80	24.40	66.10	1,030	13.56	14.9	36.6	40.9	68.9	124.0
1990...	53.70	74.60	95.60	23.20	55.50	1,160	13.74	9.3	32.6	39.4	70.9	80.0
1991...	49.10	72.70	98.00	19.70	52.20	1,100	12.27	7.1	30.8	38.4	67.8	55.0
1992...	41.60	71.30	89.00	25.80	59.50	1,130	13.15	8.6	31.8	37.7	57.6	74.0
1993...	45.20	72.60	91.20	28.60	64.40	1,160	12.84	10.0	34.0	39.0	63.4	51.0
1994...	39.90	66.70	87.20	30.90	65.60	1,170	13.01	7.6	35.0	40.4	61.4	78.0
1995...	40.50	61.80	73.10	28.00	78.20	1,130	12.78	6.5	34.4	41.6	62.4	104.0
1996...	51.90	58.70	58.40	29.90	82.20	1,090	14.75	6.6	38.1	43.3	74.9	70.0
1997...	52.90	63.10	78.90	37.90	90.30	1,100	13.36	7.7	37.7	39.9	70.3	84.0
1998...	34.40	59.60	78.80	30.60	72.30	1,120	15.41	8.0	39.3	38.0	65.5	60.0
1999[1]..	30.30	63.40	87.70	31.10	74.50	1,280	14.38	7.1	37.1	40.8	62.2	38.0
2000...	42.30	68.60	104.00	34.20	79.40	1,340	12.40	5.8	33.6	40.7	61.8	33.0

Figures below represent cents per lb for cotton, apples, and peanuts; dollars per bushel for oats, wheat, corn, barley, and soybeans; dollars per 100 lb for rice, sorghum, and potatoes; dollars per ton for cottonseed and baled hay; weighted crop year prices. The marketing year is described as follows: apples, June-May; wheat, oats, barley, hay, and potatoes, July-June; cotton, rice, peanuts, and cottonseed, Aug.-July; soybeans, Sept.-Aug.; and corn and sorghum grain, Oct.-Sept.

Year	Corn	Wheat	Upland cotton*	Oats	Barley	Rice	Soy-beans	Sor-ghum	Peanuts	Cotton-seed	Hay	Pota-toes	Apples
1940...	0.62	0.67	9.8	0.30	0.39	1.80	0.89	0.87	3.7	21.70	9.78	0.85	NA
1950...	1.52	2.00	39.9	0.79	1.19	5.09	2.47	1.88	10.9	86.60	21.10	1.50	NA
1960...	1.00	1.74	30.1	0.60	0.84	4.55	2.13	1.49	10.0	42.50	21.70	2.00	2.7
1970...	1.33	1.33	21.9	0.62	0.97	5.17	2.85	2.04	12.8	56.40	26.10	2.21	6.5
1975...	2.54	3.55	51.1	1.45	2.42	8.35	4.92	4.21	19.0	97.00	52.10	4.48	8.8
1980...	3.11	3.91	74.4	1.79	2.86	12.80	7.57	5.25	25.1	129.00	71.00	6.55	12.1
1985...	2.23	3.08	56.8	1.23	1.98	6.53	5.05	3.45	24.4	66.00	67.60	3.92	17.3
1986...	1.50	2.42	51.5	1.21	1.61	3.75	4.78	2.45	29.2	80.00	59.70	5.03	19.1
1987...	1.94	2.57	63.7	1.56	1.81	7.27	5.88	3.04	28.0	82.50	65.00	4.38	12.7
1988...	2.54	3.72	55.6	2.61	2.80	6.83	7.42	4.05	28.0	118.00	85.20	6.02	17.4
1989...	2.36	3.72	63.6	1.49	2.42	7.35	5.69	3.75	28.0	105.00	85.40	7.36	13.9
1990...	2.28	2.61	67.1	1.14	2.14	6.68	5.74	3.79	34.7	121.00	80.60	6.08	20.9
1991...	2.37	3.00	56.8	1.21	2.10	7.58	5.58	4.01	28.3	71.00	71.20	4.96	25.1
1992...	2.07	3.24	53.7	1.32	2.04	5.89	5.56	3.38	30.0	97.50	74.30	5.52	19.5
1993...	2.50	3.26	58.1	1.36	1.99	7.98	6.40	4.13	30.4	113.00	84.70	6.18	18.4
1994...	2.26	3.45	72.0	1.22	2.03	6.78	5.48	3.80	28.9	101.00	86.70	5.58	18.6
1995...	3.24	4.55	75.4	1.67	2.89	9.15	6.72	5.69	29.3	106.00	82.20	6.77	24.0
1996...	2.71	4.30	69.3	1.96	2.74	9.96	7.35	4.17	28.1	126.00	95.80	4.93	20.8
1997...	2.43	3.38	65.2	1.60	2.38	9.70	6.47	3.95	28.3	121.00	100.00	5.62	22.1
1998...	1.90	2.65	64.2	1.10	1.98	8.50	5.35	3.10	25.7	129.00	84.60	5.24	17.1
1999[1]..	1.82	2.48	45.0	1.12	2.13	5.93	4.63	2.80	25.4	89.00	76.90	5.77	21.3
2000...	1.85	2.65	56.0	1.05	2.15	5.75	4.75	3.15	25.7	106.00	83.00	4.95	17.9

*Beginning in 1964, 480-lb net weight bales. NA = Not available. (1) Revised.

Off-Farm Grain Storage Facilities, by State[1]

Source: National Agricultural Statistics Service, U.S. Dept. of Agriculture

State	No. of facilities	Total capacity (1,000 bu)	State	No. of facilities	Total capacity (1,000 bu)	State	No. of facilities	Total capacity (1,000 bu)	State	No. of facilities	Total capacity (1,000 bu)
AL.....	89	27,070	IA......	505	1,022,000	New Eng.	25	7,930	SD.....	281	147,000
AZ.....	26	19,500	KS.....	796	875,000	NJ.....	17	2,890	TN.....	210	56,600
AR....	212	235,440	KY.....	197	58,030	NM.....	27	15,460	TX.....	530	600,000
CA....	172	118,000	LA.....	41	98,100	NY.....	61	38,930	UT.....	40	19,000
CO....	130	126,180	MD.....	59	44,500	NC.....	203	65,920	VA.....	99	32,895
DE....	19	21,500	MI......	250	141,000	ND	394	244,370	WA....	280	203,365
FL.....	27	6,642	MN.....	623	494,500	OH.....	480	348,190	WI.....	361	218,000
GA....	197	46,525	MS.....	84	52,400	OK.....	264	239,500	WY....	18	9,100
ID.....	145	112,500	MO.....	441	225,970	OR.....	105	63,470	Unall.[2]..	10	924
IL.....	1,000	1,115,880	MT.....	144	59,890	PA.....	212	30,900			
IN.....	430	381,086	NE.....	527	692,391	SC.....	87	24,452	**TOTAL** .	**9,818**	**8,343,000**

(1) Data as of Dec. 1, 2000; excludes AK and HI. Off-farm capacity includes all elevators, warehouse terminals, merchant mills, other storage, and oilseed crushers which store grains, soybeans, sunflowers, or flaxseed. (2) Unallocated includes NV and WV.

World Wheat, Rice, and Corn Production, 2000

Source: UN Food and Agriculture Organization; in thousands of metric tons

COUNTRY	Wheat	Rice[1]	Corn
Afghanistan	1,469	233	115
Argentina	16,500	858	16,200
Australia	19,550	1,400	365
Austria	1,313	—	1,800
Bangladesh	1,900	35,821	2
Belgium-Lux.	1,634	—	370
Brazil	1,895	11,168	32,038
Bulgaria	2,800	7	937
Cambodia	—	3,762	95
Canada	26,804	—	6,827
Chile	1,500	113	646
China	99,370	190,168	105,231
Colombia	36	2,100	1,010
Croatia	1,080	—	800
Cuba	—	369	185
Czech Rep.	4,084	—	304
Denmark	4,700	—	—
Ecuador	22	1,520	747
Egypt	6,564	5,997	6,395
Ethiopia	1,220	—	2,600
Finland	550	—	—
France	37,559	107	16,395
Germany	21,634	—	3,241
Greece	1,770	180	1,850
Hungary	3,709	7	4,874
India	74,251	134,150	11,500
Indonesia	—	51,000	9,169
Iran	7,000	2,348	1,156
Iraq	384	130	53
Ireland	706	—	—
Italy	7,464	1,300	10,207
Japan	689	11,863	—
Kazakhstan	9,091	213	247
Kenya	105	55	1,800
Korea, North	158	1,690	1,041
Korea, South	5	7,067	79
Laos	—	2,155	77
Madagascar	91	2,300	150
Malaysia	—	2,037	57
Mexico	3,300	450	18,761
Moldova	770	—	1,091
Morocco	1,381	25	95
Myanmar	117	20,000	349
Nepal	1,184	4,030	1,445
Netherlands	1,183	—	60
New Zealand	360	—	174
Nigeria	101	3,277	5,476
Pakistan	21,079	7,000	1,351
Peru	165	1,665	1,271
Philippines	—	12,415	4,486
Poland	8,503	—	923
Portugal	429	149	905
Romania	4,320	4	4,200
Russia	36,000	440	1,800
Slovakia	1,254	—	440
South Africa	2,122	3	10,584
Spain	7,333	798	3,867
Sri Lanka	—	2,767	31
Sweden	2,530	—	—
Switzerland	548	—	209
Syria	3,105	—	180
Thailand	1	23,403	4,571
Turkey	16,500	338	2,500
Turkmenistan	1,700	37	9
Ukraine	10,159	90	3,840
United Kingdom	16,700	—	—
United States	60,512	8,669	253,208
Uruguay	310	1,175	65
Uzbekistan	2,787	175	40
Venezuela	1	737	900
Vietnam	—	32,544	1,930
Yugoslavia	1,927	—	2,944
Zimbabwe	250	—	2,108
WORLD, TOTAL	**576,317**	**598,852**	**590,791**

— production is small or nonexistent. Because not all countries are reported on this table, country totals do not add to world totals. (1) Rice paddy.

Wheat, Rice, and Corn—Exports/Imports of 10 Leading Countries, 1997-99

Source: UN Food and Agriculture Organization

(in thousands of metric tons; ranked for 1999; by marketing years)

EXPORTS

Wheat

TOP EXPORTERS	1999	1998	1997
U.S.	28,445	27,004	25,768
France	18,317	13,733	14,600
Australia	16,540	15,231	19,378
Canada	16,158	17,702	18,858
Argentina	8,797	10,371	8,791
Germany	4,666	4,932	3,862
Ukraine	4,526	1,600	812
Kazakhstan	3,104	2,501	2,792
United Kingdom	2,853	4,213	3,645
Turkey	1,865	1,109	15

Rice

TOP EXPORTERS	1999	1998	1997
Thailand	6,839	6,356	5,567
Vietnam	4,600	3,800	3,575
China	2,819	3,792	1,010
U.S.	2,668	3,113	2,296
India	2,571	4,800	2,134
Pakistan	1,791	1,972	1,767
Uruguay	699	659	649
Australia	669	552	655
Italy	677	602	632
Argentina	659	547	540

Corn

TOP EXPORTERS	1999	1998	1997
U.S.	51,975	42,125	41,792
France	8,352	7,979	7,340
Argentina	7,890	12,442	10,979
China	4,305	4,687	6,617
Hungary	1,708	2,109	1,192
Canada	889	259	263
Yugoslavia	463	463	119
South Africa	421	844	1,696
Germany	396	359	353
Romania	181	389	83

IMPORTS

Wheat

TOP IMPORTERS	1999	1998	1997
Brazil	6,895	6,395	4,850
Iran	6,156	2,770	5,942
Japan	5,973	5,758	6,315
Egypt	5,962	7,340	6,902
Italy	5,953	6,916	6,977
Russia	4,547	1,095	2,143
Algeria	4,099	3,463	3,508
Pakistan	3,240	2,520	2,500
Spain	3,234	3,308	2,974
Belgium-Lux.	3,216	2,867	2,854

Rice

TOP IMPORTERS	1999	1998	1997
Indonesia	4,748	1,895	348
Bangladesh	2,215	2,635	179
Brazil	984	1,305	816
Iran	852	2,000	637
Philippines	834	2,200	722
Iraq	781	1,000	684
Nigeria	688	1,000	731
Japan	664	499	569
Senegal	625	557	402
Malaysia	612	658	640

Corn

TOP IMPORTERS	1999	1998	1997
Japan	16,606	16,049	16,097
Korea (South)	8,115	7,111	8,313
Mexico	5,546	5,212	2,519
China	4,893	5,009	5,787
Egypt	3,585	3,043	3,059
Spain	2,935	2,616	2,503
Malaysia	2,200	1,840	2,745
Netherlands	1,843	1,734	1,769
Colombia	1,809	2,010	1,734
United Kingdom	1,267	1,301	1,473

IT'S A FACT: China, which produced 4.3 million metric tons of apples in 1990, ranking 3d behind the Soviet Union (6 mil) and U.S. (4.4 mil), boosted its production to 22.9 million metric tons—or 38% of the world's apple crop—by the year 2000. The U.S. and former Soviet Union were left way behind, with 4.8 million and 1.1 million, respectively.

World Commercial Catch of Fish, Crustaceans, and Mollusks, by Major Fishing Areas, 1994-99

Source: Food and Agriculture Organization of the United Nations (FAO); in metric tons
(in thousands of metric tons; live weight)

AREA	1994	1995	1996	1997	1998	1999
Marine						
Pacific Ocean	62,008	61,952	63,580	62,846	57,288	63,810
Atlantic Ocean	23,686	24,961	24,950	26,087	25,265	25,159
Indian Ocean	7,860	8,143	8,460	8,570	8,565	8,924
TOTAL	**93,553**	**95,056**	**96,989**	**97,503**	**91,119**	**97,894**
Inland Waters						
N. America	527	539	565	600	600	628
S. America	404	441	440	454	466	481
Europe	866	850	844	828	867	926
Former USSR	472	419	412	388	430	480
Asia	15,131	17,411	19,453	21,032	22,498	23,983
Africa	1,848	2,045	1,965	2,054	2,150	2,238
Oceania	21	23	22	24	25	26
TOTAL	**18,796**	**21,308**	**23,289**	**24,991**	**26,608**	**28,283**
GRAND TOTAL	**112,350**	**116,364**	**120,278**	**122,495**	**117,727**	**126,177**

Note: Data for marine mammals and aquatic plants are excluded.

Commercial Catch of Fish, Crustaceans, and Mollusks, for 20 Leading Countries, 1994-99[1]

Source: U.S. Dept. of Commerce, Natl. Oceanic and Atmospheric Admin., Natl. Marine Fisheries Service
(in thousands of metric tons; live weight; ranked for 1999)

COUNTRY	1999	1998	1997	1996	1995	1994	COUNTRY	1999	1998	1997	1996	1995	1994
China	41,513	39,545	36,529	33,320	29,912	25,304	Korean Rep.	2,423	2,354	2,596	2,772	2,688	2,701
Peru	8,438	4,346	7,877	9,522	8,943	12,005	Philippines	2,199	2,146	2,136	2,133	2,222	2,233
Japan	5,936	6,030	6,733	6,763	6,787	7,398	Vietnam	1,795	1,653	1,573	1,461	1,452	1,178
India	5,352	5,245	5,379	5,258	4,906	4,738	Iceland	1,740	1,686	2,210	2,064	1,616	1,560
Chile	5,325	3,558	6,083	6,909	7,591	7,839	Bangladesh	1,544	1,354	1,262	1,194	1,109	1,035
U.S.[2]	5,228	5,154	5,422	5,395	5,638	5,926	Spain	1,485	1,529	1,389	1,363	1.372	1,251
Indonesia	4,797	4,595	4,453	4,291	4,145	3,917	Denmark	1,448	1,560	1,867	1,723	2,044	1,916
Russia	4,210	4,518	4,715	4,730	4,374	3,781	Malaysia	1,407	1,287	1,281	1,239	1,245	1,182
Thailand	3,608	3,508	3,417	3,561	3,573	3,522	Mexico	1,251	1,216	1,529	1,495	1,355	1,223
Norway	3,086	3,259	3,224	2,970	2,802	2,585	Canada	1,136	1,105	1,052	976	914	1,078

(1) Includes aquaculture. (2) Includes weight of clam, oyster, scallop, and other mollusk shells. This weight is not included in U.S. landings statistics shown elsewhere.

U.S. Commercial Landings of Fish and Shellfish, 1986-2000[1]

Source: U.S. Dept. of Commerce, Natl. Oceanic and Atmospheric Admin., Natl. Marine Fisheries Service

YEAR	Landings for human food		Landings for industrial purposes[2]		TOTAL	
	mil lb	mil dollars	mil lb	mil dollars	mil lb	mil dollars
1986	3,393	$2,641	2,638	$122	6,031	$2,763
1987	3,946	2,979	2,950	136	6,896	3,115
1988	4,588	3,362	2,604	158	7,192	3,520
1989	6,204	3,111	2,259	127	8,463	3,238
1990	7,041	3,366	2,363	156	9,404	3,522
1991	7,031	3,169	2,453	139	9,484	3,308
1992	7,618	3,531	2,019	147	9,637	3,678
1993	8,214	3,317	2,253	154	10,467	3,471
1994	7,936	3,751	2,525	95	10,461	3,846
1995	7,667	3,625	2,121	145	9,788	3,770
1996	7,474	3,355	2,091	132	9,565	3,487
1997	7,244	3,285	2,598	163	9,842	3,448
1998	7,173	3,009	2,021	119	9,194	3,128
1999	6,832	3,265	2,507	202	9,339	3,467
2000	6,912	3,398	2,157	152	9,069	3,550

Note: Data does not include products of aquaculture, except oysters and clams. (1) Statistics on landings are shown in round weight for all items except univalve and bivalve mollusks such as clams, oysters, and scallops, which are shown in weight of meats (excluding the shell). All data are preliminary. (2) Processed into meal, oil, solubles, and shell products or used as bait or animal food.

U.S. Domestic Landings, by Regions, 1999-2000[1]

Source: U.S. Dept. of Commerce, Natl. Oceanic and Atmospheric Admin., Natl. Marine Fisheries Service

REGION	1999		2000	
	1,000 lb	1,000 dollars	1,000 lb	1,000 dollars
New England	583,863	$655,377	570,728	$681,092
Middle Atlantic	225,278	180,673	219,661	173,296
Chesapeake	527,407	172,012	492,110	172,210
South Atlantic	230,971	198,347	221,350	204,480
Gulf	1,945,063	757,857	1,759,993	910,685
Pacific Coast and Alaska	5,765,700	1,422,258	5,750,364	1,320,763
Great Lakes	23,843	16,009	22,245	18,508
Hawaii	36,907	64,557	32,531	68,447
TOTAL	**9,339,032**	**$3,467,090**	**$9,068,982**	**$3,549,481**

(1) Landings reported in round (live) weight items except for univalve and bivalve mollusks (e.g., clams, oysters, scallops), which are reported in weight of meats (excluding shell). Landings for Mississippi River Drainage Area states not included (not available).

EMPLOYMENT

Employment and Unemployment in the U.S., 1900-2000

Source: Bureau of Labor Statistics, U.S. Dept. of Labor

(civilian labor force, persons 16 years of age and older; annual averages; in thousands)

Year[1]	Employed	Unemployed	Unemployment rate
1900[2]	26,956	1,420	5.0%
1910[2]	34,599	2,150	5.9
1920[2]	39,208	2,132	5.2
1930[2]	44,183	4,340	8.9
1940[2]	47,520	8,120	14.6
1950	58,918	3,288	5.0
1955	62,170	2,852	4.4
1960	65,778	3,852	5.5
1965	71,088	3,366	4.5
1970	78,678	4,093	4.9
1975	85,846	7,929	8.5
1980	99,303	7,637	7.1
1985	107,150	8,312	7.2
1986	109,597	8,237	7.0
1987	112,440	7,425	6.2%
1988	114,968	6,701	5.5
1989	117,342	6,528	5.3
1990[3]	118,793	7,047	5.6
1991	117,718	8,628	6.8
1992	118,492	9,613	7.5
1993	120,259	8,940	6.9
1994[4]	123,060	7,996	6.1
1995	124,900	7,404	5.6
1996	126,708	7,236	5.4
1997[5]	129,558	6,739	4.9
1998[5]	131,463	6,210	4.5
1999[6]	133,488	5,880	4.2
2000[7]	134,337	5,655	4.0

(1) **Other early unemployment rates:** 1905, 4.3; 1915, 8.5; 1925, 3.2; 1935, 20.3; 1936, 16.9; 1937, 14.3; 1938, 19.0; 1939, 17.2. 1945, 1.9; all for 14 years of age and older. (2) Persons 14 years of age and older. (3) Beginning in 1990, data incorporate 1990 census-based population controls, adjusted for estimated undercount. (4) Beginning in 1994, not strictly comparable with prior years, because of major redesign of the survey used. (5) Not strictly comparable with 1994-96 because of revisions in population controls used in household survey. (6) Data not strictly comparable with 1998 and earlier years because of further revisions in population controls used in household survey. (7) Beginning in Jan. 2000, not strictly comparable with earlier years because of revisions to the controls used in the survey.

IT'S A FACT: Prior to any impact from Sept. 11, the U.S. unemployment rate for Aug. 2001 was up to 4.9%, the highest monthly rate since Sept. 1997.

Unemployment Insurance Data, by State, 2000

Source: Employment and Training Admin., U.S. Dept. of Labor; state programs only

STATE	Monetarily eligible claimants	First payments	Final payments	Initial claims	Benefits paid	Average weekly benefit	Employers subject to state law
AL	159,396	133,933	26,847	306,701	$195,491,708	$159.41	87,435
AK	48,201	43,557	16,962	93,452	101,215,533	189.86	16,126
AZ	93,618	69,104	23,333	157,107	154,212,642	162.51	103,011
AR	112,608	80,163	22,722	188,458	1168,300,672	210.08	59,492
CA	1,310,318	973,333	370,790	2,435,754	2,267,961,937	160.00	920,655
CO	78,121	51,952	20,506	95,406	157,662,024	255.86	131,031
CT	113,691	102,837	24,483	177,301	330,231,315	257.56	96,298
DE	28,396	26,978	4,988	58,873	71,004,291	214.85	24,678
DC	14,797	15,164	8,230	20,491	63,707,712	241.03	26,233
FL	291,584	224,212	86,357	367,325	634,749,720	220.21	380,484
GA	256,078	177,902	42,699	305,401	306,832,255	211.89	187,681
HI	33,158	25,247	7,058	70,805	95,622,651	283.67	27,932
ID	55,837	45,292	11,219	101,446	98,804,234	209.46	38,745
IL	352,447	309,386	91,236	636,532	1,155,272,020	251.58	278,082
IN	188,051	129,325	35,548	288,947	258,511,284	222.19	124,425
IA	106,275	84,455	15,626	153,331	203,843,187	238.42	68,825
KS	71,987	54,263	16,641	115,463	153,877,718	247.09	67,268
KY	117,641	111,252	18,296	241,881	263,127,972	224.78	87,397
LA	98,499	71,852	22,561	157,383	173,316,393	182.06	96,364
ME	45,372	28,203	10,569	70,379	73,412,321	202.29	38,387
MD	130,913	92,798	25,996	188,679	251,898,887	212.51	130,472
MA	215,921	172,041	53,639	327,979	745,331,280	293.45	166,440
MI	487,480	358,509	80,809	714,557	888,583,414	244.12	218,381
MN	136,654	109,278	26,734	207,638	374,256,943	290.51	128,033
MS	84,724	60,187	16,108	161,243	117,844,299	156.62	53,301
MO	204,593	137,426	34,663	342,593	313,856,788	186.22	127,558
MT	33,129	25,194	7,216	54,515	56,969,770	187.92	29,367
NE	40,856	27,838	8,179	56,314	55,649,970	188.00	44,443
NV	85,892	67,949	21,190	141,810	199,352,283	222.43	44,079
NH	21,891	13,595	817	28,460	29,928,343	217.21	39,326
NJ	304,789	245,606	109,650	468,778	1,074,571,102	289.61	255,914
NM	34,886	27,492	8,594	53,751	67,741,584	180.43	41,558
NY	508,881	424,234	186,345	842,527	1,478,421,426	247.48	467,425
NC	359,092	256,360	42,564	781,556	464,823,115	231.21	171,001
ND	18,064	11,800	4,282	27,908	38,765,227	210.01	18,684
OH	324,418	248,223	48,306	559,469	719,371,982	236.40	235,159
OK	59,316	40,872	11,846	96,123	98,302,045	214.40	74,382
OR	174,840	147,190	36,272	359,164	405,748,253	232.62	98,859
PA	505,425	396,227	96,622	952,225	1,376,499,309	264.76	256,329
PR	123,405	114,649	57,994	208,287	224,934,517	103.91	49,872
RI	46,395	37,798	12,501	83,684	135,676,689	253.48	32,609
SC	152,743	101,066	22,586	283,745	196,633,510	190.18	87,900
SD	11,209	7,791	642	18,124	15,187,464	180.86	22,344
TN	202,509	177,875	45,512	382,244	351,845,287	188.74	109,541
TX	582,054	322,307	165,241	700,580	940,473,595	227.11	386,759
UT	57,149	40,584	11,571	68,461	90,868,142	213.89	52,482
VT	22,111	17,593	2,258	33,451	41,722,949	215.55	20,805
VI	1,976	1,048	482	1,783	2,366,544	183,43	NA
VA	148,318	95,607	21,975	249,196	191,084,151	203.88	159,921
WA	295,995	205,411	57,301	477,346	827,064,448	280.94	189,933
WV	61,639	48,884	10,566	85,028	107,550,585	197.53	38,667
WI	278,483	230,458	8,808	527,241	456,092,114	233.11	121,529
WY	23,689	10,833	2,994	20,568	24,764,048	207.10	18,555
U.S.	**9,315,674**	**7,033,133**	**2,143,989**	**15,637,463**	**$19,341,633,733**	**$221.00**	**6,752.171**

NA = Not available.

Unemployment Rates, by Selected Country, 1970-2000

Source: Bureau of Labor Statistics, U.S. Dept. of Labor; civilian labor force, seasonally adjusted; Sept. 2001

Time Period	U.S.	Australia	Canada	France	Germany[1]	Italy[2]	Japan	Sweden	UK
1970	4.9	1.6	5.7	2.5	0.5	3.2	1.2	1.5	3.1
1975	8.5	4.9	6.9	4.2	3.4	3.4	1.9	1.6	4.6
1980	7.1	6.1	7.5	6.5	2.8	4.4	2.0	2.0	7.0
1981	7.6	5.8	7.6	7.6	4.0	4.9	2.2	2.5	10.5
1982	9.7	7.2	11.0	8.3	5.6	5.4	2.4	3.1	11.3
1983	9.6	10.0	11.9	8.6	6.9[3]	5.9	2.7	3.5	11.8
1984	7.5	9.0	11.3	10.0	7.1	5.9	2.8	3.1	11.7
1985	7.2	8.3	10.7	10.5	7.2	6.0	2.6	2.8	11.2
1986	7.0	8.1	9.6	10.6	6.6	7.5[3]	2.8	2.6	11.2
1987	6.2	8.1	8.8	10.8	6.3	7.9	2.9	2.2[3]	10.3
1988	5.5	7.2	7.8	10.3	6.3	7.9	2.5	1.9	8.6
1989	5.3	6.2	7.5	9.6	5.7	7.8	2.3	1.6	7.2
1990	5.6[3]	6.9	8.1	9.1	5.0	7.0	2.1	1.8	6.9
1991	6.8	9.6	10.3	9.6	5.6	6.9[3]	2.1	3.1	8.8
1992	7.5	10.8	11.2	10.4[3]	6.7	7.3	2.2	5.6	10.1
1993	6.9	10.9	11.4	11.8	7.9	10.2[3]	2.5	9.3	10.5
1994	6.1[3]	9.7	10.4	12.3	8.5	11.2	2.9	9.6	9.7
1995	5.6	8.5	9.4	11.8	8.2	11.8	3.2	9.1	8.7
1996	5.4	8.6	9.6	12.5	8.9	11.7	3.4	9.9	8.2
1997	4.9	8.6	9.1	12.4	9.9	11.9	3.4	10.1	7.0
1998	4.5	8.0	8.3	11.8	9.4	12.0	4.1	8.4	6.3
1999	4.2	7.2	6.8	11.2	8.7	11.5	4.7	7.1	6.1(P)
2000	4.0	6.6	5.8	9.7(P)	8.3(P)	10.7(P)	4.8(P)	5.9	NA
1st quarter	4.1	6.8	6.0	10.2(P)	8.4(P)	11.3(P)	4.8(P)	6.7	5.8(P)
2d quarter	4.0	6.7	5.8	9.7(P)	8.3(P)	10.8(P)	4.7(P)	6.0	5.5(P)
3d quarter	4.0	6.3	5.8	9.6(P)	8.2(P)	10.6(P)	4.7(P)	5.6	5.4(P)
4th quarter	4.0	6.5	5.7	9.2(P)	8.1(P)	10.1(P)	4.6(P)	5.2	NA

NA = Not available. P=Preliminary. **NOTE:** For the sake of comparisons, U.S. unemployment rate concepts were applied to unemployment data for other countries. Quarterly and monthly figures for France and Germany were calculated by applying annual adjustment factors to current published data and are less precise indicators of unemployment under U.S. concepts than the annual figures. (1) For former West Germany only, through 1994; from 1995 on figures are for unified Germany and not adjusted by BLS. (2) Quarterly rates are for first month of quarter. (3) As a result of revisions in survey methodology, there are breaks in the data series for the U.S. (1990, 1994), France (1992), Germany (1983), Italy (1986, 1991, 1993), and Sweden (1987); data prior to a survey change are not fully comparable to data after a survey change.

Employed Persons in the U.S., by Occupation and Sex, 1995, 2000

Source: Bureau of Labor Statistics, U.S. Dept. of Labor

(in thousands)

	TOTAL 16 years and older		MEN 16 years and older		WOMEN 16 years and older	
	1995	2000	1995	2000	1995	2000
TOTAL	**124,900**	**135,208**	**67,377**	**72,293**	**57,523**	**62,915**
Managerial and professional specialty	35,318	40,887	18,378	20,543	16,940	20,345
Executive, administrative, and managerial	17,186	19,774	9,840	10,814	7,346	8,960
Officials and administrators, public administration	710	753	371	380	339	373
Other executive, administrative, and managerial	12,151	14,089	7,471	8,291	4,680	5,797
Management-related occupations	4,325	4,932	1,998	2,143	2,327	2,789
Professional specialty	18,132	21,113	8,539	9,728	9,593	11,385
Engineers	1,934	2,093	1,771	1,886	163	207
Mathematical and computer scientists	1,195	2,074	813	1,422	382	652
Natural scientists	519	566	377	376	142	190
Health diagnosing occupations	1,002	1,038	773	757	229	281
Health assessment and treating occupations	2,762	2,966	393	425	2,369	2,541
Teachers, college and university	846	961	464	541	382	420
Teachers, except college and university	4,507	5,353	1,142	1,317	3,365	4,036
Lawyers and judges	926	926	684	651	242	275
Other professional specialty occupations	4,440	5,134	2,122	2,352	2,318	2,782
Technical, sales, and administrative support	37,417	39,442	13,310	14,288	24,107	25,154
Technicians and related support	3,909	4,385	1,900	2,118	2,009	2,267
Sales occupations	15,119	16,340	7,634	8,231	7,485	8,110
Administrative support, including clerical	18,389	18,717	3,776	3,939	14,613	14,778
Service occupations	16,930	18,278	6,774	7,245	10,155	11,034
Precision production, craft, and repair	13,524	14,882	12,323	13,532	1,201	1,351
Mechanics and repairers	4,423	4,875	4,248	4,625	175	250
Construction trades	5,098	6,120	4,978	5,960	120	160
Other precision production, craft, and repair	4,004	3,887	3,097	2,946	907	941
Operators, fabricators, and laborers	18,068	18,319	13,675	13,988	4,393	4,331
Machine operators, assemblers, and inspectors	7,907	7,319	4,958	4,622	2,949	2,697
Transportation and material moving occupations	5,171	5,557	4,682	5,003	490	554
Motor vehicle operators	3,904	4,222	3,474	3,736	429	486
Other transportation and material moving occupations	1,268	1,335	1,207	1,267	60	68
Handlers, equipment cleaners, helpers, and laborers	4,990	5,443	4,035	4,363	955	1,080
Construction laborers	780	1,015	754	977	26	38
Other handlers, equipment cleaners, etc.	4,210	4,428	3,281	3,386	929	1,042
Farming, forestry, and fishing	3,642	3,399	2,916	2,698	726	701
Farm operators and managers	NA	1,125	NA	839	NA	286
Other farming, forestry, and fishing occupations	NA	2,274	NA	1,859	NA	415

NA = Not available. **NOTE:** Beginning in Jan. 2000, data reflect revised population controls used in the household survey. Totals may not add because of independent rounding.

Elderly in the Labor Force, 1890-2000

Source: Bureau of the Census, U.S. Dept. of Commerce

The percentage of men 65 years of age and older in the labor force steadily declined between 1890 and 1990 dropping 74% in 100 years, but then increased slightly by 2000. The percentage of women 65 or older in the work force has barely changed at all.

(labor force participation rate; figs. for 1910 not available)

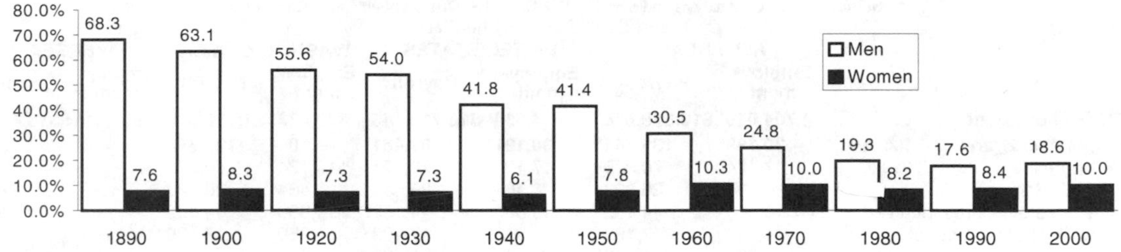

Unemployment Insurance

Source: Unemployment Insurance Service, U.S. Dept. of Labor

Unlike old-age and survivors insurance, which is entirely a federal program, unemployment insurance in the U.S. is a federal-state system that provides insured wage earners partial replacement for lost wages during a period of involuntary unemployment. The program protects most wage and salary workers. During fiscal year 2000, 127 million workers in commerce, industry, agriculture, and government were covered under the federal-state system.

Each state, as well as the District of Columbia, Puerto Rico, and the Virgin Islands, has its own law and operates its own program. The amount and duration of the weekly benefits are determined by state laws and are based on prior wages and length of employment. States are required to extend the duration of benefits when unemployment in the state rises to and remains above specified levels; costs of extended benefits are shared by the state and federal governments.

Under the Federal Unemployment Tax Act, the federal tax rate is 6.2% on the first $7,000 paid to each employee of employers with one or more employees in 20 weeks of the year or with a quarterly payroll of $1,500 or more. A credit of up to 5.4% is allowed for taxes paid under state unemployment insurance laws that meet certain criteria, for a net federal rate of 0.8%; subject employers also pay a state unemployment tax. Governmental agencies and certain nonprofit organizations are not subject to the federal tax; these employers reimburse states for benefits paid to former employees.

The secretary of labor certifies states for administrative grants to operate the program (under the Social Security Act) and for employer tax credit (under the Federal Unemployment Tax Act).

Benefits are financed solely by employer contributions, except in Alaska, New Jersey, and Pennsylvania, where employees also contribute. Benefits are paid through the states' public employment offices, at which unemployed workers must register for work and to which they must report regularly for referral to a possible job during the time when they are drawing weekly benefit payments.

During fiscal year 2000, $20.9 billion in benefits were paid under all unemployment insurance programs to 6.9 million beneficiaries. They received an average payment of $212 weekly for total unemployment, which lasted an average of 14.2 weeks.

U.S. Unemployment Rates by Selected Characteristics, 1960-2001

Source: Bureau of Labor Statistics, U.S. Dept. of Labor; seasonally adjusted, quarterly averages

						2000				2001	
	1960	1970	1980	1990	2000	I	II	III	IV	I	II
TOTAL (all civilian workers)	5.5	4.9	7.1	5.6	4.0	4.1	4.0	4.0	4.0	4.2	4.5
Men, 20 years and older	4.7	3.5	5.9	5.0	3.3	3.3	3.3	3.3	3.4	3.7	4.0
Women, 20 years and older	5.1	4.8	6.4	4.9	3.6	3.6	3.7	3.6	3.4	3.6	3.8
Both sexes, 16 to 19 years	14.7	15.3	17.8	15.5	13.1	13.3	12.5	13.5	12.9	13.7	14.0
White	5.0	4.5	6.3	4.8	3.5	3.5	3.4	3.5	3.5	3.7	3.9
Black	NA	NA	14.3	11.4	6.7	6.9	6.8	6.6	6.5	7.0	7.2
Black and other	10.2	8.2	13.1	10.1	7.7	7.8	7.7	7.6	7.5	8.1	8.2
Hispanic origin	NA	NA	10.1	8.2	5.7	5.9	5.7	5.6	5.6	6.2	6.5
Married men, spouse present	3.7	2.6	4.2	3.4	2.0	2.0	1.9	2.0	2.2	2.3	2.6
Married women, spouse present	5.2	4.9	5.8	3.8	2.7	2.7	2.7	2.8	2.5	2.6	2.9
Women who maintain families	NA	5.4	9.2	8.3	5.9	6.3	6.2	5.7	5.2	6.2	6.3
OCCUPATION											
Managerial and professional specialty	NA	NA	NA	2.1	1.7	1.7	1.7	1.8	1.7	1.8	2.0
Technical, sales, and administrative support	NA	NA	NA	4.3	3.6	3.5	3.6	3.7	3.5	3.5	4.0
Precision production, craft, and repair	NA	NA	NA	5.9	3.6	3.8	3.5	3.4	3.6	3.6	4.5
Operators, fabricators, and laborers	NA	NA	NA	8.7	6.3	6.2	6.3	6.2	6.4	7.3	7.4
Farming, forestry, and fishing	NA	NA	NA	6.4	6.0	5.3	5.9	6.0	6.7	7.6	6.9
INDUSTRY											
Nonagricultural private wage and salary workers	6.2	5.2	7.4	5.8	4.1	4.2	4.1	4.1	4.0	4.4	4.6
Goods-producing industries	7.5	6.1	9.4	7.0	4.4	4.4	4.2	4.3	4.6	5.1	5.4
Mining	9.7	3.1	6.4	4.8	4.0	3.1	3.6	4.6	4.7	3.5	5.8
Construction	13.5	9.7	14.1	11.1	6.4	6.7	5.8	6.3	6.6	6.7	6.8
Manufacturing	6.2	5.6	8.5	5.8	3.5	3.5	3.7	3.6	3.8	4.6	4.8
Durable goods	6.4	5.7	8.9	5.8	3.4	3.1	3.6	3.2	3.5	4.4	4.7
Nondurable goods	6.1	5.4	7.9	5.8	4.0	4.2	3.7	4.1	4.1	4.8	4.9
Service-producing industries	5.1	4.5	6.1	5.2	4.0	4.2	4.0	4.0	3.8	4.2	4.4
Transportation and public utilities	4.6	3.2	4.9	3.9	3.1	3.2	3.0	3.1	2.9	2.9	4.1
Wholesale and retail trade	5.9	5.3	7.4	6.4	5.0	5.2	5.1	4.9	4.8	5.1	5.3
Finance, insurance, and real estate	2.4	2.8	3.4	3.0	2.3	2.5	2.4	2.2	2.1	2.5	2.6
Services	5.1	4.7	5.9	5.0	3.8	4.0	3.8	3.8	3.6	4.1	4.2
Government workers	NA	NA	4.1	2.7	2.1	2.0	2.1	2.2	2.2	1.9	2.1
Agricultural wage/salary workers	8.3	7.5	11.0	9.8	7.6	6.0	7.7	7.7	9.0	9.9	9.0

NA = Not available.

> ***IT'S A FACT:*** The most dangerous occupation in 2000 was timber-cutting, according to the Bureau of Labor Statistics. Timber-cutters suffered fatal occupational injuries at a rate of 122 per 100,000 workers. Others who faced higher-than-average danger on the job included airplane pilots, construction laborers, truck drivers, and farm workers.

Civilian Employment of the Federal Government, May 2001

Source: Statistical Analysis and Services Division, U.S. Office of Personnel Management

(payroll in thousands of dollars)

	ALL AREAS		UNITED STATES		WASH., D.C., MSA[2]		OVERSEAS	
	Employment	Payroll	Employment	Payroll	Employment	Payroll	Employment	Payroll
TOTAL, all agencies[1]	**2,704,015***	**$10,529,616***	**2,614,584***	**$10,216,145***	**322,633***	**$1,592,373***	**89,431***	**$313,471***
Legislative Branch	**30,199***	**133,544***	**30,191***	**133,481***	**29,104***	**127,682***	**8***	**63***
Congress	17,124	70,975	17,124	70,975	17,124	70,975	—	—
U.S. Senate	6,464	26,823	6,464	26,823	6,464	26,823	—	—
House of Representatives	10,660	44,152	10,660	44,152	10,660	44,152	—	—
Architect of the Capitol	1,950	7,199	1,950	7,199	1,950	7,199	—	—
Congressional Budget Ofc	232	1,458	232	1,458	232	1,458	—	—
General Accounting Ofc	3,117	19,562	3,116	19,556	2,338	14,818	1	6
Government Printing Ofc	3,044	12,461	3,044	12,461	2,766	11,502	—	—
Library of Congress	4,268	19,530	4,261	19,473	4,240	19,395	7	57
U.S. Tax Court	241	1,438	241	1,438	241	1,438	—	—
Judicial Branch	**32,957**	**150,648**	**32,548**	**148,985**	**1,904**	**10,308**	**409**	**1,663**
Supreme Court	418	1,415	418	1,415	418	1,415	—	—
U.S. Courts	32,539	149,233	32,130	147,570	1,486	8,893	409	1,663
Executive Branch	**2,640,859***	**10,245,424***	**2,551,845***	**9,933,679***	**291,625***	**1,454,383***	**89,014***	**311,745***
Exec Ofc of the President	1,566	10,570	1,557	10,507	1,557	10,507	9	63
White House Office	369	2,061	369	2,061	369	2,061	—	—
Ofc of Vice President	15	116	15	116	15	116	—	—
Ofc of Mgmt & Budget	490	3,883	490	3,883	490	3,883	—	—
Ofc of Administration	182	1,100	182	1,100	182	1,100	—	—
Council Economic Advisors	30	170	30	170	30	170	—	—
Ofc of Policy Development	31	145	31	145	31	145	—	—
National Security Council	47	301	47	301	47	301	—	—
Ofc of Natl Drug Control	109	795	109	795	109	795	—	—
Ofc of U.S. Trade Rep	173	1,262	164	1,199	164	1,199	9	63
Executive Departments	1,593,855	6,527,928	1,512,051	6,244,115	218,226	1,082,600	81,804	283,813
State	28,054	147,512	10,586	55,906	9,295	47,026	17,468	91,606
Treasury	159,274	639,368	157,898	632,876	22,538	123,053	1,376	6,492
Defense, Total	670,568	2,390,373	616,772	2,242,590	63,119	233,318	53,796	147,783
Defense, Mil Function	645,849	2,323,444	592,127	2,175,761	62,398	231,479	53,722	147,683
Defense, Civ Function	24,719	66,929	24,645	66,829	721	1,839	74	100
Dept of the Army	227,870	629,053	205,062	563,591	19,003	40,749	22,808	65,462
Army, Mil Function	203,152	562,125	180,418	496,763	18,282	38,910	22,734	65,362
Army, Civil Function	24,718	66,928	24,644	66,828	721	1,839	74	100
Corps of Engineers	24,699	66,880	24,625	66,780	702	1,791	74	100
Dept of the Navy	182,557	739,163	174,527	706,456	24,660	99,855	8,030	32,707
Dept of the Air Force	154,438	624,766	147,651	597,306	5,123	20,737	6,787	27,460
Defense Log Agcy	23,680	89,067	22,942	85,749	2,007	8,921	738	3,318
Other Defense Activities	82,023	308,324	66,590	289,488	12,326	63,056	15,433	18,836
Justice	126,711	613,108	124,162	600,322	21,831	132,058	2,549	12,786
Interior	72,982	271,041	72,601	269,722	7,936	37,267	381	1,319
Agriculture	100,084	369,032	98,797	365,085	11,213	55,138	1,287	3,947
Commerce	39,151	171,637	38,381	168,339	20,074	103,643	770	3,298
Labor	16,016	74,725	15,973	74,554	5,364	27,622	43	171
Health & Human Services	63,323	309,912	63,064	308,371	28,227	151,882	259	1,541
Housing & Urban Dev.	10,154	55,056	10,067	54,627	3,059	18,159	87	429
Transportation	64,131	431,276	63,629	428,252	10,487	67,425	502	3,024
Energy	15,689	92,019	15,681	91,959	5,230	33,547	8	60
Education	4,581	24,268	4,576	24,247	3,195	17,579	5	21
Veterans Affairs	223,137	938,601	219,864	927,265	6,658	34,883	3,273	11,336
Independent Agencies	1,045,438*	3,706,926*	1,038,237*	3,679,057*	71,842*	361,276*	7,201*	27,869*
Bd of Gov, Fed Rsrv Sys	1,650	9,814	1,650	9,814	1,650	9,814	—	—
Environmtl Protect Agcy	17,968	98,190	17,919	97,933	6,019	35,576	49	257
Equal Employ Opp Comm	2,746	12,828	2,746	12,828	655	3,510	—	—
Federal Communic Comm	1,946	11,425	1,944	11,411	1,620	9,764	2	14
Federal Deposit Ins Corp	6,532	40,743	6,524	40,692	2,442	16,608	8	51
Fed. Emerg. Mgmt Agcy.	4,826	20,294	4,763	19,941	2,150	9,247	63	353
General Svcs Admin.	13,921	65,322	13,836	64,971	4,734	25,266	85	351
Natl Aero & Space Admin.	18,850	110,569	18,829	110,428	3,984	24,123	21	141
Natl Fnd Arts & Humanities	363	1,908	363	1,908	363	1,908	—	—
Peace Corps	1,004	3,889	584	2,511	472	2,134	420	1,378
Securities & Exch. Comm.	2,946	17,558	2,946	17,558	1,806	10,754	—	—
Small Business Adm.	3,968	18,688	3,876	18,386	823	4,666	92	302
Smithsonian Inst.	5,082*	19,825*	5,054*	19,649*	4,636*	17,879*	28*	176*
Social Security Admin	63,997	262,696	63,491	260,730	1,743	7,568	506	1,966
U.S. Postal Service	856,550	2,779,914	852,520	2,764,579	21,065	78,582	4,030	15,335

NOTE: * denotes figures that are preliminary or are based in whole or part on figures for the previous month. (1) Totals include agencies not listed. (2) Metropolitan Statistical Area.

U.S. Occupational Illnesses, by Industry and Type of Illness, 1999

Source: Bureau of Labor Statistics, U.S. Dept. of Labor

(percent distribution)

		GOODS PRODUCING				SERVICE PRODUCING				
	All private sector[1]	Agriculture[2]	Mining[3]	Construction	Manufacturing	Trans. and pub. utilities	Wholesale	Retail	Finance[4]	Service
Total [1,702,470 cases]	**100.0**	**100.0**	**100.0**	**100.0**	**100.0**	**100.0**	**100.0**	**100.0**	**100.0**	**100.0**
Nature of injury or illness:										
Sprains, strains	43.5	33.1	38.9	37.3	39.0	48.6	48.0	41.9	38.6	49.5
Bruises, contusions	9.2	9.3	10.1	7.2	8.6	10.4	8.6	10.0	6.2	9.9
Cuts, lacerations	7.8	9.6	5.7	10.2	8.9	4.3	6.1	11.9	7.0	4.7
Fractures	6.7	10.5	15.6	9.8	6.9	5.9	6.9	6.0	7.5	5.1
Heat burns	1.6	.8	1.2	1.2	1.6	.4	.6	3.2	.6	1.6
Carpal tunnel syndrome	1.6	.6	.6	.6	3.0	1.0	.9	1.2	5.5	1.4
Tendonitis	1.0	.7	—	.4	1.8	.5	.6	.7	1.5	1.0
Chemical burns	.7	.6	.5	1.0	.9	.3	.7	.5	.3	.6
Amputations	.6	.8	.8	.7	1.3	.3	.4	.3	.4	.2
Multiple traumatic injuries	3.5	3.3	6.2	4.4	3.0	3.5	3.6	3.4	4.2	3.4
Part of body affected by the injury or illness:										
Head	6.3	8.3	5.7	8.2	7.3	6.1	5.7	5.3	4.9	5.5
Neck	1.8	1.4	2.2	1.3	1.4	2.3	2.1	1.6	1.5	2.3
Trunk	37.1	30.3	35.3	32.3	34.9	40.5	42.1	35.7	30.4	40.6
Upper extremities	23.3	28.1	21.5	23.7	31.0	16.4	17.8	26.2	22.5	18.3
Lower extremities	20.6	22.8	23.9	24.3	17.5	22.8	21.8	20.6	20.4	20.0
Body systems	1.3	1.1	1.5	.8	1.2	1.5	1.0	1.0	3.1	1.7
Multiple parts	8.7	5.9	9.8	8.3	5.8	9.9	8.8	8.6	14.8	10.9
Source of injury or illness:										
Chemicals and chemical products	1.7	1.2	7.7	1.4	2.4	1.3	1.2	1.2	1.3	1.8
Containers	14.4	7.8	6.2	4.8	13.4	20.6	23.7	21.7	10.9	9.5
Furniture and fixtures	3.4	.9	.5	1.7	2.6	1.8	2.3	5.3	6.1	5.0
Machinery	6.7	7.5	13.0	5.7	11.6	2.5	6.7	7.3	6.1	3.6
Parts and materials	11.3	7.0	16.9	24.9	17.8	8.2	11.2	6.0	3.1	4.5
Worker motion or position	15.7	14.2	3.8	13.4	19.6	15.6	14.7	13.1	21.5	14.9
Floors, walkways, ground surfaces	16.0	14.8	18.4	18.0	10.4	15.5	14.5	19.5	22.3	18.2
Tools, instruments, and equipment	6.2	8.6	5.9	10.4	6.5	3.7	3.5	7.2	6.1	5.1
Vehicles	8.1	7.3	5.7	5.1	4.7	18.5	12.8	6.6	7.9	7.5
Health care patient	4.3	—	—	—	(5)	.9	—	—	.7	17.8
Event or exposure leading to injury or illness:										
Contact with objects and equipment	27.0	32.8	40.4	34.9	33.4	21.4	26.0	28.6	19.1	18.4
Struck by object	13.5	14.6	22.3	18.0	14.1	10.9	13.3	16.1	10.8	9.8
Struck against object	6.8	7.9	8.4	7.9	7.8	6.0	6.4	7.7	5.4	5.4
Caught in equipment or object	4.5	6.3	9.4	4.6	8.6	3.0	4.6	3.0	1.6	2.0
Fall to lower level	5.5	7.6	9.6	11.6	3.3	6.8	6.3	4.3	6.6	4.4
Fall on same level	11.2	8.3	7.9	7.1	8.0	9.4	8.9	16.2	16.7	14.3
Slip, trip, loss of balance—without fall	3.2	2.9	.9	3.1	2.6	3.7	3.1	3.7	2.9	3.5
Overexertion	27.0	16.1	28.7	20.7	25.3	28.4	30.9	25.6	19.6	32.4
Overexertion in lifting	15.6	9.7	10.5	12.2	13.7	15.2	19.1	17.4	12.5	17.6
Repetitive motion	4.3	2.3	1.3	1.8	8.4	2.8	3.2	2.5	9.9	3.4
Exposure to harmful substances	4.5	5.4	4.3	3.7	5.2	3.4	2.9	4.9	3.6	4.9
Transportation accidents	4.3	4.3	1.5	3.4	2.1	9.2	6.6	3.1	6.4	4.5
Fires and explosions	.2	—	.3	.4	.2	.1	.2	.3	.1	.1
Assaults and violent acts by person	1.0	.1	—	.2	.1	.3	.1	1.0	.8	3.0

NOTE: Dashes (—) indicate data are not available or do not meet publication guidelines. Because of rounding and classifications not shown, percentages may not add to 100. All injuries and illnesses reported involved days away from work. (1) Private sector includes all industries except government, but excludes farms with fewer than 11 employees. (2) Agriculture includes forestry and fishing, but excludes farms with fewer than 11 employees. (3) Data conforming to OSHA definition for mining operators in coal, metal, and nonmetal mining and for employers in railroad transportation are provided by the Mine Safety and Health Administration, U.S. Dept. of Labor, and by the Federal Railroad Administration, U.S. Dept. of Transportation. Independent mining contractors are excluded from the coal, metal, and nonmetal industries. (4) Finance includes insurance and real estate.

Fatal Occupational Injuries, 2000

Source: Bureau of Labor Statistics, U.S. Dept. of Labor

	FATALITIES Number	FATALITIES Percent
TRANSPORTATION INCIDENTS	**2,571**	**43**
Highway	1,363	23
Collision between vehicles, mobile equipment	694	12
Noncollision	356	6
Nonhighway (farm, industrial premises)	399	7
Aircraft	280	5
Worker struck by a vehicle	370	6
Water vehicle	84	1
Rail vehicle	71	1
ASSAULTS AND VIOLENT ACTS	**929**	**16**
Homicides	677	11
Shooting	533	9
Stabbing	66	1
Self-inflicted injuries	220	4
CONTACT WITH OBJECTS AND EQUIPMENT	**1,005**	**17**
Struck by object	570	10
Struck by falling object	357	6
Struck by flying object	61	1
Caught in or compressed by equipment or objects	294	5
Caught in running equipment or machinery	157	3
Caught in or crushed in collapsing materials	123	2
EXPOSURE TO HARMFUL SUBSTANCE OR ENVIRONMENTS	**480**	**8**
Contact with electric current	256	4
Contact with overhead power lines	128	2
Contact with temperature extremes	29	—
Exposure to caustic, noxious, or allergenic substances	100	2
Oxygen deficiency	93	2
Fires and explosions	177	3
Other events or exposures	**19**	—
TOTAL	**5,915**	**100**

NOTE: Totals for categories may include subcategories not shown separately. Percentages based on incidence rate per total fatalities. Dashes (—) indicate less than 0.5% or unavailable data.

U.S. Wage and Salary Workers Paid Hourly Rates, Second Quarter 2001

Source: Bureau of Labor Statistics, U.S. Dept. of Labor; unpublished tabulations from Current Population Survey

(in thousands)

SEX AND AGE	Total hourly workers	$5.15[1] or less	% of workers earning $5.15 or less	Less than $10.00	% of workers earning less than $10.00	$10.00 or more	% of workers earning $10.00 or more
Total, 16 years and older	72,455	2,315	3.2	32,590	45.0	39,865	55.0
16 to 24 years	16,478	1,300	7.9	12,473	75.7	4,005	24.3
20 to 24 years	10,116	660	6.5	6,729	66.5	3,387	33.5
25 years and older	55,977	1,015	1.8	20,117	35.9	35,860	64.1
25 to 54 years	47,849	820	1.7	16,659	34.8	31,190	65.2
25 to 34 years	15,942	372	2.3	6,379	40.0	9,563	60.0
35 to 44 years	18,174	276	1.5	5,911	32.5	12,263	67.5
45 to 54 years	13,734	171	1.2	4,370	31.8	9,364	68.2
55 years and older	8,128	196	2.4	3,458	42.5	4,670	57.5
55 to 64 years	6,387	111	1.7	2,305	36.1	4,082	63.9
65 years and older	1,741	84	4.8	1,153	66.2	588	33.8
Men, 16 years and older	36,125	802	2.2	13,583	37.6	22,542	62.4
16 to 24 years	8,477	467	5.5	5,996	70.7	2,481	29.3
20 to 24 years	5,275	166	3.1	3,178	60.2	2,097	39.8
25 years and older	27,648	334	1.2	7,587	27.4	20,061	72.6
Women, 16 years and older	36,329	1,514	4.2	19,006	52.3	17,323	47.7
16 to 24 years	8,000	833	10.4	6,476	81.0	1,524	19.1
20 to 24 years	4,841	494	10.2	3,551	73.4	1,290	26.6
25 years and older	28,329	681	2.4	12,530	44.2	15,799	55.8
RACE AND HISPANIC ORIGIN							
White							
Total, 16 years and older	59,226	1,930	3.3	26,216	44.3	33,010	55.7
Men	29,950	685	2.3	11,073	37.0	18,877	63.0
Women	29,276	1,244	4.2	15,143	51.7	14,133	48.3
Black							
Total, 16 years and older	9,985	332	3.3	5,064	50.7	4,921	49.3
Men	4,494	103	2.3	1,897	42.2	2,597	57.8
Women	5,491	229	4.2	3,167	57.7	2,324	42.3
Hispanic origin							
Total, 16 years and older	10,138	298	2.9	5,875	57.9	4,263	42.0
Men	5,915	147	2.5	3,086	52.2	2,829	47.8
Women	4,223	151	3.6	2,789	66.0	1,434	34.0
FULL- AND PART-TIME STATUS							
Full-time workers							
Total, 16 years and older	55,166	914	1.7	20,158	36.5	35,008	63.5
Men	30,316	320	1.1	9,097	30	21,219	70.0
Women	24,850	593	2.4	11,061	44.5	13,789	55.5
Part-time workers							
Total, 16 years and older	17,141	1,389	8.1	12,345	72	4,796	28.0
Men	5,730	477	8.3	4,438	77.5	1,292	22.5
Women	11,412	912	8.0	7,908	69.3	3,504	30.7

NOTE: Data refer to the sole or principal job, exclude the self-employed, and are not seasonally adjusted. Totals may not add because of independent rounding or because all subcategories are not listed. Full- or part-time status on the principal job is not identifiable for some multiple jobholders. Data for "other races" are not presented, and Hispanics are included in both white and black population groups. The data are from unpublished work tables and should not be considered as if part of an official BLS news release. (1) $5.15 = minimum wage starting Sept. 1, 1997. Source: Bureau of Labor Statistics, U.S. Dept. of Labor, unpublished tabulations from current Population Survey.

Federal Minimum Hourly Wage Rates Since 1950

Source: Bureau of Labor Statistics, U.S. Dept. of Labor

The Fair Labor Standards Act of 1938 and subsequent amendments provide for minimum wage-coverage applicable to nonprofessional workers in specified nonsupervisory employment categories.

EFFECTIVE DATE	NONFARM WORKERS			FARM WORKERS[4]
	Under laws prior to 1966[1]	Percent of avg. earnings[2]	Under 1966 and later provis.[3]	
Jan. 25, 1950	$0.75	54	NA	NA
Mar. 1, 1956	1.00	52	NA	NA
Sept. 3, 1961	1.15	50	NA	NA
Sept. 3, 1963	1.25	51	NA	NA
Feb. 1, 1967	1.40	50	$1.00	$1.00
Feb. 1, 1968	1.60	54	1.15	1.15
Feb. 1, 1969	(5)	(5)	1.30	1.30
Feb. 1, 1970	(5)	(5)	1.45	(5)
Feb. 1, 1971	(5)	(5)	1.60	(5)
May 1, 1974	2.00	46	1.90	1.60
Jan. 1, 1975	2.10	45	2.00	1.80
Jan. 1, 1976	$2.30	46	$2.20	$2.00
Jan. 1, 1977	(5)	(5)	2.30	2.20
Jan. 1, 1978	2.65	44	2.65	2.65
Jan. 1, 1979	2.90	45	2.90	2.90
Jan. 1, 1980	3.10	43	3.10	3.10
Jan. 1, 1981	3.35	42	3.35	3.35
Apr. 1, 1990	3.80[6]	35	3.80	3.80[6]
Apr. 1, 1991	4.25[6]	38	4.25	4.25[6]
Oct. 1, 1996	4.75[7]	37	4.75	4.75[7]
Sept. 1, 1997	5.15[7]	39	5.15	5.15

NA = not applicable. (1) Applies to workers covered prior to 1961 Amendments and, after Sept. 1965, to workers covered by 1961 Amendments. Rates set by 1961 Amendments were: Sept. 1961, $1.00; Sept. 1964, $1.15; and Sept. 1965, $1.25. (2) Percent of gross average hourly earnings of production workers in manufacturing. (3) Applies to workers newly covered by Amendments of 1966, 1974, and 1977, and Title IX of Education Amendments of 1972. (4) Included in coverage as of 1966, 1974, and 1977 Amendments. (5) No change in rate. (6) Training wage for workers age 16-19 in first 6 months of first job: Apr. 1, 1990, $3.35; Apr. 1, 1991, $3.62. The training wage expired Mar. 31, 1993. (7) Under 1996 legislation, a subminimum training wage of $4.25 an hour was established for employees under 20 years of age during their first 90 consecutive calendar days of employment with an employer. For workers receiving gratuities, the minimum wage remained $2.13 per hour.

Hourly Compensation Costs[1], by Selected Country, 1975-2000

Source: Bureau of Labor Statistics, U.S. Dept. of Labor
(in U.S. dollars, compensation for production workers in manufacturing)

Country/Territory	1975	1985	1990	2000
Australia	$5.62	$8.20	$13.07	$14.15
Austria	4.51	7.58	17.75	19.46
Belgium	6.41	8.97	19.17	21.11
Canada	5.96	10.95	15.95	16.16
Denmark	6.28	8.13	18.04	20.44
Finland	4.61	8.16	21.03	19.50
France	4.52	7.52	15.49	16.38
Germany[2]	6.31	9.53	21.88	22.99
Greece	1.69	3.66	6.76	—
Hong Kong[3]	0.76	1.73	3.20	5.53
Ireland	3.03	5.92	11.66	12.50
Israel	2.25	4.06	8.55	12.88
Italy	4.67	7.63	17.45	14.66
Japan	3.00	6.34	12.80	22.00
Korea, South	0.32	0.23	3.71	8.13
Luxembourg	$6.50	$7.81	$16.74	$16.69
Mexico	1.47	1.59	1.58	2.46
Netherlands	6.58	8.75	18.06	19.08
New Zealand	3.21	4.47	8.33	8.13
Norway	6.77	10.37	21.47	22.05
Portugal	1.58	1.53	3.77	4.75
Singapore	0.84	2.47	3.78	7.42
Spain	2.53	4.66	11.38	10.85
Sri Lanka	0.28	0.28	0.35	—
Sweden	7.18	9.66	20.93	20.14
Switzerland	6.09	9.66	20.86	21.24
Taiwan	0.40	1.50	3.93	5.98
United Kingdom	3.37	6.27	12.70	15.88
United States	6.36	13.01	14.91	19.86

— Data not available. (1) Compensation includes all direct pay (including bonuses, etc.), paid benefits, and for some countries, labor taxes. (2) 1975, 1985, and 1990 data are for area covered by the former West Germany. 2000 is for unified Germany. (3) Now part of China.

Top 15 U.S. Metropolitan Areas, by Average Annual Salary, 2000

Source: Bureau of Labor Statistics, U.S. Dept. of Labor

Rank	Metropolitan area	Average annual salary[1]
1.	San Jose, CA	$76,076
2.	San Francisco, CA	59,314
3.	New York, NY	56,377
4.	New Haven–Bridgeport–Stamford–Danbury–Waterbury, CT	50,585
5.	Middlesex-Somerset-Hunterdon, NJ	48,977
6.	Newark, NJ	48,733
7.	Jersey City, NJ	47,514
8.	Boulder-Longmont, CO	45,565
9.	Washington, DC–MD–VA–WV	$45,333
10.	Boston-Worcester–Lawrence–Lowell-Brockton, MA–NH	45,191
11.	Seattle–Bellevue–Everett, WA	45,171
12.	Trenton, NJ	44,576
13.	Oakland, CA	44,170
14.	Bergen–Passaic, NJ	43,789
15.	Hartford, CT	42,349

NOTE: Jacksonville, NC, recorded the **lowest average annual pay** among U.S. metropolitan areas in 2000—$21,057—followed by Yuma, AZ ($21,487), Brownsville–Harlingen–San Benito, TX ($21,561), McAllen–Edinburg–Mission, TX ($21,695), and Myrtle Beach, SC ($22,881). The nationwide metropolitan average was $36,986. (1) Data are preliminary and include workers covered by Unemployment Insurance and Unemployment Compensation for Federal Employees programs.

Average Hours and Earnings of U.S. Production Workers, 1969-2000[1]

Source: Bureau of Labor Statistics, U.S. Dept. of Labor
(annual averages)

Year	Weekly hours	Hourly earnings	Weekly earnings
1969	37.7	$3.04	$114.61
1970	37.1	3.23	119.83
1971	36.9	3.45	127.31
1972	37.0	3.70	136.90
1973	36.9	3.94	145.39
1974	36.5	4.24	154.76
1975	36.1	4.53	163.53
1976	36.1	4.86	175.45
1977	36.0	5.25	189.00
1978	35.8	5.69	203.70
1979	35.7	6.16	219.91
1980	35.3	6.66	235.10
1981	35.2	7.25	255.20
1982	34.8	7.68	267.26
1983	35.0	8.02	280.70
1984	35.2	8.32	292.86
1985	34.9	$8.57	$299.09
1986	34.8	8.76	304.85
1987	34.8	8.98	312.50
1988	34.7	9.28	322.02
1989	34.6	9.66	334.24
1990	34.5	10.01	345.35
1991	34.3	10.32	353.98
1992	34.4	10.57	363.61
1993	34.5	10.83	373.64
1994	34.7	11.12	385.86
1995	34.5	11.43	394.34
1996	34.4	11.82	406.61
1997	34.6	12.28	424.89
1998	34.6	12.78	442.19
1999	34.5	13.24	456.78
2000	34.5	13.75	474.38

(1) Private-industry production workers in mining and manufacturing; construction workers; nonsupervisory workers in services, transportation, and public utilities; wholesale or retail trade; finance, insurance, or real estate.

Median Income, by Sex, Race, Age, and Education, 1999, 2000

Source: Bureau of the Census, U.S. Dept. of Commerce

	1999	2000
MALE	**$27,275**	**$28,272**
Race		
White	28,564	29,696
Black	20,579	21,662
Hispanic origin[1]	18,234	19,833
Age		
Under 65 years	29,776	30,675
65 and over	19,079	19,168
Educational attainment		
Less than 9th grade	13,438	14,149
9th-12th grade (no diploma)	17,707	18,953
High school graduate	27,240	27,666
Some college, no degree	32,724	33,039
Associate degree	36,632	37,953
Bachelor's degree or more	47,325	49,178
FEMALE	**$15,311**	**$16,190**
Race		
White	15,362	16,218
Black	14,771	16,081
Hispanic origin[1]	11,314	12,255
Age		
Under 65 years	16,943	18,332
65 and over	10,943	10,899
Educational attainment		
Less than 9th grade	8,238	8,404
9th-12th grade (no diploma)	9,625	9,996
High school graduate	14,695	15,119
Some college, no degree	19,665	20,181
Associate degree	21,959	23,269
Bachelor's degree or more	31,642	33,365

NOTE: Includes both full-time and part-time year-round workers, 15 years old and over as of Mar. of the following year. (1) May be of any race.

Median Weekly Earnings of Wage and Salary Workers in the U.S. by Age, Sex, and Union Affiliation, 1995, 2000

Source: Bureau of Labor Statistics, U.S. Dept. of Labor

	1995				2000			
SEX AND AGE	TOTAL	Members of unions[1]	Represented by unions[2]	Non-union	TOTAL	Members of unions[1]	Represented by unions[2]	Non-union
Total, 16 years and older . .	$479	$602	$598	$447	$576	$696	$691	$542
16 to 24 years	292	375	373	287	361	437	436	355
25 years and older.	510	613	610	486	611	709	705	592
25 to 34 years	451	542	534	433	550	627	624	529
35 to 44 years	550	621	619	520	631	716	712	614
45 to 54 years	582	665	663	436	671	755	752	639
55 to 64 years	514	614	614	482	617	727	723	592
65 years and older.	389	509	506	362	442	577	565	422
Men, 16 years and older . .	538	640	638	507	646	739	737	620
16 to 24 years	303	388	388	298	376	458	457	370
25 years and older.	588	654	652	563	700	753	752	682
25 to 34 years	490	583	579	475	603	678	675	591
35 to 44 years	624	665	663	612	731	776	774	718
45 to 54 years	685	705	706	670	777	801	799	769
55 to 64 years	623	655	659	607	738	755	757	729
65 years and older.	441	615	619	400	537	613	613	514
Women, 16 years and older	406	527	523	386	491	616	613	472
16 to 24 years	275	349	345	272	342	406	405	339
25 years and older.	428	539	536	408	515	627	623	497
25 to 34 years	403	492	488	393	493	579	578	483
35 to 44 years	453	553	552	427	520	605	604	506
45 to 54 years	464	595	593	423	565	697	692	522
55 to 64 years	403	501	501	383	505	659	647	481
65 years and older.	353	435	425	333	378	485	484	365

Note: Data refer to the sole or principal job of full-time workers. Excluded are self-employed workers whose businesses are incorporated, although they technically qualify as wage and salary workers. (1) Including members of an employee association similar to a union. (2) Including members of a labor union or employee association similar to a union, and others whose jobs are covered by a union or an employee-association contract.

Work Stoppages (Strikes and Lockouts) in the U.S., 1960-2000

Source: Bureau of Labor Statistics, U.S. Dept. of Labor; involving 1,000 workers or more

Year	Number of stoppages[1]	Workers involved[1] (thousands)	Work days idle[1] (thousands)	Year	Number of stoppages[1]	Workers involved[1] (thousands)	Work days idle[1] (thousands)
1960	222	896	13,260	1985	54	324	7,079
1965	268	999	15,140	1986	69	533	11,861
1970	381	2,468	52,761	1987	46	174	4,481
1971	298	2,516	35,538	1988	40	118	4,381
1972	250	975	16,764	1989	51	452	16,996
1973	317	1,400	16,260	1990	44	185	5,926
1974	424	1,796	31,809	1991	40	392	4,584
1975	235	965	17,563	1992	35	364	3,989
1976	231	1,519	23,962	1993	35	182	3,981
1977	298	1,212	21,258	1994	45	322	5,020
1978	219	1,006	23,774	1995	31	192	5,771
1979	235	1,021	20,409	1996	37	273	4,889
1980	187	795	20,844	1997	29	339	4,497
1981	145	729	16,908	1998	34	387	5,116
1982	96	656	9,061	1999	17	73	1,996
1983	81	909	17,461	2000	39	394	20,419
1984	62	376	8,499				

(1) Numbers cover stoppages that began in the year indicated. Days of idleness include all stoppages in effect.

Work Stoppages Involving 5,000 Workers or More Beginning in 2000

Source: Bureau of Labor Statistics, U.S. Dept. of Labor

EMPLOYER; LOCATION; UNION	Began	Ended	Workers involved[1]	Estimated days idle in 2000[1]
Boeing Company; Interstate; Society of Professional Engineering Employees in Aerospace	2/9	3/19	17,000	459,000
Los Angeles janitorial maintenance contractors; Los Angeles County, CA; Service Employees	4/3	4/24	8,500	103,800
Building Owners and Managers Association; Chicago, IL; Service Employees	4/17	4/17	5,000	5,000
University of California; California; Automobile Workers	4/18	4/18	5,000	5,000
Association of National Advertisers and American Association of Advertising Agencies; Interstate; American Federation of Television and Radio Artists and Screen Actors Guild	5/1	10/30	135,000	17,280,000
Verizon Communications; Interstate; Communication Workers and Electrical Workers (IBEW)	8/6	8/23	85,000	955,000
Los Angeles County Metropolitan Transportation Authority; Los Angeles County, CA; United Transportation Union	9/16	10/16	7,400	148,000
Los Angeles County; Los Angeles County, CA; Service Employees.	10/11	10/11	47,000	47,000
Hospitals (18); California; Service Employees.	12/14	12/14	6,000	6,000

(1) Workers and days idle are rounded to the nearest 100.

Labor Union Directory

Source: Bureau of Labor Statistics, U.S. Dept. of Labor; AFL-CIO; World Almanac research as of Oct. 2001.

(*) Independent union; all others affiliated with AFL-CIO.

Actors and Artistes of America, Associated (AAAA), 165 W 46th St., Suite 500, New York, NY 10036; founded 1919; Theodore Bikel, Pres.; no individual members, 7 National Performing Arts Unions are affiliates; approx. 100,000 combined membership.

Actors' Equity Association, 165 W 46th St., New York, NY 10036; founded 1913; Patrick Quinn, Pres. (since 2000); 40,000 active members.

Air Line Pilots Association, 535 Herndon Pkwy., Herndon, VA 20170; founded 1931; Capt. Duane Woerth, Pres. (since 1999); 57,000+ members, 50 airlines.

American Federation of Labor & Congress of Industrial Organizations (AFL-CIO), 815 16th St. NW, Washington, DC 20006; founded 1955; John J. Sweeney, Pres. (since 1995); 13 mil. members.

Automobile, Aerospace & Agricultural Implement Workers of America, International Union, United (UAW), 8000 E Jefferson Ave., Detroit, MI 48214; founded 1935; Stephen P. Yokich, Pres. (since 1995); 746,000 active (500,000 ret.) members, 1,000+ locals.

Bakery, Confectionery, Tobacco Workers and Grain Millers International Union (BCTGM), 10401 Connecticut Ave., Kensington, MD 20895; founded 1886; Frank Hurt, Pres. (since 1992); 125,000 members.

Boilermakers, Iron Ship Builders, Blacksmiths, Forgers and Helpers, International Brotherhood of (IBBISB/BF&H), 753 State Ave., Suite 565, Kansas City, KS 66101; founded 1880; Charles W. Jones, Int'l Pres. (since 1983); 80,000 members, 368 locals.

Bricklayers and Allied Craftworkers, International Union of, 815 15th St. NW, Washington, DC 20005; founded 1865; John J. Flynn, Pres. (since 1999); 100,000 members, 200 locals.

Carpenters and Joiners of America, United Brotherhood of, 101 Constitution Ave., NW, Washington, DC 20001; founded 1881; Douglas J. McCarron, Gen. Pres. (since 1995); 525,000 members, 1,000 locals.

Communications Workers of America (CWA), 501 3d St. NW, Washington, DC 20001; founded 1938; Morton Bahr, Pres. (since 1985); 630,000 members, 1,400 locals.

***Education Association, National,** 1201 16th St. NW, Washington, DC 20036; founded 1857; Bob Chase, Pres. (since 1996); 2.5 mil. members, 13,500 affiliates.

Electrical Workers, International Brotherhood of (IBEW), 1125 15th St. NW, Washington, DC 20005; founded 1891; Edwin D. Hill, Pres. (since 2001); 727,836 members, 1,019 locals.

Electronic, Electrical, Salaried, Machine and Furniture Workers, International Union of (IUE), 1126 16th St. NW, Washington, DC 20036; founded 1949; Edward L. Fire, Pres. (since 1997); 120,000 members, 380 locals.

Engineers, International Union of Operating (IUOE), 1125 17th St. NW, Washington, DC 20036; founded 1896; Frank Hanley, Pres.; 360,000 members, 175 locals.

Farm Workers of America, United (UFW), 29700 Woodford-Tehachapi Rd., PO Box 62, Keene, CA 93531; founded 1962; Arturo S. Rodríguez, Pres. (since 1993); 50,000 members.

***Federal Employees, Federal District 1, National Federation of (NFFE FD1, IAMAW, AFL-CIO),** 1016 16th St. NW, Suite 300, Washington, DC 20036; founded 1917; Richard N. Brown, Pres. (1998); 120,000 members, 290 locals.

Fire Fighters, International Association of, 1750 New York Ave. NW, Washington, DC 20006; founded 1918; Harold Schaitberger, Pres. (since 2000); 250,000 members, 2,508 locals.

Firemen and Oilers, National Conference of, 1900 L St., NW, Suite 502, Washington, DC 20036; founded 1898; George J. Francisco, Jr., Pres.; 26,000 members, 133 locals.

Flight Attendants, Association of, 1275 K St. NW, Washington, DC 20005; founded 1945; Patricia A. Friend, Int'l Pres.; 46,000 members, 26 carriers.

Food and Commercial Workers International Union, United (UFCW), 1775 K St. NW, Washington, DC 20006-1598; founded 1979 following merger; Douglas H. Dority, Natl. Pres. (since 1994); 1.4 mil. members, 997 locals.

Glass, Molders, Pottery, Plastics & Allied Workers Intl. Union (GMP), 608 E Baltimore Pike, PO Box 607, Media, PA 19063; founded 1842; James H. Rankin, Pres. (since 1997); 65,000 members, 370 locals.

Government Employees, American Federation of (AFGE), 80 F St. NW, Washington, DC 20001; founded 1932; Bobby L. Harnage Sr., Pres. (since 1997); 197,000 members, 1,100 locals.

Graphic Communications International Union (GCIU), 1900 L St. NW, Washington, DC 20036; founded 1983; James J. George Tedeschi, Pres. (since 2000); 150,000 members, 321 locals.

Hotel Employees and Restaurant Employees International Union, 1219 28th St. NW, Washington, DC 20007; John W. Wilhelm, Gen. Pres. (since 1998); 350,000 members, 140 locals.

Iron Workers, International Association of Bridge, Structural, Ornamental and Reinforcing, 1750 New York Ave. NW, Suite 400, Washington, DC 20006; founded 1896; Joseph Hunt, Gen. Pres. (since 2001); 120,000 members, 242 locals.

Laborers' International Union of North America (LIUNA), 905 16th St. NW, Washington, DC 20006-1765; founded 1903; Terence M. O'Sullivan, Pres. (since 2000); 800,000 members.

Leather Goods, Plastics Novelty, and Service Workers' Union, International, 265 W 14th St., Suite 711, New York, NY 10011; 5,500 members, 80 locals.

Letter Carriers, National Association of (NALC), 100 Indiana Ave. NW, Washington, DC 20001-2144; founded 1889; Vincent R. Sombrotto, Pres. (since 1978); 312,848 members, 2,783 locals.

Locomotive Engineers, Brotherhood of (BLE), The Standard Bldg. Mezzanine, 1370 Ontario Ave., Cleveland, OH 44113-1702; founded 1863; Don M. Hans, Pres. (since 2001); 58,000 members, 600+ divisions.

Longshore & Warehouse Union, International (ILWU), 1188 Franklin St., San Francisco, CA 94109-6800; founded 1937; James Spinosa, Pres. (since 2000); 60,000 members, 58 locals, 16 units of locals.

Longshoremen's Association, International (ILA), 17 Battery Pl., Suite 1530, New York, NY 10004; John M. Bowers, Pres. (since 1987); 65,000 members.

Machinists and Aerospace Workers, International Association of (IAMAW), 9000 Machinists Pl., Upper Marlboro, MD 20772-2687; founded 1888; R. Thomas Buffenbarger, Pres. (since 1997); 780,000 members, 1,194 locals.

Maintenance of Way Employes, Brotherhood of (BMWE), 26555 Evergreen Rd., Suite 200, Southfield, MI 48076; founded 1887; M. A. "Mac" Fleming, Pres. (since 1990); 55,000 members, 790 locals.

Marine Engineers' Beneficial Assn. (MEBA), 444 N Capitol St. NW, Suite 800, Washington, DC 20001; founded 1875; Lawrence OToole, Pres. (since 1998).

Mine Workers of America, United (UMWA), 8315 Lee Highway, Fairfax, VA 22031; founded 1890; Cecil E. Roberts, Pres. (since 1995); 130,000 members, 600 locals.

Musicians of the United States and Canada, American Federation of (AFM), 1501 Broadway, Suite 600, New York, NY 10036; founded 1896; Thomas F. Lee, Pres. (since 2001); 125,000 members, 250+ locals.

Needletrades, Industrial, and Textile Employees, Union of (UNITE), 1710 Broadway, New York, NY 10019; founded 1995; Bruce S. Raynor, Pres. (since 2001); 250,000 members, 900 locals.

Newspaper Guild-Communications Workers of America (CWA), The, 501 3d St. NW, Suite 250, Washington, DC 20001-2797; founded 1933; Linda K. Foley, Pres. (since 1995); 35,000 members, 90 locals.

***Nurses Association, American (ANA),** 600 Maryland Ave. SW, Suite 100-W, Washington, DC 20024-2571; founded 1897; Mary E. Foley, Pres.; 177,000 members, 53 constituent state & territorial assns.

Office and Professional Employees International Union (OPEIU), 265 W 14th St., Suite 610, New York, NY 10011; founded 1945 (AFL Charter); Michael Goodwin, Pres. (since 1994); 130,000 members, 200 locals.

PACE International Union, AFL-CIO, CLC (PACE), 3340 Perimeter Hill Dr., PO Box 1475, Nashville, TN 37202; founded 1884; Boyd D. Young, Pres. (since 1999); 320,000 members, 1,600 locals.

Painters and Allied Trades, International Union of (IUPAT), 1750 New York Ave. NW, Washington, DC 20006; founded 1887; Michael E. Monroe, Gen. Pres.; 130,000 members, 425 locals.

Plasterers' and Cement Masons' International Association of the United States and Canada, Operative, 14405 Laurel Pl., Suite 300, Laurel, MD 20707; founded 1864; John J. Dougherty, Pres.; 40,000 members, 100 locals.

Plumbing and Pipe Fitting Industry of the United States and Canada, United Association of Journeymen and Apprentices of the, 901 Massachusetts Ave. NW, PO Box 37800, Washington, DC 20013; founded 1889; Martin J. Maddaloni, Gen. Pres. (since 1997); 307,000 members, 333 locals.

***Police, National Fraternal Order of,** 1410 Donelson Pike, A-17, Nashville, TN 37217; Steve Young, Natl. Pres. (since 2001); 290,000 members, 2,000+ affiliates.

Police Associations, International Union of, 1421 Prince St., Suite 400, Alexandria, VA 22314; Samuel Cabral, Pres. (since 1995); 80,000 members, 500 locals.

***Postal Supervisors, National Association of,** 1727 King St., Suite 400, Alexandria, VA 22314-2753; Vincent Palladino, Pres. (since 1992); 36,000 members, 400 locals.

Postal Workers Union, American (APWU), 1300 L St. NW, Washington, DC 20005; founded 1971; Moe Biller, Pres. (since 1980); 350,000 members, 1,600+ locals.

Roofers, Waterproofers & Allied Workers, United Union of, 1660 L St. NW, Suite 800, Washington, DC 20036; founded 1906; Earl J. Kruse, Pres. (since 1985); 25,000 members, 86 locals.

***Rural Letter Carriers' Association, National,** 1630 Duke St., 4th Fl., Alexandria, VA 22314; founded 1903; Gus Baffa, Pres. (since 2001); 100,000 members; 50 state org.

Seafarers International Union of North America (SIU), 5201 Auth Way and Britannia Way, Camp Springs, MD 20746; founded 1938; Michael Sacco, Pres. (since 1988); 85,000 members, 18 affiliates.

***Security, Police, and Fire Professionals of America (SPFA),** 25510 Kelly Rd., Roseville, MI 48066; founded 1948; David L. Hickey, Pres. (since 2000); 12,000 members, 160 locals.

Service Employees International Union (SEIU), 1313 L St. NW, Washington, DC 20005; founded 1921; Andrew L. Stern, Pres. (since 1996); 1.4 million members, 350 locals.

Sheet Metal Workers' International Association (SMWIA), 1750 New York Ave. NW, Washington, DC 20006; founded 1888; Michael J. Sullivan, Pres. (since 1999); 150,000 members, 194 locals.

State, County, and Municipal Employees, American Federation of (AFSCME), 1625 L St. NW, Washington, DC 20036; Gerald W. McEntee, Pres. (since 1981); 1.3 mil. members, 3,617 locals.

Steelworkers of America, United (USWA), 5 Gateway Center, Pittsburgh, PA 15222; founded 1936; Leo W. Gerard, Pres. (since 2001); 700,000+ members, 2,000 locals.

Teachers, American Federation of (AFT), 555 New Jersey Ave. NW, Washington, DC 20001; founded 1916; Sandra Feldman, Pres. (since 1997); 1 mil.+ members, 3,000 locals.

Teamsters, International Brotherhood of (IBT), 25 Louisiana Ave. NW, Washington, DC 20001; founded 1903; James P. Hoffa, Gen. Pres. (since 1999); 1.5 mil. members, 569 locals.

Television and Radio Artists, American Federation of, 260 Madison Ave., 7th fl., New York, NY 10016; founded 1937; John Connolly, Natl. Pres. (since 2001); 75,000 members, 35 locals.

Theatrical Stage Employees, Moving Picture Technicians, Artists and Allied Crafts of the United States, Its Territories, and Canada, International Alliance of (IATSE), 1515 Broadway, Suite 601, New York, NY 10036; founded 1893; Thomas C. Short, Pres. (since 1994); 95,000 members, 555+ locals.

Transit Union, Amalgamated (ATU), 5025 Wisconsin Ave. NW, 3rd Fl., Washington, DC 20016; founded 1892; James La Sala, Pres. (since 1986); 165,000 members, 285 locals.

Transportation-Communications International Union (TCU), 3 Research Place, Rockville, MD 20850; founded 1899; Robert A. Scardelletti, Pres. (since 1991); 100,000 members.

Transportation Union, United (UTU), 14600 Detroit Ave., Cleveland, OH 44107; founded 1969; Byron A. Boyd Jr., Pres. (since 2001); 135,000 members, 680 locals.

Transport Workers Union of America, 80 West End Ave., 5th Fl., New York, NY 10023; founded 1934; Sonny Hall, Int'l. Pres. (since 1993); 125,000+ members, 92 locals.

***Treasury Employees Union, National (NTEU),** 901 E St. NW, Suite 600, Washington, DC 20004; founded 1938; Colleen M. Kelley, Natl. Pres. (since 1999); 155,000 represented, 270+ chapters.

***University Professors, American Association of (AAUP),** 1012 14th St. NW, Suite 500, Washington, DC 20005; founded 1915; Jane Buck, Pres.; 44,000 members, 600 chapters.

Utility Workers Union of America (UWUA), 815 16th St. NW, Washington, DC 20006; founded 1945; Donald Wightman, Pres. (since 1996); 43,000 members, 250 locals.

U.S. Union Membership, 1930-2000

Source: Bureau of Labor Statistics, U.S. Dept. of Labor

Year	Labor force[1] (thousands)	Union members[2] (thousands)	Percentage of labor force
1930	29,424	3,401	11.6
1935	27,053	3,584	13.2
1940	32,376	8,717	26.9
1945	40,394	14,322	35.5
1950	45,222	14,267	31.5
1955	50,675	16,802	33.2
1960	54,234	17,049	31.4
1965	60,815	17,299	28.4
1970	70,920	19,381	27.3
1975	76,945	19,611	25.5
1980	90,564	19,843	21.9
1985	94,521	16,996	18.0
1986	96,903	16,975	17.5
1987	99,303	16,913	17.0
1988	101,407	17,002	16.8
1989	103,480	16,960	16.4
1990	103,905	16,740	16.1
1991	102,786	16,568	16.1
1992	103,688	16,390	15.8
1993	105,067	16,598	15.8
1994	107,989	16,748	15.5
1995	110,038	16,360	14.9
1996	111,960	16,269	14.5
1997	114,533	16,110	14.1
1998	116,730	16,211	13.9
1999	118,963	16,477	13.9
2000	120,786	16,258	13.5

(1) Does not include agricultural employment; from 1985, does not include self-employed or unemployed persons. (2) From 1930 to 1980, includes dues-paying members of traditional trade unions, regardless of employment status; after that includes employed only. From 1985, includes members of employee associations that engage in collective bargaining with employers.

TAXES

Federal Income Tax

Source: George W. Smith III, CPA, Nationally Syndicated Tax Author and Columnist

The recently enacted Economic Growth and Tax Relief Reconciliation Act of 2001 represents the largest federal income tax cut in over two decades. The legislation provides reductions that phase in over 10 years, with immediate cuts in individual tax rates.

The Economic Growth and Tax Relief Reconciliation Act of 2001 (Tax Relief Act)

On May 26, 2001, approximately four months after the Presidential Inauguration, Congress passed the massive $1.35 trillion, 10-year Tax Relief Act, and Pres. George W. Bush signed it on June 7. The measure adds more than 440 changes to the already complex Internal Revenue Tax Code.

Individual Tax Rates. The Tax Relief Act reduces marginal tax rates for individuals. Beginning July 1, 2001, the 28% rate drops to 27%, the 31% to 30%, and the 36% rate to 35%. The top 39.6% rate is lowered to 38.6% and eventually decreases to 35% in 2006. On the bottom, the 15% rate remains, but there is a new, lower 10% rate, beginning Jan. 1, 2002, which covers the first $6,000 of taxable income for those filing as single, $10,000 for heads of households, and $12,000 for married persons filing jointly.

Advance Refunds. Single taxpayers who paid federal income tax on their 2000 returns received checks for $300 in advance refund payments (less if they earned under $6,000 in taxable income). Heads of households received up to $500, a lesser amount if they earned under $10,000 of taxable income. While married couples filing jointly received up to $600, they received less if combined earned income was under $12,000. These refund payments in effect are supposed to duplicate the 10% income tax bracket benefit in 2001. The first checks were mailed in July 2001. Most mailings were completed by the end of September.

Child Credits. Beginning in 2001, the child tax credit increased from $500 to $600 for each qualifying child. The credit is scheduled to rise to $700 in 2005, $800 in 2009, and $1,000 in 2010. A credit reduces a taxpayer's income tax liability. However, for lower-income taxpayers the Tax Relief Act makes the child tax credit partially refundable although no income tax was paid.

Starting in the year 2002, the maximum expense eligible for the dependent care tax credit increases to $3,000 for one qualifying child, or other dependent incapable of self-care, with a $6,000 maximum for two or more. The income phase-out limitations also increase .

Adoption Credit. Both the maximum dollar limitation for the adoption credit and the exclusion from income of employer-provided adoption assistance increase to $10,000 per eligible child beginning in 2002. However, the credit is subject to phaseout limits based on income. After 2002, a credit can be claimed for a special needs adoption whether or not the taxpayer has qualified adoption expenses.

Education IRA. The maximum annual contribution to an education IRA savings account increases in 2002 from $500 to $2,000. For tax years beginning after Dec. 31, 2001, these accounts may be used for elementary and secondary education expenses, whether incurred in a public, private, or religious school. Qualifying distributions are not taxable to the recipient. The phase-out maximum for joint filers increases to twice that of single filers. This IRA is not tax deductible.

Student Loans. After 2001, the Tax Relief Act increases the income phase-out range for the interest deduction on student loans to $50,000-$65,000 for single taxpayers and $100,000-$130,000 for married taxpayers. Starting in 2003, these ranges will be adjusted annually for inflation. The provision also repeals the 60-month time limitation for the number of months during which interest paid is deductible.

College Education. For 2002 and 2003, a $3,000 deduction for qualified tuition and related expenses is available to individual taxpayers even if the taxpayer does not itemize. The amount increases to $4,000 for 2004 and 2005. Generally, any accredited public, nonprofit or proprietary post-secondary institution is considered an eligible education institution. These dollar amounts are phased-out for higher-income taxpayers and cannot be taken if the Hope or Lifetime Learning credits are taken for the same student. The Tax Relief Act terminates this deduction after December 31, 2005.

Marriage Penalty Relief. The standard deduction available to married taxpayers filing a joint return will gradually increase over a five year period starting in 2005 until in 2009 it reaches 200% of that allowed for single taxpayers. The higher limit of the 15% bracket for married taxpayers filing a joint return also will increase beginning in 2005.

Higher Income. The current levels for phasing out personal exemptions and the Schedule A itemized deductions for higher income taxpayers will decrease by 1/3 in 2006 and 2007, and by 2/3 in 2008 and 2009. These phase-out rules will be completely eliminated in 2010.

Estate Exclusion. The estate tax exclusion increases from $675,000 in 2001 to $1 million in 2002 and 2003. In 2004 and 2005 the exclusion increases to $1.5 million; then it goes to $2 million in 2006, and $3.5 millon in 2009. The Tax Relief Act repeals all estate taxes for the year 2010.

Note: Unless Congress acts, most of the changes enacted in the Economic Growth and Tax Relief Reconciliation Act of 2001 will go out of existence at the end of 2010, when the legislation expires.

Other Tax Matters, New Legislation, IRS Rulings

Medical Conference. A recent Internal Revenue Service (IRS) ruling allowed a taxpayer with a chronically ill dependent to take a medical deduction for expenses paid to attend a medical conference recommended by the dependent's physician.

Eye Surgery. The IRS changed its thinking and now says that the cost of certain kinds of eye surgery (radial keratomy, lasik, etc.) to improve vision is deductible, since the treatment is considered corrective, not cosmetic.

Credit Card Fees. The IRS ruled in 2001 that a taxpayer is not permitted to deduct credit card fees incurred when the card was used to pay personal income taxes.

Smoking. The IRS reversed its 20-year-old position and now allows taxpayers to deduct two treatments for quitting cigarette smoking as a medical expense: (1) participation in a smoking-cessation program, and (2) prescription drugs to alleviate the effects of nicotine withdrawal. However, over-the-counter products such as nicotine patches and chewing gum remain nondeductible.

Staying Fit. Expenses for membership in a fitness center and the like are not deductible even if there is a job requirement to stay in top physical condition.

Funeral Expenses. Funeral expenses are not deductible on an individual's income tax return. However, they are deductible on the federal estate tax return (Form 706) of the deceased.

Garage Sale. Revenues received from a garage sale usually do not result in taxable income. In most cases, the item that was sold cost more than the revenue received for it. By the same token, losses attributable to a garage sale do not result from a trade or business and, therefore, are considered personal and not deductible.

Day Camp. If both spouses work, the cost of summer day camp may qualify for the child care credit.

Divorce. Taxpayers can deduct a portion of their legal fees for a divorce if the attorney's invoice actually specifies how much of the fee was for tax advice. Legal fees paid to collect taxable alimony also are deductible.

Investment Expenses. Investors can take a miscellaneous deduction on Schedule A for investment and custodial fees, trust administration fees, investment advice, financial newspapers and reports and other expenses paid for managing their investment portfolio that produces taxable income (even a loss). However, they cannot deduct expenses for attending a convention, seminar, or similar meeting for investment purposes.

Gambling. Lottery and other gambling winnings are reported on page l, Form 1040. Gambling expenses are reported on Schedule A. Such expenses are deductible only up to the amount of the winnings reported on page 1.

Rental Income. Rental income received on homes and cottages rented for 14 days or less per year is neither taxable nor reportable on the taxpayer's return.

Parking Tickets. Penalties and fines paid to a governmental unit or department are not deductible. This includes parking and speeding tickets. This is also true for penalties for late filing of an individual's tax return and expenses incurred in a criminal case resulting in a conviction.

IT'S A FACT: In the past five years there have been over 1,900 changes to the federal tax code. Since 1986, there have been about 7,000 changes.

Tax Highlights

Medical Expenses. The page 1, Form 1040 deduction for medical insurance premiums paid in 2001 for self-employed individuals, their spouse, and dependents is 60% of the cost. Beginning in 2002 it rises to 70%, and it becomes a 100% page one deduction thereafter. The remaining portion of the premium not used in 2001 and 2002 is deductible as a medical expense on Schedule A subject to the 7.5% limitation rule. For more information call the IRS at 1-800-829-3676 and ask for Publication 535, Business Expense.

Sale of Residence. Married couples who lived in their principal residence for at least 2 years during a 5-year period ending on the date of sale and are filing a joint income tax return may exclude up to $500,000 in gain from the sale of their residence. This deduction is reusable every 2 years. Single taxpayers may exclude a gain up to $250,000. Married couples who do not share a principal residence with their spouse but continue to file a joint return also may claim up to the $250,000 exclusion for a qualifying sale or exchange of each spouse's principal residence.

Homeowners who have lived in their home fewer than 2 years and must sell because of a change in their place of work, or certain other reasons, may prorate the exclusion based on the amount of time lived there.

House Closing Points. The IRS has ruled that taxpayers need not deduct points in the year of purchase of a home. They may amortize the points over the life of the loan. This especially could help first-time home buyers who may not have sufficient deductions to itemize.

Home Office. Legislation now allows a deduction for taxpayers who set up an office at home to take care of the administrative or management side of their business. The home now can be considered a principal place of business in such cases. See the instructions on Form 8829, Expenses for Business Use of Your Home.

Mileage. The mileage allowance deduction for driving to obtain medical treatment or for automobile costs incurred in a deductible job-related move increased to 12 cents per mile for 2001, up from 10 cents in 2000. The deduction for an individual using his or her automobile in volunteer work for qualified charities remains at 14 cents per mile. The standard mileage deduction for business use of an automobile increased from 32.5 cents for 2000 to 34.5 cents for 2001. This rate also applies to a leased automobile used in business and must be used for the entire lease period.

Innocent Spouse Relief. The IRS Reform Act of 1998 provides a separate liability section for taxpayers who are divorced, legally separated, or living apart for at least 12 months. In effect, this legislation prevents them from being held liable for a spouse's tax liability.

Children's Income. Parents may elect to include on their income tax return the dividends and interest income of a dependent child under age 14 whose unearned income is more than $750 and whose gross income is less than $7,500. Form 8814, Parent's Election to Report Child's Interest and Dividends, must be attached to the parents' tax return. This election is not available if estimated tax payments were made or investments were sold in the child's name during the year.

If a dependent child with taxable income cannot file an income tax return, the parent, guardian, or other legally responsible person must file a return for the child.

An individual may not claim a dependency exemption in the year 2001 for a child who qualifies as a full-time student and is over age 23 at the end of the year, unless the child's gross income is less than $2,900.

Capital Gains. The long-term capital gains tax rate for individual taxpayers is 20% for qualified investments held more than 12 months. For taxpayers in the 15% tax bracket, the maximum long-term capital gains rate is 10%.

Starting Jan. 1, 2001, Congress lowered the capital gains rate for investments held more than 5 years to 18%; 8% if the taxpayer is in the 15% tax bracket. Further, if the taxpayer is in a tax bracket above 15%, the 5-year holding period would apply only to investments acquired after Dec. 31, 2000. For individuals in the 15% tax bracket, investments do not have to be acquired after the year 2000 to have the 5-year period begin.

Hobbies. Qualifying long-term gains for collectibles such as art, antiques, jewelry, stamps, and coins are taxed at a maximum 27% if sold after July 1, 2001.

Death Benefits. Qualified accelerated death benefits paid under a life insurance contract to terminally ill persons (certified as expected to die within 24 months) now are excludable from gross income. A similar exclusion applies to the sale or assignment of insurance death benefits to another person. Accelerated death benefits paid to a chronically ill person under a long-term care rider are tax-free up to $190 per day starting in 2000 and will be indexed for inflation.

Payroll Tax Rates. For the year 2001 the maximum wage base for withholding Social Security tax increased to $80,400, up from $76,200. The Social Security and Medicare tax rates remain at 6.2% and 1.45% respectively. There is no maximum wage base for the Medicare tax. Employer and employee each pay these taxes. Self-employed individuals pay both parts, 12.4% and 2.9%, for a total of 15.3%.

Domestic Workers. The annual threshold dollar amount in 2001 for reporting and paying Social Security and federal unemployment taxes on domestic employees, including nannies and housekeepers, is $1,300, up $100. Household workers under 18 are exempt unless household work is their principal occupation. Household employers must apply for an employer federal ID number and issue W-2 wage statements.

Filing and Payment Dates

Filing Dates. The due date for filing a timely 2001 U.S. individual income tax return is Monday, Apr. 15, 2002.

Refunds. Individuals can call the IRS toll-free at 1-800-829-4477 for a recorded message to check on the status of their expected refund. Taxpayers may have refunds deposited directly into their bank account.

Payments. Taxpayers may use their MasterCard, Discover or American Express credit cards for tax payments. To pay by credit card, call 1-888-2-PAY-TAX. There is a "convenience fee" charged by the credit card company based on the size of the payment. Estimated tax payments now can be made this way instead of filing Form 1040-ES payment vouchers. Federal income taxes may now be paid via the Internet.

Depending on the amount of tax owed, taxpayers may apply for monthly installment payments by attaching Form 9465 to their tax return. There is a nominal filing fee if the request is approved.

Estimated Taxes. Due dates for individual quarterly federal estimated tax payments for 2002 are as follows: 1st quarter, Mon., Apr. 15, 2002; 2d quarter, Mon. June 17; 3d quarter, Mon., Sept. 16; 4th quarter, Wed., Jan. 15, 2003. Different filing dates may apply for state and local quarterly estimated tax payments.

Need More Time? If individuals cannot file their 2001 tax return by Apr 15, 2002, they may apply for a 4-month extension of time. Although the extension is automatic, they must file Form 4868 no later than midnight Apr. 15, 2002.

The IRS says some eight million individuals filed for an extension of time in 2001.

Timely Postmark. The IRS must accept the postmark of qualified couriers such as UPS and FedEx as proof of timely mailing. *Caution:* When the return is mailed after Apr. 15, 2002, or after the extended due date, the IRS considers a return as filed on the date it is received by the IRS, not the date of postmark.

Statute of Limitations. Taxpayers have until Apr. 15, 2002 to file their 1998 federal tax return to claim a refund. After that date any refund for 1998 will be lost...forever.

Filing Penalties. The IRS can levy 2 potential penalties when a tax return is filed after the due date and there is a balance owing. One penalty is for failing to file a timely tax return. The second is for failure to pay the tax when due. Interest will be charged on any unpaid tax balance.

Privacy. To protect the taxpayer's privacy, Social Security numbers no longer appear on mailing labels.

IRS Services. Federal tax forms, tax legislation, relevant court decisions, and other information and resources are available from the IRS via the following:

Internet website: www.irs.gov
Telnet: iris.irs.gov
File Transfer Protocol: ftp.irs.gov
Fax: 1-703-368-9694
Forms/Publications: 1-800-829-3676

English/Spanish. The IRS provides videotaped instructions both in English and Spanish at participating libraries. Many IRS publications and tax forms including instructions are also printed in Spanish. For more information, call 1-800-TAX-FORM and ask for the free IRS Publication 1SP, Derechos del Contribuyente.

Hearing Impaired. The IRS telephone service for hearing impaired persons is available for taxpayers with access to TDD equipment. The toll-free telephone number is 1-800-829-4059.

IT'S A FACT: The IRS web site www.irs.gov recorded more than 1.5 billion hits from Jan. 1, 2001, through Apr. 16, 2001. This figure was up 57% from the previous year.

Who Must File a Tax Return?

Most U.S. citizens and resident aliens will have to file a 2001 income tax return if their gross income for the year is at least as much as the amount shown in the following table:

Filing Status	2001 Gross Income
Single	
Under 65	$ 7,450
65 or older	8,550
Married filing jointly	
Both spouses under 65	13,400
One spouse 65 or older	14,300
Both spouses 65 or older	15,200
Married filing separately	2,900
Head of household	
Under 65	9,550
65 or older	10,650
Qualifying widow(er)	
Under 65	10,500
65 or older	11,400

Regardless of the above amounts, a tax return must also be filed if:

- Taxpayer had net earnings of $400 or more from self-employment for the year.
- Taxpayer received advance earned income credit payments during the year from an employer or is entitled to receive a refundable earned income credit.
- Taxpayer paid estimated income tax payments during the year 2001 or expects an income tax refund.
- Taxpayer has losses to be carried back or forward.

Additional taxes are owed for:

—Social Security tax on unreported tips.
—Alternative minimum tax.
—Recapture of investment credit.
—Excise tax attributable to qualified retirement distributions including IRAs, annuities, and modified endowment contracts.

IT'S A FACT: Electronic tax filing hit a new high for the number of returns filed in 2001. As of Apr. 16, 2001, the IRS had received over 39 million returns electronically. This was up more than 13% from 2000 and represents about one-third of all tax returns filed for the year.

Which Tax Return Form to File?

Most U.S. citizens can use one of the following income tax forms: Form 1040, 1040A, or 1040EZ. Forms 1040A and 1040EZ are shorter and simpler to use than Form 1040.

You may be able to use the shortest tax return, Form 1040EZ, if:

- You are single or married filing jointly and do not claim any dependents.
- You are not 65 or older or blind.
- Your only income is from wages, salaries, tips, taxable scholarships or fellowships, unemployment compensation or Alaska Permanent Fund dividends.
- Your taxable income is less than $50,000 and you do not have over $400 of taxable interest income.
- You do not claim a student loan interest deduction or an education credit.
- You do not itemize deductions, claim any adjustments to income or have tax credits other than the earned income credit.
- You received no advance earned income credit payments. You did not make any estimated tax payments.

The income tax Form 1040A may be used if:

- You have income only from wages, salaries, tips, taxable scholarships or fellowships, interest and dividends, IRA distributions, pensions, annuities, unemployment compensation and/or taxable Social Security or railroad retirement benefits.
- Your taxable income is less than $50,000.
- You do not itemize deductions.
- You claim a deduction for qualified IRA contributions.
- You claim a credit for child and dependent care expenses, credit for the elderly or the disabled, the earned income credit, the adoption credit, child tax credit or education credits.
- You report employment taxes on wages paid to household employees on Schedule H.
- You take the education exclusion for interest income earned from Series EE U.S. Savings Bonds.
- You received advance earned income credit payments.
- You owe alternative minimum tax.
- You have made estimated tax payments.

You must file Form 1040 if any of these apply:

- Your taxable income is $50,000 or more. (However, you also may use Form 1040 for lower income amounts.)
- You plan to itemize deductions.
- You received any nontaxable dividends or capital gain distributions.
- You have foreign bank accounts and/or foreign trusts.
- You have taxable refunds from state or local income taxes.
- You have business, farm, or rental income or losses.
- You sold or exchanged capital assets or business property.
- You have miscellaneous income such as alimony that is not allowed on Form 1040A or 1040EZ.
- You have additional adjustments to income such as payments for alimony or moving expenses.
- You are allowed a foreign tax credit or certain other credits.
- You have other taxes to pay, such as self-employment tax or Social Security tax on tips.
- You have losses that are to be carried back or forward.
- You are required to file additional forms such as Form 2106, Employee Business Expenses; Form 2555, Foreign Earned Income; Form 3903, Moving Expenses; Form 4972, Tax on Lump-Sum Distributions.

Separately vs. jointly

- Married taxpayers can file separate income tax returns. If they later change their minds, they may amend their tax returns and file jointly. However, once a joint return is filed, they cannot later file separate returns.

Individual Income Tax Rates for Year 2001

Single

Tax Rate	Taxable Income
15%	$0 to $27,050
27.5%	$27,051 to $65,550
30.5%	$65,551 to $136,750
35.5%	$136,751 to 297,350
39.1%	More than $297,350

Married Filing Jointly or Qualifying Widow(er)

Tax Rate	Taxable Income
15%	$0 to $45,200
27.5%	$45,201 to $109,250
30.5%	$109,251 to $166,500
35.5%	$166,501 to $297,350
39.1%	More than $297,350

Married Filing Separately

Tax Rate	Taxable Income
15%	$0 to $22,600
27.5%	$22,601 to $54,625
30.5%	$54,626 to $83,250
35.5%	$83,251 to 148,675
39.1%	More than $148,675

Head of Household

Tax Rate	Taxable Income
15%	$0 to $36,250
27.5%	$36,251 to $93,650
30.5%	$93,651 to $151,650
35.5%	$151,651 to 297,350
39.1%	More than $297,350

Estates and Trusts

Tax Rate	Taxable Income
15%	$0 to $1,800
27.5%	$1,801 to $4,250
30.5%	$4,251 to $6,500
35.5%	$6,501 to $8,900
39.1%	More than $8,900

Note: Rather than incorporating a 10% tax bracket into the 2001 tax tables, Congress elected to effect this change by directing the U.S. Treasury to issue advance refund checks. The tax rates (above) reflect the average of the (older) rates of the first six months of 2001 and the new reduction in tax rates effective July 1 through Dec. 31, 2001.

"Kiddie Tax." If a child under age 14 has net investment income exceeding $1,500 for the year 2001, the excess will be taxable at the parents' top tax rate.

Exemptions

Dollar Amounts. The personal exemption amount for each taxpayer, spouse and dependent for the year 2001 is $2,900, up from $2,800 for 2000. These exemptions are adjusted each year for any cost of living increase.

Phaseout. The exemption deduction for higher income taxpayers begins to be phased out when their income exceeds certain threshold dollar amounts. Each exemption is reduced by 2% for each $2,500 ($1,250 for married filing separately) or fraction thereof by which adjusted gross income for year 2001 exceeds the following:

Threshold Dollar Amounts

Married filing jointly:	$199,450
Qualifying widow(er):	$199,450
Head of household:	$166,200
Single:	$132,950
Married filing separately:	$99,725

The exemption amount for the year 2001 is fully phased out when adjusted gross income is more than $122,500 ($61,250 for married filing separately) over the above threshold amount.

Standard Deduction

The standard deduction is a flat dollar amount that is subtracted from the adjusted gross income (AGI) of taxpayers who do not itemize deductions. The amount allowed depends on filing status and is adjusted annually for inflation.

2001 Standard Deduction Amount

Single:	$4,550
Married filing jointly or qualifying widow(er):	$7,600
Married filing separately:	$3,800
Head of household:	$6,650

These figures are not applicable if an individual can be claimed as a dependent on another person's tax return.

Standard Deduction For Dependents. An individual reported as a dependent on another person's income tax return generally may claim on his or her own tax return only the larger of $750 of earned income plus $250 not to exceed $4,550. A blind dependent may add $1,100 to this amount. Earned income includes wages, salaries, commissions and tips. Earned income also includes net profit from self-employment and any part of a scholarship or fellowship grant that must be included in gross income.

Taxpayers who are 65 or older and/or blind may claim an additional standard deduction:

2001 Additional Standard Deduction Amount

Single or head of household, 65 or older OR blind:	$1,100
Single or head of household, 65 or older AND blind:	$2,200
Married filing jointly or qualifying widow(er), 65 or older OR blind (per person):	$ 900
Married filing jointly or qualifying widow(er), 65 or older AND blind (per person):	$1,800
Married filing separately, 65 or older OR blind:	$ 900
Married filing separately, 65 or older AND blind:	$1,800

Adjustments to Income

IRA Deduction. The maximum tax-deferred Individual Retirement Arrangement (IRA) contribution for a married couple filing jointly in 2001 is $4,000, but not to exceed total earned income if less than $4,000. Each spouse can contribute up to $2,000 annually even if a spouse had little or no income. However, there are income limitations.

IRA Withdrawals. There is a 10% early withdrawal penalty for IRA distributions before age 59½ unless it qualifies for one of the following exceptions:

Distributions paid to the beneficiary after the death of the owner.

Payments paid due to the disability of the owner.

Part of a series of substantially equal periodic payments.

Made to an employee following separation from employment after age 55. This exception does not apply if a qualified distribution from a pension plan is rolled into an IRA.

Used to pay certain unreimbursed medical expenses.

Used to pay certain qualifying higher education expenses.

Used to pay certain qualified first time home buyer acquisition costs (up to $10,000).

The Roth IRA. Although contributions paid into a Roth IRA are not deductible, distributions of funds including investment earnings held in the account for 5 years or longer and distributed after age 59½ are free both of income tax and the 10% early withdrawal penalty at the time of distribution.

Any funds paid from the Roth IRA after the 5-year exclusion period to an estate or beneficiary on or after an individual's death, including funds paid to an individual who is disabled, are tax and penalty free regardless of age. This includes withdrawals of up to $10,000 if used for a first-time home purchase.

Withdrawals from a Roth IRA held less than 5 years are subject both to income tax and the 10% withdrawal penalty regardless of age. However, earnings withdrawn for "qualified higher education expenses" of the taxpayer, spouse, or any child or grandchild of the taxpayer or spouse are taxable but not subject to the early withdrawal penalty.

For more information on IRAs call the IRS at 1-800-829-3676 for a free copy of Publication 590, Individual Retirement Arrangements (IRA).

Moving Expenses. Taxpayers who change jobs or are transferred usually can deduct part of their moving expenses including travel and moving of household goods, but not meals. The standard mileage rate for automobiles used in the move increased to 12 cents per mile starting in 2001 plus parking and tolls.

In order to take a moving expense deduction, the new job must be at least 50 miles farther from the former home than the old job. Employees must work full-time for at least 39 weeks during the first 12 months after they arrive in the general area of their new job.

Itemized Deductions

If the total amount of itemized deductions is more than the standard deduction, taxpayers generally should itemize their deductions on Schedule A, Form 1040. (Only the total amount of medical expenses that exceeds 7.5% of the taxpayer's adjusted gross income is deductible.) The following examples are just a few of the deductions that may be itemized; some are subject to income limitations:

- Medicines, birth control pills, and insulin are deductible if prescribed by a doctor.
- Long-term care insurance premiums are deductible up to certain annual limits based on age. The maximum premium allowed as a medical expense deduction is: $230 if age 40 or less; $430 from 41 to 50; $860, 51 to 60; $2,290, 61 to 70, and $2,860 if over age 70. Any long-term benefits received under a qualifying policy are tax-free subject to per diem restrictions.
- Cosmetic surgery for congenital abnormality, personal injury resulting from an accident or trauma, or a disfiguring disease is allowed as a medical deduction.
- Most mortgage interest paid on a primary residence or a second home is fully deductible. However, there are limitations on mortgages in excess of $1,000,000.
- Interest paid on home equity loans is deductible but only on the first $100,000 of equity debt.
- Borrowers generally can deduct points paid on their principal home mortgage loan on Schedule A.
- Investment interest expense is deductible only to the extent of net investment income. Any investment interest expense not currently deducted is carried over to future years.
- State and local income taxes, real estate taxes, and personal property taxes are fully deductible. Sales taxes are not deductible.
- Casualty and theft losses are deductible subject to the $100 and 10% limitation rule for each occurrence.
- Taxpayers deducting individual charitable contributions of $250 or more must obtain written substantiation from the charity. If the amount is $75 or more, the charity must include a breakdown of the payment indicating how much was a (deductible) contribution and what (if any) was the (nondeductible) value of goods, meals or services received.
- Miscellaneous expenses including union and professional dues, tax preparation fees, safe-deposit box rental fees, and employee business expenses are deductible insofar as they exceed 2% of adjusted gross income.
- Unreimbursed employee business expenses including travel, automobile, telephone, and gifts are deductible on Schedule A as miscellaneous itemized deductions. Only 50% of the cost of customer meals and entertainment is deductible and it is further subject to the 2% rule.
- Employment fees paid to agencies, resume costs, postage, travel, and other expenses to look for a new job in your present occupation are deductible even if you do not get a new job.

Threshold Reduction. Many itemized deductions otherwise allowed are further reduced by the smaller of these two figures: 3% of a taxpayer's 2001 adjusted gross income in excess of the threshold amount of $132,950 ($66,475 for married taxpayers filing separately) OR 80% of the amount of these itemized deductions otherwise allowable for the year. This provision does not apply to medical expenses, investment interest expense, casualty losses, or gambling expenses. The threshold reduction is phased out over a five year period by the Tax Relief Act, starting in 2006.

Business Expenses

Business Equipment. The election to expense currently the cost of certain business machinery and other assets instead of depreciating them over a period of years is called a "Section 179 Expense Election." The maximum amount deductible for 2001 is $24,000, $4,000 more than the previous year.

Travel. Travel expenses paid for other individuals (including a spouse) traveling with the taxpayer on a business trip are not deductible unless the individual (1) is an employee, (2) has a bona fide business purpose for the travel, and (3) would otherwise be allowed to deduct the travel expense. Expenses paid for business assignments away from home in a single location that last for more than one year are no longer deductible.

Dues. Dues paid to business, social, athletic, luncheon, sporting, and country clubs, including airport and hotel clubs, are no longer deductible. However, dues paid to the Chamber of Commerce, business economic clubs and trade associations remain deductible.

Tax Credits

Adoption Credit. An adoption expense credit is available for up to $5,000 of qualified expenses for each eligible adopted person. The credit limit is per person, not per year. The adoption credit increases to $6,000 for an eligible person with special needs. The adoption credit begins to phase out when adjusted gross income (AGI) reaches $75,000. Beginning in 2002 the Tax Relief Act increases the credit to $10,000 of qualified expenses and the phase-out starting point at $150,000. Starting in 2003 the phase-out amount will be adjusted for inflation.

Earned Income Credit. The Tax Relief Act of 2001 increases the phaseout range of the earned income credit for joint filers by an additional $1,000 starting in 2002, to a maximum increase of $3,000 beginning in 2008. The $3,000 increase will be adjusted for the cost of living for years after 2008.

Lower income workers who maintain a household may be eligible for a refundable earned income credit. The credit is based on total earned income such as wages, commissions, and tips.

Congress also simplified the rules by redefining earned income, extending the definition of qualifying children to include descendents of stepchildren and eliminating the one-year residency requirement for foster children. Unless Congress decides otherwise, all these changes will be repealed after Dec. 31, 2010.

Individuals may qualify for the credit even if they are not required to file a return. However, a tax return must be filed to receive the refund. The IRS will assist individuals filing for the credit if they need help.

Education Credits. The Hope Scholarship Credit applies to qualified tuition and expenses for the first 2 years of postsecondary education in a degree or certificate program at an eligible educational institution. However, it does not apply to room and board or cost of books. The credit can be as high as $1,500 per student.

The Lifetime Learning Credit is available for taxpayers whose postsecondary education expenses are not eligible for the Hope credit. This credit equals 20% of tuition and other qualifying expenses paid for by the taxpayer, spouse or dependents. Taxpayers can deduct up to $1,000 ($5,000 of expenses x 20%) for all entitled students who are enrolled in an eligible educational institution. After 2003, this credit increases to $2,000 ($10,000 x 20%).

The Lifetime Learning Credit is allowed only for years in which the Hope credit is not used. Neither credit may be taken in any year in which funds are withdrawn from an Educational IRA for the same expenses. The credit begins to phase out when modified AGI exceeds $40,000 for singles and $80,000 on a joint return, with full phaseout at $50,000 for singles and $100,000 on joint returns. These credits are deducted from the individual's federal income tax and reported on Form 8863, Education Credits (Hope and Lifetime Learning Credits). Any excess not used is nonrefundable.

Taxable Social Security Benefits

Earnings Limitations. Age 62 to 65: Starting in the year 2001 individuals in this age group lose $1 of their Social Security benefits for every $2 of earned income over $10,680.

Age 65 or Over: As a result of The Senior Citizen's Freedom to Work Act, individuals 65 or over receiving Social Security benefits will no longer be subject to an earnings limitation.

Taxable Benefits. Up to 50% of Social Security benefits may be taxable income if the person's total income is:
over $25,000 but less than $34,000 for a single individual, head of household, qualifying widow(er), or a married person who is filing separately if spouses lived apart all year; or
over $32,000 but less than $44,000 for married individuals filing jointly.

For people with incomes exceeding these maximum amounts, 85% of Social Security benefits may become taxable. If the taxpayer is married and filing separately, and lived with a spouse at any time during the year, the percentage amounts are reduced to zero.

Most Social Security benefits will not be taxable if they are the only income received during the year 2001.

Retirement Planning

The SIMPLE Plan. A very popular retirement plan is called the Savings Incentive Match Plan for Employees (SIMPLE). This plan is available for businesses with 100 or fewer employees including self-employed individuals, and generally is easier to implement and more cost-effective to administrate than a traditional 401(k) plan.

Beginning in 2001, employees can defer up to $6,500 in compensation, a $500 increase from 2000. The Tax Relief Act increases the amount in 2002 to $7,000, annually thereafter by $1,000 until $10,000 is reached in 2005. A SIMPLE retirement plan can operate either as an IRA or as a 401(k). The tax liability on these amounts is deferred until a future date.

Profit-Sharing. Another retirement arrangement is the profit-sharing plan. This plan limits the total employer and employee contributions to the lower of 25% of compensation or $35,000 ($40,000 in 2002).

Age 70½ Plus. The owner of a traditional IRA must begin receiving distributions from the IRA by Apr. 1 of the calendar year following the year in which he or she reaches age 70½, even if the individual is not retired. However, any employee who works beyond age 70½ and is not a 5% or more owner of the business can continue to defer his or her profit sharing and pension retirement plan distributions to a later date.

Retired and Moved. States may not impose an income tax on retirement income if the person is no longer a resident of that state.

> ***IT'S A FACT:*** Only about 12% of individual returns filed during 2001 (for the year 2000) included a check mark in the "yes" box asking if the taxpayer wants $3 to go to the presidential election campaign fund. In the late 1970s close to 30% said "yes" to a similar question.

IRS Tax Audit

The IRS projects that 232.5 million individual income tax returns will be filed in 2001. Only about 1 out of every 100 of those returns will be audited. Good news…unless that one return happens to be yours!

Needless to say, the agency is very good at selecting returns that will yield additional taxes. If the IRS concludes you owe more and you disagree with the findings, you can meet with a supervisor. If you still do not agree, you can appeal to a separate Appeals Office or to the U.S. Tax Court.

For more information about income tax audits, call the IRS at 1-800-829-3676 for its free Publication 556, Examination of Returns, Appeal Rights, and Claims for Refund. Or, visit http://www.irs.gov

Your Rights as a Taxpayer

Several years ago Congress enacted the Taxpayer Bill of Rights 1. This law required the IRS to explain in easy-to-understand language any actions it proposes to take against a taxpayer, as well as to modify some of its audit and collection procedures.

Congress later passed the Taxpayer Bill of Rights 2 which created an Office of the Taxpayer Advocate within the IRS with authority to order IRS personnel to issue refund checks and meet deadlines for resolving disputes. Taxpayers Advocates can be contacted by calling 1-877-777-4778. The IRS also must pay legal fees if the taxpayer wins the case and the IRS cannot show it was "substantially justified" in pursuing the matter.

More recently, Congress created a 9-member oversight board to watch over the IRS management. The legislation shifts the burden of proof to the IRS under certain circumstances in disputes dealing with income, estate, and gift taxes. Further, it establishes procedures designed to ensure due process when the IRS seeks to collect taxes by levy.

For more information ask for IRS Publication 1, Your Rights as a Taxpayer, by calling 1-800-TAX-FORM for a free copy.

Federal Outlays to States Per Dollar of Tax Revenue Received

Source: The Tax Foundation

(figures for fiscal year 2000; ranked highest to lowest)

State	Outlay	State	Outlay	State	Outlay	State	Outlay
Dist. of Columbia	$6.49	Louisiana	$1.39	North Carolina	$1.06	Massachusetts	$0.86
New Mexico	2.03	Arkansas	1.38	Pennsylvania	1.06	New York	0.86
North Dakota	1.86	Maine	1.32	Utah	1.06	Colorado	0.85
Mississippi	1.78	Maryland	1.32	Iowa	1.04	Delaware	0.84
West Virginia	1.75	Idaho	1.30	Kansas	1.02	Wisconsin	0.83
Alaska	1.68	South Carolina	1.27	Florida	1.00	Michigan	0.81
South Dakota	1.46	Missouri	1.26	Georgia	0.99	Minnesota	0.76
Montana	1.59	Tennessee	1.20	Ohio	0.97	Illinois	0.74
Hawaii	1.56	Arizona	1.18	Texas	0.96	New Hampshire	0.71
Alabama	1.54	Rhode Island	1.18	Oregon	0.93	Nevada	0.69
Virginia	1.48	Nebraska	1.09	Indiana	0.92	New Jersey	0.66
Oklahoma	1.46	Wyoming	1.09	Washington	0.87	Connecticut	0.62
Kentucky	1.41	Vermont	1.08	California	0.86		

Tax Burden in Selected Countries[1]

Source: Organization for Economic Cooperation and Development, 1998

Country	Income tax (%)	Social Security (%)	Total payment[2] (%)	Country	Income tax (%)	Social Security (%)	Total payment[2] (%)
Denmark	34	10	43	Australia	24	2	25
Germany	21	21	42	United Kingdom	17	8	25
Turkey	24	9	33	New Zealand	20	0	20
Canada	22	6	27	Greece	2	16	18
France	14	13	27	Korea	2	5	6
United States	18	8	26	Japan	0	7	7

(1) Does not include taxes not listed, such as sales tax or VAT. (1) Totals may not add due to rounding.

State Government Individual Income Taxes

Source: Reproduced with permission from *CCH State Tax Guide,* published and copyrighted by CCH Inc., 2700 Lake Cook Road, Riverwoods, IL 60015

Below are basic state tax rates on taxable income, for 2001 unless otherwise indicated. Alaska, Florida, Nevada, South Dakota, Texas, Washington, and Wyoming did not have state income taxes and are thus not listed. For further details, see notes which follow.

Alabama
1st ......$1,000 ..... 2%
Next .....$5,000 ..... 4%
Over .....$6,000 ..... 5%

Arizona*
1st .....$20,000 ...2.87%
Next ....$30,000 ....3.2%
Next ....$50,000 ...3.74%
Next ...$200,000 ...4.72%
$300,001 and over ...5.04%

Arkansas
1st ......$2,999 ..... 1%
Next .....$3,000 ....2.5%
Next .....$3,000 ....3.5%
Next .....$6,000 ....4.5%
Next ....$10,000 ..... 6%
$25,000 or over 7%

California*
$0 to $11,496 .... 1%
$11,497 to $27,250 .... 2%
$27,251 to $43,006 .... 4%
$43,007 to $59,700 .... 6%
$59,701 to $75,450 .... 8%
Over $75,450 ....9.3%

Colorado
4.63% of federal taxable income

Connecticut
1st .....$20,000 ..... 3%
Over ....$20,000 ....4.5%

Delaware
$2,001 to $5,000 .....2.2%
Next .....$5,000 ....3.9%
Next ....$10,000 ....4.8%
Next .....$5,000 ....5.2%
Next ....$35,000 ....5.5%
Over ....$60,000 ...5.95%

District of Columbia
1st .....$10,000 ..... 5%
Next ....$20,000 ....7.5%
Over ....$30,000 ....9.3%

Georgia
1st ......$1,000 ..... 1%
Next .....$2,000 ..... 2%
Next .....$2,000 ..... 3%
Next .....$2,000 ..... 4%
Next .....$3,000 ..... 5%
Over ....$10,000 ..... 6%

Hawaii
1st ......$4,000 ....1.5%
Next .....$4,000 ....3.7%
Next .....$8,000 ....6.4%
Next .....$8,000 ....6.9%
Next .....$8,000 ....7.3%
Next .....$8,000 ....7.6%
Next ....$20,000 ....7.9%
Next ....$20,000 ....8.2%
Over ....$80,000 ....8.5%

Idaho*
1st ..... $1,000 ..... 2%
2d ....$1,000 ..... 4%
3d ....$1,000 ....4.5%
4th ......$1,000 ....5.5%
5th ......$1,000 ....6.5%
Next .....$2,500 ....7.5%
Next ....$12,500 ....7.8%
Over ...$20,000 ....8.2%

Illinois
3% of taxable net income

Indiana
3.4% of adj. gross income

Iowa
$0 to $1,185........0.36%
$1,185 to $2,370 ....0.72%
$2,370 to $4,740 ....2.43%
$4,740 to $10,665 ....4.5%
$10,665 to $17,775 ..6.12%
$17,775 to $23,700 ..6.48%
$23,700 to $35,550 ...6.8%
$35,500 to $53,325 ..7.92%
Over $53,325........8.98%

Kansas
1st .....$30,000 ....3.5%
Next ....$30,000 ...6.25%
Over....$60,000 ...6.45%

Kentucky
1st ......$3,000 ..... 2%
Next .....$1,000 ..... 3%
Next .....$1,000 ..... 4%
Next .....$3,000 ..... 5%
Over .....$8,000 ..... 6%

Louisiana*
1st .....$10,000 ..... 2%
Next ....$40,000 ..... 4%
Over....$50,000 ..... 6%

Maine
Less than $8,250 ..... 2%
$8,250 to $16,499 ....4.5%
$16,500 to $29,999 .... 7%
$33,000 or more 8.5%

Maryland
1st ......$1,000 ..... 2%
2d ....$1,000 ..... 3%
3d ....$1,000 ..... 4%
Over .....$3,000 ....4.8%

Massachusetts
Short term cap. gains . 12%
5 classes of cap. gain income 0-5%
All other income......5.6%

Michigan
4.2% of taxable income

Minnesota
$0 to $26,480.......5.35%
$26,481 to $105,200 .7.05%
Over $105,200......7.85%

Mississippi
1st ..... $5,000 ..... 3%
Next .....$5,000 ..... 4%
Over....$10,000 ..... 5%

Missouri
1st ......$1,000 ....1.5%
2d ....$1,000 ..... 2%
3d ....$1,000 ....2.5%
4th ......$1,000 ..... 3%
5th ......$1,000 ....3.5%
6th ......$1,000 ......4%
7th ......$1,000 ....4.5%
8th ......$1,000 ......5%
9th ......$1,000 ....5.5%
Over.....$9,000 ......6%

Montana
$0 to $2,199 ....2%
$2,200 to $4,299 ....3% less $22
$4,300 to $8,599 ......4% less $65
$8,600 to $12,899 ......5% less $151
$12,900 to $17,199 .....6% less $280
$17,200 to $21,499 .....7% less $452
$21,500 to $30,199 .....8% less $667
$30,200 to $43,099 .....9% less $969
$43,100 to $75,399 ....10% less $1,400
$75,000 and over .....11% less $2,154

Nebraska
1st ......$4,000 ...2.51%
Next....$26,000 ...3.49%
Next....$16,750 ...5.01%
Over....$46,750 ...6.68%

New Hampshire
5% of interest and dividends

New Jersey
1st .....$20,000 .... 1.4%
Next....$30,000 ... 1.75%
Next....$20,000 ... 2.45%
Next....$10,000 .... 3.5%
Next....$70,000 .. 5.525%
Over...$150,000 ... 6.37%

New Mexico*
Not over $8,000...... 1.7%
$8,001 to $16,000 .... 3.2%
$16,001 to $24,000 ... 4.7%
$24,001 to $40,000 .....6%
$40,001 to $64,000 .. 7.1%
$64,001 to $100,000 .. 7.9%
Over $100,000....... 8.2%

New York
1st .....$16,000 ......4%
Next.....$6,000 .... 4.5%
Next.....$4,000 ... 5.25%
Next....$14,000 .... 5.9%
Over....$40,000 ... 6.85%

North Carolina
Up to ... $21,250 ......6%
Next....$78,750 ......7%
Over...$100,000 ... 7.75%

North Dakota
1st ......$3,000 ... 2.67%
Next.....$2,000 ......4%
Next.....$3,000 ... 5.33%
Next.....$7,000 ... 6.67%
Next....$10,000 ......8%
Next.... $10,000... 9.33%
Next.... $15,000.. 10.67%
Over ... $50,000......12%

Ohio
1st...... $5,000.. 0.743%
Next..... $5,000.. 1.486%
Next..... $5,000.. 2.972%
Next..... $5,000.. 3.715%
Next.... $20,000.. 4.457%
Next.... $40,000.. 5.201%
Next.... $20,000.. 5.943%
Next... $100,000.... 6.9%
Over .. $200,000.... 7.5%

Oklahoma
1st...... $2,000.....0.5%
Next .... $3,000......1%
Next..... $2,500......2%
Next..... $2,300......3%
Next..... $2,400......4%
Next..... $2,800......5%
Next..... $6,000......6%
Remainder 6.75%

Oregon
1st...... $4,900......5%
$4,900 to $12,000......7%
Over ... $12,200......9%

Pennsylvania....... 2.8%

Rhode Island
25.5% of federal liability

South Carolina
1st ..... $2,340.... 2.5%
Next..... $2,340......3%
Next..... $2,340......4%
Next..... $2,340......5%
Next..... $2,340......6%
$11,701 and over 7%

Tennessee
6% of interest and dividends

Utah
$0 to $1,726 .... 2.3%
$1,726 to $3,450..... 3.3%
$3,450 to $5,176..... 4.2%
$5,176 to $6,900..... 5.2%
$6,900 to $8,626.......6%
Over $8,626 ....7%

Vermont
24% of federal income tax

Virginia
1st ..... $3,000......2%
Next..... $2,000......3%
Next.... $12,000......5%
Over ... $17,000.... 5.75%

West Virginia
1st .... $10,000......3%
Next.... $15,000......4%
Next.... $15,000.... 4.5%
Next.... $20,000......6%
Over ... $60,000.... 6.5%

Wisconsin*
1st $10,000.......... 4.6%
Next ... $10,000... 6.15%
Next... $130,000.... 6.5%
Over .. $150,000... 6.75%

* = Community property state in which, in general, one-half of the community income is taxable to each spouse.

Alabama: Rates shown are for married persons filing jointly. Single persons, heads of families, married persons filing separately, and estates or trusts are taxed at 2% of the first $500 of taxable income, 4% on the next $2,500, and 5% on the rest.

Arizona: Rates shown are for married persons filing jointly and unmarried heads of households. For single taxpayers and married taxpayers filing separately, rates range from 2.87% of the first $10,000 of taxable income to 5.04% of taxable income over $150,000.

California: Rates shown are the 2001 inflation-adjusted rates for residents who are filing joint returns. Rates for heads of households range from 1% on the 1st $11,500 of taxable income to $1,876.02 plus 9.3% of taxable income over $51,350. Rates for single taxpayers range from 1% on the first $5,478 to $1,660.96 plus 9.3% of taxable income over $37,725.

Colorado: Alternative minimum tax imposed. Qualified taxpayers may pay alternative tax of 0.5% of gross receipts from sales.

Connecticut: Rates shown are for married individuals filing jointly or persons filing as a surviving spouse. For: (1) unmarried individuals and married individuals filing separately, rates are 3% on the first $10,000 of CT taxable income and $300 plus 4.5% of the excess over $10,000; (2) for heads of households, rates are 3% of the first $16,000 of taxable income and $480 plus 4.5% of the excess over $16,000; and (3) for trusts or estates, rates are 4.5% of taxable income. Resident estates and trusts are subject to the 4.5% income tax rate on all of their income. Additional state minimum tax imposed on resident individuals, trusts, and estates is equal to the amount by which the CT minimum tax exceeds the CT basic income tax [the lesser of (a) 19% of adjusted federal tentative minimum tax, or (b) 5% of adjusted federal alternative minimum taxable income]. Separate provisions apply for non- and part-year resident individuals, trusts, and estates.

District of Columbia: The tax on unincorporated business is 9.975%. Minimum tax, $100.

Georgia: Rates shown are for married persons filing jointly and heads of households. Single persons pay at rates ranging from 1% on taxable net income not over $750 to 6% plus $230 on taxable net income over $7,000. Married persons filing separately pay at rates ranging from 1% on taxable net income not over $500 to 6% plus $170 on taxable net income over $5,000.

Hawaii: Rates shown are for taxpayers filing jointly and surviving spouses. For heads of households, rates range from 1.5% of taxable income up to $3,000 to $4,263 plus 8.5% of taxable income of $60,000 and over. For unmarried individuals (other than a surviving spouse or head of household), married individuals filing separately, and estates and trusts, the rates range from 1.5% of taxable income up to $2,000 to 8.5% plus $2,842 of taxable income over $40,000.

Idaho: Each person (joint returns deemed one person) filing return pays additional $10. Rates shown are for married persons filing jointly. The tax on single taxpayers and married taxpayers filing separately ranges from 2% of the first $1,000 of taxable income to 8.2% of taxable income over $20,000.

Illinois: Additional personal property replacement tax of 1.5% of net income is imposed on partnerships, trusts, and S corporations.

Indiana: Counties may impose an adjusted gross income tax on residents at 0.5%, 0.75%, or 1% and at 0.25% on nonresidents or a county option income tax at rates ranging between 0.2% and 1%, with the rate on nonresidents equal to $^1/_4$ of the rate on residents.

Iowa: Rates shown are those announced for use in preparation of 2001 estimated taxes, not final 2001 rates. An alternative minimum tax is imposed equal to 75% of the maximum state individual income tax rate for the tax year of the state alternative minimum taxable income.

Kansas: Rates shown are for married individuals filing joint returns. For single taxpayers and married taxpayers filing separately, the rate is 3.5% of the first $15,000, $525 plus 6.25% of the next $15,000, and $1,462.50 plus 6.45% of taxable income over $30,000.

Louisiana: The amount of tax due is determined from tax tables. These amounts are double (rates remain the same) for taxpayers filing joint returns.

Maine: Rates shown are for married individuals filing jointly and for surviving spouses. For unmarried or legally separated individuals who qualify as heads of household, tax rates range from 2% if taxable income is less than $6,200 and $1,268 plus 8.5% if taxable income is $24,750 or more. Additional state minimum tax is imposed equal to the amount by which the state minimum tax (27% of adjusted federal tentative minimum tax) exceeds Maine income tax liability, other than withholding tax liability.

Maryland: For a tax year beginning after 2001, income over $3,000 will be taxed at a rate of 4.75%.

Michigan: Persons with business activity allocated or apportioned to Michigan are also subject to a single business tax on an adjusted tax base.

Minnesota: Rates shown are for married taxpayers filing joint returns. For single taxpayers, rates range from 5.35% of the 1st $18,120 of taxable income to 7.85% of taxable income over $59,500. For heads of households, rates range from 5.35% of the 1st $22,300 of taxable income to 7.85% of taxable income over $89,610. A 6.4% alternative minimum tax is imposed.

Montana: Rates shown are indexed for inflation. Minimum tax, $1.

Nebraska: Rates shown are for married couples filing jointly and qualified surviving spouses. Rates for married persons filing separately range from 2.51% of the first $2,000 to $923.49 plus 6.68% of taxable income over $23,375. Rates for heads of households range from 2.51% of the first $3,800 to $1,351.46 plus 6.68% of taxable income over $35,000. Rates for single individuals range from 2.51% of the first $2,400 to $1,045.73 plus 6.68% of taxable income over $26,500.

New Jersey: Rates shown are for married persons filing jointly, heads of households, and surviving spouses. Rates for married persons filing separately, unmarried individuals, and estates and trusts range from 1.4% of the first $20,000 of taxable income to $2,651.25 plus 6.37% of taxable income over $75,000.

New Mexico: Rates shown are for married persons filing jointly and surviving spouses. For married persons filing separately, rates range from 1.7% on the first $4,000 of taxable income to $3,138 plus 8.2% on taxable income over $50,000. For heads of household, rates range from 1.7% on the first $7,000 of taxable income to $5,195 plus 8.2% on taxable income over $83,000. For single individuals, estates, and trusts, rates range from 1.7% of the first $5,500 of taxable income to $4,057.50 plus 8.2% of taxable income over $65,000. Qualified taxpayers may pay alternative tax of 0.75% of gross receipts from New Mexico sales.

New York: Rates shown are for married individuals filing jointly and surviving spouses. Separate schedules are set out for heads of households (ranging from 4% on the first $11,000 of taxable income to $1,492 plus 6.85% on taxable income over $30,000) and for unmarried individuals, married individuals filing separately, and estates and trusts (ranging from 4% of the first $8,000 of taxable income to $973 plus 6.85% of taxable income over $20,000). In addition, individuals, estates, and trusts are subject to a 6% tax on minimum taxable income. A tax table benefit recapture supplemental tax is imposed on some individuals.

North Carolina: Rates shown are for married persons filing jointly. For married persons filing separate returns, rates are 6% of income under $10,626 to $3,393.75 plus 7.75% of taxable income over $50,000. For heads of households the rate is 6% of income under $17,001 to $5,430 plus 7.75% of income over $80,000. For single individuals the rate is 6% of income under $12,751 to $4,072.50 plus 7.75% of excess over $60,000.

North Dakota: Individuals, estates, and trusts are allowed an optional method of computing the tax. The optional tax is 14% of the taxpayer's adjusted federal income tax liability for the tax year.

Ohio: Rates shown are those announced for use in preparation of 2001 estimated taxes, not final 2001 rates.

Oklahoma: Rates shown are for heads of households, married persons filing jointly, and a surviving spouse not deducting federal income taxes. Single persons, married persons filing separately, and estates and trusts not deducting federal income taxes pay at rates ranging from 0.5% on the first $1,000 of taxable income to 6.75% on taxable income over $10,000. Optional rates (ranging from 0.5% to 10%) are enacted for taxpayers who deduct federal income taxes.

Oregon: Rates shown are for year 2000 for joint returns, heads of households, or qualifying widows or widowers with a dependent child. Rates for single persons or married filing separately are 5% of the first $2,450, $123 plus 7% of the amount over $2,450, and $378 plus 9% of the amount over $6,100.

Utah: Rates shown are for married persons filing jointly or heads of households. Rates for single taxpayers, married persons filing separately, and estates and trusts range from 2.3% of adjusted federal taxable income not over $750 to $158 plus 7% of adjusted federal taxable income over $3,750.

West Virginia: For married taxpayers filing separately, rates range from 3% of taxable income not over $5,000 to $1,387.50 plus 6.5% of taxable income over $30,000.

Wisconsin: Rates shown are for married persons filing jointly. For married persons filing separately, rates range from 4.6% of the first $5,000 of taxable income to 6.75% of taxable income over $75,000. For fiduciaries and single individuals, rates range from 4.6% of the first $7,500 of taxable income to 6.75% of taxable income over $112,500. In addition, a temporary recycling surcharge is imposed on individuals, estates, partnerships, trusts, and exempt trusts, except those entities engaged only in farming, at the rate of the greater of $25 or 0.2173% of net business income. The maximum surcharge is $9,800. An individual, estate, trust, exempt trust, or partnership engaged in farming is subject to a surcharge of $25.

ENERGY

U.S. Energy Overview, 1960-2000

Source: Energy Information Administration, U.S. Dept. of Energy, *Annual Energy Review 2000;* in quadrillion Btu

	1960	1965	1970	1975	1980	1985	1990[1]	1995	1999	2000[P]
Production	**42.80**	**50.68**	**63.50**	**61.36**	**67.24**	**67.72**	**70.84**	**71.30**	**71.98**	**71.90**
Fossil fuels	39.87	47.23	59.19	54.73	59.01	57.54	58.56	57.46	57.29	57.40
Coal	10.82	13.06	14.61	14.99	18.60	19.33	22.46	22.02	23.18	22.66
Natural gas (dry)	12.66	15.78	21.67	19.64	19.91	16.98	18.36	19.10	19.13	19.74
Crude oil[2]	14.93	16.52	20.40	17.73	18.25	18.99	15.57	13.89	12.45	12.38
Natural gas plant liquids (NGPL)	1.46	1.88	2.51	2.37	2.25	2.24	2.17	2.44	2.53	2.61
Nuclear electric power	0.01	0.04	0.24	1.90	2.74	4.15	6.16	7.18	7.74	8.01
Hydroelectric pumped storage[3]	(4)	(4)	(4)	(4)	(4)	(4)	–0.04	–0.03	–0.07	–0.06
Renewable energy	2.93	3.40	4.08	4.72	5.49	6.03	6.15	6.69	7.02	6.56
Conventional hydroelectric power[5]	1.61	2.06	2.63	3.15	2.90	2.97	3.01	3.21	3.31	2.84
Geothermal	0.001	0.004	0.01	0.07	0.11	0.20	0.34	0.31	0.37	0.32
Wood, waste, alcohol[6]	1.32	1.34	1.43	1.50	2.49	2.86	2.66	3.07	3.22	3.28
Solar	NA	NA	NA	NA	NA	(*)	0.06	0.07	0.07	0.07
Wind	NA	NA	NA	NA	NA	(*)	0.03	0.03	0.05	0.05
Imports	**4.23**	**5.92**	**8.39**	**14.11**	**15.97**	**12.10**	**18.95**	**22.57**	**27.55**	**28.52**
Coal	0.01	(*)	(*)	0.02	0.03	0.0[illegible]	0.07	0.24	0.23	0.31
Natural gas	0.16	0.47	0.85	0.98	1.01	0.95	1.55	2.90	3.66	3.81
All crude oil and petroleum prods.[7]	4.00	5.40	7.47	12.95	14.66	10.61	17.12	18.88	23.13	23.78
Other[8]	0.06	0.04	0.07	0.16	0.28	0.49	0.22	0.55	0.52	0.61
Exports	**1.48**	**1.85**	**2.66**	**2.36**	**3.72**	**4.23**	**4.87**	**4.54**	**3.81**	**4.10**
Coal	1.02	1.38	1.94	1.76	2.42	2.44	2.77	2.32	1.53	1.53
Natural gas	0.01	0.03	0.07	0.07	0.05	0.06	0.09	0.16	0.16	0.24
All crude oil and petroleum prods.[7]	0.43	0.39	0.55	0.44	1.16	1.66	1.82	1.99	1.95	2.15
Other[8]	0.02	0.06	0.11	0.08	0.09	0.08	0.18	0.07	0.17	0.18
Consumption[9]	**45.12**	**54.02**	**67.86**	**72.04**	**78.43**	**76.78**	**84.34**	**90.94**	**96.87**	**98.50**
Fossil fuels	42.14	50.58	63.52	65.35	69.98	66.22	72.03	76.92	82.09	83.86
Coal	9.84	11.58	12.26	12.66	15.42	17.48	19.25	20.03	21.69	22.41
Coal coke net imports	–0.01	–0.02	–0.06	0.01	–0.04	–0.01	(*)	0.06	0.06	0.07
Natural gas[10]	12.39	15.77	21.79	19.95	20.39	17.83	19.30	22.16	22.29	23.33
Petroleum[11]	19.92	23.25	29.52	32.73	34.20	30.92	33.55	34.55	37.96	37.96
Nuclear electric power	0.01	0.04	0.24	1.90	2.74	4.15	6.16	7.16	7.74	8.01
Hydroelectric pumped storage[3]	(4)	(4)	(4)	(4)	(4)	(4)	–0.04	–0.03	–0.07	–0.06
Renewable energy	2.98	3.40	4.10	4.79	5.71	6.46	6.25	6.99	7.23	6.82
Conventional hydroelectric power[5,12]	1.66	2.06	2.65	3.22	3.12	3.40	3.15	3.48	3.51	3.11
Geothermal energy[13]	0.001	0.004	0.01	0.07	0.11	0.20	0.36	0.33	0.37	0.32
Wood, waste, alcohol[6]	1.32	1.34	1.43	1.50	2.49	2.86	2.66	3.07	3.22	3.28
Solar energy	NA	NA	NA	NA	NA	(*)	0.06	0.07	0.07	0.07
Wind energy	NA	NA	NA	NA	NA	(*)	0.03	0.03	0.05	0.05

(1) Starting in 1990, expanded coverage of nonelectric utility use of renewable energy resulted in an increase in total production and consumption figures. (2) Incl. lease condensate. (3) Total pumped storage facility production minus energy used for pumping. (4) Included in conventional hydroelectric power. (5) Starting in 1990, pumped storage is removed and expanded coverage of industrial use of hydroelectric power is included. (6) Substituted in 2000 for former "Biofuels" category; figures for 1960-99 were recalculated. Alcohol is ethanol blended into motor gasoline. (7) Incl. imports of crude oil for the Strategic Petroleum Reserve, which began in 1977. (8) Coal coke and small amts. of electricity transmitted across borders with Canada and Mexico. (9) Starting in 1990, "Consumption" includes net imports of electricity from nonrenewable energy sources. (10) Incl. supplemental gaseous fuels. (11) Petroleum products supplied, incl. natural gas plant liquids and crude oil burned as fuel. (12) Starting in 1990, includes only the part of net imports of electricity derived from hydroelectric power. (13) Incl. electricity imports from Mexico derived from geothermal energy. NA = Not available. P = preliminary. (*) = Less than 0.005 quadrillion Btu. Some figures here have been revised.

U.S. Energy Flow, 2000

Source: Energy Information Administration, U.S. Dept. of Energy, *Annual Energy Review 2000*; in quadrillion Btu

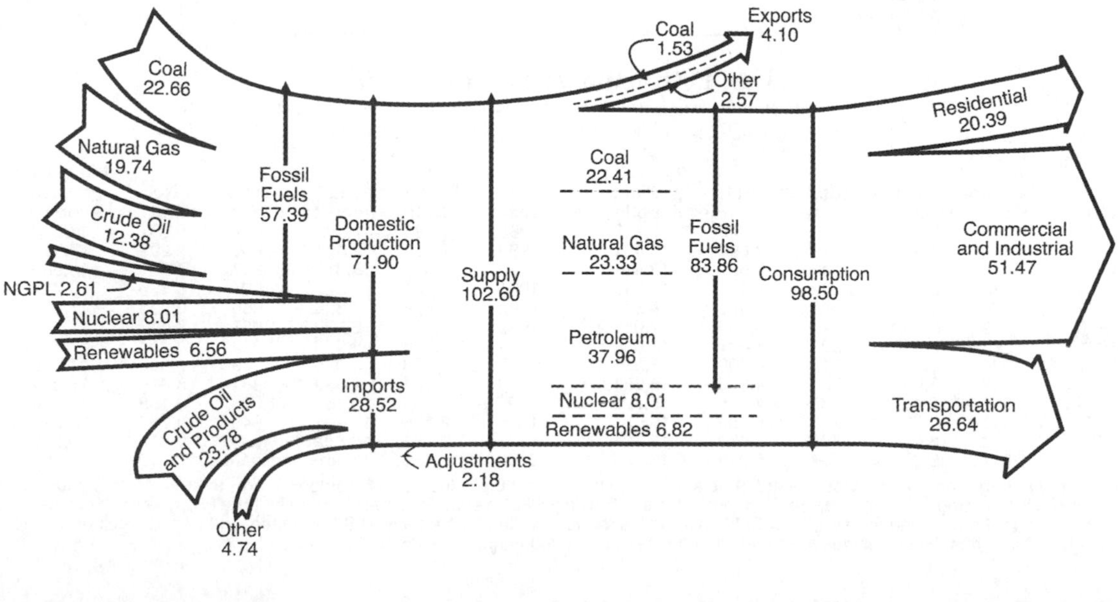

World Energy Consumption and Production Trends, 1999

Source: Energy Information Administration, U.S. Dept. of Energy, International Energy Database, Aug. 2001

The world's **consumption** of primary energy—petroleum, natural gas, coal, net hydroelectric, nuclear, geothermal, solar, wind, and wood and waste electric power, and other wood and waste (primarily for the United States)—increased from 380 quadrillion Btu in 1998 to 382 quadrillion Btu in 1999.

The 29 countries of the Organization for Economic Co-operation and Development (OECD), which includes most of the world's largest economies (the United States, Japan, and Germany), continued to dominate global energy use. OECD nations accounted for 59% of the world's primary energy consumption in 1999, up 1 percentage point from 1998.

World **production** of primary energy decreased from 383 quadrillion Btu in 1998 to 380 quadrillion Btu in 1999. World production of petroleum in 1999 was about 72 million barrels per day, or 150 quadrillion Btu, down about 1% from 1998; petroleum remained the most heavily used source of energy.

In 1999, 3 countries—the United States, Russia, and China—were the world's leading producers (38%) and consumers (41%) of energy. Russia and the United States alone supplied 30% of the world total. The United States alone accounted for 25% of the world's energy consumption. The United States consumed 35% more energy than it produced—an imbalance of 25 quadrillion Btu.

World's Major Consumers of Primary Energy, 1999

Source: Energy Information Administration, International Energy Database; quadrillion Btu

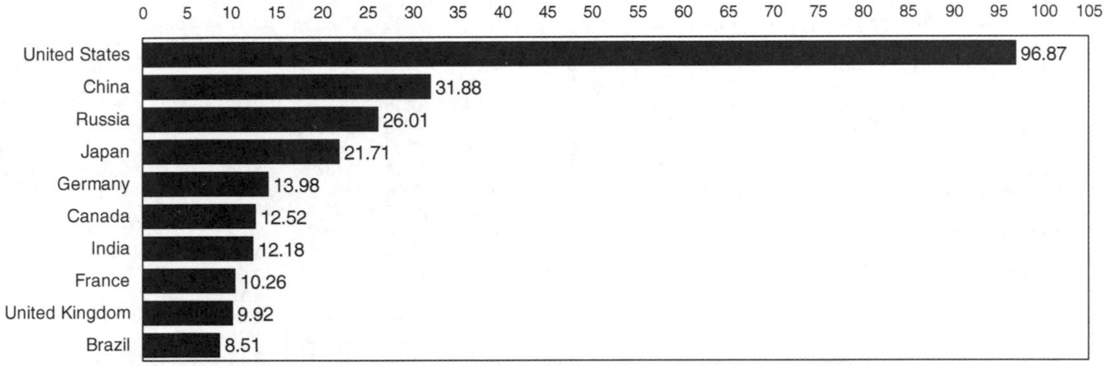

World's Major Producers of Primary Energy, 1999

Source: Energy Information Administration, International Energy Database; quadrillion Btu

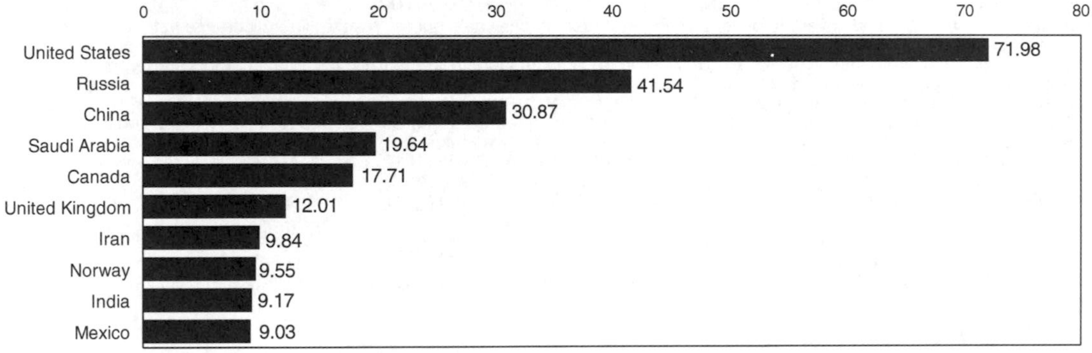

U.S. Petroleum Trade, 1975-2000

Source: Energy Information Administration, U.S. Dept. of Energy, *Monthly Energy Review,* April 2001

(in thousands of barrels per day; average for the year)

Year	Imports from Persian Gulf[1]	Total imports	Total exports	Net imports[2]	Petroleum products supplied
1975	1,165	6,056	209	5,846	16,322
1976	1,840	7,313	223	7,090	17,461
1977	2,448	8,807	243	8,565	18,431
1978	2,219	8,363	362	8,002	18,847
1979	2,069	8,456	471	7,985	18,513
1980	1,519	6,909	544	6,365	17,056
1981	1,219	5,996	595	5,401	16,058
1982	696	5,113	815	4,298	15,296
1983	442	5,051	739	4,312	15,231
1984	506	5,437	722	4,715	15,726
1985	311	5,067	781	4,286	15,726
1986	912	6,224	785	5,439	16,281
1987	1,077	6,678	764	5,914	16,665
1988	1,541	7,402	815	6,587	17,283
1989	1,861	8,061	859	7,202	17,325
1990	1,966	8,018	857	7,161	16,988
1991	1,845	7,627	1,001	6,626	16,714
1992	1,778	7,888	950	6,938	17,033
1993	1,782	8,620	1,003	7,618	17,237
1994	1,728	8,996	942	8,054	17,718
1995	1,573	8,835	949	7,886	17,725
1996	1,604	9,399	981	8,419	18,234
1997	1,755	10,162	1,003	9,158	18,620
1998	2,136	10,708	945	9,764	18,917
1999	2,464	10,852	940	9,912	19,519
2000	2,488	11,459	1,040	10,419	19,701

(1) Bahrain, Iran, Iraq, Kuwait, Qatar, Saudi Arabia, and the United Arab Emirates. (2) Net imports are total imports minus total exports. **Notes:** Beginning in Oct. 1977, imports for the Strategic Petroleum Reserves are included. U.S. geographic coverage includes the 50 states and the District of Columbia. U.S. exports include shipments to U.S. territories, and imports include receipts from U.S. territories. Figures in this table may not add, because of rounding. Some figures are revised.

Energy Consumption, Total and Per Capita, by State, 1999

Source: Energy Information Administration, U.S. Dept. of Energy, State Energy Data Report 1999

TOTAL CONSUMPTION						CONSUMPTION PER CAPITA					
Rank	State	Trillion Btu	Rank	State	Trillion Btu	Rank	State	Million Btu	Rank	State	Million Btu
1.	Texas	11,501.0	28.	Arkansas	1,203.7	1.	Alaska	1,121.5	28.	Nevada	340.1
2.	California	8,375.4	29.	Colorado	1,155.5	2.	Wyoming	879.4	29.	Oregon	334.5
3.	Ohio	4,323.4	30.	Iowa	1,121.7	3.	Louisiana	826.9	30.	Michigan	328.4
4.	New York	4,283.0	31.	Oregon	1,109.2	4.	North Dakota	577.1	31.	District of Columbia	327.1
5.	Illinois	3,882.6	32.	Kansas	1,050.0	5.	Texas	573.8	32.	South Dakota	326.0
6.	Florida	3,852.9	33.	Connecticut	839.3	6.	Arkansas	471.8	33.	Utah	325.8
7.	Pennsylvania	3,715.5	34.	West Virginia	735.4	7.	Montana	467.1	34.	Virginia	324.1
8.	Louisiana	3,615.4	35.	Alaska	694.7	8.	Kentucky	462.1	35.	Missouri	323.3
9.	Michigan	3,239.6	36.	Utah	693.9	9.	Indiana	460.3	36.	Illinois	320.1
10.	Georgia	2,798.1	37.	New Mexico	635.0	10.	Alabama	458.8	37.	North Carolina	319.8
11.	Indiana	2,735.8	38.	Nevada	615.3	11.	Mississippi	436.5	38.	New Jersey	317.9
12.	New Jersey	2,588.7	39.	Nebraska	602.0	12.	Maine	421.9	39.	Pennsylvania	309.8
13.	North Carolina	2,446.9	40.	Maine	528.6	13.	Idaho	414.1	40.	Colorado	284.9
14.	Washington	2,240.8	41.	Idaho	518.3	14.	Oklahoma	410.2	41.	New Hampshire	279.2
15.	Virginia	2,227.3	42.	Wyoming	421.8	15.	West Virginia	407.0	42.	Vermont	277.9
16.	Tennessee	2,070.5	43.	Montana	412.4	16.	Kansas	395.6	43.	Maryland	266.5
17.	Alabama	2,004.8	44.	North Dakota	365.7	17.	Iowa	390.9	44.	Rhode Island	263.5
18.	Kentucky	1,830.2	45.	New Hampshire	335.4	18.	Washington	389.3	45.	Connecticut	255.7
19.	Wisconsin	1,810.5	46.	Delaware	278.8	19.	South Carolina	384.2	46.	Arizona	255.3
20.	Missouri	1,768.0	47.	Rhode Island	261.1	20.	Ohio	384.1	47.	Florida	255.0
21.	Minnesota	1,675.3	48.	Hawaii	241.4	21.	Tennessee	377.6	48.	Massachusetts	254.1
22.	Massachusetts	1,569.1	49.	South Dakota	239.0	22.	Delaware	370.0	49.	California	252.7
23.	South Carolina	1,493.0	50.	District of Columbia	169.8	23.	New Mexico	365.0	50.	New York	235.4
24.	Maryland	1,378.2	51.	Vermont	165.0	24.	Nebraska	361.3	51.	Hawaii	203.7
25.	Oklahoma	1,377.5	**TOTAL U.S.**		**95,682.4**	25.	Georgia	359.3	**TOTAL U.S.**		**350.9**
26.	Arizona	1,219.8				26.	Minnesota	350.8			
27.	Mississippi	1,208.5				27.	Wisconsin	344.8			

Gasoline Retail Prices, U.S. City Average, 1974-2001

Source: Energy Information Administration, U.S. Dept. of Energy, *Monthly Energy Review,* Aug. 2001

(cents per gallon, including taxes)

AVERAGE	Leaded regular	Unleaded regular	Unleaded premium	All types[1]	AVERAGE	Leaded regular	Unleaded regular	Unleaded premium	All types[1]
1974	53.2	NA	NA	NA	1988	89.9	94.6	110.7	96.3
1975	56.7	NA	NA	NA	1989	99.8	102.1	119.7	106.0
1976	59.0	61.4	NA	NA	1990	114.9	116.4	134.9	121.7
1977	62.2	65.6	NA	NA	1991	NA	114.0	132.1	119.6
1978	62.6	67.0	NA	65.2	1992	NA	112.7	131.6	119.0
1979	85.7	90.3	NA	88.2	1993	NA	110.8	130.2	117.3
1980	119.1	124.5	NA	122.1	1994	NA	111.2	130.5	117.4
1981[2]	131.1	137.8	147.0[3]	135.3	1995	NA	114.7	133.6	120.5
1982	122.2	129.6	141.5	128.1	1996	NA	123.1	141.3	128.8
1983	115.7	124.1	138.3	122.5	1997	NA	123.4	141.6	129.1
1984	112.9	121.2	136.6	119.8	1998	NA	105.9	125.0	111.5
1985	111.5	120.2	134.0	119.6	1999	NA	116.5	135.7	122.1
1986	85.7	92.7	108.5	93.1	2000	NA	151.0	169.3	156.3
1987	89.7	94.8	109.3	95.7	2001 (Jan. - June)	NA	155.6	175.5	162.1

Until unleaded gas became available in 1976, leaded was the only type used in automobiles. Average retail prices (in cents per gallon) for selected years preceding those in the table above were as follows: 1950: .27; 1955: .29; 1960: .31; 1965: .31; 1970: .36. (1) Also includes types of motor gasoline not shown separately. (2) In Sept. 1981, the Bureau of Labor Statistics changed the weights in the calculation of average motor gasoline prices. Starting in Sept. 1981, gasohol is included in average for all types, and unleaded premium is weighted more heavily. (3) Based on Sept. through Dec. data only. **NOTE:** Geographic coverage for 1974-77 is 56 urban areas; for 1978 and later, 85 urban areas. NA = Not available.

Gasoline Retail Prices in Selected Countries, 1990-2000

Source: Energy Information Administration, U.S. Dept. of Energy

(average price of unleaded regular gas; dollars per gallon, including taxes)

Year	Australia	Brazil	Canada	China	Germany	Japan	Mexico	Taiwan	U.S.
1990	NA	$3.82	$1.87	NA	$2.65	$3.17	$1.00	$2.49	$1.16
1991	$1.96	2.91	1.92	NA	2.90	3.46	1.29	2.39	1.14
1992	1.89	2.92	1.73	NA	3.27	3.59	1.50	2.42	1.13
1993	1.73	2.40	1.57	NA	3.07	4.02	1.56	2.27	1.11
1994	1.84	2.80	1.45	$0.96	3.52	4.39	1.48	2.14	1.11
1995	1.95	2.16	1.53	1.03	3.96	4.43	1.12	2.23	1.15
1996	2.12	2.31	1.61	1.03	3.94	3.65	1.26	2.15	1.23
1997	2.05	2.61	1.62	1.07	3.54	3.27	1.47	2.23	1.23
1998	1.63	2.80	1.38	1.08	3.34	2.82	1.50	1.86	1.06
1999	1.72	NA	1.51	NA	3.42	3.27	1.80	1.86	1.17
2000	1.94	NA	1.86	NA	3.45	3.74	2.02	NA	1.51

NA = Not available.

Major U.S. Coal Producers, 1999

Source: Energy Information Administration

Rank	Company Name	Production (thousand short tons)	Percent of total production	Rank	Company Name	Production (thousand short tons)	Percent of total production
1	Peabody Energy Corp.	165,479	15.0	16	James River Coal Co.	12,578	1.1
2	Kennecott Energy Co.	114,506	10.4	17	RENCOAL Inc.	10,337	.9
3	Arch Coal Co.	108,752	9.9	18	Kiewit Coal Properties Inc.	8,762	.8
4	Consol Energy Inc.	69,889	6.4	19	Independence Coal Co.	8,255	.8
5	RAG Coal Internat	59,265	5.4	20	Coastal Corp.	8,172	.7
6	North American Coal Co.	31,195	2.8	21	Branham & Baker Coal Co.	7,359	.7
7	TXU Utilities Co.	28,638	2.6	22	AEP Service Corp.	7,226	.7
8	Massey Energy	26,938	2.4	23	American Coal Co.	6,534	.6
9	Vulcan Holding	23,758	2.2	24	ALCOA Inc.	6,226	.6
10	AEI Resources Inc.	21,675	2.0	25	USX Corp.	6,028	.5
11	Pacificorp	21,072	1.9	26	Westmoreland Resources	5,465	.5
12	Entech Inc.	19,838	1.8	27	Drummond Resources	5,325	.5
13	Broken Hill Proprietary Co.	15,878	1.4	28	TECO Energy Inc.	5,154.5	
14	Chevron Corp	15,228	1.4		**All other coal producers**	**268,427**	**24.4**
15	Alliance Coal	14,625	1.3		**U.S. TOTAL**	**1,100,431**	**100.0**

Note: The company is the firm owning the mineral rights to the mined coal.

Major U.S. Coal Mines, 1999

Source: Energy Information Administration

Rank	Mine Name/Company	Mine Type	State	Production (short tons)
1	Rochelle Mine Complex/Powder River Coal	Surface	Wyoming	68,865,690
2	Black Thunder/Thunder Basin Coal	Surface	Wyoming	48,670,522
3	Cordero/Cordero Mining	Surface	Wyoming	46,186,691
4	Jacobs Ranch/Jacobs Ranch Coal	Surface	Wyoming	29,069,128
5	Caballo/Caballo Coal	Surface	Wyoming	26,468,762
6	Antelope/Antelope Coal	Surface	Wyoming	22,685,237
7	Belle Ayr/RAG Coal West	Surface	Wyoming	17,893,709
8	Eagle Butte/RAG Coal West	Surface	Wyoming	17,416,240
9	Freedom-Coteau/Coteau Properties	Surface	North Dakota	16,391,229
10	Buckskin/Triton Coal	Surface	Wyoming	15,587,569
11	Coal Creek/Thunder Basin Coal	Surface	Wyoming	11,249,947
12	Spring Creek/Spring Creek Coal Co	Surface	Montana	10,995,516
13	Decker/Decker Coal	Surface	Montana	10,878,069
14	Rosebud No 6/Western Energy Co	Surface	Montana	10,621,638
15	Enlow Fork/Consol PA Coal Co	Underground	Pennsylvania	9,835,217
16	Navajo/BHP Minerals	Surface	New Mexico	9,374,060
17	Jewett/Northwestern Resources	Surface	Texas	9,215,933
18	Bailey No 1/Consol PA Coal Co.	Underground	Pennsylvania	8,518,670
19	Foidel Creek/Twenty Mile Coal	Underground	Colorado	8,500,157
20	North Rochelle/Triton Coal Co.	Surface	Wyoming	8,170,482
21	Keyenta/Peabody Western Coal	Surface	Arizona	7,251,024
22	Falkirik/Falkirk Mining	Surface	North Dakota	7,191,836
23	McKinley/Pittsburgh & Midway Coal	Surface	New Mexico	7,181,357
24	West Elk/Mountain Coal Co.	Underground	Colorado	7,078,112
25	McElroy/CONSOL	Underground	West Virginia	6,995,483
	ALL OTHER MINES			**658,139,150**
	U.S. TOTAL			**1,100,431,428**

Note: The company is the firm operating the mine.

Ten Largest Oil Fields in the United States

Source: Energy Information Administration, *Petroleum: An Energy Profile, 1999*

(as of Jan. 1, 1999, by size of total recoverable resources)

Oil Field	Location	Year Discovered	Cumulative Production	Remaining Reserves (billion barrels)	Recoverable Resources
Prudhoe Bay	Alaska	1968	9.7	3.3	13
East Texas	Texas	1930	5.3	(s)	5.4
Wilmington	California	1932	2.5	0.3	2.8
Midway-Sunset	California	1894	2.4	0.3	2.7
Kuparuk River	Alaska	1969	1.6	1	2.6
Wasson	Texas	1936	2.0	0.1	2.1
Kern River	California	1899	1.7	0.4	2.1
Yates	Texas	1926	1.4	0.6	2
Panhandle	Texas	1921	1.5	(s)	1.5
Elk Hills	California	1911	1.1	0.3	1.4

(s) = less than 100 million barrels

Production of Crude Oil, by Major States, 2000

Source: Energy Information Administration, *Petroleum Supply Annual 2000*

(thousand barrels)

Rank	State	Total	Rank	State	Total	Rank	State	Total
1	Texas	443,396	11	Colorado	18,479	21	Nebraska	2,955
2	Alaska	355,198	12	Utah	15,636	22	Indiana	2,098
3	California	271,132	13	Montana	15,427	23	Pennsylvania	1,500
4	Louisiana	105,424	14	Illinois	12,206	24	West Virginia	1,401
5	Oklahoma	69,976	15	Alabama	10,458	25	South Dakota	1,170
6	New Mexico	67,198	16	Michigan	7,907	26	Nevada	621
7	Wyoming	60,726	17	Arkansas	7,153	27	New York	210
8	Kansas	34,463	18	Ohio	6,574	28	Tennessee	346
9	North Dakota	32,718	19	Florida	4,625			
10	Mississippi	19,843	20	Kentucky	3,467		**U.S. TOTAL**	**2,130,707**

World Crude Oil and Natural Gas Reserves, Jan. 1, 2000

Sources: Energy Information Administration, U.S. Dept. of Energy, *U.S. Crude Oil and Natural Gas Liquides Reserves, Dec. 2000; Oil and Gas Journal (OGJ),* Dec. 1999; *World Oil (WO),* Aug. 2000

	Crude oil (billion barrels)		Natural gas (trillion cubic feet)	
	OGJ	***WO***	***OGJ***	***WO***
North America	**55.1**	**55.6**	**261.3**	**261.3**
Canada	4.9	5.6	63.9	63.5
Mexico	28.4	28.3	30.1	30.4
United States	21.8	21.8	167.4	167.4
Central and South America	**89.5**	**69.24**	**222.7**	**227.9**
Argentina	2.8	2.6	24.2	24.3
Bolivia	0.1	0.2	4.3	5.5
Brazil	7.4	8.1	8.0	8.2
Colombia	2.6	2.3	6.9	6.6
Ecuador	2.1	3.0	3.7	3.9
Peru	0.4	4.1	9.0	8.8
Trinidad and Tobago	0.6	0.7	19.8	21.4
Venezuela	72.6	47.1	142.5	145.8
Other	1.0	1.0	4.2	3.5
Western Europe	**18.8**	**17.6**	**159.5**	**152.7**
Denmark	1.1	0.9	3.4	2.6
Germany	0.4	0.3	12.0	9.5
Italy	0.6	0.6	8.1	7.4
Netherlands	0.1	0.1	62.5	59.8
Norway	10.8	10.0	41.4	42.9
United Kingdom	5.2	5.0	26.7	26.8
Other	0.7	0.7	5.5	3.7
Eastern Europe and Former USSR	**58.9**	**64.7**	**1,999.2**	**1,947.6**
Hungary	0.1	0.1	2.9	1.1
Kazakhstan	5.4	6.4	65.0	70.6
Romania	1.4	1.2	13.2	4.0
Russia	48.6	52.7	1,700.0	1,705.0
Other[1]	3.3	4.3	218.1	166.9
Middle East	**675.6**	**629.2**	**1,749.2**	**1,836.2**
Bahrain	0.1	NA	3.9	NA
Iran	89.7	93.1	812.3	790.0
Iraq	112.5	100.0	109.8	112.6
Kuwait	96.5	94.7	52.7	56.4
Oman	5.3	5.7	28.4	29.3
Qatar	3.7	5.4	300.0	394.0
Saudi Arabia	263.5	261.4	204.5	208.0
Syria	2.5	2.3	8.5	8.4
United Arab Emirates	97.8	63.8	212.0	209.0
Yemen	4.0	2.1	16.9	17.0
Other	(2)	0.5	0.3	11.5
Africa	**74.9**	**86.5**	**394.2**	**409.7**
Algeria	9.2	13.0	159.7	159.7
Angola	5.4	8.5	1.6	3.8
Cameroon	0.4	0.6	3.9	3.9
Congo Republic	1.5	1.7	3.2	4.3
Egypt	2.9	3.8	35.2	42.5
Libya	29.5	29.5	46.4	46.4
Nigeria	22.5	24.5	124.0	126.0
Tunisia	0.3	0.3	2.8	2.8
Other	3.1	4.7	17.4	20.3
Far East and Oceania	**44.0**	**58.7**	**363.5**	**375.4**
Australia	2.9	2.9	44.6	44.6
Brunei	1.4	1.0	13.8	9.2
China	24.0	34.1	48.3	41.3
India	4.8	3.4	22.9	16.1
Indonesia	5.0	8.4	72.3	80.8
Malaysia	3.9	4.6	81.7	85.2
New Zealand	0.1	0.1	2.5	2.1
Pakistan	0.2	0.2	21.6	22.9
Papua New Guinea	0.3	0.8	5.4	17.3
Thailand	0.3	0.3	12.5	11.1
Other	1.1	2.9	37.9	44.7
WORLD	**1,016.8**	**981.4**	**5,149.6**	**5,210.8**

(1) Albania, Azerbaijan, Belarus, Bulgaria, Czech Republic, Georgia, Kyrgyzstan, Lithuania, Poland, Slovakia, Tajikistan, Turkmenistan, Ukraine, Uzbekistan. (2) Less than 50 million barrels. **NOTE:** Data for Kuwait and Saudi Arabia include one-half of the reserves in the Neutral Zone between Kuwait and Saudi Arabia. All reserve figures except those for the former USSR and natural gas reserves in Canada are *proved reserves* recoverable with present technology and prices at the time of estimation. Former USSR figures and natural gas figures for Canada are *explored reserves*, which include proved, probable, and some partially possible. Totals may not equal sum of components as a result of independent rounding.

Nuclear Electricity Generation by Selected Country, May 2001

Source: Energy Information Administration, U.S. Dept. of Energy, *Monthly Energy Review*, Aug. 2001

(Generation for the month, in billion kilowatt-hours; E = estimate)

Argentina	0.5
Armenia	0.3
Belgium	3.5
Brazil	0.8
Bulgaria	1.8[E]
Canada	4.5
China	1.1[E]
Czech Republic	1.0
Finland	1.5
France	29.8
Germany	13.2
Hungary	1.1
India	1.6[E]
Japan	25.2
Korea, South	9.1
Lithuania	0.6
Mexico	0.4
Netherlands	0.4
Pakistan	0.2
Romania	0.5
Russia	9.6
Slovakia	1.2
Slovenia	0.1
South Africa	1.3
Spain	5.8
Sweden	5.8
Switzerland	2.5
Taiwan	2.5
Ukraine	5.4
United Kingdom[1]	6.5
United States	64.9[E]

(1) Data for the United Kingdom is the total for a 4- or 5-week reporting period, not the calendar month.

Nations Most Reliant on Nuclear Energy, 2000

Source: International Atomic Energy Agency, Sept. 2001

(Nuclear electricity generation as % of total electricity generated)

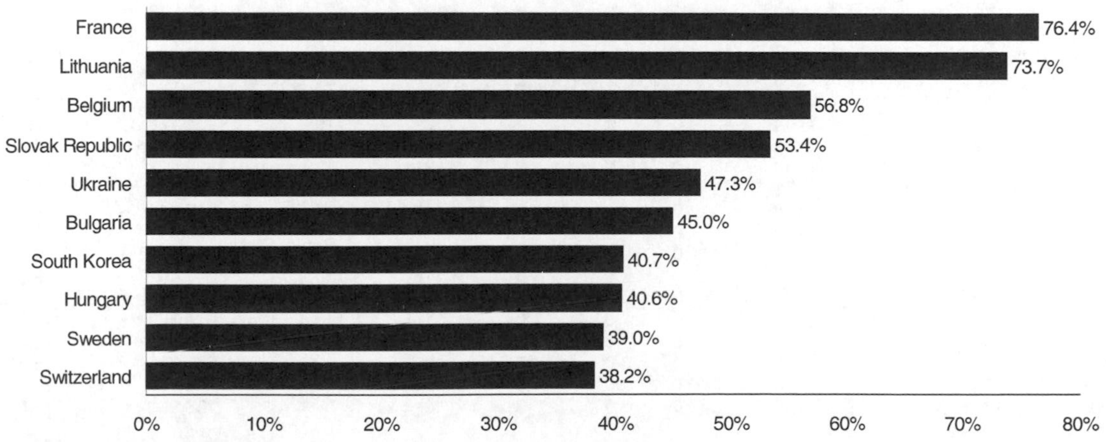

World Nuclear Power Summary, 2000

Source: International Atomic Energy Agency, Sept. 2001

Country	Reactors in operation: No. of units	Reactors in operation: Total MW(e)	Reactors under construction: No. of units	Reactors under construction: Total MW(e)	Nuclear electricity supplied in 2000: TW(e).h[1]	Nuclear electricity supplied in 2000: % of total	Total operating experience to Dec. 31, 2000: Years	Total operating experience to Dec. 31, 2000: Months
Argentina	2	935	1	692	5.73	7.26	44	7
Armenia	1	376	—	—	1.84	33.00	33	3
Belgium	7	5,712	—	—	45.40	56.75	170	7
Brazil	2	1,855	—	—	6.05	1.87	19	3
Bulgaria	6	3,538	—	—	18.18	45.00	113	2
Canada	14	9,998	—	—	68.68	11.80	433	2
China	3	2,167	8	6,420	16.00	1.19	23	5
Czech Republic	5	2,569	1	912	13.59	20.06	58	9
Finland	4	2,656	—	—	21.60	32.15	87	4
France	59	63,073	—	—	395.00	76.40	1,169	2
Germany	19	21,122	—	—	159.60	30.57	591	1
Hungary	4	1,755	—	—	14.18	40.60	62	2
India	14	2,503	2	900	14.21	3.14	181	5
Iran	—	—	2	2,111	—	—	—	—
Japan	53	43,491	3	3,190	304.87	33.82	962	8
Korea, South	16	12,990	4	3,820	103.50	40.74	169	2
Lithuania	2	2,370	—	—	8.40	73.68	30	6
Mexico	2	1,360	—	—	7.92	3.86	17	11
Netherlands	1	449	—	—	3.70	4.00	56	0
Pakistan	2	425			1.08	1.65	29	10
Romania	1	650	1	650	5.05	10.86	4	6
Russia	29	19,843	3	2,825	119.65	14.95	671	6
Slovakia	6	2,408	2	776	16.49	53.43	85	0
Slovenia	1	676	—	—	4.54	37.38	19	3
South Africa	2	1,800	—	—	12.99	6.58	32	3
Spain	9	7,512	—	—	59.30	27.63	192	2
Sweden	11	9,432	—	—	54.80	39.00	278	1
Switzerland	5	3,192	—	—	24.95	38.18	128	10
Taiwan	6	4,884	2	2,560	37.00	23.64	116	1
Ukraine	13	11,207	4	3,800	72.40	47.28	240	10
United Kingdom	35	12,968	—	—	78.30	21.94	1,238	4
United States	104	97,411	—	—	753.90	19.83	2,559	8
TOTAL	**438**	**351,327**	**33**	**28,656**	**2,448.90**	**—**	**9,819**	**11**

(1) 1 terawatt-hour [TW(e).h] = 10^6 megawatt-hour [MW(e).h]. For an average power plant, 1 TW(e).h = 0.39 megatons of coal equivalent (input) and 0.23 megatons of oil equivalent (input).

U.S. Nuclear Reactor Units and Power Plant Operations

Source: Energy Information Administration, U.S. Dept. of Energy, *Monthly Energy Review,* June 2001

	Number of reactor units									
	Licensed for operation		Construction permits							
Year	Operable	In startup	Granted	Pending	On order	Announced	Total	Total design capacity (million KWs)	Nuclear-based electricity generation (million net KW-hours)	Nuclear portion of domestic electricity generation (percent)
1977	67	2	78	49	13	2	209	203	250,883	11.8
1978	70	0	88	32	5	0	195	191	276,403	12.5
1979	69	0	90	24	3	0	185	180	255,155	11.4
1980	71	1	82	12	3	0	168	162	251,116	11.0
1981	75	0	76	11	2	0	163	157	272,674	11.9
1982	78	2	60	3	2	0	144	134	282,773	12.6
1983	81	3	53	0	2	0	138	129	293,677	12.7
1984	87	6	38	0	2	0	132	123	327,634	13.6
1985	96	3	30	0	2	0	130	121	383,691	15.5
1986	101	7	19	0	2	0	128	119	414,038	16.6
1987	107	4	14	0	2	0	127	119	455,270	17.7
1988	109	3	12	0	0	0	123	115	526,973	19.5
1989	111	1	10	0	0	0	121	113	529,402	17.8
1990	112	0	8	0	0	0	119	111	576,974	19.1
1991	111	0	8	0	0	0	119	111	612,642	19.9
1992	109	0	8	0	0	0	117	111	618,841	20.1
1993	110	0	7	0	0	0	116	110	610,367	19.1
1994	109	0	7	0	0	0	116	110	640,492	19.7
1995	109	1	6	0	0	0	116	110	673,402	20.1
1996	109	0	6	0	0	0	116	110	674,729	19.6
1997	107	0	3	0	0	0	110	102	628,644	18.0
1998	104	0	3	0	0	0	107	99	673,702	18.6
1999	104	0	0	0	0	0	104	NA	728,198	19.8
2000	104	0	0	0	0	0	104	NA	753,893	19.8

NA = Not available.

ENVIRONMENT

Greenhouse Effect and Global Warming

Source: U.S. Environmental Protection Agency

The Earth naturally absorbs incoming solar radiation and emits thermal radiation back into space. Some of the thermal radiation is trapped by so-called greenhouse gases in the atmosphere, which increases warming of the Earth's surface and atmosphere. In recent years, carbon dioxide (CO_2), a naturally occurring greenhouse gas, has been building up in the atmosphere as a result of activities such as the burning of fossil fuels (coal, oil, natural gas) and deforestation. Water vapor, methane (CH_4), nitrous oxide (N_2O), and ozone (O_3) are also naturally occurring greenhouse gases. Greenhouse gases that are mostly human-made include chlorofluorocarbons (CFCs), hydrochlorofluorocarbons (HCFCs), hydrofluorocarbons (HFCs), perfluorocarbons (PFCs), and sulfur hexafluoride (SF_6). Several nongreenhouse gases (carbon monoxide [CO], oxides of nitrogen [NOx], and nonmethane volatile organic compounds [NMVOCs]) contribute indirectly to the greenhouse effect by producing greenhouse gases during chemical transformations or by influencing the atmospheric lifetimes of greenhouse gases.

Since the beginning of the industrial revolution, atmospheric concentrations of CO_2, CH_4, and N_2O have increased by 30%, 145%, and 15%, respectively. This increasing buildup is believed by many scientists to be the major cause of higher than normal average global temperatures in the 1990s and into the 21st century; 2000 was the sixth-warmest year on record (57.6°F). The hottest year was 1998, the 2d-hottest year was 1997, and the century's 10 hottest years have all occurred since 1985. Over the 20th century, the Earth's average temperature has risen by approximately 1°F, and some scientists believe that it could rise by 2° to 6°F over the 21st century. This global warming could speed the melting of the polar ice caps, inundate coastal lowlands, and bring about major changes in crop production and in natural habitat. The United States is the world's leading producer of CO_2, followed by China, Russia, Japan, India, and Germany.

In Dec. 1997, a United Nations summit on global warming was held in Kyoto, Japan. Delegates from over 150 nations adopted an international treaty to set limits on emissions of CO_2, CH_4, N_2O, HFCs, PFCs, and SF_6. The accord, known as the Kyoto Protocol, called for a reduction in emissions 5.2% below 1990 levels by the year 2012 for all 38 industrialized countries that signed the accord. The 15 EU nations agreed to reductions of 8%, the U.S. to 7%, and Japan to 6%. Developing nations were permitted to limit their emissions voluntarily.

The U.S. signed the treaty on Nov. 12, 1998, but then-Pres. Bill Clinton did not send it to the Senate for ratification because of dim prospects for approval. Another meeting was set for July 2001 in Bonn, Germany, to work out details for monitoring and measuring emissions. The Bush adminstration declared it would pull out of the treaty, saying it placed an unfair burden on developed countries and limited economic growth. Despite U.S. opposition, delegates in Bonn agreed on binding guidelines and timetables for achieving the reduction in emissions set forth by the Kyoto Protocol. Under the Bonn agreement, high-emissions nations could meet their targets by purchasing pollution credits from nations that exceed their target reductions and gain credits for "sinks," areas such as forests and croplands that absorb CO_2 from the atmosphere.

U.S. Greenhouse Gas Emissions From Human Activities, 1990-99

Source: U.S. Environmental Protection Agency

GAS AND SOURCE	1990	1995	1996	1997	1998	1999
Carbon dioxide (CO_2)	**4,913.0**	**5,219.8**	**5,403.2**	**5,478.7**	**5,489.7**	**5,558.1**
Fossil fuel combustion	4,835.7	5,121.3	5,303.0	5,374.9	5,386.8	5,456.1
Methane (CH_4)	**644.5**	**650.5**	**638.0**	**632.0**	**624.8**	**619.6**
Coal Mining	87.9	74.6	69.3	68.8	66.5	61.8
Natural gas systems	121.2	124.2	125.8	122.7	122.1	121.8
Enteric fermentation	129.5	136.3	132.2	129.6	127.5	127.2
Nitrous oxide (N_2O)	**396.9**	**431.9**	**441.6**	**444.1**	**433.7**	**432.6**
Agricultural soil management	269.0	285.4	294.6	299.8	300.3	298.3
Hydrofluorocarbons (HFCs), perfluorocarbons (PFCs), and sulfur hexafluoride (SF_6)[1]	**83.9**	**99.0**	**115.1**	**123.3**	**138.6**	**135.7**
TOTAL U.S. EMISSIONS	**6,038.2**	**6,401.3**	**6,597.8**	**6,678.0**	**6,686.8**	**6,746.0**
NET U.S. EMISSIONS[2]	**4,978.3**	**5,382.3**	**5,576.2**	**5,696.2**	**5,703.5**	**5,755.7**

Note: Emissions are given in terms of equivalent emissions of carbon dioxide (CO_2), using units of teragrams of carbon dioxide equivalents (Tg CO_2 Eq.). Before 1999, emissions were reported using units of million metric tons of carbon equivalents (MMTCE). (1) These gases have extremely high global warming potential, and PFCs and SF_6 have long atmospheric lifetimes. (2) Total emissions minus carbon dioxide absorbed by forests or other means.

U.S. Greenhouse Gas Emissions, 1999

Source: U.S. Environmental Protection Agency

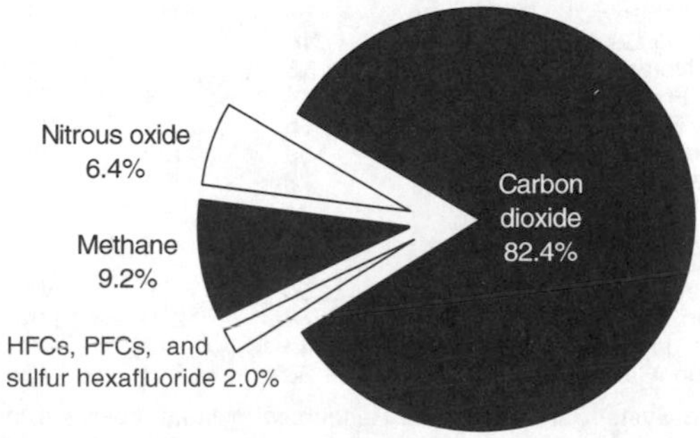

World Carbon Dioxide Emissions From the Use of Fossil Fuels, 1999

Source: Energy Information Administration

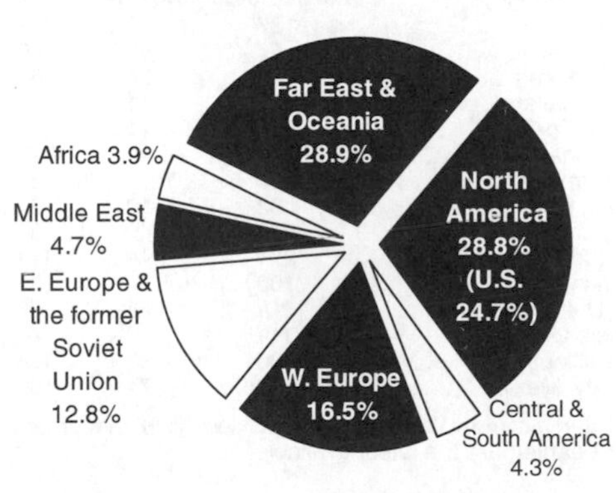

Average Global Temperatures, 1880-2000

Source: National Oceanic and Atmospheric Administration; in degrees Fahrenheit

1880-89 .. 56.65	1910-19 .. 56.57	1930-39 .. 57.00	1950-59 .. 57.06	1970-79...57.04	1990-99...57.64
1890-99 .. 56.64	1920-29 .. 56.74	1940-49 .. 57.13	1960-69 .. 57.05	1980-89...57.36	2000 .....57.60
1900-09 .. 56.52					

Toxics Release Inventory, 1998-99

Source: U.S. Environmental Protection Agency

Reported industrial releases of toxic chemicals into the environment in the U.S. by major manufacturing facilities (excluding power plants and mining facilities) decreased 5.3% from the 1998 figure and decreased by about half from the figure for 1988, the baseline year. Totals below may not add because of rounding.

Pollutant releases	1999 mil lb	1998 mil lb
Air releases	1,175	1,270
Surface water releases	254	238
Underground injection	200	210
On-site land releases	324	344
Pollutant transfers		
To recycling	2,075	2,007
To energy recovery	514	483
To treatment	241	253
To publicly owned treatment works	322	328
Other transfers	0	1
TOTAL	**3,637**	**3,511**

Top industries, total releases	1999 mil lb	1998 mil lb
Primary metals	684	719
Chemicals	671	687
Paper	226	234
Food	123	135
Transportation equipment	105	103
TOTAL	**1,952**	**2,062**
Top carcinogens, air/water/land releases		
Styrene	57	57
Dichloromethane	36	41
Formaldehyde	24	22
Acetaldehyde	13	13
Trichloroethylene	11	13
Ethylbenzene	10	10

Top 10 States, Total Releases, 1997-99

Source: U.S. Environmental Protection Agency

State	1999 mil lb	1998 mil lb	1997 mil lb
Texas	258	259	262
Pennsylvania	160	136	145
Ohio	140	144	156
Louisiana	135	175	188
Indiana	126	115	123
Illinois	96	102	127
Tennessee	88	93	107
Utah	83	99	104
Florida	77	75	95
Alabama	75	89	95

Air Pollution

Source: World Bank, *World Development Indicators 2001*

In many towns and cities exposure to air pollution is the main environmental threat to health. Winter smog—made up of soot, dust, and sulfur dioxide—has long been associated with temporary increases in deaths. Prolonged exposure to particulate pollution can lead to a host of chronic respiratory illnesses and exacerbates heart disease and other conditions. Particulate pollution has been estimated to cause 500,000 premature deaths in the world each year.

Emissions of sulfur dioxide and nitrogen oxides lead to acid rain, which spreads over long distances, upsetting the delicate chemical balance of soils, trees, and plants. Direct exposure to high levels of sulfur dioxide or acid deposition causes defoliation.

Where coal is a primary fuel, high levels of urban air pollution may result. If the coal has a high sulfur content, widespread acid deposition may result. Combustion of petroleum products is another important cause of air pollution.

In the table below, **suspended particulates** refers to smoke, soot, dust, and liquid droplets from combustion that are in the air. The level of particulates indicates the quality of the air and the level of technology and pollution controls in a given country. **Sulfur dioxide** is an air pollutant formed when fossil fuels containing sulfur are burned. Nitrogen dioxide is a poisonous, pungent gas formed when nitric oxide combines with hydrocarbons and sunlight, producing a photochemical reaction. **Nitrogen oxide** is emitted by bacteria, nitrogenous fertilizers, aerobic decomposition of organic matter in oceans and soils, combustion of fuels and biomass, and motor vehicles and industrial activities.

Data in the table are based on reports from urban monitoring sites. Annual means (measured in micrograms per cubic meter) are average concentrations observed at various sites; the resulting figures give a general indication of air quality in each city, but results should be interpreted with caution.

World Health Organization standards for acceptable air quality are 90 micrograms per cubic meter for total suspended particulates and 50 micrograms per cubic meter for sulfur dioxide and nitrogen dioxide.

Air Pollution in 30 Selected World Cities[1]

City and Country	Suspended Particulates	Sulfur Dioxide	Nitrogen Dioxide
Accra, Ghana	137	NA	NA
Ankara, Turkey	57	55	46
Athens, Greece	178	34	64
Bangkok, Thailand	223	11	23
Barcelona, Spain	117	11	43
Beijing, China	377	90	122
Berlin, Germany	50	18	26
Bucharest, Romania	82	10	71
Calcutta, India	375	49	34
Jakarta, Indonesia	271	NA	NA
Kiev, Ukraine	100	14	51
London, United Kingdom	NA	25	77
Los Angeles, U.S.A.	NA	9	74
Manila, Philippines	200	33	NA
Mexico City, Mexico	279	74	130
Moscow, Russia	100	109	NA
Nairobi, Kenya	69	NA	NA
New York City	NA	26	79
Oslo, Norway	15	8	43
Paris, France	14	14	57
Quito, Ecuador	175	31	NA
Rio de Janeiro, Brazil	139	129	NA
Rome, Italy	73	NA	NA
Seoul, South Korea	84	44	60
Sofia, Bulgaria	195	39	122
Stockholm, Sweden	9	5	29
Sydney, Australia	54	28	NA
Tehran, Iran	248	209	NA
Tokyo, Japan	49	18	68
Toronto, Canada	36	17	43

(1) Data derived from WHO's Healthy Cities Air Management Information System and the World Resources Institute, collected in 1998 or, if earlier, are the latest available.

Emissions of Principal Air Pollutants in the U.S., 1970-99

Source: U.S. Environmental Protection Agency, Office of Air Quality Planning and Standards; in thousand short tons; estimated

Source	1970	1975	1980	1985	1990	1995	1996	1997	1998	1999
Carbon monoxide	129,444	116,757	117,434	117,013	99,119	94,058	101,294	101,459	96,872	97,441
Lead	220,869	159,659	74,153	22,890	4,975	3,929	4,077	4,137	4,057	4,199
Nitrogen oxides[1]	20,928	22,632	24,384	23,198	24,170	25,051	26,053	26,352	26,020	25,393
Volatile organic compounds[1]	30,982	26,079	26,336	24,428	21,053	20,918	19,464	19,732	18,614	18,145
Particulate matter[2]	12,325	7,108	6,258	3,662	3,340	3,165	2,929	2,854	2,758	3,045
Sulfur dioxide	31,161	28,011	25,905	23,658	23,678	19,188	18,859	19,363	19,491	18,867
TOTAL[3]	**445,709**	**360,156**	**274,470**	**214,849**	**176,335**	**166,309**	**172,676**	**160,914**	**167,812**	**167,390**

(1) Ozone, a major air pollutant and the primary constituent of smog, is not emitted directly to the air but is formed by sunlight acting on emissions of nitrogen oxides and volatile organic compounds. (2) Does not include natural sources. (3) Totals are rounded, as are components of totals.

Carbon Monoxide Emission Estimates, 1970-99

Source: U.S. Environmental Protection Agency, Office of Air Quality Planning and Standards; in thousand short tons

Source	1970	1975	1980	1985	1990	1995	1996	1997	1998	1999
Fuel combustion	4,632	4,480	7,302	8,485	5,510	5,934	6,206	5,484	5,075	5,322
Industrial processes	16,899	10,770	9,250	7,215	5,852	5,790	4,759	4,932	4,955	7,590
Transportation	100,004	96,243	92,538	93,386	76,635	75,035	79,795	78,509	77,478	75,151
Miscellaneous	7,909	5,263	8,344	7,927	11,122	7,298	10,534	12,534	9,364	9,378
TOTAL[1]	**129,444**	**116,757**	**117,434**	**117,013**	**99,119**	**94,058**	**101,294**	**101,459**	**96,872**	**97,441**

(1) Totals may not add because of rounding.

Lead Emission Estimates, 1970-99

Source: U.S. Environmental Protection Agency, Office of Air Quality Planning and Standards; in short tons

Source	1970	1975	1980	1985	1990	1995	1996	1997	1998	1999
Fuel combustion[1]	10,616	10,347	4,299	515	500	490	492	493	494	501
Industrial processes	28,554	12,976	5,148	3,402	3,278	2,875	3,061	3,121	3,045	3,162
Transportation	181,698	136,336	64,706	18,973	1,197	564	525	523	518	536
TOTAL[2]	**220,869**	**159,659**	**74,153**	**22,890**	**4,975**	**3,929**	**4,077**	**4,137**	**4,057**	**4,199**

(1) Does not include transportation. (2) Totals may not add because of rounding.

Nitrogen Oxides Emission Estimates, 1970-99

Source: U.S. Environmental Protection Agency, Office of Air Quality Planning and Standards; in thousand short tons

Source	1970	1975	1980	1985	1990	1995	1996	1997	1998	1999
Fuel combustion	10,061	10,486	11,320	10,048	10,895	10,827	10,523	10,576	10,396	10,026
Industrial processes	1,215	697	666	891	892	873	903	939	950	942
Transportation	9,322	11,284	12,150	11,948	12,014	13,085	14,211	14,436	14,355	14,105
Miscellaneous	330	165	248	310	369	267	416	402	319	320
TOTAL[1]	**20,928**	**22,632**	**24,384**	**23,198**	**24,170**	**25,051**	**26,053**	**26,352**	**26,020**	**25,393**

(1) Totals may not add because of rounding.

Air Quality of Selected U.S. Metropolitan Areas, 1990-99

Source: U.S. Environmental Protection Agency, Office of Air Quality Planning and Standards

Data indicate the number of days metropolitan statistical areas failed to meet acceptable air-quality standards. All figures were revised based on new standards set in 1998.

Metropolitan statistical area	1990	1991	1992	1993	1994	1995	1996	1997	1998	1999
Atlanta, GA	42	23	20	36	15	35	25	31	50	61
Bakersfield, CA	99	113	100	97	98	105	109	55	76	88
Baltimore, MD	29	50	23	48	41	36	28	30	51	40
Boston, MA–NH	7	13	9	6	10	8	2	8	7	5
Chicago, IL	4	22	4	3	8	21	6	9	7	12
Dallas, TX	24	2	12	14	27	36	12	20	28	23
Denver, CO	9	6	11	3	1	2	0	0	5	1
Detroit, MI	11	28	8	5	11	14	13	12	17	15
El Paso, TX	19	7	10	7	11	8	7	4	6	6
Fresno, CA	62	83	69	59	55	61	70	75	67	81
Hartford, CT	13	23	15	14	18	14	5	16	10	18
Houston, TX	51	36	32	28	38	66	26	47	38	50
Las Vegas, NV–AZ	4	0	1	2	2	0	2	0	0	0
Los Angeles–Long Beach, CA	173	168	175	134	139	113	94	60	56	27
Miami, FL	1	1	3	2	1	2	1	3	8	5
Minneapolis–St. Paul, MN–WI	4	2	1	0	2	5	0	0	1	0
New Haven–Meriden, CT	17	29	10	17	14	14	8	19	10	16
New York, NY	36	49	10	19	21	19	15	23	17	24
Orange County, CA	45	35	35	25	15	9	9	3	6	1
Philadelphia, PA–NJ	39	49	24	51	26	30	22	32	37	32
Phoenix–Mesa, AZ	12	11	13	16	10	22	17	12	17	12
Pittsburgh, PA	19	21	9	13	19	25	11	21	39	23
Riverside–San Bernardino, CA	159	154	174	168	149	124	119	105	95	93
Sacramento, CA	61	46	51	20	36	41	42	15	27	38
St. Louis, MO–IL	23	32	15	9	32	34	20	15	23	29
Salt Lake City–Ogden, UT	5	20	9	5	12	4	8	1	12	2
San Diego, CA	96	67	66	58	46	48	31	14	33	16
San Francisco, CA	0	0	0	0	0	2	0	0	0	0
Seattle–Bellevue–Everett, WA	9	4	3	0	3	0	6	1	3	1
Ventura, CA	70	87	54	43	63	66	62	45	29	22
Washington, DC–MD–VA–WV	25	48	14	48	20	29	18	29	47	39

Hazardous Waste Sites in the U.S., 2001

Source: U.S. Environmental Protection Agency, *National Priorities List*, Sept. 2001

State/Territory	Total Proposed Gen	Total Proposed Fed	Total Final Gen	Total Final Fed	Total
Alabama	2	0	10	3	15
Alaska	0	0	1	6	7
Arizona	0	0	7	3	10
Arkansas	0	0	12	0	12
California	3	0	72	24	99
Colorado	2	0	12	3	17
Connecticut	1	0	14	1	16
Delaware	0	0	16	1	17
District of Columbia	0	0	0	1	1
Florida	1	0	45	6	52
Georgia	1	0	12	2	15
Hawaii	0	0	1	2	3
Idaho	4	0	4	2	10
Illinois	5	1	35	4	45
Indiana	1	0	28	0	29
Iowa	2	0	13	1	16
Kansas	1	1	9	1	12
Kentucky	0	0	14	1	15
Louisiana	2	0	12	1	15
Maine	1	0	9	3	13
Maryland	1	1	9	8	19
Massachusetts	2	0	23	8	33
Michigan	1	1	67	0	69
Minnesota	0	0	22	2	24
Mississippi	2	0	2	0	4
Missouri	1	0	22	3	26
Montana	1	0	13	0	14
Nebraska	0	0	9	1	10
Nevada	0	0	1	0	1
New Hampshire	1	0	17	1	19
New Jersey	5	0	103	8	116
New Mexico	2	0	10	1	13
New York	4	0	84	4	92
North Carolina	1	0	24	2	27
North Dakota	0	0	0	0	0
Ohio	2	2	27	3	34
Oklahoma	1	0	10	1	12
Oregon	1	0	9	2	12
Pennsylvania	3	0	90	6	99
Rhode Island	0	0	10	2	12
South Carolina	0	0	23	2	25
South Dakota	0	0	1	1	2
Tennessee	0	1	10	3	14
Texas	3	0	34	4	41
Utah	6	0	11	4	21
Vermont	0	0	9	0	9
Virginia	0	0	20	11	31
Virgin Islands	0	0	2	0	2
Washington	0	0	34	14	48
West Virginia	0	0	7	2	9
Wisconsin	2	0	38	0	40
Wyoming	0	0	1	1	2
American Samoa	0	0	0	0	0
Commonwealth of Marianas	0	0	0	0	0
Guam	0	0	1	1	2
Puerto Rico	0	0	9	0	9
Trust Territories	0	0	0	0	0
Total	**65**	**7**	**1,078**	**160**	**1,310**

Note: Gen = general superfund sites; Fed = federal facility sites.

Watersheds in the U.S.

Source: U.S. Environmental Protection Agency

A watershed is a water drainage area, or land areas bounded by ridges that catch rain and snow and drain to rivers, lakes, and groundwater within the drainage area. In a comprehensive assessment of watersheds in the continental U.S. released in Sept. 1999, the Environmental Protection Agency (EPA) concluded that 15% of the 2,262 watersheds had good water quality, 36% had moderate water quality, and 23% had less acceptable water quality. There was insufficient information to fully characterize the remaining 26%. The data indicate that polluted runoff from urban and rural areas is a major contributor to water quality problems, threatening water quality even in currently healthy watersheds.

The EPA categorized the watersheds by combining nationally available data from 15 individual databases, from both public and private sources, into a single Index of Watershed Indicators. The indicators include 7 used to assess watershed conditions (quality) and 8 used to assess vulnerability to degradation from pollution. You can find information about your own watershed on the Internet by going to the following website: http://www.epa.gov/surf3/index.html

Renewable Water Resources

Source: World Resources Institute

Globally, water supplies are abundant, but they are unevenly distributed among and within countries. In some areas, water withdrawals are so high, relative to supply, that surface water supplies are shrinking and groundwater reserves are being depleted faster than they can be replenished by precipitation.

Countries With Most Resources

Country	Cubic meters per capita	Total cubic kilometers
Iceland	606,498	168.00
Suriname	452,489	200.00
Guyana	281,542	241.00
Papua-New Guinea	174,055	801.00
Gabon	140,171	164.00
Solomon Islands	107,194	44.70
Canada	94,373	2,849.50
New Zealand	88,859	327.00
Norway	87,691	394.00
Republic of Congo	78,668	222.00

Countries With Least Resources

Country	Cubic meters per capita	Total cubic kilometers
Kuwait	11	0.02
Egypt	43	2.80
United Arab Emirates	64	0.15
Libya	100	0.60
Jordan	114	0.68
Mauritania	163	0.40
Singapore	172	0.60
Moldova	225	1.00
Turkmenistan	232	1.00
Yemen	243	4.10

Frontier Forests

Only one-fifth of the Earth's forest cover from 8,000 years ago survives unfragmented, in the large unspoiled tracts called frontier forests. These forests are big enough to provide stable habitats for a rich diversity of plant and animal species. Most surviving forests are in the far north or the tropics, and are under threat.

Percentage of Frontier Forest Under Moderate or High Threat of Destruction

Source: World Resources Institute

Europe	100
Central America	87
Africa	77
Oceania	76
Asia	60
South America	54
North America	26
World	39

IT'S A FACT: China's Shaanxi province recently banned the production, sale, or use of disposable chopsticks. According to the Chinese News Agency Xinhua, 25 million trees have been cut down in China each year to produce an annual output of 45 billion pairs of disposable chopsticks.

U.S. List of Endangered and Threatened Species

Source: Fish and Wildlife Service, U.S. Dept. of Interior; as of Aug. 2001

	ENDANGERED		THREATENED			Total species with recovery plans
Group	U.S.	Foreign	U.S.	Foreign	Total species	
Mammals	63	251	9	17	340	50
Birds	78	175	14	6	273	75
Reptiles	14	64	22	15	115	30
Amphibians	10	8	8	1	27	12
Fishes	70	11	44	0	125	95
Clams	61	2	8	0	71	56
Snails	20	1	11	0	32	27
Insects	33	4	9	0	46	28
Arachnids	12	0	0	0	12	5
Crustaceans	18	0	3	0	21	12
Animal subtotal	**379**	**516**	**128**	**39**	**1,062**	**390**
Flowering plants	565	1	141	0	707	555
Conifers and Cycads	2	0	1	2	5	2
Ferns and Allies	24	0	2	0	26	26
Lichens	2	0	0	0	2	2
Plant subtotal	**593**	**1**	**144**	**2**	**740**	**585**
GRAND TOTAL	**972**	**517**	**272**	**41**	**1,802**[1]	**975**

(1) When separate populations of a species are listed as endangered and as threatened, those species are tallied twice. Those species are the argali, chimpanzee, chinook salmon, gray wolf, green sea turtle, leopard, olive ridley sea turtle, piping plover, roseate tern, saltwater crocodile, sockeye salmon, steelhead, and Steller sea lion.

Some Endangered Animal Species

Source: Fish and Wildlife Service, U.S. Dept. of the Interior

Common name	Scientific name	Range
Armadillo, giant	Pridontes maximus	Venezuela, Guyana to Argentina
Bat, gray	Myotis grisescens	Central, southeastern U.S.
Bear, brown	Ursus arctos arctos	Palearctic
Bison, wood	Bison bison athabascae	Canada, northwestern U.S.
Bobcat, Mexican	Felis rufus escuinapae	Central Mexico
Camel, Bactrian	Camelus bactrianus	Mongolia, China
Caribou, woodland	Rangifer tarandus caribou	U.S., Canada
Cheetah	Acinonyx jubatus	Africa to India
Chimpanzee, pygmy	Pan paniscus	Congo (formerly Zaire)
Chinchilla	Chinchilla brevicaudata boliviana	Bolivia
Condor, California	Gymnogyps californianus	U.S. (AZ, CA, OR), Mexico (Baja California)
Crane, hooded	Grus monacha	Japan, Russia
Crane, whooping	Grus americana	Canada, Mexico, U.S. (Rocky Mts. to Carolinas)
Crocodile, American	Crocodylus acutus	U.S. (FL), Mexico, Caribbean Sea, Central and S America
Deer, Columbian white-tailed	Odocoileus virginianus leucurus	U.S. (OR, WA)
Dolphin, Chinese river	Lipotes vexillifer	China
Elephant, Asian	Elephas maximus	S central and southeastern Asia
Fox, northern swift	Vulpes velox hebes	U.S., Canada
Gorilla	Gorilla gorilla	Central and W Africa
Hawk, Hawaiian	Buteo solitarius	U.S. (HI)
Hyena, brown	Hyaena brunnea	Southern Africa
Kangaroo, Tasmanian forester	Macropus giganteus tasmaniensis	Australia (Tasmania)
Leopard	Panthera pardus	Africa and Asia
Lion, Asiatic	Panthera leo persica	Turkey to India
Manatee, West Indian	Trichechus manatus	Southeastern U.S., Caribbean Sea, S America
Monkey, spider	Ateles geoffroyi frontatus	Costa Rica, Nicaragua
Ocelot	Felis pardalis	U.S. (AZ, TX) to Central and S America
Orangutan	Pongo pygmaeus	Borneo, Sumatra
Ostrich, West African	Struthio camelus spatzi	W Sahara
Otter, marine	Lutra felina	Peru south to Straits of Magellan
Panda, giant	Ailuropoda melanoleuca	China
Panther, Florida	Felis concolor coryi	U.S. (LA, AR east to SC, FL)
Parakeet, golden	Aratinga guarouba	Brazil
Parrot, imperial	Amazona imperialis	West Indies (Dominica)
Penguin, Galapagos	Spheniscus mendiculus	Ecuador (Galapagos Islands)
Puma, eastern	Puma concolor couguar	Eastern N America
Python, Indian	Python molurus molurus	Sri Lanka, India
Rhinoceros, black	Diceros bicornis	Sub-Saharan Africa
Rhinoceros, northern white	Ceratotherium simum cottoni	Congo (formerly Zaire), Sudan, Uganda, Central African Republic
Salamander, Chinese giant	Andrias davidianus davidianus	Western China
Squirrel, Carolina northern flying	Glaucomys sabrinus coloratus	U.S. (NC, TN)
Stork, oriental white	Ciconia ciconia boyciana	China, Japan, Korea, Russia
Tiger	Panthera tigris	Asia
Tortoise, Galapagos	Geochelone elephantopus	Ecuador (Galapagos Islands)
Turtle, Plymouth red-bellied	Pseudemys rubriventris bangsi	U.S. (MA)
Whale, gray	Eschrichtius robustus	N Pacific Ocean
Whale, humpback	Megaptera novaeangliae	Oceania
Wolf, red	Canis rufus	Southeastern U.S. to central TX
Woodpecker, ivory-billed	Campephilus principalis	S central and southeastern U.S., Cuba
Yak, wild	Bos grunniens mutus	China (Tibet), India
Zebra, mountain	Equus zebra zebra	South Africa

Classification

Source: *Funk & Wagnalls New Encyclopedia*

In biology, classification is the identification, naming, and grouping of organisms into a formal system. The 2 fields that are most directly concerned with classification are taxonomy and systematics. Although the 2 disciplines overlap considerably, taxonomy is more concerned with nomenclature (naming) and with constructing hierarchical systems, and systematics with uncovering evolutionary relationships. Two kingdoms of living forms, Plantae and Animalia, have been recognized since Aristotle established the first taxonomy in the 4th century BC. In addition, there are the following 3 kingdoms: Protista (one-celled organisms), Monera (bacteria and blue-green algae; also known as the kingdom Procaryotae), and Fungi. The 7 basic categories of classification (from most general to most specific) are: kingdom, phylum (or division), class, order, family, genus, and species. Below are 2 examples:

ZOOLOGICAL HIERARCHY

Kingdom	Phylum	Class	Order	Family	Genus	Species name	Common name
Animalia	Chordata	Mammalia	Primates	Hominidae	Homo	Homo sapiens	Human

BOTANICAL HIERARCHY

Kingdom	Division*	Class	Order	Family	Genus	Species name	Common name
Plantae	Magnoliophyta	Magnoliopsida	Magnoliales	Magnoliaceae	Magnolia	M. virginiana	Sweet Bay

* In botany, the division is generally used in place of the phylum.

Gestation, Longevity, and Incubation of Animals

Information reviewed and updated by Ronald M. Nowak, author *Walker's Mammals of the World* (6th ed., Johns Hopkins University Press, 1999). Average longevity figures supplied by Ronald T. Reuther. These apply to animals in captivity; the potential life span of animals is rarely attained in nature. Figures on gestation and incubation are averages based on estimates.

ANIMAL	Gestation (days)	Average longevity (years)	Maximum longevity (yr-mo)
Ass	365	12	47
Baboon	187	20	45
Bear: Black	219	18	36-10
Grizzly	225	25	50
Polar	240	20	45
Beaver	105	5	50
Bison	285	15	40
Camel	406	12	50
Cat (domestic)	63	12	28
Chimpanzee	230	20	60
Chipmunk	31	6	10
Cow	284	15	30
Deer (white-tailed)	201	8	20
Dog (domestic)	61	12	20
Elephant (African)	660	35	70
Elephant (Asian)	645	40	77
Elk	250	15	26-8
Fox (red)	52	7	14
Giraffe	457	10	36-2
Goat (domestic)	151	8	18
Gorilla	258	20	54
Guinea pig	68	4	8
Hippopotamus	238	41	61
Horse	330	20	50
Kangaroo (gray)	36	7	24
Leopard	98	12	23
Lion	100	15	30
Monkey (rhesus)	166	15	37
Moose	240	12	27
Mouse (meadow)	21	3	4
Mouse (dom. white)	19	3	6
Opossum (American)	13	1	5
Pig (domestic)	112	10	27
Puma	90	12	20
Rabbit (domestic)	31	5	13
Rhinoceros (black)	450	15	45-10
Rhinoceros (white)	480	20	50
Sea lion (California)	350	12	34
Sheep (domestic)	154	12	20
Squirrel (gray)	44	10	23-6
Tiger	105	16	26-3
Wolf (maned)	63	5	15-8
Zebra (Grant's)	365	15	50

Incubation time (days)

Chicken	21
Duck	30
Goose	30
Pigeon	18
Turkey	26

Speeds of Animals

Source: *Natural History* magazine. Copyright © The American Museum of Natural History, 1974

ANIMAL	mph
Cheetah	70
Pronghorn antelope	61
Wildebeest	50
Lion	50
Thomson's gazelle	50
Quarterhorse	47.5
Elk	45
Cape hunting dog	45
Coyote	43
Gray fox	42
Hyena	40
Zebra	40
Mongolian wild ass	40
Greyhound	39.35
Whippet	35.50
Rabbit (domestic)	35
Mule deer	35
Jackal	35
Reindeer	32
Giraffe	32
White-tailed deer	30
Wart hog	30
Grizzly bear	30
Cat (domestic)	30
Human	27.89
Elephant	25
Black mamba snake	20
Six-lined race runner (lizard)	18
Wild turkey	15
Squirrel	12
Pig (domestic)	11
Chicken	9
Spider (Tegenaria atrica)	1.17
Giant tortoise	0.17
Three-toed sloth	0.15
Garden snail	0.03

Most of these measurements are for maximum speeds over approximate quarter-mile distances. Exceptions are the lion and elephant, whose speeds were clocked in the act of charging; the whippet, which was timed over a 200-yd course; the cheetah, timed over a 100-yd distance; and the black mamba, six-lined race runner, spider, giant tortoise, three-toed sloth, and garden snail, which were measured over various small distances.

Major Venomous Animals

Snakes

Asian pit viper — from 2 ft to 5 ft long; throughout Asia; reactions and mortality vary, but most bites cause tissue damage, and mortality is generally low.

Australian brown snake — 4 ft to 7 ft long; very slow onset of cardiac or respiratory distress; moderate mortality, but because death can be sudden and unexpected, it is the most dangerous of the Australian snakes; antivenom.

Barba Amarilla or fer-de-lance — up to 7 ft long; from tropical Mexico to Brazil; severe tissue damage common; moderate mortality; antivenom.

Black mamba — up to 14 ft long, fast-moving; S and C Africa; rapid onset of dizziness, difficulty breathing, erratic heartbeat; mortality high, nears 100% without antivenom.

Boomslang — less than 6 ft long; in African savannahs; rapid onset of nausea and dizziness, often followed by slight recovery and then sudden death from internal hemorrhaging; bites rare, mortality high; antivenom.

Bushmaster — up to 12 ft long; wet tropical forests of C and S America; few bites occur, but mortality rate is high.

Common or Asian cobra — 4 ft to 8 ft long; throughout southern Asia; considerable tissue damage, sometimes paralysis; mortality probably not more than 10%; antivenom.

Copperhead — less than 4 ft long; from New England to Texas; pain and swelling; very seldom fatal; antivenom seldom needed.

Coral snake — 2 ft to 5 ft long; in Americas south of Canada; bite may be painless; slow onset of paralysis, impaired breathing; mortalities rare, but high without antivenom and mechanical respiration.

Cottonmouth water moccasin — up to 5 ft long; wetlands of southern U.S. from Virginia to Texas. Rapid onset of severe pain, swelling; mortality low, but tissue destruction can be extensive; antivenom.

Death adder — less than 3 ft long; Australia; rapid onset of faintness, cardiac and respiratory distress; at least 50% mortality without antivenom.

Desert horned viper — in dry areas of Africa and western Asia; swelling and tissue damage; low mortality; antivenom.

European viper — 1 ft to 3 ft long; bleeding and tissue damage; mortality low; antivenom.

Gaboon viper — more than 6 ft long; fat; 2-in. fangs; south of the Sahara; massive tissue damage, internal bleeding; few recorded bites.

King cobra — up to 16 ft long; throughout southern Asia; rapid swelling, dizziness, loss of consciousness, difficulty breathing, erratic heartbeat; mortality varies sharply with amount of venom involved, but most bites involve nonfatal amounts; antivenom.

Krait — up to 5 ft long; in SE Asia; rapid onset of sleepiness; numbness; as much as 50% mortality even with use of antivenom.

Puff adder — up to 5 ft long; fat; south of the Sahara and throughout the Middle East; rapid large swelling, great pain, dizziness; moderate mortality, often from internal bleeding; antivenom.

Rattlesnake — 2 ft to 6 ft long; throughout W Hemisphere; rapid onset of severe pain, swelling; mortality low, but amputation of affected digits is sometimes necessary; antivenom. Mojave rattler may produce temporary paralysis.

Ringhals, or spitting, cobra — 5 ft to 7 ft long; southern Africa; squirts venom through holes in front of fangs as a defense; venom is severely irritating, can cause blindness.

Russell's viper or tic-polonga — more than 5 ft long; throughout Asia; internal bleeding; bite reports common; moderate mortality rate; antivenom.

Saw-scaled or carpet viper — as much as 2 ft long; in dry areas from India to Africa; severe bleeding, fever; high mortality, causes more human fatalities than any other snake; antivenom.

Sea snakes — throughout Pacific, Indian oceans except NE Pacific; almost painless bite, variety of muscle pain, paralysis; mortality rate low, many bites not envenomed; some antivenoms.

Sharp-nosed pit viper or one hundred pace snake — up to 5 ft long; in S Vietnam, Taiwan, and China; the most toxic of Asian pit vipers; very rapid onset of swelling and tissue damage, internal bleeding; moderate mortality; antivenom.

Taipan — up to 11 ft long; in Australia and New Guinea; rapid paralysis with severe breathing difficulty; mortality nears 100% without antivenom.

Tiger snake — 2 ft to 6 ft long; S Australia; pain, numbness, mental disturbances with rapid onset of paralysis; may be the deadliest of all land snakes, although antivenom is quite effective.

Yellow or Cape cobra — 7 ft long; in S Africa; most toxic venom of any cobra; rapid onset of swelling, breathing and cardiac difficulties; mortality is high without treatment; antivenom.

Note: Not all bites by venomous snakes are actually envenomed. Any animal bite, however, carries the danger of tetanus, and anyone suffering a venomous snake bite should seek medical attention. Antivenoms do not cure; they are only an aid in the treatment of bites. Mortality rates above are for envenomed bites; low mortality, c. 2% or less; moderate, 2%-5%; high, 5%-15%.

Lizards

Gila monster — as much as 24 in. long, with heavy body and tail; in high desert in SW U.S. and N Mexico; immediate severe pain and transient low blood pressure; no recent mortality.

Mexican beaded lizard — similar to Gila monster, Mexican west coast; reaction and mortality rate similar to Gila monster.

Insects

Ants, bees, wasps, hornets, etc. Global distribution. Usual reaction is piercing pain in area of sting. Not directly fatal, except in cases of massive multiple stings. However, many people suffer allergic reactions — swelling and rashes — and a few may die within minutes from severe sensitivity to the venom (anaphylactic shock).

Spiders, Scorpions

Atrax spider — also known as funnel web spider; several varieties, often large; in Australia; slow onset of breathing, circulation difficulties; low mortality; antivenom.

Black widow — small, round-bodied with red hourglass marking; the widow and its relatives are found in tropical and temperate zones; severe musculoskeletal pain, weakness, breathing difficulty, convulsions; may be more serious in small children; low mortality; antivenom. The **redback** spider of Australia has the hourglass marking on its back, rather than on its front, but is otherwise identical to the black widow.

Brown recluse, or fiddleback, spider — small, oblong body; throughout U.S.; pain with later ulceration at place of bite; in severe cases fever, nausea, and stomach cramps; ulceration may last months; very low mortality.

Scorpion — crablike body with stinger in tail, various sizes, many varieties throughout tropical and subtropical areas; various symptoms may include severe pain spreading from the wound, numbness, severe agitation, cramps; severe reaction may include respiratory failure; low mortality, usually in children; antivenoms.

Tarantula — large, hairy spider found around the world; the American tarantula, and probably all other tarantulas, are harmless to humans, though their bite may cause some pain and swelling.

Sea Life

Cone-shell — mollusk in small, beautiful shell; in the S Pacific and Indian oceans; shoots barbs into victims; paralysis; low mortality.

Octopus — global distribution, usually in warm waters; all varieties produce venom, but only a few can cause death; rapid onset of paralysis with breathing difficulty.

Portuguese man-of-war — jellyfishlike, with tentacles up to 70 ft long; in most warm water areas; immediate severe pain; not directly fatal, though shock may cause death in rare cases.

Sea wasp — jellyfish, with tentacles up to 30 ft long, in the S Pacific; very rapid onset of circulatory problems; high mortality because of speed of toxic reaction; antivenom.

Stingray — several varieties of differing sizes; found in tropical and temperate seas and some fresh water; severe pain, rapid onset of nausea, vomiting, breathing difficulties; wound area may ulcerate, gangrene may appear; seldom fatal.

Stonefish — brownish fish that lies motionless as a rock on bottom in shallow water; throughout S Pacific and Indian oceans; extraordinary pain, rapid paralysis; low mortality; antivenom available, amount determined by number of puncture wounds; warm water relieves pain.

Major U.S. Public Zoological Parks

Source: *World Almanac* questionnaire, 2001; budget and attendance in millions

Zoo	Budget	Attendance	Acres	Species	Some major attractions
Albuquerque (NM) Biological Park	$9.0	1.0	64	200	Polar Bears, Tropical America, Mexican Wolves, Koalas *for further information: (505) 764-6200.*
Arizona-Sonora Desert Museum (Tucson, AZ)	6.0	0.6	100	300+	Desert Loop Trail, Hummingbird Aviary, Pollination Gardens *for further information: (520) 883-2702.*
Audubon Zoological Garden (New Orleans)	28	0.8	58	360	Louisiana Swamp, Jaguar Jungle, white tigers *for further information: (800) 774-7394.*
Baltimore Zoo	10.6	0.6	161	305	Children's zoo, white rhinos, warthogs *for further information: (410) 366-7102.*
Bronx Zoo/Wildlife Conservation Park (N.Y.C.)	34.9	2.4	235	530	Congo Gorilla Forest, Wild Asia, Jungle World *for further information: (718) 367-1010.*
Brookfield Zoo (Chicago area)	47.1	2.0	216	465	Living Coast, The Swamp, Habitat Africa, Tropic World *for further information: (708) 485-0263.*
Buffalo (NY) Zoological Gardens	5.2	0.3	23.5	184	Indian Rhino Pavilion, African Predators, Gorilla Rainforest *for further information: (716) 837-3900.*
Cincinnati Zoo and Botanical Garden	18.0	1.3	85	700+	Vanishing Giants, Gorilla World, Manatee Springs *for further information: (800) 94-HIPPO.*
Cleveland Metroparks Zoo	12.0	1.3	168	643	Rainforest, African Wilderness, Australian Adventure *for further information: (216) 661-6500.*
Columbus Zoo and Aquarium (Powell, OH)	25.0	1.1	588	600	Manatee Coast, African Forest, Discovery Reef *for further information: (800) MONKEYS.*
Dallas Zoo	11.4	0.8	95	384	Chimpanzee Forest, Endangered Tiger Habitat *for further information: (214) 670-5656.*
Denver Zoo	15.3	1.7	80	715	Komodo Dragon habitat, okapi, black rhino, primates *for further information: (303) 376-4800.*
Detroit Zoological Park (Royal Oak, MI[1]	12.5	1.1	125	340	Penguinarium, Arctic Ring of Life, Great Apes of Harambee *for further information: (248) 398-0900.*
The Houston Zoo[1]	NA	1.5	55	800	Indochinese Tigers, Primates, Koala Crossing, Sun Bears *for further information: (713) 523-5888.*
Lincoln Park Zoological Gardens (Chicago)[1]	17.0	3.0	35	290	Great Ape House, Farm-in-the-Zoo, Kovler Sea Lion Pool *for further information: (312) 742-2000.*
Los Angeles Zoo	17.2	1.4	80	350	Chimpanzees of Mahale Mountains, Red Ape Rain Forest *for further information: (323) 644-6400.*
Louisville (KY) Zoo	9.2	0.7	135	134	African Petting Zoo, Islands Exhibit, Gorilla Forest *for further information: (502) 459-2181.*
Memphis (TN) Zoo	7.1	0.7	70+	500+	Cat Country, Once Upon a Farm, Primate Canyon *for further information: (901) 276-WILD.*
Miami Metrozoo	8.4	0.4	300	303	Komodo dragons, meerkats, Dr. Wilde's Rainforest Museum *for further information: (305) 251-0400.*
Milwaukee County Zoological Gardens	11.5	1.3	194	347	Birds of Prey show, Sea Lion show, ZooTrain, Zoomobile *for further information: (414) 771-3040.*
Minnesota Zoo (Apple Valley)	17.8	1.2	500	375	Coral Reef, Dolphin shows, Wells Fargo Family Farm *for further information: (800) 366-7811.*
Oklahoma City Zoological Park & Botanical Garden	10.5	0.7	110	600	Aquaticus, Cat Forest, Lion Overlook, Great EscApe *for further information: (405) 424-3344.*
Omaha's Henry Doorly Zoo	16.0	1.2	130	794	Indoor rain forest, cat complex, aquarium, indoor desert *for further information: (402) 733-8401.*
Oregon Zoo (Portland)	15.0	1.2	64	200	Penguinarium, Africa Rain Forest, Alaska Tundra, elephants *for further information: (503) 226-1561.*
Philadelphia Zoo	19.1	1.2	42	330	Primate Reserve, Amphibian and Reptile House *for further information: (215) 243-1100.*
Phoenix (AZ) Zoo	16.0	1.2	250	200	Arizona Trail, Discovery Trail, Africa Trail, Tropics Trail *for further information: (602) 273-1341.*
Point Defiance Zoo & Aquarium (Tacoma, WA) [1]	7.2	0.5	27	300	Rocky Shores, Tundra, Penguin Point, Southeast Asia *for further information: (253) 591-5337.*
Riverbanks Zoo & Garden (Columbia, SC)	6.0	0.9	170	350	Aquarium Reptile complex, Birdhouse, botanical garden *for further information: (803) 779-8717.*
St. Louis Zoo	34.7	2.9	90	681	Big Cat Country, Jungle of the Apes, Insectarium *for further information: (314) 781-0900.*
San Diego Wild Animal Park	33.0	1.7	1,800	400+	Heart of Africa walking safari, Wgasa Bush Line Railway *for further information: (619) 234-6541.*
San Diego Zoo	56.0	3.5	100	800	panda research station, Polar Bear Plunge, Gorilla Tropics *for further information: (619) 234-3153.*
San Francisco Zoo	14.0	0.9	75	250	Gorilla World, Koala Crossing, Lion House *for further information: (415) 753-7080.*
Smithsonian National Zoo (Washington, DC)	28.6	3.0	163	475	Giant pandas, Sumatran tigers, Great Cats Exchange *for further information: (202) 673-4800.*
Toledo (OH) Zoo	15.0	1.0	62	708	Hippoquarium, African Savanna, Arctic Encounter *for further information: (419) 385-5721.*
Tulsa (OK) Zoo and Living Museum	3.8	0.6	70	450	Tropical American Rain Forest, North American Living Museum *for further information: (918) 669-6600.*
Woodland Park Zoo (Seattle)	19.0	1.0	92	337	Tropical Rain Forest, Elephant Forest, African Savanna *for further information: (206) 684-4800.*
Zoo Atlanta	19.8	1.1	39	221	Gorillas of the Ford African Rain Forest, giant pandas *for further information: (404) 624-5600.*

Note: NA = Not available. (1) 2000 data.

Major Canadian Public Zoological Parks

Source: *World Almanac* questionnaire, 2001; budget in millions of dollars (Canadian), attendance in millions

Zoo	Budget	Attendance	Acres	Species	Some major attractions
Assiniboine Park Zoo (Winnipeg)	$3.5	0.4	90	375	Cold-Hardy Wildlife, Northern Cats, Free-flight Aviary *for further information: (204) 986-2327.*
Calgary Zoo	14.0	1.0	136	272	Botanical Garden, Prehistoric Park, Primate Building *for further information: (403) 232 9300.*
Granby Zoo (Quebec)[1]	8.0	0.4	85	225	Exotic Animal collection, AMAZOO water park *for further information: (877) GRANBYZOO.*
Toronto Zoo	28.0	1.2	710	460	Gorilla Rainforest, African Savanna, Polar Bears *for further information: (416) 392-5900.*

Note: NA = Not available. (1) 2000 data.

Top 50 American Kennel Club Registrations

Source: American Kennel Club, New York, NY; covers (new) dogs registered during calendar year shown

Breed	2000 Rank	2000 Number registered	1999 Rank	1999 Number registered
Labrador Retriever	1	172,841	1	154,897
Golden Retriever	2	66,300	2	62,652
German Shepherd Dog	3	57,660	3	57,256
Dachshund	4	54,773	4	50,772
Beagle	5	52,026	5	49,080
Poodle	6	45,868	6	45,852
Yorkshire Terrier	7	43,574	9	40,684
Chihuahua	8	43,096	7	42,013
Boxer	9	38,803	10	34,998
Shih Tzu	10	37,599	11	34,576
Rottweiler	11	37,355	8	41,776
Pomeranian	12	33,568	12	33,584
Miniature Schnauzer	13	30,472	14	28,649
Cocker Spaniel	14	29,393	13	29,958
Pug	15	24,373	16	21,555
Shetland Sheepdog	16	23,866	15	24,271
Miniature Pinscher	17	22,020	17	21,406
Boston Terrier	18	19,922	19	17,738
Siberian Husky	19	17,551	18	18,106
Maltese	20	17,446	20	16,358
Bulldog	21	15,215	21	13,754
Basset Hound	22	14,427	22	13,595
Doberman Pinscher	23	13,874	23	13,431
German Shorthaired Pointer	24	13,224	24	12,325
Bichons Frise	25	11,750	25	11,245
English Springer Spaniel	26	10,918	26	10,217
Pembroke Welsh Corgi	27	10,301	30	8,850
Great Dane	28	10,210	28	9,860
Pekingese	29	9,749	27	10,082
West Highland White Terrier	30	9,364	29	9,061
Brittany	31	9,230	31	8,646
Weimaraner	32	9,126	34	8,124
Lhasa Apso	33	8,122	33	8,191
Collie	34	8,042	32	8,249
Australian Shepherd	35	6,905	38	5,593
Saint Bernard	36	6,561	37	6,485
Chinese Shar-Pei	37	6,299	35	6,845
Akita	38	5,927	36	6,499
Mastiff	39	5,576	39	5,306
Cairn Terrier	40	4,887	42	4,475
Chesapeake Bay Retriever	41	4,665	41	4,594
Scottish Terrier	42	4,396	43	4,369
Papillon	43	4,128	46	3,547
Chow Chow	44	3,603	44	4,342
Great Pyrenee	45	3,569	45	3,638
Airedale Terrier	46	3,431	49	2,950
Vizsla	47	3,224	48	3,005
Alaskan Malamute	48	3,128	47	3,208
Dalmatian	49	3,084	40	4,652
Bloodhound	50	3,056	51	2,733

Cat Breeds

Source: The Cat Fanciers' Association, Manasquan, NJ

Only a small percentage of house cats in the U.S. are pedigreed or registered with one of the official registering bodies. The largest is the Cat Fanciers' Assn., Inc., with 671 member clubs. The Cat Fanciers' Assn. recognized 37 breeds as of Oct. 1, 2001 (in order of registration totals): Persian, Maine Coon, Siamese, Exotic, Abyssinian, Oriental, Birman, Scottish Fold, American Shorthair, Burmese, Tonkinese, Ocicat, Cornish Rex, Devon Rex, Norwegian Forest Cat, Russian Blue, Ragdoll, Colorpoint Shorthair, British Shorthair, Somali, Manx, Japanese Bobtail, Egyptian Mau, Turkish Angora, Chartreux, American Bobtail, American Curl, La Perm, Singapura, Siberian, Balinese, Selkirk Rex, Sphynx, Javanese, Turkish Van, American Wirehair, Bombay, Korat, Havana Brown, and European Burmese.

Trees of the U.S.

Source: American Forests, Washington, DC

Approximately 826 native and naturalized species of trees are grown in the U.S. The oldest living tree is believed to be a bristlecone pine tree in California named Methuselah, estimated to be 4,700 years old. The world's largest known living tree, the General Sherman giant sequoia in California, weighs more than 6,167 tons—as much as 41 blue whales or 740 elephants.

American Forests recognizes and lists the "National Champion" (largest known) of each U.S. tree species. Anyone can nominate candidates for the 2002-2003 National Register of Big Trees; for information, write to American Forests, PO Box 2000, Washington, DC 20013, or check their website: http://www.americanforests.org

Listed here, in alphabetical order, are ten largest National Champion trees selected by American Forests.

10 Largest National Champion Trees

Tree Type	Girth at 4.5 ft. (in.)	Height (ft.)	Crown Spread (ft.)	Total Points	Location
Giant sequoia	998	275	107	1,300	Sequoia National Park, CA
Coast redwood	950	321	80	1,291	Jedidiah Smith State Park, CA
Western redcedar	761	159	45	931	Olympic National Park, WA
Sitka spruce	707	191	96	922	Olympic National Park, WA
Coast Douglas-fir	505	281	71	804	Olympic National Forest, WA
Common baldcypress	644	83	85	748	Cat Island, LA
California-laurel	546	108	118	684	Grass Valley, CA
Sugar pine	442	232	29	681	Dorrington, CA
Port-Orford-cedar	451	219	39	680	Siskiyou National Forest, OR
Monterey cypress	537	102	116	668	Pescadero Co., CA

METEOROLOGY

National Weather Service Watches and Warnings

Source: National Weather Service, NOAA, U.S. Dept. of Commerce; *Glossary of Meteorology,* American Meteorological Society

National Weather Service forecasters issue a *Severe Thunderstorm* or *Tornado Watch* for a specific area when a severe convective storm that usually covers a relatively small geographic area or moves in a narrow path is sufficiently intense to threaten life and/or property. Examples include thunderstorms with large hail, damaging winds, and/or tornadoes. Excessive localized convective rains are not classified as severe storms but are often the product of severe local storms. Such rainfall may result in phenomena that threaten life and property, such as flash floods. Although cloud-to-ground lightning is not a criterion for severe local storms, it is acknowledged to be a leading cause of storm deaths and injuries.

A *Watch* alerts people that threatening weather is likely. Under a Watch, they should remain alert for approaching storms, activate a plan for action, and monitor ongoing events closely. A *Warning* means that severe weather is occurring or has been indicated by radar; immediate action should be taken by people in the storm's path.

Severe Thunderstorm—a thunderstorm that produces a tornado, winds of at least 50 knots (58 mph), and/or hail at least 3/4 inch in diameter. A thunderstorm with winds of at least 35 knots (40 mph) and/or hail at least ½ inch in diameter is defined as approaching severe. A *Severe Thunderstorm Watch* is issued for a specific area where such storms are most likely to develop. A *Severe Thunderstorm Warning* indicates that a severe thunderstorm has been sighted or indicated by radar.

Tornado—a violent rotating column of air (winds over 200 mph), usually pendant to a cumulonimbus cloud, with circulation reaching the ground. A tornado nearly always starts as a funnel cloud and may be accompanied by a loud roaring noise. On a local scale, it is the most destructive of all atmospheric phenomena. Tornado paths have varied in length from a few feet to more than 100 miles (avg. 5 mi); in diameter from a few feet to more than a mile (avg. 220 yd); average forward speed, 30 mph.

Cyclone—an atmospheric circulation of winds rotating counterclockwise in the northern hemisphere and clockwise in the southern hemisphere. Tornadoes, hurricanes, and the lows shown on weather maps are all examples of cyclones of various size and intensity. Cyclones are usually accompanied by precipitation or stormy weather.

Subtropical Storm—an atmospheric circulation of one-minute sustained surface winds, 34 knots (39 mph) or more. Depending on its characteristics and intensity, it can develop into a tropical storm or a hurricane.

Tropical Storm—an atmospheric circulation of one-minute sustained surface winds within a range of 34 to 63 knots (39 to 73 mph). A *Tropical Storm Watch* is an an-nouncement that a tropical storm or tropical storm conditions may pose a threat to coastal areas generally within 36 hours. A *Tropical Storm Warning* is an announcement that tropical storm conditions pose a threat along a specified segment of coastline within 24 hours.

Hurricane—a severe cyclone originating over tropical ocean waters and having one-minute sustained surface winds 64 knots (73 mph) or higher. (West of the international date line, in the western Pacific, such storms are known as *typhoons.*) The area of hurricane-force winds forms a circle or an oval, sometimes as wide as 300 mi in diameter. In the lower latitudes, hurricanes usually move west or northwest at 10 to 15 mph. When the center approaches 25° to 30° North Latitude, the direction of motion often changes to northeast, with increased forward speed.

Blizzard—a severe weather condition characterized by strong winds bearing a great amount of snow. The National Weather Service specifies winds of 35 mph or higher and sufficient falling and/or blowing snow to frequently reduce visibility to less than ¼ mi. for at least 3 hours.

Flood—Flooding takes many forms. *River Flooding:* This natural process occurs when rains, sometimes coupled with melting snow, fill river basins with too much water too quickly; torrential rains from decaying hurricanes or tropical systems can also be a major cause of river flooding. *Coastal Flooding:* Winds from tropical storms and hurricanes or intense offshore low pressure systems can drive ocean water inland and cause significant flooding. Coastal floods can also be produced by sea waves called *tsunamis,* sometimes referred to as tidal waves; these waves are produced by earthquakes or volcanic activity. *Flash Flooding:* Usually due to copious amounts of rain falling in a short time, flash flooding typically occurs within 6 hours of the rain event. Flash floods account for the majority of flood deaths in the U.S. *Urban Flooding:* Urbanization significantly increases runoff over what would occur on natural terrain, making flash flooding in these areas extremely dangerous. Streets can become swift-moving rivers, and basements can become death traps as they fill with water. *Ice Jam Flooding:* Ice can accumulate at natural or artificial obstructions and stop the flow of water. As the water flow is stopped, water builds up and flooding can occur upstream. If the jam suddenly gives way, the gush of ice and water can cause serious downstream flash flooding.

Flash Flood or Flood Watch: Flash flooding or flooding is possible within a designated area.

Flash Flood or Flood Warning: Flash flooding or flooding has been reported or is imminent; all necessary precautions should be taken immediately.

Urban and Small Stream Advisory: Small streams, streets, and low-lying areas such as railroad underpasses and urban storm drains are flooding.

National Weather Service Marine Warnings and Advisories

Small Craft Advisory alerts mariners to sustained (exceeding 2 hours) weather and/or sea conditions, either present or forecast, potentially hazardous to small boats. Although "small craft" is not defined, hazardous conditions generally include winds of 18 to 33 knots and/or dangerous wave conditions. It is the responsibility of the mariner, based on experience and on the location and size or type of boat, to determine whether conditions are hazardous to the boat. Upon receiving word of a Small Craft Advisory, the mariner should immediately obtain the latest marine forecast to determine the reason for the advisory.

Gale Warning indicates that winds within the range 34 to 47 knots, not directly associated with a tropical storm, are forecast for the area.

Tropical Storm Warning indicates that winds within the range of 34 to 63 knots are forecast in a specified coastal area to occur within 24 hours or less. Issued only for winds of tropical weather systems.

Storm Warning indicates that winds 48 knots or above, not directly associated with a tropical storm, are forecast for the area.

Hurricane Warning indicates that winds 64 knots or greater are forecast for the area within 24 hours. Issued only for winds produced by tropical weather systems.

Special Marine Warning indicates potentially hazardous weather conditions, usually of short duration (2 hours or less) and producing wind speeds of 34 knots or more, not adequately covered by existing marine warnings.

Primary sources of dissemination are commercial radio, TV, U.S. Coast Guard radio stations, and NOAA VHF-FM broadcasts. These NOAA broadcasts on 162.40 to 162.55 MHz can usually be received 20-40 mi from the transmitting antenna site, depending on terrain and quality of the receiver used. Where transmitting antennas are on high ground, the range may be somewhat greater, reaching 60 mi or more.

Monthly Normal Temperatures, Precipitation

Source: National Climatic Data Center, NESDIS, NOAA, U.S. Dept. of Commerce

The temperatures given here are based on records for the 30-year period 1961-90. For stations that did not have continuous records from the same site for the entire 30 years, the means have been adjusted to the record at the present site.

Figures are for airport stations unless otherwise indicated. * = city station. T = temperature in Fahrenheit; P = precipitation in inches; L = less than 0.05 inch.

	Jan.		Feb.		Mar.		Apr.		May		June		July		Aug.		Sept.		Oct.		Nov.		Dec.	
Station	T	P	T	P	T	P	T	P	T	P	T	P	T	P	T	P	T	P	T	P	T	P	T	P
Albany, NY	21	2.4	24	2.3	34	2.9	46	3.0	58	3.4	67	3.6	72	3.2	70	3.5	61	3.0	50	2.8	40	3.2	27	2.9
Albuquerque, NM	34	0.4	40	0.5	47	0.5	55	0.5	64	0.5	74	0.6	79	1.4	76	1.6	69	1.0	57	0.9	44	0.4	35	0.5
Anchorage, AK	15	0.8	19	0.8	26	0.7	36	0.7	47	0.7	54	1.1	58	1.7	56	2.4	48	2.7	35	2.0	21	1.1	16	1.1
Asheville, NC	36	3.3	39	3.9	47	4.6	55	3.4	63	4.4	69	4.2	73	4.5	72	4.7	66	3.9	56	3.6	48	3.6	40	3.5
Atlanta, GA	41	4.8	45	4.8	54	5.8	62	4.3	69	4.3	76	3.6	79	5.0	78	3.7	73	3.4	62	3.1	53	3.9	45	4.3
Atlantic City, NJ	31	3.5	33	3.1	42	3.6	50	3.6	60	3.3	69	2.6	75	3.8	73	4.1	66	2.9	55	2.8	46	3.6	36	3.3
Baltimore, MD	32	3.1	35	3.1	44	3.4	53	3.1	63	3.7	73	3.7	77	3.7	76	3.9	69	3.4	57	3.0	47	3.3	37	3.4
Barrow, AK	−13	0.2	−18	0.2	−15	0.2	−2	0.2	19	0.2	34	0.3	39	0.9	38	1.0	31	0.6	14	0.5	−2	0.3	−11	0.2
Birmingham, AL	42	5.1	46	4.7	54	6.2	62	5.0	69	4.9	76	3.7	80	5.3	79	3.6	73	3.9	63	2.8	53	4.3	45	5.1
Bismarck, ND	9	0.5	16	0.4	28	0.8	43	1.7	55	2.2	64	2.7	71	2.1	68	1.7	57	1.5	46	0.9	29	0.5	14	0.5
Boise, ID	29	1.5	36	1.2	43	1.3	49	1.2	58	1.1	67	1.8	74	0.4	73	0.4	63	0.8	52	0.8	40	1.5	30	1.4
Boston, MA	29	3.6	30	3.6	39	3.7	48	3.6	58	3.3	68	3.1	74	2.8	72	3.2	65	3.1	55	3.3	45	4.2	34	4.0
Buffalo, NY	24	2.7	25	2.3	34	2.7	45	2.9	57	3.1	66	3.6	71	3.1	69	4.2	62	3.5	51	3.1	41	3.8	29	3.7
Burlington, VT	16	1.8	18	1.6	31	2.2	44	2.8	56	3.1	65	3.5	71	3.7	68	4.1	59	3.3	48	2.9	37	3.1	23	2.4
Caribou, ME	9	2.4	12	1.9	25	2.4	38	2.5	51	3.1	61	2.9	66	4.0	63	4.1	54	3.5	43	3.1	31	3.6	15	3.2
Charleston, SC	48	3.5	51	3.3	58	4.3	65	2.7	73	4.0	78	6.4	82	6.8	81	7.2	76	4.7	67	2.9	58	2.5	51	3.2
Chicago, IL	21	1.5	25	1.4	37	2.7	49	3.6	59	3.3	69	3.8	73	3.7	72	4.2	64	3.8	53	2.4	40	2.9	27	2.5
Cleveland, OH	25	2.0	27	2.2	37	2.9	48	3.1	58	3.5	68	3.7	72	3.5	70	3.4	64	3.4	53	2.5	43	3.2	31	3.1
Columbus, OH	26	2.2	30	2.2	41	3.3	51	3.2	61	3.9	69	4.0	73	4.3	72	3.7	66	3.0	54	2.2	43	3.2	32	2.9
Dallas-Ft. Worth, TX	43	1.8	48	2.2	57	2.8	66	3.5	73	4.9	81	3.0	85	2.3	85	2.2	77	3.4	67	3.5	56	2.3	47	1.8
Denver, CO	30	0.5	33	0.6	39	1.3	48	1.7	57	2.4	67	1.8	74	1.9	71	1.5	62	1.2	51	1.0	39	0.9	31	0.6
Des Moines, IA	19	1.0	25	1.1	37	2.3	51	3.4	62	3.7	72	4.5	77	3.8	74	4.2	65	3.5	54	2.6	39	1.8	24	1.3
Detroit, MI	23	1.8	25	1.7	36	2.6	47	3.0	58	2.9	68	3.6	72	3.2	71	3.4	63	2.9	51	2.1	40	2.7	28	2.8
Dodge City, KS	30	0.5	35	0.6	43	1.6	55	2.0	64	3.0	74	3.1	80	3.2	78	2.7	69	1.9	57	1.3	43	0.8	32	0.6
Duluth, MN	7	1.2	12	0.8	24	1.9	39	2.3	51	3.0	60	3.8	66	3.6	64	4.0	54	3.8	44	2.5	28	1.8	13	1.2
Fairbanks, AK	−10	0.5	−4	0.4	11	0.4	31	0.3	49	0.6	60	1.4	63	1.9	57	2.0	46	1.0	25	0.9	3	0.8	−7	0.9
Fresno, CA	46	2.0	51	1.8	55	1.9	61	1.0	69	0.3	77	0.1	82	L	80	L	75	0.2	65	0.5	54	1.4	45	1.4
Galveston, TX*	53	3.3	55	2.3	62	2.2	69	2.4	76	3.6	81	4.4	83	4.0	84	4.5	80	5.9	73	2.8	64	3.4	56	3.5
Grand Junction, CO	25	0.6	34	0.5	43	0.9	52	0.7	62	0.9	72	0.5	79	0.6	76	0.8	67	0.8	55	1.0	40	0.7	29	0.6
Grand Rapids, MI	22	1.8	24	1.4	34	2.6	46	3.4	58	3.1	67	3.7	72	3.2	70	3.6	61	4.2	50	2.8	38	3.3	27	2.9
Hartford, CT	25	3.4	28	3.2	38	3.6	49	3.9	60	4.1	69	3.8	74	3.2	72	3.7	63	3.8	52	3.6	42	4.0	30	3.9
Helena, MT	20	0.6	26	0.4	34	0.7	43	1.0	53	1.8	62	1.9	69	1.1	67	1.3	55	1.2	45	0.6	32	0.5	21	0.6
Honolulu, HI	73	3.6	73	2.2	74	2.2	76	1.5	78	1.1	79	0.5	81	0.6	81	0.4	81	0.8	80	2.3	77	3.0	74	3.8
Houston, TX	50	3.2	54	3.3	61	2.7	68	4.2	75	4.7	80	4.0	83	3.3	82	3.7	78	4.9	70	3.7	61	3.4	54	3.7
Huron, SD	13	0.4	19	0.8	32	1.2	46	2.0	58	2.7	68	3.3	74	2.3	72	2.0	61	1.4	49	1.4	32	0.7	18	0.5
Indianapolis, IN	26	2.3	30	2.5	41	3.8	52	3.7	63	4.0	72	3.5	75	4.5	73	3.6	67	2.9	55	2.6	43	3.2	31	3.3
Jackson, MS	44	5.2	48	4.7	57	5.8	65	5.6	72	5.1	79	3.2	82	4.5	81	3.8	76	3.6	65	3.3	56	4.8	48	5.9
Jacksonville, FL	52	3.3	55	3.9	61	3.7	67	2.8	73	3.6	79	5.7	82	5.6	81	7.9	78	7.0	70	2.9	62	2.1	55	2.7
Juneau, AK	24	4.5	28	3.7	33	3.3	40	2.8	47	3.4	53	3.1	56	4.2	55	5.3	49	6.7	42	7.8	32	4.9	27	4.4
Kansas City, MO	26	1.1	31	1.1	43	2.5	55	3.1	64	5.0	73	4.7	79	4.4	76	4.0	68	4.9	57	3.3	43	1.9	30	1.6
Knoxville, TN	36	4.2	40	4.1	49	5.1	58	3.7	65	4.1	73	4.0	77	4.7	76	3.1	70	3.1	58	2.8	49	3.8	40	4.5
Lander, WY	20	0.5	25	0.6	34	1.2	43	2.1	53	2.3	63	1.5	71	0.8	69	0.5	58	1.1	47	1.1	31	0.8	21	0.6
Lexington, KY	31	2.9	35	3.2	45	4.4	55	3.9	64	4.5	72	3.7	76	5.0	75	3.9	68	3.2	57	2.6	46	3.4	36	4.0
Little Rock, AR	39	3.9	44	4.4	53	5.3	62	6.2	70	7.0	78	7.8	82	8.2	81	8.1	74	7.4	63	6.3	52	5.2	43	4.3
Los Angeles, CA*	58	2.9	60	3.1	61	2.6	63	1.0	66	0.2	70	L	74	L	75	0.1	74	0.5	70	0.3	63	2.0	58	2.0
Louisville, KY	32	2.9	36	3.3	46	4.7	56	4.2	65	4.6	73	3.5	77	4.5	76	3.5	70	3.2	58	2.7	47	3.7	37	3.6
Marquette, MI*	12	2.2	14	1.7	24	2.8	37	2.6	50	3.0	59	3.5	65	2.9	63	3.4	54	4.1	44	3.6	30	2.9	17	2.6
Memphis, TN	40	3.7	44	4.4	53	5.4	63	5.5	71	5.0	79	3.6	83	3.8	81	3.4	74	3.5	63	3.0	53	5.1	44	5.7
Miami, FL	67	2.0	69	2.1	72	2.4	75	2.9	79	6.2	81	9.3	83	5.7	83	7.6	82	7.6	78	5.6	74	2.7	69	1.8
Milwaukee, WI	19	1.6	23	1.5	33	2.7	44	3.5	55	2.8	65	3.2	71	3.5	69	3.5	62	3.4	50	2.4	38	2.5	24	2.3
Minneapolis, MN	12	1.0	18	0.9	31	1.9	46	2.4	59	3.4	68	4.1	74	3.5	71	3.6	61	2.7	49	2.2	33	1.6	18	1.1
Mobile, AL	50	4.8	53	5.5	61	6.4	68	4.5	75	5.7	80	5.0	82	6.9	82	7.0	78	5.9	68	2.9	60	4.1	53	5.3
Moline, IL	20	1.5	25	1.2	37	3.0	50	3.9	61	4.3	71	4.3	75	5.0	73	4.2	65	4.0	53	2.9	40	2.5	25	2.2
Nashville, TN	36	3.6	40	3.8	50	4.9	59	4.4	68	4.9	76	3.6	79	4.0	78	3.5	72	3.5	60	2.6	50	4.1	41	4.6
Newark, NJ	31	3.4	33	3.0	42	3.9	52	3.8	63	4.1	73	3.2	78	4.5	76	3.9	69	3.7	58	3.1	47	3.9	36	3.5
New Orleans, LA	51	5.1	54	6.0	62	4.9	69	4.5	75	4.6	80	5.8	82	6.1	82	6.2	78	5.5	69	3.1	61	4.4	55	5.8
New York, NY*	32	3.4	34	3.3	42	4.1	53	4.2	63	4.4	72	3.7	77	4.4	76	4.0	68	3.9	58	3.6	48	4.5	37	3.9
Norfolk, VA	39	3.8	41	3.5	49	3.7	57	3.1	66	3.8	74	3.8	78	5.1	77	4.8	72	3.9	61	3.2	53	2.9	44	3.2
Oklahoma City, OK	36	1.1	41	1.6	50	2.7	60	2.8	68	5.2	77	4.3	82	2.6	81	2.6	73	3.8	62	3.2	50	2.0	39	1.4
Omaha, NE	21	0.7	27	0.8	39	2.0	52	2.7	62	4.5	72	3.9	77	3.5	74	3.2	65	3.7	53	2.3	39	1.5	25	1.0
Philadelphia, PA	30	3.2	33	2.8	42	3.5	52	3.6	63	3.8	72	3.7	77	4.3	76	3.8	68	3.4	56	2.6	46	3.3	36	3.4
Phoenix, AZ	54	0.7	58	0.7	62	0.9	70	0.2	79	0.1	88	0.1	94	0.8	92	1.0	86	0.9	75	0.7	62	0.7	54	1.0
Pittsburgh, PA	26	2.5	29	2.4	39	3.4	50	3.2	60	3.6	68	3.7	72	3.8	71	3.2	64	3.0	52	2.4	42	2.9	32	2.9
Portland, ME	21	3.5	23	3.3	33	3.7	43	4.1	53	3.6	62	3.4	69	3.1	67	2.9	59	3.1	49	3.9	39	5.2	27	4.6
Portland, OR	40	5.4	44	3.9	47	3.6	51	2.4	57	2.1	64	1.5	68	0.6	69	1.1	63	1.8	55	2.7	46	5.3	40	6.1
Providence, RI	28	4.1	30	3.7	37	4.3	47	4.0	57	3.5	67	2.8	73	3.0	71	4.0	64	3.5	54	3.8	44	4.2	33	4.5
Raleigh, NC	39	3.6	42	3.4	50	3.7	59	2.9	67	3.7	74	3.7	78	4.4	77	4.4	71	3.3	60	2.7	51	2.9	43	3.1
Rapid City, SD	22	0.4	27	0.5	34	1.0	45	1.9	55	2.7	65	3.1	72	2.0	71	1.7	60	1.2	49	1.1	35	0.6	24	0.5
Reno, NV	33	1.1	38	1.0	43	0.7	49	0.4	57	0.7	65	0.5	72	0.3	70	0.3	60	0.4	51	0.4	40	0.9	33	1.0
Richmond, VA	37	3.2	39	3.2	48	3.6	57	3.0	66	3.8	74	3.6	78	5.0	77	4.4	70	3.3	59	3.5	50	3.2	40	3.3
St. Louis, MO	29	1.8	34	2.1	45	3.6	57	3.5	66	4.0	75	3.7	80	3.9	78	2.9	70	3.1	58	2.7	46	3.3	34	3.0
Salt Lake City, UT	28	1.1	34	1.2	42	1.9	50	2.1	59	1.8	69	0.9	78	0.8	76	0.9	65	1.3	53	1.4	41	1.3	30	1.4
San Antonio, TX	49	1.7	54	1.8	62	1.5	69	2.5	76	4.2	82	3.8	85	2.2	85	2.5	79	3.4	70	3.2	60	2.6	52	1.5
San Diego, CA	57	1.8	59	1.5	60	1.8	62	0.8	64	0.2	67	0.1	71	L	73	0.1	71	0.2	68	0.4	62	1.5	57	1.6
San Francisco, CA	49	4.4	52	3.2	53	3.1	56	1.4	58	0.2	62	0.1	63	L	64	0.1	65	0.2	61	1.2	55	2.9	49	3.1
San Juan, PR	77	2.8	77	2.1	78	2.3	79	3.8	81	5.9	82	4.0	83	4.4	83	5.3	82	5.3	82	5.7	80	5.9	78	4.7
Sault Ste. Marie, MI*	13	2.4	14	1.7	24	2.3	38	2.4	51	2.7	58	3.1	64	2.7	63	3.6	55	3.7	45	3.2	33	3.5	19	2.9
Savannah, GA	49	3.6	52	3.2	59	3.8	66	3.0	74	4.1	79	5.7	82	6.4	81	7.4	77	4.5	67	2.4	59	2.2	52	3.0
Scottsbluff, NE	25	0.5	30	0.5	36	1.1	47	1.6	56	2.8	67	2.6	74	2.1	72	1.1	61	1.1	50	0.8	36	0.6	26	0.6
Seattle, WA	41	5.4	44	4.0	47	3.8	50	2.5	56	1.8	61	1.6	65	0.9	66	1.2	61	1.9	54	3.3	46	5.7	42	6.0
Spokane, WA	27	2.0	33	1.5	39	1.5	46	1.2	54	1.4	62	1.3	69	0.7	68	0.7	59	0.7	47	1.0	35	2.2	28	2.4
Springfield, MO	31	1.8	36	2.2	46	3.9	56	4.2	65	4.4	73	5.1	78	2.9	77	3.5	69	4.6	58	3.6	46	3.8	35	3.2
Syracuse, NY	22	2.3	24	2.2	34	2.8	46	3.3	57	3.3	65	3.8	70	3.8	68	3.5	62	3.8	51	3.2	41	3.7	28	3.2
Tampa, FL	60	2.0	62	3.1	67	3.0	71	1.2	77	3.1	81	5.5	82	6.6	82	7.6	81	6.0	75	2.0	68	1.8	62	2.2
Washington, DC	31	2.7	34	2.8	43	3.2	53	3.1	62	4.0	71	3.9	76	3.5	74	3.9	67	3.4	55	3.2	45	3.3	35	3.2
Wilmington, DE	31	3.0	33	2.9	43	3.4	52	3.4	63	3.8	72	3.6	76	4.2	75	3.4	68	3.4	56	2.9	46	3.3	36	3.5

Normal High and Low Temperatures, Precipitation

Source: National Climatic Data Center, NESDIS, NOAA, U.S. Dept. of Commerce

The normal temperatures given here are based on records for the 30-year period 1961-90. The extreme temperatures (through 1990) are listed for the stations shown and may not agree with the state records shown on page 176.

Figures are for airport stations unless otherwise indicated. * = city station. Temperatures are Fahrenheit.

State	Station	NORMAL TEMPERATURE January		July		EXTREME TEMPERATURE		AVERAGE ANNUAL PRECIPITATION (inches)
		Max.	Min.	Max.	Min.	Highest	Lowest	
Alabama	Mobile	60	40	91	73	104	3	63.96
Alaska	Anchorage	21	8	65	52	85	–34	15.91
Alaska	Barrow	–7	–19	45	34	79	–56	4.49
Arizona	Phoenix	66	41	106	81	122	17	7.66
Arkansas	Little Rock	49	29	92	72	112	–5	72.10
California	Los Angeles*	68	49	84	65	112	28	14.77
California	San Diego	66	49	76	66	111	29	9.9
California	San Francisco	56	42	72	54	106	20	19.70
Colorado	Denver	43	16	88	59	104	–30	15.40
Connecticut	Hartford	33	16	85	62	102	–26	44.14
Delaware	Wilmington	39	22	86	67	102	–14	40.84
District of Columbia	Washington–National	42	27	89	71	104	–5	38.63
Florida	Jacksonville	64	41	91	72	105	7	51.32
Florida	Miami	75	59	89	76	98	30	55.91
Georgia	Atlanta	50	32	88	70	105	–8	50.77
Georgia	Savannah	60	38	91	72	105	3	49.22
Hawaii	Honolulu	80	66	88	74	94	53	22.02
Idaho	Boise	36	22	90	58	111	–25	12.11
Illinois	Chicago	29	13	84	63	104	–27	35.82
Illinois	Moline	28	11	86	65	106	–27	39.08
Indiana	Indianapolis	34	17	86	65	104	–23	39.94
Iowa	Des Moines	28	11	87	67	108	–24	33.12
Kentucky	Lexington	39	22	86	66	103	–21	44.55
Kentucky	Louisville	40	23	87	67	105	–20	44.39
Louisiana	New Orleans	61	42	91	73	102	11	61.88
Maine	Caribou	19	–2	77	55	96	–41	36.60
Maine	Portland	30	11	79	58	103	–39	44.34
Maryland	Baltimore	40	23	87	67	105	–7	40.76
Massachusetts	Boston	36	22	82	65	102	–12	41.51
Michigan	Detroit	30	16	83	61	104	–21	32.62
Michigan	Sault Ste. Marie*	21	5	76	51	98	–36	34.23
Minnesota	Duluth	16	–2	77	55	97	–39	30.00
Minnesota	Minneapolis-St. Paul	21	3	84	63	105	–34	28.32
Mississippi	Jackson	56	33	92	71	106	2	55.37
Missouri	Kansas City	35	17	89	68	109	–23	37.62
Missouri	St. Louis	38	21	89	70	107	–18	37.51
Montana	Helena	30	10	85	53	105	–42	11.60
Nebraska	Omaha	31	11	88	66	114	–23	29.86
Nebraska	Scottsbluff	38	12	90	59	109	–42	15.27
Nevada	Reno	45	21	92	51	105	–16	7.53
New Jersey	Atlantic City	40	21	85	65	106	–11	40.29
New Mexico	Albuquerque	47	22	93	64	105	–17	8.88
New York	Albany	30	11	84	60	100	–28	36.17
New York	Buffalo	30	17	80	62	99	–20	38.58
New York	New York–La Guardia	37	26	84	69	107	–3	42.12
North Carolina	Asheville	47	25	83	62	100	–16	47.59
North Carolina	Raleigh	49	29	88	68	105	–9	41.43
North Dakota	Bismarck	20	–2	84	56	109	–44	15.47
Ohio	Cleveland	32	18	82	61	104	–19	36.63
Ohio	Columbus	34	19	84	63	102	–19	38.09
Oregon	Portland	45	34	80	57	107	–3	36.30
Pennsylvania	Philadelphia	38	23	86	67	104	–7	41.41
Pennsylvania	Pittsburgh	34	19	83	62	103	–18	36.85
Rhode Island	Providence	37	19	82	63	104	–13	45.53
South Carolina	Charleston	58	38	90	73	104	6	51.53
South Dakota	Huron	24	2	87	62	112	–39	20.08
South Dakota	Rapid City	34	11	86	58	110	–30	16.64
Tennessee	Memphis	49	31	92	73	108	–13	52.10
Tennessee	Nashville	46	27	90	69	107	–17	47.30
Texas	Galveston*	58	47	87	79	101	8	42.28
Texas	Houston	61	40	93	72	107	7	46.07
Utah	Salt Lake City	36	19	92	64	107	–30	16.18
Vermont	Burlington	25	8	81	60	101	–30	34.47
Virginia	Norfolk	47	31	86	70	104	–3	44.64
Virginia	Richmond	46	26	88	68	105	–12	43.16
Washington	Seattle-Tacoma	45	35	75	55	99	0	37.19
Washington	Spokane	33	21	83	54	108	–25	16.49
Wisconsin	Milwaukee	26	12	80	62	103	–26	32.93
Wyoming	Lander	31	8	86	56	101	–37	13.01

Mean Annual Snowfall (inches) based on record through 1990: Boston, MA, 42; Sault Ste. Marie, MI, 113; Albany, NY, 65.2; Burlington, VT, 78.6; Lander, WY, 66; Juneau, AK, 105.8.

Wettest Spot: Mount Waialeale, HI, on the island of Kauai, is the rainiest place in the United States and in the world, according to the National Geographic Society; it has an average annual rainfall of 460 inches.

Below are the official temperature extremes through mid-1999. There are many unofficial claims. To qualify as official meteorological data, readings must be taken on approved instruments in a sheltered and ventilated location.

Highest Temperature: A temperature of 136° F observed at El Azizia (Al Aziziyah), near Tripoli, Libya, on Sept. 13, 1922, is generally accepted as the world's highest temperature recorded under standard conditions.

The record high in the United States was 134° F in Death Valley, CA, July 10, 1913.

Lowest Temperature: A record low temperature of –129° F was recorded at the Soviet Antarctica station of Vostok on July 21, 1983.

The record low in the United States was –80° F at Prospect Creek, AK, Jan. 23, 1971.

The lowest official temperature on the North American continent was recorded at –81° F in Feb. 1947, at an airport in the Yukon called Snag.

Annual Climatological Data, 2000

Source: National Climatic Data Center, NESDIS, NOAA, U.S. Dept. of Commerce

	Temperature °F					Precipitation[1]						Fastest[2] wind		No. of Days	
									Sleet or snow						
Station	Elev. (ft.)	Highest	Date	Lowest	Date	Total (in.)	Greatest in 24 hours	Date	Total (in.)	Greatest in 24 hours	Date	MPH	Date	Prec. .01 in. or more	Snow, sleet 1 in. or more
Albany, NY	302	90	6/25+	–9	1/29	46.92	3.37	6/05-06	83.5	13.3	4/09	43	10/12	149	17
Albuquerque, NM	5,310	99	6/15	15	1/04	8.24	0.76	10/23	10.8	6.1	12/26	51	8/08	54	2
Anchorage, AK	144	73	6/04	–10	1/16	14.37	1.02	6/28-29	52.9	5.3	1/31	41	2/02	117	20
Asheville, NC	2,200	93	7/10	9	12/20	35.59	2.29	11/08-09	14.4	4.5	1/22	45	1/20	124	5
Atlanta, GA	972	101	8/18	13	12/20	35.56	1.78	1/22-23	—	—	—	39	7/20	111	0
Atlantic City, NJ	11	93	7/10	4	3/25	47.54	5.59	8/12-13	—	—	—	32	4/18	99	—
Baltimore, MD	194	95	6/11	7	1/30	41.91	2.38	9/25-26	27.2	14.9	1/25	41	12/12	117	5
Barrow, AK	44	67	7/22	–47	2/2	5.82	0.88	7/03-04	47.3	3.2	1/13-14	55	08/10	109	12
Birmingham, AL	644	103	7/19	12	12/20	50.24	4.69	3/10	3.0	3.0	1/28	48	7/26	107	1
Bismarck, ND	1,652	104	8/11	–22	12/24	23.00	2.62	6/12-13	50.0	7.8	11/6-7	55	4/05	117	16
Boise, ID	2,872	102	8/09	14	11/17	12.04	1.06	10/11-12	16.2	2.9	11/14-15	45	12/15	89	6
Boston, MA	178	92	6/17	0	1/17	45.6	4.18	6/06-07	29.4	6.5	2/18	46	12/17	129	7
Buffalo, NY	739	88	9/01	0	1/22+	42.20	2.31	11/20	145.9	24.9	11/20	51	12/12	172	33
Burlington, VT	346	89	5/07	–14	1/23	39.26	2.04	5/08-09	117.9	14.4	4/09-10	36	4/08	175	32
Caribou, ME	628	85	6/16	–22	2/18	38.36	1.94	2/13-14	149.6	18.6	2/13-14	38	12/18	160	31
Charleston, SC	44	100	6/03	20	12/31	45.94	3.54	7/12-13	—	—	—	38	12/17	101	—
Chicago, IL	658	93	9/01	–9	12/22	33.36	2.44	9/11-12	57.8	11.1	2/18	41	8/06	125	14
Cleveland, OH	776	88	6/14+	–3	1/28	40.59	2.88	8/06	80.8	7.5	11/21	46	12/12	162	24
Columbus, OH	813	91	6/11	–5	1/25	42.89	2.67	5/28	38.4	3.5	2/03	47	12/12	140	16
Dallas-Ft. Worth, TX	572	111	9/04	20	12/12	36.25	2.20	11/05-06	—	—	—	51	2/25	84	—
Denver, CO	5,380	101	7/06	–2	12/21+	14.42	2.00	5/17-18	—	—	—	46	2/15	81	—
Des Moines, IA	910	98	9/02	–12	12/25+	26.14	2.38	6/13	—	—	—	53	6/13	104	—
Detroit, MI	641	91	6/10	–3	12/22	42.15	4.08	9/11-12	45.8	5.5	12/11	39	5/09	142	19
Duluth, MN	1,427	88	6/08	–18	12/12	31.34	2.03	11/06-07	81.6	8.0	3/14-15	40	3/25	133	29
Fairbanks, AK	434	83	6/24	–49	1/05+	10.78	0.72	1/17-18	52.5	8.7	1/17-18	26	5/13	94	19
Fresno, CA	343	106	6/15	29	1/06+	15.24	1.89	2/27-28	—	—	—	35	2/20	46	—
Grand Rapids, MI	786	90	6/10	–9	12/28	40.16	3.78	9/22-23	112.8	14.2	12/11	38	6/01	148	32
Hartford, CT	198	94	5/09	–4	1/18+	42.17	3.44	6/06-07	—	—	—	45	6/02	135	—
Helena, MT	3,895	101	7/30	–18	12/16	8.38	1.01	10/11-12	—	—	—	46	3/14	90	—
Honolulu, HI	16	90	8/16+	59	1/24	7.10	1.39	11/02-03	—	—	—	35	4/03	67	—
Houston, TX	119	109	9/04	24	1/30	47.63	7.61	5/19-20	T	T	3/04+	40	2/23	107	0
Huron, SD	1,282	102	8/14	–22	12/24	20.08	1.83	7/09	49.4	11.0	4/07	52	4/05	100	14
Indianapolis, IN	796	91	9/02	–8	1/21	40.46	2.57	10/04-05	37.5	8.1	3/11	46	9/20	122	9
Jackson, MS	294	107	8/30	15	12/31+	42.59	5.01	4/02-03	1.4	1.0	12/31	33	8/09	103	1
Jacksonville, FL	26	103	7/20	21	12/31+	39.77	4.66	9/06-07	0	0	—	43	7/11	109	0
Kansas City, MO	1050	106	9/02	–9	12/22	34.96	2.31	6/19-20	20.8	7.0	12/13	58	7/26	107	9
Knoxville, TN	992	94	8/17+	12	12/20	47.02	2.69	5/23-24	—	—	—	49	11/09	122	—
Lander, WY	5,558	99	8/01	–10	11/13	8.61	1.26	4/18-19	58.6	7.0	11/09	52	9/17	60	20
Lexington, KY	965	95	6/13	1	12/22	42.10	2.65	2/17-18	—	—	—	39	1/03	126	—
Los Angeles, CA	152	93	9/12	41	1/08	11.01	1.88	4/17-18	—	—	—	33	3/31+	32	—
Louisville, KY	481	96	8/17+	4	12/25	45.22	3.99	1/02-03	16.8	2.0	1/19	40	12/11	118	7
Marquette, MI	1,415	90	8/31	–21	2/17	29.88	1.29	11/13	251.0	17.0	11/20	—	—	176	60
Memphis, TN	335	107	8/30	12	12/22+	37.27	3.09	2/26	—	—	—	39	6/26	103	—
Miami, FL	27	95	7/16	42	12/31	61.05	12.66	10/03-04	T	T	5/15	30	9/19	136	0
Milwaukee, WI	678	89	6/10	–8	12/22	44.37	4.42	7/02	87.8	13.6	12/11	45	5/11	143	22
Minn.-St. Paul, MN	872	94	6/08	–17	12/25	30.48	2.64	7/09-10	—	—	—	45	4/05	124	—
Mobile, AL	223	105	8/29	19	12/20	45.74	3.04	3/29-30	T	T	4/24	38	11/09	95	0
Moline, IL	605	97	9/11+	–18	12/25	37.88	2.27	4/19-20	63.8	9.3	2/18	39	3/08	134	16
Nashville, TN	692	100	8/17	11	12/20	42.43	4.39	5/24-25	—	—	—	38	7/28	118	—
Newark, NJ	26	96	6/11	4	1/18	43.35	2.82	7/26-27	33.3	13.9	12/30	48	12/12	126	6
New Orleans, LA	17	101	8/30	28	12/20	38.88	4.01	11/18-19	—	—	—	39	12/16	92	—
New York, NY	37	95	6/11	5	1/18	42.48	3.57	7/26-27	30.4	13.3	12/30	41	12/12	127	7
Norfolk, VA	48	96	5/13	19	12/24	49.43	2.78	6/24	—	—	—	43	8/01	121	—
North Little Rock, AR	563	111	8/30	12	12/22	43.56	4.14	11/23-24	8.7	7.1	1/27-28	—	—	107	2
Oklahoma City, OK	1,281	108	9/03+	11	12/12	39.04	4.52	10/22-23	17.3	6.2	1/26	48	7/22+	91	6
Philadelphia, PA	60	94	6/11+	7	1/22	44.72	3.08	3/21-22	31.5	9.0	12/30	45	12/12	122	7
Phoenix, AZ	1,104	113	7/19+	32	1/03	7.87	1.92	3/05-06	0	0	—	39	7/30	33	0
Pittsburgh, PA	1,173	89	6/14	1	12/26	40.09	2.78	7/28-29	32.2	3.3	1/20	40	5/23	138	12
Portland, ME	50	90	9/01	–13	1/23	40.67	2.73	4/21-22	59.9	11.0	12/30-31	40	3/28	134	11
Portland, OR	24	98	6/27	24	11/18	30.20	1.10	12/21-22	—	—	—	43	1/16	157	—
Providence, RI	55	91	6/27+	2	1/22	46.00	2.63	12/16-17	—	—	—	41	12/12	127	—
Raleigh, NC	428	99	6/13	1	1/28	39.34	2.72	7/23-24	28.1	17.9	12/25	40	12/17	117	5
Rapid City, SD	3,166	102	8/11	–19	12/12+	15.03	1.28	7/09-10	36.2	14.0	4/19	55	2/25	70	10
Reno, NV	4,405	103	8/01	12	1/6+	5.71	0.58	1/23-24	—	—	—	45	4/28	41	—
Richmond, VA	165	94	7/10	–1	1/28	43.15	2.69	6/28-29	—	—	—	45	6/15	109	—
St. Louis, MO	562	102	8/17+	2	12/22+	37.37	3.59	6/23-24	25.7	7.6	12/13	41	5/22	105	7
Salt Lake City, UT	4,227	103	8/01	11	12/29	16.34	1.46	8/30-31	56.7	6.1	11/09	41	8/04	94	20
San Antonio, TX	809	111	9/05	23	1/05	35.85	3.66	11/02-03	T	T	4/02	39	8/21	91	0
San Diego, CA	42	89	6/25	43	1/06	6.90	1.76	2/20-21	—	T	—	38	3/05	43	—
San Francisco, CA	87	105	6/14	38	1/06	21.57	1.97	1/23-24	—	—	—	46	3/16	74	—
San Juan, PR	69	92	10/14	64	1/20	39.77	5.30	8/22-23	0	0	0	30	12/16	201	0
Sault Ste. Marie, MI	721	89	8/31	–19	1/22	22.40	1.25	11/09	—	—	—	35	1/16	153	—
Savannah, GA	49	104	7/20	19	12/21	37.44	2.74	9/04	—	—	—	36	8/10	97	—
Scottsbluff, NE	3,947	103	7/15	–9	12/11	14.79	2.10	4/18-19	36.6	4.5	2/17	47	5/17	74	19
Seattle, WA	448	88	6/27	27	11/17	28.66	1.34	2/01	—	—	—	40	1/16	150	—
Spokane, WA	2,412	99	7/31	–2	11/11	14.67	1.51	4/13-14	55.4	7	12/26+	46	3/14	98	19
Springfield, MO	1,278	102	9/03+	–10	12/22	35.36	2.06	6/16-17	26.7	14.3	12/14	43	5/24	110	6
Syracuse, NY	417	88	6/10	–10	1/17	37.01	1.39	6/13-14	156.8	9.6	12/20	48	12/12	180	42
Tampa, FL	38	95	6/16+	30	12/31	29.85	2.62	8/12-13	0	0	—	35	11/25	85	0
Washington, DC	60	94	6/11	14	12/23	40.66	2.05	9/25-26	17.4	9.3	1/25	47	7/14	117	4
Wilmington, DE	93	92	6/11	6	1/28	46.03	4.87	3/21-22	—	—	—	45	1/11	131	—

(T) Trace. (—) Data not available or incomplete. (1) Where one date is shown, it is the starting date of the storm. (2) Sustained for at least 2 minutes, not peak gust.

Record Temperatures by State

Source: National Climatic Data Center, NESDIS, NOAA, U.S. Dept. of Commerce, through 1999

State	Lowest °F	Highest °F	Latest date	Station	Approx. elevation in feet
Alabama	–27		Jan. 30, 1966	New Market	760
		112	*Sept. 5, 1925*	*Centerville*	*345*
Alaska	–80		Jan. 23, 1971	Prospect Creek	1,100
		100	*June 27, 1915*	*Fort Yukon*	*420*
Arizona	–40		Jan. 7, 1971	Hawley Lake	8,180
		128	*June 29, 1994*[1]	*Lake Havasu City*	*505*
Arkansas	–29		Feb. 13, 1905	Pond	1,250
		120	*Aug. 10, 1936*	*Ozark*	*396*
California	–45		Jan. 20, 1937	Boca	5,532
		134	*July 10, 1913*	*Greenland Ranch*	*–178*
Colorado	–61		Feb. 1, 1985	Maybell	5,920
		118	*July 11, 1888*	*Bennett*	*5,484*
Connecticut	–32		Feb. 16, 1943	Falls Village	585
		106	*July 15, 1995*	*Danbury*	*450*
Delaware	–17		Jan. 17, 1893	Millsboro	20
		110	*July 21, 1930*	*Millsboro*	*20*
Florida	–2		Feb. 13, 1899	Tallahassee	193
		109	*June 29, 1931*	*Monticello*	*207*
Georgia	–17		Jan. 27, 1940	CCC Camp F-16	1,000
		112	*July 24, 1952*	*Louisville*	*132*
Hawaii	12		May 17, 1979	Mauna Kea Obs. 111.2	13,770
		100	*Apr. 27, 1931*	*Pahala*	*850*
Idaho	–60		Jan. 18, 1943	Island Park Dam	6,285
		118	*July 28, 1934*	*Orofino*	*1,027*
Illinois	–36		Jan. 5, 1999	Congerville	635
		117	*July 14, 1954*	*East St. Louis*	*410*
Indiana	–36		Jan. 19, 1994	New Whiteland	785
		116	*July 14, 1936*	*Collegeville*	*672*
Iowa	–47		Feb. 3, 1996[1]	Elkader	770
		118	*July 20, 1934*	*Keokuk*	*614*
Kansas	–40		Feb. 13, 1905	Lebanon	1,812
		121	*July 24, 1936*[1]	*Alton (near)*	*1,651*
Kentucky	–37		Jan. 19, 1994	Shelbyville	730
		114	*July 28, 1930*	*Greensburg*	*581*
Louisiana	–16		Feb. 13, 1899	Minden	194
		114	*Aug. 10, 1936*	*Plain Dealing*	*268*
Maine	–48		Jan. 19, 1925	Van Buren	510
		105	*July 10, 1911*[1]	*North Bridgton*	*450*
Maryland	–40		Jan. 13, 1912	Oakland	2,461
		109	*July 10, 1936*[1]	*Cumberland; Frederick*	*623; 325*
Massachusetts	–35		Jan. 12, 1981	Chester	640
		107	*Aug. 2, 1975*	*Chester; New Bedford*	*640; 120*
Michigan	–51		Feb. 9, 1934	Vanderbilt	785
		112	*July 13, 1936*	*Mio*	*963*
Minnesota	–60		Feb. 2, 1996	Tower	1,430
		114	*July 6, 1936*[1]	*Moorhead*	*904*
Mississippi	–19		Jan. 30, 1966	Corinth	420
		115	*July 29, 1930*	*Holly Springs*	*600*
Missouri	–40		Feb. 13, 1905	Warsaw	700
		118	*July 14, 1954*[1]	*Warsaw; Union*	*700; 560*
Montana	–70		Jan. 20, 1954	Rogers Pass	5,470
		117	*July 5, 1937*	*Medicine Lake*	*1,950*
Nebraska	–47		Feb. 12, 1899	Camp Clarke	3,700
		118	*July 24, 1936*[1]	*Minden*	*2,169*
Nevada	–50		Jan. 8, 1937	San Jacinto	5,200
		125	*June 29, 1994*[1]	*Laughlin*	*605*
New Hampshire	–46		Jan. 28 1925	Pittsburg	1,575
		106	*July 4, 1911*	*Nashua*	*125*
New Jersey	–34		Jan. 5, 1904	River Vale	70
		110	*July 10, 1936*	*Runyon*	*18*
New Mexico	–50		Feb. 1, 1951	Gavilan	7,350
		122	*June 27, 1994*	*Waste Isolat. Pilot Plt.*	*3,418*
New York	–52		Feb. 18, 1979[1]	Old Forge	1,720
		108	*July 22, 1926*	*Troy*	*35*
North Carolina	–34		Jan. 21, 1985	Mt. Mitchell	6,525
		110	*Aug. 21, 1983*	*Fayetteville*	*213*
North Dakota	–60		Feb. 15, 1936	Parshall	1,929
		121	*July 6, 1936*	*Steele*	*1,857*
Ohio	–39		Feb. 10, 1899	Milligan	800
		113	*July 21, 1934*[1]	*Gallipolis (near)*	*673*
Oklahoma	–27		Jan. 18, 1930	Watts	958
		120	*June 27, 1994*[1]	*Tipton*	*1,350*
Oregon	–54		Feb. 10, 1933[1]	Seneca	4,700
		119	*Aug. 10, 1898*	*Pendleton*	*1,074*
Pennsylvania	–42		Jan. 5, 1904	Smethport	1,500
		111	*July 10, 1936*[1]	*Phoenixville*	*100*
Rhode Island	–25		Feb. 5, 1996	Greene	425
		104	*Aug. 2, 1975*	*Providence*	*51*
South Carolina	–19		Jan. 21, 1985	Caesars Head	3,115
		111	*June 28, 1954*[1]	*Camden*	*170*
South Dakota	–58		Feb. 17, 1936	McIntosh	2,277
		120	*July 5, 1936*	*Gannvalley*	*1,750*
Tennessee	–32		Dec. 30, 1917	Mountain City	2,471
		113	*Aug. 9, 1930*[1]	*Perryville*	*377*
Texas	–23		Feb. 8, 1933[1]	Seminole	3,275
		120	*Aug. 12, 1936*	*Seymour*	*1,291*
Utah	–69		Feb. 1, 1985	Peter's Sink	8,092
		117	*Jul. 5, 1985*	*Saint George*	*2,880*

State	Lowest °F	Highest °F	Latest date	Station	Approx. elevation in feet
Vermont	–50		Dec. 30, 1933	Bloomfield	915
		105	*July 4, 1911*	*Vernon*	*310*
Virginia	–30		Jan. 22, 1985	Mountain Lake Bio. Station	3,870
		110	*July 15, 1954*	*Balcony Falls*	*725*
Washington	–48		Dec. 30, 1968	Mazama; Winthrop	2,120; 1,755
		118	*Aug. 5, 1961*[1]	*Ice Harbor Dam*	*475*
West Virginia	–37		Dec. 30, 1917	Lewisburg	2,200
		112	*July 10, 1936*[1]	*Martinsburg*	*435*
Wisconsin	–54		Jan. 24, 1922	Danbury	908
		114	*July 13, 1936*	*Wisconsin Dells*	*900*
Wyoming	–66		Feb. 9, 1933	Riverside R.S.	6,650
		114	*July 12, 1900*	*Basin*	*3,500*

(1) Also on earlier dates at the same or other places.

IT'S A FACT: Leadville, CO, which reaches an altitude of 10,430 ft., claims to be the highest city in the U.S. Leadville's mean low temperature in January is 4.8° F. The average January high is 30.1° F.

World Temperature and Precipitation

Source: World Meteorological Organization

Average daily maximum and minimum temperatures and annual precipitation are based on records for the 30-year period 1961-90. The length of record of extreme temperatures includes all available years of data for a given location and is usually for a longer period; record temperatures may have been measured at a different location within the city. Surface elevations are supplied by the WMO and may differ from city elevation figures in other sections of *The World Almanac*. NA = Not available.

Station	Surface elevation (feet)	Temperature °F: Average daily, January Max.	January Min.	July Max.	July Min.	Extreme Max.	Extreme Min.	Average annual precipitation (inches)
Algiers, Algeria	82	61.7	42.6	87.1	65.3	NA	NA	27.0
Athens, Greece	49	56.1	44.6	88.9	73.0	NA	NA	14.6
Auckland, New Zealand	20	74.8	61.2	58.5	46.4	NA	NA	49.4
Bangkok, Thailand	66	89.6	69.8	90.9	77.0	104	51	59.0
Berlin, Germany	190	35.2	26.8	73.6	55.2	107	–4	23.3
Bogotá, Colombia	8,357	67.3	41.7	64.6	45.5	75	21	32.4
Bombay (Mumbai), India	36	85.3	66.7	86.2	77.5	110	46	85.4
Bucharest, Romania	298	34.7	22.1	83.8	60.1	105	–18	23.4
Budapest, Hungary	456	34.2	24.8	79.7	59.7	103	–10	20.3
Buenos Aires, Argentina	82	85.8	67.3	59.7	45.7	104	22	45.2
Cairo, Egypt	243	65.8	48.2	93.9	71.1	118	34	1.0
Cape Town, South Africa	138	79.0	60.3	63.3	44.6	105	28	20.5
Caracas, Venezuela	2,739	79.9	60.8	81.3	66.0	96	45	36.1
Casablanca, Morocco	203	62.8	47.1	77.7	66.7	NA	NA	16.8
Copenhagen, Denmark	16	35.6	28.4	68.9	55.0	NA	NA	NA
Damascus, Syria	2,004	54.3	32.9	97.2	61.9	NA	NA	5.6
Dublin, Ireland	279	45.7	36.5	66.0	52.5	86	8	28.8
Geneva, Switzerland	1,364	38.3	27.9	76.3	53.2	101	–3	35.6
Havana, Cuba	164	78.4	65.5	88.3	74.8	NA	NA	46.9
Hong Kong, China	203	65.5	56.5	88.7	79.9	97	32	87.2
Istanbul, Turkey	108	47.8	37.2	82.8	65.3	105	7	27.4
Jerusalem, Israel	2,483	53.4	39.4	83.8	63.0	107	26	23.2
Lagos, Nigeria	125	90.0	72.3	82.8	72.1	NA	NA	59.3
Lima, Peru	43	79.0	66.9	66.4	59.4	NA	NA	0.2
London, England	203	44.1	32.7	71.1	52.3	99	2	29.7
Manila, Philippines	79	85.8	74.8	89.1	76.8	NA	NA	49.6
Mexico City, Mexico	7,570	70.3	43.7	73.8	53.2	NA	NA	33.4
Montreal, Canada	118	21.6	5.2	79.2	59.7	100	–36	37.0
Nairobi, Kenya	5,897	77.9	50.9	71.6	48.6	NA	NA	41.9
Paris, France	213	42.8	33.6	75.2	55.2	105	–1	25.6
Prague, Czech Republic	1,197	32.7	22.5	73.9	53.2	98	–16	20.7
Reykjavik, Iceland	200	35.4	26.6	55.9	46.9	76	–3	31.5
Rome, Italy	79	53.8	35.4	88.2	62.1	NA	NA	33.0
San Salvador, El Salvador	2,037	86.5	61.3	86.2	66.4	105	45	68.3
São Paulo, Brazil	2,598	81.1	65.7	71.2	53.1	NA	NA	57.4
Shanghai, China	23	45.9	32.9	88.9	76.6	104	10	43.8
Singapore	52	85.8	73.6	87.4	75.6	NA	NA	84.6
Stockholm, Sweden	171	30.7	23.0	71.4	56.1	97	–26	21.2
Sydney, Australia	10	79.5	65.5	62.4	43.9	114	32	46.4
Tehran, Iran	3,906	45.0	30.0	98.2	75.2	109	–5	9.1
Tokyo, Japan	118	49.1	34.2	83.8	72.1	NA	NA	55.4
Toronto, Canada	567	27.5	12.0	80.2	57.6	105	–26	30.8

Hurricane and Tornado Classifications

Source: National Weather Service, NOAA, U.S. Dept. of Commerce

The Saffir-Simpson Hurricane Scale is a 1-5 rating based on a hurricane's intensity. The scale is used to give an estimate of the potential property damage and flooding expected along the coast from a hurricane landfall. Wind speed is the determining factor in the scale. The Fujita (or F) Scale, created by T. Theodore Fujita, is used to classify tornadoes. The F Scale uses rating numbers from 0 to 5, based on the amount and type of wind damage.

Saffir-Simpson Scale (Hurricanes)

Category	Wind Speed	Severity	Storm Surge[1]
1	74-95 MPH	Weak	4-5 feet
2	96-110 MPH	Moderate	6-8 feet
3	111-130 MPH	Strong	9-12 feet
4	131-155 MPH	Very Strong	13-18 feet
5	more than 155 MPH	Devastating	more than 18 feet

(1) Above normal tides.

Fujita Scale (Tornadoes)

Rank	Wind Speed	Damage	Strength
F-0	40-72 MPH	Light	Weak
F-1	73-112 MPH	Moderate	Weak
F-2	113-157 MPH	Considerable	Strong
F-3	158-206 MPH	Severe	Strong
F-4	207-260 MPH	Devastating	Violent
F-5	more than 261 MPH	Incredible	Violent

Hurricane Names in 2002

Source: National Weather Service, NOAA, U.S. Dept. of Commerce

Atlantic hurricanes — Arthur, Bertha, Cesar, Dolly, Edouard, Fran, Gustav, Hortense, Isidore, Josephine, Kyle, Lili, Marco, Nana, Omar, Paloma, Rene, Sally, Teddy, Vicky, Wilfred.

Eastern Pacific hurricanes — Alma, Boris, Cristina, Douglas, Elida, Fausto, Genevieve, Hernan, Iselle, Julio, Kenna, Lowell, Marie, Norbert, Odile, Polo, Rachel, Simon, Trudy, Vance, Winnie, Xavier, Yolanda, Zeke.

Tides and Their Causes

Source: U.S. Dept. of Commerce, Natl. Oceanic & Atmospheric Admin. (NOAA), Natl. Ocean Service (NOS)

The tides are a natural phenomenon involving the alternating rise and fall in the large fluid bodies of the earth caused by the combined gravitational attraction of the sun and moon. The combination of these two variable influences produces the complex recurrent cycle of the tides. Tides may occur in both oceans and seas, to a limited extent in large lakes, in the atmosphere, and, to a very minute degree, in the earth itself. The length of time between succeeding tides varies as the result of many factors.

The tide-generating force represents the difference between (1) the centrifugal force produced by the revolution of the earth around the common center-of-gravity of the earth-moon system and (2) the gravitational attraction of the moon acting upon the earth's overlying waters. Since, on the average, the moon is only 238,856 miles from the earth compared with the sun's much greater distance of 92,980,000 miles, this closer distance outranks the much smaller mass of the moon compared with that of the sun, and the moon's tide-raising force is, accordingly, 2.5 times that of the sun.

The effect of the tide-generating forces of the moon and sun acting tangentially to the earth's surface (the so-called "tractive force") tends to cause a maximum accumulation of the waters of the oceans at two diametrically opposite positions on the surface of the earth and to withdraw compensating amounts of water from all points 90° removed from the positions of these tidal bulges. As the earth rotates beneath the maxima and minima of these tide-generating forces, a sequence of two high tides, separated by two low tides, ideally is produced each day (semidiurnal tide).

Twice in each lunar month, when the sun, moon, and earth are directly aligned, with the moon between the earth and the sun (at new moon) or on the opposite side of the earth from the sun (at full moon), the sun and the moon exert their gravitational force in a mutual or additive fashion. The highest high tides and lowest low tides are produced at these times. These are called *spring* tides. At two positions 90° in between, the gravitational forces of the moon and sun—imposed at right angles—tend to counteract each other to the greatest extent, and the range between high and low tides is reduced. These are called *neap* tides. This semi-monthly variation between the spring and neap tides is called the *phase inequality*.

The inclination to the equator of the moon's monthly orbit and the inclination of the sun to the equator during the earth's yearly orbit produce a difference in the height of succeeding high tides and in the extent of depression of succeeding low tides that is known as the *diurnal inequality*. In most cases, this produces a so-called *mixed tide*. In extreme cases, these phenomena may result in only one high tide and one low tide each (*diurnal tide*). There are other monthly and yearly variations in the tide because of the elliptical shape of the orbits themselves.

The datum for Charting and Predictions is Mean Lower Low Water (MLLW). This became effective Nov. 1980 according to the convention of 1980, which prescribed that data on all United States coastlines would be the same; namely, Mean Higher High Water (MHHW), Mean High Water (MHW), Mean Tide Level (MTL), Mean Sea Level (MSL), Mean Low Water (MLW), Mean Lower Low Water (MLLW). Diurnal range of tide is the difference in height between MHHW and MLLW. Mean range of tide is the difference in height between MHW and MLW.

The actual range of tide in the open ocean is less than in the shoreline regions. However, as the ocean tide approaches shoal waters and its effects are augmented, the tidal range may be greatly increased. In Nova Scotia along the narrow channel of the Bay of Fundy, the range of tides, or difference between high and low waters, may reach 43½ feet or more (under spring tide conditions) as a result of resonant amplification.

At New Orleans, the periodic rise and fall of the diurnal tide is affected by the seasonal stages of the Mississippi River, being about 10 inches at low stage and zero at high. The Canadian Tide Tables for 1972 gave a maximum range of nearly 50 feet at Leaf Basin, Ungava Bay, Quebec.

In every case, actual high or low tide can vary considerably from the average, as a result of weather conditions such as strong winds, abrupt barometric pressure changes, or prolonged periods of extreme high or low pressure.

The Average Rise and Fall of Tides[1]

Places	Ft.	In.	Places	Ft.	In.	Places	Ft.	In.	Places	Ft.	In.
Baltimore, MD	1	8	Galveston, TX	1	5	Newport, RI	3	11	San Diego, CA	5	9
Boston, MA	10	4	Halifax, N.S.	4	5[2]	New York, NY	5	1	Sandy Hook, NJ	5	2
Charleston, SC	5	10	Hampton Roads, VA	2	10	Philadelphia, PA	6	9	San Francisco, CA	5	10
Cristobal, Panama	1	1	Key West, FL	1	10	Portland, ME	9	11	Seattle, WA	11	4
Eastport, ME	19	4	Mobile, AL	1	6	St. John's, Nfld.	2	7[2]	Vancouver, B.C.	10	6
Ft. Pulaski, GA	7	6	New London, CT	3	1	St. Petersburg, FL	2	3	Washington, DC	3	2

(1) Diurnal range. (2) Mean range.

Speed of Winds in the U.S.

Source: National Climatic Data Center, NESDIS, NOAA, U.S. Dept. of Commerce

Miles per hour — average, high through 2000. Wind velocities in true values.

Station	Avg.	High	Station	Avg.	High	Station	Avg.	High
Albuquerque, NM	8.9	52	Helena, MT	7.7	73	Mt. Washington, NH	35.2	231
Anchorage, AK	7.1	75	Honolulu, HI	11.3	46	New Orleans, LA	8.2	69
Atlanta, GA	9.1	60	Houston, TX	7.7	51	New York, NY(b)	9.3	40
Baltimore, MD	8.9	80	Indianapolis, IN	9.6	49	Omaha, NE	10.5	58
Bismarck, ND	10.2	55	Jacksonville, FL	7.9	57	Philadelphia, PA	9.5	73
Boston, MA	12.4	54	Kansas City, MO	10.6	58	Phoenix, AZ	6.2	43
Buffalo, NY	11.8	91	Las Vegas, NV	9.3	56	Pittsburgh, PA	9.0	58
Cape Hatteras, NC	10.9	60	Lexington, KY	9.1	47	Portland, OR	7.9	88
Casper, WY	12.8	81	Little Rock, AR	7.8	65	Rochester, NY	9.6	68
Chicago, IL	10.4	58	Los Angeles, CA	6.0	49	St. Louis, MO	9.7	52
Cleveland, OH	10.5	53	Louisville, KY	8.3	56	Salt Lake City, UT	8.8	71
Dallas-Ft. Worth, TX	10.7	73	Memphis, TN	8.8	51	San Diego, CA	7.0	56
Denver, CO	8.6	46	Miami, FL	9.2	86(a)	San Francisco, CA	8.7	47
Des Moines, IA	10.7	76	Milwaukee, WI	11.5	54	Seattle, WA	8.9	66
Detroit, MI	10.3	53	Minn.-St. Paul, MN	10.5	51	Spokane, WA	8.9	59
Hartford, CT	8.4	46	Mobile, AL	8.8	63	Washington, DC	9.4	49

(a) Highest velocity ever recorded in Miami area was 132 mph, at former station in Miami Beach in Sept. 1926. (b) Data for Central Park; Battery Place data through 1960, avg. 14.5, high 113.

IT'S A FACT: While Chicago is known as the "windy city," Boston, Buffalo, Casper, Cleveland, Dallas-Ft. Worth, Des Moines, Honolulu, Kansas City (MO), Milwaukee, Minneapolis-St. Paul, and Omaha all had higher average wind speeds in 2000.

El Niño

Source: National Weather Service, NOAA, U.S. Dept. of Commerce

El Niño is a naturally occurring climate phenomenon characterized by warmer-than-normal ocean temperatures in the equatorial eastern Pacific and along the tropical western coasts of Central and South America. The term *El Niño*, Spanish for "the Christ Child," was originally used by Ecuadorian and Peruvian fishermen to refer to a warm ocean current typically appearing around Christmastime and lasting for several months. Fish are less abundant during these warm intervals, so fishermen often take a break to repair equipment and spend time with their families. In some years, however, the water remains especially warm into May or even June. Over the years, the term has come to be reserved for those exceptionally strong, warm intervals that not only disrupt fishermen's lives but also bring heavy rains.

The first known record of El Niño is attributed to Francisco Pizarro, a Spaniard who in 1525 described unusual desert rainfall in northern Peru and its El Niño association. El Niño episodes occur generally every 2 to 6 years and typically last 12 to 18 months. Recent episodes include 1972-73, 1977-78, 1982-83, 1986-87, 1991-92, and 1997-98.

The intensity of El Niño events varies—some are strong, such as the 1982-83 and 1997-98 events; others are considerably weaker, based on intensity and area encompassed by the abnormally warm ocean temperatures. The eastward extent of the warmer than normal water varies from episode to episode. Both of these characteristics affect the patterns of temperature and precipitation variations associated with El Niño in the U.S. and elsewhere.

The 1997-98 El Niño, one of the most powerful climate events of the century, strongly impacted global weather patterns. The extremely warm temperatures in the equatorial Pacific, combined with shifts in trade winds across the tropics, contributed to wildfires in Indonesia; significant crop loss in Argentina and New Zealand; devastating floods in Chile, Peru, southern Brazil, and northern Argentina; mudslides in California; and record rains in the southeastern U.S.

El Niño has a significant influence on weather and climate patterns around the globe, and its impacts are most clearly seen in the wintertime. During El Niño years, winter temperatures in the continental U.S. tend to be warmer than normal in the northern and west coast states and cooler than normal in the Southeast. Conditions tend to be wetter than normal over central and southern California and the southwest U.S. and across much of the southern third of the contiguous 48 states, particularly along the Gulf Coast, and drier than normal over the northern portions of the Rocky Mountains and in the Ohio valley region. Globally, El Niño brings wetter than normal conditions to Peru and Chile and drier than normal conditions to Australia and Indonesia. It should be noted that El Niño is only one of a number of factors influencing seasonal variations of climate.

The opposite of El Niño is La Niña, with colder than normal sea surface temperatures in the tropical Pacific. La Niña typically brings wetter than normal conditions to the Pacific Northwest and warmer than normal temperatures to much of the southern U.S. during winter months.

El Niño and La Niña episodes are detected and monitored by observing systems, including satellites, moored buoys, and drifting buoys released by volunteer ships crossing the Pacific Ocean. Highly sophisticated numerical computer models of the global ocean and atmosphere use data from the observing systems to predict the onset and evolution of El Niño and its associated impacts. Numerous other models at research institutions worldwide also use the data from the observing systems to increase the understanding of El Niño and improve forecasting techniques.

Wind Chill Table

Source: National Weather Service, NOAA, U.S. Dept. of Commerce

Temperature and wind combine to cause heat loss from body surfaces. The following table shows that, for example, a temperature of 20° Fahrenheit, plus a wind of 20 miles per hour, causes a body heat loss equal to that in minus 10 degrees temperature with no wind. In other words, a 20-mph wind makes 20° feel like minus 10.

The top line of figures shows temperatures in degrees Fahrenheit. The column at far left shows wind speeds up to 45 mph. (Wind speeds greater than 45 mph have little additional chilling effect.)

	35	30	25	20	15	10	5	0	–5	–10	–15	–20	–25	–30	–35	–40	–45
MPH																	
5	33	27	21	16	12	7	0	–5	–10	–15	–21	–26	–31	–36	–42	–47	–52
10	22	16	10	3	–3	–9	–15	–22	–27	–34	–40	–46	–52	–58	–64	–71	–77
15	16	9	2	–5	–11	–18	–25	–31	–38	–45	–51	–58	–65	–72	–78	–85	–92
20	12	4	–3	–10	–17	–24	–31	–39	–46	–53	–60	–67	–74	–81	–88	–95	–103
25	8	1	–7	–15	–22	–29	–36	–44	–51	–59	–66	–74	–81	–88	–96	–103	–110
30	6	–2	–10	–18	–25	–33	–41	–49	–56	–64	–71	–79	–86	–93	–101	–109	–116
35	4	–4	–12	–20	–27	–35	–43	–52	–58	–67	–74	–82	–89	–97	–105	–113	–120
40	3	–5	–13	–21	–29	–37	–45	–53	–60	–69	–76	–84	–92	–100	–107	–115	–123
45	2	–6	–14	–22	–30	–38	–46	–54	–62	–70	–78	–85	–93	–102	–109	–117	–125

Heat Index

The heat index is a measure of the contribution high humidity makes, in combination with abnormally high temperatures, to reducing the body's ability to cool itself. For example, the index shows that an air temperature of 100° Fahrenheit with a relative humidity of 50% has the same effect on the human body as a temperature of 120° . Sunstroke and heat exhaustion are likely when the heat index reaches 105. This index is a measure of what hot weather "feels like" to the average person for various temperatures and relative humidities.

					Air Temperature*						
	70	**75**	**80**	**85**	**90**	**95**	**100**	**105**	**110**	**115**	**120**
Relative Humidity					**Apparent Temperature***						
0%	64	69	73	78	83	87	91	95	99	103	107
10%	65	70	75	80	85	90	95	100	105	111	116
20%	66	72	77	82	87	93	99	105	112	120	130
30%	67	73	78	84	90	96	104	113	123	135	148
40%	68	74	79	86	93	101	110	123	137	151	
50%	69	75	81	88	96	107	120	135	150		
60%	70	76	82	90	100	114	132	149			
70%	70	77	85	93	106	124	144				
80%	71	78	86	97	113	136					
90%	71	79	88	102	122						
100%	72	80	91	108							

*Degrees Fahrenheit

Ultraviolet (UV) Index Forecast

Source: National Weather Service, NOAA, U.S. Dept. of Commerce

The National Weather Service (NWS), Environmental Protection Agency (EPA), and Centers for Disease Control and Prevention (CDC) developed and began offering a UV index on June 28, 1994, in response to increasing incidence of skin cancer, cataracts, and other effects from exposure to the sun's harmful rays. The UV Index is now a regular element of NWS atmospheric forecasts.

UV Index number and forecast. The UV Index number, ranging from 0 to 10+, is an indication of the amount of UV radiation reaching the earth's surface over the one-hour period around noon. The lower the number, the less the radiation. The UV Index forecast is produced for 58 cities by the NWS Climate Prediction Center. The index number is based on several factors: latitude, day of year, time of day, total atmospheric ozone, elevation, and predicted cloud conditions. The index is valid for a radius of about 30 miles around a listed city; however, adjustments should be made for a number of factors.

Ozone. Ozone is measured by a NOAA polar orbiting satellite. The more ozone, the lower the UV radiation at the surface.

Cloudiness. Increased cloudiness lowers the Index number.

Reflectivity. Reflective surfaces intensify UV exposure. As an example, grass reflects 2.5% to 3% of UV radiation reaching the surface; sand, 20% to 30%; snow and ice, 80% to 90%; water, up to 100% (depending on reflection angle).

Elevation. At higher elevations, UV radiation travels a shorter distance to reach the surface so there is less atmosphere to absorb the rays. For every 4,000 ft. one travels above sea level, the UV Index increases by 1 unit. Snow and lack of pollutants intensify UV exposure at higher altitudes.

Latitude. The closer to the equator, the higher the UV radiation level.

Accuracy. After gathering data from 20 UV sensors (during June-Oct. 1994), the NWS determined that 32% of UV Index forecasts for that period were correct, 76% were within ±1 UV Index unit, and about 90% were within ±2 units. Unpredictable cloudiness, haze, and pollution contribute to forecast error.

SPF number. The UV Index is not linked in any way to the SPF number on suntan lotions and sunscreens. For an explanation of the SPF factor, contact the product's manufacturer or the Food and Drug Administration.

Further information. For precautions to take after learning the UV Index number, call the U.S. EPA hotline (800-296-1996) or your doctor. For questions on scientific aspects, call the NWS at 301-713-0622.

Global Measured Extremes of Temperature and Precipitation

Source: National Climatic Data Center

Highest Temperature Extremes

Continent	Highest Temp. (deg F)	Place	Elevation (Feet)	Date
Africa	136	El Azizia, Libya	367	Sept. 13, 1922
North America	134	Death Valley, CA (Greenland Ranch)	–178	July 10, 1913
Asia	129	Tirat Tsvi, Israel	–722	June 21, 1942
Australia	128	Cloncurry, Queensland	622	Jan. 16, 1889
Europe	122	Seville, Spain	26	Aug. 4, 1881
South America	120	Rivadavia, Argentina	676	Dec. 11, 1905
Oceania	108	Tuguegarao, Philippines	72	Apr. 29, 1912
Antarctica	59	Vanda Station, Scott Coast	49	Jan. 5, 1974

Lowest Temperature Extremes

Continent	Lowest Temp. (deg F)	Place	Elevation (Feet)	Date
Antarctica	–129.0	Vostok	11,220	July 21, 1983
Asia	–90.0	Oimekon, Russia	2,625	Feb. 6, 1933
Asia	–90.0	Verkhoyansk, Russia	350	Feb. 7, 1892
Greenland	–87.0	Northice	7,687	Jan. 9, 1954
North America	–81.4	Snag, Yukon, Canada	2,120	Feb. 3, 1947
Europe	–67.0	Ust'Shchugor, Russia	279	Jan.*
South America	–27.0	Sarmiento, Argentina	879	June 1, 1907
Africa	–11.0	Ifrane, Morocco	5,364	Feb. 11, 1935
Australia	–9.4	Charlotte Pass, NSW	5,758	June 29, 1994
Oceania	14.0	Haleakala Summit, Maui, HI	9,750	Jan. 2, 1961

* Exact day and year unknown.

Highest Average Annual Precipitation Extremes

Continent	Highest Avg. (Inches)	Place	Elevation (Feet)	Years of Data
South America	523.6[1,2]	Lloro, Colombia	520[3]	29
Asia	467.4[1]	Mawsynram, India	4,597	38
Oceania	460.0[1]	Mt. Waialeale, Kauai, HI	5,148	30
Africa	405.0	Debundscha, Cameroon	30	32
South America	354.0[2]	Quibdo, Colombia	120	16
Australia	340.0	Bellenden Ker, Queensland	5,102	9
North America	256.0	Henderson Lake, British Columbia	12	14
Europe	183.0	Crkvica, Bosnia-Herzegovina	3,337	22

(1) The value given is continent's highest and possibly the world's depending on measurement practices, procedures, and period of record variations. (2) The official greatest average annual precipitation for South America is 354 inches at Quibdo, Colombia. The 523.6 inches average at Lloro, Colombia (14 miles SE and at a higher elevation than Quibdo) is an estimated amount. (3) Approximate elevation.

Lowest Average Annual Precipitation Extremes

Continent	Lowest Avg. (Inches)	Place	Elevation (Feet)	Years of Data
South America	0.03	Arica, Chile	95	59
Africa	<0.1	Wadi Halfa, Sudan	410	39
Antarctica	0.8[1]	Amundsen-Scott South Pole Station	9,186	10
North America	1.2	Batagues, Mexico	16	14
Asia	1.8	Aden, Yemen	22	50
Australia	4.05	Mulka (Troudaninna), South Australia	160[2]	42
Europe	6.4	Astrakhan, Russia	45	25
Oceania	8.93	Puako, Hawaii	5	13

(1) The value given is the average amount of solid snow accumulating in one year as indicated by snow markers. The liquid content of the snow is undetermined. (2) Approximate elevation.

DISASTERS

As of Oct. 1, 2001. Listings in this chapter are selective and may not include acts of terrorism, war related disasters, or disasters with relatively low fatalities.

On Sept. 11, 2001, hijacked airliners used to attack targets in the U.S. destroyed the World Trade Center twin towers in New York City and a portion of the Pentagon, outside Washington, DC, killing thousands of people. For further details, see Some Notable Aircraft Disasters Since 1937, below. See also the front-of-the-book feature article and the Chronology of the Year's Events.

Some Notable Shipwrecks Since 1854

(Figures indicate estimated lives lost. Does not include most military disasters.)

1854, Mar.—City of Glasgow; Brit. steamer missing in N Atlantic; 480.
1854, Sept. 27—Arctic; U.S. (Collins Line) steamer sunk in collision with French steamer *Vesta* near Cape Race; 285-351.
1856, Jan. 23—Pacific; U.S. (Collins Line) steamer missing in N Atlantic; 186-286.
1858, Sept. 23—Austria; German steamer destroyed by fire in N Atlantic; 471.
1863, Apr. 27—Anglo-Saxon; Brit. steamer wrecked at Cape Race; 238.
1865, Apr. 27—Sultana; Mississippi River steamer blew up near Memphis, TN; 1,450.
1869, Oct. 27—Stonewall; steamer burned on Mississippi River below Cairo, IL; 200.
1870, Jan. 25—City of Boston; Brit. (Inman Line) steamer vanished between New York and Liverpool; 177.
1870, Oct. 19—Cambria; Brit. steamer wrecked off N Ireland; 196.
1872, Nov. 7—Mary Celeste; U.S. half-brig sailed from New York for Genoa; found abandoned; loss of life unknown.
1873, Jan. 22—Northfleet; Brit. steamer foundered off Dungeness, England; 300.
1873, Apr. 1—Atlantic; Brit. (White Star) steamer wrecked off Nova Scotia; 585.
1873, Nov. 23—Ville du Havre; French steamer sank after collision with Brit. sailing ship *Loch Earn*; 226.
1875, May 7—Schiller; German steamer wrecked off Scilly Isles; 312.
1875, Nov. 4—Pacific; U.S. steamer sank after collision off Cape Flattery; 236.
1878, Sept. 3—Princess Alice; Brit. steamer sank after collision in Thames River; 700.
1878, Dec. 18—Byzantin; French steamer sank after collision in Dardanelles; 210.
1881, May 24—Victoria; steamer capsized in Thames River, Canada; 200.
1883, Jan. 19—Cimbria; German steamer sank in collision with Brit. steamer *Sultan* in North Sea; 389.
1887, Nov. 15—Wah Yeung; Brit. steamer burned at sea; 400.
1890, Feb. 17—Duburg; Brit. steamer wrecked, China Sea; 400.
1890, Sept. 19—Ertogrul; Turkish frigate wrecked off Japan; 540.
1891, Mar. 17—Utopia; Brit. steamer sank in collision with Brit. ironclad *Anson* off Gibraltar; 562.
1895, Jan. 30—Elbe; German steamer sank in collision with Brit. steamer *Craithie* in North Sea; 332.
1895, Mar. 11—Reina Regenta; Spanish cruiser foundered near Gibraltar; 400.
1898, Feb. 15—Maine; U.S. battleship blown up in Havana Harbor; 260.
1898, July 4—La Bourgogne; French steamer sank in collision with Brit. sailing ship *Cromartyshire* off Nova Scotia; 549.
1898, Nov. 26—Portland; U.S. steamer wrecked off Cape Cod; 157.
1904, June 15—General Slocum; excursion steamer burned in East River, New York City; 1,030.
1904, June 28—Norge; Danish steamer wrecked on Rockall Island, Scotland; 620.
1906, Aug. 4—Sirio; Italian steamer wrecked off Cape Palos, Spain; 350.
1908, Mar. 23—Matsu Maru; Japanese steamer sank in collision near Hakodate, Japan; 300.
1909, Aug. 1—Waratah; Brit. steamer, Sydney to London, vanished; 300.
1910, Feb. 9—General Chanzy; French steamer wrecked off Minorca, Spain; 200.
1911, Sept. 25—Liberté; French battleship exploded at Toulon; 285.
1912, Mar. 5—Principe de Asturias; Spanish steamer wrecked off Spain; 500.
1912, Apr. 14-15—Titanic; Brit. (White Star) steamer hit iceberg in N Atlantic; 1,503.
1912, Sept. 28—Kichemaru; Japanese steamer sank off Japanese coast; 1,000.
1914, May 29—Empress of Ireland; Brit. (Canadian Pacific) steamer sunk in collision with Norwegian collier in St. Lawrence River; 1,014.
1915, May 7—Lusitania; Brit. (Cunard Line) steamer torpedoed and sunk by German submarine off Ireland; 1,198.
1915, July 24—Eastland; excursion steamer capsized in Chicago River; 812.
1916, Feb. 26—Provence; French cruiser sank in Mediterranean; 3,100.
1916, Mar. 3—Principe de Asturias; Spanish steamer wrecked near Santos, Brazil; 558.
1916, Aug. 29—Hsin Yu; Chinese steamer sank off Chinese coast; 1,000.
1917, Dec. 6—Mont Blanc, Imo; French ammunition ship and Belgian steamer collided in Halifax Harbor; 1,600.
1918, Apr. 25—Kiang-Kwan; Chinese steamer sank in collision off Hankow; 500.
1918, July 12—Kawachi; Japanese battleship blew up in Tokayama Bay; 500.
1918, Oct. 25—Princess Sophia; Canadian steamer sank off Alaskan coast; 398.
1919, Jan. 17—Chaonia; French steamer lost in Straits of Messina, Italy; 460.
1919, Sept. 9—Valbanera; Spanish steamer lost off Florida coast; 500.
1921, Mar. 18—Hong Kong; steamer wrecked in South China Sea; 1,000.
1922, Aug. 26—Niitaka; Japanese cruiser sank in storm off Kamchatka, USSR; 300.
1924, June 12—USS Mississippi; U.S. battleship; explosions in gun turret, off San Pedro, CA; 48.
1927, Oct. 25—Principessa Mafalda; Italian steamer blew up, sank off Porto Seguro, Brazil; 314.
1928, Nov. 12—Vestris; Brit. steamer sank off Virginia; 113.
1934, Sept. 8—Morro Castle; U.S. steamer, Havana to New York, burned off Asbury Park, NJ; 134.
1939, May 23—Squalus; U.S. submarine sank off Portsmouth, NH; 26.
1939, June 1—Thetis; submarine sank, Liverpool Bay; 99.
1942, Feb. 18—Truxtun and Pollux; U.S. destroyer and cargo ship ran aground, sank off Newfoundland; 204.
1942, Oct. 2—Curacao; Brit. cruiser sank after collision with liner Queen Mary; 338.
1944, Dec. 17-18—3 U.S. Third Fleet destroyers sank during typhoon in Philippine Sea; 790.
1947, Jan. 19—Himera; Greek steamer hit a mine off Athens; 392.
1947, Apr. 16—Grandcamp; French freighter exploded in Texas City, TX, harbor, starting fires; 510.
1948, Nov.—Chinese army evacuation ship exploded and sank off S Manchuria; 6,000.
1948, Dec. 3—Kiangya; Chinese refugee ship wrecked in explosion S of Shanghai; 1,100+.
1949, Sept. 17—Noronic; Canadian Great Lakes Cruiser burned at Toronto dock; 130.
1952, Apr. 26—Hobson and Wasp; U.S. destroyer and aircraft carrier collided in Atlantic; 176.
1954, May 26—Pennington; sank off Rhode Island; 103.
1954, Sept. 26—Toya Maru; Japanese ferry sank in Tsugaru Strait, Japan; 1,172.
1956, July 26—Andrea Doria and **Stockholm;** Italian liner and Swedish liner collided off Nantucket; 51.
1957, July 14—Eshghabad; Soviet ship ran aground in Caspian Sea; 270.
1960, Dec. 19—Constellation; U.S. aircraft carrier caught fire in Brooklyn Navy Yard, NY; 49.
1961, Apr. 8—Dara; British ocean liner exploded in Persian Gulf; 236.
1961, July 8—Save; Portuguese ship ran aground off Mozambique; 259.
1963, Apr. 10—Thresher; U.S. Navy atomic submarine sank in N Atlantic; 129.
1964, Feb. 10—Australian destroyer *Voyager* sank after collision with aircraft carrier *Melbourne* off New South Wales; 82.
1965, Nov. 13—Yarmouth Castle; Panamanian registered cruise ship burned and sank off Nassau; 89.
1967, July 29—Forrestal; U.S. aircraft carrier caught fire off N Vietnam; 134.
1968, Jan. 25—Dakar; Israeli submarine vanished in Mediterranean Sea; 69.
1968, late May—Scorpion; U.S. nuclear submarine sank in Atlantic near Azores; 99 (located Oct. 31).
1969, June 2—Evans; U.S. destroyer cut in half by Australian carrier *Melbourne*, S China Sea; 74.
1970, Mar. 4—Eurydice; French submarine sank in Mediterranean near Toulon; 57.

1970, Dec. 15—Namyong-Ho; South Korean ferry sank in Korea Strait; 308.
1974, May 1—Motor launch capsized off Bangladesh; 250.
1974, Sept. 26—Soviet destroyer sank in Black Sea; 200+.
1975, Nov. 10—Edmund Fitzgerald; U.S. cargo ship sank during storm on Lake Superior; 29.
1976, Oct. 20—George Prince and **Frosta;** ferryboat and Norwegian tanker collided on Mississippi R. at Luling, LA; 77.
1976, Dec. 25—Patria; Egyptian liner caught fire and sank in the Red Sea; 100.
1979, Aug. 14—23 yachts competing in Fastnet yacht race sank or abandoned during storm in S Irish Sea; 18.
1980, Sept. 9—Derbyshire; British bulk carrier sank in typhoon in Pacific Ocean near Okinawa, Japan; 44.
1981, Jan. 27—Tamponas II; Indonesian passenger ship caught fire and sank in Java Sea; 580.
1981, May 26—Nimitz; U.S. Marine combat jet crashed on deck of U.S. aircraft carrier; 14.
1983, Feb. 12—Marine Electric; coal freighter sank during storm off Chincoteague, VA; 33.
1983, May 25—10th of Ramadan; Nile steamer caught fire and sank in Lake Nasser; 357.
1986, Apr. 20—ferry sank near Barisal, Bangladesh; 262.
1986, Aug. 31—Soviet passenger ship *Admiral Nakhimov* and Soviet freighter *Pyotr Vasev* collided in Black Sea; 398.
1987, Mar. 6—British ferry capsized off Zeebrugge, Belgium; 189.
1987, Dec. 20—Philippine ferry *Dona Paz* and oil tanker *Victor* collided in Tablas Strait; 4,341.
1988, Aug. 6—Indian ferry capsized on Ganges R.; 400+.
1989, Apr. 19—USS Iowa; explosion in gun turret; 47.
1989, Aug. 20—Brit. barge *Bowbelle* struck Brit. pleasure cruiser *Marchioness* on Thames R. in central London; 56.
1989, Sept. 10—Romanian pleasure boat and Bulgarian barge collided on Danube R.; 161.
1991, Apr. 10—Auto ferry and oil tanker collided outside Livorno Harbor, Italy; 140.
1991, Dec. 14—Salem Express; ferry rammed coral reef near Safaga, Egypt; 462.
1993, Feb. 17—Neptune; ferry capsized off Port-au-Prince, Haiti; 500+.
1993, Oct. 10—West Sea Ferry; capsized in Yellow Sea near W South Korea during storm; 285.
1994, Sept. 28—Estonia; ferry sank in Baltic Sea; 1,049.
1996, May 21—Bukoba; ferry sank in Lake Victoria (Africa); 500.
1997, Feb. 20—Tamil refugee boat sank off Sri Lanka; 165.
1997, Mar. 28—Albanian refugee boat sank in Adriatic Sea after being rammed by Italian navy warship *Sibilla*; 83.
1997, Sept. 8—Pride of la Gonâve; Haitian ferry sank off Montrouis, Haiti; 200+.
1998, Apr. 4—passenger boat capsized off coast near Ibaka beach, Nigeria; 280.
1998, Sept. 2—2 passenger boats capsized on Lake Kivu, near Bukavu, Congo; 200+.
1998, Sept. 18—ferry sank S of Manila; 97.
1999, Feb. 6—Harta Rimba; cargo ship sank off Indonesia; 280+.
1999, Mar. 26—passenger boat overturned off coast, Sierra Leone; 150+.
1999, Apr. 2—passenger ferry sank off coast of Nigeria; 100+.
1999, May 1—amphibious excursion boat sank in Lake Hamilton, AR; 13.
1999, May 8—passenger ferry capsized off Bangladesh; 200+.
1999, Nov. 24—Dashun; passenger ferry capsized near Yantai, China; 280.
2000, May 3—2 ferries capsized in storm in Meghna river, Bangladesh; 72+.
2000, June 29—overloaded ferry capsized in storm off Sulawesi Island, Indonesia; 500+.
2000, Aug. 12—Kursk; Russian submarine sank in Barents Sea; 118.
2000, Sept. 26—Express Samina; Greek ferry sank off Paros, Greece; 81+.
2001, Feb. 9—Ehime Maru; Japanese trawler sunk by surfacing U.S. submarine *Greeneville*, near Hawaii; 9.

Some Notable Aircraft Disasters Since 1937

In a coordinated terrorist attack on U.S. targets, Sept. 11, 2001, 4 planes were hijacked and crashed, with all persons on board (a total of 265, including what were said to be 19 hijackers) killed:

American Airlines Flight 11, a Boeing 767-200 with 81 passengers plus 11 crew, took off from Boston's Logan Airport, 7:59 AM; planned destination was Los Angeles; crashed into Tower 1 of the World Trade Center in New York City, 8:48 AM.

United Airlines Flight 175, a Boeing 767-200, with 56 passengers plus 9 crew, took off from Boston's Logan Airport, bound for Los Angeles; crashed into Tower 2 of the World Trade Center in New York City, 9:03 AM.

American Airlines Flight 77, a Boeing 757-200, with 58 passengers plus 6 crew, left from Dulles International Airport near Washington, DC; may have been intended to hit a different target, but crashed into the Pentagon outside Washington, DC, 9:45 AM.

United Air Lines Flight 93, a Boeing 757-200 with 37 passengers plus 7 crew, bound for San Francisco, departed from Newark (NJ) Airport; apparently diverted from hijackers' target; crashed near Shanksville, in southwestern PA, 10:10 AM.

Date	Aircraft	Site of accident	Deaths
1937, May 6	German zeppelin Hindenburg	Burned at mooring, Lakehurst, NJ	36*
1944, Aug. 23	U.S. Air Force B-24 Liberator bomber	Hit school, Freckleton, England	61*
1945, July 28	U.S. Army B-25	Hit Empire State Building, New York, NY	14*
1952, Dec. 20	U.S. Air Force C-124	Fell, burned, Moses Lake, WA	87
1953, Mar. 3	Canadian Pacific Comet Jet	Karachi, Pakistan	11[1]
1953, June 18	U.S. Air Force C-124	Crashed, burned near Tokyo	129
1955, Oct. 6	United Airlines DC-4	Crashed in Medicine Bow Peak, WY	66
1955, Nov. 1	United Airlines DC-6B	Exploded, crashed near Longmont, CO	44[2]
1956, June 20	Venezuelan Super-Constellation	Crashed in Atlantic off Asbury Park, NJ	74
1956, June 30	TWA Super-Const., United DC-7	Collided over Grand Canyon, AZ	128
1960, Dec. 16	United DC-8 jet, TWA Super-Const.	Collided over New York City	134[3]
1962, Mar. 16	Flying Tiger Super-Constellation	Vanished in W Pacific	107
1962, June 3	Air France Boeing 707 jet	Crashed on takeoff from Paris	130
1962, June 22	Air France Boeing 707 jet	Crashed in storm, Guadeloupe, W.I.	113
1963, June 3	Chartered Northwest Airlines DC-7	Crashed in Pacific off British Columbia	101
1963, Nov. 29	Trans-Canada Airlines DC-8F	Crashed after takeoff from Montreal	118
1965, May 20	Pakistani Boeing 720-B	Crashed at Cairo, Egypt, airport	121
1966, Jan. 24	Air India Boeing 707 jetliner	Crashed on Mont Blanc, France-Italy	117
1966, Feb. 4	All-Nippon Boeing 727	Plunged into Tokyo Bay	133
1966, Mar. 5	BOAC Boeing 707 jetliner	Crashed on Mount Fuji, Japan	124
1966, Dec. 24	U.S. military-chartered CL-44	Crashed into village in South Vietnam	129*
1967, Apr. 20	Swiss Britannia turboprop	Crashed at Nicosia, Cyprus	126
1967, July 19	Piedmont Boeing 727, Cessna 310	Collided in air, Hendersonville, NC	82
1968, Apr. 20	S. African Airways Boeing 707	Crashed on takeoff, Windhoek, South-West Africa	122
1968, May 3	Braniff International Electra	Crashed in storm near Dawson, TX	85
1969, Mar. 16	Venezuelan DC-9	Crashed after takeoff from Maracaibo, Venezuela	155[4]
1969, Dec. 8	Olympic Airways DC-6B	Crashed near Athens in storm	93
1970, Feb. 15	Dominican DC-9	Crashed into sea on takeoff from Santo Domingo	102
1970, July 3	British chartered jetliner	Crashed near Barcelona, Spain	112
1970, July 5	Air Canada DC-8	Crashed near Toronto International Airport	108
1970, Aug. 9	Peruvian turbojet	Crashed after takeoff from Cuzco, Peru	101*
1970, Nov. 14	Southern Airways DC-9	Crashed in mountains near Huntington, WV	75[5]
1971, July 30	All-Nippon Boeing 727 and Japanese Air Force F-86	Collided over Morioka, Japan	162[6]
1971, Sept. 4	Alaska Airlines Boeing 727	Crashed into mountain near Juneau, AK	111
1972, Aug. 14	East German Ilyushin-62	Crashed on takeoff, East Berlin	156
1972, Oct. 13	Aeroflot Ilyushin-62	Crashed near Moscow	176

Date	Aircraft	Site of accident	Deaths
1972, Dec. 3	Chartered Spanish airliner	Crashed on takeoff, Canary Islands	155
1972, Dec. 29	Eastern Airlines Lockheed Tristar	Crashed on approach to Miami Intl. Airport	101
1973, Jan. 22	Chartered Boeing 707	Burst into flames during landing, Kano Airport, Nigeria	176
1973, Feb. 21	Libyan jetliner	Shot down by Israeli fighter planes over Sinai	108
1973, Apr. 10	British Vanguard turboprop	Crashed during snowstorm at Basel, Switzerland	104
1973, June 3	Soviet Supersonic TU-144	Crashed near Goussainville, France	14[7]
1973, July 11	Brazilian Boeing 707	Crashed on approach to Orly Airport, Paris	122
1973, July 31	Delta Airlines jetliner	Crashed, landing in fog at Logan Airport, Boston	89
1973, Dec. 23	French Caravelle jet	Crashed in Morocco	106
1974, Mar. 3	Turkish DC-10 jet	Crashed at Ermenonville near Paris	346
1974, Apr. 23	Pan American 707 jet	Crashed in Bali, Indonesia	107
1974, Dec. 1	TWA-727	Crashed in storm, Upperville, VA	92
1974, Dec. 4	Dutch-chartered DC-8	Crashed in storm near Colombo, Sri Lanka	191
1975, Apr. 4	Air Force Galaxy C-5A	Crashed near Saigon, S Viet., after takeoff (carrying orphans)	172
1975, June 24	Eastern Airlines 727 jet	Crashed in storm, JFK Airport, NY	113
1975, Aug. 3	Chartered 707	Hit mountainside, Agadir, Morocco	188
1976, Sept. 10	British Airways Trident, Yugoslav DC-9	Collided near Zagreb, Yugoslavia	176
1976, Sept. 19	Turkish 727	Hit mountain, S Turkey	155
1976, Oct. 13	Bolivian 707 cargo jet	Crashed in Santa Cruz, Bolivia	100[8]
1977, Mar. 27	KLM 747, Pan American 747	Collided on runway, Tenerife, Canary Islands	582[9]
1977, Nov. 19	TAP Boeing 727	Crashed on Madeira	130
1977, Dec. 4	Malaysian Boeing 737	Hijacked, then exploded in mid-air over Straits of Johore	100
1977, Dec. 13	U.S. DC-3	Crashed after takeoff at Evansville, IN	29[10]
1978, Jan. 1	Air India 747	Exploded, crashed into sea off Bombay	213
1978, Sept. 25	Boeing 727, Cessna 172	Collided in air, San Diego, CA	150
1978, Nov. 15	Chartered DC-8	Crashed near Colombo, Sri Lanka	183
1979, May 25	American Airlines DC-10	Crashed after takeoff at O'Hare Intl. Airport, Chicago	275[11]
1979, Aug. 17	Two Soviet Aeroflot jetliners	Collided over Ukraine	173
1979, Nov. 26	Pakistani Boeing 707	Crashed near Jidda, Saudi Arabia	156
1979, Nov. 28	New Zealand DC-10	Crashed into mountain in Antarctica	257
1980, Mar. 14	Polish Ilyushin 62	Crashed making emergency landing, Warsaw	87[12]
1980, Aug. 19	Saudi Arabian Tristar	Burned after emergency landing, Riyadh	301
1981, Dec. 1	Yugoslavian DC-9	Crashed into mountain in Corsica	178
1982, Jan. 13	Air Florida Boeing 737	Crashed into Potomac R. after takeoff	78
1982, July 9	Pan Am Boeing 727	Crashed after takeoff in Kenner, LA	153[13]
1983, Sept. 1	S. Korean Boeing 747	Shot down after violating Soviet airspace	269
1983, Nov. 27	Colombian Boeing 747	Crashed near Barajas Airport, Madrid	183
1985, Feb. 19	Spanish Boeing 727	Crashed into Mt. Oiz, Spain	148
1985, June 23	Air-India Boeing 747	Crashed into Atlantic Ocean S of Ireland	329
1985, Aug. 2	Delta Air Lines L-1011	Crashed at Dallas-Ft. Worth Intl. Airport	137
1985, Aug. 12	Japan Air Lines Boeing 747	Crashed into Mt. Ogura, Japan	520[14]
1985, Dec. 12	Arrow Air DC-8	Crashed after takeoff in Gander, Newfoundland	256[15]
1986, Mar. 31	Mexican Boeing 727	Crashed NW of Mexico City	166
1986, Aug. 31	Aeromexico DC-9	Collided with Piper PA-28 over Cerritos, CA	82[16]
1987, May 9	Polish Ilyushin 62M	Crashed after takeoff in Warsaw, Poland	183
1987, Aug. 16	Northwest Airlines MD-82	Crashed after takeoff in Romulus, MI	156
1987, Nov. 28	S. African Boeing 747	Crashed into Indian Ocean near Mauritius	159
1987, Nov. 29	S. Korean Boeing 707	Exploded over Thai-Burmese border	155
1988, Mar. 17	Colombian Boeing 707	Crashed into mountainside near Venezuela border	137
1988, July 3	Iranian A300 Airbus	Shot down by U.S. Navy warship *Vincennes* over Persian Gulf	290
1988, Dec. 21	Pan Am Boeing 747	Exploded and crashed in Lockerbie, Scotland	270[17]
1989, Feb. 8	U.S. Boeing 707	Crashed into mountain in Azores Islands off Portugal	144
1989, June 7	Suriname DC-8	Crashed near Paramaribo Airport, Suriname	168
1989, July 19	United Airlines DC-10	Crashed while landing in Sioux City, IA	111
1989, Sept. 19	French DC-10	Exploded in air over Niger	171
1990, Oct. 2	Chinese airline Boeing 737	Hijacked; upon landing in Guangzhou, crashed on ground	132
1991, May 26	Lauda-Air Boeing 767-300	Exploded over rural Thailand	223
1991, July 11	Nigerian DC-8	Crashed while landing at Jidda, Saudi Arabia	261
1991, Oct. 5	Indonesian military transport	Crashed after takeoff from Jakarta	137*
1992, July 31	Thai Airbus A-300-310	Crashed into mountain S. of Kathmandu, Nepal	113
1992, Oct. 4	El Al Boeing 747-200F	Crashed into 2 apartment bldgs., Amsterdam, Netherlands	120*
1994, Jan. 3	Aeroflot TU-154	Crashed and exploded after takeoff in Irkhutsk, Russia	125[18]
1994, Apr. 26	China Airlines Airbus A-300-600R	Crashed at Japan's Nagoya Airport	264
1994, June 16	China Northwest Airlines TU-154	Crashed 10 min. after takeoff	160
1994, Sept. 8	USAir Boeing 737-300	Crashed in Aliquippa, PA, near Pittsburgh Intl. Airport	132
1994, Oct. 31	American Eagle ATR-72-210	Crashed in field near Roselawn, IN	68
1995, Aug. 11	Aviateca Boeing 737	Crashed into Chichontepec volcano, El Salvador	65
1995, Dec. 20	American Airlines Boeing 757	Crashed into mountain 50 mi N of Cali, Colombia	160
1996, Jan. 8	Antonova 32 cargo jet	Crashed into central market, Kinshasa, Zaire	350+*
1996, Feb. 6	Turkish Boeing 757	Crashed into Atlantic Ocean, off Dominican Republic	189
1996, Apr. 25	T-43, a military version of a Boeing 737	Crashed into mountain near Dubrovnik, Croatia	35[19]
1996, May 11	ValuJet DC-9	Crashed into the Florida Everglades after takeoff	110
1996, July 17	Trans World Airlines Boeing 747	Exploded and crashed in Atlantic Ocean, off Long Isl., NY	230
1996, Aug. 29	Vnukovo TU-154	Crashed into mountain on Arctic island of Spitsbergen	141
1996, Oct. 2	Aeroperu Boeing 757	Crashed in Pacific after takeoff from Lima, Peru	70
1996, Oct. 31	Brazilian TAM Fokker-100	Crashed into houses in São Paulo, Brazil	98[20]
1996, Nov. 7	Nigerian Boeing 727	Crashed into a lagoon 40 mi SE of Lagos, Nigeria	143
1996, Nov. 12	Saudi Arabian Boeing 747, Kazakh Ilyushin-76 cargo plane	Collided in midair near New Delhi, India	349[21]
1996, Nov. 23	Ethiopian Boeing 767	Hijacked, then crashed in Indian Ocean off the Comoros	127
1997, Jan. 9	Comair Embraer 120	Crashed on approach into Detroit Metro. Airport	29
1997, Feb. 4	2 Sikorsky CH-53 transport helicopters	Collided in midair over northern Galilee, Israel	73
1997, May 8	China Southern Airlines Boeing 737	Crashed on approach into Shenzhen's Huangtian Airport	35
1997, July 11	Cubana de Aviación Antonov-24	Crashed into the Caribbean off SE Cuba	44
1997, Aug. 6	Korean Air Boeing 747-300	Crashed into jungle on Guam on approach into airport	228
1997, Sept. 3	Vietnamese Airlines Tupolev TU-134	Crashed on approach into Phnom Penh airport	64

Date	Aircraft	Site of accident	Deaths
1997, Sept. 14	U.S. C-141 cargo plane, German TU-154	Collided in midair off SW Africa	33
1997, Sept. 26	Indonesian Airbus A-300	Crashed near Medan, Indonesia, airport	234
1997, Oct. 10	Austral Airlines DC-9-32	Crashed and exploded near Neuvo Berlin, Uruguay	74
1997, Dec. 6	Russian AN-124 transport cargo plane	Crashed into apartment complex near Irkutsk, Siberia	67*
1997, Dec. 15	Chartered TU-154 from Tajikistan	Crashed in desert near Sharja, U.A.E., airport	85
1997, Dec. 17	Chartered Yakovlev-42 from Ukraine	Crashed in mountains near Katerini, Greece	70
1997, Dec. 19	SilkAir Boeing 737-300	Crashed in Musi River, Sumatra, Indonesia	104
1998, Jan. 14	Afghan cargo plane	Crashed into mountain, SW Pakistan	50+
1998, Feb. 2	Cebu Pacific Air DC-9-32	Crashed into mountain near Cagayan de Oro, Philippines	104
1998, Feb. 16	China Airlines Airbus 300-622R	Crashed on approach to airport, Taipei, Taiwan	203[22]
1998, Apr. 20	Air France Boeing 727-200	Crashed into mountain after takeoff from Bogotá, Colombia	53
1998, Sept. 2	Swissair MD-11	Crashed into Atlantic Ocean off Halifax, Nova Scotia	229
1998, Sept. 25	Pauknair BAE146	Crashed into hillside in Morocco	38
1998, Oct. 11	Congo Air Lines Boeing 727	Shot down by rebels in Kindu, Congo	40
1998, Dec. 11	Thai Airways Airbus A310-200	Crashed short of runway at Surat Thani airport, southern Thailand	101
1999, Feb. 3	Chartered Antonov plane	Crashed in residential area of Luanda, Angola	28
1999, Feb. 24	China Southwest Airlines TU-154	Crashed on approach to Wenzhou airport, eastern China	61
1999, Sept. 1	LAPA Boeing 737-200	Crashed on takeoff from Jorge Newbery Airport, Buenos Aires	74[23]
1999, Oct. 31	EgyptAir Boeing 767-300	Crashed off Nantucket, MA	217
1999, Dec. 25	Cubana de Aviacion Yak-42	Crashed into mountain near Valencia, Venezuela	22
2000, Jan. 14	Chartered Shorts SD-360	Crashed into Mediterranean Sea off Tripoli, Libya	22
2000, Jan. 30	Kenya Airways Airbus A310	Crashed into Atlantic Ocean after takeoff from Abidjan, Cote d'Ivoire	169
2000, Apr. 8	Marine Corps V-22 Osprey	Crashed landing at Marana, AZ	19
2000, Jan. 31	Alaska Airlines MD-83	Crashed into Pacific Ocean NW of Malibu, CA	88
2000, Apr. 19	Air Philippines Boeing 737-200	Crashed by Davao airport	131
2000, May 21	Chartered Jetstream 31	Crashed near Wilkes-Barre, PA	19
2000, July 25	Air France Concorde	Crashed into hotel after takeoff from Paris	113[25]
2000, Aug. 9	Piper Navajo and Piper Seminole	Collided over a housing development in Burlington, NJ	11
2000, Aug. 23	Gulf Air Airbus A320	Crashed into Persian Gulf near Manama, Bahrain	143
2000, Oct. 31	Singapore Airlines 747-400	Crashed immediately after takeoff, Taipei, Taiwan	81
2000, Oct. 31	Chartered Antonov 26	Exploded after takeoff in northern Angola	50
2000, Nov. 15	Chartered Antonov 24	Crashed after takeoff from Luanda, Angola	40+
2001, Jan. 25	Rutaca Airlines DC-3	Crashed shortly after takeoff from Cuidad Bolivar, Venezuela	24
2001, Jan. 27	Chartered Beechcraft King Air 200	Crashed after takeoff from Boulder, CO	10[26]
2001, Mar. 3	C23 Sherpa mil. transp.	Crashed in storm, central GA	21
2001, Mar. 29	Chartered Gulfstream III jet	Crashed into hillside on approach to Aspen, CO	18
2001, Apr. 7	M-17 helicopter	Crashed into mountain S. of Hanoi, Vietnam	16[27]
2001, July 3	Vladivostokavia Tu-154	Crashed on approach to landing at Irkutsk, Russia	145
2001, Sept. 11	2 Boeing 767s, 2 Boeing 757s	2 crashed into World Trade Center towers, NYC; 1 into Pentagon; 1 near Shanksville, PA	265

*Including those on ground and in buildings. (1) First fatal crash of commercial jet plane. (2) Caused by bomb planted by John G. Graham in insurance plot to kill his mother, a passenger. (3) Incl. all 128 aboard planes and 6 on ground. (4) Killed 84 on the plane and 71 on the ground. (5) Incl. 43 Marshall Univ. football players and coaches. (6) Airliner-fighter crash; pilot of fighter parachuted to safety, was arrested for negligence. (7) First supersonic plane crash; killed 6 crewmen and 8 on ground; there were no passengers. (8) Crew of 3 killed; 97, mostly children, killed on the ground. (9) World's worst airline disaster. (10) Incl. Univ. of Evansville basketball team. (11) Incl. 2 on the ground. Highest death toll in U.S. aviation history. (12) Incl. 22 members of U.S. boxing team. (13) Incl. 8 on the ground. (14) Worst single-plane disaster. (15) Incl. 248 members of U.S. 101st Airborne Division. (16) Incl. 15 on the ground. (17) Incl. 11 on the ground. (18) Incl. 1 on the ground. (19) Incl. U.S. Sec. of Commerce Ronald Brown. (20) Incl. 2 on the ground. (21) World's worst midair collision. (22) Incl. 6 on the ground. (23) Incl. 10 on the ground. (24) Incl. 4 on the ground. (25) World's first Concorde crash; deaths incl. 5 on the ground. (26) Incl. 7 players and staff of Oklahoma State Univ. men's basketball team. (27) Carried U.S. mil. personnel, searching for MIAs from Vietnam War.

Some Notable Railroad Disasters

Date	Location	Deaths
1876, Dec. 29	Ashtabula, OH	92
1880, Aug. 11	Mays Landing, NJ	40
1887, Aug. 10	Chatsworth, IL	81
1888, Oct. 10	Mud Run, PA	55
1889, June 12	Amagh, Ireland	80
1891, June 14	Nr. Basel, Switzerland	100
1896, July 30	Atlantic City, NJ	60
1903, Dec. 23	Laurel Run, PA	53
1904, Aug. 7	Eden, CO	96
1904, Sept. 24	New Market, TN	56
1906, Mar. 16	Florence, CO	35
1906, Oct. 28	Atlantic City, NJ	40
1906, Dec. 30	Washington, DC	53
1907, Jan. 2	Volland, KS	33
1907, Jan. 19	Fowler, IN	29
1907, Feb. 16	New York, NY	22
1907, Feb. 23	Colton, CA	26
1907, May 11	Lompoc, CA	36
1907, July 20	Salem, MI	33
1908, Sept. 25	Young's Point, MT	21
1909, Jan. 15	Dotsero, CO	21
1910, Mar. 1	Wellington, WA	96
1910, Mar. 21	Green Mountain, IA	55
1911, Aug. 25	Manchester, NY	29
1912, July 4	East Corning, NY	39
1912, July 5	Ligonier, PA	23
1914, Aug. 5	Tipton Ford, MO	43
1914, Sept. 15	Lebanon, MO	28
1915, May 22	Nr. Gretna, Scotland	227
1916, Mar. 29	Amherst, OH	27
1917, Sept. 28	Kellyville, OK	23
1917, Dec. 12	Modane, France	543
1917, Dec. 20	Shepherdsville, KY	46
1918, June 22	Ivanhoe, IN	68
1918, July 9	Nashville, TN	101
1918, Nov. 1	Brooklyn, NY	97
1919, Jan. 12	South Byron, NY	22
1919, Dec. 20	Onawa, ME	23
1921, Feb. 27	Porter, IN	37
1921, Dec. 5	Woodmont, PA	27
1922, Aug. 5	Sulphur Spring, MO	34
1922, Dec. 13	Humble, TX	22
1923, Sept. 27	Lockett, WY	31
1925, June 16	Hackettstown, NJ	50
1925, Oct. 27	Victoria, MS	21
1926, Sept. 5	Waco, CO	30
1937, July 16	Nr. Patna, India	107
1938, June 19	Saugus, MT	47
1939, Aug. 12	Harney, NV	24
1939, Dec. 22	Near Magdeburg, Germany	132
1939, Dec. 22	Near Friedrichshafen, Germany	99
1940, Apr. 19	Little Falls, NY	31
1940, July 31	Cuyahoga Falls, OH	43
1943, Aug. 29	Wayland, NY	27
1943, Sept. 6	Frankford Junction, Philad. PA	79
1943, Dec. 16	Between Rennert and Buie, NC	72
1944, Jan. 16	Leon Prov., Spain	500
1944, Mar. 2	Salerno, Italy	521
1944, July 6	High Bluff, TN	35
1944, Aug. 4	Near Stockton, GA	47
1944, Sept. 14	Dewey, IN	29
1944, Dec. 31	Bagley, UT	50
1945, Aug. 9	Michigan, ND	34
1946, Mar. 20	Aracaju, Mexico	185
1946, Apr. 25	Naperville, IL	45
1947, Feb. 18	Gallitzin, PA	24
1949, Oct. 22	Nr. Dwor, Poland	200+
1950, Feb. 17	Rockville Centre, NY	31

Date	Location	Deaths
1950, Sept. 11	Coshocton, OH	33
1950, Nov. 22	Richmond Hill, NY	79
1951, Feb. 6	Woodbridge, NJ	84
1952, Mar. 4	Nr. Rio de Janeiro, Brazil	119
1952, July 9	Rzepin, Poland	160
1952, Oct. 8	Harrow, England	112
1953, Mar. 27	Conneaut, OH	21
1955, Apr. 3	Guadalajara, Mexico	300
1956, Jan. 22	Los Angeles, CA	30
1957, Sept. 1	Kendal, Jamaica	178
1957, Sept. 29	Montgomery, W Pakistan	250
1957, Dec. 4	London, England	90
1958, May 8	Rio de Janeiro, Brazil	128
1958, Sept. 15	Elizabethport, NJ	48
1960, Nov. 14	Pardubice, Czech.	110
1962, Jan. 8	Woerden, Netherlands	91
1962, May 3	Tokyo, Japan	163
1964, July 26	Porto, Portugal	94
1970, Feb. 1	Buenos Aires, Argentina	236
1972, June 16	Vierzy, France	107
1972, July 21	Seville, Spain	76
1972, Oct. 6	Saltillo, Mexico	208
1972, Oct. 30	Chicago, IL	45
1974, Aug. 30	Zagreb, Yugoslavia	153
1975, Feb. 28	London subway train	41
1977, Jan. 18	Granville, Australia	83
1981, June 6	Bihar, India	700+
1982, Jan. 27	El Asnam, Algeria	130
1982, July 11	Tepic, Mexico	120
1983, Feb. 19	Empalme, Mexico	100
1987, July 2	Kasumbalesha Shaba, Zaire	125
1988, Dec. 12	London, England	115
1989, Jan. 15	Maizdi Khan, Bangladesh	110+
1990, Jan. 4	Sindh Prov., Pakistan	210+
1991, May 14	Shigaraki, Japan	42
1993, Sept. 22	Big Bayou Conot, AL	47
1994, Mar. 8	Nr. Durban, South Africa	63
1994, Sept. 22	Tolunda, Angola	300
1995, Aug. 20	Firozabad, India	358
1997, Mar. 3	Punjab State, Pakistan	125
1997, Mar. 31	Huarte Arakil, Spain	21
1997, Apr. 29	Hunan, China	58
1997, May 4	Rwanda	100+
1997, Sept. 14	Central India	77
1998, June 3	Eschede, Germany	102
1998, Feb. 19	Yaounde, Cameroon	100+
1999, Mar. 15	Bourbonnais, IL	11
1999, Mar. 24	Nairobi, Kenya	32+
1999, Aug. 2	Gauhati, India	285+
1999, Oct. 5	London, England	31
2000, Jan. 4	Rena, Norway	35
2000, July 28	São Paulo, Brazil	12
2000, Nov. 11	Kaprun, Austria	155
2001, Feb. 28	Great Heck, England	13
2001, Mar. 18	Nr. Des Moines, IA	1

Principal U.S. Mine Disasters Since 1900

Source: Bureau of Mines, U.S. Dept. of the Interior; Mine Safety and Health Admin., U.S. Dept. of Labor

(All are bituminous-coal mines unless otherwise noted.)

Date	Location	Deaths
1900, May 1	Scofield, UT	200
1902, May 19	Coal Creek, TN	184
1902, July 10	Johnstown, PA	112
1903, June 30	Hanna, WY	169
1904, Jan. 25	Cheswick, PA	179
1905, Feb. 26	Virginia City, AL	112
1907, Jan. 29	Stuart, WV	84
1907, Dec. 6	Monongah, WV	361
1907, Dec. 19	Jacobs Creek, PA	239
1908, Nov. 28	Marianna, PA	154
1909, Nov. 13	Cherry, IL	259
1910, Jan. 31	Primero, CO	75
1910, May 5	Palos, AL	90
1910, Nov.8	Delagua, CO	79
1911, Apr. 8	Littleton, AL	128
1911, Dec. 9	Briceville, TN	84
1912, Mar. 26	Jed, WV	83
1913, Apr. 23	Finleyville, PA	96
1913, Oct. 22	Dawson, NM	263
1914, Apr. 28	Eccles, WV	181
1915, Mar. 2	Layland, WV	112
1917, Apr. 27	Hastings, CO	121
1917, June 8	Butte, MT[1]	163
1919, June 5	Wilkes-Barre, PA[2]	92
1922, Nov. 6	Spangler, PA	77
1922, Nov. 22	Dolomite, AL	90
1923, Feb. 8	Dawson, NM	120
1923, Aug. 14	Kemmerer, WY	99
1924, Mar. 8	Castle Gate, UT	171
1924, Apr. 28	Benwood, WV	119
1926, Jan. 13	Wilburton, OK	91
1927, Apr. 30	Everettville, WV	97
1928, May 19	Mather, PA	195
1930, Nov. 5	Millfield, OH	82
1940, Jan. 10	Bartley, WV	91
1947, Mar. 25	Centralia, IL	111
1951, Dec. 21	West Frankfort, IL	119
1968, Nov. 20	Farmington, WV	78
1970, Dec. 30	Hyden, KY	38
1972, May 2	Kellogg, ID[1]	91
1976, Mar. 9	Oven Fork, KY	15
1981, Apr. 15	Redstone, CO	15
1981, Dec. 8	Whitwell, TN	13
1984, Dec. 19	Huntington, UT	27
1989, Sept. 13	Sturgis, KY	10
2001, Sept. 23	Brookwood, AL	13

Note: World's worst mine disaster killed 1,549 workers in Honkeiko Colliery in Manchuria, Apr. 25, 1942. (1) Metal mine. (2) Anthracite mine.

Some Notable U.S. Tornadoes Since 1925

Date	Location	Deaths
1925, Mar. 18	MO, IL, IN	689
1927, Apr. 12	Rock Springs, TX	74
1927, May 9	AR, Poplar Bluff, MO	92
1927, Sept. 29	St. Louis, MO	90
1930, May 6	Hill, Navarro, Ellis Co., TX	41
1932, Mar. 21	AL (series of tornadoes)	268
1936, Apr. 5	MS, GA	455
1936, Apr. 6	Gainesville, GA	203
1938, Sept. 29	Charleston, SC	32
1942, Mar. 16	Central to NE Mississippi	75
1942, Apr. 27	Rogers and Mayes Co., OK	52
1944, June 23	OH, PA, WV, MD	150
1945, Apr. 12	OK-AR	102
1947, Apr. 9	TX, OK, KS	169
1948, Mar. 19	Bunker Hill and Gillespie, IL	33
1949, Jan. 3	LA and AR	58
1952, Mar. 21	AR, MO, TN (series)	208
1953, May 11	Waco, TX	114
1953, June 8	MI, OH	142
1953, June 9	Worcester and vicinity, MA	90
1953, Dec. 5	Vicksburg, MS	38
1955, May 25	KS, MO, OK, TX	115
1957, May 20	KS, MO	48
1958, June, 4	NW Wisconsin	30
1959, Feb. 10	St. Louis, MO	21
1960, May 5, 6	Southeastern OK, AR	30
1965, Apr. 11	IN, IL, OH, MI, WI	271
1966, Mar. 3	Jackson, MS	57
1966, Mar. 3	MS, AL	61
1967, Apr. 21	IL, MI	33
1968, May 15	Midwest	71
1969, Jan. 23	MS	32
1971, Feb. 21	Mississippi delta	110
1973, May 26-27	South, Midwest (series)	47
1974, Apr. 3-4	AL, GA, TN, KY, OH	315
1977, Apr. 4	AL, MS, GA	22
1979, Apr. 10	TX, OK	60
1984, Mar. 28	NC, SC	57
1985, May 31	NY, PA, OH, Ont. (series)	75
1987, May 22	Saragosa, TX	29
1989, Nov. 15	Huntsville, AL	18
1990, Aug. 28	Northern IL	25
1991, Apr. 26	KS, OK	23
1992, Nov. 21-23	South, Midwest	26
1994, Mar. 27-28	AL, TN, GA, NC, SC (series)	52
1995, May 6-7	Southern OK, northern TX	23
1997, Mar. 1	Central AR	26
1997, May 27	Jarrell, TX	27
1998, Feb. 22-23	Central FL	42
1998, Mar. 20	Northeast GA	12

Date	Location	Deaths
1998, Mar. 24	Eastern India	145
1998, Apr. 8	AL, GA, MS	39
1998, Apr. 16	AK, KY, TN	10
1998, May 30	Spencer, SD	6
1999, Jan. 17	Western TN	8
1999, Jan. 21	AK, TN	8
1999, Apr. 3	Northwestern LA	6
1999, Apr. 9	OH, IL, IN, MO	6
1999, May 3-4	OK, KS	42
2000, Feb. 14	Southwest GA	22+
2000, Mar. 28	TX	5
2000, July 14	Alberta	11
2000, Dec. 16	AL	12

IT'S A FACT: In what may have been the worst single natural disaster in U.S. history, about 3,000 homes were swept away and at least 6,000 people were killed, Sept. 8, 1900, when a 20-foot storm surge, spawned by a hurricane, inundated Galveston Island in Texas. One reason for the high death toll is that forecasters lacked the technology to track storms accurately and provide adequate advance warning.

Some Notable Hurricanes, Typhoons, Blizzards, Other Storms

H.—hurricane; T.—typhoon

Date	Location	Deaths
1888, Mar. 11-14	Blizzard, eastern U.S.	400
1900, Sept. 8	H., Galveston, TX	6,000+
1906, Sept. 19-24	H., LA, MS	350
1906, Sept. 18	Typhoon, Hong Kong	10,000
1915, Sept. 29	H., LA	500
1926, Sept. 11-22	H., FL, AL	243
1926, Oct. 20	H., Cuba	600
1928, Sept. 6-20	H., southern FL	1,836
1930, Sept. 3	H., Dominican Republic	2,000
1935, Aug. 29-Sept. 10	H., Caribbean, southeastern U.S.	400+
1938, Sept. 21	H., Long Island, NY; New England	600
1940, Nov. 11-12	Blizzard, NE, Midwest U.S.	144
1942, Oct. 15-16	H., Bengal, India	40,000
1944, Sept. 9-16	H., NC to New England	46
1947, Dec. 26	Blizzard, NYC, N Atlantic states	55
1952, Oct. 22	Typhoon, Philippines	440
1954, Aug. 30	H. *Carol,* northeastern U.S.	68
1954, Oct. 5-18	H. *Hazel,* E Canada, U.S.; Haiti	347
1955, Aug. 12-13	H. *Connie,* NC, SC, VA, MD	43
1955, Aug. 7-21	H. *Diane,* eastern U.S.	400
1955, Sept. 19	H. *Hilda,* Mexico	200
1955, Sept. 22-28	H. *Janet,* Caribbean	500
1956, Feb. 1-29	Blizzard, W Europe	1,000
1957, June 25-30	H. *Audrey,* TX to AL	390
1958, Feb. 15-16	Blizzard, northeastern U.S.	171
1959, Sept. 17-19	T. *Sarah,* Japan, S. Korea	2,000
1959, Sept. 26-27	T. *Vera,* Honshu, Japan	4,466
1960, Sept. 4-12	H. *Donna,* Caribbean, E U.S.	148
1961, Sept. 11-14	H. *Carla,* TX	46
1961, Oct. 31	H. *Hattie,* Br. Honduras	400
1963, May 28-29	Windstorm, Bangladesh	22,000
1963, Oct. 4-8	H. *Flora,* Caribbean	6,000
1964, Oct. 4-7	H. *Hilda,* LA, MS, GA	38
1964, June 30	T. *Winnie,* N Philippines	107
1964, Sept. 5	T. *Ruby,* Hong Kong and China	735
1965, May 11-12	Windstorm, Bangladesh	17,000
1965, June 1-2	Windstorm, Bangladesh	30,000
1965, Sept. 7-12	H. *Betsy,* FL, MS, LA	74
1965, Dec. 15	Windstorm, Bangladesh	10,000
1966, June 4-10	H. *Alma,* Honduras, SE U.S.	51
1966, Sept. 24-30	H. *Inez,* Carib., FL, Mexico	293
1967, July 9	T. *Billie,* SW Japan	347
1967, Sept. 5-23	H. *Beulah,* Carib., Mex., TX	54
1967, Dec. 12-20	Blizzard, Southwest U.S.	51
1968, Nov. 18-28	T. *Nina,* Philippines	63
1969, Aug. 17-18	H. *Camille,* MS, LA	256
1970, July 30-Aug. 5	H. *Celia,* Cuba, FL, TX	31
1970, Aug. 20-21	H. *Dorothy,* Martinique	42
1970, Sept. 15	T. *Georgia,* Philippines	300
1970, Oct. 14	T. *Sening,* Philippines	583
1970, Oct. 15	T. *Titang,* Philippines	526
1970, Nov. 13	Cyclone, Bangladesh	300,000
1971, Aug. 1	T. *Rose,* Hong Kong	130
1972, June 19-29	H. *Agnes,* FL to NY	118
1972, Dec. 3	T. *Theresa,* Philippines	169
1973, June-Aug.	Monsoon rains, India	1,217
1974, June 11	Storm Dinah, Luzon Isl., Phil.	71
1974, July 11	T. *Gilda,* Japan, S. Korea	108
1974, Sept. 19-20	H. *Fifi,* Honduras	2,000
1974, Dec. 25	Cyclone leveled Darwin, Austral.	50
1975, Sept. 13-27	H. *Eloise,* Caribbean, NE U.S.	71
1976, May 20	T. *Olga,* floods, Philippines	215
1977, July 25, 31	T. *Thelma,* T. *Vera,* Taiwan	39
1978, Oct. 27	T. *Rita,* Philippines	c. 400
1979, Aug. 30 - Sept. 7	H. *David,* Caribbean, E U.S.	1,100
1980, Aug. 4-11	H. *Allen,* Caribbean, TX	272
1981, Nov. 25	T. *Irma,* Luzon Isl., Phil.	176
1983, June	Monsoon, India	900
1984, Sept. 2	T. *Ike,* S Philippines	1,363
1985, May 25	Cyclone, Bangladesh	10,000
1985, Oct. 26-Nov. 6	H. *Juan,* SE U.S.	97
1987, Nov. 25	T. *Nina,* Philippines	650
1988, Sept. 10-17	H. *Gilbert,* Caribbean, Gulf of Mex	260
1989, Sept. 16-22	H. *Hugo,* Caribbean, SE U.S.	504
1990, May 6-11	Cyclones, SE India	450
1991, Apr. 30	Cyclone, Bangladesh	139,000
1991, Nov. 5	Tropical storm, Philippines	7,000+
1992, Aug. 24-26	H. *Andrew,* southern FL, LA	23
1993, Mar. 13-14	Blizzard, eastern U.S.	200
1993, June	Monsoon, Bangladesh	2,000
1994, Nov. 8-18	Storm Gordon, Caribbean, FL	830
1995, Sept. 4-6	H. *Luis,* Caribbean	14
1995, Sept. 13-22	H. *Marilyn,* Virgin Isls., Carib.	13
1995, Oct. 2-4	H. *Opal,* S Mexico, FL, AL	59
1995, Nov. 2-3	T. *Angela,* Philippines	600+
1996, Jan. 7-8	Blizzard, northeastern U.S.	100
1996, July 8-13	H. *Bertha,* Carib., eastern U.S.	15
1996, Aug. 22	Blizzard, Himalayas, N India	239
1996, Aug. 29-Sept. 6	H. *Fran,* Carib., NC, VA, WV	28
1996, Sept. 9-10	H. *Hortense,* Caribbean	24
1996, Sept. 9	T. *Sally,* S China	114
1996, Nov. 6	Cyclone, Andhra Pradesh, India	1,000+
1996, Nov. 24-25	Ice storms, TX to MO	26
1996, Dec. 25	Tropical storm, E Malaysia	100+
1997, May 19	Cyclone, Bangladesh	108
1997, May 26	Rain storm, Philippines	29
1997, July 2	Storms, southeastern MI	16
1997, Aug. 18	Typhoon, Taiwan	24
1997, Sept. 27	Cyclone, S Bangladesh	c. 35
1997, Oct. 8-10	H. *Pauline,* SW Mexico	230
1997, Oct. 13	Cyclone, Tongi, Bangladesh	15+
1998, Feb. 4-6	Blizzard, KY, WV	10+
1998, June 9	Cyclone, Gujarat, India	1,320
1998, Aug.	Monsoon, Bangladesh	326
1998, Sept. 21-23	H. *Georges,* Caribbean, FL Keys, U.S. Gulf Coast	600+
1998, Oct. 27-29	H. *Mitch,* Honduras, Nicaragua, Guatemala, El Salvador	10,866+
1999, Sept. 4-17	H. *Floyd,* Bahamas, eastern seaboard, U.S.	69+
1999, Oct. 29	Cyclone, Eastern India	9,392
1999, Dec. 26-29	Gales, France, Switzerland, Germany	120
2000, Aug. 22-23	Typhoon *Bilis,* Taiwan	11
2000, Sept. 11	Typhoon *Saomai,* Japan	7
2000, Dec. 27	TX, OK, AR	40+
2001, June 6-17	Tropical storm *Allison,* SE U.S.	47
2001, July 30	Typhon, Taiwan	200

IT'S A FACT: On July 7, 2001, a plane tree collapsed in a storm killing at least 11 people at a concert outside Strasbourg, France.

Some Notable Floods, Tidal Waves

Date	Location	Deaths
1228	Holland	100,000
1642	China	300,000
1883, Aug. 27	Indonesia	36,000
1887	Huang He River, China	900,000
1889, May 31	Johnstown, PA	2,209
1903, June 15	Heppner, OR	325
1911	Chang Jiang River, China	100,000
1913, Mar. 25-27	OH, IN	732
1915, Aug. 17	Galveston, TX	275
1928, Mar. 13	Dam collapse, Saugus, CA	450
1928, Sept. 13	Lake Okeechobee, FL	2,000
1931, Aug.	Huang He River, China	3,700,000
1937, Jan. 22	OH, MS Valleys	250
1939	N China	200,000
1946, Apr. 1	HI, AK	159
1947, Sept. 20	Honshu Island, Japan	1,900
1951, Aug.	Manchuria	1,800
1953, Jan. 31	W Europe	2,000
1954, Aug. 17	Farahzad, Iran	2,000
1955, Oct. 7-12	India, Pakistan	1,700
1959, Nov. 1	W Mexico	2,000
1959, Dec. 2	Frejus, France	412
1960, Oct. 10	Bangladesh	6,000
1960, Oct. 31	Bangladesh	4,000
1962, Feb. 17	North Sea coast, Germany	343
1962, Sept. 27	Barcelona, Spain	445
1963, Oct. 9	Dam collapse, Vaiont, Italy	1,800
1966, Nov. 3-4	Florence, Venice, Italy	113
1967, Jan. 18-24	E Brazil	894
1967, Mar. 19	Rio de Janeiro, Brazil	436
1967, Nov. 26	Lisbon, Portugal	464
1968, Aug. 7-14	Gujarat State, India	1,000
1968, Oct. 7	NE India	780
1969, Jan. 18-26	Southern CA	100
1969, Mar. 17	Mundau Valley, Alagoas, Brazil	218
1969, Aug. 20-22	Western VA	189
1969, Sept. 15	South Korea	250
1969, Oct. 1-8	Tunisia	500
1970, May 20	Central Romania	160
1970, July 22	Himalayas, India	500
1971, Feb. 26	Rio de Janeiro, Brazil	130
1972, Feb. 26	Buffalo Creek, WV	118
1972, June 9	Rapid City, SD	236
1972, Aug. 7	Luzon Isl., Philippines	454
1972, Aug. 19-31	Pakistan	1,500
1974, Mar. 29	Tubaro, Brazil	1,000
1974, Aug. 12	Monty-Long, Bangladesh	2,500
1976, June 5	Teton Dam collapse, ID	11
1976, July 31	Big Thompson Canyon, CO	139
1976, Nov. 17	East Java, Indonesia	136
1977, July 19-20	Johnstown, PA	68
1977, Nov. 6	Toccoa, GA	39
1978, June-Sept.	N India	1,200
1979, Jan.-Feb.	Brazil	204
1979, July 17	Lomblem Isl., Indonesia	539
1979, Aug. 11	Morvi, India	15,000
1980, Feb. 13-22	Southern CA, AZ	26
1981, Apr.	N China	550
1981, July	Sichuan, Hubei Prov., China	1,300
1982, Jan. 23	Nr. Lima, Peru	600
1982, May 12	Guangdong, China	430
1982, Sept. 17-21	El Salvador, Guatemala	1,300+
1984, Aug-Sept.	South Korea	200+
1985, July 19	Dam collapse, N Italy	361
1987, Aug.-Sept.	N Bangladesh	1,000+
1988, Sept.	N India	1,000+
1990, June 14	Shadyside, OH	23
1991, Dec. 18-26	TX	18
1993, July-Aug.	Midwest	48
1994, July	GA, AL	32
1995, Jan. 30-Feb. 9	NW Europe	40
1995, July	Hunan Province, China	1,200
1995, Aug. 19	SW Morocco	136
1995, Dec. 25	KwaZulu Natal, South Africa	166
1996, Jan.	Northeastern U.S.	15+
1996, Feb. 17	Biak Isl., Indonesia	105
1996, April	Afghanistan	100+
1996, June-July	S China	950+
1996, Aug. 7	Pyrenees Mts., Spain	71
1996, Dec.-1997, Jan.	Northwestern U.S.	29
1997, Mar.	Ohio R. Valley	35
1997, July	Poland, Czech Republic	98
1997, Oct.	Israel, Egypt, Jordan	19
1997, Nov.	Spanish-Portuguese border	31+
1997, Nov.	Bardera, Somalia	1,300+
1998, Jan.	Kenya	86
1998, Feb.	California to Tijuana, Mexico	30+
1998, Mar.	SW Pakistan	300+
1998, July-Aug.	China	4,150
1998, July-Sept.	Bangladesh	1,441
1998, July 17	Papua New Guinea	3,000
1998, Aug. 24	S Texas, Mexico	16
1999, Aug. 1-4	S. Korea, Philippines, Vietnam, Thailand	188+
1999, Sept.-Oct.	NE Mexico	350+
1999, Oct.-Dec.	Central Vietnam	700+
1999, Feb. 6-11	Botswana	70+
1999, Dec.	Venezuela	9,000+
2000, Feb.-Mar.	Madagascar	150+
2000, Feb.-Mar.	Mozambique	700
2000, May 17	Timor Island	50+
2000, May 21	Colombia	21
2000, May 31	Gansu, China	36
2000, June 7	Sichaun, China	38
2000, June 8-12	Uttar Pradesh, India	43+
2000, Aug. 2	Himachal Pradesh, India	120+
2000, Aug. 2	Bhutan	200+
2000, Sept. 19-30	India, Bangladesh	1,000+
2000, Oct. 12-17	France, Britain, Italy, Switzerland	35
2001, Feb.	Mozambique	60+
2001, Aug.-Sept.	Southern Vietnam	182
2001, Aug. 1-6	Taiwan	100+
2001, Aug. 10-12	Northeastern Iran	67+
2001, Aug. 11	Phetchabun, Thailand	46+

Some Major Earthquakes

Source: Global Volcanism Network, Smithsonian Institution; U.S. Geological Survey, Dept. of the Interior; World Almanac research

Magnitude of earthquakes (Mag.) is measured on the Richter scale; each higher number represents a tenfold increase in energy. Adopted in 1935, the scale is applied to earthquakes as far back as reliable seismograms are available.

Date	Location	Deaths	Mag.
526, May 20	Antioch, Syria	250,000	NA
856	Corinth, Greece	45,000	"
1057	Chihli, China	25,000	"
1169, Feb. 11	Near Mt. Etna, Sicily	15,000[1]	"
1268	Cilicia, Asia Minor	60,000	"
1290, Sept. 27	Chihli, China	100,000	"
1293, May 20	Kamakura, Japan	30,000	"
1531, Jan. 26	Lisbon, Portugal	30,000	"
1556, Jan. 24	Shaanxi, China	830,000	"
1667, Nov.	Shemaka, Caucasia	80,000	"
1693, Jan. 11	Catania, Italy	60,000	"
1730, Dec. 30	Hokkaido, Japan	137,000	"
1737, Oct. 11	India, Calcutta	300,000	"
1755, June 7	N Persia	40,000	"
1755, Nov. 1	Lisbon, Portugal	60,000	8.75*
1783, Feb. 4	Calabria, Italy	30,000	NA
1797, Feb. 4	Quito, Ecuador	41,000	"
1811-12	New Madrid, MO (series)	NA	8.7*
1822, Sept. 5	Asia Minor, Aleppo	22,000	NA
1828, Dec. 28	Echigo, Japan	30,000	"
1868, Aug. 13-15	Peru, Ecuador	40,000	"
1875, May 16	Venezuela, Colombia	16,000	"
1886, Aug. 31	Charleston, SC	60	6.6
1896, June 15	Japan, sea wave	27,120	NA
1905, Apr. 4	Kangra, India	19,000	8.6
1906, Apr. 18-19	San Francisco, CA	503[2]	8.3
1906, Aug. 17	Valparaiso, Chile	20,000	8.6
1907, Oct. 21	Central Asia	12,000	8.1
1908, Dec. 28	Messina, Italy	83,000	7.5
1915, Jan. 13	Avezzano, Italy	29,980	7.5
1918, Oct. 11	Mona Passage, P.R.	116	7.5
1920, Dec. 16	Gansu, China	200,000	8.6
1923, Sept. 1	Yokohama, Japan	143,000	8.3
1925, Mar. 16	Yunnan, China	5,000	7.1
1927, May 22	Nan-Shan, China	200,000	8.3
1932, Dec. 25	Gansu, China	70,000	7.6
1933, Mar. 2	Japan	2,990	8.9
1933, Mar. 10	Long Beach, CA	115	6.2
1934, Jan. 15	India, Bihar-Nepal	10,700	8.4
1935, Apr. 21	Taiwan (Formosa)	3,276	7.4
1935, May 30	Quetta, India	50,000	7.5
1939, Jan. 25	Chillan, Chile	28,000	8.3
1939, Dec. 26	Erzincan, Turkey	30,000	8.0
1946, Dec. 20	Honshu, Japan	1,330	8.4

Date	Location	Deaths	Mag.
1948, June 28	Fukui, Japan	5,390	7.3
1949, Aug. 5	Pelileo, Ecuador	6,000	6.8
1950, Aug. 15	Assam, India	1,530	8.7
1953, Mar. 18	NW Turkey	1,200	7.2
1956, June 10-17	N Afghanistan	2,000	7.7
1957, July 2	N Iran	1,200	7.4
1957, Dec. 13	W Iran	1,130	7.3
1960, Feb. 29	Agadir, Morocco	12,000	5.9
1960, May 21-30	S Chile	5,000	9.5
1962, Sept. 1	NW Iran	12,230	7.3
1963, July 26	Skopje, Yugoslavia	1,100	6.0
1964, Mar. 27	Alaska	131	9.2
1966, Aug. 19	E Turkey	2,520	7.1
1968, Aug. 31	NE Iran	12,000	7.3
1970, Jan. 5	Yunnan Prov., China	15,621	7.7
1970, Mar. 28	W Turkey	1,100	7.3
1970, May 31	N Peru	66,000	7.8
1971, Feb. 9	San Fernando Val., CA	65	6.6
1972, Apr. 10	S Iran	5,054	7.1
1972, Dec. 23	Managua, Nicaragua	5,000	6.2
1974, Dec. 28	Pakistan (9 towns)	5,200	6.3
1975, Sept. 6	Turkey (Lice, etc.)	2,300	6.7
1976, Feb. 4	Guatemala	23,000	7.5
1976, May 6	NE Italy	1,000	6.5
1976, June 25	Irian Jaya, New Guinea	422	7.1
1976, July 27	Tangshan, China	255,000	8.0
1976, Aug. 16	Mindanao, Philippines.	8,000	7.8
1976, Nov. 24	NW Iran-USSR border	5,000	7.3
1977, Mar. 4	Romania	1,500	7.2
1977, Aug. 19	Indonesia	200	8.0
1977, Nov. 23	NW Argentina	100	8.2
1978, Sept. 16	NE Iran	15,000	7.8
1979, Sept. 12	Indonesia	100	8.1
1979, Dec. 12	Colombia, Ecuador	800	7.9
1980, Oct. 10	NW Algeria	3,500	7.7
1980, Nov. 23	S Italy	3,000	7.2
1981, June 11	S Iran	3,000	6.9
1981, July 28	S Iran	1,500	7.3
1982, Dec. 13	W Arabian Peninsula	2,800	6.0
1983, May 26	N Honshu, Japan	81	7.7
1983, Oct. 30	E Turkey	1,342	6.9
1985, Mar. 3	Chile	146	7.8
1985, Sept. 19	Michoacan, Mexico	9,500	8.1
1986, Oct. 10	El Salvador	1,000+	5.5
1987, Mar. 6	Colombia-Ecuador	4,000+	7.0
1988, Aug. 20	India-Nepal border	1,450	6.6
1988, Nov. 6	China-Burma border	1,000	7.3
1988, Dec. 7	Soviet Armenia	55,000	7.0
1989, Oct. 17	San Francisco Bay area	62	7.1
1990, May 30	N Peru	115	6.3
1990, June 20	W Iran	40,000+	7.7
1990, July 16	Luzon, Philippines	1,621	7.8
1991, Feb. 1	Pakistan, Afgh. border	1,200	6.8
1991, Oct. 19	N India	2,000	7.0
1992, Mar. 13, 15	E Turkey	4,000	6.2/6.0
1992, June 28	S California	1	7.5/6.6
1992, Dec. 12	Flores Isl., Indonesia	2,500	7.5
1993, July 12	off Hokkaido, Japan	200+	7.7
1992, Sept. 1	SW Nicaragua	116	7.0
1992, Oct. 12	Cairo, Egypt	450	5.9
1993, Sept. 30	Maharashtra, S India	9,748[3]	6.3
1994, Jan. 17	Northridge, CA	61	6.8
1994, Feb. 15	S Sumatra, Indon.	215	7.0
1994, June 6	Cauca, SW Colombia	1,000	6.8
1994, Aug. 19	N Algeria	164	6.0
1995, Jan. 16	Kobe, Japan	5,502	6.9
1995, May 27	Sakhalin Isl., Russia	1,989	7.5
1995, Oct. 1	SW Turkey	73	6.0
1995, Oct. 9	W coast, Mexico	c. 40+	7.6
1996, Feb. 3	SW China	200+	7.0
1996, Feb. 17	Irian Jaya, Indonesia	53	7.5
1997, Feb. 4	Turkmen.-Iran border	79	6.9
1997, Feb. 27	W Pakistan	100+	7.3
1997, Feb. 28	NW Iran	1,000+	6.1
1997, May 10	N Iran	1,560	7.5
1997, May 21	Madhya Pradesh, India	40+	6.1
1997, July 9	NE Venezuela	82	6.9
1997, Sept. 26	Central Italy	11	5.5/5.7
1997, Sept. 28	Sulawesi, Indonesia	17+	5.9
1997, Oct. 15	Illapel, Chile	8	6.8
1998, Jan. 10	Zhangbei, China	50	6.2
1998, Feb. 4, 8	Takhar province, NE Afghanistan	2,323	6.1
1998, May 22	Central Bolivia	105	6.5
1998, May 30	NE Afghanistan	4,700+	6.9
1998, June 27	Adana, Turkey	144	6.3
1998, July 9	Azores, Portugal	10	5.8
1998, Nov. 29	East Indonesia	34	7.8
1999, Jan. 25	Armenia, Colombia	1,185+	6.0
1999, Feb. 11	Central Afghanistan	60	6.0
1999, Mar. 28	Uttar Pradesh, India	87	6.8
1999, May 7	Southern Iran	26+	6.2
1999, June 16	Puebla, Mexico	16	6.7
1999, Aug. 17	Western Turkey	17,200+	7.4
1999, Sept. 7	Athens, Greece	143	5.9
1999, Sept. 21	Taichung, Taiwan	2,474	7.6
1999, Sept. 30	Oaxaca, Mexico	20	7.5
1999, Nov. 12	Duzce, Turkey	675+	7.2
2000, May 4	Eastern Indonesia	35	6.5
2000, June 4	Sumatra, Indonesia	103	7.9
2000, June 11	Central Taiwan	2	6.7
2001, Jan. 13	San Vicente, El Salvador	800+	7.6
2001, Jan. 26	Gujarat, India	20,000+	7.9
2001, Feb. 13	San Vicente, El Salvador	255	6.6
2001, Mar. 24	Hiroshima, Japan	2	6.4
2001, June 23	Arequipa, Peru	102	8.1

(*) estimated from earthquake intensity. NA = not available. (1) Once thought to have been a volcanic eruption; evidence indicates a destructive earthquake and tsunami occurred on this date. (2) With subsequent fires, death toll rose to 700; some estimates of the death toll are much higher. (3) Official death toll as released by Indian government. Other sources reported estimates of about 30,000 deaths.

Some Notable Fires Since 1835

(See also Some Notable Explosions Since 1910.)

Date	Location	Deaths
1835, Dec. 16	New York, NY, 500 bldgs. destroyed	—
1845, May	Canton, China, theater	1,670
1871, Oct. 8	Chicago, $196 million loss; 17,000 bldgs. destroyed	250
1871, Oct. 8	Peshtigo, WI, forest fire	1,182
1872, Nov. 9	Boston, 800 bldgs. destroyed	—
1876, Dec. 5	Brooklyn, NY, theater	295
1877, June 20	St. John, New Brunswick	100
1881, Dec. 8	Ring Theater, Vienna	850
1887, May 25	Opera Comique, Paris	200
1887, Sept. 4	Exeter, England, theater	200
1894, Sept. 1	MN, forest fire	413
1897, May 4	Paris, charity bazaar	150
1900, June 30	Hoboken, NJ, docks	326
1902, Sept. 20	Birmingham, AL, church	115
1903, Dec. 30	Iroquois Theater, Chicago	602
1908, Jan. 13	Rhoads Theater, Boyertown, PA	170
1908, Mar. 4	Collinwood, OH, school	176
1911, Mar. 25	Triangle Shirtwaist factory, NY, NY	146
1913, Oct. 14	Mid Glamorgan, Wales, colliery	439
1918, Apr. 13	Norman, OK, state hospital	38
1918, Oct. 12	Cloquet, MN, forest fire	400
1919, June 20	Mayagüez Theater, San Juan, P.R.	150
1923, May 17	Camden, SC, school	76
1924, Dec. 24	Babb's Switch, OK, school	35
1929, May 15	Cleveland, OH, clinic	125
1930, Apr. 21	Columbus, OH, penitentiary	320
1931, July 24	Pittsburgh, PA, home for aged	48
1934, Dec. 11	Hotel Kerns, Lansing, MI	34
1938, May 16	Atlanta, GA, Terminal Hotel	35
1940, Apr. 23	Natchez, MS, dance hall	198
1942, Nov. 28	Cocoanut Grove, Boston	491
1942, Dec. 12	St. John's, Nfld., hostel	100
1943, Sept. 7	Gulf Hotel, Houston, TX	55
1944, July 6	Ringling Circus, Hartford, CT	168
1946, June 5	LaSalle Hotel, Chicago	61
1946, Dec. 7	Winecoff Hotel, Atlanta	119
1946, Dec. 12	NY, NY, ice plant, tenement	37
1949, Apr. 5	Effingham, IL, hospital	77
1950, Jan. 7	Davenport, IA, Mercy Hospital	41
1953, Mar. 29	Largo, FL, nursing home	35
1953, Apr. 16	Chicago, metalworking plant	35
1957, Feb. 17	Warrenton, MO, home for aged	72
1958, Mar. 19	New York, NY, loft building	24
1958, Dec. 1	Chicago, parochial school	95
1958, Dec. 16	Bogotá, Colombia, store	83
1959, June 23	Stalheim, Norway, resort hotel	34
1960, Mar. 12	Pusan, Korea, chemical plant	68
1960, July 14	Guatemala City, mental hospital	225
1960, Nov. 13	Amude, Syria, movie theater	152
1961, Jan. 6	Thomas Hotel, San Francisco	20
1961, Dec. 8	Hartford, CT, hospital	16
1961, Dec. 17	Niteroi, Brazil, circus	323
1963, May 4	Diourbel, Senegal, theater	64

Date	Location	Deaths
1963, Nov. 18	Surfside Hotel, Atlantic City, NJ	25
1963, Nov. 23	Fitchville, OH, rest home	63
1963, Dec. 29	Roosevelt Hotel, Jacksonville, FL	22
1964, May 8	Manila, apartment bldg.	30
1964, Dec. 18	Fountaintown, IN, nursing home	20
1965, Mar. 1	LaSalle, Quebec, apartment	28
1965, Aug. 11-16	Watts riot fires, CA	30+
1966, Mar. 11	Numata, Japan, 2 ski resorts	31
1966, Aug. 13	Melbourne, Australia, hotel	29
1966, Oct. 17	New York, NY, bldg. (firefighters)	12
1966, Dec. 7	Erzurum, Turkey, barracks	68
1967, Feb. 7	Montgomery, AL, restaurant	25
1967, May 22	Brussels, Belgium, store	322
1967, July 16	Jay, FL, state prison	37
1968, Feb. 26	Shrewsbury, England, hospital	22
1968, May 11	Vijayawada, India, wedding hall	58
1968, Nov. 18	Glasgow, Scotland, factory	24
1969, Dec. 2	Notre Dame, Can., nursing home	54
1970, Jan. 9	Marietta, OH, nursing home	27
1970, Nov. 1	Grenoble, France, dance hall	145
1970, Dec. 20	Tucson, AZ, hotel	28
1971, Mar. 6	Burghoezli, Switzerland, psychiatric clinic	28
1971, Apr., 20	Bangkok, Thailand, hotel	24
1971, Dec., 25	Seoul, South Korea, hotel	162
1972, May 13	Osaka, Japan, nightclub	116
1972, July 5	Sherborne, England, hospital	30
1973, Feb. 6	Paris, France, school	21
1973, June 24	New Orleans, LA, bar	32
1973, Aug. 3	Isle of Man, England, amusement park	51
1973, Sept. 1	Copenhagen, Denmark, hotel	35
1973, Nov. 6	Fukui, Japan, train	28
1973, Nov. 29	Kumamoto, Japan, dept. store	107
1973, Dec. 2	Seoul, South Korea, theater	50
1974, Feb. 1	São Paulo, Brazil, bank building	189
1974, June 30	Port Chester, NY, discotheque	24
1974, Nov. 3	Seoul, S. Korea, hotel, disco	88
1975, Dec. 12	Mina, Saudi Arabia, tent city	138
1976, Oct. 24	Bronx, NY, social club	25
1977, Feb. 25	Moscow, Russia, Rossiya hotel	45
1977, May 28	Southgate, KY, nightclub	164
1977, June 9	Abidjan, Ivory Coast, nightclub	41
1977, June 26	Columbia, TN, jail	42
1977, Nov. 14	Manila, Philippines, hotel	47
1978, Jan. 28	Kansas City, Coates House Hotel	16
1978, Aug. 19	Abadan, Iran, movie theater	425+
1979, July 14	Saragossa, Spain, hotel	80
1979, Dec. 31	Chapais, Quebec, social club	42
1980, May 20	Kingston, Jamaica, nursing home	157
1980, Nov. 21	MGM Grand Hotel, Las Vegas	84
1980, Dec. 4	Stouffer Inn, Harrison, NY	26
1981, Jan. 9	Keansburg, NJ, boarding home	30
1981, Feb. 10	Las Vegas Hilton	8
1981, Feb. 14	Dublin, Ireland, discotheque	44
1982, Sept. 4	Los Angeles, apartment house	24
1982, Nov. 8	Biloxi, MS, county jail	29
1983, Feb. 13	Turin, Italy, movie theater	64
1983, Dec. 17	Madrid, Spain, discotheque	83
1984, May 11	Great Adventure Amusement Pk., NJ	8
1985, Apr. 21	Tabaco, Phil., movie theater	44
1985, Apr. 26	Buenos Aires, Argentina, hospital	79
1985, May 11	Bradford, England, soccer stadium	53
1986, Dec. 31	Puerto Rico, Dupont Plaza Hotel	96
1987, May 6-June 2	N China, forest fire	193
1987, Nov. 17	London, England, subway	30
1988, Mar. 20	Lashio, Burma, 2,000 buildings	134
1990, Mar. 25	Bronx, NY, social club	87
1991, Mar. 3	Addis Ababa, Ethiopia, munitions dump	260+
1991, Sept. 3	Hamlet, NC, processing plant	25
1991, Oct. 20-21	Oakland, Berkeley, CA, wildfire	24
1993, Apr. 19	Waco, TX, cult compound	72
1994, May 10	Bangkok, Thailand, toy factory	213
1994, July 4-10	Glenwood Springs, CO (firefighters)	14
1994, Dec. 10	Karamay, China, theater	300
1994, Nov. 2	Durunka, Egypt, burning fuel flood	500
1995, Oct. 28	Baku, Azerbaijan, subway train	300
1995, Dec. 23	Mandi Dabwali, India, school	500+
1996, Mar. 19	Quezon City, Philippines, nightclub	150+
1996, Mar. 28	Bogor, Indonesia, shopping mall	78
1996, Apr. 11	Düsseldorf, Germany, airport	16
1996, Oct. 22	Caracas, Venezuela, jail	25
1996, Nov. 20	Hong Kong, building	39
1997, Feb. 23	Baripada, India, worship site	164
1997, Apr. 15	Mina, Saudi Arabia, encampment	343
1997, June 7	Thanjavur, India, temple	60+
1997, June 13	New Delhi, India, movie theater	60
1997, July 11	Pattaya, Thailand, hotel	90
1997, Sept. 29	Home for retarded children, near Colina, Chile	30
1998, Mar. 26	Mazeras, India, school dorm.	22
1998, Dec. 3	Manila, Philippines, orphanage	28
1999, Feb. 10	Samara, Russia, police hdqtrs.	23
1999, Mar. 24	France and Italy, Mont Blanc tunnel	40
1999, June 30	Hwasung, S. Korea, camp dormitory	23
1999, Oct. 30	Inchon, S. Korea, karaoke salon	55+
2000, Mar. 10	Vitupo, Tuvalu, school	18
2000, Mar. 17	Kanungu, Uganda, church	530
2000, June 23	Queensland, Australia, hostel	15
2000, Oct. 20	Mexico City, Mexico, nightclub	20
2000, Dec. 25	Luoyang, China, shopping center	309
2001, Jan. 1	Volendam, Netherlands, cafe	10
2001, Mar. 6	Central China, school	41
2001, Mar. 26	Machakos, Kenya, school	64
2001, Aug. 6	Madras, India, home for mentally ill	27
2001, Aug. 18	Quezon City, Philippines, hotel	73
2001, Sept. 1	Tokyo, Japan nightclub	44

Some Notable Explosions Since 1910

(See also Principal U.S. Mine Disasters Since 1900.)

Date	Location	Deaths
1910, Oct. 1	Los Angeles Times Bldg.	21
1913, Mar. 7	Dynamite, Baltimore harbor	55
1915, Sept. 27	Gasoline tank car, Ardmore, OK	47
1917, Apr. 10	Munitions plant, Eddystone, PA	133
1917, Dec. 6	Halifax Harbor, Canada	1,654
1918, May 18	Chemical plant, Oakdale, PA	193
1918, July 2	Explosives, Split Rock, NY	50
1918, Oct. 4	Shell plant, Morgan Station, NJ	64
1919, May 22	Food plant, Cedar Rapids, IA	44
1920, Sept. 16	Wall Street, NY, NY, bomb	30
1921, Sept. 21	Chem. storage facility, Oppau, Ger.	561
1924, Jan. 3	Food plant, Pekin, IL	42
1927, May 18	Bath school, Lansing, MI	38
1928, April 13	Dance hall, West Plains, MO	40
1937, Mar. 18	New London, TX, school	311
1940, Sept. 12	Hercules Powder, Kenvil, NJ	55
1942, June 5	Ordnance plant, Elwood, IL	49
1944, Apr. 14	Bombay, India, harbor	700
1944, July 17	Port Chicago, CA, pier	322
1944, Oct. 21	Liquid gas tank, Cleveland	135
1947, Apr. 16	Texas City, TX, pier	576
1948, July 28	Farben works, Ludwigshafen, Ger.	184
1950, May 19	Munitions barges, S. Amboy, NJ	30
1956, Aug. 7	Dynamite trucks, Cali, Colombia	1,100
1958, Apr. 18	Sunken munitions ship, Okinawa, Japan	40
1958, May 22	Nike missiles, Leonardo, NJ	10
1959, Apr. 10	World War II bomb, Philippines	38
1959, June 28	Rail tank cars, Meldrin, GA	25
1959, Aug. 7	Dynamite truck, Roseburg, OR	13
1959, Nov. 2	Jamuri Bazar, India, explosives	46
1959, Dec. 13	2 apt. bldgs., Dortmund, Ger.	26
1960, Mar. 4	Belgian munitions ship, Havana, Cuba	100
1960, Oct. 25	Gas, Windsor, Ont., store	11
1962, Jan. 16	Gas pipeline, Edson, Alberta	8
1962, Oct. 3	Telephone Co. office, NY, NY	23
1963, Jan. 2	Packing plant, Terre Haute, IN	16
1963, Mar. 9	Dynamite plant, S. Africa	45
1963, Aug. 13	Explosives dump, Gauhaiti, India	32
1963, Oct. 31	State Fair Coliseum, Indianapolis, IN	73
1964, July 23	Bone, Algeria, harbor munitions	100
1965, Mar. 4	Gas pipeline, Natchitoches, LA	17
1965, Aug. 9	Missile silo, Searcy, AR	53
1965, Oct. 21	Bridge, Tila Bund, Pakistan	80
1965, Oct. 30	Cartagena, Colombia	48
1965, Nov. 24	Armory, Keokuk, IA	20
1967, Dec. 25	Apartment bldg., Moscow, USSR	20
1968, Apr. 6	Sports store, Richmond, IN	43
1970, Apr. 8	Subway construction, Osaka, Japan	73
1971, June 24	Tunnel, Sylmar, CA	17
1971, June 28	School, fireworks, Puebla, Mexico	13
1971, Oct. 21	Shopping center, Glasgow, Scotland	20
1973, Feb., 10	Liquid gas tank, Staten Island, NY	40
1975, Dec. 27	Coal mine, Chasnala, India	431
1976, Apr. 13	Lapua, Finland, munitions works	40
1977, Nov. 11	Freight train, Iri, South Korea	57
1977, Dec. 22	Grain elevator, Westwego, LA	35
1978, Feb. 24	Derailed tank car, Waverly, TN	12

Date	Location	Deaths
1978, July 11	Propylene tank truck, Spanish coastal campsite	150
1980, Oct. 23	School, Ortuella, Spain	64
1982, Apr. 25	Antiques exhibition, Todi, Italy	33
1982, Nov. 2	Salang Tunnel, Afghanistan	1,000+
1984, Feb. 25	Oil pipeline, Cubatao, Brazil	508
1984, June 21	Naval supply depot, Severomorsk, USSR	200+
1984, Nov. 19	Gas storage area, NE Mexico City	334
1984, Dec. 3	Chemical plant, Bhopal, India	3,849
1984, Dec. 5	Coal mine, Taipei, Taiwan	94
1985, June 25	Fireworks factory, Hallett, OK	21
1988, Apr. 10	Pakistani army ammunitions dump near Rawalpindi and Islamabad	100
1988, July 6	Oil rig, North Sea	167
1989, June 3	Gas pipeline, between Ufa, Asha, USSR	650+
1992, Mar. 3	Coal mine, Kozlu, Turkey	270+
1992, Apr. 22	Sewer, Guadalajara, Mexico	190
1992, May 9	Coal mine, Plymouth, Nova Scotia	26
1993, Feb. 26	World Trade Center, NY, NY	6
1994, July 18	Jewish community center, Buenos Aires, Argentina	100
1995, Apr. 19	Fed'l. office building, Oklahoma City	168[1]
1995, Apr. 29	Subway construction, South Korea	110
1995, Nov. 13	Military facility, Riyadh, Saudi Arabia	7
1996, Jan. 31	Bank, Colombo, Sri Lanka	53
1996, Feb. 25	Jerusalem and Ashkelon, Israel	27
1996, Mar. 3-4	Jerusalem and Tel Aviv, Israel	33
1996, June 25	U.S. military housing complex, near Dhahran, Saudi Arabia	19
		19
1996, July 24	Train, Colombo, Sri Lanka	86
1996, Nov. 10	Cemetery, Moscow, Russia	13
1996, Nov. 16	Russian military apartment, Dagestan region, Russia	68
1996, Nov. 21	Building, San Juan, Puerto Rico	29
1996, Nov. 27	Coal mine, Shanxi province, China	91+
1996, Dec. 30	Train, Assam, India	59+
1997, Jan. 18	Near courthouse, Lahore, Pakistan	25
1997, Mar. 19	Ammunition depot, Jalalabad, Afghanistan	16
1997, July 8	Train, Punjab, India	36
1997, July 9	Military airfield, S Romania	16
1997, July 30	Market, Jerusalem	15
1997, Nov. 19	Car, Hyderabad, India	23
1997, Dec. 2	Coal mine, Novokuznetsk, Siberia	68
1997, Dec. 6	Trains, southern India	10+
1998, Jan. 17	Coal mine, Sokobanja, Serbia	29
1998, Feb. 14	Oil tankers (2), Yaounde, Cameroon	120
1998, Feb. 14	17 bombs, Coimbatore, India	50
1998, Feb. 23	Train, near El Affroune, Algiers	18
1998, Mar. 5	Bus, Colombo, Sri Lanka	32
1998, Mar. 9	Train, Lahore, Pakistan	10
1998, Apr. 4	Coal mine, Donetsk, Ukraine	63
1998, Aug. 7	Bomb, U.S. Embassy, Nairobi, Kenya	213
	Bomb, U.S. Embassy, Dar-es-Salaam, Tanzania	11
1998, Aug. 15	Car bomb, Omagh, Ireland	29
1998, Aug. 16	Coal mine, Luhansk, Ukraine	24
1998, Aug. 31	Marketplace in Algiers	17
1998, Sept. 8	Two buses, Sao Paulo, Brazil	59
1998, Oct. 17	Oil pipeline, Jesse, Nigeria	700+
1999, May 16	Fuel truck, Punjab province, Pakistan	75
1999, July 27	Truck, Chongqing, China	15
1999, July 29	Gold mine, Carletonville, S. Africa	17
1999, Sept. 10	Apartment building, Moscow	94
1999, Sept. 13	Apartment building, Moscow	118
1999, Sept. 16	Apartment building, Moscow	18
1999, Sept. 26	Fireworks factory, Celaya, Mexico	56
2000, Feb. 25	Two buses with bombs, Ozamis, Philippines	41
2000, Mar. 11	Coal mine, Krasnodon, Ukraine	80
2000, Apr. 16	Airport hangar, Congo, Dem. Rep. of.	100+
2000, July 16	Oil pipeline, Warri, Nigeria	30
2000, Aug. 19	Train derailed in Nairobi, Kenya	25
2000, Aug. 20	Natural gas pipeline, Carlsbad, NM	10
2000, Sept. 9	Truck explodes in Urumqi, China	60
2000, Sept. 13	Bomb, Jakarta, Indonesia	15
2000, Sept. 19	Bomb, Islamabad, Pakistan	16
2000, Oct. 12	U.S. destroyer, Yemen	17
2001, Mar. 6	School, Wanzai County China	41
2001, Mar.15	Oil platform, Brazil	10
2001, Apr. 21	Coal mine, Shaanxi, China	51
2001, June 1	Dance club, Tel Aviv, Israel	21
2001, July 17	Coal mine, Guanxi, China	76+
2001, Aug. 19	Coal mine, Donetsk region, Ukraine	52
2001, Sept. 21	Chem. plant, Toulouse, France	29

Notable Nuclear Accidents

Oct. 7, 1957 — A fire in the Windscale plutonium production reactor N of Liverpool, England, released radioactive material; later blamed for 39 cancer deaths.

Jan. 3, 1961 — A reactor at a federal installation near Idaho Falls, ID, killed 3 workers. Radiation contained.

Oct. 5, 1966 — A sodium cooling system malfunction caused a partial core meltdown at the Enrico Fermi demonstration breeder reactor, near Detroit, MI. Radiation contained.

Jan. 21, 1969 — A coolant malfunction from an experimental underground reactor at Lucens Vad, Switzerland, released a large amount of radiation into a cavern, which was then sealed.

Mar. 22, 1975 — Fire at the Brown's Ferry reactor in Decatur, AL, caused dangerous lowering of cooling water levels.

Mar. 28, 1979 — The worst commercial nuclear accident in the U.S. occurred as equipment failures and human mistakes led to a loss of coolant and a partial core meltdown at the Three Mile Island reactor in Middletown, PA.

Feb. 11, 1981 — Eight workers were contaminated when more than 100,000 gallons of radioactive coolant fluid leaked into the containment building of TVA's Sequoyah 1 plant in Tennessee.

Apr. 25, 1981 — Some 100 workers were exposed to radiation during repairs of a nuclear plant at Tsuruga, Japan.

Jan. 6, 1986 — A cylinder of nuclear material burst after being improperly heated at a Kerr-McGee plant at Gore, OK. One worker died; 100 were hospitalized.

Apr. 26, 1986 — In the worst accident in the history of nuclear power, fires and explosions resulting from an unauthorized experiment at the Chernobyl nuclear power plant near Kiev, USSR (now in Ukraine), left at least 31 dead in the immediate aftermath and spread radioactive material over much of Europe. An estimated 135,000 people were evacuated from areas around Chernobyl, some of which were uninhabitable for years. As a result of the radiation released, tens of thousands of excess cancer deaths (as well as increased birth defects) were expected.

Sept. 30, 1999 — Japan's worst nuclear accident ever occurred at a uranium-reprocessing facility in Tokaimura, NE of Tokyo, when workers accidentally overloaded a container with uranium, thereby exposing workers and area residents to extremely high radiation levels.

> ***IT'S A FACT:*** On Dec. 24, 1984, in the worst industrial accident in history, more than 3,800 people were killed when toxic gas leaked from a storage tank in a Union Carbide insecticide factory in a heavily populated section of Bhopal, India. Up to 200,000 people suffered injuries, including severe damage to eyes, lungs, and kidneys.

Record Oil Spills

The number of tons can be multiplied by 7 to estimate roughly the number of barrels spilled; the exact number of barrels in a ton varies with the type of oil. Each barrel contains 42 gallons.

Name, place	Date	Cause	Tons
Ixtoc I oil well, S Gulf of Mexico	June 3, 1979	Blowout	600,000
Nowruz oil field, Persian Gulf	Feb. 1983	Blowout	600,000 (est.)
Atlantic Empress & Aegean Captain, off Trinidad and Tobago	July 19, 1979	Collision	300,000
Castillo de Bellver, off Cape Town, South Africa	Aug. 6, 1983	Fire	250,000
Amoco Cadiz, near Portsall, France	Mar. 16, 1978	Grounding	223,000
Torrey Canyon, off Land's End, England	Mar. 18, 1967	Grounding	119,000
Sea Star, Gulf of Oman	Dec. 19, 1972	Collision	115,000
Urquiola, La Coruna, Spain	May 12, 1976	Grounding	100,000
Hawaiian Patriot, N Pacific	Feb. 25, 1977	Fire	99,000
Othello, Tralhavet Bay, Sweden	Mar. 20, 1970	Collision	60,000-100,000

Other Notable Oil Spills

Name, place	Date	Cause	Gallons
Persian Gulf	began Jan. 23, 1991	Spillage by Iraq	130,000,000[1]
Braer, off Shetland Islands	Jan. 5, 1993	Grounding	26,000,000
Aegean Sea, off N Spain	Dec. 3, 1992	Unknown	21,500,000
Sea Empress, off SW Wales	Feb. 15, 1996	Grounding	18,000,000
World Glory, off South Africa	June 13, 1968	Hull failure	13,524,000
Exxon Valdez, Prince William Sound, AK	Mar. 24, 1989	Grounding	10,080,000
Keo, off MA	Nov. 5, 1969	Hull failure	8,820,000
Storage tank, Sewaren, NJ	Nov. 4, 1969	Tank rupture	8,400,000
Ekofisk oil field, North Sea	Apr. 22, 1977	Well blowout	8,200,000
Argo Merchant, Nantucket, MA	Dec. 15, 1976	Grounding	7,700,000
Pipeline, West Delta, LA	Oct. 15, 1967	Dragging anchor	6,720,000
Tanker off Japan	Nov. 30, 1971	Ship broke in half	6,258,000

(1) Est. by Saudi Arabia. Some estimates as low as 25 mil gal.

Historic Assassinations Since 1865

1865—Apr. 14. U.S. Pres. Abraham Lincoln shot by John Wilkes Booth, a well-known actor with Confederate sympathies, at Ford's Theater in Washington, DC; died Apr. 15.

1881—Mar. 13. Alexander II, of Russia.—July 2. U.S. Pres. James A. Garfield shot by Charles J. Guiteau, a disappointed office seeker, in Washington, DC; died Sept. 19.

1894—June 24. Pres. Sadi Carnot of France, by Italian anarchist, Sante Caserio, in Lyon.

1898—Sept. 10. Empress Elizabeth of Austria, stabbed by Italian anarchist Luigi Luccheni.

1900—July 29. Umberto I, king of Italy.

1901—Sept. 6. U.S. Pres. William McKinley in Buffalo, NY; died Sept. 14. Leon Czolgosz executed for the crime.

1908—Feb. 1. King Carlos I of Portugal and his son Luis Felipe, in Lisbon.

1913—Feb. 23. Mexican Pres. Francisco I. Madero and Vice Pres. Jose Pino Suarez.—Mar. 18. George, king of Greece.

1914—June 28. Archduke Francis Ferdinand of Austria-Hungary and his wife in Sarajevo, Bosnia (later part of Bosnia and Herzegovina), by Gavrilo Princip.

1916—Dec. 30. Grigori Rasputin, powerful Russian monk.

1918—July 12. Grand Duke Michael of Russia, at Perm.—July 16. Nicholas II, abdicated as czar of Russia; his wife, the Czarina Alexandra; their son, Czarevitch Alexis; their daughters, Grand Duchesses Olga, Tatiana, Marie, Anastasia; and 4 members of their household, executed by Bolsheviks at Ekaterinburg.

1920—May 20. Mexican Pres. Gen. Venustiano Carranza in Tlaxcalantongo.

1922—Aug. 22. Michael Collins, Irish revolutionary.—Dec. 16. Polish Pres.Gabriel Narutowicz in Warsaw.

1923—July 20. Gen. Francisco "Pancho" Villa, ex-rebel leader, in Parral, Mexico.

1928—July 17. Gen. Alvaro Obregon, president-elect of Mexico, in San Angel, Mexico.

1932—May 6. Pres. Paul Doumer of France shot by Russian émigré, Pavel Gorgulov, in Paris.

1934—July 25. In Vienna, Austrian Chancellor Engelbert Dollfuss by Nazis.

1935—Sept. 8. U.S. Sen. Huey P. Long shot in Baton Rouge, LA, by Dr. Carl Austin Weiss, who was slain by Long's bodyguards; Long died Sept. 10.

1940—Aug. 20. Leon Trotsky (Lev Bronstein), 63, exiled Russian war minister, near Mexico City, by Ramon Mercador del Rio, a Spaniard.

1948—Jan. 30. Mohandas K. Gandhi, 78, shot in New Delhi, India, by Nathuram Vinayak Godse.—Sept. 17. Count Folke Bernadotte, UN mediator for Palestine, by Jewish extremists in Jerusalem.

1951—July 20. King Abdullah ibn Hussein of Jordan.—Oct. 16. Prime Min. Liaquat Ali Khan of Pakistan shot in Rawalpindi.

1956—Sept. 21. Pres. Anastasio Somoza of Nicaragua, shot in Leon; died Sept. 29.

1957—July 26. Pres. Carlos Castillo Armas of Guatemala, in Guatemala City by one of his own guards.

1958—July 14. King Faisal of Iraq; his uncle, Crown Prince Abdullah; and July 15, Prem. Nuri as-Said, by rebels in Baghdad.

1959—Sept. 25. Prime Min. Solomon Bandaranaike of Ceylon, by Buddhist monk in Colombo.

1961—Jan. 17. Ex-Prem. Patrice Lumumba of the Congo, in Katanga Province.—May 30. Dominican dictator Rafael Leonidas Trujillo Molina, near Ciudad Trujillo.

1963—June 12. Medgar W. Evers, NAACP's Mississippi field secretary, by Byron De La Beckwith in Jackson, MS.—Nov. 2. Pres. Ngo Dinh Diem of South Vietnam and his brother, Ngo Dinh Nhu, in a military coup.—Nov. 22. U.S. Pres. John F. Kennedy shot in Dallas, TX; accused gunman Lee Harvey Oswald was murdered by Jack Ruby while awaiting trial.

1965—Jan. 21. Iranian Prem. Hassan Ali Mansour in Tehran; 4 executed.—Feb. 21. Malcolm X, black nationalist, shot in New York City.

1966—Sept. 6. Prime Min. Hendrik F. Verwoerd of South Africa stabbed to death in parliament at Cape Town.

1968—Apr. 4. Rev. Dr. Martin Luther King Jr. fatally shot in Memphis, TN; James Earl Ray convicted of crime.—June 5. Sen. Robert F. Kennedy (D, NY) shot in Los Angeles; Sirhan Sirhan, convicted of crime.

1971—Nov. 28. Prime Min. Wasfi Tal of Jordan, in Cairo, by Palestinian guerrillas.

1973—Mar. 2. U.S. Amb. Cleo A. Noel Jr., U.S. Charge d'Affaires George C. Moore, and Belgian Charge d'Affaires Guy Eid killed by Palestinian guerrillas in Khartoum, Sudan.

1974—Aug. 19. U.S. Amb. to Cyprus, Rodger P. Davies, killed by sniper's bullet in Nicosia.

1975—Feb. 11. Pres. Richard Ratsimandrava, of Madagascar, shot in Tananarive.—Mar. 25. King Faisal of Saudi Arabia shot by nephew Prince Musad Abdel Aziz, in royal palace in Riyadh.—Aug. 15. Bangladesh Pres. Sheik Mujibur Rahman killed in coup.

1976—Feb. 13. Nigerian head of state, Gen. Murtala Ramat Mohammed, by self-styled "young revolutionaries."

1977—Mar. 16. Kamal Jumblat, Lebanese Druse chieftain, shot near Beirut.—Mar. 18. Congo Pres. Marien Ngouabi shot in Brazzaville.

1978—July 9. Former Iraqi Prem. Abdul Razak Al-Naif shot in London.

1979—Feb. 14. U.S. Amb. Adolph Dubs shot by Afghan Muslim extremists in Kabul.—Aug. 27. Lord Mountbatten, World War II hero, and 2 others killed when a bomb exploded on his fishing boat off the coast of Co. Sligo, Ire. IRA claimed responsibility.—Oct. 26. South Korean Pres. Park Chung Hee and 6 bodyguards fatally shot by Kim Jae Kyu, head of South Korean CIA, and 5 aides in Seoul.

1980—Apr. 12. Liberian Pres. William R. Tolbert slain in military coup.—Sept. 17. Former Nicaraguan Pres. Anastasio Somoza Debayle shot in Paraguay.

1981—Oct. 6. Egyptian Pres. Anwar al-Sadat shot by commandos while reviewing a military parade in Cairo; 7 others killed, 28 wounded; 4 convicted as assassins and executed.

1982—Sept. 14. Lebanese Pres.-elect Bashir Gemayel killed by bomb in east Beirut.

1983—Aug. 21. Philippine opposition leader Benigno Aquino Jr. shot by gunman at Manila International Airport.

1984—Oct. 31. Indian Prime Min. Indira Gandhi shot and killed by 2 Sikh bodyguards, in New Delhi.

1986—Feb. 28. Swedish Prem. Olof Palme shot by gunman on Stockholm street.

1987—June 1. Lebanese Prem. Rashid Karami killed when bomb exploded aboard a helicopter.

1988—Apr. 16. PLO military chief Khalil Wazir (Abu Jihad) gunned down by Israeli commandos in Tunisia.

1989—Aug. 18. Colombian presidential candidate Luis Carlos Galan killed by Medellín cartel drug traffickers at campaign rally in Bogotá.—Nov. 22. Lebanese Pres. Rene Moawad killed when bomb exploded next to his motorcade.

1990—Mar. 22. Presidential candidate Bernando Jamamillo Ossa shot by gunman at an airport in Bogotá.

1991—May 21. Rajiv Gandhi, former prime min. of India, killed by bomb during election rally in Madras.

1992—June 29. Mohammed Boudiaf, pres. of Algeria, shot by gunman in Annaba.

1993—May 1. Ranasinghe Premadasa, pres. of Sri Lanka, killed by bomb in Colombo.

1994—Mar. 23. Luis Donaldo Colosio Murrieta, Mexican presidential candidate, shot by gunman Mario Aburto Martinez. — Apr. 6. Burundian Pres. Cyprien Ntaryamira and Rwandan Pres. Juvenal Habyarimana killed, with 8 others, when their plane was apparently shot down.

1995—Nov. 4. Yitzhak Rabin, prime min. of Israel, shot by gunman Yigal Amir at peace rally in Tel Aviv.

1996—Oct. 2. Andrei Lukanov, former Bulgarian prime minister, shot outside his home by an unidentified gunman.

1998—Feb. 6. Claude Erigmac, prefect of Corsica, shot in the back while walking to a concert, by two unidentified gunmen.—Apr. 26. Guatemalan Rom. Catholic Bishop Juan Gerardi

Conedera, human rights champion, found beaten to death in Guatemala City; 4 persons convicted, June 8, 2001.

1999—Mar. 23. Paraguayan Vice-Pres. Luis Maria Argaña, ambushed and shot to death, along with his driver, by four unidentified assailants.—Apr. 9. Niger's Pres. Ibrahim Bare Mainassara, ambushed and killed by dissident soldiers.—Oct. 27. Armenia's Prime Min. Vazgen Sarkissian, along with 7 others, was shot to death during a session of Parliament.

2000—Jan. 15. Serbian paramilitary leader Zeljko Raznjatovic (alias Arkan), with 2 others, shot and killed by unidentified gunman in Belgrade hotel lobby; 4 suspects later charged with the killing.— June 8. Brig. Gen. Stephen Saunders, Britain's senior military representative in Greece, shot and killed by 2 men on motorcycle, while driving a car in an Athens suburb.

2001—Jan. 16. Congolese Pres. Laurent Kabila, shot to death by a bodyguard at his presidential palace in the capital, Kinshasa.—June 1. Nepal's King Birendra, Queen Aiswarya, and (according to official report) 7 other royals fatally shot by Crown Prince Dipendra, who also fatally wounded himself.—Sept. 9. Afghan Northern Alliance (anti-Taliban) guerrilla leader Ahmed Shah Massoud, fatally injured in suicide-attack bombing in N. Afghanistan by 2 Arabs posing as journalists; died Sept. 15.—Sept. 24. Colombian culture minister Consuelo Araujo, kidnapped, later slain, by Revolutionary Armed Forces guerrillas.

Assassination Attempts

1912—Oct. 14. Former U.S. Pres. Theodore Roosevelt shot and wounded by demented man in Milwaukee, WI.

1933—Feb. 15. In Miami, FL, Joseph Zangara, anarchist, shot at Pres.-elect Franklin D. Roosevelt, but a woman seized his arm, and the bullet fatally wounded Mayor Anton J. Cermak, of Chicago, who died Mar. 6.

1944—July 20. Adolf Hitler was injured when a bomb, planted by a German officer, exploded in Hitler's headquarters. One aide was killed and 12 were injured in the explosion.

1950—Nov. 1. In an attempt to assassinate Pres. Harry Truman, 2 members of a Puerto Rican nationalist movement—Griselio Torresola and Oscar Collazo—tried to shoot their way into Blair House. Torresola was killed, and a White House policeman, Pvt. Leslie Coffelt, was fatally shot.

1970—Nov. 27. Pope Paul VI unharmed by knife-wielding assailant who attempted to attack him in Manila airport.

1972—May 15. Alabama Gov. George Wallace shot in Laurel, MD, by Arthur Bremer; seriously crippled.

1975—Sept. 5. Pres. Gerald R. Ford unharmed when a Secret Service agent grabbed a pistol aimed at him by Lynette (Squeaky) Fromme, a Charles Manson follower, in Sacramento.—Sept. 22. Pres. Ford again unharmed when Sara Jane Moore fired a revolver at him.

1980—May 29. Civil rights leader Vernon E. Jordan Jr. shot and wounded in Ft. Wayne, IN.

1981—Jan. 16. Irish political activist Bernadette Devlin McAliskey and her husband shot and seriously wounded by 3 members of a Protestant paramilitary group in Co. Tyrone, Ire.—Mar. 30. Pres. Ronald Reagan, along with Press Sec. James Brady, Secret Service agent Timothy J. McCarthy, and Washington, DC, policeman Thomas Delahanty shot and seriously wounded by John W. Hinckley Jr. in Washington, DC.—May 13. Pope John Paul II and 2 bystanders shot and wounded by Mehmet Ali Agca, an escaped Turkish murderer, in St. Peter's Square, Rome.

1982—May 12. Pope John Paul II unharmed after guards overpowered a man with a knife, in Fatima, Portugal.

1984—Oct. 12. British Prime Min. Margaret Thatcher narrowly escaped injury when a bomb, said to have been planted by the IRA, exploded at the Grand Hotel in Brighton, England, during a Conservative Party conference. Four died, including a member of Parliament.

1986—Sept. 7. Chilean Pres. Gen. Augusto Pinochet Ugarte escaped unharmed when his motorcade was attacked by rebels using rockets, bazookas, grenades, and rifles.

1995—June 26. Egyptian Pres. Hosni Mubarak unharmed when gunmen fired on his motorcade in Addis Ababa, Ethiopia. Four died, including 2 Ethiopian police officers.

1997—Feb. 12. Colombian Pres. Ernesto Samper Pizano unharmed when a bomb exploded on a runway in Barranquilla as his plane was preparing to land.—Apr. 30. Tajik Pres. Imamali Rakhmanov injured when a grenade was thrown at him. 2 others were killed.

1998—Feb. 9. Georgian Pres. Eduard A. Shevardnadze unharmed when gunmen fired on his motorcade in Tbilisi, Georgia. Three died, including 2 bodyguards and 1 assailant.

2000—Sept. 18. Armed men attempted to assassinate Côte d'Ivoire military leader Gen. Robert Guei in a predawn raid.

Notable U.S. Kidnappings Since 1924

Robert Franks, 13, in Chicago, **May 22, 1924,** by 2 youths, Richard Loeb and Nathan Leopold, who killed boy. Demand for $10,000 ignored. Loeb died in prison; Leopold paroled 1958.

Charles A. Lindbergh Jr., 20 mos. old, in Hopewell, NJ, **Mar. 1, 1932;** found dead **May 12.** Ransom of $50,000 paid to man identified as Bruno Richard Hauptmann, 35, paroled German convict who entered U.S. illegally. Hauptmann was convicted after spectacular trial at Flemington, and electrocuted in Trenton, NJ, prison, **Apr. 3, 1936.**

William A. Hamm Jr., 39, in St. Paul, **June 15, 1933.** $100,000 paid. Alvin Karpis given life, paroled in 1969.

Charles F. Urschel, in Oklahoma City, **July 22, 1933.** Released **July 31** after $200,000 paid. George "Machine Gun" Kelly and 5 others sentenced to life.

Brooke L. Hart, 22, in San Jose, CA. Thomas Thurmond and John Holmes arrested after demanding $40,000 ransom. When Hart's body was found in San Francisco Bay, **Nov. 26, 1933,** a mob attacked the jail and lynched the 2 kidnappers.

George Weyerhaeuser, 9, in Tacoma, WA, **May 24, 1935.** Returned home **June 1** after $200,000 paid. Kidnappers given 20 to 60 years.

Charles Mattson, 10, in Tacoma, WA, **Dec. 27, 1936.** Found dead **Jan. 11, 1937.** Kidnapper asked $28,000, but failed to contact for delivery.

Arthur Fried, in White Plains, NY, **Dec. 4, 1937.** Body not found. Two kidnappers executed.

Robert C. Greenlease, 6, taken from Kansas City, MO, school **Sept. 28, 1953,** held for $600,000. Body was found Oct. 7. Bonnie Brown Heady and Carl A. Hall pleaded guilty and were executed.

Peter Weinberger, 32 days old, Westbury, NY, **July 4, 1956,** for $2,000 ransom, not paid. Child found dead. Angelo John LaMarca, 31, convicted, executed.

Lee Crary, 8, in Everett, WA, **Sept. 22, 1957;** $10,000 ransom, not paid. He escaped after 3 days, led police to George E. Collins, who was convicted.

Frank Sinatra Jr., 19, from hotel room in Lake Tahoe, CA, **Dec. 8, 1963.** Released **Dec. 11** after his father paid $240,000 ransom. Three men sentenced to prison.

Barbara Jane Mackle, 20, abducted **Dec. 17, 1968,** from Atlanta, GA, motel; found unharmed 3 days later, buried in a coffin-like box 18 inches underground, after her father had paid $500,000 ransom; Gary Steven Krist sentenced to life, Ruth Eisenmann-Schier to 7 years.

Mrs. Roy Fuchs, 35, and 3 children held hostage 2 hours, **May 14, 1969,** in Long Island, NY, released after her husband, a bank manager, paid kidnappers $129,000 in bank funds; 4 men arrested, ransom recovered.

Virginia Piper, 49, abducted **July 27, 1972,** from her home in suburban Minneapolis; found unharmed near Duluth 2 days later after husband paid $1 million ransom.

Patricia "Patty" Hearst, 19, taken from her Berkeley, CA, apartment **Feb. 4, 1974.** "Symbionese Liberation Army" captors demanded her father, publisher Randolph Hearst, give millions to the area's poor. Implicated in a San Francisco bank holdup, **Apr. 15.** The FBI, **Sept. 18, 1975,** captured her and others; they were indicted on various charges. Patricia Hearst convicted of bank robbery, **Mar. 20, 1976;** released from prison under executive clemency, **Feb. 1, 1979.** In 1978, William and Emily Harris were sentenced to 10 years to life for the kidnapping; both were paroled in 1983.

J. Reginald Murphy, 40, an editor of *Atlanta* (GA) *Constitution,* kidnapped **Feb. 20, 1974;** freed **Feb. 22** after newspaper paid $700,000 ransom. William A. H. Williams arrested; most of the money recovered.

E. B. Reville, Hepzibah, GA, banker, and wife, Jean, kidnapped **Sept. 30, 1974.** Ransom of $30,000 paid. He was found alive; Jean Reville was found dead **Oct. 2.**

Jack Teich, Kings Point, NY, steel executive, seized **Nov. 12, 1974;** released **Nov. 19** after payment of $750,000.

Adam Walsh, 6, abducted from a Hollywood, FL, department store, **July 27, 1981.** Although his severed head was found 2 weeks later, his body was never recovered. John Walsh, Adam's father, became active in raising awareness about missing children.

Sidney J. Reso, oil company executive, seized **Apr. 29, 1992;** died **May 3;** Arthur D. Seale and wife, Irene, arrested **June 19.** Arthur Seale pleaded guilty, sentenced to life in prison; Irene Seale sentenced to 20-year prison term.

Polly Klaas, 12, Petaluma, CA, abducted at knife point, **Oct. 1, 1993,** during a slumber party at her home. Police arrested Richard Allen Davis on **Nov. 30;** he led them to her body, found **Dec. 4** in wooded area of Cloverdale, CA. Davis found guilty **June 18, 1996,** and sentenced to death **Sept. 26.**

Marshall I. Wais, 79, owner of 2 San Francisco steel companies, kidnapped **Nov. 19, 1996,** from his San Francisco home. Released unharmed the same day after $500,000 ransom paid; Thomas William Taylor and Michael K. Robinson arrested the same day.

The Year in Pictures

REUTERS

2002 AMERICA ATTACKED

Nineteen terrorists Sept. 11 crashed hijacked commercial airliners into the World Trade Center twin towers in New York City and the Pentagon outside Washington, DC, in the deadliest terrorist action ever carried out against the U.S. A 4th hijacked plane crashed in a field southeast of Pittsburgh. Over 5,500 people died in the attacks. Pres. George W. Bush, Sept. 15, told the nation, "We're at war." He warned that a "sustained" fight would be needed to defeat terrorists and identified Saudi-born Osama bin Laden, believed to be operating from Afghanistan, as a "prime suspect" in the attacks.

AP/WIDE WORLD PHOTOS

At 9:03 AM, 15 minutes after the Trade Center's north tower was struck, hijackers of a United Airlines flight take aim at the south tower. ►

AP/WIDE WORLD PHOTOS

◄ The impact of the United jet ignites a fireball.

AP/WIDE WORLD PHOTOS

Just 56 minutes after the United crash, the south tower crumbles to the ground. The north tower collapses 29 minutes later, at 10:28. ►

AP/WIDE WORLD PHOTOS

◄ People flee the collapse.

AP/WIDE WORLD PHOTOS

Rescuers search for survivors days after the crash. More than 5,000 people were missing. ►

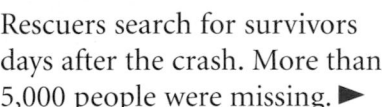

AP/WIDE WORLD PHOTOS

▲ Several thousand Chicagoans in Daley Plaza observe a moment of silence, during a national day of mourning, Sept. 14.

AP/WIDE WORLD PHOTOS

Firefighters and inspectors view damage to the Pentagon, which was also hit by a hijacked jet; about 190 people were believed killed in the plane and on the ground. ►

AP/WIDE WORLD PHOTOS

▲ BUSH INAUGURATION

George W. Bush (R) is sworn is as 43d U.S. president by Chief Justice William Rehnquist, Jan. 20, 2001, in Washington, DC. Also shown are his wife, Laura Bush, who held the Bible, and Jenna, one of their twin daughters. Elected in one of the tightest presidential races ever, Bush is a son of former Pres. George H. W. Bush, and only the second son of a president to succeed to that office.

AP/WIDE WORLD PHOTOS

◄ BACK TO WORK

Vice Pres. Dick Cheney shakes hands with his doctor before leaving a Washington, DC, hospital, Mar. 6, following surgery to repair a damaged artery. Cheney was able to promptly resume his duties, including chairmanship of the administration's energy policy task force.

JEFFORDS SWITCH ►

Sen. James Jeffords (VT) announces, May 24, that he is leaving the Republican Party to become an independent. The move shifted control of the closely divided Senate to the Democrats.

AP/WIDE WORLD PHOTOS

THE MISSING INTERN ►

Rep. Gary Condit (D, CA) leaves his Washington, DC, apartment, pursued by members of the media, July 26. Condit was questioned several times by police regarding the disappearance of 24-year-old government intern Chandra Levy (inset), missing since Apr. 30.

AP/WIDE WORLD PHOTOS

CORBIS

AP/WIDE WORLD PHOTOS

AFTER THE WHITE HOUSE

At right, former First Lady Hillary Rodham Clinton, with daughter Chelsea, in New York City Jan. 7, a few days after her swearing in as a U.S. senator.

Below, former Pres. Bill Clinton at a ceremony, July 30, welcoming him to his new offices in Harlem; also present: Rep. Charles Rangel (D, NY), whistling; actress Cicely Tyson (center); and the Boys Choir of Harlem. Clinton's exit from the White House had been clouded by controversial pardons, including that of fugitive financier Marc Rich.

AP/WIDE WORLD PHOTOS

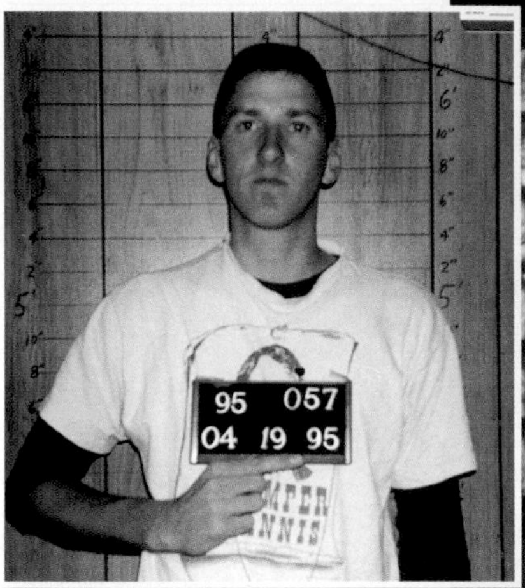

AP/WIDE WORLD PHOTOS

AP/WIDE WORLD PHOTOS

▲ McVEIGH EXECUTED

A visitor views wall of 168 photos at Oklahoma City Memorial, showing victims of the 1995 bombing at a federal office building there. Timothy McVeigh (left, shown hours after his 1995 arrest), convicted in the bombing, was executed by lethal injection in Terre Haute, IN, June 11.

AP/WIDE WORLD PHOTOS

◄ PUTIN AND BUSH

Russian Pres. Vladimir Putin (left) and Pres. Bush at the G-8 summit, July 21 in Genoa, Italy; the two leaders differed over Bush's anti-missile defense plan, but agreed to further "consultations" on the issue.

AP/WIDE WORLD PHOTOS

VIEQUES CONTROVERSY ►

A protester, Apr. 30, marches outside a San Juan, PR, prison. Demonstrators opposed to bombing exercises at the U.S. naval base on the Puerto Rican island of Vieques were being held at the prison on trespassing charges.

ENERGY CRUNCH ►

The sun sets, May 5, on power lines and turbines at San Giorgio Pass, CA, where strong winds power electric generators. Two days earlier, Pres. Bush ordered federal agencies to cut power usage, amid concerns over energy shortages in California and elsewhere in the U.S.

AFP/CORBIS

AP/WIDE WORLD PHOTOS

◄ DOT-COM WOES

Webvan trucks sit idle at an Oakland, CA, loading site, July 9, as Webvan, which took grocery orders online for delivery to homes, announced it would file for bankruptcy, just 20 months after its stock went public pegged at $4.8 billion. The collapse was one of many for "dot-com" Internet companies.

AP/WIDE WORLD PHOTOS

TAX REBATE CHECKS ►

Shopper outside a Home Depot store in West Windsor, NJ, Aug. 2. A 10-year, $1.35 trillion federal tax cut, signed into law June 7, provided for millions of tax rebate checks to be mailed out, most of them amounting to $300 per taxpayer.

KATHARINE GRAHAM ►

Publisher Katharine Graham, who made the *Washington Post* into one of the world's most influential newspapers and built a media empire around it, died July 17 at 84.

AP/WIDE WORLD PHOTOS

AP/WIDE WORLD PHOTOS

◄ DALE EARNHARDT

Champion NASCAR racer Dale Earnhardt, shown Feb. 16 at Daytona International Speedway with his son, Dale Earnhardt Jr. (left), and pole sitter Dale Jarrett, was killed 2 days later in a crash on the last lap of the Daytona 500 race.

©BETTMANN/CORBIS

▲ IMOGENE COCA

Comedian Imogene Coca died June 2 at 92. She is shown here with her early TV comedy partner, Sid Caesar.

CARROLL O'CONNOR ►

Actor Carroll O'Connor, best known as TV's Archie Bunker, died June 21 at 76. He is flanked by *All in the Family* cast members Sally Struthers (left), Rob Reiner (top left), Jean Stapleton, and Mike Evans.

©BETTMANN/CORBIS

©MITCHELL GERBER/ CORBIS

◄ JACK LEMMON

Actor Jack Lemmon died June 27 at 76. In a career spanning over 50 years he played a variety of memorable parts, and won 2 Oscars.

NATIONAL DEFENSE

Chief Commanding Officers of the U.S. Military

Chairman, Joint Chiefs of Staff
Gen. Richard B. Myers

Vice Chairman
Gen. Peter Pace

The Joint Chiefs of Staff consists of the Chairman and Vice Chairman of the Joint Chiefs of Staff; the Chief of Staff, U.S. Army; the Chief of Naval Operations; the Chief of Staff, U.S. Air Force; and the Commandant of the Marine Corps.

Army

Chief of Staff	Date of Rank
Eric K. Shinseki	Aug. 5, 1997

Other Generals	
Abrams, John N.	Sept. 14, 1998
Coburn, John G.	May 14, 1999
Franks, Tommy R.	July 6, 2000
Hendrix, John W.	Nov. 23, 1999
Keane, John M.	Jan. 22, 1999
Meigs, Montgomery C.	Nov. 10, 1998
Schoomaker, Peter J.	Oct. 24, 1997
Schwartz, Thomas A.	Aug. 31, 1998
Shelton, Henry H.	Mar. 1, 1996

Air Force

Chief of Staff	Date of Rank
Michael E. Ryan	Apr. 4, 1996

Other Generals	
Babbitt, George T., Jr.	June 1, 1997
Eberhart, Ralph E.	Aug. 1, 1997
Gamble, Patrick K.	Oct. 1, 1998
Gordon, John A.	Oct. 31, 1997
Jumper, John P.	Nov. 17, 1997
Myers, Richard B.	Sept. 1, 1997
Newton, Lloyd W.	Apr. 1, 1997
Ralston, Joseph W.	July 1, 1995
Robertson, Charles T., Jr.	Sept. 1, 1998

Navy

Chief of Naval Operations	Date of Rank
Clark, Vernon E. (surface warfare)	July 21, 2000

Other Admirals	
Blair, Dennis C. (surface warfare)	May 1, 1999
Bowman, Frank L. (submariner)	Oct. 1, 1996
Ellis, Jr., James O. (aviator)	Jan. 1, 1999
Fallon, William J. (aviator)	Nov. 1, 2000
Fargo, Thomas B. (submariner)	Dec. 1, 1999
Mies, Richard W. (submariner)	Aug. 1, 1998
Natter, Robert J. (surface warfare)	Sept. 1, 2000

Marine Corps

Commandant of the Marine Corps (CMC)	Date of Rank
Gen. James L. Jones	July 1, 1999

Other Generals	
Fulford Jr, Carlton W.	Oct. 1, 2000
Pace, Peter	Nov. 1, 2000
Williams, Michael J.	Nov. 1, 2000

Coast Guard

Commandant, with rank of Admiral	Date of Rank
James M. Loy	May 29, 1998

Vice Commandant, with rank of Vice Admiral	
Thomas H. Collins	June 9, 2000

Unified Combatant Commands Commanders in Chief

U.S. European Command, Stuttgart-Vaihingen, Germany — Gen. Joseph W. Ralston (USAF) (concurrently NATO Supreme Allied Commander, Europe)

U.S. Pacific Command, Honolulu, HI — Adm. Dennis C. Blair (USN)

U.S. Joint Forces Command, Norfolk, VA — Adm. William F. Kernan (USN) (concurrently NATO Supreme Allied Commander, Atlantic)

U.S. Special Operations Command, MacDill AFB, Florida — Gen. Charles R. Holland (USAF)

U.S. Transportation Command, Scott AFB, Illinois — Gen. Tony Robertson (USAF)

U.S. Central Command, MacDill AFB, Florida — Gen. Tommy R. Franks (USA)

U.S. Southern Command, Miami, FL — Gen. Peter Pace (USMC)

U.S. Space Command, Peterson AFB, Colorado — Gen. Ralph E. Eberhart (USAF)

U.S. Strategic Command, Offutt AFB, Nebraska — Adm. Richard W. Mies (USN)

North Atlantic Treaty Organization International Commands

NATO Headquarters:
Chairman, NATO Military Committee — Adm. Guido Venturoni (Italian Navy)

Strategic Command:
Allied Command Europe (ACE) — Gen. Joseph W. Ralston (USAF), Supreme Allied Commander Europe

Subordinate Command:
Allied Forces South Europe (AFSOUTH) — Adm. James D. Ellis Jr. (USN), Commander-in-Chief, South

Allied Forces North Europe (AFNORTH) — Gen. Sir John Deverell KCB OBE (Royal Army, UK), Commander-in-Chief, North

Strategic Command:
Allied Command Atlantic (ACLANT) — Gen. William F. Kernan (USA), Supreme Allied Commander, Atlantic

Subordinate Commands:
Western Atlantic (WESTLANT) — Adm. Robert J. Natter (USN), Commander-in-Chief, Western Atlantic

Southern Atlantic (SOUTHLANT) — Vice Adm. Americo da Silva Santos (Portuguese Navy), Commander-in-Chief, Southern Atlantic

Eastern Atlantic (EASTLANT) — Adm. Alan West KCB DSC, (Royal Navy, UK), Commander-in-Chief, Eastern Atlantic

Principal U.S. Military Training Centers

Army

Name, PO address	ZIP	Nearest city
Aberdeen Proving Ground, MD	21005	Aberdeen
Carlisle Barracks, PA	17013	Carlisle
Fort Benning, GA	31905	Columbus
Fort Bliss, TX	79916	El Paso
Fort Bragg, NC	28307	Fayetteville
Fort Gordon, GA	30905	Augusta
Fort Huachuca, AZ	85613	Sierra Vista
Fort Jackson, SC	29207	Columbia
Fort Knox, KY	40121	Radcliff
Fort Leavenworth, KS	66027	Leavenworth
Fort Lee, VA	23801	Petersburg
Fort McClellan, AL	36205	Anniston
Fort Rucker, AL	36362	Dothan
Fort Sill, OK	73503	Lawton
Fort Leonard Wood, MO	65473	Rolla
Joint Readiness Training Center, Ft. Polk, LA	71459	Leesville
National Training Center, Ft. Irwin, CA	92311	Barstow, CA
The Judge Advocate General School, VA	22901	Charlottesville

Navy

Name, PO address	ZIP	Nearest city
Naval Education & Training Ctr.	32508	Pensacola, FL
Naval Air Training Center	78419	Corpus Christi,TX
Training Command Fleet	23511	Norfolk, VA
Training Command Fleet	92113	San Diego, CA
Naval Aviation Schools Command	32508	Pensacola, FL
Naval Education & Training Ctr.	02841	Newport, RI
Naval Post Graduate School	93943	Monterey, CA
Naval Submarine School	06349	Groton, CT
Naval Training Ctr., Great Lakes	60088	N. Chicago, IL
Naval War College	02841	Newport, RI
Naval Air Tech. Training Ctr.	32508	Pensacola, FL
Fleet Antisubmarine Warfare	92147	San Diego, CA

Marine Corps

Name, PO address	ZIP	Nearest city
MCB Camp Lejeune, NC	28542	Jacksonville
MCB Camp Pendleton, CA.	92055	Oceanside
MCB Kaneohe Bay, HI	96863	Kailua
MCAGCC Twentynine Palms, CA.	92278	Palm Springs
MCCDC Quantico, VA	22134	Quantico
MCRD Parris Island, SC.	29905	Beaufort
MCRD San Diego, CA	92140	San Diego
MCAS Cherry Point, NC	28533	Havelock
MCAS Miramar, CA.	92145	San Diego
MCAS New River, NC	28545	Jacksonville
MCAS Beaufort, SC	29904	Beaufort
MCAS Yuma, AZ	85369	Yuma
MCMWTC Bridgeport, CA.	93517	Bridgeport

MCB = Marine Corps Base. MCCDC = Marine Corps Combat Development Command. MCAS = Marine Corps Air Station. MCRD = Marine Corps Recruit Depot. MCAGCC = Marine Corps Air-Ground Combat Center. MCMWTC = Marine Corps Mountain Warfare Training Center.

Air Force

Name, PO address	ZIP	Nearest city
Goodfellow AFB, TX.	76908	San Angelo
Keesler AFB, MS	39534	Biloxi
Lackland AFB, TX	78236	San Antonio
Maxwell AFB, AL.	36112	Montgomery
Sheppard AFB, TX	76311	Wichita Falls

All are Air Education and Training Command Bases.

Personal Salutes and Honors, U.S.

The U.S. national salute, 21 guns, is also the salute to a national flag. U.S. independence is commemorated by the salute to the Union—one gun for each state—fired at noon July 4, at all military posts provided with suitable artillery.

A 21-gun salute on arrival and departure, with 4 ruffles and flourishes, is rendered to the **president** of the United States, to an ex-president, and to a president-elect. The national anthem or "Hail to the Chief," as appropriate, is played for the president, and the national anthem for the others. A 21-gun salute on arrival and departure, with 4 ruffles and flourishes, also is rendered to the **sovereign or chief of state** of a foreign country or a member of a reigning royal family, and the national anthem of his or her country is played. The music is considered an inseparable part of the salute and immediately follows the ruffles and flourishes without pause. For the Honors March, generals receive the "General's March," admirals receive the "Admiral's March," and all others receive the 32-bar medley of "The Stars and Stripes Forever."

GRADE, TITLE, OR OFFICE	SALUTE (IN GUNS) Arriving	Leaving	Ruffles and flourishes	Music
Vice president of United States	19		4	Hail, Columbia
Speaker of the House	19		4	Honors March
U.S. or foreign ambassador	19		4	Nat. anthem of official
Premier or prime minister	19		4	Nat. anthem of official
Secretary of Defense, Army, Navy, or Air Force	19	19	4	Honors March
Other cabinet members, Senate president pro tempore, governor, or chief justice of U.S.	19		4	Honors March
Chairman, Joint Chiefs of Staff	19	19	4	
Army chief of staff, chief of naval operations, Air Force chief of staff, Marine commandant	19	19	4	Honors March
General of the Army, general of the Air Force, fleet admiral	19	19	4	
Generals, admirals	17	17	4	
Assistant secretaries of Defense, Army, Navy, or Air Force	17	17	4	Honors March
Chair of a committee of Congress	17		4	Honors March

OTHER SALUTES (on arrival only) include: 15 guns, with 3 ruffles and flourishes, for U.S. envoys or ministers and foreign envoys or ministers accredited to the U.S.; 15 guns, for a lieutenant general or vice admiral; 13 guns, with 2 ruffles and flourishes, for a major general or rear admiral (upper half) and for U.S. ministers resident and ministers resident accredited to the U.S.; 11 guns, with 1 ruffle and flourish, for a brigadier general or rear admiral (lower half) and for U.S. charges d'affaires and like officials accredited to the U.S.; 11 guns, no ruffles and flourishes, for consuls general accredited to the U.S.

Military Units, U.S. Army and Air Force

ARMY UNITS. Squad: In infantry usually 10 enlisted personnel under a staff sergeant. **Platoon:** In infantry 4 squads under a lieutenant. **Company:** Headquarters section and 3 platoons under a captain. (Company-size unit in the artillery is a battery; in the cavalry, a troop.) **Battalion:** Hdqts. and 4 or more companies under a lieutenant colonel. (Battalion-size unit in the cavalry is a squadron.) **Brigade:** Hdqts. and 3 or more battalions under a colonel. **Division:** Hdqts. and 3 brigades with artillery, combat support, and combat service support units under a major general. **Army Corps:** Two or more divisions with corps troops under a lieutenant general. **Field Army:** Hdqts. and 2 or more corps with field Army troops under a general.

AIR FORCE UNITS. Flight: Numerically designated flights are the lowest level unit in the Air Force. They are used primarily where there is a need for small mission elements to be incorporated into an organized unit. **Squadron:** A squadron is the basic unit in the Air Force. It is used to designate the mission units in operational commands. **Group:** The group is a flexible unit composed of 2 or more squadrons whose functions may be operational, support, or administrative in nature. **Wing:** An operational wing normally has 2 or more assigned mission squadrons in an area such as combat, flying training, or airlift. **Numbered Air Forces:** Normally an operationally oriented agency, the numbered air force is designed for the control of 2 or more wings with the same mission and/or geographical location. **Major Command:** A major subdivision of the Air Force that is assigned a major segment of the USAF mission.

The Federal Service Academies

U.S. Military Academy, West Point, NY. Founded 1802. Awards BS degree and Army commission for a 5-year service obligation. For admissions information, write Admissions Office, Bldg. 606, USMA, West Point, NY 10996.

U.S. Naval Academy, Annapolis, MD. Founded 1845. Awards BS degree and Navy or Marine Corps commission for a 5-year service obligation. For admissions information, write Candidate Guidence Office, Naval Academy, Annapolis, MD 21402-5018.

U.S. Air Force Academy, Colorado Springs, CO. Founded 1954. Awards BS degree and Air Force commission for a 6-year service obligation. For admissions information, write Registrar, U.S. Air Force Academy, CO 80840-5025.

U.S. Coast Guard Academy, New London, CT. Founded 1876. Awards BS degree and Coast Guard commission for a 5-year service obligation. For admissions information, write Director of Admissions, Coast Guard Academy, New London, CT 06320-8103.

U.S. Merchant Marine Academy, Kings Point, NY. Founded 1943. Awards BS degree, a license as a deck, engineer, or dual officer, and a U.S. Naval Reserve commission. Service obligations vary according to options taken by the graduate. For admissions information, write Admission Office, U.S. Merchant Marine Academy, Kings Point, NY 11024.

U.S. Army, Navy, Air Force, Marine Corps, and Coast Guard Insignia

Source: Dept. of the Army, Dept. of the Navy, Dept. of the Air Force, U.S. Dept. of Defense

Army

General of the Armies — Gen. John J. Pershing (1860-1948), the only person to have held this rank, in life, was authorized to prescribe his own insignia, but never wore in excess of four stars. The rank originally was established posthumously by Congress for George Washington in 1799, and he was promoted to the rank by joint resolution of Congress, approved by Pres. Gerald Ford, Oct. 19, 1976.

General of the Army — Five silver stars fastened together in a circle and the coat of arms of the United States in gold color metal with shield and crest enameled.

General	Four silver stars
Lieutenant General	Three silver stars
Major General	Two silver stars
Brigadier General	One silver star
Colonel	Silver eagle
Lieutenant Colonel	Silver oak leaf
Major	Gold oak leaf
Captain	Two silver bars
First Lieutenant	One silver bar
Second Lieutenant	One gold bar

Warrant Officers

Grade Five — Silver bar with 4 enamel silver squares
Grade Four — Silver bar with 4 enamel black squares
Grade Three — Silver bar with 3 enamel black squares
Grade Two — Silver bar with 2 enamel black squares
Grade One — Silver bar with 1 enamel black squares

Noncommissioned Officers

Sergeant Major of the Army (E-9) — Three chevrons above 3 arcs, with an American Eagle centered on the chevrons, flanked by 2 stars—one star on each side of the eagle. Also wears distinctive red and white shield collar insignia.

Command Sergeant Major (E-9) — Three chevrons above 3 arcs with a 5-pointed star with a wreath around the star between the chevrons and arcs.

Sergeant Major (E-9) — Three chevrons above 3 arcs with a 5-pointed star between the chevrons and arcs.

First Sergeant (E-8) — Three chevrons above 3 arcs with a lozenge between the chevrons and arcs.

Master Sergeant (E-8) — Three chevrons above 3 arcs.

Sergeant First Class (E-7) — Three chevrons above 2 arcs.

Staff Sergeant (E-6) — Three chevrons above 1 arc.

Sergeant (E-5) — Three chevrons.

Corporal (E-4) — Two chevrons.

Specialists

Specialist (E-4) — Eagle device only.

Other enlisted

Private First Class (E-3) — One chevron above one arc.

Private (E-2) — One chevron.

Private (E-1) — None.

Air Force

Insignia for Air Force officers are identical to those of the Army. Insignia for enlisted personnel are worn on both sleeves and consist of a star and an appropriate number of rockers. Chevrons appear above 5 rockers for the top 3 noncommissioned officer ranks, as follows (in ascending order): Master Sergeant, 1 chevron; Senior Master Sergeant, 2 chevrons; and Chief Master Sergeant, 3 chevrons. The insignia of the Chief Master Sergeant of the Air Force has 3 chevrons and a wreath around the star design.

Navy

The following stripes are worn on the lower sleeves of the Service Dress Blue uniform. They are of gold embroidery.

Rank	Insignia
Fleet Admiral*	1 two inch with 4 one-half inch
Admiral	1 two inch with 3 one-half inch
Vice Admiral	1 two inch with 2 one-half inch
Rear Admiral (upper half)	1 two inchwith 1 one-half inch
Rear Admiral (lower half)	1 two inch
Captain	4 one-half inch
Commander	3 one-half inch
Lieutenant Commander	2 one-half inch with 1 one-quarter inch between
Lieutenant	2 one-half inch
Lieutenant (j.g.)	1 one-half inch with one-quarter inch above
Ensign	1 one-half inch

Warrant Officer-W-4 — ½″stripe with 1 break
Warrant Officer W-3 — ½″stripe with 2 breaks, 2″apart
Warrant Officer W-2 — ½″stripe with 3 breaks, 2″apart
Warrant Officer W-1 — ¼″stripe with 3 breaks, 2″apart

Enlisted personnel (noncommissioned petty officers)—A rating badge worn on the upper left sleeve, consisting of a spread eagle, appropriate number of chevrons, and centered specialty mark.

*The rank of Fleet Admiral is reserved for wartime use only.

Marine Corps

Marine Corps' distinctive cap and collar ornament is the Marine Corps Emblem—a combination of the American eagle, a globe, and an anchor. Marine Corps and Army officer insignia are similar. Marine Corps enlisted insignia, although basically similar to the Army's, feature crossed rifles beneath the chevrons. Marine Corps enlisted rank insignia are as follows:

Sergeant Major of the Marine Corps (E-9) — Same as Sergeant Major (below) but with Marine Corps emblem in the center with a 5-pointed star on both sides of the emblem.

Sergeant Major (E-9) — Three chevrons above 4 rockers with a 5-pointed star in the center.

Master Gunnery Sergeant (E-9) — Three chevrons above 4 rockers with a bursting bomb insignia in the center.

First Sergeant (E-8) — Three chevrons above 3 rockers with a diamond in the middle.

Master Sergeant (E-8) — Three chevrons above 3 rockers with crossed rifles in the middle.

Gunnery Sergeant (E-7) — Three chevrons above 2 rockers with crossed rifles in the middle.

Staff Sergeant (E-6) — Three chevrons above 1 rocker with crossed rifles in the middle.

Sergeant (E-5) — Three chevrons above crossed rifles.

Corporal (E-4) — Two chevrons above crossed rifles.

Lance Corporal (E-3) — One chevron above crossed rifles.

Private First Class (E-2) — One chevron.

Private (E-1) — None.

Coast Guard

Coast Guard insignia follow Navy custom, with certain minor changes such as the officer cap insignia. The Coast Guard shield is worn on both sleeves of officers and on the right sleeve of all enlisted personnel.

U.S. Army Personnel on Active Duty[1]

Source: Dept. of the Army, U.S. Dept. of Defense

Date[2]	Total strength[3]	Commissioned officers Total	Commissioned officers Male	Commissioned officers Female[4]	Warrant officers Male[5]	Warrant officers Female	Enlisted personnel Total	Enlisted personnel Male	Enlisted personnel Female
1940	267,767	17,563	16,624	939	763	—	249,441	249,441	—
1942	3,074,184	203,137	190,662	12,475	3,285	—	2,867,762	2,867,762	—
1943	6,993,102	557,657	521,435	36,222	21,919	—	6,413,526	6,358,200	55,325
1944	7,992,868	740,077	692,351	47,726	36,893	10	7,215,888	7,144,601	71,287
1945	8,266,373	835,403	772,511	62,892	56,216	44	7,374,710	7,283,930	90,780
1946	1,889,690	257,300	240,643	16,657	9,826	18	1,622,546	1,605,847	16,699
1950	591,487	67,784	63,375	4,409	4,760	22	518,921	512,370	6,551
1955	1,107,606	111,347	106,173	5,174	10,552	48	985,659	977,943	7,716
1960	871,348	91,056	86,832	4,224	10,141	39	770,112	761,833	8,279
1965	967,049	101,812	98,029	3,783	10,285	23	854,929	846,409	8,520
1970	1,319,735	143,704	138,469	5,235	23,005	13	1,153,013	1,141,537	11,476
1975	781,316	89,756	85,184	4,572	13,214	22	678,324	640,621	37,703
1980 (Sept. 30)	772,661	85,339	77,843	7,496	13,265	113	673,944	612,593	61,351
1985 (Sept. 30)	776,244	94,103	83,563	10,540	15,296	288	666,557	598,639	67,918
1990 (Mar. 31)	746,220	91,330	79,520	11,810	15,177	470	639,713	567,015	72,698
1994	553,627	74,956	64,281	10,675	12,448	535	465,688	405,664	60,024
1995	521,036	72,646	62,250	10,396	12,053	599	435,807	377,832	57,975
1996 (May 31)	493,330	68,850	58,875	9,975	11,456	660	408,511	351,669	56,842
1997 (May 31)	487,297	67,986	58,270	9,716	11,021	719	403,072	342,817	60,255
1998	491,707	67,048	56,650	10,398	10,989	661	402,000	345,149	56,851
1999	479,100	66,613	56,952	9,661	10,767	757	388,211	329,803	58,408
2000	471,633	66,344	56,391	9,953	10,608	781	393,900	333,947	59,953

(1) Represents strength of the active Army, including Philippine Scouts, retired Regular Army personnel on extended active duty, and National Guard and Reserve personnel on extended active duty; excludes U.S. Military Academy cadets, contract surgeons, and National Guard and Reserve personnel not on extended active duty. (2) June 30, unless otherwise noted. (3) Data for 1940 to 1946 include personnel in the Army Air Forces and its predecessors (Air Service and Air Corps). (4) Includes women doctors, dentists, and Medical Service Corps officers for 1946 and subsequent years, women in the Army Nurse Corps for all years, and the Women's Army Corps and Women's Medical Specialists Corps (dietitians, physical therapists, and occupational specialists) for 1943 and subsequent years. (5) Act of Congress approved Apr. 27, 1926, directed the appointment as warrant officers of field clerks still in active service. Includes flight officers as follows: 1943, 5,700; 1944, 13,615; 1945, 31,117; 1946, 2,580.

U.S. Navy Personnel on Active Duty

Source: Dept. of the Navy, U.S. Dept. of Defense

(As of Aug. 31, 2001)

Date	Officers	Nurses	Enlisted	Officer Candi-dates	Total
1940 (June)	13,162	442	144,824	2,569	160,997
1945 (June)	320,293	11,086	2,988,207	61,231	3,380,817
1950 (June)	42,687	1,964	331,860	5,037	381,538
1960 (June)	67,456	2,103	544,040	4,385	617,984
1970 (June)	78,488	2,273	605,899	6,000	692,660
1980 (June)[1]	63,100	—	464,100	—	527,200
1990 (Sept.)	74,429	—	530,133	—	604,562
1993 (Mar.)	66,787	—	445,409	—	512,196
1994 (Apr.)	64,430	—	418,378	—	482,808
1995 (May)	61,075	—	402,626	—	463,701
1996 (June)	60,013	—	376,595	—	436,608
1997 (June)	57,341	—	340,616	—	397,957
1998 (Sept.)	55,007	—	326,196	—	381,203
1999 (June)	55,726	—	322,372	—	378,098
2000 (Oct.)	53,698	—	320,212	—	373,910
2001 (Aug.)	54,177	—	317,100	—	375,618

(1) Starting in 1980, "Nurses" are included with "Officers," and "Officer Candidates" are included with "Enlisted."

U.S. Marine Corps Personnel on Active Duty

Source: Dept. of the Marines, U.S. Dept. of Defense

(As of May 31, 2001)

Year	Officers	Enlisted	Total
1940	1,800	26,545	28,345
1945	37,067	437,613	474,680
1950	7,254	67,025	74,279
1960	16,203	154,418	170,621
1970	24,941	234,796	259,737
1980	18,198	170,271	188,469
1990	19,958	176,694	196,652
1991	19,753	174,287	194,040
1992	19,132	165,397	184,529
1993	18,878	161,205	180,083
1994	18,430	159,949	178,379
1995	18,017	153,929	171,946
1996	18,146	154,141	172,287
1997	18,089	154,240	172,329
1998	17,984	154,648	172,632
1999	17,892	155,250	173,142
2000	17,897	154,744	172,641
2001	18,072	152,559	170,631

U.S. Air Force Personnel on Active Duty

Source: Air Force Dept., U.S. Dept. of Defense

(as of May 31, 2001)

Year[1]	Strength
1918	195,023
1920	9,050
1930	13,531
1940	51,165
1941	152,125
1942	764,415
1943	2,197,114
1944	2,372,292
1945	2,282,259
1950	411,277
1960	814,213
1970	791,078
1980	557,969
1986	608,200
1990	535,233
1991	510,432
1992	470,315
1993	444,351
1994	426,327
1995	400,051
1996	389,400
1997	378,681
1998	363,479
1999	357,929
2000	357,777
2001	351,935

(1) Prior to 1947, data are for U.S. Army Air Corps and Air Service of the Signal Corps.

U.S. Coast Guard Personnel on Active Duty

Source: U.S. Coast Guard, U.S. Dept. of Defense

(midyear personnel figures)

Year	Total	Officers	Cadets	Enlisted
1970	37,689	5,512	653	31,524
1975	36,788	5,630	1,177	29,981
1980	39,381	6,463	877	32,041
1981	39,760	6,519	981	32,260
1983	39,708	6,535	811	32,362
1984	38,705	6,790	759	31,156
1985	38,595	6,775	733	31,087
1986	37,284	6,577	754	29,953
1987	38,576	6,644	859	31,073
1988	37,723	6,530	887	30,306
1989	37,453	6,614	867	29,972
1990	37,308	6,475	820	29,860
1991	38,280	7,095	900	30,285
1992	39,185	7,348	919	30,918
1993	38,832	7,724	691	30,417
1994	37,284	7,401	881	29,002
1995	36,731	7,489	841	28,401
1996	35,229	7,270	830	27,129
1997	34,717	7,079	868	26,770
1998	34,890	7,140	805	26,945
1999	35,266	7,135	880	27,251
2000	35,712	7,154	863	27,695

Chairmen of the Joint Chiefs of Staff, 1949-2001

Gen. of the Army Omar N. Bradley, USA	8/16/49 –8/14/53
Adm. Arthur W. Radford, USN	8/15/53 – 8/14/57
Gen. Nathan F. Twining, USAF	8/15/57 – 9/30/60
Gen. Lyman L. Lemnitzer, USA	10/1/60 – 10/30/62
Gen. Maxwell D. Taylor, USA	10/1/62 – 7/3/64
Gen. Earle G. Wheeler, USA	7/3/64 – 7/2/70
Adm. Thomas H. Moorer, USN	7/3/70 – 6/30/74
Gen. George S. Brown, USAF	7/1/74 – 6/20/78
Gen. David C. Jones, USAF	6/21/78 – 6/18/82
Gen. John W. Vessey Jr., USA	6/18/82 – 9/30/85
Adm. William J. Crowe, Jr., USN	10/1/85 – 9/30/89
Gen. Colin L. Powell, USA	10/1/89 – 9/30/93
Gen. John M. Shalikashvili, USA	10/1/93 – 9/30/97
Gen. Henry H. Shelton, USA	10/1/97 – 9/30/01
Gen. Richard B. Myers	10/1/01 –

Women in the U.S. Armed Forces

Source: U.S. Dept. of Defense

Women in the Army, Navy, Air Force, Marines, and Coast Guard are fully integrated with male personnel. Expansion of military women's programs began in the Department of Defense in fiscal year 1973.

Admission of women to the service academies began in the fall of 1976.

Under rules instituted in 1993, women were allowed to fly combat aircraft and serve aboard warships. Women remained restricted from service in ground combat units.

Between Apr. 1993 and July 1994, almost 260,000 positions in the armed forces were opened to women. By the mid-1990s, 80% of all jobs and more than 90% of all career fields in the military had been opened to women. As of June 30, 2000, women made up 14.4% of the armed forces.

Women Active Duty Troops in 2000

Service	% Women
Army	15.1
Navy	14.0
Marines	5.9
Air Force	19.0
Coast Guard	10.2

Women on Active Duty, All Services*: 1973-2000

Year	% Women	Year	% Women
1973	2.5	1987	10.2
1975	4.6	1993	11.6
1981	8.9	2000	14.4

*Not including the Coast Guard, which is a part of the Dept. of Transportation.

For Further Information on the U.S. Armed Forces

Army — Office of the Chief of Public Affairs, Attention: Media Relations Division—MRD, Army 1500, Wash., DC 20310-1500. **Website:** http://www.army.mil

Navy — Chief of Information, 1200 Navy Pentagon, Wash., DC 20350-1200. **Website:** http://www.navy.mil

Air Force — Office of Public Affairs, 1690 Air Force, Pentagon, Wash., DC 20330-1690. **Website:** http://www.af.mil

Marine Corps — Commandant of the Marine Corps (Code PA), Headquarters, U.S. Marine Corps, Wash. DC 20380-1775. **Website:** http://www.usmc.mil

Coast Guard — Commandant (G-IPA), U.S. Coast Guard, 2100 Second St. SW, Wash., DC 20593-0001. **Website:** http://www.uscg.mil

Additional information on all the U.S. Armed Forces branches, as well as many other related organizations, can be accessed through DefenseLINK, the official Internet site of the Dept. of Defense: http://www.defenselink.mil

African American Service in U.S. Wars

American Revolution. About 5,000 African Americans served in the Continental Army, mostly in integrated units, some in all-black combat units.

Civil War. Some 200,000 African Americans served in the Union Army; 38,000 were killed, and 22 won the Medal of Honor (the nation's highest award).

World War I. About 367,000 African Americans served in the armed forces, 100,000 in France.

World War II. Over 1 mil African Americans served in the armed forces; all-black fighter and bomber AAF units and infantry divisions gave distinguished service. (By 1954, armed forces were completely desegregated.)

Korean War. Approximately 3,100 African Americans lost their lives in the Korean combat.

Vietnam War. 274,937 African Americans served in the armed forces (1965-74); 5,681 were killed in combat.

Persian Gulf War. About 104,000 African Americans served in the Kuwaiti theater—20% of all U.S. troops, compared with 8.7% of all troops for World War II and 9.8% for Vietnam.

Defense Contracts, 2000

Source: U.S. Dept. of Defense

(in thousands of dollars)

Listed are the 50 companies (including their subsidiaries) or organizations receiving the largest dollar volume of prime contract awards from the U.S. Department of Defense during fiscal year 2000.

Lockheed Martin	$15,125,846
Boeing	12,041,420
Raytheon	6,330,613
General Dynamics	4,195,923
Northrop Grumman	3,079,615
Litton Industries	2,737,284
United Technologies	2,071,536
TRW	2,004,857
General Electric	1,609,329
Science Applications Intl.	1,522,077
Carlyle Group	1,194,713
Computer Sciences	1,164,634
Textron	1,164,465
Marconi	997,339
Honeywell International	951,255
Newport News Shipbuilding	789,900
Dyncorp	771,235
Bechtel Group, Inc.	694,717
Government of Canada	$676,881
National Amusements	619,696
Morrison Knudsen	596,635
Halliburton	595,070
BP Amoco	591,953
Energy, U.S. Dept. of	590,631
ITT Industries	553,972
Health Net	550,580
IT Group	493,335
Rockwell International	473,668
Alliant Techsystems	470,397
Longbow Limited Liability	468,091
FED EX	452,385
Maersk	426,798
Stewart & Stevenson Services	424,051
Booz Allen & Hamilton	419,576
Mitre	409,217
Federal Republic of Germany	$408,004
Raytheon Lockheed Martin	401,584
Jacobs Engineering Group	387,343
Boeing Sikorsky Comanche Team	384,751
L-3 Communications Holding	377,662
Oshkosh Truck	372,526
AT&T	352,547
Johns Hopkins Univ.	351,776
Mass. Inst. ofTechnology	347,197
Triwest Healthcore Alliance	335,878
Philipp Holzmann Aktiengesells	334,703
Aerospace Corporation	334,194
Renco Group	330,064
Electronic Data Systems	329,554
Exxon Mobil	324,816

U.S. Veteran Population

Source: U.S. Dept. of Veterans Affairs; as of September 30, 2001

(in thousands)

TOTAL VETERANS IN CIVILIAN LIFE[1]	**25,038**
Total wartime veterans[2]	**18,866**
Total Gulf War	3,096
Gulf War with service in Vietnam era	304
Gulf War with no prior wartime service	2,792
Total Vietnam era	8,221
Vietnam era with service in Korean conflict	422
Vietnam era with no prior wartime service	7,799
Total Korean conflict	3,769
Korean conflict with service in WWII	529
Korean conflict with no prior wartime service	3,240
World War II	5,033
World War I	2
Total peacetime veterans	**6,173**
Service between Vietnam era and Gulf War only	3,458
Service between Korean conflict and Vietnam era only	2,600
Other peacetime	115

NOTE: Details may not add to total shown because of rounding. (1) There are an indeterminate number of Mexican Border period veterans, 7 of whom were receiving benefits in August 2001. (2) The total for "wartime veterans" consists only of veterans from each listed war that had no prior wartime service. Figures are for U.S. veterans worldwide

Veterans Compensation and Pension Case Payments

Source: 1900-1980: Dept. of Veterans Affairs; 1990-2000: Natl. Center for Veteran Analysis and Statistics

Fiscal year	Living veteran cases	Deceased veteran cases	Total cases	Total expenditures (dollars)
1900	752,510	241,019	993,529	$138,462,130
1910	602,622	318,461	921,083	159,974,056
1920	419,627	349,916	769,543	316,418,030
1930	542,610	298,223	840,833	418,432,809
1940	610,122	239,176	849,298	429,138,465
1950	2,368,238	658,123	3,026,361	2,009,462,298
1960	3,008,935	950,802	3,959,737	3,314,761,383
1970	3,127,338	1,487,176	4,614,514	5,253,839,611
1980	3,195,395	1,450,785	4,646,180	$11,046,637,368
1990	2,746,329	837,596	3,583,925	14,674,411,000
1995	2,668,576	661,679	3,330,255	17,765,045,000
1996	2,671,026	637,232	3,308,258	17,055,809,000
1997	2,666,785	613,976	3,280,761	19,284,287,000
1998	2,668,030	594,782	3,262,812	20,164,598,000
1999	2,673,167	578,508	3,251,675	21,023,864,000
2000	2,672,407	563,754	3,236,161	21,963,216,000

Active Duty U.S. Military Personnel Strengths, Worldwide, 2000

Source: U.S. Dept. of Defense

(as of Sept. 30, 2000)

U.S. Territories & Special Locations	
U.S., 48 contiguous states	938,753
Alaska	15,757
Hawaii	33,930
Guam	3,266
Johnston Atoll	229
Puerto Rico	2,896
Transients	29,699
Afloat	101,962
REGIONAL TOTAL[1]	**1,126,521**
Europe	
Belgium	1,554
Bosnia and Herzegovina	5,708
Croatia	138
Germany	69,203
Greece	678
Greenland	125
Hungary	375
Iceland	1,636
Italy	11,190
Macedonia, F.Y.R. of	347
Netherlands	659
Portugal	1,005
Serbia	5,427
Spain	2,007
Turkey	2,006
United Kingdom	11,207
Afloat	3,767
REGIONAL TOTAL[1]	**117,411**
East Asia & Pacific	
Australia	175
Japan	40,159
Korea, South	36,565
Singapore	411
Thailand	526
Afloat	23,352
REGIONAL TOTAL[1]	**101,447**
Sub-Saharan Africa	224
TOTAL[1]	**224**
North Africa, Middle East & South Asia	
Bahrain	949
Diego Garcia	625
Egypt	499
Kuwait	4,602
Oman	251
Saudi Arabia	7,053
United Arab Emirates	402
Afloat	14,772
REGIONAL TOTAL[1]	**29,384**
Other Western Hemisphere	
Canada	156
Cuba (Guantánamo)	688
Honduras	351
Peru	425
Afloat	3,068
REGIONAL TOTAL[1]	**5,416**
Former Soviet Union	
Russia	101
TOTAL[1]	**160**
TOTAL WORLDWIDE[2]	**1,384,338**

(1) Countries and areas with fewer than 100 assigned U.S. military members not listed; regional totals include personnel stationed in those countries and areas not shown. (2) Total worldwide also includes undistributed personnel.

The Medal of Honor

The Medal of Honor is the highest military award for bravery that can be given to any individual in the United States. The first Army Medals were awarded on Mar. 25, 1863, and the first Navy Medals went to sailors and Marines on Apr. 3, 1863.

On Dec. 21, 1861, Pres. Abraham Lincoln signed into law a bill to create the Navy Medal of Honor. Lincoln later (July 14, 1862) approved a resolution providing for the presentation of Medals of Honor to enlisted men of the Army and Voluntary Forces, making it a law. The law was amended on March 3, 1863 to extend its provisions to include officers as well as enlisted men.

The Medal of Honor is awarded in the name of Congress to a person who, while a member of the armed forces, distinguishes himself or herself conspicuously by gallantry and intrepidity at the risk of life above and beyond the call of duty while engaged in an action against any enemy of the United States; while engaged in military operations involving conflict with an opposing foreign force; or while serving with friendly foreign forces engaged in an armed conflict against an opposing armed force in which the United States is not a belligerent party. The deed performed must have been one of personal bravery or self-sacrifice so conspicuous as to clearly distinguish the individual above his or her comrades and must have involved risk of life. Incontestable proof of the performance of service is required, and each recommendation for award of this decoration is considered on the standard of extraordinary merit.

Prior to World War I, the 2,625 Army Medal of Honor awards up to that time were reviewed to determine which past awards met new stringent criteria. The Army removed 911 names from the list, most of them former members of a volunteer infantry group during the Civil War who had been induced to extend their enlistments when they were promised the medal. However, in 1977 a medal was restored to Dr. Mary Walker, and in 1989 medals were restored to Buffalo Bill Cody and 7 other Indian scouts.

Since that review, Medals of Honor have been awarded in the following numbers:

World War I	124	Korean War	131
Peacetime (1920-40)	18	Vietnam War	240
World War II	441	Somalia	2

The figure for World War II includes 7 African-American soldiers who were awarded Medals of Honor (6 of them posthumously) in Jan. 1997. Previously, no black soldier had received the medal for World War II service; an Army inquiry begun in 1993 concluded that the prevailing political climate and Army practices of the time had prevented proper recognition of heroism on the part of black soldiers in that war.

Nations With Largest Armed Forces, by Active-Duty Troop Strength[1]

Source: *The Military Balance. 2000* (International Institute for Strategic Studies, published by Oxford University Press, UK)

	Troop strength (thousands) Active troops	Reserve troops	Defense expend. ($ bil)	Tanks (MBT) (army only)	Navy Cruisers/Frigates/Destroyers	Submarines	Combat aircraft (air force only) FGA	Fighters
1. CHINA	2,820.0	1,200.0	39.9	7,060	40F/20D	65	400	3,000 (est.)
2. UNITED STATES	1,371.5	1,303.3	283.1	7,900	27C/35F/52D**	74	52 tactical fighter squadrons	
3. India*	1,173.0	528.4	15.0	3,414	8F/12D**	16	18 sqn	20 sqn
4. N. Korea	1,055.0	4,700.0	2.1	3,500	3F	26	15 rgt FGA/FTR	
5. RUSSIA	1,004.1	2,400.0	56.8	21,820	7C/10F/17D**	67	575	880
6. S. Korea	672.0	4,500.0	12.1	2,330	9F/6D	19	7 tact ftr wgs	
7. Turkey	639.0	378.7	10.2	4,205	22F	14	11 sqn	7 sqn
8. Iran	545.6	350.0	5.7	1,135	3F	5	9 sqn	7 sqn
9. Pakistan*	587.0	513.0	3.5	2,285	8F	10	6 sqn	12 sqn
10. Vietnam	484.0	3,000.0	.9	1,315	6F	2	2 rgt	6 rgt
11. Egypt	450.0	254.0	3.0	5,155	10F/1D	4	7 sqn	22 sqn
12. Iraq	429.0	650.0	1.5	2,200	2F	—	130	180
13. Afghanistan	400.0	—	.27	1,000	—	0	110	80
14. Taiwan	376.0	1,657.5	15.0	739	21F/12D	4	23 sqn FGA/FTR	
15. Myanmar	343.8	—	2.0	100	—	—	3 sqn	2 sqn
16. Germany	332.8	344.7	31.1	2,815	12F/2D	14	5 wg	4 wg
17. Ethiopia	325.5	—	.4	160	—	—	53	—
18. FRANCE	317.3	419.0	37.9	1,207	1C/35F/4D**	12	7 sqn	6 sqn
19. Syria	316.0	396.0	1.0	4,850	2F	—	9-10 sqn	17 sqn
20. Ukraine	311.4	1,000.0	1.4	4,014	1C/7F	1	5 rgt	8 rgt
21. Thailand	306.0	200.0	2.6	282	14F**	—	3 sqn	3 sqn
22. Indonesia	299.0	400.0	1.5	375	17F	2	5 sqn	1 sqn
23. Brazil	291.0	1,115.0	16.0	178	14F**	5	3 sqn	2 sqn
24. Italy	265.5	72.0	22.6	1,396	1C/24F/4D**	7	8 sqn	5 sqn
25. Japan	242.6	48.6	40.4	1,070	13F/42D	16	2 sqn	10 sqn
26. Poland	240.7	406.0	3.2	1,704	2F/1D	3	9 sqn	2 sqn
27. UNITED KINGDOM	212.4	254.3	36.9	616	20F/11D**	16	10 sqn	6 sqn
28. Romania	207.0	470.0	.6	1,253	6F/1D	1	4 rgt	2 rgt
29. Eritrea	200.0	120.0	.3	80	—	—	12 FGA/FTR	
30. Morocco	196.3	150.0	1.8	644	1F	—	47	15

Nations with known strategic nuclear capability are shown in all capital letters. *India and Pakistan tested nuclear devices in 1998. MBT = main battle tank. FGA = fighter, ground attack; rgt = regiment; sqn = squadron (12-24 aircraft). **Denotes navies with aircraft carriers, as follows: United States 12, United Kingdom 3, France 2, India 1, Italy 1, Russian 1, Brazil 1, Spain 1, Thailand 1. (1) All figures are for 1999. — = not available.

Directors of the Central Intelligence Agency

In 1942, Pres. Franklin D. Roosevelt established the Office of Strategic Services (OSS); it was disbanded in 1945. In 1946, Pres. Harry Truman established the Central Intelligence Group (CIG) to operate under the National Intelligence Authority (NIA). A 1947 law replaced the NIA with the National Security Council and the CIG with the Central Intelligence Agency.

Director	Served	Appointed by President
Adm. Sidney W. Souers	1946	Truman
Gen. Hoyt S. Vandenberg	1946-1947	Truman
Adm. Roscoe H. Hillenkoetter	1947-1950	Truman
Gen. Walter Bedell Smith	1950-1953	Truman
Allen W. Dulles	1953-1961	Eisenhower
John A. McCone	1961-1965	Kennedy
Adm. William F. Raborn Jr.	1965-1966	Johnson
Richard Helms	1966-1973	Johnson
James R. Schlesinger	1973	Nixon
William E. Colby	1973-1976	Nixon
George Bush	1976-1977	Ford
Adm. Stansfield Turner	1977-1981	Carter
William J. Casey	1981-1987	Reagan
William H. Webster	1987-1991	Reagan
Robert M. Gates	1991-1993	Bush
R. James Woolsey	1993-1995	Clinton
John M. Deutch	1995-1997	Clinton
George J. Tenet	1997-	Clinton

Nuclear Arms Treaties and Negotiations: A Historical Overview

Aug. 5, 1963—Limited Test Ban Treaty signed in Moscow by U.S., USSR, and Britain; prohibited testing of nuclear weapons in space, above ground, and under water.

Jan. 27, 1967—Outer Space Treaty banned the introduction of nuclear weapons and other weapons of mass destruction into orbit around the earth, their installation on the moon or other celestial body, or their station in space.

July 1, 1968—Nuclear Nonproliferation Treaty, with U.S., USSR, and Great Britain as major signers, limited spread of nuclear material for military purposes by agreement not to assist nonnuclear nations in getting or making nuclear weapons. Extended indefinitely, May 11, 1995.

May 26, 1972—Strategic Arms Limitation Treaty (SALT I) signed in Moscow by U.S. and USSR. This short-term agreement imposed a 5-year freeze on both testing and deployment of intercontinental ballistic missiles (ICBMs) as well as submarine-launched ballistic missiles (SLBMs). In the area of defensive nuclear weapons, the separate **ABM Treaty,** signed on the same occasion, limited antiballistic missiles to 2 sites of 100 antiballistic missile launchers in each country (amended in 1974 to one site in each country). ABM Treaty amended Sept. 1997 to allow flexibility in development of shorter-range nuclear weapons.

July 3, 1974—ABM Treaty Revision (protocol on antiballistic missile systems) and **Threshold Test Ban Treaty** on limiting underground testing of nuclear weapons to 150 kilotons were signed by U.S. and USSR in Moscow.

Sept. 1977—U.S. and USSR agreed to continue to abide by SALT I, despite its expiration date.

June 18, 1979—SALT II signed in Vienna by the U.S. and USSR, constrained offensive nuclear weapons, limiting each side to 2,400 missile launchers and heavy bombers; ceiling to apply until Jan. 1, 1985. Treaty also set a subceiling of 1,320 ICBMs and SLBMs with multiple warheads on each side. SALT II never reached the Senate floor for ratification because Pres. Jimmy Carter withdrew support following Dec. 1979 Soviet invasion of Afghanistan.

Dec. 8, 1987—Intermediate-Range Nuclear Forces (INF) Treaty signed in Washington, D.C., by USSR leader Mikhail Gorbachev and U.S. Pres. Ronald Reagan, eliminating all U.S. and Soviet intermediate- and shorter-range nuclear missiles from Europe and Asia. Ratified, with conditions, by U.S. Senate on May 27, 1988; by USSR on June 1, 1988. Entered into force June 1, 1988.

July 31, 1991—Strategic Arms Reduction Treaty (START I) signed in Moscow by USSR and U.S. to reduce strategic offensive arms by about 30% in 3 phases over 7 years. START I was the first treaty to mandate reductions by the superpowers. Treaty was approved by U.S. Senate Oct. 1, 1992.

With the Soviet Union breakup in Dec. 1991, 4 former Soviet republics became independent nations with strategic nuclear weapons—Russia, Ukraine, Kazakhstan, and Belarus. The last 3 agreed in principle in 1992 to transfer their nuclear weapons to Russia and ratify START I. The Russian Supreme Soviet voted to ratify, Nov. 4, 1992, but Russia decided not to provide instruments of ratification until the other 3 republics ratified START I and acceded to the Nuclear Nonproliferation Treaty (NPT) as nonnuclear nations. By late 1994, all 3 nations had done so, and NPT entered into force on Dec. 5, 1994. In Dec. 1996, Belarus was the last of the 3 to give up its nuclear weapons.

Jan. 3, 1993—START II signed in Moscow by U.S. and Russia. Potentially the broadest disarmament pact in history, it called for both sides to reduce their long-range nuclear arsenals to about one-third of their then-current levels within a decade and disable and dismantle launching systems. The U.S. ratified START II on Jan. 26, 1996; Russia ratified it Apr. 13, 2000. On Sept. 26, 1997, the U.S. and Russia signed an agreement that would delay the dismantling of launching systems under START II to the end of 2007 (they would still be disabled by 2003). The accord was intended to facilitate Russian ratification of START II. Russia and the U.S. also agreed in writing to work toward further strategic arms cuts in a 3d round of START negotiations.

Sept. 24, 1996—Comprehensive Test Ban Treaty (CTBT) signed by U.S. and Russia. The CTBT bans all nuclear weapon tests and other nuclear explosions. It is intended to help prevent the nuclear powers from developing more advanced weapons, while limiting the ability of other states to acquire such devices. As of Oct. 2000, the CTBT had been signed by 160 nations, including China, Russia, the U.S., the U.K, and France. It had been ratified by 66, including France, Russia, and the U.K., but not the U.S. or China. The treaty will enter into force after 44 nuclear-capable states ratify it. As of Oct. 15, 2000, 30 of the 44 had done so. As of Sept. 2001, Pres. George W. Bush sought Russian assent to his plans for building an American missle defense system. These plans would be in violation of the ABM treaty and could lead to its abandonment by both sides.

Monthly Military Pay Scale[1]

Source: U.S. Dept. of Defense; effective July 1, 2001

	Years of Service							
Rank/Grade	**Under 2**	**Over 2**	**Over 4**	**Over 8**	**Over 12**	**Over 16**	**Over 20**	**Over 26**
General—0-10 (2)	$8,518.80	$8,818.50	—	$9,156.90	$9,664.20	$10,356.00	$11,049.30	$11,737.20
Lt. General—0-9	7,550.10	7,747.80	—	8,114.10	8,451.60	9,156.90	9,664.20	10,356.00
Major General—0-8	6,838.20	7,062.63	$7,252.20	7,747.80	8,114.10	8,451.60	9,156.90	*
Brig. General—0-7	5,682.30	6,068.40	6,112.50	6,514.50	6,915.90	7,747.80	*	8,322.60
Colonel—0-6	4,211.40	4,626.60	*	5,160.90	*	6,005.40	6,617.40	7,309.80
Lt. Colonel—0-5	3,368.70	3,954.90	4,280.40	*	4,831.80	5,481.60	5,790.30	*
Major—0-4	2,839.20	3,457.20	3,739.50	4,127.70	4,629.30	4,935.00	*	*
Captain—0-3	2,638.20	2,991.00	3,489.30	3,839.70	4,189.80	*	*	*
1st Lt.—0-2	2,301.00	2,620.80	3,120.30	*	*	*	*	*
2d Lt.—0-1	1,997.70	2,079.00	*	*	*	*	*	*
Chief Warrant—W-4	2,688.00	2,891.70	3,056.70	3,336.30	3,614.10	3,892.50	4,168.20	4,590.90
Warrant Officer—W-1	1,782.60	2,043.90	2,214.60	2,419.20	2,626.80	2,835.90	3,018.60	*
Sgt. Major—E-9 (3)	—	—	—	—	3,197.40	3,392.40	3,601.80	4,060.80
Master Sgt.—E-8	—	—	—	2,622.00	2,768.40	2,945.10	3,138.00	3,612.60
Sgt. 1st class—E-7	1,831.20	1,999.20	2,149.80	2,362.20	2,512.80	2,666.10	2,817.90	3,250.50
Staff Sgt.—E-6	1,575.00	1,740.30	1,891.80	2,097.30	2,248.80	2,379.60	*	*
Sergeant—E-5	1,381.80	1,549.20	1,701.00	1,888.50	1,040.30	*	*	*
Corporal—E-4	1,288.80	1,423.80	1,576.20	*	*		*	*
Pvt. 1st class—E-3	1,214.70	1,307.10	1,385.40	*	*	*	*	*
Private—E-2	1,169.10	*	*	*	*	*	*	*
Recruit—E-1 (4mos+)	1,042.80	*	*	*	*	*	*	*
Recruit—E-1 (<4mos)	964.80	*	*	*	*	*	*	*

— indicates that the pay grade could not be reached in the length of service shown. *indicates no change; pay does not increase for additional years of service at this grade. (1) Basic pay is limited for O-7 to O-10 to $11,141.70 per month, and for O-6 and below to $9,800.10 per month. (2) While serving as Chairman or Vice Chairman of the Joint Chiefs of Staff, Chief of Staff of the Army or Air Force, Chief of Naval Operations, Commandant of the Marine Corps or Coast Guard, the amount of basic pay is $12,950.70, regardless of years of service; however, the amount received is limited to $11,141.70 per month. (3) While serving as Sergeant Major of the Army, Master Chief Petty Officer of the Navy or Coast Guard, Chief Master Sergeant of the Air Force, or Sergeant Major of the Marine Corps, basic pay is $4,893.60 per month.

Casualties in Principal Wars of the U.S.

Source: U.S. Dept. of Defense, U.S. Coast Guard

Data prior to World War I are based on incomplete records in many cases. Casualty data are confined to dead and wounded personnel and, therefore, exclude personnel captured or missing in action who were subsequently returned to military control. Dash (—) indicates information is not available. off. = officers.

			CASUALTIES			
WAR	**Branch of service**	**Number serving**	**Battle deaths**	**Other deaths**	**Wounds not mortal[7]**	**Total[13]**
Revolutionary War	**Total**	—	**4,435**	—	**6,188**	10,623
1775-83	Army	184,000	4,044	—	6,004	10,048
	Navy	to	342	—	114	456
	Marines	250,000	49	—	70	119
War of 1812	**Total**	**286,730[8]**	**2,260**	—	**4,505**	**6,765**
1812-15	Army	—	1,950	—	4,000	5,950
	Navy	—	265	—	439	704
	Marines	—	45	—	66	111
Mexican War	**Total**	**78,789[8]**	**1,733**	**11,550**	**4,152**	**17,435**
1846-48	Army	—	1,721	11,550	4,102	17,373
	Navy	—	1	—	3	4
	Marines	—	11	—	47	58
	Coast Guard[12]	71 off.	—	—	—	—
Civil War						
Union forces	**Total**	**2,213,363[8]**	**140,415**	**224,097**	**281,881**	**646,392**
1861-65	Army	2,128,948	138,154	221,374	280,040	639,568
	Navy	—	2,112	2,411	1,710	6,233
	Marines	84,415	148	312	131	591
Confederate forces	**Total**	—	**74,524**	**59,297**	—	**133,821**
(estimate)[1]	Army	600,000	—	—	—	—
1863-66	Navy	to	—	—	—	—
	Marines	1,500,000	—	—	—	—
	Coast Guard[12]	219 off.	1	—	—	1
Spanish-American War	**Total**	**307,420**	**385**	**2,061**	**1,662**	**4,108**
1898	Army[3]	280,564	369	2,061	1,594	4,024
	Navy	22,875	10	0	47	57
	Marines	3,321	6	0	21	27
	Coast Guard[12]	660	0	—	—	—
World War I	**Total**	**4,743,826**	**53,513**	**63,195**	**204,002**	**320,710**
April 6, 1917 - Nov. 11, 1918	Army[4]	4,057,101	50,510	55,868	193,663	300,041
	Navy	599,051	431	6,856	819	8,106
	Marines	78,839	2,461	390	9,520	12,371
	Coast Guard	8,835	111	81	—	192
World War II	**Total**	**16,353,659**	**292,131**	**115,185**	**671,846**	**1,079,162**
Dec. 7, 1941 - Dec. 31, 1946[2]	Army[5]	11,260,000	234,874	83,400	565,861	884,135
	Navy[6]	4,183,466	36,950	25,664	37,778	100,392
	Marines	669,100	19,733	4,778	68,207	91,718
	Coast Guard	241,093	574	1,343	—	1,917
Korean War[9]	**Total**	**5,764,143**	**33,667**	**3,249**	**103,284**	**140,200**
June 25, 1950 - July 27, 1953	Army	2,834,000	27,709	2,452	77,596	107,757
	Navy	1,177,000	493	160	1,576	2,226
	Marines	424,000	4,267	339	23,744	28,353
	Air Force	1,285,000	1,198	298	368	1,864
	Coast Guard	44,143	—	—	—	—
Vietnam War[10]	**Total**	**8,752,000**	**47,393**	**10,800**	**153,363**	**211,556**
Aug. 4, 1964 - Jan. 27, 1973	Army	4,368,000	30,929	7,272	96,802	135,003
	Navy	1,842,000	1,631	931	4,178	6,740
	Marines	794,000	13,085	1,753	51,392	66,230
	Air Force	1,740,000	1,741	842	931	3,514
	Coast Guard	8,000	7	2	60	69
Persian Gulf War	**Total**	**467,939[11]**	**148**	**151**	**467**	**766**
1991	Army	246,682	98	105	—	203
	Navy	98,852	6	14	—	20
	Marines	71,254	24	26	—	50
	Air Force	50,751	20	6	—	26
	Coast Guard	400	—	—	—	—

(1) Authoritative statistics for the Confederate forces are not available. An estimated 26,000-31,000 Confederate personnel died in Union prisons. (2) Data are for Dec. 1, 1941, through Dec. 31, 1946, when hostilities were officially terminated by Presidential Proclamation; few battle deaths or wounds not mortal were incurred after Japanese acceptance of Allied peace terms on Aug. 14,1945. Numbers serving Dec. 1, 1941-Aug. 31, 1945, were: Total—14,903,213; Army—10,420,000; Navy—3,883,520; Marine Corps—599,693. (3) Number serving covers the period April 21-Aug. 13, 1898, while dead and wounded data are for the period May 1-Aug. 31, 1898. Active hostilities ceased on Aug. 13, 1898, but ratifications of the treaty of peace were not exchanged between the United States and Spain until April 11, 1899. (4) Includes Army Air Forces battle deaths and wounds not mortal, as well as casualties suffered by American forces in northern Russia to Aug. 25, 1919, and in Siberia to April 1, 1920. Other deaths covered the period April 1, 1917-Dec. 31, 1918. (5) Includes Army Air Forces. (6) Battle deaths and wounds not mortal include casualties incurred in Oct. 1941 due to hostile action. (7) Marine Corps data for World War II, the Spanish-American War, and prior wars represent the number of individuals wounded, whereas all other data in this column represent the total number (incidence) of wounds. (8) As was reported by the Commissioner of Pensions in his Annual Report for Fiscal Year 1903. (9) As a result of an ongoing Dept. of Defense review of available Korean War casualty record information, updates to previously reported figures for battle deaths and other deaths are reflected in this table. (10) Number serving covers the period Aug. 4, 1964-Jan. 27, 1973 (date of ceasefire). Includes casualties incurred in Mayaguez Incident. Wounds not mortal exclude 150,332 persons not requiring hospital care. (11) Estimated. (12) Actually the U.S. Revenue Cutter Services, predecessor to the U.S. Coast Guard. (13) Totals do not include categories for which no data are listed.

Memorable Moments in Human Spaceflight

Sources: National Aeronautics and Space Administration; Congressional Research Service; World Almanac research

Note: Boldface denotes U.S. mission by National Aeronautics and Space Administration (NASA). Other missions were sponsored by the USSR or, later, the Commonwealth of Independent States. Dates are Eastern standard time. EVA = extravehicular activity. ASTP = Apollo-Soyuz Test Project. Number of total flights by each crew member is given in parentheses when flight listed is not the first.

Dates	Mission[1]	Crew (no. of flights)	Duration (hr:min)	Remarks
4/12/61	Vostok 1	Yuri A. Gagarin	1:48	1st human orbital flight
5/5/61	**Mercury-Redstone 3**	**Alan B. Shepard Jr.**	**0:15**	**1st American in space**
7/21/61	**Mercury-Redstone 4**	**Virgil I. Grissom**	**0:15**	**Spacecraft sank, Grissom rescued**
8/6/61- 8/7/61	Vostok 2	Gherman S. Titov	25:18	1st spaceflight of more than 24 hrs
2/20/62	**Mercury-Atlas 6**	**John H. Glenn Jr.**	**4:55**	**1st American in orbit; 3 orbits**
5/24/62	**Mercury-Atlas 7**	**M. Scott Carpenter**	**4:56**	**Manual retrofire error caused 250-mi landing overshoot**
8/11/62- 8/15/62	Vostok 3	Andrian G. Nikolayev	94:22	Vostok 3 and 4 made 1st group flight
8/12/62- 8/15/62	Vostok 4	Pavel R. Popovich	70:57	On 1st orbit it came within 3 mi of Vostok 3
10/3/62	**Mercury-Atlas 8**	**Walter M. Schirra Jr.**	**9:13**	**Landed 5 mi from target**
5/15/63- 5/16/63	**Mercury-Atlas 9**	**L. Gordon Cooper**	**34:19**	**1st U.S. evaluation of effects of one day in space on a person; 22 orbits**
6/14/63- 6/19/63	Vostok 5	Valery F. Bykovsky	119:06	Vostok 5 and 6 made 2d group flight
6/16/63- 6/19/63	Vostok 6	Valentina V. Tereshkova	70:50	1st woman in space; passes within 3 mi of Vostok 5
10/12/64- 10/13/64	Voskhod 1	Vladimir M. Komarov, Konstantin P. Feoktistov, Boris B. Yegorov	24:17	1st 3-person orbital flight; 1st without space suits
3/18/65- 3/19/65	Voskhod 2	Pavel I. Belyayev, Aleksei A. Leonov	26:02	Leonov made 1st "space walk" (10 min)
3/23/65	**Gemini-Titan 3**	**Grissom (2), John W. Young**	**4:53**	**1st piloted spacecraft to change its orbital path**
6/3/65- 6/7/65	**Gemini-Titan 4**	**James A. McDivitt, Edward H. White 2d**	**97:56**	**White was 1st American to "walk in space"(36 min)**
8/21/65- 8/29/65	**Gemini-Titan 5**	**Cooper (2), Charles Conrad Jr.**	**190:55**	**Longest-duration human flight to date**
12/15/65- 12/16/65	**Gemini-Titan 6A**	**Schirra (2), Thomas P. Stafford**	**25:51**	**Completed 1st U.S. space rendezvous, with Gemini 7**
12/4/65- 12/18/65	**Gemini-Titan 7**	**Frank Borman, James A. Lovell**	**330:35**	**Longest-duration Gemini flight**
3/16/66	**Gemini-Titan 8**	**Neil A. Armstrong, David R. Scott**	**10:41**	**1st docking of one space vehicle with another; mission aborted, control malfunction; 1st Pacific landing**
6/3/66- 6/6/66	**Gemini-Titan 9A**	**Stafford (2), Eugene A. Cernan**	**72:21**	**Performed rendezvous maneuvers, including simulation of lunar module rendezvous**
7/18/66- 7/21/66	**Gemini-Titan 10**	**Young (2), Michael Collins**	**70:47**	**1st use of Agena target vehicle's propulsion systems; 1st orbital docking**
9/12/66- 9/15/66	**Gemini-Titan 11**	**Conrad (2), Richard F. Gordon Jr.**	**71:17**	**1st tethered flight; highest Earth-orbit altitude (850 mi)**
11/11/66- 11/15/66	**Gemini-Titan 12**	**Lovell (2), Edwin W. "Buzz" Aldrin Jr.**	**94:34**	**Final Gemini mission; 5 hr EVA**
4/23/67- 4/24/67	Soyuz 1	Komarov (2)	26:40	Crashed on reentry, killing Komarov
10/11/68- 10/22/68	**Apollo-Saturn 7**	**Schirra (3), Donn F. Eisele, R. Walter Cunningham**	**260:09**	**1st piloted flight of Apollo spacecraft command-service module only; live TV footage of crew**
12/21/68- 12/27/68	**Apollo-Saturn 8**	**Borman (2), Lovell (3), William A. Anders**	**147:00**	**1st lunar orbit and piloted lunar return reentry (command-service module only); views of lunar surface televised to Earth**
1/14/69- 1/17/69	Soyuz 4	Vladimir A. Shatalov	71:21	Docked with Soyuz 5
1/15/69- 1/18/69	Soyuz 5	Boris V. Volyanov, Aleksei S. Yeliseyev, Yevgeny V. Khrunov	72:54	Docked with 4; Yeliseyev and Khrunov transferred to Soyuz 4 via a spacewalk
3/3/69- 3/13/69	**Apollo-Saturn 9**	**McDivitt (2), D. Scott (2), Russell L. Schweickart**	**241:00**	**1st piloted flight of lunar module**
5/18/69- 5/26/69	**Apollo-Saturn 10**	**Stafford (3), Young (3), Cernan (2)**	**192:03**	**1st lunar module orbit of Moon, 50,000 ft from Moon surface**
7/16/69- 7/24/69	**Apollo-Saturn 11**	**Armstrong (2), Collins (2), Aldrin (2)**	**195:18**	**1st lunar landing made by Armstrong and Aldrin (720); collected 48.5 lb of soil, rock samples; lunar stay time 21:36:21**
10/11/69- 10/16/69	Soyuz 6	Georgi S. Shonin, Valery N. Kubasov	118:43	1st welding of metals in space
10/12/69- 10/17/69	Soyuz 7	Anatoly V. Flipchenko, Vladislav N. Volkov, Viktor V. Gorbatko	118:40	Space lab construction test made; Soyuz 6, 7, and 8: 1st time 3 spacecraft, 7 crew members orbited the Earth at once

Dates	Mission[1]	Crew (no. of flights)	Duration (hr:min)	Remarks
10/13/69[2]	Soyuz 8	Shatalov (2), Yeliseyev (2)	118:51	Part of space lab construction team
11/14/69-11/24/69	**Apollo-Saturn 12**	**Conrad (3), Richard F. Gordon Jr. (2), Alan L. Bean**	**244:36**	**Conrad and Bean made 2d Moon landing (11/18); collected 74.7 lb of samples, lunar stay time 31:31**
4/11/70-4/17/70	**Apollo-Saturn 13**	**Lovell (4), Fred W. Haise Jr., John L. Swigart Jr.**	**142:54**	**Aborted after service module oxygen tank ruptured; crew returned safely using lunar module**
6/1/70-6/19/70	Soyuz 9	Nikolayev (2), Vitaliy I. Sevastyanov	424:59	Longest human spaceflight to date
1/31/71-2/9/71	**Apollo-Saturn 14**	**A. Shepard (2), Stuart A. Roosa, Edgar D. Mitchell**	**216:01**	**Shepard and Mitchell made 3d Moon landing (2/3); collected 96 lb of lunar samples; lunar stay 33:31**
4/19/71[2]	Salyut 1[3]	(Occupied by Soyuz 11 crew)		1st space station
4/22/71[2]	Soyuz 10	Shatalov (3), Yeliseyev (3), Nikolay N. Rukavishnikov	47:46	1st successful docking with a space station; failed to enter space station
6/6/71-6/30/71	Soyuz 11	Georgi T. Dobrovolskiy, V. Volkov (2), Viktor I. Patsayev	570:22	Docked and entered Salyut 1 space station; orbited in Salyut 1 for 23 days, crew died during reentry from loss of pressurization
7/26/71-8/7/71	**Apollo-Saturn 15**	**D. Scott (3), James B. Irwin, Alfred M. Worden**	**295:12**	**Scott and Irwin made 4th Moon landing (7/30); 1st lunar rover use; 1st deep space walk; 170 lb of samples; 66:55 stay**
4/16/72-4/27/72	**Apollo-Saturn 16**	**Young (4), Charles M. Duke Jr., Thomas K. Mattingly 2d**	**265:51**	**Young and Duke made 5th Moon landing (4/20); collected 213 lb of lunar samples; lunar stay 71:2**
12/7/72-12/19/72	**Apollo-Saturn 17**	**Cernan (3), Ronald E. Evans, Harrison H. Schmitt**	**301:51**	**Cernan and Schmitt made 6th lunar landing (12/11); collected 243 lb of samples; record lunar stay of more than 75 hr**
5/14/73[2]	**Skylab 1[4]**	**(Occupied by Skylab 2, 3, and 4 crews)**		**1st U.S. space station**
5/25/73-6/22/73	**Skylab 2**	**Conrad (4), Joseph P. Kerwin, Paul J. Weitz**	**672:49**	**1st Amer. piloted orbiting space station; crew repaired damage caused during boost**
7/28/73-9/25/73	**Skylab 3**	**Bean (2), Owen K. Garriott, Jack R. Lousma**	**1,427:09**	**Crew systems and operational tests; exceeded pre-mission plans for scientific activities; 3 hr EVA 13:44**
11/16/73-2/8/74	**Skylab 4**	**Gerald P. Carr, Edward G. Gibson, William Pogue**	**2,017:15**	**Final Skylab mission**
7/15/75-7/21/75	Soyuz 19 (ASTP)	Leonov (2), Kubasov (2)	143:31	U.S.-USSR joint flight; crews linked up in space (7/17), conducted experiments, shared meals, held a joint news conference
7/15/75-7/24/75	**Apollo** (ASTP)	**Vance Brand, Stafford (4), Donald K. Slayton**	**217:28**	**Joint flight with Soyuz 19**
12/10/77[2]	Soyuz 26	Yuri V. Romanenko, Georgiy M. Grechko (2)	2,314:00	1st multiple docking to a space station (Soyuz 26 and 27 docked at Salyut 6)
1/10/78[2]	Soyuz 27	Vladimir A. Dzhanibekov	142:59	*See Soyuz 26*
3/2/78[2]	Soyuz 28	Aleksei A. Gubarev (2), Vladimir Remek	190:16	1st international crew launch; Remek was 1st Czech in space
4/12/81-4/14/81	**Columbia (STS-1)**	**Young (5), Robert L. Crippen**	**54:21**	**1st space shuttle to fly into Earth's orbit**
11/12/81-11/14/81	**Columbia (STS-2)**	**Joe H. Engle, Richard H. Truly**	**54:13**	**1st scientific payload; 1st reuse of space shuttle**
11/11/82-11/16/82	**Columbia (STS-5)**	**Brand (2), Robert Overmyer, William Lenoir, Joseph Allen**	**122:14**	**1st 4-person crew**
6/18/83-6/24/83	**Challenger (STS-7)**	**Crippen (2), Frederick Hauck, Sally K. Ride, John M. Fabian, Norman Thagard**	**146:24**	**Ride was 1st U.S. woman in space; 1st 5-person crew**
6/27/83[2]	Soyuz T-9	Vladimir A. Lyakhov (2), Aleksandr Pavlovich Aleksandrov	3,585:46	Docked at Salyut 7; 1st construction in space
8/30/83-9/5/83	**Challenger (STS-8)**	**Truly (2), Daniel Brandenstein, William Thornton, Guion Bluford, Dale Gardner**	**145:09**	**Bluford was 1st African-American in space**
11/28/83-12/8/83	**Columbia (STS-9)**	**Young (6), Brewster Shaw Jr., Robert Parker, Garriott (2), Byron Lichtenberg, Ulf Merbold**	**247:47**	**1st 6-person crew;1st Spacelab mission**
2/3/84-2/11/84	**Challenger (41-B)**	**Brand (3), Robert Gibson, Ronald McNair, Bruce McCandless, Robert Stewart**	**191:16**	**1st untethered EVA**
2/8/84-4/11/84	Soyuz T-10B	Leonid Kizim, Vladimir Solovyov, Oleg Atkov	1,510:43	Docked with Salyut 7; crew set space duration record of 237 days
4/3/84-10/2/84	Soyuz T-11	Yury Malyshev (2), Gennady Strekalov (3), Rakesh Sharma	4,365:48	Docked with Salyut 7; Sharma 1st Indian in space
4/6/84-4/13/84	**Challenger (41-C)**	**Crippen (3), Francis R. Scobee, George D. Nelson, Terry J. Hart, James D. van Hoften**	**167:40**	**1st in-orbit satellite repair**
7/17/84[2]	Soyuz T-12	Dzhanibekov (4), Svetlana Y. Savitskaya (2), Igor P. Volk	283:14	Docked at Salyut 7; Savitskaya was 1st woman to perform EVA
8/30/84-9/5/84	**Discovery (41-D)**	**Henry W. Hartsfield (2), Michael L. Coats, Richard M. Mullane, Steven A. Hawley, Judith A. Resnik, Charles D. Walker**	**144:56**	**1st flight of U.S. nonastronaut (Walker)**
10/5/84-10/13/84	**Challenger (41-G)**	**Crippen (4), Jon A. McBride, Kathryn D. Sullivan, Ride (2), Marc Garneau, David C. Leestma, Paul D. Scully-Power**	**197:24**	**1st 7-person crew**

Dates	Mission[1]	Crew (no. of flights)	Duration (hr:min)	Remarks
11/8/84-11/16/84	**Discovery (51-A)**	**Hauck (2); David M. Walker, Dr. Anna L. Fisher, J. Allen (2), D. Gardner (2)**	**191:45**	**1st satellite retrieval/repair**
4/12/85-4/19/85	**Discovery (51-D)**	**Karol J. Bobko, Donald E. Williams, Jake Garn, Walker (2), Jeffrey A. Hoffman, S. David Griggs, M. Rhea Seddon**	**167:55**	**Garn (R, VT) was 1st U.S. senator in space**
6/17/85-6/24/85	**Discovery (51-G)**	**Brandenstein (2), John O. Creighton, Shannon W. Lucid, Steven R. Nagel, Fabian (2), Prince Sultan Salman al-Saud, Patrick Baudry**	**169:39**	**Launched 3 satellites; Salman al-Saud was 1st Arab in space; Baudry was 1st French person on U.S. mission**
10/3/85-10/7/85	**Atlantis (51-J)**	**Bobko (3), Ronald J. Grabe, David C. Hilmers, Stewart (2), William A. Pailes**	**97:47**	**1st Atlantis flight**
10/30/85-11/6/85	**Challenger (61-A)**	**Hartsfield (3), Nagel (2), Buchli (2), Bluford (2), Bonnie J. Dunbar, Wubbo J. Ockels, Richard Furrer, Ernst Messerschmid**	**168:45**	**1st 8-person crew; 1st German Spacelab mission**
11/26/85-12/3/85	**Atlantis (61-B)**	**Shaw (2), Bryan D. O'Connor, Sherwood C. Spring, Mary L. Cleave, Jerry L. Ross, C. Walker (3), Rodolfo Neri**	**165:05**	**Space structures assembly test; Neri was 1st Mexican in space**
1/12/86-1/18/86	**Columbia (61-C)**	**R. Gibson (2), Charles F. Bolden Jr., Hawley (2), G. Nelson (2), Franklin R. Chang-Diaz, Robert J. Cenker, Bill Nelson**	**146:04**	**B. Nelson was 1st U.S. Representative in space; material and astronomy experiments conducted**
1/28/86	**Challenger (51-L)**	**Scobee (2), Michael J. Smith, Resnik (2), Ellison S. Onizuka (2), Ronald E. McNair, Gregory B. Jarvis, Christa McAuliffe**	**—**	**Exploded 73 sec after liftoff; all were killed**
2/20/86[2]	Mir[3]	—	—	Space station with 6 docking ports launched
3/13/86[2]	Soyuz T-15	Kizim (3), Solovyov (2)	3,000:01	Ferry between stations; docked at *Mir*
2/5/87-12/29/87	Soyuz TM-2	Romanenko (3), Aleksandr I. Laveikin	7,835:38	Romanenko set endurance record, since broken
7/22/87-12/29/87	Soyuz TM-3	Aleksandr Viktorenko, Aleksandr Pavlovich Aleksandrov (2), Mohammed Faris	3,847:16	Docked with *Mir*; Faris 1st Syrian in space
12/21/87-12/21/88	Soyuz TM-4	V. Titov (2), Muso Manarov, Anatoly Levchenko	8,782:39	Docked with *Mir*
6/7/88-6/17/88	Soyuz TM-5	Viktor Savinykh (3), Anatoly Solovyev, Aleksandr Panayotov Aleksandrov	236:13	Docked with *Mir*; Aleksandrov 1st Bulgarian in space
9/29/88-10/3/88	**Discovery (STS-26)**	**Hauck (3), Richard O. Covey (2), Hilmers (2), G. Nelson (2), John M. Lounge (2)**	**97:00**	**Redesigned shuttle makes 1st flight**
5/4/89-5/8/89	**Atlantis (STS-30)**	**D. Walker (2), Grabe (2), Thagard (2), Cleave (2), Mark C. Lee**	**96:56**	**Launched Venus orbiter *Magellan***
10/18/89-10/23/89	**Atlantis (STS-34)**	**Donald E. Williams (2), Michael J. McCulley, Lucid (2), Chang-Diaz (2), Ellen S. Baker**	**119:39**	**Launched Jupiter probe and orbiter *Galileo***
4/24/90-4/29/90	**Discovery (STS-31)**	**McCandless (2), Sullivan (2), Loren J. Shriver (2), Bolden (2), Hawley (3)**	**121:16**	**Launched Hubble Space Telescope**
10/6/90-10/10/90	**Discovery (STS-41)**	**Richard N. Richards (2), Robert D. Cabana, Bruce E. Melnick, William M. Shepherd (2), Thomas D. Akers**	**98:10**	**Launched *Ulysses* spacecraft to investigate interstellar space and the Sun**
4/5/91-4/11/91	**Atlantis (STS-37)**	**Nagel (3), Kenneth D. Cameron, Linda Godwin, Ross (3), Jay Apt**	**144:32**	**Launched Gamma Ray Observatory to measure celestial gamma rays**
5/18/91-10/10/91	Soyuz TM-12	Anatoly Artsebarskiy, Sergei Krikalev (2) (to *Mir*), Helen Sharman	3,471:22	Docked with *Mir;* Sharman 1st from United Kingdom in space
3/17/92-3/25/92	Soyuz TM-14	Viktorenko (3) (to *Mir*), Alexandr Kaleri (to *Mir*), Klaus-Dietrich Flade, Aleksandr Volkov (3) (from *Mir*), Krikalev (2) (from *Mir*)	3,495:11	First human CIS space mission; docked with *Mir* 3/19; Viktorenko and Kaleri to *Mir*, Volkov and Krikalev from *Mir;* Krikalev was in space313 days
5/7/92-5/16/92	**Endeavour (STS-49)**	**Brandenstein (4), Kevin C. Chilton, Melnick (2), Pierre J. Thuot (2), Richard J. Hieb (2), Kathryn Thornton (2), Akers (2)**	**213:30**	**1st 3-person EVA; satellite recovery and redeployment**
9/12/92-9/21/92	**Endeavour (STS-47)**	**R. Gibson (4), Curtis L. Brown Jr., Lee (2), Apt (2), N. Jan Davis, Mae Carol Jemison, Mamoru Mohri**	**190:30**	**Jemison was 1st black woman in space; Lee and Davis were 1st married couple to travel together in space; 1st Japanese Spacelab**
10/22/92-11/1/92	**Columbia (STS-52)**	**James D. Wetherbee (2), Michael A. Baker (2), Shepherd (3), Tamara E. Jernigan (2), Charles L. Veach (2), Steven G. MacLean**	**236:57**	**Studied influence of gravity on basic fluid and solidification processes**
4/8/93-4/17/93	**Discovery (STS-56)**	**Cameron (2), Stephen S. Oswald (2), C. Michael Foale (2), Ellen Ochoa, Kenneth D. Cockrell**	**222:08**	**2d atmospheric mission; Ochoa was 1st Hispanic woman in space**
6/21/93-7/1/93	**Endeavour (STS-57)**	**Grabe (4), Brian J. Duffy (2), G. David Low (3), Nancy J. Sherlock, Peter J. K. Wisoff, Janice E. Voss**	**239:46**	**Carried Spacelab commercial payload module**
10/18/93-11/1/93	**Columbia (STS-58)**	**John E. Blaha (4), Richard A. Searfoss, Lucid (4), David A. Wolf, William A. McArthur, Martin J. Fettman**	**336:29**	**Studied effects of microgravity**
12/2/93-12/13/93	**Endeavour (STS-61)**	**Covey (3), Kenneth D. Bowersox (2), Claude Nicollier (2), Story Musgrave (5), Akers (3), K. Thornton (3), Hoffman (4)**	**259:58**	**Hubble Space Telescope repaired; Akers set new U.S. EVA duration record (29 hr, 40 min)**
2/3/94-2/11/94	**Discovery (STS-60)**	**Bolden (3), Kenneth S. Reightier Jr. (2), Davis (2), Chang-Diaz (3), Ronald M. Sega, Krikalev (3)**	**199:10**	**Krikalev was 1st Russian on U.S. shuttle**

Dates	Mission[1]	Crew (no. of flights)	Duration (hr:min)	Remarks
4/9/94-4/20/94	**Endeavour (STS-59)**	**Sidney M. Gutierrez (2), Chilton (2), Apt (3), Michael R. Clifford (2), Godwin (2), Thomas D. Jones**	**269:50**	**Gathered data about Earth and the effects humans have on its carbon, water, and energy cycles**
7/1/94-11/4/94	Soyuz TM-19	Yuri I. Malenchenko, Talgat A. Musabayev, Merbold (2) (from *Mir*)	3,022:53	Docked with *Mir;* Merbold from *Mir*
9/9/94-9/20/94	**Discovery (STS-64)**	**Richards (4), L. Blaine Hammond Jr. (2), Jerry M. Linenger, Susan J. Helms (2), Carl J. Meade (3), Lee (3)**	**262:50**	**Performed atmospheric research; 1st untethered EVA in over 10 years**
2/3/95-2/11/95	**Discovery (STS-63)**	**Wetherbee (3), Eileen M. Collins, Bernard A. Harris (2), Foale (3), Janice E. Voss (2), V. Titov (4)**	**198:29**	***Discovery* and Russian space station rendezvous**
3/2/95-3/18/95	**Endeavour (STS-67)**	**Oswald (3), William G. Gregory, Samuel T. Durrance (2), Ronald Parise (2), Wendy B. Lawrence, Jernigan (3), John M. Grunsfeld**	**399:09**	**Shuttle data made available on the Internet; astronomy research conducted**
3/14/95-3/22/95	Soyuz TM-21	Thagard (2), Vladimir Dezhurov, Strekalov (5)	2,688[5]	Docked with *Mir* 3/16/95; Thagard was 1st Amer. on the Russ. spacecraft; Valery Polyakov returned to Earth, 3/22/95, after record stay in space (439 days)
6/27/95-7/7/95	**Atlantis (STS-71)**	**R. Gibson (5), Charles J. Precourt (2), E. Baker (3), Gregory J. Harbaugh (3), Dunbar (4), Solovyev (4) (to *Mir*), Nikolai M. Budarin (to *Mir*), Thagard (5) (from *Mir*), Strekalov (from *Mir*), Dezhurov (from *Mir*)**	**269:47**	**1st *Mir* docking; exchanged crew members with *Mir;* Thagard, with his stay on *Mir,* had spent 115 days in space**
10/20/95-11/5/95	**Columbia (STS-73)**	**Bowersox (3), Kent Rominger, K. Thornton (4), Catherine Coleman, Michael Lopez-Alegria, Fred Leslie, Albert Sacco**	**381:52**	**Most ever first-time space flyers; near-weightlessness experiments conducted in microgravity laboratory**
11/8/95-11/20/95	**Atlantis (STS-74)**	**Cameron (3), James D. Halsell Jr. (2), Chris Hadfield, Ross (5), McArthur (2)**	**196:30**	**2d *Mir* docking (11/15-11/18); erected a 15-ft permanent docking tunnel to *Mir* for future use by U.S. orbiters**
1/11/96-1/20/96	**Endeavour (STS-72)**	**Duffy (3), Brent W. Jett Jr., Winston E. Scott, Leroy Chiao (2), Daniel T. Barry, Koichi Wakata**	**214:01**	**Released NASA space probe; retrieved Japanese satellite; 13 hr EVA**
2/22/96-3/9/96	**Columbia (STS-75)**	**Andrew M. Allen (3), Scott J. Horowitz, Chang-Diaz (5), Umberto Guidoni, Hoffman (5), Maurizio Cheli, Nicollier (3)**	**377:40**	**Lost an Italian satellite when its tether was severed; microgravity experiments performed; singe marks found on 2 O-rings**
3/22/96-3/31/96	**Atlantis (STS-76)**	**Chilton (3), Searfoss (2), Sega (2), Clifford (3) Godwin (3), Lucid (5) (to *Mir*)**	**221:15**	**3d *Mir* docking (5 days); Lucid to *Mir*; 2-person EVA**
6/20/96-7/7/96	**Columbia (STS-78)**	**Terence T. Henricks (4), Kevin R. Kregel (2), Helms (3), Richard M. Linnehan, Charles E. Brady, Jean-Jacques Favier, Robert Brent Thirsk**	**405:48**	**Studied weightlessness with the Life/Microgravity Spacelab on board**
9/16/96-9/26/96	**Atlantis (STS-79)**	**Apt (4), Terry Wilcutt (2), William Readdy (3), Akers (4), Carl E. Walz (3), Lucid (5) (from *Mir*), Blaha (5) (to *Mir*)**	**243:19**	**Docked with *Mir* 9/18/96; exchanged crew members, including Lucid, who set U.S. and women's individual duration in space record (188 days)**
11/19/96-12/7/96	**Columbia (STS-80)**	**Cockrell (3), Rominger (2), Jernigan (4), Jones (3), Musgrave (6)**	**423:53**	**Longest-duration shuttle flight; Musgrave was oldest person to fly in space; 2 science satellites deployed and retrieved**
1/12/97-1/22/97	**Atlantis (STS-81)**	**M. Baker (4), Jett (2), Wisoff (3), Grunsfeld (2), Marsha Ivins (4), Linenger (2) (to *Mir*), Blaha (5) (from *Mir*)**	**243:30**	**Docked with *Mir* 1/14-1/19/97; Linenger to *Mir;* Blaha from *Mir,* spent 128 days in space**
2/11/97-2/21/97	**Discovery (STS-82)**	**Bowersox (4), Horowitz (2), Joe Tanner (2), Hawley (4), Harbaugh (4), Lee (4), Steve Smith (2)**	**238:47**	**Increased capabilities of Hubble Space Telescope; 5 EVAs used to service it**
5/15/97-5/24/97	**Atlantis (STS-84)**	**Precourt (3), E. Collins (2), Jean-François Clervoy (2), Carlos Noriega, Ed Lu, Elena Kondakova, Foale (4) (to *Mir*), Linenger (2) (from *Mir*)**	**221:20**	**Docked with *Mir* 5/16-5/21/97; Foale to *Mir;* Linenger from *Mir,* 132 days in space, 2d longest time for an American; stay on *Mir* marked by troubles incl. fire 2/23**
7/1/97-7/17/97	**Columbia (STS-94)**	**Halsell (4), Susan L. Still (2), Janice E. Voss (4), Donald A. Thomas (4), Michael Gernhardt (3), Roger Crouch (2), Greg Linteris (2)**	**376:46**	**Reflight of Microgravity Science Laboratory-1 mission (STS-83) that was aborted 4/8/97 because of problem with fuel cell**
8/5/97-2/19/98	Soyuz TM-26	Solovyev (5), Pavel Vinogradov	4,743:35	Docked with *Mir* 8/7/97; repaired damaged space station
8/7/97-8/19/97	**Discovery (STS-85)**	**Brown (4), Rominger (3), Davis (3), Robert L. Curbeam Jr., Stephen K. Robinson, Bjarni V. Tryggvason**	**284:27**	**Deployed and retrieved satellite designed to study Earth's middle atmosphere; demonstrated robotic arm**
9/25/97-10/6/97	**Atlantis (STS-86)**	**Wetherbee (4), Michael J. Bloomfield, V. Titov (4), Scott Parazynski (2), Jean-Loup Chrétien (3), Lawrence (2), Wolf (2) (to *Mir*), Foale (4) (from *Mir*)**	**236:24**	**Docked with *Mir* 9/27-10/3/97; delivered new computer to *Mir;* Wolf to *Mir;* Foale from *Mir;* stay on *Mir* marked by collision with cargo ship 6/25, worst such collision ever**
1/22/98-1/31/98	**Endeavour (STS-89)**	**Wilcutt (3), Joe F. Edwards Jr., Dunbar (5), Michael P. Anderson, James F. Reilly II, Salizhan Sharipov, Andrew Thomas (2) (to *Mir*), Wolf (2) (from *Mir*)**	**211:48**	**Docked with *Mir* 1/24-1/29/98; delivered water and cargo; Thomas to *Mir;* Wolf from *Mir,* 128 days in space**
1/29/98-8/25/98	Soyuz TM-27	Musabayev (2), Budarin (2), Leopold Eyharts	4,923:36	Docked with *Mir* 1/31/98

Dates	Mission[1]	Crew (no. of flights)	Duration (hr:min)	Remarks
4/17/98-5/3/98	Columbia (STS-90)	Searfoss (3), Scott D. Altman, Linnehan (2), Dafydd Rhys Williams, Kathryn P. Hire, Jay C. Buckey, James A. Pawelczyk	381:50	Studied effects of microgravity on the nervous systems of the crew and over 2,000 live animals; 1st surgery in space on animals meant to survive
6/2/98-6/12/98	Discovery (STS-91)	Precourt (4), Dominic L. Gorie, Lawrence (3), Chang-Diaz (6), Janet L. Kavandi, Valery Ryumin (4), A. Thomas (2) (from *Mir*)	235:53	Final docking mission with *Mir;* Thomas from *Mir,* 141 days in space
10/29/98-11/7/98	Discovery (STS-95)	Brown (5), Steven W. Lindsey (2), Parazynski (3), Robinson (2), Pedro Duque, Chiaki Mukai (2), Glenn (2)	213:44	Sen. John Glenn (D, OH), 77, was oldest person to fly in space; Duque was 1st Spaniard in space; series of experiments to study aging process performed on Glenn; Spartan 201 satellite, which studied the Sun, deployed and retrieved
12/4/98-12/15/98	Endeavour (STS-88)	Cabana (4), Frederick W. Sturckow, Nancy J. Currie (3), Ross (6), James H. Newman (3), Krivalev (4)	283:18	1st assembly of International Space Station; attached U.S.-built *Unity* connecting module with already-deployed Russian-built *Zarya* control module; 1st crew to enter ISS
5/27/99-6/6/99	Discovery (STS-96)	Rominger (4), Rick D. Husband, Ochoa (3), Jernigan (5), Barry (2), Julie Payette, Valery Ivanovich Tokarev	235:13	Transferred nearly 2 tons of supplies to ISS; small satellite STARSHINE deployed and observed by students on Earth
7/23/99-7/27/99	Columbia (STS-93)	E. Collins (3), Jeffrey S. Ashby, Hawley (5), Coleman (2), Michel Tognini (2)	118:50	Collins was 1st woman to command a space shuttle; deployed Chandra X-ray Observatory, a telescope designed to study the universe
12/19/99-12/27/99	Discovery (STS-103)	Brown (6), Scott Kelly, S. Smith (3), Foale (5), Grunsfeld (3), Nicollier (4), Clervoy (3)	191:10	Replaced equipment on and upgraded Hubble Space Telescope; 3 EVAs
2/11/00-2/22/00	Endeavour (STS-99)	Kregel (4), Gorie (2), Kavandi (2), Janice E. Voss (5), Mohri (2), Gerhard P.J. Thiele	269:38	Used radar to make most complete topographic map of Earth's surface ever produced.
5/19/00-5/29/00	Atlantis (STS-101)	Halsell (5), Horowitz (3), Helms (4), Yury Usachev (3), James S. Voss (4), Mary Ellen Weber (2), Jeffrey N. Williams	236:09	Serviced and resupplied International Space Station; boosted orbit of ISS to an altitude of about 238 mi; 1 EVA by Voss and Williams
9/8/00-9/20/00	Atlantis (STS-106)	Wilcutt (4), Altman (2), Lu (2), Richard A. Mastracchio, Daniel C. Burbank, Malenchenko (2), Boris V. Morukov	283:10	Prepared International Space Station for 1st permanent crew by connecting power, data, and communications cables, and delivering supplies; 1 EVA by all 7 crew members
10/11/00-10/24/00	Discovery (STS-92)	Duffy (4), Pamela A. Melroy, Koichi Wakata (2), Leroy Chiao (3), Wisoff (4), Lopez-Alegria (2), McArthur (3)	309:43	Installed 1st permanent framework structure on International Space Station, setting the stage for future additions; 4 EVAs
10/31/00[2]	Soyuz TM-204	Shepherd (4), Yuri Gidzenko (2),Krikalev (5)	—	Established 1st permanent manning of International Space Station with 3-person crew for a 4-month stay
11/30/00-12/11/00	Endeavour (STS-97)	Jett (3), Bloomfield (2),Tanner (3), Marc Garneau (3), Noriega (2)	259:57	Delivered 17-ton solar arrays, batteries, and radiators to I SS; 3 EVAs
2/7/01-2/20/01	Atlantis (STS-98)	Cockrell (4), Ivins (5), Jones (4), Curbeam (2), Mark L. Polansky	309:20	Installed U.S. Destiny Laboratory Module on the ISS; 3 EVAs
3/8/01-3/21/01	Discovery (STS-102)	Wetherbee (5), James M. Kelly, Helms (4) (to ISS), James S. Voss (5) (to ISS), Paul Richards, Andrew S.W. Thomas (2), Usachev (4), (to ISS) Shepherd (4) (from ISS), Gidzenko (2) (from ISS), Krikalev (5) (from ISS)	307:49	Transported 2d permanent crew (Voss, Helms, and Usachev) to International Space Station and returned 1st crew to Earth; 2 EVAs
4/19/01-5/1/01	Endeavour (STS-100)	Rominger (5), John L. Phillips, Hadfield (2), Ashby (2), Parazynski (4), Guidoni (2), Yuri V. Lonchakov	285:30	Installed the Canadarm2, a robotic arm, and delivered supplies to International Space Station; 2 EVAs
7/12/01-7/23/01	Atlantis (STS-104)	Lindsey (3), Charles O. Hobaugh, Gernhardt (4), Kavandi (3), Reilly (2)	259:58	Installed a Joint Airlock, with nitrogen and oxygen tanks to permit future spacewalks from the International Space Station; 3 EVAs
8/10/01-8/22/01	Discovery (STS-105)	Helms (5), Horowitz (4), Sturckow (2), Barry (2), Patrick G. Forrester, Culbertson (3) (to ISS), Dezhurov (2) (to ISS), Mikhail Turin, Usachev, (4), Voss (5) (from ISS), Helms (4) (from ISS)	285:13	Transported Expedition Three crew to International Space Station (Culbertson, Turin, Dezhurov) and returned Expedition Two crew to Earth; 2 EVAs

Note: As of October 1, 2001, there have been 106 space shuttle flights, 81 since the 1986 *Challenger* explosion. Active shuttles include the *Columbia* (26 flights), the *Discovery* (30), the *Atlantis* (24), and the *Endeavour* (16). (The *Challenger* completed 9 missions.) Four Soviets are known to have died in spaceflights: Komarov was killed on Soyuz 1 (1967) when the parachute lines tangled during descent; the 3-person Soyuz11 crew (1971) was asphyxiated. Seven Americans died in the *Challenger* explosion, and 3 astronauts—Virgil I. Grissom, Edward H. White, and Roger B. Chaffee—died in the Jan. 27, 1967, Apollo 1 fire on the ground at Cape Kennedy, FL. (1) For space shuttle flights, mission name is in parentheses following the name of the orbiter. (2) Launch date. (3) Space stations, such as the *Salyuts* and *Mir,* were used to house crews starting in 1971. (4) Skylab 1 deteriorated and fell from orbit without burning up upon entering the atmosphere. Pieces fell on Australia and into the Indian Ocean; no one was injured. (5) The approximate crew duration for Thagard's stay. Crew did not return together.

IT'S A FACT: NASA astronauts Thomas Jones and Robert Curbeam Jr. conducted America's 100th spacewalk on Feb. 14, 2001, outside the space shuttle *Atlantis.*

Individuals Who Have Flown in Space, 1961-2001

Source: Congressional Research Service; World Almanac research; as of Oct. 1, 2001

Country	No. of individs.	Country	No. of individs.	Country	No. of individs.	Country	No. of individs.
United States	267	Cuba	1	Japan	5	Slovakia	1
Russia/CIS	97	Czechoslovakia	1	Mexico	1	Spain	2
Afghanistan	1	France	8	Mongolia	1	Switzerland	1
Austria	1	Germany	11	Netherlands	1	Syria	1
Belgium	1	Hungary	1	Poland	1	United Kingdom	1
Bulgaria	2	India	1	Romania	1	Vietnam	1
Canada	9	Italy	3	Saudi Arabia	1	**TOTAL**	**422**

Note: All individuals flew on either a Russian/CIS-sponsored mission or on a U.S.-sponsored mission. All cosmonauts who were citizens of the USSR at the time of launch are included under "Russia/CIS." "Germany" includes former E and W Germany.

International Space Station

The International Space Station (ISS) is considered the largest cooperative scientific project in history.

16 cooperating nations: U.S., Russia, Canada, Belgium, Denmark, France, Germany, Italy, Netherlands, Norway, Spain, Sweden, Switzerland, United Kingdom, Japan, and Brazil

The station when completed:
- mass of 1,040,000 lb
- 356' x 290', with almost an acre of solar panels
- internal volume roughly equivalent to passenger cabin of a 747 jumbo jet
- 6 laboratories; living space for up to 7 people

Assembly:
- 11/20/98: U.S.owned, Russian-built *Zarya* ("sunrise") control module launched by rocket from Kazakhstan—1st step in assembly of the station
- 12/4/98: U.S.-built *Unity* connecting module launched on space shuttle *Endeavour*; shuttle crew attached *Unity* and *Zarya*
- 5/27/99: space shuttle *Discovery* launched, bringing supplies; 1st docking with ISS
- 7/26/00: Russian-built *Zvezda* ("star") service module, the primary Russian contribution to the ISS, connected with the station
- 11/2/00: 1st permanent crew arrives for 4-month stay
- 2/9/00: U.S. Destiny Laboratory Module delivered
- 4/21/01: A robotic arm, Canadarm2, delivered
- to be completed by 2005, after 44 total missions

Examples of research planned:
- growing living cells for research in an environment free of gravity
- studying the effects on humans of long-term exposure to reduced gravity
- studying large-scale long-term changes in Earth's environment by observing Earth from orbit

Summary of Worldwide Successful Launches, 1957-2000

Source: National Aeronautics and Space Administration

Year	Total[1]	Russia[2]	United States	Japan	European Space Agency	China	France	India	United Kingdom	Germany	Canada
1957-59	24	6	18	—	—	—	—	—	—	—	—
1960-69	1,035	399	614	—	2	—	4	—	1	—	—
1970-79	1,366	1,028	247	18	5	8	14	1	6	3	4
1980-89	1,431	1,132	191	26	14	16	5	9	4	7	5
1990-99	1,045	542	300	23	55	33	16	11	7	6	4
2000	81	35	29	0	12	5	0	0	0	0	0
TOTAL	**4,982**	**3,142**	**1,399**	**67**	**88**	**62**	**39**	**21**	**18**	**16**	**13**

(1) Includes launches sponsored by countries not shown. (2) Figures covering 1957-96 apply to the the Soviet Union, or, after 1991, to the Commonwealth of Independent States.

Space Junk

Source: National Aeronautics and Space Administration; North American Aerospace Defense Command

Space junk, or space debris, consists of objects that have spun off from space missions since Sputnik-1 in 1957 and orbit the Earth at high speeds, posing an increasing risk to space missions and functioning space satellites.

- There are more than 8,000 items of space junk orbiting the Earth that are officially tracked by NORAD, and in all at least 110,000 objects 1 cm or larger in size.
- Space junk includes derelict satellites, upper stages of rockets that carry payloads into orbit, specks and beads of slag from rocket exhaust, metal bolts, and garbage left by Earth-orbiting missions and space stations.
- 80% of space junk that could damage active satellites or space shuttles orbits the planet at an altitude of about 1,200 miles above Earth—what scientists call a near-Earth orbit.
- Even miniscule objects pose a threat to space missions, as they travel at speeds around 16,500 miles per hour—40 times faster than a bullet shot from a .38 caliber pistol.

Notable Proposed Space Missions

Source: National Aeronautics and Space Administration

Planned Launch date	Mission	Purpose
Dec. 2001	Space InfraRed Telescope Facility (SIRTF)	Make high-sensitivity observations of celestial sources
May 2002	Gravity Probe B (GP-B)	Attempt to prove Einstein's Theory of General Relativity by measuring minute "twisting" in space-time caused by the rotation of the Earth
Jan. 2003	The International Rosetta Mission	Rendezvous with comet 46 P/Wirtanen in 2011 to study the object's nucleus and environment
2007	Planck-Herschel Satellite	Study the origins of the Universe and "dark matter"; collect data to study whether the Universe is finite or infinite

Notable U.S. Planetary Science Missions

Source: National Aeronautics and Space Administration

Spacecraft	Launch date (Coordinated Universal Time)	Mission	Remarks
Mariner 2	Aug. 27, 1962	Venus	Passed within 22,000 mi of Venus 12/14/62; contact lost 1/3/63 at 54 million mi
Ranger 7	July 28, 1964	Moon	Yielded over 4,000 photos of lunar surface
Mariner 4	Nov. 28, 1964	Mars	Passed behind Mars 7/14/65; took 22 photos from 6,000 mi
Ranger 8	Feb. 17, 1965	Moon	Yielded over 7,000 photos of lunar surface
Surveyor 3	Apr. 17, 1967	Moon	Scooped and tested lunar soil
Mariner 5	June 14, 1967	Venus	In solar orbit; closest Venus flyby 10/19/67
Mariner 6	Feb. 24, 1969	Mars	Came within 2,000 mi of Mars 7/31/69; collected data, photos
Mariner 7	Mar. 27, 1969	Mars	Came within 2,000 mi of Mars 8/5/69
Mariner 9	May 30, 1971	Mars	First craft to orbit Mars 11/13/71; sent back over 7,000 photos
Pioneer 10	Mar. 2, 1972	Jupiter	Passed Jupiter 12/4/73; exited the planetary system 6/13/83; transmission ended 3/31/97 at 6.39 billion mi
Pioneer 11	Apr. 5, 1973	Jupiter, Saturn	Passed Jupiter 12/3/74; Saturn 9/1/79; discovered an additional ring and 2 moons around Saturn; operating in outer solar system; transmission ended 9/95
Mariner 10	Nov. 3, 1973	Venus, Mercury	Passed Venus 2/5/74; arrived Mercury 3/29/74. 1st time gravity of 1 planet (Venus) used to whip spacecraft toward another (Mercury)
Viking 1	Aug. 20, 1975	Mars	Landed on Mars 7/20/76; did scientific research, sent photos; functioned 6 years
Viking 2	Sept. 9, 1975	Mars	Landed on Mars 9/3/76; functioned 3 years
Voyager 1	Sept. 5, 1977	Jupiter, Saturn	Encountered Jupiter 3/5/79, provided evidence of Jupiter ring; passed near Saturn 11/12/80
Voyager 2	Aug. 20, 1977	Jupiter, Saturn, Uranus, Neptune	Encountered Jupiter 7/9/79; Saturn 8/25/81; Uranus 1/24/86; Neptune 8/25/89
Pioneer Venus 1	May 20, 1978	Venus	Entered Venus orbit 12/4/78; spent 14 years studying planet; ceased operating 10/19/92
Pioneer Venus 2	Aug. 8, 1978	Venus	Encountered Venus 12/9/78; probes impacted on surface
Magellan	May 4, 1989	Venus	Landed on Venus 8/10/90; orbited and mapped Venus; monitored geological activity on surface; ceased operating 10/11/94
Galileo	Oct. 18, 1989	Jupiter	Used Earth's gravity to propel it toward Jupiter; encountered Venus Feb. 1990; encountered Jupiter 12/7/95; released probe to Jovian surface; encountered moons Ganymede, Europa, Io, and Callisto
Mars Observer	Sept. 25, 1992	Mars	Communication was lost 8/21/93
Near Earth Asteroid Rendezvous (NEAR)	Feb. 17, 1996	Asteroid Eros	Rendezvoused with Eros Apr. 2000; began orbiting and studying the asteroid; communication ceased 2/28/01
Mars Global Surveyor	Nov. 7, 1996	Mars	Began orbiting Mars 9/11/97; began 2-year mapping survey of entire Martian surface 3/9/99; discovered magnetism on planet; observed Martian moon Phobos; discovered evidence of liquid water in geologically recent past 6/22/00
Mars Pathfinder	Dec. 4, 1996	Mars	Landed on Mars 7/4/97; rover Sojourner made measurements of the Martian climate and soil composition, sending thousands of surface images; ceased operating 9/27/97
Cassini	Oct. 15, 1997	Saturn	Scheduled to reach Saturn in 2004; 4-year mission to study planet's atmosphere, rings, and moons; probe will land on moon Titan
Lunar Prospector	Jan. 6, 1998	Moon	Began orbiting Moon 1/11/98; mapped abundance of 11 elements on Moon's surface; discovered evidence of water-ice at both lunar poles; made 1st precise gravity map of entire lunar surface; crashed into crater near Moon's south pole 7/31/99 to end mission
Mars Climate Orbiter	Dec. 11, 1998	Mars	Communication was lost 9/23/99
Mars Polar Lander	Jan. 3, 1999	Mars	Communication was lost 12/3/99
Stardust	Feb. 7, 1999	Comet Wild-2	Scheduled to reach comet in 2004; to gather dust samples and return them to Earth in 2006
Genesis	Aug. 8, 2001	Sun	Scheduled to travel to the Sun and collect particles from solar wind and return them to Earth in 2004.

Passenger Traffic at World Airports, 2000

Source: Airports Council International-North America

AIRPORT	Passenger Arrivals and Departures
London, UK (Heathrow)	64,607,185
Tokyo/Haneda, Japan (Tokyo Intl.)	56,402,206
Frankfurt, Germany (Rhein/Main)	49,360,620
Paris, France (Charles De Gaulle)	48,240,137
Amsterdam, Netherlands (Schiphol)	39,604,589
Seoul, South Korea (Kimpo Intl.)	36,727,124
Madrid, Spain (Barajas)	32,765,820
Hong Kong, China (Hong Kong Intl.)	32,746,737
London, UK (Gatwick)	32,056,942
Bangkok, Thailand (Bangkok Intl.)	29,621,898
Toronto, Ontario (Lester B. Pearson Intl.)	28,820,326
Singapore (Changi)	28,618,200
Tokyo, Japan (Narita)	27,389,915
Rome, Italy (Fiumicino)	25,921,886
Paris, France (Orly)	25,399,111
Sydney, Australia (Kingsford Smith)	23,553,878
Munich, Germany (Munich)	23,125,872
Zurich, Switzerland (Zurich)	22,649,539
Beijing, China (Beijing Capital Intl.)	21,659,077
Brussels, Belgium (Brussels Intl.)	21,604,478
Mexico City, Mexico (Mexico City)	21,042,610
Milan, Italy (Malpensa)	20,716,815
Osaka, Japan (Kansai Intl.)	20,472,060
Barcelona, Spain (El Prat)	19,797,135
Palma De Mallorca, Spain (Palma de Mallorca)	19,401,807
Manchester, UK (Manchester)	18,804,322
Taipei, Taiwan (Chiang Kai-Shek)	18,681,418
Stockholm, Sweden (Arlanda)	18,446,309
Copenhagen, Denmark (Copenhagen)	18,294,387
Melbourne. Australia (Melbourne)	16,442,312

Note: Excludes U.S. airports. Includes only airports participating in the Airports Council Internat. Airport Traffic Statistics collection.

Passenger Traffic at U.S. Airports, 2000

Source: Airports Council International-North America

AIRPORT	Passenger Arrivals and Departures	AIRPORT	Passenger Arrivals and Departures
Atlanta (Hartsfield Intl.—ATL)	80,171,000	Houston (George Bush Intercontinental—IAH)	35,246,000
Chicago (O'Hare—ORD)	72,136,000	Newark (EWR)	34,195,000
Los Angeles (LAX)	68,478,000	Miami (MIA)	33,570,000
Dallas/Ft. Worth (DFW)	60,687,000	New York (J. F. Kennedy Intl.—JFK)	32,779,000
San Francisco (SFO)	41,174,000	Orlando (MCO)	30,823,000
Denver (DEN)	38,749,000	St. Louis (Lambert-St. Louis Intl.—STL)	30,547,000
Las Vegas (McCarran Intl.—LAS)	36,856,000	Seattle-Tacoma (SEA)	28,404,000
Minneapolis/St. Paul (MSP)	36,688,000	Boston (Logan Intl.—BOS)	27,413,000
Phoenix (Sky Harbor Intl.—PHX)	35,890,000	New York (LaGuardia—LGA)	25,234,000
Detroit (DTW)	35,535,000	Philadelphia (PHL)	24,901,000

U.S. Scheduled Airline Traffic, 1990-2000

Source: Air Transport Association of America

(in thousands, except where otherwise noted)

	1990	1995	1999	2000
Revenue passengers enplaned	**465,600**	**547,800**	**636,000**	**665,500**
Revenue passenger miles	457,926,000	540,656,000	652,047,000	692,505,000
Available seat miles	733,375,000	807,078,000	918,419,000	956,502,000
% of seats filled with passengers	62.4	67.0	71.0	72.4
Cargo traffic (ton miles)	**12,549,000**	**16,921,000**	**21,613,000**	**23,611,000**
Financial				
Passenger revenue ($000)	$58,453,000	$69,594,000	$84,318,000	$93,572,000
Net profit ($000)	–$3,921,000	$2,314,000	$5,360,000	$2,638,000
Employees	**545,809**	**546,987**	**646,410**	**679,967**

NA=Not available.

Leading U.S. Passenger Airlines, 2000

Source: Air Transport Association of America

(in thousands)

Airline	Passengers	Airline	Passengers	Airline	Passengers	Airline	Passengers
Delta	105,591	Trans World	26,365	Atlantic Southeast	6,096	Aloha	5,177
American	86,240	America West	19,942	Mesaba	6,068	Horizon Air	5,044
United	83,854	Alaska	13,512	American Trans Air	5,940	Air Wisconsin	3,857
Southwest	72,568	American Eagle	12,176	Hawaiian	5,887	Midway	2,937
US Airways	59,772	Continental Express	7,770	Comair	5,655	Frontier	2,893
Northwest	56,835	AirTran	7,547			Spirit Air	2,836
Continental	45,139					Sun Country	2,203

U.S. Airline Safety, Scheduled Commercial Carriers, 1980-2000

Source: Air Transport Association of America

	Departures (millions)	Fatal accidents	Fatalities	Accident rate[2]		Departures (millions)	Fatal accidents	Fatalities	Accident rate[2]
1980	5.4	0	0	0.000	1990	6.9	6	39	0.087
1981	5.2	4	4	0.077	1991	6.8	4	62	0.059
1982[1]	5.0	4	234	0.060	1992	7.1	4	33	0.057
1983	5.0	4	15	0.079	1993	7.2	1	1	0.014
1984	5.4	1	4	0.018	1994	7.5	4	239	0.053
1985	5.8	4	197	0.069	1995	8.1	2	166	0.025
1986[1]	6.4	2	5	0.016	1996	8.2	3	342	0.036
1987[1]	6.6	4	231	0.046	1997	8.2	3	3	0.037
1988[1]	6.7	3	285	0.030	1998	8.3	1	1[3]	0.012
1989	6.6	8	131	0.121	1999	8.6	2	12	0.023
					2000	9.0	3	92	0.033

(1) Sabotage-caused accidents are included in the number of fatal accidents and fatalities, but not in the calculation of accident rates. (2) Fatal accidents per 100,000 departures. (3) On-ground employee fatality.

Aircraft Operating Statistics, 2000

Source: Air Transport Association of America; figures are averages for most commonly used models

	No. of seats	Speed airborne (mph)	Flight length (mi)	Fuel (gal per hr)	Operating cost per hr		No. of seats	Speed airborne (mph)	Flight length (mi)	Fuel (gal per hr)	Operating cost per hr
B747-400	379	546	4,375	3,257	$6,964	B737-400	141	411	675	803	$2,446
B747-200/300	369	521	2,951	3,664	8,615	MD-80	137	428	782	950	2,539
B747-F	0	508	2,277	3,530	7,740	B737-300	131	414	620	782	2,150
L-1011	322	503	1,576	2,524	6,565	DC-9-50	127	359	282	912	2,130
DC-10-10	309	509	2,012	2,395	4,372	A319	121	451	1,058	755	2,029
DC-10-40	285	495	1,631	2,580	6,313	B737-100/200	116	389	515	829	2,275
B-777	273	524	3,435	2,201	4,497	B717-200	114	322	492	582	1,690
MD-11	264	515	3,910	2,485	7,204	DC-9-40	112	393	519	855	1,771
DC-10-30	252	517	2,724	2,708	6,879	B737-500	109	408	584	755	2,271
A300-600	238	475	1,271	1,698	6,033	DC-9-30	98	372	512	814	2,188
B767-300ER	211	446	2,076	1,486	3,696	F-100	92	380	482	664	2,304
B767-200ER	180	487	2,191	1,450	4,103	DC-9-10	69	390	468	742	2,000
MD-90	148	431	711	815	4,392	CRJ 100	50	436	502	464	1,585
B727-200	147	434	707	1,317	2,868	CRJ 145	50	375	466	419	987
B727-F	0	431	609	1,386	4,583	ERJ-145	50	335	437	337	869
B737-800	150	459	1,155	770	2,459	ERJ-135	37	339	402	280	791
A320-100/200	146	454	1,107	811	2,324						

IT'S A FACT: Airbus Industrie's A380, scheduled to make its first flight in 2004, will be the world's largest passenger airplane, seating 555 passengers and boasting a wingspan of 261.8 ft.

Some Notable Aviation Firsts[1]

1903 — On Dec. 17, near Kitty Hawk, NC, brothers Wilbur and Orville Wright made the first human-carrying, powered flight. Each made 2 flights; the longest, about 852 ft, lasted 59 sec.
1907 — U.S. airplane manufacturing company formed by Glenn H. Curtiss.
1908 — 1st airplane passenger, Lt. Frank P. Lahm, rode with Wilbur Wright in a brief (6 min, 24 sec) flight.
1911 — 1st transportation of mail by airplane officially approved by the U.S. Postal Service began on Sept. 23. It lasted one week. In 1918, limited scheduled air mail service began. By 1921, scheduled transcontinental airmail service began between New York City and San Francisco.
1914 — 1st scheduled passenger airline service began. It operated between St. Petersburg and Tampa, FL.
1919 — 1st airline food, a basket lunch, was served as part of a commercial airline service.
1930 — Ellen Church became 1st flight attendant.
1939 — On Aug. 27, the German Heinkel He 178 made the first successful flight powered by a jet engine.
1947 — Mach 1, the sound barrier, was broken by Amer. Charles E. ("Chuck") Yeager in a Bell X-1 rocket-powered aircraft.
1947 — Largest airplane ever flown, Howard Hughes's "Spruce Goose," flew 1 mi at an altitude of 80 ft.
1953 — Jacqueline Cochran became 1st woman to fly faster than sound.
1960 — Convair B-58, 1st supersonic bomber, was introduced.
1968 — The supersonic speed of Mach 2 was accomplished for 1st time, in a Tupolev Tu-144. The plane had an approximate maximum speed of 1,200 mph.
1970 — The Tupolev Tu-144, during commercial transport, exceeded Mach 2. It reached about 1,335 mph at 53,475 ft.
1976 — The Concorde began 1st scheduled supersonic commercial service.

(1) Excludes notable around-the-world and international trips.

Some Notable Around-the-World and Intercontinental Trips

Aviator or Craft	From/To	Miles	Time	Date
Nellie Bly	New York/New York		72d 06h 11m	1889
George Francis Train	New York/New York		67d 12h 03m	1890
Charles Fitzmorris	Chicago/Chicago		60d 13h 29m	1901
J. W. Willis Sayre	Seattle/Seattle		54d 09h 42m	1903
J. Alcock-A.W. Brown [1]	Newfoundland/Ireland	1,960	16h 12m	June 14-15, 1919
2 U.S. Army airplanes	Seattle/Seattle	26,103	35d 01h 11m	1924
Richard E. Byrd, Floyd Bennett [2]	Spitsbergen (Nor.)/N. Pole	1,545	15h 30m	May 9, 1926
Amundsen-Ellsworth-Nobile Polar Expedition (in a dirigible)	Spitsbergen (Nor.)/over N. Pole to Teller, Alaska		80h	May 11-14,1926
E.S. Evans and L. Wells (*New York World*)	New York/New York	18,410[3]	28d 14h 36m 05s	June 16-July 14, 1926
Charles Lindbergh [4]	New York/Paris	3,610	33h 29m 30s	May 20-21, 1927
Amelia Earhart, W. Stultz, L. Gordon	Newfoundland/Wales		20h 40m	June 17-18, 1928
Graf Zeppelin	Friedrichshafen, Ger./Lakehurst, NJ	6,630	4d 15h 46m	Oct. 11-15, 1928
Graf Zeppelin	Friedrichshafen, Ger./Lakehurst, NJ	21,700	20d 04h	Aug. 14-Sept. 4, 1929
Wiley Post and Harold Gatty (Monoplane Winnie Mae)	New York/New York	15,474	8d 15h 51m	July 1, 1931
C. Pangborn-H. Herndon Jr. [5]	Misawa, Japan/Wenatchee, Wash.	4,458	41h 34m	Oct. 3-5, 1931
Amelia Earhart [6]	Newfoundland/Ireland	2,026	14h 56m	May 20-21, 1932
Wiley Post (Monoplane Winnie Mae)[7]	New York/New York	15,596	115h 36m 30s	July 15-22, 1933
Hindenburg Zeppelin	Lakehurst, NJ/Frankfort, Ger.		42h 53m	Aug. 9-11, 1936
Howard Hughes and 4 assistants	New York/New York	14,824	3d 19h 08m 10s	July 10-13, 1938
America, Pan American 4-engine Lockheed Constellation[8]	New York/New York	22,219	101h 32m	June 17-30, 1947
Col. Edward Eagan	New York/New York	20,559	147h 15m	Dec. 13, 1948
USAF B-50 Lucky Lady II (Capt. James Gallagher) [9]	Ft. Worth, TX/Ft. Worth, TX	23,452	94h 01m	Mar. 2, 1949
Col. D. Schilling, USAF [10]	England/Limestone, ME	3,300	10h 01m	Sept. 22, 1950
C.F. Blair Jr.	Norway/Alaska	3,300	10h 29m	May 29, 1951
Canberra Bomber [11]	N. Ireland/Newfoundland	2073	04h 34m	Aug. 26, 1952
	Newfoundland/N. Ireland	2073	03h 25m	Aug. 26, 1952
3 USAF B-52 Strato-fortresses [12]	Merced, CA/CA	24,325	45h 19m	Jan. 15-18, 1957
USSR TU-114 [13]	Moscow/New York	5,092	11h 06m	June 28, 1959
Peter Gluckmann (solo)	San Francisco/San Francisco	22,800	29d	Aug. 22-Sept. 20, 1959
Sue Snyder	Chicago/Chicago	21,219	62h 59m	June 22-24, 1960
Robert & Joan Wallick	Manila/Manila	23,129	5d 06h 17m 10s	June 2-7, 1966
Trevor K. Brougham	Darwin, Australia/Darwin	24,800	5d 05h 57m	Aug. 5-10, 1972
Arnold Palmer	Denver/Denver	22,985	57h 7m 12s	May 17-19, 1976
Boeing 747[14]	San Francisco/San Francisco	26,382	57h 25m 42s	Oct. 28-31, 1977
Richard Rutan & Jeana Yeager[15]	Edwards AFB, CA	24,986	09d 03m 44s	Dec. 14-23, 1986
Concorde	New York/New York	1,114 mph	31h 27m 49s	Aug. 15-16, 1995
Col. Douglas L. Raaberg and crew, B1 bomber[16]	Dyess AFB, Abilene, TX/ Dyess AFB	6,250	36h 13m 36s	June 3, 1995
Linda Finch[17]	Oakland, CA/Oakland, CA	26,000	73d	Mar. 17-May 28, 1997
Bertrand Piccard, Brian Jones[18]	Switzerland/Egypt	29,054.6	19d 21h 55m	Mar. 1-21, 1999

(1) Nonstop transatlantic flight. (2) Claim of reaching N. Pole in dispute; if claim is untrue, then Amundsen-Ellsworth-Nobile were the first to fly over N. Pole. (3) Includes mileage by train and auto, 4,110; by plane, 6,300; by steamship, 8,000. (4) Solo transatlantic flight in the Ryan monoplane "Spirit of St. Louis." (5) Nonstop transpacific flight. (6) First woman's transoceanic solo flight. (7) First to fly solo around N circumference of the world and first to fly twice around the world. (8) Inception of regular commercial global air service. (9) First nonstop round-the-world flight, refueled 4 times in flight. (10) Nonstop jet transatlantic flight. (11) Transatlantic round trip on same day. (12) First nonstop global flight by jet planes; refueled in flight by KC-97 aerial tankers; average speed approx. 525 mph. (13) Nonstop between Moscow and New York. (14) Speed record around the world over both Earth's poles. (15) Circled Earth nonstop without refueling. (16) Refueled in flight 6 times. Tested B-1B bomber by bombing 3 pre-arranged target sites on 3 continents. (17) Followed the intended around-the-world flight route (1937) of Amelia Earhart. (18) First to circumnavigate the globe nonstop in a balloon.

TRADE AND TRANSPORTATION

U.S. Trade With Selected Countries and Major Areas, 2000

Source: Office of Trade and Econ. Analysis, U.S. Dept. of Commerce

(in millions of dollars; countries listed by amount of total trade with U.S.)

COUNTRY	Total Trade with U.S.	U.S. Exports to	Rank[1]	U.S. Imports from	Rank[1]	U.S. trade balance with	Rank[2]
Canada	$409,779.3	$178,941.0	1	$230,838.3	1	$–51,897.3	3
Mexico	247,275.4	111,349.0	2	135,926.4	3	–24,577.4	5
Japan	211,403.8	64,924.4	3	146,479.4	2	–81,555.0	2
China	116,203.7	16,185.3	11	100,018.4	4	–83,833.2	1
Germany	87,961.2	29,448.4	5	58,512.8	5	–29,064.4	4
United Kingdom	84,915.5	41,570.4	4	43,345.1	6	–1,774.6	32
Korea (South)	68,137.7	27,830.0	6	40,307.7	8	–12,477.7	10
Taiwan	64,908.7	24,405.9	7	40,502.8	7	–16,096.8	6
France	50,161.6	20,361.5	9	29,800.1	9	–9,438.6	13
Singapore	36,984.6	17,806.3	10	19,178.3	12	–1,372.0	36
Malaysia	36,505.7	10,937.5	17	25,568.2	10	–14,630.7	7
Italy	36,103.0	11,060.3	16	25,042.7	11	–13,982.4	8
Netherlands	31,506.6	21,836.0	8	9,670.6	26	12,165.5	230
Brazil	29,173.4	15,320.9	12	13,852.5	18	1,468.3	224
Hong Kong	29,173.4	15,320.9	12	13,852.5	18	1,468.3	224
Belgium+Luxembourg	24,584.0	14,323.1	14	10,260.9	24	4,062.2	228
Venezuela	24,173.1	5,549.9	25	18,623.2	13	–13,073.2	9
Ireland	24,177.1	7,713.5	21	16,463.6	14	–8,750.1	14
Thailand	23,002.8	6,617.5	22	16,385.3	15	–9,767.8	12
Philippines	22,733.9	8,799.2	19	13,934.7	17	–5,135.5	21
Israel	20,710.3	7,745.9	20	12,964.4	19	–5,218.6	20
Saudi Arabia	20,598.8	6,234.1	24	14,364.7	16	–8,130.6	15
Switzerland	20,113.5	9,953.6	18	10,159.9	24	–206.3	64
Australia	18,920.4	12,482.3	15	6,438.1	30	6,044.2	229
India	14,353.8	3,667.2	31	10,686.6	21	–7,019.5	17
MAJOR AREA/GROUP							
North America	657,054.8	290,290.1	NA	366,764.7	NA	–76,474.7	NA
Western Europe	422,170.9	181,509.3	NA	240,661.6	NA	–59,152.3	NA
Euro Area	279,731.7	116,211.8	NA	163,519.9	NA	–47,308.1	NA
European Union (EU)	385,083.0	165,064.5	NA	220,018.5	NA	–54,954.0	NA
European Free Trade Association	28,174.1	11,770.2	NA	16,403.9	NA	–4,633.7	NA
Eastern Europe	22,043.3	5,938.7	NA	16,104.6	NA	–10,165.9	NA
Former Soviet Republics	13,719.8	3,398.9	NA	10,320.9	NA	–6,922.0	NA
OECD	420,078.6	180,584.3	NA	239,494.3	NA	–58,910.0	NA
Pacific Rim Countries	620,583.2	202,574.8	NA	418,008.4	NA	–215,433.5	NA
Asia/Middle East	57,967.2	19,005.6	NA	38,961.6	NA	–19,956.0	NA
Asia/NICS	196,061.9	84,624.2	NA	111,437.7	NA	–26,813.5	NA
Asia/South	22,119.4	4,616.4	NA	17,503.0	NA	–12,886.6	NA
ASEAN	133,393.2	46,750.3	NA	86,642.9	NA	–39,892.6	NA
APEC	1,320,628.3	506,992.1	NA	813,636.2	NA	–306,644.2	NA
South/Central America	132,630.6	59,282.7	NA	73,347.9	NA	–14,065.1	NA
Twenty Latin American Republics	367,009.8	163,754.3	NA	203,255.5	NA	–39,501.2	NA
Central American Common Market	20,863.1	9,105.4	NA	11,757.7	NA	–2,652.3	NA
Latin American Free Trade Association	334,490.6	147,980.4	NA	186,510.2	NA	–38,529.8	NA
NATO	819,218.2	355,363.0	NA	463,855.2	NA	–108,492.3	NA
OPEC	86,167.5	19,077.5	NA	67,090.0	NA	–48,012.5	NA
WORLD TOTAL	**$1,999,939.7**	**$781,917.7**	**NA**	**$1,218,022.0**	**NA**	**$–436,104.4**	**NA**

(1) Rank shown is for column to the left. (2) Rank is by size of U.S. trade deficit. NA = Not applicable. **Note:** Details may not equal totals because of rounding or incomplete enumeration.

Definitions of areas as used in the table: **North America**—Canada, Mexico. **Western Europe**—Andorra, Austria, Belgium, Bosnia and Herzegovina, Croatia, Cyprus, Denmark, Faroe Islands, Finland, France, Germany, Gibraltar, Greece, Iceland, Ireland, Italy, Liechtenstein, Luxembourg, Macedonia, Malta and Gozo, Monaco, Netherlands, Norway, Portugal, San Marino, Slovenia, Spain, Svalbard/Jan Mayen Island, Sweden, Switzerland, Turkey, United Kingdom, Vatican City, Yugoslavia. **Euro Area**—Austria, Belgium, Finland, France, Germany, Ireland, Italy, Luxemboug, Netherlands, Portugal, Spain. **EU**—(European Union) Belgium, Denmark, France, Germany, Greece, Ireland, Italy, Luxembourg, Netherlands, Portugal, Spain, United Kingdom. **EFTA**—(European Free Trade Association) Austria, Finland, Iceland, Liechtenstein, Norway, Sweden, Switzerland. **Eastern Europe**—Albania, Armenia, Azerbaijan, Belarus, Bulgaria, Czech Republic, Estonia, Georgia, Hungary, Kazakhstan, Kyrgyzstan, Latvia, Lithuania, Moldova, Poland, Romania, Russia, Slovakia, Tajikistan, Turkmenistan, Ukraine, Uzbekistan. **Former Soviet Republics**—Armenia, Azerbaijan, Belarus, Estonia, Georgia, Kazakhstan, Kyrgyzstan, Latvia, Lithuania, Moldova, Russia, Tajikistan, Turkmenistan, Ukraine, Uzbekistan. **OECD**—(Organization for Economic Cooperation & Development in Europe) Austria, Belgium, Denmark, Finland, France, Germany, Greece, Iceland, Ireland, Italy, Liechtenstein, Luxembourg, Monaco, Netherlands, Norway, Portugal, San Marino, Spain, Svalbard/Jan Mayen Island, Sweden, Switzerland, Turkey, United Kingdom. **Pacific Rim Countries/Territories**—Australia, Brunei, China, Hong Kong, Indonesia, Japan, South Korea, Macao, Malaysia, New Zealand, Papua New Guinea, Philippines, Singapore, Taiwan. **Asia/Middle East**—Bahrain, Iran, Iraq, Israel, Jordan, Kuwait, Lebanon, Oman, Qatar, Saudi Arabia, Syria, U.A.E., Yemen. **Asia/NICS**—(Newly Industrialized Countries) Hong Kong (special administrative region of China), Korea, Singapore, Taiwan. **Asia/South**—Afghanistan, Bangladesh, India, Nepal, Pakistan, Sri Lanka. **ASEAN**—(Association of Southeast Asian Nations) Brunei, Indonesia, Malaysia, Philippines, Singapore, Thailand. **APEC**—(Asia-Pacific Economic Cooperation) Australia, Brunei, Canada, Chile, China, Hong Kong, Indonesia, Japan, Korea, Malaysia, Mexico, New Zealand, Papua New Guinea, Peru, Philippines, Russia, Singapore, Taiwan, Thailand, Vietnam. **South/Central America**—Anguilla, Antigua and Barbuda, Argentina, Aruba, Bahamas, Barbados, Belize, Bermuda, Bolivia, Brazil, British Virgin Islands, Cayman Islands, Chile, Colombia, Costa Rica, Cuba, Dominica, Dominican Republic, Ecuador, El Salvador, Falkland Islands, French Guiana, Grenada, Guadeloupe, Guatemala, Guyana, Haiti, Honduras, Jamaica, Martinique, Montserrat, Netherland Antilles, Nicaragua, Panama, Paraguay, Peru, St. Kitts and Nevis, St. Lucia, St. Vincent and the Grenadines, Suriname, Trinidad and Tobago, Turks and Caicos Islands, Uruguay, Venezuela. **20 Latin American Republics**—Argentina, Bolivia, Brazil, Chile, Colombia, Costa Rica, Cuba, Dominican Republic, Ecuador, El Salvador, Guatemala, Haiti, Honduras, Mexico, Nicaragua, Panama, Paraguay, Peru, Uruguay, Venezuela. **Central American Common Market**—Costa Rica, El Salvador, Guatemala, Honduras, Nicaragua. **LAFTA**—(Latin American Free Trade Assn.) Argentina, Bolivia, Brazil, Chile, Colombia, Ecuador, Mexico, Paraguay, Peru, Uruguay, Venezuela. **NATO**—(North Atlantic Treaty Organization) Belgium, Canada, Denmark, France, Germany, Greece, Iceland, Ireland, Italy, Liechtenstein, Luxembourg, Monaco, Netherlands, Norway, Portugal, San Marino, Spain, Svalbard/Jan Mayan Island, Sweden, Switzerland, Turkey, United Kingdom. **OPEC**—(Organization of Petroleum Exporting Countries) Algeria, Indonesia, Iran, Iraq, Kuwait, Libya, Nigeria, Qatar, Saudi Arabia, United Arab Emirates, Venezuela.

U.S. Exports and Imports by Principal Commodity Groupings, 2000

Source: Office of Trade and Economic Analysis, U.S. Dept. of Commerce

(millions of dollars)

Items	Exports	Imports
TOTAL	**$781,918**	**$1,218,022**
Agricultural commodities	**51,296**	**39,186**
Animal feeds	3,780	597
Cereal flour	1,310	1,753
Coffee	9	2,350
Corn	4,695	160
Cotton, raw and linters	1,893	28
Hides and skins	1,426	109
Live animals	859	1,930
Meat and preparations	7,004	3,841
Oils/fats, vegetable	814	1,188
Rice	855	180
Soybeans	5,284	34
Sugar	4	461
Tobacco, unmanufactured	1,204	569
Vegetables and fruits	7,477	9,286
Wheat	3,374	229
Manufactured goods	**625,894**	**1,012,855**
ADP equipment; office mach	46,595	92,133
Airplane parts	15,062	5,572
Airplanes	24,777	12,412
Aluminum	3,780	6,949
Artwork/antiques	1,387	5,864
Basketware, etc.	3,309	4,840
Chemicals-cosmetics	5,292	3,539
Chemicals-dyeing	4,089	2,667
Chemicals-fertilizers	2,249	1,684
Chemicals-inorganic	5,359	6,108
Chemicals-medicinal	12,893	14,685
Chemicals-n.e.s.	12,264	5,725
Chemicals-organic	17,990	28,578
Chemicals-plastics	19,519	10,647
Clothing	8,191	64,296
Copper	1,425	4,471
Electrical machinery	89,917	108,747
Footwear	663	14,842
Furniture and bedding	4,744	18,923
Gem diamonds	1,289	12,068
General industrial mach	33,094	34,667
Glass	2,502	2,248
Glassware	865	1,918
Gold, nonmonetary	5,898	2,657
Iron and steel mill prod	5,715	15,807
Jewelry	$1,574	$6,459
Lighting, plumbing	1,384	5,104
Metal manufactures[1]	13,453	16,204
Metalworking machinery	6,191	7,726
Nickel	401	1,425
Optical goods	3,246	4,019
Paper and paperboard	10,640	15,185
Photographic equipment	4,236	6,896
Plastic articles[1]	7,607	8,034
Platinum	888	5,566
Pottery	114	1,805
Power generating mach.	32,743	33,773
Printed materials	4,776	3,680
Records/magnetic media	5,395	5,172
Rubber articles[1]	1,673	1,962
Rubber tires and tubes	2,379	4,785
Scientific instruments	30,984	22,007
Ships, boats	1,070	1,178
Silver and bullion	227	775
Spacecraft	158	217
Specialized ind. mach.	30,959	22,711
Television, VCR, etc.	27,921	70,468
Textile yearn, fabric	10,534	15,171
Toys/games/sporting goods	3,609	20,011
Travel goods	351	4,430
Vehicles	57,421	161,544
Watches/clocks/parts	348	3,481
Wood manufactures	1,842	7,228
Mineral fuels	**13,179**	**135,367**
Coal	2,162	805
Crude oil	463	89,876
Liquified propane/butane	663	1,508
Mineral fuels, other	3,320	2,194
Natural gas	411	12,594
Petroleum preparations	5,746	25,673
Selected commodities:		
Alcoholic bev., distilled	424	2,946
Cigarettes	3,304	258
Cork, wood, lumber	4,320	8,227
Crude fertilizers	1,724	1,399
Fish and preparations	2,806	9,907
Metal ores; scrap	4,234	3,817
Pulp and waste paper	4,576	3,381

(1) Those not specified elsewhere. **NOTE:** Not all products are listed in each commodity group.

Trends in U.S. Foreign Trade, 1790-2000

Source: Office of Trade and Economic Analysis, U.S. Dept. of Commerce

In 1790, U.S. exports and imports combined came to $43 million and there was a $3 million trade deficit. By 2000, U.S. exports and imports combined amounted to nearly $2 trillion, and the trade deficit, which had generally been climbing in recent years (after a century of trade surpluses), reached an all-time high, $436 billion.

(in millions of dollars)

Year	Exports	Imports	Trade Balance
1790	$20	$23	$–3
1795	48	70	–22
1800	71	91	–20
1805	96	121	–25
1810	67	85	–19
1815	53	113	–60
1820	70	74	–5
1825	91	90	1
1830	72	63	9
1835	115	137	–22
1840	124	98	25
1845	106	113	–7
1850	144	174	–29
1855	219	258	–39
1860	334	354	–20
1865	166	239	–73
1870	393	436	–43
1875	513	533	–20
1880	836	668	168
1885	742	578	165
1890	858	789	69
1895	808	732	76
1900	1,394	850	545
1905	1,519	1,118	401
1910	1,745	1,557	188
1915	2,769	1,674	1,094
1920	$8,228	$5,278	$2,950
1925	4,910	4,227	683
1930	3,843	3,061	782
1935	2,283	2,047	235
1940	4,021	2,625	1,396
1945	9,806	4,159	5,646
1950	9,997	8,954	1,043
1955	14,298	11,566	2,732
1960	19,659	15,073	4,586
1965	26,742	21,520	5,222
1970	42,681	40,356	2,325
1975	107,652	98,503	9,149
1980	220,626	244,871	–24,245
1985	213,133	345,276	–132,143
1990	394,030	495,042	–101,012
1991	421,730	485,453	–63,723
1992	448,164	532,665	–84,501
1993	465,091	580,659	–115,568
1994	512,626	683,256	–170,630
1995	584,742	743,445	–158,703
1996	625,075	795,289	–170,214
1997	689,182	870,671	–181,489
1998	682,138	911,896	–229,758
1999	695,797	1,024,618	–328,821
2000	781,918	1,218,022	–436,104

The North American Free Trade Agreement (NAFTA)

NAFTA, a comprehensive plan for free trade between the U.S., Canada, and Mexico, took effect on Jan. 1, 1994. Major provisions are:

Agriculture—Tariffs on all farm products are to be eliminated over 15 years. Domestic price-support systems may continue provided they do not distort trade.

Automobiles—After 8 years, at least 62.5% of an automobile's value must have been produced in North America for it to qualify for duty-free status. Tariffs are to be phased out over 10 years.

Banking—U.S. and Canadian banks may acquire Mexican commercial banks accounting for as much as 8% of the industry's capital. All limits on ownership end in 2004.

Disputes—Special judges have jurisdiction to resolve disagreements within strict timetables.

Energy—Mexico continues to bar foreign ownership of its oil fields but, starting in 2004, U.S. and Canadian companies can bid on contracts offered by Mexican oil and electricity monopolies.

Environment—The trade agreement cannot be used to overrule national and state environmental, health, or safety laws.

Immigration—All 3 countries must ease restrictions on the movement of business executives and professionals.

Jobs—Barriers to limit Mexican migration to U.S. remain.

Patent and copyright protection—Mexico strengthened its laws providing protection to intellectual property.

Tariffs—Tariffs on 10,000 customs goods are to be eliminated over 15 years. One-half of U.S. exports to Mexico are to be considered duty-free within 5 years.

Textiles—A "rule of origin" provision requires most garments to be made from yarn and fabric that has been produced in North America. Most tariffs are being phased out over 5 years.

Trucking—Trucks were to have free access on crossborder routes and throughout the 3 countries by 1999, but the U.S. continued to impose restrictions on Mexican trucks. On Feb. 6, 2001, an arbitration panel ruled that the U.S. restrictions were in violation of NAFTA. Pres. Bush pledged to work with Congress to bring the U.S. into compliance with NAFTA.

U.S. Trade With Mexico and Canada, 1992-2000

Source: Office of Trade and Economic Analysis, U.S. Dept. of Commerce

(U.S. exports to, imports from, Canada and Mexico in millions of dollars)

	MEXICO				CANADA		
Year	Exports	Imports	Trade Balance[1]	Year	Exports	Imports	Trade Balance[1]
1992	$40,592	$35,211	$5,381	1992	$90,594	$98,630	$–8,0361
1993	41,581	39,917	1,664	1993	100,444	111,216	–10,772
1994[2]	50,844	49,494	1,350	1994[2]	114,439	128,406	–13,968
1995	46,292	61,685	–15,393	1995	127,226	145,349	–18,123
1996	56,792	74,297	–17,506	1996	134,210	155,893	–21,682
1997	71,388	85,938	–14,549	1997	151,767	167,234	–15,467
1998	78,773	94,629	–15,857	1998	156,603	173,256	–16,653
1999	86,909	109,721	–22,812	1999	166,600	198,711	–32,111
2000	111,349	135,926	–24,577	2000	178,941	230,838	–51,897

(1) Totals may not add due to rounding. (2) NAFTA provisions began to take effect Jan. 1, 1994.

Foreign Exchange Rates, 1970-2000

Source: International Monetary Fund, Federal Reserve Board; Federal Reserve Board

(National currency units per dollar except as indicated; data are annual averages)

Year	Australia[1] (dollar)	Austria (schilling)	Belgium (franc)	Canada (dollar)	Denmark (krone)	France (franc)	Germany[2] (deutsche mark)	Greece (drachma)
1970	1.1136	25.880	49.680	1.0103	7.489	5.5200	3.6480	30.00
1975	1.3077	17.443	36.799	1.0175	5.748	4.2876	2.4613	32.29
1980	1.1400	12.945	29.237	1.1693	5.634	4.2250	1.8175	42.62
1985	0.7003	20.690	59.378	1.3655	10.596	8.9852	2.9440	138.12
1990	0.7813	11.370	33.418	1.1668	6.189	5.4453	1.6157	158.51
1995	0.7415	10.081	29.480	1.3724	5.602	4.9915	1.4331	231.66
1996	0.7829	10.587	30.962	1.3635	5.799	5.1155	1.5048	240.71
1997	0.7441	12.204	35.774	1.3846	6.604	5.8367	1.7341	273.06
1998	0.6294	12.379	36.299	1.4835	6.701	5.8995	1.7597	295.53
1999	0.6453	0.9386[3]	0.9386[3]	1.4857	6.976	0.9386[3]	0.9386[3]	305.65
2000	0.5815	0.9232[3]	0.9232[3]	1.4855	8.095	0.9232[3]	0.9232[3]	365.92

Year	India (rupee)	Ireland[1] (pound)	Italy (lira)	Japan (yen)	Malaysia (ringgit)	Mexico (new peso)	Netherlands (guilder)	Norway (krone)
1970	7.576	2.3959	623	357.60	3.0900	—	3.5970	7.1400
1975	8.409	2.2216	653	296.78	2.4030	—	2.5293	5.2282
1980	7.887	2.0577	856	226.63	2.1767	—	1.9875	4.9381
1985	12.369	1.0656	1,909	238.54	2.4830	—	3.3214	8.5972
1990	17.504	1.6585	1,198	144.79	2.7049	2.8126	1.8209	6.2597
1995	32.427	1.6038	1,628.9	94.06	2.5044	6.4194	1.6057	6.3352
1996	35.433	1.6006	1,542.9	108.78	2.5159	7.5994	1.6859	6.4498
1997	36.313	1.5180	1,703.1	120.99	2.8132	7.9185	1.9513	7.0734
1998	41.259	1.4257	1,736.2	130.91	3.9244	9.1360	1.9837	7.5451
1999	43.055	1.0668	0.9386[3]	113.91	3.8000	9.5604	0.9386[3]	7.7992
2000	45.000	0.9232[3]	0.9232[3]	107.80	3.8000	9.4590	0.9232[3]	8.8131

Year	Portugal (escudo)	Singapore (dollar)	South Korea (won)	Spain (peseta)	Sweden (krona)	Switzerland (franc)	Thailand (baht)	UK[1] (pound)
1970	28.75	3.0800	310.57	69.72	5.1700	4.3160	21.000	2.3959
1975	25.51	2.3713	484.00	57.43	4.1530	2.5839	20.379	2.2216
1980	50.08	2.1412	607.43	71.76	4.2309	1.6772	20.476	2.3243
1985	170.39	2.2002	870.02	170.04	8.6039	2.4571	27.159	1.2963
1990	142.55	1.8125	707.76	101.93	5.9188	1.3892	25.585	1.7847
1995	151.11	1.4174	771.27	124.69	7.1333	1.1825	24.915	1.5785
1996	154.24	1.4100	804.45	126.66	6.7060	1.2360	25.343	1.5617
1997	175.31	1.4848	951.29	146.41	7.6349	1.4513	31.364	1.6377
1998	180.10	1.6736	1,401.44	149.40	7.9499	1.4498	41.359	1.6564
1999	0.9386[3]	1.6950	1,188.82	0.9386[3]	8.2624	1.5022	37.814	1.6182
2000	0.9232[3]	1.7250	1,130.90	0.9232[3]	9.1735	1.6904	40.210	1.5156

(1) Value of 1 unit of foreign currency in dollars. (2) West Germany prior to 1991. (3) Euro Area member, figs. in euros per U.S. dollar. 1 EUR=13.7063 Aust. schillings, 40.3399 Bel. francs, 5.94573 Fin. markkas, 6.55957 Fr. francs, 1.95583 Ger. marks, .787564 Ir. pounds, 1936.27 It. lire, 40.3399 Lux. francs, 2.20371 Neth. guilders, 200.482 Port. escudos, 166.386 Sp. pesetas.

Foreign Direct Investment[1] in the U.S. by Selected Countries and Territories

Source: Bureau of Economic Analysis; U.S. Dept. of Commerce

(millions of dollars)

	1998	1999	2000
ALL COUNTRIES[2]	**$793,748**	**$965,632**	**$1,238,627**
Canada	**74,143**	**76,526**	**100,822**
Europe[3]	**528,601**	**670,030**	**890,611**
Austria	1,969	3,203	3,172
Belgium	10,966	10,037	14,186
Denmark	3,312	5,226	5,905
Finland	4,106	4,967	5,473
France	58,051	82,276	119,069
Germany	94,404	111,706	122,846
Ireland	12,484	15,621	23,031
Italy	4,143	4,709	6,409
Luxembourg	26,650	57,047	83,304
Netherlands	98,926	125,775	152,432
Norway	3,833	3,089	2,441
Spain	2,285	2,746	8,860
Sweden	15,139	20,843	27,389
Switzerland	48,403	53,706	81,698
United Kingdom	143,165	166,900	229,762
South and Central America[3]	**9,819**	**8,365**	**8,671**
Brazil	635	740	846
Mexico	2,432	1,730	2,471
Panama	6,504	5,475	4,004
Other W. Hemisphere[3]	**$18,034**	**$29,739**	**$34,029**
Bahamas	2,077	1,581	1,385
Bermuda	3,740	12,590	14,942
Netherlands Antilles	2,844	3,153	3,515
UK islands, Caribbean	9,009	11,082	12,513
Africa[3]	**862**	**1,547**	**2,119**
Middle East[3]	**6,346**	**4,432**	**8,373**
Israel	2,084	2,485	3,183
Kuwait	NA	916	957
Saudi Arabia	NA	946	NA
United Arab Emirates	43	13	79
Asia and Pacific[3]	**155,943**	**174,993**	**194,002**
Australia	12,883	13,230	14,487
Hong Kong	1,578	883	1,494
Japan	134,590	153,119	163,215
Korea, Republic of	974	1,853	2,696
Malaysia	100	71	36
Singapore	1,561	1,370	7,661
Taiwan	3,144	2,990	3,224
European Union (15)[4]	**NA**	**611,171**	**802,712**
OPEC[5]	**NA**	**2,129**	**6,220**

(1) The book value of foreign direct investors' equity in, and net outstanding loans to, their U.S. affiliates. A U.S. affiliate is a U.S. business enterprise in which a single foreign direct investor owns at least 10% of the voting securities or the equivalent. (2) Total includes sources not reflected in regional subtotals. (3) Totals include countries or territories not shown. (4) The European Union (15) comprises Austria, Belgium, Denmark, Finland, France, Germany, Greece, Ireland, Italy, Luxembourg, the Netherlands, Portugal, Spain, Sweden, and the United Kingdom. (5) Organization of Petroleum Exporting Countries: Algeria, Indonesia, Iran, Iraq, Kuwait, Libya, Nigeria, Qatar, Saudi Arabia, the United Arab Emirates, and Venezuela. NA = Not available.

U.S. Direct Investment[1] Abroad in Selected Countries and Territories

Source: Bureau of Economic Analysis, U.S. Dept. of Commerce

(millions of dollars)

	1990	1999	2000
ALL COUNTRIES[2]	**$424,086**	**$1,130,789**	**$1,244,654**
Canada	**67,033**	**111,051**	**126,421**
Europe	**211,194**	**588,341**	**648,731**
Austria	889	3,711	3,676
Belgium	9,050	17,347	16,409
Denmark	1,597	4,123	5,618
Finland	551	1,290	1,279
France	18,874	40,009	39,087
Germany	27,259	50,892	53,610
Greece	288	604	672
Ireland	6,880	26,084	33,369
Italy	13,117	17,914	23,622
Luxembourg	1,390	16,484	19,470
Netherlands	22,658	105,571	115,506
Norway	3,815	6,181	6,303
Portugal	598	1,463	1,784
Spain	7,704	13,244	14,561
Sweden	1,600	10,200	11,371
Switzerland	25,199	48,849	54,873
Turkey	494	1,235	1,378
United Kingdom	68,224	212,007	233,384
Other	NA	11,135	12,760
South America[3]	**23,760**	**74,743**	**79,354**
Argentina	2,956	14,175	14,489
Brazil	14,918	34,276	35,560
Chile	1,368	10,105	10,846
Colombia	1,728	3,854	4,423
Ecuador	387	1,035	838
Peru	410	2,705	3,317
Venezuela	1,490	7,342	8,423
Central America[3]	**17,719**	**68,456**	**74,754**
Costa Rica	NA	1,539	1,983
Guatemala	NA	578	904
Honduras	NA	126	115
Mexico	9,398	32,262	35,414
Panama	7,409	33,027	35,407
Other West Hemisphere[3]	**$30,113**	**$77,506**	**$85,280**
Bahamas	3,309	702	668
Barbados	NA	1,065	1,227
Bermuda	21,737	47,119	54,114
Dominican Republic	NA	956	1,126
Jamaica	604	2,311	2,596
Netherlands Antilles	–2,229	3,652	3,725
Trinidad and Tobago	508	1,329	1,331
UK islands, Caribbean	4,800	19,767	20,165
Africa[3]	**4,861**	**14,884**	**15,813**
Egypt	1,465	2,190	2,735
Nigeria	161	1,462	1,283
South Africa	956	2,905	2,826
Middle East[3]	**3,973**	**10,519**	**11,851**
Israel	756	3,051	3,426
Saudi Arabia	1,981	4,426	4,784
United Arab Emirates	519	557	573
Asia and Pacific[3]	**61,869**	**181,882**	**199,599**
Australia	14,846	34,776	35,324
China	NA	8,058	9,577
Hong Kong	6,187	20,092	23,308
India	513	1,402	1,258
Indonesia	3,226	10,495	11,605
Japan	20,997	49,438	55,606
Korea, Republic of	2,178	8,559	9,432
Malaysia	1,384	5,820	5,995
New Zealand	3,131	5,433	5,340
Philippines	1,629	3,136	2,910
Singapore	3,385	20,117	23,245
Taiwan	2,014	6,513	7,737
Thailand	1,585	6,809	7,124
European Union (15)[4]	**NA**	**520,942**	**573,416**
Eastern Europe[5]	**NA**	**9,581**	**11,009**
OPEC[6]	**NA**	**29,152**	**32,401**

(1) The book value of U.S. direct investors' equity in, and net outstanding loans to, their foreign affiliates. A foreign affiliate is a foreign business enterprise in which a single U.S. investor owns at least 10% of the voting securities or the equivalent. (2) Total includes countries not reflected in regional totals. (3) Total includes countries not shown. (4) The European Union (15) comprises Austria, Belgium, Denmark, Finland, France, Germany, Greece, Ireland, Italy, Luxembourg, the Netherlands, Portugal, Spain, Sweden, and the United Kingdom. (5) Eastern Europe consists of Albania, Armenia, Azerbaijan, Belarus, Bulgaria, Czech Rep., Estonia, Georgia, Hungary, Kazakhstan, Latvia, Lithuania, Moldova, Poland, Romania, Russia, Slovakia, Tajikstan, Turkmenistan, Ukraine, and Uzbekistan. (6) Organization of Petroleum Exporting Countries: Algeria, Indonesia, Iran, Iraq, Kuwait, Libya, Nigeria, Qatar, Saudi Arabia, the United Arab Emirates, and Venezuela. NA = not available.

U.S. International Transactions

Source: Bureau of Economic Analysis, U.S. Dept. of Commerce; revised as of July 2001
(millions of dollars)

	1965	1970	1975	1980	1985	1990	1995	1999	2000
Exports of goods, services, and income[1]	$42,722	$68,387	$157,936	$344,440	$382,749	$700,455	$991,490	$1,242,655	$1,418,568
Merchandise adjusted, excluding military[2]	26,461	42,469	107,088	224,250	215,915	389,307	575,871	684,553	772,210
Services	8,824	14,171	25,497	47,584	73,155	147,824	218,739	272,800	293,492
Income receipts on U.S. assets abroad	7,437	11,748	25,351	72,606	93,679	163,324	196,880	283,092	350,525
Imports of goods, services, and income	–32,708	–59,901	–132,745	–333,774	–484,037	–757,758	–1,086,539	–1,518,106	–1,809,099
Merchandise adjusted, excluding military[2]	–21,510	–39,866	–98,185	–249,750	–338,088	–498,337	–749,431	–1,029,987	–1,224,417
Services	–9,111	–14,520	–21,996	–41,491	–72,862	–120,019	–147,036	–189,204	–217,024
Income payments on foreign assets in the U.S.	–2,088	–5,515	–12,564	–42,532	–73,087	–139,402	–190,072	–291,603	–360,146
Unilateral transfers, net	–4,583	–6,156	–7,075	–8,349	–22,700	–34,588	–34,046	–48,913	–54,136
Capital account transactions, net	NA	NA	NA	NA	NA	NA	NA	–3,491	705
U.S. assets abroad, net(increase/ capital outflow [–])	–5,716	–9,337	–39,703	–86,967	–39,889	–74,011	–307,207	–437,067	–580,952
U.S. official reserve assets, net	1,225	2,481	–849	–8,155	–3,858	–2,158	–9,742	8,747	–290
U.S. government assets, other than official reserve assets, net	–1,605	–1,589	–3,474	–5,162	–2,821	2,307	–549	2,751	–944
U.S. private assets, net	–5,336	–10,229	–35,380	–73,651	–33,211	–74,160	–296,916	–448,565	–579,718
Foreign assets in the U.S., net (increase/capital inflow [+])	742	6,359	17,170	62,612	146,383	140,992	451,234	813,744	1,024,218
Statistical discrepancy (sum of above items with sign reversed)	–457	–219	4,417	20,886	17,494	24,911	–14,931	–48,822	696
Memorandum: Balance on current account	5,431	2,331	18,116	2,317	–123,987	–91,892	–129,095	–324,364	–444,667

NA = Not available. (1) Excludes transfers of goods and services under U.S. military grant programs. (2) Excludes exports of goods under U.S. military agency sales contracts identified in Census export documents, excludes imports of goods under direct defense expenditures identified in Census import documents, and reflects various other adjustments.

Merchant Fleets of the World by Flag of Registry, 2001

Source: Maritime Administration, U.S. Dept. of Commerce
Self- propelled oceangoing vessels of 1,000 gross deadweight tons and over, as of Jan. 1, 2001 (tonnage in thousands)

	All Vessels		Tanker		Dry Bulk Carrier		Container		Other[1]	
By Flag	No.	Tons	No.	Tons	No.	Tons	No.	Tons	No.	Tons
Panama	4,656	167,850	1,088	57,404	1,379	81,464	500	14,581	1,689	14,401
Liberia	1,498	78,743	592	44,267	370	22,674	229	6,970	307	4,832
Malta	1,395	45,852	350	21,506	437	18,238	50	872	558	5,236
Bahamas	1,021	45,025	262	27,937	140	8,120	55	1,549	564	7,419
Greece	693	43,084	280	26,166	259	14,288	46	1,797	108	833
Cyprus	1,307	36,213	172	7,910	461	20,053	125	2,807	549	5,443
Singapore	880	33,089	396	17,873	128	8,512	172	4,060	184	2,644
Norway (NIS)[2]	658	27,985	306	17,277	88	6,929	5	101	259	3,678
China	1,448	22,143	261	3,626	326	10,805	103	1,778	758	5,934
Hong Kong	336	16,418	35	1,254	193	12,355	53	1,619	55	1,190
United States	454	15,769	142	8,438	15	605	88	3,029	209	3,697
Japan	621	15,613	253	8,282	154	5,651	23	693	191	987
Marshall Islands	195	15,489	94	11,057	64	3,509	25	758	12	165
India	294	10,431	101	5,211	115	4,517	8	153	70	550
Italy	406	9,726	221	4,242	42	3,630	22	696	121	1,158
Philippines	457	9,601	66	284	166	7,614	7	78	218	1,625
Saint Vincent	760	9,588	95	1,082	129	4,602	27	179	509	3,725
Bermuda	105	9,366	27	4,841	28	3,701	16	459	34	365
Turkey	534	8,871	92	1,137	152	5,991	21	214	269	1,529
Korea (South)	480	8,347	129	1,378	99	5,055	45	835	207	1,079
All Other Flags	10,170	148,112	2,110	56,751	945	33,616	938	24,665	6,177	33,080
By Country										
Greece	3,133	142,021	799	62,923	1,345	66,469	136	3,826	853	8,803
Japan	2,712	96,845	743	37,168	805	45,369	193	5,397	971	8,911
United States	1,521	63,570	414	32,896	128	6,228	100	3,212	425	5,465
Norway	1,310	57,331	487	36,766	180	11,205	21	667	622	8,693
China	1,974	39,937	297	6,150	556	22,527	189	3,709	932	7,551
Hong Kong	588	36,319	127	14,423	248	18,444	54	1,536	159	1,916
Germany	1,791	31,620	164	4,187	110	4,826	653	16,449	864	6,158
Republic of Korea	861	25,619	209	6,871	198	14,087	103	2,608	351	2,053
Singapore	783	19,792	358	12,140	95	3,348	125	2,637	205	1,667
Taiwan	531	18,537	36	2,488	148	8,318	195	6,566	152	1,165
United Kingdom	519	17,128	149	7,589	60	4,610	90	3,301	220	1,628
Denmark	590	16,906	164	8,073	32	2,126	120	5,219	274	1,488
Russia	1,606	14,206	344	6,418	127	2,704	34	798	1,101	4,286
Italy	508	13,244	253	5,270	75	5,875	11	269	169	1,830
India	331	11,733	113	5,543	129	5,368	5	129	84	693
Saudi Arabia	104	10,485	72	9,771	1	2	5	250	26	462
Sweden	333	10,173	136	7,929	13	468	1	11	183	1,765
Turkey	597	9,122	94	1,318	155	5,771	26	248	322	1,785
Iran	140	7,122	33	4,127	47	1,913	7	180	53	902
Switzerland	239	6,752	47	1,368	46	2,181	74	2,328	72	875
All Other Countries	8,197	128,853	2,033	54,505	1,192	50,090	416	8,553	5,010	31,474
TOTAL ALL SHIPS	**28,368**	**777,315**	**7,072**	**327,923**	**5,690**	**281,929**	**2,558**	**67,893**	**13,048**	**99,570**

(1) Includes roll-on/roll-off, passenger, breakbulk ships, partial container ships, refrigerated cargo ships, barge carriers, and specialized cargo ships. (2) NIS = Norwegian International Ship Registry.

50 Busiest U.S. Ports, 1999

Source: Corps of Engineers, Dept. of the Army, U.S. Dept. of Defense

(ports ranked by tonnage handled; all figures in tons)

Rank	Port Name	Total	Domestic	Foreign	Imports	Exports
1.	South Louisiana, LA	214,196,912	119,453,922	94,742,990	29,407,479	65,335,511
2.	Houston, TX	158,828,203	56,735,755	102,092,448	69,919,172	32,173,276
3.	New York, NY and NJ	133,715,223	70,231,084	63,484,139	55,933,999	7,550,140
4.	New Orleans, LA	87,511,476	38,602,795	48,908,681	29,186,801	19,721,880
5.	Corpus Christi, TX	77,986,587	22,294,600	55,691,987	48,041,568	7,650,419
6.	Beaumont, TX	69,405,951	15,550,377	53,855,574	49,748,637	4,106,937
7.	Baton Rouge, LA	63,728,759	43,324,207	20,404,552	13,330,923	7,073,629
8.	Plaquemines, LA, Port of	62,461,023	40,611,729	21,849,294	12,838,507	9,010,787
9.	Long Beach, CA	60,882,795	18,674,640	42,208,155	28,447,669	13,760,486
10.	Valdez, AK	53,391,575	49,393,923	3,997,652	5,597	3,992,055
11.	Pittsburgh, PA	52,931,007	52,931,007	0	0	0
12.	Tampa, FL	51,518,312	32,512,833	19,005,479	6,995,638	12,009,841
13.	Lake Charles, LA	50,741,854	19,989,692	30,752,162	27,001,155	3,751,007
14.	Texas City, TX	49,502,557	18,263,045	31,239,512	28,012,342	3,227,170
15.	Mobile, AL	45,439,385	21,375,982	24,063,403	15,488,105	8,575,298
16.	Duluth-Superior, MN and WI	42,296,553	29,427,323	12,869,230	720,304	12,148,926
17.	Los Angeles, CA	42,267,055	5,609,027	36,658,028	24,492,725	12,165,303
18.	Norfolk Harbor, VA	40,781,438	10,093,797	30,687,641	8,463,891	22,223,750
19.	Philadelphia, PA	39,270,741	14,004,004	25,266,737	24,986,819	279,918
20.	Baltimore, MD	37,287,472	14,215,400	23,072,072	15,464,031	7,608,041
21.	St. Louis, MO and IL	32,650,783	32,650,783	0	0	0
22.	Portland, OR	29,339,074	12,900,369	16,438,705	4,698,328	11,740,377
23.	Pascagoula, MS	28,095,079	10,333,484	17,761,595	15,659,223	2,102,372
24.	Freeport, TX	28,076,004	5,558,866	22,517,138	20,629,944	1,887,194
25.	Paulsboro, NJ	26,845,472	8,994,963	17,850,509	17,759,373	91,136
26.	Chicago, IL	26,602,449	20,688,838	5,913,611	5,166,986	746,625
27.	Seattle, WA	25,442,979	9,054,329	16,388,650	9,031,475	7,357,175
28.	Richmond, CA	22,355,552	11,092,440	11,263,112	9,049,493	2,213,619
29.	Huntington, WV	22,313,997	22,313,997	0	0	0
30.	Boston, MA	22,170,862	9,330,341	12,840,521	12,199,093	641,428
31.	Port Everglades, FL	22,053,829	12,554,477	9,499,352	7,377,777	2,121,575
32.	Tacoma, WA	21,103,313	7,780,804	13,322,509	4,344,588	8,977,921
33.	Portland, ME	20,370,505	1,908,548	18,461,957	18,268,742	193,215
34.	Charleston, SC	19,916,241	5,029,771	14,886,470	8,222,636	6,663,834
35.	Marcus Hook, PA	19,260,811	9,399,799	9,861,012	9,829,918	31,094
36.	Jacksonville, FL	19,257,445	10,066,818	9,190,627	8,279,832	910,795
37.	Port Arthur, TX	18,307,752	6,893,486	11,414,266	9,769,508	1,644,758
38.	Savannah, GA	18,156,379	2,753,775	15,402,604	9,192,033	6,210,571
39.	Detroit, MI	16,948,335	12,332,561	4,615,774	4,257,604	358,170
40.	Memphis, TN	16,611,022	16,611,022	0	0	0
41.	Anacortes, WA	16,231,602	12,469,852	3,761,750	2,864,243	897,507
42.	San Juan, PR	15,555,483	7,964,501	7,590,982	7,015,994	574,988
43.	Cleveland, OH	15,540,366	12,438,957	3,101,409	2,781,112	320,297
44.	Indiana Harbor, IN	15,127,434	15,127,407	27	27	0
45.	Newport News, VA	14,341,901	5,607,612	8,734,289	1,382,701	7,351,588
46.	Cincinnati, OH	14,293,775	14,293,775	0	0	0
47.	Lorain, OH	12,967,688	12,693,921	273,767	273,767	0
48.	Toledo, OH	12,326,606	5,906,703	6,419,903	1,589,579	4,830,324
49.	Honolulu, HI	12,258,801	10,509,556	1,749,245	1,438,825	310,420
50.	Two Harbors, MN	11,872,419	11,571,507	300,912	0	300,912

World Trade Organization (WTO)

Following World War II, the major economic powers of the world negotiated a set of rules for reducing and limiting trade barriers and for settling trade disputes. These rules were called the General Agreement on Tariffs and Trade (GATT). Headquarters to oversee the administration of the GATT were established in Geneva, Switzerland. Periodically, rounds of multilateral trade negotiations under the GATT were carried out. The 8th round, begun in 1986 in Punta del Este, Uruguay, and dubbed the Uruguay Round, concluded on Dec. 15, 1993, when 117 countries completed a new trade-liberalization agreement. The name for the GATT was changed to the World Trade Organization (WTO), which officially came into being Jan. 1, 1995.

New Passenger Cars Imported Into the U.S., by Country of Origin,[1] 1968-2000

Source: Bureau of the Census, Foreign Trade Division

	Japan	Germany[2]	Italy	United Kingdom	Sweden	France	South Korea	Mexico	Canada	Total[3]
1968	169,849	707,972	33,843	96,787	52,515	39,551	NA	NA	500,881	1,620,452
1969	260,005	642,157	41,569	104,050	41,008	24,457	NA	NA	691,146	1,846,717
1970	381,338	674,945	42,523	76,257	57,844	37,114	NA	NA	692,783	2,013,420
1971	703,672	770,807	51,469	106,710	61,925	23,316	NA	0	802,281	2,587,484
1972	697,788	676,967	64,614	72,038	64,541	14,713	NA	9	842,300	2,485,901
1973	624,805	677,465	56,102	64,140	58,626	8,219	NA	4,469	871,557	2,437,345
1974	791,791	619,757	107,071	72,512	60,817	21,331	NA	3,914	817,559	2,572,557
1975	695,573	370,012	102,344	67,106	51,993	15,647	NA	0	733,766	2,074,653
1976	1,128,936	349,804	82,500	77,190	37,466	21,916	NA	0	825,590	2,536,749
1977	1,341,530	423,492	55,437	56,889	39,370	19,215	NA	NA	849,814	2,790,144
1978	1,563,047	416,231	69,689	54,478	56,140	28,502	NA	6	833,061	3,024,982
1979	1,617,328	495,565	72,456	46,911	65,907	27,887	NA	4	677,008	3,005,523
1980	1,991,502	338,711	46,899	32,517	61,496	47,386	NA	1	594,770	3,116,448
1981	1,911,525	234,052	21,635	12,728	68,042	42,477	NA	1	563,943	2,856,286
1982	1,801,185	259,385	9,402	13,023	89,231	50,032	NA	27	702,495	2,926,407
1983	1,871,192	239,807	5,442	17,261	114,726	40,823	NA	2	835,665	3,133,836
1984	1,948,714	335,032	8,582	19,833	114,854	37,788	NA	NA	1,073,425	3,559,427

	Japan	Germany[2]	Italy	United Kingdom	Sweden	France	South Korea	Mexico	Canada	Total[3]
1985	2,527,467	473,110	8,689	24,474	142,640	42,882	NA	13,647	1,144,805	4,397,679
1986	2,618,711	451,699	11,829	27,506	148,700	10,869	169,309	41,983	1,162,226	4,691,297
1987	2,417,509	377,542	8,648	50,059	138,565	26,707	399,856	126,266	926,927	4,589,010
1988	2,123,051	264,249	6,053	31,636	108,006	15,990	455,741	148,065	1,191,357	4,450,213
1989	2,051,525	216,881	9,319	29,378	101,571	4,885	270,609	133,049	1,151,122	4,042,728
1990	1,867,794	245,286	11,045	27,271	93,084	1,976	201,475	215,986	1,220,221	3,944,602
1991	1,762,347	171,097	2,886	14,862	62,905	1,727	186,740	249,498	1,109,248	3,612,665
1992	1,598,919	205,248	1,791	10,997	76,832	65	130,110	266,111	1,119,223	3,447,200
1993	1,501,953	180,383	1,178	20,029	58,742	23	122,943	299,634	1,371,856	3,604,361
1994	1,488,159	178,774	1,010	28,217	63,867	58	213,962	360,367	1,525,746	3,909,079
1995	1,114,360	204,932	1,031	42,450	82,593	14	131,718	462,800	1,552,691	3,624,428
1996	1,190,896	234,909	1,365	44,373	86,619	27	225,623	550,867	1,690,733	4,069,113
1997	1,387,812	300,489	1,912	43,691	79,780	67	222,568	544,075	1,731,209	4,378,295
1998	1,456,081	373,330	2,104	49,891	84,543	56	211,650	584,795	1,837,615	4,673,418
1999	1,707,277	461,061	1,697	68,394	83,399	186	372,965	639,878	2,170,427	5,639,616
2000	1,839,093	488,323	3,125	81,196	86,707	134	568,121	934,000	2,138,811	6,324,284

(1) Excludes passenger cars assembled in U.S. foreign trade zones. (2) Figures prior to 1991 are for West Germany. (3) Includes countries not shown separately.

Passenger Car Production, U.S. Plants, 1999-2000

Source: Ward's Communications

Series	1999	2000
626 (Mazda)	87,065	67,255
Cougar (Mercury)	78,078	40,176
TOTAL AUTOALLIANCE[1]	**165,143**	**107,431**
BMW Z3	48,393	38,665
TOTAL BMW	**48,393**	**38,665**
Cirrus	47,124	39,232
Sebring	—	41,116
Total Chrysler	**50,813**	**86,794**
Neon (Dodge)	165,229	179,039
Stratus	100,196	117,272
Viper	1,600	1,731
Total Dodge	**267,025**	**298,042**
Breeze	47,911	2,030
Neon (Plymouth)	63,216	43,177
Prowler	2,868	2,890
Total Plymouth	**113,995**	**48,097**
TOTAL DAIMLER CHRYSLER	**431,833**	**432,933**
Contour	114,897	17,410
Escort	114,171	—
Focus	114,682	325,720
Mustang	191,432	180,431
Taurus	382,858	441,480
Total Ford	**918,040**	**965,041**
Continental	27,121	22,589
Lincoln LS	39,266	58,791
Town Car	81,551	81,391
Total Lincoln	**147,938**	**162,771**
Mystique	38,021	5,388
Sable	106,427	114,133
Tracer	14,691	—
Total Mercury	159,139	**119,521**
TOTAL FORD MOTOR CO.	**1,225,117**	**1,247,333**
LeSabre	157,501	163,919
Park Ave.	61,009	45,418
Total Buick	**218,510**	**209,337**
Eldorado	17,638	12,043
Fleetwood Deville	88,922	118,967
Seville	38,907	28,026
Total Cadillac	**145,467**	**159,036**
Cavalier	269,564	252,028
Corvette	33,243	34,919
Malibu	251,584	260,563
Total Chevrolet	**554,391**	**547,510**
Alero	146,329	135,012
Aurora	12,260	39,921
Cutlass	20,784	—
Intrigue	98,492	59,884
Total Oldsmobile	**277,865**	**234,817**
Bonneville	48,120	66,783
Grand Am	259,471	250,435
Grand Prix	173,876	166,929
Sunfire	110,089	92,174
Total Pontiac	**591,556**	**576,321**
Saturn Ev1	318	—
Saturn LS	60,987	85,954
Saturn S	238,140	175,308
Total Saturn	**299,445**	**261,262**
TOTAL GENERAL MOTORS	**2,087,224**	**1,938,287**
Acura CL	15,804	31,440
Acura TL	78,959	83,893
Total Acura	**94,763**	**115,333**
Accord	369,181	336,034
Civic	221,956	225,723
Total Honda	**591,137**	**561,757**
TOTAL AMERICAN HONDA MOTOR CORP.	**49,967**	**49,996**
Sebring (Chrysler)	24,859	19,924
Avenger (Dodge)	15,801	1,361
Stratus (Dodge)	—	17,580
Eclipse	50,876	61,644
Galant	64,653	104,136
TOTAL MITSUBISHI	**161,931**	**221,975**
Altima	152,541	**150,129**
Sentra	15,201	—
TOTAL NISSAN	**167,742**	**150,129**
Prizm (Chevrolet)	49,967	49,996
Corolla (Toyota)	160,759	147,741
TOTAL NUMMI[2]	**210,726**	**197,737**
Legacy	93,070	107,955
TOTAL SUBARU-IZUSU	**93,070**	**107,995**
Avalon	71,227	120,252
Camry	285,613	251,625
Cavalier (Chevrolet)	6,111	1,111
TOTAL TOYOTA	**377,988**	**357,951**
GRAND TOTAL	**5,640,030**	**5,542,519**

(1) Company is a joint venture between Ford and Mazda. (2) New United Motor Manufacturing, Inc., is a joint venture between GM and Toyota.

Cars Registered in the U.S., 1900-99[1]

Source: U.S. Dept. of Transportation, Federal Highway Administration

(includes automobiles for public and private use)

Year	Cars Registered
1900	8,000
1905	77,400
1910	458,377
1915	2,332,426
1920	8,131,522
1925	17,481,001
1930	23,034,753
1935	22,567,827
1940	27,465,826
1945	25,796,985
1950	40,339,077
1955	52,144,739
1960	61,671,390
1965	75,257,588
1970	89,243,557
1975	106,705,934
1980	121,600,843
1985	127,885,193
1990	133,700,497
1991	128,299,601
1992	126,581,148
1993	127,327,189
1994	127,883,469
1995	128,386,775
1996	129,728,311
1997	129,748,704
1998	131,838,538
1999	132,432,044

(1) There were no publicly owned vehicles before 1925; statistics also exclude military vehicles for all years. Alaska and Hawaii data included since 1960.

Domestic and Imported Retail Car Sales in the U.S., 1980-99

Source: Ward's Communications

Calendar year	Domestic[1]	IMPORTS: From Japan	From Germany	From other countries	Total imports	Total U.S. sales	Import %: Total	Japan
1980	6,581,307	1,905,968	305,219	186,700	2,397,887	8,979,194	26.7	21.2
1981	6,208,760	1,858,896	282,881	185,502	2,327,279	8,536,039	27.3	21.8
1982	5,758,586	1,801,969	247,080	174,508	2,223,557	7,982,143	27.9	22.6
1983	6,795,295	1,915,621	279,748	191,403	2,386,772	9,182,067	26.0	20.9
1984	7,951,523	1,906,206	344,416	188,220	2,438,842	10,390,365	23.5	18.3
1985	8,204,542	2,217,837	423,983	195,925	2,837,745	11,042,287	25.7	20.1
1986	8,214,897	2,382,614	443,721	418,286	3,244,621	11,459,518	28.3	20.8
1987	7,080,858	2,190,405	347,881	657,465	3,195,751	10,276,609	31.1	21.3
1988	7,526,038	2,022,602	280,099	700,991	3,003,692	10,529,730	28.5	19.2
1989	7,072,902	1,897,143	248,561	553,660	2,699,364	9,772,266	27.6	19.4
1990	6,896,888	1,719,384	265,116	418,823	2,403,323	9,300,211	25.8	18.5
1991	6,136,757	1,500,309	192,776	344,814	2,037,899	8,174,656	24.9	18.4
1992	6,276,557	1,451,766	200,851	283,938	1,936,555	8,213,112	23.6	17.7
1993	6,741,667	1,328,445	186,177	261,570	1,776,192	8,517,859	20.9	15.6
1994	7,255,303	1,239,450	192,241	303,489	1,735,214	8,990,517	19.3	13.8
1995	7,128,712	981,462	207,555	317,269	1,506,257	8,634,964	17.4	11.4
1996	7,253,582	726,940	237,984	308,247	1,273,171	8,526,753	14.9	8.5
1997	6,916,769	726,104	297,028	332,173	1,355,305	8,272,074	16.4	8.8
1998	6,761,940	691,162	366,724	321,895	1,379,781	8,141,721	15.9	8.5
1999	6,979,357	757,568	466,870	494,489	1,718,927	8,698,284	21.1	8.7
2000	6,830,505	862,780	516,614	636,726	2,016,120	8,846,625	22.8	9.8

(1) Includes cars manufactured in Canada and Mexico.

Sport Utility Vehicle Sales in the U.S., 1988-2000

Source: Ward's Communications

In 1988, 960,852 sport utility vehicles (SUVs) were sold in the United States, accounting for 6.3% of all sales of light vehicles (cars, SUVs, minivans, vans, pickup trucks, and trucks under 14,000 lbs.). By 2000, sales of SUVs in the U.S. increased to 2,978,241, accounting for 35.0% of total light vehicle sales.

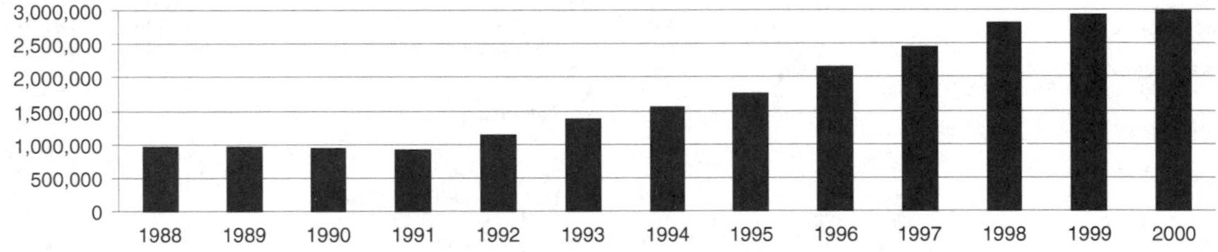

U.S. Light-Vehicle Fuel Efficiency, 1975-1999

Source: Environmental Protection Agency, Office of Mobile Sources

Since 1975, both light-duty trucks (SUVs, minivans, vans, and light trucks) and cars have generally become more fuel-efficient, but their fuel efficiency has declined in recent years. In addition, light-duty trucks, which are less fuel-efficient than cars, have come to occupy an increasing proportion of the total light vehicle market, rising from only 19% in 1975 to an estimated 46% by 1999. This increase has been a major factor in the recent decline in the fuel efficiency of the average light vehicle sold.

YEAR	Cars (MPG*)	Light-duty Trucks (MPG*)	All Light Vehicles (MPG*)	YEAR	Cars (MPG*)	Light-duty Trucks (MPG*)	All Light Vehicles (MPG*)
1975	15.8	13.7	15.3	1995	28.3	20.5	24.7
1980	23.5	18.6	22.5	1996	28.3	20.8	24.8
1985	27.0	20.6	25.0	1997	28.4	20.7	24.5
1990	27.8	20.7	25.2	1998	28.6	20.6	24.4
1994	28.0	20.8	24.6	1999	28.1	20.3	23.8

* MPG value represents city and highway fuel efficiency combined in a 55%/45% ratio.

Top-Selling Passenger Cars in the U.S. by Calendar Year,1996-2000 (Domestic and Import)

Source: Ward's Communications

2000

1. Toyota Camry 422,961
2. Honda Accord 404,515
3. Ford Taurus 382,035
4. Honda Civic 324,528
5. Ford Focus 286,166
6. Chevrolet Cavalier 236,803
7. Toyota Corolla 230,156
8. Pontiac Grand Am 214,923
9. Chevrolet Malibu 207,376
10. Saturn S 177,355
11. Chevrolet Impala 174,358
12. Ford Mustang 173,676
13. Buick Lesabre 148,633
14. Pontiac Grand Prix 148,521
15. Volkswagen Jetta 144,853
16. Dodge Intrepid 143,840
17. Buick Century 143,085
18. Nissan Altima 136,971
19. Nissan Maxima 129,235
20. Oldsmobile Alero 122,722

1999

1. Toyota Camry 448,162
2. Honda Accord 404,192
3. Ford Taurus 368,327
4. Honda Civic 318,308
5. Chevrolet Cavalier 272,122
6. Ford Escort 260,486
7. Toyota Corolla 249,128
8. Pontiac Grand Am 234,936
9. Chevrolet Malibu 218,540
10. Saturn S 207,977
11. Ford Mustang 166,915
12. Buick Century 157,035
13. Nissan Altima 153,525
14. Buick LeSabre 149,445
15. Pontiac Grand Prix 148,197
16. Dodge Intrepid 144,355
17. Ford Contour 134,487
18. Nissan Mazima 131,182
19. Volkswagen Jetta 130,054
20. Mercury Grand Marquis 122,776

1998		1997		1996	
1. Toyota Camry	429,575	1. Toyota Camry	397,156	1. Ford Taurus	401,049
2. Honda Accord	401,071	2. Honda Accord	384,609	2. Honda Accord	382,298
3. Ford Taurus	371,074	3. Ford Taurus	357,162	3. Toyota Camry	359,433
4. Honda Civic	334,562	4. Honda Civic	315,546	4. Honda Civic	286,350
5. Ford Escort	291,936	5. Chevrolet Cavalier	302,161	5. Ford Escort	284,644
6. Chevrolet Cavalier	256,099	6. Ford Escort	283,898	6. Saturn	278,574
7. Toyota Corolla	250,501	7. Saturn	250,810	7. Chevrolet Cavalier	277,222
8. Saturn	231,786	8. Chevrolet Lumina	228,451	8. Chevrolet Lumina	237,973
9. Chevrolet Malibu	223,703	9. Toyota Corolla	218,461	9. Pontiac Grand Am	222,477
10. Pontiac Grand Am	180,428	10. Pontiac Grand Am	204,078	10. Toyota Corolla	209,048

U.S. Car Sales by Vehicle Size and Type, 1983-2000

Source: Ward's Communications

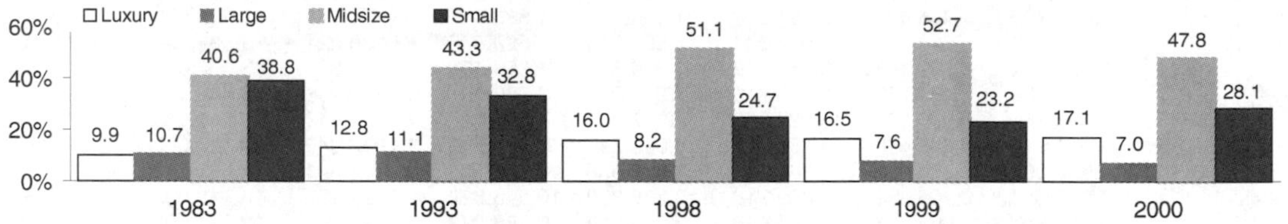

World Motor Vehicle Production, 1950-2000

Source: American Automobile Manufacturers Assn.; for 1998-2000: Automotive News Data Center and Marketing Systems GmbH
(in thousands)

Year	United States	Canada	Europe	Japan	Other	World total	U.S. % of world total
1950	8,006	388	1,991	32	160	10,577	75.7
1960	7,905	398	6,837	482	866	16,488	47.9
1970	8,284	1,160	13,049	5,289	1,637	29,419	28.2
1980	8,010	1,324	15,496	11,043	2,692	38,565	20.8
1985	11,653	1,933	16,113	12,271	2,939	44,909	25.9
1990	9,783	1,928	18,866	13,487	4,496	48,554	20.1
1991	8,811	1,888	17,804	13,245	5,180	46,928	18.8
1992	9,729	1,961	17,628	12,499	6,269	48,088	20.2
1993	10,898	2,246	15,208	11,228	7,205	46,785	23.3
1994	12,263	2,321	16,195	10,554	8,167	49,500	24.8
1995	11,985	2,408	17,045	10,196	8,349	49,983	24.0
1996	11,799	2,397	17,550	10,346	9,241	51,332	23.0
1997	12,119	2,571	17,773	10,975	10,024	53,463	22.7
1998	12,047	2,568	16,332	10,050	12,844	53,841	22.4
1999	13,107	3,042	17,603	9,985	14,050	57,787	22.7
2000	12,855	2,955	17,657	10,196	16,102	59,766	21.5

Note: Data for 1998-2000 may not be fully comparable with earlier years because derived from different source.

Leading Motor Vehicle Producers, 2000

Source: Automotive News Data Center and Marketing Systems GmbH

	Total	Passenger Cars	Trucks		Total	Passenger Cars	Trucks
United States	12,854,585	5,542,519	7,312,066	Poland	726,462	653,140	73,322
Japan	10,196,524	8,363,485	1,833,039	Sweden	511,334	396,170	115,164
Germany	5,526,619	5,131,919	394,700	Czech Republic	455,613	428,205	27,408
France	3,330,036	2,923,093	406,943	Turkey	428,901	295,430	133,471
Korea, South	3,104,666	2,602,008	502,658	Malaysia	391,720	350,100	41,620
Spain	3,052,154	2,385,639	666,515	Taiwan	367,197	255,800	111,397
Canada	2,954,795	1,550,500	1,404,295	Thailand	350,800	84,900	265,900
China	2,044,104	604,677	1,439,427	South Africa	343,191	219,813	123,378
Mexico	1,889,486	1,279,089	610,397	Argentina	339,632	238,921	100,711
United Kingdom	1,808,300	1,621,228	187,072	Australia	316,812	292,500	24,312
Italy	1,742,299	1,422,281	320,015	Netherlands	256,782	215,085	41,697
Brazil	1,671,093	1,347,923	323,170	Portugal	248,847	193,561	55,286
Russia	1,241,070	988,070	253,000	Iran	239,681	220,000	19,681
Belgium	1,033,294	912,233	121,061	Indonesia	237,602	39,500	198,102
India	842,300	715,400	126,900	**World Total[1]**	**57,787,060**	**42,246,535**	**17,519,078**

(1) Totals include countries or territories not shown.

The Most Popular Colors, by Type of Vehicle, 2000 Model Year

Source: Du Pont Automotive Products

Luxury Cars		Full Size/Intermediate Cars		Compact/Sports Cars		Light Trucks	
Color	Percent	Color	Percent	Color	Percent	Color	Percent
White Metallic	19.8	Silver	21.5	Silver	22.3	White	23.1
Silver	17.2	White	13.0	Black	14.4	Silver	14.1
Black	10.8	Black	11.5	White	11.4	Medium/Dark Blue	11.1
Light Brown	8.6	Medium/Dark Green	10.7	Light Brown	9.9	Black	10.6
White	7.1	Light Brown	8.5	Medium/Dark Green	9.7	Medium/Dark Green	8.3
Medium/Dark Blue	7.1	Medium/Dark Blue	7.0	Medium Red	8.3	Medium Red	6.2
Medium/Dark Gray	6.6	Medium Red	6.9	Bright Red	7.5	Bright Red	5.4
Medium Red	6.1	Medium/Dark Gray	4.3	Medium/Dark Blue	5.0	Gold	4.1
Gold	5.3	Bright Red	3.8	Teal	2.6	Medium/DarkGrey	3.9
Medium/Dark Green	4.3	Gold	3.5	Bright Blue	2.1	Dark Red	3.2
Other	7.1	Other	9.3	Other	6.8	Other	10.0

Licensed Drivers, by Age

Source: Federal Highway Administration, U.S. Dept. of Transportation

	1998			1999			1989	Percent change total drivers
Age	Male	Female	Total	Male	Female	Total	Total	1989-99
under 16	17,620	15,825	33,445	17,543	15,706	33,248	38,000	–12.50
16	862,038	808,143	1,670,181	753,624	704,633	1,458,257	1,463,000	–0.32
17	1,215,339	1,137,135	2,352,475	1,200,327	1,130,122	2,330,449	2,293,000	1.63
18	1,423,627	1,322,775	2,746,402	1,434,671	1,332,849	2,767,520	2,806,000	–1.37
19	1,537,666	1,442,593	2,980,260	1,563,198	1,457,470	3,020,668	3,074,000	–1.73
(19 and under)	5,056,291	4,726,471	9,782,763	4,969,362	4,640,780	9,610,142	9,674,000	–0.66
20	1,566,648	1,481,921	3,048,569	1,559,679	1,482,071	3,041,750	3,129,000	–2.79
21	1,525,147	1,455,855	2,981,002	1,561,806	1,490,535	3,052,341	3,201,000	–4.64
22	1,524,782	1,456,491	2,981,273	1,600,041	1,536,503	3,136,544	3,335,000	–5.95
23	1,583,899	1,513,310	3,097,209	1,587,646	1,520,575	3,108,221	3,536,000	–12.10
24	1,665,974	1,592,185	3,258,159	1,630,891	1,559,507	3,190,398	3,702,000	–13.82
(20-24)	7,866,450	7,499,762	15,366,213	7,940,062	7,589,191	15,529,253	16,904,000	–8.13
25-29	9,198,169	8,830,398	18,028,566	9,067,323	8,670,126	17,737,449	20,569,000	–13.77
30-34	9,741,944	9,438,468	19,180,411	9,633,540	9,279,100	18,912,641	20,514,000	–7.81
35-39	10,632,061	10,504,530	21,136,591	10,635,935	10,464,605	21,100,541	18,560,000	13.69
40-44	10,256,084	10,206,632	20,462,716	10,376,809	10,345,322	20,722,131	16,119,000	28.56
45-49	9,101,307	9,063,342	18,164,649	9,328,218	9,288,229	18,616,447	12,584,000	47.94
50-54	7,700,538	7,624,393	15,324,931	8,096,694	8,036,958	16,133,652	10,259,000	57.26
55-59	5,958,581	5,879,875	11,838,456	6,200,918	6,147,189	12,348,107	9,505,000	29.91
60-64	4,755,196	4,691,850	9,447,046	4,858,994	4,804,434	9,663,428	9,313,000	3.76
65-69	4,166,116	4,170,831	8,336,948	4,163,873	4,167,098	8,330,972	8,265,000	0.80
70-74	3,363,722	3,794,340	7,431,062	3,645,458	3,806,964	7,452,422	NA	NA
75-79	2,675,722	2,868,555	5,544,277	2,777,529	3,011,770	5,789,299	NA	NA
80-84	1,520,076	1,658,454	3,178,530	1,582,587	1,748,113	3,330,701	NA	NA
85 and over	839,480	917,538	1,757,018	889,019	1,004,218	1,893,236	NA	NA
TOTAL	**93,104,738**	**91,875,439**	**184,980,177**	**94,166,321**	**93,004,099**	**187,170,420**	**178,843,000**	**8.39**

(1) Comparisons between "licensed" drivers under age 16 in 1989 and in 1999 are not entirely valid because of a change in definition in 1990, which interpreted "licensed" drivers more strictly than before. NA = not available.

IT'S A FACT: According to the National Highway Traffic Safety Administration, there were 16,653 alcohol-related traffic fatalities in 2000, compared to 15,976 in 1999. It was the first increase in such deaths since 1995. In 2000, 40% of all traffic fatalities involved alcohol, up from the historic low of 38% in 1999.

Highway Speed Limits, by State

Source: Insurance Institute for Highway Safety

Under the National Highway System Designation Act, signed Nov. 28, 1995, by Pres. Bill Clinton, states were allowed to set their own highway speed limits, as of Dec. 8, 1995. Under federal legislation enacted in 1974 during the energy crisis, states had been, in effect, restricted to a National Maximum Speed Limit (NMSL) of 55 miles per hour (raised in 1987 to 65 mph on rural interstates). Maximum posted speed limits, in miles per hour, are given by state in the table below. Most data current as of July 1, 2001. For more information visit the Insurance Institute for Highway Safety website at http://www.hwysafety.org

STATE	Rural Interstate	Urban[1] Interstate	Limited[2] Access Roads	Other Roads
AL	70	70	65	65
AK	65	55	65	55
AZ	75	55	55	55
AR	70 (65)	55	60	55
CA	70 (55)	65	70	55
CO	75	65	65	55
CT	65	55	65	55
DE	65	55	65	55
FL	70	65	70	65
GA	70	65	65	65
HI	55	50	45	45
ID	75 (65)	65	65	65
IL	65 (55)	55	65	55
IN	65 (60)	55	55	55
IA	65	55	65	55
KS	70	70	70	65
KY	65	55	55	55
LA	70	55	70	65
ME	65	55	55	55
MD	65	65	65	55
MA	65	65	65	55
MI	70 (55)	65	70	55
MN	70	65	65	55
MS	70	70	70	65
MO	70	60	70	65
MT	75 (65)	65	70[3]	70[3]
NE	75	65	65	60
NV	75	65	70	70
NH	65	65	55	55
NJ	65	55	65	55
NM	75	55	65	55
NY	65	65	65	55
NC	70	65	65	55
ND	70	55	65	65[4]
OH	65 (55)	65	55	55
OK	75	70	70	70
OR	65 (55)	55	55	55
PA	65	55	65	55
RI	65	55	55	55
SC	70	70	60	55
SD	75	65	65	65
TN	70	70	70	55
TX	70	70	70	70
UT	75	65	55	55
VT	65	55	50	50
VA	65	55	65	55
WA	70 (60)	60	55	55
WV	70	55	65	55
WI	65	65	65	55
WY	75	60	65	65

(1) Urban interstates are determined from U.S. Census Bureau criteria, which may be adjusted by state and local governments to reflect planning and other issues. (2) Limited access roads are multiple-lane highways with restricted access via exit and entrance ramps rather than intersections. (3) Speed limit is 65 mph at night. (4) Speed limit is 55 mph at night. **NOTE:** Speeds shown in parentheses are for commercial trucks. "Night" means from one-half hour after sunset to one-half hour before sunrise.

Selected Motor Vehicle Statistics

Source: Federal Highway Administration; U.S. Dept. of Transportation; Insurance Institute for Highway Safety; Driver's license age requirements, state gas tax, and safety belt laws as of August 2001; 1999 figures where not specified.

STATE	Driver's license age requirements: Regular[1]	Driver's license age requirements: Learner's Permit	State gas tax cents/ gal.	Safety belt use law[12]	Licensed drivers per 1,000 resident pop.	Regist. motor vehicles per 1,000 pop.	Licensed drivers per motor vehicle	Gals. of fuel used per vehicle	Miles per gal.	Annual miles driven per vehicle	Vehicle miles per licensed driver
Alabama	16	15	18	P	789	906	0.87	811	17.49	14,193	16,299
Alaska	16	14	8	S	742	922	0.80	650	12.23	7,953	9,894
Arizona	16	15y, 7m	18	S	690	755	0.91	835	15.55	12,986	14,202
Arkansas	16	14	19.5	S	755	712	1.06	1,101	14.62	16,089	15,187
California	17[2]	15	18	P	628	795	0.79	642	17.74	11,382	14,405
Colorado	17	15	22	S	737	951	0.78	639	16.52	10,557	13,620
Connecticut	16y, 4[2]	16	32	P	723	843	0.86	646	16.75	10,819	12,607
Delaware	16y, 10[2]	15y, 10m	23	S	732	817	0.90	737	18.84	13,876	15,481
Dist. of Col.	18[3]	16	20	P	672	454	1.48	816	18.02	14,701	9,923
Florida	18	15	13.1	S	821	754	1.09	762	16.36	12,459	11,443
Georgia	18	15	7.5	P	702	895	0.78	864	16.40	14,178	18,069
Hawaii	16[2]	15y, 6m	16	P	635	606	1.05	581	19.44	11,302	10,783
Idaho	17[4]	14y, 6m	25	S	697	902	0.77	794	15.59	12,372	16,012
Illinois	17[2]	15	19	S	653	771	0.85	681	16.06	10,945	12,920
Indiana	18	15	15	P	649	925	0.70	733	17.38	12,746	18,163
Iowa	17[2]	14	20	P	674	1,063	0.63	682	14.01	9,554	15,057
Kansas	16	14	20	S	713	838	0.85	819	15.20	12,453	14,636
Kentucky	16y, 6m[5]	16	16.4	S	672	672	1.00	1,082	16.61	17,963	17,973
Louisiana	17[6]	155	20	P	632	802	0.79	815	14.43	11,757	14,915
Maine	16[2]	15	19	S	728	731	1.00	918	16.83	15,449	15,513
Maryland	17y, 7m[7]	15y, 9m	23.5	P	618	753	0.82	749	16.84	12,608	15,378
Massachusetts	18	16	21	S	716	864	0.83	579	16.78	9,717	11,721
Michigan	17[2]	14y, 9m	19	P	696	840	0.83	734	15.71	11,538	13,936
Minnesota	17[2]	15	20	S	609	840	0.72	782	16.39	12,821	17,685
Mississippi	16[8]	15	18.4	S[a]	646	837	0.77	950	15.84	15,055	19,506
Missouri	18	15y, 6m	17	S[a]	702	805	0.87	887	17.08	15,152	17,380
Montana	15[9]	14y, 6m	27	S	748	1,130	0.66	694	14.20	9,856	14,898
Nebraska	17	15	22.8	S	722	942	0.77	789	14.54	11,475	14,978
Nevada	16[2]	15y, 6m	24.75	S	731	642	1.14	1,021	14.65	14,962	13,155
New Hampshire	18	15y, 6m	19.5	none	765	875	0.87	736	15.38	11,318	12,948
New Jersey	17y, 6m	16	10.5	P	682	749	0.91	769	13.97	10,739	11,807
New Mexico	16y, 6m[2]	15	18.5	P	702	906	0.78	857	16.56	14,185	18,301
New York	17[2]	16[10]	29.3	P	584	591	0.99	625	18.81	11,760	11,903
North Carolina	16y, 6m	15	21.2	P	718	744	0.96	887	17.38	15,422	15,982
North Dakota	16	14	21	S	723	1,112	0.65	733	14.07	10,309	15,860
Ohio	17[2]	15y, 6m	22	S	715	909	0.79	654	15.76	10,306	13,111
Oklahoma	16	15y, 6m	17	P	689	873	0.79	846	17.16	14,521	18,407
Oregon	17	15	24	P	742	909	0.82	662	17.38	11,510	14,088
Pennsylvania	17[2]	16	25.9	S	707	751	0.94	707	16.01	11,324	12,032
Rhode Island	17y, 6[2]	16	29	S[a]	695	754	0.92	623	17.79	11,088	12,028
South Carolina	16y, 3m	15	16	S[a]	723	779	0.93	942	15.49	14,588	15,709
South Dakota	16	14	22	S	741	1,067	0.70	754	13.98	10,543	15,166
Tennessee	17[12]	15	20	S	762	807	0.94	868	16.85	14,629	15,507
Texas	16y, 6[2, 11]	15	20	P	666	702	0.95	923	16.24	14,989	15,785
Utah	17[8]	15y, 9m	24.5	S	676	741	0.91	819	17.06	13,977	15,313
Vermont	16y, 6m[2]	15	20	S[a]	835	872	0.96	748	17.74	13,267	13,850
Virginia	18[2]	15	17.5	S	688	854	0.81	780	16.14	12,588	15,627
Washington	17[2]	15	23	S	717	845	0.85	675	16.06	10,842	12,767
West Virginia	17	15	25.35	S	705	763	0.92	807	17.11	13,801	14,943
Wisconsin	16y, 9[2]	15y, 6m	25.4	S	711	812	0.88	750	17.80	13,353	15,258
Wyoming	16	15	14	S	758	1,101	0.69	1,311	11.26	14,762	21,452
AVERAGE					**686**	**793**	**0.87**	**757**	**16.44**	**12,442**	**14,379**

NOTE: Many states are moving toward graduated licensing systems that phase in full driving privileges. During the learner's phase, driving generally is not permitted unless there is an adult supervisor. In an intermediate phase, young licensees not yet having unrestricted licenses may be allowed to drive unsupervised under certain conditions but not others. (1) Unrestricted operation of private passenger car. (2) Applicants under age 18 must have completed an approved driver education course. (3) Learner's phase mandatory for all ages. Applicants under age 21 must complete a 6-month intermediate phase. (4) Applicants under age 17 must have completed an approved driver education course. (5) License holders under age 18 must complete a 4-hour course on safe driving within 1 yr. of receiving license (6) Applicants age 17 and older must have completed an educational program, but doesn't require behind-the-wheel training. (7) Initial applicants of any age must have completed an approved driver education course. (8) Applicants age 17 and older not subject to learner's permit and intermediate license requirements. (9) Applicants under age 16 must have completed an approved driver education course. (10) Driving in New York City is prohibited for all licensees age 16 and for those age 17 without driver education. (11) Effective 1/1/2002. (12) P = officer may stop vehicle for a violation (primary); S = an officer may issue seat belt citation only when vehicle is stopped for another moving violation (secondary). (a) Primary enforcement for children under a specified age: MS-8; MO-16; RI-13; SC, VT-19.

Road Mileage Between Selected U.S. Cities

	Atlanta	Boston	Chicago	Cincinnati	Cleveland	Dallas	Denver	Des Moines	Detroit	Houston
Atlanta, Ga.	...	1,037	674	440	672	795	1,398	870	699	789
Boston, Mass.	1,037	...	963	840	628	1,748	1,949	1,280	695	1,804
Chicago, Ill.	674	963	...	287	335	917	996	327	266	1,067
Cincinnati, Oh.	440	840	287	...	244	920	1,164	571	259	1,029
Cleveland, Oh.	672	628	335	244	...	1,159	1,321	652	170	1,273
Dallas Tex.	795	1,748	917	920	1,159	...	781	684	1,143	243
Denver, Col.	1,398	1,949	996	1,164	1,321	781	...	669	1,253	1,019
Detroit, Mich.	699	695	266	259	170	1,143	1,253	584	...	1,265
Houston, Tex.	789	1,804	1,067	1,029	1,273	243	1,019	905	1,265	...
Indianapolis, Ind.	493	906	181	106	294	865	1,058	465	278	987
Kansas City, Mo.	798	1,391	499	591	779	489	600	195	743	710
Los Angeles, Cal.	2,182	2,979	2,054	2,179	2,367	1,387	1,059	1,727	2,311	1,538
Memphis, Tenn.	371	1,296	530	468	712	452	1,040	599	713	561
Milwaukee, Wis.	761	1,050	87	374	422	991	1,029	361	353	1,142
Minneapolis, Minn.	1,068	1,368	405	692	740	936	841	252	671	1,157
New Orleans, La.	479	1,507	912	786	1,030	496	1,273	978	1,045	356
New York, N.Y.	841	206	802	647	473	1,552	1,771	1,119	637	1,608
Omaha, Neb.	986	1,412	459	693	784	644	537	132	716	865
Philadelphia, Pa.	741	296	738	567	413	1,452	1,691	1,051	573	1,508
Pittsburgh, Pa.	687	561	452	287	129	1,204	1,411	763	287	1,313
Portland Ore.	2,601	3,046	2,083	2,333	2,418	2,009	1,238	1,786	2,349	2,205
St. Louis, Mo.	541	1,141	289	340	529	630	857	333	513	779
San Francisco	2,496	3,095	2,142	2,362	2,467	1,753	1,235	1,815	2,399	1,912
Seattle, Wash.	2,618	2,976	2,013	2,300	2,348	2,078	1,307	1,749	2,279	2,274
Tulsa, Okla.	772	1,537	683	736	925	257	681	443	909	478
Washington, DC	608	429	671	481	346	1,319	1,616	984	506	1,375

	Indianapolis	Kansas City	Los Angeles	Louisville	Memphis	Milwaukee	Minneapolis	New Orleans	New York	Omaha
Atlanta, Ga.	493	798	2,182	382	371	761	1,068	479	841	986
Boston, Mass.	906	1,391	2,979	941	1,296	1,050	1,368	1,507	206	1,412
Chicago, Ill.	181	499	2,054	292	530	87	405	912	802	459
Cincinnati, Oh.	106	591	2,179	101	468	374	692	786	647	693
Cleveland Oh.	294	779	2,367	345	712	422	740	1,030	473	784
Dallas, Tex.	865	489	1,387	819	452	991	936	496	1,552	644
Denver, Col.	1,058	600	1,059	1,120	1,040	1,029	841	1,273	1,771	537
Detroit, Mich.	278	743	2,311	360	713	353	671	1,045	637	716
Houston, Tex.	987	710	1,538	928	561	1,142	1,157	356	1,608	865
Indianapolis, Ind.	...	485	2,073	111	435	268	586	796	713	587
Kansas City, Mo.	485	...	1,589	520	451	537	447	806	1,198	201
Los Angeles, Cal.	2,073	1,589	...	2,108	1,817	2,087	1,889	1,883	2,786	1,595
Memphis, Tenn.	435	451	1,817	367	...	612	826	390	1,100	652
Milwaukee, Wis.	268	537	2,087	379	612	...	332	994	889	493
Minneapolis, Minn.	586	447	1,889	697	826	332	...	1,214	1,207	357
New Orleans, La.	796	806	1,883	685	390	994	1,214	...	1,311	1,007
New York, N.Y.	713	1,198	2,786	748	1,100	889	1,207	1,311	...	1,251
Omaha, Neb.	587	201	1,595	687	652	493	357	1,007	1,251	...
Philadelphia, Pa.	633	1,118	2,706	668	1,000	825	1,143	1,211	100	1,183
Pittsburgh, Pa.	353	838	2,426	388	752	539	857	1,070	368	895
Portland, Ore.	2,272	1,809	959	2,320	2,259	2,010	1,678	2,505	2,885	1,654
St. Louis, Mo.	235	257	1,845	263	285	363	552	673	948	449
San Francisco	2,293	1,835	379	2,349	2,125	2,175	1,940	2,249	2,934	1,683
Seattle, Wash.	2,194	1,839	1,131	2,305	2,290	1,940	1,608	2,574	2,815	1,638
Tulsa, Okla.	631	248	1,452	659	401	757	695	647	1,344	387
Washington, DC	558	1,043	2,631	582	867	758	1,076	1,078	233	1,116

	Philadelphia	Pittsburgh	Portland	St. Louis	Salt Lake City	San Francisco	Seattle	Toledo	Tulsa	Wash., DC
Atlanta, Ga.	741	687	2,601	541	1,878	2,496	2,618	640	772	608
Boston, Mass.	296	561	3,046	1,141	2,343	3,095	2,976	739	1,537	429
Chicago, Ill.	738	452	2,083	289	1,390	2,142	2,013	232	683	671
Cincinnati, Oh.	567	287	2,333	340	1,610	2,362	2,300	200	736	481
Cleveland Oh.	413	129	2,418	529	1,715	2,467	2,348	111	925	346
Dallas, Tex.	1,452	1,204	2,009	630	1,242	1,753	2,078	1,084	257	1,319
Denver, Col.	1,691	1,411	1,238	857	504	1,235	1,307	1,218	681	1,616
Detroit, Mich.	576	287	2,349	513	1,647	2,399	2,279	59	909	506
Houston, Tex.	1,508	1,313	2,205	779	1,438	1,912	2,274	1,206	478	1,375
Indianapolis, Ind.	633	353	2,272	235	1,504	2,293	2,194	219	631	558
Kansas City, Mo.	1,118	838	1,809	257	1,086	1,835	1,839	687	248	1,043
Los Angeles, Cal.	2,706	2,426	959	1,845	715	379	1,131	2,276	1,452	2,631
Memphis, Tenn.	1,000	752	2,259	285	1,535	2,125	2,290	654	401	867
Milwaukee, Wis.	825	539	2,010	363	1,423	2,175	1,940	319	757	758
Minneapolis, Minn.	1,143	857	1,678	552	1,186	1,940	1,608	637	695	1,076
New Orleans, La.	1,211	1,070	2,505	673	1,738	2,249	2,574	986	647	1,078
New York, N.Y.	100	368	2,885	948	2,182	2,934	2,815	578	1,344	233
Omaha, Neb.	1,183	895	1,654	449	931	1,683	1,638	681	387	1,116
Philadelphia, Pa.	...	288	2,821	868	2,114	2,866	2,751	514	1,264	133
Pittsburgh, Pa.	288	...	2,535	588	1,826	2,578	2,465	228	984	221
Portland, Ore.	2,821	2,535	...	2,060	767	636	172	2,315	1,913	2,754
St. Louis, Mo.	868	588	2,060	...	1,337	2,089	2,081	454	396	793
San Francisco	2,866	2,578	636	2,089	752	...	808	2,364	1,760	2,799
Seattle, Wash.	2,751	2,465	172	2,081	836	808	...	2,245	1,982	2,684
Tulsa, Okla.	1,264	984	1,913	396	1,172	1,760	1,982	850	...	1,189
Washington, DC	133	221	2,754	793	2,047	2,799	2,684	447	1,189	...

Air Distances Between Selected World Cities in Statute Miles

Point-to-point measurements are usually from City Hall.

	Bangkok	Beijing	Berlin	Cairo	Cape Town	Caracas	Chicago	Hong Kong	Honolulu	Lima
Bangkok	...	2,046	5,352	4,523	6,300	10,555	8,570	1,077	6,609	12,244
Beijing	2,046	...	4,584	4,698	8,044	8,950	6,604	1,217	5,077	10,349
Berlin	5,352	4,584	...	1,797	5,961	5,238	4,414	5,443	7,320	6,896
Cairo	4,523	4,698	1,797	...	4,480	6,342	6,141	5,066	8,848	7,726
Cape Town	6,300	8,044	5,961	4,480	...	6,366	8,491	7,376	11,535	6,072
Caracas	10,555	8,950	5,238	6,342	6,366	...	2,495	10,165	6,021	1,707
Chicago	8,570	6,604	4,414	6,141	8,491	2,495	...	7,797	4,256	3,775
Hong Kong	1,077	1,217	5,443	5,066	7,376	10,165	7,797	...	5,556	11,418
Honolulu	6,609	5,077	7,320	8,848	11,535	6,021	4,256	5,556	...	5,947
London	5,944	5,074	583	2,185	5,989	4,655	3,958	5,990	7,240	6,316
Los Angeles	7,637	6,250	5,782	7,520	9,969	3,632	1,745	7,240	2,557	4,171
Madrid	6,337	5,745	1,165	2,087	5,308	4,346	4,189	6,558	7,872	5,907
Melbourne	4,568	5,643	9,918	8,675	6,425	9,717	9,673	4,595	5,505	8,059
Mexico City	9,793	7,753	6,056	7,700	8,519	2,234	1,690	8,788	3,789	2,639
Montreal	8,338	6,519	3,740	5,427	7,922	2,438	745	7,736	4,918	3,970
Moscow	4,389	3,607	1,006	1,803	6,279	6,177	4,987	4,437	7,047	7,862
New York	8,669	6,844	3,979	5,619	7,803	2,120	714	8,060	4,969	3,639
Paris	5,877	5,120	548	1,998	5,786	4,732	4,143	5,990	7,449	6,370
Rio de Janeiro	9,994	10,768	6,209	6,143	3,781	2,804	5,282	11,009	8,288	2,342
Rome	5,494	5,063	737	1,326	5,231	5,195	4,824	5,774	8,040	6,750
San Francisco	7,931	5,918	5,672	7,466	10,248	3,902	1,859	6,905	2,398	4,518
Singapore	883	2,771	6,164	5,137	6,008	11,402	9,372	1,605	6,726	11,689
Stockholm	5,089	4,133	528	2,096	6,423	5,471	4,331	5,063	6,875	7,166
Tokyo	2,865	1,307	5,557	5,958	9,154	8,808	6,314	1,791	3,859	9,631
Warsaw	5,033	4,325	322	1,619	5,935	5,559	4,679	5,147	7,366	7,215
Washington, DC	8,807	6,942	4,181	5,822	7,895	2,047	596	8,155	4,838	3,509

	London	Los Angeles	Madrid	Melbourne	Mexico City	Montreal	Moscow	New Delhi	New York	Paris
Bangkok	5,944	7,637	6,337	4,568	9,793	8,338	4,389	1,813	8,669	5,877
Beijing	5,074	6,250	5,745	5,643	7,753	6,519	3,607	2,353	6,844	5,120
Berlin	583	5,782	1,165	9,918	6,056	3,740	1,006	3,598	3,979	548
Cairo	2,185	7,520	2,087	8,675	7,700	5,427	1,803	2,758	5,619	1,998
Cape Town	5,989	9,969	5,308	6,425	8,519	7,922	6,279	5,769	7,803	5,786
Caracas	4,655	3,632	4,346	9,717	2,234	2,438	6,177	8,833	2,120	4,732
Chicago	3,958	1,745	4,189	9,673	1,690	745	4,987	7,486	714	4,143
Hong Kong	5,990	7,240	6,558	4,595	8,788	7,736	4,437	2,339	8,060	5,990
Honolulu	7,240	2,557	7,872	5,505	3,789	4,918	7,047	7,412	4,969	7,449
London	...	5,439	785	10,500	5,558	3,254	1,564	4,181	3,469	214
Los Angeles	5,439	...	5,848	7,931	1,542	2,427	6,068	7,011	2,451	5,601
Madrid	785	5,848	...	10,758	5,643	3,448	2,147	4,530	3,593	655
Melbourne	10,500	7,931	10,758	...	8,426	10,395	8,950	6,329	10,359	10,430
Mexico City	5,558	1,542	5,643	8,426	...	2,317	6,676	9,120	2,090	5,725
Montreal	3,254	2,427	3,448	10,395	2,317	...	4,401	7,012	331	3,432
Moscow	1,564	6,068	2,147	8,950	6,676	4,401	...	2,698	4,683	1,554
New York	3,469	2,451	3,593	10,359	2,090	331	4,683	7,318	...	3,636
Paris	214	5,601	655	10,430	5,725	3,432	1,554	4,102	3,636	...
Rio de Janeiro	5,750	6,330	5,045	8,226	4,764	5,078	7,170	8,753	4,801	5,684
Rome	895	6,326	851	9,929	6,377	4,104	1,483	3,684	4,293	690
San Francisco	5,367	347	5,803	7,856	1,887	2,543	5,885	7,691	2,572	5,577
Singapore	6,747	8,767	7,080	3,759	10,327	9,203	5,228	2,571	9,534	6,673
Stockholm	942	5,454	1,653	9,630	6,012	3,714	716	3,414	3,986	1,003
Tokyo	5,959	5,470	6,706	5,062	7,035	6,471	4,660	3,638	6,757	6,053
Warsaw	905	5,922	1,427	9,598	6,337	4,022	721	3,277	4,270	852
Washington, DC	3,674	2,300	3,792	10,180	1,885	489	4,876	7,500	205	3,840

	Rio de Janeiro	Rome	San Francisco	Singapore	Stockholm	Tehran	Tokyo	Vienna	Warsaw	Wash., DC
Bangkok	9,994	5,494	7,931	883	5,089	3,391	2,865	5,252	5,033	8,807
Beijing	10,768	5,063	5,918	2,771	4,133	3,490	1,307	4,648	4,325	6,942
Berlin	6,209	737	5,672	6,164	528	2,185	5,557	326	322	4,181
Cairo	6,143	1,326	7,466	5,137	2,096	1,234	5,958	1,481	1,619	5,822
Cape Town	3,781	5,231	10,248	6,008	6,423	5,241	9,154	5,656	5,935	7,895
Caracas	2,804	5,195	3,902	11,402	5,471	7,320	8,808	5,372	5,559	2,047
Chicago	5,282	4,824	1,859	9,372	4,331	6,502	6,314	4,698	4,679	596
Hong Kong	11,009	5,774	6,905	1,605	5,063	3,843	1,791	5,431	5,147	8,155
Honolulu	8,288	8,040	2,398	6,726	6,875	8,070	3,859	7,632	7,366	4,838
London	5,750	895	5,367	6,747	942	2,743	5,959	771	905	3,674
Los Angeles	6,330	6,326	347	8,767	5,454	7,682	5,470	6,108	5,922	2,300
Madrid	5,045	851	5,803	7,080	1,653	2,978	6,706	1,128	1,427	3,792
Melbourne	8,226	9,929	7,856	3,759	9,630	7,826	5,062	9,790	9,598	10,180
Mexico City	4,764	6,377	1,887	10,327	6,012	8,184	7,035	6,320	6,337	1,885
Montreal	5,078	4,104	2,543	9,203	3,714	5,880	6,471	4,009	4,022	489
Moscow	7,170	1,483	5,885	5,228	716	1,532	4,660	1,043	721	4,876
New York	4,801	4,293	2,572	9,534	3,986	6,141	6,757	4,234	4,270	205
Paris	5,684	690	5,577	6,673	1,003	2,625	6,053	645	852	3,840
Rio de Janeiro	...	5,707	6,613	9,785	6,683	7,374	11,532	6,127	6,455	4,779
Rome	5,707	...	6,259	6,229	1,245	2,127	6,142	477	820	4,497
San Francisco	6,613	6,259	...	8,448	5,399	7,362	5,150	5,994	5,854	2,441
Singapore	9,785	6,229	8,448	...	5,936	4,103	3,300	6,035	5,843	9,662
Stockholm	6,683	1,245	5,399	5,936	...	2,173	5,053	780	494	4,183
Tokyo	11,532	6,142	5,150	3,300	5,053	4,775	...	5,689	5,347	6,791
Warsaw	6,455	820	5,854	5,843	494	1,879	5,689	347	...	4,472
Washington, DC	4,779	4,497	2,441	9,662	4,183	6,341	6,791	4,438	4,472	...

EDUCATION

Historical Overview of U.S. Public Elementary and Secondary Schools

Source: National Center for Education Statistics, U.S. Dept. of Education

	1899-1900	1919-20	1939-40	1959-60	1969-70	1979-80	1989-90	1997-98	1998-99
Population statistics (thousands)									
Total U.S. population[1]	75,995	104,514	131,028	177,830	201,385	224,567	246,819	267,744	270,248
Population 5-17 years of age	21,573	27,571	30,151	43,881	52,386	48,041	44,947	50,490	50,915
Percentage 5-17 years of age	28.4	26.4	23.0	24.7	26.0	21.4	18.2	18.9	18.8
Enrollment (thousands)									
Elementary and secondary[2]	15,503	21,578	25,434	36,087	45,550	41,651	40,543	46,327	46,539
Kindergarten & grades 1-8	14,984	19,378	18,833	27,602	32,513	28,034	29,152	33,073	33,346
Grades 9-12	519	2,200	6,601	8,485	13,037	13,616	11,390	13,054	13,193
Percentage pop. 5-17 enrolled	71.9	78.3	84.4	82.2	87.0	86.7	90.2	91.4	91.4
Percentage in high schools	3.3	10.2	26.0	23.5	28.6	32.7	28.1	28.3	28.3
High school graduates (thousands)	62	231	1,143	1,627	2,589	2,748	2,320	2,456	2,489
School term; staff									
Average school term (in days)	144.3	161.9	175.0	178.0	178.9	178.5	*	*	*
Total instructional staff (thousands)	*	678	912	1,457	2,286	2,406	2,986	3,573	3,694
Teachers, librarians, and other nonsupervisory instructional staff (thousands)	423	657	875	1,393	2,195	2,300	2,860	3,447	3,564
Revenue and expenditures (millions)									
Total revenue	$220	$970	$2,261	$14,747	$40,267	$96,881	$208,548	$325,976	$347,330
Total expenditures	215	1,036	2,344	15,613	40,683	95,962	212,770	334,296	355,859
Current expenditures[3]	180	861	1,942	12,329[6]	34,218[6]	86,984[6]	188,229[6]	285,473[6]	302,874[6]
Capital outlay	35	154	258	2,662	4,659	6,506	17,781	36,168	39,527
Interest on school debt	*	18	131	490	1,171	1,874	3,776	7,779	6,196
Others	*	3	13	133	636	598	2,983	4,885	5,263
Salaries and pupil cost									
Avg. annual salary of instruct. staff[4]	$325	$871	$1,441	$5,174	$9,047	$16,715	$32,638	$41,243	$42,488
Expenditure per capita total pop.	2.83	9.91	17.89	88	202	427	862	1,249	1,317
Current expenditure per pupil ADA[5]	16.67	53.32	88.09	375	816	2,272	4,980	6,662	7,013

NOTE: Because of rounding, details may not add to totals. Prior to 1959-60, data do not include Alaska and Hawaii. * = Data not collected. (1) Population data for 1899-1900 are based on total population from the decennial census. From 1919-20 to 1959-60, population data are total population, including armed forces overseas, as of July 1 preceding the school year. Data for later years are for resident population that excludes armed forces overseas. (2) Data for 1899-1900 are school year enrollment; data for later years are fall enrollment. (3) In 1899-1900, includes interest on school debt. (4) Includes supervisors, principals, teachers, and nonsupervisory instructional staff. (5) ADA means average daily attendance. (6) Because of changes in the definition of "current expenditures," data for 1959-60 and later years are not entirely comparable with prior years.

Programs for the Disabled, 1990-2000

Source: Office of Special Education and Rehabilitative Services, U.S. Dept. of Education

(Number of children from 6 to 21 years old served annually in educational programs for the disabled; in thousands)

Type of Disability	1990-91	1992-93	1993-94	1994-95	1995-96	1996-97	1997-98	1998-99	1999-2000
Learning disabilities	2,144	2,366	2,428	2,510	2,602	2,674	2,754	2,817	2,834
Speech impairments	988	998	1,018	1,020	1,027	1,049	1,064	1,075	1,081
Mental retardation	551	532	554	571	586	594	603	611	600
Emotional disturbance	391	402	415	428	439	446	454	463	469
Multiple disabilities	98	103	110	90	95	99	107	108	111
Hearing impairments	59	61	65	65	68	69	70	71	71
Orthopedic impairments	49	53	57	60	63	66	67	69	71
Other health impairments	56	66	83	107	134	161	191	221	253
Visual impairments	24	24	25	25	25	26	26	26	65
Autism	NA	16	19	23	29	34	43	54	65
Deaf-blindness	2	1	1	1	1	1	1	2	2
Traumatic brain injury	NA	4	5	7	10	10	12	13	14
ALL DISABILITIES	**4,362**	**4,626**	**4,779**	**4,908**	**5,079**	**5,231**	**4,397**	**5,541**	**5,614**

NOTE: Counts are based on reports from the 50 states and the District of Columbia. Details may not add to totals because of rounding and/or incomplete enumeration. NA = not available or unreliable because of incomplete reporting.

Technology in U.S. Public Schools, 2001

Source: Quality Education Data, Inc., Denver, CO

(Number and percentage of schools in each category that have the technology indicated.)

	Elementary[1]		Middle/Jr. high[2]		Senior high[3]		K-12[4]		Special Ed./ Adult Ed.	
TOTAL SCHOOLS	**51,256**	**100%**	**14,889**	**100%**	**17,937**	**100%**	**2,436**	**100%**	**2,423**	**100%**
Schools with computers	**48,331**	**94**	**13,623**	**91**	**16,336**	**91**	**2,066**	**85**	**1,709**	**70**
By number of computers:										
1-10	3,942	8	437	3	835	5	127	5	613	25
11-20	5,867	11	855	6	1,123	6	229	9	317	13
21-50	17,143	33	3,420	23	3,323	19	641	26	445	18
51-100	14,969	29	4,517	30	3,959	22	629	26	204	8
100+	6,410	13	4,394	30	7,096	40	440	18	130	5
Schools with LANs[5]	**25,346**	**49**	**8,826**	**59**	**11,619**	**65**	**1,342**	**55**	**409**	**17**
By enrollment:										
100-299	4,869	9	1,047	7	2,161	12	535	22	206	9
300-499	9,017	18	1,837	12	1,925	11	381	16	66	3
500+	11,460	22	5,942	40	7,533	42	426	17	137	6
Schools with WANs[6]	**9,716**	**19**	**3,286**	**22**	**4,135**	**23**	**322**	**13**	**128**	**5**
By enrollment:										
100-299	1,468	3	309	2	515	2	110	5	69	3
300-499	3,381	7	617	4	489	3	89	4	21	1
500+	4,867	9	2,360	16	3,131	17	123	5	38	2

(1) Includes preschool and schools with grade spans of K-3, K-5, K-6, K-8, and K-12. (2) Includes schools with grade spans of 4-8, 7-8, and 7-9. (3) Includes vocational, technical, and alternative high schools and schools with grade spans of 7-12, 9-12, and 10-12. (4) K-12 also included under Elementary schools. (5) LAN=Local area computer network. (6) WAN=Wide area computer network.

Students Per Computer in U.S. Public Schools

Source: Quality Education Data, Inc., Denver, CO, *Technology in Public Schools, 1983-2000*

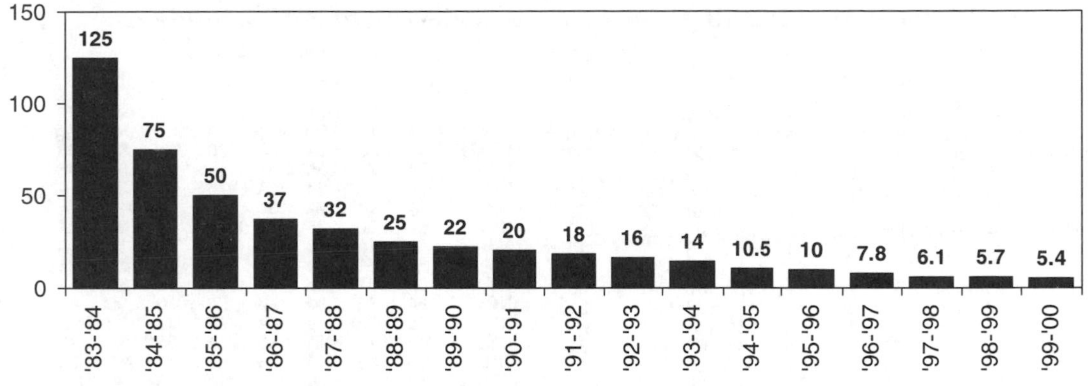

Overview of U.S. Public Schools, Fall 1999*

Source: National Center for Education Statistics, U.S. Dept. of Education; National Education Association

	Local school districts	Elementary schools[1]	Secondary schools[2]	Classroom teachers	Total enrollment	Pupils per teacher	Teacher's avg. pay[3]	Expend. per pupil[4]
AL.	128	913	410	48,614	740,732	15.2	$37,956	$5,512
AK	53	201	90	7,838	134,391	17.1	46,986	9,209
AZ.	413	1,086	403	43,892	852,612	19.4	36,302	5,235
AR	310	707	409	31,362	451,034	14.4	34,476	5,193
CA	987	6,181	2,057	287,344	6,038,589	21.0	48,923	6,045
CO	176	1,143	373	40,772	708,109	17.4	39,284	6,386
CT	166	821	212	39,907	553,933	13.9	52,100	9,620
DE	19	124	43	7,318	112,836	15.4	47,047	8,336
DC	1	132	41	4,779	77,194	16.2	48,651	10,611
FL.	67	2,145	471	130,336	2,381,396	18.3	37,824	6,443
GA	180	1,507	317	90,638	1,422,762	15.7	42,216	6,534
HI	1	196	50	10,866	185,860	17.1	41,980	6,648
ID	113	410	226	13,641	245,331	18.0	36,375	5,379
IL	896	3,169	1,021	124,815	2,027,600	16.2	48,053	7,676
IN	295	1,410	463	58,864	988,702	16.8	43,055	7,249
IA	375	1,057	437	33,480	497,301	14.9	36,479	6,548
KS	304	1,010	425	32,969	472,188	14.3	39,432	6,708
KY	176	1,015	441	41,954	648,180	15.4	37,234	6,412
LA.	75	1,030	336	50,031	756,579	16.6	34,253	6,019
ME	283	539	163	16,349	209,253	12.8	36,256	7,688
MD	24	1,071	244	50,995	846,582	16.6	44,997	7,865
MA	351	1,510	348	77,596	971,425	12.5	47,523	8,750
MI.	737	2,646	866	96,111	1,725,617	18.0	49,975	8,142
MN	407	1,252	717	56,010	854,034	15.2	40,577	7,159
MS	152	576	315	30,722	500,716	16.3	32,957	4,871
MO.	525	1,548	646	63,890	914,110	14.3	36,764	6,393
MT	455	517	365	10,353	157,556	15.2	32,930	6,768
NE	593	974	350	20,766	288,261	13.9	34,175	6,856
NV	17	354	117	17,380	325,610	18.7	40,172	5,934
NH	179	425	96	14,037	206,783	14.7	38,303	6,780
NJ.	604	1,855	438	95,883	1,289,256	13.4	53,281	10,748
NM	89	552	194	19,797	324,495	16.4	33,785	5,363
NY	707	3,081	940	202,078	2,887,776	14.3	50,920	10,514
NC	120	1,688	370	81,914	1,275,925	15.6	41,167	6,088
ND	231	343	212	8,150	112,751	13.8	30,891	5,820
OH	708	2,775	988	116,200	1,836,554	15.8	42,716	7,295
OK	544	1,219	588	41,498	627,032	15.1	34,434	5,684
OR	197	924	264	27,803	545,033	19.6	42,333	7,787
PA.	501	2,383	789	114,525	1,816,716	15.9	49,500	8,026
RI	36	258	56	11,041	156,454	14.2	48,474	9,049
SC	90	811	282	45,468	666,780	14.7	37,327	6,003
SD	176	466	288	9,384	131,037	14.0	30,265	5,613
TN	139	1,172	368	60,702	916,202	15.1	37,074	5,521
TX.	1,041	4,899	1,995	267,935	3,991,783	14.9	38,614	6,161
UT	40	509	250	21,832	480,255	22.0	36,049	4,478
VT.	287	284	71	8,474	104,559	12.3	38,651	7,984
VA.	135	1,438	399	81,073	1,133,994	14.0	40,197	6,129
WA	296	1,386	588	50,368	1,003,714	19.9	42,101	6,595
WV	55	620	193	21,082	291,811	13.8	35,764	7,176
WI.	426	1,532	533	60,778	877,753	14.4	41,646	8,062
WY	48	267	107	6,940	92,105	13.3	34,189	7,393
TOTAL U.S.	**14,928**	**64,131**[5]	**22,365**[5]	**2,906,554**	**48,857,321**	**16.1**	**$42,898**	**$7,013**

*Full-time elementary and secondary day schools only. (1) Includes schools below grade 9. (2) Includes schools with no grade lower than 7. (3) National Education Association estimate, Fall 2000. (4) Fall 1998. (5) Includes schools operated by the Bureau of Indian Affairs (BIA).

IT'S A FACT: In 2000, 57% of school-age children (ages 6 to 17 years) had access to a computer both at home and at school, while 23% had access only at school and 10% only at home.

Mathematics, Reading, and Science Achievement of U.S. Students

Source: National Assessment of Educational Progress, National Center for Education Statistics, U.S. Dept. of Education

Percent of students who scored at or above basic level in most recent national tests.[1]

	GRADE 4				Grade 8				
	Math.		Reading		Math.		Reading		Science
STATE[2]	'96	2000	'94	'98	'96	2000	'94	'98	'96
AL....	48	57	23	24	45	52	23	21	18
AK....	65	NA	NA	NA	68	52	NA	NA	31
AZ....	57	59	24	22	57	62	24	28	23
AR....	54	57	24	23	52	52	24	23	22
CA....	46	53	18	20	51	52	18	22	20
CO....	67	NA	28	34	67	NA	28	30	32
CT....	75	77	38	46	70	72	38	42	36
DE....	54	NA	23	25	55	NA	23	25	21
DC....	20	25	NA	10	20	23	NA	12	5
FL....	55	NA	23	23	54	NA	23	23	21
GA....	53	58	26	24	51	56	26	25	21
HI.....	53	55	19	17	51	52	19	19	15
IN.....	72	79	33	NA	68	76	33	NA	30
IA.....	74	78	35	35	78	NA	35	NA	36
KS....	NA	76	NA	34	NA	77	NA	35	NA
KY....	60	60	26	29	56	63	26	29	23
LA....	44	57	15	19	38	48	15	18	13
ME....	75	74	41	36	77	76	41	42	41
MD....	59	61	26	29	57	64	26	31	25
MA....	71	78	36	37	68	76	36	36	37
MI....	68	72	NA	28	67	70	NA	NA	32
MN....	76	78	33	36	75	80	33	37	37
MS....	42	45	18	18	36	41	18	19	12
MO....	66	73	31	29	64	66	31	29	28
MT...	71	73	35	37	75	72	35	38	41
NE....	70	67	34	NA	76	74	34	NA	35
NH...	NA	NA	NA	38	NA	NA	NA	NA	NA
NV....	57	60	36	21	NA	58	36	24	NA
NJ....	68	NA	33	NA	NA	NA	33	NA	NA
NM...	51	51	21	22	51	49	21	24	19
NY....	64	67	27	29	61	68	27	34	27
NC...	64	76	30	28	56	70	30	31	24
ND...	75	75	38	NA	77	77	38	NA	41
OK...	NA	70	NA	30	NA	65	NA	29	NA
OR...	65	68	NA	28	67	72	NA	33	32
PA....	68	NA	30	NA	NA	NA	30	NA	NA
RI....	61	67	32	32	60	65	32	30	26
SC....	48	60	20	22	48	54	20	22	17
TN....	58	60	27	25	53	53	27	26	22
TX....	69	77	26	29	59	69	26	28	23
UT....	69	77	30	28	70	68	30	31	32
VT....	67	74	NA	NA	72	75	NA	NA	34
VA....	62	72	26	30	58	68	26	33	27
WA...	67	NA	27	29	67	NA	27	32	27
WV...	63	67	26	29	54	62	26	27	21
WI....	74	NA	35	34	75	NA	35	33	39
WY...	64	73	32	30	68	70	32	29	34
U.S.[2]	**62**	**66**	**28**	**29**	**61**	**64**	**28**	**31**	**27**

NA = Not administered. (1) Basic level denotes a partial mastery of prerequisite knowledge and skills fundamental for proficient work at each grade. (2) Only states that chose to participate are included.

Revenues[1] for Public Elementary and Secondary Schools, by State, 2000-01

Source: National Education Association; estimated; in thousands

		Federal		State		Local and intermediate	
STATE	Total	Amount	%	Amount	%	Amount	%
Alabama	$4,389,234*	$452,405*	10.3*	$2,784,764	63.4*	$1,52,065	26.2*
Alaska	1, 231,902*	154,507*	12.5*	782,849*	63.5*	294,546*	23.9*
Arizona	4,551,781*	315,152*	6.9	2,304,498*	50.6*	1,932,130*	42.4*
Arkansas	2,777,874*	217,672*	7.8*	1,717,889*	61.8*	842,314*	30.3*
California	46,308,429*	3,521,313*	7.6*	29,563,075*	63.8*	13,224,041*	28.6*
Colorado	4,439,094*	247,840*	5.6*	1,988,846*	44.8*	2,202,407*	49.6*
Connecticut	6,353,956	299,799	4.7	2,712,957	42.7	3,341,200	52.6
Delaware	1,070,244	81,144*	7.6*	700,448*	65.4*	288,633*	27.0*
District of Columbia	791,680*	142,527*	18.0*	0	0	649,153*	82.0*
Florida	19,262,225	1,606,911*	8.3*	9,921,253*	51.5*	7,734,061*	40.2
Georgia	9,575,255*	631,877*	6.6*	4,904,331*	51.2*	4,039,018*	42.2*
Hawaii	1,397,256*	145,926*	10.4*	1,218,724*	87.2*	32,607	2.3*
Idaho	1,604,900	115,900	7.2	962,600	60.0	526,400	32.8
Illinois	15,590,487*	1,105,067*	7.1*	4,474,743*	28.7*	10,010,678*	64.2*
Indiana	8,731,931*	405,008*	4.6*	4,419,087*	50.6*	3,907,836*	44.8*
Iowa	3,657,762	151,040	4.1	1,933,454	52.9	1,573,268	43.0
Kansas	3,375,037	211,886	6.3	2,252,698	66.7	910,453	27.0
Kentucky	4,938,773*	379,743*	7.7*	3,107,763	62.9*	1,451,267*	29.4*
Louisiana	5,057,541*	577,797	11.4	2,569,914	50.8	1,909,830	37.8
Maine	1,733,290	106,902	6.2	802,155	46.3	824,233	47.6
Maryland	7,476,803*	398,387*	5.3*	2,995,166	40.1*	4,083,250*	54.6*
Massachusetts	9,873,691*	501,729*	5.1*	4,321,112*	43.8*	5,050,850*	51.2*
Michigan	15,841,426*	1,117,459*	7.1*	11,072,818*	69.9	3,651,149*	23.0*
Minnesota	7,603,474	366,421	4.8	4,505,724	59.3	22,731,329	35.9
Mississippi	2,714,220*	385,359*	14.2*	1,463,639*	53.9*	865,222*	31.9*
Missouri	6,979,365*	497,528*	7.1*	2,589,120*	37.1*	3,892,717*	55.8*
Montana	1,092,500	121,000	11.1	490,000	44.9	481,500	44.1
Nebraska	1,994,150	104,806	5.3	862,773	43.3	1,026,571*	51.5
Nevada	2,375,188	115,993*	4.9	734,843	30.9*	1,524,352	64.2*
New Hampshire	1,579,219*	62,928*	4.0*	147,218*	9.3*	1,369,074*	86.7*
New Jersey	12,992,712*	400,082*	3.1*	4,898,206*	37.7*	12,992,712*	59.2*
New Mexico	2,240,073	289,281	12.9	1,666,540	74.4	284,252	12.7
New York	31,650,180*	2,152,434*	6.8*	14,518,993*	45.9*	14,978,753*	47.3*
North Carolina	8,589,106	657,129	7.7	6,015,005	70.0	1,916,972*	22.3
North Dakota	718,959	84,171	11.7	283,612	39.4	351,177	48.8
Ohio	13,997,052	822,440	5.9	6,120,511	43.7	7,054,101	50.4
Oklahoma	4,151,405*	360,950*	8.7*	2,552,028*	61.5*	1,238,427*	29.8*
Oregon	4,385,277	289,142*	6.6*	2,547,025	58.1*	1,549,110*	35.3*
Pennsylvania	15,987,552*	880,755*	5.5*	6,519,199*	40.8*	8,587,599*	53.7*
Rhode Island	1,261,555*	49,525	3.9*	464,516*	36.8*	747,514*	59.3
South Carolina	4,916,009	386,513	7.9	2,483,734	50.5	2,045,762	41.6
South Dakota	862,886*	87,423*	10.1*	335,216*	38.8*	440,248*	51.0*

STATE	Total	Federal Amount	%	State Amount	%	Local and intermediate Amount	%
Tennessee	5,235,713*	446,453*	8.5*	2,574,893*	49.2*	2,214,367*	42.3*
Texas	29,208,423*	2,471,442*	8.5*	12,710,951*	43.5	14,026,030*	48.0*
Utah	2,561,504	181,636*	7.1*	1,602,892*	62.6*	776,976*	30.3*
Vermont	947,810*	48,054*	5.1*	820,872*	86.6*	78,884*	8.3*
Virginia	6,906,415*	381,752*	5.5*	2,622,001*	38.0*	3,902,662*	56.5*
Washington	7,714,260*	601,279*	7.8*	4,989,363*	64.7*	2,123,618*	27.5*
West Virginia	2,510,599*	235,764*	9.4*	1,571,104*	62.6*	703,732*	28.0*
Wisconsin	8,268,958	474,157	5.7	4,361,329	52.7	3,433,472	41.5
Wyoming	781,000	58,500	7.5	398,000	51.0	324,500	41.5
TOTAL U.S.	**$334,641,880**	**$23,003,983**	**6.9**	**$167,185,881**	**50.0**	**$144,452,016**	**43.2**

*Indicates NEA estimate. (1) Included as revenue receipts are all appropriations from general funds of federal, state, county, and local governments; receipts from taxes levied for school purposes; income from permanent school funds and endowments; and income from leases of school lands and miscellaneous sources (interest on bank deposits, tuition, gifts, school lunch charges, etc.).

Enrollment in Public and Private Schools, 1899-2010

Source: National Center for Education Statistics, U.S. Dept. of Education

School year[1]	Public school[2] enrollment	Private school[2] enrollment	% Private
1899-1900	15,503	1,352	8.7
1909-10	17,814	1,558	8.7
1919-20	21,578	1,699	7.9
1929-30	25,678	2,651	10.3
1939-40	25,434	2,611	10.3
1949-50	25,111	3,380	13.5
1959-60	35,182	5,675	16.1
1969-70	45,550	5,500[3]	12.1
1979-80	41,651	5,000[3]	12.0
1989-90	40,543	5,198	11.4
1999-2000[4]	46,857	6,018	11.4
2000-2001[4]	47,051	5,851	11.1
2009-2010[4]	47,178	5,836	11.0

(1) Fall enrollment. (2) In thousands. (3) Estimated. (4) Projected.

U.S. Public High School Graduation Rates, 1998-99

Source: National Center for Education Statistics, U.S. Dept. of Education

	Rate (%)[1]	Rank
Alabama	61.0	39
Alaska	66.8	35
Arizona	60.8	41
Arkansas	72.7	25
California	68.3	34
Colorado	72.0	28
Connecticut	71.8	30
Delaware	64.1	37
District of Columbia	52.9	51
Florida	57.8	47
Georgia	55.8	49
Hawaii	64.0	38
Idaho	78.0	10
Illinois	75.4	14
Indiana	71.2	31
Iowa	83.2	5
Kansas	74.5	17
Kentucky	65.7	36
Louisiana	57.0	48
Maine	75.6	13
Maryland	72.6	26
Massachusetts	75.0	15
Michigan	72.8	24
Minnesota	84.7	2
Mississippi	60.0	45
Missouri	72.4	27
Montana	78.5	8
Nebraska	87.9	1
Nevada	73.7	21
New Hampshire	73.0	23
New Jersey	79.6	7
New Mexico	60.2	44
New York	61.0	39
North Carolina	60.5	43
North Dakota	84.5	3
Ohio	69.5	32
Oklahoma	73.1	22
Oregon	74.0	20
Pennsylvania	74.9	16
Rhode Island	68.7	33
South Carolina	55.5	50
South Dakota	71.9	29
Tennessee	59.9	46
Texas	60.6	42
Utah	83.7	4
Vermont	80.5	6
Virginia	74.3	18
Washington	74.1	19
West Virginia	75.8	12
Wisconsin	78.1	9
Wyoming	76.7	11
TOTAL U.S.	**68.1**	

NOTE: Data exclude ungraded pupils and have not been adjusted for interstate migration. (1) Graduates as percentage of fall 1995 9th-grade enrollment.

Teachers' Salaries in Upper Secondary Education, Selected Countries, 1999

Source: Organization for Economic Cooperation and Development

Annual statutory teachers' salaries in public institutions in upper secondary (senior high school) education, general programs, in equivalent U.S. dollars converted using PPPs[1]

	Starting salary	Salary with 15 years' experience	Salary at top of scale
Switzerland	$46,866	$62,052	$70,548
Germany	35,546	41,745	49,445
Denmark	29,986	40,019	42,672
Belgium	29,075	41,977	50,461
Spain	29,058	33,988	43,100
Netherlands	27,133	46,148	54,720
Australia	26,658	37,138	37,577
United States	**25,405**	**36,219**	**44,394**
Austria	24,027	30,376	53,443
Korea(South)	23,613	39,265	62,135
Ireland	23,033	35,944	40,523
Norway	22,194	25,854	27,453
France	21,918	28,757	41,537
Finland	21,047	29,530	31,325
Italy	20,822	26,175	32,602
Iceland	20,775	25,795	30,954
Sweden	20,549	26,210	NA
England	19,999	33,540	33,540
Scotland	$19,765	$32,858	$32,858
Greece	19,650	23,943	28,987
Portugal	18,751	27,465	50,061
Tunisia	18,235	19,770	20,577
New Zealand	16,678	32,573	32,573
Argentina	15,789	22,266	26,759
Chile	14,644	16,214	19,597
Malaysia	13,575	21,568	29,822
Philippines	12,620	13,715	14,609
Brazil	12,598	16,103	18,556
Uruguay	10,305	12,489	15,585
Turkey	8,144	9,355	10,568
Jordan	8,096	10,652	27,347
Czech Republic	8,052	10,695	14,316
Hungary	6,908	10,355	13,217
Thailand	5,781	14,208	27,098
Peru	4,701	4,701	4,701
Indonesia	1,689	3,537	5,598

NA = Not available. (1) Purchasng power parities (PPPs) are the rates of currency conversion that equalize the purchasing power of different currencies by eliminating the differences in price levels between countries.

Population with Upper Secondary Education or Higher, Selected Countries, 1999

Source: Organization for Economic Cooperation and Development

(Percentage of the population ages 25-64 having attained at least upper secondary (senior high school) education)

United States	87	New Zealand	74	Poland[1]	54	Spain	35
Czech Republic	86	Finland	72	Ireland[1]	51	Uruguay[1]	32
Norway[1]	85	Hungary	67	Jordan	51	Zimbabwe	29
Switzerland	82	Korea	66	Greece	50	Brazil[1]	24
Germany	81	France	62	Peru[1]	46	Indonesia	22
Japan	81	United Kingdom	62	Philippines	44	Turkey	22
Denmark	80	Australia	57	Chile[1]	43	Portugal	21
Canada	79	Belgium	57	Italy	42	Mexico	20
Sweden	77	Iceland	56	Sri Lanka[1]	36	Thailand[1]	16
Austria[1]	74	Luxembourg	56	Malaysia[1]	35	Tunisia	8

(1) Year of reference 1998.

Governmental Expenditure Per Student, Selected Countries, 1998

Source: Organization for Economic Cooperation and Development

(Expenditure per student in U.S. dollars converted using PPPs[1] on public and private institutions, by level of education, based on full-time equivalents)

	Primary[2]	Secondary[3]		Primary[2]	Secondary[3]		Primary[2]	Secondary[3]
Australia	3,981	5,830	Indonesia[5]	116	497	Philippines[4]	689	726
Austria	6,065	8,163	Ireland	2,745	3,934	Poland	1,496	1,438
Belgium[3]	3,799	6,238	Israel	4,135	5,115	Portugal	3,121	4,636
Brazil[4]	837	1,076	Italy	5,653	6,458	Spain	3,267	4,274
Chile	1,500	1,713	Japan	5,075	5,890	Sweden	5,579	5,648
Czech Republic	1,645	3,182	Korea (South)	2,838	3,544	Switzerland	6,470	9,348
Denmark	6,713	7,200	Malaysia	919	1,469	Thailand	1,048	1,177
Finland	4,641	5,111	Mexico	863	1,586	Tunisia[5]	891	1,633
France	3,752	6,605	Netherlands	3,795	5,304	United Kingdom	3,329	5,230
Germany	3,531	6,209	Norway	5,761	7,343	United States	6,043	7,764
Greece	2,368	3,287	Paraguay	572	948	Uruguay	971	1,246
Hungary	2,028	2,140	Peru	479	671	Zimbabwe	768	1,179

(1) Purchasing power parities (PPPs) are the rates of currency conversion that equalize the purchasing power of different currencies by eliminating the differences in price levels between countries. (2) Primary—elementary school age. (3) Lower secondary—(junior high) and upper secondary (senior high school) combined. (4) Year of reference 1997. (5) Year of reference 1999.

U.S. Institutions of Higher Education—Charges, 1969-70 to 2000-2001

Source: National Center for Education Statistics, U.S. Dept. of Education

Figures for 1969-70 are average charges for full-time resident degree-credit students; figures for later years are average charges per full-time equivalent student. Room and board are based on full-time students. These figures are enrollment-weighted, according to the number of full-time-equivalent undergraduates, and thus vary from averages given elsewhere.

	TUITION AND FEES			BOARD RATES			DORMITORY CHARGES		
	All institutions	2-yr	4-yr	All institutions	2-yr	4-yr	All institutions	2-yr	4-yr
PUBLIC (in-state)									
1969-70	$323	$178	$427	$511	$465	$540	$369	$308	$395
1979-80	583	355	840	867	894	898	715	572	749
1989-90	1,356	756	2,035	1,635	1,581	1,728	1,513	962	1,561
1990-91	1,454	824	2,159	1,691	1,594	1,767	1,612	1,050	1,658
1991-92	1,624	937	2,410	1,780	1,612	1,852	1,731	1,074	1,789
1992-93	1,782	1,025	2,349	1,841	1,668	1,854	1,756	1,106	1,816
1993-94	1,942	1,125	2,537	1,880	1,681	1,895	1,873	1,190	1,934
1994-95	2,057	1,192	2,681	1,949	1,712	1,967	1,959	1,232	2,023
1995-96	2,179	1,239	2,848	2,020	1,681	2,045	2,057	1,297	2,121
1996-97	2,271	1,276	2,987	2,111	1,789	2,133	2,148	1,339	2,214
1997-98	2,360	1,314	3,110	2,228	1,795	2,263	2,225	1,401	2,301
1998-99	2,430	1,327	3,229	2,347	1,828	2,389	2,330	1,450	2,409
1999-2000	2,506	1,338	3,349	2,364	1,834	2,406	2,440	1,549	2,519
2000-2001	2,600	1,359	3,506	2,454	1,900	2,498	2,566	1,603	2,651
PRIVATE									
1969-70	1,533	1,034	1,809	561	546	608	436	413	503
1979-80	3,130	2,062	3,811	955	924	1,078	827	769	999
1989-90	8,147	5,196	10,348	1,948	1,811	2,339	1,923	1,663	2,411
1990-91	8,772	5,570	11,379	2,074	1,989	2,470	2,063	1,744	2,654
1991-92	9,434	5,752	12,192	2,252	2,090	2,727	2,221	1,789	2,860
1992-93	9,942	6,059	10,294	2,344	1,875	2,354	2,348	1,970	2,362
1993-94	10,572	6,370	10,952	2,434	1,970	2,445	2,490	2,067	2,506
1994-95	11,111	6,914	11,481	2,509	2,023	2,520	2,587	2,233	2,601
1995-96	11,563	7,094	12,243	2,606	2,098	2,617	2,738	2,371	2,751
1996-97	11,954	7,236	12,881	2,663	2,181	2,672	2,878	2,537	2,889
1997-98	12,921	7,464	13,344	2,762	2,785	2,761	2,954	2,672	2,964
1998-99	13,319	7,854	13,973	2,865	2,884	2,865	3,075	2,581	2,865
1999-2000	13,965	8,235	14,588	2,882	2,922	2,881	3,224	2,808	2,882
2000-2001	14,690	8,961	15,531	2,989	2,962	2,989	3,370	2,768	2,989

Top 20 Colleges and Universities in Endowment Assets, 2000[1]

Source: National Association of College and University Business Officers (NACUBO)

College/University	Endowment assets[2]
1. Harvard University	18,844,338
2. Yale University	10,084,900
3. University of Texas System	10,013,175
4. Stanford University	8,649,475
5. Princeton University	8,398,100
6. Massachusetts Institute of Technology	6,475,506
7. University of California	5,639,777
8. Emory University	5,032,683
9. Columbia University	4,263,972
10. Washington University	4,234,599
11. The Texas A&M University System and Foundations	4,205,849
12. University of Chicago	3,828,664
13. University of Michigan	3,468,372
14. Cornell University	3,436,926
15. Rice University	3,372,458
16. Northwestern University	3,368,233
17. University of Pennsylvania	3,200,812
18. University of Notre Dame	3,089,007
19. Duke University	2,663,891
20. Dartmouth College	2,490,376

NOTE: Figures are for market value of endowment assets, excluding pledges and working capital. (1) As of June 30, 2000. (2) In thousands.

U.S. Higher Education Trends: Bachelor's Degrees Conferred

Source: National Center for Education Statistics, U.S. Dept. of Education

Figures for 2001-2002 and 2009-2010 are projected.

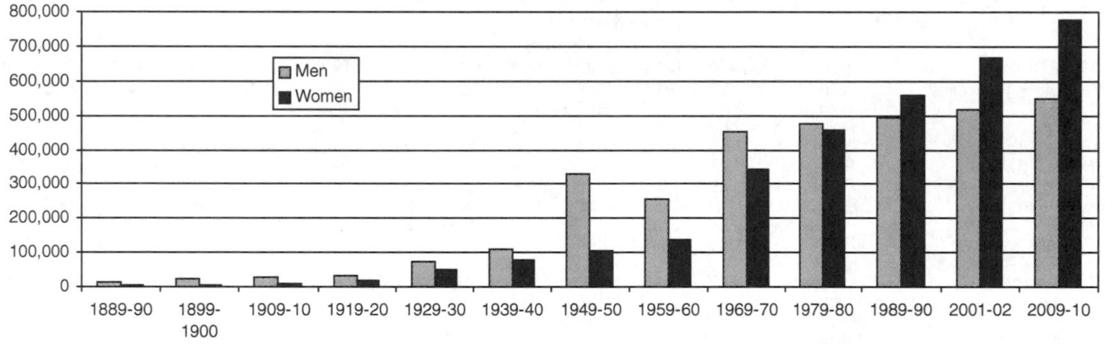

Financial Aid for College and Other Postsecondary Education

Reviewed by National Assoc. of Student Financial Aid Administrators

The cost of postsecondary education in the U.S. has increased in recent years, but financial aid, which may be in the form of grants (no repayment needed), loans, and/or work-study programs, is widely available to help families meet these expenses. Most aid is limited to family financial need as determined by standard formulas. Students interested in receiving aid are advised to apply, without making prior assumptions. Financial aid personnel at each school can provide information about programs available to students, steps to apply for them, and deadlines, all of which may vary.

First-time applicants for federal aid must file a Free Application for Federal Student Aid (FAFSA), generally as soon as possible after Jan. 1 for the academic year starting the following September. Figures provided must agree with federal income tax forms filed for the previous year. Other possible sources of aid include state governments, employers and unions, civic organizations, and the institutions themselves. There are also special federal programs that pay for postsecondary education in return for service: AmeriCorps (phone: 1-800-942-2677) and ROTC (phone: 1-800-USA-ROTC). Additional forms and certain fees may be required if a student is to be considered for institutional aid. Aid must be reapplied for annually.

A federal formula, based on information provided on the FAFSA, takes into account such factors as family after-tax income in the preceding calendar year, parental assets (excluding the parents' home) and length of time to retirement, and unusual expenses (such as very high medical expenses).

The resulting Expected Family Contribution, or EFC (which is divided among the family members—excluding parents—in college), is subtracted from the total cost of attendance for each person (including room and board or allowance for living costs) to determine financial need, and thus the maximum federal aid for which the family may be eligible. (Some institutions use a separate formula for need-based institutional aid.) Some schools guarantee to meet the full financial need of each admitted student; others try to do so but may fall short, depending on the availability of funds. Outside scholarships (even if non-need-based) are taken into account in determining need.

The aid package offered by each school may include one or more of the following resources: Federal Pell Grants, for those with greatest financial need; Federal Supplementary Educational Opportunity Grants, for those with relatively great financial need; grants from the school; federal work-study or other work programs; low-interest Perkins loans; and subsidized and unsubsidized Stafford loans. Unsubsidized Stafford loans are available without need, as are all PLUS loans to parents. Loans have varying interest rates and other requirements. Repayment of Perkins and Stafford loans does not begin until after graduation; deferments are available under certain circumstances. For PLUS loans, parents must pass a credit check and begin repayment of both principal and interest while the student is still in school.

Certain federal income tax credits—dollar for dollar reductions of the amount of tax due—are available to families who meet income and other requirements; see the chapter on Taxes.

Rules for financial aid are complex and changeable. *The Student Guide*, a comprehensive resource on financial aid from the U.S. Dept. of Education, can be found at the website http://www.ed.gov/prog_info/SFA/StudentGuide

Further information and FAFSA forms are available from the school or from the Federal Student Aid Information Center, PO Box 84, Washington, DC 20044; phone: 1-800-4-FED-AID, Mon.-Fri., 8 AM - 8 PM Eastern Time. The Information Center also has a free booklet called *The EFC Formula Book.* FAFSA forms can be obtained online at http://www.fafsa.ed.gov

Salaries of U.S. College Professors, 2000-2001

Source: American Association of University Professors

	MEN Type of institution			WOMEN Type of institution		
TEACHING LEVEL	**Public**	**Private/ Independent**	**Church-related**	**Public**	**Private/ Independent**	**Church-related**
Doctoral level						
Professor	$85,259	$109,303	$96,905	$77,379	$98,905	$89,176
Associate	61,912	71,940	66,905	57,796	66,851	61,972
Assistant	52,611	63,228	55,510	48,106	56,901	51,255
Master's level						
Professor	69,454	76,066	72,251	67,283	71,976	65,112
Associate	55,812	58,933	56,202	53,338	55,990	51,979
Assistant	45,986	48,172	45,916	44,222	45,935	43,053
General 4-year						
Professor	62,915	75,072	57,747	59,425	70,986	54,616
Associate	51,937	55,822	47,218	48,943	53,722	45,686
Assistant	43,154	45,319	39,503	41,306	44,227	38,996
2-year						
Professor	59,743	57,041	42,578	55,268	47,609	39,356
Associate	48,826	48,384	38,329	45,748	41,337	39,378
Assistant	42,462	34,459	35,465	40,736	30,716	34,585

ACT (formerly American College Testing) Mean Scores and Characteristics of College-Bound Students, 1990-2001

Source: ACT, Inc.

(for school year ending in year shown)

SCORES[1]	Unit[1]	1990	1991	1992	1993	1994	1995	1996	1997	1998	1999	2000	2001
Composite Scores	**Points**	**20.6**	**20.6**	**20.6**	**20.7**	**20.8**	**20.8**	**20.9**	**21.0**	**21.0**	**21.0**	**21.0**	**21.0**
Male	Points	21.0	20.9	20.9	21.0	20.9	21.0	21.0	21.1	21.2	21.1	21.2	21.1
Female	Points	20.3	20.4	20.5	20.5	20.7	20.7	20.8	20.8	20.9	20.9	20.9	20.9
English Score	**Points**	**20.5**	**20.3**	**20.2**	**20.3**	**20.3**	**20.2**	**20.3**	**20.3**	**20.4**	**20.5**	**20.5**	**20.5**
Male	Points	20.1	19.8	19.8	19.8	19.8	19.8	19.8	19.9	19.9	20.0	20.0	20.0
Female	Points	20.9	20.7	20.6	20.6	20.7	20.6	20.7	20.7	20.8	20.9	20.9	20.8
Math Score	**Points**	**19.9**	**20.0**	**20.0**	**20.1**	**20.2**	**20.2**	**20.2**	**20.6**	**20.8**	**20.7**	**20.7**	**20.7**
Male	Points	20.7	20.6	20.7	20.8	20.8	20.9	20.9	21.3	21.5	21.4	21.4	21.4
Female	Points	19.3	19.4	19.5	19.6	19.6	19.7	19.7	20.1	20.2	20.2	20.2	20.2
PARTICIPANTS													
Total Number	**(thousands)**	**817**	**796**	**832**	**875**	**892**	**945**	**925**	**959**	**995**	**1,019**	**1,065**	**1,070**
Male	Percent	46	45	45	45	45	44	44	44	43	43	43	43
White	Percent	79	79	79	79	79	80	79	74	76	72	72	71
Black	Percent	9	9	9	9	9	9	9	10	11	10	10	11
Composite Scores													
27 or above	Percent	12	11	12	12	13	13	13	14	14	14	14	14
18 or below	Percent	35	35	35	35	34	34	34	33	33	33	32	33

Note: Beginning with the Oct. 1989 test (1990 scores), an entirely new ACT Assessment was introduced. It is not possible to compare directly these data and data from earlier years. (1) Minimum point score, 1; maximum score, 36. Test scores and characteristics of college-bound students are based on the performance of all ACT-tested students who graduated in the spring of a given school year and who took the ACT Assessment during junior or senior year of high school.

ACT Average Composite Scores by State, 2000-01

Source: ACT, Inc.

STATE	Avg. Composite Score	% Grads Taking ACT[1]	STATE	Avg. Composite Score	% Grads Taking ACT[1]	STATE	Avg. Composite Score	% Grads Taking ACT[1]	STATE	Avg. Composite Score	% Grads Taking ACT[1]
AL	20.1	69	IL	21.6	71	MT	21.7	55	RI	21.2	5
AK	21.0	34	IN	21.4	20	NE	21.6	74	SC	19.3	28
AZ	21.5	28	IA	22.0	67	NV	21.3	39	SD	21.4	70
AR	20.1	75	KS	21.6	78	NH	22.3	7	TN	20.0	79
CA	21.4	12	KY	20.1	72	NJ	20.6	4	TX	20.3	33
CO	21.5	62	LA	19.6	80	NM	19.9	64	UT	21.4	69
CT	21.8	4	ME	21.4	6	NY	22.2	14	VT	22.2	9
DE	20.6	4	MD	20.5	11	NC	19.7	13	VA	20.6	10
DC	17.4	26	MA	21.9	8	ND	21.4	80	WA	22.4	17
FL	20.4	40	MI	21.3	69	OH	21.4	63	WV	20.2	61
GA	19.9	19	MN	22.1	66	OK	20.5	71	WI	22.2	68
HI	21.7	19	MS	18.5	89	OR	22.6	11	WY	21.5	64
ID	21.5	59	MO	21.4	70	PA	21.4	8	**U.S. AVG.**	**21.0**	**38**

(1) Based on number of high school graduates in 2001, as projected by the Western Interstate Commission for Higher Education, and number of students in the class of 2001 who took the ACT.

SAT Mean Verbal and Math Scores of College-Bound Seniors, 1975-2001

Source: The College Board

(recentered scale; for school year ending in year shown)

	1975	1980	1985	1990	1995	1996	1997	1998	1999	2000	2001
Verbal Scores	**512**	**502**	**509**	**500**	**504**	**505**	**505**	**505**	**505**	**505**	**506**
Male	515	506	514	505	505	507	507	509	509	507	509
Female	509	498	503	496	502	503	503	502	502	504	502
Math Scores	**498**	**492**	**500**	**501**	**506**	**508**	**511**	**512**	**511**	**514**	**514**
Male	518	515	522	521	525	527	530	531	531	533	533
Female	479	473	480	483	490	492	494	496	495	498	498

NOTE: In 1995, the College Board recentered the scoring scale for the SAT by reestablishing the original mean score of 500 on the 200-800 scale. Earlier scores have been adjusted to allow for this recentering.

SAT Mean Scores by State, 1990 and 1997-2001

Source: The College Board

(recentered scale; for school year ending in year shown)

STATE	1990 V	1990 M	1997 V	1997 M	1998 V	1998 M	1999 V	1999 M	2000 V	2000 M	2001 V	2001 M	% Grads Taking SAT[1]
Alabama	545	534	561	555	562	558	561	555	559	555	559	554	9
Alaska	514	501	520	517	521	520	516	514	519	515	514	510	51
Arizona	521	520	523	522	525	528	524	525	521	523	523	525	34
Arkansas	545	532	567	558	568	555	563	556	563	554	562	550	6
California	494	508	496	514	497	516	497	514	497	518	498	517	51
Colorado	533	534	536	539	537	542	536	540	534	537	539	542	31
Connecticut	506	496	509	507	510	509	510	509	508	509	509	510	82
Delaware	510	496	505	498	501	493	503	497	502	496	501	499	67
District of Columbia	483	467	490	475	488	476	494	478	494	486	482	474	56
Florida	495	493	499	499	500	501	499	498	498	500	498	499	54
Georgia	478	473	486	481	486	482	487	482	488	486	491	489	63
Hawaii	480	505	483	512	483	513	482	513	488	519	486	515	52
Idaho	542	524	544	539	545	544	542	540	540	541	543	542	17
Illinois	542	547	562	578	564	581	569	585	568	586	576	589	12
Indiana	486	486	494	497	497	500	496	498	498	501	499	501	60
Iowa	584	588	589	601	593	601	594	598	589	600	593	603	5
Kansas	566	563	578	575	582	585	578	576	574	580	577	580	9
Kentucky	548	541	548	546	547	550	547	547	548	550	550	550	12
Louisiana	551	537	560	553	562	558	561	558	562	558	564	562	7
Maine	501	490	507	504	504	501	507	503	504	500	506	500	69
Maryland	506	502	507	507	506	508	507	507	507	509	508	510	65
Massachusetts	503	498	508	508	508	508	511	511	511	513	511	515	79
Michigan	529	534	557	566	558	569	557	565	557	569	561	572	11
Minnesota	552	558	582	592	585	598	586	598	581	594	580	589	9
Mississippi	552	538	567	551	562	549	563	548	562	549	566	551	4
Missouri	548	541	567	568	570	573	572	572	572	577	577	577	8
Montana	540	542	545	548	543	546	545	546	543	546	539	539	23
Nebraska	559	562	562	564	565	571	568	571	560	571	562	568	8
Nevada	511	511	508	509	510	513	512	517	510	517	509	515	33
New Hampshire	518	510	521	518	523	520	520	518	520	519	520	516	72
New Jersey	495	498	497	508	497	508	498	510	498	513	499	513	81
New Mexico	554	546	554	545	554	551	549	542	549	543	551	542	13
New York	489	496	495	502	495	503	495	502	494	506	495	505	77
North Carolina	478	470	490	488	490	492	493	493	492	496	493	499	65
North Dakota	579	578	588	595	590	599	594	605	588	609	592	599	4
Ohio	526	522	535	536	536	540	534	568	533	539	534	539	26
Oklahoma	553	542	568	560	568	564	567	560	563	560	567	561	8
Oregon	515	509	525	524	528	528	525	525	527	527	526	526	55
Pennsylvania	497	490	498	495	497	495	498	495	498	497	500	499	71
Rhode Island	498	488	499	493	501	495	504	499	505	500	501	499	71
South Carolina	475	467	479	474	478	473	479	475	484	482	486	488	57
South Dakota	580	570	574	570	584	581	585	588	587	588	577	582	4
Tennessee	558	544	564	556	564	557	559	553	563	553	562	553	13
Texas	490	489	494	501	494	501	494	499	493	500	493	499	53
Utah	566	555	576	570	572	570	570	568	570	569	575	570	5
Vermont	507	493	508	502	508	504	514	506	513	508	511	506	69
Virginia	501	496	506	497	507	499	508	499	509	500	510	501	68
Washington	513	511	523	523	524	526	525	526	526	528	527	527	53
West Virginia	520	514	524	508	525	513	527	512	526	511	527	512	18
Wisconsin	552	559	579	590	581	594	584	595	584	597	584	596	6
Wyoming	534	538	543	543	548	546	546	551	545	545	547	545	11
NATIONAL AVERAGE	**500**	**501**	**505**	**511**	**505**	**512**	**505**	**511**	**505**	**514**	**506**	**514**	**45**

NOTE: In 1995, the College Board recentered the scoring scale for the SAT by reestablishing the original mean score of 500 on the 200-800 scale. The College Board states that comparing states or ranking them on the basis of SAT scores alone is invalid, and the College Board discourages doing so. (1) Based on number of high school graduates in 2001, as projected by the Western Interstate Commission for Higher Education, and number of students in the class of 2001 who took the SAT.

Average SAT Scores by Parental Education, 2001

Source: The College Board

(Deviation in points from mean score shown by highest level of educational attainment of test taker's parent. Mean 2001 verbal score was 506. Mean 2001 math score was 514.)

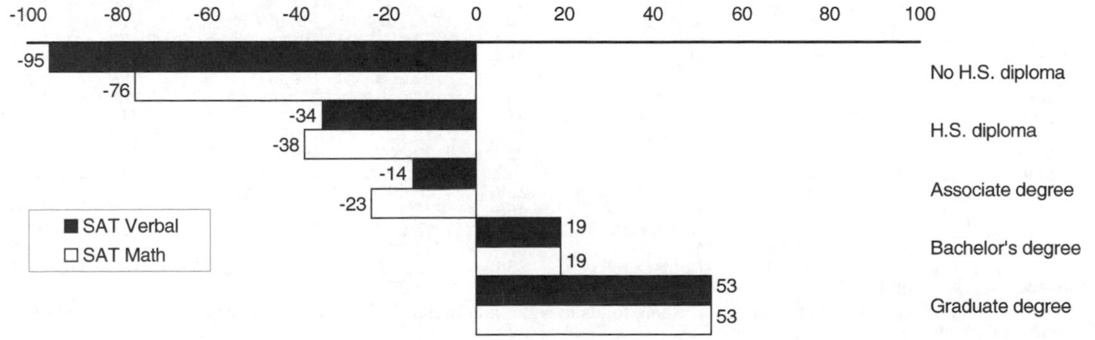

Top 50 Public Libraries in the U.S. and Canada, 2001

Source: Public Library Data Service, Statistical Report 2001, Public Library Association

Ranked at end of the 2000 fiscal year by population served.

Population served	Library name and location	No. of branches[1]	No. of holdings	Circulation	Annual acquisition expenditures
3,781,500	Los Angeles Public Library (CA)	67	6,066,546	12,986,003	$8,600,000
3,506,500	Los Angeles Public Library, County of (CA)	84	7,743,803	14,761,515	8,088,851
3,070,302	New York Public Library	80	10,258,570	12,842,278	12,596,813
2,896,016	Chicago Public Library (IL)	77	14,374,608	6,835,882	11,000,000
2,385,421	Toronto Public Library (Ontario)	97	8,540,339	25,382,467	9,048,445
2,300,664	Brooklyn Public Library (NY)	59	6,970,900	10,916,696	10,662,253
1,951,598	Queens Borough Public Library (NY)	62	9,143,990	17,239,736	11,869,373
1,919,390	Houston Public Library (TX)	37	4,476,593	6,234,526	6,447,265
1,790,391	Miami-Dade Public Library System (FL)	31	3,550,332	4,579,086	5,349,224
1,585,577	Philadelphia, The Free Library of (PA)	53	9,216,869	6,341,612	7,584,298
1,490,289	Broward County Libraries Division (FL)	36	2,582,010	7,387,149	8,365,997
1,432,700	San Antonio Public Library (TX)	18	1,859,921	4,322,695	3,482,578
1,336,449	Carnegie Library of Pittsburgh (PA)	19	6,188,697	2,609,348	3,105,861
1,285,100	San Diego Public Library (CA)	33	3,000,653	6,381,062	3,300,000
1,193,000	Hawaii State Public Library System (HI)	50	3,211,213	6,912,342	1,965,636
1,178,553	Harris County Public Library (TX)	25	2,025,196	5,062,798	1,915,689
1,088,839	Las Vegas-Clark County Library District (NV)	24	2,298,170	5,225,780	4,740,572
1,084,233	King County Library System (WA)	41	4,436,931	12,914,177	6,195,461
1,068,800	Dallas Public Library (TX)	22	5,345,887	3,738,051	3,289,532
1,041,000	San Bernardino County Library (CA)	28	1,155,140	2,793,457	1,406,920
1,037,205	Montreal, Bibliotheque de (PQ)	24	2,578,271	5,435,716	3,069,191
1,027,974	Detroit Public Library (MI)	23	2,961,367	1,228,169	2,221,474
1,001,910	Tampa-Hillsborough County Public Library (FL)	20	2,252,285	3,974,071	3,573,347
1,001,838	Providence Public Library (RI)	9	1,623,847	793,693	835,257
989,925	Fairfax County Public Library (VA)	21	2,526,793	10,813,448	5,705,199
968,532	Buffalo & Erie County Public Library (NY)	51	3,356,266	7,994,794	4,326,217
965,413	San Diego County Library (CA)	31	1,199,687	3,214,059	1,845,839
923,600	San Jose Public Library System (CA)	17	1,885,312	7,886,647	3,644,640
863,315	Tucson-Pima Public Library (AZ)	18	1,218,385	5,383,213	2,978,930
860,749	Calgary Public Library (AB)	15	2,060,560	11,448,902	3,240,391
855,000	Montgomery County Dept. of Public Libraries (MD)	22	2,505,244	10,087,585	5,278,378
846,584	Memphis/Shelby County Public Library & Info. Ctr. (TN)	22	1,989,371	3,417,439	2,060,877
843,638	St. Louis County Library District (MO)	18	2,765,732	7,800,578	4,406,800
840,443	Cincinnati & Hamilton County, The Public Lib. of (OH)	41	9,686,841	13,326,369	8,103,570
835,600	Contra Costa County Library (CA)	22	1,269,571	3,620,087	2,116,866
811,305	Orange County Library System (FL)	12	1,702,426	4,140,589	3,060,174
801,400	San Francisco Public Library (CA)	26	2,253,314	5,166,843	4,680,597
790,770	Columbus Metropolitan Library (OH)	20	2,918,193	12,239,562	7,891,218
786,100	Atlanta-Fulton Public Library (GA)	32	2,443,570	2,830,480	3,567,249
785,487	Indianapolis-Marion County Public Library (IN)	22	1,852,643	9,725,789	5,622,821
781,600	Fresno County Library (CA)	34	798,264	1,932,993	1,818,847
774,300	Prince George's County Memorial Library System (MD)	19	1,974,039	3,845,249	3,113,981
753,823	Jacksonville Public Library (FL)	14	2,330,583	4,021,675	3,229,006
735,050	Hennepin County Library (MN)	26	1,900,000	10,660,342	4,205,434
730,969	Baltimore County Public Library (MD)	16	1,771,929	9,543,864	4,898,273
717,400	Macomb County Library (MI)	0	185,824	181,271	339,794
713,968	Rochester Public Library (NY)	10	955,328	1,567,738	1,208,200
685,116	Palm Beach County Library System (FL)	14	1,150,631	5,063,446	3,799,572
664,937	Louisville Free Public Library (KY)	16	1,245,762	3,269,773	3,061,183
661,091	Charlotte & Mecklenburg County, Pub. Lib. of (NC)	22	1,726,712	6,305,590	2,987,384

(1) Main branch not included.

Number of Public Libraries and Operating Income, by State, 1998

Source: Public Libraries Survey, National Center for Education Statistics, U.S. Dept. of Education

(data for fiscal year 1998 unless otherwise indicated; operating income in thousands)

STATE	No. of libraries[1]	Operating income[2]	STATE	No. of libraries[1]	Operating income[2]	STATE	No. of libraries[1]	Operating income[2]
Alabama	273	$55,557	Kentucky	191	$62,066	Ohio	703	$609,615
Alaska	101	22,057	Louisiana	326	102,793	Oklahoma	212	49,052
Arizona	166	97,817	Maine	276	23,193	Oregon	206	84,829
Arkansas	202	33,443	Maryland	187	152,762	Pennsylvania	650	208,865
California	1,041	732,842	Massachusetts	490	174,801	Rhode Island	72	28,163
Colorado	253	130,826	Michigan	649	238,808	South Carolina	180	61,640
Connecticut	242	118,153	Minnesota	361	131,208	South Dakota	129	12,451
Delaware	31	12,062	Mississippi	240	31,058	Tennessee	288	64,700
District of Columbia	27	22,488	Missouri	377	130,992	Texas	791	248,733
Florida	450	302,993	Montana	107	13,137	Utah	103	48,694
Georgia	367	132,530	Nebraska	250	30,802	Vermont	195	11,583
Hawaii	49	21,211	Nevada	82	43,664	Virginia	317	159,357
Idaho	144	20,364	New Hampshire	238	28,647	Washington	312	196,511
Illinois	782	425,924	New Jersey	453	277,114	West Virginia	172	22,089
Indiana	422	199,302	New Mexico	94	25,647	Wisconsin	456	139,415
Iowa	556	60,640	New York	1,077	730,212	Wyoming	75	12,621
Kansas	363	61,371	North Carolina	366	125,242	**U.S. TOTAL**	**16,180**	**$6,737,820**
			North Dakota	86	7,776			

(1) Includes central libraries and branches. (2) Some totals may be underestimated because of nonresponse. (3) These libraries reported data for fiscal year 1998.

American Colleges and Universities

General Information for the 2000–2001 Academic Year

Source: Peterson's, a Thomson Learning Company, Copyright 2001

These listings **include only accredited undergraduate degree-granting institutions** in the United States and the U.S. territories **with a total enrollment of 1,000 or more.** Four-year colleges (which award a bachelor's degree as their highest undergraduate degree) are listed first, followed by two-year colleges (which generally award an associate as their highest or primary undergraduate degree). Data reported **only for institutions that provided updated information** on Peterson's Annual Survey of Undergraduate Institutions for the 2000–2001 academic year.

All institutions are coeducational except those where the ZIP code is followed directly by: (1)–men only, (2)–primarily men, (3)–women only, (4)–primarily women.

The **Tuition & Fees** column shows the annual tuition and required fees for full-time students, or, where indicated, the tuition and standard fees per credit hour (cr. hr.) for part-time students. Where tuition costs vary according to residence, the figure is given for the most local resident and is coded: (A)–area residents, (S)–state residents; all other figures apply to all students regardless of residence. Where annual expenses are expressed as a lump sum (including full-time tuition, mandatory fees, and room and board), the figure is coded: (C)–comprehensive fee. **Rm. & Board** is the average cost for one academic year.

Control: 1–independent (nonprofit), 2–independent-religious, 3–proprietary (profit-making), 4–federal, 5–state, 6–commonwealth (Puerto Rico), 7–territory (U.S. territories), 8–county, 9–district, 10–city, 11–state and local, 12–state-related. **Degree** means the highest degree offered (B–bachelor's, M–master's, F–first professional, D–doctorate). Where no letter is given, as for most two-year colleges, the highest degree offered is the associate degree.

Enrollment is the total number of matriculated undergraduate and (if applicable) graduate students.

Faculty is the total number of faculty members teaching undergraduate courses and (if available) graduate courses.

NA or a **dash** indicates category is inapplicable or data not available. **NR** indicates data not reported.

Four-Year Colleges

Name, address	Year Founded	Tuition & Fees	Rm. & Board	Control, Degree	Enroll-ment	Faculty
Abilene Christian U, Abilene, TX 79699-9100	1906	$10,910	$4,420	2-D	4,739	310
Acad of Art Coll, San Francisco, CA 94105-3410	1929	$12,060	NA	3-M	5,995	546
Adams State Coll, Alamosa, CO 81102	1921	$2,396 (S)	$5,185	5-M	6,948	170
Adelphi U, Garden City, NY 11530	1896	$15,690	$7,450	1-D	5,908	554
Adrian Coll, Adrian, MI 49221-2575	1859	$14,250	$4,990	2-B	1,082	128
Alabama Ag & Mech U, Normal, AL 35762	1875	$2,820 (S)	$4,500	5-D	5,523	307
Alabama State U, Montgomery, AL 36101-0271	1867	$2,520 (S)	$3,700	5-M	5,269	375
Albany State U, Albany, GA 31705-2717	1903	$2,398 (S)	$3,366	5-M	3,525	203
Albertus Magnus Coll, New Haven, CT 06511-1189	1925	$14,785	$6,708	2-M	2,105	147
Albion Coll, Albion, MI 49224-1831	1835	$18,952	$5,404	2-B	1,522	126
Albright Coll, Reading, PA 19612-5234	1856	$20,310	$6,040	2-B	1,728	130
Alcorn State U, Alcorn State, MS 39096-7500	1871	$2,785 (S)	$2,809	5-M	2,936	194
Alfred U, Alfred, NY 14802-1205	1836	$9,976 (S)	$7,450	1-D	2,433	208
Allegheny Coll, Meadville, PA 16335	1815	$21,570	$5,100	2-B	1,904	150
Alma Coll, Alma, MI 48801-1599	1886	$15,884	$5,726	2-B	1,409	126
Alvernia Coll, Reading, PA 19607-1799	1958	$13,435	$5,840	2-M	1,647	178
Alverno Coll, Milwaukee, WI 53234-3922 (3)	1887	$11,530	$4,610	2-M	1,872	188
Amberton U, Garland, TX 75041-5595	1971	$4,050	NA	2-M	1,648	39
American Coll of Computer & Information Scis, Birmingham, AL 35205	1988	$105	NA	3-M	9,000	27
American Intl Coll, Springfield, MA 01109-3189	1885	$13,600	$7,112	1-D	1,274	122
American Military U, Manassas Park, VA 20111 (2)	1991	$9,000	NA	3-M	2,080	101
American U, Washington, DC 20016-8001	1893	$21,399	$8,372	2-D	10,778	NA
American U of Puerto Rico, Bayamón, PR 00960-2037	1963	$3,254	NA	1-B	4,388	206
Amherst Coll, Amherst, MA 01002-5000	1821	$26,080	$6,800	1-B	1,682	220
Anderson Coll, Anderson, SC 29621-4035	1911	$10,515	$4,605	2-B	1,398	110
Anderson U, Anderson, IN 46012-3495	1917	$14,680	$4,750	2-D	2,381	190
Andrews U, Berrien Springs, MI 49104	1874	$12,916	$4,060	2-D	2,749	261
Angelo State U, San Angelo, TX 76909	1928	$2,572 (S)	$4,140	5-M	6,309	293
Anna Maria Coll, Paxton, MA 01612	1946	$14,645	$6,200	2-M	1,255	181
Appalachian State U, Boone, NC 28608	1899	$1,988 (S)	$3,810	5-D	13,227	902
Aquinas Coll, Grand Rapids, MI 49506-1799	1886	$14,876	$5,176	2-M	2,605	234
Arcadia U, Glenside, PA 19038-3295	1853	$17,830	$7,740	2-D	2,756	289
Arizona State U, Tempe, AZ 85287	1885	$2,346 (S)	$5,240	5-D	44,126	1,774
Arizona State U East, Mesa, AZ 85212	1995	$2,296 (S)	$4,540	5-M	1,939	80
Arizona State U West, Phoenix, AZ 85069-7100	1984	$2,345 (S)	NA	5-M	5,325	339
Arkansas State U, State University, AR 72467	1909	$3,160 (S)	$3,071	5-D	10,429	566
Arkansas Tech U, Russellville, AR 72801-2222	1909	$2,768 (S)	$3,336	5-M	5,172	305
Armstrong Atlantic State U, Savannah, GA 31419-1997	1935	$2,242 (S)	$4,582	5-M	5,496	357
Art Ctr Coll of Design, Pasadena, CA 91103-1999	1930	$19,900	NA	1-M	1,459	394
The Art Inst of Colorado, Denver, CO 80203	1952	$13,824	$5,994	3-B	2,090	129
The Art Inst of Phoenix, Phoenix, AZ 85021-2859 (2)	1995	$13,632	NA	3-B	1,000	47
Asbury Coll, Wilmore, KY 40390-1198	1890	$13,884	$3,566	2-M	1,357	153
Ashland U, Ashland, OH 44805-3702	1878	$16,270	$5,862	2-D	6,055	215
Assumption Coll, Worcester, MA 01609	1904	$18,120	$6,980	2-M	2,766	192
Athens State U, Athens, AL 35611-1902	1822	$2,370 (S)	NA	5-B	2,662	161
Auburn U, Auburn University, AL 36849-0002	1856	$3,154 (S)	$4,640	5-D	21,136	1,241
Auburn U Montgomery, Montgomery, AL 36124-4023	1967	$3,000 (S)	NA	5-D	4,900	307
Audrey Cohen Coll, New York, NY 10013-1919	1964	$14,750	NA	1-M	1,278	141
Augsburg Coll, Minneapolis, MN 55454-1351	1869	$16,340	$5,320	2-M	3,040	292
Augustana Coll, Rock Island, IL 61201-2296	1860	$17,913	$5,214	2-B	2,261	218
Augustana Coll, Sioux Falls, SD 57197	1860	$14,754	$4,260	2-M	1,783	169
Augusta State U, Augusta, GA 30904-2200	1925	$2,222 (S)	NA	5-M	5,070	286
Aurora U, Aurora, IL 60506-4892	1893	$12,918	$4,914	1-D	2,391	155
Austin Coll, Sherman, TX 75090-4400	1849	$15,219	$5,891	2-M	1,234	109
Austin Peay State U, Clarksville, TN 37044-0001	1927	$2,831 (S)	$3,350	5-M	7,121	463
Averett U, Danville, VA 24541-3692	1859	$14,190	$4,797	2-M	2,030	246
Avila Coll, Kansas City, MO 64145-1698	1916	$12,650	$5,000	2-M	1,412	176
Azusa Pacific U, Azusa, CA 91702-7000	1899	$15,210	$4,962	2-D	6,497	574
Babson Coll, Babson Park, MA 02457-0310	1919	$21,952	$8,746	1-M	3,397	208

Name, address	Year Founded	Tuition & Fees	Rm. & Board	Control, Degree	Enroll-ment	Faculty
Baker Coll of Auburn Hills, Auburn Hills, MI 48326-1586	1990	$5,400	NA	1-B	1,910	115
Baker Coll of Clinton Township, Clinton Township, MI 48035-4701	1990	$5,400	NA	1-B	2,560	133
Baker Coll of Flint, Flint, MI 48507-5508	1911	$5,580	NA	1-B	3,901	173
Baker Coll of Jackson, Jackson, MI 49202	1994	$5,400	NA	1-B	1,201	90
Baker Coll of Muskegon, Muskegon, MI 49442-3497	1888	$7,440	NA	1-B	2,727	145
Baker Coll of Owosso, Owosso, MI 48867-4400	1984	$7,200	NA	1-B	1,905	103
Baker Coll of Port Huron, Port Huron, MI 48060-2597	1990	$7,220	NA	1-M	1,230	95
Baker U, Baldwin City, KS 66006-0065	1858	$12,300	$4,750	2-M	2,817	NA
Baldwin-Wallace Coll, Berea, OH 44017-2088	1845	$15,340	$5,460	2-M	4,831	348
Ball State U, Muncie, IN 47306-1099	1918	$3,830 (S)	$4,830	5-D	19,004	1,084
Bard Coll, Annandale-on-Hudson, NY 12504	1860	$24,950	$7,440	1-D	1,430	191
Barnard Coll, New York, NY 10027-6598 (3)	1889	$23,056	$9,358	1-B	2,268	291
Barry U, Miami Shores, FL 33161-6695	1940	$16,600	$6,400	2-D	8,650	730
Barton Coll, Wilson, NC 27893-7000	1902	$11,462	$4,270	2-B	1,202	73
Bates Coll, Lewiston, ME 04240-6028	1855	$32,650 (C)	NA	1-B	1,694	189
Bayamón Central U, Bayamón, PR 00960-1725	1970	$3,570	NA	2-M	3,269	212
Baylor U, Waco, TX 76798	1845	$11,938	$5,238	2-D	13,719	795
Becker Coll, Worcester, MA 01609	1784	$12,520	$6,420	1-B	1,163	84
Belhaven Coll, Jackson, MS 39202-1789	1883	$10,860	$4,010	2-M	1,630	193
Bellarmine U, Louisville, KY 40205-0671	1950	$13,760	$4,160	2-M	2,175	244
Bellevue U, Bellevue, NE 68005-3098	1965	$3,238	NA	1-M	3,445	133
Belmont U, Nashville, TN 37212-3757	1951	$12,330	$5,437	2-M	2,976	418
Beloit Coll, Beloit, WI 53511-5596	1846	$21,550	$4,882	1-B	1,254	112
Bemidji State U, Bemidji, MN 56601-2699	1919	$3,575 (S)	$3,880	5-M	4,608	224
Benedict Coll, Columbia, SC 29204	1870	$8,272	$4,390	2-B	2,208	142
Benedictine Coll, Atchison, KS 66002-1499	1859	$12,630	$4,790	2-M	1,369	84
Benedictine U, Lisle, IL 60532-0900	1887	$14,880	$5,500	2-D	2,868	215
Bentley Coll, Waltham, MA 02452-4705	1917	$18,910	$8,600	1-M	5,728	227
Berea Coll, Berea, KY 40404	1855	$205	$4,099	1-B	1,590	156
Berklee Coll of Music, Boston, MA 02215-3693	1945	$16,916	$8,890	1-B	3,361	466
Bernard M. Baruch Coll of the City U of New York, New York, NY 10010-5585	1919	$3,350 (S)	NA	11-D	15,698	837
Berry Coll, Mount Berry, GA 30149-0159	1902	$12,500	$5,488	2-M	2,033	NA
Bethel Coll, Mishawaka, IN 46545-5591	1947	$12,650	$4,150	2-M	1,647	149
Bethel Coll, St. Paul, MN 55112-6999	1871	$15,935	$5,700	2-M	2,953	277
Bethune-Cookman Coll, Daytona Beach, FL 32114-3099	1904	$8,988	$5,782	2-B	2,745	220
Biola U, La Mirada, CA 90639-0001	1908	$16,630	$5,272	2-D	3,928	298
Birmingham-Southern Coll, Birmingham, AL 35254	1856	$16,269	$5,652	2-M	1,477	138
Black Hills State U, Spearfish, SD 57799	1883	$3,629 (S)	$2,883	5-M	4,171	183
Bloomfield Coll, Bloomfield, NJ 07003-9981	1868	$10,950	$5,350	2-B	1,771	205
Bloomsburg U of Pennsylvania, Bloomsburg, PA 17815-1905	1839	$4,668 (S)	$4,032	5-M	7,548	374
Bluefield State Coll, Bluefield, WV 24701-2198	1895	$2,288 (S)	NA	5-B	2,648	198
Bluffton Coll, Bluffton, OH 45817-1196	1899	$14,306	$5,122	2-M	1,063	101
Boise State U, Boise, ID 83725-0399	1932	$2,451 (S)	$3,570	5-D	16,282	917
Boston Coll, Chestnut Hill, MA 02467-3800	1863	$23,270	$8,510	2-D	13,551	1,203
Boston U, Boston, MA 02215	1839	$25,044	$8,450	1-D	28,318	3,277
Bowdoin Coll, Brunswick, ME 04011	1794	$25,890	$6,760	1-B	1,609	182
Bowie State U, Bowie, MD 20715-9465	1865	$3,778 (S)	$4,744	5-M	4,700	310
Bowling Green State U, Bowling Green, OH 43403	1910	$5,184 (S)	$5,768	5-D	18,096	929
Bradley U, Peoria, IL 61625-0002	1897	$14,580	$5,460	1-M	5,951	505
Brandeis U, Waltham, MA 02454-9110	1948	$26,166	$7,189	1-D	4,753	461
Brewton-Parker Coll, Mt. Vernon, GA 30445-0197	1904	$6,340	$3,100	2-B	1,416	183
Briarcliffe Coll, Bethpage, NY 11714	1966	$10,400	NA	3-B	1,790	136
Bridgewater Coll, Bridgewater, VA 22812-1599	1880	$14,970	$6,970	2-B	1,223	89
Bridgewater State Coll, Bridgewater, MA 02325-0001	1840	$2,895 (S)	$4,887	5-M	8,839	252
Brigham Young U, Provo, UT 84602-1001	1875	$2,940	$4,650	2-D	32,554	1,827
Brigham Young U–Hawaii Campus, Laie, HI 96762-1294	1955	$2,988	$5,275	2-B	2,358	185
Brooklyn Coll of the City U of New York, Brooklyn, NY 11210-2889	1930	$3,393 (S)	NA	11-M	15,039	992
Brown U, Providence, RI 02912	1764	$26,184	$7,346	1-D	7,723	737
Bryant Coll, Smithfield, RI 02917-1284	1863	$17,330	$7,250	1-M	3,373	195
Bryn Mawr Coll, Bryn Mawr, PA 19010-2899 (3)	1885	$24,160	$8,340	1-D	1,784	167
Bucknell U, Lewisburg, PA 17837	1846	$23,839	$5,596	1-M	3,582	311
Buena Vista U, Storm Lake, IA 50588	1891	$17,176	$4,795	2-M	1,381	130
Butler U, Indianapolis, IN 46208-3485	1855	$18,230	$6,140	1-F	4,166	416
Cabrini Coll, Radnor, PA 19087-3698	1957	$16,900	$7,560	2-M	2,062	166
Caldwell Coll, Caldwell, NJ 07006-6195	1939	$13,200	$6,250	2-M	2,094	170
California Baptist U, Riverside, CA 92504-3206	1950	$10,380	$4,966	2-M	2,043	157
California Coll for Health Scis, National City, CA 91950-6605	1978	$399	NA	3-M	5,742	17
California Coll of Arts & Crafts, San Francisco, CA 94107	1907	$19,460	$6,051	1-M	1,213	326
California Inst of Technology, Pasadena, CA 91125-0001	1891	$21,120	$6,543	1-D	1,968	317
California Inst of the Arts, Valencia, CA 91355-2340	1961	$20,930	NA	1-M	1,224	274
California Lutheran U, Thousand Oaks, CA 91360-2787	1959	$17,000	$6,460	2-M	2,766	246
California Polytechnic State U, San Luis Obispo, San Luis Obispo, CA 93407	1901	$2,135 (S)	$6,246	5-M	16,877	1,107
California State Polytechnic U, Pomona, Pomona, CA 91768-2557	1938	$1,772 (S)	$6,113	5-M	18,424	1,183
California State U, Bakersfield, Bakersfield, CA 93311-1099	1970	$1,875 (S)	$4,345	5-M	5,594	357
California State U, Chico, Chico, CA 95929-0722	1887	$2,030 (S)	$6,216	5-M	15,909	959
California State U, Dominguez Hills, Carson, CA 90747-0001	1960	$840	NA	5-M	12,848	762
California State U, Fresno, Fresno, CA 93740-8027	1911	$1,746 (S)	$5,876	5-D	19,114	1,110
California State U, Fullerton, Fullerton, CA 92834-9480	1957	$1,810 (S)	NA	5-M	28,381	1,711
California State U, Long Beach, Long Beach, CA 90840	1949	$1,723 (S)	$5,800	5-M	30,918	1,587
California State U, Los Angeles, Los Angeles, CA 90032-8530	1947	$1,722 (S)	$6,201	5-D	19,593	1,081
California State U, Monterey Bay, Seaside, CA 93955-8001	1994	$1,893 (S)	$4,100	5-M	2,267	NA
California State U, Sacramento, Sacramento, CA 95819-6048	1947	$1,867 (S)	$5,510	5-M	24,530	1,492
California State U, San Marcos, San Marcos, CA 92096-0001	1990	$1,706 (S)	NA	5-M	5,500	389
California State U, Stanislaus, Turlock, CA 95382	1957	$1,877 (S)	$6,480	5-M	7,062	421
California U of Pennsylvania, California, PA 15419-1394	1852	$4,937 (S)	$4,662	5-M	5,899	344
Calumet Coll of Saint Joseph, Whiting, IN 46394-2195	1951	$6,750	NA	2-B	1,004	81

Name, address	Year Founded	Tuition & Fees	Rm. & Board	Control, Degree	Enroll-ment	Faculty
Calvin Coll, Grand Rapids, MI 49546-4388	1876	$14,040	$4,890	2-M	4,309	343
Cambridge Coll, Cambridge, MA 02138-5304	1971	$8,550	NA	1-M	2,700	50
Cameron U, Lawton, OK 73505-6377	1908	$2,090 (S)	$2,746	5-M	5,002	264
Campbellsville U, Campbellsville, KY 42718-2799	1906	$9,050	$4,320	2-M	1,601	155
Campbell U, Buies Creek, NC 27506	1887	$11,539	$4,122	2-D	3,490	432
Canisius Coll, Buffalo, NY 14208-1098	1870	$16,447	$6,730	2-M	4,814	456
Capital U, Columbus, OH 43209-2394	1830	$16,880	$5,170	2-F	3,897	470
Cardinal Stritch U, Milwaukee, WI 53217-3985	1937	$11,800	$4,820	2-D	5,994	164
Caribbean U, Bayamón, PR 00960-0493	1969	$2,510	NA	1-M	1,115	158
Carleton Coll, Northfield, MN 55057-4001	1866	$24,570	$4,950	1-B	1,936	214
Carlow Coll, Pittsburgh,PA 15213-3165 (4)	1929	$13,342	$5,280	2-M	1,925	194
Carnegie Mellon U, Pittsburgh, PA 15213-3891	1900	$23,022	$7,028	1-D	8,514	874
Carroll Coll, Helena, MT 59625-0002	1909	$12,816	$5,168	2-B	1,251	122
Carroll Coll, Waukesha, WI 53186-5593	1846	$15,580	$4,740	2-M	2,902	237
Carson-Newman Coll, Jefferson City, TN 37760	1851	$11,920	$3,980	2-M	2,230	203
Carthage Coll, Kenosha, WI 53140-1994	1847	$17,350	$5,210	2-M	2,222	148
Case Western Reserve U, Cleveland, OH 44106	1826	$20,260	$5,815	1-D	9,304	557
Castleton State Coll, Castleton, VT 05735	1787	$5,180 (S)	$5,346	5-M	1,632	160
Catawba Coll, Salisbury, NC 28144-2488	1851	$13,330	$4,980	2-M	1,342	119
The Catholic U of America, Washington, DC 20064	1887	$19,930	$8,073	2-D	5,493	672
Cedar Crest Coll, Allentown, PA 18104-6196 (3)	1867	$17,790	$6,465	2-B	1,566	150
Cedarville U, Cedarville, OH 45314-0601	1887	$11,562	$4,929	2-M	2,857	212
Centenary Coll, Hackettstown, NJ 07840-2100	1867	$15,370	$6,250	2-M	1,338	141
Centenary Coll of Louisiana, Shreveport, LA 71134-1188	1825	$14,600	$4,500	2-M	1,035	107
Ctr for Creative Studies-Coll of Art & Design, Detroit, MI 48202-4034	1926	$16,526	NA	1-B	1,086	192
Central Coll, Pella, IA 50219-1999	1853	$14,754	$5,242	2-B	1,336	120
Central Connecticut State U, New Britain, CT 06050-4010	1849	$3,972 (S)	$5,824	5-M	12,252	837
Central Methodist Coll, Fayette, MO 65248-1198	1854	$11,760	$4,330	2-M	1,232	97
Central Michigan U, Mount Pleasant, MI 48859	1892	$3,775 (S)	$4,828	5-D	26,775	1,063
Central Missouri State U, Warrensburg, MO 64093	1871	$3,210 (S)	$4,230	5-M	10,936	530
Central State U, Wilberforce, OH 45384	1887	$3,573 (S)	$5,031	5-M	1,129	108
Central Washington U, Ellensburg, WA 98926-7463	1891	$3,162 (S)	$4,821	5-M	8,050	512
Centre Coll, Danville, KY 40422-1394	1819	$22,450 (C)	NA	2-B	1,057	98
Chadron State Coll, Chadron, NE 69337	1911	$2,361 (S)	$3,492	5-M	2,686	123
Chaminade U of Honolulu, Honolulu, HI 96816-1578	1955	$12,160	$5,720	2-M	2,489	318
Champlain Coll, Burlington, VT 05402-0670	1878	$11,005	$7,965	1-B	2,530	185
Chapman U, Orange, CA 92866	1861	$21,314	$7,246	2-F	4,341	455
Charleston Southern U, Charleston, SC 29423-8087	1964	$11,346	$4,362	2-M	2,603	235
Charter Oak State Coll, New Britain, CT 06053-2142	1973	$71/credit (S)	NA	5-B	1,453	66
Chatham Coll, Pittsburgh, PA 15232-2826 (3)	1869	$18,236	$6,246	1-M	1,004	73
Chestnut Hill Coll, Philadelphia, PA 19118-2693 (3)	1924	$19,800	$6,834	2-D	1,561	283
Cheyney U of Pennsylvania, Cheyney, PA 19319	1837	$4,447 (S)	$4,983	5-M	1,496	97
Chicago State U, Chicago, IL 60628	1867	$3,151 (S)	$5,825	5-M	7,580	NA
Christian Brothers U, Memphis, TN 38104-5581	1871	$14,330	$4,470	2-M	2,156	185
Christopher Newport U, Newport News, VA 23606-2998	1960	$3,056 (S)	$5,350	5-M	5,249	335
The Citadel, The Military Coll of South Carolina, Charleston, SC 29409 (2)	1842	$4,288 (S)	$4,350	5-M	3,872	186
City Coll of the City U of New York, New York, NY 10031-9198	1847	$3,309 (S)	NA	11-D	10,362	810
City U, Bellevue, WA 98004-6442	1973	$6,920	NA	1-M	5,410	1,095
Claflin U, Orangeburg, SC 29115	1869	$7,538	$4,116	2-B	1,315	93
Claremont McKenna Coll, Claremont, CA 91711	1946	$22,390	$7,420	1-B	1,002	139
Clarion U of Pennsylvania, Clarion, PA 16214	1867	$4,359 (S)	$2,420	5-M	6,192	369
Clark Atlanta U, Atlanta, GA 30314	1865	$11,120	$5,974	2-D	5,010	NA
Clarke Coll, Dubuque, IA 52001-3198	1843	$14,136	$5,172	2-M	1,118	100
Clarkson U, Potsdam, NY 13699	1896	$21,000	$7,781	1-D	2,877	185
Clark U, Worcester, MA 01610-1477	1887	$24,620	$4,350	1-D	2,910	232
Clayton Coll & State U, Morrow, GA 30260-0285	1969	$2,266 (S)	NA	5-B	4,455	393
Clemson U, Clemson, SC 29634	1889	$3,780 (S)	$4,325	5-D	17,465	1,025
Cleveland State U, Cleveland, OH 44115	1964	$4,110 (S)	$5,200	5-D	15,293	906
Coastal Carolina U, Conway, SC 29528-6054	1954	$3,500 (S)	$5,240	5-M	4,653	295
Coe Coll, Cedar Rapids, IA 52402-5070	1851	$18,415	$5,200	2-M	1,301	121
Colby Coll, Waterville, ME 04901-8840	1813	$32,750 (C)	NA	1-B	1,814	201
Colegio Universitario del Este, Carolina, PR 00984-2010	1949	$3,780	NA	1-B	7,077	439
Coleman Coll, La Mesa, CA 91942-1532	1963	$25,400	NA	1-M	1,060	101
Colgate U, Hamilton, NY 13346-1386	1819	$25,740	$6,330	1-M	2,781	281
Coll Misericordia, Dallas, PA 18612-1098	1924	$16,010	$6,470	2-M	1,765	183
Coll of Aeronautics, Flushing, NY 11369-1037 (2)	1932	$10,550	NA	1-B	1,301	60
Coll of Charleston, Charleston, SC 29424-0001	1770	$3,630 (S)	$4,260	5-B	11,129	753
Coll of Mount St. Joseph, Cincinnati, OH 45233-1670	1920	$14,000	$5,100	2-M	2,216	233
Coll of Mount Saint Vincent, Riverdale, NY 10471-1093	1911	$16,330	$7,300	1-M	1,512	149
The Coll of New Jersey, Ewing, NJ 08628	1855	$5,991 (S)	$6,504	5-M	6,859	669
The Coll of New Rochelle, New Rochelle, NY 10805-2308 (4)	1904	$12,080	$6,250	1-M	6,787	152
Coll of Notre Dame of Maryland, Baltimore, MD 21210-2476 (3)	1873	$15,875	$6,800	2-M	3,178	90
Coll of Saint Benedict, Saint Joseph, MN 56374-2091 (3)	1887	$17,241	$5,272	2-B	2,024	181
Coll of St. Catherine, St. Paul, MN 55105-1789 (3)	1905	$16,442	$4,690	2-M	4,487	474
Coll of Saint Elizabeth, Morristown, NJ 07960-6989 (3)	1899	$14,610	$6,850	2-M	1,772	175
The Coll of Saint Rose, Albany, NY 12203-1419	1920	$13,160	$6,550	1-M	4,231	338
The Coll of St. Scholastica, Duluth, MN 55811-4199	1912	$16,290	$4,952	2-M	1,984	168
Coll of Santa Fe, Santa Fe, NM 87505-7634	1947	$16,110	$5,072	1-M	1,618	257
Coll of Staten Island of the City U of NewYork, Staten Island, NY 10314-6600	1955	$3,316 (S)	NA	11-M	11,115	735
Coll of the Holy Cross, Worcester, MA 01610-2395	1843	$23,815	$7,540	2-B	2,826	268
Coll of the Ozarks, Point Lookout, MO 65726	1906	$150	$2,500	2-B	1,404	123
The Coll of William & Mary, Williamsburg, VA 23187-8795	1693	$4,687 (S)	$5,096	5-D	7,530	718
The Coll of Wooster, Wooster, OH 44691-2363	1866	$21,520	$5,680	2-B	1,837	174
Collins Coll: A School of Design &Technology,Tempe, AZ 85281-5206	1978	$18,800	NA	3-B	1,577	70
Colorado Christian U, Lakewood, CO 80226-7499	1914	$10,400	$5,330	2-M	2,082	237
The Colorado Coll, Colorado Springs, CO 80903-3294	1874	$23,165	$5,808	1-M	1,942	202

Name, address	Year Founded	Tuition & Fees	Rm. & Board	Control, Degree	Enroll-ment	Faculty
Colorado School of Mines, Golden, CO 80401-1887	1874	$5,412 (S)	$4,900	5-D	3,211	355
Colorado State U, Fort Collins, CO 80523-0015	1870	$3,858 (S)	$5,280	5-D	23,098	993
Colorado Tech U, Colorado Springs, CO 80907-3896	1965	$8,043	NA	3-D	1,851	118
Columbia Coll, New York, NY 10027	1754	$25,922	$7,939	1-B	3,913	NA
Columbia Coll, Caguas, PR 00726	1966	$5,018	NA	3-B	1,043	77
Columbia Coll, Columbia, SC 29203-5998 (3)	1854	$15,060	$4,990	2-M	1,398	171
Columbia Coll Chicago, Chicago, IL 60605-1996	1890	$11,780	NA	1-M	9,056	1,188
Columbia Union Coll, Takoma Park, MD 20912-7796	1904	$13,305	$4,619	2-M	1,030	48
Columbia U, School of General Studies, New York, NY 10027-6939	1754	$24,944	$9,000	1-B	1,117	632
Columbia U, The Fu Foundation School of Engineering & Applied Sci, New York, NY 10027	1864	$25,922	$7,939	1-D	1,248	NA
Columbus Coll of Art & Design, Columbus, OH 43215-1758	1879	$15,010	$6,100	1-B	1,546	173
Columbus State U, Columbus, GA 31907-5645	1958	$3,040 (S)	$4,560	5-M	737	335
Concord Coll, Athens, WV 24712-1000	1872	$2,620 (S)	$4,150	5-B	2,955	200
Concordia Coll, Moorhead, MN 56562	1891	$14,136	$3,900	2-B	2,826	253
Concordia U, Irvine, CA 92612-3299	1972	$15,700	$5,600	2-M	1,334	118
Concordia U, River Forest, IL 60305-1499	1864	$13,843	$5,365	2-D	1,900	194
Concordia U, St. Paul, MN 55104-5494	1893	$14,752	$5,160	2-M	1,711	239
Concordia U, Seward, NE 68434-1599	1894	$12,470	$4,136	2-M	1,270	115
Concordia U, Portland, OR 97211-6099	1905	$15,500	$3,820	2-M	1,040	110
Concordia U Wisconsin, Mequon, WI 53097-2402	1881	$12,985	$4,820	2-M	4,603	163
Connecticut Coll, New London, CT 06320-4196	1911	$31,985 (C)	NA	1-M	1,856	192
Converse Coll, Spartanburg, SC 29302-0006 (3)	1889	$15,840	$4,830	1-M	1,513	94
Coppin State Coll, Baltimore, MD 21216-3698	1900	$3,973 (S)	$5,274	5-M	3,765	202
Cornell U, Ithaca, NY 14853-0001	1865	$10,922 (S)	$8,086	1-D	18,995	1,753
Cornerstone U, Grand Rapids, MI 49525-5897	1941	$12,005	$4,830	2-B	1,848	136
Covenant Coll, Lookout Mountain, GA 30750	1955	$17,070	$5,000	2-M	1,158	64
Creighton U, Omaha, NE 68178-0001	1878	$14,910	$5,782	2-D	6,235	886
The Culinary Inst of America, Hyde Park, NY 12538-1499	1946	$18,055	NA	1-B	2,028	131
Cumberland Coll, Williamsburg, KY 40769-1372	1889	$9,920	$4,276	2-M	1,702	99
Cumberland U, Lebanon, TN 37087-3554	1842	$9,900	$3,500	1-M	1,206	100
Curry Coll, Milton, MA 02186-9984	1879	$17,960	$6,870	1-M	2,330	393
Daemen Coll, Amherst, NY 14226-3592	1947	$12,920	$6,100	1-M	1,825	182
Dakota State U, Madison, SD 57042-1799	1881	$3,805 (S)	$2,840	5-M	1,807	102
Dallas Baptist U, Dallas, TX 75211-9299	1965	$9,150	$3,774	2-M	4,032	296
Dalton State Coll, Dalton, GA 30720-3797	1963	$1,508 (S)	NA	5-B	3,139	126
Daniel Webster Coll, Nashua, NH 03063-1300	1965	$16,570	$6,340	1-B	1,099	87
Dartmouth Coll, Hanover, NH 03755	1769	$25,653	$6,723	1-D	5,386	647
Davenport U, Dearborn, MI 48126-3799	1962	$6,984	NA	1-M	3,158	190
Davenport U, Grand Rapids, MI 49503	1866	$9,408	NA	1-M	2,214	127
Davenport U, Kalamazoo, MI 49006-2791	1866	$8,940	NA	1-B	1,081	110
Davenport U, Lansing, MI 48933-2197	1979	$8,364	NA	1-B	1,140	105
Davenport U, Warren, MI 48092-5209	1962	$6,768	NA	1-M	2,561	114
David N. Myers Coll, Cleveland, OH 44115-1096	1848	$8,760	NA	1-M	1,227	86
Davidson Coll, Davidson, NC 28036-1719	1837	$25,095	$6,571	2-B	1,679	166
Delaware State U, Dover, DE 19901-2277	1891	$3,650 (S)	$5,710	5-M	3,103	254
Delaware Valley Coll, Doylestown, PA 18901-2697	1896	$16,448	$6,340	1-M	1,887	194
Delta State U, Cleveland, MS 38733-0001	1924	$2,696 (S)	$2,990	5-D	3,875	302
Denison U, Granville, OH 43023	1831	$22,210	$6,300	1-B	2,108	187
DePaul U, Chicago, IL 60604-2287	1898	$15,420	$6,675	2-D	20,548	1,629
DePauw U, Greencastle, IN 46135-1772	1837	$20,510	$6,324	2-B	2,225	234
DeSales U, Center Valley, PA 18034-9568	1964	$14,790	$6,090	2-M	2,549	114
Des Moines U Osteopathic Medical Ctr, Des Moines, IA 50312-4104	1898	$12,675	NA	1-F	1,153	106
DeVry Coll of Technology, North Brunswick, NJ 08902-3362	1969	$8,315	NA	3-B	3,779	162
DeVry Inst of Technology, Phoenix, AZ 85021-2995	1967	$8,315	NA	3-B	3,705	117
DeVry Inst of Technology, Fremont, CA 94555	1998	$9,315	NA	3-B	2,149	49
DeVry Inst of Technology, Long Beach, CA 90806	1984	$8,315	NA	3-B	2,877	117
DeVry Inst of Technology, Pomona, CA 91768-2642	1983	$8,315	NA	3-B	3,674	143
DeVry Inst of Technology, Alpharetta, GA 30004	1997	$8,315	NA	3-B	1,513	61
DeVry Inst of Technology, Decatur, GA 30030-2198	1969	$8,315	NA	3-B	2,916	118
DeVry Inst of Technology, Addison, IL 60101-6106	1982	$8,250	NA	3-B	4,006	198
DeVry Inst of Technology, Chicago, IL 60618-5994	1931	$8,315	NA	3-B	4,095	145
DeVry Inst of Technology, Tinley Park, IL 60477	2000	$8,315	NA	3-B	1,010	53
DeVry Inst of Technology, Kansas City, MO 64131-3698	1931	$8,250	NA	3-B	2,708	137
DeVry Inst of Technology, Long Island City, NY 11101	1998	$9,315	NA	3-B	1,652	62
DeVry Inst of Technology, Columbus, OH 43209-2705	1952	$8,315	NA	3-B	3,570	122
DeVry Inst of Technology, Irving, TX 75063-2439	1969	$8,250	NA	3-B	3,462	168
Dickinson Coll, Carlisle, PA 17013-2896	1773	$24,450	$6,450	1-B	2,115	193
Dickinson State U, Dickinson, ND 58601-4896	1918	$2,378 (S)	$2,716	5-B	2,012	113
Dillard U, New Orleans, LA 70122-3097	1869	$9,200	$5,350	2-B	1,953	165
Doane Coll, Crete, NE 68333-2430	1872	$12,826	$3,900	2-M	2,135	125
Dominican Coll, Orangeburg, NY 10962-1210	1952	$13,910	$7,400	1-M	1,636	NA
Dominican U, River Forest, IL 60305-1099	1901	$14,820	$5,030	2-M	2,317	205
Dominican U of California, San Rafael, CA 94901-2298	1890	$17,606	$8,440	2-M	1,369	199
Dordt Coll, Sioux Center, IA 51250-1697	1955	$13,350	$3,800	2-M	1,420	124
Dowling Coll, Oakdale, NY 11769-1999	1955	$14,090	NA	1-D	6,885	430
Drake U, Des Moines, IA 50311-4516	1881	$17,090	$4,870	1-D	5,126	293
Drew U, Madison, NJ 07940-1493	1867	$24,018	$6,782	2-D	2,412	157
Drexel U, Philadelphia, PA 19104-2875	1891	$16,150	$7,842	1-D	13,128	911
Drury U, Springfield, MO 65802-3791	1873	$11,204	$4,304	1-M	1,762	146
Duke U, Durham, NC 27708-0586	1838	$25,630	$7,387	2-D	12,192	2,179
Duquesne U, Pittsburgh, PA 15282-0001	1878	$16,520	$6,504	2-D	9,667	847
D'Youville Coll, Buffalo, NY 14201-1084	1908	$11,930	$5,720	1-M	2,397	175
Earlham Coll, Richmond, IN 47374-4095	1847	$21,070	$4,936	2-B	1,104	105
East Carolina U, Greenville, NC 27858-4353	1907	$2,257 (S)	$4,220	5-D	18,750	1,091
East Central U, Ada, OK 74820-6899	1909	$1,962 (S)	$1,167	5-M	4,110	265
Eastern Coll, St. Davids, PA 19087-3696	1952	$15,150	$6,490	2-M	2,400	290
Eastern Connecticut State U, Willimantic, CT 06226-2295	1889	$4,146 (S)	$5,850	5-M	5,145	351

Name, address	Year Founded	Tuition & Fees	Rm. & Board	Control, Degree	Enroll-ment	Faculty
Eastern Illinois U, Charleston, IL 61920-3099	1895	$3,688 (S)	$5,400	5-M	10,635	664
Eastern Kentucky U, Richmond, KY 40475-3102	1906	$2,542 (S)	$3,796	5-M	14,428	604
Eastern Mennonite U, Harrisonburg, VA 22802-2462	1917	$14,150	$5,120	2-F	1,398	127
Eastern Michigan U, Ypsilanti, MI 48197	1849	$3,900 (S)	$5,016	5-D	23,181	1,099
Eastern Nazarene Coll, Quincy, MA 02170-2999	1918	$14,062	$4,550	2-M	1,387	65
Eastern New Mexico U, Portales, NM 88130	1934	$1,944 (S)	$3,690	5-M	3,564	198
Eastern Washington U, Cheney, WA 99004-2431	1882	$3,012 (S)	$4,558	5-M	8,597	636
East Stroudsburg U of Pennsylvania, East Stroudsburg, PA 18301-2999	1893	$4,492 (S)	$3,938	5-M	5,802	271
East Tennessee State U, Johnson City, TN 37614	1911	$2,759 (S)	$3,818	5-D	11,063	730
East Texas Baptist U, Marshall, TX 75670-1498	1912	$8,450	$3,126	2-B	1,402	109
Eckerd Coll, St. Petersburg, FL 33711	1958	$18,785	$5,110	2-B	1,572	136
Edgewood Coll, Madison, WI 53711-1997	1927	$12,450	$4,516	2-M	2,077	190
Edinboro U of Pennsylvania, Edinboro, PA 16444	1857	$4,569 (S)	$4,104	5-M	7,212	395
Elizabeth City State U, Elizabeth City, NC 27909-7806	1891	$1,684 (S)	$4,008	5-B	1,920	139
Elizabethtown Coll, Elizabethtown, PA 17022-2298	1899	$19,100	$5,600	2-M	1,825	192
Elmhurst Coll, Elmhurst, IL 60126-3296	1871	$15,000	$5,500	2-M	2,611	267
Elmira Coll, Elmira, NY 14901	1855	$22,540	$7,280	1-B	1,581	107
Elon U, Elon College, NC 27244	1889	$13,781	$4,660	2-M	4,138	270
Embry-Riddle Aeronautical U, Prescott, AZ 86301-3720 (2)	1978	$10,420	$5,050	1-B	1,689	105
Embry-Riddle Aeronautical U, Daytona Beach, FL 32114-3900 (2)	1926	$11,690	$5,390	1-M	4,803	254
Embry-Riddle Aeronautical U, Extended Campus, Daytona Beach, FL 32114-3900 (2)	1970	$2,064	NA	1-M	8,044	2,836
Emerson Coll, Boston, MA 02116-4624	1880	$20,014	$9,020	1-D	4,074	304
Emmanuel Coll, Boston, MA 02115	1919	$16,412	$7,390	2-M	1,524	78
Emory U, Atlanta, GA 30322-1100	1836	$24,532	$7,868	2-D	11,294	NA
Emporia State U, Emporia, KS 66801-5087	1863	$2,744 (S)	$3,774	5-D	5,614	338
Endicott Coll, Beverly, MA 01915-2096	1939	$15,116	$7,698	1-M	1,958	143
Evangel U, Springfield, MO 65802-2191	1955	$9,810	$3,790	2-M	1,538	106
The Evergreen State Coll, Olympia, WA 98505	1967	$2,997 (S)	$5,244	5-M	4,125	306
Excelsior Coll, Albany, NY 12203-5159	1970	$1,150	NA	1-M	18,041	NA
Fairfield U, Fairfield, CT 06430-5195	1942	$21,435	$7,630	2-M	5,188	490
Fairleigh Dickinson U, Florham-Madison Campus, Madison, NJ 07940-1099	1942	$16,346	$6,842	1-M	3,451	262
Fairleigh Dickinson U, Teaneck–Hackensack Campus, Teaneck, NJ 07666	1942	$16,346	$6,842	1-D	5,931	466
Fairmont State Coll, Fairmont, WV 26554	1865	$2,316 (S)	$4,084	5-B	6,496	440
Fashion Inst of Technology, New York, NY 10001-5992	1944	$3,314 (S)	$7,535	11-M	10,813	883
Faulkner U, Montgomery, AL 36109-3398	1942	$7,800	$3,950	2-F	2,660	57
Fayetteville State U, Fayetteville, NC 28301-4298	1867	$1,770 (S)	$3,800	5-D	4,373	234
Felician Coll, Lodi, NJ 07644-2198	1942	$11,690	$2,980	2-M	1,531	152
Ferris State U, Big Rapids, MI 49307	1884	$4,386 (S)	$5,258	5-F	9,847	662
Fitchburg State Coll, Fitchburg, MA 01420-2697	1894	$3,018 (S)	$4,680	5-M	5,715	303
Flagler Coll, St. Augustine, FL 32085-1027	1968	$6,320	$3,910	1-B	1,830	152
Florida Ag & Mech U, Tallahassee, FL 32307-3200	1887	$2,269 (S)	$4,214	5-D	12,126	NA
Florida Atlantic U, Boca Raton, FL 33431-0991	1961	$2,396 (S)	$4,993	5-D	21,229	1,210
Florida Gulf Coast U, Fort Myers, FL 33965-6565	1991	$2,201 (S)	$7,000	5-M	3,664	180
Florida Inst of Technology, Melbourne, FL 32901-6975	1958	$18,450	$5,590	1-D	4,248	236
Florida Intl U, Miami, FL 33199	1965	$2,410 (S)	NA	5-D	31,945	1,067
Florida Metro U–Orlando Coll, South, Orlando, FL 32809	NR	$6,338	NA	3-M	1,500	NA
Florida Metro U–Tampa Coll, Tampa, FL 33614-5899	1890	$7,206	NA	3-M	1,218	54
Florida State U, Tallahassee, FL 32306	1857	$2,379 (S)	$5,150	5-D	33,951	NA
Fontbonne Coll, St. Louis, MO 63105-3098	1917	$11,843	$5,200	2-M	2,059	180
Fordham U, New York, NY 10458	1841	$20,660	$8,310	2-D	13,650	1,125
Fort Hays State U, Hays, KS 67601-4099	1902	$2,178 (S)	$2,375	5-M	5,506	NA
Fort Lewis Coll, Durango, CO 81301-3999	1911	$2,330 (S)	$5,172	5-B	4,289	244
Fort Valley State U, Fort Valley, GA 31030-3298	1895	$1,742 (S)	$1,801	5-D	2,561	148
Framingham State Coll, Framingham, MA 01701-9101	1839	$2,830 (S)	$4,154	5-M	6,626	217
Franciscan U of Steubenville, Steubenville, OH 43952-1763	1946	$12,970	$5,070	2-M	2,154	158
Francis Marion U, Florence, SC 29501-0547	1970	$3,600 (S)	$3,720	5-M	3,569	207
Franklin & Marshall Coll, Lancaster, PA 17604-3003	1787	$24,866	$5,994	1-B	1,892	187
Franklin Coll of Indiana, Franklin, IN 46131-2598	1834	$14,245	$4,590	2-B	1,020	107
Franklin Pierce Coll, Rindge, NH 03461-0060	1962	$18,985	$6,250	1-M	1,489	150
Franklin U, Columbus, OH 43215-5399	1902	$5,983	NA	1-M	5,087	255
Freed-Hardeman U, Henderson, TN 38340-2399	1869	$8,990	$4,620	2-M	1,886	116
Fresno Pacific U, Fresno, CA 93702-4709	1944	$14,254	$4,530	2-M	1,686	317
Friends U, Wichita, KS 67213	1898	$11,660	$3,540	1-M	3,247	225
Frostburg State U, Frostburg, MD 21532-1099	1898	$4,132 (S)	$5,214	5-M	5,348	344
Furman U, Greenville, SC 29613	1826	$19,156	$5,144	1-M	3,272	273
Gallaudet U, Washington, DC 20002-3625	1864	$8,420	$7,340	1-D	1,661	221
Gannon U, Erie, PA 16541-0001	1925	$14,112	$5,670	2-D	3,377	298
Gardner-Webb U, Boiling Springs, NC 28017	1905	$12,010	$4,760	2-M	3,194	131
George Fox U, Newberg, OR 97132-2697	1891	$17,610	$5,550	2-D	2,635	217
George Mason U, Fairfax, VA 22030-4444	1957	$3,768 (S)	$5,240	5-D	23,408	1,401
Georgetown Coll, Georgetown, KY 40324-1696	1829	$11,590	$4,400	2-M	1,736	150
Georgetown U, Washington, DC 20057	1789	$24,168	$9,103	2-D	12,427	895
The George Washington U, Washington, DC 20052	1821	$25,040	$8,538	1-D	20,527	1,546
Georgia Coll & State U, Milledgeville, GA 31061	1889	$2,358 (S)	$4,312	5-M	5,076	345
Georgia Inst of Technology, Atlanta, GA 30332-0001	1885	$3,308 (S)	$5,234	5-D	14,804	728
Georgian Court Coll, Lakewood, NJ 08701-2697 (3)	1908	$13,092	$4,700	2-M	2,486	209
Georgia Southern U, Statesboro, GA 30460	1906	$2,500 (S)	$4,154	5-D	14,184	745
Georgia Southwestern State U, Americus, GA 31709-4693	1906	$2,398 (S)	$3,690	5-M	2,622	151
Georgia State U, Atlanta, GA 30303-3083	1913	$3,132 (S)	NA	5-D	23,619	1,276
Gettysburg Coll, Gettysburg, PA 17325-1483	1832	$24,875	$5,956	2-B	2,218	252
Global U of the Assemblies of God, Springfield, MO 65804	1948	$1,800	NA	2-M	2,756	450
Golden Gate U, San Francisco, CA 94105-2968	1853	$8,832	NA	1-D	5,379	846
Goldey-Beacom Coll, Wilmington, DE 19808-1999	1886	$9,398	NA	1-M	1,400	49
Gonzaga U, Spokane, WA 99258	1887	$17,560	$5,500	2-D	4,347	434
Gordon Coll, Wenham, MA 01984-1899	1889	$17,270	$5,200	2-M	1,617	116

Name, address	Year Founded	Tuition & Fees	Rm. & Board	Control, Degree	Enroll-ment	Faculty
Goshen Coll, Goshen, IN 46526-4794	1894	$13,140	$4,640	2-B	1,084	136
Goucher Coll, Baltimore, MD 21204-2794	1885	$21,290	$7,650	1-M	1,973	156
Governors State U, University Park, IL 60466-0975	1969	$2,520 (S)	NA	5-M	6,116	201
Grace Coll, Winona Lake, IN 46590-1294	1948	$10,998	$4,685	2-M	1,059	80
Graceland U, Lamoni, IA 50140	1895	$12,350	$4,100	2-M	3,192	95
Grambling State U, Grambling, LA 71245	1901	$2,301 (S)	$2,636	5-D	5,070	224
Grand Canyon U, Phoenix, AZ 85061-1097	1949	$9,660	$4,286	2-M	3,615	190
Grand Valley State U, Allendale, MI 49401-9403	1960	$4,272 (S)	$5,030	5-M	18,579	1,063
Grand View Coll, Des Moines, IA 50316-1599	1896	$12,886	$4,166	2-B	1,380	126
Greenville Coll, Greenville, IL 62246-0159	1892	$12,974	$4,994	2-M	1,179	121
Grinnell Coll, Grinnell, IA 50112-1690	1846	$20,500	$5,820	1-B	1,344	142
Grove City Coll, Grove City, PA 16127-2104	1876	$7,710	$4,206	2-B	2,332	158
Guilford Coll, Greensboro, NC 27410-4173	1837	$16,970	$5,610	2-B	1,246	105
Gustavus Adolphus Coll, St. Peter, MN 56082-1498	1862	$18,330	$4,605	2-B	2,560	226
Gwynedd-Mercy Coll, Gwynedd Valley, PA 19437-0901	1948	$14,400 (C)	NA	2-M	1,941	185
Hamilton Coll, Clinton, NY 13323-1296	1812	$26,100	$6,440	1-B	1,767	193
Hamline U, St. Paul, MN 55104-1284	1854	$16,275	$5,445	2-D	3,380	275
Hampshire Coll, Amherst, MA 01002	1965	$26,125	$6,814	1-B	1,172	114
Hampton U, Hampton, VA 23668	1868	$16,200 (C)	NA	1-D	5,743	356
Hannibal-LaGrange Coll, Hannibal, MO 63401-1999	1858	$8,830	$3,226	2-B	1,104	89
Hanover Coll, Hanover, IN 47243-0108	1827	$11,770	$4,930	2-B	1,142	99
Harding U, Searcy, AR 72149-0001	1924	$8,425	$4,336	2-M	4,337	281
Hardin-Simmons U, Abilene, TX 79698-0001	1891	$9,950	$3,405	2-F	2,304	190
Hartwick Coll, Oneonta, NY 13820-4020	1797	$25,085	$6,530	1-B	1,419	166
Harvard U, Cambridge, MA 02138	1636	$25,128	$7,982	1-D	17,850	2,336
Hastings Coll, Hastings, NE 68901-7696	1882	$13,666	$4,188	2-M	1,130	109
Haverford Coll, Haverford, PA 19041-1392	1833	$24,940	$7,910	1-B	1,135	112
Hawaii Pacific U, Honolulu, HI 96813-2785	1965	$8,920	$8,350	1-M	8,228	707
Heidelberg Coll, Tiffin, OH 44883-2462	1850	$17,278	$5,974	2-M	1,562	122
Henderson State U, Arkadelphia, AR 71999-0001	1890	$2,795 (S)	$3,152	5-M	3,430	220
Hendrix Coll, Conway, AR 72032-3080	1876	$12,665	$4,625	2-B	1,137	93
Heritage Coll, Toppenish, WA 98948-9599	1982	$5,430	NA	1-M	1,127	130
High Point U, High Point, NC 27262-3598	1924	$12,440	$5,770	2-M	2,788	212
Hillsdale Coll, Hillsdale, MI 49242-1298	1844	$13,840	$5,700	1-B	1,138	131
Hiram Coll, Hiram, OH 44234-0067	1850	$18,440	$6,254	2-B	1,199	100
Hobart & William Smith Colls, Geneva, NY 14456-3397	1822	$25,192	$6,808	1-B	1,854	151
Hofstra U, Hempstead, NY 11549	1935	$15,042	$7,240	1-D	13,144	1,226
Hollins U, Roanoke, VA 24020-1603 (3)	1842	$17,210	$6,415	1-M	1,033	99
Holy Family Coll, Philadelphia, PA 19114-2094	1954	$12,760	NA	2-M	2,546	236
Hood Coll, Frederick, MD 21701-8575 (3)	1893	$18,620	$6,900	1-M	1,708	92
Hope Coll, Holland, MI 49422-9000	1866	$16,644	$5,224	2-B	3,015	275
Houghton Coll, Houghton, NY 14744	1883	$15,180	$5,400	2-B	1,409	97
Houston Baptist U, Houston, TX 77074-3298	1960	$11,142	$4,080	2-M	2,416	191
Howard Payne U, Brownwood, TX 76801-2715	1889	$9,000	$3,820	2-B	1,480	131
Howard U, Washington, DC 20059-0002	1867	$9,330	$4,304	1-D	10,700	1,305
Humboldt State U, Arcata, CA 95521-8299	1913	$1,861 (S)	$5,845	5-M	7,433	551
Hunter Coll of the City U of NewYork, NewYork, NY 10021-5085	1870	$3,343 (S)	NA	11-M	20,012	1,253
Husson Coll, Bangor, ME 04401-2999	1898	$9,580	$5,150	1-M	1,892	151
The Illinois Inst of Art, Chicago, IL 60654	1916	$12,780	NA	3-B	1,250	80
Illinois Inst of Technology, Chicago, IL 60616-3793	1890	$18,100	$5,428	1-D	6,003	531
Illinois State U, Normal, IL 61790-2200	1857	$4,482 (S)	$4,544	5-D	20,755	1,126
Illinois Wesleyan U, Bloomington, IL 61702-2900	1850	$20,410	$5,150	1-B	2,102	183
Immaculata Coll, Immaculata, PA 19345-0500 (3)	1920	$13,950	$6,800	2-D	3,278	220
Indiana State U, Terre Haute, IN 47809-1401	1865	$3,564 (S)	$4,604	5-D	11,051	NA
Indiana U Bloomington, Bloomington, IN 47405	1820	$4,405 (S)	$5,608	5-D	37,076	NA
Indiana U East, Richmond, IN 47374-1289	1971	$3,206 (S)	NA	5-B	2,335	NA
Indiana U Kokomo, Kokomo, IN 46904-9003	1945	$3,212 (S)	NA	5-M	2,682	NA
Indiana U Northwest, Gary, IN 46408-1197	1959	$3,237 (S)	NA	5-M	4,649	NA
Indiana U of Pennsylvania, Indiana, PA 15705-1087	1875	$4,617 (S)	$3,966	5-D	13,410	738
Indiana U–Purdue U Fort Wayne, Fort Wayne, IN 46805-1499	1917	$2,946 (S)	NA	5-M	10,532	637
Indiana U–Purdue U Indianapolis, Indianapolis, IN 46202-2896	1969	$3,881 (S)	$5,098	5-D	27,525	NA
Indiana U South Bend, South Bend, IN 46634-7111	1922	$3,300 (S)	NA	5-M	7,252	NA
Indiana U Southeast, New Albany, IN 47150-6405	1941	$3,248 (S)	NA	5-M	6,427	NA
Indiana Wesleyan U, Marion, IN 46953-4999	1920	$12,250	$4,740	2-M	6,969	166
Inter American U of Puerto Rico, Aguadilla Campus, Aguadilla, PR 00605	1957	$2,978	NA	1-B	3,370	201
Inter American U of Puerto Rico, Arecibo Campus, Arecibo, PR 00614-4050	1957	$2,990	NA	1-M	3,926	234
Inter American U of Puerto Rico, Barranquitas Campus, Barranquitas, PR 00794	1957	$3,000	NA	1-B	1,710	90
Inter American U of Puerto Rico, Bayamón Campus, Bayamón, PR 00957	1912	$3,236	NA	1-B	4,960	280
Inter American U of Puerto Rico, Fajardo Campus, Fajardo, PR 00738-7003	1965	$3,600	NA	1-B	1,971	163
Inter American U of Puerto Rico, Guayama Campus, Guayama, PR 00785	1958	$3,700	NA	1-B	1,246	132
Inter American U of Puerto Rico, Metro Campus, San Juan, PR 00919-1293	1960	$3,830	NA	1-D	10,586	643
Inter American U of Puerto Rico, Ponce Campus, Mercedita, PR 00715-1602	1962	$3,900	NA	1-B	4,123	212
Inter American U of Puerto Rico, San Germán Campus, San Germán, PR 00683-5008	1912	$3,030	$2,400	1-D	5,309	307
Intl Acad of Design & Technology, Tampa, FL 33634-7350	1984	$10,260	NA	3-B	1,359	84
Intl Acad of Design & Technology, Chicago, IL 60602-9736	1977	$10,890	NA	3-B	1,739	129
Intl Coll, Naples, FL 34119	1990	$7,230	NA	1-M	1,026	66
Intl Fine Arts Coll, Miami, FL 33132-1121	1965	$13,025	NA	3-M	1,100	93
Iona Coll, New Rochelle, NY 10801-1890	1940	$15,870	$8,835	2-M	4,469	344
Iowa State U of Sci & Technology, Ames, IA 50011	1858	$3,132 (S)	$4,432	5-D	26,845	1,662

Name, address	Year Founded	Tuition & Fees	Rm. & Board	Control, Degree	Enroll-ment	Faculty
Ithaca Coll, Ithaca, NY 14850-7020	1892	$19,192	$8,284	1-M	6,170	534
Jackson State U, Jackson, MS 39217	1877	$2,788 (S)	$3,492	5-D	6,820	412
Jacksonville State U, Jacksonville, AL 36265-1602	1883	$2,640 (S)	$3,080	5-M	8,002	370
Jacksonville U, Jacksonville, FL 32211-3394	1934	$15,510	$5,400	1-M	2,052	231
James Madison U, Harrisonburg, VA 22807	1908	$4,000 (S)	$5,290	5-D	15,326	934
Jamestown Coll, Jamestown, ND 58405	1883	$7,550	$3,300	2-B	1,195	75
John Brown U, Siloam Springs, AR 72761-2121	1919	$11,492	$4,478	2-M	1,517	120
John Carroll U, University Heights, OH 44118-4581	1886	$17,837	$6,312	2-M	4,384	402
John F. Kennedy U, Orinda, CA 94563-2603	1964	$10,746	NA	1-D	1,567	930
John Jay Coll of Criminal Justice of the City U of New York, New York, NY 10019-1093	1964	$3,312 (S)	NA	11-M	10,472	628
Johns Hopkins U, Baltimore, MD 21218-2699	1876	$24,930	$8,185	1-D	5,285	1,104
Johnson & Wales U, Providence, RI 02903-3703	1914	$13,845	$5,970	1-D	9,172	358
Johnson C. Smith U, Charlotte, NC 28216-5398	1867	$11,605	$4,035	1-B	1,585	119
Johnson State Coll, Johnson, VT 05656-9405	1828	$5,047 (S)	$5,346	5-M	1,450	181
Jones Intl U, Englewood, CO 80112	1995	$4,290	NA	1-M	1,796	38
Judson Coll, Elgin, IL 60123-1498	1963	$13,890	$5,360	2-B	1,111	48
Juniata Coll, Huntingdon, PA 16652-2119	1876	$19,360	$5,290	2-B	1,291	111
Kalamazoo Coll, Kalamazoo, MI 49006-3295	1833	$19,764	$5,961	2-B	1,322	104
Kansas State U, Manhattan, KS 66506	1863	$2,333 (S)	$4,240	5-D	21,929	NA
Kean U, Union, NJ 07083	1855	$4,612 (S)	NA	5-M	11,468	371
Keene State Coll, Keene, NH 03435	1909	$5,304 (S)	$5,086	5-M	4,573	377
Kennesaw State U, Kennesaw, GA 30144-5591	1963	$2,306 (S)	NA	5-M	13,392	627
Kent State U, Kent, OH 44242-0001	1910	$5,285 (S)	$4,764	5-D	21,924	1,205
Kentucky State U, Frankfort, KY 40601	1886	$2,440 (S)	$3,740	12-M	2,254	121
Kenyon Coll, Gambier, OH 43022-9623	1824	$26,080	$4,370	1-B	1,599	155
Kettering U, Flint, MI 48504-4898	1919	$16,110	$4,340	1-M	3,341	157
King's Coll, Wilkes-Barre, PA 18711-0801	1946	$16,700	$6,890	2-M	2,233	184
Knox Coll, Galesburg, IL 61401	1837	$21,174	$5,436	1-B	1,199	126
Kutztown U of Pennsylvania, Kutztown, PA 19530-0730	1866	$4,667 (S)	$4,522	5-M	8,068	430
Lafayette Coll, Easton, PA 18042-1798	1826	$22,929	$7,106	2-B	2,279	225
Lake Forest Coll, Lake Forest, IL 60045-2399	1857	$21,190	$5,000	1-M	1,270	126
Lake Superior State U, Sault Sainte Marie, MI 49783-1626	1946	$4,014 (S)	$5,078	5-M	3,118	234
Lamar U, Beaumont, TX 77710	1923	$2,776 (S)	$3,591	5-D	8,568	416
Lander U, Greenwood, SC 29649-2099	1872	$3,955 (S)	$3,950	5-M	2,935	189
La Roche Coll, Pittsburgh, PA 15237-5898	1963	$12,000	$6,474	2-M	1,875	193
La Salle U, Philadelphia, PA 19141-1199	1863	$18,400	$7,676	2-D	5,567	454
La Sierra U, Riverside, CA 92515-8247	1922	$14,910	$4,191	2-D	1,466	NA
Lawrence Technological U, Southfield, MI 48075-1058	1932	$11,300	NA	1-M	4,117	NA
Lawrence U, Appleton, WI 54912-0599	1847	$21,855	$4,791	1-B	1,289	153
Lebanon Valley Coll, Annville, PA 17003-1400	1866	$18,420	$5,680	2-M	2,100	176
Lee U, Cleveland, TN 37320-3450	1918	$6,862	$4,020	2-M	3,361	257
Lehigh U, Bethlehem, PA 18015-3094	1865	$24,000	$6,440	1-D	6,509	457
Lehman Coll of the City U of New York, Bronx, NY 10468-1589	1931	$3,320 (S)	NA	11-M	9,074	694
Le Moyne Coll, Syracuse, NY 13214-1399	1946	$15,870	$6,760	2-M	3,130	237
LeMoyne-Owen Coll, Memphis, TN 38126-6595	1862	$7,500	$4,200	2-B	1,013	79
Lenoir-Rhyne Coll, Hickory, NC 28603	1891	$13,356	$4,920	2-M	1,483	159
Lesley U, Cambridge, MA 02138-2790 (4)	1909	$16,475	$7,520	1-D	5,630	37
LeTourneau U, Longview, TX 75607-7001	1946	$12,840	$5,420	2-M	2,981	167
Lewis & Clark Coll, Portland, OR 97219-7899	1867	$21,520	$6,100	1-F	3,014	321
Lewis-Clark State Coll, Lewiston, ID 83501-2698	1893	$4,720 (S)	$3,050	5-B	2,702	127
Lewis U, Romeoville, IL 60446	1932	$13,320	$6,000	2-M	4,304	NA
Liberty U, Lynchburg, VA 24502	1971	$8,950	$4,800	2-D	6,192	256
Limestone Coll, Gaffney, SC 29340	1845	$10,100	$4,800	1-B	1,967	71
Lincoln Memorial U, Harrogate, TN 37752-1901	1897	$9,600	$3,900	1-M	1,753	125
Lincoln U, Jefferson City, MO 65102	1866	$3,598 (S)	$3,790	5-M	3,347	158
Lincoln U, Lincoln University, PA 19352	1854	$5,621 (S)	$5,256	12-M	2,008	131
Lindenwood U, St. Charles, MO 63301-1695	1827	$11,250	$5,600	2-M	6,056	136
Lindsey Wilson Coll, Columbia, KY 42728-1298	1903	$9,798	$4,680	2-M	1,415	98
Linfield Coll, McMinnville, OR 97128-6894	1849	$18,600	$5,350	2-B	1,534	145
Lipscomb U, Nashville, TN 37204-3951	1891	$10,268	$5,080	2-F	2,545	217
Lock Haven U of Pennsylvania, Lock Haven, PA 17745-2390	1870	$4,548 (S)	$4,264	5-M	4,105	231
Loma Linda U, Loma Linda, CA 92350	1905	$14,220	$8,370	2-D	3,110	2,330
Long Island U, Brooklyn Campus, Brooklyn, NY 11201-8423	1926	$16,920	$5,730	1-D	7,782	1,040
Long Island U, C.W. Post Campus, Brookville, NY 11548-1300	1954	$16,840	$6,790	1-D	13,147	330
Long Island U, Southampton Coll, Southampton, NY 11968-4198	1963	$16,950	$7,990	1-M	2,725	223
Longwood Coll, Farmville, VA 23909-1800	1839	$4,003 (S)	$4,820	5-M	3,961	239
Loras Coll, Dubuque, IA 52004-0178	1839	$15,329	$5,475	2-M	1,725	170
Louisiana State U & Ag & Mech Coll, Baton Rouge, LA 70803	1860	$3,395 (S)	$4,270	5-D	30,861	1,395
Louisiana State U Health Scis Ctr, New Orleans, LA 70112-2223	1931	$3,659 (S)	NA	5-D	2,720	3,000
Louisiana State U in Shreveport, Shreveport, LA 71115-2399	1965	$2,300 (S)	NA	5-M	4,108	228
Louisiana Tech U, Ruston, LA 71272	1894	$2,748 (S)	$3,195	5-D	10,363	442
Lourdes Coll, Sylvania, OH 43560-2898	1958	$9,624	NA	2-B	1,270	117
Loyola Coll in Maryland, Baltimore, MD 21210-2699	1852	$21,515	$6,040	2-D	6,073	443
Loyola Marymount U, Los Angeles, CA 90045-2659	1911	$20,834	$7,100	2-F	7,515	749
Loyola U Chicago, Chicago, IL 60611-2196	1870	$18,726	$7,266	2-D	12,605	2,049
Loyola U New Orleans, New Orleans, LA 70118-6195	1912	$16,700	$6,806	2-F	5,279	386
Lubbock Christian U, Lubbock, TX 79407-2099	1957	$9,824	$3,588	2-M	1,617	131
Luther Coll, Decorah, IA 52101-1045	1861	$18,080	$3,914	2-B	2,621	221
Lycoming Coll, Williamsport, PA 17701-5192	1812	$18,540	$5,145	2-B	1,402	103
Lynchburg Coll, Lynchburg, VA 24501-3199	1903	$18,105	$4,400	2-M	1,932	190
Lyndon State Coll, Lyndonville, VT 05851-0919	1911	$5,028 (S)	$5,346	5-M	1,179	122
Lynn U, Boca Raton, FL 33431-5598	1962	$19,250	$6,800	1-D	2,037	168
Macalester Coll, St. Paul, MN 55105-1899	1874	$21,614	$5,932	2-B	1,794	212
Macon State Coll, Macon, GA 31206-5144	1968	$1,438 (S)	NA	5-B	4,118	215
Madonna U, Livonia, MI 48150-1173	1947	$7,120	$4,852	2-M	3,819	266
Malone Coll, Canton, OH 44709-3897	1892	$13,030	$5,450	2-M	2,162	192
Manchester Coll, North Manchester, IN 46962-1225	1889	$14,610	$5,300	2-M	1,106	83

Name, address	Year Founded	Tuition & Fees	Rm. & Board	Control, Degree	Enroll-ment	Faculty
Manhattan Coll, Riverdale, NY 10471	1853	$19,500	$7,550	2-M	2,943	244
Manhattanville Coll, Purchase, NY 10577-2132	1841	$20,410	$8,320	1-M	2,200	198
Mansfield U of Pennsylvania, Mansfield, PA 16933	1857	$4,766 (S)	$4,198	5-M	3,113	190
Marian Coll, Indianapolis, IN 46222-1997	1851	$14,836	$5,046	2-B	1,411	145
Marian Coll of Fond du Lac, Fond du Lac, WI 54935-4699	1936	$12,979	$4,482	2-M	2,016	81
Marietta Coll, Marietta, OH 45750-4000	1835	$18,240	$5,196	1-M	1,262	118
Marist Coll, Poughkeepsie, NY 12601-1387	1929	$15,746	$7,828	1-M	5,498	508
Marquette U, Milwaukee, WI 53201-1881	1881	$17,336	$6,090	2-D	10,887	1,057
Marshall U, Huntington, WV 25755	1837	$2,620 (S)	$4,772	5-D	13,643	737
Mars Hill Coll, Mars Hill, NC 28754	1856	$12,800	$4,500	2-B	1,224	143
Martin Luther Coll, New Ulm, MN 56073	1995	$5,195	$2,340	2-B	1,026	102
Mary Baldwin Coll, Staunton, VA 24401 (4)	1842	$15,310	$7,450	2-M	1,451	125
Marygrove Coll, Detroit, MI 48221-2599 (4)	1905	$10,650	$5,200	2-M	5,644	72
Maryland Inst, Coll of Art, Baltimore, MD 21217-4191	1826	$20,080	NA	1-M	1,302	220
Marylhurst U, Marylhurst, OR 97036-0261	1893	$10,770	$5,928	2-M	1,071	477
Marymount Manhattan Coll, New York, NY 10021-4597	1936	$14,180	NA	1-B	2,497	352
Marymount U, Arlington, VA 22207-4299	1950	$14,420	$6,350	2-M	3,422	344
Maryville U of Saint Louis, St. Louis, MO 63141-7299	1872	$13,000	$5,600	1-M	3,055	281
Mary Washington Coll, Fredericksburg, VA 22401-5358	1908	$3,246 (S)	$5,448	5-M	4,282	322
Marywood U, Scranton, PA 18509-1598	1915	$16,487	$6,900	2-D	2,859	237
Massachusetts Coll of Art, Boston, MA 02115-5882	1873	$3,883 (S)	$7,192	5-M	2,315	207
Massachusetts Coll of Liberal Arts, North Adams, MA 01247-4100	1894	$3,357 (S)	$4,290	5-M	1,520	121
Massachusetts Coll of Pharmacy & Health Scis, Boston, MA 02115-5896	1823	$17,343	$8,400	1-D	1,886	128
Massachusetts Inst of Technology, Cambridge, MA 02139-4307	1861	$26,050	$7,175	1-D	10,090	1,763
The Master's Coll & Seminary, Santa Clarita, CA 91321-1200	1927	$14,300	$5,800	2-F	1,435	108
McKendree Coll, Lebanon, IL 62254-1299	1828	$12,300	$4,570	2-B	1,993	178
McMurry U, Abilene, TX 79697	1923	$10,425	$4,379	2-B	1,344	110
McNeese State U, Lake Charles, LA 70609	1939	$2,471 (S)	$2,620	5-M	7,603	381
MCP Hahnemann U, Philadelphia, PA 19102-1192	1848	$11,155	$8,526	1-D	2,451	193
Medaille Coll, Buffalo, NY 14214-2695	1875	$11,960	$5,500	1-M	1,618	184
Medgar Evers Coll of the City U of New York, Brooklyn, NY 11225-2298	1969	$3,282 (S)	NA	11-B	4,700	424
Medical Coll of Georgia, Augusta, GA 30912	1828	$2,974 (S)	NA	5-D	1,930	764
Medical U of South Carolina, Charleston, SC 29425-0002	1824	$5,180 (S)	NA	5-D	2,346	1,248
Mercer U, Macon, GA 31207-0003	1833	$17,028	$5,378	2-D	6,908	511
Mercy Coll, Dobbs Ferry, NY 10522-1189	1951	$8,500	$7,700	1-M	10,000	665
Mercyhurst Coll, Erie, PA 16546	1926	$14,340	$5,364	2-M	3,222	204
Meredith Coll, Raleigh, NC 27607-5298 (3)	1891	$9,840	$4,260	2-M	2,595	267
Merrimack Coll, North Andover, MA 01845-5800	1947	$16,775	$7,750	2-M	2,587	215
Mesa State Coll, Grand Junction, CO 81501	1925	$2,123 (S)	$5,048	5-M	4,892	301
Messiah Coll, Grantham, PA 17027	1909	$16,130	$5,770	2-B	2,797	242
Methodist Coll, Fayetteville, NC 28311-1420	1956	$13,410	$5,080	2-M	2,134	NA
Metro State Coll of Denver, Denver, CO 80217-3362	1963	$1,926 (S)	NA	5-B	17,688	1,025
Metro State U, St. Paul, MN 55106-5000	1971	$2,942 (S)	NA	5-M	5,623	520
Miami U, Oxford, OH 45056	1809	$6,403 (S)	$5,590	12-D	16,290	975
Michigan State U, East Lansing, MI 48824	1855	$5,772 (S)	$4,472	5-D	43,366	2,634
Michigan Technological U, Houghton, MI 49931-1295	1885	$4,666 (S)	$4,917	5-D	6,336	415
MidAmerica Nazarene U, Olathe, KS 66062-1899	1966	$11,066	$5,074	2-M	1,714	162
Middlebury Coll, Middlebury, VT 05753-6002	1800	$32,765 (C)	NA	1-D	2,284	221
Middle Tennessee State U, Murfreesboro, TN 37132	1911	$2,808 (S)	$3,436	5-D	19,121	943
Midland Lutheran Coll, Fremont, NE 68025-4200	1883	$13,940	$3,760	2-B	1,025	90
Midwestern State U, Wichita Falls, TX 76308	1922	$3,112 (S)	$4,024	5-M	5,809	328
Miles Coll, Birmingham, AL 35208	1905	$4,630	$2,940	2-B	1,390	56
Millersville U of Pennsylvania, Millersville, PA 17551-0302	1855	$4,804 (S)	$4,900	5-M	7,378	416
Millikin U, Decatur, IL 62522-2084	1901	$17,359	$5,594	2-B	2,307	240
Millsaps Coll, Jackson, MS 39210-0001	1890	$15,814	$6,339	2-M	1,277	98
Mills Coll, Oakland, CA 94613-1000 (3)	1852	$19,056	$7,588	1-D	1,070	154
Milwaukee School of Engineering, Milwaukee, WI 53202-3109 (2)	1903	$19,845	$4,530	1-M	2,620	235
Minnesota State U, Mankato, Mankato, MN 56001	1868	$3,208 (S)	$3,344	5-M	12,085	645
Minnesota State U Moorhead, Moorhead, MN 56563-0002	1885	$3,520 (S)	$3,500	5-M	NA	272
Minot State U, Minot, ND 58707-0002	1913	$2,454 (S)	$3,080	5-M	3,081	222
Mississippi Coll, Clinton, MS 39058	1826	$9,674	$4,176	2-F	3,423	235
Mississippi State U, Mississippi State, MS 39762	1878	$3,430 (S)	$3,990	5-D	16,561	1,056
Mississippi U for Women, Columbus, MS 39701-9998	1884	$2,656 (S)	$2,790	5-M	2,814	206
Mississippi Valley State U, Itta Bena, MS 38941-1400	1946	$2,746 (S)	$2,825	5-M	2,687	136
Missouri Baptist Coll, St. Louis, MO 63141-8698	1964	$9,980	$4,750	2-M	2,806	127
Missouri Southern State Coll, Joplin, MO 64801-1595	1937	$2,496 (S)	$3,610	5-B	5,785	294
Missouri Valley Coll, Marshall, MO 65340-3197	1889	$11,950	$5,000	2-B	1,641	86
Missouri Western State Coll, St. Joseph, MO 64507-2294	1915	$3,026 (S)	$3,476	5-B	5,089	304
Molloy Coll, Rockville Centre, NY 11571-5002	1955	$13,250	NA	1-M	2,490	309
Monmouth Coll, Monmouth, IL 61462-1998	1853	$16,380	$4,500	2-B	1,069	99
Monmouth U, West Long Branch, NJ 07764-1898	1933	$16,326	$6,900	1-M	5,636	441
Montana State U–Billings, Billings, MT 59101-9984	1927	$3,052 (S)	$4,500	5-M	4,296	227
Montana State U–Bozeman, Bozeman, MT 59717	1893	$3,079 (S)	$4,650	5-D	11,658	735
Montana State U–Northern, Havre, MT 59501-7751	1929	$2,692 (S)	$3,800	5-M	1,525	120
Montana Tech of The U of Montana, Butte, MT 59701-8997	1895	$3,006 (S)	$4,278	5-M	2,067	143
Montclair State U, Upper Montclair, NJ 07043-1624	1908	$4,864 (S)	$6,454	5-D	13,502	846
Montreat Coll, Montreat, NC 28757-1267	1916	$11,730	$4,614	2-M	1,203	57
Moody Bible Inst, Chicago, IL 60610-3284	1886	$1,225	$5,500	2-F	1,470	101
Moravian Coll, Bethlehem, PA 18018-6650	1742	$19,400	$6,120	2-F	1,315	221
Morehead State U, Morehead, KY 40351	1922	$2,710 (S)	$3,800	5-M	8,322	457
Morehouse Coll, Atlanta, GA 30314 (1)	1867	$12,432	$7,382	1-B	2,970	235
Morgan State U, Baltimore, MD 21251	1867	$4,388 (S)	$5,780	5-D	6,269	NA
Morningside Coll, Sioux City, IA 51106-1751	1894	$13,606	$4,794	2-M	1,049	112
Morris Brown Coll, Atlanta, GA 30314-4140	1881	$10,426	$5,262	2-B	2,787	186
Mountain State U, Beckley, WV 25802-2830	1933	$4,080	$1,824	1-M	2,110	130
Mount Aloysius Coll, Cresson, PA 16630-1999	1939	$10,980	$4,830	2-B	1,297	112
Mount Holyoke Coll, South Hadley, MA 01075 (3)	1837	$25,370	$7,410	1-M	2,089	228

Name, address	Year Founded	Tuition & Fees	Rm. & Board	Control, Degree	Enroll-ment	Faculty
Mount Ida Coll, Newton Centre, MA 02459-3310	1899	$15,830	$8,950	1-B	1,471	176
Mount Marty Coll, Yankton, SD 57078-3724	1936	$10,634	$4,272	2-M	1,125	59
Mount Mary Coll, Milwaukee, WI 53222-4597 (3)	1913	$12,646	$4,380	2-M	1,246	154
Mount Mercy Coll, Cedar Rapids, IA 52402-4797	1928	$13,850	$4,600	2-B	1,363	121
Mount Olive Coll, Mount Olive, NC 28365	1951	$9,210	$4,000	2-B	1,713	122
Mount Saint Mary Coll, Newburgh, NY 12550-3494	1960	$11,960	$5,980	1-M	2,019	199
Mount St. Mary's Coll, Los Angeles, CA 90049-1599 (4)	1925	$17,666	$7,070	2-M	1,973	269
Mount Saint Mary's Coll & Seminary, Emmitsburg, MD 21727-7799	1808	$17,400	$6,860	2-F	1,859	129
Mt. Sierra Coll, Monrovia, CA 91016	NR	$9,552	NA	3-B	1,222	55
Mount Union Coll, Alliance, OH 44601-3993	1846	$15,550	$4,560	2-B	2,334	158
Mount Vernon Nazarene Coll, Mount Vernon, OH 43050-9500	1964	$11,926	$4,203	2-M	1,961	168
Muhlenberg Coll, Allentown, PA 18104-5586	1848	$21,050	$5,650	2-B	2,470	221
Muskingum Coll, New Concord, OH 43762	1837	$12,665	$5,100	2-M	1,871	113
Naropa U, Boulder, CO 80302-6697	1974	$14,096	NA	1-M	1,037	120
Natl-Louis U, Evanston, IL 60201-1796	1886	$13,095	$5,519	1-D	4,183	297
Natl U, La Jolla, CA 92037-1011	1971	$7,755	NA	1-M	22,858	1,938
Nazareth Coll of Rochester, Rochester, NY 14618-3790	1924	$14,666	$6,470	1-M	2,975	192
Nebraska Wesleyan U, Lincoln, NE 68504-2796	1887	$13,704	$4,006	2-B	1,699	166
Neumann Coll, Aston, PA 19014-1298	1965	$14,320	$6,760	2-M	1,625	171
New Coll of California, San Francisco, CA 94102-5206	1971	$9,240	NA	1-M	1,088	90
New Jersey City U, Jersey City, NJ 07305-1597	1927	$4,642 (S)	$5,600	5-M	8,267	520
New Jersey Inst of Technology, Newark, NJ 07102-1982	1881	$6,730 (S)	$7,300	5-D	8,261	577
Newman U, Wichita, KS 67213-2097	1933	$10,298	$3,950	2-M	1,967	201
New Mexico Highlands U, Las Vegas, NM 87701	1893	$1,992 (S)	$3,970	5-M	3,396	136
New Mexico Inst of Mining & Technology, Socorro, NM 87801	1889	$2,499 (S)	$3,704	5-D	1,548	131
New Mexico State U, Las Cruces, NM 88003-8001	1888	$2,790 (S)	$3,892	5-D	14,958	934
New School Bachelor of Arts, New School U, New York, NY 10011-8603	1919	$13,808	NA	1-D	1,395	442
New York Inst of Technology, Old Westbury, NY 11568-8000	1955	$13,700	$7,310	1-F	8,965	772
New York U, New York, NY 10012-1019	1831	$24,336	$9,226	1-D	37,150	3,713
Niagara U, Niagara University, NY 14109	1856	$14,560	$6,660	2-M	3,146	275
Nicholls State U, Thibodaux, LA 70310	1948	$2,368 (A)	$3,002	5-M	7,326	276
Nichols Coll, Dudley, MA 01571-5000	1815	$15,650	$7,810	1-M	1,363	42
Norfolk State U, Norfolk, VA 23504	1935	$2,856 (S)	$5,267	5-D	6,668	465
North Carolina Central U, Durham, NC 27707-3129	1910	$1,975 (S)	$3,837	5-F	5,476	388
North Carolina State U, Raleigh, NC 27695	1887	$2,746 (S)	$5,274	5-D	28,619	1,588
North Central Coll, Naperville, IL 60566-7063	1861	$16,095	$5,472	2-M	2,532	227
North Central U, Minneapolis, MN 55404-1322	1930	$8,554	$2,270	2-B	1,163	88
North Dakota State U, Fargo, ND 58105	1890	$3,010 (S)	$3,542	5-D	9,894	589
Northeastern Illinois U, Chicago, IL 60625-4699	1961	$2,664 (S)	NA	5-M	10,941	565
Northeastern State U, Tahlequah, OK 74464-2399	1846	$2,375 (S)	$2,724	5-D	8,351	424
Northeastern U, Boston, MA 02115-5096	1898	$19,590	$9,135	1-D	17,990	1,097
Northern Arizona U, Flagstaff, AZ 86011	1899	$2,346 (S)	$4,391	5-D	19,964	1,077
Northern Illinois U, De Kalb, IL 60115-2854	1895	$4,275 (S)	$5,036	5-D	23,248	1,249
Northern Kentucky U, Highland Heights, KY 41099	1968	$2,328 (A)	$3,656	5-F	12,080	816
Northern Michigan U, Marquette, MI 49855-5301	1899	$3,234 (S)	$4,976	5-M	8,082	336
Northern State U, Aberdeen, SD 57401-7198	1901	$3,312 (S)	$2,663	5-M	3,283	109
North Georgia Coll & State U, Dahlonega, GA 30597-1001	1873	$2,354 (S)	$3,612	5-M	3,525	168
North Greenville Coll, Tigerville, SC 29688-1892	1892	$8,000	$4,570	2-B	1,282	114
North Park U, Chicago, IL 60625-4895	1891	$16,910	$5,870	2-D	2,387	121
Northwest Coll, Kirkland, WA 98083-0579	1934	$10,748	$5,490	2-M	1,096	82
Northwestern Coll, Orange City, IA 51041-1996	1882	$13,000	$3,650	2-B	1,243	108
Northwestern Coll, St. Paul, MN 55113-1598	1902	$14,982	$4,834	2-B	2,084	151
Northwestern Oklahoma State U, Alva, OK 73717-2799	1897	$1,993 (S)	$2,500	5-M	2,019	136
Northwestern State U of Louisiana, Natchitoches, LA 71497	1884	$2,493 (S)	$2,896	5-D	9,011	310
Northwestern U, Evanston, IL 60208	1851	$24,648	$7,471	1-D	15,663	2,528
Northwest Missouri State U, Maryville, MO 64468-6001	1905	$3,330 (S)	$4,150	5-M	6,442	249
Northwest Nazarene U, Nampa, ID 83686-5897	1913	$14,240	$4,140	2-M	1,316	86
Northwood U, Midland, MI 48640-2398	1959	$12,063	$5,592	1-M	3,823	102
Northwood U, Texas Campus, Cedar Hill, TX 75104-1204	1966	$12,063	$5,160	1-B	1,014	24
Norwich U, Northfield, VT 05663	1819	$15,626	$5,890	1-M	2,703	272
Notre Dame Coll, Manchester, NH 03104-2299	1950	$15,647	$6,213	2-M	1,120	58
Notre Dame de Namur U, Belmont, CA 94002-1997	1851	$16,850	$7,860	2-M	1,670	194
Nyack Coll, Nyack, NY 10960-3698	1882	$12,740	$5,800	2-B	2,160	150
Oakland City U, Oakland City, IN 47660-1099	1885	$11,266	$4,100	2-D	1,789	56
Oakland U, Rochester, MI 48309-4401	1957	$4,279 (S)	$4,833	5-D	15,235	755
Oakwood Coll, Huntsville, AL 35896	1896	$9,058	$2,646	2-B	1,767	168
Oberlin Coll, Oberlin, OH 44074	1833	$25,530	$6,364	1-M	2,955	280
Occidental Coll, Los Angeles, CA 90041-3314	1887	$23,850	$6,880	1-M	1,726	179
Oglethorpe U, Atlanta, GA 30319-2797	1835	$18,390	$5,560	1-M	1,267	122
Ohio Dominican Coll, Columbus, OH 43219-2099	1911	$10,910	$5,220	2-B	2,085	114
Ohio Northern U, Ada, OH 45810-1599	1871	$21,435	$5,265	2-F	3,227	239
The Ohio State U, Columbus, OH 43210	1870	$4,761 (S)	$5,807	5-D	47,952	3,464
The Ohio State U at Lima, Lima, OH 45804-3576	1960	$3,351 (S)	NA	5-B	118	91
The Ohio State U–Mansfield Campus, Mansfield, OH 44906-1599	1958	$3,351 (S)	NA	5-B	1,583	64
The Ohio State U–Newark Campus, Newark, OH 43055-1797	1957	$3,351 (S)	NA	5-B	2,025	89
Ohio U, Athens, OH 45701-2979	1804	$5,085 (S)	$5,922	5-D	19,667	1,146
Ohio U–Chillicothe, Chillicothe, OH 45601-0629	1946	$3,033 (S)	NA	5-B	1,726	133
Ohio U–Eastern, St. Clairsville, OH 43950-9724	1957	$3,033 (S)	NA	5-B	1,118	114
Ohio U–Lancaster, Lancaster, OH 43130-1097	1968	$3,033 (S)	NA	5-M	1,585	NA
Ohio U–Zanesville, Zanesville, OH 43701-2695	1946	$3,048 (S)	NA	5-M	1,357	96
Ohio Wesleyan U, Delaware, OH 43015	1842	$21,880	$6,610	2-B	1,880	176
Oklahoma Baptist U, Shawnee, OK 74804	1910	$9,440	$3,510	2-M	2,017	158
Oklahoma Christian U, Oklahoma City, OK 73136-1100	1950	$9,700	NA	2-M	1,808	134
Oklahoma City U, Oklahoma City, OK 73106-1402	1904	$10,170	$9,300	2-F	4,041	321
Oklahoma Panhandle State U, Goodwell, OK 73939-0430	1909	$1,911 (S)	$2,580	5-B	1,178	85
Oklahoma State U, Stillwater, OK 74078	1890	$2,587 (S)	$4,716	5-D	21,252	1,054
Old Dominion U, Norfolk, VA 23529	1930	$3,916 (S)	$5,232	5-D	18,969	937

Name, address	Year Founded	Tuition & Fees	Rm. & Board	Control, Degree	Enroll-ment	Faculty
Olivet Nazarene U, Bourbonnais, IL 60914-2271	1907	$12,708	$4,696	2-M	2,859	108
Oral Roberts U, Tulsa, OK 74171-0001	1963	$12,260	$5,228	2-D	3,552	263
Oregon Health & Sci U, Portland, OR 97201-3098	1974	$5,747 (S)	NA	12-D	1,849	836
Oregon Inst of Technology, Klamath Falls, OR 97601-8801	1947	$3,459 (S)	$4,898	5-M	2,842	169
Oregon State U, Corvallis, OR 97331	1868	$3,014 (S)	$5,508	5-D	16,788	1,873
Otterbein Coll, Westerville, OH 43081	1847	$16,911	$5,289	2-M	2,917	205
Ouachita Baptist U, Arkadelphia, AR 71998-0001	1886	$10,000	$4,100	2-B	1,714	158
Our Lady of Holy Cross Coll, New Orleans, LA 70131-7399	1916	$6,090	NA	2-M	1,233	110
Our Lady of the Lake U of San Antonio, San Antonio, TX 78207-4689	1895	$11,966	$4,466	2-D	3,474	269
Pace U, New York City Campus, New York, NY 10038	1906	$16,230	$6,900	1-D	7,911	563
Pace U, Pleasantville/Briarcliff Campus, Pleasantville, NY 10570	NR	$16,230	$6,900	1-D	3,533	340
Pacific Lutheran U, Tacoma, WA 98447	1890	$16,800	$5,300	2-M	3,515	281
Pacific Union Coll, Angwin, CA 94508-9707	1882	$15,075	$4,530	2-M	1,456	113
Pacific U, Forest Grove, OR 97116-1797	1849	$17,800	$4,903	1-D	2,148	271
Palm Beach Atlantic Coll, West Palm Beach, FL 33416-4708	1968	$11,120	$4,470	2-M	2,456	185
Palmer Coll of Chiropractic, Davenport, IA 52803-5287	1897	$16,290	NA	1-F	1,731	100
Park U, Parkville, MO 64152-4358	1875	$4,950	$4,850	2-M	1,367	100
Parsons School of Design, New School U, New York, NY 10011-8878	1896	$21,930	$9,083	1-M	2,815	709
Peirce Coll, Philadelphia, PA 19102-4699 (4)	1865	$9,100	NA	1-B	2,334	295
Pennsylvania Coll of Technology, Williamsport, PA 17701-5778	1965	$7,422 (S)	NA	12-B	5,320	405
The Pennsylvania State U Abington Coll, Abington, PA 19001-3918	1950	$6,722 (S)	NA	12-B	3,049	197
The Pennsylvania State U Altoona Coll, Altoona, PA 16601-3760	1939	$6,742 (S)	$4,910	12-B	3,765	255
The Pennsylvania State U at Erie, The Behrend Coll, Erie, PA 16563	1948	$6,852 (S)	$4,910	12-M	3,791	246
The Pennsylvania State U Berks Campus of the Berks–Lehigh Valley Coll, Reading, PA 19610-6009	1924	$6,742 (S)	$4,910	12-B	2,197	149
The Pennsylvania State U Harrisburg Campus of the Capital Coll, Middletown, PA 17057-4898	1966	$6,832 (S)	$4,910	12-D	3,258	258
The Pennsylvania State U Schuylkill Campus of the Capital Coll, Schuylkill Haven, PA 17972-2208	1934	$6,626 (S)	$4,910	12-B	1,124	81
The Pennsylvania State U U Park Campus, University Park, PA 16802-1503	1855	$6,852 (S)	$4,910	12-D	40,571	2,371
Pepperdine U, Malibu, CA 90263-0002	1937	$24,050	$7,290	2-D	6,317	312
Peru State Coll, Peru, NE 68421	1867	$2,378 (S)	$3,526	5-M	1,664	100
Pfeiffer U, Misenheimer, NC 28109-0960	1885	$12,066	$4,734	2-M	1,494	124
Philadelphia Biblical U, Langhorne, PA 19047-2990	1913	$10,355	$5,073	2-M	1,434	118
Philadelphia U, Philadelphia, PA 19144-5497	1884	$15,458	$6,882	1-M	3,316	381
Piedmont Coll, Demorest, GA 30535-0010	1897	$9,500	$4,400	2-M	1,728	167
Pikeville Coll, Pikeville, KY 41501	1889	$7,800	$3,340	2-F	1,156	68
Pittsburg State U, Pittsburg, KS 66762	1903	$2,260 (S)	$3,990	5-M	6,418	328
Plattsburgh State U of New York, Plattsburgh, NY 12901-2681	1889	$4,065 (S)	$5,116	5-M	6,153	401
Plymouth State Coll, Plymouth, NH 03264-1595	1871	$5,314 (S)	$5,206	5-M	4,041	314
Point Loma Nazarene U, San Diego, CA 92106-2899	1902	$14,340	$5,990	2-M	2,741	270
Point Park Coll, Pittsburgh, PA 15222-1984	1960	$12,996	$5,718	1-M	2,619	263
Polytechnic U, Brooklyn Campus, Brooklyn, NY 11201-2990	1854	$21,780	$5,250	1-D	3,059	295
Polytechnic U of Puerto Rico, Hato Rey, PR 00919	1966	$5,115	NA	1-M	5,299	252
Pomona Coll, Claremont, CA 91711	1887	$24,170	$8,170	1-B	1,565	191
Pontifical Catholic U of Puerto Rico, Ponce, PR 00717-0777	1948	$3,880	$2,660	2-D	7,448	246
Portland State U, Portland, OR 97207-0751	1946	$3,525 (S)	$6,150	5-D	18,889	853
Prairie View A&M U, Prairie View, TX 77446-0188	1878	$2,693 (S)	NA	5-M	6,609	385
Pratt Inst, Brooklyn, NY 11205-3899	1887	$20,084	$7,800	1-M	4,199	718
Presbyterian Coll, Clinton, SC 29325	1880	$17,342	$5,156	2-B	1,147	120
Princeton U, Princeton, NJ 08544-1019	1746	$25,430	$7,206	1-D	6,547	914
Providence Coll, Providence, RI 02918	1917	$18,750	$7,625	2-M	5,336	340
Purchase Coll, State U of New York, Purchase, NY 10577-1400	1967	$3,997 (S)	$5,654	5-M	4,078	365
Purdue U, West Lafayette, IN 47907	1869	$3,872 (S)	$5,800	5-D	37,871	1,870
Purdue U Calumet, Hammond, IN 46323-2094	1951	$3,116 (A)	NA	5-M	NA	483
Purdue U North Central, Westville, IN 46391-9528	1967	$3,338 (S)	NA	5-M	3,459	229
Queens Coll, Charlotte, NC 28274-0002	1857	$11,360	$5,890	2-M	1,596	96
Queens Coll of the City U of New York, Flushing, NY 11367-1597	1937	$3,403 (S)	NA	11-M	15,686	1,076
Quincy U, Quincy, IL 62301-2699	1860	$14,700	$4,780	2-M	1,188	111
Quinnipiac U, Hamden, CT 06518-1940	1929	$17,780	$8,170	1-F	6,477	400
Radford U, Radford, VA 24142	1910	$2,950 (S)	$4,938	5-M	8,837	544
Ramapo Coll of New Jersey, Mahwah, NJ 07430-1680	1969	$5,638 (S)	$7,044	5-M	5,195	326
Randolph-Macon Coll, Ashland, VA 23005-5505	1830	$18,350	$4,205	2-B	1,171	138
Reed Coll, Portland, OR 97202-8199	1908	$25,020	$6,820	1-M	1,385	126
Regis Coll, Weston, MA 02493 (3)	1927	$17,500	$8,100	2-M	1,138	130
Regis U, Denver, CO 80221-1099	1877	$17,570	$6,700	2-M	11,000	108
Reinhardt Coll, Waleska, GA 30183-2981	1883	$8,800	$4,716	2-B	1,078	150
Rensselaer Polytechnic Inst, Troy, NY 12180-3590	1824	$25,538	$8,308	1-D	8,022	465
Rhode Island School of Design, Providence, RI 02903-2784	1877	$22,305	$6,600	1-F	2,086	399
Rhodes Coll, Memphis, TN 38112-1690	1848	$19,503	$5,671	2-M	1,554	156
Rice U, Houston, TX 77251-1892	1912	$16,400	$6,850	1-D	4,171	NA
The Richard Stockton Coll of New Jersey, Pomona, NJ 08240-0195	1969	$4,784 (S)	$5,516	5-M	6,312	344
Rider U, Lawrenceville, NJ 08648-3001	1865	$17,450	$7,080	1-M	5,274	457
Rivier Coll, Nashua, NH 03060-5086	1933	$15,520	$6,100	2-M	2,300	209
Roanoke Coll, Salem, VA 24153-3794	1842	$17,765	$5,722	2-B	1,677	174
Robert Morris Coll, Chicago, IL 60605	1913	$11,550	NA	1-B	4,938	346
Robert Morris Coll, Moon Township, PA 15108-1189	1921	$10,034	$6,320	1-D	4,660	364
Roberts Wesleyan Coll, Rochester, NY 14624-1997	1866	$14,220	$4,758	2-M	1,596	141
Rochester Inst of Technology, Rochester, NY 14623-5698	1829	$18,252	$6,996	1-D	13,274	1,091
Rockford Coll, Rockford, IL 61108-2393	1847	$16,800	$5,430	1-M	1,315	148
Rockhurst U, Kansas City, MO 64110-2561	1910	$13,800	$4,850	2-M	2,727	212
Rogers State U, Claremore, OK 74017-3252	1909	$1,649 (S)	$2,063	5-M	2,622	110
Roger Williams U, Bristol, RI 02809	1956	$17,980	$8,200	1-F	4,115	312
Rollins Coll, Winter Park, FL 32789-4499	1885	$22,868	$7,000	1-M	2,320	227
Roosevelt U, Chicago, IL 60605-1394	1945	$13,070	$6,040	1-D	7,359	601
Rose-Hulman Inst of Technology, Terre Haute, IN 47803-3920 (2)	1874	$20,814	$5,175	1-M	1,725	132
Rosemont Coll, Rosemont, PA 19010-1699 (3)	1921	$15,270	$7,030	2-M	1,178	183

Name, address	Year Founded	Tuition & Fees	Rm. & Board	Control, Degree	Enroll-ment	Faculty
Rowan U, Glassboro, NJ 08028-1701	1923	$5,347 (S)	$5,776	5-D	9,364	661
Rutgers, The State U of New Jersey, Camden, Camden, NJ 08102-1401	1927	$6,204 (S)	$6,519	5-F	5,125	365
Rutgers, The State U of New Jersey, Newark, Newark, NJ 07102	1892	$6,204 (S)	$6,519	5-D	9,352	643
Rutgers, The State U of New Jersey, New Brunswick, New Brunswick, NJ 08901-1281	1766	$6,204 (S)	$6,519	5-D	35,237	2,142
Sacred Heart U, Fairfield, CT 06432-1000	1963	$15,464	$7,614	2-M	5,684	466
Saginaw Valley State U, University Center, MI 48710	1963	$3,607 (S)	$5,015	5-M	8,622	32
St. Ambrose U, Davenport, IA 52803-2898	1882	$13,890	$5,160	2-D	3,011	319
Saint Anselm Coll, Manchester, NH 03102-1310	1889	$18,950	$7,000	2-B	1,985	165
St. Augustine Coll, Chicago, IL 60640-3501	1980	$6,740	NA	1-B	1,543	131
Saint Augustine's Coll, Raleigh, NC 27610-2298	1867	$7,900	$4,960	2-B	1,465	111
St. Bonaventure U, St. Bonaventure, NY 14778-2284	1858	$15,140	$5,800	2-M	2,801	221
St. Cloud State U, St. Cloud, MN 56301-4498	1869	$3,067 (S)	$3,468	5-D	15,140	740
St. Edward's U, Austin, TX 78704-6489	1885	$11,896	$5,000	2-M	3,824	325
St. Francis Coll, Brooklyn Heights, NY 11201-4398	1884	$8,990	NA	2-B	2,336	201
Saint Francis U, Loretto, PA 15940-0600	1847	$16,519	$7,075	2-M	2,129	120
St. John Fisher Coll, Rochester, NY 14618-3597	1948	$14,990	$6,300	2-M	2,739	404
Saint John's U, Collegeville, MN 56321 (1)	1857	$17,241	$5,129	2-F	2,020	190
St. John's U, Jamaica, NY 11439	1870	$15,930	$8,950	2-D	18,621	1,134
Saint Joseph Coll, West Hartford, CT 06117-2700 (3)	1932	$17,430	$7,140	2-M	1,823	83
St. Joseph's Coll, New York, Brooklyn, NY 11205-3688	1916	$9,030	NA	1-B	1,283	139
St. Joseph's Coll, Suffolk Campus, Patchogue, NY 11772-2399	1916	$9,622	NA	1-M	3,236	317
Saint Joseph's U, Philadelphia, PA 19131-1395	1851	$19,965	$7,856	2-D	6,961	447
St. Lawrence U, Canton, NY 13617-1455	1856	$23,955	$7,475	1-M	2,061	191
Saint Leo U, Saint Leo, FL 33574-6665	1889	$12,090	$6,300	2-M	1,624	138
Saint Louis U, St. Louis, MO 63103-2097	1818	$18,438	$6,140	2-D	13,873	1,800
Saint Martin's Coll, Lacey, WA 98503-7500	1895	$14,930	$4,768	2-M	1,498	73
Saint Mary-of-the-Woods Coll, Saint Mary-of-the-Woods, IN 47876 (3)	1840	$14,840	$5,540	2-M	1,470	61
Saint Mary's Coll, Notre Dame, IN 46556 (3)	1844	$18,300	$6,174	2-B	1,455	181
Saint Mary's Coll of California, Moraga, CA 94556	1863	$18,255	$7,560	2-D	4,142	517
St. Mary's Coll of Maryland, St. Mary's City, MD 20686-3001	1840	$7,549 (S)	$6,555	5-B	1,547	182
Saint Mary's U of Minnesota, Winona, MN 55987-1399	1912	$14,355	$4,620	2-D	4,991	139
St. Mary's U of San Antonio, San Antonio, TX 78228-8507	1852	$11,880	$5,251	2-D	4,137	322
Saint Michael's Coll, Colchester, VT 05439	1904	$18,782	$7,253	2-M	2,687	193
St. Norbert Coll, De Pere, WI 54115-2099	1898	$16,820	$5,162	2-M	2,140	163
St. Olaf Coll, Northfield, MN 55057-1098	1874	$19,400	$4,500	2-B	3,014	315
Saint Peter's Coll, Jersey City, NJ 07306-5997	1872	$15,606	$6,446	2-M	3,282	289
St. Thomas Aquinas Coll, Sparkill, NY 10976	1952	$12,820	$7,350	1-M	2,185	150
St. Thomas U, Miami, FL 33054-6459	1961	$14,020	$4,400	2-F	2,231	123
Saint Vincent Coll, Latrobe, PA 15650-2690	1846	$15,781	$5,168	2-B	1,206	108
Saint Xavier U, Chicago, IL 60655-3105	1847	$14,530	$5,744	2-M	4,602	339
Salem Coll, Winston-Salem, NC 27108-0548 (3)	1772	$13,945	$8,240	2-M	1,027	88
Salem State Coll, Salem, MA 01970-5353	1854	$3,583 (S)	$2,860	5-M	13,417	423
Salisbury State U, Salisbury, MD 21801-6837	1925	$4,312 (S)	$5,990	5-M	6,421	426
Salve Regina U, Newport, RI 02840-4192	1934	$17,550	$7,750	2-D	2,246	250
Samford U, Birmingham, AL 35229-0002	1841	$10,738	$4,720	2-D	4,379	392
Sam Houston State U, Huntsville, TX 77341	1879	$2,228 (S)	$3,672	5-D	12,385	521
San Diego State U, San Diego, CA 92182	1897	$1,776 (S)	$7,110	5-D	31,609	1,821
San Francisco State U, San Francisco, CA 94132-1722	1899	$1,826 (S)	$6,930	5-D	26,826	1,712
San Jose State U, San Jose, CA 95192-0001	1857	$1,939 (S)	$6,248	5-M	26,698	NA
Santa Clara U, Santa Clara, CA 95053	1851	$20,577	$8,034	2-D	7,356	636
Sarah Lawrence Coll, Bronxville, NY 10708	1926	$26,668	$8,460	1-M	1,449	209
Savannah Coll of Art & Design, Savannah, GA 31402-3146	1978	$16,200	$6,750	1-M	4,923	234
Savannah State U, Savannah, GA 31404	1890	$2,494 (S)	$4,144	5-M	2,500	NA
School of the Art Inst of Chicago, Chicago, IL 60603-3103	1866	$20,220	NA	1-M	2,578	459
School of Visual Arts, New York, NY 10010-3994	1947	$16,110	NA	3-M	5,312	848
Seattle Pacific U, Seattle, WA 98119-1997	1891	$15,381	$5,895	2-D	3,491	251
Seattle U, Seattle, WA 98122	1891	$17,010	$6,075	2-D	5,790	433
Seton Hall U, South Orange, NJ 07079-2697	1856	$19,400	$8,060	2-D	9,920	802
Seton Hill Coll, Greensburg, PA 15601 (4)	1883	$15,225	$5,200	2-M	1,366	111
Shawnee State U, Portsmouth, OH 45662-4344	1986	$3,162 (S)	$4,588	5-B	3,280	267
Shaw U, Raleigh, NC 27601-2399	1865	$7,430	$4,648	2-F	2,527	291
Shenandoah U, Winchester, VA 22601-5195	1875	$16,300	$6,000	2-D	2,428	282
Shepherd Coll, Shepherdstown, WV 25443-3210	1871	$2,508 (S)	$4,112	5-B	4,603	321
Shippensburg U of Pennsylvania, Shippensburg, PA 17257-2299	1871	$4,746 (S)	$4,274	5-M	7,011	343
Shorter Coll, Rome, GA 30165	1873	$9,470	$4,900	2-M	1,925	195
Siena Coll, Loudonville, NY 12211-1462	1937	$15,800	$6,680	2-M	3,314	260
Siena Heights U, Adrian, MI 49221-1796	1919	$12,400	$4,502	2-M	1,917	NA
Simmons Coll, Boston, MA 02115 (4)	1899	$20,890	$8,410	1-D	3,340	410
Simpson Coll, Indianola, IA 50125-1297	1860	$15,150	$5,040	2-B	1,912	115
Simpson Coll & Graduate School, Redding, CA 96003-8606	1921	$10,540	$4,680	2-M	1,186	86
Skidmore Coll, Saratoga Springs, NY 12866-1632	1903	$25,475	$7,260	1-M	2,503	201
Slippery Rock U of Pennsylvania, Slippery Rock, PA 16057	1889	$4,684 (S)	$3,988	5-D	6,952	383
Smith Coll, Northampton, MA 01063 (3)	1871	$23,400	$8,160	1-D	3,113	274
Sonoma State U, Rohnert Park, CA 94928-3609	1960	$2,002 (S)	$6,471	5-M	7,402	542
South Carolina State U, Orangeburg, SC 29117-0001	1896	$3,724 (S)	NA	5-D	4,525	NA
South Dakota School of Mines & Technology, Rapid City, SD 57701-3995	1885	$4,161 (S)	$3,224	5-D	2,307	129
South Dakota State U, Brookings, SD 57007	1881	$3,588 (S)	$2,922	5-D	8,662	530
Southeastern Coll of the Assemblies of God, Lakeland, FL 33801-6099	1935	$6,563	$3,850	2-B	1,232	71
Southeastern Louisiana U, Hammond, LA 70402	1925	$2,490 (S)	$3,006	5-M	14,525	483
Southeastern Oklahoma State U, Durant, OK 74701-0609	1909	$1,670 (S)	$1,246	5-M	3,776	222
Southeastern U, Washington, DC 20024-2788	1879	$8,265	NA	1-M	1,057	86
Southeast Missouri State U, Cape Girardeau, MO 63701-4799	1873	$3,390 (S)	$4,718	5-M	8,948	388
Southern Adventist U, Collegedale, TN 37315-0370	1892	$11,040	$3,886	2-M	2,046	167
Southern Arkansas U–Magnolia, Magnolia, AR 71753	1909	$2,454 (S)	$2,930	5-M	2,977	201
Southern Connecticut State U, New Haven, CT 06515-1355	1893	$3,850 (S)	$6,011	5-M	12,127	824

Name, address	Year Founded	Tuition & Fees	Rm. & Board	Control, Degree	Enroll-ment	Faculty
Southern Illinois U Carbondale, Carbondale, IL 62901-6806	1869	$4,113 (S)	$4,104	5-D	22,552	1,071
Southern Illinois U Edwardsville, Edwardsville, IL 62026-0001	1957	$3,001 (S)	$4,598	5-F	12,193	746
Southern Methodist U, Dallas, TX 75275	1911	$19,620	$7,177	2-D	10,064	787
Southern Nazarene U, Bethany, OK 73008	1899	$9,380	$4,316	2-M	1,950	137
Southern New Hampshire U, Manchester, NH 03106-1045	1932	$15,850	$6,792	1-D	5,363	107
Southern Oregon U, Ashland, OR 97520	1926	$3,369 (S)	$5,649	5-M	5,493	228
Southern Polytechnic State U, Marietta, GA 30060-2896	1948	$2,210 (S)	$4,308	5-M	3,397	207
Southern U & Ag & Mech Coll, Baton Rouge, LA 70813	1880	$2,286 (S)	$3,410	5-D	9,449	574
Southern U at New Orleans, New Orleans, LA 70126-1009	1959	$1,874 (S)	NA	5-M	5,000	NA
Southern Utah U, Cedar City, UT 84720-2498	1897	$2,066 (S)	$2,866	5-M	5,963	327
Southern Wesleyan U, Central, SC 29630-1020	1906	$12,104	$4,290	2-M	1,803	153
Southwest Baptist U, Bolivar, MO 65613-2597	1878	$9,746	$3,010	2-M	2,701	242
Southwestern Adventist U, Keene, TX 76059	1894	$9,510	$4,550	2-M	1,187	97
Southwestern Assemblies of God U, Waxahachie, TX 75165-2397	1927	$6,480	$3,990	2-M	1,782	70
Southwestern Coll, Winfield, KS 67156-2499	1885	$12,260	$4,320	2-M	1,143	100
Southwestern Oklahoma State U, Weatherford, OK 73096-3098	1901	$2,089 (S)	$2,506	5-F	4,862	230
Southwestern U, Georgetown, TX 78626	1840	$15,750	$6,318	2-B	1,312	145
Southwest Missouri State U, Springfield, MO 65804-0094	1905	$3,748 (S)	$4,032	5-M	17,703	956
Southwest State U, Marshall, MN 56258-1598	1963	$3,428 (S)	$3,588	5-M	4,707	230
Southwest Texas State U, San Marcos, TX 78666	1899	$3,296 (S)	$4,564	5-D	22,423	659
Spalding U, Louisville, KY 40203-2188	1814	$11,496	$2,930	2-D	1,481	140
Spelman Coll, Atlanta, GA 30314-4399 (3)	1881	$11,135	$6,880	1-B	1,897	147
Spring Arbor U, Spring Arbor, MI 49283-9799	1873	$12,416	$4,580	2-M	2,558	72
Springfield Coll, Springfield, MA 01109-3797	1885	$17,300	$6,035	1-D	3,946	330
Spring Hill Coll, Mobile, AL 36608-1791	1830	$16,254	$5,768	2-M	1,484	126
Stanford U, Stanford, CA 94305-9991	1891	$24,441	$8,030	1-D	18,549	1,671
State U of New York at Binghamton, Binghamton, NY 13902-6000	1946	$4,463 (S)	$5,772	5-D	12,473	784
State U of New York at Farmingdale, Farmingdale, NY 11735	1912	$4,085 (S)	$6,400	5-B	5,045	295
State U of New York at New Paltz, New Paltz, NY 12561	1828	$3,985 (S)	$5,368	5-D	7,720	554
State U of New York at Oswego, Oswego, NY 13126	1861	$4,011 (S)	$6,350	5-M	8,159	361
State U of New York Coll at Brockport, Brockport, NY 14420-2997	1867	$4,103 (S)	$5,800	5-M	8,524	601
State U of New York Coll at Buffalo, Buffalo, NY 14222-1095	1867	$3,909 (S)	$5,170	5-M	11,402	574
State U of New York Coll at Cortland, Cortland, NY 13045	1868	$4,104 (S)	$5,750	5-M	7,178	485
State U of New York Coll at Fredonia, Fredonia, NY 14063	1826	$4,225 (S)	$5,330	5-M	5,086	416
State U of New York Coll at Geneseo, Geneseo, NY 14454-1401	1871	$4,221 (S)	$4,890	5-M	5,477	340
State U of New York Coll at Old Westbury, Old Westbury, NY 11568-0210	1965	$3,946 (S)	$5,345	5-B	2,992	233
State U of New York Coll at Oneonta, Oneonta, NY 13820-4015	1889	$4,123 (S)	$5,456	5-M	5,584	382
State U of New York Coll at Potsdam, Potsdam, NY 13676	1816	$4,099 (S)	$6,100	5-M	4,231	327
State U of New York Coll of Agriculture & Technology at Cobleskill, Cobleskill, NY 12043	1916	$3,971 (S)	$6,100	5-B	2,301	155
State U of New York Coll of Environmental Sci & Forestry, Syracuse, NY 13210	1911	$3,702 (S)	$8,310	5-D	1,749	125
State U of New York Coll of Technology at Canton, Canton, NY 13617	1906	$4,400 (A)	$4,800	5-B	2,126	116
State U of New York Empire State Coll, Saratoga Springs, NY 12866-4391	1971	$3,555 (S)	NA	5-M	8,009	329
State U of New York Inst of Technology at Utica/Rome, Utica, NY 13504-3050	1966	$4,045 (S)	$6,030	5-M	2,660	164
State U of New York Upstate Medical U, Syracuse, NY 13210-2334	1950	$3,810 (S)	$6,515	5-D	1,173	695
State U of West Georgia, Carrollton, GA 30118	1933	$2,394 (S)	$3,854	5-D	8,959	410
Stephen F. Austin State U, Nacogdoches, TX 75962	1923	$2,168 (S)	$4,370	5-D	11,284	670
Stetson U, DeLand, FL 32720-3781	1883	$18,385	$6,070	1-F	3,199	253
Stevens Inst of Technology, Hoboken, NJ 07030	1870	$22,150	$7,492	1-D	3,714	212
Stillman Coll, Tuscaloosa, AL 35403-9990	1876	$5,880	$3,764	2-B	1,458	86
Stonehill Coll, Easton, MA 02357	1948	$17,320	$8,166	2-M	2,649	240
Stony Brook U, State U of New York, Stony Brook, NY 11794	1957	$4,228 (S)	$6,524	5-D	19,924	1,737
Strayer U, Washington, DC 20005-2603	1892	$8,100	NA	3-M	12,096	370
Suffolk U, Boston, MA 02108-2770	1906	$16,616	$9,990	1-D	6,982	748
Sullivan U, Louisville, KY 40205	1864	$10,080	NA	3-M	4,000	70
Sul Ross State U, Alpine, TX 79832	1920	$2,792 (S)	$3,690	5-M	2,010	113
Susquehanna U, Selinsgrove, PA 17870	1858	$20,440	$5,770	2-B	1,829	164
Swarthmore Coll, Swarthmore, PA 19081-1397	1864	$25,200	$7,804	1-B	1,428	201
Syracuse U, Syracuse, NY 13244-0003	1870	$20,816	$8,750	1-D	14,478	1,395
Tarleton State U, Stephenville, TX 76402	1899	$3,440 (S)	$3,802	5-M	7,667	452
Taylor U, Upland, IN 46989-1001	1846	$15,820	$4,740	2-B	1,886	146
Teikyo Post U, Waterbury, CT 06723-2540	1890	$14,125	$6,300	1-B	1,356	190
Temple U, Philadelphia, PA 19122-6096	1884	$6,948 (S)	$6,482	12-D	28,355	2,309
Tennessee State U, Nashville, TN 37209-1561	1912	$2,697 (S)	$3,600	5-D	8,640	534
Tennessee Technological U, Cookeville, TN 38505	1915	$2,704 (S)	$3,650	5-D	8,410	502
Texas A&M Intl U, Laredo, TX 78041-1900	1969	$2,455 (S)	NA	5-M	3,016	160
Texas A&M U, College Station, TX 77843	1876	$3,374 (S)	$5,164	5-D	44,026	2,123
Texas A&M U at Galveston, Galveston, TX 77553-1675	1962	$2,915 (S)	$3,977	5-B	1,363	135
Texas A&M U–Commerce, Commerce, TX 75429-3011	1889	$2,711 (S)	$4,550	5-D	7,483	418
Texas A&M U–Kingsville, Kingsville, TX 78363	1925	$2,062 (S)	$3,744	5-D	5,946	359
Texas A&M U–Texarkana, Texarkana, TX 75505-5518	1971	$1,848 (S)	NA	5-M	1,233	73
Texas Christian U, Fort Worth, TX 76129-0002	1873	$13,125	$4,290	2-D	7,775	578
Texas Lutheran U, Seguin, TX 78155-5999	1891	$12,570	$4,880	2-B	1,506	122
Texas Southern U, Houston, TX 77004-4584	1947	$2,030 (S)	$4,016	5-D	6,886	292
Texas Tech U, Lubbock, TX 79409	1923	$3,400 (S)	$5,079	5-D	24,558	998
Texas Wesleyan U, Fort Worth, TX 76105-1536	1890	$10,010	$3,964	2-F	3,133	245
Texas Woman's U, Denton, TX 76204 (4)	1901	$2,072 (S)	$3,872	5-D	8,404	762
Thiel Coll, Greenville, PA 16125-2181	1866	$11,464	$5,674	2-B	1,039	105
Thomas Edison State Coll, Trenton, NJ 08608-1176	1972	$2,500 (S)	NA	5-M	8,137	593
Thomas Jefferson U, Philadelphia, PA 19107	1824	$17,500	NA	1-M	2,260	98
Thomas More Coll, Crestview Hills, KY 41017-3495	1921	$12,580	$3,756	2-M	1,417	128
Tiffin U, Tiffin, OH 44883-2161	1888	$11,130	$5,100	1-M	1,558	148
Tougaloo Coll, Tougaloo, MS 39174	1869	$7,555	$3,200	2-B	1,000	90
Touro Coll, New York, NY 10010	1971	$9,950	NA	1-D	7,791	937

Name, address	Year Founded	Tuition & Fees	Rm. & Board	Control, Degree	Enroll-ment	Faculty
Towson U, Towson, MD 21252-0001	1866	$4,984 (S)	$6,104	5-D	16,729	1,214
Transylvania U, Lexington, KY 40508-1797	1780	$15,270	$5,530	2-B	1,083	99
Trevecca Nazarene U, Nashville, TN 37210-2877	1901	$10,848	$4,904	2-D	1,709	154
Trinity Coll, Hartford, CT 06106-3100	1823	$25,440	$7,160	1-M	2,246	261
Trinity Coll, Washington, DC 20017-1094 (3)	1897	$14,440	$6,700	2-M	1,325	141
Trinity Intl U, Deerfield, IL 60015-1284	1897	$14,390	$5,100	2-D	2,152	68
Trinity U, San Antonio, TX 78212-7200	1869	$15,804	$6,330	2-M	2,571	264
Tri-State U, Angola, IN 46703-1764	1884	$13,700	$4,950	1-B	1,267	87
Troy State U, Troy, AL 36082	1887	$3,126 (S)	$4,274	5-M	6,299	NA
Troy State U Dothan, Dothan, AL 36304-0368	1961	$3,084 (S)	NA	5-M	1,958	127
Troy State U Montgomery, Montgomery, AL 36103-4419	1965	$2,760 (S)	NA	5-M	3,090	215
Truman State U, Kirksville, MO 63501-4221	1867	$3,712 (S)	$4,552	5-M	6,009	411
Tufts U, Medford, MA 02155	1852	$25,714	$7,680	1-D	8,933	1,097
Tulane U, New Orleans, LA 70118-5669	1834	$25,390	$6,908	1-D	12,160	880
Tusculum Coll, Greeneville, TN 37743-9997	1794	$13,100	$4,300	2-M	1,681	36
Tuskegee U, Tuskegee, AL 36088	1881	$10,084	$5,328	1-D	2,826	262
Union Coll, Schenectady, NY 12308-2311	1795	$24,963	$6,639	1-M	2,439	218
The Union Inst, Cincinnati, OH 45206-1925	1969	$6,528	NA	1-D	1,812	30
Union U, Jackson, TN 38305-3697	1823	$11,920	$3,830	2-M	2,392	216
United States Air Force Acad, USAF Academy, CO 80840-5025 (2)	1954	$0 (C)	NA	4-B	4,319	531
United States Intl U, San Diego, CA 92131-1799	1952	$13,611	$5,775	1-D	1,335	146
U del Turabo, Turabo, PR 00778-3030	1972	$3,324	NA	1-M	8,065	410
U Metroa, Río Piedras, PR 00928-1150	1980	$3,324	NA	1-M	5,857	358
U at Buffalo, The State U of New York, Buffalo, NY 14260	1846	$4,715 (S)	$6,054	5-D	24,830	1,949
The U of Akron, Akron, OH 44325-0001	1870	$4,496 (S)	$5,350	5-D	21,556	1,615
The U of Alabama, Tuscaloosa, AL 35487	1831	$3,014 (S)	$3,800	5-D	19,277	1,036
The U of Alabama at Birmingham, Birmingham, AL 35294	1969	$3,363 (S)	$6,471	5-D	14,951	782
The U of Alabama in Huntsville, Huntsville, AL 35899	1950	$3,284 (S)	$4,300	5-D	6,563	427
U of Alaska Anchorage, Anchorage, AK 99508-8060	1954	$1,800 (S)	NA	5-M	14,765	982
U of Alaska Fairbanks, Fairbanks, AK 99775	1917	$3,510 (S)	$4,610	5-D	7,131	555
U of Alaska Southeast, Juneau, AK 99801	1972	$2,602 (S)	NA	5-M	2,799	NA
The U of Arizona, Tucson, AZ 85721	1885	$2,348 (S)	$5,888	5-D	34,560	1,391
U of Arkansas, Fayetteville, AR 72701-1201	1871	$3,660 (S)	$4,358	5-D	15,346	864
U of Arkansas at Little Rock, Little Rock, AR 72204-1099	1927	$2,996 (S)	NA	5-D	10,541	701
U of Arkansas at Monticello, Monticello, AR 71656	1909	$2,530 (S)	$2,580	5-M	2,323	136
U of Arkansas at Pine Bluff, Pine Bluff, AR 71601-2799	1873	$3,090 (S)	$4,100	5-M	3,042	230
U of Baltimore, Baltimore, MD 21201-5779	1925	$4,504 (S)	NA	5-D	4,674	332
U of Bridgeport, Bridgeport, CT 06601	1927	$15,012	$7,070	1-D	2,686	324
U of California, Berkeley, Berkeley, CA 94720-1500	1868	$4,047 (S)	$8,670	5-D	31,347	1,773
U of California, Davis, Davis, CA 95616	1905	$4,072 (S)	$6,800	5-D	26,121	NA
U of California, Irvine, Irvine, CA 92697	1965	$4,058 (S)	$6,724	5-D	20,211	989
U of California, Los Angeles, Los Angeles, CA 90095	1919	$3,698 (S)	$8,565	5-D	36,890	3,435
U of California, Riverside, Riverside, CA 92521-0102	1954	$3,862 (S)	$7,200	5-D	13,063	718
U of California, San Diego, La Jolla, CA 92093	1959	$3,848 (S)	$7,425	5-D	20,212	1,045
U of California, Santa Barbara, Santa Barbara, CA 93106	1909	$3,836 (S)	$7,577	5-D	19,962	933
U of Central Arkansas, Conway, AR 72035-0001	1907	$3,402 (S)	$3,290	5-D	8,481	496
U of Central Florida, Orlando, FL 32816	1963	$2,385 (S)	$5,436	5-D	33,713	1,828
U of Central Oklahoma, Edmond, OK 73034-5209	1890	$1,965 (S)	$2,905	5-M	14,173	719
U of Charleston, Charleston, WV 25304-1099	1888	$13,200	$4,637	1-M	1,280	131
U of Chicago, Chicago, IL 60637-1513	1891	$25,239	$8,070	1-D	12,531	1,860
U of Cincinnati, Cincinnati, OH 45221-0091	1819	$5,337 (S)	$6,399	5-D	27,327	1,174
U of Colorado at Boulder, Boulder, CO 80309	1876	$3,188 (S)	$5,538	5-D	28,910	2,216
U of Colorado at Colorado Springs, Colorado Springs, CO 80933-7150	1965	$2,743 (S)	$5,893	5-D	6,588	465
U of Colorado at Denver, Denver, CO 80217-3364	1912	$2,723 (S)	NA	5-D	14,113	868
U of Colorado Health Scis Ctr, Denver, CO 80262	1883	$8,304 (S)	NA	5-D	2,059	1,700
U of Connecticut, Storrs, CT 06269	1881	$5,596 (S)	$6,062	5-D	19,393	1,067
U of Dallas, Irving, TX 75062-4736	1955	$14,994	$5,600	2-D	3,300	239
U of Dayton, Dayton, OH 45469-1300	1850	$16,070	$5,080	2-D	10,310	797
U of Delaware, Newark, DE 19716	1743	$4,993 (S)	$5,312	12-D	20,276	1,233
U of Denver, Denver, CO 80208	1864	$20,556	$6,438	1-D	9,444	841
U of Detroit Mercy, Detroit, MI 48219-0900	1877	$14,332	$5,470	2-D	6,212	407
U of Evansville, Evansville, IN 47722-0002	1854	$16,404	$5,270	2-M	2,797	184
The U of Findlay, Findlay, OH 45840-3653	1882	$16,020	$5,960	2-M	4,510	347
U of Florida, Gainesville, FL 32611	1853	$2,256 (S)	$5,440	5-D	45,114	1,809
U of Georgia, Athens, GA 30602	1785	$3,276 (S)	$5,080	5-D	31,288	2,038
U of Great Falls, Great Falls, MT 59405	1932	$10,260	NA	2-M	1,072	98
U of Guam, Mangilao, GU 96923	1952	$2,250 (S)	$2,905	7-M	3,748	230
U of Hartford, West Hartford, CT 06117-1599	1877	$19,696	$7,840	1-D	6,897	721
U of Hawaii at Hilo, Hilo, HI 96720-4091	1970	$1,466 (S)	$4,989	5-B	2,874	215
U of Hawaii at Manoa, Honolulu, HI 96822	1907	$3,157 (S)	$4,933	5-D	17,263	1,136
U of Houston, Houston, TX 77204	1927	$2,444 (S)	$4,513	5-D	32,123	1,611
U of Houston–Clear Lake, Houston, TX 77058-1098	1974	$4,402 (S)	NA	5-M	3,634	514
U of Houston–Downtown, Houston, TX 77002-1001	1974	$2,414 (S)	NA	5-B	8,951	471
U of Houston–Victoria, Victoria, TX 77901-4450	1973	$2,124 (S)	NA	5-M	1,698	94
U of Idaho, Moscow, ID 83844-2282	1889	$2,476 (S)	$4,064	5-D	11,635	600
U of Illinois at Chicago, Chicago, IL 60607-7128	1946	$4,800 (S)	$5,856	5-D	24,541	1,549
U of Illinois at Springfield, Springfield, IL 62794-9243	1969	$3,395 (S)	NA	5-M	3,942	243
U of Illinois at Urbana–Champaign, Urbana, IL 61801	1867	$5,028 (S)	$5,844	5-D	37,965	2,538
U of Indianapolis, Indianapolis, IN 46227-3697	1902	$14,630	$5,225	2-D	3,601	368
The U of Iowa, Iowa City, IA 52242-1316	1847	$3,204 (S)	$4,597	5-D	28,311	1,714
U of Kansas, Lawrence, KS 66045	1866	$6,425 (S)	$4,114	5-D	28,329	1,743
U of Kentucky, Lexington, KY 40506-0032	1865	$3,446 (S)	$3,782	5-D	23,114	NA
U of La Verne, La Verne, CA 91750-4443	1891	$16,940	$5,400	1-D	3,004	396
U of Louisiana at Lafayette, Lafayette, LA 70504	1898	$2,275 (S)	$2,726	5-D	15,742	665
U of Louisiana at Monroe, Monroe, LA 71209-0001	1931	$2,719 (S)	$3,660	5-D	9,400	516
U of Louisville, Louisville, KY 40292-0001	1798	$3,447 (S)	$3,500	5-D	19,771	1,857
U of Maine, Orono, ME 04469	1865	$4,764 (S)	$5,360	5-D	10,282	624

Name, address	Year Founded	Tuition & Fees	Rm. & Board	Control, Degree	Enroll-ment	Faculty
The U of Maine at Augusta, Augusta, ME 04330-9410	1965	$3,575 (S)	NA	5-B	5,617	228
U of Maine at Farmington, Farmington, ME 04938-1990	1863	$3,996 (S)	$4,614	5-B	2,413	157
U of Maine at Presque Isle, Presque Isle, ME 04769-2888	1903	$3,520 (S)	$4,140	5-B	1,427	123
U of Mary, Bismarck, ND 58504-9652	1959	$8,685 (A)	$3,600	2-M	2,119	155
U of Mary Hardin-Baylor, Belton, TX 76513	1845	$9,300	$3,931	2-M	2,590	189
U of Maryland, Baltimore County, Baltimore, MD 21250-5398	1963	$5,490 (S)	$5,850	5-D	10,759	821
U of Maryland, Coll Park, College Park, MD 20742	1856	$5,136 (S)	$6,328	5-D	33,189	2,020
U of Maryland Eastern Shore, Princess Anne, MD 21853-1299	1886	$3,994 (S)	$5,130	5-D	3,297	280
U of Maryland U Coll, Adelphi, MD 20783	1947	$4,584 (S)	NA	5-D	18,276	788
U of Massachusetts Amherst, Amherst, MA 01003	1863	$5,370 (S)	$4,895	5-D	24,416	1,264
U of Massachusetts Boston, Boston, MA 02125-3393	1964	$4,222 (S)	NA	5-D	13,346	887
U of Massachusetts Dartmouth, North Dartmouth, MA 02747-2300	1895	$4,129 (S)	$5,143	5-D	7,122	477
U of Massachusetts Lowell, Lowell, MA 01854-2881	1894	$4,255 (S)	$4,994	5-D	12,189	553
The U of Memphis, Memphis, TN 38152	1912	$3,087 (S)	$3,908	5-D	19,986	1,321
U of Miami, Coral Gables, FL 33124	1925	$22,528	$7,934	1-D	13,963	1,122
U of Michigan, Ann Arbor, MI 48109	1817	$6,513 (S)	$5,780	5-D	37,595	2,633
U of Michigan–Dearborn, Dearborn, MI 48128-1491	1959	$4,440 (S)	NA	5-M	8,484	443
U of Michigan–Flint, Flint, MI 48502-1950	1956	$3,916 (S)	NA	5-M	6,316	407
U of Minnesota, Crookston, Crookston, MN 56716-5001	1966	$5,070 (S)	$4,100	5-B	2,464	NA
U of Minnesota, Duluth, Duluth, MN 55812-2496	1947	$5,017 (S)	$4,338	5-F	9,089	441
U of Minnesota, Morris, Morris, MN 56267-2134	1959	$5,567 (S)	$4,102	5-B	1,944	114
U of Minnesota, Twin Cities Campus, Minneapolis, MN 55455-0213	1851	$4,877 (S)	$4,914	5-D	45,481	3,079
U of Mississippi, University, MS 38677	1844	$3,153 (S)	$3,790	5-D	12,105	594
U of Mississippi Medical Ctr, Jackson, MS 39216-4505	1955	$2,478 (S)	NA	5-D	1,694	714
U of Missouri–Columbia, Columbia, MO 65211	1839	$4,581 (S)	$4,545	5-D	23,309	1,664
U of Missouri–Kansas City, Kansas City, MO 64110-2499	1929	$5,050 (S)	$4,950	5-D	12,698	884
U of Missouri–Rolla, Rolla, MO 65409-0910	1870	$4,805 (S)	$4,838	5-D	4,626	380
U of Missouri–St. Louis, St. Louis, MO 63121-4499	1963	$4,944 (S)	$4,620	5-D	15,397	1,051
U of Mobile, Mobile, AL 36663-0220	1961	$8,300	$2,550	2-M	1,918	150
The U of Montana–Missoula, Missoula, MT 59812-0002	1893	$3,136 (S)	$4,525	5-D	12,406	645
U of Montevallo, Montevallo, AL 35115	1896	$3,480 (S)	$3,390	5-M	3,014	205
U of Nebraska at Kearney, Kearney, NE 68849-0001	1903	$2,873 (S)	$3,620	5-M	6,506	400
U of Nebraska at Omaha, Omaha, NE 68182	1908	$2,970 (S)	NA	5-D	13,479	800
U of Nebraska–Lincoln, Lincoln, NE 68588	1869	$3,450 (S)	$4,310	5-D	22,268	1,099
U of Nebraska Medical Ctr, Omaha, NE 68198	1869	$4,022 (S)	NA	5-D	2,695	786
U of Nevada, Las Vegas, Las Vegas, NV 89154-9900	1957	$2,386 (S)	$5,800	5-D	22,043	1,247
U of Nevada, Reno, Reno, NV 89557	1874	$2,454 (S)	$5,650	5-D	13,149	680
U of New England, Biddeford, ME 04005-9526	1831	$16,250	$6,420	1-F	3,323	210
U of New Hampshire, Durham, NH 03824	1866	$7,395 (S)	$5,154	5-D	13,426	706
U of New Hampshire at Manchester, Manchester, NH 03102-8597	1967	$5,064 (S)	NA	5-B	1,086	108
U of New Haven, West Haven, CT 06516-1916	1920	$15,520	$6,960	1-D	4,348	615
U of New Mexico, Albuquerque, NM 87131-2039	1889	$2,794 (S)	$4,860	5-D	23,545	1,313
U of New Orleans, New Orleans, LA 70148	1958	$2,882 (S)	$3,300	5-D	16,218	840
U of North Alabama, Florence, AL 35632-0001	1830	$2,915 (S)	$3,506	5-M	5,601	307
The U of North Carolina at Asheville, Asheville, NC 28804-3299	1927	$2,063 (S)	$4,300	5-M	3,234	293
The U of North Carolina at Chapel Hill, Chapel Hill, NC 27599	1789	$2,768 (S)	$5,630	5-D	24,892	2,973
The U of North Carolina at Charlotte, Charlotte, NC 28223-0001	1946	$2,138 (S)	$4,354	5-D	17,241	1,002
The U of North Carolina at Greensboro, Greensboro, NC 27412-5001	1891	$2,201 (S)	$5,422	5-D	12,831	660
The U of North Carolina at Pembroke, Pembroke, NC 28372-1510	1887	$1,860 (S)	$3,680	5-M	3,445	224
The U of North Carolina at Wilmington, Wilmington, NC 28403-3201	1947	$2,360 (S)	$4,862	5-M	9,885	608
U of North Dakota, Grand Forks, ND 58202	1883	$3,572 (S)	$3,614	5-D	11,031	581
U of Northern Colorado, Greeley, CO 80639	1890	$2,783 (S)	$4,996	5-D	12,163	591
U of Northern Iowa, Cedar Falls, IA 50614	1876	$3,130 (S)	$4,149	5-D	14,106	860
U of North Florida, Jacksonville, FL 32224-2645	1965	$1,941 (S)	$4,990	5-D	12,550	666
U of North Texas, Denton, TX 76203	1890	$3,055 (S)	$4,202	5-D	27,054	1,270
U of Notre Dame, Notre Dame, IN 46556	1842	$23,357	$5,920	2-D	10,800	NA
U of Oklahoma, Norman, OK 73019-0390	1890	$2,581 (S)	$4,610	5-D	21,607	1,138
U of Oklahoma Health Scis Ctr, Oklahoma City, OK 73190	1890	$2,978 (S)	NA	5-D	2,831	441
U of Oregon, Eugene, OR 97403	1872	$3,825 (S)	$5,564	5-D	17,801	1,125
U of Pennsylvania, Philadelphia, PA 19104	1740	$25,170	$7,826	1-D	18,145	1,761
U of Phoenix–Colorado Campus, Lone Tree, CO 80124[1]	NR	$7,740	NA	3-D	66,534	8,163
U of Phoenix–Fort Lauderdale Campus, Plantation, FL 33324[1]	NR	$7,740	NA	3-D	66,534	8,163
U of Phoenix–Grand Rapids Campus, Grand Rapids, MI 49546[1]	NR	$7,740	NA	3-D	66,534	8,163
U of Phoenix–Hawaii Campus, Honolulu, HI 96813[1]	NR	$7,740	NA	3-D	66,534	8,163
U of Phoenix–Jacksonville Campus, Jacksonville, FL 32216[1]	NR	$7,740	NA	3-D	66,534	8,163
U of Phoenix-Louisiana Campus, Metairie, LA 70001[1]	NR	$7,740	NA	3-D	66,534	8,163
U of Phoenix–Maryland Campus, Columbia, MD 21045[1]	NR	$7,740	NA	3-D	66,534	8,163
U of Phoenix–Metro Detroit Campus, Troy, MI 48098[1]	NR	$7,740	NA	3-D	66,534	8,163
U of Phoenix–Nevada Campus, Las Vegas, NV 89106[1]	NR	$7,740	NA	3-D	66,534	8,163
U of Phoenix–New Mexico Campus, Albuquerque, NM 87109[1]	NR	$7,740	NA	3-D	66,534	8,163
U of Phoenix–Northern California Campus, Pleasanton, CA 94588[1]	NR	$7,740	NA	3-D	66,534	8,163
U of Phoenix–Ohio Campus, Independence, OH 44131[1]	NR	$7,740	NA	3-D	66,534	8,163
U of Phoenix–Oklahoma City Campus, Oklahoma City, OK 73116[1]	NR	$7,740	NA	3-D	66,534	8,163
U of Phoenix–Oregon Campus, Portland, OR 97223[1]	NR	$7,740	NA	3-D	66,534	8,163
U of Phoenix–Orlando Campus, Maitland, FL 32751[1]	NR	$7,740	NA	3-D	66,534	8,163
U of Phoenix–Philadelphia Campus, Wayne, PA 19087[1]	NR	$7,740	NA	3-D	66,534	8,163
U of Phoenix–Phoenix Campus, Phoenix, AZ 85040-1958[1]	1976	$7,740	NA	3-D	66,534	8,163
U of Phoenix–Pittsburgh Campus, Pittsburgh, PA 15276[1]	NR	$7,740	NA	3-D	66,534	8,163
U of Phoenix–Puerto Rico Campus, Guaynabo, PR 00970-3870[1]	NR	$7,740	NA	3-D	66,534	8,163
U of Phoenix–Sacramento Campus, Sacramento, CA 95833[1]	NR	$7,740	NA	3-D	66,534	8,163
U of Phoenix–Saint Louis Campus, Maryland Heights, MO 63043-4822[1]	NR	$7,740	NA	3-D	66,534	8,163
U of Phoenix–San Diego Campus, San Diego, CA 92123[1]	NR	$7,740	NA	3-D	66,534	8,163
U of Phoenix–Southern Arizona Campus, Tucson, AZ 85712[1]	NR	$7,740	NA	3-D	66,534	8,163
U of Phoenix–Southern California Campus, Fountain Valley, CA 92708[1]	NR	$7,740	NA	3-D	66,534	8,163
U of Phoenix–Southern Colorado Campus, Colorado Springs, CO 80919[1]	NR	$7,740	NA	3-D	66,534	8,163

Name, address	Year Founded	Tuition & Fees	Rm. & Board	Control, Degree	Enroll-ment	Faculty
U of Phoenix–Tampa Campus, Tampa, FL 33637[1]	NR	$7,740	NA	3-D	66,534	8,163
U of Phoenix–Tulsa Campus, Tulsa, OK 74146[1]	NR	$7,740	NA	3-D	66,534	8,163
U of Phoenix–Utah Campus, Salt Lake City, UT 84123[1]	NR	$7,740	NA	3-D	66,534	8,163
U of Phoenix–Washington Campus, Seattle, WA 98188[1]	NR	$7,740	NA	3-D	66,534	8,163
U of Pittsburgh, Pittsburgh, PA 15260	1787	$7,002 (S)	$5,936	12-D	26,329	1,959
U of Pittsburgh at Bradford, Bradford, PA 16701-2812	1963	$6,876 (S)	$5,150	12-B	1,204	111
U of Pittsburgh at Greensburg, Greensburg, PA 15601-5860	1963	$6,936 (S)	$5,540	12-B	1,587	100
U of Pittsburgh at Johnstown, Johnstown, PA 15904-2990	1927	$6,984 (S)	$5,400	12-B	3,031	199
U of Portland, Portland, OR 97203-5798	1901	$18,450	$5,398	2-M	2,926	282
U of Puerto Rico at Humacao, Humacao, PR 00791	1962	$1,245 (S)	NA	6-B	4,592	278
U of Puerto Rico at Ponce, Ponce, PR 00732-7186	1970	$1,910 (S)	NA	6-B	4,150	200
U of Puerto Rico at Utuado, Utuado, PR 00641-2500	1979	$1,315 (S)	NA	6-B	1,620	90
U of Puerto Rico, Cayey U Coll, Cayey, PR 00736	1967	$1,245	NA	6-B	4,089	229
U of Puerto Rico, Mayagüez Campus, Mayagüez, PR 00681-9000	1911	$1,160 (S)	NA	6-D	12,414	761
U of Puerto Rico, Medical Scis Campus, San Juan, PR 00936-5067 (4)	1950	$1,900 (S)	NA	6-D	2,739	762
U of Puerto Rico, Río Piedras, San Juan, PR 00931	1903	$790 (S)	$4,180	6-D	21,539	1,293
U of Puget Sound, Tacoma, WA 98416	1888	$21,425	$5,510	1-M	2,867	263
U of Redlands, Redlands, CA 92373-0999	1907	$20,836	$7,590	1-M	1,803	196
U of Rhode Island, Kingston, RI 02881	1892	$5,154 (S)	$6,688	5-D	14,362	682
U of Richmond, University of Richmond, VA 23173	1830	$20,140	$4,530	1-F	3,583	360
U of Rio Grande, Rio Grande, OH 45674	1876	$8,619 (A)	$5,169	1-M	2,107	132
U of Rochester, Rochester, NY 14627-0250	1850	$23,730	$7,740	1-D	8,336	1,371
U of St. Francis, Joliet, IL 60435-6169	1920	$14,290	$5,350	2-M	2,742	148
U of Saint Francis, Fort Wayne, IN 46808-3994	1890	$12,495	$4,800	2-M	1,587	161
U of St. Thomas, St. Paul, MN 55105-1096	1885	$17,313	$5,407	2-D	11,288	778
U of St. Thomas, Houston, TX 77006-4696	1947	$12,412	$5,550	2-D	4,073	232
U of San Diego, San Diego, CA 92110-2492	1949	$19,128	$8,440	2-D	6,943	606
U of San Francisco, San Francisco, CA 94117-1080	1855	$19,060	$8,242	2-D	6,207	716
U of Sci & Arts of Oklahoma, Chickasha, OK 73018-0001	1908	$1,923 (S)	$2,390	5-B	1,409	96
The U of Scranton, Scranton, PA 18510	1888	$18,660	$8,112	2-M	4,615	381
U of Sioux Falls, Sioux Falls, SD 57105-1699	1883	$12,100	$3,790	2-M	1,272	90
U of South Alabama, Mobile, AL 36688-0002	1963	$2,670 (S)	$3,114	5-D	11,673	917
U of South Carolina, Columbia, SC 29208	1801	$3,918 (S)	$4,588	5-D	23,728	1,389
U of South Carolina Aiken, Aiken, SC 29801-6309	1961	$3,598 (S)	$4,090	5-M	3,278	228
U of South Carolina Spartanburg, Spartanburg, SC 29303-4999	1967	$3,714 (S)	$4,040	5-M	3,709	246
U of South Dakota, Vermillion, SD 57069-2390	1862	$3,678 (S)	$3,037	5-D	7,488	362
U of Southern California, Los Angeles, CA 90089	1880	$24,124	$7,610	1-D	29,194	1,959
U of Southern Colorado, Pueblo, CO 81001-4901	1933	$2,369 (S)	$5,164	5-M	5,791	204
U of Southern Indiana, Evansville, IN 47712-3590	1965	$3,075 (S)	$5,182	5-M	9,012	545
U of Southern Maine, Portland, ME 04104-9300	1878	$4,309 (S)	$5,202	5-D	10,820	704
U of Southern Mississippi, Hattiesburg, MS 39406	1910	$2,970 (S)	$4,212	5-D	14,510	664
U of South Florida, Tampa, FL 33620-9951	1956	$2,388 (S)	$4,494	5-D	36,015	2,009
The U of Tampa, Tampa, FL 33606-1490	1931	$16,102	$5,418	1-M	3,560	288
The U of Tennessee, Knoxville, TN 37996	1794	$3,612 (S)	$4,490	5-D	25,890	1,248
The U of Tennessee at Chattanooga, Chattanooga, TN 37403-2598	1886	$2,834 (S)	NA	5-M	8,319	616
The U of Tennessee at Martin, Martin, TN 38238-1000	1900	$3,330 (S)	$3,800	5-M	5,877	309
The U of Texas at Arlington, Arlington, TX 76019	1895	$2,864 (A)	$3,795	5-D	20,424	916
The U of Texas at Austin, Austin, TX 78712-1111	1883	$3,575 (S)	$5,113	5-D	49,996	2,580
The U of Texas at Brownsville, Brownsville, TX 78520-4991	1973	$1,722 (A)	NA	5-M	NA	399
The U of Texas at Dallas, Richardson, TX 75083-0688	1969	$3,287 (S)	$5,799	5-D	10,945	531
The U of Texas at El Paso, El Paso, TX 79968-0001	1913	$2,541 (S)	$5,165	5-D	15,224	866
The U of Texas at San Antonio, San Antonio, TX 78249-0617	1969	$3,455 (S)	$5,994	5-D	18,830	948
The U of Texas at Tyler, Tyler, TX 75799-0001	1971	$2,336 (S)	NA	5-M	3,594	380
The U of Texas Health Sci Ctr at Houston, Houston, TX 77225-0036	1972	$3,463 (S)	NA	5-D	3,143	1,080
The U of Texas Medical Branch at Galveston, Galveston, TX 77555	1891	$1,832 (S)	NA	5-D	1,927	123
The U of Texas of the Permian Basin, Odessa, TX 79762-0001	1969	$2,530 (S)	$3,950	5-M	2,272	146
The U of Texas–Pan American, Edinburg, TX 78539-2999	1927	$2,363 (S)	$5,531	5-D	12,759	791
The U of Texas Southwestern Medical Ctr at Dallas, Dallas, TX 75390	1943	$1,941 (S)	NA	5-D	1,505	103
The U of the Arts, Philadelphia, PA 19102-4944	1870	$17,850	NA	1-M	2,063	394
U of the District of Columbia, Washington, DC 20008-1175	1976	$2,070 (S)	NA	9-M	5,358	442
U of the Incarnate Word, San Antonio, TX 78209-6397	1881	$12,670	$5,010	2-D	3,702	383
U of the Pacific, Stockton, CA 95211-0197	1851	$20,725	$6,378	1-D	5,609	575
U of the Sacred Heart, San Juan, PR 00914-0383	1935	$4,820	NA	2-M	5,234	374
U of the Scis in Philadelphia, Philadelphia, PA 19104-4495	1821	$16,100	$7,600	1-D	2,262	225
U of the South, Sewanee, TN 37383-1000	1857	$20,130	$5,610	2-D	1,494	150
U of the Virgin Islands, Charlotte Amalie, VI 00802-9990	1962	$4,946 (S)	$5,830	7-M	2,268	239
U of Toledo, Toledo, OH 43606-3398	1872	$4,681 (S)	$4,798	5-D	19,491	961
U of Tulsa, Tulsa, OK 74104-3189	1894	$13,810	$4,810	2-D	4,158	414
U of Utah, Salt Lake City, UT 84112-1107	1850	$2,897 (S)	$4,669	5-D	26,031	1,157
U of Vermont, Burlington, VT 05405	1791	$8,288 (S)	$5,806	5-D	10,118	687
U of Virginia, Charlottesville, VA 22903	1819	$4,335 (S)	$4,767	5-D	22,411	1,236
The U of Virginia's Coll at Wise, Wise, VA 24293	1954	$3,330 (S)	$5,080	5-B	1,447	101
U of Washington, Seattle, WA 98195	1861	$3,761 (S)	$6,060	5-D	36,134	3,035
The U of West Alabama, Livingston, AL 35470	1835	$2,894 (S)	$2,822	5-M	1,924	108
U of West Florida, Pensacola, FL 32514-5750	1963	$2,398 (S)	$4,668	5-D	8,479	441
U of Wisconsin–Eau Claire, Eau Claire, WI 54702-4004	1916	$3,252 (S)	$3,435	5-M	10,549	482
U of Wisconsin–Green Bay, Green Bay, WI 54311-7001	1968	$3,306 (S)	NA	5-M	5,507	304
U of Wisconsin–La Crosse, La Crosse, WI 54601-3742	1909	$2,594 (S)	$3,360	5-M	9,013	481
U of Wisconsin–Madison, Madison, WI 53706-1380	1848	$3,780 (S)	$4,500	5-D	41,219	NA
U of Wisconsin–Milwaukee, Milwaukee, WI 53201-0413	1956	$3,785 (S)	NA	5-D	23,578	1,333
U of Wisconsin–Oshkosh, Oshkosh, WI 54901	1871	$3,024 (S)	$3,234	5-M	10,756	509
U of Wisconsin–Parkside, Kenosha, WI 53141-2000	1968	$3,094 (S)	$4,230	5-M	4,856	180
U of Wisconsin–Platteville, Platteville, WI 53818-3099	1866	$3,132 (S)	$3,338	5-M	5,558	NA
U of Wisconsin–River Falls, River Falls, WI 54022-5001	1874	$3,138 (S)	$3,452	5-M	6,743	216
U of Wisconsin–Stevens Point, Stevens Point, WI 54481-3897	1894	$3,165 (S)	$3,616	5-M	8,858	435
U of Wisconsin–Stout, Menomonie, WI 54751	1891	$3,286 (S)	$3,530	5-M	7,702	389
U of Wisconsin–Superior, Superior, WI 54880-4500	1893	$2,974 (S)	$3,426	5-M	2,660	140

Name, address	Year Founded	Tuition & Fees	Rm. & Board	Control, Degree	Enroll-ment	Faculty
U of Wisconsin–Whitewater, Whitewater, WI 53190-1790	1868	$3,144 (S)	$3,324	5-M	10,540	465
U of Wyoming, Laramie, WY 82071	1886	$2,575 (S)	$4,568	5-D	11,057	643
U System Coll for Lifelong Learning, Concord, NH 03301	1972	$4,158 (S)	NA	11-B	NA	498
Urbana U, Urbana, OH 43078-2091	1850	$11,548	$5,000	2-M	1,372	87
Ursinus Coll, Collegeville, PA 19426-1000	1869	$23,770	$6,140	2-B	1,290	143
Ursuline Coll, Pepper Pike, OH 44124-4398 (4)	1871	$13,500	$4,560	2-M	1,252	166
Utah State U, Logan, UT 84322	1888	$2,403 (S)	$4,040	5-D	21,490	674
Utica Coll of Syracuse U, Utica, NY 13502-4892	1946	$17,104	$6,660	1-M	2,189	194
Valdosta State U, Valdosta, GA 31698	1906	$2,992 (S)	$4,214	5-D	8,687	495
Valley City State U, Valley City, ND 58072	1890	$3,173 (S)	$2,892	5-B	1,090	77
Valparaiso U, Valparaiso, IN 46383-6493	1859	$17,636	$4,660	2-F	3,654	328
Vanderbilt U, Nashville, TN 37240-1001	1873	$24,712	$5,364	1-D	10,092	953
Vanguard U of Southern California, Costa Mesa, CA 92626-6597	1920	$13,778	$5,060	2-M	1,660	129
Vassar Coll, Poughkeepsie, NY 12604	1861	$24,940	$6,940	1-M	2,400	298
Vermont Tech Coll, Randolph Center, VT 05061-0500	1866	$5,830 (S)	$5,346	5-B	1,145	113
Villa Julie Coll, Stevenson, MD 21153	1952	$11,500	NA	1-M	2,259	253
Villanova U, Villanova, PA 19085-1699	1842	$22,825	$8,050	2-D	10,003	812
Virginia Coll at Birmingham, Birmingham, AL 35209	1989	$8,000	NA	3-B	1,143	51
Virginia Commonwealth U, Richmond, VA 23284-9005	1838	$3,650 (S)	$4,955	5-D	24,066	1,897
Virginia Military Inst, Lexington, VA 24450 (2)	1839	$5,072 (S)	$4,564	5-B	1,300	143
Virginia Polytechnic Inst & State U, Blacksburg, VA 24061	1872	$3,640 (S)	$3,954	5-D	25,643	1,491
Virginia State U, Petersburg, VA 23806-0001	1882	$3,228 (S)	$5,310	5-M	4,353	278
Virginia Union U, Richmond, VA 23220-1170	1865	$9,580	$4,250	2-D	1,700	104
Virginia Wesleyan Coll, Norfolk, VA 23502-5599	1961	$15,035	$5,750	2-B	1,421	117
Viterbo U, La Crosse, WI 54601-4797	1890	$13,050	$4,530	2-M	2,106	173
Wagner Coll, Staten Island, NY 10301-4495	1883	$19,200	$6,800	1-M	2,013	NA
Wake Forest U, Winston-Salem, NC 27109	1834	$22,410	$6,340	2-D	6,124	520
Walla Walla Coll, College Place, WA 99324-1198	1892	$14,568	$4,176	2-M	1,795	177
Walsh Coll of Accountancy & Business Administration, Troy, MI 48007-7006	1922	$5,264	NA	1-M	2,981	130
Walsh U, North Canton, OH 44720-3396	1958	$12,410	$5,600	2-M	1,545	182
Wartburg Coll, Waverly, IA 50677-0903	1852	$15,765	$4,500	2-B	1,600	147
Washburn U of Topeka, Topeka, KS 66621	1865	$3,052 (S)	$3,410	10-F	5,917	492
Washington & Jefferson Coll, Washington, PA 15301-4801	1781	$19,650	$5,160	1-B	1,241	110
Washington & Lee U, Lexington, VA 24450-0303	1749	$17,965	$5,690	1-F	2,100	237
Washington Coll, Chestertown, MD 21620-1197	1782	$21,300	$5,740	1-M	1,225	125
Washington State U, Pullman, WA 99164	1890	$3,894 (S)	$4,826	5-D	21,122	1,230
Washington U in St. Louis, St. Louis, MO 63130-4899	1853	$24,745	$7,724	1-D	12,118	1,192
Wayland Baptist U, Plainview, TX 79072-6998	1908	$7,850	$3,121	2-M	5,093	288
Waynesburg Coll, Waynesburg, PA 15370-1222	1849	$11,950	$2,460	2-M	1,692	126
Wayne State Coll, Wayne, NE 68787	1910	$2,513 (S)	$3,330	5-M	3,518	200
Wayne State U, Detroit, MI 48202	1868	$3,970 (S)	NA	5-D	30,408	1,744
Weber State U, Ogden, UT 84408-1001	1889	$2,118 (S)	$3,878	5-M	16,051	738
Webster U, St. Louis, MO 63119-3194	1915	$13,080	$5,658	1-D	16,730	1,930
Wellesley Coll, Wellesley, MA 02481 (3)	1870	$24,174	$7,480	1-B	2,267	316
Wentworth Inst of Technology, Boston, MA 02115-5998	1904	$13,000	$6,800	1-B	3,152	239
Wesleyan U, Middletown, CT 06459-0260	1831	$26,180	$6,630	1-D	3,158	308
Wesley Coll, Dover, DE 19901-3875	1873	$11,719	$5,266	2-M	1,490	81
West Chester U of Pennsylvania, West Chester, PA 19383	1871	$4,630 (S)	$4,650	5-M	12,274	779
Western Carolina U, Cullowhee, NC 28723	1889	$2,260 (S)	$3,284	5-D	6,699	529
Western Connecticut State U, Danbury, CT 06810-6885	1903	$3,910 (S)	$5,668	5-M	5,806	334
Western Illinois U, Macomb, IL 61455-1390	1899	$3,972 (S)	$4,706	5-M	13,093	688
Western Intl U, Phoenix, AZ 85021-2718	1978	$7,920	NA	3-M	2,690	200
Western Kentucky U, Bowling Green, KY 42101-3576	1906	$2,534 (S)	$3,813	5-M	15,479	938
Western Maryland Coll, Westminster, MD 21157-4390	1867	$19,600	$5,350	1-M	3,147	150
Western Michigan U, Kalamazoo, MI 49008-5202	1903	$4,094 (S)	$5,073	5-D	28,657	1,104
Western Montana Coll of The U of Montana, Dillon, MT 59725-3598	1893	$2,795 (S)	$3,810	5-B	1,160	60
Western New England Coll, Springfield, MA 01119-2654	1919	$15,504	$7,050	1-F	NA	336
Western New Mexico U, Silver City, NM 88062-0680	1893	$1,768 (S)	$2,938	5-M	2,580	145
Western Oregon U, Monmouth, OR 97361-1394	1856	$3,342 (S)	$5,043	5-M	4,729	281
Western State Coll of Colorado, Gunnison, CO 81231	1901	$2,980 (S)	$5,652	5-B	2,366	128
Western Washington U, Bellingham, WA 98225-5996	1893	$3,102 (S)	$5,100	5-M	11,708	650
Westfield State Coll, Westfield, MA 01086	1838	$2,916 (S)	$4,556	5-M	5,008	251
West Liberty State Coll, West Liberty, WV 26074	1837	$2,420 (S)	$3,340	5-B	2,606	162
Westminster Coll, New Wilmington, PA 16172-0001	1852	$16,270	$4,530	2-M	1,599	138
Westminster Coll, Salt Lake City, UT 84105-3697	1875	$13,730	$4,570	1-M	2,403	244
Westmont Coll, Santa Barbara, CA 93108-1099	1937	$20,964	$7,068	2-B	1,278	138
West Texas A&M U, Canyon, TX 79016-0001	1909	$2,075 (S)	$3,474	5-M	6,775	307
West Virginia State Coll, Institute, WV 25112-1000	1891	$2,464 (S)	$3,800	5-B	4,823	150
West Virginia U, Morgantown, WV 26506	1867	$2,836 (S)	$5,152	5-D	21,987	1,648
West Virginia U Inst of Technology, Montgomery, WV 25136	1895	$3,064 (S)	$4,210	5-M	2,326	167
West Virginia Wesleyan Coll, Buckhannon, WV 26201	1890	$18,050	$4,350	2-M	1,622	150
Westwood Coll of Technology–Denver North, Denver, CO 80221-3653	1953	$8,919	NA	3-B	1,925	69
Wheaton Coll, Wheaton, IL 60187-5593	1860	$15,540	$5,260	2-D	2,827	244
Wheaton Coll, Norton, MA 02766	1834	$24,450	$6,920	1-B	1,474	144
Wheeling Jesuit U, Wheeling, WV 26003-6295	1954	$16,230	$5,250	2-M	1,515	96
Wheelock Coll, Boston, MA 02215 (4)	1888	$17,410	$6,950	1-M	1,077	139
Whitman Coll, Walla Walla, WA 99362-2083	1859	$21,742	$6,090	1-B	1,424	154
Whittier Coll, Whittier, CA 90608-0634	1887	$20,724	$6,736	1-F	2,203	133
Whitworth Coll, Spokane, WA 99251-0001	1890	$16,928	$5,500	2-M	2,012	110
Wichita State U, Wichita, KS 67260	1895	$2,757 (S)	$4,120	5-D	14,810	517
Widener U, Chester, PA 19013-5792	1821	$17,950	$7,325	1-D	5,680	379
Wilkes U, Wilkes-Barre, PA 18766-0002	1933	$17,220	$7,438	1-F	3,570	275
Willamette U, Salem, OR 97301-3931	1842	$22,822	$5,930	2-F	2,405	230
William Carey Coll, Hattiesburg, MS 39401-5499	1906	$9,630	$1,890	2-M	1,980	172
William Jewell Coll, Liberty, MO 64068-1843	1849	$13,800	$4,250	2-B	1,153	163
William Paterson U of New Jersey, Wayne, NJ 07470-8420	1855	$5,150 (S)	$6,350	5-M	9,126	875
William Penn U, Oskaloosa, IA 52577-1799	1873	$12,770	$4,140	2-B	1,353	55

Name, address	Year Founded	Tuition & Fees	Rm. & Board	Control, Degree	Enrollment	Faculty
Williams Coll, Williamstown, MA 01267	1793	$24,740	$6,730	1-M	2,066	250
William Woods U, Fulton, MO 65251-1098	1870	$13,390	$5,600	2-M	1,479	95
Wilmington Coll, New Castle, DE 19720-6491	1967	$6,290	NA	1-D	7,000	584
Wilmington Coll, Wilmington, OH 45177	1870	$14,566	$5,220	2-B	1,262	80
Wingate U, Wingate, NC 28174-0159	1896	$13,050	$5,200	2-M	1,283	101
Winona State U, Winona, MN 55987-5838	1858	$3,300 (S)	$3,600	5-M	7,318	357
Winston-Salem State U, Winston-Salem, NC 27110-0003	1892	$1,805 (S)	$3,628	5-B	2,717	263
Winthrop U, Rock Hill, SC 29733	1886	$4,282 (S)	$4,150	5-M	6,062	415
Wofford Coll, Spartanburg, SC 29303-3663	1854	$17,730	$5,235	2-B	1,087	109
Woodbury U, Burbank, CA 91504-1099	1884	$17,458	$6,756	1-M	1,342	215
Worcester Polytechnic Inst, Worcester, MA 01609-2280	1865	$23,452	$7,592	1-D	3,874	345
Worcester State Coll, Worcester, MA 01602-2597	1874	$2,508 (S)	$5,000	5-M	5,303	258
Wright State U, Dayton, OH 45435	1964	$4,335 (S)	$4,700	5-D	14,773	689
Xavier U, Cincinnati, OH 45207	1831	$15,880	$6,940	2-D	6,523	556
Xavier U of Louisiana, New Orleans, LA 70125-1098	1925	$9,700	$5,100	2-F	3,797	216
Yale U, New Haven, CT 06520	1701	$25,220	$7,660	1-D	11,032	NA
Yeshiva U, New York, NY 10033-3201	1886	$17,295	$5,520	1-D	5,481	NA
York Coll of Pennsylvania, York, PA 17405-7199	1787	$6,978	$4,860	1-M	5,254	361
York Coll of the City U of NewYork, Jamaica, NY 11451-0001	1967	$3,292 (S)	NA	11-B	5,389	424
Youngstown State U, Youngstown, OH 44555-0001	1908	$3,994 (S)	$4,800	5-D	11,787	782

(1) All the information given for these schools applies to the University of Phoenix system as a whole.

Two-Year Colleges

Figures for Room & Board are given where applicable.

Name, address	Year Founded	Tuition & Fees	Rm. & Board	Control, Degree	Enrollment	Faculty
Adirondack Comm Coll, Queensbury, NY 12804	1960	$2,476 (S)	—	11	3,151	244
Aiken Tech Coll, Aiken, SC 29802-0696	1972	$1,210 (S)	—	11	2,339	131
Aims Comm Coll, Greeley, CO 80632-0069	1967	$1,305 (A)	—	9	7,171	342
Alabama Southern Comm Coll, Monroeville, AL 36461	1965	$1,344 (S)	—	5	1,600	107
Alamance Comm Coll, Graham, NC 27253-8000	1959	$940 (S)	—	5	3,428	218
Alexandria Tech Coll, Alexandria, MN 56308-3707	1961	$2,512 (S)	—	5	1,897	100
Allan Hancock Coll, Santa Maria, CA 93454-6399	1920	$390 (S)	—	11	8,851	497
Allegany Coll of Maryland, Cumberland, MD 21502-2596	1961	$2,720 (A)	—	11	2,591	193
Allen County Comm Coll, Iola, KS 66749-1607	1923	$1,568 (S)	$3,000	11	1,968	151
Allentown Business School, Allentown, PA 18103-3880	1869	$18,000	—	3	1,561	77
Alpena Comm Coll, Alpena, MI 49707-1495	1952	$2,000 (A)	—	11	1,871	119
Alvin Comm Coll, Alvin, TX 77511-4898	1949	$656 (S)	—	11	3,665	219
Angelina Coll, Lufkin, TX 75902-1768	1968	$640 (A)	$2,600	11	4,418	292
Anne Arundel Comm Coll, Arnold, MD 21012-1895	1961	$1,920 (A)	—	11	11,758	699
Anoka-Hennepin Tech Coll, Anoka, MN 55303	1967	$1,663 (A)	—	5	2,024	103
Anoka-Ramsey Comm Coll, Cambridge Campus, Cambridge, MN 55008-5706	NA	$85/credit (S)	—	5	1,345	55
Arapahoe Comm Coll, Littleton, CO 80160-9002	1965	$1,386 (S)	—	5	7,446	414
Arizona Western Coll, Yuma, AZ 85366-0929	1962	$1,140 (S)	$3,550	11	7,002	96
Arkansas State U–Beebe, Beebe, AR 72012-1000	1927	$1,272 (S)	$2,800	5	3,186	97
The Art Inst of Atlanta, Atlanta, GA 30328	1949	$12,624	—	3-B	2,237	127
The Art Inst of Dallas, Dallas, TX 75231-9959	1978	$12,870	—	3	1,532	94
The Art Inst of Fort Lauderdale, Fort Lauderdale, FL 33316-3000	1968	$311	$4,486	3-B	3,145	110
The Art Inst of Houston, Houston, TX 77056-4115	1978	$13,905	—	3	1,679	121
The Art Inst of Philadelphia, Philadelphia, PA 19103-5198	1966	$13,905	—	3-B	2,672	230
The Art Inst of Pittsburgh, Pittsburgh, PA 15219	1921	$14,050	—	3-B	2,500	115
The Art Inst of Seattle, Seattle, WA 98121-1642	1982	$8,910	—	3	2,623	189
Asheville-Buncombe Tech Comm Coll, Asheville, NC 28801-4897	1959	$772 (S)	—	5	4,940	502
Ashland Comm Coll, Ashland, KY 41101-3683	1937	$1,230 (S)	—	5	2,252	153
Asnuntuck Comm Coll, Enfield, CT 06082-3800	1972	$1,886 (S)	—	5	1,850	116
Athens Tech Coll, Athens, GA 30601-1500	1958	$1,074 (S)	—	5	2,747	97
Atlantic Cape Comm Coll, Mays Landing, NJ 08330-2699	1964	$1,959 (A)	—	8	5,162	254
Austin Comm Coll, Austin, TX 78752-4390	1972	$1,631 (A)	—	9	26,856	1,469
Bainbridge Coll, Bainbridge, GA 31717	1972	$1,356 (S)	—	5	1,316	68
Baltimore City Comm Coll, Baltimore, MD 21215-7893	1947	$1,990 (S)	—	5	5,883	399
Barstow Coll, Barstow, CA 92311-6699	1959	$264 (S)	—	11	3,330	109
Barton County Comm Coll, Great Bend, KS 67530-9283	1969	$1,472 (A)	$2,904	11	4,612	199
Bates Tech Coll, Tacoma, WA 98405-4895	NA	$2,300 (S)	—	5	16,162	290
Beaufort County Comm Coll, Washington, NC 27889-1069	1967	$908 (S)	—	5	1,377	136
Bellevue Comm Coll, Bellevue, WA 98007-6484	1966	$1,734 (S)	—	5	10,321	572
Bellingham Tech Coll, Bellingham, WA 98225	NA	$2,029 (S)	—	5	3,791	162
Belmont Tech Coll, St. Clairsville, OH 43950-9735	1971	$2,200 (S)	—	5	1,522	101
Bergen Comm Coll, Paramus, NJ 07652-1595	1965	$1,664 (A)	—	8	11,993	647
Berkeley Coll, West Paterson, NJ 07424-3353	1931	$12,945	$8,100	3	1,994	127
Berkeley Coll, New York, NY 10017-4604	1936	$12,945	—	3-B	1,640	134
Berkshire Comm Coll, Pittsfield, MA 01201-5786	1960	$2,490 (A)	—	5	2,496	162
Bessemer State Tech Coll, Bessemer, AL 35021-0308	1966	$1,576 (S)	—	5	1,354	105
Bevill State Comm Coll, Sumiton, AL 35148	1969	$1,662 (S)	—	5	3,554	280
Big Bend Comm Coll, Moses Lake, WA 98837-3299	1962	$1,641 (S)	$4,097	5	2,082	113
Bishop State Comm Coll, Mobile, AL 36603-5898	1965	$1,248 (S)	—	5	3,393	176
Bismarck State Coll, Bismarck, ND 58506-5587	1939	$2,418 (S)	$2,800	5	2,744	202
Black Hawk Coll, Moline, IL 61265-5899	1946	$1,650 (A)	—	11	5,633	339
Blackhawk Tech Coll, Janesville, WI 53547-5009	1968	$1,980 (S)	—	9	3,784	293
Black River Tech Coll, Pocahontas, AR 72455	1972	$1,032 (A)	—	5	1,243	70
Blinn Coll, Brenham, TX 77833-4049	1883	$39/sem. hr. (A)	—	11	12,025	474
Blue Mountain Comm Coll, Pendleton, OR 97801-1000	1962	$1,692 (S)	—	11	1,816	256
Blue Ridge Comm Coll, Flat Rock, NC 28731-9624	1969	$880 (S)	—	11	1,485	241
Blue Ridge Comm Coll, Weyers Cave, VA 24486-0080	1967	$1,210 (S)	—	5	2,873	157
Blue River Comm Coll, Blue Springs, MO 64015	NA	$1,590 (A)	—	11	2,019	150
Borough of Manhattan Comm Coll of the City U of New York, New York, NY 10007-1097	1963	$2,500 (S)	—	11	15,875	1,020

Name, address	Year Founded	Tuition & Fees	Rm. & Board	Control, Degree	Enroll-ment	Faculty
Bossier Parish Comm Coll, Bossier City, LA 71111-5801	1967	$1,360 (S)	—	5	3,754	158
Bowling Green State U–Firelands Coll, Huron, OH 44839-9791	1968	$3,202 (S)	—	5	1,471	92
Brazosport Coll, Lake Jackson, TX 77566-3199	1968	$666 (A)	—	11	3,855	187
Bristol Comm Coll, Fall River, MA 02720-7395	1965	$2,484 (S)	—	5	6,060	279
Bronx Comm Coll of the City U of New York, Bronx, NY 10453	1959	$2,610 (S)	—	11	6,928	683
Brookdale Comm Coll, Lincroft, NJ 07738-1597	1967	$2,160 (A)	—	8	11,552	627
Brooks Coll, Long Beach, CA 90804-3291	1971	$10,910	$5,080	3	1,100	100
Broome Comm Coll, Binghamton, NY 13902-1017	1946	$2,548 (S)	—	11	5,665	148
Broward Comm Coll, Fort Lauderdale, FL 33301-2298	1960	$1,305 (S)	—	5	30,333	775
Brown Inst, Mendota Heights, MN 55120	1946	$9,550	—	3	1,800	96
Bucks County Comm Coll, Newtown, PA 18940-1525	1964	$2,583 (A)	—	8	8,469	550
Bunker Hill Comm Coll, Boston, MA 02129	1973	$1,560 (S)	—	5	6,385	327
Burlington County Coll, Pemberton, NJ 08068-1599	1966	$1,800 (A)	—	8	5,972	344
Butler County Comm Coll, El Dorado, KS 67042-3280	1927	$1,068 (S)	$3,550	11	7,810	423
Butler County Comm Coll, Butler, PA 16003-1203	1965	$1,440 (A)	—	8	3,011	65
Butte Coll, Oroville, CA 95965-8399	1966	$488 (S)	—	9	10,213	582
Cabrillo Coll, Aptos, CA 95003-3194	1959	$294 (S)	—	9	13,147	552
Caldwell Comm Coll & Tech Inst, Hudson, NC 28638-2397	1964	$912 (S)	—	5	3,161	401
Calhoun Comm Coll, Decatur, AL 35609-2216	1965	$3,648 (S)	—	5	7,981	387
Cambria County Area Comm Coll, Johnstown, PA 15907-0068	NA	$1,815 (A)	—	11	1,450	145
Camden County Coll, Blackwood, NJ 08012-0200	1967	$2,112 (A)	—	11	12,131	748
Cañada Coll, Redwood City, CA 94061-1099	1968	$298 (S)	—	11	5,332	275
Cape Cod Comm Coll, West Barnstable, MA 02668-1599	1961	$2,310 (S)	—	5	3,950	331
Cape Fear Comm Coll, Wilmington, NC 28401-3993	1959	$956 (S)	—	5	5,470	429
Capital Comm Coll, Hartford, CT 06105-2354	1946	$2,128 (S)	—	5	3,050	143
Carl Albert State Coll, Poteau, OK 74953-5208	1934	$1,324 (S)	$2,272	5	2,169	160
Carroll Comm Coll, Westminster, MD 21157	1993	$1,728 (A)	—	11	2,488	136
Carteret Comm Coll, Morehead City, NC 28557-2989	1963	$1,160 (S)	—	5	1,506	93
Casper Coll, Casper, WY 82601-4699	1945	$1,172 (S)	$2,750	9	3,971	239
Catawba Valley Comm Coll, Hickory, NC 28602-9699	1960	$904 (S)	—	11	3,731	242
Cayuga County Comm Coll, Auburn, NY 13021-3099	1953	$2,805 (S)	—	11	2,518	162
Cecil Comm Coll, North East, MD 21901-1999	1968	$1,615 (A)	—	8	1,293	176
Cedar Valley Coll, Lancaster, TX 75134-3799	1977	$610 (A)	—	5	2,832	155
Central Alabama Comm Coll, Alexander City, AL 35011-0699	1965	$1,728 (S)	—	5	1,609	193
Central Carolina Comm Coll, Sanford, NC 27330-9000	1962	$1,086 (S)	—	11	4,084	426
Central Carolina Tech Coll, Sumter, SC 29150-2499	1963	$1,200 (A)	—	5	2,528	121
Central Comm Coll–Columbus Campus, Columbus, NE 68602-1027	1968	$1,440 (S)	$2,880	11	2,176	91
Central Comm Coll–Grand Island Campus, Grand Island, NE 68802-4903	1976	$1,440 (S)	—	11	2,956	114
Central Comm Coll–Hastings Campus, Hastings, NE 68902-1024	1966	$1,440 (S)	$2,880	11	2,368	91
Central Florida Comm Coll, Ocala, FL 34478-1388	1957	$1,484 (S)	—	11	5,282	229
Central Georgia Tech Coll, Macon, GA 31206-3628	1966	$999 (S)	—	5	3,787	217
Centralia Coll, Centralia, WA 98531-4099	1925	$1,776 (S)	—	5	3,766	197
Central Lakes Coll, Brainerd, MN 56401-3904	1938	$2,313 (S)	—	5	2,857	140
Central Maine Tech Coll, Auburn, ME 04210-6498	1964	$2,409 (S)	$3,760	5	1,248	52
Central Ohio Tech Coll, Newark, OH 43055-1767	1971	$1,608 (S)	—	5	1,832	158
Central Oregon Comm Coll, Bend, OR 97701-5998	1949	$1,989 (A)	$5,034	9	4,079	263
Central Piedmont Comm Coll, Charlotte, NC 28235-5009	1963	$808 (S)	—	11	15,038	1,235
Central Texas Coll, Killeen, TX 76540-1800	1967	$672 (A)	$2,742	11	14,636	609
Central Virginia Comm Coll, Lynchburg, VA 24502-2498	1966	$1,225 (S)	—	5	3,940	103
Central Wyoming Coll, Riverton, WY 82501-2273	1966	$1,524 (S)	$2,670	11	1,644	41
Century Comm & Tech Coll, White Bear Lake, MN 55110	1970	$2,486 (S)	—	5	6,914	283
Cerro Coso Comm Coll, Ridgecrest, CA 93555-9571	1973	$330 (S)	—	5	10,474	286
Chabot Coll, Hayward, CA 94545-5001	1961	$264 (S)	—	5	12,925	531
Chaffey Coll, Rancho Cucamonga, CA 91737-3002	1883	$286 (S)	—	9	16,236	540
Chattahoochee Tech Coll, Marietta, GA 30060	1961	$1,128 (S)	—	5	2,739	137
Chattahoochee Valley Comm Coll, Phenix City, AL 36869-7928	1974	$1,800 (S)	—	5	1,729	126
Chattanooga State Tech Comm Coll, Chattanooga, TN 37406-1097	1965	$1,266 (S)	—	5	8,162	626
Chemeketa Comm Coll, Salem, OR 97309-7070	1955	$1,710 (S)	—	11	9,779	699
Chesapeake Coll, Wye Mills, MD 21679-0008	1965	$1,890 (A)	—	11	2,186	166
Chipola Jr Coll, Marianna, FL 32446-3065	1947	$1,536 (S)	—	5	1,970	69
Chippewa Valley Tech Coll, Eau Claire, WI 54701-6162	1912	$1,946 (S)	—	9	3,254	400
Cincinnati State Tech & Comm Coll, Cincinnati, OH 45223-2690	1966	$3,245 (S)	—	5	6,675	822
Cisco Jr Coll, Cisco, TX 76437-9321	1940	$1,096 (A)	$2,600	11	2,642	98
City Coll of San Francisco, San Francisco, CA 94112-1821	1935	$360 (S)	—	11	68,737	1,593
City Colls of Chicago, Harold Washington Coll, Chicago, IL 60601-2449	1962	$1,529 (A)	—	11	8,434	231
City Colls of Chicago, Harry S Truman Coll, Chicago, IL 60640-5616	1956	$1,250 (A)	—	11	6,051	696
City Colls of Chicago, Kennedy-King Coll, Chicago, IL 60621-3733	1935	$1,529 (A)	—	11	6,575	77
City Colls of Chicago, Malcolm X Coll, Chicago, IL 60612-3145	1911	$1,604 (A)	—	11	8,791	134
City Colls of Chicago, Olive-Harvey Coll, Chicago, IL 60628-1645	1970	$50/cr. hr. (A)	—	11	3,165	133
City Colls of Chicago, Richard J. Daley Coll, Chicago, IL 60652-1242	1960	$1,475 (A)	—	11	6,448	140
City Colls of Chicago, Wilbur Wright Coll, Chicago, IL 60634-1591	1934	$1,600 (A)	—	11	5,800	231
Clackamas Comm Coll, Oregon City, OR 97045-7998	1966	$1,476 (S)	—	9	6,369	502
Clarendon Coll, Clarendon, TX 79226-0968	1898	$1,024 (A)	$2,250	11	1,000	61
Clark Coll, Vancouver, WA 98663-3598	1933	$1,776 (A)	—	5	7,952	487
Clark State Comm Coll, Springfield, OH 45501-0570	1962	$3,156 (S)	—	5	2,808	251
Clatsop Comm Coll, Astoria, OR 97103-3698	1958	$1,476 (S)	—	8	1,796	198
Cleveland Comm Coll, Shelby, NC 28152	1965	$956 (S)	—	5	2,225	148
Cleveland Inst of Electronics, Cleveland, OH 44114-3636 (2)	1934	$1,495	—	3	3,044	5
Cleveland State Comm Coll, Cleveland, TN 37320-3570	1967	$1,435 (S)	—	5	3,056	180
Clinton Comm Coll, Clinton, IA 52732-6299	1946	$1,950 (S)	—	11	1,163	75
Clinton Comm Coll, Plattsburgh, NY 12901-9573	1969	$2,566 (S)	—	11	1,697	136
Clover Park Tech Coll, Lakewood, WA 98499	NA	$2,127 (A)	—	5	8,286	224
Clovis Comm Coll, Clovis, NM 88101-8381	1971	$548 (A)	—	5	2,899	161
Coahoma Comm Coll, Clarksdale, MS 38614-9799	1949	$1,200 (S)	$2,844	11	1,244	55
Coastal Bend Coll, Beeville, TX 78102-2197	1965	$898 (A)	$1,920	8	3,249	176
Coastal Carolina Comm Coll, Jacksonville, NC 28546-6899	1964	$910 (S)	—	11	3,772	238
Coastal Georgia Comm Coll, Brunswick, GA 31520-3644	1961	$1,406 (S)	—	5	2,127	94

Name, address	Year Founded	Tuition & Fees	Rm. & Board	Control, Degree	Enroll-ment	Faculty
Coastline Comm Coll, Fountain Valley, CA 92708-2597	1976	$380 (S)	—	11	9,236	349
Cochise Coll, Douglas, AZ 85607-9724	1962	$898 (S)	$2,806	11	4,166	253
Coconino Comm Coll, Flagstaff, AZ 86003	1991	$990 (S)	—	5	3,689	205
Coffeyville Comm Coll, Coffeyville, KS 67337-5063	1923	$1,440 (S)	$2,925	11	1,680	62
Colby Comm Coll, Colby, KS 67701-4099	1964	$1,440 (S)	$3,098	11	2,160	53
Coll of Alameda, Alameda, CA 94501-2109	1970	$264 (S)	—	11	5,500	166
Coll of DuPage, Glen Ellyn, IL 60137-6599	1967	$1,846 (A)	—	11	28,862	1,537
Coll of Eastern Utah, Price, UT 84501-2699	1937	$1,466 (S)	$3,300	5	2,704	182
Coll of Lake County, Grayslake, IL 60030-1198	1967	$1,770 (A)	—	9	12,457	874
Coll of Marin, Kentfield, CA 94904	1926	$354 (S)	—	11	8,458	464
Coll of San Mateo, San Mateo, CA 94402-3784	1922	$0 (S)	—	11	10,872	476
Coll of Southern Maryland, La Plata, MD 20646-0910	1958	$2,016 (A)	—	11	6,400	379
Coll of The Albemarle, Elizabeth City, NC 27906-2327	1960	$777 (S)	—	5	2,071	122
Coll of the Canyons, Santa Clarita, CA 91355-1899	1969	$382 (S)	—	11	10,260	469
Coll of the Desert, Palm Desert, CA 92260-9305	1959	$11/unit (A)	—	11	10,420	320
Coll of the Mainland, Texas City, TX 77591-2499	1967	$718 (A)	—	11	3,400	581
Coll of the Redwoods, Eureka, CA 95501-9300	1964	$284 (S)	$5,435	11	6,828	485
Coll of the Siskiyous, Weed, CA 96094-2899	1957	$350 (S)	$4,122	11	3,235	192
Collin County Comm Coll District, Plano, TX 75093-8309	1985	$874 (A)	—	11	12,998	701
Colorado Northwestern Comm Coll, Rangely, CO 81648-3598	1962	$1,998 (S)	$4,490	5	1,979	187
Columbia Basin Coll, Pasco, WA 99301-3397	1955	$1,755 (S)	—	5	5,837	609
Columbia-Greene Comm Coll, Hudson, NY 12534-0327	1969	$2,460 (S)	—	11	1,598	106
Columbia State Comm Coll, Columbia, TN 38402-1315	1966	$1,314 (S)	—	5	4,299	285
Columbus State Comm Coll, Columbus, OH 43216-1609	1963	$2,086 (S)	—	5	17,662	966
Comm Coll of Allegheny County, Pittsburgh, PA 15233-1894	1966	$1,781 (A)	—	8	16,148	1,168
Comm Coll of Aurora, Aurora, CO 80011-9036	1983	$1,785 (S)	—	5	4,300	215
Comm Coll of Beaver County, Monaca, PA 15061-2588	1966	$2,002 (A)	—	5	2,187	104
Comm Coll of Denver, Denver, CO 80217-3363	1970	$1,700 (S)	—	5	6,511	372
Comm Coll of Philadelphia, Philadelphia, PA 19130-3991	1964	$2,490 (A)	—	11	42,000	1,182
Comm Coll of Rhode Island, Warwick, RI 02886-1807	1964	$1,806 (S)	—	5	15,583	676
Comm Coll of Southern Nevada, North Las Vegas, NV 89030-4296	1971	$1,404 (S)	—	5	30,132	2,223
Comm Coll of Vermont, Waterbury, VT 05676-0120	1970	$3,004 (S)	—	5	4,599	508
Compton Comm Coll, Compton, CA 90221-5393	1927	$384 (S)	—	11	7,003	347
Connors State Coll, Warner, OK 74469-9700	1908	$1,508 (S)	$1,205	5	1,979	109
Contra Costa Coll, San Pablo, CA 94806-3195	1948	$332 (S)	—	11	7,074	222
Copiah-Lincoln Comm Coll, Wesson, MS 39191-0457	1928	$1,000 (S)	$1,900	11	1,840	113
Corning Comm Coll, Corning, NY 14830-3297	1956	$2,875 (S)	—	11	4,696	187
Cosumnes River Coll, Sacramento, CA 95823-5799	1970	$332 (S)	—	9	15,002	425
County Coll of Morris, Randolph, NJ 07869-2086	1966	$2,310 (A)	—	8	7,951	471
Cowley County Comm Coll & Area Voc–Tech School, Arkansas City, KS 67005	1922	$1,440 (S)	$2,990	11	3,840	184
Crafton Hills Coll, Yucaipa, CA 92399-1799	1972	$11/unit (S)	—	11	5,200	217
Craven Comm Coll, New Bern, NC 28562-4984	1965	$777 (S)	—	5	2,351	166
Crowder Coll, Neosho, MO 64850-9160	1963	$1,440 (A)	$3,700	11	1,719	176
Cuesta Coll, San Luis Obispo, CA 93403-8106	1964	$400 (S)	—	9	9,732	401
Cumberland County Coll, Vineland, NJ 08362-0517	1963	$2,041 (A)	—	11	2,709	151
Cuyahoga Comm Coll, Cleveland, OH 44115-2878	1963	$1,845 (A)	—	11	20,321	1,142
Cuyamaca Coll, El Cajon, CA 92019-4304	1978	$592 (S)	—	5	6,842	342
Cypress Coll, Cypress, CA 90630-5897	1966	$310 (S)	—	11	13,790	590
Dabney S. Lancaster Comm Coll, Clifton Forge, VA 24422	1964	$1,339 (S)	—	5	1,489	160
Dakota County Tech Coll, Rosemount, MN 55068	NA	$2,738 (S)	—	5	2,901	199
Danville Area Comm Coll, Danville, IL 61832-5199	1946	$2,040 (A)	—	11	2,846	121
Danville Comm Coll, Danville, VA 24541-4088	1967	$1,184 (S)	—	5	3,750	180
Darton Coll, Albany, GA 31707-3098	1965	$1,650 (S)	—	5	2,805	124
Davenport U, Midland, MI 48642	1907	$6,102	—	1-B	1,500	144
Davidson County Comm Coll, Lexington, NC 27293-1287	1958	$777 (S)	—	11	2,303	212
Daytona Beach Comm Coll, Daytona Beach, FL 32120-2811	1958	$1,417 (S)	—	5	10,417	893
De Anza Coll, Cupertino, CA 95014-5793	1967	$357 (S)	—	11	26,451	775
DeKalb Tech Coll, Clarkston, GA 30021-2397	1961	$1,323 (S)	—	5	3,112	854
Delaware County Comm Coll, Media, PA 19063-1094	1967	$1,746 (A)	—	11	8,943	565
Delaware Tech & Comm Coll, Jack F. Owens Campus, Georgetown, DE 19947	1967	$1,566 (S)	—	5	3,735	200
Delaware Tech & Comm Coll, Stanton/Wilmington Campus, Newark, DE 19713	1968	$1,676 (S)	—	5	6,976	471
Delaware Tech & Comm Coll, Terry Campus, Dover, DE 19901	1972	$1,676 (S)	—	5	2,096	150
Delgado Comm Coll, New Orleans, LA 70119-4399	1921	$1,411 (S)	—	5	12,784	734
Del Mar Coll, Corpus Christi, TX 78404-3897	1935	$818 (A)	—	11	9,968	694
Delta Coll, University Center, MI 48710	1961	$1,955 (A)	—	9	9,599	515
Denmark Tech Coll, Denmark, SC 29042-0327	1948	$1,700 (S)	$3,096	5	1,240	70
Des Moines Area Comm Coll, Ankeny, IA 50021-8995	1966	$2,022 (S)	—	11	10,998	616
Diablo Valley Coll, Pleasant Hill, CA 94523-1544	1949	$360 (S)	—	11	21,464	700
Diné Coll, Tsaile, AZ 86556	1968	$620	$2,940	4	1,870	152
Dixie State Coll of Utah, St. George, UT 84770-3876	1911	$1,480 (S)	$2,850	5-B	6,350	259
Dodge City Comm Coll, Dodge City, KS 67801-2399	1935	$1,144 (S)	$3,240	11	2,259	163
Doña Ana Branch Comm Coll, Las Cruces, NM 88003-8001	1973	$840 (A)	—	11	4,638	272
Dunwoody Inst, Minneapolis, MN 55403 (2)	1914	$7,050	—	1	1,059	77
Durham Tech Comm Coll, Durham, NC 27703-5023	1961	$904 (S)	—	5	5,302	514
Dutchess Comm Coll, Poughkeepsie, NY 12601-1595	1957	$2,420 (S)	—	11	6,582	415
Dyersburg State Comm Coll, Dyersburg, TN 38024	1969	$1,576 (S)	—	5	2,278	172
East Arkansas Comm Coll, Forrest City, AR 72335-2204	1974	$984 (A)	—	5	1,358	101
East Central Coll, Union, MO 63084-0529	1968	$1,288 (A)	—	9	3,190	173
East Central Comm Coll, Decatur, MS 39327-0129	1928	$1,200 (S)	$3,370	11	2,256	124
Eastern Arizona Coll, Thatcher, AZ 85552-0769	1888	$748 (A)	$3,380	11	6,226	272
Eastern Maine Tech Coll, Bangor, ME 04401-4206	1966	$1,848 (S)	$3,500	5	1,277	129
Eastern New Mexico U–Roswell, Roswell, NM 88202-6000	1958	$709 (A)	—	5	2,919	262
Eastern Oklahoma State Coll, Wilburton, OK 74578-4999	1908	$1,430 (S)	$2,400	5	1,954	54
Eastern Wyoming Coll, Torrington, WY 82240-1699	1948	$1,608 (S)	$2,730	11	1,420	142
Eastfield Coll, Mesquite, TX 75150-2099	1970	$700 (A)	—	11	8,021	542

Name, address	Year Founded	Tuition & Fees	Rm. & Board	Control, Degree	Enroll-ment	Faculty
East Georgia Coll, Swainsboro, GA 30401-2699	1973	$1,429 (S)	—	5	1,309	54
East Los Angeles Coll, Monterey Park, CA 91754-6001	1945	$286 (S)	—	11	17,197	450
ECPI Coll of Technology, Hampton, VA 23666	1966	$7,145	—	3	2,263	130
ECPI Coll of Technology, Virginia Beach, VA 23462	1966	$8,400	—	3	2,416	130
ECPI Tech Coll, Richmond, VA 23236	1966	$8,400	—	3	1,137	80
Edgecombe Comm Coll, Tarboro, NC 27886-9399	1968	$904 (S)	—	11	1,886	217
Edison Comm Coll, Fort Myers, FL 33906-6210	1962	$1,402 (S)	—	11	8,929	417
Edison State Comm Coll, Piqua, OH 45356-9253	1973	$1,848 (S)	—	5	2,830	298
Edmonds Comm Coll, Lynnwood, WA 98036-5999	1967	$1,761 (S)	—	11	8,157	419
Elaine P. Nunez Comm Coll, Chalmette, LA 70043-1249	1992	$1,360 (S)	—	5	1,868	105
El Centro Coll, Dallas, TX 75202-3604	1966	$700 (A)	—	8	4,333	310
Elizabethtown Comm Coll, Elizabethtown, KY 42701-3081	1964	$1,230 (S)	—	5	3,513	190
El Paso Comm Coll, El Paso, TX 79998-0500	1969	$1,286 (S)	—	8	19,551	1,146
Erie Comm Coll, City Campus, Buffalo, NY 14203-2698	1971	$2,655 (A)	—	11	2,220	178
Erie Comm Coll, North Campus, Williamsville, NY 14221-7095	1946	$2,655 (A)	—	11	4,665	326
Erie Comm Coll, South Campus, Orchard Park, NY 14127-2199	1974	$2,655 (A)	—	11	3,122	325
Essex County Coll, Newark, NJ 07102-1798	1966	$2,022 (A)	—	8	8,868	530
Eugenio María de Hostos Comm Coll of the City U of New York, Bronx, NY 10451	1968	$2,552 (S)	—	11	3,118	341
Everett Comm Coll, Everett, WA 98201-1327	1941	$1,634 (S)	—	5	7,700	307
Fayetteville Tech Comm Coll, Fayetteville, NC 28303-0236	1961	$789 (S)	—	5	8,310	761
Feather River Comm Coll District, Quincy, CA 95971-9124	1968	$382 (S)	—	11	1,200	87
Finger Lakes Comm Coll, Canandaigua, NY 14424-8395	1965	$2,520 (S)	—	11	4,667	258
Fiorello H. LaGuardia Comm Coll of the City U of New York, Long Island City, NY 11101-3071	1970	$2,622 (A)	—	11	11,997	732
Fisher Coll, Boston, MA 02116-1500	1903	$14,485	$7,350	1-B	1,510	57
Flathead Valley Comm Coll, Kalispell, MT 59901-2622	1967	$1,748 (A)	—	11	1,822	168
Florence-Darlington Tech Coll, Florence, SC 29501-0548	1963	$1,520 (A)	—	5	3,814	222
Florida Comm Coll at Jacksonville, Jacksonville, FL 32202-4030	1963	$1,775 (S)	—	5	20,838	1,102
Florida Culinary Inst, West Palm Beach, FL 33407	NA	$21,000	—	3	1,152	21
Florida Keys Comm Coll, Key West, FL 33040-4397	1965	$1,445 (S)	—	5	1,711	138
Florida Natl Coll, Hialeah, FL 33012	1982	$8,090	—	3	1,164	37
Floyd Coll, Rome, GA 30162-1864	1970	$1,932 (S)	—	5	2,085	126
Foothill Coll, Los Altos Hills, CA 94022-4599	1958	$398 (S)	—	11	18,500	587
Forsyth Tech Comm Coll, Winston-Salem, NC 27103-5197	1964	$678 (S)	—	5	5,874	573
Fort Scott Comm Coll, Fort Scott, KS 66701	1919	$1,350 (S)	$2,800	11	1,677	173
Fox Valley Tech Coll, Appleton, WI 54912-2277	1967	$1,968 (S)	—	11	6,291	1,170
Frank Phillips Coll, Borger, TX 79008-5118	1948	$1,035 (A)	$2,290	11	1,045	97
Frederick Comm Coll, Frederick, MD 21702-2097	1957	$2,018 (A)	—	11	4,343	240
Fresno City Coll, Fresno, CA 93741-0002	1910	$288 (S)	—	9	21,829	832
Front Range Comm Coll, Westminster, CO 80030-2105	1968	$1,547 (S)	—	5	12,483	781
Fullerton Coll, Fullerton, CA 92832-2095	1913	$357 (S)	—	11	20,555	821
Fulton-Montgomery Comm Coll, Johnstown, NY 12095-3790	1964	$2,692 (S)	—	11	1,886	116
Gadsden State Comm Coll, Gadsden, AL 35902-0227	1985	$1,344 (S)	$2,350	5	4,729	288
Gainesville Coll, Gainesville, GA 30503-1358	1964	$1,316 (S)	—	5	3,032	112
Galveston Coll, Galveston, TX 77550-7496	1967	$560 (S)	—	11	2,255	105
Garden City Comm Coll, Garden City, KS 67846-6399	1919	$1,408 (S)	$3,450	8	2,039	135
Garland County Comm Coll, Hot Springs, AR 71913	1973	$980 (A)	—	11	2,300	105
Gaston Coll, Dallas, NC 28034-1499	1963	$794 (S)	—	11	4,044	370
Gateway Comm Coll, Phoenix, AZ 85034-1795	1968	$1,230 (S)	—	11	7,895	259
Gateway Comm Coll, New Haven, CT 06511-5918	1992	$1,886 (S)	—	5	4,157	237
Gateway Tech Coll, Kenosha, WI 53144-1690	1911	$1,845 (S)	—	11	5,741	238
Gavilan Coll, Gilroy, CA 95020-9599	1919	$308 (S)	—	11	5,710	164
Genesee Comm Coll, Batavia, NY 14020-9704	1966	$2,772 (S)	—	11	4,521	324
George Corley Wallace State Comm Coll, Selma, AL 36702-1049	1966	$1,800 (S)	—	5	1,445	93
Georgia Military Coll, Milledgeville, GA 31061-3398	1879	$10,553	$3,450	11	3,465	193
Georgia Perimeter Coll, Decatur, GA 30034-3897	1964	$1,406 (S)	—	5	13,708	905
Germanna Comm Coll, Locust Grove, VA 22508-2102	1970	$1,264 (S)	—	5	5,294	135
Glendale Comm Coll, Glendale, AZ 85302-3090	1965	$994 (A)	—	11	20,091	847
Glendale Comm Coll, Glendale, CA 91208-2894	1927	$308 (S)	—	11	15,629	715
Glen Oaks Comm Coll, Centreville, MI 49032-9719	1965	$1,653 (A)	—	11	1,491	90
Gloucester County Coll, Sewell, NJ 08080	1967	$2,432 (A)	—	8	4,669	218
Golden West Coll, Huntington Beach, CA 92647-2748	1966	$372 (S)	—	11	12,784	440
Gordon Coll, Barnesville, GA 30204-1762	1852	$1,420 (S)	$2,688	5	2,838	139
Grand Rapids Comm Coll, Grand Rapids, MI 49503-3201	1914	$1,474 (A)	—	9	13,400	490
Grays Harbor Coll, Aberdeen, WA 98520-7599	1930	$1,700 (S)	—	5	2,816	220
Grayson County Coll, Denison, TX 75020-8299	1964	$1,199 (A)	$2,600	11	3,289	100
Great Basin Coll, Elko, NV 89801-3348	1967	$1,500 (S)	—	5-B	3,283	213
Greenfield Comm Coll, Greenfield, MA 01301-9739	1962	$1,814 (S)	—	5	2,144	152
Green River Comm Coll, Auburn, WA 98092-3699	1965	$1,779 (S)	—	5	6,548	362
Griffin Tech Coll, Griffin, GA 30223	1965	$661 (S)	—	5	3,217	110
Grossmont Coll, El Cajon, CA 92020-1799	1961	$297 (S)	—	11	16,175	653
Guam Comm Coll, Guam Main Facility, GU 96921-3069	1977	$1,310 (S)	—	7	2,000	94
Guilford Tech Comm Coll, Jamestown, NC 27282-0309	1958	$923 (S)	—	11	6,305	416
Gulf Coast Comm Coll, Panama City, FL 32401-1058	1957	$1,473 (S)	—	5	5,341	669
Hagerstown Comm Coll, Hagerstown, MD 21742-6590	1946	$2,310 (A)	—	8	2,516	187
Halifax Comm Coll, Weldon, NC 27890-0809	1967	$902 (S)	—	11	2,166	68
Harford Comm Coll, Bel Air, MD 21015-1698	1957	$1,950 (A)	—	11	4,821	297
Harrisburg Area Comm Coll, Harrisburg, PA 17110-2999	1964	$2,115 (A)	—	11	10,574	650
Harry M. Ayers State Tech Coll, Anniston, AL 36202-1647	1966	$1,632 (S)	—	5	1,102	39
Hawaii Comm Coll, Hilo, HI 96720-4091	1954	$1,082 (S)	$3,818	5	2,285	149
Hawkeye Comm Coll, Waterloo, IA 50704-8015	1967	$2,370 (S)	—	11	4,263	233
Haywood Comm Coll, Clyde, NC 28721-9453	1964	$880 (S)	—	11	1,724	190
Hazard Comm Coll, Hazard, KY 41701-2403	1968	$1,230 (S)	—	5	2,218	60
Heald Coll, Schools of Business & Technology, Hayward, CA 94545-1557	1863	$7,560	—	1	1,077	75
Heartland Comm Coll, Normal, IL 61761	1990	$1,290 (A)	—	11	4,205	249
Henderson Comm Coll, Henderson, KY 42420-4623	1963	$1,140 (S)	—	5	1,170	98

Name, address	Year Founded	Tuition & Fees	Rm. & Board	Control, Degree	Enroll-ment	Faculty
Henry Ford Comm Coll, Dearborn, MI 48128-1495	1938	$1,992 (A)	—	9	12,123	997
Herkimer County Comm Coll, Herkimer, NY 13350	1966	$2,430 (S)	—	11	2,591	114
Hesser Coll, Manchester, NH 03103-7245	1900	$9,015	$4,900	3-B	2,766	250
Hibbing Comm Coll, Hibbing, MN 55746-3300	1916	$2,453 (S)	—	5	3,050	82
Highland Comm Coll, Freeport, IL 61032-9341	1962	$1,480 (A)	—	11	2,595	188
Highland Comm Coll, Highland, KS 66035	1858	$870 (S)	—	11	2,673	226
Highline Comm Coll, Des Moines, WA 98198-9800	1961	$1,641 (S)	—	5	6,062	398
Hillsborough Comm Coll, Tampa, FL 33631-3127	1968	$1,442 (S)	—	5	16,846	1,280
Hinds Comm Coll, Raymond, MS 39154-1100	1917	$1,070 (S)	$1,850	11	12,126	905
Hocking Coll, Nelsonville, OH 45764-9588	1968	$2,151 (S)	—	5	6,435	229
Holmes Comm Coll, Goodman, MS 39079-0369	1928	$1,210 (A)	—	11	3,067	125
Holyoke Comm Coll, Holyoke, MA 01040-1099	1946	$1,790 (S)	—	5	5,754	274
Honolulu Comm Coll, Honolulu, HI 96817-4598	1920	$1,042 (S)	—	5	4,487	216
Hopkinsville Comm Coll, Hopkinsville, KY 42241-2100	1965	$1,250 (S)	—	5	3,186	177
Horry-Georgetown Tech Coll, Conway, SC 29528-6066	1965	$1,492 (A)	—	11	3,587	250
Housatonic Comm Coll, Bridgeport, CT 06604-4704	1965	$1,886 (S)	—	5	3,902	203
Houston Comm Coll System, Houston, TX 77270-7849	1971	$840 (A)	—	11	40,929	2,574
Howard Coll, Big Spring, TX 79720	1945	$748 (A)	$2,764	11	2,135	146
Howard Comm Coll, Columbia, MD 21044-3197	1966	$2,673 (A)	—	11	5,452	369
Hudson County Comm Coll, Jersey City, NJ 07306	1974	$2,801 (A)	—	11	4,874	347
Hudson Valley Comm Coll, Troy, NY 12180-6096	1953	$2,530 (S)	—	11	8,116	511
Hutchinson Comm Coll & Area Voc School, Hutchinson, KS 67501-5894	1928	$1,504 (S)	$3,110	11	3,466	282
Illinois Eastern Comm Colls, Frontier Comm Coll, Fairfield, IL 62837-2601	1976	$1,344 (A)	—	11	1,519	161
Illinois Eastern Comm Colls, Lincoln Trail Coll, Robinson, IL 62454	1969	$1,344 (A)	—	11	1,279	75
Illinois Eastern Comm Colls, Olney Central Coll, Olney, IL 62450	1962	$1,344 (A)	—	11	1,648	86
Illinois Eastern Comm Colls, Wabash Valley Coll, Mount Carmel, IL 62863	1960	$1,344 (A)	—	11	3,421	88
Illinois Valley Comm Coll, Oglesby, IL 61348-9692	1924	$1,717 (A)	—	9	4,582	189
Imperial Valley Coll, Imperial, CA 92251-0158	1922	$264 (S)	—	11	7,009	310
Independence Comm Coll, Independence, KS 67301-0708	1925	$44/sem. hr. (S)	$3,600	5	1,132	48
Indian Hills Comm Coll, Ottumwa, IA 52501-1398	1966	$1,860 (S)	$3,060	11	3,787	148
Indian River Comm Coll, Fort Pierce, FL 34981-5596	1960	$1,500 (S)	—	5	13,186	1,010
Instituto Comercial de Puerto Rico Jr Coll, San Juan, PR 00919-0304	1946	$3,822	—	3	1,530	102
Interboro Inst, New York, NY 10019-3602	1888	$7,025	—	3	1,124	46
Inver Hills Comm Coll, Inver Grove Heights, MN 55076-3224	1969	$2,519 (S)	—	5	4,286	220
Iowa Central Comm Coll, Fort Dodge, IA 50501-5798	1966	$1,614 (S)	$3,800	11	4,295	280
Iowa Lakes Comm Coll, Estherville, IA 51334-2295	1967	$2,416 (S)	$3,200	11	2,863	123
Iowa Western Comm Coll, Council Bluffs, IA 51502	1966	$2,370 (S)	—	9	4,484	222
Itasca Comm Coll, Grand Rapids, MN 55744	1922	$3,219 (S)	—	5	1,033	88
Itawamba Comm Coll, Fulton, MS 38843	1947	$960 (S)	$1,900	11	3,500	102
Ivy Tech State Coll–Central Indiana, Indianapolis, IN 46206-1763	1963	$1,986 (S)	—	5	6,734	341
Ivy Tech State Coll–Columbus, Columbus, IN 47203-1868	1963	$1,986 (S)	—	5	3,286	266
Ivy Tech State Coll–Eastcentral, Muncie, IN 47302-9448	1968	$1,986 (S)	—	5	3,349	291
Ivy Tech State Coll–Kokomo, Kokomo, IN 46903-1373	1968	$1,986 (S)	—	5	1,813	146
Ivy Tech State Coll–Lafayette, Lafayette, IN 47905-5266	1968	$1,986 (S)	—	5	3,014	174
Ivy Tech State Coll–North Central, South Bend, IN 46601	1968	$1,986 (S)	—	5	3,475	228
Ivy Tech State Coll–Northeast, Fort Wayne, IN 46805-1430	1969	$1,986 (S)	—	5	3,378	298
Ivy Tech State Coll–Northwest, Gary, IN 46409-1499	1963	$1,986 (S)	—	5	4,675	103
Ivy Tech State Coll–Southcentral, Sellersburg, IN 47172-1829	1968	$1,986 (S)	—	5	1,933	141
Ivy Tech State Coll–Southeast, Madison, IN 47250-1883	1963	$1,986 (S)	—	5	1,418	109
Ivy Tech State Coll–Southwest, Evansville, IN 47710-3398	1963	$1,986 (S)	—	5	3,442	218
Ivy Tech State Coll–Wabash Valley, Terre Haute, IN 47802	1966	$1,986 (S)	—	5	3,321	190
Ivy Tech State Coll–Whitewater, Richmond, IN 47374-1220	1963	$1,986 (S)	—	5	1,262	115
Jackson Comm Coll, Jackson, MI 49201-8399	1928	$1,800 (A)	—	8	6,600	363
Jackson State Comm Coll, Jackson, TN 38301-3797	1967	$1,445 (S)	—	5	3,739	211
James H. Faulkner State Comm Coll, Bay Minette, AL 36507	1965	$2,016 (S)	$1,162	5	2,784	144
James Sprunt Comm Coll, Kenansville, NC 28349-0398	1964	$918 (S)	—	5	1,276	104
Jamestown Comm Coll, Jamestown, NY 14701-1999	1950	$97/term (S)	—	11	4,041	266
Jefferson Coll, Hillsboro, MO 63050-2441	1963	$1,410 (A)	—	11	3,997	228
Jefferson Comm Coll, Louisville, KY 40202-2005	1968	$1,280 (S)	—	5	9,536	436
Jefferson Comm Coll, Watertown, NY 13601	1961	$2,562 (S)	—	11	3,652	173
Jefferson Comm Coll, Steubenville, OH 43952-3598	1966	$1,740 (A)	—	11	1,395	124
Jefferson State Comm Coll, Birmingham, AL 35215-3098	1965	$1,800 (S)	—	5	5,652	323
John A. Logan Coll, Carterville, IL 62918-9900	1967	$1,380 (A)	—	11	5,130	42
Johnson County Comm Coll, Overland Park, KS 66210-1299	1967	$1,200 (A)	—	11	16,400	702
Johnston Comm Coll, Smithfield, NC 27577-2350	1969	$787 (S)	—	5	3,061	303
John Tyler Comm Coll, Chester, VA 23831-5316	1967	$977 (S)	—	5	5,371	258
John Wood Comm Coll, Quincy, IL 62301-9147	1974	$1,680 (A)	—	9	2,167	181
Joliet Jr Coll, Joliet, IL 60431-8938	1901	$1,380 (A)	—	11	11,333	597
Jones County Jr Coll, Ellisville, MS 39437-3901	1928	$848 (S)	$1,902	11	4,363	175
J. Sargeant Reynolds Comm Coll, Richmond, VA 23285-5622	1972	$1,377 (S)	—	5	10,091	581
Kalamazoo Valley Comm Coll, Kalamazoo, MI 49003-4070	1966	$1,403 (A)	—	11	9,319	394
Kansas City Kansas Comm Coll, Kansas City, KS 66112-3003	1923	$1,204 (S)	—	11	5,238	414
Kapiolani Comm Coll, Honolulu, HI 96816-4421	1957	$1,052 (S)	—	5	6,760	317
Keiser Coll, Fort Lauderdale, FL 33309	1977	$8,500	—	3	2,850	56
Keiser Coll, Melbourne, FL 32901-1461	1989	$7,980	—	3	3,041	196
Kellogg Comm Coll, Battle Creek, MI 49017-3397	1956	$1,655 (A)	—	11	5,300	356
Kent State U, Salem Campus, Salem, OH 44460-9412	1966	$3,004 (S)	—	5	1,050	93
Kent State U, Trumbull Campus, Warren, OH 44483-1998	1954	$3,004 (S)	—	5	2,223	133
Kent State U, Tuscarawas Campus, New Philadelphia, OH 44663-9403	1962	$2,948 (S)	—	5-B	1,678	163
Keystone Coll, La Plume, PA 18440	1868	$12,250	$6,300	1-B	1,231	157
Kilgore Coll, Kilgore, TX 75662-3299	1935	$744 (A)	$2,800	11	4,000	222
Kingsborough Comm Coll of the City U of New York, Brooklyn, NY 11235	1963	$2,600 (S)	—	11	14,801	914
Kingwood Coll, Kingwood, TX 77339-3801	1984	$684 (A)	—	11	4,564	216
Kirkwood Comm Coll, Cedar Rapids, IA 52406-2068	1966	$1,950 (S)	—	11	11,645	645

Name, address	Year Founded	Tuition & Fees	Rm. & Board	Control, Degree	Enroll-ment	Faculty
Kirtland Comm Coll, Roscommon, MI 48653-9699	1966	$1,803 (A)	—	9	1,175	95
Kishwaukee Coll, Malta, IL 60150	1967	$1,560 (A)	—	11	3,663	207
Labette Comm Coll, Parsons, KS 67357-4299	1923	$1,350 (S)	$2,500	11	1,408	208
Lake Area Tech Inst, Watertown, SD 57201	1964	$2,916 (A)	—	5	1,132	65
Lake Land Coll, Mattoon, IL 61938-9366	1966	$1,560 (A)	—	11	5,901	197
Lakeland Comm Coll, Kirtland, OH 44094-5198	1967	$1,791 (A)	—	11	7,812	597
Lake Michigan Coll, Benton Harbor, MI 49022-1899	1946	$59/cr. hr. (A)	—	9	3,406	240
Lake Region State Coll, Devils Lake, ND 58301-1598	1941	$2,078 (A)	$2,818	5	1,219	66
Lakeshore Tech Coll, Cleveland, WI 53015-1414	1967	$2,240 (S)	—	11	2,742	555
Lake-Sumter Comm Coll, Leesburg, FL 34788-8751	1962	$1,409 (S)	—	11	2,751	154
Lake Superior Coll, Duluth, MN 55811	1995	$2,455 (S)	—	5	2,768	156
Lake Tahoe Comm Coll, South Lake Tahoe, CA 96150-4524	1975	$261 (S)	—	11	3,053	186
Lake Washington Tech Coll, Kirkland, WA 98034-8506	1949	$2,400 (A)	—	9	4,934	225
Lamar State Coll–Orange, Orange, TX 77630-5899	1969	$1,834 (S)	—	5	1,939	114
Lane Comm Coll, Eugene, OR 97405-0640	1964	$1,375 (S)	—	11	10,626	605
Laney Coll, Oakland, CA 94607-4893	1953	$364 (S)	—	11	9,516	315
Laramie County Comm Coll, Cheyenne, WY 82007-3299	1968	$1,524 (S)	$4,064	8	3,394	246
Laredo Comm Coll, Laredo, TX 78040-4395	1946	$766 (A)	$3,641	11	7,322	321
Lawson State Comm Coll, Birmingham, AL 35221-1798	1949	$1,556 (S)	—	5	1,820	105
Lee Coll, Baytown, TX 77522-0818	1934	$586 (A)	—	9	5,906	355
Leeward Comm Coll, Pearl City, HI 96782-3393	1968	$1,047 (S)	—	5	6,000	236
Lenoir Comm Coll, Kinston, NC 28502-0188	1960	$786 (S)	—	5	2,205	163
Lewis & Clark Comm Coll, Godfrey, IL 62035-2466	1970	$1,602 (A)	—	9	6,629	382
Lima Tech Coll, Lima, OH 45804-3597	1971	$2,189 (S)	—	5	2,521	193
Lincoln Coll, Normal, IL 61761	1865	$11,100	—	1	1,010	35
Lincoln Land Comm Coll, Springfield, IL 62794-9256	1967	$1,426 (A)	—	9	6,700	398
Linn-Benton Comm Coll, Albany, OR 97321	1966	$1,544 (S)	—	11	4,419	529
Long Beach City Coll, Long Beach, CA 90808-1780	1927	$380 (S)	—	5	27,001	984
Longview Comm Coll, Lee's Summit, MO 64081-2105	1969	$1,590 (A)	—	11	6,571	208
Lorain County Comm Coll, Elyria, OH 44035	1963	$2,247 (A)	—	11	7,106	460
Lord Fairfax Comm Coll, Middletown, VA 22645-0047	1969	$943 (S)	—	5	3,510	152
Los Angeles Harbor Coll, Wilmington, CA 90744-2397	1949	$264 (S)	—	11	7,467	270
Los Angeles Valley Coll, Valley Glen, CA 91401-4096	1949	$22/unit (S)	—	11	17,786	510
Louisiana State U at Alexandria, Alexandria, LA 71302-9121	1960	$1,147 (S)	—	5	2,400	107
Louisiana State U at Eunice, Eunice, LA 70535-1129	1967	$1,164 (S)	—	5	2,901	108
Lower Columbia Coll, Longview, WA 98632-0310	1934	$1,411 (S)	—	5	4,697	207
Luzerne County Comm Coll, Nanticoke, PA 18634-9804	1966	$1,800 (A)	—	8	5,920	461
Macomb Comm Coll, Warren, MI 48093-3896	1954	$1,735 (A)	—	9	22,001	808
Manatee Comm Coll, Bradenton, FL 34206-7046	1957	$1,359 (S)	—	5	7,834	326
Manchester Comm Coll, Manchester, CT 06045-1046	1963	$1,814 (S)	—	5	5,135	306
Maple Woods Comm Coll, Kansas City, MO 64156-1299	1969	$1,590 (A)	—	11	5,536	187
Marion Tech Coll, Marion, OH 43302-5694	1971	$2,410 (S)	—	12	1,656	105
Massachusetts Bay Comm Coll, Wellesley Hills, MA 02481	1961	$2,010 (S)	—	5	4,676	332
Massasoit Comm Coll, Brockton, MA 02302-3996	1966	$2,040 (S)	—	5	6,706	394
Maui Comm Coll, Kahului, HI 96732	1967	$1,068 (S)	—	5	2,657	94
Maysville Comm Coll, Maysville, KY 41056	1967	$1,230 (S)	—	5	1,241	111
McDowell Tech Comm Coll, Marion, NC 28752-9724	1964	$880 (S)	—	5	1,078	58
McHenry County Coll, Crystal Lake, IL 60012-2761	1967	$1,544 (A)	—	11	5,086	277
McLennan Comm Coll, Waco, TX 76708-1499	1965	$1,260 (A)	—	8	5,731	314
Mendocino Coll, Ukiah, CA 95482-0300	1973	$342 (S)	—	11	5,200	193
Merced Coll, Merced, CA 95348-2898	1962	$382 (S)	—	11	7,814	421
Mercer County Comm Coll, Trenton, NJ 08690-1004	1966	$2,310 (A)	—	11	7,751	473
Meridian Comm Coll, Meridian, MS 39307	1937	$1,055 (S)	$2,420	11	3,007	194
Merritt Coll, Oakland, CA 94619-3196	1953	$268 (S)	—	11	6,000	201
Mesabi Range Comm & Tech Coll, Virginia, MN 55792-3448	1918	$2,538 (S)	—	5	1,304	67
Mesa Comm Coll, Mesa, AZ 85202-4866	1965	$1,050 (A)	—	11	24,000	852
Metro Comm Coll, Omaha, NE 68103-0777	1974	$1,350 (S)	—	11	11,534	803
Miami-Dade Comm Coll, Miami, FL 33132-2296	1960	$1,191 (S)	—	11	47,081	2,046
Miami U–Hamilton Campus, Hamilton, OH 45011-3399	1968	$3,050 (S)	—	5-M	3,139	225
Miami U–Middletown Campus, Middletown, OH 45042-3497	1966	$3,049 (S)	—	5-B	2,775	158
Middle Georgia Coll, Cochran, GA 31014-1599	1884	$1,742 (S)	$3,392	5	1,941	119
Middlesex Comm Coll, Middletown, CT 06457-4889	1966	$1,888 (S)	—	5	2,309	120
Middlesex Comm Coll, Bedford, MA 01730-1655	1970	$1,826 (S)	—	5	7,500	468
Middlesex County Coll, Edison, NJ 08818-3050	1964	$2,430 (A)	—	8	10,500	552
Midland Coll, Midland, TX 79705-6399	1969	$1,800 (A)	$1,500	11	4,832	194
Mid-South Comm Coll, West Memphis, AR 72301	1993	$894 (A)	—	5	1,072	66
Mid-State Tech Coll, Wisconsin Rapids, WI 54494-5599	1917	$116/credit (A)	—	11	1,927	96
Milwaukee Area Tech Coll, Milwaukee, WI 53233-1443	1912	$2,476 (S)	—	9	72,211	1,759
Mineral Area Coll, Park Hills, MO 63601-1000	1922	$1,500 (A)	—	9	2,851	186
Minnesota State Coll–Southeast Tech, Winona, MN 55987	1992	$89/cr. hr. (S)	—	5	2,000	84
Minnesota West Comm & Tech Coll, Pipestone, MN 56164	1967	$2,408	—	5	2,044	127
MiraCosta Coll, Oceanside, CA 92056-3899	1934	$296 (S)	—	5	9,892	369
Mission Coll, Santa Clara, CA 95054-1897	1977	$370 (S)	—	11	10,500	382
Mississippi County Comm Coll, Blytheville, AR 72316-1109	1975	$946 (A)	—	5	1,922	93
Mississippi Delta Comm Coll, Moorhead, MS 38761-0668	1926	$1,250 (S)	—	9	2,619	130
Mississippi Gulf Coast Comm Coll, Perkinston, MS 39573-0548	1911	$890 (S)	$1,746	9	8,765	367
Mitchell Comm Coll, Statesville, NC 28677-5293	1852	$880 (S)	—	5	1,830	133
Moberly Area Comm Coll, Moberly, MO 65270-1304	1927	$1,310 (A)	—	11	2,938	175
Mohave Comm Coll, Kingman, AZ 86401	1971	$720 (S)	—	5	5,847	349
Mohawk Valley Comm Coll, Utica, NY 13501-5394	1946	$2,720 (S)	$5,024	11	5,957	273
Monroe Coll, Bronx, NY 10468-5407	1933	$7,160	—	3-B	3,373	149
Monroe Comm Coll, Rochester, NY 14623-5780	1961	$2,656 (S)	—	11	15,315	1,041
Monroe County Comm Coll, Monroe, MI 48161-9047	1964	$1,244 (A)	—	8	3,555	201
Montana State U–Great Falls Coll of Technology, Great Falls, MT 59405	1969	$2,267 (S)	—	5	1,191	83
Montcalm Comm Coll, Sidney, MI 48885-0300	1965	$1,636 (A)	—	11	1,545	90
Monterey Peninsula Coll, Monterey, CA 93940-4799	1947	$306 (S)	—	5	15,475	388
Montgomery Coll, Rockville, MD 20850	NA	$2,652 (A)	—	11	20,823	1,038

Name, address	Year Founded	Tuition & Fees	Rm. & Board	Control, Degree	Enroll-ment	Faculty
Montgomery Coll, Conroe, TX 77384	1995	$684 (A)	—	11	5,312	236
Montgomery County Comm Coll, Blue Bell, PA 19422-0796	1964	$2,400 (A)	—	8	9,099	693
Moorpark Coll, Moorpark, CA 93021-1695	1967	$286 (S)	—	8	13,595	613
Moraine Park Tech Coll, Fond du Lac, WI 54936-1940	1967	$2,043 (S)	—	11	6,921	1,058
Moraine Valley Comm Coll, Palos Hills, IL 60465-0937	1967	$1,470 (A)	—	11	12,972	738
Morgan Comm Coll, Fort Morgan, CO 80701-4399	1967	$3,632 (S)	—	5	1,524	217
Morton Coll, Cicero, IL 60804-4398	1924	$1,488 (A)	—	11	4,430	213
Motlow State Comm Coll, Lynchburg, TN 37352-8500	1969	$1,451 (S)	—	5	3,331	215
Mott Comm Coll, Flint, MI 48503-2089	1923	$2,270 (A)	—	9	8,659	424
Mountain View Coll, Dallas, TX 75211-6599	1970	$700 (A)	—	11	5,128	264
Mt. San Jacinto Coll, San Jacinto, CA 92583-2399	1963	$22/unit (A)	—	11	10,086	415
Mount Wachusett Comm Coll, Gardner, MA 01440-1000	1963	$830 (S)	—	5	3,367	203
Murray State Coll, Tishomingo, OK 73460-3130	1908	$46/cr. hr. (S)	—	5	1,825	73
Muscatine Comm Coll, Muscatine, IA 52761-5396	1929	$3,900 (S)	—	5	1,170	95
Muskingum Area Tech Coll, Zanesville, OH 43701-2626	1969	$2,580 (S)	—	11	2,064	116
Napa Valley Coll, Napa, CA 94558-6236	1942	$264 (S)	—	11	5,998	304
Nash Comm Coll, Rocky Mount, NC 27804-0488	1967	$588 (S)	—	5	2,150	110
Nashville State Tech Inst, Nashville, TN 37209-4515	1970	$1,419 (S)	—	5	7,315	393
Nassau Comm Coll, Garden City, NY 11530-6793	1959	$2,320 (S)	—	11	19,621	1,342
Navarro Coll, Corsicana, TX 75110-4899	1946	$986 (A)	$3,096	11	4,048	242
Neosho County Comm Coll, Chanute, KS 66720-2699	1936	$1,440 (S)	$3,250	11	1,500	96
Newbury Coll, Brookline, MA 02445	1962	$14,300	$7,400	1-B	2,152	89
New England Inst of Technology, Warwick, RI 02886-2244	1940	$11,795	—	1-B	2,603	208
New England Inst of Technology at Palm Beach, West Palm Beach, FL 33407	1983	$17,000	—	3	1,175	55
New Hampshire CommTech Coll, Nashua/Claremont, Nashua, NH 03063	1967	$3,520 (S)	—	5	1,282	118
New Mexico Jr Coll, Hobbs, NM 88240-9123	1965	$332 (A)	$3,325	11	3,200	120
New Mexico State U–Alamogordo, Alamogordo, NM 88311-0477	1958	$864 (A)	—	5	1,920	84
New River Comm Coll, Dublin, VA 24084-1127	1969	$1,183 (S)	—	5	3,692	192
The New York Coll for Wholistic Health Education & Research, Syosset, NY 11791-4413	1981	$10,020	—	1-M	1,000	65
Niagara County Comm Coll, Sanborn, NY 14132-9460	1962	$2,685 (A)	—	11	4,656	273
Nicolet Area Tech Coll, Rhinelander, WI 54501-0518	1968	$2,680 (S)	—	11	1,415	84
Normandale Comm Coll, Bloomington, MN 55431-4399	1968	$2,447 (S)	—	5	6,531	190
Northampton County Area Comm Coll, Bethlehem, PA 18020-7599	1967	$2,430 (A)	$4,920	11	5,666	511
North Arkansas Coll, Harrison, AR 72601	1974	$1,008 (A)	—	11	1,817	119
North Central State Coll, Mansfield, OH 44901-0698	1961	$1,550 (S)	—	5	2,760	182
Northcentral Tech Coll, Wausau, WI 54401-1899	1912	$2,049 (S)	—	9	3,609	226
North Central Texas Coll, Gainesville, TX 76240-4699	1924	$648 (A)	—	8	4,845	250
North Country Comm Coll, Saranac Lake, NY 12983-0089	1967	$2,495 (S)	—	11	1,086	112
North Dakota State Coll of Sci, Wahpeton, ND 58076	1903	$1,920 (S)	$3,500	5	2,425	146
Northeast Alabama Comm Coll, Rainsville, AL 35986-0159	1963	$1,620 (S)	—	5	1,659	55
Northeast Comm Coll, Norfolk, NE 68702-0469	1973	$1,478 (S)	—	11	4,520	202
Northeastern Jr Coll, Sterling, CO 80751-2399	1941	$2,313 (S)	$5,949	5	3,633	286
Northeastern Oklahoma Ag & Mech Coll, Miami, OK 74354-6434	1919	$46/cr. hr. (S)	—	5	2,000	100
Northeast Iowa Comm Coll, Calmar, IA 52132-0480	1966	$2,460 (S)	—	11	3,480	168
Northeast Mississippi Comm Coll, Booneville, MS 38829	1948	$1,030 (S)	$1,900	5	2,962	126
Northeast State Tech Comm Coll, Blountville, TN 37617-0246	1966	$1,455 (A)	—	5	4,126	223
Northeast Texas Comm Coll, Mount Pleasant, TX 75456-1307	1985	$950 (A)	$2,700	11	2,052	114
Northeast Wisconsin Tech Coll, Green Bay, WI 54307-9042	1913	$1,970 (S)	—	11	6,670	215
Northern Virginia Comm Coll, Annandale, VA 22003-3796	1965	$39/cr. hr. (S)	—	5	37,073	1,508
North Florida Comm Coll, Madison, FL 32340-1602	1958	$1,560 (S)	—	5	1,012	44
North Harris Coll, Houston, TX 77073-3499	1972	$684 (A)	—	11	9,127	497
North Hennepin Comm Coll, Minneapolis, MN 55445-2231	1966	$2,086 (S)	—	5	5,300	200
North Idaho Coll, Coeur d'Alene, ID 83814-2199	1933	$1,218 (A)	—	11	4,825	331
North Iowa Area Comm Coll, Mason City, IA 50401-7299	1918	$2,215 (S)	$3,320	11	2,803	238
North Lake Coll, Irving, TX 75038-3899	1977	$562 (A)	—	8	8,000	666
Northland Comm & Tech Coll, Thief River Falls, MN 56701	1965	$2,674 (S)	—	5	1,881	92
Northland Pioneer Coll, Holbrook, AZ 86025-0610	1974	$750 (S)	—	11	5,086	305
North Shore Comm Coll, Danvers, MA 01923-4093	1965	$1,752 (S)	—	5	6,285	506
NorthWest Arkansas Comm Coll, Bentonville, AR 72712	1989	$1,290 (A)	—	11	4,058	239
Northwest Coll, Powell, WY 82435-1898	1946	$1,620 (S)	$3,058	11	1,569	161
Northwestern Michigan Coll, Traverse City, MI 49686-3061	1951	$2,128 (A)	$4,600	11	3,965	87
Northwestern Tech Coll, Rock Springs, GA 30739	1966	$1,292 (A)	—	5	1,500	79
Northwest Indian Coll, Bellingham, WA 98226	1978	$2,088 (S)	—	4	1,689	63
Northwest Mississippi Comm Coll, Senatobia, MS 38668-1701	1927	$1,200 (S)	$2,000	11	5,000	200
Northwest-Shoals Comm Coll, Muscle Shoals, AL 35662	1963	$1,680 (S)	—	5	3,826	262
Northwest State Comm Coll, Archbold, OH 43502-9542	1968	$2,280 (S)	—	5	2,720	149
Norwalk Comm Coll, Norwalk, CT 06854-1655	1961	$1,886 (S)	—	5	5,377	346
Oakland Comm Coll, Bloomfield Hills, MI 48304-2266	1964	$1,550 (A)	—	11	23,188	892
Oakton Comm Coll, Des Plaines, IL 60016-1268	1969	$1,415 (A)	—	9	10,133	647
Ocean County Coll, Toms River, NJ 08754-2001	1964	$76/credit (A)	—	8	7,143	322
Odessa Coll, Odessa, TX 79764-7127	1946	$976 (A)	$2,000	11	4,777	271
Ohio U–Southern Campus, Ironton, OH 45638-2214	1956	$2,805 (S)	—	5-M	2,500	120
Ohlone Coll, Fremont, CA 94539-5884	1967	$382 (S)	—	11	10,500	434
Okaloosa-Walton Comm Coll, Niceville, FL 32578-1295	1963	$1,316 (S)	—	11	7,802	241
Oklahoma City Comm Coll, Oklahoma City, OK 73159-4419	1969	$1,308 (S)	—	5	8,909	425
Oklahoma State U, Oklahoma City, Oklahoma City, OK 73107-6120	1961	$1,757 (S)	—	5	4,011	223
Oklahoma State U, Okmulgee, Okmulgee, OK 74447-3901	1946	$2,995 (S)	—	5	2,401	130
Olympic Coll, Bremerton, WA 98337-1699	1946	$1,770 (S)	—	5	5,613	320
Onondaga Comm Coll, Syracuse, NY 13215-2099	1962	$2,786 (A)	—	11	7,848	457
Orangeburg-Calhoun Tech Coll, Orangeburg, SC 29118-8299	1968	$1,296 (A)	—	11	1,861	112
Orange Coast Coll, Costa Mesa, CA 92628-5005	1947	$376 (A)	—	11	21,942	1,143
Orange County Comm Coll, Middletown, NY 10940-6437	1950	$2,345 (S)	—	11	5,587	336
Otero Jr Coll, La Junta, CO 81050-3415	1941	$1,900 (S)	$2,780	5	1,360	73
Owensboro Comm Coll, Owensboro, KY 42303-1899	1986	$1,250 (A)	—	5	2,252	124
Owens Comm Coll, Findlay, OH 45840	1983	$1,820 (S)	—	5	1,985	158
Ozarks Tech Comm Coll, Springfield, MO 65801	1990	$1,680 (A)	—	9	6,343	264

Name, address	Year Founded	Tuition & Fees	Rm. & Board	Control, Degree	Enroll-ment	Faculty
Paducah Comm Coll, Paducah, KY 42002-7380	1932	$725 (S)	—	5	3,200	151
Palm Beach Comm Coll, Lake Worth, FL 33461-4796	1933	$1,450 (S)	—	5	17,326	942
Palomar Coll, San Marcos, CA 92069-1487	1946	$330 (A)	—	11	25,701	1,285
Palo Verde Coll, Blythe, CA 92225-1118	1947	$264 (S)	—	11	2,278	69
Panola Coll, Carthage, TX 75633-2397	1947	$930 (A)	$2,860	11	1,424	93
Paradise Valley Comm Coll, Phoenix, AZ 85032-1200	1985	$1,300 (A)	—	11	7,000	330
Paris Jr Coll, Paris, TX 75460-6298	1924	$1,106 (A)	$2,650	11	2,850	147
Parkland Coll, Champaign, IL 61821-1899	1967	$1,590 (A)	—	9	8,026	491
Pasadena City Coll, Pasadena, CA 91106-2041	1924	$356 (S)	—	11	24,000	1,296
Pasco-Hernando Comm Coll, New Port Richey, FL 34654-5199	1972	$1,401 (S)	—	5	5,165	242
Passaic County Comm Coll, Paterson, NJ 07505-1179	1968	$2,432 (S)	—	8	4,633	357
Pellissippi State Tech Comm Coll, Knoxville, TN 37933-0990	1974	$1,464 (S)	—	5	7,859	412
Peninsula Coll, Port Angeles, WA 98362-2779	1961	$1,855 (S)	$5,403	5	4,618	221
The Pennsylvania State U Delaware County Campus of the Commonwealth Coll, Media, PA 19063-5596	1966	$6,636 (S)	—	12-B	1,679	123
The Pennsylvania State U DuBois Campus of the Commonwealth Coll, DuBois, PA 15801-3199	1935	$6,626 (S)	—	12-B	1,012	89
The Pennsylvania State U Fayette Campus of the Commonwealth Coll, Uniontown, PA 15401-0519	1934	$6,626 (S)	—	12-B	1,127	90
The Pennsylvania State U Hazleton Campus of the Commonwealth Coll, Hazleton, PA 18201-1291	1934	$6,636 (S)	$4,910	12-B	1,377	97
The Pennsylvania State U Mont Alto Campus of the Commonwealth Coll, Mont Alto, PA 17237-9703	1929	$6,636 (S)	$4,910	12-B	1,307	98
The Pennsylvania State U Shenango Campus of the Commonwealth Coll, Sharon, PA 16146-1537	1965	$6,636 (S)	—	12-B	1,032	91
The Pennsylvania State U Worthington Scranton Campus of the Commonwealth Coll, Dunmore, PA 18512-1699	1923	$6,626 (S)	—	12-B	1,716	113
The Pennsylvania State U York Campus of the Commonwealth Coll, York, PA 17403-3298	1926	$6,626 (S)	—	12-B	2,006	132
Penn Valley Comm Coll, Kansas City, MO 64111	1969	$1,590 (A)	—	11	4,886	248
Pensacola Jr Coll, Pensacola, FL 32504-8998	1948	$984 (A)	—	5	4,820	568
Petit Jean Coll, Morrilton, AR 72110	1961	$1,192 (A)	—	5	1,236	65
Phillips Comm Coll of the U of Arkansas, Helena, AR 72342-0785	1965	$984 (A)	—	11	2,100	70
Phoenix Coll, Phoenix, AZ 85013-4234	1920	$41/cr. hr. (A)	—	11	12,386	104
Piedmont Comm Coll, Roxboro, NC 27573-1197	1970	$814 (S)	—	5	1,934	105
Piedmont Tech Coll, Greenwood, SC 29648-1467	1966	$1,375 (A)	—	5	4,104	233
Piedmont Virginia Comm Coll, Charlottesville, VA 22902-7589	1972	$1,172 (S)	—	5	4,277	248
Pierce Coll, Lakewood, WA 98498-1999	1967	$1,194 (S)	—	5	13,000	579
Pikes Peak Comm Coll, Colorado Springs, CO 80906-5498	1968	$1,854 (S)	—	5	9,997	550
Pima Comm Coll, Tucson, AZ 85709-1010	1966	$850 (S)	—	11	28,466	1,621
Pitt Comm Coll, Greenville, NC 27835-7007	1961	$920 (S)	—	11	5,363	280
Polk Comm Coll, Winter Haven, FL 33881-4299	1964	$1,723 (S)	—	5	5,611	322
Porterville Coll, Porterville, CA 93257-6058	1927	$382 (S)	—	5	5,418	140
Portland Comm Coll, Portland, OR 97280-0990	1961	$1,988 (S)	—	11	43,790	1,385
Potomac State Coll of West Virginia U, Keyser, WV 26726-2698	1901	$2,194 (S)	$4,196	5	1,109	95
Prairie State Coll, Chicago Heights, IL 60411-8226	1958	$1,404 (A)	—	11	5,188	315
Pratt Comm Coll & Area Voc School, Pratt, KS 67124-8317	1938	$1,472 (S)	—	11	1,331	49
Prince George's Comm Coll, Largo, MD 20774-2199	1958	$2,416 (A)	—	8	11,563	646
Pueblo Comm Coll, Pueblo, CO 81004-1499	1933	$1,875 (S)	—	5	4,626	329
Pulaski Tech Coll, North Little Rock, AR 72118	1945	$1,530 (S)	—	5	4,306	145
Queensborough Comm Coll of the City U of New York, Bayside, NY 11364	1958	$2,616 (S)	—	11	10,827	774
Quincy Coll, Quincy, MA 02169-4522	1958	$2,880	—	10	3,764	69
Quinebaug Valley Comm Coll, Danielson, CT 06239-1440	1971	$1,886 (S)	—	5	1,347	82
Quinsigamond Comm Coll, Worcester, MA 01606-2092	1963	$1,890 (S)	—	5	5,617	371
Randolph Comm Coll, Asheboro, NC 27204-1009	1962	$1,167 (S)	—	5	1,950	164
Ranken Tech Coll, St. Louis, MO 63113 (2)	1907	$7,155	—	1	1,423	67
Rappahannock Comm Coll, Glenns, VA 23149-2616	1970	$953 (S)	—	12	1,810	79
Raritan Valley Comm Coll, Somerville, NJ 08876-1265	1965	$2,330 (S)	—	8	5,751	316
Redlands Comm Coll, El Reno, OK 73036-5304	1938	$1,448 (S)	—	5	2,173	107
Red Rocks Comm Coll, Lakewood, CO 80228-1255	1969	$1,595 (S)	—	5	7,394	276
Reedley Coll, Reedley, CA 93654-2099	1926	$330 (S)	—	11	10,078	258
Rend Lake Coll, Ina, IL 62846-9801	1967	$1,260 (A)	—	5	4,331	147
Renton Tech Coll, Renton, WA 98056	1942	$1,810 (S)	—	5	5,424	253
Richard Bland Coll of The Coll of William & Mary, Petersburg, VA 23805-7100	1961	$1,802 (S)	—	5	1,267	64
Richland Comm Coll, Decatur, IL 62521-8513	1971	$1,505 (A)	—	9	3,243	233
Richmond Comm Coll, Hamlet, NC 28345-1189	1964	$688 (S)	—	5	1,335	100
Ricks Coll, Rexburg, ID 83460-4107	1888	$2,100	$3,668	2	9,000	437
Rio Salado Coll, Tempe, AZ 85281-6950	1978	$1,240 (A)	—	11	11,386	1,001
Riverland Comm Coll, Austin, MN 55912	1940	$2,594 (S)	—	5	3,175	158
Riverside Comm Coll, Riverside, CA 92506-1299	1916	$350 (S)	—	11	27,110	1,589
Roane State Comm Coll, Harriman, TN 37748-5011	1971	$1,584 (S)	—	5	5,099	407
Roanoke-Chowan Comm Coll, Ahoskie, NC 27910	1967	$917 (S)	—	5	1,014	62
Rochester Comm & Tech Coll, Rochester, MN 55904-4999	1915	$2,545 (S)	—	5	4,404	225
Rockingham Comm Coll, Wentworth, NC 27375-0038	1964	$1,004 (S)	—	5	1,964	106
Rockland Comm Coll, Suffern, NY 10901-3699	1959	$2,454 (S)	—	11	6,262	412
Rock Valley Coll, Rockford, IL 61114-5699	1964	$43/cr. hr. (A)	—	9	8,000	260
Rogue Comm Coll, Grants Pass, OR 97527-9298	1970	$1,620 (S)	—	11	4,024	505
Rose State Coll, Midwest City, OK 73110-2799	1968	$1,100 (S)	—	11	7,350	412
Roxbury Comm Coll, Roxbury Crossing, MA 02120-3400	1973	$1,656 (S)	—	5	2,382	120
Sacramento City Coll, Sacramento, CA 95822-1386	1916	$332 (S)	—	11	19,531	554
Saint Charles County Comm Coll, St. Peters, MO 63376-0975	1986	$1,500 (A)	—	5	5,565	311
St. Cloud Tech Coll, St. Cloud, MN 56303-1240	1948	$2,325 (S)	—	5	2,912	162
St. Johns River Comm Coll, Palatka, FL 32177-3897	1958	$1,315 (S)	—	5	3,459	157
St. Louis Comm Coll at Florissant Valley, St. Louis, MO 63135-1499	1963	$1,344 (A)	—	9	7,365	350
St. Louis Comm Coll at Forest Park, St. Louis, MO 63110-1316	1962	$1,344 (A)	—	9	6,749	314
St. Louis Comm Coll at Meramec, Kirkwood, MO 63122-5720	1963	$1,008 (A)	—	9	12,518	570
St. Paul Tech Coll, St. Paul, MN 55102-1800	1919	$2,222 (S)	—	12	4,785	124

Name, address	Year Founded	Tuition & Fees	Rm. & Board	Control, Degree	Enroll-ment	Faculty
St. Petersburg Jr Coll, St. Petersburg, FL 33731-3489	1927	$1,209 (S)	—	11	19,900	1,656
St. Philip's Coll, San Antonio, TX 78203-2098	1898	$973 (A)	—	9	8,293	472
Salem Comm Coll, Carneys Point, NJ 08069-2799	1972	$1,876 (A)	—	8	1,166	75
Salish Kootenai Coll, Pablo, MT 59855-0117	1977	$2,487 (A)	—	1-B	1,042	55
Salt Lake Comm Coll, Salt Lake City, UT 84130-0808	1948	$1,636 (S)	—	5	21,596	1,130
Sampson Comm Coll, Clinton, NC 28329-0318	1965	$781 (S)	—	11	1,296	100
San Antonio Coll, San Antonio, TX 78212-4299	1925	$938 (S)	—	11	21,000	948
Sandhills Comm Coll, Pinehurst, NC 28374-8299	1963	$908 (S)	—	11	2,851	224
San Diego Mesa Coll, San Diego, CA 92111-4998	1964	$352 (S)	—	11	23,294	585
San Diego Miramar Coll, San Diego, CA 92126-2999	1969	$382 (S)	—	11	6,659	231
San Jacinto Coll Central Campus, Pasadena, TX 77501-2007	1961	$788 (S)	—	11	10,507	533
San Jacinto Coll North Campus, Houston, TX 77049-4599	1974	$724 (A)	—	11	4,946	274
San Jacinto Coll South Campus, Houston, TX 77089-6099	1979	$724 (A)	—	11	6,263	251
San Joaquin Delta Coll, Stockton, CA 95207-6370	1935	$380 (S)	—	9	18,526	213
San Juan Coll, Farmington, NM 87402-4699	1958	$360 (S)	—	8	3,921	278
Santa Ana Coll, Santa Ana, CA 92706-3398	1915	$301 (S)	—	5	29,268	2,354
Santa Barbara City Coll, Santa Barbara, CA 93109-2394	1908	$386 (S)	—	11	14,230	615
Santa Fe Comm Coll, Gainesville, FL 32606-6200	1966	$1,452 (S)	—	11	12,796	874
Santa Fe Comm Coll, Santa Fe, NM 87505-4887	1983	$646 (A)	—	11	3,076	335
Santa Monica Coll, Santa Monica, CA 90405-1628	1929	$388 (S)	—	11	27,985	1,207
Santa Rosa Jr Coll, Santa Rosa, CA 95401-4395	1918	$264 (S)	—	11	25,233	1,767
Santiago Canyon Coll, Orange, CA 92869	2000	$301 (S)	—	5	9,249	2,354
Sauk Valley Comm Coll, Dixon, IL 61021	1965	$1,440 (A)	—	9	2,386	152
Savannah Tech Coll, Savannah, GA 31405	1929	$966 (S)	—	5	2,048	145
Schenectady County Comm Coll, Schenectady, NY 12305-2294	1968	$2,455 (S)	—	11	3,473	207
Scott Comm Coll, Bettendorf, IA 52722-6804	1966	$1,950 (S)	—	11	3,847	235
Scottsdale Comm Coll, Scottsdale, AZ 85256-2626	1969	$1,240 (A)	—	11	10,397	509
Seminole Comm Coll, Sanford, FL 32773-6199	1966	$1,387 (S)	—	11	7,865	332
Seward County Comm Coll, Liberal, KS 67905-1137	1969	$1,440 (S)	$3,200	11	2,325	208
Shasta Coll, Redding, CA 96049-6006	1948	$292 (S)	—	11	10,901	491
Shelton State Comm Coll, Tuscaloosa, AL 35405	1979	$1,560 (S)	—	5	6,211	215
Sheridan Coll, Sheridan, WY 82801-1500	1948	$1,464 (S)	$3,370	11	2,670	207
Shoreline Comm Coll, Seattle, WA 98133-5696	1964	$1,797 (S)	—	5	7,000	415
Sinclair Comm Coll, Dayton, OH 45402-1460	1887	$1,325 (A)	—	11	19,026	974
Skagit Valley Coll, Mount Vernon, WA 98273-5899	1926	$1,629 (S)	—	5	6,767	318
Skyline Coll, San Bruno, CA 94066-1698	1969	$298 (S)	—	11	8,573	300
Snead State Comm Coll, Boaz, AL 35957-0734	1898	$1,470 (S)	$1,766	5	1,579	88
Snow Coll, Ephraim, UT 84627-1203	1888	$1,354 (S)	$3,800	5	2,999	142
Solano Comm Coll, Suisun, CA 94585-3197	1945	$315 (S)	—	11	10,076	374
Somerset Comm Coll, Somerset, KY 42501-2973	1965	$1,238 (A)	—	5	2,387	151
South Arkansas Comm Coll, El Dorado, AR 71731-7010	1975	$1,018 (A)	—	5	2,700	63
Southeast Arkansas Coll, Pine Bluff, AR 71603	1991	$1,290 (S)	—	5	1,955	89
Southeast Comm Coll, Cumberland, KY 40823-1099	1960	$1,230 (S)	—	5	2,529	119
Southeast Comm Coll, Lincoln Campus, Lincoln, NE 68520-1299	1973	$1,800 (S)	—	9	5,431	548
Southeastern Baptist Theological Seminary, Wake Forest, NC 27588-1889	1950	$140/credit (S)	—	2-D	1,702	53
Southeastern Comm Coll, North Campus, West Burlington, IA 52655-0180	1968	$1,950 (S)	$2,700	11	1,922	94
Southeastern Illinois Coll, Harrisburg, IL 62946-4925	1960	$1,248 (A)	$2,600	5	3,272	208
Southeast Tech Inst, Sioux Falls, SD 57107-1301	1968	$2,836 (S)	—	5	2,283	149
Southern Maine Tech Coll, South Portland, ME 04106	1946	$2,540 (S)	$4,200	5	2,334	208
Southern State Comm Coll, Hillsboro, OH 45133-9487	1975	$2,487 (S)	—	5	1,845	109
Southern Union State Comm Coll, Wadley, AL 36276	1922	$1,440 (S)	—	5	4,500	217
Southern U at Shreveport, Shreveport, LA 71107	1964	$1,200 (S)	—	5	1,324	98
South Florida Comm Coll, Avon Park, FL 33825-9356	1965	$1,407 (S)	—	5	2,076	203
South Mountain Comm Coll, Phoenix, AZ 85040	1979	$970 (A)	—	11	3,091	195
South Piedmont Comm Coll, Polkton, NC 28135-0126	1962	$660 (S)	—	5	1,699	146
South Texas Comm Coll, McAllen, TX 78501	1993	$1,412 (A)	—	9	11,319	554
Southwestern Coll, Chula Vista, CA 91910-7299	1961	$364 (S)	—	11	18,403	749
Southwestern Comm Coll, Creston, IA 50801	1966	$2,528 (S)	—	5	1,204	75
Southwestern Comm Coll, Sylva, NC 28779	1964	$778 (S)	—	5	1,669	227
Southwestern Illinois Coll, Belleville, IL 62221-5899	1946	$1,128 (A)	—	9	13,923	791
Southwestern Michigan Coll, Dowagiac, MI 49047-9793	1964	$1,767 (A)	—	11	3,047	211
Southwestern Oregon Comm Coll, Coos Bay, OR 97420-2912	1961	$1,779	$5,230	11	3,057	600
Southwest Georgia Tech Coll, Thomasville, GA 31792	1963	$1,350 (S)	—	5	1,400	125
Southwest Mississippi Comm Coll, Summit, MS 39666	1918	$1,050 (S)	$1,900	11	1,651	94
Southwest Missouri State U–West Plains, West Plains, MO 65775	1963	$2,320 (S)	$4,000	5	1,525	89
Southwest Tennessee Comm Coll, Memphis, TN 38101-0780	NA	$1,321 (S)	—	5	12,194	763
Southwest Virginia Comm Coll, Richlands, VA 24641-1510	1968	$40/cr. hr. (S)	—	5	4,191	387
Southwest Wisconsin Tech Coll, Fennimore, WI 53809-9778	1967	$2,208 (S)	—	11	2,200	112
Spoon River Coll, Canton, IL 61520-9801	1959	$1,760 (A)	—	5	1,861	148
Springfield Tech Comm Coll, Springfield, MA 01105-1296	1967	$2,130 (S)	—	5	6,898	342
Stanly Comm Coll, Albemarle, NC 28001-7458	1971	$918 (S)	—	5	1,532	106
Stark State Coll of Technology, Canton, OH 44720-7299	1970	$2,635 (S)	—	11	4,565	467
State U of New York Coll of Agriculture & Technology at Morrisville, Morrisville, NY 13408-0901	1908	$3,765 (S)	$5,440	5-B	3,033	148
State U of New York Coll of Technology at Alfred, Alfred, NY 14802	1908	$4,030 (S)	$5,358	5-B	2,733	275
State U of New York Coll of Technology at Delhi, Delhi, NY 13753	1913	$3,765 (S)	$5,470	5-B	2,013	140
Suffolk County Comm Coll, Selden, NY 11784-2899	1959	$2,674 (A)	—	11	18,044	1,333
Sullivan County Comm Coll, Loch Sheldrake, NY 12759	1962	$2,656 (S)	—	11	1,552	98
Surry Comm Coll, Dobson, NC 27017-0304	1965	$908 (A)	—	5	2,996	96
Sussex County Comm Coll, Newton, NJ 07860	1981	$2,340 (A)	—	11	2,286	188
Tacoma Comm Coll, Tacoma, WA 98466	1965	$1,694 (S)	—	5	6,283	309
Taft Coll, Taft, CA 93268-2317	1922	$370 (S)	$2,720	11	1,989	85
Tallahassee Comm Coll, Tallahassee, FL 32304-2895	1966	$1,288 (S)	—	11	10,816	423
Tarrant County Coll District, Fort Worth, TX 76102-6599	1967	$852 (A)	—	8	25,968	1,341
Tech Career Insts, New York, NY 10001-2705	1909	$7,230	—	3	4,018	210
Terra State Comm Coll, Fremont, OH 43420-9670	1968	$1,944 (S)	—	5	2,454	156
Texarkana Coll, Texarkana, TX 75599-0001	1927	$710 (A)	—	11	4,010	171
Texas State Tech Coll, Sweetwater, TX 79556-4108	1970	$146/credit (S)	$2,360	5	1,224	142

Name, address	Year Founded	Tuition & Fees	Rm. & Board	Control, Degree	Enroll-ment	Faculty
Texas State Tech Coll–Harlingen, Harlingen, TX 78550-3697	1967	$1,665 (S)	$3,048	5	3,237	281
Texas State Tech Coll–Waco Campus, Waco, TX 76705-1695	1965	$1,770 (A)	—	5	3,928	255
Three Rivers Comm Coll, Poplar Bluff, MO 63901-2393	1966	$1,694 (A)	—	11	2,500	67
Tidewater Comm Coll, Norfolk, VA 23510	1968	$891 (S)	—	5	20,760	1,200
Tomball Coll, Tomball, TX 77375-4036	1988	$684 (A)	—	11	5,880	293
Tompkins Cortland Comm Coll, Dryden, NY 13053-0139	1968	$2,834 (S)	—	11	2,674	195
Tri-County Comm Coll, Murphy, NC 28906-7919	1964	$671 (S)	—	5	1,151	55
Tri-County Tech Coll, Pendleton, SC 29670-0587	1962	$1,700 (A)	—	5	3,612	384
Trident Tech Coll, Charleston, SC 29423-8067	1964	$1,300 (A)	—	11	10,246	523
Trinidad State Jr Coll, Trinidad, CO 81082-2396	1925	$2,113 (S)	$3,362	5	2,543	173
Trinity Valley Comm Coll, Athens, TX 75751-2765	1946	$456 (A)	—	11	4,447	116
Triton Coll, River Grove, IL 60171-9983	1964	$1,766 (A)	—	5	16,927	579
Truckee Meadows Comm Coll, Reno, NV 89512-3901	1971	$1,116 (S)	—	5	9,930	635
Truett-McConnell Coll, Cleveland, GA 30528	1946	$6,600	$3,300	2	2,033	178
Tulsa Comm Coll, Tulsa, OK 74135-6198	1968	$1,582 (S)	—	5	16,203	1,200
Tyler Jr Coll, Tyler, TX 75711-9020	1926	$830 (A)	$2,800	11	8,240	364
Ulster County Comm Coll, Stone Ridge, NY 12484	1961	$2,596 (S)	—	11	2,671	195
Union County Coll, Cranford, NJ 07016-1528	1933	$2,559 (A)	—	11	8,655	398
Universal Tech Inst, Houston, TX 77073-5598	NA	$15,100	—	13	1,400	65
The U of Akron–Wayne Coll, Orrville, OH 44667-9192	1972	$3,270 (S)	—	5	1,515	150
U of Alaska Anchorage, Kenai Peninsula Coll, Soldotna, AK 99669-9798	1964	$2,398 (S)	—	5	1,779	107
U of Alaska Anchorage, Matanuska-Susitna Coll, Palmer, AK 99645-2889	1958	$1,906 (S)	—	5	1,600	NA
U of Arkansas Comm Coll at Batesville, Batesville, AR 72503	NA	$874 (A)	—	5	1,024	67
U of Arkansas Comm Coll at Hope, Hope, AR 71801-0140	1966	$1,004 (A)	—	5	1,235	68
U of Cincinnati Raymond Walters Coll, Cincinnati, OH 45236-1007	1967	$4,998 (S)	—	5	3,600	277
U of Kentucky, Lexington Comm Coll, Lexington, KY 40506-0235	1965	$2,400 (S)	—	5	7,214	379
U of New Mexico–Gallup, Gallup, NM 87301-5603	1968	$960 (S)	—	5-B	2,618	159
U of New Mexico–Valencia Campus, Los Lunas, NM 87031-7633	1981	$1,368 (S)	—	5	1,544	93
U of Northwestern Ohio, Lima, OH 45805-1498	1920	$8,556	—	1-B	2,053	83
U of South Carolina Beaufort, Beaufort, SC 29902-4601	1959	$2,100 (S)	—	5	1,070	64
U of South Carolina Sumter, Sumter, SC 29150-2498	1966	$2,200 (S)	—	5	1,173	82
U of Wisconsin–Fox Valley, Menasha, WI 54952	1933	$2,476 (S)	—	5	1,512	58
U of Wisconsin–Marathon County, Wausau, WI 54401-5396	1933	$2,640 (S)	$3,300	5	1,339	79
U of Wisconsin–Waukesha, Waukesha, WI 53188-2799	1966	$2,280 (S)	—	5	2,018	80
Utah Valley State Coll, Orem, UT 84058-5999	1941	$2,002 (S)	—	5-B	20,946	976
Valencia Comm Coll, Orlando, FL 32802-3028	1967	$1,480 (S)	—	5	27,565	934
Vance-Granville Comm Coll, Henderson, NC 27536-0917	1969	$679 (S)	—	5	4,434	311
Ventura Coll, Ventura, CA 93003-3899	1925	$352 (S)	—	11	11,310	574
Vernon Regional Jr Coll, Vernon, TX 76384-4092	1970	$960 (A)	$2,243	11	2,095	128
Victor Valley Coll, Victorville, CA 92392-5849	1961	$274 (S)	—	5	15,374	325
Vincennes U, Vincennes, IN 47591-5202	1801	$2,541 (S)	$4,438	5	4,755	383
Virginia Highlands Comm Coll, Abingdon, VA 24212-0828	1967	$1,259 (S)	—	5	3,867	135
Virginia Western Comm Coll, Roanoke, VA 24038	1966	$1,178 (S)	—	5	8,020	300
Vista Comm Coll, Berkeley, CA 94704-5102	1974	$330 (S)	—	11	4,500	153
Volunteer State Comm Coll, Gallatin, TN 37066-3188	1970	$1,435 (S)	—	5	6,567	381
Wake Tech Comm Coll, Raleigh, NC 27603-5696	1958	$786 (S)	—	11	9,679	895
Wallace State Comm Coll, Hanceville, AL 35077-2000	1966	$1,008 (S)	$1,350	5	4,770	339
Walla Walla Comm Coll, Walla Walla, WA 99362-9267	1967	$1,768 (S)	—	5	5,916	312
Walters State Comm Coll, Morristown, TN 37813-6899	1970	$1,433 (S)	—	5	6,163	253
Washington State Comm Coll, Marietta, OH 45750-9225	1971	$2,500 (S)	—	5	1,911	132
Washtenaw Comm Coll, Ann Arbor, MI 48106	1965	$1,726 (A)	—	11	11,089	654
Waubonsee Comm Coll, Sugar Grove, IL 60554-9799	1966	$1,468 (A)	—	9	7,602	397
Wayne Comm Coll, Goldsboro, NC 27533-8002	1957	$781 (S)	—	11	2,791	215
Wayne County Comm Coll District, Detroit, MI 48226-3010	1967	$1,770 (A)	—	11	10,000	286
Weatherford Coll, Weatherford, TX 76086-5699	1869	$970 (A)	$2,599	11	2,566	100
Westark Coll, Fort Smith, AR 72913-3649	1928	$1,260 (A)	—	11-B	5,286	285
West Central Tech Coll, Carrollton, GA 30116	1968	$963 (S)	—	5	2,056	101
Westchester Business Inst, White Plains, NY 10602	1915	$12,975	—	3	1,072	84
Westchester Comm Coll, Valhalla, NY 10595-1698	1946	$2,653 (S)	—	11	10,819	641
Western Iowa Tech Comm Coll, Sioux City, IA 51102-5199	1966	$2,190 (S)	$1,800	5	4,338	87
Western Nebraska Comm Coll, Scottsbluff, NE 69361	1926	$1,440 (S)	$3,140	11	2,500	144
Western Nevada Comm Coll, Carson City, NV 89703-7316	1971	$1,440 (S)	—	5	5,117	354
Western Oklahoma State Coll, Altus, OK 73521-1397	1926	$1,484 (S)	$1,600	5	2,296	99
Western Piedmont Comm Coll, Morganton, NC 28655-4511	1964	$1,000 (S)	—	5	2,200	133
Western Texas Coll, Snyder, TX 79549-6105	1969	$1,100 (A)	$2,300	11	1,180	55
Western Wyoming Comm Coll, Rock Springs, WY 82902-0428	1959	$1,344 (S)	$2,820	11	2,525	193
West Georgia Tech Coll, LaGrange, GA 30240	NA	$1,224 (S)	—	5	1,234	75
West Hills Comm Coll, Coalinga, CA 93210-1399	1932	$288 (S)	$4,911	5	3,463	170
Westmoreland County Comm Coll, Youngwood, PA 15697-1895	1970	$1,560 (S)	—	8	5,506	389
West Shore Comm Coll, Scottville, MI 49454-0277	1967	$1,674 (S)	—	9	1,338	55
West Virginia Northern Comm Coll, Wheeling, WV 26003-3699	1972	$1,632 (S)	—	5	2,464	122
West Virginia U at Parkersburg, Parkersburg, WV 26101-9577	1971	$1,436 (S)	—	5-B	3,291	176
Wharton County Jr Coll, Wharton, TX 77488-3298	1946	$855 (A)	$2,180	11	4,449	222
Whatcom Comm Coll, Bellingham, WA 98226-8003	1970	$1,680 (S)	—	5	5,000	224
Wilkes Comm Coll, Wilkesboro, NC 28697	1965	$918 (S)	—	5	2,127	309
Wilson Tech Comm Coll, Wilson, NC 27893-3310	1958	$791 (S)	—	5	1,596	99
Wisconsin Indianhead Tech Coll, New Richmond Campus, New Richmond, WI 54017-1738	1972	$1,968 (S)	—	9	1,495	65
Wisconsin Indianhead Tech Coll, Rice Lake Campus, Rice Lake, WI 54868	1941	$1,896 (S)	—	9	1,598	79
Wisconsin Indianhead Tech Coll, Superior Campus, Superior, WI 54880-5207	1912	$2,143 (S)	—	9	1,113	76
Wor-Wic Comm Coll, Salisbury, MD 21804	1976	$1,764 (A)	—	11	2,329	127
Yakima Valley Comm Coll, Yakima, WA 98907-2520	1928	$1,716 (S)	$4,800	5	3,946	292
York Tech Coll, Rock Hill, SC 29730-3395	1961	$1,236 (A)	—	5	3,597	230
York Tech Inst, York, PA 17402-9017	NA	$10,538	—	13	1,100	52
Yuba Coll, Marysville, CA 95901-7699	1927	$672 (S)	—	11	12,348	595

Some Notable Movies, Sept. 2000 – Aug. 2001

Movies	Stars	Director
A.I.	Haley Joel Osment, Frances O'Connor, William Hurt, Jude Law	Steven Spielberg
Almost Famous	Billy Crudup, Frances McDormand, Patrick Fugit, Kate Hudson	Cameron Crowe
Along Came a Spider	Morgan Freeman, Monica Potter	Lee Tamahori
American Pie 2	Jason Biggs, Mena Suvari, Chris Klein, Alyson Hannigan, Tara Reid	James B. Rogers
America's Sweethearts	John Cusack, Catherine Zeta-Jones, Julia Roberts, Billy Crystal	Joe Roth
Atlantis: The Lost Empire	Michael J. Fox, James Garner, Cree Summer, Don Novello	Gary Trousdale, Kirk Wise
Before Night Falls	Javier Bardem, Johnny Depp	Julian Schnabel
Billy Elliott	Julie Walters, Jamie Bell, Gary Lewis	Stephen Daldry
Bridget Jones's Diary	Renee Zellweger, Colin Firth, Hugh Grant, Gemma Jones, Jim Broadbent	Sharon Maguire
Cast Away	Tom Hanks, Helen Hunt	Robert Zemeckis
Cats & Dogs	Jeff Goldblum, Elizabeth Perkins, Tobey Maguire, Alec Baldwin, Susan Sarandon	Lawrence Guterman
Charlie's Angels	Cameron Diaz, Drew Barrymore, Lucy Liu, Bill Murray	Joseph McGinty Nichol
Chocolat	Juliette Binoche, Carrie-Ann Moss, Lena Olin, Johnny Depp, Judi Dench, Alfred Molina	Lasse Hallstrom
Contender, The	Joan Allen, Jeff Bridges, Gary Oldman, Christian Slater	Rod Lurie
Crouching Tiger, Hidden Dragon	Chow Yun-Fat, Michelle Yeoh, Ziyi Zhang, Pei-pei Cheng	Ang Lee
Dancer in the Dark	Björk, Catherine Deneuve, David Morse, Peter Stormare	Lars Von Trier
Dr. Dolittle 2	Eddie Murphy, Kristen Wilson, Jeffrey Jones, Kevin Pollak	Steve Carr
Dr. Seuss' How the Grinch Stole Christmas	Jim Carrey, Taylor Momsen, Christine Baranski, Moly Shannon	Ron Howard
Emperor's New Groove, The	David Spade, John Goodman, Eartha Kitt	Mark Dindal
Family Man, The	Nicolas Cage, Téa Leoni, Don Cheadle, Jeremy Piven	Brett Ratner
Fast And The Furious, The	Vin Diesel, Paul Walker, Michelle Rodriguez, Jordana Brewster	Rob Cohen
Hannibal	Julianne Moore, Anthony Hopkins, Gary Oldman, Ray Liotta, Giancarlo Giannini	Ridley Scott
Jurassic Park III	Téa Leoni, William H. Macy, Sam Neill	Joe Johnston
Lara Croft: Tomb Raider	Angelina Jolie, Jon Voight, Iain Glen, Daniel Craig	Simon West
Legally Blonde	Reese Witherspoon, Luke Wilson	Robert Luketic
Meet the Parents	Ben Stiller, Robert De Niro, Teri Polo, Blythe Danner, Owen Wilson	Jay Roach
Memento	Guy Pierce, Carrie-Ann Moss, Joe Pantoliano	Christopher Nolan
Miss Congeniality	Sandra Bullock, Michael Caine, Benjamin Bratt, William Shatner	Donald Petrie
Mummy Returns, The	Brendan Fraser, Rachel Weisz, Arnold Vosloo, John Hannah	Stephen Sommers
O Brother, Where Art Thou?	George Clooney, John Turturro, Tim Blake Nelson	Joel Coen
Others, The	Nicole Kidman, Elaine Cassidy, Christopher Eccleston	Alejandro Amenábar
Pearl Harbor	Ben Affleck, Kate Beckinsdale, Josh Hartnett, Jon Voight	Michael Bay
Planet of the Apes	Mark Wahlberg, Helena Bonham Carter, Tim Roth, Michael Clarke Duncan	Tim Burton
Pollock	Ed Harris, Marcia Gay Harden	Ed Harris
Princess Diaries, The	Julie Andrews, Anne Hathaway, Hector Elizondo, Heather Matarazzo	Garry Marshall
Quills	Geoffrey Rush, Kate Winslet, Joaquin Phoenix, Michael Caine	Philip Kaufman
Remember the Titans	Denzel Washington, Will Patton, Wood Harris, Ryan Hurst	Boaz Yakin
Requiem For A Dream	Jared Leto, Jennifer Connelly, Ellen Burstyn, Marlon Wayans	Darren Aronofsky
Rugrats in Paris: The Movie	Christine Cavanaugh, Elizabeth Daily, Cheryl Chase, Tara Strong, Kath Soucie, Cree Summer	Stig Bergqvist, Paul Demeyer
Rush Hour 2	Jackie Chan, Chris Tucker, Ziyi Zhang	Brett Ratner
Save the Last Dance	Julia Stiles, Sean Patrick Thomas, Kerry Washington, Fredro Starr	Thomas Carter
Scary Movie 2	Marlon Wayans, Shawn Wayans, Anna Faris	Keenen Ivory Wayans
Score, The	Robert De Niro, Edward Norton, Angela Bassett, Marlon Brando	Frank Oz
Shadow of the Vampire	John Malkovich, Willem Dafoe, Cary Elwes	E. Elias Merhige
Shrek	Mike Myers, Cameron Diaz, Eddie Murphy, John Lithgow	Andrew Adamson, Vicky Jenson
Spy Kids	Antonio Banderas, Carla Gugino, Daryl Sabara, Alexa Vega, Alan Cumming	Robert Rodriguez
Traffic	Catherine Zeta-Jones, Michael Douglas, Don Cheadle, Benicio Del Toro, Erika Christensen, Luis Guzman	Steven Soderbergh
Unbreakable	Bruce Willis, Samuel L. Jackson, Robin Wright, Spencer Treat Clark	M. Night Shyamalan
What Women Want	Mel Gibson, Helen Hunt, Marisa Tomei	Nancy Meyers
You Can Count On Me	Laura Linney, Mark Ruffalo, Matthew Broderick, Rory Culkin	Kenneth Lonergan

50 Top-Grossing Movies, 2000

Source: *Variety*, box-office grosses in the U.S. and Canada during calendar year 2000

Rank	Title	Gross (millions)
1.	Dr. Seuss' How the Grinch Stole Christmas	$253.4
2.	Mission: Impossible 2	215.4
3.	Gladiator	186.7
4.	The Perfect Storm	182,6
5.	Meet the Parents	161.3
6.	X-Men	157.3
7.	Scary Movie	157.0
8.	What Lies Beneath	155.4
9.	Dinosaur	137.7
10.	Erin Brokovich	125.6
11.	Nutty Professor II: The Klumps	123.3
12.	Charlie's Angels	122.8
13.	Big Momma's House	117.6
14.	What Women Want	115.8
15.	Remember the Titans	113.7
16.	The Patriot	113.3
17.	Cast Away	$109.7
18.	Chicken Run	106.8
19.	Gone in Sixty Seconds	101.6
20.	Me, Myself & Irene	90.6
21.	Space Cowboys	90.2
22.	Unbreakable	90.0
23.	Scream 3	89.1
24.	U-571	77.1
25.	Hollow Man	73.2
26.	Rugrats in Paris: The Movie	71.2
27.	Shaft	70.3
28.	Disney's The Kid	69.7
29.	Road Trip	68.5
30.	Bring It On	68.4
31.	Rules of Engagement	61.3
32.	The Cell	61.3
33.	Mission to Mars	60.9
34.	Coyote Ugly	$60.8
35.	Stuart Little	60.6
36.	The Green Mile	60.1
37.	Snow Day	60.0
38.	American Beauty	59.1
39.	102 Dalmatians	58.3
40.	Fantasia 2000	58.1
41.	The Whole Nine Yards	57.3
42.	Next Friday	57.2
43.	Shanghai Noon	56.9
44.	Romeo Must Die	56.0
45.	The Cider House Rules	53.3
46.	Final Destination	53.3
47.	Vertical Limit	51.9
48.	The Road to Eldorado	50.9
49.	The Emperor's New Groove	50.8
50.	The Hurricane	50.2

National Film Registry, 1989-2000

Source: National Film Registry, Library of Congress

"Culturally, historically, or esthetically significant" films placed on the registry. * = selected in 2000.

Adam's Rib (1949)
The Adventures of Robin Hood (1938)
The African Queen (1951)
All About Eve (1950)
All That Heaven Allows (1955)
All Quiet on the Western Front (1930)
An American in Paris (1951)
American Graffiti (1973)
A Movie (1958)
Annie Hall (1977)
The Apartment (1960)
Apocalypse Now (1979)*
A Star is Born (1954)*
A Streetcar Named Desire (1951)
The Awful Truth (1937)
Badlands (1973)
The Band Wagon (1953)
The Bank Dick (1940)
The Battle of San Pietro (1945)
Ben-Hur (1926)
The Best Years of Our Lives (1946)
Big Business (1929)
The Big Parade (1925)
The Big Sleep (1946)
The Birth of a Nation (1915)
The Black Pirate (1926)
Blacksmith Scene (1893)
Blade Runner (1982)
The Blood of Jesus (1941)
Bonnie and Clyde (1967)
Bride of Frankenstein (1935)
The Bridge on the River Kwai (1957)
Bringing Up Baby (1938)
Broken Blossoms (1919)
Cabaret (1972)
Carmen Jones (1954)
Casablanca (1942)
Castro Street (1966)
Cat People (1942)
Chan Is Missing (1982)
The Cheat (1915)
Chinatown (1974)
Chulas Fronteras (1976)
Citizen Kane (1941)
The City (1939)
City Lights (1931)
Civilization (1916)
The Conversation (1974)
The Cool World (1963)
Cops (1922)
A Corner in Wheat (1909)
The Crowd (1928)
Czechoslovakia 1968 (1968)
David Holzman's Diary (1968)
The Day the Earth Stood Still (1951)
Dead Birds (1964)
The Deer Hunter (1978)
Destry Rides Again (1939)
Detour (1946)
Dodsworth (1936)
The Docks of New York (1928)
Dog Star Man (1964)
Don't Look Back (1967)
Do the Right Thing (1989)
Double Indemnity (1944)
Dracula (1931)*
Dr. Strangelove (or, How I Learned to Stop Worrying and Love the Bomb) (1964)
Duck Amuck (1953)
Duck Soup (1933)
Easy Rider (1969)
Eaux D'Artifice (1953)
El Norte (1983)
The Emperor Jones (1933)
E.T.: The Extra-Terrestrial (1982)
The Exploits of Elaine (1914)
The Fall of the House of Usher (1928)*
Fantasia (1940)
Fatty's Tintype Tangle (1915)
Five Easy Pieces (1970)*
Flash Gordon serial (1936)
Footlight Parade (1933)
Force of Evil (1948)
The Forgotten Frontier (1931)
42nd Street (1933)
The Four Horsemen of the Apocalypse (1921)
Frankenstein (1931)
Frank Film (1973)
Freaks (1932)
The Freshman (1925)
From the Manger to the Cross (1912)
Fury (1936)
The General (1927)
Gerald McBoing Boing (1951)
Gertie the Dinosaur (1914)
Gigi (1958)
The Godfather (1972)
The Godfather, Part II (1974)
The Gold Rush (1925)
Gone With the Wind (1939)
GoodFellas (1990)*
The Graduate (1967)
The Grapes of Wrath (1940)
Grass (1925)
The Great Dictator (1940)
The Great Train Robbery (1903)
Greed (1924)
Gun Crazy (1949)
Gunga Din (1939)
Harlan County, U.S.A. (1976)
Harold and Maude (1972)
The Heiress (1949)
Hell's Hinges (1916)
High Noon (1952)
High School (1968)
Hindenburg Disaster Newsreel Footage (1937)
His Girl Friday (1940)
The Hitch-Hiker (1953)
Hospital (1970)
The Hospital (1971)
How Green Was My Valley (1941)
How the West Was Won (1962)
The Hustler (1961)
I Am a Fugitive From a Chain Gang (1932)
The Immigrant (1917)
In the Land of the Head-Hunters aka In the Land of the War Canoes (1914)
Intolerance (1916)
Invasion of the Body Snatchers (1956)
It Happened One Night (1934)
It's a Wonderful Life (1946)
The Italian (1915)
Jammin' the Blues (1944)
Jazz on a Summer's Day (1959)
The Jazz Singer (1927)
Killer of Sheep (1977)
King: A Filmed Record . . . Montgomery to Memphis (1970)
King Kong (1933)
The Kiss (1896)
Kiss Me Deadly (1955)
Knute Rockne, All American (1940)
Koyaanisqatsi (1983)*
The Lady Eve (1941)
Lambchops (1929)
The Land Beyond the Sunset (1912)*
Lassie Come Home (1943)
The Last of the Mohicans (1920)
The Last Picture Show (1972)
Laura (1944)
Lawrence of Arabia (1962)
The Learning Tree (1969)
Let's All Go to the Lobby (1957)*
Letter From an Unknown Woman (1948)
The Life and Death of 9413—A Hollywood Extra (1928)
Life and Times of Rosie the Riveter (1980)
The Life of Emile Zola (1937)*
Little Caesar (1930)*
The Little Fugitive (1953)
Little Miss Marker (1934)
The Living Desert (1953)*
The Lost World (1925)
Louisiana Story (1948)
Love Finds Andy Hardy (1938)
Love Me Tonight (1932)
Magical Maestro (1952)
The Magnificent Ambersons (1942)
The Maltese Falcon (1941)
The Manchurian Candidate (1962)
Manhattan (1921)
March of Time: Inside Nazi Germany—1938 (1938)
Marty (1955)
M*A*S*H (1970)
Master Hands (1936)
Mean Streets (1973)
Meet Me in St. Louis (1944)
Meshes of the Afternoon (1943)
Midnight Cowboy (1969)
Mildred Pierce (1945)
Modern Times (1936)
Modesta (1956)
Morocco (1930)
Motion Painting No. 1 (1947)
Mr. Smith Goes to Washington (1939)
Multiple Sidosis (1970)*
The Music Box (1932)
My Darling Clementine (1946)
My Man Godfrey (1936)
The Naked Spur (1953)
Nanook of the North (1922)
Nashville (1975)
Network (1976)*
A Night at the Opera (1935)
The Night of the Hunter (1955)
Night of the Living Dead (1968)
Ninotchka (1939)
North by Northwest (1959)
Nothing but a Man (1964)
One Flew Over the Cuckoo's Nest (1975)
On the Waterfront (1954)
The Outlaw Josey Wales (1976)
Out of the Past (1947)
The Ox-Bow Incident (1943)
Pass the Gravy (1928)
Paths of Glory (1957)
Peter Pan (1924)*
Phantom of the Opera (1925)
The Philadelphia Story (1940)
Pinocchio (1940)
A Place in the Sun (1951)
The Plow That Broke the Plains (1936)
Point of Order (1964)
The Poor Little Rich Girl (1917)
Porky in Wackyland (1938)*
Powers of Ten (1978)
President McKinley Inauguration Footage (1901)*
Primary (1960)
The Prisoner of Zenda (1937)
The Producers (1968)
Psycho (1960)
The Public Enemy (1931)
Pull My Daisy (1959)
Raging Bull (1980)
Raiders of the Lost Ark (1981)
Rear Window (1954)
Rebel Without a Cause (1955)
Red River (1948)
Regeneration (1915)*
Republic Steel Strike Riots Newsreel Footage (1937)
Return of the Secaucus 7 (1980)
Ride the High Country (1962)
Rip Van Winkle (1896)
The River (1937)
Road to Morocco (1942)
Roman Holiday (1953)
Safety Last (1923)
Salesman (1969)
Salomé (1922)*
Salt of the Earth (1954)
Scarface (1932)
The Searchers (1956)
Seventh Heaven (1927)
Shadow of a Doubt (1943)
Shadows (1959)
Shaft (1971)*
Shane (1953)
She Done Him Wrong (1933)
Sherlock, Jr. (1924)
Sherman's March (1986)*
Shock Corridor (1963)
The Shop Around the Corner (1940)
Show Boat (1936)
Singin' in the Rain (1952)
Sky High (1922)
Snow White (1933)
Snow White and the Seven Dwarfs (1937)
Some Like It Hot (1959)
Stagecoach (1939)
Star Wars (1977)
Steamboat Willie (1928)
Sullivan's Travels (1941)
Sunrise (1927)
Sunset Boulevard (1950)
Sweet Smell of Success (1957)
Tabu (1933)
Tacoma Narrows Bridge Collapse (1940)
The Tall T (1957)*
Taxi Driver (1976)
The Ten Commandments (1956)
Tevye (1939)
The Thief of Bagdad (1924)
The Thin Man (1934)
To Be or Not To Be (1942)
To Fly (1976)
To Kill a Mockingbird (1962)
Tootsie (1982)
Topaz (1943-45)
Top Hat (1935)
Touch of Evil (1958)
Trance and Dance in Bali (1939)
The Treasure of the Sierra Madre (1948)
Trouble in Paradise (1932)
Tulips Shall Grow (1942)
Twelve O'Clock High (1949)
2001: A Space Odyssey (1968)
Verbena Tragica (1939)
Vertigo (1958)

Westinghouse Works 1904 (1904)
West Side Story (1961)
What's Opera, Doc? (1957)
Where Are My Children? (1916)
Why We Fight (Series/1943-45)*
The Wild Bunch (1969)
Will Success Spoil Rock Hunter? (1957)*
The Wind (1928)
Wings (1927)
Within Our Gates (1920)
The Wizard of Oz (1939)
Woman of the Year (1942)
A Woman Under the Influence (1974)
Woodstock (1970)
Yankee Doodle Dandy (1942)
Zapruder Film (1963)

100 Best American Movies of All Time

Source: American Film Institute

Compiled in 1998 based on ballots sent to 1,500 figures, mostly from the film world. Criteria for judging included historical significance, critical recognition and awards, and popularity. The year each film was first released is in parentheses.

1. Citizen Kane (1941)
2. Casablanca (1942)
3. The Godfather (1972)
4. Gone With the Wind (1939)
5. Lawrence of Arabia (1962)
6. The Wizard of Oz (1939)
7. The Graduate (1967)
8. On the Waterfront (1954)
9. Schindler's List (1993)
10. Singin' in the Rain (1952)
11. It's a Wonderful Life (1946)
12. Sunset Boulevard (1950)
13. The Bridge on the River Kwai (1957)
14. Some Like It Hot (1959)
15. Star Wars (1977)
16. All About Eve (1950)
17. The African Queen (1951)
18. Psycho (1960)
19. Chinatown (1974)
20. One Flew Over the Cuckoo's Nest (1975)
21. The Grapes of Wrath (1940)
22. 2001: A Space Odyssey (1968)
23. The Maltese Falcon (1941)
24. Raging Bull (1980)
25. E.T.: The Extra-Terrestrial (1982)
26. Dr. Strangelove (1964)
27. Bonnie and Clyde (1967)
28. Apocalypse Now (1979)
29. Mr. Smith Goes to Washington (1939)
30. Treasure of the Sierra Madre (1948)
31. Annie Hall (1977)
32. The Godfather, Part II (1974)
33. High Noon (1952)
34. To Kill a Mockingbird (1962)
35. It Happened One Night (1934)
36. Midnight Cowboy (1969)
37. The Best Years of Our Lives (1946)
38. Double Indemnity (1944)
39. Doctor Zhivago (1965)
40. North by Northwest (1959)
41. West Side Story (1961)
42. Rear Window (1954)
43. King Kong (1933)
44. The Birth of a Nation (1915)
45. A Streetcar Named Desire (1951)
46. A Clockwork Orange (1971)
47. Taxi Driver (1976)
48. Jaws (1975)
49. Snow White and the Seven Dwarfs (1937)
50. Butch Cassidy and the Sundance Kid (1969)
51. The Philadelphia Story (1940)
52. From Here to Eternity (1953)
53. Amadeus (1984)
54. All Quiet on the Western Front (1930)
55. The Sound of Music (1965)
56. M*A*S*H (1970)
57. The Third Man (1949)
58. Fantasia (1940)
59. Rebel Without a Cause (1955)
60. Raiders of the Lost Ark (1981)
61. Vertigo (1958)
62. Tootsie (1982)
63. Stagecoach (1939)
64. Close Encounters of the Third Kind (1977)
65. The Silence of the Lambs (1991)
66. Network (1976)
67. The Manchurian Candidate (1962)
68. An American in Paris (1951)
69. Shane (1953)
70. The French Connection (1971)
71. Forrest Gump (1994)
72. Ben-Hur (1959)
73. Wuthering Heights (1939)
74. The Gold Rush (1925)
75. Dances With Wolves (1990)
76. City Lights (1931)
77. American Graffiti (1973)
78. Rocky (1976)
79. The Deer Hunter (1978)
80. The Wild Bunch (1969)
81. Modern Times (1936)
82. Giant (1956)
83. Platoon (1986)
84. Fargo (1996)
85. Duck Soup (1933)
86. Mutiny on the Bounty (1935)
87. Frankenstein (1931)
88. Easy Rider (1969)
89. Patton (1970)
90. The Jazz Singer (1927)
91. My Fair Lady (1964)
92. A Place in the Sun (1951)
93. The Apartment (1960)
94. Goodfellas (1990)
95. Pulp Fiction (1994)
96. The Searchers (1956)
97. Bringing Up Baby (1938)
98. Unforgiven (1992)
99. Guess Who's Coming to Dinner (1967)
100. Yankee Doodle Dandy (1942)

100 Most Thrilling American Movies

Source: American Film Institute

Compiled in 2001 based on ballots sent to 1,500 leading figures of the American film community. Choices based on the "total adrenaline-inducing impact of a film's artistry and craft" regardless of genre. The year each film was first released is in parentheses.

1. Psycho (1960)
2. Jaws (1975)
3. The Exorcist (1973)
4. North by Northwest (1959)
5. The Silence of the Lambs (1991)
6. Alien (1979)
7. The Birds (1963)
8. The French Connection (1971)
9. Rosemary's Baby (1968)
10. Raiders of the Lost Ark (1981)
11. The Godfather (1972)
12. King Kong (1933)
13. Bonnie and Clyde (1967)
14. Rear Window (1954)
15. Deliverance (1972)
16. Chinatown (1974)
17. The Manchurian Candidate (1962)
18. Vertigo (1958)
19. The Great Escape (1963)
20. High Noon (1952)
21. A Clockwork Orange (1971)
22. Taxi Driver (1976)
23. Lawrence of Arabia (1962)
24. Double Indemnity (1944)
25. Titanic (1997)
26. The Maltese Falcon (1941)
27. Star Wars (1977)
28. Fatal Attraction (1987)
29. The Shining (1980)
30. The Deer Hunter (1978)
31. Close Encounters of the Third Kind (1977)
32. Strangers on a Train (1951)
33. The Fugitive (1993)
34. The Night of the Hunter (1955)
35. Jurassic Park (1993)
36. Bullitt (1968)
37. Casablanca (1942)
38. Notorious (1946)
39. Die Hard (1988)
40. 2001: A Space Odyssey (1968)
41. Dirty Harry (1971)
42. The Terminator (1984)
43. The Wizard of Oz (1939)
44. E.T. : The Extra-Terrestrial (1982)
45. Saving Private Ryan (1998)
46. Carrie (1976)
47. Invasion of the Body Snatchers (1956)
48. Dial M For Murder (1954)
49. Ben-Hur (1959)
50. Marathon Man (1976)
51. Raging Bull (1980)
52. Rocky (1976)
53. Pulp Fiction (1994)
54. Butch Cassidy and the Sundance Kid (1969)
55. Wait Until Dark (1967)
56. Frankenstein (1931)
57. All the President's Men (1976)
58. The Bridge on the River Kwai (1957)
59. Planet of the Apes (1968)
60. The Sixth Sense (1999)
61. Cape Fear (1962)
62. Spartacus (1960)
63. What Ever Happened to Baby Jane? (1962)
64. Touch of Evil (1958)
65. The Dirty Dozen (1967)
66. The Matrix (1999)
67. The Treasure of the Sierra Madre (1948)
68. Halloween (1978)
69. The Wild Bunch (1969)
70. Dog Day Afternoon (1975)
71. Goldfinger (1964)
72. Platoon (1986)
73. Laura (1944)
74. Blade Runner (1982)
75. The Third Man (1949)
76. Thelma & Louise (1991)
77. Terminator 2: Judgment Day (1991)
78. Gaslight (1944)
79. The Magnificent Seven (1960)
80. Rebecca (1940)
81. The Omen (1976)
82. The Day the Earth Stood Still (1951)
83. The Phantom of the Opera (1925)
84. Poltergeist (1982)
85. Dracula (1931)
86. The Picture of Dorian Gray (1945)
87. The Thing from Another World (1951)
88. 12 Angry Men (1957)
89. The Guns of Navarone (1961)
90. The Poseidon Adventure (1972)
91. Braveheart (1995)
92. Body Heat (1981)
93. Night of the Living Dead (1968)
94. The China Syndrome (1979)
95. Full Metal Jacket (1987)
96. Blue Velvet (1986)
97. Safety Last (1923)
98. Blood Simple (1984)
99. Speed (1994)
100. The Adventures of Robin Hood (1938)

All-Time Top-Grossing American Movies Through 2000

Source: *Variety* magazine

Rank	Title/Date	Gross[1]
1.	Titanic (1997)	$600.8
2.	Star Wars: Episode IV—A New Hope (1977)	461.0
3.	Star Wars: Episode I—The Phantom Menace (1999)	431.0
4.	E.T.: The Extra-Terrestrial (1982)	399.8
5.	Jurassic Park (1993)	357.1
6.	Forrest Gump (1994)	329.7
7.	The Lion King (1994)	312.9
8.	Return of the Jedi (1983)	309.2
9.	Independence Day (1996)	306.2
10.	The Sixth Sense (1999)	293.5
11.	The Empire Strikes Back (1980)	290.3
12.	Home Alone (1990)	285.8
13.	Dr. Seuss' The Grinch Who Stole Christmas (2000)	260.0
14.	Jaws (1975)	260.0
15.	Batman (1989)	251.2
16.	Men in Black (1997)	250.0
17.	Toy Story 2 (1999)	$245.9
18.	Raiders of the Lost Ark (1981)	242.4
19.	Twister (1996)	241.7
20.	Ghostbusters (1984)	238.6
21.	Beverly Hills Cop (1984)	234.8
22.	Cast Away (2000)	229.2
23.	The Lost World: Jurassic Park (1997)	229.1
24.	Mrs. Doubtfire (1993)	219.2
25.	Ghost (1990)	217.6
26.	Aladdin (1992)	217.4
27.	Saving Private Ryan (1998)	216.3
28.	Mission: Impossible 2 (2000)	215.4
29.	Back to the Future (1985)	208.2
30.	Austin Powers: The Spy Who Shagged Me (1999)	206.0
31.	Terminator 2 (1991)	204.8
32.	The Exorcist (1973)	204.7
33.	Armageddon (1998)	201.6
34.	Gone With the Wind (1939)	$198.6
35.	Indiana Jones and the Last Crusade (1989)	197.2
36.	Toy Story (1995)	191.8
37.	Gladiator (2000)	186.9
38.	Dances With Wolves (1990)	184.2
39.	Batman Forever (1995)	184.0
40.	The Fugitive (1993)	183.9
41.	The Perfect Storm (2000)	182.6
42.	Grease (1978)	181.5
43.	Liar, Liar (1997)	181.4
44.	Mission: Impossible (1996)	181.0
45.	What Women Want (2000)	180.9
46.	Indiana Jones and the Temple of Doom (1984)	179.9
47.	Pretty Woman (1990)	178.4
48.	Tootsie (1982)	177.2
49.	Top Gun (1986)	176.8
50.	There's Something About Mary (1998)	176.5

(1) Gross is in millions of absolute dollars based on box office sales in the U.S. and Canada. Rising ticket prices favor newer films, but older films have the advantage of reissues.

WORLD ALMANAC EDITORS' PICKS

The World Almanac staff ranked the following as the funniest (American) films they had ever seen:

1. Airplane!
2. Some Like It Hot
3. Annie Hall
4. Young Frankenstein
5. Raising Arizona
6. Trading Places
7. Dirty Rotten Scoundrels
8. Tootsie
9. National Lampoon's Animal House
10. When Harry Met Sally . . .

Most Popular Movie Videos, 2000

Source: Alexander & Associates/Video Flash, New York, NY

Top 10 Rentals, 2000

1. The Sixth Sense
2. The Green Mile
3. The Matrix
4. Double Jeopardy
5. American Pie
6. Runaway Bride
7. American Beauty
8. Sleepy Hollow
9. The Bone Collector
10. The General's Daughter

All-Time Top 10 Rentals[1]

1. Pretty Woman
2. Top Gun
3. The Little Mermaid
5. Home Alone
4. Ghost
6. The Lion King
7. Beauty and the Beast
8. Terminator 2: Judgment Day
9. Forrest Gump
10. Aladdin

Top 10 Sales, 2000

1. Tarzan (Disney 1999)
2. Toy Story 2
3. Stuart Little
4. Star Wars Episode One (stand. ed.)
5. The Matrix
6. The Little Mermaid II
7. The Sixth Sense
8. Pokemon: The First Movie
9. Big Daddy
10. Erin Brockovich

All-Time Top 10 Sales[2]

1. The Lion King
2. Aladdin
3. Beauty and the Beast
4. Snow White and the Seven Dwarfs
5. Forrest Gump
6. Toy Story
7. 101 Dalmatians (animated)
8. The Little Mermaid
9. Jurassic Park
10. Pocohontas

(1) Rented Mar. 1, 1987-Dec. 29, 2000. (2) Sold Feb. 16, 1988-Dec. 28, 2000.

Top 50 Record Long-Run Broadway Plays[1]

Source: The League of American Theatres and Producers, Inc., New York, NY

Title	Performances
Cats	7,485
A Chorus Line	6,137
*Les Miserables	5,975
Oh! Calcutta! (revival)	5,959
*The Phantom of the Opera	5,678
Miss Saigon	4,092
42nd Street	3,486
Grease (original)	3,388
Fiddler on the Roof	3,242
Life With Father	3,224
Tobacco Road	3,182
*Beauty and the Beast	3,004
Hello Dolly	2,844
My Fair Lady	2,717
Annie	2,377
Man of La Mancha	2,328
Abie's Irish Rose	2,327
*Rent	2,232
Oklahoma!	2,212
Smokey Joe's Cafe	2,037
*Chicago (revival)	2,003
Pippin	1,944
South Pacific	1,925
The Magic Show	1,920
Gemini	1,819
Deathtrap	1,793
Harvey	1,775
Dancin'	1,774
La Cage aux Folles	1,761
Hair	1,750
The Wiz	1,672
Born Yesterday	1,642
Crazy for You	1,622
Ain't Misbehavin'	1,604
*The Lion King	1,593
The Best Little Whorehouse in Texas	1,584
Mary, Mary	1,572
Evita	1,567
The Voice of theTurtle	1,557
Jekyll & Hyde	1,543
Barefoot in the Park	1,530
Dreamgirls	1,521
Mame	1,508
Grease (revival)	1,505
Same Time, Next Year	1,453
Arsenic and Old Lace	1,444
The Sound of Music (orig.)	1,443
How to Succeed in Business Without Really Trying (orig.)	1,417
Me and My Girl	1,417
Hellzapoppin	1,404

* Still running Sept. 2, 2001. (1) Number of performances through Sept. 2, 2001.

Broadway Season Statistics, 1959-2001

Source: The League of American Theatres and Producers, Inc., New York, NY

Season	Gross (mil $)	Attendance (mil)	Playing Weeks	New Productions
1959-1960	46	7.9	1,156	58
1960-1961	44	7.7	1,210	48
1961-1962	44	6.8	1,166	53
1962-1963	44	7.4	1,134	54
1963-1964	40	6.8	1,107	63
1964-1965	50	8.2	1,250	67
1965-1966	54	9.6	1,295	68
1966-1967	55	9.3	1,269	69
1967-1968	59	9.5	1,259	74
1968-1969	58	8.6	1,209	67
1969-1970	53	7.1	1,047	62
1970-1971	55	7.4	1,107	49
1971-1972	52	6.5	1,157	55
1972-1973	45	5.4	889	55
1973-1974	46	5.7	907	43
1974-1975	57	6.6	1,101	54

Season	Gross (mil $)	Attendance (mil)	Playing Weeks	New Productions
1975-1976	71	7.3	1,136	55
1976-1977	93	8.8	1,349	54
1977-1978	114	9.6	1,433	42
1978-1979	134	9.6	1,542	50
1979-1980	146	9.6	1,540	61
1980-1981	197	11.0	1,544	60
1981-1982	223	10.1	1,455	48
1982-1983	209	8.4	1,258	50
1983-1984	227	7.9	1,097	36
1984-1985	209	7.3	1,078	33
1985-1986	190	6.5	1,041	34
1986-1987	208	7.1	1,039	41
1987-1988	253	8.1	1,113	30
1988-1989	262	8.1	1,108	33
1989-1990	282	8.0	1,070	40
1990-1991	267	7.3	971	28
1991-1992	293	7.4	905	37
1992-1993	328	7.9	1,019	34
1993-1994	356	8.1	1,066	39
1994-1995	406	9.0	1,120	33
1995-1996	436	9.5	1,146	38
1996-1997	499	10.6	1,349	37
1997-1998	558	11.5	1,442	33
1998-1999	588	11.7	1,441	39
1999-2000	603	11.4	1,464	37
2000-2001	666	11.9	1,484	28

Some Notable Non-Profit Theater Companies in the U.S

Source: Theatre Communications Group, Inc.

Theater Company	City	State
A Contemporary Theatre	Seattle	WA
Actors Theatre of Louisville	Louisville	KY
Alabama Shakespeare Festival	Montgomery	AL
Alley Theatre	Houston	TX
Alliance Theatre Company	Atlanta	GA
American Conservatory Theatre	San Francisco	CA
American Repertory Theatre	Cambridge	MA
Arena Stage	Washington	DC
Arizona Theatre Company	Tucson	AZ
Berkeley Repertory Theatre	Berkeley	CA
Center Stage	Baltimore	MD
Chicago Shakespeare Theater	Chicago	IL
Children's Theatre Company, The	Minneapolis	MN
Cincinnati Playhouse in the Park	Cincinnati	OH
Cleveland Play House, The	Cleveland	OH
Denver Center Theatre Company	Denver	CO
Ford's Theatre	Washington	DC
Geffen Playhouse	Los Angeles	CA
The Globe Theater	San Diego	CA
Goodman Theatre	Chicago	IL
Goodspeed Musicals	East Haddam	CT
Guthrie Theater, The	Minneapolis	MN
Hartford Stage Company	Hartford	CT
Huntington Theatre Company	Boston	MA
Joseph Papp Public Theater	New York	NY
La Jolla Playhouse	La Jolla	CA
Lincoln Center Theater	New York	NY
Long Wharf Theatre	New Haven	CT
Manhattan Theatre Club	New York	NY
Mark Taper Forum	Los Angeles	CA
McCarter Theatre	Princeton	NJ
Milwaukee Repertory Theater	Milwaukee	WI
Oregon Shakespeare Festival	Ashland	OR
Pasadena Playhouse	Pasadena	CA
Pittsburgh Public Theater	Pittsburgh	PA
Repertory Theater of St. Louis	St. Louis	MO
Roundabout Theater Company	New York	NY
Seattle Children's Theater	Seattle	WA
Seattle Repertory Theatre	Seattle	WA
Shakespeare Theatre, The	Washington	DC
South Coast Repertory	Costa Mesa	CA
Steppenwolf Theatre Company	Chicago	IL
Trinity Repertory Company	Providence	RI
Village Theater	Issaquah	WA
Walnut Street Theater	Philadelphia	PA

U.S. Symphony Orchestras[1]

Source: American Symphony Orchestra League, 33 West 60th St., New York, NY 10023; data as of Nov. 1, 2000

Symphony Orchestra[2]	Music Director[3]
Alabama Symphony (AL)	Richard Westerfield
American (NY)	Leon Botstein
Atlanta (GA)	Robert Spano
Austin (TX)	Peter Bay
Baltimore (MD)	Yuri Temirkanov
Boston (MA)	Seiji Ozawa
Boulder Philharmonic (CO)	Theodore Kuchar
Brooklyn Philharmonic (NY)	Robert Spano
Buffalo Philharmonic (NY)	JoAnne Falletta
Charlotte (NC)	Peter McCoppin
Chicago (IL)	Daniel Barenboim
Chicago Sinfonietta (IL)	Paul Freeman
Cincinnati (OH)	Jesus Lopez-Cobos
Cleveland Orchestra (OH)	Christoph von Dohnanyi
Colorado (CO)	Marin Alsop
Colorado Springs (CO)	Lawrence Leighton Smith
Columbus (OH)	Alessandro Siciliani
Dallas (TX)	Andrew Litton
Dayton Philharmonic (OH)	Neal Gittleman
Detroit (MI)	Neeme Jarvi
EOS Orchestra (NY)	Jonathan Sheffer
Florida Orchestra (Tampa)	Jahja Ling
Florida Philharmonic (Ft. Lauderdale)	James Judd
Florida Symphonic Pops Boca Raton)	Crafton Beck
Florida West Coast (FL)	Leif Bjaland
Fort Wayne Philharmonic (IN)	Edvard Tchivzhel
Fort Worth (TX)	Miguel Harth-Bedoya
Grand Rapids (MI)	David Lockington
Grant Park (Chicago, IL)	Carlos Kalmar
Hartford (CT)	Joseph Silverstein
Honolulu (HI)	Samuel Wong
Houston (TX)	Hans Graf
Indianapolis (IN)	Raymond Leppard
Jacksonville (FL)	Fabio Mechetti
Kansas City (MO)	Anne Manson
Knoxville (TN)	Kirk Trevor
Long Beach (CA)	JoAnn Falletta
Los Angeles Chamber (CA)	Jeffrey Kahane
Los Angeles Philharmonic (CA)	Esa-Pekka Salonen
Louisiana Philharmonic (New Orleans)	Klauspeter Seibel
Louisville Orchestra (KY)	Uriel Segal
Memphis (TN)	David Loebel
Milwaukee (WI)	Andreas Delfs
Minnesota (Minneapolis)	Eiji Oue
Naples Philharmonic (FL)	Christopher Seaman
Nashville Symphony (TN)	Kenneth D. Schermerhorn
National (Washington, DC)	Leonard Slatkin
New Haven (CT)	Jung-Ho Pak
New Jersey (Newark)	Zdenek Macal
New Mexico (Albuquerque)	David Lockington
New York Philharmonic (NYC)	Kurt Masur
North Carolina Symphony (Raleigh)	Gerhardt Zimmermann
Oklahoma City Philharmonic (OK)	Joel A. Levine
Omaha Symphony (NE)	Victor Yampolsky
Oregon Symphony (Portland)	James DePreist
Pacific Symphony (Santa Ana, CA)	Carl St. Clair
Philadelphia (PA)	Wolfgang Sawallisch
Philharmonia Baroque (San Francisco, CA)	Nicholas McGegan
Phoenix Symphony (AZ)	Hermann Michael
Pittsburgh (PA)	Mariss Jansons
Portland (ME)	Toshiyuki Shimada
Rhode Island Philharmonic (RI)	Larry Rachleff
Richmond Symphony (VA)	Mark Russell Smith
Rochester Philharmonic Orch. (NY)	Christopher Seaman
St. Louis (MO)	Hans Vonk
St. Paul Chamber Orchestra (MN)	Andreas Delfs
San Antonio (TX)	Christopher Wilkins
San Francisco (CA)	Michael Tilson Thomas
San Jose (CA)	Leonid Grin
Savannah (GA)	Philip B. Greenberg
Seattle (WA)	Gerard Schwarz
Spokane (WA)	Fabio Mechetti
Syracuse (NY)	Daniel Hege
Toledo (OH)	Andrew Massey
Tucson (AZ)	George Hanson
Tulsa Philharmonic (OK)	Kenneth Jean
Utah (Salt Lake City)	Keith Lockhart
Virginia Symphony (VA)	JoAnn Falletta
West Virginia (Charleston)	Thomas B. Conlin

(1) Includes only orchestras with annual expenses $2 mil or greater. (2) If only place name is given, add Symphony Orchestra. (3) General title; listed is highest-ranking member of conducting personnel.

U.S. Opera Companies With Budgets of $1 Million or More

Source: OPERA America, 1156 15th Street NW, Washington, DC 20005-1704; July 2001

Academy of Vocal Arts Opera Theatre (Philadelphia, PA); K. James McDowell, dir.
Arizona Opera (Tucson); David Speers, gen. dir.
Aspen Opera Theater Center (CO); Robert Harth, pres./ceo
Atlanta Opera (GA); Alfred Kennedy, exec. dir.
Austin Lyric Opera (TX); Joseph McClain, gen. dir.
Baltimore Opera Company (MD); Michael Harrison, gen. dir.
Boston Lyric Opera Company (MA); Janice Mancini Del Sesto, gen. dir.
Brooklyn Academy of Music (NY); Karen Brooks Hopkins, pres.
Central City Opera (Denver, CO); Pelham Pearce, gen. dir.
Cincinnati Opera (OH); Patricia Beggs, mng. dir.
Cleveland Opera (OH); David Bamberger, gen. dir.
Connecticut Opera (Hartford); Willie Anthony Waters, gen./art. dir.
Dallas Opera (TX); Anthony Whitworth-Jones, gen. dir.
Dayton Opera (OH); Mark Light, pres.
Des Moines Metro Opera, Inc. (IA); Jerilee Mace, exec. dir.
Florentine Opera Company (Milwaukee, WI); Dennis Hanthorn, gen. dir.
Florida Grand Opera (Miami, FL); Robert Heuer, gen. dir./ceo
Fort Worth Opera (TX); Keith Wolfe, man. dir.
Glimmerglass Opera (Cooperstown, NY); Esther Nelson, gen. dir.
Goodspeed Musicals (East Haddam, CT); Michael Price, exec. dir.
Hawaii Opera Theatre (Honolulu); Henry Akina, gen./art. dir.
Houston Grand Opera (TX); David Gockley, gen. dir.
Indianapolis Opera (IN); John C. Pickett, exec. dir.
Kentucky Opera (Louisville); Deborah S. Sandler, gen. dir.
Knoxville Opera Company (TN); Francis Graffeo, gen. dir.
Los Angeles Opera (CA); Plácido Domingo, art. dir.
Lyric Opera of Chicago (IL); William Mason, gen. dir.
Lyric Opera of Kansas City (MO); Evan R. Luskin, gen. dir.
Metro Lyric Opera (Allenhurst, NJ); Era M. Tognoli, gen./art. dir.
Metropolitan Opera (New York, NY); Joseph Volpe, gen. mgr.
Michigan Opera Theatre (Detroit); David DiChiera, gen. dir.
Minnesota Opera (Minneapolis); Kevin Smith, pres./ceo
Nashville Opera Association (TN); Carol Penterman, ceo/exec. dir.
New Jersey State Opera (Newark); Alfredo Silipigni, art. dir.
New Orleans Opera Association (LA); Robert Lyall, gen. dir.
New York City Opera (NY); Paul Kellogg, gen. dir.
Opera Carolina (Charlotte, NC); James Meena, gen. dir.
Opera Colorado (Denver); Peter Russell, pres./gen. dir.
Opera/Columbus (OH); William F. Russell, gen. dir.
Opera Company of Philadelphia (PA); Robert B. Driver, gen. dir.
OperaDelaware (Wilmington); Leland P. Kimball III, gen. dir.
Opera Festival of New Jersey (Princeton); Karen Tiller, gen. dir.
Opera Memphis (TN); Michael Ching, gen./art. dir.
Opera Omaha (NE); Jane Hill, exec. dir.
Opera Orchestra of New York (NY); Eve Queler, mus. dir.
Opera Pacific (Irvine, CA); Martin G. Hubbard, exec. dir.
Opera Theatre of Saint Louis (MO); Charles MacKay, gen. dir.
Opera San José (CA); Irene Dalis, gen. dir.
Orlando Opera (FL); Robert Swedberg, gen. dir.
Palm Beach Opera (FL); Herbert P. Benn, gen. dir.
Pittsburgh Opera (PA); Mark Weinstein, gen. dir.
Portland Opera (OR); Robert Bailey, gen. dir.
San Diego Civic Light Opera Association (CA); Brian Wells, prod'g. art. dir.
San Diego Opera (CA); Ian D. Campbell, gen. dir.
San Francisco Opera (CA); Pamela Rosenberg, gen. dir.
Santa Barbara Civic Light Opera (CA); Paul Iannaccone, exec. prod.
Santa Fe Opera (NM); Richard Gaddes, gen. dir.
Sarasota Opera (FL); Susan T. Danis, exec. dir.
Seattle Opera (WA); Speight Jenkins, gen. dir.
Skylight Opera Theatre (Milwaukee, WI); Christopher Libby, mng. dir.
Tulsa Opera (OK); Carol I. Crawford, gen. dir.
Utah Festival Opera Company (Logan); Michael Ballam, gen. dir.
Utah Opera (Salt Lake City); Anne Ewers, gen. dir.
Virginia Opera (Norfolk); Peter Mark, gen./art. dir.
Washington Opera (DC); Walter Arnheim, exec. dir.
West Virginia Symphony Orchestra (Charleston); Paul A. Helfrich, exec. dir.
Wolf Trap Opera Company (Vienna, VA); Kim Pensinger Witman, gen. dir.

Some Notable U.S. Dance Companies

Source: DanceUSA

Organization	City	State
Alabama Ballet	Birmingham	AL
Alvin Ailey American Dance Theater	New York	NY
American Ballet Theatre	New York	NY
American Repertory Ballet Company	New Brunswick	NJ
American Repertory Dance Company	Los Angeles	CA
Aspen Santa Fe Ballet	Aspen	CO
The Atlanta Ballet, Inc.	Atlanta	GA
Ballet Austin	Austin	TX
Ballet Concierto de Puerto Rico	Santurce	PR
Ballet Florida	W. Palm Beach	FL
Ballet Hispanico of New York	New York	NY
Ballet Internationale, Inc.	Indianapolis	IN
Ballet Memphis	Cordova	TN
Ballet San Jose Silicon Valley	San Jose	CA
Ballet West	Salt Lake City	UT
BalletMet Columbus	Columbus	OH
Betty Salamun's DANCECIRCUS	Milwaukee	WI
Bill T. Jones/Arnie Zane Dance Company	New York	NY
Boston Ballet	Boston	MA
Caribbean Dance Company of the Virgin Islands	St. Croix	USVI
Carolyn Dorfman Dance Company	Union	NJ
Charleston Ballet Theatre	Charleston	SC
Chen & Dancers	New York	NY
Cincinnati Ballet	Cincinnati	OH
Collage Dance Theatre	Los Angeles	CA
Contemporary Dance/Fort Worth	Fort Worth	TX
Cunningham Dance Foundation	New York	NY
Dance Alloy	Pittsburgh	PA
Dance Institute of Washington	Washington	DC
Dance Theatre of Harlem	New York	NY
Dayton Ballet	Dayton	OH
Dayton Contemporary Dance Company	Dayton	OH
Demetrius Klein Dance Company	Lake Worth	FL
Diavolo Dance Theater	Los Angeles	CA
Donald Byrd/The Group	Brooklyn	NY
Doug Varone & Dancers/DOVA, Inc.	New York	NY
EIKO & KOMA	New York	NY
Felice Lesser Dance Theater	New York	NY
Flamenco Vivo Carlota Santana	New York	NY
Fort Worth Dallas Ballet	Fort Worth	TX
Garth Fagan Dance	Rochester	NY
Gina Gibney Dance Inc.	New York	NY
Houston Ballet	Houston	TX
Hubbard Street Dance Chicago	Chicago	IL
James Sewell Ballet	Minneapolis	MN
Joe Goode Performance Group	San Francisco	CA
Joffrey Ballet of Chicago	Chicago	IL
June Watanabe in Company	San Rafael	CA
Kansas City Ballet	Kansas City	MO
Ko-Thi Dance Company	Milwaukee	WI
Lar Lubovitch Dance Company	New York	NY
Lily Cai Chinese Dance Company	San Francisco	CA
Limón Dance Company	New York	NY
Liz Lerman Dance Exchange	Takoma Park	MD
Malashock Dance & Company	San Diego	CA
Margaret Jenkins Dance Company	San Francisco	CA
Mark Morris Dance Group	Brooklyn	NY
Meredith Monk/The House Foundation	New York	NY
Milwaukee Ballet	Milwaukee	WI
Monte/Brown Dance	New York	NY
Montgomery Ballet	Montgomery	AL
Nai-Ni Chen Dance Company	Fort Lee	NJ
Nashville Ballet	Nashville	TN
New York City Ballet	New York	NY
North Carolina Dance Theater	Charlotte	NC
OPC/San Francisco	San Francisco	CA
Ohio Ballet	Akron	OH
Pacific Northwest Ballet	Seattle	WA
Parsons Dance Company.	New York	NY
Paul Taylor Dance Foundation	New York	NY
Paula Josa-Jones/Performance Works	Chilmark	MA
Pittsburgh Ballet Theatre	Pittsburgh	PA
Richmond Ballet	Richmond	VA
San Francisco Ballet	San Francisco	CA
Stephen Petronio Company	New York	NY
Tennessee Children's Dance Ensemble	Knoxville	TN
Trinity Irish Dance Co.	Chicago	IL
Troika Ranch	Brooklyn	NY
Tulsa Ballet Theatre	Tulsa	OK
Urban Bush Women	Brooklyn	NY
The Washington Ballet	Washington	DC

Some Notable Museums

This unofficial list of the largest museums in the U.S. by budget was compiled with the assistance of the American Association of Museums, a national association representing the concerns of the museum community. Association members also include zoos, aquariums, arboretums, botanical gardens, and planetariums, but these are not included in *The World Almanac* listing. See also Major U.S. Public Zoological Parks and Major Canadian Public Zoological Parks.

Museum	City	State
American Museum of Natural History	New York	NY
Amon Carter Museum of Western Art	Ft. Worth	TX
The Art Institute of Chicago	Chicago	IL
Autry Museum of Western Heritage	Los Angeles	CA
Brooklyn Museum of Art	Brooklyn	NY
Busch-Reisinger Museum	Cambridge	MA
California Academy of Science	San Francisco	CA
California Science Center	Los Angeles	CA
Carnegie Museums of Pittsburgh	Pittsburgh	PA
Chicago Historical Society	Chicago	IL
Children's Museum of Indianapolis	Indianapolis	IN
Cincinnati Art Museum	Cincinnati	OH
Cincinnati Museum Center	Cincinnati	OH
Cleveland Museum of Art	Cleveland	OH
Colonial Williamsburg	Williamsburg	VA
Corning Museum of Glass	Corning	NY
Dallas Museum of Art	Dallas	TX
Denver Art Museum	Denver	CO
Denver Museum of Nature and Science	Denver	CO
Detroit Institute of Arts	Detroit	MI
Exploratorium	San Francisco	CA
The Field Museum of Natural History	Chicago	IL
Fine Arts Museum of San Francisco	San Francisco	CA
Franklin Institute	Philadelphia	PA
The Frick Collection	New York	NY
Harvard University Art Museum	Cambridge	MA
Henry F. Dupont Winterthur Museum	Winterthur	DE
Henry Ford Museum/Greenfield Village	Dearborn	MI
High Museum of Art	Atlanta	GA
Houston Museum of Natural Science	Houston	TX
Jamestown-Yorktown Foundation	Williamsburg	VA
Jewish Museum	New York	NY
L.A. County Museum of Art	Los Angeles	CA
Liberty Science Center, Liberty State Park	Jersey City	NJ
Maryland Academy of Sciences	Baltimore	MD
Maryland Science Center	Baltimore	MD
Mashantucket Pequot Museum and Research Center	Mashantucket	CT
Metropolitan Museum of Art	New York	NY
Milwaukee Public Museum	Milwaukee	WI
Minneapolis Institute of Art	Minneapolis	MN
Museum of African American History	Detroit	MI
Museum of Contemporary Art	Los Angeles	CA
Museum of Fine Arts	Boston	MA
Museum of Fine Arts	Houston	TX
Museum of Modern Art	New York	NY
Museum of New Mexico	Santa Fe	NM
Museum of Science	Boston	MA
Mystic Seaport Museum	Mystic	CT
National Air & Space Museum	Washington	DC
National Baseball Hall of Fame and Museum, Inc.	Cooperstown	NY
National Gallery of Art	Washington	DC
National Museum of American History-Smithsonian Inst.	Washington	DC
National Museum of Natural History	Washington	DC
Nelson-Atkins Museum of Art	Kansas City	MO
New York Historical Society	New York	NY
New York State Museum	Albany	NY
The Newseum	Arlington	VA
Peabody Essex Museum	Salem	MA
Pennsylvania Historical & Museum Commission	Harrisburg	PA
Philadelphia Museum of Art	Philadelphia	PA
Public Museum of Grand Rapids	Grand Rapids	MI
Rock & Roll Hall of Fame and Museum Inc.	Cleveland	OH
San Diego Museum of Art	San Diego	CA
San Francisco Museum of Modern Art	San Francisco	CA
Science Museum of Minnesota	Saint Paul	MN
Scottsdale Museum of Contemp. Art	Scottsdale	AZ
St. Louis Science Center	St. Louis	MO
Toledo Museum of Art	Toledo	OH
U.S. Holocaust Memorial Museum	Washington	DC
Univ. of Pennsylvania Museum	Philadelphia	PA
Virginia Museum of Fine Arts	Richmond	VA
Wadsworth Atheneum	Hartford	CT
Walker Art Center	Minneapolis	MN
Whitney Museum of American Art	New York	NY

Best-Selling U.S. Magazines, 2000

Source: Audit Bureau of Circulations, Schaumburg, IL

General magazines, exclusive of groups and comics; also excluding magazines that failed to file reports to ABC by press time. Based on total average paid circulation during the 6 months ending Dec. 31, 2000.

	Magazine	Circulation
1.	Reader's Digest	12,566,047
2.	TV Guide	9,948,792
3.	National Geographic Magazine	7,828,642
4.	Better Homes and Gardens	7,617,985
5.	Family Circle	5,002,042
6.	Good Housekeeping	4,558,524
7.	Woman's Day	4,244,383
8.	Ladies' Home Journal	4,101,550
9.	Time	4,056,150
10.	McCall's	4,005,958
11.	People Weekly	3,552,287
12.	Home and Away	3,314,650
13.	Sports Illustrated	3,205,241
14.	Playboy	3,151,580
15.	Newsweek	3,144,695
16.	Prevention	3,110,642
17.	Via Magazine	2,630,852
18.	The American Legion Magazine	2,602,005
19.	Cosmopolitan	2,592,887
20.	Southern Living	2,537,485
21.	Maxim	2,458,150
22.	Martha Stewart Living	2,436,422
23.	Seventeen	2,374,803
24.	Redbook	2,269,605
25.	YM	2,202,979
26.	O The Oprah Magazine	2,162,668
27.	Glamour	2,147,263
28.	National Enquirer	2,075,063
29.	U.S. News & World Report	2,070,511
30.	'Teen	2,057,623
31.	Smithsonian	2,051,045
32.	Parents	2,004,929
33.	Money	1,906,352
34.	V.F.W. Magazine	1,824,672
35.	Field & Stream	1,752,937
36.	Ebony	1,728,986
37.	Country Living	1,673,792
38.	Star	1,630,092
39.	Men's Health	1,629,568
40.	Shape	1,618,130
41.	Woman's World	1,604,003
42.	Teen People	1,600,504
43.	In Style	1,584,691
44.	Golf Digest	1,563,476
45.	Popular Science	1,554,698
46.	American Rifleman	1,552,512
47.	First for Women	1,542,566
48.	Entertainment Weekly	1,520,463
49.	Parenting Magazine	1,460,041
50.	Cooking Light	1,453,558
51.	Sunset	1,448,005
52.	Golf Magazine	1,405,020
53.	Car and Driver	1,402,657
54.	Outdoor Life	1,351,394
55.	Health	1,339,754
56.	Self	1,294,091
57.	Motor Trend	1,285,178
58.	Bon Appetit	1,280,105
59.	PC World	1,265,443
60.	Boys' Life	1,259,656
61.	Family Life	1,259,656
62.	Rolling Stone	1,254,148
63.	Popular Mechanics	1,238,681
64.	Family Fun	1,232,544
65.	PC Magazine	1,228,362
66.	Scholastic Parent and Child	1,224,098
67.	Endless Vacation	1,219,393
68.	ESPN The Magazine	1,175,526
69.	Vogue	1,174,183
70.	The American Hunter	1,173,549
71.	The Family Handyman	1,156,280
72.	Fitness	1,121,229
73.	Life	1,120,338
74.	Sesame Street Magazine	1,119,405
75.	Soap Opera Digest	1,109,433
76.	The Elks Magazine	1,103,316
77.	Mademoiselle	1,100,185
78.	Vanity Fair	1,050,684
79.	Country Home	1,045,729
80.	Scouting	1,037,249
81.	Michigan Living	1,030,169
82.	Home	1,020,938
83.	Kiplinger's Personal Finance Magazine	1,018,924
84.	Discover	1,005,981
85.	Essence	1,004,452
86.	Yahoo Internet Life	1,003,771
87.	Ziff Davis Smart Business	1,001,591
88.	Sport	1,000,830
89.	Consumers Digest	1,000,033
90.	American Homestyle & Gardening	982,205
91.	Victoria	973,629
92.	Travel & Leisure	960,485
93.	Today's Homeowner	951,747
94.	Business Week	949,860
95.	Marie Claire	948,321
96.	Gourmet	946,345
97.	Elle	945,897
98.	Jet	944,073
99.	Child	925,326
100.	GQ	898,508

Some Notable New Books, 2000

Source: List published by American Library Association, Chicago, IL, 2001, for books published in 2000

Fiction

Margaret Atwood, *The Blind Assassin*
Frederich Busch, *Don't Tell Anyone*
Michael Chabon, *Amazing Adventures of Kavalier & Clay*
J. M. Coetzee, *Disgrace*
Jim Crace, *Being Dead*
Helen DeWitt, *Last Samurai*
Laura Kalpakian, *Delinquent Virgin*
Thomas King, *Truth and Bright Water*
Matthew Kneale, *English Passengers*
Antonya Nelson, *Living to Tell*
Michael Ondaatje, *Anil's Ghost*
Tom Paine, *Scar Vegas*
Zadie Smith, *White Teeth*
Joy Williams, *The Quick and the Dead*

Poetry

Seamus Heaney, *Beowulf: A New Translation*
Stanley Kunitz, *Collected Poems of Stanley Kunitz*
Les Murray, *Learning Human: Selected Poems*

Nonfiction

Jacques Barzun, *From Dawn to Decadence: 500 Years of Western Cultural Life, 1500 to the Present*
Dave Eggers, *A Heartbreaking Work of Staggering Genius*
Fergus Fleming, *Barrow's Boys*
Alice Yaeger Kaplan, *Collaborator: The Trial and Execution of Robert Brasillach*
Matt Ridley, *Genome: The Autobiography of a Species in 23 Chapters*
Elaine Sciolino, *Persian Mirrors: The Elusive Face of Iran*
Nicholas Shakespeare, *Bruce Chatwin*
Colin Thubron, *In Siberia*

Young Adults
Nonfiction

Lance Armstrong with Sally Jenkins, *It's Not About the Bike . . . My Journey Back to Life*
Susan D. Bachrach, *The Nazi Olympics: Berlin 1936*
Susan Campbell Bartoletti, *Kids on Strike!*
Wendy Beckett, *My Favorite Things: 75 Works of Art from Around the World*
Dennis Brindell Fradin and Judith Bloom Fradin, *Ida B. Wells: Mother of the Civil Rights Movement*
Betsy Franco, editor, *You Hear Me?: Poems and Writings by Teenage Boys*
Russell Freedman, *Give me Liberty!: The Story of the Declaration of Independence*
Savion Glover and Bruce Weber, *Savion: My Life in Tap*
Lori Gottlieb, *Stick Figure: A Diary of My Former Self*
Jon Katz, *Geeks: How Two Boys Rode the Internet Out of Idaho*
Tom Lalicki, *Spellbinder: The Life of Harry Houdini*
Shannon Lanier and Jane Feldman, *Jefferson's Children: The Story of One American Family*
Ellen Levine, *Darkness over Denmark: The Danish Resistance and the Rescue of the Jews*
Alber Marrin, *Sitting Bull and His World*
Gary Paulsen, *The Beet Fields*
Judith St. George, *In the Line of Fire: Presidents' Lives at Stake*
Ann Turner, *Learning to Swim*
Loung Ung, *First They Killed My Father: A Daughter of Cambodia Remembers*
Judd Winick, *Pedro and Me*

Young Adults
Fiction

David Almond, *Kit's Wilderness*
Laurie Halse Anderson, *Fever 1793*
Kathi Appelt, *Kissing Tennessee and Other Stories from the Stardust Dance*
Adam Bagdasarian, *Forgotten Fire*
Cat Bauer, *Harley, Like a Person*
Joan Bauer, *Hope was Here*
Gary Blackwood, *Shakespeare's Scribe*
Martha Brooks, *Being with Henry*
Meg Cabot, *The Princess Diaries*
Dia Calhoun, *Aria of the Sea*
Tracy Chevalier, *Girl with a Pearl Earring*
Carolyn Coman, *Many Stones*
Sharon Creech, *The Wanderer*
Michael Crichton, *Timeline*
Gillian Cross, *Tightrope*
Sarah Dessen, *Dreamland*
Carl Deuker, *Night Hoops*
Anna Fienberg, *Borrowed Light*
Adrian Fogelin, *Crossing Jordan*
Patricia Reilly Giff, *Nory Ryan's Song*
Mel Glenn, *Split Image*
Dianne E. Gray, *Holding up the Earth*
Kent Haruf, *Plainsong*
Catherine Ryan Hyde, *Pay It Forward*
Anne Isaacs, *Torn Thread*
Kathleen Karr, *The Boxer*
Cristina Kessler, *No Condition is Permanent*
E. L. Konigsburg, *Silent to the Bone*
Amy Goldman Koss, *The Girls*
Iain Lawrence, *Ghost Boy*
Benjamin Lebert, *Crazy*
Mary Logue, *Dancing with an Alien*
Chris Lynch, *Gold Dust*
Juliet Marillier, *Daughter of the Forest*
Gerald Morris, *The Savage Damsel and the Dwarf*
Rita Murphy, *Night Flying*
Walter Dean Myers, *145th Street Short Stories*
Jerrie Oughton, *Perfect Family*
Richard Peck, *A Year Down Yonder*
Julie Anne Peters, *Define "Normal"*
Rodman Philbrick, *The Last Book in the Universe*
Randall Beth Platt, *The Likes of Me*
Carol Plum-Ucci, *The Body of Christopher Creed*
Louise Plummer, *A Dance for Three*
Louise Rennison, *Angus, Thongs, and Full-Frontal Snogging: Confessions of Georgia Nicolson*
Pam Munoz Ryan, *Esperanza Rising*
Virginia Frances Schwartz, *Send One Angel Down*
Jerry Spinelli, *Stargirl*
Terry Trueman, *Stuck in Neutral*
Rich Wallace, *Playing Without a Ball*
Ken Wells, *Meely LaBauve*
Gloria Whelan, *Homeless Bird*
Ruth White, *Memories of Summer*
Lori Aurelia Williams, *When Kambia Elaine Flew in from Neptune*
Ellen Wittlinger, *What's in a Name*
Jacqueline Woodson, *Miracle's Boys*
Jane Yolen and Robert J. Harris, *Queen's Own Fool*

Some Notable New Books for Children, 2000

Source: List published by American Library Association, Chicago, IL, 2001, for books published in 2000

Younger Readers

David A. Adler, *America's Champion Swimmer: Gertrude Ederle*
Kate Banks, *Night Worker*
Lynne Barasch, *Radio Rescue*
Don Brown, *Uncommon Traveler: Mary Kingsley in Africa*
Doreen Cronin, *Click, Clack, Moo: Cows That Type*
Ian Falconer, *Olivia*
Bob Graham, *Max*
Elissa Haden Guest, *Iris and Walter*
Kevin Henkes, *Wemberly Worried*
Elizabeth Fitzgerald Howard, *Virgie Goes to School with Us Boys*
Simon James (Editor), *Days Like This: A Collection of Small Poems*
Reeve Lindbergh (Editor), *In Every Tiny Grain of Sand: A Child's Books of Prayers and Praise*

Angela Shelf Medearis, *Seven Spools of Thread: A Kwanzaa Story*
Mary Pope Osborne, *Kate and the Beanstalk*
Lisa Westberg Peters, *Cold Little Duck, Duck, Duck*
Gary Soto, *Chato and the Party Animals*
Ernest Lawrence Thayer, *Casey at the Bat: A Ballad of the Republic Sung in the Year 1888*
Shelley Moore Thomas, *Good Night, Good Knight*
Uzo Unobagha, *Off to the Sweet Shores of Africa and Other Talking Drum Rhymes*
Jane Yolen, *How Do Dinosaurs Say Good Night?*

Middle Readers

Nic Bishop, *Digging for Bird-Dinosaurs: An Expedition to Madagascar*
Joseph Bruchac, *Crazy Horse's Vision*
Lesa Cline-Ransome, *Satchel Paige*
Lynn Curlee, *Liberty*
Kate DiCamillo, *Because of Winn-Dixie*
Jack Gantos, *Joey Pigza Loses Control*
Beverly Gherman, *Norman Rockwell: Storyteller with a Brush*
James Cross Giblin, *The Amazing Life of Benjamin Franklin*
Patricia Reilly Giff, *Nory Ryan's Song*
Alan Govenar (Editor), *Osceola: Memories of a Sharecropper's Daughter*
Jan and Sandra Jordan Greenberg, *Frank O. Gehry: Outside In*
Ann M. Martin and Laura Godwin , *The Doll People*
Megan McDonald, *Judy Moody*
Christopher Myers, *Wings*
Laurie Myers, *Surviving Brick Johnson*
Andrea Davis Pinkney, *Let It Shine: Stories of Black Women Freedom Fighters*
Doreen Rappaport, *Freedom River*
Anne Rockwell, *Only Passing Through: The Story of Sojourner Truth*
Judith St. George, *So You Want to be President?*
Sylvia Waugh, *Space Race*
Sophie Webb, *My Season with Penguins: An Antarctic Journal*

Older Readers-JHS

David Almond, *Kit's Wilderness*
Marc Aronson, *Sir Walter Raleigh and the Quest for El Dorado*
Joan Bauer, *Hope Was Here*
Barbara Brenner (Editor), *Voices: Poetry and Art From Around the World*
Daniella Carmi (Translated from Hebrew by Yael Lotan.), *Samir and Yonatan*
Sharon Creech, *The Wanderer*
Joan Dash, *The Longitude Prize*
Dennis Brindell Fradin and Judith Bloom Fradin, *Ida B. Wells: Mother of the Civil Rights Movement*
Lorenz Graham, *How God Fix Jonah*
Lynn Joseph, *The Color of My Words*
Iain Lawrence, *Ghost Boy*
Christian Lehmann (translated from French by William Rodarmor), *Ultimate Game*
J. Patrick Lewis, *Freedom Like Sunlight: Praisesongs for Black Americans*
Janet Taylor Lisle, *The Art of Keeping Cool*
Chris Lynch, *Gold Dust*
David Macaulay, *Building Big*
Jim Murphy, *Blizzard! The Storm That Changed America*
William Nicholson, *The Wind Singer*
Richard Peck, *A Year Down Yonder*
Philip Pullman, *The Amber Spyglass*
Gloria Whelan, *Homeless Bird*
Judd Winick, *Pedro and Me: Friendship, Loss, and What I Learned*

All Ages

Aesop's Fables, Illustrated by Jerry Pinkney
L. Frank Baum, *The Wonderful Wizard of Oz: A Commemorative Pop-Up*
Mary Casanova, *The Hunter: A Chinese Folktale*
Cynthia DeFelice, *Cold Feet*
Ray Hicks as told to Lynn Salsi, *The Jack Tales*
Paul B. Janeczko (Editor), *Stone Bench in an Empty Park*
Francisco Jimenez, *Christmas Gift: El regalo de Navidad*
Eric A. Kimmel, *Gershon's Monster: A Story for the Jewish New Year*
J.K. Rowling, *Harry Potter and the Goblet of Fire*

> ***IT'S A FACT:*** A total of 23.3 million Harry Potter books were sold in the U.S. in 2000.

Best-Selling Books, 2000

Source: *Publishers Weekly*

Rankings are based on copies "shipped and billed" in 2000, minus returns through early 2001.

Fiction

1. *The Brethren,* John Grisham
2. *The Mark: The Beast Rules the World,* Jerry B. Jenkins and Tim LaHaye
3. *The Bear and the Dragon,* Tom Clancy
4. *The Indwelling: The Beast Takes Possession,* Jerry B. Jenkins and Tim LaHaye
5. *The Last Precinct,* Patricia Cornwell
6. *Journey,* Danielle Steel
7. *The Rescue,* Nicholas Sparks
8. *Rose Are Red,* James Patterson
9. *Cradle and All,* James Patterson
10. *The House on Hope Street,* Danielle Steel
11. *The Wedding,* Danielle Steel
12. *Drowning Ruth,* Christina Schwarz
13. *Before I Say Good-Bye,* Mary Higgins Clark
14. *Deck the Halls,* Mary and Carol Higgins Clark
15. *Gap Creek,* Robert Morgan

Nonfiction

1. *Who Moved My Cheese?,* Spencer Johnson
2. *Guinness World Records 2001,* Guinness World Records Ltd.
3. *Body for Life,* Bill Phillips
4. *Tuesdays with Morrie,* Mitch Albomy
5. *The Beatles Anthology,* The Beatles
6. *The OReilly Factor,* Bill OReilly. Broadway
7. *Relationship Rescue,* Philip C. McGraw, Ph. D
8. *The Millionaire Mind,* Thomas J. Stanley
9. *Ten Things I Wish I'd Known Before I Went Out into the Real World,* Maria Shriver
10. *Eating Well for Optimum Health,* Andrew Weil, M.D. Knopf
11. *The Prayer of Jabez,* Dr. Bruce Wilkinson
12. *Flags of Our Fathers,* James Bradley with Ron Powers
13. *A Short Guide to a Happy Life,* Anna Quindlen
14. *On Writing,* Stephen King. Scribner
15. *Nothing Like It in the World,* Stephen E. Ambrose

Trade Paperbacks

1. *A Child Called "It,"* Dave Pelzer
2. *Left Behind,* Jerry B. Jenkins and Tim LaHaye
3. *The Poisonwood Bible,* Barbara Kingsolver
4. *Chicken Soup for the Couple's Soul,* Jack Canfield, Mark Victor Hansen, Mark & Chrissy Donnelly and Barbara De Angelis
5. *Apollyon,* Jerry B. Jenkins and Tim LaHaye
6. *Tribulation Force,* Jerry B. Jenkins and Tim LaHaye
7. *Chicken Soup for the Teenage Soul III,* Jack Canfield, Mark Victor Hansen, Kimberly Kirberger
8. *While I Was Gone,* Sue Miller
9. *House of Sand and Fog,* Andre Dubus III
10. *Talking Dirty with the Queen of Clean,* Linda Cobb
11. *The Lost Boy,* Dave Pelzer
12. *The Worst Case Scenario Survival Handbook,* Joshua Piven and David Borgenicht
13. *Chicken Soup for the Golfer's Soul,* Jack Canfield, Mark Victor Hansen, Jeff Aubery, and Mark & Chrissy Donnelly
14. *Nicolae,* Jerry B. Jenkins and Tim LaHaye
15. *The Millionaire Next Door,* William Danko and Thomas Stanley

Almanacs, Atlases, and Annuals

1. *The World Almanac and Book of Facts 2001,* Edited by Ken Park
2. *Guinness World Records 2000,* Mark C. Young
3. *The World Almanac and Book of Facts 2000,* Edited by Ken Park
4. *J.K. Lasser's Your Income Tax 2001*
5. *The Ernst & Young Tax Guide 2001*

Mass Market

1. *The Testament,* John Grisham
2. *The Brethren,* John Grisham
3. *Hannibal,* Thomas Harris
4. *The Green Mile,* Stephen King
5. *Heart of the Sea,* Nora Roberts
6. *Tears of the Moon,* Nora Roberts
7. *Black Notice,* Patricia Cornwell
8. *Irresistible Forces,* Danielle Steel
9. *Timeline,* Michael Crichton
11. *The Girl Who Loved Tom Gordon,* Stephen King
12. *We'll Meet Again,* Mary Higgins Clark
13. *Pop Goes the Weasel,* James Patterson
14. *River's End,* Nora Roberts
15. *False Memory,* Dean Koontz

Leading U.S. Daily Newspapers, 2000

Source: 2001 *Editor & Publisher International Yearbook*

(Circulation as of Sept. 30, 2000; m = morning, e = evening)

As of Feb. 1, 2001, the number of U.S. daily newspapers had dropped to 1,480, for a net loss of 3 since Feb. 1, 2000. Average daily circulation for the 6 months ending Sept. 30, 2000, was 55,772,847, down 206,485 from the same period in 1999, for a decrease of about 0.4%. The overall number of Sunday papers increased by 12 to 917. Average Sunday circulation for the 6 months ending Sept. 30, 2000, fell 473,582, or about 0.8%, from 59,894,381 to 59,420,999.

Newspaper		Circulation
1. New York (NY) *Wall Street Journal*	(m)	1,762,751
2. Arlington (VA) *USA Today*	(m)	1,692,666
3. New York (NY) *Times*	(m)	1,097,180
4. Los Angeles (CA) *Times*	(m)	1,033,399
5. Washington (DC) *Post*	(m)	762,009
6. New York (NY) *Daily News*	(m)	704,463
7. Chicago (IL) *Tribune*	(m)	661,699
8. Long Island (NY) *Newsday*	(m)	576,345
9. Houston (TX) *Chronicle*	(m)	546,799
10. Dallas (TX) *Morning News*	(m)	495,597
11. Chicago (IL) *Sun-Times*	(m)	471,031
12. Boston (MA) *Globe*	(m)	464,472
13. San Francisco (CA) *Chronicle*	(m)	457,028
14. Phoenix (AZ) *Arizona Republic*	(m)	445,322
15. New York (NY) *Post*	(m)	443,951
16. Denver (CO) *Rocky Mountain News*	(m)	426,465
17. Denver (CO) *Post*	(m)	420,033
18. Newark (NJ) *Star-Ledger*	(m)	407,537
19. Philadelphia (PA) *Inquirer*	(m)	400,385
20. San Diego (CA) *Union-Tribune*	(all day)	370,395
21. Detroit (MI) *Free Press*	(m)	365,579
22. Cleveland(OH) *Plain Dealer*	(m)	364,708
23. Orange County (CA) *Register*	(m)	358,654
24. Portland (OR) *Oregonian*	(all day)	348,468
25. Miami (FL) *Herald*	(m)	343,877
27. Minneapolis (MN) *Star Tribune*	(m)	336,476
26. St. Petersburg (FL) *Times*	(m)	325,633
28. Baltimore (MD) *Sun*	(m)	315,306
29. Atlanta (GA) *Constitution*	(m)	311,342
30. Los Angeles (CA) *Investor's Business Daily*	(m)	303,596
31. St. Louis (MO) *Post-Dispatch*	(m)	294,434
32. Sacramento (CA) *Bee*	(m)	289,751
33. San Jose (CA) *Mercury News*	(m)	286,679
34. Milwaukee (WI) *Journal Sentinel*	(m)	278,377
35. Kansas City (MO) *Star*	(m)	267,664
36. New Orleans (LA) *Times-Picayune*	(m)	264,001
37. Orlando (FL) *Sentinel*	(all day)	260,802
38. Boston (MA) *Herald*	(m)	257,761
39. Fort Lauderdale (FL) *Sun-Sentinel*	(m)	256,690
40. Indianapolis (IN) *Star*	(m)	248,144
41. Columbus (OH) *Dispatch*	(m)	244,177
42. Charlotte (NC) *Observer*	(m)	240,594
43. Pittsburgh (PA) *Post-Gazette*	(m)	240,245
44. Detroit (MI) *News*	(e)	237,518
45. Louisville (KY) *Courier-Journal*	(m)	231,630
46. Buffalo (NY) *News*	(all day)	226,342
47. Seattle (WA) *Times*	(m)	225,687
48. San Antonio (TX) *Express-News*	(m)	221,246
49. Fort Worth (TX) *Star-Telegram*	(m)	220,096
50. Omaha (NE) *World-Herald*	(all day)	214,651
51. Tampa (FL) *Tribune*	(m)	213,032
52. St. Paul (MN) *Pioneer Press*	(m)	205,798
53. Hartford (CT) *Courant*	(m)	202,509
54. Los Angeles (CA) *Daily News*	(m)	200,387
55. Oklahoma City (OK) *Daily Oklahoman*	(m)	198,576
56. Norfolk (VA) *Virginian-Pilot*	(m)	197,574
57. Richmond (VA) *Times-Dispatch*	(m)	196,432
58. Cincinnati (OH) *Enquirer*	(m)	195,360
59. Austin (TX) *American-Statesman*	(m)	187,789
60. Nashville (TN) *Tennessean*	(m)	186,793
61. Walnut Creek (CA) *Contra Costa Times*	(m)	182,682
62. Seattle (WA) *Post-Intelligencer*	(m)	175,794
63. Rochester (NY) *Democrat and Chronicle*	(m)	173,398
64. Jacksonville (FL) *Times-Union*	(m)	172,734
65. West Palm Beach (FL) *Palm Beach Post*	(m)	172,523
66. Little Rock (AR) *Democrat-Gazette*	(m)	172,214
67. Riverside (CA) *Press-Enterprise*	(m)	166,935
68. Providence (RI) *Journal*	(m)	162,358
69. Memphis (TN) *Commercial Appeal*	(m)	161,274
70. Raleigh (NC) *News & Observer*	(m)	161,175
71. Neptune (NJ) *Asbury Park Press*	(all day)	160,069
72. Las Vegas (NV) *Review-Journal*	(m)	158,970
73. Fresno (CA) *Bee*	(m)	156,915
74. Des Moines (IA) *Register*	(m)	155,698
75. Philadelphia (PA) *Daily News*	(m)	154,145
76. Birmingham (AL) *News*	(m)	148,851
77. Arlington Heights (IL) *Daily Herald*	(m)	145,902
78. Tulsa (OK) *World*	(m)	145,697
79. Bergen County (NJ) *Record*	(m)	145,595
80. White Plains (NY) *Journal News*	(m)	143,685
81. Grand Rapids (MI) *Press*	(e)	141,303
82. Akron (OH) *Beacon Journal*	(m)	140,137
83. Toledo (OH) *Blade*	(m)	137,972
84. Salt Lake City (UT) *Tribune*	(m)	134,542
85. Dayton (OH) *Daily News*	(m)	134,393
86. Tacoma (WA) *News Tribune*	(m)	127,629
87. Allentown (PA) *Morning Call*	(m)	127,175
88. Columbia (SC) *State*	(m)	118,783
89. Los Angeles (CA) *La Opinion*	(m)	117,558
90. Knoxville (TN) *News-Sentinel*	(m)	116,564
91. Albuquerque (NM) *Journal*	(m)	108,931
92. Lexington (KY) *Herald-Leader*	(m)	108,550
93. Honolulu (HI) *Advertiser*	(m)	106,590
94. Charleston (SC) *Post & Courier*	(m)	105,868
95. Sarasota (FL) *Herald-Tribune*	(m)	105,672
96. Spokane (WA) *Spokesman-Review*	(m)	105,550
97. Worcester (MA) *Telegram & Gazette*	(m)	103,565
98. Washington (DC) *Times*	(m)	102,957
99. Jackson (MS) *Clarion-Ledger*	(m)	101,886
100. Daytona Beach (FL) *News-Journal*	(m)	100,356

Leading Canadian Daily Newspapers, 2000

Source: 2001 Editor & Publisher International Yearbook

(Circulation as of Sept. 30, 2000; m = morning)

Newspaper	Circulation
Toronto (ON) *Star* (m)	460,989
Toronto (ON) *Globe and Mail* (m)	368,857
Toronto (ON) *National Post* (m)	320,224
Montreal (QC) *Le Journal* (m)	259,081
Toronto (ON) *Sun* (m)	228,596
Vancouver (BC) *Sun* (m)	187,170
Montreal (QC) *La Presse* (m)	183,178
Vancouver (BC) *Province* (m)	157,896
Montreal (QC) *Gazette* (m)	136,463
Edmonton (AB) *Journal* (m)	136,193

Top 15 News/Information Websites, August 2001

Source: Media Metrix, Inc.

Rank		Visitors[1]
1.	www.about.com	15,404
2.	www.msnbc.com	13,427
3.	www.cnn.com	10,283
4.	www.weather.com	4,595
5.	www.nytimes.com	5,335
6.	www.cbs.com sites*	4,250
7.	ABC NEWS*	4,092
8.	www.usatoday.com	3,672
9.	www.discovery.com	3,631
10.	www.washingtonpost.com	2,935
11.	www.weatherbug.com	2,928
12.	www.timeinc.net	2,558
13.	www.slate.com	2,392
14.	www.time.com	2,299
15.	LA TIMES*	2,102

(1) Number of unique visitors in thousands who visited website at least once in Aug. 2001. *Represents an aggregation of commonly owned/branded domain names.

U.S. Commercial Radio Stations, by Format, 1994-2001[1]

Source: M Street Corporation, Littleton, NH © 2001; counts are for Aug. of each year

Stations, by primary format	2001	1999	1998	1997	1996	1995	1994
1. Country	2,190	2,306	2,368	2,491	2,525	2,613	2,642
2. News/Talk	1,139	1,159	1,131	1,111	1,116	1,036	1,197
3. Oldies	785	766	799	755	738	710	714
4. Adult Contemporary (AC)	709	775	844	902	952	1,052	923
5. Spanish	574	536	493	474	463	427	401
6. Adult Standards	569	595	561	551	499	470	435
7. Top-40	468	401	379	358	333	318	358
8. Soft Adult Contemporary	375	382	368	346	337	347	345
9. Hot AC	369	325	281	260	283	256	242
10. Religion (Teaching, Variety)	356	363	356	404	424	418	426
11. Classic Rock	338	314	282	240	349	306	264
11. Sports	338	256	251	220	156	148	196
13. Rock	282	280	266	262	273	301	309
14. Classic Hits	265	222	192	172			
15. Black Gospel	264	257	238	208	166	147	122
16. Southern Gospel	255	269	273	255	248	239	204
Off Air	113	96	102	143	279	308	309
Changing formats/not available	3	3	3	2	4	19	6
TOTAL STATIONS	**10,561**	**10,444**	**10,292**	**10,207**	**9,991**	**9,889**	**9,778**

(1) Data for 2000 unavailable.

Top-Grossing North American Concert Tours, 1985-2000

Source: Pollstar, Fresno, CA

Artist (Year)	Total gross[1]	Cities/ Shows
1. The Rolling Stones (1994)	$121.2	43/60
2. Pink Floyd (1994)	103.5	39/59
3. The Rolling Stones (1989)	98.0	33/60
4. The Rolling Stones (1997)	89.3	26/33
5. Tina Turner (2000)	80.2	88/95
6. U2 (1997)	79.9	37/46
7. The Eagles (1994)	79.4	32/54
8. 'N Sync (2000)	76.4	64/86
9. The New Kids on the Block (1990)	74.1	122/152
10. Dave Matthews Band (2000)	68.2	43/63
11. U2 (1992)	$67.0	61/73
12. The Rolling Stones (1999)	64.7	26/34
13. The Eagles (1995)	63.3	46/58
14. KISS (2000)	62.7	120/128
15. Bruce Springsteen & The E Street Band (1999)	61.4	18/54
16. Barbra Streisand (1994)	58.9	6/22
17. The Grateful Dead (1994)	52.4	29/84
18. 'N Sync (1999)	51.5	108/121
19. Tim McGraw/Faith Hill (2000)	48.8	64/66
20. Dave Matthews Band (1999)	48.5	47/62

(1) In millions. Not adjusted for inflation.

Sales of Recorded Music and Music Videos, by Genre and Format, 1995-2000

Source: Recording Industry Assn. of America, Washington, DC

Breakdown is by percentage of all recorded music sold, ranked for 2000.

GENRE	2000	1999	1998	1997	1996	1995
Rock	24.8%	25.2%	25.7%	32.5%	32.6%	33.5%
Rap	12.9	10.8	9.7	10.1	8.9	6.7
Pop	11.0	10.3	10.0	9.4	9.3	10.1
Country	10.7	10.8	14.1	14.4	14.7	16.7
R&B	9.7	10.5	12.8	11.2	12.1	11.3
Other	8.3	9.1	7.9	5.7	5.2	7.0
Religious	4.8	5.1	6.3	4.5	4.3	3.1
Jazz	2.9	3.0	1.9	2.8	3.3	3.0
Classical	2.7	3.5	3.3	2.8	3.4	2.9
Oldies	0.9	0.7	0.7	0.8	0.8	1.0
Soundtracks	0.7%	0.8%	1.7%	1.2%	0.8%	0.9%
Children's	0.6	0.4	0.4	0.9	0.7	0.5
New Age	0.5	0.5	0.6	0.8	0.7	0.7
FORMAT						
Compact disc (CD)	89.3	83.2	74.8	70.2	68.4	65.0
Cassette	4.9	8.0	14.8	18.2	19.3	25.1
Singles (all types)	2.5	5.4	6.8	9.3	9.3	7.5
Music video	0.8	0.9	1.0	0.6	1.0	0.9
LP	0.5	0.5	0.7	0.7	0.6	0.5

Note: Totals may not equal 100% because of "Don't know/no answer" responses to survey.

Sales of Recorded Music and Music Videos, by Units Shipped and Value, 1991-2000

Source: Recording Industry Assn. of America, Washington, DC

(in millions, net after returns)

FORMAT	1991	1992	1993	1994	1995	1996	1997	1998	1999	2000	% change 1999-2000
Compact disc (CD)											
Units shipped	333.3	407.5	495.4	662.1	722.9	778.9	753.1	847.0	938.9	942.5	0.4
Dollar value	4,337.7	5,326.5	6,511.4	8,464.5	9,377.4	9,934.7	9,915.1	11,416.0	12,816.3	13,214.5	3.1
CD single											
Units shipped	5.7	7.3	7.8	9.3	21.5	43.2	66.7	56.0	55.9	34.2	-38.8
Dollar value	35.1	45.1	45.8	56.1	110.9	184.1	272.7	213.2	222.4	142.7	-35.8
Cassette											
Units shipped	360.1	366.4	339.5	345.4	272.6	225.3	172.6	158.5	123.6	76.0	-38.5
Dollar value	3,019.6	3,116.3	2,915.8	2,976.4	2,303.6	1,905.3	1,522.7	1,419.9	1,061.6	626.0	-41.0
Cassette single											
Units shipped	69.0	84.6	85.6	81.1	70.7	59.9	42.2	26.4	14.2	1.3	-91.0
Dollar value	230.4	298.8	298.5	274.9	236.3	189.3	133.5	94.4	48.0	4.6	-90.3
LP/EP											
Units shipped	4.8	2.3	1.2	1.9	2.2	2.9	2.7	3.4	2.9	2.2	-24.6
Dollar value	29.4	13.5	10.6	17.8	25.1	36.8	33.3	34.0	31.8	27.7	-12.7
Vinyl single											
Units shipped	22.0	19.8	15.1	11.7	10.2	10.1	7.5	5.4	5.3	4.8	-8.1
Dollar value	63.9	66.4	51.2	47.2	46.7	47.5	35.6	25.7	27.9	26.3	-5.4
Music video											
Units shipped	6.1	7.6	11.0	11.2	12.6	16.9	18.6	27.2	19.8	18.2	-8.0
Dollar value	118.1	157.4	213.3	231.1	220.3	236.1	323.9	508.0	376.7	281.9	-25.2
DVD											
Units shipped	—	—	—	—	—	—	—	0.5	2.5	3.3	35.2
Dollar value	—	—	—	—	—	—	—	12.2	66.3	80.3	21.1
TOTAL UNITS	**801.0**	**895.5**	**955.6**	**1,122.7**	**1,112.7**	**1,137.2**	**1,063.4**	**1,124.3**	**1,160.6**	**1,079.3**	-7.0
TOTAL VALUE	**7,834.2**	**9,024.0**	**10,046.6**	**12,068.0**	**12,320.3**	**12,533.8**	**12,236.8**	**13,723.5**	**14,584.5**	**14,323.0**	**-1.8**

Multi-Platinum and Platinum Awards for Recorded Music and Music Videos, 2000

Source: Recording Industry Assn. of America, Washington, DC

To achieve platinum status, an album must reach a minimum sale of 1 mil units in LPs, tapes, and CDs, with a manufacturer's dollar volume of at least $2 mil based on one-third of the suggested retail list price for each record, tape, or CD sold. To achieve multi-platinum status, an album must reach a minimum sale of at least 2 mil units in LPs, tapes, and CDs, with a manufacturer's dollar volume of at least $4 mil based on one-third of the list price.

Singles must sell 1 mil units to achieve a platinum award and 2 mil to achieve a multi-platinum award. EP singles count as 2 units. Double-CD sets count as 2 units. Music videos (long form) must sell 100,000 units to qualify for a platinum award and must sell more than 200,000 units for a multi-platinum award. Video singles, which must have a maximum running time of 15 minutes and no more than 2 songs per title, must sell 50,000 units to qualify for a platinum award and at least 100,000 units to qualify for a multi-platinum award.

Awards listed were for albums and singles released in 2000 and for music videos released at any time.

Albums, Multi-Platinum

(numbers in parentheses = millions sold)

Alma Caribena – Caribbean Soul, Gloria Estefan (2)
The Better Life, Three Doors Down (2)
Country Grammar, Nelly (4)
The Heat, Toni Braxton (2)
The History of Rock, Kid Rock (2)
Infest, Papa Roach (2)
Mad Season, Matchbox 20 (2)
The Marshall Mathers LP, Eminem (7)
Music, Madonna (2)
My Name is Joe, Joe (2)
No Strings Attached, 'N Sync (9)
Oops!...I Did It Again, Britney Spears (7)
Revelation, 98 Degrees (2)
Son by Four, Son by Four (2)
Stankonia, Outkast (2)
Whitney: The Greatest Hits, Whitney Houston (2)
Who Let the Dogs Out, Baha Men (2)

Albums, Platinum

Anarchy, Busta Rhymes
Arrasando, Thalia
The Better Life, 3 Doors Down
BTNHResurrection, Bone Thugs 'n Harmony
Can't Take Me Home, Pink
Caricias, Rocio Durcal
Charlie's Angels, Soundtrack
Coyote Ugly, Soundtrack
Crush, Bon Jovi
Grammy Nominees 2000, Various
The Greatest Hits, Whitney Houston
Greatest Hits, Lenny Kravitz
Hooray for Boobies, Bloodhound Gang
I Hope You Dance, Lee Ann Womack
Inside Job, Don Henley
J.E. Heartbreak, Jagged Edge
Latest Greatest Straitest Hits, George Strait
Let's Get Ready, Mystikal
Life Story, Black Rob
Maroon, Barenaked Ladies
Mer de Noms, A Perfect Circle
Mission: Impossible 2, Soundtrack
The Notorious K.I.M., Lil' Kim
Nutty Professor 2: The Klumps, Soundtrack
One Voice, Billy Gillman
Por El Pasado, Grupo Bryndis
Return of Saturn, No Doubt
Riding with The King, Eric Clapton and B.B. King
Romeo Must Die, Soundtrack
Rule 3:36, Ja Rule
Ryde or Die, Volume 2, Various
Sacred Arias, Andrea Bocelli
Secreto De Amor, Joan Sebastian
The Sickness, Disturbed
Sittin' Fat Down South, Lil' Troy
Skull and Bones, Cypress Hill
Snoop Dogg Presents Tha Eastsidaz, Snoop Dogg
Timeless: Live in Concert, Barbara Streisand
Totally Hits 2, Various
Voodoo, D'Angelo
The Wall Live 1980-81: Is There Anybody Out There, Pink Floyd

Singles, Platinum

Music, Madonna
Maria, Maria, Santana
Incomplete, Sisqo
From the Bottom of My Broken Heart, Britney Spears

Music Videos, Multi-Platinum

(numbers in parentheses = millions sold)

Live From Madison Square Garden, 'N Sync (2)

Music Videos, Platinum

The Greatest Hits, Whitney Houston
Supernatural Live, Santana

Top-Selling Video Games, 2000

Source: The NPD TRSTS Video Game Tracking Service, The NPD Group, Inc., Port Washington, NY; ranked by units sold

Platform, Title

1. Nintendo Gameboy Color, Pokémon Silver
2. Nintendo Gameboy Color, Pokémon Gold
3. Nintendo Gameboy, Pokémon Yellow
4. N64 Gameboy, Pokémon Stadium
5. Sony Playstation, Tony Hawk's Pro Skater 2
6. N64, Legend of Zelda: Majora's Mask
7. Sony Playstation, Tony Hawk's Pro Skater
8. Sony Playstation, Gran Turismo 2
9. Nintendo Gameboy, Pokémon Blue
10. Nintendo Gameboy, Pokémon Red
11. Sony Playstation, WWF Smackdown!
12. N64, Tony Hawk's Pro Skater
13. Nintendo Gameboy, Pokémon Trading Card Game
14. Nintendo Gameboy Color, Super Mario Brothers Deluxe
15. Sony Playstation, Madden NFL 2001

U.S. Television Set Owners

Source: Nielsen Media Research; January 1, 2001

Of the 102.2 million homes (98.2% of U.S. households) that owned at least one TV set in 2000:

99% had color televisions
35% had 2 TV sets
41% had 3 or more TV sets
86% had a VCR
69% received basic cable
32% received premium cable

Some Television Addresses, Phone Numbers, Internet Sites

ABC–American Broadcasting Co.
77 W 66th St.
New York, NY 10023 (212) 456-7777
Website: http://www.abc.com

CBS–Columbia Broadcasting System
524 W. 57th St.
New York, NY 10019 (212) 975-4321
Website: http://www.cbs.com

NBC–National Broadcasting Co.
30 Rockefeller Plaza
New York, NY 10112 (212) 664-4444
Website: http://www.nbc.com

Fox Television
205 E 67th St.
New York, NY 10021 (212) 452-5555
Website: http://www.fox.com

PBS–Public Broadcasting Service
1320 Braddock Place
Alexandria, VA 22314 (703) 739-5000
Website: http://www.pbs.org

CABLE

A&E–Arts & Entertainment Network
235 E 45th St.
New York, NY 10017 (212) 210-1400
Website: http://www.aande.com

AMC–American Movie Classics
Rainbow Media Holdings, Inc.
200 Jericho Triangle
Jericho, NY 11753 (516) 803-4300
Website: http://www.amctv.com

BET–Black Entertainment Television
2000 M Street NW, Suite 602
Washington, DC 20036 (202) 533-1990
Website: http://www.bet.com

CNBC–Consumer News and Business Channel
2200 Fletcher Ave.
Fort Lee, NJ 07024 (201) 585-2622
Website: http://www.cnbc.com

CNN–Cable News Network
One CNN Center, Box 105366
Atlanta, GA 30348-5366 (404) 827-1500
Website: http://www.cnn.com

C-SPAN–Cable-Satellite Public Affairs Network
400 N Capitol St. NW, Suite 650
Washington, DC 20001 (202) 737-3220
Website: http://www.c-span.org

DIS–The Disney Channel
3800 W Alameda Ave.
Burbank, CA 91505 (818) 569-7500
Website: http://www.disneychannel.com

ESPN–ESPN, Inc.
ESPN Plaza, 935 Middle St.
Bristol, CT 06010 (860) 585-2000
Website: http://espn.com

LIF–Lifetime
309 W 49th St.
New York, NY 10019 (212) 424-7000
Website: http://www.lifetimetv.com

MSNBC
1 MSNBC Plaza
Secaucus, NJ 07094 (201) 583-5000
Website: http://www.msnbc.com

MTV–Music Television
MTV Networks, Inc.
1515 Broadway
New York, NY 10036 (212) 258-8000
Website: http://www.mtv.com

NICK–Nickelodeon/Nick at Nite
MTV Networks, Inc.
1515 Broadway
New York, NY 10036 (212) 258-8000
Websites: http://www.nick.com
http://www.nick-at-nite.com

TBS–Turner Broadcasting System
Turner Entertainment Group
One CNN Center, Box 105366
Atlanta, GA 30348-5366
(404) 827-1700
Website: http://www.turner.com

TDC–The Discovery Channel
Discovery Communications
7700 Wisconsin Ave., Suite 700
Bethesda, MD 20814 (301) 986-0444
Website: http://www.discovery.com

USA–USA Network
USA Networks
152 W. 57th St.
New York, NY 10019 (212) 314-7200
Website: http://www.usanetwork.com

WORLD ALMANAC EDITORS' PICKS

The World Almanac staff ranked the following as their favorite animated cartoons of all time:

1. The Simpsons
2. Bugs Bunny
3. Scooby Doo
4. South Park
5. The Jetsons
6. Charlie Brown
7. Tom & Jerry
8. Rocky and Bullwinkle
9. Heckle and Jeckle
10. The Flintstones

Number of Cable TV Systems,[1] 1975-2001

Source: *Television and Cable Factbook*, Warren Publishing, Inc., Washington, DC; estimates as of Jan. 1

Year	Systems	Year	Systems	Year	Systems	Year	Systems	Year	Systems
1975	3,506	1981	4,375	1987	7,900	1992	11,073	1997	10,943
1976	3,681	1982	4,825	1988	8,500	1993	11,108	1998	10,845
1977	3,832	1983	5,600	1989	9,050	1994	11,214	1999	10,700
1978	3,875	1984	6,200	1990	9,575	1995	11,215	2000	10,500
1979	4,150	1985	6,600	1991	10,704	1996	11,220	2001	10,929
1980	4,225	1986	7,500						

(1) The satellite-signal-receiving hardware, cable lines, and cable boxes that provide cable programming to homes within a geographic area.

Top 20 Cable TV Networks, 2001

Source: *Cable Television Developments,* Natl. Cable Television Assn., Jan.-Apr. 2001; ranked by number of subscribers

Network[1]	Affiliates	Subscribers (mil)
1. TBS Superstation (1976)	11,668	83.3
2. The Discovery Channel (1985)	NA	82.5
3. ESPN (1979)	NA	82.1
4. USA Network (1980)	NA	81.9
5. TNT (Turner Network Television) (1988)	10,637	81.8
CNN (1980)	11,528	81.8
7. Nickelodeon (1979)/Nick at Nite (1985)	11,788	81.3
C-SPAN (1979)	7,047	81.3
9. A&E Television Networks (1984)	12,000[2]	81.0
TNN (The Nashville Network) (1983)	NA	81.0
11. FOX Family Channel (1998[3])	13,700	80.9
12. Lifetime Television (1984)	11,000	80.6
13. The Weather Channel (1982)	12,763	80.2
14. QVC Network (1986)	7,511	79.6
15. MTV: Music Television (1981)	9,176	79.2
16. The Learning Channel (1980)	NA	78.9
17. CNBC (1989)	5,000	78.1
18. AMC (American Movie Classics) (1984)	NA	77.9
CNN Headline News (1982)	7,039	77.9
20. ESPN2 (1993)	NA	76.9
VH1 (Music First) (1985)	5,457	76.9

NA = Not available. **Note:** Data include noncable affiliates. (1) Date in parentheses is year service began. (2) U.S. and Canada. (3) Began 1977 as the Family Channel; relaunched as FOX Family Channel, 1998.

U.S. Households With Cable Television, 1977-2000

Source: Nielsen Media Research

Year	Basic cable subscribers	As % of households with TVs	Year	Basic cable subscribers	As % of households with TVs	Year	Basic cable subscribers	As % of households with TVs
1977	12,168,450	16.6	1985	39,872,520	46.2	1993	58,834,440	62.5
1978	13,391,910	17.9	1986	42,237,140	48.1	1994	60,483,600	63.4
1979	14,814,380	19.4	1987	44,970,880	50.5	1995	62,956,470	65.7
1980	17,671,490	22.6	1988	48,636,520	53.8	1996	64,654,160	66.7
1981	23,219,200	28.3	1989	52,564,470	57.1	1997	65,929,420	67.3
1982	29,340,570	35.0	1990	54,871,330	59.0	1998	67,011,180	67.4
1983	34,113,790	40.5	1991	55,786,390	60.6	1999	67,592,000	68.0
1984	37,290,870	43.7	1992	57,211,600	61.5	2000	68,544,000	68.0

Average U.S. Television Viewing Time, October 2000

Source: Nielsen Media Research (hours: minutes per week)

Group	Age	Total per week[1]	Early Morning M-F 7-10 AM	Daytime M-F 10 AM-4 PM	Primetime M-Sat. 8-11 PM & Sun. 7-11 PM	Late Night M-F 11:30 PM-1 AM	Saturday 7 AM-1 PM	Sunday 1-7 PM
Women	18+	33:16	2:07	4:55	9:34	2:13	0:47	1:33
	18-24	21:50	0:59	3:47	7:17	1:43	0:32	1:04
	25-54	31:35	2:05	4:08	9:12	2:15	0:48	1:30
	55+	41:10	2:38	6:49	11:50	2:21	0:51	1:49
Working Women		28:34	1:35	2:58	8:58	2:07	0:45	1:29
Men	18+	30:14	1:29	3:14	9:13	2:15	0:44	1:55
	18-24	21:10	0:50	2:44	5:18	1:51	0:31	1:18
	25-54	29:04	1:21	2:40	9:02	2:21	0:44	1:55
	55+	37:28	2:09	4:46	11:28	2:13	0:49	2:12
Teens	12-17	19:40	0:41	1:41	5:56	1:14	0:43	1:13
Children	2-11	20:30	0:50	3:05	4:50	0:42	1:10	1:05
ALL PEOPLE		**29:04**	**1:44**	**3:45**	**8:24**	**1:55**	**0:49**	**1:34**

(1) Total Day = Mon-Sun 24 hours. All times are Eastern Time.

TV Viewing Shares, Broadcast Years 1990-2000

Source: *Cable TV Facts,* Cable Advertising Bureau, New York, NY

	All Television Households[2]							All Cable Households[2]							Pay Cable Households[2]						
	'90	'95	'96	'97	'98	'99	'00	'90	'95	'96	'97	'98	'99	'00	'90	'95	'96	'97	'98	'99	'00
Network Affiliates[3]	55	48	46	43	41	46	44	46	41	40	38	36	41	40	43	38	36	35	34	39	37
Indep. TV Stations[4]	20	22	21	20	20	11	12	16	17	17	17	16	8	9	16	17	18	17	17	8	9
Public TV Stations	3	3	3	3	3	3	3	3	3	3	3	3	2	2	2	2	3	2	2	2	2
Basic Cable[5]	21	30	33	36	40	44	46	32	42	43	46	49	54	55	30	41	43	46	49	55	55
Pay Cable	6	6	6	13	7	6	6	10	8	8	8	9	7	7	18	15	14	13	12	10	11

(1) Broadcast year (season) ends in May of the year shown, began the previous Sept. (2) Share figures refer to percentage of the viewing audience for all television viewing, 24 hours/day. As a result of multiset use and rounding of numbers, share figures add to more than 100. (3) Includes CBS, NBC, ABC, and FOX. (4) Includes WB, UPN, and PAX. (5) Includes ad-supported cable and all other cable (non-pay and non-ad-supported channels).

Favorite Syndicated Programs, 2000-2001

Source: Nielsen Media Research, Oct. 2, 2000-Aug. 27, 2001

Average audience percentages, or ratings, are estimates of the percentage of TV-owning households watching a program.

Rank	Program	Avg. audience (%)
1.	Wheel of Fortune	9.8
2.	Jeopardy	8.1
3.	World Wrestling Federation	7.2
4.	MMN Home Team Baseball	6.7
5.	Judge Judy	6.0
6.	ESPN NFL Regular Season	5.9
	Entertainment Tonight	5.9
	Oprah Winfrey Show	5.9
	Century 16	5.9
10.	Friends	5.5
11.	ESPN NFL Regular Season 2	5.4
12.	Frasier	4.8
	Wheel of Fortune (weekend)	4.8
14.	Seinfeld	4.7
15.	WCW Wrestling	4.3
16.	Seinfeld (weekend)	4.2
17.	Warner Bros. Volume 32	4.0
18.	Live! with Regis & Kelly	3.8
19.	Entertainment Tonight (weekend)	3.6
	Judge Joe Brown	3.6
	Buena Vista III	3.6
	Jerry Springer	3.6
	Warner Bros. Volume 31	3.6

TV Parental Guidelines

On Dec. 19, 1996, representatives of the television industry announced the creation of TV Parental Guidelines, a rating system intended to give parents advance information about the content of programs. The guidelines, modeled after the Motion Picture Ratings System and developed by a broad spectrum of industry representatives, began to appear on broadcast and cable television programs in Jan. 1997. On July 10, 1997, most of the television industry, after negotiations with advocacy groups, agreed to add the labels D, L, S, and V to the existing ratings. The added labels, which went into effect by Oct. 1, provide more specific information about the degree of violence, coarse language, and sexually suggestive content. Some of the networks that did not add the labels Oct. 1 began to add their own parental advisories to shows.

There are two categories of ratings, one for children's programs and one for programs not specifically designed for children. The ratings are as follows:

The following categories apply to programs designed solely for children:

TV-Y **All Children.** *This program is designed to be appropriate for all children.* Whether animated or live action, the themes and elements in this program are specifically designed for a very young audience, including children ages 2-6. This program is not expected to frighten younger children.

TV-Y7 **Directed to Older Children.** *This program is designed for children age 7 and above.* It may be more appropriate for children who have acquired the developmental skills needed to distinguish between make-believe and reality. Themes and elements in this program may include mild fantasy or comedic violence, or may frighten children under the age of 7. Therefore, parents may wish to consider the suitability of this program for their very young children. Programs containing fantasy violence that may be more intense or more combative than other programs in this category are designated as **TV-Y7-FV**.

The following categories apply to programs designed for the entire audience:

TV-G **General Audience.** *Most parents would find this program suitable for all ages.* Although this rating does not signify a program designed specifically for children, most parents may let younger children watch this program unattended. It contains little or no violence, no strong language, and little or no sexual dialogue or situations.

TV-PG **Parental Guidance Suggested.** *This program contains material that parents may find unsuitable for younger children.* Many parents may want to watch it with their younger children. The theme itself may call for parental guidance and/or the program contains one or more of the following: moderate violence (V), some sexual situations (S), infrequent coarse language (L), or some suggestive dialogue (D).

TV-14 **Parents Strongly Cautioned.** *This program contains some material that many parents would find unsuitable for children under 14 years of age.* Parents are strongly urged to exercise greater care in monitoring this program and are cautioned against letting children under the age of 14 watch unattended. This program contains one or more of the following: intense violence (V), intense sexual situations (S), strong coarse language (L), or intensely suggestive dialogue (D).

TV-MA **Mature Audience Only.** *This program is specifically designed to be viewed by adults and therefore may be unsuitable for children under 17.* This program contains one or more of the following: graphic violence (V), explicit sexual activity (S), or crude, indecent language (L).

When a program is broadcast, the appropriate icon should appear in the upper left corner of the picture frame for the first 15 seconds. If the program is longer than 1 hour, the icon should be repeated at the beginning of the 2d hour. Guidelines are also displayed in TV listings in a number of newspapers and magazines.

Favorite Prime-Time Television Programs, 2000-2001

Source: Nielsen Media Research

Data are for regularly scheduled network programs in 2000-2001 season through May 23; ranked by average audience percentage. Average audience percentages, or ratings, are estimates of the percentage of all TV-owning households that are watching a particular program. Audience share percentages are estimates of the percentage of those watching TV that are tuned into a particular program. The top 50 programs are listed (there are 2 programs tied for 49th place).

Rank	Program	Average Audience	Audience Share
1.	Survivor II	17.4	27
2.	E.R.	15.2	25
3.	Millionaire - Wed.	13.7	22
4.	Millionaire - Tue	13.0	21
5.	Friends	12.9	21
6.	NFL Monday Night Football	12.7	22
7.	Millionaire - Sun	12.6	18
	Everybody Loves Raymond	12.6	19
9.	Law And Order	12.3	20
10.	Practice, The	12.0	19
11.	CSI	11.7	19
12.	Millionaire - Thu	11.6	18
	West Wing	11.6	18
14.	Will & Grace	11.4	17
15.	60 Minutes	11.1	19
16.	Cursed	10.9	17
	Becker	10.9	16
18.	Temptation Island	10.7	16
	Frasier	10.7	16
20.	Just Shoot Me	10.5	16
	Millionaire - Fri	10.5	18
22.	NFL Monday Showcase	10.2	16
23.	Judging Amy	9.9	16
24.	Touched By An Angel	9.8	15
25.	NYPD Blue	9.7	16
26.	Law And Order: SVU	9.4	17
27.	Weber Show	9.2	14
	JAG	9.2	15
29.	King Of Queens	8.9	14
30.	Yes, Dear	8.7	13
	District, The	8.7	16
	CBS Sunday Movie	8.7	14
33.	My Wife & Kids	8.6	14
	Family Law	8.6	14
35.	Weakest Link	8.5	14
36.	Simpsons	8.3	13
	Dharma & Greg	8.3	13
	20/20 - Fri	8.3	15
	Providence	8.3	15
	60 Minutes II	8.3	13
41.	Drew Carey Show	8.2	13
	Three Sisters	8.2	13
43.	Malcolm in the Middle	8.1	12
	Dateline Fri	8.1	14
45.	Mole, The	8.0	12
	Ally Mcbeal	8.0	12
	Primetime Thursday	8.0	13
48.	X-Files	7.9	12
49.	ABC Monday Night Movie	7.8	12
	Third Watch	7.8	13

All-Time Top Television Programs

Source: Nielsen Media Research, Jan. 1961-Apr. 30, 2000

Estimates exclude unsponsored or joint network telecasts or programs under 30 minutes long. Ranked by rating (percentage of TV-owning households tuned in to the program).

Rank	Program	Telecast date	Network	Rating (%)	Avg. households (in thousands)
1.	M*A*S*H (last episode)	2/28/83	CBS	60.2	50,150
2.	Dallas (Who Shot J.R.?)	11/21/80	CBS	53.3	41,470
3.	Roots-Pt. 8	1/30/77	ABC	51.1	36,380
4.	Super Bowl XVI	1/24/82	CBS	49.1	40,020
5.	Super Bowl XVII	1/30/83	NBC	48.6	40,480
6.	XVII Winter Olympics - 2d Wed.	2/23/94	CBS	48.5	45,690
7.	Super Bowl XX	1/26/86	NBC	48.3	41,490
8.	Gone With the Wind-Pt. 1	11/7/76	NBC	47.7	33,960
9.	Gone With the Wind-Pt. 2	11/8/76	NBC	47.4	33,750
10.	Super Bowl XII	1/15/78	CBS	47.2	34,410
11.	Super Bowl XIII	1/21/79	NBC	47.1	35,090
12.	Bob Hope Christmas Show	1/15/70	NBC	46.6	27,260
	Super Bowl XVIII	1/22/84	CBS	46.4	38,800
13.	Super Bowl XIX	1/20/85	ABC	46.4	39,390
15.	Super Bowl XIV	1/20/80	CBS	46.3	35,330
	Super Bowl XXX	1/28/96	NBC	46.0	44,150
16.	ABC Theater (The Day After)	11/20/83	ABC	46.0	38,550
	Roots-Pt. 6	1/28/77	ABC	45.9	32,680
18.	The Fugitive	8/29/67	ABC	45.9	25,700
20.	Super Bowl XXI	1/25/87	CBS	45.8	40,030
21.	Roots-Pt. 5	1/27/77	ABC	45.7	32,540
	Super Bowl XXVIII	1/30/94	NBC	45.5	42,860
22.	Cheers (last episode)	5/20/93	NBC	45.5	42,360
24.	Ed Sullivan	2/9/64	CBS	45.3	23,240
25.	Super Bowl XXVII	1/31/93	NBC	45.1	41,990
26.	Bob Hope Christmas Show	1/14/71	NBC	45.0	27,050
27.	Roots-Pt. 3	1/25/77	ABC	44.8	31,900
28.	Super Bowl XXXII	1/25/98	NBC	44.5	43,630
	Super Bowl XI	1/9/77	NBC	44.4	31,610
29.	Super Bowl XV	1/25/81	NBC	44.4	34,540
31.	Super Bowl VI	1/16/72	CBS	44.2	27,450
	XVII Winter Olympics - 2d Fri.	2/25/94	CBS	44.1	41,540
32.	Roots-Pt. 2	1/24/77	ABC	44.1	31,400
34.	Beverly Hillbillies	1/8/64	CBS	44.0	22,570
	Roots-Pt. 4	1/26/77	ABC	43.8	31,190
35.	Ed Sullivan	2/16/64	CBS	43.8	22,445
37.	Super Bowl XXIII	1/22/89	NBC	43.5	39,320
38.	Academy Awards	4/7/70	ABC	43.4	25,390
39.	Super Bowl XXXI	1/26/97	FOX	43.3	42,000
	Super Bowl XXXIV	1/30/00	ABC	43.3	43,620
41.	Thorn Birds-Pt. 3	3/29/83	ABC	43.2	35,990
42.	Thorn Birds-Pt. 4	3/30/83	ABC	43.1	35,900
43.	CBS NFC Championship	1/10/82	CBS	42.9	34,960
44.	Beverly Hillbillies	1/15/64	CBS	42.8	21,960
45.	Super Bowl VII	1/14/73	NBC	42.7	27,670

IT'S A FACT: According to preliminary Nielsen figures, the "America: A Tribute to Heroes" telethon Sept. 21, 2001, drew a 65 share of U.S. households, averaging 59.3 million U.S. viewers. Broadcast on more than 30 networks and seen in an estimated 210 countries, the star-studded show raised more than $150 million in pledges to the United Way for victims of the Sept. 11 terrorist attacks and their families.

Top-Rated TV Shows of Each Season, 1950-51 to 2000-2001

Source: Nielsen Media Research; regular series programs, Sept.-May season

Season	Program	Rating[1]	TV-owning households (in thousands)
1950-51	Texaco Star Theatre	61.6	10,320
1951-52	Godfrey's Talent Scouts	53.8	15,300
1952-53	I Love Lucy	67.3	20,400
1953-54	I Love Lucy	58.8	26,000
1954-55	I Love Lucy	49.3	30,700
1955-56	$64,000 Question	47.5	34,900
1956-57	I Love Lucy	43.7	38,900
1957-58	Gunsmoke	43.1	41,920
1958-59	Gunsmoke	39.6	43,950
1959-60	Gunsmoke	40.3	45,750
1960-61	Gunsmoke	37.3	47,200
1961-62	Wagon Train	32.1	48,555
1962-63	Beverly Hillbillies	36.0	50,300
1963-64	Beverly Hillbillies	39.1	51,600
1964-65	Bonanza	36.3	52,700
1965-66	Bonanza	31.8	53,850
1966-67	Bonanza	29.1	55,130
1967-68	Andy Griffith	27.6	56,670
1968-69	Rowan & Martin Laugh-In	31.8	58,250
1969-70	Rowan & Martin Laugh-In	26.3	58,500
1970-71	Marcus Welby, MD	29.6	60,100
1971-72	All in the Family	34.0	62,100
1972-73	All in the Family	33.3	64,800
1973-74	All in the Family	31.2	66,200
1974-75	All in the Family	30.2	68,500
1975-76	All in the Family	30.1	69,600
1976-77	Happy Days	31.5	71,200
1977-78	Laverne & Shirley	31.6	72,900
1978-79	Laverne & Shirley	30.5	74,500
1979-80	60 Minutes	28.2	76,300
1980-81	Dallas	31.2	79,900
1981-82	Dallas	28.4	81,500
1982-83	60 Minutes	25.5	83,300
1983-84	Dallas	25.7	83,800
1984-85	Dynasty	25.0	84,900
1985-86	Cosby Show	33.8	85,900
1986-87	Cosby Show	34.9	87,400
1987-88	Cosby Show	27.8	88,600
1988-89	Roseanne	25.5	90,400
1989-90	Roseanne	23.4	92,100
1990-91	Cheers	21.6	93,100
1991-92	60 Minutes	21.7	92,100
1992-93	60 Minutes	21.6	93,100
1993-94	Home Improvement	21.9	94,200
1994-95	Seinfeld	20.5	95,400
1995-96	E.R.	22.0	95,900
1996-97	E.R.	21.2	97,000
1997-98	Seinfeld	22.0	98,000
1998-99	E.R.	17.8	99,400
1999-2000	Who Wants to Be a Millionaire	18.6	100,800
2000-2001	Survivor II	17.4	102,200

(1) Rating is percent of TV-owning households tuned in to the program. Data prior to 1988-89 exclude Alaska and Hawaii.

100 Leading U.S. Advertisers, 1999

Source: Competitive Media Reporting and Publishers Information Bureau, New York, © copyright 2000
(in thousands of dollars)

Rank	Advertiser	Ad Spending
1.	General Motors Corp.	$2,921,275.2
2.	Procter & Gamble Co	1,738,686.1
3.	DaimlerChrysler AG	1,511,578.9
4.	Philip Morris Cos Inc.	1,371,333.8
5.	Ford Motor Co.	1,191,082.2
6.	Time Warner Inc	1,075,013.2
7.	Walt Disney Co	889,797.2
8.	Johnson & Johnson.	852,038.9
9.	AT&T Corp	828,708.4
10.	MCI Worldcom Inc.	759,898.3
11.	Pfizer Inc	737,974.9
12.	Toyota Motor Corp	709,833.8
13.	Sears Roebuck & Co.	687,293.5
14.	National Amusements Inc	682,780.0
15.	Sony Corp	662,232.3
16.	News Corp Ltd	661,187.9
17.	Diageo PLC	659,392.1
18.	Unilever	640,774.5
19.	McDonald's Corp.	630,343.3
20.	Federated Dept Stores Inc	620,161.3
21.	Honda Motor Co Ltd	561,958.2
22.	Tricon Global Restaurants Inc	551,749.2
23.	Nissan Motor Co Ltd	528,830.0
24.	Ford Motor Co Dlr Assn	524,538.3
25.	Seagram Co Ltd	502,070.7
26.	Sprint Corp	499,151.3
27.	General Mills Inc	482,317.2
28.	Circuit City Stores Inc	481,988.9
29.	May Dept Stores Co	471,542.7
30.	Loreal SA	460,931.1
31.	Ford Motor Co Loc Dlr.	448,568.5
32.	Target Corp	427,163.8
33.	General Motors Corp Loc Dlr.	415,522.0
34.	US Govt.	407,470.7
35.	JC Penney Co Inc	374,313.3
36.	Volkswagen AG.	370,526.7
37.	General Electric Co.	$355,710.5
38.	Pepsico Inc	354,967.0
39.	Coca-Cola Co	354,446.2
40.	Anheuser-Busch Co Inc.	350,531.6
41.	American Express Co.	343,405.9
42.	K Mart Corp.	338,894.5
43.	Verizon Communications	335,357.5
44.	American Home Pdts Corp	330,540.8
45.	Nestle SA	328,736.6
46.	IBM Corp.	321,731.5
47.	Kellogg Co	320,715.0
48.	SmithKline Beecham Plc.	320,596.1
49.	General Motors Corp Dlr Assn	312,094.2
50.	Wal-Mart Stores Inc	305,204.2
51.	Valassis Communications Inc.	305,010.8
52.	Bristol-Myers Squibb Co	301,998.0
53.	Mars Inc	297,606.3
54.	DaimlerChrysler AG Dlr Assn	290,557.9
55.	Toyota Motor Corp Loc Dlr.	288,231.7
56.	Microsoft Corp.	286,423.4
57.	Gap Inc	278,617.9
58.	Visa USA Inc.	268,436.8
59.	DaimlerChrysler AG Loc Dlr.	264,948.9
60.	Morgan Stanley Dean Witter Dscvr & Co.	260,306.5
61.	Schering-Plough Corp.	257,764.3
62.	Bayer AG Group	256,937.9
63.	Glaxo Wellcome Plc	255,163.9
64.	Best Buy Co Inc	253,058.3
65.	Home Depot Inc	249,281.3
66.	Quaker Oats Co	249,075.6
67.	Toyota Motor Corp Dlr Assn	$248,734.8
68.	Clorox Co	243,944.1
69.	SBC Communications Inc.	240,871.8
70.	Novartis AG.	237,863.0
71.	Gateway Inc	231,384.4
72.	Berkshire Hathaway Inc	224,995.8
73.	Nike Inc.	224,312.3
74.	Kimberly-Clark Corp	224,288.0
75.	Not Itemized-Real Estate Dev	221,726.2
76.	Wendy's Intl Inc.	220,016.7
77.	SC Johnson & Son Inc	214,825.6
78.	Reckitt Benckiser PLC	211,016.1
79.	Mattel Inc	203,985.3
80.	Dillard Inc	200,105.2
81.	Sara Lee Corp.	199,677.6
82.	Merck & Co Inc	198,798.7
83.	Hasbro Inc.	194,228.0
84.	Charles Schwab Corp.	190,799.5
85.	Nabisco Group Holdings Corp	187,137.3
86.	Mazda Motor Corp	187,058.9
87.	Mitsubishi Motors Corp	185,857.5
88.	Campbell Soup Co	184,757.9
89.	Alltel	181,029.6
90.	Mastercard Intl Inc	180,872.6
91.	Pharmacia Corp	177,656.0
92.	Nissan Motor Co Ltd Dlr Assn	177,316.9
93.	Astrazeneca PLC	176,187.2
94.	Ralston Purina Co.	172,182.9
95.	Intel Corp	171,769.3
96.	Hershey Foods Corp.	170,083.1
97.	MacAndrews & Forbes Holdings Inc	169,335.6
98.	First Union Corp	168,927.1
99.	FMR Corp	168,490.3
100.	Hewlett-Packard Co	168,414.2
	TOTAL	**$43,135,029.0**

U.S. Ad Spending by Top Categories, 1999

Source: Competitive Media Reporting and Publishers Information Bureau, New York, © copyright 2000
(in thousands of dollars, Jan.-Dec. 1999)

Category	Total	Magazines	Sunday Magazines	Newspapers	Network Television	Spot Television	Syndicated Television	Cable Networks	Radio
Automotive, Access & Equip	$10,454,795.8	1,835,927.3	32,178.2	1,544,638.2	2,573,474.8	3,451,710.2	211,398.5	784,641.7	20,826.9
Retail	8,501,141.7	657,730.0	146,274.1	4,006,854.2	822,194.5	2,241,942.4	87,576.6	505,002.0	33,567.9
Media & Advertising	4,978,059.1	873,616.6	21,039.8	1,806,501.0	896,274.7	752,231.7	150,874.8	419,942.9	57,577.6
Financial	3,965,350.1	750,212.8	31,874.1	998,998.4	981,838.4	458,723.1	62,176.7	662,212.8	19,313.8
Drugs & Proprietary Remedies	3,903,115.6	858,833.4	104,938.5	47,116.0	1,650,110.0	275,531.0	345,725.2	574,529.7	46,331.8
Telecommunications	3,344,281.6	233,696.7	4,188.7	812,485.6	870,683.3	665,859.9	235,260.2	504,970.5	17,136.7
Automotive Dealers & Services	3,311,770.5	9,885.0	1,882.5	2,531,810.0	63,660.6	660,315.0	2,613.1	41,393.7	210.6
Restaurants	3,111,872.4	29,154.4	4,647.1	76,993.9	1,189,395.4	1,358,220.4	133,619.3	307,482.5	12,359.4
Public Transportation, Hotels & Resorts	2,800,402.9	694,704.7	46,695.7	1,122,379.7	192,299.6	446,016.1	27,298.8	245,765.5	25,242.8
Department Stores	2,641,114.8	90,474.9	13,772.1	1,871,244.0	259,753.0	275,343.2	59,572.7	57,851.4	13,103.5
Misc Services & Amusements	2,573,037.9	238,977.6	33,009.7	935,778.2	71,643.8	1,039,573.0	48,305.2	173,805.8	31,944.6
Direct Response Companies	2,511,166.6	1,142,442.4	315,762.6	217,759.1	144,873.8	104,465.3	96,797.4	486,804.0	2,262.0
Computers, Software	2,114,277.3	985,713.9	5,349.5	144,723.8	539,895.6	130,407.9	21,008.2	282,343.5	4,834.9
Ins. & Real Estate	1,926,440.0	269,750.1	23,577.9	631,149.1	301,803.1	405,948.9	47,167.0	205,611.6	41,432.3
Beverages	1,384,038.2	165,916.1	11,193.1	7,872.5	656,363.0	251,404.8	83,523.3	195,873.1	11,892.3
Personal Hygiene & Health	1,365,754.6	288,012.3	3,768.7	1,014.6	588,543.0	94,920.3	164,382.5	220,775.5	4,337.7
Prepared Foods	1,332,927.5	198,147.3	16,584.6	9,986.7	453,443.9	295,180.7	123,042.7	229,726.0	6,815.6
Cosmetics & Beauty	1,309,497.6	581,971.8	7,569.7	17,989.9	449,758.1	86,024.8	77,819.7	82,610.1	5,753.5
Dairy, Produce, Meat & Bakery Goods	1,301,380.8	286,355.1	21,239.0	9,406.6	309,589.9	379,631.0	94,935.0	194,793.3	5,430.9
Confectionery & Snacks	1,191,371.1	159,467.1	5,406.4	2,204.0	508,151.6	98,129.3	167,247.7	235,892.7	14,872.3
Government, Politics & Organizations	1,033,761.4	128,368.8	21,997.3	213,414.3	230,031.6	289,768.0	32,657.6	104,515.4	13,008.4
Discount Department & Variety Stores	1,011,183.1	50,820.3	9,095.6	350,034.8	268,006.7	238,021.9	35,588.5	58,844.7	770.6
Games, Toys & Hobbycraft	876,535.4	81,046.8	42.5	2,452.6	296,605.8	60,936.0	78,329.8	355,920.4	1,201.5
Audio & Video Equipment & Supplies	861,403.2	276,377.1	36,405.9	24,687.5	250,906.5	69,089.3	43,594.6	144,665.3	15,677.0
Beer & Wine	825,372.9	64,068.0	3,103.0	6,536.5	433,807.0	134,398.4	34,145.4	145,325.3	3,989.3

NOTED PERSONALITIES

This chapter contains the following sections:

Widely Known Americans of the Present

Political leaders, journalists, other widely known living persons. As of Sept. 2001. Excludes many in categories listed elsewhere in Noted Personalities, such as Writers of the Present and Entertainment Personalities of the Present, or in the Sports section.

Spencer Abraham, b 6/12/52 (East Lansing, MI), energy sec.
Roger Ailes, b 5/15/40 (Warren, OH), TV exec.
Madeleine K. Albright, b 5/15/37 (Prague, Czech.), former sec. of state.
Lamar Alexander, b 7/3/40 (Maryville, TN), former TN gov., presid. candidate.
Stephen E. Ambrose, b 1/10/36 (Decatur, IL), historian.
Walter H. Annenberg, b 3/13/08 (Milwaukee), publisher, philanthropist.
Roone Arledge, b 7/8/31 (Forest Hills, NY), TV exec.
Richard K. Armey, b 7/7/40 (Cando, ND), House majority leader.
Neil Armstrong, b 8/5/30 (Wapakoneta, OH), former astronaut.
John Ashcroft, b 5/9/42 (Chicago), attorney general.
Bruce Babbitt, b 6/27/38 (Los Angeles), former AZ gov, interior sec.
F. Lee Bailey, b 6/10/33 (Waltham, MA), attorney.
Russell Baker, b 8/14/25 (Loudoun Co., VA), columnist.
Dave Barry, b 7/3/47 (Armonk, NY), humorist.
Marion Barry, b 3/6/36 (Itta Bena, MS), former Wash., DC, mayor.
Gary Bauer, b 1956 (Covington, KY), political activist, former presid. candidate.
Lloyd Bentsen, b 2/11/21 (Mission, TX), former senator, treasury sec., vice-presid. nominee.
Samuel "Sandy" Berger, b 10/28/45 (Sharon, CT), former national security adviser.
Jeff Bezos, b 1/12/64 (Albuquerque, NM), founder and CEO of Amazon.com.
Joseph R. Biden Jr., b 11/20/42 (Scranton, PA), senator (DE).
James H. Billington, b 6/1/29 (Bryn Mawr, PA), librarian of Congress.
Julian Bond, b 1/14/40 (Nashville), civil rights leader.
David Bonior, b 6/6/45 (Detroit), House minority whip.
Daniel Boorstin, b 10/1/14 (Atlanta), historian, former librarian of Congress.
Barbara Boxer, b 11/11/40 (Brooklyn, NY), senator (CA).
Bill Bradley, b 7/28/43 (Crystal City, MO), former senator (NJ), basketball player, presid. candidate.
Ed Bradley, b 6/22/41 (Philadelphia), TV journalist.
James Brady, b 9/17/44 (Grand Rapids, MI), former presid. press sec.; gun control advocate.
Jimmy Breslin, b 10/17/30 (Jamaica, NY), columnist, author.
Stephen Breyer, b 8/15/38 (San Francisco), Sup. Ct. justice.
David Brinkley, b 7/10/20 (Wilmington, NC), former TV journalist.
David Broder, b 9/11/29 (Chicago Heights, IL), journalist.
Tom Brokaw, b 2/6/40 (Webster, SD), TV journalist.
Joyce Brothers, b 9/20/28 (NYC), psychologist.
Edmund G. ("Jerry") Brown Jr., b 4/7/38 (San Francisco), Oakland mayor; former CA gov., pres. candidate.
Willie Brown, b 3/20/34 (Mineola, TX), San Francisco mayor.
Pat Buchanan, b 11/2/38 (Wash., DC), journalist, former presid. candidate.
Art Buchwald, b 10/20/25 (Mt. Vernon, NY), humorist.
William F. Buckley Jr., b 11/24/25 (NYC), columnist, author.
Warren Buffett, b 8/30/30 (Omaha), investor.
Dan Burton, b 6/21/38 (Indianapolis), U.S. representative.
Barbara Bush, b 6/8/25 (Rye, NY), former first lady.
Barbara Bush, b 11/25/81 (Dallas, TX), daughter of Pres. George W. Bush.
George H. W. Bush, b 6/12/24 (Milton, MA), former president.
George W. Bush, b 7/6/46 (New Haven, CT), U.S. president.
Jeb Bush, b 2/11/53 (Houston), FL governor.
Jenna Bush, b 11/25/81(Dallas, TX), daughter of Pres. George W. Bush.
Laura Bush, b 11/4/46 (Midland, TX), first lady.
Robert Byrd, b 11/20/17 (N. Wilkesboro, NC), senator (WV), former majority leader.
Andrew Card, b 5/10/47 (Brockton, MA), White House chief of staff.
Tucker Carlson, b 5/16/69 (San Francisco), journalist, TV commentator.
Jimmy Carter, b 10/1/24 (Plains, GA), former president.
Rosalynn Carter, b 8/18/27 (Plains, GA), former first lady.
James Carville Jr., b 10/25/44 (Fort Benning, GA), political consultant.
Steve Case, b 8/21/58 (Honolulu, HI), AOL Time Warner exec.
Elaine Chao, b 3/26/53 (Taipei, Taiwan), labor sec.
Dick Cheney, b 1/30/41 (Lincoln, NE), U.S. vice president.
Lynne Cheney, b 8/14/41 (Casper, WY), political commentator, wife of Dick Cheney.
Julia Child, b 8/15/12 (Pasadena, CA), TV chef, author.
Noam Chomsky, b 12/7/28 (Philadelphia), linguist; activist.
Connie Chung, b 8/20/46 (Wash., DC), TV journalist.
Liz Claiborne, b 3/31/29 (Brussels, Belg.), fashion designer.
Bill Clinton, b 8/19/46 (Hope, AR), former U.S. president.
Chelsea Clinton, b 2/27/80 (Little Rock, AR), daughter of former Pres. Clinton and Hillary Rodham Clinton.
Hillary Rodham Clinton, b 10/26/47 (Chicago), senator (NY), former first lady.
Johnnie L. Cochran Jr., b 10/2/37 (Shreveport, LA), attorney.
Gary Condit, b 4/21/48 (Salina, OK), U.S. representative (CA).
Ward Connerly, b 6/15/39 (Leesville, LA), anti-affirmative-action activist.
Bob Costas, b 3/22/52 (NYC), TV journalist.
Katie Couric, b 1/7/57 (Wash., DC), TV journalist.
Walter Cronkite, b 11/4/16 (St. Joseph, MO), former TV journalist.
Andrew Cuomo, b 12/6/57 (NYC), HUD sec.
Mario Cuomo, b 6/15/32 (Queens, NY), former NY gov.
Richard M. Daley, b 4/24/42 (Chicago), Chicago mayor.
Thomas Daschle, b 12/9/47 (Aberdeen, SD), Senate majority leader.
Gray Davis, b 12/26/42 (NYC), CA governor.
Tom DeLay, b 4/8/47 (Laredo, TX), House majority whip.
Alan Dershowitz, b 9/1/38 (Brooklyn, NY), attorney.
Barry Diller, b 2/2/42 (San Francisco), TV exec.
Christopher Dodd, b 5/27/44 (Willimantic, CT), senator.
Elizabeth Hanford Dole, b 7/29/36 (Salisbury, NC), former Red Cross pres., transp. sec., labor sec., presid. contender.
Robert Dole, b 7/22/23 (Russell, KS), former Senate majority leader, presid. nominee.
Pete Domenici, b 5/7/32 (Albuquerque, NM), senator.
Sam Donaldson, b 3/11/34 (El Paso, TX), TV journalist.
Elizabeth Drew, b 11/16/35 (Cincinnati), journalist.
Michael S. Dukakis, b 11/3/33 (Boston), former MA gov., presid. nominee.
Roger Ebert, b 6/18/42 (Urbana, IL), film critic.
Marian Wright Edelman, b 6/6/39 (Bennettsville, SC), children's rights advocate.
John Edwards, b 6/10/53 (Robbins, NC), senator.
Edward Egan, Cardinal, b 4/2/32 (Oak Park, IL), Catholic archbishop of New York.
Michael Eisner, b 3/7/42 (NYC), Disney Co. exec.
John Engler, b 10/12/48 (Mount Pleasant, MI), MI gov.
Donald Evans, b 7/27/46 (Houston), commerce sec.
Rev. Jerry Falwell, b 8/11/33 (Lynchburg, VA), TV evangelist, religious educator.
Louis Farrakhan, b 5/11/33 (NYC), Nation of Islam leader.
Russell Feingold, b 3/2/53 (Janesville, WI), senator.
Dianne Feinstein, b 6/22/33 (San Francisco), senator.
Geraldine Ferraro, b 8/26/35 (Newburgh, NY), former U.S. representative, vice-presid. nominee.

Larry Flynt, b 11/1/42 (Magoffin Co., KY), publisher.
Shelby Foote, b 11/17/16 (Greenville, MS), historian.
Malcolm "Steve" Forbes Jr., b 7/18/47 (Morristown, NJ), publisher, former presid. contender.
Betty Ford, b 4/8/18 (Chicago), former first lady.
Gerald R. Ford, b 7/14/13 (Omaha), former president.
John Hope Franklin, b 1/2/15 (Rentiesville, OK), historian.
Betty Friedan, b 2/4/21 (Peoria, IL), author, feminist.
Milton Friedman, b 7/31/12 (Brooklyn, NY), economist.
Francis Fukuyama, b 10/27/52 (Chicago), social scientist.
John Kenneth Galbraith, b 10/15/08 (Iona Station, Ont.), economist.
Bill Gates, b 10/28/55 (Seattle), Microsoft exec.
Henry Louis Gates Jr., b 9/16/50 (Keyser, WV), scholar.
David Geffen, b 2/21/43 (Brooklyn, NY), entertainment exec.
Richard Gephardt, b 1/31/41 (St. Louis, MO), House minority leader.
Louis Gerstner, b 3/1/42 (Mineola, NY), IBM exec.
Newt Gingrich, b 6/17/43 (Harrisburg, PA), former House Speaker.
Ruth Bader Ginsburg, b 3/15/33 (Bklyn, NY), Sup. Ct. justice.
Rudolph Giuliani, b 5/28/44 (Bklyn, NY) NYC mayor.
John Glenn, b 7/18/21 (Cambridge, OH), former senator, astronaut.
Ellen Goodman, b 4/11/41 (Newton, MA), columnist.
Doris Kearns Goodwin, b 1/4/43 (Rockville Centre, NY), historian, TV commentator.
Berry Gordy, b 11/28/29 (Detroit), Motown founder.
Al Gore Jr., b 3/31/48 (Wash., DC), former U.S. vice president, presid. candidate.
Tipper Gore, b 8/19/48 (Wash., DC), wife of Al Gore.
Stephen Jay Gould, b 9/10/41 (NYC), biologist, author.
Rev. Billy Graham, b 11/7/18 (Charlotte, NC), evangelist.
Phil Gramm, b 7/8/42 (Ft. Benning, GA), senator (TX), former presid. contender.
Jeff Greenfield, b 6/10/43 (NYC), TV journalist.
Alan Greenspan, b 3/6/26 (NYC), Fed chairman.
Andrew Grove, b 9/2/36 (Budapest, Hungary), Intel exec.
Bryant Gumbel, b 9/29/48 (New Orleans), TV journalist.
David Halberstam, b 4/10/34 (NYC), journalist, author.
Pete Hamill, b 6/24/35 (Brooklyn, NY), journalist, author.
Paul Harvey, b 9/4/18 (Tulsa, OK), radio journalist.
J. Dennis Hastert, b 1/2/42 (Aurora, IL), House Speaker.
Orrin Hatch, b 3/22/34 (Homestead Park, PA), senator (UT).
Hugh Hefner, b 4/9/26 (Chicago), publisher.
Jesse Helms, b 10/18/21 (Monroe, NC), senator.
Leona Helmsley, b c1920 (NYC), real estate exec.
Heloise, b 4/15/51 (Waco, TX), advice columnist.
Anita Hill, b 7/10/56 (Morris, OK), legal scholar, complainant against Clarence Thomas.
Christopher Hitchens, b 4/13/49 (Portsmouth, England), journalist, author.
James P. Hoffa, b 5/19/41, (Detroit), Teamsters Union head.
Richard Holbrooke, b 4/24/41 (NYC), former U.S. rep. to UN.
David Horowitz, b 1/10/39 (NYC), columnist, author.
Karen Hughes, b 1957 (Paris, France), Bush communications director.
H. Wayne Huizenga, b 12/29/39 (Evergreen Park, IL), entrepreneur, sports exec.
Kay Bailey Hutchison, b 7/22/43 (Galveston, TX), senator.
Henry J. Hyde, b 4/18/24 (Chicago), U.S. representative.
Lee Iacocca, b 10/15/24 (Allentown, PA), former auto exec.
Carl Icahn, b 1936 (Queens, NY), financier.
Don Imus, b 7/23/40 (?) (Riverside, CA), talk-show host.
Patricia Ireland, b 10/19/45 (Oak Park, IL), feminist leader.
Molly Ivins, b 1944 (Texas), columnist.
Rev. Jesse Jackson, b 10/8/41 (Greenville, SC), civil rights leader, former presid. contender.
James Jeffords, b 5/11/34 (Rutland, VT), senator.
Steven Jobs, b 2/24/55 (San Francisco), Apple Computer exec.
Lady Bird Johnson, b 12/22/12 (Karnack, TX), former first lady.
Vernon E. Jordan Jr., b 8/15/35 (Atlanta), attorney, former presid. adviser, civil rights leader.
Donna Karan, b 10/2/48 (Forest Hills, NY), fashion designer.
John R. Kasich, b 5/13/52 (McKees Rocks, PA), U.S. representative (OH).
Jeffrey Katzenberg, b 1950 (NYC), entertainment exec.
Garrison Keillor, b 8/7/42 (Anoka, MN), author, broadcaster.
Jack Kemp, b 7/13/35 (Los Angeles), former vice-presid. nominee, HUD sec., pro football quarterback.
Anthony Kennedy, b 7/23/36 (Sacramento, CA), Sup. Ct. justice.
Caroline Kennedy Schlossberg, b 11/27/57 (Boston), author, daughter of Pres. Kennedy.
Edward M. Kennedy, b 2/22/32 (Brookline, MA), senator.
Bob (Joseph Robert) Kerrey, b 8/27/43 (Lincoln, NE), former senator.
John Kerry, b 12/11/43 (Denver), senator (MA).
Jack Kevorkian, b 5/26/28 (Pontiac, MI), physican, assisted-suicide activist.
Coretta Scott King, b 4/27/27 (Marion, AL), civil rights leader, widow of Martin Luther King Jr.
Larry King, b 11/19/33 (Brooklyn, NY), TV journalist.
Michael Kinsley, b 3/9/51 (Detroit), editor, political commenator.
Jeane J. Kirkpatrick, b 11/19/26 (Duncan, OK), political scientist, former ambassador to UN.
Henry Kissinger, b 5/27/23 (Fuerth, Germany), former sec. of state, national security adviser; Nobel Peace Prize winner.
Calvin Klein, b 11/19/42 (NYC), fashion designer.
Philip H. Knight, b 2/24/38 (Portland, OR), CEO of Nike.
Edward I. Koch, b 12/12/24 (NYC), former NYC mayor.
C. Everett Koop, b 10/14/16 (Brooklyn, NY), former surgeon general.
Ted Koppel, b 2/8/40 (Lancashire, England), TV journalist.
William Kristol, b 12/23/52 (NYC), editor, columnist.
Brian Lamb, b 10/9/41 (Lafayette, IN), cable TV exec., journalist.
Ann Landers, b 7/4/18 (Sioux City, IA), advice columnist.
Estee Lauder, b 9/1/08 (NYC), founder, cosmetics and fragrance firm.
Matt Lauer, b 12/30/57 (NYC), TV journalist.
Ralph Lauren, b 10/14/39 (Bronx, NY), fashion designer.
Patrick Leahy, b 3/31/40 (Montpelier, VT), senator.
Norman Lear, b 7/27/22 (New Haven, CT), TV producer, political activist.
Jim Lehrer, b 5/19/34 (Wichita, KS), TV journalist, author.
James Levine, b 6/23/43 (Cincinnati) conductor.
Monica Lewinsky, b 7/23/73 (San Francisco), former White House intern, key figure in White House scandal.
Joseph Lieberman, b 2/24/42 (Stamford, CT), senator, former vice presid. candidate.
Rush Limbaugh, b 1/12/51 (Cape Girardeau, MO), radio talk-show host.
Gary Locke, b 1/21/50 (Seattle), WA gov.
Trent Lott, b 10/9/41 (Grenada, MS), Senate minority leader.
Shannon Lucid, b 1/14/43 (Shanghai, China), astronaut.
Richard G. Lugar, b 4/4/32 (Indianapolis), senator.
Connie Mack, b 10/29/40 (Philadelphia), senator (FL).
Melquiades Martinez, b 10/23/46 (Sagua la Grande, Cuba), housing and urban development sec.
Janet Maslin, b 8/12/49 (NYC), film critic, author.
Mary Matalin, b 8/19/53 (Chicago), political commentator.
Chris Matthews, b 1945 (Philadelphia), TV journalist.
John McCain, b 8/29/36 (Panama Canal Zone), senator (AZ); former presid. contender.
George McGovern, b 7/19/22 (Avon, SD), former senator, presid. nominee.
John McLaughlin, b 3/29/27 (Providence, RI), TV journalist.
Robert S. McNamara, b 6/9/16 (San Francisco), former defense sec., World Bank head.
Kweisi Mfume, b 10/24/48 (Baltimore), civil rights leader, former U.S. representative.
Kate Michelman, b 8/4/42 (New Jersey), abortion-rights activist.
Kate Millett, b 9/14/34 (St. Paul, MN), author, feminist.
Norman Mineta, b 11/12/31 (San Jose, CA), transportation sec.
George Mitchell, b 8/20/33, (Waterville, ME), former Senate majority leader, N. Ireland peace negotiator.
Walter Mondale, b 1/5/28 (Ceylon, MN), former vice pres., senator, presid. nominee.
Bill Moyers, b 6/5/34 (Hugo, OK), TV journalist, author.
Daniel P. Moynihan, b 3/16/27 (Tulsa, OK), former senator, author.
Robert S. Mueller III, b 8/7/44 (NYC), FBI director.
Rupert Murdoch, b 3/11/31 (Melbourne, Aust.), media exec.
Ralph Nader, b 2/27/34 (Winsted, CT), consumer advocate, former presid. candidate.
John Negroponte, b 7/21/39 (London, Eng.), U.S. representative to UN.
Don Nickles, b 12/6/48 (Ponca City, OK), Senate minority whip.
Oliver North, b 10/7/43 (San Antonio, TX), radio talk-show host, former National Security Council aide.
Eleanor Holmes Norton, b 6/13/37 (Wash., DC), U.S. House delegate.
Gale Norton, b 3/11/54 (Wichita, KS), interior sec.
Robert Novak, b 2/26/31 (Joliet, IL), journalist.
Sam Nunn, b 9/8/38 (Perry, GA), former senator.
Sandra Day O'Connor, b 3/26/30 (El Paso, TX), Sup. Ct. justice.
Paul O'Neill, b 12/4/35 (Pittsburgh), treasury sec.
Michael Ovitz, b 12/4/46 (Encino, CA), entertainment exec.
Camille Paglia, b 1947 (Endicott, NY), scholar, author.
Roderick R. Paige, b 6/17/33 (Monticello, MS), education sec.
Leon F. Panetta, b 6/28/38 (Monterey, CA), former White House chief of staff, U.S. representative.
Rosa Parks, b 2/4/13 (Tuskegee, AL), civil rights activist.
George Pataki, b 6/24/45 (Peekskill, NY), NY gov.
Jane Pauley, b 10/31/50 (Indianapolis), TV journalist.
H. Ross Perot, b 6/27/30 (Texarkana, TX), entrepreneur, former presid. nominee.
George Plimpton, b 3/18/27 (NYC), author, editor.

Alvin F. Poussaint, b 5/15/34 (NYC), child psychiatrist.
Colin Powell, b 4/5/37 (NYC), sec. of state; former national security adviser, Joint Chiefs of Staff chairman.
Anthony Principi, b 4/16/44 (Bronx, NY), sec. of veterans affairs.
Dan Quayle, b 2/4/47 (Indianapolis), former U.S. vice pres., senator, presid. contender.
Anna Quindlen, b 7/8/53 (Philadelphia), author, columnist.
Dan Rather, b 10/31/31 (Wharton, TX), TV journalist.
Nancy Reagan, b 7/6/23 (NYC), former first lady.
Ronald Reagan, b 2/6/11 (Tampico, IL), former president.
Sumner Redstone, b 5/27/23 (Boston), media exec.
Ralph Reed, b 6/24/61 (Portsmouth, VA), political adviser.
William Rehnquist, b 10/1/24 (Milwaukee), Sup. Ct. chief justice.
Robert B. Reich, b 6/24/46 (Scranton, PA), economist, former labor sec.
Janet Reno, b 7/21/38 (Miami, FL), former attorney general.
Condoleezza Rice, b 11/14/54 (Birmingham, AL), national security advisor.
Ann Richards, b 9/3/33 (Waco, TX), former TX gov.
Bill Richardson, b 11/15/47 (Pasadena, CA), former energy sec., UN ambassador, congressman.
Sally K. Ride, b 5/26/51 (Encino, CA), former astronaut.
Tom (Thomas Joseph) Ridge, b 8/26/45 (Munhall, PA), director, Office of Homeland Security; former PA gov.
Richard Riordan, b 1930 (Flushing, NY), Los Angeles mayor.
Cokie Roberts, b 12/27/43 (New Orleans), TV journalist.
Rev. Oral Roberts, b 1/24/18 (nr. Ada, OK), TV evangelist, educator.
Rev. Pat Robertson, b 3/22/30 (Lexington, VA), religious broadcasting exec, former presid. contender.
David Rockefeller, b 6/12/15 (NYC), banker.
John D. "Jay" Rockefeller 4th, b 6/18/37 (NYC), senator, former WV gov.
Laurance S. Rockefeller, b 5/26/10 (NYC), philanthropist.
Fred Rogers, b 3/20/28 (Latrobe, PA), children's TV personality.
Andy Rooney, b 1/14/19 (Albany, NY), TV commentator.
Charlie Rose, b 1/5/42 (Henderson, NC), TV journalist.
Karl Rove, b 12/25/50 (Denver) political consultant.
Louis Rukeyser, b 1/30/33 (NYC), TV journalist, financial analyst.
Donald Rumsfeld, b 7/9/32 (Chicago), defense sec.
Tim Russert, b 5/7/50 (Buffalo, NY), TV journalist.
William Safire, b 12/17/29 (NYC), columnist.
Diane Sawyer, b 12/22/45 (Glasgow, KY), TV journalist.
Antonin Scalia, b 3/11/36 (Trenton, NJ), Sup. Ct. justice.
Phyllis Schlafly, b 8/15/24 (St. Louis, MO) political activist.
Arthur Schlesinger Jr., b 10/15/17 (Columbus, OH), historian.
Patricia Schroeder, b 7/30/40 (Portland, OR), former U.S. representative.
Rev. Robert Schuller, b 9/16/26 (Alton, IA), TV evangelist.
Charles Schumer, b. 11/23/50 (Brooklyn, NY), senator.
H. Norman Schwarzkopf, b 8/22/34 (Trenton, NJ), former military leader.
Allan H. ("Bud") Selig, b 7/30/34 (Milwaukee), baseball comm.
Donna E. Shalala, b 2/14/41 (Cleveland), former sec. of health and human services.
Al Sharpton, b 10/3/54 (NYC), activist, civil rights leader.
Bernard Shaw, b 1940 (Chicago) TV journalist.
Maria Shriver, b 11/6/55 (Chicago), TV journalist.
George P. Shultz, b 12/13/20 (NYC), former sec. of state, other cabinet posts.
O. J. Simpson, b 7/9/47 (San Francisco), former football star, murder defendant.
Liz Smith, b 2/2/23 (Ft. Worth, TX), gossip columnist.
David H. Souter, b 9/17/39 (Melrose, MA), Sup. Ct. justice.
George Soros, b 8/12/30 (Budapest, Hungary), financier, philanthropist.
Arlen Specter, b 2/12/30 (Wichita, KS), senator (PA).
Kenneth Starr, b 7/21/46 (Vernon, TX), former Whitewater independent counsel.
Shelby Steele, b 1/1/46 (Chicago), scholar, critic.
George Steinbrenner, b 7/4/30 (Rocky River, OH), NY Yankees owner.
Gloria Steinem, b 3/25/34 (Toledo, OH), author, feminist.
George Stephanopoulos, b 2/10/61 (Fall River, MA), TV journalist, former presid. adviser.
David J. Stern, b 9/22/42 (NYC), basketball comm.
John Paul Stevens, b 4/20/20 (Chicago), Sup. Ct. justice.
Martha Stewart, b 8/3/41 (Nutley, NJ), homemaking adviser, entrepreneur.
Arthur Ochs Sulzberger Jr., b 9/22/51 (Mt. Kisco, NY), newspaper publisher.
John H. Sununu, b 7/2/39 (Havana, Cuba), political commentator, former White House chief of staff.
John J. Sweeney, b 5/5/34 (NYC), AFL-CIO pres.
Paul Tagliabue, b 11/24/40 (Jersey City, NJ), football comm.
George Tenet, b 1/5/53 (Queens, NY), CIA director.
Clarence Thomas, b 6/23/48 (Savannah, GA), Sup. Ct. justice.
Helen Thomas, b 8/4/20 (Winchester, KY), journalist.
R. David Thomas, b 7/2/32 (Atlantic City, NJ), Wendy's founder.
Fred Thompson, b 8/19/42 (Sheffield, AL), senator.
Hunter S. Thompson, b 7/18/37 (Louisville, KY), journalist.
Tommy G. Thompson, b 11/19/41 (Elroy, WI), sec. of health and human services, former WI gov.
J. Strom Thurmond, b 12/5/02 (Edgefield, SC), senator.
Laurence Tisch, b 3/15/23 (NYC), entertainment exec.
Margaret Truman, b 2/17/24 (Independence, MO), author, daughter of Pres. Truman.
Donald Trump, b 1946 (NYC), real estate exec.
Ted Turner, b 11/19/38 (Cincinnati), TV exec, philanthropist.
Peter Ueberroth, b 9/2/37 (Chicago), sports & travel exec.
Jack Valenti, b 9/5/21 (Houston), movie industry exec.
Abigail Van Buren, b 7/4/18 (Sioux City, IA), advice columnist.
Ann Veneman, b 6/29/49 (Sacramento, CA), agriculture sec.
Jesse Ventura, b 7/15/51 (Minneapolis), MN governor, former wrestler.
Mike Wallace, b 5/9/18 (Brookline, MA), TV journalist.
Barbara Walters, b 9/25/31 (Boston), TV journalist.
J. C. Watts Jr. b 11/18/57 (Eufaula, OK), U.S. representative, Republican Conference chair.
Andrew Weil, b 6/8/42 (Philadelphia), health adviser.
Caspar Weinberger, b 8/18/17 (San Francisco), business exec, former defense sec., other cabinet posts.
Harvey Weinstein, b 3/19/52 (NYC), movie executive.
Jack Welch, b 1935 (Salem, MA), former General Electric CEO.
Jann Wenner, b 1/7/46 (NYC), publisher.
Cornel West, b 6/23/53 (Tulsa, OK), scholar, critic.
Ruth Westheimer, b. 1928 (Germany), human sexuality expert.
Christine Todd Whitman, b 9/26/46 (NYC), EPA head, former NJ gov.
Elie Wiesel, b 9/30/28 (Sighet, Romania), scholar, author, Nobel Peace Prize winner.
L. Douglas Wilder, b 1/17/31 (Richmond, VA), former VA gov.
George Will, b 5/4/41 (Champaign, IL), journalist, author.
Jody Williams, b 10/9/50 (Brattleboro, VT), anti-landmine activist, Nobel Peace Prize winner.
Pete Wilson, b 8/23/33 (Lake Forest, IL), former CA gov.
Paul Wolfowitz, b 12/22/43 (NYC), deputy defense secretary.
Bob Woodward, b 3/26/43 (Geneva, IL), journalist, author.
Paula Zahn, 2/24/56 (Omaha, NE), TV journalist.
Mortimer Zuckerman, b 6/4/37 (Montreal, Quebec), publisher, columnist.

African-Americans of the Past

See also other categories.

Ralph David Abernathy, 1926-90, organizer, 1957, pres., 1968, Southern Christian Leadership Conf.
Crispus Attucks, c1723-70, leader of group of colonists that clashed with British soldiers in 1770 Boston Massacre.
Benjamin Banneker, 1731-1806, inventor, astronomer, mathematician, gazetteer.
Daisy Bates, 1920?-99, Arkansas, civil rights leader who fought for school integration.
James P. Beckwourth, 1798-c1867, western fur trader, scout; Beckwourth Pass in N California named for him.
Mary McCleod Bethune, 1875-1955, adviser to FDR and Truman; founder, pres., Bethune-Cookman College.
Henry Blair, 19th cent., pioneer inventor; obtained patents for a corn-planter, 1834, and cotton-planter, 1836.
Edward Bouchet, 1852-1918, first black to earn a PhD at a U.S. university (Yale, 1876).
Tom Bradley, 1917-98, first African-American mayor of L.A.
Sterling A. Brown, 1901-89, poet, literature professor; helped establish African-American literary criticism.
William Wells Brown, 1815-84, memoirist, ex-slave; first African American to publish a novel, 1853.
Ralph Bunche, 1904-71, first black to win the Nobel Peace Prize, 1950; undersecretary of the UN, 1950.
Stokely Carmichael (Kwame Toure), 1941-98, black power activist.
George Washington Carver, 1864-1943, botanist, chemist, and educator; transformed the economy of the South.
Charles Waddell Chesnutt, 1858-1932, author known for his short stories, such as in *The Conjure Woman* (1899).
Eldridge Cleaver, 1935-98, revolutionary social critic; former "minister of information" for Black Panthers; *Soul on Ice.*
James Cleveland, 1931-91, composer, musician, singer; first black gospel artist to appear at Carnegie Hall.
Countee Cullen, 1903-46, poet, prominent in the Harlem Renaissance of the 1920s; *The Black Christ.*
Benjamin O. Davis Sr., 1877-1970, first African-American general, 1940, in U.S. Army.
William L. Dawson, 1886-1970, Illinois congressman, first black chairman of a major U.S. House committee.

Aaron Douglas, 1900-79, "father of black American art."
Frederick Douglass, 1817-95, author, editor, orator, diplomat; edited abolitionist weekly *The North Star.*
St. Clair Drake, 1911-90, black studies pioneer, *Black Metropolis* (1945, with Horace R. Cayton).
Charles Richard Drew, 1904-50, physician, pioneered in development of blood banks.
William Edward Burghardt (W.E.B.) Du Bois, 1868-1963, historian, sociologist; an NAACP founder, 1909.
Paul Laurence Dunbar, 1872-1906, poet, novelist; won fame with *Lyrics of Lowly Life,* 1896.
Jean Baptiste Point du Sable, c1750-1818, pioneer trader and first settler of Chicago, 1779.
Medgar Evers, 1925-63, Mississippi civil rights leader; campaigned to register black voters; assassinated.
James Farmer, 1920-99, civil rights leader; founded Congress of Racial Equality.
Henry O. Flipper, 1856-1940, first African-American to graduate, 1877, from West Point.
Marcus Garvey, 1887-1940, founded Universal Negro Improvement Assn., 1911.
Ewart Guinier, 1911-90, trade unionist; first chairman of Harvard Univ.'s Dept. of African American Studies.
Prince Hall, 1735-1807, activist; founded black Freemasonry; served in American Revolutionary war.
Jupiter Hammon, c1720-1800, poet; first African-American to have his works published, 1761.
Lorraine Hansberry, 1930-65, playwright; won New York Drama Critics Circle Award, 1959; *A Raisin in the Sun.*
William H. Hastie, 1904-76, first black federal judge, appointed 1937; governor of Virgin Islands, 1946-49.
Matthew A. Henson, 1866-1955, member of Peary's 1909 expedition to the North Pole; placed U.S. flag at the pole.
Chester Himes, 1909-84, novelist; *Cotton Comes to Harlem.*
William A. Hinton, 1883-1959, physician, developed tests for syphilis; first black prof., 1949, at Harvard Med. School.
Charles Hamilton Houston, 1895-1950, lawyer, Howard University instructor, champion of minority rights,
Langston Hughes, 1902-67, poet, lyric writer, author; a major influence in 1920s Harlem Renaissance.
Daniel James Jr., 1920-78, first black 4-star general, 1975; commander, North American Air Defense Command.
Henry Johnson, 1897-1929, first American decorated by France in WW1 with the Croix de Guerre.
James Weldon Johnson, 1871-1938, poet, novelist, diplomat; lyricist for *Lift Every Voice and Sing.*
Barbara Jordan, 1936-96, congresswoman, orator, educator.; first black woman to win a seat in the Texas senate, 1966.
Ernest Everett Just, 1883-1941, marine biologist; studied egg development; author, *Biology of Cell Surfaces,* 1941.
Rev. Martin Luther King Jr., 1929-68, civil rights leader; led 1956 Montgomery, AL, boycott; founder, pres., Southern Christian Leadership Conference, 1957; Nobel laureate (1964); assassinated.
Lewis H. Latimer, 1848-1928, associate of Edison; supervised installation of first electric street lighting in NYC.
Mickey Leland, 1944-89, U.S. representative from Texas, 1978 until death; chairman of Congressional Black Caucus.
Henry Lewis, 1932-1996, (U.S.) conductor; first black conductor and musical director of major American orchestra.
Malcolm X (Little), 1925-65, Black Muslim, black nationalist leader; promoted black pride; assassinated.
Thurgood Marshall, 1908-93, first black U.S. solicitor general, 1965; first black justice of U.S. Sup. Ct., 1967-91.
Jan Matzeliger, 1852-89, invented lasting machine, patented 1883, which revolutionized the shoe industry.
Benjamin Mays, 1895-1984, educator, civil rights leader; headed Morehouse College, 1940-67.
Ronald McNair, 1950-86, physicist, astronaut; killed in *Challenger* explosion.
Dorie Miller, 1919-43, Navy hero of Pearl Harbor attack.
Elijah Muhammad, 1897-1975, founded Nation of Islam, 1931.
Pedro Alonzo Niño, navigator of Columbus's Niña, 1492.
Frederick D. Patterson, 1901-88, founder of United Negro College Fund, 1944.
Harold R. Perry, 1916-91, first black American Roman Catholic bishop in the 20th cent.
Adam Clayton Powell Jr., 1908-72, early civil rights leader, congressman, 1945-69.
Joseph H. Rainey, 1832-87, first black elected to U.S. House, 1869, from South Carolina.
A. Philip Randolph, 1889-1979, organized Brotherhood of Sleeping Car Porters, 1925; an organizer of 1941 and 1963 March on Washington movements.
Hiram R. Revels, 1822-1901, first African-American U.S. senator, elected in Mississippi, served 1870-71.
Norbert Rillieux, 1806-94; invented a vacuum pan evaporator, 1846, revolutionizing sugar-refining industry.
Paul Robeson, 1898-1976, actor, singer, civil rights activist; ostracized by conservatives in the 1950s.
Jackie Robinson, 1919-72, first African-American in major league baseball, 1947, and the Baseball Hall of Fame, 1962.
Carl T. Rowan, 1925-2000, reporter, columnist, author.
Bayard Rustin, 1910-87, an organizer of the 1963 March on Washington; exec. director, A. Philip Randolph Institute.
Peter Salem, at the Battle of Bunker Hill, June 17, 1775, shot and killed British commander Maj. John Pitcairn.
Carl Stokes, 1927-1996, first black mayor of a major American city (Cleveland), 1967-72.
Willard Townsend, 1895-1957, organized the United Transport Service Employees (redcaps), 1935.
Sojourner Truth, 1797-1883, born Isabella Baumfree; preacher, abolitionist; worked for black educ. opportunity.
Harriet Tubman, 1823-1913, Underground Railroad conductor, nurse and spy for Union Army in the Civil War.
Nat Turner, 1800-31, led most significant of more than 200 slave revolts in U.S., in Southampton, VA; hanged.
Booker T. Washington, 1856-1915, founder, 1881, and first pres. of Tuskegee Institute; *Up From Slavery.*
Harold Washington, 1922-87, first black mayor of Chicago.
Robert C. Weaver, 1907-97, first African-American appointed to cabinet; secretary of HUD.
Phillis Wheatley, c1753-84, poet; 2d American woman and first black woman to be published, 1770.
Walter White, 1893-1955, exec. sec., NAACP, 1931-55.
Roy Wilkins, 1901-81, exec. director, NAACP, 1955-77.
Daniel Hale Williams, 1858-1931, surgeon; performed one of first two open-heart operations, 1893.
Granville T. Woods, 1856-1910, invented third-rail system now used in subways, and automatic air brake.
Carter G. Woodson, 1875-1950, historian; founded Assn. for the Study of Negro Life and History.
Frank Yerby, 1916-91, first best-selling African-American novelist; *The Foxes of Harrow.*
Coleman A. Young, 1918-97, first Afr.-Amer. mayor of Detroit, 1974-93.

Architects and Some of Their Achievements

Max Abramovitz, b 1908, Avery Fisher Hall, NYC; U.S. Steel Bldg. (now USX Towers), Pittsburgh, PA.
Henry Bacon, 1866-1924, Lincoln Memorial, Wash., DC.
Pietro Belluschi, 1899-1994, Juilliard School, Lincoln Center, Pan Am, now MetLife, Bldg. (with Walter Gropius), NYC.
Marcel Breuer, 1902-81, Whitney Museum of American Art (with Hamilton Smith), NYC.
Charles Bulfinch, 1763-1844, State House, Boston; Capitol (part), Wash., DC.
Gordon Bunshaft, 1909-90, Lever House, Park Ave, NYC; Hirshhorn Museum, Wash., DC.
Daniel H. Burnham, 1846-1912, Union Station, Wash. DC; Flatiron Bldg., NYC.
Irwin Chanin, 1892-1988, theaters, skyscrapers, NYC.
Lucio Costa, 1902-98, master plan for city of Brasilia, with Oscar Niemeyer.
Ralph Adams Cram, 1863-1942, Cath. of St. John the Divine, NYC; U.S. Military Acad. (part), West Point, NY.
R. Buckminster Fuller, 1895-1983, U.S. Pavilion (geodesic domes), Expo 67, Montreal.
Frank O. Gehry, b 1929, Guggenheim Museum, Bilbao, Spain; Experience Music Project, Seattle, WA.
Cass Gilbert, 1859-1934, Custom House, Woolworth Bldg., NYC; Supreme Court Bldg., Wash., DC.
Bertram G. Goodhue, 1869-1924, Capitol, Lincoln, NE; St. Thomas's Church, St. Bartholomew's Church, NYC.
Michael Graves, b 1934, Portland Bldg., Portland, OR; Humana Bldg., Lexington, KY.
Walter Gropius, 1883-1969, Pan Am Bldg. (now MetLife Bldg.) (with Pietro Belluschi), NYC.
Lawrence Halprin, b 1916, Ghirardelli Sq., San Francisco; Nicollet Mall, Minneapolis; FDR Memorial, Wash., DC.
Peter Harrison, 1716-75, Touro Synagogue, Redwood Library, Newport, RI.
Wallace K. Harrison, 1895-1981, Metropolitan Opera House, Lincoln Center, NYC.
Thomas Hastings, 1860-1929, NY Public Library (with John Carrère), Frick Mansion, NYC.
James Hoban, 1762-1831, White House, Wash., DC.
Raymond Hood, 1881-1934, Rockefeller Center (part), Daily News, NYC; Tribune, Chicago, IL.
Richard M. Hunt, 1827-95, Metropolitan Museum (part), NYC; National Observatory, Wash., DC.
Helmut Jahn, b 1940, United Airlines Terminal, O'Hare Airport, Chicago.
William Le Baron Jenney, 1832-1907, Home Insurance (demolished 1931), Chicago, IL.

Philip C. Johnson, b 1906, AT&T headquarters (now 550 Madison Ave.), NYC; Transco Tower, Houston, TX.
Albert Kahn, 1869-1942, General Motors Bldg., Detroit, MI.
Louis Kahn, 1901-74, Salk Laboratory, La Jolla, CA; Yale Art Gallery, New Haven, CT.
Christopher Grant LaFarge, 1862-1938, Roman Catholic Chapel, West Point, NY.
Benjamin H. Latrobe, 1764-1820, Capitol (part), Wash., DC; State Capitol Bldg., Richmond, VA.
Le Corbusier, (Charles-Edouard Jeanneret), 1887-1965, Salvation Army Hostel and Swiss Dormitory, both Paris; master plan for cities of Algiers and Buenos Aires.
William Lescaze, 1896-1969, Philadelphia Savings Fund Society; Borg-Warner Bldg., Chicago.
Maya Lin, b 1959, Vietnam Veterans Memorial, Wash., DC.
Charles Rennie Mackintosh, 1868-1928, Glasgow School of Art; Hill House, Helensburgh.
Bernard R. Maybeck, 1862-1957, Hearst Hall, Univ. of CA, Berkeley; First Church of Christ Scientist, Berkeley, CA.
Charles F. McKim, 1847-1909, Public Library, Boston; Columbia Univ. (part), NYC.
Charles M. McKim, b 1920, KUHT-TV Transmitter Bldg., Lutheran Church of the Redeemer, Houston, TX.
Richard Meier, b 1934, Getty Center Museum, Los Angeles, CA; High Museum of Art, Atlanta, GA.
Ludwig Mies van der Rohe, 1886-1969, Seagram Bldg., (with Philip C. Johnson), NYC; National Gallery, Berlin.
Robert Mills, 1781-1855, Washington Monument, Wash., DC.
Charles Moore, 1925-93, Sea Ranch, near San Francisco; Piazza d'Italia, New Orleans, LA.
Richard J. Neutra, 1892-1970, Mathematics Park, Princeton, NJ; Orange Co. Courthouse, Santa Ana, CA.
Oscar Niemeyer, b 1907, government buildings, Brasilia Palace Hotel, all Brasilia.
Gyo Obata, b 1923, Natl. Air & Space Museum, Smithsonian Inst., Wash., DC; Dallas-Ft. Worth Airport.
Frederick L. Olmsted, 1822-1903, Central Park, NYC; Fairmount Park, Philadelphia, PA.
I(eoh) M(ing) Pei, b 1917, East Wing, Natl. Gallery of Art, Wash., DC; Pyramid, The Louvre, Paris; Rock & Roll Hall of Fame and Museum, Cleveland, OH.
Cesar Pelli, b 1926, World Financial Center, Carnegie Hall Tower, NYC; Petronas Twin Towers, Malaysia.
William Pereira, 1909-85, Cape Canaveral; Transamerica Bldg., San Francisco, CA.
John Russell Pope, 1874-1937, National Gallery, Wash., DC.
John Portman, b 1924, Peachtree Center, Atlanta, GA.
George Browne Post, 1837-1913, NY Stock Exchange; Capitol, Madison, WI.
James Renwick Jr., 1818-95, Grace Church, St. Patrick's Cath., NYC.; Corcoran (now Renwick) Gallery, Wash., DC.
Henry H. Richardson, 1838-86, Trinity Church, Boston, MA.
Kevin Roche, b 1922, Oakland Museum, Oakland, CA; Fine Arts Center, University of Massachusetts, Amherst.
James Gamble Rogers, 1867-1947, Columbia-Presbyterian Medical Center, NYC; Northwestern Univ., Evanston, IL.
John Wellborn Root, 1887-1963, Palmolive Bldg., Chicago; Hotel Statler, Wash., DC.
Paul Rudolph, 1918-97, Jewitt Art Center, Wellesley Colllege, MA; Art & Architecture Bldg., Yale Univ., New Haven, CT.
Eero Saarinen, 1910-61, Gateway to the West Arch, St. Louis, MO; Trans World Flight Center, NYC.
Louis Skidmore, 1897-1962, Atomic Energy Commission town site, Oak Ridge, TN; Terrace Plaza Hotel, Cincinnati, OH.
Clarence S. Stein, 1882-1975, Temple Emanu-El, NYC.
Edward Durell Stone, 1902-78, U.S. Embassy, New Delhi, India; (H. Hartford) Gallery of Modern Art, NYC.
Louis H. Sullivan, 1856-1924, Auditorium Bldg., Chicago, IL.
Richard Upjohn, 1802-78, Trinity Church, NYC.
Max O. Urbahn, 1912-95, Vehicle Assembly Bldg., Cape Canaveral, FL.
Robert Venturi, b 1925, Gordon Wu Hall, Princeton, NJ; Mielparque Nikko Kirifuri Resort, Japan.
Ralph T. Walker, 1889-1973, NY Telephone Bldg. (now NYNEX); IBM Research Lab, Poughkeepsie, NY.
Roland A. Wank, 1898-1970, Cincinnati Union Terminal, OH; head architect (1933-44), Tennessee Valley Authority.
Stanford White, 1853-1906, Washington Arch in Washington Square Park, first Madison Square Garden, NYC.
Frank Lloyd Wright, 1867-1959, Imperial Hotel, Tokyo; Guggenheim Museum, NYC; Marin County Civic Center, San Rafael; Kaufmann "Fallingwater" house, Bear Run, PA.; Taliesen West, Scottsdale, AZ.
William Wurster, 1895-1973, Ghirardelli Sq., San Francisco; Cowell College, UC, Berkeley, CA.
Minoru Yamasaki, 1912-86, World Trade Center, NYC.

Artists, Photographers, and Sculptors of the Past

Artists are painters unless otherwise indicated.

Berenice Abbott, 1898-1991, (U.S.) photographer. Documentary of New York City, *Changing New York* (1939).
Ansel Easton Adams, 1902-84, (U.S.) photographer. Landscapes of the American Southwest.
Washington Allston, 1779-1843, (U.S.) landscapist. *Belshazzar's Feast.*
Albrecht Altdorfer, 1480-1538, (Ger.) landscapist.
Andrea del Sarto, 1486-1530, (It.) frescoes. *Madonna of the Harpies.*
Fra Angelico, c1400-55, (It.) Renaissance muralist. *Madonna of the Linen Drapers' Guild.*
Diane Arbus, 1923-71, (U.S.) photographer. Disturbing images.
Alexsandr Archipenko, 1887-1964, (U.S.) sculptor. *Boxing Match, Medranos.*
Eugène Atget, 1856-1927, (Fr.) photographer. Paris life.
John James Audubon, 1785-1851, (U.S.) *Birds of America.*
Hans Baldung-Grien, 1484-1545, (Ger.) *Todentanz.*
Ernst Barlach, 1870-1938, (Ger.) Expressionist sculptor. *Man Drawing a Sword.*
Frederic-Auguste Bartholdi, 1834-1904, (Fr.) *Liberty Enlightening the World, Lion of Belfort.*
Fra Bartolommeo, 1472-1517, (It.) *Vision of St. Bernard.*
Aubrey Beardsley, 1872-98, (Br.) illustrator. *Salome, Lysistrata, Morte d'Arthur, Volpone.*
Max Beckmann, 1884-1950, (Ger.) Expressionist. *The Descent From the Cross.*
Gentile Bellini, 1426-1507, (It.) Renaissance. *Procession in St. Mark's Square.*
Giovanni Bellini, 1428-1516, (It.) *St. Francis in Ecstasy.*
Jacopo Bellini, 1400-70, (It.) *Crucifixion.*
George Wesley Bellows, 1882-1925, (U.S.) sports artist, portraitist, landscapist. *Stag at Sharkey's, Edith Clavell.*
Thomas Hart Benton, 1889-1975, (U.S.) American regionalist. *Threshing Wheat, Arts of the West.*
Gianlorenzo Bernini, 1598-1680, (It.) Baroque sculpture. *The Assumption.*
Albert Bierstadt, 1830-1902, (U.S.) landscapist. *The Rocky Mountains, Mount Corcoran.*
George Caleb Bingham, 1811-79, (U.S.) *Fur Traders Descending the Missouri.*
William Blake, 1752-1827, (Br.) engraver. *Book of Job, Songs of Innocence, Songs of Experience.*
Rosa Bonheur, 1822-99, (Fr.) *The Horse Fair.*
Pierre Bonnard, 1867-1947, (Fr.) Intimist. *The Breakfast Room, Girl in a Straw Hat.*
Gutzon Borglum, 1871-1941, (U.S.) sculptor. Mt. Rushmore Memorial.
Hieronymus Bosch, 1450-1516, (Flem.) religious allegories. *The Crowning With Thorns.*
Sandro Botticelli, 1444-1510, (It.) Renaissance. *Birth of Venus, Adoration of the Magi, Guiliano de'Medici.*
Margaret Bourke-White, 1906-71, (U.S.) photographer, photojournalist. WW2, USSR, rural South during the Depression.
Mathew Brady, c1823-96, (U.S.) photographer. Official photographer of the Civil War.
Constantin Brancusi, 1876-1957, (Romanian-Fr.) Nonobjective sculptor. *Flying Turtle, The Kiss.*
Georges Braque, 1882-1963, (Fr.) Cubist. *Violin and Palette.*
Pieter Bruegel the Elder, c1525-69, (Flem.) *The Peasant Dance, Hunters in the Snow, Magpie on the Gallows.*
Pieter Bruegel the Younger, 1564-1638, (Flem.) *Village Fair, The Crucifixion.*
Edward Burne-Jones, 1833-98, (Br.) Pre-Raphaelite artist-craftsman. *The Mirror of Venus.*
Alexander Calder, 1898-1976, (U.S.) sculptor. *Lobster Trap and Fish Tail.*
Julia Cameron, 1815-79, (Br.) photographer. Considered one of the most important portraitists of the 19th cent.
Robert Capa (Andrei Friedmann), 1913-54, (Hung.-U.S.) photographer. War photojournalist; invasion of Normandy.
Michelangelo Merisi da Caravaggio, 1573-1610, (It.) Baroque. *The Supper at Emmaus.*
Emily Carr, 1871-1945, (Can.) landscapist. *Blunden Harbour, Big Raven, Rushing Sea of Undergrowth.*
Carlo Carrà, 1881-1966, (It.) Metaphysical school. *Lot's Daughters, The Enchanted Room.*
Mary Cassatt, 1844-1926, (U.S.) Impressionist. *The Cup of Tea, Woman Bathing, The Boating Party.*
George Catlin, 1796-1872, (U.S.) American Indian life. *Gallery of Indians, Buffalo Dance.*
Benvenuto Cellini, 1500-71, (It.) Mannerist sculptor, goldsmith. *Perseus and Medusa.*
Paul Cézanne, 1839-1906, (Fr.) *Card Players, Mont-Sainte-Victoire With Large Pine Trees.*
Marc Chagall, 1887-1985, (Russ.) Jewish life and folklore. *I and the Village, The Praying Jew.*

Jean Simeon Chardin, 1699-1779, (Fr.) still lifes. *The Kiss, The Grace.*
Frederick Church, 1826-1900, (U.S.) Hudson River school. *Niagara, Andes of Ecuador.*
Giovanni Cimabue, 1240-1302, (It.) Byzantine mosaicist. *Madonna Enthroned With St. Francis.*
Claude Lorrain (Claude Gellé), 1600-82, (Fr.) ideal-landscapist. *The Enchanted Castle.*
Thomas Cole, 1801-48, (U.S.) Hudson River school. *The Ox-Bow, In the Catskills.*
John Constable, 1776-1837, (Br.) landscapist. *Salisbury Cathedral From the Bishop's Grounds.*
John Singleton Copley, 1738-1815, (U.S.) portraitist. *Samuel Adams, Watson and the Shark.*
Lovis Corinth, 1858-1925, (Ger.) Expressionist. *Apocalypse.*
Jean-Baptiste-Camille Corot, 1796-1875, (Fr.) landscapist. *Souvenir de Mortefontaine, Pastorale.*
Correggio, 1494-1534, (It.) Renaissance muralist. *Mystic Marriages of St. Catherine.*
Gustave Courbet, 1819-77, (Fr.) Realist. *The Artist's Studio.*
Lucas Cranach the Elder, 1472-1553, (Ger.) Protestant Reformation portraitist. *Luther.*
Imogen Cunningham, 1883-1976, (U.S.) photographer, portraitist. Plant photography.
Nathaniel Currier, 1813-88, and **James M. Ives,** 1824-95, (both U.S.) lithographers. *A Midnight Race on the Mississippi, American Forest Scene—Maple Sugaring.*
John Steuart Curry, 1897-1946, (U.S.) Americana, murals. *Baptism in Kansas.*
Salvador Dalí, 1904-89, (Sp.) Surrealist. *Persistence of Memory, The Crucifixion.*
Honoré Daumier, 1808-79, (Fr.) caricaturist. *The Third-Class Carriage.*
Jacques-Louis David, 1748-1825, (Fr.) Neoclassicist. *The Oath of the Horatii.*
Arthur Davies, 1862-1928, (U.S.) Romantic landscapist. *Unicorns, Leda and the Dioscuri.*
Willem de Kooning, 1904-1997, (Dutch-U.S.) abstract expressionist. *Excavation, Woman I, Door to the River.*
Edgar Degas, 1834-1917, (Fr.) *The Ballet Class.*
Eugène Delacroix, 1798-1863, (Fr.) Romantic. *Massacre at Chios, Liberty Leading the People.*
Paul Delaroche, 1797-1856, (Fr.) historical themes. *Children of Edward IV.*
Luca Della Robbia, 1400-82, (It.) Renaissance terracotta artist. *Cantoria* (singing gallery), Florence cathedral.
Donatello, 1386-1466, (It.) Renaissance sculptor. *David, Gattamelata.*
Jean Dubuffet, 1902-85, (Fr.) painter, sculptor, printmaker. *Group of Four Trees.*
Marcel Duchamp, 1887-1968, (Fr.) Dada artist. *Nude Descending a Staircase, No. 2.*
Raoul Dufy, 1877-1953, (Fr.) Fauvist. *Chateau and Horses.*
Asher Brown Durand, 1796-1886, (U.S.) Hudson River school. *Kindred Spirits.*
Albrecht Dürer, 1471-1528, (Ger.) Renaissance painter, engraver, woodcuts. *St. Jerome in His Study, Melencolia I.*
Anthony van Dyck, 1599-1641, (Flem.) Baroque portraitist. *Portrait of Charles I Hunting.*
Thomas Eakins, 1844-1916, (U.S.) Realist. *The Gross Clinic.*
Alfred Eisenstaedt, 1898-1995, (Ger.-U.S.) photographer, photojournalist. Famous photo, V-J Day, Aug. 14, 1945.
Peter Henry Emerson, 1856-1936, (Br.) photographer. Promoted photography as an independent art form.
Jacob Epstein, 1880-1959, (Br.) religious and allegorical sculptor. *Genesis, Ecce Homo.*
Jan van Eyck, c1390-1441, (Flem.) naturalistic panels. *Adoration of the Lamb.*
Roger Fenton, 1819-68, (Br.) photographer. Crimean War.
Anselm Feuerbach, 1829-80, (Ger.) Romantic Classicist. *Judgment of Paris, Iphigenia.*
John Bernard Flannagan, 1895-1942, (U.S.) animal sculptor. *Triumph of the Egg.*
Jean-Honoré Fragonard, 1732-1806, (Fr.) Rococo. *The Swing.*
Daniel Chester French, 1850-1931, (U.S.) *The Minute Man of Concord;* seated *Lincoln,* Lincoln Memorial, Wash., DC.
Caspar David Friedrich, 1774-1840, (Ger.) Romantic landscapes. *Man and Woman Gazing at the Moon.*
Thomas Gainsborough, 1727-88, (Br.) portraitist. *The Blue Boy, The Watering Place, Orpin the Parish Clerk.*
Alexander Gardner, 1821-82, (U.S.) photographer. Civil War; railroad construction; Great Plains Indians.
Paul Gauguin, 1848-1903, (Fr.) Post-impressionist. *The Tahitians, Spirit of the Dead Watching.*
Lorenzo Ghiberti, 1378-1455, (It.) Renaissance sculptor. Gates of Paradise baptistery doors, Florence.
Alberto Giacometti, 1901-66, (Swiss) attenuated sculptures of solitary figures. *Man Pointing.*
Giorgione, c1477-1510, (It.) Renaissance. *The Tempest.*
Giotto di Bondone, 1267-1337, (It.) Renaissance. *Presentation of Christ in the Temple.*
François Girardon, 1628-1715, (Fr.) Baroque sculptor of classical themes. *Apollo Tended by the Nymphs.*
Vincent van Gogh, 1853-90, (Dutch) *The Starry Night, L'Arlesienne, Bedroom at Arles, Self-Portrait.*
Arshile Gorky, 1905-48, (U.S.) Surrealist. *The Liver Is the Cock's Comb.*
Francisco de Goya y Lucientes, 1746-1828, (Sp.) *The Naked Maja, The Disasters of War* (etchings).
El Greco, 1541-1614, (Sp.) *View of Toledo, Assumption of the Virgin.*
Horatio Greenough, 1805-52, (U.S.) Neo-classical sculptor.
Matthias Grünewald, 1480-1528, (Ger.) mystical religious themes. *The Resurrection.*
Frans Hals, c1580-1666, (Dutch) portraitist. *Laughing Cavalier, Gypsy Girl.*
Austin Hansen, 1910-96, (U.S.) photographer. Harlem, NY, life.
Childe Hassam, 1859-1935, (U.S.) Impressionist. *Southwest Wind, July 14 Rue Daunon.*
Edward Hicks, 1780-1849, (U.S.) folk painter. *The Peaceable Kingdom.*
Lewis Wickes Hine, 1874-1940, (U.S.) photographer. Studies of immigrants, children in industry.
Hans Hofmann, 1880-1966, (U.S.) early abstract Expressionist. *Spring, The Gate.*
William Hogarth, 1697-1764, (Br.) caricaturist. *The Rake's Progress.*
Katsushika Hokusai, 1760-1849, (Jpn.) printmaker. *Crabs.*
Hans Holbein the Elder, 1460-1524, (Ger.) late Gothic. *Presentation of Christ in the Temple.*
Hans Holbein the Younger, 1497-1543, (Ger.) portraitist. *Henry VIII, The French Ambassadors.*
Winslow Homer, 1836-1910, (U.S.) naturalist painter, marine themes. *Marine Coast, High Cliff.*
Edward Hopper, 1882-1967, (U.S.) realistic urban scenes. *Nighthawks, House by the Railroad.*
Horst P. Horst, 1906-99, (Ger.) fashion, celebrity photographer.
Jean-Auguste-Dominique Ingres, 1780-1867, (Fr.) Classicist. *Valpincon Bather.*
George Inness, 1825-94, (U.S.) luminous landscapist. *Delaware Water Gap.*
William Henry Jackson, 1843-1942, (U.S.) photographer. American West, building of Union Pacific Railroad.
Donald Judd, 1928-94, (U.S.) sculptor, major Minimalist.
Frida Kahlo, 1907-54, (Mex.) painter; *Self-Portrait With Monkey.*
Vasily Kandinsky, 1866-1944, (Russ.) Abstractionist. *Capricious Forms, Improvisation 38 (second version).*
Paul Klee, 1879-1940, (Swiss) Abstractionist. *Twittering Machine, Pastoral, Death and Fire.*
Gustav Klimt, 1862-1918, (Austrian) cofounder of Vienna Secession Movement, *The Kiss.*
Oscar Kokoschka, 1886-1980, (Austrian) Expressionist. *View of Prague, Harbor of Marseilles.*
Kathe Kollwitz, 1867-1945, (Ger.) printmaker, social justice themes. *The Peasant War.*
Gaston Lachaise, 1882-1935, (U.S.) figurative sculptor. *Standing Woman.*
John La Farge, 1835-1910, (U.S.) muralist. *Red and White Peonies, The Ascension.*
Sir Edwin (Henry) Landseer, 1802-73, (Br.) painter, sculptor. *Shoeing, Rout of Comus.*
Dorothea Lange, 1895-1965, (U.S.) photographer. Depression photographs, migrant farm workers.
Fernand Léger, 1881-1955, (Fr.) machine art. *The Cyclists.*
Leonardo da Vinci, 1452-1519, (It.) *Mona Lisa, Last Supper, The Annunciation.*
Emanuel Leutze, 1816-68, (U.S.) historical themes. *Washington Crossing the Delaware.*
Roy Lichtenstein, 1923-97, (U.S.) pop artist.
Jacques Lipchitz, 1891-1973, (Fr.) Cubist sculptor. *Harpist.*
Filippino Lippi, 1457-1504, (It.) Renaissance.
Fra Filippo Lippi, 1406-69, (It.) Renaissance. *Coronation of the Virgin, Madonna and Child With Angels.*
Morris Louis, 1912-62, (U.S.) abstract Expressionist. *Signa, Stripes, Alpha-Phi.*
Aristide Maillol, 1861-1944, (Fr.) sculptor. *L'Harmonie.*
Édouard Manet, 1832-83, (Fr.) forerunner of Impressionism. *Luncheon on the Grass, Olympia.*
Andrea Mantegna, 1431-1506, (It.) Renaissance frescoes. *Triumph of Caesar.*
Franz Marc, 1880-1916, (Ger.) Expressionist. *Blue Horses.*
John Marin, 1870-1953, (U.S.) Expressionist seascapes. *Maine Island.*
Reginald Marsh, 1898-1954, (U.S.) satirical artist. *Tattoo and Haircut.*
Masaccio, 1401-28, (It.) Renaissance. *The Tribute Money.*
Henri Matisse, 1869-1954, (Fr.) Fauvist. *Woman With the Hat.*
Michelangelo Buonarroti, 1475-1564, (It.) *Pietà, David, Moses, The Last Judgment,* Sistine Chapel ceiling.

Jean-Francois Millet, 1814-75, (Fr.) painter of peasant subjects. *The Gleaners, The Man With a Hoe.*
Joan Miró, 1893-1983, (Sp.) Exuberant colors, playful images. Catalan landscape, *Dutch Interior.*
Amedeo Modigliani, 1884-1920, (It.) *Reclining Nude.*
Piet Mondrian, 1872-1944, (Dutch) Abstractionist. *Composition With Red, Yellow and Blue.*
Claude Monet, 1840-1926, (Fr.) Impressionist. *The Bridge at Argenteuil, Haystacks.*
Henry Moore, 1898-1986, (Br.) sculptor of large-scale, abstract works. *Reclining Figure* (several).
Gustave Moreau, 1826-98, (Fr.) Symbolist. *The Apparition, Dance of Salome.*
James Wilson Morrice, 1865-1924, (Can.) landscapist. *The Ferry, Quebec, Venice, Looking Over the Lagoon.*
William Morris, 1834-1896, (Br.) decorative artist, leader of the Arts and Crafts movement.
Grandma Moses, 1860-1961, (U.S.) folk painter. *Out for the Christmas Trees, Thanksgiving Turkey.*
Edvard Munch, 1863-1944, (Nor.) Expressionist. *The Cry.*
Bartolome Murillo, 1618-82, (Sp.) Baroque religious artist. *Vision of St. Anthony, The Two Trinities.*
Eadweard Muybridge, 1830-1904, (Br.-U.S.) photographer. Studies of motion, *Animal Locomotion.*
Nadar (Gaspar-Félix Tournachon), 1820-1910, (Fr.) photographer, caricaturist, portraitist. Invented photo-essay.
Barnett Newman, 1905-70, (U.S.) abstract Expressionist. *Stations of the Cross.*
Isamu Noguchi, 1904-88, (U.S.) abstract sculptor, designer. *Kouros, BirdC(MU),* sculptural gardens.
Georgia O'Keeffe, 1887-1986, (U.S.) Southwest motifs. *Cow's Skull: Red, White, and Blue, The Shelton With Sunspots.*
José Clemente Orozco, 1883-1949, (Mex.) frescoes. *House of Tears, Pre-Columbian Golden Age.*
Timothy H. O'Sullivan, 1840-82, (U.S.) Civil War photographer.
Charles Willson Peale, 1741-1827, (U.S.) Amer. Revolutionary portraitist. *The Staircase Group,* U.S. presidents.
Rembrandt Peale, 1778-1860, (U.S.) portraitist. Thomas Jefferson.
Pietro Perugino, 1446-1523, (It.) Renaissance. *Delivery of the Keys to St. Peter.*
Pablo Picasso, 1881-1973, (Sp.) painter, sculptor. *Guernica; Dove; Head of a Woman; Head of a Bull, Metamorphosis.*
Piero della Francesca, c1415-92, (It.) Renaissance. *Duke of Urbino, Flagellation of Christ.*
Camille Pissarro, 1830-1903, (Fr.) Impressionist. *Boulevard des Italiens, Morning, Sunlight; Bather in the Woods.*
Jackson Pollock, 1912-56, (U.S.) abstract Expressionist. *Autumn Rhythm.*
Nicolas Poussin, 1594-1665, (Fr.) Baroque pictorial classicism. *St. John on Patmos.*
Maurice B. Prendergast, c1860-1924, (U.S.) Post-impressionist water colorist. *Umbrellas in the Rain.*
Pierre-Paul Prud'hon, 1758-1823, (Fr.) Romanticist. *Crime Pursued by Vengeance and Justice.*
Pierre Cecile Puvis de Chavannes, 1824-98, (Fr.) muralist. *The Poor Fisherman.*
Raphael Sanzio, 1483-1520, (It.) Renaissance. *Disputa, School of Athens, Sistine Madonna.*
Man Ray, 1890-1976, (U.S.) Dada artist. *Observing Time, The Lovers, Marquis de Sade.*
Odilon Redon, 1840-1916, (Fr.) Symbolist painter, lithographer. *In the Dream, Vase of Flowers.*
Rembrandt van Rijn, 1606-69, (Dutch) *The Bridal Couple, The Night Watch.*
Frederic Remington, 1861-1909, (U.S.) painter, sculptor. Portrayer of the American West, *Bronco Buster.*
Pierre-Auguste Renoir, 1841-1919, (Fr.) Impressionist. *The Luncheon of the Boating Party, Dance in the Country.*
Joshua Reynolds, 1723-92, (Br.) portraitist. *Mrs. Siddons as the Tragic Muse.*
Diego Rivera, 1886-1957, (Mex.) frescoes. *The Fecund Earth.*
Henry Peach Robinson, 1830-1901 (Br.) photographer. A leader of "high art" photography.
Norman Rockwell, 1894-1978, (U.S.) painter, illustrator. *Saturday Evening Post* covers.
Auguste Rodin, 1840-1917, (Fr.) sculptor. *The Thinker.*
Mark Rothko, 1903-70, (U.S.) abstract Expressionist. *Light, Earth and Blue.*
Georges Rouault, 1871-1958, (Fr.) Expressionist. *Three Judges.*
Henri Rousseau, 1844-1910, (Fr.) primitive exotic themes. *The Snake Charmer.*
Theodore Rousseau, 1812-67, (Swiss-Fr.) landscapist. *Under the Birches, Evening.*
Peter Paul Rubens, 1577-1640, (Flem.) Baroque. *Mystic Marriage of St. Catherine.*
Jacob van Ruisdael, c1628-82, (Dutch) landscapist. *Jewish Cemetery.*
Charles M. Russell, 1866-1926, (U.S.) Western life.
Salomon van Ruysdael, c1600-70, (Dutch) landscapist. *River With Ferry-Boat.*
Albert Pinkham Ryder, 1847-1917, (U.S.) seascapes and allegories. *Toilers of the Sea.*
Augustus Saint-Gaudens, 1848-1907, (U.S.) memorial statues. *Farragut, Mrs. Henry Adams (Grief).*
Andrea Sansovino, 1460-1529, (It.) Renaissance sculptor. *Baptism of Christ.*
Jacopo Sansovino, 1486-1570, (It.) Renaissance sculptor. *St. John the Baptist.*
John Singer Sargent, 1856-1925, (U.S.) Edwardian society portraitist. The Wyndham Sisters, Madam X.
George Segal, 1924-2000, (U.S.) sculptor of life-sized figures realistically depicting daily life.
Georges Seurat, 1859-91, (Fr.) Pointillist. *Sunday Afternoon on the Island of La Grande Jatte.*
Gino Severini, 1883-1966, (It.) Futurist and Cubist. *Dynamic Hieroglyph of the Bal Tabarin.*
Ben Shahn, 1898-1969, (U.S.) social and political themes. Sacco and Vanzetti series, *Seurat's Lunch, Handball.*
Charles Sheeler, 1883-1965, (U.S.) abstractionist.
David Alfaro Siqueiros, 1896-1974, (Mex.) political muralist. *March of Humanity.*
David Smith, 1906-65, (U.S.) welded metal sculpture. *Hudson River Landscape, Zig, Cubi* series.
Edward Steichen, 1879-1973, (U.S.) photographer. Credited with transforming photography into an art form.
Alfred Stieglitz, 1864-1946, (U.S.) photographer, editor; helped create acceptance of photography as art.
Paul Strand, 1890-1976, (U.S.) photographer. People, nature, landscapes.
Gilbert Stuart, 1755-1828, (U.S.) portraitist. George Washington, Thomas Jefferson, James Madison.
Thomas Sully, 1783-1872, (U.S.) portraitist. *Col. Thomas Handasyd Perkins, The Passage of the Delaware.*
William Henry Fox Talbot, 1800-77, (Br.) photographer. *Pencil of Nature,* early photographically illustrated book.
George Tames, 1919-94, (U.S.) photographer. Chronicled presidents, political leaders.
Yves Tanguy, 1900-55, (Fr.) Surrealist. *Rose of the Four Winds, Mama, Papa Is Wounded!*
Giovanni Battista Tiepolo, 1696-1770, (It.) Rococo frescoes. *The Crucifixion.*
Jacopo Tintoretto, 1518-94, (It.) Mannerist. *The Last Supper.*
Titian, c1485-1576, (It.) Renaissance. *Venus and the Lute Player, The Bacchanal.*
Jose Rey Toledo, 1916-94, (U.S.) Native American artist. Captured the essence of tribal dances on canvas.
Henri de Toulouse-Lautrec, 1864-1901, (Fr.) *At the Moulin Rouge.*
John Trumbull, 1756-1843, (U.S.) historical themes. *The Declaration of Independence.*
J(oseph) M(allord) W(illiam) Turner, 1775-1851, (Br.) Romantic landscapist. *Snow Storm.*
Paolo Uccello, 1397-1475, (It.) Gothic-Renaissance. *The Rout of San Romano.*
Maurice Utrillo, 1883-1955, (Fr.) Impressionist. *Sacre-Coeur de Montmartre.*
John Vanderlyn, 1775-1852, (U.S.) Neo-classicist. *Ariadne Asleep on the Island of Naxos.*
Diego Velázquez, 1599-1660, (Sp.) Baroque. *Las Meninas, Portrait of Juan de Pareja.*
Jan Vermeer, 1632-75, (Dutch) interior genre subjects. *Young Woman With a Water Jug.*
Paolo Veronese, 1528-88, (It.) devotional themes, vastly peopled canvases. *The Temptation of St. Anthony.*
Andrea del Verrocchio, 1435-88, (It.) Floren. sculptor. *Colleoni.*
Maurice de Vlaminck, 1876-1958, (Fr.) Fauvist landscapist. *Red Trees.*
Andy Warhol, 1928-87, (U.S.) Pop Art. *Campbell's Soup Cans, Marilyn Diptych.*
Antoine Watteau, 1684-1721, (Fr.) Rococo painter of "scenes of gallantry." *The Embarkation for Cythera.*
George Frederic Watts, 1817-1904, (Br.) painter and sculptor of grandiose allegorical themes. *Hope.*
Benjamin West, 1738-1820, (U.S.) realistic historical themes. *Death of General Wolfe.*
Edward Weston, 1886-1958, (U.S.) photographer. Landscapes of American West.
James Abbott McNeill Whistler, 1834-1903, (U.S.) *Arrangement in Grey and Black, No. 1: The Artist's Mother.*
Archibald M. Willard, 1836-1918, (U.S.) *The Spirit of '76.*
Grant Wood, 1891-1942, (U.S.) Midwestern regionalist. *American Gothic, Daughters of Revolution.*
Ossip Zadkine, 1890-1967, (Russ.) School of Paris sculptor. *The Destroyed City, Musicians, Christ.*

Business Leaders and Philanthropists of the Past

Elizabeth Arden (F. N. Graham), 1884-1966, (U.S.) Canadian-born founder of cosmetics empire.
Philip D. Armour, 1832-1901, (U.S.) industrialist; streamlined meatpacking.
John Jacob Astor, 1763-1848, (U.S.) German-born fur trader, banker, real estate magnate; at death, richest in U.S.
Francis W. Ayer, 1848-1923, (U.S.) ad industry pioneer.
August Belmont, 1816-90, (U.S.) German-born financier.
James B. (Diamond Jim) Brady, 1856-1917, (U.S.) financier, philanthropist, legendary bon vivant.
Adolphus Busch, 1839-1913, (U.S.) German-born businessman; established brewery empire.
Asa Candler, 1851-1929, (U.S.) founded Coca-Cola Co.
Andrew Carnegie, 1835-1919, (U.S.) Scottish-born industrialist; philanthropist; founded Carnegie Steel Co.
Tom Carvel, 1908-89, (Gr.-U.S.) founded ice cream chain.
William Colgate, 1783-1857, (Br.-U.S.) Br.-born businessman, philanthropist; founded soap-making empire.
Jay Cooke, 1821-1905, (U.S.) financier; sold $1 billion in Union bonds during Civil War.
Peter Cooper, 1791-1883, (U.S.) industrialist, inventor, philanthropist; founded Cooper Union (1859).
Ezra Cornell, 1807-74, (U.S.) businessman, philanthropist; headed Western Union, established university.
Erastus Corning, 1794-1872, (U.S.) financier; headed N.Y. Central.
Charles Crocker, 1822-88, (U.S.) railroad builder, financier.
Samuel Cunard, 1787-1865, (Can.) pioneered trans-Atlantic steam navigation.
Marcus Daly, 1841-1900, (U.S.) Irish-born copper magnate.
W. Edwards Deming, 1900-93, (U.S.) quality-control expert who revolutionized Japanese manufacturing.
Walt Disney, 1901-66, (U.S.) pioneer in cinema animation; built entertainment empire.
Herbert H. Dow, 1866-1930, (U.S.) founder of chemical co.
James Duke, 1856-1925, (U.S.) founded American Tobacco, Duke Univ.
Eleuthere I. du Pont, 1771-1834, (Fr.-U.S.) gunpowder manufacturer; founded one of the largest business empires.
Thomas C. Durant, 1820-85, (U.S.) railroad official, financier.
William C. Durant, 1861-1947, (U.S.) industrialist; formed General Motors.
George Eastman, 1854-1932, (U.S.) inventor; manufacturer of photographic equipment.
Marshall Field, 1834-1906, (U.S.) merchant; founded Chicago's largest department store.
Harvey Firestone, 1868-1938, (U.S.) founded tire company.
Avery Fisher, 1906-94, (U.S.) industrialist, philanthropist, founded Fisher electronics.
Henry M. Flagler, 1830-1913, (U.S.) financier; helped form Standard Oil; developed Florida as resort state.
Malcolm Forbes, 1919-90, (U.S.) magazine publisher.
Henry Ford, 1863-1947, (U.S.) auto maker; developed first popular low-priced car.
Henry Ford 2d, 1917-87, (U.S.) headed auto company founded by grandfather.
Henry C. Frick, 1849-1919, (U.S.) steel and coke magnate; had prominent role in development of U.S. Steel.
Jakob Fugger (Jakob the Rich), 1459-1525, (Ger.) headed leading banking, trading house, in 16th-cent. Europe.
Alfred C. Fuller, 1885-1973, (U.S.) Canadian-born businessman; founded brush company.
Elbert H. Gary, 1846-1927, (U.S.) one of the organizers of U.S. Steel; chaired board of directors, 1903-27.
Jean Paul Getty, 1892-1976, (U.S.) founded oil empire.
Amadeo Giannini, 1870-1949, (U.S.) founded Bank of America.
Stephen Girard, 1750-1831, (U.S.) French-born financier, philanthropist; richest man in U.S. at his death.
Leonard H. Goldenson, 1905-99, (U.S.) turned ABC into major TV network.
Jay Gould, 1836-92, (U.S.) railroad magnate, financier.
Hetty Green, 1834-1916, (U.S.) financier, the "witch of Wall St."; richest woman in U.S. in her day.
William Gregg, 1800-67, (U.S.) launched textile industry in S.
Meyer Guggenheim, 1828-1905, (U.S.) Swiss-born merchant, philanthropist; built merchandising, mining empires.
Armand Hammer, 1898-1990, (U.S.) headed Occidental Petroleum; promoted U.S.-Soviet ties.
Edward H. Harriman, 1848-1909, (U.S.) railroad financier, administrator; headed Union Pacific.
Henry J. Heinz, 1844-1919, (U.S.) founded food empire.
James J. Hill, 1838-1916, (U.S.) Canadian-born railroad magnate, financier; founded Great Northern Railway.
Conrad N. Hilton, 1888-1979, (U.S.) hotel chain founder.
Howard Hughes, 1905-76, (U.S.) industrialist, aviator, movie maker.
H. L. Hunt, 1889-1974, (U.S.) oil magnate.
Collis P. Huntington, 1821-1900, (U.S.) railroad magnate.
Henry E. Huntington, 1850-1927, (U.S.) railroad builder, philanthropist.
Walter L. Jacobs, 1898-1985, (U.S.) founder of the first rental car agency, which later became Hertz.
Howard Johnson, 1896-1972, (U.S.) founded restaurants.
Henry J. Kaiser, 1882-1967, (U.S.) industrialist; built empire in steel, aluminum.
Minor C. Keith, 1848-1929, (U.S.) railroad magnate; founded United Fruit Co.
Will K. Kellogg, 1860-1951, (U.S.) businessman, philanthropist; founded breakfast food co.
Richard King, 1825-85, (U.S.) cattleman; founded half-million-acre King Ranch in Texas.
William S. Knudsen, 1879-1948, (U.S.) Danish-born auto industry executive.
Samuel H. Kress, 1863-1955, (U.S.) businessman, art collector, philanthropist; founded "dime store" chain.
Ray A. Kroc, 1902-84, (U.S.) founded McDonald's fast-food chain.
Alfred Krupp, 1812-87, (Ger.) armaments magnate.
William Levitt, 1907-94, (U.S.) industrialist, "suburb maker".
Thomas Lipton, 1850-1931, (Scot.) merchant, tea empire.
James McGill, 1744-1813, (Scot.-Can.) founded university.
Andrew W. Mellon, 1855-1937, (U.S.) financier, industrialist; benefactor of National Gallery of Art.
Charles E. Merrill, 1885-1956, (U.S.) financier; developed firm of Merrill Lynch.
John Pierpont Morgan, 1837-1913, (U.S.) most powerful figure in finance and industry at the turn of the cent.
Akio Morita, 1921-99, (Japan) co-founded Sony Corp.
Malcolm Muir, 1885-1979, (U.S.) created *Business Week* magazine; headed *Newsweek,* 1937-61.
Samuel Newhouse, 1895-1979, (U.S.) publishing and broadcasting magnate; built communications empire.
Aristotle Onassis, 1906-75, (Gr.) shipping magnate.
William S. Paley, 1901-90, (U.S.) built CBS communic. empire.
George Peabody, 1795-1869, (U.S.) merchant, financier, philanthropist.
James C. Penney, 1875-1971, (U.S.) businessman; developed department store chain.
William C. Procter, 1862-1934, (U.S.) headed soap co.
John D. Rockefeller, 1839-1937, (U.S.) industrialist; established Standard Oil.
John D. Rockefeller Jr., 1874-1960, (U.S.) philanthropist; established foundation; provided land for UN.
Meyer A. Rothschild, 1743-1812, (Ger.) founded international banking house.
Thomas Fortune Ryan, 1851-1928, (U.S.) financier; a founder of American Tobacco.
Edmond J. Safra, 1932-99, (U.S.) founded Republic National Bank of New York.
David Sarnoff, 1891-1971, (U.S.) broadcasting pioneer; established first radio network, NBC.
Richard Sears, 1863-1914, (U.S.) founded mail-order co.
Werner von Siemens, 1816-92, (Ger.) industrialist; inventor.
Alfred P. Sloan, 1875-1966, (U.S.) industrialist, philanthropist; headed General Motors.
A. Leland Stanford, 1824-93, (U.S.) railroad official, philanthropist; founded university.
Nathan Straus, 1848-1931, (U.S.) German-born merchant, philanthropist; headed Macy's.
Levi Strauss, c1829-1902, (U.S.) pants manufacturer.
Clement Studebaker, 1831-1901, (U.S.) wagon, carriage (maker).
Gustavus Swift, 1839-1903, (U.S.) pioneer meatpacker.
Gerard Swope, 1872-1957, (U.S.) industrialist, economist; headed General Electric.
James Walter Thompson, 1847-1928, (U.S.) ad executive.
Alice Tully, 1902-93, (U.S.) philanthropist, arts patron.
Theodore N. Vail, 1845-1920, (U.S.) organized Bell Telephone system; headed AT&T.
Cornelius Vanderbilt, 1794-1877, (U.S.) financier; established steamship, railroad empires.
Henry Villard, 1835-1900, (U.S.) German-born railroad executive, financier.
George Westinghouse, 1846-1914, (U.S) inventor, manufacturer; organized Westinghouse Electric Co., 1886.
Charles R. Walgreen, 1873-1939, (U.S.) founded drugstore chain.
DeWitt Wallace, 1889-1981, (U.S.) and **Lila Wallace,** 1889-1984, (U.S.) cofounders of *Reader's Digest* magazine.
Sam Walton, 1918-92, (U.S.) founder of Wal-Mart stores.
John Wanamaker, 1838-1922, (U.S.) pioneered department-store merchandising.
Aaron Montgomery Ward, 1843-1913, (U.S.) established first mail-order firm.
Thomas J. Watson, 1874-1956, (U.S.) IBM head, 1914-56.
John Hay Whitney, 1905-82, (U.S.) publisher, sportsman, philanthropist.
Charles E. Wilson, 1890-1961, (U.S.) auto industry exec., public official.
Frank W. Woolworth, 1852-1919, (U.S.) created 5 & 10 chain.
William Wrigley Jr., 1861-1932, (U.S.) founded Wrigley chewing gum company.

American Cartoonists

Reviewed by Lucy Shelton Caswell, Professor and Curator, Cartoon Research Library, Ohio State University

Scott Adams, b 1957, Dilbert.
Charles Addams, 1912-88, macabre cartoons.
Brad Anderson, b 1924, Marmaduke.
Sergio Aragones, b 1937, *MAD Magazine.*
Peter Arno, 1904-68, *The New Yorker.*
Tex Avery, 1908-80, animator, Bugs Bunny, Porky Pig.
George Baker, 1915-75, The Sad Sack.
Carl Barks, b 1901, Donald Duck comic books.
C. C. Beck, 1910-89, Captain Marvel.
Jim Berry, b 1932, Berry's World.
Herb Block (Herblock), b 1909, political cartoonist.
George Booth, b 1926, *The New Yorker.*
Berkeley Breathed, b 1957, Bloom County.
Dik Browne, 1917-89, Hi & Lois, Hagar the Horrible.
Marjorie Buell, 1904-93, Little Lulu.
Ernie Bushmiller, 1905-82, Nancy.
Milton Caniff, 1907-88, Terry & the Pirates, Steve Canyon.
Al Capp, 1909-79, Li'l Abner.
Roz Chast, b 1954, *The New Yorker.*
Paul Conrad, 1924, political cartoonist.
Roy Crane, 1901-77, Captain Easy, Buz Sawyer.
Robert Crumb, b 1943, underground cartoonist.
Shamus Culhane, 1908-96, animator.
Jay N. Darling (Ding), 1876-1962, political cartoonist.
Jack Davis, b 1926, *MAD Magazine.*
Jim Davis, b 1945, Garfield.
Billy DeBeck, 1890-1942, Barney Google.
Rudolph Dirks, 1877-1968, The Katzenjammer Kids.
Walt Disney, 1901-66, produced animated cartoons, created Mickey Mouse, Donald Duck.
Steve Ditko, b 1927, Spider-Man.
Mort Drucker, b 1929, *MAD Magazine.*
Will Eisner, b 1917, The Spirit.
Jules Feiffer, b 1929, political cartoonist.
Bud Fisher, 1884-1954, Mutt & Jeff.
Ham Fisher, 1900-55, Joe Palooka.
Max Fleischer, 1883-1972, Betty Boop.
Hal Foster, 1892-1982, Tarzan, Prince Valiant.
Fontaine Fox, 1884-1964, Toonerville Folks.
Isadore "Friz" Freleng, 1905-95, animator, Yosemite Sam, Porky Pig, Sylvester and Tweety Bird.
Rube Goldberg, 1883-1970, Boob McNutt.
Chester Gould, 1900-85, Dick Tracy.
Harold Gray, 1894-1968, Little Orphan Annie.
Matt Groening, b 1954, Life in Hell, The Simpsons.
Cathy Guisewite, b 1950, Cathy.
Bill Hanna, 1910-2001, & **Joe Barbera,** b 1911, animators, Tom & Jerry, Yogi Bear, Flintstones.
Johnny Hart, b 1931, BC, Wizard of Id.
Oliver Harrington, 1912-95, Bootsie.
Alfred Harvey, 1913-94, created Casper the Friendly Ghost.
Jimmy Hatlo, 1898-1963, Little Iodine.
John Held Jr., 1889-1958, Jazz Age.
George Herriman, 1881-1944, Krazy Kat.
Harry Hershfield, 1885-1974, Abie the Agent.
Al Hirschfeld, b 1903, *N.Y. Times* theater caricaturist.
Burne Hogarth, 1911-96, Tarzan.
Helen Hokinson, 1900-49, *The New Yorker.*
Nicole Hollander, b 1939, Sylvia.
Lynn Johnston, b 1947, For Better or For Worse.
Chuck Jones, b 1912, animator, Bugs Bunny, Porky Pig.
Mike Judge, b. 1962, Beavis and Butt-head, King of the Hill.
Bob Kane, b 1916-98, Batman.
Bil Keane, b 1922, The Family Circus.
Walt Kelly, 1913-73, Pogo.
Hank Ketcham, 1920-2001, Dennis the Menace.
Ted Key, b 1912, Hazel.
Frank King, 1883-1969, Gasoline Alley.
Jack Kirby, 1917-94, Fantastic Four, The Incredible Hulk.
Rollin Kirby, 1875-1952, political cartoonist.
B(ernard) Kliban, 1935-91, cat books.
Edward Koren, b 1935, *The New Yorker.*
Harvey Kurtzman, 1921-93, *MAD Magazine.*
Walter Lantz, 1900-94, Woody Woodpecker.
Gary Larson, b 1950, The Far Side.
Mell Lazarus, b 1929, Momma, Miss Peach.
Stan Lee, b 1922, Marvel Comics.
David Levine, b 1926, *N.Y. Review of Books* caricatures.
Doug Marlette, b 1949, political cartoonist, Kudzu.
Don Martin, 1931-2000, *MAD Magazine.*
Bill Mauldin, b 1921, political cartoonist.
Jeff MacNelly, 1947-2000, political cartoonist, Shoe.
Winsor McCay, 1872-1934, Little Nemo.
John T. McCutcheon, 1870-1949, political cartoonist.
George McManus, 1884-1954, Bringing Up Father.
Dale Messick, b 1906, Brenda Starr.
Norman Mingo, 1896-1980, Alfred E. Neuman.
Bob Montana, 1920-75, Archie.
Dick Moores, 1909-86, Gasoline Alley.
Willard Mullin, 1902-78, sports cartoonist; Dodgers "Bum," Mets "Kid."
Russell Myers, b 1938, Broom Hilda.
Thomas Nast, 1840-1902, political cartoonist; Republican elephant.
Pat Oliphant, b 1935, political cartoonist.
Frederick Burr Opper, 1857-1937, Happy Hooligan.
Richard Outcault, 1863-1928, Yellow Kid, Buster Brown.
Trey Parker, b 1969?, animator, co-creator of *South Park.*
Mike Peters, b 1943, cartoonist, Mother Goose & Grimm.
George Price, 1901-95, *The New Yorker.*
Antonio Prohias, 1921(?)-98, Spy vs. Spy.
Alex Raymond, 1909-56, Flash Gordon, Jungle Jim.
Forrest (Bud) Sagendorf, 1915-94, Popeye.
Art Sansom, 1920-91, The Born Loser.
Charles Schulz, 1922-2000, Peanuts.
Elzie C. Segar, 1894-1938, Popeye.
Joe Shuster, 1914-92, & **Jerry Siegel,** 1914-96, Superman.
Sidney Smith, 1887-1935, The Gumps.
Otto Soglow, 1900-75, Little King.
Art Spiegelman, b 1948, Raw, Maus.
William Steig, b 1907, *The New Yorker.*
Matt Stone, b 1971?, animator, co-creator of South Park.
Paul Szep, b 1941, political cartoonist.
James Swinnerton, 1875-1974, Little Jimmy, Canyon Kiddies.
Paul Terry, 1887-1971, animator of Mighty Mouse.
Bob Thaves, b 1924, Frank and Ernest.
James Thurber, 1894-61, *The New Yorker.*
Garry Trudeau, b 1948, Doonesbury.
Mort Walker, b 1923, Beetle Bailey.
Bill Watterson, b 1958, Calvin and Hobbes.
Russ Westover, 1887-1966, Tillie the Toiler.
Signe Wilkinson, b 1950, political cartoonist.
Frank Willard, 1893-1958, Moon Mullins.
J. R. Williams, 1888-1957, The Willets Family, Out Our Way.
Gahan Wilson, b 1930, *The New Yorker.*
Tom Wilson, b 1931, Ziggy.
Art Young, 1866-1943, political cartoonist.
Chic Young, 1901-73, Blondie.

Economists, Educators, Historians, and Social Scientists of the Past

For Psychologists see Scientists of the Past.

Brooks Adams, 1848-1927, (U.S.) historian, political theoretician; *The Law of Civilization and Decay.*
Henry Adams, 1838-1918, (U.S.) historian, autobiographer; *History of the United States of America, The Education of Henry Adams.*
Francis Bacon, 1561-1626, (Eng.) philosopher, essayist, and statesman; championed observation and induction.
George Bancroft, 1800-91, (U.S.) historian; wrote 10-volume *History of the United States.*
Jack Barbash, 1911-94, (U.S.) labor economist who helped create the AFL-CIO.
Henry Barnard, 1811-1900, (U.S.) public school reformer.
Charles A. Beard, 1874-1948, (U.S.) historian; *The Economic Basis of Politics.*
Bede (the Venerable), c673-735, (Br.) scholar, historian; *Ecclesiastical History of the English People.*
Ruth Benedict, 1887-1948, (U.S.) anthropologist; studied Indian tribes of the Southwest.
Sir Isaiah Berlin, 1909-97, (Br.) philosopher, historian; *The Age of Enlightenment.*
Louis Blanc, 1811-82, (Fr.) Socialist leader and historian.
Sarah G. Blanding, 1899-1985, (U.S.) head of Vassar College, 1946-64.
Leonard Bloomfield, 1887-1949, (U.S.) linguist; *Language.*
Franz Boas, 1858-1942, (U.S.) German-born anthropologist; studied American Indians.
Van Wyck Brooks, 1886-1963, (U.S.) historian; critic of New England culture, especially literature.
Edmund Burke, 1729-97, (Ir.) British parliamentarian and political philosopher; *Reflections on the Revolution in France.*
Nicholas Murray Butler, 1862-1947, (U.S.) educator; headed Columbia Univ., 1902-45; Nobel Peace Prize, 1931.

Joseph Campbell, 1904-87, (U.S.) author, editor, teacher; wrote books on mythology, folklore.
Thomas Carlyle, 1795-1881, (Sc.) historian, critic; *Sartor Resartus, Past and Present, The French Revolution.*
Edward Channing, 1856-1931, (U.S.) historian; wrote 6-volume *History of the United States.*
Henry Steele Commager, 1902-98, (U.S.) historian, educator; wrote *The Growth of the American Republic.*
John R. Commons, 1862-1945, (U.S.) economist, labor historian; *Legal Foundations of Capitalism.*
James B. Conant, 1893-1978, (U.S.) educator, diplomat; *The American High School Today.*
Benedetto Croce, 1866-1952, (It.) philosopher, statesman, and historian; *Philosophy of the Spirit.*
Bernard A. De Voto, 1897-1955, (U.S.) historian; wrote trilogy on American West; edited Mark Twain manuscripts.
Melvil Dewey, 1851-1931, (U.S.) devised decimal system of library-book classification.
Emile Durkheim, 1858-1917, (Fr.) a founder of modern sociology; *The Rules of Sociological Method.*
Charles Eliot, 1834-1926, (U.S.) educator, Harvard president.
Friedrich Engels, 1820-95, (Ger.) political writer; with Marx wrote the *Communist Manifesto.*
Irving Fisher, 1867-1947, (U.S.) economist; contributed to the development of modern monetary theory.
John Fiske, 1842-1901, (U.S.) historian and lecturer; popularized Darwinian theory of evolution.
Charles Fourier, 1772-1837, (Fr.) utopian socialist.
Giovanni Gentile, 1875-1944, (It.) philosopher, educator; reformed Italian educational system.
Sir James George Frazer, 1854-1941, (Br.) anthropologist; studied myth in religion; *The Golden Bough.*
Henry George, 1839-97, (U.S.) economist, reformer; led single-tax movement.
Edward Gibbon, 1737-94, (Br.) historian; *The History of the Decline and Fall of the Roman Empire.*
Francesco Guicciardini, 1483-1540, (It.) historian; *Storia d'Italia,* principal historical work of the 16th cent.
Thomas Hobbes, 1588-1679, (Eng.) philosopher, political theorist; *Leviathan.*
Richard Hofstadter, 1916-70, (U.S.) historian; *The Age of Reform.*
John Holt, 1924-85, (U.S.) educator and author.
John Maynard Keynes, 1883-1946, (Br.) economist; principal advocate of deficit spending.
Russell Kirk, 1918-94, (U.S.), social philosopher; *The Conservative Mind.*
Alfred L. Kroeber, 1876-1960, (U.S.) cultural anthropologist; studied Indians of North and South America.
Christopher Lasch, 1932-94, (U.S.) social critic, historian; *The Culture of Narcissism.*
James L. Laughlin, 1850-1933, (U.S.) economist; helped establish Federal Reserve System.
Lucien Lévy-Bruhl, 1857-1939, (Fr.) philosopher; studied the psychology of primitive societies; *Primitive Mentality.*
John Locke, 1632-1704, (Eng.) philosopher and political theorist; *Two Treatises of Government.*
Thomas B. Macaulay, 1800-59, (Br.) historian, statesman.
Niccolò Machiavelli, 1469-1527, (It.) writer, statesman. *The Prince.*
Bronislaw Malinowski, 1884-1942, (Pol.) considered the father of social anthropology.
Thomas R. Malthus, 1766-1834, (Br.) economist; famed for *Essay on the Principle of Population.*
Horace Mann, 1796-1859, (U.S.) pioneered modern public school system.
Karl Mannheim, 1893-1947, (Hung.) sociologist, historian; *Ideology and Utopia.*
Karl Marx, 1818-83, (Ger.) political theorist, proponent of Communism; *Communist Manifesto, Das Kapital.*
Giuseppe Mazzini, 1805-72, (It.) political philosopher.
William H. McGuffey, 1800-73, (U.S.) whose *Reader* was a mainstay of 19th-cent. U.S. public education.
George H. Mead, 1863-1931, (U.S.) philosopher, social psychologist.
Margaret Mead, 1901-78, (U.S.) cultural anthropologist; popularized field; *Coming of Age in Samoa.*
Alexander Meiklejohn, 1872-1964, (U.S.) Br.-born educator; championed academic freedom and experimental curricula.
James Mill, 1773-1836, (Sc.) philosopher, historian, economist; a proponent of utilitarianism.
Perry G. Miller, 1905-63, (U.S.) historian; interpreted 17th-cent. New England.
Theodor Mommsen, 1817-1903, (Ger.) historian; *The History of Rome.*
Ashley Montagu, 1905-99, (Eng.) anthropologist; *The Natural Superiority of Women.*
Charles-Louis Montesquieu, 1689-1755, (Fr.) social philosopher; *The Spirit of Laws.*
Maria Montessori, 1870-1952, (It.) educator, physician; started Montessori method of student self-motivation.
Samuel Eliot Morison, 1887-1976, (U.S.) historian; chronicled voyages of early explorers.
Lewis Mumford, 1895-1990, (U.S.) sociologist, critic; *The Culture of Cities.*
Gunnar Myrdal, 1898-1987, (Swed.) economist, social scientist; *Asian Drama: An Inquiry Into the Poverty of Nations.*
Joseph Needham, 1900-95, (Br.) scientific historian; *Science and Civilization in China.*
Allan Nevins, 1890-1971, (U.S.) historian, biographer; *The Ordeal of the Union.*
José Ortega y Gasset, 1883-1955, (Sp.) philosopher; advocated control by elite, *The Revolt of the Masses.*
Robert Owen, 1771-1858, (Br.) political philosopher, reformer; pioneer in cooperative movement.
Thomas (Tom) Paine, 1737-1809, (U.S.) political theorist, writer. *Common Sense.*
Vilfredo Pareto, 1848-1923, (It.) economist, sociologist.
Francis Parkman, 1823-93, (U.S.) historian; *France and England in North America.*
Elizabeth P. Peabody, 1804-94, (U.S.) education pioneer; founded 1st kindergarten in U.S., 1860.
William Prescott, 1796-1859, (U.S.) early American historian; *The Conquest of Peru.*
Pierre Joseph Proudhon, 1809-65, (Fr.) social theorist; father of anarchism; *The Philosophy of Property.*
François Quesnay, 1694-1774, (Fr.) economic theorist.
David Ricardo, 1772-1823, (Br.) economic theorist; advocated free international trade.
Jean-Jacques Rousseau, 1712-78, (Fr.) social philosopher; the father of romantic sensibility; *Confessions.*
Edward Sapir, 1884-1939, (Ger.-U.S.) anthropologist; studied ethnology and linguistics of U.S. Indian groups.
Ferdinand de Saussure, 1857-1913, (Swiss) a founder of modern linguistics.
Hjalmar Schacht, 1877-1970, (Ger.) economist.
Joseph Schumpeter, 1883-1950, (Czech.-U.S.) economist, sociologist.
Elizabeth Seton, 1774-1821, (U.S.) nun; est. parochial school education in U.S.; first native-born American saint.
George Simmel, 1858-1918, (Ger.) sociologist, philosopher; helped establish German sociology.
Adam Smith, 1723-90, (Br.) economist; advocated laissez-faire economy, free trade; *The Wealth of Nations.*
Jared Sparks, 1789-1866, (U.S.) historian, educator, editor; *The Library of American Biography.*
Oswald Spengler, 1880-1936, (Ger.) philosopher and historian; *The Decline of the West.*
William G. Sumner, 1840-1910, (U.S.) social scientist, economist; laissez-faire economy, Social Darwinism.
Hippolyte Taine, 1828-93, (Fr.) historian; basis of naturalistic school; *The Origins of Contemporary France.*
A(lan) J(ohn) P(ercivale) Taylor, 1906-89, (Br.) historian; *The Origins of the Second World War.*
Nikolaas Tinbergen, 1907-88, (Dutch-Br.) ethologist; pioneer in study of animal behavior.
Alexis de Tocqueville, 1805-59, (Fr.) political scientist, historian; *Democracy in America.*
Francis E. Townsend, 1867-1960, (U.S.) led old-age pension movement, 1933.
Arnold Toynbee, 1889-1975, (Br.) historian; *A Study of History,* sweeping analysis of hist. of civilizations.
George Trevelyan, 1838-1928, (Br.) historian, statesman; favored "literary" over "scientific" history; *History of England.*
Barbara Tuchman, 1912-89, (U.S.) author of popular history books, *The Guns of August, The March of Folly.*
Frederick J. Turner, 1861-1932, (U.S.) historian, educator; *The Frontier in American History.*
Thorstein B. Veblen, 1857-1929, (U.S.) economist, social philosopher; *The Theory of the Leisure Class.*
Giovanni Vico, 1668-1744, (It.) historian, philosopher; regarded by many as first modern historian; *New Science.*
Izaak Walton, 1593-1683, (Eng.) wrote biographies; political-philosophical study of fishing, *The Compleat Angler.*
Sidney J., 1859-1947, and **Beatrice,** 1858-1943, **Webb,** (Br.) leading figures in Fabian Society and Labor Party.
Max Weber, 1864-1920, (Ger.) sociologist; *The Protestant Ethic and the Spirit of Capitalism.*
Emma Hart Willard, 1787-1870, (U.S.) pioneered higher education for women.
C. Vann Woodward, 1908-99, (U.S.) historian; *The Strange Career of Jim Crow.*

American Journalists of the Past

Reviewed by Dean Mills, Dean, Missouri School of Journalism

See also African-Americans, Business Leaders, Cartoonists, Writers of the Past.

Franklin P. Adams (F.P.A.), 1881-1960, humorist; wrote column "The Conning Tower."
Martin Agronsky, 1915-99, broadcast journalist; developed Agronsky & Company.
Joseph W. Alsop, 1910-89, and **Stewart Alsop**, 1914-74, Washington-based political analysts, columnists.
Brooks Atkinson, 1894-1984, theater critic.
James Gordon Bennett, 1795-1872, editor and publisher; founded *NY Herald.*
James Gordon Bennett, 1841-1918, succeeded father, financed expeditions, founded afternoon paper.
Elias Boudinot, d 1839, founding editor of first Native American newspaper in U.S., *Cherokee Phoenix* (1828-34).
Margaret Bourke-White, 1904-71, photojournalist.
Arthur Brisbane, 1864-1936, editor; helped introduce "yellow journalism" with sensational, simply written articles.
Heywood Broun, 1888-1939, author, columnist; founded American Newspaper Guild.
Herb Caen, 1916-97, longtime columnist for *San Francisco Chronicle* and *Examiner.*
John Campbell, 1653-1728, published *Boston News-Letter*, first continuing newspaper in the American colonies.
Jimmy Cannon, 1909-73, syndicated sports columnist.
John Chancellor, 1927-96, TV journalist; anchored *NBC Nightly News.*
Harry Chandler, 1864-1944, *Los Angeles Times* publisher, 1917-41; made it a dominant force.
Marquis Childs, 1903-90, reporter and columnist for *St. Louis Post-Dispatch* and United Feature syndicate.
Craig Claiborne, 1920-2000, NY Times food editor and critic; key in internationalizing American taste.
Elizabeth Cochrane (Nellie Bly), pioneer woman journalist, investig. reporter, noted for series on trip around the world.
Charles Collingwood, 1917-85, CBS news correspondent, foreign affairs reporter, documentary host.
Howard Cosell, 1920-95, TV and radio sportscaster.
Gardner Cowles, 1861-1946, founded newspaper chain.
Cyrus Curtis, 1850-1933, publisher of *Saturday Evening Post, Ladies Home Journal, Country Gentleman.*
Charles Anderson Dana, 1819-97, editor, publisher; made *NY Sun* famous for its news reporting.
Elmer (Holmes) Davis, 1890-1958, *NY Times* editorial writer; radio commentator.
Richard Harding Davis, 1864-1916, war correspondent, travel writer, fiction writer.
Benjamin Day, 1810-89, published *NY Sun* beginning in 1833, introducing penny press to the U.S.
Frederick Douglass, 1817-95, ex-slave, social reformer, newspaper editor.
Finley Peter Dunne, 1867-1936, humorist, social critic, wrote "Mr. Dooley" columns.
Mary Baker Eddy, 1821-1910, founded Christian Science movement and *Christian Science Monitor.*
Rowland Evans Jr., 1921-2001, Washington columnist and commentator.
Marshall Field III, 1893-1956, retail magnate, *Chicago Sun* founder.
Doris Fleeson, 1901-70, war correspondent, columnist.
James Franklin, 1697-1735, printer, pioneer journalist, publisher of *New England Courant* and *Rhode Island Gazette.*
Fred W. Friendly, 1915-98, radio, TV reporter, announcer, producer, executive, collaborator with Edward R. Murrow.
Margaret Fuller, 1810-50, social reformer, transcendentalist, critic and foreign correspondent for *NY Tribune.*
Frank E. Gannett, 1876-1957, founded newspaper chain.
William Lloyd Garrison, 1805-79, abolitionist; publisher of *The Liberator.*
Elizabeth Meriwether Gilmer (Dorothy Dix), 1861-1951, reporter, pioneer of the advice column genre.
Edwin Lawrence Godkin, 1831-1902, founder of *The Nation*, editor of *N.Y. Evening Post.*
Katharine Graham,1917-2001, publisher of the *Washington Post.*
Sheilah Graham, 1904-89, Hollywood gossip columnist.
Horace Greeley, 1811-72, editor and politician; founded *NY Tribune.*
Meg Greenfield, 1930-1999, *Newsweek* columnist, editorial page editor Wash. Post.
Gilbert Hovey Grosvenor, 1875-1966, longtime editor of *National Geographic* magazine.
John Gunther, 1901-70, *Chicago Daily News* foreign correspondent, author.
Sarah Josepha Buell Hale, 1788-1879, first female magazine editor, (Ladies' Magazine, later Godey's Lady's Book)
Benjamin Harris, 1673-1716, publisher (1690) of *Publick Occurrences*, 1st newspaper in the American colonies; suppressed after one issue.
William Randolph Hearst, 1863-1951, founder of Hearst newspaper chain and one of the pioneer yellow journalists.
Gabriel Heatter, 1890-1972, radio commentator.
John Hersey, 1914-98, foreign correspondent for *Time, Life,* and *The New Yorker*; author.
Marguerite Higgins, 1920-66, reporter, war correspondent.
Hedda Hopper, 1885-1966, Hollywood gossip columnist.
Roy Howard, 1883-1964, editor, executive, Scripps-Howard papers and United Press (later United Press International).
Chet (Chester Robert) Huntley, 1911-74, co-anchor of NBC's *Huntley-Brinkley Report.*
Ralph Ingersoll, 1900-85, editor, *Fortune, Time, Life* exec.
H. V. (Hans von) Kaltenborn, 1878-1965, radio commentator, reporter.
Murray Kempton, 1917-97, reporter, columnist for magazines and newspapers, including *NY Post.*
John S. Knight, 1894-1981, editor, publisher; founded Knight newspaper group, which merged into Knight-Ridder.
Joseph Kraft, 1942-86, foreign policy columnist.
Arthur Krock, 1886-1974, *NY Times* political writer, Washington bureau chief.
Charles Kuralt, 1934-97, TV anchor and host of CBS "On the Road" feature stories about life in the U.S.
David Lawrence, 1888-1973, reporter, columnist, publisher; founded *U.S. News & World Report.*
Frank Leslie, 1821-80, engraver and publisher of newspapers and magazines, notably *Leslie's Illustrated Newspaper.*
Alexander Liberman, 1912-99, editorial director for Conde Nast magazines.
A(bbott) J(oseph) Liebling, 1904-63, foreign correspondent, critic, principally with *The New Yorker.*
Walter Lippmann, 1889-1974, political analyst, social critic, columnist, author.
Peter Lisagor, 1915-76, Washington bureau chief, *Chicago Daily News;* broadcast commentator.
David Ross Locke, 1833-88, humorist, satirist under pseudonym P.V. Nasby; owned *Toledo (Ohio) Blade.*
Elijah Parish Lovejoy, 1802-37, abolitionist editor in St. Louis and in Alton, IL; killed by proslavery mob.
Clare Booth Luce, 1903-87, war correspondent for *Life;* diplomat, playwright.
Henry R. Luce, 1898-1967, founded *Time, Fortune, Life, Sports Illustrated.*
C(harles) K(enny) McClatchy, 1858-1936 founder of McClatchy newspaper chain.
Samuel McClure, 1857-1949, founder (1893) of *McClure's Magazine*, famous for its investigative reporting.
Anne O'Hare McCormick, 1889-1954, foreign correspondent, first woman on *NY Times* editorial board.
Robert R. McCormick, 1880-1955, editor, publisher, executive of *Chicago Tribune* and *NY Daily News.*
Dwight Macdonald, 1906-1982, reporter, social critic for *The New Yorker, The Nation, Esquire.*
Ralph McGill, 1893-1969, crusading editor and publisher of *Atlanta Constitution.*
O(scar) O(dd) McIntyre, 1884-1938, feature writer, syndicated columnist concentrating on everyday life in New York City.
Don Marquis, 1878-1937, humor columnist for *NY Sun* and *N.Y. Tribune;* wrote "archy and mehitabel" stories.
Robert Maynard, 1937-97, first African-American editor and then owner of major U.S. paper, the *Oakland Tribune.*
Joseph Medill, 1823-99, longtime *editor of Chicago Tribune.*
H(enry) L(ouis) Mencken, 1880-1956, reporter, editor, columnist with *Baltimore Sun* papers; anti-establishment viewpoint.
Edwin Meredith, 1876-1928, founder of magazine company.
Frank A. Munsey, 1854-1925, owner, editor, and publisher of newspapers and magazines, including *Munsey's Magazine.*
Edward R. Murrow, 1908-65, broadcast reporter, executive; reported from Britain in WW2; hosted *See It Now, Person to Person.*
William Rockhill Nelson, 1841-1915, cofounder, editor, and publisher, *Kansas City Star.*
Adolph S. Ochs, 1858-1935, publisher; built *NY Times* into a leading newspaper.
Louella Parsons, 1881-1972, Hollywood gossip columnist.
Drew (Andrew Russell) Pearson, 1879-1969, investigative reporter and columnist.
(James) Westbrook Pegler, 1894-1969, reporter, columnist.
Shirley Povich, 1905-98, sports columnist.
Joseph Pulitzer, 1847-1911, *NY World* publisher; founded Columbia Journalism School, Pulitzer Prizes.
Joseph Pulitzer II, 1885-1955, longtime *St. Louis Post-Dispatch* editor, publisher; built it into major paper.
Ernie (Ernest Taylor) Pyle, 1900-45, reporter, war correspondent; killed in WW2.
Henry Raymond, 1820-69, cofounder, editor, *NY Times.*
Harry Reasoner, 1923-91, TV reporter, anchor.

John Reed, 1887-1920, reporter, foreign correspondent famous for coverage of Bolshevik Revolution.
Whitelaw Reid, 1837-1912, longtime editor, *NY Tribune*.
James Reston, 1909-95 *NY Times* political reporter, columnist.
Frank Reynolds, 1923-83, TV reporter, anchor.
(Henry) Grantland Rice, 1880-1954, sportswriter.
Jacob Riis, 1849-1914, reporter, photographer; exposed slum conditions in *How the Other Half Lives*.
Max Robinson, 1939-88, TV journalist, first African-American to anchor network news, 1978.
Harold Ross, 1892-1951, founder, editor, The *New Yorker*.
Mike Royko, 1932-97, columnist for *Chicago Sun-Times* and *Chicago Tribune*.
(Alfred) Damon Runyon, 1884-1946, sportswriter, columnist; stories collected in *Guys and Dolls*.
John B. Russwurm, 1799-1851, cofounded (1827) nation's first black newspaper, *Freedom's Journal,* in NYC.
Adela Rogers St. Johns, 1894-1988, reporter, sportswriter for Hearst newspapers.
Harrison Salisbury, 1908-93, reporter, foreign correspondent; a Soviet specialist.
E(dward) W(lyllis) Scripps, 1854-1926, founded first large U.S. newspaper chain, pioneered syndication.
Eric Sevareid, 1912-92, war correspondent, radio newscaster, TV commentator.
William L. Shirer, 1904-93, broadcaster, foreign correspondent; wrote *The Rise and Fall of the Third Reich*.
Red (Walter) Smith, 1905-82, sportswriter.
Edgar P. Snow, 1905-71, correspondent, expert on Chinese Communist movement.
Lawrence Spivak, 1900-94, co-creator, moderator, producer of *Meet the Press*.
(Joseph) Lincoln Steffens, 1866-1936, muckraking journalist.
I(sidor) F(einstein) Stone, 1907-89, one-man editor of *I.F. Stone's Weekly*.
Arthur Hays Sulzberger, 1891-1968, longtime publisher of *N.Y. Times*.
C(yrus) L(eo) Sulzberger, 1912-93, *N.Y. Times* foreign correspondent and columnist.
David Susskind, 1920-87, TV producer, public affairs talk-show host (*Open End*).
John Cameron Swayze, 1906-95, newscaster, anchor of *Camel News Caravan*.
Herbert Bayard Swope, 1882-1958, war correspondent and editor of *N.Y. World*.
Ida Tarbell, 1857-1944, muckraking journalist.
Isaiah Thomas, 1750-1831, printer, publisher, cofounder of revolutionary journal, *Massachusetts Spy*.
Lowell Thomas, 1892-1981, radio newscaster, world traveler.
Dorothy Thompson, 1894-1961, foreign correspondent, columnist, radio commentator.
Ida Bell Wells-Barnett, 1862-1931, African-American reporter, editor, anti-lynching crusader.
William Allen White, 1868-1944, editor, publisher; made *Emporia* (KS) *Gazette* known worldwide.
Walter Winchell, 1897-1972, reporter, columnist, broadcaster of celebrity news.
John Peter Zenger, 1697-1746, printer and journalist; acquitted in precedent-setting libel suit (1735).

Military and Naval Leaders of the Past

Reviewed by Alan C. Aimone, USMA Library

Creighton Abrams, 1914-74, (U.S.) commanded forces in Vietnam, 1968-72.
Alexander the Great, 356-323 B.C., (Maced.) conquered Persia and much of the world known to Europeans.
Harold Alexander, 1891-1969, (Br.) led Allied invasion of Italy, 1943, WW2.
Ethan Allen, 1738-89, (U.S.) headed Green Mountain Boys; captured Ft. Ticonderoga, 1775, Amer. Rev.
Edmund Allenby, 1861-1936, (Br.) in Boer War, WW1; led Egyptian expeditionary force, 1917-18.
Benedict Arnold, 1741-1801, (U.S.) victorious at Saratoga; tried to betray West Point to British, Amer. Rev.
Henry "Hap" Arnold, 1886-1950, (U.S.) commanded Army Air Force in WW2.
John Barry, 1745-1803, (U.S.) won numerous sea battles during Amer. Rev.
Belisarius, c505-565, (Byzant.) won remarkable victories for Byzantine Emperor Justinian I.
Pierre Beauregard, 1818-93, (U.S.) Confed. general, ordered bombardment of Ft. Sumter that began Civil War.
Gebhard von Blücher, 1742-1819, (Ger.) helped defeat Napoleon at Waterloo.
Napoleon Bonaparte, 1769-1821, (Fr.) defeated Russia and Austria at Austerlitz, 1805; invaded Russia, 1812; defeated at Waterloo, 1815.
Edward Braddock, 1695-1755, (Br.) commanded forces in French and Indian War.
Omar N. Bradley, 1893-1981, (U.S.) headed U.S. ground troops in Normandy invasion, 1944, WW2.
John Burgoyne, 1722-92, (Br.) defeated at Saratoga, Amer. Rev.
Julius Caesar, 100-44 BC (Rom.) general and politician; conquered N Gaul; overthrew Roman Republic.
Claire Lee Chennault, 1893-1958, (U.S.) headed Flying Tigers in WW2.
Mark W. Clark, 1896-1984, (U.S.) helped plan N African invasion in WW2; commander of UN forces, Korean War.
Karl von Clausewitz, 1780-1831, (Pruss.) military theorist.
Lucius D. Clay, 1897-1978, (U.S.) led Berlin airlift, 1948-49.
Henry Clinton, 1738-95, (Br.) commander of forces in Amer. Rev., 1778-81.
Cochise, c1815-74, (Nat. Am.) chief of Chiricahua band of Apache Indians in Southwest.
Charles Cornwallis, 1738-1805, (Br.) victorious at Brandywine, 1777; surrendered at Yorktown, Amer. Rev.
Hernan Cortes, 1485-1547, (Sp.) led Spanish conquistadors in the defeat of the Aztec empire, 1519-28.
Crazy Horse, 1849-77, (Nat. Am.) Sioux war chief victorious at battle of Little Bighorn.
George Armstrong Custer, 1839-76, (U.S.) U.S. army officer defeated and killed at battle of Little Bighorn.
Moshe Dayan, 1915-81, (Isr.) directed campaigns in the 1967, 1973 Arab-Israeli wars.
Stephen Decatur, 1779-1820, (U.S.) naval hero of Barbary wars, War of 1812.
Anton Denikin, 1872-1947, (Russ.) led White forces in Russian civil war.
George Dewey, 1837-1917, (U.S.) destroyed Spanish fleet at Manila, 1898, Span.-Amer. War.
Karl Doenitz, 1891-1980, (Ger.) submarine com. in chief and naval commander, WW2.
Hugh C. Dowding, 1883-1970, (Br.) headed RAF, 1936-40, WW2.
Jubal Early, 1816-94, (U.S.) Confed. general, led raid on Washington, 1864, Civil War.
Dwight D. Eisenhower, 1890-1969, (U.S.) commanded Allied forces in Europe, WW2.
David Farragut, 1801-70, (U.S.) Union admiral, captured New Orleans, Mobile Bay, Civil War.
Ferdinand Foch, 1851-1929, (Fr.) headed victorious Allied armies, 1918, WW1.
Nathan Bedford Forrest, 1821-77, (U.S.) Confed. general, led raids against Union supply lines, Civil War.
Frederick the Great, 1712-86, (Pruss.) led Prussia in Seven Years War.
Horatio Gates, 1728-1806, (U.S.) commanded army at Saratoga, Amer. Rev.
Genghis Khan, 1162-1227, (Mongol) unified Mongol tribes and subjugated much of Asia, 1206-21.
Geronimo, 1829-1909, (Nat. Am.) leader of Chiricahua band of Apache Indians.
Charles G. Gordon, 1833-85, (Br.) led forces in China, Crimean War; killed at Khartoum.
Ulysses S. Grant, 1822-85, (U.S.) headed Union army, Civil War, 1864-65; forced Lee's surrender, 1865.
Nathanael Greene, 1742-86, (U.S.) defeated British in Southern campaign, 1780-81.
Heinz Guderian, 1888-1953, (Ger.) tank theorist, led panzer forces in Poland, France, Russia, WW2.
Che (Ernesto) Guevara, 1928-67, (Arg.) guerrilla leader; prominent in Cuban revolution; killed in Bolivia.
Gustavus Adolphus, 1594-1632, (Swed.) King; military tactician; reformer; led forces in Thirty Years' War.
Douglas Haig, 1861-1928, (Br.) led British armies in France, 1915-18, WW1.
William F. Halsey, 1882-1959, (U.S.) defeated Japanese fleet at Leyte Gulf, 1944, WW2.
Hannibal, 247-183 B.C., (Carthag.) invaded Rome, crossing Alps, in Second Punic War, 218-201 B.C.
Sir Arthur Travers Harris, 1895-1984, (Br.) led Britain's WW2 bomber command.
Richard Howe, 1726-99, (Br.) commanded navy in Amer. Rev., 1776-78; June 1 victory against French, 1794.
William Howe, 1729-1814, (Br.) commanded forces in Amer. Rev., 1776-78.
Isaac Hull, 1773-1843, (U.S.) sunk British frigate Guerriere, War of 1812.
Thomas (Stonewall) Jackson, 1824-63, (U.S.) Confed. general, led Shenandoah Valley campaign, Civil War.
Joseph Joffre, 1852-1931, (Fr.) headed Allied armies, won Battle of the Marne, 1914, WW1.
Chief Joseph, c1840-1904, (Nat. Am.) chief of the Nez Percé, led his tribe across 3 states seeking refuge in Canada; surrendered about 30 mi from Canadian border.

John Paul Jones, 1747-92, (U.S.) commanded Bonhomme Richard in victory over Serapis, Amer. Rev., 1779.
Stephen Kearny, 1794-1848, (U.S.) headed Army of the West in Mexican War.
Albert Kesselring, 1885-1960 (Ger.) field marshal who led the defense of Italy in WW2.
Ernest J. King, 1878-1956, (U.S.) key WW2 naval strategist.
Horatio H. Kitchener, 1850-1916, (Br.) led forces in Boer War; victorious at Khartoum; organized army in WW1.
Henry Knox, 1750-1806, (U.S.) general in Amer. Rev.; first sec. of war under U.S. Constitution.
Lavrenti Kornilov, 1870-1918, (Russ.) commander-in-chief, 1917; led counter-revolutionary march on Petrograd.
Thaddeus Kosciusko, 1746-1817, (Pol.) aided Amer. Rev.
Walter Krueger, 1881-1967, (U.S.) led Sixth Army in WW2 in Southwest Pacific.
Mikhail Kutuzov, 1745-1813, (Russ.) fought French at Borodino, Napoleonic Wars, 1812; abandoned Moscow; forced French retreat.
Marquis de Lafayette, 1757-1834, (Fr.) fought in, secured French aid for Amer. Rev.
T(homas) E. Lawrence (of Arabia), 1888-1935, (Br.) organized revolt of Arabs against Turks in WW1.
Henry (Light-Horse Harry) Lee, 1756-1818, (U.S.) cavalry officer in Amer. Rev.
Robert E. Lee, 1807-70, (U.S.) Confed. general defeated at Gettysburg, Civil War; surrendered to Grant, 1865.
Curtis LeMay, 1906-90, (U.S.) Air Force commander in WW2, Korean War, and Vietnam War.
Lyman Lemnitzer, 1899-1988, (U.S.) WW2 hero, later general, chairman of Joint Chiefs of Staff.
James Longstreet, 1821-1904, (U.S.) aided Lee at Gettysburg, Civil War.
Maurice, Count of Nassau, 1567-1625, (Dutch) military innovator; led forces in Thirty Years' War.
Douglas MacArthur, 1880-1964, (U.S.) commanded forces in SW Pacific in WW2; headed occupation forces in Japan, 1945-51; UN commander in Korean War.
Erich von Manstein, 1887-1973, (Ger.) served WW1–2, planned inv. of France (1940), convicted of war crimes.
Carl Gustaf Mannerheim, 1867-1951, (Finn.) army officer and pres. of Finland 1944-46.
Francis Marion, 1733-95, (U.S.) led guerrilla actions in South Carolina during Amer. Rev.
Duke of Marlborough, 1650-1722, (Br.) led forces against Louis XIV in War of the Spanish Succession.
George C. Marshall, 1880-1959, (U.S.) chief of staff in WW2; authored Marshall Plan.
George B. McClellan, 1826-85, (U.S.) Union general, commanded Army of the Potomac, 1861-62, Civil War.
George Meade, 1815-72, (U.S.) commanded Union forces at Gettysburg, Civil War.
Billy Mitchell, 1879-1936, (U.S.) WW1 air-power advocate; court-martialed for insubordination, later vindicated.
Helmuth von Moltke, 1800-91, (Ger.) victorious in Austro-Prussian, Franco-Prussian wars.
Louis de Montcalm, 1712-59, (Fr.) headed troops in Canada, French and Indian War; defeated at Quebec, 1759.
Bernard Law Montgomery, 1887-1976, (Br.) stopped German offensive at Alamein, 1942, WW2; helped plan Normandy.
Daniel Morgan, 1736-1802, (U.S.) victorious at Cowpens, 1781, Amer. Rev.
Louis Mountbatten, 1900-79, (Br.) Supreme Allied Commander of SE Asia, 1943-46, WW2.
Joachim Murat, 1767-1815, (Fr.) led cavalry at Marengo, Austerlitz, and Jena, Napoleonic Wars.
Horatio Nelson, 1758-1805, (Br.) naval commander, destroyed French fleet at Trafalgar.
Michel Ney, 1769-1815, (Fr.) commanded forces in Switz., Aust., Russ., Napoleonic Wars; defeated at Waterloo.
Chester Nimitz, 1885-1966, (U.S.) commander of naval forces in Pacific in WW2.
George S. Patton, 1885-1945, (U.S.) led assault on Sicily, 1943, Third Army invasion of Europe, WW2.
Oliver Perry, 1785-1819, (U.S.) won Battle of Lake Erie in War of 1812.
John Pershing, 1860-1948, (U.S.) commanded Mexican border campaign, 1916, Amer. Expeditionary Force, WW1.
Henri Philippe Pétain, 1856-1951, (Fr.) defended Verdun, 1916; headed Vichy government in WW2.
George E. Pickett, 1825-75, (U.S.) Confed. general famed for "charge" at Gettysburg, Civil War.
Charles Portal, 1893-1971, (Br.) chief of staff, Royal Air Force, 1940-45, led in Battle of Britain.
Hyman Rickover, 1900-86, (U.S.) father of nuclear navy.
Matthew Bunker Ridgway, 1895-1993, (U.S.) commanded Allied ground forces in Korean War.
Erwin Rommel, 1891-1944, (Ger.) headed Afrika Korps, WW2.
Gerd von Rundstedt, 1875-1953, (Ger.) supreme commander in West, 1942-45, WW2.
Aleksandr Samsonov, 1859-1914, (Russ.) led invasion of E Prussia, WW1, defeated at Tannenberg, 1914.
Winfield Scott, 1786-1866, (U.S.) hero of War of 1812; headed forces in Mexican War, took Mexico City.
Philip Sheridan, 1831-88, (U.S.) Union cavalry officer, headed Army of the Shenandoah, 1864-65, Civil War.
William T. Sherman, 1820-91, (U.S.) Union general, sacked Atlanta during "march to the sea," 1864, Civil War.
Carl Spaatz, 1891-1974, (U.S.) directed strategic bombing against Germany, later Japan, in WW2.
Raymond Spruance, 1886-1969, (U.S.) victorious at Midway Island, 1942, WW2.
Joseph W. Stilwell, 1883-1946, (U.S.) headed forces in the China, Burma, India theater in WW2.
J.E.B. Stuart, 1833-64, (U.S.) Confed. cavalry commander, Civil War.
Aleksandr Suvorov, 1729-1800, (Rus.) commanded Allied Russian and Austrian armies against Ottoman Turks in Russo-Turkish War.
George H. Thomas, 1816-70, (U.S.) saved Union army at Chattanooga, 1863; won at Nashville, 1864, Civil War.
Semyon Timoshenko, 1895-1970, (USSR) defended Moscow, Stalingrad, WW2; led winter offensive, 1942-43.
Alfred von Tirpitz, 1849-1930, (Ger.) responsible for submarine blockade in WW1.
Sebastien Le Prestre de Vauban, 1633-1707, (Fr.) innovative military engineer and theorist.
Jonathan M. Wainwright, 1883-1953, (U.S.) forced to surrender on Corregidor, 1942, WW2.
George Washington, 1732-99, (U.S.) led Continental army, 1775-83, Amer. Rev.
Archibald Wavell, 1883-1950, (Br.) commanded forces in N and E Africa, and SE Asia in WW2.
Anthony Wayne, 1745-96, (U.S.) captured Stony Point, 1779, Amer. Rev.
Duke of Wellington, 1769-1852, (Br.) defeated Napoleon at Waterloo, 1815.
James Wolfe, 1727-59, (Br.) captured Quebec from French, 1759, French and Indian War.
Isoroku Yamamoto, 1884-1943, (Jpn.) com. in chief of Japanese fleet and naval planner before and during WW2.
Georgi Zhukov, 1895-1974, (Russ.) defended Moscow, 1941, led assault on Berlin, 1945, WW2.

Philosophers and Religious Figures of the Past

For other Greeks and Romans, see Historical Figures chapter.

Lyman Abbott, 1835-1922, (U.S.) clergyman, reformer; advocate of Christian Socialism.
Pierre Abelard, 1079-1142, (Fr.) philosopher, theologian, teacher; used dialectic method to support Christian beliefs.
Mortimer Adler, 1902-2001, (U.S.) philosopher, helped create "Great Books" program.
Felix Adler, 1851-1933, (U.S.) German-born founder of the Ethical Culture Society.
(St.) Anselm, c1033-1109, (It.) philosopher-theologian, church leader; "ontological argument" for God's existence.
(St.) Thomas Aquinas, 1225-74, (It.) preeminent medieval philosopher-theologian; *Summa Theologica*.
Aristotle, 384-322 BC, (Gr.) pioneering wide-ranging philosopher, logician, ethician, naturalist.
(St.) Augustine, 354-430, (N Africa) philosopher, theologian, bishop; *Confessions, City of God, On the Trinity.*
J. L. Austin, 1911-60, (Br.) ordinary-language philosopher.
Averroes (Ibn Rushd), 1126-98, (Sp.) Islamic philosopher, physician.
Avicenna (Ibn Sina), 980-1037, (Iran.) Islamic philosopher, scientist.
A(lfred) J(ules) Ayer, 1910-89, (Br.) philosopher; logical positivist; *Language, Truth, and Logic.*
Roger Bacon, c1214-94, (Eng.) philosopher and scientist.
Bahaullah (Mirza Husayn Ali), 1817-92, (Pers.) founder of Bahá'í faith.
Karl Barth, 1886-1968, (Swiss) theologian; a leading force in 20th-cent. Protestantism.
Thomas à Becket, 1118-70, (Eng.) archbishop of Canterbury; opposed Henry II; murdered by King's men.
(St.) Benedict, c480-547, (It.) founded the Benedictines.
Jeremy Bentham, 1748-1832, (Br.) philosopher, reformer; enunciated utilitarianism.
Henri Bergson, 1859-1941, (Fr.) philosopher of evolution.
George Berkeley, 1685-1753, (Ir.) idealist philosopher, churchman.
John Biddle, 1615-62, (Eng.) founder of English Unitarianism.
Jakob Boehme, 1575-1624, (Ger.) theosophist and mystic.

Dietrich Bonhoeffer, 1906-1945 (Ger.) Lutheran theologian, pastor; executed as opponent of Nazis.
William Brewster, 1567-1644, (Eng.) headed Pilgrims.
Emil Brunner, 1889-1966, (Swiss) Protestant theologian.
Giordano Bruno, 1548-1600, (It.) philosopher, pantheist.
Martin Buber, 1878-1965, (Ger.) Jewish philosopher, theologian; *I and Thou.*
Buddha (Siddhartha Gautama**),** c563-c483 BC, (Indian) phil-osopher; founded Buddhism.
John Calvin, 1509-64, (Fr.) theologian; a key figure in the Protestant Reformation.
Rudolph Carnap, 1891-1970, (U.S.) German-born analytic philosopher; a founder of logical positivism.
William Ellery Channing, 1780-1842, (U.S.) clergyman; early spokesman for Unitarianism.
Auguste Comte, 1798-1857, (Fr.) philosopher; originated positivism.
Confucius, 551-479 BC, (Chin.) founder of Confucianism.
John Cotton, 1584-1652, (Eng.) Puritan theologian.
Thomas Cranmer, 1489-1556, (Eng.) churchman; wrote much of *Book of Common Prayer.*
René Descartes, 1596-1650, (Fr.) philosopher, mathematician; "father of modern philosophy." *Discourse on Method, Meditations on First Philosophy.*
John Dewey, 1859-1952, (U.S.) philosopher, educator; instrumentalist theory of knowledge; helped inaugurate progressive education movement.
Denis Diderot, 1713-84, (Fr.) philosopher, encyclopedist.
John Duns Scotus, c1266-1308, (Sc.) Franciscan philosopher and theologian.
Mary Baker Eddy, 1821-1910, (U.S.) founder of Christian Science; *Science and Health.*
Jonathan Edwards, 1703-58, (U.S.) preacher, theologian; "Sinners in the Hands of an Angry God."
(Desiderius) Erasmus, c1466-1536, (Dutch) Renaissance humanist; *On the Freedom of the Will.*
Johann Fichte, 1762-1814, (Ger.) idealist philosopher.
Michel Foucault, 1926-84, (Fr.) structuralist philosopher, historian.
George Fox, 1624-91, (Br.) founder of Society of Friends.
(St.) Francis of Assisi, 1182-1226, (It.) founded Franciscans.
al-Ghazali, 1058-1111, Islamic philosopher.
Georg W. F. Hegel, 1770-1831, (Ger.) idealist philosopher; *Phenomenology of Mind.*
Martin Heidegger, 1889-1976, (Ger.) existentialist philosopher; affected many fields; *Being and Time.*
Johann G. Herder, 1744-1803, (Ger.) philosopher, cultural historian; a founder of German Romanticism.
Thomas Hobbes, 1588-1679, (Eng.) philosopher, political theorist; *Leviathan.*
David Hume, 1711-76, (Sc.) empiricist philosopher; *Enquiry Concerning Human Understanding.*
Jan Hus, 1369-1415, (Czech.) religious reformer.
Edmund Husserl, 1859-1938, (Ger.) philosopher; founded the phenomenological movement.
Thomas Huxley, 1825-95, (Br.) philosopher, educator.
William Inge, 1860-1954, (Br.) theologian; explored mystic aspects of Christianity.
William James, 1842-1910, (U.S.) philosopher, psychologist; pragmatist; studied religious experience.
Karl Jaspers, 1883-1969, (Ger.) existentialist philosopher.
Joan of Arc, 1412-1431, (Fr.) national heroine and a patron saint of France; key figure in the Hundred Years' War.
Immanuel Kant, 1724-1804, (Ger.) philosopher; founder of modern critical philosophy; *Critique of Pure Reason.*
Thomas à Kempis, c1380-1471, (Ger.) monk, devotional writer; *Imitation of Christ.* attributed to him.
Soren Kierkegaard, 1813-55, (Dan.) religious philosopher; pre-existentialist; *Either/Or, The Sickness Unto Death.*
John Knox, 1505-72, (Sc.) leader of the Protestant Reformation in Scotland.
Lao-Tzu, 604-531 BC, (Chin.) philosopher; considered the founder of the Taoist religion.
Gottfried von Leibniz, 1646-1716, (Ger.) rationalistic philosopher, logician, mathematician.
John Locke, 1632-1704, (Eng.) political theorist, empiricist philosopher; *Essay Concerning Human Understanding.*
(St.) Ignatius Loyola, 1491-1556, (Sp.) founder of the Jesuits; *Spiritual Exercises.*
Martin Luther, 1483-1546, (Ger.) leader of the Protestant Reformation, founded Lutheran church.
Jean-Francois Lyotard, 1924-98, (Fr.) postmodern philosopher, lecturer; *The Post-Modern Condition.*
Maimonides, 1135-1204, (Sp.) major Jewish philosopher.
Gabriel Marcel, 1889-1973, (Fr.) Roman Catholic existentialist philosopher, dramatist, and critic,
Jacques Maritain, 1882-1973, (Fr.) Neo-Thomist philosopher.
Cotton Mather, 1663-1728, (U.S.) defender of orthodox Puritanism; founded Yale, 1701.
Philipp Melanchthon, 1497-1560, (Ger.) theologian, humanist; an important voice in the Reformation.
Maurice Merleau-Ponty, 1908-61, (Fr.) existentialist philosopher; *Phenomenology of Perception.*
Thomas Merton, 1915-68, (U.S.) Trappist monk, spiritual writer; *The Seven Storey Mountain.*
John Stuart Mill, 1806-73, (Br.) philosopher, economist; libertarian political theorist; *Utilitarianism.*
Muhammad, c570-632, (Arab) the prophet of Islam.
Dwight Moody, 1837-99, (U.S.) evangelist.
G(eorge) E(dward) Moore, 1873-1958, (Br.) philosopher; *Principia Ethica,* "A Defense of Common Sense."
Elijah Muhammad, 1897-1975, (U.S.) leader of the Black Muslim sect.
Heinrich Muhlenberg, 1711-87, (Ger.) organized the Lutheran Church in America.
John H. Newman, 1801-90, (Br.) Roman Catholic convert, cardinal; led Oxford Movement; *Apologia pro Vita Sua.*
Reinhold Niebuhr, 1892-1971, (U.S.) Protestant theologian.
Richard Niebuhr, 1894-1962 (U.S.) Protestant theologian.
Friedrich Nietzsche, 1844-1900, (Ger.) philosopher; *The Birth of Tragedy, Beyond Good and Evil, Thus Spake Zarathustra.*
Blaise Pascal, 1623-62, (Fr.) philosopher, mathematician; *Pensées.*
(St.) Patrick, c389-c461, (Br.) brought Christianity to Ireland.
(St.) Paul, ?-c67, a key proponent of Christianity; his epistles are first Christian theological writing.
Norman Vincent Peale, 1898-1993, (U.S.) minister, author; *The Power of Positive Thinking.*
C(harles) S. Peirce, 1839-1914, (U.S.) philosopher, logician; originated concept of pragmatism, 1878.
Plato, c428-347 BC, (Gr.) philosopher; wrote classic Socratic dialogues; argued for universal truths and independent reality of ideas or forms; *Republic.*
Plotinus, 205-70, (Rom.) a founder of neo-Platonism; *Enneads.*
W(illard) V(an) O(rman) Quine, 1908-2001, (U.S.) philosopher, logician; "On What There Is."
Josiah Royce, 1855-1916, (U.S.) idealist philosopher
Bertrand Russell, 1872-1970, (Br.) philosopher, logician; one of the founders of modern logic; a prolific popular writer.
Charles T. Russell, 1852-1916, (U.S.) founder of Jehovah's Witnesses.
Gilbert Ryle, 1900-76, (Br.) analytic philosopher; *The Concept of Mind.*
George Santayana, 1863-1952, (U.S.) philosopher, writer, critic; *The Sense of Beauty, The Realms of Being.*
Jean-Paul Sartre, 1905-80, (Fr.) philosopher, novelist, playwright. *Nausea, No Exit, Being and Nothingness.*
Friedrich von Schelling, 1775-1854, (Ger.) philosopher of romantic movement.
Friedrich Schleiermacher, 1768-1834, (Ger.) theologian; a founder of modern Protestant theology.
Arthur Schopenhauer, 1788-1860, (Ger.) philosopher; *The World as Will and Idea.*
Albert Schweitzer, 1875-1965, (Ger.) theologian, social philosopher, medical missionary.
Joseph Smith, 1805-44, (U.S.) founded Latter-Day Saints (Mormon) movement, 1830.
Socrates, 469-399 BC, (Gr.) influential philosopher immortalized by Plato.
Herbert Spencer, 1820-1903, (Br.) philosopher of evolution.
Baruch de Spinoza, 1632-77, (Dutch) rationalist philosopher; *Ethics.*
Billy Sunday, 1862-1935, (U.S.) evangelist.
Pierre Teilhard de Chardin, 1881-1955, (Fr.) Jesuit priest, paleontologist, philosopher-theologian; *The Divine Milieu.*
Daisetz Teitaro Suzuki, 1870-1966, (Jpn.) Buddhist scholar.
(St.) Theresa of Lisieux, 1873-97, (Fr.) Carmelite nun revered for everyday sanctity; *The Story of a Soul.*
Emanuel Swedenborg, 1688-1772, (Swed.) philosopher, mystic.
Paul Tillich, 1886-1965, (U.S.) German-born philosopher and theologian; brought depth psychology to Protestantism.
John Wesley, 1703-91, (Br.) theologian, evangelist; founded Methodism.
Alfred North Whitehead, 1861-1947, (Br.) philosopher, mathematician; *Process and Reality.*
William of Occam, c1285-c1349 (Eng.) medieval scholastic philosopher; nominalist.
Roger Williams, c1603-83, (U.S.) clergyman; championed religious freedom and separation of church and state.
Ludwig Wittgenstein, 1889-1951, (Austrian) philosopher; major influence on contemporary language philosophy; *Tractatus Logico-Philosophicus, Philosophical Investigations.*
John Woolman, 1720-72, (U.S.) Quaker social reformer, abolitionist, writer; *The Journal.*
John Wycliffe, 1320-84, (Eng.) theologian, reformer.
(St.) Francis Xavier, 1506-52, (Sp.) Jesuit missionary, "Apostle of the Indies."
Brigham Young, 1801-77, (U.S.) Mormon leader after Smith's assassination; colonized Utah.
Huldrych Zwingli, 1484-1531, (Swiss) theologian; led Swiss Protestant Reformation.

Political Leaders of the Past

(U.S. presidents, vice presidents, Supreme Ct. justices, signers of Decl. of Indep. listed elsewhere.)

Abu Bakr, 573-634, Muslim leader, first caliph, chosen successor to Muhammad.
Dean Acheson, 1893-1971, (U.S.) sec. of state; architect of cold war foreign policy.
Samuel Adams, 1722-1803, (U.S.) patriot, Boston Tea Party firebrand.
Konrad Adenauer, 1876-1967, (Ger.) first West German chancellor.
Emilio Aguinaldo, 1869-1964, (Philip.) revolutionary; fought against Spain and the U.S.
Akbar, 1542-1605, greatest Mogul emperor of India.
Carl Albert, 1908-2000 (U.S.) House rep. from OK, Speaker, 1971-76.
Salvador Allende Gossens, 1908-1973, (Chilean) Marxist pres. 1970-73; ousted and died in coup.
Hafez al Assad, 1930-2000 (Syr.), Syrian ruler from 1970.
Herbert H. Asquith, 1852-1928, (Br.) liberal prime min.; instituted major social reform.
Atahualpa, ?-1533, Inca (ruling chief) of Peru.
Kemal Ataturk, 1881-1938, (Turk.) founded modern Turkey.
Clement Attlee, 1883-1967, (Br.) Labour party leader, prime min.; enacted natl. health, nationalized many industries.
Stephen F. Austin, 1793-1836, (U.S.) led Texas colonization.
Mikhail Bakunin, 1814-76, (Rus.) revolutionary; leading exponent of anarchism.
Arthur J. Balfour, 1848-1930, (Br.) foreign sec. under Lloyd George; issued Balfour Declaration backing Zionism.
Bernard M. Baruch, 1870-1965, (U.S.) financier, govt. adviser.
Fulgencio Batista y Zaldívar, 1901-73, (Cub.) Cuban pres. (1940-44, 1952-59), overthrown by Castro.
Lord Beaverbrook, 1879-1964, (Br.) financier, statesman, newspaper owner.
Menachem Begin, 1913-92, (Isr.) Israeli prime min., shared 1978 Nobel Peace Prize.
Eduard Benes, 1884-1948, (Czech.) pres. during interwar and post-WW2 eras.
David Ben-Gurion, 1886-1973, (Isr.) first prime min. of Israel, 1948-53, 1955-63.
Thomas Hart Benton, 1782-1858, (U.S.) Missouri senator; championed agrarian interests and westward expansion.
Aneurin Bevan, 1897-1960, (Br.) Labour party leader.
Ernest Bevin, 1881-1951, (Br.) Labour party leader, foreign minister; helped lay foundation for NATO.
Otto von Bismarck, 1815-98, (Ger.) statesman known as the Iron Chancellor; uniter of Germany, 1870.
James G. Blaine, 1830-93, (U.S.) Republican politician, diplomat; influential in Pan-American movement.
Léon Blum, 1872-1950, (Fr.) socialist leader, writer; headed first Popular Front government.
Simón Bolívar, 1783-1830, (Venez.) S. Amer. Revolutionary who liberated much of the continent from Spanish rule.
William E. Borah, 1865-1940, (U.S.) isolationist senator; helped block U.S. membership in League of Nations.
Cesare Borgia, 1476-1507, (It.) soldier, politician; an outstanding figure of the Italian Renaissance.
Willy Brandt, 1913-92, (Ger.) statesman, chancellor of West Germany, 1969-74; promoted East/West peace, *Ostpolitik.*
Leonid Brezhnev, 1906-82, (USSR) Soviet leader, 1964-82.
Aristide Briand, 1862-1932, (Fr.) foreign min.; chief architect of Locarno Pact and anti-war Kellogg-Briand Pact.
William Jennings Bryan, 1860-1925, (U.S.) Democratic, populist leader, orator; 3 times lost race for presidency.
Ralph Bunche, 1904-71, (U.S.) a founder and key diplomat of United Nations for more than 20 years.
John C. Calhoun, 1782-1850, (U.S.) political leader; champion of states' rights and a symbol of the Old South.
Robert Castlereagh, 1769-1822, (Br.) foreign sec.; guided Grand Alliance against Napoleon.
Camillo Benso Cavour, 1810-61, (It.) statesman; largely responsible for uniting Italy under the House of Savoy.
Nicolae Ceausescu, 1918-89, (Roman.) Communist leader, head of state 1967-89; executed.
Austen Chamberlain, 1863-1937, (Br.) statesman; helped finalize Locarno Treaties, both 1925.
Neville Chamberlain, 1869-1940, (Br.) Conservative prime min. whose appeasement of Hitler led to Munich Pact.
Chiang Kai-shek, 1887-1975, (Chin.) Nationalist Chinese pres. whose government was driven from mainland to Taiwan.
Winston Churchill, 1874-1965, (Br.) prime min., soldier, author; guided Britain through WW2.
Galeazzo Ciano, 1903-44, (It.) fascist foreign minister; helped create Rome-Berlin Axis, executed by Mussolini.
Henry Clay, 1777-1852, (U.S.) "The Great Compromiser," one of the most influential pre-Civil War political leaders.
Georges Clemenceau, 1841-1929, (Fr.) twice prem., Wilson's antagonist at Paris Peace Conference after WW1.
DeWitt Clinton, 1769-1828, (U.S.) political leader; responsible for promoting idea of the Erie Canal.
Robert Clive, 1725-74, (Br.) first administrator of Bengal; laid foundation for British Empire in India.
Jean Baptiste Colbert, 1619-83, (Fr.) statesman; influential under Louis XIV, created the French navy.
Bettino Craxi, 1934-2000, (It.) Italy's first post-WWII Socialist premier.
Oliver Cromwell, 1599-1658, (Br.) Lord Protector of England, led parliamentary forces during Civil War.
Curzon of Kedleston, 1859-1925, (Br.) viceroy of India, foreign sec.; major force in post-WW1 world.
Édouard Daladier, 1884-1970, (Fr.) Radical Socialist politician, arrested by Vichy, interned by Germans until 1945.
Richard J. Daley, 1902-1976, (U.S.) Chicago mayor.
Georges Danton, 1759-94, (Fr.) leading French Rev. figure.
Jefferson Davis, 1808-89, (U.S.) pres. of the Confederacy.
Charles G. Dawes, 1865-1951, (U.S.) statesman, banker; advanced plan to stabilize post-WW1 German finances.
Alcide De Gasperi, 1881-1954, (It.) prime min.; founder of Christian Democratic party.
Charles De Gaulle, 1890-1970, (Fr.) general, statesman; first pres. of the Fifth Republic.
Deng Xiaoping, 1904-97, (Chin.) "paramount leader" of China; backed economic modernization.
Eamon De Valera, 1882-1975, (Ir.-U.S.) statesman; led fight for Irish independence.
Thomas E. Dewey, 1902-71, (U.S.) NY governor; twice loser in try for presidency.
Ngo Dinh Diem, 1901-63, (Viet.) South Vietnamese pres.; assassinated in government takeover.
Everett M. Dirksen, 1896-1969, (U.S.) Senate Republican minority leader, orator.
Benjamin Disraeli, 1804-81, (Br.) prime min.; considered founder of modern Conservative party.
Engelbert Dollfuss, 1892-1934, (Austrian) chancellor; assassinated by Austrian Nazis.
Andrea Doria, 1466-1560, (It.) Genoese admiral, statesman; called "Father of Peace" and "Liberator of Genoa."
Stephen A. Douglas, 1813-61, (U.S.) Democratic leader, orator; opposed Lincoln for the presidency.
Alexander Dubcek, 1921-92, (Czech.) statesman whose attempted liberalization was crushed, 1968.
John Foster Dulles, 1888-1959, (U.S.) sec. of state under Eisenhower, cold war policy-maker.
Friedrich Ebert, 1871-1925, (Ger.) Social Democratic movement leader; 1st pres., Weimar Republic, 1919-25.
Sir Anthony Eden, 1897-1977, (Br.) foreign sec., prime min. during Suez invasion of 1956.
Ludwig Erhard, 1897-1977, (Ger.) economist, West German chancellor; led nation's economic rise after WW2.
Amintore Fanfani, 1908-99, (It.) six-time premier of Italy.
Joao Baptista de Figueiredo, 1918-99, (Braz.) president of Brazil, restored the nation's democracy.
Hamilton Fish, 1808-93, (U.S.) sec. of state, successfully mediated disputes with Great Britain, Latin America.
James V. Forrestal, 1892-1949, (U.S.) sec. of navy, first sec. of defense.
Francisco Franco, 1892-1975, (Sp.) leader of rebel forces during Spanish Civil War and longtime ruler of Spain.
Benjamin Franklin, 1706-90, (U.S.) printer, publisher, author, inventor, scientist, diplomat.
Louis de Frontenac, 1620-98, (Fr.) governor of New France (Canada); encouraged explorations, fought Iroquois.
J. William Fulbright, 1905-95, (U.S.) U.S. senator; leading figure in U.S. foreign policy during cold war years.
Hugh Gaitskell, 1906-63, (Br.) Labour party leader; major force in reversing its stand for unilateral disarmament.
Albert Gallatin, 1761-1849, (U.S.) sec. of treasury; instrumental in negotiating end of War of 1812.
Léon Gambetta, 1838-82, (Fr.) statesman, politician; one of the founders of the Third Republic.
Indira Gandhi, 1917-84, (In.) daughter of Jawaharlal Nehru, prime min. of India, 1966-77, 1980-84; assassinated.
Mohandas K. Gandhi, 1869-1948, (In.) political leader, ascetic; led movement against British rule; assassinated.
Giuseppe Garibaldi, 1807-82, (It.) patriot, soldier; a leader in the Risorgimento, Italian unification movement.
William E. Gladstone, 1809-98, (Br.) prime min. 4 times; dominant force of Liberal party from 1868 to 1894.
Paul Joseph Goebbels, 1897-1945, (Ger.) Nazi propagandist, master of mass psychology.
Barry Goldwater, 1909-98 (U.S.) conservative U.S. senator and 1964 Republican presid. nominee.
Klement Gottwald, 1896-1953, (Czech.) Communist leader; ushered Communism into his country.

Alexander Hamilton, 1755-1804, (U.S.) first treasury sec.; champion of strong central government.
Dag Hammarskjold, 1905-61, (Swed.) statesman; UN sec.-general.
Hassan II, King, 1929-99, (Moroc.), ruler of Morocco,1962-99.
John Hay, 1838-1905, (U.S.) sec. of state; primarily associated with Open Door Policy toward China.
Patrick Henry, 1736-99, (U.S.) major revolutionary figure, remarkable orator.
Édouard Herriot, 1872-1957, (Fr.) Radical Socialist leader; twice prem., pres. of National Assembly.
Theodor Herzl, 1860-1904, (Hung.) founded modern Zionism.
Heinrich Himmler, 1900-45, (Ger.) head of Nazi SS and Gestapo.
Paul von Hindenburg, 1847-1934, (Ger.) field marshal, WW1; 2d pres. of Weimar Republic, 1925-34.
Adolf Hitler, 1889-1945, (Ger.) dictator; built Nazism, launched WW2, presided over the Holocaust.
Ho Chi Minh, 1890-1969, (Viet.) N Vietnamese pres., Vietnamese Communist leader.
Harry L. Hopkins, 1890-1946, (U.S.) New Deal administrator; closest adviser to FDR during WW2.
Edward M. House, 1858-1938, (U.S.) diplomat; confidential adviser to Woodrow Wilson.
Samuel Houston, 1793-1863, (U.S.) leader of struggle to win control of Texas from Mexico.
Cordell Hull, 1871-1955, (U.S.) sec. of state, 1933-44; initiated reciprocal trade to lower tariffs, helped organize UN.
Hubert H. Humphrey, 1911-78, (U.S.) Minnesota Democrat; senator; vice pres., pres. candidate.
Hussein, King, 1935-99 (Jordan), peacemaker; ruler of Jordan, 1952-99.
Jinnah, Muhammad Ali, 1876-1948, (Pak.) founder, first governor-general of Pakistan.
Benito Juarez, 1806-72, (Mex.) rallied his country against foreign threats, sought to create democratic, federal republic.
Constantine Karamanlis, 1907-98, (Gr.) Greek prime min. (1955-63, 1974-80); restored democracy; later president.
Frank B. Kellogg, 1856-1937, (U.S.) sec. of state; negotiated Kellogg-Briand Pact to outlaw war.
Robert F. Kennedy, 1925-68, (U.S.) attorney general, senator; assassinated while seeking presidency.
Aleksandr Kerensky, 1881-1970, (Russ.) headed provisional government after Feb. 1917 revolution.
Ayatollah Ruhollah Khomeini, 1900-89, (Iranian), religious-political leader, spearheaded overthrow of shah, 1979.
Nikita Khrushchev, 1894-1971, (USSR) prem., first sec. of Communist party; initiated de-Stalinization.
Kim Il Sung, 1912-94, (Korean) N Korean dictator, 1948-94.
Lajos Kossuth, 1802-94, (Hung.) principal figure in 1848 Hungarian revolution.
Pyotr Kropotkin, 1842-1921, (Russ.) anarchist; championed the peasants but opposed Bolshevism.
Kublai Khan, c1215-94, Mongol emperor; founder of Yüan dynasty in China.
Béla Kun, 1886-c1939, (Hung.) member of 3d Communist Internat.; tried to foment worldwide revolution.
Robert M. LaFollette, 1855-1925, (U.S.) Wisconsin public official; leader of progressive movement.
Fiorello La Guardia, 1882-1947, (U.S.) colorful NYC reform mayor.
Pierre Laval, 1883-1945, (Fr.) politician, Vichy foreign min.; executed for treason.
Andrew Bonar Law, 1858-1923, (Br.) Conservative party politician; led opposition to Irish home rule.
Vladimir Ilyich Lenin (Ulyanov), 1870-1924, (Russ.) revolutionary; founded Bolshevism; Soviet leader 1917-24.
Ferdinand de Lesseps, 1805-94, (Fr.) diplomat, engineer; conceived idea of Suez Canal.
Rene Levesque, 1922-87, (Can.) prem. of Quebec, 1976-85; led unsuccessful separartist campaign.
Maxim Litvinov, 1876-1951, (Pol.-Russ.) revolutionary, commissar of foreign affairs; favored cooperation with West.
Liu Shaoqi, c1898-1974, (Chin.) Communist leader; fell from grace during Cultural Revolution.
David Lloyd George, 1863-1945, (Br.) Liberal party prime min.; laid foundations for modern welfare state.
Henry Cabot Lodge, 1850-1924, (U.S.) Republican senator; led opposition to participation in League of Nations.
Huey P. Long, 1893-1935, (U.S.) Louisiana political demagogue, governor; assassinated.
Rosa Luxemburg, 1871-1919, (Ger.) revolutionary; leader of the German Social Democratic party and Spartacus party.
J. Ramsay MacDonald, 1866-1937, (Br.) first Labour party prime min. of Great Britain.
Harold Macmillan, 1895-1986, (Br.) prime min. of Great Britain, 1957-63.
Joseph R. McCarthy, 1908-57, (U.S.) senator, extremist in searching out alleged Communists and pro-Communists.
Makarios III, 1913-77, (Cypriot) Greek Orthodox archbishop; first pres. of Cyprus.
Mao Zedong, 1893-1976, (Chin.) chief Chinese Marxist theorist, revolutionary, political leader; led Chinese revolution establishing his nation as Communist state.
Jean Paul Marat, 1743-93, (Fr.) revolutionary, politician; identified with radical Jacobins; assassinated.
José Martí, 1853-95, (Cub.) patriot, poet; leader of Cuban struggle for independence.
Jan Masaryk, 1886-1948, (Czech.) foreign min.; died by mysterious alleged suicide following Communist coup.
Thomas G. Masaryk, 1850-1937, (Czech.) statesman, philosopher; first pres. of Czechoslovak Republic.
Jules Mazarin, 1602-61, (Fr.) cardinal, statesman; prime min. under Louis XIII and queen regent Anne of Austria.
Giusseppe Mazzini, 1805-72, (It.), reformer dedicated to Risorgimento movement for renewal of Italy.
Tom Mboya, 1930-69, (Kenyan) political leader; instrumental in securing independence for Kenya.
Cosimo I de' Medici, 1519-74, (It.) Duke of Florence, grand duke of Tuscany.
Lorenzo de' Medici, the Magnificent, 1449-92, (It.) merchant prince; a towering figure in Italian Renaissance.
Catherine de Médicis, 1519-89, (Fr.) queen consort of Henry II, regent of France; influential in Catholic-Huguenot wars.
Golda Meir, 1898-1978, (Isr.) a founder of the state of Israel and prime min., 1969-74.
Klemens W. N. L. Metternich, 1773-1859, (Austrian) statesman; arbiter of post-Napoleonic Europe.
François Mitterrand, 1916-96, (Fr.) pres. of France, 1981-95.
Mobutu Sese Seko, 1930-97, (Zaire) longtime ruler of Zaire (now Congo) (1965-97); exiled after rebellion.
Guy Mollet, 1905-75, (Fr.) socialist politician, resistance leader.
Henry Morgenthau Jr., 1891-1967, (U.S.) sec. of treasury; fund-raiser for New Deal and U.S. WW2 activities.
Gouverneur Morris, 1752-1816, (U.S.) statesman, diplomat. financial expert, helped plan decimal coinage.
Benito Mussolini, 1883-1945, (It.) leader of the Italian fascist state; assassinated.
Imre Nagy, c1896-1958, (Hung.) Communist prem.; assassinated after Soviets crushed 1956 uprising.
Gamal Abdel Nasser, 1918-70, (Egypt.) leader of Arab unification, 2d Egyptian pres.
Jawaharlal Nehru, 1889-1964, (In.) prime min.; guided India through its early years of independence.
Kwame Nkrumah, 1909-72, (Ghan.) 1st prime min., 1957-60, and pres., 1960-66, of Ghana.
Frederick North, 1732-92, (Br.) prime min.; his inept policies led to loss of American colonies.
Daniel O'Connell, 1775-1847, (Ir.) political leader; known as The Liberator.
Julius K. Nyerere, 1923?-99, (Tanz.) founding father, 1st pres., 1962-85, of Tanzania.
Omar, c581-644, Muslim leader; 2d caliph, led Islam to become an imperial power.
Thomas P. (Tip) O'Neill Jr., 1912-94, (U.S.) U.S. congressman, Speaker of the House, 1977-86.
Ignace Paderewski, 1860-1941, (Pol.) statesman, pianist; composer, briefly prime min., an ardent patriot.
Viscount Palmerston, 1784-1865, (Br.) Whig-Liberal prime min., foreign min.; embodied British nationalism.
Andreas George Papandreou, 1919-1996, (Gk.) leftist politician, served 2 times as prem. (1981-89, 1993-96).
Georgios Papandreou, 1888-1968, (Gk.) Republican politician; served 3 times as prime min.
Franz von Papen, 1879-1969, (Ger.) politician; major role in overthrow of Weimar Republic and rise of Hitler.
Charles Stewart Parnell, 1846-1891, (Ir.) nationalist leader; "uncrowned king of Ireland."
Lester Pearson, 1897-1972, (Can.) diplomat, Liberal party leader, prime min.
Robert Peel, 1788-1850, (Br.) reformist prime min., founder of Conservative party.
Eva (Evita) Perón, 1919-52 (Arg.) highly influential 2d wife of Juan Perón.
Juan Perón, 1895-1974, (Arg.) dynamic pres. of Argentina (1946-55, 1973-74).
Joseph Pilsudski, 1867-1935, (Pol.) statesman; instrumental in reestablishing Polish state in the 20th cent.
Charles Pinckney, 1757-1824, (U.S.) founding father; his Pinckney plan largely incorporated into Constitution.
Christian Pineau, 1905-95, (Fr.) leader of French Resistance during WW2; French foreign min., 1956-58.
William Pitt, the Elder, 1708-78, (Br.) statesman; the "Great Commoner," transformed Britain into imperial power.
William Pitt, the Younger, 1759-1806, (Br.) prime min. during French Revolutionary wars.
Georgi Plekhanov, 1857-1918, (Russ.) revolutionary, social philosopher; called "father of Russian Marxism."
Raymond Poincaré, 1860-1934, (Fr.) 9th pres. of the Republic; advocated harsh punishment of Germany after WW1.

Pol Pot, 1925-98, (Camb.) leader of Khmer Rouge; ruled Cambodia, 1975-79; responsible for mass deaths.
Georges Pompidou, 1911-74, (Fr.) Gaullist political leader; pres. 1969-74.
Grigori Potemkin, 1739-91, (Russ.) field marshal; favorite of Catherine II.
Yitzhak Rabin, 1922-95, (Isr.) military, political leader; prime min. of Israel, 1974-77, 1992-95; assassinated.
Edmund Randolph, 1753-1813, (U.S.) attorney; prominent in drafting, ratification of constitution.
John Randolph, 1773-1833, (U.S.) Southern planter; strong advocate of states' rights.
Jeannette Rankin, 1880-1973, (U.S.) pacifist; first woman member of U.S. Congress.
Walter Rathenau, 1867-1922, (Ger.) industrialist, statesman.
Sam Rayburn, 1882-1961, (U.S.) Democratic leader; representative for 47 years, House Speaker for 17.
Paul Reynaud, 1878-1966, (Fr.) statesman; prem. in 1940 at the time of France's defeat by Germany.
Syngman Rhee, 1875-1965, (Korean) first pres. of S Korea.
Cecil Rhodes, 1853-1902, (Br.) imperialist, industrial magnate; established Rhodes scholarships in his will.
Cardinal de Richelieu, 1585-1642, (Fr.) statesman, known as "red eminence;" chief minister to Louis XIII.
Maximilien Robespierre, 1758-94, (Fr.) leading figure in French Revolution and Reign of Terror.
Nelson Rockefeller, 1908-79, (U.S.) Republican governor of NY, 1959-73; U.S. vice pres., 1974-77.
George W. Romney, 1907-95, (U.S.) auto exec.; 3-term Republican governor of Michigan.
Eleanor Roosevelt, 1884-1962, (U.S.) influential First Lady, humanitarian, UN diplomat.
Elihu Root, 1845-1937, (U.S.) lawyer, statesman, diplomat; leading Republican supporter of the League of Nations.
Dean Rusk, 1909-95, (U.S.) statesman; sec. of state, 1961-69.
John Russell, 1792-1878, (Br.) Liberal prime min. during the Irish potato famine.
Anwar al-Sadat, 1918-81, (Egypt.) pres., 1970-1981, promoted peace with Israel; Nobel laureate; assassinated.
António de Salazar, 1889-1970, (Port.) longtime dictator.
José de San Martin, 1778-1850, S Amer. revolutionary; protector of Peru.
Eisaku Sato, 1901-75, (Jpn.) prime min.; presided over Japan's post-WW2 emergence as major world power.
Abdul Aziz Ibn Saud, c1880-1953, (Saudi Arabia) king of Saudi Arabia, 1932-53.
Robert Schuman, 1886-1963, (Fr.) statesman; founded European Coal and Steel Community.
Carl Schurz, 1829-1906, (U.S.) German-American political leader, journalist, orator, dedicated reformer.
Kurt Schuschnigg, 1897-1977, (Austrian) chancellor; unsuccessful in stopping Austria's annexation by Germany.
William H. Seward, 1801-72, (U.S.) anti-slavery activist; as U.S. sec. of state purchased Alaska.
Carlo Sforza, 1872-1952, (It.) foreign min., anti-fascist.
Sitting Bull, c1831-90, (Nat. Am.) Sioux leader in Battle of Little Bighorn over George A. Custer, 1876.
Alfred E. Smith, 1873-1944, (U.S.) NY Democratic governor; first Roman Catholic to run for presidency.
Margaret Chase Smith, 1897-1995, (U.S.) congresswoman, senator; 1st woman elected to both houses of Congress.
Jan C. Smuts, 1870-1950, (S. African) statesman, philosopher, soldier, prime min.
Paul Henri Spaak, 1899-1972, (Belg.) statesman, socialist leader.
Joseph Stalin, 1879-1953, (USSR) Soviet dictator, 1924-53; instituted forced collectivization, massive purges, and labor camps, causing millions of deaths.
Edwin M. Stanton, 1814-69, (U.S.) sec. of war, 1862-68.
Edward R. Stettinius Jr., 1900-49, (U.S.) industrialist, sec. of state who coordinated aid to WW2 allies.
Adlai E. Stevenson, 1900-65, (U.S.) Democratic leader, diplomat, Illinois governor, presidenial candidate.
Henry L. Stimson, 1867-1950, (U.S.) statesman; served in 5 administrations, foreign policy adviser in 30s and 40s.
Gustav Stresemann, 1878-1929, (Ger.) chancellor, foreign minister; strove to regain friendship for post-WW1 Germany.
Sukarno, 1901-70, (Indon.) dictatorial first pres. of the Indonesian republic.
Sun Yat-sen, 1866-1925, (Chin.) revolutionary; leader of Kuomintang, regarded as the father of modern China.
Robert A. Taft, 1889-1953, (U.S.) conservative Senate leader, called "Mr. Republican."
Charles de Talleyrand, 1754-1838, (Fr.) statesman, diplomat; the major force of the Congress of Vienna of 1814-15.
U Thant, 1909-74 (Bur.) statesman, UN sec.-general.
Norman M. Thomas, 1884-1968, (U.S.) social reformer; 6 times Socialist party presidential candidate.
Josip Broz Tito, 1892-1980, (Yug.) pres. of Yugoslavia from 1953, WW2 guerrilla chief, postwar rival of Stalin.
Palmiro Togliatti, 1893-1964, (It.) major Italian Communist leader.
Hideki Tojo, 1885-1948, (Jpn.) statesman, soldier; prime min. during most of WW2.
François Toussaint L'Ouverture, c1744-1803, (Haitian) patriot, martyr; thwarted French colonial aims.
Leon Trotsky, 1879-1940, (Russ.) revolutionary, founded Red Army, expelled from party in conflict with Stalin; assassinated.
Pierre Elliott Trudeau, 1919-2000, (Can.) longtime liberal prime minister of Canada, 1968-79, 1980-84; achieved native Canadian constitution.
Rafael L. Trujillo Molina, 1891-1961, (Dom.) dictator of Dominican Republic, 1930-61; assassinated.
Moise K. Tshombe, 1919-69, (Cong.) pres. of secessionist Katanga, prem. of Congo.
William M. Tweed, 1823-78, (U.S.) politicial boss of Tammany Hall, NYC's Democratic political machine.
Walter Ulbricht, 1893-1973, (Ger.) Communist leader of German Democratic Republic.
Arthur H. Vandenberg, 1884-1951, (U.S.) senator; proponent of bipartisan anti-Communist foreign policy.
Eleutherios Venizelos, 1864-1936, (Gk.) most prominent Greek statesman of early 20th cent.
Hendrik F. Verwoerd, 1901-66, (S. African) prime min.; rigorously applied apartheid policy despite protest.
George Wallace, 1919-98, (U.S.) former segregationist governor of Alabama and presid. candidate.
Robert Walpole, 1676-1745, (Br.) statesman; generally considered Britain's first prime min.
Daniel Webster, 1782-1852, (U.S.) orator, politician; advocate of business interests during Jacksonian agrarianism.
Chaim Weizmann, 1874-1952, (Russ.-Isr.) Zionist leader, scientist; first Israeli pres.
Wendell L. Willkie, 1892-1944, (U.S.) Republican who tried to unseat FDR when he ran for his 3d term.
Harold Wilson, 1916-95, (Br.) Labour party leader; prime min., 1964-70, 1974-76.
Emiliano Zapata, c1879-1919, (Mex.) revolutionary; major influence on modern Mexico.
Todor Zhivkov, 1911-98, (Bulg.) Communist ruler of Bulgaria from 1954 until ousted in a 1989 coup.
Zhou Enlai, 1898-1976, (Chin.) diplomat, prime min.; a leading figure of the Chinese Communist party.

Scientists of the Past

Revised by Peter Barker, Prof. & Chair, Dept. of the Hist. of Science, Univ. of Oklahoma

For pre-modern scientists see also Philosophers and Religious Figures of the Past and Historical Figures chapter.

Albertus Magnus, c1200-1280, (Ger.) theologian, philosopher; helped found medieval study of natural science.
Alhazen (Ibn al-Haytham), c965-ca.1040, mathematician, astronomer; optical theorist.
Andre-Marie Ampère, 1775-1836, (Fr.) mathematician, chemist; founder of electrodynamics.
John V. Atanasoff, 1903-95, (U.S.) physicist; co-invented Atanasoff-Berry Computer (1939-41), regarded in law as the original "automatic electronic digital computer".
Amedeo Avogadro, 1776-1856, (It.) chemist, physicist; proposed that equal volumes of gas contain equal numbers of molecules, permitting determination of molecular weights.
John Bardeen, 1908-91, (U.S.) double Nobel laureate in physics (transistor, 1956; superconductivity, 1972).
A. H. Becquerel, 1852-1908, (Fr.) physicist; discovered radioactivity in uranium (1896).
Alexander Graham Bell, 1847-1922, (U.S.) inventor; first to patent and commercially exploit the telephone (1876).
Daniel Bernoulli, 1700-82, (Swiss) mathematician; developed fluid dynamics and kinetic theory of gases.
Clifford Berry, 1918-1963, (U.S.) collaborated with Atanasoff on the ABC computer (1939-41).
Jöns Jakob Berzelius, 1779-1848, (Swed.) chemist; developed modern chemical symbols and formulas, discovered selenium and thorium.
Henry Bessemer, 1813-98, (Br.) engineer; invented Bessemer steel-making process.
Bruno Bettelheim, 1903-90, (Austrian-U.S.) psychoanalyst specializing in autistic and other disturbed children; *Uses of Enchantment* (1976).
Louis Blériot, 1872-1936, (Fr.) engineer; monoplane pioneer, first Channel flight (1909).
Franz Boas, 1858-1942, (Ger.-U.S.) founded modern anthropology; studied Pacific Coast tribes.
Niels Bohr, 1885-1962, (Dan.) atomic and nuclear physicist; founded quantum mechanics.

Max Born, 1882-1970, (Ger.) atomic and nuclear physicist; helped develop quantum mechanics.
Satyendranath Bose, 1894-1974, (Indian) physicist; forerunner of modern quantum theory for integral-spin particles.
Louis de Broglie, 1892-1987, (Fr.) physicist; proposed quantum wave-particle duality.
Robert Bunsen, 1811-99, (Ger.) chemist; pioneered spectroscopic analysis; discovered rubidium, caesium.
Luther Burbank, 1849-1926, (U.S.) naturalist; developed plant breeding into a modern science.
Vannevar Bush, 1890-1974, (U.S.) electrical engineer; developed differential analyzer, an early analogue computer; headed WWII Office of Scientific Res. and Dev.
Marvin Camras, 1916-95, (U.S.) inventor, electrical engineer; invented magnetic tape recording.
Alexis Carrel, 1873-1944, (Fr.) surgeon, biologist; developed methods of suturing blood vessels and transplanting organs.
Rachel Carson, 1907-64, (U.S.) marine biologist, environmentalist; *Silent Spring* (1962).
George Washington Carver, c1864-1943, (U.S.) agricultural scientist, nutritionist; improved and pioneered new uses for peanuts and sweet potatoes.
James Chadwick, 1891-1974, (Br.) physicist; discovered the neutron (1932); led British Manhattan Project group in U.S. (1943-45).
Albert Claude, 1898-1983, (Belg.-U.S.) a founder of modern cell biology; determined role of mitochondria.
Nicolaus Copernicus, 1473-1543, (Pol.) first modern astronomer to propose sun as center of the planets' motions.
Jacques Yves Cousteau, 1910-1997, (Fr.) oceanographer; co-inventor, with E. Gagnan, of the Aqualung (1943).
Seymour Cray, 1925-96, (U.S.) computer industry pioneer; developed supercomputers.
Marie, 1867-1934 (Pol.-Fr.) and **Pierre Curie**, 1859-1906, (Fr.) physical chemists; pioneer investigators of radioactivity, discovered radium and polonium (1898).
Gottlieb Daimler, 1834-1900, (Ger.) engineer, inventor; pioneer automobile manufacturer.
John Dalton, 1766-1844, (Br.) chemist, physicist; formulated atomic theory, made first table of atomic weights.
Charles Darwin, 1809-82, (Br.) naturalist; established theory of organic evolution; *Origin of Species* (1859).
Lee De Forest, 1873-1961, (U.S.) inventor of triode, pioneer in wireless telegraphy, sound pictures, television.
Max Delbruck, 1906-81, (Ger.-U.S.) founded molecular biology.
Rudolf Diesel, 1858-1913, (Ger.) mechanical engineer; patented Diesel engine (1892).
Theodosius Dobzhansky, 1900-75, (Russ.-U.S.) biologist; reconciled genetics and natural selection contributing to "modern synthesis" in evolution.
Christian Doppler, 1803-53, (Austrian) physicist; showed change in wave frequency caused by motion of source, now known as Doppler effect.
J. Presper Eckert Jr., 1919-95, (U.S.) co-inventor, with Mauchly, of the ENIAC computer (1943-45).
Thomas A. Edison, 1847-1931, (U.S.) inventor; held more than 1,000 patents, including incandescent electric lamp.
Paul Ehrlich, 1854-1915, (Ger.) medical researcher in immunology and bacteriology; pioneered antitoxin production.
Albert Einstein, 1879-1955, (Ger.-U.S.) theoretical physicist; founded relativity theory, replacing Newton's theories of space, time, and gravity. Proved $E=mc^2$ (1905).
John F. Enders, 1897-1985, (U.S.) virologist, helped discover vaccines against polio, measles, mumps and chicken pox.
Erik Erikson, 1902-94, (U.S.) psychoanalyst, author; theory of developmental stages of life, *Childhood and Society* (1950).
Leonhard Euler, 1707-83, (Swiss) mathematician, physicist; pioneer of calculus, revived ideas of Fermat.
Gabriel Fahrenheit, 1686-1736, (Ger.) physicist; improved thermometers and introduced Fahrenheit temperature scale.
Michael Faraday, 1791-1867, (Br.) chemist, physicist; discovered electrical induction and invented dynamo (1831).
Philo T. Farnsworth, 1906-71, (U.S.) inventor; built first television system (San Francisco), 1928.
Pierre de Fermat, 1601-65, (Fr.) mathematician; founded modern theory of numbers.
Enrico Fermi, 1901-54, (It.-U.S.) nuclear physicist; demonstrated first controlled chain reaction (Chicago, 1942).
Richard Feynman, 1918-88, (U.S.) theoretical physicist, author; founder of Quantum Electrodynamics (QED).
Alexander Fleming, 1881-1955, (Br.) bacteriologist; discovered penicillin (1928).
Jean B. J. Fourier, 1768-1830, (fr.) introduced method of analysis in math and physics known as Fourier Series.
Sigmund Freud, 1856-1939, (Austrian) psychiatrist; founder of psychoanalysis. *Interpretation of Dreams* (1901).
Erich Fromm, 1900-1980, (U.S.) psychoanalyst. *Man for Himself* (1947).
Galileo Galilei, 1564-1642, (It.) physicist; used telescope to vindicate Copernicus, founded modern science of motion.
Luigi Galvani, 1737-98, (It.) physiologist; studied electricity in living organisms.
Carl Friedrich Gauss, 1777-1855, (Ger.) math. physicist; completed work of Fermat and Euler in number theory.
Joseph Gay-Lussac, 1778-1850, (Fr.) chemist, physicist; investigated behavior of gases, discovered boron.
Josiah W. Gibbs, 1839-1903, (U.S.) theoretical physicist, chemist; founded chemical thermodynamics.
Robert H. Goddard, 1882-1945, (U.S.) physicist; invented liquid fuel rocket (1926).
George W. Goethals, 1858-1928, (U.S.) chief engineer who completed Panama Canal (1907-14).
William C. Gorgas, 1854-1920, (U.S.) physician; pioneer in prevention of yellow fever and malaria.
Ernest Haeckel, 1834-1919, (Ger.) zoologist, evolutionist; early Darwinist, introduced concept of "ecology."
Otto Hahn, 1879-1968, (Ger.) chemist; with Meitner discovered nuclear fission (1938).
Edmund Halley, 1656-1742, (Br.) astronomer; predicted return of 1682 comet ("Halley's Comet") in 1759.
William Harvey, 1578-1657, (Br.) physician, anatomist; discovered circulation of the blood (1628).
Werner Heisenberg, 1901-76, (Ger.) physicist; developed matrix mechanics and uncertainty principle (1927).
Hermann von Helmholtz, 1821-94, (Ger.) physicist, physiologist; formulated principle of conservation of energy.
William Herschel, 1738-1822, (Ger.-Br.) astronomer; discovered Uranus (1781).
Heinrich Hertz, 1857-94, (Ger.) physicist; discovered radio waves and photo-electric effect (1886-7).
David Hilbert, 1862-1943, (Ger.) mathematician; contributed to algebra, calculus and foundational studies (formalism).
Edwin P. Hubble, 1889-1953, (U.S.) astronomer; discovered observational evidence of expanding universe.
Alexander von Humboldt, 1769-1859, (Ger.) naturalist, author; explored S America, created ecology.
Edward Jenner, 1749-1823, (Br.) physician; pioneered vaccination, introduced term "virus."
James Joule, 1818-89, (Br.) physicist; found relation between heat and mechanical energy (conservation of energy).
Carl Jung, 1875-1961, (Swiss) psychiatrist; founder of analytical psychology.
Sister Elizabeth Kenny, 1886-1952, (Austral.) nurse; developed treatment for polio.
Johannes Kepler, 1571-1630, (Ger.) astronomer; discovered laws of planetary motion.
Al-Khawarizmi, early 9th cent., (Arab.), mathematician; regarded as founder of algebra.
Robert Koch, 1843-1910 (Ger.) bacteriologist; isolated bacterial causes of tuberculosis and other diseases.
Georges Köhler, 1946-95, (Ger.) immunologist; with Cesar Milstein he developed monoclonal antibody technique.
Jacques Lacan, 1901-81, (Fr.) controversial influential psychoanalyst.
Joseph Lagrange, 1736-1813, (Fr.) geometer, astronomer; showed that gravity of earth and moon cancels creating stable points in space around them.
Jean B. Lamarck, 1744-1829, (Fr.) naturalist; forerunner of Darwin in evolutionary theory.
Pierre Simon de Laplace, 1749-1827, (Fr.) astronomer, physicist; proposed nebular origin for solar system.
Antoine Lavoisier, 1743-94, (Fr.) a founder of mod. chemistry.
Ernest O. Lawrence, 1901-58, (U.S.) physicist; invented the cyclotron.
Jerome Lejeune, 1927-94, (Fr.) geneticist; discovered chromosomal cause of Down syndrome (1959).
Louis 1903-72, and **Mary Leakey**, 1913-96, (Br.) early hominid paleoanthropologists; discovered remains in Africa.
Anton van Leeuwenhoek, 1632-1723, (Dutch) founder of microscopy.
Kurt Lewin, 1890-1947, (Ger.-U.S.) social psychologist; studied human motivation and group dynamics.
Justus von Liebig, 1803-73, (Ger.) founded quantitative organic chemistry.
Joseph Lister, 1827-1912, (Br.) physician; pioneered antiseptic surgery.
Konrad Lorenz, 1903-89, (Austrian) ethologist; pioneer in study of animal behavior.
Percival Lowell, 1855-1916, (U.S.) astronomer; predicted the existence of Pluto.
Louis, 1864-1948, and **Auguste Lumière**, 1862-1954, (Fr.) invented cinematograph and made first motion picture (1895).
Guglielmo Marconi, 1874-1937, (It.) physicist; developed wireless telegraphy.
John W. Mauchly, 1907-80, (U.S.) co-inventor, with Eckert, of computer ENIAC (1943-45).
James Clerk Maxwell, 1831-79, (Br.) physicist; unified electricity and magnetism; electromagnetic theory of light.
Maria Goeppert Mayer, 1906-72, (Ger.-U.S.) physicist; developed shell model of atomic nuclei.

Barbara McClintock, 1902-92, (U.S.) geneticist; showed that some genetic elements are mobile.
Lise Meitner, 1878-1968, (Austrian) co-discoverer, with Hahn, of nuclear fission (1938).
Gregor J. Mendel, 1822-84, (Austrian) botanist, monk; his experiments became the foundation of modern genetics.
Dmitri Mendeleyev, 1834-1907, (Russ.) chemist; established Periodic Table of the Elements.
Franz Mesmer, 1734-1815, (Ger.) physician; introduced hypnotherapy.
Albert A. Michelson, 1852-1931, (U.S.) physicist; invented interferometer.
Robert A. Millikan, 1868-1953, (U.S.) physicist; measured electronic charge.
Thomas Hunt Morgan, 1866-1945, (U.S.) geneticist, embryologist; established role of chromosomes in heredity.
Isaac Newton, 1642-1727, (Br.) natural philosopher; discovered laws of gravitation, motion; with Leibniz, founded calculus.
Robert N. Noyce, 1927-90, (U.S.) invented microchip.
J. Robert Oppenheimer, 1904-67, (U.S.) physicist; scientific director of Manhattan project.
Wilhelm Ostwald, 1853-1932, (Ger.) chemist, philosopher; main founder of modern physical chemistry.
Louis Pasteur, 1822-95, (Fr.) chemist; showed that germs cause disease and fermentation, originated pasteurization.
Linus C. Pauling, 1901-94, (U.S.) chemist; studied chemical bonds; campaigned for nuclear disarmament.
Jean Piaget, 1896-1980, (Swiss) psychologist; four-stage theory of intellectual development in children.
Max Planck, 1858-1947, (Ger.) physicist; introduced quantum hypothesis (1900).
Walter S. Reed, 1851-1902, (U.S.) army physician; proved mosquitoes transmit yellow fever.
Theodor Reik, 1888-1969, (Austrian-U.S.) psychoanalyst, major Freudian disciple.
Bernhard Riemann, 1826-66, (Ger.) mathematician; developed non-Euclidean geometry used by Einstein.
Wilhelm Roentgen, 1845-1923, (Ger.) physicist; discovered X-rays (1895).
Carl Rogers, 1902-87, (U.S.) psychotherapist, author; originated nondirective therapy.
Ernest Rutherford, 1871-1937, (Br.) physicist; pioneer investigator of radioactivity, identified the atomic nucleus.
Albert B. Sabin, 1906-93, (Russ.-U.S.), developed oral polio live-virus vaccine.
Carl Sagan, 1934-96, (U.S.) astronomer, author.
Jonas Salk, 1914-95, (U.S.) developed first successful polio vaccine, widely used in U.S. after 1955.
Giovanni Schiaparelli, 1835-1910, (It.) astronomer; reported canals on Mars.
Erwin Schrödinger, 1887-1961, (Austrian) physicist; developed wave equation for quantum systems.
Glenn T. Seaborg, 1912-99, (U.S.) chemist, Nobel Prize winner (1951); codiscoverer of plutonium.
Harlow Shapley, 1885-1972, (U.S.) astronomer; mapped galactic clusters and position of Sun in our own galaxy.
B(urrhus) F(rederick) Skinner, 1904-89, (U.S.) psychologist; leading advocate of behaviorism.
Roger W. Sperry, 1913-94, (U.S.) neuorobiologist; established different functions of right and left sides of brain.
Benjamin Spock, 1903-98, (U.S.) pediatrician, child care expert; *Common Sense Book of Baby and Child Care.*
Charles P. Steinmetz, 1865-1923, (Ger.-U.S.) electrical engineer; developed basic ideas on alternating current.
Leo Szilard, 1898-1964, (Hung.-U.S.) physicist; helped on Manhattan project, later opposed nuclear weapons.
Nikola Tesla, 1856-1943, (Serb.-U.S.) invented a number of electrical devices including a.c. dynamos, transformers and motors.
William Thomson (Lord Kelvin), 1824-1907, (Br.) physicist; aided in success of transatlantic telegraph cable (1865); proposed Kelvin absolute temperature scale.
Alan Turing, 1912-54, (Br.) mathematician; helped develop basis for computers.
Rudolf Virchow, 1821-1902, (Ger.) pathologist; pioneered the modern theory that diseases affect the body through cells.
Alessandro Volta, 1745-1827, (It.) physicist; electricity pioneer.
Werner von Braun, 1912-77, (Ger.-U.S.) developed rockets for warfare and space exploration.
John Von Neumann, 1903-57, (Hung.-U.S.) mathematician; originated game theory; basic design for modern computers.
Alfred Russell Wallace, 1823-1913, (Br.) naturalist; proposed concept of evolution independently of Darwin.
John B. Watson, 1878-1958, (U.S.) psychologist; a founder of behaviorism.
James E. Watt, 1736-1819, (Br.) mechanical engineer, inventor; invented modern steam engine (1765).
Alfred L. Wegener, 1880-1930, (Ger.) meteorologist, geophysicist; postulated continental drift.
Norbert Wiener, 1894-1964, (U.S.) mathematician; founder of cybernetics.
Sewall Wright, 1889-1988, (U.S.) evolutionary theorist; helped found population genetics.
Wilhelm Wundt, 1832-1920, (Ger.) founder of experimental psychology.
Ferdinand von Zeppelin, 1838-1917, (Ger.) soldier, aeronaut, airship designer.

Social Reformers, Activists, and Humanitarians of the Past

Jane Addams, 1860-1935, (U.S.) cofounder of Hull House; won Nobel Peace Prize, 1931.
Susan B. Anthony, 1820-1906, (U.S.) a leader in temperance, anti-slavery, and woman suffrage movements.
Thomas Barnardo, 1845-1905, (Br.) social reformer; pioneered in care of destitute children.
Clara Barton, 1821-1912, (U.S.) organized American Red Cross.
Henry Ward Beecher, 1813-87, (U.S.) clergyman, abolitionist.
Amelia Bloomer, 1818-94, (U.S.) suffragette, social reformer.
William Booth, 1829-1912, (Br.) founded Salvation Army.
John Brown, 1800-59, (U.S.) abolitionist who led murder of 5 pro-slavery men, was hanged.
Frances Xavier (Mother) Cabrini, 1850-1917, (It.-U.S.) Italian-born nun; founded charitable institutions; first American canonized as a saint, 1946.
Carrie Chapman Catt, 1859-1947, (U.S.) suffragette.
Cesar Chavez, 1927-93, (U.S.) labor leader; helped establish United Farm Workers of America.
Clarence Darrow, 1857-1938, (U.S.) lawyer; defender of "underdog," opponent of capital punishment.
Dorothy Day, 1897-1980, (U.S.) founder of Catholic Worker movement.
Eugene V. Debs, 1855-1926, (U.S.) labor leader; led Pullman strike, 1894; 4-time Socialist presidential candidate.
Dorothea Dix, 1802-87, (U.S.) crusader for mentally ill.
Thomas Dooley, 1927-61, (U.S.) "jungle doctor," noted for efforts to supply medical aid to developing countries.
Marjory Stoneman Douglas, 1890-1998, (U.S.) writer and environmentalist; campaigned to save Florida Everglades.
William Lloyd Garrison, 1805-79, (U.S.) abolitionist.
Emma Goldman, 1869-1940, (Russ.-U.S.) published anarchist *Mother Earth,* birth-control advocate.
Samuel Gompers, 1850-1924, (U.S.) labor leader.
Michael Harrington, 1928-89, (U.S.) exposed poverty in affluent U.S. in *The Other America,* 1963.
Sidney Hillman, 1887-1946, (U.S.) labor leader; helped organize CIO.
Samuel G. Howe, 1801-76, (U.S.) social reformer; changed public attitudes toward the handicapped.
Helen Keller, 1880-1968, (U.S.) crusader for better treatment for the handicapped; deaf and blind herself.
Maggie Kuhn, 1905-95, (U.S.) founded Gray Panthers, 1970.
William Kunstler, 1919-95, (U.S.) civil liberties attorney.
John L. Lewis, 1880-1969, (U.S.) labor leader; headed United Mine Workers, 1920-60.
Karl Menninger, 1893-1990, (U.S.) with brother William founded Menninger Clinic and Menninger Foundation.
Lucretia Mott, 1793-1880, (U.S.) reformer, pioneer feminist.
Philip Murray, 1886-1952, (U.S.) Scottish-born labor leader.
Florence Nightingale, 1820-1910, (Br.) founder of modern nursing.
Emmeline Pankhurst, 1858-1928, (Br.) woman suffragist.
Walter Reuther, 1907-70, (U.S.) labor leader; headed UAW.
Jacob Riis, 1849-1914, (U.S.) crusader for urban reforms.
Margaret Sanger, 1883-1966, (U.S.) social reformer; pioneered the birth-control movement.
Earl of Shaftesbury (A. A. Cooper), 1801-85, (Br.) social reformer.
Elizabeth Cady Stanton, 1815-1902, (U.S.) woman suffrage pioneer.
Lucy Stone, 1818-93, (U.S.) feminist, abolitionist.
Mother Teresa of Calcutta, 1910-97, (Alban.) nun; founded order to care for sick, dying poor; 1979 Nobel Peace Prize.
Philip Vera Cruz, 1905-94, (Filipino-U.S.) helped to found the United Farm Workers Union.
William Wilberforce, 1759-1833, (Br.) social reformer; prominent in struggle to abolish the slave trade.
Frances E. Willard, 1839-98, (U.S.) temperance, women's rights leader.
Mary Wollstonecraft, 1759-97, (Br.) wrote *Vindication of the Rights of Women.*

Writers of the Present

Name (Birthplace)	Birthdate
Chinua Achebe (Ogidi, Nigeria)	11/16/30
Richard Adams (Newbury, Eng.)	5/10/20
Edward Albee (Wash., DC)	3/12/28
Isabel Allende (Chile)	8/2/42
Jorge Amado (Bahia, Brazil)	8/10/12
Martin Amis (Oxford, Eng.)	8/25/49
Maya Angelou (St. Louis, MO)	4/4/28
Piers Anthony (Oxford, Eng.)	8/6/34
Jeffrey Archer (Somerset, Eng.)	4/15/40
Oscar Arias Sanchez (Heredia, Costa Rica)	9/13/41
John Ashbery (Rochester, NY)	1927
Margaret Atwood (Ottawa, Ont.)	11/18/39
Louis Auchincloss (Lawrence, NY)	9/27/17
Paul Auster (Newark, NJ)	2/3/47
Russell Banks (Newton, MA)	3/28/40
John Barth (Cambridge, MD)	5/27/30
Ann Beattie (Wash., DC)	9/7/47
Saul Bellow (Lachine, Que.)	6/10/15
Peter Benchley (NYC)	5/8/40
John Berendt (Syracuse, NY)	12/5/39
Thomas Berger (Cincinnati, OH)	7/20/24
Judy Blume (Elizabeth, NJ)	2/12/38
T. Coraghessan Boyle (Peekskill, NY)	12/2/47
Ray Bradbury (Waukegan, IL)	8/22/20
Barbara Taylor Bradford (Leeds, Eng.)	5/10/33
Rita Mae Brown (Hanover, PA)	11/28/44
Christopher Buckley (NYC)	1952
Hortense Calisher (NYC)	12/20/11
Ethan Canin (Ann Arbor, MI)	7/19/60
Camilo Jose Cela (Galicia, Spain)	5/11/16
Michael Chabon (Washington, DC)	1963
Sandra Cisneros (Chicago, IL)	12/20/54
Tom Clancy (Baltimore, MD)	4/12/47
Mary Higgins Clark (NYC)	12/24/31
Arthur C. Clarke (Minehead, Eng.)	12/16/17
Beverly Cleary (McMinnville, OR)	4/12/16
Jackie Collins (London, Eng.)	10/4/41?
Evan S. Connell (Kansas City, MO)	8/17/24
Pat Conroy (Atlanta, GA)	10/26/45
Robin Cook (NYC)	5/4/40
Patricia Cornwell (Miami, FL)	6/9/56
Harry Crews (Alma, GA)	6/6/35
Michael Crichton (Chicago, IL)	10/23/42
Michael Cunningham (Ohio)	1952
Don DeLillo (NYC)	11/20/36
Joan Didion (Sacramento, CA)	12/5/34
E. L. Doctorow (NYC)	1/6/31
Takako Doi (Hyogo, Jap.)	11/30/28
Rita Dove (Akron, OH)	8/28/52
Roddy Doyle (Dublin, Ireland)	5/58
John Gregory Dunne (Hartford, CT)	5/25/32
Umberto Eco (Alessandria, Italy)	1/5/32
Bret Easton Ellis (Los Angeles, CA)	1964
James Ellroy (Los Angeles)	3/4/48
Louise Erdrich (Little Falls, MN)	7/6/54
Laura Esquivel (Mexico City, Mexico)	1950
Howard Fast (NYC)	11/11/14
Ken Follet (Cardiff, Wales)	6/5/49
Dario Fo (San Giano, Italy)	3/26/26
Horton Foote (Wharton, TX)	3/14/16
Richard Ford (Jackson, MS)	2/16/44
Frederick Forsyth (Ashford, Eng.)	1938
John Fowles (Leigh-on-Sea, Eng.)	3/31/26
Paula Fox (NYC)	4/22/23
Dick Francis (Lawrenny, S. Wales)	10/31/20
Michael Frayn (London, Eng.)	9/8/33
Marilyn French (NYC)	11/21/29
Brian Friel (Omagh, Ire.)	11/9/29
Carlos Fuentes (Mexico City, Mex.)	11/11/28
Ernest J. Gaines (Oscar, LA)	1/15/33
Gabriel Garcia Marquez (Aracata, Colombia)	3/6/28
Frank Gilroy (NYC)	10/13/25
Gail Godwin (Birmingham, AL)	6/18/37
William Goldman (Chicago, IL)	8/12/31
Nadine Gordimer (Springs, S. Africa)	11/20/23
Mary Gordon (Long Island, NY)	12/8/49
Sue Grafton (Louisville, KY)	4/24/40
Günter Grass (Danzig, Ger.)	10/16/27
Shirley Ann Grau (New Orleans, LA)	7/8/29
John Grisham (Jonesboro, AR)	2/8/55
John Guare (NYC)	2/5/38
Arthur Hailey (Luton, Eng.)	4/5/20
David Hare (St. Leonards, Sussex, Eng.)	6/5/47
Robert Hass (San Francisco, CA)	3/1/41
Vaclav Havel (Prague, Czech.)	10/5/36
Seamus Heaney (N. Ireland)	1939
Mark Helprin (NYC)	6/28/47
Tony Hillerman (Sacred Heart, OK)	5/27/25
S. E. Hinton (Tulsa, OK)	1948
Alice Hoffman (NYC)	3/16/52
John Irving (Exeter, NH)	3/2/42
John Jakes (Chicago, IL)	3/31/32
P. D. James (Oxford, Eng.)	8/3/20
Erica Jong (NYC)	3/26/42
Garrison Keillor (Anoka, MN)	8/7/42
Thomas Keneally (Sydney, Austral.)	10/7/35
William Kennedy (Albany, NY)	1/16/28
Ken Kesey (La Junta, CO)	9/17/35
Jamaica Kincaid (St. Johns, Antigua)	5/25/39
Stephen King (Portland, ME)	9/21/47
Barbara Kingsolver (Annapolis, MD)	4/8/55
Maxine Hong Kingston (Stockton, CA)	10/27/40
Galway Kinnell (Providence, RI)	2/1/27
John Knowles (Fairmont, WV)	9/16/26
Kenneth Koch (Cincinnati, OH)	2/27/25
Dean Koontz (Everett, PA)	7/9/45
Judith Krantz (NYC)	1/9/28
Maxine Kumin (Philadelphia, PA)	6/6/25
John Le Carré (Poole, Eng.)	10/19/31
Ursula K. Le Guin (Berkeley, CA)	10/21/29
Madeleine L'Engle (NYC)	11/29/18
Elmore Leonard (New Orleans, LA)	10/11/25
Doris Lessing (Kermanshah, Persia)	10/22/19
Ira Levin (NYC)	8/27/29
Alison Lurie (Chicago, IL)	9/3/26
Naguib Mahfouz (Cairo, Egypt)	12/11/11
Norman Mailer (Long Branch, NJ)	1/31/23
David Mamet (Chicago, IL)	11/30/47
Bobbie Ann Mason (nr. Mayfield, KY)	5/1/40
Peter Matthiessen (NYC)	5/22/27
Ed McBain (NYC)	10/15/26
Cormac McCarthy (Providence, RI)	7/20/33
Frank McCourt (Brooklyn, NY)	1930
Colleen McCullough (Wellington, N.S.W.)	6/1/37
Alice McDermot (NYC)	6/27/53
Thomas McGuane (Wyandotte, MI)	12/11/39
Terry McMillan (Port Huron, MI)	10/18/51
Larry McMurtry (Wichita Falls, TX)	6/3/36
John McPhee (Princeton, NJ)	3/8/31
Arthur Miller (NYC)	10/17/15
Czeslaw Milosz (Seteiniai, Lithuania)	6/30/11
Toni Morrison (Lorain, OH)	2/18/31
Walter Mosley (Los Angeles, CA)	1952
Alice Munro (Wingham, Ont.)	7/10/31
Haruki Murakami (Kyoto, Japan)	1/12/49
V. S. Naipaul (Port-of-Spain, Trin.)	8/17/32
Joyce Carol Oates (Lockport, NY)	6/16/38
Edna O'Brien (Tuamgraney, Ir.)	12/15/30
Tim O'Brien (Austin, MN)	10/1/46
Kenzaburo Oe (Ose, Shikoku, Japan)	1/31/35
Cynthia Ozick (NYC)	4/17/28
Grace Paley (NYC)	12/11/22
Marge Piercy (Detroit, MI)	3/31/36
Robert Pinsky (Long Branch, NJ)	10/20/40
Harold Pinter (London, Eng.)	10/10/30
Chaim Potok (NYC)	2/17/29
Reynolds Price (Macon, NC)	2/1/33
Richard Price (NYC)	10/12/49
E. Annie Proulx (Norwich, CT)	8/22/35
Thomas Pynchon (Glen Cove, NY)	5/8/37
David Rabe (Dubuque, IA)	3/10/40
Ishmael Reed (Chattanooga, TN)	2/22/38
Ruth Rendell (England)	2/17/30
Anne Rice (New Orleans, LA)	10/14/41
Adrienne Rich (Baltimore, MD)	5/16/29
Philip Roth (Newark, NJ)	3/19/33
J.K. Rowling (Bristol, Eng.)	7/31/66
Salman Rushdie (Bombay, India)	6/19/47
J. D. Salinger (NYC)	1/1/19
Jose Saramago (Azinhaga, Portugal)	1922
David Sedaris (Binghamton, NY)	12/26/56
Vikram Seth (Calcutta, India)	6/20/52
Sidney Sheldon (Chicago, IL)	2/11/17
Sam Shepard (Ft. Sheridan, IL)	11/5/43
Carol Shields (Oak Park, IL)	6/2/35
Claude Simon (Tananarive, Madagascar)	1913
Neil Simon (NYC)	7/4/27
Jane Smiley (Los Angeles, CA)	9/26/49
Aleksandr Solzhenitsyn (Kislovodsk, Russia)	12/11/18
Susan Sontag (NYC)	1/16/33
Wole Soyinka (Abeokuta, Nigeria)	7/13/34
Mickey Spillane (Brooklyn, NY)	3/9/18
Danielle Steel (NYC)	8/14/47
Richard Stern (NYC)	2/25/28

Name (Birthplace)	Birthdate
Mary Stewart (Sunderland, Eng.)	9/17/16
Tom Stoppard (Zlin, Czech.)	7/13/37
William Styron (Newport News, VA)	6/11/25
Wislawa Szymborska (Kornik, Poland)	7/2/23
Amy Tan (Oakland, CA)	2/19/52
Paul Theroux (Medford, MA)	4/10/41
Scott F. Turow (Chicago, IL)	4/2/49
Anne Tyler (Minneapolis, MN)	10/25/41
John Updike (Shillington, PA)	3/18/32
Leon Uris (Baltimore, MD)	8/3/24
Mario Vargas Llosa (Arequipa, Peru)	3/28/36
Gore Vidal (West Point, NY)	10/3/25
Paula Vogel (Wash., DC)	11/16/51
Kurt Vonnegut Jr. (Indianapolis, IN)	11/11/22
Derek Walcott (Castries, Saint Lucia)	1930
Alice Walker (Eatonton, GA)	2/9/44
Robert James Waller (Rockford, IA)	8/1/39
Joseph Wambaugh (East Pittsburgh, PA)	1/22/37
Wendy Wasserstein (NYC)	10/18/50
August Wilson (Pittsburgh, PA)	4/27/45
Lanford Wilson (Lebanon, MO)	4/13/37
Tom Wolfe (Richmond, VA)	3/2/31
Tobias Wolff (Birmingham, AL)	6/19/45
Herman Wouk (NYC)	5/27/15

Writers of the Past

See also Journalists of the Past, and Greeks and Romans in Historical Figures chapter.

Alice Adams, 1926-99, (U.S.) novelist, short-story writer. *Superior Woman.*

James Agee, 1909-55, (U.S.) novelist. *A Death in the Family.*

Conrad Aiken, 1889-1973, (U.S.) poet, critic. *Ushant.*

Louisa May Alcott, 1832-88, (U.S.) novelist. *Little Women.*

Sholom Aleichem, 1859-1916, (Russ.) Yiddish writer. *Tevye's Daughter, The Old Country.*

Vicente Aleixandre, 1898-1984, (Sp.) poet. *La destrucción o el amor, Dialogolos del conocimiento.*

Horatio Alger, 1832-1899, (U.S.) "rags-to-riches" books.

Eric Ambler, 1909-98, (Br.) suspense novelist. *A Coffin for Dimitrios.*

Kingsley Amis, 1922-95, (Br.) novelist, critic. *Lucky Jim.*

Hans Christian Andersen, 1805-75, (Dan.) author of fairy tales. *The Ugly Duckling.*

Maxwell Anderson, 1888-1959, (U.S.) playwright. *What Price Glory?, High Tor, Winterset, Key Largo.*

Sherwood Anderson, 1876-1941, (U.S.) short-story writer. "Death in the Woods;" *Winesburg, Ohio.*

Reinaldo Arenas, 1943-1990, (Cuba) short-story writer, novelist. *Before Night Falls.*

Matthew Arnold, 1822-88, (Br.) poet, critic. "Thrysis," "Dover Beach," "Culture and Anarchy."

Isaac Asimov, 1920-92, (U.S.) versatile writer, espec. of science-fiction. *I Robot.*

Miguel Angel Asturias, 1899-1974, (Guatemala) novelist. *El Señor Presidente.*

W(ystan) H(ugh) Auden, 1907-73, (Br.) poet, playwright, literary critic. "The Age of Anxiety."

Jane Austen, 1775-1817, (Br.) novelist. *Pride and Prejudice, Sense and Sensibility, Emma, Mansfield Park.*

Isaac Babel, 1894-1941, (Russ.) short-story writer, playwright. *Odessa Tales, Red Cavalry.*

James Baldwin, 1924-87, author, playwright. *The Fire Next Time, Blues for Mister Charlie.*

Honoré de Balzac, 1799-1850, (Fr.) novelist. *Le Père Goriot, Cousine Bette, Eugénie Grandet.*

James M. Barrie, 1860-1937, (Br.) playwright, novelist. *Peter Pan, Dear Brutus, What Every Woman Knows.*

Charles Baudelaire, 1821-67, (Fr.) poet. *Les Fleurs du Mal.*

L(yman) Frank Baum, 1856-1919, (U.S.) *Wizard of Oz* series.

Simone de Beauvoir, 1908-86, (Fr.) novelist, essayist. *The Second Sex, Memoirs of a Dutiful Daughter.*

Samuel Beckett, 1906-89, (Ir.) novelist, playwright. *Waiting for Godot, Endgame* (plays); *Murphy, Watt, Molloy* (novels).

Brendan Behan, 1923-64, (Ir.) playwright. *The Quare Fellow, The Hostage, Borstal Boy.*

Robert Benchley, 1889-1945, (U.S.) humorist.

Stephen Vincent Benét, 1898-1943, (U.S.) poet, novelist. *John Brown's Body.*

John Berryman, 1914-72, (U.S.) poet. *Homage to Mistress Bradstreet.*

Ambrose Bierce, 1842-1914, (U.S.) short-story writer, journalist. *In the Midst of Life, The Devil's Dictionary.*

Elizabeth Bishop, 1911-79, (U.S.) poet. *North and South—A Cold Spring.*

William Blake, 1757-1827, (Br.) poet, artist. *Songs of Innocence, Songs of Experience.*

Giovanni Boccaccio, 1313-75, (It.) poet. *Decameron.*

Heinrich Böll, 1917-85, (Ger.) novelist, short-story writer. *Group Portrait With Lady.*

Jorge Luis Borges, 1900-86, (Arg.) short-story writer, poet, essayist. *Labyrinths.*

James Boswell, 1740-95, (Sc.) biographer. *The Life of Samuel Johnson.*

Pierre Boulle, (1913-94), (Fr.) novelist. *The Bridge Over the River Kwai, Planet of the Apes.*

Paul Bowles, 1910-99, (U.S.) novelist, short-story writer. The Sheltering Sky

Anne Bradstreet, c1612-72, (U.S.) poet. *The Tenth Muse Lately Sprung Up in America.*

Bertolt Brecht, 1898-1956, (Ger.) dramatist, poet. *The Threepenny Opera, Mother Courage and Her Children.*

Charlotte Brontë, 1816-55, (Br.) novelist. *Jane Eyre.*

Emily Brontë, 1818-48, (Br.) novelist. *Wuthering Heights.*

Elizabeth Barrett Browning, 1806-61, (Br.) poet. *Sonnets From the Portuguese, Aurora Leigh.*

Joseph Brodsky, 1940-96, (Russ.-U.S.) poet. *A Part of Speech, Less Than One, To Urania.*

Robert Browning, 1812-89, (Br.) poet. "My Last Duchess," "Fra Lippo Lippi," *The Ring and The Book.*

Pearl S. Buck, 1892-1973, (U.S.) novelist. *The Good Earth.*

Mikhail Bulgakov, 1891-1940, (Russ.) novelist, playwright. *The Heart of a Dog, The Master and Margarita.*

John Bunyan, 1628-88, (Br.) writer. *Pilgrim's Progress.*

Anthony Burgess, 1917-93, (Br.) author. *A Clockwork Orange.*

Frances Hodgson Burnett, 1849-1924, (Br.-U.S.) novelist. *The Secret Garden.*

Robert Burns, 1759-96, (Sc.) poet. "Flow Gently, Sweet Afton," "My Heart's in the Highlands," "Auld Lang Syne."

Edgar Rice Burroughs, 1875-1950, (U.S.) "Tarzan" books.

William S. Burroughs, 1914-97, (U.S.) novelist. *Naked Lunch.*

George Gordon, Lord Byron, 1788-1824, (Br.) poet. *Don Juan, Childe Harold, Manfred, Cain.*

Italo Calvino, 1923-85, (It.) novelist, short-story writer. *If on a Winter's Night a Traveler.*

Albert Camus, 1913-60, (Fr.) writer. *The Stranger, The Fall.*

Karel Capek, 1890-1938, (Czech.) playwright, novelist, essayist. *R.U.R. (Rossum's Universal Robots).*

Truman Capote, 1924-84, (U.S.) author. *Other Voices, Other Rooms, Breakfast at Tiffany's, In Cold Blood.*

Lewis Carroll (Charles Dodgson), 1832-98, (Br.) writer, mathematician. *Alice's Adventures in Wonderland.*

Giacomo Casanova, 1725-98, (It.) adventurer, memoirist.

Willa Cather, 1873-1947, (U.S.) novelist. *O Pioneers!, My Ántonia, Death Comes for the Archbishop.*

Miguel de Cervantes Saavedra, 1547-1616, (Sp.) novelist, dramatist, poet. *Don Quixote.*

Raymond Chandler, 1888-1959, (U.S.) writer of detective fiction. Philip Marlowe series.

Geoffrey Chaucer, c1340-1400, (Br.) poet. *The Canterbury Tales, Troilus and Criseyde.*

John Cheever, 1912-82, (U.S.) novelist, short-story writer. *The Wapshot Scandal,* "The Country Husband."

Anton Chekhov, 1860-1904, (Russ.) short-story writer, dramatist. *Uncle Vanya, The Cherry Orchard, The Three Sisters.*

G(ilbert) K(eith) Chesterton, 1874-1936, (Br.) critic, novelist, relig. apologist. Father Brown series of mysteries.

Kate Chopin, 1851-1904, (U.S.) writer. *The Awakening.*

Agatha Christie, 1890-1976, (Br.) mystery writer; created Miss Marple, Hercule Poirot; *And Then There Were None., Murder on the Orient Express, Murder of Roger Ackroyd.*

James Clavell, 1924-94, (Br.-U.S.) novelist. *Shogun, King Rat.*

Jean Cocteau, 1889-1963, (Fr.) writer, visual artist, filmmaker. *The Beauty and the Beast, Les Enfants Terribles.*

Samuel Taylor Coleridge, 1772-1834, (Br.) poet, critic. "Kubla Khan," "The Rime of the Ancient Mariner."

(Sidonie) Colette, 1873-1954, (Fr.) novelist. *Claudine, Gigi.*

Wilkie Collins, 1824-89, (Br.) Novelist. *The Moonstone.*

Joseph Conrad, 1857-1924, (Br.) novelist. *Lord Jim, Heart of Darkness, The Nigger of the Narcissus.*

James Fenimore Cooper, 1789-1851, (U.S.) novelist. *Leatherstocking Tales, The Last of the Mohicans.*

Pierre Corneille, 1606-84, (Fr.) dramatist. *Medeé, Le Cid.*

Hart Crane, 1899-1932, (U.S.) poet. "The Bridge."

Stephen Crane, 1871-1900, (U.S.) novelist, short-story writer. *The Red Badge of Courage,* "The Open Boat."

E. E. Cummings, 1894-1962, (U.S.) poet. *Tulips and Chimneys.*

Roald Dahl, 1916-90, (Br.-U.S.) writer. *Charlie and the Chocolate Factory, James and the Giant Peach.*

Gabriele D'Annunzio, 1863-1938, (It.) poet, novelist, dramatist. *The Child of Pleasure, The Intruder, The Victim.*

Dante Alighieri, 1265-1321, (It.) poet. *The Divine Comedy.*

Robertson Davies, 1913-95, (Can.) novelist, playwright, essayist. Salterton, Deptford, and Cornish trilogies.

Daniel Defoe, 1660-1731, (Br.) writer. *Robinson Crusoe, Moll Flanders, Journal of the Plague Year.*
Charles Dickens, 1812-70, (Br.) novelist. *David Copperfield, Oliver Twist, Great Expectations, A Tale of Two Cities.*
James Dickey, 1923-1997, (U.S.) poet, novelist. *Deliverance.*
Emily Dickinson, 1830-86, (U.S.) lyric poet. "Because I could not stop for Death . . .," "Success is counted sweetest . . ."
Isak Dinesen (Karen Blixen), 1885-1962, (Dan.) author. *Out of Africa, Seven Gothic Tales, Winter's Tales.*
John Donne, 1573-1631, (Br.) poet, divine. *Songs and Sonnets.*
José Donoso, 1924-96, (Chil.) surreal novelist and short-story writer. *The Obscene Bird of Night.*
John Dos Passos, 1896-1970, (U.S.) novelist. *U.S.A.*
Fyodor Dostoyevsky, 1821-81, (Russ.) novelist. *Crime and Punishment, The Brothers Karamazov, The Possessed.*
Arthur Conan Doyle, 1859-1930, (Br.) novelist. Sherlock Holmes mystery stories.
Theodore Dreiser, 1871-1945, (U.S.) novelist. *An American Tragedy, Sister Carrie.*
John Dryden, 1631-1700, (Br.) poet, dramatist, critic. *All for Love, Mac Flecknoe, Absalom and Achitophel.*
Alexandre Dumas, 1802-70, (Fr.) novelist, dramatist. *The Three Musketeers, The Count of Monte Cristo.*
Alexandre Dumas (fils), 1824-95, (Fr.) dramatist, novelist. *La Dame aux Camélias, Le Demi-Monde.*
Lawrence Durrell, 1912-90, (Br.) novelist, poet. *Alexandria Quartet.*
Ilya G. Ehrenburg, 1891-1967, (Russ.) writer. *The Thaw.*
George Eliot (Mary Ann Evans or Marian Evans), 1819-80, (Br.) novelist. *Silas Marner, Middlemarch.*
T(homas) S(tearns) Eliot, 1888-1965, (Br.) poet, critic. *The Waste Land,* "The Love Song of J. Alfred Prufrock."
Stanley Elkin, 1930-95, (U.S.) novelist, short story writer. *George Mills.*
Ralph Ellison, 1914-94, (U.S.), writer. *Invisible Man.*
Ralph Waldo Emerson, 1803-82, (U.S.) poet, essayist. "Brahma," "Nature," "The Over-Soul," "Self-Reliance."
James T. Farrell, 1904-79, (U.S.) novelist. *Studs Lonigan.*
William Faulkner, 1897-1962, (U.S.) novelist. *Sanctuary, Light in August, The Sound and the Fury, Absalom, Absalom!*
Edna Ferber, 1887-1968, (U.S.) novelist, short-story writer, playwright. *So Big, Cimarron, Show Boat.*
Henry Fielding, 1707-54, (Br.) novelist. *Tom Jones.*
F(rancis) Scott Fitzgerald, 1896-1940, (U.S.) short-story writer, novelist. *The Great Gatsby, Tender Is the Night.*
Gustave Flaubert, 1821-80, (Fr.) novelist. *Madame Bovary.*
Ian Fleming, 1908-64, (Br.) novelist; James Bond spy thrillers.
Ford Madox Ford, 1873-1939, (Br.) novelist, critic, poet. *The Good Soldier.*
C(ecil) S(cott) Forester, 1899-1966, (Br.) writer. Horatio Hornblower books.
E(dward) M(organ) Forster, 1879-1970, (Br.) novelist. *A Passage to India, Howards End.*
Anatole France, 1844-1924, (Fr.) writer. *Penguin Island, My Friend's Book, The Crime of Sylvestre Bonnard.*
Robert Frost, 1874-1963, (U.S.) poet. "Birches," "Fire and Ice," "Stopping by Woods on a Snowy Evening."
William Gaddis, 1922-98, (U.S.) novelist. *The Recognitions.*
John Galsworthy, 1867-1933, (Br.) novelist, dramatist. *The Forsyte Saga.*
Erle Stanley Gardner, 1889-1970, (U.S.) mystery writer; created Perry Mason.
Jean Genet, 1911-86, (Fr.) playwright, novelist. *The Maids.*
Kahlil Gibran, 1883-1931, (Lebanese-U.S.) mystical novelist, essayist, poet. *The Prophet.*
André Gide, 1869-1951, (Fr.) writer. *The Immoralist, The Pastoral Symphony, Strait Is the Gate.*
Allen Ginsberg, 1926-1997, (U.S.) Beat poet. "Howl."
Jean Giraudoux, 1882-1944, (Fr.) novelist, dramatist. *Electra, The Madwoman of Chaillot, Ondine, Tiger at the Gate.*
Johann Wolfgang von Goethe, 1749-1832, (Ger.) poet, dramatist, novelist. *Faust, Sorrows of Young Werther.*
Nikolai Gogol, 1809-52, (Russ.) short-story writer, dramatist, novelist. *Dead Souls, The Inspector General.*
William Golding, 1911-93, (Br.) novelist. *Lord of the Flies.*
Oliver Goldsmith, 1728-74, (Br.-Ir.) dramatist, novelist. *The Vicar of Wakefield, She Stoops to Conquer.*
Maxim Gorky, 1868-1936, (Russ.) dramatist, novelist. *The Lower Depths.*
Robert Graves, 1895-1985, (Br.) poet, classical scholar, novelist. *I, Claudius; The White Goddess.*
Thomas Gray, 1716-71, (Br.) poet. "Elegy Written in a Country Churchyard," "The Progress of Poesy."
Julien Green, 1900-98, (U.S.-Fr.) expatriate American, French novelist. *Moira, Each Man in His Darkness.*
Graham Greene, 1904-91, (Br.) novelist. *The Power and the Glory, The Heart of the Matter, The Ministry of Fear.*
Zane Grey, 1872-1939, (U.S.) writer of Western stories.
Jakob Grimm, 1785-1863, (Ger.) philologist, folklorist; with brother **Wilhelm,** 1786-1859, collected *Grimm's Fairy Tales.*
Alex Haley, 1921-92, (U.S.) author. *Roots.*
Dashiell Hammett, 1894-1961, (U.S.) detective-story writer; created Sam Spade. *The Maltese Falcon, The Thin Man.*
Knut Hamsun, 1859-1952 (Nor.) novelist. *Hunger.*
Thomas Hardy, 1840-1928, (Br.) novelist, poet. *The Return of the Native, Tess of the D'Urbervilles, Jude the Obscure.*
Joel Chandler Harris, 1848-1908, (U.S.) Uncle Remus stories.
Moss Hart, 1904-61, (U.S.) playwright. *Once in a Lifetime, You Can't Take It With You, The Man Who Came to Dinner.*
Bret Harte, 1836-1902, (U.S.) short-story writer, poet. *The Luck of Roaring Camp.*
Jaroslav Hasek, 1883-1923, (Czech.) writer, playwright. *The Good Soldier Schweik.*
John Hawkes, 1925-98, (U.S.) experimental fiction writer. *The Goose on the Grave, Blood Oranges.*
Nathaniel Hawthorne, 1804-64, (U.S.) novelist, short-story writer. *The Scarlet Letter,* "Young Goodman Brown."
Heinrich Heine, 1797-1856, (Ger.) poet. *Book of Songs.*
Joseph Heller, 1923-99, (U.S.) novelist. Catch-22.
Lillian Hellman, 1905-84, (U.S.) playwright, author of memoirs. *The Little Foxes, An Unfinished Woman, Pentimento.*
Ernest Hemingway, 1899-1961, (U.S.) novelist, short-story writer. *A Farewell to Arms, For Whom the Bell Tolls.*
O. Henry (W. S. Porter), 1862-1910, (U.S.) short-story writer. "The Gift of the Magi."
George Herbert, 1593-1633, (Br.) poet. "The Altar," "Easter Wings."
Zbigniew Herbert, 1924-98, (Pol.) poet. "Apollo and Marsyas."
Robert Herrick, 1591-1674, (Br.) poet. "To the Virgins to Make Much of Time."
James Herriot (James Alfred Wight), 1916-95, (Br.) novelist, veterinarian. *All Creatures Great and Small.*
John Hersey, 1914-93, (U.S.) novelist, journalist. *Hiroshima, A Bell for Adano.*
Hermann Hesse, 1877-1962, (Ger.) novelist, poet. *Death and the Lover, Steppenwolf, Siddhartha.*
James Hilton, 1900-54, (Br.) novelist. *Lost Horizon.*
Oliver Wendell Holmes, 1809-94, (U.S.) poet, novelist. *The Autocrat of the Breakfast-Table.*
Gerard Manley Hopkins, 1844-89, (Br.) poet. "Pied Beauty."
A(lfred) E. Housman, 1859-1936, (Br.) poet. *A Shropshire Lad.*
William Dean Howells, 1837-1920, (U.S.) novelist, critic. *The Rise of Silas Lapham.*
Langston Hughes, 1902-67, (U.S.) poet, playwright. *The Weary Blues, One-Way Ticket, Shakespeare in Harlem.*
Ted Hughes, 1930-98, (Br.) British poet laureate, 1984-98. *Crow, The Hawk in the Rain.*
Victor Hugo, 1802-85, (Fr.) poet, dramatist, novelist. *Notre Dame de Paris, Les Misérables.*
Zora Neale Hurston, 1903-60, (U.S.) novelist, folklorist. *Their Eyes Were Watching God, Mules and Men.*
Aldous Huxley, 1894-1963, (Br.) writer. *Brave New World.*
Henrik Ibsen, 1828-1906, (Nor.) dramatist, poet. *A Doll's House, Ghosts, The Wild Duck, Hedda Gabler.*
William Inge, 1913-73, (U.S.) playwright. *Picnic; Come Back, Little Sheba; Bus Stop.*
Eugene Ionesco, 1910-94, (Fr.) surrealist dramatist. *The Bald Soprano, The Chairs.*
Washington Irving, 1783-1859, (U.S.) writer. "Rip Van Winkle," "The Legend of Sleepy Hollow."
Christopher Isherwood, 1904-1986, (Br.) novelist, playwright. *The Berlin Stories.*
Shirley Jackson, 1919-65, (U.S.) writer. "The Lottery."
Henry James, 1843-1916, (U.S.) novelist, short-story writer, critic. *The Portrait of a Lady, The Ambassadors, Daisy Miller.*
Robinson Jeffers, 1887-1962, (U.S.) poet, dramatist. *Tamar and Other Poems, Medea.*
Samuel Johnson, 1709-84, (Br.) author, scholar, critic. *Dictionary of the English Language, Vanity of Human Wishes.*
Ben Jonson, 1572-1637, (Br.) dramatist, poet. *Volpone.*
James Joyce, 1882-1941, (Ir.) writer. *Ulysses, Dubliners, A Portrait of the Artist as a Young Man, Finnegans Wake.*
Ernst Junger, 1895-1998, (Ger.) novelist, essayist. *The Peace, On the Marble Cliff.*
Franz Kafka, 1883-1924, (Ger.) novelist, short-story writer. *The Trial, The Castle, The Metamorphosis.*
George S. Kaufman, 1889-1961, (U.S.) playwright. *The Man Who Came to Dinner, You Can't Take It With You, Stage Door.*
Nikos Kazantzakis, 1883-1957, (Gk.) novelist. *Zorba the Greek, A Greek Passion.*
Alfred Kazin, 1915-98 (U.S.) author, critic, teacher. *On Native Grounds.*
John Keats, 1795-1821, (Br.) poet. "Ode on a Grecian Urn," "Ode to a Nightingale," "La Belle Dame Sans Merci."
Jack Kerouac, 1922-1969, (U.S.), author, Beat poet. *On the Road, The Dharma Bums,* "Mexico City Blues."
Joyce Kilmer, 1886-1918, (U.S.) poet. "Trees."
Rudyard Kipling, 1865-1936, (Br.) author, poet. "The White Man's Burden," "Gunga Din," *The Jungle Book.*
Jean de la Fontaine, 1621-95, (Fr.) poet. *Fables choisies.*

Pär Lagerkvist, 1891-1974, (Swed.) poet, dramatist, novelist. *Barabbas, The Sybil.*
Selma Lagerlöf, 1858-1940, (Swed.) novelist. *Jerusalem, The Ring of the Lowenskolds.*
Alphonse de Lamartine, 1790-1869, (Fr.) poet, novelist, statesman. *Méditations poétiques.*
Charles Lamb, 1775-1834, (Br.) essayist. *Specimens of English Dramatic Poets, Essays of Elia.*
Giuseppe di Lampedusa, 1896-1957, (It.) novelist. *The Leopard.*
William Langland, c1332-1400, (Eng.) poet. *Piers Plowman.*
Ring Lardner, 1885-1933, (U.S.) short-story writer, humorist.
Louis L'Amour, 1908-88, (U.S.) western author, screenwriter. *Hondo, The Cherokee Trail.*
D(avid) H(erbert) Lawrence, 1885-1930, (Br.) novelist. *Sons and Lovers, Women in Love, Lady Chatterley's Lover.*
Halldor Laxness, 1902-98, (Icelandic) novelist. *Iceland's Bell.*
Mikhail Lermontov, 1814-41, (Russ.) novelist, poet. "Demon," *Hero of Our Time.*
Alain-René Lesage, 1668-1747, (Fr.) novelist. *Gil Blas de Santillane.*
Gotthold Lessing, 1729-81, (Ger.) dramatist, philosopher, critic. *Miss Sara Sampson, Minna von Barnhelm.*
C(live) S(taples) Lewis, 1898-1963, (Br.) critic, novelist, religious writer. *Allegory of Love; The Lion, the Witch and the Wardrobe; Out of the Silent Planet.*
Sinclair Lewis, 1885-1951, (U.S.) novelist. *Babbitt, Main Street, Arrowsmith, Dodsworth.*
Vachel Lindsay, 1879-1931, (U.S.) poet. *General William Booth Enters Into Heaven, The Congo.*
Hugh Lofting, 1886-1947, (Br.) writer. Dr. Doolittle series.
Jack London, 1876-1916, (U.S.) novelist, journalist. *Call of the Wild, The Sea-Wolf, White Fang.*
Henry Wadsworth Longfellow, 1807-82, (U.S.) poet. *Evangeline, The Song of Hiawatha.*
Amy Lowell, 1874-1925, (U.S.) poet, critic. "Lilacs."
James Russell Lowell, 1819-91, (U.S.) poet, editor. *Poems, The Biglow Papers.*
Robert Lowell, 1917-77, (U.S.) poet. "Lord Weary's Castle."
Archibald MacLeish, 1892-1982, (U.S.) poet. *Conquistador.*
Bernard Malamud, 1914-86, (U.S.) short-story writer, novelist. "The Magic Barrel," *The Assistant, The Fixer.*
Stéphane Mallarmé, 1842-98, (Fr.) poet. *Poésies.*
Sir Thomas Malory, ?-1471, (Br.) writer. *Morte d'Arthur.*
Andre Malraux, 1901-76, (Fr.) novelist. *Man's Fate.*
Osip Mandelstam, 1891-1938, (Russ.) poet. *Stone, Tristia.*
Thomas Mann, 1875-1955, (Ger.) novelist, essayist. *Buddenbrooks, The Magic Mountain,* "Death in Venice."
Katherine Mansfield, 1888-1923, (Br.) short-story writer. "Bliss."
Christopher Marlowe, 1564-93, (Br.) dramatist, poet. *Tamburlaine the Great,* Dr. *Faustus, The Jew of Malta.*
Andrew Marvell, 1621-78, (Br.) poet. "To His Coy Mistress."
John Masefield, 1878-1967, (Br.) poet. "Sea Fever," "Cargoes," *Salt Water Ballads.*
Edgar Lee Masters, 1869-1950, (U.S.) poet, biographer. *Spoon River Anthology.*
W(illiam) Somerset Maugham, 1874-1965, (Br.) author. *Of Human* Bondage, *The Moon and Sixpence.*
Guy de Maupassant, 1850-93, (Fr.) novelist, short-story writer. "A Life," "Bel-Ami," "The Necklace."
François Mauriac, 1885-1970, (Fr.) novelist, dramatist. *Viper's Tangle, The Kiss to the Leper.*
Vladimir Mayakovsky, 1893-1930, (Russ.) poet, dramatist. *The Cloud in Trousers.*
Mary McCarthy, 1912-89, (U.S.) critic, novelist, memoirist. *Memories of a Catholic Girlhood.*
Carson McCullers, 1917-67, (U.S.) novelist. *The Heart Is a Lonely Hunter, Member of the Wedding.*
Herman Melville, 1819-91, (U.S.) novelist, poet. *Moby-Dick, Typee, Billy Budd, Omoo.*
George Meredith, 1828-1909, (Br.) novelist, poet. *The Ordeal of Richard Feverel, The Egoist.*
Prosper Mérimée, 1803-70, (Fr.) author. *Carmen.*
James Merrill, 1926-95, (U.S.) poet. *Divine Comedies.*
James Michener, 1907-97, (U.S.) novelist. *Tales of the South Pacific.*
Edna St. Vincent Millay, 1892-1950, (U.S.) poet. *The Harp Weaver and Other Poems.*
Henry Miller, 1891-1980, (U.S.) erotic novelist. *Tropic of Cancer.*
A(lan) A(lexander) Milne, 1882-1956, (Br.) author. *Winnie-the-Pooh.*
John Milton, 1608-74, (Br.) poet, writer. *Paradise Lost, Comus, Lycidas, Areopagitica.*
Mishima Yukio (Hiraoka Kimitake), 1925-70, (Jpn.) writer. *Confessions of a Mask.*
Gabriela Mistral, 1889-1957, (Chil.) poet. *Sonnets of Death.*
Margaret Mitchell, 1900-49, (U.S.) novelist. *Gone With the Wind.*
Jean Baptiste Molière, 1622-73, (Fr.) dramatist. *Le Tartuffe, Le Misanthrope, Le Bourgeois Gentilhomme.*
Ferenc Molnár, 1878-1952, (Hung.) dramatist, novelist. *Liliom, The Guardsman, The Swan.*
Michel de Montaigne, 1533-92, (Fr.) essayist. *Essais.*
Eugenio Montale, 1896-1981, (It.) poet.
Brian Moore, 1921-99, (Ir.-U.S.) novelist. *The Lonely Passion of Judith Hearne.*
Clement C. Moore, 1779-1863, (U.S.) poet, educator. "A Visit From Saint Nicholas."
Marianne Moore, 1887-1972, (U.S.) poet.
Alberto Moravia, 1907-90, (It.) novelist, short-story writer. *The Time of Indifference.*
Sir Thomas More, 1478-1535, (Br.) writer, statesman, saint. *Utopia.*
Wright Morris, 1910-98 (U.S.) novelist. *My Uncle Dudley.*
Murasaki Shikibu, c978-1026, (Jpn.) novelist. *The Tale of Genji.*
Iris Murdoch, 1919-99 (Br.), novelist, philosopher. *The Sea, The Sea.*
Alfred de Musset, 1810-57, (Fr.) poet, dramatist. *La Confession d'un Enfant du Siècle.*
Vladimir Nabokov, 1899-1977, (Russ.-U.S.) novelist. *Lolita, Pale Fire.*
Ogden Nash, 1902-71, (U.S.) poet of light verse.
Pablo Neruda, 1904-73, (Chil.) poet. *Twenty Love Poems and One Song of Despair, Toward the Splendid City.*
Patrick O'Brian, 1914-2000, (Br.) historical novelist. *Master and Commander, Blue at the Mizzen.*
Sean O'Casey, 1884-1964, (Ir.) dramatist. *Juno and the Paycock, The Plough and the Stars.*
Frank O'Connor (Michael Donovan), 1903-66, (Ir.) short-story writer. "Guests of a Nation."
Flannery O'Connor, 1925-64, (U.S.) novelist, short-story writer. *Wise Blood,* "A Good Man Is Hard to Find."
Clifford Odets, 1906-63, (U.S.) playwright. *Waiting for Lefty, Awake and Sing, Golden Boy, The Country Girl.*
John O'Hara, 1905-70, (U.S.) novelist, short-story writer. *From the Terrace, Appointment in Samarra, Pal Joey.*
Omar Khayyam, c1028-1122, (Per.) poet. *Rubaiyat.*
Eugene O'Neill, 1888-1953, (U.S.) playwright. *Emperor Jones, Anna Christie, Long Day's Journey Into Night.*
George Orwell, 1903-50, (Br.) novelist, essayist. *Animal Farm, Nineteen Eighty-Four.*
John Osborne, 1929-95, (Br.) dramatist, novelist. *Look Back in Anger, The Entertainer.*
Wilfred Owen, 1893-1918 (Br.) poet. "Dulce et Decorum Est."
Dorothy Parker, 1893-1967, (U.S.) poet, short-story writer. *Enough Rope, Laments for the Living.*
Boris Pasternak, 1890-1960, (Russ.) poet, novelist. *Doctor Zhivago.*
Octavio Paz, 1914-98, (Mex.) poet, essayist. *The Labyrinth of Solitude, They Shall Not Pass!, The Sun Stone.*
Samuel Pepys, 1633-1703, (Br.) public official, diarist.
S(idney) J(oseph) Perelman, 1904-79, (U.S.) humorist. *The Road to Miltown, Under the Spreading Atrophy.*
Charles Perrault, 1628-1703, (Fr.) writer. *Tales From Mother Goose* (*Sleeping Beauty, Cinderella*).
Petrarch (Francesco Petrarca), 1304-74, (It.) poet. *Africa, Trionfi, Canzoniere.*
Luigi Pirandello, 1867-1936, (It.) novelist, dramatist. *Six Characters in Search of an Author.*
Sylvia Plath, 1932-63, (U.S.) author, poet. *The Bell Jar.*
Edgar Allan Poe, 1809-49, (U.S.) poet, short-story writer, critic. "Annabel Lee," "The Raven," "The Purloined Letter."
Alexander Pope, 1688-1744, (Br.) poet. *The Rape of the Lock, The Dunciad, An Essay on Man.*
Katherine Anne Porter, 1890-1980, (U.S.) novelist, short-story writer. *Ship of Fools.*
Ezra Pound, 1885-1972, (U.S.) poet. *Cantos.*
Anthony Powell, 1905-2000, (Br.) novelist. *A Dance to the Music of Time* series.
J(ohn) B. Priestley, 1894-1984, (Br.) novelist, dramatist. *The Good Companions.*
Marcel Proust, 1871-1922, (Fr.) novelist. *Remembrance of Things Past.*
Aleksandr Pushkin, 1799-1837, (Russ.) poet, novelist. *Boris Godunov, Eugene Onegin.*
Mario Puzo, 1920-99, (U.S.) novelist. *The Godfather.*
François Rabelais, 1495-1553, (Fr.) writer. *Gargantua.*
Jean Racine, 1639-99, (Fr.) dramatist. *Andromaque, Phèdre, Bérénice, Britannicus.*
Ayn Rand, 1905-82, (Russ.-U.S.) novelist, moral theorist. *The Fountainhead, Atlas Shrugged.*
Terence Rattigan, 1911-77, (Br.) playwright. *Separate Tables, The Browning Version.*
Erich Maria Remarque, 1898-1970, (Ger.-U.S.) novelist. *All Quiet on the Western Front.*
Samuel Richardson, 1689-1761, (Br.) novelist. *Pamela; or Virtue Rewarded.*
Rainer Maria Rilke, 1875-1926, (Ger.) poet. *Life and Songs, Duino Elegies, Poems From the Book of Hours.*
Arthur Rimbaud, 1854-91, (Fr.) poet. *A Season in Hell.*

Edwin Arlington Robinson, 1869-1935, (U.S.) poet. "Richard Cory," "Miniver Cheevy," *Merlin.*
Theodore Roethke, 1908-63, (U.S.) poet. *Open House, The Waking, The Far Field.*
Romain Rolland, 1866-1944, (Fr.) novelist, biographer. *Jean-Christophe.*
Pierre de Ronsard, 1524-85, (Fr.) poet. *Sonnets pour Hélène, La Franciade.*
Christina Rossetti, 1830-94, (Br.) poet. "When I Am Dead, My Dearest."
Dante Gabriel Rossetti, 1828-82, (Br.) poet, painter. "The Blessed Damozel."
Edmond Rostand, 1868-1918, (Fr.) poet, dramatist. *Cyrano de Bergerac.*
Damon Runyon, 1880-1946, (U.S.) short-story writer, journalist. *Guys and Dolls, Blue Plate Special.*
John Ruskin, 1819-1900, (Br.) critic, social theorist. *Modern Painters, The Seven Lamps of Architecture.*
Antoine de Saint-Exupéry, 1900-44, (Fr.) writer. *Wind, Sand and Stars, The Little Prince.*
Saki, or H(ector) H(ugh) Munro, 1870-1916, (Br.) writer. *The Chronicles of Clovis.*
George Sand (Amandine Lucie Aurore Dupin), 1804-76, (Fr.) novelist. *Indiana, Consuelo.*
Carl Sandburg, 1878-1967, (U.S.) poet. *The People, Yes; Chicago Poems, Smoke and Steel, Harvest Poems.*
William Saroyan, 1908-81, (U.S.) playwright, novelist. *The Time of Your Life, The Human Comedy.*
Nathalie Sarraute, 1900-99, (Fr.) Nouveau Roman novelist. Tropismes.
May Sarton, 1914-95, (Belg.-U.S.) poet, novelist. *Encounter in April, Anger.*
Dorothy L. Sayers, 1893-1957, (Br.) mystery writer; created Lord Peter Wimsey.
Richard Scarry, 1920-94, (U.S.) author of children's books. *Richard Scarry's Best Story Book Ever.*
Friedrich von Schiller, 1759-1805, (Ger.) dramatist, poet, historian. *Don Carlos, Maria Stuart, Wilhelm Tell.*
Sir Walter Scott, 1771-1832, (Sc.) novelist, poet. *Ivanhoe.*
Jaroslav Seifert, 1902-86, (Czech.) poet.
Dr. Seuss (Theodor Seuss Geisel), 1904-91, (U.S.) children's book author and illustrator. *The Cat in the Hat.*
William Shakespeare, 1564-1616, (Br.) dramatist, poet. *Romeo and Juliet, Hamlet, King Lear, Julius Caesar,* sonnets.
Karl Shapiro, 1913-2000, (U.S.) poet. "Elegy for a Dead Soldier".
George Bernard Shaw, 1856-1950, (Ir.-Br.) playwright, critic. *St. Joan, Pygmalion, Major Barbara, Man and Superman.*
Mary Wollstonecraft Shelley, 1797-1851, (Br.) novelist, feminist. *Frankenstein. The Last Man.*
Percy Bysshe Shelley, 1792-1822, (Br.) poet. *Prometheus Unbound, Adonais,* "Ode to the West Wind," "To a Skylark."
Richard B. Sheridan, 1751-1816, (Br.) dramatist. *The Rivals, School for Scandal.*
Robert Sherwood, 1896-1955, (U.S.) playwright, biographer. *The Petrified Forest, Abe Lincoln in Illinois.*
Mikhail Sholokhov, 1906-84, (Russ.) writer. *The Silent Don.*
Upton Sinclair, 1878-1968, (U.S.) novelist. *The Jungle.*
Isaac Bashevis Singer, 1904-91, (Pol.-U.S.) novelist, short-story writer, in Yiddish. *The Magician of Lublin.*
C(harles) P(ercy) Snow, 1905-80, (Br.) novelist, scientist. *Strangers and Brothers, Corridors of Power.*
Stephen Spender, 1909-95, (Br.) poet, critic, novelist. *Twenty Poems,* "Elegy for Margaret."
Edmund Spenser, 1552-99, (Br.) poet. *The Faerie Queen.*
Johanna Spyri, 1827-1901, (Swiss) children's author. *Heidi.*
Christina Stead, 1903-83, (Austral.) novelist, short-story writer. *The Man Who Loved Children.*
Richard Steele, 1672-1729, (Br.) essayist, playwright, began the *Tatler* and *Spectator. The Conscious Lovers.*
Gertrude Stein, 1874-1946, (U.S.) writer. *Three Lives.*
John Steinbeck, 1902-68, (U.S.) novelist. *The Grapes of Wrath, Of Mice and Men, The Winter of Our Discontent.*
Stendhal (Marie Henri Beyle), 1783-1842, (Fr.) novelist. *The Red and the Black, The Charterhouse of Parma.*
Laurence Sterne, 1713-68, (Br.) novelist. *Tristram Shandy.*
Wallace Stevens, 1879-1955, (U.S.) poet. *Harmonium, The Man With the Blue Guitar, Notes Toward a Supreme Fiction.*
Robert Louis Stevenson, 1850-94, (Br.) novelist, poet, essayist. *Treasure Island, A Child's Garden of Verses.*
Bram Stoker, 1845-1910, (Br.) writer. *Dracula.*
Rex Stout, 1886-1975, (U.S.) mystery writer; created Nero Wolfe.
Harriet Beecher Stowe, 1811-96, (U.S.) novelist. *Uncle Tom's Cabin.*
Lytton Strachey, 1880-1932, (Br.) biographer, critic. *Eminent Victorians. Queen Victoria, Elizabeth and Essex.*
August Strindberg, 1849-1912, (Swed.) dramatist, novelist. *The Father, Miss Julie, The Creditors.*
Jonathan Swift, 1667-1745, (Br.) satirist, poet. *Gulliver's Travels,* "A Modest Proposal."
Algernon C. Swinburne, 1837-1909, (Br.) poet, dramatist. *Atalanta in Calydon.*
John M. Synge, 1871-1909, (Ir.) poet, dramatist. *Riders to the Sea, The Playboy of the Western World.*
Rabindranath Tagore, 1861-1941, (In.) author, poet. *Sadhana, The Realization of Life, Gitanjali.*
Booth Tarkington, 1869-1946, (U.S.) novelist. *Seventeen.*
Peter Taylor, 1917-94, (U.S.) novelist. A *Summons to Memphis.*
Sara Teasdale, 1884-1933, (U.S.) poet. *Helen of Troy and Other Poems, Rivers to the Sea.*
Alfred, Lord Tennyson, 1809-92, (Br.) poet. *Idylls of the King, In Memoriam,* "The Charge of the Light Brigade."
William Makepeace Thackeray, 1811-63, (Br.) novelist. *Vanity Fair, Henry Esmond, Pendennis.*
Dylan Thomas, 1914-53, (Welsh) poet. *Under Milk Wood, A Child's Christmas in Wales.*
Henry David Thoreau, 1817-62, (U.S.) writer, philosopher, naturalist. *Walden,* "Civil Disobedience."
James Thurber, 1894-1961, (U.S.) humorist; "The Secret Life of Walter Mitty," *My Life and Hard Times.*
J(ohn) R(onald) R(euel) Tolkien, 1892-1973, (Br.) writer. *The Hobbit, Lord of the Rings* trilogy.
Leo Tolstoy, 1828-1910, (Russ.) novelist, short-story writer. *War and Peace, Anna Karenina,* "The Death of Ivan Ilyich."
Anthony Trollope, 1815-82, (Br.) novelist. *The Warden, Barchester Towers,* the Palliser novels.
Ivan Turgenev, 1818-83, (Russ.) novelist, short-story writer. *Fathers and Sons, First Love, A Month in the Country.*
Amos Tutuola, 1920-97, (Nigerian) novelist. *The Palm-Wine Drunkard, My Life in the Bush of Ghosts.*
Mark Twain (Samuel Clemens), 1835-1910, (U.S.) novelist, humorist. *The Adventures of Huckleberry Finn, Tom Sawyer; Life on the Mississippi.*
Sigrid Undset, 1881-1949, (Nor.) novelist, poet. *Kristin Lavransdatter.*
Paul Valéry, 1871-1945, (Fr.) poet, critic. *La Jeune Parque, The Graveyard by the Sea.*
Jules Verne, 1828-1905, (Fr.) novelist. *Twenty Thousand Leagues Under the Sea.*
François Villon, 1431-63?, (Fr.) poet. *The Lays, The Grand Testament.*
Voltaire (F.M. Arouet), 1694-1778, (Fr.) writer of "philosophical romances"; philosopher, historian; *Candide.*
Robert Penn Warren, 1905-89, (U.S.) novelist, poet, critic. *All the King's Men.*
Evelyn Waugh, 1903-66, (Br.) novelist. *The Loved One, Brideshead Revisited, A Handful of Dust.*
H(erbert) G(eorge) Wells, 1866-1946, (Br.) novelist. *The Time Machine, The Invisible Man, The War of the Worlds.*
Eudora Welty, 1909-2001, (U.S.) Southern short story writer. "Why I Live at the P.O," "The Ponder Heart."
Rebecca West, 1893-1983, (Br.) novelist, critic, journalist. *Black Lamb and Grey Falcon.*
Edith Wharton, 1862-1937, (U.S.) novelist. *The Age of Innocence, The House of Mirth, Ethan Frome.*
E(lwyn) B(rooks) White, 1899-1985, (U.S.) essayist, novelist. *Charlotte's Web, Stuart Little.*
Patrick White, 1912-90, (Austral.) novelist. *The Tree of Man.*
T(erence) H(anbury) White, 1906-64, (Br.) author. *The Once and Future King, A Book of Beasts.*
Walt Whitman, 1819-92, (U.S.) poet. *Leaves of Grass.*
John Greenleaf Whittier, 1807-92, (U.S.) poet, journalist. *Snow-Bound.*
Oscar Wilde, 1854-1900, (Ir.) novelist, playwright. *The Picture of Dorian Gray, The Importance of Being Earnest.*
Laura Ingalls Wilder, 1867-1957, (U.S.) novelist. Little House on the Prairie series of children's books.
Thornton Wilder, 1897-1975, (U.S.) playwright. *Our Town, The Skin of Our Teeth, The Matchmaker.*
Tennessee Williams, 1911-83, (U.S.) playwright. *A Streetcar Named Desire, Cat on a Hot Tin Roof, The Glass Menagerie.*
William Carlos Williams, 1883-1963, (U.S.) poet, physician. *Tempers, Al Que Quiere! Paterson,* "This Is Just to Say."
Edmund Wilson, 1895-1972, (U.S.) critic, novelist. *Axel's Castle, To the Finland Station.*
P(elham) G(renville) Wodehouse, 1881-1975, (Br.-U.S.) humorist. The "Jeeves" novels, *Anything Goes.*
Thomas Wolfe, 1900-38, (U.S.) novelist. *Look Homeward, Angel; You Can't Go Home Again.*
Virginia Woolf, 1882-1941, (Br.) novelist, essayist. *Mrs. Dalloway, To the Lighthouse, A Room of One's Own.*
William Wordsworth, 1770-1850, (Br.) poet. "Tintern Abbey," "Ode: Intimations of Immortality," *The Prelude.*
Richard Wright, 1908-60, novelist, short-story writer. *Native Son, Black Boy, Uncle Tom's Children.*
Elinor Wylie, 1885-1928, (U.S.) poet. *Nets to Catch the Wind.*
William Butler Yeats, 1865-1939, (Ir.) poet, playwright. "The Second Coming," *The Wild Swans at Coole.*
Émile Zola, 1840-1902, (Fr.) novelist. *Nana, Thérésè Raquin.*

Poets Laureate

There is no record of the origin of the office of Poet Laureate of England. Henry III (1216-72) reportedly had a Versificator Regis, or King's Poet, paid 100 shillings a year. Other poets said to have filled the role include Geoffrey Chaucer (d 1400), Edmund Spenser (d 1599), Ben Jonson (d 1637), and Sir William d'Avenant (d 1668).

The first official English poet laureate was John Dryden, appointed 1668, for life (as was customary). Then came Thomas Shadwell, in 1689; Nahum Tate, 1692; Nicholas Rowe, 1715; Rev. Laurence Eusden, 1718; Colley Cibber, 1730; William Whitehead, 1757; Rev. Thomas Warton, 1785; Henry James Pye, 1790; Robert Southey, 1813; William Wordsworth, 1843; Alfred, Lord Tennyson, 1850; Alfred Austin, 1896; Robert Bridges, 1913; John Masefield, 1930; C. Day Lewis, 1968; Sir John Betjeman, 1972; Ted Hughes, 1984; Andrew Motion, 1999.

In U.S., appointment is by Librarian of Congress and is not for life: Robert Penn Warren, 1986; Richard Wilbur, 1987; Howard Nemerov, 1988; Mark Strand, 1990; Joseph Brodsky, 1991; Mona Van Duyn, 1992; Rita Dove, 1993; Robert Hass, 1995; Robert Pinsky, 1997; Stanley Kunitz, 2000; Billy Collins, 2001.

Composers of Classical and Avant Garde Music

Carl Philipp Emanuel Bach, 1714-88, (Ger.) Cantatas, passions, numerous keyboard and instrumental works.
Johann Christian Bach, 1735-82, (Ger.) Concertos, operas, sonatas.
Johann Sebastian Bach, 1685-1750, (Ger.) St. Matthew Passion, The Well-Tempered Clavier.
Samuel Barber, 1910-81, (U.S.) Adagio for Strings, Vanessa.
Béla Bartók, 1881-1945, (Hung.) Concerto for Orchestra, The Miraculous Mandarin.
Amy Beach (Mrs. H. H. A. Beach), 1867-1944, (U.S.) The Year's at the Spring, Fireflies, The Chambered Nautilus.
Ludwig van Beethoven, 1770-1827, (Ger.) Concertos (Emperor), sonatas (Moonlight, Pathetique), 9 symphonies.
Vincenzo Bellini, 1801-35, (It.) I Puritani, La Sonnambula, Norma.
Alban Berg, 1885-1935, (Austrian) Wozzeck, Lulu.
Hector Berlioz, 1803-69, (Fr.) Damnation of Faust, Symphonie Fantastique, Requiem.
Leonard Bernstein, 1918-90, (U.S.) Chichester Psalms, Jeremiah Symphony, Mass.
Georges Bizet, 1838-75, (Fr.) Carmen, Pearl Fishers.
Ernest Bloch, 1880-1959, (Swiss-U.S.) Macbeth (opera), Schelomo, Voice in the Wilderness.
Luigi Boccherini, 1743-1805, (It.) Chamber music and guitar pieces.
Alexander Borodin, 1833-87, (Russ.) Prince Igor, In the Steppes of Central Asia, Polovtzian Dances.
Pierre Boulez, b 1925, (Fr.) LeVisage nuptial, Edats/Multiple, Domaines.
Johannes Brahms, 1833-97, (Ger.) Liebeslieder Waltzes, Acad. Festival Overture, chamber music, 4 symphonies.
Benjamin Britten, 1913-76, (Br.) Peter Grimes, Turn of the Screw, A Ceremony of Carols, War Requiem.
Anton Bruckner, 1824-96, (Austrian) 9 symphonies.
Dietrich Buxtehude, 1637-1707, (Dan.) Organ works, vocal music.
William Byrd, 1543-1623, (Br.) Masses, motets.
John Cage, 1912-92, (U.S.) Winter Music, Fontana Mix.
Emmanuel Chabrier, 1841-94, (Fr.) Le Roi Malgré Lui, Espana.
Gustave Charpentier, 1860-1956, (Fr.) Louise.
Frédéric Chopin, 1810-49, (Pol.) Mazurkas, waltzes, etudes, nocturnes, polonaises, sonatas.
Aaron Copland, 1900-90, (U.S.) Appalachian Spring, Fanfare for the Common Man, Lincoln Portrait.
Claude Debussy, 1862-1918, (Fr.) Pelleas et Melisande, La Mer, Prelude to the Afternoon of a Faun.
Gaetano Donizetti, 1797-1848, (It.) Elixir of Love, Lucia di Lammermoor, Daughter of the Regiment.
Paul Dukas, 1865-1935, (Fr.) Sorcerer's Apprentice.
Antonin Dvorak, 1841-1904, (Czech.) Songs My Mother Taught Me, Symphony in E Minor (From the New World).
Edward Elgar, 1857-1934, (Br.) Enigma Variations, Pomp and Circumstance.
Manuel de Falla, 1876-1946, (Sp.) El Amor Brujo, La Vida Breve, The Three-Cornered Hat.
Gabriel Faurè, 1845-1924, (Fr.) Requiem, Elègie for Cello and Piano.
Cesar Franck, 1822-90, (Belg.) Symphony in D minor, Violin Sonata.
George Gershwin, 1898-1937, (U.S.) Rhapsody in Blue, An American in Paris, Porgy and Bess.
Philip Glass, b 1937, (U.S.) Einstein on the Beach, The Voyage.
Mikhail Glinka, 1804-57, (Russ.) A Life for the Tsar, Ruslan and Ludmilla.
Christoph W. Gluck, 1714-87, (Ger.) Alceste, Iphigènie en Tauride.
Charles Gounod, 1818-93, (Fr.) Faust, Romeo and Juliet.
Edvard Grieg, 1843-1907, (Nor.) Peer Gynt Suite, Concerto in A minor for piano.
George Frideric Handel, 1685-1759, (Ger.-Br.) Messiah, Water Music.
Howard Hanson, 1896-1981, (U.S.) Symphonies No. 1 (Nordic) and No. 2 (Romantic).
Roy Harris, 1898-1979, (U.S.) Symphonies.
(Franz) Joseph Haydn, 1732-1809, (Austrian) Symphonies (Clock, London, Toy), chamber music, oratorios.
Paul Hindemith, 1895-1963, (U.S.) Mathis der Maler.
Gustav Holst, 1874-1934, (Br.) The Planets.
Arthur Honegger, 1892-1955, (Fr.) Judith, Le Roi David, Pacific 231.
Alan Hovhaness, 1911-2000, (U.S.) Symphonies, Magnificat.
Engelbert Humperdinck, 1854-1921, (Ger.) Hansel and Gretel.
Charles Ives, 1874-1954, (U.S.) Concord Sonata, symphonies.
Aram Khachaturian, 1903-78, (Russ.) Ballets, piano pieces, Sabre Dance.
Zoltán Kodaly, 1882-1967, (Hung.) Háry János, Psalmus Hungaricus.
Fritz Kreisler, 1875-1962, (Austrian) Caprice Viennois, Tambourin Chinois.
Edouard Lalo, 1823-92, (Fr.) Symphonie Espagnole.
Ruggero Leoncavallo, 1857-1919, (It.) Pagliacci.
Franz Liszt, 1811-86, (Hung.) 20 Hungarian rhapsodies, symphonic poems.
Edward MacDowell, 1861-1908, (U.S.) To a Wild Rose.
Gustav Mahler, 1860-1911, (Austrian) Das Lied von der Erde; 9 complete symphonies.
Pietro Mascagni, 1863-1945, (It.) Cavalleria Rusticana.
Jules Massenet, 1842-1912, (Fr.) Manon, Le Cid, Thaïs.
Felix Mendelssohn, 1809-47, (Ger.) A Midsummer Night's Dream, Songs Without Words, violin concerto.
Gian-Carlo Menotti, b 1911, (It.-U.S.) The Medium, The Consul, Amahl and the Night Visitors.
Claudio Monteverdi, 1567-1643, (It.) Opera, masses, madrigals.
Modest Moussorgsky, 1839-81, (Russ.) Boris Godunov, Pictures at an Exhibition.
Wolfgang Amadeus Mozart, 1756-91, (Austrian) Chamber music, concertos, operas (Magic Flute, Marriage of Figaro), 41 symphonies.
Jacques Offenbach, 1819-80, (Fr.) Tales of Hoffmann.
Carl Orff, 1895-1982, (Ger.) Carmina Burana.
Johann Pachelbel, 1653-1706, (Ger.) Canon and Fugue in D major.
Ignacy Paderewski, 1860-1941, (Pol.) Minuet in G.
Niccolò Paganini, 1782-1840, (It.) Caprices for violin solo.
Giovanni Palestrina, c1525-94, (It.) Masses, madrigals.
Krzystof Pendercki, b 1933, (Pol.) Psalmus, Polymorphia, De natura sonoris.
Francis Poulenc, 1899-1963, (Fr.) Dialogues des Carmèlites.
Mel Powell, 1923-98, (U.S.) *Duplicates: A Concerto for Two Pianos and Orchestra, Cantilena Concertante.*
Sergei Prokofiev, 1891-1953, (Russ.) Classical Symphony, Love for Three Oranges, Peter and the Wolf.
Giacomo Puccini, 1858-1924, (It.) La Boheme, Manon Lescaut, Tosca, Madama Butterfly.
Henry Purcell, 1659-95, (Eng.) Dido and Aeneas.
Sergei Rachmaninoff, 1873-1943, (Russ.) Concertos, preludes (Prelude in C sharp minor), symphonies.
Maurice Ravel, 1875-1937, (Fr.) Bolèro, Daphnis et Chloè, Piano Concerto in D for Left Hand Alone.
Nikolai Rimsky-Korsakov, 1844-1908, (Russ.) Golden Cockerel, Scheherazade, Flight of the Bumblebee.
Gioacchino Rossini, 1792-1868, (It.) Barber of Seville, Othello, William Tell.
Camille Saint-Saëns, 1835-1921, (Fr.) Carnival of Animals (The Swan), Samson and Delilah, Danse Macabre.
Alessandro Scarlatti, 1660-1725, (It.) Cantatas, oratorios, operas.
Domenico Scarlatti, 1685-1757, (It.) Harpsichord works.
Alfred Schnittke, 1934-98 (Sov.-Ger.) *Life With an Idiot.*
Arnold Schoenberg, 1874-1951, (Austrian) Pelleas and Melisande, Pierrot Lunaire, Verklärte Nacht.
Franz Schubert, 1797-1828, (Austrian) Chamber music (Trout Quintet), lieder, symphonies (Unfinished).
Robert Schumann, 1810-56, (Ger.) Die Frauenliebe und Leben, Träumerei.
Dimitri Shostakovich, 1906-75, (Russ.) Symphonies, Lady Macbeth of the District Mzensk.
Jean Sibelius, 1865-1957, (Finn.) Finlandia.
Bedrich Smetana, 1824-84, (Czech.) The Bartered Bride.
Karlheinz Stockhausen, b 1928, (Ger.) KontraPunkte, Kontakte for Electronic Instruments.

Richard Strauss, 1864-1949, (Ger.) Salome, Elektra, Der Rosenkavalier, Thus Spake Zarathustra.
Igor Stravinsky, 1882-1971, (Russ.) Noah and the Flood, The Rake's Progress, The Rite of Spring.
Toru Takemitsu, 1930-96, (Jpn.) Requiem for Strings, Dorian Horizon.
Peter I. Tchaikovsky, 1840-93, (Russ.) Nutcracker, Swan Lake, The Sleeping Beauty.
Virgil Thomson, 1896-1989, (U.S.) Opera, film music, Four Saints in Three Acts.
Dmitri Tiomkin, 1894-1979, (Russ.-U.S.) film scores, including *High Noon.*
Sir Michael Tippett, 1905-98, (Br.) *A Child of Our Time, The Midsummer Marriage, The Knot Garden.*
Ralph Vaughan Williams, 1872-1958, (Eng.) Fantasiz on a Theme by Thomas Tallis, symphonies, vocal music.
Giuseppe Verdi, 1813-1901, (It.) Aida, Rigoletto, Don Carlo, Il Trovatore, La Traviata, Falstaff, Macbeth.
Heitor Villa-Lobos, 1887-1959, (Brazil) Bachianas Brasileiras.
Antonio Vivaldi, 1678-1741, (It.) Concerto grossos (The Four Seasons).
Richard Wagner, 1813-83, (Ger.) Rienzi, Tannhäuser, Lohengrin, Tristan und Isolde.
Carl Maria von Weber, 1786-1826, (Ger.) Der Freischutz.

Composers of Operettas, Musicals, and Popular Music

Richard Adler, b 1921, (U.S.) *Pajama Game; Damn Yankees.*
Milton Ager, 1893-1979, (U.S.) I Wonder What's Become of Sally; Hard Hearted Hannah; Ain't She Sweet?
Arthur Altman, 1910-94, (U.S.) *All or Nothing at All.*
Leroy Anderson, 1908-75, (U.S.) Sleigh Ride, Blue Tango, Syncopated Clock.
Paul Anka, b 1941, (Can.) My Way; *Tonight Show* theme.
Harold Arlen, 1905-86, (U.S.) Stormy Weather; Over the Rainbow; Blues in the Night; That Old Black Magic.
Burt Bacharach, b 1928, (U.S.) Raindrops Keep Fallin' on My Head; Walk on By; What the World Needs Now Is Love.
Ernest Ball, 1878-1927, (U.S.) Mother Machree; When Irish Eyes Are Smiling.
Irving Berlin, 1888-1989, (U.S.) *Annie Get Your Gun; Call Me Madam;* God Bless America; White Christmas.
Leonard Bernstein, 1918-90, (U.S.) *On the Town; Wonderful Town; Candide; West Side Story.*
Eubie Blake, 1883-1983, (U.S.) *Shuffle Along;* I'm Just Wild About Harry.
Jerry Bock, b 1928, (U.S.) *Mr. Wonderful; Fiorello; Fiddler on the Roof; The Rothschilds.*
Carrie Jacobs Bond, 1862-1946, (U.S.) I Love You Truly.
Nacio Herb Brown, 1896-1964, (U.S.) Singing in the Rain; You Were Meant for Me; All I Do Is Dream of You.
Hoagy Carmichael, 1899-1981, (U.S.) Stardust; Georgia on My Mind; Old Buttermilk Sky.
George M. Cohan, 1878-1942, (U.S.) Give My Regards to Broadway; You're a Grand Old Flag; Over There.
Cy Coleman, b 1929, (U.S.) *Sweet Charity;* Witchcraft.
John Frederick Coots, 1897-?, (U.S.) Santa Claus Is Coming to Town; You Go to My Head; For All We Know.
Noel Coward, 1899-1973, (Br.) *Bitter Sweet;* Mad Dogs and Englishmen; Mad About the Boy.
Neil Diamond, b 1941, (U.S.) I'm a Believer; Sweet Caroline.
Walter Donaldson, 1893-1947, (U.S.) My Buddy; Carolina in the Morning; Makin' Whoopee.
Vernon Duke, 1903-69, (U.S.) April in Paris.
Bob Dylan, b 1941, (U.S.) Blowin' in the Wind.
Gus Edwards, 1879-1945, (U.S.) School Days; By the Light of the Silvery Moon; In My Merry Oldsmobile.
Sherman Edwards, 1919-81, (U.S.) See You in September; Wonderful! Wonderful!
Duke Ellington, 1899-1974, (U.S.) Sophisticated Lady; Satin Doll; It Don't Mean a Thing; Solitude.
Sammy Fain, 1902-89, (U.S.) I'll Be Seeing You; Love Is a Many-Splendored Thing.
Fred Fisher, 1875-1942, (U.S.) Peg O' My Heart; Chicago.
Stephen Collins Foster, 1826-64, (U.S.) My Old Kentucky Home; Old Folks at Home, Beautiful Dreamer.
Rudolf Friml, 1879-1972, (Czech-U.S.) *The Firefly; Rose Marie; Vagabond King; Bird of Paradise.*
John Gay, 1685-1732, (Br.) *The Beggar's Opera.*
George Gershwin, 1898-1937, (U.S.) Someone to Watch Over Me; I've Got a Crush on You; Embraceable You.
Morton Gould, 1913-96, (U.S.) Fall River Suite, Holocaust Suite, Spirituals for Orchestra, Stringmusic.
Ferde Grofe, 1892-1972, (U.S.) Grand Canyon Suite.
Marvin Hamlisch, b 1944, (U.S.) The Way We Were; Nobody Does It Better; *A Chorus Line.*
Ray Henderson, 1896-1970, (U.S.) *George White's Scandals;* That Old Gang of Mine; Five Foot Two, Eyes of Blue.
Victor Herbert, 1859-1924, (Ir.-U.S.) *Mlle. Modiste; Babes in Toyland; The Red Mill; Naughty Marietta; Sweethearts.*
Jerry Herman, b 1933, (U.S.) *Hello Dolly; Mame.*
Brian Holland, b 1941, **Lamont Dozier,** b 1941, **Eddie Holland,** b 1939, (all U.S.) Heat Wave; Stop! In the Name of Love; Baby, I Need Your Loving.
Antonio Carlos Jobim, 1927-94, (Brazil) *The Girl From Ipanema; Desafinado; One Note Samba.*
Billy (William Martin) Joel, b 1949, (U.S.) *Just the Way You Are; Honesty;* Piano Man.
Scott Joplin, 1868-1917, (U.S.) Maple Leaf Rag; *Treemonisha.*
John Kander, b 1927, (U.S.) *Cabaret; Chicago; Funny Lady.*
Jerome Kern, 1885-1945, (U.S.) *Sally; Sunny; Show Boat.*
Carole King, b 1942, (U.S.) Will You Love Me Tomorrow?; Natural Woman; One Fine Day; Up on the Roof.
Burton Lane, 1912-1997, (U.S.) *Finian's Rainbow.*
Franz Lehar, 1870-1948, (Hung.) *Merry Widow.*
Jerry Leiber, & **Mike Stoller,** both b 1933, (both U.S.) Hound Dog; Searchin'; Yakety Yak; Love Me Tender.
Mitch Leigh, b 1928, (U.S.) *Man of La Mancha.*
John Lennon, 1940-80, & **Paul McCartney,** b 1942, (both Br.) I Want to Hold Your Hand; She Loves You.
Andrew Lloyd Webber, b 1948, (Br.) *Jesus Christ Superstar; Evita; Cats; The Phantom of the Opera.*
Frank Loesser, 1910-69, (U.S.) *Guys and Dolls; Where's Charley?; The Most Happy Fella; How to Succeed....*
Frederick Loewe, 1901-88, (Austrian-U.S.) *Brigadoon; Paint Your Wagon; My Fair Lady; Camelot.*
Henry Mancini, 1924-94, (U.S.) Moon River; Days of Wine and Roses; Pink Panther Theme.
Barry Mann, b 1939, & **Cynthia Weil,** b 1937, (both U.S.) You've Lost That Loving Feeling.
Jimmy McHugh, 1894-1969, (U.S.) Don't Blame Me; I'm in the Mood for Love; I Feel a Song Coming On.
Alan Menken, b 1949, (U.S.) *Little Shop of Horrors, Beauty and the Beast.*
Joseph Meyer, 1894-1987, (U.S.) If You Knew Susie; California, Here I Come; Crazy Rhythm.
Chauncey Olcott, 1858-1932, (U.S.) Mother Machree.
Jerome "Doc" Pomus, 1925-91, (U.S.) Save the Last Dance for Me; A Teenager in Love.
Cole Porter, 1893-1964, (U.S.) *Anything Goes; Kiss Me Kate; Can Can; Silk Stockings.*
Smokey Robinson, b 1940, (U.S.) Shop Around; My Guy; My Girl; Get Ready.
Richard Rodgers, 1902-79, (U.S.) *Oklahoma!; Carousel; South Pacific; The King and I; The Sound of Music.*
Sigmund Romberg, 1887-1951, (Hung.) *Maytime; The Student Prince; Desert Song; Blossom Time.*
Harold Rome, 1908-93, (U.S.) *Pins and Needles; Call Me Mister; Wish You Were Here; Fanny; Destry Rides Again.*
Vincent Rose, b 1880-1944, (U.S.) Avalon; Whispering; Blueberry Hill.
Harry Ruby, 1895-1974, (U.S.) Three Little Words; Who's Sorry Now?
Arthur Schwartz, 1900-84, (U.S.) *The Band Wagon;* Dancing in the Dark; By Myself; That's Entertainment.
Neil Sedaka, b 1939, (U.S.) Breaking Up Is Hard to Do.
Paul Simon, b 1942, (U.S.) Sounds of Silence; I Am a Rock; Mrs. Robinson; Bridge Over Troubled Waters.
Stephen Sondheim, b 1930, (U.S.) *A Little Night Music; Company; Sweeney Todd; Sunday in the Park With George.*
John Philip Sousa, 1854-1932, (U.S.) *El Capitan;* Stars and Stripes Forever.
Oskar Straus, 1870-1954, (Austrian) *Chocolate Soldier.*
Johann Strauss, 1825-99, (Austrian) *Gypsy Baron; Die Fledermaus;* waltzes: Blue Danube; Artist's Life.
Charles Strouse, b 1928, (U.S.) *Bye Bye, Birdie; Annie.*
Jule Styne, 1905-94, (Br.-U.S.) *Gentlemen Prefer Blondes; Bells Are Ringing; Gypsy; Funny Girl.*
Arthur S. Sullivan, 1842-1900, (Br.) *H.M.S. Pinafore; Pirates of Penzance; The Mikado.*
Deems Taylor, 1885-1966, (U.S.) *Peter Ibbetson.*
Harry Tobias, 1905-94, (U.S.) *I'll Keep the Lovelight Burning.*
Egbert van Alstyne, 1882-1951, (U.S.) In the Shade of the Old Apple Tree; Memories; Pretty Baby.
Jimmy Van Heusen, 1913-90, (U.S.) Moonlight Becomes You; Swinging on a Star; All the Way; Love and Marriage.
Albert von Tilzer, 1878-1956, (U.S.) I'll Be With You in Apple Blossom Time; Take Me Out to the Ball Game.
Harry von Tilzer, 1872-1946, (U.S.) Only a Bird in a Gilded Cage; On a Sunday Afternoon.
Fats Waller, 1904-43, (U.S.) Honeysuckle Rose; Ain't Misbehavin'.
Harry Warren, 1893-1981, (U.S.) You're My Everything; We're in the Money; I Only Have Eyes for You.
Jimmy Webb, b 1946, (U.S.) Up, Up and Away; By the Time I Get to Phoenix; Didn't We?; Wichita Lineman.
Kurt Weill, 1900-50, (Ger.-U.S.) *Threepenny Opera; Lady in the Dark; Knickerbocker Holiday; One Touch of Venus.*
Percy Wenrich, 1887-1952, (U.S.) When You Wore a Tulip; Moonlight Bay; Put On Your Old Gray Bonnet.

Richard A. Whiting, 1891-1938, (U.S.) Till We Meet Again; Sleepytime Gal; Beyond the Blue Horizon; My Ideal.
John Williams, b 1932, (U.S.) *Jaws; E.T.; Star Wars* series; *Raiders of the Lost Ark* series.
Meredith Willson, 1902-84, (U.S.) *The Music Man.*
Stevie Wonder, b 1950, (U.S.) You Are the Sunshine of My Life; Signed, Sealed, Delivered, I'm Yours.
Vincent Youmans, 1898-1946, (U.S.) *Two Little Girls in Blue; Wildflower; No, No, Nanette; Hit the Deck; Rainbow; Smiles.*

Lyricists

Howard Ashman, 1950-91, (U.S.) Little Shop of Horrors; The Little Mermaid.
Johnny Burke, 1908-84, (U.S.) Misty; Imagination.
Irving Caesar, 1895-1996, (U.S.) Swanee; Tea for Two; Just a Gigolo.
Sammy Cahn, 1913-93, (U.S.) High Hopes; Love and Marriage; The Second Time Around; It's Magic.
Leonard Cohen, b 1934, (Can.) Suzanne; Stranger Song.
Betty Comden, b 1919, (U.S.) and **Adolph Green,** b 1915, (U.S.) The Party's Over; Just in Time; New York, New York.
Hal David, b 1921, (U.S.) What the World Needs Now Is Love.
Buddy De Sylva, 1895-1950, (U.S.) When Day Is Done; Look for the Silver Lining; April Showers.
Howard Dietz, 1896-1983, (U.S.) Dancing in the Dark; You and the Night and the Music; That's Entertainment.
Al Dubin, 1891-1945, (U.S.) Tiptoe Through the Tulips; Anniversary Waltz; Lullaby of Broadway.
Fred Ebb, b 1936, (U.S.) Cabaret; Zorba; Woman of the Year.
Dorothy Fields, 1905-74, (U.S.) On the Sunny Side of the Street; Don't Blame Me; The Way You Look Tonight.
Ira Gershwin, 1896-1983, (U.S.) The Man I Love; Fascinating Rhythm; S'Wonderful; Embraceable You.
William S. Gilbert, 1836-1911, (Br.) The Mikado; H.M.S. Pinafore; Pirates of Penzance.
Gerry Goffin, b 1939, (U.S.) Will You Love Me Tomorrow; Take Good Care of My Baby; Up on the Roof.
Mack Gordon, 1905-59, (Pol.-U.S.) You'll Never Know; The More I See You; Chattanooga Choo-Choo.
Oscar Hammerstein II, 1895-1960, (U.S.) Ol' Man River; Oklahoma; Carousel.
E. Y. (Yip) Harburg, 1898-1981, (U.S.) Brother, Can You Spare a Dime; April in Paris; Over the Rainbow.
Lorenz Hart, 1895-1943, (U.S.) Isn't It Romantic; Blue Moon; Lover; Manhattan; My Funny Valentine.
DuBose Heyward, 1885-1940, (U.S.) Summertime.
Gus Kahn, 1886-1941, (U.S.) Memories; Ain't We Got Fun.
Alan J. Lerner, 1918-86, (U.S.) Brigadoon; My Fair Lady; Camelot; Gigi; On a Clear Day You Can See Forever.
Johnny Mercer, 1909-76, (U.S.) Blues in the Night; Come Rain or Come Shine; Laura; That Old Black Magic.
Bob Merrill, 1921-98, (U.S.) People; (How Much Is That) Doggie in the Window.
Jack Norworth, 1879-1959, (U.S.) Take Me Out to the Ball Game; Shine On Harvest Moon.
Mitchell Parish, 1901-93, (U.S.) Stairway to the Stars; Stardust.
Andy Razaf, 1895-1973, (U.S.) Honeysuckle Rose; Ain't Misbehavin'; S'posin'.
Leo Robin, 1900-84, (U.S.) Thanks for the Memory; Hooray for Love; Diamonds Are a Girl's Best Friend.
Paul Francis Webster, 1907-84, (U.S.) Secret Love; The Shadow of Your Smile; Love Is a Many-Splendored Thing.
Jack Yellen, 1892-1991, (U.S.) Down by the O-Hi-O; Ain't She Sweet; Happy Days Are Here Again.

Blues and Jazz Artists of the Past

Julian "Cannonball" Adderley, 1928-75, alto sax
Nat Adderley, 1931-2000, cornet, trumpet, composer
Louis "Satchmo" Armstrong, 1900-71, trumpet, singer; "scat" vocals
Mildred Bailey, 1907-51, blues singer
Chet Baker, 1929-88, trumpet
Count Basie, 1904-84, orchestra leader, piano
Sidney Bechet, 1897-1959, early innovator, soprano sax
Bix Beiderbecke, 1903-31, cornet, piano, composer
Tex Beneke, 1914-2000, tenor sax, vocalist, band leader
Tommy Benford, 1906-94, drummer
Bunny Berigan, 1909-42, trumpet, singer
Barney Bigard, 1906-80, clarinet
Ed Blackwell, 1929-92, drummer
Jimmy Blanton, 1921-42, bass
Charles "Buddy" Bolden, 1868-1931, cornet; formed first jazz band.
Lester Bowie, 1941-99, trumpet, composer, band leader
Big Bill Broonzy, 1893-1958, blues singer, guitar
Clifford Brown, 1930-56, trumpet
Don Byas, 1912-72, tenor sax
Charlie Byrd, 1925-99, guitarist; popularized bossa nova
Cab Calloway, 1907-94, band leader
Harry Carney, 1910-74, baritone sax
Betty Carter, 1930-98, jazz singer
Sidney Catlett, 1910-51, drums
Doc Cheatham, 1905-97, trumpet
Don Cherry, 1937-95, lyrical jazz trumpet
Charlie Christian, 1919-42, guitar
Kenny Clarke, 1914-85, modern drums
Buck Clayton, 1911-91, trumpet, arranger
James Cleveland, 1931-91, gospel singer
Al Cohn, 1925-88, tenor sax, composer
Cozy Cole, 1909-81, drums
Johnny Coles, 1926-96, trumpet
John Coltrane, 1926-67, tenor sax innovator
Eddie Condon, 1904-73, guitar, band leader; Dixieland
Tadd Dameron, 1917-65, piano, composer
Eddie "Lockjaw" Davis, 1921-86, tenor sax
Miles Davis, 1926-91, trumpet; pioneer of cool jazz
Wild Bill Davison, 1906-89, cornet, early Chicago jazz
Paul Desmond, 1924-77, alto sax
Vic Dickenson, 1906-84, trombone, composer
Willie Dixon, 1915-92, songwriter, blues, "You Shook Me"
Warren "Baby" Dodds, 1898-1959, Dixieland drummer
Johnny Dodds, 1892-1940, clarinet
Jimmy Dorsey, 1904-57, clarinet, alto sax; band leader
Tommy Dorsey, 1905-56, trombone; band leader
Roy Eldridge, 1911-89, trumpet, drums, singer
Duke Ellington, 1899-1974, piano, band leader, composer
Bill Evans, 1929-80, piano
Gil Evans, 1912-88, composer, arranger, piano
Tal Farlow, 1921-98, jazz guitarist
Ella Fitzgerald, 1918-1996, jazz vocalist, "first lady of song"
"Red" Garland, 1923-84, piano
Erroll Garner, 1921-77, piano, composer, "Misty"
Stan Getz, 1927-91, tenor sax
Dizzy Gillespie, 1917-93, trumpet, composer; bop developer
Benny Goodman, 1909-86, clarinet; band, combo leader
Dexter Gordon, 1923-90, tenor sax, bop-derived style
Stéphane Grappelli, 1908-97, violin
Bobby Hackett, 1915-76, trumpet, cornet
W. C. Handy, 1873-1958, composer, "St. Louis Blues"
Coleman Hawkins, 1904-69, tenor sax, "Body and Soul"
Fletcher Henderson, 1898-1952, orchestra leader, arranger
Woody Herman, 1913-87, clarinet, alto sax, band leader
Jay C. Higginbotham, 1906-73, trombone
Earl "Fatha" Hines, 1905-83, piano, songwriter
Al Hirt, 1922-99, trumpet
Johnny Hodges, 1906-70, alto sax
Billie Holiday, 1915-59, blues singer, "Strange Fruit"
John Lee Hooker, 1917-2001, blues guitarist, singer
Sam "Lightnin" Hopkins, 1912-82, blues singer, guitarist
Howlin' Wolf, 1910-1976, blues singer, harmonica, guitar
Mahalia Jackson, 1911-72, gospel singer
Elmore James, 1918-63, blues songwriter, singer, guitarist
Blind Lemon Jefferson, 1897-1930, blues singer, guitar
Little Willie John, 1937-68, singer, songwriter
Bunk Johnson, 1879-1949, cornet, trumpet
J.J. Johnson, 1924-2001, modern jazz trombone
James P. Johnson, 1891-1955, piano, composer
Robert Johnson, 1912-38, blues songwriter, singer, guitarist
Jo Jones, 1911-85, drums
Philly Joe Jones, 1923-85, drums
Thad Jones, 1923-86, trumpet, cornet
Scott Joplin, 1868-1917, ragtime composer
Louis Jordan, 1908-75, singer, alto sax
Stan Kenton, 1912-79, orchestra leader, composer, piano
Albert King, 1923-92, blues guitarist
Gene Krupa, 1909-73, drums, band and combo leader
Scott LaFaro, 1936-61, bass
Huddie Ledbetter (Lead Belly), 1888-1949, blues singer, guitar
John Lewis, 1920-2001, pianist, Modern Jazz Quartet founder
Mel Lewis, 1929-90, drummer, orchestra leader
Jimmie Lunceford, 1902-47, band leader, sax
Jimmy McPartland, 1907-91, trumpet
Carmen McRae, 1920-94, jazz singer
Glenn Miller, 1904-44, trombone, dance band leader
Charles Mingus, 1922-79, bass, composer, combo leader
Thelonious Monk, 1920-82, piano, composer, combo leader; bop developer
Wes Montgomery, 1925-68, guitar
"Jelly Roll" Morton, 1885-1941, composer, piano, singer
Bennie Moten, 1894-1935, piano
Gerry Mulligan, 1927-96, baritone sax, songwriter, "cool school"
Turk Murphy, 1915-87, trombone, band leader
Theodore "Fats" Navarro, 1923-50, trumpet
Red Nichols, 1905-65, cornet, combo leader

King Oliver, 1885-1938, cornet, band leader; Louis Armstrong
Sy Oliver, 1910-88, Swing Era arranger, composer, conductor
Kid Ory, 1886-1973, trombone, "Muskrat Ramble"
Charlie "Bird" Parker, 1920-55, alto sax, noted jazz improviser
Joe Pass, 1929-94, guitarist
Art Pepper, 1925-82, alto sax
Oscar Pettiford, 1922-60, a leading bop-era bassist
Bud Powell, 1924-66, piano; modern jazz pioneer
Louis Prima, 1911-78, singer, band leader.
Tito Puente, 1923-2000, jazz percussionist, band leader
Don Pullen, 1942-95, piano; percussive pianist
Sun Ra, 1915?-93, bandleader, pianist, composer
Gertrude "Ma" Rainey, 1886-1939, blues singer
Don Redman, 1900-64, composer, arranger
Django Reinhardt, 1910-53, guitar; influenced Amer. jazz
Buddy Rich, 1917-87, drums, band leader
Red Rodney, 1928-94, trumpeter
Frank Rosollino, 1926-78, trombone
Jimmy Rowles, 1918-96, jazz composer, accompanist
Jimmy Rushing, 1903-72, blues singer
Pee Wee Russell, 1906-69, clarinet
Zoot Sims, 1925-85, tenor, alto sax, clarinet
Zutty Singleton, 1898-1975, Dixieland drummer
Bessie Smith, 1894-1937, blues singer
Clarence "Pinetop" Smith, 1904-29, piano, singer; pioneer of boogie woogie
Willie "The Lion" Smith, 1897-1973, stride style pianist
Muggsy Spanier, 1906-67, cornet, band leader
Billy Strayhorn, 1915-67, composer, piano
Sonny Stitt, 1924-82, alto, tenor sax
Art Tatum, 1910-56, piano; technical virtuoso
Art Taylor, 1929-95, jazz drummer, bandleader
Jack Teagarden, 1905-64, trombone, singer
Mel Torme, 1925-99, "Velvet Fog", singer
Dave Tough, 1908-48, drums
Lennie Tristano, 1919-78, piano, composer
Joe Turner, 1911-85, blues singer
Sarah Vaughan, 1924-90, singer
Joe Venuti, 1904-78, first great jazz violinist
T-Bone Walker, 1910-75, guitarist; electric blues guitar
Thomas "Fats" Waller, 1904-43, piano, singer, composer
Dinah Washington, 1924-63, singer
Grover Washington Jr., 1943-99, jazz sax, composer
Ethel Waters, 1896-1977, jazz and blues singer
Muddy Waters, 1915-83, blues singer, songwriter
Johnny Watson, 1935-96, rhythm and blues guitarist
Chick Webb, 1902-39, band leader, drums
Ben Webster, 1909-73, tenor sax
Junior Wells, 1934-98, blues singer, harmonica
Paul Whiteman, 1890-1967, jazz orchestra leader
Charles "Cootie" Williams, 1908-85, trumpet, band leader
Mary Lou Williams, 1914-81, piano, composer
John Lee "Sonny Boy" Williamson, 1914-48, blues singer, harmonica virtuoso
Sonny Boy Williamson ("Rice" Miller), 1900?-65, Delta bluesman, singer, songwriter, harmonica
Teddy Wilson, 1912-86, piano, composer
Kai Winding, 1922-83, trombone, composer
Jimmy Yancey, 1894-1951, piano
Lester "Pres" Young, 1909-59, tenor sax, composer

Noted Country Music Artists of the Past and Present

Roy Acuff, 1903-92, fiddler, singer, songwriter; "Wabash Cannon Ball"
Alabama (Randy Owen, 1949- ; Jeff Cook, 1949- ; Teddy Gentry, 1952- ; Mark Herndon, 1955-) "Feels So Right"
Eddy Arnold, 1918- , singer, guitarist, the "Tennessee Plowboy"
Chet Atkins, 1924-2001, guitarist, composer, producer, helped create the "Nashville sound"
Gene Autry, 1907-98, first great singing movie cowboy; "Back in the Saddle Again"
Garth Brooks, 1962- , singer, songwriter; "Friends in Low Places"
Brooks & Dunn (Kix Brooks, 1955- ; Ronnie Dunn, 1953-) "Hard Workin' Man"
Boudleaux and Felice Bryant (Boudleau, 1920-87; Felice, 1925-), songwriting team; "Hey Joe"
Glen Campbell, 1936- , singer, instrumentalist, TV host; "Gentle on My Mind," "Rhinestone Cowboy"
Mary Chapin Carpenter, 1958- , singer, songwriter; "I Feel Lucky"
Carter Family (original members,"Mother" Maybelle 1909-78; A.P., 1891-1960, Sara, 1898-1979) "Wildwood Flower"
Johnny Cash, 1932- , singer, songwriter; "I Walk the Line," "Ring of Fire," "Folsom Prison Blues"
Patsy Cline, 1932-63, singer; "Walkin' After Midnight," "Crazy," "Sweet Dreams"
John Denver, 1943-97, singer, songwriter; "Rocky Mountain High"
Dixie Chicks (Natalie Maines, 1974- ; Martie Seidel, 1969- ; Emily Erwin Robison, 1972-) "Wide Open Spaces," "Fly"
Dale Evans (Lucille Wood Smith), 1912-2001, singer, actress, married Roy Rogers
Flatt & Scruggs (Lester Flatt, 1914-79; Earl Scruggs, 1924-), guitar-banjo duo and soloists; "Foggy Mountain Breakdown"
Red Foley, 1910-68, singer; "Chattanoogie Shoe Shine Boy"
Tennessee Ernie Ford, 1919-91, singer, TV host; "Sixteen Tons"
Lefty Frizzell, 1928-75, singer, guitarist; "Long Black Veil"
Vince Gill, 1957- , singer, songwriter; "When I Call Your Name"
Merle Haggard, 1937- , singer, songwriter; "Okie from Muskogee"
Emmylou Harris, 1947- , singer, songwriter, folk-country crossover artist; "If I Could Only Win Your Love"
Faith Hill, 1967-, singer, songwriter, married Tim McGraw; "Wild One," "This Kiss," "Breathe"
Waylon Jennings, 1937– , singer, songwriter, "outlaw country" pioneer; "Luckenbach, Texas"
George Jones, 1931- , singer, songwriter; "Why Baby Why"
The Judds (Naomi, 1946- ; Wynonna, 1964-), mother-daughter duo; Wynonna also a solo act
Alison Krauss, 1971– , bluegrass fiddler, singer, bandleader; "When You Say Nothing at All"
Kris Kristofferson, 1936- , singer, songwriter, actor; "Me and Bobby McGee"
Brenda Lee, 1944- , singer; "I'm Sorry"
Patty Loveless, 1957- , singer, songwriter; "How Can I Help You Say Goodbye"
Lyle Lovett, 1957- , singer, songwriter, bandleader, actor; "Cowboy Man"
Loretta Lynn, 1935- , singer, songwriter; "Coal Miner's Daughter"
Kathy Mattea, 1959- , singer, songwriter; "Eighteen Wheels and a Dozen Roses"
Reba McEntire, 1955- , singer, songwriter, actress; "Whoever's in New England"
Tim McGraw, 1967- , singer; "It's Your Love," with wife, Faith Hill
Roger Miller, 1936-92, singer, songwriter; "King of the Road"
Ronnie Milsap, 1944- , singer, songwriter; "There's No Gettin' Over Me"
Bill Monroe, 1911-96, singer, songwriter, mandolin player, "father of bluegrass music"; "Mule Skinner Blues"
Patsy Montana, 1908-96, yodeling/singing cowgirl; "I Want to Be a Cowboy's Sweetheart"
Willie Nelson, 1933- , singer, songwriter, actor; "On the Road Again"
Mark O'Connor, 1961- , fiddler, country-classical crossover composer
Dolly Parton, 1946- , singer, songwriter, actress; "Dollywood" theme park; "Here You Come Again," "9 to 5"
Minnie Pearl, 1912-96, comedienne, Grand Ole Opry star
Charley Pride, 1938- , singer, 1st African-American country star; "Kiss an Angel Good Mornin'"
Jim Reeves, 1923-64, singer, songwriter; "Four Walls"
Charlie Rich, 1932-95, singer, songwriter called the "Silver Fox"; "The Most Beautiful Girl"
LeAnn Rimes, 1982- , singer; "Blue"
Tex Ritter, 1905-74, singer, songwriter; "Jingle, Jangle, Jingle"
Marty Robbins, 1925-82, singer, songwriter; "A White Sport Coat and a Pink Carnation"
Jimmie Rodgers, 1897-1933, singer, songwriter; "T for Texas"
Kenny Rogers, 1938- , singer, songwriter, actor; "The Gambler"
Roy Rogers (Leonard Slye), 1911-98, singer, actor, "King of the Cowboys," sang with Sons of the Pioneers.
Fred Rose, 1898-1954, songwriter, singer, producer; "Blue Eyes Cryin' in the Rain"
Ricky Skaggs, 1954- , singer, songwriter, bandleader; "Don't Cheat in Our Hometown"
George Strait, 1952- , singer, bandleader; "Ace in the Hole"
Merle Travis, 1917-83, singer, guitarist, songwriter; "Divorce Me C.O.D."
Randy Travis, 1959- , singer, songwriter; "Forever and Ever, Amen"
Ernest Tubb, 1914-84, singer, songwriter, guitarist; "Walking the Floor Over You"
Shania Twain, 1965- , singer, songwriter; "You're Still the One"
Conway Twitty, 1933-93, singer, songwriter; "Hello Darlin' "
Dottie West, 1932-91, singer, songwriter; "Here Comes My Baby"
Hank Williams Jr., 1949- , singer, songwriter; "Bocephus"; "All My Rowdy Friends (Have Settled Down)"
Hank Williams Sr., 1923-53, singer, songwriter; "Your Cheatin' Heart"
Bob Wills, 1905-75, Western Swing fiddler, singer, bandleader, songwriter; "New San Antonio Rose"
Tammy Wynette, 1942-98, singer; "Stand By Your Man"
Trisha Yearwood, 1964- , singer, songwriter; "How Do I Live"
Dwight Yoakam, 1957- , singer, songwriter, actor; "Ain't That Lonely Yet"

Dance Figures of the Past

Source: Reviewed by Gary Parks, Reviews editor, *Dance* magazine

Alvin Ailey, 1931-89, (U.S.) modern dancer, choreographer; melded modern dance and Afro-Caribbean techniques.
Frederick Ashton, 1904-88, (Br.) ballet choreographer; director of Great Britain's Royal Ballet, 1963-70.
Fred Astaire, 1899-1987, (U.S.) dancer, actor; teamed with dancer/actress **Ginger Rogers** (1911-95) in movie musicals.
George Balanchine, 1904-83, (Russ.-U.S.) ballet choreographer, teacher; most influential exponent of the neoclassical style; founded, with Lincoln Kirstein, School of American Ballet and New York City Ballet.
Carlo Blasis, 1803-78, (It.) ballet dancer, choreographer, writer; his teaching methods are standards of classical dance.
August Bournonville, 1805-79, (Dan.) ballet dancer, choreographer, teacher; exuberant, light style.
Gisella Caccialanza, 1914-97, (U.S.) ballerina, charter member of Balanchine's American Ballet.
Enrico Cecchetti, 1850-1928, (It.) ballet dancer, leading dancer of Russia's Imperial Ballet; his technique was basis for Britain's Imperial Soc. of Teachers of Dancing.
Gower Champion, 1921-80, (U.S.) dancer, choreographer, director; with his wife **Marge,** b 1923, (U.S.) choreographed, danced in Broadway musicals and films.
John Cranko, 1927-73, (S. African) choreographer; created narrative ballets based on literary works.
Agnes de Mille, 1909-93, (U.S.) ballerina, choreographer; known for using American themes, she choreographed the ballet *Rodeo* and the musical *Oklahoma*.
Dame Ninette DeValois, 1898-2001, (Br.) choreographer, founding director London's Royal Ballet; *The Rake's Progress.*
Sergei Diaghilev, 1872-1929, (Russ.) impresario; founded Les Ballet Russes; saw ballet as an art unifying dance, drama, music, and decor.
Alexandra Danilova, 1903-97, (Russ.) ballerina; noted teacher at the School of American Ballet.
Isadora Duncan, 1877-1927, (U.S.) expressive dancer who united free movement with serious music; one of the founders of modern dance.
Fanny Elssler, 1810-84, (Austrian) ballerina of the Romantic era; known for dramatic skill, sensual style.
Michel Fokine, 1880-1942, (Russ.) ballet dancer, choreographer, teacher; rejected strict classicism in favor of dramatically expressive style.
Margot Fonteyn, 1919-91, (Br.) prima ballerina, Royal Ballet of Great Britain; famed performance partner of Rudolf Nureyev.
Bob Fosse, 1927-87, (U.S.) jazz dancer, choreographer, director; Broadway musicals and film.
Serge Golovine, 1924-98, (Fr.) ballet dancer with Grand Ballet du Marquis de Cuevas; choreographer.
Martha Graham, 1893-1991, (U.S.) modern dancer, choreographer; created and codified her own dramatic technique.
Martha Hill, 1901-95, (U.S.) educator; leading figure in modern dance; founded American Dance Festival.
Doris Humphrey, 1895-1958, (U.S.) modern dancer, choreographer, writer, teacher.
Robert Joffrey, 1930-88, (U.S.) ballet dancer, choreographer; cofounded with **Gerald Arpino,** b 1928, (U.S.), the Joffrey Ballet.
Kurt Jooss, 1901-79, (Ger.) choreographer, teacher; created expressionist works using modern and classical techniques.
Tamara Karsavina, 1885-1978, (Russ.) prima ballerina of Russia's Imperial Ballet and Diaghilev's Ballets Russes; partner of Nijinsky.
Nora Kaye, 1920-87, (U.S.) ballerina with Metropolitan Opera Ballet and Ballet Theater (now American Ballet Theatre).
Lincoln Kirstein, 1907-96 (U.S.) brought ballet as an art form to U.S.; founded, with George Balanchine, School of American Ballet and New York City Ballet.
Serge Lifar, 1905-86, (Russ.-Fr.) prem. danseur, choreographer; director of dance at Paris Opera, 1930-45, 1947-58.
José Limón, 1908-72, (Mex.-U.S.) modern dancer, choreographer, teacher; developed technique based on Humphrey.
Catherine Littlefield, 1908-51, (U.S.) ballerina, choreographer, teacher; pioneer of American ballet.
Léonide Massine, 1896-1979, (Russ.-U.S.) ballet dancer, choreographer; his "symphonic ballet" used concert music previously thought unsuitable for dance.
Kenneth MacMillan, 1929-92, (Br.) dancer, choreographer; directed Royal Ballet of Great Britain 1970-77.
Vaslav Nijinsky, 1890-50, (Russ.) prem. danseur, choreographer; leading member of Diaghilev's Ballets Russes; his ballets were revolutionary for their time.
Alwin Nikolais, 1910-93, (U.S.) modern choreographer; created dance theater utilizing mixed media effects.
Jean-George Noverre, 1727-1810, (Fr.) ballet choreographer, teacher, writer; "Shakespeare of the Dance."
Rudolf Nureyev, 1938-93, (Russ.) prem. danseur, choreographer; leading male dancer of his generation; director of dance at Paris Opera, 1983-89.
Ruth Page, 1903-91, (U.S.) ballerina, choreographer; danced and directed ballet at Chicago Lyric Opera.
Anna Pavlova, 1881-1931, (Russ.) prima ballerina; toured with her own company to world acclaim.
Marius Petipa, 1818-1910, (Fr.) ballet dancer, choreographer; ballet master of the Imperial Ballet; established Russian classicism as leading style of late 19th cent.
Pearl Primus, 1919-95, (Trinidad-U.S.) modern dancer, choreographer, scholar; combined African, Caribbean, and African-American styles.
Jerome Robbins, 1918-98, (U.S.) choreographer, director, dancer; *The King and I, West Side Story, Fiddler on the Roof; Gypsy.*
Bill (Bojangles) Robinson, 1878-1949, (U.S.) famed tap dancer; called King of Tapology on stage and screen.
Ruth St. Denis, 1877-1968, (U.S.) influential interpretive dancer, choreographer, teacher.
Ted Shawn, 1891-1972, (U.S.) modern dancer, choreographer; formed dance company and school with Ruth St. Denis; established Jacob's Pillow Dance Festival.
Marie Taglioni, 1804-84, (It.) ballerina, teacher; in title role of *La Sylphide* established image of the ethereal ballerina.
Antony Tudor, 1908-87, (Br.) choreographer, teacher; exponent of the "psychological ballet."
Galina Ulanova, 1910-98, (Russ.) revered ballerina with Bolshoi Ballet.
Agrippina Vaganova, 1879-1951, (Russ.) ballet teacher, director; codified Soviet ballet technique that developed virtuosity; called "queen of variations."
Mary Wigman, 1886-1973, (Ger.) modern dancer, choreographer, teacher; influenced European expressionist dance.

Opera Singers of the Past

Frances Alda, 1883-1952, (NZ) soprano
Paul Althouse, 1889-1954, (U.S.) tenor
Pasquale Amato, 1878-1942, (It.) baritone
Marian Anderson, 1897-1993, (U.S.) contralto
Jussi Björling, 1911-60, (Swed.) tenor
Lucrezia Bori, 1887-1960, (It.) soprano
Maria Callas, 1923-77, (U.S.) soprano
Emma Calvé, 1858-1942, (Fr.) soprano
Enrico Caruso, 1873-1921, (It.) tenor
Feodor Chaliapin, 1873-1938, (Russ.) bass
Boris Christoff, 1914-93, (Bulg.) bass
Richard Crooks, 1900-72, (U.S.) tenor
Giuseppe De Luca, 1876-1950, (It.) baritone
Edouard De Reszke, 1853-1917, (Pol.) bass
Jean De Reszke, 1850-1925, (Pol.) tenor
Emmy Destinn, 1878-1930, (Czech.) soprano
Todd Duncan, 1903-98, (U.S.) baritone
Emma Eames, 1865-1952, (U.S.) soprano
Geraldine Farrar, 1882-1967, (U.S.) soprano
Kirsten Flagstad, 1895-1962, (Nor.) soprano
Olive Fremstad, 1871-1951, (Swed.-U.S.) soprano
Amelita Galli-Curci, 1882-1963, (It.) soprano
Mary Garden, 1874-1967, (Br.) soprano
Beniamino Gigli, 1890-1957, (It.) tenor
Tito Gobbi, 1913-84, (It.) baritone
Frieda Hempel, 1885-1955, (Ger.) soprano
Maria Jeritza, 1887-1982, (Czech.) soprano
Alexander Kipnis, 1891-1978, (Russ.-U.S.) bass
Lilli Lehmann, 1848-1929, (Ger.) soprano
Lotte Lehmann, 1888-1976, (Ger.-U.S.) soprano
Jenny Lind, 1820-87, (Swed.) soprano
John McCormack, 1884-1945, (Ir.) tenor
Blanche Marchesi, 1863-1940, (Fr.) soprano
Nellie Melba, 1861-1931, (Austral.) soprano.
Lauritz Melchior, 1890-1973, (Dan.) tenor
Zinka Milanov, 1906-89, (Yugo.) soprano
Lillian Nordica, 1857-1914, (U.S.) soprano
Adelina Patti, 1843-1919, (It.) soprano
Peter Pears, 1910-86, (Eng.) tenor
Jan Peerce, 1904-84, (U.S.) tenor
Ezio Pinza, 1892-1957, (It.) bass
Lily Pons, 1898-1976, (Fr.) soprano
Rosa Ponselle, 1897-1981, (U.S.) soprano
Hermann Prey, 1929-98, (Ger.) baritone.
Marcella Sembrich, 1858-1935, (Pol.) soprano
Eleanor Steber, 1916-90, (U.S.) soprano
Ferrucio Tagliavini, 1913-95, (It.) tenor
Luisa Tetrazzini, 1871-1940, (It.) soprano
Lawrence Tibbett, 1896-1960, (U.S.) baritone
Richard Tucker, 1913-75, (U.S.) tenor
Pauline Viardot, 1821-1910, (Fr.) mezzo-soprano
Leonard Warren, 1911-60, (U.S.) baritone

Rock and Roll, Rhythm and Blues, and Rap Artists

Titles in quotation marks are singles; others are albums.

AC/DC: "Back in Black"
Bryan Adams: "Cuts Like a Knife"
***Aerosmith (2001):** "Sweet Emotion"
Alice In Chains: "Heaven Beside You"
***The Allman Brothers Band (1995):** "Ramblin' Man"
***The Animals (1994):** "House of the Rising Sun"
Paul Anka: "Lonely Boy"
Fiona Apple: "Criminal"
The Association: "Cherish"
Frankie Avalon: "Venus"
The B-52s: "Love Shack"
Bachman Turner Overdrive: "Takin' Care of Business"
Backstreet Boys: "I Want it That Way"
Bad Company: "Can't Get Enough"
Erykah Badu: "On and On"
***La Vern Baker (1991):** "I Cried a Tear"
***Hank Ballard and the Midnighters (1990):** "Work With Me, Annie"
***The Band (1994):** "The Weight"
Barenaked Ladies: "One Week"
***The Beach Boys (1988):** "Good Vibrations"
Beastie Boys: "(You Gotta) Fight for Your Right (to Party)"
***The Beatles (1988):** *Sgt. Pepper's Lonely Hearts Club Band*
Beck: "Loser"
***The Bee Gees (1997):** "Stayin' Alive"
Ben Folds Five: "Brick"
***Chuck Berry (1986):** "Johnny B. Goode"
Pat Benatar: "Hit Me With Your Best Shot"
The Big Bopper: "Chantilly Lace"
Björk: "Human Behavior"
The Black Crowes: "Hard to Handle"
Black Sabbath: "Paranoid"
***Bobby "Blue" Bland (1992):** "Turn On Your Love Light"
Mary J. Blige: *My Life*
Blind Faith: "Can't Find My Way Home"
Blink-182: "All the Small Things"
Blondie: "Heart of Glass"
Blood, Sweat, and Tears: "Spinning Wheel"
Blues Traveler: "Run-Around"
Gary "U.S." Bonds: "Quarter to Three"
Bon Jovi: "Livin' on a Prayer"
***Booker T. and the M.G.'s (1992):** "Green Onions"
Earl Bostic: "Flamingo"
Boston: "More Than A Feeling"
***David Bowie (1996):** "Space Oddity"
Boyz II Men: "I'll Make Love to You"
Toni Braxton: "Un-Break My Heart"
***James Brown (1986):** "Papa's Got a Brand New Bag"
***Ruth Brown (1993):** "Lucky Lips"
Jackson Browne: "Doctor My Eyes"
***Buffalo Springfield (1997):** "For What It's Worth"
***Jimmy Buffet:** "Margaritaville"
***Solomon Burke (2001):** "Over and Over (Huggin' and Lovin')"
Bush: "Glycerine"
***The Byrds (1991):** "Turn! Turn! Turn!"
Mariah Carey: "Vision of Love"
The Cars: "Shake It Up"
***Johnny Cash (1992):** "I Walk the Line"
***Ray Charles (1986):** "Georgia on My Mind"
Cheap Trick: "Surrender"
Chicago: "Saturday in the Park"
Chubby Checker: "The Twist"
***Eric Clapton (2000):** "Layla"
The Clash: "Rock the Casbah"
***The Coasters (1987):** "Yakety Yak"
***Eddie Cochran (1987):** "Summertime Blues"
Joe Cocker: "With a Little Help From My Friends"
Collective Soul: "The World I Know"
Phil Collins: "Against All Odds"
***Sam Cooke (1986):** "You Send Me"
Coolio: "Gangsta's Paradise"
Alice Cooper: "School's Out"
Elvis Costello: "Alison"
Counting Crows: "Mr. Jones"
***Cream (1993):** "Sunshine of Your Love"
Creed: "Arms Wide Open"
***Creedence Clearwater Revival (1993):** "Proud Mary"
***Crosby, Stills, and Nash (1997):** "Suite: Judy Blue Eyes"
Sheryl Crow: "All I Want to Do"
The Cure: "Boys Don't Cry"
The Crystals: "Da Doo Ron Ron"
Cypress Hill: "Insane in the Brain"
Danny and the Juniors: "At the Hop"
***Bobby Darin (1990):** "Splish Splash"
Spencer Davis Group: "Gimme Some Lovin' "
Deep Purple: "Smoke on the Water"
Def Leppard: "Photograph"
Depeche Mode: "Strange Love"
***Bo Diddley (1987):** "Who Do You Love?"
***Dion and the Belmonts (1989):** "A Teenager in Love"
Celine Dion: "Because You Loved Me"
Dire Straits: "Money for Nothing"
DMX: "What's My Name"
***Fats Domino (1986):** "Blueberry Hill"
Donovan: "Mellow Yellow"
The Doobie Brothers: "What a Fool Believes"
***The Doors (1993):** "Light My Fire"
Dr. Dre: "Nothin' But a 'G' Thang"
***The Drifters (1988):** "Save the Last Dance for Me"
Duran Duran: "Hungry Like the Wolf"
***Bob Dylan (1988):** "Like a Rolling Stone"
***The Eagles (1998):** "Hotel California"
***Earth, Wind, and Fire (2000):** "Shining Star"
***Duane Eddy (1994):** "Rebel-Rouser"
Missy Elliott: "Sock It 2 Me"
Emerson, Lake, and Palmer: "Lucky Man"
Eminem: "The Real Slim Shady"
En Vogue: "Hold On"
The Eurythmics: "Sweet Dreams (Are Made of This)"
Everclear: "Father Of Mine"
***The Everly Brothers (1986):** "Wake Up, Little Susie"
The Five Satins: "In the Still of the Night"
***The Flamingos (2001):** "I Only Have Eyes for You"
***Fleetwood Mac (1998):** *Rumours*
The Foo Fighters: "I'll Stick Around"
Foreigner: "Double Vision"
***The Four Seasons (1990):** "Sherry"
***The Four Tops (1990):** "I Can't Help Myself (Sugar Pie, Honey Bunch)"
***Aretha Franklin (1987):** "Respect"
Peter Gabriel: "Shock the Monkey"
Marvin Gaye (1987): "I Heard It Through the Grapevine"
Genesis: "No Reply at All"
Grand Funk Railroad: "We're an American Band"
Grand Master Flash and the Furious Five: "The Message"
***The Grateful Dead (1994):** "Uncle John's Band"
Macy Gray: "I Try"
***Al Green (1995):** "Let's Stay Together"
Green Day: "Time of Your Life"
The Guess Who: "American Woman"
Guns N' Roses: "Sweet Child o' Mine"
***Bill Haley and His Comets (1987):** "Rock Around the Clock"
Hall and Oates: "Kiss on My List"
Hanson: "MMMBop"
Juliana Hatfield: "Spin the Bottle"
Heart: "Barracuda"
Lauryn Hill: "Doo-Wop (That Thing)"
Hole: "Doll Parts"
Hootie and the Blowfish: *Cracked Rear View*
Whitney Houston: "I Will Always Love You"
***The Impressions (1991):** "For Your Precious Love"
Indigo Girls: "Closer to Fine"
INXS: "Need You Tonight"
***The Isley Brothers (1992):** "It's Your Thing"
***The Jackson Five (1997):** "ABC"
Janet Jackson: *Rhythm Nation*
***Michael Jackson (2000):** *Thriller*
***Etta James (1993):** "At Last"
Tommy James & The Shondells: "Crimson and Clover"
Jane's Addiction: "Jane Says"
Jay and the Americans: "This Magic Moment"
Jay-Z: "Can I Live"
***Jefferson Airplane (1996):** "White Rabbit"
Jethro Tull: *Aqualung*
Joan Jett: "I Love Rock 'n' Roll"
Jewel: "You Were Meant for Me"
***Billy Joel (1999):** "Piano Man"
***Elton John (1994):** "Candle in the Wind"
***Little Willie John (1996):** "Sleep"
***Janis Joplin (1995):** "Me and Bobby McGee"
Journey: "Don't Stop Believin'"
K.C. and the Sunshine Band: "Get Down Tonight"
R. Kelly: "I Can't Sleep Baby (If I)"
Kid Rock: "Cowboy"

*B.B. King (1987): "The Thrill Is Gone"
Carole King: *Tapestry*
*The Kinks (1990): "You Really Got Me"
Kiss: "Rock 'n' Roll All Night"
*Gladys Knight and the Pips (1996): "Midnight Train to Georgia"
Korn: "Blind"
Lenny Kravitz: "Are You Gonna Go My Way?"
*Led Zeppelin (1995): "Stairway to Heaven"
Brenda Lee: "I'm Sorry"
*John Lennon (1994): "Imagine"
*Jerry Lee Lewis (1986): "Whole Lotta Shakin' Going On"
Lil' Kim: "No Matter What They Say"
Limp Bizkit: "Break Stuff"
Little Anthony and the Imperials: "Tears on My Pillow"
*Little Richard (1986): "Tutti Frutti"
Live: "Lightning Crashes"
L. L. Cool J: "Mama Said Knock You Out"
*The Lovin' Spoonful (2000): "Summer in the City"
*Frankie Lymon and the Teenagers (1993): "Why Do Fools Fall in Love?"
Lynyrd Skynyrd: "Free Bird"
Madonna: "Material Girl"
*The Mamas and the Papas (1998): "Monday, Monday"
Aimee Mann: "Save Me"
Marilyn Manson: "Beautiful People"
*Bob Marley (1994): *Exodus*
*Martha and the Vandellas (1995): "Dancin' in the Streets"
The Marvelettes: "Please, Mr. Postman"
Matchbox 20: "Push"
Dave Matthews Band: "Don't Drink the Water"
*Curtis Mayfield (1999): "Superfly"
*Paul McCartney (1999): "Band on the Run"
Don McLean: "American Pie"
*Clyde McPhatter (1987): "A Lover's Question"
Meat Loaf: "Paradise by the Dashboard Light"
John (Cougar) Mellencamp: "Jack and Diane"
Men at Work: "Who Can It Be Now?"
Metallica: "Enter Sandman"
George Michael: "Faith"
*Joni Mitchell (1997): "Big Yellow Taxi"
Moby: "Bodyrock"
The Monkees: "I'm a Believer"
Moody Blues: "Nights in White Satin"
*The Moonglows (2000): "Blue Velvet"
Alanis Morissette: "Ironic"
*Van Morrison (1993): "Brown-Eyed Girl"
*Ricky Nelson (1987): "Hello, Mary Lou"
Nine Inch Nails: "Closer"
Nirvana: *Nevermind*
The Notorious B.I.G.: "Mo Money Mo Problems"
'N Sync: "Bye, Bye, Bye"
Oasis: "Wonderwall"
The Offspring: "Pretty Fly (for a White Guy)"
*Roy Orbison (1987): "Oh, Pretty Woman"
Ozzy Osbourne: "Crazy Train"
*Parliament/Funkadelic (1997): "One Nation Under a Groove"
Pearl Jam: "Jeremy"
*Carl Perkins (1987): "Blue Suede Shoes"
Peter, Paul, and Mary: "Leaving on a Jet Plane"
Tom Petty and the Heartbreakers: "Refugee"
Liz Phair: *Exile in Guyville*
Phish: "Sample in a Jar"
*Wilson Pickett (1991): "Land of 1,000 Dances"
*Pink Floyd (1996): *The Wall*
*The Platters (1990): "The Great Pretender"
Poco: "Crazy Love"
The Police: "Every Breath You Take"
Iggy Pop: "Lust for Life"
*Elvis Presley (1986): "Love Me Tender"
The Pretenders: "Brass in Pocket"
*Lloyd Price (1998): "Stagger Lee"
Prince (The Artist): "Purple Rain"
Procol Harum: "A Whiter Shade of Pale"
Public Enemy: "Fight the Power"
Puff Daddy and the Family: *No Way Out*
*Queen (2000): "Bohemian Rhapsody"
Radiohead: "Creep"
Rage Against the Machine: "Bulls on Parade"
*Bonnie Raitt (2000): "Something to Talk About"
The Ramones: "I Wanna Be Sedated"
*Otis Redding (1989): "(Sittin' on) the Dock of the Bay"
Red Hot Chili Peppers: "Under the Bridge"
*Jimmy Reed (1991): "Ain't That Loving You, Baby?"
Lou Reed: "Walk on the Wild Side"
R.E.M.: "Losing My Religion"
REO Speedwagon: "Can't Fight This Feeling"
Busta Rhymes: "What's It Gonna Be?"
The Righteous Brothers: "You've Lost That Lovin' Feelin' "
Johnny Rivers: "Poor Side of Town"
*Smokey Robinson and the Miracles (1987): "Shop Around"
*The Rolling Stones (1989): "Satisfaction"
The Ronettes: "Be My Baby"
Linda Ronstadt: "You're No Good"
Run-D.M.C.: "Raisin' Hell"
Rush: "Tom Sawyer"
Sade: "Smooth Operator"
Salt-N-Pepa: "Shoop"
*Sam and Dave (1992): "Soul Man"
*Santana (1998): "Black Magic Woman"
Seal: "Kiss From a Rose"
Neil Sedaka: "Breaking Up Is Hard to Do"
The Sex Pistols: "Anarchy in the U.K."
Tupac Shakur: "How Do U Want It"
*Del Shannon (1999): "Runaway"
*The Shirelles (1996): "Soldier Boy"
Carly Simon: "You're So Vain"
*Paul Simon (2001): "50 Ways to Leave Your Lover"
*Simon and Garfunkel (1990): "Bridge Over Troubled Water"
Sisqo: "Thong Song"
*Sly and the Family Stone (1993): "Everyday People"
Smashing Pumpkins: "Today"
Patti Smith: "Because the Night"
Will Smith: "Gettin' Jiggy With It"
The Smiths: "This Charming Man"
Snoop Dogg: "Gin and Juice"
Sonic Youth: "Bull in the Heather"
Soundgarden: "Black Hole Sun"
Britney Spears: "Hit Me Baby One More Time"
Spice Girls: "Wannabe"
*Dusty Springfield (1999): "I Only Want to Be With You"
*Bruce Springsteen (1999): "Born to Run"
Squeeze (2001): "Tempted"
*Staple Singers (1999): "I'll Take You There"
*Steely Dan (2001): "Rikki Don't Lose That Number"
Steppenwolf: "Born to Be Wild"
*Rod Stewart (1994): "Maggie Mae"
Sting: "If You Love Somebody, Set Them Free"
Styx: "Come Sail Away"
Sublime: "What I Got"
The Sugar Hill Gang: "Rapper's Delight"
Donna Summer: "Bad Girls"
*The Supremes (1988): "Stop! In the Name of Love"
Talking Heads: "Once in a Lifetime"
*James Taylor (2001): "You've Got a Friend"
*The Temptations (1989): "My Girl"
Three Dog Night: "Joy to the World"
TLC: "Waterfalls"
T. Rex: "Bang a Gong (Get It On)"
*Big Joe Turner (1987): "Shake, Rattle & Roll"
*Ike and Tina Turner (1991): "Proud Mary"
*Tina Turner: "What's Love Got to Do With It?"
The Turtles: "Happy Together"
U2: "With or Without You"
Usher: "You Make Me Wanna"
*Ritchie Valens (2001): "La Bamba"
Van Halen: "Running With the Devil"
Stevie Ray Vaughan: "Crossfire"
*The Velvet Underground (1996): "Sweet Jane"
*Gene Vincent[1] (1998): "Be-Bop-A-Lula"
Tom Waits: "Downtown Train"
The Wallflowers: "One Headlight"
Dionne Warwick: "I Say a Little Prayer"
*Muddy Waters (1987): "I Can't Be Satisfied"
Mary Wells: "My Guy"
*The Who (1990): *Tommy*
*Jackie Wilson (1987): "That's Why"
*Stevie Wonder (1989): "You Are the Sunshine of My Life"
Wu-Tang Clan: "Protect Ya Neck"
*The Yardbirds (1992): "For Your Love"
Yes: "Roundabout"
*Neil Young (1995): "Down by the River"
*The Young Rascals/The Rascals (1997): "Good Lovin' "
*Frank Zappa[1]/Mothers of Invention (1995): *Sheik Yerbouti*
ZZ Top: "Legs"

* Inducted into Rock and Roll Hall of Fame as performer between 1986 and 2001; year is in parentheses. (1) Only individual performer is in Rock and Roll Hall of Fame.

Entertainment Personalities of the Present

Living actors, musicians, dancers, singers, producers, directors, radio-TV performers.

Name	Birthplace	Birthdate
Abbado, Claudio	Milan, Italy	6/26/33
Abdul, Paula	San Fernando, CA	6/19/62
Abraham, F. Murray	Pittsburgh, PA	10/24/39
Adams, Bryan	Kingston, Ontario	11/5/59
Adams, Don	New York, NY	4/19/26
Adams, Edie	Kingston, PA	4/16/29
Adams, Mason	New York, NY	2/26/19
Adjani, Isabelle	Paris, France	6/27/55
Affleck, Ben	Berkeley, CA	8/15/72
Agar, John	Chicago, IL	1/31/21
Agutter, Jenny	London, England	12/20/52
Aiello, Danny	New York, NY	6/20/33
Aimee, Anouk	Paris, France	4/27/34
Albanese, Licia	Bari, Italy	7/22/13
Alberghetti, Anna Maria	Pesaro, Italy	5/15/36
Albert, Eddie	Rock Island, IL	4/22/08
Albert, Marv	New York, NY	6/12/43
Alda, Alan	New York, NY	1/28/36
Alexander, Jane	Boston, MA	10/28/39
Alexander, Jason	Newark, NJ	9/23/59
Allen, Debbie	Houston, TX	1/16/50
Allen, Joan	Rochelle, IL	8/20/56
Allen, Karen	Carrollton, IL	10/5/51
Allen, Steve	New York, NY	12/26/21
Allen, Tim	Denver, CO	6/13/53
Allen, Woody	Brooklyn, NY	12/1/35
Alley, Kirstie	Wichita, KS	1/12/51
Allman, Gregg	Nashville, TN	12/7/47
Allyson, June	New York, NY	10/7/17
Alonso, Maria Conchita	Cienfuegos, Cuba	6/29/57
Alpert, Herb	Los Angeles, CA	3/31/35
Altman, Robert	Kansas City, MO	2/20/25
Almodóvar, Pedro	Calzada de Calatrava, Spain	9/25/51
Ames, Ed	Boston, MA	7/9/27
Amos, John	Newark, NJ	12/27/42
Amos, Tori	North Carolina	8/22/64
Anderson, Gillian	Chicago, IL	8/9/68
Anderson, Harry	Newport, RI	10/14/49
Anderson, Ian	Dunfermline, Scotland	8/10/47
Anderson, Kevin	Illinois	1/13/60
Anderson, Loni	St. Paul, MN	8/5/46
Anderson, Lynn	Grand Forks, ND	9/26/47
Anderson, Melissa Sue	Berkeley, CA	9/26/62
Anderson, Richard	Long Branch, NJ	8/8/26
Anderson, Richard Dean	Minneapolis, MN	1/23/50
Andersson, Bibi	Stockholm, Sweden	11/11/35
Andress, Ursula	Bern, Switzerland	3/19/36
Andrews, Anthony	London, England	1/12/48
Andrews, Julie	Walton, England	10/1/35
Andrews, Patty	Minneapolis, MN	2/16/20
Aniston, Jennifer	Sherman Oaks, CA	2/11/69
Anka, Paul	Ottawa, Ontario	7/30/41
Ann-Margret	Stockholm, Sweden	4/28/41
Antonioni, Michelangelo	Ferrara, Italy	9/29/12
Apple, Fiona	New York, NY	9/13/77
Applegate, Christina	Los Angeles, CA	11/25/72
Archer, Anne	Los Angeles, CA	8/25/47
Arkin, Adam	Brooklyn, NY	8/19/56
Arkin, Alan	New York, NY	3/26/34
Arnaz, Desi, Jr.	Los Angeles, CA	1/19/53
Arnaz, Lucie	Los Angeles, CA	7/17/51
Arness, James	Minneapolis, MN	5/26/23
Arnold, Eddy	Henderson, TN	5/15/18
Arnold, Tom	Ottumwa, IA	3/6/59
Arquette, Patricia	New York, NY	4/8/68
Arquette, Rosanna	New York, NY	8/10/59
Arroyo, Martina	New York, NY	2/2/37
Arthur, Beatrice	New York, NY	5/13/23
Ashley, Elizabeth	Ocala, FL	8/30/41
Asner, Ed	Kansas City, MO	11/15/29
Assante, Armand	New York, NY	10/4/49
Astin, John	Baltimore, MD	3/30/30
Atkinson, Rowan	Newcastle-Upon-Tyne, Eng.	1/6/55
Attenborough, Richard	Cambridge, England	8/29/23
Auberjonois, Rene	New York, NY	6/1/40
Austin, Patti	New York, NY	8/10/48
Autry, Alan	Shreveport, LA	7/31/52
Avalon, Frankie	Philadelphia, PA	9/18/39
Aykroyd, Dan	Ottawa, Ontario	7/1/52
Azaria, Hank	Forest Hills, NY	4/25/64
Aznavour, Charles	Paris, France	5/22/24
Babyface	Indianapolis, IN	4/10/59
Bacall, Lauren	New York, NY	9/16/24
Bacon, Kevin	Philadelphia, PA	7/8/58
Badu, Erykah	Dallas, TX	2/26/71
Baez, Joan	Staten Island, NY	1/9/41
Bain, Conrad	Lethbridge, Alberta	2/4/23
Baio, Scott	Brooklyn, NY	9/22/61
Baker, Anita	Toledo, OH	1/26/58
Baker, Carroll	Johnstown, PA	5/28/31
Baker, Diane	Hollywood, CA	2/25/38
Baker, Joe Don	Groesbeck, TX	2/12/36
Baker, Kathy	Midland, TX	6/8/50
Bakula, Scott	St. Louis, MO	10/9/55
Baldwin, Alec	Massapequa, NY	4/3/58
Baldwin, Daniel	Massapequa, NY	10/5/60
Baldwin, Stephen	Massapequa, NY	5/12/66
Baldwin, William	Massapequa, NY	2/21/63
Bale, Cristian	Pembrokeshire, Wales	1/30/74
Ballard, Kaye	Cleveland, OH	11/20/26
Bancroft, Anne	New York, NY	9/17/31
Banderas, Antonio	Málaga, Spain	8/10/60
Banks, Tyra	Los Angeles, CA	12/4/73
Bannon, Jack	Los Angeles, CA	6/14/40
Baranski, Christine	Buffalo, NY	5/2/52
Bardem, Javier	Canary Islands	5/1/69
Bardot, Brigitte	Paris, France	9/28/34
Barker, Bob	Darrington, WA	12/12/23
Barkin, Ellen	New York, NY	4/16/55
Barrie, Barbara	Chicago, IL	5/23/31
Barry, Gene	New York, NY	6/14/19
Barrymore, Drew	Los Angeles, CA	2/22/75
Bartoli, Cecilia	Rome, Italy	6/4/66
Baryshnikov, Mikhail	Riga, Latvia	1/28/48
Basinger, Kim	Athens, GA	12/8/53
Bass, Lance	Mississippi	5/4/79
Bassett, Angela	New York, NY	8/16/58
Bassey, Shirley	Cardiff, Wales	1/8/37
Bateman, Jason	Rye, NY	1/14/69
Bateman, Justine	Rye, NY	2/19/66
Bates, Alan	Allestree, England	2/17/34
Bates, Kathy	Memphis, TN	6/28/48
Battle, Kathleen	Portsmouth, OH	8/13/48
Baxter, Meredith	Los Angeles, CA	6/21/47
Bean, Orson	Burlington, VT	7/22/28
Beatty, Ned	Louisville, KY	7/6/37
Beatty, Warren	Richmond, VA	3/30/37
Beck (Hansen)	Los Angeles, CA	7/8/70
Beck, Jeff	Surrey, England	6/24/44
Beck, John	Chicago, IL	1/28/43
Beckinsale, Kate	London, England	7/26/73
Bedelia, Bonnie	New York, NY	3/25/48
Begley, Ed, Jr.	Los Angeles, CA	9/16/49
Belafonte, Harry	New York, NY	3/1/27
Bel Geddes, Barbara	New York, NY	10/31/22
Bell, Art	Pahrump, NV	6/17/45
Bello, Maria	Norristown, PA	4/18/67
Belmondo, Jean-Paul	Neuilly-sur-Seine, France	4/9/33
Belushi, Jim	Chicago, IL	6/15/54
Belzer, Richard	Bridgeport, CT	8/4/44
Benatar, Pat	Brooklyn, NY	1/10/53
Benedict, Dirk	Helena, MT	3/1/45
Benigni, Roberto	Misericordia, Italy	10/27/52
Bening, Annette	Topeka, KS	5/29/58
Benjamin, Richard	New York, NY	5/22/38
Bennett, Tony	New York, NY	8/3/26
Benson, George	Pittsburgh, PA	3/22/43
Benson, Robby	Dallas, TX	1/21/56
Berenger, Tom	Chicago, IL	5/31/50
Bergen, Candice	Beverly Hills, CA	5/9/46
Bergen, Polly	Knoxville, TN	7/14/30
Bergman, Ingmar	Uppsala, Sweden	7/14/18
Berle, Milton	New York, NY	7/12/08
Berlinger, Warren	Brooklyn, NY	8/31/37
Berman, Lazar	Leningrad, Russia	2/26/30
Berman, Shelley	Chicago, IL	2/3/26
Bernard, Crystal	Dallas, TX	9/30/64
Bernhard, Sandra	Flint, MI	6/6/55
Bernsen, Corbin	N. Hollywood, CA	9/7/54
Berry, Chuck	St. Louis, MO	10/18/26
Berry, Halle	Cleveland, OH	8/14/68
Berry, Ken	Moline, IL	11/3/33
Bertinelli, Valerie	Wilmington, DE	4/23/60
Bialik, Mayim	San Diego, CA	12/12/75
Bertolucci, Bernardo	Parma, Italy	3/16/41
Biggs, Jason	Pompton Plains, NJ	5/12/78
Bikel, Theodore	Vienna, Austria	5/2/24
Billingsley, Barbara	Los Angeles, CA	12/22/22
Binoche, Juliette	Paris, France	4/9/64
Birney, David	Washington, DC	4/23/39
Bishop, Joey	Bronx, NY	2/3/18
Bisset, Jacqueline	Weybridge, England	9/13/44
Bissett, Josie	Seattle, WA	10/5/69

Name	Birthplace	Birthdate
Björk (Gudmundsdottir)	Rheinberg, Iceland	11/21/65
Black, Clint	Katy, TX	2/4/62
Black, Karen	Park Ridge, IL	7/1/42
Blades, Ruben	Panama City, Panama	7/16/48
Blair, Janet	Altoona, PA	4/23/21
Blair, Linda	St. Louis, MO	1/22/59
Blair, Selma	Southfield, MI	6/23/72
Blake, Robert	Nutley, NJ	9/18/33
Blanchett, Cate	Melbourne, Australia	1969
Bledsoe, Tempestt	Chicago, IL	8/1/73
Bleeth, Yasmine	New York, NY	6/16/68
Blethyn, Brenda	Kent, England	2/20/46
Blige, Mary J.	Bronx, NY	1/11/71
Bloom, Claire	London, England	2/15/31
Blyth, Ann	Mt. Kisco, NY	8/16/28
Bochco, Steven	New York, NY	12/16/43
Bogdanovich, Peter	Kingston, NY	7/30/39
Bogosian, Eric	Boston, MA	4/24/53
Bologna, Joseph	Brooklyn, NY	12/30/38
Bolton, Michael	New Haven, CT	2/26/53
Bonet, Lisa	San Francisco, CA	11/16/67
Bonham Carter, Helena	London, England	5/23/66
Bon Jovi, Jon	Sayreville, NJ	3/2/62
Bono (Vox)	Dublin, Ireland	5/10/60
Boone, Debby	Hackensack, NJ	9/22/56
Boone, Pat	Jacksonville, FL	6/1/34
Boreanaz, David	Buffalo, NY	5/16/71
Borgnine, Ernest	Hamden, CT	1/24/17
Bosson, Barbara	Charleroi, PA	11/1/39
Bosco, Philip	Jersey City, NJ	9/26/30
Bosley, Tom	Chicago, IL	10/1/27
Bostwick, Barry	San Mateo, CA	2/24/45
Bottoms, Timothy	Santa Barbara, CA	8/30/51
Bowie, David	London, England	1/8/47
Boxleitner, Bruce	Elgin, IL	5/12/50
Boy George	London, England	6/14/61
Boyle, Lara Flynn	Davenport, IA	3/24/70
Boyle, Peter	Philadelphia, PA	10/18/33
Bracco, Lorraine	Brooklyn, NY	10/2/55
Bracken, Eddie	New York, NY	2/7/20
Branagh, Kenneth	Belfast, N. Ireland	12/10/60
Brandauer, Klaus Maria	Steiermark, Austria	6/22/44
Brando, Marlon	Omaha, NE	4/3/24
Brandy (Norwood)	McComb, MS	2/11/79
Braschi, Nicoletta	Gesena, Italy	1960
Bratt, Benjamin	San Francisco, CA	12/16/63
Braugher, Andre	Chicago, Il	7/1/62
Braxton, Toni	Severn, MD	10/7/66
Brennan, Eileen	Los Angeles, CA	9/3/35
Brenner, David	Philadelphia, PA	2/4/45
Brewer, Teresa	Toledo, OH	5/7/31
Bridges, Beau	Hollywood, CA	12/9/41
Bridges, Jeff	Los Angeles, CA	12/4/49
Brightman, Sarah	Berkhamstead, England	8/14/60
Brimley, Wilford	Salt Lake City, UT	9/27/34
Brinkley, Christie	Malibu, CA	2/2/54
Broderick, Matthew	New York, NY	3/21/62
Brolin, James	Los Angeles, CA	7/18/40
Bronson, Charles	Ehrenfeld, PA	11/3/22
Brooks, Albert	Beverly Hills, CA	7/22/47
Brooks, Foster	Louisville, KY	5/11/12
Brooks, Garth	Tulsa, OK	2/7/62
Brooks, James L	North Bergen, NJ	5/9/40
Brooks, Mel	New York, NY	6/28/26
Brosnan, Pierce	Co. Meath, Ireland	5/16/53
Brown, Blair	Washington, DC	1948
Brown, Bobby	Boston, MA	2/5/69
Brown, Bryan	Sydney, Australia	6/23/47
Brown, James	Pulaski, TN (?)	6/17/28(?)
Browne, Jackson	Heidelberg, Germany	10/9/48
Browne, Roscoe Lee	Woodbury, NJ	5/2/25
Brubeck, Dave	Concord, CA	12/6/20
Bryson, Peabo	Greenville, SC	4/13/51
Buckley, Betty	Ft. Worth, TX	7/3/47
Buffett, Jimmy	Pascagoula, MS	12/25/46
Bujold, Genevieve	Montreal, Quebec	7/1/42
Bullock, Sandra	Arlington, VA	7/26/64
Bumbry, Grace	St. Louis, MO	1/4/37
Burghoff, Gary	Bristol, CT	5/24/40
Burke, Delta	Orlando, FL	7/30/56
Burnett, Carol	San Antonio, TX	4/26/33
Burns, Edward	Valley Stream, NY	1/29/68
Burrows, Darren E.	Winfield, KS	9/12/66
Burstyn, Ellen	Detroit, MI	12/7/32
Burton, LeVar	Landstuhl, W Germany	2/16/57
Burton, Tim	Burbank, CA	8/25/58
Buscemi, Steve	Brooklyn, NY	12/13/57
Busey, Gary	Goose Creek, TX	6/29/44
Busfield, Timothy	Lansing, MI	6/12/57
Butler, Brett	Montgomery, AL	1/30/58
Buttons, Red	New York, NY	2/5/19
Buzzi, Ruth	Westerly, RI	7/24/36
Byrne, David	Dumbarton, Scotland	5/14/52
Byrne, Gabriel	Dublin, Ireland	5/12/50
Caan, James	New York, NY	3/26/39
Caballe, Montserrat	Barcelona, Spain	4/12/33
Caesar, Sid	Yonkers, NY	9/8/22
Cage, Nicolas	Long Beach, CA	1/7/64
Cain, Dean	Mt. Clemens, MI	7/31/66
Caine, Michael	London, England	3/14/33
Caldwell, Sarah	Maryville, MO	3/6/24
Caldwell, Zoe	Melbourne, Australia	9/14/33
Cameron, James	Kapuskasiny, Ontario	8/16/54
Cameron, Kirk	Panorama City, CA	10/12/70
Camp, Hamilton	London, England	10/30/34
Campanella, Joseph	New York, NY	11/21/27
Campbell, Bruce	Royal Oak, MI	6/22/58
Campbell, Glen	Billstown, AR	4/22/36
Campbell, Naomi	London, England	5/22/70
Campbell, Neve	Toronto, Ontario	10/3/73
Campion, Jane	Wellington, New Zealand	1955
Cannell, Stephen J.	Los Angeles, CA	2/5/42
Cannon, Dyan	Tacoma, WA	1/4/37
Capshaw, Kate	Ft. Worth, TX	11/3/53
Cardinale, Claudia	Tunis, Tunisia	4/15/39
Carey, Drew	Cleveland, OH	5/23/58
Carey, Mariah	Huntington, NY	3/27/70
Cariou, Len	Winnipeg, Canada	9/30/39
Carlin, George	New York, NY	5/12/37
Carlisle Hart, Kitty	New Orleans, LA	9/3/15
Carlyle, Robert	Glasgow, Scotland	4/14/61
Carmen, Eric	Cleveland, OH	8/11/49
Carney, Art	Mt. Vernon, NY	11/4/18
Carpenter, John	Carthage, NY	1/16/48
Carpenter, Mary Chapin	Princeton, NJ	2/21/58
Caron, Leslie	Boulogne, France	7/1/31
Carr, Vikki	El Paso, TX	7/19/41
Carradine, David	Hollywood, CA	10/8/36
Carradine, Keith	San Mateo, CA	8/8/49
Carreras, Jose	Barcelona, Spain	12/5/46
Carrere, Tia	Honolulu, HI	1/2/66
Carrey, Jim	Toronto, Ontario	1/17/62
Carroll, Diahann	Bronx, NY	7/17/35
Carroll, Pat	Shreveport, LA	5/5/27
Carson, Johnny	Corning, IA	10/23/25
Carson, Lisa Nicole	Brooklyn, NY	7/12/69
Carter, Benny	New York, NY	8/8/07
Carter, Dixie	McLemoresville, TN	5/25/39
Carter, Jack	New York, NY	6/24/23
Carter, June	Maces Spring, VA	6/23/29
Carter, Lynda	Phoenix, AZ	7/24/51
Carter, Nell	Birmingham, AL	9/13/48
Carter, Ron	Royal Oak Twp, MI	5/4/37
Carter, Nick	Jamestown, NY	1/28/80
Cartwright, Nancy	Dayton, OH	10/25/59
Caruso, David	Forest Hills, NY	1/17/56
Carvey, Dana	Missoula, MT	4/2/55
Cash, Johnny	Kingsland, AR	2/26/32
Cash, Rosanne	Memphis, TN	5/24/55
Cassidy, David	New York, NY	4/12/50
Castellaneta, Dan	Chicago, IL	1958
Cates, Phoebe	New York, NY	7/16/63
Cathbert, Lacey	Purvis, MS	9/30/82
Cattrall, Kim	Liverpool, England	8/21/56
Cavett, Dick	Gibbon, NE	11/19/36
Chamberlain, Richard	Beverly Hills, CA	3/31/35
Chan, Jackie	Hong Kong	4/7/54
Channing, Carol	Seattle, WA	1/31/23
Channing, Stockard	New York, NY	2/13/44
Chaplin, Geraldine	Santa Monica, CA	7/31/44
Chapman, Tracy	Cleveland, OH	3/30/64
Charisse, Cyd	Amarillo, TX	3/8/21
Charles, Ray	Albany, GA	9/23/30
Charo	Murcia, Spain	1/15/51
Chase, Chevy	New York, NY	10/8/43
Chasez, Joshua (J.C.)	Washington, DC	8/8/76
Cheadle, Don	Kansas City, MO	11/29/64
Checker, Chubby	Philadelphia, PA	10/3/41
Cher	El Centro, CA	5/20/46
Chiklis, Michael	Lowell, MA	8/30/63
Cho, Margaret	San Francisco	12/5/68
Chong, Rae Dawn	Vancouver, Canada	2/28/62
Chong, Thomas	Edmonton, Alberta	5/24/38
Chow Yun-Fat	Hong Kong	5/18/55
Christensen, Helena	Copenhagen, Denmark	12/25/68
Christie, Julie	Assam, India	4/14/40
Christopher, William	Evanston, IL	10/20/32
Church, Thomas Haden	El Paso, TX	6/17/61
Clapton, Eric	Surrey, England	3/30/45
Clark, Dick	Mt. Vernon, NY	11/30/29
Clark, Petula	Ewell, Surrey, England	11/15/32

Name	Birthplace	Birthdate
Clark, Roy	Meherrin, VA	4/15/33
Clay, Andrew Dice	Brooklyn, NY	9/29/58
Clayburgh, Jill	New York, NY	4/30/44
Cleese, John	Weston-Super-Mare, Eng.	10/27/39
Cliburn, Van	Shreveport, LA	7/12/34
Clooney, George	Lexington, KY	5/6/61
Clooney, Rosemary	Maysville, KY	5/23/28
Close, Glenn	Greenwich, CT	3/19/47
Coburn, James	Laurel, NE	8/31/28
Coen, Ethan	St. Louis Park, MN	9/21/57
Coen, Joel	St. Louis Park, MN	11/29/54
Cole, Gary	Park Ridge, IL	9/20/57
Cole, Natalie	Los Angeles, CA	2/6/50
Cole, Olivia	Memphis, TN	11/26/42
Cole, Paula	Manchester, CT	4/5/68
Coleman, Dabney	Austin, TX	1/3/32
Coleman, Gary	Zion, IL	2/8/68
Coleman, Ornette	Fort Worth, TX	3/9/30
Collins, Joan	London, England	5/23/33
Collins, Judy	Seattle, WA	5/1/39
Collins, Pauline	Exmouth, England	9/3/40
Collins, Phil	London, England	1/30/51
Collins, Stephen	Des Moines, IA	10/1/47
Colvin, Shawn	Vermillion, SD	1/10/56
Combs, Sean "Puffy"	Harlem, NY	11/9/69
Comden, Betty	Brooklyn, NY	5/3/19
Connelly, Jennifer	Catskill Mountains, NY	12/12/70
Connery, Sean	Edinburgh, Scotland	8/25/30
Connick, Harry, Jr.	New Orleans, LA	9/11/67
Conniff, Ray	Attleboro, MA	11/6/16
Connors, Mike	Fresno, CA	8/15/25
Conrad, Robert	Chicago, IL	3/1/35
Constantine, Michael	Reading, PA	5/22/27
Conti, Tom	Paisley, Scotland	11/22/41
Conway, Tim	Willoughby, OH	12/15/33
Cook, Barbara	Atlanta, GA	10/25/27
Cooke, Alistair	Manchester, England	11/20/08
Coolidge, Rita	Nashville, TN	5/1/45
Coolio	Los Angeles, CA	8/1/63
Cooper, Alice	Detroit, MI	2/4/48
Cooper, Jackie	Los Angeles, CA	9/15/21
Copperfield, David	Metuchen, NJ	9/16/56
Coppola, Francis Ford	Detroit, MI	4/7/39
Corbin, Barry	Lamesa, TX	10/16/40
Cord, Alex	New York, NY	8/3/31
Corea, Chick	Chelsea, MA	6/12/41
Corelli, Franco	Ancona, Italy	4/8/23
Corey, Jeff	New York, NY	8/10/14
Corley, Pat	Dallas, TX	6/1/30
Cosby, Bill	Philadelphia, PA	7/12/37
Costas, Bob	New York, NY	3/22/52
Costello, Elvis	London, England	8/25/54
Costner, Kevin	Compton, CA	1/18/55
Courtenay, Tom	Hull, England	2/25/37
Cox, Courteney	Birmingham, AL	6/15/64
Cox, Nikki	Los Angeles, CA	6/2/78
Cox, Ronny	Cloudcroft, NM	8/23/38
Coyote, Peter	New York, NY	10/10/42
Crain, Jeanne	Barstow, CA	5/25/25
Crawford, Cindy	DeKalb, IL	2/20/66
Crawford, Michael	Salisbury, England	1/19/42
Crenna, Richard	Los Angeles, CA	11/30/26
Crespin, Regine	Marseilles, France	2/23/26
Cronyn, Hume	London, Ontario	7/18/11
Crosby, David	Los Angeles, CA	8/14/41
Cross, Ben	London, England	12/16/47
Crouse, Lindsay	New York, NY	5/12/48
Crow, Sheryl	Kennett, MO	2/11/63
Crowe, Cameron	Palm Springs, CA	7/13/57
Crowe, Russell	New Zealand	4/7/64
Crowell, Rodney	Houston, TX	8/17/50
Crudup, Billy	New York, NY	6/8/68
Cruise, Tom	Syracuse, NY	7/3/62
Cruz, Penelope	Madrid, Spain	4/28/74
Crystal, Billy	Long Beach, NY	3/14/47
Culkin, Macaulay	New York, NY	8/26/80
Cullum, John	Knoxville, TN	3/2/30
Culp, Robert	Oakland, CA	8/16/30
Cummings, Constance	Seattle, WA	5/15/10
Curry, Tim	Cheshire, England	4/19/46
Curtin, Jane	Cambridge, MA	9/6/47
Curtis, Jamie Lee	Los Angeles, CA	11/22/58
Curtis, Keene	Salt Lake City, UT	2/15/23
Curtis, Tony	New York, NY	6/3/25
Cusack, Joan	Evanston, IL	10/11/62
Cusack, John	Evanston, IL	6/28/66
Cyrus, Billy Ray	Flatwoods, KY	8/25/61
Dafoe, Willem	Appleton, WI	7/22/55
Dahl, Arlene	Minneapolis, MN	8/11/28
Dale, Jim	Rothwell, England	8/15/35
Dalton, Abby	Las Vegas, NV	8/15/32
Dalton, Timothy	Colwyn Bay, Wales	3/21/44
Daltrey, Roger	London, England	3/1/44
Daly, Timothy	Suffern, NY	3/1/58
Daly, Tyne	Madison, WI	2/21/47
Damon, Matt	Cambridge, MA	10/8/70
Damone, Vic	Brooklyn, NY	6/12/28
Danes, Claire	New York, NY	4/12/79
D'Angelo	'Richmond, VA	2/11/74
D'Angelo, Beverly	Columbus, OH	11/15/54
Dangerfield, Rodney	Babylon, NY	11/22/21
Daniels, Charlie	Wilmington, NC	10/28/36
Daniels, Jeff	Georgia	2/19/55
Daniels, William	Brooklyn, NY	3/31/27
Danner, Blythe	Philadelphia, PA	2/3/44
Danson, Ted	San Diego, CA	12/29/47
Danza, Tony	New York, NY	4/21/51
Darby, Kim	Hollywood, CA	7/8/48
David, Larry	Brooklyn, NY	1947
Davidson, John	Pittsburgh, PA	12/13/41
Davis, Ann B.	Schenectady, NY	5/5/26
Davis, Clifton	Chicago, IL	10/4/45
Davis, Geena	Wareham, MA	1/21/57
Davis, Judy	Perth, Australia	1955
Davis, Mac	Lubbock, TX	1/21/42
Davis, Ossie	Cogdell, GA	12/18/17
Dawber, Pam	Farmington Hills, MI	10/18/51
Dawson, Richard	Hampshire, England	11/20/32
Day, Doris	Cincinnati, OH	4/3/24
Day, Laraine	Roosevelt, UT	10/13/20
Day-Lewis, Daniel	London, England	4/29/57
Dean, Jimmy	Plainview, TX	8/10/28
Dearie, Blossom	E. Durham, NY	4/28/26
DeCarlo, Yvonne	Vancouver, BC	9/1/22
Dee, Frances	Los Angeles, CA	11/26/07
Dee, Ruby	Cleveland, OH	10/27/23
Dee, Sandra	Bayonne, NJ	4/23/42
DeFranco, Buddy	Camden, NJ	2/17/23
DeGeneres, Ellen	Metairie, LA	1/26/58
DeHaven, Gloria	Los Angeles, CA	7/23/25
De Havilland, Olivia	Tokyo, Japan	7/1/16
Delaney, Kim	Philadelphia, PA	11/29/64
Delany, Dana	New York, NY	3/11/56
DeLaurentiis, Dino	Torre Annunziata, Italy	8/8/19
Delon, Alain	Sceaux, France	11/8/35
Del Toro, Benicio	Santurce, Puerto Rico	2/19/67
DeLuise, Dom	Brooklyn, NY	8/1/33
Demme, Jonathan	Rockville Centre, NY	2/22/44
DeMornay, Rebecca	Santa Rosa, CA	11/29/61
Dench, Judi	York, England	12/9/34
Deneuve, Catherine	Paris, France	10/22/43
De Niro, Robert	New York, NY	8/17/43
Dennehy, Brian	Bridgeport, CT	7/9/38
Denver, Bob	New Rochelle, NY	1/9/35
DePalma, Brian	Newark, NJ	9/11/40
Depardieu, Gerard	Chateauroux, France	12/27/48
Depp, Johnny	Owensboro, KY	6/9/63
Derek, Bo	Long Beach, CA	11/20/56
Dern, Bruce	Chicago, IL	6/4/36
Dern, Laura	Santa Monica, CA	2/1/67
Devane, William	Albany, NY	9/5/37
DeVito, Danny	Neptune, NJ	11/17/44
DeWitt, Joyce	Wheeling, WV	4/23/49
Dey, Susan	Pekin, IL	12/10/52
Diamond, Neil	Brooklyn, NY	1/24/41
Diaz, Cameron	San Diego, CA	8/30/72
DiCaprio, Leonardo	Los Angeles, CA	11/11/74
Dick, Andy	Charleston, SC	12/21/66
Dickinson, Angie	Kulm, ND	9/30/31
Diddley, Bo	McComb, MS	12/20/28
Diesel, Vin	New York, NY	7/18/67
Diggs, Taye	Rochester, NY	1/2/71
Diller, Phyllis	Lima, OH	7/17/17
Dillman, Bradford	San Francisco, CA	4/14/30
Dion, Celine	Charlemagne, Quebec	3/30/68
Dillon, Matt	New Rochelle, NY	2/18/64
Dobson, Kevin	New York, NY	3/18/44
Dogg, Snoop	Long Beach, CA	10/20/72
Doherty, Shannen	Memphis, TN	4/21/71
Dolenz, Mickey	Los Angeles, CA	3/8/45
Domingo, Placido	Madrid, Spain	1/21/41
Domino, Fats	New Orleans, LA	2/26/28
Donahue, Phil	Cleveland, OH	12/21/35
D'Onofrio, Vincent	Brooklyn, NY	6/30/59
Donovan (Leitch)	Glasgow, Scotland	2/10/46
Dorn, Michael	Luling, TX	12/5/52
Dorough, Howie	Orlando, FL	8/22/73
Dotrice, Roy	Guernsey, England	5/26/23
Douglas, Kirk	Amsterdam, NY	12/9/16
Douglas, Michael	New Brunswick, NJ	9/25/44

Name	Birthplace	Birthdate
Dow, Tony	Holywood, CA	3/17/45
Down, Lesley-Ann	London, England	3/17/54
Downey, Robert, Jr.	New York, NY	4/4/65
Downey, Roma	Derry, Northern Ireland	5/6/60
Downs, Hugh	Akron, OH	2/14/21
Drescher, Fran	Queens, NY	9/30/57
Drew, Ellen	Kansas City, MO	11/23/15
Dreyfuss, Richard	Brooklyn, NY	10/29/47
Driver, Minnie	London, England	1/31/71
Dryer, Fred	Hawthorne, CA	7/6/46
Duchovny, David	New York, NY	8/7/60
Duffy, Julia	Minneapolis, MN	6/27/51
Duffy, Patrick	Townsend, MT	3/17/49
Dukakis, Olympia	Lowell, MA	6/20/31
Duke, Patty	New York, NY	12/14/46
Dukes, David	San Francisco, CA	6/6/45
Dullea, Keir	Cleveland, OH	5/30/36
Dunaway, Faye	Bascom, FL	1/14/41
Duncan, Sandy	Henderson, TX	2/20/46
Dunham, Katherine	Joliet, IL	6/22/10
Dunne, Griffin	New York, NY	6/8/55
Dunst, Kirsten	New Jersey	4/30/82
Durbin, Deanna	Winnipeg, Manitoba	12/4/21
Durning, Charles	Highland Falls, NY	2/28/23
Dussault, Nancy	Pensacola, FL	6/30/36
Dutton, Charles S.	Baltimore, MD	1/30/51
Duvall, Robert	San Diego, CA	1/5/31
Duvall, Shelley	Houston, TX	7/7/49
Dylan, Bob	Duluth, MN	5/24/41
Dylan, Jakob	New York, NY	12/9/69
Dysart, Richard	Augusta, ME	3/30/29
Easton, Sheena	Bellshill, Scotland	4/27/59
Eastwood, Clint	San Francisco, CA	5/31/30
Ebert, Roger	Urbana, IL	6/18/42
Ebsen, Buddy	Belleville, IL	4/2/08
Eden, Barbara	Tucson, AZ	8/23/34
Edwards, Anthony	Santa Barbara, CA	7/19/63
Edwards, Blake	Tulsa, OK	7/26/22
Edwards, Ralph	Merino, CO	6/13/13
Eichhorn, Lisa	Reading, PA	2/4/52
Eikenberry, Jill	New Haven, CT	1/21/47
Ekberg, Anita	Malmo, Sweden	9/29/31
Ekland, Britt	Stockholm, Sweden	10/6/42
Elam, Jack	Miami, AZ	11/13/16
Electra, Carmen	Cincinnati, OH	4/20/73
Elfman, Jenna	Los Angeles, CA	9/30/71
Elizondo, Hector	New York, NY	12/22/36
Elliott, Bob	Boston, MA	3/26/23
Elliott, Chris	New York, NY	1960
Elliott, Sam	Sacramento, CA	8/9/44
Elvira	Manhattan, KS	9/17/51
Eminem	St. Joseph, MO	10/17/72
Enberg, Dick	Auburn Hills, MI	1/5/35
Englund, Robert	Hollywood, CA	6/6/48
Enya	Gweedore, Ireland	5/17/61
Ephron, Nora	New York, NY	5/19/41
Estefan, Gloria	Havana, Cuba	9/1/57
Estevez, Emilio	New York, NY	5/12/62
Estrada, Erik	New York, NY	3/16/49
Etheridge, Melissa	Leavenworth, KS	5/29/61
Evans, Linda	Hartford, CT	11/18/42
Evans, Robert	New York, NY	6/29/30
Everett, Chad	South Bend, IN	6/11/36
Everett, Rupert	Norfolk, England	5/29/59
Everly, Don	Brownie, KY	2/1/37
Everly, Phil	Chicago, IL	1/19/39
Evigan, Greg	South Amboy, NJ	10/14/53
Fabares, Shelley	Santa Monica, CA	1/19/42
Fabian (Forte)	Philadelphia, PA	2/6/43
Fabio	Milan, Italy	3/15/61
Fabray, Nanette	San Diego, CA	10/27/20
Fairchild, Morgan	Dallas, TX	2/3/50
Falana, Lola	Philadelphia, PA	9/11/46
Falco, Edie	Brooklyn, NY	1965?
Falk, Peter	New York, NY	9/16/27
Farentino, James	Brooklyn, NY	2/24/38
Fargo, Donna	Mt. Airy, NC	11/10/45
Farina, Dennis	Chicago, IL	2/29/44
Farr, Jamie	Toledo, OH	7/1/34
Farrell, Eileen	Willimantic, CT	2/13/20
Farrell, Mike	St. Paul, MN	2/6/39
Farrow, Mia	Los Angeles, CA	2/9/45
Fatone, Joey	New York, NY	1/29/76
Faustino, David	California	3/3/74
Fawcett, Farrah	Corpus Christi, TX	2/2/47
Feinstein, Michael	Columbus, OH	9/7/56
Feldon, Barbara	Pittsburgh, PA	3/12/41
Feliciano, Jose	Lares, Puerto Rico	9/10/45
Feldshuh, Tovah	New York, NY	12/27/53

Name	Birthplace	Birthdate
Fenn, Sherilyn	Detroit, MI	2/1/65
Ferrell, Conchata	Charleston, WV	3/28/43
Ferrell, Will	Irvine, CA	7/16/68
Ferrer, Mel	Elberon, NJ	8/25/17
Fiedler, John	Platteville, WI	2/3/25
Field, Sally	Pasadena, CA	11/6/46
Fiennes, Joseph	Salisbury, England	5/27/70
Fiennes, Ralph	Suffolk, England	12/22/62
Fincher, David	Denver, CO	1962
Finney, Albert	Salford, England	5/9/36
Fiorentino, Linda	Philadelphia, PA	3/9/60
Firth, Colin	Grayshott, England	9/10/60
Firth, Peter	Yorkshire, England	10/27/53
Fischer-Dieskau, Dietrich	Berlin, Germany	5/28/25
Fishburne, Laurence	Augusta, GA	7/30/61
Fisher, Carrie	Beverly Hills, CA	10/21/56
Fisher, Eddie	Philadelphia, PA	8/10/28
Fitzgerald, Geraldine	Dublin, Ireland	11/24/13
Flack, Roberta	Black Mountain, NC	2/10/39
Flanagan, Fionnula	Dublin, Ireland	12/10/41
Fleming, Rhonda	Hollywood, CA	8/10/23
Fletcher, Louise	Birmingham, AL	7/22/34
Flockhart, Calista	Freeport, IL	11/11/64
Foch, Nina	Leyden, Netherlands	4/20/24
Fogelberg, Dan	Peoria, IL	8/13/51
Fogerty, John	Berkeley, CA	5/28/45
Foley, Dave	Toronto, Ontario	1/4/63
Fonda, Bridget	Los Angeles, CA	1/27/64
Fonda, Jane	New York, NY	12/21/37
Fonda, Peter	New York, NY	2/23/40
Fontaine, Joan	Tokyo, Japan	10/22/17
Ford, Faith	Alexandria, LA	9/14/64
Ford, Glenn	Quebec, Canada	5/1/16
Ford, Harrison	Chicago, IL	7/13/42
Forman, Milos	Caslav, Czechoslovakia	2/18/32
Forsythe, John	Penns Grove, NJ	1/29/18
Foster, Jodie	New York, NY	11/19/62
Fox, James	London, England	5/19/39
Fox, Matthew	Crowheart, WY	7/14/66
Fox, Michael J.	Edmonton, Alberta	6/9/61
Fox, Vivica A.	Indianapolis, IN	7/30/64
Foxworth, Robert	Houston, TX	11/1/41
Foxworthy, Jeff	Atlanta, GA	9/6/57
Foxx, Jamie	Terrell, TX	12/13/67
Frampton, Peter	Kent, England	4/22/50
Franciosa, Anthony	New York, NY	10/25/28
Francis, Anne	Ossining, NY	9/16/30
Francis, Connie	Newark, NJ	12/12/38
Franken, Al	New York, NY	5/21/51
Frankenheimer, John	Malba, NY	2/19/30
Franklin, Aretha	Memphis, TN	3/25/42
Franklin, Bonnie	Santa Monica, CA	1/6/44
Franz, Dennis	Maywood, IL	10/28/44
Fraser, Brendan	Indianapolis, IN	12/3/67
Freeman, Al, Jr.	San Antonio, TX	3/21/34
Freeman, Mona	Baltimore, MD	6/9/26
Freeman, Morgan	Memphis, TN	6/1/37
French, Dawn	Holyhead, Wales	10/11/57
Fricker, Brenda	Dublin, Ireland	2/17/45
Friedkin, William	Chicago, IL	8/29/35
Frost, David	Tenterden, England	4/7/39
Fry, Stephen	London, England	8/24/57
Fuentes, Daisy	Havana, Cuba	11/17/66
Funicello, Annette	Utica, NY	10/22/42
Furlong, Edward	Glendale, CA	8/2/77
Gabor, Zsa Zsa	Budapest, Hungary	2/6/17
Gabriel, John	Niagara Falls, NY	5/25/31
Gabriel, Peter	London, England	2/13/50
Galway, James	Belfast, Ireland	12/8/39
Gandolfini, James	Westwood, NJ	9/18/61
Garagiola, Joe	St. Louis, MO	2/12/26
Garcia, Andy	Havana, Cuba	4/12/56
Garofalo, Janeane	New Jersey	9/28/64
Garfunkel, Art	New York, NY	11/5/41
Garland, Beverly	Santa Cruz, CA	10/17/26
Garner, James	Norman, OK	4/7/28
Garr, Teri	Lakewood, OH	12/11/45
Garrett, Betty	St. Joseph, MO	5/23/19
Garth, Jennie	Champaign, IL	4/3/72
Gatlin, Larry	Seminole, TX	5/2/48
Gavin, John	Los Angeles, CA	4/8/28
Gayle, Crystal	Paintsville, KY	1/9/51
Gaynor, Mitzi	Chicago, IL	9/4/30
Gazzara, Ben	New York, NY	8/28/30
Geary, Anthony	Coalville, UT	5/29/47
Geary, Cynthia	Jackson, MS	3/21/66
Gedda, Nicolai	Stockholm, Sweden	7/11/25
Gellar, Sarah Michelle	New York, NY	4/14/77
Gere, Richard	Philadelphia, PA	8/31/49
Getty, Estelle	New York, NY	7/25/24

Name	Birthplace	Birthdate
Ghostley, Alice	Eve, MO	8/14/26
Giannini, Giancarlo	Spezia, Italy	8/1/42
Gibb, Barry	Isle of Man, England	9/1/46
Gibb, Maurice	Manchester, England	12/22/49
Gibb, Robin	Manchester, England	12/22/49
Gibbons, Leeza	South Carolina	3/26/57
Gibbs, Marla	Chicago, IL	6/14/31
Gibson, Deborah	New York, NY	8/31/70
Gibson, Henry	Germantown, PA	9/21/35
Gibson, Mel	Peekskill, NY	1/3/56
Gibson, Thomas	Charleston, SC	7/3/62
Gifford, Frank	Santa Monica, CA	8/16/30
Gifford, Kathie Lee	Paris, France	8/16/53
Gilbert, Sara	Santa Monica, CA	1/29/75
Gilbert, Melissa	Los Angeles, CA	5/8/64
Gilberto, Astrud	Salvador, Brazil	3/30/40
Gill, Vince	Norman, OK	4/12/57
Gillette, Anita	Baltimore, MD	8/16/38
Gilley, Mickey	Natchez, MS	3/9/36
Gilliam, Terry	Minneapolis, MN	11/22/40
Gilpin, Peri	Waco, TX	5/27/63
Ginty, Robert	New York, NY	11/14/48
Givens, Robin	New York, NY	11/27/64
Glaser, Paul Michael	Cambridge, MA	3/25/42
Glenn, Scott	Pittsburgh, PA	1/26/42
Gless, Sharon	Los Angeles, CA	5/31/43
Glover, Crispin	New York, NY	9/20/64
Glover, Danny	San Francisco, CA	7/22/47
Glover, Savion	Newark, NJ	1973
Godard, Jean Luc	Paris, France	12/3/30
Goldberg, Whoopi	New York, NY	11/13/49
Goldblum, Jeff	Pittsburgh, PA	10/22/52
Goldthwait, Bobcat	Syracuse, NY	5/1/62
Goldwyn, Tony	Los Angeles, CA	5/20/60
Gooding, Cuba, Jr.	Bronx, NY	1/2/68
Goodman, John	St. Louis, MO	6/20/52
Gordon-Levitt, Joseph	Los Angeles, CA	2/17/81
Gorme, Eydie	Bronx, NY	8/16/32
Gorshin, Frank	Pittsburgh, PA	4/5/34
Gossett, Louis, Jr.	Brooklyn, NY	5/27/36
Gould, Elliott	Brooklyn, NY	8/29/38
Gould, Harold	Schenectady, NY	12/10/23
Goulet, Robert	Lawrence, MA	11/26/33
Gowdy, Curt	Green River, WY	7/31/19
Graham, Heather	Milwaukee, WI	1/29/70
Grammer, Kelsey	St. Thomas, Virgin Isl.	2/20/55
Granger, Farley	San Jose, CA	7/1/25
Grant, Amy	Augusta, GA	12/25/60
Grant, Hugh	London, England	9/9/60
Grant, Lee	New York, NY	10/31/29
Graves, Peter	Minneapolis, MN	3/18/26
Gray, Linda	Santa Monica, CA	9/12/40
Gray, Macy	Canton, OH	1970?
Gray, Spaulding	Barrington, RI	6/5/41
Grayson, Kathryn	Winston-Salem, NC	2/9/22
Green, Adolph	New York, NY	12/2/15
Green, Al	Forrest City, AR	4/13/46
Green, Seth	Philadelphia, PA	2/8/74
Green, Tom	Pembroke, Canada	7/30/71
Greene, Shecky	Chicago, IL	4/8/26
Greenwood, Bruce	Quebec, Canada	8/12/56
Greer, Jane	Washington, DC	9/9/24
Gregory, Cynthia	Los Angeles, CA	7/8/46
Gregory, Dick	St. Louis, MO	10/12/32
Gregory, James	Bronx, NY	12/23/11
Grey, Jennifer	New York, NY	3/22/60
Grey, Joel	Cleveland, OH	4/11/32
Grier, David Alan	Detroit, MI	6/30/55
Grier, Pam	Winston-Salem, NC	5/26/49
Griffin, Merv	San Mateo, CA	7/6/25
Griffith, Andy	Mount Airy, NC	6/1/26
Griffith, Melanie	New York, NY	8/9/57
Grimes, Tammy	Lynn, MA	1/30/34
Grizzard, George	Roanoke Rapids, NC	4/1/28
Grodin, Charles	Pittsburgh, PA	4/21/35
Grosbard, Ulu	Antwerp, Belgium	1/19/29
Gross, Michael	Chicago, IL	6/21/47
Guest, Christopher	New York, NY	2/5/48
Guillaume, Robert	St. Louis, MO	11/30/37
Gumbel, Greg	New Orleans, LA	5/3/46
Guthrie, Arlo	New York, NY	7/10/47
Guttenberg, Steve	New York, NY	8/24/58
Guy, Buddy	Lettsworth, LA	7/30/36
Guy, Jasmine	Boston, MA	3/10/64
Hackett, Buddy	Brooklyn, NY	8/31/24
Hackman, Gene	San Bernardino, CA	1/30/30
Hagen, Uta	Gottingen, Germany	6/12/19
Haggard, Merle	Bakersfield, CA	4/6/37
Hagman, Larry	Weatherford, TX	9/21/31
Haid, Charles	San Francisco, CA	6/2/44
Haines, Connie	Savannah, GA	1/20/22
Hale, Barbara	DeKalb, IL	4/18/22
Hall, Arsenio	Cleveland, OH	2/12/55
Hall, Daryl	Pottstown, PA	10/11/48
Hall, Deidre	Milwaukee, WI	10/31/48
Hall, Monty	Winnipeg, Manitoba	8/25/25
Hall, Tom T.	Olive Hill, KY	5/25/36
Halliwell, Geri	Hertfordshire, England	8/6/72
Hamill, Mark	Oakland, CA	9/25/51
Hamilton, George	Memphis, TN	8/12/39
Hamilton, Linda	Salisbury, MD	9/26/56
Hamlin, Harry	Pasadena, CA	10/30/51
Hammer	Oakland, CA	3/29/63
Hammond, Darrell	Melbourne, FL	10/8/60
Hampton, Lionel	Louisville, KY	4/20/08
Hancock, Herbie	Chicago, IL	4/12/40
Hanks, Tom	Oakland, CA	7/9/56
Hannah, Daryl	Chicago, IL	12/3/60
Hanson, Curtis	Los Angeles, CA	3/24/45
Hanson, Isaac	Tulsa, OK	11/17/80
Hanson, Taylor	Tulsa, OK	3/14/83
Hanson, Zac	Arlington, VA	10/22/85
Hardison, Kadeem	New York, NY	7/24/66
Harewood, Dorian	Dayton, OH	8/6/51
Harmon, Mark	Burbank, CA	9/2/51
Harper, Jessica	Chicago, IL	10/10/49
Harper, Tess	Mammoth Springs, AR	8/15/50
Harper, Valerie	Suffern, NY	8/22/40
Harrelson, Woody	Midland, TX	7/23/61
Harrington, Pat	New York, NY	8/13/29
Harris, Barbara	Evanston, IL	7/25/35
Harris, Ed	Englewood, NJ	11/28/50
Harris, Emmylou	Birmingham, AL	4/2/47
Harris, Julie	Grosse Pte. Park, MI	12/2/25
Harris, Neil Patrick	Albuquerque, NM	6/15/73
Harris, Richard	Co. Limerick, Ireland	10/1/33
Harris, Rosemary	Ashby, England	9/19/30
Harrison, George	Liverpool, England	2/25/43
Harrison, Gregory	Avalon, CA	5/31/50
Harry, Deborah	Miami, FL	7/1/45
Hart, Mary	Madison, SD	11/8/51
Hart, Melissa Joan	Sayville, NY	4/18/76
Hartley, Hal	Lindenhurst, NY	11/3/59
Hartley, Mariette	New York, NY	6/21/40
Hartman, David	Pawtucket, RI	5/19/35
Hartman, Lisa	Houston, TX	6/1/56
Hartnett, Josh	San Francisco, CA	7/21/78
Hasselhoff, David	Baltimore, MD	7/17/52
Hatcher, Teri	Sunnyvale, CA	12/8/64
Hauer, Rutger	Breukelen, Netherlands	1/23/44
Haver, June	Rock Island, IL	6/10/26
Havoc, June	Seattle, WA	11/8/16
Hawke, Ethan	Austin, TX	11/6/70
Hawn, Goldie	Washington, DC	11/21/45
Hayden, Melissa	Toronto, Ontario	4/25/23
Hayek, Salma	Coatzacoalcos, Mexico	9/2/68
Hayes, Isaac	Covington, TN	8/20/42
Hayes, Sean	Glen Ellyn, IL	6/26/70
Hays, Robert	Bethesda, MD	7/24/47
Heard, John	Washington, DC	3/7/45
Hearn, George	Memphis, TN	1935
Heaton, Patricia	Bay Village, OH	3/4/59
Heche, Anne	Aurora, OH	5/25/69
Heckart, Eileen	Columbus, OH	3/29/19
Hedren, Tippi	New Ulm, MN	1/19/35
Helfgott, David	Melbourne, Australia	5/19/47
Helmond, Katherine	Galveston, TX	7/5/34
Hemingway, Mariel	Mill Valley, CA	11/21/61
Hemmings, David	Guildford, England	11/18/41
Hemsley, Sherman	Philadelphia, PA	2/1/38
Henderson, Florence	Dale, IN	2/14/34
Henderson, Skitch	Halstad, MN	1/27/18
Henley, Don	Gilmer, TX	7/22/47
Henner, Marilu	Chicago, IL	4/6/52
Henry, Buck	New York, NY	12/9/30
Hepburn, Katharine	Hartford, CT	5/12/07
Herman, Pee-Wee	Peekskill, NY	8/27/52
Herrmann, Edward	Washington, DC	7/21/43
Hershey, Barbara	Los Angeles, CA	2/5/48
Hesseman, Howard	Lebanon, OR	2/27/40
Heston, Charlton	Evanston, IL	10/4/24
Hetfield, James	Los Angeles, CA	8/3/63
Hewett, Christopher	Sussex, England	4/5/22
Hewitt, Jennifer Love	Waco, TX	2/21/79
Hildegarde	Adell, WI	2/1/06
Hill, Arthur	Melfort, Sask	8/1/22
Hill, Faith	Jackson, MS	9/21/67
Hill, George Roy	Minneapolis, MN	12/20/22
Hill, Lauryn	South Orange, NJ	5/25/75
Hill, Steven	Seattle, WA	2/24/22
Hiller, Wendy	Stockport, England	8/15/12

Name	Birthplace	Birthdate
Hillerman, John	Denison, TX	12/30/32
Hines, Gregory	New York, NY	2/14/46
Hines, Roy	Boston, MA	3/13/26
Hines, Jerome	Hollywood, CA	11/8/21
Hingle, Pat	Miami, FL	7/19/24
Hirsch, Judd	New York, NY	3/15/35
Ho, Don	Kakaako, Oahu, HI	8/13/30
Hoffman, Dustin	Los Angeles, CA	8/8/37
Hoffman, Philip Seymour	Fairport, NY	7/23/67
Hogan, Paul	New South Wales, Australia	10/8/39
Holbrook, Hal	Cleveland, OH	2/17/25
Holder, Geoffrey	Trinidad	8/1/30
Holliday, Polly	Jasper, AL	8/2/37
Holliman, Earl	Delhi, LA	9/11/28
Holly, Lauren	Bristol, PA	10/28/63
Holm, Celeste	New York, NY	4/29/19
Holm, Ian	Goodmayes, England	9/12/31
Holmes, Katie	Toledo, OH	12/18/78
Hooks, Jan	Decatur, GA	4/23/57
Hope, Bob	London, England	5/29/03
Hopkins, Anthony	Port Talbot, South Wales	12/31/37
Hopkins, Bo	Greenville, SC	2/2/42
Hopkins, Telma	Louisville, KY	10/28/48
Hopper, Dennis	Dodge City, KS	5/17/36
Horne, Lena	Brooklyn, NY	6/30/17
Horne, Marilyn	Bradford, PA	1/16/34
Hornsby, Bruce	Williamsburg, VA	11/23/54
Horsley, Lee	Muleshoe, TX	5/15/55
Hoskins, Bob	Suffolk, England	10/26/42
Houston, Whitney	E Orange, NJ	8/9/63
Howard, Ken	El Centro, CA	3/28/44
Howard, Ron	Duncan, OK	3/1/54
Howell, C. Thomas	Los Angeles, CA	12/7/66
Howes, Sally Ann	London, England	7/20/30
Hudson, Kate	Los Angeles, CA	4/19/79
Hughes, Barnard	Bedford Hills, NY	7/16/15
Hulce, Tom	Whitewater, WI	12/6/53
Humperdinck, Engelbert	Madras, India	5/3/36
Hunt, Helen	Los Angeles, CA	6/15/63
Hunt, Linda	Morristown, NJ	4/2/45
Hunter, Holly	Conyers, GA	3/20/58
Hunter, Kim	Detroit, MI	11/12/22
Hunter, Tab	New York, NY	7/11/31
Hurley, Elizabeth	Hampshire, England	6/10/65
Hurt, John	Chesterfield, England	1/22/40
Hurt, Mary Beth	Marshalltown, IA	9/26/46
Hurt, William	Washington, DC	3/20/50
Hussey, Ruth	Providence, RI	10/30/14
Huston, Anjelica	Santa Monica, CA	7/8/51
Hutton, Betty	Battle Creek, MI	2/26/21
Hutton, Lauren	Charleston, SC	11/17/44
Hutton, Timothy	Malibu, CA	8/16/60
Hyman, Earle	Rocky Mount, NC	10/11/26
Ian, Janis	New York, NY	4/7/51
Ice Cube	Los Angeles, CA	6/15/69
Ice-T	Newark, NJ	2/16/58
Idle, Eric	Durham, England	3/29/43
Idol, Billy	London, England	11/30/55
Iglesias, Enrique	Madrid, Spain	5/8/75
Iglesias, Julio	Madrid, Spain	9/23/43
Iman	Mogadishu, Somalia	7/25/55
Imbruglia, Natalie	Australia	2/4/75
Imus, Don	Riverside, CA	7/23/40
Ireland, Kathy	Santa Barbara, CA	3/8/63
Ingram, James	Akron, OH	2/16/56
Irons, Jeremy	Cowes, England	9/19/48
Irving, Amy	Palo Alto, CA	9/10/53
Irving, George S.	Springfield, MA	11/1/22
Irwin, Bill	Santa Monica, CA	4/11/50
Ivey, Judith	El Paso, TX	9/4/51
Ivory, James	Berkeley, CA	6/7/28
Jackee	Winston-Salem, NC	8/14/56
Jackman, Hugh	Sydney, Australia	10/12/68
Jackson, Anne	Allegheny, PA	9/3/25
Jackson, Glenda	Liverpool, England	5/9/36
Jackson, Janet	Gary, IN	5/16/66
Jackson, Jermaine	Gary, IN	12/11/54
Jackson, Jonathan	Orlando, FL	5/11/82
Jackson, Joshua	Vancouver, Brit. Columbia	6/11/78
Jackson, Kate	Birmingham, AL	10/29/48
Jackson, La Toya	Gary, IN	5/29/56
Jackson, Michael	Gary, IN	8/29/58
Jackson, Samuel L.	Chattanooga, TN	12/21/48
Jacobi, Derek	London, England	10/22/38
Jagger, Mick	Dartford, England	7/26/43
James, Etta	Los Angeles, CA	1938
James, Kevin	Stony Brook, NY	4/26/65
Janis, Conrad	New York, NY	2/11/28
Janney, Allison	Dayton, OH	11/19/60

Name	Birthplace	Birthdate
Janssen, Famke	Amsterdam, Netherlands	11/5/65
Jardine, Al	Lima, OH	9/3/42
Jarmusch, Jim	Akron, OH	1/22/53
Jarreau, Al	Milwaukee, WI	3/12/40
Jarrette, Keith	Allentown, PA	5/8/45
Jeffreys, Anne	Goldsboro, NC	1/26/23
Jennings, Waylon	Littlefield, TX	6/15/37
Jeter, Michael	Lawrenceburg, TN	8/20/52
Jett, Joan	Philadelphia, PA	9/22/60
Jewel (Kilcher)	Payson, UT	5/23/74
Jewison, Norman	Toronto, Ontario	7/21/26
Jillian, Ann	Cambridge, MA	1/29/50
Jillette, Penn	Greenfield, MA	3/5/55
Joel, Billy	Bronx, NY	5/9/49
John, Elton	Middlesex, England	3/25/47
Johns, Glynis	Durban, S Africa	10/5/23
Johnson, Arte	Benton Harbor, MI	1/20/29
Johnson, Beverly	Buffalo, NY	10/13/52
Johnson, Don	Flatt Creek, MO	12/15/49
Johnson, Van	Newport, RI	8/25/16
Johnston, Bruce	Chicago, IL	6/24/44
Johnston, Kristen	Washington, DC	9/20/67
Jolie, Angelina	Los Angeles, CA	6/4/75
Jones, Charlie	Ft. Smith, AR	11/9/30
Jones, Davy	Manchester, England	12/30/45
Jones, Dean	Morgan City, AL	1/25/35
Jones, Elvin	Pontiac, MI	9/9/27
Jones, George	Saratoga, TX	9/12/31
Jones, Grace	Spanishtown, Jamaica	5/19/52
Jones, Jack	Hollywood, CA	1/14/38
Jones, James Earl	Tate Co., MS	1/17/31
Jones, Jennifer	Tulsa, OK	3/2/19
Jones, Quincy	Chicago, IL	3/14/33
Jones, Shirley	Smithton, PA	3/31/34
Jones, Tom	Pontypridd, Wales	6/7/40
Jones, Tommy Lee	San Saba, TX	9/15/46
Jourdan, Louis	Marseilles, France	6/19/19
Jovovich, Milla	Kiev, Ukraine	12/19/75
Judd, Ashley	Los Angeles, CA	4/19/68
Judd, Naomi	Ashland, KY	1/11/46
Judd, Wynonna	Ashland, KY	5/30/64
Jump, Gordon	Dayton, OH	4/1/32
Kanaly, Steve	Burbank, CA	3/14/46
Kane, Carol	Cleveland, OH	6/18/52
Kaplan, Gabe	Brooklyn, NY	3/31/45
Karlen, John	New York, NY	5/28/33
Karn, Richard	Seattle, WA	2/17/56
Karras, Alex	Gary, IN	7/15/35
Kasem, Casey	Detroit, MI	4/27/33
Kavner, Julie	Los Angeles, CA	9/7/51
Kazan, Elia	Istanbul, Turkey	9/7/09
Kazan, Lainie	New York, NY	5/15/42
Keach, Stacy	Savannah, GA	6/2/41
Keaton, Diane	Santa Ana, CA	1/5/46
Keaton, Michael	Pittsburgh, PA	9/9/51
Keel, Howard	Gillespie, IL	4/13/17
Keener, Catherine	Miami FL	1961
Keeshan, Bob	Lynbrook, NY	6/27/27
Keitel, Harvey	Brooklyn, NY	5/13/39
Keith, David	Knoxville, TN	5/8/54
Keith, Penelope	Sutton, Surrey, Eng.	4/2/40
Kellerman, Sally	Long Beach, CA	6/2/37
Kelly, R(obert)	Illinois	1967?
Kennedy, George	New York, NY	2/18/25
Kennedy, Jayne	Washington, DC	11/27/51
Kenny G	Seattle, WA	6/5/56
Kent, Allegra	Los Angeles, CA	8/11/37
Kercheval, Ken	Wolcottville, IN	7/15/35
Kerns, Joanna	San Francisco, CA	2/12/53
Kerr, Deborah	Helensburgh, Scotland	9/30/21
Kessel, Barney	Muskogee, OK	10/17/23
Khan, Chaka	Great Lakes, IL	3/23/53
Kidder, Margot	Yellowknife, N.W.T.	10/17/48
Kidman, Nicole	Honolulu, HI	6/20/67
Kilborn, Craig	Hastings, MN	8/24/62
Kilmer, Val	Los Angeles, CA	12/31/59
Kimbrough, Charles	St. Paul, MN	5/23/36
King, Alan	Brooklyn, NY	12/26/27
King, B. B.	Itta Bena, MS	9/16/25
King, Carole	Brooklyn, NY	2/9/42
King, Larry	Brooklyn, NY	11/19/33
King, Perry	Alliance, OH	4/30/48
Kingsley, Ben	Yorkshire, England	12/31/43
Kinnear, Greg	Logansport, IN	6/17/63
Kinney, Kathy	Stevens Point, WI	11/3/54
Kinski, Nastassja	Berlin, W. Germany	1/24/60
Kirby, Bruno	New York, NY	4/28/49
Kirkland, Gelsey	Bethlehem, PA	12/29/53
Kirkpatrick, Chris	Pennsylvania	10/17/71
Kitt, Eartha	North, SC	1/17/27

Name	Birthplace	Birthdate
Klein, Robert	New York, NY	2/8/42
Kline, Kevin	St. Louis, MO	10/24/47
Klugman, Jack	Philadelphia, PA	4/27/22
Knight, Gladys	Atlanta, GA	5/28/44
Knight, Shirley	Goessel, KS	7/5/36
Knight, Wayne	Cartersville, GA	8/7/55
Knotts, Don	Morgantown, WV	7/21/24
Knowles, Beyoncé	Houston, TX	9/4/80
Konitz, Lee	Chicago, IL	10/13/27
Kopell, Bernie	New York, NY	6/21/33
Korman, Harvey	Chicago, IL	2/15/27
Kotto, Yaphet	New York, NY	11/15/37
Krakowski, Jane	Parsippany, NJ	1969
Kramer, Stanley	New York, NY	9/29/13
Kristofferson, Kris	Brownsville, TX	6/22/36
Kudrow, Lisa	Encino, CA	5/30/63
Kurtz, Swoosie	Omaha, NE	9/6/44
LaBelle, Patti	Philadelphia, PA	5/24/44
Ladd, Cheryl	Huron, SD	7/12/51
Ladd, Diane	Meridian, MS	11/29/32
Lagassé, Emeril	Fall River, MA	10/15/59?
Lahti, Christine	Detroit, MI	4/5/50
Laine, Cleo	Middlesex, England	10/28/27
Laine, Frankie	Chicago, IL	3/30/13
Lake, Ricki	New York, NY	9/21/68
Lamas, Lorenzo	Santa Monica, CA	1/20/58
Lambert, Christopher	New York, NY	3/29/57
Landau, Martin	New York, NY	6/20/34
Landis, John	Chicago, IL	8/3/50
Lane, Diane	New York, NY	1/22/63
Lane, Nathan	Jersey City, NJ	2/3/56
lang, k.d.	Consort, Alberta	11/2/61
Lang, Stephen	New York, NY	7/11/52
Lange, Hope	Redding Ridge, CT	11/28/31
Lange, Jessica	Cloquet, MN	4/20/49
Langella, Frank	Bayonne, NJ	1/1/40
Langford, Frances	Lakeland, FL	4/4/13
Lansbury, Angela	London, England	10/16/25
LaPaglia, Anthony	Adelaide, Australia	1/31/59
Laredo, Ruth	Detroit, MI	11/20/37
Larroquette, John	New Orleans, LA	11/25/47
LaSalle, Eriq	Hartford, CT	6/23/63
Lauper, Cyndi	New York, NY	6/20/53
Laurie, Piper	Detroit, MI	1/22/32
Lavin, Linda	Portland, ME	10/15/37
Law, Jude	London, England	12/29/72
Lawless, Lucy	Mount Albert, New Zealand	3/28/68
Lawrence, Carol	Melrose Park, IL	9/5/34
Lawrence, Joey	Montgomery, PA	4/20/76
Lawrence, Martin	Frankfurt, Germany	4/16/65
Lawrence, Steve	Brooklyn, NY	7/8/35
Lawrence, Vicki	Inglewood, CA	3/26/49
Leach, Robin	London, England	8/29/41
Leachman, Cloris	Des Moines, IA	4/4/26
Lear, Norman	New Haven, CT	7/27/22
Leary, Denis	Boston, MA	8/18/57
Learned, Michael	Washington, DC	4/9/39
LeBlanc, Matt	Newton, MA	7/25/67
LeBon, Simon	Bushey, England	10/27/58
Ledger, Heath	Perth, Australia	4/4/79
Lee, Ang	Taiwan	10/23/54
Lee, Brenda	Atlanta, GA	12/11/44
Lee, Christopher	London, England	5/27/22
Lee, Michele	Los Angeles, CA	6/24/42
Lee, Pamela Anderson	Comox, Canada	7/1/67
Lee, Peggy	Jamestown, ND	5/26/20
Lee, Spike	Atlanta, GA	3/20/57
Leeves, Jane	London, England	4/18/62
Legrand, Michel	Paris, France	2/24/32
Leguizamo, John	Bogota, Colombia	7/22/65
Leibman, Ron	New York, NY	10/11/37
Leigh, Janet	Merced, CA	7/6/27
Leigh, Jennifer Jason	Los Angeles, CA	2/5/62
Leighton, Laura	Iowa City, IA	3/14/69
Lennox, Annie	Aberdeen, Scotland	12/25/54
Leno, Jay	New Rochelle, NY	4/28/50
Leonard, Robert Sean	Westwood, NJ	2/28/69
Leoni, Tea	New York, NY	2/25/66
Leslie, Joan	Detroit, MI	1/26/25
Leto, Jared	Bossier City, LA	12/26/71
Letterman, David	Indianapolis, IN	4/12/47
Levine, James	Cincinnati, OH	6/23/43
Levinson, Barry	Baltimore, MD	6/2/32
Levy, Eugene	Hamilton, Canada	12/17/46
Lewis, Al	New York, NY	4/30/10
Lewis, Huey	New York, NY	7/5/51
Lewis, Jerry	Newark, NJ	3/16/26
Lewis, Jerry Lee	Ferriday, LA	9/29/35
Lewis, John	La Grange, IL	5/30/20
Lewis, Juliette	San Fernando Valley, CA	6/21/73
Lewis, Richard	New York, NY	6/29/47
Li, Jet	Beijing, China	4/26/63
Light, Judith	Trenton, NJ	2/9/50
Lightfoot, Gordon	Orillia, Ontario	11/17/38
Lil' Kim	Brooklyn, NY	7/11/75
Linden, Hal	New York, NY	3/20/31
Linkletter, Art	Saskatchewan, Canada	7/17/12
Linn-Baker, Mark	St. Louis, MO	6/17/53
Linney, Laura	New York, NY	2/5/64
Liotta, Ray	Newark, NJ	12/18/55
Lithgow, John	Rochester, NY	10/19/45
Little, Rich	Ottawa, Ontario	11/26/38
Little Richard	Macon, GA	12/5/32
Littrell, Brian	Lexington, KY	2/20/75
Liu, Lucy	New York, NY	12/2/67
L. L. Cool J.	New York, NY	1/14/68
Lloyd, Christopher	Stamford, CT	10/22/38
Lloyd, Emily	England	9/29/70
Lloyd Webber, Andrew	London, England	3/22/48
Locke, Sondra	Shelbyville, TN	5/28/47
Lockhart, June	New York, NY	6/25/25
Locklear, Heather	Los Angeles, CA	9/25/61
Loggia, Robert	New York, NY	1/3/30
Loggins, Kenny	Everett, WA	1/17/47
Lollobrigida, Gina	Subiaco, Italy	7/4/27
Lom, Herbert	Prague, Czechoslovakia	1/9/17
Long, Nia	New York, NY	10/30/70
Long, Shelley	Ft. Wayne, IN	8/23/49
Lopez, Jennifer	Bronx, NY	7/24/70
Loren, Sophia	Rome, Italy	9/20/34
Loring, Gloria	New York, NY	12/10/46
Loudon, Dorothy	Boston, MA	9/17/33
Louis-Dreyfus, Julia	New York, NY	1/13/61
Love, Courtney	San Francisco, CA	7/9/64
Love, Mike	Los Angeles, CA	3/15/41
Lovett, Lyle	Klein, TX	11/1/57
Lovitz, Jon	Tarzana, CA	7/21/57
Loveless, Patty	Pikeville, KY	1/4/57
Lowe, Rob	Charlottesville, VA	3/17/64
Lucas, George	Modesto, CA	5/14/44
Lucci, Susan	Scarsdale, NY	12/23/48
Luckinbill, Laurence	Ft. Smith, AR	11/21/34
Ludwig, Christa	Berlin, Germany	3/16/28
Lumet, Sidney	Philadelphia, PA	6/25/24
LuPone, Patti	Northport, NY	4/21/49
Lynch, David	Missoula, MT	1/20/46
Lynley, Carol	New York, NY	2/13/42
Lynn, Loretta	Butcher Hollow, KY	4/14/35
Lynne, Shelby	Quantico, VA	10/22/68
Lyonne, Natasha	New York, NY	4/4/79
Ma, Yo Yo	Paris, France	10/7/55
Maazel, Lorin	Paris, France	3/6/30
MacArthur, James	Los Angeles, CA	12/8/37
MacCorkindale, Simon	Cambridge, England	2/12/52
MacDowell, Andie	Gaffney, SC	4/21/58
MacGraw, Ali	Pound Ridge, NY	4/1/38
MacLachlan, Kyle	Yakima, WA	2/22/59
MacLaine, Shirley	Richmond, VA	4/24/34
MacLeod, Gavin	Mt. Kisco, NY	2/28/30
MacNee, Patrick	London, England	2/6/22
MacNeil, Cornell	Minneapolis, MN	9/24/22
MacNicol, Peter	Dallas, TX	4/10/54
MacPherson, Elle	Sydney, Australia	3/29/64
Macchio, Ralph	Long Island, NY	11/4/62
Macy, Bill	Revere, MA	5/18/22
Macy, William H.	Miami, FL	3/13/50
Madden, John	Austin, MN	4/10/36
Madigan, Amy	Chicago, IL	9/11/51
Madonna (Ciccone)	Bay City, MI	8/16/58
Maher, Bill	Rivervale, NJ	1/20/56
Mahoney, John	Manchester, England	6/20/40
Majors, Lee	Wyandotte, MI	4/23/40
Malden, Karl	Chicago, IL	3/22/13
Malick, Terrence	Ottawa, IL	11/30/43
Malick, Wendie	Buffalo, NY	12/13/50
Malkovich, John	Christopher, IL	12/9/53
Malone, Dorothy	Chicago, IL	1/30/25
Mamet, David	Chicago, IL	11/30/47
Manchester, Melissa	Bronx, NY	2/15/51
Mandel, Howie	Toronto, Ontario	11/29/55
Mandrell, Barbara	Houston, TX	12/25/48
Mangione, Chuck	Rochester, NY	11/29/40
Manilow, Barry	New York, NY	6/17/46
Mann, Herbie	New York, NY	4/16/30
Manoff, Dinah	New York, NY	1/25/58
Manson, Marilyn	Canton, OH	1/5/69
Mantegna, Joe	Chicago, IL	11/13/47
Marceau, Marcel	Strasbourg, France	3/22/23
Margulies, Julianna	Spring Valley, NY	6/8/66
Marin, Cheech	Los Angeles, CA	7/13/46

Name	Birthplace	Birthdate
Marinaro, Ed	New York, NY	3/31/50
Markova, Alicia	London, England	12/1/10
Marriner, Neville	Lincoln, England	4/15/24
Marsalis, Branford	New Orleans, LA	8/26/60
Marsalis, Wynton	New Orleans, LA	10/18/61
Marsh, Jean	London, England	7/1/34
Marshall, Garry	New York, NY	11/13/34
Marshall, Penny	New York, NY	10/15/43
Marshall, Peter	Huntington, WV	3/30/27
Martin, Dick	Detroit, MI	1/30/23
Martin, Kellie	Riverside, CA	10/16/75
Martin, Ricky	San Juan, Puerto Rico	12/24/71
Martin, Steve	Waco, TX	1945
Martin, Tony	San Francisco, CA	12/25/13
Martins, Peter	Copenhagen, Denmark	10/27/46
Mason, Jackie	Sheboygan, WI	6/9/31
Mason, Marsha	St. Louis, MO	4/3/42
Masterson, Mary Stuart	Los Angeles, CA	6/28/66
Mastrantonio, Mary Elizabeth	Lombard, IL	11/17/58
Masur, Kurt	Brieg, Germany	7/18/27
Masur, Richard	New York, NY	11/20/48
Mathers, Jerry	Sioux City, IA	6/2/48
Matheson, Tim	Glendale, CA	12/31/47
Mathis, Johnny	San Francisco, CA	9/30/35
Matlin, Marlee	Morton Grove, IL	8/24/65
Matthews, Dave	Johannesburg, South Africa	1/9/67
May, Elaine	Philadelphia, PA	4/21/32
Mayo, Virginia	St. Louis, MO	11/30/20
Mazursky, Paul	Brooklyn, NY	4/25/30
McArdle, Andrea	Philadelphia, PA	11/5/63
McBride, Patricia	Teaneck, NJ	8/23/42
McCallum, David	Glasgow, Scotland	9/19/33
McCambridge, Mercedes	Joliet, IL	3/17/18
McCarthy, Andrew	Westfield, NJ	11/29/62
McCarthy, Jenny	Chicago, IL	11/1/72
McCarthy, Kevin	Seattle, WA	2/15/14
McCartney, Paul	Liverpool, England	6/18/42
McCarver, Tim	Memphis, TN	10/16/41
McClanahan, Rue	Healdton, OK	2/21/36
McConaughey, Matthew	Uvalde, Texas	11/4/69
McCoo, Marilyn	Jersey City, NJ	9/30/43
McCormack, Eric	Toronto, Canada	4/18/63
McCormack, Mary	Plainsfield, NJ	4/8/69
McDaniel, James	Washington, DC	3/25/58
McDermott, Dylan	Waterbury, CT	10/26/62
McDonald, Audra	Berlin, Germany	1970
McDonnell, Mary	Wilkes-Barre, PA	1952
McDormand, Frances	Illinois	6/23/57
McDowell, Malcolm	Leeds, England	6/13/43
McEntire, Reba	McAlester, OK	3/28/55
McFerrin, Bobby	New York, NY	3/11/50
McGavin, Darren	Spokane, WA	5/7/22
McGillis, Kelly	Newport Beach, CA	7/9/57
McGoohan, Patrick	New York, NY	3/19/28
McGovern, Elizabeth	Evanston, IL	7/18/61
McGovern, Maureen	Youngstown, OH	7/27/49
McGraw, Tim	Delhi, LA	5/1/67
McGregor, Ewan	Crieff, Scotland	3/31/71
McGuire, Al	New York, NY	9/7/31
McKean, Michael	New York, NY	10/17/47
McKechnie, Donna	Pontiac, MI	11/16/42
McKellen, Ian	Burnley, England	5/25/39
McLachlan, Sarah	Halifax, Nova Scotia	1/28/68
McLean, A.J.	West Palm Beach, FL	1/9/78
McMahon, Ed	Detroit, MI	3/6/23
McNichol, Kristy	Los Angeles, CA	9/11/62
McPartland, Marian	Stough, England	3/20/20
McRaney, Gerald	Collins, MS	8/19/48
Meadows, Jayne	Wu Chang, China	9/27/20
Meara, Anne	New York, NY	9/20/29
Meat Loaf	Dallas, TX	9/27/47
Mehta, Zubin	Bombay, India	4/29/36
Mellencamp, John	Seymour, IN	10/7/51
Mendes, Sergio	Niteroi, Brazil	2/11/41
Mercer, Marian	Akron, OH	11/26/35
Merchant, Natalie	Jamestown, NY	10/26/63
Merrill, Dina	New York, NY	12/9/25
Merrill, Robert	Brooklyn, NY	6/4/19
Messing, Debra	Brooklyn, NY	8/15/68
Metcalf, Laurie	Carbondale, IL	6/16/55
Michael, George	Watford, England	6/26/63
Michaels, Al	New York, NY	11/12/44
Michaels, Lorne	Toronto, Canada	11/17/44
Midler, Bette	Honolulu, HI	12/1/45
Midori	Osaka, Japan	10/25/71
Milano, Alyssa	New York, NY	12/19/72
Miles, Sarah	Ingatestone, England	12/31/41
Miles, Vera	near Boise City, OK	8/23/29
Miller, Ann	Houston, TX	4/12/19

Name	Birthplace	Birthdate
Miller, Dennis	Pittsburgh, PA	11/3/53
Miller, Mitch	Rochester, NY	7/4/11
Miller, Penelope Ann	Los Angeles, CA	1/13/64
Mills, Donna	Chicago, IL	12/11/42
Mills, John	Suffolk, England	2/22/08
Milner, Martin	Detroit, MI	12/28/27
Milnes, Sherrill	Downers Grove, IL	1/10/35
Milsap, Ronnie	Robinsville, NC	1/16/44
Minghella, Anthony	Isle of Wight, England	1/6/54
Minnelli, Liza	Los Angeles, CA	3/12/46
Mirren, Helen	London, England	7/2/46
Mitchell, Joni	McLeod, Alberta	11/7/43
Moby	New York, NY	9/11/65
Modine, Matthew	Loma Linda, CA	3/22/59
Moffat, Donald	Plymouth, England	12/26/30
Moffo, Anna	Wayne, PA	6/27/27
Molinaro, Al	Kenosha, WI	6/24/19
Moll, Richard	Pasadena, CA	1/13/43
Monica (Arnold)	College Park, GA	10/24/80
Montalban, Ricardo	Mexico City, Mexico	11/25/20
Moody, Ron	London, England	1/8/24
Moore, Demi	Roswell, NM	11/11/62
Moore, Dudley	London, England	4/19/35
Moore, Julianne	Boston, MA	12/30/60
Moore, Mary Tyler	Brooklyn, NY	12/29/36
Moore, Melba	New York, NY	10/29/45
Moore, Roger	London, England	10/14/27
Moore, Terry	Los Angeles, CA	1/1/29
Moranis, Rick	Toronto, Ontario	4/18/53
Moreau, Jeanne	Paris, France	1/23/28
Moreno, Rita	Humacao, PR	12/11/31
Morgan, Harry	Detroit, MI	4/10/15
Moriarty, Michael	Detroit, MI	4/5/41
Morissette, Alanis	Ottawa, Ontario	6/1/74
Morita, Pat	Isleton, CA	6/28/32
Morris, Garrett	New Orleans, LA	2/1/37
Morris, Howard	New York, NY	9/4/25
Morrison, Van	Belfast, N. Ireland	8/31/45
Morrissey	Manchester, England	5/22/59
Morrow, Rob	New Rochelle, NY	9/21/62
Morse, David	Hamilton, MA	10/11/53
Morse, Robert	Newton, MA	5/18/31
Morton, Joe	New York, NY	10/18/47
Moses, William	Los Angeles, CA	11/17/59
Moss, Carrie-Ann	Vancouver, Canada	8/21/67
Moss, Kate	London, England	1/16/74
Mr. T	Chicago, IL	5/21/52
Mueller-Stahl, Armin	Tilsit, E. Prussia	12/17/20
Muldaur, Diana	New York, NY	8/19/38
Mulgrew, Kate	Dubuque, IA	4/29/55
Mull, Martin	Chicago, IL	8/18/43
Mullaly, Megan	Los Angeles, CA	11/12/58
Mulroney, Dermot	Alexandria, VA	10/31/63
Muniz, Frankie	Ridgewood, NJ	12/5/85
Munsel, Patrice	Spokane, WA	5/14/25
Murphy, Ben	Jonesboro, AR	3/6/42
Murphy, Eddie	Brooklyn, NY	4/3/61
Murphy, Michael	Los Angeles, CA	5/5/38
Murray, Anne	Springhill, Nova Scotia	6/20/45
Murray, Bill	Evanston, IL	9/21/50
Murray, Don	Hollywood, CA	7/31/29
Musburger, Brent	Portland, OR	5/26/39
Muti, Riccardo	Naples, Italy	7/28/41
Myers, Mike	Toronto, Ontario	5/25/63
Nabors, Jim	Sylacauga, AL	6/12/33
Nash, Graham	Blackpool, England	2/2/42
Naughton, James	Middletown, CT	7/6/46
Neal, Patricia	Packard, KY	1/20/26
Nealon, Kevin	Bridgeport, CT	11/18/53
Neeson, Liam	Ballymena, N. Ireland	6/7/52
Neill, Sam	Ulster, N. Ireland	9/14/47
Nelligan, Kate	London, Ontario	3/16/51
Nelson, Craig T.	Spokane, WA	4/4/46
Nelson, Ed	New Orleans, LA	12/21/28
Nelson, Judd	Portland, ME	11/28/59
Nelson, Tracy	Santa Monica, CA	10/25/63
Nelson, Willie	Abbott, TX	4/30/33
Nero, Peter	New York, NY	5/22/34
Nesmith, Mike	Dallas, TX	12/30/42
Nettleton, Lois	Oak Park, IL	8/16/31
Neuwirth, Bebe	Princeton, NJ	12/31/58
Neville, Aaron	New Orleans, LA	1/24/41
Newhart, Bob	Oak Park, IL	9/5/29
Newman, Paul	Cleveland, OH	1/26/25
Newman, Randy	Los Angeles, CA	11/28/43
Newton, Wayne	Norfolk, VA	4/3/42
Newton-John, Olivia	Cambridge, England	9/26/47
Nicholas, Denise	Detroit, MI	7/12/44
Nicholas, Fayard	Philadelphia, PA	10/20/14
Nichols, Mike	Berlin, Germany	11/6/31

Name	Birthplace	Birthdate
Nicholson, Jack	Neptune, NJ	4/28/37
Nicks, Stevie	Phoenix, AZ	5/26/48
Nielsen, Leslie	Regina, Sask	2/11/26
Nilsson, Birgit	Karup, Sweden	5/17/18
Nimoy, Leonard	Boston, MA	3/26/31
Nolte, Nick	Omaha, NE	2/8/40
Noone, Peter	Manchester, England	11/5/47
Norman, Jessye	Augusta, GA	9/15/45
Norris, Chuck	Ryan, OK	3/10/40
North, Sheree	Los Angeles, CA	1/17/33
Norton, Edward	Columbia, MD	1969
Noth, Christopher	Madison, WI	11/13/57
Novak, Kim	Chicago, IL	2/13/33
Nuyen, France	Marseille, France	7/31/39
Oates, John	New York, NY	4/7/48
O'Brian, Hugh	Rochester, NY	4/19/25
O'Brien, Conan	Brookline, MA	4/18/63
O'Brien, Margaret	San Diego, CA	1/15/37
Ocean, Billy	Fyzabad, Trinidad	1/21/50
O'Connor, Carroll	New York, NY	8/2/24
O'Connor, Donald	Chicago, IL	8/28/25
O'Connor, Sinead	Dublin, Ireland	12/8/66
Odetta	Birmingham, AL	12/31/30
O'Donnell, Chris	Winnetka, IL	6/26/70
O'Donnell, Rosie	Commack, NY	3/21/62
O'Hara, Catherine	Toronto, Canada	3/4/54
O'Hara, Maureen	Dublin, Ireland	8/17/20
O'Herlihy, Dan	Wexford, Ireland	5/1/19
Oldman, Gary	London, England	3/21/58
Olin, Ken	Chicago, IL	7/30/54
Olin, Lena	Stockholm, Sweden	3/22/55
Olmos, Edward James	E. Los Angeles, CA	2/24/47
Olsen, Ashley	California	6/13/86
Olsen, Mary-Kate	California	6/13/86
Olsen, Merlin	Logan, UT	9/15/40
O'Neal, Ryan	Los Angeles, CA	4/20/41
O'Neal, Tatum	Los Angeles, CA	11/5/63
O'Neill, Ed	Youngstown, OH	4/12/46
Ontkean, Michael	Vancouver, B.C.	1/24/46
Orbach, Jerry	New York, NY	10/20/35
Orlando, Tony	New York, NY	4/3/44
Ormond, Julia	Epsom, England	1/4/65
Osbourne, Ozzy	Birmingham, England	12/3/48
O'Shea, Milo	Dublin, Ireland	6/2/26
Oslin, K.T.	Crosset, AR	1942
Osment, Haley Joel	Los Angeles, CA	4/10/88
Osmond, Donny	Ogden, UT	12/9/57
Osmond, Marie	Ogden, UT	10/13/59
O'Toole, Annette	Houston, TX	4/1/53
O'Toole, Peter	Connemara, Ireland	8/2/32
Owens, Buck	Sherman, TX	8/12/29
Oz, Frank	Herford, England	5/25/44
Ozawa, Seiji	Shenyang, China	9/1/35
Paar, Jack	Canton, OH	5/1/18
Pacino, Al	New York, NY	4/25/40
Packer, Billy	Wellsville, NY	2/25/40
Page, Bettie	Kingsport, TN	4/22/23
Page, Jimmy	Heston, England	1/9/44
Page, Patti	Claremore, OK	11/8/27
Paget, Debra	Denver, CO	8/19/33
Paige, Janis	Tacoma, WA	9/16/22
Palance, Jack	Lattimer, PA	2/18/20
Palin, Michael	Sheffield, England	5/5/43
Palmer, Betsy	East Chicago, IN	11/1/29
Palmer, Geoffrey	London, England	6/4/27
Palmer, Robert	Bately, England	1/19/49
Palminteri, Chazz	Bronx, NY	5/15/51
Paltrow, Gwyneth	Los Angeles, CA	9/28/73
Papas, Irene	Chiliomedion, Greece	3/9/26
Paquin, Anna	Wellington, New Zealand	6/24/82
Parker, Alan	London, England	2/14/44
Parker, Eleanor	Cedarville, OH	6/26/22
Parker, Fess	Ft. Worth, TX	8/16/25
Parker, Jameson	Baltimore, MD	11/18/47
Parker, Jean	Deer Lodge, MT	8/11/15
Parker, Mary-Louise	Fort Jackson, SC	8/2/64
Parker, Sarah Jessica	Nelsonville, OH	3/25/65
Parsons, Estelle	Lynn, MA	11/20/27
Parton, Dolly	Sevierville, TN	1/19/46
Patinkin, Mandy	Chicago, IL	11/30/52
Patric, Jason	Queens, NY	6/17/66
Patton, Will	Charleston, SC	6/14/54
Paul, Adrian	London, England	5/29/59
Pavarotti, Luciano	Modena, Italy	10/12/35
Paxton, Bill	Fort Worth, TX	5/17/55
Paycheck, Johnny	Greenfield, OH	5/31/41
Peck, Gregory	La Jolla, CA	4/5/16
Peet, Amanda	New York, NY	1/11/72
Pendergrass, Teddy	Philadelphia, PA	3/26/50

Name	Birthplace	Birthdate
Penn, Arthur	Philadelphia, PA	9/27/22
Penn, Sean	Burbank, CA	8/17/60
Penny, Joe	London, England	9/14/56
Perez, Rosie	Brooklyn, NY	9/6/64
Perkins, Elizabeth	New York, NY	11/18/60
Perlman, Itzhak	Tel Aviv, Israel	8/31/45
Perlman, Rhea	Brooklyn, NY	3/31/48
Perlman, Ron	New York, NY	4/13/50
Perrine, Valerie	Galveston, TX	9/3/43
Perry, Luke	Fredericktown, OH	10/11/66
Perry, Mathew	Williamstown, MA	8/19/69
Persoff, Nehemiah	Jerusalem, Israel	8/14/20
Pesci, Joe	Newark, NJ	2/9/43
Peters, Bernadette	New York, NY	2/28/48
Peters, Brock	New York, NY	7/2/27
Peters, Roberta	New York, NY	5/4/30
Peterson, Oscar	Montreal, Quebec	8/15/25
Peterson, Wolfgang	Emden, Germany	3/14/41
Petty, Tom	Gainesville, FL	10/20/53
Pfeiffer, Michelle	Santa Ana, CA	4/29/57
Philbin, Regis	New York, NY	8/25/34
Phillippe, Ryan	New Castle, DE	9/10/75
Phillips, Lou Diamond	Philippines	2/17/62
Phillips, Mackenzie	Alexandria, VA	11/10/59
Phillips, Michelle	Long Beach, CA	6/4/44
Phoenix, Joaquin	Puerto Rico	10/28/74
Pickett, Wilson	Prattville, AL	3/18/41
Pierce, David Hyde	Albany, NY	4/3/59
Pinchot, Bronson	New York, NY	5/20/59
Pinkett Smith, Jada	Baltimore, MD	8/18/71
Pirner, David	Green Bay, WI	4/16/64
Piscopo, Joe	Passaic, NJ	6/17/51
Pitt, Brad	Shawnee, OK	12/18/64
Plant, Robert	W. Bromwich, England	8/20/48
Pleshette, Suzanne	New York, NY	1/31/37
Plowright, Joan	Brigg, England	10/28/29
Plummer, Amanda	New York, NY	3/23/57
Plummer, Christopher	Toronto, Ontario	12/13/27
Poitier, Sidney	Miami, FL	2/20/27
Polanski, Roman	Paris, France	8/18/33
Pollack, Sydney	Lafayette, IN	7/1/34
Ponti, Carlo	Milan, Italy	12/11/13
Pop, Iggy	Ann Arbor, MI	4/21/47
Portman, Natalie	Jerusalem, Israel	6/9/81
Posey, Parker	Baltimore, MD	11/8/64
Post, Markie	Palo Alto, CA	11/4/50
Poston, Tom	Columbus, OH	10/17/27
Potts, Annie	Nashville, TN	10/28/52
Povich, Maury	Washington, DC	1/17/39
Powell, Jane	Portland, OR	4/1/28
Powers, Stefanie	Hollywood, CA	11/2/42
Prentiss, Paula	San Antonio, TX	3/4/39
Presley, Priscilla	New York, NY	5/24/46
Preston, Billy	Houston, TX	9/9/46
Previn, Andre	Berlin, Germany	4/6/29
Price, Leontyne	Laurel, MS	2/10/27
Price, Ray	Perryville, TX	1/12/26
Pride, Charley	Sledge, MS	3/18/38
Priestley, Jason	Vancouver, Brit. Columbia	8/28/69
Prince (The Artist)	Minneapolis, MN	6/7/58
Principal, Victoria	Fukuoka, Japan	1/3/50
Prinze, Freddie, Jr.	Albuquerque, NM	3/8/76
Prosky, Robert	Philadelphia, PA	12/13/30
Pryce, Jonathan	Wales	6/1/47
Pryor, Richard	Peoria, IL	12/1/40
Pulliam, Keshia Knight	Newark, NJ	4/9/79
Pullman, Bill	Hornell, NY	12/17/54
Purcell, Sarah	Richmond, IN	10/8/48
Quaid, Dennis	Houston, TX	4/9/54
Quaid, Randy	Houston, TX	10/1/50
Queen Latifah	East Orange, NJ	3/18/70
Quinn, Aidan	Chicago, IL	3/8/59
Quinn, Martha	Albany, NY	5/11/59
Rachins, Alan	Cambridge, MA	10/10/47
Rae, Charlotte	Milwaukee, WI	4/22/26
Raffi	Cairo, Egypt	7/8/48
Rainer, Luise	Vienna, Austria	1/12/10
Raitt, Bonnie	Burbank, CA	11/8/49
Ramey, Samuel	Colby, KS	3/28/42
Ramone, Dee Dee	Berlin, Germany	9/18/52
Ramone, Johnny	Long Island, NY	10/8/51
Ramone, Tommy	Budapest, Hungary	1/29/52
Randall, Tony	Tulsa, OK	2/26/20
Randolph, John	New York, NY	6/1/15
Randolph, Joyce	Detroit, MI	10/21/25
Raphael, Sally Jessy	Easton, PA	2/25/43
Rashad, Phylicia	Houston, TX	6/17/48
Ratzenberger, John	Bridgeport, CT	4/6/47
Rawls, Lou	Chicago, IL	12/1/36

Name	Birthplace	Birthdate
Reagan, Ronald	Tampico, IL	2/6/11
Reddy, Helen	Melbourne, Australia	10/25/41
Redford, Robert	Santa Monica, CA	8/18/37
Redgrave, Lynn	London, England	3/8/43
Redgrave, Vanessa	London, England	1/30/37
Reed, Jerry	Atlanta, GA	3/20/37
Reed, Lou	Long Island, NY	3/2/43
Reed, Rex	Ft. Worth, TX	10/2/38
Reese, Della	Detroit, MI	7/6/31
Reeve, Christopher	New York, NY	9/25/52
Reeves, Keanu	Beirut, Lebanon	9/2/64
Regalbuto, Joe	New York, NY	8/24/49
Reid, Tim	Norfolk, VA	12/19/44
Reilly, Charles Nelson	New York, NY	1/13/31
Reilly, John C.	Chicago, IL	5/24/65
Reiner, Carl	Bronx, NY	3/20/22
Reiner, Rob	Bronx, NY	3/6/45
Reinhold, Judge	Wilmington, DE	5/21/56
Reinking, Ann	Seattle, WA	11/10/50
Reiser, Paul	New York, NY	3/30/57
Reitman, Ivan	Czechoslovakia	10/27/46
Remini, Leah	Brooklyn, NY	6/15/70
Resnik, Regina	New York, NY	8/30/24
Reynolds, Burt	Waycross, GA	2/11/36
Reynolds, Debbie	El Paso, TX	4/1/32
Reznor, Trent	Mercer, PA	5/17/65
Rhames, Ving	New York, NY	5/12/61
Rhymes, Busta	Brooklyn, NY	5/20/72
Ribisi, Giovanni	Los Angeles, CA	12/17/74
Ricci, Christina	Santa Monica, CA	2/12/80
Richards, Denise	Downers Grove, IL	2/17/72
Richards, Keith	Kent, England	12/18/43
Richards, Michael	Culver City, CA	7/21/49
Richardson, Ian	Edinburgh, Scotland	4/7/34
Richardson, Kevin	Lexington, KY	10/3/72
Richardson, Miranda	Lancashire, England	3/3/58
Richardson, Natasha	London, England	5/11/63
Richardson, Patricia	Bethesda, MD	2/23/51
Richie, Lionel	Tuskegee, AL	6/20/50
Rickles, Don	New York, NY	5/8/26
Rickman, Alan	Hammersmith, England	2/21/46
Riegert, Peter	New York, NY	4/11/47
Rigg, Diana	Doncaster, England	7/20/38
Rimes, LeAnn	Jackson, MS	8/28/82
Ringwald, Molly	Roseville, CA	2/18/68
Ritter, John	Burbank, CA	9/17/48
Rivera, Chita	Washington, DC	1/23/33
Rivera, Geraldo	New York, NY	7/4/43
Rivers, Joan	Brooklyn, NY	6/8/37
Roach, Max	Elizabeth City, NC	1/10/24
Robbins, Tim	W. Covina, CA	10/16/58
Roberts, Doris	St. Louis, MO	11/4/25
Roberts, Eric	Biloxi, MS	4/18/56
Roberts, Julia	Smyrna, GA	10/28/67
Roberts, Pernell	Waycross, GA	5/18/30
Roberts, Tony	New York, NY	10/22/39
Robertson, Cliff	La Jolla, CA	9/9/25
Robertson, Dale	Harrah, OK	7/14/23
Robinson, Smokey	Detroit, MI	2/19/40
Roche, Eugene	Boston, MA	9/22/28
Rochon, Lela	Los Angeles, CA	4/17/64
Rock, Chris	South Carolina	2/7/66
Rodgers, Jimmy	Camas, WA	9/18/33
Rodriquez, Johnny	Sabinal, TX	12/10/51
Rogers, Fred	Latrobe, PA	3/20/28
Rogers, Kenny	Houston, TX	8/21/38
Rogers, Mimi	Coral Gables, FL	1/27/56
Rogers, Wayne	Birmingham, AL	4/7/33
Rollins, Sonny	New York, NY	9/7/29
Romano, Ray	New York, NY	12/21/57
Ronstadt, Linda	Tucson, AZ	7/15/46
Rooney, Mickey	Brooklyn, NY	9/23/20
Rose, Axl	Lafayette, IN	2/6/62
Rose Marie	New York, NY	8/15/25
Roseanne	Salt Lake City, UT	11/3/52
Ross, Diana	Detroit, MI	3/26/44
Ross, Katharine	Hollywood, CA	1/29/42
Rossdale, Gavin (Bush)	London, England	10/30/67
Ross, Marion	Albert Lea, MN	10/25/28
Rossellini, Isabella	Rome, Italy	6/18/52
Rostropovich, Mstislav	Baku, Azerbaijan	3/12/27
Roth, David Lee	Bloomington, IN	10/10/55
Roth, Tim	London, England	5/14/61
Rotten, Johnny	England	1/31/56
Rourke, Mickey	Schenectady, NY	7/16/53
Routledge, Patricia	Birkenhead, England	2/17/29
Rowlands, Gena	Cambria, WI	6/19/30
Rudner, Rita	Coconut Grove, FL	9/17/56
Ruehl, Mercedes	Queens, NY	2/28/48
Ruffalo, Mark	Kenosha, WI	1968
Rush, Barbara	Denver, CO	1/4/30
Rush, Geoffrey	Toowoomba, Australia	1951
Russell, Jane	Bemidji, MN	6/21/21
Russell, Ken	Southampton, England	7/3/27
Russell, Keri	Fountain Valley, CA	3/23/76
Russell, Kurt	Springfield, MA	3/17/51
Russell, Mark	Buffalo, NY	8/23/32
Russell, Leon	Lawton, OK	4/2/41
Russell, Nipsey	Atlanta, GA	10/13/24
Russell, Theresa	San Diego, CA	3/20/57
Russo, Rene	Burbank, CA	2/17/54
Rutherford, Ann	Toronto, Ontario	11/2/20
Ruttan, Susan	Oregon City, OR	9/16/50
Ryan, Meg	Fairfield, CT	11/19/61
Ryan, Roz	Detroit, MI	7/7/51
Rydell, Bobby	Philadelphia, PA	4/26/42
Ryder, Winona	Winona, MN	10/29/71
Sabato, Antonio, Jr.	Italy	2/29/72
Sade	Ibadan, Nigeria	1/16/59
Sagal, Katie	Los Angeles, CA	1956
Saget, Bob	Philadelphia, PA	5/17/56
Sahl, Mort	Montreal, Quebec	5/11/27
Saint, Eva Marie	Newark, NJ	7/4/24
St. James, Susan	Los Angeles, CA	8/14/46
St. John, Jill	Los Angeles, CA	8/19/40
Sajak, Pat	Chicago, IL	10/26/47
Saks, Gene	New York, NY	11/8/21
Sales, Soupy	Franklinton, NC	1/8/26
Samms, Emma	London, England	8/28/60
Sandler, Adam	Brooklyn, NY	9/9/66
Sands, Julian	Yorkshire, England	1/15/58
Sanford, Isabel	New York, NY	8/29/17
San Giacomo, Laura	Hoboken, NJ	11/14/62
Santana, Carlos	Autlan, Mexico	7/20/47
Sarandon, Susan	New York, NY	10/4/46
Sarnoff, Dorothy	New York, NY	5/25/17
Sartain, Gailard	Tulsa, OK	9/18/46
Savage, Ben	Chicago, IL	9/13/80
Savage, Fred	Highland Park, IL	7/9/76
Sawa, Devon	Vancouver, Canada	9/7/78
Saxon, John	Brooklyn, NY	8/5/35
Sayles, John	Schenectady, NY	9/28/50
Scaggs, Boz	Dallas, TX	6/8/44
Scales, Prunella	Surrey, England	1933
Scalia, Jack	Brooklyn, NY	11/10/51
Schallert, William	Los Angeles, CA	7/6/22
Scheider, Roy	Orange, NJ	11/10/32
Schell, Maria	Vienna, Austria	1/15/26
Schell, Maximilian	Vienna, Austria	12/8/30
Schenkel, Chris	Bippus, IN	8/21/23
Schiffer, Claudia	Rheinbach, Germany	8/25/70
Schneider, John	Mt. Kisco, NY	4/8/54
Schneider, Rob	San Francisco, CA	10/31/64
Schreiber, Liev	San Francisco, CA	10/4/67
Schroder, Rick	Staten Island, NY	4/13/70
Schwarzenegger, Arnold	Graz, Austria	7/30/47
Schwarzkopf, Elisabeth	Jarotschin, Poland	12/9/15
Schwimmer, David	Queens, NY	11/12/67
Sciorra, Annabella	New York, NY	3/24/64
Scofield, Paul	Hurst, Pierpont, England	1/21/22
Scolari, Peter	New Rochelle, IL	9/12/54
Scorsese, Martin	New York, NY	11/17/42
Scott, Lizabeth	Scranton, PA	9/29/22
Scott, Martha	Jamesport, MO	9/22/14
Scott, Ridley	Durham, England	11/30/37
Scott-Heron, Gil	Chicago, IL	4/1/49
Scott Thomas, Kristin	Cornwall, England	1960
Scotto, Renata	Savona, Italy	2/24/35
Scully, Vin	New York, NY	11/29/27
Seagal, Steven	Lansing, MI	4/10/51
Secor, Kyle	Tacoma, WA	5/31/60
Sedaka, Neil	New York, NY	3/13/39
Seeger, Pete	New York, NY	5/3/19
Segal, George	Great Neck, NY	2/13/34
Seidelman, Susan	Philadelphia, PA	12/11/52
Seinfeld, Jerry	New York, NY	4/29/54
Sellecca, Connie	New York, NY	5/25/55
Selleck, Tom	Detroit, MI	1/29/45
Severinsen, Doc	Arlington, OR	7/7/27
Sevigny, Chloë	Springfield, MA	11/18/74
Sewell, Rufus	London, England	10/29/67
Seymour, Jane	Middlesex, England	2/15/51
Shackelford, Ted	Oklahoma City, OK	6/23/46
Shaffer, Paul	Thunder Bay, Ontario	11/28/49
Shandling, Garry	Chicago, IL	11/29/49
Shankar, Ravi	Benares, India	4/7/20
Shannon, Molly	Shaker Heights, OH	9/16/64
Sharif, Omar	Alexandria, Egypt	4/10/32
Shatner, William	Montreal, Quebec	3/22/31
Shaughnessy, Charles	London, England	2/9/55
Shaver, Helen	St. Thomas, Ontario	2/24/51

Name	Birthplace	Birthdate
Shaw, Artie	New York, NY	5/23/10
Shea, John	N. Conway, NH	4/14/49
Shearer, Harry	Los Angeles, CA	12/23/43
Shearer, Moira	Scotland	1/17/26
Shearing, George	London, England	8/13/19
Sheedy, Ally	New York, NY	6/12/62
Sheen, Charlie	Los Angeles, CA	9/3/65
Sheen, Martin	Dayton, OH	8/3/40
Sheindlin, Judge Judy	Brooklyn, NY	1942?
Shelley, Carole	London, England	8/16/39
Shepard, Sam	Ft. Sheridan, IL	11/5/43
Shepherd, Cybill	Memphis, TN	2/18/49
Sheridan, Nicollette	Northington, England	11/21/63
Shields, Brooke	New York, NY	5/31/65
Shire, Talia	New York, NY	4/25/46
Short, Bobby	Danville, IL	9/15/24
Short, Martin	Hamilton, Ontario	3/26/50
Show, Grant	Detroit, MI	4/27/62
Shue, Andrew	South Orange, NJ	2/20/67
Shue, Elisabeth	Wilmington, DE	6/10/63
Shull, Richard B.	Evanston, IL	2/24/29
Shyamalan, M. Night	Pondicherry, India	8/6/70
Siepi, Cesare	Milan, Italy	2/10/23
Sikking, James B.	Los Angeles, CA	3/5/34
Sills, Beverly	Brooklyn, NY	5/25/29
Silver, Ron	New York, NY	7/2/46
Silverman, Jonathan	Los Angeles, CA	8/5/66
Silverstone, Alicia	San Francisco, CA	10/4/76
Simmons, Gene	Haifa, Israel	8/25/49
Simmons, Jean	London, England	1/31/29
Simmons, Richard	New Orleans, LA	7/12/48
Simon, Carly	New York, NY	6/25/45
Simon, Paul	Newark, NJ	10/13/41
Simone, Nina	Tyron, NC	2/21/33
Sinatra, Nancy	Jersey City, NJ	6/8/40
Sinbad	Benton Harbor, MI	11/10/56
Sinise, Gary	Blue Island, IL	3/7/55
Singleton, John	Los Angeles, CA	1/6/68
Singleton, Penny	Philadelphia, PA	9/15/08
Sizemore, Tom	Detroit, MI	9/29/64
Skerritt, Tom	Detroit, MI	8/25/33
Skye, Ione	Hertfordshire, England	9/4/70
Slater, Christian	New York, NY	8/19/69
Slater, Helen	Massapequa, NY	12/14/63
Slezak, Erika	Hollywood, CA	8/5/46
Slick, Grace	Chicago, IL	10/30/39
Smirnoff, Yakov	Odessa, Ukraine	1/24/51
Smith, Allison	New York, NY	12/9/69
Smith, Jaclyn	Houston, TX	10/26/47
Smith, Keely	Norfolk, VA	3/9/35
Smith, Kevin	Red Bank, NJ	8/2/70
Smith, Maggie	Ilford, England	12/28/34
Smith, Will	Philadelphia, PA	9/25/68
Smits, Jimmy	New York, NY	7/9/55
Smothers, Dick	New York, NY	11/20/39
Smothers, Tom	New York, NY	2/2/37
Snipes, Wesley	Orlando, FL	7/31/63
Snyder, Tom	Milwaukee, WI	5/12/36
Soderbergh, Steven	Baton Rouge, LA	1/14/63
Somers, Suzanne	San Bruno, CA	10/16/46
Sommer, Elke	Berlin, Germany	11/5/41
Sorbo, Kevin	Mound, MN	9/24/58
Sorvino, Mira	Tenafly, NJ	9/28/70
Sorvino, Paul	Brooklyn, NY	4/13/39
Soul, David	Chicago, IL	8/28/43
Spacek, Sissy	Quitman, TX	12/25/49
Spacey, Kevin	S. Orange, NJ	7/26/59
Spade, David	Birmingham, MI	7/22/65
Spader, James	Boston, MA	2/7/60
Spano, Joe	San Francisco, CA	7/7/46
Spears, Britney	Kentwood, LA	12/2/81
Spector, Phil	Bronx, NY	12/25/40
Spelling, Aaron	Dallas, TX	4/22/28
Spelling, Tori	Los Angeles, CA	5/16/73
Spencer, John	New York, NY	12/20/46
Spielberg, Steven	Cincinnati, OH	12/18/47
Spiner, Brent	Houston, TX	2/2/49
Springer, Jerry	London, England	2/13/44
Springfield, Rick	Sydney, Australia	8/23/49
Springsteen, Bruce	Freehold, NJ	9/23/49
Stack, Robert	Los Angeles, CA	1/13/19
Stafford, Jo	Coalinga, CA	11/12/18
Stahl, Richard	Detroit, MI	1/4/32
Stallone, Sylvester	New York, NY	7/6/46
Stamos, John	Cypress, CA	8/19/63
Stamp, Terence	Stepney, England	7/22/39
Stang, Arnold	New York, NY	9/28/25
Stanton, Harry Dean	West Irvine, KY	7/14/26
Stapleton, Jean	New York, NY	1/19/23
Stapleton, Maureen	Troy, NY	6/21/25
Starr, Ringo	Liverpool, England	7/7/40

Name	Birthplace	Birthdate
Steenburgen, Mary	Newport, AR	2/8/53
Stefani, Gwen	Anaheim, CA	10/3/69
Steiger, Rod	W. Hampton, NY	4/14/25
Stein, Ben	Washington, DC	11/25/44
Stephens, James	Mt. Kisco, NY	5/18/51
Stern, Daniel	Stamford, CT	5/28/57
Stern, Howard	New York, NY	1/12/54
Sternhagen, Frances	Washington, DC	1/13/30
Stevens, Andrew	Memphis, TN	6/10/55
Stevens, Cat	London, England	7/21/48
Stevens, Connie	Brooklyn, NY	8/8/38
Stevens, Rise	New York, NY	6/11/13
Stevens, Stella	Yazoo City, MS	10/1/36
Stevenson, Parker	Philadelphia, PA	6/4/52
Stewart, French	Albuquerque, NM	2/20/64
Stewart, Jon	New York, NY	11/28/62
Stewart, Patrick	Mirfield, England	7/13/40
Stewart, Rod	London, England	1/10/45
Stiers, David Ogden	Peoria, IL	10/31/42
Stiles, Julia	New York, NY	3/28/81
Stiller, Ben	New York, NY	11/30/65
Stiller, Jerry	New York, NY	6/8/27
Stills, Stephen	Dallas, TX	1/3/45
Sting	Newcastle, England	10/2/51
Stipe, Michael	Decatur, GA	1/4/60
Stockwell, Dean	Hollywood, CA	3/5/36
Stoltz, Eric	American Samoa	9/30/61
Stone, Dee Wallace	Kansas City, KS	12/14/48
Stone, Oliver	New York, NY	9/15/46
Stone, Sharon	Meadville, PA	3/10/58
Stookey, Paul	Baltimore, MD	12/30/37
Storch, Larry	New York, NY	1/8/23
Storm, Gale	Bloomington, TX	4/5/22
Stowe, Madeleine	Los Angeles, CA	8/18/58
Straight, Beatrice	Old Westbury, NY	8/2/18
Strait, George	Pearsall, TX	5/18/52
Strasser, Robin	New York, NY	5/7/45
Stratas, Teresa	Toronto, Ontario	5/26/38
Strathairn, David	San Francisco, CA	1/26/49
Strauss, Peter	New York, NY	2/20/47
Streep, Meryl	Summit, NJ	6/22/49
Streisand, Barbra	Brooklyn, NY	4/24/42
Stringfield, Sherry	Colorado Springs, CO	6/24/67
Stritch, Elaine	Detroit, MI	2/2/26
Stroman, Susan	Wilmington, DE	10/17/54
Struthers, Sally	Portland, OR	7/28/48
Stuart, Gloria	Santa Monica, CA	7/4/10
Stuarti, Enzo	Rome, Italy	3/3/25
Sullivan, Susan	New York, NY	11/18/44
Sumac, Yma	Ichocan, Peru	9/10/27
Summer, Donna	Boston, MA	12/31/48
Sutherland, Donald	St. John, New Brunswick	7/17/34
Sutherland, Joan	Sydney, Australia	11/7/26
Sutherland, Kiefer	London, England	12/20/66
Suvari, Mena	Newport, RI	2/9/79
Swank, Hilary	Bellingham, WA	7/30/74
Swayze, Patrick	Houston, TX	8/18/54
Swit, Loretta	Passaic, NJ	11/4/37
Takei, George	Los Angeles, CA	4/20/39
Tallchief, Maria	Fairfax, OK	1/24/25
Tamblyn, Russ	Los Angeles, CA	12/30/34
Tarantino, Quentin	Knoxville, TN	3/27/63
Taylor, Billy	Greenville, SC	7/24/21
Taylor, Buck	Hollywood, CA	5/13/38
Taylor, Elizabeth	London, England	2/27/32
Taylor, James	Boston, MA	3/12/48
Taylor, Rip	Washington, DC	1/13/30
Taylor, Rod	Sydney, Australia	1/11/29
Taymor, Julie	Newton, MA	12/15/52
Te Kanawa, Kiri	Gisborne, New Zealand	3/6/44
Tebaldi, Renata	Pesaro, Italy	2/1/22
Teller	Philadelphia, PA	2/14/48
Temple Black, Shirley	Santa Monica, CA	4/23/28
Tennant, Victoria	London, England	9/30/50
Tennille, Toni	Montgomery, AL	5/8/43
Tesh, John	Garden City, NY	7/9/52
Tharp, Twyla	Portland, IN	7/1/41
Thaxter, Phyllis	Portland, ME	11/20/21
Theron, Charlize	South Africa	8/7/75
Thicke, Alan	Kirkland Lake, Ontario	3/1/47
Thiessen, Tiffani-Amber	Long Beach, CA	1/23/74
Thomas, Jay	New Orleans, LA	7/12/48
Thomas, Jonathan Taylor	Bethlehem, PA	9/8/81
Thomas, Marlo	Detroit, MI	11/21/38
Thomas, Michael Tilson	Hollywood, CA	12/21/44
Thomas, Philip Michael	Columbus, OH	5/26/49
Thomas, Richard	New York, NY	6/13/51
Thompson, Emma	London, England	4/15/59
Thompson, Jack	Sydney, Australia	8/31/40

Name	Birthplace	Birthdate
Thompson, Lea	Rochester, MN	5/31/61
Thompson, Sada	Des Moines, IA	9/27/29
Thorne-Smith, Courtney	San Francisco, CA	11/8/68
Thornton, Billy Bob	Hot Springs, AR	8/4/55
Thurman, Uma	Boston, MA	4/29/70
Tiegs, Cheryl	Minnesota	9/25/47
Tillis, Mel	Tampa, FL	8/8/32
Tilly, Jennifer	Los Angeles, CA	9/6/61
Tilly, Meg	Texada, B.C.	2/14/60
Timberlake, Justin	Memphis, TN	1/31/81
Todd, Richard	Dublin, Ireland	6/11/19
Tomei, Marisa	New York, NY	12/4/64
Tomlin, Lily	Detroit, MI	9/1/39
Tork, Peter	Washington, DC	2/13/44
Torn, Rip	Temple, TX	2/6/31
Townsend, Robert	Chicago, IL	2/6/57
Townshend, Peter	Chiswick, England	5/19/45
Travanti, Daniel J.	Kenosha, WI	3/7/40
Travers, Mary	Louisville, KY	11/9/36
Travis, Nancy	New York, NY	9/21/61
Travis, Randy	Marshville, NC	5/4/59
Travolta, John	Englewood, NJ	2/18/54
Trebek, Alex	Sudbury, Ontario	7/22/40
Tritt, Travis	Marietta, GA	2/9/63
Tucci, Stanley	Katonah, NY	1/11/60
Tucker, Chris	Atlanta, GA	8/31/72
Tucker, Michael	Baltimore, MD	2/6/44
Tucker, Tanya	Seminole, TX	10/10/58
Tune, Tommy	Wichita Falls, TX	2/28/39
Turlington, Christy	San Francisco, CA	1/2/69
Turner, Janine	Lincoln, NE	12/6/62
Turner, Kathleen	Springfield, MO	6/19/54
Turner, Tina	Brownsville, TN	11/26/39
Turturro, John	Brooklyn, NY	2/28/57
Twain, Shania	Windsor, Ontario	8/28/65
Twiggy	London, England	9/19/46
Tyler, Liv	Portland, ME	7/1/77
Tyler, Steven	Boston, MA	3/26/48
Tyson, Cicely	New York, NY	12/19/33
Uecker, Bob	Milwaukee, WI	1/26/35
Uggams, Leslie	New York, NY	5/25/43
Ullman, Tracey	Slough, England	12/30/59
Ullmann, Liv	Tokyo, Japan	12/16/38
Ulrich, Skeet	North Carolina	1/20/70
Underwood, Blair	Tacoma, WA	8/25/64
Urich, Robert	Toronto, Ohio	12/19/47
Usher (Raymond IV)	Chattanooga,TN	10/14/79
Ustinov, Peter	London, England	4/16/21
Vaccaro, Brenda	Brooklyn, NY	11/18/39
Vale, Jerry	New York, NY	7/8/31
Valente, Caterina	Paris, France	1/14/31
Valli, Frankie	Newark, NJ	5/3/37
Van Ark, Joan	New York, NY	6/16/43
Vance, Courtney B.	Detroit, MI	3/12/60
Vandross, Luther	New York, NY	4/20/51
Van Damme, Jean-Claude	Brussels, Belgium	10/18/60
Van Der Beek, James	Chesire, CT	3/8/77
Van Doren, Mamie	Rowena, SD	2/6/36
Van Dyke, Dick	West Plains, MO	12/13/25
Van Dyke, Jerry	Danville, IL	7/27/31
Van Halen, Eddie	Nijmegen, Netherlands	1/26/57
Van Patten, Dick	New York, NY	12/9/28
Van Peebles, Mario	Mexico	1/15/57
Van Sant, Gus	Louisville, KY	7/24/52
Vaughn, Robert	New York, NY	11/22/32
Vaughn, Vince	Minneapolis, MN	3/28/70
Vedder, Eddie	Evanston, IL	12/23/65
Vereen, Ben	Miami, FL	10/10/46
Verrett, Shirley	New Orleans, LA	5/31/31
Vickers, Jon	Prince Albert, Sask.	10/26/26
Vigoda, Abe	New York, NY	2/24/21
Vincent, Jan-Michael	Denver, CO	7/15/44
Vinton, Bobby	Canonsburg, PA	4/16/35
Vitale, Dick	East Rutherford, NJ	6/9/40
Voight, Jon	Yonkers, NY	12/29/38
Von Stade, Frederica	Somerville, NJ	6/1/45
Von Sydow, Max	Lund, Sweden	4/10/29
Von Trier, Lars	Copenhagen, Denmark	4/30/56
Wagner, Jack	Washington, MO	10/3/59
Wagner, Lindsay	Los Angeles, CA	6/22/49
Wagner, Robert	Detroit, MI	2/10/30
Wahl, Ken	Chicago, IL	2/14/56
Wahlberg, Mark	Dorchester, MA	6/5/71
Wain, Bea	Bronx, NY	4/30/17
Waite, Ralph	White Plains, NY	6/22/29
Waits, Tom	Pomona, CA	12/7/49
Walden, Robert	New York, NY	9/25/43

Name	Birthplace	Birthdate
Walken, Christopher	New York, NY	3/31/43
Wallace, Marcia	Creston, IA	11/1/42
Wallach, Eli	Brooklyn, NY	12/7/15
Walter, Jessica	New York, NY	1/31/44
Ward, Fred	San Diego, CA	1943
Ward, Sela	Meridian, MS	8/11/56
Ward, Simon	London, England	10/19/41
Warden, Jack	Newark, NJ	9/18/20
Warfield, Marsha	Chicago, IL	3/5/54
Warner, Malcolm-Jamal	Jersey City, NJ	8/18/70
Warren, Lesley Ann	New York, NY	8/16/46
Warrick, Ruth	St. Joseph, MO	6/29/16
Warwick, Dionne	East Orange, NJ	12/12/41
Washington, Denzel	Mt. Vernon, NY	12/28/54
Waters, John	Baltimore, MD	4/22/46
Waters, Roger	Great Bookham, England	9/9/44
Waterston, Sam	Cambridge, MA	11/15/40
Watson, Emily	London, England	1/14/67
Watts, Andre	Nuremberg, Germany	6/20/46
Wayans, Damon	New York, NY	9/4/60
Wayans, Keenen Ivory	New York, NY	6/8/58
Weathers, Carl	New Orleans, LA	1/14/48
Weaver, Dennis	Joplin, MO	6/4/24
Weaver, Fritz	Pittsburgh, PA	1/19/26
Weaver, Sigourney	New York, NY	10/8/49
Weir, Peter	Sydney, Australia	8/8/44
Weitz, Bruce	Norwalk, CT	5/27/43
Welch, Raquel	Chicago, IL	9/5/40
Weld, Tuesday	New York, NY	8/27/43
Wells, Kitty	Nashville, TN	8/30/19
Wendt, George	Chicago, IL	10/17/48
West, Adam	Walla Walla, WA	9/19/29
Wettig, Patricia	Cincinnati, OH	12/4/51
Whalley-Kilmer, Joanne	Manchester, England	8/25/64
Wheaton, Wil	Burbank, CA	7/29/72
Whitaker, Forest	Longview, TX	7/15/61
White, Barry	Galveston, TX	9/12/44
White, Betty	Oak Park, IL	1/17/22
White, Jaleel	Los Angeles, CA	11/27/76
White, Vanna	N. Myrtle Beach, SC	2/18/57
Whitford, Bradley	Wisconsin	10/10/59
Whiting, Margaret	Detroit, MI	7/22/24
Whitman, Stuart	San Francisco, CA	2/1/26
Whitmore, James	White Plains, NY	10/1/21
Widmark, Richard	Sunrise, MN	12/26/14
Wiest, Dianne	Kansas City, MO	3/28/48
Wilder, Billy	Vienna, Austria	6/22/06
Wilder, Gene	Milwaukee, WI	6/11/35
Williams, Andy	Wall Lake, IA	12/3/30
Williams, Barry	Santa Monica, CA	9/30/54
Williams, Billy Dee	New York, NY	4/6/37
Williams, Cindy	Van Nuys, CA	8/22/47
Williams, Esther	Los Angeles, CA	8/8/23
Williams, Hal	Columbus, OH	12/14/38
Williams, Hank, Jr.	Shreveport, LA	5/26/49
Williams, JoBeth	Houston, TX	1949
Williams, Lucinda	Lake Charles, LA	1/26/53
Williams, Michelle	Kalcspell, MT	9/9/80
Williams, Montel	Baltimore, MD	7/3/56
Williams, Paul	Omaha, NE	9/19/40
Williams, Robin	Chicago, IL	7/21/52
Williams, Treat	Rowayton, CT	12/1/51
Williams, Vanessa	New York, NY	3/18/63
Williamson, Kevin	Bern, NC	3/14/65
Williamson, Nicol	Hamilton, Scotland	9/14/38
Willis, Bruce	W. Germany	3/19/55
Wilson, Brian	Hawthorne, CA	6/20/42
Wilson, Cassandra	Jackson, MS	12/4/55
Wilson, Demond	Valdosta, GA	10/13/46
Wilson, Elizabeth	Grand Rapids, MI	4/4/25
Wilson, Luke	Dallas, TX	9/21/71
Wilson, Nancy	Chillicothe, OH	2/20/37
Windom, William	New York, NY	9/28/23
Winfield, Paul	Los Angeles, CA	5/22/41
Winfrey, Oprah	Kosciusko, MS	1/29/54
Winger, Debra	Cleveland, OH	5/16/55
Winkler, Henry	New York, NY	10/30/45
Winningham, Mare	Phoenix, AZ	5/6/59
Winslet, Kate	Reading, England	10/5/75
Winter, Johnny	Beaumont,TX	2/23/44
Winters, Jonathan	Dayton, OH	11/11/25
Winters, Shelley	St. Louis, MO	8/18/22
Winwood, Steve	Birmingham, England	5/12/48
Wiseman, Joseph	Montreal, Quebec	5/15/18
Withers, Jane	Atlanta, GA	4/12/26
Witherspoon, Reese	Nashville, TN	4/22/76
Witt, Alicia	Worcester, MA	8/21/75
Wolf, Scott	Boston, MA	6/4/68
Wonder, Stevie	Saginaw, MI	5/13/50
Wong, Faye	Beijing, China	8/8/69
Woo, John	Guangzhou, China	5/1/46

Name	Birthplace	Birthdate
Wood, Elijah	Cedar Rapids, IA	1/28/81
Woodard, Alfre	Tulsa, OK	11/2/53
Woods, James	Vernal, NJ	4/18/47
Woodward, Edward	Croyden, England	6/1/30
Woodward, Joanne	Thomasville, GA	2/27/30
Wopat, Tom	Lodi, WI	9/9/50
Worth, Irene	Nebraska	6/23/16
Wray, Fay	Alberta, Canada	9/10/07
Wright, Martha	Seattle, WA	3/23/26
Wright, Max	Detroit, MI	8/2/43
Wright, Steven	New York, NY	12/6/55
Wright, Teresa	New York, NY	10/27/18
Wright Penn, Robin	Dallas, TX	4/8/66
Wyatt, Jane	Campgaw, NJ	8/10/11
Wyle, Noah	Hollywood, CA	6/4/71
Wyman, Bill	London, England	10/24/36
Wyman, Jane	St. Joseph, MO	1/4/14
Yankovic, Weird Al	Lynwood, CA	10/23/59
Yanni	Kalamata, Greece	11/4/54
Yarborough, Glenn	Milwaukee, WI	1/12/30
Yarrow, Peter	New York, NY	5/31/38
Yearwood, Trisha	Monticello, GA	9/19/64
Yoakam, Dwight	Pikesville, KY	10/23/56
York, Michael	Fulmer, England	3/27/42
York, Susannah	London, England	1/9/42
Young, Alan	Northumberland, England	11/19/19
Young, Burt	New York, NY	4/30/40
Young, Neil	Toronto, Ontario	11/12/45
Young, Sean	Louisville, KY	11/20/59
Zane, Billy	Chicago, IL	2/24/66
Zeffirelli, Franco	Florence, Italy	2/12/23
Zellweger, Renee	Katy, TX	1969
Zemeckis, Robert	Chicago, IL	5/14/51
Zerbe, Anthony	Long Beach, CA	5/20/36
Zeta-Jones, Catherine	Swansea, Wales	9/25/69
Zimbalist, Efrem, Jr.	New York, NY	11/30/23
Zimbalist, Stephanie	Encino, CA	10/8/56
Zimmer, Kim	Grand Rapids, MI	2/2/55
Zukerman, Pinchas	Tel Aviv, Israel	7/16/48
Zuniga, Daphne	San Francisco, CA	10/28/62

Entertainment Personalities of the Past

See also other lists for some deceased entertainers not included here.

Name	Born	Died
Aaliyah	1979	2001
Abbott, Bud	1895	1974
Abbott, George	1887	1995
Acuff, Roy	1903	1992
Adams, Joey	1911	1999
Adams, Maude	1872	1953
Adler, Jacob P	1855	1926
Adler, Luther	1903	1984
Adoree, Renee	1898	1933
Aherne, Brian	1902	1986
Ailey, Alvin	1931	1989
Akins, Claude	1918	1994
Albertson, Frank	1909	1964
Albertson, Jack	1907	1981
Alda, Robert	1914	1986
Alexander, Ben	1911	1969
Allen, Fred	1894	1956
Allen, Gracie	1906	1964
Allen, Mel	1913	1996
Allen, Peter	1944	1992
Allgood, Sara	1883	1950
Ameche, Don	1908	1993
Ames, Leon	1903	1993
Amsterdam, Morey	1909?	1996
Anderson, Judith	1897	1992
Anderson, Marian	1902	1993
Andre the Giant	1946	1993
Andrews, Dana	1909	1992
Andrews, Laverne	1913	1967
Andrews, Maxine	1918	1995
Angeli, Pier	1933	1971
Anita Louise	1915	1970
Arbuckle, Fatty (Roscoe)	1887	1933
Arden, Eve	1908	1990
Arlen, Richard	1900	1976
Arliss, George	1868	1946
Armetta, Henry	1888	1945
Armstrong, Louis	1900	1971
Arnaz, Desi	1917	1986
Arnold, Edward	1890	1956
Arquette, Cliff	1905	1974
Arthur, Jean	1900	1991
Ashcroft, Peggy	1907	1991
Astaire, Fred	1899	1987
Astor, Mary	1906	1987
Atkins, Chet	1924	2001
Atwill, Lionel	1885	1946
Auer, Mischa	1905	1967
Aumont, Jean-Pierre	1911	2001
Austin, Gene	1900	1972
Autry, Gene	1907	1998
Axton, Hoyt	1938	1999
Ayres, Lew	1908	1996
Backus, Jim	1913	1989
Bailey, Pearl	1918	1990
Bainter, Fay	1892	1968
Baker, Josephine	1906	1975
Balanchine, George	1904	1983
Ball, Lucille	1911	1989
Balsam, Martin	1919	1996
Bancroft, George	1882	1956
Bankhead, Tallulah	1903	1968
Banks, Leslie	1890	1952
Bara, Theda	1890	1955
Barnes, Binnie	1903	1998
Barnum, Phineas T.	1810	1891
Barrymore, Ethel	1879	1959
Barrymore, John	1882	1942
Barrymore, Lionel	1878	1954
Barrymore, Maurice	1848	1905
Bartel, Paul	1938	2000
Barthelmess, Richard	1897	1963
Bartholomew, Freddie	1924	1992
Bartok, Eva	1926	1998
Barty, Billy	1924	2001
Basehart, Richard	1914	1984
Basie, Count	1904	1984
Bates, Clayton (Peg Leg)	1907	1998
Bates, Florence	1888	1954
Bavier, Francis	1902	1989
Baxter, Anne	1923	1985
Baxter, Warner	1889	1951
Beatty, Clyde	1904	1965
Beaumont, Hugh	1909	1982
Beavers, Louise	1902	1962
Beery, Noah, Sr.	1884	1946
Beery, Noah, Jr.	1913	1994
Beery, Wallace	1889	1949
Begley, Ed	1901	1970
Bellamy, Ralph	1904	1991
Belushi, John	1949	1982
Benaderet, Bea	1906	1968
Bendix, William	1906	1964
Bennett, Constance	1904	1965
Bennett, Joan	1910	1990
Bennett, Michael	1943	1987
Benny, Jack	1894	1974
Benzell, Mimi	1924	1970
Beradino, John	1917	1996
Berg, Gertrude	1899	1966
Bergen, Edgar	1903	1978
Bergman, Ingrid	1915	1982
Berkeley, Busby	1895	1976
Bernardi, Herschel	1923	1986
Bernhardt, Sarah	1844	1923
Bernie, Ben	1893	1943
Bessell, Ted	1939	1996
Bickford, Charles	1889	1967
Big Bopper, The	1930	1959
Bing, Rudolf	1902	1997
Bissell, Whit	1909	1996
Bixby, Bill	1934	1993
Bjoerling, Jussi	1911	1960
Blackmer, Sidney	1895	1973
Blackstone, Harry	1885	1965
Blake, Amanda	1931	1989
Blaine, Vivian	1921	1995
Blanc, Mel	1908	1989
Blocker, Dan	1928	1972
Blondell, Joan	1909	1979
Blore, Eric	1888	1959
Blue, Ben	1901	1975
Blyden, Larry	1925	1975
Bogarde, Dirk	1920	1999
Bogart, Humphrey	1899	1957
Boland, Mary	1880	1965
Boles, John	1895	1969
Bolger, Ray	1904	1987
Bond, Ward	1903	1960
Bondi, Beulah	1892	1981
Bono, Sonny	1935	1998
Boone, Richard	1917	1981
Booth, Edwin	1833	1893
Booth, Junius Brutus	1796	1852
Booth, Shirley	1898	1992
Borge, Victor	1909	2001
Bow, Clara	1905	1965
Bowes, Maj. Edward	1874	1946
Bowman, Lee	1914	1979
Brown, Les	1912	2001
Boxcar Willie	1931	1999
Boyd, Stephen	1928	1977
Boyd, William	1898	1972
Boyer, Charles	1899	1978
Brady, Alice	1893	1939
Brand, Neville	1921	1992
Brazzi, Rossano	1916	1994
Brennan, Walter	1894	1974
Brent, George	1904	1979
Brett, Jeremy	1935	1995
Brice, Fanny	1891	1951
Bridges, Lloyd	1913	1998
Broderick, Helen	1891	1959
Brown, Joe E.	1892	1973
Bruce, Lenny	1925	1966
Bruce, Nigel	1895	1953
Bruce, Virginia	1910	1982
Brynner, Yul	1915	1985
Buchanan, Edgar	1903	1979
Buñuel, Luis	1900	1983
Buono, Victor	1938	1982
Burke, Billie	1885	1970
Burnette, Smiley	1911	1967
Burns, George	1896	1996
Burr, Raymond	1917	1993
Burton, Richard	1925	1984
Busch, Mae	1897	1946
Bushman, Francis X.	1883	1966
Butterworth, Charles	1896	1946
Byington, Spring	1893	1971
Cabot, Bruce	1904	1972
Cabot, Sebastian	1918	1977
Cagney, James	1899	1986
Calhern, Louis	1895	1956
Calhoun, Rory	1923	1999
Callas, Maria	1923	1977
Calloway, Cab	1907	1994
Cambridge, Godfrey	1933	1976
Campbell, Mrs. Patrick	1865	1940
Candy, John	1950	1994
Cantin, Has	1911	1993
Cantor, Eddie	1892	1964
Capra, Frank	1897	1991
Carey, Harry	1878	1947
Carey, Macdonald	1913	1994
Carle, Frankie	1903	2001
Carpenter, Karen	1950	1983
Carradine, John	1906	1988
Carrillo, Leo	1880	1961
Carroll, Leo G.	1892	1972
Carroll, Madeleine	1906	1987
Carroll, Nancy	1905	1965
Carson, Jack	1910	1963

Name	Born	Died
Caruso, Enrico	1873	1921
Casadesus, Gaby	1901	1999
Casals, Pablo	1876	1973
Cass, Peggy	1924	1999
Cassidy, Jack	1927	1976
Cassavetes, John	1929	1989
Castle, Irene	1893	1969
Castle, Vernon	1887	1918
Caulfield, Joan	1922	1991
Chaliapin, Feodor	1873	1938
Champion, Gower	1919	1980
Chandler, Jeff	1918	1961
Chaney, Lon	1883	1930
Chaney, Lon, Jr.	1905	1973
Chapin, Harry	1942	1981
Chaplin, Charles	1889	1977
Chapman, Graham	1941	1989
Chase, Ilka	1905	1978
Chatterton, Ruth	1893	1961
Cherrill, Virginia	1908	1996
Chevalier, Maurice	1888	1972
Clair, René	1898	1981
Clark, Bobby	1888	1960
Clark, Dane	1913	1998
Clark, Fred	1914	1968
Clift, Montgomery	1920	1966
Cline, Patsy	1932	1963
Clyde, Andy	1892	1967
Cobain, Kurt	1967	1994
Cobb, Lee J.	1911	1976
Coburn, Charles	1877	1961
Coca, Imogene	1908	2001
Cochran, Steve	1917?	1965
Cody, Buffalo Bill	1846	1917
Cody, Iron Eyes	1907	1999
Cohan, George M.	1878	1942
Cohen, Myron	1902	1986
Colbert, Claudette	1903	1996
Cole, Nat "King"	1919	1965
Collins, Ray	1890	1965
Colman, Ronald	1891	1958
Columbo, Russ	1908	1934
Como, Perry	1912	2001
Connors, Chuck	1921	1992
Conrad, William	1920	1994
Conried, Hans	1917	1982
Conte, Richard	1911	1975
Convy, Bert	1933	1991
Conway, Tom	1904	1967
Coogan, Jackie	1914	1984
Cook, Elisha, Jr.	1904	1995
Cooke, Sam	1935	1964
Cooper, Gary	1901	1961
Cooper, Gladys	1888	1971
Cooper, Melville	1896	1973
Corby, Ellen	1913	1999
Corio, Ann	1914	1999
Cornell, Katharine	1893	1974
Correll, Charles ("Andy")	1890	1972
Costello, Dolores	1905	1979
Costello, Lou	1906	1959
Cotten, Joseph	1905	1994
Coward, Noel	1899	1973
Cox, Wally	1924	1973
Crabbe, Buster	1908	1983
Crane, Bob	1928	1978
Crawford, Broderick	1911	1986
Crawford, Joan	1904	1977
Crews, Laura Hope	1880	1942
Crisp, Donald	1880	1974
Croce, Jim	1942	1973
Crosby, Bing	1903	1977
Crothers, Scatman	1910	1986
Cugat, Xavier	1900	1990
Cukor, George	1899	1983
Cullen, Bill	1920	1990
Cummings, Robert	1908	1990
Currie, Finlay	1878	1968
Curtis, Ken	1916	1991
Cushing, Peter	1913	1994
Dailey, Dan	1914	1978
Dandridge, Dorothy	1923	1965
Daniell, Henry	1894	1963
Daniels, Bebe	1901	1971
Darin, Bobby	1936	1973
Darnell, Linda	1921	1965
Darwell, Jane	1879	1967
Da Silva, Howard	1909	1986
Davenport, Harry	1866	1949
Davies, Marion	1897	1961
Davis, Bette	1908	1989
Davis, Joan	1907	1961
Davis, Sammy Jr.	1925	1990
Day, Dennis	1917	1988
Dean, James	1931	1955
Defore, Don	1917	1993
Dekker, Albert	1905	1968
Del Rio, Dolores	1908	1983
Demarest, William	1892	1983
DeMille, Agnes	1905	1993
DeMille, Cecil B.	1881	1959
Denison, Michael	1915	1998
Denning, Richard	1914	1998
Dennis, Sandy	1937	1992
Denny, Reginald	1891	1967
Denver, John	1943	1997
Derek, John	1926	1998
DeSica, Vittorio	1901	1974
Devine, Andy	1905	1977
Dewhurst, Colleen	1924	1991
De Wilde, Brandon	1942	1972
De Wolfe, Billy	1907	1974
Diamond, Selma	1920	1985
Dietrich, Marlene	1901	1992
Digges, Dudley	1879	1947
Disney, Walt	1901	1966
Dix, Richard	1894	1949
Dmytryk, Edward	1908	1999
Donahue, Troy	1936	2001
Donat, Robert	1905	1958
Donlevy, Brian	1901?	1972
Dors, Diana	1931	1984
Douglas, Melvyn	1901	1981
Douglas, Paul	1907	1959
Dove, Billie	1900	1998
Downey, Morton, Jr.	1933	2001
Doyle, David	1929	1997
Drake, Alfred	1914	1992
Draper, Ruth	1889	1956
Dresser, Louise	1881	1965
Dressler, Marie	1869	1934
Drew, Mrs. John	1820	1897
Dru, Joanne	1923	1996
Duchin, Eddy	1909	1951
Duff, Howard	1917	1990
Dumbrille, Douglass	1890	1974
Dumont, Margaret	1889	1965
Duncan, Isadora	1878	1927
Dunn, James	1905	1967
Dunne, Irene	1898	1990
Dunnock, Mildred	1904	1991
Durante, Jimmy	1893	1980
Duryea, Dan	1907	1968
Duse, Eleanora	1858	1924
Eagels, Jeanne	1894	1929
Eckstine, Billy	1914	1993
Eddy, Nelson	1901	1967
Edelman, Herb	1933	1996
Edwards, Cliff	1897	1971
Edwards, Gus	1879	1945
Edwards, Vince	1928	1996
Egan, Richard	1923	1987
Ellington, Duke	1899	1974
Elliot, Cass	1941	1974
Elman, Mischa	1891	1967
Errol, Leon	1881	1951
Evans, Dale	1912	2001
Evans, Edith	1888	1976
Evans, Maurice	1901	1989
Ewell, Tom	1909	1994
Fadiman, Clifton	1904	1999
Fairbanks, Douglas	1883	1939
Fairbanks, Douglas, Jr.	1909	2000
Farley, Chris	1964	1997
Farmer, Frances	1914	1970
Farnsworth, Richard	1920	2000
Farnum, Dustin	1870	1929
Farnum, William	1876	1953
Farrar, Geraldine	1882	1967
Farrell, Charles	1901	1990
Farrell, Glenda	1904	1971
Fassbinder, Rainer Werner	1946	1982
Fay, Frank	1897	1961
Faye, Alice	1912	1998
Fazenda, Louise	1895	1962
Feld, Fritz	1900	1993
Feldman, Marty	1933	1982
Fell, Norman	1924	1998
Fellini, Federico	1920	1993
Fenneman, George	1919	1997
Ferrer, Jose	1912	1992
Fetchit, Stepin	1898	1985
Fiedler, Arthur	1894	1979
Field, Betty	1918	1973
Fields, Gracie	1898	1979
Fields, W.C.	1879	1946
Fields, Totie	1931	1978
Finch, Peter	1916	1977
Fine, Larry	1902	1975
Firkusny, Rudolf	1912	1994
Fiske, Minnie Maddern	1865	1932
Fitzgerald, Barry	1888	1961
Flagstad, Kirsten	1895	1962
Fleming, Art	1924	1995
Fleming, Eric	1925	1966
Flippen, Jay C.	1900	1971
Flynn, Errol	1909	1959
Flynn, Joe	1925	1974
Foley, Red	1910	1968
Fonda, Henry	1905	1982
Fontaine, Frank	1920	1978
Fontanne, Lynn	1887	1983
Fonteyn, Margot	1919	1991
Ford, John	1895	1973
Ford, Paul	1901	1976
Ford, Tennessee Ernie	1919	1991
Ford, Wallace	1899	1966
Forrest, Helen	1918	1999
Fosse, Bob	1927	1987
Foster, Phil	1914	1985
Foster, Preston	1901	1970
Foxx, Redd	1922	1991
Foy, Eddie	1857	1928
Franchi, Sergio	1933?	1990
Francis, Arlene	1908	2001
Francis, Kay	1903	1968
Franciscus, James	1934	1991
Frann, Mary	1943	1998
Frawley, William	1893	1966
Frederick, Pauline	1885	1938
French, Victor	1934	1989
Friganza, Trixie	1870	1955
Frisco, Joe	1890	1958
Froman, Jane	1907	1980
Fuller, Samuel	1912	1997
Funt, Allen	1914	1999
Furness, Betty	1916	1994
Gabin, Jean	1904	1976
Gable, Clark	1901	1960
Gabor, Eva	1920	1995
Garbo, Greta	1905	1990
Garcia, Jerry	1942	1995
Gardenia, Vincent	1922	1992
Gardner, Ava	1922	1990
Garfield, John	1913	1952
Garland, Judy	1922	1969
Garson, Greer	1904	1996
Gassman, Vittorio	1922	2000
Gaye, Marvin	1939	1984
Gaynor, Janet	1906	1984
Geer, Will	1902	1978
George, Gladys	1900	1954
Gibb, Andy	1958	1988
Gibson, Hoot	1892	1962
Gielgud, John	1904	2000
Gilbert, Billy	1894	1971
Gilbert, John	1895	1936
Gilford, Jack	1907	1990
Gillette, William	1855	1937
Gingold, Hermione	1897	1987
Gish, Dorothy	1898	1968
Gish, Lillian	1893	1993
Gleason, Jackie	1916	1987
Gleason, James	1886	1959
Gluck, Alma	1884	1938
Gobel, George	1919	1991
Goddard, Paulette	1905	1990
Godfrey, Arthur	1903	1983
Godunov, Alexander	1949	1995
Goldwyn, Samuel	1882	1974
Gomez, Thomas	1905	1971
Goodman, Benny	1909	1986
Gorcey, Leo	1915	1969
Gordon, Gale	1906	1995
Gordon, Ruth	1896	1985
Gosden, Freeman ("Amos")	1899	1982
Gottschalk, Ferdinand	1869	1944
Gottschalk, Louis	1829	1869
Gould, Glenn	1932	1982

Name	Born	Died
Gould, Morton	1913	1996
Grable, Betty	1916	1973
Graham, Martha	1894	1991
Graham, Virginia	1912	1998
Grahame, Gloria	1925	1981
Granger, Stewart	1913	1993
Grant, Cary	1904	1986
Granville, Bonita	1923	1988
Greco, Jose	1918	2001
Greene, Lorne	1915	1987
Greenstreet, Sydney	1879	1954
Griffith, David Wark	1874	1948
Griffith, Hugh	1912	1980
Guardino, Harry	1925	1995
Guinness, Sir Alec	1914	2000
Guthrie, Woody	1912	1967
Gwenn, Edmund	1875	1959
Gwynne, Fred	1926	1993
Hale, Alan	1892	1950
Hale, Alan, Jr.	1918	1990
Haley, Bill	1925	1981
Haley, Jack	1899	1979
Hall, Huntz	1919	1999
Hamilton, Margaret	1902	1985
Hammerstein, Oscar	1847	1919
Hardwicke, Cedric	1893	1964
Hardy, Oliver	1892	1957
Harlow, Jean	1911	1937
Harris, Phil	1904	1995
Harrison, Rex	1908	1990
Hart, William S.	1870	1946
Hartman, Phil	1948	1998
Harvey, Laurence	1928	1973
Hawkins, Jack	1910	1973
Hawkins, Screamin' Jay	1929	2000
Hayakawa, Sessue	1890	1973
Hayden, Sterling	1916	1986
Hayes, Gabby	1885	1969
Hayes, Helen	1900	1993
Hayes, Peter Lind	1915	1998
Hayward, Leland	1902	1971
Hayward, Louis	1909	1985
Hayward, Susan	1917	1975
Hayworth, Rita	1918	1987
Head, Edith	1907	1981
Healy, Ted	1896	1937
Heflin, Van	1910	1971
Heifetz, Jascha	1901	1987
Held, Anna	1873	1918
Hemingway, Margaux	1955	1996
Hendrix, Jimi	1942	1970
Henie, Sonja	1912	1969
Henreid, Paul	1908	1992
Henson, Jim	1936	1990
Hepburn, Audrey	1929	1993
Hersholt, Jean	1886	1956
Hickey, William	1928	1997
Hickson, Joan	1906	1998
Hill, Benny	1925	1992
Hirt, Al	1922	1999
Hitchcock, Alfred	1899	1980
Hobson, Valerie	1917	1998
Hodiak, John	1914	1955
Holden, Fay	1894	1973
Holden, William	1918	1981
Holliday, Judy	1922	1965
Holloway, Sterling	1905	1992
Holly, Buddy	1936	1959
Holt, Jack	1888	1951
Holt, Tim	1918	1973
Homolka, Oscar	1898	1978
Hooker, John Lee	1917	2001
Hoon, Shannon	1967	1995
Hopkins, Miriam	1902	1972
Hopper, DeWolf	1858	1935
Hopper, William	1915	1970
Horowitz, Vladimir	1904	1989
Horton, Edward Everett	1886	1970
Houdini, Harry	1874	1926
Houseman, John	1902	1988
Howard (Horwitz), Curly	1903	1952
Howard, Eugene	1881	1965
Howard, Joe	1867	1961
Howard, Leslie	1890	1943
Howard (Horwitz), Moe	1897	1975
Howard (Horwitz), Shemp	1895	1955
Howard, Tom	1885	1955
Howard, Trevor	1916	1988
Howard, Willie	1885	1949
Hudson, Rock	1925	1985

Name	Born	Died
Hull, Henry	1890	1977
Hull, Josephine	1886	1957
Humphrey, Doris	1895	1958
Hunter, Jeffrey	1925	1969
Hunter, Ross	1921	1996
Husing, Ted	1901	1962
Huston, John	1906	1987
Huston, Walter	1884	1950
Hutchence, Michael	1960	1997
Hutton, Jim	1934	1979
Hutton, Robert	1920	1994
Hyde-White, Wilfrid	1903	1991
Ingram, Rex	1895	1969
Iturbi, Jose	1895	1980
Ireland, Jill	1936	1990
Ireland, John	1915	1992
Irving, Henry	1838	1905
Ives, Burl	1909	1995
Jack, Wolfman	1938	1995
Jackson, Joe	1875	1942
Jackson, Mahalia	1911	1972
Jackson, Milt	1922	1999
Jaeckel, Richard	1926	1997
Jaffe, Sam	1891	1984
Jagger, Dean	1903	1991
James, Dennis	1917	1997
James, Harry	1916	1983
Janis, Elsie	1889	1956
Jannings, Emil	1886	1950
Janssen, David	1930	1980
Jenkins, Allen	1900	1974
Jessel, George	1898	1981
Johnson, Ben	1918	1996
Johnson, Celia	1908	1982
Johnson, Chic	1892	1962
Johnson, J.J.	1924	2001
Jolson, Al	1886	1950
Jones, Brian	1942	1969
Jones, Buck	1889	1942
Jones, Carolyn	1933	1983
Jones, Henry	1912	1999
Jones, Spike	1911	1965
Joplin, Janis	1943	1970
Jory, Victor	1902	1982
Joslyn, Allyn	1905	1981
Julia, Raul	1940	1994
Kahn, Madeline	1942	1999
Kane, Helen	1910	1966
Kanin, Garson	1912	1999
Karloff, Boris	1887	1969
Karns, Roscoe	1893	1970
Kaufman, Andy	1949	1984
Kaye, Danny	1913	1987
Kaye, Stubby	1918	1997
Kean, Charles	1811	1868
Kean, Mrs. Charles	1806	1880
Kean, Edmund	1787	1833
Keaton, Buster	1895	1966
Keeler, Ruby	1910	1993
Keith, Brian	1921	1997
Kellaway, Cecil	1894	1973
Kelley, DeForest	1920	1999
Kelly, Emmett	1898	1979
Kelly, Gene	1912	1996
Kelly, Grace	1929	1982
Kelly, Jack	1927	1992
Kelly, Nancy	1921	1985
Kelly, Patsy	1910	1981
Kelton, Pert	1907	1968
Kendall, Kay	1926	1959
Kennedy, Arthur	1914	1990
Kennedy, Edgar	1890	1948
Kibbee, Guy	1886	1956
Kilbride, Percy	1888	1964
Kiley, Richard	1922	1999
Kirby, George	1923	1995
Kirby, Durward	1912	2000
Klemperer, Werner	1919	2000
Knight, Ted	1923	1986
Kostelanetz, Andre	1901	1980
Kovacs, Ernie	1919	1962
Kramer, Stanley	1913	2001
Kruger, Otto	1885	1974
Kubrick, Stanley	1928	1999
Kulp, Nancy	1921	1991
Kurosawa, Akira	1910	1998
Ladd, Alan	1913	1964
Lahr, Bert	1895	1967

Name	Born	Died
Lake, Arthur	1905	1987
Lake, Veronica	1919	1973
Lamarr, Hedy	1913	2000
Lamas, Fernando	1915	1982
Lamour, Dorothy	1914	1996
Lancaster, Burt	1913	1994
Lanchester, Elsa	1902	1986
Lane, Pricilla	1917	1995
Landis, Carole	1919	1948
Landis, Jessie Royce	1904	1972
Landon, Michael	1936	1991
Lang, Fritz	1890	1976
Langdon, Harry	1884	1944
Langtry, Lillie	1853	1929
Lanza, Mario	1921	1959
LaRue, Lash (Alfred)	1917	1996
Lauder, Harry	1870	1950
Laughton, Charles	1899	1962
Laurel, Stan	1890	1965
Lawford, Peter	1923	1984
Lawrence, Gertrude	1898	1952
Lean, David	1908	1991
Lee, Bernard	1908	1981
Lee, Bruce	1940	1973
Lee, Canada	1907	1952
Lee, Gypsy Rose	1914	1970
LeGallienne, Eva	1899	1991
Lehmann, Lotte	1888	1976
Leigh, Vivien	1913	1967
Leighton, Margaret	1922	1976
Lemmon, Jack	1925	2001
Lennon, John	1940	1980
Lenya, Lotte	1898	1981
Leonard, Eddie	1870	1941
Leonard, Sheldon	1907	1997
LeRoy, Mervyn	1900	1987
Levant, Oscar	1906	1972
Levene, Sam	1905	1980
Levenson, Sam	1911	1980
Lewis, Joe E.	1902	1971
Lewis, Shari	1934	1998
Lewis, Ted	1892	1971
Liberace	1919	1987
Lillie, Beatrice	1894	1989
Lind, Jenny	1820	1887
Lindfors, Viveca	1920	1995
Lindley, Audra	1918	1997
Linville, Larry	1939	2000
Little, Cleavon	1939	1992
Llewelyn, Desmond	1914	1999
Lloyd, Harold	1893	1971
Lloyd, Marie	1870	1922
Lockhart, Gene	1891	1957
Logan, Ella	1913	1969
Lombard, Carole	1909	1942
Lombardo, Guy	1902	1977
Long, Richard	1927	1974
Lopez, Vincent	1895	1975
Lord, Jack	1920?	1998
Lorne, Marion	1888	1968
Lorre, Peter	1904	1964
Lovejoy, Frank	1912	1962
Lowe, Edmund	1890	1971
Loy, Myrna	1905	1993
Lubitsch, Ernst	1892	1947
Ludden, Allen	1918	1981
Lugosi, Bela	1882	1956
Lukas, Paul	1894	1971
Lundigan, William	1914	1975
Lunt, Alfred	1892	1977
Lupino, Ida	1918	1995
Lymon, Frankie	1942	1968
Lynde, Paul	1926	1982
Lynn, Diana	1926	1971
MacDonald, Jeanette	1903	1965
Mack, Ted	1904	1976
MacLane, Barton	1902	1969
MacMurray, Fred	1908	1991
MacRae, Gordon	1921	1986
Macready, George	1909	1973
Madison, Guy	1922	1996
Magnani, Anna	1908	1973
Mancini, Henry	1924	1994
Main, Marjorie	1890	1975
Malle, Louis	1932	1995
Mansfield, Jayne	1932	1967
Mantovani, Annunzio	1905	1980
Marais, Jean	1913	1998
March, Fredric	1897	1975
March, Hal	1920	1970

Name	Born	Died
Marchand, Nancy	1928	2000
Marley, Bob	1945	1981
Marshall, Brenda	1915	1992
Marshall, E.G.	1910	1998
Marshall, Herbert	1890	1966
Martin, Dean	1917	1995
Martin, Mary	1913	1990
Martin, Ross	1920	1981
Marvin, Lee	1924	1987
Marx, Arthur (Harpo)	1888	1964
Marx, Herbert (Zeppo)	1901	1979
Marx, Julius (Groucho)	1890	1977
Marx, Leonard (Chico)	1886	1961
Marx, Milton (Gummo)	1893	1977
Mason, James	1909	1984
Massey, Daniel	1933	1998
Massey, Raymond	1896	1983
Mastroianni, Marcello	1924	1996
Matthau, Walter	1920	2000
Mature, Victor	1916	1999
Maxwell, Marilyn	1921	1972
Mayer, Louis B.	1885	1957
Mayfield, Curtis	1942	1999
Maynard, Ken	1895	1973
Mazurki, Mike	1909	1990
McCartney, Linda	1941	1998
McClure, Doug	1935	1995
McCormack, John	1884	1945
McCrea, Joel	1905	1990
McDaniel, Hattie	1895	1952
McDowall, Roddy	1928	1998
McFarland, George "Spanky"	1928	1993
McHugh, Frank	1899	1981
McIntire, John	1907	1991
McLaglen, Victor	1883	1959
McMahon, Horace	1907	1971
McNeill, Don	1907	1979
McQueen, Butterfly	1911	1995
McQueen, Steve	1930	1980
Meadows, Audrey	1924	1996
Medford, Kay	1920	1980
Meek, Donald	1880	1946
Meeker, Ralph	1920	1989
Melba, Nellie	1861	1931
Melchior, Lauritz	1890	1973
Menjou, Adolphe	1890	1963
Menken, Helen	1902	1966
Menuhin, Yehudi	1916	1999
Mercouri, Melina	1925	1994
Mercury, Freddie	1946	1991
Meredith, Burgess	1909	1997
Merman, Ethel	1908	1984
Merrick, David	1911	2000
Merrill, Gary	1915	1990
Mifune, Toshiro	1920	1997
Milland, Ray	1905	1986
Miller, Glenn	1904	1944
Miller, Marilyn	1898	1936
Miller, Roger	1936	1992
Mills, Harry	1913	1982
Minnevitch, Borrah	1903	1955
Mineo, Sal	1939	1976
Mingus, Charles	1922	1979
Miranda, Carmen	1913	1955
Mitchell, Cameron	1918	1994
Mitchell, Thomas	1892	1962
Mitchum, Robert	1917	1997
Mix, Tom	1880	1940
Monica, Corbett	1930	1998
Monroe, Marilyn	1926	1962
Monroe, Vaughn	1911	1973
Montand, Yves	1921	1991
Montez, Maria	1917	1951
Montgomery, Elizabeth	1933	1995
Montgomery, George	1916	2000
Montgomery, Robert	1904	1981
Moore, Clayton	1914	1999
Moore, Colleen	1900	1988
Moore, Grace	1901	1947
Moore, Garry	1914	1993
Moore, Victor	1876	1962
Moorehead, Agnes	1906	1974
Morgan, Dennis	1910	1994
Morgan, Frank	1890	1949
Morgan, Helen	1900	1941
Morgan, Henry	1915	1994
Morley, Robert	1908	1992
Morris, Chester	1901	1970
Morris, Greg	1934	1996
Morris, Wayne	1914	1959
Morrison, Jim	1943	1971

Name	Born	Died
Morrow, Vic	1932	1982
Mostel, Zero	1915	1977
Mowbray, Alan	1897	1969
Mulhare, Edward	1923	1997
Mulligan, Gerry	1927	1996
Mulligan, Richard	1932	2000
Muni, Paul	1895	1967
Munshin, Jules	1915	1970
Murphy, Audie	1924	1971
Murphy, George	1902	1992
Murray, Arthur	1895	1991
Murray, Kathryn	1906	1999
Murray, Mae	1885	1965
Nagel, Conrad	1896	1970
Naish, J. Carroll	1900	1973
Naldi, Nita	1898	1961
Nance, Jack	1943	1997
Natwick, Mildred	1908	1994
Negri, Pola	1897	1987
Nelson, Harriet (Hilliard)	1914	1994
Nelson, Ozzie	1906	1975
Nelson, Rick	1940	1985
Nesbit, Evelyn	1885	1967
Newley, Anthony	1931	1999
Newton, Robert	1905	1956
Nicholas, Harold	1924	2000
Nijinsky, Vaslav	1890	1950
Nilsson, Anna Q.	1893	1974
Niven, David	1909	1983
Nolan, Lloyd	1902	1985
Normand, Mabel	1894	1930
Notorious B.I.G.	1972	1997
Novarro, Ramon	1899	1968
Nureyev, Rudolf	1938	1993
Oakie, Jack	1903	1978
Oakley, Annie	1860	1926
Oates, Warren	1928	1982
Oberon, Merle	1911	1979
O'Brien, Edmond	1915	1985
O'Brien, Pat	1899	1983
O'Connell, Arthur	1908	1981
O'Connell, Helen	1921	1993
O'Connor, Carroll	1924	2001
O'Connor, Una	1880	1959
O'Keefe, Dennis	1908	1968
Oland, Warner	1880	1938
Olcott, Chauncey	1860	1932
Oliver, Edna May	1883	1942
Olivier, Laurence	1907	1989
Olsen, Ole	1892	1963
O'Neill, James	1849	1920
Orbison, Roy	1936	1988
Ormandy, Eugene	1899	1985
O'Sullivan, Maureen	1911	1998
Ouspenskaya, Maria	1876	1949
Owen, Reginald	1887	1972
Paderewski, Ignace	1860	1941
Page, Geraldine	1924	1987
Pakula, Alan	1928	1998
Pallette, Eugene	1889	1954
Palmer, Lilli	1914	1986
Pangborn, Franklin	1894	1958
Parks, Bert	1914	1992
Parks, Larry	1914	1975
Pasternack, Josef A.	1881	1940
Pastor, Tony (Vaudevillian)	1837	1908
Pastor, Tony (Bandleader)	1907	1969
Patti, Adelina	1843	1919
Patti, Carlotta	1840	1889
Patrick, Gail	1911	1980
Pavlova, Anna	1885	1931
Payne, John	1912	1989
Pearl, Minnie	1912	1996
Peerce, Jan	1904	1984
Pendleton, Nat	1899	1967
Penner, Joe	1905	1941
Peppard, George	1928	1994
Perkins, Anthony	1932	1992
Perkins, Carl	1932	1998
Perkins, Marlin	1905	1986
Peters, Jean	1926	2000
Peters, Susan	1921	1952
Phillips, John	1935	2001
Phoenix, River	1970	1993
Piaf, Edith	1915	1963
Pickens, Slim	1919	1983
Pickford, Mary	1893	1979
Picon, Molly	1898	1992

Name	Born	Died
Pidgeon, Walter	1897	1984
Pinza, Ezio	1892	1957
Pitts, Zasu	1898	1963
Plato, Dana	1964	1999
Pleasence, Donald	1919	1995
Pons, Lily	1904	1976
Ponselle, Rosa	1897	1981
Porter, Nyree Dawn	1940	2001
Powell, Dick	1904	1963
Powell, Eleanor	1912	1982
Powell, William	1892	1984
Power, Tyrone	1913	1958
Preminger, Otto	1905	1986
Presley, Elvis	1935	1977
Preston, Robert	1918	1987
Price, Vincent	1911	1993
Prima, Louis	1911	1978
Prinze, Freddie	1954	1977
Prowse, Juliet	1936	1996
Puente, Tito	1923	2000
Pyle, Denver	1920	1997
Quayle, Anthony	1913	1989
Questel, Mae	1908	1998
Quinn, Anthony	1915	2001
Quintero, José	1924	1999
Rabb, Ellis	1930	1998
Rabbit, Eddie	1941	1998
Radner, Gilda	1946	1989
Raft, George	1895	1980
Rains, Claude	1890	1967
Ralston, Esther	1902	1994
Ramone, Joey	1951	2001
Rampal, Jean-Pierre	1922	2000
Rathbone, Basil	1892	1967
Ratoff, Gregory	1897	1960
Ray, Aldo	1926	1991
Ray, Johnnie	1927	1990
Rayburn, Gene	1917	1999
Raye, Martha	1916	1994
Raymond, Gene	1908	1998
Redding, Otis	1941	1967
Redgrave, Michael	1908	1985
Reed, Donna	1921	1986
Reed, Oliver	1938	1999
Reed, Robert	1932	1992
Reeves, George	1914	1959
Reeves, Steve	1926	2000
Reinhardt, Max	1873	1943
Remick, Lee	1935	1991
Renaldo, Duncan	1904	1980
Rennie, Michael	1909	1971
Renoir, Jean	1894	1979
Rettig, Tommy	1941	1996
Reynolds, Marjorie	1923	1997
Rich, Charlie	1932	1995
Richardson, Ralph	1902	1983
Riddle, Nelson	1921	1985
Ripperton, Minnie	1947	1979
Ritchard, Cyril	1898	1977
Ritter, Tex	1907	1974
Ritter, Thelma	1905	1969
Ritz, Al	1901	1965
Ritz, Harry	1906	1986
Ritz, Jimmy	1903	1985
Robards, Jason	1922	2000
Robbins, Jerome	1918	1998
Robbins, Marty	1925	1982
Robeson, Paul	1898	1976
Robinson, Bill	1878	1949
Robinson, Edward G.	1893	1973
Rochester (E. Anderson)	1905	1977
Roddenberry, Gene	1921	1991
Rodgers, Jimmie	1897	1933
Rogers, Buddy	1904	1999
Rogers, Ginger	1911	1995
Rogers, Roy	1911	1998
Rogers, Will	1879	1935
Roland, Gilbert	1905	1994
Rolle, Esther	1920?	1998
Rollins, Howard	1950	1996
Roman, Ruth	1924	1999
Romero, Cesar	1907	1994
Rooney, Pat	1880	1962
Rose, Billy	1899	1966
Rossellini, Roberto	1906	1977
Rowan, Dan	1922	1987
Rubinstein, Artur	1887	1982
Ruggles, Charles	1886	1970
Russell, Gail	1924	1961

Name	Born	Died
Russell, Lillian	1861	1922
Russell, Rosalind	1911	1976
Rutherford, Margaret	1892	1972
Ryan, Irene	1903	1973
Ryan, Robert	1909	1973
Sargent, Dick	1933	1994
St. Cyr, Lili	1917	1999
St. Denis, Ruth	1877	1968
Sakall, S.Z.	1884	1955
Sale (Chic), Charles	1885	1936
Sanders, George	1906	1972
Savalas, Telly	1924	1994
Schildkraut, Joseph	1895	1964
Schipa, Tito	1889	1965
Schnabel, Artur	1882	1951
Scott, George C.	1927	1999
Scott, Hazel	1920	1981
Scott, Randolph	1898	1987
Scott, Zachary	1914	1965
Scott-Siddons, Mrs.	1843	1896
Seberg, Jean	1938	1979
Seeley, Blossom	1892	1974
Segovia, Andres	1893	1987
Selena	1971	1995
Sellers, Peter	1925	1980
Selznick, David O.	1902	1965
Sennett, Mack	1884	1960
Senor Wences	1896	1999
Serling, Rod	1924	1975
Shakur, Tupac	1971	1996
Shaw, Robert (actor)	1927	1978
Shaw, Robert (conductor)	1916	1999
Shawn, Ted	1891	1972
Shean, Al	1868	1949
Shearer, Norma	1902	1983
Sheridan, Ann	1915	1967
Shore, Dinah	1917	1994
Shubert, Lee	1875	1953
Siddons, Mrs. Sarah	1755	1831
Sidney, Sylvia	1910	1999
Signoret, Simone	1921	1985
Silverheels, Jay	1912	1980
Silvers, Phil	1912	1985
Sim, Alastair	1900	1976
Sims, Irene	1930	2001
Sinatra, Frank	1915	1998
Sinclair, Madge	1938	1995
Siskel, Gene	1946	1999
Sitka, Emil	1914	1998
Sjostrom, Victor	1879	1960
Skelton, Red	1913	1997
Skinner, Otis	1858	1942
Smith, Alexis	1921	1992
Smith, Buffalo Bob	1917	1998
Smith, C. Aubrey	1863	1948
Smith, Kate	1907	1986
Snow, Hank	1914	1999
Solti, George	1912	1997
Sondergaard, Gale	1899	1985
Sothern, Ann	1909	2001
Sousa, John Philip	1854	1932
Sparks, Ned	1884	1957
Springfield, Dusty	1939	1999
Stander, Lionel	1908	1994
Stanley, Kim	1925	2001
Stanwyck, Barbara	1907	1990
Stern, Isaac	1920	2001
Stevens, Craig	1918	2000
Stevens, Inger	1934	1970
Stevens, Mark	1916	1994
Stevenson, McLean	1929	1996
Stewart, James	1908	1997
Stickney, Dorothy	1896	1998
Stokowski, Leopold	1882	1977
Stone, Lewis	1879	1953
Stone, Milburn	1904	1980
Strasberg, Lee	1901	1999
Strasberg, Susan	1938	1999
Sturges, Preston	1898	1959
Sullavan, Margaret	1911	1960
Sullivan, Barry	1912	1994
Sullivan, Ed	1902	1974
Sullivan, Francis L.	1903	1956
Summerville, Slim	1892	1946
Swanson, Gloria	1899	1983
Swarthout, Gladys	1904	1969
Switzer, Carl "Alfalfa"	1926	1959
Talbot, Lyle	1904	1996
Talmadge, Norma	1893	1957
Tamiroff, Akim	1899	1972
Tandy, Jessica	1909	1994
Tanguay, Eva	1878	1947
Tati, Jacques	1908	1982
Taylor, Deems	1885	1966
Taylor, Dub	1907	1994
Taylor, Estelle	1899	1958
Taylor, Laurette	1887	1946
Taylor, Robert	1911	1969
Terry, Ellen	1847	1928
Thalberg, Irving	1899	1936
Thomas, Danny	1912	1991
Thomas, John Charles	1892	1960
Thorndike, Sybil	1882	1976
Tibbett, Lawrence	1896	1960
Tierney, Gene	1920	1991
Tiny Tim	1932?	1996
Tippett, Sir Michael	1905	1998
Todd, Michael	1909	1958
Tomlinson, David	1917	2000
Tone, Franchot	1903	1968
Torme, Mel	1925	1999
Toscanini, Arturo	1867	1957
Tracy, Lee	1898	1968
Tracy, Spencer	1900	1967
Traubel, Helen	1903	1972
Travers, Henry	1874	1965
Treacher, Arthur	1894	1975
Tree, Herbert Beerbohm	1853	1917
Trevor, Claire	1909	2000
Truex, Ernest	1890	1973
Truffaut, Francois	1932	1984
Tucker, Forrest	1919	1986
Tucker, Richard	1913	1975
Tucker, Sophie	1884	1966
Turner, Lana	1920	1995
Turpin, Ben	1874	1940
Twelvetrees, Helen	1908	1959
Twitty, Conway	1933	1993
Valens, Ritchie	1941	1959
Valentino, Rudolph	1895	1926
Vallee, Rudy	1901	1986
Van, Bobby	1928	1980
Vance, Vivian	1912	1979
Van Fleet, Jo	1922	1996
Varney, Jim	1949	2000
Vaughan, Sarah	1924	1990
Veidt, Conrad	1893	1943
Velez, Lupe	1908	1944
Vera-Ellen	1926	1981
Verdon, Gwen	1925	2000
Villechaize, Herve	1943	1993
Vincent, Gene	1935	1971
Vicious, Sid	1958	1979
Vinson, Helen	1907	1999
Von Stroheim, Erich	1885	1957
Von Zell, Harry	1906	1981
Walker, Junior	1942	1995
Walker, Nancy	1922	1992
Walker, Robert	1918	1951
Wallenda, Karl	1905	1978
Walsh, J. T.	1943	1998
Walsh, Raoul	1887	1980
Walston, Ray	1914	2001
Walter, Bruno	1876	1962
Ward, Helen	1916	1998
Waring, Fred	1900	1984
Warner, H. B.	1876	1958
Washington, Dinah	1924	1963
Waters, Ethel	1896	1977
Waxman, Al	1935	2001
Wayne, David	1914	1995
Wayne, John	1907	1979
Webb, Clifton	1891	1966
Webb, Jack	1920	1982
Weems, Ted	1901	1963
Weissmuller, Johnny	1904	1984
Welk, Lawrence	1903	1992
Welles, Orson	1915	1985
Wellman, William	1896	1975
Werner, Oskar	1922	1984
West, Mae	1892	1980
Weston, Jack	1924	1996
Whale, James	1889	1957
Wheeler, Bert	1895	1968
White, Jesse	1919	1997
White, Pearl	1889	1938
Whiteman, Paul	1891	1967
Whitty, May	1865	1948
Wickes, Mary	1910	1995
Wilde, Cornel	1918	1989
Wilding, Michael	1912	1979
Williams, Bert	1877	1922
Williams, Guy	1924	1989
Williams, Hank Sr.	1923	1953
Wills, Bob	1905	1975
Wills, Chill	1903	1978
Wilson, Carl	1946	1998
Wilson, Dennis	1944	1983
Wilson, Dooley	1894	1953
Wilson, Flip	1933	1998
Wilson, Marie	1917	1972
Windsor, Marie	1919	2000
Winninger, Charles	1884	1969
Withers, Grant	1904	1959
Wong, Anna May	1907	1961
Wood, Natalie	1938	1981
Wood, Peggy	1892	1978
Woolley, Monty	1888	1963
Wyler, William	1902	1981
Wynette, Tammy	1942	1998
Wynn, Ed	1886	1966
Wynn, Keenan	1916	1986
Yankovic, Frank	1915	1998
York, Dick	1929	1992
Young, Clara Kimball	1890	1960
Young, Gig	1913	1978
Young. Loretta	1913	1999
Young, Robert	1907	1998
Young, Roland	1887	1953
Youngman, Henny	1906	1998
Zanuck, Darryl F.	1902	1979
Zappa, Frank	1940	1993
Zinneman, Fred	1907	1997
Ziegfeld, Florenz	1869	1932
Zukor, Adolph	1873	1976

Original Names of Selected Entertainers

EDIE ADAMS: Elizabeth Edith Enke
EDDIE ALBERT: Edward Albert Heimberger
ALAN ALDA: Alphonso D'Abruzzo
JASON ALEXANDER: Jay Greenspan
FRED ALLEN: John Sullivan
WOODY ALLEN: Allen Konigsberg
JUNE ALLYSON: Ella Geisman
JULIE ANDREWS: Julia Wells
EVE ARDEN: Eunice Quedens
BEATRICE ARTHUR: Bernice Frankel
JEAN ARTHUR: Gladys Greene
FRED ASTAIRE: Frederick Austerlitz
BABYFACE: Kenneth Edmonds
LAUREN BACALL: Betty Joan Perske
ERYKAH BADU: Erica Wright
ANNE BANCROFT: Anna Maria Italiano
GENE BARRY: Eugene Klass
PAT BENATAR: Patricia Andrejewski
TONY BENNETT: Anthony Benedetto
IRVING BERLIN: Israel Baline
JACK BENNY: Benjamin Kubelsky
JOEY BISHOP: Joseph Gottlieb
THE BIG BOPPER: Jiles Perry "J.P." Richardson
BONO (VOX): Paul Hewson
VICTOR BORGE: Borge Rosenbaum
DAVID BOWIE: David Robert Jones
BOY GEORGE: George Alan O'Dowd
FANNY BRICE: Fanny Borach
CHARLES BRONSON: Charles Buchinski
ALBERT BROOKS: Albert Einstein
MEL BROOKS: Melvin Kaminsky
GEORGE BURNS: Nathan Birnbaum
ELLEN BURSTYN: Edna Gilhooley

RICHARD BURTON: Richard Jenkins
RED BUTTONS: Aaron Chwatt
NICOLAS CAGE: Nicholas Coppola
MICHAEL CAINE: Maurice Micklewhite
MARIA CALLAS: Maria Kalogeropoulos
DIAHANN CARROLL: Carol Diahann Johnson
JACKIE CHAN: Chan Kwong-Sung
CYD CHARISSE: Tula Finklea
RAY CHARLES: Ray Charles Robinson
CHUBBY CHECKER: Ernest Evans
CHER: Cherilyn Sarkisian
PATSY CLINE: Virginia Patterson Hensley
LEE J. COBB: Leo Jacoby
CLAUDETTE COLBERT: Lily Chauchoin
ALICE COOPER: Vincent Furnier
DAVID COPPERFIELD: David Kotkin
HOWARD COSELL: Howard Cohen
ELVIS COSTELLO: Declan McManus
LOU COSTELLO: Louis Cristillo
PETER COYOTE: Peter Cohon
MICHAEL CRAWFORD: Michael Dumble-Smith
TOM CRUISE: Thomas Mapother IV
TONY CURTIS: Bernard Schwartz
VIC DAMONE: Vito Farinola
RODNEY DANGERFIELD: Jacob Cohen
BOBBY DARIN: Walden Robert Cassotto
DORIS DAY: Doris von Kappelhoff
YVONNE DE CARLO: Peggy Middleton
SANDRA DEE: Alexandra Zuck
JOHN DENVER: Henry John Deutschendorf Jr.
BO DEREK: Mary Cathleen Collins
DANNY DEVITO: Daniel Michaeli
ANGIE DICKINSON: Angeline Brown
BO DIDDLEY: Elias Bates
PHYLLIS DILLER: Phyllis Driver
TROY DONAHUE: Merle Johnson Jr.
KIRK DOUGLAS: Issur Danielovitch
MELVYN DOUGLAS: Melvyn Hesselberg
BOB DYLAN: Robert Zimmerman
BARBARA EDEN: Barbara Huffman
ELVIRA: Cassandra Peterson
EMINEM: Marshall Mathers
ENYA: Eithne Ni Bhraonian
DALE EVANS: Frances Smith
CHAD EVERETT: Raymond Cramton
DOUGLAS FAIRBANKS: Douglas Ullman
MORGAN FAIRCHILD: Patsy McClenny
JAMIE FARR: Jameel Farah
ALICE FAYE: Alice Jeanne Leppert
STEPIN FETCHIT: Lincoln Perry
W.C. FIELDS: William Claude Dukenfield
BARRY FITZGERALD: William Shields
JOAN FONTAINE: Joan de Havilland
JODIE FOSTER: Alicia Christian Foster
REDD FOXX: John Sanford
ANTHONY FRANCIOSA: Anthony Papaleo
ARLENE FRANCIS: Arlene Kazanjian
CONNIE FRANCIS: Concetta Franconero
GRETA GARBO: Greta Gustafsson
VINCENT GARDENIA: Vincent Scognamiglio
JOHN GARFIELD: Julius Garfinkle
JUDY GARLAND: Frances Gumm
JAMES GARNER: James Bumgarner
CRYSTAL GAYLE: Brenda Gayle Webb
KATHIE LEE GIFFORD: Kathie Epstein
WHOOPI GOLDBERG: Caryn Johnson
EYDIE GORME: Edith Gormezano
STEWART GRANGER: James Stewart
CARY GRANT: Archibald Leach
LEE GRANT: Lyova Rosenthal
JOEL GREY: Joe Katz
ROBERT GUILLAUME: Robert Williams
BUDDY HACKETT: Leonard Hacker
HAMMER: Stanley Kirk Burrell
JEAN HARLOW: Harlean Carpentier
REX HARRISON: Reginald Carey
LAURENCE HARVEY: Larushka Skikne
HELEN HAYES: Helen Brown
SUSAN HAYWARD: Edythe Marriner
RITA HAYWORTH: Margarita Cansino
PEE-WEE HERMAN: Paul Reubenfeld
WILLIAM HOLDEN: William Beedle
BILLIE HOLIDAY: Eleanora Fagan
JUDY HOLLIDAY: Judith Tuvim
HARRY HOUDINI: Ehrich Weiss
LESLIE HOWARD: Leslie Stainer
HOWLIN' WOLF: Chester Burnett
ROCK HUDSON: Roy Scherer Jr. (later Fitzgerald)
ENGELBERT HUMPERDINCK: Arnold Dorsey
KIM HUNTER: Janet Cole
BETTY HUTTON: Betty Thornberg
ICE CUBE: O'Shea Jackson
ICE-T: Tracy Morrow
BILLY IDOL: William Broad
DAVID JANSSEN: David Meyer
JAY-Z: Shawn Carter
ANN JILLIAN: Anne Nauseda
ELTON JOHN: Reginald Dwight
DON JOHNSON: Donald Wayne
AL JOLSON: Asa Yoelson
JENNIFER JONES: Phylis Isley
TOM JONES: Thomas Woodward
LOUIS JOURDAN: Louis Gendre
WYNONNA JUDD: Christina Ciminella
BORIS KARLOFF: William Henry Pratt
DANNY KAYE: David Kaminsky
DIANE KEATON: Diane Hall
MICHAEL KEATON: Michael Douglas
CHAKA KHAN: Yvette Stevens
CAROLE KING: Carole Klein
LARRY KING: Larry Zeigler
BEN KINGSLEY: Krishna Banji
NASTASSJA KINSKI: Nastassja Naksyznyski
TED KNIGHT: Tadeus Wladyslaw Konopka
CHERYL LADD: Cheryl Stoppelmoor
VERONICA LAKE: Constance Ockleman
HEDY LAMARR: Hedwig Kiesler
DOROTHY LAMOUR: Mary Leta Dorothy Slaton
MICHAEL LANDON: Eugene Orowitz
MARIO LANZA: Alfredo Cocozza
QUEEN LATIFAH: Dana Owens
STAN LAUREL: Arthur Jefferson
STEVE LAWRENCE: Sidney Leibowitz
BRENDA LEE: Brenda Mae Tarpley
GYPSY ROSE LEE: Rose Louise Hovick
MICHELLE LEE: Michelle Dusiak
PEGGY LEE: Norma Egstrom
JANET LEIGH: Jeanette Morrison
VIVIEN LEIGH: Vivian Hartley
HUEY LEWIS: Hugh Cregg
JERRY LEWIS: Joseph Levitch
LIL' KIM: Kimberly Denise Jones
CAROLE LOMBARD: Jane Peters
JACK LORD: John Joseph Ryan
SOPHIA LOREN: Sophia Scicolone
PETER LORRE: Laszio Lowenstein
MYRNA LOY: Myrna Williams
BELA LUGOSI: Bela Ferenc Blasko
MOMS MABLEY: Loretta Mary Aitken
SHIRLEY MACLAINE: Shirley Beaty
ELLE MACPHERSON: Eleanor Gow
LEE MAJORS: Harvey Lee Yeary 2d
KARL MALDEN: Mladen Sekulovich
BARRY MANILOW: Barry Alan Pincus
JAYNE MANSFIELD: Vera Jane Palmer
MARILYN MANSON: Brian Warner
FREDRIC MARCH: Frederick Bickel
PETER MARSHALL: Pierre LaCock
WALTER MATTHAU: Walter Matuschanskayasky
DEAN MARTIN: Dino Crocetti
MEAT LOAF: Marvin Lee Aday
FREDDIE MERCURY: Frederick Bulsara
ETHEL MERMAN: Ethel Zimmerman
GEORGE MICHAEL: Georgios Panayiotou
RAY MILLAND: Reginald Truscott-Jones
ANN MILLER: Lucille Collier
JONI MITCHELL: Roberta Joan Anderson
MOBY: Richard Melville Hall
MARILYN MONROE: Norma Jean Mortenson (later Baker)
YVES MONTAND: Ivo Livi
RON MOODY: Ronald Moodnick
DEMI MOORE: Demetria Guynes
GARRY MOORE: Thomas Garrison Morfit
RITA MORENO: Rosita Alverio
HARRY MORGAN: Harry Bratsburg
MR. T: Lawrence Tero
PAUL MUNI: Muni Weisenfreund
MIKE NICHOLS: Michael Igor Peschowsky
CHUCK NORRIS: Carlos Ray
NOTORIOUS B.I.G.: Christopher Wallace
HUGH O'BRIAN: Hugh Krampke
MAUREEN O'HARA: Maureen Fitzsimons
PATTI PAGE: Clara Ann Fowler
JACK PALANCE: Walter Palanuik
BERT PARKS: Bert Jacobson
MINNIE PEARL: Sarah Ophelia Cannon
BERNADETTE PETERS: Bernadette Lazzaro
EDITH PIAF: Edith Gassion
SLIM PICKENS: Louis Lindley
MARY PICKFORD: Gladys Smith
STEFANIE POWERS: Stefania Federkiewicz
PAULA PRENTISS: Paula Ragusa
ROBERT PRESTON: Robert Preston Meservey
PRINCE (THE ARTIST): Prince Rogers Nelson
DEE DEE RAMONE: Douglas Colvin
JOEY RAMONE: Jeffrey Hyman
JOHNNY RAMONE: John Cummings
TOMMY RAMONE: Tom Erdelyi
TONY RANDALL: Leonard Rosenberg
JOHNNIE RAY: John Alvin
MARTHA RAYE: Margaret O'Reed
DONNA REED: Donna Belle Mullenger
DELLA REESE: Delloreese Patricia Early
BUSTER RHYMES: Trevor Smith Jr.
JOAN RIVERS: Joan Sandra Molinsky
EDWARD G. ROBINSON: Emmanuel Goldenberg
GINGER ROGERS: Virginia McMath
ROY ROGERS: Leonard Franklin Slye
MICKEY ROONEY: Joe Yule Jr.
JOHNNY ROTTEN: John Lydon
LILLIAN RUSSELL: Helen Leonard
MEG RYAN: Margaret Hyra
WINONA RYDER: Winona Horowitz
SADE: Helen Folsad Abu
SOUPY SALES: Milton Hines
SUSAN SARANDON: Susan Tomaling
SEAL: Samuel Sealhenry
RANDOLPH SCOTT: George Randolph Crane
JANE SEYMOUR: Joyce Frankenberg
OMAR SHARIF: Michael Shalhoub
CHARLIE SHEEN: Carlos Irwin Estevez
MARTIN SHEEN: Ramon Estevez
BEVERLY SILLS: Belle Silverman
TALIA SHIRE: Talia Coppola
PHIL SILVERS: Philip Silversmith
SINBAD: David Atkins
"BUFFALO BOB" SMITH: Robert Schmidt
SNOOP DOGGY DOG: Calvin Broadus
ANN SOTHERN: Harriette Lake
ROBERT STACK: Robert Modini
BARBARA STANWYCK: Ruby Stevens
JEAN STAPLETON: Jeanne Murray
RINGO STARR: Richard Starkey
CONNIE STEVENS: Concetta Ingolia
STING: Gordon Sumner
DONNA SUMMER: La Donna Gaines
RIP TAYLOR: Charles Elmer Jr.
ROBERT TAYLOR: Spangler Brugh
DANNY THOMAS: Muzyad Yakhoob, later Amos Jacobs
TINY TIM: Herbert Khaury
RIP TORN: Elmore Rual Torn Jr.
RANDY TRAVIS: Randy Traywick
SOPHIE TUCKER: Sophia Kalish
TINA TURNER: Annie Mae Bullock
TWIGGY: Leslie Hornby
CONWAY TWITTY: Harold Lloyd Jenkins
RUDOLPH VALENTINO: Rudolpho D'Antonguolla
FRANKIE VALLI: Frank Castelluccio
SID VICIOUS: John Simon Ritchie
JOHN WAYNE: Marion Morrison
CLIFTON WEBB: Webb Hollenbeck
RAQUEL WELCH: Raquel Tejada
GENE WILDER: Jerome Silberman
SHELLEY WINTERS: Shirley Schrift
STEVIE WONDER: Stevland Morris
JANE WYMAN: Sarah Jane Fulks
GIG YOUNG: Byron Barr
LORETTA YOUNG: Gretchen Michaels

CITIES OF THE U.S.

Source: Bureau of Labor Statistics: employment; Bureau of Economic Analysis: per capita personal income. All other data from Census 2000, U.S. Census Bureau.

Included here are the 100 most populous U.S. cities, based on the 2000 Census. Population rank indicated by figure in parentheses. Most data are for the city proper; employment figures are for 2000, income figures for 1999. Some statistics, where noted, apply to the whole Metropolitan Statistical Area (MSA). Inc.=incorporated; est.=established.

Note: Websites are as of Sept. 2001 and subject to change.

Akron, Ohio

Population: 217,074 (81); **Pop. density:** 3,496 per sq. mi; **Pop. change (1990-2000):** –2.7%. **Area:** 62.1 sq. mi. **Employment:** 107,664 employed; 5.6% unemployed. **Per capita income (MSA):** $24,579; change (1998-99): 3.6%

History: settled 1825; inc. as city 1865; located on Ohio-Erie Canal and is a port of entry; polymer center of the Americas.

Transportation: 1 airport; major trucking industry; Conrail, Amtrak; metro transit system. **Communications:** 7 radio stations. **Medical facilities:** 4 hosp.; specialized children's treatment center. **Educational facilities:** 4 univ. and colleges; 68 pub. schools. **Further information:** Akron Regional Development Board, One Cascade Plaza, Akron, OH 44308.

Websites: http://www.ci.akron.oh.us
http://www.ardb.org

Albuquerque, New Mexico

Population: 448,607 (35); **Pop. density:** 2,484 per sq. mi; **Pop. change (1990-2000):** 16.6%. **Area:** 180.6 sq. mi. **Employment:** 233,859 employed; 3.1% unemployed. **Per capita income (MSA):** $25,619; change (1998-99): 3.4%.

History: founded 1706 by the Spanish; inc. 1890.

Transportation: 1 intl. airport; 1 railroad; 11 bus service/charters. **Communications:** 14 TV, 42 radio stations. **Medical facilities:** 6 major hosp. **Educational facilities:** 1 univ., 13 colleges. **Further information:** Albuquerque Convention & Visitors Bureau, PO Box 26866, Albuquerque, NM 87125-6866.

Websites: http://www.abqcvb.org
http://www.cabq.org

Anaheim, California

Population: 328,014 (55); **Pop. density:** 6,708 per sq. mi; **Pop. change (1990-2000):** 23.1%. **Area:** 48.9 sq. mi. **Employment:** 161,960 employed; 2.9% unemployed. **Per capita income (MSA):** $33,805; change (1998-99): 5.1%.

History: founded 1857; inc. 1870; now known as home of The Disneyland Resort, the Mighty Ducks of Anaheim, and the Anaheim Angels.

Transportation: 4 railroads; Greyhound buses (MSA). **Communications:** 1 TV, 2 radio stations (MSA). **Medical facilities:** 6 hosp.; 5 medical centers (MSA). **Educational facilities:** 10 univ. and colleges; 39 elem., 10 junior high, 11 high schools (MSA). **Further information:** Chamber of Commerce, 100 South Anaheim Blvd., Ste. 300, Anaheim, CA 92805.

Website: http://www.anaheim.net

Anchorage, Alaska

Population: 260,283 (65); **Pop. density:** 153 per sq. mi; **Pop. change (1990-2000):** 15.0%. **Area:** 1,697.2 sq. mi. **Employment:** 161,960 employed; 2.9% unemployed. **Per capita income (MSA):** $33,813; change (1998-99): 2.5%.

History: founded 1914 as a construction camp for railroad; HQ of Alaska Defense Command, WWII; severely damaged in earthquake 1964, but now rebuilt and currently population center of Alaska.

Transportation: 1 intl. airport; 1 railroad; transit system, 1 port. **Communications:** 7 TV, 22 radio stations. **Medical facilities:** 4 hosp. **Educational facilities:** 3 univ., 1 college, 89 pub. schools. **Further information:** Chamber of Commerce, 441 W. 5th Ave., Ste. 300, Anchorage, AK 99501-2309.

Websites: http://www.ci.anchorage.ak.us
http://www.anchoragechamber.org

Arlington, Texas

Population: 332,969 (53); **Pop. density:** 3,476 per sq. mi; **Pop. change (1990-2000):** 27.2%. **Area:** 95.8 sq. mi. **Employment:** 187,399 employed; 2.7% unemployed. **Per capita income (MSA):** $28,035; change (1998-99): 3.7%.

History: settled in 1840s between Dallas and Ft. Worth; inc. 1884.

Transportation: Dallas/Ft. Worth airport is 20 min. away; 11 railway lines; intercity transport system in planning stage. **Communications:** 11 TV, 44 radio stations. **Medical facilities:** 2 hosp. **Educational facilities:** 1 univ., 1 junior college; 60 pub. schools. **Further information:** The Arlington Chamber, 316 W. Main St., Arlington, TX 76010.

Websites: http://www.ci.arlington.tx.us
http://www.chamber.arlingtontx.com

Atlanta, Georgia

Population: 416,474 (39); **Pop. density:** 3,162 per sq. mi; **Pop. change (1990-2000):** 5.7%. **Area:** 131.7 sq. mi. **Employment:** 215,316 employed; 5.1% unemployed. **Per capita income (MSA):** $32,486; change (1998-99): 4.7%.

History: founded as "Terminus" 1837; renamed Atlanta 1845; inc. 1847; played major role in Civil War; became permanent state capital 1877; birthplace of civil rights movement; host to 1996 Centennial Olympic Games.

Transportation: 1 intl. airport; 3 railroad lines; MARTA bus and rapid rail service. **Communications:** 14 TV, 56 radio stations; 29 cable TV cos. **Medical facilities:** 61 hosp.; VA hosp.; U.S. Centers for Disease Control and Prevention; American Cancer Society. **Educational facilities:** 43 colleges, univ., seminaries, junior colleges; 813 pub. schools (metro area). **Further information:** Metro Atlanta Chamber of Commerce, 235 Andrew Young Intl. Blvd. NW, Atlanta, GA 30303.

Websites: http://www.atlantasmartcity.com
http://www.metroatlantachamber.com

Augusta, Georgia

Population: 199,775 (89); **Pop. density:** 661 per sq. mi. **Pop. change (1990-2000):** 347.5%. **Area:** 302.1 sq. mi. **Employment:** 75,761 employed; 5.8% unemployed. **Per capita income (MSA):** $23,549; change (1998-99): 3.1%.

History: founded 1736 as colonial trading post; one of the few pre-Civil War manufacturing centers in the South. Augusta National Golf Club, home of Masters Tournament, founded 1933.

Transportation: 1 regional, 1 local airport; Savannah River; 1 interstate highway. **Communications:** 4 TV, 20+ radio stations; 2 major cable TV providers. **Medical facilities:** 9 hosp., including Eisenhower Army Medical Center. **Educational facilities:** 1 univ., 1 medical college, 1 tech. college; 36 elem., 10 middle, 9 pub. high schools. **Further information:** Augusta Metropolitan Convention and Visitors Bureau, PO Box 1331, Augusta, GA 30903.

Websites: http://www.augustaga.org
http://www.augustagausa.com

Aurora, Colorado

Population: 276,393 (61); **Pop. density:** 1,940 per sq. mi; **Pop. change (1990-2000):** 24.4%. **Area:** 142.5 sq. mi. **Employment:** 158,371 employed; 2.2% unemployed. **Per capita income (MSA):** $36,058; change (1998-99): 6.3%.

History: located 5 mi east of Denver; early growth stimulated by presence of military bases; fast-growing trade center.

Transportation: adjacent to Denver Intl. Airport; 1 airport; bus system. **Communications:** 1 TV station. **Medical facilities:** 2 private hosp. **Educational facilities:** 1 univ., 4 community and junior colleges, 2 technical colleges; 86 pub. schools. **Further information:** Aurora Planning Dept., 1470 S. Havana St., Rm. 608, Aurora, CO 80012.

Websites: http://www.auroragov.org
http://www.aurorachamber.org

Austin, Texas

Population: 656,562 (16); **Pop. density:** 2,611 per sq. mi; **Pop. change (1990-2000):** 41.0%. **Area:** 251.5 sq. mi. **Employment:** 381,777 employed; 2.2% unemployed. **Per capita income (MSA):** $31,794; change (1998-99): 7.8%.

History: first permanent settlement 1835; capital of Rep. of Texas 1839; named after Stephen Austin; inc. 1840.

Transportation: 1 intl. airport; 4 railroads. **Communications:** 7 TV, 20 radio stations. **Medical facilities:** 11 hosp. **Educational facilities:** 7 univ. and colleges. **Further information:** Chamber of Commerce, PO Box 1967, Austin, TX 78767.

Websites: http://www.ci.austin.tx.us
http://www.austinchamber.org

IT'S A FACT: Austin is home to North America's largest urban bat population; up to 1.5 million Mexican free-tail bats fly there at night.

Bakersfield, California

Population: 247,057 (69); **Pop. density:** 2,184 per sq. mi; **Pop. change (1990-2000):** 41.3%. **Area:** 113.1 sq. mi. **Employment:** 92,072 employed; 8.3% unemployed. **Per capita income (MSA):** $19,886; change (1998-99): 0.8%.

History: named after Col. Thomas Baker, an early settler; inc. 1898.

Transportation: 1 airport; 3 railroads; Amtrak; Greyhound buses; local bus system. **Communications:** 5 TV, 34 radio stations. **Medical facilities:** 7 major hosp.; 9 convalescent, 1 psychiatric, 3 physical rehab., 5 urgent care facilities; 3 clinics. **Educational facilities:** 7 univ., 1 community college, 11 vocational schools, 1 adult school, 1 college of law, 139 elem., 39 junior high, 25 high schools. **Further information:** Greater Bakersfield Chamber of Commerce, 1725 Eye St., PO Box 1947, Bakersfield, CA 93303.

Website: http://www.bakersfieldchamber.org

Baltimore, Maryland

Population: 651,154 (17); **Pop. density:** 8,058 per sq. mi; **Pop. change (1990-2000):** –11.5%. **Area:** 80.8 sq. mi. **Employment:** 269,277 employed; 8.1% unemployed. **Per capita income (MSA):** $31,434; change (1998-99): 5.4%.

History: founded by Maryland legislature 1729; inc. 1797; bombing of Ft. McHenry (1814) inspired Francis Scott Key to write "Star-Spangled Banner"; birthplace of America's railroads 1828; rebuilt after fire 1904; site of National Aquarium 1981.

Transportation: 1 major airport; 3 railroads; bus system; subway system; light rail system; Inner Harbor water taxi system; 2 underwater tunnels. **Communications:** 5 TV, 26 radio stations. **Medical facilities:** 31 hosp.; 2 major medical centers. **Educational facilities:** over 30 univ. and colleges; 185 pub. schools. **Further information:** Greater Baltimore Committee, 111 S. Calvert St., Ste. 1700, Baltimore, MD 21202-6180.

Websites: http://www.ci.baltimore.md.us
http://www.baltimore.org

Baton Rouge, Louisiana

Population: 227,818 (74); **Pop. density:** 2,966 per sq. mi; **Pop. change (1990-2000):** 3.8%. **Area:** 76.8 sq. mi. **Employment:** 114,251 employed; 4.9% unemployed. **Per capita income (MSA):** $25,316; change (1998-99): 2.9%.

History: claimed by Spain at time of Louisiana Purchase 1803; est. independence by rebellion 1810; inc. as town 1817; became state capital 1849; Union-held most of Civil War.

Transportation: 1 airport, 5 airlines; 1 bus line; 3 railroad trunk lines. **Communications:** 5 TV, 19 radio stations. **Medical facilities:** 5 hosp. **Educational facilities:** 2 univ.; 105 pub., 52 nonpublic schools. **Further information:** The Chamber of Greater Baton Rouge, PO Box 3217, Baton Rouge, LA 70821.

Websites: http://www.baton-rouge.com/BatonRouge
http://www.brchamber.org

Birmingham, Alabama

Population: 242,820 (71); **Pop. density:** 1,620 per sq. mi; **Pop. change (1990-2000):** –8.7%. **Area:** 149.9 sq. mi. **Employment:** 123,444 employed; 5.0% unemployed. **Per capita income (MSA):** $27,896; change (1998-99): 4.4%.

History: settled 1871 at the intersection of 2 major railroads within proximity of elements required for iron and steel production.

Transportation: 1 intl. airport; 4 major rail freight lines, Amtrak; 1 bus line; 75 truck line terminals; 5 air cargo cos.; 7 barge lines; 5 interstate highways. **Communications:** 7 TV, 32 radio stations; 1 educational TV, 1 educational radio station. **Medical facilities:** 16, including the Univ. of Alabama at Birmingham Medical Center; VA hosp. **Educational facilities:** 1 pub., 2 private univ.; 4 private colleges, 1 private law school. **Further information:** Birmingham Area Chamber of Commerce, 2027 First Ave. N, Birmingham, AL 35203.

Websites: http://www.birminghamchamber.com
http://www.ci.bham.al.us

Boston, Massachusetts

Population: 589,141 (20); **Pop. density:** 12,172 per sq. mi; **Pop. change (1990-2000):** 2.6%. **Area:** 48.4 sq. mi. **Employment:** 283,782 employed; 2.9% unemployed. **Per capita income (MSA):** $36,285; change (1998-99): 6.6%.

History: settled 1630 by John Winthrop; capital of Mass. Bay Colony; figured strongly in Am. Revolution, earning distinction as the "Cradle of Liberty"; inc. 1822.

Transportation: 1 major airport; 2 railroads; city rail and subway system; 3 underwater tunnels; port. **Communications:** 12 TV, 21 radio stations. **Medical facilities:** 31 hosp.; 8 major medical research centers. **Educational facilities:** 30 univ. and colleges. **Further information:** Greater Boston Convention and Visitors Bureau, 2 Copley Pl., Suite 105, Boston, MA 02116.

Websites: http://www.bostonusa.com
http://www.gbcc.org

Buffalo, New York

Population: 292,648 (58); **Pop. density:** 7,208 per sq. mi; **Pop. change (1990-2000):** –10.8%. **Area:** 40.6 sq. mi. **Employment:** 127,264 employed; 8.1% unemployed. **Per capita income (MSA):** $26,710; change (1998-99): 4.4%.

History: founded 1790 by the Dutch; raided twice by British, War of 1812; served as western terminus for Erie Canal, became a center for trade and manufacturing; inc. 1832; last stop on the Underground Railroad; key point for Canada-U.S. political, trade, and social relations.

Transportation: 1 intl. airport; 4 Class I railroads; Amtrak metro rail system; water service to Great Lakes-St. Lawrence Seaway system and Atlantic seaboard. **Communications:** 8 TV, 23 radio stations. **Medical facilities:** 14 hosp., 37 research centers. **Educational facilities:** 12 colleges and univ.; 111 pub. and private schools. **Further information:** Buffalo Niagara Partnership, 300 Main Place Tower, Buffalo, NY 14202-3797.

Websites: http://www.ci.buffalo.ny.us/city/index.html
http://buffaloniagara.org

Charlotte, North Carolina

Population: 540,828 (26); **Pop. density:** 2,232 per sq. mi; **Pop. change (1990-2000):** 36.6%. **Area:** 242.3 sq. mi. **Employment:** 283,996 employed; 2.7% unemployed. **Per capita income (MSA):** $30,340; change (1998-99): 4.6%.

History: settled by Scotch-Irish immigrants 1740s; inc. 1768 and named after Queen Charlotte, George III's wife; scene of first major U.S. gold discovery 1799.

Transportation: 1 airport; 2 major railway lines; 1 bus line; 388 trucking firms. **Communications:** 7 TV, 26 radio stations. **Medical facilities:** 10 hosp., 1 medical center. **Educational facilities:** 4 univ., 8 colleges, 86 elem. schools, 28 middle schools, 14 high schools. **Further information:** Chamber of Commerce, PO Box 32785, Charlotte, NC 28232.

Website: http://www.charlottechamber.com

Chesapeake, Virginia

Population: 199,184 (91); **Pop. density:** 585 per sq. mi; **Pop. change (1990-2000):** 31.1%. **Area:** 340.7 sq. mi. **Employment:** 104,691 employed; 2.2% unemployed. **Per capita income (MSA):** $24,979; change (1998-99): 4.0%.

History: region settled in 1620s with first English colonies on banks of Elizabeth River; home to Great Dismal Swamp Canal, first envisioned by George Washington in 1763; Battle of Great Bridge fought here Dec. 1775; inc. as a city 1963.

Transportation: Amtrak, freight rail service; bus service; 1 intl. airport, 2 regional airports; deepwater ports. **Communications:** 9 TV, 48 radio stations. **Medical facilities:** 1 hosp. **Educational facilities:** 9 colleges and univ.; 49 pub. schools and educational centers. **Further information:** Hampton Roads Chamber of Commerce, Chesapeake Div., 400 Volvo Pky., Chesapeake, VA 23320.

Website: http://www.chesapeake.va.us

Chicago, Illinois

Population: 2,896,016 (3); **Pop. density:** 12,752 per sq. mi; **Pop. change (1990-2000):** 4.0%. **Area:** 227.1 sq. mi. **Employment:** 1,266,759 employed; 5.6% unemployed. **Per capita income (MSA):** $34,743; change (1998-99): 4.0%.

History: site acquired from Indians 1795; significant white settlement began with opening of Erie Canal 1825; chartered as city 1837; boomed with arrival of railroads from east and canal to Mississippi R.; about one-third of city destroyed by fire 1871; major grain and livestock market.

Transportation: 3 airports; major railroad system, trucking industry. **Communications:** 9 TV, 31 radio stations. **Medical facilities:** over 123 hosp. **Educational facilities:** 95 insts. of higher learning. **Further information:** Chicagoland Chamber of Commerce, 1 IBM Plaza, Ste. 2800, Chicago, IL 60611.

Websites: http://www.ci.chi.il.us
http://www.chicagolandchamber.org

IT'S A FACT: Baltimore was home of the first U.S. umbrella factory (1828) and first ice cream freezer (1848); the Ouija board was invented there (1892), and the first U.S. stage coach route ran from Baltimore to Philadelphia (1773).

Cincinnati, Ohio

Population: 331,285 (54); **Pop. density:** 4,247 per sq. mi; **Pop. change (1990-2000):** –9.0%. **Area:** 78.0 sq. mi. **Employment:** 164,282 employed; 5.1% unemployed. **Per capita income (MSA):** $30,105; change (1998-99): 4.4%.

History: founded 1788 and named after the Society of Cincinnati, an organization of Revolutionary War officers; chartered as village 1802; inc. as city 1819.

Transportation: 1 intl. airport; 3 railroads; 1 bus system. **Communications:** 7 TV, 25 radio stations. **Medical facilities:** 27 hosp.; Children's Hosp. Medical Center; VA hosp. **Educational facilities:** 4 univ., 11 colleges, 8 technical & 2-year colleges. **Further information:** Chamber of Commerce, 300 Carew Tower, 441 Vine St., Cincinnati, OH 45202.

Websites: http://www.cincinnatichamber.com
http://www.cincinnatiusa.org

Cleveland, Ohio

Population: 478,403 (33); **Pop. density:** 6,165 per sq. mi; **Pop. change (1990-2000):** –5.4%. **Area:** 77.6 sq. mi. **Employment:** 188,011 employed; 8.7% unemployed. **Per capita income (MSA):** $30,472; change (1998-99): 3.9%.

History: surveyed in 1796; given recognition as village 1815, inc. as city 1836; annexed Ohio City 1854.

Transportation: 1 intl. airport; rail service; major port; rapid transit system. **Communications:** 9 TV, 21 radio stations. **Medical facilities:** 14 hosp. **Educational facilities:** 8 univ. and colleges; 127 pub. schools. **Further information:** Greater Cleveland Growth Assn., Tower City Center, 50 Pub. Square, Suite 200, Cleveland, OH 44113-2291.

Websites: http://www.cleveland.oh.us
http://www.clevelandgrowth.com

Colorado Springs, Colorado

Population: 360,890 (48); **Pop. density:** 1,943 per sq. mi; **Pop. change (1990-2000):** 28.4%. **Area:** 185.7 sq. mi. **Employment:** 185,764 employed; 3.2% unemployed. **Per capita income (MSA):** $27,255; change (1998-99): 4.5%.

History: city founded in 1871 at the foot of Pike's Peak; inc. 1872.

Transportation: 1 municipal airport; 1 bus line. **Communications:** 9 TV, 28 radio stations. **Medical facilities:** 5 hosp. **Educational facilities:** 11 univ., 5 colleges. **Further information:** Chamber of Commerce, PO Box B, Colorado Springs, CO 80901.

Websites: http://www.coloradosprings-travel.com/cscvb
http://www.coloradospringschamber.org

Columbus, Ohio

Population: 711,470 (15); **Pop. density:** 3,383 per sq. mi; **Pop. change (1990-2000):** 12.4%. **Area:** 210.3 sq. mi. **Employment:** 386,107 employed; 2.8% unemployed. **Per capita income (MSA):** $29,777; change (1998-99): 4.4%.

History: first settlement 1797; laid out as new capital 1812 with current name; became city 1834.

Transportation: 6 airports; 2 railroads; 2 intercity bus lines. **Communications:** 8 TV, 29 radio stations. **Medical facilities:** 17 hosp. **Educational facilities:** 11 univ. and colleges; 8 technical/2-year schools; 129 pub. schools (67 elem., 21 middle, 14 high, 27 magnet). **Further information:** Greater Columbus Chamber of Commerce, 37 N. High St., Columbus, OH 43215.

Website: http://www.columbus.org

Corpus Christi, Texas

Population: 277,454 (60); **Pop. density:** 1,795 per sq. mi; **Pop. Change (1990-2000):** 7.8%. **Area:** 154.6 sq. mi. **Employment:** 121,666 employed; 6.2% unemployed. **Per capita income (MSA):** $21,936; change (1998-99): 2.6%.

History: settled 1839 and inc. 1852.

Transportation: 1 intl. airport; 2 bus lines, metro bus system; 3 freight railroads. **Communications:** 6 TV, 17 radio stations. **Medical facilities:** 14 hosp. including a children's center. **Educational facilities:** 1 univ., 1 college. **Further information:** Corpus Christi Regional Economic Development Corp., PO Box 2724, Corpus Christi, TX 78403.

Website: http://www.ccredc.com

Dallas, Texas

Population: 1,188,580 (8); **Pop. density:** 3,470 per sq. mi; **Pop. change (1990-2000):** 18.0%. **Area:** 342.5 sq. mi. **Employment:** 652,205 employed; 4.0% unemployed. **Per capita income (MSA):** $34,690; change (1998-99): 4.8%.

History: first settled 1841; platted 1846; inc. 1871; developed as the financial and commercial center of Southwest; headquarters of regional Federal Reserve Bank; major center for distribution and high-tech manufacturing.

Transportation: 1 intl. airport, 1 regional airport; Amtrak; transit system. **Communications:** 17 TV, 61 radio stations. **Medical facilities:** 15 general hosp.; major medical center. **Educational facilities:** 220 pub. schools, 11 univ. and colleges, 3 community college campuses. **Further information:** Greater Dallas Chamber, Resource Center, 1201 Elm St., Ste. 2000, Dallas, TX 75270.

Websites: http://www.dallaschamber.org
http://www.ci.dallas.tx.us

Denver, Colorado

Population: 554,636 (25); **Pop. density:** 3,616 per sq. mi; **Pop. change (1990-2000):** 18.6%. **Area:** 153.4 sq. mi. **Employment:** 269,413 employed; 3.0% unemployed. **Per capita income (MSA):** $36,058; change (1998-99): 6.3%.

History: settled 1858 by gold prospectors and miners; inc. 1861; became territorial capital 1867; growth spurred by gold and silver boom; became financial, industrial, cultural center of Rocky Mt. region.

Transportation: 1 intl. airport, 3 corporate reliever airports; 5 rail freight lines, Amtrak; 1 bus line. **Communications:** 14 TV, 29 radio stations. **Medical facilities:** 20 hosp. **Educational facilities:** 15 four-yr. colleges and univ.; 8 two-yr. and community colleges. **Further information:** Denver Metro Chamber of Commerce, 1445 Market St., Denver, CO 80202-1729.

Website: http://www.denverchamber.org

Des Moines, Iowa

Population: 198,682 (93); **Pop. density:** 2,621 per sq. mi; **Pop. change (1990-2000):** 2.8%. **Area:** 75.8 sq. mi. **Employment:** 117,938 employed; 2.6% unemployed. **Per capita income (MSA):** $31,118; change (1998-99): 4.5%.

History: Fort Des Moines built 1843; settled and inc. 1851; chartered as city 1857.

Transportation: 1 intl. airport; 4 bus lines; 4 railroads; metro bus system. **Communications:** 6 TV, 21 radio stations. **Medical facilities:** 6 hosp. **Educational facilities:** 2 univ., 6 colleges. **Further information:** Greater Des Moines Partnership, 700 Locust St., Ste. 100, Des Moines, IA 50309.

Websites: http://www.desmoinesmetro.com
http://www.ci.des-moines.ia.us

Detroit, Michigan

Population: 951,270 (10); **Pop. density:** 6,854 per sq. mi; **Pop. change (1990-2000):** –7.5%. **Area:** 138.8 sq. mi. **Employment:** 376,024 employed; 6.6% unemployed. **Per capita income (MSA):** $31,472; change (1998-99): 5.1%.

History: founded by French 1701; controlled by British 1760; acquired by U.S. 1796; destroyed by fire 1805; inc. as city 1824; capital of state 1837-47; auto manufacturing began 1899.

Transportation: 1 intl. airport; 5 railroads; major intl. port; pub. transit system. **Communications:** 11 TV, 37 radio stations. **Medical facilities:** 23 hosp.; 2 major medical centers. **Educational facilities:** 5 univ. and colleges. **Further information:** Detroit Regional Chamber, One Woodward Ave., PO Box 33840, Detroit, MI 48232-0840.

Website: http://www.detroitchamber.com

El Paso, Texas

Population: 563,662 (23); **Pop. density:** 2,263 per sq. mi; **Pop. change (1990-2000):** 9.4%. **Area:** 249.1 sq. mi. **Employment:** 235,681 employed; 7.9% unemployed. **Per capita income (MSA):** $17,216; change (1998-99): 2.6%.

History: first settled 1827; inc. 1873; arrival of railroad 1881 boosted city's population and industries.

Transportation: 1 intl. airport; 3 rail providers; 2 interstate highways; 4 intl. ports of entry. **Communications:** 12 TV, 21 radio stations. **Medical facilities:** 6 hosp.; 3 rehabilitation, 11 specialty centers. **Educational facilities:** 5 univ., 2 colleges (1 grad. only). **Further information:** Greater El Paso Chamber of Commerce, 10 Civic Center Plaza, El Paso, TX 79901.

Website: http://www.elpaso.org

Fort Wayne, Indiana

Population: 205,727 (84); **Pop. density:** 2,604 per sq. mi; **Pop. Change (1990-2000):** 18.9%. **Area:** 79.0 sq. mi. **Employment:** 94,284 employed; 3.8% unemployed. **Per capita income (MSA):** $27,355; change (1998-99): 3.5%.

History: French fort 1680; U.S. fort 1794; settled by 1832; inc. 1840 prior to Wabash-Erie canal completion 1843.

Transportation: 1 airport; 3 railroads; 6 bus lines. **Communications:** 5 TV, 21 radio stations. **Medical facilities:** 5 regional hosp.; VA hosp. **Educational facilities:** 5 univ., 4 colleges, 3 bus. schools; 80 pub. schools. **Further information:** Chamber of Commerce, 826 Ewing Street, Fort Wayne, IN 46802-2182.

Websites: http://www.ft-wayne.in.us
http://www.fwchamber.org

Fort Worth, Texas

Population: 534,694 (27); **Pop. density:** 1,828 per sq. mi; **Pop. change (1990-2000):** 19.5%. **Area:** 292.5 sq. mi. **Employment:** 265,060 employed; 4.2% unemployed. **Per capita income (MSA):** $28,035; change (1998-99): 3.7%.

History: established as military post 1849; inc. 1873; oil discovered 1917.

Transportation: 2 intl. airport; 10 major railroads, Amtrak; local bus service; 2 transcontinental, 2 intrastate bus lines. **Communications:** 14 TV, 11 local radio stations. **Medical facilities:** 25 hosp.; 1 children's hosp.; 4 government hosp. **Educational facilities:** 8 univ. and colleges. **Further information:** Chamber of Commerce, 777 Taylor St. #900, Fort Worth, TX 76102.

Website: http://www.fortworthchamber.com

Fremont, California

Population: 203,413 (85); **Pop. density:** 2,652 per sq. mi; **Pop. change (1990-2000):** 17.3%. **Area:** 76.7 sq. mi. **Employment:** 108,777 employed; 2.0% unemployed. **Per capita income (MSA):** $35,666; change (1998-99): 6.9%.

History: area first settled by Spanish 1769; inc. 1956 with consolidation of 5 communities.

Transportation: intracity bus line; Bay Area Rapid Transit System (southern terminal). **Communications:** 1 radio station. **Medical facilities:** 1 hosp.; 2 major medical facilities; 18 clinics. **Educational facilities:** 1 community college; 42 pub. schools. **Further information:** Chamber of Commerce, 39488 Stevenson Place, Suite 100, Fremont, CA 94539.

Website: http://www.fremontbusiness.com

Fresno, California

Population: 427,652 (37); **Pop. density:** 4,096 per sq. mi; **Pop. change (1990-2000):** 20.7%. **Area:** 104.4 sq. mi. **Employment:** 174,282 employed; 12.9% unemployed. **Per capita income (MSA):** $20,776; change (1998-99): 4.5%.

History: founded 1872; inc. as city 1885.

Transportation: 1 municipal airport; Amtrak; 1 bus line; intracity bus system. **Communications:** 13 TV, 23 radio stations. **Medical facilities:** 17 general hosp. **Educational facilities:** 9 colleges; 102 pub. schools. **Further information:** Greater Fresno Area Chamber of Commerce, PO Box 1469, Fresno, CA 93716-1469.

Websites: http://www.fresnochamber.com
http://fresno-online.com/cvb

Garland, Texas

Population: 215,768 (82); **Pop. density:** 3,778 per sq. mi; **Pop. change (1990-2000):** 19.4%. **Area:** 57.1 sq. mi. **Employment:** 122,050 employed; 2.7% unemployed. **Per capita income (MSA):** $34,690; change (1998-99): 4.8%.

History: settled 1850s; inc. 1891.

Transportation: 30 min. from Dallas/Ft. Worth Intl. Airport; 2 railroads. **Communications:** 14 local TV (Dallas/Ft. Worth), 25+ radio stations. **Medical facilities:** 2 hosp.; 348 beds. **Educational facilities:** 3 univ., 2 community colleges; 62 pub. schools. **Further information:** Chamber of Commerce, 914 S. Garland Ave., Garland, TX 75040.

Website: http://www.garlandchamber.com

Glendale, Arizona

Population: 218,812 (80); **Pop. density:** 3,928 per sq. mi; **Pop. change (1990-2000):** 47.7%. **Area:** 55.7 sq. mi. **Employment:** 122,050 employed; 2.7% unemployed. **Per capita income (MSA):** $27,617; change (1998-99): 4.3%.

History: est. 1892; inc. 1910.

Transportation: 1 local airport, 30 min. from Phoenix Sky Harbor Intl. Airport. **Communications:** 12 TV stations, 40 radio stations. **Medical facilities:** 3 hosp. **Educational facilities:** 12 institutes of higher education, 9 pub. school districts. **Further information:** Chamber of Commerce, PO Box 249, 7105 N. 59th Ave., Glendale, AZ 85311.

Website: http://www.glendaleazchamber.org

Glendale, California

Population: 194,973 (98); **Pop. density:** 6,372 per sq. mi; **Pop. change (1990-2000):** 8.3%. **Area:** 30.6 sq. mi. **Employment:** 93,377 employed, 5.1% unemployed. **Per capita income (MSA):** $28,276; change (1998-99): 3.6%.

History: became a town in 1887; inc. 1906.

Transportation: near Los Angeles Intl. airport; 1 local airport; commuter trains, Amtrak; bus system. **Communications:** 21 TV, 70 radio stations. **Medical facilities:** 3 hosp; other facilities. **Educational facilities:** 1 community college; 26 pub. schools. **Further information:** City of Glendale Public Information Officer, 613 E. Broadway, Glendale, CA 91206.

Website: http://www.ci.glendale.ca.us

Grand Rapids, Michigan

Population: 197,800 (94); **Pop. density:** 4,434 per sq. mi; **Pop. change (1990-2000):** 4.6%. **Area:** 44.6 sq. mi. **Employment:** 113,127 employed, 4.4% unemployed. **Per capita income (MSA):** $27,616; change (1998-99): 3.9%.

History: originally site of Ottawa Indian village; trading post 1826; became lumbering center and incorporated city 1850.

Transportation: 1 intl. airport; 3 rail carriers; Amtrak, Greyhound bus line; transit bus system. **Communications:** 8 TV, 39 radio stations. **Medical facilities:** 13 hosp. **Educational facilities:** 16 colleges; 19 pub. schools, 17 charter schools. **Further information:** Right Place Program, 111 Pearl St. NW, Grand Rapids, MI 49503

Website: http://www.rightplace.org

Greensboro, North Carolina

Population: 223,891 (77); **Pop. density:** 2,138 per sq. mi; **Pop. change (1990-2000):** 22.0%. **Area:** 104.7 sq. mi. **Employment:** 113,137 employed; 3.1% unemployed. **Per capita income (MSA):** $28,896; change (1998-99): 3.9%.

History: settled 1749; site of Revolutionary War conflict 1781 between Generals Nathanael Greene and Cornwallis; inc. 1807.

Transportation: 1 regional airport; 2 railroads; Trailways/Greyhound bus service. **Communications:** all cable TV stations; 11 radio stations. **Medical facilities:** 4 hosp. **Educational facilities:** 2 univ., 3 colleges; 94 pub. schools. **Further information:** Chamber of Commerce, PO Box 3246, Greensboro, NC 27402.

Websites: http://www.ci.greensboro.nc.us
http://www.greensboro.org

Hialeah, Florida

Population: 226,419 (75); **Pop. density:** 11,793 per sq. mi; **Pop. change (1990-2000):** 20.4%. **Area:** 19.2 sq. mi. **Employment:** 99,361 employed; 5.5% unemployed. **Per capita income (MSA):** $24,733; change (1998-99): 3.2%.

History: founded 1917, inc. 1925; industrial and residential city NW of Miami; Hialeah Park Horse Racing Track.

Transportation: 5 mi from Miami Intl. Airport; access to Port of Miami; Amtrak; 2 rail freight lines; Metrorail, Metrobus systems. **Communications:** 5 TV, 7 radio stations. **Medical facilities:** 4 hosp. (30 more in the area). **Educational facilities:** 8 univ. and colleges, 25 pub., 39 private schools. **Further information:** Hialeah-Dade Development, Inc., 501 Palm Ave., Hialeah, FL 33010.

Websites: http://www.ci.hialeah.fl.us
http://www.hddi.org

Honolulu, Hawaii

Population: 371,657 (46); **Pop. density:** 4,336 per sq. mi; **Pop. change (1990-2000):** 1.7%. **Area:** 85.7 sq. mi. **Employment (MSA):** 407,601 employed, 3.8% unemployed. **Per capita income (MSA):** $29,465; change (1998-99): 2.9%.

History: harbor entered by Europeans 1778; declared capital of kingdom by King Kamehameha III 1850; Pearl Harbor naval base attacked by Japanese Dec. 7, 1941.

Transportation: 1 major airport; large, active port for passengers and cargo. **Communications:** 10 TV, 33 radio stations. **Medical facilities:** 15 major medical centers. **Educational facilities:** 4 univ., 6 colleges; 257 pub. schools, 91 private schools. **Further information:** Hawaii Visitors and Convention Bureau, 2270 Kalakaua Avenue, Honolulu, HI 96815.

Websites: http://www.co.honolulu.hi.us
http://www.gohawaii.com

Houston, Texas

Population: 1,953,631 (4); **Pop. density:** 3,372 per sq. mi; **Pop. change (1990-2000):**19.8%. **Area:** 579.4 sq. mi. **Employment:** 974,799 employed; 5.1% unemployed. **Per capita income (MSA):** $32,386; change (1998-99): 2.9%.

History: founded 1836; inc. 1837; capital of Repub. of Texas 1837-39; developed rapidly after construction of channel to Gulf of Mexico 1914; world center of oil and natural gas technology.

Transportation: 3 commercial airports; 2 mainline railroads; major bus transit system; major intl. port. **Communications:** 16 TV, 55 radio stations. **Medical facilities:** 63 hosp.; major medical center. **Educational facilities:** 34 univ. and colleges. **Further information:** Greater Houston Partnership, 1200 Smith St., Houston, TX 77002-4400.

Websites: http://www.houston.org
http://www.ci.houston.tx.us

Indianapolis, Indiana

Population: 791,926 (12); **Pop. density:** 2,191 per sq. mi; **Pop. change (1990-2000):** 6.7%. **Area:** 361.5 sq. mi. **Employment:** 402,436 employed; 3.0% unemployed. **Per capita income (MSA):** $30,523; change (1998-99): 4.2%.

History: settled 1820; became capital 1825.

Transportation: 1 intl. airport; 5 railroads; 3 interstate bus lines. **Communications:** 10 TV, 27 radio stations. **Medical facilities:** 17 hosp.; 1 major medical and research center. **Educational facilities:** 8 univ. and colleges; major pub. library system. **Further information:** Chamber of Commerce, 320 N. Meridian St., Suite 200, Indianapolis, IN 46204.

Websites: http://www.ci.indianapolis.in.us
http://www.indychamber.com

Irving, Texas

Population: 191,615 (100); **Pop. density:** 2,851 per sq. mi; **Pop. change (1990-2000):** 23.6%. **Area:** 67.2 sq. mi. **Employment:** 113,495 employed, 2.9% unemployed. **Per capita income (MSA):** $34,690; change (1998-99): 4.8% .

History: founded 1903; inc. 1914; remained a small city until the 1950s.

Transportation: 1 intl. airport, 1 bus system, 1 RR express (train service). **Communications:** 9 TV, 7 radio stations. **Medical facilities:** 2 hosp. **Educational facilities:** 1 univ., 2 colleges. **Further information:** Greater Irving-Las Colinas Chamber of Commerce, 3333 N. MacArthur Blvd., Ste. 100, Irving, TX 75062.

Websites: http://www.irvingchamber.com
http://www.ci.irving.tx.us

Jacksonville, Florida

Population: 735,617 (14); **Pop. density:** 971 per sq. mi; **Pop. change (1990-2000):** 15.8%. **Area:** 757.7 sq. mi. **Employment:** 379,835 employed; 3.3% unemployed. **Per capita income (MSA):** $27,625; change (1998-99): 1.1%.

History: settled 1816 as Cowford; renamed after Andrew Jackson 1822; inc. 1832; rechartered 1851; scene of conflicts in Seminole and Civil wars.

Transportation: 1 intl. airport; 3 railroads; 2 interstate bus lines; 2 seaports. **Communications:** 7 TV, 34 radio stations. **Medical facilities:** 11 hosp. **Educational facilities:** 7 univ., 5 colleges, 2 community colleges; 233 pub. schools, 155 private schools. **Further information:** Chamber of Commerce, 3 Independent Drive, Jacksonville, FL 32202.

Websites: http://www.expandinjax.com
http://www.myjaxchamber.com

Jersey City, New Jersey

Population: 240,055 (72); **Pop. density:** 16,111 per sq. mi; **Pop. change (1990-2000):** 5.0%. **Area:** 14.9 sq. mi. **Employment:** 103,110 employed; 7.1% unemployed. **Per capita income (MSA):** $27,662; change (1998-99): 3.8%.

History: site bought from Indians 1630; chartered as town by British 1668; scene of Revolutionary War conflict 1779; chartered under present name 1838; important station on Underground Railroad.

Transportation: Intercity bus and subway system; ferry service to Manhattan. **Communications:** see New York, NY. **Medical facilities:** 4 hosp. **Educational facilities:** 3 colleges. **Further information:** Hudson County Chamber of Commerce, 253 Washington St., Jersey City, NJ 07302.

Website: http://www.jerseycitynet.com

Kansas City, Missouri

Population: 441,545 (36); **Pop. density:** 1,408 per sq. mi; **Pop. change (1990-2000):** 1.5%. **Area:** 313.5 sq. mi. **Employment:** 255,344 employed; 4.0% unemployed. **Per capita income (MSA):** $30,225; change (1998-99): 5.2%.

History: settled by 1838 at confluence of the Missouri and Kansas rivers; inc. 1851.

Transportation: 1 intl. airport; a major rail center; more than 300 motor freight carriers; 7 barge lines. **Communications:** 9 TV, 43 radio stations. **Medical facilities:** 50 hosp.; 2 VA hosp. **Educational facilities:** 22 univ. and colleges. **Further information:** Greater Kansas City Chamber of Commerce, 911 Main St., Ste. 2600, Kansas City, MO 64105.

Websites: http://www.kansascity.com
http://www.kcchamber.com

Las Vegas, Nevada

Population: 478,434 (32); **Pop. density:** 4,223 per sq. mi; **Pop. change (1990-2000):** 85.2%. **Area:** 113.3 sq. mi. **Employment:** 233,299 employed; 4.1% unemployed. **Per capita income (MSA):** $29,486; change (1998-99): 4.1%.

History: occupied by Mormons 1855-57; bought by railroad 1903; city of Las Vegas inc. 1911; gambling legalized 1931.

Transportation: 1 intl. airport; 2 railroads; bus system. **Communications:** 7 TV, 30 radio stations. **Medical facilities:** 12 hosp. **Educational facilities:** 1 univ., 5 colleges; 217 pub. schools in area. **Further information:** Las Vegas Chamber of Commerce, 3720 Howard Hughes Parkway, Las Vegas, NV 89109.

Website: http://www.lvchamber.com

Lexington, Kentucky

Population: 260,512 (64); **Pop. density:** 916 per sq. mi; **Pop. change (1990-2000):** 15.6%. **Area:** 284.5 sq. mi. **Employment:** 143,992 employed; 1.8% unemployed. **Per capita income (MSA):** $28,161; change (1998-99): 4.0%.

History: site was founded and named in 1775 by hunters who heard of the Revolutionary War battle at Lexington, Mass.; settled 1779; chartered 1782; inc. as a city 1832.

Transportation: 8 comm. airlines; 2 railroads; city buses. **Communications:** 5 TV, 20 radio stations. **Medical facilities:** 5 general, 5 specialized hosp. **Educational facilities:** 2 univ., 4 colleges. **Further information:** Greater Lexington Chamber of Commerce, 330 E. Main St., Lexington, KY 40507.

Website: http://www.lexchamber.com

Lincoln, Nebraska

Population: 225,581 (76); **Pop. density:** 3,023 per sq. mi; **Pop. change (1990-2000):** 17.5%. **Area:** 74.6 sq. mi. **Employment:** 127,445 employed; 2.7% unemployed. **Per capita income (MSA):** $28,493; change (1998-99): 4.2%.

History: originally called Lancaster; chosen state capital 1867, renamed after Abraham Lincoln; inc. 1869.

Transportation: 1 airport; Greyhound; Amtrak, 2 railroads. **Communications:** 2 TV, 15 radio stations. **Medical facilities:** 5 hosp. including VA, rehabilitation facilities. **Educational facilities:** 3 univ., 3 voc.-tech./business colleges; 53 pub., 15 private schools. **Further information:** Chamber of Commerce, PO Box 83006, Lincoln, NE 68501-3006.

Websites: http://www.lincoln.org
http://www.lcoc.com

Long Beach, California

Population: 461,522 (34); **Pop. density:** 9,157 per sq. mi; **Pop. change (1990-2000):** 7.5%. **Area:** 50.4 sq. mi. **Employment:** 211,293 employed; 5.0% unemployed. **Per capita income (MSA):** $28,276; change (1998-99): 3.6%.

History: settled as early as 1784 by Spanish; by 1884 present site developed on harbor; inc. 1888; oil discovered 1921.

Transportation: 1 airport; 3 railroads; major intl. port; 4 bus co. with 40 bus lines, light rail service. **Communications:** 1 radio station, 1 CATV franchise. **Medical facilities:** 5 hosp. **Educational facilities:** 1 univ., 1 community college (2 campuses); 87 pub. schools in district. **Further information:** Long Beach City Hall, 333 W. Ocean Blvd., Long Beach, CA 90802

Websites: http://www.ci.long-beach.ca.us
http://www.lbchamber.com

Los Angeles, California

Population: 3,694,820 (2); **Pop. density:** 7,876 per sq. mi; **Pop. change (1990-2000):** 6.0%. **Area:** 469.1 sq. mi. **Employment:** 1,790,620 employed; 6.1% unemployed. **Per capita income (MSA):** $28,276; change (1998-99): 3.6%.

History: founded by Spanish 1781; captured by U.S. 1846; inc. 1850; grew rapidly after coming of rairoads, 1876 & 1885, Hollywood a district of L.A.

Transportation: 1 intl. airport; 3 railroads; major freeway system; intracity transit system. **Communications:** 21 TV, 70 radio stations. **Medical facilities:** 822 hosp. and clinics in metro. area. **Educational facilities:** 192 univ. and colleges (incl. junior, community, and other); 1,678 pub. schools; 1,470 private schools. **Further information:** Los Angeles Area Chamber of Commerce, 350 S. Bixel St., PO Box 513696, Los Angeles, CA 90051-1696.

Websites: http://www.ci.la.ca.us
http://www.lachamber.org

Louisville, Kentucky

Population: 256,231 (66); **Pop. density:** 4,126 per sq. mi; **Pop. change (1990-2000):** –4.8%. **Area:** 62.1 sq. mi. **Employment:** 125,955 employed; 3.7% unemployed. **Per capita income (MSA):** $29,342; change (1998-99): 4.6%.

History: settled 1778; named for Louis XVI of France; inc. 1828; base for Union forces in Civil War.

Transportation: 1 municipal airport, 1 private-craft airport; 1 terminal, 4 trunk-line railroads; metro bus line, Greyhound station; 5 barge lines. **Communications:** 6 TV, 21 radio stations, 2 educational. **Medical facilities:** 23 hosp. **Educational facilities:** 10 univ. and colleges, 32 business and vocational schools. **Further information:** Greater Louisville, Inc. Metro Chamber of Commerce, 600 W. Main St., Louisville, KY 40202.

Website: http://www.greaterlouisville.com

Lubbock, Texas

Population: 199,564 (90); **Pop. density:** 1,738 per sq. mi; **Pop. change (1990-2000):** 7.2%. **Area:** 114.8 sq. mi. **Employment:** 101,972 employed; 2.6% unemployed. **Per capita income (MSA):** $24,459; change (1998-99): 3.0%.
History: settled 1879; laid out 1891; inc. 1909 through merger of two towns.
Transportation: 1 intl. airport; 2 railroads, bus line. **Communications:** 9 TV, 25 radio stations. **Medical facilities:** 7 hosp. **Educational facilities:** 3 univ., 1 junior college; 51 pub. schools. **Further information:** Chamber of Commerce, 1301 Broadway, Lubbock, TX 79401.
Websites: http://www.ci.lubbock.tx.us
http://www.lubbockchamber.com

Madison, Wisconsin

Population: 208,054 (83); **Pop. density:** 3,028 per sq. mi; **Pop. change (1990-2000):** 8.8%. **Area:** 68.7 sq. mi. **Employment:** 126,881 employed; 1.8% unemployed. **Per capita income (MSA):** $31,999; change (1998-99): 4.6%.
History: first white settlement 1832; selected as site for state capital, named after James Madison, 1836; chartered 1856.
Transportation: 1 airport, 10 airlines; 1 intracity, 3 intercity bus systems; 3 freight rail lines. **Communications:** 8 TV, 24 radio stations, 3 cable providers. **Medical facilities:** 6 hosp., 92 clinics. **Educational facilities:** 7 colleges and univ., including main branch of Univ. of Wisconsin; 29 elem. schools, 12 middle schools, 5 high schools. **Further information:** Greater Madison Chamber of Commerce, PO Box 71, Madison, WI 53701-0071.
Websites: http://www.ci.madison.wi.us
http://www.greatermadisonchamber.com

Memphis, Tennessee

Population: 650,100 (18); **Pop. density:** 2,328 per sq. mi; **Pop. change (1990-2000):** 6.5%. **Area:** 279.3 sq. mi. **Employment:** 301,486 employed; 4.8% unemployed. **Per capita income (MSA):** $28,828; change (1998-99): 3.7%.
History: French, Spanish, and U.S. forts by 1797; settled by 1819; inc. as town 1826, as city 1840; surrendered charter to state 1879 after yellow fever epidemics; rechartered as city 1893.
Transportation: 1 intl. airport; 5 railroads; 1 bus system. **Communications:** 7 TV, 32 radio stations. **Medical facilities:** 19 hosp. **Educational facilities:** 17 univ. and colleges; 209 pub., 76 private schools. **Further information:** Memphis Area Chamber of Commerce, 22 N. Front St., Ste. 200, PO Box 224, Memphis, TN 38101-0224.
Websites: http://www.ci.memphis.tn.us
http://www.memphischamber.com

Mesa, Arizona

Population: 396,375 (42); **Pop. density:** 3,171 per sq. mi; **Pop. change (1990-2000):** 37.6%. **Area:** 125.0 sq. mi. **Employment:** 197,233 employed; 2.2% unemployed. **Per capita income (MSA):** $27,617; change (1998-99): 4.3%.
History: founded by Mormons 1878; inc. 1883; 13 mi. from Phoenix; population boomed fivefold 1960-80.
Transportation: near Sky Harbor Intl. Airport in Phoenix; 2 railroads; bus line. **Medical facilities:** 4 major hosp. **Educational facilities:** 1 univ., 3 colleges; 70 pub. schools. **Further information:** Convention and Visitor's Bureau, 120 N. Center, Mesa, AZ 85201.
Websites: http://www.ci.mesa.az.us
http://www.mesacvb.com

Miami, Florida

Population: 362,470 (47); **Pop. density:** 10,153 per sq. mi; **Pop. change (1990-2000):** 1.1%. **Area:** 35.7 sq. mi. **Employment:** 167,648 employed; 7.7% unemployed. **Per capita income (MSA):** $24,733; change (1998-99): 3.2%.
History: site of fort 1836; settlement began 1870; inc. 1896, modern city developed into financial and recreation center; land speculation in 1920s added to city's growth, as did Cuban, Central and South American, and Haitian immigration since 1960.
Transportation: 1 intl. airport; seaport; Amtrak, transit rail system; 2 bus lines; 65 truck lines. **Communications:** 9 commercial, 2 educational TV stations; 41 radio stations. **Medical facilities:** 8 hosp.; VA hosp. **Educational facilities:** 6 univ. and colleges. **Further information:** Miami-Dade Dept. of Planning, Development, and Regulation, Research Div., 111 NW 1st St., Ste. 1220, Miami, FL 33128.
Websites: http://www.ci.miami.fl.us
http://www.greatermiami.com

Milwaukee, Wisconsin

Population: 596,974 (19); **Pop. density:** 6,212 per sq. mi; **Pop. change (1990-2000):** –5.0%. **Area:** 96.1 sq. mi. **Employment:** 248,725 employed; 6.7% unemployed. **Per capita income (MSA):** 31,805; change (1998-99): 4.6%.
History: Indian trading post by 1674; settlement began 1835; inc. as city 1848; famous beer industry.
Transportation: 1 intl. airport; 3 railroads; major port; 4 bus lines. **Communications:** 12 TV, 37 radio stations. **Medical facilities:** 8 hosp.; major medical center. **Educational facilities:** 7 univ. and colleges, 160 pub. schools. **Further information:** Metropolitan Milwaukee Association of Commerce, 756 N. Milwaukee St., Milwaukee, WI 53202.
Websites: http://www.ci.mil.wi.us
http://www.milwaukee.org

Minneapolis, Minnesota

Population: 382,618 (45); **Pop. density:** 6,969 per sq. mi; **Pop. change (1990-2000):** 3.9%. **Area:** 54.9 sq. mi. **Employment:** 199,623 employed; 3.2% unemployed. **Per capita income (MSA):** $35,250; change (1998-99): 4.5%.
History: site visited by Hennepin 1680; included in area of military reservations 1819; inc. 1867.
Transportation: 1 intl. airport; 5 railroads. **Communications:** 7 TV, 30 radio stations. **Medical facilities:** 7 hosp., incl. leading heart hosp. at Univ. of Minnesota. **Educational facilities:** 10 univ. and colleges; 121 pub., 28 private schools. **Further information:** City of Minneapolis Office of Pub. Affairs, 323M City Hall, 350 S. 5th St., Minneapolis, MN 55415.
Website: http://www.ci.minneapolis.mn.us

Mobile, Alabama

Population: 198,915 (92); **Pop. density:** 1,687 per sq. mi; **Pop. change (1990-2000):** 1.3%. **Area:** 117.9 sq. mi. **Employment:** 96,871 employed; 5.9% unemployed. **Per capita income (MSA):** $21,814; change (1998-99): 2.9%.
History: settled by French 1711; occupied by U.S. 1813; inc. as city 1814; only seaport of Alabama.
Transportation: 4 rail freight lines, Amtrak; 6 airlines; 55 truck lines; leading river system. **Communications:** 8 TV, 27 radio stations. **Medical facilities:** 7 hosp. **Educational facilities:** 3 univ., 5 colleges. **Further information:** Chamber of Commerce, PO Box 2187, Mobile, AL 36652.
Websites: http://www.ci.mobile.al.us
http://www.mobilechamber.com

Montgomery, Alabama

Population: 201,568 (87); **Pop. density:** 1,297 per sq. mi; **Pop. change (1990-2000):** 7.7%. **Area:** 155.4 sq. mi. **Employment:** 96,945 employed; 3.8% unemployed. **Per capita income (MSA):** $25,637; change (1998-99): 5.0%.
History: inc. as town 1819, as city 1837; became state capital 1846; first capital of Confederacy 1861.
Transportation: 3 airlines; 2 railroads; 2 bus lines; Alabama R. navigable to Gulf of Mexico. **Communications:** 4 TV, 2 CATV, 1 public TV, 16 radio stations. **Medical facilities:** 3 major hosp.; VA and 32 clinics. **Educational facilities:** 8 colleges and univ.; 35 pub., 35 private schools. **Further information:** Montgomery Area Chamber of Commerce, PO Box 79, Montgomery, AL 36101.
Website: http://www.montgomerychamber.com

Nashville, Tennessee

Population: 569,891 (22); **Pop. density:** 1,204 per sq. mi; **Pop. change (1990-2000):** 11.6%. **Area:** 473.3 sq. mi. **Employment:** 296,992 employed; 2.9% unemployed. **Per capita income (MSA):** $30,510; change (1998-99): 4.1%.
History: settled 1779; first chartered 1806; became permanent state capital 1843; home of Grand Ole Opry.
Transportation: 1 airport; 1 railroad; bus line; transit system of buses and trolleys. **Communications:** 11 TV, 34 radio stations. **Medical facilities:** 14 hosp.; VA and speech-hearing center. **Educational facilities:** 17 universities and colleges, 129 pub. schools. **Further information:** Chamber of Commerce, 211 Commerce St., Ste 100, Nashville, TN 37201.
Website: http://www.nashvillechamber.com

Newark, New Jersey

Population: 273,546 (63); **Pop. density:** 11,494 per sq. mi; **Pop. change (1990-2000):** –0.6%. **Area:** 23.8 sq. mi. **Employment:** 100,983 employed; 8.1% unemployed. **Per capita income (MSA):** $38,715; change (1998-99): 4.5%.
History: settled by Puritans 1666; used as supply base by Washington 1776; inc. as town 1833, as city 1836.
Transportation: 1 intl. airport; 1 intl. seaport, 3 railroads; bus system; subways. **Communications:** 5 TV, 5 radio stations within city limits, 1 daily newspaper, 8 weekly papers. **Medical facilities:** 5 hosp. **Educational facilities:** 5 univ. and colleges; 71 pub. schools, 40 private schools. **Further information:** Newark Public Information Office, City of Newark, 920 Broad St., Newark, NJ 07102.
Websites: http://www.ci.newark.nj.us
http://www.rbp.org

New Orleans, Louisiana

Population: 484,674 (31); **Pop. density:** 2,684 per sq. mi; **Pop. change (1990-2000):** –2.5%. **Area:** 180.6 sq. mi. **Employment:** 183,166 employed, 5.7% unemployed. **Per capita income (MSA):** $25,960; change (1998-99): 2.2%.

History: founded by French 1718; became major seaport on Mississippi R.; acquired by U.S. as part of Louisiana Purchase 1803; inc. as city 1805; Battle of New Orleans was last battle of War of 1812.

Transportation: 2 airports; major railroad center; major intl. port. **Communications:** 8 TV, 26 radio stations. **Medical facilities:** 25 hosp.; 2 major research centers. **Educational facilities:** 12 univ. and 9 colleges. **Further information:** New Orleans Metropolitan Convention & Visitors Bureau, Inc., 1520 Sugar Bowl Dr., New Orleans, LA 70112.

Website: http://www.neworleanscvb.com

New York, New York

Population: 8,008,278 (1); **Pop. density:** 26,404 per sq. mi; **Pop. change (1990-2000):** 9.4%. **Area:** 303.3 sq. mi. **Employment:** 3,357,363 employed; 5.7% unemployed. **Per capita income (MSA):** $38,814; change (1998-99): 5.9%.

History: trading post established 1625; British took control from Dutch 1664 and named city New York; briefly U.S. capital; Washington inaugurated as president 1789; under new charter, 1898, city expanded to include 5 boroughs: The Bronx, Brooklyn, Queens, and Staten Island, as well as Manhattan.

Transportation: 3 intl. airports serve area; 2 rail terminals; major subway network that includes 25 routes; 235 bus routes; ferry system; 4 underwater tunnels. **Communications:** 17 TV, 67 radio stations. **Medical facilities:** 82 hosp.; 8 academic medical centers. **Educational facilities:** 100 univ. and colleges; 1,189 pub. schools. **Further information:** Convention and Visitors Bureau, 810 Seventh Ave., New York, NY 10019.

Websites: http://www.ci.nyc.ny.us
http://www.nycvisit.com

Norfolk, Virginia

Population: 234,403 (73); **Pop. density:** 4,365 per sq. mi; **Pop. change (1990-2000):** –10.3%. **Area:** 53.7 sq. mi. **Employment:** 80,129 employed, 4.2% unemployed. **Per capita income (MSA):** $24,979; change (1998-99): 4.0%.

History: founded 1682; burned by patriots to prevent capture by British during Revolutionary War; rebuilt and inc. as town 1805, as city 1845; site of world's largest naval base.

Transportation: 1 intl. airport; 2 railroads; Amtrak; bus system. **Communications:** 13 TV, 6 city-access TV, 27 radio stations. **Medical facilities:** 6 hosp. **Educational facilities:** 3 univ., 1 college, 1 medical school; 59 pub. schools. **Further information:** Norfolk Convention and Visitors Bureau, 232 E. Main St., Norfolk, VA 23510.

Websites: http://www.norfolk.va.us
http://www.norfolkcvb.com

Oakland, California

Population: 399,484 (41); **Pop. density:** 7,121 per sq. mi; **Pop. change (1990-2000):** 7.3%. **Area:** 56.1 sq. mi. **Employment:** 183,613 employed; 4.7% unemployed. **Per capita income (MSA):** $35,666; change (1998-99): 6.9%.

History: area settled by Spanish 1820; inc. as city under present name 1854.

Transportation: 1 intl. airport; western terminus for 2 railroads; underground, 75-mi underwater subway. **Communications:** 1 TV, 3 radio stations in city. **Medical facilities:** 10 hosp. in MSA. **Educational facilities:** 8 East Bay colleges and univ.; 81 pub. schools. **Further information:** Oakland Metropolitan Chamber of Commerce, 475 14th St., Oakland, CA 94612-1903.

Websites: http://www.oaklandchamber.com
http://www.oaklandnet.com

Oklahoma City, Oklahoma

Population: 506,132 (29); **Pop. density:** 834 per sq. mi; **Pop. change (1990-2000):** 13.8%. **Area.** 607.0 sq. mi. **Employment:** 246,353 employed; 2.7% unemployed. **Per capita income (MSA):** $24,437; change (1998-99): 3.8%.

History: settled during land rush in Midwest 1889; inc. 1890; became capital 1910; oil discovered 1928.

Transportation: 1 intl. airport; 2 railroads; pub. transit system; 3 major bus lines. **Communications:** 8 TV, 22 radio stations. **Medical facilities:** 21 hosp. **Educational facilities:** 17 univ. and colleges; 87 pub., 30 private schools. **Further information:** Chamber of Commerce, Economic Development Division, 123 Park Ave., Oklahoma City, OK 73102.

Websites: http://www.okcchamber.com
http://www.okccvb.org

Omaha, Nebraska

Population: 390,007 (44); **Pop. density:** 3,371 per sq. mi; **Pop. change (1990-2000):** 16.1%. **Area:** 115.7 sq. mi. **Employment:** 199,838 employed; 3.5% unemployed. **Per capita income (MSA):** $30,692; change (1998-99): 5.3%.

History: founded 1854; inc. 1857; large food-processing, telecommunications, information-processing center; home of more than 20 insurance companies.

Transportation: 12 major airlines; 3 major railroads; intercity bus line. **Communications:** 8 TV, 22 radio stations. **Medical facilities:** 16 hosp.; institute for cancer research. **Educational facilities:** 5 univ., 4 colleges; 243 pub., 78 private schools. **Further information:** Greater Omaha Chamber of Commerce, 1301 Harney St., Omaha, NE 68102.

Websites: http://www.ci.omaha.ne.us
http://www.accessomaha.com

Philadelphia, Pennsylvania

Population: 1,517,550 (5) **Pop. density:** 11,233 per sq. mi; **Pop. change (1990-2000):** –4.3%. **Area:** 135.1 sq. mi. **Employment:** 590,114 employed; 6.1% unemployed. **Per capita income (MSA):** $32,627; change (1998-99): 4.2%. .

History: first settled by Swedes 1638; Swedes surrendered to Dutch 1654; settled by English and Scottish Quakers 1678; named Philadelphia 1682; chartered 1701; Continental Congresses convened 1774, 1775; Declaration of Independence signed here 1776; national capital 1790-1800; state capital 1683-1799.

Transportation: 1 major airport; 3 railroads; major freshwater port; subway, el, rail commuter, bus, and streetcar system. **Communications:** 2 major daily newspapers, 11 TV, 45 radio stations. **Medical facilities:** 47 hosp. **Educational facilities:** 25 degree-granting institutions; 10 community college campuses. **Further information:** Greater Philadelphia Chamber of Commerce, Business Information Center, 200 South Broad St., Suite 700, Philadelphia PA 19102.

Websites: http://www.phila.gov
http://www.gpcc.com

Phoenix, Arizona

Population: 1,321,045 (6); **Pop. density:** 2,782 per sq. mi; **Pop. change (1990-2000):** 34.3%. **Area:** 474.9 sq. mi. **Employment:** 699,905 employed; 2.9% unemployed. **Per capita income (MSA):** $27,617; change (1998-99): 4.3%.

History: settled 1870; inc. as city 1881; became territorial capital 1889.

Transportation: 1 intl. airport; 5 railroads; transcontinental bus line; pub. transit system. **Communications:** 15 TV, 45 radio stations. **Medical facilities:** 28 hosp., 1 medical research center. **Educational facilities:** 88 institutions of higher learning; 186 pub. schools (143 elem., 19 junior high, 24 high schools). **Further information:** Greater Phoenix Chamber of Commerce, 201 N. Central Ave., 27th fl., Phoenix, AZ 85073.

Websites: http://www.ci.phoenix.az.us
http://www.phoenixchamber.com

Pittsburgh, Pennsylvania

Population: 334,563 (52); **Pop. density:** 6,017 per sq. mi; **Pop. change (1990-2000):** –9.5%. **Area:** 55.6 sq. mi. **Employment:** 152,981 employed; 4.1% unemployed. **Per capita income (MSA):** $29,587; % change (1998-99): 5.6%.

History: settled around Ft. Pitt 1758; inc. as city 1816; has one of the largest inland ports; by Civil War, already a center for iron production.

Transportation: 1 intl. airport; 20 railroads; 2 bus lines; trolley/subway system. **Communications:** 6 TV, 26 radio stations. **Medical facilities:** 35 hosp.; VA installation. **Educational facilities:** 3 univ., 6 colleges; 86 pub. schools. **Further information:** Pittsburgh Regional Alliance, Regional Enterprise Tower, 36th Floor, 425 Sixth Ave., Pittsburgh, PA 15219.

Website: http://www.pittsburghregion.org

Plano, Texas

Population: 222,030 (78); **Pop. density:** 3,101 per sq. mi; **Pop. change (1990-2000):** 72.5%. **Area:** 71.6 sq. mi. **Employment:** 138,842 employed; 1.7% unemployed. **Per capita income (MSA):** $34,690; change (1998-99): 4.8%.

History: settled 1846; inc. as city 1873.

Transportation: 1 bus line. **Communications:** 2 TV, 1 radio stations. **Medical facilities:** 6 medical facilities. **Educational facilities:** 1 institute of higher learning, 58 pub. schools. **Further information:** Plano Chamber of Commerce, PO Drawer 940287, Plano, TX 75094-0287.

Website: http://www.planocc.org

Portland, Oregon

Population: 529,121 (28); **Pop. density:** 3,940 per sq. mi; **Pop. change (1990-2000):** 21.0%. **Area:** 134.3 sq. mi. **Employment:** 267,373 employed; 4.6% unemployed. **Per capita income (MSA):** $30,672; change (1998-99): 5.3%.

History: settled by pioneers 1845; developed as trading center, aided by California Gold Rush 1849; city chartered 1851.

Transportation: 1 intl. airport; 2 major rail freight lines, Amtrak; 2 intercity bus lines; 27-mi. frontage freshwater port; mass transit bus and rail system. **Communications:** 9 TV, 27 radio stations. **Medical facilities:** 12 hosp.; VA hosp. **Educational facilities:** 25 univ. and colleges, 1 community college. **Further information:** Portland Metropolitan Chamber of Commerce, 221 N.W. 2d Ave., Portland, OR 97209.

Website: http://www.pdxchamber.org

Raleigh, North Carolina

Population: 276,093 (62); **Pop. density:** 2,409 per sq. mi; **Pop. change (1990-2000):** 32.8%. **Area:** 114.6 sq. mi. **Employment:** 171,363 employed; 1.8% unemployed. **Per capita income (MSA):** $32,054; change (1998-99): 5.0%.

History: named after Sir Walter Raleigh; site chosen for capital 1788; laid out 1792; inc. 1795; occupied by Gen. Sherman 1865.

Transportation: 1 intl. airport, 14 airlines, 8 commuter airlines; 3 railroads; 2 bus lines. **Communications:** 8 TV, 31 radio stations. **Medical facilities:** 6 hosp. **Educational facilities:** 6 univ. and colleges; 1 community college; 1067 pub. schools (county). **Further information:** Chamber of Commerce, 800 S. Salisbury St., PO Box 2978, Raleigh, NC 27602.

Websites: http://www.raleigh.acn.net
http://www.raleighchamber.org

Richmond, Virginia

Population: 197,790 (95); **Pop. density:** 3,291 per sq. mi; **Pop. change (1990-2000):** –2.6%. **Area:** 60.1 sq. mi. **Employment:** 92,425 employed, 2.9% unemployed. **Per capita income (MSA):** $30,593; change (1998-99): 4.9%.

History: first settled 1607; became capital of Commonwealth of Virginia, 1779; attacked by British under Benedict Arnold 1781; inc. as city 1782; capital of Confederate States of America, 1861-65.

Transportation: 1 intl. airport; 3 railroads; 2 intracity bus lines; deepwater terminal accessible to oceangoing ships. **Communications:** 6 TV, 28 radio stations. **Medical facilities:** Medical Coll. of Virginia renowned for heart and kidney transplants; 8 hosp. **Educational facilities:** 20 univ. and colleges incl. 7 branches; 173 pub., 49 private schools. **Further information:** Chamber of Commerce, PO Box 12280, Richmond, VA 23241.

Websites: http://www.ci.richmond.va.us
http://www.grcc.com

Riverside, California

Population: 255,166 (67); **Pop. density:** 3,267 per sq. mi; **Pop. change (1990-2000):** 12.7%. **Area:** 78.1 sq. mi. **Employment:** 147,084 employed; 5.4% unemployed. **Per capita income (MSA):** $22,060; change (1998-99): 4.0%.

History: founded 1870; inc. 1886; known for its citrus industry; home of the parent navel orange tree and the historic Mission Inn.

Transportation: municipal airport, intl. airport nearby; rail freight lines, commuter line; trolley/bus system; interstate freeways. **Communications:** 15 TV, 47 radio stations. **Medical facilities:** 3 hosp.; many clinics. **Educational facilities:** 3 univ., 1 community college. **Further information:** Chamber of Commerce, 3985 University Ave., Riverside, CA 92501.

Websites: http://www.ci.riverside.ca.us
http://www.riverside-chamber.com

Rochester, New York

Population: 219,773 (79); **Pop. density:** 6,139 per sq. mi; Pop. change (1990-2000): –5.1%. **Area:** 35.8 sq. mi. **Employment:** 104,481 employed; 6.7% unemployed. **Per capita income (MSA):** $28,162; change (1998-99): 3.3%.

History: first permanent settlement 1812; inc. as village 1817, as city 1834; developed as Erie Canal town.

Transportation: 1 intl. airport; Amtrak; 2 bus lines; intracity transit service; Port of Rochester. **Communications:** 6 TV, 19 radio stations. **Medical facilities:** 8 general hosp. **Educational facilities:** 11 colleges, 3 community colleges. **Further information:** Greater Rochester Metro Chamber of Commerce, 55 St. Paul St., Rochester, NY 14604-1391.

Websites: http://www.rnychamber.com
http://www.ci.rochester.ny.us

Sacramento, California

Population: 407,018 (40); **Pop. density:** 4,187 per sq. mi; **Pop. change (1990-2000):** 10.2%. **Area:** 97.2 sq. mi. **Employment:** 193,518 employed; 5.2% unemployed. **Per capita income (MSA):** $28,718; change (1998-99): 5.3%.

History: settled 1839; important trading center during California Gold Rush 1840s; became state capital 1854.

Transportation: international, executive, and cargo airports; 2 mainline transcontinental rail carriers; bus and light rail system; Port of Sacramento. **Communications:** 8 TV, 34 radio stations; 3 cable TV cos. **Medical facilities:** 12 major hosp. **Educational facilities:** 7 colleges and univ., 5 private colleges and univ., 5 community colleges, 81 pub. schools. **Further information:** Sacramento Metropolitan Chamber of Commerce, 917 7th St., Sacramento, CA 95814.

Websites: http://www.ci.sacramento.ca.us
http://www.sacog.org

St. Louis, Missouri

Population: 348,189 (49); **Pop. density:** 5,625 per sq. mi; **Pop. change (1990-2000):** –12.2%. **Area:** 61.9 sq. mi. **Employment:** 146,338 employed; 6.6% unemployed. **Per capita income (MSA):** $30,382; change (1998-99): 3.6%.

History: founded 1764 as a fur trading post by French; acquired by U.S. 1803; chartered as city 1822; became independent city 1876; lies on Mississippi R., near confluence with Missouri R.

Transportation: 2 intl. airports; 2d largest rail center, 10 trunk-line railroads; 2d largest inland port; Amtrak; Greyhound; bus & light rail; 13 barge lines. **Communications:** 8 TV, 17 radio stations. **Medical facilities:** 9 hosp., 2 teaching hosp. **Educational facilities:** 7 univ., 12 colleges and seminaries, 69 elem., 22 middle, 16 high schools. **Further information:** St. Louis Planning & Urban Design Agency, 1015 Locust St., Ste. 1200, St. Louis, MO 63101.

Website: http://stlouis.missouri.org

St. Paul, Minnesota

Population: 287,151 (59); **Pop. density:** 5,438 per sq. mi; **Pop. change (1990-2000):** 5.5%. **Area:** 52.8 sq. mi. **Employment:** 135,522 employed; 3.5% unemployed. **Per capita income (MSA):** $35,250; change (1998-99): 4.5%.

History: founded in early 1840s as "Pig's Eye Landing"; became capital of the Minnesota territory 1849 and chartered as St. Paul 1854.

Transportation: 1 intl., 1 business airport; 6 major rail lines; 2 interstate bus lines; pub. transit system. **Communications:** 9 TV, 47 radio stations. **Medical facilities:** 6 hosp. **Educational facilities:** 5 univ., 4 colleges; 1 technical, 1 law school, 1 art and design college; 65 public, 39 private schools. **Further information:** St. Paul Area Chamber of Commerce, 332 Minnesota St., Ste. N-205, St. Paul, MN 55101.

Websites: http://www.ci.stpaul.mn.us
http://www.stpaulcvb.org

St. Petersburg, Florida

Population: 248,232 (68); **Pop. density:** 4,165 per sq. mi; **Pop. change (1990-2000):** 4.0%. **Area:** 59.6 sq. mi. **Employment:** 133,821 employed; 2.9% unemployed. **Per capita income (MSA):** $28,145; change (1998-99): 4.2%.

History: founded 1888; inc. 1892.

Transportation: 2 intl. airports; Amtrak bus connection; county-wide public bus system; 1 cruise port. **Communications:** 17 TV, 41 radio stations in area. **Medical facilities:** 4 major hosp.; VA hosp. **Educational facilities:** 1 univ., 2 colleges, 1 law school; 144 pub. schools (county-wide). **Further information:** St. Petersburg Area Chamber of Commerce, PO Box 1371, St. Petersburg, FL 33731.

Website: http://www.stpete.com

San Antonio, Texas

Population: 1,144,646 (9); **Pop. density:** 2,808 per sq. mi; **Pop. change (1990-2000):** 22.3%. **Area:** 407.6 sq. mi. **Employment:** 511,662 employed; 3.8% unemployed. **Per capita income (MSA):** $24,716; change (1998-99): 3.5%.

History: first Spanish garrison 1718; Battle at the Alamo fought here 1836; city subsequently captured by Texans; inc. 1837.

Transportation: 1 intl. airport; 4 railroads; 3 bus lines; pub. transit system. **Communications:** 9 TV, 42 radio stations. **Medical facilities:** 36 hosp.; major medical center. **Educational facilities:** 18 univ. and colleges; 16 pub. school districts. **Further information:** Chamber of Commerce, 602 E. Commerce, PO Box 1628, San Antonio, TX 78296.

Websites: http://www.ci.sat.tx.us
http://www.sachamber.org

IT'S A FACT: When a reporter from New York called St. Paul "another Siberia, unfit for human habitation," the Chamber of Commerce instituted a Winter Carnival to show that things were lively there all year round. A centerpiece of the first carnival (1886) was an elaborate ice castle.

San Diego, California

Population: 1,223,400 (7); **Pop. density:** 3,772 per sq. mi; **Pop. change (1990-2000):** 10.2%. **Area:** 324.3 sq. mi. **Employment:** 624,560 employed; 3.0% unemployed. **Per capita income (MSA):** $29,489; change (1998-99): 6.2%.

History: claimed by the Spanish 1542; first mission est. 1769; scene of conflict during Mexican-American War 1846; inc. 1850.

Transportation: 1 major airport; 1 railroad; major freeway system; bus system; trolley system. **Communications:** 9 TV, 25 radio stations, 2 cable providers. **Medical facilities:** 16 hosp. **Educational facilities:** 25 colleges and univ.; 176 pub. schools. **Further information:** San Diego Regional Chamber of Commerce, 402 W. Broadway, Ste. 1000, San Diego, CA 92101.

Websites: http://www.sannet.gov
http://www.sdchamber.org

San Francisco, California

Population: 776,733 (13); **Pop. density:** 16,632 per sq. mi; **Pop. change (1990-2000):** 7.3%. **Area:** 46.7 sq. mi. **Employment:** 423,010 employed; 2.8% unemployed. **Per capita income (MSA):** $49,695; change (1998-99): 7.9%.

History: nearby Farallon Islands sighted by Spanish 1542; city settled by 1776; claimed by U.S. 1846; became a major city during California Gold Rush 1849; inc. as city 1850; earthquake devastated city 1906.

Transportation: 1 major airport; intracity railway system; 2 railway transit systems; bus and railroad service; ferry system; 1 underwater tunnel. **Communications:** 10 TV; 15 radio stations. **Medical facilities:** 16 medical centers. **Educational facilities:** 16 univ. and colleges, 111 pub. schools, 5 charter schools. **Further information:** Convention & Visitors Bureau, 201 3d St., Ste. 900, San Francisco, CA 94103.

Websites: http://www.ci.sf.ca.us
http://www.sfchamber.com

San Jose, California

Population: 894,943 (11); **Pop. density:** 5,117 per sq. mi; **Pop. change (1990-2000):** 14. 4%. **Area:** 174.9 sq. mi. **Employment:** 497,361 employed; 2.3% unemployed. **Per capita income (MSA):** $46,649; change (1998-99): 15.5%.

History: founded by the Spanish 1777 between San Francisco and Monterey; state cap. 1849-51; inc. 1850.

Transportation: 1 intl. airport; 2 railroads; light rail system; bus system. **Communications:** 4 TV, 14 radio stations. **Medical facilities:** 6 hosp. **Educational facilities:** 3 univ. and colleges. **Further information:** Chamber of Commerce, 310 S. First St., San Jose, CA 95113.

Websites: http://www.ci.san-jose.ca.us
http://www.sjchamber.com

Santa Ana, California

Population: 337,977 (51); **Pop. density:** 12,471 per sq. mi; **Pop. change (1990-2000):** 15.1%. **Area:** 27.1 sq. mi. **Employment:** 160,664 employed; 4.6% unemployed. **Per capita income (MSA):** $33,805; change (1998-99): 5.1%

History: founded 1769; inc. as city 1869.

Transportation: 1 airport; 5 major freeways including main Los Angeles-San Diego artery; Amtrak. **Communications:** 14 TV, 28 radio stations. **Medical facilities:** 4 hosp. **Educational facilities:** 1 community college. **Further information:** Santa Ana Chamber of Commerce, 2020 N. Broadway, 2d floor, Santa Ana, CA 92706.

Website: http://www.santaanachamber.com

Scottsdale, Arizona

Population: 202,705 (86); **Pop. density:** 1,100 per sq. mi; **Pop. change (1990-2000):** 55.8%. **Area:** 184.2 sq. mi. **Employment:** 102,278 employed; 1.9% unemployed. **Per capita income (MSA):** $27,617; change (1998-99): 4.3%

History: founded 1888 by Army Chaplain Winfield Scott; inc. June 25, 1951; Frank Lloyd Wright built winter home here (Taliesin West); slogan "West's Most Western Town," by Mayor Malcolm White adopted 1951.

Transportation: 1 intl. airport in area, 1 local airport; regional bus system; local bus system; taxi system. **Communications:** 12 TV, 45 radio stations. **Medical facilities:** 2 general hospitals; Mayo Clinic. **Educational facilities:** 1 univ. nearby, 1 community college; 3 unified school districts. **Further information:** Scottsdale Chamber of Commerce, 7343 Scottsdale Rd., Scottsdale, AZ 85251.

Websites: http://www.ci.scottsdale.az.us
http://www.scottsdalechamber.com

Seattle, Washington

Population: 563,374 (24); **Pop. density:** 6,715 per sq. mi; **Pop. change (1990-2000):** 9.1%. **Area:** 83.9 sq. mi. **Employment:** 342,569 employed; 4.2% unemployed. **Per capita income (MSA):** $39,880; change (1998-99): 8.9%.

History: settled 1851; inc. 1869; suffered severe fire 1889; played prominent role during Alaska Gold Rush 1897; growth followed opening of Panama Canal 1914; center of aircraft industry WWII.

Transportation: 1 intl. airport; 2 railroads; ferries serve Puget Sound, Alaska, Canada. **Communications:** 7 TV, 39 radio stations. **Medical facilities:** 40 hosp. **Educational facilities:** 7 univ., 6 colleges, 11 community colleges. **Further information:** Greater Seattle Chamber of Commerce, 1301 5th Ave., Ste. 2400, Seattle, WA 98101-2611.

Websites: http://www.ci.seattle.wa.us
http://www.seattlechamber.com

Shreveport, Louisiana

Population: 200,145 (88); **Pop. density:** 1,941 per sq. mi; **Pop. change (1990-2000):** 0.8%. **Area:** 103.1 sq. mi. **Employment:** 88,594 employed; 5.4% unemployed. **Per capita income (MSA):** $24,053; change (1998-99): 3.5%.

History: founded 1833 near site of a 160-mi logjam cleared by Capt. Henry Shreve; inc. 1839; oil discovered 1905.

Transportation: 2 airports; 3 bus lines. **Communications:** 6 TV, 20 radio stations. **Medical facilities:** 16 hosp. **Educational facilities:** 4 univ., 1 college; 74 pub. schools. **Further information:** Chamber of Commerce, PO Box 20074, 400 Edwards St., Shreveport, LA 71120.

Website: http://www.shreveportchamber.org

Spokane, Washington

Population: 195,629 (97); **Pop. density:** 3,385 per sq. mi; **Pop. change (1990-2000):** 10.4%. **Area:** 57.8 sq. mi. **Employment:** 94,108 employed, 6.3% unemployed. **Per capita income (MSA):** $24,368; change (1998-99): 4.3%.

History: settled 1872; inc. as village of Spokane Falls 1881, destroyed in fire 1889; reinc. as city of Spokane 1891.

Transportation: 1 intl. airport; 2 railroads; bus system. **Communications:** 5 TV, 25 radio stations. **Medical facilities:** 6 major hosp. **Educational facilities:** 9 univ. and colleges; 14 pub. school districts, 16 high schools. **Further information:** Spokane Regional Chamber of Commerce, 801 W. Riverside Ave., Spokane, WA 99201.

Website: http://www.spokanechamber.org

Stockton, California

Population: 243,771 (70); **Pop. density:** 4,457 per sq. mi; **Pop. change (1990-2000):** 15.6%. **Area:** 54.7 sq. mi. **Employment:** 96,289 employed; 10.4% unemployed. **Per capita income (MSA):** $21,544; change (1998-99): 3.7%.

History: site purchased 1842; settled 1847; inc. 1850; chief distributing point for agric. products of San Joaquin Valley.

Transportation: 1 airport; deepwater inland seaport; 4 railroads; 2 bus lines, county bus system. **Communications:** 5 TV stations. **Medical facilities:** 4 hosp.; regional burn, cancer, heart centers. **Educational facilities:** 6 univ. and colleges; 58 pub. schools. **Further information:** Chamber of Commerce, 445 W. Weber Ave., Ste. 220, Stockton, CA 95203.

Websites: http://www.stocktongov.com
http://www.stocktonchamber.org

Tacoma, Washington

Population: 193,556 (99); **Pop. density:** 3,863 per sq. mi; **Pop. change (1990-2000):** 9.6%. **Area:** 50.1 sq. mi. **Employment:** 93,257 employed, 6.1% unemployed. **Per capita income (MSA):** $25,289; change (1998-99) 3.4%.

History: first European explorer of area was British Capt. George Vancouver 1792; colonized by Hudson's Bay Co. at Ft. Nisqually 1833; inc. 1884.

Transportation: 1 intl. airport; 3 railroads; transit system; Port of Tacoma. **Communications:** 6 TV stations. **Medical facilities:** 7 hosp.; Army Medical Center; VA facility. **Educational facilities:** 3 univ., 4 colleges. **Further information:** Tacoma-Pierce County Chamber of Commerce, PO Box 1933, Tacoma, WA 98401.

Websites: http://www.cityoftacoma.org
http://www.tacomachamber.org

Tampa, Florida

Population: 303,447 (57); **Pop. density:** 2,707 per sq. mi; **Pop. change (1990-2000):** 8.4%. **Area:** 112.1 sq. mi. **Employment:** 173,311 employed; 3.2% unemployed. **Per capita income (MSA):** $28,145; change (1998-99): 4.2%.

History: U.S. army fort on site 1824; inc. 1849; Ybor City National Historical Landmark district.

Transportation: 1 intl. airport; Port of Tampa; CSX rail, Amtrak Rail; bus system; downtown trolley. **Communications:** 14 TV, 57 radio stations. **Medical facilities:** 21 hosp. **Educational facilities:** 6 univ. and colleges; 172 pub. schools. **Further information:** Chamber of Commerce, 401 E. Jackson St., PO Box 420, Tampa, FL 33601.

Website: http://www.tampachamber.com

Toledo, Ohio

Population: 313,619 (56); **Pop. density:** 3,892 per sq. mi; **Pop. change (1990-2000):** –5.8%. **Area:** 80.6 sq. mi. **Employment:** 151,050 employed; 5.7% unemployed. **Per capita income (MSA):** $27,087; change (1998-99): 3.9%.

History: site of Ft. Industry 1794; Battles of Ft. Meigs and Ft. Timbers 1812; figured in "Toledo War" 1835-36 between Ohio and Michigan over borders; inc. 1837.

Transportation: 5 major airlines; 4 railroads; 53 motor freight lines; 7 interstate bus lines. **Communications:** 7 TV, 22 radio stations. **Medical facilities:** 7 major hosp. complexes. **Educational facilities:** 6 univ. and colleges. **Further information:** Toledo Area Chamber of Commerce, 300 Madison Ave., Ste. 200, Toledo, OH 43604.

Website: http://www.toledochamber.com

Tucson, Arizona

Population: 486,699 (30); **Pop. density:** 2,500 per sq. mi; **Pop. change (1990-2000):** 20.1%. **Area:** 194,7 sq. mi. **Employment:** 231,046 employed; 3.2% unemployed. **Per capita income (MSA):** $23,911; change (1998-99): 4.7%.

History: settled 1775 by Spanish as a presidio; acquired by U.S. in Gadsden Purchase 1853; inc. 1877.

Transportation: 1 intl. airport; 2 railroads; 6 bus systems, 1 trolley. **Communications:** 10 TV, 34 radio stations. **Medical facilities:** 12 hosp. **Educational facilities:** 3 univ., 1 community college; 184 pub. schools. **Further information:** Tucson Metropolitan Chamber of Commerce, PO Box 991, Tucson, AZ 85702.

Websites: http://www.ci.tucson.az.us
http://www.tucsonchamber.org

Tulsa, Oklahoma

Population: 393,049 (43); **Pop. density:** 2,152 per sq. mi; **Pop. change (1990-2000):** 7.0%. **Area:** 182.6 sq. mi. **Employment:** 213,264 employed; 3.0% unemployed. **Per capita income (MSA):** $27,654; change (1998-99): 1.6%.

History: settled in 1836 by Creek Indians; modern town founded 1882 and inc. 1898; oil discovered early 20th century; emerging as telecommunications hub.

Transportation: 1 intl. airport; 5 rail lines; 5 bus lines; transit bus system. **Communications:** 130 TV, 31 radio stations. **Medical facilities:** 10 hosp. **Educational facilities:** 8 univ. and colleges; 85 pub., 39 private schools. **Further information:** Tulsa Metro Chamber, 616 S. Boston Ave., Ste. 100, Tulsa, OK 74119-1298.

Website: http://www.tulsachamber.com

Virginia Beach, Virginia

Population: 425,257 (38); **Pop. density:** 1,713 per sq. mi; **Pop. change (1990-2000):** 8.2%. **Area:** 248.3 sq. mi. **Employment:** 208,588 employed; 2.2% unemployed. **Per capita income (MSA):** $24,979; change (1998-99): 4.0%.

History: area founded by Capt. John Smith 1607; formed by merger with Princess Anne Co. 1963.

Transportation: 1 airport; 2 railroads; 1 bus line; pub. transit system. **Communications:** 8 TV, 44 radio stations. **Medical facilities:** 2 hosp. **Educational facilities:** 1 univ., 2 colleges; 84 pub. schools. **Further information:** Virginia Beach Dept. of Economic Development, One Columbus Center, Ste. 300, Virginia Beach, VA 23462.

Website: http://www.yesvirginiabeach.com

Washington, District of Columbia

Population: 572,059 (21); **Pop. density:** 9,317 per sq. mi; **Pop. change (1990-2000):** –5.7%. **Area:** 61.4 sq. mi. **Employment:** 262,763 employed, 5.8% unemployed. **Per capita income (MSA):** $38,403; change (1998-99): 5.5%.

History: U.S. capital; site at Potomac R. chosen by George Washington 1790 on land ceded from VA and MD (portion S of Potomac returned to VA 1846); Congress first met 1800; inc. 1802; sacked by British, War of 1812; one of the most important Civil War battles was fought at Ft. Stevens.

Transportation: 3 intl. airports in area; Amtrak, 6 other passenger & cargo rail lines; Metrobus/Metrorail transit system; bus line. **Communications:** 5 TV, 61 radio stations. **Medical facilities:** 16 hosp. **Educational facilities:** 10 univ. and colleges. **Further information:** DC Chamber of Commerce, 1213 K Street NW, Washington, DC 20005.

Websites: http://www.ci.washington.dc.us
http://www.dcchamber.org

Wichita, Kansas

Population: 344,284 (50); **Pop. density:** 2,535 per sq. mi; **Pop. change (1990-2000):** 13.2%. **Area:** 135.8 sq. mi. **Employment:** 170,286 employed; 4.6% unemployed. **Per capita income (MSA):** $26,916; change (1998-99): 1.5%.

History: founded 1864; inc. 1871.

Transportation: 2 airports; 3 major rail freight lines; 2 bus lines. **Communications:** 80 TV, 34 radio stations. **Medical facilities:** 7 hosp., 2 psychiatric rehab. centers. **Educational facilities:** 3 univ., 1 medical school; 96 pub. schools. **Further information:** Chamber of Commerce, 350 W. Douglas Ave., Wichita, KS 67202.

Websites: http://www.wichitakansas.org
http://webs.witchita.edu/cedbn

Yonkers, New York

Population: 196,086 (96); **Pop. density:** 10,833 per sq. mi; **Pop. change (1990-2000):** 4.3%. **Area:** 18.1 sq. mi. **Employment:** 86,024 employed; 4.1% unemployed. **Per capita income (MSA):** $38,814; change (1998-99): 5.9%.

History: founded 1641 by the Dutch; inc. as town 1855; chartered as city 1872; directly north of NYC.

Transportation: intracity bus system; rail service. **Communications:** see New York, NY. **Medical facilities:** 3 hosp. **Educational facilities:** 1 college; 32 pub. schools. **Further information:** Chamber of Commerce, 20 S. Broadway, 12th fl., Yonkers, NY 10701.

Websites: http://www.cityofyonkers.com
http://www.yonkerschamber.com

Fastest-Growing Big Cities*

City	2000 population	1990 population	% change
1. Las Vegas, NV	478,434	258,295	85.2
2. Plano, TX	222,030	128,713	72.5
3. Scottsdale, AZ	202,705	130,069	55.8
4. Glendale, AZ	218,812	148,134	47.7
5. Bakersfield, CA	247,057	174,820	41.3
6. Austin, TX	656,562	465,622	41.0
7. Mesa, AZ	396,375	288,091	37.6
8. Charlotte, NC	540,828	395,934	36.6
9. Phoenix, AZ	1,321,045	983,403	34.3
10. Raleigh, NC	276,093	207,951	32.8

Fastest-Shrinking Big Cities*

City	2000 population	1990 population	% change
1. St. Louis, MO	348,189	396,685	–12.2
2. Baltimore, MD	651,154	736,014	–11.5
3. Buffalo, NY	292,648	328,123	–10.8
4. Norfolk, VA	234,403	261,229	–10.3
5. Pittsburgh, PA	334,563	369,879	–9.5
6. Cincinnati, OH	331,285	364,040	–9.0
7. Birmingham, AL	242,820	265,968	–8.7
8. Detroit, MI	951,270	1,027,974	–7.5
9. Toledo, OH	313,619	332,943	–5.8
10. Washington, DC	572,059	606,900	–5.7

*Among those with populations of 200,000 or more, based on 2000 U.S. Census.

Percent of Population by Race and Hispanic Origin for the 10 Largest Cities

City	White	Black or African-Amer.	Amer. Indian, Alaska Native	Asian	Hawaiian & Other Pacific Isl.	Some other race	Two or more races	Hispanic or Latino (of any race)
1. New York, NY	44.7	26.6	0.5	9.8	0.1	13.4	4.9	27.0
2. Los Angeles, CA	46.9	11.2	0.8	10.0	0.2	25.7	5.2	46.5
3. Chicago, IL	42.0	36.8	0.4	4.3	0.1	13.6	2.9	26.0
4. Houston, TX	49.3	25.3	0.4	5.3	0.1	16.5	3.1	37.4
5. Philadelphia, PA	45.0	43.2	0.3	4.5	0.0	4.8	2.2	8.5
6. Phoenix, AZ	71.1	5.1	2.0	2.0	0.1	16.4	3.3	34.1
7. San Diego, CA	60.2	7.9	0.6	13.6	0.5	12.4	4.8	25.4
8. Dallas, TX	50.8	25.9	0.5	2.7	0.0	17.2	2.7	35.6
9. San Antonio, TX	67.7	6.8	0.8	1.6	0.1	19.3	3.7	58.7
10. Detroit, MI	12.3	81.6	0.3	1.0	0.0	2.5	2.3	5.0

STATES AND OTHER AREAS OF THE U.S.

Sources: Population: U.S. Commerce Dept., Bureau of the Census—Census 2000: April 1, 2000 (including armed forces stationed in the state). Area: Bureau of the Census, Geography Division; forested land: Agriculture Dept., Forest Service. Lumber production: Bureau of the Census, Industry Division; mineral production: Dept. of Interior, Office of Mineral Information; commercial fishing: Commerce Dept., Natl. Marine Fisheries Service; value of construction: McGraw-Hill Information Systems Co., F.W. Dodge Division. Personal per capita income: Commerce Dept., Bureau of Economic Analysis; sales tax: CCH Inc.; unemployment: Labor Dept., Bureau of Labor Statistics. Tourism: Tourism Industries/ITA, Tourism Works for America Report. Lottery figures (not all states have a lottery): *La Fleur's Lottery World*. Finance: Federal Deposit Insurance Corp. Federal employees: Labor Dept., Office of Personnel Management. Energy: Energy Dept., Energy Information Administration. Other information from sources in individual states.

Note: Population density is for land area only. Categories under racial or distribution may not add to 100% due to rounding. "Nat. American" includes American Indians and Alaska Natives (including Eskimos and Aleuts). Hispanic population may be any race and is dispersed among racial categories, besides being listed separately. Nonfuel mineral values for some states exclude small amounts to avoid disclosing proprietary data. Categories under employment distribution are not all-inclusive. Famous Persons lists may include some nonnatives associated with the state as well as persons born there. Website addresses listed may not be official state sites and are not endorsed by *The World Almanac;* all website addresses are subject to change.

Alabama

Heart of Dixie, Camellia State

People. Population (2000): 4,447,100; rank: 23; **net change** (1999-2000): 10.1%. **Pop. density**: 87.6 per sq mi. **Racial distribution** (2000): 71.1% white; 26.0% black; 0.7% Asian; 0.5% Native American/Nat. AK; <0.1% Hawaiian/Pacific Islander; 0.7% other race; 2 or more races, 1.0%. **Hispanic pop.** (any race): 1.7%.

Geography. Total area: 52,419 sq mi; rank: 30. **Land area:** 50,744 sq mi; rank: 28. **Acres forested:** 21,974,000. **Location:** East South Central state extending N-S from Tenn. to the Gulf of Mexico; E of the Mississippi River. **Climate:** long, hot summers; mild winters; generally abundant rainfall. **Topography:** coastal plains, including Prairie Black Belt, give way to hills, broken terrain; highest elevation, 2,407 ft. **Capital:** Montgomery.

Economy. Chief industries: pulp & paper, chemicals, electronics, apparel, textiles, primary metals, lumber and wood products, food processing, fabricated metals, automotive tires, oil and gas exploration. **Chief manuf. goods:** electronics, cast iron & plastic pipe, fabricated steel products, ships, paper products, chemicals, steel, mobile homes, fabrics, poultry processing, soft drinks, furniture, tires. **Chief crops**: cotton, greenhouse & nursery, peanuts, sweet potatoes, potatoes and other vegetables. **Livestock**: (Jan. 2001) 1.4 mil cattle/calves; (Dec. 2000): 165,000 hogs/pigs; 15.6 mil chickens (excl. broilers); 971.2 mil broilers. **Timber/lumber** (2000): pine, hardwoods; 2.7 bil bd. ft. **Nonfuel minerals** (est. 2000): $1.1 bil; mostly portland cement, crushed stone, lime, sand & gravel, and masonry. **Commercial fishing** (1999): $50 mil. **Chief port:** Mobile. **Principal internat. airports at:** Birmingham, Huntsville. **Value of construction** (1997): $4.8 bil. **Gross state product** (1999): $115.1 bil. **Employment distrib.** (May 2001): 24.7% services; 23.2% trade; 18.3% govt.; 18.1% mfg. **Per cap. pers. income** (2000): $23,471. **Sales tax** (2001): 4%. **Unemployment** (2000): 4.6%. **Tourism expends.** (1997): $4.7 bil.

Finance. FDIC-insured commercial banks (2000): 158. **Deposits:** $129.9 bil. **FDIC-insured savings institutions** (2000): 12. **Assets:** $2.2 bil.

Federal govt. Fed. civ. employees (Mar. 2000): 36,203. **Avg. salary:** $50,329. **Notable fed. facilities:** George C. Marshall NASA Space Center; Gunter Annex & Maxwell AFB; Ft. Rucker; Ft. McClellan; Natl. Fertilizer Develop. Center; Navy Station & U.S. Corps of Engineers; Redstone Arsenal.

Energy. Electricity production (2000, kWh, by source): Coal: 76 bil; Petroleum: 241 mil; Gas: 3.6 bil; Hydroelectric: 5.8 bil; Nuclear: 31.4 bil.

State data. Motto: We dare defend our rights. **Flower:** Camellia. **Bird:** Yellowhammer. **Tree:** Southern Longleaf pine. **Song:** Alabama. **Entered union** Dec. 14, 1819; rank, 22d. **State fair:** Regional and county fairs held in Sept. and Oct.; no state fair.

History. Alabama was inhabited by the Creek, Cherokee, Chickasaw, Alabama, and Choctaw peoples when the Europeans arrived. The first Europeans were Spanish explorers in the early 1500s. The French made the first permanent settlement on Mobile Bay, 1702. France later gave up the entire region to England under the Treaty of Paris, 1763. Spanish forces took control of the Mobile Bay area, 1780, and it remained Spanish until U.S. troops seized the area, 1813. Most of present-day Alabama was held by the Creeks until Gen. Andrew Jackson broke their power, 1814, and they were removed to Oklahoma Territory. The state seceded, 1861, and the Confederate states were organized Feb. 4, at Montgomery, the first capital; it was readmitted, 1868.

Tourist attractions. First White House of the Confederacy, Civil Rights Memorial, Alabama Shakespeare Festival, all Montgomery; Ivy Green, Helen Keller's birthplace, Tuscumbia; Civil Rights Museum, statue of Vulcan, Birmingham; Carver Museum, Tuskegee; W. C. Handy Home & Museum, Florence; Alabama Space and Rocket Center, Huntsville; Moundville State Monument, Moundville; Pike Pioneer Museum, Troy; USS *Alabama* Memorial Park, Mobile; Russell Cave Natl. Monument, near Bridgeport: a detailed record of occupancy by humans from about 10,000 BC to AD 1650.

Famous Alabamians. Hank Aaron, Tallulah Bankhead, Hugo L. Black, Paul "Bear" Bryant, George Washington Carver, Nat King Cole, William C. Handy, Bo Jackson, Helen Keller, Coretta Scott King, Harper Lee, Joe Louis, Willie Mays, John Hunt Morgan, Jim Nabors, Jesse Owens, Condoleezza Rice, George Wallace, Booker T. Washington, Hank Williams.

Tourist information. Bureau of Tourism and Travel, 401 Adams Avenue, Suite 126, Montgomery, AL 36104.

Toll-free travel information. 1-800-ALABAMA out of state.

Website. http://alaweb.asc.edu

Tourism website. http://www.touralabama.org

Alaska

The Last Frontier (unofficial)

People. Population (2000): 626,932; rank: 48; **net change** (1999-2000): 14%. **Pop. density**: 1.1 per sq mi. **Racial distribution** (2000): 69.3% white; 3.5% black; 4.0% Asian; 15.6% Native American/Nat. AK; 0.5% Hawaiian/Pacific Islander; 1.6% other race; 2 or more races, 5.4%. **Hispanic pop.** (any race): 4.1%.

Geography. Total area: 663,267 sq mi; rank: 1. **Land area:** 571,951 sq mi; rank: 1. **Acres forested:** 129,131,000. **Location:** NW corner of North America, bordered on E by Canada. **Climate:** SE, SW, and central regions, moist and mild; far north extremely dry. Extended summer days, winter nights, throughout. **Topography:** includes Pacific and Arctic mountain systems, central plateau, and Arctic slope. Mt. McKinley, 20,320 ft, is the highest point in North America. **Capital:** Juneau.

Economy. Chief industries: petroleum, tourism, fishing, mining, forestry, transportation, aerospace. **Chief manuf. goods:** fish products, lumber & pulp, furs. **Agriculture: Chief crops:** greenhouse products, barley, oats, hay, potatoes, lettuce, aquaculture. **Livestock**: (Jan. 2001) 10,500 cattle/calves; (Dec. 2000) 800 hogs/pigs. **Timber/lumber:** (2000): spruce, yellow cedar, hemlock; NA;. **Nonfuel minerals** (est. 2000): $1.1 bil; mostly zinc, lead, silver, gold, sand & gravel. **Commercial fishing** (1999): $1.1 bil. **Chief ports:** Anchorage, Dutch Harbor, Kodiak, Seward, Skagway, Juneau, Sitka, Valdez, Wrangell. **Principal internat. airports at:** Anchorage, Fairbanks, Juneau. **Value of construction** (1997): $1 bil. **Gross state product** (1999): $26.3 bil. **Employment distrib.** (May 2001): 26.2% services; 20.2% trade; 26.1% govt.; 4.4% mfg. **Per cap. pers. income** (2000): $30,064. **Sales tax:** none. **Unemployment** (2000): 6.6%. **Tourism expends.** (1997): $1.4 bil.

Finance. FDIC-insured commercial banks (2000): 6. **Deposits:** $4.1 bil. **FDIC-insured savings institutions** (2000): 2. **Assets:** $304 mil.

Federal govt. Fed. civ. employees (Mar. 2000): 11,170. **Avg. salary:** $47, 316.

Energy. Electricity production (2000, kWh, by source): Coal: 183 mil; Petroleum: 385 mil; Gas: 3.2 bil; Hydroelectric: 823 mil.

State data. Motto: North to the future. **Flower:** Forget-Me-Not. **Bird:** Willow ptarmigan. **Tree:** Sitka spruce. **Song:** Alaska's Flag. **Entered union** Jan. 3, 1959; rank, 49th. **State fair** at Palmer; late Aug.-early Sept.

History. Early inhabitants were the Tlingit-Haida people and tribes of the Athabascan family. The Aleut and Inuit (Eskimo), who arrived about 4,000 years ago from Siberia, lived in the coastal areas. Vitus Bering, a Danish explorer working for Russia, was the first European to land in Alaska, 1741. The first permanent Russian settlement was established on Kodiak Island, 1784. In 1799, the Russian-American Co. controlled the region, and the first chief manager, Aleksandr Baranov, set up headquarters at Archangel, near present-day Sitka. Sec. of State William H. Seward bought Alaska from Russia for $7.2 mil in 1867, a bargain some called "Seward's Folly." In 1896, gold was discovered in the Klondike region, and the famed gold rush began. Alaska became a territory in 1912.

Tourist attractions. Inside Passage; Portage Glacier; Mendenhall Glacier; Ketchikan Totems; Glacier Bay Natl. Park and Preserve; Denali Natl. Park, one of N. America's great wildlife sanctuaries, surrounding Mt. McKinley, N. America's highest peak; Mt. Roberts Tramway, Juneau; Pribilof Islands fur seal rookeries; restored St. Michael's Russian Orthodox Cathedral, Sitka; Katmai Natl. Park & Preserve.

Famous Alaskans. Tom Bodett, Susan Butcher, Ernest Gruening, Jewel (Kilcher), Gov. Tony Knowles, Sydney Laurence, Libby Riddles, Jefferson "Soapy" Smith.

Tourist information. Alaska Division of Tourism, PO Box 110801, Juneau, AK 99811-0801; 1-907-465-2012.

Website. http://www.state.ak.us

Tourism website.http://www.dced.state.ak.us/tourism

Arizona

Grand Canyon State

People. Population (2000): 5,130,632; rank: 20; **net change** (1999-2000): 40%. **Pop. density**: 45.2 per sq mi. **Racial distribution** (2000): 75.5% white; 3.1% black; 1.8% Asian; 5.0% Native American/Nat. AK; 0.1% Hawaiian/Pacific Islander; 11.6% other race; 2 or more races, 2.9%. **Hispanic pop.** (any race): 25.3%.

Geography. Total area: 113,998 sq mi; rank: 6. **Land area:** 113,635 sq mi; rank: 6. **Acres forested:** 19,596,000. **Location:** in the southwestern U.S. **Climate:** clear and dry in the southern regions and northern plateau; high central areas have heavy winter snows. **Topography:** Colorado plateau in the N, containing the Grand Canyon; Mexican Highlands running diagonally NW to SE; Sonoran Desert in the SW. **Capital:** Phoenix.

Economy. Chief industries: manufacturing, construction, tourism, mining, agriculture. **Chief manuf. goods:** electronics, printing & publishing, foods, prim. & fabric. metals, aircraft and missiles, apparel. **Chief crops:** cotton, lettuce, cauliflower, broccoli, sorghum, barley, corn, wheat, citrus fruits. **Livestock:** (Jan. 2001) 850,000 cattle/calves; 132,000 sheep/lambs; (Dec. 2000) 9,000 hogs/pigs. **Timber/lumber** (2000): pine, fir, spruce; 68 mil bd. ft. **Nonfuel minerals** (est. 2000): $2.6 bil; mostly copper, sand & gravel, cement, molybdenum, crushed stone. **Principal internat. airports at:** Phoenix, Tucson. **Value of construction** (1997): $10 bil. **Gross state product** (1999): $143.7 bil. **Employment distrib.** (May 2001): 31.7% services; 23.4% trade; 16.4% govt.; 9.4% mfg. **Per cap. pers. income** (2000): $25,578. **Sales tax** (2001): 5%. **Unemployment** (2000): 3.9%. **Tourism expends.** (1997): $8.7 bil. **Lottery** (2000): total sales: $255.6 mil; net income: $75.7 mil.

Finance. FDIC-insured commercial banks (2000): 45. **Deposits**: $30.7 bil. **FDIC-insured savings institutions** (2000): 4. **Assets:** $826 mil.

Federal govt. Fed. civ. employees (Mar. 2000): 28,169. **Avg. salary:** $44,036. **Notable fed. facilities:** Luke, Davis-Monthan AF bases; Ft. Huachuca Army Base; Yuma Proving Grounds.

Energy. Electricity production (2000, kWh, by source): Coal: 40.3 bil; Petroleum: 191 mil; Gas: 8.3 bil; Hydroelectric: 8.6 bil; Nuclear: 30.4 bil.

State data. Motto: Ditat Deus (God enriches). **Flower:** Blossom of the Saguaro cactus. **Bird:** Cactus wren. **Tree:** Paloverde. **Song:** Arizona. **Entered union** Feb. 14, 1912; rank, 48th. **State fair** at Phoenix; late Oct.-early Nov.

History. Anasazi, Mogollon, and Hohokam civilizations inhabited the area c 300 BC-AD 1300, later Pueblo peoples; Navajo and Apache came c 15th cent. Marcos de Niza, a Franciscan, and Estevanico, a former black slave, explored, 1539; Spanish explorer Francisco Vásquez de Coronado visited, 1540. Eusebio Francisco Kino, a Jesuit missionary, taught Indians 1692-1711, and left missions. Tubac, a Spanish fort, became the first European settlement, 1752. Spain ceded Arizona to Mexico, 1821. The U.S. took over, 1848, after the Mexican War. The area below the Gila River was obtained from Mexico in the Gadsden Purchase, 1853. Arizona became a territory, 1863. Apache wars ended with Geronimo's surrender, 1886.

Tourist attractions. The Grand Canyon of the Colorado; Painted Desert; Petrified Forest Natl. Park; Canyon de Chelly; Meteor Crater; London Bridge, Lake Havasu City; Biosphere 2, Oracle; Navajo Natl. Monument; Sedona.

Famous Arizonans. Bruce Babbitt, Cochise, Alice Cooper, Geronimo, Barry Goldwater, Zane Grey, Carl Hayden, George W. P. Hunt, Helen Jacobs, Bil Keane, Percival Lowell, William H. Pickering, John J. Rhodes, Morris Udall, Stewart Udall, Frank Lloyd Wright.

Tourist information. Arizona Office of Tourism, Ste. 4015, 2702 N. 3rd St., Phoenix, AZ 85004, 1-800-520-3433.

Website. http://www.state.az.us

Tourism website. http://www.arizonaguide.com

Arkansas

The Natural State, The Razorback State

People. Population (2000): 2,673,400; rank: 33; **net change** (1999-2000): 13.7%. **Pop. density**: 51.3 per sq mi. **Racial distribution** (2000): 80.0% white; 15.7% black; 0.8% Asian; 0.8% Native American/Nat. AK; 0.1% Hawaiian/Pacific Islander; 1.5% other race; 2 or more races, 1.3%. **Hispanic pop.** (any race): 3.2%.

Geography. Total area: 53,179 sq mi; rank: 29. **Land area:** 52,068 sq mi; rank: 27. **Acres forested:** 17,864,000. **Location:** in the west south-central U.S. **Climate:** long, hot summers, mild winters; generally abundant rainfall. **Topography:** eastern delta and prairie, southern lowland forests, and the northwestern highlands, which include the Ozark Plateaus. **Capital:** Little Rock.

Economy. Chief industries: manufacturing, agriculture, tourism, forestry. **Chief manuf. goods:** food products, chemicals, lumber, paper, plastics, electric motors, furniture, auto components, airplane parts, apparel, machinery, steel. **Chief crops:** rice, soybeans, cotton, tomatoes, grapes, apples, commercial vegetables, peaches, wheat. **Livestock:** (Jan. 2001) 1.8 mil cattle/calves; (Dec. 2000) 685,000 hogs/pigs; 23.1 mil chickens (excl. broilers); 1.2 bil broilers. **Timber/lumber** (2000): oak, hickory, gum, cypress, pine; 2.6 bil bd. ft. **Nonfuel minerals** (est. 2000): $506 mil; mostly bromine, crushed stone, portland cement, sand & gravel. **Chief ports:** Little Rock, Pine Bluff, Osceola, Helena, Fort Smith, Van Buren, Camden, Dardanelle, North Little Rock, West Memphis, Crossett, McGehee, Morrilton. **Value of construction** (1997): $3 bil. **Gross state product** (1999): $64.8 bil. **Employment distrib.** (May 2001): 24.2% services; 23.2% trade; 16.6% govt.; 20.7% mfg. **Per cap. pers. income** (2000): $22,257. **Sales tax** (2001): 5.125%. **Unemployment** (2000): 4.4%. **Tourism expends.** (1997): $3.3 bil.

Finance. FDIC-insured commercial banks (2000): 185. **Deposits:** $21.5 bil. **FDIC-insured savings institutions** (2000): 9. **Assets:** $3.3 bil.

Federal govt. Fed. civ. employees (Mar. 2000): 10,959. **Avg. salary:** $42,810. **Notable fed. facilities:** Nat'l. Center for Toxicological Research, Jefferson; Pine Bluff Arsenal, Little Rock AFB.

Energy. Electricity production (2000, kWh, by source): Coal: 23.9 bil; Petroleum: 202 mil; Gas: 3.3 bil; Hydroelectric: 2.1 bil; Nuclear: 11.7 bil.

State data. Motto: Regnat Populus (The people rule). **Flower:** Apple blossom. **Bird:** Mockingbird. **Tree:** Pine. **Song:** Arkansas. **Entered union** June 15, 1836; rank, 25th. **State fair** at Little Rock; late Sept.-early Oct.

IT'S A FACT: The American Robert McCulloch bought the London Bridge from the British government in 1962 for $2,460,000 and had it moved by ship and trucks to Lake Havasu City, AZ. On Oct. 10, 1971, the reassembled bridge was officially opened.

History. Quapaw, Caddo, Osage, Cherokee, and Choctaw peoples lived in the area at the time of European contact. The first European explorers were de Soto, 1541; Marquette and Jolliet, 1673; and La Salle, 1682. The first settlement was by the French under Henri de Tonty, 1686, at Arkansas Post. In 1762, the area was ceded by France to Spain, then given back again, 1800, and was part of the Louisiana Purchase, 1803. It was made a territory, 1819. Arkansas seceded in 1861, only after the Civil War began; more than 10,000 Arkansans fought on the Union side.

Tourist attractions. Hot Springs Natl. Park (water ranging from 95° F-147° F); Eureka Springs; Ozark Folk Center, Blanchard Caverns, both near Mountain View; Crater of Diamonds (only U.S. diamond mine) near Murfreesboro; Toltec Mounds Archeological State Park, Little Rock; Buffalo Natl. River; Mid-America Museum, Hot Springs; Pea Ridge National Military Park, Pead Ridge; Tanyard Springs, Morrilton; Wiederkehr Wine Village, Wiederkehr Village.

Famous Arkansans. Daisy Bates, Dee Brown, Paul "Bear" Bryant, Glen Campbell, Johnny Cash, Hattie Caraway, Bill Clinton, "Dizzy" Dean, Orval Faubus, James W. Fulbright, John Grisham, John H. Johnson, Douglas MacArthur, John L. McClellan, James S. McDonnell, Scottie Pippen, Dick Powell, Brooks Robinson, Billy Bob Thornton, Winthrop Rockefeller, Mary Steenburgen, Edward Durell Stone, Sam Walton, Archibald Yell.

Tourist Information. Arkansas Dept. of Parks & Tourism, One Capitol Mall, Little Rock, AR 72201

Toll-free travel information. 1-800-NATURAL.

Website. http://www.state.ar.us

Tourism website. http://www.arkansas.com

California

Golden State

People. Population (2000): 33,871,648; rank: 1; **net change** (1999-2000): 13.8%. **Pop. density**: 217.2 per sq mi. **Racial distribution** (2000): 59.5; white; 6.7% black; 10.9% Asian; 1.0% Native American/Nat. AK; 0.3% Hawaiian/Pacific Islander; 16.8% other race; 2 or more races, 4.7% **Hispanic pop.** (any race): 32.4%.

Geography. Total area: 163,696 sq mi; rank: 3. **Land area:** 155,959 sq mi; rank: 3. **Acres forested:** 37,263,000. **Location:** on western coast of the U.S. **Climate:** moderate temperatures and rainfall along the coast; extremes in the interior. **Topography:** long mountainous coastline; central valley; Sierra Nevada on the east; desert basins of the southern interior; rugged mountains of the north. **Capital:** Sacramento.

Economy. Chief industries: agriculture, tourism, apparel, electronics, telecommunications, entertainment. **Chief manuf. goods:** electronic and electrical equip., computers, industrial machinery, transportation equip. and instruments, food. **Chief farm products**: milk and cream, grapes, cotton, flowers, oranges, rice, nursery products, hay, tomatoes, lettuce, strawberries, almonds, asparagus. **Livestock**: (Jan. 2001) 5.2 mil cattle/calves; 840,000 sheep/lambs; (Dec. 2000) 150,000 hogs/pigs; 30 mil chickens (excl. broilers); **Timber/lumber** (2000): fir, pine, redwood, oak; 3.3 bil bd. ft. **Nonfuel minerals** (est. 2000): $3.4 bil; mostly portland cement, sand & gravel, boron, crushed stone, gold. **Commercial fishing** (1999): $150.3 mil. **Chief ports:** Long Beach, Los Angeles, San Diego, Oakland, San Francisco, Sacramento, Stockton. **Principal internat. airports at:** Fresno, Los Angeles, Oakland, Ontario, Sacramento, San Diego, San Francisco, San Jose. **Value of construction** (1997): $36.7 bil. **Gross state product** (1999): $1.2 tril. **Employment distrib.** (May 2001): 32.0% services; 22.7% trade; 16.0% govt.; 13.0% mfg. **Per cap. pers. income** (2000): $32,275. **Sales tax** (2001): 7%. **Unemployment** (2000): 4.9%. **Tourism expends.** (1997): $65.8 bil. **Lottery** (2000): total sales: $2.6 bil; net income: $907.3 mil.

Finance. FDIC-insured commercial banks (2000): 304. **Deposits**: $238.5 bil. **FDIC-insured savings institutions** (2000): 46. **Assets:** $353.5 bil.

Federal govt. Fed. civ. employees (Mar. 2000): 141,377. **Avg. salary:** $50,639. **Notable fed. facilities:** Vandenberg, Beale, Travis, McClellan AF bases; San Francisco Mint.

Energy. Electricity production (2000, kWh, by source): Petroleum: 145 mil; Gas: 12.3 bil; Hydroelectric: 37.8 bil; Nuclear: 35.2 bil.; **Other:** 145 mil.

State data. Motto: Eureka (I have found it). **Flower:** Golden poppy. **Bird:** California valley quail. **Tree:** California redwood. **Song:** I Love You, California. **Entered union** Sept. 9, 1850; rank, 31st. **State fair** at Sacramento; late Aug.-early Sept.

History. Early inhabitants included more than 100 different Native American tribes with multiple dialects. The first European explorers were Cabrillo, 1542, and Drake, 1579. The first settlement was the Spanish Alta California mission at San Diego, 1769, first in a string founded by Franciscan Father Junípero Serra. U.S. traders and settlers arrived in the 19th cent. and staged the Bear Flag revolt, 1846, in protest against Mexican rule; later that year U.S. forces occupied California. At the end of the Mexican War, Mexico ceded the territory to the U.S., 1848; that same year gold was discovered, and the famed gold rush began.

Tourist attractions. The *Queen Mary*, Long Beach; Palomar Mountain; Disneyland, Anaheim; Getty Center, Los Angeles; Tournament of Roses and Rose Bowl, Pasadena; Universal Studios, Hollywood; Long Beach Aquarium of the Pacific; Golden State Museum, Sacramento; San Diego Zoo; Yosemite Valley; Lassen and Sequoia-Kings Canyon natl. parks; Lake Tahoe; Mojave and Colorado deserts; San Francisco Bay; Napa Valley; Monterey Peninsula; oldest living things on earth believed to be a stand of Bristlecone pines in the Inyo National Forest, est. 4,700 years old; world's tallest tree, 365-ft "National Geographic Society" coast redwood, in Humboldt Redwoods State Park.

Famous Californians. Edmund G. (Pat) Brown, Jerry Brown, Luther Burbank, Ted Danson, Cameron Diaz, Leonardo DiCaprio, Joe DiMaggio, John C. Fremont, Robert Frost, Tom Hanks, Helen Hunt, Bret Harte, William Randolph Hearst, Jack Kemp, Monica Lewinsky, Jack London, George Lucas, Mark McGwire, Aimee Semple McPherson, Marilyn Monroe, John Muir, Richard M. Nixon, George S. Patton Jr., Ronald Reagan, Sally K. Ride, William Saroyan, Father Junípero Serra, O.J. Simpson, Kevin Spacey, Leland Stanford, John Steinbeck, Shirley Temple, Earl Warren, Tiger Woods.

California Division of Tourism. P.O. Box 1499, Sacramento, CA 95812-1499.

Toll-free travel information. 1-800-GOCALIF.

Website. http://www.state.ca.us/s

Tourism website. http://gocalif.ca.gov

Colorado

Centennial State

People. Population (2000): 4,301,261; rank: 24; **net change** (1999-2000): 30.6%. **Pop. density**: 41.5 per sq mi. **Racial distribution** (2000): 82.8% white; 3.8% black; 2.2% Asian; 1.0% Native American/Nat. AK; 0.1% Hawaiian/Pacific Islander; 7.2% other race; 2 or more races, 2.8% **Hispanic pop.** (any race): 17.1%.

Geography. Total area: 104,094 sq mi; rank: 8. **Land area:** 103,718 sq mi; rank: 8. **Acres forested:** 21,338,000. **Location:** in W central U.S. **Climate:** low relative humidity, abundant sunshine, wide daily, seasonal temp. ranges; alpine conditions in the high mountains. **Topography:** eastern dry high plains; hilly to mountainous central plateau; western Rocky Mountains of high ranges, with broad valleys, deep, narrow canyons. **Capital:** Denver.

Economy. Chief industries: manufacturing, construction, government, tourism, agriculture, aerospace, electronics equipment. **Chief manuf. goods:** computer equip. & instruments, foods, machinery, aerospace products. **Chief crops:** corn, wheat, hay, sugar beets, barley, potatoes, apples, peaches, pears, dry edible beans, sorghum, onions, oats, sunflowers, vegetables. **Livestock**: (Jan. 2001) 3.2 mil cattle/calves; 420,000 sheep/lambs; (Dec. 2000) 840,000 hogs/pigs; 4.2 mil chickens (excl. broilers). **Timber/lumber** (2000): oak, ponderosa pine, Douglas fir; 105 mil bd. ft.. **Nonfuel minerals** (est. 2000): $566 mil; mostly sand & gravel, portland cement, crushed stone, gold, helium. **Principal internat. airport at:** Denver. **Value of construction** (1997): $9.2 bil. **Gross state product** (1999): $154 bil. **Employment distrib.** (May 2001): 31.4% services; 23.9% trade; 15.4% govt.; 9.0% mfg. **Per cap. pers. income** (2000): $32,949. **Sales tax** (2001): 2.9%. **Unemployment** (2000): 2.7%. **Tourism expends.** (1997): $8.8 bil. **Lottery** (2000): total sales: $371 mil; net income: $89.3 mil.

Finance. FDIC-insured commercial banks (2000): 181. **Deposits:** $37.3 bil. **FDIC-insured savings institutions** (2000): 10. **Assets:** $838 mil.

Federal govt. Fed. civ. employees (Mar. 2000): 33,138. **Avg. salary:** $51,098. **Notable fed. facilities:** U.S. Air Force Academy; U.S. Mint; Ft. Carson; Natl. Renewable Energy Labs; U.S. Rail Transportation Test Center; N. American Aerospace Defense Command; Consolidated Space Operations Ctr.; Denver Federal Center; Natl. Center for Atmospheric Research; Natl. Instit. for Standards in Technology; Natl. Oceanic and Atmospheric Administration.

Energy. Electricity production (2000, kWh, by source): Coal: 34.8 bil; Petroleum: 92 mil; Gas: 3.6 bil; Hydroelectric: 1.4 bil.

State data. Motto: Nil Sine Numine (Nothing Without Providence). **Flower:** Rocky Mountain columbine. **Bird:** Lark bunting. **Tree:** Colorado blue spruce. **Song:** Where the Columbines Grow. **Entered union** Aug. 1, 1876; rank 38th. **State fair** at Pueblo; mid-Aug. - early Sept.

History. Early civilization centered around the Mesa Verde c 2,000 years ago, later, Ute, Pueblo, Cheyenne, and Arapaho peoples lived in the area. The region was claimed by Spain, but passed to France. The U.S. acquired eastern Colorado in the Louisiana Purchase, 1803. Lt. Zebulon M. Pike explored the area, 1806, discovering the peak that bears his name. After the Mexican War, 1846-48, U.S. immigrants settled in the east, former Mexicans in the south. Gold was discovered in 1858, causing a population boom. Displaced Native Americans protested, resulting in the so-called Sand Creek Massacre, 1864, where more than 200 Cheyenne and Arapaho were killed. All Native Americans were later removed to Oklahoma Territory.

Tourist attractions. Rocky Mountain Natl. Park; Aspen Ski Resort; Garden of the Gods, Colorado Springs; Great Sand Dunes, Dinosaur, Black Canyon of the Gunnison, and Colorado natl. monuments; Pikes Peak and Mt. Evans highways; Mesa Verde Natl. Park (ancient Anasazi Indian cliff dwellings); Grand Mesa Natl. Forest; mining towns of Central City, Silverton, Cripple Creek; Burlington's Old Town; Bent's Fort, outside La Junta; Georgetown Loop Historic Mining Railroad Park, Cumbres & Toltec Scenic Railroad; limited stakes gaming in Central City, Blackhawk, Cripple Creek, Ignacio, and Towaoe.

Famous Coloradans. Tim Allen, Frederick Bonfils, Henry Brown, Molly Brown, William N. Byers, M. Scott Carpenter, Lon Chaney, Jack Dempsey, Mamie Eisenhower, Douglas Fairbanks, Barney Ford, Scott Hamilton, Chief Ourey, "Baby Doe" Tabor, Lowell Thomas, Byron R. White, Paul Whiteman.

State Chamber of Commerce. 1776 Lincoln, Ste. 1200, Denver, CO 80203. Phone: 303-831-7411

Tourist information. Colorado Travel and Tourism Authority, P. O. Box 3524, Englewood, CO 80155.

Toll-free travel information. 1-800-COLORADO.

Website. http://www.state.co.us

Tourism website. http://www.colorado.com

Connecticut

Constitution State, Nutmeg State

People. Population (2000): 3,405,565; rank: 29; **net change** (1999-2000): 3.6%. **Pop. density**: 702.9 per sq mi. **Racial distribution** (2000): 81.6% white; 9.1% black; 2.4% Asian; 0.3% Native American/Nat. AK; <0.1% Hawaiian/Pacific Islander; 4.3% other race; 2 or more races, 2.2%. **Hispanic pop.** (any race): 9.4%.

Geography. Total area: 5,543 sq mi; rank: 48. **Land area:** 4,845 sq mi; rank: 48. **Acres forested:** 1,819,000. **Location:** New England state in NE corner of the U.S. **Climate:** moderate; winters avg. slightly below freezing; warm, humid summers. **Topography:** western upland, the Berkshires, in the NW, highest elevations; narrow central lowland N-S; hilly eastern upland drained by rivers. **Capital:** Hartford.

Economy. Chief industries: manufacturing, retail trade, government, services, finances, insurance, real estate. **Chief manuf. goods:** aircraft engines and parts, submarines, helicopters, machinery and computer equipment, electronics and electrical equipment, medical instruments, pharmaceuticals. **Chief crops:** nursery stock, Christmas trees, mushrooms, vegetables, sweet corn, tobacco, apples. **Livestock:** (Jan. 2001) 63,000 cattle/calves; (Dec. 2000) 4,000 hogs/pigs; 3.9 mil chickens (excl. broilers). **Timber/lumber** (2000): oak, birch, beech, maple; 52 mil bd. ft. **Nonfuel minerals** (est. 2000): $99.5 mil; mostly crushed stone, sand & gravel, dimension stone, clays, gemstones. **Commercial fishing** (1999): $38 mil. **Chief ports:** New Haven, Bridgeport, New London. **Principal internat. airport at:** Windsor Locks. **Value of construction** (1997): $3.8 bil. **Gross state product** (1999): $152 bil. **Employment distrib.** (May 2001): 31.8% services; 21.6% trade; 14.4% govt.; 15.1% mfg. **Per cap. pers. income** (2000): $40,640. **Sales tax** (2001): 6%. **Unemployment** (2000): 2.3%. **Tourism expends.** (1997): $4.5 bil. **Lottery** (2000): total sales: $837.5 mil; net income: $254.4 mil.

Finance. FDIC-insured commercial banks (2000): 23. **Deposits:** $2.8 bil. **FDIC-insured savings institutions** (2000): 46. **Assets**: $46.2 bil.

Federal govt. Fed. civ. employees (Mar. 2000): 7,130. **Avg. salary:** $51,290. **Notable fed. facilities:** U.S. Coast Guard Academy; U.S. Navy Submarine Base.

Energy. Electricity production (2000, kWh, by source): Petroleum: 2.2 bil; Gas: 657 mil; Hydroelectric: 400 mil; Nuclear: 16.4 bil.; **Other:** 436 mil.

State data. Motto: Qui Transtulit Sustinet (He who transplanted still sustains). **Flower:** Mountain laurel. **Bird:** American robin. **Tree:** White oak. **Song:** Yankee Doodle. **Fifth** of the 13 original states to ratify the Constitution, Jan. 9, 1788. **State Fair:** largest fair at Durham, late Sept.; no state fair.

History. At the time of European contact, inhabitants of the area were Algonquian peoples, including the Mohegan and Pequot. Dutch explorer Adriaen Block was the first European visitor, 1614. By 1634, settlers from Plymouth Bay had started colonies along the Connecticut River; in 1637 they defeated the Pequots. The Colony of Connecticut was chartered by England, 1662, adding New Haven, 1665. In the American Revolution, Connecticut Patriots fought in most major campaigns, while Connecticut privateers captured British merchant ships.

Tourist attractions. Mark Twain House, Hartford; Yale University's Art Gallery, Peabody Museum, both in New Haven; Mystic Seaport; Mystic Marine Life Aquarium; P. T. Barnum Museum, Bridgeport; Gillette Castle, Hadlyme; U.S.S. *Nautilus* Memorial, Groton (1st nuclear-powered submarine); Mashantucket Pequot Museum & Research Center, Foxwoods Resort & Casino, both in Ledyard, Mohegan Sun, Uncasville; Lake Compounce, Bristol.

Famous "Nutmeggers." Ethan Allen, Phineas T. Barnum, Samuel Colt, Jonathan Edwards, Nathan Hale, Katharine Hepburn, Isaac Hull, Robert Mitchum, J. Pierpont Morgan, Israel Putnam, Wallace Stevens, Harriet Beecher Stowe, Mark Twain, Noah Webster, Eli Whitney.

Tourist information. Dept. of Economic and Community Development, 505 Hudson St., Hartford, CT 06106.

Toll-free travel information. 1-800-CTBOUND

Website. http://www.state.ct.us

Tourism website. http://www.ctbound.org

Delaware

First State, Diamond State

People. Population (2000): 783,600; rank: 45; **net change** (1999-2000): 17.6%. **Pop. density**: 401 per sq mi. **Racial distribution** (2000): 74.6% white; 19.2% black; 2.1% Asian; 0.3% Native American/Nat. AK; <0.1% Hawaiian/Pacific Islander; 2.0% other race; 2 or more races, 1.7%. **Hispanic pop.** (any race): 4.8%.

Geography. Total area: 2,489 sq mi; rank: 49. **Land area:** 1,954 sq mi; rank: 49. **Acres forested:** 398,000. **Location:** occupies the Delmarva Peninsula on the Atlantic coastal plain. **Climate:** moderate. **Topography:** Piedmont plateau to the N, sloping to a near sea-level plain. **Capital:** Dover.

Economy. Chief industries: chemicals, agriculture, finance, poultry, shellfish, tourism, auto assembly, food processing, transportation equipment. **Chief manuf. goods:** nylon, apparel, luggage, foods, autos, processed meats and vegetables, railroad & aircraft equipment. **Chief crops:** soybeans, potatoes, corn, mushrooms, lima beans, green peas, barley, cucumbers, wheat, corn, grain sorghum, greenhouse & nursery. **Livestock:** (Jan. 2001) 27,000 cattle/calves; (Dec. 2000) 29,000 hogs/pigs; 1.6 mil chickens (excl. broilers); 251.7 mil broilers. **Timber/lumber** (2000): hardwoods and softwoods (except for southern yellow pine);15 mil bd. ft. **Nonfuel minerals** (est. 2000): $12 mil; mostly magnesium compounds, sand & gravel, gemstones. **Commercial fishing** (1999): $6.9 mil. **Chief ports:** Wilmington. **Principal internat. airport at:** Philadelphia/Wilmington. **Value of construction** (1997): $935 mil. **Gross state product** (1999): $35.7 bil. **Employment distrib.** (May 2001): 28.9% services; 21.8% trade; 13.5% govt.; 13.5% mfg. **Per cap. pers. income** (2000): $31,255. **Sales tax:** none. **Unemployment** (2000): 4%. **Tourism expends.** (1997): $1 bil. **Lottery** (2000): total sales: $556.5 mil; net income: $238.2 mil.

Finance. FDIC-insured commercial banks (2000): 32. **Deposits:** $77.8 bil. **FDIC-insured savings institutions** (2000): 6. **Assets:** $12.7 bil.

Federal govt. Fed. civ. employees (Mar. 2000): 2,478. **Avg. salary:** $45,033. **Notable fed. facilities:** Dover Air Force Base, Federal Wildlife Refuge, Bombay Hook.

Energy. Electricity production (2000, kWh, by source): Coal: 3.3 bil; Petroleum: 380 mil; Gas: 419 mil.

State data. Motto: Liberty and independence. **Flower:** Peach blossom. **Bird:** Blue hen chicken. **Tree:** American holly. **Song:** Our Delaware. **First** of original 13 states to ratify the Constitution, Dec. 7, 1787. **State fair** at Harrington; end of July.

History. The Lenni Lenape (Delaware) people lived in the region at the time of European contact. Henry Hudson located the Delaware R., 1609, and in 1610, English explorer Samuel Argall entered Delaware Bay, naming the area after Virginia's governor, Lord De La Warr. The Dutch first settled near present Lewes, 1631, but the colony was destroyed by Indians. Swedes settled at Fort Christina (now Wilmington), 1638. Dutch settled anew, 1651, near New Castle and seized the Swedish settlement, 1655, only to lose all Delaware and New Netherland to the British, 1664. After 1682, Delaware became part of Pennsylvania, and in 1704 it was granted its own assembly. In 1776, it adopted a constitution as the state of Delaware. Although it remained in the Union during the Civil War, Delaware retained slavery until abolished by the 13th Amendment in 1865.

Tourist attractions. Ft. Christina Monument, site of founding of New Sweden, Holy Trinity (Old Swedes) Church, erected 1698, the oldest Protestant church in the U.S. still in use, Wilmington; Hagley Museum, Winterthur Museum and Gardens, both near Wilmington; historic district, New Castle; John Dickinson "Penman of the Revolution" home, Dover; Rehoboth Beach, "nation's summer capital," Rehoboth; Dover Downs Intl. Speedway.

Famous Delawareans. Thomas F. Bayard, Henry Seidel Canby, E. I. du Pont, John P. Marquand, Howard Pyle, Caesar Rodney.

Chamber of Commerce. 1200 N. Orange St., Ste. 200, Wilmington, DE 19899-0671.

Toll-free travel information. 1-800-2VISITDE.

Website. http://www.delaware.gov

Tourism website. http://www.visitdelaware.net

Florida

Sunshine State

People. Population (2000): 15,982,378; rank: 4; **net change** (1999-2000): 23.5%. **Pop. density**: 296.4 per sq mi. **Racial distribution** (2000): 78.0% white; 14.6% black; 1.7% Asian; 0.3% Native American/Nat. AK; 0.1% Hawaiian/Pacific Islander; 3.0% other race; 2 or more races, 2.4%. **Hispanic pop.** (any race): 16.8%.

Geography. Total area: 65,755 sq mi; rank: 22. **Land area:** 53,927 sq mi; rank: 26. **Acres forested:** 16,549,000. **Location:** peninsula jutting southward 500 mi between the Atlantic and the Gulf of Mexico. **Climate:** subtropical N of Bradenton-Lake Okeechobee-Vero Beach line; tropical S of line. **Topography:** land is flat or rolling; highest point is 345 ft in the NW. **Capital:** Tallahassee.

Economy. Chief industries: tourism, agriculture, manufacturing, construction, services, international trade. **Chief manuf. goods:** electric & electronic equipment, transportation equipment, food, printing & publishing, chemicals, instruments, industrial machinery. **Chief crops:** citrus fruits, vegetables, melons, greenhouse and nursery products, potatoes, sugarcane, strawberries. **Livestock:** (Jan. 2001) 1.8 mil cattle/calves; (Dec. 2000) 40,000 hogs/pigs; 12.8 mil chickens (excl. broilers); 122.2 mil broilers. **Timber/lumber** (2000): pine, cypress, cedar; 751 mil bd. ft. **Nonfuel minerals** (est. 2000): $1.9 bil; mostly phosphate rock, crushed stone, portland and masonry cement, sand & gravel. **Commercial fishing** (1999): $197.7 mil. **Chief ports:** Pensacola, Tampa, Manatee, Miami, Port Everglades, Jacksonville, St. Petersburg, Canaveral. **Principal internat. airports at:** Daytona Beach, Ft. Lauderdale/Hollywood, Ft. Myers, Jacksonville, Key West, Melbourne, Miami, Orlando, Panama City, St. Petersburg/Clearwater, Sarasota/Bradenton, Tampa, West Palm Beach. **Value of construction** (1997): $25.2 bil. **Gross state product** (1999): $442.9 bil. **Employment distrib.** (May 2001): 38.1% services; 24.6% trade; 13.8% govt.; 6.6% mfg. **Per cap. pers. income** (2000): $28,145. **Sales tax** (2001): 6%. **Unemployment** (2000): 3.6% **Tourism expends.** (1997): $52.1 bil. **Lottery** (2000): total sales: $2.2 bil; net income: $907.6 mil.

Finance. FDIC-insured commercial banks (2000): 265. **Deposits:** $47.7 bil. **FDIC-insured savings institutions** (2000): 46. **Assets:** $23.7 bil.

Federal govt. Fed. civ. employees (Mar. 2000): 60,220. **Avg. salary:** $48,494. **Notable fed. facilities:** John F. Kennedy Space Center, NASA-Kennedy Space Center's Spaceport USA; Eglin Air Force Base; Pensacola Naval Training Center; MacDill Air Force Base, Tampa.

Energy. Electricity production (2000, kWh, by source): Coal: 66.5 bil; Petroleum: 33.8 bil; Gas: 35.9 bil; Hydroelectric: 83 mil; Nuclear: 32.3 bil.

State data. Motto: In God we trust. **Flower:** Orange blossom. **Bird:** Mockingbird. **Tree:** Sabal palmetto palm. **Song:** Old Folks at Home. **Entered union** Mar. 3, 1845; rank, 27th. **State fair** at Tampa; early Feb.

History. The original inhabitants of Florida included the Timucua, Apalachee, and Calusa peoples. Later the Seminole migrated from Georgia to Florida, becoming dominant there in the early 18th cent. The first European to see Florida was Ponce de León, 1513. France established a colony, Fort Caroline, on the St. John River, 1564. Spain settled St. Augustine, 1565, and Spanish troops massacred most of the French. Britain's Sir Francis Drake burned St. Augustine, 1586. In 1763, Spain ceded Florida to Great Britain, which held the area briefly, 1763-83, before returning it to Spain. After Andrew Jackson led a U.S. invasion, 1818, Spain ceded Florida to the U.S., 1819. The Seminole War, 1835-42, resulted in removal of most Native Americans to Oklahoma Territory. Florida seceded from the Union, 1861, and was readmitted in 1868.

Tourist attractions. Miami Beach; St. Augustine, oldest permanent European settlement in U.S.; Castillo de San Marcos, St. Augustine; Walt Disney World's Magic Kingdom, EPCOT Center, and Disney-MGM Studios, Animal Kingdom all near Orlando; Sea World, Universal Studios, near Orlando; Spaceport USA, Kennedy Space Center; Everglades Natl. Park; Ringling Museum of Art, Ringling Museum of the Circus, both in Sarasota; Cypress Gardens, Winter Haven; Busch Gardens, Tampa; U.S. Astronaut Hall of Fame, Mariana Caverns; Church St. Station, Orlando; Silver Springs, Ocala.

Famous Floridians. Edna Buchanan, Jeb Bush, Marjory Stoneman Douglas, Henry M. Flagler, Carl Hiaasen, James Weldon Johnson, MacKinlay Kantor, John D. MacDonald, Chief Osceola, Claude Pepper, Henry B. Plant, A. Philip Randolph, Marjorie Kinnan Rawlings, Janet Reno, Joseph W. Stilwell, Charles P. Summerall, Ben Vereen.

Tourist information. Visit Florida, P.O. Box 1100, Tallahassee, FL 32302-1100, 1-850-488-5607.

Toll-free number. 1-888-735-2872 (1-888-7FLA-USA)

Website. http://www.state.fl.us

Tourism website. http://www.flausa.com

Georgia

Empire State of the South, Peach State

People. Population (2000): 8,186,453; rank: 10; **net change** (1999-2000): 26.4%. **Pop. density**: 141.4 per sq mi. **Racial distribution** (2000): 65.1% white; 28.7% black; 2.1% Asian; 0.3% Native American/Nat. AK; 0.1% Hawaiian/Pacific Islander; 2.4% other race; 2 or more races, 1.4%. **Hispanic pop.** (any race): 5.3% .

Geography. Total area: 59,425 sq mi; rank: 24. **Land area:** 57,906 sq mi; rank: 21. **Acres forested:** 24,137,000. **Location:** South Atlantic state. **Climate:** maritime tropical air masses dominate in summer; polar air masses in winter; E central area drier. **Topography:** most southerly of the Blue Ridge Mts. cover NE and N central; central Piedmont extends to the fall line of rivers; coastal plain levels to the coast flatlands. **Capital:** Atlanta.

Economy. Chief industries: services, manufacturing, retail trade. **Chief manuf. goods:** textiles, apparel, food, and kindred products, pulp & paper products. **Chief crops:** peanuts, cotton, corn, tobacco, hay, soybeans. **Livestock:** (Jan. 2001) 1.3 mil cattle/calves; (Dec. 2000) 380,000 hogs/pigs; 29.8 mil chickens (excl. broilers); 1.2 bil broilers. **Timber/lumber** (2000): pine, hardwood; 3.2 bil bd. ft. **Nonfuel minerals** (est. 2000): $1.7 bil; crushed stone, portland cement, sand & gravel. **Commercial fishing** (1999): $21.1 mil. **Chief ports:** Savannah, Brunswick. **Principal internat. airports at:** Atlanta, Savannah. **Value of construction** (1997): $13.6 bil. **Gross state product** (1999): $276 bil. **Employment distrib.** (May 2001): 28.7% services; 24.8% trade; 15.2% govt.; 14.1% mfg. **Per cap. pers. income** (2000): $27,940. **Sales tax** (2001): 4%. **Unemployment** (2000): 3.7%. **Tourism expends.** (1997): $12.6 bil. **Lottery** (2000): total sales: $2.2 bil; net income: $682.4 mil.

Finance. FDIC-insured commercial banks (2000): 337. **Deposits:** $109.3 bil. **FDIC-insured savings institutions** (2000): 24. **Assets:** $7.2 bil.

Federal govt. Fed. civ. employees (Mar. 2000): 62,345. **Avg. salary:** $46,787. **Notable fed. facilities:** Dobbins AFB; Ft. Benning; Ft. Gordon; Ft. Gillem; Ft. Stewart; King's Bay Naval Base; Moody Air Force Base; Navy Supply Corps School; Ft. McPherson; Fed. Law Enforcement Training Ctr., Glynco, Robins AFB; Centers for Disease Control.

Energy. Electricity production (2000, kWh, by source): Coal: 77.8 bil; Petroleum: 633 mil; Gas: 1.8 bil; Hydroelectric: 2.3 bil; Nuclear: 32.5 bil.

State data. Motto: Wisdom, justice and moderation. **Flower:** Cherokee rose. **Bird:** Brown thrasher. **Tree:** Live oak. **Song:** Georgia On My Mind. **Fourth** of the 13 original states to ratify the Constitution, Jan. 2, 1788. **State fair** at Macon, 3d week in Oct.

History. Creek and Cherokee peoples were early inhabitants of the region. The earliest known European settlement was the Spanish mission of Santa Catalina, 1566, on Saint Catherines Island. Gen. James Oglethorpe established a colony at Savannah, 1733, for the poor and religiously persecuted. Oglethorpe defeated a Spanish army from Florida at Bloody Marsh, 1742. In the American Revolution, Georgians seized the Savannah armory, 1775, and sent the munitions to the Continental Army. They fought seesaw campaigns with Cornwallis's British troops, twice liberating Augusta and forcing final evacuation by the British from Savannah, 1782. The Cherokee were removed to Oklahoma Territory, 1832-38, and thousands died on the long march, known as the Trail of Tears. Georgia seceded from the Union, 1861, and was invaded by Union forces, 1864, under Gen. William T. Sherman, who took Atlanta, Sept. 2, and proceeded on his famous "march to the sea," ending in Dec., in Savannah. Georgia was readmitted, 1870.

Tourist attractions. State Capitol, Stone Mt. Park, Six Flags Over Georgia, Kennesaw Mt. Natl. Battlefield Park, Martin Luther King Jr. Natl. Historic Site, Underground Atlanta, Jimmy Carter Library & Museum, all Atlanta; Chickamauga and Chattanooga Natl. Military Park, near Dalton; Chattahoochee Natl. Forest; alpine village of Helen; Dahlonega, site of America's first gold rush; Brasstown Bald Mt.; Lake Lanier; Franklin D. Roosevelt's Little White House, Warm Springs; Callaway Gardens, Pine Mt.; Andersonville Natl. Historic Site; Okefenokee Swamp, near Waycross; Jekyll Island; St. Simons Island; Cumberland Island Natl. Seashore; historic riverfront district, Savannah.

Famous Georgians. Kim Basinger, Griffin Bell, James Bowie, James Brown, Erskine Caldwell, Jimmy Carter, Ray Charles, Lucius D. Clay, Ty Cobb, James Dickey, John C. Fremont, Newt Gingrich, Joel Chandler Harris, "Doc" Holliday, Alan Jackson, Martin Luther King Jr., Gladys Knight, Sidney Lanier, Little Richard, Juliette Gordon Low, Margaret Mitchell, Sam Nunn, Flannery O'Connor, Otis Redding, Burt Reynolds, Julia Roberts, Jackie Robinson, Clarence Thomas, Travis Tritt, Ted Turner, Carl Vinson, Alice Walker, Herschel Walker, Joseph Wheeler, Joanne Woodward, Trisha Yearwood, Andrew Young.

Chamber of Commerce. 235 International Blvd., Atlanta, GA 30303; (404) 880-9000.

Toll-free travel information. 1-800-VISITGA.

Website. http://www.state.ga.us

Tourism website. http://www.georgia.org/tourism

Hawai'i

Aloha State

People. Population (2000): 1,211,537; rank: 42; **net change** (1999-2000): 9.3%. **Pop. density**: 188.6 per sq mi. **Racial distribution** (2000): 24.3% white; 1.8% black; 41.6% Asian; 0.3% Native American/Nat. AK; 9.4% Hawaiian/Pacific Islander; 1.3% other race; 2 or more races, 21.4%. **Hispanic pop.** (any race): 7.2%.

Geography. Total area: 10,931 sq mi; rank: 43. **Land area:** 6,423 sq mi; rank: 47. **Acres forested:** 1,748,000. **Location:** Hawaiian Islands lie in the North Pacific, 2,397 mi SW from San Francisco. **Climate:** subtropical, with wide variations in rainfall; Waialeale, on Kaua'i, wettest spot in U.S. (annual rainfall 460 in.) **Topography:** islands are tops of a chain of submerged volcanic mountains; active volcanoes: Mauna Loa, Kilauea. **Capital:** Honolulu.

Economy. Chief industries: tourism, defense, sugar, pineapples. **Chief manuf. goods:** processed sugar, canned pineapple, clothing, foods, printing & publishing. **Chief crops:** sugar, pineapples, macadamia nuts, fruits, coffee, vegetables, floriculture. **Livestock:** (Jan. 2001) 150,000 cattle/calves; (Dec. 2000) 26,000 hogs/pigs; 722,000 chickens (excl. broilers); 1 mil broilers. **Timber/Lumber (2000):** NA; **Nonfuel minerals** (est. 2000): $91.4 mil; mostly crushed stone, portland and masonry cement, sand and gravel, gemstones. **Commercial fishing** (1998): $65 mil. **Chief ports:** Honolulu, Hilo, Kailua. **Principal internat. airports at:** Hilo, Honolulu, Kailua, Kahului. **Value of construction** (1997): $1.7 bil. **Gross state product** (1999): $41 bil. **Employment distrib.** (May 2001): 33.6% services; 25.1% trade; 20.2% govt.; 3.2% mfg. **Per cap. pers. income** (2000): $28,221. **Sales tax** (2001): 4%. **Unemployment** (2000): 4.3%. **Tourism expenditures** (1997): $14.2 bil.

Finance. FDIC-insured commercial banks (2000): 8. **Deposits:** $17.5 bil. **FDIC-insured savings institutions** (2000): 2. **Assets:** $6.5 bil.

Federal govt. Fed. civ. employees (Mar. 2000): 19,084. **Avg. salary:** $45,157. **Notable fed. facilities:** Pearl Harbor Naval Shipyard; Hickam AFB; Schofield Barracks; Ft. Shafter; Marine Corps Base-Kaneohe Bay; Barbers Point NAS; Wheeler AFB; Prince Kuhio Federal Building.

Energy. Electricity production (2000, kWh, by source): Petroleum: 6.3 bil; Hydroelectric: 19 mil.

State data. Motto: The life of the land is perpetuated in righteousness. **Flower:** Yellow hibiscus. **Bird:** Hawaiian goose. **Tree:** Kukui (Candlenut). **Song:** Hawai'i Pono'i. **Entered union** Aug. 21, 1959; rank, 50th. **State fair:** at O'ahu, late June.

History. Polynesians from islands 2,000 mi to the south settled the Hawaiian Islands, probably between AD 300 and AD 600. The first European visitor was British captain James Cook, 1778. Between 1790 and 1810, the islands were united politically under the leadership of a native king, Kamehameha I, whose four successors—all bearing the name Kamehameha—ruled the kingdom from his death, 1819, until the end of the dynasty, 1872. Missionaries arrived, 1820, bringing Western culture. King Kamehameha III and his chiefs created the first constitution and a legislature that set up a public school system. Sugar production began, 1835, and it became the dominant industry. In 1893, Queen Liliuokalani was deposed, and a republic was instituted, 1894, headed by Sanford B. Dole. Annexation by the U.S. came in 1898. The Japanese attack on Pearl Harbor, Dec. 7, 1941, brought the U.S. into World War II.

Tourist attractions. Hawaii Volcanoes, Haleakala natl. parks; Natl. Memorial Cemetery of the Pacific, Waikiki Beach, Diamond Head, Honolulu; U.S.S. *Arizona* Memorial, Pearl Harbor; Hanauma Bay; Polynesian Cultural Center, Laie; Nu'uanu Pali; Waimea Canyon; Wailoa and Wailuku River state parks.

Famous Islanders. Bernice Pauahi Bishop, Tia Carrera, Father Damien de Veuster, Don Ho, Duke Kahanamoku, King Kamehameha, Brook Mahealani Lee, Daniel K. Inouye, Jason Scott Lee, Queen Liliuokalani, Bette Midler, Ellison Onizuka.

Chamber of Commerce of Hawaii. 1132 Bishop St., Suite 200, Honolulu, HI 96813; phone: (808) 545-4300.

Toll-free travel information. 1-800-GOHAWAII.

Website. http://www.hawaii.gov

Tourism website. http://www.gohawaii.com

Idaho

Gem State

People. Population (2000): 1,293,953; rank: 39; **net change** (1999-2000): 28.5%. **Pop. density**: 15.6 per sq mi. **Racial distribution** (2000): 91.0% white; 0.4% black; 0.9% Asian; 1.4% Native American/Nat. AK; 0.1% Hawaiian/Pacific Islander; 4.2% other race; 2 or more races, 2.0%. **Hispanic pop.** (any race): 7.9%.

Geography. Total area: 83,570 sq mi; rank: 14. **Land area:** 82,747 sq mi; rank: 11. **Acres forested:** 21,621,000. **Location:** northwestern Mountain state bordering on British Columbia. **Climate:** tempered by Pacific westerly winds; drier, colder, continental climate in SE; altitude an important factor. **Topography:** Snake R. plains in the S; central region of mountains, canyons, gorges (Hells Canyon, 7,900 ft, deepest in N. America); subalpine northern region. **Capital:** Boise.

Economy. Chief industries: manufacturing, agriculture, tourism, lumber, mining, electronics. **Chief manuf. goods:** electronic components, computer equipment, processed foods, lumber and wood products, chemical products, primary metals, fabricated metal products, machinery. **Chief crops:** potatoes, peas, dry beans, sugar beets, alfalfa seed, lentils, wheat, hops, barley, plums and prunes, mint, onions, corn, cherries, apples, hay. **Livestock:** (Jan. 2001) 2 mil cattle/calves; 275,000 sheep/lambs; (Dec. 2000) 24,000 hogsyellow, white pine; Douglas fir; white spruce; /pigs; 1.2 mil chickens (excl. broilers). **Timber/lumber** (2000): 1.9 bil bd. ft. **Nonfuel minerals** (est. 2000): $398 mil; mostly phosphate rock, silver, sand & gravel, molybdenum, lead. **Chief port:** Lewiston. **Value of construction** (1997) $1.8 bil. **Gross state product** (1999): $34 bil. **Employment distrib.** (May 2001): 26.1% services; 25.4% trade; 19.3% govt.; 13.3% mfg. **Per cap. pers. income** (2000): $24,180. **Sales tax** (2001: 5%. **Unemployment** (2000): 4.9%. **Tourism expenditures** (1997): $1.9 bil. **Lottery** (2000): total sales: $86.5 mil; net income: $8.2 mil.

Finance. FDIC-insured commercial banks (2000): 18. **Deposits:** $2.1 bil. **FDIC-insured savings institutions** (2000): 2. **Assets:** $590 mil.

Federal govt. Fed. civ. employees (Mar. 2000): 7,336. **Avg. salary:** $46,236. **Notable fed. facilities:** Idaho Natl. Engineering Lab; Mt. Home Air Force Base.

Energy. Electricity production (2000, kWh, by source): Petroleum: 3 mil; Hydroelectric: 10.1 bil.

State data. Motto: Esto Perpetua (It is perpetual). **Flower:** Syringa. **Bird:** Mountain bluebird. **Tree:** White pine. **Song:** Here We Have Idaho. **Entered union** July 3, 1890; rank, 43d. **State fair** at Boise, late Aug.; at Blackfoot, early Sept.

History. Early inhabitants were Shoshone, Northern Paiute, Bannock, and Nez Percé peoples. White exploration of the region began with Lewis and Clark, 1805-6. Next came fur traders, setting up posts, 1809-34, and missionaries, 1830s-50s. Mormons made their first permanent settlement at Franklin, 1860. Idaho's gold rush began the same year and brought thousands of permanent settlers. Most remarkable of the Indian wars was the 1,700-mi trek, 1877, of Chief Joseph and the Nez Percé, pursued by U.S. troops through 3 states and caught just short of the Canadian border. The Idaho territory was organized, 1863. Idaho adopted a progressive constitution and became a state, 1890.

Tourist attractions. Hells Canyon, deepest gorge in N. America; World Center for Birds of Prey; Craters of the Moon; Sun Valley, in Sawtooth Mts.; Crystal Falls Cave; Shoshone Falls; Lava Hot Springs; Lake Pend Oreille; Lake Coeur d'Alene; Sawtooth Natl. Recreation Area; River of No Return Wilderness Area; Redfish Lake.

Famous Idahoans. William E. Borah, Frank Church, Fred T. Dubois, Ezra Pound, Chief Joseph, Sacagawea, Picabo Street, Lana Turner.

Tourist information. Department of Commerce, 700 W. State St., Boise, ID 83720.

Toll-free travel information. 1-800-842-5858.

Website. http://www.state.id.us

Tourism website. http://www.visitid.org

Illinois

Prairie State

People. Population (2000): 12,419,293; rank: 5; **net change** (1999-2000): 8.6%. **Pop. density**: 223.4 per sq mi. **Racial distribution** (2000): 73.5% white; 15.1% black; 3.4% Asian; 0.2% Native American/Nat. AK; <0.1% Hawaiian/Pacific Islander; 5.8% other race; 2 or more races, 1.9%. **Hispanic pop.** (any race): 12.3%.

Geography. Total area: 57,914 sq mi; rank: 25. **Land area:** 55,584 sq mi; rank: 24. **Acres forested:** 4,266,000. **Location:** East North Central state; western, southern, and eastern boundaries formed by Mississippi, Ohio, and Wabash rivers, respectively. **Climate:** temperate; typically cold, snowy winters, hot summers. **Topography:** prairie and fertile plains throughout; open hills in the southern region. **Capital:** Springfield.

Economy. Chief industries: services, manufacturing, travel, wholesale and retail trade, finance, insurance, real estate, construction, health care, agriculture. **Chief manuf. goods:** machinery, electric and electronic equipment, prim. & fabric. metals, chemical products, printing & publishing, food and kindred products. **Chief crops:** corn, soybeans, wheat, sorghum, hay. **Livestock:** (Jan. 2001) 1.5 mil cattle/calves; 75,000 sheep/lambs; (Dec. 2000) 4.2 mil hogs/pigs; 4 mil chickens (excl. broilers). **Timber/lumber** (2000): oak, hickory, maple, cottonwood; 121 mil bd. ft. **Nonfuel minerals** (est. 2000): $907 mil; mostly crushed stone, portland cement, sand & gravel, lime. **Commercial fishing** (1999): $50,000. **Chief ports:** Chicago. **Principal internat. airport at:** Chicago. **Value of construction** (1997): $12.5 bil. **Gross state product** (1998): $445.6 bil. **Employment distrib.** (May 2001): 31.0% services; 22.6% trade; 13.7% govt.; 15.2% mfg. **Per cap. pers. income** (2000): $32,259. **Sales tax** (2001): 6.25%. **Unemployment** (2000): 4.4%. **Tourism expends.** (1997): $19.6 bil. **Lottery** (2000): total sales: $1.5 bil; net income: $533.2 mil.

Finance. FDIC-insured commercial banks (2000): 711. **Deposits:** $249.4 bil. **FDIC-insured savings institutions** (2000): 117. **Assets:** $39.5 bil.

Federal govt. Fed. civ. employees (Mar. 2000): 40,696. **Avg. salary:** $52,350. **Notable fed. facilities:** Fermi Natl. Accelerator Lab; Argonne Natl. Lab; Rock Island Arsenal; Great Lakes, Naval Training Station, Scott AFB.

Energy. Electricity production (2000, kWh, by source): Coal: 32.1 bil; Petroleum: 128 mil; Gas: 342 mil; Hydroelectric: 53 mil; Nuclear: 82.5 bil.; **Other:** 64 mil.

State data. Motto: State sovereignty—national union. **Flower:** Native violet. **Bird:** Cardinal. **Tree:** White oak. **Song:** Illinois. **Entered union** Dec. 3, 1818; rank, 21st. **State fair** at Springfield, mid-Aug.; DuQuoin, late Aug.

History. Seminomadic Algonquian peoples, including the Peoria, Illinois, Kaskaskia, and Tamaroa, lived in the region at the time of European contact. Fur traders were the first Europeans in Illinois, followed shortly by Jolliet and Marquette, 1673, and La Salle, 1680, who built a fort near present-day Peoria. The first settlements were French, at Cahokia, near present-day St. Louis, 1699, and Kaskaskia, 1703. France ceded the area to Britain, 1763, and in 1778, American Gen. George Rogers Clark took Kaskaskia from the British without a shot. Defeat of Native American tribes in Black Hawk War, 1832, and growth of railroads brought change to the area. In 1787, it became part of the Northwest Territory. Post-Civil War Illinois became a center for the labor movement as bitter strikes, such as the Haymarket Square riot, occurred in 1885-86.

Tourist attractions. Chicago museums and parks; Lincoln shrines at Springfield, New Salem, Sangamon County; Cahokia Mounds, Collinsville; Starved Rock State Park; Crab Orchard Wildlife Refuge; Mormon settlement at Nauvoo; Fts. Kaskaskia, Chartres, Massac (parks); Shawnee Natl. Forest, Southern Illinois; Illinois State Museum, Springfield; Dickson Mounds Museum, between Havana and Lewistown.

Famous Illinoisans. Jane Addams, Saul Bellow, Jack Benny, Ray Bradbury, Gwendolyn Brooks, William Jennings Bryan, St. Frances Xavier Cabrini, Hillary Rodham Clinton, Clarence Darrow, John Deere, Stephen A. Douglas, James T. Farrell, George W. Ferris, Marshall Field, Betty Friedan, Benny Goodman, Ulysses S. Grant, Dennis Hastert, Ernest Hemingway, Wild Bill Hickok, Henry J. Hyde, Abraham Lincoln, Vachel Lindsay, Edgar Lee Masters, Oscar Mayer, Cyrus McCormick, Ronald Reagan, Carl Sandburg, Adlai Stevenson, Frank Lloyd Wright, Philip Wrigley.

Tourist information. Illinois Dept. of Commerce and Community Affairs, 620 E. Adams St., Springfield, IL 62701.

Toll-free travel information: 1-800-2-CONNECT.

Website. http://www.state.il.us

Tourism website. http://www.enjoyillinois.com

Indiana

Hoosier State

People. Population (2000): 6,080,485; rank: 14; **net change** (1999-2000): 9.7%. **Pop. density**: 169.5 per sq mi. **Racial distribution** (2000): 87.5% white; 8.4% black; 1.0% Asian; 0.3% Native American/Nat. AK; <0.1% Hawaiian/Pacific Islander; 1.6% other race; 2 or more races, 1.2%. **Hispanic pop.** (any race): 3.5%.

Geography. Total area: 36,418 sq mi; rank: 38. **Land area:** 35,867 sq mi; rank: 38. **Acres forested:** 4,439,000. **Location:** East North Central state; Lake Michigan on N border. **Climate:** 4 distinct seasons with a temperate climate. **Topography:** hilly southern region; fertile rolling plains of central region; flat, heavily glaciated north; dunes along Lake Michigan shore. **Capital:** Indianapolis.

Economy: Chief industries: manufacturing, services, agriculture, government, wholesale and retail trade, transportation and public utilities. **Chief manuf. goods:** primary metals, transportation equipment, motor vehicles & equip., industrial machinery & equipment, electronic & electric equipment. **Chief crops:** corn, soybeans, wheat, nursery and greenhouse products, vegetables, popcorn, fruit, hay, tobacco, mint. **Livestock:** (Jan. 2001) 880,000 mil cattle/calves; 66,000 sheep/lambs; (Dec. 2000) 3.4 mil hogs/pigs; 28.7 mil chickens (excl. broilers). **Timber/lumber** (2000): oak, tulip, beech, sycamore; 358 mil bd. ft. **Nonfuel minerals** (est. 2000): $729 mil; mostly crushed stone, portland and masonry cement, sand & gravel, lime. **Commercial fishing** (1997): $327,000. **Chief ports:** Burns Harbor, Portage; Southwind Maritime, Mt. Vernon; Clark Maritime, Jeffersonville. **Principal internat. airports at:** Indianapolis, Ft. Wayne. **Value of construction** (1997): $9.2 bil. **Gross state product** (1999):

$182.2 bil. **Employment distrib.** (May 2001): 25.4% services; 23.5% trade; 13.9% govt.; 22.2% mfg. **Per cap. pers. income** (2000): $27,011. **Sales tax** (2001): 5%. **Unemployment** (2000): 3.2%. **Tourism expends.** (1997): $5.6 bil. **Lottery** (2000): total sales: $582.5 mil; net income: $165.2 mil.

Finance. FDIC-insured commercial banks (2000): 153. **Deposits:** $54.6 bil. **FDIC-insured savings institutions** (2000): 66. **Assets:** $16.9 bil.

Federal govt. Fed. civ. employees (Mar. 2000): 21,884. **Avg. salary:** $43,145. **Notable fed. facilities:** Naval Air Warfare Center; Ft. Benjamin Harrison; Del. Grissom AFB; Naval Surface Warfare Center.

Energy. Electricity production (2000, kWh, by source): Coal: 116.9 bil; Petroleum: 838 mil; Gas: 671 mil; Hydroelectric: 588 mil.

State data. Motto: Crossroads of America. **Flower:** Peony. **Bird:** Cardinal. **Tree:** Tulip poplar. **Song:** On the Banks of the Wabash, Far Away. **Entered union** Dec. 11, 1816; rank, 19th. **State fair** at Indianapolis; mid-Aug.

History. When the Europeans arrived, Miami, Potawatomi, Kickapoo, Piankashaw, Wea, and Shawnee peoples inhabited the area. A French trading post was built, 1731-32, at Vincennes. La Salle visited the present South Bend area, 1679 and 1681. The first French fort was built near present-day Lafayette, 1717. France ceded the area to Britain, 1763. During the American Revolution, American Gen. George Rogers Clark captured Vincennes, 1778, and defeated British forces, 1779. At war's end, Britain ceded the area to the U.S. Miami Indians defeated U.S. troops twice, 1790, but were beaten, 1794, at Fallen Timbers by Gen. Anthony Wayne. At Tippecanoe, 1811, Gen. William H. Harrison defeated Tecumseh's Indian confederation. The Delaware, Potawatomi, and Miami were moved farther west, 1820-1850.

Tourist attractions. Lincoln Log Cabin Historic Site, near Charleston; George Rogers Clark Park, Vincennes; Wyandotte Cave; Tippecanoe Battlefield Memorial Park; Benjamin Harrison home; Indianapolis 500 raceway and museum, all Indianapolis; Indiana Dunes, near Chesterton; National College Football Hall of Fame, South Bend; Hoosier Nat'l. Forest, south-central Indiana.

Famous "Hoosiers." Larry Bird, Ambrose Burnside, Hoagy Carmichael, Jim Davis, James Dean, Eugene V. Debs, Theodore Dreiser, Paul Dresser, Jeff Gordon, Benjamin Harrison, Gil Hodges, Michael Jackson, David Letterman, John Mellencamp, Jane Pauley, Cole Porter, Dan Quayle, Gene Stratton Porter, Ernie Pyle, James Whitcomb Riley, Oscar Robertson, Red Skelton, Booth Tarkington, Kurt Vonnegut, Lew Wallace, Wendell L. Willkie, Wilbur Wright.

Chamber of Commerce. One North Capital, Suite 200, Indianapolis, IN 46204.

Toll-free travel information. 1-888-ENJOYIN.

Website. http://www.ai.org

Tourism website. http://www.enjoyindiana.com

Iowa

Hawkeye State

People. Population (2000): 2,926,324; rank: 30; **net change** (1999-2000): 5.4%. **Pop. density**: 52.4 per sq mi. **Racial distribution** (2000): 93.9% white; 2.1% black; 1.3% Asian; 0.3% Native American/Nat. AK; <0.1% Hawaiian/Pacific Islander; 1.3% other race; 2 or more races, 1.1%. **Hispanic pop.** (any race): 2.8%.

Geography. Total area: 56,272 sq mi; rank: 26. **Land area:** 55,869 sq mi; rank: 23. **Acres forested:** 2,050,000. **Location:** West North Central state bordered by Mississippi R. on the E and Missouri R. on the W. **Climate:** humid, continental. **Topography:** Watershed from NW to SE; soil especially rich and land level in the N central counties. **Capital:** Des Moines.

Economy. Chief industries: agriculture, communications, construction, finance, insurance, trade, services, manufacturing. **Chief manuf. goods:** processed food products, tires, farm machinery, electronic products, appliances, household furniture, chemicals, fertilizers, auto accessories. **Chief crops:** silage and grain corn, soybeans, oats, hay. **Livestock:** (Jan. 2001) 3.7 mil cattle/calves; 270,000 sheep/lambs; (Dec. 2000) 15.4 mil hogs/pigs; 37.8 mil chickens (excl. broilers); **Timber/lumber** (2000): red cedar; 88 mil bd. ft. **Nonfuel minerals** (est. 2000): $510 mil; mostly crushed stone, portland cement, sand & gravel, gypsum, lime. **Principal internat. airport at:** Des Moines. **Value of construction** (1997): $3.2 bil. **Gross state product** (1999): $85 bil. **Employment distrib.** (May 2001): 26.7% services; 24.3% trade; 16.3% govt.; 17.6% mfg. **Per cap. pers. income** (2000): $26,723. **Sales tax** (2001): 5%. **Unemployment** (2000): 2.6%. **Tourism expends.** (1997): $3.7 bil. **Lottery** (2000): total sales: $178.2 mil; net income: $44.4 mil.

Finance. FDIC-insured commercial banks (2000): 431. **Deposits:** $34.3 bil. **FDIC-insured savings institutions** (2000): 25. **Assets:** $4.4 bil.

Federal govt. Fed. civ. employees (Mar. 2000): 6,950 **Avg. salary:** $44,449.

Energy. Electricity production (2000, kWh, by source): Coal: 33.5 bil; Petroleum: 91 mil; Gas: 321 mil; Hydroelectric: 890 mil; Nuclear: 4.5 bil.

State data. Motto: Our liberties we prize, and our rights we will maintain. **Flower:** Wild rose. **Bird:** Eastern goldfinch. **Tree:** Oak. **Rock:** Geode. **Entered union** Dec. 28, 1846; rank, 29th. **State fair** at Des Moines; mid-Aug.

History. Early inhabitants were Mound Builders who dwelt on Iowa's fertile plains. Later, Woodland tribes including the Iowa and Yankton Sioux lived in the area. The first Europeans, Marquette and Jolliet, gave France its claim to the area, 1673. In 1762, France ceded the region to Spain, but Napoleon took it back, 1800. It became part of the U.S. through the Louisiana Purchase, 1803. Native American Sauk and Fox tribes moved into the area from states farther east but relinquished their land in defeat, after the 1832 uprising led by the Sauk chieftain Black Hawk. By mid-19th cent. they were forced to move on to Kansas. Iowa became a territory in 1838, and entered as a free state, 1846, strongly supporting the Union.

Tourist attractions. Herbert Hoover birthplace and library, West Branch; Effigy Mounds Natl. Monument, prehistoric Indian burial site, Marquette; Amana Colonies; Grant Wood's paintings and memorabilia, Davenport Municipal Art Gallery; Living History Farms, Des Moines; Adventureland, Altoona; Boone & Scenic Valley Railroad, Boone; Greyhound Parks, in Dubuque and Council Bluffs; Prairie Meadows horse racing, Altoona; riverboat cruises and casino gambling, Mississippi and Missouri Rivers; Iowa Great Lakes, Okoboji.

Famous Iowans. Tom Arnold, Johnny Carson, Marquis Childs, Buffalo Bill Cody, Mamie Dowd Eisenhower, Bob Feller, George Gallup, Susan Glaspell, James Norman Hall, Harry Hansen, Herbert Hoover, Ann Landers, Glenn Miller, Lillian Russell, Billy Sunday, James A. Van Allen, Carl Van Vechten, Henry Wallace, John Wayne, Meredith Willson, Grant Wood.

Tourist information. Division of Tourism, Iowa Dept. of Economic Development, 200 E. Grand Ave., Des Moines, IA 50309.

Toll-free travel information. 1-800-476-6035

Website. http://www.state.ia.us

Tourism website. http://www.traveliowa.com

Kansas

Sunflower State

People. Population (2000): 2,688,418; rank: 32; **net change** (1999-2000): 8.5%. **Pop. density**: 32.9 per sq mi. **Racial distribution** (2000): 86.1% white; 5.7% black; 1.7% Asian; 0.9% Native American/Nat. AK; 0.1% Hawaiian/Pacific Islander; 3.4% other race; 2 or more races, 2.1%. **Hispanic pop.** (any race): 7.0%.

Geography. Total area: 82,277 sq mi; rank: 15. **Land area:** 81,815 sq mi; rank: 13. **Acres forested:** 1,359,000. **Location:** West North Central state, with Missouri R. on E. **Climate:** temperate but continental, with great extremes between summer and winter. **Topography:** hilly Osage Plains in the E; central region level prairie and hills; high plains in the W. **Capital:** Topeka.

Economy. Chief industries: manufacturing, finance, insurance, real estate, services. **Chief manuf. goods:** transportation equipment, machinery & computer equipment, food and kindred products, printing & publishing. **Chief crops:** wheat, sorghum, corn, hay, soybeans, sunflowers. **Livestock:** (Jan. 2001) 6.7 mil cattle/calves; 110,000 sheep/lambs; (Dec. 2000) 1.6 mil hogs/pigs; 2 mil chickens (excl. broilers). **Timber/lumber:** (2000): oak, walnut; 14 mil bd. ft. **Nonfuel minerals** (est. 2000): $624 mil; mostly portland cement, salt, crushed stone, helium, sand & gravel. **Chief ports:** Kansas City. **Principal internat. airport at:** Kansas City. **Value of construction** (1997): $3.8 bil. **Gross state product** (1999): $81 bil. **Employment distrib.** (May 2001): 26.0% services; 23.7% trade; 18.1% govt.; 15.4% mfg. **Per cap. pers. income** (2000): $27,816. **Sales tax** (2001): 4.9%. **Unemployment** (2000): 3.7%. **Tourism expends.** (1997): $3.1 bil. **Lottery** (2000): total sales: $192.6 mil; net income: $56.8 mil.

IT'S A FACT: Found in offices all over the world, the indispensable Post-it® note—available in more than 25 sizes, 50 standard shapes, and 50 colors—is manufactured in the small town of Cynthiana, KY (2000 population: 6,258).

Finance. FDIC-insured commercial banks (2000): 376. **Deposits**: $31 bil. **FDIC-insured savings institutions** (2000): 17. **Assets** : $11.5 bil.

Federal govt. Fed. civ. employees (Mar. 2000): 14,669. **Avg. salary:** $45,588. **Notable fed. facilities:** Fts. Riley, Leavenworth; Leavenworth Federal Penitentiary; Colmery-O'Neal Veterans Hospital.

Energy. Electricity production (2000, kWh, by source): Coal: 32.2 bil; Petroleum: 407 mil; Gas: 2.8 bil; Nuclear: 9.1 bil.

State data. Motto: Ad Astra per Aspera (To the stars through difficulties). **Flower:** Native sunflower. **Bird:** Western meadowlark. **Tree:** Cottonwood. **Song:** Home on the Range. **Entered union** Jan. 29, 1861; rank, 34th. **State fair** at Hutchinson; begins Friday after Labor Day.

History. When Coronado first explored the area, Wichita, Pawnee, Kansa, and Osage peoples lived there. These Native Americans—hunters who also farmed—were joined on the Plains by the nomadic Cheyenne, Arapaho, Comanche, and Kiowa about 1800. French explorers established trading between 1682 and 1739, and the U.S. took over most of the area in the Louisiana Purchase, 1803. After 1830, thousands of eastern Native Americans were removed to Kansas. Kansas became a territory, 1854. Violent incidents between pro- and antislavery settlers caused the territory to be known as "Bleeding Kansas." It eventually entered the Union as a free state, 1861. Railroad construction after the war made Abilene and Dodge City terminals of large cattle drives from Texas.

Tourist attractions. Eisenhower Center, Abilene; Agricultural Hall of Fame and Natl. Center, Bonner Springs; Dodge City-Boot Hill & Frontier Town; Old Cowtown Museum, Wichita; Ft. Scott and Ft. Larned, restored 1800s cavalry forts; Kansas Cosmosphere and Space Center, Hutchinson; Woodlands Racetrack, Kansas City; U.S. Cavalry Museum, Ft. Riley; NCAA Visitors Center, Shawnee; Heartland Park Raceway, Topeka.

Famous Kansans. Kirstie Alley, Ed Asner, Roscoe "Fatty" Arbuckle, Gwendolyn Brooks, John Brown, George Washington Carver, Wilt Chamberlain, Walter P. Chrysler, Glenn Cunningham, John Stuart Curry, Robert Dole, Amelia Earhart, Wyatt Earp, Dwight D. Eisenhower, Ron Evans, Maurice Greene, Wild Bill Hickok, Cyrus Holliday, Dennis Hopper, William Inge, Don Johnson, Walter Johnson, Nancy Landon Kassebaum, Buster Keaton, Emmett Kelly, Alf Landon, Edgar Lee Masters, Hattie McDaniel, Oscar Micheaux, Carry Nation, Georgia Neese-Gray, Charlie Parker, Gordon Parks, Jim Ryun, Barry Sanders, Vivian Vance, William Allen White, Jess Willard.

Tourist information. Kansas Dept. of Commerce & Housing, Travel and Tourism Div., 700 SW Harrison, Suite 1300, Topeka, KS 66601; 1-913-296-2009.

Toll-free travel information. 1-800-2KANSAS.

Website. http://www.ink.org

Tourism website. http://www.travelks.com

Kentucky

Bluegrass State

People. Population (2000): 4,041,769; rank: 25; **net change** (1999-2000): 9.7%. **Pop. density**: 101.7 per sq mi. **Racial distribution** (2000): 90.1% white; 7.3% black; 0.7% Asian; 0.2% Native American/Nat. AK; <0.1% Hawaiian/Pacific Islander; 0.6% other race; 2 or more races, 1.1%. **Hispanic pop.** (any race): 1.5%.

Geography. Total area: 40,409 sq mi; rank: 37. **Land area:** 39,728 sq mi; rank: 36. **Acres forested:** 12,714,000. **Location:** East South Central state, bordered on N by Illinois, Indiana, Ohio; on E by West Virginia and Virginia; on S by Tennessee; on W by Missouri. **Climate:** moderate, with plentiful rainfall. **Topography:** mountainous in E; rounded hills of the Knobs in the N; Bluegrass, heart of state; wooded rocky hillsides of the Pennyroyal; Western Coal Field; the fertile Purchase in the SW. **Capital:** Frankfort.

Economy. Chief industries: manufacturing, services, finance, insurance and real estate, retail trade, public utilities. **Chief manuf. goods:** transportation & industrial machinery, apparel, printing & publishing, food products, electric & electronic equipment. **Chief crops:** tobacco, corn, soybeans. **Livestock:** (Jan. 2001) 2.3 mil cattle/calves; (Dec. 2000) 430,000 hogs/pigs; 5.8 mil chickens (excl. broilers); (Dec. 1998) 188.8 mil broilers. **Timber/lumber** (2000): hardwoods, pines; 816 mil bd. ft. **Nonfuel minerals** (est. 2000): $497 mil; mostly crushed stone, lime, portland cement, sand & gravel, clays. **Chief ports:** Paducah, Louisville, Covington, Owensboro, Ashland, Henderson County, Lyon County, Hickman-Fulton County. **Principal internat. airports at:** Covington/Cincinnati, Louisville. **Value of construction** (1997): $4.8 bil. **Gross state product** (1999): $114 bil. **Employment distrib.** (May 2001): 26.4% services; 23.6% trade; 17.2% govt.; 17.0% mfg. **Per cap. pers. income** (2000): $24,294. **Sales tax** (2001): 6%. **Unemployment** (2000): 4.1%. **Tourism expends.** (1997): $4.7 bil. **Lottery** (2000): total sales: $583.7 mil; net income: $156 mil.

Finance. FDIC-insured commercial banks (2000): 233. **Deposits:** $39.5 bil. **FDIC-insured savings institutions** (2000): 34. **Assets:** $3.1 bil.

Federal govt. Fed. civ. employees (Mar. 2000): 19,407. **Avg. salary:** $41,471. **Notable fed. facilities:** U.S. Gold Bullion Depository, Fort Knox; Federal Correctional Institution, Lexington.

Energy. Electricity production (2000, kWh, by source): Coal: 77.9 bil; Petroleum: 119 mil; Gas: 307 mil; Hydroelectric: 2.3 bil.

State data. Motto: United we stand, divided we fall. **Flower:** Goldenrod. **Bird:** Cardinal. **Tree:** Tulip Poplar. **Song:** My Old Kentucky Home. **Entered union** June 1, 1792; rank, 15th. **State fair** at Louisville, late Aug.

History. The area was predominantly hunting grounds for Shawnee, Wyandot, Delaware, and Cherokee peoples. Explored by Americans Thomas Walker and Christopher Gist, 1750-51, Kentucky was the first area west of the Alleghenies settled by American pioneers. The first permanent settlement was Harrodsburg, 1774. Daniel Boone blazed the Wilderness Trail through the Cumberland Gap and founded Ft. Boonesborough, 1775. Conflicts with Native Americans, spurred by the British, were unceasing until, during the American Revolution, Gen. George Rogers Clark captured British forts in Indiana and Illinois, 1778. In 1792, Virginia dropped its claims to the region, and it became the 15th state. Although officially a Union state, Kentuckians had divided loyalties during the Civil War and were forced to choose sides; its slaves were freed only after the adoption of the 13th Amendment to the U.S. Constitution, 1865.

Tourist attractions. Kentucky Derby; Louisville; Land Between the Lakes Natl. Recreation Area, Kentucky Lake and Lake Barkley; Mammoth Cave Natl. Park; Echo River, 360 ft below ground; Lake Cumberland; Lincoln's birthplace, Hodgenville; My Old Kentucky Home State Park, Bardstown; Cumberland Gap Natl. Historical Park, Middlesboro; Kentucky Horse Park, Lexington; Shaker Village, Pleasant Hill.

Famous Kentuckians. Muhammad Ali, John James Audubon, Alben W. Barkley, Daniel Boone, Louis D. Brandeis, John C. Breckinridge, Kit Carson, Albert B. "Happy" Chandler, Henry Clay, Jefferson Davis, D. W. Griffith, "Casey" Jones, Abraham Lincoln, Mary Todd Lincoln, Thomas Hunt Morgan, Carry Nation, Col. Harland Sanders, Diane Sawyer, Jesse Stuart, Adlai Stevenson, Zachary Taylor, Hunter S. Thompson, Robert Penn Warren, Whitney Young Jr.

Tourist Information. Kentucky Dept. of Travel, 500 Mero St., #2200, Frankfort, KY 40601.

Toll-free travel information. 1-800-225-TRIP.

Website. http://www.state.ky.us

Tourism website. http://www.kentuckytourism.com

Louisiana

Pelican State

People. Population (2000): 4,468,976; rank: 22; **net change** (1999-2000): 5.9%. **Pop. density**: 102.6 per sq mi. **Racial distribution** (2000): 63.9% white; 32.5% black; 1.2% Asian; 0.6% Native American/Nat. AK; <0.1% Hawaiian/Pacific Islander; 0.7% other race; 2 or more races, 1.1%. **Hispanic pop.** (any race): 2.4%.

Geography. Total area: 51,840 sq mi; rank: 31. **Land area:** 43,562 sq mi; rank: 33. **Acres forested:** 13,864,000. **Location:** West South Central state on the Gulf Coast. **Climate:** subtropical, affected by continental weather patterns. **Topography:** lowlands of marshes and Mississippi R. flood plain; Red R. Valley lowlands; upland hills in the Florida Parishes; average elevation, 100 ft. **Capital:** Baton Rouge.

Economy. Chief industries: wholesale and retail trade, tourism, manufacturing, construction, transportation, communication, public utilities, finance, insurance, real estate, mining. **Chief manuf. goods:** chemical products, foods, transportation equipment, electronic equipment, petroleum products, lumber, wood, and paper. **Chief crops:** soybeans, sugarcane, rice, corn, cotton, sweet potatoes, pecans, sorghum, aquaculture. **Livestock:** (Jan. 2001) 860,000 cattle/calves; (Dec. 2000) 29,000 hogs/pigs; 2.6 mil chickens (excl. broilers). **Timber/lumber** (2000): pines, hardwoods, oak; 1.3 bil bd. ft. **Nonfuel minerals** (est. 2000): $404 mil; mostly salt, sulfur, sand & gravel, crushed stone, clays. **Commercial fishing** (2000): $302.7 mil. **Chief ports:** New Orleans, Baton Rouge, Lake Charles, Port of S. Louisiana (La Place), Shreveport, Plaquemine, St. Bernard, Alexandria. **Principal internat. airport at:** New Orleans. **Value of construction** (1997): $4.7 bil. **Gross state product** (1999): $129 bil. **Employment distrib.** (May 2001): 28.0% services; 23.7% trade; 19.1% govt.; 9.2% mfg. **Per cap. pers. income** (2000): $23,334. **Sales tax** (2001): 4%. **Unemployment** (2000): 5.5%. **Tourism expends.** (1997): $7.3 bil. **Lottery** (2000): total sales: $276.4 mil; net income: $99.9 mil.

Finance. FDIC-insured commercial banks (2000): 149. **Deposits**: $41.9 bil. **FDIC-insured savings institutions** (2000): 33. **Assets:** $4.5 bil.

Federal govt. Fed. civ. employees (Mar. 2000): 19,767. **Avg. salary:** $44,728. **Notable federal facilities:** Strategic Petroleum Reserve, Michoud Assembly Plant, Southeast U.S. Agricultural Research Ctr., U.S. Army Corps of Engineers, all New Orleans; Ft. Polk military bases, Barksdale; U.S. Public Service Hospital, Carville; Naval Air Station, Chalmette; V.A. Hospital, Pineville.

Energy. Electricity production (2000, kWh, by source): Coal: 14.4 bil; Petroleum: 616 mil; Gas: 26.6 bil; Nuclear: 15.8 bil.

State data. Motto: Union, justice, and confidence. **Flower:** Magnolia. **Bird:** Eastern brown pelican. **Tree:** Cypress. **Song:** Give Me Louisiana. **Entered union** Apr. 30, 1812; rank, 18th. **State fair** at Shreveport; Oct.

History. Caddo, Tunica, Choctaw, Chitimacha, and Chawash peoples lived in the region at the time of European contact. Europeans Cabeza de Vaca and Panfilo de Narvaez first visited, 1530. The region was claimed for France by La Salle, 1682. The first permanent settlement was by the French at Biloxi, now in Mississippi, 1699. France ceded the region to Spain, 1762, took it back, 1800, and sold it to the U.S., 1803, in the Louisiana Purchase. During the American Revolution, Spanish Louisiana aided the Americans. Admitted as a state in 1812, Louisiana was the scene of the Battle of New Orleans, 1815.

Louisiana Creoles are descendants of early French and/or Spanish settlers. About 4,000 Acadians, French settlers in Nova Scotia, Canada, were forcibly transported by the British to Louisiana in 1755 (an event commemorated in Longfellow's "Evangeline") and settled near Bayou Teche; their descendants became known as Cajuns. Another group, the Islenos, were descendants of Canary Islanders brought to Louisiana by a Spanish governor in 1770. Traces of Spanish and French survive in local dialects.

Tourist attractions. Mardi Gras, French Quarter, Superdome, Dixieland jazz, Aquarium of the Americas, Audubon Zoo & Gardens, all New Orleans; Battle of New Orleans site; Longfellow-Evangeline Memorial Park, St. Martinville; Kent House Museum, Alexandria; Hodges Gardens, Natchitoches, USS Kidd Memorial, Baton Rouge.

Famous Louisianans. Louis Armstrong, Pierre Beauregard, Judah P. Benjamin, Braxton Bragg, Kate Chopin, Harry Connick Jr., Ellen DeGeneres, Lillian Hellman, Grace King, Bob Livingston, Huey Long, Wynton Marsalis, Leonidas K. Polk, Anne Rice, Henry Miller Shreve, Britney Spears, Edward D. White Jr.

Tourist information. Louisiana Office of Tourism, PO Box 94291, Baton Rouge, LA 70804-9291.

Toll-free travel information. 1-800-677-4082.

Website. http://www.state.la.us

Tourism website. http://www.louisianatravel.com

Maine

Pine Tree State

People. Population (2000): 1,274,923; rank: 40; **net change** (1999-2000): 3.8%. **Pop. density**: 41.3 per sq mi. **Racial distribution** (2000): 96.9% white; 0.5% black; 0.7% Asian; 0.6% Native American/Nat. AK; <0.1% Hawaiian/Pacific Islander; 0.2% other race; 2 or more races, 1.0%. **Hispanic pop.** (any race): 0.7%.

Geography. Total area: 35,385 sq mi; rank: 39. **Land area:** 30,862 sq mi; rank: 39. **Acres forested:** 17,533,000. **Location:** New England state at northeastern tip of U.S. **Climate:** Southern interior and coastal, influenced by air masses from the S and W; northern clime harsher, avg. over 100 in. snow in winter. **Topography:** Appalachian Mts. extend through state; western borders have rugged terrain; long sand beaches on southern coast; northern coast mainly rocky promontories, peninsulas, fjords. **Capital:** Augusta.

Economy. Chief industries: manufacturing, agriculture, fishing, services, trade, government, finance, insurance, real estate, construction. **Chief manuf. goods:** paper & wood products, transportation equipment. **Chief crops:** potatoes, aquaculture products. **Livestock:** (Jan. 2001) 97,000 cattle/calves; (Dec. 2000) 6,500 hogs/pigs; 5.7 mil chickens (excl. broilers). **Timber/lumber** (2000): pine, spruce, fir; 1.2 bil bd. ft. **Nonfuel minerals** (est. 2000): $102 mil; mostly sand & gravel, portland and masonry cement, crushed stone, peat. **Commercial fishing** (1999): $265.2 mil. **Chief ports:** Searsport, Portland, Eastport. **Principal internat. airports at:** Bangor, Portland. **Value of construction** (1997): $1.1 bil. **Gross state product** (1999): $34 bil. **Employment distrib.** (May 2001): 30.9% services; 24.8% trade; 16.5% govt.; 13.4% mfg. **Per cap. pers. income (2000)**: $25,623. **Sales tax** (2001): 5%. **Unemployment** (2000): 3.5%. **Tourism expends.** (1997): $2 bil. **Lottery** (2000): total sales: $148 mil; net income: $39.6 mil.

Finance. FDIC-insured commercial banks (2000): 15. **Deposits**: $5.4 bil. **FDIC-insured savings institutions** (2000): 26. **Assets**: $7.3 bil.

Federal govt. Fed. civ. employees (Mar. 2000): 7,893. **Avg. salary:** $45,496. **Notable fed. facilities:** Kittery Naval Shipyard; Brunswick Naval Air Station.

Energy. Electricity production (2000, kWh, by source): Petroleum: 2 mil; Hydroelectric: 4 mil.

State data. Motto: Dirigo (I direct). **Flower:** White pine cone and tassel. **Bird:** Chickadee. **Tree:** Eastern white pine. **Song:** State of Maine Song. **Entered union** Mar. 15, 1820; rank, 23d. **State fair:** at Bangor, late July; at Skowhegan, mid-Aug.

History. When the Europeans arrived, Maine was inhabited by Algonquian peoples including the Abnaki, Penobscot, and Passamaquoddy. Maine's rocky coast was believed to have been explored by the Cabots, 1498-99. French settlers arrived, 1604, at the St. Croix River, English, c 1607, on the Kennebec; both settlements failed. Maine was made part of Massachusetts, 1691. In the American Revolution, a Maine regiment fought at Bunker Hill. A British fleet destroyed Falmouth (now Portland), 1775, but the British ship *Margaretta* was captured near Machiasport. In 1820, Maine broke off and became a separate state.

Tourist attractions. Acadia Natl. Park, Bar Harbor, on Mt. Desert Island; Old Orchard Beach; Portland's Old Port; Kennebunkport; Common Ground Country Fair; Portland Headlight; Baxter State Pk.; Freeport/L. L. Bean.

Famous "Down Easters." Leon Leonwood (LL) Bean, James G. Blaine, Cyrus H. K. Curtis, Hannibal Hamlin, Sarah Jewett, Stephen King, Henry Wadsworth Longfellow, Sir Hiram and Hudson Maxim, Edna St. Vincent Millay, George Mitchell, Edmund Muskie, Judd Nelson, Edwin Arlington Robinson, Joan Benoit Samuelson, Liv Tyler, Kate Douglas Wiggin, Ben Ames Williams.

Chamber of Commerce and Industry. Maine Chamber & Business Alliance, 7 Community Dr., Augusta, ME 04330.

Toll-free travel information. 1-888-MAINE45 (from within the United States and Canada).

Website. http://www.state.me.us

Tourism website. http://www.visitmaine.com

Maryland

Old Line State, Free State

People. Population (2000): 5,296,486; rank: 19; **net change** (1999-2000): 10.8%. **Pop. density**: 541.9 per sq mi. **Racial distribution** (2000): 64.0% white; 27.9% black; 4.0% Asian; 0.3% Native American/Nat. AK; <0.1% Hawaiian/Pacific Islander; 1.8% other race; 2 or more races, 2.0%. **Hispanic pop.** (any race): 4.3%.

Geography. Total area: 12,407 sq mi; rank: 42. **Land area:** 9,774 sq mi; rank: 42. **Acres forested:** 2,700,000. **Location:** South Atlantic state stretching from the Ocean to the Allegheny Mts. **Climate:** continental in the west; humid subtropical in the east. **Topography:** Eastern Shore of coastal plain and Maryland Main of coastal plain, piedmont plateau, and the Blue Ridge, separated by the Chesapeake Bay. **Capital:** Annapolis.

Economy. Chief industries: manufacturing, biotechnology and information technology, services, tourism. **Chief manuf. goods:** electric and electronic equipment; food and kindred products, chemicals and allied products, printed materials. **Chief crops:** greenhouse and nursery products, soybeans, corn. **Livestock:** (Jan. 2001) 235,000 cattle/calves; (Dec. 2000) 58,000 hogs/pigs; 4.5 mil chickens (excl. broilers); 294.4 mil broilers. **Timber/lumber:** (2000): hardwoods; 293 mil bd. ft. **Nonfuel minerals** (est. 2000): $357 mil; mostly crushed stone, portland cement, sand & gravel, masonry cement, dimension stone. **Commercial fishing** (1999): $63.8 mil. **Chief port:** Baltimore. **Principal internat. airport at:** Baltimore. **Value of construction** (1997): $5.9 bil. **Gross state product** (1999): $174.7 bil. **Employment distrib.** (May 2001): 35.4% services; 22.3% trade; 18.3% govt.; 7.1% mfg. **Per cap. pers. income** (2000): $33,872. **Sales tax** (2001): 5%. **Unemployment** (2000): 3.9%. **Tourism expends.** (1997): $6.9 bil. **Lottery** (2000): total sales: $1.2 bil; net income: $406.8 mil.

Finance. FDIC-insured commercial banks (2000): 74. **Deposits:** $35.6 bil. **FDIC-insured savings institutions** (2000): 63. **Assets:** $8.5 bil.

Federal govt. Fed. civ. employees (Mar. 2000): 101,425. **Avg. salary:** $58,911. **Notable fed. facilities:** U.S. Naval Academy; Natl. Agriculture Research Center; Ft. George G. Meade, Aberdeen Proving Ground; Goddard Space Flight Center; Natl. Institutes of Health; Natl. Institute of Standards & Technology; Food & Drug Administration; Bureau of the Census.

Energy. Electricity production (2000, kWh, by source): Coal: 20.2 bil; Petroleum: 1.4 bil; Gas: 1.9 bil; Hydroelectric: 1.7 bil; Nuclear: 6.3 bil.

State data. Motto: Fatti Maschii, Parole Femine (Manly deeds, womanly words). **Flower:** Black-eyed Susan. **Bird:** Baltimore oriole. **Tree:** White oak. **Song:** Maryland, My Maryland. **Seventh** of the original 13 states to ratify Constitution, Apr. 28, 1788. **State fair** at Timonium; late Aug.-early Sept.

History. Europeans encountered Algonquian-speaking Nanticoke and Piscataway and Iroquois-speaking Susquehannock when they first visited the area. Italian explorer Verrazano visited the Chesapeake region in the early 16th cent. English Capt. John Smith explored and mapped the area, 1608. William Claiborne set up a trading post on Kent Island in Chesapeake Bay, 1631. King Charles I granted land to Cecilius Calvert, Lord Baltimore, 1632; Calvert's brother Leonard, with about 200 settlers, founded St. Marys, 1634. The bravery of Maryland troops in the American Revolution, as at the Battle of Long Island, won the state its nickname "The Old Line State." In the War of 1812, when a British fleet tried to take Ft. McHenry, Marylander Francis Scott Key wrote "The Star Spangled Banner," 1814. Although a slave-holding state, Maryland remained with the Union during the Civil War and was the site of the battle of Antietam, 1862, which halted Gen. Robert E. Lee's march north.

Tourist attractions. The Preakness at Pimlico track, Baltimore; The Maryland Million at Laurel Race Course; Ocean City; restored Ft. McHenry, near which Francis Scott Key wrote "The Star-Spangled Banner"; Edgar Allan Poe house, Ravens Football at Memorial Stadium, Camden Yards, Natl. Aquarium, Harborplace, all Baltimore; Antietam Battlefield, near Hagerstown; South Mountain Battlefield; U.S. Naval Academy, Annapolis; Maryland State House, Annapolis, 1772, the oldest still in legislative use in the U.S.

Famous Marylanders. John Astin, Benjamin Banneker, Tom Clancy, Jonathan Demme, Francis Scott Key, H. L. Mencken, Kwersi Mfume, Ogden Nash, Charles Willson Peale, William Pinkney, Edgar Allan Poe, Cal Ripken, Jr., Babe Ruth, Upton Sinclair, Roger B. Taney, John Waters, Montel Williams.

Maryland Dept. of Business & Economic Development. 217 E. Redwood St., Baltimore, MD 21202; (410) 767-6870.

Toll-free travel information. 1-800-MDISFUN.

Website. http://www.state.md.us

Tourism website. http//www.mdisfun.org

Massachusetts

Bay State, Old Colony

People. Population (2000): 6,349,097; rank: 13; **net change** (1999-2000): 5.5%. **Pop. density**: 809.8 per sq mi. **Racial distribution** (2000): 84.5% white; 5.4% black; 3.8% Asian; 0.2% Native American/Nat. AK; <0.1% Hawaiian/Pacific Islander; 3.7% other race; 2 or more races, 2.3%. **Hispanic pop.** (any race): 6.8%.

Geography. Total area: 10,555 sq mi; rank: 44. **Land area:** 7,840 sq mi; rank: 45. **Acres forested:** 3,203,000. **Location:** New England state along Atlantic seaboard. **Climate:** temperate, with colder and drier clime in western region. **Topography:** jagged indented coast from Rhode Island around Cape Cod; flat land yields to stony upland pastures near central region and gentle hilly country in west; except in west, land is rocky, sandy, and not fertile. **Capital:** Boston.

Economy. Chief industries: services, trade, manufacturing. **Chief manuf. goods:** electric and electronic equipment, instruments, industrial machinery and equipment, printing and publishing, fabricated metal products. **Chief crops:** cranberries, greenhouse, nursery, vegetables. **Livestock:** (Jan. 2001) 50,000 cattle/calves; (Dec. 2000) 20,000 hogs/pigs; 357,000 chickens (excl. broilers). **Timber/lumber** (2000): white pine, oak, other hard woods; NA; **Nonfuel minerals** (est. 2000): $210 mil; mostly crushed stone, sand & gravel, dimension stone, lime, clays. **Commercial fishing** (1999): $260.2 mil. **Chief ports:** Boston, Fall River, New Bedford, Salem, Gloucester, Plymouth. **Principal internat. airport at:** Boston. **Value of construction** (1997): $10.3 bil. **Gross state product** (1999): $262.6 bil. **Employment distrib.** (May 2001): 36.9% services; 22.4% trade; 12.7% govt.; 12.7% mfg. **Per cap. pers. income** (2000): $37,992. **Sales tax** (2001): 5%. **Unemployment** (2000): 2.6%. **Tourism expends.** (1997): $10.8 bil. **Lottery** (2000): total sales: $3.7 bil; net income: $853.2 mil.

Finance. FDIC-insured commercial banks (2000): 44. **Deposits:** $72.6 bil. **FDIC-insured savings institutions** (2000): 185. **Assets**: $62.3 bil.

Federal govt. Fed. civ. employees (Mar. 2000): 24,905. **Avg. salary:** $50,963. **Notable fed. facilities:** Thomas P. O'Neill Jr. Federal Bldg., J.W. McCormack Bldg., John Fitzgerald Kennedy Federal Bldg., Q.M. Laboratory, Natick.

Energy. Electricity production (2000, kWh, by source): Coal: 1.1 bil; Petroleum: 125 mil; Gas: 335 mil; Hydroelectric: 70 mil.

State data. Motto: Ense Petit Placidam Sub Libertate Quietem (By the sword we seek peace, but peace only under liberty). **Flower:** Mayflower. **Bird:** Chickadee. **Tree:** American elm. **Song:** All Hail to Massachusetts. **Sixth** of the original 13 states to ratify Constitution, Feb. 6, 1788. **State Fair** at Topsfield, early Oct.

History. Early inhabitants were the Algonquian, Nauset, Wampanoag, Massachuset, Pennacook, Nipmuc, and Pocumtuc peoples. Pilgrims settled in Plymouth, 1620, giving thanks for their survival with the first Thanksgiving Day, 1621. About 20,000 new settlers arrived, 1630-40. Native American relations with the colonists deteriorated leading to King Philip's War, 1675-76, which the colonists won, ending Native American resistance. Demonstrations against British restrictions set off the Boston Massacre, 1770, and the Boston Tea Party, 1773. The first bloodshed of American Revolution was at Lexington, 1775.

Tourist attractions. Provincetown artists' colony; Cape Cod; Plymouth Rock, Plymouth Plantation, Mayflower II, all Plymouth; Freedom Trail, Museum of Fine Arts, Children's Museum, Museum of Science, New England Aquarium, JFK Library, Boston Ballet, Boston Pops, Boston Symphony Orchestra, all Boston; Tanglewood, Jacob's Pillow Dance Festival, Hancock Shaker Village Berkshire Railway Museum, all in the Berkshires; Salem; Old Sturbridge Village; Deerfield Historic District; Walden Pond; Naismith Memorial Basketball Hall of Fame, Springfield.

Famous "Bay Staters." John Adams, John Quincy Adams, Samuel Adams, Louisa May Alcott, Horatio Alger, Susan B. Anthony, Crispus Attucks, Clara Barton, Alexander Graham Bell, Stephen Breyer, George Bush, John Cheever, E. E. Cummings, Emily Dickinson, Charles Eliot, Ralph Waldo Emerson, William Lloyd Garrison, Edward Everett Hale, John Hancock, Nathaniel Hawthorne, Oliver Wendell Holmes, Winslow Homer, Elias Howe, John F. Kennedy, Jack Lemmon, James Russell Lowell, Cotton Mather, Samuel F. B. Morse, Edgar Allan Poe, Paul Revere, Dr. Seuss (Theodore Seuss Geisel), Henry David Thoreau, Barbara Walters, James McNeil Whistler, John Greenleaf Whittier.

Tourist information. Massachusetts Office of Travel & Tourism, 100 Cambridge St., 13th Floor, Boston, MA 02202.

Toll-free travel information. 1-800-227-MASS.

Website. http://www.state.ma.us

Tourism website. http://www.massvacation.com

Michigan

Great Lakes State, Wolverine State

People. Population (2000): 9,938,444; rank: 8; **net change** (1999-2000): 6.9%. **Pop. density**: 175 per sq mi. **Racial distribution** (2000): 80.2% white; 14.2% black; 1.8% Asian; 0.6% Native American/Nat. AK; <0.1% Hawaiian/Pacific Islander; 1.3% other race; 2 or more races, 1.9%. **Hispanic pop.** (any race): 3.3%.

Geography. Total area: 96,716 sq mi; rank: 11. **Land area:** 56,804 sq mi; rank: 22. **Acres forested:** 18,253,000. **Location:** East North Central state bordering on 4 of the 5 Great Lakes, divided into an Upper and Lower Peninsula by the Straits of Mackinac, which link lakes Michigan and Huron. **Climate:** well-defined seasons tempered by the Great Lakes. **Topography:** low rolling hills give way to northern tableland of hilly belts in Lower Peninsula; Upper Peninsula is level in the east, with swampy areas; western region is higher and more rugged. **Capital:** Lansing.

Economy. Chief industries: manufacturing, services, tourism, agriculture, forestry/lumber. **Chief manuf. goods:** automobiles, transportation equipment, machinery, fabricated metals, food products, plastics, office furniture. **Chief crops:** corn, wheat, soybeans, dry beans, hay, potatoes, sweet corn, apples, cherries, sugar beets, blueberries, cucumbers, Niagra grapes. **Livestock:** (Jan. 2001) 980,000 cattle/calves; 71,000 sheep/lambs; (Dec. 2000) 950,000 hogs/pigs; 7.6 mil chickens (excl. broilers); 750,000 broilers. **Timber/lumber** (2000): maple, oak, aspen; 681 mil bd. ft. **Nonfuel minerals** (est. 2000): $1.7 bil; mostly portland cement, iron ore, sand & gravel, magnesium compounds, crushed stone. **Commercial fishing** (1999): $9.3 mil. **Chief ports:** Detroit, Saginaw River, Escanaba, Muskegon, Sault Ste. Marie, Port Huron, Marine City. **Principal internat. airports at:** Detroit, Flint, Grand Rapids, Kalamazoo, Lansing, Saginaw. **Value of construction** (1997): $10.7 bil. **Gross state product** (1999): $308.3 bil **Employment distrib.** (May 2001): 28.5% services; 23.4% trade; 14.9% govt.; 20.2% mfg. **Per cap. pers. income** (2000): $29,612. **Sales tax** (2001): 6%. **Unemployment** (2000): 3.6%. **Tourism expends.** (1997): $10.1 bil. **Lottery** (2000): total sales: $1.7 bil; net income: $607.5 mil.

Finance. FDIC-insured commercial banks (2000): 168. **Deposits**: $97.7 bil. **FDIC-insured savings institutions** (2000): 22. **Assets:** $31.5 bil.

Federal govt. Fed. civ. employees (Mar. 2000): 21,348. **Avg. salary:** $51,553. **Notable fed. facilities:** Isle Royal, Sleeping Bear Dunes national parks.

Energy. Electricity production (2000, kWh, by source): Coal: 66.4 bil; Petroleum: 957 mil; Gas: 2.4 bil; Hydroelectric: 234 mil; Nuclear: 18.9 bil.

State data. Motto: Si Quaeris Peninsulam Amoenam, Circumspice (If you seek a pleasant peninsula, look about you). **Flower:** Apple blossom. **Bird:** Robin. **Tree:** White pine. **Song:** Michigan, My Michigan. **Entered union** Jan. 26, 1837; rank, 26th. **State fair** at Detroit, late Aug.–early Sept.; Upper Peninsula (Escanaba), mid-Aug.

History. Early inhabitants were the Ojibwa, Ottawa, Miami, Potawatomi, and Huron. French fur traders and missionaries visited the region, 1616, set up a mission at Sault Ste. Marie, 1641, and a settlement there, 1668. French settlements were taken over, 1763, by the British, who crushed a Native American uprising led by Ottawa chieftain Pontiac that same year. Treaty of Paris ceded territory to U.S., 1783, but British remained until 1796. The British seized Ft. Mackinac and Detroit, 1812. After Oliver H. Perry's Lake Erie victory and William H. Harrison's victory near the Thames River, 1813, the British retreated to Canada. The opening of the Erie Canal, 1825, and new land laws and Native American cessions led the way for a flood of settlers.

Tourist attractions. Henry Ford Museum, Greenfield Village, both in Dearborn; Michigan Space Center, Jackson; Tahquamenon *(Hiawatha)* Falls; DeZwaan windmill and Tulip Festival, Holland; "Soo Locks," St. Mary's Falls Ship Canal, Sault Ste. Marie, Kalamazoo Aviation History Museum; Mackinac Island; Kellogg's Cereal City USA, Battle Creek; Museum of African-American History, Motown Historical Museum, both Detroit.

Famous People. Ralph Bunche, Paul de Kruif, Thomas A. Edison, Edna Ferber, Gerald R. Ford, Henry Ford, Aretha Franklin, Edgar Guest, Lee Iacocca, Robert Ingersoll, Magic Johnson, Casey Kasem, Will Kellogg, Ring Lardner, Elmore Leonard, Charles Lindbergh, Joe Louis, Madonna,Terry McMillan, Jack Paar, Pontiac, Diana Ross, Glenn Seaborg, Tom Selleck, Sinbad (David Adkins), John Smoltz, Lily Tomlin, Stewart Edward White, Malcolm X.

State Chamber of Commerce. 600 S. Walnut, Lansing, MI 48933. Phone: 517-371-2100

Toll-free travel information. 1-888-78GREAT.

Website. http://www.michigan.gov

Tourism website. http://travel.michigan.org

Minnesota

North Star State, Gopher State

People. Population (2000): 4,919,479; rank: 21; **net change** (1999-2000): 12.4%. **Pop. density**: 61.8 per sq mi. **Racial distribution** (2000): 89.4% white; 3.5% black; 2.9% Asian; 1.1% Native American/Nat. AK; <0.1% Hawaiian/Pacific Islander; 1.3% other race; 2 or more races, 1.7%. **Hispanic pop.** (any race): 2.9%.

Geography. Total area: 86,939 sq mi; rank: 12. **Land area:** 79,610 sq mi; rank: 14. **Acres forested:** 16,718,000. **Location:** West North Central state bounded on the E by Wisconsin and Lake Superior, on the N by Canada, on the W by the Dakotas, and on the S by Iowa. **Climate:** northern part of state lies in the moist Great Lakes storm belt; the western border lies at the edge of the semi-arid Great Plains. **Topography:** central hill and lake region covering approx. half the state; to the NE, rocky ridges and deep lakes; to the NW, flat plain; to the S, rolling plains and deep river valleys. **Capital:** St. Paul.

Economy. Chief industries: agribusiness, forest products, mining, manufacturing, tourism. **Chief manuf. goods:** food, chemical and paper products, industrial machinery, electric and electronic equipment, computers, printing & publishing, scientific and medical instruments, fabricated metal products, forest products. **Chief crops:** corn, soybeans, wheat, sugar beets, hay, barley, potatoes, sunflowers. **Livestock:** (Jan. 2001) 2.6 mil cattle/calves; 170,000 sheep/lambs; (Dec. 2000) 5.8 mil hogs/pigs; 15.8 mil chickens (excl. broilers); 44.2 mil broilers. **Timber/lumber** (2000): needleleaves and hardwoods; 303 mil bd. ft. **Nonfuel minerals** (est. 2000): $1.6 bil; mostly iron ore, sand & gravel, crushed stone, dimension stone. **Commercial fishing** (1998): $197,000. **Chief ports:** Duluth, St. Paul, Minneapolis. **Principal internat. airport at:** Minneapolis-St. Paul. **Value of construction** (1997): $6.2 bil. **Gross state product** (1998): $173 bil. **Employment distrib.** (May 2001): 29.6% services; 23.7% trade; 14.7% govt.; 16.1% mfg. **Per cap. pers. income** (2000): $32,101. **Sales tax** (2001): 6.5%. **Unemployment** (2000): 3.3%. **Tourism expends.** (1997): $6.2 bil. **Lottery** (2000): total sales: $397.3 mil; net income: $86.4 mil.

Finance. FDIC-insured commercial banks (2000): 492. **Deposits**: $122.6 bil. **FDIC-insured savings institutions** (2000): 22. **Assets**: $2.9 bil.

Federal govt. Fed. civ. employees (Mar. 2000): 12,926. **Avg. salary:** $49,685.

Energy. Electricity production (2000, kWh, by source): Coal: 29.6 bil; Petroleum: 432 mil; Gas: 432 mil; Hydroelectric: 629 mil; Nuclear: 13 bil.; **Other:** 416 mil.

State data. Motto: L'Etoile du Nord (The star of the north). **Flower:** Pink and white lady's-slipper. **Bird:** Common loon. **Tree:** Red pine. **Song:** Hail! Minnesota. **Entered union** May 11, 1858; rank, 32d. **State fair** at St. Paul/Minneapolis; late Aug.-early Sept.

History. Dakota Sioux were early inhabitants of the area, and in the 16th cent., the Ojibwa began moving in from the east. French fur traders Médard Chouart and Pierre Esprit Radisson entered the region in the mid-17th cent. In 1679, French explorer Daniel Greysolon, sieur Duluth, claimed the entire region in the name of France. Britain took the area east of the Mississippi, 1763. The U.S. took over that portion after the American Revolution and in 1803, gained the western area in the Louisiana Purchase. The U.S. built Ft. St. Anthony (now Ft. Snelling), 1819, and in 1837, bought Native American lands, spurring an influx of settlers from the east. In 1849, the Territory of Minnesota was created. Sioux Indians staged a bloody uprising, the Battle of Woods Lake, 1862, and were driven from the state.

IT'S A FACT: Tonka toy metal trucks were first manufactured in Mound, MN, in 1947 and took their name—which means "great" in Sioux—from nearby Lake Minnetonka.

Tourist attractions. Minneapolis Institute of Arts, Walker Art Center, Minneapolis Sculpture Garden, Minnehaha Falls (inspiration for Longfellow's *Hiawatha),* Guthrie Theater, Minneapolis; Ordway Theater, St. Paul; Voyageurs Natl. Park; Mayo Clinic, Rochester; St. Paul Winter Carnival; North Shore (of Lake Superior).

Famous Minnesotans. Warren Burger, Ethan and Joel Coen, William O. Douglas, Bob Dylan, F. Scott Fitzgerald, Al Franken, Judy Garland, Cass Gilbert, Hubert Humphrey, Garrison Keillor, Sister Elizabeth Kenny, Jessica Lange, Sinclair Lewis, Paul Manship, Roger Maris, E. G. Marshall, William and Charles Mayo, Eugene McCarthy, Walter F. Mondale, Prince (Nelson Rodgers), Charles Schulz, Harold Stassen, Thorstein Veblen, Jesse Ventura.

Chamber of Commerce. 30 East 7th St., Suite 1700, St. Paul, MN 55101-4901.

Toll-free travel information. 1-800-657-3700.

Website. http://www.state.mn.us

Tourism website. http://www.exploreminnesota.com

Mississippi

Magnolia State

People. Population (2000): 2,844,658; rank: 31; **net change** (1999-2000): 10.5%. **Pop. density**: 60.6 per sq mi. **Racial distribution** (2000): 61.4% white; 36.3% black; 0.7% Asian; 0.4% Native American/Nat. AK; <0.1% Hawaiian/Pacific Islander; 0.5% other race; 2 or more races, 0.7%. **Hispanic pop.** (any race): 1.4%.

Geography. Total area: 48,430 sq mi; rank: 32. **Land area:** 46,907 sq mi; rank: 31. **Acres forested:** 17,000,000. **Location:** East South Central state bordered on the W by the Mississippi R. and on the S by the Gulf of Mexico. **Climate:** semi-tropical, with abundant rainfall, long growing season, and extreme temperatures unusual. **Topography:** low, fertile delta between the Yazoo and Mississippi rivers; loess bluffs stretching around delta border; sandy gulf coastal terraces followed by piney woods and prairie; rugged, high sandy hills in extreme NE followed by Black Prairie Belt, Pontotoc Ridge, and flatwoods into the north central highlands. **Capital:** Jackson.

Economy. Chief industries: warehousing & distribution, services, manufacturing, government, wholesale and retail trade. **Chief manuf. goods:** chemicals & plastics, food & kindred products, furniture, lumber & wood products, electrical machinery, transportation equipment. **Chief crops:** cotton, rice, soybeans. **Livestock:** (Jan. 2001) 1.1 mil cattle/calves; (Dec. 2000) 315,000 hogs/pigs; 10.3 mil chickens (excl. broilers); 735.1 mil broilers. **Timber/lumber** (2000): pine, oak, hardwoods; 3.0 bil bd. ft. **Nonfuel minerals** (est. 2000): $157 mil; mostly sand & gravel, portland cement, clays, dimension stone. **Commercial fishing** (1999): $48.5 mil. **Chief ports:** Pascagoula, Vicksburg, Gulfport, Natchez, Greenville. **Principal internat. airport at:** Jackson. **Value of construction** (1997): $2.6 bil. **Gross state product** (1998): $64.2 bil. **Employment distrib.** (May 2001): 23.8% services; 22.5% trade; 20.7% govt.; 19.2% mfg. **Per cap. pers. income** (2000): $20,993. **Sales tax** (2001): 7%. **Unemployment** (2000): 5.7%. **Tourism expends.** (1997): $3.8 bil.

Finance. FDIC-insured commercial banks (2000): 101. **Deposits:** $26.4 bil. **FDIC-insured savings institutions** (2000): 8. **Assets:** $1 bil.

Federal govt. Fed. civ. employees (Mar. 2000): 16,591. **Avg. salary:** $44,618. **Notable fed. facilities:** Columbus, Keesler AF bases; Meridian Naval Air Station, John C. Stennis Space Center; U.S. Army Corps of Engineers Waterway Experiment Station.

Energy. Electricity production (2000, kWh, by source): Coal: 13.6 bil; Petroleum: 2.9 bil; Gas: 6.4 bil; Nuclear: 10.7 bil.

State data. Motto: Virtute et Armis (By valor and arms). **Flower:** Magnolia. **Bird:** Mockingbird. **Tree:** Magnolia. **Song:** Go, Mississippi! **Entered union** Dec. 10, 1817; rank, 20th. **State fair** at Jackson; early Oct.

History. Early inhabitants of the region were Choctaw, Chickasaw, and Natchez peoples. Hernando de Soto explored the area, 1540, and sighted the Mississippi River, 1541. Robert La Salle traced the river from Illinois to its mouth and claimed the entire valley for France, 1682. The first settlement was the French Ft. Maurepas, near Ocean Springs, 1699. The area was ceded to Britain, 1763; American settlers followed. During the American Revolution, Spain seized part of the area, remaining even after the U.S. acquired title at the end of the conflict; Spain finally moved out, 1798. The Territory of Mississippi was formed, 1798. Mississippi seceded, 1861. Union forces captured Corinth and Vicksburg and destroyed Jackson and much of Meridian. Mississippi was readmitted to the Union in 1870.

Tourist attractions. Vicksburg Natl. Military Park and Cemetery, other Civil War sites; Hattiesburg; Natchez Trace; Indian mounds; Antebellum homes; pilgrimages in Natchez and some 25 other cities; Smith Robertson Museum, Mynelle Gardens, both Jackson; Mardi Gras and Shrimp Festival, both in Biloxi; Gulf Islands Natl. Seashore; Casinos on the Mississippi River; the Mississippi Coast.

Famous Mississippians. Dana Andrews, Margaret Walker Alexander, Jimmy Buffett, Hodding Carter III, Bo Diddley, William Faulkner, Brett Favre, Shelby Foote, Morgan Freeman, John Grisham, Fannie Lou Hamer, Jim Henson, Faith Hill, John Lee Hooker, Robert Johnson, James Earl Jones, B. B. King, L. Q. C. Lamar, Trent Lott, Gerald McRaney, Willie Morris, Walter Payton, Elvis Presley, Leontyne Price, Charley Pride, LeAnn Rimes, Muddy Waters, Eudora Welty, Tennessee Williams, Oprah Winfrey, Johnny Winter, Richard Wright, Tammy Wynette.

Tourist Information. Dept. of Economic & Community Development. PO Box 849, Jackson, MS 39205-0849.

Toll-free travel information. 1-800-927-6378.

Website. http://www.state.ms.us

Tourism website. http://www.visitmississippi.org

Missouri

Show Me State

People. Population (2000): 5,595,211; rank: 17; **net change** (1999-2000): 9.3%. **Pop. density**: 81.2 per sq mi. **Racial distribution** (2000): 84.9% white; 11.2% black; 1.1% Asian; 0.4% Native American/Nat. AK; 0.1% Hawaiian/Pacific Islander; 0.8% other race; 2 or more races, 1.5%. **Hispanic pop.** (any race): 2.1%.

Geography. Total area: 69,704 sq mi; rank: 21. **Land area:** 68,886 sq mi; rank: 18. **Acres forested:** 14,007,000. **Location:** West North Central state near the geographic center of the conterminous U.S.; bordered on the E by the Mississippi R., on the NW by the Missouri R. **Climate:** continental, susceptible to cold Canadian air, moist, warm gulf air, and drier SW air. **Topography:** rolling hills, open, fertile plains, and well-watered prairie N of the Missouri R.; south of the river land is rough and hilly with deep, narrow valleys; alluvial plain in the SE; low elevation in the west. **Capital:** Jefferson City.

Economy. Chief industries: agriculture, manufacturing, aerospace, tourism. **Chief manuf. goods:** transportation equipment, food and related products, electrical and electronic equipment, chemicals. **Chief crops:** soybeans, corn, wheat, hay. **Livestock:** (Jan. 2001) 4.3 mil cattle/calves; 73,000 sheep/lambs; (Dec. 2000) 2.9 mil hogs/pigs; 8.2 mil chickens (excl. broilers); 249.7 mil broilers. **Timber/lumber** (2000): oak, hickory; 721 mil bd. ft. **Nonfuel minerals** (est. 2000): $1.3 bil; mostly crushed stone, lead, portland cement, lime, zinc. **Chief ports:** St. Louis, Kansas City. **Principal internat. airports at:** Kansas City, St. Louis. **Value of construction** (1997): $5.9 bil. **Gross state product** (1999): $170.5 bil. **Employment distrib.** (May 2001): 28.7% services; 23.5% trade; 15.6% govt.; 14.0% mfg. **Per cap. pers. income** (2000): $27,445. **Sales tax** (2001): 4.225%. **Unemployment** (2000): 3.5%. **Tourism expends.** (1997): $8.4 bil. **Lottery** (2000): total sales: $507.9 mil; net income: $154 mil.

Finance. FDIC-insured commercial banks (2000): 362. **Deposits:** $52.4 bil. **FDIC-insured savings institutions** (2000): 39. **Assets:** $7.1 bil.

Federal govt. Fed. civ. employees (Mar. 2000): 32,295. **Avg. salary:** $44,478. **Notable fed. facilities:** Federal Reserve banks; Ft. Leonard Wood; Jefferson Barracks; Whiteman AFB.

Energy. Electricity production (1999 kWh, by source): Coal: 62.1 bil; Petroleum: 237 mil; Gas: 2.9 bil; Hydroelectric: 413 mil; Nuclear: 10 bil.; **Other:** 73 mil.

State data. Motto: Salus Populi Suprema Lex Esto (The welfare of the people shall be the supreme law). **Flower:** Hawthorn. **Bird:** Bluebird. **Tree:** Dogwood. **Song:** Missouri Waltz. **Entered union** Aug. 10, 1821; rank, 24th. **State fair** at Sedalia; 3d week in Aug.

History. Early inhabitants of the region were Algonquian Sauk, Fox, and Illinois and Siouan Osage, Missouri, Iowa, and Kansa peoples. Hernando de Soto visited 1541. French hunters and lead miners made the first settlement c 1735, at Ste. Genevieve. The territory was ceded to Spain by the French, 1763, then returned to France, 1800. The U.S. acquired Missouri as part of the Louisiana Purchase, 1803. The influx of white settlers drove Native American tribes to the Kansas and Oklahoma territories; most were gone by 1836. The fur trade and the Santa Fe Trail provided prosperity; St. Louis became the gateway for pioneers heading West. Missouri entered the Union as a slave state, 1821. Though it remained with the Union, pro- and anti-slavery forces battled there during the Civil War.

Tourist attractions. Silver Dollar City, Branson; Mark Twain Area, Hannibal; Pony Express Museum, St. Joseph; Harry S. Truman Library, Independence; Gateway Arch, St. Louis; Worlds of Fun, Kansas City; Lake of the Ozarks; Churchill Mem., Fulton; State Capitol, Jefferson City.

Famous Missourians. Maya Angelou, Robert Altman, Burt Bacharach, Josephine Baker, Scot Bakula, Thomas Hart Benton, Tom Berenger, Yogi Berra, Chuck Berry, William Burroughs, George Caleb Bingham, Daniel Boone, Omar Bradley, Kate Capshaw, Dale Carnegie, George Washington Carver, Bob Costas, Walter Cronkite, Walt Disney, T. S. Eliot, Richard Gephardt, John Goodman, Betty Grable, Edwin Hubble, Jesse James, Marianne Moore, Reinhold Niebuhr, J. C. Penney, John J. Pershing, Brad Pitt, Joseph Pulitzer, Ginger Rogers, Bess Truman, Harry S. Truman, Kathleen Turner, Tina Turner, Mark Twain, Dick Van Dyke, Tennessee Williams, Lanford Wilson, Shelley Winters, Jane Wyman.

Chamber of Commerce. 428 E. Capitol, Jefferson City, MO 65101.

Toll-free travel information. 1-888-877-1234, ext. 124.

Website. http://www.ecodev.state.mo.us

Tourism website. http://www.missouritourism.com

Montana

Treasure State

People. Population (2000): 902,195; rank: 44; **net change** (1999-2000): 12.9%. **Pop. density**: 6.2 per sq mi. **Racial distribution** (2000): 90.6% white; 0.3% black; 0.5% Asian; 6.2% Native American/Nat. AK; 0.1% Hawaiian/Pacific Islander; 0.6% other race; 2 or more races, 1.7%. **Hispanic pop.** (any race): 2.0%.

Geography. Total area: 147,042 sq mi; rank: 4. **Land area:** 145,552 sq mi; rank: 4. **Acres forested:** 22,512,000. **Location:** Mountain state bounded on the E by the Dakotas, on the S by Wyoming, on the SSW by Idaho, and on the N by Canada. **Climate:** colder, continental climate with low humidity. **Topography:** Rocky Mts. in western third of the state; eastern two-thirds gently rolling northern Great Plains. **Capital:** Helena.

Economy. Chief industries: agriculture, timber, mining, tourism, oil and gas. **Chief manuf. goods:** food products, wood & paper products, primary metals, printing & publishing, petroleum and coal products. **Chief crops:** wheat, barley, sugar beets, hay, oats. **Livestock:** (Jan. 2001) 2.6 mil cattle/calves; 360,000 sheep/lambs; (Dec. 2000) 155,000 hogs/pigs; 460,000 chickens (excl. broilers). **Timber/lumber** (2000): Douglas fir, pines, larch; 1.3 bil bd. ft. **Nonfuel minerals** (est. 2000): $582 mil; mostly copper, gold, portland cement, palladium, sand and gravel. **Principal internat. airports at:** Billings, Missoula. **Value of construction** (1997): $827 mil. **Gross state product** (1999): $20.6 bil. **Employment distrib.** (May 2001): 30.6% services; 26.2% trade; 20.4% govt.; 6.1% mfg. **Per cap. pers. income** (2000): $22,569. **Sales tax:** none. **Unemployment** (2000): 4.9%. **Tourism expends.** (1997): $1.8 bil. **Lottery** (2000): total sales: $29.9 mil; net income: $5.8 mil.

Finance. FDIC-insured commercial banks (2000): 84. **Deposits**: $8.8 bil. **FDIC-insured savings institutions** (2000): 5. **Assets**: $1.5 bil.

Federal govt. Fed. civ. employees (Mar. 2000): 7,780. **Avg. salary:** $44,987. **Notable fed. facilities:** Malmstrom AFB; Ft. Peck, Hungry Horse, Libby, Yellowtail dams; numerous missile silos.

Energy. Electricity production (2000, kWh, by source): Coal: 14.9 bil; Petroleum: 15 mil; Gas: 13 mil; Hydroelectric: 6.3 bil.

State data. Motto: Oro y Plata (Gold and silver). **Flower:** Bitterroot. **Bird:** Western meadowlark. **Tree:** Ponderosa pine. **Song:** Montana. **Entered union** Nov. 8, 1889; rank, 41st. **State fair** at Great Falls; late July-early Aug.

History. Cheyenne, Blackfoot, Crow, Assiniboin, Salish (Flatheads), Kootenai, and Kalispel peoples were early inhabitants of the area. French explorers visited the region, 1742. The U.S. acquired the area partly through the Louisiana Purchase, 1803, partly through explorations of Lewis and Clark, 1805-6. Fur traders and missionaries established posts early 19th cent. Gold was discovered, 1863, and the Montana territory was established, 1864. Indian uprisings reached their peak with the Battle of Little Bighorn, 1876. Chief Joseph and the Nez Percé tribe surrendered here, 1877, after long trek across the state. Mining activity and the coming of the Northern Pacific Railway, 1883, brought population growth. Copper wealth from the Butte pits resulted in the turn of the century "War of Copper Kings" as factions fought for control of "the richest hill on earth."

Tourist attractions. Glacier Natl. Park; Yellowstone Natl. Park; Museum of the Rockies, Bozeman; Museum of the Plains Indian, Blackfeet Reservation, near Browning; Little Bighorn Battlefield Natl. Monument and Custer Natl. Cemetery; Flathead Lake; Helena; Lewis and Clark Caverns State Park, near Whitehall; Lewis and Clark Interpretive Center, Great Falls.

Famous Montanans. Dana Carvey, Gary Cooper, Marcus Daly, Chet Huntley, Will James, Myrna Loy, David Lynch, Mike Mansfield, Brent Musburger, Jeannette Rankin, Charles M. Russell, Lester Thurow.

Chamber of Commerce. 2030 11th Ave., PO Box 1730, Helena, MT 59624.

Toll-free travel information. 1-800-VISITMT.

Website. http://www.discoveringmontana.com

Tourism website. http://www.visitmt.com

Nebraska

Cornhusker State

People. Population (2000): 1,711,263; rank: 38; **net change** (1999-2000): 8.4%. **Pop. density**: 22.3 per sq mi. **Racial distribution** (2000): 89.6% white; 4.0% black; 1.3% Asian; 0.9% Native American/Nat. AK; 0.1% Hawaiian/Pacific Islander; 2.8% other race; 2 or more races, 1.4%. **Hispanic pop.** (any race): 5.5%.

Geography. Total area: 77,354 sq mi; rank: 16. **Land area:** 76,872 sq mi; rank: 15. **Acres forested:** 722,000. **Location:** West North Central state with the Missouri R. for a NE and E border. **Climate:** continental semi-arid. **Topography:** till plains of the central lowland in the eastern third rising to the Great Plains and hill country of the north central and NW. **Capital:** Lincoln.

Economy. Chief industries: agriculture, manufacturing. **Chief manuf. goods:** processed foods, industrial machinery, printed materials, electric and electronic equipment, primary and fabricated metal products, transportation equipment. **Chief crops:** corn, sorghum, soybeans, hay, wheat, dry beans, oats, potatoes, sugar beets. **Livestock:** (Jan. 2001) 6.6 mil cattle/calves; 114,000 sheep/lambs; (Dec. 2000) 3.1 mil hogs/pigs; 13.9 mil chickens (excl. broilers); 11.5 mil broilers. **Timber/lumber** (2000): oak, hickory, and elm; 34 mil bd. ft. **Nonfuel minerals** (est. 2000): $170 mil; mostly portland and masonry cement, sand & gravel, crushed stone, lime. **Chief ports:** Omaha, Sioux City, Brownville, Blair, Plattsmouth, Nebraska City. **Value of construction** (1997): $2 bil. **Gross state product** (1999): $53.7 bil. **Employment distrib.** (May 2001): 28.3% services; 23.7% trade; 17.2% govt.; 13.0% mfg. **Per cap. pers. income** (2000): $27,829. **Sales tax** (2001): 5%. **Unemployment** (2000): 3%. **Tourism expends.** (1997): $2.3 bil. **Lottery** (2000): total sales: $68.2 mil; net income: $16.1 mil.

Finance. FDIC-insured commercial banks (2000): 276. **Deposits**: $23.7 bil. **FDIC-insured savings institutions** (2000): 15. **Assets**: $15.1 bil.

Federal govt. Fed. civ. employees (Mar. 2000): 7,515. **Avg. salary:** $45,942. **Notable fed. facilities:** Offutt AFB.

Energy. Electricity production (2000, kWh, by source): Coal: 18.3 bil; Petroleum: 57 mil; Gas: 435 mil; Hydroelectric: 1.5 bil; Nuclear: 8.6 bil.

State data. Motto: Equality before the law. **Flower:** Goldenrod. **Bird:** Western meadowlark. **Tree:** Cottonwood. **Song:** Beautiful Nebraska. **Entered union** Mar. 1, 1867; rank, 37th. **State fair** at Lincoln; Aug.- Sept.

History. When the Europeans first arrived, Pawnee, Ponca, Omaha, and Oto peoples lived in the region. Spanish and French explorers and fur traders visited the area prior to its acquisition in the Louisiana Purchase, 1803. Lewis and Clark passed through, 1804-6. The first permanent settlement was Bellevue, near Omaha, 1823. The region was gradually settled, despite the 1834 Indian Intercourse Act, which declared Nebraska Indian country and excluded white settlement. Conflicts with settlers eventually forced Native Americans to move to reservations. Many Civil War veterans settled under free land terms of the 1862 Homestead Act; as agriculture grew, struggles followed between homesteaders and ranchers.

Tourist attractions. State Museum (Elephant Hall), State Capitol, both Lincoln; Stuhr Museum of the Prairie Pioneer, Grand Island; Museum of the Fur Trade, Chadron; Henry Doorly Zoo, Joslyn Art Museum, both Omaha; Ashfall Fossil Beds, Strategic Air Command Museum, Ashland; Boys Town, west of Omaha; Arbor Lodge State Park, Nebraska City; Buffalo Bill Ranch State Hist. Park, North Platte; Pioneer Village, Minden; Oregon Trail landmarks; Scotts Bluff Natl. Monument; Chimney Rock Historic Site; Ft. Robinson; Hastings Museum, Hastings.

Famous Nebraskans. Grover Cleveland Alexander, Fred Astaire, Marlon Brando, Charles W. Bryan, William Jennings Bryan, Warren Buffett, Johnny Carson, Willa Cather, Dick Cavett, Dick Cheney, William F. "Buffalo Bill" Cody, Loren Eiseley, Rev. Edward J. Flanagan, Henry Fonda, Gerald R. Ford, Bob Gibson, Rollin Kirby, Harold Lloyd, Malcolm X, J. Sterling Morton, John Neihardt, Nick Nolte, George Norris, Tom Osborne, John J. Pershing, Roscoe Pound, Chief Red Cloud, Mari Sandoz, Robert Taylor, Darryl F. Zanuck.

Chamber of Commerce and Industry. 1320 Lincoln Mall, Ste. 201, Lincoln, NE 68508; 402-474-4422

Toll-free travel information. 1-800-228-4307.

Website. http://www.state.ne.us

Tourism website. http://www.visitnebraska.org

Nevada

Sagebrush State, Battle Born State, Silver State

People. Population (2000): 1,998,257; rank: 35; **net change** (1999-2000): 66.3%. **Pop. density**: 18.2 per sq mi. **Racial distribution** (2000): 75.2% white; 6.8% black; 4.5% Asian; 1.3% Native American/Nat. AK; 0.4% Hawaiian/Pacific Islander; 8.0% other race; 2 or more races, 3.8%. **Hispanic pop.** (any race): 19.7%.

Geography. Total area: 110,561 sq mi; rank: 7. **Land area:** 109,826 sq mi; rank: 7. **Acres forested:** 8,938,000. **Location:** Mountain state bordered on N by Oregon and Idaho, on E by Utah and Arizona, on SE by Arizona, and on SW and W by California. **Climate:** semi-arid and arid. **Topography:** rugged N-S mountain ranges; highest elevation, Boundary Peak, 13,140 ft; southern area is within the Mojave Desert; lowest elevation, Colorado River at southern tip of state, 479 ft. **Capital:** Carson City.

Economy. Chief industries: gaming, tourism, mining, manufacturing, government, retailing, warehousing, trucking. **Chief manuf. goods:** food products, plastics, chemicals, aerospace products, lawn and garden irrigation equipment, seismic and machinery-monitoring devices. **Chief crops:** hay, alfalfa seed, potatoes, onions, garlic, barley, wheat. **Livestock:** (Jan. 2001) 520,000 cattle/calves; 95,000 sheep/lambs; (Dec. 2000) 7,500 hogs/pigs. **Timber/lumber:** (2000): piñon, juniper, other pines; < .5 mil bd. ft. **Nonfuel minerals** (est. 2000): $2.8 bil; mostly gold, lime, silver, sand & gravel, diatomite. **Principal internat. airports at:** Las Vegas, Reno. **Value of construction** (1997): $6.7 bil. **Gross state product** (1999): $69.8 bil. **Employment distrib.** (May 2001): 42.9% services; 21.4% trade; 11.7% govt.; 4.3% mfg. **Per cap. pers. income** (2000): $30,529. **Sales tax** (2001): 6.5%. **Unemployment** (2000): 4.1%. **Tourism expends.** (1997): $18.5 bil.

Finance. FDIC-insured commercial banks (2000): 32. **Deposits**: $13.3 bil. **FDIC-insured savings institutions** (2000): 2. **Assets**: $942 mil.

Federal govt. Fed. civ. employees (Mar. 2000): 7,234. **Avg. salary:** $49,514. **Notable fed. facilities:** Nevada Test Site; Hawthorne Army Ammunition Plant, Nellis Air Force Base and Gunnery Range; Fallon Naval Air Station; Palomino Valley Wild Horse and Burro Placement Center.

Energy. Electricity production (2000, kWh, by source): Coal: 19 bil; Petroleum: 47 mil; Gas: 7.9 bil; Hydroelectric: 2.4 bil.

State data. Motto: All for our country. **Flower:** Sagebrush. **Bird:** Mountain bluebird. **Trees:** Single-leaf piñon and bristlecone pine. **Song:** Home Means Nevada. **Entered union** Oct. 31, 1864; rank, 36th. **State fair** at Reno; late Aug.

History. Shoshone, Paiute, Bannock, and Washoe peoples lived in the area at the time of European contact. Nevada was first explored by Spaniards, 1776. Hudson's Bay Co. trappers explored the north and central region, 1825; trader Jedediah Smith crossed the state, 1826-27. The area was acquired by the U.S., 1848, at the end of the Mexican War. The first settlement, Mormon Station, now Genoa, was established, 1849. Discovery of the Comstock Lode, rich in gold and silver, 1859, spurred a population boom. In the early 20th cent., Nevada adopted progressive measures such as the initiative, referendum, recall, and woman suffrage.

Tourist attractions. Legalized gambling at: Lake Tahoe, Reno, Las Vegas, Laughlin, Elko County, and elsewhere. Hoover Dam; Lake Mead; Great Basin Natl. Park; Valley of Fire State Park; Virginia City; Red Rock Canyon Natl. Conservation Area; Liberace Museum, the Las Vegas Strip, Guinness World of Records Museum, Lost City Museum, Overton, Lamoille Canyon, Pyramid Lake, all Las Vegas. Skiing near Lake Tahoe.

Famous Nevadans. Andre Agassi, Walter Van Tilburg Clark, George Ferris, Sarah Winnemucca Hopkins, Paul Laxalt, Dat So La Lee, John William Mackay, Anne Martin, Pat McCarran, Key Pittman, William Morris Stewart.

Tourist information. Commission on Tourism, 5151 S. Carson St., Carson City, NV 89701.

Toll-free travel information. 1-800-NEVADA8.

Website. http://www.silverstate.nv.us

Tourism website. http://www.travelnevada.com

New Hampshire

Granite State

People. Population (2000): 1,235,786; rank: 41; **net change** (1999-2000): 11.4%. **Pop. density**: 137.8 per sq mi. **Racial distribution** (2000): 96.0% white; 0.7% black; 1.3% Asian; 0.2% Native American/Nat. AK; <0.1% Hawaiian/Pacific Islander; 0.6% other race; 2 or more races, 1.1%. **Hispanic pop.** (any race): 1.7%.

Geography. Total area: 9,350 sq mi; rank: 46. **Land area:** 8,968 sq mi; rank: 44. **Acres forested:** 4,981,000. **Location:** New England state bounded on S by Massachusetts, on W by Vermont, on N and NW by Canada, on E by Maine and the Atlantic Ocean. **Climate:** highly varied, due to its nearness to high mountains and ocean. **Topography:** low, rolling coast followed by countless hills and mountains rising out of a central plateau. **Capital:** Concord.

Economy. Chief industries: tourism, manufacturing, agriculture, trade, mining. **Chief manuf. goods:** machinery, electrical and electronic products, plastics, fabricated metal products. **Chief crops:** dairy products, nursery & greenhouse products, hay, vegetables, fruit, maple syrup & sugar products. **Livestock**: (Jan. 2001) 42,000 cattle/calves; (Dec. 2000) 4,000 hogs/pigs; 262,000 chickens (excl. broilers). **Timber/lumber** (2000): white pine, hemlock, oak, birch; 340 mil bd. ft. **Nonfuel minerals** (est. 2000): $59.2 mil; mostly sand & gravel, crushed and dimension stone, gemstones. **Commercial fishing** (1999): $12.5 mil. **Chief ports:** Portsmouth, Hampton, Rye. **Value of construction** (1997): $1.3 bil. **Gross state product** (1999): $44.2 bil. **Employment distrib.** (May 2001): 31.1% services; 26.5% trade; 13.1% govt.; 16.5% mfg. **Per cap. pers. income** (2000): $33,332. **Sales tax:** none. **Unemployment** (2000): 2.8%. **Tourism expends.** (1997): $1.9 bil. **Lottery** (2000): total sales: $190.9 mil; net income: $61.5 mil.

Finance. FDIC-insured commercial banks (2000): 16. **Deposits:** $16.4 bil. **FDIC-insured savings institutions** (2000): 19. **Assets**: $9.3 bil.

Federal govt. Fed. civ. employees (Mar. 2000): 3,094. **Avg. salary:** $55,698.

Energy. Electricity production (2000, kWh, by source): Coal: 3.9 bil; Petroleum: 382 mil; Gas: 77 mil; Hydroelectric: 327 mil; Nuclear: 7.9 bil.

State data. Motto: Live free or die. **Flower:** Purple lilac. **Bird:** Purple finch. **Tree:** White birch. **Song:** Old New Hampshire. **Ninth** of the original 13 states to ratify the Constitution, June 21, 1788. **State Fair:** Many agricultural fairs statewide, July through Sept.; no State fair.

History. Algonquian-speaking peoples, including the Pennacook, lived in the region when the Europeans arrived. The first explorers to visit the area were England's Martin Pring, 1603, and France's Champlain, 1605. The first settlement was Odiorne's Point (now port of Rye), 1623. Native American conflicts were ended, 1759, by Robert Rogers' Rangers. Before the American Revolution, New Hampshire residents seized a British fort at Portsmouth, 1774, and drove the royal governor out, 1775. New Hampshire became the first colony to adopt its own constitution, 1776. Three regiments served in the Continental Army, and scores of privateers raided British shipping.

Tourist attractions. Mt. Washington, highest peak in Northeast; Lake Winnipesaukee; White Mt. National Forest; Crawford, Franconia—famous for the Old Man of the Mountain, described by Hawthorne as the Great Stone Face, Pinkham notches, all White Mt. region; the Flume, a spectacular gorge; the aerial tramway, Cannon Mt.; Strawbery

Banke, Portsmouth; Shaker Village, Canterbury; Saint-Gaudens, natl. historic site, Cornish; Mt. Monadnock.

Famous New Hampshirites. Salmon P. Chase, Ralph Adams Cram, Mary Baker Eddy, Daniel Chester French, Robert Frost, Horace Greeley, Sarah Buell Hale, Franklin Pierce, Augustus Saint-Gaudens, Adam Sandler, Alan Shepard, David H. Souter, Daniel Webster.

Tourist information. Division of Travel & Tourism Development, PO Box 1856, Concord, NH 03302-1856.

Toll-free travel information. 1-800-FUNINNH ext. 169.

Website. http://www.state.nh.us

Tourism website. http://www.visitnh.gov

New Jersey

Garden State

People. Population (2000): 8,414,350; rank: 9; **net change** (1999-2000): 8.9%. **Pop. density**: 1,134.5 per sq mi. **Racial distribution** (2000): 72.6% white; 13.6% black; 5.7% Asian; 0.2% Native American/Nat. AK; <0.1% Hawaiian/Pacific Islander; 5.4% other race; 2 or more races, 2.5%. **Hispanic pop.** (any race): 13.3%.

Geography. Total area: 8,721 sq mi; rank: 47. **Land area:** 7,417 sq mi; rank: 46. **Acres forested:** 2,007,000. **Location:** Middle Atlantic state bounded on N and E by New York and Atlantic Ocean, on S and W by Delaware and Pennsylvania. **Climate:** moderate, with marked difference bet. NW and SE extremities. **Topography:** Appalachian Valley in the NW also has highest elevation, High Pt., 1,801 ft; Appalachian Highlands, flat-topped NE-SW mountain ranges; Piedmont Plateau, low plains broken by high ridges (Palisades) rising 400-500 ft; Coastal Plain, covering three-fifths of state in SE, rises from sea level to gentle slopes. **Capital:** Trenton.

Economy. Chief industries: pharmaceuticals/drugs, telecommunications, biotechnology, printing & publishing. **Chief manuf. goods:** chemicals, electronic equipment, food. **Chief crops:** nursery/greenhouse, tomatoes, blueberries, peaches, peppers, cranberries, soybeans. **Livestock:** (Jan. 2001) 48,000 cattle/calves; (Dec. 2000) 14,000 hogs/pigs; 2.3 mil chickens (excl. broilers). **Timber/lumber:** (2000): pine, cedar, mixed hardwoods; 19 mil bd. ft. **Nonfuel minerals** (est. 2000): $286 mil; mostly crushed stone, sand & gravel, clays, peat. **Commercial fishing** (1999): $98 mil. **Chief ports:** Newark, Elizabeth, Hoboken, Camden. **Principal internat. airports at:** Atlantic City, Newark. **Value of construction** (1997): $8.3 bil. **Gross state product** (1999): $331.5 bil. **Employment distrib.** (May 2001): 33.3% services; 23.3% trade; 14.8% govt.; 11.2% mfg. **Per cap. pers. income** (2000): $36,983. **Sales tax** (2001): 6%. **Unemployment** (2000): 3.8%. **Tourism expends.** (1997): $13.9 bil. **Lottery** (2000): total sales: $1.8 bil; net income: $719.4 mil.

Finance. FDIC-insured commercial banks (2000): 81. **Deposits**: $80 bil. **FDIC-insured savings institutions** (2000): 71. **Assets**: $42.3 bil.

Federal govt. Fed. civ. employees (Mar. 2000): 25,782. **Avg. salary:** $55,478. **Notable fed. facilities:** McGuire AFB; Fort Dix; Fort Monmouth; Picatinny Arsenal; Lakehurst Naval Air Engineering Center.

Energy. Electricity production (2000, kWh, by source): Coal: 5.1 bil; Petroleum: 261 mil; Gas: 1.6 bil; Hydroelectric: -141 mil; Nuclear: 18.2 bil.

State data. Motto: Liberty and prosperity. **Flower:** Purple violet. **Bird:** Eastern goldfinch. **Tree:** Red oak. **Third** of the original 13 states to ratify the Constitution, Dec. 18, 1787. **State fair** at Cherry Hill; late July-early Aug.

History. The Lenni Lenape (Delaware) peoples lived in the region and had mostly peaceful relations with European colonists, who arrived after the explorers Verrazano, 1524, and Hudson, 1609. The first permanent European settlement was Dutch, at Bergen (now Jersey City), 1660. When the British took New Netherland, 1664, the area between the Delaware and Hudson Rivers was given to Lord John Berkeley and Sir George Carteret. During the American Revolution, New Jersey was the scene of nearly 100 battles, large and small, including Trenton, 1776; Princeton, 1777; Monmouth, 1778.

Tourist attractions. 127 mi of beaches; Miss America Pageant, Atlantic City; Grover Cleveland birthplace, Caldwell; Cape May Historic District; Edison Natl. Historic Site, W. Orange; Six Flags Great Adventure, Jackson; Liberty State Park, Jersey City; Meadowlands Sports Complex, E. Rutherford; Pine Barrens wilderness area; Princeton University; numerous Revolutionary War historical sites; State Aquarium, Camden.

Famous New Jerseyans. Jason Alexander, Count Basie, Judy Blume, Bill Bradley, Jon Bon Jovi, Aaron Burr, Grover Cleveland, Danny DeVito, James Fenimore Cooper, Stephen Crane, Thomas Edison, Albert Einstein, Allen Ginsberg, Alexander Hamilton, Ed Harris, Whitney Houston, Buster Keaton, Joyce Kilmer, Norman Mailer, George McClellan, Jack Nicholson, Thomas Paine, Dorothy Parker, Joe Pesci, Molly Pitcher, Paul Robeson, Philip Roth, Antonin Scalia, Wally Schirra, H. Norman Schwarzkopf, Frank Sinatra, Bruce Springsteen, Martha Stewart, Meryl Streep, John Travolta, Walt Whitman, William Carlos Williams, Woodrow Wilson.

Chamber of Commerce. 50 W. State St., Trenton, NJ 08608.

Toll-free travel information. 1-800-VISITNJ.

Website. http://www.state.nj.us

Tourism website. http://www.visitnj.org

New Mexico

Land of Enchantment

People. Population (2000): 1,819,046; rank: 36; **net change** (1999-2000): 20.1%. **Pop. density**: 15 per sq mi. **Racial distribution** (2000): 66.8% white; 1.9% black; 1.1% Asian; 9.5% Native American/Nat. AK; 0.1% Hawaiian/Pacific Islander; 17.0% other race; 2 or more races, 3.6%. **Hispanic pop.** (any race): 42.1%.

Geography. Total area: 121,589 sq mi; rank: 5. **Land area:** 121,356 sq mi rank: 5. **Acres forested:** 15,296,000. **Location:** southwestern state bounded by Colorado on the N, Oklahoma, Texas, and Mexico on the E and S, and Arizona on the W. **Climate:** dry, with temperatures rising or falling 5× F with every 1,000 ft elevation. **Topography:** eastern third, Great Plains; central third, Rocky Mts. (85% of the state is over 4,000-ft elevation); western third, high plateau. **Capital:** Santa Fe.

Economy. Chief industries: government, services, trade. **Chief manuf. goods:** foods, machinery, apparel, lumber, printing, transportation equipment, electronics, semiconductors. **Chief crops:** hay, onions, chiles, greenhouse nursery, pecans, cotton. **Livestock:** (Jan. 2001) 1.6 mil cattle/calves; 255,000 sheep/lambs; (Dec. 2000) 3,000 hogs/pigs. **Timber/lumber** (2000): ponderosa pine, Douglas fir; 100 mil bd. ft. **Nonfuel minerals** (est. 2000): $812 mil; mostly copper, potash, sand & gravel, portland cement, crushed stone. **Principal internat. airport at:** Albuquerque. **Value of construction** (1997): $1.9 bil. **Gross state product** (1999): $51 bil. **Employment distrib.** (May 2001): 29.5% services; 23.0% trade; 24.4% govt.; 5.6% mfg. **Per cap. pers. income** (2000): $22,203. **Sales tax** (2001): 5%. **Unemployment** (2000): 4.9%. **Tourism expends.** (1997): $3.3 bil. **Lottery** (2000): total sales: $110.8 mil; net income: $24.5 mil.

Finance. FDIC-insured commercial banks (2000): 54. **Deposits**: $11.3 bil. **FDIC-insured savings institutions** (2000): 10. **Assets**: $3.2 bil.

Federal govt. Fed. civ. employees (Mar. 2000): 20,936. **Avg. salary:** $45,817. **Notable fed. facilities:** Kirtland, Cannon, Holloman AF bases; Los Alamos Scientific Laboratory; White Sands Missile Range; Natl. Solar Observatory; Natl. Radio Astronomy Observatory, Sandia National Laboratories.

Energy. Electricity production (2000, kWh, by source): Coal: 28.6 bil; Petroleum: 29 mil; Gas: 3.5 bil; Hydroelectric: 230 mil.

State data. Motto: Crescit Eundo (It grows as it goes). **Flower:** Yucca. **Bird:** Roadrunner. **Tree:** Piñon. **Song:** O, Fair New Mexico; Asi Es Nuevo Mexico. **Entered union** Jan. 6, 1912; rank, 47th. **State fair** at Albuquerque; mid-Sept.

History. Early inhabitants were peoples of the Mogollon and Anasazi civilizations, followed by the Pueblo peoples, Anasazi descendants. The nomadic Navajo and Apache tribes arrived c 15th cent. Franciscan Marcos de Niza and a former black slave, Estevanico, explored the area, 1539, seeking gold. First settlements were at San Juan Pueblo, 1598, and Santa Fe, 1610. Settlers alternately traded and fought with the Apache, Comanche, and Navajo. Trade on the Santa Fe Trail to Missouri started, 1821. The Mexican War was declared in May 1846; Gen. Stephen Kearny took Santa Fe without firing a shot, Aug. 18, 1846, declaring New Mexico part of the U.S. All Hispanic New Mexicans and Pueblo became U.S. citizens by terms of the 1848 treaty ending the war, but Congress denied the area statehood and created the territory of New Mexico, 1850. Pancho Villa raided Columbus, 1916, and U.S. troops were sent to the area. The world's first atomic bomb was exploded near Alamogordo, south of Santa Fe, 1945.

Tourist attractions. Carlsbad Caverns Natl. Park, with the largest natural underground chamber in the world; Santa Fe, oldest capital in U.S.; White Sands Natl. Monument, the largest gypsum deposit in the world; Chaco Culture National Historical Park; Acoma Pueblo, the "sky city," built atop a 357-ft mesa; Taos; Taos Art Colony; Taos Ski Valley; Ute Lake State Park; Shiprock.

Famous New Mexicans. Ben Abruzzo, Maxie Anderson, Billy (the Kid) Bonney, Kit Carson, Bob Foster, Peter Hurd, Tony Hillerman, Archbishop Jean Baptiste Lamy, Nancy Lopez, Bill Mauldin, Georgia O'Keeffe, Kim Stanley, Al Unser, Bobby Unser, Lew Wallace.

Tourist information. New Mexico Dept. of Tourism, PO Box 20002, Santa Fe, NM 87503.

Toll-free travel information. 1-800-733-6396, ext. 0643

Website. http://www.state.nm.us

Tourism website. http://www.newmexico.org

New York

Empire State

People. Population (2000): 18,976,457; rank: 3; **net change** (1999-2000): 5.5%. **Pop. density**: 401.9 per sq mi. **Racial distribution** (2000): 67.9% white; 15.9% black; 5.5% Asian; 0.4% Native American/Nat. AK; 0.1% Hawaiian/Pacific Islander; 7.1% other race; 2 or more races, 3.1%. **Hispanic pop.** (any race): 15.1%.

Geography. Total area: 54,556 sq mi; rank: 27. **Land area:** 47,214 sq mi; rank: 30. **Acres forested:** 18,713,000. **Location:** Middle Atlantic state, bordered by the New England states, Atlantic Ocean, New Jersey and Pennsylvania, Lakes Ontario and Erie, and Canada. **Climate:** variable; the SE region moderated by the ocean. **Topography:** highest and most rugged mountains in the NE Adirondack upland; St. Lawrence-Champlain lowlands extend from Lake Ontario NE along the Canadian border; Hudson-Mohawk lowland follows the flows of the rivers N and W, 10-30 mi wide; Atlantic coastal plain in the SE; Appalachian Highlands, covering half the state westward from the Hudson Valley, include the Catskill Mts., Finger Lakes; plateau of Erie-Ontario lowlands. **Capital:** Albany.

Economy. Chief industries: manufacturing, finance, communications, tourism, transportation, services. **Principal manufactured goods:** books & periodicals, clothing & apparel, pharmaceuticals, machinery, instruments, toys & sporting goods, electronic equipment, automotive & aircraft components. **Chief crops:** apples, grapes, strawberries, cherries, pears, onions, potatoes, cabbage, sweet corn, green beans, cauliflower, field corn, hay, wheat, oats, dry beans. **Products:** milk, cheese, maple syrup, wine. **Livestock:** (Jan. 2001) 1.4 mil cattle/calves; 60,000 sheep/lambs; (Dec. 2000) 80,000 hogs/pigs; 5.6 mil chickens (excl. broilers); 2 mil broilers. **Timber/lumber** (2000): birch, sugar and red maple, basswood, hemlock, pine, oak, ash; 578 mil bd. ft. **Nonfuel minerals** (est. 2000): $970 mil; mostly crushed stone, portland cement, salt, sand & gravel, zinc. **Commercial fishing** (1999): $76 mil. **Chief ports:** New York, Buffalo, Albany. **Principal internat. airports at:** Albany, Buffalo, New York, Newburgh, Rochester, Syracuse. **Value of construction** (1997): $15.8 bil. **Gross state product** (2000): $754.6 bil. **Employment distrib.** (May 2001): 35.7% services; 20.2% trade; 16.8% govt.; 9.7% mfg. **Per cap. pers. income** (2000): $34,547. **Sales tax** (2001): 4%. **Unemployment** (2000): 4.6%. **Tourism expends.** (1997): $32.9 bil. **Lottery** (2000): total sales: $3.6 bil; net income: $1.4 bil.

Finance. FDIC-insured commercial banks (2000): 148. **Deposits:** $790.7 bil. **FDIC-insured savings institutions** (2000): 82. **Assets**: $132.4 bil.

Federal govt. Fed. civ. employees (Mar. 2000): 57,456. **Avg. salary:** $49,330. **Notable fed. facilities:** West Point Military Academy; Merchant Marine Academy; Ft. Drum; Rome Labs.; Watervliet Arsenal.

Energy. Electricity production (2000, kWh, by source): Coal: 4.1 bil; Petroleum: 11 bil; Gas: 8.9 bil; Hydroelectric: 18.8 bil; Nuclear: 29.9 bil.

State data. Motto: Excelsior (Ever upward). **Flower:** Rose. **Bird:** Bluebird. **Tree:** Sugar maple. **Song:** I Love New York. **Eleventh** of the original 13 states to ratify the Constitution, July 26, 1788. **State fair** at Syracuse; late Aug.-early Sept.

History. Algonquians including the Mahican, Wappinger, and Lenni Lenape inhabited the region, as did the Iroquoian Mohawk, Oneida, Onondaga, Cayuga, and Seneca tribes, who established the League of the Five Nations. In 1609, Henry Hudson visited the river named for him, and Champlain explored the lake named for him. The first permanent settlement was Dutch, near present-day Albany, 1624. New Amsterdam was settled, 1626, at the S tip of Manhattan Island. A British fleet seized New Netherland, 1664. Ninety-two of the 300 or more engagements of the American Revolution were fought in New York, including the Battle of Bemis Heights-Saratoga, 1777, a turning point of the war. Completion of Erie Canal, 1825, established the state as a gateway to the West. The first woman's rights convention was held in Seneca Falls, 1848.

Tourist attractions. New York City; Adirondack and Catskill Mts.; Finger Lakes; Great Lakes; Thousand Islands; Niagara Falls; Saratoga Springs; Philipsburg Manor, Sunnyside (Washington Irving's home), the Dutch Church of Sleepy Hollow, all in Tarrytown area; Corning Glass Center and Steuben factory, Corning; Fenimore House, Natl. Baseball Hall of Fame and Museum, both in Cooperstown; Ft. Ticonderoga overlooking Lakes George and Champlain; Empire State Plaza, Albany; Lake Placid; Franklin D. Roosevelt Natl. Historic Site, including the Roosevelt Library, Hyde Park; Long Island beaches; Theodore Roosevelt estate, Sagamore Hill, Oyster Bay; Turning Stone Casino.

Famous New Yorkers. Woody Allen, Susan B. Anthony, James Baldwin, Lucille Ball, L. Frank Baum, Humphrey Bogart, Mel Brooks, Benjamin Cardozo, De Witt Clinton, Peter Cooper, Aaron Copland, George Eastman, Millard Fillmore, Lou Gehrig, George and Ira Gershwin, Ruth Bader Ginsburg, Rudolph Giuliani, Jackie Gleason, Julia Ward Howe, Charles Evans Hughes, Washington Irving, Henry and William James, John Jay, Michael Jordan, Edward Koch, Fiorello LaGuardia, Herman Melville, J. Pierpont Morgan Jr., Eddie Murphy, Joyce Carol Oates, Carroll O'Connor, Rosie O'Donnell, Eugene O'Neill, Colin Powell, Nancy Reagan, John D. Rockefeller, Nelson Rockefeller, Eleanor Roosevelt, Franklin D. Roosevelt, Theodore Roosevelt, J. D. Salinger, Jerry Seinfeld, Paul Simon, Alfred E. Smith, Elizabeth Cady Stanton, Barbra Streisand, Donald Trump, William (Boss) Tweed, Martin Van Buren, Gore Vidal, Edith Wharton, Walt Whitman.

Tourist information. Empire State Development, Travel Information Center, 1 Commerce Plaza, Albany, NY 12245.

Toll-free travel information. 1-800-CALLNYS from U.S. states and territories and Canada; 1-518-474-4116 from other areas.

Website. http://www.empire.state.ny.us

Tourism website. http://www.iloveny.state.ny.us

North Carolina

Tar Heel State, Old North State

People. Population (2000): 8,049,313; rank: 11; **net change** (1999-2000): 21.4%. **Pop. density**: 165.2 per sq mi. **Racial distribution** (2000): 72.10% white; 21.6% black; 1.4% Asian; 1.2% Native American/Nat. AK; 0.1% Hawaiian/Pacific Islander; 2.3% other race; 2 or more races, 1.3%. **Hispanic pop.** (any race): 4.7%.

Geography. Total area: 53,819 sq mi; rank: 28. **Land area:** 48,711 sq mi; rank: 29. **Acres forested:** 19,278,000. **Location:** South Atlantic state bounded by Virginia, South Carolina, Georgia, Tennessee, and the Atlantic Ocean. **Climate:** sub-tropical in SE, medium-continental in mountain region; tempered by the Gulf Stream and the mountains in W. **Topography:** coastal plain and tidewater, two-fifths of state, extending to the fall line of the rivers; piedmont plateau, another two-fifths, of gentle to rugged hills; southern Appalachian Mts. contains the Blue Ridge and Great Smoky Mts. **Capital:** Raleigh.

Economy. Chief industries: manufacturing, agriculture, tourism. **Chief manuf. goods:** food products, textiles, industrial machinery and equipment, electrical and electronic equipment, furniture, tobacco products, apparel. **Chief crops:** tobacco, cotton, soybeans, corn, food grains, wheat, peanuts, sweet potatoes. **Livestock:** (Jan. 2001) 950,000 cattle/calves; (Dec. 2000) 9.4 mil hogs/pigs; 17.6 mil chickens (excl. broilers); 674.6 mil broilers. **Timber/lumber** (2000): yellow pine, oak, hickory, poplar, maple; 2.5 bil bd. ft. **Nonfuel minerals** (est. 1999): $779 mil; mostly crushed stone, phosphate rock, sand & gravel, clays. **Commercial fishing** (1999): $97.3 mil. **Principal internat. airports at:** Charlotte, Greensboro, Raleigh/Durham, Wilmington. **Chief ports:** Morehead City, Wilmington. **Value of construction** (1997): $14 bil. **Gross state product** (1999): $258.6 bil. **Employment distrib.** (May 2001): 26.6% services; 22.8% trade; 16.2% govt.; 18.8% mfg. **Per cap. pers. income** (2000): $27,194. **Sales tax** (2001): 4%. **Unemployment** (2000): 3.6%. **Tourism expends.** (1997): $10.7 bil.

Finance. FDIC-insured commercial banks (2000): 75. **Deposits**: $645.4 bil. **FDIC-insured savings institutions** (2000): 43. **Assets:** $6.2 bil.

Federal govt. Fed. civ. employees (Mar. 2000): 30,294. **Avg. salary:** $44,344. **Notable fed. facilities:** Ft. Bragg; Camp LeJeune Marine Base; U.S. EPA Research and Development Labs, Cherry Point Marine Corps Air Station; Natl. Humanities Center; Natl. Inst. of Environmental Health Science; Natl. Center for Health Statistics Lab, Research Triangle Park.

Energy. Electricity production (2000, kWh, by source): Coal: 71.1 bil; Petroleum: 471 mil; Gas: 838 mil; Hydroelectric: 2.3 bil; Nuclear: 39.1 bil.

State data. Motto: Esse Quam Videri (To be rather than to seem). **Flower:** Dogwood. **Bird:** Cardinal. **Tree:** Pine. **Song:** The Old North State. **Twelfth** of the original 13 states to ratify the Constitution, Nov. 21, 1789. **State fair** at Raleigh; mid-Oct.

History. Algonquian, Siouan, and Iroquoian peoples lived in the region at the time of European contact. The first English colony in America was the first of 2 established by Sir Walter Raleigh on Roanoke Island, 1585 and 1587. The first group returned to England; the second, the "Lost Colony," disappeared without a trace. Permanent settlers came from Virginia, c 1660. Roused by British repression, the colonists drove out the royal governor, 1775. The province's congress was the first to vote for independence; ten regiments were furnished to the Continental Army. Cornwallis's forces were defeated at Kings Mountain, 1780, and forced out after Guilford Courthouse, 1781. The state seceded in 1861, and provided more troops to the Confederacy than any other state; readmitted in 1868.

Tourist attractions. Cape Hatteras and Cape Lookout natl. seashores; Great Smoky Mts.; Guilford Courthouse and Moore's Creek parks; 66 American Revolution battle sites; Bennett Place, near Durham, where Gen. Joseph Johnston surrendered the last Confederate army to Gen. William Sherman; Ft. Raleigh, Roanoke Island, where Virginia Dare, first child of English parents in the New World, was born Aug. 18, 1587; Wright Brothers Natl. Memorial, Kitty Hawk; Battleship *North Carolina*, Wilmington; NC Zoo, Asheboro; NC Symphony, NC Museum, Raleigh; Carl Sandburg Home, Hendersonville, Biltmore House & Gardens, Asheville.

Famous North Carolinians. David Brinkley, Robert Byrd, Shirley Caesar, John Coltrane, Rick Dees, Elizabeth Dole, Ava Gardner, Richard J. Gatling, Billy Graham, Andy Griffith, O. Henry, Andrew Jackson, Andrew Johnson, Michael Jordan, Wm. Rufus King, Charles Kuralt, Meadowlark Lemon, Dolley Madison, Theolonius Monk, Edward R. Murrow, Arnold Palmer, Richard Petty, James K. Polk, Charlie Rose, Carl Sandburg, Enos Slaughter, Dean Smith, James Taylor, Thomas Wolfe, Orville and Wilbur Wright.

Tourist information. North Carolina Division of Tourism, Film & Sports Development, 301 N. Wilmington St., Raleigh, NC 27601.

Toll-free travel information. 1-800-VISITNC.

Website. http://www.ncgov.com

Tourism website. http://www.visitnc.com

North Dakota

Peace Garden State

People. Population (2000): 642,200; rank: 47; **net change** (1999-2000): 0.5%. **Pop. density**: 9.3 per sq mi. **Racial distribution** (2000): 92.4% white; 0.6% black; 0.6% Asian; 4.9% Native American/Nat. AK; <0.1% Hawaiian/Pacific Islander; 0.4% other race; 2 or more races, 1.2%. **Hispanic pop.** (any race): 1.2%.

Geography. Total area: 70,700 sq mi; rank: 19. **Land area:** 68,976 sq mi; rank: 17. **Acres forested:** 462,000. **Location:** West North Central state, situated exactly in the middle of North America, bounded on the N by Canada, on the E by Minnesota, on the S by South Dakota, on the W by Montana. **Climate:** continental, with a wide range of temperature and moderate rainfall. **Topography:** Central Lowland in the E comprises the flat Red River Valley and the Rolling Drift Prairie; Missouri Plateau of the Great Plains on the W. **Capital:** Bismarck.

Economy. Chief industries: agriculture, mining, tourism, manufacturing, telecommunications, energy, food processing. **Chief manuf. goods:** farm equipment, processed foods, fabricated metal, high-tech. electronics. **Chief crops:** spring wheat, durum, barley, flaxseed, oats, potatoes, dry edible beans, honey, soybeans, sugar beets, sunflowers, hay. **Livestock**: (Jan. 2001) 2 mil cattle/calves; 138,000 sheep/lambs; (Dec. 2000) 185,000 hogs/pigs. **Timber/lumber:** (2000): oak, ash, cottonwood, aspen; 1 mil bd. ft. **Nonfuel minerals** (est. 2000): $42 mil; mostly sand & gravel, lime, crushed stone, clays, gemstones. **Principal internat. airport at:** Fargo. **Value of construction** (1997): $788 mil. **Gross state product** (1999): $17 bil. **Employment distrib.** (May 2001): 28.3% services; 25.1% trade; 22.1% govt.; 7.7% mfg. **Per cap. pers. income** (2000): $25,068. **Sales tax** (2001): 5%. **Unemployment** (2000): 3%. **Tourism expends.** (1997): $1.1 bil.

Finance. FDIC-insured commercial banks (2000): 110. **Deposits:** $13.3 bil. **FDIC-insured savings institutions** (2000): 3. **Assets:** $912 mil.

Federal govt. Fed. civ. employees (Mar. 2000): 4,970. **Avg. salary:** $42,719. **Notable fed. facilities:** Strategic Air Command Base; Northern Prairie Wildlife Research Center; Garrison Dam; Theodore Roosevelt Natl. Park; Grand Forks Energy Research Center; Ft. Union Natl. Historic Site.

Energy. Electricity production (2000, kWh, by source): Coal: 29 bil; Petroleum: 48 mil; Gas: <0.5; Hydroelectric: 2.1 bil.

State data. Motto: Liberty and union, now and forever, one and inseparable. **Flower:** Wild prairie rose. **Bird:** Western meadowlark. **Tree:** American elm. **Song:** North Dakota Hymn. **Entered union** Nov. 2, 1889; rank, 39th. **State fair** at Minot; July.

History. At the time of European contact, the Ojibwa, Yanktonai and Teton Sioux, Mandan, Arikara, and Hidatsa peoples lived in the region. Pierre de Varennes was the first French fur trader in the area, 1738, followed later by the English. The U.S. acquired half the territory in the Louisiana Purchase, 1803. Lewis and Clark built Ft. Mandan, near present-day Stanton, 1804-5, and wintered there. In 1818, American ownership of the other half was confirmed by agreement with Britain. The first permanent settlement was at Pembina, 1812. Missouri River steamboats reached the area, 1832, the first railroad, 1873, bringing many homesteaders. The "bonanza farm" craze of the 1870s-80s attracted many settlers. The state was first to hold a national Presidential primary, 1912.

Tourist attractions. North Dakota Heritage Center, Bismarck; Bonanzaville, Fargo; Ft. Union Trading Post Natl. Historic Site; Lake Sakakawea; Intl. Peace Garden; Theodore Roosevelt Natl. Park, including Elkhorn Ranch, Badlands; Ft. Abraham Lincoln State Park and Museum, near Mandan; Dakota Dinosaur Museum, Dickinson; Knife River Indian Villages-National Historic Site.

Famous North Dakotans. Maxwell Anderson, Angie Dickinson, John Bernard Flannagan, Phil Jackson, Louis L'Amour, Peggy Lee, Eric Sevareid, Vilhjalmur Stefansson, Lawrence Welk.

Greater North Dakota Association (Chamber of Commerce). PO Box 2639, 2000 Schafer St., Bismarck, ND 58501.

Toll-free travel information. 1-800-HELLO-ND

Website. http://www.discovernd.com

Tourism website. http//www.ndtourism.com

Ohio

Buckeye State

People. Population (2000): 11,353,140; rank: 7; **net change** (1999-2000): 4.7%. **Pop. density**: 277.3 per sq mi. **Racial distribution** (2000): 85.0% white; 11.5% black; 1.2% Asian; 0.2% Native American/Nat. AK; <0.1% Hawaiian/Pacific Islander; 0.8% other race; 2 or more races, 1.4%. **Hispanic pop.** (any race): 1.9%.

Geography. Total area: 44,825 sq mi; rank: 34. **Land area:** 40,948 sq mi; rank: 35. **Acres forested:** 7,863,000. **Location:** East North Central state bounded on the N by Michigan and Lake Erie; on the E and S by Pennsylvania, West Virginia, and Kentucky; on the W by Indiana. **Climate:** temperate but variable; weather subject to much precipitation. **Topography:** generally rolling plain; Allegheny plateau in E; Lake Erie plains extend southward; central plains in the W. **Capital:** Columbus.

Economy. Chief industries: manufacturing, trade, services. **Chief manuf. goods:** transportation equipment, machinery, primary and fabricated metal products. **Chief crops:** corn, hay, winter wheat, oats, soybeans. **Livestock:** (Jan. 2001) 1.2 mil cattle/calves; 142,000 sheep/lambs; (Dec. 2000) 1.5 mil hogs/pigs; 35.8 mil chickens (excl. broilers); 50.5 mil broilers. **Timber/lumber** (2000): oak, ash, maple, walnut, beech; 393 mil bd. ft. **Nonfuel minerals** (est. 2000): $1.1 bil; mostly crushed stone, sand & gravel, salt, lime, portland cement. **Commercial fishing** (1999): $2.1 mil. **Chief ports:** Toledo, Conneaut, Cleveland, Ashtabula. **Principal internat. airports at:** Akron, Cincinnati, Cleveland, Columbus, Dayton. **Value of construction** (1997): $14.7 bil. **Gross**

state product (1999): $362 bil. **Employment distrib.** (May 2001): 28.5% services; 24.1% trade; 14.0% govt.; 18.8% mfg. **Per cap. pers. income** (2000): $28,400. **Sales tax** (2001): 5%. **Unemployment** (2000): 4.1%. **Tourism expends.** (1997): $11.7 bil. **Lottery** (2000): total sales: $2.2 bil; net income: $592.9 mil.

Finance. FDIC-insured commercial banks (2000): 211. **Deposits:** $229.8 bil. **FDIC-insured savings institutions** (2000): 128. **Assets:** $72.3 bil.

Federal govt. Fed. civ. employees (Mar. 2000): 42,376. **Avg. salary:** $51,679. **Notable fed. facilities:** Wright Patterson AFB; Defense Construction Supply Center; Lewis Research Ctr.; Portsmouth Gaseous Diffusion Plant.

Energy. Electricity production (2000, kWh, by source): Coal: 125.1 bil; Petroleum: 343 mil; Gas: 427 mil; Hydroelectric: 582 mil; Nuclear: 16.8 bil.

State data. Motto: With God, all things are possible. **Flower:** Scarlet carnation. **Bird:** Cardinal. **Tree:** Buckeye. **Song:** Beautiful Ohio. **Entered union** Mar. 1, 1803; rank, 17th. **State fair** at Columbus; Aug.

History. Wyandot, Delaware, Miami, and Shawnee peoples sparsely occupied the area when the first Europeans arrived. La Salle visited the region, 1669, and France claimed the area, 1682. Around 1730, traders from Pennsylvania and Virginia entered the area; the French and their Native American allies sought to drive them out. France ceded its claim, 1763, to Britain. During the American Revolution, George Rogers Clark seized British posts and held the region, until Britain gave up its claim, 1783, in the Treaty of Paris. The region became U.S. territory after the American Revolution. First organized settlement was at Marietta, 1788. Indian warfare ended with Anthony Wayne's victory at Fallen Timbers, 1794. In the War of 1812, Oliver Hazard Perry's victory on Lake Erie and William Henry Harrison's invasion of Canada, 1813, ended British incursions.

Tourist attractions. Mound City Group Natl. Monuments, a group of 24 prehistoric Indian burial mounds; Neil Armstrong Air and Space Museum, Wapakoneta; Air Force Museum, Dayton; Pro Football Hall of Fame, Canton; King's Island amusement park, Mason; Lake Erie Islands, Cedar Point amusement park, both Sandusky; birthplaces, homes of, and memorials to U.S. Pres.s W. H. Harrison, Grant, Garfield, Hayes, McKinley, Harding, Taft, Benjamin Harrison; Amish Region, Tuscarawas/Holmes counties; German Village, Columbus; Sea World, Aurora; Jack Nicklaus Sports Center, Mason; Bob Evans Farm, Rio Grande; Rock and Roll Hall of Fame and Museum, Cleveland.

Famous Ohioans. Sherwood Anderson, Neil Armstrong, George Bellows, Ambrose Bierce, Erma Bombeck, Drew Carey, Hart Crane, George Custer, Clarence Darrow, Paul Laurence Dunbar, Thomas Edison, Clark Gable, John Glenn, Zane Grey, Bob Hope, William Dean Howells, Toni Morrison, Jack Nicklaus, Jesse Owens, Pontiac, Eddie Rickenbacker, John D. Rockefeller Sr. and Jr., Roy Rogers, Pete Rose, Arthur Schlesinger Jr., Gen. William Sherman, Steven Spielberg, Gloria Steinem, Harriet Beecher Stowe, Charles Taft, Robert A. Taft, William H. Taft, Tecumseh, James Thurber, Orville and Wilbur Wright.

Chamber of Commerce. PO Box 15159. 230 E. Town St., Columbus, OH 43215-0159.

Toll-free travel information. 1-800-BUCKEYE.

Website. http://www.state.oh.us

Tourism website. http://www.ohiotourism.com

Oklahoma

Sooner State

People. Population (2000): 3,450,654; rank: 27; **net change** (1999-2000): 9.7%. **Pop. density**: 50.3 per sq mi. **Racial distribution** (2000): 76.2% white; 7.6% black; 1.4% Asian; 7.9% Native American/Nat. AK; 0.1% Hawaiian/Pacific Islander; 2.4% other race; 2 or more races, 4.5%. **Hispanic pop.** (any race): 5.2%.

Geography. Total area: 69,898 sq mi; rank: 20. **Land area:** 68,667 sq mi; rank: 19. **Acres forested:** 7,539,000. **Location:** West South Central state bounded on the N by Colorado and Kansas; on the E by Missouri and Arkansas; on the S and W by Texas and New Mexico. **Climate:** temperate; southern humid belt merging with colder northern continental; humid eastern and dry western zones. **Topography:** high plains predominate in the W, hills and small mountains in the E; the east central region is dominated by the Arkansas R. Basin, and the Red R. Plains, in the S. **Capital:** Oklahoma City.

Economy. Chief industries: manufacturing, mineral and energy exploration and production, agriculture, services. **Chief manuf. goods:** nonelectrical machinery, transportation equipment, food products, fabricated metal products. **Chief crops:** wheat, cotton, hay, peanuts, grain sorghum, soybeans, corn, pecans. **Livestock:** (Jan. 2001) 5.1 mil cattle/calves; 55,000 sheep/lambs; (Dec. 2000) 2.3 mil hogs/pigs; 5.3 mil chickens (excl. broilers); 216.4 mil broilers. **Timber/lumber:** (2000): ine, oak, hickory; NA; **Nonfuel minerals** (est. 2000): $453 mil; mostly crushed stone, portland cement, sand & gravel, helium. **Chief ports:** Catoosa, Muskogee. **Principal internat. airports at:** Oklahoma City, Tulsa. **Value of construction** (1997): $3.1 bil. **Gross state product** (1999): $86.4 bil. **Employment distrib.** (May 2001): 29.0% services; 22.8% trade; 19.4% govt.; 11.9% mfg. **Per cap. pers. income** (2000): $23,517. **Sales tax** (2001): 4.5%. **Unemployment** (2000): 3%. **Tourism expends.** (1997): $6.5 bil.

Finance. FDIC-insured commercial banks (2000): 286. **Deposits:** $33.4 bil. **FDIC-insured savings institutions** (2000): 8. **Assets:** $6.5 bil.

Federal govt. Fed. civ. employees (Mar. 2000): 32,212. **Avg. salary:** $44,482. **Notable fed. facilities:** Federal Aviation Agency and Tinker AFB, Oklahoma City; Ft. Sill, Lawton; Altus AFB; Vance AFB.

Energy. Electricity production (2000, kWh, by source): Coal: 32.6 bil; Petroleum: 49 mil; Gas: 16.3 bil; Hydroelectric: 2.1 bil.

State data. Motto: Labor Omnia Vincit (Labor conquers all things). **Flower:** Mistletoe. **Bird:** Scissor-tailed flycatcher. **Tree:** Redbud. **Song:** Oklahoma! **Entered union** Nov. 16, 1907; rank, 46th. **State fair** at Oklahoma City; last 2 full weeks of Sept.

History. The region was sparsely inhabited by Native American tribes when Coronado, the first European, arrived in 1541; in the 16th and 17th cent., French traders visited. Part of the Louisiana Purchase, 1803, Oklahoma was established as Indian Territory (but not given territorial government). It became home to the "Five Civilized Tribes"—Cherokee, Choctaw, Chickasaw, Creek, and Seminole—after the forced removal of Indians from the eastern U.S., 1828-46. The land was also used by Comanche, Osage, and other Plains Indians. As white settlers pressed west, land was opened for homesteading by runs and lottery, the first run on Apr. 22, 1889. The most famous run was to the Cherokee Outlet, 1893.

Tourist attractions. Cherokee Heritage Center, Tahlequah; White Water Bay and Frontier City theme pks., both Oklahoma City; Will Rogers Memorial, Claremore; Natl. Cowboy Hall of Fame and Remington Park Race Track, both Oklahoma City; Ft. Gibson Stockade, near Muskogee; Ouachita Natl. Forest; Tulsa's art deco district; Wichita Mts. Wildlife Refuge, Lawton; Woolaroc Museum & Wildlife Preserve, Bartlesville; Sequoyah's Home Site, near Sallisaw; Philbrook Museum of Art and Gilcrease Museum, both Tulsa.

Famous Oklahomans. Troy Aikman, Carl Albert, Gene Autry, Johnny Bench, Garth Brooks, William "Hopalong Cassidy" Boyd, Lon Chaney, Walter Cronkite, L. Gordon Cooper, Jerome "Dizzy" Dean, Ralph Ellison, John Hope Franklin, James Garner, Geronimo, Woody Guthrie, Paul Harvey, Ron Howard, Gen. Patrick J. Hurley, Ben Johnson, Jeane Kirkpatrick, Louis L'Amour, Shannon Lucid, Mickey Mantle, Reba McEntire, Wiley Post, Tony Randall, Oral Roberts, Will Rogers, Barry Switzer, Maria Tallchief, Jim Thorpe, J.C. Watts Jr.

Chamber of Commerce. Chamber of Commerce, 330 NE 10th, Oklahoma City, OK 73104.

Tourism Dept. PO Box 60789, Oklahoma City, OK 73146-0789.

Toll-free travel information. 1-800-652-OKLA.

Website. http://www.state.ok.us

Tourism website. http://www.travelok.com

Oregon

Beaver State

People. Population (2000): 3,421,399; rank: 28; **net change** (1999-2000): 20.4%. **Pop. density**: 35.6 per sq mi. **Racial distribution** (2000): 86.6% white; 1.6% black; 3.0% Asian; 1.3% Native American/Nat. AK; 0.2% Hawaiian/Pacific Islander; 4.2% other race; 2 or more races, 3.1%. **Hispanic pop.** (any race): 8.0%.

Geography. Total area: 98,381 sq mi; rank: 9. **Land area:** 95,997 sq mi; rank: 10. **Acres forested:** 27,997,000. **Location:** Pacific state, bounded on N by Washington; on E by Idaho; on S by Nevada and California; on W by the Pacific. **Climate:** coastal mild and humid climate; continental dryness and extreme temperatures in the interior. **Topography:** Coast Range of rugged mountains; fertile Willamette R. Valley to E and S; Cascade Mt. Range of volcanic peaks E of the valley; plateau E of Cascades, remaining two-thirds of state. **Capital:** Salem.

Economy. Chief industries: manufacturing, services, trade, finance, insurance, real estate, government, construction. **Chief manuf. goods:** electronics & semiconductors, lumber & wood products, metals, transportation equipment, processed food, paper. **Chief crops:** greenhouse, hay, wheat, grass seed, potatoes, onions, Christmas trees, pears, mint. **Livestock:** (Jan. 2001) 1.4 mil cattle/calves; 245,000 sheep/lambs; (Dec. 2000) 32,000 hogs/pigs; 3.7 mil chickens (excl. broilers). **Timber/lumber** (2000): Douglas fir, hemlock, ponderosa pine; 6.3 bil bd. ft. **Nonfuel minerals** (est. 2000): $439 mil; mostly sand & gravel, crushed stone, portland cement, diatomite, pumice and pumicite. **Commercial fishing** (1999): $68 mil. **Chief ports:** Portland, Astoria, Coos Bay. **Principal internat. airports at:** Portland, Medford. **Value of construction** (1997): $6 bil. **Gross state product** (1999): $109.7 bil. **Employment distrib.** (May 2001): 27.6% services; 24.5% trade; 16.7% govt.; 14.8% mfg. **Per cap. pers. income** (2000): $28,350. **Sales tax:** none. **Unemployment** (2000): 4.9%. **Tourism expends.** (1997): $5.2 bil. **Lottery** (2000): total sales: $760.2 mil; net income: $315.2 mil.

Finance. FDIC-insured commercial banks (2000): 43. **Deposits:** $6.5 bil. **FDIC-insured savings institutions** (2000): 5. **Assets:** $2.2 bil.

Federal govt. Fed. civ. employees (Mar. 2000): 17,044. **Avg. salary:** $47,906. **Notable fed. facilities:** Bonneville Power Administration.

Energy. Electricity production (2000, kWh, by source): Coal: 3.8 bil; Petroleum: 53 mil; Gas: 4.4 bil; Hydroelectric: 37.7 bil.

State data. Motto: She flies with her own wings. **Flower:** Oregon grape. **Bird:** Western meadowlark. **Tree:** Douglas fir. **Song:** Oregon, My Oregon. **Entered union** Feb. 14, 1859; rank, 33d. **State fair** at Salem; 12 days ending with Labor Day.

History. More than 100 Native American tribes inhabited the area at the time of European contact, including the Chinook, Yakima, Cayuse, Modoc, and Nez Percé. Capt. Robert Gray sighted and sailed into the Columbia River, 1792; Lewis and Clark, traveling overland, wintered at its mouth, 1805-6; John Jacob Astor established a trading post in the Columbia River region, 1811. Settlers arrived in the Williamette Valley, 1834. In 1843, the first large wave of settlers arrived via the Oregon Trail. Early in the 20th cent., the "Oregon System"—political reforms that included the initiative, referendum, recall, direct primary, and woman suffrage—was adopted.

Tourist attractions. John Day Fossil Beds Natl. Monument; Columbia River Gorge; Timberline Lodge, Mt. Hood Natl. Forest; Crater Lake Natl. Park; Oregon Dunes Natl. Recreation Area; Ft. Clatsop Natl. Memorial; Oregon Caves Natl. Monument; Oregon Museum of Science and Industry, Portland; Shakespearean Festival, Ashland; High Desert Museum, Bend; Multnomah Falls; Diamond Lake; "Spruce Goose," Evergreen Aviation Museum, McMinnville.

Famous Oregonians. Ernest Bloch, Ernest Haycox, Chief Joseph, Ken Kesey, Phil Knight, Ursula K. Le Guin, Edwin Markham, Tom McCall, Dr. John McLoughlin, Joaquin Miller, Bob Packwood, Linus Pauling, Steve Prefontaine, John Reed, Alberto Salazar, Mary Decker Slaney, William Simon U'Ren.

Tourist information. Economic Development Department, 775 Summer St. NE, Salem, OR 97310.

Toll-free travel information. 1-800-547-7842.

Website. http://www.state.or.us

Tourism website. http://www.traveloregon.com

Pennsylvania

Keystone State

People. Population (2000): 12,281,054; rank: 6; **net change** (1999-2000): 3.4%. **Pop. density**: 274 per sq mi. **Racial distribution** (2000): 85.4% white; 10.0% black; 1.8% Asian; 0.1% Native American/Nat. AK; <0.1% Hawaiian/Pacific Islander; 1.5% other race; 2 or more races, 1.2%. **Hispanic pop.** (any race): 3.2%.

Geography. Total area: 46,055 sq mi; rank: 33. **Land area:** 44,817 sq mi; rank: 32. **Acres forested:** 16,969,000. **Location:** Middle Atlantic state, bordered on the E by the Delaware R.; on the S by the Mason-Dixon Line; on the W by West Virginia and Ohio; on the N/NE by Lake Erie and New York. **Climate:** continental with wide fluctuations in seasonal temperatures. **Topography:** Allegheny Mts. run SW to NE, with Piedmont and Coast Plain in the SE triangle; Allegheny Front a diagonal spine across the state's center; N and W rugged plateau falls to Lake Erie Lowland. **Capital:** Harrisburg.

Economy. Chief industries: agribusiness, advanced manufacturing, health care, travel & tourism, depository institutions, biotechnology, printing & publishing, research & consulting, trucking & warehousing, transportation by air, engineering & management, legal services. **Chief manuf. goods:** fabricated metal products; industrial machinery & equipment, transportation equipment, rubber & plastics, electronic equipment, chemicals & pharmaceuticals, lumber & wood products, stone, clay, & glass products. **Chief crops:** corn, hay, mushrooms, apples, potatoes, winter wheat, oats, vegetables, tobacco, grapes, peaches. **Livestock:** (Jan. 2001) 1.6 mil cattle/calves; 81,000 sheep/lambs; (Dec. 2000) 1 mil hogs/pigs; 30.6 mil chickens (excl. broilers); 135.3 mil broilers. **Timber/lumber** (2000): pine, oak, maple; 1.2 bil bd. ft. **Nonfuel minerals** (est. 2000): $1.3 bil; mostly crushed stone, portland and masonry cement, lime, sand & gravel. **Chief ports:** Philadelphia, Pittsburgh, Erie. **Principal internat. airports at:** Allentown, Harrisburg, Philadelphia, Pittsburgh, Wilkes-Barre/Scranton. **Value of construction** (1997): $10.1 bil. **Gross state product** (1999): $383 bil. **Employment distrib.** (May 2001): 33.2% services; 22.5% trade; 12.8% govt.; 15.7% mfg. **Per cap. pers. income** (2000): $29,539. **Sales tax** (2001): 6%. **Unemployment** (2000): 4.2%. **Tourism expends.** (1997): $13.7 bil. **Lottery** (2000): total sales: $1.7 bil; net income: $672 mil.

Finance. FDIC-insured commercial banks (2000): 187. **Deposits:** $138.7 bil. **FDIC-insured savings institutions** (2000): 116. **Assets:** $75.8 bil.

Federal govt. Fed. civ. employees (Mar. 2000): 61,448. **Avg. salary:** $45,927. **Notable fed. facilities:** Carlisle Barracks; Army War College; Naval Inventory Control Point, Phila. and Mechanicsbrg; Defense Personnel Supply Center, Phila.; Defense Distribution Center, New Cumberland; Tobyhanna Army Depot; Letterkenny Army Depot; NAS Willow Grove; 911th Air Wing, Pittsburgh; Naval Surface Warfare Center, Phila.; Charles E. Kelly Support Facility.

Energy. Electricity production (2000, kWh, by source): Coal: 36.3 bil; Petroleum: 1.6 bil; Gas: 230 mil; Hydroelectric: 1.2 bil; Nuclear: 57.3 bil.

State data. Motto: Virtue, liberty and independence. **Flower:** Mountain laurel. **Bird:** Ruffed grouse. **Tree:** Hemlock. **Song:** Pennsylvania. **Second** of the original 13 states to ratify the Constitution, Dec. 12, 1787. **State fair** at Harrisburg; 2d week in Jan. at State Farm Show Building.

History. At the time of European contact, Lenni Lenape (Delaware), Shawnee and Iroquoian Susquehannocks, Erie, and Seneca occupied the region. Swedish explorers established the first permanent settlement, 1643, on Tinicum Island. In 1655, the Dutch seized the settlement but lost it to the British, 1664. The region was given by Charles II to William Penn, 1681. Philadelphia ("brotherly love") was the capital of the colonies during most of the American Revolution, and of the U.S., 1790-1800. Philadelphia was taken by the British, 1777; Washington's troops encamped at Valley Forge in the bitter winter of 1777-78. The Declaration of Independence, 1776, and the Constitution, 1787, were signed in Philadelphia. The Civil War battle of Gettysburg, July 1-3, 1863, marked a turning point, favoring Union forces.

Tourist attractions. Independence Natl. Historic Park, Franklin Institute Science Museum, Philadelphia Museum of Art, all in Philadelphia; Valley Forge Natl. Historic Park; Gettysburg Natl. Military Park; Pennsylvania Dutch Country; Hershey; Duquesne Incline, Carnegie Institute, Heinz Hall, all in Pittsburgh; Pocono Mts.; Pennsylvania's Grand Canyon, Tioga County; Allegheny Natl. Forest; Laurel Highlands; Presque Isle State Park; Fallingwater, Ligonier; Johnstown; SteamTown U.S.A., Scranton; State Flagship Niagara, Erie; Oil Heritage Region, Northwest PA.

Famous Pennsylvanians. Marian Anderson, Maxwell Anderson, George Blanda, James Buchanan, Andrew Carnegie, Rachel Carson, Perry Como, Thomas Eakins, Stephen Foster, Benjamin Franklin, Robert Fulton, Martha Graham, Milton Hershey, Gene Kelly, Grace Kelly (Princess Grace of Monaco), George C. Marshall, Dan Marino, John J. McCloy, Margaret Mead, Andrew W. Mellon, Joe Montana, Stan Musial, Joe Namath, John O'Hara, Arnold Palmer, Robert E. Peary, Mike Piazza, Mary Roberts Rinehart, Betsy Ross, Will Smith, Jimmy Stewart, Jim Thorpe, Johnny Unitas, John Updike, Honus Wagner, Andy Warhol, Benjamin West.

Chamber of Business and Industry. 417 Walnut St., Harrisburg, PA 17120; 800-VISITPA.

Toll-free travel information. 1-800-VISITPA.

Website. http://www.state.pa.us

Tourism website. http://www.experiencepa.com

Rhode Island

Little Rhody, Ocean State

People. Population (2000): 1,048,319; rank: 43; **net change** (1999-2000): 4.5%. **Pop. density**: 1,003.2 per sq mi. **Racial distribution** (2000): 85.0% white; 4.5% black; 2.3% Asian; 0.5% Native American/Nat. AK; 0.1% Hawaiian/Pacific Islander; 5.0% other race; 2 or more races, 2.7%. **Hispanic pop.** (any race): 8.7%.

Geography. Total area: 1,545 sq mi; rank: 50. **Land area:** 1,045 sq mi; rank: 50. **Acres forested:** 401,000. **Location:** New England state. **Climate:** invigorating and changeable. **Topography:** eastern lowlands of Narragansett Basin; western uplands of flat and rolling hills. **Capital:** Providence.

Economy. Chief industries: services, manufacturing. **Chief manuf. goods:** costume jewelry, toys, machinery, textiles, electronics. **Chief crops:** nursery products, turf & vegetable production. **Livestock:** (Jan. 2001) 6,000 cattle/calves; (Dec. 2000) 3,000 hogs/pigs; 65,000 chickens (excl. broilers). **Timber/lumber:** (2000): oak; 11 mil bd. ft. **Nonfuel minerals** (est. 2000): $23.7 mil; mostly sand & gravel, crushed stone, gemstones. **Commercial fishing** (1999): $79.3 mil. **Chief ports:** Providence, Quonset Point, Newport. **Value of construction** (1997): $773 mil. **Gross state product** (1999): $32.6 bil. **Employment distrib.** (May 2001): 34.6% services; 23.1% trade; 13.4% govt.; 14.7% mfg. **Per cap. pers. income** (2000): $29,685. **Sales tax** (2001): 7%. **Unemployment** (2000): 4.1%. **Tourism expends.** (1997): $1.2 bil. **Lottery** (2000): total sales: $864.1 mil; net income: $150.2 mil.

Finance. FDIC-insured commercial banks (2000): 7. **Deposits:** $114.5 bil. **FDIC-insured savings institutions** (2000): 6. **Assets:** $1.9 bil.

Federal govt. Fed. civ. employees (Mar. 2000): 5,771. **Avg. salary:** $53,138. **Notable fed. facilities:** Naval War College; Naval Underwater Warfare Center; Natl. Marine Fisheries Laboratory; EPA Environmental Research Laboratory.

Energy. Electricity production (2000, kWh, by source): Petroleum: 9 mil.

State data. Motto: Hope. **Flower:** Violet. **Bird:** Rhode Island red. **Tree:** Red maple. **Song:** Rhode Island. **Thirteenth** of original 13 states to ratify the Constitution, May 29, 1790. **State fair** at Richmond; mid-Aug.

History. When the Europeans arrived Narragansett, Niantic, Nipmuc, and Wampanoag peoples lived in the region. Verrazano visited the area, 1524. The first permanent settlement was founded at Providence, 1636, by Roger Williams, who was exiled from the Massachusetts Bay Colony; Anne Hutchinson, also exiled, settled Portsmouth, 1638. Quaker and Jewish immigrants seeking freedom of worship began arriving, 1650s-60s. The colonists broke the power of the Narragansett in the Great Swamp Fight, 1675, the decisive battle in King Philip's War. British trade restrictions angered colonists, and they burned the British customs vessel *Gaspee*, 1772. The colony became the first to formally renounce all allegiance to King George III, May 4, 1776. Initially opposed to joining the Union, Rhode Island was the last of the 13 colonies to ratify the Constitution, 1790.

Tourist attractions. Newport mansions; yachting races including Newport to Bermuda; Block Island; Touro Synagogue, oldest in U.S., Newport; first Baptist Church in America, Providence; Slater Mill Historic Site, Pawtucket; Gilbert Stuart birthplace, Saunderstown.

Famous Rhode Islanders. Ambrose Burnside, George M. Cohan, Nelson Eddy, Jabez Gorham, Nathanael Greene, Christopher and Oliver La Farge, Matthew C. and Oliver Hazard Perry, Gilbert Stuart.

Tourist Information. Rhode Island Economic Development Corporation, One W. Exchange St., Providence, RI 02903.

Toll-free travel information. 1-800-556-2484.

Website. http://www.state.ri.us

Tourism website. http://visitrhodeisland.com

South Carolina

Palmetto State

People. Population (2000): 4,012,012; rank: 26; **net change** (1999-2000): 15.1%. **Pop. density**: 133.2 per sq mi. **Racial distribution** (2000): 67.2% white; 29.5% black; 0.9% Asian; 0.3% Native American/Nat. AK; <0.1% Hawaiian/Pacific Islander; 1.0% other race; 2 or more races, 1.0%. **Hispanic pop.** (any race): 2.4%.

Geography. Total area: 32,020 sq mi; rank: 40. **Land area:** 30,109 sq mi; rank: 40. **Acres forested:** 12,257,000. **Location:** South Atlantic state, bordered by North Carolina on the N; Georgia on the SW and W; the Atlantic Ocean on the E, SE, and S. **Climate:** humid subtropical. **Topography:** Blue Ridge province in NW has highest peaks; piedmont lies between the mountains and the fall line; coastal plain covers two-thirds of the state. **Capital:** Columbia.

Economy. Chief industries: tourism, agriculture, manufacturing. **Chief manuf. goods:** textiles, chemicals and allied products, machinery and fabricated metal products, apparel and related products. **Chief crops:** tobacco, cotton, soybeans, corn, wheat, peaches, tomatoes. **Livestock:** (Jan. 2001) 445,000 cattle/calves; (Dec. 2000) 290,000 hogs/pigs; 6.9 mil chickens (excl. broilers); 191.3 mil broilers. **Timber/lumber** (2000): pine, oak; 1.5 bil bd. ft. **Nonfuel minerals** (est. 2000): $560 mil; mostly portland and masonry, cement, crushed stone, gold, sand & gravel, clays. **Commercial fishing** (1999: $29.3 mil. **Chief ports:** Charleston, Georgetown, Beaufort/ Port Royal. **Principal internat. airports at:** Charleston, Greenville/Spartanburg, Myrtle Beach. **Value of construction** (1997): $6 bil. **Gross state product** (2000): $107 bil. **Employment distrib.** (May 2001): 25.4% services; 24.0% trade; 16.9% govt.; 18.0% mfg. **Per cap. pers. income** (2000): $24,321. **Sales tax** (2001): 5%. **Unemployment** (2000): 3.9%. **Tourism expends.** (1997): $6.5 bil.

Finance. FDIC-insured commercial banks (2000): 79. **Deposits:** $18.4 bil. **FDIC-insured savings institutions** (2000): 29. **Assets:** $8 bil.

Federal govt. Fed. civ. employees (Mar. 2000): 15,880. **Avg. salary:** $44,672. **Notable fed. facilities:** Polaris Submarine Base; Barnwell Nuclear Power Plant; Ft. Jackson; Parris Island; Savannah River Plant.

Energy. Electricity production (2000, kWh, by source): Coal: 38.3 bil.; Petroleum: 265 mil; Gas: 188 mil; Hydroelectric: 416 mil; Nuclear: 50.9 bil.

State data. Motto: Dum Spiro Spero (While I breathe, I hope). **Flower:** Yellow jessamine. **Bird:** Carolina wren. **Tree:** Palmetto. **Song:** Carolina. **Eighth** of the original 13 states to ratify the Constitution, May 23, 1788. **State fair** at Columbia; mid-Oct.

History. At the time of European settlement, Cherokee, Catawba, and Muskogean peoples lived in the area. The first English colonists settled near the Ashley River, 1670, and moved to the site of Charleston, 1680. The colonists seized the government, 1775, and the royal governor fled. The British took Charleston, 1780, but were defeated at Kings Mountain that same year, and at Cowpens and Eutaw Springs, 1781. In the 1830s, South Carolinians, angered by federal protective tariffs, adopted the Nullification Doctrine, holding that a state can void an act of Congress. The state was the first to secede from the Union, 1860, and Confederate troops fired on and forced the surrender of U.S. troops at Ft. Sumter, in Charleston Harbor, launching the Civil War. South Carolina was readmitted,1868.

Tourist attractions. Historic Charleston; Ft. Sumter Natl. Monument, in Charleston Harbor; Charleston Museum, est. 1773, oldest museum in U.S.; Middleton Place, Magnolia Plantation, Cypress Gardens, Drayton Hall, all near Charleston; other gardens at Brookgreen, Edisto, Glencairn; Myrtle Beach; Hilton Head Island; Revolutionary War battle sites; Andrew Jackson State Park & Museum; South Carolina State Museum, Columbia; Riverbanks Zoo, Columbia.

Famous South Carolinians. Charles Bolden, James F. Byrnes, John C. Calhoun, Joe Fraizer, DuBose Heyward, Ernest F. Hollings, Andrew Jackson, Jesse Jackson,"Shoeless" Joe Jackson, James Longstreet, Francis Marion, Andie McDowell, Ronald McNair, Charles Pinckney, John Rutledge, Thomas Sumter, Strom Thurmond, John B. Watson.

Tourist information. S. Carolina Dept. of Parks, Recreation, & Tourism; 803-734-0122.

Toll-free travel information. 1-800-346-3634.

Website. http://www.myscgov.com

Tourism website. http://www.discoversouthcarolina.com

South Dakota

Coyote State, Mount Rushmore State

People. Population (2000): 754,844; rank: 46; **net change** (1999-2000): 8.5%. **Pop. density**: 9.9 per sq mi. **Racial distribution** (2000): 88.7% white; 0.6% black; 0.6% Asian; 8.3% Native American/Nat. AK; <0.1% Hawaiian/Pacific Islander; 0.5% other race; 2 or more races, 1.3%. **Hispanic pop.** (any race): 1.4%.

Geography. Total area: 77,116 sq mi; rank: 17. **Land area:** 75,885 sq mi; rank: 16. **Acres forested:** 1,690,000. **Location:** West North Central state bounded on the N by North

Dakota; on the E by Minnesota and Iowa; on the S by Nebraska; on the W by Wyoming and Montana. **Climate:** characterized by extremes of temperature, persistent winds, low precipitation and humidity. **Topography:** Prairie Plains in the E; rolling hills of the Great Plains in the W; the Black Hills, rising 3,500 ft, in the SW corner. **Capital:** Pierre.

Economy. Chief industries: agriculture, services, manufacturing. **Chief manuf. goods:** food and kindred products, machinery, electric and electronic equipment. **Chief crops:** corn, soybeans, oats, wheat, sunflowers, sorghum. **Livestock:** (Jan. 2001) 4.1 mil cattle/calves; 420,000 sheep/lambs; (Dec. 2000) 1.4 mil hogs/pigs; 2.5 mil chickens (excl. broilers). **Timber/lumber** (2000): ponderosa pine; NA; **Nonfuel minerals** (est. 2000): $260 mil; mostly gold, portland cement, sand & gravel, crushed and dimension stone. **Value of construction** (1997): $742 mil. **Gross state product** (1999): $21.6 bil. **Employment distrib.** (May 2001): 27.8% services; 24.5% trade; 18.6% govt.; 12.3% mfg. **Per cap. pers. income** (2000): $26,115. **Sales tax** (2001): 4%. **Unemployment** (2000): 2.3%. **Tourism expends.** (1997) $1.1 bil. **Lottery** (2000): total sales: $580.7 mil; net income: $101 mil.

Finance. FDIC-insured commercial banks (2000): 97. **Deposits:** $13.5 bil. **FDIC-insured savings institutions** (2000): 4. **Assets:** $1.1 bil.

Federal govt. Fed. civ. employees (Mar. 2000): 6,768. **Avg. salary:** $41,880. **Notable fed. facilities:** Ellsworth AFB, Corp of Engineers, Nat'l Park Service.

Energy. Electricity production (2000, kWh, by source): Coal: 3.6 bil; Petroleum: 53 mil; Gas: 259 mil; Hydroelectric: 5.7 bil.

State data. Motto: Under God, the people rule. **Flower:** Pasqueflower. **Bird:** Chinese ring-necked pheasant. **Tree:** Black Hills spruce. **Song:** Hail, South Dakota. **Entered union** Nov. 2, 1889; rank, 40th. **State fair** at Huron; late Aug.-early Sept.

History. At the time of first European contact, Mandan, Hidatsa, Arikara and Sioux lived in the area. The French Verendrye brothers explored the region, 1742-43. The U.S. acquired the area, 1803, in the Louisiana Purchase. Lewis and Clark passed through the area, 1804-6. In 1817 a trading post was opened at Fort Pierre, which later became the site of the first European settlement in South Dakota. Gold was discovered, 1874, in the Black Hills on the great Sioux reservation; the "Great Dakota Boom" began in 1879. Conflicts with Native Americans led to the Great Sioux Agreement, 1889, which established reservations and opened up more land for white settlement. The massacre of Native American families at Wounded Knee, 1890, ended Sioux resistance.

Tourist attractions. Black Hills; Mt. Rushmore; Needles Highway; Harney Peak, tallest E. of Rockies; Deadwood, 1876 Gold Rush town; Custer State Park; Jewel Cave Natl. Monument; Badlands Natl. Park "moonscape"; "Great Lakes of S. Dakota"; Ft. Sisseton; Great Plains Zoo & Museum, Sioux Falls; Corn Palace, Mitchell; Wind Cave Natl. Park; Crazy Horse Memorial, mountain carving in progress.

Famous South Dakotans. Sparky Anderson, Black Elk, Bob Barker, Tom Brokaw, Crazy Horse, Thomas Daschle, Myron Floren, Mary Hart, Cheryl Ladd, Dr. Ernest O. Lawrence, George McGovern, Billy Mills, Allen Neuharth, Pat O'Brien, Sitting Bull.

Tourist information. Department of Tourism, Capitol Lake Plaza, 711 E. Wells Ave., c/o 500 E. Capitol Ave., Pierre, SD 57501-5070.

Toll-free travel information. 1-800-SDAKOTA.

Website. http://www.state.sd.us

Tourism website. http://www.travelsd.com

Tennessee

Volunteer State

People. Population (2000): 5,689,283; rank: 16; **net change** (1999-2000): 16.7%. **Pop. density**: 138 per sq mi. **Racial distribution** (2000): 80.2% white; 16.4% black; 1.0% Asian; 0.3% Native American/Nat. AK; <0.1% Hawaiian/Pacific Islander; 1.0% other race; 2 or more races, 1.1%. **Hispanic pop.** (any race): 2.2%.

Geography. Total area: 42,143 sq mi; rank: 36. **Land area:** 41,217 sq mi; rank: 34. **Acres forested:** 13,612,000. **Location:** East South Central state bounded on the N by Kentucky and Virginia; on the E by North Carolina; on the S by Georgia, Alabama, and Mississippi; on the W by Arkansas and Missouri. **Climate:** humid continental to the N; humid subtropical to the S. **Topography:** rugged country in the E; the Great Smoky Mts. of the Unakas; low ridges of the Appalachian Valley; the flat Cumberland Plateau; slightly rolling terrain and knobs of the Interior Low Plateau, the largest region; Eastern Gulf Coastal Plain to the W, laced with streams; Mississippi Alluvial Plain, a narrow strip of swamp and flood plain in the extreme W. **Capital:** Nashville.

Economy. Chief industries: manufacturing, trade, services, tourism, finance, insurance, real estate. **Chief manuf. goods:** chemicals, food, transportation equipment, industrial machinery & equipment, fabricated metal products, rubber/plastic products, paper & allied products, printing & publishing. **Chief crops:** tobacco, cotton, lint, soybeans, grain, corn. **Livestock:** (Jan. 2001) 2.2 mil cattle/calves; (Dec. 2000) 230,000 hogs/pigs; 2.2 mil chickens (excl. broilers); 150.8 mil broilers. **Timber/lumber** (2000): red oak, white oak, yellow poplar, hickory; 954 mil bd. ft. **Nonfuel minerals** (est. 2000): $770 mil; mostly crushed stone, zinc, portland cement, sand & gravel, clays. **Chief ports:** Memphis, Nashville, Chattanooga, Knoxville. **Principal internat. airports at:** Memphis, Nashville. **Value of construction** (1997): $8.2 bil. **Gross state product** (1999): $170.1 bil. **Employment distrib.** (May 2001): 27.6% services; 23.8% trade; 14.5% govt.; 17.8% mfg. **Per cap. pers. income** (2000): $26,239. **Sales tax** (2001): 6%. **Unemployment** (2000): 3.9%. **Tourism expends.** (1997): $9 bil.

Finance. FDIC-insured commercial banks (2000): 197. **Deposit:** $61.9 bil. **FDIC-insured savings institutions** (2000): 24. **Assets:** $5.2 bil.

Federal govt. Fed. civ. employees (Mar. 2000): 35,309 **Avg. salary:** $48,308. **Notable fed. facilities:** Tennessee Valley Authority; Oak Ridge Nat'l. Laboratories; Arnold Engineering Development Center; Ft. Campbell Army Base; Millington Naval Station.

Energy. Electricity production (2000, kWh, by source): Coal: 60.1 bil; Petroleum: 542 mil; Gas: 127 mil; Hydroelectric: 5.1 bil; Nuclear: 25.8 bil.

State data. Motto: Agriculture and commerce. **Flower:** Iris. **Bird:** Mockingbird. **Tree:** Tulip poplar. **Song:** The Tennessee Waltz. **Entered union** June 1, 1796; rank, 16th. **State fair** at Nashville; mid-Sept.

History. When the first European explorers arrived, Creek and Yuchi peoples lived in the area; the Cherokee moved into the region in the early 18th cent. Spanish explorers first visited the area, 1541. English traders crossed the Great Smokies from the east while France's Marquette and Jolliet sailed down the Mississippi on the west, 1673. The first permanent settlement was by Virginians on the Watauga River, 1769. During the American Revolution, the colonists helped win the Battle of Kings Mountain (NC), 1780, and joined other eastern campaigns. The state seceded from the Union, 1861, and saw many Civil War engagements, but 30,000 soldiers fought for the Union. Tennessee was readmitted in 1866, the only former Confederate state not to have a postwar military government.

Tourist attractions. Reelfoot Lake; Lookout Mountain, Chattanooga; Fall Creek Falls; Great Smoky Mountains Natl. Park; Lost Sea, Sweetwater; Cherokee Natl. Forest; Cumberland Gap Natl. Park; Andrew Jackson's home, the Hermitage, near Nashville; homes of Pres.s Polk and Andrew Johnson; American Museum of Science and Energy, Oak Ridge; Parthenon, Grand Old Opry, Opryland USA, all Nashville; Dollywood theme park, Pigeon Forge; Tennessee Aquarium, Chattanooga; Graceland, home of Elvis Presley, Memphis; Alex Haley Home and Museum, Henning; Casey Jones Home and Museum, Jackson.

Famous Tennesseans. Roy Acuff, Davy Crockett, David Farragut, Ernie Ford, Aretha Franklin, Morgan Freeman, Al Gore Jr., Alex Haley, William C. Handy, Sam Houston, Cordell Hull, Andrew Jackson, Andrew Johnson, Casey Jones, Estes Kefauver, Grace Moore, Dolly Parton, Minnie Pearl, James Polk, Elvis Presley, Dinah Shore, Bessie Smith, Hank Williams, Jr., Alvin York.

Tourist information. Dept. of Tourist Development, 5th Floor, Rachel Jackson Bldg., 320 6th Ave. N., Nashville, TN 37202.

Toll-free travel information. 1-800-Go2TENN

Website. http://www.state.tn.us

Tourism website. http://www.tnvacation.com

IT'S A FACT: As a tribute to its musical heritage, the state of Tennessee actually has not one, but 5 official state songs: "My Homeland, Tennessee"; "When It's Iris Time in Tennessee"; "My Tennessee"; "Tennessee Waltz"; and "Rocky Top."

Texas

Lone Star State

People. Population (2000): 20,851,820; rank: 2; **net change** (1999-2000): 22.8%. **Pop. density**: 79.6 per sq mi. **Racial distribution** (2000): 71.0% white; 11.5% black; 2.7% Asian; 0.6% Native American/Nat. AK; 0.1% Hawaiian/Pacific Islander; 11.7% other race; 2 or more races, 2.5%. **Hispanic pop.** (any race): 32.0%.

Geography. Total area: 268,581 sq mi; rank: 2. **Land area:** 261,797 sq mi; rank: 2. **Acres forested:** 19,193,000. **Location:** Southwestern state, bounded on the SE by the Gulf of Mexico; on the SW by Mexico, separated by the Rio Grande; surrounding states are Louisiana, Arkansas, Oklahoma, New Mexico. **Climate:** extremely varied; driest region is the Trans-Pecos; wettest is the NE. **Topography:** Gulf Coast Plain in the S and SE; North Central Plains slope upward with some hills; the Great Plains extend over the Panhandle, are broken by low mountains; the Trans-Pecos is the southern extension of the Rockies. **Capital:** Austin.

Economy. Chief industries: manufacturing, trade, oil and gas extraction, services. **Chief manuf. goods:** industrial machinery and equipment, foods, electrical and electronic products, chemicals and allied products, apparel. **Chief crops:** cotton, grains (wheat), sorghum grain, vegetables, citrus and other fruits, greenhouse/nursery, pecans, peanuts. **Chief farm products:** milk, eggs **Livestock**: (Jan. 2001) 13.7 mil cattle/calves; 1.1 mil sheep/lambs; (Dec. 2000) 920,000 hogs/pigs; 25.4 mil chickens (excl. broilers); 507.9 mil broilers. **Timber/lumber** (2000): pine, cypress; 1.6 bil bd. ft. **Nonfuel minerals** (est. 2000): $2.1 bil; mostly portland cement, crushed stone, sand & gravel, lime, salt. **Commercial fishing** (1999): $209.2 mil. **Chief ports:** Houston, Galveston, Brownsville, Beaumont, Port Arthur, Corpus Christi. **Principal internat. airports at:** Amarillo, Austin, Corpus Christi, Dallas/Ft. Worth, El Paso, Harlingen, Houston, Lubbock, Odessa, San Antonio. **Value of construction** (1997): $27.2 bil. **Gross state product** (1999): $687.3 bil. **Employment distrib.** (May 2001): 29.2% services; 23.6% trade; 16.4% govt.; 11.2% mfg. **Per cap. pers. income** (2000): $27,871. **Sales tax** (2001): 6.25%. **Unemployment** (2000): 4.2%. **Tourism expends.** (1997): $29.2 bil. **Lottery** (2000): total sales: $2.7 bil; net income: $747.1 mil.

Finance. FDIC-insured commercial banks (2000): 710. **Deposits:** $133.2 bil. **FDIC-insured savings institutions** (2000): 49. **Assets:** $67 bil.

Federal govt. Fed. civ. employees (Mar. 2000): 102,361. **Avg. salary:** $46,714. **Notable fed. facilities:** Fort Hood, Kelly AFB, and Ft. Sam Houston.

Energy. Electricity production (2000, kWh, by source): Coal: 132.8 bil; Petroleum: 1.2 bil; Gas: 119.6 bil; Hydroelectric: 832 mil; Nuclear: 37.6 bil.

State data. Motto: Friendship. **Flower:** Bluebonnet. **Bird:** Mockingbird. **Tree:** Pecan. **Song:** Texas, Our Texas. **Entered union** Dec. 29, 1845; rank, 28th. **State fair** at Dallas; mid-Oct.

History. At the time of European contact, Native American tribes in the region were numerous and diverse in culture. Coahuiltecan, Karankawa, Caddo, Jumano, and Tonkawa peoples lived in the area, and during the 19th cent., the Apache, Comanche, Cherokee, and Wichita arrived. Spanish explorer Pineda sailed along the Texas coast, 1519; Cabeza de Vaca and Coronado visited the interior, 1541. Spaniards made the first settlement at Ysleta, near El Paso, 1682. Americans moved into the land early in the 19th cent. Mexico, of which Texas was a part, won independence from Spain, 1821; Santa Anna became dictator in 1835; Texans rebelled. Santa Anna wiped out defenders of the Alamo, 1836; Sam Houston's Texans defeated Santa Anna at San Jacinto, and independence was proclaimed that same year. The Republic of Texas, with Sam Houston as its first president, functioned as a nation until 1845, when it was admitted to the Union.

Tourist attractions. Padre Island Natl. Seashore; Big Bend, Guadalupe Mts. natl. parks; The Alamo; Ft. Davis; Six Flags Amusement Park; Sea World and Fiesta Texas, both in San Antonio; San Antonio Missions Natl. Historical Park; Cowgirl Hall of Fame, Fort Worth; Lyndon B. Johnson Natl. Historical Park, marking his birthplace, boyhood home, and ranch, near Johnson City; Lyndon B. Johnson Library and Museum, Austin; Texas State Aquarium, Corpus Christi; Kimball Art Museum, Fort Worth; George Bush Library, College Station.

Famous Texans. Lance Armstrong, Stephen F. Austin, Lloyd Bentsen, James Bowie, Carol Burnett, George Bush, George W. Bush, Joan Crawford, J. Frank Dobie, Dwight D. Eisenhower, Morgan Fairchild, Farrah Fawcett, Sam Houston, Howard Hughes, Molly Ivins, Kay Bailey Hutchinson, Lyndon B. Johnson, Tommy Lee Jones, Janis Joplin, Barbara Jordan, Mary Martin, Chester Nimitz, Sandra Day O'Connor, H. Ross Perot, Katherine Ann Porter, Dan Rather, Sam Rayburn, Ann Richards, Sissy Spacek, Kenneth Starr, George Strait.

Chamber of Commerce. 900 Congress, Suite 501, Austin, TX 78701.

Toll-free travel information. 1-800-8888TEX.

Website. http://www.state.tx.us

Tourism website. http://www.traveltex.com

Utah

Beehive State

People. Population (2000): 2,233,169; rank: 34; **net change** (1999-2000): 29.6%. **Pop. density**: 27.2 per sq mi. **Racial distribution** (2000): 89.2% white; 0.8% black; 1.7% Asian; 1.3% Native American/Nat. AK; 0.7% Hawaiian/Pacific Islander; 4.2% other race; 2 or more races, 2.1%. **Hispanic pop.** (any race): 9.0%.

Geography. Total area: 84,899 sq mi; rank: 13. **Land area:** 82,144 sq mi; rank: 12. **Acres forested:** 16,234,000. **Location:** Middle Rocky Mountain state; its southeastern corner touches Colorado, New Mexico, and Arizona, and is the only spot in the U.S. where 4 states join. **Climate:** arid; ranging from warm desert in SW to alpine in NE. **Topography:** high Colorado plateau is cut by brilliantly colored canyons of the SE; broad, flat, desert-like Great Basin of the W; the Great Salt Lake and Bonneville Salt Flats to the NW; Middle Rockies in the NE run E-W; valleys and plateaus of the Wasatch Front. **Capital:** Salt Lake City.

Economy. Chief industries: services, trade, manufacturing, government, transportation, utilities. **Chief manuf. goods:** medical instruments, electronic components, food products, fabricated metals, transportation equipment, steel and copper. **Chief crops:** hay, corn, wheat, barley, apples, potatoes, cherries, onions, peaches, pears. **Livestock:** (Jan. 2001) 910,000 cattle/calves; 390,000 sheep/lambs; (Dec. 2000) 550,000 hogs/pigs; 3.8 mil chickens (excl. broilers). **Timber/lumber:** (2000): aspen, spruce, pine; 105 mil bd. ft. **Nonfuel minerals** (est. 2000): $1.4 bil; mostly copper, gold, magnesium metal, portland cement, sand & gravel. **Principal internat. airport at:** Salt Lake City. **Value of construction** (1997): $5.3 bil. **Gross state product** (1999): $62.6 bil. **Employment distrib.** (May 2001): 29.4% services; 23.3% trade; 17.2% govt.; 11.9% mfg. **Per cap. pers. income** (2000): $23,907. **Sales tax** (2001): 4.75%. **Unemployment** (2000): 3.2%. **Tourism expends.** (1997): $3.7 bil.

Finance. FDIC-insured commercial banks (2000): 56. **Deposits:** $69.7 bil. **FDIC-insured savings institutions** (2000): 5. **Assets:** $1.5 bil.

Federal govt. Fed. civ. employees (Mar. 2000): 24,166. **Avg. salary:** $42,710. **Notable fed. facilities:** Hill AFB; Tooele Army Depot; IRS Western Service Center.

Energy. Electricity production (2000, kWh, by source): Coal: 33.7 bil; Petroleum: 33 mil; Gas: 913 mil; Hydroelectric: 766 mil.; **Other:** 152 mil.

State data. Motto: Industry. **Flower:** Sego lily. **Bird:** Seagull. **Tree:** Blue spruce. **Song:** Utah, We Love Thee. **Entered union** Jan. 4, 1896; rank, 45th. **State fair** at Salt Lake City; Sept.

History. Ute, Gosiute, Southern Paiute, and Navajo peoples lived in the region at the time of European contact. Spanish Franciscans visited the area, 1776; American fur traders followed. Permanent settlement began with the arrival of the Mormons, 1847; they made the arid land bloom and created a prosperous economy. The State of Deseret was organized in 1849, and asked admission to the Union. In 1850, Congress established the region as the territory of Utah, and Brigham Young was appointed governor. The Union and Pacific Railroads met near Promontory, May 10, 1869, creating the first transcontinental railroad. Statehood was not achieved until 1896, after a long period of controversy over the Mormon Church's doctrine of polygamy, which it discontinued in 1890.

Tourist attractions. Temple Square, Mormon Church headquarters, Salt Lake City; Great Salt Lake; Zion National Park, Canyonlands, Bryce Canyon, Arches, and Capitol Reef natl. parks; Dinosaur, Rainbow Bridge, Timpanogos Cave, and Natural Bridges natl. monuments; Lake Powell; Flaming Gorge Natl. Recreation Area.

Famous Utahans. Maude Adams, Rosanne Barr, Ezra Taft Benson, John Moses Browning, Mariner Eccles, Philo Farnsworth, James Fletcher, David M. Kennedy, J. Willard Marriott, Merlin Olsen, Osmond family, Ivy Baker Priest, George Romney, Wallace Stegner, Brigham Young, Loretta Young.

Tourist information. Utah Travel Council, Council Hall, Salt Lake City, UT 84114; 801-538-1030.

Toll-free travel information. 1-800-200-1160 or 1-800 UTAH-FUN.

Website. http://www.utah.gov

Tourism website. http://www.utah.com

Vermont

Green Mountain State

People. Population (2000): 608,827; rank: 49; **net change** (1999-2000): 8.2%. **Pop. density**: 65.8 per sq mi. **Racial distribution** (2000): 96.8% white; 0.5% black; 0.9% Asian; 0.4% Native American/Nat. AK; <0.1% Hawaiian/Pacific Islander; 0.2% other race; 2 or more races, 1.2%. **Hispanic pop.** (any race): 0.9%.

Geography. Total area: 9,614 sq mi; rank: 45. **Land area:** 9,250 sq mi; rank: 43. **Acres forested:** 4,538,000. **Location:** northern New England state. **Climate:** temperate, with considerable temperature extremes; heavy snowfall in mountains. **Topography:** Green Mts. N-S backbone 20-36 mi wide; avg. altitude 1,000 ft. **Capital:** Montpelier.

Economy. Chief industries: manufacturing, tourism, agriculture, trade, finance, insurance, real estate, government. **Chief manuf. goods:** machine tools, furniture, scales, books, computer components, speciality foods. **Chief crops:** dairy products, apples, maple syrup, greenhouse/nursery, vegetables and small fruits. **Livestock:** (Jan. 2001) 295,000 cattle/calves; (Dec. 2000) 3,000 hogs/pigs; 246,000 chickens (excl. broilers). **Timber/lumber** (2000): pine, spruce, fir, hemlock; 233 mil bd. ft. **Nonfuel minerals** (est. 2000): $43 mil; mostly dimension stone, crushed stone, sand & gravel, talc & pyrophyllite, gemstones. **Principal internat. airport at:** Burlington. **Value of construction** (1997): $622 mil. **Gross state product** (1999): $17.2 bil. **Employment distrib.** (May 2001): 30.7% services; 22.9% trade; 16.6% govt.; 16.3% mfg. **Per cap. pers. income** (2000): $26,901. **Sales tax** (2001): 5%. **Unemployment** (2000): 2.9%. **Tourism expends.** (1997): $1.4 bil. **Lottery** (2000): total sales: $75 mil; net income: $19.4 mil.

Finance. FDIC-insured commercial banks (2000): 18. **Deposits:** $6.3 bil. **FDIC-insured savings institutions** (2000): 5. **Assets:** $1 bil.

Federal govt. Fed. civ. employees (Mar. 2000): 2,853. **Avg. salary:** $44,862.

Energy. Electricity production (2000, kWh, by source): Petroleum: 71 mil; Gas: 91 mil; Hydroelectric: 388 mil; Nuclear: 4.5 bil.; **Other:** 188 mil.

State data. Motto: Freedom and unity. **Flower:** Red clover. **Bird:** Hermit thrush. **Tree:** Sugar maple. **Song:** These Green Mountains. **Entered union** Mar. 4, 1791; rank, 14th. **State fair** at Rutland; early Sept.

History. Before the arrival of the Europeans, Abnaki and Mahican peoples lived in the region. Champlain explored the lake that bears his name, 1609. The first American settlement was Ft. Dummer, 1724, near Brattleboro. During the American Revolution, Ethan Allen and the Green Mountain Boys captured Ft. Ticonderoga (NY), 1775; John Stark defeated part of Burgoyne's forces near Bennington, 1777. In the War of 1812, Thomas MacDonough defeated a British fleet on Lake Champlain off Plattsburgh (NY), 1814.

Tourist attractions. Shelburne Museum; Rock of Ages Quarry, Graniteville; Vermont Marble Exhibit, Proctor; Bennington Battle Monument; Pres. Calvin Coolidge homestead, Plymouth; Maple Grove Maple Museum, St. Johnsbury; Ben & Jerry's Factory, Waterbury.

Famous Vermonters. Ethan Allen, Chester A. Arthur, Calvin Coolidge, John Deere, George Dewey, John Dewey, Stephen A. Douglas, Dorothy Canfield Fisher, James Fisk, James Jeffords, Rudy Vallee.

Chamber of Commerce. PO Box 37, Montpelier, VT 05601.

Tourist information. Vermont Dept. of Tourism and Marketing, 6 Baldwin St., Drawer 33, Montpelier, VT 05633-1301.

Toll-free travel information. 1-800-VERMONT

Website. http://www.state.vt.us

Tourism website. http://www.1-800-vermont.com

Virginia

Old Dominion

People. Population (2000): 7,078,515; rank: 12; **net change** (1999-2000): 14.4%. **Pop. density**: 178.8 per sq mi. **Racial distribution** (2000): 72.3% white; 19.6% black; 3.7% Asian; 0.3% Native American/Nat. AK; 0.1% Hawaiian/Pacific Islander; 2.0% other race; 2 or more races, 2.0%. **Hispanic pop.** (any race): 4.7%.

Geography. Total area: 42,774 sq mi; rank: 35. **Land area:** 39,594 sq mi; rank: 37. **Acres forested:** 15,858,000. **Location:** South Atlantic state bounded by the Atlantic Ocean on the E and surrounded by North Carolina, Tennessee, Kentucky, West Virginia, and Maryland. **Climate:** mild and equable. **Topography:** mountain and valley region in the W, including the Blue Ridge Mts.; rolling piedmont plateau; tidewater, or coastal plain, including the eastern shore. **Capital:** Richmond.

Economy. Chief industries: services, trade, government, manufacturing, tourism, agriculture. **Chief manuf. goods:** food processing, transportation equipment, printing, textiles, electronic & electrical equipment, industrial machinery & equipment, lumber & wood products, chemicals, rubber & plastics, furniture. **Chief crops:** tobacco, grain corn, soybeans, winter wheat, peanuts, lint & seed cotton. **Livestock:** (Jan. 2001) 1.7 mil cattle/calves; 61,000 sheep/lambs; (Dec. 2000) 425,000 hogs/pigs; 4.6 mil chickens (excl. broilers); (Dec. 1998) 268.7 mil broilers. **Timber/lumber** (2000): pine and hardwoods; 1.6 bil bd. ft. **Nonfuel minerals** (est. 2000): $692 mil; mostly crushed stone, sand & gravel, portland cement, lime, clays. **Commercial fishing** (1999): $108.3 mil. **Chief ports:** Hampton Roads, Richmond, Alexandria. **Principal internat. airports at:** Arlington, Norfolk, Loudon, Richmond, Newport News. **Value of construction** (1997): $10.1 bil. **Gross state product** (1999): $242 bil. **Employment distrib.** (May 2001): 32.6% services; 21.8% trade; 17.5% govt.; 10.6% mfg. **Per cap. pers. income** (2000): $31,162. **Sales tax** (2001): 4.5%. **Unemployment** (2000): 2.2%. **Tourism expends.** (1997): $11.6 bil. **Lottery** (2000): total sales: $973 mil; net income: $323.5 mil.

Finance. FDIC-insured commercial banks (2000): 143. **Deposits:** $42.9 bil. **FDIC-insured savings institutions** (2000): 19. **Assets:** $31.6 bil.

Federal govt. Fed. civ. employees (Mar. 2000): 114,436 **Avg. salary:** $55,322. **Notable fed. facilities:** Pentagon; Norfolk Naval Station, Norfolk Naval Air Station; Naval Shipyard; Marine Corps Base; Langley AFB; NASA at Langley.

Energy. Electricity production (2000, kWh, by source): Coal: 33.7 bil; Petroleum: 2.3 bil; Gas: 1.8 bil; Hydroelectric: -681 mil; Nuclear: 28.3 bil.

State data. Motto: Sic Semper Tyrannis (Thus always to tyrants). **Flower:** Dogwood. **Bird:** Cardinal. **Tree:** Dogwood. **Song Emeritus:** Carry Me Back to Old Virginia. **Tenth** of the original 13 states to ratify the Constitution, June 25, 1788. **State fair** at Richmond; late Sept.-early Oct.

History. Living in the area at the time of European contact were the Cherokee and Susquehanna and the Algonquians of the Powhatan Confederacy. English settlers founded Jamestown, 1607. Virginians took over much of the government from royal governor Dunmore, 1775, forcing him to flee. Virginians under George Rogers Clark freed the Ohio-Indiana-Illinois area of British forces. Benedict Arnold burned Richmond and Petersburg for the British, 1781. That same year, Britain's Cornwallis was trapped at Yorktown and surrendered, ending the American Revolution. Virginia seceded from the Union, 1861, and Richmond became the capital of the Confederacy. Hampton Roads, off the Virginia coast, was the site of the famous naval battle of the USS *Monitor* and CSS *Virginia* (Merrimac), 1862. Virginia was readmitted, 1870.

Tourist attractions. Colonial Williamsburg; Busch Gardens, Williamsburg; Wolf Trap Farm, near Falls Church; Arlington Natl. Cemetery; Mt. Vernon, home of George Washington; Jamestown Festival Park; Yorktown; Jefferson's Monticello, Charlottesville; Robert E. Lee's birthplace, Stratford Hall, and grave, Lexington; Appomattox; Shenandoah Natl. Park; Blue Ridge Parkway; Virginia Beach; Paramount's King's Dominion, near Richmond.

IT'S A FACT: Before joining the original 13 colonies to become the 14th state, Vermont was an independent republic from 1777 to 1791—complete with its own coinage and postal service.

Famous Virginians. Richard E. Byrd, James B. Cabell, Henry Clay, Jubal Early, Jerry Falwell, William Henry Harrison, Patrick Henry, A.P. Hill, Thomas Jefferson, Joseph E. Johnston, Robert E. Lee, Meriwether Lewis and William Clark, James Madison, John Marshall, George Mason, James Monroe, George Pickett, Pocahontas, Edgar Allan Poe, John Randolph, Walter Reed, John Smith, J.E.B. Stuart, William Styron, Zachary Taylor, John Tyler, Maggie Walker, Booker T. Washington, George Washington, L. Douglas Wilder, Woodrow Wilson.

Chamber of Commerce. 9 South Fifth St., Richmond, VA 23219.

Toll-free travel information. 1-800-321-3244.

Website. http://www.state.va.us

Tourism website. http://www.virginia.org

Washington

Evergreen State

People. Population (2000): 5,894,121; rank: 15; **net change** (1999-2000): 21.1%. **Pop. density**: 88.6 per sq mi. **Racial distribution** (2000): 81.8% white; 3.2% black; 5.5% Asian; 1.6% Native American/Nat. AK; 0.4% Hawaiian/Pacific Islander; 3.9% other race; 2 or more races, 3.6%. **Hispanic pop.** (any race): 7.5%.

Geography. Total area: 71,300 sq mi; rank: 18. **Land area:** 66,544 sq mi; rank: 20. **Acres forested:** 20,483,000. **Location:** Pacific state bordered by Canada on the N; Idaho on the E; Oregon on the S; and the Pacific Ocean on the W. **Climate:** mild, dominated by the Pacific Ocean and protected by the Cascades. **Topography:** Olympic Mts. on NW peninsula; open land along coast to Columbia R.; flat terrain of Puget Sound Lowland; Cascade Mts. region's high peaks to the E; Columbia Basin in central portion; highlands to the NE; mountains to the SE. **Capital:** Olympia.

Economy. Chief industries: advanced technology, aerospace, biotechnology, intl. trade, forestry, tourism, recycling, agriculture & food processing. **Chief manuf. goods:** computer software, aircraft, pulp & paper, lumber and plywood, aluminum, processed fruits and vegetables, machinery, electronics. **Chief crops:** apples, potatoes, hay, farm forest products. **Livestock:** (Jan. 2001) 1.2 mil cattle/calves; 54,000 sheep/lambs; (Dec. 2000) 27,000 hogs/pigs; 6.7 mil chickens (excl. broilers). **Timber/lumber** (2000): Douglas fir, hemlock, cedar, pine; 44.8 bil bd. ft. **Nonfuel minerals** (est. 2000): $691 mil; mostly sand & gravel, magnesium metal, crushed stone, portland cement, gold. **Commercial fishing** (1999): $98.5 mil. **Chief ports:** Seattle, Tacoma, Vancouver, Kelso-Longview. **Principal internat. airports at:** Seattle/Tacoma, Spokane, Boeing Field. **Value of construction** (1997): $8.5 bil. **Gross state product** (1999): $209.3 bil. **Employment distrib.** (May 2001): 29.2% services; 23.9% trade; 17.8% govt.; 12.3% mfg. **Per cap. pers. income** (2000): $31,528. **Sales tax** (2001): 6.5%. **Unemployment** (2000): 5.2%. **Tourism expends.** (1997): $7.7 bil. **Lottery** (2000): total sales: $452.8 mil; net income: $59.8. mil.

Finance. FDIC-insured commercial banks (2000): 79. **Deposits:** $12.3 bil. **FDIC-insured savings institutions** (2000): 23. **Assets:** $57.3 bil.

Federal govt. Fed. civ. employees (Mar. 2000): 41,775. **Avg. salary:** $49,125. **Notable fed. facilities:** Bonneville Power Admin.; Ft. Lewis; McChord AFB; Hanford Nuclear Reservation; Bremerton Naval Shipyards.

Energy. Electricity production (2000, kWh, by source): Coal: 3.3 bil; Petroleum: 229 mil; Gas: 2.5 bil; Hydroelectric: 79.9 bil; Nuclear: 8.6 bil.; **Other:** 362 mil.

State data. Motto: Alki (By and by). **Flower:** Western rhododendron. **Bird:** Willow goldfinch. **Tree:** Western hemlock. **Song:** Washington, My Home. **Entered union** Nov. 11, 1889; rank, 42d. **State fairs:** 5 area fairs, in Aug. and Sept.; no state fair.

History. At the time of European contact, many Native American tribes lived in the area, including the Nez Percé, Spokan, Yakima, Cayuse, Okanogan, Walla Walla, and Colville peoples, who lived in the interior region, and the Nooksak, Chinook, Nisqually, Clallam, Makah, Quinault, and Puyallup peoples, who inhabited the coastal area. Spain's Bruno Hezeta sailed the coast, 1775. In 1792, British naval officer George Vancouver mapped Puget Sound area, and that same year, American Capt. Robert Gray sailed up the Columbia River. Canadian fur traders set up Spokane House, 1810. Americans under John Jacob Astor established a post at Ft. Okanogan, 1811, and missionary Marcus Whitman settled near Walla Walla, 1836. Final agreement on the border of Washington and Canada was made with Britain, 1846, and Washington became part of the Oregon Territory, 1848. Gold was discovered, 1855.

Tourist attractions. Seattle Waterfront, Seattle Center and Space Needle, Museum of Flight, all Seattle; Mt. Rainier, Olympic, and North Cascades natl. parks; Mt. St. Helens; Puget Sound; San Juan Islands; Grand Coulee Dam; Columbia R. Gorge Natl. Scenic Area; Spokane's Riverfront Park.

Famous Washingtonians. Raymond Carver, Kurt Cobain, Bing Crosby, William O. Douglas, Bill Gates, Jimi Hendrix, Henry M. Jackson, Gary Larson, Mary McCarthy, Robert Motherwell, Edward R. Murrow, Theodore Roethke, Ann Rule, Hilary Swank, Julia Sweeney, Adam West, Marcus Whitman, Minoru Yamasaki.

Tourist information. WA State Tourism Division, PO Box 42500, Olympia, WA 98504-2500; 360-725-5052

Website. http://access.wa.gov

Tourism website. http://www.tourism.wa.gov

West Virginia

Mountain State

People. Population (2000): 1,808,344; rank: 37; **net change** (1999-2000): 0.8%. **Pop. density**: 75.1 per sq mi. **Racial distribution** (2000): 95.0% white; 3.2% black; 0.5% Asian; 0.2% Native American/Nat. AK; <0.1% Hawaiian/Pacific Islander; 0.2% other race; 2 or more races, 0.9%. **Hispanic pop.** (any race): 0.7%.

Geography. Total area: 24,230 sq mi; rank: 41. **Land area:** 24,078 sq mi; rank: 41. **Acres forested:** 12,128,000. **Location:** South Atlantic state bounded on the N by Ohio, Pennsylvania, Maryland; on the S and W by Virginia, Kentucky, Ohio; on the E by Maryland and Virginia. **Climate:** humid continental climate except for marine modification in the lower panhandle. **Topography:** ranging from hilly to mountainous; Allegheny Plateau in the W, covers two-thirds of the state; mountains here are the highest in the state, over 4,000 ft. **Capital:** Charleston.

Economy. Chief industries: manufacturing, services, mining, tourism. **Chief manuf. goods:** machinery, plastic & hardwood prods., fabricated metals, chemicals, aluminum, automotive parts, steel. **Chief crops:** apples, peaches, hay, tobacco, corn, wheat, oats. **Chief farm products:** dairy products, eggs. **Livestock:** (Jan. 2001) 400,000 cattle/calves; 35,000 sheep/lambs; (Dec. 2000) 10,000 hogs/pigs; 1.9 mil chickens (excl. broilers); 89.5 mil broilers. **Timber/lumber** (2000): oak, yellow poplar, hickory, walnut, cherry; 804 mil bd. ft. **Nonfuel minerals** (est. 2000): $182 mil; mostly crushed stone, portland cement, sand & gravel, lime, salt. **Chief port:** Huntington. **Value of construction** (1997): $1.2 bil. **Gross state product** (1999): $40.7 bil. **Employment distrib.** (May 2001): 31.1% services; 22.3% trade; 19.3% govt.; 10.6% mfg. **Per cap. pers. income** (2000): $21,915. **Sales tax** (2001): 6%. **Unemployment** (2000): 5.5%. **Tourism expends.** (1997): $1.6 bil. **Lottery** (2000): total sales: $448 mil; net income: $139.6 mil.

Finance. FDIC-insured commercial banks (2000): 70. **Deposits:** $13.8 bil. **FDIC-insured savings institutions** (2000): 7. **Assets:** $884 mil.

Federal govt. Fed. civ. employees (Mar. 2000): 11,755. **Avg. salary:** $45,770. **Notable fed. facilities:** National Radio Astronomy Observatory; Bureau of Public Debt Bldg.; Harpers Ferry Natl. Park; Correctional Institution for Women; FBI Identification Center.

Energy. Electricity production (2000, kWh, by source): Coal: 88.3 bil; Petroleum: 254 mil; Gas: 42 mil; Hydroelectric: 338 mil; **Other:** 14 mil.

State data. Motto: Montani Semper Liberi (Mountaineers are always free). **Flower:** Big rhododendron. **Bird:** Cardinal. **Tree:** Sugar maple. **Songs:** The West Virginia Hills; This Is My West Virginia; West Virginia, My Home, Sweet Home. **Entered union** June 20, 1863; rank, 35th. **State fair** at Lewisburg (Fairlea); late Aug.

History. Sparsely inhabited at the time of European contact, the area was primarily Native American hunting grounds. British explorers Thomas Batts and Robert Fallam reached the New River, 1671. Early American explorers included George Washington, 1753, and Daniel Boone. In the fall of 1774, frontiersmen defeated an allied Indian uprising at Point Pleasant. The area was part of Virginia and often objected to rule by the eastern part of the state. When Virginia seceded in 1861, the Wheeling Convention repudiated the act and created a new state, Kanawha, later renamed West Virginia. It was admitted to the Union 1863.

Tourist attractions. Harpers Ferry Natl. Historic Park; Science and Cultural Center, Charleston; White Sulphur (in

Greenbrier) and Berkeley Springs mineral water spas; New River Gorge, Fayetteville; Winter Place, Exhibition Coal Mine, both Beckley; Monongahela Natl. Forest; Fenton Glass, Williamstown; Viking Glass, New Martinsville; Blenko Glass, Milton; Sternwheel Regatta, Charleston; Mountain State Forest Festival; Snowshoe Ski Resort, Slaty Fork; Canaan State Park, Davis; Mountain State Arts & Crafts Fair, Ripley; Ogle Bay, Wheeling; White water rafting, several locations.

Famous West Virginians. Newton D. Baker, Pearl Buck, John W. Davis, Thomas "Stonewall" Jackson, Don Knotts, Dwight Whitney Morrow, Michael Owens, Mary Lou Retton, Walter Reuther, Cyrus Vance, Jerry West, Charles "Chuck" Yeager.

Tourist information. Dept. of Commerce, West Virginia Division of Tourism, State Capitol, Charleston WV 25305.

Toll-free travel information. 1-800-CALLWVA.

Website. http://www.state.wv.us

Tourism website. http://www.callwva.com

Wisconsin

Badger State

People. Population (2000): 5,363,675; rank: 18; **net change** (1999-2000): 9.6%. **Pop. density**: 98.8 per sq mi. **Racial distribution** (2000): 88.9% white; 5.7% black; 1.7% Asian; 0.9% Native American/Nat. AK; <0.1% Hawaiian/Pacific Islander; 1.6% other race; 2 or more races, 1.2%. **Hispanic pop.** (any race): 3.6%.

Geography. Total area: 65,498 sq mi; rank: 23. **Land area:** 54,310 sq mi; rank: 25. **Acres forested:** 15,513,000. **Location:** East North Central state, bounded on the N by Lake Superior and Upper Michigan; on the E by Lake Michigan; on the S by Illinois; on the W by the St. Croix and Mississippi rivers. **Climate:** long, cold winters and short, warm summers tempered by the Great Lakes. **Topography:** narrow Lake Superior Lowland plain met by Northern Highland, which slopes gently to the sandy crescent Central Plain; Western Upland in the SW; 3 broad parallel limestone ridges running N-S are separated by wide and shallow lowlands in the SE. **Capital:** Madison.

Economy. Chief industries: services, manufacturing, trade, government, agriculture, tourism. **Chief manuf. goods:** food products, motor vehicles & equip., paper products, medical instruments and supplies, printing, plastics. **Chief crops:** corn, hay, soybeans, potatoes, cranberries, sweet corn, peas, oats, snap beans. **Chief products:** milk, butter, cheese, canned and frozen vegetables. **Livestock:** (Jan. 2001) 3.4 mil cattle/calves; 80,000 sheep/lambs; (Dec. 2000) 620,000 hogs/pigs; 5.8 mil chickens (excl. broilers); 33.8 mil broilers. **Timber/lumber** (2000): maple, birch, oak, evergreens; 629 mil bd. ft. **Nonfuel minerals** (est. 2000): $349 mil; mostly crushed and dimension stone, sand & gravel, lime. **Commercial fishing** (1999): $4.2 mil. **Chief ports:** Superior, Ashland, Milwaukee, Green Bay, Kewaunee, Pt. Washington, Manitowoc, Sheboygan, Marinette, Kenosha. **Principal internat. airports at:** Green Bay, Milwaukee. **Value of construction** (1997): $6.1 bil. **Gross state product** (1999): $166 bil. **Employment distrib.** (May 2001): 27.5% services; 22.3% trade; 14.5% govt.; 21.1% mfg. **Per cap. pers. income** (2000): $28,232. **Sales tax** (2001): 5%. **Unemployment** (2000): 3.5%. **Tourism expends.** (1997): $5.8 bil. **Lottery** (2000): total sales: $406.7 mil; net income: $188.4 mil.

Finance. FDIC-insured commercial banks (2000): 315. **Deposits:** $58.8 bil. **FDIC-insured savings institutions** (2000): 39. **Assets:** $19.1 bil.

Federal govt. Fed. civ. employees (Mar. 2000): 10,967. **Avg. salary:** $45,252. **Notable fed. facilities:** Ft. McCoy.

Energy. Electricity production (2000, kWh, by source): Coal: 40.7 bil; Petroleum: 190 mil; Gas: 907 mil; Hydroelectric: 1.7 bil; Nuclear: 11.5 bil; **Other:** 263 mil.

State data. Motto: Forward. **Flower:** Wood violet. **Bird:** Robin. **Tree:** Sugar maple. **Song:** On, Wisconsin! **Entered union** May 29, 1848; rank, 30th. **State fair** at State Fair Park, West Allis; July-Aug.

History. At the time of European contact, Ojibwa, Menominee, Winnebago, Kickapoo, Sauk, Fox, and Potawatomi peoples inhabited the region. Jean Nicolet was the first European to see the Wisconsin area, arriving in Green Bay, 1634; French missionaries and fur traders followed. The British took over, 1763. The U.S. won the land after the American Revolution, but the British were not ousted until after the War of 1812. Lead miners came next, then farmers. In 1816, the U.S. government built a fort at Prairie du Chien on Wisconsin's border with Iowa. Native Americans in the area rebelled against the seizure of their tribal lands in the Black Hawk War of 1832, but treaties from 1829 to 1848, transferred all land titles in Wisconsin to the U.S. government. Railroads were started in 1851, serving growing wheat harvests and iron mines. Some 96,000 soldiers served the Union cause during the Civil War.

Tourist attractions. Old Wade House and Carriage Museum, Greenbush; Villa Louis, Prairie du Chien; Circus World Museum, Baraboo; Wisconsin Dells; Old World Wisconsin, Eagle; Door County peninsula; Chequamegon and Nicolet national forests; Lake Winnebago; House on the Rock, Dodgeville; Monona Terrace, Madison.

Famous Wisconsinites. Don Ameche, Carrie Chapman Catt, Willem Dafoe, Edna Ferber, King Camp Gillette, Harry Houdini, Robert La Follette, Alfred Lunt, Pat O'Brien, Georgia O'Keeffe, William H. Rehnquist, John Ringling, Donald K. "Deke" Slayton, Spencer Tracy, Thorstein Veblen, Orson Welles, Laura Ingalls Wilder, Thornton Wilder, Frank Lloyd Wright.

Tourist information. Wisconsin Dept. of Tourism, 201 W. Washington Ave., PO Box 7976, Madison, WI 53707-7976.

Toll-free travel information. 1-800-432-TRIP.

Website. http://www.wisconsin.gov

Tourism website. http://www.travelwisconsin.com

Wyoming

Equality State, Cowboy State

People. Population (2000): 493,782; rank: 50; **net change** (1999-2000): 8.9%. **Pop. density**: 5.1 per sq mi. **Racial distribution** (2000): 92.1% white; 0.8% black; 0.6% Asian; 2.3% Native American/Nat. AK; 0.1% Hawaiian/Pacific Islander; 2.5% other race; 2 or more races, 1.8%. **Hispanic pop.** (any race): 6.4%.

Geography. Total area: 97,814 sq mi; rank: 10. **Land area:** 97,100 sq mi; rank: 9. **Acres forested:** 9,966,000. **Location:** Mountain state lying in the high western plateaus of the Great Plains. **Climate:** semi-desert conditions throughout; true desert in the Big Horn and Great Divide basins. **Topography:** the eastern Great Plains rise to the foothills of the Rocky Mts.; the Continental Divide crosses the state from the NW to the SE. **Capital:** Cheyenne.

Economy. Chief industries: mineral extraction, oil, natural gas, tourism and recreation, agriculture. **Chief manuf. goods:** refined petroleum, wood, stone, clay products, foods, electronic devices, sporting apparel, and aircraft. **Chief crops:** wheat, beans, barley, oats, sugar beets, hay. **Livestock:** (Jan. 2001) 1.6 mil cattle/calves; 530,000 sheep/lambs; (Dec. 2000) 108,000 hogs/pigs; 17,000 chickens (excl. broilers). **Timber/lumber** (2000): ponderosa & lodgepole pine, Douglas fir, Engelmann spruce; 172 mil bd. ft. **Nonfuel minerals** (est. 2000): $922 mil; mostly soda ash, clays, helium, portland cement, crushed stone. **Principal internat. airport at:** Casper. **Value of construction** (1997): $655 mil. **Gross state product** (1999): $17.5 bil. **Employment distrib.** (May 2001): 23.1% services; 23.1% trade; 24.7% govt.; 4.6% mfg. **Per cap. pers. income** (2000): $27,230. **Sales tax** (2001): 4%. **Unemployment** (2000): 3.9%. **Tourism expends.** (1997): $1.4 bil.

Finance. FDIC-insured commercial banks (2000): 46. **Deposits:** $5.7 bil. **FDIC-insured savings institutions** (2000): 4. **Assets:** $387 mil.

Federal govt. Fed. civ. employees (Mar. 2000): 4,354. **Avg. salary:** $43,757. **Notable fed. facilities:** Warren AFB.

Energy. Electricity production (2000, kWh, by source): Coal: 43 bil; Petroleum: 35 mil; Gas: 204 mil; Hydroelectric: 1 bil.

State data. Motto: Equal Rights. **Flower:** Indian Paintbrush. **Bird:** Western Meadowlark. **Tree:** Plains Cottonwood. **Song:** Wyoming. **Entered union** July 10, 1890; rank, 44th. **State fair** at Douglas; late Aug.

History. Shoshone, Crow, Cheyenne, Oglala Sioux, and Arapaho peoples lived in the area at the time of European contact. France's François and Louis La Verendrye were the first Europeans to see the region, 1743. John Colter, an American, was first to traverse Yellowstone area, 1807-8. Trappers and fur traders followed in the 1820s. Forts Laramie and Bridger became important stops on the pioneer trails to the West Coast. Population grew after the Union Pacific crossed the state, 1868. Women won the vote, for the first time in the U.S., from the Territorial Legislature, 1869. Disputes between large land owners and small ranchers culminated in the Johnson County Cattle War, 1892; federal troops were called in to restore order.

Tourist attractions. Yellowstone Natl. Park, the first U.S. national park, est. 1872; Grand Teton Natl. Park; Natl. Elk Refuge; Devils Tower Natl. Monument; Fort Laramie Natl. Historic Site and nearby pioneer trail ruts; Buffalo Bill Historical Center, Cody; Cheyenne Frontier Days, Cheyenne.

Famous Wyomingites. James Bridger, William F. "Buffalo Bill" Cody, Curt Gowdy, Esther Hobart Morris, Nellie Tayloe Ross.

Tourist information. Division of Tourism & State Marketing, I-25 at College Dr., Cheyenne, WY 82002.

Toll-free travel information. 1-800-CALLWYO.

Website. http://www.state.wy.us

Tourism website. http://www.wyomingtourism.org

District of Columbia

People. Population (2000): 572,059; **net change** (1999-2000): -5.7%. **Pop. density**: 9,378 per sq mi. **Racial distribution** (2000): 30.8% white; 60.0% black; 2.7% Asian; 0.3% Native American/Nat. AK; 0.1% Hawaiian/Pacific Islander; 3.8% other race; 2 or more races, NA. **Hispanic pop.** (any race): 7.9%.

Geography. Total area: 68 sq mi; rank: 50. **Land area:** 61 sq mi; rank: 51. **Location:** at the confluence of the Potomac and Anacostia rivers, flanked by Maryland on the N, E, and SE and by Virginia on the SW. **Climate:** hot humid summers, mild winters. **Topography:** low hills rise toward the N away from the Potomac R. and slope to the S; highest elevation, 410 ft, lowest Potomac R., 1 ft.

Economy. Chief industries: government, service, tourism. **Value of construction** (1997): $673 mil. **Gross state product** (1999): $55.8 bil. **Employment distrib.** (May 2001): 47.1% services; 7.5% trade; 34.0% govt.; 1.8% mfg. **Per cap. pers. income** (2000): $37,383. **Sales tax** (2001): 5.75%. **Unemployment** (2000): 5.8%. **Tourism expenditures** (1997): $5.4 bil. **Lottery** (2000) total sales: $215.8 mil; net income: $69.5 mil.

Finance. FDIC-insured commercial banks & trust companies (2000): 6. **Deposits:** $614 mil. **FDIC-insured savings institutions** (2000): 1. **Assets:** $255 mil.

Federal govt. No. of federal employees (Mar. 2000): 143,751. **Avg. salary:** $65,176.

Energy. Electricity production (2000, kWh, by source): Petroleum: 95 mil; **Other:** 28 mil.

District data. Motto: Justitia omnibus (Justice for all). **Flower:** American beauty rose. **Tree:** Scarlet oak. **Bird:** Wood thrush.

History. The District of Columbia, coextensive with the city of Washington, is the seat of the U.S. federal government. It lies on the west central edge of Maryland on the Potomac River, opposite Virginia. Its area was originally 100 sq mi taken from the sovereignty of Maryland and Virginia. Virginia's portion south of the Potomac was given back to that state in 1846.

The 23d Amendment (1961) granted residents the right to vote for president and vice president for the first time since 1800 and gave them 3 members in the Electoral College. The first such votes were cast in Nov. 1964.

Congress, which has legislative authority over the District under the Constitution, established in 1874 a government of 3 commissioners appointed by the president. The Reorganization Plan of 1967 substituted a single appointive commissioner (also called mayor), assistant, and 9-member City Council. Funds were still appropriated by Congress; residents had no vote in local government, except to elect school board members. In Sept. 1970, Congress approved legislation giving the District one delegate to the House of Representatives, who can vote in committee but not on the floor. The first delegate was elected 1971.

In May 1974, voters approved a congressionally drafted charter giving them the right to elect their own mayor and a 13-member city council; the first took office Jan. 2, 1975. The district won the right to levy taxes; Congress retained power to veto council actions and approve the city budget.

Proposals for a "federal town" for the deliberations of the Continental Congress were made in 1783, 4 years before the adoption of the Constitution. Rivalry between Northern and Southern delegates over the site appeared in the First Congress, 1789. John Adams, presiding officer of the Senate, cast the deciding vote of that body for Germantown, PA. In 1790 Congress compromised by making Philadelphia the temporary capital for 10 years. The Virginia members of the House wanted a permanent capital on the eastern bank of the Potomac, while the Southerners opposed having the nation assume the war debts of the 13 original states as provided under the Assumption Bill, fathered by Alexander Hamilton. Hamilton and Jefferson arranged a compromise: the Virginia men voted for the Assumption Bill, and the Northerners conceded the capital to the Potomac. Pres. Washington chose the site in Oct. 1790 and persuaded landowners to sell their holdings to the government. The capital was named Washington.

Washington appointed Pierre Charles L'Enfant, a Frenchman, to plan the capital on an area not more than 10 mi square. The L'Enfant plan, for streets 100 to 110 ft. wide and one avenue 400 ft. wide and a mile long, seemed grandiose and foolhardy, but Washington endorsed it. When L'Enfant ordered a wealthy landowner to remove his new manor house because it obstructed a vista, and demolished it when the owner refused, Washington stepped in and dismissed the architect. Andrew Ellicott, who was working on surveying the area, finished the official map and design of the city. Ellicott was assisted by Benjamin Banneker, a distinguished black architect and astronomer.

On Sept. 18, 1793, Pres. Washington laid the cornerstone of the north wing of the Capitol. On June 3, 1800, Pres. John Adams moved to Washington, and on June 10, Philadelphia ceased to be the temporary capital. The City of Washington was incorporated in 1802; the District of Columbia was created as a municipal corporation in 1874, embracing Washington, Georgetown, and Washington County.

Tourist attractions: *See* Washington, DC, Capital of the U.S.

Tourist information. Washington, DC Convention and Visitors Association, 1212 New York Ave. NW, #600, Washington, DC 20005; phone: 202-789-7000.

Website. http://dcpages.ari.net

Tourism website. http://www.washington.org

OUTLYING U.S. AREAS

American Samoa

People. Population (2000 est.): 65,446. **Population growth rate** (2000 est.): 2.5%. **Pop. density (2000): 849.9 per sq mi. Major ethnic group:** Samoan (Polynesian), Caucasian, Tongan. **Languages:** Samoan, English.

Land area: 77 sq. mi. **Total area:** 90 sq mi. **Capital:** Pago Pago, Island of Tutuila. **Motto:** Samoa Muamua le Atua (In Samoa, God Is First). **Song:** Amerika Samoa. **Flower:** Paogo (Ula-fala). **Plant:** Ava.

Public education. Student-teacher ratio (1995): 20.

Boasting spectacular scenery and delightful South Seas climate, American Samoa is the most southerly of all lands under U.S. sovereignty. It is an unincorporated territory consisting of 7 small islands of the Samoan group: **Tutuila, Aunu'u, Manu'a Group (Ta'u, Olosega, Ofu), Rose,** and **Swains Island.** The islands are 2,300 mi SW of Honolulu.

Economy. Chief industries: tuna processing, trade, services, tourism. **Chief crops:** vegetables, nuts, melons and other fruits. **Livestock** (1999): 140 cattle; 35,301 hogs/pigs; 60,463 chickens. **Commercial fishing** (2000): $2 mil. **Principal airport at:** Pago Pago.

Finance. FDIC-insured commercial banks (2000): 1. **Deposits:** $61 mil.

Energy. Electricity production (1998): 125 bil kWh.

A tripartite agreement between Great Britain, Germany, and the U.S. in 1899 gave the U.S. sovereignty over the eastern islands of the Samoan group; these islands became American Samoa. Local chiefs ceded Tutuila and Aunu'u to the U.S. in 1900, and the Manu'a group and Rose in 1904; Swains Island was annexed in 1925. Samoa (Western), comprising the larger islands of the Samoan group, was a New Zealand mandate and UN Trusteeship until it became independent Jan. 1, 1962 (now called Samoa).

Tutuila and Aunu'u have an area of 53 sq mi. Ta'u has an area of 17 sq mi, and the islets of Ofu and Olosega, 5 sq mi with a population of a few thousand. Swains Island has nearly 2 sq mi and a population of about 100.

About 70% of the land is bush and mountains. Chief exports are fish products. Taro, breadfruit, yams, coconuts, pineapples, oranges, and bananas are also produced.

From 1900 to 1951, American Samoa was under the jurisdiction of the U.S. Navy. Since 1951, it has been under the Interior Dept. On Jan. 3, 1978, the first popularly elected Samoan governor and lieutenant governor were inaugurated. Previously, the governor was appointed by the Secretary of the Interior. American Samoa has a bicameral legislature and elects a delegate to the House of Representatives, with no vote except in committees.

The American Samoans are of Polynesian origin. They are nationals of the U.S.; approximately 20,000 live in Hawaii, 65,000 in California and Washington.

Website. http://www.samoanet.com

Tourism website: http://www.amsamoa.com

Guam

Where America's Day Begins

People. Population (2000 est.): 154,623. **Population growth rate** (2000 est.): 1.7%. **Pop. density** (2000 est.): 736.3 per sq mi. **Major ethnic groups** Chamorro, Filipino, Caucasian, Chinese, Japanese, Korean. (Native Guamanians, ethnically Chamorros, are basically of Indonesian stock, with a mixture of Spanish and Filipino; in addition to the official language, they speak the native Chamorro). **Languages:** English, Chamorro, Japanese. **Migration** (1990): About 52% of population were born elsewhere; of these, 48% in Asia, 40% in U.S.

Geography. Total area: 217 sq mi. **Land area:** 210 sq. mi. **Location:** largest and southernmost of the Mariana Islands in the West Pacific, 3,700 mi W of Hawaii. **Climate:** tropical, with temperatures from 70° to 90° F; avg. annual rainfall, about 70 in. **Topography:** coralline limestone plateau in the N; southern chain of low volcanic mountains sloping gently to the W, more steeply to coastal cliffs on the E; general elevation, 500 ft; highest point, Mt. Lamlam, 1,334 ft. **Capital:** Hagatna.

Economy. Chief industries: tourism, U.S. military, construction, banking, printing & publishing. **Chief manuf. goods:** textiles, foods. **Chief crops:** cabbages, eggplants, cucumber, long beans, tomatoes, bananas, coconuts, watermelon, yams, cantaloupe, papayas, maize, sweet potatoes. **Livestock** (1998): NA cattle; 2,287 hogs/pigs; 11,540 chickens. **Commercial fishing** (2000): $1.3 mil. **Chief port:** Apra Harbor. **Principal internat. airport at:** Hagatna. **Construction sales** (1997): $506 mil. **Employment distrib.** (1995): 31% govt.; 21% trade; 33% serv. **Per capita income** (1996 est.): $19,000. **Unemployment** (1994): 6.7%. **Tourism expends.** (1995): $4.9 bil.

Finance. FDIC-insured commercial banks (2000): 2. **Deposits:** $656 mil. **FDIC-insured savings institutions** (2000): 2. **Assets:** $280 mil.

Federal govt. Federal employees (1990): 7,200. **Notable fed. facilities:** Anderson AFB; naval, air, and port bases.

Public education. Student-teacher ratio (1995): 18.3.

Misc. data. Flower: Puti Tai Nobio (Bougainvillea). **Bird:** Toto (Fruit dove). **Tree:** Ifit (Intsiabijuga). **Song:** Stand Ye Guamanians.

History. Guam was probably settled by voyagers from the Indonesian-Philippine archipelago by 3d cent. BC. Pottery, rice cultivation, and megalithic technology show strong East Asian cultural influence. Centralized, village clan-based communities engaged in agriculture and offshore fishing. The estimated population by the early 16th cent. was 50,000-75,000. Magellan arrived in the Marianas Mar. 6, 1521. They were colonized in 1668 by Spanish missionaries, who named them the Mariana Islands in honor of Maria Anna, queen of Spain. When Spain ceded Guam to the U.S., it sold the other Marianas to Germany. Japan obtained a League of Nations mandate over the German islands in 1919; in Dec. 1941 it seized Guam, which was retaken by the U.S. in July-August 1944.

Guam is a self-governing organized unincorporated U.S. territory. The Organic Act of 1950 provided for a governor, elected to a 4-year term, and a 21-member unicameral legislature, elected biennially by the residents, who are American citizens. In 1970, the first governor was elected. In 1972, a U.S. law gave Guam one delegate to the U.S. House of Representatives who has a voice but no vote, except in committees.

Guam's quest to change its status to a U.S. Commonwealth began in the late 1970s. The Guam Commission on Self-Determination, created in 1984, developed a draft Commonwealth Act. In 1993, legislation proposing a change of status was submitted to the U.S. Congress. In 1994, the U.S. Congress passed legislation transferring 3,200 acres of land on Guam from federal to local control.

Tourist attractions. Tropical climate, oceanic marine environment; annual mid-Aug. Merizo Water Festival; Tarzan Falls; beaches; water sports; duty-free port shopping.

Website. http://www.gov.gu

Tourism website. http://www.visitguam.org

Commonwealth of the Northern Mariana Islands

People. Population (2000 est.): 71,912.**Population growth rate** (2000 est.): 3.8%. **Pop. density** (2000): 401.7 per sq mi. **Major ethnic Groups:** Chamorro, Carolinians and other Micronesians, Caucasian, Japanese, Chinese, Korean. **Languages:** English, Chamorro, Carolinian.

Total area: 189 sq. mi. **Land area:** 179 sq. mi. Located in the perpetually warm climes between Guam and the Tropic of Cancer, the 14 islands of the Northern Marianas form a 300-mi. long archipelago. The indigenous population in 1990 was concentrated on the 3 largest of the 6 inhabited islands: **Saipan,** the seat of government and commerce (38,896), **Rota** (2,295), and **Tinian** (2,118).

Economy. Chief industries: trade, services, and tourism. **Chief manuf. goods:** apparel, stone, clay and glass products. **Chief crops:** melons, vegetables, horticulture, fruits and nuts. **Livestock**: (1998): 1,789 cattle; 831 hogs/pigs; 29,409 chickens. **Commercial fishing (2000): $938,365 Construction sales (1997): $88 mil. Employment distrib.** (1992): 53% trade; 33% serv.; 8% const.; 6% manuf.

Education. Pupil-teacher ratio (1995): 20.9.

The people of the Northern Marianas are predominantly of Chamorro cultural extraction, although Carolinians and immigrants from other areas of E. Asia and Micronesia have also settled in the islands. English is among the several languages commonly spoken. Pursuant to the Covenant of 1976, which established the Northern Marianas as a commonwealth in political union with the U.S., most of the indigenous population and many domiciliaries of these islands achieved U.S. citizenship on Nov. 3, 1986, when the U.S. terminated its administration of the UN trusteeship as it affected the Northern Marianas. From July 18, 1947, the U.S. had administered the Northern Marianas under a trusteeship agreement with the UN Security Council.

The Northern Mariana Islands has been self-governing since 1978, when a constitution drafted and adopted by the people became effective and a popularly elected bicameral legislature (2-year term), with offices of governor (4-year term) and lieut. governor, was inaugurated.

Tourism website: http://www.visitmarianas.com

Commonwealth of Puerto Rico

(Estado Libre Asociado de Puerto Rico)

People. Population (2000): 3,808,610 (about 2.7 mil more Puerto Ricans reside in the mainland U.S.); **Population growth rate:** (2000 est.) .56%; **net change** (1999-2000): 8.1%. **Pop. density**: 1,112.0 per sq mi. **Urban** (1990): 66.8%. **Racial distribution** (2000): 80.5% white; 8% black; 0.2% Asian; 0.4% Native American/Nat. AK; 3.8% Other. **Hispanic pop.** (any race): 98.8%. **Languages:** Spanish and English are joint official languages.

Geography. Total area: 5,324 sq. mi. **Land area:** 3,425 sq mi. **Location:** island lying between the Atlantic to the N and the Caribbean to the S; it is easternmost of the West Indies group called the Greater Antilles, of which Cuba, Hispaniola, and Jamaica are the larger islands. **Climate:** mild, with a mean temperature of 77° F. **Topography:** mountainous throughout three-fourths of its rectangular area, surrounded by a broken coastal plain; highest peak, Cerro de Punto, 4,390 ft. **Capital:** San Juan.

Economy. Chief industries: manufacturing, service. **Chief manuf. goods:** pharmaceuticals, apparel, electronics & other electric equipment, industrial machinery. **Gross domestic product:** (1999 est.) $38.1 bil. **Chief crops:** coffee, plantains, pineapples, tomatoes, sugarcane, bananas, mangos, ornamental plants. **Livestock** (1998): 386,980 cattle; 101,619 hogs/pigs; 12.6 mil. chickens. **Nonfuel minerals** (1996): $31.1 mil, mostly portland cement, crushed stone. **Commercial fishing** (2000): $6.4 mil. **Chief ports/river shipping:** San Juan, Ponce, Mayagüez. **Principal airports at:** San Juan, Ponce, Mayagüez, Aguadilla. **Construction sales** (1997): $4 bil. **Employment distrib.** (2000): 28.2% govt.; 21.5% trade; 21.3% serv.; 13.9% mfg. **Per capita in-**

come (1999 est.): $9,800. **Unemployment** (2000): 10.1%. **Tourism expends.** (1999): $2.1 mil.

Finance. FDIC-insured commercial banks (2000): 12. **Deposits:** $29.7 bil.

Federal govt. Fed. civ. employees (1997): 13,874. **Notable fed. facilities:** U.S. Naval Station at Roosevelt Roads; P.R. National Guard Training Area at Camp Santiago, and at Ft. Allen, Juana Diaz; Sabana SECA Communications Center (U.S. Navy); U.S. Army Station at Ft. Buchanan.

Energy. Electricity production (1998): 17.8 bil kWh.

Public education. Student-teacher ratio (1995): 16. **Min. teachers' salary** (1997): $1,500 monthly.

Misc. data. Motto: Joannes Est Nomen Eius (John is his name). **Flower:** Maga. **Bird:** Reinita. **Tree:** Ceiba. **National anthem:** La Borinqueña.

History. Puerto Rico (or Borinquen, after the original Arawak Indian name, Boriquen) was visited by Columbus on his second voyage, Nov. 19, 1493. In 1508, the Spanish arrived.

Sugarcane was introduced, 1515, and slaves were imported 3 years later. Gold mining petered out, 1570. Spaniards fought off a series of British and Dutch attacks; slavery was abolished, 1873. Under the treaty of Paris, Puerto Rico was ceded to the U.S. after the Spanish-American War, 1898. In 1952 the people voted in favor of Commonwealth status.

The Commonwealth of Puerto Rico is a self-governing part of the U.S. with a primarily Hispanic culture. The island's citizens have virtually the same control over their internal affairs as do the 50 states of the U.S. However, they do not vote in national general elections, only in national primaries.

Puerto Rico is represented in the U.S. House of Representatives by a delegate who has a voice but no vote, except in committees.

No federal income tax is collected from residents on income earned from local sources in Puerto Rico. Nevertheless, as part of the U.S. legal system, Puerto Rico is subject to the provisions of the U.S. Constitution; most federal laws apply as they do in the 50 states.

Puerto Rico's famous "Operation Bootstrap," begun in the late 1940s, succeeded in changing the island from "The Poorhouse of the Caribbean" to an area with the highest per capita income in Latin America. This program encouraged manufacturing and development of the tourist trade by selective tax exemption, low-interest loans, and other incentives. Despite the marked success of Puerto Rico's development efforts over an extended period of time, per capita income in Puerto Rico is low in comparison to that of the U.S.

Tourist attractions. Ponce Museum of Art; Forts El Morro and San Cristobal; Old Walled City of San Juan; Arecibo Observatory; Cordillera Central and state parks; El Yunque Rain Forest; San Juan Cathedral; Porta Coeli Chapel and Museum of Religious Art, Interamerican Univ., San Germán; Condado Convention Center; Casa Blanca, Ponce de León family home, Puerto Rican Family Museum of 16th and 17th centuries, and Fine Arts Center all in San Juan.

Cultural facilities and events. Festival Casals classical music concerts, mid-June; Puerto Rico Symphony Orchestra at Music Conservatory; Botanical Garden and Museum of Anthropology, Art, and History at the University of Puerto Rico; Institute of Puerto Rican Culture, at the Dominican Convent; and many popular festivals.

Famous Puerto Ricans. Julia de Burgos, Marta Casals Istomin, Pablo Casals, José Celso Barbosa, Orlando Cepeda, Roberto Clemente, José de Diego, José Feliciano, Doña Felisa Rincón de Gautier, Luis A. Ferré, José Ferrer, Commodore Diégo E. Hernández, Miguel Hernández Agosto, Rafael Hernández (El Jibarito), Rafael Hernández Colón, Raúl Julía, René Marqués, Ricky Martin, Concha Meléndez, Rita Moreno, Luis Muñoz Marín, Luis Palés Matos, Adm. Horacio Rivero.

Chamber of Commerce. 100 Tetuán, PO Box S-3789, San Juan, PR 00902.

Website. http://fortaleza.govpr.org

Tourism website. http://www.prtourism.com

Virgin Islands

St. John, St. Croix, St. Thomas

People. Population (2000 est.): 120,917. **Population growth rate** (2000): 1.1%. **Pop. density (2000 est.): 902.4 per sq mi. Major ethnic groups:** West Indian, French, Hispanic. **Languages:** English (official), Spanish, Creole.

Geography. Total area: 171 sq mi. **Land area:** 134 sq mi. **Location:** 3 larger and 50 smaller islands and cays in the S and W of the V.I. group (British V.I. colony to the N and E), which is situated 70 mi E of Puerto Rico, located W of the Anegada Passage, a major channel connecting the Atlantic Ocean and the Caribbean Sea. **Climate:** subtropical; the sun tempered by gentle trade winds; humidity is low; average temperature, 78° F. **Topography:** St. Thomas is mainly a ridge of hills running E and W, and has little tillable land; St. Croix rises abruptly in the N but slopes to the S to flatlands and lagoons; St. John has steep, lofty hills and valleys with little level tillable land. **Capital:** Charlotte Amalie, St. Thomas.

Economy. Chief industries: tourism, rum, alumina, petroleum refining, watches, textiles, electronics, printing & publishing. **Chief manuf. goods:** rum, textiles, pharmaceuticals, perfumes, stone, glass & clay products. **Chief crops:** vegetables, horticulture, fruits and nuts. **Livestock** (1998): 6,636 cattle; 1,436 hogs/pigs; 3,538 chickens. **Minerals:** sand, gravel. **Chief ports:** Cruz Bay, St. John; Frederiksted and Christiansted, St. Croix; Charlotte Amalie, St. Thomas. **Principal internat. airports on:** St. Thomas, St. Croix. **Construction sales** (1997): $185 mil. **Employment distrib.** (1992): 50% trade; 43% serv. **Per capita income** (1999 est.): $15,000. **Unemployment** (1999): 4.9%. **Tourism expends.** (1995): $792 mil.

Finance. FDIC-insured commercial banks (2000): 2. **Deposits:** $113 mil.

Energy. Electricity production (1998): 1bil kWh.

Public education. Student-teacher ratio (1995): 14.

Misc. data. Flower: Yellow elder or yellow trumpet, local designation Ginger Thomas. **Bird:** Yellow breast. **Song:** Virgin Islands March.

History. The islands were visited by Columbus in 1493. Spanish forces, 1555, defeated the Caribes and claimed the territory; by 1596 the native population was annihilated. First permanent settlement in the U.S. territory, 1672, by the Danes; U.S. purchased the islands, 1917, for defense purposes.

The Virgin Islands has a republican form of government, headed by a governor and lieut. governor elected, since 1970, by popular vote for 4-year terms. There is a 15-member unicameral legislature, elected by popular vote for a 2-year term. Residents of the V.I. have been U.S. citizens since 1927. Since 1973 they have elected a delegate to the U.S. House of Representatives, who has a voice but no vote, except in committees.

Tourist attractions. Magens Bay, St. Thomas; duty-free shopping; Virgin Islands Natl. Park, beaches, Indian relics, and evidence of colonial Danes.

Tourist information. Dept. of Economic Development & Agriculture: St. Thomas, PO Box 6400, St. Thomas, VI 00801; St. Croix, PO Box 4535, Christiansted, St. Croix 00820.

Website. http://www.usvi.net

Other Islands

Navassa lies between Jamaica and Haiti, 100 mi south of Guantanamo Bay, Cuba, in the Caribbean; it covers about 2 sq mi, is reserved by the U.S. for a lighthouse, and is uninhabited. It is administered by the U.S. Coast Guard.

Wake Atoll, and its neighboring atolls, **Wilkes** and **Peale,** lie in the Pacific Ocean on the direct route from Hawaii to Hong Kong, about 2,300 mi W of Honolulu and 1,290 mi E of Guam. The group is 4.5 mi long, 1.5 mi wide, and totals less than 3 sq mi in land area. The U.S. flag was hoisted over Wake Atoll, July 4, 1898; formal possession taken Jan. 17, 1899. Wake was administered by the U.S. Air Force, 1972-94. The population consists of about 200 persons.

Midway Atoll, acquired in 1867, consists of 2 atolls, **Sand** and **Eastern,** in N Pacific 1,150 mi. NW of Honolulu, with an area of about 2 sq mi, administered by the U.S. Navy. There is no indigenous population; total pop. is about 450. **Johnston Atoll,** 717 mi WSW of Honolulu, area 1 sq mi, is operated by the Defense Nuclear Agency, and the Fish and Wildlife Service, U.S. Dept. of the Interior; its population is about 1,200. **Kingman Reef,** 920 mi S of Hawaii, is under Navy control. **Howland, Jarvis,** and **Baker Islands,** 1,400-1,650 mi SW of Honolulu, uninhabited since World War II, are under the Interior Dept. **Palmyra** is an atoll about 1,000 mi S of Hawaii, 5 sq mi. Privately owned, it is under the Interior Dept.

WASHINGTON, DC, CAPITAL OF THE U.S.

Most attractions are free. All times are subject to change. For more details call the Washington, DC, Convention and Visitors Association at 202-789-7000, or check out the website at: http://www.washington.org

Bureau of Engraving and Printing

The **Bureau of Engraving and Printing** of the U.S. Treasury Dept. is the headquarters for the making of U.S. paper money. Free 35-minute self-guided tours (tickets required) Mon.-Fri., 9 AM-2 PM year-round; extended hours, June-Aug., 5 PM-6:40 PM. Closed federal holidays. 14th and C Sts. SW. Phone: 202-874-3019.

Website. http://www.moneyfactory.com

Capitol

The **United States Capitol** was originally designed by Dr. William Thornton, an amateur architect, who submitted a plan in 1793 that won him $500 and a city lot.

The south, or House, wing was completed in 1807 under the direction of Benjamin H. Latrobe.

The present Senate and House wings and the iron dome were designed and constructed by Thomas U. Walter, 4th architect of the Capitol, between 1851 and 1863.

The present cast iron dome at its greatest exterior measures 135 ft 5 in., and it is topped by the bronze Statue of Freedom that stands 19$^{1}/_{2}$ ft and weighs 14,985 lb. On its base are the words *E Pluribus Unum* (Out of Many, One).

The Capitol is open from 9 AM to 8 PM, March-Aug., and 9 AM to 4:30 PM, Sept.-Feb., daily. It is closed Jan. 1, Thanksgiving Day, and Dec. 25. Tours through the Capitol, including the House and Senate galleries, are conducted Mon.-Sat.

To observe debate in the House or Senate while Congress is in session, individuals living in the U.S. may obtain tickets to the visitor's galleries from their U.S. representative or senator. Visitors from other countries may obtain passes at the Capitol. Between Constitution & Independence Ave., at Pennsylvania Ave. Phone: 202-225-6827.

Website. http://www.aoc.gov

Federal Bureau of Investigation

The **Federal Bureau of Investigation** offers guided one-hour tours of its headquarters, beginning with a videotape presentation. Visitors learn about the history of the FBI and see such things as the weapons confiscated from famous gangsters, photos of the most-wanted fugitives, the DNA laboratory, goods forfeited and seized in narcotics operations, and a sharpshooting demonstration.

Tours are conducted Mon.-Fri., 8:45 AM-4:15 PM, except Jan. 1, Dec. 25, and other federal holidays. Tickets may be obtained at the FBI on day of tour or through a U.S. representative or senator. J. Edgar Hoover Bldg., Pennsylvania Ave., between 9th and 10th Sts. NW. Phone: 202-324-3000.

Website. http://www.fbi.gov

Folger Shakespeare Library

The **Folger Shakespeare Library,** on Capitol Hill, is a research institution holding rare books and manuscripts of the Renaissance period and the largest collection of Shakespearean materials in the world, including 79 copies of the First Folio. The library's museum and performing arts programs are presented in the Elizabethan Theatre, which resembles an innyard theater of Shakespeare's day.

Exhibit may be visited Mon.-Sat., 10 AM-4 PM., 201 E. Capitol St., SE , Phone: 202-544-4600.

Website. http://www.folger.edu

Holocaust Memorial Museum

The **U.S. Holocaust Memorial Museum** opened on Apr. 21, 1993. The museum documents, through permanent and temporary displays, interactive videos, and special lectures, the events of the Holocaust beginning in 1933 and continuing World War II. The permanent exhibition is not recommended for children under the age of 11.

The museum is open daily, 10 AM-5:30 PM, except Yom Kippur and Dec. 25, and extended hours (8 AM-10 PM) Apr. 3-Sept. 2. A limited number of free tickets are available on day of visit; advance tickets may be ordered for a small fee. 100 Raoul Wallenberg Pl. SW. Phone: 202-488-0400.

Website. http://www.ushmm.org/

Jefferson Memorial

Dedicated in 1943, the **Thomas Jefferson Memorial** stands on the south shore of the Tidal Basin in West Potomac Park. It is a circular stone structure, with Vermont marble on the exterior and Georgia white marble inside, and combines architectural elements of the dome of the Pantheon in Rome and the rotunda designed by Jefferson for the University of Virginia.

The memorial, on the south edge of the Tidal Basin, is open daily, 8 AM-midnight. An elevator and curb ramps for the handicapped are in service. Phone: 202-426-6841.

Website. http://www.nps.gov/jefm/index2.htm

John F. Kennedy Center

The **John F. Kennedy Center for the Performing Arts,** designated by Congress as the National Cultural Center and the official memorial in Washington, DC, to Pres. John F. Kennedy, opened Sept. 8, 1971. Designed by Edward Durell Stone, the center includes an opera house, a concert hall, several theaters, 2 restaurants, and a library.

Free tours are available daily, 10 AM-1 PM. 2700 F St. NW. Phone: 202-416-8340, or 1-800-444-1324.

Website. http://www.kennedy-center.org

Korean War Veterans Memorial

Dedicated on July 27, 1995, the **Korean War Veterans Memorial** honors all Americans who served in the Korean War. Situated at the west end of the Mall, across the reflecting pool from the Vietnam Memorial, the triangular-shaped stone and steel memorial features a multiservice formation of 19 troops clad in ponchos with the wind at their back, ready for combat. A granite wall, with images of the men and women who served, juts into a pool of water, the Pool of Remembrance, and is inscribed with the words *Freedom Is Not Free*.

The $18 mil memorial, which was funded by private donations, is open 8 AM-midnight. Independence Ave. at Lincoln Memorial. Phone: 202-619-7222.

Website. http://www.nps.gov/kwvm/index.htm

Library of Congress

Established by and for Congress in 1800, the **Library of Congress** has extended its services over the years to other government agencies and other libraries, to scholars, and to the general public, and it now serves as the national library. It contains more than 80 million items in 470 languages.

The library's exhibit halls are open to the public Mon.-Fri., 8:30 AM-9:30 PM; Sat., 8:30 AM-6:30 PM. The library is closed Jan. 1 and Dec. 25. 101 Independence Ave., SE. Phone: 202-707-8000.

Website. http://www.loc.gov

Lincoln Memorial

Designed by Henry Bacon, the **Lincoln Memorial** in West Potomac Park, on the axis of the Capitol and the Washington Monument, consists of a large marble hall enclosing a heroic statue of Abraham Lincoln in meditation sitting on a large armchair. The memorial was dedicated on May 30, 1922. The statue was designed by Daniel Chester French and sculpted by French and the Piccirilli brothers. Murals and ornamentation on the bronze ceiling beams are by Jules Guerin. The text of the Gettysburg Address is in the south chamber; that of Lincoln's Second Inaugural speech is in the north chamber. Each is engraved on a stone tablet.

The memorial is open 24 hr daily. An elevator for the handicapped is in service. W. Potomac Park at 23rd St. NW. Phone: 202-619-7222.

Website. http://www.nps.gov/linc/home.htm

National Archives and Records

Original copies of the Declaration of Independence, the Constitution, and the Bill of Rights are on permanent display in the **National Archives** Exhibition Hall. The National Archives also holds other valuable U.S. government records and historic maps, photographs, and manuscripts.

IT'S A FACT: On an average day, the U.S. Treasury Department's Bureau of Engraving and Printing produces 37 million notes, with a face value of approximately $696 million.

Central Research and Microfilm Research Rooms are also available to the public for genealogical research.

The Exhibition Hall is open daily, 10 AM-5:30 PM; closed Dec. 25. 7th & Pennsylvania Ave. NW. Phone: 202-501-5000.

Website. http://www.nara.gov

National Gallery of Art

The **National Gallery of Art**, situated on the north side of the Mall facing Constitution Avenue, was established by Congress, Mar. 24, 1937, and opened Mar. 17, 1941. The original West building was designed by John Russell Pope. The East building, opened in 1978, was designed by I. M. Pei. The National Gallery is separate from, but maintains a relationship with, the Smithsonian Institution.

Open daily, 10 AM-5 PM; Sunday, 11 AM-6 PM. Closed Jan. 1 and Dec. 25. 4th & Constitution Ave NW. Phone: 202-737-4215.

Website. http://www.nga.gov

Franklin Delano Roosevelt Memorial

Opened May 2, 1997, by Pres. Bill Clinton, the **FDR Memorial** features 9 bronze sculptural ensembles depicting FDR, Eleanor Roosevelt (the first First Lady to be honored in a national memorial), and events from the Great Depression and World War II. This 7.5-acre memorial is located near the Tidal Basin in a park-like setting and includes waterfalls, quiet pools, and reddish Dakota granite upon which some of Pres. Roosevelt's well-known words are carved. The monument is wheelchair accessible.

Grounds, staffed daily, 8 AM-midnight, except Dec. 25. 1850 W. Basin Dr. SW. Phone: 202-426-6841.

Website. http://www.nps.gov/fdrm/

Smithsonian Institution

The **Smithsonian Institution,** established in 1846, is the world's largest museum complex and consists of 14 museums and the National Zoo. It holds some 100 mil. artifacts and specimens in its trust. Nine museums are on the National Mall between the Washington Monument and the Capitol; 5 other museums and the zoo are elsewhere in Washington (the Cooper-Hewitt Museum and the National Museum of the American Indian, also administered by the Smithsonian, are in New York City). The **Smithsonian Information Center** is located in "the Castle" on the Mall. Also on the Mall are the **National Museum of American History,** the **National Museum of Natural History,** the **National Air and Space Museum,** the **Hirshhorn Museum and Sculpture Garden,** the **Arthur M. Sackler Gallery,** the **National Museum of African Art,** the **Freer Gallery of Art,** and the **Arts and Industries Building.** Near the Sackler Gallery is the **Enid A. Haupt Garden.** Located nearby are the **National Postal Museum,** the **National Museum of American Art,** the **National Portrait Gallery,** and the **Renwick Gallery.** Farther away, at 1901 Fort Place SE, is the **Anacostia Museum.**

Most museums are open daily, except Dec. 25, 10 AM-5:30 PM. Phone: 202-357-2700.

Website. http://www.si.edu

Vietnam Veterans Memorial

Originally dedicated on Nov. 13, 1982, the **Vietnam Veterans Memorial** is a recognition of the men and women who served in the armed forces in the Vietnam War. On a V-shaped black-granite wall, designed by Maya Ying Lin, are inscribed the names of the more than 58,000 Americans who lost their lives or remain missing.

Since 1982, 2 additions have been made to the Memorial. The 1st, dedicated on Nov. 11, 1984, is the Frederick Hart sculpture *Three Servicemen*. On Nov. 11, 1993, the Vietnam Women's Memorial was dedicated, honoring the more than 11,500 women who served in Vietnam. The bronze sculpture, portraying 3 women helping a wounded male soldier, was designed by Glenna Goodacre.

The memorial is open 8am-12am daily. Constitution Ave. & Bacon Dr. NW. Phone: 202-426-6841.

Website. http://www.thevirtualwall.org

Washington Monument

The **Washington Monument,** dedicated in 1885, is a tapering shaft, or obelisk, of white marble, 555 ft, 5 1/8 inches in height and 55 ft, 1½ in. square at base. Eight small windows, 2 on each side, are located at the 500-ft level, where points of interest are indicated.

Open daily (except Dec. 25), 9 AM-4:30 PM; 8 AM-midnight, Apr.-Labor Day. Free timed passes are available; passes are available in advance for a small fee. 15th & Constitution Ave. NW. Phone: 202-426-6841.

Website. http://www.nps.gov/wash

White House

The **White House,** the President's residence, stands on 18 acres on the south side of Pennsylvania Ave., between the Treasury and the old Executive Office Building. The walls are of sandstone, quarried at Aquia Creek, VA. The exterior walls were painted, causing the building to be termed the "White House." On Aug. 24, 1814, during Madison's administration, the house was burned by the British. James Hoban rebuilt it by Oct. 1817.

The White House is normally open for free self-guided tours Tues.-Sat., 10 AM-noon (passes, necessary mid-March-mid-Sept., are available at White House Visitor's Center, 8 AM-noon, located at 1450 Pennsylvania Ave., NW). Only the public rooms on the ground floor and state floor may be visited. Free reserved tickets for guided congressional tours can be obtained 8 to 10 weeks in advance from your local U.S. representative or senator. 1600 Pennsylvania Ave. Phone: 202-456-7041.

Website. http://www.whitehouse.gov

Attractions Near Washington, DC

Arlington National Cemetery

Arlington National Cemetery, on the former Custis estate in Arlington, VA, is the site of the **Tomb of the Unknowns** and is the final resting place of Pres. John Fitzgerald Kennedy, who was buried there on Nov. 25, 1963. His wife, Jacqueline Bouvier Kennedy Onassis, was buried at the same site on May 23, 1994. An eternal flame burns over the grave site. In an adjacent area is the grave of Pres. Kennedy's brother Sen. Robert F. Kennedy (NY), interred on June 8, 1968. Many other famous Americans are also buried at Arlington, as well as more than 200,000 American soldiers from every major war.

North of the National Cemetery, approximately 350 yd, stands the **U.S. Marine Corps War Memorial**, also known as Iwo Jima. The memorial is a bronze statue of the raising of the U.S. flag on Mt. Suribachi, Feb. 23, 1945, during World War II, executed by Felix de Weldon from the photograph by Joe Rosenthal.

On the southern side of the Memorial Bridge, near the cemetery entrance, a memorial honoring the women in the military was dedicated, Oct. 18, 1997. The **Women in Military Service for America Memorial** is a half-circle granite monument, 30 ft. high and 226 ft. in diameter, with the Great Seal of the United States in the center.

Open daily, 8 AM-5 PM (8 AM-7 PM., Apr.-Sept.), Arlington, VA. Phone: 703-607-8052.

Website. http://www.http://www.arlingtoncemetery.org

Mount Vernon

Mount Vernon, George Washington's estate, is on the south bank of the Potomac R., 16 mi below Washington, DC, in northern Virginia. The present house is an enlargement of one apparently built on the site by Augustine Washington, who lived there 1735-38. His son Lawrence came there in 1743, and renamed the plantation Mount Vernon in honor of Admiral Vernon, under whom he had served in the West Indies. Lawrence Washington died in 1752 and was succeeded as proprietor by his half-brother, George Washington. The estate has been restored to its 18th-century appearance and includes many original furnishings. Washington and his wife, Martha, are buried on the grounds.

Open 365 days, 8 AM-5 PM, Apr.-Aug., 9 AM-5 PM, Sept., Oct., Mar.; 9 AM-4 PM, Nov.-Feb. Phone: 703-780-2000, or 1-800-429-1520. Admission: adults $8, seniors (62+) $7.50, children (6-11) $4, age 5 and under free.

Website. http://www.mountvernon.org

The Pentagon

The **Pentagon,** headquarters of the Department of Defense, is one of the world's largest office buildings. Situated in Arlington, VA, it has housed more than 23,000 employees in offices occupying 3,707,745 sq ft. The building was, however, severely damaged when struck by a plane Sept. 11, 2001.

Free tours (about every 2 hrs) have generally been available Mon.-Fri. (excluding federal holidays), starting at 9 AM; check for current details. Arlington, VA (I-395 South to Boundary Channel Drive exit). Phone: 703-695-1776.

Website. http://www.defenselink.mil/pubs/pentagon

UNITED STATES POPULATION

Census 2000: The Results Start Rolling In

by William G. Barron, Jr., acting director, U.S. Census Bureau, Dept. of Commerce

As New Year's revelers were preparing to ring in 2001, U.S. Census Bureau officials at the Commerce Department could do some celebrating of their own. On Dec. 28, 2000, they released the Census 2000 state population counts as required by Article I, Section 2 of the Constitution. These population figures are used to reapportion the 435 seats in the U.S. House of Representatives among the states. Thus, the 2000 census, described by then Census Bureau Director Kenneth Prewitt as the "most accurate headcount in the nation's history," had borne its first fruit. And the net national undercount rate—the rate at which people were missed—had declined from 1.6% in the 1990 census to 1.2% in Census 2000.

March 2001 brought the next major milestone: the delivery to each state's governor and legislative leaders of Census 2000 state population counts cross-tabulated by above/below voting age and by race and Hispanic/non-Hispanic origin down to the census block level. As required by law, the Census Bureau provides these data to the states to be used to redraw federal, state, and local legislative districts to reflect population shifts since the last census, so that constituents can have equal representation in legislatures.

The Census Bureau continued to release Census 2000 data during the year. Next came detailed tabulations based on the questions asked of all people and every housing unit. Topics covered included sex, age, race, Hispanic or Latino origin, household relationship, household and family characteristics, occupancy status of the housing unit, and tenure (i.e., whether the home is owned or rented). For most subjects, data were released down to the block level.

Here is a look at some of the highlights of the Census 2000 results released thus far.

Southward and Westward Shift Continues

Census 2000 showed that the resident population of the United States on April 1, 2000, was 281,421,906, an increase of 13.2% over the 248,709,873 people counted during the 1990 census. This increase of 32.7 million people was the largest total census-to-census population increase in U.S. history, exceeding the jump of 28.0 million people between 1950 and 1960, which occurred during the "baby boom." (In percentage terms, it was the largest since 1960-70).

During the past decade, the fastest-growing region in the United States was the West at 19.7%; the region added 10.4 million people in the 1990s, for a total of 63.2 million. The fastest-growing states in the nation were all located in the West: Nevada (an increase of 66.3%), Arizona (40.0%), Colorado (30.6%), Utah (29.6%), and Idaho (28.5%). California recorded the largest numeric increase of any state, 4.1 million people.

The South was the 2d-fastest-growing region (17.3%), adding a total of 14.8 million people in the 1990s. Georgia was the region's fastest growing state (up 26.4%), making the 1990s the only decade in the 20th century in which this distinction did not belong to Florida. Texas, which grew by 3.9 million, and Florida, up 3.0 million, showed the largest numeric increases in the South.

The Midwest grew by 7.9%, adding 4.7 million people. Minnesota (up 12.4%) was the region's bellwether for the 3d straight decade. Illinois, up 988,000, and Michigan, up 643,000, recorded the largest numeric increases. The region also was home to the state with the nation's smallest population growth: North Dakota (up just 0.5%). It is worth noting that growth in Nebraska, Iowa, and Missouri, while below the national average, was the fastest in many decades.

Population in the Northeast increased the least, by 2.8 million, or 5.5%, with New Hampshire (up 11.4%) growing the fastest in that region for the 4th straight decade. New York, up 986,000, and New Jersey, up 684,000, gained the most numerically.

Counties with large population increases generally were in or near major metropolitan areas such as Atlanta, Phoenix, Las Vegas, Houston, and Washington, DC. Maricopa County, AZ (Phoenix), had the largest population gain: 950,000 people. Douglas County, CO, near Denver, had the largest percentage growth in population: 191%. Counties in Florida, northern Georgia, North Carolina, Tennessee, southwest Missouri, and eastern, central, and southern Texas also experienced rapid growth.

A band of counties that lost population—in some cases more than 10%—stretched across the Great Plains states from the Mexican to the Canadian borders. A 2d band of slow growth included much of the interior Northeast and Appalachia, extending from Maine through western Pennsylvania and West Virginia to eastern Kentucky.

Las Vegas was the fastest growing metropolitan area, with a growth rate of 83%, followed by Naples, FL, which grew 65%.

Reflecting the movement of the population during the 1990s and continuing a decades-old southwesterly trend, the nation's center of population moved 12.1 miles south and 32.5 miles west during the decade, to a point about 3 miles east of Edgar Springs, MO.

Racial and Ethnic Diversity

Census 2000 results showed a racially diverse America. For Census 2000, respondents were given the option of selecting one or more race categories to indicate their racial identities. About 6.8 million people, or 2.4% of the nation's population, took advantage of the option to select more than one category. Of this group, 93% reported exactly 2 races. The most common combination was "White and Some other race," reported by 32% of the 2-or-more-races population.

The remaining 274.6 million people reported only one race. Of the entire population, 75.1% reported they were White only; 12.3%, Black or African American only; 0.9%, American Indian and Alaska Native only; 3.6%, Asian only; 0.1%, Native Hawaiian and Other Pacific Islander only; and 5.5%, Some other race only. Among Asians (reporting one race only), the largest groups were Chinese (23% of Asians), Filipino (18.1%), and Indian (16.4%).

A separate question on the census form collected information on Hispanic or Latino origin. Hispanics, who may be of any race, totaled 35.3 million, or about 13% of the total population. Among Latinos who specified a detailed origin, 20.6 million were Mexican; 3.4 million, Puerto Rican; 1.7 million, Central American; 1.4 million, South American; 1.2 million, Cuban; and 0.8 million, Dominican.

The Latino population rose nearly 13 million, or 57.9%, between the 1990 and 2000 censuses. In 2000, one-half of Hispanics lived in California and Texas. The counties with the highest concentrations of Latinos were located along the nation's southwestern border; among the 50 counties nationwide where Hispanics comprised a majority of the total population, 43 were in either Texas or New Mexico.

An Aging Nation

The median age (meaning half are older and half younger) of the U.S. population in 2000 was 35.3 years, the highest it has ever been.

The increase in the median age—which was 32.9 years in 1990—reflects the aging of the "baby boomers" (those born from 1946 to 1964). So it's not surprising that the most rapid increase in size of any age group was the 49% jump in the population 45-to-54-years-old. This increase, to 37.7 million in 2000, was fueled mainly by the entry into this age group of the first half of the "baby boom" generation.

IT'S A FACT: The U.S. POPClock at the Census Bureau keeps a running estimate of the total U.S. population. Taking the 2000 Census as a starting point, the clock assumed the following standards as of Aug. 31, 2001: one birth every 8 seconds; one death every 14 seconds; a net gain of one international migrant every 34 seconds; a net gain of one U.S. citizen returning every 3,202 seconds. That results in an overall net gain of one person every 11 seconds. To find out the projected current population, visit http://www.census.gov/main/www/popclock.html

Ironically, while the median age increased by nearly 2½ years between 1990 and 2000, the growth of the population age 65 and over (12% to 35.0 million) was the lowest recorded rate of growth in any decade for this age group. This also marked the first time in the history of the census that the 65-and-over population actually increased at a slower rate than the overall population. The slower growth reflected the relatively low number of births in the late 1920s and early 1930s. Meanwhile, the number of children under 18 increased 14% to 72.3 million, but the number of young adults ages 18 to 34 actually declined 4%.

Wider Range of Living Arrangements

The average size of the nation's households declined from 2.63 persons in 1990 to 2.59 persons in 2000. This decline reflected the changing living arrangements that took place during the decade. For example, the most common type of household, those consisting of married-couple families, declined from 55% to 52% of all households, for a total of 54.5 million households in 2000. And the 2d most common, those consisting of people living alone, rose from 25% to 26% of all households for a total of 27.2 million.

Families maintained by women with no husband present remained steady as a proportion of all households, at 12%, totaling 12.9 million in 2000. Those maintained by men with no wife present increased from 3% to 4%, totaling 4.4 million households in the latest census.

Another 5.5 million households consisted of unmarried partners: 4.9 million of the opposite sex, 0.6 million of the same sex. These unmarried-partner households comprised 5% of all households in 2000, up from less than 4% in 1990, and 2% in 1980.

Family patterns differ from one part of the nation to another. For example, counties with the highest proportion of married-couple households were located in Utah and Idaho and in the West Central section of the country, from west Texas northwards. In contrast, the counties in the coastal states extending from the lower Mississippi Valley up the Atlantic seaboard to New England had the lowest proportions of these households. Counties in the lower Mississippi Valley had relatively high percentages of family households maintained by women with no husband present.

Living alone can be characteristic of people in different stages of life. For example, it includes young adults starting out on their own with their first jobs, as well as elderly people who are widowed and living apart from their children. Hence, the 10 large cities with the highest proportion of one-person households included both the college town of Cambridge, MA, and Fort Lauderdale, FL, where nearly 3 out of 10 one-person households were elderly.

"Sample" Data Coming in 2002

The Census Bureau had not finished painting the demographic portrait of the nation as it appeared in 2000. In fact, it had just begun.

By the time 2002 comes to a close, it will be possible to paint a richer, more detailed portrait. That is because, during the course of the year, the Census Bureau will start releasing data collected on the Census 2000 "long form," which was delivered to a representative sample of the nation's housing units (in all, roughly 1 in 6).

Data will be provided on marital status, educational attainment, ancestry, language other than English spoken at home, nativity, disability, grandparents as caregivers, veterans' status, means and length of commute to work, industry, occupation, and income. The long-form information also will include data on the nation's housing, including age, type and size of home, plumbing and kitchen facilities, number of vehicles, and housing costs.

Statistics on these characteristics will be provided down to the block group level. Then, in late 2002 through early 2003, data on these same characteristics will be released state-by-state for detailed race and Hispanic or Latino categories and American Indian and Alaska Native tribes and ancestry groups down to the census tract (neighborhood) level. During the same time frame, the Census Bureau will issue its Public Use Microdata Sample files, which consist of a sample of complete long-form questionnaires, with identifying information removed (i.e., names and addresses); these files will enable users to generate more complex cross-tabulations.

All of the Census 2000 information can be found on the Internet at http://www.census.gov, as well as in CD-ROM, DVD, and printed report format. A new search feature on the Census Bureau's Web site—"American FactFinder"—permits users to quickly access the census results they need for specific geographic areas.

Release of the final Census 2000 results in 2003 will close another chapter in the history of the nation's statistical description. The next chapter is already being outlined, as planning for the 2010 census is now under way. At the Census Bureau, the job of measuring America is never-ending.

The Census: Looking Back

The U.S. Census is conducted every 10 years as mandated by the U.S. Constitution, Article I, Section 2. The primary purpose is to apportion seats in the U.S. House of Representatives and determine state legislative district boundaries. The data are also critical for a vast array of government programs at every level, and for providing demographic information to individuals and businesses.

The first U.S. census, which counted 3.9 million people, was conducted in 1790, shortly after George Washington became president. It counted the number of free white males age 16 and over, and under 16 (to measure how many men might be available for military service), the number of free white females, all other free persons (including any American Indians who paid taxes), and slaves. It took 18 months to collect the data, often on unofficial sheets of paper supplied by U.S. marshals. In contrast to today's pledge of confidentiality, the 1790 census was displayed "at two of the most public places." The 1790 census resulted in an increase of 41 seats (65 to 106) in the House of Representatives.

As the nation expanded, so did the scope of the census data. The first inquiry on manufactures was made in 1810. Questions on agriculture, mining, and fisheries were added in 1840. In 1850, the census included inquiries on social issues—taxation, churches, pauperism, and crime.

The 1880 census contained so many questions that it took the full 10 years between censuses to publish all the results. Because of this delay, Congress limited the 1900 census to questions on population, manufactures, agriculture, and mortality. (Many of the dropped topics reappeared in later censuses.)

For many years, the undertaking of each census had to be authorized by a specific act of Congress. In 1954, Congress specified the laws under which the Census Bureau operates in Title 13 of the U.S. Code. This title delineates the basic scope of the census, the requirements for the public to provide information as well as for the Bureau to keep information confidential, and the penalties for violating any of these obligations. The secretary of commerce (and through that individual, the Census Bureau) is now directed by law to take censuses of population, housing, agriculture, irrigation, manufactures, mineral industries, other businesses (wholesale trade, retail trade, services), construction, transportation, and governments at stated intervals, and may take surveys related to any of these subjects.

U.S. marshals supervised their assistants' enumeration of the first 9 censuses and reported to the president (1790), the secretary of state (1800-1840), or the secretary of the interior (1850-1870). There was no continuity of personnel from one census to the next. However, in 1902, Congress authorized the president to set up a permanent Census Office in the Interior Dept. In 1903, the agency was transferred to the new Dept. of Commerce and Labor, and when the department split in 1913, the Bureau of the Census was placed in the Commerce Dept.

The Census Bureau began using statistical sampling techniques in the 1940s, computers in the 1950s, and mail enumeration in the 1960s, all as part of an effort to publish more data sooner and at a lower cost, and with less burden on the public.

U.S. Population by Official

STATE	1790[1]	1800[1]	1810[1]	1820[1]	1830[1]	1840	1850	1860	1870	1880	1890	1900
AL	......	1	9	128	310	590,756	771,623	964,201	996,992	1,262,505	1,513,401	1,828,697
AK	......	......	......	......	......	......	......	......	......	33,426	32,052	63,592
AZ	......	......	......	......	......	......	......	......	9,658	40,440	88,243	122,931
AR	......	......	1	14	30	97,574	209,897	435,450	484,471	802,525	1,128,211	1,311,564
CA	......	......	......	......	......	......	92,597	379,994	560,247	864,694	1,213,398	1,485,053
CO	......	......	......	......	......	......	......	34,277	39,864	194,327	413,249	539,700
CT	238	251	262	275	298	309,978	370,792	460,147	537,454	622,700	746,258	908,420
DE	59	64	73	73	77	78,085	91,532	112,216	125,015	146,608	168,493	184,735
DC	......	8	16	23	30	33,745	51,687	75,080	131,700	177,624	230,392	278,718
FL	......	......	......	......	35	54,477	87,445	140,424	187,748	269,493	391,422	528,542
GA	83	163	252	341	517	691,392	906,185	1,057,286	1,184,109	1,542,180	1,837,353	2,216,331
HI	......	......	......	......	......	......	......	......	......	......	......	154,001
ID	......	......	......	......	......	......	......	......	14,999	32,610	88,548	161,772
IL	......	......	12	55	157	476,183	851,470	1,711,951	2,539,891	3,077,871	3,826,352	4,821,550
IN	......	6	25	147	343	685,866	988,416	1,350,428	1,680,637	1,978,301	2,192,404	2,516,462
IA	......	......	......	......	......	43,112	192,214	674,913	1,194,020	1,624,615	1,912,297	2,231,853
KS	......	......	......	......	......	......	......	107,206	364,399	996,096	1,428,108	1,470,495
KY	74	221	407	564	688	779,828	982,405	1,155,684	1,321,011	1,648,690	1,858,635	2,147,174
LA	......	......	77	153	216	352,411	517,762	708,002	726,915	939,946	1,118,588	1,381,625
ME	97	152	229	298	399	501,793	583,169	628,279	626,915	648,936	661,086	694,466
MD	320	342	381	407	447	470,019	583,034	687,049	780,894	934,943	1,042,390	1,188,044
MA	379	423	472	523	610	737,699	994,514	1,231,066	1,457,351	1,783,085	2,238,947	2,805,346
MI	......	......	5	9	32	212,267	397,654	749,113	1,184,059	1,636,937	2,093,890	2,420,982
MN	......	......	......	......	......	......	6,077	172,023	439,706	780,773	1,310,283	1,751,394
MS	......	8	31	75	137	375,651	606,526	791,305	827,922	1,131,597	1,289,600	1,551,270
MO	......	......	20	67	140	383,702	682,044	1,182,012	1,721,295	2,168,380	2,679,185	3,106,665
MT	......	......	......	......	......	......	......	......	20,595	39,159	142,924	243,329
NE	......	......	......	......	......	......	......	28,841	122,993	452,402	1,062,656	1,066,300
NV	......	......	......	......	......	......	......	6,857	42,491	62,266	47,355	42,335
NH	142	184	214	244	269	284,574	317,976	326,073	318,300	346,991	376,530	411,588
NJ	184	211	246	278	321	373,306	489,555	672,035	906,096	1,131,116	1,444,933	1,883,669
NM	......	......	......	......	......	......	61,547	93,516	91,874	119,565	160,282	195,310
NY	340	589	959	1,373	1,919	2,428,921	3,097,394	3,880,735	4,382,759	5,082,871	6,003,174	7,268,894
NC	394	478	556	639	736	753,419	869,039	992,622	1,071,361	1,399,750	1,617,949	1,893,810
ND	......	......	......	......	......	......	......	......	2,405[2]	36,909	190,983	319,146
OH	......	45	231	581	938	1,519,467	1,980,329	2,339,511	2,665,260	3,198,062	3,672,329	4,157,545
OK	......	......	......	......	......	......	......	......	......	......	258,657	790,391
OR	......	......	......	......	......	......	12,093	52,465	90,923	174,768	317,704	413,536
PA	434	602	810	1,049	1,348	1,724,033	2,311,786	2,906,215	3,521,951	4,282,891	5,258,113	6,302,115
RI	69	69	77	83	97	108,830	147,545	174,620	217,353	276,531	345,506	428,556
SC	249	346	415	503	581	594,398	668,507	703,708	705,606	995,577	1,151,149	1,340,316
SD	......	......	......	......	......	......	......	4,837[2]	11,776[2]	98,268	348,600	401,570
TN	36	106	262	423	682	829,210	1,002,717	1,109,801	1,258,520	1,542,359	1,767,518	2,020,616
TX	......	......	......	......	......	......	212,592	604,215	818,579	1,591,749	2,235,527	3,048,710
UT	......	......	......	......	......	......	11,380	40,273	86,786	143,963	210,779	276,749
VT	85	154	218	236	281	291,948	314,120	315,098	330,551	332,286	332,422	343,641
VA	692	808	878	938	1,044	1,025,227	1,119,348	1,219,630	1,225,163	1,512,565	1,655,980	1,854,184
WA	......	......	......	......	......	......	1,201	11,594	23,955	75,116	357,232	518,103
WV	56	79	105	137	177	224,537	302,313	376,688	442,014	618,457	762,794	958,800
WI	......	......	......	......	......	30,945	305,391	775,881	1,054,670	1,315,497	1,693,330	2,069,042
WY	......	......	......	......	......	......	......	......	9,118	20,789	62,555	92,531
U.S.	**3,929**	**5,308**	**7,240**	**9,638**	**12,866[3]**	**17,068,953[3]**	**23,191,876**	**31,443,321**	**38,558,371**	**50,189,209**	**62,979,766**	**76,212,168**

Note: Where possible, population shown is that of the 2000 area of the state. Members of the Armed Forces overseas or other U.S. nationals abroad are not included. Totals revised to include corrections of initial tabulated counts. (1) Totals for 1790 through 1830 are in thousands. (2) 1860 figure is for Dakota Territory; 1870 figures are for parts of Dakota Territory. (3) Includes persons (5,318 in 1830 and 6,100 in 1840) on public ships in the service of the U.S. not credited to any region, division, or state.

Estimated Population of American Colonies, 1630-1780

Source: Bureau of the Census, U.S. Dept. of Commerce; in thousands

Colony	1630	1650	1670	1690	1700	1720	1740	1750	1770	1780
TOTAL	**4.6**	**50.4**	**111.9**	**210.4**	**250.9**	**466.2**	**905.6**	**1,170.8**	**2,148.1**	**2,780.4**
Maine (counties)[1]	0.4	1.0	...	...	...	...	...	...	31.3	49.1
New Hampshire[2]	0.5	1.3	1.8	4.2	5.0	9.4	23.3	27.5	62.4	87.8
Vermont[3]	...	...	...	...	...	...	...	...	10.0	47.6
Plymouth and Massachusetts[1,2,4]	0.9	15.6	35.3	56.9	55.9	91.0	151.6	188.0	235.3	268.6
Rhode Island[2]	...	0.8	2.2	4.2	5.9	11.7	25.3	33.2	58.2	52.9
Connecticut[2]	...	4.1	12.6	21.6	26.0	58.8	89.6	111.3	183.9	206.7
New York[2]	0.4	4.1	5.8	13.9	19.1	36.9	63.7	76.7	162.9	210.5
New Jersey[2]	...	...	1.0	8.0	14.0	29.8	51.4	71.4	117.4	139.6
Pennsylvania[2]	...	...	...	11.4	18.0	31.0	85.6	119.7	240.1	327.3
Delaware[2]	...	0.2	0.7	1.5	2.5	5.4	19.9	28.7	35.5	45.4
Maryland[2]	...	4.5	13.2	24.0	29.6	66.1	116.1	141.1	202.6	245.5
Virginia[2]	2.5	18.7	35.3	53.0	58.6	87.8	180.4	231.0	447.0	538.0
North Carolina[2]	...	...	3.8	7.6	10.7	21.3	51.8	73.0	197.2	270.1
South Carolina[2]	...	...	0.2	3.9	5.7	17.0	45.0	64.0	124.2	180.0
Georgia[2]	...	...	...	...	...	...	2.0	5.2	23.4	56.1
Kentucky[5]	...	...	...	...	...	...	...	...	15.7	45.0
Tennessee[6]	...	...	...	...	...	...	...	...	1.0	10.0

(1) For 1660-1750, Maine counties are included with Massachusetts. Maine was part of Massachusetts until it became a separate state in 1820. (2) One of the original 13 states. (3) Admitted to statehood in 1791. (4) Plymouth became a part of the Province of Massachusetts in 1691. (5) Admitted to statehood in 1792. (6) Admitted to statehood in 1796.

Census, 1790-2000

1910	1920	1930	1940	1950	1960	1970	1980	1990	2000
2,138,093	2,348,174	2,646,248	2,832,961	3,061,743	3,266,740	3,444,354	3,894,025	4,040,587	4,447,100
64,356	55,036	59,278	72,524	128,643	226,167	302,583	401,851	550,043	626,932
204,354	334,162	435,573	499,261	749,587	1,302,161	1,775,399	2,716,546	3,665,228	5,130,632
1,574,449	1,752,204	1,854,482	1,949,387	1,909,511	1,786,272	1,923,322	2,286,357	2,350,725	2,673,400
2,377,549	3,426,861	5,677,251	6,907,387	10,586,223	15,717,204	19,971,069	23,667,764	29,760,021	33,871,648
799,024	939,629	1,035,791	1,123,296	1,325,089	1,753,947	2,209,596	2,889,735	3,294,394	4,301,261
1,114,756	1,380,631	1,606,903	1,709,242	2,007,280	2,535,234	3,032,217	3,107,564	3,287,116	3,405,565
202,322	223,003	238,380	266,505	318,085	446,292	548,104	594,338	666,168	783,600
331,069	437,571	486,869	663,091	802,178	763,956	756,668	638,432	606,900	572,059
752,619	968,470	1,468,211	1,897,414	2,771,305	4,951,560	6,791,418	9,746,961	12,937,926	15,982,378
2,609,121	2,895,832	2,908,506	3,123,723	3,444,578	3,943,116	4,587,930	5,462,982	6,478,216	8,186,453
191,874	255,881	368,300	422,770	499,794	632,772	769,913	964,691	1,108,229	1,211,537
325,594	431,866	445,032	524,873	588,637	667,191	713,015	944,127	1,006,749	1,293,953
5,638,591	6,485,280	7,630,654	7,897,241	8,712,176	10,081,158	11,110,285	11,427,409	11,430,602	12,419,293
2,700,876	2,930,390	3,238,503	3,427,796	3,934,224	4,662,498	5,195,392	5,490,214	5,544,159	6,080,485
2,224,771	2,404,021	2,470,939	2,538,268	2,621,073	2,757,537	2,825,368	2,913,808	2,776,755	2,926,324
1,690,949	1,769,257	1,880,999	1,801,028	1,905,299	2,178,611	2,249,071	2,364,236	2,477,574	2,688,418
2,289,905	2,416,630	2,614,589	2,845,627	2,944,806	3,038,156	3,220,711	3,660,324	3,685,296	4,041,769
1,656,388	1,798,509	2,101,593	2,363,880	2,683,516	3,257,022	3,644,637	4,206,116	4,219,973	4,468,976
742,371	768,014	797,423	847,226	913,774	969,265	993,722	1,125,043	1,227,928	1,274,923
1,295,346	1,449,661	1,631,526	1,821,244	2,343,001	3,100,689	3,923,897	4,216,933	4,781,468	5,296,486
3,366,416	3,852,356	4,249,614	4,316,721	4,690,514	5,148,578	5,689,170	5,737,093	6,016,425	6,349,097
2,810,173	3,668,412	4,842,325	5,256,106	6,371,766	7,823,194	8,881,826	9,262,044	9,295,297	9,938,444
2,075,708	2,387,125	2,563,953	2,792,300	2,982,483	3,413,864	3,806,103	4,075,970	4,375,099	4,919,479
1,797,114	1,790,618	2,009,821	2,183,796	2,178,914	2,178,141	2,216,994	2,520,770	2,573,216	2,844,658
3,293,335	3,404,055	3,629,367	3,784,664	3,954,653	4,319,813	4,677,623	4,916,766	5,117,073	5,595,211
376,053	548,889	537,606	559,456	591,024	674,767	694,409	786,690	799,065	902,195
1,192,214	1,296,372	1,377,963	1,315,834	1,325,510	1,411,330	1,485,333	1,569,825	1,578,385	1,711,263
81,875	77,407	91,058	110,247	160,083	285,278	488,738	800,508	1,201,833	1,998,257
430,572	443,083	465,293	491,524	533,242	606,921	737,681	920,610	1,109,252	1,235,786
2,537,167	3,155,900	4,041,334	4,160,165	4,835,329	6,066,782	7,171,112	7,365,011	7,730,188	8,414,350
327,301	360,350	423,317	531,818	681,187	951,023	1,017,055	1,303,302	1,515,069	1,819,046
9,113,614	10,385,227	12,588,066	13,479,142	14,830,192	16,782,304	18,241,391	17,558,165	17,990,455	18,976,457
2,206,287	2,559,123	3,170,276	3,571,623	4,061,929	4,556,155	5,084,411	5,880,095	6,628,637	8,049,313
577,056	646,872	680,845	641,935	619,636	632,446	617,792	652,717	638,800	642,200
4,767,121	5,759,394	6,646,697	6,907,612	7,946,627	9,706,397	10,657,423	10,797,603	10,847,115	11,353,140
1,657,155	2,028,283	2,396,040	2,336,434	2,233,351	2,328,284	2,559,463	3,025,487	3,145,585	3,450,654
672,765	783,389	953,786	1,089,684	1,521,341	1,768,687	2,091,533	2,633,156	2,842,321	3,421,399
7,665,111	8,720,017	9,631,350	9,900,180	10,498,012	11,319,366	11,800,766	11,864,720	11,881,643	12,281,054
542,610	604,397	687,497	713,346	791,896	859,488	949,723	947,154	1,003,464	1,048,319
1,515,400	1,683,724	1,738,765	1,899,804	2,117,027	2,382,594	2,590,713	3,120,729	3,486,703	4,012,012
583,888	636,547	692,849	642,961	652,740	680,514	666,257	690,768	696,004	754,844
2,184,789	2,337,885	2,616,556	2,915,841	3,291,718	3,567,089	3,926,018	4,591,023	4,877,185	5,689,283
3,896,542	4,663,228	5,824,715	6,414,824	7,711,194	9,579,677	11,198,655	14,225,513	16,986,510	20,851,820
373,351	449,396	507,847	550,310	688,862	890,627	1,059,273	1,461,037	1,722,850	2,233,169
355,956	352,428	359,611	359,231	377,747	389,881	444,732	511,456	562,758	608,827
2,061,612	2,309,187	2,421,851	2,677,773	3,318,680	3,966,949	4,651,448	5,346,797	6,187,358	7,078,515
1,141,990	1,356,621	1,563,396	1,736,191	2,378,963	2,853,214	3,413,244	4,132,353	4,866,692	5,894,121
1,221,119	1,463,701	1,729,205	1,901,974	2,005,552	1,860,421	1,744,237	1,950,186	1,793,477	1,808,344
2,333,860	2,632,067	2,939,006	3,137,587	3,434,575	3,951,777	4,417,821	4,705,642	4,891,769	5,363,675
145,965	194,402	225,565	250,742	290,529	330,066	332,416	469,557	453,588	493,782
92,228,496	**106,021,537**	**123,202,624**	**132,164,569**	**151,325,798**	**179,323,175**	**203,302,031**	**226,542,203**	**248,709,873**	**281,421,906**

U.S. Center of Population, 1790-2000

Source: Bureau of the Census, U.S. Dept. of Commerce

The U.S. Center of Population is considered here as the center of population gravity, or that point upon which the U.S. would balance if it were a rigid plane without weight and the population distributed thereon, with each individual assumed to have equal weight and to exert an influence on a central point proportional to his or her distance from that point. The 2000 center is 12.1 miles south and 32.5 miles west of the1990 center of population, and is more than 1,000 miles from the 1790 center.

YEAR	N Lat °	'	"	W Long °	'	"	APPROXIMATE LOCATION
1790	39	16	30	76	11	12	23 miles east of Baltimore, MD
1800	39	16	6	76	56	30	18 miles west of Baltimore, MD
1810	39	11	30	77	37	12	40 miles northwest by west of Washington, DC (in VA)
1820	39	5	42	78	33	0	16 miles east of Moorefield, WV[1]
1830	38	57	54	79	16	54	19 miles west-southwest of Moorefield, WV[1]
1840	39	2	0	80	18	0	16 miles south of Clarksburg, WV[1]
1850	38	59	0	81	19	0	23 miles southeast of Parkersburg, WV[1]
1860	39	0	24	82	48	48	20 miles south by east of Chillicothe, OH
1870	39	12	0	83	35	42	48 miles east by north of Cincinnati, OH
1880	39	4	8	84	39	40	8 miles west by south of Cincinnati, OH (in KY)
1890	39	11	56	85	32	53	20 miles east of Columbus, IN
1900	39	9	36	85	48	54	6 miles southeast of Columbus, IN
1910	39	10	12	86	32	20	In the city of Bloomington, IN
1920	39	10	21	86	43	15	8 miles south-southeast of Spencer, Owen Co., IN
1930	39	3	45	87	8	6	3 miles northeast of Linton, Greene Co., IN
1940	38	56	54	87	22	35	2 miles southeast by east of Carlisle, Haddon township, Sullivan Co., IN
1950 (incl. Alaska & Hawaii)	38	48	15	88	22	8	3 miles northeast of Louisville, Clay Co., IL
1960	38	35	58	89	12	35	6½miles northwest of Centralia, Clinton Co., IL
1970	38	27	47	89	42	22	5 miles east southeast of Mascoutah, St. Clair Co., IL
1980	38	8	13	90	34	26	¼ mile west of De Soto, Jefferson Co., MO
1990	37	52	20	91	12	55	9.7 miles northwest of Steelville, MO
2000	37	41	49	91	48	34	2.8 miles east of Edgar Springs, MO

(1) West Virginia was set off from Virginia on Dec. 31, 1862, and was admitted as a state on June 20, 1863.

Congressional Apportionment

Source: Bureau of the Census, U.S. Dept. of Commerce

	2000	1990	1980	1970	1950	1900	1850
AL	7	7	7	7	9	9	7
AK	1	1	1	1	NA	NA	NA
AZ	8	6	5	4	2	NA	NA
AR	4	4	4	4	6	7	2
CA	53	52	45	43	30	8	2
CO	7	6	6	5	4	3	NA
CT	5	6	6	6	6	5	4
DE	1	1	1	1	1	1	1
FL	25	23	19	15	8	3	1
GA	13	11	10	10	10	11	8
HI	2	2	2	2	NA	NA	NA
ID	2	2	2	2	2	1	NA
IL	19	20	22	24	25	25	9
IN	9	10	10	11	11	13	11
IA	5	5	6	6	8	11	2
KS	4	4	5	5	6	8	NA
KY	6	6	7	7	8	11	10
LA	7	7	8	8	8	7	4
ME	2	2	2	2	3	4	6
MD	8	8	8	8	7	6	6
MA	10	10	11	12	14	14	11
MI	15	16	18	19	18	12	4
MN	8	8	8	8	9	9	2
MS	4	5	5	5	6	8	5
MO	9	9	9	10	11	16	7
MT	1	1	2	2	2	1	NA
NE	3	3	3	3	4	6	NA
NV	3	2	2	1	1	1	NA
NH	2	2	2	2	2	2	3
NJ	13	13	14	15	14	10	5
NM	3	3	3	2	2	NA	NA
NY	29	31	34	39	43	37	33
NC	13	12	11	11	12	10	8
ND	1	1	1	1	2	2	NA
OH	18	19	21	23	23	21	21
OK	5	6	6	6	6	5	NA
OR	5	5	5	4	4	2	1
PA	19	21	23	25	30	32	25
RI	2	2	2	2	2	2	2
SC	6	6	6	6	6	7	6
SD	1	1	1	2	2	2	NA
TN	9	9	9	9	9	10	10
TX	32	30	27	24	22	16	2
UT	3	3	3	2	2	1	NA
VT	1	1	1	1	1	2	3
VA	11	11	10	10	10	10	13
WA	9	9	8	7	7	3	NA
WV	3	3	4	4	6	5	NA
WI	8	9	9	9	10	11	3
WY	1	1	1	1	1	1	NA
TOTAL	**435**	**435**	**435**	**435**	**435**	**391**	**237**

The Constitution, in Article 1, Section 2, provided for a census of the population every 10 years to establish a basis for apportionment of representatives among the states. This apportionment largely determines the number of electoral votes allotted to each state.

The number of representatives of each state in Congress is determined by the state's population, but each state is entitled to one representative regardless of population. A congressional apportionment has been made after each decennial census except that of 1920. Prior to 1970, 3/5 the number of slaves who added to the total free population. Indians "not taxed" were excluded until 1940.

Under provisions of a law that became effective Nov. 15, 1941, representatives are apportioned by the method of equal proportions. In the application of this method, the apportionment is made so that the average population per representative has the least possible variation between one state and any other.

The first House of Representatives, in 1789, had 65 members, as provided by the Constitution. Of these, the largest numbers are from Virginia (19), Massachusetts (14), and Pennsylvania (13).

As the population grew, the number of representatives was increased, but the total membership has been fixed at 435 since the apportionment based on the 1910 census.

U.S. Area and Population, 1790-2000

Source: Bureau of the Census, U.S. Dept. of Commerce

	AREA			POPULATION			
						Increase over preceding census	
Census date	Gross Area	Land Area	Water Area	Number	Per sq mi of land	Number	%
1790 (Aug. 2)	891,364	864,746	26,618	3,929,214	4.5	—	—
1800 (Aug. 4)	891,364	864,746	26,618	5,308,483	6.1	1,379,269	35.1
1810 (Aug. 6)	1,722,685	1,681,828	40,857	7,239,881	4.3	1,931,398	36.4
1820 (June 1)	1,792,552	1,749,462	43,090	9,638,453	5.5	2,398,572	33.1
1830 (June 1)	1,792,552	1,749,462	43,090	12,866,020[2]	7.4	3,227,567	33.5
1840 (June 1)	1,792,552	1,749,462	43,090	17,068,953[2]	9.8	4,203,433	32.7
1850 (June 1)	2,991,655	2,940,042	51,613	23,191,876	7.9	6,122,423	35.9
1860 (June 1)	3,021,295	2,969,640	51,655	31,443,321	10.6	8,251,445	35.6
1870 (June 1)	3,612,299	3,540,705	71,594	38,558,371	10.9	7,115,050	22.6
1880 (June 1)	3,612,299	3,540,705	71,594	50,189,209	14.2	11,630,838	30.2
1890 (June 1)	3,612,299	3,540,705	71,594	62,979,766	17.8	12,790,557	25.5
1900 (June 1)	3,618,770	3,547,314	71,456	76,212,168	21.5	13,232,402	21.0
1910 (Apr. 15)	3,618,770	3,547,045	71,725	92,228,496	26.0	16,016,328	21.0
1920 (Jan. 1)	3,618,770	3,546,931	71,839	106,021,537	29.9	13,793,041	15.0
1930 (Apr. 1)	3,618,770	3,551,608	67,162	123,202,624	34.7	17,181,087	16.2
1940 (Apr. 1)	3,618,770	3,554,608	64,162	132,164,569	37.2	8,961,945	7.3
1950 (Apr. 1)	3,618,770	3,552,206	66,564	151,325,798	42.6	19,161,229	14.5
1960 (Apr. 1)	3,618,770	3,540,911	77,859	179,323,175	50.6	27,997,377	18.5
1970 (Apr. 1)	3,618,770	3,536,855	81,915	203,302,031	57.5	23,978,856	13.4
1980 (Apr. 1)	3,618,770	3,539,289	79,481	226,542,203	64.0	23,240,172	11.4
1990 (Apr. 1)	3,717,796[1]	3,536,278	181,518[1]	248,709,873	70.3	22,167,670	9.8
2000 (Apr. 1)	3,794,085	3,537,440	256,648[1]	281,421,906	79.6	32,712,033	13.2

(1) Includes inland, coastal, and Great Lakes. Data before 1990 cover inland water only. (2) The U.S. total includes persons (5,318 in 1830 and 6,100 in 1840) on public ships in the service of the U.S. not credited to any region, division, or state. **NOTE**: Percent changes are computed on the basis of change in population since the preceding census date, so the period covered is not always exactly 10 years. Population density figures given for various years represent the area within the boundaries of the U.S. that was under the jurisdiction on the date in question—including, in some cases, considerable areas not organized or settled and not actually covered by the census. In 1870, for example, Alaska was not covered by the census. Population figures shown here may reflect corrections made to the initial tabulated census counts.

Population, by Sex, Race, Residence, and Median Age, 1790-2000

Source: Bureau of the Census, U.S. Dept. of Commerce
(in thousands, except as indicated)

	SEX		RACE				RESIDENCE		MEDIAN AGE (years)		
				Black							
	Male	Female	White	Number	Percent	Other[5]	Urban	Rural	All races	White	Black
Conterminous U.S.[1]											
1790 (Aug. 2)	NA	NA	3,172	757	19.3	NA	202	3,728	NA	NA	NA
1810 (Aug. 6)	NA	NA	5,862	1,378	19.0	NA	525	6,714	NA	16.0	NA
1820 (Aug. 7)	4,897	4,742	7,867	1,772	18.4	NA	693	8,945	16.7	16.6	17.2
1840 (June 1)	8,689	8,381	14,196	2,874	16.8	NA	1,845	15,224	17.8	17.9	17.6
1860 (June 1)	16,085	15,358	26,923	4,442	14.1	79	6,217	25,227	19.4	19.7	17.5
1870 (June 1)	19,494	19,065	33,589	4,880	12.7	89	9,902	28,656	20.2	20.4	18.5
1880 (June 1)	25,519	24,637	43,403	6,581	13.1	172	14,130	36,026	20.9	21.4	18.0
1890 (June 1)	32,237	30,711	55,101	7,489	11.9	358	22,106	40,841	22.0	22.5	17.8
1900 (June 1)	38,816	37,178	66,809	8,834	11.6	351	30,160	45,835	22.9	23.4	19.4
1920 (Jan. 1)	53,900	51,810	94,821	10,463	9.9	427	54,158	51,553	25.3	25.5	22.3
1930 (Apr. 1)	62,137	60,638	110,287	11,891	9.7	597	68,955	53,820	26.5	26.9	23.5
1940 (Apr. 1)	66,062	65,608	118,215	12,866	9.8	589	74,424	57,246	29.0	29.5	25.3
United States											
1950 (Apr. 1)	74,833	75,864	135,150	15,045	9.9	1,131	96,467	54,230	30.2	30.7	26.2
1960 (Apr. 1)	88,331	90,992	158,832	18,872	10.5	1,620	125,269	54,054	29.5	30.3	23.5
1970 (Apr. 1)[2]	98,912	104,300	177,749	22,580	11.1	2,883	149,647	53,565	28.1	28.9	22.4
1980 (Apr. 1)[3]	110,053	116,493	194,713	26,683	11.8	5,150	167,051	59,495	30.0	30.9	24.9
1985 (July 1, est.)	115,730	122,194	202,031	28,569	12.0	7,324	NA	NA	31.4	32.3	26.6
1990 (Apr. 1)	121,239	127,470	199,686	29,986	12.1	9,233	187,053	61,656	32.9	34.4	28.1
1991 (July 1, est.)	122,984	129,122	210,979	31,107	12.3	10,020	NA	NA	33.1	34.1	28.1
1992 (July 1, est.)	124,506	130,496	212,885	31,670	12.4	10,446	NA	NA	33.4	34.4	28.5
1993 (July 1, est.)	125,938	131,858	214,760	32,168	12.5	10,867	NA	NA	33.7	34.7	28.7
1994 (July 1, est.)	127,216	133,076	216,413	32,653	12.5	11,227	NA	NA	34.0	35.0	29.0
1995 (July 1, est.)	128,569	134,321	218,149	33,095	12.6	11,646	NA	NA	34.3	35.3	29.2
1996 (July 1, est.)	129,746	135,434	219,686	33,514	12.6	11,979	NA	NA	34.6	35.7	29.5
1997 (July 1, est.)	131,018	136,618	221,334	33,947	12.7	12,355	NA	NA	34.9	36.0	29.7
1998 (July 1, est.)	132,263	137,766	222,932	34,370	12.7	12,727	NA	NA	35.3	36.3	29.9
1999 (July 1, est.)	133,352	139,526	224,692	34,903	12.8	13,283	NA	NA	35.5	36.6	30.1
2000 (Apr. 1)[4]	138,054	143,368	211,461	34,658	12.3	13,118	NA	NA	35.3	NA	NA

NA=Not available. **NOTE:** Urban and rural definitions may change from census to census. Figures have been adjusted to be consistent with the 1990 urban and rural definitions. (1) Excludes Alaska and Hawaii. (2) The revised 1970 resident population count is 203,302,031, which incorporates changes due to errors found after tabulations were completed. The race and sex data shown here reflect the official 1970 census count; the residence data come from the tabulated count. (3) The race data shown for Apr. 1, 1980, have been modified. (4) Race data for 2000 do not include the 2.4% of the population that reported being of more than one race. (5) "Other" consists of American Indians, Alaska Natives, Asians, and Pacific Islanders.

IT'S A FACT: The gap in numbers between males and females in the U.S. narrowed during the 1990s, as immigration and falling death rates boosted the male population. The number of males grew by 14%, the number of females by 13%, leaving a ratio of 96.3 males for every female in 2000. Alaska had the most males per female (107 per 100), followed by Nevada (103.9); the District of Columbia had the fewest (89.0), followed by Rhode Island (92.5).

U.S. Population by Race and Hispanic or Latino Origin, 1990-2000

	Census 2000		1990 Census		% inc, 1990 - 2000	
	One race only	One race or more[3]	Number	% of total pop.	Using one race only for 2000	Using one race only or in combination for 2000[4]
RACE[1]						
Total U.S. population[2]	281,421,906	281,421,906	248,709,873	100.0	13.2	13.2
White	211,460,626	216,930,975	199,686,070	80.3	5.9	8.6
Black or African American	34,658,190	36,419,434	29,986,060	12.1	15.6	21.5
American Indian and Alaska Native	2,475,956	4,119,301	1,959,234	0.8	26.4	110.3
Asian	10,242,998	11,898,828	6,908,638	2.8	48.3	72.2
Native Hawaiian and other Pac. Isl.	398,835	874,414	365,024	0.1	9.3	139.5
Some other race	15,359,073	18,521,486	9,804,847	3.9	56.6	88.9
HISPANIC OR LATINO AND RACE						
Total U.S. population[2]	281,421,906	281,421,906	248,709,873	100.0	13.2	13.2
Hispanic or Latino (of any race)[2]	35,305,818	35,305,818	22,354,059	9.0	57.9	57.9
Not Hispanic or Latino[2]	246,116,088	246,116,088	226,355,814	91.0	8.7	8.7
White	194,552,774	198,177,900	188,128,296	75.6	3.4	5.3
Black or African American	33,947,837	35,383,751	29,216,293	11.7	16.2	21.1
American Indian and Alaska Native	2,068,883	3,444,700	1,793,773	0.7	15.3	92.0
Asian	10,123,169	11,579,494	6,642,481	2.7	52.4	74.3
Native Hawaiian and other Pac. Isl.	353,509	748,149	325,878	0.1	8.5	129.6
Some other race	467,770	1,770,645	249,093	0.1	87.8	610.8

(1) Because individuals could report only one race in 1990 and could report more than one race in 2000, and because of other changes in the census questionnaire, the race data for 1990 and 2000 are not directly comparable. (2) The data for total U.S. population, Hispanic or Latino population, and total Not Hispanic or Latino population are not affected by the changes cited in (1). Hispanic or Latino persons may be of any race. (3) Alone or in combination with one or more of the other five races listed. (4) Columns 5 and 6 provide, respectively, a "minimum-maximum" range for the percent increase in population for each race between 1990 and 2000.

Population by State, 1990-2000

Source: Bureau of the Census, U.S. Dept. of Commerce

Rank	State	1990 population	2000 population	Percentage change 1990-2000
1.	California	29,760,021	33,871,648	13.8
2.	Texas	16,986,510	20,851,820	22.8
3.	New York	17,990,455	18,976,457	5.5
4.	Florida	12,937,926	15,982,378	23.5
5.	Illinois	11,430,602	12,419,293	8.6
6.	Pennsylvania	11,881,643	12,281,054	3.4
7.	Ohio	10,847,115	11,353,140	4.7
8.	Michigan	9,295,297	9,938,444	6.9
9.	New Jersey	7,730,188	8,414,350	8.9
10.	Georgia	6,478,216	8,186,453	26.4
11.	North Carolina	6,628,637	8,049,313	21.4
12.	Virginia	6,187,358	7,078,515	14.4
13.	Massachusetts	6,016,425	6,349,097	5.5
14.	Indiana	5,544,159	6,080,485	9.7
15.	Washington	4,866,692	5,894,121	21.1
16.	Tennessee	4,877,185	5,689,283	16.7
17.	Missouri	5,117,073	5,595,211	9.3
18.	Wisconsin	4,891,769	5,363,675	9.6
19.	Maryland	4,781,468	5,296,486	10.8
20.	Arizona	3,665,228	5,130,632	40.0
21.	Minnesota	4,375,099	4,919,479	12.4
22.	Louisiana	4,219,973	4,468,976	5.9
23.	Alabama	4,040,587	4,447,100	10.1
24.	Colorado	3,294,394	4,301,261	30.6
25.	Kentucky	3,685,296	4,041,769	9.7
26.	South Carolina	3,486,703	4,012,012	15.1
27.	Oklahoma	3,145,585	3,450,654	9.7
28.	Oregon	2,842,321	3,421,399	20.4
29.	Connecticut	3,287,116	3,405,565	3.6
30.	Iowa	2,776,755	2,926,324	5.4
31.	Mississippi	2,573,216	2,844,658	10.5
32.	Kansas	2,477,574	2,688,418	8.5
33.	Arkansas	2,350,725	2,673,400	13.7
34.	Utah	1,722,850	2,233,169	29.6
35.	Nevada	1,201,833	1,998,257	66.3
36.	New Mexico	1,515,069	1,819,046	20.1
37.	West Virginia	1,793,477	1,808,344	0.8
38.	Nebraska	1,578,385	1,711,263	8.4
39.	Idaho	1,006,749	1,293,953	28.5
40.	Maine	1,227,928	1,274,923	3.8
41.	New Hampshire	1,109,252	1,235,786	11.4
42.	Hawaii	1,108,229	1,211,537	9.3
43.	Rhode Island	1,003,464	1,048,319	4.5
44.	Montana	799,065	902,195	12.9
45.	Delaware	666,168	783,600	17.6
46.	South Dakota	696,004	754,844	8.5
47.	North Dakota	638,800	642,200	0.5
48.	Alaska	550,043	626,932	14.0
49.	Vermont	562,758	608,827	8.2
50.	District of Columbia	606,900	572,059	–5.7
51.	Wyoming	453,588	493,782	8.9
	Total Resident Population	**248,709,873**	**281,421,906**	**13.2**

Density of Population by State, 1930-2000

Source: Bureau of the Census, U.S. Dept. of Commerce

(per square mile, land area only)

STATE	1930	1960	1980	1990	2000
AL	51.8	64.2	76.6	79.6	87.6
AK*	.1	0.4	0.7	1.0	1.1
AZ	3.8	11.5	23.9	32.3	45.2
AR	35.2	34.2	43.9	45.1	51.3
CA	36.2	100.4	151.4	190.8	217.2
CO	10.0	16.9	27.9	31.8	41.5
CT	328.0	520.6	637.8	678.4	702.9
DE	120.5	225.2	307.6	340.8	401.0
DC	7,981.5	12,523.9	10,132.3	9,882.8	9,378.0
FL	27.1	91.5	180.0	239.6	296.4
GA	49.7	67.8	94.1	111.9	141.4
HI*	57.5	98.5	150.1	172.5	188.6
ID	5.4	8.1	11.5	12.2	15.6
IL	136.4	180.4	205.3	205.6	223.4
IN	89.4	128.8	152.8	154.6	169.5
IA	44.1	49.2	52.1	49.7	52.4
KS	22.9	26.6	28.9	30.3	32.9
KY	65.2	76.2	92.3	92.8	101.7
LA	46.5	72.2	94.5	96.9	102.6
ME	25.7	31.3	36.3	39.8	41.3
MD	165.0	313.5	428.7	489.2	541.9
MA	537.4	657.3	733.3	767.6	809.8
MI	84.9	137.7	162.6	163.6	175.0
MN	32.0	43.1	51.2	55.0	61.8
MS	42.4	46.0	53.4	54.9	60.6
MO	52.4	62.6	71.3	74.3	81.2
MT	3.7	4.6	5.4	5.5	6.2
NE	18.0	18.4	20.5	20.5	22.3
NV	.8	2.6	7.3	10.9	18.2
NH	51.6	67.2	102.4	123.7	137.8
NJ	537.3	805.5	986.2	1,042.0	1,134.5
NM	3.5	7.8	10.7	12.5	15.0
NY	262.6	350.6	370.6	381.0	401.9
NC	64.5	93.2	120.4	136.1	165.2
ND	9.7	9.1	9.4	9.3	9.3
OH	161.6	236.6	263.3	264.9	277.3
OK	34.6	33.8	44.1	45.8	50.3
OR	9.9	18.4	27.4	29.6	35.6
PA	213.8	251.4	264.3	265.1	274.0
RI	649.8	819.3	897.8	960.3	1,003.2
SC	56.8	78.7	103.4	115.8	133.2
SD	9.1	9.0	9.1	9.2	9.9
TN	62.4	86.2	111.6	118.3	138.0
TX	22.1	36.4	54.3	64.9	79.6
UT	6.2	10.8	17.8	21.0	27.2
VT	38.8	42.0	55.2	60.8	65.8
VA	60.7	99.6	134.7	156.3	178.8
WA	23.3	42.8	62.1	73.1	88.6
WV	71.8	77.2	80.8	74.5	75.1
WI	53.7	72.6	86.5	90.1	98.8
WY	2.3	3.4	4.9	4.7	5.1
U.S.	**41.2**	**50.6**	**64.0**	**70.3**	**79.6**

* For purposes of comparison, Alaska and Hawaii are included in above tabulation for 1930, even though not states then.

25 Largest Counties, by Population, 1990-2000

Source: Bureau of the Census, U.S. Dept of Commerce

County	2000 Population	1990 Population	Percentage change, 1990-2000
Los Angeles, CA	9,519,338	8,863,164	7.4
Cook, IL	5,376,741	5,105,067	5.3
Harris, TX	3,400,578	2,818,199	20.7
Maricopa, AZ	3,072,149	2,122,101	44.8
Orange, CA	2,846,289	2,410,556	18.1
San Diego, CA	2,813,833	2,498,016	12.6
Kings, NY	2,465,326	2,300,664	7.2
Miami-Dade, FL	2,253,362	1,937,094	16.3
Queens, NY	2,229,379	1,951,598	14.2
Dallas, TX	2,218,899	1,852,810	19.8
Wayne, MI	2,061,162	2,111,687	–2.4
King, WA	1,737,034	1,507,319	15.2
San Bernardino, CA	1,709,434	1,418,380	20.5
Santa Clara, CA	1,682,585	1,497,577	12.4
Broward, FL	1,623,018	1,255,488	29.3
Riverside, CA	1,545,387	1,170,413	32.0
New York, NY	1,537,195	1,487,536	3.3
Philadelphia, PA	1,517,550	1,585,577	–4.3
Middlesex, MA	1,465,396	1,398,468	4.8
Tarrant, TX	1,446,219	1,170,103	23.6
Alameda, CA	1,443,741	1,279,182	12.9
Suffolk, NY	1,419,369	1,321,864	7.4
Cuyahoga, OH	1,393,978	1,412,140	–1.3
Bexar, TX	1,392,931	1,185,394	17.5
Clark, NV	1,375,765	741,459	85.5

Note on least populated counties: The following are the smallest counties by 2000 population: Loving County, TX (64); Kalawao County, HI (147); King County, TX (356); Kenedy County, TX (414); Arthur County, NE (444); Petroleum County, MT (493); McPherson County, NE (533); San Juan County, CO (558); Blaine County, NE (583); and Loup County, NE (712).

Metropolitan Areas, 1990-2000

Source: Bureau of the Census, U.S. Dept. of Commerce

(CMSAs and MSAs of more than 600,000 persons listed by Census 2000 population counts)

Metropolitan statistical areas (MSAs) are defined for federal statistical use by the Office of Management and Budget (OMB), with technical assistance from the Bureau of the Census. Most individual metropolitan areas with populations over 1 million may, under specified circumstances, be subdivided into component Primary Metropolitan Statistical Areas (PMSAs), in which case the area as a whole is designated a Consolidated Metropolitan Statistical Area (CMSA).

Effective June 30, 1999, the Office of Management and Budget designated 261 MSAs, 76 PMSAs, and 19 CMSAs for the U.S. and Puerto Rico.

CMSAs and MSAs	Population 2000	Population 1990	Percent Change 1990-2000
New York–Northern New Jersey–Long Island, NY–NJ–CT–PA CMSA	21,199,865	19,549,649	8.4
Los Angeles–Riverside–Orange County, CA CMSA	16,373,645	14,531,529	12.7
Chicago–Gary–Kenosha, IL–IN–WI CMSA	9,157,540	8,239,820	11.1
Washington–Baltimore, DC–MD–VA–WV CMSA	7,608,070	6,727,050	13.1
San Francisco–Oakland–San Jose, CA CMSA	7,039,362	6,253,311	12.6
Philadelphia–Wilmington–Atlantic City, PA–NJ–DE–MD CMSA	6,188,463	5,892,937	5.0
Boston–Worcester–Lawrence, MA–NH–ME–CT CMSA	5,819,100	5,455,403	6.7
Detroit–Ann Arbor–Flint, MI CMSA	5,456,428	5,187,171	5.2
Dallas–Fort Worth, TX CMSA	5,221,801	4,037,282	29.3
Houston–Galveston–Brazoria, TX CMSA	4,669,571	3,731,131	25.2
Atlanta, GA MSA	4,112,198	2,959,950	38.9
Miami–Fort Lauderdale, FL CMSA	3,876,380	3,192,582	21.4
Seattle–Tacoma–Bremerton, WA CMSA	3,554,760	2,970,328	19.7
Phoenix–Mesa, AZ MSA	3,251,876	2,238,480	45.3
Minneapolis–St. Paul, MN–WI MSA	2,968,806	2,538,834	16.9
Cleveland–Akron, OH CMSA	2,945,831	2,859,644	3.0
San Diego, CA MSA	2,813,833	2,498,016	12.6
St. Louis, MO–IL MSA	2,603,607	2,492,525	4.5
Denver–Boulder–Greeley, CO CMSA	2,581,506	1,980,140	30.4
San Juan–Caguas–Arecibo, PR CMSA	2,450,292	2,270,808	7.9
Tampa–St. Petersburg–Clearwater, FL MSA	2,395,997	2,067,959	15.9
Pittsburgh, PA MSA	2,358,695	2,394,811	–1.5
Portland–Salem, OR–WA CMSA	2,265,223	1,793,476	26.3
Cincinnati–Hamilton, OH–KY–IN CMSA	1,979,202	1,817,571	8.9
Sacramento–Yolo, CA CMSA	1,796,857	1,481,102	21.3
Kansas City, MO–KS MSA	1,776,062	1,582,875	12.2
Milwaukee–Racine, WI CMSA	1,689,572	1,607,183	5.1
Orlando, FL MSA	1,644,561	1,224,852	34.3
Indianapolis, IN MSA	1,607,486	1,380,491	16.4
San Antonio, TX MSA	1,592,383	1,324,749	20.2
Norfolk–Virginia Beach–Newport News, VA–NC MSA	1,569,541	1,443,244	8.8
Las Vegas, NV–AZ MSA	1,563,282	852,737	83.3
Columbus, OH MSA	1,540,157	1,345,450	14.5
Charlotte–Gastonia–Rock Hill, NC–SC MSA	1,499,293	1,162,093	29.0
New Orleans, LA MSA	1,337,726	1,285,270	4.1
Salt Lake City–Ogden, UT MSA	1,333,914	1,072,227	24.4
Greensboro–Winston-Salem–High Point, NC MSA	1,251,509	1,050,304	19.2
Austin–San Marcos, TX MSA	1,249,763	846,227	47.7
Nashville, TN MSA	1,231,311	985,026	25.0
Providence–Fall River–Warwick, RI–MA MSA	1,188,613	1,134,350	4.8
Raleigh–Durham–Chapel Hill, NC MSA	1,187,941	855,545	38.9
Hartford, CT MSA	1,183,110	1,157,585	2.2
Buffalo–Niagara Falls, NY MSA	1,170,111	1,189,288	–1.6
Memphis, TN–AR–MS MSA	1,135,614	1,007,306	12.7
West Palm Beach–Boca Raton, FL MSA	1,131,184	863,518	31.0
Jacksonville, FL MSA	1,100,491	906,727	21.4
Rochester, NY MSA	1,098,201	1,062,470	3.4
Grand Rapids–Muskegon–Holland, MI MSA	1,088,514	937,891	16.1
Oklahoma City, OK MSA	1,083,346	958,839	13.0
Louisville, KY–IN MSA	1,025,598	948,829	8.1
Richmond–Petersburg, VA MSA	996,512	865,640	15.1
Greenville–Spartanburg–Anderson, SC MSA	962,441	830,563	15.9
Dayton–Springfield, OH MSA	950,558	951,270	–0.1
Fresno, CA MSA	922,516	755,580	22.1
Birmingham, AL MSA	921,106	840,140	9.6
Honolulu, HI MSA	876,156	836,231	4.8
Albany–Schenectady–Troy, NY MSA	875,583	861,424	1.6
Tucson, AZ MSA	843,746	666,880	26.5
Tulsa, OK MSA	803,235	708,954	13.3
Syracuse, NY MSA	732,117	742,177	–1.4
Omaha, NE–IA MSA	716,998	639,580	12.1
Albuquerque, NM MSA	712,738	589,131	21.0
Knoxville, TN MSA	687,249	585,960	17.3
El Paso, TX MSA	679,622	591,610	14.9
Bakersfield, CA MSA	661,645	543,477	21.7
Allentown–Bethlehem–Easton, PA MSA	637,958	595,081	7.2
Harrisburg–Lebanon–Carlisle, PA MSA	629,401	587,986	7.0
Scranton–Wilkes-Barre–Hazleton, PA MSA	624,776	638,466	–2.1
Toledo, OH MSA	618,203	614,128	0.7
Baton Rouge, LA MSA	602,894	528,264	14.1

Final 2000 census figures showed that the nation in that year had 50 metropolitan areas of at least 1 mil people, including 9 that had reached that size since 1990. The 50 areas had 161.5 mil people, or 57.5% of the U.S. population, in 2000.

Almost 226 mil people resided in metropolitan areas in 2000, an increase of 27.6 mil (13.9%) since 1990. The population outside metropolitan areas totaled 55.4 mil in 2000, up 5.1 mil (10.2%) from 1990. The metropolitan population in 2000 was 80.3% of the U.S. total, compared with 79.8% in 1990 and 76.2% in 1980.

Population of 100 Largest U.S. Cities, 1850-2000

Source: Bureau of the Census, U.S. Dept. of Commerce (100 most populous cities ranked by Census 2000 population counts)

Rank	City	2000	1990	1980	1970	1950	1900	1850
1.	New York, NY	8,008,278	7,322,564	7,071,639	7,895,563	7,891,957	3,437,202	696,115
2.	Los Angeles, CA	3,694,820	3,485,398	2,968,528	2,811,801	1,970,358	102,479	1,610
3.	Chicago, IL	2,896,016	2,783,726	3,005,072	3,369,357	3,620,962	1,698,575	29,963
4.	Houston, TX	1,953,631	1,630,553	1,595,138	1,233,535	596,163	44,633	2,396
5.	Philadelphia, PA	1,517,550	1,585,577	1,688,210	1,949,996	2,071,605	1,293,697	121,376
6.	Phoenix, AZ	1,321,045	983,403	789,704	584,303	106,818	5,544	...
7.	San Diego, CA	1,223,400	1,110,549	875,538	697,471	334,387	17,700	...
8.	Dallas, TX	1,188,580	1,006,877	904,599	844,401	434,462	42,638	...
9.	San Antonio, TX	1,144,646	935,933	785,940	654,153	408,442	53,321	3,488
10.	Detroit, MI	951,270	1,027,974	1,203,368	1,514,063	1,849,568	285,704	21,019
11.	San Jose, CA	894,943	782,248	629,400	459,913	95,280	21,500	...
12.	Indianapolis, IN[1]	791,926	741,952	700,807	736,856	427,173	169,164	8,091
13.	San Francisco, CA	776,733	723,959	678,974	715,674	775,357	342,782	34,776
14.	Jacksonville, FL[1]	735,617	635,230	540,920	504,265	204,517	28,429	1,045
15.	Columbus, OH	711,470	632,910	565,021	540,025	375,901	125,560	17,882
16.	Austin, TX	656,562	465,622	345,890	253,539	132,459	22,258	629
17.	Baltimore, MD	651,154	736,014	786,741	905,787	949,708	508,957	169,054
18.	Memphis, TN	650,100	610,337	646,174	623,988	396,000	102,320	8,841
19.	Milwaukee, WI	596,974	628,088	636,297	717,372	637,392	285,315	20,061
20.	Boston, MA	589,141	574,283	562,994	641,071	801,444	560,892	136,881
21.	Washington, DC	572,059	606,900	638,432	756,668	802,178	278,718	40,001
22.	Nashville, TN[1]	569,891	510,784	455,651	426,029	174,307	80,865	10,165
23.	El Paso, TX	563,662	515,342	425,259	322,261	130,485	15,906	...
24.	Seattle, WA	563,374	516,259	493,846	530,831	467,591	80,671	...
25.	Denver, CO	554,636	467,610	492,686	514,678	415,786	133,859	...
26.	Charlotte, NC	540,828	395,934	315,474	241,420	134,042	18,091	1,065
27.	Fort Worth, TX	534,694	447,619	385,164	393,455	278,778	26,688	...
28.	Portland, OR	529,121	437,319	368,148	379,967	373,628	90,426	...
29.	Oklahoma City, OK	506,132	444,719	404,014	368,164	243,504	10,037	...
30.	Tucson, AZ	486,699	405,390	330,537	262,933	45,454	7,531	...
31.	New Orleans, LA	484,674	496,938	557,927	593,471	570,445	287,104	116,375
32.	Las Vegas, NV	478,434	258,295	164,674	125,787	24,624	...	...
33.	Cleveland, OH	478,403	505,616	573,822	750,879	914,808	381,768	17,034
34.	Long Beach, CA	461,522	429,433	361,498	358,879	250,767	2,252	...
35.	Albuquerque, NM	448,607	384,736	332,920	244,501	96,815	6,238	...
36.	Kansas City, MO	441,545	435,146	448,028	507,330	456,622	163,752	...
37.	Fresno, CA	427,652	354,202	217,491	165,655	91,669	12,470	...
38.	Virginia Beach, VA	425,257	393,069	262,199	172,106	5,390	...	...
39.	Atlanta, GA	416,474	394,017	425,022	495,039	331,314	89,872	2,572
40.	Sacramento, CA	407,018	369,365	275,741	257,105	137,572	29,282	6,820
41.	Oakland, CA	399,484	372,242	339,337	361,561	384,575	66,960	...
42.	Mesa, AZ	396,375	288,091	152,404	63,049	16,790	722	...
43.	Tulsa, OK	393,049	367,302	360,919	330,350	182,740	1,390	...
44.	Omaha, NE	390,007	335,795	313,939	346,929	251,117	102,555	...
45.	Minneapolis, MN	382,618	368,383	370,951	434,400	521,718	202,718	...
46.	Honolulu, HI[2]	371,657	365,272	365,048	324,871	248,034	39,306	...
47.	Miami, FL	362,470	358,548	346,681	334,859	249,276	1,681	...
48.	Colorado Springs, CO	360,890	281,140	215,105	135,517	45,472	21,085	...
49.	St. Louis, MO	348,189	396,685	452,801	622,236	856,796	575,238	77,860
50.	Wichita, KS	344,284	304,011	279,838	276,554	168,279	24,671	...
51.	Santa Ana, CA	337,977	293,742	204,023	155,710	45,533	4,933	...
52.	Pittsburgh, PA	334,563	369,879	423,959	520,089	676,806	321,616	46,601
53.	Arlington, TX	332,969	261,721	160,113	90,229	7,692	1,079	...
54.	Cincinnati, OH	331,285	364,040	385,409	453,514	503,998	325,902	115,435
55.	Anaheim, CA	328,014	266,406	219,494	166,408	14,556	1,456	...
56.	Toledo, OH	313,619	332,943	354,635	383,062	303,616	131,822	3,829
57.	Tampa, FL	303,447	280,015	271,577	277,714	124,681	15,839	...
58.	Buffalo, NY	292,648	328,123	357,870	462,768	580,132	352,387	42,261
59.	St. Paul, MN	287,151	272,235	270,230	309,866	311,349	163,065	1,112
60.	Corpus Christi, TX	277,454	257,453	232,134	204,525	108,287	4,703	...
61.	Aurora, CO	276,393	222,103	158,588	74,974	11,421	202	...
62.	Raleigh, NC	276,093	207,951	150,255	122,830	65,679	13,643	4,518
63.	Newark, NJ	273,546	275,221	329,248	381,930	438,776	246,070	38,894
64.	Lexington, KY	260,512	225,366	204,165	108,137	55,534	26,369	8,159
65.	Anchorage, AK	260,283	226,338	174,431	48,081	11,254	...	...
66.	Louisville, KY	256,231	269,063	298,694	361,706	369,129	204,731	43,194
67.	Riverside, CA	255,166	226,505	170,591	140,089	46,764	7,973	...
68.	St. Petersburg, FL	248,232	238,629	238,647	216,159	96,738	1,575	...
69.	Bakersfield, CA	247,057	174,820	105,611	69,515	34,784	4,836	...
70.	Stockton, CA	243,771	210,943	148,283	109,963	70,853	17,506	...
71.	Birmingham, AL	242,820	265,968	284,413	300,910	326,037	38,415	...
72.	Jersey City, NJ	240,055	228,537	223,532	260,350	299,017	206,433	6,856
73.	Norfolk, VA	234,403	261,229	266,979	307,951	213,513	46,624	14,326
74.	Baton Rouge, LA	227,818	219,531	220,394	165,921	125,629	11,269	3,905
75.	Hialeah, FL	226,419	188,004	145,254	102,452	19,676	...	...
76.	Lincoln, NE	225,581	191,972	171,932	149,518	98,884	40,169	...
77.	Greensboro, NC	223,891	183,521	155,642	144,076	74,389	10,035	...
78.	Plano, TX	222,030	128,713	72,331	17,872	2,126	1,304	...
79.	Rochester, NY	219,773	231,636	241,741	295,011	332,488	162,608	36,403
80.	Glendale, AZ	218,812	148,134	96,988	36,228	8,179	...	...
81.	Akron, OH	217,074	223,019	237,177	275,425	274,605	42,728	3,266
82.	Garland, TX	215,768	180,650	138,857	81,437	10,571	819	...
83.	Madison, WI	208,054	191,262	170,616	171,809	96,056	19,164	1,525
84.	Fort Wayne, IN	205,727	173,072	172,391	178,269	133,607	45,115	4,282
85.	Fremont, CA	203,413	173,339	131,945	100,869	...	...	...
86.	Scottsdale, AZ	202,705	130,069	88,364	67,823	2,032	...	...

Rank	City	2000	1990	1980	1970	1950	1900	1850
87.	Montgomery, AL	201,568	187,106	177,857	133,386	106,525	30,346	8,728
88.	Shreveport, LA	200,145	198,525	206,989	182,064	127,206	16,013	1,728
89.	Augusta, GA[1]	199,775	44,639	47,532	59,864	71,508	39,441	9,448
90.	Lubbock, TX	199,564	186,206	174,361	149,101	71,747	...	...
91.	Chesapeake, VA	199,184	151,976	114,486	89,580	...	...	...
92.	Mobile, AL	198,915	196,278	200,452	190,026	129,009	38,469	20,515
93.	Des Moines, IA	198,682	193,187	191,003	201,404	177,965	62,139	...
94.	Grand Rapids, MI	197,800	189,126	181,843	197,649	176,515	87,565	2,686
95.	Richmond, VA	197,790	203,056	219,214	249,332	230,310	85,050	27,570
96.	Yonkers, NY	196,086	188,082	195,351	204,297	152,798	47,931	...
97.	Spokane, WA	195,629	177,196	171,300	170,516	161,721	36,848	...
98.	Glendale, CA	194,973	180,038	139,060	133,000	96,000	...	...
99.	Tacoma, WA	193,556	176,664	158,501	154,407	143,673	37,714	...
100.	Irving, TX	191,615	155,037	109,943	97,457	2,615	...	...

(1) Indianapolis, IN; Jacksonville, FL; Nashville, TN; and Augusta, GA, are parts of consolidated city-county governments. Populations of other incorporated places in the county have been excluded from the population totals shown here. For years that predate the establishment of a consolidated city-county government, city population is shown. (2) Locations in Hawaii are called "census designated places (CDPs)." Although these areas are not incorporated, they are recognized for census purposes as large urban places. Honolulu CDP is coextensive with Honolulu Judicial District within the city and county of Honolulu.

Mobility, by Selected Characteristics, 1999-2000

Source: Bureau of the Census, U.S. Dept. of Commerce
(numbers in thousands)

	Total no. of movers[1]	MOVED TO: Same county	Diff. county, same state	Diff. state	Abroad
Marital status					
Married, spouse present	13,143	6,892	2,812	2,860	580
Married, spouse absent	707	348	139	127	92
Widowed	944	521	198	200	24
Divorced	3,771	2,246	748	724	55
Separated	1,231	790	220	195	27
Never married	13,393	7,562	2,819	2,439	574
Educational attainment[2]					
Less than 9th grade	1,469	854	274	219	123
Grades 9-12, no diploma	2,164	1,347	430	329	59
High school grad.	7,152	4,079	1,413	1,455	206
Some college or AA degree	6,051	3,390	1,372	1,148	141
Bachelor's degree	4,581	2,199	1,083	1,058	241
Prof. or graduate degree	1,912	843	401	520	147

	Total no. of movers[1]	MOVED TO: Same county	Diff. county, same state	Diff. state	Abroad
Income[3]					
Under $5,000	4,107	2,144	814	883	264
$5,000-$9,999	4,142	2,420	846	738	138
$10,000-$19,999	7,308	4,365	1,448	1,316	179
$20,000-$29,999	5,292	3,129	1,099	941	124
$30,000-$39,999	3,465	1,878	824	683	78
$40,000-$49,999	1,893	973	493	392	36
$50,000-$59,999	1,143	582	266	266	29
$60,000-$74,999	1,056	576	237	210	33
$75,000-$99,999	723	331	175	188	29
$100,000 and over	713	323	152	193	45
Ownership status					
Owner	17,150	9,396	4,042	1,189	478
Renter	26,238	15,003	4,772	1,917	1,268
TOTAL[4]	**43,388**	**24,399**	**8,814**	**8,428**	**1,746**

(1) People who moved to a new residence in 12-month period ending in Mar. 2000. (2) People 25 years and older. (3) People 15 years and older. (4) People 1 year and older.

U.S. Population, by Age, Sex, and Household, 2000

Source: Bureau of the Census, U.S. Dept. of Commerce; 2000 Census

	Number	%
Total population	**281,421,906**	
AGE		
Under 5 years	19,175,798	6.8
5 to 9 years	20,549,505	7.3
10 to 14 years	20,528,072	7.3
15 to 19 years	20,219,890	7.2
20 to 24 years	18,964,001	6.7
25 to 34 years	39,891,724	14.2
35 to 44 years	44,148,527	16.0
45 to 54 years	37,677,952	13.4
55 to 59 years	13,469,237	4.8
60 to 64 years	10,805,447	3.8
65 to 74 years	18,390,986	6.5
75 to 84 years	12,361,180	4.4
85 years and over	4,239,587	1.5
18 years and over	209,128,094	74.3
Male	100,994,367	35.9
Female	108,133,727	38.4
21 years and over	196,899,193	70.0

	Number	%
62 years and over	41,256,029	14.7
65 years and over	34,991,753	12.4
SEX		
Male	138,053,563	49.1
Female	143,368,343	50.9
HOUSEHOLDS BY TYPES		
Total Households	105,480,101	
Family households (families)	71,787,347	68.1
Married-couple families	54,493,232	51.7
Female householder, no husband present	12,900,103	12.2
Nonfamily households	33,692,754	31.9
Householder living alone	27,230,075	25.8
Householder 65 years and over	9,722,857	9.2
Persons living in households	273,643,273	NA
Persons per household	2.59	NA
Persons living in group quarters	7,778,633	NA
Institutionalized persons	4,059,039	NA
Other persons in group quarters	3,719,594	NA

NA = not applicable.

U.S. Population Abroad, by Selected Country, 1999

Source: U.S. Dept. of State

Area	Resident U.S. citizens[1]	Area	Resident U.S. citizens[1]	Area	Resident U.S. citizens[1]	Area	Resident U.S. citizens[1]
Argentina	27,600	Egypt	10,890	Italy	168,970	Saudi Arabia	35,990
Australia	102,800	France	101,750	Japan	70,350	South Korea	30,000
Belgium	35,330	Germany	210,880	Mexico	1,036,300	Spain	94,510
Brazil	40,640	Greece	72,500	Netherlands	23,707	Switzerland	12,110
Canada	687,700	Hong Kong	48,220	Panama	19,700	United Kingdom	224,000
Costa Rica	19,800	Ireland	46,980	Philippines	105,000	Venezuela	25,000
Dominican Republic	82,000	Israel	18,000	Portugal	2,170	**Total[2]**	**3,784,690**

Note: Figures do not include U.S. military or government personnel or their dependents. (1) Estimated. (2) Includes other areas not shown separately.

U.S. Foreign-Born Population

Source: Bureau of the Census, U.S. Dept. of Commerce

Percentage of Population That Is Foreign-Born, 1900-2000

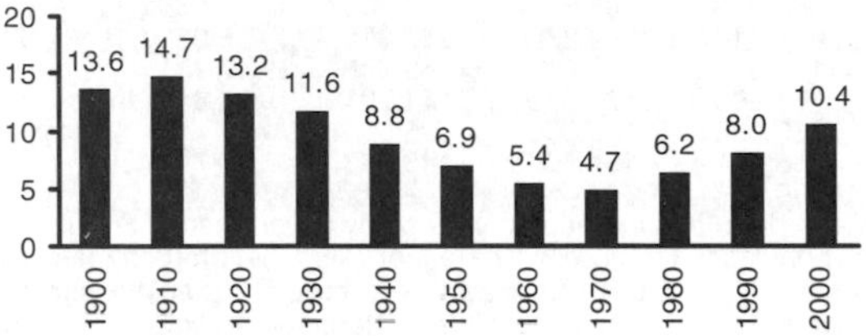

U.S. Foreign Born Population by Regional Origin, 1995-2000

Region	2000 (in thousands)	1995 (in thousands)
Europe	4,355	3,937
Under 18	250	232
Asia	7,246	6,121
Under 18	657	767
Latin America	14,477	11,777
Under 18	1,684	1,481
Other	2,301	2,658
Under 18	245	275
ALL REGIONS	28,379	24,493
Under 18	2,837	2,726

Foreign-Born Population: Top Countries of Origin, 1920, 1960, 1999

Source: Bureau of the Census, U.S. Dept. of Commerce

(in thousands)

Country	1920 Number	Percent	Country	1960 Number	Percent	Country	1999 Number	Percent
Germany	1,686	12.1	Italy	1,256	12.9	Mexico	7,197	27.2
Italy	1,610	11.6	Germany	989	10.2	Philipines	1,455	5.5
Soviet Union	1,400	10.1	Canada	953	9.8	China and Hong Kong	985	3.7
Poland	1,139	8.2	Great Britain	764	7.9	Vietnam	966	3.7
Canada	1,138	8.2	Poland	747	7.7	Cuba	943	3.6
Great Britain	1,135	8.2	Soviet Union	690	7.1	India	839	3.2
Ireland	1,037	7.5	Mexico	575	5.9	El Salvador	761	2.9
Sweden	625	4.5	Ireland	338	3.5	Dominican Republic	679	2.6
Austria	575	4.1	Austria	304	3.1	Great Britian	655	2.5
Mexico	486	3.5	Hungary	245	2.5	Korea	611	2.3

Immigrants Admitted, by Top 30 Metropolitan Areas of Intended Residence, 1998

Source: Immigration and Naturalization Service, U.S. Dept. of Justice

(fiscal year 1998)

Metropolitan Statistical Area	Number	Percentage	Metropolitan Statistical Area	Number	Percentage
TOTAL immigrants admitted to U.S.	**660,477**	**100.0**	San Diego, CA	9,836	1.5
New York, NY	81,417	12.3	Dallas, TX	9,641	1.5
Los Angeles–Long Beach, CA	60,220	9.1	Newark, NJ	9,551	1.4
Chicago, IL	31,033	4.7	Seattle–Bellevue–Everett, WA	9,497	1.4
Miami, FL	29,242	4.4	Philadelphia, PA–NJ	9,197	1.4
Washington, DC–MD–VA	25,639	3.9	Bergen–Passaic, NJ	8,645	1.3
San Francisco, CA	14,553	2.2	Nassau–Suffolk, NY	7,932	1.2
Orange County, CA	14,291	2.2	Atlanta, GA	7,504	1.1
Oakland, CA	13,499	2.0	Jersey City, NJ	5,860	0.9
Houston, TX	13,189	2.0	Minneapolis–St. Paul, MN–WI	5,719	0.9
Boston–Lawrence–Lowell–Brockton, MA	12,854	1.9	Middlesex–Somerset– Hunterdon, NJ	5,457	0.8
San Jose, CA	11,811	1.8	West Palm Beach–Boca Raton, FL	4,951	0.7
Riverside–San Bernardino, CA	10,227	1.5	Portland–Vancouver, OR–WA	4,751	0.7
Fort Lauderdale, FL	9,954	1.5	Fresno, CA	4,517	0.7
Detroit, MI	9,852	1.5	Honolulu, HI	4,214	0.6
			Sacramento, CA	4,212	0.6

Immigrants Admitted, by State of Intended Residence, 1998

Source: Immigration and Naturalization Service, U.S. Dept. of Justice

(fiscal year 1998)

STATE	Immigrants	STATE	Immigrants	STATE	Immigrants	STATE	Immigrants
AL	1,608	IA	1,655	NJ	35,091	VT	513
AK	1,008	KS	3,184	NM	2,199	VA	15,686
AZ	6,211	KY	2,017	NY	96,559	WA	16,920
AR	914	LA	2,193	NC	6,415	WV	375
CA	170,126	ME	709	ND	472	WI	3,724
CO	6,513	MD	15,561	OH	7,697	WY	159
CT	7,780	MA	15,869	OK	2,273	Other:	
DE	1,063	MI	13,943	OR	5,909	Guam	1,835
DC	2,377	MN	6,981	PA	11,942	N Mariana Isls.	103
FL	59,965	MS	701	RI	1,976	Puerto Rico	3,251
GA	10,445	MO	3,588	SC	2,125	Virgin Isls.	979
HI	5,465	MT	299	SD	356	Armed Service Posts	88
ID	1,504	NE	1,267	TN	2,806	Other or unknown	6,030
IL	33,163	NV	6,106	TX	44,428	**TOTAL**	**660,477**
IN	3,981	NH	1,010	UT	3,360		

The Elderly U.S. Population, 1900-2000

Source: Bureau of the Census, U.S. Dept. of Commerce

Year[1]	65 AND OVER Number[2]	65 AND OVER Percent	85 AND OVER Number[2]	85 AND OVER Percent
1900	3,080	4.1	122	0.2
1910	3,949	4.3	167	0.2
1920	4,933	4.7	210	0.2
1930	6,634	5.4	272	0.2
1940	9,019	6.8	365	0.3
1950	12,269	8.1	577	0.4
1960	16,560	9.2	929	0.5
1970	19,980	9.8	1,409	0.7
1980	25,550	11.3	2,240	1.0
1990	31,079	12.5	3,021	1.2
1995	33,619	12.8	3,685	1.4
2000	34,992	12.4	4,240	1.5

NOTE: Figures for 1900 to 1950 exclude Alaska and Hawaii. (1) Date of Census. (2) Resident population, in thousands.

Projections of Total Population, by Age, 2010-2100

Source: Bureau of the Census, U.S. Dept. of Commerce

Age	2010 Population[1]	2010 Percentage Distribution	2025 Population[1]	2025 Percentage Distribution	2050 Population[1]	2050 Percentage Distribution	2100 Population[1]	2100 Percentage Distribution
TOTAL	299,862	**100.0**	**337,815**	**100.0**	**403,687**	**100.0**	**570,954**	**100.0**
Under 5 years	20,099	6.7	22,551	6.7	26,914	6.7	36,068	6.3
5-14 years	39,346	13.1	44,486	13.2	52,869	13.1	71,807	12.6
15-24 years	42,819	14.3	43,614	12.9	52,769	13.1	72,620	12.7
25-34 years	38,851	13.0	42,872	12.7	50,458	12.5	68,775	12.0
35-44 years	39,443	13.2	43,234	12.8	49,588	12.3	67,912	11.9
45-54 years	44,161	14.7	38,291	11.3	45,445	11.3	63,787	11.2
55-64 years	35,429	11.8	40,125	11.9	43,644	10.8	58,822	10.3
65 years and over	39,715	13.2	62,641	18.5	81,999	20.3	131,163	23.0
85 years and over	5,786	1.9	7,441	2.2	19,352	4.8	37,030	6.5
100 years and over	129	0.0	313	0.1	1,095	0.3	5,323	0.9

NOTE: Assumptions were based on July 1 estimates of U.S. population consistent with the 1990 decennial census, as enumerated. All figures shown are for July 1 of the given year, exclude Armed Forces overseas, and are middle series population projections. For the series shown, different assumptions were made regarding fertility rates (lifetime births per woman), life expectancy, and immigr ation in the coming decades.Yearly net immigration was assumed to be 820,000. Percentage distribution may not equal 100, because of over-lapping categories shown and rounding. (1) In thousands.

U.S. Households, 1960-2000[1]

Source: Bureau of the Census, U.S. Dept. of Commerce
(as of Mar.)

YEAR	Total	Married-couple households	% of Total	Unmarried-couple households
1960	52,799	39,254	74	439
1970	63,401	44,728	71	523
1980	80,776	49,112	61	1,589
1981	82,368	49,294	60	1,808
1982	83,527	49,630	59	1,863
1983	83,918	49,908	59	1,891
1984	85,407	50,090	59	1,988
1985	86,789	50,350	58	1,983
1986	88,458	50,933	58	2,220
1987	89,479	51,537	58	2,334
1988	91,124	51,675	57	2,588
1989	92,830	52,100	56	2,764
1990	93,347	52,317	56	2,856
1991	94,312	52,147	55	3,039
1992	95,669	52,457	55	3,308
1993	96,426	53,090	55	3,510
1994	97,107	53,171	55	3,661
1995	98,990	53,858	54	3,668
1996	99,627	53,567	54	3,958
1997	101,018	53,604	53	4,130
1998	102,528	54,317	53	4,236
1999	103,874	54,770	53	4,486
2000	105,480	54,493	52	5,476

(1) All household numbers in thousands.

Young Adults Living at Home[1] in the U.S., 1960-2000

Source: Bureau of the Census, U.S. Dept. of Commerce
(numbers in thousands)

18-24 years old	Male Total	Male Percent	Female Total	Female Percent
2000	13,291	57	13,242	47
1999	12,936	58	13,031	49
1998	12,633	59	12,568	48
1997	12,534	60	12,452	48
1996	12,402	59	12,441	48
1995	12,545	58	12,613	47
1994	12,683	60	12,792	46
1993	12,049	59	12,260	47
1992	12,083	60	12,351	48
1991	12,275	60	12,627	49
1990	12,450	58	12,860	48
1985	13,695	60	14,149	48
1980	14,278	54	14,844	43
1970	10,398	54	11,959	41
1960	6,842	52	7,876	35

25-34 years old	Male Total	Male Percent	Female Total	Female Percent
2000	18,563	13	19,222	8
1999	18,924	14	19,551	9
1998	19,526	15	19,828	8
1997	20,039	15	20,217	9
1996	20,390	16	20,528	9
1995	20,589	15	20,800	8
1994	20,873	16	21,073	9
1993	20,856	16	21,007	9
1992	21,125	15	21,368	9
1991	21,319	15	21,586	9
1990	21,462	15	21,779	8
1985	20,184	13	20,673	8
1980	18,107	10	18,689	7
1970	11,929	9	12,637	7
1960	10,896	11	11,587	7

(1) Includes young adults living in their parent(s)' home and unmarried college students living in a dormitory.

Grandchildren Living in the Home of Their Grandparents, 1970-2000

Source: Bureau of the Census, U.S. Dept. of Commerce (numbers in thousands)

		Grandchildren living with grandparents				
			WITH PARENT(S) PRESENT			
YEAR	Total children under 18	Total	Both parents present	Mother only present	Father only present	Without parent(s) present
1970	69,276	2,214	363	817	78	957
1980	63,369	2,306	310	922	86	988
1990	64,137	3,155	467	1,563	191	935
1991	65,093	3,320	559	1,674	151	937
1992	65,965	3,253	502	1,740	144	867
1993	66,893	3,368	475	1,647	229	1,017
1994	69,508	3,735	436	1,764	175	1,359
1995	70,254	3,965	427	1,876	195	1,466
1996	70,908	4,060	467	1,943	220	1,431
1997	70,983	3,894	554	1,785	247	1,309
1998	71,377	3,989	503	1,827	241	1,417
1999	71,703	3,919	535	1,803	250	1,331
2000	72,012	3,842	531	1,732	220	1,359

Living Arrangements of Children, 1970-2000

Source: Bureau of the Census, U.S. Dept. of Commerce

(excludes persons under 18 years of age who maintained households or resided in group quarters)

		Percentage of children who live with—							
			MOTHER ONLY						
Race, Hispanic origin, and year	Number (1,000)	BOTH PARENTS	Total	Divorced	Married Spouse absent	Single[1]	Widowed	FATHER ONLY	NEITHER PARENT
White									
1970	58,790	90	8	3	3	Z	2	1	2
1980	52,242	83	14	7	4	1	2	2	2
1990	51,390	79	16	8	4	3	1	3	2
1997	55,868	75	18	8	4	5	1	4	3
1998	56,118	74	18	8	4	5	1	5	3
1999	56,265	74	18	NA	NA	NA	NA	4	3
2000	56,455	75	17	NA	NA	NA	NA	4	3
Black									
1970	9,422	59	30	5	16	4	4	2	10
1980	9,375	42	44	11	16	13	4	2	12
1990	10,018	38	51	10	12	27	2	4	8
1997	11,369	35	52	9	11	31	1	5	8
1998	11,407	36	51	9	9	32	1	4	9
1999	11,425	35	51	NA	NA	NA	NA	4	10
2000	11,412	38	49	NA	NA	NA	NA	4	9
Hispanic[2]									
1970	4,006[3]	78	NA	NA	NA	NA	NA	NA	NA
1980	5,459	75	20	6	8	4	2	2	4
1990	7,174	67	27	7	10	8	2	3	3
1997	10,525	64	27	7	7	12	1	4	5
1998	10,857	64	27	6	8	12	1	4	5
1999	11,236	63	27	NA	NA	NA	NA	5	5
2000	11,613	65	25	NA	NA	NA	NA	4	5

NA = Not available. Z = Less than 0.5%. (1) Never married. (2) Hispanic persons may be of any race. (3) All persons under 18 years old.

Block Grants for Welfare (Temporary Assistance for Needy Families), 2000

Source: Admin. for Children and Families, Off. of Planning, Research, and Evaluation, U.S. Dept. of Health and Human Services

State (for 2000)	Total Federal and State TANF Expenditures[1]	Average Monthly Number of Families	Average Monthly Number of Recipients	Average Monthly Number of Children[2]	Average Monthly Expenditure per Family	Average Monthly Expenditure per Recipient
Alabama	$92,200	19,083	56,408	36,703	$402.62	$136.21
Alaska	93,427	7,347	23,838	15,099	1,059.73	326.60
Arizona	254,820	33,723	87,217	66,346	629.69	243.47
Arkansas	112,328	12,354	29,313	22,042	757.73	319.34
California	5,776,830	501,019	1,307,941	1,003,235	960.85	368.06
Colorado	204,624	11,154	28,837	21,788	1,528.73	591.32
Connecticut	424,587	28,095	66,407	46,692	1,259.40	532.81
Delaware	54,298	6,058	12,879	8,432	746.92	351.33
District of Columbia .	133,869	17,563	46,893	34,271	635.20	237.90
Florida	747,633	67,355	152,709	120,545	924.99	407.98
Georgia	382,763	53,269	140,967	108,947	598.79	226.27
Hawaii.	110,529	14,969	44,425	30,996	615.31	207.33
Idaho	43,706	1,275	2,309	1,885	2,855.69	1,577.73
Illinois	875,026	88,493	264,066	197,923	824.01	276.14
Indiana	310,144	35,872	99,073	71,696	720.49	260.87
Iowa	157,252	20,025	53,267	36,119	654.42	246.01
Kansas	151,184	12,585	31,620	22,808	1,001.09	398.44
Kentucky.	204,951	38,542	88,747	64,300	443.13	192.45
Louisiana	126,336	27,820	86,619	58,718	378.44	121.54
Maine	91,509	10,864	28,191	19,147	701.94	270.50
Maryland	276,909	29,340	72,724	53,565	786.49	317.30
Massachusetts	585,786	44,189	101,452	72,427	1,104.71	481.17

State (for 2000)	Total Federal and State TANF Expenditures[1]	Average Monthly Number of Families	Average Monthly Number of Recipients	Average Monthly Number of Children[2]	Average Monthly Expenditure per Family	Average Monthly Expenditure per Recipient
Michigan	1,198,333	74,709	207,463	152,926	1,336.67	481.34
Minnesota	382,014	39,040	115,749	80,348	815.43	275.03
Mississippi	62,076	14,970	33,801	26,855	345.56	153.05
Missouri	310,512	46,776	124,773	89,326	553.20	207.38
Montana	44,270	4,555	14,249	8,702	809.99	258.90
Nebraska	72,870	9,538	24,037	17,470	636.63	252.63
Nevada	55,937	6,259	15,906	12,024	744.81	293.06
New Hampshire	73,219	5,841	14,035	9,637	1,044.66	434.74
New Jersey	309,510	51,630	130,317	97,259	499.56	197.92
New Mexico	149,090	23,655	72,343	50,007	525,23	171.74
New York	3,235,946	258,702	723,793	491,254	1,042.36	372.57
North Carolina	434,747	45,725	99,553	76,341	792.32	363.92
North Dakota	32,561	2,901	8,706	5,815	935.33	311.68
Ohio	986,714	97,969	245,085	179,956	839.31	335.50
Oklahoma	133,771	14,364	35,472	27,592	776.06	314.26
Oregon	256,047	16,918	41,889	30,024	1,261.22	509.37
Pennsylvania	891,326	89,899	239,125	175,247	826.23	310.62
Rhode Island	151,405	16,324	45,161	31,085	772.91	279.38
South Carolina	124,381	16,059	37,285	29,041	645.43	278.00
South Dakota	21,035	2,802	6,755	5,480	625.65	259.50
Tennessee	270,477	56,476	151,438	107,434	399.10	148.84
Texas	743,149	127,880	342,383	252,107	484.28	180.88
Utah	89,449	8,410	22,292	16,027	886.36	334.39
Vermont	59,035	6,043	16,119	10,322	814.09	305.20
Virginia	218,886	31,864	72,573	53,158	572.45	251.34
Washington	530,635	57,008	153,057	104,750	775.67	288.91
West Virginia	134,042	12,146	32,262	22,105	919.69	346.24
Wisconsin	417,982	16,719	38,056	32,346	2,083.42	915.29
Wyoming	14,787	604	1,183	955	2,041.56	1,041,41
2000 Totals[2]	**22,614,919**	**2,236,776**	**5,890,761**	**4,309,274**	**842.54**	**319.92**
1999 Totals[2]	**18,462,703**	**2,673,610**	**7,187,658**	**5,318,722[3]**	**575.46**	**214.06**
1998 Totals[2]	**20,595,662**	**3,199,700**	**8,790,149**	**6,329,970[3]**	**536.40**	**195.25**

NOTE: Data are preliminary and cover the fiscal year. Under 1996 legislation, the Aid to Families with Dependent Children (AFDC) program was converted to this state block-grant program. (1) In thousands. FY 2000 represents combined spending in FY 2000 from all Federal TANF grants for Fiscal Years 1997-2000. Data provided in prior years for FYs 1997-1999 were for the specified year only. (2)Totals include territories not listed. (3) Based on sample data for some states.

Adults Receiving TANF[1] (Welfare) Funds, by Employment Status, 1998-99

Source: Admin. for Children and Families, Off. of Planning, Research, and Evaluation, U.S Dept. of Health and Human Services

STATE	Adults	% Employed	STATE	Adults	% Employed	STATE	Adults	% Employed
AL	9,960	20.1	LA	27,438	22.4	OR	14,048	6.9
AK	8,663	37.5	ME	11,676	28.4	PA	87,233	27.7
AZ	23,030	34.1	MD	28,503	7.3	Puerto Rico	39,254	6.9
AR	6,574	13.8	MA	39,484	19.7	RI	15,994	27.5
CA	505,957	42.8	MI	68,702	44.7	SC	10,197	23.3
CO	10,345	24.0	MN	36,131	14.6	SD	1,692	21.5
CT	25,770	41.6	MS	8,656	11.0	TN	41,590	27.5
DE	4,091	30.1	MO	36,309	12.1	TX	84,085	6.4
DC	15,142	24.7	MT	4,731	14.7	UT	7,499	27.7
FL	47,222	22.8	NE	9,811	15.4	VT	6,576	27.0
GA	36,431	16.1	NV	5,216	22.5	VA	24,028	30.0
Guam	2,667	2.3	NH	5,085	21.5	Virgin Islands	968	1.1
HI	14,629	24.4	NJ	44,944	13.3	WA	59,769	34.2
ID	613	28.2	NM	26,160	22.1	WV	11,949	9.0
IL	100,384	43.1	NY	270,105	17.5	WI	8,510	11.7
IN	33,749	21.4	NC	32,513	13.2	WY	407	15.2
IA	18,947	31.5	ND	2,379	16.1	**U.S.**	**2,068,024**	**27.6**
KS	9,207	19.9	OH	81,195	26.9			
KY	28,219	17.5	OK	13,590	25.3			

(1) TANF = the state block grant program known asTemporary Assistance for Needy Families.

Poverty Rate

Source: Bureau of the Census, U.S. Dept. of Commerce

The poverty rate is the proportion of the population whose income falls below the government's official poverty level, which is adjusted each year for inflation. The national poverty rate was 11.8% in 1999, a decrease from the 1998 rate of 12.7% and the 1990 rate of 13.5%. The 1999 data showed 16.9% of children lived in poverty; the poverty rate among people 65 and over was 9.7%.

Poverty Level by Family Size, 1980-99

Source: Bureau of the Census, U.S. Dept. of Commerce

	1999	1998	1990	1980
1 person	$8,501	$8,316	$6,652	$4,186
Under 65 years	8,667	8,480	6,800	4,284
65 years and over	7,990	7,818	6,268	3,950
2 persons	10,869	10,634	8,509	5,361
Householder under 65 years	11,214	10,972	8,794	5,537
Householder 65 years and over	10,075	9,862	7,905	4,982

	1999	1998	1990	1980
3 persons	$13,290	$13,003	$10,419	$6,570
4 persons	17,029	16,660	13,359	8,415
5 persons	20,127	19,680	15,792	9,967
6 persons	22,727	22,228	17,839	11,272
7 persons	25,912	25,257	20,241	12,761
8 persons	28,967	28,166	22,582	14,199
9 persons or more	34,417	33,339	26,848	16,896

Persons Below Poverty Level, 1960-99

Source: Bureau of the Census, U.S. Dept. of Commerce

YEAR	Number below poverty level (in millions) All races[1]	White	Black	Hispanic origin[2]	Percentage below poverty level All races[1]	White	Black	Hispanic origin[2]	Avg. income cutoffs for family of 4 at poverty level[3]
1960	39.9	28.3	NA	NA	22.2	17.8	NA	NA	$3,022
1970	25.4	17.5	7.5	NA	12.6	9.9	33.5	NA	3,968
1980	29.3	19.7	8.6	3.5	13.0	10.2	32.5	25.7	8,414
1990	33.6	22.3	9.8	6.0	13.5	10.7	31.9	28.1	13,359
1991	35.7	23.7	10.2	6.3	14.2	11.3	32.7	28.7	13,924
1992	38.0	25.3	10.8	7.6	14.8	11.9	33.4	29.6	14,335
1993	39.3	26.2	10.9	8.1	15.1	12.2	33.1	30.6	14,763
1994	38.1	25.4	10.2	8.4	14.5	11.7	30.6	30.7	15,141
1995	36.4	24.4	9.9	8.6	13.8	11.2	29.3	30.3	15,569
1996	36.5	24.7	9.7	8.7	13.7	11.2	28.4	29.4	16,036
1997	35.6	24.4	9.1	8.3	13.3	11.0	26.5	27.1	16,400
1998	34.5	23.5	9.1	8.1	12.7	10.5	26.1	25.6	16,660
1999	32.3	21.9	8.4	7.4	11.8	9.8	23.6	22.8	17,029

NA = Not available. **NOTE:** Because of a change in the definition of poverty, data prior to 1980 are not directly comparable to data since 1980. (1) Includes other races not shown separately. (2) Persons of Hispanic origin may be of any race. (3) Figures for 1960-80 represent only nonfarm families.

Poverty by Family Status, Sex, and Race, 1986-99

Source: Bureau of the Census, U.S. Dept. of Commerce

(numbers in thousands)

	1999 No.	1999 %[1]	1998 No.	1998 %[1]	1995 No.	1995 %[1]	1990 No.	1990 %[1]	1986 No.	1986 %[1]
TOTAL POOR	**32,258**	**11.8**	**34,476**	**12.7**	**36,425**	**13.8**	**33,585**	**13.5**	**32,370**	**13.6**
In families	**23,396**	**10.2**	**25,370**	**11.2**	**27,501**	**12.3**	**25,232**	**12.0**	**24,754**	**12.0**
Head of household	6,676	9.3	7,186	10.0	7,532	10.8	7,098	10.7	7,023	10.9
Related children	11,510	16.3	12,845	18.3	13,999	20.2	12,715	19.9	12,257	19.8
Unrelated individuals	**8,305**	**19.1**	**8,478**	**19.9**	**8,247**	**20.9**	**7,446**	**20.7**	**6,846**	**21.6**
In families, female householder, no husband present	**11,607**	**30.4**	**12,907**	**33.1**	**14,205**	**36.5**	**12,578**	**37.2**	**11,944**	**38.3**
Head of household	3,531	27.8	3,831	29.9	4,057	32.4	3,768	33.4	3,613	34.6
Related children	6,602	41.9	7,627	46.1	8,364	50.3	7,363	53.4	6,943	54.4
Unrelated female individuals	**4,907**	**21.7**	**5,013**	**22.6**	**4,865**	**23.5**	**4,589**	**24.0**	**4,311**	**25.1**
All other families	**11,789**	**6.2**	**12,463**	**6.6**	**13,296**	**7.2**	**12,654**	**7.1**	**12,811**	**7.3**
Head of household	3,145	5.3	3,355	5.7	3,475	6.1	3,330	6.0	3,410	6.3
Related children	4,908	9.0	5,218	9.7	5,635	10.7	5,352	10.7	5,313	10.8
Unrelated male individuals	**3,398**	**16.3**	**3,465**	**17.0**	**3,382**	**18.0**	**2,857**	**16.9**	**2,536**	**17.5**
TOTAL WHITE POOR	**21,922**	**9.8**	**23,454**	**10.5**	**24,423**	**11.2**	**22,326**	**10.7**	**22,183**	**11.0**
In families	**15,141**	**8.1**	**16,549**	**8.9**	**17,593**	**9.6**	**15,916**	**9.0**	**16,393**	**9.4**
Head of household	4,377	7.3	4,829	8.0	4,994	8.5	4,622	8.1	4,811	8.6
Related children	7,123	12.9	7,935	14.4	8,474	15.5	7,696	15.1	7,714	15.3
Female householder, no spouse present	**1,883**	**22.5**	**2,123**	**24.9**	**2,200**	**26.6**	**2,010**	**26.8**	**2,041**	**28.2**
Unrelated individuals	**6,375**	**17.6**	**6,386**	**18.0**	**6,336**	**19.0**	**5,739**	**18.6**	**5,198**	**19.2**
TOTAL BLACK POOR	**8,360**	**23.6**	**9,091**	**26.1**	**9,872**	**29.3**	**9,837**	**31.9**	**8,983**	**31.1**
In families	**6,688**	**22.7**	**7,259**	**24.7**	**8,189**	28.5	**8,160**	**31.0**	**7,410**	**29.7**
Head of household	1,898	21.9	1,981	23.4	2,127	26.4	2,193	29.3	1,987	28.0
Related children	3,644	32.7	4,073	36.4	4,644	41.5	4,412	44.2	4,039	42.7
Female householder, no spouse present	**1,499**	**39.3**	**1,557**	**40.8**	**1,701**	**45.1**	**1,648**	**48.1**	**1,488**	**50.1**
Unrelated individuals	**1,552**	**27.6**	**1,752**	**32.5**	**1,551**	**32.6**	**1,491**	**35.1**	**1,431**	**38.5**

(1) Percentage of total U.S. population in each category who fell below poverty level and are enumerated here. For example, of all persons in families in 1999, 10.2%, or 23,396,000, were poor.

Persons in Poverty, by State, 1989, 1999

Source: Bureau of the Census, U.S. Dept. of Commerce

STATE	1999 Percentage	1989 Percentage	STATE	1999 Percentage	1989 Percentage	STATE	1999 Percentage	1989 Percentage
AL	14.8	18.3	KY	12.8	19.0	OH	11.6	12.5
AK	8.5	9.0	LA	19.1	23.6	OK	13.4	16.7
AZ	14.3	15.7	ME	10.5	10.8	OR	13.8	12.4
AR	14.7	19.1	MD	7.2	8.3	PA	10.3	11.1
CA	14.6	12.5	MA	10.2	8.9	RI	10.7	9.6
CO	8.7	11.7	MI	10.3	13.1	SC	12.7	15.4
CT	8.3	6.8	MN	8.8	10.2	SD	9.3	15.9
DE	10.3	8.7	MS	16.9	25.2	TN	12.7	15.7
DC	18.6	16.9	MO	10.7	13.3	TX	15.0	18.1
FL	12.8	12.7	MT	16.1	16.1	UT	7.3	11.4
GA	13.2	14.7	NE	11.6	11.1	VT	9.8	9.9
HI	10.9	8.3	NV	10.9	10.2	VA	8.4	10.2
ID	13.5	13.3	NH	8.8	6.4	WA	9.2	10.9
IL	10.0	11.9	NJ	8.2	7.6	WV	16.8	19.7
IN	8.0	10.7	NM	20.5	20.6	WI	8.7	10.7
IA	8.3	11.5	NY	15.4	13.0	WY	11.1	11.9
KS	10.9	11.5	NC	13.8	13.0			
			ND	14.1	14.4	**U.S. TOTAL**	**11.8**	**12.8**

U.S. Places of 5,000 or More Population—With ZIP and Area Codes

Source: U.S. Bureau of the Census, Dept. of Commerce; NeuStar Inc.

The following is a list of places of 5,000 or more inhabitants recognized by the Bureau of the Census, U.S. Dept. of Commerce, based on 2000 Census results. Also given are 1990 census populations. This list includes **places that are incorporated** under the laws of their respective states as cities, boroughs, towns, and villages, as well as boroughs in Alaska and towns in the 6 New England states, New York, and Wisconsin.

Places that the Census Bureau designates as **"census designated places" (CDPs)** are also included; these are marked (c). CDP boundaries can change from one census to another. Hawaii is the only state that has no incorporated places recognized by the Census Bureau; all places shown for Hawaii are CDPs.

This list also includes, in *italics*, **minor civil divisions (MCDs)**, for Connecticut, Maine, Massachusetts, New Hampshire, Rhode Island, and Vermont. MCDs are not incorporated and not recognized as CDPs, but are often the primary political or administrative divisions of a county.

An **asterisk** (*) denotes that the ZIP code given is for general delivery; named streets and/or P.O. boxes within the community may differ; consult local postmaster. **Area codes** are given in parentheses. Some regions have 2 or more area codes intermixed; these are known as **overlays**. States where this occurs are noted. When 2 or more area codes are listed for one place, consult local operators for assistance. Area codes based on latest information as of Sept. 2001. For a listing in numerical order of all area codes in the U.S., Canada, and the Caribbean, see the Telecommunications chapter.

For some places listed, no area code and/or ZIP code is available. — = Not available.

Alabama

ZIP	Place	Area Code	2000	1990
35007	Alabaster	(205)	22,619	14,619
*35950	Albertville	(256)	17,247	14,507
*35010	Alexander City	(256)	15,008	14,917
36420	Andalusia	(334)	8,794	9,269
*36201	Anniston	(256)	24,276	26,638
35016	Arab	(251)	7,174	6,321
*35611	Athens	(256)	18,967	16,901
*36502	Atmore	(334)	7,676	8,046
35954	Attalla	(256)	6,592	6,859
*36830	Auburn	(251)	42,987	33,830
36507	Bay Minette	(251)	7,820	7,168
*35020	Bessemer	(205)	29,672	33,581
*35203	Birmingham	(205)	242,820	265,347
*35957	Boaz	(256)	7,411	6,928
*36426	Brewton	(251)	5,498	5,885
35243	Cahaba Heights (c)	(205)	5,203	4,778
35220	Center Point (c)	(205)	22,784	22,658
36671	Chickasaw	(251)	6,364	6,651
*35045	Clanton	(205)	7,800	7,669
*35055	Cullman	(256)	13,995	13,367
36526	Daphne	(251)	16,581	11,291
*35601	Decatur	(256)	53,929	49,917
36732	Demopolis	(334)	7,540	7,512
*36302	Dothan	(334)	57,737	54,131
*36330	Enterprise	(334)	21,178	20,119
*36027	Eufaula	(334)	13,908	13,220
35064	Fairfield	(205)	12,381	12,200
*36532	Fairhope	(251)	12,480	9,189
*35630	Florence	(256)	36,264	36,426
*36535	Foley	(251)	7,590	4,937
35214	Forestdale (c)	(205)	10,509	10,395
*35967	Fort Payne	(256)	12,938	11,838
36362	Fort Rucker (c)	(334)	6,052	7,593
35068	Fultondale	(205)	6,595	6,400
*35901	Gadsden	(256)	38,978	42,523
35071	Gardendale	(205)	11,626	9,251
35905	Glencoe	(256)	5,152	4,687
35235	Grayson Valley (c)	(205)	5,447	—
36037	Greenville	(334)	7,228	7,847
36542	Gulf Shores	(251)	5,044	3,261
*35976	Guntersville	(256)	7,395	7,038
35570	Hamilton	(205)	6,786	6,171
35640	Hartselle	(256)	12,019	11,114
35080	Helena	(205)	10,296	4,303
35259	Homewood	(205)	25,043	23,644
*35244	Hoover	(205)	62,742	39,988
35023	Hueytown	(205)	15,364	15,280
*35801	Huntsville	(256)	158,216	159,880
35210	Irondale	(205)	9,813	9,458
36545	Jackson	(251)	5,419	5,819
36265	Jacksonville	(256)	8,404	10,283
*35501	Jasper	(205)	14,052	13,553
—	Lake Purdy (c)		5,799	1,840
36863	Lanett	(334)	7,897	8,985
35094	Leeds	(205)	10,455	10,009
*35758	Madison	(256)	29,329	14,792
35228	Midfield	(205)	5,626	5,559
36054	Millbrook	(334)	10,386	6,046
*36601	Mobile	(251)	198,915	199,973
*36460	Monroeville	(251)	6,862	6,993
*36104	Montgomery	(334)	201,568	190,350
35004	Moody	(205)	8,053	4,921
—	Moores Mill (c)	(256)	5,178	3,362
35253	Mountain Brook	(205)	20,604	19,810
*35661	Muscle Shoals	(256)	11,924	9,611
*35476	Northport	(205)	19,435	17,297
35121	Oneonta	(205)	5,576	4,844
*36801	Opelika	(334)	23,498	22,122
36467	Opp	(334)	6,607	7,011
36203	Oxford	(256)	14,592	9,537
*36360	Ozark	(334)	15,119	13,030
35124	Pelham	(205)	14,369	9,356
*35125	Pell City	(205)	9,565	7,945
*36867	Phenix City	(334)	28,265	25,311
36272	Piedmont	(256)	5,120	5,347
35126	Pinson (c)	(205)	5,033	10,987
35127	Pleasant Grove	(205)	9,983	8,458
*36067	Prattville	(334)	24,303	19,816
36610	Prichard	(251)	28,633	34,320
35906	Rainbow City	(256)	8,428	7,667
36274	Roanoke	(334)	6,563	6,362
*35653	Russellville	(256)	8,971	7,812
36201	Saks (c)	(256)	10,698	11,138
36571	Saraland	(251)	12,288	11,784
36572	Satsuma	(251)	5,687	5,194
*35768	Scottsboro	(256)	14,762	13,786
*36701	Selma	(334)	20,512	23,755
35660	Sheffield	(256)	9,652	10,380
36877	Smiths (c)	(334)	21,756	3,456
35901	Southside	(256)	7,036	5,580
*36527	Spanish Fort	(251)	5,423	3,732
*35150	Sylacauga	(256)	12,616	12,520
*35160	Talladega	(256)	15,143	18,175
35217	Tarrant	(205)	7,022	8,046
*36582	Theodore (c)	(251)	6,811	6,509
36619	Tillman's Corner (c)	(251)	15,685	17,988
*36081	Troy	(334)	13,935	13,051
35173	Trussville	(205)	12,924	8,283
*35401	Tuscaloosa	(205)	77,906	77,866
35674	Tuscumbia	(256)	7,856	8,413
36083	Tuskegee	(334)	11,846	12,257
*36854	Valley	(334)	9,198	9,556
35266	Vestavia Hills	(205)	24,476	19,550
*36092	Wetumpka	(334)	5,726	4,670

Alaska (907)

ZIP	Place	2000	1990
*99501	Anchorage	260,283	226,338
99559	Bethel	5,471	4,674
*99708	College (c)	11,402	11,249
99702	Eielson AFB (c)	5,400	5,251
*99701	Fairbanks	30,224	30,843
*99801	Juneau	30,711	26,751
—	Kalifornsky (c)	5,846	285
99611	Kenai	6,942	6,327
*99901	Ketchikan	7,922	8,263
—	Knik-Fairview (c)	7,049	272
*99615	Kodiak	6,334	6,365
—	Lakes (c)	6,706	—
99835	Sitka	8,835	8,588
*99654	Wasilla	5,469	4,028

Arizona

ZIP	Place	Area Code	2000	1990
*85220	Apache Junction	(480)	31,814	18,092
85323	Avondale	(623)	35,883	17,595
85653	Avra Valley (c)	(520)	5,038	3,403
—	Big Park (c)	(928)	5,245	3,024
85603	Bisbee	(520)	6,090	6,288
85326	Buckeye	(623)	6,537	4,436
*86442	Bullhead City	(928)	33,769	21,951
86322	Camp Verde	(928)	9,451	6,243
*85222	Casa Grande	(520)	25,224	19,076
85740	Casas Adobes (c)	(520)	54,011	—
85738	Catalina (c)	(520)	7,025	4,864
—	Catalina Foothills (c)	(520)	53,794	—
*85225	Chandler	(480)	176,581	89,862
86503	Chinle (c)	(928)	5,366	5,059
86323	Chino Valley	(928)	7,835	4,837
85228	Coolidge	(520)	7,786	6,934
86326	Cottonwood	(928)	9,179	5,918
86326	Cottonwood-Verde Village (c)	(928)	10,610	7,037
86327	Dewey-Humboldt (c)	(928)	6,295	3,640
*85607	Douglas	(520)	14,312	13,908
—	Drexel Heights (c)	(520)	23,849	—
85335	El Mirage	(623)	7,609	5,001
85231	Eloy	(520)	10,375	7,211
*86004	Flagstaff	(928)	52,894	45,857
85232	Florence	(520)	17,054	7,321
85726	Flowing Wells (c)	(520)	15,050	14,013
—	Fortuna Foothills (c)	(928)	20,478	7,737
*85268	Fountain Hills	(480)	20,235	10,030

ZIP	Place	Area Code	2000	1990
*85299	Gilbert	(480)	109,697	29,149
*85301	Glendale	(623)	218,812	147,070
*85501	Globe	(928)	7,486	6,062
—	Gold Camp (c)		6,029	—
85338	Goodyear	(623)	18,911	6,258
*85622	Green Valley (c)	(520)	17,283	13,231
85283	Guadalupe	(480)	5,228	5,458
*86401	Kingman	(928)	20,069	13,208
*86403	Lake Havasu City	(928)	41,938	24,363
85653	Marana	(520)	13,556	2,565
*85201	Mesa	(480)	396,375	289,199
*86440	Mohave Valley (c)	(928)	13,694	6,962
—	New Kingman-Butler (c)	(928)	14,810	11,627
*85027	New River (c)	(602)	10,740	—
*85621	Nogales	(520)	20,878	19,489
85737	Oro Valley	(520)	29,700	9,024
86040	Page	(928)	6,809	6,598
85253	Paradise Valley	(480)	13,664	11,903
*85541	Payson	(928)	13,620	8,377
*85345	Peoria	(623)	108,364	51,080
*85034	Phoenix	(602)	1,321,045	988,015
—	Picture Rocks (c)	(520)	8,139	4,026
*86301	Prescott	(928)	33,938	26,592
*86314	Prescott Valley	(928)	23,535	8,904
*85546	Safford	(928)	9,232	7,359
85349	San Luis	(928)	15,322	4,212
*85251	Scottsdale	(480)	202,705	130,099
*86336	Sedona	(928)	10,192	7,720
*85901	Show Low	(928)	7,695	5,020
*85635	Sierra Vista	(520)	37,775	32,983
85635	Sierra Vista Southeast (c)	(520)	14,348	9,237
85350	Somerton	(928)	7,266	5,293
85713	South Tucson	(520)	5,490	5,171
*85351	Sun City (c)	(623)	38,309	38,126
*85351	Sun City West (c)	(623)	26,344	15,997
85248	Sun Lakes (c)	(480)	11,936	6,578
*85374	Surprise	(623)	30,848	7,122
—	Tanque Verde (c)	(520)	16,195	—
*85285	Tempe	(480)	158,625	141,993
—	Three Points (c)	(520)	5,273	2,175
86045	Tuba City (c)	(928)	8,225	7,323
*85726	Tucson	(520)	486,699	415,444
—	Tucson Estates (c)	(520)	9,755	2,662
85941	Whiteriver (c)	(928)	5,220	3,775
*85390	Wickenburg	(928)	5,082	4,515
86047	Winslow	(928)	9,520	9,279
*85364	Yuma	(928)	77,515	56,966

Arkansas

ZIP	Place	Area Code	2000	1990
71923	Arkadelphia	(870)	10,912	10,014
*72501	Batesville	(870)	9,445	9,187
*72714	Bella Vista (c)	(501)	16,582	9,083
*72015	Benton	(501)	21,906	18,177
72712	Bentonville	(501)	19,730	11,257
*72315	Blytheville	(870)	18,272	22,523
*72022	Bryant	(501)	9,764	5,940
72023	Cabot	(501)	15,261	8,319
*71701	Camden	(870)	13,154	14,701
72830	Clarksville	(501)	7,719	5,833
*72032	Conway	(501)	43,167	26,481
71635	Crossett	(870)	6,097	6,282
71832	De Queen	(870)	5,765	4,633
71639	Dumas	(870)	5,238	5,520
72065	East End (c)	(501)	5,623	—
*71730	El Dorado	(870)	21,530	23,146
*72701	Fayetteville	(501)	58,047	42,247
*72335	Forrest City	(870)	14,774	13,364
*72901	Fort Smith	(501)	80,268	72,798
72936	Greenwood	(501)	7,112	3,984
*72601	Harrison	(870)	12,152	9,936
72543	Heber Springs	(501)	6,432	5,628
72342	Helena	(870)	6,323	7,491
*71801	Hope	(870)	10,616	9,768
*71901	Hot Springs	(501)	35,750	33,095
*71909	Hot Springs Village (c)	(501)	8,397	6,361
*72076	Jacksonville	(501)	29,916	29,101
*72401	Jonesboro	(870)	55,515	46,535
*72201	Little Rock	(501)	183,133	175,727
72745	Lowell	(501)	5,013	1,224
*71753	Magnolia	(870)	10,858	11,151
72104	Malvern	(501)	9,021	9,236
72360	Marianna	(870)	5,181	6,033
72364	Marion	(870)	8,901	4,405
72113	Maumelle	(501)	10,557	6,714
71953	Mena	(501)	5,637	5,475
*71655	Monticello	(870)	9,146	8,119
72110	Morrilton	(501)	6,550	6,551
*72653	Mountain Home	(870)	11,012	9,027
72112	Newport	(870)	7,811	7,459
*72114	North Little Rock	(501)	60,433	61,829
72370	Osceola	(870)	8,875	9,165
*72450	Paragould	(870)	22,017	18,540
*71601	Pine Bluff	(870)	55,085	57,140
72455	Pocahontas	(870)	6,518	6,151
*72756	Rogers	(501)	38,829	24,692
*72801	Russellville	(501)	23,682	21,260
*72143	Searcy	(501)	18,928	15,180
72120	Sherwood	(501)	21,511	18,890
72761	Siloam Springs	(501)	10,843	8,151
*72764	Springdale	(501)	45,798	29,945
72160	Stuttgart	(870)	9,745	10,420
71854	Texarkana	(870)	26,448	22,631
72472	Trumann	(870)	6,889	6,346
*72956	Van Buren	(501)	18,986	14,899
71671	Warren	(870)	6,442	6,455
72390	West Helena	(870)	8,689	10,137
*72301	West Memphis	(870)	27,666	28,259
72396	Wynne	(870)	8,615	8,187

California

ZIP	Place	Area Code	2000	1990
92301	Adelanto	(760)	18,130	6,815
*91376	Agoura Hills	(818)	20,537	20,396
*94501	Alameda	(510)	72,259	73,979
94507	Alamo (c)	(925)	15,626	12,277
94706	Albany	(510)	16,444	16,327
*91802	Alhambra	(323)/(626)	85,804	82,087
92656	Aliso Viejo (c)	(949)	40,166	7,612
90249	Alondra Park (c)	(310)	8,622	12,215
*91901	Alpine (San Diego) (c)	(619)	13,143	9,695
*91003	Altadena (c)	(626)	42,610	42,658
95945	Alta Sierra (c)	(530)	6,522	5,709
95127	Alum Rock (c)	(408)	13,479	—
94589	American Canyon	(707)	9,774	7,734
*92803	Anaheim	(909)	328,014	266,406
96007	Anderson	(530)	9,022	8,299
*94509	Antioch	(925)	90,532	62,195
*92307	Apple Valley	(760)	54,239	46,079
*95003	Aptos (c)	(831)	9,396	9,061
*91006	Arcadia	(626)	53,054	48,284
*95521	Arcata	(707)	16,651	15,211
95825	Arden-Arcade (c)	(916)	96,025	92,040
*93420	Arroyo Grande	(805)	15,851	14,432
*90701	Artesia	(562)	16,380	15,464
93203	Arvin	(661)	12,956	9,286
94577	Ashland (c)	(510)	20,793	16,590
*93422	Atascadero	(805)	26,411	23,138
94027	Atherton	(650)	7,194	7,163
95301	Atwater	(209)	23,113	22,282
*95603	Auburn	(530)	12,462	10,653
95201	August (c)	(209)	7,808	6,376
93204	Avenal	(559)	14,674	9,770
91746	Avocado Heights (c)	(626)	15,148	14,232
91702	Azusa	(626)	44,712	41,203
*93302	Bakersfield	(661)	247,057	176,264
91706	Baldwin Park	(626)	75,837	69,330
92220	Banning	(951)	23,562	20,572
*92312	Barstow	(760)	21,119	21,472
94565	Bay Point (c)	(925)	21,534	17,453
—	Bayview-Montalvin (c)	(510)	5,004	3,988
93402	Baywood-Los Osos (c)	(805)	14,351	14,377
95903	Beale AFB (c)	(530)	5,115	6,912
92223	Beaumont	(951)	11,384	9,685
90201	Bell	(323)	36,664	34,365
*90706	Bellflower	(323)	72,878	61,815
90202	Bell Gardens	(213)/(323)/(562)	44,054	42,315
94002	Belmont	(650)	25,123	24,165
94510	Benicia	(707)	26,865	24,437
*94704	Berkeley	(510)	102,743	102,724
92201	Bermuda Dunes (c)	(760)	6,229	4,571
*90210	Beverly Hills	(213)/(310)/(323)	33,784	31,971
92314	Big Bear City (c)	(909)	5,779	4,920
92315	Big Bear Lake	(909)	5,438	5,351
94526	Blackhawk-Camino Tassajara (c)	(925)	10,048	6,199
92316	Bloomington (c)	(909)	19,318	15,116
*92225	Blythe	(760)	12,155	10,835
93637	Bonadelle Ranchos-Madera Ranchos (c)	(559)	7,300	5,705
*91902	Bonita (c)	(619)	12,401	12,542
92021	Bostonia (c)	(619)	15,169	13,670
95416	Boyes Hot Springs (c)	(707)	6,665	5,973
92227	Brawley	(760)	22,052	18,923
*92822	Brea	(562)/(714)	35,410	32,873
94513	Brentwood	(925)	23,302	7,563
—	Bret Harte (c)	(209)	5,161	—
*90622	Buena Park	(714)	78,282	68,784
*91510	Burbank (Los Angeles)	(818)	100,316	93,649
—	Burbank (Santa Clara) (c)	(408)	5,239	4,902
*94010	Burlingame	(650)	28,158	26,666
*91372	Calabasas	(818)	20,033	16,577
*92231	Calexico	(760)	27,109	18,633
*93504	California City	(760)	8,385	5,955
92320	Calimesa	(909)	7,139	6,654
92233	Calipatria	(760)	7,289	2,701
94515	Calistoga	(707)	5,190	4,468
*93010	Camarillo	(805)	57,077	52,297
93428	Cambria (c)	(805)	6,232	5,382
95682	Cameron Park (c)	(530)	14,549	11,897
*95008	Campbell	(408)	38,138	36,088
92055	Camp Pendleton North (c)	(949)	8,197	10,373
92055	Camp Pendleton South (c)	(949)	8,854	11,299
92587	Canyon Lake	(951)	9,952	9,991
95010	Capitola	(831)	10,033	10,171
*92008	Carlsbad	(760)	78,247	63,292
*95608	Carmichael (c)	(916)	49,742	48,702
*93013	Carpinteria	(805)	14,194	13,747
*90745	Carson	(310)	89,730	83,995
92077	Casa de Oro-Mt. Helix (c)	(619)	18,874	30,727

ZIP	Place	Area Code	2000	1990
*94546	Castro Valley (c)	(510)	57,292	48,619
95012	Castroville (c)	(831)	6,724	5,272
*92235	Cathedral City	(760)	42,647	30,085
95307	Ceres	(209)	34,609	26,413
90703	Cerritos	(562)	51,488	53,244
91724	Charter Oak (c)	(626)	9,027	8,858
94541	Cherryland (c)	(510)	13,837	11,088
92223	Cherry Valley (c)	(909)	5,891	5,945
*95926	Chico	(530)	59,954	39,970
*91708	Chino	(909)	67,168	59,682
91709	Chino Hills	(909)	66,787	37,868
93610	Chowchilla	(559)	11,127	5,930
*91910	Chula Vista	(619)	173,556	135,160
91702	Citrus (c)	(626)	10,581	9,481
*95621	Citrus Heights	(916)	85,071	107,439
91711	Claremont	(909)	33,998	32,610
94517	Clayton	(925)	10,762	7,317
95422	Clearlake	(707)	13,142	11,804
95425	Cloverdale	(707)	6,831	4,924
*93612	Clovis	(559)	68,468	50,323
92236	Coachella	(760)	22,724	16,896
93210	Coalinga	(559)	11,668	8,212
92324	Colton	(909)	47,662	40,213
95932	Colusa	(530)	5,402	4,934
90022	Commerce	(323)/(562)	12,568	12,135
*90221	Compton	(310)	93,493	90,454
*94520	Concord	(925)	121,780	111,308
93212	Corcoran	(559)	14,458	13,360
96021	Corning	(530)	6,741	5,870
*91718	Corona	(951)	124,966	75,943
*92138	Coronado	(619)	24,100	26,540
*94925	Corte Madera	(415)	9,100	8,272
*92628	Costa Mesa	(714)/(949)	108,724	96,357
94931	Cotati	(707)	6,471	5,714
92679	Coto de Caza (c)	(949)	13,057	2,853
94556	Country Club (c)	(209)	9,462	9,325
*91722	Covina	(626)	46,837	43,332
92325	Crestline (c)	(909)	10,218	8,594
90201	Cudahy	(323)	24,208	22,817
*90230	Culver City	(230)/(310)/(323)	38,816	38,793
*95014	Cupertino	(408)	50,546	39,967
90630	Cypress	(714)	46,229	42,655
*94015	Daly City	(415)/(650)	103,621	92,088
92629	Dana Point	(949)	35,110	31,896
*94526	Danville	(925)	41,715	31,306
*95616	Davis	(530)	60,308	46,322
90250	Del Aire (c)	(310)	9,012	8,040
*93215	Delano	(661)	38,824	22,762
95315	Delhi (c)	(209)	8,022	3,280
*92240	Desert Hot Springs	(760)	16,582	11,668
91765	Diamond Bar	(909)	56,287	53,672
93618	Dinuba	(559)	16,844	12,743
94514	Discovery Bay (c)	(925)	8,981	5,351
95620	Dixon	(707)	16,103	10,417
*90241	Downey	(562)	107,323	91,444
*91009	Duarte	(626)	21,486	20,716
94568	Dublin	(925)	29,973	23,229
95938	Durham (c)	(530)	5,220	4,784
93219	Earlimart (c)	(661)	6,583	5,881
90220	East Compton (c)	(310)	9,286	7,967
—	East Foothills (c)		8,133	14,898
92343	East Hemet (c)	(909)	14,823	17,611
90638	East La Mirada (c)	(562)	9,538	9,367
90022	East Los Angeles (c)	(323)/(562)	124,283	126,379
94303	East Palo Alto	(650)	29,506	23,451
91117	East Pasadena (c)		6,045	5,910
93257	East Porterville (c)	(559)	6,730	5,790
—	East San Gabriel (c)	(626)	14,512	12,736
93523	Edwards AFB (c)	(661)	5,909	7,423
*92020	El Cajon	(619)	94,869	88,918
*92244	El Centro	(760)	37,835	31,405
94530	El Cerrito	(510)	23,171	22,869
95762	El Dorado Hills (c)	(916)	18,016	6,395
94018	El Granada (c)	(650)	5,724	4,426
*95624	Elk Grove (c)	(916)	59,984	17,483
*91734	El Monte	(626)	115,965	106,162
*93446	El Paso de Robles	(805)	24,297	18,583
93030	El Rio (c)	(805)	6,193	6,419
90245	El Segundo	(310)	16,033	15,223
*94802	El Sobrante (c)	(510)	12,260	9,852
*94617	Emeryville	(510)	6,882	5,740
*92024	Encinitas	(760)	58,014	55,406
95320	Escalon	(209)	5,963	4,437
*92025	Escondido	(760)	133,559	108,648
*95501	Eureka	(707)	26,128	27,025
93221	Exeter	(559)	9,168	7,276
*94930	Fairfax	(415)	7,319	6,931
94533	Fairfield	(707)	96,178	78,650
95628	Fair Oaks (Sacramento) (c)	(916)	28,008	26,867
—	Fairview (c)		9,470	9,045
*92028	Fallbrook (c)	(760)	29,100	22,095
93223	Farmersville	(559)	8,737	6,235
*93015	Fillmore	(805)	13,643	11,992
93622	Firebaugh	(209)	5,743	4,429
90001	Florence-Graham (c)	(323)	60,197	57,147
95828	Florin (c)	(916)	27,653	24,330
*95630	Folsom	(916)	51,884	29,802
*92334	Fontana	(909)	128,929	87,535
95841	Foothill Farms (c)	(916)	17,426	17,135
92610	Foothill Ranch (c)	(949)	10,899	—
95437	Fort Bragg	(707)	7,026	6,078
95540	Fortuna	(707)	10,497	8,788
94404	Foster City	(650)	28,803	28,176
*92728	Fountain Valley	(714)	54,978	53,691
95019	Freedom (c)	(831)	6,000	8,361
*94537	Fremont	(510)	203,413	173,339
*93706	Fresno	(559)	427,652	354,091
*92834	Fullerton	(714)	126,003	114,144
95632	Galt	(209)	19,472	8,889
*90247	Gardena	(310)	57,746	51,481
95205	Garden Acres (c)	(209)	9,747	8,547
*92842	Garden Grove	(714)	165,196	142,965
*95020	Gilroy	(408)	41,464	31,487
92509	Glen Avon (c)	(951)	14,853	12,663
*91209	Glendale	(323)/(626)/(818)	194,973	180,038
*91741	Glendora	(626)	49,415	47,832
93561	Golden Hills (c)	(661)	7,434	5,423
95670	Gold River (c)	(916)	8,023	—
*93116	Goleta (c)	(805)	55,204	—
93926	Gonzales	(831)	7,525	4,660
92324	Grand Terrace	(909)	11,626	10,946
95746	Granite Bay (c)	(916)	19,388	—
*95945	Grass Valley	(530)	10,922	9,048
93927	Greenfield (Monterey)	(831)	12,583	7,464
95948	Gridley	(530)	5,382	4,631
93433	Grover Beach	(805)	13,067	11,602
93434	Guadalupe	(805)	5,659	5,479
91745	Hacienda Heights (c)	(626)	53,122	52,354
94019	Half Moon Bay	(650)	11,842	8,886
*93230	Hanford	(559)	41,686	30,463
90716	Hawaiian Gardens	(323)	14,779	13,639
*90250	Hawthorne	(213)/(310)/(323)	84,112	71,349
*94544	Hayward	(510)	140,030	114,705
95448	Healdsburg	(707)	10,722	9,469
*92546	Hemet	(951)	58,812	43,366
94547	Hercules	(510)	19,488	16,829
90254	Hermosa Beach	(310)	18,566	18,219
*92340	Hesperia	(760)	62,582	50,418
92346	Highland	(909)	44,605	34,439
94010	Hillsborough	(650)	10,825	10,667
*95023	Hollister	(831)	34,413	19,318
92250	Holtville	(760)	5,612	4,820
91720	Home Gardens (c)	(951)	9,461	7,780
*92647	Huntington Beach	(714)	189,594	181,519
90255	Huntington Park	(323)	61,348	56,129
93234	Huron	(559)	6,306	4,766
92251	Imperial	(760)	7,560	4,113
*91932	Imperial Beach	(619)	26,992	26,512
*92201	Indio	(760)	49,116	36,850
*90301	Inglewood	(213)/(310)/(323)	112,580	109,602
—	Interlaken (c)	(831)	7,328	6,404
95640	Ione	(209)	7,129	6,516
*92619	Irvine	(714)/(949)	143,072	110,330
93117	Isla Vista (c)	(805)	18,344	20,395
91935	Jamul (c)	(619)	5,920	2,258
94914	Kentfield (c)	(415)	6,351	6,030
93630	Kerman	(559)	8,551	5,448
93930	King City	(831)	11,094	7,634
93631	Kingsburg	(559)	9,199	7,245
*91011	La Canada Flintridge	(818)	20,318	19,378
*91224	La Crescenta-Montrose (c)	(818)	18,532	16,968
90045	Ladera Heights (c)	(310)	6,568	6,316
94549	Lafayette	(925)	23,908	23,366
—	Laguna (c)		34,309	9,828
*92652	Laguna Beach	(949)	23,727	23,170
*92654	Laguna Hills	(949)	31,178	22,719
*92607	Laguna Niguel	(949)	61,891	44,723
—	Laguna West-Lakeside (c)		8,414	—
*92654	Laguna Woods	(949)	16,507	—
*90631	La Habra	(562)/(949)	58,974	51,263
90631	La Habra Heights	(562)	5,712	6,226
92352	Lake Arrowhead (c)	(909)	8,934	6,539
*92531	Lake Elsinore	(951)	28,928	19,733
92630	Lake Forest	(714)	58,707	56,036
92530	Lakeland Village (c)	(951)	5,626	5,159
93535	Lake Los Angeles (c)	(661)	11,523	7,977
92040	Lakeside (c)	(619)	19,560	39,412
*90714	Lakewood	(562)	79,345	73,553
*91941	La Mesa	(619)	54,749	52,911
*90638	La Mirada	(562)/(714)	46,783	40,452
93241	Lamont (c)	(661)	13,296	11,517
*93539	Lancaster	(661)	118,718	97,300
90623	La Palma	(562)/(714)	15,408	15,392
—	La Presa (c)		32,721	—
*91747	La Puente	(626)	41,063	36,955
92253	La Quinta	(760)	23,694	11,215
95401	La Riviera (c)	(916)	10,273	10,986
95403	Larkfield-Wikiup (c)	(707)	7,479	6,779
*94939	Larkspur	(415)	12,014	11,068
92688	Las Flores (c)	(949)	5,625	—
95330	Lathrop	(209)	10,445	6,841
91750	La Verne	(909)	31,638	30,843
*90260	Lawndale	(310)	31,711	27,331
*91945	Lemon Grove	(619)	24,918	23,984
93245	Lemoore	(559)	19,712	13,622
93245	Lemoore Station (c)	(559)	5,749	0
90304	Lennox (c)	(310)	22,950	22,757
95648	Lincoln	(916)	11,205	7,248
95901	Linda (c)	(530)	13,474	13,033
93247	Lindsay	(559)	10,297	8,338

ZIP	Place	Area Code	2000	1990
95062	Live Oak (Santa Cruz) (c)	(831)	16,628	15,212
95953	Live Oak (Sutter)	(530)	6,229	4,320
*94550	Livermore	(925)	73,345	56,741
95334	Livingston	(209)	10,473	7,317
*95240	Lodi	(209)	56,999	51,874
92354	Loma Linda	(909)	18,681	18,470
90717	Lomita	(213)	20,046	19,442
*93436	Lompoc	(805)	41,103	37,649
*90801	Long Beach	(310)/(562)	461,522	429,321
95650	Loomis	(916)	6,260	5,705
*90720	Los Alamitos	(562)/(949)	11,536	11,788
*94022	Los Altos	(650)	27,693	26,599
94022	Los Altos Hills	(650)	7,902	7,514
*90086	Los Angeles	(213)/(310)/(323)/(818)	3,694,820	3,485,557
93635	Los Banos	(209)	25,869	14,519
*95030	Los Gatos	(408)	28,592	27,357
94903	Lucas Valley-Marinwood (c)	(415)	6,357	5,982
90262	Lynwood	(213)/(310)/(323)	69,845	61,945
93250	Mc Farland	(661)	9,618	7,005
95521	McKinleyville (c)	(707)	13,599	10,749
*93638	Madera	(559)	43,207	29,283
93637	Madera Acres (c)	(559)	7,741	5,245
95954	Magalia (c)	(530)	10,569	8,987
*90265	Malibu	(310)	12,575	11,730
93546	Mammoth Lakes	(760)	7,093	4,785
*90266	Manhattan Beach	(310)	33,852	32,063
*95336	Manteca	(209)	49,258	40,773
93933	Marina	(831)	25,101	26,512
*90291	Marina del Rey (c)	(310)	8,176	7,431
94553	Martinez	(925)	35,866	31,800
95901	Marysville	(530)	12,268	12,324
—	Mayflower Village (c)		5,081	4,978
90270	Maywood	(323)	28,083	27,893
92254	Mecca (c)	(619)	5,402	1,966
93640	Mendota	(559)	7,890	6,821
*94025	Menlo Park	(650)	30,785	28,403
92359	Mentone (c)	(909)	7,803	5,675
*95340	Merced	(209)	63,893	56,155
94030	Millbrae	(650)	20,718	20,414
*94941	Mill Valley	(415)	13,600	13,029
*95035	Milpitas	(408)	62,698	50,690
91752	Mira Loma (c)	(951)	17,617	15,786
93641	Mira Monte (c)	(805)	7,177	7,744
*92690	Mission Viejo	(949)	93,102	79,464
*95350	Modesto	(209)	188,856	164,746
*91017	Monrovia	(626)	36,929	35,733
91763	Montclair	(909)	33,049	28,434
90640	Montebello	(323)	62,150	59,564
*93150	Montecito	(805)	10,000	—
*93940	Monterey	(831)	29,674	31,954
*91754	Monterey Park	(323)/(626)/(818)	60,051	60,738
*93021	Moorpark	(805)	31,415	25,494
*94556	Moraga	(925)	16,290	15,987
*92552	Moreno Valley	(951)	142,381	118,779
*95037	Morgan Hill	(408)	33,556	23,928
*93442	Morro Bay	(805)	10,350	9,664
*94041	Mountain View	(650)	70,708	67,365
*92564	Murrieta	(951)	44,282	18,557
92405	Muscoy (c)	(714)	8,919	7,541
*94558	Napa	(707)	72,585	61,865
*91950	National City	(619)	54,260	54,249
94560	Newark	(510)	42,471	37,861
95360	Newman	(209)	7,093	4,158
*92658	Newport Beach	(949)	70,032	66,643
93444	Nipomo (c)	(805)	12,626	7,109
91760	Norco	(951)	24,157	23,302
95603	North Auburn (c)	(530)	11,847	10,301
94025	North Fair Oaks (c)	(650)	15,440	13,912
95660	North Highlands (c)	(916)	44,187	42,105
*90650	Norwalk	(562)	103,298	94,279
*94947	Novato	(415)	47,630	47,585
95361	Oakdale	(209)	15,503	11,978
*94617	Oakland	(510)	399,484	372,242
94561	Oakley	(925)	25,619	18,374
93445	Oceano (c)	(805)	7,260	6,169
*92056	Oceanside	(760)	161,029	128,090
93308	Oildale (c)	(661)	27,885	26,553
*93023	Ojai	(805)	7,862	7,613
95961	Olivehurst (c)	(530)	11,061	9,738
*91761	Ontario	(909)	158,007	133,179
95060	Opal Cliffs (c)	(831)	6,458	5,940
*92863	Orange	(714)	128,821	110,658
93646	Orange Cove	(559)	7,722	5,604
95662	Orangevale (c)	(916)	26,705	26,266
93457	Orcutt (c)	(805)	28,830	—
94563	Orinda	(925)	17,599	16,642
95963	Orland	(530)	6,281	5,052
93647	Orosi (c)	(559)	7,318	5,486
*95965	Oroville	(530)	13,004	11,885
95965	Oroville East (c)	(530)	8,680	8,462
*93030	Oxnard	(805)	170,358	142,560
94044	Pacifica	(650)	38,390	37,670
93950	Pacific Grove	(831)	15,522	16,117
95968	Palermo (c)	(530)	5,720	5,260
*93590	Palmdale	(661)	116,670	73,314
*92260	Palm Desert	(760)	41,155	23,252
*92262	Palm Springs	(760)	42,807	40,144
*94303	Palo Alto	(650)	58,598	55,900
90274	Palos Verdes Estates	(310)	13,340	13,512
*95969	Paradise	(530)	26,408	25,401

ZIP	Place	Area Code	2000	1990
90723	Paramount	(562)	55,266	47,669
95823	Parkway-So. Sacramento (c)	(916)	36,468	31,903
93648	Parlier	(559)	11,145	7,938
*91109	Pasadena	(323)/(626)/(818)	133,936	131,586
	Paso Robles. *See El Paso de Robles*			
95363	Patterson	(209)	11,606	8,626
92509	Pedley (c)	(951)	11,207	8,869
*92572	Perris	(951)	36,189	21,500
*94952	Petaluma	(707)	54,548	43,166
—	Phoenix Lake-Cedar Ridge (c)		5,123	3,569
90660	Pico Rivera	(562)	63,428	59,177
94611	Piedmont	(510)	10,952	10,602
94564	Pinole	(510)	19,039	17,460
*93449	Pismo Beach	(805)	8,551	7,669
94565	Pittsburg	(925)	56,769	47,607
*92871	Placentia	(714)	46,488	41,259
95667	Placerville	(530)	9,610	8,286
94523	Pleasant Hill	(925)	32,837	31,583
*94566	Pleasanton	(925)	63,654	50,570
*91769	Pomona	(909)	149,473	131,700
*93257	Porterville	(559)	39,615	29,521
*93041	Port Hueneme	(805)	21,845	20,322
92679	Portola Hills (c)	(949)	6,391	2,677
*92064	Poway	(858)	48,044	43,396
93907	Prunedale (c)	(831)	16,432	7,393
*93551	Quartz Hill (c)	(661)	9,890	9,626
92065	Ramona (c)	(760)	15,691	13,040
*95670	Rancho Cordova (c)	(916)	55,060	48,731
*91729	Rancho Cucamonga	(909)	127,743	101,409
92270	Rancho Mirage	(760)	13,249	9,778
90275	Rancho Palos Verdes	(310)	41,145	41,667
91941	Rancho San Diego (c)	(619)	20,155	6,977
92688	Rancho Santa Margarita (c)	(949)	47,214	11,390
96080	Red Bluff	(530)	13,147	12,363
*96049	Redding	(530)	80,865	66,176
*92373	Redlands	(909)	63,591	62,667
*90277	Redondo Beach	(310)	63,261	60,167
*94063	Redwood City	(650)	75,402	66,072
93654	Reedley	(559)	20,756	15,791
*92377	Rialto	(909)	91,873	72,395
*94802	Richmond	(510)	99,216	86,019
*93556	Ridgecrest	(760)	24,927	28,295
95003	Rio del Mar (c)	(831)	9,198	8,919
95673	Rio Linda (c)	(916)	10,466	9,481
95366	Ripon	(209)	10,146	7,455
95367	Riverbank	(209)	15,826	8,591
*92502	Riverside	(951)	255,166	226,546
*95677	Rocklin	(916)	36,330	18,806
94572	Rodeo (c)	(510)	8,717	7,589
*94928	Rohnert Park	(707)	42,236	36,326
90274	Rolling Hills Estates	(310)	7,676	7,789
93560	Rosamond (c)	(661)	14,349	7,430
—	Rosedale (c)	(805)	8,445	4,673
95401	Roseland (c)	(707)	6,369	8,779
91770	Rosemead	(626)	53,505	51,638
95826	Rosemont (c)	(916)	22,904	22,851
*95678	Roseville	(916)	79,921	44,685
90720	Rossmoor (c)	(714)	10,298	9,893
91748	Rowland Heights (c)	(818)	48,553	42,647
92519	Rubidoux (c)	(951)	29,180	24,367
92382	Running Springs (c)	(909)	5,125	4,195
*95814	Sacramento	(916)	407,018	369,365
94574	Saint Helena	(707)	5,950	4,990
95368	Salida (c)	(209)	12,560	4,499
*93907	Salinas	(831)	151,060	108,777
*94960	San Anselmo	(415)	12,378	11,735
*92401	San Bernardino	(909)	185,401	170,036
94066	San Bruno	(650)	40,165	38,961
*93001	San Buenaventura (Ventura)	(805)	100,916	92,557
94070	San Carlos	(650)	27,718	26,382
*92674	San Clemente	(949)	49,936	41,100
*92138	San Diego	(619)/(858)	1,223,400	1,110,623
92065	San Diego Country Estates (c)	(760)	9,262	6,874
91773	San Dimas	(909)	34,980	32,398
*91341	San Fernando	(818)	23,564	22,580
*94142	San Francisco	(415)	776,733	723,959
*91778	San Gabriel	(626)	39,804	37,120
93657	Sanger	(559)	18,931	16,839
*92581	San Jacinto	(951)	23,779	17,614
*95113	San Jose	(408)	894,943	782,224
*92690	San Juan Capistrano	(949)	33,826	26,183
*94577	San Leandro	(510)	79,452	68,223
94580	San Lorenzo (c)	(510)	21,898	19,987
*93401	San Luis Obispo	(805)	44,174	41,958
*92069	San Marcos	(760)	54,977	38,974
*91109	San Marino	(626)	12,945	12,959
*94402	San Mateo	(650)	92,482	85,619
94806	San Pablo	(510)	30,215	25,158
*94915	San Rafael	(415)	56,063	48,410
94583	San Ramon	(925)	44,722	35,303
*92711	Santa Ana	(714)/(949)	337,977	293,827
*93102	Santa Barbara	(805)	92,325	85,571
*95050	Santa Clara	(408)	102,361	93,613
*91380	Santa Clarita	(661)	151,088	120,050
*95060	Santa Cruz	(831)	54,593	49,711
90670	Santa Fe Springs	(562)	17,438	15,520
*93454	Santa Maria	(805)	77,423	61,552
*90401	Santa Monica	(310)	84,084	86,905
*93060	Santa Paula	(805)	28,598	25,062
*95402	Santa Rosa	(707)	147,595	113,261

ZIP	Place	Area Code	2000	1990
*92071	Santee	(619)	52,975	52,902
*95070	Saratoga	(408)	29,843	28,061
*94965	Sausalito	(415)	7,330	7,152
*95066	Scotts Valley	(831)	11,385	8,667
90740	Seal Beach	(714)	24,157	25,098
93955	Seaside	(831)	31,696	38,826
*95472	Sebastopol	(707)	7,774	7,008
93662	Selma	(559)	19,444	14,757
—	Shackelford (c)		5,170	—
93263	Shafter	(661)	12,736	9,404
*96019	Shasta Lake	(916)	9,008	8,821
*91025	Sierra Madre	(626)	10,578	10,762
90806	Signal Hill	(562)	9,333	8,371
*93065	Simi Valley	(805)	111,351	100,218
92075	Solana Beach	(858)	12,979	12,956
93960	Soledad	(831)	11,263	13,426
*93463	Solvang	(805)	5,332	4,741
95476	Sonoma	(707)	9,128	8,168
95073	Soquel (c)	(831)	5,081	9,188
91733	South El Monte	(626)	21,144	20,850
90280	South Gate	(323)/(562)	96,375	86,284
*96151	South Lake Tahoe	(530)	23,609	21,586
95965	South Oroville (c)	(530)	7,695	7,463
*91030	South Pasadena	(213)/(323)/(626)/(818)	24,292	23,936
*94080	South San Francisco	(650)	60,552	54,312
91770	South San Gabriel (c)	(626)	7,595	7,700
91744	South San Jose Hills (c)	(626)	20,218	17,814
90605	South Whittier (c)	(562)	55,193	49,514
95991	South Yuba City (c)	(530)	12,651	8,816
*91977	Spring Valley (c)	(619)	26,663	55,331
94309	Stanford (c)	(650)	13,315	18,097
90680	Stanton	(714)	37,403	30,491
*95208	Stockton	(209)	243,771	210,943
95375	Strawberry (c)	(209)	5,302	4,377
94585	Suisun City	(707)	26,118	22,704
*92586	Sun City (c)	(951)	17,773	14,930
*94086	Sunnyvale	(408)	131,760	117,324
*96130	Susanville	(530)	13,541	12,130
93268	Taft	(661)	6,400	5,902
94941	Tamalpais-Homestead Valley (c)	(415)	10,691	9,601
94806	Tara Hills (c)	(510)	5,332	4,998
*93581	Tehachapi	(661)	10,957	6,182
*92589	Temecula	(951)	57,716	27,177
91780	Temple City	(626)	33,377	31,153
95965	Thermalito (c)	(530)	6,045	5,646
*91359	Thousand Oaks	(805)	117,005	104,381
92276	Thousand Palms (c)	(760)	5,120	4,122
94920	Tiburon	(415)	8,666	7,554
*90503	Torrance	(310)	137,946	133,107
*95376	Tracy	(209)	56,929	33,558
*96161	Truckee	(916)	13,864	8,848
*93274	Tulare	(559)	43,994	33,249
*95380	Turlock	(209)	55,810	42,224
*92781	Tustin	(714)/(949)	67,504	50,689
92705	Tustin Foothills (c)	(714)	24,044	24,358
*92277	Twentynine Palms	(760)	14,764	11,821
92278	Twentynine Palms Base (c)	(760)	8,413	10,606
95060	Twin Lakes (c)	(831)	5,533	5,379
95482	Ukiah	(707)	15,497	14,632
94587	Union City	(510)	66,869	53,762
*91785	Upland	(909)	68,393	63,374
*95687	Vacaville	(707)	88,625	71,476
91744	Valinda (c)	(626)	21,776	18,735
*94590	Vallejo	(707)	116,760	109,199
92343	Valle Vista (c)	(909)	10,488	8,751
92082	Valley Center (c)	((760)	7,323	1,711
93437	Vandenberg AFB (c)	(805)	6,151	9,846
93436	Vandenberg Village (c)	(805)	5,802	5,971
	Ventura. *See San Buenaventura*			
*92393	Victorville	(760)	64,029	50,103
90043	View Park-Windsor Hills (c)	(310)	10,958	11,769
92861	Villa Park	(714)	5,999	6,299
—	Vincent (c)		15,097	13,713
—	Vineyard (c)		10,109	—
*93291	Visalia	(559)	91,565	75,659
*92083	Vista	(760)	89,857	71,861
—	Waldon (c)		5,133	—
*91788	Walnut	(909)	30,004	29,105
*94596	Walnut Creek	(925)	64,296	60,569
90255	Walnut Park (c)	(213)	16,180	14,722
93280	Wasco	(661)	21,263	12,412
95386	Waterford	(209)	6,924	4,771
*95076	Watsonville	(831)	44,265	31,099
90044	West Athens (c)	(310)	9,101	8,859
90502	West Carson (c)	(323)	21,138	20,143
90247	West Compton (c)	(310)	5,435	5,451
*91790	West Covina	(626)	105,080	96,226
90069	West Hollywood	(310)/(323)	35,716	36,118
*91359	Westlake Village	(805)	8,368	7,455
*92685	Westminster	(714)	88,207	78,293
—	West Modesto (c)		6,096	—
90047	Westmont (c)	(323)	31,623	31,044
91746	West Puente Valley (c)	(626)	22,589	20,254
*95691	West Sacramento	(916)	31,615	28,898
*90606	West Whittier-Los Nietos (c)	(562)	25,129	24,164
*90605	Whittier	(562)	83,680	77,671
92595	Wildomar (c)	(951)	14,064	10,411
95490	Willits	(707)	5,073	5,027
90222	Willowbrook (c)	(323)	34,138	32,772
95988	Willows	(530)	6,220	5,988
95492	Windsor	(707)	22,744	12,002
—	Winter Gardens (c)		19,771	—
95694	Winters	(530)	6,125	4,639
95388	Winton (c)	(209)	8,832	7,559
92502	Woodcrest (c)	(951)	8,342	7,796
93286	Woodlake	(559)	6,651	5,678
*95695	Woodland	(530)	49,151	40,230
94062	Woodside	(650)	5,352	5,034
*92885	Yorba Linda	(714)	58,918	52,422
96097	Yreka	(530)	7,290	6,948
*95991	Yuba City	(530)	36,758	27,385
92399	Yucaipa	(909)	41,207	32,819
*92286	Yucca Valley	(760)	16,865	16,539

Colorado

Area code (720) overlays area code (303). See introductory note.

ZIP	Place	Area Code	2000	1990
*80840	Air Force Academy (c)	(719)	7,526	9,062
81101	Alamosa	(719)	7,960	7,579
80401	Applewood (c)	(303)	7,123	11,069
*80004	Arvada	(303)	102,153	89,261
*81611	Aspen	(970)	5,914	5,049
*80017	Aurora	(303)	276,393	222,103
81620	Avon	(970)	5,561	1,798
—	Berkley (c)		10,743	—
80908	Black Forest (c)	(719)	13,247	8,143
*80302	Boulder	(303)	94,673	85,127
80601	Brighton	(303)	20,905	14,203
*80020	Broomfield	(303)	38,272	24,638
80723	Brush	(970)	5,117	4,165
*81212	Canon City	(719)	15,431	12,687
81623	Carbondale	(970)	5,196	3,004
—	Castle Pines (c)	(303)	5,958	—
80104	Castle Rock	(303)	20,224	8,710
80120	Castlewood (c)	(303)	25,567	24,392
80110	Cherry Hills Village	(303)	5,958	5,245
81220	Cimarron Hills (c)	(719)	15,194	11,160
81520	Clifton (c)	(970)	17,345	12,671
*80903	Colorado Springs	(719)	360,890	280,430
80120	Columbine (c)	(303)	24,095	23,969
*80022	Commerce City	(303)	20,991	16,466
81321	Cortez	(970)	7,977	7,284
*81625	Craig	(970)	9,189	8,091
81416	Delta	(970)	6,400	3,789
*80202	Denver	(303)	554,636	467,610
80022	Derby (c)	(303)	6,423	6,043
*81301	Durango	(970)	13,922	12,439
80214	Edgewater	(303)	5,445	4,613
81632	Edwards (c)	(970)	8,257	—
*80110	Englewood	(303)	31,727	29,396
80516	Erie	(303)	6,291	1,258
*80517	Estes Park	(970)	5,413	3,184
80620	Evans	(970)	9,514	5,876
*80439	Evergreen (c)	(303)	9,216	7,582
80221	Federal Heights	(303)	12,065	9,342
80913	Fort Carson (c)	(719)	10,566	11,309
*80525	Fort Collins	(970)	118,652	87,491
80621	Fort Lupton	(970)	6,787	5,159
80701	Fort Morgan	(970)	11,034	9,068
80817	Fountain	(719)	15,197	10,754
81521	Fruita	(970)	6,478	4,045
81504	Fruitvale (c)	(303)	6,936	5,222
*81601	Glenwood Springs	(970)	7,736	6,561
*80401	Golden	(303)	17,159	13,127
*81501	Grand Junction	(970)	41,986	32,893
*80631	Greeley	(970)	76,930	60,454
*80111	Greenwood Village	(303)	11,035	7,589
80501	Gunbarrel (c)	(303)	9,435	9,388
*81230	Gunnison	(970)	5,409	4,636
80163	Highlands Ranch (c)	(303)	70,931	10,181
80127	Ken Caryl (c)	(303)	30,887	24,391
80026	Lafayette	(303)	23,197	14,708
81050	La Junta	(719)	7,568	7,678
*80226	Lakewood	(303)	144,126	126,475
81052	Lamar	(719)	8,869	8,343
*80126	Littleton	(303)	40,340	33,711
*80501	Longmont	(303)	71,093	51,976
80027	Louisville	(303)	18,937	12,363
*80538	Loveland	(970)	50,608	37,357
*81401	Montrose	(970)	12,344	8,854
80233	Northglenn	(303)	31,575	27,195
80649	Orchard Mesa (c)	(303)	6,456	5,977
*80134	Parker	(303)	23,558	5,450
*81003	Pueblo	(719)	102,121	98,640
81007	Pueblo West (c)	(719)	16,899	4,386
81503	Redlands (c)	(970)	8,043	9,355
81650	Rifle	(970)	6,784	4,858
81201	Salida	(719)	5,504	4,737
80911	Security-Widefield (c)	(719)	29,845	23,822
80110	Sheridan	(303)	5,600	4,976
80221	Sherrelwood (c)	(303)	17,657	16,636
80122	Southglenn (c)	(303)	43,520	43,087
*80477	Steamboat Springs	(970)	9,815	6,695
80751	Sterling	(970)	11,360	10,362
—	Stonegate (c)		6,284	—
80906	Stratmoor (c)	(719)	6,650	5,854
80027	Superior	(303)	9,011	255
—	The Pinery (c)	(303)	7,253	4,885

ZIP	Place	Area Code	2000	1990
80229	Thornton	(303)	82,384	55,031
81082	Trinidad	(719)	9,078	8,580
81251	Twin Lakes (c)	(719)	6,301	—
80229	Welby (c)	(303)	12,973	10,218
80030	Westminster	(303)	100,940	74,619
*80033	Wheat Ridge	(303)	32,913	29,419
80550	Windsor	(970)	9,896	5,062
*80863	Woodland Park	(719)	6,515	4,610
80132	Woodmoor (c)	(719)	7,177	3,858

Connecticut

See introductory note.

ZIP	Place	Area Code	2000	1990
06401	Ansonia	(203)	18,554	18,403
06001	*Avon*	(860)	15,832	13,937
06403	*Beacon Falls*	(203)	5,246	5,083
06037	*Berlin*	(860)	18,215	16,787
06524	*Bethany*	(203)	5,040	—
06801	*Bethel*	(203)	18,067	17,541
06002	*Bloomfield*	(860)	19,587	19,483
06043	*Bolton*	(860)	5,017	—
06405	*Branford*	(203)	28,683	27,603
06405	Branford Center (c)	(203)	5,735	5,688
*06602	Bridgeport	(203)	139,529	141,686
*06010	Bristol	(860)	60,062	60,640
06804	*Brookfield*	(203)	15,664	14,113
06234	*Brooklyn*	(860)	7,173	6,681
06013	*Burlington*	(860)	8,190	7,026
06019	*Canton*	(860)	8,840	8,268
06040	Central Manchester (c)	(860)	30,595	30,934
06410	*Cheshire*	(203)	28,543	25,684
06410	Cheshire Village (c)	(203)	5,789	5,759
06413	*Clinton*	(860)	13,094	12,767
06415	*Colchester*	(860)	14,551	10,980
06340	Conning Towers-Nautilus Park (c)	(860)	10,241	10,013
06238	*Coventry*	(860)	11,504	10,063
06416	*Cromwell*	(860)	12,871	12,286
*06810	Danbury	(203)	74,848	65,585
06820	*Darien*	(203)	19,607	18,196
06418	Derby	(203)	12,391	12,199
06422	*Durham*	(860)	6,627	5,732
06423	*East Haddam*	(860)	8,333	6,676
06424	*East Hampton*	(860)	13,352	10,428
*06101	*East Hartford*	(860)	49,575	50,452
06512	*East Haven*	(203)	28,189	26,144
06333	*East Lyme*	(860)	18,118	15,340
06612	*Easton*	(203)	7,272	6,303
06088	*East Windsor*	(860)	9,818	10,081
06029	*Ellington*	(860)	12,921	11,197
*06082	*Enfield*	(860)	45,212	45,532
06426	*Essex*	(860)	6,505	5,904
*06430	*Fairfield*	(203)	57,340	53,418
*06032	*Farmington*	(860)	23,641	20,608
06033	*Glastonbury*	(860)	31,876	27,901
06033	Glastonbury Center (c)	(860)	7,157	7,082
06035	*Granby*	(860)	10,347	9,369
*06830	*Greenwich*	(203)	61,101	58,441
06351	*Griswold*	(860)	10,807	10,384
*06340	Groton	(860)	10,010	9,837
06340	*Groton*	(860)	39,907	45,144
06437	*Guilford*	(203)	21,398	19,848
06438	*Haddam*	(860)	7,157	6,769
*06514	*Hamden*	(203)	56,913	52,434
*06101	Hartford	(860)	121,578	139,739
06791	*Harwinton*	(860)	5,283	5,228
06248	*Hebron*	(860)	8,610	7,079
06037	Kensington (c)	(860)	8,541	8,306
06239	*Killingly*	(860)	16,472	15,889
06419	*Killingworth*	(860)	6,018	4,814
06249	*Lebanon*	(860)	6,907	6,041
06339	*Ledyard*	(860)	14,687	14,913
06759	*Litchfield*	(860)	8,316	8,365
06443	*Madison*	(203)	17,858	15,485
*06040	*Manchester*	(860)	54,740	51,618
06250	*Mansfield*	(860)	20,720	21,103
06447	*Marlborough*	(860)	5,709	5,535
*06450	Meriden	(203)	58,244	59,479
06762	*Middlebury*	(203)	6,451	6,145
06457	Middletown	(860)	43,167	42,762
06460	Milford	(203)	52,305	48,168
06468	*Monroe*	(203)	19,247	16,896
06353	*Montville*	(860)	18,546	16,673
06770	Naugatuck	(203)	30,989	30,625
*06050	New Britain	(860)	71,538	75,491
06840	*New Canaan*	(203)	19,395	17,864
06812	*New Fairfield*	(203)	13,953	12,911
06057	*New Hartford*	(860)	6,088	5,769
*06511	New Haven	(203)	123,626	130,474
*06101	*Newington*	(860)	29,306	29,208
06320	New London	(860)	25,671	28,540
06776	*New Milford*	(860)	27,121	23,629
06470	*Newtown*	(203)	25,031	20,779
06471	*North Branford*	(203)	13,906	12,996
06473	*North Haven*	(203)	23,035	22,247
*06856	Norwalk	(203)	82,951	78,331
06360	Norwich	(860)	36,117	37,391
06779	Oakville (c)	(860)	8,618	8,741
06371	*Old Lyme*	(860)	7,406	6,535
06475	*Old Saybrook*	(860)	10,367	9,552
06477	*Orange*	(203)	13,233	12,830
06478	*Oxford*	(203)	9,821	8,685
06379	Pawcatuck (c)	(860)	5,474	5,289
06374	*Plainfield*	(860)	14,619	14,363
06062	*Plainville*	(860)	17,328	17,392
06782	*Plymouth*	(860)	11,634	11,822
06480	*Portland*	(860)	8,732	8,418
06712	*Prospect*	(203)	8,707	7,775
06260	*Putnam*	(860)	9,002	9,031
06260	Putnam District(c)	(860)	6,746	6,835
06896	*Redding*	(203)	8,270	7,927
06877	Ridgefield (c)	(203)	7,212	6,363
06877	*Ridgefield*	(203)	23,643	20,919
06066	Rockville (c)	(860)	7,708	—
06067	*Rocky Hill*	(860)	17,966	16,554
06483	*Seymour*	(203)	15,454	14,288
06484	Shelton	(203)	38,101	35,418
06082	Sherwood Manor (c)	(860)	5,689	6,357
06070	*Simsbury*	(860)	23,234	22,023
06070	Simsbury Center (c)	(860)	5,603	5,577
06071	*Somers*	(860)	10,417	9,108
06488	*Southbury*	(203)	18,567	15,818
06489	*Southington*	(860)	39,728	38,518
06074	*South Windsor*	(860)	24,412	22,090
06082	Southwood Acres (c)	(860)	8,067	8,963
06075	*Stafford*	(860)	11,307	11,091
*06904	Stamford	(203)	117,083	108,056
06378	*Stonington*	(860)	17,906	16,919
06268	Storrs (c)	(860)	10,996	12,198
*06602	*Stratford*	(203)	49,976	49,389
06078	*Suffield*	(860)	13,552	11,427
06786	Terryville (c)	(860)	5,360	5,426
06787	*Thomaston*	(860)	7,503	6,947
06277	*Thompson*	(860)	8,878	8,668
06082	Thompsonville (c)	(860)	8,125	8,458
06084	*Tolland*	(860)	13,146	11,001
06790	Torrington	(860)	35,202	33,687
06611	*Trumbull*	(203)	34,243	32,016
06066	*Vernon*	(860)	28,063	29,841
06492	*Wallingford*	(203)	43,026	40,822
06492	Wallingford Center (c)	(203)	17,509	17,827
*06702	Waterbury	(203)	107,271	108,961
06385	*Waterford*	(860)	19,152	17,930
06795	*Watertown*	(860)	21,661	20,456
06498	*Westbrook*	(860)	6,292	5,414
*06101	*West Hartford*	(860)	63,589	60,110
06516	West Haven	(203)	52,360	54,021
06883	*Weston*	(203)	10,037	8,648
*06880	*Westport*	(203)	25,749	24,410
*06101	*Wethersfield*	(860)	26,271	25,651
06226	Willimantic (c)	(860)	15,823	14,746
06279	*Willington*	(860)	5,959	5,979
06897	*Wilton*	(203)	17,633	15,989
06094	*Winchester*	(860)	10,664	11,524
06280	*Windham*	(860)	22,857	22,039
06095	*Windsor*	(860)	28,237	27,817
06096	*Windsor Locks*	(860)	12,043	12,358
06098	Winsted (c)	(860)	7,321	8,254
06716	*Wolcott*	(203)	15,215	13,700
06525	*Woodbridge*	(203)	8,983	7,924
06798	*Woodbury*	(203)	9,198	8,131
06281	*Woodstock*	(860)	7,221	6,008

Delaware (302)

ZIP	Place	2000	1990
19701	Bear (c)	17,593	—
19713	Brookside (c)	14,806	15,307
19703	Claymont (c)	9,220	9,800
*19901	Dover	32,135	27,630
19809	Edgemoor (c)	5,992	5,853
19805	Elsmere	5,800	5,935
19702	Glasgow (c)	12,840	—
19707	Hockessin (c)	12,902	—
19709	Middletown	6,161	3,834
19963	Milford	6,732	6,032
*19711	Newark	28,547	26,463
—	North Star (c)	8,277	—
19800	Pike Creek (c)	19,751	10,163
19973	Seaford	6,699	5,689
19977	Smyrna	5,679	5,231
*19899	Wilmington	72,664	71,529
19720	Wilmington Manor (c)	8,262	8,568

District of Columbia (202)

ZIP	Place	2000	1990
*20090	Washington	572,059	606,900

Florida

Area code (321) overlays area code (407). Area code (754) overlays (954). Area code (786) overlays (305). See introductory note

ZIP	Place	Area Code	2000	1990
*32615	Alachua	(386)	6,098	4,667
*32714	Altamonte Springs	(407)	41,200	35,167
—	Andover (c)		8,489	6,251
33572	Apollo Beach (c)	(813)	7,444	6,025
*32712	Apopka	(407)	26,642	13,611
*34266	Arcadia	(863)	6,604	6,488
32233	Atlantic Beach	(904)	13,368	11,636
33823	Auburndale	(863)	11,032	8,846
*33160	Aventura	(305)	25,267	14,914
*33825	Avon Park	(863)	8,542	8,078

ZIP	Place	Area Code	2000	1990
32857	Azalea Park (c)	(407)	11,073	8,926
*33830	Bartow	(863)	15,340	14,716
33154	Bay Harbor Islands	(305)	5,146	4,703
—	Bay Hill (c)	(407)	5,177	5,346
34667	Bayonet Point (c)	(727)	23,577	21,860
33505	Bayshore Gardens (c)	(941)	17,350	17,062
33589	Beacon Square (c)	(727)	7,263	6,265
34233	Bee Ridge (c)	(941)	8,744	6,406
32073	Bellair-Meadowbrook Terrace (c)	(904)	16,539	15,606
33430	Belle Glade	(561)	14,906	16,177
*32802	Belle Isle	(407)	5,531	5,272
*34420	Belleview (c)	(352)	21,201	19,386
*34461	Beverly Hills (c)	(352)	8,317	6,163
33043	Big Pine Key (c)		5,032	4,206
*33509	Bloomingdale (c)	(813)	19,839	13,912
—	Boca Del Mar (c)		21,832	17,754
*33431	Boca Raton	(561)	74,764	61,486
*34135	Bonita Springs	(941)	32,797	13,600
—	Boyette (c)		5,895	—
*33436	Boynton Beach	(561)	60,389	46,284
*34206	Bradenton	(941)	49,504	43,769
*33509	Brandon (c)	(813)	77,895	57,985
32503	Brent (c)	(850)	22,257	21,624
33317	Broadview Park (c)	(954)	6,798	6,109
33313	Broadview-Pompano Park (c)	(954)	5,314	5,230
*34601	Brooksville	(352)	7,264	7,589
33142	Brownsville (c)	(305)	14,393	15,607
32404	Callaway	(850)	14,233	12,253
32920	Cape Canaveral	(321)	8,829	8,014
*33909	Cape Coral	(941)	102,286	74,991
33055	Carol City (c)	(305)	59,443	53,331
*32707	Casselberry	(407)	22,629	20,736
—	Cedar Grove	(850)	5,367	1,479
33401	Century Village (c)	(305)	7,616	8,363
—	Cheval (c)		7,602	—
—	Citrus Park (c)		20,226	—
*32966	Citrus Ridge (c)		12,015	—
*33758	Clearwater	(727)	108,787	98,669
*34711	Clermont	(352)	9,333	6,910
33440	Clewiston	(863)	6,460	6,085
*32922	Cocoa	(321)	16,412	17,710
*32931	Cocoa Beach	(321)	12,482	12,123
32922	Cocoa West (c)	(321)	5,921	6,160
*33097	Coconut Creek	(954)	43,566	27,269
33064	Collier Manor-Cresthaven (c)	(954)	7,741	7,322
33801	Combee Settlement (c)	(863)	5,436	5,463
32809	Conway (c)	(407)	14,394	13,159
33328	Cooper City	(954)	27,939	21,335
*33114	Coral Gables	(305)	42,249	40,091
*33075	Coral Springs	(954)	117,549	78,864
33157	Coral Terrace (c)	(305)	24,380	23,255
—	Country Club (c)		36,310	3,408
—	Country Walk (c)	(305)	10,653	—
*32536	Crestview	(850)	14,766	9,886
33803	Crystal Lake (c)	(863)	5,341	5,300
33157	Cutler (c)	(305)	17,390	16,201
33157	Cutler Ridge (c)	(305)	24,781	21,268
33884	Cypress Gardens (c)	(863)	8,844	9,188
33919	Cypress Lake (c)	(941)	12,072	10,491
*33525	Dade City	(352)	6,188	5,633
33004	Dania Beach	(954)	20,061	—
33329	Davie	(954)	75,720	47,143
*32114	Daytona Beach	(386)	64,112	61,991
32713	De Bary	(386)	15,559	9,327
*33441	Deerfield Beach	(954)	64,583	46,997
*32433	DeFuniak Springs	(850)	5,089	5,200
*32720	De Land	(386)	20,904	16,622
*33444	Delray Beach	(561)	60,020	47,184
*32738	Deltona	(407)	69,543	49,429
*32541	Destin	(850)	11,119	8,090
—	Doctor Phillips (c)	(407)	9,548	7,963
—	Doral (c)	(305)	20,438	3,126
*34698	Dunedin	(727)	35,691	34,427
33610	East Lake (c)		29,394	—
33610	East Lake-Orient Park (c)	(813)	5,703	6,171
—	East Perrine (c)		7,079	—
*32132	Edgewater	(386)	18,668	15,351
32542	Eglin AFB (c)	(850)	8,082	8,347
—	Egypt Lake-Leto (c)		32,782	—
34680	Elfers (c)	(727)	13,161	12,356
*34295	Englewood (c)	(941)	16,196	15,025
32534	Ensley (c)	(850)	18,752	16,362
33928	Estero (c)	(941)	9,503	3,177
*32726	Eustis	(352)	15,106	12,856
32804	Fairview Shores (c)	(305)	13,898	13,192
*32034	Fernandina Beach	(904)	10,549	8,765
32730	Fern Park (c)	(407)	8,318	8,294
32514	Ferry Pass (c)	(850)	27,176	26,301
33034	Florida City	(305)	7,843	5,978
32960	Florida Ridge (c)	(561)	15,217	12,218
32714	Forest City (c)	(407)	12,612	10,638
*33310	Fort Lauderdale	(954)	152,397	149,238
33841	Fort Meade	(863)	5,691	5,151
*33902	Fort Myers	(941)	48,208	44,947
*33931	Fort Myers Beach (c)	(941)	6,561	9,284
*33922	Fort Myers Shores (c)	(941)	5,793	5,460
*34981	Fort Pierce	(561)	37,516	36,830
33452	Fort Pierce North (c)	(561)	7,386	5,833
34982	Fort Pierce South (c)	(561)	5,672	5,320
*32548	Fort Walton Beach	(850)	19,973	21,407
—	Fountainbleau (c)		59,549	—
*32043	Fruit Cove (c)	(904)	16,077	5,904
34230	Fruitville (c)	(941)	12,741	9,808
—	Fussels Corner (c)		5,313	3,840
*32602	Gainesville	(352)	95,447	91,482
33534	Gibsonton (c)	(813)	8,752	7,706
32960	Gifford (c)	(561)	7,599	6,278
33138	Gladeview (c)	(954)	14,468	15,637
33143	Glenvar Heights (c)	(305)	16,243	14,823
34116	Golden Gate (c)	(941)	20,951	14,148
33055	Golden Glades (c)	(305)	32,623	25,474
—	Golden Lakes (c)	(561)	6,694	3,867
32733	Goldenrod (c)	(407)	12,871	12,362
32560	Gonzalez (c)	(850)	11,365	7,669
33170	Goulds (c)	(305)	7,453	7,284
—	Greater Carrollwood (c)		33,519	—
—	Greater Northdale (c)		20,461	16,318
—	Greater Sun Center (c)		16,321	—
33454	Greenacres	(561)	27,569	18,683
32043	Green Cove Springs	(904)	5,378	4,497
*32561	Gulf Breeze	(850)	5,665	5,530
33581	Gulf Gate Estates (c)	(941)	11,647	11,622
33737	Gulfport	(727)	12,527	11,709
*33844	Haines City	(863)	13,174	11,683
*33009	Hallandale	(305)/(954)	34,282	30,997
—	Hamptons at Boca Raton (c)		11,306	11,686
34442	Hernando (c)	(352)	8,253	2,103
*33010	Hialeah	(305)	226,419	188,008
33016	Hialeah Gardens	(305)	19,297	7,727
*33455	Hobe Sound (c)	(561)	11,376	11,507
*34689	Holiday (c)	(727)	21,904	19,360
32125	Holly Hill	(386)	12,119	11,141
*33022	Hollywood	(954)	139,357	121,720
*33030	Homestead	(305)	31,909	26,694
34447	Homosassa Springs (c)	(352)	12,458	6,271
*34668	Hudson (c)	(727)	12,765	7,344
—	Hunters Creek (c)	(407)	9,369	—
*34142	Immokalee (c)	(941)	19,763	14,120
32937	Indian Harbour Beach	(321)	8,152	6,933
32963	Indian River Estates (c)	(561)	5,793	4,858
33785	Indian Rocks Beach	(727)	5,072	3,963
34956	Indiantown (c)	(561)	5,588	4,794
*34450	Inverness	(352)	6,789	5,797
—	Inverness Highlands South (c)		5,781	—
33880	Inwood (c)	(863)	6,925	6,824
—	Iona(c)	(941)	11,756	9,565
33036	Islamorada, Village of Islands	(305)	6,846	1,220
33162	Ives Estates (c)	(305)	17,586	13,531
*32203	Jacksonville	(904)	735,617	635,230
*32250	Jacksonville Beach	(904)	20,990	17,839
33880	Jan Phyl Village (c)	(863)	5,633	5,308
33568	Jasmine Estates (c)	(727)	18,213	17,136
*34957	Jensen Beach (c)	(561)	11,100	9,884
*33458	Jupiter	(561)	39,328	26,753
33183	Kendale Lakes (c)	(305)	56,901	48,524
33256	Kendall (c)	(305)	75,226	87,271
—	Kendall West (c)		38,034	—
33149	Key Biscayne	(305)	10,507	8,854
33037	Key Largo (c)	(305)	11,886	11,336
—	Keystone (c)		14,627	—
*33040	Key West	(305)	25,478	24,832
*33573	Kings Point (c)	(305)	12,207	12,422
*34744	Kissimmee	(407)	47,814	30,337
*32159	Lady Lake	(352)	11,828	8,071
—	Lake Butter (c)		7,062	—
*32055	Lake City	(386)	9,980	9,626
*33804	Lakeland	(863)	78,452	70,576
33801	Lakeland Highlands (c)	(863)	12,557	9,972
32569	Lake Lorraine (c)	(850)	7,106	6,779
33054	Lake Lucerne (c)	(305)	9,132	9,478
33612	Lake Magdalene (c)	(813)	28,755	15,973
*32746	Lake Mary	(407)	11,458	5,929
33403	Lake Park	(561)	8,721	6,704
—	Lakes by the Bay (c)		9,055	5,615
32073	Lakeside (c)	(904)	30,927	29,137
*33853	Lake Wales	(863)	10,194	9,670
34951	Lakewood Park (c)	(561)	10,458	7,211
*33461	Lake Worth	(561)	35,133	28,564
—	Lake Worth Corridor (c)		18,663	—
34639	Land O'Lakes (c)	(813)	20,971	7,892
33465	Lantana	(561)	9,437	8,392
*33770	Largo	(727)	69,371	65,910
33313	Lauderdale Lakes	(954)	31,705	27,341
33313	Lauderhill	(954)	57,585	49,015
34272	Laurel (c)	(941)	8,393	8,245
*34461	Lecanto (c)	(352)	5,161	1,243
*34748	Leesburg	(352)	15,956	14,783
*33936	Lehigh Acres (c)	(941)	33,430	13,611
33033	Leisure City (c)	(305)	22,152	19,379
33074	Lighthouse Point	(954)	10,767	10,378
*32060	Live Oak	(386)	6,480	6,332
32860	Lockhart (c)	(407)	12,944	11,636
34228	Longboat Key	(941)	7,603	5,937
*32750	Longwood	(407)	13,745	13,316
*33549	Lutz (c)	(813)	17,081	10,552
32444	Lynn Haven	(850)	12,451	9,270
—	McGregor (c)		7,136	6,504
*32751	Maitland	(407)	12,019	8,932

ZIP	Place	Area Code	2000	1990
33550	Mango (c)	(813)	8,842	8,700
33050	Marathon	(305)	10,255	8,857
*34145	Marco Island	(941)	14,879	—
33093	Margate	(954)	53,909	42,985
*32446	Marianna	(850)	6,230	6,292
—	Meadow Woods (c)		11,286	4,876
—	Medulla (c)	(863)	6,637	3,977
*32901	Melbourne	(321)	71,382	60,034
32666	Melrose Park (c)	(954)	7,114	6,477
33561	Memphis (c)	(941)	7,264	6,760
*32953	Merritt Island (c)	(321)	36,090	32,886
*33101	Miami	(305)	362,470	358,648
*33152	Miami Beach	(305)	87,933	92,639
33014	Miami Lakes (c)	(305)	22,676	12,750
33153	Miami Shores	(305)	10,380	10,084
33266	Miami Springs	(305)	13,712	13,268
32976	Micco (c)	(561)	9,498	8,757
*32068	Middleburg (c)	(904)	10,338	6,223
*32570	Milton	(850)	7,045	7,216
32754	Mims (c)	(321)	9,147	9,412
34755	Minneola	(352)	5,435	1,515
33023	Miramar	(954)	72,739	40,663
*32757	Mount Dora	(352)	9,418	7,294
32526	Myrtle Grove (c)	(850)	17,211	17,402
*34102	Naples	(941)	20,976	19,505
—	Naples Manor (c)	(941)	5,186	4,574
34102	Naples Park (c)	(941)	6,741	8,002
32266	Neptune Beach	(904)	7,270	6,816
*34653	New Port Richey	(727)	16,117	14,044
33552	New Port Richey East (c)	(727)	9,916	9,683
*32168	New Smyrna Beach	(386)	20,048	16,549
*32578	Niceville	(850)	11,684	10,509
33269	Norland (c)	(305)	22,995	22,109
33308	North Andrews Gardens (c)	(954)	9,656	9,002
33141	North Bay Village	(305)	6,733	5,383
33918	North Fort Myers (c)	(941)	40,214	30,027
33068	North Lauderdale	(954)	32,264	26,473
33261	North Miami	(305)	59,880	50,001
33160	North Miami Beach	(305)	40,786	35,361
33408	North Palm Beach	(561)	12,064	11,538
*34287	North Port	(941)	22,797	11,973
34234	North Sarasota (c)	(941)	6,738	6,702
33307	Oakland Park	(305)	30,966	26,326
33860	Oak Ridge (c)	(407)	22,349	15,388
*34478	Ocala	(352)	45,943	42,045
32548	Ocean City (c)	(850)	5,594	5,422
34761	Ocoee	(407)	24,391	12,778
33163	Ojus (c)	(305)	16,642	15,519
*34972	Okeechobee	(863)	5,376	4,943
34677	Oldsmar	(813)	11,910	8,361
33265	Olympia Heights (c)	(305)	13,452	37,792
*33054	Opa-Locka	(305)	14,951	15,283
33054	Opa-Locka North (c)	(305)	6,224	6,568
*32763	Orange City	(386)	6,604	5,372
*32073	Orange Park	(904)	9,081	9,488
*32802	Orlando	(407)	185,951	164,674
32861	Orlo Vista (c)	(407)	6,047	5,990
*32174	Ormond Beach	(386)	36,301	29,721
32074	Ormond By-The-Sea (c)	(386)	8,430	8,157
*32765	Oviedo	(407)	26,316	11,114
32571	Pace (c)	(850)	7,393	6,277
33476	Pahokee	(561)	5,985	6,822
*32177	Palatka	(386)	10,033	10,447
*32905	Palm Bay	(321)	79,413	62,543
33480	Palm Beach	(561)	10,468	9,814
33408	Palm Beach Gardens	(561)	35,058	24,139
*34990	Palm City (c)	(561)	20,097	3,925
*32135	Palm Coast	(386)	32,732	14,287
*34221	Palmetto	(941)	12,571	9,268
33157	Palmetto Estates (c)	(305)	13,675	12,293
*34683	Palm Harbor (c)	(727)	59,248	50,256
*33601	Palm River-Clair Mel (c)	(813)	17,589	13,691
33460	Palm Springs	(561)	11,699	9,763
33012	Palm Springs North (c)	(305)	5,460	5,300
32082	Palm Valley (c)	(904)	19,860	9,960
*32401	Panama City	(850)	36,417	34,396
32417	Panama City Beach	(850)	7,671	4,051
33060	Parkland	(954)	13,835	3,773
—	Pelican Bay (c)		5,686	—
33021	Pembroke Park	(954)	6,299	4,933
33029	Pembroke Pines	(954)	137,427	65,566
*32502	Pensacola	(850)	56,255	59,198
*32347	Perry	(850)	6,847	7,151
32859	Pine Castle (c)	(407)	8,803	8,276
33156	Pinecrest	(305)	19,055	—
32858	Pine Hills (c)	(407)	41,764	35,322
—	Pine Island Ridge (c)	(954)	5,199	5,244
*33781	Pinellas Park	(727)	45,658	43,571
—	Pine Ridge (c)		5,490	—
33168	Pinewood (c)	(305)	16,523	15,518
33318	Plantation	(954)	82,934	66,814
*33566	Plant City	(813)	29,915	22,754
*34758	Poinciana (c)	(407)	13,647	—
*33060	Pompano Beach	(954)	78,191	72,411
33064	Pompano Beach Highlands (c)	(954)	6,505	17,915
*33952	Port Charlotte (c)	(941)	46,451	41,535
32129	Port Orange	(904)	45,823	35,399
32927	Port St. John (c)	(321)	12,112	8,933
*34981	Port St. Lucie	(561)	88,769	55,761
—	Port St. Lucie-River Park (c)		5,175	4,874
34992	Port Salerno (c)	(561)	10,141	7,786
*33032	Princeton (c)	(305)	10,090	7,073
*33950	Punta Gorda	(941)	14,344	10,637
*32351	Quincy	(850)	6,982	7,452
33156	Richmond Heights (c)	(305)	8,479	8,583
—	Richmond West (c)		28,082	—
—	Ridge Wood Heights (c)		5,028	4,851
*33569	Riverview (c)	(813)	12,035	6,478
33419	Riviera Beach	(561)	29,884	27,646
*32955	Rockledge	(321)	20,170	16,023
33947	Rotonda (c)	(941)	6,574	3,576
33411	Royal Palm Beach	(561)	21,523	15,532
33570	Ruskin (c)	(813)	8,321	6,046
34695	Safety Harbor	(727)	17,203	15,120
*32084	Saint Augustine	(904)	11,592	11,695
—	Saint Augustine South (c)		5,035	4,218
*34769	Saint Cloud	(407)	20,074	12,684
*33736	Saint Pete Beach	(727)	9,929	9,200
*33733	Saint Petersburg	(727)	248,232	240,318
33912	San Carlos Park (c)	(941)	16,317	11,785
33432	Sandalfoot Cove (c)	(305)	16,582	14,214
*32771	Sanford	(407)	38,291	32,387
33957	Sanibel	(941)	6,064	5,468
*34230	Sarasota	(941)	52,715	50,897
33577	Sarasota Springs (c)	(941)	15,875	16,088
32937	Satellite Beach	(321)	9,577	9,889
33055	Scott Lake (c)	(305)	14,401	14,588
*32958	Sebastian	(561)	16,181	10,248
*33870	Sebring	(863)	9,667	8,841
*33584	Seffner (c)	(813)	5,467	5,371
*33770	Seminole	(813)	10,890	9,251
34610	Shady Hills (c)	(727)	7,798	—
*34242	Siesta Key (c)	(941)	7,150	7,772
34472	Silver Springs Shores (c)	(352)	6,690	6,421
32809	Sky Lake (c)	(407)	5,651	6,202
32703	South Apopka (c)	(407)	5,800	6,360
33505	South Bradenton (c)	(941)	21,587	20,398
32121	South Daytona	(386)	13,177	12,488
—	Southeast Arcadia (c)		6,064	4,145
34277	Southgate (c)	(941)	7,455	7,324
34233	South Gate Ridge (c)	(941)	5,655	5,924
—	South Highpoint (c)		8,839	—
33243	South Miami	(305)	10,741	10,404
33157	South Miami Heights (c)	(305)	33,522	30,030
33707	South Pasadena	(727)	5,778	5,644
32937	South Patrick Shores (c)	(321)	8,913	10,249
34230	South Sarasota (c)	(941)	5,314	5,298
33595	South Venice (c)	(941)	13,539	11,951
32401	Springfield	(850)	8,810	8,719
*34601	Spring Hill (c)	(352)	69,078	31,117
32091	Starke	(904)	5,593	5,226
*34994	Stuart	(561)	14,633	11,936
—	Sugarmill Woods (c)	(352)	6,409	4,073
33160	Sunny Isles Beach	(305)	15,315	—
33345	Sunrise	(954)	85,779	65,683
33283	Sunset (c)	(305)	17,150	15,810
33144	Sweetwater	(305)	14,226	13,909
*32301	Tallahassee	(850)	150,624	124,773
33320	Tamarac	(954)	55,588	44,822
33144	Tamiami (c)	(305)	54,788	33,845
*33601	Tampa	(813)	303,447	280,015
34689	Tarpon Springs	(727)	21,003	17,874
32778	Tavares	(352)	9,700	7,488
33687	Temple Terrace	(813)	20,918	16,444
33469	Tequesta	(561)	5,273	4,499
33186	The Crossings (c)		23,557	—
—	The Hammocks (c)		47,379	—
32159	The Villages (c)		8,333	—
33592	Thonotosassa (c)	(813)	6,091	—
—	Three Lakes (c)		6,955	—
33025	Timber Pines (c)		5,840	3,182
*32780	Titusville	(321)	40,670	39,394
32685	Town 'n' Country (c)	(813)	72,523	60,946
33706	Treasure Island	(727)	7,450	7,266
32867	Union Park (c)	(407)	10,191	6,890
33024	University (c)		30,736	—
34201	University Park (c)	(941)	26,538	—
32401	Upper Grand Lagoon (c)	(850)	10,889	7,855
32580	Valparaiso	(850)	6,408	6,316
*33594	Valrico (c)	(813)	6,582	—
—	Vamo (c)	(941)	5,285	3,325
*34285	Venice	(941)	17,764	17,052
33595	Venice Gardens (c)	(941)	7,466	7,701
*32960	Vero Beach	(561)	17,705	17,350
32960	Vero Beach South (c)	(561)	20,362	16,973
33901	Villas (c)	(941)	11,346	9,898
32507	Warrington (c)	(850)	15,207	16,040
32791	Wekiva Springs (c)	(407)	23,169	23,026
33414	Wellington	(561)	38,216	20,670
33543	Wesley Chapel (c)	(813)	5,691	—
—	West and East Lealman (c)		21,753	—
33626	Westchase (c)	(813)	11,116	—
33155	Westchester (c)	(305)	30,271	29,883
—	Westgate-Belvedere Homes (c)		8,134	6,880
33138	West Little River (c)	(305)	32,498	33,575
32912	West Melbourne	(321)	9,824	8,398
33144	West Miami	(305)	5,863	5,727
33326	Weston	(954)	49,286	—

ZIP	Place	Area Code	2000	1990
*33416	West Palm Beach	(561)	82,103	67,764
32505	West Pensacola (c)	(850)	21,939	22,107
33157	West Perrine (c)	(305)	8,600	—
—	West Samoset (c)		5,507	3,819
—	West Vero Corridor (c)		7,695	—
33168	Westview (c)	(305)	9,692	9,668
33165	Westwood Lakes (c)	(305)	12,005	11,522
—	Whisper Walk (c)		5,135	3,037
—	Williamsburg (c)	(407)	6,736	3,093
33305	Wilton Manors	(954)	12,697	11,804
33803	Winston (c)	(813)	9,024	9,118
*34787	Winter Garden	(407)	14,351	9,863
*33880	Winter Haven	(863)	26,487	24,725
*32789	Winter Park	(407)	24,090	24,260
*32707	Winter Springs	(407)	31,666	22,151
32547	Wright (c)	(850)	21,697	18,945
34972	Yeehaw Junction (c)	(407)	21,778	—
*32097	Yulee (c)	(904)	8,392	6,915
*33540	Zephyrhills	(813)	10,833	8,220
—	Zephyrhills West (c)		5,242	4,249

Georgia

Area code (470) overlays area codes (404), (678), and (770). Area code (678) overlays (770). See introductory note.

ZIP	Place	Area Code	2000	1990
*30101	Acworth	(770)	13,422	4,519
31620	Adel	(229)	5,307	5,093
*31706	Albany	(229)	76,939	78,804
*30004	Alpharetta	(770)	34,854	13,002
31709	Americus	(229)	17,013	16,516
*30603	Athens-Clarke County[1]	(706)	101,489	86,522
*30301	Atlanta	(404)	416,474	393,929
30011	Auburn	(770)	6,904	3,139
*30903	Augusta-Richmond County[2]	(706)	199,775	186,616
30168	Austell	(770)	5,359	4,173
*31717	Bainbridge	(229)	11,722	10,803
30204	Barnesville	(770)	5,972	4,747
30032	Belvedere Park (c)	(404)	18,945	18,089
31723	Blakely	(229)	5,696	5,595
*31520	Brunswick	(912)	15,600	16,433
*30518	Buford	(404)	10,668	8,771
31728	Cairo	(229)	9,239	9,035
*30701	Calhoun	(706)	10,667	7,135
31730	Camilla	(229)	5,669	5,124
30032	Candler-McAfee (c)	(404)	28,294	29,491
*30114	Canton	(770)	7,709	4,817
*30117	Carrollton	(770)	19,843	16,029
*30120	Cartersville	(770)	15,925	12,037
30125	Cedartown	(770)	9,470	7,976
30366	Chamblee	(404)	9,552	7,668
30021	Clarkston	(404)	7,231	5,385
30337	College Park	(404)	20,382	20,645
*31908	Columbus	(706)	186,291	178,683
30529	Commerce	(770)	5,292	4,108
30288	Conley (c)	(404)	6,188	5,528
*30013	Conyers	(404)	10,689	7,380
*31015	Cordele	(229)	11,608	10,833
—	Country Club Estates (c)		7,594	7,500
*30014	Covington	(770)	11,547	9,860
*30132	Dallas	(770)	5,056	2,810
*30720	Dalton	(706)	27,912	22,218
31742	Dawson	(229)	5,058	5,295
*30030	Decatur (DeKalb)	(404)	18,147	17,304
31520	Dock Junction (c)	(912)	6,951	7,094
30362	Doraville	(404)	9,862	7,626
*31533	Douglas	(912)	10,639	10,464
*30134	Douglasville	(404)	20,065	11,635
30333	Druid Hills (c)	(404)	12,741	12,174
*31021	Dublin	(478)	15,857	16,312
*30096	Duluth	(404)	22,122	9,821
30356	Dunwoody (c)	(404)	32,808	26,302
31023	Eastman	(478)	5,440	5,153
30364	East Point	(404)	39,595	34,595
31024	Eatonton	(706)	6,764	6,479
30809	Evans (c)	(706)	17,727	13,713
30213	Fairburn	(770)	5,464	4,013
30060	Fair Oaks (c)	(404)	8,443	6,996
30535	Fairview (c)	(706)	6,601	6,444
*30214	Fayetteville	(404)	11,148	5,827
31750	Fitzgerald	(229)	8,758	8,901
*30297	Forest Park	(404)	21,447	16,958
31905	Fort Benning South (c)	(706)	11,737	14,617
30742	Fort Oglethorpe	(706)	6,940	5,880
*31313	Fort Stewart (c)	(912)	11,205	13,774
31030	Fort Valley	(478)	8,005	8,198
*30501	Gainesville	(770)	25,578	17,885
31418	Garden City	(912)	11,289	7,410
31754	Georgetown (c)	(912)	10,599	5,554
30316	Gresham Park (c)	(404)	9,215	9,000
*30223	Griffin	(770)	23,451	21,325
30813	Grovetown	(706)	6,089	3,596
30354	Hapeville	(404)	6,180	5,483
*31313	Hinesville	(912)	30,392	21,596
—	Irondale (c)		7,727	3,352
*31546	Jesup	(912)	9,279	8,958
*30144	Kennesaw	(404)	21,675	8,936
31548	Kingsland	(912)	10,506	6,089
30728	La Fayette	(706)	6,702	6,655
*30240	LaGrange	(706)	25,998	25,574
*30045	Lawrenceville	(404)	22,397	17,250
*30047	Lilburn	(404)	11,307	9,295
30052	Loganville	(770)	5,435	3,180
30126	Mableton (c)	(404)	29,733	25,725
30253	McDonough	(770)	8,493	2,929
*31201	Macon	(478)	97,255	107,365
*30060	Marietta	(404)	58,748	44,129
30917	Martinez (c)	(706)	27,749	33,731
—	Midway-Hardwick (c)		5,135	4,910
31061	Milledgeville	(478)	18,757	17,727
*30655	Monroe	(770)	11,407	9,759
*31768	Moultrie	(229)	14,387	14,865
30087	Mountain Park (c)	(404)	11,753	11,025
*30263	Newnan	(770)	16,242	12,497
*30071	Norcross	(404)	8,410	5,947
30319	North Atlanta (c)	(404)	38,579	27,812
30033	North Decatur (c)	(404)	15,270	13,936
30033	North Druid Hills (c)	(404)	18,852	14,170
30032	Panthersville (c)	(404)	11,791	9,874
30269	Peachtree City	(404)	31,580	19,027
31069	Perry	(478)	9,602	9,452
31322	Pooler	(912)	6,239	4,649
30127	Powder Springs	(404)	12,481	6,862
30074	Redan (c)	(404)	33,841	24,376
31324	Richmond Hill	(912)	6,959	2,934
*30274	Riverdale	(404)	12,478	9,495
*30161	Rome	(706)	34,980	30,425
*30077	Roswell	(404)	79,334	47,986
31558	Saint Marys	(912)	13,761	8,204
31522	Saint Simons (c)	(912)	13,381	12,026
31082	Sandersville	(478)	6,144	6,290
30358	Sandy Springs (c)	(404)	85,781	67,842
*31402	Savannah	(912)	131,510	137,812
30079	Scottdale (c)	(404)	9,803	8,636
—	Skidaway Island (c)	(912)	6,914	4,495
*30080	Smyrna	(404)	40,999	32,453
*30078	Snellville	(404)	15,351	12,084
*30458	Statesboro	(912)	22,698	20,770
30281	Stockbridge	(404)	9,853	3,359
*30086	Stone Mountain	(404)	7,145	6,544
30518	Sugar Hill	(404)	11,399	4,519
30024	Suwanee	(770)	8,725	2,412
30401	Swainsboro	(478)	6,943	7,361
31791	Sylvester	(229)	5,990	6,023
30286	Thomaston	(706)	9,411	9,127
*31792	Thomasville	(229)	18,162	17,554
30824	Thomson	(706)	6,828	6,862
*31794	Tifton	(229)	15,060	14,215
*30577	Toccoa	(706)	9,323	8,720
*30084	Tucker (c)	(404)	26,532	25,781
30291	Union City	(404)	11,621	9,347
*31603	Valdosta	(229)	43,724	40,038
*30474	Vidalia	(912)	10,491	11,118
30339	Vinings (c)	(404)	9,677	7,417
*31088	Warner Robins	(478)	48,804	43,861
*31501	Waycross	(912)	15,333	16,410
30830	Waynesboro	(706)	5,813	5,669
—	Whitemarsh Island (c)		5,824	2,824
31410	Wilmington Island (c)	(912)	14,213	11,230
30680	Winder	(770)	10,201	7,373
*30188	Woodstock	(770)	10,050	4,361

(1) Athens merged with Clarke County in 1991. The 2000 and 1990 populations are for all of Clarke County except for Winterville and Bogart, which are part of the county but are also separate incorporated places. (2) Augusta merged with Richmond County in 1996. The 2000 and 1990 populations are for all of Richmond County except for Blythe and Hephzibah, which are part of the county but are also separate incorporated places.

Hawaii (808)

ZIP	Place	2000	1990
—	Ahuimanu (c)	8,506	8,387
96701	Aiea (c)	9,019	8,906
96706	Ewa Beach (c)	14,650	14,315
—	Haiku-Pauwela (c)	6,578	4,509
—	Halawa (c)	13,891	13,408
96778	Hawaiian Paradise Park (c)	7,051	3,389
96853	Hickam Housing (c)	5,471	6,553
*96720	Hilo (c)	40,759	37,808
96725	Holualoa (c)	6,107	3,834
*96820	Honolulu (c)	371,657	377,059
*96732	Kahului (c)	20,146	16,889
96734	Kailua (Hawaii) (c)	9,870	9,126
96863	Kailua (Honolulu) (c)	36,513	36,818
96740	Kalaoa (c)	6,794	4,490
96744	Kaneohe (c)	34,970	35,448
—	Kaneohe Station (c)	11,827	11,662
96746	Kapaa (c)	9,472	8,149
96753	Kihei (c)	16,749	11,107
*96761	Lahaina (c)	9,118	9,073
96766	Lihue (c)	5,674	5,536
96792	Maili (c)	5,943	6,059
96792	Makaha (c)	7,753	7,990
96706	Makakilo (c)	13,156	9,828
96768	Makawao (c)	6,327	5,405
96789	Mililani Town (c)	28,608	29,359
96792	Nanakuli (c)	10,814	9,575

ZIP	Place	2000	1990
96761	Napili-Honokowai (c)	6,788	4,332
96782	Pearl City (c)	30,976	30,993
96788	Pukalani (c)	7,380	5,879
96786	Schofield Barracks (c)	14,428	19,597
—	Village Park (c)	9,625	7,407
96786	Wahiawa (c)	16,151	17,386
96792	Waianae (c)	10,506	8,758
—	Waihee-Waiehu (c)	7,310	4,004
96753	Wailea-Makena (c)	5,671	3,799
96793	Wailuku (c)	12,296	10,688
—	Waimalu (c)	29,371	29,967
96796	Waimea (c)	7,028	5,972
96797	Waipahu (c)	33,108	31,435
96797	Waipio (c)	11,672	11,812
96786	Waipio Acres (c)	5,298	5,304

Idaho (208)

ZIP	Place	2000	1990
83401	Ammon	6,187	5,002
83221	Blackfoot	10,419	9,646
*83707	Boise	185,787	126,685
83318	Burley	9,316	8,702
*83605	Caldwell	25,967	18,586
83202	Chubbuck	9,700	7,794
*83814	Coeur d'Alene	34,514	24,561
83616	Eagle	11,085	3,327
83617	Emmett	5,490	4,601
83714	Garden City	10,624	6,369
83333	Hailey	6,200	3,575
83835	Hayden	9,159	4,888
*83402	Idaho Falls	50,730	43,973
83338	Jerome	7,780	6,529
*83654	Kuna	5,382	1,955
83501	Lewiston	30,904	28,082
*83642	Meridian	34,919	9,596
83843	Moscow	21,291	18,398
83647	Mountain Home	11,143	7,913
83648	Mountain Home AFB (c)	8,894	5,936
*83653	Nampa	51,867	28,365
83661	Payette	7,054	5,672
*83201	Pocatello	51,466	46,117
*83854	Post Falls	17,247	7,349
83440	Rexburg	17,257	14,298
83350	Rupert	5,645	5,455
83864	Sandpoint	6,835	5,561
*83301	Twin Falls	34,469	27,634
83672	Weiser	5,343	4,571

Illinois

Area code (224) overlays area code (847). See introductory note.

ZIP	Place	Area Code	2000	1990
60101	Addison	(630)	35,914	32,053
60102	Algonquin	(847)	23,276	11,764
60803	Alsip	(708)	19,725	18,227
62002	Alton	(618)	30,496	33,060
62906	Anna	(618)	5,136	4,805
60002	Antioch	(847)	8,788	6,105
*60005	Arlington Heights	(847)	76,031	75,463
*60505	Aurora	(630)	142,990	99,672
*60010	Barrington	(847)	10,168	9,538
60103	Bartlett	(630)	36,706	19,395
61607	Bartonville	(309)	6,310	6,555
60510	Batavia	(630)	23,866	17,076
60085	Beach Park	(847)	10,072	9,492
62618	Beardstown	(217)	5,766	5,270
*62220	Belleville	(618)	41,410	42,806
60104	Bellwood	(708)	20,535	20,241
61008	Belvidere	(815)	20,820	16,059
60106	Bensenville	(630)	20,703	17,767
62812	Benton	(618)	6,880	7,216
60163	Berkeley	(708)	5,245	5,137
60402	Berwyn	(708)	54,016	45,426
62010	Bethalto	(618)	9,454	9,507
60108	Bloomingdale	(630)	21,675	16,614
*61701	Bloomington	(309)	64,808	51,889
60406	Blue Island	(708)	23,463	21,203
*60440	Bolingbrook	(630)	56,321	40,843
60538	Boulder Hill (c)	(630)	8,169	8,894
60914	Bourbonnais	(815)	15,256	13,929
60915	Bradley	(815)	12,784	10,954
60408	Braidwood	(815)	5,203	3,584
60455	Bridgeview	(708)	15,335	14,402
60153	Broadview	(708)	8,264	8,538
60513	Brookfield	(708)	19,085	18,876
60089	Buffalo Grove	(847)	42,909	36,417
60459	Burbank	(708)	27,902	27,600
60521	Burr Ridge	(630)	10,408	8,247
62206	Cahokia	(618)	16,391	17,550
60409	Calumet City	(708)	39,071	37,840
60643	Calumet Park	(708)	8,516	8,418
61520	Canton	(309)	15,288	13,959
*62901	Carbondale	(618)	20,681	27,033
62626	Carlinville	(217)	5,685	5,416
62821	Carmi	(618)	5,422	5,735
*60188	Carol Stream	(630)	40,438	31,759
60110	Carpentersville	(847)	30,586	23,049
60013	Cary	(847)	15,531	10,025
62801	Centralia	(618)	14,136	14,476
62206	Centreville	(618)	5,951	7,489
*61821	Champaign	(217)	67,518	63,502
60410	Channahon	(815)	7,344	4,266
61920	Charleston	(217)	21,039	20,398
62629	Chatham	(217)	8,583	6,074
62233	Chester	(618)	5,185	8,204
*60607	Chicago	(312)/(773)	2,896,016	2,783,726
*60411	Chicago Heights	(708)	32,776	32,966
60415	Chicago Ridge	(708)	14,127	13,643
61523	Chillicothe	(309)	5,996	5,959
60804	Cicero	(708)	85,616	67,436
60514	Clarendon Hills	(630)	7,610	6,994
61727	Clinton	(217)	7,485	7,437
62234	Collinsville	(618)	24,707	22,424
61241	Colona	(309)	5,173	2,237
62236	Columbia	(618)	7,922	5,524
60478	Country Club Hills	(708)	16,169	15,431
60525	Countryside	(708)	5,991	5,961
60435	Crest Hill	(815)	13,329	10,999
60445	Crestwood	(708)	11,251	10,823
60417	Crete	(708)	7,346	6,773
61610	Creve Coeur	(309)	5,448	5,938
*60014	Crystal Lake	(815)	38,000	24,692
*61832	Danville	(217)	33,904	33,828
60561	Darien	(630)	22,860	20,556
*62525	Decatur	(217)	81,860	83,900
60015	Deerfield	(847)	18,420	17,327
60115	DeKalb	(815)	39,018	35,076
*60018	Des Plaines	(847)	58,720	53,414
61021	Dixon	(815)	15,941	15,134
60419	Dolton	(708)	25,614	23,956
*60515	Downers Grove	(630)	48,724	47,464
62832	Du Quoin	(618)	6,448	6,697
62024	East Alton	(618)	6,830	7,063
61244	East Moline	(309)	20,333	20,147
61611	East Peoria	(309)	22,638	21,378
*62201	East St. Louis	(618)	31,542	40,944
62025	Edwardsville	(618)	21,491	14,582
62401	Effingham	(217)	12,384	11,927
*60120	Elgin	(847)	94,487	77,014
*60009	Elk Grove Village	(847)	34,727	33,429
60126	Elmhurst	(630)	42,762	42,029
60707	Elmwood Park	(708)	25,405	23,206
*60201	Evanston	(847)	74,239	73,233
60805	Evergreen Park	(708)	20,821	20,874
62837	Fairfield	(618)	5,421	5,439
62208	Fairview Heights	(618)	15,034	14,768
62839	Flora	(618)	5,086	5,054
60422	Flossmoor	(708)	9,301	8,651
60130	Forest Park	(708)	15,688	14,918
60020	Fox Lake	(847)	9,178	7,539
60423	Frankfort	(815)	10,391	7,180
—	Frankfort Square (c)	(815)	7,766	6,227
60131	Franklin Park	(847)	19,434	18,485
61032	Freeport	(815)	26,443	25,840
60030	Gages Lake (c)	(847)	10,415	8,349
*61401	Galesburg	(309)	33,706	33,530
61254	Geneseo	(309)	6,480	5,990
60134	Geneva	(630)	19,515	12,625
62034	Glen Carbon	(618)	10,425	7,774
60022	Glencoe	(847)	8,762	8,499
60139	Glendale Heights	(630)	31,765	27,915
*60137	Glen Ellyn	(630)	26,999	24,919
60025	Glenview	(847)	41,847	38,436
60425	Glenwood	(708)	9,000	9,289
62035	Godfrey	(618)	16,286	15,675
—	Goodings Grove (c)	(815)	17,084	14,054
62040	Granite City	(618)	31,301	32,766
60030	Grayslake	(847)	18,506	7,388
62246	Greenville	(618)	6,955	5,108
60031	Gurnee	(847)	28,834	13,715
60103	Hanover Park	(630)	38,278	32,918
62946	Harrisburg	(618)	9,860	9,318
60033	Harvard	(815)	7,996	5,975
60426	Harvey	(708)	30,000	29,771
60656	Harwood Heights	(708)	8,297	7,680
60047	Hawthorn Woods	(847)	6,002	4,423
60429	Hazel Crest	(708)	14,816	13,334
62948	Herrin	(618)	11,298	10,857
60457	Hickory Hills	(708)	13,926	13,021
62249	Highland	(618)	8,438	7,546
60035	Highland Park	(847)	31,365	30,575
60162	Hillside	(708)	8,155	7,672
*60521	Hinsdale	(630)	17,349	16,029
*60195	Hoffman Estates	(847)	49,495	46,363
60430	Homewood	(708)	19,543	19,278
60942	Hoopeston	(217)	5,965	5,871
60142	Huntley	(847)	5,730	2453
60067	Inverness	(847)	6,749	6,516
60042	Island Lake	(847)	8,153	4,449
60143	Itasca	(630)	8,302	6,947
*62650	Jacksonville	(217)	18,940	19,327
62052	Jerseyville	(618)	7,984	7,382
60050	Johnsburg	(815)	5,391	—
*60436	Joliet	(815)	106,221	77,217

ZIP	Place	Area Code	2000	1990
60458	Justice	(708)	12,193	11,137
60901	Kankakee	(815)	27,491	27,541
61443	Kewanee	(309)	12,944	12,969
60525	La Grange	(708)	15,608	15,362
60526	La Grange Park	(708)	13,295	12,861
60044	Lake Bluff	(847)	6,056	5,486
60045	Lake Forest	(847)	20,059	17,836
60102	Lake in the Hills	(847)	23,152	5,882
60046	Lake Villa	(847)	5,864	2,857
60047	Lake Zurich	(847)	18,104	14,927
60438	Lansing	(708)	28,332	28,131
61301	La Salle	(815)	9,796	9,717
60439	Lemont	(630)	13,098	7,359
*60048	Libertyville	(847)	20,742	19,174
62656	Lincoln	(217)	15,369	15,418
60069	Lincolnshire	(847)	6,108	4,928
60645	Lincolnwood	(847)	12,359	11,365
60046	Lindenhurst	(847)	12,539	8,044
60532	Lisle	(630)	21,182	19,584
62056	Litchfield	(217)	6,815	6,883
60441	Lockport	(815)	15,191	9,401
60148	Lombard	(630)	42,322	39,408
60047	Long Grove	(847)	6,735	4,747
*61130	Loves Park	(815)	20,044	15,457
60411	Lynwood	(708)	7,377	6,535
60534	Lyons (Cook)	(708)	10,255	9,828
*60050	McHenry	(815)	21,501	16,343
61115	Machesney Park	(815)	20,759	19,042
61455	Macomb	(309)	18,558	19,952
60950	Manteno	(815)	6,414	3,709
60152	Marengo	(815)	6,355	4,768
62959	Marion	(618)	16,035	14,597
60426	Markham (Cook)	(708)	12,620	13,136
62258	Mascoutah	(618)	5,659	5,511
60443	Matteson	(708)	12,928	11,378
61938	Mattoon	(217)	18,291	18,441
60153	Maywood	(708)	26,987	27,139
*60160	Melrose Park	(708)	23,171	20,859
61342	Mendota	(815)	7,272	7,017
62960	Metropolis	(618)	6,482	6,734
60445	Midlothian	(708)	14,315	14,372
61264	Milan	(309)	5,348	5,753
60448	Mokena	(708)	14,583	6,128
*61265	Moline	(309)	43,768	43,080
61462	Monmouth	(309)	9,841	9,489
60538	Montgomery	(630)	5,471	4,487
60450	Morris	(815)	11,928	10,274
61550	Morton	(309)	15,198	13,799
60053	Morton Grove	(847)	22,451	22,373
62863	Mount Carmel	(618)	7,982	8,287
60056	Mount Prospect	(847)	56,265	53,168
62864	Mount Vernon	(618)	16,269	17,082
60060	Mundelein	(847)	30,935	21,224
62966	Murphysboro	(618)	13,295	9,176
*60540	Naperville	(630)	128,358	85,806
60451	New Lenox	(815)	17,771	9,698
60714	Niles	(847)	30,068	28,375
61761	Normal	(309)	45,386	40,023
60634	Norridge	(708)	14,582	14,459
60542	North Aurora	(630)	10,585	6,010
*60062	Northbrook	(708)	33,435	32,565
60064	North Chicago	(847)	35,918	34,978
60093	Northfield	(847)	5,389	4,924
60164	Northlake	(708)	11,878	12,505
60546	North Riverside	(708)	6,688	6,180
60521	Oak Brook	(630)	8,702	9,087
60452	Oak Forest	(708)	28,051	26,202
*60303	Oak Lawn	(708)	55,245	56,182
*60303	Oak Park	(708)	52,524	53,648
62269	O'Fallon	(618)	21,910	16,064
62450	Olney	(618)	8,631	8,873
60477	Orland Hills	(708)	6,779	5,510
*60462	Orland Park	(708)	51,077	35,720
60543	Oswego	(630)	13,326	3,949
61350	Ottawa	(815)	18,307	17,574
*60067	Palatine	(847)	65,479	41,554
60463	Palos Heights	(708)	11,260	11,478
60465	Palos Hills	(708)	17,665	17,803
62557	Pana	(217)	5,614	5,796
61944	Paris	(217)	9,077	9,105
60085	Park City	(847)	6,637	4,677
60466	Park Forest	(708)	23,462	24,656
60068	Park Ridge	(847)	37,775	37,075
61554	Pekin	(309)	33,857	32,254
*61601	Peoria	(309)	112,936	113,508
61603	Peoria Heights	(309)	6,635	6,930
61354	Peru	(815)	9,835	9,302
62274	Pinckneyville	(618)	5,464	3,372
60544	Plainfield	(815)	13,038	4,557
60545	Plano	(630)	5,633	5,104
61764	Pontiac	(815)	11,864	11,428
62040	Pontoon Beach	(618)	5,620	4,013
61356	Princeton	(815)	7,501	7,197
60070	Prospect Heights	(847)	17,081	15,236
*62301	Quincy	(217)	40,366	39,682
61866	Rantoul	(217)	12,857	17,212
60471	Richton Park	(708)	12,533	10,523
60827	Riverdale	(708)	15,055	13,671
60305	River Forest	(708)	11,635	11,669
60171	River Grove	(708)	10,668	9,961
60546	Riverside	(708)	8,895	8,774
60472	Robbins	(708)	6,635	7,498
62454	Robinson	(618)	6,822	6,740
61068	Rochelle	(815)	9,424	8,769
61071	Rock Falls	(815)	9,580	9,669
*61125	Rockford	(815)	150,115	142,815
*61201	Rock Island	(309)	39,684	40,630
61072	Rockton	(815)	5,296	2,928
60008	Rolling Meadows	(847)	24,604	22,598
60446	Romeoville	(815)	21,153	14,101
61072	Roscoe	(815)	6,244	2,079
60172	Roselle	(630)	23,115	20,803
60073	Round Lake	(847)	5,842	3,550
60073	Round Lake Beach	(847)	25,859	16,406
60073	Round Lake Park	(847)	6,038	4,045
*60174	Saint Charles	(630)	27,896	22,636
62881	Salem	(618)	7,909	7,470
60548	Sandwich	(815)	6,509	5,607
60411	Sauk Village	(708)	10,411	10,734
*60194	Schaumburg	(847)	75,386	68,586
60176	Schiller Park	(847)	11,850	11,189
*62269	Shiloh	(618)	7,643	2,655
60436	Shorewood	(815)	7,686	6,264
61282	Silvis	(309)	7,269	6,926
*60077	Skokie	(847)	63,348	59,432
61080	South Beloit	(815)	5,397	4,072
60177	South Elgin	(847)	16,100	7,474
60473	South Holland	(708)	22,147	22,105
*62703	Springfield	(217)	111,454	105,412
61362	Spring Valley	(815)	5,398	5,246
62088	Staunton	(618)	5,030	4,806
60475	Steger	(708)	9,682	9,251
61081	Sterling	(815)	15,451	15,142
60402	Stickney	(708)	6,148	5,678
60165	Stone Park	(708)	5,127	4,383
60107	Streamwood	(630)	36,407	31,197
61364	Streator	(815)	14,190	14,121
60501	Summit	(708)	10,637	9,971
62221	Swansea	(618)	10,579	8,201
60178	Sycamore	(815)	12,020	9,896
62568	Taylorville	(217)	11,427	11,133
60477	Tinley Park	(708)	48,401	37,115
62294	Troy	(618)	8,524	6,194
60466	University Park	(708)	6,662	6,204
*61801	Urbana	(217)	36,395	36,383
62471	Vandalia	(618)	6,975	6,114
60061	Vernon Hills	(847)	20,120	15,319
60181	Villa Park	(630)	22,075	22,279
60555	Warrenville	(630)	13,363	11,389
61571	Washington	(309)	10,841	10,136
62204	Washington Park	(618)	5,345	7,431
62298	Waterloo	(618)	7,614	5,030
60970	Watseka	(815)	5,670	5,424
60084	Wauconda	(847)	9,448	6,294
*60085	Waukegan	(847)	87,901	69,481
60154	Westchester	(708)	16,824	17,301
*60185	West Chicago	(630)	23,469	14,808
60118	West Dundee	(847)	5,428	3,728
60558	Western Springs	(708)	12,493	11,956
62896	West Frankfort	(618)	8,196	8,526
60559	Westmont	(630)	24,554	21,402
*60187	Wheaton	(630)	55,416	51,441
60090	Wheeling	(847)	34,496	29,911
60514	Willowbrook	(630)	8,967	8,651
60480	Willow Springs	(708)	5,027	4,509
60091	Wilmette	(847)	27,651	26,694
60481	Wilmington	(815)	5,134	4,743
60190	Winfield	(630)	8,718	7,096
60093	Winnetka	(847)	12,419	12,210
60096	Winthrop Harbor	(847)	6,670	6,240
60097	Wonder Lake (c)	(815)	7,463	6,664
60191	Wood Dale	(630)	13,535	12,394
60517	Woodridge	(630)	30,934	26,359
62095	Wood River	(618)	11,296	11,490
60098	Woodstock	(815)	20,151	14,368
60482	Worth	(708)	11,047	11,208
60560	Yorkville	(630)	6,189	3,974
60099	Zion	(847)	22,866	19,783

Indiana

ZIP	Place	Area Code	2000	1990
46001	Alexandria	(765)	6,260	5,709
*46011	Anderson	(765)	59,734	59,518
46703	Angola	(260)	7,344	5,851
46706	Auburn	(260)	12,074	9,386
46123	Avon	(317)	6,248	—
47006	Batesville	(812)	6,033	4,720
47421	Bedford	(812)	13,768	13,817
46107	Beech Grove	(317)	14,880	13,383
*47408	Bloomington	(812)	69,291	62,735
46714	Bluffton	(260)	9,536	9,104
47601	Boonville	(812)	6,834	6,686

ZIP	Place	Area Code	2000	1990
47834	Brazil	(812)	8,188	7,640
47025	Bright (c)	(812)	5,405	3,945
46112	Brownsburg	(317)	14,520	7,751
*46032	Carmel	(317)	37,733	25,380
46303	Cedar Lake	(219)	9,279	8,885
47111	Charlestown	(812)	5,993	5,889
46304	Chesterton	(219)	10,488	9,118
47129	Clarksville (Clark)	(812)	21,400	19,838
47842	Clinton	(765)	5,126	5,040
46725	Columbia City	(260)	7,077	5,883
*47201	Columbus	(812)	39,059	33,948
47331	Connersville	(765)	15,411	15,550
47933	Crawfordsville	(765)	15,243	13,584
46307	Crown Point	(219)	19,806	17,728
46229	Cumberland	(317)	5,500	4,557
46122	Danville	(317)	6,418	4,345
46733	Decatur	(260)	9,528	8,642
46514	Dunlap (c)	(574)	5,887	5,705
46311	Dyer	(219)	13,895	10,923
46312	East Chicago	(219)	32,414	33,892
*46515	Elkhart	(574)	51,874	44,661
47429	Ellettsville	(812)	5,078	3,275
46036	Elwood	(765)	9,737	9,494
*47708	Evansville	(812)	121,582	126,272
46038	Fishers	(317)	37,835	7,189
*46802	Fort Wayne	(260)	205,727	195,680
46041	Frankfort	(765)	16,662	14,754
46131	Franklin	(317)	19,463	12,932
46738	Garrett	(260)	5,803	5,349
*46401	Gary	(219)	102,746	116,646
46933	Gas City	(765)	5,940	6,311
*46526	Goshen	(574)	29,383	23,794
46530	Granger (c)	(574)	28,284	20,241
46135	Greencastle	(765)	9,880	8,984
46140	Greenfield	(317)	14,600	11,657
47240	Greensburg	(812)	10,260	9,286
*46142	Greenwood	(317)	36,037	26,507
46319	Griffith	(219)	17,334	17,914
*46320	Hammond	(219)	83,048	84,236
47348	Hartford City	(765)	6,928	6,960
46322	Highland	(219)	23,546	23,696
46342	Hobart	(219)	25,363	24,440
47542	Huntingburg	(812)	5,598	5,236
46750	Huntington	(260)	17,450	16,389
*46206	Indianapolis	(317)	791,926	731,278
*47546	Jasper	(812)	12,100	10,030
*47130	Jeffersonville	(812)	27,362	24,016
46755	Kendallville	(260)	9,616	7,984
*46902	Kokomo	(765)	46,113	44,996
*47901	Lafayette	(765)	56,397	45,933
—	Lakes of the Four Seasons (c)	(219)	7,291	6,556
46405	Lake Station	(219)	13,948	13,899
*46350	La Porte	(574)	21,621	21,507
46226	Lawrence	(317)	38,915	26,849
46052	Lebanon	(765)	14,222	12,059
47441	Linton	(812)	5,774	5,814
46947	Logansport	(574)	19,684	16,865
46356	Lowell	(219)	7,505	6,430
47250	Madison	(812)	12,004	12,006
*46952	Marion	(765)	31,320	32,607
46151	Martinsville	(765)	11,698	11,677
*46401	Merrillville	(219)	30,560	27,257
*46360	Michigan City	(219)	32,900	33,822
*46544	Mishawaka	(574)	46,557	42,635
47960	Monticello	(574)	5,723	5,237
46158	Mooresville	(317)	9,273	5,779
47620	Mount Vernon	(812)	7,478	7,217
*47302	Muncie	(765)	67,430	71,170
46321	Munster	(219)	21,511	19,949
46550	Nappanee	(574)	6,710	5,474
*47150	New Albany	(812)	37,603	36,322
47362	New Castle	(765)	17,780	17,753
46774	New Haven	(260)	12,406	11,234
*46060	Noblesville	(317)	28,590	17,655
46962	North Manchester	(260)	6,260	6,383
47265	North Vernon	(812)	6,515	5,129
47130	Oak Park (c)	(812)	5,379	5,630
46970	Peru	(765)	12,994	12,843
46168	Plainfield	(317)	18,396	14,953
46563	Plymouth	(574)	9,840	8,291
46368	Portage	(219)	33,496	29,062
47371	Portland	(260)	6,437	6,483
47670	Princeton	(812)	8,175	8,127
47978	Rensselaer	(219)	5,294	5,045
*47374	Richmond	(765)	39,124	38,705
46975	Rochester	(574)	6,414	5,969
46173	Rushville	(765)	5,995	5,533
46373	Saint John	(219)	8,382	4,921
47167	Salem	(812)	6,172	5,619
46375	Schererville	(219)	24,851	20,155
47170	Scottsburg	(812)	6,040	5,334
47172	Sellersburg	(812)	6,071	5,936
47274	Seymour	(812)	18,101	15,605
46176	Shelbyville	(765)	17,951	15,347
*46624	South Bend	(574)	107,789	105,511
46383	South Haven (c)	(219)	5,619	6,112
46224	Speedway	(317)	12,881	13,092
47586	Tell City	(812)	7,845	8,088
*47808	Terre Haute	(812)	59,614	57,475
46072	Tipton	(765)	5,251	4,751
*46383	Valparaiso	(219)	27,428	24,414
47591	Vincennes	(812)	18,701	19,867
46992	Wabash	(260)	11,743	12,127
*46580	Warsaw	(574)	12,415	10,968
47501	Washington	(812)	11,380	10,864
46074	Westfield	(317)	9,293	3,304
*46580	West Lafayette	(765)	28,778	26,144
46394	Whiting	(219)	5,137	5,155
47394	Winchester	(765)	5,037	5,095
46077	Zionsville	(317)	8,775	6,207

Iowa

ZIP	Place	Area Code	2000	1990
50511	Algona	(515)	5,741	6,015
50009	Altoona	(515)	10,345	7,242
*50010	Ames	(515)	50,731	47,198
52205	Anamosa	(319)	5,494	5,100
50021	Ankeny	(515)	27,117	18,482
50022	Atlantic	(712)	7,257	7,432
52722	Bettendorf	(563)	31,275	28,139
*50036	Boone	(515)	12,803	12,392
52601	Burlington	(319)	26,839	27,208
51401	Carroll	(712)	10,106	9,579
50613	Cedar Falls	(319)	36,145	34,298
*52401	Cedar Rapids	(319)	120,758	108,772
52544	Centerville	(641)	5,924	5,936
50616	Charles City	(641)	7,812	7,878
51012	Cherokee	(712)	5,369	6,026
51632	Clarinda	(712)	5,690	5,104
50428	Clear Lake	(641)	8,161	8,183
*52732	Clinton	(563)	27,772	29,201
50325	Clive	(515)	12,855	7,446
52241	Coralville	(319)	15,123	10,347
*51501	Council Bluffs	(712)	58,268	54,315
50801	Creston	(641)	7,597	7,911
*52802	Davenport	(563)	98,359	95,333
52101	Decorah	(563)	8,172	8,063
51442	Denison	(712)	7,339	6,604
*50318	Des Moines	(515)	198,682	193,189
*50274	De Witt	(563)	5,049	4,514
*52001	Dubuque	(563)	57,686	57,538
51334	Estherville	(712)	6,656	6,720
52556	Fairfield	(641)	9,509	9,955
50501	Fort Dodge	(515)	25,136	26,057
52627	Fort Madison	(319)	10,715	11,614
51534	Glenwood	(712)	5,358	4,960
50111	Grimes	(515)	5,098	2,653
50112	Grinnell	(641)	9,105	8,902
*51537	Harlan	(712)	5,282	5,148
52233	Hiawatha	(319)	6,480	5,354
50644	Independence	(319)	6,014	5,972
50125	Indianola	(515)	12,998	11,340
*52240	Iowa City	(319)	62,220	59,735
50126	Iowa Falls	(641)	5,193	5,435
50131	Johnston	(515)	8,649	4,702
52632	Keokuk	(319)	11,427	12,451
50138	Knoxville	(641)	7,731	8,232
51031	Le Mars	(712)	9,237	8,454
52057	Manchester	(563)	5,257	5,137
52060	Maquoketa	(563)	6,112	6,130
52302	Marion	(319)	26,294	20,422
50158	Marshalltown	(641)	26,009	25,178
*50401	Mason City	(641)	29,172	29,040
52641	Mount Pleasant	(319)	8,751	7,959
52761	Muscatine	(563)	22,697	22,881
50201	Nevada	(515)	6,658	6,009
50208	Newton	(641)	15,579	14,799
52317	North Liberty	(319)	5,367	2,926
50211	Norwalk	(515)	6,884	5,726
50662	Oelwein	(319)	6,692	6,691
51041	Orange City	(712)	5,582	4,940
52577	Oskaloosa	(641)	10,938	10,600
52501	Ottumwa	(641)	24,998	24,488
50219	Pella	(641)	9,832	9,270
50220	Perry	(515)	7,633	6,652
*50317	Pleasant Hill	(515)	5,070	3,671
*51566	Red Oak	(712)	6,197	6,264
51601	Shenandoah	(712)	5,546	5,572
51250	Sioux Center	(712)	6,002	5,074
*51101	Sioux City	(712)	85,013	80,505
51301	Spencer	(712)	11,317	11,066
50588	Storm Lake	(712)	10,076	8,769
*50318	Urbandale	(515)	29,072	23,775
52349	Vinton	(319)	5,102	5,103
52353	Washington	(319)	7,047	7,074
*50701	Waterloo	(319)	68,747	66,467
50263	Waukee	(515)	5,126	2,512
50677	Waverly	(319)	8,968	8,539
50595	Webster City	(515)	8,176	7,894
*50265	West Des Moines	(515)	46,403	31,702

Kansas

ZIP	Place	Area Code	2000	1990
67410	Abilene	(785)	6,543	6,242
67002	Andover	(316)	6,698	4,204
67005	Arkansas City	(620)	11,963	12,762
66002	Atchison	(913)	10,232	10,656
67010	Augusta	(316)	8,423	7,848
66952	Bel Aire	(316)	5,836	3,695
66012	Bonner Springs	(913)	6,768	6,413
66720	Chanute	(620)	9,411	9,488
67337	Coffeyville	(620)	11,021	12,917
67701	Colby	(785)	5,450	5,510
66901	Concordia	(785)	5,714	6,152
67037	Derby	(316)	17,807	14,691
67801	Dodge City	(620)	25,176	21,129
67042	El Dorado	(316)	12,057	11,495
66801	Emporia	(620)	26,760	25,512
66442	Fort Riley North (c)	(785)	8,114	12,848
66701	Fort Scott	(620)	8,297	8,362
67846	Garden City	(620)	28,451	24,097
66030	Gardner	(913)	9,396	4,277
67530	Great Bend	(620)	15,345	15,427
67601	Hays	(785)	20,013	18,632
67060	Haysville	(316)	8,502	8,364
*67501	Hutchinson	(620)	40,787	39,308
67301	Independence	(620)	9,846	10,030
66749	Iola	(620)	6,302	6,351
66441	Junction City	(785)	18,886	20,642
*66102	Kansas City	(913)	146,866	151,521
66043	Lansing	(913)	9,199	7,120
*66044	Lawrence	(785)	80,098	65,608
66048	Leavenworth	(913)	35,420	38,495
66209	Leawood	(913)	27,656	19,693
66214	Lenexa	(913)	40,238	34,110
*67901	Liberal	(620)	19,666	16,573
67460	McPherson	(620)	13,770	12,422
*66502	Manhattan	(785)	44,831	43,081
66202	Merriam	(913)	11,008	11,819
66203	Mission	(913)	9,727	9,504
67110	Mulvane	(316)	5,155	4,683
67114	Newton	(316)	17,190	16,700
*66061	Olathe	(913)	92,962	63,402
66067	Ottawa	(785)	11,921	10,667
66204	Overland Park	(913)	149,080	111,790
66071	Paola	(913)	5,011	4,698
67219	Park City	(316)	5,814	5,081
67357	Parsons	(620)	11,514	11,919
66762	Pittsburg	(620)	19,243	17,789
66208	Prairie Village	(913)	22,072	23,186
67124	Pratt	(620)	6,570	6,687
66205	Roeland Park	(913)	6,817	7,706
*67401	Salina	(785)	45,679	42,299
66203	Shawnee	(913)	47,996	37,962
*66601	Topeka	(785)	122,377	119,883
67880	Ulysses	(620)	5,960	5,474
67152	Wellington	(620)	8,647	8,517
*67202	Wichita	(316)	344,284	304,017
67156	Winfield	(620)	12,206	11,931

Kentucky

ZIP	Place	Area Code	2000	1990
41001	Alexandria	(859)	8,286	5,592
*41101	Ashland	(606)	21,981	23,622
40004	Bardstown	(502)	10,374	6,712
41073	Bellevue	(859)	6,480	6,997
40403	Berea	(859)	9,851	9,129
*42101	Bowling Green	(270)	49,296	41,688
40261	Buechel (c)	(502)	7,272	7,081
41005	Burlington (c)	(859)	10,779	6,070
*42718	Campbellsville	(270)	10,498	9,592
42330	Central City	(270)	5,893	4,979
*40701	Corbin	(606)	7,742	7,644
*41011	Covington	(859)	43,370	43,646
41031	Cynthiana	(859)	6,258	6,497
*40422	Danville	(859)	15,477	14,454
41074	Dayton	(859)	5,966	6,576
40243	Douglass Hills	(502)	5,718	5,431
41017	Edgewood	(859)	9,400	8,143
*42701	Elizabethtown	(270)	22,542	18,167
41018	Elsmere	(859)	8,139	6,847
41018	Erlanger	(859)	16,676	15,979
40118	Fairdale (c)	(502)	7,658	6,563
40291	Fern Creek (c)	(502)	17,870	16,406
41139	Flatwoods	(606)	7,605	7,799
*41042	Florence	(859)	23,551	18,586
42223	Fort Campbell North (c)	(270)	14,338	18,861
40121	Fort Knox (c)	(270)	12,377	21,495
41017	Fort Mitchell	(859)	8,089	7,438
41075	Fort Thomas	(859)	16,495	16,032
41011	Fort Wright	(859)	5,681	6,404
*40601	Frankfort	(502)	27,741	26,535
*42134	Franklin	(270)	7,996	7,607
40324	Georgetown	(502)	18,080	11,414
*42141	Glasgow	(270)	13,019	12,777
40330	Harrodsburg	(859)	8,014	7,335
*42420	Henderson	(270)	27,373	25,945
41076	Highland Heights	(859)	6,554	4,223
40228	Highview (c)	(502)	15,161	14,814
40229	Hillview	(502)	7,037	6,119
*42240	Hopkinsville	(270)	30,089	29,809
41051	Independence	(859)	14,982	10,444
40269	Jeffersontown	(502)	26,633	23,223
40031	La Grange	(502)	5,676	3,901
40342	Lawrenceburg	(502)	9,014	5,911
40033	Lebanon	(270)	5,718	5,695
*42754	Leitchfield	(270)	6,139	4,965
*40507	Lexington	(859)	260,512	225,366
40741	London	(606)	5,692	5,757
40232	Louisville	(502)	256,231	269,555
40252	Lyndon	(502)	9,369	8,037
42431	Madisonville	(270)	19,307	18,693
42066	Mayfield	(270)	10,349	9,935
41056	Maysville	(606)	8,993	8,113
40965	Middlesborough	(606)	10,384	11,328
40253	Middletown	(502)	5,744	5,016
42633	Monticello	(606)	5,981	5,357
40351	Morehead	(606)	5,914	8,357
40353	Mount Sterling	(859)	5,876	5,362
40047	Mount Washington	(502)	8,485	5,256
42071	Murray	(270)	14,950	14,442
40218	Newburg (c)	(502)	20,636	21,647
*41071	Newport	(859)	17,048	18,871
*40356	Nicholasville	(859)	19,680	13,603
—	Oakbrook (c)		7,726	4,113
42262	Oak Grove	(502)	7,064	2,863
40259	Okolona(c)	(502)	17,807	18,902
*42301	Owensboro	(270)	54,067	53,577
*42003	Paducah	(270)	26,307	27,256
*40361	Paris	(859)	9,183	8,730
*41501	Pikeville	(606)	6,295	6,324
40268	Pleasure Ridge Park (c)	(502)	25,776	25,131
42445	Princeton	(270)	6,536	6,940
*40160	Radcliff	(502)	21,961	19,778
*40475	Richmond	(859)	27,152	21,183
42276	Russellville	(270)	7,149	7,454
40216	Saint Dennis (c)	(502)	9,177	10,326
*40206	Saint Matthews	(502)	15,852	15,691
*40066	Shelbyville	(502)	10,085	6,155
40165	Shepherdsville	(502)	8,334	4,805
40256	Shively	(502)	15,157	15,535
*42501	Somerset	(606)	11,352	10,735
41015	Taylor Mill	(859)	6,913	5,530
40272	Valley Station (c)	(502)	22,946	22,840
40383	Versailles	(859)	7,511	7,269
41016	Villa Hills	(859)	7,948	7,370
40769	Williamsburg	(606)	5,143	5,493
40390	Wilmore	(859)	5,905	4,215
*40391	Winchester	(859)	16,724	15,799

Louisiana

ZIP	Place	Area Code	2000	1990
*70510	Abbeville	(337)	11,887	11,769
*71301	Alexandria	(318)	46,342	49,049
70032	Arabi (c)	(504)	8,093	8,787
70094	Avondale (c)	(504)	5,441	5,813
*70714	Baker	(225)	13,793	13,087
*71220	Bastrop	(318)	12,988	13,916
*70821	Baton Rouge	(225)	227,818	219,531
70360	Bayou Cane (c)	(985)	17,046	15,876
70037	Belle Chasse (c)	(504)	9,848	8,512
*70427	Bogalusa	(985)	13,365	14,280
*71111	Bossier City	(318)	56,461	52,721
70517	Breaux Bridge	(337)	7,281	6,694
70094	Bridge City (c)	(504)	8,323	8,327
70518	Broussard	(337)	5,874	3,213
70811	Brownfields (c)	(225)	5,222	5,229
71291	Brownsville-Bawcomville (c)	(318)	7,616	7,397
70520	Carencro	(337)	6,120	5,518
*70043	Chalmette (c)	(504)	32,069	31,860
71291	Claiborne (c)	(318)	9,830	8,300
*70433	Covington	(985)	8,483	7,691
*70526	Crowley	(337)	14,225	13,983
70345	Cut Off (c)	(985)	5,635	5,325
*70726	Denham Springs	(225)	8,757	8,381
70634	De Ridder	(337)	9,808	10,475
70047	Destrehan (c)	(985)	11,260	8,031
70346	Donaldsonville	(225)	7,605	7,949
—	Eden Isle (c)		6,261	3,768
70072	Estelle (c)	(504)	15,880	14,091
70535	Eunice	(337)	11,499	11,162
71459	Fort Polk South (c)	(337)	11,000	10,911
70538	Franklin	(337)	8,354	9,004
70354	Galliano (c)	(985)	7,356	4,294
70820	Gardere (c)	(225)	8,992	7,209
*70737	Gonzales	(225)	8,156	7,208
*70053	Gretna	(504)	17,423	17,208
*70401	Hammond	(985)	17,639	15,871
70123	Harahan	(504)	9,885	9,927
*70058	Harvey (c)	(504)	22,226	21,222
*70360	Houma	(985)	32,393	30,495
70544	Jeanerette	(337)	5,997	6,205
70502	Jefferson (c)	(504)	11,843	14,521
70546	Jennings	(337)	10,986	11,305
70548	Kaplan	(337)	5,177	4,535

ZIP	Place	Area Code	2000	1990
*70062	Kenner	(504)	70,517	72,033
70445	Lacombe (c)	(985)	7,518	6,523
*70501	Lafayette	(337)	110,257	101,865
*70601	Lake Charles	(337)	71,757	70,580
71254	Lake Providence	(318)	5,104	5,380
*70068	Laplace (c)	(985)	27,684	24,194
70373	Larose (c)	(985)	7,306	5,772
*71446	Leesville	(337)	6,753	7,638
70070	Luling (c)	(985)	11,512	2,803
*70471	Mandeville	(985)	10,489	7,474
71052	Mansfield	(318)	5,582	5,389
71351	Marksville	(318)	5,537	5,526
*70072	Marrero (c)	(504)	36,165	36,671
70075	Meraux (c)	(504)	10,192	8,849
70812	Merrydale (c)	(225)	10,427	10,395
*70009	Metairie (c)	(504)	146,136	149,428
*71055	Minden	(318)	13,027	13,661
*71207	Monroe	(318)	53,107	54,909
*70380	Morgan City	(985)	12,703	14,531
70612	Moss Bluff (c)	(337)	10,535	8,039
*71457	Natchitoches	(318)	17,865	16,609
*70560	New Iberia	(337)	32,623	31,828
*70140	New Orleans	(504)	484,674	496,938
71463	Oakdale	(318)	8,137	6,837
70808	Oak Hills Place (c)	(225)	7,996	5,479
—	Old Jefferson (c)		5,631	4,531
*70570	Opelousas	(337)	22,860	19,091
70392	Patterson	(985)	5,130	5,166
*71360	Pineville	(318)	13,829	15,308
*70764	Plaquemine	(225)	7,064	7,101
70454	Ponchatoula	(985)	5,180	5,499
70767	Port Allen	(225)	5,278	6,277
70601	Prien (c)	(337)	7,215	6,448
70394	Raceland (c)	(985)	10,224	5,564
70578	Rayne	(337)	8,552	8,502
71037	Red Chute (c)	(318)	5,984	5,431
70084	Reserve (c)	(985)	9,111	8,847
70123	River Ridge (c)	(504)	14,588	14,800
*71270	Ruston	(318)	20,546	20,071
70582	Saint Martinville	(337)	6,989	7,226
70087	Saint Rose (c)	(504)	6,540	6,259
70395	Schriever (c)	(985)	5,880	4,958
70583	Scott	(337)	7,870	4,912
70817	Shenandoah (c)		17,070	13,429
*71102	Shreveport	(318)	200,145	198,518
*70458	Slidell	(985)	25,695	24,124
71075	Springhill	(318)	5,439	5,668
*70663	Sulphur	(337)	20,512	20,125
*71282	Tallulah	(318)	9,189	8,526
70056	Terrytown (c)	(504)	25,430	23,787
*70301	Thibodaux	(985)	14,431	14,125
70053	Timberlane (c)	(504)	11,405	12,614
70809	Village Saint George (c)	(225)	6,993	6,242
70586	Ville Platte	(337)	8,145	9,037
70092	Violet (c)	(504)	8,555	8,574
70094	Waggaman (c)	(504)	9,435	9,405
*71291	West Monroe	(318)	13,250	14,096
*70094	Westwego	(504)	10,763	11,218
71483	Winnfield	(318)	5,749	6,138
71295	Winnsboro	(318)	5,344	5,755
—	Woodmere (c)		13,058	—
70791	Zachary	(225)	11,275	9,036

Maine (207)

See introductory note.

ZIP	Place	2000	1990
*04210	Auburn	23,203	24,309
*04330	Augusta	18,560	21,325
*04401	Bangor	31,473	33,181
04530	Bath	9,266	9,799
04915	Belfast	6,381	6,355
03901	*Berwick*	6,353	5,995
*04005	Biddeford	20,942	20,710
04412	Brewer	8,987	9,021
04011	Brunswick (c)	14,816	14,683
04011	*Brunswick*	21,172	20,906
04093	*Buxton*	7,452	6,494
04843	*Camden*	5,254	5,060
04107	*Cape Elizabeth*	9,068	8,854
04736	Caribou	8,312	9,415
04021	*Cumberland*	7,159	5,836
03903	*Eliot*	5,954	5,329
04605	Ellsworth	6,456	5,975
04937	*Fairfield*	6,573	6,718
04105	*Falmouth*	10,310	7,610
04938	*Farmington*	7,410	7,436
04032	*Freeport*	7,800	6,905
04345	Gardiner	6,198	6,746
04038	*Gorham*	14,141	11,856
04039	*Gray*	6,820	5,904
04444	*Hampden*	6,327	5,974
04079	*Harpswell*	5,239	5,012
04730	Houlton (c)	5,270	5,627
04730	*Houlton*	6,476	6,613
04043	*Kennebunk*	10,476	8,004
03904	*Kittery*	9,543	9,372
04027	*Lebanon*	5,083	—
*04240	Lewiston	35,690	39,757
04457	*Lincoln*	5,221	5,587
04250	*Lisbon*	9,077	9,457
04462	*Millinocket*	5,203	6,956
04462	Millinocket (c)	5,190	6,922
04963	*Oakland*	5,959	5,595
04064	*Old Orchard Beach*	8,856	7,789
04064	Old Orchard Beach (c)	8,856	7,789
04468	Old Town	8,130	8,317
04473	*Orono*	9,112	10,573
04473	Orono (c)	8,253	9,789
*04101	Portland	64,249	64,157
04769	Presque Isle	9,511	10,550
04841	Rockland	7,609	7,972
04276	*Rumford*	6,472	7,078
04072	Saco	16,822	15,181
04073	Sanford (c)	10,133	10,296
04073	*Sanford*	20,806	20,463
*04074	*Scarborough*	16,970	12,518
04976	Skowhegan (c)	6,696	6,990
04976	*Skowhegan*	8,824	8,725
03908	*South Berwick*	6,671	5,877
*04101	South Portland	23,324	23,163
04084	*Standish*	9,285	7,678
04086	Topsham (c)	6,271	6,147
04086	*Topsham*	9,100	8,746
04087	*Waterboro*	6,214	4,510
*04901	Waterville	15,605	17,173
04090	*Wells*	9,400	7,778
*04092	Westbrook	16,142	16,121
04062	*Windham*	14,904	13,020
04901	Winslow (c)	7,743	5,436
04901	*Winslow*	7,743	7,997
04364	*Winthrop*	6,232	5,968
04096	*Yarmouth*	8,360	7,862
03909	*York*	12,854	9,818

Maryland

Area code (240) overlays area code (301). Area code (443) overlays (410). See introductory note.

ZIP	Place	Area Code	2000	1990
21001	Aberdeen	(410)	13,842	13,087
20607	Accokeek (c)	(301)	7,349	4,477
20783	Adelphi (c)	(301)	14,998	13,524
20762	Andrews AFB (c)	(410)	7,925	10,228
*21401	Annapolis	(410)	35,838	33,195
21227	Arbutus (c)	(410)	20,116	19,750
21012	Arnold (c)	(410)	23,422	20,261
20916	Aspen Hill (c)	(301)	50,228	45,494
21220	Ballenger Creek (c)	(410)	13,518	5,546
*21203	Baltimore	(410)	651,154	736,014
*21014	Bel Air	(410)	10,080	8,942
21050	Bel Air North (c)	(410)	25,798	14,880
21014	Bel Air South (c)	(410)	39,711	26,421
*20705	Beltsville (c)	(301)	15,690	14,476
—	Bennsville (c)		7,325	—
*20814	Bethesda (c)	(301)	55,277	62,936
20710	Bladensburg	(301)	7,661	8,064
*20715	Bowie	(301)	50,269	37,642
21220	Bowleys Quarters (c)	(410)	6,314	5,595
21225	Brooklyn Park (c)	(410)	10,938	10,987
20866	Burtonsville (c)	(301)	7,305	5,853
20619	California (c)	(410)	9,307	7,626
20705	Calverton (c)	(301)	12,610	12,046
21613	Cambridge	(410)	10,911	11,514
20748	Camp Springs (c)	(301)	17,968	16,392
21401	Cape St. Clair (c)	(410)	8,022	7,878
21234	Carney (c)	(410)	28,264	25,578
21228	Catonsville (c)	(410)	39,820	35,233
20657	Chesapeake Ranch Estates-Drum Point (c)	(301)	11,503	5,423
20784	Cheverly	(301)	6,433	6,023
*20814	Chevy Chase (c)	(301)	9,381	8,559
20783	Chillum (c)	(301)	34,252	31,309
20735	Clinton (c)	(301)	26,064	19,987
20904	Cloverly (c)	(301)	7,835	7,904
21030	Cockeysville (c)	(410)	19,388	18,668
20914	Colesville (c)	(301)	19,810	18,819
*20740	College Park	(301)	24,657	23,714
*21045	Columbia (c)	(410)/(301)	88,254	75,883
20743	Coral Hills (c)	(410)	10,720	11,032
—	Cresaptown-Bel Air (c)		5,884	4,586
21114	Crofton (c)	(410)	20,091	12,781
*21502	Cumberland	(301)	21,518	23,712
20872	Damascus (c)	(301)	11,430	9,817
*20874	Darnestown (c)	(301)	6,378	—
*20747	District Heights	(301)	5,958	6,711
21222	Dundalk (c)	(410)	62,306	65,800
21601	Easton	(410)	11,708	9,372
20737	East Riverdale (c)	(301)	14,961	14,187
21219	Edgemere (c)	(410)	9,248	9,226
21040	Edgewood (c)	(410)	23,378	23,903
21784	Eldersburg (c)	(410)	27,741	9,720
21227	Elkridge (c)	(410)	22,042	12,953
*21921	Elkton	(410)	11,893	9,073

ZIP	Place	Area Code	2000	1990
*21043	Ellicott City (c)	(410)	56,397	41,396
21221	Essex (c)	(410)	39,078	40,872
20904	Fairland (c)	(301)	21,738	19,828
21047	Fallston (c)	(410)	8,427	5,730
21061	Ferndale (c)	(410)	16,056	16,355
—	Forest Glen (c)		7,344	—
20747	Forestville (c)	(301)	12,707	16,731
20755	Fort Meade (c)	(301)	9,882	12,509
*20744	Fort Washington (c)	(301)	23,845	24,032
*21701	Frederick	(301)	52,767	40,186
20744	Friendly (c)	(301)	10,938	9,028
21532	Frostburg	(301)	7,873	8,069
*20877	Gaithersburg	(301)	52,613	39,676
21055	Garrison (c)	(410)	7,969	5,045
*20874	Germantown (c)	(301)	55,419	41,145
20706	Glenarden	(301)	6,318	5,025
*21061	Glen Burnie (c)	(410)	38,922	37,305
20769	Glenn Dale (c)	(301)	12,609	9,689
—	Goddard (c)		5,554	4,576
—	Greater Landover (c)		22,900	—
20772	Greater Upper Marlboro (c)		18,720	11,528
*20770	Greenbelt	(301)	21,456	20,561
21122	Green Haven (c)	(410)	17,415	14,416
21771	Green Valley (c)	(301)	12,262	9,424
*21740	Hagerstown	(301)	36,687	35,306
21740	Halfway (c)	(301)	10,065	8,873
21074	Hampstead	(410)	5,060	2,608
21211	Hampton (c)		5,004	4,926
21078	Havre de Grace	(410)	11,331	8,952
20748	Hillcrest Heights (c)	(301)	16,359	17,136
*20780	Hyattsville	(301)	14,733	13,864
20794	Jessup (c)	(410)	7,865	6,537
21085	Joppatowne (c)	(410)	11,391	11,084
—	Kemp Mill (c)		9,956	—
20772	Kettering (c)	(301)	11,008	9,901
—	Lake Arbor (c)		8,533	—
21122	Lake Shore (c)	(410)	13,065	13,269
20787	Langley Park (c)	(301)	16,214	17,474
20706	Lanham-Seabrook (c)	(301)	18,190	16,792
21227	Lansdowne-Baltimore Highlands(c)		15,724	15,509
20646	La Plata	(301)	6,551	5,841
20772	Largo (c)	(301)	8,408	9,475
*20707	Laurel	(301)	19,960	19,086
20653	Lexington Park (c)	(410)	11,021	9,943
—	Linganore-Bartonsville (c)		12,529	4,079
21090	Linthicum (c)	(410)	7,539	7,547
21207	Lochearn (c)	(410)	25,269	25,240
21037	Londontowne (c)	(410)	7,595	6,992
*21093	Lutherville-Timonium (c)	(410)	15,814	16,442
20748	Marlow Heights (c)	(301)	6,059	5,885
20772	Marlton (c)	(301)	7,798	5,523
20707	Maryland City (c)	(301)	6,814	6,813
21093	Mays Chapel (c)	(410)	11,427	10,132
21220	Middle River (c)	(410)	23,958	24,616
21207	Milford Mill (c)	(410)	26,527	22,547
20717	Mitchellville (c)	(301)	9,611	12,593
20886	Montgomery Village (c)	(301)	38,051	32,315
21771	Mount Airy	(301)/(410)	6,425	3,730
20712	Mount Rainier	(301)	8,498	7,954
20784	New Carrollton	(301)	12,589	12,002
20815	North Bethesda (c)	(301)	38,610	29,656
20895	North Kensington (c)	(301)	8,940	8,607
20707	North Laurel (c)	(301)	20,468	15,008
20878	North Potomac (c)	(301)	23,044	18,456
*21842	Ocean City	(410)	7,173	5,146
21811	Ocean Pines (c)	(410)	10,496	4,251
21113	Odenton (c)	(410)	20,534	12,833
*20832	Olney (c)	(301)	31,438	23,019
21206	Overlea (c)	(410)	12,148	12,137
21117	Owings Mills (c)	(410)	20,193	9,474
*20750	Oxon Hill-Glassmanor (c)	(301)	35,355	35,794
21234	Parkville (c)	(410)	31,118	31,617
21401	Parole (c)	(410)	14,031	10,054
*21122	Pasadena(c)	(410)	12,093	10,012
21128	Perry Hall (c)	(410)	28,705	22,723
21282	Pikesville (c)	(410)	29,123	24,815
20837	Poolesville	(301)	5,151	3,796
*20850	Potomac (c)	(301)	44,822	45,634
21227	Pumphrey (c)	(410)	5,317	5,483
21133	Randallstown (c)	(301)	30,870	26,277
—	Redland (c)	(301)	16,998	16,145
21136	Reisterstown (c)	(410)	22,438	19,314
*20737	Riverdale Park	(301)	6,690	4,843
21017	Riverside (c)		6,128	—
21122	Riviera Beach (c)	(410)	12,695	11,376
*20850	Rockville	(301)	47,388	44,830
20772	Rosaryville (c)	(301)	12,322	8,976
21237	Rosedale (c)	(410)	19,199	18,703
—	Rossmoor (c)		7,569	6,182
21221	Rossville (c)	(410)	11,515	9,492
20602	Saint Charles (c)	(301)	33,379	28,717
*21801	Salisbury	(410)	23,743	20,592
20763	Savage-Guilford (c)	(410)	12,918	9,669
21144	Severn (c)	(410)	35,076	24,499
21146	Severna Park (c)	(410)	28,507	25,879
20764	Shady Side (c)	(301)	5,559	4,107
*20907	Silver Spring (c)	(301)	76,540	76,046
21061	South Gate (c)	(410)	28,672	27,564
20895	South Kensington (c)	(301)	7,887	8,777
20707	South Laurel (c)	(301)	20,479	18,591
21666	Stevensville (c)	(410)	5,880	1,862
*20752	Suitland-Silver Hills (c)	(301)	33,515	35,111
*20913	Takoma Park	(301)	17,299	16,724
21787	Taneytown	(410)	5,128	3,695
*20748	Temple Hills (c)	(301)	7,792	6,865
21788	Thurmont	(301)	5,588	3,398
*21202	Towson (c)	(410)	51,793	49,445
—	Travilah (c)	(301)	7,442	—
*20602	Waldorf (c)	(301)	22,312	15,058
20743	Walker Mill (c)	(301)	11,104	10,920
21793	Walkersville	(301)	5,192	4,145
*21157	Westminster	(410)	16,731	13,060
20902	Wheaton-Glenmont (c)	(301)	57,694	53,720
21162	White Marsh (c)	(410)	8,485	8,183
20903	White Oak (c)	(301)	20,973	18,671
21207	Woodlawn (c) (Baltimore)	(410)	36,079	32,907
21284	Woodlawn (c) (Prince George's)	(410)	6,251	5,329
—	Woodmore (c)	(240)/(301)	6,077	2,874

Massachusetts

Area code (339) overlays area code (781). Area code (351) overlays (978). Area code (774) overlays (508). Area code (857) overlays (617). See introductory note.

ZIP	Place	Area Code	2000	1990
02351	*Abington*	(781)	14,605	13,817
01720	*Acton*	(978)	20,331	17,872
02743	*Acushnet*	(508)	10,161	9,554
01220	*Adams*	(413)	8,809	9,445
01220	Adams (c)	(413)	5,784	6,356
01001	Agawam	(413)	28,144	27,323
01913	*Amesbury*	(978)	16,450	14,997
01913	Amesbury (c)	(978)	12,327	12,109
*01002	*Amherst*	(413)	34,874	35,228
*01002	Amherst Center (c)	(413)	17,050	17,824
01810	Andover (c)	(978)	7,900	8,242
01810	*Andover*	(978)	31,247	29,151
*02205	*Arlington*	(781)	42,389	44,630
01430	*Ashburnham*	(978)	5,546	5,433
01721	*Ashland*	(508)	14,674	12,066
01331	*Athol*	(978)	11,299	11,451
01331	Athol (c)	(978)	8,370	8,732
02703	Attleboro	(508)	42,068	38,383
01501	*Auburn*	(508)	15,901	15,005
01432	*Ayer*	(978)	7,287	6,871
02630	Barnstable Town	(508)	47,821	40,949
01005	*Barre*	(978)	5,113	1,094
01730	*Bedford*	(781)	12,595	12,996
01007	*Belchertown*	(413)	12,968	10,579
02019	*Bellingham*	(508)	15,314	14,877
02478	*Belmont*	(781)	24,194	24,720
02779	*Berkley*	(508)	5,749	4,237
01915	Beverly	(978)	39,862	38,195
*01821	*Billerica*	(978)	38,981	37,609
01504	*Blackstone*	(508)	8,804	8,023
—	Bliss Corner (c)		5,466	4,908
*02205	Boston	(617)	589,141	574,283
02532	*Bourne*	(508)	18,721	16,064
01921	*Boxford*	(978)	7,921	6,266
*02205	*Braintree*	(781)	33,828	33,836
02631	*Brewster*	(508)	10,094	8,440
*02324	Bridgewater (c)	(508)	6,664	7242
02324	*Bridgewater*	(508)	25,185	21,249
*02303	Brockton	(508)	94,304	92,788
*02446	*Brookline*	(617)	57,107	54,718
01803	*Burlington*	(781)	22,876	23,302
*02139	Cambridge	(617)	101,355	95,802
02021	*Canton*	(781)	20,775	18,530
02330	*Carver*	(508)	11,163	10,590
01507	*Charlton*	(508)	11,263	9,576
02633	*Chatham*	(508)	6,625	6,579
01824	*Chelmsford*	(978)	33,858	32,383
02150	Chelsea	(617)	35,080	28,710
*01020	Chicopee	(413)	54,653	56,632
01510	*Clinton*	(978)	13,435	13,222
01510	Clinton (c)	(978)	7,884	7,943
01778	Cochituate (c)	(508)	6,768	6,046
02025	*Cohasset*	(781)	7,261	7,075
01742	*Concord*	(978)	16,993	17,076
*01226	*Dalton*	(413)	6,892	7,155
01923	*Danvers*	(978)	25,212	24,174
02714	*Dartmouth*	(508)	30,666	27,244
*02026	*Dedham*	(781)	23,464	23,782
02638	*Dennis*	(508)	15,973	13,864
02715	*Dighton*	(508)	6,175	5,631
01516	*Douglas*	(508)	7,045	5,438
02030	*Dover*	(508)	5,558	4,915
01826	*Dracut*	(978)	28,562	25,594
01571	*Dudley*	(508)	10,036	9,540
*02332	*Duxbury*	(781)	14,248	13,895
02333	*East Bridgewater*	(508)	12,974	11,104
02536	East Falmouth (c)	(508)	6,615	5,577
02642	*Eastham*	(508)	5,453	4,462
01027	*Easthampton*	(413)	15,994	15,537
01028	*East Longmeadow*	(413)	14,100	13,367

ZIP	Place	Area Code	2000	1990
02334	*Easton*	(508)	22,299	19,807
02149	Everett	(617)	38,037	35,701
02719	*Fairhaven*	(508)	16,159	16,132
*02722	Fall River	(508)	91,938	92,703
*02540	*Falmouth*	(508)	32,660	27,960
01420	Fitchburg	(978)	39,102	41,194
02035	*Foxborough*	(508)	16,246	14,637
02035	Foxborough (c)	(508)	5,509	5,706
*01701	*Framingham*	(508)	66,910	64,989
02038	Franklin	(508)	29,560	22,095
02702	*Freetown*	(508)	8,472	8,522
01440	Gardner	(978)	20,770	20,125
01833	*Georgetown*	(978)	7,377	6,384
*01930	Gloucester	(978)	30,273	28,716
01519	*Grafton*	(508)	14,894	13,035
01033	*Granby*	(413)	6,132	5,565
01230	*Great Barrington*	(413)	7,527	7,725
01301	*Greenfield*	(413)	18,168	18,666
01301	Greenfield (c)	(413)	13,716	14,016
01450	*Groton*	(978)	9,547	7,511
01834	*Groveland*	(978)	6,038	5,214
02338	*Halifax*	(781)	7,500	6,526
01936	*Hamilton*	(978)	8,315	7,280
01036	*Hampden*	(413)	5,171	—
02339	*Hanover*	(781)	13,164	11,912
02341	*Hanson*	(781)	9,495	9,028
01451	*Harvard*	(978)	5,981	12,329
02645	*Harwich*	(508)	12,386	10,275
*01830	Haverhill	(978)	58,969	51,418
*02018	Hingham (c)	(781)	5,352	5,454
02043	*Hingham*	(781)	19,882	19,821
02343	*Holbrook*	(781)	10,785	11,041
01520	*Holden*	(508)	15,621	14,628
01746	*Holliston*	(508)	13,801	12,926
*01040	Holyoke	(413)	39,838	43,704
01747	*Hopedale*	(508)	5,907	5,666
01748	*Hopkinton*	(508)	13,346	9,191
01749	*Hudson*	(978)	18,113	17,233
01749	Hudson (c)	(978)	14,388	14,267
02045	*Hull*	(781)	11,050	10,466
02601	Hyannis (c)	(508)	11,050	14,120
01938	*Ipswich*	(978)	12,987	11,873
02364	Kingston (c)	(781)	5,380	4,774
02364	*Kingston*	(781)	11,780	9,045
02347	*Lakeville*	(508)	9,821	7,785
01523	*Lancaster*	(978)	7,380	6,661
*01842	Lawrence	(978)	72,043	70,207
01238	*Lee*	(413)	5,985	5,849
01524	*Leicester*	(508)	10,471	10,191
01240	*Lenox*	(413)	5,077	5,069
01453	Leominster	(978)	41,303	38,145
*02205	*Lexington*	(781)	30,355	28,974
01773	*Lincoln*	(781)	8,056	7,666
01460	*Littleton*	(978)	8,184	7,051
*01028	*Longmeadow*	(413)	15,633	15,467
*01853	Lowell	(978)	105,167	103,439
01056	*Ludlow*	(413)	21,209	18,820
01462	*Lunenburg*	(978)	9,401	9,117
*01901	Lynn	(781)	89,050	81,245
01940	*Lynnfield*	(781)	11,542	11,049
02148	Malden	(781)	56,340	53,884
01944	*Manchester-by-the-Sea*	(978)	5,228	5,286
*02048	*Mansfield*	(508)	22,414	16,568
02048	Mansfield Center (c)	(508)	7,320	7,170
01945	*Marblehead*	(781)	20,377	19,971
02738	*Marion*	(508)	5,123	4,496
01752	Marlborough	(508)	36,255	31,813
02050	*Marshfield*	(781)	24,324	21,531
02649	*Mashpee*	(508)	12,946	7,884
02739	*Mattapoisett*	(508)	6,268	5,850
01754	*Maynard*	(978)	10,433	10,325
02052	Medfield (c)	(508)	6,670	5,985
02052	*Medfield*	(508)	12,273	10,531
*02155	Medford	(781)	55,765	57,407
02053	*Medway*	(508)	12,448	9,931
02176	Melrose	(781)	27,134	28,150
01756	*Mendon*	(508)	5,286	—
01860	*Merrimac*	(978)	6,138	5,166
01844	Methuen	(978)	43,789	39,990
02346	*Middleborough*	(508)	19,941	17,867
02346	Middleborough Center (c)	(508)	6,913	6,837
01949	*Middleton*	(978)	7,744	4,921
01757	*Milford*	(508)	26,799	25,355
01757	Milford (c)	(508)	24,230	23,339
01527	*Millbury*	(508)	12,784	12,228
02054	*Millis*	(508)	7,902	7,613
02186	*Milton*	(617)	26,062	25,725
01057	*Monson*	(413)	8,359	7,776
01351	*Montague*	(413)	8,489	8,316
*02584	*Nantucket*	(508)	9,520	6,012
01760	*Natick*	(508)	32,170	30,510
*02205	*Needham*	(781)	28,911	27,557
*02740	New Bedford	(508)	93,768	99,922
01951	*Newbury*	(978)	6,717	5,623
01950	Newburyport	(978)	17,189	16,317
*02205	Newton	(617)	83,829	82,585
02056	*Norfolk*	(508)	10,460	9,259
01247	North Adams	(413)	14,681	16,797
01059	North Amherst (c)	(413)	6,019	6,239
*01060	Northampton	(413)	28,978	11,929
01845	*North Andover*	(978)	27,202	29,289
*02760	*North Attleborough*	(508)	27,143	22,792
02760	North Attleborough Center (c)	(508)	16,796	16,178
01532	Northborough (c)	(508)	6,257	5,761
01532	*Northborough*	(508)	14,013	13,371
01534	*Northbridge*	(508)	13,182	12,002
01864	*North Reading*	(978)	13,837	25,038
02060	North Scituate (c)	(781)	5,065	4,891
02766	*Norton*	(508)	18,036	14,265
02061	*Norwell*	(781)	9,765	9,279
02062	*Norwood*	(781)	28,587	28,700
02065	Ocean Bluff-Brant Rock (c)	(781)	5,100	4,541
01364	*Orange*	(978)	7,518	7,312
02653	*Orleans*	(508)	6,341	5,838
01540	Oxford (c)	(508)	5,899	5,969
01540	*Oxford*	(508)	13,352	12,588
01069	*Palmer*	(413)	12,497	12,054
*01960	Peabody	(978)	48,129	47,264
02359	*Pembroke*	(781)	16,927	14,544
01463	*Pepperell*	(978)	11,142	10,098
01866	Pinehurst (c)	(978)	6,941	6,614
*01201	Pittsfield	(413)	45,793	48,622
02762	*Plainville*	(508)	7,683	6,871
*02360	Plymouth (c)	(508)	7,658	7,258
*02360	*Plymouth*	(508)	51,701	45,608
*02205	Quincy	(617)	88,025	84,985
02368	*Randolph*	(781)	30,963	30,093
02767	*Raynham*	(508)	11,739	9,867
01867	*Reading*	(781)	23,708	22,539
02769	*Rehoboth*	(508)	10,172	8,656
02151	Revere	(781)	47,283	42,786
02370	*Rockland*	(781)	17,670	16,123
01966	Rockport (c)	(978)	5,606	5,448
01966	*Rockport*	(978)	7,767	7,482
01969	*Rowley*	(978)	5,500	4,452
01543	*Rutland*	(508)	6,353	4,936
*01970	Salem	(978)	40,407	38,091
01952	*Salisbury*	(978)	7,827	6,882
02563	*Sandwich*	(508)	20,136	15,489
01906	*Saugus*	(781)	26,078	25,549
02066	Scituate (c)	(781)	5,069	5,180
02066	*Scituate*	(781)	17,863	16,786
02771	*Seekonk*	(508)	13,425	13,046
02067	*Sharon*	(781)	17,408	15,517
02067	Sharon (c)	(781)	5,941	5,893
01464	*Shirley*	(978)	6,373	6,118
01545	*Shrewsbury*	(508)	31,640	24,146
*02722	*Somerset*	(508)	18,234	17,655
*02205	Somerville	(617)	77,478	76,210
01002	South Amherst (c)	(413)	5,039	5,053
01073	*Southampton*	(413)	5,387	—
01772	*Southborough*	(508)	8,781	6,628
01550	*Southbridge*	(508)	17,214	17,816
01550	Southbridge (c)	(508)	12,878	13,631
01075	*South Hadley*	(413)	17,196	16,685
01077	*Southwick*	(413)	8,835	7,667
02664	South Yarmouth (c)	(508)	11,603	10,358
01562	Spencer (c)	(508)	6,032	6,306
01562	*Spencer*	(508)	11,691	11,645
*01101	Springfield	(413)	152,082	156,983
01564	*Sterling*	(978)	7,257	6,481
02180	*Stoneham*	(781)	22,219	22,203
02072	*Stoughton*	(781)	27,149	26,777
01775	*Stow*	(978)	5,902	5,328
01566	*Sturbridge*	(508)	7,837	7,775
01776	*Sudbury*	(978)	16,841	14,358
01590	*Sutton*	(508)	8,250	6,824
01907	*Swampscott*	(781)	14,412	13,650
02777	*Swansea*	(508)	15,901	15,411
02780	Taunton	(508)	55,976	49,832
01468	*Templeton*	(978)	6,799	6,438
01876	*Tewksbury*	(978)	28,851	27,266
01983	*Topsfield*	(978)	6,141	5,754
01469	*Townsend*	(978)	9,198	8,496
01879	*Tyngsborough*	(978)	11,081	8,642
01568	*Upton*	(508)	5,642	4,677
01569	*Uxbridge*	(508)	11,156	10,415
01880	*Wakefield*	(781)	24,804	24,825
*02081	Walpole (c)	(508)	5,867	5,495
02081	*Walpole*	(508)	22,824	20,223
*02205	Waltham	(781)	59,226	57,878
01082	Ware (c)	(413)	6,174	6,533
01082	*Ware*	(413)	9,707	9,808
02571	*Wareham*	(508)	20,335	19,232
*02205	Watertown	(781)	32,986	33,284
01778	*Wayland*	(508)	13,100	11,874
01570	*Webster*	(508)	16,415	16,196
01570	Webster (c)	(508)	11,600	11,849
*02205	*Wellesley*	(781)	26,613	26,615
01581	*Westborough*	(508)	17,997	14,133
01583	*West Boylston*	(508)	7,481	6,611
02379	*West Bridgewater*	(508)	6,634	6,389
01742	West Concord (c)	(978)	5,632	5,761
*01085	Westfield	(413)	40,072	38,372

ZIP	Place	Area Code	2000	1990
01886	*Westford*	(978)	20,754	16,392
01473	*Westminster*	(978)	6,907	6,191
02493	*Weston*	(781)	11,469	10,200
02790	*Westport*	(508)	14,183	13,852
*01089	*West Springfield*	(413)	27,899	27,537
02090	*Westwood*	(781)	14,117	12,557
02673	West Yarmouth (c)	(508)	6,460	5,409
*02205	*Weymouth*	(781)	53,988	54,063
01588	Whitinsville (c)	(508)	6,340	5,639
02382	*Whitman*	(781)	13,882	13,240
01095	*Wilbraham*	(413)	13,473	12,635
01267	*Williamstown*	(413)	8,424	8,220
01887	*Wilmington*	(978)	21,363	17,651
01475	*Winchendon*	(978)	9,611	8,805
01890	*Winchester*	(781)	20,810	20,267
02152	*Winthrop*	(617)	18,303	18,127
*01801	Woburn	(781)	37,258	35,943
*01613	Worcester	(508)	172,648	169,759
02093	*Wrentham*	(508)	10,554	9,006
02675	*Yarmouth*	(508)	24,807	21,174
02675	Yarmouth Port (c)	(508)	5,395	4,271

Michigan

Area code (947) overlays area code (248). See introductory note.

ZIP	Place	Area Code	2000	1990
49221	Adrian	(517)	21,574	22,097
49224	Albion	(517)	9,144	10,066
49401	Allendale (c)	(616)	11,555	6,950
48101	Allen Park	(313)	29,376	31,092
48801	Alma	(989)	9,275	9,034
49707	Alpena	(989)	11,304	11,354
*48106	Ann Arbor	(734)	114,024	109,608
*48321	Auburn Hills	(248)	19,837	17,076
*49016	Battle Creek	(616)	53,364	53,516
*48707	Bay City	(989)	36,817	38,936
48505	Beecher (c)	(810)	12,793	14,465
48809	Belding	(616)	5,877	5,969
*49022	Benton Harbor	(616)	11,182	12,818
49022	Benton Heights (c)	(616)	5,458	5,465
48072	Berkley	(248)	15,531	16,960
48025	Beverly Hills	(248)	10,437	10,610
49307	Big Rapids	(231)	10,849	12,603
*48012	Birmingham	(248)	19,291	19,997
48301	Bloomfield (c)	(248)	43,021	42,137
48722	Bridgeport (c)	(989)	7,849	8,569
*48116	Brighton	(810)	6,701	5,686
48601	Buena Vista (c)	(989)	7,845	8,196
*48501	Burton	(810)	30,308	27,437
49601	Cadillac	(231)	10,000	10,104
*48185	Canton (c)	(734)	76,366	57,047
48724	Carrollton (c)	(989)	6,602	6,521
48015	Center Line	(586)	8,531	9,026
48813	Charlotte	(517)	8,389	8,083
49721	Cheboygan	(231)	5,295	4,997
48017	Clawson	(248)	12,732	13,874
*48046	Clinton (c)	(517)	95,648	85,866
49036	Coldwater	(517)	12,697	9,607
49321	Comstock Park (c)	(616)	10,674	6,530
49508	Cutlerville (c)	(616)	15,114	11,228
48423	Davison	(810)	5,536	5,693
*48120	Dearborn	(313)	97,775	89,286
*48127	Dearborn Heights	(313)	58,264	60,838
*48231	Detroit	(313)	951,270	1,027,974
49047	Dowagiac	(616)	6,147	6,418
49506	East Grand Rapids	(616)	10,764	10,807
*48826	East Lansing	(517)	46,525	50,677
48021	Eastpointe	(586)	34,077	35,283
49001	Eastwood (c)	(616)	6,265	6,340
48827	Eaton Rapids	(517)	5,330	4,695
48229	Ecorse	(313)	11,229	12,180
49829	Escanaba	(906)	13,140	13,659
49022	Fair Plain (c)	(616)	7,828	8,051
*48333	Farmington	(248)	10,423	10,170
48333	Farmington Hills	(248)	82,111	74,614
48430	Fenton	(810)	10,582	8,434
48220	Ferndale	(248)	22,105	25,084
48134	Flat Rock	(734)	8,488	7,290
*48501	Flint	(810)	124,943	140,925
48433	Flushing	(810)	8,348	8,542
49506	Forest Hills (c)	(616)	20,942	16,690
48026	Fraser	(586)	15,297	13,899
48623	Freeland (c)	(989)	5,147	1,421
*48135	Garden City	(734)	30,047	31,846
49837	Gladstone	(906)	5,032	4,565
48439	Grand Blanc	(810)	8,242	7,760
49417	Grand Haven	(616)	11,168	11,951
48837	Grand Ledge	(517)	7,813	7,562
*49501	Grand Rapids	(616)	197,800	189,126
*49418	Grandville	(616)	16,263	15,624
48838	Greenville	(616)	7,935	8,101
48138	Grosse Ile (c)	(734)	10,894	9,781
*48231	Grosse Pointe	(313)	5,670	5,681
48230	Grosse Pointe Farms	(313)	9,764	10,092
48230	Grosse Pointe Park	(313)	12,443	12,857
48230	Grosse Pointe Woods	(313)	17,080	17,715
48212	Hamtramck	(313)	22,976	18,372
48225	Harper Woods	(313)	14,254	14,903
48625	Harrison (c)	(989)	24,461	24,685
48840	Haslett (c)	(517)	11,283	10,230
49058	Hastings	(616)	7,095	6,549
48030	Hazel Park	(248)	18,963	20,051
48203	Highland Park	(313)	16,746	20,121
49242	Hillsdale	(517)	8,233	8,175
*49423	Holland	(616)	35,048	30,745
48442	Holly	(248)	6,135	5,595
48842	Holt (c)	(517)	11,315	11,744
49931	Houghton	(906)	7,010	7,498
*48844	Howell	(517)	9,232	8,147
49426	Hudsonville	(616)	7,160	6,170
48070	Huntington Woods	(248)	6,151	6,419
48141	Inkster	(313)/(734)	30,115	30,772
48846	Ionia	(616)	10,569	10,349
49801	Iron Mountain	(906)	8,154	8,525
49938	Ironwood	(906)	6,293	6,849
49849	Ishpeming	(906)	6,686	7,200
*49204	Jackson	(517)	36,316	37,425
*49428	Jenison (c)	(616)	17,211	17,882
*49001	Kalamazoo	(616)	77,145	80,277
49518	Kentwood	(616)	45,255	37,826
49802	Kingsford	(906)	5,549	5,480
48144	Lambertville (c)	(734)	9,299	7,860
*48901	Lansing	(517)	119,128	127,321
48446	Lapeer	(810)	9,072	7,759
48146	Lincoln Park	(313)	40,008	41,832
*48150	Livonia	(734)	100,545	100,850
49431	Ludington	(231)	8,357	8,507
48071	Madison Heights	(248)	31,101	32,196
49660	Manistee	(231)	6,586	6,734
49855	Marquette	(906)	19,661	21,977
49068	Marshall	(616)	7,459	6,941
48040	Marysville	(810)	9,684	8,515
48854	Mason	(517)	6,714	6,768
48122	Melvindale	(313)	10,735	11,216
49858	Menominee	(906)	9,131	9,398
*48640	Midland	(989)	41,685	38,053
*48381	Milford	(248)	6,272	5,500
*48161	Monroe	(734)	22,076	22,902
*48046	Mount Clemens	(586)	17,312	18,405
*48804	Mount Pleasant	(989)	25,946	23,299
*49440	Muskegon	(231)	40,105	39,809
49444	Muskegon Heights	(231)	12,049	13,176
*48047	New Baltimore	(586)	7,405	5,798
49120	Niles	(616)	12,204	12,458
49505	Northview (c)	(616)	14,730	13,712
48167	Northville	(248)	6,459	6,226
49441	Norton Shores	(231)	22,527	21,755
*48376	Novi	(248)	47,386	32,998
48237	Oak Park	(248)	29,793	30,468
*48805	Okemos (c)	(517)	22,805	20,216
48867	Owosso	(989)	15,713	16,322
49770	Petoskey	(231)	6,080	6,056
48170	Plymouth	(734)	9,022	9,560
48170	Plymouth Township (c)	(734)	27,798	23,646
*48343	Pontiac	(248)	66,337	71,136
*49081	Portage	(616)	44,897	41,042
*48061	Port Huron	(810)	32,338	33,694
*48231	Redford (c)	(313)	51,622	54,387
48218	River Rouge	(313)	9,917	11,314
48192	Riverview	(734)	13,272	13,894
*48308	Rochester	(248)	10,467	7,130
48306	Rochester Hills	(248)	68,825	61,766
48174	Romulus	(313)/(734)	22,979	22,897
48066	Roseville	(586)	48,129	51,412
*48068	Royal Oak	(248)	60,062	65,410
*48605	Saginaw	(989)	61,799	69,512
48604	Saginaw Township North (c)	(989)	24,994	23,018
48603	Saginaw Township South (c)	(989)	13,801	13,987
48079	Saint Clair	(810)	5,802	5,116
*48080	Saint Clair Shores	(313)	63,096	68,107
48879	Saint Johns	(989)	7,485	7,392
49085	Saint Joseph	(616)	8,789	9,214
48176	Saline	(734)	8,034	6,663
49783	Sault Sainte Marie	(906)	16,542	14,689
49455	Shelby (c)	(231)	65,159	48,655
48609	Shields (c)	(989)	6,590	6,634
*48037	Southfield	(248)	78,296	75,727
48195	Southgate	(734)	30,136	30,771
49090	South Haven	(616)	5,021	5,563
48178	South Lyon	(248)	10,036	6,479
48161	South Monroe (c)	(734)	6,370	5,266
49015	Springfield	(616)	5,189	5,582
*48311	Sterling Heights	(586)	124,471	117,810
49091	Sturgis	(616)	11,285	10,130
48473	Swartz Creek	(810)	5,102	4,851
48180	Taylor	(313)/(734)	65,868	70,811
49286	Tecumseh	(517)	8,574	7,462
48182	Temperance (c)	(734)	7,757	6,542
49093	Three Rivers	(616)	7,328	7,464
*49684	Traverse City	(231)	14,532	15,155
48183	Trenton	(734)	19,584	20,586
*48099	Troy	(248)	80,959	72,884
49504	Walker	(616)	21,842	17,279
*48390	Walled Lake	(248)	6,713	6,278
*48090	Warren	(586)	138,247	144,864
*48329	Waterford (c)	(248)	73,150	66,692

ZIP	Place	Area Code	2000	1990
48917	Waverly (c)	(517)	16,194	15,614
48184	Wayne	(734)	19,051	19,899
*48325	West Bloomfield Township (c)	(248)	64,862	54,843
*48185	Westland	(313)/(734)	86,602	84,724
49019	Westwood (c)	(616)	9,122	8,957
48189	Whitmore Lake (c)	(734)	6,574	3,251
48393	Wixom	(248)	13,263	8,550
48183	Woodhaven	(734)	12,530	11,631
48192	Wyandotte	(734)	28,006	30,938
49509	Wyoming	(616)	69,368	63,891
*48197	Ypsilanti	(734)	22,362	24,846
49464	Zeeland	(616)	5,805	5,417

Minnesota

ZIP	Place	Area Code	2000	1990
56007	Albert Lea	(507)	18,356	18,310
56308	Alexandria	(320)	8,820	8,029
55304	Andover	(763)	26,588	15,216
*55303	Anoka	(612)/(763)	18,076	17,192
55124	Apple Valley	(952)	45,527	34,598
55112	Arden Hills	(651)	9,652	9,199
55912	Austin	(507)	23,314	21,926
56425	Baxter	(218)	5,555	3,695
*56601	Bemidji	(218)	11,917	11,165
55309	Big Lake	(763)	6,063	3,113
55014	Blaine	(651)/(763)	44,942	38,975
*55420	Bloomington	(952)	85,172	86,335
56401	Brainerd	(218)	13,178	12,353
55429	Brooklyn Center	(763)	29,172	28,887
55443	Brooklyn Park	(763)	67,388	56,381
55313	Buffalo	(763)	10,097	7,302
*55337	Burnsville	(651)/(952)	60,220	51,288
55008	Cambridge	(763)	5,520	5,094
55316	Champlin	(763)	22,193	16,849
55317	Chanhassen	(952)	20,321	11,736
55318	Chaska	(952)	17,449	11,339
55720	Cloquet	(218)	11,201	10,885
55421	Columbia Heights	(612)	18,520	18,910
55433	Coon Rapids	(763)	61,607	52,978
55340	Corcoran	(763)	5,630	5,199
55016	Cottage Grove	(651)	30,582	22,935
56716	Crookston	(218)	8,192	8,119
55428	Crystal	(763)	22,698	23,788
*56501	Detroit Lakes	(218)	7,348	7,141
*55806	Duluth	(218)	86,918	85,493
*55121	Eagan	(651)/(952)	63,557	47,409
55005	East Bethel	(763)	10,941	8,050
56721	East Grand Forks	(218)	7,501	8,658
*55344	Eden Prairie	(612)/(952)	54,901	39,311
55424	Edina	(952)	47,425	46,075
55330	Elk River	(763)	16,447	11,143
56031	Fairmont	(507)	10,889	11,265
55113	Falcon Heights	(651)	5,572	5,380
55021	Faribault	(507)	20,818	17,085
55024	Farmington	(651)/(952)	12,365	5,940
*56537	Fergus Falls	(218)	13,471	12,362
55025	Forest Lake	(651)	6,798	5,833
55432	Fridley	(763)	27,449	28,335
55336	Glencoe	(320)	5,453	4,648
55427	Golden Valley	(763)	20,281	20,971
*55744	Grand Rapids	(218)	7,764	7,976
*55304	Ham Lake	(763)	12,710	8,924
55033	Hastings	(651)	18,204	15,478
55810	Hermantown	(218)	7,448	6,761
*55746	Hibbing	(218)	17,071	18,046
*55343	Hopkins	(952)	17,145	16,529
55038	Hugo	(651)	6,363	4,417
55350	Hutchinson	(320)	13,080	11,459
56649	Internationa Falls	(218)	6,703	8,325
*55349	Inver Grove Heights	(651)	29,751	22,477
55042	Lake Elmo	(651)	6,863	5,900
55044	Lakeville	(952)	43,128	24,854
55014	Lino Lakes	(651)	16,791	8,807
55355	Litchfield	(320)	6,562	6,041
55117	Little Canada	(651)	9,771	8,971
56345	Little Falls	(320)	7,719	7,371
—	Lower Red Lake UT	(218)	5,057	—
55115	Mahtomedi	(651)	7,563	5,633
*56001	Mankato	(507)	32,427	31,459
55311	Maple Grove	(763)	50,365	38,736
55109	Maplewood	(651)	34,947	30,954
56258	Marshall	(507)	12,735	12,023
55118	Mendota Heights	(651)	11,434	9,388
*55440	Minneapolis	(612)/(763)/(952)	382,618	368,383
55345	Minnetonka	(952)	51,301	48,370
56265	Montevideo	(320)	5,346	5,499
*55362	Monticello	(763)	7,868	5,045
*56560	Moorhead	(218)	32,177	32,295
56267	Morris	(320)	5,068	5,613
55364	Mound	(952)	9,435	9,634
55112	Mounds View	(763)	12,738	12,541
55112	New Brighton	(651)	22,206	22,207
54427	New Hope	(763)	20,873	21,853
56073	New Ulm	(507)	13,594	13,132
55056	North Branch	(651)/(763)	8,023	4,267
55057	Northfield	(507)	17,147	14,684
56001	North Mankato	(507)	11,798	10,662
55109	North Saint Paul	(651)	11,929	12,376
55128	Oakdale	(651)	26,653	18,377
—	Oak Grove	(763)	6,903	5,488
55323	Orono	(952)	7,538	7,285
—	Otsego	(763)	6,389	5,219
55060	Owatonna	(507)	22,434	19,386
*55446	Plymouth	(763)	65,894	50,889
55372	Prior Lake	(952)	15,917	11,482
55303	Ramsey	(763)	18,510	12,408
55066	Red Wing	(651)	16,116	15,134
56283	Redwood Falls	(507)	5,459	4,859
55423	Richfield	(612)	34,439	35,710
55422	Robbinsdale	(763)	14,123	14,396
*55901	Rochester	(507)	85,806	70,729
55068	Rosemount	(651)/(952)	14,619	8,622
55113	Roseville	(651)	33,690	33,485
55418	Saint Anthony	(612)	8,012	7,727
*56301	Saint Cloud	(320)	59,107	48,812
55426	Saint Louis Park	(952)	44,126	43,787
*55374	Saint Michael	(763)	9,099	2,506
*55101	Saint Paul	(651)	287,151	272,235
55071	Saint Paul Park	(651)	5,070	4,965
56082	Saint Peter	(507)	9,747	9,481
56377	Sartell	(320)	9,641	5,409
56379	Sauk Rapids	(320)	10,213	7,823
56378	Savage	(952)	21,115	9,906
55379	Shakopee	(612)	20,568	11,739
55126	Shoreview	(651)	25,924	24,587
55331	Shorewood	(952)	7,400	5,913
55075	South Saint Paul	(651)	20,167	20,197
55432	Spring Lake Park	(763)	6,772	6,532
55976	Stewartville	(507)	5,411	4,520
*55082	Stillwater	(651)	15,143	13,882
56701	Thief River Falls	(218)	8,410	8,010
55127	Vadnais Heights	(651)	13,069	11,041
*55792	Virginia	(218)	9,157	9,432
55387	Waconia	(952)	6,814	3,498
56387	Waite Park	(320)	6,568	5,020
56093	Waseca	(507)	8,493	8,385
—	West Crow Wing UT		5,144	—
55118	West Saint Paul	(651)	19,405	19,248
*55110	White Bear Lake	(651)	24,325	24,622
56201	Willmar	(320)	18,351	17,531
55987	Winona	(507)	27,069	25,435
55125	Woodbury	(651)	46,463	20,075
56187	Worthington	(507)	11,283	9,977

Mississippi

ZIP	Place	Area Code	2000	1990
39730	Aberdeen	(662)	6,415	6,837
38821	Amory	(662)	6,956	7,093
38606	Batesville	(662)	7,113	6,403
*39520	Bay Saint Louis	(228)	8,209	8,063
*39530	Biloxi	(228)	50,644	46,319
38829	Booneville	(662)	8,625	7,955
*39042	Brandon	(601)	16,436	11,089
*39601	Brookhaven	(601)	9,861	10,243
39272	Byram (c)	(601)	7,386	—
39046	Canton	(601)	12,911	11,723
38614	Clarksdale	(662)	20,645	21,180
*38732	Cleveland	(662)	13,841	15,384
*39056	Clinton	(601)	23,347	21,847
39429	Columbia	(601)	6,603	6,815
*39701	Columbus	(662)	25,944	23,799
*38834	Corinth	(662)	14,054	11,820
39059	Crystal Springs	(601)	5,873	5,643
39525	Diamondhead (c)	(228)	5,912	2,661
39532	D'Iberville	(228)	7,608	6,566
39074	Forest	(601)	5,987	5,062
39553	Gautier	(228)	11,681	10,088
*38701	Greenville	(662)	41,633	45,226
*38930	Greenwood	(662)	18,425	18,906
*38901	Grenada	(662)	14,879	10,864
39564	Gulf Hills (c)	(228)	5,900	5,004
*39501	Gulfport	(228)	71,127	64,045
*39401	Hattiesburg	(601)	44,779	45,325
38632	Hernando		6,812	3,125
*38635	Holly Springs	(662)	7,957	7,261
38637	Horn Lake	(662)	14,099	9,069
38751	Indianola	(662)	12,066	11,809
*39205	Jackson	(601)	184,256	202,062
39090	Kosciusko	(662)	7,372	6,986
*39440	Laurel	(601)	18,393	18,827
38756	Leland	(662)	5,502	6,366
39560	Long Beach	(228)	17,320	15,804
39339	Louisville	(662)	7,006	7,165
*39648	McComb	(601)	13,337	11,797
*39110	Madison	(601)	14,692	7,471
*39302	Meridian	(601)	39,968	41,036
*39563	Moss Point	(228)	15,851	17,837
*39120	Natchez	(601)	18,464	19,460
38652	New Albany	(662)	7,607	6,775
*39564	Ocean Springs	(228)	17,225	15,221
38654	Olive Branch	(662)	21,054	3,567
38655	Oxford	(662)	11,756	10,026
*39567	Pascagoula	(228)	26,200	25,899
39571	Pass Christian	(228)	6,579	5,557
39288	Pearl	(601)	21,961	19,588
39465	Petal	(601)	7,579	7,883

ZIP	Place	Area Code	2000	1990
39350	Philadelphia	(601)	7,303	6,758
39466	Picayune	(601)	10,535	10,633
38863	Pontotoc	(662)	5,253	4,570
39218	Richland	(601)	6,027	4,014
*39157	Ridgeland	(601)	20,173	11,714
38663	Ripley	(662)	5,478	5,371
39533	Saint Martin (c)	(228)	6,676	6,349
38668	Senatobia	(601)	6,682	4,772
38671	Southaven	(662)	28,977	18,705
*39759	Starkville	(662)	21,869	18,458
*38801	Tupelo	(662)	34,211	30,685
*39180	Vicksburg	(601)	26,407	26,886
39576	Waveland	(228)	6,674	5,369
39367	Waynesboro	(601)	5,197	5,143
—	West Hattiesburg (c)	(601)	6,305	5,450
39773	West Point	(662)	12,145	8,489
38967	Winona	(662)	5,482	5,965
39194	Yazoo City	(662)	14,550	12,427

Missouri

Area code (557) overlays area code (314). Area code (975) overlays (816). See introductory note.

ZIP	Place	Area Code	2000	1990
63123	Affton (c)	(314)	20,535	21,106
63010	Arnold	(636)	19,965	18,828
65605	Aurora	(417)	7,014	6,459
*63011	Ballwin	(636)	31,283	27,054
63012	Barnhart (c)	(314)	6,108	4,911
63137	Bellefontaine Neighbors	(314)	11,271	10,918
64012	Belton	(816)	21,730	18,145
63134	Berkeley	(314)	10,063	12,250
63031	Black Jack	(314)	6,792	6,131
*64015	Blue Springs	(816)	48,080	40,103
65613	Bolivar	(417)	9,143	6,845
65233	Boonville	(660)	8,202	7,095
*65615	Branson	(417)	6,050	3,706
63144	Brentwood	(314)	7,693	8,150
63044	Bridgeton	(314)	15,550	17,732
64429	Cameron	(816)	8,312	6,782
*63701	Cape Girardeau	(573)	35,349	34,475
64834	Carl Junction	(417)	5,294	4,123
64836	Carthage	(417)	12,668	10,747
63830	Caruthersville	(573)	6,760	7,389
*63017	Chesterfield	(636)	46,802	42,325
64601	Chillicothe	(660)	8,968	8,799
63105	Clayton	(314)	12,825	13,926
64735	Clinton	(660)	9,311	8,703
*65201	Columbia	(573)	84,531	69,133
63128	Concord (c)	(314)	16,689	19,859
63126	Crestwood	(314)	11,863	11,229
63141	Creve Coeur	(314)	16,500	12,289
*63135	Dellwood	(314)	5,255	5,245
63020	De Soto	(636)	6,375	5,993
63131	Des Peres	(636)	8,592	8,395
63841	Dexter	(573)	7,356	7,506
63011	Ellisville	(636)	9,104	7,183
63025	Eureka	(636)	7,676	4,683
64024	Excelsior Springs	(816)	10,847	10,373
63640	Farmington	(573)	13,924	11,596
63135	Ferguson	(314)	22,406	22,290
63028	Festus	(636)	9,660	8,105
*63033	Florissant	(314)	50,497	51,038
65473	Fort Leonard Wood (c)	(573)	13,666	15,863
65251	Fulton	(573)	12,128	10,033
64118	Gladstone	(816)	26,365	26,243
65254	Glasgow Village (c)	(573)	5,234	5,199
63122	Glendale	(314)	5,767	5,945
64029	Grain Valley	(816)	5,160	1,898
64030	Grandview	(816)	24,881	24,973
63401	Hannibal	(573)	17,757	18,004
64701	Harrisonville	(816)	8,946	7,696
63042	Hazelwood	(314)	26,206	15,512
*64050	Independence	(816)	113,288	112,301
63755	Jackson	(573)	11,947	9,256
*65101	Jefferson City	(573)	39,636	35,517
63136	Jennings	(314)	15,469	15,841
*64801	Joplin	(417)	45,504	41,175
*64108	Kansas City	(816)	441,545	434,829
64060	Kearney	(816)	5,472	1,790
63857	Kennett	(573)	11,260	10,941
63501	Kirksville	(660)	16,988	17,152
63122	Kirkwood	(314)	27,324	28,318
63124	Ladue (St. Louis Co.)	(314)	8,645	8,795
63367	Lake Saint Louis	(636)	10,169	7,536
65536	Lebanon	(417)	12,155	9,983
*64063	Lee's Summit	(816)	70,700	46,418
63125	Lemay (c)	(314)	17,215	18,005
*64068	Liberty	(816)	26,232	20,459
63552	Macon	(660)	5,538	5,571
63011	Manchester	(636)	19,161	6,506
63143	Maplewood	(314)	9,228	9,962
65340	Marshall	(660)	12,433	12,711
65706	Marshfield	(417)	5,720	4,374
63043	Maryland Heights	(314)	25,756	25,440
64468	Maryville	(816)	10,581	10,663
63129	Mehlville (c)	(314)	28,822	27,557
65265	Mexico	(573)	11,320	11,290
65270	Moberly	(660)	11,945	12,839
65708	Monett	(417)	7,396	6,529
63026	Murphy (c)	(636)	9,048	9,342
64850	Neosho	(417)	10,505	9,254
64772	Nevada	(417)	8,607	8,597
65714	Nixa	(417)	12,124	4,893
63121	Normandy	(314)	5,153	4,480
64075	Oak Grove	(816)	5,535	4,565
63129	Oakville (c)	(314)	35,309	31,750
63366	O'Fallon	(636)	46,169	17,427
63132	Olivette	(314)	7,438	7,573
63114	Overland	(314)	16,838	17,987
65721	Ozark	(417)	9,665	4,401
63069	Pacific	(636)	5,482	4,350
63601	Park Hills	(573)	7,861	7,866
63775	Perryville	(573)	7,667	6,933
64080	Pleasant Hill	(816)	5,582	3,827
*63901	Poplar Bluff	(573)	16,651	16,841
64083	Raymore	(816)	11,146	5,592
64133	Raytown	(816)	30,388	30,601
65738	Republic	(417)	8,438	6,290
64085	Richmond	(816)	6,116	5,738
63117	Richmond Heights	(314)	9,602	10,448
*65401	Rolla	(573)	16,367	14,090
63074	Saint Ann	(314)	13,607	14,449
*63301	Saint Charles	(636)	60,321	50,634
63114	Saint John	(314)	6,871	7,502
*64501	Saint Joseph	(816)	73,990	71,852
*63166	Saint Louis	(314)	348,189	396,685
63376	Saint Peters	(636)	51,381	40,660
63126	Sappington (c)	(314)	7,287	10,917
*65301	Sedalia	(660)	20,339	19,800
63119	Shrewsbury	(314)	6,644	6,416
63801	Sikeston	(573)	16,992	17,641
64089	Smithville	(816)	5,514	2,525
63138	Spanish Lake (c)	(314)	21,337	20,322
*65801	Springfield	(417)	151,580	140,494
63080	Sullivan	(573)	6,351	5,661
63127	Sunset Hills	(314)	8,267	4,915
63006	Town and Country	(314)	10,894	10,944
64683	Trenton	(660)	6,216	6,129
63379	Troy	(314)	6,737	3,811
63084	Union	(636)	7,757	6,196
63130	University City	(314)	37,428	40,087
63088	Valley Park	(636)	6,518	4,165
64093	Warrensburg	(660)	16,340	15,244
63383	Warrenton	(636)	5,281	3,564
63090	Washington	(636)	13,243	11,367
64870	Webb City	(417)	9,812	7,538
63119	Webster Groves	(314)	23,230	22,992
63304	Weldon Spring	(636)	5,270	1,470
63385	Wentzville	(636)	6,896	4,640
65775	West Plains	(417)	10,866	9,214
*63011	Wildwood	(314)	32,884	16,742

Montana (406)

ZIP	Place	2000	1990
59711	Anaconda	9,417	10,356
59714	Belgrade	5,728	3,422
*59101	Billings	89,847	81,125
*59718	Bozeman	27,509	22,660
*59701	Butte	34,606	33,336
59901	Evergreen (c)	6,215	4,109
*59401	Great Falls	56,690	55,125
59501	Havre	9,621	10,201
*59601	Helena	25,780	24,609
—	Helena Valley Southeast (c)	7,141	4,601
—	Helena Valley West Central (c)	6,983	6,327
*59901	Kalispell	14,223	11,917
59044	Laurel	6,255	5,686
59457	Lewistown	5,813	6,097
59047	Livingston	6,851	6,701
59301	Miles City	8,487	8,461
*59801	Missoula	57,053	42,918
59801	Orchard Homes (c)	5,199	10,317
59937	Whitefish	5,032	4,368

Nebraska

ZIP	Place	Area Code	2000	1990
69301	Alliance	(308)	8,959	9,765
68310	Beatrice	(402)	12,496	12,352
*68108	Bellevue	(402)	44,382	39,240
*68008	Blair	(402)	7,512	6,860
69337	Chadron	(308)	5,634	5,588
68108	Chalco (c)	(402)	10,736	7,337
68601	Columbus	(402)	20,971	19,480
68333	Crete	(402)	6,028	4,841
68022	Elkhorn	(402)	6,062	1,398
*68025	Fremont	(402)	25,174	23,680
69341	Gering	(308)	7,751	7,946
*68802	Grand Island	(308)	42,940	39,487
*68901	Hastings	(402)	24,064	22,837
68949	Holdrege	(308)	5,636	5,671
*68847	Kearney	(308)	27,431	24,396
68128	La Vista	(402)	11,699	9,992
68850	Lexington	(308)	10,011	6,600
*68501	Lincoln	(402)	225,581	191,972
69001	McCook	(308)	7,994	8,112
68410	Nebraska City	(402)	7,228	6,547

ZIP	Place	Area Code	2000	1990
*68701	Norfolk	(402)	23,516	21,476
*69101	North Platte	(308)	23,878	22,605
68113	Offutt AFB (c)	(402)	8,901	—
*68005	Omaha	(402)	390,007	344,463
*68046	Papillion	(402)	16,363	13,892
68048	Plattsmouth	(402)	6,887	6,415
68127	Ralston	(402)	6,314	6,236
68661	Schuyler	(402)	5,371	4,052
*69361	Scottsbluff	(308)	14,732	13,711
68434	Seward	(402)	6,319	5,641
69162	Sidney	(308)	6,282	5,959
68776	South Sioux City	(402)	11,925	9,677
68787	Wayne	(402)	5,583	5,142
68467	York	(402)	8,081	7,940

Nevada

ZIP	Place	Area Code	2000	1990
*89005	Boulder City	(702)	14,966	12,567
*89701	Carson City	(775)	52,457	40,443
89403	Dayton (c)	(775)	5,907	2,217
*89801	Elko	(775)	16,708	14,836
—	Enterprise (c)		14,676	6,412
*89406	Fallon	(775)	7,536	6,430
89408	Fernley (c)	(775)	8,543	5,164
89410	Gardnerville Ranchos (c)	(775)	11,054	7,455
*89015	Henderson	(702)	175,381	64,948
*89450	Incline Village-Crystal Bay (c)	(775)	9,952	7,119
*89125	Las Vegas	(702)	478,434	258,877
*89028	Laughlin (c)	(702)	7,076	4,791
89506	Lemmon Valley-Golden Valley (c)	(702)	6,855	—
*89024	Mesquite	(702)	9,389	1,871
89040	Moapa Valley (c)	(702)	5,784	3,444
89191	Nellis AFB (c)	(702)	8,896	8,377
*89030	North Las Vegas	(702)	115,488	47,849
*89041	Pahrump (c)	(775)	24,631	7,424
89109	Paradise (c)	(775)	186,070	124,682
*89501	Reno	(775)	180,480	134,230
89436	Spanish Springs (c)	(775)	9,018	—
*89431	Sparks	(775)	66,346	53,367
89815	Spring Creek (c)	(702)	10,548	5,866
—	Spring Valley (c)	(702)	117,390	51,726
89110	Sunrise Manor (c)	(702)	156,120	95,362
89433	Sun Valley (c)	(775)	19,461	11,391
89101	Winchester (c)	(702)	26,958	23,365
*89445	Winnemucca	(775)	7,174	6,473

New Hampshire (603)

See introductory note.

ZIP	Place	2000	1990
03031	*Amherst*	10,769	9,068
03811	*Atkinson*	6,178	5,188
03825	*Barrington*	7,475	6,164
03110	*Bedford*	18,274	12,563
03220	*Belmont*	6,716	5,796
03570	Berlin	10,331	11,824
03304	*Bow*	7,138	5,500
03743	Claremont	13,151	13,902
*03301	Concord	40,687	36,006
03818	*Conway*	8,604	7,940
03038	Derry (c)	22,661	20,446
03038	*Derry*	34,021	29,603
*03820	Dover	26,884	25,042
03824	Durham (c)	9,024	9,236
03824	*Durham*	12,664	11,818
03042	*Epping*	5,476	5,162
03833	Exeter (c)	9,759	9,556
03833	*Exeter*	14,058	12,481
03835	*Farmington*	5,774	5,739
03235	Franklin	8,405	8,304
03246	*Gilford*	6,803	5,867
03045	*Goffstown*	16,929	14,621
03841	*Hampstead*	8,297	6,732
*03842	Hampton (c)	9,126	7,989
*03842	*Hampton*	14,937	12,278
03755	Hanover Compact (c)	8,162	6,538
03755	*Hanover*	10,850	9,212
03049	*Hollis*	7,015	5,705
03106	*Hooksett*	11,721	9,002
03229	*Hopkinton*	5,399	4,806
03051	Hudson (c)	7,814	7,626
03051	*Hudson*	22,928	19,530
03452	*Jaffrey*	5,476	5,361
03431	Keene	22,563	22,430
03848	*Kingston*	5,862	5,591
*03246	Laconia	16,411	15,743
*03766	Lebanon	12,568	12,183
03052	*Litchfield*	7,360	5,516
03561	*Littleton*	5,845	5,827
03053	Londonderry (c)	11,417	10,114
03053	*Londonderry*	23,236	19,781
*03103	Manchester	107,006	99,332
03253	*Meredith*	5,943	4,837
03054	*Merrimack*	25,119	22,156
03055	Milford (c)	8,293	8,015
03055	*Milford*	13,535	11,795
*03060	Nashua	86,605	79,662
03857	Newmarket (c)	5,124	4,917
03857	*Newmarket*	8,027	7,157
03773	*Newport*	6,269	6,110
03076	*Pelham*	10,914	9,408
03275	*Pembroke*	6,897	6,561
03458	*Peterborough*	5,883	5,239
03102	Pinardville (c)	5,779	4,654
03865	*Plaistow*	7,747	7,316
03264	*Plymouth*	5,892	5,811
*03801	Portsmouth	20,784	25,925
03077	*Raymond*	9,674	8,713
03461	*Rindge*	5,451	4,941
*03867	Rochester	28,461	26,630
03870	*Rye*	5,182	—
03079	*Salem*	28,112	25,746
03873	*Sandown*	5,143	—
03874	*Seabrook*	7,934	6,503
03878	Somersworth	11,477	11,249
03106	South Hooksett (c)	5,282	3,638
03885	*Stratham*	6,355	4,955
03275	Suncook (c)	5,362	5,214
03446	*Swanzey*	6,800	6,236
03281	*Weare*	7,776	6,193
03087	*Windham*	10,709	9,000
03894	*Wolfeboro*	6,083	4,807

New Jersey

Area code (551) overlays area code (201). Area code (848) overlays (732). Area code (862) overlays (973). See introductory note.

ZIP	Place	Area Code	2000	1990
08201	Absecon	(609)	7,638	7,298
07401	Allendale	(201)	6,699	5,900
07712	Asbury Park	(732)	16,930	16,799
08034	Ashland (c)		8,375	—
*08401	Atlantic City	(609)	40,517	37,986
08106	Audubon	(856)	9,182	9,205
07001	Avenel (c)	(732)	17,552	15,504
—	Barclay-Kingston (c)		10,728	—
08007	Barrington	(856)	7,084	6,792
07002	Bayonne	(201)	61,842	61,464
08722	Beachwood	(732)	10,375	9,324
07109	Belleville (c)	(973)	35,928	34,213
*08031	Bellmawr	(856)	11,262	12,603
07719	Belmar	(732)	6,045	5,877
07621	Bergenfield	(201)	26,247	24,458
07922	Berkeley Heights (c)	(908)	13,407	11,980
08009	Berlin	(856)	6,149	5,672
07924	Bernardsville	(908)	7,345	6,597
07003	Bloomfield (c)	(973)	47,683	45,061
07403	Bloomingdale	(973)	7,610	7,530
07603	Bogota	(201)	8,249	7,824
07005	Boonton	(973)	8,496	8,343
08805	Bound Brook	(732)	10,155	9,487
08302	Bridgeton	(856)	22,771	18,942
08203	Brigantine	(609)	12,594	11,354
08015	Browns Mills (c)	(609)	11,257	11,429
07828	Budd Lake (c)	(973)	8,100	7,272
08016	Burlington	(609)	9,736	9,835
07405	Butler	(973)	7,420	7,392
*07006	Caldwell	(973)	7,584	7,542
*08101	Camden	(856)	79,904	87,492
07072	Carlstadt	(201)	5,917	5,510
08069	Carney's Point (c)	(856)	6,914	8,443
07008	Carteret	(732)	20,709	19,025
07009	Cedar Grove (c) (Essex)	(973)	12,300	12,053
07928	Chatham	(973)	8,460	8,007
08002	Cherry Hill Mall (c)	(856)	13,238	—
07066	Clark (c)	(732)/(908)	14,597	14,629
08312	Clayton	(856)	7,139	6,155
07010	Cliffside Park	(201)	23,007	20,393
*07015	Clifton	(973)	78,672	71,984
07624	Closter	(201)	8,383	8,094
08108	Collingswood	(856)	14,326	15,289
07067	Colonia (c)	(732)	17,811	18,238
07016	Cranford (c)	(908)	22,578	22,633
07626	Cresskill	(201)	7,746	7,558
08759	Crestwood Village (c)	(732)	8,392	8,030
08810	Dayton (c)	(732)	6,235	4,321
*07801	Dover	(973)	18,188	15,115
07628	Dumont	(201)	17,503	17,187
08812	Dunellen	(732)	6,823	6,528
08816	East Brunswick (c)	(732)	46,756	43,548
*07019	East Orange	(973)	69,824	73,552
07073	East Rutherford	(201)/(973)	8,716	7,902
*07724	Eatontown	(732)	14,008	13,800
08043	Echelon (c)	(856)	10,440	—
07020	Edgewater	(201)	7,677	5,001
*08818	Edison (c)	(732)/(908)	97,687	88,680
*07207	Elizabeth	(908)	120,568	110,002
07407	Elmwood Park	(201)	18,925	17,623
07630	Emerson	(201)	7,197	6,930
07631	Englewood	(201)	26,203	24,850
07632	Englewood Cliffs	(201)	5,322	5,634
08002	Erlton-Ellisburg (c)		8,168	—
08618	Ewing (c)	(609)	35,707	34,185
07004	Fairfield (Essex) (c)	(973)	7,063	7,615
07704	Fair Haven	(732)	5,937	5,270

ZIP	Place	Area Code	2000	1990
07410	Fair Lawn	(201)/(973)	31,637	30,548
07022	Fairview (Bergen)	(201)	13,255	10,733
07023	Fanwood	(908)	7,174	7,115
08518	Florence-Roebling (c)	(609)	8,200	8,564
07932	Florham Park	(973)	8,857	8,521
08863	Fords (c)	(732)	15,032	14,392
08640	Fort Dix (c)	(609)	7,464	10,205
07024	Fort Lee	(201)	35,461	31,997
07416	Franklin (Sussex)	(973)	5,160	4,977
07417	Franklin Lakes	(201)	10,422	9,873
07728	Freehold	(732)	10,976	10,742
07026	Garfield	(201)	29,786	26,727
08028	Glassboro	(856)	19,068	15,614
07028	Glen Ridge	(973)	7,271	7,076
07452	Glen Rock	(201)	11,546	10,883
08030	Gloucester City	(856)	11,484	12,649
—	Greentree (c)		11,536	—
07093	Guttenberg	(201)	10,807	8,268
*07602	Hackensack	(201)	42,677	37,049
07840	Hackettstown	(908)	10,403	8,120
08033	Haddonfield	(856)	11,659	11,633
08035	Haddon Heights	(856)	7,547	7,860
*07508	Haledon	(973)	8,252	6,951
08037	Hammonton	(609)	12,604	12,208
07029	Harrison	(973)	14,424	13,425
07604	Hasbrouck Heights	(201)	11,662	11,488
*07506	Hawthorne	(973)	18,218	17,084
07422	Highland Lake (c)	(973)	5,051	4,550
08904	Highland Park (Middlesex)	(732)	13,999	13,279
07732	Highlands	(732)	5,097	4,849
08520	Hightstown	(609)	5,216	5,126
07642	Hillsdale	(201)	10,087	9,750
07205	Hillside (c)	(908)/(973)	21,747	21,044
07030	Hoboken	(201)	38,577	33,397
08753	Holiday City-Berkeley (c)	(732)	13,884	14,293
07843	Hopatcong	(973)	15,888	15,586
07111	Irvington (c)	(973)	60,695	59,774
08830	Iselin (c)	(732)	16,698	16,141
08831	Jamesburg	(732)	6,025	5,294
*07303	Jersey City	(201)	240,055	228,517
07734	Keansburg	(732)	10,732	11,069
07032	Kearny	(201)/(973)	40,513	34,874
08824	Kendall Park (c)	(908)	9,006	7,127
07033	Kenilworth	(908)	7,675	7,574
07735	Keyport	(732)	7,568	7,586
07405	Kinnelon	(973)	9,365	8,470
07871	Lake Mohawk (c)	(973)	9,755	8,930
08701	Lakewood (c)	(732)	36,065	26,095
08879	Laurence Harbor (c)	(732)	6,227	6,361
*08733	Leisure Village West-Pine Lake Park (c)	(732)	11,085	10,139
07605	Leonia	(201)	8,914	8,365
07035	Lincoln Park	(973)	10,930	10,978
07738	Lincroft (c)	(732)	6,255	6,193
07036	Linden	(732)/(908)	39,394	36,701
08021	Lindenwold	(856)	17,414	18,734
08221	Linwood	(609)	7,172	6,866
07424	Little Falls (c)	(973)	10,855	11,294
07643	Little Ferry	(201)	10,800	9,989
07739	Little Silver	(732)	6,170	5,721
07039	Livingston (c)	(973)	27,391	26,609
07644	Lodi	(201)/(973)	23,971	22,355
07740	Long Branch	(732)	31,340	28,658
07071	Lyndhurst (c)	(201)	19,383	18,262
08641	McGuire AFB (c)	(609)	6,478	7,580
07940	Madison	(973)	16,530	15,850
08859	Madison Park (c)	(732)	6,929	7,490
08736	Manasquan	(732)	6,310	5,369
08835	Manville	(908)	10,343	10,567
07040	Maplewood (c)	(973)	23,868	21,756
08402	Margate City	(609)	8,193	8,431
08053	Marlton (c)	(856)	10,260	10,228
07747	Matawan	(732)	8,910	9,239
07607	Maywood	(201)	9,523	9,536
07945	Mendham	(973)	5,097	4,890
08619	Mercerville-Hamilton Sq. (c)	(609)	26,419	26,873
08840	Metuchen	(732)	12,840	12,804
08846	Middlesex	(732)	13,717	13,055
07432	Midland Park	(201)	6,947	7,047
07041	Millburn (c)	(973)	19,765	18,630
08850	Milltown (Middlesex)	(732)	7,000	6,968
08332	Millville	(856)	26,847	25,992
*07042	Montclair (c)	(973)	38,977	37,729
07645	Montvale	(201)	7,034	6,946
08057	Moorestown-Lenola (c)	(856)	13,860	13,242
07751	Morganville (c)	(732)	11,255	—
07950	Morris Plains	(973)	5,236	5,219
*07960	Morristown	(973)	18,544	16,189
07092	Mountainside	(908)	6,602	6,657
08087	Mystic Island (c)	(609)	8,694	7,400
07753	Neptune City	(732)	5,218	4,997
*07102	Newark	(973)	273,546	275,221
*08901	New Brunswick	(732)	48,573	41,711
07646	New Milford	(201)	16,400	15,990
07974	New Providence	(908)	11,907	11,439
07860	Newton	(973)	8,244	7,521
07031	North Arlington	(201)	15,181	13,790
08902	North Brunswick Twp. (c)	(732)	36,287	31,287
07006	North Caldwell	(973)	7,375	6,706
08225	Northfield	(609)	7,725	7,305
07508	North Haledon	(973)	7,920	7,987
07060	North Plainfield	(908)	21,103	18,820
07648	Norwood	(201)	5,751	4,858
07110	Nutley (c)	(973)	27,362	27,099
07436	Oakland	(201)	12,466	11,997
*08050	Ocean Acres (c)	(609)	13,155	5,587
08226	Ocean City	(609)	15,378	15,512
07757	Oceanport	(732)	5,807	6,146
08857	Old Bridge (c)	(732)	22,833	22,151
07675	Old Tappan	(201)	5,482	4,254
07649	Oradell	(201)	8,047	8,024
*07051	Orange (c)	(973)	32,868	29,925
07650	Palisades Park	(201)	17,073	14,536
08065	Palmyra	(856)	7,091	7,056
*07652	Paramus	(201)	25,737	25,004
07656	Park Ridge	(201)	8,708	8,102
07055	Passaic	(973)	67,861	58,041
*07510	Paterson	(973)	149,222	140,891
08066	Paulsboro	(856)	6,160	6,577
08110	Pennsauken (c)	(856)	35,737	34,738
08070	Pennsville (c)	(856)	11,657	12,218
*08861	Perth Amboy	(732)	47,303	41,967
08865	Phillipsburg	(908)	15,166	15,757
08021	Pine Hill	(856)	10,880	9,854
08071	Pitman	(856)	9,331	9,365
*07061	Plainfield	(908)	47,829	46,577
08232	Pleasantville	(609)	19,012	16,027
08742	Point Pleasant	(732)	19,306	18,177
08742	Point Pleasant Beach	(732)	5,314	5,112
07442	Pompton Lakes	(973)	10,640	10,539
*08540	Princeton	(609)	14,203	12,016
—	Princeton Meadows (c)	(609)	13,436	—
07508	Prospect Park	(973)	5,779	5,053
07065	Rahway	(732)	26,500	25,325
08057	Ramblewood (c)	(856)	6,003	6,181
07446	Ramsey	(201)	14,351	13,228
—	Ramtown (c)	(732)	5,932	—
08869	Raritan	(908)	6,338	5,798
07701	Red Bank	(732)	11,844	10,636
07657	Ridgefield	(201)	10,830	9,996
07660	Ridgefield Park	(201)	12,873	12,454
*07451	Ridgewood	(201)/(973)	24,936	24,152
07456	Ringwood	(973)	12,396	12,623
07661	River Edge	(201)	10,946	10,603
07675	River Vale (c)	(201)	9,449	9,410
07662	Rochelle Park (c)	(201)	5,528	5,587
07866	Rockaway	(973)	6,473	6,243
07068	Roseland	(973)	5,298	4,847
07203	Roselle	(908)	21,274	20,314
07204	Roselle Park	(908)	13,281	12,805
07760	Rumson	(732)	7,137	6,701
08078	Runnemede	(856)	8,533	9,042
07070	Rutherford	(201)	18,110	17,790
07663	Saddle Brook (c)	(201)/(973)	13,155	13,296
08079	Salem	(856)	5,857	6,883
*08872	Sayreville	(732)	40,377	34,998
07076	Scotch Plains (c)	(732)/(908)	22,732	21,150
*07094	Secaucus	(201)	15,931	14,061
08083	Somerdale	(856)	5,192	5,440
*08873	Somerset (c)	(732)	23,040	22,070
08244	Somers Point	(609)	11,614	11,216
08876	Somerville	(908)	12,423	11,632
08879	South Amboy	(732)	7,913	7,851
07079	South Orange (c)	(973)	16,964	16,390
07080	South Plainfield	(732)/(908)	21,810	20,489
08882	South River	(732)	15,322	13,692
08884	Spotswood	(732)	7,880	7,983
—	Springdale (c)		14,409	—
07081	Springfield (c)	(908)/(973)	14,429	13,420
07762	Spring Lake Heights	(732)	5,227	5,341
08084	Stratford	(856)	7,271	7,614
07747	Strathmore (c)	(732)	6,740	7,060
07876	Succasunna-Kenvil (c)	(201)	12,569	11,781
*07901	Summit	(908)	21,131	19,757
07666	Teaneck (c)	(201)	39,260	37,825
07670	Tenafly	(201)	13,806	13,326
07724	Tinton Falls	(732)	15,053	12,361
*08753	Toms River (c)	(732)	86,327	7,524
*07512	Totowa	(973)	9,892	10,177
*08650	Trenton	(609)	85,403	88,675
08520	Twin Rivers (c)	(609)	7,422	7,715
07083	Union (Union) (c)	(908)	54,405	50,024
07735	Union Beach	(732)	6,649	6,156
07087	Union City	(201)	67,088	58,012
07458	Upper Saddle River	(201)	7,741	7,198
08406	Ventnor City	(609)	12,910	11,005
07044	Verona (c)	(973)	13,533	13,597
08251	Villas (c)	(609)	9,064	8,136
*08360	Vineland	(856)	56,271	54,780
07463	Waldwick	(201)	9,622	9,757
07057	Wallington	(201)/(973)	11,583	10,828
07465	Wanaque	(201)/(973)	10,266	9,711
07882	Washington	(908)	6,712	6,474

ZIP	Place	Area Code	2000	1990
07675	Washington Twp. (Bergen) (c)	(201)	8,938	9,245
07060	Watchung	(908)	5,613	5,110
*07470	Wayne (c)	(973)	54,069	47,025
07007	West Caldwell (c)	(973)	11,233	10,422
*07091	Westfield	(732)/(908)	29,644	28,870
07728	West Freehold (c)	(732)	12,498	11,166
07764	West Long Branch	(732)	8,258	7,690
07480	West Milford (c)	(973)	26,410	25,430
07093	West New York	(201)	45,768	38,125
07052	West Orange (c)	(973)	44,943	39,103
07424	West Paterson	(973)	10,987	10,982
07675	Westwood	(201)	10,999	10,446
07885	Wharton	(973)	6,298	5,405
08610	White Horse (c)	(609)	9,373	9,397
07886	White Meadow Lake (c)	(973)	9,052	8,002
08260	Wildwood	(609)	5,436	4,484
08094	Williamstown (c)	(856)	11,812	10,891
07095	Woodbridge (c)	(732)	18,309	17,434
08096	Woodbury	(856)	10,307	10,904
07675	Woodcliff Lake	(201)	5,745	5,303
07075	Wood-Ridge	(201)/(973)	7,644	7,506
07481	Wyckoff (c)	(201)	16,508	15,372
08620	Yardville-Groveville (c)	(609)	9,208	9,248
07726	Yorketown (c)	(609)	6,712	6,313

New Mexico

Area code (575) goes into effect Mar. 3, 2002. Before then use area code (505). See introductory note.

ZIP	Place	Area Code	2000	1990
*88310	Alamogordo	(505)	35,582	27,596
*87101	Albuquerque	(575)	448,607	384,915
88021	Anthony (c)	(505)	7,904	5,160
*88210	Artesia	(505)	10,692	10,610
87410	Aztec	(505)	6,378	5,480
87002	Belen	(575)	6,901	6,547
87004	Bernalillo	(575)	6,611	5,864
87413	Bloomfield	(505)	6,417	5,214
*88220	Carlsbad	(505)	25,625	24,952
88021	Chaparral (c)	(505)	6,117	2,962
*88101	Clovis	(505)	32,667	30,954
87048	Corrales	(575)	7,334	5,453
*88030	Deming	(505)	14,116	11,422
—	El Cerro-Monterey Park (c)	(575)	5,483	—
—	Eldorado at Santa Fe (c)	(575)	5,799	2,260
*87532	Espanola	(505)	9,688	8,389
*87401	Farmington	(505)	37,844	33,997
*87301	Gallup	(505)	20,209	19,157
87020	Grants	(505)	8,806	8,626
*88240	Hobbs	(505)	28,657	29,121
87417	Kirtland (c)	(505)	6,190	3,552
*88001	Las Cruces	(505)	74,267	62,360
87701	Las Vegas	(505)	14,565	14,753
87544	Los Alamos (c)	(575)	11,909	11,455
87002	Los Chaves (c)	(575)	5,033	3,872
87031	Los Lunas	(575)	10,034	6,013
87107	Los Ranchos de Albuquerque	(575)	5,092	5,075
88260	Lovington	(505)	9,471	9,322
87107	North Valley (c)	(575)	11,923	12,507
88130	Portales	(505)	11,131	10,690
87740	Raton	(505)	7,282	7,372
*87124	Rio Rancho	(575)	51,765	32,512
*88201	Roswell	(505)	45,293	44,260
*88345	Ruidoso	(505)	7,698	4,600
*87501	Santa Fe	(575)	62,203	56,537
87420	Shiprock (c)	(505)	8,156	7,687
*88061	Silver City	(505)	10,545	10,683
87801	Socorro	(505)	8,877	8,159
87105	South Valley (c)	(575)	39,060	35,701
88063	Sunland Park	(505)	13,309	8,179
87901	Truth or Consequences	(505)	7,289	6,221
88401	Tucumcari	(505)	5,989	6,827
87544	White Rock (c)	(575)	6,045	6,192
87327	Zuni Pueblo (c)	(505)	6,367	5,857

WORLD ALMANAC EDITORS' PICKS

The World Almanac staff ranked the following as the oddest or funniest names for U.S. towns. (Except for the first, they do not appear in the Census Bureau list of Places of 5,000 or More.)

1. Truth or Consequences, NM
2. Intercourse, PA
3. Hot Coffee, MS
4. Chicken, AK
5. Rambo Riviera, AR

New York

Area code (347) overlays area code (718).
Area codes (646) and (917) overlay (212). See introductory note.

ZIP	Place	Area Code	2000	1990
10901	Airmont	(845)	7,799	7,674
*12201	Albany	(518)	95,658	100,031
11507	Albertson (c)	(516)	5,200	5,166
14411	Albion	(585)	1,438	5,863
*11701	Amityville	(516)/(631)	9,441	9,286
12010	Amsterdam	(518)	18,355	20,714
12603	Arlington (c)	(845)	12,481	11,948
*13021	Auburn	(315)	28,574	31,258
11702	Babylon	(631)	12,615	12,249
11510	Baldwin (c)	(516)	23,455	22,719
11510	Baldwin Harbor (c)	(516)	8,147	7,899
13027	Baldwinsville	(315)	7,053	6,591
12020	Ballston Spa	(518)	5,556	5,194
*14020	Batavia	(585)	16,256	16,310
14810	Bath	(607)	5,641	5,801
11705	Bayport (c)	(631)	8,662	7,702
11706	Bay Shore (c)	(631)	23,852	21,279
11709	Bayville	(516)	7,135	7,193
11751	Baywood (c)	(631)	7,571	7,351
12508	Beacon	(845)	13,808	13,243
11710	Bellmore (c)	(516)	16,441	16,438
11714	Bethpage (c)	(516)	16,543	15,761
*13902	Binghamton	(607)	47,380	53,008
10913	Blauvelt (c)	(845)	5,207	4,838
11716	Bohemia (c)	(631)	9,871	9,556
11717	Brentwood (c)	(631)	53,917	45,218
10510	Briarcliff Manor	(914)	7,696	7,070
14610	Brighton (c)	(585)	35,584	34,455
14420	Brockport	(585)	8,103	8,749
10708	Bronxville	(914)	6,543	6,028
*14240	Buffalo	(716)	292,648	328,175
11933	Calverton (c)	(631)	5,704	4,759
*14424	Canandaigua	(585)	11,264	10,725
13617	Canton	(315)	5,882	6,379
11514	Carle Place (c)	(516)	5,247	5,107
10512	Carmel Hamlet (c)	(845)	5,650	4,800
11516	Cedarhurst	(516)	6,164	5,716
11720	Centereach (c)	(631)	27,285	26,720
11934	Center Moriches (c)	(631)	6,655	5,987
11721	Centerport (Suffolk) (c)	(631)	5,446	5,333
11722	Central Islip (c)	(516)	31,950	26,028
10514	Chappaqua (c)	(914)	9,468	—
14225	Cheektowaga (c)	(716)	79,988	84,387
10977	Chestnut Ridge	(845)	7,829	7,517
12047	Cohoes	(518)	15,521	16,825
12205	Colonie	(518)	7,916	8,019
11725	Commack (c)	(631)	36,367	36,124
10920	Congers (c)	(845)	8,303	8,003
11726	Copiague (c)	(631)	21,922	20,769
11727	Coram (c)	(631)	34,923	30,111
14830	Corning	(607)	10,842	11,938
13045	Cortland	(607)	18,740	19,801
*10520	Croton-on-Hudson	(914)	7,606	7,018
11729	Deer Park (c)	(631)	28,316	28,840
12054	Delmar (c)	(518)	8,292	8,360
14043	Depew	(716)	16,629	17,673
11746	Dix Hills (c)	(631)	26,024	25,849
10522	Dobbs Ferry	(914)	10,622	9,940
14048	Dunkirk	(716)	13,131	13,989
14052	East Aurora(585)/	(716)	6,673	6,647
10709	Eastchester (c)	(914)	18,564	18,537
11735	East Farmingdale	(516)/(631)	5,400	4,510
12302	East Glenville (c)	(518)	6,064	6,518
11576	East Hills	(516)	6,842	6,746
11730	East Islip (c)	(631)	14,078	14,325
11758	East Massapequa (c)	(516)	19,565	19,550
11554	East Meadow (c)	(516)	37,461	36,909
11731	East Northport (c)	(631)	20,845	20,411
11772	East Patchogue (c)	(631)	20,824	20,195
14445	East Rochester	(585)	6,650	6,932
11518	East Rockaway	(516)	10,414	10,152
11786	East Shoreham (c)	(631)	5,809	5,461
*14901	Elmira	(607)	30,940	33,724
11003	Elmont (c)	(516)	32,657	28,612
11731	Elwood (c)	(631)	10,916	10,916
*13760	Endicott	(607)	13,038	13,531
13762	Endwell (c)	(607)	11,706	12,602
13219	Fairmount (c)	(315)	10,795	12,266
14450	Fairport	(585)	5,740	5,943
—	Fairview (c)	(845)	5,421	4,811
11735	Farmingdale	(516)	8,399	8,022
11738	Farmingville (c)	(631)	16,458	14,842
*11001	Floral Park	(516)	15,967	15,947
13603	Fort Drum (c)	(315)	12,123	11,578
11768	Fort Salonga (c)	(631)	9,634	9,176
11010	Franklin Square (Nassau) (c)	(516)	29,342	28,205
14063	Fredonia	(716)	10,706	10,436
11520	Freeport	(516)	43,783	39,894
13069	Fulton	(315)	11,855	12,929
*11530	Garden City	(516)	21,672	21,675
11040	Garden City Park (c)	(516)	7,554	7,437
14624	Gates-North Gates (c)	(585)	15,138	14,995
14454	Geneseo	(585)	7,579	7,187
14456	Geneva	(315)	13,617	14,143
11542	Glen Cove	(516)	26,622	24,149
12801	Glens Falls	(518)	14,354	15,023

ZIP	Place	Area Code	2000	1990
12801	Glens Falls North (c)	(518)	8,061	7,978
12078	Gloversville	(518)	15,413	16,656
10924	Goshen	(845)	5,676	5,255
13642	Gouverneur	(315)	7,418	4,604
*11021	Great Neck	(516)	9,538	8,745
11020	Great Neck Plaza	(516)	6,433	5,897
14616	Greece (c)	(585)	14,614	15,632
11740	Greenlawn (c)	(631)	13,286	13,208
*10583	Greenville (Westchester) (c)	(914)	8,648	9,528
14075	Hamburg	(716)	10,116	10,442
11946	Hampton Bays (c)	(631)	12,236	7,893
10528	Harrison	(914)	24,154	23,308
10530	Hartsdale (c)	(914)	9,830	9,587
10706	Hastings-on-Hudson	(914)	7,648	8,000
*11788	Hauppauge (c)	(631)	20,100	19,750
10927	Haverstraw	(845)	10,117	9,438
10532	Hawthorne (c)	(845)	5,083	4,764
*11551	Hempstead	(516)	56,554	45,982
13350	Herkimer	(315)	7,498	7,945
11557	Hewlett (c)	(516)	7,060	6,620
*11802	Hicksville (c)	(516)	41,260	40,174
12528	Highland (c)	(845)	5,060	4,492
10977	Hillcrest (c)	(845)	7,106	6,447
14468	Hilton	(585)	5,856	5,216
11741	Holbrook (c)	(631)	27,512	25,273
11742	Holtsville (c)	(631)	17,006	14,972
14843	Hornell	(607)	9,019	9,877
*14845	Horseheads	(607)	6,452	6,802
12534	Hudson	(518)	7,524	8,034
12839	Hudson Falls	(518)	6,927	7,651
11743	Huntington (c)	(631)	18,403	18,243
11746	Huntington Station (c)	(631)	29,910	28,247
13357	Ilion	(315)	8,610	8,888
11096	Inwood (c)	(516)	9,325	7,767
14617	Irondequoit (c)	(585)	52,354	52,322
10533	Irvington	(914)	6,631	6,348
11751	Islip (c)	(631)	20,575	18,924
11752	Islip Terrace (c)	(631)	5,641	5,530
*14850	Ithaca	(607)	29,287	29,541
*14702	Jamestown	(716)	31,730	34,681
10535	Jefferson Valley-Yorktown (c)	(914)	14,891	14,118
11753	Jericho (Nassau) (c)	(516)	13,045	13,141
13790	Johnson City	(607)	15,535	16,578
12095	Johnstown	(518)	8,511	9,058
14217	Kenmore	(716)	16,426	17,180
11754	Kings Park (c)	(631)	16,146	17,773
11024	Kings Point	(516)	5,076	4,843
*12401	Kingston	(845)	23,456	23,095
10950	Kiryas Joel	(845)	13,138	7,437
14218	Lackawanna	(716)	19,064	20,585
10512	Lake Carmel (c)	(845)	8,663	8,489
11755	Lake Grove	(631)	10,250	9,612
10547	Lake Mohegan (c)	(914)	5,979	—
11779	Lake Ronkonkoma (c)	(631)	19,701	18,997
11552	Lakeview (c)	(516)	5,607	5,476
14086	Lancaster	(716)	11,188	11,940
10538	Larchmont	(914)	6,485	6,181
11559	Lawrence	(516)	6,522	6,513
11756	Levittown (c)	(516)	53,067	53,286
11757	Lindenhurst	(631)	27,819	26,879
13365	Little Falls	(315)	5,188	5,829
*14094	Lockport	(716)	22,279	24,426
11561	Long Beach	(516)	35,462	33,510
11563	Lynbrook	(516)	19,911	19,208
10541	Mahopac (c)	(845)	8,478	7,755
12953	Malone	(518)	6,075	6,777
11565	Malverne	(516)	8,934	9,054
10543	Mamaroneck	(914)	18,752	17,325
11030	Manhasset (c)	(516)	8,362	7,718
11050	Manorhaven	(516)	6,138	5,672
11949	Manorville (c)	(631)	11,131	6,198
11758	Massapequa (c)	(516)	22,652	22,018
11762	Massapequa Park	(516)	17,499	18,044
13662	Massena	(315)	11,209	11,716
11950	Mastic (c)	(631)	15,436	13,778
11951	Mastic Beach (c)	(631)	11,543	10,293
13211	Mattydale (c)	(315)	6,367	6,418
—	Mechanicstown (c)	(845)	6,061	—
12118	Mechanicville	(518)	5,019	5,249
11763	Medford (c)	(631)	21,985	21,274
14103	Medina	(585)/(716)	6,415	6,686
11747	Melville (c)	(631)	14,533	12,586
11566	Merrick (c)	(516)	22,764	23,042
11953	Middle Island (c)	(631)	9,702	7,848
*10940	Middletown	(845)	25,388	24,160
11764	Miller Place (c)	(631)	10,580	9,315
11501	Mineola	(516)	19,234	19,005
10950	Monroe	(845)	7,780	6,672
10952	Monsey (c)	(845)	14,504	13,986
12701	Monticello	(845)	6,512	6,597
10970	Mount Ivy (c)	(845)	6,536	6,013
10549	Mount Kisco	(914)	9,983	9,108
11766	Mount Sinai (c)	(631)	8,734	8,023
*10551	Mount Vernon	(914)	68,381	67,153

ZIP	Place	Area Code	2000	1990
12590	Myers Corner (c)	(845)	5,546	5,599
10954	Nanuet (c)	(845)	16,707	14,065
11767	Nesconset (c)	(631)	11,992	10,712
14513	Newark	(315)	9,682	9,849
*12550	Newburgh	(845)	28,259	26,454
11590	New Cassel (c)	(516)	13,298	10,257
10956	New City (c)	(845)	34,038	33,673
*11040	New Hyde Park	(516)	9,523	9,728
12561	New Paltz	(845)	6,034	5,470
*10802	New Rochelle	(914)	72,182	67,265
*12550	New Windsor (c)	(845)	9,077	8,898
*10001	New York	(212)/(718)	8,008,278	7,322,564
*14302	Niagara Falls	(716)	55,593	61,840
11701	North Amityville (c)	(631)	16,572	13,849
11703	North Babylon (c)	(631)	17,877	18,081
11706	North Bay Shore (c)	(631)	14,992	12,799
11710	North Bellmore (c)	(516)	20,079	19,707
11713	North Bellport (c)	(631)	9,007	8,182
11757	North Lindenhurst (c)	(631)	11,767	10,563
11758	North Massapequa (c)	(516)	19,152	19,365
11566	North Merrick (c)	(516)	11,844	12,113
11040	North New Hyde Park (c)	(516)	14,542	14,359
11772	North Patchogue (c)	(631)	7,825	7,374
11768	Northport	(631)	7,606	7,572
13212	North Syracuse	(315)	6,862	7,363
14120	North Tonawanda	(716)	33,262	34,989
11580	North Valley Stream (c)	(516)	15,789	14,574
11793	North Wantagh (c)	(516)	12,156	12,276
13815	Norwich	(607)	7,355	7,613
10960	Nyack	(845)	6,737	6,558
11769	Oakdale (c)	(631)	8,075	7,875
11572	Oceanside (c)	(516)	32,733	32,423
13669	Ogdensburg	(315)	12,364	13,521
11804	Old Bethpage (c)	(516)	5,400	5,610
14760	Olean	(585)/(716)	15,347	16,946
13421	Oneida	(315)	10,987	10,850
13820	Oneonta	(607)	13,292	13,954
12550	Orange Lake (c)	(845)	6,085	5,196
10562	Ossining	(914)	24,010	22,582
13126	Oswego	(315)	17,954	19,195
11771	Oyster Bay (c)	(516)	6,826	6,687
11772	Patchogue	(631)	11,919	11,060
10965	Pearl River (c)	(845)	15,553	15,314
10566	Peekskill	(914)	22,441	19,536
10803	Pelham	(914)	6,400	5,443
10803	Pelham Manor	(914)	5,466	6,413
14527	Penn Yan	(315)	5,219	5,248
11714	Plainedge (c)	(516)	9,195	8,739
11803	Plainview (c)	(516)	25,637	26,207
*12901	Plattsburgh	(518)	18,816	21,255
10570	Pleasantville	(914)	7,172	6,592
10573	Port Chester	(914)	27,867	24,728
11777	Port Jefferson	(631)	7,837	7,455
11776	Port Jefferson Station (c)	(631)	7,527	7,232
12771	Port Jervis	(845)	8,860	9,060
11050	Port Washington (c)	(516)	15,215	15,387
13676	Potsdam	(315)	9,425	10,251
*12601	Poughkeepsie	(845)	29,871	28,844
12144	Rensselaer	(518)	7,761	8,255
11961	Ridge (c)	(631)	13,380	11,734
11901	Riverhead (c)	(631)	10,513	8,814
*14692	Rochester	(585)	219,773	230,356
*11571	Rockville Centre	(516)	24,568	24,727
11778	Rocky Point (c)	(631)	10,185	8,596
*13440	Rome	(315)	34,950	44,350
11779	Ronkonkoma (c)	(631)	20,029	20,391
11575	Roosevelt (c)	(516)	15,854	15,030
11577	Roslyn Heights (c)	(516)	6,295	6,405
12303	Rotterdam (c)	(518)	20,536	21,228
10580	Rye	(914)	14,955	14,936
10573	Rye Brook	(914)	8,602	7,765
11780	Saint James (c)	(631)	13,268	12,703
14779	Salamanca	(716)	6,097	6,566
13454	Salisbury (c)	(315)	12,341	12,226
12983	Saranac Lake	(518)	5,041	5,377
12866	Saratoga Springs	(518)	26,186	25,001
11782	Sayville (c)	(631)	16,735	16,550
10583	Scarsdale	(914)	17,823	16,987
*12301	Schenectady	(518)	61,821	65,566
10940	Scotchtown (c)	(845)	8,954	8,765
12302	Scotia	(518)	7,957	7,359
11579	Sea Cliff	(516)	5,066	5,054
11783	Seaford (c)	(516)	15,791	15,597
11507	Searingtown (c)	(516)	5,034	5,020
11784	Selden (c)	(631)	21,861	20,608
13148	Seneca Falls	(315)	6,861	7,370
11733	Setauket-East Setauket (c)	(516)	15,931	13,634
11967	Shirley (Suffolk) (c)	(631)	25,395	22,936
10591	Sleepy Hollow[1]	(914)	9,212	8,152
11787	Smithtown (c)	(631)	26,901	25,638
13209	Solvay	(315)	6,845	6,717
11789	Sound Beach (c)	(631)	9,807	9,102
11735	South Farmingdale (c)	(516)	15,061	15,377
14850	South Hill (c)	(607)	6,003	5,423

ZIP	Place	Area Code	2000	1990
11746	South Huntington (c)	(631)	9,465	9,624
14094	South Lockport (c)	(716)	8,552	7,112
11971	Southold (c)	(631)	5,465	5,192
14904	Southport (c)	(607)	7,396	7,753
11581	South Valley Stream (c)	(516)	5,638	5,328
10977	Spring Valley	(845)	25,464	21,802
*11790	Stony Brook (c)	(631)	13,727	13,726
10980	Stony Point (c) (Rockland)	(845)	11,744	10,587
10901	Suffern	(845)	11,006	11,055
11791	Syosset (c)	(516)	18,544	18,967
*13220	Syracuse	(315)	147,306	163,860
10983	Tappan (c)	(845)	6,757	6,867
10591	Tarrytown	(914)	11,090	10,739
11776	Terryville (c)	(631)	10,589	10,275
10594	Thornwood (c)	(914)	5,980	7,025
*14150	Tonawanda	(716)	16,136	17,284
*14150	Tonawanda (c)	(716)	61,729	65,284
*12180	Troy	(518)	49,170	54,269
10707	Tuckahoe	(914)	6,211	6,302
11553	Uniondale (c)	(516)	23,011	20,328
*13504	Utica	(315)	60,651	68,637
10595	Valhalla (c)	(914)	5,379	—
10989	Valley Cottage (c)	(845)	9,269	9,007
*11582	Valley Stream	(516)	36,368	33,946
—	Viola (c)		5,931	4,504
11792	Wading River (c)	(631)	6,668	5,317
12586	Walden	(845)	6,164	5,836
11793	Wantagh (c)	(516)	18,971	18,567
10990	Warwick	(845)	6,412	5,984
10992	Washingtonville	(845)	5,851	4,906
13165	Waterloo	(315)	5,111	5,116
*13601	Watertown	(315)	26,705	29,429
12189	Watervliet	(518)	10,207	11,061
14580	Webster	(585)	5,216	5,464
14895	Wellsville	(585)	5,171	5,241
*11704	West Babylon (c)	(631)	43,452	42,410
11590	Westbury (Nassau)	(516)	14,263	13,060
14905	West Elmira (c)	(607)	5,136	5,218
12801	West Glens Falls (c)	(518)	6,721	5,964
10993	West Haverstraw	(845)	10,295	9,183
11552	West Hempstead (c)	(516)	18,713	17,689
11743	West Hills (c)	(631)	5,607	5,849
11795	West Islip (c)	(631)	28,907	28,419
12203	Westmere (c)	(518)	7,188	6,750
*10996	West Point (c)	(845)	7,138	8,024
11796	West Sayville (c)	(631)	5,003	4,680
14224	West Seneca (c)	(716)	45,943	47,866
13219	Westvale (c)	(315)	5,166	5,952
11798	Wheatley Heights (c)	(631)	5,013	5,027
*10602	White Plains	(914)	53,077	48,718
14231	Williamsville	(716)	5,573	5,583
11596	Williston Park	(516)	7,261	7,516
11797	Woodbury (c)	(516)	9,010	8,008
11598	Woodmere (c)	(516)	16,447	15,578
11798	Wyandach (c)	(631)	10,546	8,950
11980	Yaphank (c)	(631)	5,025	4,637
*10702	Yonkers	(914)	196,086	188,082
10598	Yorktown Heights (c)	(914)	7,972	7,690

(1) North Tarrytown changed its name to Sleepy Hollow on Dec. 12, 1996.

North Carolina

Area code (980) overlays area code (704). Area code (984) overlays (919). See introductory note.

ZIP	Place	Area Code	2000	1990
*28001	Albemarle	(704)	15,680	14,940
27502	Apex	(919)	20,212	4,789
27263	Archdale	(336)	9,014	6,975
*27203	Asheboro	(336)	21,672	16,362
*28802	Asheville	(828)	68,889	63,379
28012	Belmont	(704)	8,705	8,434
28016	Bessemer City	(704)	5,119	4,698
28711	Black Mountain	(828)	7,511	7,156
28607	Boone	(828)	13,472	12,949
28712	Brevard	(828)	6,789	5,452
*27215	Burlington	(336)	44,917	39,498
27509	Butner (c)	(919)	5,792	4,679
27510	Carrboro	(919)	16,782	12,134
*27511	Cary	(919)	94,536	44,394
*27514	Chapel Hill	(919)	48,715	38,711
*28204	Charlotte	(704)	540,828	419,558
28021	Cherryville	(704)	5,361	4,756
27520	Clayton	(919)	6,973	4,756
27012	Clemmons	(336)	13,827	5,982
*28328	Clinton	(910)	8,600	8,385
*28025	Concord	(704)	55,977	29,591
28613	Conover	(828)	6,604	5,311
28031	Cornelius	(704)	11,969	2,581
28036	Davidson	(704)	7,139	4,046
*28334	Dunn	(910)	9,196	9,258
*27701	Durham	(919)	187,035	138,894
*27288	Eden	(336)	15,908	15,238
27932	Edenton	(252)	5,394	5,268
*27909	Elizabeth City	(252)	17,188	16,087
27244	Elon College	(336)	6,738	4,394
*28302	Fayetteville	(910)	121,015	75,850
28043	Forest City	(828)	7,549	7,475
28307	Fort Bragg (c)	(910)	29,183	34,744
27526	Fuquay-Varina	(919)	7,898	4,447
27529	Garner	(919)	17,757	14,716
*28052	Gastonia	(704)	66,277	54,725
*27530	Goldsboro	(919)	39,043	40,736
27253	Graham	(336)	12,833	10,368
*27420	Greensboro	(336)	223,891	185,125
*27834	Greenville	(252)	60,476	46,274
28540	Half Moon (c)	(910)	6,645	6,306
28345	Hamlet	(910)	6,018	6,722
28532	Havelock	(252)	22,442	20,300
27536	Henderson	(252)	16,095	15,655
*28739	Hendersonville	(828)	10,420	7,284
*28603	Hickory	(828)	37,222	28,474
*27260	High Point	(336)	85,839	69,428
27278	Hillsborough	(919)	5,446	4,263
27540	Holly Springs	(919)	9,192	1,203
28348	Hope Mills	(910)	11,237	8,272
*28070	Huntersville	(704)	24,960	3,014
28079	Indian Trail	(704)	11,905	1,942
*28540	Jacksonville	(910)	66,715	78,031
—	James City (c)	(252)	5,420	4,279
*28081	Kannapolis	(704)	36,910	31,592
*27284	Kernersville	(336)	17,126	11,860
27948	Kill Devil Hills	(252)	5,897	4,238
27021	King	(336)	5,952	4,059
—	Kings Grant (c)		7,738	—
28086	Kings Mountain	(704)	9,693	8,768
*28502	Kinston	(252)	23,688	25,295
27545	Knightdale	(919)	5,958	1,884
*28352	Laurinburg	(910)	15,874	16,131
28645	Lenoir	(828)	16,793	16,337
27023	Lewisville	(336)	8,826	6,433
*27292	Lexington	(336)	19,953	16,583
*28092	Lincolnton	(704)	9,965	6,955
*28358	Lumberton	(910)	20,795	18,656
28403	Masonboro (c)	(910)	11,812	7,010
*28105	Matthews	(704)	22,127	13,756
27302	Mebane	(919)	7,284	4,754
28227	Mint Hill	(704)	14,922	13,637
*28110	Monroe	(704)	26,228	18,623
*28115	Mooresville	(704)	18,823	9,563
28557	Morehead City	(252)	7,691	6,473
*28655	Morganton	(828)	17,310	15,085
27560	Morrisville	(919)	5,208	1,022
27030	Mount Airy	(336)	8,484	7,156
28120	Mount Holly	(704)	9,618	7,710
—	Murraysville (c)		7,279	—
—	Myrtle Grove (c)		7,125	4,275
*28562	New Bern	(252)	23,128	20,728
28658	Newton	(828)	12,560	11,134
*28465	Oak Island	(910)	6,571	—
—	Ogden (c)		5,481	3,228
27565	Oxford	(919)	8,338	7,965
*28374	Pinehurst	(910)	9,706	5,825
28399	Piney Green (c)	(910)	11,658	8,999
*27611	Raleigh	(919)	276,093	218,859
*27320	Reidsville	(336)	14,485	14,085
27870	Roanoke Rapids	(252)	16,957	15,722
*28379	Rockingham	(910)	9,672	9,399
*27801	Rocky Mount	(252)	55,893	53,078
27573	Roxboro	(336)	8,696	7,332
28704	Royal Pines (c)		5,334	4,418
28601	Saint Stephens (c)	(828)	9,439	8,734
*28144	Salisbury	(704)	26,462	23,626
*27330	Sanford	(919)	23,220	18,881
27576	Selma	(919)	5,914	4,600
*28150	Shelby	(704)	19,477	15,460
27344	Siler City	(919)	6,966	4,808
—	Silver Lake (c)		5,788	4,071
27577	Smithfield	(919)	11,510	10,180
*28387	Southern Pines	(910)	10,918	9,213
28052	South Gastonia (c)	(704)	5,433	5,487
28390	Spring Lake	(910)	8,098	7,552
*28677	Statesville	(704)	23,320	20,647
27358	Summerfield	(336)	7,018	2,051
27886	Tarboro	(252)	11,138	11,037
*27360	Thomasville	(336)	19,788	15,915
27370	Trinity	(336)	6,690	5,469
*27587	Wake Forest	(919)	12,588	5,832
27889	Washington	(252)	9,583	9,160
28786	Waynesville	(828)	9,232	7,282
28104	Weddington	(704)	6,696	3,803
28472	Whiteville	(910)	5,148	5,340
27892	Williamston	(252)	5,843	5,870
*28402	Wilmington	(910)	75,838	55,530
*27893	Wilson	(252)	44,405	38,400
*27102	Winston-Salem	(336)	185,776	162,292

North Dakota (701)

ZIP	Place	2000	1990
*58501	Bismarck	55,532	49,272
58301	Devils Lake	7,222	7,782
*58601	Dickinson	16,010	16,097
*58102	Fargo	90,599	74,084

ZIP	Place	2000	1990
*58201	Grand Forks	49,321	49,417
*58401	Jamestown	15,527	15,571
58554	Mandan	16,718	15,177
*58701	Minot	36,567	34,544
*58701	Minot AFB (c)	7,599	9,095
58072	Valley City	6,826	7,163
*58075	Wahpeton	8,586	8,751
58078	West Fargo	14,940	12,287
*58801	Williston	12,512	13,136

Ohio

Area code (234) overlays area code (330). Area code (380) overlays (614). Area code (567) overlays (419). See introductory note.

ZIP	Place	Area Code	2000	1990
45810	Ada	(419)	5,582	5,428
*44309	Akron	(330)	217,074	223,019
44601	Alliance	(330)	23,253	23,376
44001	Amherst	(440)	11,797	10,332
44805	Ashland	(419)	21,249	20,079
*44004	Ashtabula	(440)	20,962	21,633
45701	Athens	(740)	21,342	21,265
44202	Aurora	(330)	13,556	9,192
44515	Austintown (c)	(330)	31,627	32,371
44011	Avon	(440)	11,446	7,337
44012	Avon Lake	(440)	18,145	15,066
44203	Barberton	(330)	27,899	27,623
44140	Bay Village	(440)	16,087	17,000
44122	Beachwood	(216)	12,186	10,644
45434	Beavercreek	(937)	37,984	33,626
—	Beckett Ridge (c)		8,663	4,505
44146	Bedford	(216)/(440)	14,214	14,822
44146	Bedford Heights	(216)/(440)	11,375	12,131
45305	Bellbrook	(937)	7,009	6,511
43311	Bellefontaine	(937)	13,069	12,126
44811	Bellevue	(419)	8,193	8,157
45714	Belpre	(740)	6,660	6,796
44017	Berea	(440)	18,970	19,051
43209	Bexley	(614)	13,203	13,088
43004	Blacklick Estates (c)	(614)	9,518	10,080
45242	Blue Ash	(513)	12,513	11,923
44513	Boardman (c)	(330)	37,215	38,596
43402	Bowling Green	(419)	29,636	28,303
44141	Brecksville	(440)	13,382	11,818
45211	Bridgetown North (c)	(513)	12,569	11,748
44147	Broadview Heights	(440)	15,967	12,219
44144	Brooklyn	(216)	11,586	11,706
44142	Brook Park	(216)/(440)	21,218	22,865
45309	Brookville	(937)	5,289	4,621
44212	Brunswick	(330)	33,388	28,218
43506	Bryan	(419)	8,333	8,348
44820	Bucyrus	(419)	13,224	13,496
43725	Cambridge	(740)	11,520	11,748
44405	Campbell	(330)	9,460	10,038
44614	Canal Fulton	(330)	5,061	4,157
44406	Canfield	(330)	7,374	5,409
*44711	Canton	(330)	80,806	84,161
45005	Carlisle	(937)	5,121	4,872
45822	Celina	(419)	10,303	9,945
*45441	Centerville (Montgomery)	(937)	23,024	21,082
44024	Chardon	(440)	5,156	4,446
45211	Cheviot	(513)	9,015	9,616
45601	Chillicothe	(740)	21,796	21,923
*45202	Cincinnati	(513)	331,285	364,114
43113	Circleville	(740)	13,485	11,666
45315	Clayton	(937)	13,347	713
*44101	Cleveland	(216)	478,403	505,616
44118	Cleveland Heights	(216)	49,958	54,052
43410	Clyde	(419)	6,064	6,087
44408	Columbiana	(330)	5,635	4,961
*43216	Columbus	(614)	711,470	632,945
44030	Conneaut	(440)	12,485	13,241
44410	Cortland	(330)	6,830	5,652
43812	Coshocton	(740)	11,682	12,193
45238	Covedale (c)	(513)	6,360	6,669
44827	Crestline	(419)	5,088	4,934
*44222	Cuyahoga Falls	(330)	49,374	48,950
*45401	Dayton	(937)	166,179	182,011
45236	Deer Park	(513)	5,982	6,181
43512	Defiance	(419)	16,465	16,787
43015	Delaware	(740)	25,243	19,966
45833	Delphos	(419)	6,944	7,093
45247	Dent (c)	(513)	7,612	6,416
44622	Dover (Tuscarawas)	(330)	12,210	11,329
45663	Dry Run (c)	(614)	6,553	5,389
*43016	Dublin	(614)/(740)	31,392	16,366
44112	East Cleveland	(216)	27,217	33,096
44094	Eastlake	(440)	20,255	21,161
43920	East Liverpool	(330)	13,089	13,654
45320	Eaton	(937)	8,133	7,396
*44035	Elyria	(440)	55,953	56,746
45322	Englewood	(937)	12,235	11,402
*44117	Euclid	(216)	52,717	54,875
45324	Fairborn	(937)	32,052	31,300
*45011	Fairfield	(513)	42,097	39,709
44334	Fairlawn	(330)	7,307	5,779
44126	Fairview Park	(440)	17,572	18,028
*45839	Findlay	(419)	38,967	35,703
45224	Finneytown (c)	(513)	13,492	13,096
45405	Forest Park	(513)	19,463	18,621
45230	Forestville (c)	(513)	10,978	9,185
44830	Fostoria	(419)	13,931	14,971
45005	Franklin	(513)	11,396	11,026
43420	Fremont	(419)	17,375	17,619
43230	Gahanna	(614)	32,636	23,898
44833	Galion	(419)	11,341	11,859
44125	Garfield Heights	(216)	30,734	31,739
44041	Geneva	(440)	6,595	6,597
44420	Girard	(330)	10,902	11,304
43212	Grandview Heights	(614)	6,695	7,010
44232	Green	(330)	22,817	19,179
45331	Greenville	(937)	13,294	12,863
45253	Groesbeck (c)	(513)	7,202	6,684
43123	Grove City	(614)	27,075	19,661
*45011	Hamilton	(513)	60,690	61,438
45030	Harrison	(513)	7,487	7,520
43056	Heath	(740)	8,527	7,231
44134	Highland Heights	(440)	8,082	6,249
43026	Hilliard	(614)/(740)	24,230	11,794
45133	Hillsboro	(937)	6,368	6,235
44484	Howland Center (c)	(330)	6,481	6,732
44425	Hubbard	(330)	8,284	8,248
45424	Huber Heights	(937)	38,212	38,696
*44236	Hudson	(330)	22,439	5,159
44839	Huron	(419)	7,958	7,067
44131	Independence (Cuyahoga)	(216)/(440)	7,109	6,500
45638	Ironton	(740)	11,211	12,751
45640	Jackson	(740)	6,184	6,167
*44240	Kent	(330)	27,906	28,835
43326	Kenton	(419)	8,336	8,356
43606	Kenwood (c)	(513)	7,423	7,469
45429	Kettering	(937)	57,502	60,569
44094	Kirtland	(440)	6,670	5,881
44107	Lakewood	(216)	56,646	59,718
43130	Lancaster	(740)	35,335	34,507
45039	Landen (c)	(513)	12,766	9,263
45036	Lebanon (Warren)	(513)	16,962	10,461
*45802	Lima	(419)	40,081	45,553
43228	Lincoln Village (c)	(614)	9,482	9,958
43138	Logan	(740)	6,704	6,725
43140	London	(614)/(740)	8,771	7,807
*44052	Lorain	(440)	68,652	71,245
44641	Louisville	(330)	8,904	8,087
45140	Loveland	(513)	11,677	10,122
44124	Lyndhurst	(216)/(440)	15,279	15,982
44056	Macedonia	(330)	9,224	7,509
—	Mack South (c)		5,837	5,767
45243	Madeira	(513)	8,923	9,141
*44901	Mansfield	(419)	49,346	50,627
44137	Maple Heights	(216)	26,156	27,089
45750	Marietta	(740)	14,515	15,026
*43302	Marion	(740)	35,318	34,075
43935	Martins Ferry	(740)	7,226	8,003
43040	Marysville	(937)	15,942	10,362
45040	Mason	(513)	22,016	11,450
*44646	Massillon	(330)	31,325	30,969
43537	Maumee	(419)	15,237	15,561
44124	Mayfield Heights	(440)	19,386	19,847
*44256	Medina	(330)	25,139	19,231
*44060	Mentor	(440)	50,278	47,491
44060	Mentor-on-the-Lake	(216)	8,127	8,271
*45343	Miamisburg	(937)	19,489	17,834
44130	Middleburg Heights	(216)/(440)	15,542	14,702
*45042	Middletown	(513)	51,605	46,758
45150	Milford	(513)	6,284	5,660
45050	Monroe	(513)	7,133	5,380
45242	Montgomery	(513)	10,163	9,733
—	Montrose-Ghent (c)		5,261	4,906
45439	Moraine	(937)	6,897	5,989
45231	Mount Healthy	(513)	7,149	7,580
43050	Mount Vernon	(740)	14,375	14,550
44262	Munroe Falls	(330)	5,314	5,359
43545	Napoleon	(419)	9,318	8,884
45764	Nelsonville	(740)	5,230	4,563
*43055	Newark	(740)	46,279	44,396
45344	New Carlisle	(937)	5,735	6,049
44663	New Philadelphia	(330)	17,056	15,698
44444	Newton Falls	(330)	5,002	4,866
44446	Niles	(330)	20,932	21,128
45239	Northbrook (c)	(513)	11,076	11,471
44720	North Canton	(330)	16,369	14,904
45239	North College Hill	(513)	10,082	11,002
45251	Northgate (c)	(513)	8,016	7,864
44057	North Madison (c)	(440)	8,451	8,699
44070	North Olmsted	(440)	34,113	34,204
45502	Northridge (c) (Clark)	(937)	6,853	5,939
45414	Northridge (c) (Montgomery)	(937)	8,487	9,448
44039	North Ridgeville	(440)	22,338	21,564
44133	North Royalton	(440)	28,648	23,197
43619	Northwood	(419)	5,471	5,506
44203	Norton	(330)	11,523	11,477
44857	Norwalk	(419)	16,238	14,731
45212	Norwood	(513)	21,675	23,674

ZIP	Place	Area Code	2000	1990
*45873	Oakwood	(973)	9,215	8,957
44074	Oberlin	(440)	8,195	8,191
44138	Olmsted Falls	(440)	7,962	6,741
44862	Ontario	(419)	5,303	4,026
*45054	Oregon	(419)	19,355	18,334
44667	Orrville	(330)	8,551	7,955
45056	Oxford	(513)	21,943	19,013
44077	Painesville	(440)	17,503	15,769
44129	Parma	(216)/(440)	85,655	87,876
44130	Parma Heights	(216)/(440)	21,659	21,448
43062	Pataskala	(740)	10,249	3,046
44124	Pepper Pike	(216)/(440)	6,040	6,185
44646	Perry Heights (c)	(330)	8,900	9,055
*43551	Perrysburg	(419)	16,945	12,551
43147	Pickerington	(614)/(740)	9,792	5,668
45356	Piqua	(937)	20,738	20,612
—	Pleasant Run (c)		5,267	4,964
44319	Portage Lakes (c)	(330)	9,870	13,373
43452	Port Clinton	(419)	6,391	7,106
45662	Portsmouth	(740)	20,909	22,676
43065	Powell	(614)	6,247	2,154
44266	Ravenna	(330)	11,771	12,069
45215	Reading	(513)	11,292	12,038
43068	Reynoldsburg	(614)/(740)	32,069	25,748
44143	Richmond Heights	(216)/(440)	10,944	9,611
44270	Rittman	(330)	6,314	6,147
45431	Riverside	(937)	23,545	1,471
44116	Rocky River	(440)	20,735	20,410
43460	Rossford	(419)	6,406	5,861
43950	Saint Clairsville	(740)	5,057	5,136
45885	Saint Marys	(419)	8,342	8,441
44460	Salem	(330)	12,197	12,233
*44870	Sandusky	(419)	27,844	29,764
44870	Sandusky South (c)	(419)	6,599	6,336
44131	Seven Hills	(216)/(440)	12,080	12,339
44122	Shaker Heights	(216)	29,405	30,955
*45241	Sharonville	(513)	13,804	13,121
44054	Sheffield Lake	(440)	9,371	9,825
44875	Shelby	(419)	9,821	9,610
44878	Shiloh (c)	(419)	11,272	11,607
45365	Sidney	(937)	20,211	18,710
45236	Silverton	(513)	5,178	5,859
44139	Solon	(440)	21,802	18,548
44121	South Euclid	(216)	23,537	23,866
45066	Springboro	(513)	12,380	6,574
45246	Springdale	(513)	10,563	10,621
*45501	Springfield	(937)	65,358	70,487
*43952	Steubenville	(740)	19,015	22,125
44224	Stow	(330)	32,139	27,998
44241	Streetsboro	(330)	12,311	9,932
44136	Strongsville	(440)	43,858	35,308
44471	Struthers	(330)	11,756	12,284
—	Summerside (c)		5,523	4,573
43560	Sylvania	(419)	18,670	17,489
44278	Tallmadge	(330)	16,390	14,870
45243	The Village of Indian Hill	(513)	5,907	5,383
44883	Tiffin	(419)	18,135	18,604
45371	Tipp City	(937)	9,221	6,483
*43601	Toledo	(419)	313,619	332,943
43964	Toronto	(740)	5,676	6,127
45067	Trenton	(513)	8,746	6,189
45426	Trotwood	(937)	27,420	29,358
45373	Troy	(937)	21,999	19,478
44087	Twinsburg	(330)	17,006	9,606
44683	Uhrichsville	(740)	5,662	5,604
45322	Union	(937)	5,574	5,531
44122	University Heights	(216)	14,146	14,787
43221	Upper Arlington	(614)	33,686	34,128
43351	Upper Sandusky	(419)	6,533	5,906
43078	Urbana	(937)	11,613	11,353
45377	Vandalia	(937)	14,603	13,872
45891	Van Wert	(419)	10,690	10,922
44089	Vermilion	(440)	10,927	11,127
*44281	Wadsworth	(330)	18,437	15,718
45895	Wapakoneta	(419)	9,474	9,214
*44481	Warren	(330)	46,832	50,793
44122	Warrensville Heights	(216)	15,109	15,884
43160	Washington	(740)	13,524	13,080
43567	Wauseon	(419)	7,091	6,322
45692	Wellston	(740)	6,078	6,049
45449	West Carrollton City	(937)	13,818	14,403
*43081	Westerville	(614)	35,318	30,269
44145	Westlake	(440)	31,719	27,018
45694	Wheelersburg (c)	(740)	6,471	5,113
43213	Whitehall	(614)	19,201	20,572
45239	White Oak (c)	(513)	13,277	12,430
44092	Wickliffe	(440)	13,484	14,558
44890	Willard	(419)	6,806	6,210
*44094	Willoughby	(440)	22,621	20,510
44094	Willoughby Hills	(440)	8,595	8,427
*44095	Willowick	(440)	14,361	15,269
45177	Wilmington	(937)	11,921	11,199
45459	Woodbourne-Hyde Park (c)	(937)	7,910	7,837
44691	Wooster	(330)	24,811	22,427
43085	Worthington	(614)	14,125	14,869
45433	Wright-Patterson AFB (c)	(937)	6,656	8,579
45215	Wyoming	(513)	8,261	8,128
45385	Xenia	(937)	24,164	24,836
*44501	Youngstown	(330)	82,026	95,732
*43701	Zanesville	(740)	25,586	26,778

Oklahoma

ZIP	Place	Area Code	2000	1990
*74820	Ada	(405)	15,691	15,765
*73521	Altus	(405)	21,447	21,910
73717	Alva	(580)	5,288	5,495
73005	Anadarko	(405)	6,645	6,586
*73401	Ardmore	(405)	23,711	23,079
*74003	Bartlesville	(918)	34,748	34,256
73008	Bethany	(405)	20,307	20,075
74008	Bixby	(918)	13,336	9,502
74631	Blackwell	(405)	7,668	7,538
*74012	Broken Arrow	(918)	74,859	58,082
74015	Catoosa	(918)	5,449	2,954
*73018	Chickasha	(405)	15,850	14,988
73020	Choctaw	(405)	9,377	8,545
*74017	Claremore	(918)	15,873	13,280
73601	Clinton	(405)	8,833	9,298
74429	Coweta	(918)	7,139	6,159
74023	Cushing	(918)	8,371	7,218
73115	Del City	(405)	22,128	23,928
*73533	Duncan	(405)	22,505	21,732
*74701	Durant	(405)	13,549	12,929
*73034	Edmond	(405)	68,315	52,310
*73644	Elk City	(405)	10,510	10,428
73036	El Reno	(405)	16,212	15,414
*73701	Enid	(405)	47,045	45,309
74033	Glenpool	(918)	8,123	6,688
*74344	Grove	(918)	5,131	4,020
73044	Guthrie	(405)	9,925	10,440
73942	Guymon	(405)	10,472	7,803
74437	Henryetta	(918)	6,096	5,872
74743	Hugo	(405)	5,536	5,978
74745	Idabel	(405)	6,952	6,957
74037	Jenks	(918)	9,557	7,484
*73501	Lawton	(405)	92,757	80,561
*74501	McAlester	(918)	17,783	16,739
*74354	Miami	(918)	13,704	13,142
73140	Midwest City	(405)	54,088	52,267
73153	Moore	(405)	41,138	40,318
*74401	Muskogee	(918)	38,310	37,708
73064	Mustang	(405)	13,156	10,434
73065	Newcastle	(405)	5,434	4,214
73068	Noble	(405)	5,260	4,710
*73069	Norman	(405)	95,694	80,071
*73125	Oklahoma City	(405)	506,132	444,724
74447	Okmulgee	(918)	13,022	13,441
74055	Owasso	(918)	18,502	11,151
73075	Pauls Valley	(405)	6,256	6,150
73077	Perry	(405)	5,230	4,978
*74601	Ponca City	(405)	25,919	26,359
74953	Poteau	(918)	7,939	7,210
74361	Pryor Creek	(918)	8,659	8,327
73080	Purcell	(405)	5,571	4,784
74955	Sallisaw	(918)	7,989	7,122
74063	Sand Springs	(918)	17,451	15,339
*74066	Sapulpa	(918)	19,166	18,074
*74868	Seminole	(405)	6,899	7,071
*74801	Shawnee	(405)	28,692	26,017
74070	Skiatook	(918)	5,396	4,910
*74074	Stillwater	(405)	39,065	36,676
*74464	Tahlequah	(918)	14,458	10,586
74873	Tecumseh	(405)	6,098	5,750
73156	The Village	(405)	10,157	10,353
*74103	Tulsa	(918)	393,049	367,302
74301	Vinita	(918)	6,472	5,804
*74467	Wagoner	(918)	7,669	6,894
73123	Warr Acres	(405)	9,735	9,288
73096	Weatherford	(405)	9,859	10,124
*73801	Woodward	(405)	11,853	12,340
*73099	Yukon	(405)	21,043	20,935

Oregon

Area code (971) overlays area code (503). See introductory note.

ZIP	Place	Area Code	2000	1990
97321	Albany	(541)	40,852	33,523
*97006	Aloha (c)	(503)	41,741	34,284
97601	Altamont (c)	(541)	19,603	18,591
97520	Ashland	(541)	19,522	16,252
97103	Astoria	(503)	9,813	10,069
97814	Baker City	(541)	9,860	9,140
*97005	Beaverton	(503)	76,129	53,307
*97701	Bend	(541)	52,029	23,740
97415	Brookings	(541)	5,447	4,400
97013	Canby	(503)	12,790	8,990
97225	Cedar Hills (c)	(503)	8,949	9,294
97291	Cedar Mill (c)	(503)	12,597	9,697
97502	Central Point	(541)	12,493	7,512
97058	City of the Dalles	(541)	12,156	11,021
97015	Clackamas (c)	(503)	5,177	2,578
97420	Coos Bay	(541)	15,374	15,076
97113	Cornelius	(503)	9,652	6,148
*97333	Corvallis	(541)	49,322	44,757
97424	Cottage Grove	(541)	8,445	7,403
97338	Dallas	(503)	12,459	9,422

ZIP	Place	Area Code	2000	1990
*97440	Eugene	(541)	137,893	112,733
97024	Fairview	(503)	7,561	2,588
97439	Florence	(541)	7,263	5,171
97116	Forest Grove	(503)	17,708	13,559
97301	Four Corners (c)	(503)	13,922	12,156
97223	Garden Home-Whitford (c)	(503)	6,931	6,652
97027	Gladstone	(503)	11,438	10,152
*97526	Grants Pass	(541)	23,003	17,503
97470	Green (c)	(541)	6,174	5,076
*97030	Gresham	(503)	90,205	68,285
97303	Hayesville (c)	(503)	18,222	14,318
97838	Hermiston	(541)	13,154	10,047
*97123	Hillsboro	(503)	70,186	37,598
97031	Hood River	(541)	5,831	4,632
97351	Independence	(503)	6,035	4,425
97222	Jennings Lodge (c)	(503)	7,036	6,530
97307	Keizer	(503)	32,203	21,884
*97601	Klamath Falls	(541)	19,462	17,737
97850	La Grande	(541)	12,327	11,766
*97034	Lake Oswego	(503)	35,278	30,576
97739	La Pine (c)	(541	)5,799	—
97355	Lebanon	(541)	12,950	10,950
97367	Lincoln City	(541)	7,437	5,903
97128	McMinnville	(503)	26,499	17,894
97741	Madras	(541)	5,078	3,443
*97501	Medford	(541)	63,154	47,021
97862	Milton-Freewater	(541)	6,470	5,533
97269	Milwaukie	(503)	20,490	18,670
97038	Molalla	(503)	5,647	3,651
97361	Monmouth	(503)	7,741	6,288
97132	Newberg	(503)	18,064	13,086
97365	Newport	(541)	9,532	8,437
97459	North Bend	(541)	9,544	9,614
97268	Oak Grove (c)	(503)	12,808	12,576
—	Oak Hills (c)		9,050	6,450
—	Oatfield (c)		15,750	15,348
97914	Ontario	(541)	10,985	9,394
97045	Oregon City	(503)	25,754	14,698
97801	Pendleton	(541)	16,354	15,142
*97208	Portland	(503)	529,121	485,975
97754	Prineville	(541)	7,356	5,355
97225	Raleigh Hills (c)	(503)	5,865	6,066
97756	Redmond	(541)	13,481	7,165
—	Redwood (c)		5,844	3,702
—	Rockcreek (c)		9,404	8,282
97470	Roseburg	(541)	20,017	18,389
97470	Roseburg North (c)	(541)	5,473	6,831
97051	Saint Helens	(503)	10,019	7,535
*97309	Salem	(503)	136,924	107,793
97055	Sandy	(503)	5,385	4,154
97138	Seaside	(503)	5,900	5,359
97140	Sherwood	(503)	11,791	3,093
97381	Silverton	(503)	7,414	5,635
*97477	Springfield	(541)	52,864	44,664
97383	Stayton	(503)	6,816	5,011
—	Sunnyside (c)	(503)	6,791	4,423
97479	Sutherlin	(541)	6,669	5,020
97386	Sweet Home	(541)	8,016	6,850
97540	Talent	(541)	5,589	3,274
97281	Tigard	(503)	41,223	29,435
97060	Troutdale	(503)	13,777	7,852
97062	Tualatin	(503)	22,791	14,664
97225	West Haven-Sylvan (c)	(503)	7,147	6,009
97068	West Linn	(503)	22,261	16,389
*97225	West Slope (c)	(503)	6,442	7,959
97503	White City (c)	(541)	5,466	5,891
97070	Wilsonville	(503)	13,991	7,510
97071	Woodburn	(503)	20,100	13,404

Pennsylvania

Area code (267) overlays area code (215). Area code (484) overlays (610). Area code (878) overlays (412). See introductory note.

ZIP	Place	Area Code	2000	1990
15001	Aliquippa	(724)	11,734	13,374
*18105	Allentown (Lehigh)	(610)	106,632	105,301
*16603	Altoona	(814)	49,523	51,881
19002	Ambler	(215)	6,426	6,609
15003	Ambridge	(724)	7,769	8,133
18403	Archbald	(570)	6,220	6,291
19003	Ardmore (c)	(610)	12,616	12,646
15210	Arlington Heights (c)	(412)	5,132	4,768
15068	Arnold	(724)	5,667	6,113
19407	Audubon (c)	(610)	6,549	6,328
15202	Avalon	(412)	5,294	5,784
—	Back Mountain (c)		26,690	—
15234	Baldwin	(412)	19,999	21,923
18013	Bangor	(610)	5,319	5,383
15010	Beaver Falls	(724)	9,920	10,687
16823	Bellefonte	(814)	6,395	6,358
15202	Bellevue	(412)	8,770	9,126
18603	Berwick	(570)	10,774	10,976
15102	Bethel Park	(412)	33,556	33,823
*18016	Bethlehem	(610)	71,329	71,427
19508	Birdsboro	(610)	5,064	4,222
18447	Blakely	(570)	7,027	7,222
17815	Bloomsburg	(570)	12,375	12,439
19422	Blue Bell (c)	(215)/(610)	6,395	6,091
19061	Boothwyn (c)	(610)	5,206	5,069
16701	Bradford	(814)	9,175	9,625
15227	Brentwood	(412)	10,466	10,823
15017	Bridgeville	(412)	5,341	5,445
19007	Bristol	(215)	9,923	10,405
19015	Brookhaven	(610)	7,985	8,570
19008	Broomall (c)	(610)	11,046	10,930
*16001	Butler	(724)	15,121	15,714
15419	California	(724)	5,274	5,748
*17011	Camp Hill	(717)	7,636	7,831
15317	Canonsburg	(724)	8,607	9,200
18407	Carbondale	(570)	9,804	10,664
17013	Carlisle	(717)	17,970	18,419
15106	Carnegie	(412)	8,389	9,278
15108	Carnot-Moon (c)	(412)	10,637	10,187
15234	Castle Shannon	(412)	8,556	9,135
18032	Catasauqua	(610)	6,588	6,662
17201	Chambersburg	(717)	17,862	16,647
*19013	Chester	(610)	36,854	41,856
15025	Clairton	(412)	8,491	9,656
16214	Clarion	(814)	6,185	6,457
18411	Clarks Summit	(570)	5,126	5,433
16830	Clearfield	(814)	6,631	6,633
19018	Clifton Heights	(610)	6,779	7,111
19320	Coatesville	(610)	10,838	11,038
19426	Collegeville	(610)	8,032	4,227
19023	Collingdale	(610)	8,664	9,175
17109	Colonial Park (c) (Dauphin)	(717)	13,259	13,777
17512	Columbia	(717)	10,311	10,701
15425	Connellsville	(724)	9,146	9,229
19428	Conshohocken	(610)	7,589	8,064
15108	Coraopolis	(412)	6,131	6,747
16407	Corry	(814)	6,834	7,216
15205	Crafton	(412)	6,706	7,188
19021	Croydon (c)	(215)	9,993	9,967
19023	Darby	(610)	10,299	11,140
19036	Darby Twp. (c)	(610)	9,622	10,955
19333	Devon-Berwyn (c)	(610)	5,067	5,019
18519	Dickson City	(570)	6,205	6,276
15033	Donora	(724)	5,653	5,928
15216	Dormont	(412)	9,305	9,772
19335	Downingtown	(610)	7,589	7,749
18901	Doylestown	(215)	8,227	8,575
19026	Drexel Hill (c)	(610)	29,364	29,744
15801	Du Bois	(814)	8,123	8,286
18512	Dunmore	(570)	14,018	15,403
15110	Duquesne	(412)	7,332	8,525
19401	East Norriton (c)	(610)	13,211	13,324
*18042	Easton	(610)	26,263	26,276
18301	East Stroudsburg	(570)	9,888	8,781
17402	East York (c)	(717)	8,782	8,487
15005	Economy	(724)	9,363	9,305
16412	Edinboro	(814)	6,950	7,736
17022	Elizabethtown	(717)	11,887	9,952
16117	Ellwood City	(724)	8,688	8,894
18049	Emmaus	(610)	11,313	11,157
17025	Enola (c)	(717)	5,627	5,961
17522	Ephrata	(717)	13,213	12,133
*16501	Erie	(814)	103,717	108,718
18643	Exeter	(570)	5,955	5,691
19030	Fairless Hills (c)	(215)	8,365	9,026
16121	Farrell	(724)	6,050	6,835
19053	Feasterville-Trevose (c)	(215)	6,525	6,696
16063	Fernway (c)	(724)	12,188	9,072
19032	Folcroft	(610)	6,978	7,506
19033	Folsom (c)	(610)	8,072	8,173
15221	Forest Hills	(412)	6,831	7,335
15238	Fox Chapel	(412)	5,436	5,319
16323	Franklin	(814)	7,212	7,329
15143	Franklin Park	(412)	11,364	10,109
18052	Fullerton (c)	(610)	14,268	13,127
17325	Gettysburg	(717)	7,490	7,025
19036	Glenolden	(610)	7,476	7,260
19038	Glenside (c)	(215)	7,914	8,704
15601	Greensburg	(724)	15,889	16,318
16125	Greenville	(724)	6,380	6,734
16127	Grove City	(412)	8,024	8,240
15101	Hampton Twp. (c) (Allegheny)	(412)	17,526	15,568
17331	Hanover	(717)	14,535	14,399
19438	Harleysville (c)	(215)	8,795	7,405
*17105	Harrisburg	(717)	48,950	52,376
15065	Harrison Twp. (c) (Allegheny)	(412)	10,934	11,763
19040	Hatboro	(215)	7,393	7,382
18201	Hazleton	(570)	23,329	24,730
18055	Hellertown	(610)	5,606	5,662
16148	Hermitage	(724)	16,157	15,260
17033	Hershey (c)	(717)	12,771	11,860
16648	Hollidaysburg	(814)	5,368	5,624
16001	Homeacre-Lyndora (c)	(724)	6,685	7,511
19044	Horsham (c)	(215)	14,779	15,051
16652	Huntingdon	(814)	6,918	6,843
15701	Indiana	(724)	14,895	15,174
15644	Jeannette	(724)	10,654	11,221
15025	Jefferson Hills	(412)	9,666	—
*15907	Johnstown	(814)	23,906	28,124
15108	Kennedy Twp. (c)	(412)	7,504	7,152
19348	Kennett Square	(610)	5,273	5,218
19406	King of Prussia (c)	(610)	18,511	18,406

ZIP	Place	Area Code	2000	1990
18704	Kingston	(570)	13,855	14,507
19443	Kulpsville (c)	(215)	8,005	5,183
19530	Kutztown	(610)	5,067	4,704
*17604	Lancaster	(717)	56,348	55,551
19446	Lansdale	(215)	16,071	16,362
19050	Lansdowne	(610)	11,044	11,712
15650	Latrobe	(724)	8,994	9,265
17540	Leacock-Leola-Bareville (c)	(717)	6,625	5,685
*17042	Lebanon	(717)	24,461	24,800
18235	Lehighton	(610)	5,537	5,914
*19055	Levittown (c)	(215)	53,966	55,362
17837	Lewisburg	(570)	5,620	5,785
17044	Lewistown (Mifflin)	(717)	8,998	9,341
17112	Linglestown (c)	(717)	6,414	5,862
19353	Lionville-Marchwood (c)	(610)	6,298	6,468
17543	Lititz	(717)	9,029	8,280
17745	Lock Haven	(570)	9,149	9,230
17011	Lower Allen (c)	(717)	6,619	6,329
15068	Lower Burrell	(724)	12,608	12,251
15237	McCandless Twp. (c)	(412)	29,022	28,781
*15134	McKeesport	(412)	24,040	26,016
15136	McKees Rocks	(412)	6,622	7,691
19002	Maple Glen (c)	(215)	7,042	5,881
16335	Meadville	(814)	13,685	14,318
17055	Mechanicsburg	(717)	9,042	9,452
*19063	Media	(610)	5,533	5,957
17057	Middletown (Dauphin)	(717)	9,242	9,254
18017	Middletown (c) (Northampton)	(610)	7,378	6,866
17551	Millersville	(717)	7,774	8,099
17847	Milton	(570)	6,650	6,746
15061	Monaca	(724)	6,286	6,739
15062	Monessen	(724)	8,669	9,901
18936	Montgomeryville (c)	(215)	12,031	9,114
18507	Moosic	(570)	5,575	5,397
19067	Morrisville (Bucks)	(215)	10,023	9,765
18707	Mountain Top (c)	(570)	15,269	—
17851	Mount Carmel	(570)	6,390	7,196
17552	Mount Joy	(717)	6,765	6,398
15228	Mount Lebanon (c)	(412)	33,017	34,414
15120	Munhall	(412)	12,264	13,158
15146	Municipality of Monroeville	(412)	29,349	29,169
15668	Municipality of Murrysville	(724)	18,872	17,240
18634	Nanticoke	(570)	10,955	12,267
18064	Nazareth	(610)	6,023	5,713
19086	Nether Providence Twp. (c)	(610)	13,456	12,730
15066	New Brighton	(724)	6,641	6,854
*16108	New Castle	(724)	26,309	28,334
17070	New Cumberland	(717)	7,349	7,665
17557	New Holland	(717)	5,092	4,484
15068	New Kensington	(724)	14,701	15,894
*19403	Norristown	(610)	31,282	30,754
18067	Northampton	(610)	9,405	8,717
15104	North Braddock	(412)	6,410	7,036
15137	North Versailles (c)	(412)	11,125	13,294
16421	Northwest Harborcreek (c)	(814)	8,658	7,485
19074	Norwood (Delaware)	(610)	5,985	6,162
15139	Oakmont (Allegheny)	(412)	6,911	6,961
15238	O'Hara Twp. (c)	(412)	8,856	9,096
16301	Oil City	(814)	11,504	11,949
18518	Old Forge	(570)	8,798	8,834
19075	Oreland (c)	(215)	5,509	5,695
18071	Palmerton	(610)	5,248	5,394
17078	Palmyra	(717)	7,096	6,910
19301	Paoli (c)	(610)	5,425	5,277
16801	Park Forest Village (c)	(814)	8,830	6,703
17331	Parkville (c)	(717)	6,593	5,009
17112	Paxtonia (c)	(570)	5,254	4,862
15235	Penn Hills (c)	(412)	46,809	57,632
19096	Penn Wynne (c)	(610)	5,382	5,807
18944	Perkasie	(215)	8,828	7,878
*19104	Philadelphia	(215)	1,517,550	1,585,577
19460	Phoenixville	(610)	14,788	15,066
*15233	Pittsburgh	(412)	334,563	369,879
*18640	Pittston	(570)	8,104	9,389
15236	Pleasant Hills	(412)	8,397	8,884
15239	Plum	(412)	26,940	25,609
18651	Plymouth	(570)	6,507	7,134
19462	Plymouth Meeting (c)	(610)	5,593	6,241
*19464	Pottstown	(610)	21,859	21,831
17901	Pottsville	(570)	15,549	16,603
17109	Progress (c)	(717)	9,647	9,654
19076	Prospect Park	(610)	6,594	6,764
15767	Punxsutawney	(814)	6,271	6,782
18951	Quakertown	(215)	8,931	8,982
19087	Radnor Twp. (c)	(610)	30,878	27,676
*19612	Reading	(610)	81,207	78,380
17356	Red Lion	(717)	6,149	6,130
18954	Richboro (c)	(215)	6,678	5,141
19078	Ridley Park	(610)	7,196	7,592
15136	Robinson Twp. (Allegheny) (c)	(412)	12,289	10,830
15237	Ross Twp. (c)	(412)	32,551	35,102
15857	Saint Marys	(814)	14,502	14,020
19464	Sanatoga (c)	(610)	7,734	3,723
18840	Sayre	(570)	5,813	5,791
17972	Schuylkill Haven	(570)	5,548	5,610
15106	Scott Twp. (c)	(412)	17,288	20,413
*18505	Scranton	(570)	76,415	81,805
17870	Selinsgrove	(570)	5,383	5,384
15116	Shaler Twp. (c)	(412)	29,757	33,694
17872	Shamokin	(570)	8,009	9,184
16146	Sharon	(724)	16,328	17,533
19079	Sharon Hill	(610)	5,468	5,771
17976	Shenandoah	(570)	5,624	6,221
19607	Shillington	(610)	5,059	5,062
17404	Shiloh (c)	(717)	10,192	5,315
17257	Shippensburg	(717)	5,586	5,331
15501	Somerset	(814)	6,762	6,454
18964	Souderton	(215)	6,730	5,957
15129	South Park Twp. (c)	(814)	14,340	14,292
17701	South Williamsport	(570)	6,412	6,496
19064	Springfield (c) (Delaware)	(610)	23,677	25,326
*16804	State College	(814)	38,420	38,981
17113	Steelton	(717)	5,858	5,152
—	Stonybrook-Wilshire (c)		5,414	4,887
15136	Stowe Twp. (c)	(412)	6,706	9,202
18360	Stroudsburg	(570)	5,756	5,312
16323	Sugarcreek	(814)	5,331	5,532
17801	Sunbury	(570)	10,610	11,591
19081	Swarthmore	(610)	6,170	6,157
15218	Swissvale	(412)	9,653	10,637
18704	Swoyersville	(570)	5,157	5,630
18252	Tamaqua	(570)	7,174	7,943
18517	Taylor	(570)	6,475	6,941
16354	Titusville	(814)	6,146	6,434
19401	Trooper (c)	(610)	6,061	7,370
15145	Turtle Creek	(412)	6,076	6,556
16686	Tyrone	(814)	5,528	5,743
15401	Uniontown (Fayette)	(724)	12,422	12,034
19063	Upper Providence Twp. (c)	(610)	10,509	9,477
15241	Upper Saint Clair (c)	(412)	20,053	19,023
15690	Vandergrift	(724)	5,455	5,904
19013	Village Green-Green Ridge (c)	(610)	8,279	9,026
16365	Warren	(814)	10,259	11,122
15301	Washington (Washington)	(724)	15,268	15,864
17268	Waynesboro	(717)	9,614	9,578
17315	Weigelstown (c)	(717)	10,117	8,665
*19380	West Chester	(610)	17,861	18,041
19380	West Goshen (c)	(610)	8,472	8,948
*15122	West Mifflin	(412)	22,464	23,644
15905	Westmont	(814)	5,523	5,789
19401	West Norriton (c)	(610)	14,901	15,209
18643	West Pittston	(570)	5,072	5,590
15229	West View	(412)	7,277	7,734
15227	Whitehall (Allegheny)	(412)	14,444	14,451
15131	White Oak	(412)	8,437	8,761
*18703	Wilkes-Barre	(570)	43,123	47,523
15221	Wilkinsburg	(412)	19,196	21,080
15145	Wilkins Twp. (c)	(412)	6,917	7,487
*17701	Williamsport	(570)	30,706	31,933
19090	Willow Grove (c) (Montgomery)	(215)	16,234	16,325
17584	Willow Street (c)	(717)	7,258	5,817
15025	Wilson	(412)	7,682	7,830
19094	Woodlyn (c)	(610)	10,036	10,151
19038	Wyndmoor (c)	(215)	5,601	5,682
19610	Wyomissing	(610)	8,587	7,332
19050	Yeadon	(610)	11,762	11,980
*17405	York	(717)	40,862	42,192

Rhode Island (401)

See introductory note.

ZIP	Place	2000	1990
02806	*Barrington*	16,819	15,849
02809	*Bristol*	22,469	21,625
02830	*Burrillville*	15,796	16,230
02863	Central Falls	18,928	17,637
02813	*Charlestown*	7,859	6,478
02816	*Coventry*	33,668	31,083
*02904	Cranston	79,269	76,060
02864	*Cumberland*	31,840	29,038
02864	Cumberland Hill (c)	7,738	6,379
02818	*East Greenwich*	12,948	11,865
02914	East Providence	48,688	50,380
02822	*Exeter*	6,045	5,461
02814	*Glocester*	9,948	9,227
02828	Greenville (c)	8,626	8,303
02833	*Hopkinton*	7,836	6,873
02835	Jamestown	5,622	4,999
02919	*Johnston*	28,195	26,542
02881	Kingston (c)	5,446	6,504
02865	*Lincoln*	20,898	18,045
02842	*Middletown*	17,334	19,460
02882	*Narragansett*	16,361	15,004
02840	Newport	26,475	28,227
02843	Newport East (c)	11,463	11,080
02852	*North Kingstown*	26,326	23,786
02908	*North Providence*	32,411	32,090
02896	*North Smithfield*	10,618	10,497
*02860	Pawtucket	72,958	72,644
02871	*Portsmouth*	17,149	16,857
*02904	Providence	173,618	160,728
02812	*Richmond*	7,222	5,351
02857	*Scituate*	10,324	9,796
02917	*Smithfield*	20,613	19,163
02879	*South Kingstown*	27,921	24,612
02878	Tiverton (c)	7,282	7,259

ZIP	Place	2000	1990
02878	*Tiverton*	15,260	14,312
02864	Valley Falls (c)	11,599	11,175
*02879	Wakefield-Peacedale (c)	8,468	7,134
02885	*Warren*	11,360	11,385
*02886	Warwick	85,808	85,427
02891	Westerly (c)	17,682	16,477
02891	*Westerly*	22,966	21,605
02817	*West Greenwich*	5,085	—
02893	*West Warwick*	29,581	29,268
02895	Woonsocket	43,224	43,877

South Carolina

ZIP	Place	Area Code	2000	1990
29620	Abbeville	(864)	5,840	5,778
*29801	Aiken	(803)	25,337	20,386
*29621	Anderson	(864)	25,514	26,385
29812	Barnwell	(803)	5,035	5,255
—	Batesburg-Leesville	(803)	5,517	6,107
*29902	Beaufort	(843)	12,950	9,576
29841	Belvedere (c)	(803)	5,631	6,133
29512	Bennettsville	(843)	9,425	10,095
29611	Berea (c)	(864)	14,158	13,535
29902	Burton (c)	(843)	7,180	6,917
29020	Camden	(803)	6,682	6,696
29033	Cayce	(803)	12,150	10,824
—	Centerville (c)	(573)	5,181	4,866
*29402	Charleston	(843)	96,650	88,256
29520	Cheraw	(843)	5,524	5,553
29706	Chester	(803)	6,476	7,158
*29631	Clemson	(864)	11,939	11,145
29325	Clinton	(864)	8,091	9,603
*29201	Columbia	(803)	116,278	110,734
*29526	Conway	(843)	11,788	9,819
*29532	Darlington	(843)	6,720	7,310
29204	Dentsville (c)	(803)	13,009	11,839
29536	Dillon	(843)	6,316	6,829
*29640	Easley	(864)	17,754	15,179
—	Five Forks (c)		8,064	—
*29501	Florence	(843)	30,248	29,913
29206	Forest Acres	(803)	10,558	7,181
*29715	Fort Mill	(803)	7,587	4,930
29644	Fountain Inn	(864)	6,017	4,388
*29341	Gaffney	(864)	12,968	13,149
29605	Gantt (c)	(864)	13,962	13,891
29576	Garden City (c)	(843)	9,357	6,305
*29442	Georgetown	(843)	8,950	9,517
29445	Goose Creek	(843)	29,208	24,692
*29602	Greenville	(864)	56,002	58,256
*29646	Greenwood	(864)	22,071	20,807
*29650	Greer	(864)	16,843	10,322
29406	Hanahan	(843)	12,937	13,176
*29550	Hartsville	(843)	7,556	8,372
*29928	Hilton Head Island	(843)	33,862	23,694
29621	Homeland Park (c)	(864)	6,337	6,569
29063	Irmo	(803)	11,039	11,284
29456	Ladson (c)	(843)	13,264	13,540
29560	Lake City	(843)	6,478	7,153
*29720	Lancaster	(803)	8,177	8,914
29902	Laurel Bay (c)	(843)	6,625	4,972
29360	Laurens	(864)	9,916	9,694
*29072	Lexington	(803)	9,793	4,046
29566	Little River (c)	(843)	7,027	3,470
29078	Lugoff (c)	(803)	6,278	3,211
29571	Marion	(843)	7,042	7,658
29662	Mauldin	(864)	15,224	11,662
29461	Moncks Corner	(843)	5,952	5,599
*29465	Mount Pleasant	(843)	47,609	30,108
29574	Mullins	(843)	5,029	5,910
29576	Murrells Inlet (c)	(843)	5,519	3,334
*29575	Myrtle Beach	(803)	22,759	24,848
29108	Newberry	(803)	10,580	10,543
*29841	North Augusta	(803)	17,574	15,684
*29410	North Charleston	(843)	79,641	70,304
*29582	North Myrtle Beach	(843)	10,974	8,731
29565	Oak Grove (c)	(803)	8,183	7,173
*29115	Orangeburg	(803)	12,765	13,772
—	Parker (c)		10,760	11,072
—	Powderville (c)		5,362	—
29072	Red Bank (c)	(803)	8,811	5,950
29020	Red Hill (c)	(843)	10,509	6,112
*29730	Rock Hill	(803)	49,765	42,112
29417	Saint Andrews (c)	(843)	21,814	25,692
29609	Sans Souci (c)	(864)	7,836	7,612
*29678	Seneca	(864)	7,652	7,726
29210	Seven Oaks (c)	(803)	15,755	15,722
*29681	Simpsonville	(864)	14,352	11,744
29577	Socastee (c)	(843)	14,295	10,426
*29306	Spartanburg	(864)	39,673	43,479
*29483	Summerville	(843)	27,752	22,519
*29150	Sumter	(803)	39,643	40,977
29687	Taylors (c)	(864)	20,125	19,619
29379	Union	(864)	8,793	9,840
29607	Wade Hampton (c)	(864)	20,458	20,014
29488	Walterboro	(843)	5,153	5,595
29611	Welcome (c)	(864)	6,390	6,560
*29169	West Columbia	(803)	13,064	10,974
29206	Woodfield (c)	(803)	9,238	8,862
29745	York	(803)	6,985	6,709

South Dakota (605)

ZIP	Place	2000	1990
*57401	Aberdeen	24,658	24,995
57005	Brandon	5,693	3,545
57006	Brookings	18,504	16,270
57350	Huron	11,893	12,448
57042	Madison	6,540	6,257
57301	Mitchell	14,558	13,798
57501	Pierre	13,876	12,906
*57701	Rapid City	59,607	54,523
57701	Rapid Valley (c)	7,043	5,968
*57101	Sioux Falls	123,975	100,836
57783	Spearfish	8,606	6,966
57785	Sturgis	6,442	5,537
57069	Vermillion	9,765	10,034
57201	Watertown	20,237	17,623
57078	Yankton	13,528	12,703

Tennessee

ZIP	Place	Area Code	2000	1990
37701	Alcoa	(865)	7,734	6,400
*37303	Athens	(423)	13,220	12,054
38184	Bartlett	(901)	40,543	27,038
37660	Bloomingdale (c)	(423)	10,350	10,953
38008	Bolivar	(731)	5,802	5,969
*37027	Brentwood	(615)	23,445	16,392
*37621	Bristol	(423)	24,821	23,421
38012	Brownsville	(731)	10,748	10,017
*37401	Chattanooga	(423)	155,554	152,393
37642	Church Hill	(423)	5,916	5,208
*37040	Clarksville	(615)	103,455	75,542
*37311	Cleveland	(423)	37,192	32,236
*37716	Clinton	(865)	9,409	8,960
37315	Collegedale	(423)	6,514	5,048
*38017	Collierville	(901)	31,872	14,501
37663	Colonial Heights (c)	(423)	7,067	6,716
*38401	Columbia	(615)	33,055	28,583
*38501	Cookeville	(615)	23,923	21,744
38019	Covington	(901)	8,463	7,487
*38555	Crossville	(615)	8,981	6,930
37321	Dayton	(423)	6,180	5,671
*37055	Dickson	(615)	12,244	10,487
*38024	Dyersburg	(731)	17,452	16,321
37411	East Brainerd (c)	(423)	14,132	11,594
37412	East Ridge	(423)	20,640	21,101
*37643	Elizabethton	(423)	13,372	13,087
37650	Erwin	(423)	5,610	5,318
37062	Fairview	(615)	5,800	4,210
37922	Farragut	(865)	17,720	12,802
37334	Fayetteville	(615)	6,994	7,158
*37064	Franklin	(615)	41,842	20,098
37066	Gallatin	(615)	23,230	18,794
*38138	Germantown	(901)	37,348	33,159
*37072	Goodlettsville	(615)	13,780	11,219
*37743	Greeneville	(423)	15,198	13,532
37215	Green Hill (c)	(615)	7,068	6,763
37748	Harriman	(865)	6,744	7,119
37341	Harrison (c)	(423)	7,630	7,191
38340	Henderson	(731)	5,670	4,760
*37075	Hendersonville	(615)	40,620	32,188
38343	Humboldt	(731)	9,467	9,651
*38301	Jackson	(731)	59,643	49,145
37760	Jefferson City	(865)	7,760	5,875
*37601	Johnson City	(423)	55,469	50,354
*37662	Kingsport	(423)	44,905	40,457
37763	Kingston	(423)	5,264	4,552
*37950	Knoxville	(865)	173,890	169,761
37766	La Follette	(423)	7,926	7,201
38002	Lakeland	(901)	6,862	1,204
37086	La Vergne	(615)	18,687	7,496
38464	Lawrenceburg	(615)	10,796	10,397
*37087	Lebanon	(615)	20,235	15,208
*37771	Lenoir City	(865)	6,819	6,147
37091	Lewisburg	(615)	10,413	9,879
38351	Lexington	(731)	7,393	5,810
37352	Lynchburg	(615)	5,740	4,721
38201	McKenzie	(731)	5,295	5,168
*37110	McMinnville	(615)	12,749	11,194
*37355	Manchester	(615)	8,294	7,709
38237	Martin	(731)	10,515	8,588
*37804	Maryville	(865)	23,120	19,208
*38101	Memphis	(901)	650,100	618,652
37343	Middle Valley (c)	(423)	11,854	12,255
38358	Milan	(731)	7,664	7,512
37072	Millersville	(615)	5,308	2,575
*38053	Millington	(901)	10,433	17,866
*37813	Morristown	(423)	24,965	22,513
*37122	Mount Juliet	(615)	12,366	5,389
*37130	Murfreesboro	(615)	68,816	44,922
*37202	Nashville	(615)	569,891	488,366
*37821	Newport	(423)	7,242	7,123
*37830	Oak Ridge	(865)	27,387	27,310
37363	Ooltewah (c)	(423)	5,681	4,903
38242	Paris	(731)	9,763	9,332
*37862	Pigeon Forge	(865)	5,083	3,027
37148	Portland	(615)	8,458	5,539
38478	Pulaski	(615)	7,871	7,916
37415	Red Bank	(423)	12,418	12,320

ZIP	Place	Area Code	2000	1990
38063	Ripley	(731)	7,844	6,634
37854	Rockwood	(865)	5,774	5,348
38372	Savannah	(731)	6,917	6,547
*37862	Sevierville	(865)	11,757	7,178
37865	Seymour (c)	(865)	8,850	7,026
*37160	Shelbyville	(615)	16,105	14,042
37377	Signal Mountain	(423)	7,429	7,034
37167	Smyrna	(615)	25,569	14,720
*37379	Soddy-Daisy	(423)	11,530	8,240
37311	South Cleveland (c)	(423)	6,216	5,372
37172	Springfield	(615)	14,329	11,227
37174	Spring Hill	(931)	7,715	1,464
37874	Sweetwater	(423)	5,586	5,066
37388	Tullahoma	(615)	17,994	16,761
*38261	Union City	(731)	10,876	10,513
37188	White House	(615)	7,220	2,987
37398	Winchester	(615)	7,329	6,305

Texas

Area codes (281) and (832) overlay area code (713). Area code (682) overlays (817). Area codes (972) and (469) overlay (214). See introductory note.

ZIP	Place	Area Code	2000	1990
*79604	Abilene	(915)	115,930	106,707
—	Abram-Perezville (c)		5,444	3,999
75001	Addison	(214)	14,166	8,783
78516	Alamo	(956)	14,760	8,352
78209	Alamo Heights	(210)	7,319	6,502
77039	Aldine (c)	(713)	13,979	11,133
*78332	Alice	(361)	19,010	19,788
*75002	Allen	(214)	43,554	19,315
*79830	Alpine	(915)	5,786	5,622
—	Alton North (c)		5,051	—
*77511	Alvin	(713)	21,413	19,220
*79105	Amarillo	(806)	173,627	157,571
78750	Anderson Mill (c)		8,953	9,468
79714	Andrews	(915)	9,652	10,678
*77515	Angleton	(979)	18,130	17,140
*78336	Aransas Pass	(361)	8,138	7,180
*76004	Arlington	(817)	332,969	261,717
77346	Atascocita (c)	(281)	35,757	—
75751	Athens	(903)	11,297	10,982
75551	Atlanta	(214)	5,745	6,118
*78712	Austin	(512)	656,562	472,020
*76020	Azle	(817)	9,600	8,868
77518	Bacliff (c)	(409)	6,962	5,549
75180	Balch Springs	(214)	19,375	17,406
78602	Bastrop	(512)	5,340	4,044
*77414	Bay City	(979)	18,667	18,170
*77520	Baytown	(713)	66,430	63,843
*77707	Beaumont	(409)	113,866	114,323
*76021	Bedford	(817)	47,152	43,762
*78102	Beeville	(361)	13,129	13,547
*77401	Bellaire	(713)	15,642	13,844
76715	Bellmead	(254)	9,214	8,336
76513	Belton	(254)	14,623	12,463
76126	Benbrook	(817)	20,208	19,564
*79720	Big Spring	(915)	25,233	23,093
*78006	Boerne	(830)	6,178	4,361
75418	Bonham	(903)	9,990	6,688
*79007	Borger	(806)	14,302	15,675
76230	Bowie	(940)	5,219	4,990
76825	Brady	(915)	5,523	5,946
76424	Breckenridge	(254)	5,868	5,665
*77833	Brenham	(979)	13,507	11,952
—	Briar (c)		5,350	3,899
77611	Bridge City	(409)	8,651	8,010
79316	Brownfield	(806)	9,488	9,560
*78520	Brownsville	(956)	139,722	107,027
*76801	Brownwood	(915)	18,813	18,387
78717	Brushy Creek (c)	(903)	15,371	5,833
*77801	Bryan	(979)	65,660	55,002
76354	Burkburnett	(940)	10,927	10,145
*76028	Burleson	(817)	20,976	16,113
76520	Cameron	(254)	5,634	5,635
—	Cameron Park (c)		5,961	3,802
79835	Canutillo (c)	(915)	5,129	4,442
79015	Canyon	(806)	12,875	11,365
78130	Canyon Lake (c)	(830)	16,870	9,975
78834	Carrizo Springs	(830)	5,655	5,745
*75006	Carrolton	(214)	109,576	82,169
75633	Carthage	(903)	6,664	6,496
*75104	Cedar Hill	(214)	32,093	19,988
*78613	Cedar Park	(512)	26,049	5,161
75935	Center	(936)	5,678	4,950
77530	Channelview (c)	(713)	29,685	25,564
79201	Childress	(940)	6,778	5,055
—	Cinco Ranch (c)	(281)	11,196	—
*76031	Cleburne	(817)	26,005	22,205
*77327	Cleveland	(713)	7,605	7,124
77015	Cloverleaf (c)	(713)	23,508	18,230
77531	Clute	(979)	10,424	9,467
76834	Coleman	(915)	5,127	5,410
*77840	College Station	(979)	67,890	52,443
76034	Colleyville	(817)	19,636	12,724
*75428	Commerce	(903)	7,669	6,825
*77301	Conroe	(936)	36,811	27,675
78109	Converse	(210)	11,508	8,887
75019	Coppell	(214)	35,958	16,881
76522	Copperas Cove	(254)	29,592	24,079
76205	Corinth	(940)	11,325	3,944
*78469	Corpus Christi	(361)	277,454	257,428
*75110	Corsicana	(903)	24,485	22,911
75835	Crockett	(936)	7,141	7,024
76036	Crowley	(817)	7,467	6,974
78839	Crystal City	(830)	7,190	8,263
77954	Cuero	(361)	6,571	6,700
79022	Dalhart	(806)	7,237	6,246
*75221	Dallas	(214)	1,188,580	1,007,618
77535	Dayton	(936)	5,709	5,042
76234	Decatur	(214)	5,201	4,245
77536	Deer Park	(713)	28,520	27,424
*78840	Del Rio	(830)	33,867	30,705
*75020	Denison	(903)	22,773	21,505
*76201	Denton	(940)	80,537	66,270
*75115	De Soto	(214)	37,646	30,544
75941	Diboll	(936)	5,470	4,341
77539	Dickinson	(713)	17,093	11,692
78537	Donna	(956)	14,768	12,652
79029	Dumas	(806)	13,747	12,871
*75138	Duncanville	(214)	36,081	35,008
76135	Eagle Mountain (c)	(817)	6,599	5,847
*78852	Eagle Pass	(830)	22,413	20,651
*78539	Edinburg	(956)	48,465	31,091
77957	Edna	(361)	5,899	5,436
—	Eidson Road (c)		9,348	—
77437	El Campo	(979)	10,945	10,511
78621	Elgin	(512)	5,700	4,846
*79910	El Paso	(915)	563,662	515,342
78543	Elsa	(956)	5,549	5,242
*75119	Ennis	(214)	16,045	13,869
*76039	Euless	(817)	46,005	38,149
76140	Everman	(817)	5,836	5,672
79838	Fabens (c)	(915)	8,043	5,599
78355	Falfurrias	(361)	5,297	5,788
75381	Farmers Branch	(214)	27,508	24,250
78114	Floresville	(830)	5,868	5,247
*75067	Flower Mound	(214)	50,702	15,527
76119	Forest Hill	(817)	12,949	11,482
75126	Forney	(214)	5,588	4,070
79906	Fort Bliss (c)	(915)	8,264	13,915
76544	Fort Hood (c)	(254)	33,711	35,580
79735	Fort Stockton	(915)	7,846	8,524
*76161	Fort Worth	(817)	534,694	447,619
78624	Fredericksburg	(830)	8,911	6,934
*77541	Freeport	(979)	12,708	11,389
77545	Fresno	(281)	6,603	3,182
*77546	Friendswood	(713)	29,037	22,814
*75034	Frisco	(214)	33,714	6,138
*76240	Gainesville	(940)	15,538	14,256
77547	Galena Park	(713)	10,592	10,033
*77550	Galveston	(409)	57,247	59,067
*75040	Garland	(214)	215,768	180,635
76528	Gatesville	(254)	15,591	11,492
*78626	Georgetown	(512)	28,339	14,840
78942	Giddings	(979)	5,105	4,093
75647	Gladewater	(903)	6,078	6,027
75115	Glenn Heights	(214)	7,224	4,564
78629	Gonzales	(830)	7,202	6,527
76450	Graham	(940)	8,716	8,986
*76048	Granbury	(817)	5,718	4,045
*75051	Grand Prairie	(214)	127,427	99,606
*76051	Grapevine	(817)	42,059	29,407
—	Greatwood (c)		6,640	—
*75401	Greenville	(903)	23,960	23,071
77619	Groves	(409)	15,733	16,744
75147	Gun Barrel City	(903)	5,145	3,526
76117	Haltom City	(817)	39,018	32,856
76548	Harker Heights	(254)	17,308	12,932
*78550	Harlingen	(956)	57,564	48,746
*75652	Henderson	(903)	11,273	11,139
79045	Hereford	(806)	14,597	14,745
76643	Hewitt	(254)	11,085	8,983
78557	Hidalgo	(956)	7,322	3,292
75205	Highland Park	(214)	8,842	8,739
77562	Highlands (c)	(713)	7,089	6,632
75067	Highland Village	(214)	12,173	7,027
76645	Hillsboro	(254)	8,232	7,072
77563	Hitchcock	(409)	6,386	5,868
—	Homestead Meadows South (c)		6,807	—
78861	Hondo	(830)	7,897	6,018
*79927	Horizon City	(915)	5,233	2,308
*77052	Houston	(281)/(713)/(832)	1,953,631	1,654,348
*77338	Humble	(713)	14,579	12,060
*77340	Huntsville	(936)	35,078	30,628
*76053	Hurst	(817)	36,273	33,574
78362	Ingleside	(361)	9,388	5,696
76367	Iowa Park	(940)	6,431	6,072
*75015	Irving	(214)	191,615	155,037
77029	Jacinto City	(713)	10,302	9,343
75766	Jacksonville	(214)	13,868	12,765
75951	Jasper	(409)	8,247	7,160
77040	Jersey Village	(713)	6,880	4,826

ZIP	Place	Area Code	2000	1990
78729	Jollyville (c)	(512)	15,813	15,206
*77449	Katy	(713)	11,775	8,004
75142	Kaufman	(214)	6,490	5,251
76059	Keene	(817)	5,003	3,944
*76248	Keller	(817)	27,345	13,683
76060	Kennedale	(817)	5,850	4,096
79745	Kermit	(915)	5,714	6,875
*78028	Kerrville	(830)	20,425	17,384
*75662	Kilgore	(903)	11,301	11,066
*76540	Killeen	(254)	86,911	63,535
*78363	Kingsville	(361)	25,575	25,276
78219	Kirby	(210)	8,673	8,326
78640	Kyle	(512)	5,314	2,225
78236	Lackland AFB (c)	(210)	7,123	9,352
76705	Lacy-Lakeview	(254)	5,764	3,617
78559	La Feria	(956)	6,115	4,360
—	La Homa (c)		10,433	1,403
75065	Lake Dallas	(940)	6,166	3,656
77566	Lake Jackson	(979)	26,386	22,771
78734	Lakeway	(512)	8,002	4,044
77568	La Marque	(409)	13,682	14,120
79331	Lamesa	(806)	9,952	10,809
76550	Lampasas	(512)	6,786	6,382
*75146	Lancaster	(214)	25,894	22,117
*77571	La Porte	(713)	31,880	27,923
*78041	Laredo	(956)	176,576	122,893
*77573	League City	(713)	45,444	30,159
*78641	Leander	(512)	7,596	3,354
78268	Leon Valley	(210)	9,239	9,581
*79336	Levelland	(806)	12,866	13,986
*75067	Lewisville	(214)	77,737	46,521
77575	Liberty	(936)	8,033	7,690
79339	Littlefield	(806)	6,507	6,489
78233	Live Oak	(210)	9,156	10,023
77351	Livingston	(936)	5,433	5,019
78644	Lockhart	(512)	11,615	9,205
*75606	Longview	(903)	73,344	70,311
*79408	Lubbock	(806)	199,564	186,206
*75901	Lufkin	(936)	32,709	30,210
78648	Luling	(830)	5,080	4,661
77657	Lumberton	(409)	8,731	6,640
*78501	McAllen	(956)	106,414	84,021
*75070	McKinney	(214)	54,369	21,283
76063	Mansfield	(817)	28,031	15,615
76661	Marlin	(254)	6,628	6,386
*75670	Marshall	(903)	23,935	23,682
78368	Mathis	(361)	5,034	5,423
78570	Mercedes	(956)	13,649	12,694
*75149	Mesquite	(214)	124,523	101,484
76667	Mexia	(254)	6,563	6,933
*79701	Midland	(915)	94,996	89,343
76065	Midlothian	(214)	7,480	5,040
*76067	Mineral Wells	(940)	16,946	14,935
*78572	Mission	(956)	45,408	28,653
—	Mission Bend (c)		30,831	24,945
*77489	Missouri City	(713)	52,913	36,143
79756	Monahans	(915)	6,821	8,101
*75455	Mount Pleasant	(903)	13,935	12,291
*75961	Nacogdoches	(936)	29,914	30,872
77868	Navasota	(936)	6,789	6,296
77627	Nederland	(409)	17,422	16,192
*78130	New Braunfels	(830)	36,494	27,334
—	New Territory (c)	(281)	13,861	—
*76161	North Richland Hills	(817)	55,635	45,895
—	Nurillo (c)		5,056	—
*79761	Odessa	(915)	90,943	89,699
*77630	Orange	(409)	18,643	19,370
77465	Palacios	(361)	5,153	4,418
*75801	Palestine	(903)	17,598	18,042
—	Palmview South (c)		6,219	—
*79065	Pampa	(806)	17,887	19,959
*75460	Paris	(903)	25,898	24,799
*77501	Pasadena	(713)	141,674	119,604
*77581	Pearland	(713)	37,640	18,927
78061	Pearsall	(830)	7,157	6,924
78721	Pecan Grove (c)		13,551	9,502
79772	Pecos	(915)	9,501	12,069
79070	Perryton	(806)	7,774	7,619
*78660	Pflugerville	(512)	16,335	4,444
78577	Pharr	(956)	46,660	32,921
*79072	Plainview	(806)	22,336	21,698
*75074	Plano	(214)	222,030	127,885
78064	Pleasanton	(830)	8,266	7,678
*77640	Port Arthur	(409)	57,755	58,551
78374	Portland	(361)	14,827	12,224
77979	Port Lavaca	(361)	12,035	10,886
77651	Port Neches	(409)	13,601	12,908
78580	Raymondville	(956)	9,733	8,880
76028	Rendon (c)	(817)	9,022	7,658
*75080	Richardson	(214)	91,802	74,840
76118	Richland Hills	(817)	8,132	7,978
*77469	Richmond	(713)	11,081	10,042
78043	Rio Bravo		5,553	—
78582	Rio Grande City	(956)	11,923	10,725
76219	River Oaks	(817)	6,985	6,580
76701	Robinson	(254)	7,845	7,111
78380	Robstown	(361)	12,727	12,849

ZIP	Place	Area Code	2000	1990
76567	Rockdale	(512)	5,439	5,235
*78382	Rockport	(361)	7,385	5,619
*75087	Rockwall	(214)	17,976	10,486
78584	Roma	(956)	9,617	8,059
77471	Rosenberg	(713)	24,043	20,183
*78681	Round Rock	(512)	61,136	30,923
*75088	Rowlett	(214)	44,503	23,260
75785	Rusk	(903)	5,085	4,366
75048	Sachse	(214)	9,751	5,346
76179	Saginaw	(817)	12,374	8,551
*76902	San Angelo	(915)	88,439	84,462
*78265	San Antonio	(210)	1,144,646	976,514
78586	San Benito	(956)	23,444	20,125
79849	San Elizario (c)	(915)	11,046	4,385
78589	San Juan	(956)	26,229	12,561
*78666	San Marcos	(512)	34,733	28,738
*77510	Santa Fe	(409)	9,548	8,429
78154	Schertz	(210)	18,694	10,597
77586	Seabrook	(713)	9,443	6,685
75159	Seagoville	(214)	10,823	8,969
77474	Sealy	(979)	5,248	4,541
*78155	Seguin	(830)	22,011	18,692
79360	Seminole	(915)	5,910	6,342
—	Shady Hollow (c)		5,140	—
*75090	Sherman	(903)	35,082	31,584
77656	Silsbee	(409)	6,393	6,368
78387	Sinton	(361)	5,676	5,549
79364	Slaton	(806)	6,109	6,078
*79549	Snyder	(915)	10,783	12,195
79910	Socorro	(915)	27,152	22,995
77587	South Houston	(713)	15,833	14,207
76092	Southlake	(817)	21,519	7,082
*77373	Spring (c)	(713)	36,385	33,111
*77477	Stafford	(713)	15,681	8,395
76401	Stephenville	(254)	14,921	13,502
*77478	Sugar Land	(713)	63,328	33,712
*75482	Sulphur Springs	(903)	14,551	14,062
79556	Sweetwater	(915)	11,415	11,967
76574	Taylor	(512)	13,575	11,472
*76501	Temple	(254)	54,514	46,150
*75160	Terrell	(214)	13,606	12,490
78209	Terrell Hills	(210)	5,019	4,592
*75501	Texarkana	(903)	34,782	32,294
*77590	Texas City	(409)	41,521	40,822
75056	The Colony	(214)	26,531	22,113
77387	The Woodlands (c)	(713)	55,649	29,205
—	Timberwood Park (c)	(210)	5,889	2,578
*77375	Tomball	(713)	9,089	6,370
76262	Trophy Club	(817)	6,350	3,922
79088	Tulia	(806)	5,117	4,699
*75702	Tyler	(903)	83,650	75,450
*78148	Universal City	(830)	14,849	13,057
76308	University Park	(214)	23,324	22,259
*78801	Uvalde	(830)	14,929	14,729
*76384	Vernon	(940)	11,660	12,001
*77901	Victoria	(361)	60,603	55,076
*77662	Vidor	(409)	11,440	10,935
*76702	Waco	(254)	113,726	103,590
75501	Wake Village	(903)	5,129	4,761
76148	Watauga	(817)	21,908	20,009
*75165	Waxahachie	(214)	21,426	17,984
*76086	Weatherford	(817)	19,000	14,804
77598	Webster	(281)	9,083	4,678
78728	Wells Branch (c)		11,271	7,094
*78596	Weslaco	(956)	26,935	22,739
—	West Livingston (c)		6,612	—
79764	West Odessa (c)	(915)	17,799	16,568
77005	West University Place	(713)	14,211	12,920
77488	Wharton	(979)	9,237	9,011
75791	Whitehouse	(903)	5,346	4,018
75693	White Oak	(903)	5,624	5,136
76108	White Settlement	(817)	14,831	15,472
*76307	Wichita Falls	(940)	104,197	96,259
78239	Windcrest	(210)	5,105	5,331
—	Windemere (c)		6,868	3,207
76712	Woodway	(254)	8,733	8,695
75098	Wylie	(214)	15,132	8,716
77995	Yoakum	(361)	5,731	5,611

Utah

ZIP	Place	Area Code	2000	1990
84004	Alpine	(385)	7,146	3,492
84003	American Fork	(385)	21,941	15,722
*84010	Bountiful	(385)	41,301	37,544
84302	Brigham City	(435)	17,411	15,644
84109	Canyon Rim (c)	(801)	10,428	10,527
*84720	Cedar City	(435)	20,527	13,443
84014	Centerville	(385)	14,585	11,500
*84015	Clearfield	(385)	25,974	21,435
84015	Clinton	(385)	12,585	7,945
84121	Cottonwood Heights (c)	(801)	27,569	28,766
84121	Cottonwood West (c)	(801)	18,727	17,476
84020	Draper	(801)	25,220	7,143
84109	East Millcreek (c)	(801)	21,385	21,184
84025	Farmington	(385)	12,081	9,049
84029	Grantsville	(435)	6,015	4,500

ZIP	Place	Area Code	2000	1990
84032	Heber	(801)	7,291	4,782
84003	Highland	(385)	8,172	5,007
84117	Holladay	(801)	14,561	14,095
84737	Hurricane	(435)	8,250	3,915
84319	Hyrum	(435)	6,316	4,829
84037	Kaysville	(385)	20,351	13,961
84118	Kearns (c)	(801)	33,659	28,374
*84041	Layton	(385)	58,474	41,784
84043	Lehi	(385)	19,028	8,475
84042	Lindon	(385)	8,363	3,818
—	Little Cottonwood Creek Valley (c)	(801)	7,221	5,042
*84321	Logan	(435)	42,670	32,771
84044	Magna (c)	(801)	22,770	17,829
84664	Mapleton	(801)	5,809	3,572
84047	Midvale	(801)	27,029	11,886
84109	Millcreek (c)	(801)	30,377	32,230
84117	Mount Olympus (c)	(801)	7,103	7,413
84157	Murray	(801)	34,024	31,274
84341	North Logan	(435)	6,163	3,775
84404	North Ogden	(385)	15,026	11,593
84054	North Salt Lake	(801)	8,749	6,464
*84401	Ogden	(385)	77,226	63,943
—	Oquirrh (c)	(801)	10,390	7,593
*84057	Orem	(801)	84,324	67,561
*84060	Park City	(801)	7,371	4,468
84651	Payson	(385)	12,716	9,510
84062	Pleasant Grove	(385)	23,468	13,476
84404	Pleasant View	(385)	5,632	3,597
84501	Price	(435)	8,402	8,712
*84601	Provo	(385)	105,166	86,835
84701	Richfield	(435)	6,847	5,593
84403	Riverdale	(385)	7,656	6,419
84065	Riverton	(801)	25,011	11,261
84067	Roy	(385)	32,885	24,560
*84770	Saint George	(435)	49,663	28,572
*84101	Salt Lake City	(801)	181,743	159,928
*84070	Sandy	(801)	88,418	75,240
84335	Smithfield	(435)	7,261	5,566
84095	South Jordan	(801)	29,437	12,215
84403	South Ogden	(385)	14,377	12,105
84165	South Salt Lake	(801)	22,038	10,129
84660	Spanish Fork	(385)	20,246	11,272
84663	Springville	(385)	20,424	13,950
84098	Summit Park (c)		6,597	—
84015	Sunset	(385)	5,204	5,128
84075	Syracuse	(385)	9,398	4,658
84107	Taylorsville	(801)	57,439	51,550
84074	Tooele	(435)	22,502	13,887
84337	Tremonton	(435)	5,592	4,262
*84078	Vernal	(435)	7,714	6,640
84780	Washington	(435)	8,186	4,198
84403	Washington Terrace	(385)	8,551	8,189
*84084	West Jordan	(801)	68,336	42,915
84015	West Point	(385)	6,033	4,258
84170	West Valley City	(801)	108,896	86,969
84070	White City (c)	(801)	5,988	6,506
84087	Woods Cross	(385)	6,419	5,384

Vermont (802)

See introductory note.

ZIP	Place	2000	1990
05641	Barre	9,291	9,482
05641	*Barre*	7,602	7,411
05201	Bennington (c)	9,168	9,532
05201	*Bennington*	15,737	16,451
*05301	*Brattleboro*	12,005	12,241
*05301	Brattleboro (c)	8,289	8,612
*05401	Burlington	38,889	39,127
*05446	*Colchester*	16,986	14,731
05451	*Essex*	18,626	16,498
*05452	Essex Junction	8,591	8,396
05047	*Hartford*	10,367	9,404
05465	*Jericho*	5,015	1,405
05849	*Lyndon*	5,448	5,371
05753	Middlebury (c)	6,252	6,007
*05753	*Middlebury*	8,183	8,034
05468	*Milton*	9,479	8,404
*05602	Montpelier	8,035	8,247
05661	*Morristown*	5,139	4,733
05855	Newport	5,005	4,434
05663	*Northfield*	5,791	5,610
05101	*Rockingham*	5,309	5,484
*05701	Rutland	17,292	18,230
05478	Saint Albans	7,650	7,339
05478	*Saint Albans*	5,086	4,606
05819	Saint Johnsbury (c)	6,319	6,424
05819	*Saint Johnsbury*	7,571	7,608
05482	*Shelburne*	6,944	5,871
*05401	South Burlington	15,814	12,809
05156	*Springfield*	9,078	9,579
05488	*Swanton*	6,203	5,636
05495	*Williston*	7,650	4,887
05404	Winooski	6,561	6,649

Virginia

Area code (571) overlays area code (703). See introductory note.

ZIP	Place	Area Code	2000	1990
*24210	Abingdon	(276)	7,780	7,003
*22313	Alexandria	(703)	128,283	111,182
22003	Ananndale (c)	(703)	54,994	50,975
22554	Aquia Harbour (c)	(703)	7,856	6,308
*22210	Arlington (c)	(703)	189,453	170,897
23005	Ashland	(804)	6,619	5,864
*22041	Bailey's Crossroads (c)	(703)	23,166	19,507
24523	Bedford	(540)	6,299	6,177
22306	Belle Haven (c)	(757)	6,269	6,427
23234	Bellwood (c)	(804)	5,974	6,178
23234	Bensley (c)	(804)	5,435	5,093
*24060	Blacksburg	(540)	39,573	34,590
24605	Bluefield	(276)	5,078	5,363
23235	Bon Air (c)	(804)	16,213	16,413
22812	Bridgewater	(540)	5,203	3,918
*24203	Bristol	(276)	17,367	18,426
24416	BuenaVista	(540)	6,349	6,406
—	Bull Run (c)	(703)	11,337	5,525
*22150	Burke (c)	(703)	57,737	57,734
24018	Cave Spring (c)	(540)	24,941	24,053
*20120	Centreville (c)	(703)	48,661	26,585
*20151	Chantilly (c)	(703)	41,041	29,337
*22906	Charlottesville	(434)	45,049	40,475
*23320	Chesapeake	(757)	199,184	151,982
*23831	Chester (c)	(804)	17,890	14,986
*24073	Christiansburg	(540)	16,947	15,004
24078	Collinsville (c)	(276)	7,777	7,280
23834	Colonial Heights	(804)	16,897	16,064
24426	Covington	(540)	6,303	7,198
22701	Culpeper	(540)	9,664	8,581
22193	Dale City (c)	(540)	55,971	47,170
*24541	Danville	(434)	48,411	53,056
23228	Dumbarton (c)	(804)	6,674	8,526
22027	Dunn Loring (c)	(703)	7,861	6,509
23222	East Highland Park (c)	(804)	12,488	11,850
23847	Emporia	(434)	5,665	5,479
23803	Ettrick (c)	(804)	5,627	5,290
*22030	Fairfax	(703)	21,498	19,894
*22046	Falls Church	(703)	10,377	9,522
23901	Farmville	(434)	6,845	6,505
24551	Forest (c)	(434)	8,006	5,624
22060	Fort Belvoir (c)	(703)	7,176	8,590
22308	Fort Hunt (c)	(703)	12,923	12,989
23801	Fort Lee (c)	(804)	7,269	6,895
22310	Franconia (c)	(703)	31,907	19,882
23851	Franklin	(757)	8,346	7,864
*22404	Fredericksburg	(540)	19,279	19,027
22630	Front Royal	(540)	13,589	11,880
24333	Galax	(276)	6,837	6,699
*23060	Glen Allen (c)	(804)	12,562	9,010
23062	Gloucester Point (c)	(804)	9,429	8,509
22066	Great Falls (c)	(703)	8,549	6,945
22306	Groveton (c)	(703)	21,296	19,997
*23670	Hampton	(757)	146,437	133,811
*22801	Harrisonburg	(540)	40,468	30,707
*20170	Herndon	(703)	21,655	16,139
23075	Highland Springs (c)	(804)	15,137	13,823
24019	Hollins (c)	(540)	14,309	13,305
23860	Hopewell	(804)	22,354	23,101
22303	Huntington (c)	(703)	8,325	7,489
22306	Hybla Valley (c)	(703)	16,721	15,491
22043	Idylwood (c)	(703)	16,005	14,710
22042	Jefferson (c)	(703)	27,422	25,782
22041	Lake Barcroft (c)	(703)	8,906	8,686
22963	Lake Monticello (c)	(434)	6,852	2,331
22191	Lake Ridge (c)	(540)	30,404	23,862
23228	Lakeside (c)	(804)	11,157	12,081
23060	Laurel (c)	(804)	14,875	13,011
*20175	Leesburg	(703)	28,311	16,202
24450	Lexington	(540)	6,867	6,959
22312	Lincolnia (c)	(703)	15,788	13,041
—	Linton Hall (c)		8,620	—
*22079	Lorton (c)	(703)	17,786	15,385
*24506	Lynchburg	(434)	65,269	66,049
*22101	McLean (c)	(703)	38,929	38,168
24572	Madison Heights (c)	(434)	11,584	11,700
*20110	Manassas	(703)	35,135	27,957
20113	Manassas Park	(703)	10,290	6,734
22030	Mantua (c)	(703)	7,485	6,804
24354	Marion	(276)	6,349	6,630
*24112	Martinsville	(276)	15,416	16,162
*23111	Mechanicsville (c)	(804)	30,464	22,027
*22116	Merrifield (c)	(703)	11,170	8,399
—	Montclair (c)		15,728	11,399
23231	Montrose (c)	(804)	7,018	6,405
22121	Mount Vernon (c)	(703)	28,582	27,485
22122	Newington (c)	(703)	19,784	17,965
*23607	Newport News	(757)	180,150	171,439
*23501	Norfolk	(757)	234,403	261,250
22151	North Springfield (c)	(703)	9,173	8,996
22124	Oakton (c)	(703)	29,348	24,610
*23804	Petersburg	(804)	33,740	37,027
22043	Pimmit Hills (c)	(703)	6,152	6,019
23662	Poquoson	(757)	11,566	11,005
*23707	Portsmouth	(757)	100,565	103,910

ZIP	Place	Area Code	2000	1990
24301	Pulaski	(540)	9,473	9,985
22134	Quantico Station (c)	(703)	6,571	7,425
*24141	Radford	(540)	15,859	15,940
*20190	Reston (c)	(703)	56,407	48,556
*23232	Richmond	(804)	197,790	202,798
*24022	Roanoke	(540)	94,911	96,509
24281	Rose Hill (c)	(276)	15,058	12,675
24153	Salem	(540)	24,747	23,797
22044	Seven Corners (c)	(703)	8,701	7,280
*23430	Smithfield	(757)	6,324	4,686
24592	South Boston	(434)	8,491	6,997
*22150	Springfield (c)	(703)	30,417	23,706
*24402	Staunton	(540)	23,853	24,461
24477	Stuarts Draft (c)	(540)	8,367	5,087
23162	Sudley (c)	(540)	7,719	7,321
*23434	Suffolk	(757)	63,677	52,143
24502	Timberlake (c)	(434)	10,683	10,314
22172	Triangle (c)	(703)	5,500	4,740
23229	Tuckahoe (c)	(804)	43,242	42,629
22101	Tysons Corner (c)	(703)	18,540	13,124
*22180	Vienna	(703)	14,453	14,852
24179	Vinton	(540)	7,782	7,643
*23450	Virginia Beach	(757)	425,257	393,089
*20186	Warrenton	(540)	6,670	4,882
22980	Waynesboro	(540)	19,520	18,549
22110	West Gate (c)	(703)	7,493	6,565
22152	West Springfield (c)	(703)	28,378	28,126
*23185	Williamsburg	(757)	11,998	11,409
*22601	Winchester	(540)	23,585	21,947
24592	Wolf Trap (c)	(703)	14,001	13,133
*22191	Woodbridge (c)	(540)	31,941	26,401
—	Wyndham (c)		6,176	—
24382	Wytheville	(276)	7,804	8,036
22110	Yorkshire (c)	(703)	6,732	5,699

Washington

ZIP	Place	Area Code	2000	1990
98520	Aberdeen	(360)	16,461	16,565
98036	Alderwood Manor (c)	(425)	15,329	22,945
98221	Anacortes	(360)	14,557	11,451
98223	Arlington	(360)	11,713	4,037
98335	Artondale (c)	(253)	8,630	7,141
*98002	Auburn	(253)	40,314	33,650
98110	Bainbridge Island	(206)	20,308	—
98315	Bangor Trident Base (c)	(360)	7,253	3,702
98604	Battle Ground	(360)	9,296	3,758
*98009	Bellevue	(425)	109,569	95,213
*98225	Bellingham	(360)	67,171	52,179
98390	Bonney Lake	(360)	9,687	7,494
*98011	Bothell	(425)	30,150	12,575
*98337	Bremerton	(360)	37,259	38,142
98036	Brier	(425)	6,383	5,633
98178	Bryn Mawr-Skyway (c)	(206)	13,977	12,514
98166	Burien	(206)	31,881	27,507
98233	Burlington	(360)	6,757	4,349
—	Camano (c)		13,347	—
98607	Camas	(360)	12,534	6,762
98055	Cascade-Fairwood (c)	(425)	34,580	30,107
98531	Centralia	(360)	14,742	12,101
98532	Chehalis	(360)	7,057	6,527
99004	Cheney	(509)	8,832	7,723
99403	Clarkston	(509)	7,337	6,753
—	Clarkston Heights-Vineland (c)		6,117	2,832
99324	College Place	(509)	7,818	6,308
98072	Cottage Lake (c)	(206)	24,330	—
99218	Country Homes (c)	(509)	5,203	5,126
98042	Covington	(253)	13,783	—
98198	Des Moines	(206)	29,267	20,830
99213	Dishman (c)	(509)	10,031	9,671
—	East Hill-Meridian (c)		29,308	42,696
98366	East Port Orchard (c)	(360)	5,116	5,409
98056	East Renton Highlands (c)	(425)	13,264	13,218
98802	East Wenatchee	(509)	5,757	3,886
98801	East Wenatchee Bench (c)	(509)	13,658	12,539
98371	Edgewood	(253)	9,089	8,702
*98020	Edmonds	(425)	39,515	30,743
98387	Elk Plain (c)		15,697	12,197
98926	Ellensburg	(509)	15,414	12,360
98022	Enumclaw	(360)	11,116	7,243
98823	Ephrata	(509)	6,808	5,349
*98201	Everett	(425)	91,488	70,937
99218	Fairwood (c)	(509)	6,764	5,807
*98002	Federal Way	(253)	83,259	67,535
98685	Felida (c)	(360)	5,683	3,109
98248	Ferndale	(360)	8,758	5,398
99336	Finley (c)	(509)	5,770	4,897
98466	Fircrest	(253)	5,868	5,270
98597	Five Corners (c)		12,207	6,776
98433	Fort Lewis (c)	(253)	19,089	22,224
98373	Frederickson (c)	(206)	5,758	3,502
*98329	Gig Harbor	(253)	6,465	3,236
98338	Graham (c)	(253)	8,739	—
98930	Grandview	(509)	8,377	7,169
99016	Green Acres (c)	(509)	5,158	4,626
98660	Hazel Dell North (c)	(360)	9,261	6,924
98665	Hazel Dell South (c)	(360)	6,605	5,796
98025	Hobart (c)		6,251	—
—	Hockinson (c)	(360)	5,136	—

ZIP	Place	Area Code	2000	1990
98550	Hoquiam	(360)	9,097	8,972
98011	Inglewood-Finn Hill (c)	(425)	22,661	29,132
*98027	Issaquah	(425)	11,212	7,786
98626	Kelso	(360)	11,895	11,767
98028	Kenmore (c)	(425)	18,678	8,917
*99336	Kennewick	(509)	54,693	42,148
*98031	Kent	(253)/(425)	79,524	37,960
98033	Kingsgate (c)	(425)	12,222	14,259
*98033	Kirkland	(425)	45,054	40,059
98509	Lacey	(360)	31,226	19,279
98155	Lake Forest North (c)	(206)	13,142	8,002
98002	Lakeland North (c)	(253)	15,085	14,402
98002	Lakeland South (c)	(253)	11,436	9,027
—	Lake Morton-Berrydale (c)		9,659	—
98665	Lake Shore (c)	(360)	6,670	6,268
98258	Lake Stevens	(425)	6,361	3,435
98259	Lakewood	(253)	58,211	55,937
—	Lea Hill (c)		10,871	6,876
98632	Longview	(360)	34,660	31,499
98264	Lynden	(360)	9,020	5,709
*98046	Lynnwood	(425)	33,847	28,637
98290	Maltby (c)	(360)	8,267	—
98038	Maple Valley	(425)	14,209	1,211
98012	Martha Lake (c)	(425)	12,633	10,155
*98270	Marysville	(360)	25,315	12,248
98040	Mercer Island	(206)	22,036	20,816
98444	Midland (c)	(253)	7,414	5,587
98082	Mill Creek	(425)	11,525	7,180
—	Mill Plain (c)	(360)	7,400	—
98354	Milton	(253)	5,795	4,995
98661	Minnehaha (c)	(360)	7,689	9,661
98272	Monroe	(360)	13,795	4,275
98837	Moses Lake	(509)	14,953	11,235
98043	Mountlake Terrace	(425)	20,362	19,320
*98273	Mount Vernon	(360)	26,232	17,647
—	Mount Vista (c)		5,770	—
98275	Mukilteo	(425)	18,019	11,575
98059	Newcastle	(425)	7,737	4,649
98166	Normandy Park	(206)	6,392	6,794
—	North Creek (c)		25,742	23,236
98270	North Marysville (c)	(425)	21,161	18,711
98277	Oak Harbor	(360)	19,795	17,176
*98501	Olympia	(360)	42,514	33,729
99214	Opportunity (c)	(509)	25,065	22,326
98662	Orchards (c)	(360)	17,852	—
*99327	Othello	(509)	5,847	4,638
99027	Otis Orchards-East Farms (c)	(360)	6,318	5,811
98047	Pacific	(253)	5,527	4,622
—	Paine Field-Lake Stickney (c)		24,383	18,670
98444	Parkland (c)	(253)	24,053	20,882
98366	Parkwood (c)	(360)	7,213	6,853
*99301	Pasco	(509)	32,066	20,337
—	Picnic Point-North Lynnwood (c)		22,953	—
*98362	Port Angeles	(360)	18,397	17,710
*98366	Port Orchard	(360)	7,693	4,984
98368	Port Townsend	(360)	8,334	7,001
98370	Poulsbo	(360)	6,813	4,848
98390	Prairie Ridge (c)		11,688	8,278
*99163	Pullman	(509)	24,675	23,478
*98371	Puyallup	(253)	33,011	23,878
98848	Quincy	(509)	5,044	3,738
*98052	Redmond	(425)	45,256	35,800
*98058	Renton	(425)	50,052	41,688
99352	Richland	(509)	38,708	32,315
98188	Riverton-Boulevard Park (c)	(206)	11,188	15,337
98686	Salmon Creek (c)	(360)	16,767	11,989
*98074	Sammamish	(425)	34,104	—
*98148	Seatac	(206)	25,496	22,760
*98101	Seattle	(206)/(425)	563,374	516,259
—	Seattle Hill-Silver Firs (c)		35,311	—
98284	Sedro-Woolley	(360)	8,658	6,333
98942	Selah	(509)	6,310	5,113
98584	Shelton	(360)	8,442	7,241
*98133	Shoreline	(206)	53,025	46,979
*98315	Silverdale (c)	(360)	15,816	7,660
*98290	Snohomish	(360)	8,494	6,499
98373	South Hill (c)		31,623	12,963
98387	Spanaway (c)	(253)	21,588	15,001
*99210	Spokane	(509)	195,629	177,165
98388	Steilacoom	(253)	6,049	5,728
*98371	Summit (c)	(253)	8,041	6,312
98390	Sumner	(253)	8,504	7,535
98944	Sunnyside	(509)	13,905	11,238
*98402	Tacoma	(253)	193,556	176,664
98501	Tanglewilde-Thompson Place (c)	(360)	5,670	6,061
—	Terrace Heights (c)		6,447	4,223
98948	Toppenish	(509)	8,946	7,419
98138	Tukwila	(206)	17,181	14,506
98501	Tumwater	(360)	12,698	9,976
*98901	Union Gap	(509)	5,621	3,120
—	Union Hill-Novelty Hill (c)		11,265	—
98467	University Place	(253)	29,933	26,724
*98661	Vancouver	(360)	143,560	62,065
*98013	Vashon (c)	(206)	10,123	—
99037	Veradale (c)	(509)	9,387	7,836
99362	Walla Walla	(509)	29,686	26,482

ZIP	Place	Area Code	2000	1990
—	Waller (c)		9,200	6,415
—	Walnut Grove (c)		7,164	3,906
98671	Washougal	(360)	8,595	4,764
*98801	Wenatchee	(509)	27,856	21,746
—	West Lake Sammamish (c)		5,937	6,087
98258	West Lake Stevens (c)	(425)	18,071	12,453
99353	West Richland	(509)	8,385	3,962
99181	West Valley (c)		10,433	6,594
98166	White Center (c)	(206)	20,975	20,531
98072	Woodinville	(425)	9,194	7,628
*98903	Yakima	(509)	71,845	58,427

West Virginia (304)

ZIP	Place	2000	1990
*25801	Beckley	17,254	18,274
24701	Bluefield	11,451	12,756
26330	Bridgeport	7,306	6,837
26201	Buckhannon	5,725	5,909
*25301	Charleston	53,421	57,287
*26507	Cheat Lake (c)	6,396	3,992
*26301	Clarksburg	16,743	17,970
25301	Cross Lanes (c)	10,353	10,878
25064	Dunbar	8,154	8,697
26241	Elkins	7,032	7,494
*26554	Fairmont	19,097	20,210
26354	Grafton	5,489	5,524
*25704	Huntington	51,475	54,844
25526	Hurricane	5,222	4,461
26726	Keyser	5,303	5,870
*25401	Martinsburg	14,972	14,073
*26505	Morgantown	26,809	25,879
26041	Moundsville	9,998	10,753
26155	New Martinsville	5,984	6,705
25143	Nitro	6,824	6,851
25901	Oak Hill	7,589	6,812
*26101	Parkersburg	33,099	33,862
—	Pea Ridge (c)	6,363	6,535
24740	Princeton	6,347	7,043
25177	Saint Albans	11,567	12,241
25303	South Charleston	13,390	13,645
25569	Teays Valley (c)	12,704	8,436
26105	Vienna	10,861	10,862
26062	Weirton	20,411	22,124
26003	Wheeling	31,419	34,882

Wisconsin

ZIP	Place	Area Code	2000	1990
54301	Allouez	(920)	15,443	14,431
54720	Altoona	(715)	6,698	5,889
54409	Antigo	(715)	8,560	8,284
*59411	Appleton	(920)	70,087	65,695
54806	Ashland	(715)	8,620	8,695
54304	Ashwaubenon	(920)	17,634	16,376
53913	Baraboo	(608)	10,711	9,203
53916	Beaver Dam	(920)	15,169	14,196
54311	Bellevue Town (c)	(920)	11,828	7,541
*53511	Beloit	(608)	35,775	35,571
54923	Berlin	(920)	5,305	5,371
*53045	Brookfield	(262)	38,649	35,184
53209	Brown Deer	(414)	12,170	12,236
53105	Burlington	(262)	9,936	8,851
53012	Cedarburg	(262)	10,908	10,086
54729	Chippewa Falls	(715)	12,925	12,749
53110	Cudahy	(414)	18,429	18,659
53532	De Forest	(608)	7,368	4,882
53018	Delafield	(262)	6,472	5,347
53115	Delavan	(262)	7,956	6,073
54115	De Pere	(920)	20,559	16,594
*54703	Eau Claire	(715)	61,704	56,806
53121	Elkhorn	(262)	7,305	5,337
53122	Elm Grove	(262)	6,249	6,261
53714	Fitchburg	(608)	20,501	15,648
*54935	Fond du Lac	(920)	42,203	37,755
53538	Fort Atkinson	(920)	11,621	10,213
53217	Fox Point	(414)	7,012	7,238
53132	Franklin	(414)	29,494	21,855
53022	Germantown	(262)	18,260	13,658
53209	Glendale	(414)	13,367	14,088
53024	Grafton	(262)	10,312	9,340
*54303	Green Bay	(920)	102,313	96,466
53129	Greendale	(414)	14,405	15,128
53220	Greenfield	(414)	35,476	33,403
53130	Hales Corners	(414)	7,765	7,623
53027	Hartford	(262)	10,905	8,188
53029	Hartland	(262)	7,905	6,906
54636	Holmen	(608)	6,200	3,220
54303	Howard	(920)	13,546	9,874
54016	Hudson	(715)	8,775	6,378
*53545	Janesville	(608)	59,498	52,210
53549	Jefferson	(920)	7,338	6,078
54130	Kaukauna	(920)	12,983	11,982
*53140	Kenosha	(262)	90,352	80,426
54136	Kimberly	(920)	6,146	5,406
*54601	La Crosse	(608)	51,818	51,140
53147	Lake Geneva	(262)	7,148	5,979
54140	Little Chute	(920)	10,476	9,207
53558	McFarland	(608)	6,416	5,232
*53714	Madison	(608)	208,054	190,766
*54220	Manitowoc	(920)	34,053	32,521
54143	Marinette	(715)	11,749	11,843
54449	Marshfield	(715)	18,800	19,293
54952	Menasha	(920)	16,331	14,711
*53051	Menomonee Falls	(262)	32,647	26,840
54751	Menomonie	(715)	14,937	13,547
53097	Mequon	(414)	21,823	18,885
54452	Merrill	(715)	10,146	9,860
53562	Middleton	(608)	15,770	13,785
*53201	Milwaukee	(414)	596,974	628,088
53716	Monona	(608)	8,018	8,637
53566	Monroe	(608)	10,843	10,241
53572	Mount Horeb	(608)	5,860	4,182
53149	Mukwonago	(262)	6,162	4,495
53150	Muskego	(414)	21,397	16,813
*54956	Neenah	(920)	24,507	23,219
*53186	New Berlin	(262)	38,220	33,592
54961	New London	(920)	7,085	6,658
54017	New Richmond	(715)	6,310	5,106
53154	Oak Creek	(414)	28,456	19,513
53066	Oconomowoc	(262)	12,382	10,993
54650	Onalaska	(608)	14,839	12,201
53575	Oregon	(608)	7,514	4,519
*54901	Oshkosh	(920)	62,916	55,006
53072	Pewaukee	(262)	11,783	5,287
53818	Platteville	(608)	9,989	9,862
53158	Pleasant Prairie	(262)	16,136	12,037
54467	Plover	(715)	10,520	8,176
53073	Plymouth	(920)	7,781	6,769
53901	Portage	(608)	9,728	8,640
53074	Port Washington	(262)	10,467	9,338
53821	Prairie du Chien	(608)	6,018	5,657
*53401	Racine	(262)	81,855	84,298
53959	Reedsburg	(608)	7,827	5,834
54501	Rhinelander	(715)	7,735	7,382
54401	Rib Mountain (c)	(715)	6,059	4,634
54868	Rice Lake	(715)	8,320	7,998
53581	Richland Center	(608)	5,114	5,018
54971	Ripon	(920)	6,828	7,241
54022	River Falls	(715)	12,560	10,610
53235	Saint Francis	(414)	8,662	9,245
54166	Shawano	(715)	8,298	7,598
*53081	Sheboygan	(920)	50,792	49,587
53085	Sheboygan Falls	(920)	6,772	5,823
53211	Shorewood	(414)	13,763	14,116
53172	South Milwaukee	(414)	21,256	20,958
54656	Sparta	(608)	8,648	7,788
54481	Stevens Point	(715)	24,551	23,002
53589	Stoughton	(608)	12,354	8,786
54235	Sturgeon Bay	(920)	9,437	9,176
53177	Sturtevant	(262)	5,287	3,803
53590	Sun Prairie	(608)	20,369	15,352
54880	Superior	(715)	27,368	27,134
53089	Sussex	(262)	8,828	5,039
54660	Tomah	(608)	8,419	7,572
53181	Twin Lakes	(262)	5,124	3,989
54241	Two Rivers	(920)	12,639	13,030
53593	Verona	(608)	7,052	5,374
*53094	Watertown	(920)	21,598	19,142
*53186	Waukesha	(262)	64,825	56,894
53597	Waunakee	(608)	8,995	5,897
54981	Waupaca	(715)	5,676	4,946
53963	Waupun	(920)	10,718	8,844
*54403	Wausau	(715)	38,426	37,060
53213	Wauwatosa	(414)	47,271	49,366
53214	West Allis	(414)	61,254	63,221
*53095	West Bend	(262)	28,152	24,470
54476	Weston (c)	(715)	12,079	9,714
53217	Whitefish Bay	(414)	14,163	14,272
53190	Whitewater	(262)	13,437	12,636
53185	Wind Lake (c)	(262)	5,202	3,748
*54494	Wisconsin Rapids	(715)	18,435	18,245

Wyoming (307)

ZIP	Place	2000	1990
*82609	Casper	49,644	46,765
*82009	Cheyenne	53,011	50,008
82414	Cody	8,835	7,897
82633	Douglas	5,288	5,076
*82930	Evanston	11,507	10,904
*82716	Gillette	19,646	17,545
82935	Green River	11,808	12,711
*83002	Jackson	8,647	4,708
82520	Lander	6,867	7,023
*82072	Laramie	27,204	26,687
82435	Powell	5,373	5,292
82301	Rawlins	8,538	9,380
82501	Riverton	9,310	9,202
*82901	Rock Springs	18,708	19,050
82801	Sheridan	15,804	13,904
82240	Torrington	5,776	5,651
82401	Worland	5,250	5,742

Populations and Areas of Counties and States

Source: U.S. Bureau of the Census, Dept. of Commerce; World Almanac research

Land areas, state population figures, and 2000 county population figures are from the 2000 U.S. census; 1990 county population figures are from the 1990 U.S. census. County areas may not add to total state areas because of rounding.

Alabama

(67 counties, 50,744 sq. mi. land; pop. 4,447,100)

County	County seat or courthouse	2000 Pop.	1990 Pop.	Land area sq. mi.
Autauga	Prattville	43,671	34,222	596
Baldwin	Bay Minette	140,415	98,280	1,596
Barbour	Clayton	29,038	25,417	885
Bibb	Centreville	20,826	16,598	623
Blount	Oneonta	51,024	39,248	646
Bullock	Union Springs	11,714	11,042	625
Butler	Greenville	21,399	21,892	777
Calhoun	Anniston	112,249	116,032	608
Chambers	Lafayette	36,583	36,876	597
Cherokee	Centre	23,988	19,543	553
Chilton	Clanton	39,593	32,458	694
Choctaw	Butler	15,922	16,018	914
Clarke	Grove Hill	27,867	27,240	1,238
Clay	Ashland	14,254	13,252	605
Cleburne	Heflin	14,123	12,730	553
Coffee	Elba	43,615	40,240	679
Colbert	Tuscumbia	54,984	51,666	595
Conecuh	Evergreen	14,089	14,054	851
Coosa	Rockford	12,202	11,063	652
Covington	Andalusia	37,631	36,478	1,034
Crenshaw	Luverne	13,665	13,635	610
Cullman	Cullman	77,483	67,613	738
Dale	Ozark	49,129	49,633	561
Dallas	Selma	46,365	48,130	981
De Kalb	Fort Payne	64,452	54,651	778
Elmore	Wetumpka	65,874	49,210	621
Escambia	Brewton	38,440	35,518	947
Etowah	Gadsden	103,459	99,840	535
Fayette	Fayette	18,495	17,962	628
Franklin	Russellville	31,223	27,814	636
Geneva	Geneva	25,764	23,647	576
Greene	Eutaw	9,974	10,153	646
Hale	Greensboro	17,185	15,498	644
Henry	Abbeville	16,310	15,374	562
Houston	Dothan	88,787	81,331	580
Jackson	Scottsboro	53,926	47,796	1,079
Jefferson	Birmingham	662,047	651,520	1,113
Lamar	Vernon	15,904	15,715	605
Lauderdale	Florence	87,966	79,661	669
Lawrence	Moulton	34,803	31,513	693
Lee	Opelika	115,092	87,146	609
Limestone	Athens	65,676	54,135	568
Lowndes	Hayneville	13,473	12,658	718
Macon	Tuskegee	24,105	24,928	611
Madison	Huntsville	276,700	238,912	805
Marengo	Linden	22,539	23,084	977
Marion	Hamilton	31,214	29,830	741
Marshall	Guntersville	82,231	70,832	567
Mobile	Mobile	399,843	378,643	1,233
Monroe	Monroeville	24,324	23,968	1,026
Montgomery	Montgomery	223,510	209,085	790
Morgan	Decatur	111,064	100,043	582
Perry	Marion	11,861	12,759	719
Pickens	Carrollton	20,949	20,699	881
Pike	Troy	29,605	27,595	671
Randolph	Wedowee	22,380	19,881	581
Russell	Phenix City	49,756	46,860	641
Saint Clair	Ashville & Pell City	64,742	49,811	634
Shelby	Columbiana	143,293	99,363	795
Sumter	Livingston	14,798	16,174	905
Talladega	Talladega	80,321	74,109	740
Tallapoosa	Dadeville	41,475	38,826	718
Tuscaloosa	Tuscaloosa	164,875	150,500	1,324
Walker	Jasper	70,713	67,670	794
Washington	Chatom	18,097	16,694	1,081
Wilcox	Camden	13,183	13,568	889
Winston	Double Springs	24,843	22,053	614

Alaska

(27 divisions, 571,951 sq. mi. land; pop. 626,932)

Census Division	2000 Pop.	1990 Pop.	Land area sq. mi.
Aleutians East Borough	2,697	2,464	6,988
Aleutians West Census Area	5,465	9,478	4,397
Anchorage Borough	260,283	226,338	1,697
Bethel Census Area	16,006	13,660	40,633
Bristol Bay Borough	1,258	1,410	505
Denali Borough	1,893	1,682	12,750
Dillingham Census Area	4,922	4,010	18,675
Fairbanks North Star Borough	82,840	77,720	7,366
Haines Borough	2,392	2,117	2,344
Juneau Borough	30,711	26,752	2,717
Kenai Peninsula Borough	49,691	40,802	16,013
Ketchikan Gateway Borough	14,070	13,828	1,233
Kodiak Island Borough	13,913	13,309	6,560
Lake and Peninsula Borough	1,823	1,666	23,782
Matanuska-Susitna Borough	59,322	39,683	24,682
Nome Census Area	9,196	8,288	23,001
North Slope Borough	7,385	5,986	88,817
Northwest Arctic Borough	7,208	6,106	35,898
Prince of Wales-Outer Ketchikan Census Area	6,146	6,278	7,411
Sitka Borough	8,835	8,588	2,874
Skagway-Hoonah-Angoon Census Area	3,436	3,679	7,896
Southeast Fairbanks Census Area	6,174	5,925	24,815
Valdez-Cordova Census Area	10,195	9,920	34,319
Wade Hampton Census Area	7,028	5,789	17,194
Wrangell-Petersburg Census Area	6,684	7,042	5,835
Yakutat Borough	808	725	7,650
Yukon-Koyukuk Census Area	6,551	6,798	145,900

Arizona

(15 counties, 113,635 sq. mi. land; pop. 5,130,632)

County	County seat or courthouse	2000 Pop.	1990 Pop.	Land area sq. mi.
Apache	Saint Johns	69,423	61,591	11,205
Cochise	Bisbee	117,755	97,624	6,169
Coconino	Flagstaff	116,320	96,591	18,617
Gila	Globe	51,335	40,216	4,768
Graham	Safford	33,489	26,554	4,629
Greenlee	Clifton	8,547	8,008	1,847
La Paz	Parker	19,715	13,844	4,500
Maricopa	Phoenix	3,072,149	2,122,101	9,203
Mohave	Kingman	155,032	93,497	13,312
Navajo	Holbrook	97,470	77,674	9,953
Pima	Tucson	843,746	666,957	9,186
Pinal	Florence	179,727	116,397	5,370
Santa Cruz	Nogales	38,381	29,676	1,238
Yavapai	Prescott	167,517	107,714	8,123
Yuma	Yuma	160,026	106,895	5,514

Arkansas

(75 counties, 52,068 sq. mi. land; pop. 2,673,400)

County	County seat or courthouse	2000 Pop.	1990 Pop.	Land area sq. mi.
Arkansas	DeWitt & Stuttgart	20,749	21,653	988
Ashley	Hamburg	24,209	24,319	921
Baxter	Mountain Home	38,386	31,186	554
Benton	Bentonville	153,406	97,530	846
Boone	Harrison	33,948	28,297	591
Bradley	Warren	12,600	11,793	651
Calhoun	Hampton	5,744	5,826	628
Carroll	Berryville & Eureka Springs	25,357	18,623	630
Chicot	Lake Village	14,117	15,713	644
Clark	Arkadelphia	23,546	21,437	865
Clay	Corning & Piggott	17,609	18,107	639
Cleburne	Heber Springs	24,046	19,411	553
Cleveland	Rison	8,571	7,781	595
Columbia	Magnolia	25,603	25,691	766
Conway	Morrilton	20,336	19,151	556
Craighead	Jonesboro & Lake City	82,148	68,956	711
Crawford	Van Buren	53,247	42,493	595
Crittenden	Marion	50,866	49,939	610
Cross	Wynne	19,526	19,225	616
Dallas	Fordyce	9,210	9,614	667
Desha	Arkansas City	15,341	16,798	765
Drew	Monticello	18,723	17,369	828
Faulkner	Conway	86,014	60,006	647
Franklin	Charleston & Ozark	17,771	14,897	610
Fulton	Salem	11,642	10,037	618
Garland	Hot Springs	88,068	73,397	677
Grant	Sheridan	16,464	13,948	632
Greene	Paragould	37,331	31,804	578
Hempstead	Hope	23,587	21,621	729
Hot Spring	Malvern	30,353	26,115	615
Howard	Nashville	14,300	13,569	587
Independence	Batesville	34,233	31,192	764
Izard	Melbourne	13,249	11,364	581
Jackson	Newport	18,418	18,944	634
Jefferson	Pine Bluff	84,278	85,487	885
Johnson	Clarksville	22,781	18,221	662
Lafayette	Lewisville	8,559	9,643	527
Lawrence	Walnut Ridge	17,774	17,455	587
Lee	Marianna	12,580	13,053	602
Lincoln	Star City	14,492	13,690	561
Little River	Ashdown	13,628	13,966	532
Logan	Booneville & Paris	22,486	20,557	710
Lonoke	Lonoke	52,828	39,268	766
Madison	Huntsville	14,243	11,618	837
Marion	Yellville	16,140	12,001	598
Miller	Texarkana	40,443	38,467	624
Mississippi	Blytheville & Osceola	51,979	57,525	898

County	County seat or courthouse	2000 Pop.	1990 Pop.	Land area sq. mi.
Monroe	Clarendon	10,254	11,333	607
Montgomery	Mount Ida	9,245	7,841	781
Nevada	Prescott	9,955	10,101	620
Newton	Jasper	8,608	7,666	823
Ouachita	Camden	28,790	30,574	732
Perry	Perryville	10,209	7,969	551
Phillips	Helena	26,445	28,830	693
Pike	Murfreesboro	11,303	10,086	603
Poinsett	Harrisburg	25,614	24,664	758
Polk	Mena	20,229	17,347	859
Pope	Russellville	54,469	45,883	812
Prairie	Des Arc & De Valls Bluff	9,539	9,518	646
Pulaski	Little Rock	361,474	349,773	771
Randolph	Pocahontas	18,195	16,558	652
Saint Francis	Forrest City	29,329	28,497	634
Saline	Benton	83,529	64,183	723
Scott	Waldron	10,996	10,205	894
Searcy	Marshall	8,261	7,841	667
Sebastian	Fort Smith & Greenwood	115,071	99,590	536
Sevier	De Queen	15,757	13,637	564
Sharp	Ash Flat	17,119	14,109	604
Stone	Mountain View	11,499	9,775	607
Union	El Dorado	45,629	46,719	1,039
Van Buren	Clinton	16,192	14,008	712
Washington	Fayetteville	157,715	113,409	950
White	Searcy	67,165	54,676	1,034
Woodruff	Augusta	8,741	9,520	587
Yell	Danville & Dardanelle	21,139	17,759	928

California

(58 counties, 155,959 sq. mi. land; pop. 33,871,648)

County	County seat or courthouse	2000 Pop.	1990 Pop.	Land area sq. mi.
Alameda	Oakland	1,443,741	1,304,347	738
Alpine	Markleeville	1,208	1,113	739
Amador	Jackson	35,100	30,039	593
Butte	Oroville	203,171	182,120	1,639
Calaveras	San Andreas	40,554	31,998	1,020
Colusa	Colusa	18,804	16,275	1,151
Contra Costa	Martinez	948,816	803,731	720
Del Norte	Crescent City	27,507	23,460	1,008
El Dorado	Placerville	156,299	125,995	1,711
Fresno	Fresno	799,407	667,479	5,963
Glenn	Willows	26,453	24,798	1,315
Humboldt	Eureka	126,518	119,118	3,572
Imperial	El Centro	142,361	109,303	4,175
Inyo	Independence	17,945	18,281	10,203
Kern	Bakersfield	661,645	544,981	8,141
Kings	Hanford	129,461	101,469	1,391
Lake	Lakeport	58,309	50,631	1,258
Lassen	Susanville	33,828	27,598	4,557
Los Angeles	Los Angeles	9,519,338	8,863,052	4,061
Madera	Madera	123,109	88,090	2,136
Marin	San Rafael	247,289	230,096	520
Mariposa	Mariposa	17,130	14,302	1,451
Mendocino	Ukiah	86,265	80,345	3,509
Merced	Merced	210,554	178,403	1,929
Modoc	Alturas	9,449	9,678	3,944
Mono	Bridgeport	12,853	9,956	3,044
Monterey	Salinas	401,762	355,660	3,322
Napa	Napa	124,279	110,765	754
Nevada	Nevada City	92,033	78,510	958
Orange	Santa Ana	2,846,289	2,410,668	789
Placer	Auburn	248,399	172,796	1,404
Plumas	Quincy	20,824	19,739	2,554
Riverside	Riverside	1,545,387	1,170,413	7,207
Sacramento	Sacramento	1,223,499	1,066,789	966
San Benito	Hollister	53,234	36,697	1,389
San Bernardino	San Bernardino	1,709,434	1,418,380	20,053
San Diego	San Diego	2,813,833	2,498,016	4,200
San Francisco	San Francisco	776,733	723,959	47
San Joaquin	Stockton	563,598	480,628	1,399
San Luis Obispo	San Luis Obispo	246,681	217,162	3,304
San Mateo	Redwood City	707,161	649,623	449
Santa Barbara	Santa Barbara	399,347	369,608	2,737
Santa Clara	San Jose	1,682,585	1,497,577	1,291
Santa Cruz	Santa Cruz	255,602	229,734	445
Shasta	Redding	163,256	147,036	3,785
Sierra	Downieville	3,555	3,318	953
Siskiyou	Yreka	44,301	43,531	6,287
Solano	Fairfield	394,542	339,469	829
Sonoma	Santa Rosa	458,614	388,222	1,576
Stanislaus	Modesto	446,997	370,522	1,494
Sutter	Yuba City	78,930	64,409	603
Tehama	Red Bluff	56,039	49,625	2,951
Trinity	Weaverville	13,022	13,063	3,179
Tulare	Visalia	368,021	311,932	4,824
Tuolumne	Sonora	54,501	48,456	2,235
Ventura	Ventura	753,197	669,016	1,845
Yolo	Woodland	168,660	141,212	1,013
Yuba	Marysville	60,219	58,234	631

Colorado

(63 counties, 103,718 sq. mi. land; pop. 4,301,261)

County	County seat or courthouse	2000 Pop.	1990 Pop.	Land area sq. mi.
Adams	Brighton	363,857	265,038	1,192
Alamosa	Alamosa	14,966	13,617	723
Arapahoe	Littleton	487,967	391,572	803
Archuleta	Pagosa Springs	9,898	5,345	1,350
Baca	Springfield	4,517	4,556	2,556
Bent	Las Animas	5,998	5,048	1,514
Boulder	Boulder	291,288	225,339	742
Chaffee	Salida	16,242	12,684	1,013
Cheyenne	Cheyenne Wells	2,231	2,397	1,781
Clear Creek	Georgetown	9,322	7,619	395
Conejos	Conejos	8,400	7,453	1,287
Costilla	San Luis	3,663	3,190	1,227
Crowley	Ordway	5,518	3,946	789
Custer	Westcliffe	3,503	1,926	739
Delta	Delta	27,834	20,980	1,142
Denver	Denver	554,636	467,549	153
Dolores	Dove Creek	1,844	1,504	1,067
Douglas	Castle Rock	175,766	60,391	840
Eagle	Eagle	41,659	21,928	1,688
Elbert	Kiowa	19,872	9,646	1,851
El Paso	Colorado Springs	516,929	397,014	2,126
Fremont	Canon City	46,145	32,273	1,533
Garfield	GlenwoodSprings	43,791	29,974	2,947
Gilpin	Central City	4,757	3,070	150
Grand	Hot Sulphur Springs	12,442	7,966	1,847
Gunnison	Gunnison	13,956	10,273	3,239
Hinsdale	Lake City	790	467	1,118
Huerfano	Walsenburg	7,862	6,009	1,591
Jackson	Walden	1,577	1,605	1,613
Jefferson	Golden	527,056	438,430	772
Kiowa	Eads	1,622	1,688	1,771
Kit Carson	Burlington	8,011	7,140	2,161
Lake	Leadville	7,812	6,007	377
La Plata	Durango	43,941	32,284	1,692
Larimer	Fort Collins	251,494	186,136	2,601
Las Animas	Trinidad	15,207	13,765	4,772
Lincoln	Hugo	6,087	4,529	2,586
Logan	Sterling	20,504	17,567	1,839
Mesa	Grand Junction	116,255	93,145	3,328
Mineral	Creede	831	558	876
Moffat	Craig	13,184	11,357	4,742
Montezuma	Cortez	23,830	18,672	2,037
Montrose	Montrose	33,432	24,423	2,241
Morgan	Fort Morgan	27,171	21,939	1,285
Otero	La Junta	20,311	20,185	1,263
Ouray	Ouray	3,742	2,295	540
Park	Fairplay	14,523	7,174	2,201
Phillips	Holyoke	4,480	4,189	688
Pitkin	Aspen	14,872	12,661	970
Prowers	Lamar	14,483	13,347	1,640
Pueblo	Pueblo	141,472	123,051	2,389
Rio Blanco	Meeker	5,986	6,051	3,221
Rio Grande	Del Norte	12,413	10,770	912
Routt	Steamboat Springs	19,690	14,088	2,362
Saguache	Saguache	5,917	4,619	3,168
San Juan	Silverton	558	745	387
San Miguel	Telluride	6,594	3,653	1,287
Sedgwick	Julesburg	2,747	2,690	548
Summit	Breckenridge	23,548	12,881	608
Teller	Cripple Creek	20,555	12,468	557
Washington	Akron	4,926	4,812	2,521
Weld	Greeley	180,936	131,821	3,992
Yuma	Wray	9,841	8,954	2,366

Connecticut

(8 counties, 4,845 sq. mi. land; pop. 3,405,565)

County	County seat or courthouse	2000 Pop.	1990 Pop.	Land area sq. mi.
Fairfield	Bridgeport	882,567	827,645	626
Hartford	Hartford	857,183	851,783	735
Litchfield	Litchfield	182,193	174,092	920
Middlesex	Middletown	155,071	143,196	369
New Haven	New Haven	824,008	804,219	606
New London	New London	259,088	254,957	666
Tolland	Rockville	136,364	128,699	410
Windham	Putnam	109,091	102,525	513

Delaware

(3 counties, 1,954 sq. mi. land; pop. 783,600)

County	County seat or courthouse	2000 Pop.	1990 Pop.	Land area sq. mi.
Kent	Dover	126,697	110,993	590
New Castle	Wilmington	500,265	441,946	426
Sussex	Georgetown	156,638	113,229	938

District of Columbia

(61 sq. mi. land; pop. 572,059)

Has no counties; coextensive with city of Washington.

Florida

(67 counties, 53,927 sq. mi. land; pop. 15,982,378)

County	County seat or courthouse	2000 Pop.	1990 Pop.	Land area sq. mi.
Alachua	Gainesville	217,955	181,596	874
Baker	Macclenny	22,259	18,486	585
Bay	Panama City	148,217	126,994	764
Bradford	Starke	26,088	22,515	293
Brevard	Titusville	476,230	398,978	1,018
Broward	Fort Lauderdale	1,623,018	1,255,531	1,205
Calhoun	Blountstown	13,017	11,011	567
Charlotte	Punta Gorda	141,627	110,975	694
Citrus	Inverness	118,085	93,513	584
Clay	Green Cove Springs	140,814	105,986	601
Collier	Naples	251,377	152,099	2,025
Columbia	Lake City	56,513	42,613	797
De Soto	Arcadia	32,209	23,865	637
Dixie	Cross City	13,827	10,585	704
Duval	Jacksonville	778,879	672,971	774
Escambia	Pensacola	294,410	262,445	662
Flagler	Bunnell	49,832	28,701	485
Franklin	Apalachicola	11,057	8,967	544
Gadsden	Quincy	45,087	41,116	516
Gilchrist	Trenton	14,437	9,667	349
Glades	Moore Haven	10,576	7,591	774
Gulf	Port Saint Joe	13,332	11,504	555
Hamilton	Jasper	13,327	10,930	515
Hardee	Wauchula	26,938	19,499	637
Hendry	La Belle	36,210	25,773	1,153
Hernando	Brooksville	130,802	101,115	478
Highlands	Sebring	87,366	68,432	1,028
Hillsborough	Tampa	998,948	834,054	1,051
Holmes	Bonifay	18,564	15,778	482
Indian River	Vero Beach	112,947	90,208	503
Jackson	Marianna	46,755	41,375	916
Jefferson	Monticello	12,902	11,296	598
Lafayette	Mayo	7,022	5,578	543
Lake	Tavares	210,528	152,104	953
Lee	Fort Myers	440,888	335,113	804
Leon	Tallahassee	239,452	192,493	667
Levy	Bronson	34,450	25,912	1,118
Liberty	Bristol	7,021	5,569	836
Madison	Madison	18,733	16,569	692
Manatee	Bradenton	264,002	211,707	741
Marion	Ocala	258,916	194,835	1,579
Martin	Stuart	126,731	100,900	556
Miami-Dade	Miami	2,253,362	1,937,194	1,946
Monroe	Key West	79,589	78,024	997
Nassau	Fernandina Beach	57,663	43,941	652
Okaloosa	Crestview	170,498	143,777	936
Okeechobee	Okeechobee	35,910	29,627	774
Orange	Orlando	896,344	677,491	907
Osceola	Kissimmee	172,493	107,728	1,322
Palm Beach	West Palm Beach	1,131,184	863,503	1,974
Pasco	New Port Richey	344,765	281,131	745
Pinellas	Clearwater	921,482	851,659	280
Polk	Bartow	483,924	405,382	1,874
Putnam	Palatka	70,423	65,070	722
Saint Johns	Saint Augustine	123,135	83,829	609
Saint Lucie	Fort Pierce	192,695	150,171	572
Santa Rosa	Milton	117,743	81,961	1,017
Sarasota	Sarasota	325,957	277,776	572
Seminole	Sanford	365,196	287,521	308
Sumter	Bushnell	53,345	31,577	546
Suwannee	Live Oak	34,844	26,780	688
Taylor	Perry	19,256	17,111	1,042
Union	Lake Butler	13,442	10,252	240
Volusia	De Land	443,343	370,737	1,103
Wakulla	Crawfordville	22,863	14,202	607
Walton	De Funiak Springs	40,601	27,759	1,058
Washington	Chipley	20,973	16,919	580

Georgia

(159 counties, 57,906 sq. mi. land; pop. 8,186,453)

County	County seat or courthouse	2000 Pop.	1990 Pop.	Land area sq. mi.
Appling	Baxley	17,419	15,744	509
Atkinson	Pearson	7,609	6,213	338
Bacon	Alma	10,103	9,566	285
Baker	Newton	4,074	3,615	343
Baldwin	Milledgeville	44,700	39,530	258
Banks	Homer	14,422	10,308	234
Barrow	Winder	46,144	29,721	162
Bartow	Cartersville	76,019	55,915	459
Ben Hill	Fitzgerald	17,484	16,245	252
Berrien	Nashville	16,235	14,153	452
Bibb	Macon	153,887	150,137	250
Bleckley	Cochran	11,666	10,430	217
Brantley	Nahunta	14,629	11,077	444
Brooks	Quitman	16,450	15,398	494
Bryan	Pembroke	23,417	15,438	442
Bulloch	Statesboro	55,983	43,125	682
Burke	Waynesboro	22,243	20,579	830
Butts	Jackson	19,522	15,326	187
Calhoun	Morgan	6,320	5,013	280
Camden	Woodbine	43,664	30,167	630
Candler	Metter	9,577	7,744	247
Carroll	Carrollton	87,268	71,422	499
Catoosa	Ringgold	53,282	42,464	162
Charlton	Folkston	10,282	8,496	781
Chatham	Savannah	232,048	216,774	438
Chattahoochee	Cusseta	14,882	16,934	249
Chattooga	Summerville	25,470	22,236	313
Cherokee	Canton	141,903	90,204	424
Clarke	Athens	101,489	87,594	121
Clay	Fort Gaines	3,357	3,364	195
Clayton	Jonesboro	236,517	181,436	143
Clinch	Homerville	6,878	6,160	809
Cobb	Marietta	607,751	447,745	340
Coffee	Douglas	37,413	29,592	599
Colquitt	Moultrie	42,053	36,645	552
Columbia	Appling	89,288	66,031	290
Cook	Adel	15,771	13,456	229
Coweta	Newnan	89,215	53,853	443
Crawford	Knoxville	12,495	8,991	325
Crisp	Cordele	21,996	20,011	274
Dade	Trenton	15,154	13,183	174
Dawson	Dawsonville	15,999	9,429	211
Decatur	Bainbridge	28,240	25,517	597
De Kalb	Decatur	665,865	546,174	268
Dodge	Eastman	19,171	17,607	500
Dooly	Vienna	11,525	9,901	393
Dougherty	Albany	96,065	96,321	330
Douglas	Douglasville	92,174	71,120	199
Early	Blakely	12,354	11,854	511
Echols	Statenville	3,754	2,334	404
Effingham	Springfield	37,535	25,687	479
Elbert	Elberton	20,511	18,949	369
Emanuel	Swainsboro	21,837	20,546	686
Evans	Claxton	10,495	8,724	185
Fannin	Blue Ridge	19,798	15,992	386
Fayette	Fayetteville	91,263	62,415	197
Floyd	Rome	90,565	81,251	513
Forsyth	Cumming	98,407	44,083	226
Franklin	Carnesville	20,285	16,650	263
Fulton	Atlanta	816,006	648,776	529
Gilmer	Ellijay	23,456	13,368	427
Glascock	Gibson	2,556	2,357	144
Glynn	Brunswick	67,568	62,496	422
Gordon	Calhoun	44,104	35,067	356
Grady	Cairo	23,659	20,279	458
Greene	Greensboro	14,406	11,793	388
Gwinnett	Lawrenceville	588,448	352,910	433
Habersham	Clarkesville	35,902	27,622	278
Hall	Gainesville	139,277	95,434	394
Hancock	Sparta	10,076	8,908	473
Haralson	Buchanan	25,690	21,966	282
Harris	Hamilton	23,695	17,788	464
Hart	Hartwell	22,997	19,712	232
Heard	Franklin	11,012	8,628	296
Henry	McDonough	119,341	58,741	323
Houston	Perry	110,765	89,208	377
Irwin	Ocilla	9,931	8,649	357
Jackson	Jefferson	41,589	30,005	342
Jasper	Monticello	11,426	8,453	370
Jeff Davis	Hazlehurst	12,684	12,032	333
Jefferson	Louisville	17,266	17,408	528
Jenkins	Millen	8,575	8,247	350
Johnson	Wrightsville	8,560	8,329	304
Jones	Gray	23,639	20,739	394
Lamar	Barnesville	15,912	13,038	185
Lanier	Lakeland	7,241	5,531	187
Laurens	Dublin	44,874	39,988	812
Lee	Leesburg	24,757	16,250	356
Liberty	Hinesville	61,610	52,745	519
Lincoln	Lincolnton	8,348	7,442	211
Long	Ludowici	10,304	6,202	401
Lowndes	Valdosta	92,115	75,981	504
Lumpkin	Dahlonega	21,016	14,573	284
McDuffie	Thomson	21,231	20,119	260
McIntosh	Darien	10,847	8,634	433
Macon	Oglethorpe	14,074	13,114	403
Madison	Danielsville	25,730	21,050	284
Marion	Buena Vista	7,144	5,590	367
Meriwether	Greenville	22,534	22,411	503
Miller	Colquitt	6,383	6,280	283
Mitchell	Camilla	23,932	20,275	512
Monroe	Forsyth	21,757	17,113	396
Montgomery	Mount Vernon	8,270	7,379	245
Morgan	Madison	15,457	12,883	350
Murray	Chatsworth	36,506	26,147	344
Muscogee	Columbus	186,291	179,280	216
Newton	Covington	62,001	41,808	276
Oconee	Watkinsville	26,225	17,618	186
Oglethorpe	Lexington	12,635	9,763	441
Paulding	Dallas	81,678	41,611	313
Peach	Fort Valley	23,668	21,189	151
Pickens	Jasper	22,983	14,432	232
Pierce	Blackshear	15,636	13,328	343
Pike	Zebulon	13,688	10,224	218

County	County seat or courthouse	2000 Pop.	1990 Pop.	Land area sq. mi.
Polk	Cedartown	38,127	33,815	311
Pulaski	Hawkinsville	9,588	8,108	247
Putnam	Eatonton	18,812	14,137	345
Quitman	Georgetown	2,598	2,210	152
Rabun	Clayton	15,050	11,648	371
Randolph	Cuthbert	7,791	8,023	429
Richmond	Augusta	199,775	189,719	324
Rockdale	Conyers	70,111	54,091	131
Schley	Ellaville	3,766	3,590	168
Screven	Sylvania	15,374	13,842	648
Seminole	Donalsonville	9,369	9,010	238
Spalding	Griffin	58,417	54,457	198
Stephens	Toccoa	25,435	23,436	179
Stewart	Lumpkin	5,252	5,654	459
Sumter	Americus	33,200	30,232	485
Talbot	Talbotton	6,498	6,524	393
Taliaferro	Crawfordville	2,077	1,915	195
Tattnall	Reidsville	22,305	17,722	484
Taylor	Butler	8,815	7,642	377
Telfair	McRae	11,794	11,000	441
Terrell	Dawson	10,970	10,653	335
Thomas	Thomasville	42,737	38,943	548
Tift	Tifton	38,407	34,998	265
Toombs	Lyons	26,067	24,072	367
Towns	Hiawassee	9,319	6,754	167
Treutlen	Soperton	6,854	5,994	201
Troup	La Grange	58,779	55,532	414
Turner	Ashburn	9,504	8,703	286
Twiggs	Jeffersonville	10,590	9,806	360
Union	Blairsville	17,289	11,993	323
Upson	Thomaston	27,597	26,300	325
Walker	La Fayette	61,053	58,310	447
Walton	Monroe	60,687	38,586	329
Ware	Waycross	35,483	35,471	902
Warren	Warrenton	6,336	6,078	286
Washington	Sandersville	21,176	19,112	680
Wayne	Jesup	26,565	22,356	645
Webster	Preston	2,390	2,263	210
Wheeler	Alamo	6,179	4,903	298
White	Cleveland	19,944	13,006	242
Whitfield	Dalton	83,525	72,462	290
Wilcox	Abbeville	8,577	7,008	380
Wilkes	Washington	10,687	10,597	471
Wilkinson	Irwinton	10,220	10,228	447
Worth	Sylvester	21,967	19,744	570

Hawaii

(5 counties, 6,423 sq. mi. land; pop. 1,211,537)

County	County seat or courthouse	2000 Pop.	1990 Pop.	Land area sq. mi.
Hawaii	Hilo	148,677	120,317	4,028
Honolulu	Honolulu	876,156	836,231	600
Kalawao[1]		147	130	13
Kauai	Lihue	58,463	51,177	622
Maui	Wailuku	128,094	100,374	1,159

(1) Administered by state government.

Idaho

(44 counties, 82,747 sq. mi. land; pop. 1,293,953)

County	County seat or courthouse	2000 Pop.	1990 Pop.	Land area sq. mi.
Ada	Boise	300,904	205,775	1,055
Adams	Council	3,476	3,254	1,365
Bannock	Pocatello	75,565	66,026	1,113
Bear Lake	Paris	6,411	6,084	971
Benewah	Saint Maries	9,171	7,937	776
Bingham	Blackfoot	41,735	37,583	2,095
Blaine	Hailey	18,991	13,552	2,645
Boise	Idaho City	6,670	3,509	1,902
Bonner	Sandpoint	36,835	26,622	1,738
Bonneville	Idaho Falls	82,522	72,207	1,868
Boundary	Bonners Ferry	9,871	8,332	1,269
Butte	Arco	2,899	2,918	2,233
Camas	Fairfield	991	727	1,075
Canyon	Caldwell	131,441	90,076	590
Caribou	Soda Springs	7,304	6,963	1,766
Cassia	Burley	21,416	19,532	2,566
Clark	Dubois	1,022	762	1,765
Clearwater	Orofino	8,930	8,505	2,461
Custer	Challis	4,342	4,133	4,925
Elmore	Mountain Home	29,130	21,205	3,078
Franklin	Preston	11,329	9,232	665
Fremont	Saint Anthony	11,819	10,937	1,867
Gem	Emmett	15,181	11,844	563
Gooding	Gooding	14,155	11,633	731
Idaho	Grangeville	15,511	13,768	8,485
Jefferson	Rigby	19,155	16,543	1,095
Jerome	Jerome	18,342	15,138	600
Kootenai	Coeur d'Alene	108,685	69,795	1,245
Latah	Moscow	34,935	30,617	1,077
Lemhi	Salmon	7,806	6,899	4,564
Lewis	Nez Perce	3,747	3,516	479
Lincoln	Shoshone	4,044	3,308	1,206
Madison	Rexberg	27,467	23,674	472
Minidoka	Rupert	20,174	19,361	760
Nez Perce	Lewiston	37,410	33,754	849
Oneida	Malad City	4,125	3,492	1,200
Owyhee	Murphy	10,644	8,392	7,678
Payette	Payette	20,578	16,434	408
Power	American Falls	7,538	7,086	1,406
Shoshone	Wallace	13,771	13,931	2,634
Teton	Driggs	5,999	3,439	450
Twin Falls	Twin Falls	64,284	53,580	1,925
Valley	Cascade	7,651	6,109	3,678
Washington	Weiser	9,977	8,550	1,456

Illinois

(102 counties, 55,584 sq. mi. land; pop. 12,419,293)

County	County seat or courthouse	2000 Pop.	1990 Pop.	Land area sq. mi.
Adams	Quincy	68,277	66,090	857
Alexander	Cairo	9,590	10,626	236
Bond	Greenville	17,633	14,991	380
Boone	Belvidere	41,786	30,806	281
Brown	Mount Sterling	6,950	5,836	306
Bureau	Princeton	35,503	35,688	869
Calhoun	Hardin	5,084	5,322	254
Carroll	Mount Carroll	16,674	16,805	444
Cass	Virginia	13,695	13,437	376
Champaign	Urbana	179,669	173,025	997
Christian	Taylorville	35,372	34,418	709
Clark	Marshall	17,008	15,921	502
Clay	Louisville	14,560	14,460	469
Clinton	Carlyle	35,535	33,944	474
Coles	Charleston	53,196	51,644	508
Cook	Chicago	5,376,741	5,105,044	946
Crawford	Robinson	20,452	19,464	444
Cumberland	Toledo	11,253	10,670	346
De Kalb	Sycamore	88,969	77,932	634
De Witt	Clinton	16,798	16,516	398
Douglas	Tuscola	19,922	19,464	417
Du Page	Wheaton	904,161	781,689	334
Edgar	Paris	19,704	19,595	624
Edwards	Albion	6,971	7,440	222
Effingham	Effingham	34,264	31,704	479
Fayette	Vandalia	21,802	20,893	716
Ford	Paxton	14,241	14,275	486
Franklin	Benton	39,018	40,319	412
Fulton	Lewiston	38,250	38,080	866
Gallatin	Shawneetown	6,445	6,909	324
Greene	Carrollton	14,761	15,317	543
Grundy	Morris	37,535	32,337	420
Hamilton	McLeansboro	8,621	8,499	435
Hancock	Carthage	20,121	21,373	795
Hardin	Elizabethtown	4,800	5,189	178
Henderson	Oquawka	8,213	8,096	379
Henry	Cambridge	51,020	51,159	823
Iroquois	Watseka	31,334	30,787	1,116
Jackson	Murphysboro	59,612	61,067	588
Jasper	Newton	10,117	10,609	494
Jefferson	Mount Vernon	40,045	37,020	571
Jersey	Jerseyville	21,668	20,539	369
Jo Daviess	Galena	22,289	21,821	601
Johnson	Vienna	12,878	11,347	345
Kane	Geneva	404,119	317,471	520
Kankakee	Kankakee	103,833	96,255	677
Kendall	Yorkville	54,544	39,413	321
Knox	Galesburg	55,836	56,393	716
Lake	Waukegan	644,356	516,418	448
La Salle	Ottawa	111,509	106,913	1,135
Lawrence	Lawrenceville	15,452	15,972	372
Lee	Dixon	36,062	34,392	725
Livingston	Pontiac	39,678	39,301	1,044
Logan	Lincoln	31,183	30,798	618
McDonough	Macomb	32,913	35,244	589
McHenry	Woodstock	260,077	183,241	604
McLean	Bloomington	150,433	129,180	1,184
Macon	Decatur	114,706	117,206	581
Macoupin	Carlinville	49,019	47,679	864
Madison	Edwardsville	258,941	249,238	725
Marion	Salem	41,691	41,561	572
Marshall	Lacon	13,180	12,846	386
Mason	Havana	16,038	16,269	539
Massac	Metropolis	15,161	14,752	239
Menard	Petersburg	12,486	11,164	314
Mercer	Aledo	16,957	17,290	561
Monroe	Waterloo	27,619	22,422	388
Montgomery	Hillsboro	30,652	30,728	704
Morgan	Jacksonville	36,616	36,397	569
Moultrie	Sullivan	14,287	13,930	336
Ogle	Oregon	51,032	45,957	759
Peoria	Peoria	183,433	182,827	620
Perry	Pinckneyville	23,094	21,412	441
Piatt	Monticello	16,365	15,548	440
Pike	Pittsfield	17,384	17,577	830
Pope	Golconda	4,413	4,373	371
Pulaski	Mound City	7,348	7,523	201

County	County seat or courthouse	2000 Pop.	1990 Pop.	Land area sq. mi.
Putnam	Hennepin	6,086	5,730	160
Randolph	Chester	33,893	34,583	578
Richland	Olney	16,149	16,545	360
Rock Island	Rock Island	149,374	148,723	427
Saint Clair	Belleville	256,082	262,852	664
Saline	Harrisburg	26,733	26,551	383
Sangamon	Springfield	188,951	178,386	868
Schuyler	Rushville	7,189	7,498	437
Scott	Winchester	5,537	5,644	251
Shelby	Shelbyville	22,893	22,261	759
Stark	Toulon	6,332	6,534	288
Stephenson	Freeport	48,979	48,052	564
Tazewell	Pekin	128,485	123,692	649
Union	Jonesboro	18,293	17,619	416
Vermilion	Danville	83,919	88,257	899
Wabash	Mount Carmel	12,937	13,111	223
Warren	Monmouth	18,735	19,181	543
Washington	Nashville	15,148	14,965	563
Wayne	Fairfield	17,151	17,241	714
White	Carmi	15,371	16,522	495
Whiteside	Morrison	60,653	60,186	685
Will	Joliet	502,266	357,313	837
Williamson	Marion	61,296	57,733	423
Winnebago	Rockford	278,418	252,913	514
Woodford	Eureka	35,469	32,653	528

Indiana

(92 counties, 35,867 sq. mi. land; pop. 6,080,485)

County	County seat or courthouse	2000 Pop.	1990 Pop.	Land area sq. mi.
Adams	Decatur	33,625	31,095	339
Allen	Fort Wayne	331,849	300,836	657
Bartholomew	Columbus	71,435	63,657	407
Benton	Fowler	9,421	9,441	406
Blackford	Hartford City	14,048	14,067	165
Boone	Lebanon	46,107	38,147	423
Brown	Nashville	14,957	14,080	312
Carroll	Delphi	20,165	18,809	372
Cass	Logansport	40,930	38,413	413
Clark	Jeffersonville	96,472	87,774	375
Clay	Brazil	26,556	24,705	358
Clinton	Frankfort	33,866	30,974	405
Crawford	English	10,743	9,914	306
Daviess	Washington	29,820	27,533	431
Dearborn	Lawrenceburg	46,109	38,835	305
Decatur	Greensburg	24,555	23,645	373
De Kalb	Auburn	40,285	35,324	363
Delaware	Muncie	118,769	119,659	393
Dubois	Jasper	39,674	36,616	430
Elkhart	Goshen	182,791	156,198	464
Fayette	Connersville	25,588	26,015	215
Floyd	New Albany	70,823	64,404	148
Fountain	Covington	17,954	17,808	396
Franklin	Brookville	22,151	19,580	386
Fulton	Rochester	20,511	18,840	369
Gibson	Princeton	32,500	31,913	489
Grant	Marion	73,403	74,169	414
Greene	Bloomfield	33,157	30,410	542
Hamilton	Noblesville	182,740	108,936	398
Hancock	Greenfield	55,391	45,527	306
Harrison	Corydon	34,325	29,890	485
Hendricks	Danville	104,093	75,717	408
Henry	New Castle	48,508	48,139	393
Howard	Kokomo	84,964	80,827	293
Huntington	Huntington	38,075	35,427	383
Jackson	Brownstown	41,335	37,730	509
Jasper	Rensselaer	30,043	24,823	560
Jay	Portland	21,806	21,512	384
Jefferson	Madison	31,705	29,797	361
Jennings	Vernon	27,554	23,661	377
Johnson	Franklin	115,209	88,109	320
Knox	Vincennes	39,256	39,884	516
Kosciusko	Warsaw	74,057	65,294	538
Lagrange	Lagrange	34,909	29,477	380
Lake	Crown Point	484,564	475,594	497
La Porte	La Porte	110,106	107,066	598
Lawrence	Bedford	45,922	42,836	449
Madison	Anderson	133,358	130,669	452
Marion	Indianapolis	860,454	797,159	396
Marshall	Plymouth	45,128	42,182	444
Martin	Shoals	10,369	10,369	336
Miami	Peru	36,082	36,897	376
Monroe	Bloomington	120,563	108,978	394
Montgomery	Crawfordsville	37,629	34,436	505
Morgan	Martinsville	66,689	55,920	406
Newton	Kentland	14,566	13,551	402
Noble	Albion	46,275	37,877	411
Ohio	Rising Sun	5,623	5,315	87
Orange	Paoli	19,306	18,409	400
Owen	Spencer	21,786	17,281	385
Parke	Rockville	17,241	15,410	445
Perry	Cannelton	18,899	19,107	381
Pike	Petersburg	12,837	12,509	336
Porter	Valparaiso	146,798	128,932	418
Posey	Mount Vernon	27,061	25,968	409
Pulaski	Winamac	13,755	12,780	434
Putnam	Greencastle	36,019	30,315	480
Randolph	Winchester	27,401	27,148	453
Ripley	Versailles	26,523	24,616	446
Rush	Rushville	18,261	18,129	408
Saint Joseph	South Bend	265,559	247,052	457
Scott	Scottsburg	22,960	20,991	190
Shelby	Shelbyville	43,445	40,307	413
Spencer	Rockport	20,391	19,490	399
Starke	Knox	23,556	22,747	309
Steuben	Angola	33,214	27,446	309
Sullivan	Sullivan	21,751	18,993	447
Switzerland	Vevay	9,065	7,738	221
Tippecanoe	Lafayette	148,955	130,598	500
Tipton	Tipton	16,577	16,119	260
Union	Liberty	7,349	6,976	162
Vanderburgh	Evansville	171,922	165,058	235
Vermillion	Newport	16,788	16,773	257
Vigo	Terre Haute	105,848	106,107	403
Wabash	Wabash	34,960	35,069	413
Warren	Williamsport	8,419	8,176	365
Warrick	Boonville	52,383	44,920	384
Washington	Salem	27,223	23,717	514
Wayne	Richmond	71,097	71,951	404
Wells	Bluffton	27,600	25,948	370
White	Monticello	25,267	23,265	505
Whitley	Columbia City	30,707	27,651	336

Iowa

(99 counties, 55,869 sq. mi. land; pop. 2,926,324)

County	County seat or courthouse	2000 Pop.	1990 Pop.	Land area sq. mi.
Adair	Greenfield	8,243	8,409	569
Adams	Corning	4,482	4,866	424
Allamakee	Waukon	14,675	13,855	640
Appanoose	Centerville	13,721	13,743	496
Audubon	Audubon	6,830	7,334	443
Benton	Vinton	25,308	22,429	716
Black Hawk	Waterloo	128,012	123,798	567
Boone	Boone	26,224	25,186	571
Bremer	Waverly	23,325	22,813	438
Buchanan	Independence	21,093	20,844	571
Buena Vista	Storm Lake	20,411	19,965	575
Butler	Allison	15,305	15,731	580
Calhoun	Rockwell City	11,115	11,508	570
Carroll	Carroll	21,421	21,423	569
Cass	Atlantic	14,684	15,128	564
Cedar	Tipton	18,187	17,444	580
Cerro Gordo	Mason City	46,447	46,733	568
Cherokee	Cherokee	13,035	14,098	577
Chickasaw	New Hampton	13,095	13,295	505
Clarke	Osceola	9,133	8,287	431
Clay	Spencer	17,372	17,585	569
Clayton	Elkader	18,678	19,054	779
Clinton	Clinton	50,149	51,040	695
Crawford	Denison	16,942	16,775	714
Dallas	Adel	40,750	29,755	586
Davis	Bloomfield	8,541	8,312	503
Decatur	Leon	8,689	8,338	532
Delaware	Manchester	18,404	18,035	578
Des Moines	Burlington	42,351	42,614	416
Dickinson	Spirit Lake	16,424	14,909	381
Dubuque	Dubuque	89,143	86,403	608
Emmet	Estherville	11,027	11,569	396
Fayette	West Union	22,008	21,843	731
Floyd	Charles City	16,900	17,058	501
Franklin	Hampton	10,704	11,364	582
Fremont	Sidney	8,010	8,226	511
Greene	Jefferson	10,366	10,045	568
Grundy	Grundy Center	12,369	12,029	503
Guthrie	Guthrie Center	11,353	10,935	591
Hamilton	Webster City	16,438	16,071	577
Hancock	Garner	12,100	12,638	571
Hardin	Eldora	18,812	19,094	569
Harrison	Logan	15,666	14,730	697
Henry	Mount Pleasant	20,336	19,226	434
Howard	Cresco	9,932	9,809	473
Humboldt	Dakota City	10,381	10,756	434
Ida	Ida Grove	7,837	8,365	432
Iowa	Marengo	15,671	14,630	586
Jackson	Maquoketa	20,296	19,950	636
Jasper	Newton	37,213	34,795	730
Jefferson	Fairfield	16,181	16,310	435
Johnson	Iowa City	111,006	96,119	614
Jones	Anamosa	20,221	19,444	575
Keokuk	Sigourney	11,400	11,624	579
Kossuth	Algona	17,163	18,591	973
Lee	Fort Madison & Keokuk	38,052	38,687	517
Linn	Cedar Rapids	191,701	168,767	717
Louisa	Wapello	12,183	11,592	402
Lucas	Chariton	9,422	9,070	431
Lyon	Rock Rapids	11,763	11,952	588

County	County seat or courthouse	2000 Pop.	1990 Pop.	Land area sq. mi.
Madison	Winterset	14,019	12,483	561
Mahaska	Oskaloosa	22,335	21,532	571
Marion	Knoxville	32,052	30,001	554
Marshall	Marshalltown	39,311	38,276	572
Mills	Glenwood	14,547	13,202	437
Mitchell	Osage	10,874	10,928	469
Monona	Onawa	10,020	10,034	693
Monroe	Albia	8,016	8,114	433
Montgomery	Red Oak	11,771	12,076	424
Muscatine	Muscatine	41,722	39,907	439
O'Brien	Primghar	15,102	15,444	573
Osceola	Sibley	7,003	7,267	399
Page	Clarinda	16,976	16,870	535
Palo Alto	Emmetsburg	10,147	10,669	564
Plymouth	Le Mars	24,849	23,388	864
Pocahontas	Pocahontas	8,662	9,525	578
Polk	Des Moines	374,601	327,140	569
Pottawattamie	Council Bluffs	87,704	82,628	954
Poweshiek	Montezuma	18,815	19,033	585
Ringgold	Mount Ayr	5,469	5,420	538
Sac	Sac City	11,529	12,324	576
Scott	Davenport	158,668	150,973	458
Shelby	Harlan	13,173	13,230	591
Sioux	Orange City	31,589	29,903	768
Story	Nevada	79,981	74,252	573
Tama	Toledo	18,103	17,419	721
Taylor	Bedford	6,958	7,114	534
Union	Creston	12,309	12,750	424
Van Buren	Keosauqua	7,809	7,676	485
Wapello	Ottumwa	36,051	35,696	432
Warren	Indianola	40,671	36,033	572
Washington	Washington	20,670	19,612	569
Wayne	Corydon	6,730	7,067	526
Webster	Fort Dodge	40,235	40,342	715
Winnebago	Forest City	11,723	12,122	400
Winneshiek	Decorah	21,310	20,847	690
Woodbury	Sioux City	103,877	98,276	873
Worth	Northwood	7,909	7,991	400
Wright	Clarion	14,334	14,269	581

Kansas

(105 counties, 81,815 sq. mi. land; pop. 2,688,418)

County	County seat or courthouse	2000 Pop.	1990 Pop.	Land area sq. mi.
Allen	Iola	14,385	14,638	503
Anderson	Garnett	8,110	7,803	583
Atchison	Atchison	16,774	16,932	432
Barber	Medicine Lodge	5,307	5,874	1,134
Barton	Great Bend	28,205	29,382	894
Bourbon	Fort Scott	15,379	14,966	637
Brown	Hiawatha	10,724	11,128	571
Butler	El Dorado	59,482	50,580	1,428
Chase	Cottonwood Falls	3,030	3,021	776
Chautauqua	Sedan	4,359	4,407	642
Cherokee	Columbus	22,605	21,374	587
Cheyenne	Saint Francis	3,165	3,243	1,020
Clark	Ashland	2,390	2,418	975
Clay	Clay Center	8,822	9,158	644
Cloud	Concordia	10,268	11,023	716
Coffey	Burlington	8,865	8,404	630
Comanche	Coldwater	1,967	2,313	788
Cowley	Winfield	36,291	36,915	1,126
Crawford	Girard	38,242	35,582	593
Decatur	Oberlin	3,472	4,021	894
Dickinson	Abilene	19,344	18,958	848
Doniphan	Troy	8,249	8,134	392
Douglas	Lawrence	99,962	81,798	457
Edwards	Kinsley	3,449	3,787	622
Elk	Howard	3,261	3,327	647
Ellis	Hays	27,507	26,004	900
Ellsworth	Ellsworth	6,525	6,586	716
Finney	Garden City	40,523	33,070	1,302
Ford	Dodge City	32,458	27,463	1,099
Franklin	Ottawa	24,784	21,994	574
Geary	Junction City	27,947	30,453	385
Gove	Gove	3,068	3,231	1,071
Graham	Hill City	2,946	3,543	898
Grant	Ulysses	7,909	7,159	575
Gray	Cimarron	5,904	5,396	869
Greeley	Tribune	1,534	1,774	778
Greenwood	Eureka	7,673	7,847	1,140
Hamilton	Syracuse	2,670	2,388	996
Harper	Anthony	6,536	7,124	801
Harvey	Newton	32,869	31,028	539
Haskell	Sublette	4,307	3,886	577
Hodgeman	Jetmore	2,085	2,177	860
Jackson	Holton	12,657	11,525	656
Jefferson	Oskaloosa	18,426	15,905	536
Jewell	Mankato	3,791	4,251	909
Johnson	Olathe	451,086	355,021	477
Kearny	Lakin	4,531	4,027	871
Kingman	Kingman	8,673	8,292	863
Kiowa	Greensburg	3,278	3,660	722
Labette	Oswego	22,835	23,693	649
Lane	Dighton	2,155	2,375	717
Leavenworth	Leavenworth	68,691	64,371	463
Lincoln	Lincoln	3,578	3,653	719
Linn	Mound City	9,570	8,254	599
Logan	Oakley	3,046	3,081	1,073
Lyon	Emporia	35,935	34,732	851
McPherson	McPherson	29,554	27,268	900
Marion	Marion	13,361	12,888	943
Marshall	Marysville	10,965	11,705	903
Meade	Meade	4,631	4,247	978
Miami	Paola	28,351	23,466	577
Mitchell	Beloit	6,932	7,203	700
Montgomery	Independence	36,252	38,816	645
Morris	Council Grove	6,104	6,198	697
Morton	Elkhart	3,496	3,480	730
Nemaha	Seneca	10,717	10,446	718
Neosho	Erie	16,997	17,035	572
Ness	Ness City	3,454	4,033	1,075
Norton	Norton	5,953	5,947	878
Osage	Lyndon	16,712	15,248	704
Osborne	Osborne	4,452	4,867	892
Ottawa	Minneapolis	6,163	5,634	721
Pawnee	Larned	7,233	7,555	754
Phillips	Phillipsburg	6,001	6,590	886
Pottawatomie	Westmoreland	18,209	16,128	844
Pratt	Pratt	9,647	9,702	735
Rawlins	Atwood	2,966	3,404	1,070
Reno	Hutchinson	64,790	62,389	1,254
Republic	Belleville	5,835	6,482	716
Rice	Lyons	10,761	10,610	727
Riley	Manhattan	62,843	67,139	610
Rooks	Stockton	5,685	6,039	888
Rush	LaCrosse	3,551	3,842	718
Russell	Russell	7,370	7,835	885
Saline	Salina	53,597	49,301	720
Scott	Scott City	5,120	5,289	718
Sedgwick	Wichita	452,869	403,662	999
Seward	Liberal	22,510	18,743	640
Shawnee	Topeka	169,871	160,976	550
Sheridan	Hoxie	2,813	3,043	895
Sherman	Goodland	6,760	6,926	1,056
Smith	Smith Center	4,536	5,078	895
Stafford	Saint John	4,789	5,365	792
Stanton	Johnson	2,406	2,333	680
Stevens	Hugoton	5,463	5,048	728
Sumner	Wellington	25,946	25,841	1,182
Thomas	Colby	8,180	8,258	1,075
Trego	WaKeeney	3,319	3,694	888
Wabaunsee	Alma	6,885	6,603	797
Wallace	Sharon Springs	1,749	1,821	914
Washington	Washington	6,483	7,073	898
Wichita	Leoti	2,531	2,758	719
Wilson	Fredonia	10,332	10,289	574
Woodson	Yates Center	3,788	4,116	501
Wyandotte	Kansas City	157,882	162,026	151

Kentucky

(120 counties, 39,728 sq. mi. land; pop. 4,041,769)

County	County seat or courthouse	2000 Pop.	1990 Pop.	Land area sq. mi.
Adair	Columbia	17,244	15,360	407
Allen	Scottsville	17,800	14,628	346
Anderson	Lawrenceburg	19,111	14,571	203
Ballard	Wickliffe	8,286	7,902	251
Barren	Glasgow	38,033	34,001	491
Bath	Owingsville	11,085	9,692	279
Bell	Pineville	30,060	31,506	361
Boone	Burlington	85,991	57,589	246
Bourbon	Paris	19,360	19,236	291
Boyd	Catlettsburg	49,752	51,096	160
Boyle	Danville	27,697	25,590	182
Bracken	Brooksville	8,279	7,766	203
Breathitt	Jackson	16,100	15,703	495
Breckinridge	Hardinsburg	18,648	16,312	572
Bullitt	Shepherdsville	61,236	47,567	299
Butler	Morgantown	13,010	11,245	428
Caldwell	Princeton	13,060	13,232	347
Calloway	Murray	34,177	30,735	386
Campbell	Newport	88,616	83,866	152
Carlisle	Bardwell	5,351	5,238	192
Carroll	Carrollton	10,155	9,292	130
Carter	Grayson	26,889	24,340	411
Casey	Liberty	15,447	14,211	446
Christian	Hopkinsville	72,265	68,941	721
Clark	Winchester	33,144	29,496	254
Clay	Manchester	24,556	21,746	471
Clinton	Albany	9,634	9,135	197
Crittenden	Marion	9,384	9,196	362
Cumberland	Burkesville	7,147	6,784	306
Daviess	Owensboro	91,545	87,189	462
Edmonson	Brownsville	11,644	10,357	303
Elliott	Sandy Hook	6,748	6,455	234
Estill	Irvine	15,307	14,614	254
Fayette	Lexington	260,512	225,366	285
Fleming	Flemingsburg	13,792	12,292	351
Floyd	Prestonsburg	42,441	43,586	394
Franklin	Frankfort	47,687	44,143	210
Fulton	Hickman	7,752	8,271	209

County	County seat or courthouse	2000 Pop.	1990 Pop.	Land area sq. mi.
Gallatin	Warsaw	7,870	5,393	99
Garrard	Lancaster	14,792	11,579	231
Grant	Williamstown	22,384	15,737	260
Graves	Mayfield	37,028	33,550	556
Grayson	Leitchfield	24,053	21,050	504
Green	Greensburg	11,518	10,371	289
Greenup	Greenup	36,891	36,796	346
Hancock	Hawesville	8,392	7,864	189
Hardin	Elizabethtown	94,174	89,240	628
Harlan	Harlan	33,202	36,574	467
Harrison	Cynthiana	17,983	16,248	310
Hart	Munfordville	17,445	14,890	416
Henderson	Henderson	44,829	43,044	440
Henry	New Castle	15,060	12,823	289
Hickman	Clinton	5,262	5,566	244
Hopkins	Madisonville	46,519	46,126	551
Jackson	McKee	13,495	11,955	346
Jefferson	Louisville	693,604	665,123	385
Jessamine	Nicholasville	39,041	30,508	173
Johnson	Paintsville	23,445	23,248	262
Kenton	Covington	151,464	142,005	162
Knott	Hindman	17,649	17,906	352
Knox	Barbourville	31,795	29,676	388
Larue	Hodgenville	13,373	11,679	263
Laurel	London	52,715	43,438	436
Lawrence	Louisa	15,569	13,998	419
Lee	Beattyville	7,916	7,422	210
Leslie	Hyden	12,401	13,642	404
Letcher	Whitesburg	25,277	27,000	339
Lewis	Vanceburg	14,092	13,029	484
Lincoln	Stanford	23,361	20,096	336
Livingston	Smithland	9,804	9,062	316
Logan	Russellville	26,573	24,416	556
Lyon	Eddyville	8,080	6,624	216
McCracken	Paducah	65,514	62,879	251
McCreary	Whitley City	17,080	15,603	428
McLean	Calhoun	9,938	9,628	254
Madison	Richmond	70,872	57,508	441
Magoffin	Salyersville	13,332	13,077	309
Marion	Lebanon	18,212	16,499	346
Marshall	Benton	30,125	27,205	305
Martin	Inez	12,578	12,526	231
Mason	Maysville	16,800	16,666	241
Meade	Brandenburg	26,349	24,170	309
Menifee	Frenchburg	6,556	5,092	204
Mercer	Harrodsburg	20,817	19,148	251
Metcalfe	Edmonton	10,037	8,963	291
Monroe	Tompkinsville	11,756	11,401	331
Montgomery	Mount Sterling	22,554	19,561	199
Morgan	West Liberty	13,948	11,648	381
Muhlenberg	Greenville	31,839	31,318	475
Nelson	Bardstown	37,477	29,710	423
Nicholas	Carlisle	6,813	6,725	197
Ohio	Hartford	22,916	21,105	594
Oldham	La Grange	46,178	33,263	189
Owen	Owenton	10,547	9,035	352
Owsley	Booneville	4,858	5,036	198
Pendleton	Falmouth	14,390	12,062	281
Perry	Hazard	29,390	30,283	342
Pike	Pikeville	68,736	72,584	788
Powell	Stanton	13,237	11,686	180
Pulaski	Somerset	56,217	49,489	662
Robertson	Mount Olivet	2,266	2,124	100
Rockcastle	Mount Vernon	16,582	14,803	318
Rowan	Morehead	22,094	20,353	281
Russell	Jamestown	16,315	14,716	254
Scott	Georgetown	33,061	23,867	285
Shelby	Shelbyville	33,337	24,824	384
Simpson	Franklin	16,405	15,145	236
Spencer	Taylorsville	11,766	6,801	186
Taylor	Campbellsville	22,927	21,146	270
Todd	Elkton	11,971	10,940	376
Trigg	Cadiz	12,597	10,361	443
Trimble	Bedford	8,125	6,090	149
Union	Morganfield	15,637	16,557	345
Warren	Bowling Green	92,522	77,720	545
Washington	Springfield	10,916	10,441	301
Wayne	Monticello	19,923	17,468	459
Webster	Dixon	14,120	13,955	335
Whitley	Williamsburg	35,865	33,326	440
Wolfe	Campton	7,065	6,503	223
Woodford	Versailles	23,208	19,955	191

Louisiana

(64 parishes, 43,562 sq. mi. land; pop. 4,468,976)

Parish	Parish seat or courthouse	2000 Pop.	1990 Pop.	Land area sq. mi.
Acadia	Crowley	58,861	55,882	655
Allen	Oberlin	25,440	21,226	765
Ascension	Donaldsonville	76,627	58,214	292
Assumption	Napoleonville	23,388	22,753	339
Avoyelles	Marksville	41,481	39,159	832
Beauregard	De Ridder	32,986	30,083	1,160
Bienville	Arcadia	15,752	16,232	811
Bossier	Benton	98,310	86,088	839
Caddo	Shreveport	252,161	248,253	882
Calcasieu	Lake Charles	183,577	168,134	1,071
Caldwell	Columbia	10,560	9,806	529
Cameron	Cameron	9,991	9,260	1,313
Catahoula	Harrisonburg	10,920	11,065	704
Claiborne	Homer	16,851	17,405	755
Concordia	Vidalia	20,247	20,828	696
De Soto	Mansfield	25,494	25,668	877
East Baton Rouge	Baton Rouge	412,852	380,105	455
East Carroll	Lake Providence	9,421	9,709	421
East Feliciana	Clinton	21,360	19,211	453
Evangeline	Ville Platte	35,434	33,274	664
Franklin	Winnsboro	21,263	22,387	624
Grant	Colfax	18,698	17,526	645
Iberia	New Iberia	73,266	68,297	575
Iberville	Plaquemine	33,320	31,049	619
Jackson	Jonesboro	15,397	15,859	570
Jefferson	Gretna	455,466	448,306	307
Jefferson Davis	Jennings	31,435	30,722	652
Lafayette	Lafayette	190,503	164,762	270
Lafourche	Thibodaux	89,974	85,860	1,085
La Salle	Jena	14,282	13,662	624
Lincoln	Ruston	42,509	41,745	471
Livingston	Livingston	91,814	70,523	648
Madison	Tallulah	13,728	12,463	624
Morehouse	Bastrop	31,021	31,938	794
Natchitoches	Natchitoches	39,080	37,254	1,255
Orleans	New Orleans	484,674	496,938	181
Ouachita	Monroe	147,250	142,191	611
Plaquemines	Pointe a la Hache	26,757	25,575	845
Pointe Coupee	New Roads	22,763	22,540	557
Rapides	Alexandria	126,337	131,556	1,323
Red River	Coushatta	9,622	9,526	389
Richland	Rayville	20,981	20,629	558
Sabine	Many	23,459	22,646	865
Saint Bernard	Chalmette	67,229	66,631	465
Saint Charles	Hahnville	48,072	42,437	284
Saint Helena	Greensburg	10,525	9,874	408
Saint James	Convent	21,216	20,879	246
Saint John the Baptist	Edgard	43,044	39,996	219
Saint Landry	Opelousas	87,700	80,312	929
Saint Martin	Saint Martinville	48,583	44,097	740
Saint Mary	Franklin	53,500	58,086	613
Saint Tammany	Covington	191,268	144,500	854
Tangipahoa	Amite	100,588	85,709	790
Tensas	Saint Joseph	6,618	7,103	602
Terrebonne	Houma	104,503	96,982	1,255
Union	Farmerville	22,803	20,796	878
Vermilion	Abbeville	53,807	50,055	1,174
Vernon	Leesville	52,531	61,961	1,328
Washington	Franklinton	43,926	43,185	670
Webster	Minden	41,831	41,989	595
West Baton Rouge	Port Allen	21,601	19,419	191
West Carroll	Oak Grove	12,314	12,093	359
West Feliciana	Saint Francisville	15,111	12,915	406
Winn	Winnfield	16,894	16,498	950

Maine

(16 counties, 30,862 sq. mi. land; pop. 1,274,923)

County	County seat or courthouse	2000 Pop.	1990 Pop.	Land area sq. mi.
Androscoggin	Auburn	103,793	105,259	470
Aroostook	Houlton	73,938	86,936	6,672
Cumberland	Portland	265,612	243,135	836
Franklin	Farmington	29,467	29,008	1,698
Hancock	Ellsworth	51,791	46,948	1,588
Kennebec	Augusta	117,114	115,904	868
Knox	Rockland	39,618	36,310	366
Lincoln	Wiscasset	33,616	30,357	456
Oxford	South Paris	54,755	52,602	2,078
Penobscot	Bangor	144,919	146,601	3,396
Piscataquis	Dover-Foxcroft	17,235	18,653	3,966
Sagadahoc	Bath	35,214	33,535	254
Somerset	Skowhegan	50,888	49,767	3,927
Waldo	Belfast	36,280	33,018	730
Washington	Machias	33,941	35,308	2,568
York	Alfred	186,742	164,587	991

Maryland

(23 counties, 1 ind. city, 9,774 sq. mi. land; pop. 5,296,486)

County	County seat or courthouse	2000 Pop.	1990 Pop.	Land area sq. mi.
Allegany	Cumberland	74,930	74,946	425
Anne Arundel	Annapolis	489,656	427,239	416
Baltimore	Towson	754,292	692,134	599
Calvert	Prince Frederick	74,563	51,372	215
Caroline	Denton	29,772	27,035	320
Carroll	Westminster	150,897	123,372	449
Cecil	Elkton	85,951	71,347	348
Charles	La Plata	120,546	101,154	461
Dorchester	Cambridge	30,674	30,236	558

County	County seat or courthouse	2000 Pop.	1990 Pop.	Land area sq. mi.
Frederick	Frederick	195,277	150,208	663
Garrett	Oakland	29,846	28,138	648
Harford	Bel Air	218,590	182,132	440
Howard	Ellicott City	247,842	187,328	252
Kent	Chestertown	19,197	17,842	279
Montgomery	Rockville	873,341	762,875	496
Prince George's	Upper Marlboro	801,515	722,705	485
Queen Anne's	Centreville	40,563	33,953	372
Saint Mary's	Leonardtown	86,211	75,974	361
Somerset	Princess Anne	24,747	23,440	327
Talbot	Easton	33,812	30,549	269
Washington	Hagerstown	131,923	121,393	458
Wicomico	Salisbury	84,644	74,339	377
Worcester	Snow Hill	46,543	35,028	473
Independent City				
Baltimore		651,154	736,014	81

Massachusetts

(14 counties, 7,840 sq. mi. land; pop. 6,349,097)

County	County seat or courthouse	2000 Pop.	1990 Pop.	Land area sq. mi.
Barnstable	Barnstable	222,230	186,605	396
Berkshire	Pittsfield	134,953	139,352	931
Bristol	Taunton	534,678	506,325	556
Dukes	Edgartown	14,987	11,639	104
Essex	Salem	723,419	670,080	501
Franklin	Greenfield	71,535	70,086	702
Hampden	Springfield	456,228	456,310	618
Hampshire	Northampton	152,251	146,568	529
Middlesex	East Cambridge	1,465,396	1,398,468	823
Nantucket	Nantucket	9,520	6,012	48
Norfolk	Dedham	650,308	616,087	400
Plymouth	Plymouth	472,822	435,276	661
Suffolk	Boston	689,807	663,906	59
Worcester	Worcester	750,963	709,711	1,513

Michigan

(83 counties, 56,804 sq. mi. land; pop. 9,938,444)

County	County seat or courthouse	2000 Pop.	1990 Pop.	Land area sq. mi.
Alcona	Harrisville	11,719	10,145	674
Alger	Munising	9,862	8,972	918
Allegan	Allegan	105,665	90,509	827
Alpena	Alpena	31,314	30,605	574
Antrim	Bellaire	23,110	18,185	477
Arenac	Standish	17,269	14,906	367
Baraga	L'Anse	8,746	7,954	904
Barry	Hastings	56,755	50,057	556
Bay	Bay City	110,157	111,723	444
Benzie	Beulah	15,998	12,200	321
Berrien	Saint Joseph	162,453	161,378	571
Branch	Coldwater	45,787	41,502	507
Calhoun	Marshall	137,985	135,982	709
Cass	Cassopolis	51,104	49,477	492
Charlevoix	Charlevoix	26,090	21,468	417
Cheboygan	Cheboygan	26,448	21,398	716
Chippewa	Sault Sainte Marie	38,543	34,604	1,561
Clare	Harrison	31,252	24,952	567
Clinton	Saint Johns	64,753	57,893	571
Crawford	Grayling	14,273	12,260	558
Delta	Escanaba	38,520	37,780	1,170
Dickinson	Iron Mountain	27,472	26,831	766
Eaton	Charlotte	103,655	92,879	576
Emmet	Petoskey	31,437	25,040	468
Genesee	Flint	436,141	430,459	640
Gladwin	Gladwin	26,023	21,896	507
Gogebic	Bessemer	17,370	18,052	1,102
Grand Traverse	Traverse City	77,654	64,273	465
Gratiot	Ithaca	42,285	38,982	570
Hillsdale	Hillsdale	46,527	43,431	599
Houghton	Houghton	36,016	35,446	1,012
Huron	Bad Axe	36,079	34,951	837
Ingham	Mason	279,320	281,912	559
Ionia	Ionia	61,518	57,024	573
Iosco	Tawas City	27,339	30,209	549
Iron	Crystal Falls	13,138	13,175	1,166
Isabella	Mount Pleasant	63,351	54,624	574
Jackson	Jackson	158,422	149,756	707
Kalamazoo	Kalamazoo	238,603	223,411	562
Kalkaska	Kalkaska	16,571	13,497	561
Kent	Grand Rapids	574,335	500,631	856
Keweenaw	Eagle River	2,301	1,701	541
Lake	Baldwin	11,333	8,583	567
Lapeer	Lapeer	87,904	74,768	654
Leelanau	Leland	21,119	16,527	348
Lenawee	Adrian	98,890	91,476	751
Livingston	Howell	156,951	115,645	568
Luce	Newberry	7,024	5,763	903
Mackinac	Saint Ignace	11,943	10,674	1,022
Macomb	Mount Clemens	788,149	717,400	480
Manistee	Manistee	24,527	21,265	544
Marquette	Marquette	64,634	70,887	1,821
Mason	Ludington	28,274	25,537	495
Mecosta	Big Rapids	40,553	37,308	556
Menominee	Menominee	25,326	24,920	1,044
Midland	Midland	82,874	75,651	521
Missaukee	Lake City	14,478	12,147	567
Monroe	Monroe	145,945	133,600	551
Montcalm	Stanton	61,266	53,059	708
Montmorency	Atlanta	10,315	8,936	548
Muskegon	Muskegon	170,200	158,983	509
Newaygo	White Cloud	47,874	38,206	842
Oakland	Pontiac	1,194,156	1,083,592	873
Oceana	Hart	26,873	22,455	540
Ogemaw	West Branch	21,645	18,681	564
Ontonagon	Ontonagon	7,818	8,854	1,312
Osceola	Reed City	23,197	20,146	566
Oscoda	Mio	9,418	7,842	565
Otsego	Gaylord	23,301	17,957	515
Ottawa	Grand Haven	238,314	187,768	566
Presque Isle	Rogers City	14,411	13,743	660
Roscommon	Roscommon	25,469	19,776	521
Saginaw	Saginaw	210,039	211,946	809
Saint Clair	Port Huron	164,235	145,607	724
Saint Joseph	Centreville	62,422	58,913	504
Sanilac	Sandusky	44,547	39,928	964
Schoolcraft	Manistique	8,903	8,302	1,178
Shiawassee	Corunna	71,687	69,770	539
Tuscola	Caro	58,266	55,498	812
Van Buren	Paw Paw	76,263	70,060	611
Washtenaw	Ann Arbor	322,895	282,937	710
Wayne	Detroit	2,061,162	2,111,687	614
Wexford	Cadillac	30,484	26,360	565

Minnesota

(87 counties, 79,610 sq. mi. land; pop. 4,919,479)

County	County seat or courthouse	2000 Pop.	1990 Pop.	Land area sq. mi.
Aitkin	Aitkin	15,301	12,425	1,819
Anoka	Anoka	298,084	243,641	424
Becker	Detroit Lakes	30,000	27,881	1,310
Beltrami	Bemidji	39,650	34,384	2,505
Benton	Foley	34,226	30,185	408
Big Stone	Ortonville	5,820	6,285	497
Blue Earth	Mankato	55,941	54,044	752
Brown	New Ulm	26,911	26,984	611
Carlton	Carlton	31,671	29,259	860
Carver	Chaska	70,205	47,915	357
Cass	Walker	27,150	21,791	2,018
Chippewa	Montevideo	13,088	13,228	583
Chisago	Center City	41,101	30,521	418
Clay	Moorhead	51,229	50,422	1,045
Clearwater	Bagley	8,423	8,309	995
Cook	Grand Marais	5,168	3,868	1,451
Cottonwood	Windom	12,167	12,694	640
Crow Wing	Brainerd	55,099	44,249	997
Dakota	Hastings	355,904	275,210	570
Dodge	Mantorville	17,731	15,731	440
Douglas	Alexandria	32,821	28,674	634
Faribault	Blue Earth	16,181	16,937	714
Fillmore	Preston	21,122	20,777	861
Freeborn	Albert Lea	32,584	33,060	708
Goodhue	Red Wing	44,127	40,690	758
Grant	Elbow Lake	6,289	6,246	546
Hennepin	Minneapolis	1,116,200	1,032,431	557
Houston	Caledonia	19,718	18,497	558
Hubbard	Park Rapids	18,376	14,939	922
Isanti	Cambridge	31,287	25,921	439
Itasca	Grand Rapids	43,992	40,863	2,665
Jackson	Jackson	11,268	11,677	702
Kanabec	Mora	14,996	12,802	525
Kandiyohi	Willmar	41,203	38,761	796
Kittson	Hallock	5,285	5,767	1,097
Koochiching	International Falls	14,355	16,299	3,102
Lac qui Parle	Madison	8,067	8,924	765
Lake	Two Harbors	11,058	10,415	2,099
Lake of the Woods	Baudette	4,522	4,076	1,297
Le Sueur	Le Center	25,426	23,239	449
Lincoln	Ivanhoe	6,429	6,890	537
Lyon	Marshall	25,425	24,789	714
McLeod	Glencoe	34,898	32,030	492
Mahnomen	Mahnomen	5,190	5,044	556
Marshall	Warren	10,155	10,993	1,772
Martin	Fairmont	21,802	22,914	709
Meeker	Litchfield	22,644	20,846	609
Mille Lacs	Milaca	22,330	18,670	574
Morrison	Little Falls	31,712	29,604	1,125
Mower	Austin	38,603	37,385	712
Murray	Slayton	9,165	9,660	704
Nicollet	Saint Peter	29,771	28,076	452
Nobles	Worthington	20,832	20,098	715
Norman	Ada	7,442	7,975	876
Olmsted	Rochester	124,277	106,470	653
Otter Tail	Fergus Falls	57,159	50,714	1,980
Pennington	Thief River Falls	13,584	13,306	617
Pine	Pine City	26,530	21,264	1,411
Pipestone	Pipestone	9,895	10,491	466

County	County seat or courthouse	2000 Pop.	1990 Pop.	Land area sq. mi.
Polk	Crookston	31,369	32,589	1,970
Pope	Glenwood	11,236	10,745	670
Ramsey	Saint Paul	511,035	485,760	156
Red Lake	Red Lake Falls	4,299	4,525	432
Redwood	Redwood Falls	16,815	17,254	880
Renville	Olivia	17,154	17,673	983
Rice	Faribault	56,665	49,183	498
Rock	Luverne	9,721	9,806	483
Roseau	Roseau	16,338	15,026	1,663
Saint Louis	Duluth	200,528	198,232	6,225
Scott	Shakopee	89,498	57,846	357
Sherburne	Elk River	64,417	41,945	436
Sibley	Gaylord	15,356	14,366	589
Stearns	Saint Cloud	133,166	119,324	1,345
Steele	Owatonna	33,680	30,729	430
Stevens	Morris	10,053	10,634	562
Swift	Benson	11,956	10,724	744
Todd	Long Prairie	24,426	23,363	942
Traverse	Wheaton	4,134	4,463	574
Wabasha	Wabasha	21,610	19,744	525
Wadena	Wadena	13,713	13,154	535
Waseca	Waseca	19,526	18,079	423
Washington	Stillwater	201,130	145,860	392
Watonwan	Saint James	11,876	11,682	435
Wilkin	Breckenridge	7,138	7,516	751
Winona	Winona	49,985	47,828	626
Wright	Buffalo	89,986	68,710	661
Yellow Medicine	Granite Falls	11,080	11,684	758

Mississippi

(82 counties, 46,907 sq. mi. land; pop. 2,844,658)

County	County seat or courthouse	2000 Pop.	1990 Pop.	Land area sq. mi.
Adams	Natchez	34,340	35,356	460
Alcorn	Corinth	34,558	31,722	400
Amite	Liberty	13,599	13,328	730
Attala	Kosciusko	19,661	18,481	735
Benton	Ashland	8,026	8,046	407
Bolivar	Cleveland & Rosedale	40,633	41,875	876
Calhoun	Pittsboro	15,069	14,908	587
Carroll	Carrollton & Vaiden	10,769	9,237	628
Chickasaw	Houston & Okolona	19,440	18,085	502
Choctaw	Ackerman	9,758	9,071	419
Claiborne	Port Gibson	11,831	11,370	487
Clarke	Quitman	17,955	17,313	691
Clay	West Point	21,979	21,120	409
Coahoma	Clarksdale	30,622	31,665	554
Copiah	Hazlehurst	28,757	27,592	777
Covington	Collins	19,407	16,527	414
De Soto	Hernando	107,199	67,910	478
Forrest	Hattiesburg	72,604	68,314	467
Franklin	Meadville	8,448	8,377	565
George	Lucedale	19,144	16,673	478
Greene	Leakesville	13,299	10,220	713
Grenada	Grenada	23,263	21,555	422
Hancock	Bay Saint Louis	42,967	31,760	477
Harrison	Gulfport	189,601	165,365	581
Hinds	Jackson & Raymond	250,800	254,441	869
Holmes	Lexington	21,609	21,604	756
Humphreys	Belzoni	11,206	12,134	418
Issaquena	Mayersville	2,274	1,909	413
Itawamba	Fulton	22,770	20,017	532
Jackson	Pascagoula	131,420	115,243	727
Jasper	Bay Springs & Paulding	18,149	17,114	676
Jefferson	Fayette	9,740	8,653	519
Jefferson Davis	Prentiss	13,962	14,051	408
Jones	Ellisville & Laurel	64,958	62,031	694
Kemper	De Kalb	10,453	10,356	766
Lafayette	Oxford	38,744	31,826	631
Lamar	Purvis	39,070	30,424	497
Lauderdale	Meridian	78,161	75,555	704
Lawrence	Monticello	13,258	12,458	431
Leake	Carthage	20,940	18,436	583
Lee	Tupelo	75,755	65,579	450
Leflore	Greenwood	37,947	37,341	592
Lincoln	Brookhaven	33,166	30,278	586
Lowndes	Columbus	61,586	59,308	502
Madison	Canton	74,674	53,794	717
Marion	Columbia	25,595	25,544	542
Marshall	Holly Springs	34,993	30,361	706
Monroe	Aberdeen	38,014	36,582	764
Montgomery	Winona	12,189	12,387	407
Neshoba	Philadelphia	28,684	24,800	570
Newton	Decatur	21,838	20,291	578
Noxubee	Macon	12,548	12,604	695
Oktibbeha	Starkville	42,902	38,375	458
Panola	Batesville & Sardis	34,274	29,996	684
Pearl River	Poplarville	48,621	38,714	811
Perry	New Augusta	12,138	10,865	647
Pike	Magnolia	38,940	36,882	409
Pontotoc	Pontotoc	26,726	22,237	497
Prentiss	Booneville	25,556	23,278	415
Quitman	Marks	10,117	10,490	405
Rankin	Brandon	115,327	87,161	775
Scott	Forest	28,423	24,137	609
Sharkey	Rolling Fork	6,580	7,066	428
Simpson	Mendenhall	27,639	23,953	589
Smith	Raleigh	16,182	14,798	636
Stone	Wiggins	13,622	10,750	445
Sunflower	Indianola	34,369	35,129	694
Tallahatchie	Charleston & Sumner	14,903	15,210	644
Tate	Senatobia	25,370	21,432	404
Tippah	Ripley	20,826	19,523	458
Tishomingo	Iuka	19,163	17,683	424
Tunica	Tunica	9,227	8,164	455
Union	New Albany	25,362	22,085	415
Walthall	Tylertown	15,156	14,352	404
Warren	Vicksburg	49,644	47,880	587
Washington	Greenville	62,977	67,935	724
Wayne	Waynesboro	21,216	19,517	810
Webster	Walthall	10,294	10,222	422
Wilkinson	Woodville	10,312	9,678	677
Winston	Louisville	20,160	19,433	607
Yalobusha	Coffeeville & Water Valley	13,051	12,033	467
Yazoo	Yazoo City	28,149	25,506	919

Missouri

(114 counties, 1 ind. city, 68,886 sq. mi. land; pop. 5,595,211)

County	County seat or courthouse	2000 Pop.	1990 Pop.	Land area sq. mi.
Adair	Kirksville	24,977	24,577	567
Andrew	Savannah	16,492	14,632	435
Atchison	Rockport	6,430	7,457	545
Audrain	Mexico	25,853	23,599	693
Barry	Cassville	34,010	27,547	779
Barton	Lamar	12,541	11,312	594
Bates	Butler	16,653	15,025	848
Benton	Warsaw	17,180	13,859	706
Bollinger	Marble Hill	12,029	10,619	621
Boone	Columbia	135,454	112,379	685
Buchanan	Saint Joseph	85,998	83,083	410
Butler	Poplar Buff	40,867	38,765	698
Caldwell	Kingston	8,969	8,380	429
Callaway	Fulton	40,766	32,809	839
Camden	Camdenton	37,051	27,495	655
Cape Girardeau	Jackson	68,693	61,633	579
Carroll	Carrollton	10,285	10,748	695
Carter	Van Buren	5,941	5,515	508
Cass	Harrisonville	82,092	63,808	699
Cedar	Stockton	13,733	12,093	476
Chariton	Keytesville	8,438	9,202	756
Christian	Ozark	54,285	32,644	563
Clark	Kahoka	7,416	7,547	507
Clay	Liberty	184,006	153,411	396
Clinton	Plattsburg	18,979	16,595	419
Cole	Jefferson City	71,397	63,579	391
Cooper	Boonville	16,670	14,835	565
Crawford	Steelville	22,804	19,173	743
Dade	Greenfield	7,923	7,449	490
Dallas	Buffalo	15,661	12,646	542
Daviess	Gallatin	8,016	7,865	567
De Kalb	Maysville	11,597	9,967	424
Dent	Salem	14,927	13,702	754
Douglas	Ava	13,084	11,876	815
Dunklin	Kennett	33,155	33,112	546
Franklin	Union	93,807	80,603	923
Gasconade	Hermann	15,342	14,006	521
Gentry	Albany	6,861	6,854	492
Greene	Springfield	240,391	207,949	675
Grundy	Trenton	10,432	10,536	436
Harrison	Bethany	8,850	8,469	725
Henry	Clinton	21,997	20,044	702
Hickory	Hermitage	8,940	7,335	399
Holt	Oregon	5,351	6,034	462
Howard	Fayette	10,212	9,631	466
Howell	West Plains	37,238	31,447	928
Iron	Ironton	10,697	10,726	551
Jackson	Independence	654,880	633,234	605
Jasper	Carthage	104,686	90,465	640
Jefferson	Hillsboro	198,099	171,380	657
Johnson	Warrensburg	48,258	42,514	830
Knox	Edina	4,361	4,482	506
Laclede	Lebanon	32,513	27,158	766
Lafayette	Lexington	32,960	31,107	629
Lawrence	Mount Vernon	35,204	30,236	613
Lewis	Monticello	10,494	10,233	505
Lincoln	Troy	38,944	28,892	630
Linn	Linneus	13,754	13,885	620
Livingston	Chillicothe	14,558	14,592	535
McDonald	Pineville	21,681	16,938	540
Macon	Macon	15,762	15,345	804
Madison	Fredericktown	11,800	11,127	497
Maries	Vienna	8,903	7,976	528
Marion	Palmyra	28,289	27,682	438
Mercer	Princeton	3,757	3,723	454

County	County seat or courthouse	2000 Pop.	1990 Pop.	Land area sq. mi.
Miller	Tuscumbia	23,564	20,700	592
Mississippi	Charleston	13,427	14,442	413
Moniteau	California	14,827	12,298	417
Monroe	Paris	9,311	9,104	646
Montgomery	Montgomery City	12,136	11,355	537
Morgan	Versailles	19,309	15,574	597
New Madrid	New Madrid	19,760	20,928	678
Newton	Neosho	52,636	44,445	626
Nodaway	Maryville	21,912	21,709	877
Oregon	Alton	10,344	9,470	791
Osage	Linn	13,062	12,018	606
Ozark	Gainesville	9,542	8,598	742
Pemiscot	Caruthersville	20,047	21,921	493
Perry	Perryville	18,132	16,648	475
Pettis	Sedalia	39,403	35,437	685
Phelps	Rolla	39,825	35,248	673
Pike	Bowling Green	18,351	15,969	673
Platte	Platte City	73,781	57,867	420
Polk	Bolivar	26,992	21,826	637
Pulaski	Waynesville	41,165	41,307	547
Putnam	Unionville	5,223	5,079	518
Ralls	New London	9,626	8,476	471
Randolph	Huntsville	24,663	24,370	482
Ray	Richmond	6,689	21,968	569
Reynolds	Centerville	13,509	6,661	811
Ripley	Doniphan	283,883	12,303	629
Saint Charles	Saint Charles	9,652	212,751	560
Saint Clair	Osceola	6,689	8,457	677
Sainte Genevieve	Sainte Genevieve	17,842	16,037	502
Saint Francois	Farmington	55,641	48,904	449
Saint Louis	Clayton	1,016,315	993,508	508
Saline	Marshall	23,756	23,523	756
Schuyler	Lancaster	4,170	4,236	308
Scotland	Memphis	4,983	4,822	438
Scott	Benton	40,422	39,376	421
Shannon	Eminence	8,324	7,613	1,004
Shelby	Shelbyville	6,799	6,942	501
Stoddard	Bloomfield	29,705	28,895	827
Stone	Galena	28,658	19,078	463
Sullivan	Milan	7,219	6,326	651
Taney	Forsyth	39,703	25,561	632
Texas	Houston	23,003	21,476	1,179
Vernon	Nevada	20,454	19,041	834
Warren	Warrenton	24,525	19,534	431
Washington	Potosi	23,344	20,380	760
Wayne	Greenville	13,259	11,543	761
Webster	Marshfield	31,045	23,753	593
Worth	Grant City	2,382	2,440	267
Wright	Hartville	17,955	16,758	682
Independent City				
Saint Louis		348,189	396,685	62

Montana

(56 counties, 145,552 sq. mi. land; pop. 902,195)

County	County seat or courthouse	2000 Pop.	1990 Pop.	Land area sq. mi.
Beaverhead	Dillon	9,202	8,424	5,542
Big Horn	Hardin	12,671	11,337	4,995
Blaine	Chinook	7,009	6,728	4,226
Broadwater	Townsend	4,385	3,318	1,191
Carbon	Red Lodge	9,552	8,080	2,048
Carter	Ekalaka	1,360	1,503	3,340
Cascade	Great Falls	80,357	77,691	2,698
Chouteau	Fort Benton	5,970	5,452	3,973
Custer	Miles City	11,696	11,697	3,783
Daniels	Scobey	2,017	2,266	1,426
Dawson	Glendive	9,059	9,505	2,373
Deer Lodge	Anaconda	9,417	10,356	737
Fallon	Baker	2,837	3,103	1,620
Fergus	Lewistown	11,893	12,083	4,339
Flathead	Kalispell	74,471	59,218	5,098
Gallatin	Bozeman	67,831	50,484	2,606
Garfield	Jordan	1,279	1,589	4,668
Glacier	Cut Bank	13,247	12,121	2,995
Golden Valley	Ryegate	1,042	912	1,175
Granite	Philipsburg	2,830	2,548	1,727
Hill	Havre	16,673	17,654	2,896
Jefferson	Boulder	10,049	7,939	1,657
Judith Basin	Stanford	2,329	2,282	1,870
Lake	Polson	26,507	21,041	1,494
Lewis & Clark	Helena	55,716	47,495	3,461
Liberty	Chester	2,158	2,295	1,430
Lincoln	Libby	18,837	17,481	3,613
McCone	Circle	1,977	2,276	2,643
Madison	Virginia City	6,851	5,989	3,587
Meagher	White Sulphur Springs	1,932	1,819	2,392
Mineral	Superior	3,884	3,315	1,220
Missoula	Missoula	95,802	78,687	2,598
Musselshell	Roundup	4,497	4,106	1,867
Park	Livingston	15,694	14,515	2,802
Petroleum	Winnett	493	519	1,654
Phillips	Malta	4,601	5,163	5,140
Pondera	Conrad	6,424	6,433	1,625
Powder River	Broadus	1,858	2,090	3,297
Powell	Deer Lodge	7,180	6,620	2,326
Prairie	Terry	1,199	1,383	1,737
Ravalli	Hamilton	36,070	25,010	2,394
Richland	Sidney	9,667	10,716	2,084
Roosevelt	Wolf Point	10,620	10,999	2,356
Rosebud	Forsyth	9,383	10,505	5,012
Sanders	Thompson Falls	10,227	8,669	2,762
Sheridan	Plentywood	4,105	4,732	1,677
Silver Bow	Butte	34,606	33,941	718
Stillwater	Columbus	8,195	6,536	1,795
Sweet Grass	Big Timber	3,609	3,154	1,855
Teton	Choteau	6,445	6,271	2,273
Toole	Shelby	5,267	5,046	1,911
Treasure	Hysham	861	874	979
Valley	Glasgow	7,675	8,239	4,921
Wheatland	Harlowton	2,259	2,246	1,423
Wibaux	Wibaux	1,068	1,191	889
Yellowstone	Billings	129,352	113,419	2,635

Nebraska

(93 counties, 76,872 sq. mi. land; pop. 1,711,263)

County	County seat or courthouse	2000 Pop.	1990 Pop.	Land area sq. mi.
Adams	Hastings	31,151	29,625	563
Antelope	Neligh	7,452	7,965	857
Arthur	Arthur	444	462	715
Banner	Harrisburg	819	852	746
Blaine	Brewster	583	675	711
Boone	Albion	6,259	6,667	687
Box Butte	Alliance	12,158	13,130	1,075
Boyd	Butte	2,438	2,835	540
Brown	Ainsworth	3,525	3,657	1,221
Buffalo	Kearney	42,259	37,447	968
Burt	Tekamah	7,791	7,868	493
Butler	David City	8,767	8,601	584
Cass	Plattsmouth	24,334	21,318	559
Cedar	Hartington	9,615	10,131	740
Chase	Imperial	4,068	4,381	895
Cherry	Valentine	6,148	6,307	5,961
Cheyenne	Sidney	9,830	9,494	1,196
Clay	Clay Center	7,039	7,123	573
Colfax	Schuyler	10,441	9,139	413
Cuming	West Point	10,203	10,117	572
Custer	Broken Bow	11,793	12,270	2,576
Dakota	Dakota City	20,253	16,742	264
Dawes	Chadron	9,060	9,021	1,396
Dawson	Lexington	24,365	19,940	1,013
Deuel	Chappell	2,098	2,237	440
Dixon	Ponca	6,339	6,143	476
Dodge	Fremont	36,160	34,500	534
Douglas	Omaha	463,585	416,444	331
Dundy	Benkelman	2,292	2,582	920
Fillmore	Geneva	6,634	7,103	576
Franklin	Franklin	3,574	3,938	575
Frontier	Stockville	3,099	3,101	975
Furnas	Beaver City	5,324	5,553	718
Gage	Beatrice	22,993	22,794	855
Garden	Oshkosh	2,292	2,460	1,704
Garfield	Burwell	1,902	2,141	570
Gosper	Elwood	2,143	1,928	458
Grant	Hyannis	747	769	776
Greeley	Greeley	2,714	3,006	569
Hall	Grand Island	53,534	48,925	546
Hamilton	Aurora	9,403	8,862	544
Harlan	Alma	3,786	3,810	553
Hayes	Hayes Center	1,068	1,222	713
Hitchcock	Trenton	3,111	3,750	710
Holt	O'Neill	11,551	12,599	2,413
Hooker	Mullen	783	793	721
Howard	Saint Paul	6,567	6,057	569
Jefferson	Fairbury	8,333	8,759	573
Johnson	Tecumseh	4,488	4,673	376
Kearney	Minden	6,882	6,629	516
Keith	Ogallala	8,875	8,584	1,061
Keya Paha	Springview	983	1,029	773
Kimball	Kimball	4,089	4,108	952
Knox	Center	9,374	9,564	1,108
Lancaster	Lincoln	250,291	213,641	839
Lincoln	North Platte	34,632	32,508	2,564
Logan	Stapleton	774	878	571
Loup	Taylor	712	683	570
McPherson	Tryon	533	546	859
Madison	Madison	35,226	32,655	573
Merrick	Central City	8,204	8,062	485
Morrill	Bridgeport	5,440	5,423	1,424
Nance	Fullerton	4,038	4,275	441
Nemaha	Auburn	7,576	7,980	409
Nuckolls	Nelson	5,057	5,786	575
Otoe	Nebraska City	15,396	14,252	616
Pawnee	Pawnee City	3,087	3,317	432

County	County seat or courthouse	2000 Pop.	1990 Pop.	Land area sq. mi.
Perkins	Grant	3,200	3,367	883
Phelps	Holdrege	9,747	9,715	540
Pierce	Pierce	7,857	7,827	574
Platte	Columbus	31,662	29,820	678
Polk	Osceola	5,639	5,655	439
Red Willow	McCook	9,531	11,705	717
Richardson	Falls City	1,756	9,937	553
Rock	Bassett	13,843	2,019	1,008
Saline	Wilber	122,595	12,715	575
Sarpy	Papillion	19,830	102,583	241
Saunders	Wahoo	9,531	18,285	754
Scotts Bluff	Gering	36,951	36,025	739
Seward	Seward	16,496	15,450	575
Sheridan	Rushville	6,198	6,750	2,441
Sherman	Loup City	3,318	3,718	566
Sioux	Harrison	1,475	1,549	2,067
Stanton	Stanton	6,455	6,244	430
Thayer	Hebron	6,055	6,635	575
Thomas	Thedford	729	851	713
Thurston	Pender	7,171	6,936	394
Valley	Ord	4,647	5,169	568
Washington	Blair	18,780	16,607	390
Wayne	Wayne	9,851	9,364	443
Webster	Red Cloud	4,061	4,279	575
Wheeler	Bartlett	886	948	575
York	York	14,598	14,428	576

Nevada

(16 counties, 1 ind. city, 109,826 sq. mi. land; pop. 1,998,257)

County	County seat or courthouse	2000 Pop.	1990 Pop.	Land area sq. mi.
Churchill	Fallon	23,982	17,938	4,929
Clark	Las Vegas	1,375,765	741,368	7,910
Douglas	Minden	41,259	27,637	710
Elko	Elko	45,291	33,463	17,179
Esmeralda	Goldfield	971	1,344	3,589
Eureka	Eureka	1,651	1,547	4,176
Humboldt	Winnemucca	16,106	12,844	9,648
Lander	Battle Mountain	5,794	6,266	5,494
Lincoln	Pioche	4,165	3,775	10,634
Lyon	Yerington	34,501	20,001	1,994
Mineral	Hawthorne	5,071	6,475	3,756
Nye	Tonopah	32,485	17,781	18,147
Pershing	Lovelock	6,693	4,336	6,037
Storey	Virginia City	3,399	2,526	263
Washoe	Reno	339,486	254,667	6,342
White Pine	Ely	9,181	9,264	8,876
Independent City				
Carson City		52,457	40,443	143

New Hampshire

(10 counties, 8,968 sq. mi. land; pop. 1,235,786)

County	County seat or courthouse	2000 Pop.	1990 Pop.	Land area sq. mi.
Belknap	Laconia	56,325	49,216	401
Carroll	Ossipee	43,666	35,410	934
Cheshire	Keene	73,825	70,121	707
Coos	Lancaster	33,111	34,828	1,800
Grafton	Woodsville	81,743	74,929	1,713
Hillsborough	Nashua	380,841	335,838	876
Merrimack	Concord	136,225	120,240	934
Rockingham	Brentwood	277,359	245,845	695
Strafford	Dover	112,233	104,233	369
Sullivan	Newport	40,458	38,592	537

New Jersey

(21 counties, 7,417 sq. mi. land; pop. 8,414,350)

County	County seat or courthouse	2000 Pop.	1990 Pop.	Land area sq. mi.
Atlantic	Mays Landing	252,552	224,327	561
Bergen	Hackensack	884,118	825,380	234
Burlington	Mount Holly	423,394	395,066	805
Camden	Camden	508,932	502,824	222
Cape May	Cape May Court House	102,326	95,089	255
Cumberland	Bridgeton	146,438	138,053	489
Essex	Newark	793,633	777,964	126
Gloucester	Woodbury	254,673	230,082	325
Hudson	Jersey City	608,975	553,099	47
Hunterdon	Flemington	121,989	107,852	430
Mercer	Trenton	350,761	325,759	226
Middlesex	New Brunswick	750,162	671,712	310
Monmouth	Freehold	615,301	553,192	472
Morris	Morristown	470,212	421,330	469
Ocean	Toms River	510,916	433,203	636
Passaic	Paterson	489,049	470,872	185
Salem	Salem	64,285	65,294	338
Somerset	Somerville	297,490	240,222	305
Sussex	Newton	144,166	130,936	521
Union	Elizabeth	522,541	493,819	103
Warren	Belvidere	102,437	91,675	358

New Mexico

(33 counties, 121,356 sq. mi. land; pop. 1,819,046)

County	County seat or courthouse	2000 Pop.	1990 Pop.	Land area sq. mi.
Bernalillo	Albuquerque	556,678	480,577	1,166
Catron	Reserve	3,543	2,563	6,928
Chaves	Roswell	61,382	57,849	6,071
Cibola	Grants	25,595	23,794	4,539
Colfax	Raton	14,189	12,925	3,757
Curry	Clovis	45,044	42,207	1,406
DeBaca	Fort Sumner	2,240	2,252	2,325
Dona Ana	Las Cruces	174,682	135,510	3,807
Eddy	Carlsbad	51,658	48,605	4,182
Grant	Silver City	31,002	27,676	3,966
Guadalupe	Santa Rosa	4,680	4,156	3,030
Harding	Mosquero	810	987	2,125
Hidalgo	Lordsburg	5,932	5,958	3,446
Lea	Lovington	55,511	55,765	4,393
Lincoln	Carrizozo	19,411	12,219	4,831
Los Alamos	Los Alamos	18,343	18,115	109
Luna	Deming	25,016	18,110	2,965
McKinley	Gallup	74,798	60,686	5,449
Mora	Mora	5,180	4,264	1,931
Otero	Alamogordo	62,298	51,928	6,627
Quay	Tucumcari	10,155	10,823	2,875
Rio Arriba	Tierra Amarilla	41,190	34,365	5,858
Roosevelt	Portales	18,018	16,702	2,449
Sandoval	Bernalillo	89,908	63,319	3,709
San Juan	Aztec	113,801	91,605	5,514
San Miguel	Las Vegas	30,126	25,743	4,717
Santa Fe	Santa Fe	129,292	98,928	1,909
Sierra	Truth or Consequences	13,270	9,912	4,180
Socorro	Socorro	18,078	14,764	6,646
Taos	Taos	29,979	23,118	2,203
Torrance	Estancia	16,911	10,285	3,345
Union	Clayton	4,174	4,124	3,830
Valencia	Los Lunas	66,152	45,235	1,068

New York

(62 counties, 47,214 sq. mi. land; pop. 18,976,457)

County	County seat or courthouse	2000 Pop.	1990 Pop.	Land area sq. mi.
Albany	Albany	294,565	292,812	523
Allegany	Belmont	49,927	50,470	1,030
Bronx[1]	Bronx	1,332,650	1,203,789	42
Broome	Binghamton	200,536	212,160	707
Cattaraugus	Little Valley	83,955	84,234	1,310
Cayuga	Auburn	81,963	82,313	693
Chautauqua	Mayville	139,750	141,895	1,062
Chemung	Elmira	91,070	95,195	408
Chenango	Norwich	51,401	51,768	894
Clinton	Plattsburgh	79,894	85,969	1,039
Columbia	Hudson	63,094	62,982	636
Cortland	Cortland	48,599	48,963	500
Delaware	Delhi	48,055	47,352	1,446
Dutchess	Poughkeepsie	280,150	259,462	802
Erie	Buffalo	950,265	968,584	1,044
Essex	Elizabethtown	38,851	37,152	1,797
Franklin	Malone	51,134	46,540	1,631
Fulton	Johnstown	55,073	54,191	496
Genesee	Batavia	60,370	60,060	494
Greene	Catskill	48,195	44,739	648
Hamilton	Lake Pleasant	5,379	5,279	1,720
Herkimer	Herkimer	64,427	65,809	1,411
Jefferson	Watertown	111,738	110,943	1,272
Kings[1]	Brooklyn	2,465,326	2,300,664	71
Lewis	Lowville	26,944	26,796	1,275
Livingston	Geneseo	64,328	62,372	632
Madison	Wampsville	69,441	69,166	656
Monroe	Rochester	735,343	713,968	659
Montgomery	Fonda	49,708	51,981	405
Nassau	Mineola	1,334,544	1,287,873	287
New York[1]	New York	1,537,195	1,487,536	23
Niagara	Lockport	219,846	220,756	523
Oneida	Utica	235,469	250,836	1,213
Onondaga	Syracuse	458,336	468,973	780
Ontario	Canandaigua	100,224	95,101	644
Orange	Goshen	341,367	307,571	816
Orleans	Albion	44,171	41,846	391
Oswego	Oswego	122,377	121,785	953
Otsego	Cooperstown	61,676	60,390	1,003
Putnam	Carmel	95,745	83,941	231
Queens[1]	Jamaica	2,229,379	1,951,598	109
Rensselaer	Troy	152,538	154,429	654
Richmond[1]	Saint George	443,728	378,977	58
Rockland	New City	286,753	265,475	174
Saint Lawrence	Canton	111,931	111,974	2,686
Saratoga	Ballston Spa	200,635	181,276	812
Schenectady	Schenectady	146,555	149,285	206
Schoharie	Schoharie	31,582	31,840	622
Schuyler	Watkins Glen	19,224	18,662	329
Seneca	Waterloo	33,342	33,683	325
Steuben	Bath	98,726	99,088	1,393
Suffolk	Riverhead	1,419,369	1,321,339	912
Sullivan	Monticello	73,966	69,277	970

County	County seat or courthouse	2000 Pop.	1990 Pop.	Land area sq. mi.
Tioga	Owego	51,784	52,337	519
Tompkins	Ithaca	96,501	94,097	476
Ulster	Kingston	177,749	165,380	1,126
Warren	Lake George	63,303	59,209	869
Washington	Hudson Falls	61,042	59,330	835
Wayne	Lyons	93,765	89,123	604
Westchester	White Plains	923,459	874,866	433
Wyoming	Warsaw	43,424	42,507	593
Yates	Penn Yan	24,621	22,810	338

(1) New York City comprises 5 counties: Bronx, Kings (Brooklyn), New York (Manhattan), Queens, and Richmond (Staten Island).

North Carolina

(100 counties, 48,711 sq. mi. land; pop. 8,049,313)

County	County seat or courthouse	2000 Pop.	1990 Pop.	Land area sq. mi.
Alamance	Graham	130,800	108,213	430
Alexander	Taylorsville	33,603	27,544	260
Alleghany	Sparta	10,677	9,590	235
Anson	Wadesboro	25,275	23,474	532
Ashe	Jefferson	24,384	22,209	426
Avery	Newland	17,167	14,867	247
Beaufort	Washington	44,958	42,283	828
Bertie	Windsor	19,773	20,388	699
Bladen	Elizabethtown	32,278	28,663	875
Brunswick	Bolivia	73,143	50,985	855
Buncombe	Asheville	206,330	174,357	656
Burke	Morganton	89,148	75,740	507
Cabarrus	Concord	131,063	98,935	364
Caldwell	Lenoir	77,415	70,709	472
Camden	Camden	6,885	5,904	241
Carteret	Beaufort	59,383	52,407	520
Caswell	Yanceyville	23,501	20,662	425
Catawba	Newton	141,685	118,412	400
Chatham	Pittsboro	49,329	38,979	683
Cherokee	Murphy	24,298	20,170	455
Chowan	Edenton	14,526	13,506	173
Clay	Hayesville	8,775	7,155	215
Cleveland	Shelby	96,287	84,958	465
Columbus	Whiteville	54,749	49,587	935
Craven	New Bern	91,436	81,812	708
Cumberland	Fayetteville	302,963	274,713	653
Currituck	Currituck	18,190	13,736	262
Dare	Manteo	29,967	22,746	384
Davidson	Lexington	147,246	126,688	552
Davie	Mocksville	34,835	27,859	265
Duplin	Kenansville	49,063	39,995	818
Durham	Durham	223,314	181,844	290
Edgecombe	Tarboro	55,606	56,692	505
Forsyth	Winston-Salem	306,067	265,855	410
Franklin	Louisburg	47,260	36,414	492
Gaston	Gastonia	190,365	174,769	356
Gates	Gatesville	10,516	9,305	341
Graham	Robbinsville	7,993	7,196	292
Granville	Oxford	48,498	38,341	531
Greene	Snow Hill	18,974	15,384	265
Guilford	Greensboro	421,048	347,431	649
Halifax	Halifax	57,370	55,516	725
Harnett	Lillington	91,025	67,833	595
Haywood	Waynesville	54,033	46,948	554
Henderson	Hendersonville	89,173	69,747	374
Hertford	Winton	22,601	22,317	353
Hoke	Raeford	33,646	22,856	391
Hyde	Swan Quarter	5,826	5,411	613
Iredell	Statesville	122,660	93,205	576
Jackson	Sylva	33,121	26,835	491
Johnston	Smithfield	121,965	81,306	792
Jones	Trenton	10,381	9,361	472
Lee	Sanford	49,040	41,370	257
Lenoir	Kinston	59,648	57,274	400
Lincoln	Lincolnton	63,780	50,319	299
McDowell	Marion	42,151	35,681	442
Macon	Franklin	29,811	23,504	516
Madison	Marshall	19,635	16,953	449
Martin	Williamston	25,593	25,078	461
Mecklenburg	Charlotte	695,454	511,211	526
Mitchell	Bakersville	15,687	14,433	221
Montgomery	Troy	26,822	23,359	492
Moore	Carthage	74,769	59,000	698
Nash	Nashville	87,420	76,677	540
New Hanover	Wilmington	160,307	120,284	199
Northampton	Jackson	22,086	21,004	536
Onslow	Jacksonville	150,355	149,838	767
Orange	Hillsborough	118,227	93,662	400
Pamlico	Bayboro	12,934	11,368	337
Pasquotank	Elizabeth City	34,897	31,298	227
Pender	Burgaw	41,082	28,855	871
Perquimans	Hertford	11,368	10,447	247
Person	Roxboro	35,623	30,180	392
Pitt	Greenville	133,798	108,480	652
Polk	Columbus	18,324	14,458	238
Randolph	Asheboro	130,454	106,546	787
Richmond	Rockingham	46,564	44,511	474
Robeson	Lumberton	123,339	105,170	949
Rockingham	Wentworth	91,928	86,064	566
Rowan	Salisbury	130,340	110,605	511
Rutherford	Rutherfordton	62,899	56,956	564
Sampson	Clinton	60,161	47,297	945
Scotland	Laurinburg	35,998	33,763	319
Stanly	Albemarle	58,100	51,765	395
Stokes	Danbury	44,711	37,224	452
Surry	Dobson	71,219	61,704	537
Swain	Bryson City	12,968	11,268	528
Transylvania	Brevard	29,334	25,520	378
Tyrrell	Columbia	4,149	3,856	390
Union	Monroe	123,677	84,210	637
Vance	Henderson	42,954	38,892	254
Wake	Raleigh	627,846	426,311	832
Warren	Warrenton	19,972	17,265	429
Washington	Plymouth	13,723	13,997	348
Watauga	Boone	42,695	36,952	313
Wayne	Goldsboro	113,329	104,666	553
Wilkes	Wilkesboro	65,632	59,393	757
Wilson	Wilson	73,814	66,061	371
Yadkin	Yadkinville	36,348	30,488	336
Yancey	Burnsville	17,774	15,419	312

North Dakota

(53 counties, 68,976 sq. mi. land; pop. 642,200)

County	County seat or courthouse	2000 Pop.	1990 Pop.	Land area sq. mi.
Adams	Hettinger	2,593	3,174	988
Barnes	Valley City	11,775	12,545	1,492
Benson	Minnewaukan	6,964	7,198	1,381
Billings	Medora	888	1,108	1,151
Bottineau	Bottineau	7,149	8,011	1,669
Bowman	Bowman	3,242	3,596	1,162
Burke	Bowbells	2,242	3,002	1,104
Burleigh	Bismarck	69,416	60,131	1,633
Cass	Fargo	123,138	102,874	1,765
Cavalier	Langdon	4,831	6,064	1,488
Dickey	Ellendale	5,757	6,107	1,131
Divide	Crosby	2,283	2,899	1,260
Dunn	Manning	3,600	4,005	2,010
Eddy	New Rockford	2,757	2,951	630
Emmons	Linton	4,331	4,830	1,510
Foster	Carrington	3,759	3,983	635
Golden Valley	Beach	1,924	2,108	1,002
Grand Forks	Grand Forks	66,109	70,683	1,438
Grant	Carson	2,841	3,549	1,659
Griggs	Cooperstown	2,754	3,303	709
Hettinger	Mott	2,715	3,445	1,132
Kidder	Steele	2,753	3,332	1,351
La Moure	La Moure	4,701	5,383	1,147
Logan	Napoleon	2,308	2,847	993
McHenry	Towner	5,987	6,528	1,874
McIntosh	Ashley	3,390	4,021	975
McKenzie	Watford City	5,737	6,383	2,742
McLean	Washburn	9,311	10,457	2,110
Mercer	Stanton	8,644	9,808	1,045
Morton	Mandan	25,303	23,700	1,926
Mountrail	Stanley	6,631	7,021	1,824
Nelson	Lakota	3,715	4,410	982
Oliver	Center	2,065	2,381	724
Pembina	Cavalier	8,585	9,238	1,119
Pierce	Rugby	4,675	5,052	1,018
Ramsey	Devils Lake	12,066	12,681	1,185
Ransom	Lisbon	5,890	5,921	863
Renville	Mohall	2,610	3,160	875
Richland	Wahpeton	17,998	18,148	1,437
Rolette	Rolla	13,674	12,772	902
Sargent	Forman	4,366	4,549	859
Sheridan	McClusky	1,710	2,148	972
Sioux	Fort Yates	4,044	3,761	1,094
Slope	Amidon	767	907	1,218
Stark	Dickinson	22,636	22,832	1,338
Steele	Finley	2,258	2,420	712
Stutsman	Jamestown	21,908	22,241	2,221
Towner	Cando	2,876	3,627	1,025
Traill	Hillsboro	8,477	8,752	862
Walsh	Grafton	12,389	13,840	1,282
Ward	Minot	58,795	57,921	2,013
Wells	Fessenden	5,102	5,864	1,271
Williams	Williston	19,761	21,129	2,070

Ohio

(88 counties, 40,048 sq. mi. land; pop. 11,353,140)

County	County seat or courthouse	2000 Pop.	1990 Pop.	Land area sq. mi.
Adams	West Union	27,330	25,371	584
Allen	Lima	108,473	109,755	404
Ashland	Ashland	52,523	47,507	424
Ashtabula	Jefferson	102,728	99,880	702
Athens	Athens	62,223	59,549	507
Auglaize	Wapakoneta	46,611	44,585	401
Belmont	Saint Clairsville	70,226	71,074	537
Brown	Georgetown	42,285	34,966	492

County	County seat or courthouse	2000 Pop.	1990 Pop.	Land area sq. mi.
Butler	Hamilton	332,807	291,479	467
Carroll	Carrollton	28,836	26,521	395
Champaign	Urbana	38,890	36,019	429
Clark	Springfield	144,742	147,538	400
Clermont	Batavia	177,977	150,094	452
Clinton	Wilmington	40,543	35,444	411
Columbiana	Lisbon	112,075	108,276	532
Coshocton	Coshocton	36,655	35,427	564
Crawford	Bucyrus	46,966	47,870	402
Cuyahoga	Cleveland	1,393,978	1,412,140	458
Darke	Greenville	53,309	53,617	600
Defiance	Defiance	39,500	39,350	411
Delaware	Delaware	109,989	66,929	442
Erie	Sandusky	79,551	76,781	255
Fairfield	Lancaster	122,759	103,468	505
Fayette	Washington Court House	28,433	27,466	407
Franklin	Columbus	1,068,978	961,437	540
Fulton	Wauseon	42,084	38,498	407
Gallia	Gallipolis	31,069	30,954	469
Geauga	Chardon	90,895	81,087	404
Greene	Xenia	147,886	136,731	415
Guernsey	Cambridge	40,792	39,024	522
Hamilton	Cincinnati	845,303	866,228	407
Hancock	Findlay	71,295	65,536	531
Hardin	Kenton	31,945	31,111	470
Harrison	Cadiz	15,856	16,085	404
Henry	Napoleon	29,210	29,108	417
Highland	Hillsboro	40,875	35,728	553
Hocking	Logan	28,241	25,533	423
Holmes	Millersburg	38,943	32,849	423
Huron	Norwalk	59,487	56,238	493
Jackson	Jackson	32,641	30,230	420
Jefferson	Steubenville	73,894	80,298	410
Knox	Mount Vernon	54,500	47,473	527
Lake	Painesville	227,511	215,500	228
Lawrence	Ironton	62,319	61,834	455
Licking	Newark	145,491	128,300	687
Logan	Bellefontaine	46,005	42,310	458
Lorain	Elyria	284,664	271,126	493
Lucas	Toledo	455,054	462,361	340
Madison	London	40,213	37,078	465
Mahoning	Youngstown	257,555	264,806	415
Marion	Marion	66,217	64,274	404
Medina	Medina	151,095	122,354	422
Meigs	Pomeroy	23,072	22,987	429
Mercer	Celina	40,924	39,443	463
Miami	Troy	98,868	93,184	407
Monroe	Woodsfield	15,180	15,497	456
Montgomery	Dayton	559,062	573,809	462
Morgan	McConnelsville	14,897	14,194	418
Morrow	Mount Gilead	31,628	27,749	406
Muskingum	Zanesville	84,585	82,068	665
Noble	Caldwell	14,058	11,336	399
Ottawa	Port Clinton	40,985	40,029	255
Paulding	Paulding	20,293	20,488	416
Perry	New Lexington	34,078	31,557	410
Pickaway	Circleville	52,727	48,248	502
Pike	Waverly	27,695	24,249	441
Portage	Ravenna	152,061	142,585	492
Preble	Eaton	42,337	40,113	425
Putnam	Ottawa	34,726	33,819	484
Richland	Mansfield	128,852	126,137	497
Ross	Chillicothe	73,345	69,330	688
Sandusky	Fremont	61,792	61,963	409
Scioto	Portsmouth	79,195	80,327	612
Seneca	Tiffin	58,683	59,733	551
Shelby	Sidney	47,910	44,915	409
Stark	Canton	378,098	367,585	576
Summit	Akron	542,899	514,990	413
Trumbull	Warren	225,116	227,795	616
Tuscarawas	New Philadelphia	90,914	84,090	568
Union	Marysville	40,909	31,969	437
Van Wert	Van Wert	29,659	30,464	410
Vinton	McArthur	12,806	11,098	414
Warren	Lebanon	158,383	113,973	400
Washington	Marietta	63,251	62,254	635
Wayne	Wooster	111,564	101,461	555
Williams	Bryan	39,188	36,956	422
Wood	Bowling Green	121,065	113,269	617
Wyandot	Upper Sandusky	22,908	22,254	406

Oklahoma

(77 counties, 68,667 sq. mi. land; pop. 3,450,654)

County	County seat or courthouse	2000 Pop.	1990 Pop.	Land area sq. mi.
Adair	Stillwell	21,038	18,421	576
Alfalfa	Cherokee	6,105	6,416	867
Atoka	Atoka	13,879	12,778	978
Beaver	Beaver	5,857	6,023	1,814
Beckham	Sayre	19,799	18,812	902
Blaine	Watonga	11,976	11,470	928
Bryan	Durant	36,534	32,089	909
Caddo	Anadarko	30,150	29,550	1,278
Canadian	El Reno	87,697	74,409	900
Carter	Ardmore	45,621	42,919	824
Cherokee	Tahlequah	42,521	34,049	751
Choctaw	Hugo	15,342	15,302	774
Cimarron	Boise City	3,148	3,301	1,835
Cleveland	Norman	208,016	174,253	536
Coal	Coalgate	6,031	5,780	518
Comanche	Lawton	114,996	111,486	1,069
Cotton	Walters	6,614	6,651	637
Craig	Vinita	14,950	14,104	761
Creek	Sapulpa	67,367	60,915	956
Custer	Arapaho	26,142	26,897	987
Delaware	Jay	37,077	28,070	741
Dewey	Taloga	4,743	5,551	1,000
Ellis	Arnett	4,075	4,497	1,229
Garfield	Enid	57,813	56,735	1,058
Garvin	Pauls Valley	27,210	26,605	807
Grady	Chickasha	45,516	41,747	1,101
Grant	Medford	5,144	5,689	1,001
Greer	Mangum	6,061	6,559	639
Harmon	Hollis	3,283	3,793	538
Harper	Buffalo	3,562	4,063	1,039
Haskell	Stigler	11,792	10,940	577
Hughes	Holdenville	14,154	13,014	807
Jackson	Altus	28,439	28,764	803
Jefferson	Waurika	6,818	7,010	759
Johnston	Tishomingo	10,513	10,032	645
Kay	Newkirk	48,080	48,056	919
Kingfisher	Kingfisher	13,926	13,212	903
Kiowa	Hobart	10,227	11,347	1,015
Latimer	Wilburton	10,692	10,333	722
Le Flore	Poteau	48,109	43,270	1,586
Lincoln	Chandler	32,080	29,216	958
Logan	Guthrie	33,924	29,011	744
Love	Marietta	8,831	7,788	515
McClain	Purcell	27,740	22,795	570
McCurtain	Idabel	34,402	33,433	1,852
McIntosh	Eufaula	19,456	16,779	620
Major	Fairview	7,545	8,055	957
Marshall	Madill	13,184	10,829	371
Mayes	Pryor	38,369	33,366	656
Murray	Sulphur	12,623	12,042	418
Muskogee	Muskogee	69,451	68,078	814
Noble	Perry	11,411	11,045	732
Nowata	Nowata	10,569	9,992	565
Okfuskee	Okemah	11,814	11,551	625
Oklahoma	Oklahoma City	660,448	599,611	709
Okmulgee	Okmulgee	39,685	36,490	697
Osage	Pawhuska	44,437	41,645	2,251
Ottawa	Miami	33,194	30,561	471
Pawnee	Pawnee	16,612	15,575	569
Payne	Stillwater	68,190	61,507	686
Pittsburg	McAlester	43,953	40,950	1,306
Pontotoc	Ada	35,143	34,119	720
Pottawatomie	Shawnee	65,521	58,760	788
Pushmataha	Antlers	11,667	10,997	1,397
Roger Mills	Cheyenne	3,436	4,147	1,142
Rogers	Claremore	70,641	55,170	675
Seminole	Wewoka	24,894	25,412	633
Sequoyah	Sallisaw	38,972	33,828	674
Stephens	Duncan	43,182	42,299	874
Texas	Guymon	20,107	16,419	2,037
Tillman	Frederick	9,287	10,384	872
Tulsa	Tulsa	563,299	503,341	570
Wagoner	Wagoner	57,491	47,883	563
Washington	Bartlesville	48,996	48,066	417
Washita	Cordell	11,508	11,441	1,003
Woods	Alva	9,089	9,103	1,287
Woodward	Woodward	18,486	18,976	1,242

Oregon

(36 counties, 95,997 sq. mi. land; pop. 3,421,399)

County	County seat or courthouse	2000 Pop.	1990 Pop.	Land area sq. mi.
Baker	Baker City	16,741	15,317	3,068
Benton	Corvallis	78,153	70,811	676
Clackamas	Oregon City	338,391	278,850	1,868
Clatsop	Astoria	35,630	33,301	827
Columbia	Saint Helens	43,560	37,557	657
Coos	Coquille	62,779	60,273	1,600
Crook	Prineville	19,182	14,111	2,979
Curry	Gold Beach	21,137	19,327	1,627
Deschutes	Bend	115,367	74,976	3,018
Douglas	Roseburg	100,399	94,649	5,037
Gilliam	Condon	1,915	1,717	1,204
Grant	Canyon City	7,935	7,853	4,529
Harney	Burns	7,609	7,060	10,134

County	County seat or courthouse	2000 Pop.	1990 Pop.	Land area sq. mi.
Hood River	Hood River	20,411	16,903	522
Jackson	Medford	181,269	146,387	2,785
Jefferson	Madras	19,009	13,676	1,781
Josephine	Grants Pass	75,726	62,649	1,640
Klamath	Klamath Falls	63,775	57,702	5,944
Lake	Lakeview	7,422	7,186	8,136
Lane	Eugene	322,959	282,912	4,554
Lincoln	Newport	44,479	38,889	980
Linn	Albany	103,069	91,227	2,292
Malheur	Vale	31,615	26,038	9,887
Marion	Salem	284,834	228,483	1,184
Morrow	Heppner	10,995	7,625	2,032
Multnomah	Portland	660,486	583,887	435
Polk	Dallas	62,380	49,541	741
Sherman	Moro	1,934	1,918	823
Tillamook	Tillamook	24,262	21,570	1,102
Umatilla	Pendleton	70,548	59,249	3,215
Union	La Grande	24,530	23,598	2,037
Wallowa	Enterprise	7,226	6,911	3,145
Wasco	The Dalles	23,791	21,683	2,381
Washington	Hillsboro	445,342	311,554	724
Wheeler	Fossil	1,547	1,396	1,715
Yamhill	McMinnville	84,992	65,551	716

Pennsylvania

(67 counties, 44,817 sq. mi. land; pop. 12,281,054)

County	County seat or courthouse	2000 Pop.	1990 Pop.	Land area sq. mi.
Adams	Gettysburg	91,292	78,274	520
Allegheny	Pittsburgh	1,281,666	1,336,449	730
Armstrong	Kittanning	72,392	73,478	654
Beaver	Beaver	181,412	186,093	434
Bedford	Bedford	49,984	47,919	1,015
Berks	Reading	373,638	336,523	859
Blair	Hollidaysburg	129,144	130,542	526
Bradford	Towanda	62,761	60,967	1,151
Bucks	Doylestown	597,635	541,174	607
Butler	Butler	174,083	152,013	789
Cambria	Ebensburg	152,598	163,062	688
Cameron	Emporium	5,974	5,913	397
Carbon	Jim Thorpe	58,802	56,803	381
Centre	Bellefonte	135,758	124,812	1,108
Chester	West Chester	433,501	376,389	756
Clarion	Clarion	41,765	41,699	602
Clearfield	Clearfield	83,382	78,097	1,147
Clinton	Lock Haven	37,914	37,182	891
Columbia	Bloomsburg	64,151	63,202	486
Crawford	Meadville	90,366	86,166	1,013
Cumberland	Carlisle	213,674	195,257	550
Dauphin	Harrisburg	251,798	237,813	525
Delaware	Media	550,864	547,658	184
Elk	Ridgway	35,112	34,878	829
Erie	Erie	280,843	275,575	802
Fayette	Uniontown	148,644	145,351	790
Forest	Tionesta	4,946	4,802	428
Franklin	Chambersburg	129,313	121,082	772
Fulton	McConnellsburg	14,261	13,837	438
Greene	Waynesburg	40,672	39,550	576
Huntingdon	Huntingdon	45,586	44,164	874
Indiana	Indiana	89,605	89,994	829
Jefferson	Brookville	45,932	46,083	655
Juniata	Mifflintown	22,821	20,625	392
Lackawanna	Scranton	213,295	219,097	459
Lancaster	Lancaster	470,658	422,822	949
Lawrence	New Castle	94,643	96,246	360
Lebanon	Lebanon	120,327	113,744	362
Lehigh	Allentown	312,090	291,130	347
Luzerne	Wilkes-Barre	319,250	328,149	891
Lycoming	Williamsport	120,044	118,710	1,235
McKean	Smethport	45,936	47,131	982
Mercer	Mercer	120,293	121,003	672
Mifflin	Lewistown	46,486	46,197	412
Monroe	Stroudsburg	138,687	95,681	609
Montgomery	Norristown	750,097	678,193	483
Montour	Danville	18,236	17,735	131
Northampton	Easton	267,066	247,110	374
Northumberland	Sunbury	94,556	96,771	460
Perry	New Bloomfield	43,602	41,172	554
Philadelphia	Philadelphia	1,517,550	1,585,577	135
Pike	Milford	46,302	28,032	547
Potter	Coudersport	18,080	16,717	1,081
Schuylkill	Pottsville	150,336	152,585	778
Snyder	Middleburg	37,546	36,680	331
Somerset	Somerset	80,023	78,218	1,075
Sullivan	Laporte	6,556	6,104	450
Susquehanna	Montrose	42,238	40,380	823
Tioga	Wellsboro	41,373	41,126	1,134
Union	Lewisburg	41,624	36,176	317
Venango	Franklin	57,565	59,381	675
Warren	Warren	43,863	45,050	883
Washington	Washington	202,897	204,584	857
Wayne	Honesdale	47,722	39,944	729
Westmoreland	Greensburg	369,993	370,321	1,025
Wyoming	Tunkhannock	28,080	28,076	397
York	York	381,751	339,574	904

Rhode Island

(5 counties, 1,045 sq. mi. land; pop. 1,048,319)

County	County seat or courthouse	2000 Pop.	1990 Pop.	Land area sq. mi.
Bristol	Bristol	50,648	48,859	25
Kent	East Greenwich	167,090	161,143	170
Newport	Newport	85,433	87,194	104
Providence	Providence	621,602	596,270	413
Washington	West Kingston	123,546	109,998	333

South Carolina

(46 counties, 30,110 sq. mi. land; pop. 4,012,012)

County	County seat or courthouse	2000 Pop.	1990 Pop.	Land area sq. mi.
Abbeville	Abbeville	26,167	23,862	508
Aiken	Aiken	142,552	120,991	1,073
Allendale	Allendale	11,211	11,727	408
Anderson	Anderson	165,740	145,177	718
Bamberg	Bamberg	16,658	16,902	393
Barnwell	Barnwell	23,478	20,293	548
Beaufort	Beaufort	120,937	86,425	587
Berkeley	Moncks Corner	142,651	128,658	1,098
Calhoun	Saint Matthews	15,185	12,753	380
Charleston	Charleston	309,969	295,159	919
Cherokee	Gaffney	52,537	44,506	393
Chester	Chester	34,068	32,170	581
Chesterfield	Chesterfield	42,768	38,575	799
Clarendon	Manning	32,502	28,450	607
Colleton	Walterboro	38,264	34,377	1,056
Darlington	Darlington	67,394	61,851	561
Dillon	Dillon	30,722	29,114	405
Dorchester	Saint George	96,413	83,060	575
Edgefield	Edgefield	24,595	18,360	502
Fairfield	Winnsboro	23,454	22,295	687
Florence	Florence	125,761	114,344	800
Georgetown	Georgetown	55,797	46,302	815
Greenville	Greenville	379,616	320,127	790
Greenwood	Greenwood	66,271	59,567	456
Hampton	Hampton	21,386	18,186	560
Horry	Conway	196,629	144,053	1,134
Jasper	Ridgeland	20,678	15,487	656
Kershaw	Camden	52,647	43,599	726
Lancaster	Lancaster	61,351	54,516	549
Laurens	Laurens	69,567	58,132	715
Lee	Bishopville	20,119	18,437	410
Lexington	Lexington	216,014	167,526	699
McCormick	McCormick	9,958	8,868	360
Marion	Marion	35,466	33,899	489
Marlboro	Bennettsville	28,818	29,716	480
Newberry	Newberry	36,108	33,172	631
Oconee	Walhalla	66,215	57,494	625
Orangeburg	Orangeburg	91,582	84,804	1,106
Pickens	Pickens	110,757	93,896	497
Richland	Columbia	320,677	286,321	756
Saluda	Saluda	19,181	16,441	452
Spartanburg	Spartanburg	253,791	226,793	811
Sumter	Sumter	104,646	101,276	665
Union	Union	29,881	30,337	514
Williamsburg	Kingstree	37,217	36,815	934
York	York	164,614	131,497	682

South Dakota

(66 counties, 75,885 sq. mi. land; pop. 754,844)

County	County seat or courthouse	2000 Pop.	1990 Pop.	Land area sq. mi.
Aurora	Plankinton	3,058	3,135	708
Beadle	Huron	17,023	18,253	1,259
Bennett	Martin	3,574	3,206	1,185
Bon Homme	Tyndall	7,260	7,089	563
Brookings	Brookings	28,220	25,207	794
Brown	Aberdeen	35,460	35,580	1,713
Brule	Chamberlain	5,364	5,485	819
Buffalo	Gannvalley	2,032	1,759	471
Butte	Belle Fourche	9,094	7,914	2,249
Campbell	Mound City	1,782	1,965	736
Charles Mix	Lake Andes	9,350	9,131	1,098
Clark	Clark	4,143	4,403	958
Clay	Vermillion	13,537	13,186	412
Codington	Watertown	25,897	22,698	688
Corson	McIntosh	4,181	4,195	2,473
Custer	Custer	7,275	6,179	1,558
Davison	Mitchell	18,741	17,503	435
Day	Webster	6,267	6,978	1,029
Deuel	Clear Lake	4,498	4,522	624
Dewey	Timber Lake	5,972	5,523	2,303
Douglas	Armour	3,458	3,746	434
Edmunds	Ipswich	4,367	4,356	1,146
Fall River	Hot Springs	7,453	7,353	1,740
Faulk	Faulkton	2,640	2,744	1,000
Grant	Milbank	7,847	8,372	683
Gregory	Burke	4,792	5,359	1,016
Haakon	Philip	2,196	2,624	1,813
Hamlin	Hayti	5,540	4,974	507
Hand	Miller	3,741	4,272	1,437
Hanson	Alexandria	3,139	2,994	435
Harding	Buffalo	1,353	1,669	2,671

County	County seat or courthouse	2000 Pop.	1990 Pop.	Land area sq. mi.
Hughes	Pierre	16,481	14,817	741
Hutchinson	Olivet	8,075	8,262	813
Hyde	Highmore	1,671	1,696	861
Jackson	Kadoka	2,930	2,811	1,869
Jerauld	Wessington Springs	2,295	2,425	530
Jones	Murdo	1,193	1,324	971
Kingsbury	De Smet	5,815	5,925	838
Lake	Madison	11,276	10,550	563
Lawrence	Deadwood	21,802	20,655	800
Lincoln	Canton	24,131	15,427	578
Lyman	Kennebec	3,895	3,638	1,640
McCook	Salem	5,832	5,688	575
McPherson	Leola	2,904	3,228	1,137
Marshall	Britton	4,576	4,844	838
Meade	Sturgis	24,253	21,878	3,471
Mellette	White River	2,083	2,137	1,306
Miner	Howard	2,884	3,272	570
Minnehaha	Sioux Falls	148,281	123,809	810
Moody	Flandreau	6,595	6,507	520
Pennington	Rapid City	88,565	81,343	2,776
Perkins	Bison	3,363	3,932	2,872
Potter	Gettysburg	2,693	3,190	866
Roberts	Sisseton	10,016	9,914	1,101
Sanborn	Woonsocket	2,675	2,833	569
Shannon	(Attached to Fall River)	12,466	9,902	2,094
Spink	Redfield	7,454	7,981	1,504
Stanley	Fort Pierre	2,772	2,453	1,443
Sully	Onida	1,556	1,589	1,007
Todd	(Attached to Tripp)	9,050	8,352	1,388
Tripp	Winner	6,430	6,924	1,614
Turner	Parker	8,849	8,576	617
Union	Elk Point	12,584	10,189	460
Walworth	Selby	5,974	6,087	708
Yankton	Yankton	21,652	19,252	522
Ziebach	Dupree	2,519	2,220	1,962

Tennessee

(95 counties, 41,217 sq. mi. land; pop. 5,689,283)

County	County seat or courthouse	2000 Pop.	1990 Pop.	Land area sq. mi.
Anderson	Clinton	71,330	68,250	338
Bedford	Shelbyville	37,586	30,411	474
Benton	Camden	16,537	14,524	395
Bledsoe	Pikeville	12,367	9,669	406
Blount	Maryville	105,823	85,962	559
Bradley	Cleveland	87,965	73,712	329
Campbell	Jacksboro	39,854	35,079	480
Cannon	Woodbury	12,826	10,467	266
Carroll	Huntingdon	29,475	27,514	599
Carter	Elizabethton	56,742	51,505	341
Cheatham	Ashland City	35,912	27,140	303
Chester	Henderson	15,540	12,819	289
Claiborne	Tazewell	29,862	26,137	434
Clay	Celina	7,976	7,238	236
Cocke	Newport	33,565	29,141	434
Coffee	Manchester	48,014	40,343	429
Crockett	Alamo	14,532	13,378	265
Cumberland	Crossville	46,802	34,736	682
Davidson	Nashville	569,891	510,786	502
Decatur	Decaturville	11,731	10,472	334
De Kalb	Smithville	17,423	14,360	305
Dickson	Charlotte	43,156	35,061	490
Dyer	Dyersburg	37,279	34,854	511
Fayette	Somerville	28,806	25,559	705
Fentress	Jamestown	16,625	14,669	499
Franklin	Winchester	39,270	34,923	555
Gibson	Trenton	48,152	46,315	603
Giles	Pulaski	29,447	25,741	611
Grainger	Rutledge	20,659	17,095	280
Greene	Greeneville	62,909	55,832	622
Grundy	Altamont	14,332	13,362	361
Hamblen	Morristown	58,128	50,480	161
Hamilton	Chattanooga	307,896	285,536	542
Hancock	Sneedville	6,786	6,739	222
Hardeman	Bolivar	28,105	23,377	668
Hardin	Savannah	25,578	22,633	578
Hawkins	Rogersville	53,563	44,565	487
Haywood	Brownsville	19,797	19,437	533
Henderson	Lexington	25,522	21,844	520
Henry	Paris	31,115	27,888	562
Hickman	Centerville	22,295	16,754	613
Houston	Erin	8,088	7,018	200
Humphreys	Waverly	17,929	15,813	532
Jackson	Gainesboro	10,984	9,297	309
Jefferson	Dandridge	44,294	33,016	274
Johnson	Mountain City	17,499	13,766	298
Knox	Knoxville	382,032	335,749	508
Lake	Tiptonville	7,954	7,129	163
Lauderdale	Ripley	27,101	23,491	470
Lawrence	Lawrenceburg	39,926	35,303	617
Lewis	Hohenwald	11,367	9,247	282
Lincoln	Fayetteville	31,340	28,157	570
Loudon	Loudon	39,086	31,255	229
McMinn	Athens	49,015	42,383	430
McNairy	Selmer	24,653	22,422	560
Macon	Lafayette	20,386	15,906	307
Madison	Jackson	91,837	77,982	557
Marion	Jasper	27,776	24,683	498
Marshall	Lewisburg	26,767	21,539	375
Maury	Columbia	69,498	54,812	613
Meigs	Decatur	11,086	8,033	195
Monroe	Madisonville	38,961	30,541	635
Montgomery	Clarksville	134,768	100,498	539
Moore	Lynchburg	5,740	4,696	129
Morgan	Wartburg	19,757	17,300	522
Obion	Union City	32,450	31,717	545
Overton	Livingston	20,118	17,636	433
Perry	Linden	7,631	6,612	415
Pickett	Byrdstown	4,945	4,548	163
Polk	Benton	16,050	13,643	435
Putnam	Cookeville	62,315	51,373	401
Rhea	Dayton	28,400	24,344	316
Roane	Kingston	51,910	47,227	361
Robertson	Springfield	54,433	41,492	476
Rutherford	Murfreesboro	182,023	118,570	619
Scott	Huntsville	21,127	18,358	532
Sequatchie	Dunlap	11,370	8,863	266
Sevier	Sevierville	71,170	51,050	592
Shelby	Memphis	897,472	826,330	755
Smith	Carthage	17,712	14,143	314
Stewart	Dover	12,370	9,479	458
Sullivan	Blountville	153,048	143,596	413
Sumner	Gallatin	130,449	103,281	529
Tipton	Covington	51,271	37,568	459
Trousdale	Hartsville	7,259	5,920	114
Unicoi	Erwin	17,667	16,549	186
Union	Maynardville	17,808	13,694	224
Van Buren	Spencer	5,508	4,846	273
Warren	McMinnville	38,276	32,992	433
Washington	Jonesborough	107,198	92,336	326
Wayne	Waynesboro	16,842	13,935	734
Weakley	Dresden	34,895	31,972	580
White	Sparta	23,102	20,090	377
Williamson	Franklin	126,638	81,021	583
Wilson	Lebanon	88,809	67,675	571

Texas

(254 counties, 261,797 sq. mi. land; pop. 20,851,820)

County	County seat or courthouse	2000 Pop.	1990 Pop.	Land area sq. mi.
Anderson	Palestine	55,109	48,024	1,071
Andrews	Andrews	13,004	14,338	1,501
Angelina	Lufkin	80,130	69,884	802
Aransas	Rockport	22,497	17,892	252
Archer	Archer City	8,854	7,973	910
Armstrong	Claude	2,148	2,021	914
Atascosa	Jourdanton	38,628	30,533	1,232
Austin	Bellville	23,590	19,832	653
Bailey	Muleshoe	6,594	7,064	827
Bandera	Bandera	17,645	10,562	792
Bastrop	Bastrop	57,733	38,263	888
Baylor	Seymour	4,093	4,385	871
Bee	Beeville	32,359	25,135	880
Bell	Belton	237,974	191,073	1,060
Bexar	San Antonio	1,392,931	1,185,394	1,247
Blanco	Johnson City	8,418	5,972	711
Borden	Gail	729	799	899
Bosque	Meridian	17,204	15,125	989
Bowie	Boston	89,306	81,665	888
Brazoria	Angleton	241,767	191,707	1,386
Brazos	Bryan	152,415	121,862	586
Brewster	Alpine	8,866	8,653	6,193
Briscoe	Silverton	1,790	1,971	900
Brooks	Falfurrias	7,976	8,204	943
Brown	Brownwood	37,674	34,371	944
Burleson	Caldwell	16,470	13,625	666
Burnet	Burnet	34,147	22,677	996
Caldwell	Lockhart	32,194	26,392	546
Calhoun	Port Lavaca	20,647	19,053	512
Callahan	Baird	12,905	11,859	899
Cameron	Brownsville	335,227	260,120	906
Camp	Pittsburg	11,549	9,904	198
Carson	Panhandle	6,516	6,576	923
Cass	Linden	30,438	29,982	937
Castro	Dimmitt	8,285	9,070	898
Chambers	Anahuac	26,031	20,088	599
Cherokee	Rusk	46,659	41,049	1,052
Childress	Childress	7,688	5,953	710
Clay	Henrietta	11,006	10,024	1,098
Cochran	Morton	3,730	4,377	775
Coke	Robert Lee	3,864	3,424	899
Coleman	Coleman	9,235	9,710	1,260
Collin	McKinney	491,675	264,036	848
Collingsworth	Wellington	3,206	3,573	919
Colorado	Columbus	20,390	18,383	963
Comal	New Braunfels	78,021	51,832	561

County	County seat or courthouse	2000 Pop.	1990 Pop.	Land area sq. mi.
Comanche	Comanche	14,026	13,381	938
Concho	Paint Rock	3,966	3,044	991
Cooke	Gainesville	36,363	30,777	874
Coryell	Gatesville	74,978	64,226	1,052
Cottle	Paducah	1,904	2,247	901
Crane	Crane	3,996	4,652	786
Crockett	Ozona	4,099	4,078	2,807
Crosby	Crosbyton	7,072	7,304	900
Culberson	Van Horn	2,975	3,407	3,812
Dallam	Dalhart	6,222	5,461	1,505
Dallas	Dallas	2,218,899	1,852,691	880
Dawson	Lamesa	14,985	14,349	902
Deaf Smith	Hereford	18,561	19,153	1,497
Delta	Cooper	5,327	4,857	277
Denton	Denton	432,976	273,644	889
DeWitt	Cuero	20,013	18,840	909
Dickens	Dickens	2,762	2,571	904
Dimmit	Carrizo Springs	10,248	10,433	1,331
Donley	Clarendon	3,828	3,696	930
Duval	San Diego	13,120	12,918	1,793
Eastland	Eastland	18,297	18,488	926
Ector	Odessa	121,123	118,934	901
Edwards	Rocksprings	2,162	2,266	2,120
Ellis	Waxahachie	111,360	85,167	940
El Paso	El Paso	679,622	591,610	1,013
Erath	Stephenville	33,001	27,991	1,086
Falls	Marlin	18,576	17,712	769
Fannin	Bonham	31,242	24,804	891
Fayette	La Grange	21,804	20,095	950
Fisher	Roby	4,344	4,842	901
Floyd	Floydada	7,771	8,497	992
Foard	Crowell	1,622	1,794	707
Fort Bend	Richmond	354,452	225,421	875
Franklin	Mount Vernon	9,458	7,802	286
Freestone	Fairfield	17,867	15,818	877
Frio	Pearsall	16,252	13,472	1,133
Gaines	Seminole	14,467	14,123	1,502
Galveston	Galveston	250,158	217,396	398
Garza	Post	4,872	5,143	896
Gillespie	Fredericksburg	20,814	17,204	1,061
Glasscock	Garden City	1,406	1,447	901
Goliad	Goliad	6,928	5,980	854
Gonzales	Gonzales	18,628	17,205	1,068
Gray	Pampa	22,744	23,967	928
Grayson	Sherman	110,595	95,019	934
Gregg	Longview	111,379	104,948	274
Grimes	Anderson	23,552	18,843	794
Guadalupe	Seguin	89,023	64,873	711
Hale	Plainview	36,602	34,671	1,005
Hall	Memphis	3,782	3,905	903
Hamilton	Hamilton	8,229	7,733	836
Hansford	Spearman	5,369	5,848	920
Hardeman	Quanah	4,724	5,283	695
Hardin	Kountze	48,073	41,320	894
Harris	Houston	3,400,578	2,818,101	1,729
Harrison	Marshall	62,110	57,483	899
Hartley	Channing	5,537	3,634	1,462
Haskell	Haskell	6,093	6,820	903
Hays	San Marcos	97,589	65,614	678
Hemphill	Canadian	3,351	3,720	910
Henderson	Athens	73,277	58,543	874
Hidalgo	Edinburg	569,463	383,545	1,570
Hill	Hillsboro	32,321	27,146	962
Hockley	Levelland	22,716	24,199	908
Hood	Granbury	41,100	28,981	422
Hopkins	Sulphur Springs	31,960	28,833	782
Houston	Crockett	23,185	21,375	1,231
Howard	Big Spring	33,627	32,343	903
Hudspeth	Sierra Blanca	3,344	2,915	4,571
Hunt	Greenville	76,596	64,343	841
Hutchinson	Stinnett	23,857	25,689	887
Irion	Mertzon	1,771	1,629	1,051
Jack	Jacksboro	8,763	6,981	917
Jackson	Edna	14,391	13,039	829
Jasper	Jasper	35,604	31,102	937
Jeff Davis	Fort Davis	2,207	1,946	2,264
Jefferson	Beaumont	252,051	239,389	904
Jim Hogg	Hebbronville	5,281	5,109	1,136
Jim Wells	Alice	39,326	37,679	865
Johnson	Cleburne	126,811	97,165	729
Jones	Anson	20,785	16,490	931
Karnes	Karnes City	15,446	12,455	750
Kaufman	Kaufman	71,313	52,220	786
Kendall	Boerne	23,743	14,589	662
Kenedy	Sarita	414	460	1,457
Kent	Jayton	859	1,010	902
Kerr	Kerrville	43,653	36,304	1,106
Kimble	Junction	4,468	4,122	1,251
King	Guthrie	356	354	912
Kinney	Brackettville	3,379	3,119	1,363
Kleberg	Kingsville	31,549	30,274	871
Knox	Benjamin	4,253	4,837	849
Lamar	Paris	48,499	43,949	917
Lamb	Littlefield	14,709	15,072	1,016
Lampasas	Lampasas	17,762	13,521	712
La Salle	Cotulla	5,866	5,254	1,489
Lavaca	Hallettsville	19,210	18,690	970
Lee	Giddings	15,657	12,854	629
Leon	Centerville	15,335	12,665	1,072
Liberty	Liberty	70,154	52,726	1,160
Limestone	Groesbeck	22,051	20,946	909
Lipscomb	Lipscomb	3,057	3,143	932
Live Oak	George West	12,309	9,556	1,036
Llano	Llano	17,044	11,631	935
Loving	Mentone	67	107	673
Lubbock	Lubbock	242,628	222,636	899
Lynn	Tahoka	6,550	6,758	892
McCulloch	Brady	8,205	8,778	1,069
McLennan	Waco	213,517	189,123	1,042
McMullen	Tilden	851	817	1,113
Madison	Madisonville	12,940	10,931	470
Marion	Jefferson	10,941	9,984	381
Martin	Stanton	4,746	4,956	915
Mason	Mason	3,738	3,423	932
Matagorda	Bay City	37,957	36,928	1,114
Maverick	Eagle Pass	47,297	36,378	1,280
Medina	Hondo	39,304	27,312	1,328
Menard	Menard	2,360	2,252	902
Midland	Midland	116,009	106,611	900
Milam	Cameron	24,238	22,946	1,017
Mills	Goldthwaite	5,151	4,531	748
Mitchell	Colorado City	9,698	8,016	910
Montague	Montague	19,117	17,274	931
Montgomery	Conroe	293,768	182,201	1,044
Moore	Dumas	20,121	17,865	900
Morris	Daingerfield	13,048	13,200	255
Motley	Matador	1,426	1,532	989
Nacogdoches	Nacogdoches	59,203	54,753	947
Navarro	Corsicana	45,124	39,926	1,008
Newton	Newton	15,072	13,569	933
Nolan	Sweetwater	15,802	16,594	912
Nueces	Corpus Christi	313,645	291,145	836
Ochiltree	Perryton	9,006	9,128	918
Oldham	Vega	2,185	2,278	1,501
Orange	Orange	84,966	80,509	356
Palo Pinto	Palo Pinto	27,026	25,055	953
Panola	Carthage	22,756	22,035	801
Parker	Weatherford	88,495	64,785	904
Parmer	Farwell	10,016	9,863	882
Pecos	Fort Stockton	16,809	14,675	4,764
Polk	Livingston	41,133	30,687	1,057
Potter	Amarillo	113,546	97,841	909
Presidio	Marfa	7,304	6,637	3,856
Rains	Emory	9,139	6,715	232
Randall	Canyon	104,312	89,673	914
Reagan	Big Lake	3,326	4,514	1,175
Real	Leakey	3,047	2,412	700
Red River	Clarksville	14,314	14,317	1,050
Reeves	Pecos	13,137	15,852	2,636
Refugio	Refugio	7,828	7,976	770
Roberts	Miami	887	1,025	924
Robertson	Franklin	16,000	15,511	855
Rockwall	Rockwall	43,080	25,604	129
Runnels	Ballinger	11,495	11,294	1,051
Rusk	Henderson	47,372	43,735	924
Sabine	Hemphill	10,469	9,586	490
San Augustine	San Augustine	8,946	7,999	528
San Jacinto	Coldspring	22,246	16,372	571
San Patricio	Sinton	67,138	58,749	692
San Saba	San Saba	6,186	5,401	1,134
Schleicher	Eldorado	2,935	2,990	1,311
Scurry	Snyder	16,361	18,634	903
Shackelford	Albany	3,302	3,316	914
Shelby	Center	25,224	22,034	794
Sherman	Stratford	3,186	2,858	923
Smith	Tyler	174,706	151,309	928
Somervell	Glen Rose	6,809	5,360	187
Starr	Rio Grande City	53,597	40,518	1,223
Stephens	Breckenridge	9,674	9,010	895
Sterling	Sterling City	1,393	1,438	923
Stonewall	Aspermont	1,693	2,013	919
Sutton	Sonora	4,077	4,135	1,454
Swisher	Tulia	8,378	8,133	900
Tarrant	Fort Worth	1,446,219	1,170,103	863
Taylor	Abilene	126,555	119,655	916
Terrell	Sanderson	1,081	1,410	2,358
Terry	Brownfield	12,761	13,218	890
Throckmorton	Throckmorton	1,850	1,880	912
Titus	Mount Pleasant	28,118	24,009	411
Tom Green	San Angelo	104,010	98,458	1,522
Travis	Austin	812,280	576,407	989
Trinity	Groveton	13,779	11,445	693
Tyler	Woodville	20,871	16,646	923
Upshur	Gilmer	35,291	31,370	588
Upton	Rankin	3,404	4,447	1,242
Uvalde	Uvalde	25,926	23,340	1,557
Val Verde	Del Rio	44,856	38,721	3,170
Van Zandt	Canton	48,140	37,944	849
Victoria	Victoria	84,088	74,361	883
Walker	Huntsville	61,758	50,917	787

County	County seat or courthouse	2000 Pop.	1990 Pop.	Land area sq. mi.
Waller	Hempstead	32,663	23,374	514
Ward	Monahans	10,909	13,115	835
Washington	Brenham	30,373	26,154	609
Webb	Laredo	193,117	133,239	3,357
Wharton	Wharton	41,188	39,955	1,090
Wheeler	Wheeler	5,284	5,879	914
Wichita	Wichita Falls	131,664	122,378	628
Wilbarger	Vernon	14,676	15,121	971
Willacy	Raymondville	20,082	17,705	597
Williamson	Georgetown	249,967	139,551	1,123
Wilson	Floresville	32,408	22,650	807
Winkler	Kermit	7,173	8,626	841
Wise	Decatur	48,793	34,679	905
Wood	Quitman	36,752	29,380	650
Yoakum	Plains	7,322	8,786	800
Young	Graham	17,943	18,126	922
Zapata	Zapata	12,182	9,279	997
Zavala	Crystal City	11,600	12,162	1,298

Utah

(29 counties, 82,144 sq. mi. land; pop. 2,233,169)

County	County seat or courthouse	2000 Pop.	1990 Pop.	Land area sq. mi.
Beaver	Beaver	6,005	4,765	2,590
Box Elder	Brigham City	42,745	36,485	5,723
Cache	Logan	91,391	70,183	1,165
Carbon	Price	20,422	20,228	1,478
Daggett	Manila	921	690	698
Davis	Farmington	238,994	187,941	304
Duchesne	Duchesne	14,371	12,645	3,238
Emery	Castle Dale	10,860	10,332	4,452
Garfield	Panguitch	4,735	3,980	5,174
Grand	Moab	8,485	6,620	3,682
Iron	Parowan	33,779	20,789	3,298
Juab	Nephi	8,238	5,817	3,392
Kane	Kanab	6,046	5,169	3,992
Millard	Fillmore	12,405	11,333	6,589
Morgan	Morgan	7,129	5,528	609
Piute	Junction	1,435	1,277	758
Rich	Randolph	1,961	1,725	1,029
Salt Lake	Salt Lake City	898,387	725,956	737
San Juan	Monticello	14,413	12,621	7,820
Sanpete	Manti	22,763	16,259	1,588
Sevier	Richfield	18,842	15,431	1,910
Summit	Coalville	29,736	15,518	1,871
Tooele	Tooele	40,735	26,601	6,930
Uintah	Vernal	25,224	22,211	4,477
Utah	Provo	368,536	263,590	1,998
Wasatch	Heber City	15,215	10,089	1,177
Washington	Saint George	90,354	48,560	2,427
Wayne	Loa	2,509	2,177	2,460
Weber	Ogden	196,533	158,330	576

Vermont

(14 counties, 9,250 sq. mi. land; pop. 608,827)

County	County seat or courthouse	2000 Pop.	1990 Pop.	Land area sq. mi.
Addison	Middlebury	35,974	32,953	770
Bennington	Bennington	36,994	35,845	676
Caledonia	Saint Johnsbury	29,702	27,846	651
Chittenden	Burlington	146,571	131,761	539
Essex	Guildhall	6,459	6,405	665
Franklin	Saint Albans	45,417	39,980	637
Grand Isle	North Hero	6,901	5,318	83
Lamoille	Hyde Park	23,233	19,735	461
Orange	Chelsea	28,226	26,149	689
Orleans	Newport	26,277	24,053	698
Rutland	Rutland	63,400	62,142	933
Washington	Montpelier	58,039	54,928	689
Windham	Newfane	44,216	41,588	789
Windsor	Woodstock	57,418	54,055	971

Virginia

(95 counties, 40 ind. cities, 39,594 sq. mi. land; pop. 7,078,515)

County	County seat or courthouse	2000 Pop.	1990 Pop.	Land area sq. mi.
Accomack	Accomac	38,305	31,703	455
Albemarle	Charlottesville	79,236	68,177	723
Alleghany	Covington	12,926	12,815	445
Amelia	Amelia Court House	11,400	8,787	357
Amherst	Amherst	31,894	28,578	475
Appomattox	Appomattox	13,705	12,300	334
Arlington	Arlington	189,453	170,895	26
Augusta	Staunton	65,615	54,557	970
Bath	Warm Springs	5,048	4,799	532
Bedford	Bedford	60,371	45,553	755
Bland	Bland	6,871	6,514	359
Botetourt	Fincastle	30,496	24,992	543
Brunswick	Lawrenceville	18,419	15,987	566
Buchanan	Grundy	26,978	31,333	504
Buckingham	Buckingham	15,623	12,873	581
Campbell	Rustburg	51,078	47,499	504
Caroline	Bowling Green	22,121	19,217	533
Carroll	Hillsville	29,245	26,519	476
Charles City	Charles City	6,926	6,282	183
Charlotte	Charlotte Court House	12,472	11,688	475
Chesterfield	Chesterfield	259,903	209,599	426
Clarke	Berryville	12,652	12,101	177
Craig	New Castle	5,091	4,372	331
Culpeper	Culpeper	34,262	27,791	381
Cumberland	Cumberland	9,017	7,825	298
Dickenson	Clintwood	16,395	17,620	332
Dinwiddie	Dinwiddie	24,533	22,279	504
Essex	Tappahannock	9,989	8,689	258
Fairfax	Fairfax	969,749	818,310	395
Fauquier	Warrenton	55,139	48,700	650
Floyd	Floyd	13,874	11,965	381
Fluvanna	Palmyra	20,047	12,429	287
Franklin	Rocky Mount	47,286	39,549	692
Frederick	Winchester	59,209	45,723	415
Giles	Pearisburg	16,657	16,366	357
Gloucester	Gloucester	34,780	30,131	217
Goochland	Goochland	16,863	14,163	284
Grayson	Independence	17,917	16,278	443
Greene	Standardsville	15,244	10,297	157
Greensville	Emporia	11,560	8,553	295
Halifax	Halifax	37,355	36,030	819
Hanover	Hanover	86,320	63,306	473
Henrico	Richmond	262,300	217,878	238
Henry	Collinsville	57,930	56,942	382
Highland	Monterey	2,536	2,635	416
Isle of Wight	Isle of Wight	29,728	25,053	316
James City	Williamsburg	48,102	34,779	143
King and Queen	King and Queen Court House	6,630	6,289	316
King George	King George	16,803	13,527	180
King William	King William	13,146	10,913	275
Lancaster	Lancaster		10,896	133
Lee	Jonesville	11,567	24,496	437
Loudoun	Leesburg	23,589	86,185	520
Louisa	Louisa	169,599	20,325	497
Lunenburg	Lunenburg	25,627	11,419	432
Madison	Madison	13,146	11,949	321
Mathews	Mathews	12,520	8,348	86
Mecklenburg	Boydton	32,380	29,241	624
Middlesex	Saluda	9,932	8,653	130
Montgomery	Christiansburg	83,629	73,913	388
Nelson	Lovingston	14,445	12,778	472
New Kent	New Kent	13,462	10,466	210
Northampton	Eastville	13,093	13,061	207
Northumberland	Heathsville	12,259	10,524	192
Nottoway	Nottoway	15,725	14,993	315
Orange	Orange	25,881	21,421	342
Page	Luray	23,177	21,690	311
Patrick	Stuart	19,407	17,473	483
Pittsylvania	Chatham	61,745	55,672	971
Powhatan	Powhatan	22,377	15,328	261
Prince Edward	Farmville	19,720	17,320	353
Prince George	Prince George	33,047	27,390	266
Prince William	Manassas	280,813	214,954	338
Pulaski	Pulaski	35,127	34,496	321
Rappahannock	Washington	6,983	6,622	267
Richmond	Warsaw	8,809	7,273	191
Roanoke	Salem	85,778	79,278	251
Rockbridge	Lexington	20,808	18,350	600
Rockingham	Harrisonburg	67,725	57,482	851
Russell	Lebanon	30,308	28,667	475
Scott	Gate City	23,403	23,204	537
Shenandoah	Woodstock	35,075	31,636	512
Smyth	Marion	33,081	32,370	452
Southampton	Courtland	17,482	17,022	600
Spotsylvania	Spotsylvania	90,395	57,397	401
Stafford	Stafford	92,446	62,255	270
Surry	Surry	6,829	6,145	279
Sussex	Sussex	12,504	10,248	491
Tazewell	Tazewell	44,598	45,960	520
Warren	Front Royal	31,584	26,142	214
Washington	Abingdon	51,103	45,887	563
Westmoreland	Montross	16,718	15,480	229
Wise	Wise	40,123	39,573	404
Wythe	Wytheville	27,599	25,471	463
York	Yorktown	56,297	42,434	106
Independent Cities				
Alexandria		128,283	111,183	15
Bedford		6,299	6,176	7
Bristol		17,367	18,426	13
Buena Vista		6,349	6,406	7
Charlottesville		45,049	40,470	10
Chesapeake		199,184	151,982	341
Clifton Forge		4,289	4,679	3
Colonial Heights		16,897	16,064	7
Covington		6,303	7,352	6
Danville		48,411	53,056	43
Emporia		5,665	5,556	7
Fairfax		21,498	19,945	6
Falls Church		10,377	9,464	2

County	County seat or courthouse	2000 Pop.	1990 Pop.	Land area sq. mi.
Franklin		8,346	8,392	8
Fredericksburg		19,279	19,033	11
Galax		6,837	6,745	8
Hampton		146,437	133,773	52
Harrisonburg		40,468	30,707	18
Hopewell		22,354	23,101	10
Lexington		6,867	6,959	2
Lynchburg		65,269	66,120	49
Manassas		35,135	27,757	10
Manassas Park		10,290	6,798	2
Martinsville		15,416	16,162	11
Newport News		180,150	171,477	68
Norfolk		234,403	261,250	54
Norton		3,904	4,247	8
Petersburg		33,740	37,071	23
Poquoson		11,566	11,005	16
Portsmouth		100,565	103,910	33
Radford		15,859	15,940	10
Richmond		197,790	202,713	60
Roanoke		94,911	96,487	43
Salem		24,747	23,835	15
Staunton		23,853	24,581	20
Suffolk		63,677	52,143	400
Virginia Beach		425,257	393,089	248
Waynesboro		19,520	18,549	15
Williamsburg		11,998	11,600	9
Winchester		23,585	21,947	9

Washington

(39 counties, 66,544 sq. mi. land; pop. 5,894,121)

County	County seat or courthouse	2000 Pop.	1990 Pop.	Land area sq. mi.
Adams	Ritzville	16,428	13,603	1,925
Asotin	Asotin	20,551	17,605	635
Benton	Prosser	142,475	112,560	1,703
Chelan	Wenatchee	66,616	52,250	2,921
Clallam	Port Angeles	64,525	56,210	1,739
Clark	Vancouver	345,238	238,053	628
Columbia	Dayton	4,064	4,024	869
Cowlitz	Kelso	92,948	82,119	1,139
Douglas	Waterville	32,603	26,205	1,821
Ferry	Republic	7,260	6,295	2,204
Franklin	Pasco	49,347	37,473	1,242
Garfield	Pomeroy	2,397	2,248	711
Grant	Ephrata	74,698	54,798	2,681
Grays Harbor	Montesano	67,194	64,175	1,917
Island	Coupeville	71,558	60,195	208
Jefferson	Port Townsend	25,953	20,406	1,814
King	Seattle	1,737,034	1,507,305	2,126
Kitsap	Port Orchard	231,969	189,731	396
Kittitas	Ellensburg	33,362	26,725	2,297
Klickitat	Goldendale	19,161	16,616	1,872
Lewis	Chehalis	68,600	59,358	2,408
Lincoln	Davenport	10,184	8,864	2,311
Mason	Shelton	49,405	38,341	961
Okanogan	Okanogan	39,564	33,350	5,268
Pacific	South Bend	20,984	18,882	933
Pend Oreille	Newport	11,732	8,915	1,400
Pierce	Tacoma	700,820	586,203	1,679
San Juan	Friday Harbor	14,077	10,035	175
Skagit	Mount Vernon	102,979	79,545	1,735
Skamania	Stevenson	9,872	8,289	1,656
Snohomish	Everett	606,024	465,628	2,089
Spokane	Spokane	417,939	361,333	1,764
Stevens	Colville	40,066	30,948	2,478
Thurston	Olympia	207,355	161,238	727
Wahkiakum	Cathlamet	3,824	3,327	264
Walla Walla	Walla Walla	55,180	48,439	1,271
Whatcom	Bellingham	166,814	127,780	2,120
Whitman	Colfax	40,740	38,775	2,159
Yakima	Yakima	222,581	188,823	4,296

West Virginia

(55 counties, 24,078 sq. mi. land; pop. 1,808,344)

County	County seat or courthouse	2000 Pop.	1990 Pop.	Land area sq. mi.
Barbour	Philippi	15,557	15,699	341
Berkeley	Martinsburg	75,905	59,253	321
Boone	Madison	25,535	25,870	503
Braxton	Sutton	14,702	12,998	513
Brooke	Wellsburg	25,447	26,992	89
Cabell	Huntington	96,784	96,827	282
Calhoun	Grantsville	7,582	7,885	281
Clay	Clay	10,330	9,983	342
Doddridge	West Union	7,403	6,994	320
Fayette	Fayetteville	47,579	47,952	664
Gilmer	Glenville	7,160	7,669	340
Grant	Petersburg	11,299	10,428	477
Greenbrier	Lewisburg	34,453	34,693	1,021
Hampshire	Romney	20,203	16,498	642
Hancock	New Cumberland	32,667	35,233	83
Hardy	Moorefield	12,669	10,977	583
Harrison	Clarksburg	68,652	69,371	416
Jackson	Ripley	28,000	25,938	466
Jefferson	Charles Town	42,190	35,926	210
Kanawha	Charleston	200,073	207,619	903
Lewis	Weston	16,919	17,223	382
Lincoln	Hamlin	22,108	21,382	437
Logan	Logan	37,710	43,032	454
McDowell	Welch	27,329	35,233	535
Marion	Fairmont	56,598	57,249	310
Marshall	Moundsville	35,519	37,356	307
Mason	Point Pleasant	25,957	25,178	432
Mercer	Princeton	62,980	64,980	420
Mineral	Keyser	27,078	26,697	328
Mingo	Williamson	28,253	33,739	423
Monongalia	Morgantown	81,866	75,509	361
Monroe	Union	14,583	12,406	473
Morgan	Berkeley Springs	14,943	12,128	229
Nicholas	Summersville	26,562	26,775	649
Ohio	Wheeling	47,427	50,871	106
Pendleton	Franklin	8,196	8,054	698
Pleasants	St. Marys	7,514	7,546	131
Pocahontas	Marlinton	9,131	9,008	940
Preston	Kingwood	29,334	29,037	648
Putnam	Winfield	51,589	42,835	346
Raleigh	Beckley	79,220	76,819	607
Randolph	Elkins	28,262	27,803	1,040
Ritchie	Harrisville	10,343	10,233	454
Roane	Spencer	15,446	15,120	484
Summers	Hinton	12,999	14,204	361
Taylor	Grafton	16,089	15,144	173
Tucker	Parsons	7,321	7,728	419
Tyler	Middlebourne	9,592	9,796	258
Upshur	Buckhannon	23,404	22,867	355
Wayne	Wayne	42,903	41,636	506
Webster	Webster Springs	9,719	10,729	556
Wetzel	New Martinsville	17,693	19,258	359
Wirt	Elizabeth	5,873	5,192	233
Wood	Parkersburg	87,986	86,915	367
Wyoming	Pineville	25,708	28,990	501

Wisconsin

(72 counties, 54,310 sq. mi. land; pop. 5,363,675)

County	County seat or courthouse	2000 Pop.	1990 Pop.	Land area sq. mi.
Adams	Friendship	18,643	15,682	648
Ashland	Ashland	16,866	16,307	1,044
Barron	Barron	44,963	40,750	863
Bayfield	Washburn	15,013	14,008	1,476
Brown	Green Bay	226,778	194,594	529
Buffalo	Alma	13,804	13,584	684
Burnett	Siren	15,674	13,084	822
Calumet	Chilton	40,631	34,291	320
Chippewa	Chippewa Falls	55,195	52,360	1,010
Clark	Neillsville	33,557	31,647	1,216
Columbia	Portage	52,468	45,088	774
Crawford	Prairie du Chien	17,243	15,940	573
Dane	Madison	426,526	367,085	1,202
Dodge	Juneau	85,897	76,559	882
Door	Sturgeon Bay	27,961	25,690	483
Douglas	Superior	43,287	41,758	1,309
Dunn	Menomonie	39,858	35,909	852
Eau Claire	Eau Claire	93,142	85,183	638
Florence	Florence	5,088	4,590	488
Fond du Lac	Fond du Lac	97,296	90,083	723
Forest	Crandon	10,024	8,776	1,014
Grant	Lancaster	49,597	49,266	1,148
Green	Monroe	33,647	30,339	584
Green Lake	Green Lake	19,105	18,651	354
Iowa	Dodgeville	22,780	20,150	763
Iron	Hurley	6,861	6,153	757
Jackson	Black River Falls	19,100	16,588	987
Jefferson	Jefferson	74,021	67,783	557
Juneau	Mauston	24,316	21,650	768
Kenosha	Kenosha	149,577	128,181	273
Kewaunee	Kewaunee	20,187	18,878	343
La Crosse	La Crosse	107,120	97,904	453
Lafayette	Darlington	16,137	16,074	634
Langlade	Antigo	20,740	19,505	873
Lincoln	Merrill	29,641	26,993	883
Manitowoc	Manitowoc	82,887	80,421	592
Marathon	Wausau	125,834	115,400	1,545
Marinette	Marinette	43,384	40,548	1,402
Marquette	Montello	15,832	12,321	455
Menominee	Keshena	4,562	4,075	358
Milwaukee	Milwaukee	940,164	959,212	242
Monroe	Sparta	40,899	36,633	901
Oconto	Oconto	35,634	30,226	998
Oneida	Rhinelander	36,776	31,679	1,125
Outagamie	Appleton	160,971	140,510	640
Ozaukee	Port Washington	82,317	72,894	232
Pepin	Durand	7,213	7,107	232

County	County seat or courthouse	2000 Pop.	1990 Pop.	Land area sq. mi.
Pierce	Ellsworth	36,804	32,765	576
Polk	Balsam Lake	41,319	34,773	917
Portage	Stevens Point	67,182	61,405	806
Price	Phillips	15,822	15,600	1,253
Racine	Racine	188,831	175,034	333
Richland	Richland Center	17,924	17,521	586
Rock	Janesville	152,307	139,510	720
Rusk	Ladysmith	15,347	15,079	913
Saint Croix	Hudson	63,155	50,251	722
Sauk	Baraboo	55,225	46,975	838
Sawyer	Hayward	16,196	14,181	1,256
Shawano	Shawano	40,664	37,157	893
Sheboygan	Sheboygan	112,646	103,877	514
Taylor	Medford	19,680	18,901	975
Trempealeau	Whitehall	27,010	25,263	734
Vernon	Viroqua	28,056	25,617	795
Vilas	Eagle River	21,033	17,707	874
Walworth	Elkhorn	93,759	75,000	555
Washburn	Shell Lake	16,036	13,772	810
Washington	West Bend	117,493	95,328	431
Waukesha	Waukesha	360,767	304,715	556
Waupaca	Waupaca	51,731	46,104	751
Waushara	Wautoma	23,154	19,385	626
Winnebago	Oshkosh	156,763	140,320	439
Wood	Wisconsin Rapids	75,555	73,605	793

Wyoming

(23 counties, 97,100 sq. mi. land; pop. 493,782)

County	County seat or courthouse	2000 Pop.	1990 Pop.	Land area sq. mi.
Albany	Laramie	32,014	30,797	4,273
Big Horn	Basin	11,461	10,525	3,137
Campbell	Gillette	33,698	29,370	4,797
Carbon	Rawlins	15,639	16,659	7,896
Converse	Douglas	12,052	11,128	4,255
Crook	Sundance	5,887	5,294	2,859
Fremont	Lander	35,804	33,662	9,182
Goshen	Torrington	12,538	12,373	2,225
Hot Springs	Thermopolis	4,882	4,809	2,004
Johnson	Buffalo	7,075	6,145	4,166
Laramie	Cheyenne	81,607	73,142	2,686
Lincoln	Kemmerer	14,573	12,625	4,069
Natrona	Casper	66,533	61,226	5,340
Niobrara	Lusk	2,407	2,499	2,626
Park	Cody	25,786	23,178	6,942
Platte	Wheatland	8,807	8,145	2,085
Sheridan	Sheridan	26,560	23,562	2,523
Sublette	Pinedale	5,920	4,843	4,883
Sweetwater	Green River	37,613	38,823	10,425
Teton	Jackson	18,251	11,173	4,008
Unita	Evanston	19,742	18,705	2,082
Washakie	Worland	8,289	8,388	2,240
Weston	Newcastle	6,644	6,518	2,398

Population of Outlying Areas

Source: Bureau of the Census, U.S. Dept. of Commerce; World Almanac research

Census 2000 figures are given for all populations and land areas of Puerto Rican municipios (a municipio is the governmental unit that is the primary legal subdivision of Puerto Rico; the Census Bureau treats the municipio as the statistical equivalent of a county); all other land area figures are from the 1990 census. Because only selected areas are shown, the population and land area figures may not equal the total reported. ZIP codes with an asterisk (*) are general delivery ZIP codes. Consult the local postmaster for more specific delivery information. Wake Atoll, Johnston Atoll, and Midway Atoll receive mail through APO and FPO addresses.

Commonwealth of Puerto Rico

ZIP code	Municipio	2000 Pop.	Land area sq. mi.
00601	Adjuntas	19,143	67
00602	Aguada	42,042	31
*00605	Aguadilla	64,685	37
00703	Aguas Buenas	29,032	31
00705	Aibonito	26,493	31
00610	Añasco	28,348	39
*00613	Arecibo	100,131	126
00714	Arroyo	19,117	15
00617	Barceloneta	22,322	19
00794	Barranquitas	28,909	34
*00958	Bayamón	224,044	44
00623	Cabo Rojo	46,911	70
*00726	Caguas	140,502	59
00627	Camuy	35,244	46
00729	Canóvanas	43,335	33
*00984	Carolina	186,076	45
*00963	Cataño	30,071	5
*00737	Cayey	47,370	52
00735	Ceiba	18,004	29
00638	Ciales	19,811	67
00739	Cidra	42,753	36
00769	Coamo	37,597	78
00782	Comerío	20,002	28
00783	Corozal	36,867	43
00775	Culebra	1,868	12
00646	Dorado	34,017	23
00738	Fajardo	40,712	30
00650	Florida	12,367	15
00653	Guánica	21,888	37
*00785	Guayama	44,301	65
00656	Guayanilla	23,072	42
*00970	Guaynabo	100,053	27
00778	Gurabo	36,743	28
00659	Hatillo	38,925	42
00660	Hormigüeros	16,614	11
*00791	Humacao	59,035	45
00662	Isabela	44,444	55
00664	Jayuya	17,318	45
00795	Juana Díaz	50,531	60
00777	Juncos	36,452	27
00667	Lajas	26,261	60
00669	Lares	34,415	61
00670	Las Marías	11,061	46
00771	Las Piedras	34,485	34
00772	Loíza	32,537	19
00773	Luquillo	19,817	26
00674	Manatí	45,409	45
00606	Maricao	6,449	37
00707	Maunabo	12,741	21
*00681	Mayagüez	98,434	78
00676	Moca	39,697	50
00687	Morovis	29,965	39
00718	Naguabo	23,753	52
00719	Naranjito	29,709	27
00720	Orocovis	23,844	63
00723	Patillas	20,152	47
00624	Peñuelas	26,719	44
*00732	Ponce	186,475	115
00678	Quebradillas	25,450	23
00677	Rincón	14,767	14
00745	Río Grande	52,362	61
00637	Sabana Grande	25,935	36
00751	Salinas	31,113	69
00683	San Germán	37,105	55
*00902	San Juan	434,374	48
00754	San Lorenzo	40,997	53
00685	San Sebastián	44,204	70
00757	Santa Isabel	21,665	34
*00954	Toa Alta	63,929	27
*00950	Toa Baja	94,085	23
*00976	Trujillo Alto	75,728	21
00641	Utuado	35,336	113
00692	Vega Alta	37,910	28
*00694	Vega Baja	61,929	46
00765	Vieques	9,106	51
00766	Villalba	27,913	35
00767	Yabucoa	39,246	55
00698	Yauco	46,384	68
TOTAL		**3,808,610**	**3,425**

Commonwealth of the Northern Mariana Islands

ZIP code	Municipality	2000 Pop.	Land area sq. mi.
96950	Northern Islands	6	60
96951	Rota	3,283	33
96950	Saipan	62,392	47
96952	Tinian	3,540	39
TOTAL		**69,221**	**179**

Other U.S. External Territories

ZIP code	Location	2000 Pop.	Land area sq. mi.
	American Samoa		
96799	American Samoa	57,291	77
	Guam		
96919	Agaña Hts.	3,940	1
96928	Agat	5,656	10
96922	Asan	2,090	6
*96913	Barrigada	8,652	9
96924	Chalan-Pago-Ordot	5,923	6
96912	Dededo	42,980	30
*96913	Hagatña	1,100	1
96917	Inarajan	3,052	6
96923	Mangilao	13,313	6
96916	Merizo	2,163	19
96927	Mongmong-Toto-Maite	5,845	10
96925	Piti	1,666	6
96915	Santa Rita	7,500	2
96926	Sinajana	2,853	7
96930	Talofofo	3,215	17
*96913	Tamuning	18,012	1
96918	Umatac	887	17
96929	Yigo	19,474	35
96914	Yona	6,484	20
TOTAL		**154,805**	**210**
	Virgin Islands		
00820	Saint Croix	53,234	83
*00820	Christiansted	2,637	
*00841	Frederiksted	732	
*00830	Saint John	4,197	20
*00801	Saint Thomas	51,181	31
*00801	Charlotte Amalie	11,004	
TOTAL		**108,612**	**134**

LANGUAGE

New Words in English

The following words and definitions were provided by Merriam-Webster Inc., publishers of *Merriam-Webster's Collegiate Dictionary, Tenth Edition*. The words or meanings are among those that the Merriam-Webster editors decided had achieved enough currency in English to be added in the 2000 or 2001 copyright revision of the dictionary.

day trader: a speculator who seeks profit from the intraday fluctuation in the price of a security or commodity by completing double trades of buying and selling or selling and covering during a single session of the market

dot-com: a company that markets its products or services online via a Web site

duh: 1: used to express actual or feigned ignorance or stupidity; **2:** used derisively to indicate that something just stated is all too obvious or self-evident

electronica: dance music featuring extensive use of synthesizers, electronic percussion, and samples of recorded music or sound

eye candy: something superficially attractive to look at

fashionista: a designer, promoter, or follower of the latest fashions

fibromyalgia: a painful rheumatic condition of uncertain cause that is characterized by diffuse or localized pain, tenderness, and stiffness of skeletal muscles and associated connective tissue and that is usually accompanied by fatigue

flatline: 1a: to register on an electronic monitor as having no brain waves or heartbeat; **b:** die; **2a:** to be in a state of no progress or advancement; **b:** to come to an end

foosball: a table game resembling soccer in which the ball is moved by manipulating rods to which small figures of players are attached

gated community: a residential area protected by a private security force, enclosed by physical barriers, and entered through a controlled gate

Global Positioning System: a navigational system using satellite signals to fix the location of a radio receiver on or above the earth's surface; also: the radio receiver so used

gotcha: an unexpected usually disconcerting challenge, revelation, or catch

habanero: a very hot roundish chili pepper that is usually orange when mature

half-pipe: a U-shaped high-sided ramp or runway used in snowboarding, skateboarding, or in-line skating

jones: 1: slang: habit, addiction; especially: addiction to heroin; **2:** slang: heroin; **3:** slang: an avid desire or appetite for something; craving

max out: 1: to be at the upper limit; **2:** to use up all the available credit on

nutraceutical: a foodstuff (as a fortified food or dietary supplement) that provides health benefits

sick building syndrome: a set of symptoms (as headache, fatigue, and eye irritation) typically affecting workers in modern airtight office buildings that is believed to be caused by indoor pollutants (as formaldehyde fumes or microorganisms)

skanky: 1: slang: repugnantly filthy or squalid; **2:** slang: of low or sleazy character

soccer mom: a typically suburban mother who accompanies her children to their soccer games and is considered as part of a significant voting bloc or demographic group

sophomore (adjective): being the second in a series

24-7: for twenty-four hours seven days a week

urban legend: an often lurid story or anecdote that is based on hearsay and widely circulated as true

vision quest: a solitary vigil by an adolescent American Indian boy to seek spiritual power and learn through a vision the identity of his usually animal or bird guardian spirit

wakeboard: a short board with foot bindings on which a rider is towed by a motorboat across its wake and especially up off the crest for aerial maneuvers

warp speed: the highest possible speed

Eponyms

(words named for people)

algorithm—a step-by-step procedure for solving a problem; after al-Khwarizmi, a 9th-century mathematician working in Baghdad.

Bloody Mary—a vodka and tomato juice drink; after the nickname of Mary I, Queen of England (1553-58), notorious for persecution of Protestants

bloomers—full, loose trousers that are gathered at the knee; after Amelia Bloomer, an American social reformer who advocated (1851) such clothing

bobbies—in Great Britain, police officers; named after Sir Robert Peel, the statesman who organized the London police force in 1850

bowdlerize—to delete written matter considered indelicate; after Thomas Bowdler, English editor of an expurgated Shakespeare (1825)

boycott—to avoid trade or dealings with, as a protest; after Charles C. Boycott, an English land agent in County Mayo, Ireland, who was ostracized in 1880 for refusing to reduce rents

Braille—a system of writing for the blind; after Louis Braille, the French teacher of the blind who invented it (1853)

cardigan—a knitted sweater or jacket fastened up the front; after J.T. Brudenell, 7th Earl of Cardigan (1797-1868), who introduced such a garment to his military regiment

Casanova—a man who is a promiscuous and unscrupulous lover; after Giovanni Giacomo Casanova (1725-98), an Italian adventurer

chauvinist—excessively patriotic; after Nicolas Chauvin, a character in a 19th-cent. play who is devoted to Napoleon

derby—a stiff felt hat with a dome-shaped crown and rather narrow rolled brim; after Edward Stanley, 12th earl of Derby, who in 1780 founded the Derby horse race, to which these hats are worn

diesel—a type of internal combustion engine or a vehicle driven by it; after Rudolf Diesel (1858-1913), who built the first successful diesel engine

gerrymander—to draw an election district in such a way as to favor a political party; after Elbridge Gerry, who created (1812) just such an election district (shaped like a salamander) during his governorship of Massachusetts

guillotine—a machine for beheading; after Joseph Guillotin, a French physician who proposed its use in 1789 as more humane than hanging

leotard—a close-fitting garment for the torso, worn by dancers, acrobats, and the like; after Julius Leotard, a 19th-cent. French aerial gymnast

maverick—one who refuses to go along with a group or party; after Samuel A. Maverick (1803-1870), a Texas cattleman who declined to brand his calves.

sandwich—2 or more slices of bread with a filling in between; after John Montagu, 4th earl of Sandwich (1718-92), who supposedly ate food in this form so that he would not have to leave the gaming table

silhouette—an outline image; from Étienne de Silhouette (1709-67), a close-fisted French finance minister

> ***IT'S A FACT:*** The word "bee," as in "spelling bee," refers to a community gathering focused on a single activity. The word in this usage may be derived from the Middle English "bene," meaning "prayer" or "favor."

National Spelling Bee

The Scripps Howard National Spelling Bee, conducted by Scripps Howard Newspapers and other leading newspapers since 1939, was instituted by the Louisville (KY) *Courier-Journal* in 1925. Children under 16 years old and not beyond 8th grade are eligible to compete for cash prizes at the finals, held annually in Washington, DC. The 2001 winners were: 1st place, Sean Conley, Anoka, MN; 2d place, Kristin Hawkins, Sterling,VA.

Here are the last words given, and spelled correctly, in each of the years 1980-2001 at the national spelling bee.

1980 elucubrate
1981 sarcophagus
1982 psoriasis
1983 purim
1984 luge
1985 milieu
1986 odontalgia
1987 staphylococci
1988 elegiacal
1989 spoliator
1990 fibranne
1991 antipyretic
1992 lyceum
1993 kamikaze
1994 antediluvian
1995 xanthosis
1996 vivisepulture
1997 euonym
1998 chiaroscurist
1999 logorrhea
2000 demarche
2001 succedaneum

Foreign Words and Phrases

(L=Latin; G=Greek; F=French; Y=Yiddish; I=Italian; S=Spanish)

ad hoc (L; ad HOK): for the end or purpose at hand
ad nauseam (L; ad NAWZ-ee-um): to a sickening degree
aficionado (S; ah-fish-ee-uh-NAH-do) one who is fervently devoted to a certain activity or interest
alma mater L; AHL-mah MAH-ter): one's school or college
aloha (Hawaiian; ah-LOH-hah): love to you: greetings; farewell
apropos (F; ah-pruh-POH): relevant
au contraire (F; oh kon-TRAIR): on the contrary
bête noire (F; BET NWAHR): a thing or person viewed with particular dislike or fear
bon appétit (F; BOH nap-uh-teet): have a good meal!
bona fide (L; BOH nuh-fid): genuine; in good faith
carte blanche (F; kahrt BLANNSH): full discretionary power
cause célèbre (F; kawz suh-LEB-ruh): a notorious incident
chutzpah (Y; KHOOT-spuh): nerve bordering on arrogance
coup de grâce (F; kooh duh GRAHS): the final blow
coup d'état (F; kooh day TAH): overthrow of a government by a small group
crème de la crème (F; KREM duh luh KREM): best of the best
cum laude/magna cum laude/summa cum laude (L; kuhm LOUD-ay; MAGN-a ...; SOO-ma ...): with praise or honor/with great praise or honor/with the highest praise or honor
de facto (L; day FAK-toh): in fact, though not by right
déjà vu (F; DAY-zhah VOOH): the sensation that something happening has happened before
de jure (L; dee JOOR-ee, day YOOR-ay): in accordance with right or law; officially
de rigueur (F; duh ree-GUR): necessary according to convention or etiquette
détente (F; day-TAHNT): an easing of strained relations
éminence grise (F; ay-meh-NAHNN-suh GREEZ): one who wields power behind the scenes
enfant terrible (F; ahn-FAHN te-REE-bluh): one whose unconventional behavior causes embarrassment
en masse (F; ahn MAHS): in a large body
esprit de corps (F; es-PREE duh KAWR): group spirit; feeling of camaraderie
eureka (G; Yoor-EE-kuh): I have found it!
ex post facto (L; eks pohst FAK-toh): retroactive(ly)
fait accompli (F; fayt uh-kom-PLEE): an accomplished fact
faux pas (F; fowe PAH): a social blunder
hoi polloi (G; hoy puh-LOY): the masses
in loco parentis (L; in LOH-koh puh-REN-tis): in place of parent
in memoriam (L; in muh-MAWR-ee-uhm): in memory of
in situ (L; in SEE-tooh): in the original place or position
je ne sais quoi (F; zhuh nuh say KWAH): I don't know what; the little something that eludes description
joie de vivre (F; zhwah duh VEEV-ruh): zest for life
mano a mano (S; MAH-noh ah MAH-noh): hand to hand; in direct combat
mea culpa (L; MAY-uh CUL-puh): through my fault
modus operandi (L; MOH-duhs op-uh-RAN-dee): method of operation
noblesse oblige (F; noh-BLES oh-BLEEZH): the obligation of nobility to help the less fortunate
non compos mentis (L; non KOM-puhs MEN-tis): not of sound mind
nouveau riche (F; nooh-voh REESH): a person newly rich, perhaps one who spends money conspicuously
persona non grata (L; per-SOH-nah non GRAH-tah): unwelcome person
postmortem (L; pohst-MORE-tuhm): after death; autopsy; analysis after an event
prima donna (I; pree-muh DAH-nuh): a principal female opera singer; temperamental person
pro bono (L; proh BOH-noh): (legal work) donated for the public good
qua (L) (kwah): in the capacity or character of
que sera sera (S; keh sair-AH sair-AH): what will be will be
quid pro quo (L; kwid proh KWOH): something given or received for something else
raison d'être (F; RAY-zohnn DET-ruh): reason for being
sans souci (F; SAHNN sooh-SEE): without worry
savoir faire (F; sav-wahr-FAIR): dexterity in social affairs
schlemiel (Y; shleh-MEEL): an unlucky, bungling person
semper fidelis (L; SEM-puhr fee-DAY-lis): always faithful
status quo (L; STAY-tus QWOH): the existing order of things
terra firma (L; TER-uh FUR-muh): solid ground
tour de force (F; TOOR duh FAWRS): feat accomplished through great skill
tristesse (F; tree-STES): melancholy
verbatim (L; ver-BAY-tuhm): word for word
vis-à-vis (F; vee-ZUH-VEE): compared with; with regard to
zeitgeist (German; ZITE-guyste): the general intellectual, moral, and cultural climate of an era

Murder for a Jar of Red Rum

The phrase above is a **palindrome**. A palindrome, from Greek words for "run back again," reads the same backward as forward. Famous palindromes range from Adam's possible introduction to his wife ("Madam, I'm Adam"), to what Napoleon might have said ("Able was I ere I saw Elba") to a slogan that could have applied to Theodore Roosevelt ("A man, a plan, a canal, Panama!"). Some are one word only (e.g., "radar") or simple phrases (e.g. "Ma has a ham") Some are more involved (e.g., "Doc, note: I dissent. A fast never prevents a fatness. I diet on cod.")

Some Common Abbreviations and Acronyms

Acronyms are pronounceable words formed from first letters (or syllables) of other words. Some abbreviations below (e.g., AIDS, NATO) are thus acronyms. Some acronyms are words coined as abbreviations and written in lower case (e.g., "radar," "yuppie"). Acronyms do not have periods; usage for other abbreviations varies, but periods have become less common. Capitalization usage may vary from what is shown here. Italicized words preceding parenthetical definitions below are Latin unless otherwise noted. See also other chapters, including Computers and Internet; Weights and Measures.

AA=Alcoholics Anonymous
AAA=American Automobile Association
AC=alternating current
AD=*anno Domini* (in the year of the Lord)
AFL-CIO=American Federation of Labor and Congress of Industrial Organizations
AIDS=acquired immune deficiency syndrome
AM=*ante meridiem* (before noon)
AMA=American Medical Association
anon=anonymous
APO=army post office
ASAP=as soon as possible
ASCAP=American Society of Composers, Authors, and Publishers
ASPCA=American Society for Prevention of Cruelty to Animals
ATM=automated teller machine
AWOL=absent without leave
BA=Bachelor of Arts
bbl=barrel(s)
BC=before Christ
BCE=before Common Era
bpd=barrels per day
BS=Bachelor of Science
Btu=British thermal unit(s)
bu=bushel(s)
C= Celsius, centigrade
c=*circa* (about), copyright
CE=Common Era
CEO=chief executive officer
CFO=chief financial officer
cm=centimeter(s)
COD=cash (or collect) on delivery
Col.=Colonel
COLA=cost of living allowance
CPA=certified public accountant
Cpl.=Corporal
CPR=cardiopulmonary resuscitation
DA=district attorney
DAR=Daughters of the American Revolution
DC=direct current
DD=Doctor of Divinity
DDS=Doctor of Dental Science (or Surgery)
DNA=deoxyribonucleic acid
DNR=do not resuscitate
DOA=dead on arrival
DWI=driving while intoxicated
ed.=edited, edition, editor
e.g.=*exempli gratia* (for example)
EKG=electrocardiogram
EPA=Environmental Protection Agency
ESP=extrasensory perception
et al.=*et alii* (and others)
etc.=*et cetera* (and so forth)
EU=European Union
F=Fahrenheit
FBI=Federal Bureau of Investigation
FICA=Federal Insurance Contributions Act (Social Security)
FOB=free on board
FY=fiscal year
FYI=for your information

GB=gigabyte(s)
GDP=gross domestic product
GIGO=garbage in, garbage out
GNP=gross national product
GOP=Grand Old Party (Republican Party)
Hon.=the Honorable
HOV=high-occupancy vehicle
ht=height
HVAC=heating, ventilating, and air-conditioning
i.e.=*id est* (that is)
IMF=International Monetary Fund
IQ=intelligence quotient
IRA=individual retirement account; Irish Republican Army
IRS=Internal Revenue Service
ISBN=International Standard Book Number
JD=*Juris Doctor* (doctor of laws)
K=Kelvin
k=karat
KB=kilobyte(s); kg=kilogram(s); km=kilometer(s); kw=kilowatt(s)
kwh=kilowatt-hour(s)
l=liter(s)
lb=*libra* (pound or pounds)
Lieut. or Lt.=Lieutenant
LLB=*Legum Baccalaureus* (bachelor of laws)
m=meter(s)
MA=Master of Arts
MB=megabyte(s)
MD=*Medicinae Doctor* (doctor of medicine)
MFN=most favored nation
MIA=missing in action
ml=milliliter(s); mm=millimeter(s)
mph=miles per hour
MS=Master of Science
MSG=monosodium glutamate
Msgr.=Monsignor
MVP=most valuable player
NAACP=National Association for the Advancement of Colored People
NASA=National Aeronautics and Space Administration
NAFTA=North American Free Trade Agreement
NATO=North Atlantic Treaty Organization
NB=*nota bene* (note carefully)
NCAA=National Collegiate Athletic Association
NOW=National Organization for Women
op=*opus* (work)
OPEC=Organization of Petroleum Exporting Countries
p, pp=page(s)
PAC=political action committee
PC=personal computer
PhD=*Philosophiae Doctor* (doctor of philosophy)
PIN=Personal Identification Number
PM=*post meridiem* (afternoon)
PO=post office
POW=prisoner of war
PS=*post scriptum* (postscript)
pt=part(s), pint(s), point(s)
Pvt.=Private
q.v.=*quod vide* (which see)
radar=radio detecting and ranging
REM=rapid eye movement
Rev.=Reverend
RFD=rural free delivery
RIP=*requiescat in pace* (May he/she rest in peace)
RN=registered nurse
RNA=ribonucleic acid
ROTC=Reserve Officers' Training Corps
rpm=revolutions per minute
RR=railroad
RSVP=*répondez s'il vous plaît* (Fr.) (Please reply)
SASE=self-addressed stamped envelope
Sgt.=Sergeant
SIDS=sudden infant death syndrome
S.J.=Society of Jesus (Jesuits)
SRO=standing room only
St.=Saint, Street
TGIF=Thank God It's Friday
UFO=unidentified flying object
UHF=ultrahigh frequency
UNESCO=United Nations Educational, Social, and Cultural Organization
UNICEF=United Nations (International) Children's (Emergency) Fund
UPC=Universal Product Code
USS=United States ship
v (or vs)=*versus* (against)
VCR=videocassette recorder
VHF=very high frequency
VISTA=Volunteers in Service to America
W=watt(s)
wasp=white Anglo-Saxon Protestant
WHO=World Health Organization
yd=yard(s)
yuppie=young urban professional
ZIP=zone improvement plan (U.S. Postal Service)

Names of the Days

ENGLISH	RUSSIAN	HEBREW	FRENCH	ITALIAN	SPANISH	GERMAN	JAPANESE
Sunday	Voskresenye	Yom rishon	Dimanche	Domenica	Domingo	Sonntag	Nichiyo\bi
Monday	Ponedelnik	Yom sheni	Lundi	Lunedì	Lunes	Montag	Getsuyo\bi
Tuesday	Vtornik	Yom shlishi	Mardi	Martedì	Martes	Dienstag	Kayo\bi
Wednesday	Sreda	Yom ravii	Mercredi	Mercoledì	Miércoles	Mittwoch	Suiyo\bi
Thursday	Chetverg	Yom hamishi	Jeudi	Giovedì	Jueves	Donnerstag	Mokuyo\bi
Friday	Pyatnitsa	Yom shishi	Vendredi	Venerdì	Viernes	Freitag	Kin-yo\bi
Saturday	Subbota	Shabbat	Samedi	Sabato	Sábado	Samstag	Doyo\bi

Names for Animal Young

bunny: rabbit
calf: cattle, elephant, antelope, rhino, hippo, whale, others
cheeper: grouse, partridge, quail
chick, chicken: fowl
cockerel: rooster
codling, sprag: codfish
colt: horse (male)
cub: lion, bear, shark, fox, others
cygnet: swan
duckling: duck
eaglet: eagle
elver: eel
eyas: hawk, others
fawn: deer
filly: horse (female)
fingerling: fish generally
flapper: wild fowl
fledgling: birds generally
foal: horse, zebra, others
fry: fish generally
gosling: goose
heifer: cow
joey: kangaroo, others
kid: goat
kit: fox, beaver, rabbit, cat
kitten, kitty, catling: cats, other small mammals
lamb, lambkin, cosset, hog: sheep
leveret: hare
nestling: birds generally
owlet: owl
parr, smolt, grilse: salmon
piglet, shoat, farrow, suckling: pig
polliwog, tadpole: frog
poult: turkey
pullet: hen
pup: dog, seal, sea lion, fox
puss, pussy: cat
spike, blinker, tinker: mackerel
squab: pigeon
squeaker: pigeon, others
whelp: dog, tiger, beasts of prey
yearling: cattle, sheep, horse, others

Top 10 First Names of Americans by Decade of Birth

Source: Compiled by Dr. Cleveland Kent Evans, Bellevue University, Bellevue, NE; based on Social Security Administration records

Dr. Evans, a noted onomastician, or expert in name forms and origins, prepared these lists with data from his own research, as well as data supplied to him by the Social Security Administration.

BOYS:

1880-1889	John, William, Charles, George, James, Frank, Joseph, Harry, Henry, Edward
1890-1899	John, William, George, James, Charles, Joseph, Frank, Robert, Harry, Henry
1900-1909	John, William, James, George, Joseph, Charles, Robert, Frank, Edward, Henry
1910-1919	John, William, James, Robert, Joseph, Charles, George, Edward, Frank, Walter
1920-1929	John, Robert, James, William, Charles, George, Joseph, Richard, Edward, Donald
1930-1939	Robert, James, John, William, Richard, Charles, Donald, George, Thomas, Joseph
1940-1949	James, Robert, John, William, Richard, David, Charles, Thomas, Michael, Ronald
1950-1959	Michael, James, Robert, John, David, William, Steven, Richard, Thomas, Mark
1960-1969	Michael, John, David, James, Robert, Mark, Steven, William, Jeffrey, Richard
1970-1979	Michael, Christopher, Jason, David, James, John, Brian, Robert, Steven, William
1980-1989	Michael, Christopher, Matthew, Joshua, David, Daniel, James, John, Robert, Brian
1990-1999	Michael, Christopher, Matthew, Joshua, Nicholas, Jacob, Andrew, Daniel, Brandon, Tyler

GIRLS:

1880-1889	Mary, Anna, Elizabeth, Catherine, Margaret, Emma, Bertha, Minnie, Florence, Clara
1890-1899	Mary, Anna, Margaret, Helen, Catherine, Elizabeth, Florence, Ruth, Rose, Ethel
1900-1909	Mary, Helen, Margaret, Anna, Ruth, Catherine, Elizabeth, Dorothy, Marie, Mildred
1910-1919	Mary, Helen, Dorothy, Margaret, Ruth, Catherine, Mildred, Anna, Elizabeth, Frances
1920-1929	Mary, Dorothy, Betty, Helen, Margaret, Ruth, Virginia, Catherine, Doris, Frances
1930-1939	Mary, Betty, Barbara, Shirley, Patricia, Dorothy, Joan, Margaret, Carol, Nancy
1940-1949	Mary, Linda, Barbara, Patricia, Carol, Sandra, Nancy, Sharon, Judith, Susan
1950-1959	Deborah, Mary, Linda, Patricia, Susan, Barbara, Karen, Nancy, Donna, Catherine
1960-1969	Lisa, Deborah, Mary, Karen, Michelle, Susan, Kimberly, Lori, Teresa, Linda
1970-1979	Jennifer, Michelle, Amy, Melissa, Kimberly, Lisa, Angela, Heather, Kelly, Sarah
1980-1989	Jessica, Jennifer, Ashley, Sarah, Amanda, Stephanie, Nicole, Melissa, Katherine, Megan
1990-1999	Ashley, Jessica, Sarah, Brittany, Emily, Kaitlyn, Samantha, Megan, Brianna, Katherine

Origins of Popular American Given Names

Source: Dr. Cleveland Kent Evans, Bellevue University, Bellevue, NE

Boys

Andrew: Gr. *andreios*, "man, manly"
Austin: Eng. form of Lat. *Augustinus*, "magnificent"
Brandon: Eng. place name, "gorse-covered hill"
Brian: Irish, perhaps Celtic *Brigonos*, "high, noble"
Charles: Ger. *ceorl*, "free man"
Christopher: Gr. *Khristophoros*, "bearing Christ [in one's heart]"
Daniel: Heb. "God is my judge"
David: Heb. *Dodavehu*, perhaps "darling"
Donald: Scots Gaelic *Domhnall*, "world rule"
Edward: Old Eng. *Eadweard*, "wealth-guard"
Frank: Ger. "Frenchman"
George: Gr. *georgos*, "soil tiller, farmer"
Harry: Middle Eng. form of Henry
Henry: Ger. *Haimric*, "home-power"
Jacob: Heb. *Yaakov*, "God protects" or "supplanter"
James: Late Lat. *Iacomus*, form of Jacob
Jason: Gr. *Iason*, "healer"
Jeffrey: Norman Fr. , from Ger. *Gaufrid*, "land-peace," or *Gisfrid*, "pledge-peace"
John: Heb. *Yohanan*, "God is gracious"
Joseph: Heb. *Yosef*, "[God] shall add"
Joshua: Heb. *Yoshua*, "God saves"
Mark: Lat. *Marcus*, perhaps "of Mars, the war god"
Matthew: Heb. *Mattathia*, "gift of God"
Michael: Heb. "Who could ever be like God?"
Nicholas: Gr. *Nikolaos*, "victory-people"
Richard: Ger. "power-hardy"
Robert: Ger. *Hrodberht*, "fame-bright"
Ronald: Scots form of Old Norse *Rögnvaldr*, "advice-ruler"
Steven: Gr. *stephanos*, "crown, garland"
Thomas: Aramaic "twin"
Tyler: Old Eng. *tigeler*, "tile layer"
Walter: Ger. *Waldheri*, "rule-army"
William: Ger. *Wilhelm*, "will-helmet"
Zachary: Eng. form of Heb. *Zechariah*, "God has remembered"

Girls

Alice: Old Fr. form of Ger. *Adalheidis*, "noble kind"
Amanda: 17th-cent. invention from Lat., "lovable"
Amy: Old Fr. *Amee*, "beloved"
Angela: Gr. *angelos*, "messenger [of God]"
Anna: Lat. and Gr. form of Hannah
Ashley: Eng. place name, "ash grove"
Barbara: Gr. *barbarus*, "foreign"
Bertha, Ger. *behrt*, "bright"
Betty: 18th-cent. pet form of Elizabeth
Brianna: modern fem. form of Brian
Brittany: place name, Fr. province settled by Britons
Caitlin: Irish form of Katherine
Carol: form of Charles
Clara: Lat. *clarus*, "famous"
Deborah: Heb. "bee"
Donna: Ital. "lady"
Doris: Gr. "woman of the Dorian tribe," name of a sea nymph
Dorothy: Gr. *Dorothea*, "gift of God"
Elizabeth: Heb. *Elisheba*, perhaps "God is my oath" or "God is good fortune"
Emily: Roman *Aemilia*, possibly from Lat. *aemulus*, "rival"
Emma: Ger. *ermen*, "whole, entire"
Ethel: Old Eng. *aethel*, "noble"
Florence: Lat. *florens*, "flourishing"
Frances: fem. form of Francis, "a Frenchman"
Haley: Eng. place name, "hay clearing"
Hannah: Heb. "He has favored me"
Heather: Middle Eng. *hathir*, "heather"
Helen: Gr. *Helene*, possibly "sunbeam"
Jennifer: Cornish form of Welsh *Gwenhwyfar*, "fair-smooth"
Jessica: Shakespearean invention, probably fem. form of Jesse, Heb. "God exists"
Joan: Middle Eng. fem. form of John
Judith: Hebrew "Jewish woman"
Kaitlyn: modern American spelling of Caitlin
Karen: Danish form of Katherine
Katherine: from *Aikaterine*, Egyptian name later modified to resemble Gr. *katharos*, "pure"
Kayla: modern invention; or Yiddish form of Kelila, Heb. "crown of laurel"
Kelly: Irish Gaelic *Ceallagh*, perhaps "churchgoer" or "bright-headed"
Kimberly: Eng. place name, "Cyneburgh's clearing"
Linda: Sp. "pretty" or Ger. "tender"
Lisa: pet form of Elizabeth
Lori: pet form of either Lorraine (French "land of Lothar's people") or Laura (Latin "laurel")
Madison: Middle Eng. surname, "son of Madeline or Maud"
Margaret: Gr. *margaron*, "pearl"
Maria: Lat. form of Mary
Marie: Fr. form of Mary
Mary: Eng. form of Heb. *Maryam*, perhaps "seeress" or "wished-for child"
Megan: Welsh form of Margaret
Melissa: Gr. "bee"
Michelle: Fr. fem. form of Michael
Mildred: Old Eng. *Mildthryth*, "mild-strength"
Minnie: Pet form of Wilhelminia, fem. form of William
Nancy: medieval Eng. pet form of Agnes, Gr. *hagnos*, "holy"; later also used as pet form for Ann
Nicole: Fr. fem. form of Nicholas
Patricia: Lat. *Patricius*, "belonging to the noble class"
Rose: Ger. *hros*, "horse," or Lat. *rosa*, "rose"
Ruth: Heb., perhaps "companion"
Samantha: colonial American invention, probably combining Sam from Samuel [Heb. "name of God"] with -antha from Gr. *anthos*, "flower"
Sandra: short form of Alessandra, Ital. fem. of Alexander, Gr. "defend-man"
Sarah: Heb., "princess"
Sharon: Biblical place name, Hebrew "plain"
Shirley: Eng. place name, "bright clearing" or "shire meadow"
Stephanie: Fr. fem. form of Steven
Susan: Eng. form of Heb. *Shoshana*, "lily"
Taylor: Anglo-Norman *taillour*, "tailor"
Teresa: Spanish, perhaps "woman from Therasia"
Virginia: Lat., "virgin-like"

Pen Names

Shalom Aleichem (Solomon J. Rabinowitz)
Woody Allen (Allen Stewart Konigsberg)
Currer, Ellis, and Acton Bell (Charlotte, Emily, and Anne Brontë)
John le Carré (David John Moore Cornwell)
Lewis Carroll (Charles Lutwidge Dodgson)
Colette (Sidonie Gabrielle Colette)
Isak Dinesen (Karen Blixen)
Elia (Charles Lamb)
George Eliot (Mary Ann or Marian Evans)
Maksim Gorky (Aleksey Maksimovich Peshkov)
O. Henry (William Sydney Porter)
James Herriot (James Alfred Wight)
P. D. James (Phyllis Dorothy James White)
[John] Ross Macdonald (Kenneth Millar)
André Maurois (Émile Herzog)
Molière (Jean Baptiste Poquelin)
Frank O'Connor (Michael Donovan)
George Orwell (Eric Arthur Blair)
Mary Renault (Mary Challans)
Ellery Queen (Frederic Dannay and Manfred B. Lee)
Françoise Sagan (Françoise Quoirez)
Saki (Hector Hugh Munro)
George Sand (Amandine Lucie Aurore Dupin)
Dr. Seuss (Theodor Seuss Geisel)
Stendhal (Marie Henri Beyle)
Mark Twain (Samuel Clemens)
Voltaire (François Marie Arouet)
Tom Wolfe (Thomas Kennerly Jr.)

Forms of Address

	Address	Salutation
GOVERNMENT		
President of the U.S.	The President, The White House, Washington, DC 20500; also, The President and Mrs. ____ or The President and Mr. ____	Dear Sir or Madam; Mr. President or Madam President; Dear Mr. President or Dear Madam President
U.S. Vice President	The Vice President, The White House, Washington, DC 20500; also, The Vice President and Mrs. ____ or The Vice President and Mr. ____	Dear Sir or Madam; Mr. Vice President or Madam Vice President; Dear Mr. Vice President or Dear Madam Vice President
Chief Justice	The Hon. *Firstname Surname*, Chief Justice of the U.S., The Supreme Court, Washington, DC 20543	Dear Sir or Madam; Dear Mr. or Madam Chief Justice
Associate Justice	The Hon. Justice *Firstname Surname,* The Supreme Court, Washington, DC 20543	Dear Sir or Madam; Dear Justice *Surname*
Judge	The Hon. *Firstname Surname*, Associate Judge, U.S. District Court	Dear Judge *Surname*
Attorney General	The Hon. *Firstname Surname,* Attorney General, Dept. of Justice, Constitution Ave. & 10th St. NW, Washington, DC 20530	Dear Sir or Madam; Dear Mr. or Ms. Attorney General
Cabinet Officer	The Hon. *Firstname Surname*, Secretary of ____	Dear Mr. or Madam Secretary; or Dear Mr. or Ms. *Surname*
Senator	The Hon. or Sen. *Firstname Surname*, U.S. Senate, Washington, DC 20510	Dear Mr. or Madam Senator, or Dear Mr. or Ms. *Surname*
Representative	The Hon. or Rep. *Firstname Surname*, House of Representatives, Washington, DC 20515	Dear Mr. or Madam *Surname*
Speaker of the House	The Hon. Speaker of the House of Representatives, House of Representatives, Washington, DC 20515	Dear Mr. or Madam Speaker
Ambassador, U.S.	The Hon. *Firstname Surname*, American Ambassador[1]	Sir or Madam; Dear Mr. or Madam Ambassador
Ambassador, Foreign	His or Her Excellency[2] *Firstname Surname*, Ambassador of ____	Excellency[2] ; Dear Mr. or Madam Ambassador
Governor	The Hon. *Firstname Surname*, Governor of *State*; or in some states, His or Her Excellency, the Governor of *State*	Sir or Madam; Dear Governor *Surname*
Mayor	The Hon. *Firstname Surname*, Mayor of *City*	Sir or Madam; Dear Mayor *Surname*
MILITARY PERSONNEL		
All Titles	Full or abbreviated rank + full name + comma + abbreviation for branch of service. *Example*: Adm. John Smith, USN	Dear *Rank Surname*
RELIGIOUS		
Clergy, Protestant	The Reverend *Firstname Surname*[3]	Dear Ms. or Mr. *Surname*
Pope	His Holiness Pope *Name* or His Holiness the Pope	Your Holiness or Most Holy Father
Priest	The Reverend *Firstname Surname* or The Reverend Father *Surname*	Reverend Father, Dear Father *Surname*, or Dear Father
Rabbi	Rabbi *Firstname Surname*	Dear Rabbi *Surname*
ROYALTY AND NOBILITY		
King/Queen	His or Her Majesty, King or Queen of *Country*	Sir or Madam, or May it please Your Majesty

(1) If in Canada or Latin America, The Ambassador of the United States of America. (2) An American ambassador is not properly addressed as His or Her Excellency. (3) A member of the Protestant clergy who has a doctorate may be so addressed; for example, The Reverend Firstname Surname, DD, and Dear Dr. Surname.

Commonly Misspelled English Words

accidentally
accommodate
acknowledgment
acquainted
all right
already
amateur
appearance
appropriate
bureau
business
character
commitment
committee
conscientious
conscious
convenience
deceive
defendant
describe
description
desirable
despair
desperate
eliminate
embarrass
environment
existence
fascinating
February
finally
fluorine
foreign
forty
government
grammar
harass
humorous
hurrying
incidentally
independent
indispensable
inoculate
irresistible
judgment
laboratory
license
lightning
liquefy
maintenance
marriage
millennium
miniature
misspelled
mysterious
necessary
noticeable
occasionally
occurrence
opportunity
optimistic
parallel
performance
permanent
perseverance
privilege
receive
receipt
rhythm
ridiculous
separate
seize
similar
sincerely
supersede
transferred
Wednesday
weird

Commonly Confused English Words

adverse: unfavorable
averse: opposed

affect: to influence
effect: to bring about

allusion: an indirect reference
illusion: an unreal impression

appraise: to set a value on
apprise: to inform

capital: the seat of government
capitol: building where a legislature meets

complement: to make complete; something that completes
compliment: to praise; praise

denote: to mean
connote: to suggest beyond the explicit meaning

discreet: prudent
discrete: separate, distinct

disinterested: impartial
uninterested: without interest

elicit: to draw or bring out
illicit: illegal

emigrate: to leave for another place of residence
immigrate: to come to another place of residence

grisly: inspiring horror or great fear
grizzly: sprinkled or streaked with gray

historic: important in history
historical: relating to history

imminent: ready to take place
eminent: standing out

imply: to suggest but not explicitly; to entail
infer: to assume or understand information not relayed explicitly

include: used when the items following are part of a whole
comprise: used when the items following are all of a whole

incredible: unbelievable
incredulous: skeptical

ingenious: clever
ingenuous: innocent

oral: spoken, as opposed to written
verbal: relating to language

prostrate: stretched out face down
prostate: relating to prostate gland

The Principal Languages of the World

Source: From Ethnologue Volume 1, Languages of the World, 14th edition, Edited by Barbara F. Grimes, © 2000 by SIL International. Used by permission.

The following tables count only "first language" speakers.

Languages Spoken by the Most People

Language	Speakers (millions)
Chinese, Mandarin	874
Hindi	366
English	341
Spanish	322-58
Bengali	207
Portuguese	176
Russian	167
Japanese	125
German, Standard	100
Korean	78
French	77
Chinese, Wu	77
Javanese	75
Chinese, Yue	71
Telugu	69

Languages Spoken by at Least 2 Million People

A "Hub" country is the country of origin, not necessarily the country where the most speakers reside (e.g., Portugal is the "hub" country of Portuguese, although more Portuguese speakers live in Brazil). "Cts." means number of countries where the language is spoken as a first language. "Sps." means minimum number of speakers in millions.

Language	Hub	Cts.	Sps.
Chinese, Mandarin	China	16	874
Hindi	India	17	366
English	United Kingdom	104	341
Spanish	Spain	43	322-58
Bengali	Bangladesh	9	207
Portuguese	Portugal	33	176
Russian	Russia	30	167
Japanese	Japan	26	125
German, standard	Germany	40	100
Korean	Korea, South	31	78
French	France	53	77
Chinese, Wu	China	1	77
Javanese	Indonesia	4	75
Chinese, Yue	China	20	71
Telugu	India	7	69
Marathi	India	3	68
Vietnamese	Vietnam	20	68
Tamil	India	15	66
Italian	Italy	29	62
Turkish	Turkey	35	61
Urdu	Pakistan	21	60
Ukrainian	Ukraine	25	47
Gujarati	India	17	46
Arabic, Egyptian, spoken	Egypt	9	46
Chinese, Jinyu	China	1	45
Chinese, Min Nan	China	9	45
Polish	Poland	21	44
Chinese, Xiang	China	1	36
Malayalam	India	9	35
Kannada	India	1	35
Chinese, Hakka	China	16	33
Oriya	India	2	32
Burmese	Myanmar	5	32
Panjabi, Western	Pakistan	7	30-45
Sunda	Indonesia	1	27
Panjabi, Eastern	India	11	27
Romanian	Romania	17	26
Bhojpuri	India	3	26
Azerbaijani, South	Iran	8	24
Farsi, Western	Iran	26	24
Hausa	Nigeria	13	24
Maithili	India	2	24
Arabic, Algerian, spoken	Algeria	6	22
Serbo-Croatian	Yugoslavia	23	21
Thai	Thailand	5	20-25
Yoruba	Nigeria	5	20
Dutch	Netherlands	14	20
Awadhi	India	2	20
Chinese, Gan	China	1	20
Sindhi	Pakistan	7	19
Arabic, Moroccan, spoken	Morocco	8	19
Arabic, Saidi, spoken	Egypt	1	18
Igbo	Nigeria	1	18
Uzbek, Northern	Uzbekistan	12	18
Malay	Malaysia	8	18
Indonesian	Indonesia	6	17-30
Tagalog	Philippines	8	17
Amharic	Ethiopia	4	17
Nepali	Nepal	4	16
Arabic, Sudanese, spoken	Sudan	5	16-19
Arabic, N. Levantine, spoken	Syria	15	15
Saraiki	Pakistan	3	15-30
Cebuano	Philippines	2	15
Assamese	India	3	15
Thai, Northeastern	Thailand	1	15-23
Hungarian	Hungary	11	14
Chittagonian	Bangladesh	2	14
Haryanvi	India	1	13
Sinhala	Sri Lanka	7	13
Madura	Indonesia	2	13
Arabic, Mesop., spoken	Iraq	5	13
Greek	Greece	35	12
Marwari	India	2	12
Czech	Czech Republic	9	12
Magahi	India	1	11
Chhattisgarhi	India	1	11
Zhuang, Northern	China	1	10
Belarusan	Belarus	16	10
Deccan	India	1	10
Chinese, Min Bei	China	2	10
Arabic, Najdi, spoken	Saudi Arabia	7	9
Zulu	South Africa	6	9
Pashto, Southern	Afghanistan	6	9
Somali	Somalia	12	9-10
Arabic, Tunisian, spoken	Tunisia	5	9
Swedish	Sweden	7	9
Malagasy	Madagascar	3	9
Bulgarian	Bulgaria	11	9
Pashto, Northern	Pakistan	5	9
Lombard	Italy	3	8
Ilocano	Philippines	2	8
Oromo, West-Central	Ethiopia	2	8
Kazakh	Kazakhstan	13	8
Tatar	Russia	19	7
Haitian-Creole French	Haiti	8	7
Fulfulde, Nigerian	Nigeria	3	7
Hiligaynon	Philippines	2	7
Uyghur	China	16	7
Shona	Zimbabwe	4	7
Khmer, Central	Cambodia	6	7
Kurmanji	Turkey	25	7-8
Akan	Ghana	1	7
Azerbaijani, North	Azerbaijan	9	7
Arabic, Sanaani, spoken	Yemen	1	7
Napoletano-Calabrese	Italy	1	7
Farsi, Eastern	Afghanistan	2	7
Rwanda	Rwanda	5	7
Arabic, Hijazi spoken	Saudi Arabia	2	6
Luba-Kasai	Dem. Rep. of Congo	1	6
Thai, Northern	Thailand	2	6
Finnish	Finland	7	6
Arabic, N. Mesopotamian, spoken	Iraq	4	6
Afrikaans	South Africa	10	6
Arabic, S. Levantine, spoken	Jordan	8	6
Armenian	Armenia	29	6
Rundi	Burundi	4	6
Santali	India	4	6
Alemannisch	Switzerland	5	6
Catalan-Valencian-Balear	Spain	18	6
Turkmen	Turkmenistan	13	6
Xhosa	South Africa	3	6
Kanauji	India	1	6
Arabic, Taizzi-Adeni, spoken	Yemen	5	6
Minangkabau	Indonesia	1	6
Kurdi	Iraq	3	6
Sylhetti	Bangladesh	2	5
Slovak	Slovakia	8	5
Swahili	Tanzania	12	5
Thai, Southern	Thailand	1	5
Tigrigna	Ethiopia	3	5
Hebrew	Israel	8	5
Nyanja	Malawi	6	5
Danish	Denmark	8	5

Language	Hub	Cts.	Sps.
Guarani, Paraguayan	Paraguay	2	5
Gikuyu	Kenya	1	5
Moore	Burkina Faso	6	5
Sukuma	Tanzania	1	5
Norwegian, Bokmaal	Norway	6	5
Lithuanian	Lithuania	19	4
Oromo, Eastern	Ethiopia	1	4
Tswana	Botswana	4	4
Arabic, Libyan, spoken	Libya	3	4
Sotho, Southern	Lesotho	3	4
Umbundu	Angola	2	4
Kashmiri	India	3	4
Konkani	India	1	4
Galician	Spain	2	4
Georgian	Georgia	13	4
Luri	Iran	3	4
Tajiki	Tajikistan	7	4
Sicilian	Italy	1	4
Kituba	Dem. Rep. of Congo	1	4
Zhuang, Southern	China	1	4
Bali	Indonesia	1	3
Kabyle	Algeria	3	3
Gilaki	Iran	1	3
Aceh	Indonesia	1	3
Kanuri, Central	Nigeria	6	3
Emiliano-Romagnolo	Italy	2	3
Mazanderani	Iran	1	3
Wolof	Senegal	7	3
Yiddish, Eastern	Israel	20	3
Shan	Myanmar	3	3
Luo	Kenya	2	3
Luyia	Kenya	2	3
Tachelhit	Morocco	3	3
Malay, Pattani	Thailand	1	3
Tamazight, Central Atlas	Morocco	3	3
Quechua, South Bolivian	Bolivia	2	3
Balochi, Southern	Pakistan	4	3
Ganda	Uganda	2	3
Albanian, Tosk	Albania	9	3
Kongo	Dem. Rep. of Congo	3	3
Oromo, Borana-Arsi-Guji	Ethiopia	3	3
Bugis	Indonesia	2	3
Lao	Laos	5	3
Banjar	Indonesia	2	3
Mbundu, Loanda	Angola	1	3
Piedmontese	Italy	3	3
Tsonga	South Africa	4	3
Mongolian, Peripheral	China	2	3
Sotho, Northern	South Africa	2	3
Kamba	Kenya	1	2
Garhwali	India	1	2
Dogri-Kangri	India	1	2
Mundari	India	3	2
Venetian	Italy	3	2
Lambadi	India	1	2
Bemba	Zambia	5	2
Sasak	Indonesia	1	2
Aymara, Central	Bolivia	4	2
Karen, Sgaw	Myanmar	2	2
Albanian, Gheg	Yugoslavia	7	2
Kirghiz	Kyrgyzstan	7	2
SW-Caribbean-Creole English	Jamaica	7	2
Betawi	Indonesia	1	2
Macedonian	Macedonia	7	2
Tumbuka	Malawi	3	2
Rajbangsi	India	3	2
Batak Toba	Indonesia	1	2
Arabic, Gulf, spoken	Iraq	9	2
Waray-Waray	Philippines	1	2
Mongolian, Halh	Mongolia	4	2
Malagasy, Southern	Madagascar	1	2
Konkani, Goanese	India	3	2
Kalenjin	Kenya	1	2
Bicolano, Central	Philippines	1	2
Bagri	India	2	2
Zarma	Niger	5	2
Baoule	Côte d'Ivoire	1	2
Kumauni	India	2	2
Lomwe	Mozambique	2	2
Tarifit	Morocco	4	2
Saxon, Upper	Germany	1	2
Kurux	India	2	2
Makhuwa	Mozambique	2	2
Maninka, Kankan	Guinea	3	2
Tiv	Nigeria	2	2
Bamanankan	Mali	7	2
Ewe	Ghana	2	2
Pulaar	Senegal	6	2
Hassaniyya	Mauritania	6	2
Arakanese	Myanmar	3	2
Slovenian	Slovenia	10	2
Jula	Burkina Faso	3	2
Bouyei	China	2	2
Brahui	Pakistan	4	2
Fuuta Jalon	Guinea	6	2

IT'S A FACT: The Cambodian alphabet is the world's largest alphabet, with 74 letters. The world's shortest alphabet, used in the Solomon Islands, has only 11.

American Manual Alphabet

In the American Manual Alphabet, each letter of the alphabet is represented by a position of the fingers. This system was originally developed in France by Abbe Charles Michel De I'Epee in the late 1700s. It was brought to the United States by Laurent Clerce (1785-1869), a Frenchman who taught deaf or hearing-impaired people.

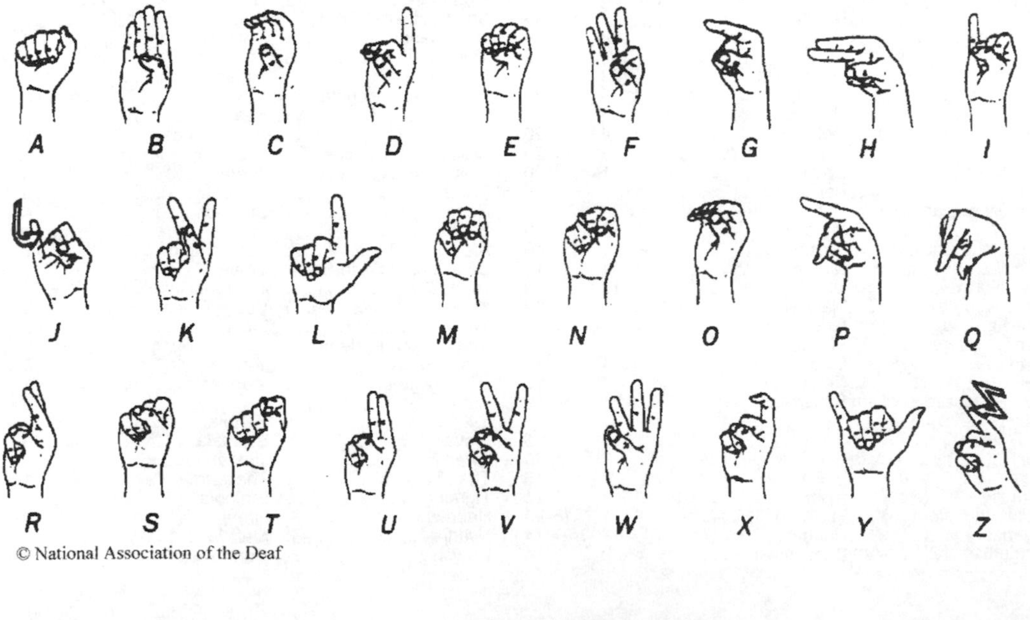

© National Association of the Deaf

WORLD EXPLORATION AND GEOGRAPHY

Early Explorers of the Western Hemisphere

Reviewed by Susan Skomal, PhD, American Anthropological Assn., and Paul B. Frederic, PhD, prof. of geography, Univ. of Maine.

In the light of recent discoveries and re-analysis of older finds, theories about how the first people arrived in the western hemisphere are being reconsidered. It was once thought that they all came across a "land bridge" from Siberia to Alaska, spreading through the Americas 12,000 to 14,000 years ago. Genetic, skeletal, and linguistic evidence indeed indicates that current Native Americans are descended from peoples from Asia. Skeletal remains of Kennewick Man found (1996) in Washington state (dated to between 9,200 and 9,600 years old) and "Luzia" from Brazil (estimated to be at least 11,500 years old), however, suggest the earlier arrival of a people with markedly different physical characteristics and uncertain origin.

In 1997, archaeologists confirmed evidence of habitation at least 12,500 years ago at Monte Verde (Chile), predating a site in Clovis, NM, by over 1,000 years. In addition, 40 skeletons from the same period as Luzia were found in 1999 at Lagoa Santa in Brazil. Because a glacier covered most of N America from 20,000 to 13,000 years ago or later, those who settled in S America may have traveled in vessels along the west coast, sailed directly from Australia, or perhaps even spread from N to S America before the ice came.

Norsemen (Norwegian Vikings sailing out of Iceland and Greenland), led by Leif Ericson, are credited by most scholars with having been the first Europeans to reach America, with at least 5 voyages occurring about AD 1000 to areas they called Helluland, Markland, and Vinland—possibly what are known today as Labrador, Nova Scotia or Newfoundland, and New England. L'Anse aux Meadows, on the N tip of Newfoundland, is the only documented settlement.

Sustained contact between the hemispheres began with the first voyage of Christopher Columbus (born Cristoforo Colombo, c 1451, in or near Genoa, Italy). Columbus made trips to the New World while sailing for the Spanish.

He left Palos, Spain, Aug. 3, 1492, with 88 (est.) men and landed at San Salvador (Watling Islands, Bahamas), Oct. 12, 1492. His fleet consisted of 3 vessels—the *Niña, Pinta,* and *Santa María*. Stops also were made on Cuba and Hispaniola. A 2d expedition left Cadiz, Spain, Sept. 25, 1493, with 17 ships and 1,500 men, reaching the Lesser Antilles Nov. 3. His 3d voyage brought him from Sanlucar, Spain (May 30, 1498, with 6 ships), to the N coast of S America. A 4th voyage reached the mainland of Central America, after leaving Cadiz, Spain, May 9, 1502. Columbus died in 1506 convinced he had reached Asia by sailing west.

In N America, John and Sebastian Cabot, Italian explorers sailing for the English, reached Newfoundland and possibly Nova Scotia in 1497. John's 2d voyage (1498), seeking a new trade route to Asia, resulted in the loss of his entire fleet. During this period exploration was dominated by Spain and Portugal. In 1497 and 1499 Amerigo Vespucci (for whom the Americas are named), an Italian explorer sailing for Spain, passed along the N and E coasts of S America. He was the first to argue that the newly discovered lands were a continent other than Asia. The basic geography of the hemisphere became well understood by the early 1800s, as explorers from many countries helped fill in the map.

Year	Explorer	Nationality (sponsor, if different)	Area reached or explored
c1000	Leif Ericson	Norse	Newfoundland
1492-1502	Christopher Columbus	Italian (Spanish)	West Indies, S. and C. America
1497	John and Sebastian Cabot	Italian (English)	Atlantic Canada
1497-98	Vasco de Gama	Portuguese	Cape of Good Hope (Africa), India
1497-99	Amerigo Vespucci	Italian (Spanish)	E and N Coast of S. America
1499	Alonso de Ojeda	Spanish	N South American coast, Venezuela
1500, Feb.	Vicente Yañez Pinzon	Spanish	S. American coast, Amazon R.
1500, Apr.	Pedro Álvarez Cabral	Portuguese	Brazil
1500-02	Gaspar Corte-Real	Portuguese	Labrador
1501	Rodrigo de Bastidas	Spanish	Central America
1513	Vasco Nunez de Balboav	Spanish	Panama, Pacific Ocean
1513	Juan Ponce de Leon	Spanish	Florida, Yucatán Peninsula
1515	Juan de Solis	Spanish	Río de la Plata
1519	Alonso de Pineda	Spanish	Mouth of Mississippi R.
1519	Hernando Cortesv	Spanish	Mexico
1519-20	Ferdinand Magellan	Portuguese (Spanish)	Straits of Magellan, Tierra del Fuego
1524	Giovanni da Verrazano	Italian (French)	Atlantic coast, inc. New York harbor
1528	Cabeza de Vaca	Spanish	Texas coast and interior
1532	Francisco Pizarro	Spanish	Peru
1534	Jacques Cartier	French	Canada, Gulf of St. Lawrence
1536	Pedro de Mendoza	Spanish	Buenos Aires
1539	Francisco de Ulloa	Spanish	California coast
1539-41	Hernando de Soto	Spanish	Mississippi R., near Memphis
1539	Marcos de Niza	Italian (Spanish)	SW United States
1540	Francisco de Coronado	Spanish	SW United States
1540	Hernando Alarcon	Spanish	Colorado R.
1540	Garcia de Lopez Cardenas	Spanish	Colorado, Grand Canyon
1541	Francisco de Orellana	Spanish	Amazon R.
1542	Juan Rodriguez Cabrillo	Portuguese (Spanish)	W Mexico, San Diego harbor
1565	Pedro Menéndez de Aviles	Spanish	St. Augustine, FL
1576	Sir Martin Frobisher	English	Frobisher's Bay, Canada
1577-80	Sir Francis Drake	English	California coast
1582	Antonio de Espejo	Spanish	Southwest U.S. (New Mexico)
1584	Amadas & Barlow (for Raleigh)	English	Virginia
1585-87	Sir Walter Raleigh's men	English	Roanoke Isl., NC
1595	Sir Walter Raleigh	English	Orinoco R.
1603-09	Samuel de Champlain	French	Canadian interior, Lake Champlain
1607	Capt. John Smith	English	Atlantic coast
1609-10	Henry Hudson	English (Dutch)	Hudson R., Hudson Bay
1634	Jean Nicolet	French	Lake Michigan, Wisconsin
1673	Jacques Marquette, Louis Jolliet	French	Mississippi R., S to Arkansas
1682	Robert Cavelier, sieur de La Salle	French	Mississippi R., S to Gulf of Mexico
1727-29	Vitus Bering	Danish (Russian)	Bering Strait and Alaska
1789	Sir Alexander Mackenzie	Canadian	NW Canada
1804-06	Meriwether Lewis and William Clark	American	Missouri R., Rocky Mts., Columbia R.

IT'S A FACT: During his four voyages to the New World between 1492 and 1504, Christopher Columbus reached parts of the Caribbean, as well as Central America and the north coast of South America, but he never actually set foot on the continent of North America.

Arctic Exploration

Early Explorers

1587 — John Davis (Eng.). Davis Strait to Sanderson's Hope, 72°12′ N.

1596 — Willem Barents and Jacob van Heemskerck (Holland). Discovered Bear Isl., touched NW tip of Spitsbergen, 79°49′ N, rounded Novaya Zemlya, wintered at Ice Haven.

1607 — Henry Hudson (Eng.). North along Greenland's E coast to Cape Hold-with-Hope, 73°30′, then N of Spitsbergen to 80°23′. Explored Hudson's Touches (Jan Mayen).

1616 — William Baffin and Robert Bylot (Eng.). Baffin Bay to Smith Sound.

1728 — Vitus Bering (Russ.). Sailed through strait (Bering) proving Asia and America are separate.

1733-40 — Great Northern Expedition (Russ.). Surveyed Siberian Arctic coast.

1741 — Vitus Bering (Russ.). Sighted Alaska, named Mount St. Elias. His lieutenant, Chirikof, explored coast.

1771 — Samuel Hearne (Hudson's Bay Co.). Overland from Prince of Wales Fort (Churchill) on Hudson Bay to mouth of Coppermine R.

1778 — James Cook (Brit.). Through Bering Strait to Icy Cape, AK, and North Cape, Siberia.

1789 — Alexander Mackenzie (North West Co., Brit.). Montreal to mouth of Mackenzie River.

1806 — William Scoresby (Brit.). N of Spitsbergen to 81°30′.

1820-23 — Ferdinand von Wrangel (Russ.). Surveyed Siberian Arctic coast. His exploration joined James Cook's at North Cape, confirming separation of the continents.

1878-79 — (Nils) Adolf Erik Nordenskjöld (Swed.). The first to navigate the Northeast Passage—an ocean route connecting Europe's North Sea, along the Arctic coast of Asia and through the Bering Sea, to the Pacific Ocean.

1881 — The U.S. steamer *Jeannette*, led by Lt. Cmdr. George W. DeLong, was trapped in ice and crushed, June 1881. DeLong and 11 others died; 12 survived.

1888 — Fridtjof Nansen (Nor.) crossed Greenland icecap.

1893-96 — Nansen in *Fram* drifted from New Siberian Isls. to Spitsbergen; tried polar dash in 1895, reached Franz Josef Land, 86°14′ N.

1897 — Salomon A. Andrée (Sweden) and 2 others started in balloon from Spitsbergen, July 11, to drift across pole to U.S., and disappeared. Aug. 6, 1930, their bodies were found on White Isl., 82°57′ N, 29°52′ E.

1903-6 — Roald Amundsen (Nor.) first sailed the Northwest Passage—an ocean route linking the Atlantic Ocean to the Pacific via Canada's marine waterways.

North Pole Exploration

Robert E. Peary explored Greenland's coast, 1891-92; tried for North Pole, 1893. In 1900 he reached N limit of Greenland and 83°50′ N; in 1902 he reached 8°06′ N; in 1906 he went from Ellesmere Isl. to 87°06′ N. He sailed in the *Roosevelt,* July 1908, to winter off Cape Sheridan, Grant Land. The dash for the North Pole began Mar. 1 from Cape Columbia, Ellesmere Isl. Peary reportedly reached the pole, 90°N, Apr. 6, 1909; however, subsequent research suggests that he may have miscalculated and fallen short of his goal by c. 30-60 mi. Peary had several supporting groups carrying supplies until the last group turned back at 87°47′ N. Peary, Matthew Henson, and 4 Eskimos proceeded with dog teams and sleds. They were said to have crossed the pole several times, then built an igloo there and remained 36 hours. Started south, Apr. 7 at 4 PM, for Cape Columbia.

1914 — Donald MacMillan (U.S.). Northwest, 200 mi, from Axel Heiberg Isl. to seek Peary's Crocker Land.

1915-17 — Vihjalmur Stefansson (Can.). Discovered Borden, Brock, Meighen, and Lougheed Isls.

1918-20 — Amundsen sailed the Northeast Passage.

1925 — Amundsen and Lincoln Ellsworth (U.S.) reached 87°44′ N in attempt to fly to North Pole from Spitsbergen.

1926 — Richard E. Byrd and Floyd Bennett (U.S.) reputedly flew over North Pole, May 9. (Claim to have reached the Pole is in dispute, however.)

1926 — Amundsen, Ellsworth, and Umberto Nobile (It.) flew from Spitsbergen over North Pole May 12, to Teller, AK, in dirigible *Norge.*

1928 — Nobile crossed North Pole in airship, May 24; crashed, May 25. Amundsen died attempting a rescue.

North Pole Exploration Records

On Aug. 3, 1958, the *Nautilus,* under Comdr. William R. Anderson, became the first ship to cross the North Pole beneath the Arctic ice.

In Aug. 1960, the nuclear-powered U.S. submarine *Seadragon* (Comdr. George P. Steele 2d) made the first E-W underwater transit through the Northwest Passage. Traveling submerged for the most part, it took 6 days to make the 850-mi trek from Baffin Bay to the Beaufort Sea.

On Aug. 16, 1977, the Soviet nuclear icebreaker *Arktika* reached the North Pole, becoming the first surface ship to break through the Arctic ice pack.

On Apr. 30, 1978, Naomi Uemura (Jap.) became the first person to reach the North Pole alone, traveling by dog sled in a 54-day, 600-mi trek over the frozen Arctic.

In Apr. 1982, Sir Ranulph Fiennes and Charles Burton, Brit. explorers, reached the North Pole and became the first to circle the earth from pole to pole. They had reached the South Pole 16 months earlier. The 52,000-mi trek took 3 years, involved 23 people, and cost an estimated $18 mil.

On May 2, 1986, 6 explorers reached the North Pole assisted only by dogs. They became the first to reach the pole without aerial logistics support since at least 1909. The explorers, Amer. Will Steger, Paul Schurke, Anne Bancroft, and Geoff Carroll, and Can. Brent Boddy and Richard Weber, completed the 500-mi journey in 56 days.

On June 15, 1995, Weber and Russ. Mikhail Malakhov became the first pair to make it to the pole and back without any mechanical assistance. The 940-mi trip, made entirely on skis, took 121 days.

Antarctic Exploration

Antarctica has been approached since 1773-75, when Capt. James Cook (Brit.) reached 71° 10′ S. Many sea and landmarks bear names of early explorers. Fabian von Bellingshausen (Russ.) discovered Peter I and Alexander I Isls., 1819-21. Nathaniel Palmer (U.S.) traveled throughout Palmer Peninsula, 60°W, 1820, without realizing that this was a continent. Capt. John Davis (U.S.) made the first known landing on the continent on Feb. 7, 1821. Later, in 1823, James Weddell (Brit.) found Weddell Sea, 74° 15′ S, the southernmost point that had been reached.

First to announce existence of the continent of Antarctica was Charles Wilkes (U.S.), who followed the coast for 1,500 mi, 1840. Adelie Coast, 140° E, was found by Dumont d'Urville (Fr.), 1840. Ross Ice Shelf was found by James Clark Ross (Brit.), 1841-42.

1895 — Leonard Kristensen (Nor.) landed a party on the coast of Victoria Land. They were the first ashore on the main continental mass. C. E. Borchgrevink, a member of that party, returned in 1899 with a Brit. expedition, first to winter on Antarctica.

1902-4 — Robert F. Scott (Brit.) explored Edward VII Peninsula to 82°17′ S, 146°33′ E from McMurdo Sound.

1908-9 — Ernest Shackleton (Brit.) introduced the use of Manchurian ponies in Antarctic sledging. He reached 88°23′ S, discovering a route on to the plateau by way of the Beardmore Glacier and pioneering the way to the pole.

1911 — Roald Amundsen (Nor.) with 4 men and dog teams reached the South Pole, Dec. 14.

1912 — Scott reached the pole from Ross Isl., Jan. 18, with 4 companions. None of Scott's party survived. Their bodies and expedition notes were found, Nov. 12.

1928 — First person to use an airplane over Antarctica was Sir George Hubert Wilkins (Austral.).

1929 — Richard E. Byrd (U.S.) established Little America on Bay of Whales. On 1,600-mi airplane flight begun Nov. 28, he crossed South Pole, Nov. 29, with 3 others.

1934-35 — Byrd led 2d expedition to Little America, explored 450,000 sq mi, wintered alone at weather station, 80°08′ S.

1934-37 — John Rymill led British Graham Land expedition; discovered Palmer Penin. is part of mainland.

1935 —Lincoln Ellsworth (U.S.) flew S along E Coast of Palmer Penin., then crossed continent to Little America, making 4 landings on unprepared terrain in bad weather.

1939-41 — U.S. Antarctic Service Expedition built West Base on Ross Ice Shelf under Paul Siple, and East Base on Palmer Peninsula under Richard Black. U.S. Navy plane flights discovered about 150,000 sq mi of new land.

1940 — Byrd charted most of coast between Ross Sea and Palmer Penin.

1946-47 — U.S. Navy undertook Operation Highjump, commanded by Byrd, included 13 ships and 4,000 men. Airplanes photomapped coastline and penetrated beyond pole.

1946-48 — Ronne Antarctic Research Expedition Comdr., Finn Ronne, USNR, determined the Antarctic to be only one continent with no strait between Weddell Sea and Ross Sea; explored 250,000 sq mi of land by flights to 79° S. Mrs. Ronne and Mrs. H. Darlington were the first women to winter on Antarctica.

1955-57 — U.S. Navy's Operation Deep Freeze led by Adm. Byrd. Supporting U.S. scientific efforts for the International Geophysical Year (IGY), the operation was commanded by Rear Adm. George Dufek. It established 5 coastal stations fronting the Indian, Pacific, and Atlantic oceans and also 3 interior stations; explored more than 1,000,000 sq mi in Wilkes Land.

1957-58 — During the IGY, July 1957 through Dec. 1958, scientists from 12 countries conducted Antarctic research at a network of some 60 stations on Antarctica.

Dr. Vivian E. Fuchs led a 12-person Trans-Antarctic Expedition on the first land crossing of Antarctica. Starting from the Weddell Sea, they reached Scott Station, Mar. 2, 1958, after traveling 2,158 mi in 98 days.

1958 — A group of 5 U.S. scientists led by Edward C. Thiel, seismologist, moving by tractor from Ellsworth Station on Weddell Sea, identified a huge mountain range, 5,000 ft above the ice sheet and 9,000 ft above sea level. The range, originally seen by a Navy plane, was named the Dufek Massif, for Rear Adm. George Dufek.

1959 — Argentina, Australia, Belgium, Chile, France, Japan, New Zealand, Norway, South Africa, U.S.S.R., U.K., and U.S. signed a treaty suspending territorial claims for 30 yrs. and reserving the continent, S of 60° S, for research.

1961-62 — Scientists discovered the Bentley Trench, running from Ross Ice Shelf into Marie Byrd Land, near the end of the Ellsworth Mts., toward the Weddell Sea.

1962 — First nuclear power plant began operation at McMurdo Sound.

1963 — On Feb. 22, a U.S. plane made the longest nonstop flight ever in the South Pole area, covering 3,600 mi in 10 hr. The flight was from McMurdo Station S past the pole to Shackleton Mts., SE to the "Area of Inaccessibility," and back to McMurdo Station.

1964 — A Brit. survey team was landed by helicopter on Cook Island, the first recorded visit since 1775.

1964 — New Zealanders mapped the mountain area from Cape Adare W some 400 mi to Pennell Glacier.

1985 — Igor A. Zotikov, a Russian researcher, discovered sediments in the Ross Ice Shelf that seem to support the continental drift theory. Research by the Ocean Drilling Project off the Queen Maud Land coast indicated that the ice sheets of E Antarctica are 37 million yrs. old.

1989 — Victoria Murden and Shirley Metz became both the first women and the first Americans to reach the South Pole overland when they arrived with 9 others on Jan. 17, 1989. The 51-day trek on skis covered 740 mi.

1991 — 24 nations approved a protocol to the 1959 Antarctica Treaty, Oct. 4. New conservation provisions, including banning oil and other mineral exploration for 50 yrs.

1995 — On Dec. 22, a Norwegian, Borge Ousland, reached the South Pole in the fastest time on skis: 44 days.

1996-97 — Ousland became 1st person to traverse Antarctica alone; reached South Pole Dec. 19, 1996; traveled 1,675 mi in 64 days, ending Jan. 18, 1997.

Volcanoes

Sources: *Volcanoes of the World*, Geoscience Press; Global Volcanism Network, Smithsonian Institution

Roughly 540 volcanoes are known to have erupted during historical times. Nearly 75% of these historically active volcanoes lie along the so-called Ring of Fire, running along the W coast of the Americas from the southern tip of Chile to Alaska, down the E coast of Asia from Kamchatka to Indonesia, and continuing from New Guinea to New Zealand. The Ring of Fire marks the boundary between the mobile tectonic plates underlying the Pacific Ocean and those of the surrounding continents. Other active regions occur along rift zones, where plates pull apart, as in Iceland, or where molten material moves up from the mantle over local "hot spots," as in Hawaii. The vast majority of the earth's volcanism occurs at submarine rift zones. For more information on volcanoes, see the website at http://www.nmnh.si.edu/gvp

IT'S A FACT: One of the world's most active volcanoes, Hawaii's Kilauea has been erupting continuously since January 3, 1983. By 2000, its lava flows had destroyed more than 180 homes, a church, and the Hawaii Volcanoes National Park visitor center.

Notable Volcanic Eruptions

Approximately 7,000 years ago, Mazama, a 9,900-ft volcano in southern Oregon, erupted violently, ejecting large amounts of ash and pumice and voluminous pyroclastic flows. The ash spread over the entire northwestern U.S. and as far away as Saskatchewan, Can. During the eruption, the top of the mountain collapsed, leaving a caldera 6 mi across and about a half mile deep, which filled with rainwater to form what is now called Crater Lake.

In AD 79, Vesuvio, or Vesuvius, a 4,190-ft volcano overlooking Naples Bay, became active after several centuries of apparent quiescence. On Aug. 24 of that year, a heated mud and ash flow swept down the mountain, engulfing the cities of Pompeii, Herculaneum, and Stabiae with debris more than 60 ft deep. About 10% of the population of the 3 towns were killed.

In 1883, an eruption similar to the Mazama eruption occurred on the island of Krakatau. At least 2,000 people died in pyroclastic flows on Aug. 26. The next day, the 2,640-ft peak of the volcano collapsed to 1,000 ft below sea level, sinking most of the island and killing over 3,000. A tsunami (tidal wave) generated by the collapse killed more than 31,000 people in Java and Sumatra, and eventually reached England. Ash from the eruption colored sunsets around the world for 2 years. A similar, even more powerful eruption had taken place 68 years earlier at Mt. Tambora on the Indonesian island of Sumbawa.

Date	Volcano	Deaths (est.)
Aug. 24, AD 79	Mt. Vesuvius, Italy	16,000
1586	Kelut, Java, Indon.	10,000
Dec. 15, 1631	Mt. Vesuvius, Italy	4,000
Aug. 12, 1772	Mt. Papandayan, Java, Indon.	3,000
June 8, 1783	Laki, Iceland	9,350
May 21, 1792	Mt. Unzen, Japan	14,500
Apr. 10-12, 1815	Mt. Tambora, Sumbawa, Indon	92,000[1]
Aug. 26-28, 1883	Krakatau, Indon.	36,000
Apr. 24, 1902	Santa María, Guatemala	1,000[2]
May 8, 1902	Mt. Pelée, Martinique	28,000
Jan. 30, 1911	Mt. Taal, Phil.	1,400
May 19, 1919	Mt. Kelut, Java, Indon.	5,000
Jan. 17-21, 1951	Mt. Lamington, New Guinea	3,000
May 18, 1980	Mt. St. Helens, U.S.	57
Mar. 28, 1982	El Chichon, Mex.	1,880
Nov. 13, 1985	Nevado del Ruiz, Colombia	23,000
Aug. 21, 1986	Lake Nyos, Cameroon	1,700
June 15, 1991	Mt. Pinatubo, Luzon, Phil.	800

(1) Of these, 10,000 were directly related to the eruption; an additional 82,000 were the result of starvation and disease brought on by the event. (2) An additional 3,000 deaths due to a malaria outbreak are sometimes attributed to the eruption.

Notable Active Volcanoes

Active volcanoes display a wide range of activity. In this table, years are given for last display of eruptive activity, as of mid-2001; list does not include submarine volcanoes. An eruption may involve explosive ejection of new or old fragmental material, escape of liquid lava, or both. Volcanoes are listed by height, which does not reflect eruptive magnitude.

Name (latest eruption)		Height (ft)
Africa		
Mt. Cameroon (2000)	Cameroon	13,435
Nyiragongo (1994)	Congo	11,400
Nyamuragira (2001)	Congo	10,028
Ol Doinyo Lengai (2000)	Tanzania	9,469
Fogo (1995)	Cape Verde Isls.	9,281
Karthala (1991)	Comoros.	8,000
Piton de la Fournaise (2001)	Réunion Isl., Indian O.	5,981
Lake Nyos (1986)	Cameroon	3,011
Erta-Ale (1995)	Ethiopia	1,650
Antarctica		
Erebus (1998)	Ross Isl	12,450
Deception Island (1970)	S. Shetland Isl.	1,890
Asia-Oceania		
Kliuchevskoi (1999)	Kamchatka, Russia	15,863
Kerinci (1999)	Sumatra, Indon.	12,467
Fuji (1708)	Honshu, Japan	12,388
Tolbachik (1976)	Kamchatka, Russia	12,080
Semeru (2000)	Java, Indon.	12,060
Slamet (1999)	Java, Indon.	11,247
Raung (2000)	Java, Indon.	10,932
Shiveluch (2001)	Kamchatka, Russia	10,771
On-take (1980)	Honshu, Japan	10,049
Mayon (2001)	Luzon, Phil.	9,991
Merapi (2001)	Java, Indon.	9,550
Bezymianny (2000)	Kamchatka, Russia	9,455
Ruapehu (1996)	New Zealand	9,175
Peuet Sague (2000)	Sumatra, Indon.	9,120
Heard (1993)	Indian Ocean	9,006
Baitoushan (1702)	China/Korea	9,003
Asama (1990)	Honshu, Japan	8,300
Niigata Yake-yama (1989)	Honshu, Japan	8,111
Canlaon (1996)	Negros Isls., Phil.	8,070
Alaid (1996)	Kuril Isl., Russia	7,674
Ulawun (2001)	Papua New Guinea	7,532
Ngauruhoe (1977)	New Zealand	7,515
Chokai (1974)	Honshu, Japan	7,300
Galunggung (1984)	Java, Indon.	7,113
Azuma (1977)	Honshu, Japan	6,700
Bagana (1995)	Papua New Guinea	6,558
Sangeang Api (1988)	Lesser Sunda Isl., Indon.	6,351
Nasu (1963)	Honshu, Japan	6,210
Tiatia (1973)	Kuril Isl., Russia	6,013
Soputan (2000)	Sulawesi, Indon.	5,994
Bandai (1888)	Honshu, Japan	5,968
Manam (2000)	Papua New Guinea	5,928
Kuju (1996)	Kyushu, Japan	5,866
Karangetang-Api Siau (2001)	Sangihe, Indon.	5,853
Kelut (1990)	Java, Indon.	5,679
Adatara (1996)	Honshu, Japan	5,636
Gamalama (1994)	Halmahera, Indon.	5,627
Kirishima (1992)	Kyushu, Japan	5,577
Gamkonora (1987)	Halmahera, Indon.	5,364
Pinatubo (1995)	Luzon, Phil.	5,249
Aso (1996)	Kyushu, Japan	5,223
Lokon-Empung (2001)	Sulawesi, Indon.	5,187
Bulusan (1995)	Luzon, Phil.	5,115
Sarychev Peak (1989)	Kuril Isl., Russia	4,960
Karkar (1979)	Papua New Guinea	4,920
Akan (1998)	Hokkaido, Japan	4,917
Akademia Nauk (1996)	Kamchatka, Russia	4,875
Karymsky (2000)	Kamchatka, Russia	4,875
Lopevi (2000)	Vanuatu	4,755
Akita-Yake-yama (1997)	Japan	4,482
Unzen (1996)	Kyushu, Japan	4,462
Ambrym (2000)	Vanuatu	4,376
Langila (2000)	Papua New Guinea	4,363
Awu (1992)	Sangihe Isl., Indon.	4,350
Sakura-jima (2000)	Kyushu, Japan	3,665
Komaga-take (2000)	Hokkaido, Japan	3,740
Dukono (1995)	Halmahera, Indonesia	3,566
Miyake-jima (2001)	Izu Isls., Japan	2,673
Krakatau (2000)	Indonesia	2,667
Suwanose-jima (1997)	Kyushu, Japan	2,621
Gaua (1982)	Vanuatu	2,614
Oshima (1990)	Izu Isls., Japan	2,487
Usu (2000)	Hokkaido, Japan	2,398
Rabaul (2000)	Papua New Guinea	2,257
Pagan (1993)	N. Mariana Isl.	1,870
Yasur (2000)	Tanna Island, Vanuatu.	1,184
White Island (2000)	Bay of Plenty, New Zealand	1,053
Taal (1977)	Luzon, Phil.	984
McDonald Islands (2000)	Indian Ocn., Australia	610
Central America—Caribbean		
Acatenango (1972)	Guatemala	12,992
Tacaná (1986)	Guatemala	12,400
Santa María (2000)	Guatemala	12,375
Fuego (2000)	Guatemala	12,346
Irazú (1994)	Costa Rica	11,260
Turrialba (1866)	Costa Rica	10,958
Póas (1994)	Costa Rica	8,884
Pacaya (2000)	Guatemala	8,373
San Miguel (1986)	El Salvador	6,994
Rincón de la Vieja (1999)	Costa Rica	6,286
San Cristobal (2001)	Nicaragua	5,725
Arenal (2000)	Costa Rica	5,436
Concepción (1999)	Nicaragua	5,282
Soufrière Guadeloupe (1977)	Guadeloupe	4,813
Pelee (1932)	Martinique	4,583
Momotombo (1905)	Nicaragua	4,127
Soufrière St. Vincent (1979)	St. Vincent	3,865
Soufrière Hills (2001)	Montserrat	3,001
Masaya (2001)	Nicaragua	2,083
South America		
Llullaillaco (1877)	Argentina-Chile	22,057
Guallatiri (1960)	Chile	19,918
Cotopaxi (1940)	Ecuador	19,347
El Misti (1870?)	Peru	19,101
Tupungatito (1986)	Chile	18,504
Láscar (2000)	Chile	18,346
Ruiz (1991)	Colombia	17,457
Sangay (1998)	Ecuador	17,021
Irruputuncu (1995)	Chile	16,939
Guagua Pichincha (2000)	Ecuador	15,696
Puracé (1977)	Colombia	15,601
Galeras (2000)	Colombia	14,029
Llaima (1998)	Chile	10,253
Villarrica (2000)	Chile	9,340
Cerro Hudson (1991)	Chile	8,580
Fernandina (1995)	Galapagos Isls., Ecuad.	4,905
Mid-Pacific		
Mauna Loa (1984)	Hawaii, HI	13,680
Kilauea (2001)	Hawaii, HI	4,009
Mid-Atlantic Ridge		
Jan Mayen (1985)	N. Atlantic Ocn., Norway	7,470
Grímsvötn (1998)	Iceland	5,659
Hekla (2000)	Iceland	4,892
Krafla (1984)	Iceland	2,145
Europe		
Etna (2001)	Italy	11,053
Vesuvius (1944)	Italy	4,203
Stromboli (2000)	Italy	3,038
Santorini (1950)	Greece	1,850
North America		
Pico de Orizaba (1687)	Mexico	18,555
Popocatépetl (2001)	Mexico	17,930
Rainier (1894?)	Washington	14,410
Wrangell (1907?)	Alaska	14,163
Shasta (1786)	California	14,162
Colima (1999)	Mexico	12,361
Redoubt (1990)	Alaska	10,197
Iliamna (1953)	Alaska	10,016
Shishaldin (1999)	Aleutian Isl., AK	9,373
Pavlof (1997)	Alaska	8,264
St. Helens (1991)	Washington	8,363
Veniaminof (1995)	Alaska	8,225
El Chichón (1982)	Mexico	7,300
Novarupta (Katmai) (1912)	Alaska	6,715
Makushin (1980)	Aleutian Isl., AK	6,680
Great Sitkin (1974)	Aleutian Isl., AK	5,710
Cleveland (2001)	Aleutian Isl., AK	5,675
Gareloi (1989)	Aleutian Isl., AK	5,161
Atka (1998)	Aleutian Isl., AK	5,029
Korovin (1998)	Aleutian Isl., AK	4,852
Akutan (1992)	Aleutian Isl., AK	4,275
Kiska (1990)	Aleutian Isl., AK	4,275
Augustine (1986)	Alaska	3,999
Okmok (1997)	Aleutian Isl., AK	3,520
Seguam (1993)	Aleutian Isl., AK	3,458

Mountains

Height of Mount Everest

Mt. Everest, the world's highest mountain, was considered 29,002 ft when Edmund Hillary and Tenzing Norgay became the first climbers to scale it, in 1953. This triangulation figure had been accepted since 1850. In 1954 the Surveyor General of the Republic of India set the height at 29,028 ft, plus or minus 10 ft because of snow; this figure was also accepted by the National Geographic Society.

In 1999, a team of climbers sponsored by Boston's Museum of Science and the National Geographic Society measured the height at the summit using sophisticated satellite-based technology. This new measurement, of 29,035 ft., was accepted by the National Geographic Society and other authorities, including the U.S. National Imagery and Mapping Agency.

IT'S A FACT: More than 1,000 people have succeeded in reaching the summit of Mount Everest since 1953, while more than 150 people have died in the attempt. Records were broken in May 2001, when the first blind person, the youngest yet (age 15), and the oldest yet (age 64) reached the top.

United States, Canada, Mexico

Name	Place	Height (ft)
McKinley	AK	20,320
Logan	Yukon	19,551
Pico de Orizaba	Mexico	18,555
St. Elias	AK-Yukon	18,008
Popocatépetl	Mexico	17,930
Foraker	AK	17,400
Iztaccihuatl	Mexico	17,343
Lucania	Yukon	17,147
King	Yukon	16,971
Steele	Yukon	16,644
Bona	AK	16,550
Blackburn	AK	16,390
Kennedy	AK	16,286
Sanford	AK	16,237
Vancouver	AK-Yukon	15,979
South Buttress	AK	15,885
Wood	Yukon	15,885
Churchill	AK	15,638
Fairweather	AK-BC	15,300
Zinantecatl (Toluca)	Mexico	15,016
Hubbard	AK-Yukon	15,015
Bear	AK	14,831
Walsh	Yukon	14,780
East Buttress	AK	14,730
Matlalcueyetl	Mexico	14,636
Hunter	AK	14,573
Alverstone	AK-Yukon	14,565
Browne Tower	AK	14,530
Whitney	CA	14,494
Elbert	CO	14,433
Massive	CO	14,421
Harvard	CO	14,420
Rainier	WA	14,410
University Peak	AK	14,410
Williamson	CA	14,375
La Plata Peak	CO	14,361
Blanca Peak	CO	14,345
Uncompahgre Peak	CO	14,309
Crestone Peak	CO	14,294
Lincoln	CO	14,286
Grays Peak	CO	14,270
Antero	CO	14,269
Torreys Peak	CO	14,267
Castle Peak	CO	14,265
Quandary Peak	CO	14,265
Evans	CO	14,264
Longs Peak	CO	14,255
McArthur	Yukon	14,253
Wilson	CO	14,246
White Mt. Peak	CA	14,246
North Palisade	CA	14,242
Cameron	CO	14,238
Shavano	CO	14,229
Belford	CO	14,197
Princeton	CO	14,197
Crestone Needle	CO	14,197
Yale	CO	14,196
Bross	CO	14,172
Kit Carson	CO	14,165
Wrangell	AK	14,163
Shasta	CA	14,162
El Diente Peak	CO	14,159
Point Success	WA	14,158
Maroon Peak	CO	14,156
Tabeguache	CO	14,155
Oxford	CO	14,153
Sill	CA	14,153
Sneffels	CO	14,150
Democrat	CO	14,148
Capitol Peak	CO	14,130
Liberty Cap	WA	14,112
Pikes Peak	CO	14,110
Snowmass	CO	14,092
Russell	CA	14,088
Eolus	CO	14,083
Windom	CO	14,082
Columbia	CO	14,073
Augusta	AK	14,070

South America

Peak, country	Height (ft)
Aconcagua, Argentina	22,834
Ojos del Salado, Arg.-Chile	22,572
Bonete, Argentina	22,546
Tupungato, Argentina-Chile	22,310
Pissis, Argentina	22,241
Mercedario, Argentina	22,211
Huascaran, Peru	22,205
Llullaillaco, Argentina-Chile	22,057
El Libertador, Argentina	22,047
Cachi, Argentina	22,047
Incahuasi, Argentina-Chile	21,720
Yerupaja, Peru	21,709
Galan, Argentina	21,654
El Muerto, Argentina-Chile	21,457
Sajama, Bolivia	21,391
Nacimiento, Argentina	21,302
Illimani, Bolivia	21,201
Coropuna, Peru	21,083
Laudo, Argentina	20,997
Ancohuma, Bolivia	20,958
Ausangate, Peru	20,945
Toro, Argentina-Chile	20,932
Illampu, Bolivia	20,873
Tres Cruces, Argentina-Chile	20,853
Huandoy, Peru	20,852
Parinacota, Bolivia-Chile	20,768
Tortolas, Argentina-Chile	20,745
Ampato, Peru	20,702
El Condor, Argentina	20,669
Salcantay, Peru	20,574
Chimborazo, Ecuador	20,561
Huancarhuas, Peru	20,531
Famatina, Argentina	20,505
Pumasillo, Peru	20,492
Solo, Argentina	20,492
Polleras, Argentina	20,456
Pular, Chile	20,423
Chani, Argentina	20,341
Aucanquilcha, Chile	20,295
Juncal, Argentina-Chile	20,276
Negro, Argentina	20,184
Quela, Argentina	20,128
Condoriri, Bolivia	20,095
Palermo, Argentina	20,079
Solimana, Peru	20,068
San Juan, Argentina-Chile	20,049
Sierra Nevada, Arg.-Chile	20,023
Antofalla, Argentina	20,013
Marmolejo, Argentina-Chile	20,013
Chachani, Peru	19,931

The highest point in the West Indies is in the Dominican Republic, Pico Duarte (10,417 ft).

Africa

Peak, country	Height (ft)
Kilimanjaro, Tanzania	19,340
Kenya, Kenya	17,058
Margherita Pk., Uganda-Congo	16,763
Ras Dashan, Ethiopia	15,158
Meru, Tanzania	14,979
Karisimbi, Congo-Rwanda	14,787
Elgon, Kenya-Uganda	14,178
Batu, Ethiopia	14,131
Guna, Ethiopia	13,881
Gughe, Ethiopia	13,780
Toubkal, Morocco	13,661
Cameroon, Cameroon	13,435

Australia, New Zealand, SE Asian Islands

Peak, country/island	Height (ft)
Jaya, New Guinea	16,500
Trikora, New Guinea	15,585
Mandala, New Guinea	15,420
Wilhelm, New Guinea	14,793
Kinabalu, Malaysia	13,455
Kerinci, Sumatra, Indon.	12,467
Cook, New Zealand	12,349
Semeru, Java, Indon.	12,060
Kosciusko, Australia	7,310

Europe

Peak, country	Height (ft)
Alps	
Mont Blanc, Fr.-It.	15,771
Monte Rosa (highest peak of group), Switz.	15,203
Dom, Switz.	14,911
Liskamm, It., Switz.	14,852
Weisshorn, Switz.	14,780
Taschhorn, Switz.	14,733
Matterhorn, It., Switz.	14,690
Dent Blanche, Switz.	14,293
Nadelhorn, Switz.	14,196
Grand Combin, Switz.	14,154
Lenzpitze, Switz.	14,088
Finsteraarhorn, Switz.	14,022
Castor, Switz.	13,865
Zinalrothorn, Switz.	13,849
Hohberghom, Switz.	13,842
Alphubel, Switz.	13,799
Rimpfischhom, Switz.	13,776
Aletschorn, Switz.	13,763
Strahlhorn, Switz.	13,747
Dent D'Herens, Switz.	13,686
Breithorn, It., Switz.	13,665
Bishorn, Switz.	13,645
Jungfrau, Switz.	13,642
Ecrins, Fr.	13,461
Monch, Switz.	13,448
Pollux, Switz.	13,422
Schreckhorn, Switz.	13,379
Ober Gabelhorn, Switz.	13,330
Gran Paradiso, It.	13,323
Bernina, It., Switz.	13,284
Fiescherhorn, Switz.	13,283
Grunhorn, Switz.	13,266
Lauteraarhorn, Switz.	13,261
Durrenhorn, Switz.	13,238
Allalinhorn, Switz.	13,213
Weissmies, Switz.	13,199
Lagginhorn, Switz.	13,156
Zupo, Switz.	13,120
Fletschhorn, Switz.	13,110
Adlerhorn, Switz.	13,081
Gletscherhorn, Switz.	13,068
Schalihorn, Switz.	13,040
Scerscen, Switz.	13,028
Eiger, Switz.	13,025
Jagerhorn, Switz.	13,024
Rottalhorn, Switz.	13,022
Pyrenees	
Aneto, Sp.	11,168
Posets, Sp.	11,073
Perdido, Sp.	11,007
Vignemale, Fr.-Sp.	10,820
Long, Sp.	10,479
Estats, Sp.	10,304
Montcalm, Sp.	10,105
Caucasus (Europe-Asia)	
Elbrus, Russia	18,510
Shkhara, Georgia	17,064
Dykh Tau, Russia	17,054
Kashtan Tau, Russia	16,877
Janqi, Georgia	16,565
Kazbek, Georgia	16,558

Asia (Mainland)

Peak	Place	Height (ft)
Everest	Nepal-Tibet	29,035
K2 (Godwin Austen)	Kashmir	28,250
Kanchenjunga	India-Nepal	28,208
Lhotse I (Everest)	Nepal-Tibet	27,923
Makalu I	Nepal-Tibet	27,824
Lhotse II (Everest)	Nepal-Tibet	27,560
Dhaulagiri	Nepal	26,810
Manaslu I	Nepal	26,760
Cho Oyu	Nepal-Tibet	26,750
Nanga Parbat	Kashmir	26,660
Annapurna I	Nepal	26,504
Gasherbrum	Kashmir	26,470
Broad	Kashmir	26,400
Gosainthan	Tibet	26,287
Annapurna II	Nepal	26,041
Gyachung Kang	Nepal-Tibet	25,910
Disteghil Sar	Kashmir	25,868
Himalchuli	Nepal	25,801
Nuptse (Everest)	Nepal-Tibet	25,726
Masherbrum	Kashmir	25,660
Nanda Devi	India	25,645
Rakaposhi	Kashmir	25,550
Kamet	India-Tibet	25,447
Namcha Barwa	Tibet	25,445
Gurla Mandhata	Tibet	25,355
Ulugh Muz Tagh	Xinjiang- Tibet	25,340
Kungur	Xinjiang	25,325
Tirich Mir	Pakistan	25,230
Makalu II	Nepal-Tibet	25,120
Minya Konka	China	24,900
Kula Gangri	Bhutan-Tibet	24,784
Changtzu (Everest)	Nepal-Tibet	24,780
Muz Tagh Ata	Xinjiang	24,757
Skyang Kangri	Kashmir	24,750
Ismail Semani Peak	Tajikistan	24,590
Jongsang Peak	India-Nepal	24,472
Jengish Chokusu	Xinjiang-Kyrgyzstan	24,406
Sia Kangri	Kashmir	24,350
Haramosh Peak	Pakistan	24,270
Istoro Nal	Pakistan	24,240
Tent Peak	India-Nepal	24,165
Chomo Lhari	Bhutan-Tibet	24,040
Chamlang	Nepal	24,012
Kabru	India-Nepal	24,002
Alung Gangri	Tibet	24,000
Baltoro Kangri	Kashmir	23,990
Mussu Shan	Xinjiang	23,890
Mana	India	23,860
Baruntse	Nepal	23,688
Nepal Peak	India-Nepal	23,500
Amne Machin	China	23,490
Gauri Sankar	Nepal-Tibet	23,440
Badrinath	India	23,420
Nunkun	Kashmir	23,410
Lenin Peak	Tajikistan	23,405
Pyramid	India-Nepal	23,400
Api	Nepal	23,399
Pauhunri	India-Tibet	23,385
Trisul	India	23,360
Kangto	India-Tibet	23,260
Nyenchhe Thanglha	Tibet	23,255
Trisuli	India	23,210
Pumori	Nepal-Tibet	23,190
Dunagiri	India	23,184
Lombo Kangra	Tibet	23,165
Saipal	Nepal	23,100
Macha Pucchare	Nepal	22,958
Numbar	Nepal	22,817
Kanjiroba	Nepal	22,580
Ama Dablam	Nepal	22,350
Cho Polu	Nepal	22,093
Lingtren	Nepal-Tibet	21,972
Khumbutse	Nepal-Tibet	21,785
Hlako Gangri	Tibet	21,266
Mt. Grosvenor	China	21,190
Thagchhab Gangri	Tibet	20,970
Damavand	Iran	18,606
Ararat	Turkey	16,804

Antarctica

Peak	Height (ft)
Vinson Massif	16,864
Tyree	16,290
Shinn	15,750
Gardner	15,375
Epperly	15,100
Kirkpatrick	14,855
Elizabeth	14,698
Markham	14,290
Bell	14,117
Mackellar	14,098
Anderson	13,957
Bentley	13,934
Kaplan	13,878
Andrew Jackson	13,750
Sidley	13,720
Ostenso	13,710
Minto	13,668
Miller	13,650
Long Gables	13,620
Dickerson	13,517
Giovinetto	13,412
Wade	13,400
Fisher	13,386
Fridtjof Nansen	13,350
Wexler	13,202
Lister	13,200
Shear	13,100
Odishaw	13,008
Donaldson	12,894
Ray	12,808
Sellery	12,779
Waterman	12,730
Anne	12,703
Press	12,566
Falla	12,549
Rucker	12,520
Goldthwait	12,510
Morris	12,500
Erebus	12,450
Campbell	12,434
Don Pedro Christophersen	12,355
Lysaght	12,326
Huggins	12,247
Sabine	12,200
Astor	12,175
Mohl	12,172
Frankes	12,064
Jones	12,040
Gjelsvik	12,008
Coman	12,000

Some Notable U.S. Mountains

Name	Place	Height (ft)
Gannett Peak	WY	13,804
Grand Teton	WY	13,766
Kings	UT	13,528
Cloud	WY	13,175
Wheeler	NM	13,161
Boundary	NV	13,140
Granite	MT	12,799
Borah	ID	12,662
Humphreys	AZ	12,633
Adams	WA	12,277
San Gorgonio	CA	11,502
Hood	OR	11,239
Lassen	CA	10,457
Granite	CA	10,321
Guadalupe	TX	8,749
Olympus	WA	7,965
Harney	SD	7,242
Mitchell	NC	6,684
Clingmans Dome	NC-TN	6,643
Washington	NH	6,288
Rogers	VA	5,729
Marcy	NY	5,344
Katahdin	ME	5,268
Spruce Knob	WV	4,861
Mansfield	VT	4,393
Black Mountain	KY	4,145

Important Islands and Their Areas

Reviewed by Laurel Duda, Marine Biological Laboratory/Woods Hole Oceanographic Inst. Library.

Figures are for total areas in square miles. Figure in parentheses shows rank among the world's 10 largest individual islands. Because some islands have not been surveyed accurately, some areas shown are estimates. Some "islands" listed are island groups. Only the largest islands in a group are listed individually. Only islands over 10 sq. miles in area are listed.

Antarctica

Adelaide	1,400
Alexander	16,700
Berkner	18,500
Roosevelt	2,900

Arctic Ocean

Akimiski, Nunavut	1,159
Amund Ringnes, Nun.	2,029
Axel Heiberg, Nun.	16,671
Baffin, **(5)**	195,928
Banks, Northwest Territories	27,038
Bathurst, Nun.	6,194
Bolshevik, Russia	4,368
Bolshoy Lyakhovsky, Russia	1,776
Borden, NWT., Nun.	1,079
Bylot, Nun.	4,273
Coats, Nun.	2,123
Cornwallis, Nun.	2,701
Devon, Nun.	21,331
Disko, Greenland	3,312
Ellef Ringnes, Nun.	4,361
Ellesmere, Nun. **(10)**	75,767
Faddayevskiy, Russia	1,930
Franz Josef Land, Russia	8,000
Iturup (Etorofu), Russia	2,596
King William, Nun.	5,062
Komsomolets, Russia	3,477
Mackenzie King, NWT	1,949
Mansel, Nun.	1,228
Melville, NWT, Nun.	16,274
Milne Land, Greenland	1,400
New Siberian Islands, Russia	14,500
Kotelnyy, Russia	4,504
Novaya Zemlya, Russia (2 isls.)	31,730
Oktyabrskoy, Russia	5,471
Prince Charles, NWT	3,676
Prince of Wales, Nun.	12,872
Prince Patrick, NWT	6,119
Somerset, Nun.	9,570
Southampton, Nun.	15,913
Svalbard (tot. group)	23,957
Nordaustlandet	5,410
Spitsbergen	15,060
Traill, Greenland	1,300
Victoria, NWT, Nun. **(9)**	83,897
Wrangel, Russia	2,800

Atlantic Ocean

Anticosti, Canada	3,068
Ascension, UK	34
Azores, Portugal (tot. group)	868
Faial	67
San Miguel	291
Bahama Isls., Bahama (tot. group)	5,382
Andros, Bahamas	2,300
Bermuda Islands, UK	20
Bioko Isl., Equatorial Guinea	785
Block Islands, RI, US	21
Canary Islands, Spain (tot. group)	2,807
Fuerteventura	688
Gran Canaria	592
Tenerife	795
Cape Breton, Canada	3,981
Cape Verde Islands	1,557
Caviana, Para, Brazil	1,918
Channel Islands, UK (tot. group)	75
Guernsey	24
Jersey	45
Faroe Islands, Denmark	540
Falkland Islands, UK (tot. group)	4,700
East Falkland	2,550
West Falkland	1,750
Great Britain, UK **(8)**	84,200
Greenland, Denmark **(1)**	840,000
Gurupa, Para, Brazil	1,878
Hebrides, Scotland	2,744
Iceland	39,699
Ireland (tot. group)	32,589
Irish Republic	27,137
Northern Ireland	5,452
Isle of Man, UK	227
Isle of Wight, England	147
Long Island, NY, US	1,320
Madeira Islands, Portugal	306

Atlantic Ocean

Marajo, Brazil	15,444
Martha's Vineyard, MA, US	89
Mount Desert, ME, US	104
Nantucket, MA, US	45
Newfoundland, Canada	42,031
Orkney Islands, Scotland	390
Prince Edward, Canada	2,185
St. Helena, UK	47
Shetland Islands, Scotland	587
Skye, Scotland	670
South Georgia, UK	1,450
Tierra del Fuego, Chile, Arg.	18,800
Tristan da Cunha, UK	40

Baltic Sea

Aland Islands, Finland	590
Bornholm, Denmark	227
Gotland, Sweden	1,159

Caribbean Sea

Antigua	108
Aruba, Netherlands	75
Barbados	166
Cuba	42,804
Isle of Youth	926
Cayman Islands	100
Curacao, Netherlands	171
Dominica	290
Guadeloupe, France	687
Hispaniola (Haiti and Dominican Rep)	29,389
Jamaica	4,244
Martinique, France	436
Puerto Rico, US	3,339
Tobago	116
Trinidad	1,864
Virgin Islands, UK	59
Virgin Islands, US	134

East Indies

Bali, Indonesia	2,171
Bangka, Indonesia	4,375
Borneo, Indonesia-Malaysia-Brunei **(3)**	280,100
Bougainville, Papua New Guinea	3,880
Buru, Indonesia	3,670
Celebes, Indonesia	69,000
Flores, Indonesia	5,500
Halmahera, Indonesia	6,865
Java (Jawa), Indonesia	48,900
Madura, Indonesia	2,113
Moluccas, Indonesia	32,307
New Britain, Papua New Guinea	14,093
New Guinea, Indon.-PNG **(2)**	306,000
New Ireland, PNG	3,707
Seram, Indonesia	6,621
Sumba, Indonesia	4,306
Sumbawa, Indonesia	5,965
Sumatra, Indonesia **(6)**	165,000
Timor, Indonesia	13,094
Yos Sudarsa, Indonesia	4,500

Indian Ocean

Andaman Isls., India	2,500
Kerguelen	2,247
Madagascar **(4)**	226,658
Mauritius	720
Pemba, Tanzania	380
Reunion, France	970
Seychelles	176
Sri Lanka	25,332
Zanzibar, Tanzania	640

Mediterranean Sea

Balearic Isls., Spain	1,927
Corfu, Greece	229
Corsica, France	3,369
Crete, Greece	3,189
Cyprus	3,572
Elba, Italy	86
Euboea, Greece	1,411
Malta	95
Rhodes, Greece	540
Sardinia, Italy	9,301
Sicily, Italy	9,926

Pacific Ocean

Admiralty, AK, US	1,709
Aleutian Isls., AK, US (tot. group)	6,912
Adak	275
Amchitka	116
Attu	350
Kanaga	142
Kiska	106
Tanaga	195
Umnak	686
Unalaska	1,051
Unimak	1,571
Baranof, AK, US	1,636
Chichagof, AK, US	2,062
Chiloe, Chile	3,241
Christmas, Kiribati	94
Diomede, Big, Russia	11
Easter Isl., Chile	69
Fiji (tot. group)	7,056
Vanua Levu	2,242
Viti Levu	4,109
Galapagos Isls., Ecuador	3,043
Graham Isl., British Columbia	2,456
Guadalcanal, Solomon Isls.	2,180
Guam, US	210
Hainan, China	13,000
Hawaiian Isls., HI, US (tot. group)	6,428
Hawaii	4,028
Oahu	600
Hong Kong, China	31
Hoste, Chile	1,590
Japan (tot. group)	145,850
Hokkaido	30,144
Honshu **(7)**	87,805
Kyushu	14,114
Okinawa	459
Shikoku	7,049
Kangaroo, South Australia	1,680
Kodiak, AK, US	3,485
Kupreanof, AK, US	1,084
Marquesas Isls., France	492
Marshall Isls.	70
Melville, Northern Territory, Aus.	2,240
Micronesia	271
New Caledonia, France	6,530
New Zealand (tot. group)	104,454
Chatham Isls.	372
North	44,204
South	58,384
Stewart	674
North Mariana Isls., US	179
Nunivak, AK, US	1,600
Palau	188
Philippines (tot. group)	115,860
Leyte	2,787
Luzon	40,680
Mindanao	36,775
Mindoro	3,690
Negros	4,907
Palawan	4,554
Panay	4,446
Samar	5,050
Prince of Wales, AK, US	2,770
Revillagigedo, AK, US	1,134
Riesco, Chile	1,973
St. Lawrence, AK, US	1,780
Sakhalin, Russia	29,500
Samoa Isls. (tot. group)	1,177
American Samoa, US	77
Tutuila, US	55
Savail, Samoa	659
Upolu, Samoa	432
Santa Catalina, CA, US	75
Santa Ines, Chile	1,407
Tahiti, France	402
Taiwan, China (tot. group)	13,969
Jinmen Dao (Quemoy)	56
Tasmania, Australia	26,178
Tonga Isls.	290
Vancouver Isl., Brit. Columbia Canada	12,079
Vanuatu	4,707
Wellington, Chile	2,549

Persian Gulf

Bahrain	217

Areas and Average Depths of Oceans, Seas, and Gulfs

Geographers and mapmakers recognize 4 major bodies of water: the Pacific, the Atlantic, the Indian, and the Arctic oceans. The Atlantic and Pacific oceans are considered divided at the equator into the N and S Atlantic and the N and S Pacific. The Arctic Ocean is the name for waters N of the continental landmasses in the region of the Arctic Circle.

	Area (sq mi)	Avg. depth (ft)
Pacific Ocean	64,186,300	12,925
Atlantic Ocean	33,420,000	11,730
Indian Ocean	28,350,500	12,598
Arctic Ocean	5,105,700	3,407
South China Sea	1,148,500	4,802
Caribbean Sea	971,400	8,448
Mediterranean Sea	969,100	4,926
Bering Sea	873,000	4,893
Gulf of Mexico	582,100	5,297
Sea of Okhotsk	537,500	3,192
Sea of Japan	391,100	5,468
Hudson Bay	281,900	305
East China Sea	256,600	620
Andaman Sea	218,100	3,667
Black Sea	196,100	3,906
Red Sea	174,900	1,764
North Sea	164,900	308
Baltic Sea	147,500	180
Yellow Sea	113,500	121
Persian Gulf	88,800	328
Gulf of California	59,100	2,375

Principal Ocean Depths

Source: National Imagery and Mapping Agency, U.S. Dept. of Defense

Name of area	Location (lat.)	Location (long.)	Depth (meters)	Depth (fathoms)	Depth (ft)
Pacific Ocean					
Mariana Trench	11° 22′ N	142° 36′ E	10,924	5,973	35,840
Tonga Trench	23° 16′ S	174° 44′ W	10,800	5,906	35,433
Philippine Trench	10° 38′ N	126° 36′ E	10,057	5,499	32,995
Kermadec Trench	31° 53′ S	177° 21′ W	10,047	5,494	32,963
Bonin Trench	24° 30′ N	143° 24′ E	9,994	5,464	32,788
Kuril Trench	44° 15′ N	150° 34′ E	9,750	5,331	31,988
Izu Trench	31° 05′ N	142° 10′ E	9,695	5,301	31,808
New Britain Trench	06° 19′ S	153° 45′ E	8,940	4,888	29,331
Yap Trench	08° 33′ N	138° 02′ E	8,527	4,663	27,976
Japan Trench	36° 08′ N	142° 43′ E	8,412	4,600	27,599
Peru-Chile Trench	23° 18′ S	71° 14′ W	8,064	4,409	26,457
Palau Trench	07° 52′ N	134° 56′ E	8,054	4,404	26,424
Aleutian Trench	50° 51′ N	177° 11′ E	7,679	4,199	25,194
New HebridesTrench	20° 36′ S	168° 37′ E	7,570	4,139	24,836
North Ryukyu Trench	24° 00′ N	126° 48′ E	7,181	3,927	23,560
Mid. America Trench	14° 02′ N	93° 39′ W	6,662	3,643	21,857
Atlantic Ocean					
Puerto Rico Trench	19° 55′ N	65° 27′ W	8,605	4,705	28,232
S SandwichTrench	55° 42′ S	25° 56′ W	8,325	4,552	27,313
Romanche Gap	0° 13′ S	18° 26′ W	7,728	4,226	25,354
Cayman Trench	19° 12′ N	80° 00′ W	7,535	4,120	24,721
Brazil Basin	09° 10′ S	23° 02′ W	6,119	3,346	20,076
Indian Ocean					
Java Trench	10° 19′ S	109° 58′ E	7,125	3,896	23,376
Ob' Trench	09° 45′ S	67° 18′ E	6,874	3,759	22,553
Diamantina Trench	35° 50′ S	105° 14′ E	6,602	3,610	21,660
Vema Trench	09° 08′ S	67° 15′ E	6,402	3,501	21,004
Agulhas Basin	45° 20′ S	26° 50′ E	6,195	3,387	20,325
Arctic Ocean					
Eurasia Basin	82° 23′ N	19° 31′ E	5,450	2,980	17,881
Mediterranean Sea					
Ionian Basin	36° 32′ N	21° 06′ E	5,150	2,816	16,896

Note: Greater depths have been reported in some areas but are not officially confirmed by research vessels.

Latitude, Longitude, and Altitude of World Cities

Source: National Imagery Mapping Agency, U.S. Dept. of Defense

City	Lat. °	Lat. ′	Long. °	Long. ′	Alt. (ft)
Athens, Greece	37	59 N	23	44 E	300
Bangkok, Thailand	13	45 N	100	31 E	0
Beijing, China	39	56 N	116	24 E	600
Berlin, Germany	52	31 N	13	25 E	110
Bogotá, Colombia	04	36 N	74	05 W	8,660
Bombay (Mumbai), India	18	58 N	72	50 E	27
Buenos Aires, Argentina	34	36 S	58	28 W	0
Cairo, Egypt	30	03 N	31	15 E	381
Jakarta, Indonesia	06	10 S	106	48 E	26
Jerusalem, Israel	31	46 N	35	14 E	2,500
Johannesburg, So. Afr.	26	12 S	28	05 E	5,740
Kathmandu, Nepal	27	43 N	85	19 E	4,500
Kiev, Ukraine	50	26 N	30	31 E	587
London, UK (Greenwich)	51	30 N	00	00	245
Manila, Philippines	14	35 N	121	00 E	0
Mexico City, Mexico	19	24 N	99	09 W	7,347
Moscow, Russia	55	45 N	37	35 E	394
New Delhi, India	28	36 N	77	12 E	770
Panama City, Panama	08	58 N	79	32 W	0
Paris, France	48	52 N	02	20 E	300
Quito, Ecuador	00	13 S	78	30 W	9,222
Rio de Janeiro, Brazil	22	43 S	43	13 W	30
Rome, Italy	41	53 N	12	30 E	95
Santiago, Chile	33	27 S	70	40 W	4,921
Seoul, South Korea	37	34 N	127	00 E	34
Sydney, Australia	33	53 S	151	12 E	25
Tehran, Iran	35	40 N	51	26 E	3,937
Tokyo, Japan	35	42 N	139	46 E	30
Warsaw, Poland	52	15 N	21	00 E	360
Wellington, New Zealand	41	18 S	174	47 E	0

Latitude, Longitude, and Altitude of U.S. and Canadian Cities

Source: U.S. geographic positions, U.S. altitudes provided by Geological Survey, U.S. Dept. of the Interior. Canadian geographic positions and altitudes provided by the Canada Flight Supplement, Natural Resources Canada.

City	Lat. N °	′	″	Long. W °	′	″	Elev. (ft)
Abilene, TX	32	26	55	99	43	58	1,718
Akron, OH	41	4	53	81	31	9	1,050
Albany, NY	42	39	9	73	45	24	20
Albuquerque, NM	35	5	4	106	39	2	4,955
Alert, N.W.T.	82	31	04	62	16	50	100
Allentown, PA	40	36	30	75	29	26	350
Amarillo, TX	35	13	19	101	49	51	3,685
Anchorage, AK	61	13	5	149	54	1	101
Ann Arbor, MI	42	16	15	83	43	35	880
Asheville, NC	35	36	3	82	33	15	2,134
Ashland, KY	38	28	42	82	38	17	558
Atlanta, GA	33	44	56	84	23	17	1,050
Atlantic City, NJ	39	21	51	74	25	24	8
Augusta, GA	33	28	15	81	58	30	414
Augusta, ME	44	18	38	69	46	48	45
Austin, TX	30	16	1	97	44	34	501
Bakersfield, CA	35	22	24	119	1	4	408
Baltimore, MD	39	17	25	76	36	45	100
Bangor, ME	44	48	4	68	46	42	158
Baton Rouge, LA	30	27	2	91	9	16	53
Battle Creek, MI	42	19	16	85	10	47	820
Bay City, MI	43	35	40	83	53	20	595
Beaumont, TX	30	5	9	94	6	6	20
Belleville, Ont.	44	11	32	77	18	34	320
Bellingham, WA	48	45	35	122	29	13	100
Berkeley, CA	37	52	18	122	16	18	150
Billings, MT	45	47	0	108	30	0	3,124
Biloxi, MS	30	23	45	88	53	7	25
Binghamton, NY	42	5	55	75	55	6	865
Birmingham, AL	33	31	14	86	48	9	600
Bismarck, ND	46	48	30	100	47	0	1,700
Bloomington, IL	40	29	3	88	59	37	829
Boise, ID	43	36	49	116	12	9	2,730
Boston, MA	42	21	30	71	3	37	20
Bowling Green, KY	36	59	25	86	26	37	510
Brandon, Man.	49	54	35	99	57	03	1,343
Brantford, Ont.	43	07	53	80	20	33	815
Brattleboro, VT	42	51	3	72	33	30	240
Bridgeport, CT	41	10	1	73	12	19	10
Brockton, MA	42	5	0	71	1	8	112
Buffalo, NY	42	53	11	78	52	43	585
Burlington, Ont.	43	26	33	79	51	03	640
Burlington, VT	44	28	33	73	12	45	113
Butte, MT	46	0	14	112	32	2	5,549
Calgary, Alta.	51	06	50	114	01	13	3,557
Cambridge, MA	42	22	30	71	6	22	30
Canton, OH	40	47	56	81	22	43	1,100
Carson City, NV	39	9	50	119	45	59	4,730
Cedar Rapids, IA	42	0	30	91	38	38	730
Central Islip, NY	40	47	26	73	12	8	88
Champaign, IL	40	6	59	88	14	36	740
Charleston, SC	32	46	35	79	55	52	118
Charleston, WV	38	20	59	81	37	58	606
Charlotte, NC	35	13	37	80	50	36	850
Charlottetown, P.E.I.	46	17	24	63	07	16	160
Chattanooga, TN	35	2	44	85	18	35	685
Cheyenne, WY	41	8	24	104	49	11	6,067
Chicago, IL	41	51	0	87	39	0	596
Churchill, Man.	58	44	14	94	03	26	94
Cincinnati, OH	39	9	43	84	27	25	683
Cleveland, OH	41	29	58	81	41	44	690
Colorado Springs, CO	38	50	2	104	49	15	6,008
Columbia, MO	38	57	6	92	20	2	758
Columbia, SC	34	0	2	81	2	6	314
Columbus, GA	32	27	39	84	59	16	300
Columbus, OH	39	57	40	82	59	56	800
Concord, NH	43	12	29	71	32	17	288
Corpus Christi, TX	27	48	1	97	23	46	35
Dallas, TX	32	47	0	96	48	0	463
Dawson, Yukon	64	02	35	139	07	40	1,214
Dayton, OH	39	45	32	84	11	30	750
Daytona Beach, FL	29	12	38	81	1	23	10
Decatur, IL	39	50	25	88	57	17	670
Denver, CO	39	44	21	104	59	3	5,260
Des Moines, IA	41	36	2	93	36	32	803
Detroit, MI	42	19	53	83	2	45	585
Dodge City, KS	37	45	10	100	1	0	2,550
Dubuque, IA	42	30	2	90	39	52	620
Duluth, MN	46	47	0	92	6	23	610
Durham, NC	35	59	38	78	53	56	394

City	Lat. N °	′	″	Long. W °	′	″	Elev. (ft)
Eau Claire, WI	44	48	41	91	29	54	850
Edmonton, Alta.	53	34	21	113	31	14	2,200
Elizabeth, NJ	40	39	50	74	12	40	38
El Paso, TX	31	45	31	106	29	11	3,695
Enid, OK	36	23	44	97	52	41	1,246
Erie, PA	42	7	45	80	5	7	650
Eugene, OR	44	3	8	123	5	8	419
Eureka, CA	40	48	8	124	9	45	44
Evansville, IN	37	58	29	87	33	21	388
Fairbanks, AK	64	50	16	147	42	59	440
Fall River, MA	41	42	5	71	9	20	200
Fargo, ND	46	52	38	96	47	22	900
Flagstaff, AZ	35	11	53	111	39	2	6,900
Flint, MI	43	0	45	83	41	15	750
Ft. Smith, AR	35	23	9	94	23	54	446
Ft. Wayne, IN	41	7	50	85	7	44	781
Ft. Worth, TX	32	43	31	97	19	14	670
Fredericton, N.B.	45	52	10	66	31	54	67
Fresno, CA	36	44	52	119	46	17	296
Gadsden, AL	34	0	51	86	0	24	554
Gainesville, FL	29	39	5	82	19	30	183
Gallup, NM	35	31	41	108	44	31	6,508
Galveston, TX	29	18	4	94	47	51	10
Gary, IN	41	35	36	87	20	47	600
Grand Junction, CO	39	3	50	108	33	0	4,597
Grand Rapids, MI	42	57	48	85	40	5	610
Great Falls, MT	47	30	1	111	18	0	3,334
Green Bay, WI	44	31	9	88	1	11	594
Greensboro, NC	36	4	21	79	47	32	770
Greenville, SC	34	51	9	82	23	39	966
Guelph, Ont.	43	33	0	80	16	0	1,100
Gulfport, MS	30	22	2	89	5	34	25
Halifax, N.S.	44	52	51	63	30	31	477
Hamilton, OH	39	23	58	84	33	41	600
Hamilton, Ont.	43	10	19	79	55	53	780
Harrisburg, PA	40	16	25	76	53	5	320
Hartford, CT	41	45	49	72	41	8	40
Helena, MT	46	35	34	112	2	7	4,090
Hilo, HI	19	43	47	155	5	24	38
Honolulu, HI	21	18	25	157	51	30	18
Houston, TX	29	45	47	95	21	47	40
Huntsville, AL	34	43	49	86	35	10	641
Indianapolis, IN	39	46	6	86	9	29	717
Iowa City, IA	41	39	40	91	31	48	685
Jackson, MI	42	14	45	84	24	5	940
Jackson, MS	32	17	55	90	11	5	294
Jacksonville, FL	30	19	55	81	39	21	12
Jersey City, NJ	40	43	41	74	4	41	83
Johnstown, PA	40	16	42	76	19	0	521
Joplin, MO	37	5	3	94	30	47	990
Juneau, AK	58	18	7	134	25	11	50
Kalamazoo, MI	42	17	30	85	35	14	755
Kansas City, KS	39	6	51	94	37	38	750
Kansas City, MO	39	5	59	94	34	42	740
Kenosha, WI	42	35	5	87	49	16	610
Key West, FL	24	33	19	81	46	58	8
Kingston, Ont.	44	13	31	76	35	49	305
Kitchener, Ont.	43	27	32	80	23	04	1,040
Knoxville, TN	35	57	38	83	55	15	889
Lafayette, IN	40	25	0	86	52	31	567
Lancaster, PA	40	2	16	76	18	21	368
Lansing, MI	42	43	57	84	33	20	830
Laredo, TX	27	30	22	99	30	26	414
Las Vegas, NV	36	10	30	115	8	11	2,000
Lawrence, MA	42	42	25	71	9	49	50
Lethbridge, Alta.	49	37	49	112	47	59	3,047
Lexington, KY	37	59	19	84	28	40	955
Lihue, HI	21	58	52	159	22	16	206
Lima, OH	40	44	33	84	6	19	875
Lincoln, NE	40	48	0	96	40	0	1,150
Little Rock, AR	34	44	47	92	17	22	350
London, Ont.	42	57	31	81	13	33	875
Los Angeles, CA	34	3	8	118	14	34	330
Louisville, KY	38	15	15	85	45	34	462
Lowell, MA	42	38	0	71	19	0	102
Lubbock, TX	33	34	40	101	51	17	3,195

City	Lat. N °	′	″	Long. W °	′	″	Elev. (ft)
Macon, GA	32	50	26	83	37	57	400
Madison, WI	43	4	23	89	24	4	863
Manchester, NH	42	59	44	71	27	19	175
Marshall, TX	32	32	41	94	22	2	410
Medicine Hat, Alta.	50	01	08	110	43	15	2,352
Memphis, TN	35	8	58	90	2	56	254
Meriden, CT	41	32	17	72	48	27	190
Miami, FL	25	46	26	80	11	38	11
Milwaukee, WI	43	2	20	87	54	23	634
Minneapolis, MN	44	58	48	93	15	49	815
Minot, ND	48	13	57	101	17	45	1,555
Mobile, AL	30	41	39	88	2	35	16
Moncton, N.B.	46	06	44	64	40	57	232
Montgomery, AL	32	22	0	86	18	0	250
Montpelier, VT	44	15	36	72	34	33	525
Montréal, Que.	45	41	06	73	55	52	221
Moose Jaw, Sask.	50	19	48	105	33	29	1,892
Muncie, IN	40	11	36	85	23	11	952
Nashville, TN	36	9	57	86	47	4	440
Natchez, MS	31	33	37	91	24	11	230
Newark, NJ	40	44	8	74	10	22	95
New Britain, CT	41	39	40	72	46	48	200
New Haven, CT	41	18	29	72	55	43	40
New Orleans, LA	29	57	16	90	4	30	11
New York, NY	40	42	51	74	0	23	55
Niagara Falls, Ont.	43	07	0	79	04	0	589
Nome, AK	64	30	4	165	24	23	25
Norfolk, VA	36	50	48	76	17	8	10
North Bay, Ont.	46	26	0	79	28	0	1,200
Oakland, CA	37	48	16	122	16	11	42
Ogden, UT	41	13	23	111	58	23	4,299
Oklahoma City, OK	35	28	3	97	30	58	1,195
Omaha, NE	41	15	31	95	56	15	1,040
Orlando, FL	28	32	17	81	22	46	106
Ottawa, Ont.	45	19	09	76	01	20	382
Paducah, KY	37	5	0	88	36	0	345
Pasadena, CA	34	8	52	118	8	37	865
Paterson, NJ	40	55	0	74	10	20	70
Pensacola, FL	30	25	16	87	13	1	32
Peoria, IL	40	41	37	89	35	20	470
Peterborough, Ont.	44	13	48	78	21	48	628
Philadelphia, PA	39	57	8	75	9	51	40
Phoenix, AZ	33	26	54	112	4	24	1,090
Pierre, SD	44	22	6	100	21	2	1,484
Pittsburgh, PA	40	26	26	79	59	46	770
Pittsfield, MA	42	27	0	73	14	45	1,039
Pocatello, ID	42	52	17	112	26	41	4,464
Pt. Arthur, TX	29	53	55	93	55	43	10
Portland, ME	43	39	41	70	15	21	25
Portland, OR	45	31	25	122	40	30	50
Portsmouth, NH	43	4	18	70	45	47	21
Portsmouth, VA	36	50	7	76	17	55	10
Prince Rupert, B.C.	54	17	10	130	26	41	116
Providence, RI	41	49	26	71	24	48	80
Provo, UT	40	14	2	111	39	28	4,549
Pueblo, CO	38	15	16	104	36	31	4,662
Québec City, Que.	46	47	36	71	23	29	244
Racine, WI	42	43	34	87	46	58	630
Raleigh, NC	35	46	19	78	38	20	350
Rapid City, SD	44	4	50	103	13	50	3,247
Reading, PA	40	20	8	75	55	38	266
Regina, Sask.	50	25	55	104	39	57	1,894
Reno, NV	39	31	47	119	48	46	4,498
Richmond, VA	37	33	13	77	27	38	190
Roanoke, VA	37	16	15	79	56	30	940
Rochester, MN	44	1	18	92	28	11	990
Rochester, NY	43	9	17	77	36	57	515
Rockford, IL	42	16	16	89	5	38	715
Sacramento, CA	38	34	54	121	29	36	20
Saginaw, MI	43	25	10	83	57	3	595
St. Catharines, Ont.	43	11	30	79	10	18	321
St. Cloud, MN	45	33	39	94	9	44	1,040
St. John, N.B.	45	18	58	65	53	25	357
St. John's, Nfld.	47	37	07	52	45	07	461
St. Joseph, MO	39	46	7	94	50	47	850
St. Louis, MO	38	37	38	90	11	52	455
St. Paul, MN	44	56	40	93	5	35	780
St. Petersburg, FL	27	46	14	82	40	46	44

City	Lat. N °	′	″	Long. W °	′	″	Elev. (ft)
Salem, OR	44	56	35	123	2	2	154
Salina, KS	38	50	25	97	36	40	1,225
Salt Lake City, UT	40	45	39	111	53	25	4,266
San Antonio, TX	29	25	26	98	29	36	650
San Bernardino, CA	34	6	30	117	17	20	1,200
San Diego, CA	32	42	55	117	9	23	40
San Francisco, CA	37	46	30	122	25	6	63
San Jose, CA	37	20	22	121	53	38	87
San Juan, P.R.	18	28	6	66	6	22	8
Santa Barbara, CA	34	25	15	119	41	50	50
Santa Cruz, CA	36	58	27	122	1	47	20
Santa Fe, NM	35	41	13	105	56	14	6,989
Sarasota, FL	27	20	10	82	31	51	27
Saskatoon, Sask.	52	10	15	106	41	59	1,653
Sault Ste. Marie, Ont.	46	29	06	84	30	34	630
Savannah, GA	32	5	0	81	6	0	42
Schenectady, NY	42	48	51	73	56	24	245
Seattle, WA	47	36	23	122	19	51	350
Sheboygan, WI	43	45	3	87	42	52	630
Sherbrooke, Que.	45	26	17	71	41	26	792
Sheridan, WY	44	47	50	106	57	20	3,742
Shreveport, LA	32	31	30	93	45	0	209
Sioux City, IA	42	30	0	96	24	0	1,117
Sioux Falls, SD	43	33	0	96	42	0	1,442
South Bend, IN	41	41	0	86	15	0	725
Spartanburg, SC	34	56	58	81	55	56	816
Spokane, WA	47	39	32	117	25	30	2,000
Springfield, IL	39	48	6	89	38	37	610
Springfield, MA	42	6	5	72	35	25	70
Springfield, MO	37	12	55	93	17	53	1,300
Springfield, OH	39	55	27	83	48	32	1,000
Stamford, CT	41	3	12	73	32	21	35
Steubenville, OH	40	22	11	80	38	3	1,060
Stockton, CA	37	57	28	121	17	23	15
Sudbury, Ont.	46	37	30	80	47	56	1,140
Superior, WI	46	43	15	92	6	14	642
Sydney, N.S.	46	09	41	60	02	52	203
Syracuse, NY	43	2	53	76	8	52	400
Tacoma, WA	47	15	11	122	26	35	380
Tallahassee, FL	30	26	17	84	16	51	188
Tampa, FL	27	56	50	82	27	31	48
Terre Haute, IN	39	28	0	87	24	50	501
Texarkana, TX	33	25	30	94	2	51	324
Thunder Bay, Ont.	48	22	19	89	19	26	653
Timmins, Ont	48	34	11	81	22	36	967
Toledo, OH	41	39	50	83	33	19	615
Topeka, KS	39	2	54	95	40	40	1,000
Toronto, Ont.	43	37	39	79	23	46	251
Trenton, NJ	40	13	1	74	44	36	54
Trois-Rivières, Que.	46	21	10	72	40	46	198
Troy, NY	42	43	42	73	41	32	35
Tucson, AZ	32	13	18	110	55	33	2,390
Tulsa, OK	36	9	14	95	59	33	804
Urbana, IL	40	6	38	88	12	26	725
Utica, NY	43	6	3	75	13	59	415
Vancouver, B.C.	49	11	42	123	10	55	14
Victoria, B.C.	48	38	49	123	25	33	63
Waco, TX	31	32	57	97	8	47	405
Walla Walla, WA	46	3	53	118	20	31	1,000
Washington, DC	38	53	42	77	2	12	25
Waterloo, IA	42	29	34	92	20	34	850
West Palm Beach, FL	26	42	54	80	3	13	21
Wheeling, WV	40	3	50	80	43	16	672
Whitehorse, Yukon	60	42	36	135	04	06	2,305
White Plains, NY	41	2	2	73	45	48	220
Wichita, KS	37	41	32	97	20	14	1,305
Wilkes-Barre, PA	41	14	45	75	52	54	550
Wilmington, DE	39	44	45	75	32	49	100
Wilmington, NC	34	13	32	77	56	42	50
Windsor, Ont.	42	16	29	82	57	30	622
Winnipeg, Man.	49	54	39	97	14	36	783
Winston-Salem, NC	36	5	59	80	14	40	912
Worcester, MA	42	15	45	71	48	10	480
Yakima, WA	46	36	8	120	30	17	1,066
Yellowknife, N.W.T.	62	27	46	114	26	25	675
Youngstown, OH	41	5	59	80	38	59	861
Yuma, AZ	32	43	31	114	37	25	160
Zanesville, OH	39	56	25	82	0	48	710

Principal World Rivers

Reviewed by Laurel Duda, Marine Biological Laboratory, Woods Hole Oceanogr. Inst. Library. For N American rivers, see separate table.

River	Outflow	Length (mi)
Africa		
Chari	Lake Chad	500
Congo	Atlantic Ocean	2,900
Gambia	Atlantic Ocean	700
Kasai	Congo River	1,000
Limpopo	Indian Ocean	1,100
Lualaba	Congo River	1,100
Niger	Gulf of Guinea	2,590
Nile	Mediterranean	4,160
Okavango	Okavango Delta	1,000
Orange	Atlantic Ocean	1,300
Senegal	Atlantic Ocean	1,020
Ubangi	Congo River	660
Zambezi	Indian Ocean	1,700
Asia		
Amu Darya	Aral Sea	1,550
Amur	Tatar Strait	1,780
Angara	Yenisey River	1,151
Brahmaputra	Bay of Bengal	1,800
Chang	East China Sea	3,964
Euphrates	Shatt al-Arab	1,700
Ganges	Bay of Bengal	1,560
Godavari	Bay of Bengal	900
Hsi (see Xi)		
Huang	Yellow Sea	3,395
Indus	Arabian Sea	1,800
Irrawaddy	Andaman Sea	1,337
Jordan	Dead Sea	200
Kolyma	Arctic Ocean	1,323
Krishna	Bay of Bengal	800
Kura	Caspian Sea	848
Lena	Laptev Sea	2,734
Mekong	South China Sea	2,700
Narbada (see Narmada)		
Narmada	Arabian Sea	800
Ob	Gulf of Ob	2,268
Ob-Irtysh	Gulf of Ob	3,362
Salween	Gulf of Martaban	1,500
Songhua	Amur River	1,150
Sungari	Amur River	1,197
Sutlej	Indus River	900
Syr	Aral Sea	1,370
Tarim	Lop Nor Basin	1,261
Tigris	Shatt al-Arab	1,180
Xi	South China Sea	1,200
Yamuna	Ganges River	855
Yangtze (see Chang)		
Yellow (see Huang)		
Yenisey	Kara Seav	2,543
Australia		
Murray-Darling	Indian Ocean	2,310
Murrumbidgee	Murray River	981
Europe		
Bug, Northern	Wisla	481
Bug, Southern	Dnieper River	532
Danube	Black Sea	1,776
Don	Sea of Azov	1,224
Dnieper	Black Sea	1,420
Dniester	Black Sea	877
Drava	Danube River	447
Dvina, North	White Sea	824
Dvina, West	Gulf of Riga	634
Ebro	Mediterranean	565
Elbe	North Sea	724
Garonne	Bay of Biscay	357
Kama	Volga River	1,122
Loire	Bay of Biscay	634
Mame	Seine River	326
Meuse	North Sea	580
Oder	Baltic Sea	567
Oka	Volga River	932
Pechora	Barents Sea	1,124
Po	Adriatic Sea	405
Rhine	North Sea	820
Rhone	Gulf of Lions	505
Seine	English Channel	496
Shannon	Atlantic Ocean	230
Tagus	Atlantic Ocean	626
Thames	North Sea	210
Tiber	Tyrrhenian Sea	252
Tisza	Danube River	600
Ural	Caspian Sea	1,575
Volga	Caspian Sea	2,290
Weser	North Sea	454
Wisla	Gulf of Gdansk	675
South America		
Amazon	Atlantic Ocean	4,000
Araguaia	Tocantins River	1,100
Iça (see Putumayo)		
Iguaça	Parana River	808
Japura	Amazon River	1,750
Madeira	Amazon River	2,013
Magdalena	Caribbean Sea	956
Negro	Amazon River	1,400
Orinoco	Atlantic Ocean	1,600
Paraguay	Parana River	1,584
Parana	Rio de la Plata	2,485
Pilcomayo	Paraguay River	1,000
Purus	Amazon River	2,100
Putumayo	Amazon River	1,000
Rio de la Plata	Atlantic Ocean	150
Rio Roosevelt	Aripuana	400
Sao Francisco	Atlantic Ocean	1,988
Tocantins	Para River	1,677
Ucayali	Marañón River	910
Uruguay	Rio de la Plata	1,000
Xingu	Amazon River	1,300

IT'S A FACT: The Rio Grande River, which forms the border between Texas and Mexico, dried up 500 yards short of its mouth in the Gulf of Mexico in the spring of 2001. Increased water use in the U.S. and Mexico and an eight-year drought were to blame.

Major Rivers in North America

Reviewed by Laurel Duda, Marine Biological Laboratory, Woods Hole Oceanographic Inst. Library

River	Source or upper limit of length	Outflow	Length (mi)
Alabama	Gilmer County, GA	Mobile River	729
Albany	Lake St. Joseph, Ontario	James Bay	610
Allegheny	Potter County, PA	Ohio River	325
Altamaha-Ocrnulgee	Junction of Yellow and South Rivers, Newton County, GA	Atlantic Ocean	392
Apalachicola-Chattahoochee	Towns County, GA	Gulf of Mexico	524
Arkansas	Lake County, CO	Mississippi River	1,459
Assiniboine	Eastern Saskatchewan	Red River	450
Attawapiskat	Attawapiskat, Ontario	James Bay	465
Back (NWT)	Contwoyto Lake	Chantrey Inlet, Arctic Ocean	605
Big Black (MS)	Webster County, MS	Mississippi River	330
Brazos	Junction of Salt and Double Mountain Forks, Stonewall County, TX	Gulf of Mexico	950
Canadian	Las Animas County, CO	Arkansas River	906
Cedar (IA)	Dodge County, MN	Iowa River	329
Cheyenne	Junction of Antelope Creek and Dry Fork, Converse County, WY	Missouri River	290
Churchill, Man.	Methy Lake, Saskatchewan	Hudson Bay	1,000
Cimarron	Colfax County, NM	Arkansas River	600
Colorado (AZ)	Rocky Mountain Natl. Park, CO (90 mi in Mexico)	Gulf of California	1,450
Colorado (TX)	West Texas	Matagorda Bay	862
Columbia	Columbia Lake, British Columbia	Pacific Ocean, bet. OR and WA	1,243
Columbia, Upper	Columbia Lake, British Columbia	To mouth of Snake River	890
Connecticut	Third Connecticut Lake, NH	Long Island Sound, CT	407
Coppermine (NWT)	Lac de Gras	Coronation Gulf, Arctic Ocean	525
Cumberland	Letcher County, KY	Ohio River	720
Delaware	Schoharie County, NY	Liston Point, Delaware Bay	390
Fraser	Near Mount Robson (on Continental Divide)	Strait of Georgia	850
Gila	Catron County, NM	Colorado River	649
Green (UT-WY)	Junction of Wells and Trail Creeks, Sublette County, WY	Colorado River	730
Hamilton (Lab.)	Lake Ashuanipi	Atlantic Ocean	532
Hudson	Henderson Lake, Essex County, NY	Upper NY Bay	306
Illinois	St. Joseph County, IN	Mississippi River	420
James (ND-SD)	Wells County, ND	Missouri River	710
James (VA)	Junction of Jackson and Cowpasture Rivers, Botetourt County, VA	Hampton Roads	340
Kanawha-New	Junction of North and South Forks of New River, NC	Ohio River	352
Kentucky	Junction of North and Middle Forks, Lee County, KY	Ohio River	259

River	Source or upper limit of length	Outflow	Length (mi)
Klamath	Lake Ewauna, Klamath Falls, OR	Pacific Ocean	250
Kootenay	Kootenay Lake, British Columbia	Columbia River	485
Koyukuk	Endicott Mountains, AK	Yukon River	470
Kuskokwim	Alaska Range	Kuskokwim Bay	724
Liard	Southern Yukon, AK	Mackenzie River	693
Little Missouri	Crook County, WY	Missouri River	560
Mackenzie	Great Slave Lake, N.W.T.	Arctic Ocean	1,060
Milk	Junction of North and South Forks, Alberta	Missouri River	625
Minnesota	Big Stone Lake, MN	Mississippi River	332
Mississippi	Lake Itasca, MN	Gulf of Mexico	2,340
Mississippi-Missouri-Red Rock	Source of Red Rock, Beaverhead Co., MT	Gulf of Mexico	3,710
Missouri	Junction of Jefferson, Madison, and Gallatin Rivers, Gallatin County, MT	Mississippi River	2,315
Missouri-Red Rock	Source of Red Rock, Beaverhead Co., MT	Mississippi River	2,540
Mobile-Alabama-Coosa	Gilmer County, GA	Mobile Bay	774
Nelson (Man.)	Lake Winnipeg	Hudson Bay	410
Neosho	Morris County, KS	Arkansas River, OK	460
Niobrara	Niobrara County, WY	Missouri River, NE	431
North Canadian	Union County, NM	Canadian River, OK	800
North Platte	Junction of Grizzly and Little Grizzly Creeks, Jackson County, CO	Platte River, NE	618
Ohio	Junction of Allegheny and Monongahela Rivers, Pittsburgh, PA	Mississippi River	981
Ohio-Allegheny	Potter County, PA	Mississippi River	1,310
Osage	East-central Kansas	Missouri River	500
Ottawa	Lake Capimitchigama	St. Lawrence River	790
Ouachita	Polk County, AR	Black River	605
Peace	Stikine Mountains, B.C.	Slave River	1,210
Pearl	Neshoba County, MS	Gulf of Mexico	411
Pecos	Mora County, NM	Rio Grande	926
Pee Dee-Yadkin	Watauga County, NC	Winyah Bay	435
Pend Oreille-Clark Fork	Near Butte, MT	Columbia River	531
Platte	Junction of North and South Platte Rivers, NE	Missouri River	310
Porcupine	Ogilvie Mountains, AK	Yukon River, AK	569
Potomac	Garrett County, MD	Chesapeake Bay	383
Powder	Junction of South and Middle Forks, WY	Yellowstone River	375
Red (OK-TX-LA)	Curry County, NM	Mississippi River	1,290
Red River of the North	Junction of Otter Tail and Bois de Sioux Rivers, Wilkin County, MN	Lake Winnipeg	545
Republican	Junction of North Fork and Arikaree River, NE	Kansas River	445
Rio Grande	San Juan County, CO	Gulf of Mexico	1,900
Roanoke	Junction of N and S Forks, Montgomery Co., VA	Albemarle Sound	380
Rock (IL-WI)	Dodge County, WI	Mississippi River	300
Sabine	Junction of S and Caddo Forks, Hunt County, TX	Sabine Lake	380
Sacramento	Siskiyou County, CA	Suisun Bay	377
St. Francis	Iron County, MO	Mississippi River	425
St. John	Northwestern Maine	Bay of Fundy	418
St. Lawrence	Lake Ontario	Gulf of St. Lawrence, Atlantic Ocean	800
Saguenay	Lake St. John, Quebec	St. Lawrence River	434
Salmon (ID)	Custer County, ID	Snake River	420
San Joaquin	Junction of S and Middle Forks, Madera Co., CA	Suisun Bay	350
San Juan	Silver Lake, Archuleta County, CO	Colorado River	360
Santee-Wateree-Catawba	McDowell County, NC	Atlantic Ocean	538
Saskatchewan, North	Rocky Mountains	Saskatchewan R.	800
Saskatchewan, South	Rocky Mountains	Saskatchewan R.	865
Savannah	Junction of Seneca andTugaloo Rivers, Anderson County, SC	Atlantic Ocean, GA-SC	314
Severn (Ont.)	Sandy Lake	Hudson Bay	610
Smoky Hill	Cheyenne County, CO	Kansas River, KS	540
Snake	Teton County, WY	Columbia River, WA	1,038
South Platte	Junction of S and Middle Forks, Park County, CO	Platte River	424
Susitna	Alaska Range	Cook Inlet	313
Susquehanna	Huyden Creek, Otsego County, NY	Chesapeake Bay	447
Tallahatchie	Tippah County, MS	Yazoo River	301
Tanana	Wrangell Mountains, AK	Yukon River	659
Tennessee	Junction of French Broad and Holston Rivers	Ohio River	652
Tennessee-French Broad	Courthouse Creek, Transylvania County, NC	Ohio River	886
Tombigbee	Prentiss County, MS	Mobile River	525
Trinity	North of Dallas, TX	Galveston Bay	360
Wabash	Darke County, OH	Ohio River	512
Washita	Hemphill County, TX	Red River, OK	500
White (AR-MO)	Madison County, AR	Mississippi River	722
Willamette	Douglas County, OR	Columbia River	309
Wind-Bighorn	Junction of Wind and Little Wind Rivers, Fremont Co., WY (Source of Wind R. isTogwotee Pass,Teton Co., WY)	Yellowstone River	338
Wisconsin	Lac Vieux Desert, Vilas County, WI	Mississippi River	430
Yellowstone	Park County, WY	Missouri River	682
Yukon	McNeil R., Yukon Territory	Bering Sea	1,979

Highest and Lowest Continental Altitudes

Source: National Geographic Society

Continent	Highest point	Elev. (ft)	Lowest point	ft below sea level
Asia	Mount Everest, Nepal-Tibet	29,035	Dead Sea, Israel-Jordan	1,348
South America	Mount Aconcagua, Argentina	22,834	Valdes Peninsula, Argentina	131
North America	Mount McKinley, AK	20,320	Death Valley, California	282
Africa	Kilimanjaro, Tanzania	19,340	Lake Assal, Djibouti	512
Europe	Mount Elbrus, Russia	18,510	Caspian Sea, Russia, Azerbaijan	92
Antarctica	Vinson Massif	16,864	Bentley Subglacial Trench	8,327[1]
Australia	Mount Kosciusko, New South Wales	7,310	Lake Eyre, South Australia	52

(1) Estimated level of the continental floor. Lower points that have yet to be discovered may exist further beneath the ice.

➤ ***IT'S A FACT:*** Russia's Lake Baykal contains about 20% of the Earth's surface fresh water—as much as the Great Lakes combined—and is home to about 100,000 nerpa (*Phoca sibirica*), the world's only freshwater seal.

Major Natural Lakes of the World

Source: Geological Survey, U.S. Dept. of the Interior

A lake is generally defined as a body of water surrounded by land. By this definition some bodies of water that are called seas, such as the Caspian Sea and the Aral Sea, are really lakes. In the following table, the word *lake* is omitted when it is part of the name.

Name	Continent	Area (sq mi)	Length (mi)	Maximum depth (ft)	Elevation (ft)
Caspian Sea[1]	Asia-Europe	143,244	760	3,363	–92
Superior	North America	31,700	350	1,330	600
Victoria	Africa	26,828	250	270	3,720
Huron	North America	23,000	206	750	579
Michigan	North America	22,300	307	923	579
Aral Sea[1]	Asia	13,000[2]	260	220	125
Tanganyika	Africa	12,700	420	4,823	2,534
Baykal	Asia	12,162	395	5,315	1,493
Great Bear	North America	12,096	192	1,463	512
Nyasa (Malawi)	Africa	11,150	360	2,280	1,550
Great Slave	North America	11,031	298	2,015	513
Erie	North America	9,910	241	210	570
Winnipeg	North America	9,417	266	60	713
Ontario	North America	7,340	193	802	245
Balkhash[1]	Asia	7,115	376	85	1,115
Ladoga	Europe	6,835	124	738	13
Maracaibo	South America	5,217	133	115	sea level
Onega	Europe	3,710	145	328	108
Eyre[1]	Australia	3,600[3]	90	4	–52
Titicaca	South America	3,200	122	922	12,500
Nicaragua	North America	3,100	102	230	102
Athabasca	North America	3,064	208	407	700
Reindeer	North America	2,568	143	720	1,106
Tonle Sap	Asia	2,500[3]	...	45	...
Turkana (Rudolf)	Africa	2,473	154	240	1,230
Issyk Kul[1]	Asia	2,355	115	2,303	5,279
Torrens[1]	Australia	2,230[3]	130	...	92
Vanern	Europe	2,156	91	328	144
Nettilling	North America	2,140	67	...	95
Winnipegosis	North America	2,075	141	38	830
Albert	Africa	2,075	100	168	2,030
Nipigon	North America	1,872	72	540	1,050
Gairdner[1]	Australia	1,840[3]	90	...	112
Urmia[1]	Asia	1,815	90	49	4,180
Manitoba	North America	1,799	140	12	813
Chad	Africa	839[4]	175	24	787

(1) Salt lake. (2) Approximate figure, could be less. The diversion of feeder rivers since the 1960s has devastated the Aral—once the world's 4th largest lake (26,000 sq. miles). By 2000, the Aral had effectively become three lakes, with the total area shown. (3) Approximate figure, subject to great seasonal variation. (4) Once 4th largest lake in Africa (about 10,000 sq. miles in the 1960s), Chad had shrunk more than 90% by 2001 due to irrigation and long-term drought.

The Great Lakes

Source: National Ocean Service, U.S. Dept. of Commerce

The Great Lakes form the world's largest body of fresh water (in surface area), and with their connecting waterways are the largest inland water transportation unit. Draining the great North Central basin of the U.S., they enable shipping to reach the Atlantic via their outlet, the St. Lawrence R., and to reach the Gulf of Mexico via the Illinois Waterway, from Lake Michigan to the Mississippi R. A 3d outlet connects with the Hudson R. and then the Atlantic via the New York State Barge Canal System. Traffic on the Illinois Waterway and the N.Y. State Barge Canal System is limited to recreational boating and small shipping vessels.

Only one of the lakes, Lake Michigan, is wholly in the U.S.; the others are shared with Canada. Ships move from the shores of Lake Superior to Whitefish Bay at the E end of the lake, then through the Soo (Sault Ste. Marie) locks, through the St. Mary's R. and into Lake Huron. To reach Gary and the Port of Indiana and South Chicago, IL, ships move W from Lake Huron to Lake Michigan through the Straits of Mackinac. Lake Superior is 601 ft above low water datum at Rimouski, Quebec, on the International Great Lakes Datum (1985). From Duluth, MN, to the E end of Lake Ontario is 1,156 mi.

	Superior	Michigan	Huron	Erie	Ontario
Length in mi	350	307	206	241	193
Breadth in mi	160	118	183	57	53
Deepest soundings in ft	1,333	923	750	210	802
Volume of water in cu mi	2,935	1,180	850	116	393
Area (sq mi) water surface—U.S.	20,600	22,300	9,100	4,980	3,460
Canada	11,100	.....	13,900	4,930	3,880
Area (sq mi) entire drainage basin—U.S.	16,900	45,600	16,200	18,000	15,200
Canada	32,400	.....	35,500	4,720	12,100
TOTAL AREA (sq mi) U.S. and Canada	**81,000**	**67,900**	**74,700**	**32,630**	**34,850**
Low water datum above mean water level at Rimouski, Quebec, avg. level in ft (1985)	601.10	577.50	577.50	569.20	243.30
Latitude, N	46° 25′	41° 37′	43° 00′	41° 23′	43° 11′
	49° 00′	46° 06′	46° 17′	42° 52′	44° 15′
Longitude, W	84° 22′	84° 45′	79° 43′	78° 51′	76° 03′
	92° 06′	88° 02′	84° 45′	83° 29′	79° 53′
National boundary line in mi	282.8	None	260.8	251.5	174.6
United States shoreline (mainland only) mi	863	1,400	580	431	300

Famous Waterfalls

Source: National Geographic Society

The earth has thousands of waterfalls, some of considerable magnitude. Their relative importance is determined not only by height but also by volume of flow, steadiness of flow, crest width, whether the water drops sheerly or over a sloping surface, and whether it descends in one leap or in a succession of leaps. A series of low falls flowing over a considerable distance is known as a **cascade**.

Estimated mean annual flow, in cubic feet per second, of major waterfalls are as follows: Niagara, 212,200; Paulo Afonso, 100,000; Urubupunga, 97,000; Iguazu, 61,000; Patos-Maribondo, 53,000; Victoria, 35,400; and Kaieteur, 23,400.

Height = total drop in feet in one or more leaps. #=falls of more than one leap; *= falls that diminish greatly seasonally; **= falls that reduce to a trickle or are dry for part of each year. If the river names are not shown, they are same as the falls.
R. = river; (C) = cascade type.

Name and location	Height (ft)
Africa	
Angola	
Ruacana, Cuene R.	406
Ethiopia	
Fincha	508
Lesotho	
Maletsunyane*	630
Zimbabwe-Zambia	
Victoria, Zambezi R.*	343
South Africa	
Augrabies, Orange R.*	480
Tugela#	2,014
Tanzania-Zambia	
Kalambo*	726
Asia	
India	
Cauvery*	330
Jog (Gersoppa),Sharavathi R.*	830
Japan	
Kegon, Daiya R.*	330
Australia	
New South Wales	
Wentworth	614
Wollomombi	1,100
Queensland	
Tully	885
Wallaman, Stony Cr.#	1,137
New Zealand	
Helena	890
Sutherland, Arthur R.#	1,904
Europe	
Austria	
Gastein#	492
Gavarnie*	1,385
Great Britain	
Scotland	
Glomach	370
Wales	
Rhaiadr	240
Italy	
Frua, Toce R. (C)	470
Norway	
Mardalsfossen (Northern)	1,535
Mardalsfossen (Southern)#	2,149
Skjeggedal, Nybuai R.#**	1,378
Skykje**	984
Vetti, Morka-Koldedola R.	900
Sweden	
Handol#	427
Switzerland	
Giessbach (C)	984
Reichenbach#	656
Simmen#	459
Staubbach	984
Trummelbach#	1,312
North America	
Canada	
Alberta	
Panther, Nigel Cr.	600
British Columbia	
Della#	1,443
Takakkaw, Daly Glacier#	1,200
Quebec	
Montmorency	274
Canada—United States	
Niagara: American	182
Horseshoe	173
United States	
California	
Feather, Fall R.*	640
Yosemite National Park	
Bridalveil*	620
Illilouette*	370
Nevada, Merced R.*	594
Ribbon**	1,612
Silver Strand, Meadow Br.**	1,170
Vernal, Merced R. *	317
Yosemite#**	2,425
Colorado	
Seven, South Cheyenne Cr.#	300
Hawaii	
Akaka, Kolekole Str.	442
Idaho	
Shoshone, Snake R.**	212
Kentucky	
Cumberland	68
Maryland	
Great, Potomac R. (C) *	71
Minnesota	
Minnehaha**	53
New Jersey	
Passaic	70
New York	
Taughannock*	215
Oregon	
Multnomah#	620
Tennessee	
Fall Creek	256
Washington	
Mt. Rainier Natl. Park	
Sluiskin, Paradise R.	300
Snoqualmie**	268
Wisconsin	
Big Manitou, Black R. (C)*	165
Wyoming	
Yellowstone Natl. Pk. Tower	132
Yellowstone (upper)*	109
Yellowstone (lower)*	308
Mexico	
El Salo	218
South America	
Argentina-Brazil	
Iguazu	230
Brazil	
Glass	1,325
Patos-Maribondo, Grande R.	115
Paulo Afonso, Sao Francisco R.	275
Urubupunga, Parana R.	39
Colombia	
Catarata de Candelas, Cusiana R.	984
Tequendama, Bogota R.*	427
Ecuador	
Agoyan, Pastaza R.*	200
Guyana	
Kaieteur, Potaro R.	741
Great, Kamarang R.	1,600
Marina, Ipobe R.#	500
Venezuela	
Angel#*	3,212
Cuquenan	2,000

Notable Deserts of the World

Arabian (Eastern), 70,000 sq mi in Egypt between the Nile R. and Red Sea, extending southward into Sudan
Atacama, 600-mi-long area rich in nitrate and copper deposits in N Chile
Chihuahuan, 140,000 sq mi in TX, NM, AZ, and Mexico
Dasht-e Kauir, approx. 300 mi long by approx. 100 mi wide in N central Iran
Dasht-e Lut, 20,000 sq mi in E Iran
Death Valley, 3,300 sq mi in CA and NV
Gibson, 120,000 sq mi in the interior of W Australia
Gobi, 500,000 sq mi in Mongolia and China
Great Sandy, 150,000 sq mi in W Australia
Great Victoria, 150,000 sq mi in SW Australia
Kalahari, 225,000 sq mi in S Africa
Kara Kum, 120,000 sq mi in Turkmenistan
Kyzyl Kum, 100,000 sq mi in Kazakhstan and Uzbekistan
Libyan, 450,000 sq mi in the Sahara, extending from Libya through SW Egypt into Sudan
Mojave, 15,000 sq mi in southern CA
Namib, long narrow area (varies from 30-100 mi wide) extending 800 mi along SW coast of Africa
Nubian, 100,000 sq mi in the Sahara in NE Sudan
Patagonia, 300,000 sq mi in S Argentina
Painted Desert, section of high plateau in northern AZ extending 150 mi
Rub al-Khali (Empty Quarter), 250,000 sq mi in the S Arabian Peninsula
Sahara, 3,500,000 sq mi in N Africa, extending westward to the Atlantic. Largest desert in the world
Sonoran, 70,000 sq mi in southwestern AZ and southeastern CA extending into NW Mexico
Syrian, 100,000-sq-mi arid wasteland extending over much of N Saudi Arabia, E Jordan, S Syria, and W Iraq
Taklimakan, 140,000 sq mi in Xinjiang Prov., China
Thar (Great Indian), 100,000-sq-mi arid area extending 400 mi along India-Pakistan border

WORLD HISTORY

Chronology of World History

Prehistory: Our Ancestors Emerge

Revised by Susan Skomal, Ph.D., American Anthropological Association

Homo sapiens. The precise origins of *Homo sapiens,* the species to which all humans belong, are subject to broad speculation based on a small, but increasing, number of fossils, on genetic and anatomical studies, and on interpretation of the geological record. Most scientists at least agree that humans evolved from apelike primate ancestors in a process that began millions of years ago.

Current theories trace the first hominid (humanlike primate) to Africa, where at least 2 lines of hominids appeared 5 to 7 million years before the present (BP). In one line was *Australopithecus*, a social animal that lived from perhaps 5 million to 3 million years BP, then apparently died out. In the other, human line was *Homo habilis*, a large-brained specimen that walked upright and had a dextrous hand. *Homo habilis* appeared some 2.5 million years BP, lived in semipermanent camps, had a food-gathering economy, and probably produced stone tools.

Homo erectus, the nearest ancestor to humans, appeared in Africa perhaps 2 million years BP and began spreading into Asia and Europe soon after. It had a fairly large brain and a skeletal structure similar to that of modern humans. *Homo erectus* hunted, learned to control fire, and may have had some primitive language skills. Brain development to *Homo sapiens,* then to the subspecies *Homo sapiens sapiens,* occurred between 500,000 and 50,000 years BP in Africa. All modern humans are members of the subspecies *Homo sapiens sapiens.*

Humans have roamed widely over the globe throughout their development. There is increasing evidence that migration from Asia to Australia via the Timor Straits took place as early as 100,000 BP. Evidence of hominids in Siberia dates as early as 300,000 BP. First confirmation for the crossing from Asia to the Americas, by land bridge, dates to the end of the last Ice Age, at 12,500 BP. A growing body of evidence in N and S America, however, suggests that humans sailed even earlier from Asia to the New World, either along coastal routes or directly across the Pacific.

Earliest cultures. A variety of cultural modes—in toolmaking, diet, shelter, and possibly social arrangements and spiritual expression—arose as humans adapted to different geographic and climatic zones and the database of knowledge grew. Sites from all over the world show seasonal migration patterns and efficient exploitation of a wide range of plant and animal foods.

Archaeologists recognize 5 basic toolmaking traditions as arising and often coexisting from more than 2.5 million years ago to the near past: (1) the *chopper tradition*—also known as the Oldowan—found in Africa, producing crude chopping tools and simple flake tools; (2) the *biface* or hand-ax tradition, found in Africa, W and S Europe, and S Asia, producing pointed hand axes chipped on both faces for cutting; (3) the *flake tradition*, found in Africa and Europe, producing small cutting and flaking tools; (4) the *blade tradition*, a more efficient technology characteristic of the Upper Paleolithic, found across Eurasia to Siberia and N Africa, producing many usable blades from a single stone; and (5) the *microlith tradition*, found throughout the inhabited world, producing specialized small tools for use as projectile points, in carving softer materials, and in making more complex tools.

Sketchy evidence remains for the stages in increasing control over the environment. Fire was used for heating and cooking by 465,000 BP in W France. Fire-hardened wooden spears, weighted and set with small stone blades, were fashioned by big-game hunters 400,000 years ago in Germany. Scraping tools found at certain sites (200,000-30,000 BP in Europe, N Africa, the Middle East, and Cent. Asia) suggest the treatment of skins for clothing. By the time Australia was settled, human ancestors had learned to navigate in boats over open water. The earliest bone tools found to date were developed 80,000 years ago in the Congo basin by fishermen, who created sophisticated fishing tackle to catch giant catfish.

Early human ancestors included artists and musicians. About 60,000 years ago the earliest immigrants to Australia carved and painted abstract designs on rocks. Painting and decoration flourished, along with stone and ivory sculpture, from 30,000 BP in Europe; more than 200 caves, mainly in S France and N Spain, show remarkable examples of naturalistic wall painting. Other examples have been found in Africa. Proto-religious rites are suggested by these works, and by evidence of ritual burial. A variety of musical instruments, including bone flutes with precisely bored holes, have been found in Paleolithic (early Stone Age) sites going back as far as 40,000-80,000 years BP.

Neolithic advances. Some time after 10,000 BC, among widely separated communities, a series of dramatic technological and social changes occurred, marking the Neolithic, or New Stone, Age. As the world climate became drier and warmer, humans learned to cultivate plants. This in turn encouraged growth of permanent settlements. Animals were domesticated. Manufacture of pottery and cloth began. These techniques permitted a dramatic increase in world population and social complexity, and accelerated humankind's ability to manipulate the environment.

Sites in N, Cent., and S America, SE Europe, and the Middle East show roughly contemporaneous (10,000-8000 BC) evidence of one or more Neolithic traits. Dates near 6000-3000 BC have been given for E and S Asian, W European, and sub-Saharan African Neolithic remains. The variety of crops—field grains, rice, maize, and roots—and varying mix of other characteristics suggest that this adaptation occurred independently in all these regions.

History Begins: 4000-1000 BC

Near Eastern cradle. If history began with writing, the first chapter opened in Mesopotamia, the Tigris-Euphrates river valley. The Sumerians used clay tablets with pictographs to keep records after 4000 BC. A **cuneiform** (wedge-shaped) script evolved by 3000 BC as a full syllabic alphabet. Neighboring peoples adapted the script for their own use.

Sumerian life centered, from 4000 BC, on large cities (Eridu, Ur, Uruk, Nippur, Kish, and Lagash) organized around temples and priestly bureaucracies, with surrounding plains watered by vast irrigation works and worked with traction plows. Sailboats, wheeled vehicles, potter's wheels, and kilns were used. Copper was smelted and tempered from c 4000 BC; bronze was produced not long after. Ores, as well as precious stones and metals, were obtained through long-distance ship and caravan trade. Iron was used from c 2000 BC. Improved ironworking, developed partly by the Hittites, became widespread by 1200 BC.

Sumerian political primacy passed among cities and their kingly dynasties. Semitic-speaking peoples, with cultures derived from the Sumerian, founded a succession of dynasties that ruled in Mesopotamia and neighboring areas for most of 1,800 years; among them were the **Akkadians** (first under Sargon I, c 2350 BC), the Amorites (whose laws, codified by **Hammurabi,** c 1792-1750 BC, have biblical parallels), and the Assyrians, with interludes of rule by the Hittites, Kassites, and Mitanni.

Mesopotamian learning, maintained by scribes and preserved in vast libraries, was practically oriented. Advances in mathematics related mostly to construction, commerce, and administration. Lists of astronomical phenomena, plants, animals, and stones were maintained; medical texts listed ailments and herbal cures. The Sumerians worshiped anthropomorphic gods representing natural forces. Sacrifices were made at **ziggurats**—huge stepped temples.

The Syria-Palestine area, site of some of the earliest urban remains (Jericho, 7000 BC), and of the recently uncovered **Ebla** civilization (fl 2500 BC), experienced Egyptian cultural and political influence along with Mesopotamian. The

Phoenician coast was an active commercial center. A phonetic alphabet was invented here before 1600 BC. It became the ancestor of many other alphabets.

Egypt. Agricultural villages along the Nile River were united by around 3300 BC into 2 kingdoms, Upper and Lower Egypt, which were unified (c 3100 BC) under the pharaoh Menes. A bureaucracy supervised construction of canals and monuments (**pyramids** starting 2700 BC). Control over Nubia to the S was asserted from 2600 BC. Brilliant Old Kingdom Period achievements in architecture, sculpture, and painting, which reached their height during the 3d and 4th Dynasties, set the standards for subsequent Egyptian civilization. **Hieroglyphic writing** appeared by 3200 BC, recording a sophisticated literature that included religious writings, philosophies, history, and science. An ordered hierarchy of gods, including totemistic animal elements, was served by a powerful priesthood in Memphis. The pharaoh was identified with the falcon god Horus. Other trends included belief in an afterlife and short-lived quasi-monotheistic reforms introduced by the pharaoh **Akhenaton** (c 1379-1362 BC).

After a period of dominance by Semitic Hyksos from Asia (c 1700-1550 BC), the New Kingdom established an empire in Syria. Egypt became increasingly embroiled in Asiatic wars and diplomacy. Conquered by Persia in 525 BC, it eventually faded away as an independent culture.

India. An urban civilization with a so-far-undeciphered writing system stretched across the Indus Valley and along the Arabian Sea c 3000-1500 BC. Major sites are Harappa and **Mohenjo-Daro** in Pakistan, well-planned geometric cities with underground sewers and vast granaries. The entire region may have been ruled as a single state. Bronze was used, and arts and crafts were well developed. Religious life apparently took the form of fertility cults. Indus civilization was probably in decline when it was destroyed by **Aryans who arrived** from the NW, speaking an Indo-European language from which most languages of Pakistan, N India, and Bangladesh descend. Led by a warrior aristocracy whose legendary deeds are in the **Rig Veda**, the Aryans spread E and S, bringing their sky gods, priestly (Brahman) ritual, and the beginnings of the caste system; local customs and beliefs were assimilated by the conquerors.

Europe. On Crete, the Bronze Age **Minoan civilization** emerged c 2500 BC. A prosperous economy and richly decorative art was supported by seaborne commerce. Mycenae and other cities in mainland Greece and Asia Minor (e.g., **Troy**) preserved elements of the culture until c 1200 BC. Cretan Linear A script (c 2000-1700 BC) remains undeciphered; Linear B script (c 1300-1200 BC) records an early Greek dialect. Unclear is the possible connection between Mycenaean monumental stonework and the megalithic monuments of W Europe, Iberia, and Malta (c 4000-1500 BC).

China. Proto-Chinese neolithic cultures had long covered N and SE China when the first large political state was organized in the N by the **Shang dynasty** (c 1523 BC). Shang kings called themselves Sons of Heaven, and they presided over a cult of human and animal sacrifice to ancestors and nature gods. The Chou dynasty, starting c 1027 BC, expanded the area of the Son of Heaven's dominion, but feudal states exercised most temporal power. A writing system with 2,000 characters was already in use under the Shang, with **pictographs** later supplemented by phonetic characters. Many of its principles and symbols, despite changes in spoken Chinese, were preserved in later writing systems. Technical advances allowed urban specialists to create fine ceramic and jade products, and bronze casting after 1500 BC was the most advanced in the world. Bronze artifacts have recently been discovered in N Thailand dating from 3600 BC, hundreds of years before similar Middle Eastern finds.

Americas. Olmecs settled (1500 BC) on the Gulf coast of Mexico and soon developed the first known civilization in the western hemisphere. Temple cities and huge stone sculpture date from 1200 BC. A rudimentary calendar and writing system existed. Olmec religion, centering on a jaguar god, and Olmec art forms influenced all later Meso-American cultures.

Paleontology: The History of Life

All dates are approximate, and are subject to change based on new fossil finds or new dating techniques, but the sequence of events is generally accepted. Dates are in years before the present.

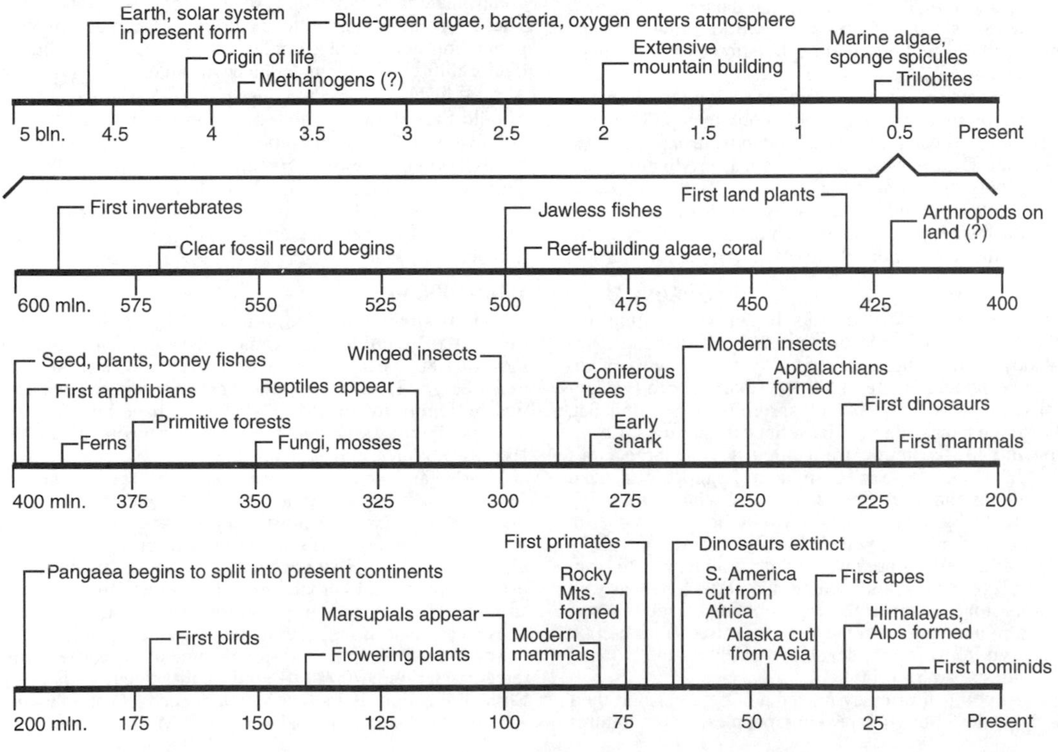

Major Gods & Goddesses of Ancient Egypt

Name	Relations	Sphere or Position	Emblem/Attribute
Ra (Re)/Atum/Amon	Self-created	The sun, creation	Hawk
Thoth (Djeheuty)	Son of Ra	The moon, wisdom, writing	Ibis/baboon
Ptah	Creator of Atum	Creation, craftsmen	----
Osiris	Brother of Set(h) & Isis	The underworld (dead), fertility, resurrection, vegetation	Bull
Isis	Sister/consort of Osiris	The underworld (dead)	----
Set(h)	Brother of Osiris	Evil, trickery, chaos	Boar, pig
Horus	Son of Osiris & Isis/ Ra & Hathor	The earth	Falcon
Hathor	Consort of Ra	Motherhood, love	Cow
Anubis	Son of Osiris	Embalmer & judge of the dead	Jackal/dog

Classical Era of Old World Civilizations: 1000 BC-400 BC

Greece. After a period of decline during the Dorian Greek invasions (1200-1000 BC), Greece and the Aegean area developed a unique civilization. Drawing upon Mycenaean traditions, Mesopotamian learning (weights and measures, lunisolar calendar, astronomy, musical scales), the Phoenician alphabet (modified for Greek), and Egyptian art, the revived **Greek city-states** saw a rich elaboration of intellectual life. The two great epic poems attributed to Homer, the *Iliad* and the *Odyssey,* were probably composed around the 8th cent. BC. Long-range commerce was aided by metal coinage (introduced by the Lydians in Asia Minor before 700 BC); colonies were founded around the Mediterranean (Cumae in Italy in 760 BC; Massalia in France c 600 BC) and Black Sea shores.

Philosophy, starting with Ionian speculation on the nature of matter (Thales, c 634-546 BC), continued by other "Pre-Socratics" (e.g., Heraclitus, c 535-415 BC; Parmenides, born c 515 BC), reached a high point in Athens in the rationalist idealism of **Plato** (c 428-347 BC), a disciple of **Socrates** (c 469-399 BC; executed for alleged impiety), and in **Aristotle** (384-322 BC), a pioneer in many fields, from natural sciences to logic, ethics, and metaphysics. The **arts** were highly valued. Architecture culminated in the **Parthenon** (438 BC) by Phidias (fl 490-430 BC). Poetry (Sappho, c 610-580 BC; Pindar, c 518-438 BC) and **drama** (Aeschylus, 525-456 BC; Sophocles, c 496-406 BC; Euripides, c 484-406 BC) thrived. Male beauty and strength, a chief artistic theme, were enhanced at the gymnasium and celebrated at the national games at Olympia.

Ruled by local tyrants or **oligarchies**, the Greeks were not politically united, but managed to resist inclusion in the Persian Empire—Persian king Darius was defeated at Marathon (490 BC), his son Xerxes at Salamis (480 BC), and the Persian army at Plataea (479 BC). Local warfare was common; the **Peloponnesian Wars** (431-404 BC) ended in Sparta's victory over Athens. Greek political power subsequently waned, but Greek cultural forms spread throughout the ancient world.

Hebrews. Nomadic Hebrew tribes entered Canaan before 1200 BC, settling among other Semitic peoples speaking the same language. They brought from the desert a **monotheistic** faith said to have been revealed to Abraham in Canaan c 1800 BC and Moses at Mt. Sinai c 1250 BC, after the Hebrews' escape from bondage in Egypt. David (r 1000-961 BC) and Solomon (r 961-922 BC) united them in a kingdom that briefly dominated the area. **Phoenicians** to the N founded Mediterranean colonies (Carthage, c 814 BC) and sailed into the Atlantic.

A temple in Jerusalem became the national religious center, with sacrifices performed by a hereditary priesthood. Polytheistic influences, especially of the fertility cult of Baal, were opposed by **prophets** (Elijah, Amos, Isaiah).

Divided into **two kingdoms** after Solomon, the Hebrews were unable to resist the revived Assyrian empire, which conquered Israel, the N kingdom, in 722 BC. Judah, the S kingdom, was conquered in 586 BC by the Babylonians under Nebuchadnezzar II. With the fixing of most of the biblical canon by the mid-4th cent. BC and the emergence of rabbis, Judaism successfully survived the loss of Hebrew autonomy. A Jewish kingdom was revived under the Hasmoneans (168-42 BC).

China. During the **Eastern Chou** dynasty (770-256 BC), Chinese culture spread E to the sea and S to the Yangtze R. Large feudal states on the periphery of the empire contended for preeminence, but continued to recognize the Son of Heaven (king), who retained a purely ritual role enriched with courtly music and dance. In the Age of Warring States (403-221 BC), when the first sections of the **Great Wall** were built, the Ch'in state in the W gained supremacy and finally united all of China.

Iron tools entered China c 500 BC, and casting techniques were advanced, aiding agriculture. Peasants owned their land and owed civil and military service to nobles. China's cities grew in number and size; barter remained the chief trade medium.

Intellectual ferment among noble scribes and officials produced the Classical Age of Chinese literature and philosophy. **Confucius** (551-479 BC) urged a restoration of a supposedly harmonious social order of the past through proper conduct in accordance with one's station and through filial and ceremonial piety. The *Analects* attributed to him are revered throughout E Asia.

Among other thinkers of this period, Mencius (d 289 BC) added the view that the Mandate of Heaven can be removed from an unjust dynasty. The Legalists sought to curb the supposed natural wickedness of people through new institutions and harsh laws; they aided the Ch'in rise to power. The Naturalists emphasized the balance of opposites—yin, yang—in the world. **Taoists** sought mystical knowledge through meditation and disengagement.

India. The political and cultural center of India shifted from the Indus to the Ganges River Valley. Buddhism, Jainism, and mystical revisions of orthodox Vedism all developed c 500-300 BC. The *Upanishads,* last part of the *Veda,* urged escape from the physical world. Vedism remained the preserve of the Brahman caste.

In contrast, **Buddhism**, founded by Siddarta Gautama (c 563-c 483 BC)—Buddha ("Enlightened One")—appealed to merchants in the urban centers and took hold at first (and most lastingly) on the geographic fringes of Indian civilization. The classic Indian epics were composed in this era: the **Ramayana** perhaps c 300 BC, the **Mahabharata** over a period starting around 400 BC.

N India was divided into a large number of monarchies and aristocratic republics, probably derived from tribal groupings, when the Magadha kingdom was formed in Bihar c 542 BC. It soon became the dominant power. The **Maurya dynasty**, founded by Chandragupta c 321 BC, expanded the kingdom, uniting most of N India in a centralized bureaucratic empire. The third Mauryan king, **Asoka** (reigned c 274-236 BC), conquered most of the subcontinent. He converted to Buddhism and inscribed its tenets on pillars throughout India. He downplayed the caste system and tried to end expensive sacrificial rites.

Before its final decline in India, Buddhism developed into a popular worship of heavenly Bodhisattvas ("enlightened beings"), and it produced a refined architecture (the Great Stupa [shrine] at Sanchi, AD 100) and sculpture (Gandhara reliefs, AD 1-400).

The Seven Wonders of the Ancient World

These ancient works of art and architecture were considered awe-inspiring by the Greek and Roman world of the first few centuries BC. Later classical writers disagreed as to which works belonged, but the following were usually included:

The Pyramids of Egypt: The only surviving ancient Wonder, these monumental structures of masonry, located at Giza on the W bank of the Nile R above Cairo, were built from c 2700 to 2500 BC as royal tombs. Three—Khufu (Cheops), Khafra (Chephren), and Menkaura (Mycerimus)—were often grouped as the first Wonder of the World. The largest, the Great Pyramid of Khufu, is a solid mass of limestone blocks covering 13 acres. It is estimated to contain 2.3 million blocks of stone, the stones themselves averaging 2½ tons and some weighing 30 tons. Its construction reputedly took 100,000 laborers 20 years.

The Hanging Gardens of Babylon: These gardens were laid out on a brick terrace 400 ft square and 75 ft above the ground. To irrigate the plants, screws were turned to lift water from the Euphrates R. The gardens were probably built by King Nebuchadnezzar II about 600 BC. The Walls of Babylon, long, thick, and made of colorfully glazed brick, were also considered by some among the Seven Wonders.

The Pharos (Lighthouse) of Alexandria: This structure was designed about 270 BC, during the reign of Ptolemy II, by the Greek architect Sostratos. Estimates of its height range from 200 to 600 ft.

The Colossus of Rhodes: A bronze statue of the sun god Helios, the Colossus was worked on for 12 years in the third cent. BC by the sculptor Chares. It was probably 120 ft high. A symbol of the city of Rhodes at its height, the statue stood on a promontory overlooking the harbor.

The Temple of Artemis (Diana) at Ephesus: This largest and most complex temple of ancient times was built about 550 BC and was made of marble except for its tile-covered wooden roof. It was begun in honor of a non-Hellenic goddess who later became identified with the Greek goddess of the same name. Ephesus was one of the greatest of the Ionian cities.

The Mausoleum at Halicarnassus: The source of our word *mausoleum*, this marble tomb was built in what is now SE Turkey by Artemisia for her husband Mausolus, king of Caria in Asia Minor, who died in 353 BC. About 135 ft high, the tomb was adorned with the works of 4 sculptors.

The Statue of Zeus (Jupiter) at Olympia: This statue of the king of the gods showed him seated on a throne. His flesh was made of ivory, his robe and ornaments of gold. Reputedly 40 ft high, the statue was made by Phidias and was placed in the great temple of Zeus in the sacred grove of Olympia about 457 BC.

Persia. Aryan peoples (Persians, Medes) dominated the area of present Iran by the beginning of the 1st millennium BC. The prophet **Zoroaster** (born c 628 BC) introduced a dualistic religion in which the forces of good (Ahura Mazda, "Lord of Wisdom") and evil (Ahriam) battle for dominance; individuals are judged by their actions and earn damnation or salvation. Zoroaster's hymns (*Gathas*) are included in the *Avesta*, the Zoroastrian scriptures. A version of this faith became the established religion of the Persian Empire and probably influenced later monotheistic religions.

Africa. Nubia, periodically occupied by Egypt since about 2600 BC, ruled Egypt c 750-661 BC and survived as an independent Egyptianized kingdom (**Kush;** capital Meroe) for 1,000 years. The Iron Age Nok culture flourished c 500 BC- AD 200 on the Benue Plateau of **Nigeria.**

Americas. The Chavin culture controlled N Peru c 900 BC to 200 BC. Its ceremonial centers, featuring the jaguar god, survived long after. Its architecture, ceramics, and textiles had influenced other Peruvian cultures. **Mayan civilization** began to develop in Central America as early as 1500 BC.

Great Empires Unite the Civilized World: 400 BC-AD 400

Persia and Alexander the Great. Cyrus, ruler of a small kingdom in Persia from 559 BC, united the Persians and Medes within 10 years and conquered Asia Minor and Babylonia in another 10. His son Cambyses, followed by **Darius** (r 522-486 BC), added vast lands to the E and N as far as the Indus Valley and Central Asia, as well as Egypt and Thrace. The whole empire was ruled by an international bureaucracy and army, with Persians holding the chief positions. The resources and styles of all the subject civilizations were exploited to create a rich syncretic art.

The kingdom of Macedon, which under Philip II dominated the Greek world and Egypt, was passed on to his son **Alexander** in 336 BC. Within 13 years, Alexander had conquered all the Persian dominions. Imbued by his tutor Aristotle with Greek ideals, Alexander encouraged Greek colonization, and Greek-style cities were founded. After his death in 323 BC, wars of succession divided the empire into 3 parts—**Macedon,** Egypt (ruled by the **Ptolemies**), and the **Seleucid** Empire.

In the ensuing 300 years (the **Hellenistic Era**), a cosmopolitan Greek-oriented culture permeated the ancient world from W Europe to the borders of India, absorbing native elites everywhere.

Hellenistic philosophy stressed the private individual's search for happiness. The Cynics followed Diogenes (c 372-287 BC), who stressed self-sufficiency and restriction of desires and expressed contempt for luxury and social convention. Zeno (c 335-c 263 BC) and the **Stoics** exalted reason, identified it with virtue, and counseled an ascetic disregard for misfortune. The **Epicureans** tried to build lives of moderate pleasure without political or emotional involvement. Hellenistic arts imitated life realistically, especially in sculpture and literature (comedies of Menander, 342-292 BC).

The sciences thrived, especially at Alexandria, where the Ptolemies financed a great library and museum. Fields of study included mathematics (**Euclid's** geometry, c 300 BC); astronomy (heliocentric theory of Aristarchus, 310-230 BC; Julian calendar, 45 BC; Ptolemy's *Almagest*, c AD 150); geography (world map of Eratosthenes, 276-194 BC); hydraulics (**Archimedes,** 287-212 BC); medicine (Galen, AD 130-200); and chemistry. Inventors refined uses for siphons, valves, gears, springs, screws, levers, cams, and pulleys.

A restored Persian empire under the **Parthians** (northern Iranian tribesmen) controlled the eastern Hellenistic world from 250 BC to AD 229. The Parthians and the succeeding Sassanian dynasty (c AD 224-651) fought with Rome periodically. The **Sassanians** revived Zoroastrianism as a state religion and patronized a nationalistic artistic and scholarly renaissance.

Rome. The city of Rome was founded, according to legend, by Romulus in 753 BC. Through military expansion and colonization, and by granting citizenship to conquered tribes, the city annexed all of Italy S of the Po in the 100-year period before 268 BC. The Latin and other Italic tribes were annexed first, followed by the **Etruscans** (founders of a great civilization, N of Rome) and the Greek colonies in the S. With a large standing army and reserve forces of several hundred thousand, Rome was able to defeat **Carthage** in the 3 **Punic Wars** (264-241, 218-201, 149-146 BC), despite the invasion of Italy (218 BC) by **Hannibal,** thus gaining Sicily and territory in Spain and N Africa.

Rome exploited local disputes to conquer Greece and Asia Minor in the 2d cent. BC, and Egypt in the 1st (after the defeat and suicide of **Antony and Cleopatra**, 30 BC). All the Mediterranean civilized world up to the disputed Parthian border was now Roman and remained so for 500 years. Less civilized regions were added to the Empire: Gaul (conquered by **Julius Caesar**, 58-51 BC), Britain (AD 43), and Dacia NE of the Danube (AD 107).

The original aristocratic republican government, with democratic features added in the 5th and 4th cent. BC, deteriorated under the pressures of empire and class conflict (**Gracchus** brothers, social reformers, murdered in 133 BC and 121 BC; slave revolts in 135 BC and 73 BC). After a series of civil wars (Marius vs. Sulla 88-82 BC, Caesar vs. **Pompey** 49-45 BC, triumvirate vs. Caesar's assassins 44-43 BC, Antony vs. Octavian 32-30 BC), the empire came under the rule of a deified monarch (first emperor, **Augustus**, 27 BC-AD 14). Provincials (nearly all granted citizenship by Caracalla, AD 212) came to dominate the army and civil service. Traditional Roman law, systematized and interpreted by independent jurists, and local self-rule in provincial cities were supplanted by a vast tax-collecting bureaucracy in the 3d and 4th cent. The legal rights of women, children, and slaves were strengthened.

Roman innovations in **civil engineering** included water mills, windmills, and rotary mills and use of cement that hardened under water. Monumental architecture (baths, theaters, temples) relied on the arch and the dome. The network of roads (some still standing) stretched 53,000 mi, passing through mountain tunnels as long as 3.5 mi. Aqueducts brought water to cities; underground sewers removed waste.

Roman art and literature were to a large extent derivative of Greek models. Innovations were made in sculpture (naturalistic busts, equestrian statues), decorative wall painting (as at Pompeii), satire (Juvenal, AD 60-127), history (Tacitus, AD 56-120), prose romance (Petronius, d AD 66). Gladiatorial contests dominated public amusements, which were supported by the state.

India. The **Gupta** monarchs reunited N India c AD 320. Their peaceful and prosperous reign saw a revival of Hindu religious thought and Brahman power. The old Vedic traditions were combined with devotion to many indigenous deities (who were seen as manifestations of Vedic gods). **Caste lines** were reinforced, and Buddhism gradually disappeared. The art (often erotic), architecture, and literature of the period, patronized by the Gupta court, are considered among India's finest achievements (Kalidasa, poet and dramatist, fl. c AD 400). Mathematical innovations included use of the zero and decimal numbers. Invasions by White Huns from the NW destroyed the empire c 550.

Rich cultures also developed in S India during this period. Emotional Tamil religious poetry contributed to the Hindu revival. The Pallava kingdom controlled much of S India c 350-880 and helped to spread Indian civilization to SE Asia.

China. The Ch'in ruler Shih Huang Ti (r 221-210 BC), known as the First Emperor, centralized political authority in China, standardized the written language, laws, weights, measures, and coinage, and conducted a census, but tried to destroy most philosophical texts. The **Han dynasty** (202 BC-AD 220) instituted the Mandarin bureaucracy, which lasted for 2,000 years. Local officials were selected by examination in the Confucian classics and trained at the imperial university and at provincial schools. The invention of **paper** facilitated this bureaucratic system. Agriculture was promoted, but the peasants bore most of the tax burden. Irrigation was improved, water clocks and sundials were used, astronomy and mathematics thrived, and landscape painting was perfected.

With the expansion S and W (to nearly the present borders of today's China), trade was opened with India, SE Asia, and the Middle East, over sea and caravan routes. Indian missionaries brought Mahayana Buddhism to China by the 1st cent. AD and spawned a variety of sects. Taoism was revived and merged with popular superstitions. Taoist and Buddhist monasteries and convents multiplied in the turbulent centuries after the collapse of the Han dynasty.

Major Gods & Goddesses of the Classical World

Greek	Roman	Relations	Sphere or Position
Aphrodite	Venus	Daughter of Zeus & Dione	Love
Apollo	——	Son of Zeus & Leto	Healing, poetry, light
Ares	Mars	Son of Zeus & Hera	War
Artemis	Diana	Daughter of Zeus & Leto	Hunting, chastity
Athena	Minerva	Daughter of Zeus & Metis	Wisdom, crafts, war
Cronus	Saturn	Father of Zeus	Titans' ruler
Demeter	Ceres	Sister of Zeus	Agriculture, fertility
Dionysus	Bacchus	Son of Zeus & Semele	Wine, fertility, ecstasy
Eros	Cupid	Son of Ares & Aphrodite	Love
Hades	Pluto	Brother of Zeus	The underworld, death
Hephaestus	Vulcan	Son of Zeus & Hera	Fire
Hera	Juno	Wife & sister of Zeus	Earth
Hermes	Mercury	Son of Zeus & Maia	Travel, commerce, gods' messenger
Hestia	Vesta	Sister of Zeus	The hearth
Pan	——	Son of Hermes & a wood nymph	Forests, flocks, shepherds
Persephone	Proserpina	Daughter of Zeus & Demeter	Grain
Poseidon	Neptune	Brother of Zeus	The sea
Rhea	Ops	Mother of Zeus	The earth
Uranus	Uranus	Father of Titans (elder gods)	The heavens
Zeus	Jupiter	Son of Cronus & Rhea	Ruler of the gods

Monotheism Spreads: AD 1-750

Roman Empire. Polytheism was practiced in the Roman Empire, and religions indigenous to particular Middle Eastern nations became international. Roman citizens worshiped **Isis** of Egypt, **Mithras** of Persia, **Demeter** of Greece, and the great mother **Cybele** of Phrygia. Their cults centered on mysteries (secret ceremonies) and the promise of an afterlife, symbolized by the death and rebirth of the god. The Jews of the empire preserved their monotheistic religion, Judaism, the world's oldest (c 1300 BC) continuous religion. Its teachings are contained in the Bible (the Old Testament). First-cent. Judaism embraced several sects, including the **Sadducees**, mostly drawn from the Temple priesthood, who were culturally Hellenized; the **Pharisees**, who upheld the full range of traditional customs and practices as of equal weight to literal scriptural law and elaborated synagogue worship; and the **Essenes**, an ascetic, millennarian sect. Messianic fervor led to repeated, unsuccessful rebellions against Rome (66-70, 135). As a result, the Temple in Jerusalem was destroyed and the population decimated; this event marked the beginning of the Diaspora (living in exile). To preserve the faith, a program of codification of law was begun at the academy of Yavneh. The work continued for some 500 years in Palestine and in Babylonia, ending in the final redaction (c 600) of the **Talmud**, a huge collection of legal and moral debates, rulings, liturgy, biblical exegesis, and legendary materials.

Christianity, which emerged as a distinct sect by the 2d half of the 1st cent., is based on the teachings of **Jesus**, whom believers considered the Savior (Messiah or Christ) and son of God. Missionary activities of the Apostles and such early leaders as **Paul of Tarsus** spread the faith. Intermittent persecution, as in Rome under Nero in AD 64, on grounds of suspected disloyalty, failed to disrupt the Christian communities. Each congregation, generally urban and of plebeian character, was tightly organized under a leader (bishop), elders (presbyters or priests), and assistants (deacons). The four **Gospels** (accounts of the life and teachings of Jesus) and the Acts of the Apostles were written down in the late 1st and early 2d cent. and circulated along with letters of Paul and other Christian leaders. An authoritative canon of these writings was not fixed until the 4th cent.

A school for priests was established at Alexandria in the 2d cent. Its teachers (**Origen** c 182-251) helped define doctrine and promote the faith in Greek-style philosophical works. Neoplatonism was given Christian coloration in the writings of Church Fathers such as **Augustine** (354-430). Christian hermits began to associate in monasteries, first in Egypt (St. Pachomius c 290-345), then in other eastern lands, then in the W (**St. Benedict's rule**, 529). Devotion to saints, especially Mary, mother of Jesus, spread. Under **Constantine** (r 306-37), Christianity became in effect the established religion of the Empire. Pagan temples were expropriated, state funds were used to build churches and support the hierarchy, and laws were adjusted in accordance with Christian ideas. Pagan worship was banned by the end of the 4th cent., and severe restrictions were placed on Judaism.

The newly established church was rocked by doctrinal disputes, often exacerbated by regional rivalries. Chief heresies (as defined by church councils, backed by imperial authority) were **Arianism**, which denied the divinity of Jesus; the **Monophysite** position denying the human nature of Christ; **Donatism**, which regarded as invalid any sacraments administered by sinful clergy; and **Pelagianism,** which denied the necessity of unmerited divine aid (grace) for salvation.

Islam. The earliest Arab civilization emerged by the end of the 2d millennium BC in the watered highlands ofYemen. Seaborne and caravan trade in frankincense and myrrh connected the area with the Nile and Fertile Crescent. The Minaean, Sabean (Sheba), and Himyarite states successively held sway. By Muhammad's time (7th cent. AD), the region was a province of Sassanian Persia. In the N, the Nabataean kingdom at Petra and the kingdom of Palmyra were Aramaicized, Romanized, and finally absorbed, as neighboring Judea had been, into the Roman Empire. Nomads shared the central region with a few trading towns and oases. Wars between tribes and raids on communities were common and were celebrated in a poetic tradition that by the 6th cent. helped establish a classic literary Arabic.

About 610, **Muhammad**, a 40-year-old Arab of Mecca, emerged as a prophet to his people. He proclaimed a revelation from the one true God, calling on contemporaries to abandon idolatry and restore the faith of Abraham. He introduced his religion as "Islam," meaning "submission" to the one God, Allah, as a continuation of the biblical faith of Abraham, Moses, and Jesus, all respected as prophets in this system. His teachings, recorded in the **Koran** (al-Qur'an in Arabic), in many ways were inclusive of Abrahamic monotheistic ideas known to the Jews and Christians in Arabia. A key aspect of the Abrahamic connection was insistence on justice in society, which led to severe opposition among the aristocrats in Mecca. As conditions worsened for Muhammad and his followers, he decided in 622 to make a ***hijra*** (emigration) to Medina, 200 mi to the N. This event marks the beginning of the Muslim lunar calendar. Hostilities between Mecca and Medina increased, and in 629 Muhammad conquered Mecca. By the time of his death in 632, nearly all the Arabian peninsula accepted his political and religious leadership.

After his death the majority of Muslims recognized the leadership of the **caliph** ("successor") Abu Bakr (632-34), followed by Umar (634-44), Uthman (644-56), and Ali (656-60). A minority, the **Shiites**, insisted instead on the leadership of Ali, Muhammad's cousin and son-in-law. By 644, **Muslim rule** over Arabia was confirmed. Muslim armies had threatened the Byzantine and Persian empires, which were weakened by wars and disaffection among subject peoples (including Coptic and Syriac Christians opposed to the Byzantine Orthodox establishment). Syria, Palestine, Egypt, Iraq, and Persia fell to Muslim armies. The new administration assimilated existing systems in the region; hence the conquered peoples participated in running of the empire. The Koran recognized the so-called Peoples of the Book, i.e., Christians, Jews, and Zoroastrians, as tolerated monotheists, and Muslim policy was relatively tolerant to minorities living as "protected" peoples. An expanded tax system, based on conquests of the Persian and Byzantine empires, provided revenue to organize campaigns against neighboring non-Muslim regions.

Disputes over succession, and pious opposition to injustices in society, led to a number of oppositional movements, which also led to the factionalization of Muslim community. The **Shiites** supported leadership candidates descended from Muhammad, believing them to be carriers of some kind of divine authority. The **Kharijites** supported an egalitarian system derived from the Koran, opposing and even engaging in battle against those who did not agree with them.

Under the **Umayyads** (661-750) and **Abbasids** (750-1256), territorial expansion led Muslim armies across N Africa and into Spain (711). Muslim armies in the W were stopped at Tours (France) in 732 by the Frankish ruler **Charles Martel**. Asia Minor, the Indus Valley, and Transoxiana were conquered in the E. The conversion of conquered peoples to Islam was gradual. In many places the official Arabic language supplanted the local tongues. But in the eastern regions the Arab rulers and their armies adopted Persian cultures and language as part of their Muslim identity.

Major Norse Gods & Goddesses

Name	Relations	Sphere or Position	Emblem/Attribute
Odin	Father of the Aesir (gods)	War and death, poetry, wisdom, magic	Spear, mead, ring/One-eyed
Thor	Son of Odin	Thunder, lightning, rain; champion of the gods	Hammer, belt
Njord	Father of Freyja & Freyr	Wind and sea, wealth and prosperity	----
Frigg	Wife of Odin	Marriage and motherhood, home	----
Freyja (Freya)	Daughter of Njord	Fertility, birth, crops	Necklace
Freyr	Son of Njord	Agriculture, sun, rain	Magic ship, golden boar
Tyr	Son of Odin ?	Justice, war	Spear/One-handed
Heimdall	Son of nine giantesses	Watchman of the gods; keen sight & hearing	Horn
Balder (Baldur)	Son of Odin	Light, purity	----
Loki	Son of giants; father of Hel (goddess of death), Jormungand (serpent encompassing the world), Fenrir (the wolf).	Malicious trickster	----

New Peoples Enter World History: 400-900

Barbarian invasions. Germanic tribes infiltrated S and E from their Baltic homeland during the 1st millennium BC, reaching S Germany by 100 BC and the Black Sea by AD 214. Organized into large federated tribes under elected kings, most resisted Roman domination and raided the empire in time of civil war (Goths took Dacia in 214, raided Thrace in 251-69). Germanic troops and commanders dominated the Roman armies by the end of the 4th cent. **Huns**, invaders from Asia, entered Europe in 372, driving more Germans into the W empire. Emperor Valens allowed Visigoths to cross the Danube in 376. Huns under Attila (d 453) raided Gaul, Italy, and the Balkans.

The W empire, weakened by overtaxation and social stagnation, was overrun in the 5th cent. Gaul was effectively lost in 406-7, Spain in 409, Britain in 410, Africa in 429-39. Rome was sacked in 410 by Visigoths under Alaric and in 455 by Vandals. The last western emperor, Romulus Augustulus, was deposed in 476 by the Germanic chief Odovacar.

Celts. Celtic cultures, which in pre-Roman times covered most of W Europe, were confined almost entirely to the British Isles after the Germanic invasions. **St. Patrick** completed (c 457-92) the conversion of Ireland. A strong monastic tradition took hold. Irish monastic missionaries in Scotland, England, and the continent (Columba c 521-97; Columban c 543-615) helped restore Christianity after the Germanic invasions. Monasteries became centers of classic and Christian learning and presided over the recording of a Christianized Celtic mythology, elaborated by secular writers and bards. An intricate decorative art style developed, especially in book illumination (Lindisfarne Gospels, c 700; Book of Kells, 8th cent.).

Successor states. The Visigothic kingdom in Spain (from 419) and much of France (to 507) saw continuation of Roman administration, language, and law (Breviary of Alaric, 506) until its destruction by the Muslims (711). The Vandal kingdom in Africa (from 429) was conquered by the Byzantines in 533. Italy was ruled successively by an Ostrogothic kingdom under Byzantine suzerainty (489-554), direct Byzantine government, and German Lombards (568-774). The Lombards divided the peninsula with the Byzantines and papacy under the dynamic reformer **Pope Gregory the Great** (590-604) and successors.

King Clovis (r 481-511) united the Franks on both sides of the Rhine and, after his conversion to Christianity, defeated the Arian heretics, Burgundians (after 500), and Visigoths (507) with the support of native clergy and the papacy. Under the **Merovingian** kings, a feudal system emerged: Power was fragmented among hierarchies of military landowners. Social stratification, which in late Roman times had acquired legal, hereditary sanction, was reinforced. The Carolingians (747-987) expanded the kingdom and restored central power. **Charlemagne** (r 768-814) conquered nearly all the Germanic lands, including Lombard Italy, and was crowned Emperor by Pope Leo III in Rome in 800. A centuries-long decline in commerce and arts was reversed under Charlemagne's patronage. He welcomed Jews to his kingdom, which became a center of Jewish learning (Rashi, 1040-1105). He sponsored the Carolingian Renaissance of learning under the Anglo-Latin scholar Alcuin (c 732-804), who reformed church liturgy.

Byzantine Empire. Under **Diocletian** (r 284-305) the empire had been divided into 2 parts to facilitate administration and defense. **Constantine** founded (330) **Constantinople** (at old Byzantium) as a fully Christian city. Commerce and taxation financed a sumptuous, orientalized court, a class of hereditary bureaucratic families, and magnificent urban construction (Hagia Sophia, 532-37). The city's fortifications and naval innovations repelled assaults by Goths, Huns, Slavs, Bulgars, Avars, Arabs, and Scandinavians. Greek replaced Latin as the official language by c 700. Byzantine art, a solemn, sacral, and stylized variation of late classical styles (mosaics at the Church of San Vitale, Ravenna, Italy 526-48), was a starting point for medieval art in E and W Europe.

Justinian (r 527-65) reconquered parts of Spain, N Africa, and Italy, codified Roman law (Codex Justinianus [529] was medieval Europe's chief legal text), closed the Platonic Academy at Athens, and ordered all pagans to convert. Lombards in Italy and Arabs in Africa retook most of his conquests. The Isaurian dynasty from Anatolia (from 717) and the Macedonian dynasty (867-1054) restored military and commercial power. The Iconoclast controversy (726-843) over the permissibility of images helped alienate the Eastern Church from the papacy.

Abbasid Empire. Baghdad (est. 762), became seat of the **Abbasid dynasty** (est 750), while Ummayads continued to rule in Spain. A brilliant cosmopolitan civilization emerged, inaugurating a Muslim-Arab golden age. Arabic was the lingua franca of the empire; intellectual sources from Persian, Sanskrit, Greek, and Syriac were rendered into Arabic. Christians and Jews equally participated in this translation movement, which also involved interaction between Jewish legal thought and Islamic law, as much as between Christian theology and Muslim scholasticism. Persian-style court life, with art and music, flourished at the court of **Harun al-Rashid** (786-809), celebrated in the masterpiece known to English readers as *The Arabian Nights*. The sciences, medicine, and mathematics were pursued at Baghdad, Cordova, and Cairo (est. 969). The culmination of this intellectual synthesis in Islamic civilization came with the scientific and philosophical works of **Avicenna** (Ibn Sina, 980-1037), **Averroes** (Ibn Rushd, 1126-98), and **Maimonides** (1135-1204), a Jew who wrote in Arabic. This intellectual tradition was translated into Latin and opened a new period in Christian thought.

The decentralization of the Abbasid empire, from 874, led to establishment of various Muslim dynasties under different ethnic groups. Persians, Berbers, and Turks ruled different regions, retaining connection with the Abbasid caliph at the religious level. The Abbasid period also saw various religious movements against the orthodox position held by governing authorities. This situation in religion led to establishment of different legal, theological, and mystical schools of thought. The most influential mass movement was **Sufism**, which aimed at the reaching out of the average individual in quest of a spiritual path. Al-Ghazali (1058-1111) is credited with reconciling personal Sufism with orthodox Sunni tradition.

Africa. Immigrants from Saba in S Arabia helped set up the **Axum** kingdom in Ethiopia in the 1st cent. (their language, Ge'ez, is preserved by the Ethiopian Church). In the 3d cent., when the kingdom became Christianized, it defeated Kushite Meroe and expanded its influence into Yemen. Axum was the center of a vast ivory trade and controlled the Red Sea coast until c 1100. Arab conquest in Egypt cut Axum's political and economic ties with Byzantium.

The Iron Age entered W Africa by the end of the 1st millennium BC. **Ghana**, the first known sub-Saharan state, ruled in the upper Senegal-Niger region c 400-1240, controlling the trade of gold from mines in the S to trans-Sahara caravan routes to the N. The **Bantu** peoples, probably of W African origin, began to spread E and S perhaps 2,000 years ago, displacing the Pygmies and Bushmen of central and S Africa during a 1,500-year period.

Japan. The advanced Neolithic Yayoi period, when irrigation, rice farming, and iron and bronze casting techniques were introduced from China or Korea, persisted to c AD 400. The myriad Japanese states were then united by the **Yamato** clan, under an emperor who acted as chief priest of the animistic Shinto cult. Japanese political and military intervention by the 6th cent. in Korea, then under strong Chinese influence, quickened a Chinese cultural invasion of Japan, bringing Buddhism, the Chinese language (which long remained a literary and governmental medium), Chinese ideographs, and Buddhist styles in painting, sculpture, literature, and architecture (7th cent., Horyu-ji temple at Nara). The Taika Reforms (646) tried unsuccessfully to centralize Japan according to Chinese bureaucratic and Buddhist philosophical values.

A nativist reaction against the Buddhist **Nara period** (710-94) ushered in the **Heian period** (794-1185) centered at the new capital, Kyoto. Japanese elegance and simplicity modified Chinese styles in architecture, scroll painting, and literature; the writing system was also simplified. The courtly novel *Tale of Genji* (1010-20) testifies to the enhanced role of women.

Southeast Asia. The historic peoples of SE Asia began arriving some 2,500 years ago from China and Tibet, displacing scattered aborigines. Their agriculture relied on rice and yams. Indian cultural influences were strongest; literacy and Hindu and Buddhist ideas followed the S India-China trade route. From the S tip of Indochina, the kingdom of **Funan** (1st-7th cent.) traded as far W as Persia. It was absorbed by Chenla, itself conquered by the **Khmer Empire** (600-1300). The Khmers, under Hindu god-kings (Suryavarman II, 1113-c 1150), built the monumental Angkor Wat temple center for the royal phallic cult. The **Nam-Viet** kingdom in Annam, dominated by China and Chinese culture for 1,000 years, emerged in the 10th cent., growing at the expense of the Khmers, who also lost ground in the NW to the new, highly organized **Thai** kingdom. On Sumatra, the **Srivijaya** Empire controlled vital sea lanes (7th to 10th cent.). A Buddhist dynasty, the Sailendras, ruled central **Java** (8th-9th cent.), building at Borobudur one of the largest stupas in the world.

China. The Sui dynasty (581-618) ushered in a period of commercial, artistic, and scientific achievement in China, continuing under the **Tang** dynasty (618-906). Inventions like the magnetic compass, gunpowder, the abacus, and printing were introduced or perfected. Medical innovations included cataract surgery. The state, from its cosmopolitan capital, Chang-an, supervised foreign trade, which exchanged Chinese silks, porcelains, and art for spices, ivory, etc., over Central Asian caravan routes and sea routes reaching Africa. A golden age of poetry bequeathed valuable works to later generations (Tu Fu, 712-70; Li Po, 701-62). Landscape painting flourished. Commercial and industrial expansion continued under the **Northern Sung** dynasty (960-1126), facilitated by paper money and credit notes. But commerce never achieved respectability; government monopolies expropriated successful merchants. The population, long stable at 50 million, doubled in 200 years with the introduction of early-ripening rice and the double harvest. In art, native Chinese styles were revived.

Americas. From 300 to 600 a Native American empire stretched from the Valley of Mexico to Guatemala, centering on the huge city **Teotihuacán** (founded 100 BC). To the S, in Guatemala, a high **Mayan** civilization developed (150-900) around hundreds of rural ceremonial centers. The Mayans improved on Olmec writing and the calendar and pursued astronomy and mathematics (using the idea of zero). In South America, a widespread pre-Inca culture grew from **Tiahuanacu,** Bolivia, near Lake Titicaca (Gateway of the Sun, c 700).

Christian Europe Regroups and Expands: 900-1300

Scandinavians. Pagan Danish and Norse (Viking) adventurers, traders, and pirates raided the coasts of the British Isles (Dublin, est. c 831), France, and even the Mediterranean for over 200 years beginning in the late 8th cent. Inland settlement in the W was limited to Great Britain (King Canute, 994-1035) and Normandy, settled (911) under Rollo, as a fief of France. Vikings also reached Iceland (874), Greenland (c 986), and North America (**Leif Ericson** and others, c 1000). Norse traders (**Varangians**) developed Russian river commerce from the 8th to the 11th cent. and helped set up a state at Kiev in the late 9th cent. Conversion to Christianity occurred in the 10th cent., reaching Sweden 100 years later. In the 11th cent. Norman bands conquered S Italy and Sicily, and Duke **William of Normandy** conquered (1066) England, bringing feudalism and the French language, essential elements in later English civilization.

Central and East Europe. Slavs began to expand from about AD 150 in all directions in Europe, and by the 7th cent. they reached as far S as the Adriatic and Aegean seas. In the Balkan Peninsula they dislocated Romanized local populations or assimilated newcomers (Bulgarians, a Turkic people). The first Slavic states were Moravia (628) in Central Europe and the Bulgarian state (680) in the Balkans. Missions of St. Methodius and Cyril (whose Greek-based cyrillic alphabet is still used by some S and E Slavs) converted (863) Moravia.

The Eastern Slavs, part-civilized under the overlordship of the Turkish-Jewish **Khazar** trading empire (7th-10th cent.), gravitated toward Constantinople by the 9th cent. The **Kievan state** adopted (989) Eastern Christianity under Prince Vladimir. King Boleslav I (992-1025) began **Poland's** long history of eastern conquest. The Magyars (**Hungarians**), in present-day Hungary since 896, accepted (1001) Latin Christianity.

Germany. The German kingdom that emerged after the breakup of Charlemagne's W Empire remained a confederation of largely autonomous states. Otto I, a Saxon who was king from 936, established the **Holy Roman Empire**—a union of Germany and N Italy—in alliance with Pope John XII, who crowned (962) him emperor; he defeated (955) the Magyars. Imperial power was greatest under the **Hohenstaufens** (1138-1254), despite the growing opposition of the papacy, which ruled central Italy, and the Lombard League cities. Frederick II (1194-1250) improved administration and patronized the arts; after his death, German influence was removed from Italy.

Christian Spain. From its N mountain redoubts, Christian rule slowly migrated S through the 11th cent., when Muslim unity collapsed. After the capture (1085) of **Toledo**, the kingdoms of Portugal, Castile, and Aragon undertook repeated crusades of reconquest, finally completed in 1492. Elements of Islamic civilization persisted in recaptured areas, influencing all Western Europe.

Crusades. Pope Urban II called (1095) for a crusade to restore Asia Minor to Byzantium and to regain the Holy Land from the Turks. Some ten crusades (lasting until 1291) succeeded only in founding four temporary Frankish states in the Levant. The 4th crusade sacked (1204) Constantinople. In Rhineland (1096), England (1290), and France (1306), Jews were massacred or expelled, and wars were launched against Christian heretics (**Albigensian** crusade in France, 1229). Trade in eastern luxuries expanded, led by the Venetian naval empire.

Economy. The agricultural base of European life benefited from improvements in **plow design** (c 1000) and by draining of lowlands and clearing of forests, leading to a rural population increase. Towns grew in N Italy, Flanders, and N Germany (Hanseatic League). Improvements in **loom design** permitted factory textile production. **Guilds** dominated urban trades from the 12th cent. Banking (centered in Italy, 12th-15th cent.) facilitated long-distance trade.

The Church. The split between the Eastern and Western churches was formalized in 1054. Western and Central Europe was divided into 500 bishoprics under one united hierarchy, but conflicts between secular and church authorities were frequent (German **Investiture Controversy**, 1075-1122). Clerical power was first strengthened through the international monastic reform begun at Cluny in 910. Popular religious enthusiasm often expressed itself in heretical movements (Waldensians from 1173), but was channeled by the **Dominican** (1215) and **Franciscan** (1223) friars into the religious mainstream.

Arts. Romanesque architecture (11th-12th cent.) expanded on late Roman models, using the rounded arch and massed stone to support enlarged basilicas. Painting and sculpture followed Byzantine models. The literature of **chivalry** was exemplified by the epic (*Chanson de Roland*, c 1100) and by courtly love poems of the troubadours of Provence and minnesingers of Germany. **Gothic** architecture emerged in France (choir of St. Denis, c 1040) and spread as French cultural influence came to predominate in Europe. Rib vaulting and pointed arches were used to combine soaring heights with delicacy, and they freed walls for display of stained glass. Exteriors were covered with painted relief sculpture and embellished with elaborate architectural detail.

Learning. Law, medicine, and philosophy were advanced at independent **universities** (Bologna, late 11th cent.), originally corporations of students and masters. Twelfth-cent. translations of Greek classics, especially Aristotle, encouraged an analytic approach. Scholastic philosophy, from Anselm (1033-1109) to **Aquinas** (1225-74), attempted to understand revelation through reason.

Apogee of Central Asian Power; Islam Grows: 1250-1500

Turks. Turkic peoples, of Central Asian ancestry, were a military threat to the Byzantine and Persian Empires from the 6th cent. After several waves of invasions, during which most of the Turks adopted Islam, the **Seljuk Turks** took (1055) Baghdad. They ruled Persia, Iraq and, after 1071, Asia Minor, where massive numbers of Turks settled. The empire was divided in the 12th cent. into smaller states ruled by Seljuks, Kurds (**Saladin,** c 1137-93), and Mamluks (a military caste of former Turk, Kurd, and Circassian slaves), which governed Egypt and the Middle East until the Ottoman era (c 1290-1922).

Osman I (r c 1290-1326) and succeeding sultans united Anatolian Turkish warriors in a militaristic state that waged holy war against Byzantium and Balkan Christians. Most of the Balkans had been subdued, and Anatolia united, when Constantinople fell (1453). By the mid-16th cent., Hungary, the Middle East, and N Africa had been conquered. The Turkish advance was stopped at Vienna (1529) and at the naval battle of Lepanto (1571) by Spain, Venice, and the papacy.

The Ottoman state was governed in accordance with orthodox Muslim law. Greek, Armenian, and Jewish communities were segregated and were ruled by religious leaders responsible for taxation; they dominated trade. State offices and most army ranks were filled by slaves through a system of child conscription among Christians.

India. Mahmud of Ghazni (971-1030) led repeated Turkish raids into N India. Turkish power was consolidated in 1206 with the start of the **Sultanate at Delhi**. Centralization of state power under the early Delhi sultans went far beyond traditional Indian practice. Muslim rule of most of the subcontinent lasted until the British conquest some 600 years later.

Mongols. Genghis Khan (c 1167-1227) first united the feuding Mongol tribes, and built their armies into an effective offensive force around a core of highly mobile cavalry. He and his immediate successors created the largest land empire in history; by 1279 it stretched from the E coast of Asia to the Danube, from the Siberian steppes to the Arabian Sea. East-West trade and contacts were facilitated (Marco Polo, c 1254-1324). The W Mongols were Islamized by 1295; successor states soon lost their Mongol character by assimilation. They were briefly reunited under the Turk Tamerlane (1336-1405).

Kublai Khan ruled China from his new capital Beijing (est. c 1264). Naval campaigns against Japan (1274, 1281) and Java (1293) were defeated, the latter by the Hindu-Buddhist maritime kingdom of Majapahit. The **Yuan** dynasty used Mongols and other foreigners (including Europeans) in official posts and tolerated the return of Nestorian Christianity (suppressed 841-45) and the spread of Islam in the S and W. A native reaction expelled the Mongols in 1367-68.

Russia. The Kievan state in Russia, weakened by the decline of Byzantium and the rise of the Catholic Polish-Lithuanian state, was overrun (1238-40) by the Mongols. Only the northern trading republic of Novgorod remained independent. The grand dukes of Moscow emerged as leaders of a coalition of princes that eventually (by 1481) defeated the Mongols. After the fall of Constantinople in 1453, the **Tsars** (Caesars) at Moscow (from Ivan III, r 1462-1505) set up an independent Russian Orthodox Church. Commerce failed to revive. The isolated Russian state remained agrarian, with the peasant class falling into serfdom.

Persia. A revival of Persian literature, making use of the Arab alphabet and literary forms, began in the 10th cent. (epic of Firdausi, 935-1020). An art revival, influenced by Chinese styles introduced after the Mongols came to power in Iran, began in the 13th cent. Persian cultural and political forms, and often the Persian language, were used for centuries by Turkish and Mongol elites from the Balkans to India. Persian mystics from Rumi (1207-73) to Jami (1414-92) promoted **Sufism** in their poetry.

Africa. Two militant Islamic Berber dynasties emerged from the Sahara to carve out empires from the Sahel to central Spain—the **Almoravids** (c 1050-1140) and the fanatical **Almohads** (c 1125-1269). The Ghanaian empire was replaced in the upper Niger by Mali (c 1230-1340), whose Muslim rulers imported Egyptians to help make **Timbuktu** a center of commerce (in gold, leather, and slaves) and learning. The Songhay empire (to 1590) replaced Mali. To the S, forest kingdoms produced refined artworks (Ife terra cotta, **Benin** bronzes). Other Muslim states in Nigeria (Hausas) and Chad originated in the 11th cent. and continued in some form until the 19th-cent. European conquest. Less-developed Bantu kingdoms existed across central Africa.

Some 40 Muslim Arab-Persian trading colonies and city-states were established all along the E African coast from the 10th cent. (Kilwa, Mogadishu). The interchange with Bantu peoples produced the **Swahili** language and culture. Gold, palm oil, and slaves were brought from the interior, stimulating the growth of the Monamatapa kingdom of the Zambezi (15th cent.). The Christian Ethiopian empire (from 13th cent.) continued the traditions of Axum.

Southeast Asia. Islam was introduced into Malaya and the Indonesian islands by Arab, Persian, and Indian traders. Coastal Muslim cities and states (starting before 1300) soon dominated the interior. Chief among these was the **Malacca** state (c 1400-1511), on the Malay peninsula.

Arts and Statecraft Thrive in Europe: 1350-1600

Italian Renaissance and Humanism. Distinctive Italian achievements in the arts in the late Middle Ages (**Dante,** 1265-1321; Giotto, 1276-1337) led to the vigorous new styles of the Renaissance (14th-16th cent.). Patronized by the rulers of the quarreling petty states of Italy (**Medicis** in Florence and the papacy, c 1400-1737), the plastic arts perfected realistic techniques, including **perspective** (Masaccio, 1401-28, **Leonardo,** 1452-1519). Classical motifs were used in architecture, and increased talent and expense were put into secular buildings. The Florentine dialect was refined as a national literary language (**Petrarch,** 1304-74). Greek refugees from the E strengthened the respect of humanist scholars for the classic sources. Soon an international movement aided by the spread of **printing** (Gutenberg, c 1397(?)-1468), **humanism** was optimistic about the power of human reason (Erasmus of Rotterdam, 1466-1536, **More's** *Utopia*, 1516) and valued individual effort in the arts and in politics (**Machiavelli,** 1469-1527).

France. The French monarchy, strengthened in its repeated struggles with powerful nobles (Burgundy, Flanders, Aquitaine) by alliances with the growing commercial towns, consolidated bureaucratic control under Philip IV (r 1285-1314) and extended French influence into Germany and Italy (popes at Avignon, France, 1309-1417). The **Hundred Years War** (1337-1453) ended English dynastic claims in France (battles of Crécy, 1346, and Poitiers, 1356; Joan of Arc executed, 1431). A French Renaissance, dating from royal invasions (1494, 1499) of Italy, was encouraged at the court of Francis I (r 1515-47), who centralized taxation and law. French vernacular literature consciously asserted its independence (La Pléiade, 1549).

England. The evolution of England's unique political institutions began with the **Magna Carta** (1215), by which King John guaranteed the privileges of nobles and church against the monarchy and assured jury trial. After the **Wars of the Roses** (1455-85), the **Tudor dynasty** reasserted royal prerogatives (Henry VIII, r 1509-47), but the trend toward independent departments and ministerial government also continued. English trade (wool exports from c 1340) was protected by the nation's growing maritime power (**Spanish Armada** destroyed, 1588).

English replaced French and Latin in the late 14th cent. in law and literature (**Chaucer,** c 1340-1400) and English translation of the Bible began (Wycliffe, 1380s). **Elizabeth I** (r 1558-1603) presided over a confident flowering of poetry (Spenser, 1552-99), drama (**Shakespeare**, 1564-1616), and music.

German Empire. From among a welter of minor feudal states, church lands, and independent cities, the **Habsburgs** assembled a far-flung territorial domain, based in Austria from 1276. Family members held the title of Holy Roman Emperor from 1438 to the Empire's dissolution in 1806, but failed to centralize its domains, leaving Germany disunited for centuries. Resistance to Turkish expansion brought Hungary under Austrian control from the 16th cent. The Netherlands, Luxembourg, and Burgundy were added in 1477, curbing French expansion.

The Flemish painting tradition of naturalism, technical proficiency, and bourgeois subject matter began in the 15th cent. (**Jan Van Eyck,** c 1390-1441), the earliest northern manifestation of the Renaissance. Albrecht **Dürer** (1471-1528) typified the merging of late Gothic and Italian trends in 16th-cent. German art. Imposing civic architecture flourished in the prosperous commercial cities.

Spain. Despite the unification of Castile and Aragon in 1479, the 2 countries retained separate governments, and the nobility, especially in Aragon and Catalonia, retained many privileges. Spanish lands in Italy (Naples, Sicily) and the Netherlands entangled the country in European wars through the mid-17th cent., while explorers, traders, and conquerors built up a Spanish empire in the Americas and the Philippines. From the late 15th cent., a **golden age** of literature and art produced works of social satire (plays of Lope de Vega, 1562-1635; **Cervantes,** 1547-1616), as well as spiritual intensity (**El Greco,** 1541-1614; **Velazquez,** 1599-1660).

Black Death. The bubonic plague reached Europe from the E in 1348, killing as much as half the population by 1350. Labor scarcity forced a rise in wages and brought greater freedom to the peasantry, making possible **peasant uprisings** (Jacquerie in France, 1358; Wat Tyler's rebellion in England, 1381). In the *ciompi* revolt (1378), Florentine wage earners demanded a say in economic and political power.

Explorations. Organized European maritime exploration began, seeking to evade the Venice-Ottoman monopoly of E trade and to promote Christianity. Beginning in 1418, expeditions from Portugal explored the W coast of Africa, until Vasco da Gama rounded the Cape of Good Hope in 1497 and reached India. A Portuguese trading empire was consolidated by the seizure of Goa (1510) and Malacca (1551). Japan was reached in 1542. The voyages of Christopher **Columbus** (1492-1504) uncovered a world new to Europeans, which Spain hastened to subdue. Navigation schools in Spain and Portugal, the development of large sailing ships (carracks), and the invention (c 1475) of the rifle aided European penetration.

Mughals and Safavids. E of the Ottoman Empire, 2 Muslim dynasties ruled unchallenged in the 16th and 17th cent. The Mughal dynasty of India, founded by Persianized Turkish invaders from the NW under Babur, dates from their 1526 conquest of the Delhi Sultanate. The dynasty ruled most of India for more than 200 years, surviving nominally until 1857. **Akbar** (r 1556-1605) consolidated administration at his glorious court, where the Urdu language (Persian-influenced Hindi) developed. Trade relations with Europe increased. Under Shah Jahan (1629-58), a secularized art fusing Hindu and Muslim element flourished in miniature painting and in architecture (**Taj Mahal**). **Sikhism** (founded c 1519) combined elements of both faiths. Suppression of Hindus and Shi'ite Muslims in S India in the late 17th cent. weakened the empire.

Fanatical devotion to the Shi'ite sect characterized the Safavids (1502-1736) of Persia and led to hostilities with the Sunni Ottomans for more than a century. The prosperity and the strength of the empire are evidenced by the mosques at its capital city, **Isfahan**. The Safavids enhanced Iranian national consciousness.

China. The **Ming** emperors (1368-1644), the last native dynasty in China, wielded unprecedented personal power, while the Confucian bureaucracy began to suffer from inertia. European trade (Portuguese monopoly through **Macao** from 1557) was strictly controlled. Jesuit scholars and scientists (Matteo Ricci, 1552-1610) introduced some Western science; their writings familiarized the West with China. Chinese technological inventiveness declined from this era, but the arts thrived, especially painting and ceramics.

Japan. After the decline of the first hereditary shogunate (chief generalship) at **Kamakura** (1185-1333), fragmentation of power accelerated, as did the consequent social mobility. Under Kamakura and the Ashikaga shogunate (1338-1573), the daimyos (lords) and samurai (warriors) grew more powerful and promoted a martial ideology. Japanese pirates and traders plied the China coast. Popular Buddhist movements included the nationalist Nichiren sect (from c 1250) and **Zen** (brought from China, 1191), which stressed meditation and a disciplined esthetic (tea ceremony, gardening, martial arts, No drama).

IT'S A FACT: Close to half of the 130 ships of the Spanish Armada of 1588 were destroyed—but only about 4 in battle with the British. The rest were wrecked or foundered in storms after the fleet was repelled and forced to sail northward around the British Isles.

Reformed Europe Expands Overseas: 1500-1700

Reformation begun. Theological debate and protests against real and perceived clerical corruption existed in the medieval Christian world, expressed by such dissenters as John **Wycliffe** (c 1320-84) and his followers, the Lollards, in England, and **Huss** (burned as a heretic, 1415) in Bohemia.

Martin **Luther** (1483-1546) preached that faith alone leads to salvation, without the mediation of clergy or good works. He attacked the authority of the pope, rejected priestly celibacy, and recommended individual study of the Bible (which he translated c 1525). His 95 Theses (1517) led to his excommunication (1521). John **Calvin** (1509-64) said that God's elect were predestined for salvation and that good conduct and success were signs of election. Calvin in Geneva and John Knox (1505-72) in Scotland established theocratic states.

Henry VIII asserted English national authority and secular power by breaking away (1534) from the Catholic Church. Monastic property was confiscated, and some Protestant doctrines given official sanction.

Religious wars. A century and a half of religious wars began with a S German peasant uprising (1524), repressed with Luther's support. Radical sects—democratic, pacifist, millennarian—arose (Anabaptists ruled Münster in 1534-35) and were suppressed violently. Civil war in France from 1562 between **Huguenots** (Protestant nobles and merchants) and Catholics ended with the 1598 **Edict of Nantes,** tolerating Protestants (revoked 1685). Habsburg attempts to restore Catholicism in Germany were resisted in 25 years of fighting; the 1555 Peace of Augsburg guarantee of religious independence to local princes and cities was confirmed only after the **Thirty Years War** (1618-48), when much of Germany was devastated by local and foreign armies (Sweden, France).

A Catholic Reformation, or **Counter Reformation**, met the Protestant challenge, clearly defining an official theology at the Council of Trent (1545-63). The **Jesuit** order (Society of Jesus), founded in 1534 by Ignatius Loyola (1491-1556), helped reconvert large areas of Poland, Hungary, and S Germany and sent missionaries to the New World, India, and China, while the Inquisition helped suppress heresy in Catholic countries. A revival of piety appeared in the devotional literature (Teresa of Avila, 1515-82) and grandiose Baroque art (Bernini, 1598-1680) of Roman Catholic countries.

Scientific Revolution. The late nominalist thinkers (Ockham, c 1300-49) of Paris and Oxford challenged Aristotelian orthodoxy, allowing for a freer scientific approach. At the same time, metaphysical values, such as the Neoplatonic faith in an orderly, mathematical cosmos, still motivated and directed inquiry. Nicolaus **Copernicus** (1473-1543) promoted the heliocentric theory, which was confirmed when Johannes Kepler (1571-1630) discovered the mathematical laws describing the orbits of the planets. The traditional Christian-Aristotelian belief that heavens and earth were fundamentally different collapsed when **Galileo** (1564-1642) discovered moving sunspots, irregular moon topography, and moons around Jupiter. He and Sir Isaac **Newton** (1642-1727) developed a mechanics that unified cosmic and earthly phenomena. Newton and Gottfried von Leibniz (1646-1716) invented calculus, and René Descartes (1596-1650) invented analytic geometry.

An explosion of **observational science** included the discovery of blood circulation (Harvey, 1578-1657) and microscopic life (Leeuwenhoek, 1632-1723) and advances in anatomy (Vesalius, 1514-64, dissected corpses) and chemistry (Boyle, 1627-91). Scientific research institutes were founded: Florence (1657), London (**Royal Society**, 1660), Paris (1666). Inventions proliferated (Savery's steam engine, 1696).

Arts. Mannerist trends of the High Renaissance (**Michelangelo**, 1475-1564) exploited virtuosity, grace, novelty, and exotic subjects and poses. The notion of artistic genius was promoted, in contrast to the anonymous medieval artisan. Private connoisseurs entered the art market. These trends were elaborated in the 17th cent. **Baroque** era on a grander scale. Dynamic movement in painting and sculpture was emphasized by sharp lighting effects, use of rich materials (colored marble, gilt), and realistic details. Curved facades, broken lines, rich, deep-cut detail, and ceiling decoration characterized Baroque architecture, especially in Germany. Monarchs, princes, and prelates, usually Catholic, used Baroque art to enhance and embellish their authority, as in royal portraits (Velazquez, 1599-1660; Van Dyck, 1599-1641).

National styles emerged. In France, a taste for rectilinear order and serenity (Poussin, 1594-1665), linked to the new rational philosophy, was expressed in classical forms. The influence of **classical values** in French literature (tragedies of **Racine**, 1639-99) gave rise to the "battle of the Ancients and Moderns." New forms included the essay (**Montaigne**, 1533-92) and novel (*Princesse de Cleves*, La Fayette, 1678).

Dutch painting of the 17th cent. was unique in its wide social distribution. The Flemish tradition of undemonstrative realism reached its peak in **Rembrandt** (1606-69) and Jan Vermeer (1632-75).

Economy. European economic expansion was stimulated by the new trade with the East, by New World gold and silver, and by a doubling of population (50 million in 1450, 100 million in 1600). New business and financial techniques were developed and refined, such as joint-stock companies, insurance, and letters of credit and exchange. The Bank of Amsterdam (1609) and the Bank of England (1694) broke the old monopoly of private banking families. The rise of a business mentality was typified by the spread of clock towers in cities in the 14th cent. By the mid-15th cent., portable clocks were available; the first watch was invented in 1502.

By 1650, most governments had adopted the **mercantile system**, in which they sought to amass metallic wealth by protecting their merchants' foreign and colonial trade monopolies. The rise in prices and the new coin-based economy undermined the craft guild and feudal manorial systems. Expanding industries (clothweaving, mining) benefited from technical advances. Coal replaced disappearing wood as the chief fuel; it was used to fuel new 16th-cent. blast furnaces making cast iron.

New World. The **Aztecs** united much of the Meso-American culture area in a militarist empire by 1519, from their capital, Tenochtitlán (pop. 300,000), which was the center of a cult requiring ritual human sacrifice. Most of the civilized areas of South America were ruled by the centralized Inca Empire (1476-1534), stretching 2,000 mi from Ecuador to NW Argentina. Lavish and sophisticated traditions in pottery, weaving, sculpture, and architecture were maintained in both regions.

These empires, beset by revolts, fell in 2 short campaigns to gold-seeking Spanish forces based in the Antilles and Panama. Hernan **Cortes** took Mexico (1519-21); Francisco **Pizarro,** Peru (1532-35). From these centers, land and sea expeditions claimed most of North and South America for Spain. The Indian high cultures did not survive the impact of Christian missionaries and the new upper class of whites and mestizos. In turn, New World silver and such Indian products as potatoes, tobacco, corn, peanuts, chocolate, and rubber exercised a major economic influence on Europe. Although the Spanish administration intermittently concerned itself with the welfare of Indians, the population remained impoverished at most levels. European diseases reduced the native population.

Brazil, which the Portuguese reached in 1500 and settled after 1530, and the Caribbean colonies of several European nations developed a plantation economy where sugarcane, tobacco, cotton, coffee, rice, indigo, and lumber were grown by slaves. From the early 16th to late 19th cent., 10 million Africans were transported to **slavery** in the New World.

Netherlands. The urban, Calvinist N provinces of the Netherlands rebelled (1568) against Habsburg Spain and founded an oligarchic mercantile republic. Their strategic control of the Baltic grain market enabled them to exploit Mediterranean food shortages. Religious refugees—French and Belgian Protestants, Iberian Jews—added to the cosmopolitan commercial talent pool. After Spain absorbed Portugal in 1580, the Dutch seized Portuguese possessions and created a vast, though short-lived commercial empire in Brazil, the Antilles, Africa, India, Ceylon, Malacca, Indonesia, and Taiwan and challenged or supplanted Portuguese traders in China and Japan. Revolution in 1640 restored Portuguese independence.

England. Anglicanism became firmly established under **Elizabeth I** after a brief Catholic interlude under "Bloody Mary" (1553-58). But religious and political conflicts led to a rebellion (1642) by Parliament. Roundheads (Puritans) defeated Cavaliers (Royalists); Charles I was beheaded (1649). The new Commonwealth was ruled as a military dictatorship by Oliver **Cromwell**, who also brutally crushed (1649-51) an Irish rebellion. Conflicts within the Puritan camp (democratic Levelers defeated, 1649) aided the Stuart restoration (1660), but Parliament was strengthened and the peaceful **"Glorious Revolution"** (1688) advanced political and religious liberties (writings of **Locke,** 1632-1704). British privateers (Drake, 1540-96) challenged Spanish control of the New World and penetrated Asian trade routes (Madras taken, 1639). North American colonies (Jamestown, 1607; Plymouth, 1620) provided an outlet for religious dissenters from Europe.

France. Emerging from the religious civil wars in 1628, France regained military and commercial great power status (under the ministries of **Richelieu**, Mazarin, and Colbert). Under **Louis XIV** (reigned 1643-1715), royal absolutism triumphed over nobles and local *parlements* (defeat of Fronde, 1648-53). Permanent colonies were founded in Canada (1608), the Caribbean (1626), and India (1674).

Sweden. Sweden seceded from the Scandinavian Union in 1523. The thinly populated agrarian state (with copper, iron, and timber exports) was united by the Vasa kings, whose conquests by the mid-17th cent. made Sweden the dominant Baltic power. The empire collapsed in the Great Northern War (1700-21).

Poland. After the union with Lithuania in 1447, Poland ruled vast territories from the Baltic to the Black Sea, resisting German and Turkish incursions. Catholic nobles failed to gain the loyalty of their Orthodox Christian subjects in the E; commerce and trades were practiced by German and Jewish immigrants. The bloody 1648-49 Cossack uprising began the kingdom's dismemberment.

China. A new dynasty, the **Manchus,** invaded from the NE, seized power in 1644, and expanded Chinese control to its greatest extent in Central and SE Asia. Trade and diplomatic contact with Europe grew, carefully controlled by China. New crops (sweet potato, maize, peanut) allowed an economic and population growth (pop. 300 million, in 1800). Traditional arts and literature were pursued with increased sophistication (*Dream of the Red Chamber*, novel, mid-18th cent.).

Japan. Tokugawa Ieyasu, shogun from 1603, finally unified and pacified feudal Japan. Hereditary daimyos and samurai monopolized government office and the professions. An urban merchant class grew, literacy spread, and a cultural renaissance occurred (**haiku,** a verse innovation of the poet Basho, 1644-94). Fear of European domination led to persecution of Christian converts from 1597 and to stringent isolation from outside contact from 1640.

Philosophy, Industry, and Revolution: 1700-1800

Science and Reason. Greater faith in human reason and empirical observation as a source of truth and a means to improve the physical and social environment, espoused since the Renaissance (Francis Bacon, 1561-1626), was bolstered by scientific discoveries in spite of theological opposition (Galileo's forced retraction, 1633). René **Descartes** (1596-1650) used a rationalistic approach modeled on geometry and introspection to discover "self-evident" truths as a foundation of knowledge. Sir Isaac **Newton** emphasized induction from experimental observation. Baruch de **Spinoza** (1632-77), who called for political and intellectual freedom, developed a systematic rationalistic philosophy in his classic work *Ethics*.

French philosophers assumed leadership of the **Enlightenment** in the 18th cent. Montesquieu (1689-1755) used British history to support his notions of limited government.

Voltaire's (1694-1778) diaries and novels of exotic travel illustrated the intellectual trends toward secular ethics and relativism. Jean-Jacques **Rousseau's** (1712-1778) radical concepts of the **social contract** and of the inherent goodness of the common man gave impetus to antimonarchical republicanism. The *Encyclopedia* (1751-72, edited by Diderot and d'Alembert), designed as a monument to reason, was largely devoted to practical technology.

In England, ideals of political and religious liberty were connected with empiricist philosophy and science in the followers of Locke. But British empiricism, especially as developed by the skeptical David **Hume** (1711-76), radically reduced the role of reason in philosophy, as did the evolutionary approach to law and politics of Edmund Burke (1729-97) and the utilitarian ethics of Jeremy Bentham (1748-1832). Adam Smith (1723-90) and other **physiocrats** called for a rationalization of economic activity by removing artificial barriers to a supposedly natural free exchange of goods.

German writers participated in the new philosophical trends popularized by Christian von Wolff (1679-1754). Immanuel **Kant's** (1724-1804) transcendental idealism, unifying an empirical epistemology with a priori moral and logical concepts, directed German thought away from skepticism. Italian contributions included work on electricity (Galvani, 1737-98; Volta, 1745-1827), the pioneer historiography of Vico (1668-1744), and writings on penal reform (Beccaria, 1738-94). Benjamin Franklin (1706-90) was celebrated in Europe for his varied achievements.

The growth of the **press** (*Spectator*, 1711-12) and the wide distribution of realistic but sentimental **novels** attested to the increase of a large bourgeois public.

Arts. Rococo art, characterized by extravagant decorative effects, asymmetries copied from organic models, and artificial pastoral subjects, was favored by the continental aristocracy for most of the cent. (Watteau, 1684-1721) and had musical analogies in the ornamentalized polyphony of late Baroque. The **Neoclassical** art after 1750, associated with the new scientific archaeology, was more streamlined and was infused with the supposed moral and geometric rectitude of the Roman Republic (David, 1748-1825). In England, **town planning** on a grand scale began.

Industrial Revolution in England. Agricultural improvements, such as the sowing drill (1701) and livestock breeding, were implemented on the large fields provided by enclosure of common lands by private owners. Profits from agriculture and from colonial and foreign trade (1800 volume, £54 million) were channeled through hundreds of banks and the **Stock Exchange** (est 1773) into new industrial processes.

The Newcomen steam pump (1712) aided coal mining. Coal fueled the new efficient steam engines patented by James Watt in 1769, and coke-smelting produced cheap, sturdy iron for machinery by the 1730s. The **flying shuttle** (1733) and **spinning jenny** (c 1764) were used in the large new cotton textile factories, where women and children were much of the work force. Goods were transported cheaply over **canals** (2,000 mi; built 1760-1800).

American Revolution. The British colonies in North America attracted a mass immigration of religious dissenters and poor people throughout the 17th and 18th cent., coming from the British Isles, Germany, the Netherlands, and other countries. The population reached 3 million non-natives by the 1770s. The small native population was greatly reduced by European diseases and by wars with and between the various colonies. British attempts to control colonial trade and to tax the colonists to pay for the costs of colonial administration and defense clashed with traditions of local self-government and eventually provoked the colonies to rebellion.

Central and East Europe. The monarchs of the three states that dominated E Europe—Austria, Prussia, and Russia—accepted the advice and legitimation of philosophes in creating more modern, centralized institutions in their kingdoms, which were enlarged by the division (1772-95) of Poland.

Under **Frederick II** (r 1740-86) Prussia, with its efficient modern army, doubled in size. State monopolies and tariff protection fostered industry, and some legal reforms were introduced. Austria's heterogeneous realms were unified under **Maria Theresa** (r 1740-80) and **Joseph II** (r 1780-90). Reforms in education, law, and religion were enacted, and the Austrian serfs were freed (1781). With its defeat in the Seven Years' War in 1763, Austria failed to regain Silesia, which had been seized by Prussia, but it was compensated by expansion to the E and S (Hungary, Slavonia, 1699; Galicia, 1772).

Russia, whose borders continued to expand in all directions, adopted some Western bureaucratic and economic policies under **Peter I** (r 1682-1725) and **Catherine II** (r 1762-96). Trade and cultural contacts with the West multiplied from the new Baltic Sea capital, **St. Petersburg** (est 1703).

French Revolution. The growing French middle class lacked political power and resented aristocratic tax privileges, especially in light of the successful American Revolution. Peasants lacked adequate land and were burdened with feudal obligations to nobles. War with Britain led to the loss of French Canada and drained the treasury, finally forcing the king to call the **Estates-General** in 1789 (first time since 1614), in an atmosphere of food riots (poor crop in 1788).

Aristocratic resistance to absolutism was soon overshadowed by the reformist Third Estate (middle class), which proclaimed itself the **National Constituent Assembly** June 17 and took the "Tennis Court oath" on June 20 to secure a constitution. The storming of the **Bastille** on July 14, 1789, by Parisian artisans was followed by looting and seizure of aristocratic property throughout France. Assembly reforms included abolition of class and regional privileges, a Declaration of Rights, suffrage by taxpayers (75% of males), and the **Civil Constitution of the Clergy** providing for election and loyalty oaths for priests. A republic was declared Sept. 22, 1792, in spite of royalist pressure from Austria and Prussia, which had declared war in April (joined by Britain the next year). Louis XVI was beheaded Jan. 21, 1793, and Queen Marie Antoinette was beheaded Oct. 16, 1793.

Royalist uprisings in La Vendée and military reverses led to institution of a **reign of terror** in which tens of thousands of opponents of the Revolution and criminals were executed. Radical reforms in the **Convention** period (Sept. 1793-Oct. 1795) included the abolition of colonial slavery, economic measures to aid the poor, support of public education, and a short-lived de-Christianization.

Division among radicals (execution of Hebert, Danton, and Robespierre, 1794) aided the ascendancy of a moderate **Directory**, which consolidated military victories. **Napoleon Bonaparte** (1769-1821), a popular young general, exploited political divisions and participated in a coup Nov. 9, 1799, making himself first consul (dictator).

India. Sikh and Hindu rebels (Rajputs, Marathas) and Afghans destroyed the power of the Mughals during the 18th cent. After France's defeat (1763) in the Seven Years' War, Britain was the primary European trade power in India. Its control of inland **Bengal and Bihar** was recognized (1765) by the Mughal shah, who granted the **British East India Co**. (under Clive, 1725-74) the right to collect land revenue there. Despite objections from Parliament (1784 India Act), the company's involvement in local wars and politics led to repeated acquisitions of new territory. The company exported Indian textiles, sugar, and indigo.

Change Gathers Steam: 1800-40

French ideals and empire spread. Inspired by the ideals of the French Revolution, and supported by the expanding French armies, new republican regimes arose near France: the **Batavian** Republic in the Netherlands (1795-1806), the **Helvetic** Republic in Switzerland (1798-1803), the **Cisalpine** Republic in N Italy (1797-1805), the **Ligurian** Republic in Genoa (1797-1805), and the **Parthenopean** Republic in S Italy (1799). A Roman Republic existed briefly in 1798 after Pope Pius VI was arrested by French troops. In Italy and Germany, new nationalist sentiments were stimulated both in imitation of and in reaction to developments in France (anti-French and anti-Jacobin peasant uprisings in Italy, 1796-99).

From 1804, when Napoleon declared himself emperor, to 1812, a succession of military victories (Austerlitz, 1805; Jena, 1806) extended his control over most of Europe, through puppet states (**Confederation of the Rhine** united W German states for the first time and **Grand Duchy of Warsaw** revived Polish national hopes), expansion of the empire, and alliances.

Among the lasting reforms initiated under Napoleon's absolutist reign were: establishment of the Bank of France, centralization of tax collection, codification of law along Roman models (Code Napoléon), and reform and extension of secondary and university education. In an 1801 concordat, the papacy recognized the effective autonomy of the French Catholic Church.

Napoleon's continental successes were offset by British victory under Adm. Horatio Nelson in the **Battle of Trafalgar** (1805).

In all, some 400,000 French soldiers were killed in the Napoleonic Wars, along with about 600,000 foreign troops.

Last gasp of old regime. The disastrous 1812 invasion of Russia exposed Napoleon's overextension. After Napoleon's 1814 exile at Elba, his armies were defeated (1815) at **Waterloo**, by British and Prussian troops.

At the **Congress of Vienna**, the monarchs and princes of Europe redrew their boundaries, to the advantage of Prussia (in Saxony and the Ruhr), Austria (in Illyria and Venetia), and Russia (in Poland and Finland). British conquest of Dutch and French colonies (S Africa, Ceylon, Mauritius) was recognized, and France, under the restored Bourbons, retained its expanded 1792 borders. The settlement brought 50 years of international peace to Europe.

But the Congress was unable to check the advance of liberal ideals and of nationalism among the smaller European nations. The 1825 **Decembrist uprising** by liberal officers in Russia was easily suppressed. But an independence movement in **Greece**, stirred by commercial prosperity and a cultural revival, succeeded in expelling Ottoman rule by 1831, with the aid of Britain, France, and Russia.

A constitutional monarchy was secured in France by the **1830 Revolution**; Louis Philippe became king. The revolutionary contagion spread to **Belgium**, which gained its independence (1830) from the Dutch monarchy, to **Poland**, whose rebellion was defeated (1830-31) by Russia, and to Germany.

Romanticism. A new style in intellectual and artistic life began to replace Neoclassicism and Rococo after the mid-18th cent. By the early 19th cent., this style, Romanticism, had prevailed in the European world.

Rousseau had begun the reaction against rationalism; in education (*Émile*, 1762) he stressed subjective spontaneity over regularized instruction. German writers (Lessing, 1729-81; Herder, 1744-1803) favorably compared the German folk song to classical forms and began a cult of Shakespeare, whose passion and "natural" wisdom was a model for the romantic *Sturm und Drang* (Storm and Stress) movement. **Goethe's** *Sorrows of Young Werther* (1774) set the model for the tragic, passionate genius.

A new interest in **Gothic architecture** in England after 1760 (Walpole, 1717-97) spread through Europe, associated with an aesthetic Christian and mystic revival (**Blake,** 1757-1827). Celtic, Norse, and German mythology and folk tales were revived or imitated (Macpherson's Ossian translation, 1762; Grimm's Fairy Tales, 1812-22). The medieval revival (Scott's *Ivanhoe*, 1819) led to a new interest in history, stressing national differences and organic growth (**Carlyle,** 1795-1881; Michelet, 1798-1874), corresponding to theories of natural evolution (Lamarck's *Philosophie Zoologique*, 1809; Lyell's *Geology*, 1830-33). A reaction against classicism characterized the English **romantic poets** (beginning with **Wordsworth,** 1770-1850). Revolution and war fed an emphasis on freedom and conflict, expressed by both poets (**Byron**, 1788-1824; **Hugo**, 1802-85) and philosophers (**Hegel**, 1770-1831).

Wild gardens replaced the formal French variety, and painters favored rural, stormy, and mountainous landscapes (**Turner**, 1775-1851; **Constable**, 1776-1837). Clothing became freer, with wigs, hoops, and ruffles discarded. Originality and genius were expected in the life as well as the work of inspired artists (Murger's *Scenes from Bohemian Life*, 1847-49). Exotic locales and themes (as in Gothic horror stories) were used in art and literature (Delacroix, 1798-1863; **Poe**, 1809-49).

Music exhibited the new dramatic style and a breakdown of classical forms (**Beethoven,** 1770-1827). The use of folk melodies and modes aided the growth of distinct national traditions (Glinka in Russia, 1804-57).

Latin America. Francois **Toussaint L'Ouverture** led a successful slave revolt in Haiti, which subsequently became the first Latin American state to achieve independence (1804). The mainland Spanish colonies won their independence (1810-24), under such leaders as Simon **Bolivar** (1783-1830). Brazil became an independent empire (1822) under the Portuguese prince regent. A new class of military officers divided power with large landholders and the church.

United States. Heavy immigration and exploitation of ample natural resources fueled rapid economic growth. The spread of the franchise, public education, and antislavery sentiment were signs of a widespread democratic ethic.

China. Failure to keep pace with Western arms technology exposed China to greater European influence and hampered efforts to bar imports of opium, which had damaged Chinese society and drained wealth overseas. In the **Opium War** (1839-42), Britain forced China to expand trade opportunities and to cede Hong Kong.

Triumph of Progress: 1840-80

Idea of Progress. As a result of the cumulative scientific, economic, and political changes of the preceding eras, the idea took hold among literate people in the West that continuing growth and improvement was the usual state of human and natural life.

Darwin's statement of the **theory of evolution** and survival of the fittest (*Origin of Species*, 1859), defended by intellectuals and scientists against theological objections, was taken as confirmation that progress was the natural direction of life. The controversy helped define popular ideas of the dedicated scientist and ever-expanding human knowledge of and control over the world (Foucault's demonstration of earth's rotation, 1851; **Pasteur's** germ theory, 1861).

Liberals following Ricardo (1772-1823) in their faith that unrestrained competition would bring continuous economic expansion sought to adjust political life to the new social realities and believed that unregulated competition of ideas would yield truth (**Mill,** 1806-73). In England, successive reform bills (1832, 1867, 1884) gave representation to the new industrial towns and extended the franchise to the middle and lower classes and to Catholics, Dissenters, and Jews. On both sides of the Atlantic, reformists tried to improve conditions for the mentally ill (**Dix,** 1802-87), women (Anthony, 1820-1906), and prisoners. Slavery was barred in the British Empire (1833), the U.S. (1865), and Brazil (1888).

IT'S A FACT: Tennessean William Walker was once president of Nicaragua. The adventurer and soldier of fortune led a revolution there, declaring himself president on July 12, 1856. Forced to flee on May 1, 1857, he returned to Central America twice more before being captured by the British Navy, turned over to authorities in Honduras, court-martialed, and executed in 1860.

IT'S A FACT: On May 15, 1862, Maj. Gen. Benjamin F. Butler, commander of the Union forces occupying New Orleans, issued General Orders No. 28—the "Woman's Order"—declaring women who insulted uniformed soldiers "by word, gesture, or movement" subject to arrest as prostitutes. His order earned him the nickname "Beast," and widespread scorn in Europe and the South.

Socialist theories based on ideas of human perfectibility or progress were widely disseminated. Utopian socialists such as Saint-Simon (1760-1825) envisaged an orderly, just society directed by a technocratic elite. A model factory town, New Lanark, Scotland, was set up by utopian Robert Owen (1771-1858), and communal experiments were tried in the U.S. (most notably, Brook Farm, Mass., 1841-47). Bakunin's (1814-76) anarchism represented the opposite utopian extreme of total freedom. Karl **Marx** (1818-83) posited the inevitable triumph of socialism in industrial countries through a dialectical process of class conflict.

Spread of industry. The technical processes and managerial innovations of the English industrial revolution spread to Europe (especially Germany) and the U.S., causing an explosion of industrial production, demand for raw materials, and competition for markets. Inventors, both trained and self-educated, provided the means for larger-scale production (Bessemer steel, 1856; sewing machine, 1846). Many inventions were shown at the 1851 London Great Exhibition at the **Crystal Palace,** the theme of which was universal prosperity.

Local specialization and long-distance trade were aided by a revolution in transportation and communication. Railroads were first introduced in the 1820s in England and the U.S. More than 150,000 mi of track had been laid worldwide by 1880, with another 100,000 mi laid in the next decade. Steamships were improved (*Savannah* crossed Atlantic, 1819). The **telegraph**, perfected by 1844 (Morse), connected the Old and New Worlds by cable in 1866 and quickened the pace of international commerce and politics. The first commercial **telephone** exchange went into operation in the U.S. in 1878.

The new class of industrial workers, uprooted from their rural homes, lacked job security and suffered from dangerous overcrowded conditions at work and at home. Many responded by organizing **trade unions** (legalized in England, 1824; France, 1884). The U.S. Knights of Labor had 700,000 members by 1886. The First International (1864-76) tried to unite workers internationally around a Marxist program. The quasi-Socialist Paris Commune uprising (1871) was violently suppressed. Factory Acts to reduce child labor and regulate conditions were passed (1833-50 in England). Social security measures were introduced by the Bismarck regime (1883-89) in Germany.

Revolutions of 1848. Among the causes of the continent-wide revolutions were an international collapse of credit and resulting unemployment, bad harvests in 1845-47, and a cholera epidemic. The new urban proletariat and expanding bourgeoisie demanded a greater political role. Republics were proclaimed in France, Rome, and Venice. Nationalist feelings reached fever pitch in the Habsburg empire, as Hungary declared independence under Kossuth, as a Slav Congress demanded equality, and as Piedmont tried to drive Austria from Lombardy. A national liberal assembly at Frankfurt called for German unification.

But riots fueled bourgeois fears of socialism (**Marx and Engels**, *Communist Manifesto*, 1848), and peasants remained conservative. The old establishment—the Papacy, the Habsburgs with the help of the Czarist Russian army — was able to rout the revolutionaries by 1849. The French Republic succumbed to a renewed monarchy by 1852 (Emperor Napoleon III).

Great nations unified. Using the "blood and iron" tactics of Bismarck from 1862, Prussia controlled N Germany by 1867 (war with Denmark, 1864; Austria, 1866). After defeating France in 1870 (annexation of Alsace-Lorraine), it won the allegiance of S German states. A new **German Empire** was proclaimed (1871). **Italy**, inspired by Giuseppe Mazzini (1805-72) and Giuseppe Garibaldi (1807-82), was unified by the reformed Piedmont kingdom through uprisings, plebiscites, and war.

The **U.S.**, its area expanded after the 1846-48 Mexican War, defeated (1861-65) a secession attempt by slave states. in the **Civil War.** Canadian provinces were united in an autonomous **Dominion of Canada** (1867). Control in **India** was removed from the East India Co. and centralized under British administration after the 1857-58 Sepoy rebellion, laying the groundwork for the modern Indian State. Queen Victoria was named Empress of India (1876).

Europe dominates Asia. The Ottoman Empire began to collapse in the face of Balkan nationalisms and European imperial incursions in N Africa (**Suez Canal**, 1869). The Turks had lost control of most of both regions by 1882. Russia completed its expansion S by 1884 (despite the temporary setback of the **Crimean War** with Turkey, Britain, and France, 1853-56), taking Turkestan, all the Caucasus, and Chinese areas in the E and sponsoring Balkan Slavs against the Turks. A succession of reformist and reactionary regimes presided over a slow modernization (serfs freed, 1861). Persian independence suffered as Russia and British India competed for influence.

China was forced to sign a series of unequal treaties with European powers and Japan. Overpopulation and an inefficient dynasty brought misery and caused rebellions (Taiping, Muslims) leaving tens of millions dead. **Japan** was forced by the U.S. (Commodore Perry's visits, 1853-54) and Europe to end its isolation. The Meiji restoration (1868) gave power to a Westernizing oligarchy. Intensified empire-building gave Burma to Britain (1824-85) and Indochina to France (1862-95). Christian missionary activity followed imperial and trade expansion in Asia.

Respectability. The fine arts were expected to reflect and encourage the good morals and manners among the Victorians. Prudery, exaggerated delicacy, and familial piety were heralded by **Bowdler's** expurgated edition (1818) of Shakespeare. Government-supported mass education sought to inculcate a work ethic as a means to escape poverty (**Horatio Alger,** 1832-99).

The official **Beaux Arts** school in Paris set an international style of imposing public buildings (Paris Opera, 1861-74; Vienna Opera, 1861-69) and uplifting statues (Bartholdi's Statue of Liberty, 1884). Realist painting, influenced by photography (Daguerre, 1837), appealed to a new mass audience with social or historical narrative (Wilkie, 1785-1841; Poynter, 1836-1919) or with serious religious, moral, or social messages (pre-Raphaelites, Millet's *Angelus*, 1858) often drawn from ordinary life. The **Impressionists** (Monet, 1840-1926; Pissarro, 1830-1903; Renoir, 1841-1919) rejected the formalism, sentimentality, and precise techniques of academic art in favor of a spontaneous, undetailed rendering of the world through careful representation of the effect of natural light on objects.

Realistic **novelists** presented the full panorama of social classes and personalities, but retained sentimentality and moral judgment (**Dickens,** 1812-70; **Eliot,** 1819-80; **Tolstoy,** 1828-1910; **Balzac,** 1799-1850).

Veneer of Stability: 1880-1900

Imperialism triumphant. The vast **African** interior, visited by European explorers (Barth, 1821-65; Livingstone, 1813-73), was conquered by the European powers in rapid, competitive thrusts from their coastal bases after 1880, mostly for domestic political and international strategic reasons. W African Muslim kingdoms (Fulani), Arab slave traders (Zanzibar), and Bantu military confederations (Zulu) were alike subdued. Only Christian Ethiopia (defeat of Italy, 1896) and Liberia resisted successfully. France (W Africa) and Britain ("Cape to Cairo," **Boer War,** 1899-1902) were the major beneficiaries. The ideology of "the white man's burden" (Kipling, *Barrack Room Ballads*, 1892) or of a "civilizing mission" (France) justified the conquests.

W European foreign capital investment soared to nearly $40 billion by 1914, but most was in E Europe (France, Germany), the Americas (Britain), and the Europeans' colonies. The foundation of the modern interdependent world economy was laid, with cartels dominating raw material trade.

An industrious world. Industrial and technological proficiency characterized the 2 new great powers—Germany and the U.S. Coal and iron deposits enabled Germany to reach 2d or 3d place status in iron, steel, and shipbuilding by the 1900s. German electrical and chemical industries were world leaders. The U.S. post-Civil War boom (interrupted by "panics"—1884, 1893, 1896) was shaped by massive immigration from S and E Europe from 1880, government subsidy of railroads, and huge private monopolies (Standard Oil, 1870; U.S. Steel, 1901). The **Spanish-American War**, 1898 (Philippine Insurrection, 1899-1902), and the **Open Door policy** in China (1899) made the U.S. a world power.

England led in **urbanization**, with **London** the world capital of finance, insurance, and shipping. Sewer systems (Paris, 1850s), electric subways (London, 1890), parks, and bargain department stores helped improve living standards for most of the urban population of the industrial world.

Westernization of Asia. Asian reaction to European economic, military, and religious incursions took the form of imitation of Western techniques and adoption of Western ideas of progress and freedom. The Chinese "self-strengthening" movement of the 1860s and 1870s included rail, port, and arsenal improvements and metal and textile mills. Reformers such as **K'ang Yu-wei** (1858-1927) won liberalizing reforms in 1898, right after the European and Japanese "scramble for concessions."

A universal education system in Japan and importation of foreign industrial, scientific, and military experts aided Japan's unprecedented rapid modernization after 1868, under the authoritarian Meiji regime. Japan's victory in the **Sino-Japanese War** (1894-95) put Formosa and Korea in its power.

In India, the British alliance with the remaining princely states masked reform sentiment among the Westernized urban elite; higher education had been conducted largely in English for 50 years. The **Indian National Congress**, founded in 1885, demanded a larger government role for Indians.

***Fin-de-siècle* sophistication. Naturalist** writers pushed realism to its extreme limits, adopting a quasi-scientific attitude and writing about formerly taboo subjects such as sex, crime, extreme poverty, and corruption (Flaubert, 1821-80; Zola, 1840-1902; Hardy, 1840-1928). Unseen or repressed psychological motivations were explored in the clinical and theoretical works of Sigmund **Freud** (1856-1939) and in works of fiction (**Dostoyevsky,** 1821-81; James, 1843-1916; Schnitzler, 1862-1931; others).

A contempt for bourgeois life or a desire to shock a complacent audience was shared by the French **symbolist** poets (Verlaine, 1844-96; Rimbaud, 1854-91), by neopagan English writers (Swinburne, 1837-1909), by continental dramatists (**Ibsen,** 1828-1906), and by satirists (**Wilde,** 1854-1900). The German philosopher Friedrich **Nietzsche** (1844-1900) was influential in his elitism and pessimism.

Postimpressionist art neglected long-cherished conventions of representation (Cézanne, 1839-1906) and showed a willingness to learn from primitive and non-European art (Gauguin, 1848-1903; Japanese prints).

Racism. Gobineau (1816-82) gave a pseudobiological foundation to modern racist theories, which spread in Europe in the latter 19th cent., along with **Social Darwinism**, the belief that societies are and should be organized as a struggle for survival of the fittest. The medieval period was interpreted as an era of natural Germanic rule (Chamberlain, 1855-1927), and notions of racial superiority were associated with German national aspirations (Treitschke, 1834-96). **Anti-Semitism**, with a new racist rationale, became a significant political force in Germany (Anti-Semitic Petition, 1880), Austria (Lueger, 1844-1910), and France (Dreyfus **case**, 1894-1906).

Last Respite: 1900-9

Alliances. While the peace of Europe (and its dependencies) continued to hold (1907 **Hague Conference** extended the rules of war and international arbitration procedures), imperial rivalries, protectionist trade practices (in Germany and France), and the escalating arms race (British *Dreadnought* battleship launched; Germany widens Kiel canal, 1906) exacerbated minor disputes (German-French Moroccan "crises," 1905, 1911).

Security was sought through alliances: **Triple Alliance** (Germany, Austria-Hungary, Italy; renewed in 1902 and 1907); Anglo-Japanese Alliance (1902), Franco-Russian Alliance (1899), **Entente Cordiale** (Britain, France, 1904), Anglo-Russian Treaty (1907), German-Ottoman friendship.

Ottomans decline. The inefficient, corrupt Ottoman government was unable to resist further loss of territory. Nearly all European lands were lost in 1912 to Serbia, Greece, Montenegro, and Bulgaria. Italy took Libya and the Dodecanese islands the same year, and Britain took Kuwait (1899) and the Sinai (1906). The **Young Turk** revolution in 1908 forced the sultan to restore a constitution, and it introduced some social reform, industrialization, and secularization.

British Empire. British trade and cultural influence remained dominant in the empire, but constitutional reforms presaged its eventual dissolution: The colonies of **Australia** were united in 1901 under a self-governing commonwealth. **New Zealand** acquired dominion status in 1907. The old Boer republics joined Cape Colony and Natal in the self-governing **Union of South Africa** in 1910.

The 1909 Indian Councils Act enhanced the role of elected province legislatures in **India**. The Muslim League (founded 1906) sought separate communal representation.

East Asia. Japan exploited its growing industrial power to expand its empire. Victory in the 1904-5 war against Russia (naval battle of Tsushima, 1905) assured Japan's domination of **Korea** (annexed 1910) and Manchuria (Port Arthur taken, 1905).

In China, central authority began to crumble (empress died, 1908). Reforms (Confucian exam system ended 1905, modernization of the army, building of railroads) were inadequate, and secret societies of reformers and nationalists, inspired by the Westernized **Sun Yat-sen** (1866-1925) fomented periodic uprisings in the S.

Siam, whose independence had been guaranteed by Britain and France in 1896, was split into spheres of influence by those countries in 1907.

Russia. The population of the Russian Empire approached 150 million in 1900. Reforms in education, in law, and in local institutions (*zemstvos*) and an industrial boom starting in the 1880s (oil, railroads) created the beginnings of a modern state, despite the autocratic tsarist regime. Liberals (1903 Union of Liberation), Socialists (Social Democrats founded 1898, Bolsheviks split off 1903), and populists (Social Revolutionaries founded 1901) were periodically repressed, and national minorities were persecuted (anti-Jewish pogroms, 1903, 1905-6).

An industrial crisis after 1900 and harvest failures aggravated poverty among urban workers, and the 1904-5 defeat by Japan (which checked Russia's Asian expansion) sparked **the Revolution of 1905-6.** A **Duma** (parliament) was created, and an agricultural reform (under Stolypin, prime minister 1906-11) created a large class of land-owning peasants (kulaks).

IT'S A FACT: But for a deadly volcanic eruption that killed 28,000 people on the island of Martinique in May 1902, there might have been a "Nicaragua Canal." A nervous U.S. Congress, flustered by news of volcanic activity, decided in June to favor a route across Panama instead.

The world shrinks. Developments in transportation and communication and mass population movements helped create an awareness of an interdependent world. Early **automobiles** (Daimler, Benz, 1885) were experimental or were designed as luxuries. Assembly-line mass production (Ford Motor Co., 1903) made the invention practicable, and by 1910 nearly 500,000 motor vehicles were registered in the U.S. alone. **Heavier-than-air flights** began in 1903 in the U.S. (Wright brothers), preceded by glider, balloon, and model plane advances in several countries. Trade was advanced by improvements in **ship design** (gyrocompass, 1910), speed (*Lusitania* crossed Atlantic in 5 days, 1907), and reach (Panama Canal begun, 1904).

The first transatlantic **radio** telegraphic transmission occurred in 1901, 6 years after Marconi discovered radio. Radio transmission of human speech had been made in 1900. Telegraphic transmission of photos was achieved in 1904, lending immediacy to news reports. **Phonographs**, popularized by Caruso's recordings (starting 1902), made for quick international spread of musical styles (ragtime). **Motion pictures**, perfected in the 1890s (Dickson, Lumière brothers), became a popular and artistic medium after 1900; newsreels appeared in 1909.

Emigration from crowded European centers soared in the decade: 9 million migrated to the U.S., and millions more went to Siberia, Canada, Argentina, Australia, South Africa, and Algeria. Some 70 million Europeans emigrated in the cent. before 1914. Several million Chinese, Indians, and Japanese migrated to SE Asia, where their urban skills often enabled them to take a predominant economic role.

Social reform. The social and economic problems of the poor were kept in the public eye by realist fiction writers (Dreiser's *Sister Carrie*, 1900; Gorky's *Lower Depths*, 1902; Sinclair's *The Jungle*, 1906), journalists (U.S. **muckrakers**—Steffens, Tarbell), and artists (Ashcan school). Frequent labor strikes and occasional assassinations by anarchists or radicals (Empress Elizabeth of Austria, 1898; King Umberto I of Italy, 1900; U.S. Pres. McKinley, 1901; Russian Interior Minister Plehve, 1904; Portugal's King Carlos, 1908) added to social tension and fear of revolution.

But democratic reformism prevailed. In Germany, Bernstein's (1850-1932) **revisionist Marxism**, downgrading revolution, was accepted by the powerful Social Democrats and trade unions. The British Fabian Society (the Webbs, Shaw) and the Labour Party (founded 1906) worked for reforms such as Social Security and union rights (1906), while woman suffragists grew more militant. U.S. **progressives** fought big business (Pure Food and Drug Act, 1906). In France, the 10-hour work day (1904) and separation of church and state (1905) were reform victories, as was universal suffrage in Austria (1907).

Arts. An unprecedented period of experimentation, centered in France, produced several new **painting** styles: Fauvism exploited bold color areas (Matisse, *Woman With Hat*, 1905); expressionism reflected powerful inner emotions (the Brücke group, 1905); cubism combined several views of an object on one flat surface (Picasso's *Demoiselles*, 1906-7); futurism tried to depict speed and motion (Italian Futurist Manifesto, 1910). **Architects** explored new uses of steel structures, with facades either neoclassical (Adler and Sullivan in U.S.); curvilinear Art Nouveau (Gaudi's Casa Mila, 1905-10); or functionally streamlined (Wright's Robie House, 1909).

Music and dance shared the experimental spirit. Ruth St. Denis (1877-1968) and Isadora Duncan (1878-1927) pioneered modern dance, while Sergei Diaghilev in Paris revitalized classic ballet from 1909. Composers explored atonal music (Debussy, 1862-1918) and dissonance (Schoenberg, 1874-1951) or revolutionized classical forms (Stravinsky, 1882-1971), often showing jazz or folk music influences.

War and Revolution: 1910-19

War threatens. Germany under Wilhelm II sought a political and imperial role consonant with its industrial strength, challenging Britain's world supremacy and threatening France, which was still resenting the loss (1871) of Alsace-Lorraine. Austria wanted to curb an expanded Serbia (after 1912) and the threat it posed to its own Slav lands. Russia feared Austrian and German political and economic aims in the Balkans and Turkey.

An accelerated arms race resulted from these circumstances. The German standing army rose to more than 2 million men by 1914. Russia and France had more than a million each, and Austria and the British Empire nearly a million each. Dozens of enormous battleships were built by the powers after 1906.

The **assassination of Austrian Archduke Franz Ferdinand** by a Serbian, June 28, 1914, was the pretext for war. The system of alliances made the conflict Europe-wide; Germany's invasion of Belgium to outflank France forced Britain to enter the war. Patriotic fervor was nearly unanimous among all classes in most countries.

World War I. German forces were stopped in France in one month. The rival armies dug **trench networks**. Artillery and improved machine guns prevented either side from any lasting advance despite repeated assaults (600,000 dead at **Verdun**, Feb.-July 1916). Poison gas, used by Germany in 1915, proved ineffective. The entrance of more than 1 million U.S. troops tipped the balance after mid-1917, forcing Germany to sue for peace the next year. The formal armistice was signed on Nov. 11, 1918.

In the E, the Russian armies were thrown back (battle of **Tannenberg**, Aug. 20, 1914), and the war grew unpopular in Russia. An allied attempt to relieve Russia through Turkey failed (**Gallipoli,** 1915). The **Russian Revolution** (1917) abolished the monarchy. The new Bolshevik regime signed the capitulatory Brest-Litovsk peace in March 1918. Italy entered the war on the allied side in May 1915 but was pushed back by Oct. 1917. A renewed offensive with Allied aid in Oct.-Nov. 1918 forced Austria to surrender.

The British Navy successfully blockaded Germany, which responded with submarine U-boat attacks; **unrestricted submarine warfare** against neutrals after Jan. 1917 helped bring the U.S. into the war. Other battlefields included Palestine and Mesopotamia, both of which Britain wrested from the Turks in 1917, and the African and Pacific colonies of Germany, most of which fell to Britain, France, Australia, Japan, and South Africa.

Settlement. At the **Paris Peace Conference** (Jan.-June 1919), concluded by the **Treaty of Versailles**, and in subsequent negotiations and local wars (Russian-Polish War, 1920), the map of Europe was redrawn with a nod to U.S. Pres. Wilson's principle of self-determination. Austria and Hungary were separated, and much of their land was given to Yugoslavia (formerly Serbia), Romania, Italy, and the newly independent Poland and Czechoslovakia. Germany lost territory in the W, N, and E, while Finland and the Baltic states were detached from Russia. Turkey lost nearly all its Arab lands to British-sponsored Arab states or to direct French and British rule. Belgium's sovereignty was recognized.

From 1916, the civilian populations and economies of both sides were mobilized to an unprecedented degree. Hardships intensified among fighting nations in 1917 (French mutiny crushed in May). More than 10 million soldiers died in the war.

A huge **reparations** burden and partial demilitarization were imposed on Germany. Pres. Wilson obtained approval for a League of Nations, but the U.S. Senate refused to allow the U.S. to join.

Russian revolution. Military defeats and high casualties caused a contagious lack of confidence in Tsar Nicholas, who was forced to abdicate Mar. 1917. A liberal provisional government failed to end the war, and massive desertions, riots, and fighting between factions followed. A moderate socialist government under Aleksandr Kerensky was overthrown (Nov. 1917) in a violent coup by the **Bolsheviks** in Petrograd under **Lenin,** who later disbanded the elected Constituent Assembly.

The Bolsheviks brutally suppressed all opposition and ended the war with Germany in Mar. 1918. **Civil war** broke out in the summer between the Red Army, including the Bolsheviks and their supporters, and monarchists, anarchists, nationalities (Ukrainians, Georgians, Poles), and others. Small U.S., British, French, and Japanese units also opposed the Bolsheviks (1918-19; Japan in Vladivostok to 1922). The civil war, anarchy, and pogroms devastated the country until the 1920 Red Army victory. The wartime total monopoly of political, economic, and police power by the Communist Party leadership was retained.

Other European revolutions. An unpopular monarchy in **Portugal** was overthrown in 1910. The new republic took severe anticlerical measures in 1911.

After a century of Home Rule agitation, during which **Ireland** was devastated by famine (1 million dead, 1846-47) and emigration, republican militants staged an unsuccessful uprising in Dublin during Easter 1916. The execution of the leaders and mass arrests by the British won popular support for the rebels. The Irish Free State, comprising all but the 6 N counties, achieved dominion status in 1922.

In the aftermath of the world war, radical revolutions were attempted in Germany (**Spartacist** uprising, Jan. 1919), **Hungary** (Kun regime, 1919), and elsewhere. All were suppressed or failed for lack of support.

Chinese revolution. The Manchu Dynasty was overthrown and a republic proclaimed in Oct. 1911. First Pres. Sun Yat-sen resigned in favor of strongman Yuan Shih-k'ai. Sun organized the parliamentarian **Kuomintang** party.

Students launched protests on May 4, 1919, against League of Nations concessions in China to Japan. Nationalist, liberal, and socialist ideas and political groups spread. The **Communist Party** was founded in 1921. A Communist regime took power in Mongolia with Soviet support in 1921.

India restive. Indian objections to British rule erupted in nationalist riots as well as in the nonviolent tactics of Mahatma **Gandhi** (1869-1948). Nearly 400 unarmed demonstrators were shot at **Amritsar** in Apr. 1919. Britain approved limited self-rule that year.

Mexican revolution. Under the long Diaz dictatorship (1877-1911) the economy advanced, but Indian and mestizo lands were confiscated, and concessions to foreigners (mostly U.S.) damaged the middle class. A **revolution in 1910** led to civil wars and U.S. intervention (1914, 1916-17). Land reform and a more democratic constitution (1917) were achieved.

The Aftermath of War: 1920-29

U.S. Easy credit, technological ingenuity, and war-related industrial decline in Europe caused a long economic boom, in which ownership of the new products—**autos, phones, radios**—became democratized. Prosperity, an increase in women workers, woman suffrage (1920), and drastic change in fashion (flappers, mannish bob for women, clean-shaven men) created a wide perception of social change, despite prohibition of alcoholic beverages (1919-33). Union membership and strikes increased. Fear of radicals led to Palmer raids (1919-20) and the Sacco/Vanzetti case (1921-27).

Europe sorts itself out. Germany's liberal **Weimar constitution** (1919) could not guarantee a stable government in the face of rightist violence (Rathenau assassinated, 1922) and Communist refusal to cooperate with Socialists. Reparations and Allied occupation of the Rhineland caused staggering inflation that destroyed middle-class savings, but economic expansion resumed after mid-decade, aided by U.S. loans. A sophisticated, **innovative culture** developed in architecture and design (Bauhaus, 1919-28), film (Lang, *M*, 1931), painting (Grosz), music (Weill, *Threepenny Opera*, 1928), theater (Brecht, *A Man's a Man*, 1926), criticism (Benjamin), philosophy (Jung), and fashion. This culture was considered decadent and socially disruptive by rightists.

England elected its first Labour governments (Jan. 1924, June 1929). A 10-day general strike in support of coal miners failed in May 1926. In **Italy**, strikes, political chaos, and violence by small Fascist bands culminated in the Oct. 1922 Fascist March on Rome, which established Mussolini's dictatorship. Strikes were outlawed (1926), and Italian influence was pressed in the Balkans (Albania a protectorate, 1926). A conservative dictatorship was also established in **Portugal** in a 1926 military coup.

Czechoslovakia, the only stable democracy to emerge from the war in Central or East Europe, faced opposition from Germans (in the Sudetenland), Ruthenians, and some Slovaks. As the industrial heartland of the old Habsburg empire, it remained fairly prosperous. With French backing, it formed the Little Entente with Yugoslavia (1920) and **Romania** (1921) to block Austrian or Hungarian irredentism. Hungary remained dominated by the landholding classes and expansionist feeling. Croats and Slovenes in **Yugoslavia** demanded a federal state until King Alexander I proclaimed (1929) a royal dictatorship. Poland faced nationality problems as well (Germans, Ukrainians, Jews); Pilsudski ruled as dictator from 1926. The Baltic states were threatened by traditionally dominant ethnic Germans and by Soviet-supported Communists.

An economic collapse and famine in **Russia** (1921-22) claimed 5 million lives. The New Economic Policy (1921) allowed land ownership by peasants and some private commerce and industry. Stalin was absolute ruler within 4 years of Lenin's death (1924). He inaugurated a brutal collectivization program (1929-32) and used foreign Communist parties for Soviet state advantage.

Internationalism. Revulsion against World War I led to pacifist agitation, to the Kellogg-Briand Pact renouncing aggressive war (1928), and to **naval disarmament** pacts (Washington, 1922; London, 1930). But the League of Nations was able to arbitrate only minor disputes (Greece-Bulgaria, 1925).

Middle East. Mustafa Kemal (**Ataturk**) led **Turkish** nationalists in resisting Italian, French, and Greek military advances (1919-23). The sultanate was abolished (1922), and elaborate reforms were passed, including secularization of law and adoption of the Latin alphabet. Ethnic conflict led to persecution of **Armenians** (more than 1 million dead in 1915, 1 million expelled), Greeks (forced Greek-Turk population exchange, 1923), and Kurds (1925 uprising).

With evacuation of the Turks from **Arab** lands, the puritanical Wahabi dynasty of E Arabia conquered (1919-25) what is now Saudi Arabia. British, French, and Arab dynastic and nationalist maneuvering resulted in the creation of 2 more Arab monarchies in 1921—Iraq and Transjordan (both under British control)—and 2 French mandates—Syria and Lebanon. Jewish immigration into British-mandated **Palestine**, inspired by the Zionist movement, was resisted by Arabs, at times violently (1921, 1929 massacres).

Reza Khan ruled **Persia** after his 1921 coup (shah from 1925), centralized control, and created the trappings of a modern secular state.

China. The Kuomintang under **Chiang Kai-shek** (1887-1975) subdued the warlords by 1928. The Communists were brutally suppressed after their alliance with the Kuomintang was broken in 1927. Relative peace thereafter allowed for industrial and financial improvements, with some Russian, British, and U.S. cooperation.

Arts. Nearly all bounds of subject matter, style, and attitude were broken in the arts of the period. **Abstract** art first took inspiration from natural forms or narrative themes (Kandinsky from 1911) and then worked free of any representational aims (Malevich's suprematism, 1915-19; Mondrian's geometric style from 1917). The **Dada** movement (from 1916) mocked artistic pretension with absurd collages and constructions (Arp, Tzara, from 1916). Paradox, illusion, and psychological taboos were exploited by **surrealists** by the latter 1920s (Dali, Magritte). Architectural schools celebrated industrial values, whether vigorous abstract constructivism (Tatlin, *Monument to 3rd International*, 1919) or the machined, streamlined **Bauhaus** style, which was extended to many design fields (Helvetica typeface).

Prose writers explored revolutionary narrative modes related to dreams (Kafka's *Trial*, 1925), internal monologue

(Joyce's **Ulysses**, 1922), and word play (Stein's *Making of Americans*, 1925). Poets and novelists wrote of modern alienation (Eliot's ***Waste Land***, 1922) and aimlessness (Lost Generation).

Sciences. Scientific specialization prevailed by the 20th cent. Advances in knowledge and technological aptitude increased with the geometric rise in the number of practitioners. Physicists challenged common-sense views of causality, observation, and a mechanistic universe, putting science further beyond popular grasp (**Einstein's** general theory of relativity, 1916; Bohr's quantum mechanics, 1913; Heisenberg's uncertainty principle, 1927).

Rise of Totalitarians: 1930-39

Depression. A worldwide financial panic and economic depression began with the Oct. 1929 U.S. stock market crash and the May 1931 failure of the Austrian Credit-Anstalt. A credit crunch caused international bankruptcies and **unemployment**: 12 million jobless by 1932 in the U.S., 5.6 million in Germany, 2.7 million in England. Governments responded with **tariff restrictions** (Smoot-Hawley Act, 1930; Ottawa Imperial Conference, 1932), which dried up world trade. Government public works programs were vitiated by deflationary budget balancing.

Germany. Years of agitation by violent extremists were brought to a head by the Depression. Nazi leader Adolf **Hitler** was named chancellor in Jan. 1933 and given dictatorial power by the Reichstag in March. Opposition parties were disbanded, strikes banned, and all aspects of economic, cultural, and religious life were brought under central government and Nazi party control and manipulated by sophisticated propaganda. Severe persecution of Jews began (**Nuremberg Laws,** Sept. 1935). Many Jews, political opponents, and others were sent to concentration camps (Dachau, 1933), where thousands died or were killed. Public works, renewed conscription (1935), arms production, and a 4-year plan (1936) all but ended unemployment.

Hitler's expansionism started with reincorporation of the Saar (1935), occupation of the **Rhineland** (Mar. 1936), and annexation of Austria (Mar. 1938). At **Munich** (Sept. 1938) an indecisive Britain and France sanctioned German dismemberment of Czechoslovakia.

Russia. Urbanization and education advanced. Rapid industrialization was achieved through successive **5-year plans** starting in 1928, using severe labor discipline and mass forced labor. Industry was financed by a decline in living standards and exploitation of agriculture, which was almost totally collectivized by the early 1930s (*kolkhoz*, collective farm; *sovkhoz*, state farm, often in newly worked lands). Successive **purges** increased the role of professionals and management at the expense of workers. Millions perished in a series of manufactured disasters: extermination (1929-34) of kulaks (peasant landowners), severe famine (1932-33), party purges and show trials (Great Purge, 1936-38), suppression of nationalities, and poor conditions in labor camps.

Spain. An industrial revolution during World War I created an urban proletariat, which was attracted to socialism and anarchism; Catalan nationalists challenged central authority. The 5 years after King Alfonso left Spain in Apr. 1931 were dominated by tension between intermittent leftist and anticlerical governments and clericals, monarchists, and other rightists. Anarchist and Communist rebellions were crushed, but a July 1936 extreme right rebellion led by Gen. Francisco **Franco** and aided by Nazi Germany and Fascist Italy succeeded, after a 3-year **civil war** (more than 1 million dead in battles and atrocities). The war polarized international public opinion.

Italy. Despite propaganda for the ideal of the Corporate State, few domestic reforms were attempted. An entente with Hungary and Austria (Mar. 1934), a pact with Germany and Japan (Nov. 1937), and intervention by 50,000-75,000 troops in Spain (1936-39) sealed Italy's identification with the fascist bloc (anti-Semitic laws after Mar. 1938). Ethiopia was conquered (1935-36), and Albania annexed (Jan. 1939) in conscious imitation of ancient Rome.

East Europe. Repressive regimes fought for power against an active opposition (liberals, socialists, Communists, peasants, Nazis). Minority groups and Jews were restricted within national boundaries that did not coincide with ethnic population patterns. In the destruction of **Czechoslovakia**, Hungary occupied S Slovakia (Nov. 1938) and Ruthenia (Mar. 1939), and a pro-Nazi regime took power in the rest of Slovakia. Other boundary disputes (e.g., Poland-Lithuania, Yugoslavia-Bulgaria, Romania-Hungary) doomed attempts to build joint fronts against Germany or Russia. Economic depression was severe.

East Asia. After a period of liberalism in **Japan**, nativist militarists dominated the government with peasant support. Manchuria was seized (Sept. 1931-Feb. 1932), and a puppet state was set up (Manchukuo). Adjacent Jehol (Inner Mongolia) was occupied in 1933. China proper was invaded in July 1937; large areas were conquered by Oct. 1938. Hundreds of thousands of rapes, murders, and other atrocities were attributed to the Japanese.

In **China** Communist forces left Kuomintang-besieged strongholds in the S in a Long March (1934-35) to the N. The Kuomintang-Communist civil war was suspended in Jan. 1937 in the face of threatening Japan.

The democracies. The Roosevelt Administration, in office Mar. 1933, embarked on an extensive program of **New Deal** social reform and economic stimulation, including protection for labor unions (heavy industries organized), Social Security, public works, wage-and-hour laws, and assistance to farmers. Isolationist sentiment (1937 Neutrality Act) prevented U.S. intervention in Europe, but military expenditures were increased in 1939.

French political instability and polarization prevented resolution of economic and international security questions. The **Popular Front** government under Leon Blum (June 1936-Apr. 1938) passed social reforms (40-hour week) and raised arms spending. National coalition governments, which ruled Britain from Aug. 1931, brought some economic recovery but failed to define a consistent international policy until Chamberlain's government (from May 1937), which practiced deliberate **appeasement** of Germany and Italy.

India. Twenty years of agitation for autonomy and then for independence (Gandhi's **salt march**, 1930) achieved some constitutional reform (extended provincial powers, 1935) despite Muslim-Hindu strife. Social issues assumed prominence with peasant uprisings (1921), strikes (1928), Gandhi's efforts for untouchables (1932 "fast unto death"), and social and agrarian reform by the provinces after 1937.

Arts. The streamlined, geometric design motifs of Art Deco (from 1925) prevailed through the 1930s. **Abstract art** flourished (Moore sculptures from 1931) alongside a new **realism** related to social and political concerns (Socialist Realism, the official Soviet style from 1934; Mexican muralist Rivera, 1886-1957; and Orozco, 1883-1949), which were also expressed in fiction and poetry (Steinbeck's *Grapes of Wrath*, 1939; Sandburg's *The People, Yes*, 1936). Modern architecture (International Style, 1932) was unchallenged in its use of artificial materials (concrete, glass), lack of decoration, and monumentality (Rockefeller Center, 1929-40). U.S.-made films captured a worldwide audience with their larger-than-life fantasies *(Gone With the Wind, The Wizard of Oz,* both 1939).

War, Hot and Cold: 1940-49

War in Europe. The Nazi-Soviet nonaggression pact (Aug. 1939) freed Germany to attack Poland (Sept.). Britain and France, which had guaranteed Polish independence, declared war on Germany. Russia seized E Poland (Sept.), attacked Finland (Nov.), and took the Baltic states (July 1940). Mobile German forces staged *blitzkrieg* attacks during Apr.-June 1940, conquering neutral Denmark, Norway, and the Low Countries and defeating France; 350,000 British and French troops were evacuated at **Dunkirk** (May). The **Battle of Britain** (June-Dec. 1940) denied Germany air superiority. German-Italian campaigns won the Balkans by Apr. 1941. Three million Axis troops **invaded Russia** in June 1941, marching through Ukraine to the Caucasus, and through White Russia and the Baltic republics to Moscow and Leningrad.

Russian winter counterthrusts (1941-42 and 1942-43) stopped the German advance (**Stalingrad,** Sept. 1942-Feb. 1943). With British and U.S. Lend-Lease aid and sustaining great casualties, the Russians drove the Axis from all E Europe and the Balkans in the next 2 years. Invasions of N Africa (Nov. 1942), Italy (Sept. 1943), and **Normandy** (launched on D-Day, June 6, 1944) brought U.S., British, Free French, and allied troops to Germany by spring 1945. Germany surrendered May 7, 1945.

War in Asia-Pacific. Japan occupied Indochina in Sept. 1940, dominated Thailand in Dec. 1941, and attacked Hawaii (**Pearl Harbor**), the Philippines, Hong Kong, and Malaya on Dec. 7, 1941 (precipitating U.S. entrance into the war). Indonesia was attacked in Jan. 1942, and Burma was conquered in Mar. 1942. The Battle of **Midway** (June 1942) turned back the Japanese advance. "Island-hopping" battles (**Guadalcanal,** Aug. 1942-Jan. 1943; **Leyte Gulf,** Oct. 1944; **Iwo Jima,** Feb.-Mar. 1945; **Okinawa,** Apr. 1945) and massive bombing raids on Japan from June 1944 wore out Japanese defenses. U.S. atom bombs, dropped Aug. 6 and 9 on **Hiroshima** and Nagasaki, forced Japan to agree, on Aug. 14, to surrender; formal surrender was on Sept. 2, 1945.

Atrocities. The war brought 20th-cent. cruelty to its peak. The Nazi regime systematically killed an estimated 5-6 million Jews, including some 3 million who died in death camps (e.g., **Auschwitz**). Gypsies, political opponents, sick and retarded people, and others deemed undesirable were also murdered by the Nazis, as were vast numbers of Slavs, especially leaders.

Civilian deaths. German bombs killed 70,000 British civilians. More than 100,000 Chinese civilians were killed by Japanese forces in the capture and occupation of Nanking. Severe retaliation by the Soviet army, E European partisans, Free French, and others took a heavy toll. U.S. and British bombing of Germany killed hundreds of thousands, as did U.S. bombing of Japan (80,000-200,000 at Hiroshima alone). Some 45 million people lost their lives in the war.

Settlement. The **United Nations** charter was signed in San Francisco on June 26, 1945, by 50 nations. The International Tribunal at **Nuremberg** convicted 22 German leaders for war crimes in Sept. 1946; 23 Japanese leaders were convicted in Nov. 1948. Postwar border changes included large gains in territory for the USSR, losses for Germany, a shift to the W in Polish borders, and minor losses for Italy. Communist regimes, supported by Soviet troops, took power in most of E Europe, including Soviet-occupied Germany (GDR proclaimed Oct. 1949). Japan lost all overseas lands.

Recovery. Basic political and social changes were imposed on Japan and W Germany by the western allies (Japan constitution adopted, Nov. 1946; W German basic law, May 1949). U.S. **Marshall Plan** aid ($12 billion, 1947-51) spurred W European economic recovery after a period of severe inflation and strikes in Europe and the U.S. The British Labour Party introduced a national health service and nationalized basic industries in 1946.

Cold War. Western fears of further Soviet advances (Cominform formed in Oct. 1947; Czechoslovakia coup, Feb. 1948; Berlin blockade, Apr. 1948-Sept. 1949) led to the formation of **NATO**. Civil War in Greece and Soviet pressure on Turkey led to U.S. aid under the **Truman Doctrine** (Mar. 1947). Other anti-Communist security pacts were the Organization of American States (Apr. 1948) and the SE Asia Treaty Organization (Sept. 1954). A new wave of **Soviet purges** and repression intensified in the last years of Stalin's rule, extending to E Europe (Slansky trial in Czechoslovakia, 1951). Only Yugoslavia resisted Soviet control (expelled by Cominform, June 1948; U.S. aid, June 1949).

China, Korea. Communist forces emerged from World War II strengthened by the Soviet takeover of industrial Manchuria. In 4 years of fighting, the Kuomintang was driven from the mainland; the People's Republic was proclaimed Oct. 1, 1949. Korea was divided by USSR and U.S. occupation forces. Separate republics were proclaimed in the 2 zones in Aug.-Sept. 1948.

India. India and Pakistan became independent dominions on Aug. 15, 1947. Millions of Hindu and Muslim refugees were created by the partition; riots (1946-47) took hundreds of thousands of lives; Mahatma **Gandhi** was assassinated in Jan. 1948. Burma became completely independent in Jan. 1948; Ceylon took dominion status in Feb.

Middle East. The UN approved partition of Palestine into Jewish and Arab states. **Israel** was proclaimed a state, May 14, 1948. Arabs rejected partition, but failed to defeat Israel in war (May 1948-July 1949). Immigration from Europe and the Middle East swelled Israel's Jewish population. British and French forces left Lebanon and Syria in 1946. Transjordan occupied most of Arab Palestine.

Southeast Asia. Communists and others fought against restoration of French rule in Indochina from 1946; a non-Communist government was recognized by France in Mar. 1949, but fighting continued. Both Indonesia and the Philippines became independent; the former in 1949 after 4 years of war with Netherlands, the latter in 1946. Philippine economic and military ties with the U.S. remained strong; a Communist-led peasant rising was checked in 1948.

Arts. New York became the center of the world art market; **abstract expressionism** was the chief mode (Pollock from 1943, de Kooning from 1947). Literature and philosophy explored **existentialism** (Camus's *The Stranger*, 1942; Sartre's *Being and Nothingness*, 1943). Non-Western attempts to revive or create regional styles (Senghor's Négritude, Mishima's novels) only confirmed the emergence of a universal culture. Radio and phonograph records spread American popular music (swing, bebop) around the world.

The American Decade: 1950-59

Polite decolonization. The peaceful decline of European political and military power in Asia and Africa accelerated in the 1950s. Nearly all of **N Africa** was freed by 1956, but France fought a bitter war to retain Algeria, with its large European minority, until 1962. **Ghana**, independent in 1957, led a parade of new black African nations (more than 2 dozen by 1962), which altered the political character of the UN. Ethnic disputes often exploded in the new nations after decolonization (UN troops in Cyprus, 1964; **Nigerian civil war**, 1967-70). Leaders of the new states, mostly sharing socialist ideologies, tried to create an Afro-Asian bloc (Bandung Conference, 1955), but Western economic influence and U.S. political ties remained strong (Baghdad Pact, 1955).

Trade. World trade volume soared, in an atmosphere of monetary stability assured by international accords (**Bretton Woods,** 1944). In Europe, economic integration advanced (**European Economic Community,** 1957; European Free Trade Association, 1960). Comecon (1949) coordinated the economies of Soviet-bloc countries.

U.S. Economic growth produced an abundance of consumer goods (9.3 million motor vehicles sold, 1955). Suburban housing tracts changed life patterns for middle and working classes (Levittown, 1947-51). Pres. Dwight **Eisenhower's** landslide election victories (1952, 1956) reflected consensus politics. Senate condemnation of Senator Joseph **McCarthy** (Dec. 1954) curbed the political abuse of anti-Communism. A system of alliances and military bases bolstered U.S. influence on all continents. Trade and payments surpluses were balanced by overseas investments and foreign aid ($50 billion, 1950-59).

USSR. In the "thaw" after Stalin's death in 1953, relations with the West improved (evacuation of Vienna, Geneva summit conference, both 1955). Repression of scientific and cultural life eased, and many prisoners were freed or rehabilitated culminating in **de-Stalinization** (1956). **Nikita Khrushchev's** leadership aimed at consumer sector growth, but farm production lagged, despite the virgin lands program (from 1954). Soviet crushing of the 1956 Hungarian revolution, the 1960 U-2 spy plane episode, and other incidents renewed East-West tension and domestic curbs.

East Europe. Resentment of Russian domination and Stalinist repression combined with nationalist, economic, and religious factors to produce periodic violence. E Berlin workers rioted (1953), Polish workers rioted in Poznan (June 1956), and a broad-based **revolution** broke out in **Hungary** (Oct. 1956). All were suppressed by Soviet force or threats (at least 7,000 dead in Hungary). But Poland was allowed to restore private ownership of farms, and a degree of personal and economic freedom returned to Hungary. Yugoslavia experimented with worker self-management and a market economy.

Korea. The 1945 division of Korea along the 38th parallel left industry in the N, which was organized into a militant regime and armed by the USSR. The S was politically disunited. More than 60,000 N Korean troops invaded the S on June 25, 1950. The U.S., backed by the UN Security Council, sent troops. UN troops reached the Chinese border in Nov. Some 200,000 Chinese troops crossed the Yalu R. and drove back UN forces. By spring 1951 battle lines had become stabilized near the original 38th parallel border, but heavy fighting continued. Finally, an armistice was signed on July 27, 1953. U.S. troops remained in the S, and U.S. economic and military aid continued. The war stimulated rapid economic recovery in Japan.

China. Starting in 1952, industry, agriculture, and social institutions were forcibly collectivized. In a massive purge, as many as several million people were executed as Kuomintang supporters or as class and political enemies. The **Great Leap Forward** (1958-60) unsuccessfully tried to force the pace of development by substituting labor for investment.

Indochina. Ho Chi Minh's forces, aided by the USSR and the new Chinese Communist government, fought French and pro-French Vietnamese forces to a standstill and captured the strategic **Dienbienphu** camp in May 1954. The Geneva Agreements divided Vietnam in half pending elections (never held) and recognized Laos and Cambodia as independent. The U.S. aided the anti-Communist Republic of Vietnam in the S.

Middle East. Arab revolutions placed leftist, militantly nationalist regimes in power in Egypt (1952) and Iraq (1958). But Arab unity attempts failed (United Arab Republic joined Egypt, Syria, Yemen, 1958-61). Arab refusal to recognize Israel (Arab League economic blockade began Sept. 1951) led to a permanent state of war, with repeated incidents (Gaza, 1955). Israel occupied Sinai, and Britain and France took (Oct. 1956) the Suez Canal, but were replaced by the UN Emergency Force. The Mossadegh government in Iran nationalized (May 1951) the British-owned oil industry in May, but was overthrown (Aug. 1953) in a U.S.-aided coup.

Latin America. Argentinian dictator Juan **Perón,** in office 1946, enforced land reform, some nationalization, welfare state measures, and curbs on the Roman Catholic Church, and crushed opposition. A Sept. 1955 coup deposed Perón. The 1952 revolution in Bolivia brought land reform, nationalization of tin mines, and improvement in the status of Indians, who nevertheless remained poor. The Batista regime in Cuba was overthrown (Jan. 1959) by Fidel **Castro,** who imposed a Communist dictatorship, aligned Cuba with the USSR, but improved education and health care. A U.S.-backed anti-Castro invasion (**Bay of Pigs**, Apr. 1961) was crushed. Self-government advanced in the British Caribbean.

Technology. Large outlays on research and development in the U.S. and the USSR focused on military applications (H-bomb in U.S., 1952; USSR, 1953; Britain, 1957; intercontinental missiles, late 1950s). Soviet launching of the **Sputnik** satellite (Oct. 4, 1957) spurred increases in U.S. science education funds (National Defense Education Act).

Literature and film. Alienation from social and literary conventions reached an extreme in the theater of the absurd (Beckett's *Waiting for Godot,* 1952), the "new novel" (Robbe-Grillet's *Voyeur,* 1955), and avant-garde film (Antonioni's *L'Avventura,* 1960). U.S. beatniks (Kerouac's *On the Road,* 1957) and others rejected the supposed conformism of Americans (Riesman's *The Lonely Crowd,* 1950).

Rising Expectations: 1960-69

Economic boom. The longest sustained economic boom on record spanned almost the entire decade in the capitalist world; the closely watched GNP figure doubled (1960-70) in the U.S., fueled by Vietnam War–related budget deficits. The **General Agreement on Tariffs and Trade** (1967) stimulated W European prosperity, which spread to peripheral areas (Spain, Italy, E Germany). Japan became a top economic power. Foreign investment aided the industrialization of Brazil. There were limited Soviet economic reform attempts.

Reform and radicalization. Pres. John F. **Kennedy**, inaugurated 1961, emphasized youthful idealism and vigor; his assassination Nov. 22, 1963, was a national trauma. A series of political and social reform movements took root in the U.S., later spreading to other countries. Blacks demonstrated nonviolently and with partial success against segregation and poverty (1963 March on Washington; 1964 **Civil Rights Act**), but some urban ghettos erupted in extensive riots (Watts, 1965; Detroit, 1967; Martin Luther King assassination, Apr. 4, 1968). New concern for the poor (Harrington's *Other America*, 1963) helped lead to Pres. Lyndon Johnson's **"Great Society"** programs (Medicare, Water Quality Act, Higher Education Act, all 1965). Concern with the **environment** surged (Carson's *Silent Spring*, 1962). **Feminism** revived as a cultural and political movement (Friedan's *Feminine Mystique*, 1963; National Organization for Women founded 1966), and a movement for homosexual rights emerged (Stonewall riot in NYC, 1969). Pope John XXIII called the **Second Vatican Council** (1962-65), which liberalized Roman Catholic liturgy and some other aspects of Catholicism.

Opposition to U.S. involvement in Vietnam, especially among university students (**Moratorium** protest, Nov. 1969), turned violent (Weatherman Chicago riots, Oct. 1969). **New Left** and Marxist theories became popular, and membership in radical groups (Students for a Democratic Society, Black Panthers) increased. Maoist groups, especially in Europe, called for total transformation of society. In France, students sparked a nationwide strike affecting 10 million workers in May-June 1968, but an electoral reaction barred revolutionary change.

Arts and styles. The boundary between fine and popular arts was blurred to some extent by Pop Art (Warhol) and rock musicals (*Hair*, 1968). Informality and exaggeration prevailed in fashion (beards, miniskirts). A nonpolitical "counterculture" developed, rejecting traditional bourgeois life goals and personal habits, and use of marijuana and hallucinogens spread (**Woodstock** festival, Aug. 1969). Indian influence was felt in religion (Ram Dass) and fashion, and The **Beatles,** who brought unprecedented sophistication to rock music, became for many a symbol of the decade.

Science. Achievements in space (**humans on the moon,** July 1969) and electronics (lasers, integrated circuits) encouraged a faith in scientific solutions to problems in agriculture ("green revolution"), medicine (heart transplants, 1967), and other areas. Harmful technology, it was believed, could be controlled (1963 nuclear weapon test ban treaty, 1968 nonproliferation treaty).

IT'S A FACT: Astronaut Eugene Cernan (*Apollo 17*) was the last man to walk on the moon. Cernan and Harrison Schmitt spent a record 75 hrs. (Dec. 11-14, 1972) on the lunar surface, more than 3 times as long as astronauts Neil Armstrong and Edwin Aldrin spent on there in the famous first lunar expedition in 1969.

China. Mao's revolutionary militancy caused disputes with the USSR under "revisionist" Khrushchev, starting in 1960. The 2 powers exchanged fire in 1969 border disputes. China used force to capture (1962) areas disputed with India. The **"Great Proletarian Cultural Revolution"** tried to impose a utopian egalitarian program in China and spread revolution abroad; political struggle, often violent, convulsed China in 1965-68.

Indochina. Communist-led guerrillas aided by N Vietnam fought from 1960 against the S Vietnam government of Ngo Dinh Diem (killed 1963). The U.S. military role increased after the 1964 **Tonkin Gulf** incident. U.S. forces peaked at 543,400 in Apr. 1969. Massive numbers of N Vietnamese troops also fought. Laotian and Cambodian neutrality were threatened by Communist insurgencies, with N Vietnamese aid, and U.S. intrigues.

Third World. A bloc of authoritarian leftist regimes among the newly independent nations emerged in political opposition to the U.S.-led Western alliance and came to dominate the conference of nonaligned nations (Belgrade, 1961; Cairo, 1964; Lusaka, 1970). Soviet political ties and military bases were established in Cuba, Egypt, Algeria, Guinea, and other countries whose leaders were regarded as revolutionary heroes by opposition groups in pro-Western or colonial countries. Some leaders were ousted in coups by pro-Western groups—Zaire's Patrice Lumumba (killed 1961), Ghana's Kwame Nkrumah (exiled 1966), and Indonesia's Sukarno (effectively ousted in 1965 after a Communist coup failed).

Middle East. Arab-Israeli tension erupted into a brief war June 1967. Israel emerged from the war as a major regional power. Military shipments before and after the war brought much of the Arab world into the Soviet political sphere. Most Arab states broke U.S. diplomatic ties, while Communist countries cut their ties to Israel. Intra-Arab disputes continued: Egypt and Saudi Arabia supported rival factions in a bloody Yemen civil war 1962-70; Lebanese troops fought Palestinian commandos 1969.

East Europe. To stop the large-scale exodus of citizens, E German authorities built (Aug. 1961) a **fortified wall across Berlin.** Soviet sway in the Balkans was weakened by Albania's support of China (USSR broke ties in Dec. 1961) and Romania's assertion (1964) of industrial and foreign policy autonomy. Liberalization (spring 1968) in Czechoslovakia was crushed with massive force by troops of 5 Warsaw Pact countries. W German treaties (1970) with the USSR and Poland facilitated the transfer of German technology and confirmed postwar boundaries.

Disillusionment: 1970-79

U.S.: Caution and neoconservatism. A relatively sluggish economy, energy shortages, and environmental problems contributed to a **"limits of growth"** philosophy. Suspicion of science and technology killed or delayed major projects (supersonic transport dropped, 1971; Seabrook nuclear power plant protests, 1977-78) and was fed by the Three Mile Island nuclear reactor accident (Mar. 1979).

There were signs of growing mistrust of big government and less support for new social policies. School busing and racial quotas were opposed (Bakke decision, June 1978); the proposed Equal Rights Amendment for women languished; civil rights legislation aimed at protecting homosexuals was opposed (Dade County referendum, June 1977).

Completion of Communist forces' takeover of **South Vietnam** (evacuation of U.S. civilians, Apr. 1975), revelations of Central Intelligence Agency misdeeds (Rockefeller Commission report, June 1975), and **Watergate** scandals (Nixon resigned in Aug. 1974) reduced faith in U.S. moral and material capacity to influence world affairs. Revelations of Soviet crimes (Solzhenitsyn's *Gulag Archipelago,* 1974) and Soviet intervention in Africa helped foster a revival of anti-Communist sentiment.

Economy sluggish. The 1960s boom faltered in the 1970s; a severe recession in the U.S. and Europe (1974-75) followed a huge oil price hike (Dec. 1973). Monetary instability (U.S. cut ties to gold in Aug. 1971), the decline of the dollar, and protectionist moves by industrial countries (1977-78) threatened trade. Business investment and spending for research declined. Severe inflation plagued many countries (25% in Britain, 1975; 18% in U.S., 1979).

China picks up pieces. After the 1976 deaths of Mao Zedong and Zhou Enlai, struggle for the leadership succession was won by pragmatists. A nationwide purge of orthodox Maoists was carried out, and the **Gang of Four,** led by Mao's widow, Chiang Ching, arrested. The new leaders freed more than 100,000 political prisoners and reduced public adulation of Mao. Political and trade ties were expanded with Japan, Europe, and the U.S. in the late 1970s, as relations worsened with the USSR, Cuba, and Vietnam (4-week invasion by China, 1979). Ideological guidelines in industry, science, education, and the armed forces, which the ruling faction said had caused chaos and decline, were reversed (bonuses to workers, Dec. 1977; exams for college entrance, Oct. 1977). Severe restrictions on cultural expression were eased.

Europe. European unity moves (EEC-EFTA trade accord, 1972) faltered as economic problems appeared (Britain floated pound, 1972; France floated franc, 1974). Germany and Switzerland curbed guest workers from southern Europe. Greece and Turkey quarreled over Cyprus and Aegean oil rights.

All non-Communist Europe was under democratic rule after free elections were held (June 1976) in **Spain** 7 months after the death of Franco. The conservative, colonialist regime in **Portugal** was overthrown in Apr. 1974. In **Greece** the 7-year-old military dictatorship yielded power in 1974. Northern Europe, though ruled mostly by Socialists (**Swedish** Socialists unseated in 1976 after 44 years in power), turned more conservative. The **British** Labour government imposed (1975) wage curbs and suspended nationalization schemes. Terrorism in **Germany** (1972 Munich Olympics killings) led to laws curbing some civil liberties. **French** "new philosophers" rejected leftist ideologies, and the shaky Socialist-Communist coalition lost a 1978 election bid.

Religion and politics. The improvement in **Muslim** countries' political fortunes by the 1950s (with the exception of Central Asia under Soviet and Chinese rule) and the growth of Arab oil wealth were followed by a resurgence of traditional religious fervor. Libyan dictator Muammar al-Qaddafi mixed Islamic laws with socialism and called for Muslim return to Spain and Sicily. The illegal Muslim Brotherhood in **Egypt** was accused of violence, while extreme groups bombed (1977) theaters to protest secular values.

In **Turkey**, the National Salvation Party was the first Islamic group to share (1974) power since secularization in the 1920s. In **Iran, Ayatollah Ruhollah Khomeini,** led a revolution that deposed the secular shah (Jan. 1979) and created an Islamic republic there. Religiously motivated Muslims took part in an insurrection in Saudi Arabia that briefly seized (1979) the Grand Mosque in Mecca. Muslim puritan opposition to **Pakistan** Pres. Zulfikar Ali-Bhutto helped lead to his overthrow in July 1977. Muslim solidarity, however, could not prevent Pakistan's eastern province (**Bangladesh**) from declaring (Dec. 1971) independence after a bloody civil war.

Muslim and Hindu resentment of coerced sterilization in **India** helped defeat the Gandhi government, which was replaced (Mar. 1977) by a coalition including religious Hindu parties. Muslims in the S **Philippines**, aided by Libya, rebelled against central rule from 1973.

The Buddhist Soka Gakkai movement launched (1964) the Komeito party in **Japan,** which became a major opposition party in 1972 and 1976 elections.

IT'S A FACT: Since the Vietnam War, an estimated 30,000-40,000 people in Vietnam have been killed by land mines and unexploded ordnance left over from the war years.

Evangelical Protestant groups grew in the U.S. A revival of interest in Orthodox Christianity occurred among **Russian** intellectuals (Solzhenitsyn). The secularist **Israeli** Labor party, after decades of rule, was ousted in 1977 by conservatives led by Menachem Begin; religious militants founded settlements on the disputed West Bank, part of biblically promised Israel. U.S. Reform Judaism revived many previously discarded traditional practices.

Old-fashioned religious wars raged intermittently in **Northern Ireland** (Catholic vs. Protestant, 1969-) and **Lebanon** (Christian vs. Muslim, 1975-), while religious militancy complicated the Israel-Arab dispute (1973 Israel-Arab war). Despite a 1979 **peace treaty between Egypt and Israel,** increased militancy on the West Bank impeded further progress.

Latin America. Repressive conservative regimes strengthened their hold on most of the continent, with a violent coup against the elected (Sept. 1973) Allende government in **Chile**, a 1976 military coup in **Argentina**, and coups against reformist regimes in **Bolivia** (1971, 1979) and **Peru** (1976). In Central America increasing liberal and leftist militancy led to the ouster (1979) of the Somoza regime of **Nicaragua** and to civil conflict in **El Salvador**.

Indochina. Communist victories in Vietnam, Cambodia, and Laos by May 1975 led to new turmoil. The **Pol Pot regime** ordered millions of city-dwellers to resettle in rural areas, in a program of forced labor, combined with terrorism, that cost more than 1 million lives (1975-79) and caused hundreds of thousands of ethnic Chinese and others to flee Vietnam ("boat people," 1979). The Vietnamese invasion of Cambodia swelled the refugee population and contributed to widespread starvation in that devastated country.

Russian expansion. Soviet influence, checked in some countries (troops ousted by Egypt, 1972), was projected farther afield, often with the use of Cuban troops (Angola, 1975-89; Ethiopia, 1977-88) and aided by a growing navy, a merchant fleet, and international banking ability. **Détente** with the West—1972 Berlin pact, 1972 strategic arms pact (**SALT**)—gave way to a more antagonistic relationship in the late 1970s, exacerbated by the Soviet invasion (1979) of **Afghanistan.**

Africa. The last remaining European colonies were granted independence (**Spanish Sahara**, 1976; **Djibouti,** 1977) and, after 10 years of civil war and many negotiation sessions, a black government took over (1979) in Zimbabwe (Rhodesia); white domination remained in **South Africa**. Great power involvement in local wars (Russia in **Angola, Ethiopia**; France in **Chad, Zaire, Mauritania**) and the use of tens of thousands of Cuban troops were denounced by some African leaders. Ethnic or tribal clashes made Africa a locus of sustained warfare during the late 1970s.

Arts. Traditional modes of painting, architecture, and music received increased popular and critical attention in the 1970s. These more conservative styles coexisted with modernist works in an atmosphere of increased variety and tolerance.

Revitalization of Capitalism, Demand for Democracy: 1980-89

USSR, Eastern Europe. A troublesome 1980-85 for the USSR was followed by 5 years of astonishing change: the surrender of the Communist monopoly, the remaking of the Soviet state, and the beginning of the disintegration of the Soviet empire. After the deaths of Leonid **Brezhnev** (1982) and 2 successors (Andropov in 1984 and Chernenko in 1985), the harsh treatment of dissent and restriction of emigration, and the Soviet invasion (Dec. 1979) of Afghanistan, Gen. Sec. Mikhail **Gorbachev** (in office 1985-1991) promoted *glasnost* and *perestroika*—economic, political, and social reform. Supported by the Communist Party (July 1988), he signed (Dec. 1987) the INF disarmament treaty, and he pledged (1988) to cut the military budget. Military withdrawal from Afghanistan was completed in Feb. 1989, the process of democratization went ahead unhindered in Poland and Hungary, and the Soviet people chose (Mar. 1989) part of the new Congress of People's Deputies from competing candidates. By decade's end the **Cold War** appeared to be fading away.

In **Poland, Solidarity**, the labor union founded (1980) by Lech **Walesa**, was outlawed in 1982 and then legalized in 1988, after years of unrest. Poland's first free election since the Communist takeover brought Solidarity victory (June 1989); Tadeusz Mazowiecki, a Walesa adviser, became (Aug. 1989) prime minister in a government with the Communists. In the fall of 1989 the failure of Marxist economies in **Hungary**, **East Germany**, **Czechoslovakia, Bulgaria,** and **Romania** brought the collapse of the Communist monopoly and a demand for democracy. In a historic step, the **Berlin Wall** was opened in Nov. 1989.

U.S. "The Reagan Years" (1981-88) brought the **longest economic boom** yet in U.S. history via budget and tax cuts, deregulation, "junk bond" financing, leveraged buyouts, and mergers and takeovers. However, there was a stock market crash (Oct. 1987), and federal budget deficits and the trade deficit increased. Foreign policy showed a **strong anti-Communist stance**, via increased defense spending, aid to anti-Communists in Central America, invasion of Cuba-threatened Grenada, and championing of the MX missile system and "Star Wars" missile defense program. Four Reagan-Gorbachev summits (1985-88) climaxed in the INF treaty (1987), as the Cold War began to wind down. The Iran-contra affair (North's TV testimony, July 1987) was a major political scandal. Homelessness and drug abuse (especially "crack" cocaine) were growing social problems. In 1988, Vice Pres. George Bush was elected to succeed Ronald Reagan as president.

Middle East. The Middle East remained militarily unstable, with sharp divisions along economic, political, racial, and religious lines. In **Iran**, the Islamic revolution of 1979 created a strong anti-U.S. stance (hostage crisis, Nov. 1979-Jan. 1981). In Sept. 1980, **Iraq** repudiated its border agreement with Iran and began major hostilities that led to an 8-year war in which millions were killed.

Libya's support for international terrorism induced the U.S. to close (May 1981) its diplomatic mission there and embargo (Mar. 1982) Libyan oil. The U.S. accused Libyan leader Muammar al-Qaddafi of aiding (Dec. 1985) terrorists in Rome and of Vienna airport attacks, and retaliated by bombing Libya (Apr. 1986).

Israel affirmed (July 1980) all Jerusalem as its capital, destroyed (1981) an Iraqi atomic reactor, and invaded (1982) Lebanon, forcing the PLO to agree to withdraw. A **Palestinian uprising**, including women and children hurling rocks and bottles at troops, began (Dec. 1987) in Israeli-occupied Gaza and spread to the West Bank; troops responded with force, killing 300 by the end of 1988, with 6,000 more in detention camps.

Israeli withdrawal from **Lebanon** began in Feb. 1985 and ended in June 1985, as Lebanon continued torn by military and political conflict. Artillery duels (Mar.-Apr. 1989) between Christian East Beirut and Muslim West Beirut left 200 dead and 700 wounded. At decade's end, violence still dominated.

Latin America. In **Nicaragua**, the leftist Sandinista National Liberation Front, in power after the 1979 civil war, faced problems as a result of Nicaragua's military aid to leftist guerrillas in El Salvador and U.S. backing of antigovernment contras. The U.S. CIA admitted (1984) having directed the mining of Nicaraguan ports, and the U.S. sent humanitarian (1985) and military (1986) aid. Profits from secret arms sales to Iran were found (1987) diverted to contras. Cease-fire talks between the Sandinista government and contras came in 1988, and elections were held in Feb. 1990.

In **El Salvador**, a military coup (Oct. 1979) failed to halt extreme right-wing violence and left-wing terrorism. Archbishop Oscar Romero was assassinated in Mar. 1980; from Jan. to June some 4,000 civilians reportedly were killed in the civil unrest. In 1984, newly elected Pres. José Napoleon Duarte worked to stem human rights abuses, but violence continued.

In **Chile**, Gen. Augusto Pinochet yielded the presidency after a democratic election (Dec. 1989), but remained as head of the army. He had ruled the country since 1973, imposing harsh measures against leftists and dissidents; at the same time he introduced economic programs that restored prosperity to Chile.

Africa. 1980-85 marked a rapid decline in the economies of virtually all African countries, a result of accelerating desertification, the world economic recession, heavy indebtedness to overseas creditors, rapid population growth, and political instability. Some 60 million Africans faced prolonged hunger in 1981; much of Africa had one of the worst droughts ever in 1983, and by year's end **150 million faced near-famine**. "Live Aid," a marathon rock concert, was presented in July 1985, and the U.S. and Western nations sent aid in Sept. 1985. Economic hardship fueled political unrest and coups. Wars in Ethiopia and Sudan and military strife in several other nations continued. AIDS took a heavy toll.

South Africa. Anti-apartheid sentiment gathered force in South Africa as demonstrations and violent police response grew. White voters approved (Nov. 1983) the first constitution to give Coloureds and Asians a voice, while still excluding blacks (70% of the population). The U.S. imposed economic sanctions in Aug. 1985, and 11 Western nations followed in September. P. W. **Botha**, 1980s president, was succeeded by F. W. **de Klerk**, in Sept. 1989, who promised "evolutionary" change via negotiation with the black population.

China. During the 1980s the Communist government and paramount leader **Deng Xiaoping** pursued **far-reaching changes**, expanding commercial and technical ties to the industrialized world and increasing the role of market forces in stimulating urban development. Apr. 1989 brought new demands for political reforms; student demonstrators camped out in Tiananmen Sq., Beijing, in a massive peaceful protest. Some 100,000 students and workers marched, and at least 20 other cities saw protests. In response, martial law was imposed; army troops crushed the demonstration in and around Tiananmen Square on June 3-4, with death toll estimates at 500-7,000, up to 10,000 dissidents arrested, 31 people tried and executed. The conciliatory Communist Party chief was ousted; the Politburo adopted (July) reforms against official corruption.

Japan. Japan's relations with other nations, especially the U.S., were dominated by **trade imbalances favoring Japan**. In 1985 the U.S. trade deficit with Japan was $49.7 billion, one-third of the total U.S. trade deficit. After Japan was found (Apr. 1986) to sell semiconductors and computer memory chips below cost, the U.S. was assured a "fair share" of the market, but charged (Mar. 1987) Japan with failing to live up to the agreement.

European Community. With the addition of Greece, Portugal, and Spain, the EC became a common market of more than **300 million people**, the West's largest trading entity. Margaret **Thatcher** became the first British prime minister in the 20th century to win a 3d consecutive term (1987). France elected (1981) its first socialist president, François **Mitterrand**, who was reelected in 1988. Italy elected (1983) its first socialist premier, Bettino **Craxi**.

International terrorism. With the 1979 overthrow of the shah of Iran, terrorism became a prominent tactic. It increased through the 1980s, but with fewer high-profile attacks after 1985. In 1979-81, Iranian militants held 52 U.S. hostages in Iran for 444 days; in 1983 a TNT-laden suicide terrorist blew up U.S. Marine headquarters in Beirut, killing 241 Americans, and a truck bomb blew up a French paratroop barracks, killing 58. The *Achille Lauro* cruise ship was hijacked in 1986, and an American passenger killed; the U.S. subsequently intercepted the Egyptian plane flying the terrorists to safety. Incidents rose to 700 in 1985, and to 1,000 in 1988. **Assassinated leaders** included Egypt's Pres. Anwar al-**Sadat** (1981), India's Prime Min. Indira **Gandhi** (1984), and Lebanese Premier Rashid **Karami** (1987).

Post–Cold War World: 1990-99

Soviet Empire breakup. The world community witnessed the extraordinary spectacle of a superpower's disintegration when the **Soviet Union** broke apart into 15 independent states. The 1980s had already seen internal reforms and a decline of Communist power both within the Soviet Union and in Eastern Europe. The Soviet breakup began in earnest with the declarations of independence adopted by the Baltic republics of **Lithuania, Latvia,** and **Estonia** during an abortive coup against reformist leader Mikhail **Gorbachev** (Aug. 1991). The other republics soon took the same step. In Dec. 1991, **Russia, Ukraine**, and **Belarus** declared the Soviet Union dead; Gorbachev resigned, and the Soviet Parliament went out of existence. The Warsaw Pact and the Council for Mutual Economic Assistance (Comecon) were disbanded. Most of the former republics joined in a loose confederation called the **Commonwealth of Independent States**. **Russia** remained the predominant country after the breakup, but its people soon suffered severe economic hardship as the nation, under Pres. Boris **Yeltsin,** moved to revamp the economy and to adopt a free market system. In Oct. 1993, **anti-Yeltsin forces** occupied the Parliament building and were ousted by the army; about 140 people died in the fighting.

The Muslim republic of **Chechnya** declared independence from the rest of Russia, but this was met with an invasion by Russian troops (Dec. 1994). After almost 21 months of vicious fighting, a cease-fire took hold in 1996, and the Russians withdrew. In 1999 Russia forcibly suppressed Muslim insurgents in Dagestan and entered neighboring Chechnya, again fighting to gain control over separatist rebels there. Yeltsin resigned office Dec. 31, 1999, to be replaced by Vladimir **Putin** (elected in his own right, Mar. 2000).

Europe. Yugoslavia broke apart, and hostilities ensued among the republics along ethnic and religious lines. **Croatia, Slovenia**, and **Macedonia** declared independence (1991), followed by **Bosnia-Herzegovina** (1992). **Serbia** and **Montenegro** remained as the republic of Yugoslavia. Bitter fighting followed, especially in Bosnia, where Serbs reportedly engaged in **"ethnic cleansing"** of the Muslim population; a peace plan (Dayton accord), brokered by the United States, was signed by **Bosnia, Serbia**, and **Croatia** (Dec. 1995), with **NATO** responsible for policing its implementation. In spring 1999, NATO conducted a bombing campaign aimed at stopping Yugoslavia from its campaign to drive out ethnic Albanians from the Kosovo region; a peace accord was reached in June under which NATO peacekeeping troops entered Kosovo.

The two **Germanys** were reunited after 45 years (Oct. 1990). The union was greeted with jubilation, but stresses became apparent when free market principles were applied to the aging East German industries, resulting in many plant closings and rising unemployment. German chancellor Helmut **Kohl**, a Christian Democrat, lost power after 16 years, in Sept. 1998 elections; Gerhard **Schroeder**, a Social Democrat, took over. Czechoslovakia broke apart peacefully (Jan. 1993), becoming the **Czech Republic** and **Slovakia.** In **Poland**, Lech **Walesa** was elected president (Dec. 1991) but was defeated in his bid for a 2d term (Nov. 1995).

NATO approved the **Partnership for Peace** Program (Jan. 1994) coordinating the defense of **Eastern** and **Central European** countries; Russia joined the program later that year. NATO signed a pact with **Russia** (1997) providing for NATO expansion into the former Soviet-bloc countries; a similar treaty was set up with **Ukraine**. The **Czech Republic**, **Hungary**, and **Poland** became members in Jan.

1999; in that year **NATO** celebrated its 50th anniversary. Efforts toward European unity continued with adoption of a single market (Jan. 1993) and conversion of the European Community to the **European Union** as the Maestricht Treaty took effect (Nov. 1993). Agreement was reached for 11 EU members to participate in Economic and Monetary Union, adopting a common currency **(euro)** for some purposes in Jan. 1999, with the euro to go into common circulation in 2002.

An intraparty revolt forced Margaret **Thatcher** out as prime minister of **Great Britain**, to be succeeded by John **Major** (Nov. 1990); 7 years later, Major suffered an overwhelming defeat at the hands of the new Labour Party leader, Tony **Blair** (May 1997). The divorce of Prince **Charles and** Princess **Diana**, followed by the death of Diana in a car accident (Aug. 1997), made headlines around the world. Talks on **peace** in **Northern Ireland** that included participation of Sinn Fein, political arm of the IRA, led to a ground-breaking peace plan, approved in an all-Ireland vote (May 1998). In Dec. 1999, Northern Ireland was granted home rule under a power-sharing cabinet. In **Scotland** voters overwhelmingly approved establishment of a regional legislature (1997), and in **Wales** voters narrowly approved establishment of a local assembly (1997). In a historic innovation, the Church of England **ordained 32 women** as priests (Mar. 1994).

Middle East. In Aug. 1990, **Iraq's Saddam Hussein** ordered his troops to invade **Kuwait.** The UN approved military action in response (Nov. 1990), and U.S. Pres. George **Bush** put together an international military force. Allied planes bombed Iraq (Jan. 1991) and launched a land attack, crushing the invasion (Feb. 1991). After Iraq accepted a cease-fire (Apr. 1991), U.S. troops withdrew, but "no-fly" zones were set up over northern Iraq to protect the Kurds and over southern Iraq to protect Shiite Muslims. The **UN** imposed **sanctions** on Iraq for failure to abide by the cease-fire. Iraq's reported failure to cooperate with UN arms inspectors seeking to eliminate "weapons of mass destruction" led to repeated air strikes by the U.S. and Britain.

The last Western hostages were freed in **Lebanon,** June 1992. **Israel** and the **Palestine Liberation Organization** signed a peace accord (Sept. 1993) providing for Palestinian self-government in the West Bank and Gaza Strip. Prime Min. Yitzhak **Rabin** and Foreign Min. Shimon **Peres** of Israel and Yasir **Arafat** of the PLO received the Nobel Peace Prize for their efforts (1994). Six Arab nations relaxed their boycott against Israel (1994), and Israel and **Jordan** signed a peace treaty (Oct. 1994). **Rabin was assassinated** (Nov. 1995) by an Israeli opponent of the peace process. After new elections (May 1996), Benjamin Netanyahu as prime minister adopted a harder line in peace negotiations. **Arafat** was elected to the presidency of the Palestinian Authority (Jan. 1996). A long-delayed interim agreement (the Wye Memorandum) on Israel military withdrawal from part of the West Bank was reached Oct. 1998. A Labour government under Ehud **Barak** took power after May 1999 elections, but further progress in peace negotiations proved elusive.

King **Hussein** of Jordan died (Feb. 1999), to be succeeded by his son Abdullah.

Asia and the Pacific. Hong Kong was returned to **China** (July 1997) after being a British colony for 156 years. China, which emerged in the decade as a major developing economic power, had agreed to follow a policy of "one country, two systems" in Hong Kong. The territory of **Macao** reverted to Chinese sovereignty (Dec. 1999) after over 400 years of Portuguese rule; it retained its capitalist economic system. **Jiang Zemin**, general secretary of the Chinese Communist Party, assumed the additional post of president of China (Mar. 1993) and emerged as the key leader after the death of paramount leader **Deng Xiaoping** (Feb. 1997). China released from prison—and exiled—some well-known dissidents but continued to be criticized for detentions and other alleged widespread **human rights abuses**. In Nov. 1999 the U.S. and China signed a landmark pact normalizing trade relations.

After years of prosperity, **Thailand, Indonesia**, and **South Korea** in 1997 began to suffer economic reverses that had a worldwide ripple effect. These countries received billion-dollar IMF bailout packages. In **Indonesia**, protests over mismanagement led to the resignation of Pres. **Suharto** (May 1998) after 32 years of nearly autocratic rule. Abdurraham Wahid was elected (Oct. 1999) in the country's first fully democratic elections. In a referendum (Aug. 1999), **East Timor** voted overwhelmingly for independence from Indonesia; pro-Indonesian militias then rampaged through the territory, but a multinational peacekeeping force was allowed in (Sept. 1999) to help restore order. In **South Korea**, former dissident **Kim Dae Jung** was elected president (Dec. 1997). Two previous presidents, Roh Tae Woo and Chun Doo Hwan, were convicted of crimes committed in office but were given amnesty by the new president.

In Japan members of a religious cult, released the nerve gas sarin on 5 Tokyo subway cars, killing 12 people and injuring more than 5,500 (Mar. 1995). Tamil rebels continued their armed conflict in **Sri Lanka**. In **Afghanistan** the **Taliban**, an extreme Islamic fundamentalist group, gained control of Kabul (Sept. 1996) and, eventually, most of the country. In **North Korea**, longtime dictator **Kim Il Sung** died (July 1994), to be succeeded by his son, **Kim Jong Il**. In the same year the country signed an agreement with the U.S. setting a timetable for North Korea to eliminate its nuclear program. The country also suffered a severe drought, and widespread starvation was feared.

India was beset by riots following destruction of a mosque by Hindu militants (Dec. 1992); Indian army troops repeatedly clashed with pro-independence demonstrators in the disputed Muslim region of **Kashmir**, exacerbating relations with **Pakistan**. Uneasy relations between India and Pakistan reached a new level when both nations conducted nuclear tests in 1998. Conflict in Pakistan between government and the military led to a bloodless coup (Oct. 1999).

Africa. South Africa was transformed as the white-dominated government abandoned **apartheid** and the country made the transition to a nonracial democratic government. Pres. F. W. **de Klerk** released Nelson **Mandela** from prison (Feb. 1990), after he had been held by the government for 27 years, and lifted a ban on the African National Congress. The white government repealed its apartheid laws (1990, 1991). **Mandela** was elected **president** (Apr. 1994), and a new constitution became law (Dec. 1996). Thabo **Mbeki**, the ANC's candidate to succeed Mandela, was overwhelmingly elected president in June 1999. In **Nigeria**, Gen. Olusegun **Obasanjo** was elected president (Feb. 1999), to become the country's first civilian leader in 15 years.

The decades-long rule of **Mobutu** Sese Seko in **Zaire** came to an end (May 1997) at the hands of rebel forces led by Laurent **Kabila**; an ailing Mobutu fled the country and soon after died. Kabila changed the country's name back to **Democratic Republic of the Congo**; conditions remained unstable. After the presidents of **Burundi** and **Rwanda** were killed in an airplane crash (Apr. 1994), violence erupted in Rwanda between Hutu and Tutsi factions; tens of thousands were slain. The conflict spread to refugee camps in neighboring Zaire and Burundi. Factional fighting also erupted in **Somalia** after Pres. Muhammad Siad Barre was ousted (Jan. 1991). The UN sent a U.S.-led **peacekeeping force**, but it was unsuccessful in restoring order. Some soldiers of the peacekeeping force were killed, including 23 Pakistanis (June 1993) and 18 U.S. Rangers (Oct. 1993). The UN ended its mission (Mar. 1995) with no durable government in place. **Liberia** endured factional fighting that lasted almost 5 years and claimed over 150,000 lives; a cease-fire was concluded in Aug. 1995. The World Health Organization reported (1995) that Africa accounted for 70% of **AIDS** cases worldwide.

A 16-year civil war appeared to end in **Angola** (May 1991) when the government signed a peace accord with the rebel UNITA faction. But despite the inauguration of a national unity government (Apr. 1997), insurgents continued to fight and gain territory. **Namibia** officially became independent in Mar. 1990. Claimed by South Africa since 1919 and placed under UN authority in 1971, it had long been a focus of colonial rivalries. In **Algeria,** the army cancelled a 2d round of parliamentary elections (Jan. 1992) after the Islamic party won a first round. Islamic fundamentalists then began a terrorist campaign that, along with killings by pro-government squads, eventually claimed thousands of lives. A peace plan was worked out with the militants in 1999.

North America. The **North American Free Trade Agreement** (NAFTA), liberalizing trade between the United States, Canada, and Mexico, went into effect Jan. 1, 1994. In **Canada**, the Progressive Conservative Party suffered a crushing defeat in general elections (Oct. 1993), and liberal Jean **Chrétien** became prime minister. The map of Canada was altered in Apr. 1999 to create a new territory, **Nunavut,** out of an area that had been part of Northwest Territories.

In the **United States**, in the 1992 presidential election, Democrat Bill **Clinton** defeated Pres. George Bush, but in 1994 congressional elections Republicans gained control of Congress. Congress passed legislation under which federal protection for welfare recipients was ended and funds turned over to the states for their programs. Clinton reached agreement with Congress on measures to eliminate the federal budget deficit. Clinton won reelection in 1996; the new administration was plagued by scandals but remained popular amid continued economic prosperity. In Dec. 1998 **Clinton** was **impeached** by the U.S. House on charges related to the Monica Lewinsky scandal; he was **acquitted** by the Senate in Feb. 1999.

The U.S. Army and Navy were torn by sexual scandals involving abuse of women personnel. The **United States** suffered embarrassment with the discovery of espionage by CIA agents (Aldrich Ames, Harold Nicholson).

In **Mexico,** Ernesto **Zedillo** of the ruling PRI party was elected president (July 1994) after the party's first candidate was assassinated. The country soon faced a crisis affecting the value of the peso, but recovered with the help of a bailout package from the U.S. A peasant revolt spearheaded by the **Zapatista National Liberation Army** erupted in the state of Chiapas (Jan. 1994) and was suppressed.

Central America. In **Haiti**, Jean-Bertrand **Aristide** was elected president (Dec. 1990) but was ousted in a military coup after 9 months in office. The UN approved a U.S.-led invasion to restore the elected leader; shortly before troops arrived, a delegation headed by former U.S. Pres. Jimmy Carter arranged (Sept. 1994) for the junta to step aside for Aristide. In **Nicaragua**, Violetta Chamarro defeated Daniel **Ortega** in the presidential election (Feb. 1990), thus ousting the Sandinistas. In **Panama**, U.S. troops invaded and overthrew the government of Manuel **Noriega** (Dec. 1989), who was wanted on drug charges; Noriega was captured Jan. 1990. On Dec. 31, 1999, Panama assumed full control of the **Panama Canal**, in accord with a treaty with the U.S. In **El Salvador** (1992) and **Guatemala** (1996) the governments signed agreements with rebel factions aimed at ending long-running civil conflicts.

South America. Alberto **Fujimori** was elected president of **Peru** in June 1990 and, despite his suppression of the constitution (1992), was reelected in 1995. Peru succeeded in capturing (Sept. 1992) the leader of the **Shining Path** guerrilla movement. Leftist guerrillas took hostages at an ambassador's residence in Lima (Dec. 1996); one hostage was killed during a government assault rescuing the rest (Apr. 1997). Peronist Pres. Carlos Saúl **Menem** served as **Argentina**'s president for much of the decade (elected 1989, reelected 1995), imposing stringent economic measures; he was succeeded in 1999 by Fernando de la **Rúa**.

Former Chilean Pres. Gen. Augusto **Pinochet** continued to head the army until Mar. 1998; he was arrested in London (Oct. 1998) on human rights charges but was judged medically unfit for trial and returned to Chile (Mar. 2000).

In **Brazil**, Fernando Henrique **Cardoso** was elected president (Oct. 1994) and reelected in 1998 amid a growing economic slump; the IMF announced a $42 billion aid package (Nov. 1998). The first UN Conference on Environment and Development, or **Earth Summit**, was held (June 1992) in **Rio de Janeiro,** with delegates from 178 nations.

Terrorism and Crime. Terrorism, often linked to Mideastern sources and with the U.S. as object, continued. A terrorist bomb exploded in a garage beneath New York City's **World Trade Center**, killing 6 people (Feb. 1993). Bombings of a U.S. military training center (Nov. 1995) and a barracks holding U.S. airmen (June 1996), both in **Saudi Arabia,** killed 7 and 19, respectively. Bombs exploded outside **U.S. embassies** in Kenya and Tanzania, Aug. 1998, killing over 220 people; the U.S. retaliated with missiles fired at alleged terrorist-linked sites in Afghanistan and Sudan. The Alfred P. Murrah Federal Building in **Oklahoma City**, OK, was destroyed by a bomb that killed 168 people (Apr. 1995).

Science. The powerful **Hubble Space Telescope** was launched in Apr. 1990; flaws in its mirrors and solar panels were repaired by space-walking astronauts (Dec. 1993). The U.S. space shuttle ***Atlantis*** docked with the orbiting Russian space station ***Mir*** (June 1995) for the first time, in the first of several joint missions in a spirit of post-Cold-War cooperation. The last Russian crew of the aging *Mir* space station departed in Aug. 1999. In Nov. 1998 the first component for a new **International Space Station** was launched into space from Kazakhstan. Two U.S. unmanned space probes sent to explore **Mars** were lost (1999) before they could send back any information.

Scottish scientist Ian Wilmut announced (Feb. 1997) the **cloning** of a sheep, nicknamed Dolly—the first mammal successfully cloned from a cell from an adult animal.

World Population Growth: AD 1-2001

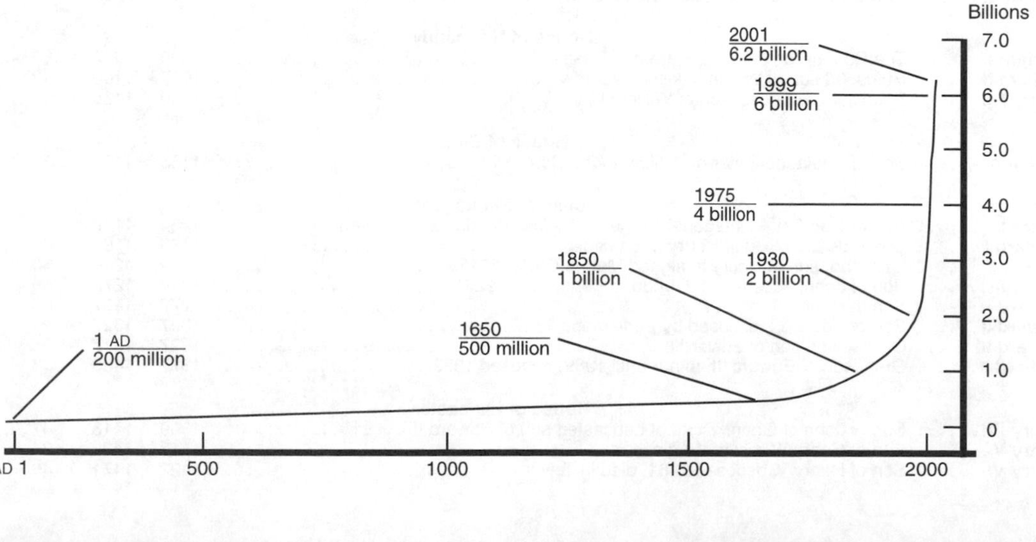

HISTORICAL FIGURES

Ancient Greeks and Romans

Greeks

Aeschines, orator, 389-314 BC
Aeschylus, dramatist, 525-456 BC
Aesop, fableist, c620-c560 BC
Alcibiades, politician, 450-404 BC
Anacreon, poet, c582-c485 BC
Anaxagoras, philosopher, c500-428 BC
Anaximander, philosopher, 611-546 BC
Anaximenes, philosopher, c570-500 BC
Antiphon, speechwriter, c480-411 BC
Apollonius, mathematician, c265-170 BC
Archimedes, math., 287-212 BC
Aristophanes, dramatist, c448-380 BC
Aristotle, philosopher, 384-322 BC
Athenaeus, scholar, fl. c200
Callicrates, architect, fl. 5th cent. BC
Callimachus, poet, c305-240 BC
Cratinus, comic dramatist, 520-421 BC
Democritus, philosopher, c460-370 BC
Demosthenes, orator, 384-322 BC
Diodorus, historian, fl. 20 BC
Diogenes, philosopher, 372-c287 BC
Dionysius, historian, d. c7 BC
Empedocles, philosopher, c490-430 BC
Epicharmus, dramatist, c530-440 BC
Epictetus, philosopher, c55-c135
Epicurus, philosopher, 341-270 BC
Eratosthenes, scientist, 276-194 BC
Euclid, mathematician, fl. c300 BC
Euripides, dramatist, c484-406 BC
Galen, physician, 130-200
Heraclitus, philosopher, c540-c475 BC
Herodotus, historian, c484-420 BC
Hesiod, poet, 8th cent. BC
Hippocrates, physician, c460-377 BC
Homer, poet, fl. c700 BC(?)
Isocrates, orator, 436-338 BC
Menander, dramatist, 342-292 BC
Parmenides, philosopher, b c515 BC
Pericles, statesman, c495-429 BC
Phidias, sculptor, c500-435 BC
Pindar, poet, c518-c438 BC
Plato, philosopher, c428-347 BC
Plutarch, biographer, c46-120
Polybius, historian, c200-c118 BC
Praxiteles, sculptor, 400-330 BC
Pythagoras, phil., math., c580-c500 BC
Sappho, poet, c610-c580 BC
Simonides, poet, 556-c468 BC
Socrates, philosopher, 469-399 BC
Solon, statesman, 640-560 BC
Sophocles, dramatist, c496-406 BC
Strabo, geographer, c63 BC-AD 24
Thales, philosopher, c634-546 BC
Themistocles, politician, c524-c460 BC
Theocritus, poet, c310-250 BC
Theophrastus, phil., c372-c287 BC
Thucydides, historian, fl. 5th cent. BC
Timon, philosopher, c320-c230 BC
Xenophon, historian, c434-c355 BC
Zeno, philosopher, c335-c263 BC

Romans

Ammianus, historian, c330-395
Apuleius, satirist, c124-c170
Boethius, scholar, c480-524
Caesar, Julius, leader, 100-44 BC
Catiline, politician, c108-62 BC
Cato (Elder), statesman, 234-49 BC
Catullus, poet, c84-54 BC
Cicero, orator, 106-43 BC
Claudian, poet, c370-c404
Ennius, poet, 239-170 BC
Gellius, author, c130-c165
Horace, poet, 65-8 BC
Juvenal, satirist, 60-127
Livy, historian, 59 BC-AD 17
Lucan, poet, 39-65
Lucilius, poet, c180-c102 BC
Lucretius, poet, c99-c55 BC
Martial, epigrammatist, c38-c103
Nepos, historian, c100-c25 BC
Ovid, poet, 43 BC-AD 17
Persius, satirist, 34-62
Plautus, dramatist, c254-c184 BC
Pliny the Elder, scholar, 23-79
Pliny the Younger, author, 62-113
Quintilian, rhetorician, c35-c97
Sallust, historian, 86-34 BC
Seneca, philosopher, 4 BC-AD 65
Silius, poet, c25-101
Statius, poet, c45-c96
Suetonius, biographer, c69-c122
Tacitus, historian, 56-120
Terence, dramatist, 185-c159 BC
Tibullus, poet, c55-c19 BC
Vergil, poet, 70-19 BC
Vitruvius, architect, fl. 1st cent. BC

Rulers of England and Great Britain

ENGLAND

Name		Reign Began	Died	Death Age	Years Reigned
	Saxons and Danes				
Egbert	King of Wessex, won allegiance of all English	829	839	—	10
Ethelwulf	Son, King of Wessex, Sussex, Kent, Essex	839	858	—	19
Ethelbald	Son of Ethelwulf, displaced father in Wessex	858	860	—	2
Ethelbert	2d son of Ethelwulf, united Kent and Wessex	860	866	—	6
Ethelred I	3d son, King of Wessex, fought Danes	866	871	—	5
Alfred	The Great, 4th son, defeated Danes, fortified London	871	899	52	28
Edward	The Elder, Alfred's son, united English, claimed Scotland	899	924	55	25
Athelstan	The Glorious, Edward's son, King of Mercia, Wessex	924	940	45	16
Edmund	3d son of Edward, King of Wessex, Mercia	940	946	25	6
Edred	4th son of Edward	946	955	32	9
Edwy	The Fair, eldest son of Edmund, King of Wessex	955	959	18	3
Edgar	The Peaceful, 2d son of Edmund, ruled all English	959	975	32	17
Edward	The Martyr, eldest son of Edgar, murdered by stepmother	975	978	17	4
Ethelred II	The Unready, 2d son of Edgar, married Emma of Normandy	978	1016	48	37
Edmund II	Ironside, son of Ethelred II, King of London	1016	1016	27	0
Canute	The Dane, gave Wessex to Edmund, married Emma	1016	1035	40	19
Harold I	Harefoot, natural son of Canute	1035	1040	—	5
Hardecanute	Son of Canute by Emma, Danish King	1040	1042	24	2
Edward	The Confessor, son of Ethelred II (canonized 1161)	1042	1066	62	24
Harold II	Edward's brother-in-law, last Saxon King	1066	1066	44	0
	House of Normandy				
William I	The Conqueror, defeated Harold at Hastings	1066	1087	60	21
William II	Rufus, 3d son of William I, killed by arrow	1087	1100	43	13
Henry I	Beauclerc, youngest son of William I	1100	1135	67	35
	House of Blois				
Stephen	Son of Adela, daughter of William I, and Count of Blois	1135	1154	50	19
	House of Plantagenet				
Henry II	Son of Geoffrey Plantagenet (Angevin) by Matilda, daughter of Henry I	1154	1189	56	35
Richard I	Coeur de Lion, son of Henry II, crusader	1189	1199	42	10
John	Lackland, son of Henry II, signed Magna Carta, 1215	1199	1216	50	17
Henry III	Son of John, acceded at 9, under regency until 1227	1216	1272	65	56
Edward I	Son of Henry III	1272	1307	68	35
Edward II	Son of Edward I, deposed by Parliament, 1327	1307	1327	43	20
Edward III	Of Windsor, son of Edward II	1327	1377	65	50
Richard II	Grandson of Edward III, minor until 1389, deposed 1399	1377	1400	33	22
	House of Lancaster				
Henry IV	Son of John of Gaunt, Duke of Lancaster, son of Edward III	1399	1413	47	13
Henry V	Son of Henry IV, victor of Agincourt	1413	1422	34	9
Henry VI	Son of Henry V, deposed 1461, died in Tower	1422	1471	49	39

Name		Reign Began	Died	Death Age	Years Reigned
	House of York				
Edward IV	Great-great-grandson of Edward III, son of Duke ofYork	1461	1483	40	22
Edward V	Son of Edward IV, murdered in Tower of London	1483	1483	13	0
Richard III	Brother of Edward IV, fell at Bosworth Field	1483	1485	32	2
	House of Tudor				
Henry VII	Son of Edmund Tudor, Earl of Richmond, whose father had married the widow of Henry V; descended from Edward III through his mother, Margaret Beaufort via John of Gaunt. By marriage with daughter of Edward IV he united Lancaster and York	1485	1509	53	24
Henry VIII	Son of Henry VII, by Elizabeth, daughter of Edward IV	1509	1547	56	38
Edward VI	Son of Henry VIII, by Jane Seymour, his 3d queen. Ruled under regents. Was forced to name Lady Jane Grey his successor. Council of State proclaimed her queen July 10, 1553. Mary Tudor won Council, was proclaimed queen July 19, 1553. Mary had Lady Jane Grey beheaded for treason, Feb. 1554	1547	1553	16	6
Mary I	Daughter of Henry VIII, by Catherine of Aragon	1553	1558	43	5
Elizabeth I	Daughter of Henry VIII, by Anne Boleyn	1558	1603	69	44
	GREAT BRITAIN				
	House of Stuart				
James I	James VI of Scotland, son of Mary, Queen of Scots. *First to call himself King of Great Britain. This became official with the Act of Union, 1707*	1603	1625	59	22
Charles I	Only surviving son of James I; beheaded Jan. 30, 1649	1625	1649	48	24
	Commonwealth, 1649–1660				
	Council of State, 1649; Protectorate, 1653[1]				
The Cromwells	Oliver Cromwell, Lord Protector	1653	1658	59	5
	Richard Cromwell, son, Lord Protector, resigned May 25, 1659	1658	1712	86	1
	House of Stuart (Restored)				
Charles II	Eldest son of Charles I, died without issue	1660	1685	55	25
James II	2d son of Charles I. Deposed 1688. Interregnum Dec. 11, 1688, to Feb. 13, 1689	1685	1701	68	3
William III	Son of William, Prince of Orange, by Mary, daughter of Charles I	1689	1702	51	13
and Mary II	Eldest daughter of James II and wife of William III	1689	1694	33	6
Anne	2d daughter of James II	1702	1714	49	12
	House of Hanover				
George I	Son of Elector of Hanover, by Sophia, granddaughter of James I	1714	1727	67	13
George II	Only son of George I, married Caroline of Brandenburg	1727	1760	77	33
George III	Grandson of George II, married Charlotte of Mecklenburg	1760	1820	81	59
George IV	Eldest son of George III, Prince Regent, from Feb. 1811	1820	1830	67	10
William IV	3d son of George III, married Adelaide of Saxe-Meiningen	1830	1837	71	7
Victoria	Daughter of Edward, 4th son of George III; married (1840) Prince Albert of Saxe-Coburg and Gotha, who became Prince Consort	1837	1901	81	63
	House of Saxe-Coburg and Gotha				
Edward VII	Eldest son of Victoria, married Alexandra, Princess of Denmark	1901	1910	68	9
	House of Windsor[2]				
George V	2d son of Edward VII, married Princess Mary of Teck	1910	1936	70	25
Edward VIII	Eldest son of George V; acceded Jan. 20, 1936, abdicated Dec. 11	1936	1972	77	1
George VI	2d son of George V; married Lady Elizabeth Bowes-Lyon	1936	1952	56	15
Elizabeth II	Elder daughter of George VI, acceded Feb. 6, 1952	1952			

= age/birth date not certain. (1) The Cromwells ruled Britain following overthow of the monarchy in 1649. (2) Name adopted by proclamation of George V, July 17, 1917.

Rulers of Scotland

Kenneth I MacAlpin was the first Scot to rule both Scots and Picts, AD 846.

Duncan I was the first general ruler, 1034. Macbeth seized the kingdom 1040, was slain by Duncan's son, Malcolm III MacDuncan (Canmore), 1057.

Malcolm married Margaret, Saxon princess who had fled from the Normans. Queen Margaret introduced English language and English monastic customs. She was canonized, 1250. Her son Edgar, 1097, moved the court to Edinburgh. His brothers Alexander I and David I succeeded. Malcolm IV, the Maiden, 1153, grandson of David I, was followed by his brother, William the Lion, 1165, whose son was Alexander II, 1214. The latter's son, Alexander III, 1249, defeated the Norse and regained the Hebrides. When he died, 1286, his granddaughter, Margaret, child of Eric of Norway and grandniece of Edward I of England, known as the Maid of Norway, was chosen ruler, but died 1290, aged 8.

John Baliol, 1292-1296. (Interregnum, 10 years.)

Robert Bruce (The Bruce), 1306-1329, victor at Bannockburn, 1314.

David II, only son of Robert Bruce, ruled 1329-1371.

Robert II, 1371-1390, grandson of Robert Bruce, son of Walter, the Steward of Scotland, was called The Steward, first of the so-called Stuart line.

Robert III, son of Robert II, 1390-1406.

James I, son of Robert III, 1406-1437.

James II, son of James I, 1437-1460.

James III, eldest son of James II, 1460-1488.

James IV, eldest son of James III, 1488-1513.

James V, eldest son of James IV, 1513-1542.

Mary, daughter of James V, born 1542, became queen when one week old; was crowned 1543. Married, 1558, Francis, son of Henry II of France, who became king 1559, died 1560. Mary ruled Scots 1561 until abdication, 1567. She also married Henry Stewart, Lord Darnley (1565), and James, Earl of Bothwell (1567). Imprisoned by Elizabeth I, Mary was beheaded 1587.

James VI, 1566-1625, son of Mary and Lord Darnley, became King of England on death of Elizabeth in 1603. Although the thrones were thus united, the legislative union of Scotland and England was not effected until the Act of Union, May 1, 1707.

IT'S A FACT: The real Macbeth did not murder Duncan in bed, as Shakespeare's villain did, but killed him on the battlefield. When Macbeth was slain in turn, he was buried in a place of honor on the island of Iona.

Prime Ministers of Great Britain

Designations in parentheses describe each government;
W=Whig; T=Tory; Cl=Coalition; P=Peelite; L=Liberal; C=Conservative[1]; La=Labour.

Sir Robert Walpole (W)[2]	1721-1742
Earl of Wilmington (W)	1742-1743
Henry Pelham (W)	1743-1754
Duke of Newcastle (W)	1754-1756
Duke of Devonshire (W)	1756-1757
Duke of Newcastle (W)	1757-1762
Earl of Bute (T)	1762-1763
George Grenville (W)	1763-1765
Marquess of Rockingham (W)	1765-1766
William Pitt the Elder (Earl of Chatham) (W)	1766-1768
Duke of Grafton (W)	1768-1770
Frederick North (Lord North) (T)	1770-1782
Marquess of Rockingham (W)	1782
Earl of Shelburne (W)	1782-1783
Duke of Portland (Cl)	1783
William Pitt the Younger (T)	1783-1801
Henry Addington (T)	1801-1804
William Pitt the Younger (T)	1804-1806
William Wyndham Grenville, Baron Grenville (W)	1806-1807
Duke of Portland (T)	1807-1809
Spencer Perceval (T)	1809-1812
Earl of Liverpool (T)	1812-1827
George Canning (T)	1827
Viscount Goderich (T)	1827-1828
Duke of Wellington (T)	1828-1830
Earl Grey (W)	1830-1834
Viscount Melbourne (W)	1834
Sir Robert Peel (C)	1834-1835
Viscount Melbourne (W)	1835-1841
Sir Robert Peel (C)	1841-1846
Lord (later Earl) John Russell (W)	1846-1852
Earl of Derby (C)	1852
Earl of Aberdeen (P)	1852-1855
Viscount Palmerston (Li)	1855-1858
Earl of Derby (C)	1858-1859
Viscount Palmerston (Li)	1859-1865
Earl Russell (Li)	1865-1866
Earl of Derby (C)	1866-1868
Benjamin Disraeli (C)	1868
William E. Gladstone (Li)	1868-1874
Benjamin Disraeli (C)	1874-1880
William E. Gladstone (Li)	1880-1885
Marquess of Salisbury (C)	1885-1886
William E. Gladstone (Li)	1886
Marquess of Salisbury (C)	1886-1892
William E. Gladstone (Li)	1892-1894
Earl of Rosebery (Li)	1894-1895
Marquess of Salisbury (C)	1895-1902
Arthur J. Balfour (C)	1902-1905
Sir Henry Campbell Bannerman (Li)	1905-1908
Herbert H. Asquith (Li)	1908-1915
Herbert H. Asquith (Cl)	1915-1916
David Lloyd George (Cl)	1916-1922
Andrew Bonar Law (C)	1922-1923
Stanley Baldwin (C)	1923-1924
James Ramsay MacDonald (La)	1924
Stanley Baldwin (C)	1924-1929
James Ramsay MacDonald (La)	1929-1931
James Ramsay MacDonald (Cl)	1931-1935
Stanley Baldwin (Cl)	1935-1937
Neville Chamberlain (Cl)	1937-1940
Winston Churchill (Cl)	1940-1945
Winston Churchill (C)	1945
Clement Attlee (La)	1945-1951
Sir Winston Churchill (C)	1951-1955
Sir Anthony Eden (C)	1955-1957
Harold Macmillan (C)	1957-1963
Sir Alec Douglas-Home (C)	1963-1964
Harold Wilson (La)	1964-1970
Edward Heath (C)	1970-1974
Harold Wilson (La)	1974-1976
James Callaghan (La)	1976-1979
Margaret Thatcher (C)	1979-1990
John Major (C)	1990-1997
Tony Blair (La)	1997-

(1) The Conservative Party was formed in 1834, an outgrowth and, in some respects, a continuation of the Tory party. (2) Walpole is commonly regarded as the first prime minister of Britain, though the title was not commonly used until later in the century and did not become official until 1905.

Historical Periods of Japan

Yamato	c. 300-592	Conquest of Yamato plain c. AD 300.
Asuka	592-710	Accession of Empress Suiko, 592.
Nara	710-794	Completion of Heijo (Nara), 710; the capital moves to Nagaoka, 784.
Heian	794-1185	Completion of Heian (Kyoto), 794.
Fujiwara	858-1160	Fujiwara-no-Yoshifusa becomes regent, 858.
Taira	1160-1185	Taira-no-Kiyomori assumes control, 1160; Minamoto-no-Yoritomo victor over Taira, 1185.
Kamakura	1192-1333	Yoritomo becomes shogun, 1192.
Namboku	1334-1392	Restoration of Emperor Godaigo, 1334; Southern Court established by Godaigo at Yoshino, 1336.
Ashikaga	1338-1573	Ashikaga Takauji becomes shogun, 1338.
Muromachi	1392-1573	Unification of Southern and Northern Courts, 1392.
Sengoku	1467-1600	Beginning of the Onin war, 1467.
Momoyama	1573-1603	Oda Nobunaga enters Kyoto, 1568; Nobunaga deposes last Ashikaga shogun, 1573; Tokugawa Ieyasu victor at Sekigahara, 1600.
Edo	1603-1867	Ieyasu becomes shogun, 1603.
Meiji	1868-1912	Enthronement of Emperor Mutsuhito (Meiji), 1867; Meiji Restoration and Charter Oath, 1868.
Taisho	1912-1926	Accession of Emperor Yoshihito, 1912.
Showa	1926-1989	Accession of Emperor Hirohito, 1926.
Heisei	1989-	Accession of Emperor Akihito, 1989.

Rulers of France: Kings, Queens, Presidents

Caesar to Charlemagne

Julius Caesar subdued the Gauls, native tribes of Gaul (France), 58 to 51 BC. The Romans ruled 500 years. The Franks, a Teutonic tribe, reached the Somme from the East c. AD 250. By the 5th century the Merovingian Franks ousted the Romans. In 451, with the help of Visigoths, Burgundians, and others, they defeated Attila and the Huns at Chalons-sur-Marne.

Childeric I became leader of the Merovingians 458. His son Clovis I (Chlodwig, Ludwig, Louis), crowned 481, founded the dynasty. After defeating the Alemanni (Germans) 496, he was baptized a Christian and made Paris his capital. His line ruled until Childeric III was deposed, 751.

The West Merovingians were called Neustrians, the eastern Austrasians. Pepin of Herstal (687-714), major domus, or head of the palace, of Austrasia, took over Neustria as dux (leader) of the Franks. Pepin's son, Charles, called Martel (the Hammer), defeated the Saracens at Tours-Poitiers, 732; was succeeded by his son, Pepin the Short, 741, who deposed Childeric III and ruled as king until 768.

His son, Charlemagne, or Charles the Great (742-814), became king of the Franks, 768, with his brother Carloman, who died 771. Charlemagne ruled France, Germany, parts of Italy, Spain, and Austria, and enforced Christianity. Crowned Emperor of the Romans by Pope Leo III in St. Peter's, Rome, Dec. 25, 800. Succeeded by son, Louis I the Pious, 814. At death, 840, Louis left empire to sons, Lothair (Roman emperor); Pepin I (king of Aquitaine); Louis II (of Germany); Charles the Bald (France). They quarreled and, by the peace of Verdun, 843, divided the empire.

The date preceding each entry is year of accession.

The Carolingians

843 Charles I (the Bald); Roman Emperor, 875
877 Louis II (the Stammerer), son
879 Louis III (died 882) and Carloman, brothers
885 Charles II (the Fat); Roman Emperor, 881
888 Eudes (Odo), elected by nobles
898 Charles III (the Simple), son of Louis II, defeated by
922 Robert, brother of Eudes, killed in war
923 Rudolph (Raoul), Duke of Burgundy
936 Louis IV, son of Charles III
954 Lothair, son, aged 13, defeated by Capet
986 Louis V (the Sluggard), left no heirs

The Capets

987 Hugh Capet, son of Hugh the Great
996 Robert II (the Wise), his son
1031 Henry I, his son
1060 Philip I (the Fair), son
1108 Louis VI (the Fat), son
1137 Louis VII (the Younger), son
1180 Philip II (Augustus), son, crowned at Reims
1223 Louis VIII (the Lion), son
1226 Louis IX, son, crusader; Louis IX (1214-1270) reigned 44 years, arbitrated disputes with English King Henry III; led crusades, 1248 (captured in Egypt 1250) and 1270, when he died of plague in Tunis. Canonized 1297 as St. Louis.
1270 Philip III (the Hardy), son
1285 Philip IV (the Fair), son, king at 17
1314 Louis X (the Headstrong), son. His posthumous son, John I, lived only 7 days
1316 Philip V (the Tall), brother of Louis X
1322 Charles IV (the Fair), brother of Louis X

House of Valois

1328 Philip VI (of Valois), grandson of Philip III
1350 John II (the Good), his son, retired to England
1364 Charles V (the Wise), son
1380 Charles VI (the Beloved), son
1422 Charles VII (the Victorious), son. In 1429 Joan of Arc (Jeanne d'Arc) promised Charles to oust the English, who occupied northern France. Joan won at Orleans and Patay and had Charles crowned at Reims, July 17, 1429. Joan was captured May 24, 1430, and executed May 30, 1431, at Rouen for heresy. Charles ordered her rehabilitation, effected 1455.
1461 Louis XI (the Cruel), son, civil reformer
1483 Charles VIII (the Affable), son
1498 Louis XII, great-grandson of Charles V
1515 Francis I, of Angouleme, nephew, son-in-law. Francis I (1494-1547) reigned 32 years, fought 4 big wars, was patron of the arts, aided Cellini, del Sarto, Leonardo da Vinci, Rabelais, embellished Fontainebleau.
1547 Henry II, son, killed at a joust in a tournament. He was the husband of Catherine de Medicis (1519-1589) and the lover of Diane de Poitiers (1499-1566). Catherine was born in Florence, daughter of Lorenzo de Medici. By her marriage to Henry II she became the mother of Francis II, Charles IX, Henry III, and Queen Margaret (Reine Margot), wife of Henry IV. She persuaded Charles IX to order the massacre of Huguenots on the Feast of St. Bartholomew, Aug. 24, 1572, six days after her daughter was married to Henry of Navarre.
1559 Francis II, son. In 1548, Mary, Queen of Scots since infancy, was betrothed when 6 to Francis, aged 4. They were married 1558. Francis died 1560, aged 16; Mary ruled Scotland, abdicated 1567.
1560 Charles IX, brother
1574 Henry III, brother, assassinated

House of Bourbon

1589 Henry IV, of Navarre, assassinated. Henry IV made enemies when he gave tolerance to Protestants by Edict of Nantes, 1598. He was grandson of Queen Margaret of Navarre, literary patron. He married Margaret of Valois, daughter of Henry II and Catherine de Medicis; was divorced; in 1600 married Marie de Medicis, who became Regent of France, 1610-1617, for her son, Louis XIII, but was exiled by Richelieu, 1631.
1610 Louis XIII (the Just), son. Louis XIII (1601-1643) married Anne of Austria. His ministers were Cardinals Richelieu and Mazarin.
1643 Louis XIV (The Grand Monarch), son. Louis XIV was king 72 years. He exhausted a prosperous country in wars for thrones and territory. By revoking the Edict of Nantes (1685) he caused the emigration of the Huguenots. He said: "I am the state."
1715 Louis XV, great-grandson. Louis XV married a Polish princess;' lost Canada to the English. His favorites, Mme. Pompadour and Mme. Du Barry, influenced policies. Noted for saying "After me, the deluge."
1774 Louis XVI, grandson; married Marie Antoinette, daughter of Empress Maria Therese of Austria. King and queen beheaded by Revolution, 1793. Their son, called Louis XVII, died in prison, never ruled.

First Republic

1792 National Convention of the French Revolution
1795 Directory, under Barras and others
1799 Consulate, Napoleon Bonaparte, first consul. Elected consul for life, 1802.

First Empire

1804 Napoleon I (Napoleon Bonaparte), emperor. Josephine (de Beauharnais), empress, 1804-1809; Marie Louise, empress, 1810-1814. Her son, Francois (1811-1832), titular King of Rome, later Duke de Reichstadt and "Napoleon II," never ruled. Napoleon abdicated 1814, died 1821.

Bourbons Restored

1814 Louis XVIII, king; brother of Louis XVI
1824 Charles X, brother; reactionary; deposed by the July Revolution, 1830

House of Orleans

1830 Louis-Philippe, the "citizen king"

Second Republic

1848 Louis Napoleon Bonaparte, president, nephew of Napoleon I.

Second Empire

1852 Napoleon III (Louis Napoleon Bonaparte), emperor; Eugenie (de Montijo), empress. Lost Franco-Prussian war, deposed 1870. Son, Prince Imperial (1856-1879), died in Zulu War. Eugenie died 1920.

Third Republic—Presidents

1871 Thiers, Louis Adolphe (1797-1877)
1873 MacMahon, Marshal Patrice M. de (1808-1893)
1879 Grevy, Paul J. (1807-1891)
1887 Sadi-Carnot, M. (1837-1894), assassinated
1894 Casimir-Perier, Jean P. P. (1847-1907)
1895 Faure, François Felix (1841-1899)
1899 Loubet, Emile (1838-1929)
1906 Fallieres, C. Armand (1841-1931)
1913 Poincare, Raymond (1860-1934)
1920 Deschanel, Paul (1856-1922)
1920 Millerand, Alexandre (1859-1943)
1924 Doumergue, Gaston (1863-1937)
1931 Doumer, Paul (1857-1932), assassinated
1932 Lebrun, Albert (1871-1950), resigned 1940
1940 Vichy govt. under German armistice: Henri Philippe Petain (1856-1951), Chief of State, 1940-1944.

Provisional govt. after liberation: Charles de Gaulle (1890-1970), Oct. 1944-Jan. 21, 1946; Felix Gouin (1884-1977), Jan. 23, 1946; Georges Bidault (1899-1983), June 24, 1946.

Fourth Republic—Presidents

1947 Auriol, Vincent (1884-1966)
1954 Coty, Rene (1882-1962)

Fifth Republic—Presidents

1959 De Gaulle, Charles Andre J. M. (1890-1970)
1969 Pompidou, Georges (1911-1974)
1974 Giscard d'Estaing, Valery (1926-)
1981 Mitterrand, François (1916-1996)
1995 Chirac, Jacques (1932-)

Rulers of Middle Europe; Rise and Fall of Dynasties; Rulers of Germany

Carolingian Dynasty

Charles the Great, or Charlemagne, ruled France, Italy, and Middle Europe; established Ostmark (later Austria); crowned Roman emperor by pope in Rome, AD 800; died 814.

Louis I (Ludwig) the Pious, son; crowned by Charlemagne 814; died 840.

Louis II, the German, son; succeeded to East Francia (Germany) 843-876.

Charles the Fat, son; inherited East Francia and West Francia (France) 876, reunited empire, crowned emperor by pope 881, deposed 887.

Arnulf, nephew, 887-899. Partition of empire.

Louis the Child, 899-911, last direct descendant of Charlemagne.

Conrad I, duke of Franconia, first elected German king, 911-918, founded House of Franconia.

Saxon Dynasty; First Reich

Henry I, the Fowler, duke of Saxony, 919-936.

Otto I, the Great, 936-973, son; crowned Holy Roman Emperor by pope, 962.

Otto II, 973-983, son; failed to oust Greeks and Arabs from Sicily.

Otto III, 983-1002, son; crowned emperor at 16.

Henry II, the Saint, duke of Bavaria, 1002-1024, great-grandson of Otto the Great.

House of Franconia

Conrad II, 1024-1039, elected king of Germany.

Henry III, the Black, 1039-1056, son; deposed 3 popes; annexed Burgundy.

Henry IV, 1056-1106, son; regency by his mother, Agnes of Poitou. Banned by Pope Gregory VII, he did penance at Canossa.

Henry V, 1106-1125, son; last of Salic House.

Lothair, duke of Saxony, 1125-1137. Crowned emperor in Rome, 1134.

House of Hohenstaufen

Conrad III, duke of Swabia, 1138-1152. In 2d Crusade.

Frederick I, Barbarossa, 1152-1190; Conrad's nephew.

Henry VI, 1190-1196, took lower Italy from Normans. Son became king of Sicily.

Philip of Swabia, 1197-1208, brother.

Otto IV, of House of Welf, 1198-1215; deposed.

Frederick II, 1215-1250, son of Henry VI; king of Sicily; crowned king of Jerusalem in 5th Crusade.

Conrad IV, 1250-1254, son; lost lower Italy to Charles of Anjou.

Conradin, 1252-1268, son, king of Jerusalem and Sicily, beheaded. Last Hohenstaufen.

Interregnum, 1254-1273, Rise of the Electors.

Transition

Rudolph I of Hapsburg, 1273-1291, defeated King Ottocar II of Bohemia. Bequeathed duchy of Austria to eldest son, Albert.

Adolph of Nassau, 1292-1298, killed in war with Albert of Austria.

Albert I, king of Germany, 1298-1308, son of Rudolph.

Henry VII, of Luxemburg, 1308-1313, crowned emperor in Rome. Seized Bohemia, 1310.

Louis IV of Bavaria (Wittelsbach), 1314-1347. Also elected was Frederick of Austria, 1314-1330 (Hapsburg). Abolition of papal sanction for election of Holy Roman Emperor.

Charles IV, of Luxemburg, 1347-1378, grandson of Henry VII, German emperor and king of Bohemia, Lombardy, Burgundy; took Mark of Brandenburg.

Wenceslaus, 1378-1400, deposed.

Rupert, Duke of Palatine, 1400-1410.

Sigismund, 1411-1437.

Hungary

Stephen I, house of Arpad, 997-1038. Crowned king 1000; converted Magyars; canonized 1083. After several centuries of feuds Charles Robert of Anjou became Charles I, 1308-1342.

Louis I, the Great, son, 1342-1382; joint ruler of Poland with Casimir III, 1370. Defeated Turks.

Mary, daughter, 1382-1395, ruled with husband. Sigismund of Luxemburg, 1387-1437, also king of Bohemia. As brother of Wenceslaus he succeeded Rupert as Holy Roman Emperor, 1410.

Albert, 1438-1439, son-in-law of Sigismund; also Roman emperor as Albert II *(see under Hapsburg).*

Ulaszlo I of Poland, 1440-1444.

Ladislaus V, posthumous son of Albert II, 1444-1457. John Hunyadi (Hunyadi Janos), governor (1446-1452), fought Turks, Czechs; died 1456.

Matthias I (Corvinus), son of Hunyadi, 1458-1490. Shared rule of Bohemia, captured Vienna, 1485, annexed Austria, Styria, Carinthia.

Ulaszlo II (king of Bohemia), 1490-1516.

Louis II, son, aged 10, 1516-1526. Wars with Suleiman, Turk. In 1527 Hungary split between Ferdinand I, Archduke of Austria, bro.-in-law of Louis II, and John Zapolya of Transylvania. After Turkish invasion, 1547, Hungary split between Ferdinand, Prince John Sigismund (Transylvania), and the Turks.

House of Hapsburg

Albert V of Austria, Hapsburg, crowned king of Hungary, Jan. 1438, Roman emperor, March 1438, as Albert II; died 1439.

Frederick III, cousin, 1440-1493. Fought Turks.

Maximilian I, son, 1493-1519. Assumed title of Holy Roman Emperor (German), 1493.

Charles V, grandson, 1519-1556. King of Spain with mother co-regent; crowned Roman emperor at Aix, 1520. Confronted Luther at Worms; attempted church reform and religious conciliation; abdicated 1556.

Ferdinand I, king of Bohemia, 1526, of Hungary, 1527; disputed. German king, 1531. Crowned Roman emperor on abdication of brother Charles V, 1556.

Maximilian II, son, 1564-1576.

Rudolph II, son, 1576-1612.

Matthias, brother, 1612-1619, king of Bohemia and Hungary.

Ferdinand II of Styria, king of Bohemia, 1617, of Hungary, 1618, Roman emperor, 1619. Bohemian Protestants deposed him, elected Frederick V of Palatine, starting Thirty Years War.

Ferdinand III, son, king of Hungary, 1625, Bohemia, 1627, Roman emperor, 1637. Peace of Westphalia, 1648, ended war. Leopold I, 1658-1705; Joseph I, 1705-1711; Charles VI, 1711-1740.

Maria Theresa, daughter, 1740-1780, Archduchess of Austria, queen of Hungary; ousted pretender, Charles VII, crowned 1742; in 1745 obtained election of her husband Francis I as Roman emperor and co-regent (d. 1765). Fought Seven Years' War with Frederick II of Prussia. Mother of Marie Antoinette.

Joseph II, son, 1765-1790, Roman emperor, reformer; powers restricted by Empress Maria Theresa until her death, 1780. First partition of Poland. Leopold II, 1790-1792.

Francis II, son, 1792-1835. Fought Napoleon. Proclaimed first hereditary emperor of Austria, 1804. Forced to abdicate as Roman emperor, 1806; last use of title. Ferdinand I, son, 1835-1848, abdicated during revolution.

Austro-Hungarian Monarchy

Francis Joseph I, nephew, 1848-1916, emperor of Austria, king of Hungary. Dual monarchy of Austria-Hungary formed, 1867. After assassination of heir, Archduke Francis Ferdinand, June 28, 1914, Austrian diplomacy precipitated World War I.

Charles I, grand-nephew, 1916-1918, last emperor of Austria and king of Hungary. Abdicated Nov. 11-13, 1918, died 1922.

Rulers of Prussia

Nucleus of Prussia was the Mark of Brandenburg. First margrave Albert the Bear (Albrecht), 1134-1170. First Hohenzollern margrave was Frederick, burgrave of Nuremberg, 1417-1440.

Frederick William, 1640-1688, the Great Elector. Son, Frederick III, 1688-1713, crowned King Frederick of Prussia, 1701.

Frederick William I, son, 1713-1740.

Frederick II, the Great, son, 1740-1786, annexed Silesia, part of Austria.

Frederick William II, nephew, 1786-1797.

Frederick William III, son, 1797-1840. Napoleonic wars.

Frederick William IV, son, 1840-1861. Uprising of 1848 and first parliament and constitution.

Second and Third Reich

William I, 1861-1888, brother. Annexation of Schleswig and Hanover; Franco-Prussian war, 1870-1871, proclamation of German Reich, Jan. 18, 1871, at Versailles; William, German emperor (Deutscher Kaiser), Bismarck, chancellor.

Frederick III, son, 1888.

William II, son, 1888-1918. Led Germany in World War I, abdicated as German emperor and king of Prussia, Nov. 9, 1918. Died in exile in Netherlands, June 4, 1941. Minor rulers of Bavaria, Saxony, Wurttemberg also abdicated.

Germany proclaimed republic at Weimar, July 1, 1919. Presidents included: Frederick Ebert, 1919-1925; Paul von Hindenburg-Beneckendorff, 1925, reelected 1932, d. Aug. 2, 1934. Adolf Hitler, chancellor, chosen successor as Leader-Chancellor (Fuehrer-Reichskanzler) of Third Reich. Annexed Austria, Mar. 1938. Precipitated World War II, 1939-1945. Suicide Apr. 30, 1945.

Germany After 1945

Following World War II, Germany was split between democratic West and Soviet-dominated East. West German chancellors: Konrad Adenauer, 1949-1963; Ludwig Erhard, 1963-1966; Kurt Georg Kiesinger, 1966-1969; Willy Brandt, 1969-1974; Helmut Schmidt, 1974-1982; Helmut Kohl, 1982-1990. East German Communist party leaders: Walter Ulbricht, 1946-1971; Erich Honecker, 1971-1989; Egon Krenz, 1989-1990.

Germany reunited Oct. 3, 1990. Post-reunification chancellors: Helmut Kohl, 1990-1998; Gerhard Schröder, 1998- .

Rulers of Poland

House of Piasts

Miesko I, 962?-992; Poland Christianized 966. Expansion under 3 Boleslavs: I, 992-1025, son, crowned king 1024; II, 1058-1079, great-grandson, exiled after killing bishop Stanislav who became chief patron saint of Poland; III, 1106-1138, nephew, divided Poland among 4 sons, eldest suzerain.

1138-1306, feudal division. 1226 founding in Prussia of military order Teutonic Knights. 1226 invasion by Tartars/Mongols.

Vladislav I, 1306-1333, reunited most Polish territories, crowned king 1320. Casimir III the Great, 1333-1370, son, developed economic, cultural life, foreign policy.

House of Anjou

Louis I, 1370-1382, nephew/was also Louis I of Hungary.

Jadwiga, 1384-1399, daughter, married 1386 Jagiello, Grand Duke of Lithuania.

House of Jagiellonians

Vladislav II, 1386-1434, Christianized Lithuania, founded personal union between Poland and Lithuania. Defeated 1410 Teutonic Knights at Grunwald.

Vladislav III, 1434-1444, son, simultaneously king of Hungary. Fought Turks, killed 1444 in battle of Varna.

Casimir IV, 1446-1492, brother, competed with Hapsburgs, put son Vladislav on throne of Bohemia, later also of Hungary (Ulaszlo II).

Sigismund I, 1506-1548, son, patronized science and arts, his and son's reign "Golden Age."

Sigismund II, 1548-1572, son, established 1569 real union of Poland and Lithuania (lasted until 1795).

Elective Kings

Polish nobles in 1572 proclaimed Poland a republic headed by king to be elected by whole nobility.

Stephen Batory, 1576-1586, duke of Transylvania, married Ann, sister of Sigismund II August. Fought Russians.

Sigismund III Vasa, 1587-1632, nephew of Sigismund II. 1592-1598 also king of Sweden. His generals fought Russians, Turks.

Vladislav II Vasa, 1632-1648, son. Fought Russians.

John II Casimir Vasa, 1648-1668, brother. Fought Cossacks, Swedes, Russians, Turks, Tatars (the "Deluge"). Abdicated 1668.

John III Sobieski, 1674-1696. Won Vienna from besieging Turks, 1683.

Stanislav II, 1764-1795, last king. Encouraged reforms; 1791 1st modern Constitution in Europe. 1772, 1793, 1795 Poland partitioned among Russia, Prussia, Austria. Unsuccessful insurrection against foreign invasion 1794 under Kosciusko, American-Polish general.

1795-1918: Poland Under Foreign Rule

1807-1815 Grand Duchy of Warsaw created by Napoleon I, Frederick August of Saxony grand duke.

1815 Congress of Vienna proclaimed part of Poland "Kingdom" in personal union with Russia.

Polish uprisings: 1830 against Russia; 1846, 1848 against Austria; 1863 against Russia—all repressed.

1918-1939: Second Republic

1918-1922 Head of State Jozef Pilsudski. Presidents: Gabriel Narutowicz 1922, assassinated; Stanislav Wojciechowski 1922-1926, had to abdicate after Pilsudski's coup d'état; Ignacy Moscicki, 1926-1939, ruled (with Pilsudski until his death, 1935) as virtual dictator.

1939-1945: Poland Under Foreign Occupation

Nazi and Soviet invasion Sept. 1939. Polish government-in-exile, first in France, then in England. Vladislav Raczkiewicz president; Gen. Vladislav Sikorski, then Stanislav Mikolajczyk, prime ministers. Soviet-sponsored Polish Committee of National Liberation proclaimed at Lublin July 1944, transformed into government Jan. 1, 1945.

Poland After 1945

In the late 1940s, Poland came increasingly under Soviet control. Communist party ruled in Poland until Aug. 1989, when democratic Solidarity party, led by Lech Walesa, gained control of government. Walesa was elected president in 1990. Solidarity lost the presidency in 1995, but the government remained democratic.

IT'S A FACT: The Danish monarchy stretches back to Viking times in an uninterrupted line of 50 kings and 2 queens, a record among the world's nations.

Rulers of Denmark, Sweden, Norway

Denmark

Earliest rulers invaded Britain; King Canute, who ruled in London 1016-1035, was most famous. The Valdemars furnished kings until the 15th century. In 1282 the Danes won the first national assembly, Danehof, from King Erik V.

Most redoubtable medieval character was Margaret, daughter of Valdemar IV, born 1353, married at 10 to King Haakon VI of Norway. In 1376 she had her first infant son Olaf made king of Denmark. After his death, 1387, she was regent of Denmark and Norway. In 1388 Sweden accepted her as sovereign. In 1389 she made her grand-nephew, Duke Erik of Pomerania, titular king of Denmark, Sweden, and Norway, with herself as regent. In 1397 she effected the Union of Kalmar of the three kingdoms and had Erik VII crowned. In 1439 the three kingdoms deposed him and elected, 1440, Christopher of Bavaria king (Christopher III). On his death, 1448, the union broke up.

Succeeding rulers were unable to enforce their claims as rulers of Sweden until 1520, when Christian II conquered Sweden. He was thrown out 1522, and in 1523 Gustavus Vasa united Sweden. Denmark continued to dominate Norway until the Napoleonic wars, when Frederick VI, 1808-1839, joined the Napoleonic cause after Britain had destroyed the Danish fleet, 1807. In 1814 he was forced to cede Norway to Sweden and Helgoland to Britain, receiving Lauenburg. Successors Christian VIII, 1839; Frederick VII, 1848; Christian IX, 1863; Frederick VIII, 1906; Christian X, 1912; Frederick IX, 1947; Margrethe II, 1972.

Sweden

Early kings ruled at Uppsala, but did not dominate the country. Sverker, c1130-c1156, united the Swedes and Goths. In 1435 Sweden obtained the Riksdag, or parliament. After the Union of Kalmar, 1397, the Danes either ruled or harried the country until Christian II of Denmark conquered it anew, 1520. This led to a rising under Gustavus Vasa, who ruled Sweden 1523-1560, and established an independent kingdom. Charles IX, 1599-1611, crowned 1604, conquered Moscow. Gustavus II Adolphus, 1611-1632, was called the Lion of the North. Later rulers: Christina, 1632; Charles X Gustavus, 1654; Charles XI, 1660; Charles XII (invader of Russia and Poland, defeated at Poltava, June 28, 1709), 1697; Ulrika Eleanora, sister, elected queen 1718; Frederick I (of Hesse), her husband, 1720; Adolphus Frederick, 1751; Gustavus III, 1771; Gustavus IV Adolphus, 1792; Charles XIII, 1809. (Union with Norway began 1814.) Charles XIV John, 1818 (he was Jean Bernadotte, Napoleon's Prince of Ponte Corvo, elected 1810 to succeed Charles XIII); he founded the present dynasty: Oscar I, 1844; Charles XV, 1859; Oscar II, 1872; Gustavus V, 1907; Gustav VI Adolf, 1950; Carl XVI Gustaf, 1973.

Norway

Overcoming many rivals, Harald Haarfager, 872-930, conquered Norway, Orkneys, and Shetlands; Olaf I, great-grandson, 995-1000, brought Christianity into Norway, Iceland, and Greenland. In 1035 Magnus the Good also became king of Denmark. Haakon V, 1299-1319, had married his daughter to Erik of Sweden. Their son, Magnus, became ruler of Norway and Sweden at 6. His son, Haakon VI, married Margaret of Denmark; their son Olaf IV became king of Norway and Denmark, followed by Margaret's regency and the Union of Kalmar, 1397.

In 1450 Norway became subservient to Denmark. Christian IV, 1588-1648, founded Christiania, now Oslo. After Napoleonic wars, when Denmark ceded Norway to Sweden, a strong nationalist movement forced recognition of Norway as an independent kingdom united with Sweden under the Swedish kings, 1814-1905. In 1905 the union was dissolved and Prince Charles of Denmark became Haakon VII. He died Sept. 21, 1957; succeeded by son, Olav V. Olav V died Jan. 17, 1991; succeeded by son, Harald V.

Rulers of the Netherlands and Belgium

The Netherlands (Holland)

William Frederick, Prince of Orange, led a revolt against French rule, 1813; crowned king, 1815. Belgium seceded Oct. 4, 1830, after a revolt. The secession was ratified by the two kingdoms by treaty, Apr. 19, 1839.

Succession: William II, son, 1840; William III, son, 1849; Wilhelmina, daughter of William III and his 2d wife Princess Emma of Waldeck, 1890; Wilhelmina abdicated, Sept. 4, 1948, in favor of daughter, Juliana. Juliana abdicated, Apr. 30, 1980, in favor of daughter, Beatrix.

Belgium

A national congress elected Prince Leopold of Saxe-Coburg as king; he took the throne July 21, 1831, as Leopold I.

Succession: Leopold II, son, 1865; Albert I, nephew of Leopold II, 1909; Leopold III, son of Albert, 1934; Prince Charles, Regent 1944; Leopold returned 1950, yielded powers to son Baudouin, Prince Royal, Aug. 6, 1950, abdicated July 16, 1951. Baudouin I took throne July 17, 1951, died July 31, 1993; succeeded by brother, Albert II.

Roman Rulers

From Romulus to the end of the Empire in the West. Rulers in the East sat in Constantinople and, for a brief period, in Nicaea, until the capture of Constantinople by the Turks in 1453, when Byzantium was succeeded by the Ottoman Empire.

BC **The Kingdom**
753 Romulus (Quirinus)
716 Numa Pompilius
673 Tullus Hostilius
640 Ancus Marcius
616 L. Tarquinius Priscus
578 Servius Tullius
534 L. Tarquinius Superbus

The Republic
509 Consulate established
509 Quaestorship instituted
498 Dictatorship introduced
494 Plebeian Tribunate created
494 Plebeian Aedileship created
444 Consular Tribunate organized
435 Censorship instituted
366 Praetorship established
366 Curule Aedileship created
362 Military Tribunate elected
326 Proconsulate introduced
311 Naval Duumvirate elected
217 Dictatorship of Fabius Maximus
133 Tribunate of Tiberius Gracchus
123 Tribunate of Gaius Gracchus
82 Dictatorship of Sulla
60 First Triumvirate formed (Caesar, Pompeius, Crassus)
46 Dictatorship of Caesar
43 Second Triumvirate formed (Octavianus, Antonius, Lepidus)

The Empire
27 Augustus (Gaius Julius Caesar Octavianus)
AD
14 Tiberius I
37 Gaius Caesar (Caligula)
41 Claudius I
54 Nero
68 Galba
69 Galba; Otho, Vitellius
69 Vespasianus
79 Titus
81 Domitianus
96 Nerva
98 Trajanus
117 Hadrianus
138 Antoninus Pius
161 Marcus Aurelius and Lucius Verus
169 Marcus Aurelius (alone)
180 Commodus
193 Pertinax; Julianus I
193 Septimius Severus
211 Caracalla and Geta
212 Caracalla (alone)
217 Macrinus
218 Elagabalus (Heliogabalus)
222 Alexander Severus
235 Maximinus I (the Thracian)
238 Gordianus I and Gordianus II; Pupienus and Balbinus
238 Gordianus III
244 Philippus (the Arabian)
249 Decius
251 Gallus and Volusianus
253 Aemilianus
253 Valerianus and Gallienus
258 Gallienus (alone)
268 Claudius Gothicus
270 Quintillus
270 Aurelianus
275 Tacitus
276 Florianus
276 Probus
282 Carus
283 Carinus and Numerianus
286 Diocletianus and Maximianus
305 Galerius and Constantius I
306 Galerius, Maximinus II, Severus I
307 Galerius, Maximinus II, Constantinus I, Licinius, Maxentius
311 Maximinus II, Constantinus I, Licinius, Maxentius
314 Maximinus II, Constantinus I, Licinius
314 Constantinus I and Licinius
324 Constantinus I (the Great)
337 Constantinus II, Constans I, Constantius II
340 Constantius II and Constans I
350 Constantius II
361 Julianus II (the Apostate)
363 Jovianus

West (Rome) and East (Constantinople)
364 Valentinianus I (West) and Valens (East)
367 Valentinianus I with Gratianus (West) and Valens (East)
375 Gratianus with Valentinianus II (West) and Valens (East)
378 Gratianus with Valentinianus II (West),Theodosius I (East)
383 Valentinianus II (West) and Theodosius I (East)
394 Theodosius I (the Great)
395 Honorius (West) and Arcadius (East)
408 Honorius (West) and Theodosius II (East)
423 Valentinianus III (West) and Theodosius II (East)
450 Valentinianus III (West) and Marcianus (East)
455 Maximus (West), Avitus (West); Marcianus (East)
456 Avitus (West), Marcianus (East)
457 Majorianus (West), Leo I (East)
461 Severus II (West), Leo I (East)
467 Anthemius (West), Leo I (East)
472 Olybrius (West), Leo I (East)
473 Glycerius (West), Leo I (East)
474 Julius Nepos (West), Leo II (East)
475 Romulus Augustulus (West) and Zeno (East)
476 End of Empire in West; Odovacar, King, drops title of Emperor; murdered by King Theodoric of Ostrogoths, 493

Rulers of Modern Italy

After the fall of Napoleon in 1814, the Congress of Vienna, 1815, restored Italy as a political patchwork, comprising the Kingdom of Naples and Sicily, the Papal States, and smaller units. Piedmont and Genoa were awarded to Sardinia, ruled by King Victor Emmanuel I of Savoy.

United Italy emerged under the leadership of Camillo, Count di Cavour (1810-1861), Sardinian prime minister. Agitation was led by Giuseppe Mazzini (1805-1872) and Giuseppe Garibaldi (1807-1882), soldier; Victor Emmanuel I abdicated 1821. After a brief regency for a brother, Charles Albert was king 1831-1849, abdicating when defeated by the Austrians at Novara. Succeeded by Victor Emmanuel II, 1849-1861.

In 1859 France forced Austria to cede Lombardy to Sardinia, which gave rights to Savoy and Nice to France. In 1860 Garibaldi led 1,000 volunteers in a spectacular campaign, took Sicily and expelled the King of Naples. In 1860 the House of Savoy annexed Tuscany, Parma, Modena, Romagna, the Two Sicilys, the Marches, and Umbria. Victor Emmanuel assumed the title of King of Italy at Turin Mar. 17, 1861.

In 1866, Victor Emmanuel allied with Prussia in the Austro-Prussian War, and with Prussia's victory received Venetia. On Sept. 20, 1870, his troops under Gen. Raffaele Cadorna entered Rome and took over the Papal States, ending the temporal power of the Roman Catholic Church.

Succession: Umberto I, 1878, assassinated 1900; Victor Emmanuel III, 1900, abdicated 1946, died 1947; Humbert II, 1946, ruled a month. In 1921 Benito Mussolini (1883-1945) formed the Fascist party; he became prime minister Oct. 31, 1922. He entered World War II as an ally of Hitler. He was deposed July 25, 1943.

At a plebiscite June 2, 1946, Italy voted for a republic; Premier Alcide de Gasperi became chief of state June 13, 1946. On June 28, 1946, the Constituent Assembly elected Enrico de Nicola, Liberal, provisional president. Successive presidents: Luigi Einaudi, elected May 11, 1948; Giovanni Gronchi, Apr. 29, 1955; Antonio Segni, May 6, 1962; Giuseppe Saragat, Dec. 28, 1964; Giovanni Leone, Dec. 29, 1971; Alessandro Pertini, July 9, 1978; Francesco Cossiga, July 9, 1985; Oscar Luigi Scalfaro, May 28, 1992, Carlo Azeglio Ciampi, May 18, 1999.

Rulers of Spain

From 8th to 11th centuries Spain was dominated by the Moors (Arabs and Berbers). The Christian reconquest established small kingdoms (Asturias, Aragon, Castile, Catalonia, Leon, Navarre, and Valencia). In 1474 Isabella, b. 1451, became Queen of Castile & Leon. Her husband, Ferdinand, b. 1452, inherited Aragon 1479, with Catalonia, Valencia, and the Balearic Islands, became Ferdinand V of Castile. By Isabella's request Pope Sixtus IV established the Inquisition, 1478. Last Moorish kingdom, Granada, fell 1492. Columbus opened New World of colonies, 1492. Isabella died 1504, succeeded by her daughter, Juana "the Mad," but Ferdinand ruled until his death 1516.

Charles I, b. 1500, son of Juana, grandson of Ferdinand and Isabella, and of Maximilian I of Hapsburg; succeeded later as Holy Roman Emperor, Charles V, 1520; abdicated 1556. Philip II, son, 1556-1598, inherited only Spanish throne; conquered Portugal, fought Turks, sent Armada vs. England. Married to Mary I of England, 1554-1558. Succession: Philip III, 1598-1621; Philip IV, 1621-1665; Charles II, 1665-1700, left Spain to Philip of Anjou, grandson of Louis XIV, who as Philip V, 1700-1746, founded Bourbon dynasty; Ferdinand VI, 1746-1759; Charles III, 1759-1788; Charles IV, 1788-1808, abdicated.

Napoleon now dominated politics and made his brother Joseph King of Spain 1808, but the Spanish ousted him in 1813. Ferdinand VII, 1808, 1814-1833, lost American colonies; succeeded by daughter Isabella II, aged 3, with wife Maria Christina of Naples regent until 1843. Isabella deposed by revolution 1868. Elected king by the Cortes, Amadeo of Savoy, 1870; abdicated 1873. First republic, 1873-74. Alfonso XII, son of Isabella, 1875-85. His posthumous son was Alfonso XIII, with his mother, Queen Maria Christina regent; Spanish-American war, Spain lost Cuba, gave up Puerto Rico, Philippines, Sulu Is., Marianas. Alfonso took throne 1902, aged 16, married British Princess Victoria Eugenia of Battenberg. Dictatorship of Primo de Rivera, 1923-30, precipitated revolution of 1931. Alfonso agreed to leave without formal abdication. Monarchy abolished; the second republic established, with socialist backing. Niceto Alcala Zamora was president until 1936, when Manuel Azaña was chosen.

In July 1936, the army in Morocco revolted against the government and General Francisco Franco led the troops into Spain. The revolution succeeded by Feb. 1939, when Azaña resigned. Franco became chief of state, with provisions that if he was incapacitated, the Regency Council by two-thirds vote could propose a king to the Cortes, which needed to have a two-thirds majority to elect him.

Alfonso XIII died in Rome Feb. 28, 1941, aged 54. His property and citizenship had been restored.

A law restoring the monarchy was approved in a 1947 referendum. Prince Juan Carlos, b. 1938, grandson of Alfonso XIII, was designated by Franco and the Cortes (Parliament) in 1969 as future king and chief of state. Franco died in office, Nov. 20, 1975; Juan Carlos proclaimed king, Nov. 22.

Leaders in the South American Wars of Liberation

Simon Bolivar (1783-1830), Jose Francisco de San Martin (1778-1850), and Francisco Antonio Gabriel Miranda (1750-1816) are among the heroes of the early 19th-century struggles of South American nations to free themselves from Spain. All three, and their contemporaries, operated in periods of factional strife, during which soldiers and civilians suffered.

Miranda, a Venezuelan, who had served with the French in the American Revolution and commanded parts of the French Revolutionary armies in the Netherlands, attempted to start a revolt in Venezuela in 1806 and failed. In 1810, with British and American backing, he returned and was briefly a dictator, until the British withdrew their support. In 1812 he was overcome by the royalists in Venezuela and taken prisoner, dying in a Spanish prison in 1816.

San Martin was born in Argentina and during 1789-1811 served in campaigns of the Spanish armies in Europe and Africa. He first joined the independence movement in Argentina in 1812 and in 1817 invaded Chile with 4,000 men over the mountain passes. Here he and Gen. Bernardo O'Higgins (1778-1842) defeated the Spaniards at Chacabuco, 1817; O'Higgins was named Liberator and became first director of Chile, 1817-23. In 1821 San Martin occupied Lima and Callao, Peru, and became protector of Peru.

Bolivar, the greatest leader of South American liberation from Spain, was born in Venezuela, the son of an aristocratic family. He first served under Miranda in 1812 and in 1813 captured Caracas, where he was named Liberator. Forced out next year by civil strife, he led a campaign that captured Bogota in 1814. In 1817 he was again in control of Venezuela and was named dictator. He organized Nueva Granada with the help of General Francisco de Paula Santander (1792-1840). By joining Nueva Granada, Venezuela, and the area that is now Panama and Ecuador, the republic of Colombia was formed, with Bolivar president. After numerous setbacks he decisively defeated the Spaniards in the second battle of Carabobo, Venezuela, June 24, 1821.

In May 1822, Gen. Antonio Jose de Sucre, Bolivar's lieutenant, took Quito. Bolivar went to Guayaquil to confer with San Martin, who resigned as protector of Peru and withdrew from politics. With a new army of Colombians and Peruvians Bolivar defeated the Spaniards in a battle at Junin in 1824 and cleared Peru.

De Sucre organized Charcas (Upper Peru) as Republica Bolivar (now Bolivia) and acted as president in place of Bolivar, who wrote its constitution. De Sucre defeated the Spanish faction of Peru at Ayacucho, Dec. 19, 1824.

Continued civil strife finally caused the Colombian federation to break apart. Santander turned against Bolivar, but the latter defeated him and banished him. In 1828 Bolivar gave up the presidency he had held precariously for 14 years. He became ill from tuberculosis and died Dec. 17, 1830. He is buried in the national pantheon in Caracas.

Rulers of Russia; Leaders of the USSR and Russian Federation

First ruler to consolidate Slavic tribes was Rurik, leader of the Russians who established himself at Novgorod, AD 862. He and his immediate successors had Scandinavian affiliations. They moved to Kiev after 972 and ruled as Dukes of Kiev. In 988 Vladimir was converted and adopted the Byzantine Greek Orthodox service, later modified by Slav influences. Important as organizer and lawgiver was Yaroslav, 1019-1054, whose daughters married kings of Norway, Hungary, and France. His grandson, Vladimir II (Monomakh), 1113-1125, was progenitor of several rulers, but in 1169 Andrew Bogolubski overthrew Kiev and began the line known as Grand Dukes of Vladimir.

Of the Grand Dukes of Vladimir, Alexander Nevsky, 1246-1263, had a son, Daniel, first to be called Duke of Muscovy (Moscow), who ruled 1263-1303. His successors became Grand Dukes of Muscovy. After Dmitri III Donskoi defeated the Tatars in 1380, they also became Grand Dukes of all Russia. Tatar independence and considerable territorial expansion were achieved under Ivan III, 1462-1505.

Tsars of Muscovy—Ivan III was referred to in church ritual as Tsar. He married Sofia, niece of the last Byzantine emperor. His successor, Basil III, died in 1533 when Basil's son Ivan was only 3. He became Ivan IV, "the Terrible"; crowned 1547 as Tsar of all the Russias, ruled until 1584. Under the weak rule of his son, Feodor I, 1584-1598, Boris Godunov had control. The dynasty died, and after years of tribal strife and intervention by Polish and Swedish armies, the Russians united under 17-year-old Michael Romanov, distantly related to the first wife of Ivan IV. He ruled 1613-1645 and established the Romanov line. Fourth ruler after Michael was Peter I.

Tsars, or Emperors, of Russia (Romanovs)—Peter I, 1682-1725, known as Peter the Great, took title of Emperor in 1721. His successors and dates of accession were: Catherine, his widow, 1725; Peter II, his grandson, 1727; Anne, Duchess of Courland, 1730, daughter of Peter the Great's brother, Tsar Ivan V; Ivan VI, 1740, great-grandson of Ivan V, child, kept in prison and murdered 1764; Elizabeth, daughter of Peter I, 1741; Peter III, grandson of Peter I, 1761, deposed 1762 for his consort, Catherine II, former princess of Anhalt Zerbst (Germany) who is known as Catherine the Great; Paul I, her son, 1796, killed 1801; Alexander I, son of Paul, 1801, defeated Napoleon; Nicholas I, his brother, 1825; Alexander II, son of Nicholas, 1855, assassinated 1881 by terrorists; Alexander III, son, 1881. Nicholas II, son, 1894-1917, last Tsar of Russia, was forced to abdicate by the Revolution that followed losses to Germany in WWI. The Tsar, the Empress, the Tsarevich (Crown Prince), and the Tsar's 4 daughters were murdered by the Bolsheviks in Yekaterinburg, July 16, 1918.

Provisional Government—Prince Georgi Lvov and Alexander Kerensky, premiers, 1917.

Union of Soviet Socialist Republics

Bolshevik Revolution, Nov. 7, 1917, removed Kerensky from power; council of People's Commissars formed, Lenin (Vladimir Ilyich Ulyanov) became premier. Lenin died Jan. 21, 1924. Aleksei Rykov (executed 1938) and V. M. Molotov held the office, but actual ruler was Joseph Stalin (Joseph Vissarionovich Djugashvili), general secretary of the Central Committee of the Communist Party. Stalin became president of the Council of Ministers (premier) May 7, 1941, died Mar. 5, 1953. Succeeded by Georgi M. Malenkov, as head of the Council and premier, and Nikita S. Khrushchev, first secretary of the Central Committee. Malenkov resigned Feb. 8, 1955, became deputy premier, was dropped July 3, 1957. Marshal Nikolai A. Bulganin became premier Feb. 8, 1955; was demoted and Khrushchev became premier Mar. 27, 1958.

Khrushchev was ousted Oct. 14-15, 1964, replaced by Leonid I. Brezhnev as first secretary of the party and by Aleksei N. Kosygin as premier. On June 16, 1977, Brezhnev also took office as president. He died Nov. 10, 1982; 2 days later the Central Committee elected former KGB head Yuri V. Andropov president. Andropov died Feb. 9, 1984; on Feb. 13, Konstantin U. Chernenko chosen by Central Committee as its general secretary. Chernenko died Mar. 10, 1985; on Mar. 11, he was succeeded as general secretary by Mikhail Gorbachev, who replaced Andrei Gromyko as president on Oct. 1, 1988. Gorbachev resigned Dec. 25, 1991, and the Soviet Union officially disbanded the next day. A loose Commonwealth of Independent States, made up of most of the 15 former Soviet constituent republics, was created.

Post-Soviet Russia

After adopting a degree of sovereignty, the Russian Republic had held elections in June 1991. Boris Yeltsin was sworn in July 10, 1991, as Russia's first elected president. With the Dec. 1991 dissolution of the Soviet Union, Russia (officially Russian Federation) became a founding member of the Commonwealth of Independent States. On Dec. 31, 1999, Yeltsin stepped down as president; he named Vladimir Putin his interim successor. Putin won a presidential election Mar. 26, 2000, and was sworn in May 7.

IT'S A FACT: It is said that the wealthiest private person in medieval Russia was a woman, Marfa Boretskaia, a 15th-century mayoress of the city-state of Novgorod. Russian women at this time could inherit, own, and control certain property independent of their husbands.

Governments of China

(Until 221 BC and frequently thereafter, China was not a unified state. Where dynastic dates overlap, the rulers or events referred to appeared in different areas of China.)

Hsia	1994 BC – c1523 BC
Shang	c1523 – c1028
Western Chou	c1027 – 770
Eastern Chou	770 – 256
Warring States	403 – 222
Ch'in (first unified empire)	221 – 206
Han	202 BC – AD 220
Western Han (expanded Chinese state beyond the Yellow and Yangtze River valleys)	202 BC – AD 9
Hsin (Wang Mang, usurper)	AD 9 – 23
Eastern Han (expanded Chinese state into Indochina and Turkestan)	25 – 220
Three Kingdoms (Wei, Shu, Wu)	220 – 265
Chin (western)	265 – 317
(eastern)	317 – 420
Northern Dynasties (followed several short-lived governments by Turks, Mongols, etc.)	386 – 581
Southern Dynasties (capital: Nanjing)	420 – 589
Sui (reunified China)	581 – 618
Tang (a golden age of Chinese culture; capital: Xian)	618 – 906
Five Dynasties (Yellow River basin)	902 – 960
Ten Kingdoms (southern China)	907 – 979
Liao (Khitan Mongols; capital at site of Beijing)	947 – 1125
Sung	960 – 1279
Northern Sung (reunified central and southern China)	960 – 1126
Western Hsai (non-Chinese rulers in northwest)	990 – 1227
Chin (Tatars; drove Sung out of central China)	1115 – 1234
Yuan (Mongols; Kublai Khan est. capital at site of Beijing, c. 1264)	1271 – 1368
Ming (China reunified under Chinese rule; capital: Nanjing, then Beijing in 1420)	1368 – 1644
Ch'ing (Manchus, descendents of Tatars)	1644 – 1911
Republic (disunity; provincial rulers, warlords)	1912 – 1949
People's Republic of China	1949 –

Leaders of China Since 1949

Mao Zedong	Chairman, Central People's Administrative Council, Communist Party (CPC), 1949-1976
Zhou Enlai	Premier, foreign minister, 1949-1976
Deng Xiaoping	Vice Premier, 1952-1966, 1973-1976, 1977-1980; "paramount leader," 1978-1997
Liu Shaoqi	President, 1959-1969
Hua Guofeng	Premier, 1976-1980; CPC Chairman, 1976-1981
Zhao Ziyang	Premier, 1980-1988; CPC General Secretary, 1987-1989
Hu Yaobang	CPC Chairman, 1981-1982; CPC General Secretary, 1982-1987
Li Xiannian	President, 1983-1988
Yang Shangkun	President, 1988-1993
Li Peng	Premier, 1988-98
Jiang Zemin	CPC General Secretary, 1989-; President, 1993-
Zhu Rongi	Premier, 1998-

AFGHANISTAN
ALBANIA
ALGERIA
ANDORRA
ANGOLA
ANTIGUA AND BARBUDA
ARGENTINA
ARMENIA
AUSTRALIA
AUSTRIA
AZERBAIJAN
THE BAHAMAS
BAHRAIN
BANGLADESH
BARBADOS
BELARUS
BELGIUM
BELIZE
BENIN
BHUTAN
BOLIVIA
BOSNIA AND HERZEGOVINA
BOTSWANA
BRAZIL
BRUNEI
BULGARIA
BURKINA FASO
BURUNDI
CAMBODIA
CAMEROON
CANADA
CAPE VERDE
CENTRAL AFRICAN REPUBLIC
CHAD
CHILE
CHINA
COLOMBIA
COMOROS
CONGO, DEM. REP. OF THE
CONGO REPUBLIC
COSTA RICA
CÔTE D'IVOIRE
CROATIA
CUBA
CYPRUS
CZECH REPUBLIC
DENMARK
DJIBOUTI
DOMINICA
DOMINICAN REPUBLIC
ECUADOR
EGYPT
EL SALVADOR
EQUATORIAL GUINEA
ERITREA

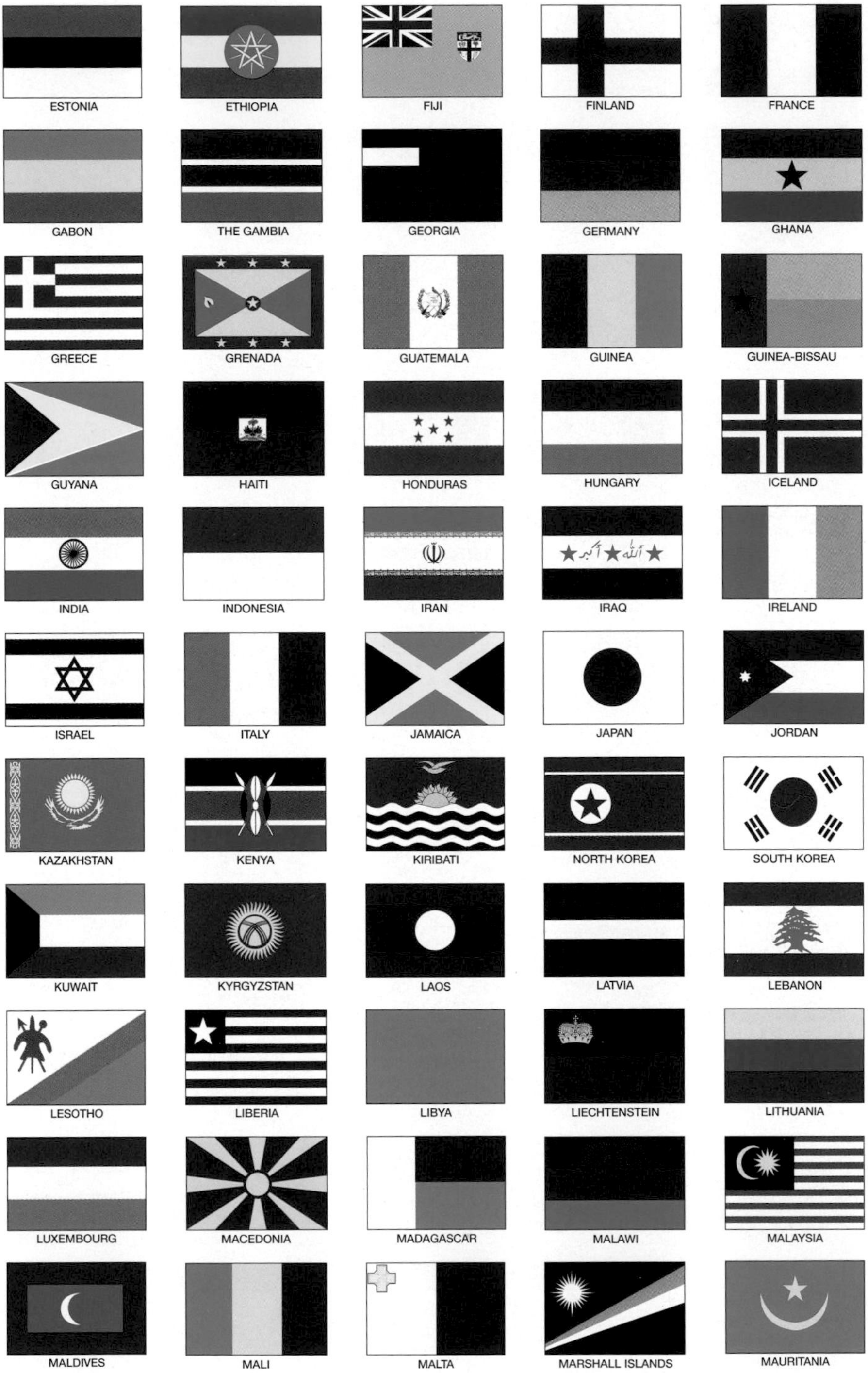

ESTONIA ETHIOPIA FIJI FINLAND FRANCE
GABON THE GAMBIA GEORGIA GERMANY GHANA
GREECE GRENADA GUATEMALA GUINEA GUINEA-BISSAU
GUYANA HAITI HONDURAS HUNGARY ICELAND
INDIA INDONESIA IRAN IRAQ IRELAND
ISRAEL ITALY JAMAICA JAPAN JORDAN
KAZAKHSTAN KENYA KIRIBATI NORTH KOREA SOUTH KOREA
KUWAIT KYRGYZSTAN LAOS LATVIA LEBANON
LESOTHO LIBERIA LIBYA LIECHTENSTEIN LITHUANIA
LUXEMBOURG MACEDONIA MADAGASCAR MALAWI MALAYSIA
MALDIVES MALI MALTA MARSHALL ISLANDS MAURITANIA

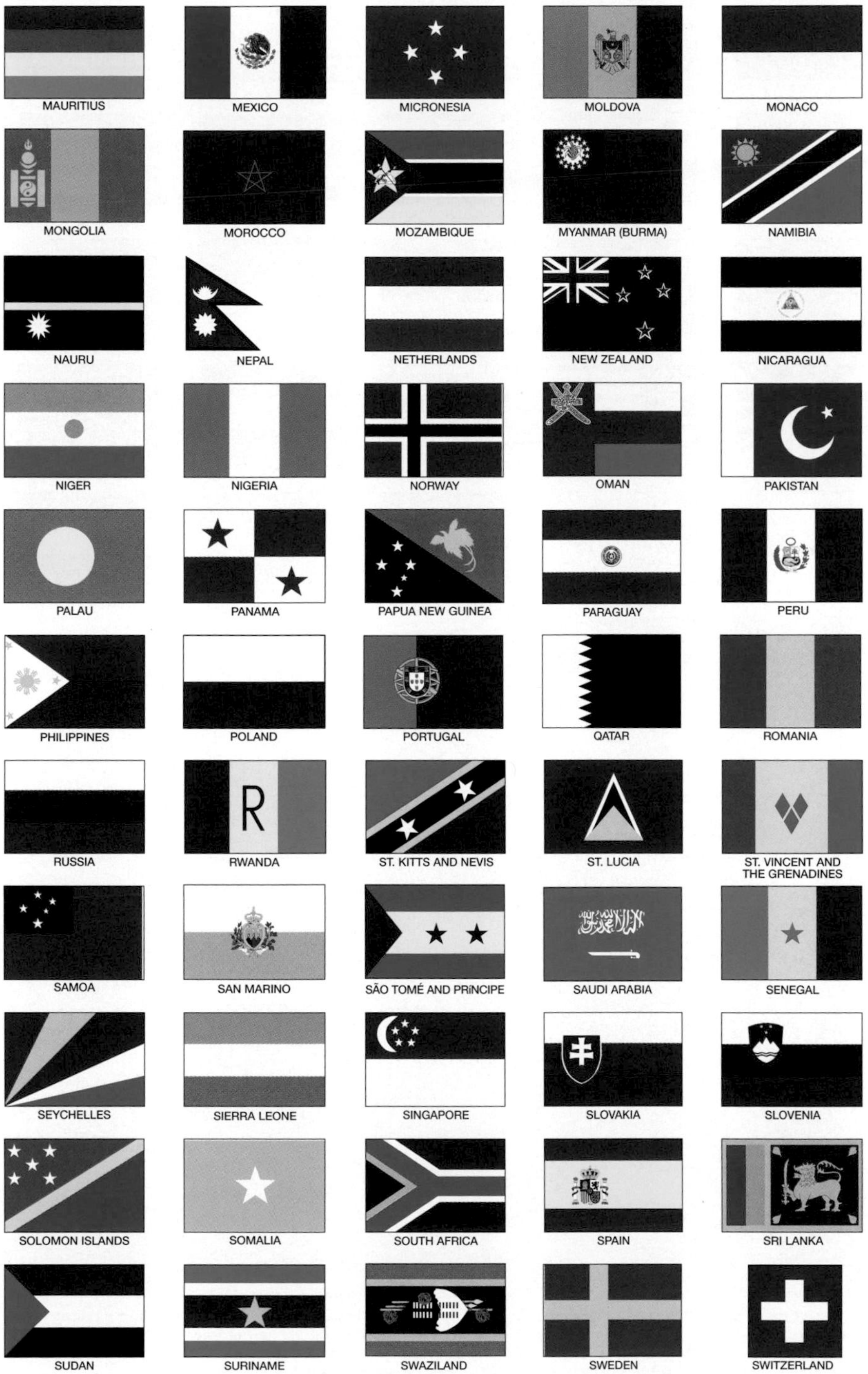
MAURITIUS
MEXICO
MICRONESIA
MOLDOVA
MONACO
MONGOLIA
MOROCCO
MOZAMBIQUE
MYANMAR (BURMA)
NAMIBIA
NAURU
NEPAL
NETHERLANDS
NEW ZEALAND
NICARAGUA
NIGER
NIGERIA
NORWAY
OMAN
PAKISTAN
PALAU
PANAMA
PAPUA NEW GUINEA
PARAGUAY
PERU
PHILIPPINES
POLAND
PORTUGAL
QATAR
ROMANIA
RUSSIA
RWANDA
ST. KITTS AND NEVIS
ST. LUCIA
ST. VINCENT AND THE GRENADINES
SAMOA
SAN MARINO
SÃO TOMÉ AND PRíNCIPE
SAUDI ARABIA
SENEGAL
SEYCHELLES
SIERRA LEONE
SINGAPORE
SLOVAKIA
SLOVENIA
SOLOMON ISLANDS
SOMALIA
SOUTH AFRICA
SPAIN
SRI LANKA
SUDAN
SURINAME
SWAZILAND
SWEDEN
SWITZERLAND

SYRIA

TAIWAN

TAJIKISTAN

TANZANIA

THAILAND

TOGO

TONGA

TRINIDAD AND TOBAGO

TUNISIA

TURKEY

TURKMENISTAN

TUVALU

UGANDA

UKRAINE

UNITED ARAB EMIRATES

UNITED KINGDOM

UNITED STATES

URUGUAY

UZBEKISTAN

VANUATU

VATICAN CITY

VENEZUELA

VIETNAM

YEMEN

YUGOSLAVIA

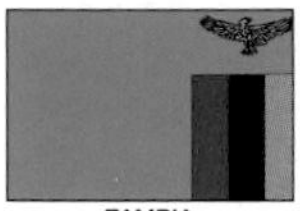

ZAMBIA

ZIMBABWE

INTERNATIONAL TIME ZONES

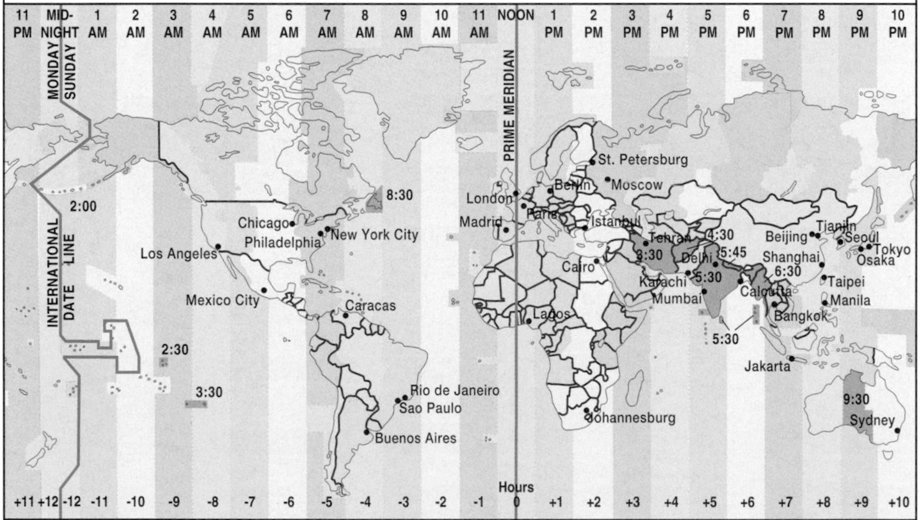

The world is divided into 24 time zones, each 15° longitude wide. The longitudinal meridian passing through Greenwich, England, is the starting point, and is called the *prime meridian.* The 12th zone is divided by the 180th meridian (International Date Line). When the line is crossed going west, the date is advanced one day; when crossed going east, the date becomes a day earlier.

© MAPQUEST.COM

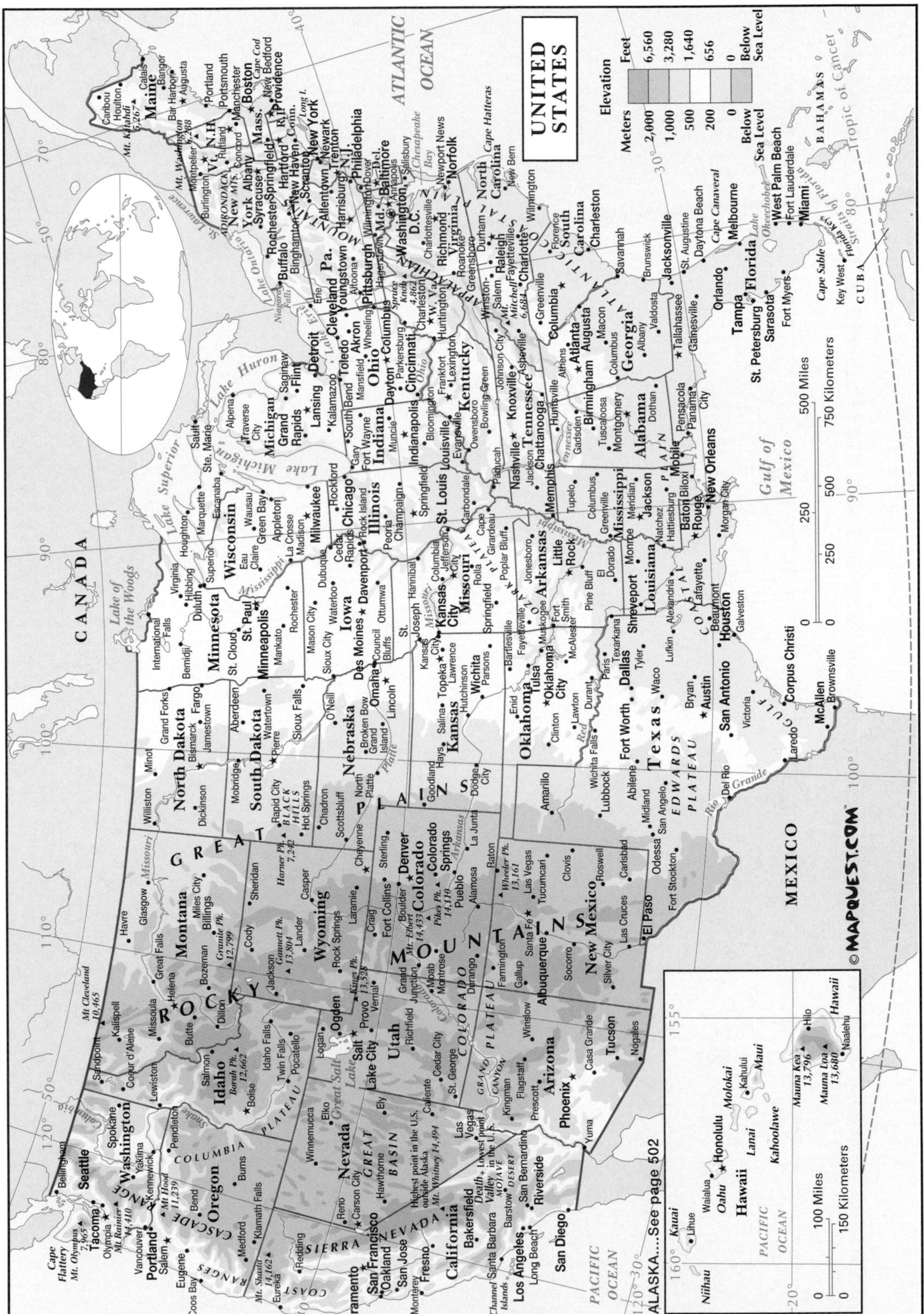
UNITED STATES
Elevation
Meters Feet
2,000 6,560
1,000 3,280
500 1,640
200 656
0 0
Below Sea Level Below Sea Level
CANADA
MEXICO
CUBA
BAHAMAS
ATLANTIC OCEAN
PACIFIC OCEAN
Gulf of Mexico
Tropic of Cancer
GREAT PLAINS
ROCKY MOUNTAINS
APPALACHIAN MOUNTAINS
ATLANTIC COASTAL PLAIN
GULF COASTAL PLAIN
Washington
Oregon
California
Nevada
Idaho
Montana
Wyoming
Utah
Arizona
Colorado
New Mexico
North Dakota
South Dakota
Nebraska
Kansas
Oklahoma
Texas
Minnesota
Iowa
Missouri
Arkansas
Louisiana
Wisconsin
Illinois
Michigan
Indiana
Ohio
Kentucky
Tennessee
Mississippi
Alabama
Georgia
Florida
South Carolina
North Carolina
Virginia
W. Va.
Pa.
New York
Maine
Hawaii
Oahu
Kauai
Niihau
Molokai
Lanai
Maui
Kahoolawe
ALASKA...See page 502
0 250 500 Miles
0 250 500 750 Kilometers
0 100 Miles
0 150 Kilometers
© MAPQUEST.COM

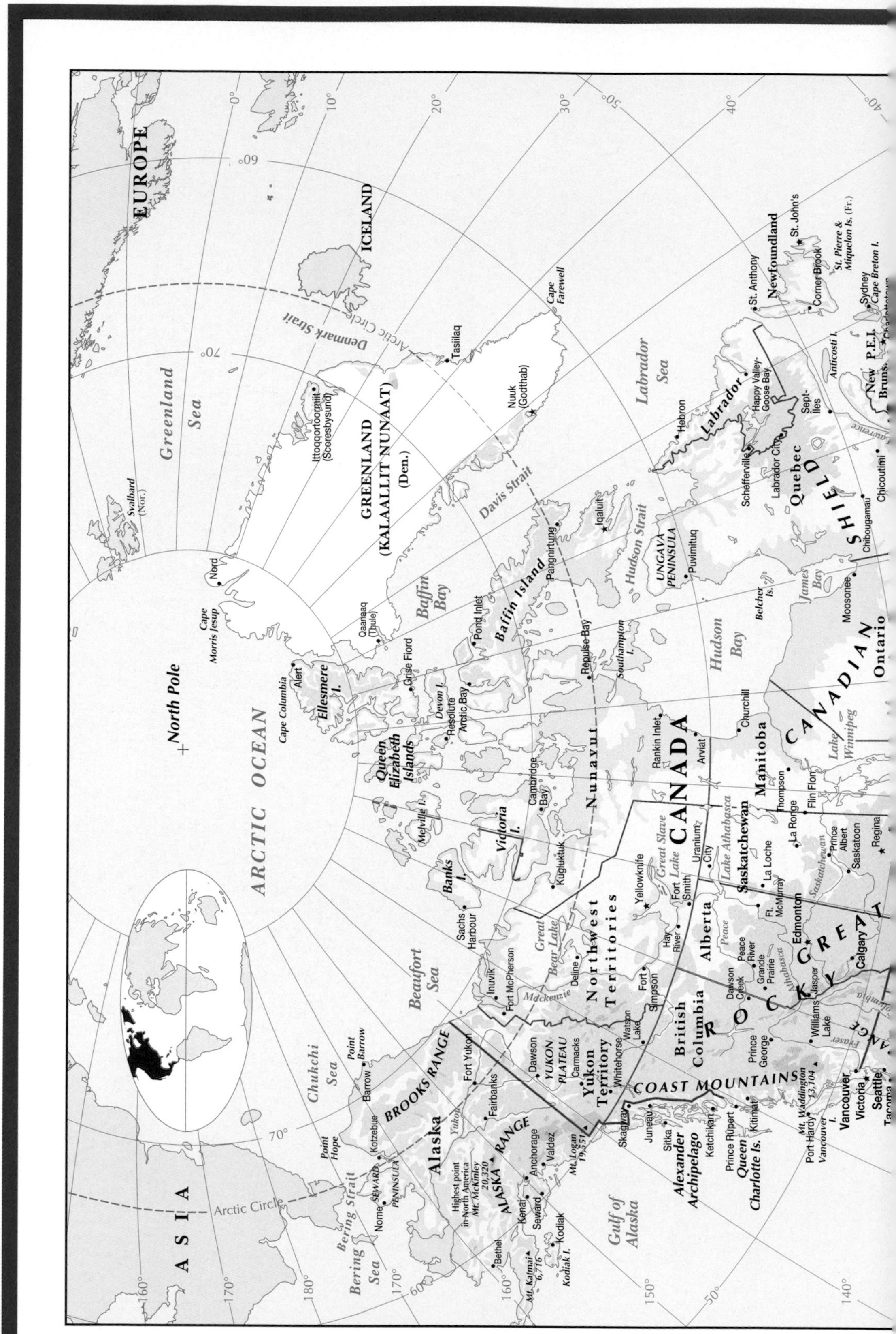

ARCTIC OCEAN
North Pole
EUROPE
ICELAND
ASIA
GREENLAND
(KALAALLIT NUNAAT)
(Den.)
Greenland Sea
Svalbard (Nor.)
Denmark Strait
Arctic Circle
Labrador Sea
Davis Strait
Baffin Bay
Baffin Island
Hudson Strait
Hudson Bay
James Bay
UNGAVA PENINSULA
CANADA
Nunavut
Northwest Territories
Yukon Territory
British Columbia
Alberta
Saskatchewan
Manitoba
Ontario
Quebec
Newfoundland
Labrador
CANADIAN SHIELD
ROCKY
COAST MOUNTAINS
BROOKS RANGE
ALASKA RANGE
YUKON PLATEAU
Alaska
Beaufort Sea
Chukchi Sea
Bering Sea
Bering Strait
Gulf of Alaska
Alexander Archipelago
Queen Charlotte Is.
Ellesmere I.
Queen Elizabeth Islands
Devon I.
Melville I.
Banks I.
Victoria I.
Southampton I.
Belcher Is.
Anticosti I.
Kodiak I.
Vancouver I.
Great Bear Lake
Great Slave Lake
Lake Athabasca
Lake Winnipeg
Mackenzie
Yukon
Peace
Athabasca
Saskatchewan
Fraser
Cape Columbia
Cape Morris Jesup
Cape Farewell
Point Barrow
Point Hope
SEWARD PENINSULA
Nord
Ittoqqortoormiit (Scoresbysund)
Tasiilaq
Nuuk (Godthab)
Qaanaaq (Thule)
Alert
Grise Fiord
Resolute
Arctic Bay
Pond Inlet
Pangnirtung
Iqaluit
Repulse Bay
Rankin Inlet
Arviat
Churchill
Cambridge Bay
Kugluktuk
Sachs Harbour
Inuvik
Fort McPherson
Deline
Yellowknife
Fort Smith
Hay River
Fort Simpson
Uranium City
La Loche
Ft. McMurray
Edmonton
Calgary
Peace River
Grande Prairie
Dawson Creek
Jasper
Prince George
Williams Lake
Prince Rupert
Kitimat
Ketchikan
Port Hardy
Vancouver
Victoria
Seattle
Tacoma
Whitehorse
Watson Lake
Dawson
Carmacks
Skagway
Juneau
Sitka
Fort Yukon
Fairbanks
Barrow
Kotzebue
Nome
Bethel
Anchorage
Valdez
Kenai
Seward
Kodiak
Mt. Katmai 6,716
Mt. McKinley 20,320
Highest point in North America
Mt. Logan 19,551
Mt. Waddington 13,104
La Ronge
Prince Albert
Saskatoon
Regina
Flin Flon
Thompson
Moosonee
Chibougamau
Chicoutimi
Sept-Iles
Schefferville
Labrador City
Happy Valley-Goose Bay
Hebron
Puvirnituq
St. Anthony
Corner Brook
St. John's
St. Pierre & Miquelon Is. (Fr.)
Sydney
Cape Breton I.
P.E.I.
New Bruns.

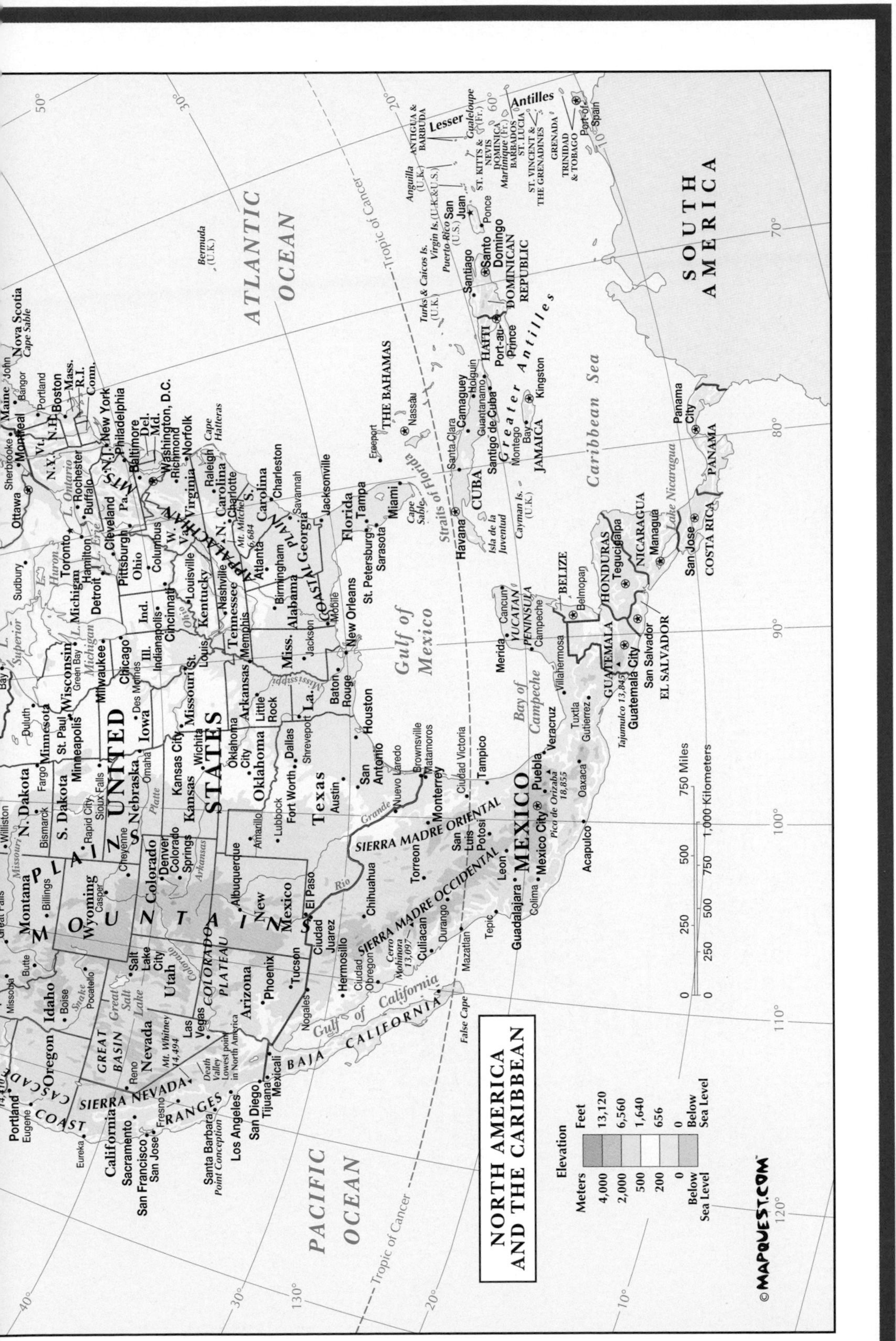
NORTH AMERICA AND THE CARIBBEAN
Elevation
Meters Feet
4,000 13,120
2,000 6,560
500 1,640
200 656
0 0
Below Sea Level Below Sea Level
0 250 500 750 Miles
0 250 500 750 1,000 Kilometers
© MAPQUEST.COM
PACIFIC OCEAN
ATLANTIC OCEAN
Gulf of Mexico
Caribbean Sea
Bay of Campeche
Gulf of California
Straits of Florida
Tropic of Cancer
UNITED STATES
MEXICO
SOUTH AMERICA
CUBA
THE BAHAMAS
HAITI
DOMINICAN REPUBLIC
JAMAICA
BELIZE
GUATEMALA
HONDURAS
EL SALVADOR
NICARAGUA
COSTA RICA
PANAMA
Greater Antilles
Lesser Antilles
ROCKY MOUNTAINS
GREAT PLAINS
GREAT BASIN
COLORADO PLATEAU
SIERRA NEVADA
CASCADE RANGE
COAST RANGES
APPALACHIAN MTS.
COASTAL PLAIN
SIERRA MADRE ORIENTAL
SIERRA MADRE OCCIDENTAL
BAJA CALIFORNIA
YUCATAN PENINSULA
Oregon
Idaho
Montana
N. Dakota
S. Dakota
Wyoming
Nevada
Utah
Colorado
California
Arizona
New Mexico
Texas
Oklahoma
Kansas
Nebraska
Minnesota
Iowa
Missouri
Arkansas
La.
Wisconsin
Ill.
Miss.
Alabama
Tennessee
Kentucky
Ind.
Michigan
Ohio
Georgia
Florida
S. Carolina
N. Carolina
Virginia
W. Va.
Pa.
N.Y.
N.J.
Del.
Md.
Washington, D.C.
Vt.
N.H.
Mass.
R.I.
Conn.
Maine
Nova Scotia
Ottawa
Montreal
Toronto
Sudbury
New York
Boston
Philadelphia
Baltimore
Chicago
Detroit
Houston
Dallas
New Orleans
Miami
Los Angeles
San Francisco
San Diego
Denver
Phoenix
Mexico City
Guadalajara
Monterrey
Havana
Kingston
Port-au-Prince
Santo Domingo
San Juan
Guatemala City
Tegucigalpa
San Salvador
Managua
San Jose
Panama City
Belmopan
Port-of-Spain
Bermuda (U.K.)
Mt. Whitney 14,494
Death Valley Lowest point in North America
Pico de Orizaba 18,855
Tajumulco 13,845
Mt. Mitchell 6,684
Cerro Mohinora 13,097

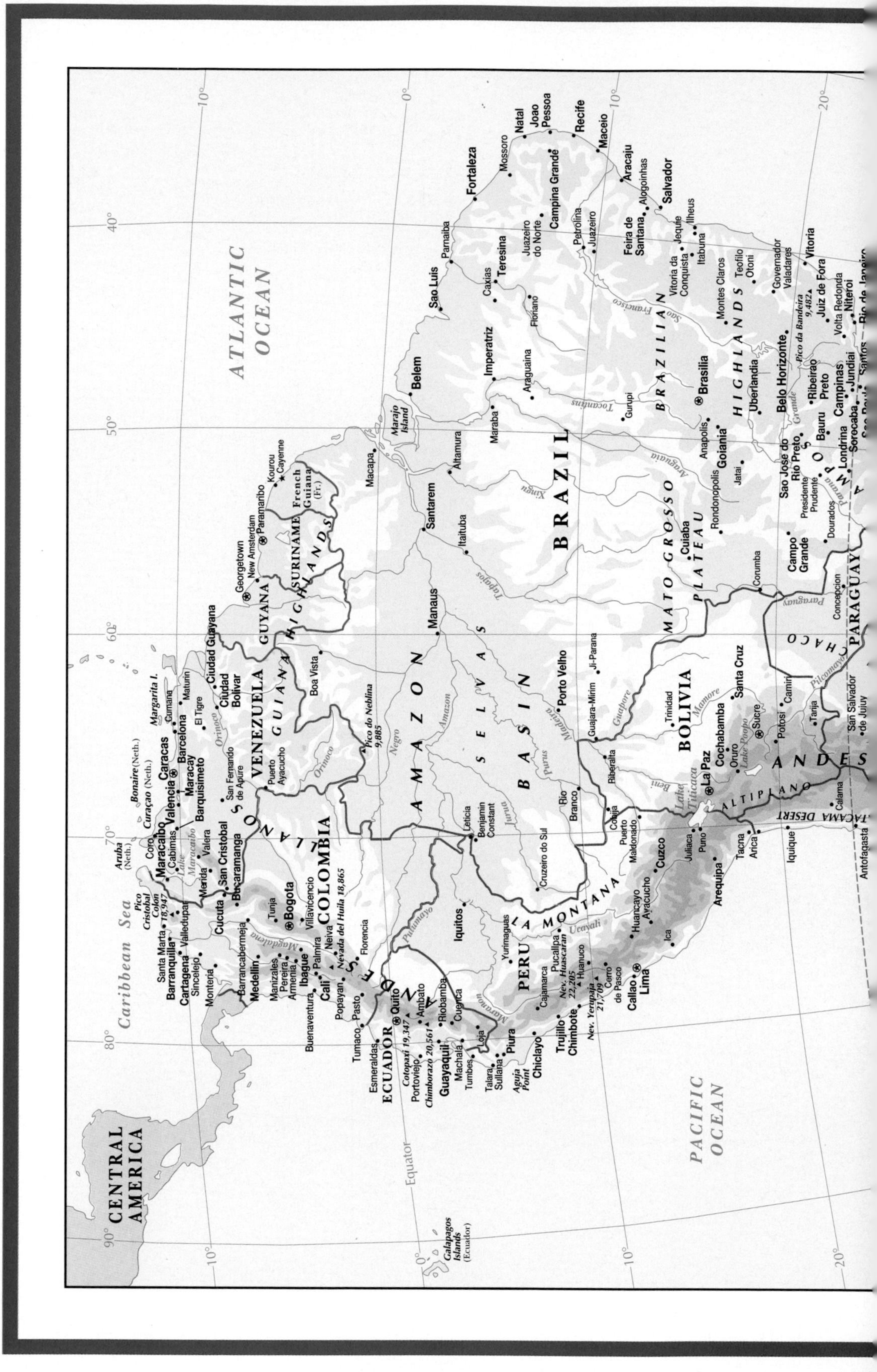

CENTRAL AMERICA
Caribbean Sea
ATLANTIC OCEAN
PACIFIC OCEAN
Equator
Galapagos Islands (Ecuador)
Aruba (Neth.)
Bonaire (Neth.)
Curaçao (Neth.)
Margarita I.
Pico Cristobal Colon 18,947
Santa Marta
Barranquilla
Cartagena
Sincelejo
Monteria
Valledupar
Maracaibo
Cabimas
Lake Maracaibo
Coro
Valencia
Caracas
Maracay
Barcelona
Cumana
Maturin
Barquisimeto
Merida
Valera
San Cristobal
Cucuta
Bucaramanga
Barrancabermeja
Medellin
Manizales
Pereira
Armenia
Ibague
Tunja
Bogota
Villavicencio
Neiva
Nevada del Huila 18,865
Cali
Palmira
Buenaventura
Popayan
Pasto
Florencia
Tumaco
COLOMBIA
VENEZUELA
LLANOS
El Tigre
Ciudad Guayana
Ciudad Bolivar
San Fernando de Apure
Puerto Ayacucho
Orinoco
GUYANA
Georgetown
New Amsterdam
SURINAME
Paramaribo
French Guiana (Fr.)
Kourou
Cayenne
GUIANA HIGHLANDS
Boa Vista
Pico do Neblina 9,885
Negro
ECUADOR
Esmeraldas
Quito
Cotopaxi 19,347
Ambato
Chimborazo 20,561
Riobamba
Portoviejo
Guayaquil
Cuenca
Machala
Loja
Tumbes
Talara
Sullana
Piura
Aguja Point
Chiclayo
Trujillo
Chimbote
Cajamarca
Nev. Huascaran 22,205
Huanuco
Nev. Yerupaja 21,709
Cerro de Pasco
Callao
Lima
Ica
Huancayo
Ayacucho
Cuzco
Juliaca
Puno
Arequipa
Tacna
Arica
Iquique
Antofagasta
Calama
PERU
Iquitos
Yurimaguas
Pucallpa
Ucayali
Putumayo
Maranon
LA MONTANA
Leticia
Benjamin Constant
Cruzeiro do Sul
Rio Branco
Cobija
Puerto Maldonado
Riberalta
Guajara-Mirim
Ji-Parana
Porto Velho
Madeira
Purus
Jurua
Beni
Mamore
Guapore
AMAZON BASIN
SELVAS
Amazon
Manaus
Santarem
Itaituba
Tapajos
Macapa
Marajo Island
Belem
Altamura
Xingu
BOLIVIA
Trinidad
La Paz
Lake Titicaca
Oruro
Lake Poopo
Cochabamba
Santa Cruz
Sucre
Potosi
Camiri
Tarija
San Salvador de Jujuy
Pilcomayo
ALTIPLANO
ANDES
ATACAMA DESERT
CHACO
PARAGUAY
Paraguay
Concepcion
MATO GROSSO PLATEAU
Corumba
Cuiaba
Rondonopolis
Jatai
Campo Grande
Dourados
Presidente Prudente
Sao Jose do Rio Preto
Bauru
Londrina
Sorocaba
Campinas
Jundiai
Ribeirao Preto
Uberlandia
Goiania
Anapolis
Brasilia
Araguaia
Tocantins
Gurupi
Maraba
Araguaina
Imperatriz
BRAZIL
BRAZILIAN HIGHLANDS
Belo Horizonte
Montes Claros
Teofilo Otoni
Governador Valadares
Pico da Bandeira 9,482
Juiz de Fora
Volta Redonda
Niteroi
Rio de Janeiro
Santos
Vitoria
Vitoria da Conquista
Jequie
Itabuna
Ilheus
Feira de Santana
Alagoinhas
Salvador
Aracaju
Maceio
Recife
Joao Pessoa
Natal
Campina Grande
Mossoro
Fortaleza
Juazeiro do Norte
Petrolina
Juazeiro
Sao Francisco
Teresina
Caxias
Floriano
Parnaiba
Sao Luis
Grande
Parana

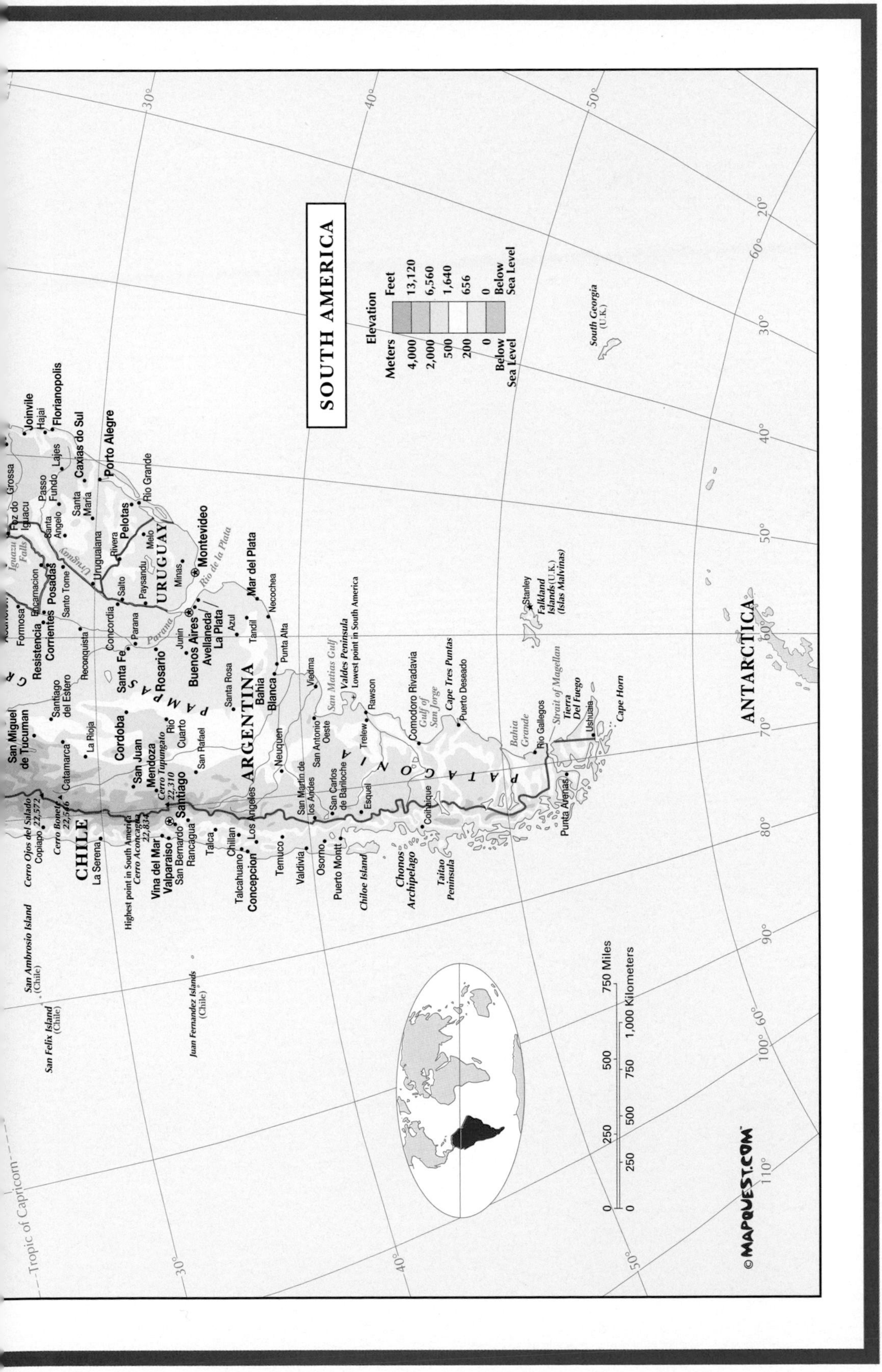
SOUTH AMERICA
Elevation
Meters
Feet
4,000
13,120
2,000
6,560
500
1,640
200
656
0
0
Below Sea Level
Below Sea Level
CHILE
ARGENTINA
URUGUAY
PAMPAS
PATAGONIA
ANTARCTICA
Tropic of Capricorn
San Felix Island (Chile)
San Ambrosio Island (Chile)
Juan Fernandez Islands (Chile)
Cerro Ojos del Salado 22,572
Cerro Bonete 22,546
Highest point in South America Cerro Aconcagua 22,834
Cerro Tupungato 22,310
Copiapo
La Serena
Vina del Mar
Valparaiso
Santiago
San Bernardo
Rancagua
Talca
Chillan
Talcahuano
Concepcion
Los Angeles
Temuco
Valdivia
Osorno
Puerto Montt
Chiloe Island
Chonos Archipelago
Taitao Peninsula
Coihaique
Punta Arenas
San Miguel de Tucuman
Catamarca
La Rioja
Santiago del Estero
Cordoba
San Juan
Mendoza
Rio Cuarto
San Rafael
Neuquen
San Martin de los Andes
San Carlos de Bariloche
Esquel
San Antonio Oeste
Trelew
Rawson
Comodoro Rivadavia
Gulf of San Jorge
Cape Tres Puntas
Puerto Deseado
Bahia Grande
Rio Gallegos
Strait of Magellan
Tierra Del Fuego
Ushuaia
Cape Horn
San Matias Gulf
Valdes Peninsula
Lowest point in South America
Viedma
Bahia Blanca
Santa Rosa
Punta Alta
Tandil
Azul
Necochea
Mar del Plata
La Plata
Avellaneda
Buenos Aires
Junin
Rosario
Santa Fe
Parana
Concordia
Reconquista
Resistencia
Corrientes
Formosa
Encarnacion
Posadas
Santo Tome
Salto
Paysandu
Minas
Montevideo
Rio de la Plata
Uruguaiana
Uruguay
Rivera
Melo
Pelotas
Rio Grande
Santa Maria
Santo Angelo
Passo Fundo
Lajes
Caxias do Sul
Porto Alegre
Foz do Iguacu
Iguazu Falls
Grossa
Joinvile
Hajai
Florianopolis
Stanley
Falkland Islands (U.K.) (Islas Malvinas)
South Georgia (U.K.)
0
250
500
750 Miles
0
250
500
750
1,000 Kilometers
© MAPQUEST.COM

EUROPE
Elevation
Meters
Feet
4,000
13,120
2,000
6,560
500
1,640
200
656
0
0
Below Sea Level
Below Sea Level
GREENLAND
(KALAALLIT NUNAAT)
(Denmark)
Isafjordhur
Akureyri
Keflavik
ICELAND
Reykjavik
Seydhisfjordhur
Arctic Circle
Norwegian Sea
Torshavn
Faroe Islands
(Den.)
Shetland Islands
(U.K.)
Orkney Islands
Hebrides
Thurso
Inverness
Scotland
Aberdeen
Dundee
Glasgow
Ayr
Edinburgh
ATLANTIC OCEAN
Londonderry
Northern Ireland
Belfast
Galway
UNITED KINGDOM
Newcastle
IRELAND
Dublin
Limerick
Cork
Waterford
Liverpool
Manchester
Leeds
Kingston upon Hull
Sheffield
Birmingham
Coventry
Norwich
Wales
Swansea
Cardiff
Bristol
England
Thames
London
Plymouth
Land's End
Portsmouth
Dover
English Channel
Channel Is.
(U.K.)
Brest
Le Havre
Caen
Rouen
Rennes
Paris
Seine
Le Mans
Orleans
Nantes
Loire
Tours
Bay of Biscay
Limoges
FRANCE
Dijon
Clermont-Ferrand
Lyon
Saint-Etienne
Bordeaux
Garonne
Grenoble
Geneva
Mt. Blanc
15,771
Toulouse
Montpellier
Rhone
Avignon
Marseille
Nice
Toulon
MONACO
Donostia-San Sebastian
Pamplona
PYRENEES
Pico de Aneto
11,168
ANDORRA
A Coruña
Gijon
Santander
Bilbao
Vigo
Leon
Vitoria-Gasteiz
Braga
Porto
Duero
Valladolid
Ebro
Coimbra
IBERIAN
Salamanca
Zaragoza
PORTUGAL
Lisbon
Setubal
Tagus
Madrid
Badajoz
Toledo
SPAIN
Barcelona
Tarragona
Valencia
Castellon de la Plana
PENINSULA
Cape St. Vincent
Cordoba
Seville
Majorca
Minorca
Palma de Mallorca
Balearic Is.
(Sp.)
Alicante
Murcia
Cartagena
Cadiz
Malaga
Granada
Almeria
Strait of Gibraltar
Gibraltar
(U.K.)
AFRICA
0
250
500 Miles
0
250
500
750 Kilometers
Namsos
Trondheim
Molde
Alesund
NORWAY
SWEDEN
Ostersund
Sundsvall
Narvik
Bodo
Bergen
Oslo
Haugesund
Stavanger
Drammen
Skien
Kristiansand
Karlstad
Borlange
Uppsala
Orebro
Stockholm
Vanern
Norrkoping
Linkoping
Vattern
Jonkoping
Goteborg
North Sea
Alborg
Jutland
Arhus
Vaxjo
Halmstad
Copenhagen
Helsingborg
Esbjerg
DENMARK
Odense
Malmo
Bornholm
(Den.)
Kiel
Lubeck
Rostock
Hamburg
Gdansk
Groningen
NETHERLANDS
Bremen
Elbe
Szczecin
Amsterdam
The Hague
Hannover
Rotterdam
Bielefeld
Antwerp
Essen
Madgeburg
Berlin
NORTHERN
Bydgoszcz
Brussels
Lille
Cologne
GERMANY
Kassel
Oder
Poznan
BELGIUM
Liege
Bonn
Erfurt
Leipzig
Dresden
POLAND
Wroclaw
LUXEMBOURG
Luxembourg
Wiesbaden
Frankfurt
Chemnitz
Liberec
Walbrzych
Mannheim
Prague
Plzen
Ostrava
Saarbrucken
Nancy
Nurnberg
CZECH REP.
Strasbourg
Rhine
Stuttgart
Regensburg
Brno
Bratislava
Augsburg
Munich
Linz
Danube
Vienna
Basel
Zurich
Bern
Salzburg
Innsbruck
SWITZERLAND
LIECHTENSTEIN
AUSTRIA
Gyor
Graz
Klagenfurt
Matterhorn
14,690
Bergamo
Udine
SLOVENIA
Milan
Verona
Trieste
Ljubljana
Zagreb
Pecs
Torino
Venice
Rijeka
CROATIA
Genoa
Parma
Po
Bologna
SAN MARINO
Florence
Pisa
Ancona
APENNINES
Perugia
Split
Corsica
(Fr.)
Elba
Tiber
Ajaccio
Adriatic Sea
Dubrovnik
VATICAN CITY
Rome
ITALY
Foggia
Sassari
Sardinia
(It.)
Naples
Vesuvius
4,202
Bari
Salerno
Taranto
Cagliari
Tyrrhenian Sea
Ionian Sea
Mediterranean Sea
Palermo
Messina
Etna
11,053
Reggio di Calabria
Sicily
(It.)
Catania
MALTA
Valletta

Novaya Zemlya
Barents Sea
North Cape
Vardo
Murmansk
Ivalo
LAPLAND
Kiruna
Apatity
KOLA PENINSULA
White Sea
Rovaniemi
Arkhangelsk
Naryan-Mar
Pechora
Pechora
ASIA
URAL MOUNTAINS
Ukhta
RUSSIA
Lulea
Oulu
Skelleftea
Belomorsk
Syktyvkar
Berezniki
FINLAND
Dvina
Kotlas
Vaasa
Kuopio
Lake Onega
Jyvaskyla
Lake Ladoga
Petrozavodsk
Perm
Tampere
Pori
Lahti
Kirov
Izhevsk
Kama
Ufa
Turku
Kotka
Helsinki
St. Petersburg
Cherepovets
Vologda
Naberezhnye Chelny
Aland Is. (Fin.)
Gulf of Finland
Tallinn
ESTONIA
Tartu
Rybinsk
Yaroslavl
Kostroma
Yoshkar Ola
Kazan
Velikiy Novgorod
Nizhniy Novgorod
Cheboksary
Ivanovo
Sterlitzmak
Pskov
Ulyanovsk
Orsk
Tver
Vladimir
Tolyatti
Orenburg
Riga
LATVIA
Moscow
Saransk
Samara
Ural
Liepaja
Daugavpils
Kaluga
Ryazan
Klaipeda
Vitsyebsk
LITHUANIA
Kaunas
Vilnius
(RUSSIA)
Kaliningrad
Smolensk
Tula
Penza
Volga
Orsha
PLAIN
Mahilyow
Lipetsk
Tambov
Saratov
EUROPEAN
Minsk
Bryansk
Hrodna
Babruysk
KAZAKHSTAN
Bialystok
BELARUS
Voronezh
Homyel
Kursk
Warsaw
Brest
Pinsk
Chernihiv
Sumy
Belgorod
Lodz
Radom
Volgograd
Lublin
Kiev (Kyiv)
Kharkiv
Kielce
Poltava
Katowice
Zhytomyr
Cherkasy
Luhansk
UKRAINE
Astrakhan
Krakow
Lviv
Vinnytsia
Dnieper
Donetsk
Horlivka
Don
Dnipropetrovsk
CARPATHIAN MOUNTAINS
SLOVAKIA
Chernivtsi
Rostov-na-Donu
Banska Bystrica
Kosice
Zaporizhzhia
Kryvyi Rih
MOLDOVA
Mariupol
Miskolc
Budapest
Prut
Chisinau
Mykolaiv
Sea of Azov
Caspian
Debrecen
Iasi
Odesa
Stavropol
Oradea
Groznyy
Makhachkala
Kecskemet
Cluj-Napoca
CRIMEA PENINSULA
Krasnodar
Mt. Elbrus 18,510
Nalchik
Vladikavkaz
Szeged
ROMANIA
Galati
Simferopol
Highest point in Europe
CAUCASUS MTS.
Sea
Timisoara
Brasov
Novi Sad
Ploiesti
Sevastopol
Belgrade
Bucharest
Constanta
Serbia
Craiova
Danube
Black Sea
Ruse
YUGOSLAVIA
Pleven
Varna
Nis
BULGARIA
Sofia
Stara Zagora
Burgas
Podgorica
BALKAN
Skopje
Plovdiv
Shkoder
F.Y.R. MAC.
Istanbul
Durres
PENINSULA
Kavala
TURKEY
Tirana
ALBANIA
Thessaloniki
Olympus 9,570
Vlore
Dardanelles
Larisa
Aegean Sea
Ioannina
Volos
Corfu
GREECE
ASIA
Patras
Athens
Corinth
Peloponnesus
Kalamata
Cyclades
Sparta
Rhodes (Gr.)
Sea of Crete
Crete
Hania
(Gr.)
Iraklion
©MAPQUEST.COM
20°
30°
40°
50°
60°
70°

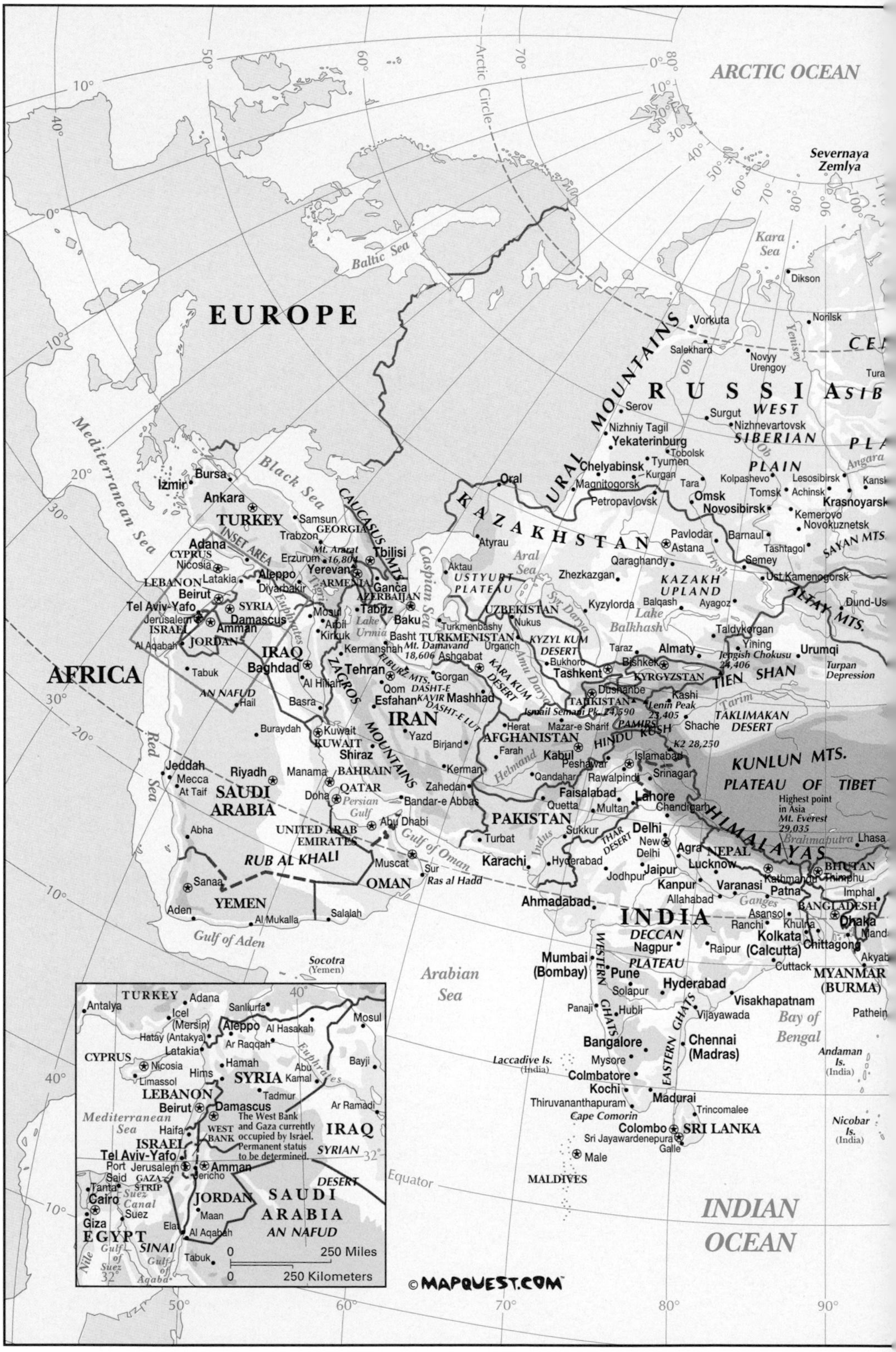
ARCTIC OCEAN
Severnaya Zemlya
Arctic Circle
Kara Sea
Dikson
Norilsk
Vorkuta
Salekhard
Novyy Urengoy
Ob
Yenisey
EUROPE
Baltic Sea
URAL MOUNTAINS
RUSSIA
Serov
Surgut
WEST SIBERIAN PLAIN
Nizhnevartovsk
Nizhniy Tagil
Yekaterinburg
Tobolsk
Tyumen
Chelyabinsk
Kurgan
Magnitogorsk
Oral
Petropavlovsk
Omsk
Novosibirsk
Tara
Kolpashevo
Tomsk
Lesosibirsk
Achinsk
Krasnoyarsk
Kemerovo
Novokuznetsk
Angara
KAZAKHSTAN
Astana
Pavlodar
Barnaul
Tashtagol
SAYAN MTS.
Semey
Ust Kamenogorsk
Irtysh
Qaraghandy
KAZAKH UPLAND
Zhezkazgan
Atyrau
Aral Sea
Aktau
USTYURT PLATEAU
Kyzylorda
Balqash
Ayagoz
Lake Balkhash
ALTAY MTS.
Dund-Us
Syr Darya
UZBEKISTAN
Nukus
Turkmenbashy
TURKMENISTAN
Urganch
KYZYL KUM DESERT
Taldykorgan
Yining
Taraz
Almaty
Bishkek
Urumqi
Jengish Chokusu 24,406
Turpan Depression
Bukhoro
Tashkent
KYRGYZSTAN
TIEN SHAN
Ashgabat
KARA KUM DESERT
Amu Darya
Dushanbe
TAJIKISTAN
Kashi
Ismail Semani Pk. 24,590
Lenin Peak 23,405
Tarim
TAKLIMAKAN DESERT
PAMIRS
Shache
Mazar-e Sharif
HINDU KUSH
K2 28,250
Herat
AFGHANISTAN
Farah
Kabul
Islamabad
Srinagar
KUNLUN MTS.
PLATEAU OF TIBET
Helmand
Peshawar
Qandahar
Rawalpindi
Faisalabad
Lahore
Quetta
Multan
Chandigarh
Highest point in Asia Mt. Everest 29,035
PAKISTAN
Sukkur
Turbat
Indus
THAR DESERT
Delhi
New Delhi
HIMALAYAS
Brahmaputra
Lhasa
Karachi
Hyderabad
Agra
NEPAL
Lucknow
Jodhpur
Jaipur
Kanpur
Varanasi
Kathmandu
BHUTAN
Thimphu
Imphal
Patna
Ahmadabad
Allahabad
Ganges
BANGLADESH
INDIA
Asansol
Ranchi
Khulna
Dhaka
Kolkata (Calcutta)
Chittagong
DECCAN
Nagpur
PLATEAU
Raipur
Cuttack
Akyab
Mumbai (Bombay)
Pune
WESTERN GHATS
Solapur
Hyderabad
MYANMAR (BURMA)
Visakhapatnam
Panaji
Hubli
Vijayawada
Bay of Bengal
Pathein
EASTERN GHATS
Bangalore
Chennai (Madras)
Laccadive Is. (India)
Mysore
Coimbatore
Andaman Is. (India)
Kochi
Madurai
Thiruvananthapuram
Trincomalee
Cape Comorin
Colombo
SRI LANKA
Sri Jayawardenepura
Galle
Nicobar Is. (India)
Male
MALDIVES
INDIAN OCEAN
Arabian Sea
Socotra (Yemen)
Mediterranean Sea
Black Sea
CAUCASUS MTS.
Bursa
Izmir
Ankara
TURKEY
Samsun
GEORGIA
Trabzon
Tbilisi
Adana
INSET AREA
CYPRUS
Nicosia
Erzurum
Mt. Ararat 16,804
Yerevan
ARMENIA
Ganca
AZERBAIJAN
Caspian Sea
LEBANON
Latakia
Aleppo
Diyarbakir
Beirut
SYRIA
Tabriz
Baku
Tel Aviv-Yafo
Jerusalem
ISRAEL
Damascus
Amman
Mosul
Arbil
Kirkuk
Lake Urmia
JORDAN
Al Aqabah
IRAQ
Kermanshah
Basht
Mt. Damavand 18,606
Baghdad
ELBURZ MTS.
Tehran
Gorgan
AFRICA
Tabuk
Al Hillah
Qom
DASHT-E KAVIR
Mashhad
AN NAFUD
Hail
Basra
Esfahan
ZAGROS MOUNTAINS
IRAN
DASHT-E LUT
Buraydah
Kuwait
KUWAIT
Yazd
Birjand
Shiraz
Red Sea
Jeddah
Riyadh
Manama
BAHRAIN
Kerman
Mecca
At Taif
SAUDI ARABIA
Doha
QATAR
Zahedan
Persian Gulf
Bandar-e Abbas
Abha
UNITED ARAB EMIRATES
Abu Dhabi
Gulf of Oman
RUB AL KHALI
Muscat
Sur
Ras al Hadd
Sanaa
OMAN
YEMEN
Aden
Al Mukalla
Salalah
Gulf of Aden
TURKEY
Adana
Antalya
Sanliurfa
Icel (Mersin)
Aleppo
Mosul
Al Hasakah
Hatay (Antakya)
Ar Raqqah
Latakia
CYPRUS
Nicosia
Hamah
Euphrates
Bayji
Hims
Abu Kamal
Limassol
SYRIA
LEBANON
Tadmur
Beirut
Damascus
Ar Ramadi
Mediterranean Sea
Haifa
WEST BANK
The West Bank and Gaza currently occupied by Israel. Permanent status to be determined.
IRAQ
ISRAEL
SYRIAN
Tel Aviv-Yafo
Port Said
Jerusalem
Amman
GAZA STRIP
Jericho
DESERT
Tanta
Cairo
Suez Canal
JORDAN
SAUDI ARABIA
Suez
Maan
Giza
Elat
AN NAFUD
EGYPT
Al Aqabah
SINAI
Nile
Gulf of Suez
Gulf of Aqaba
Tabuk
0 250 Miles
0 250 Kilometers
© MAPQUEST.COM

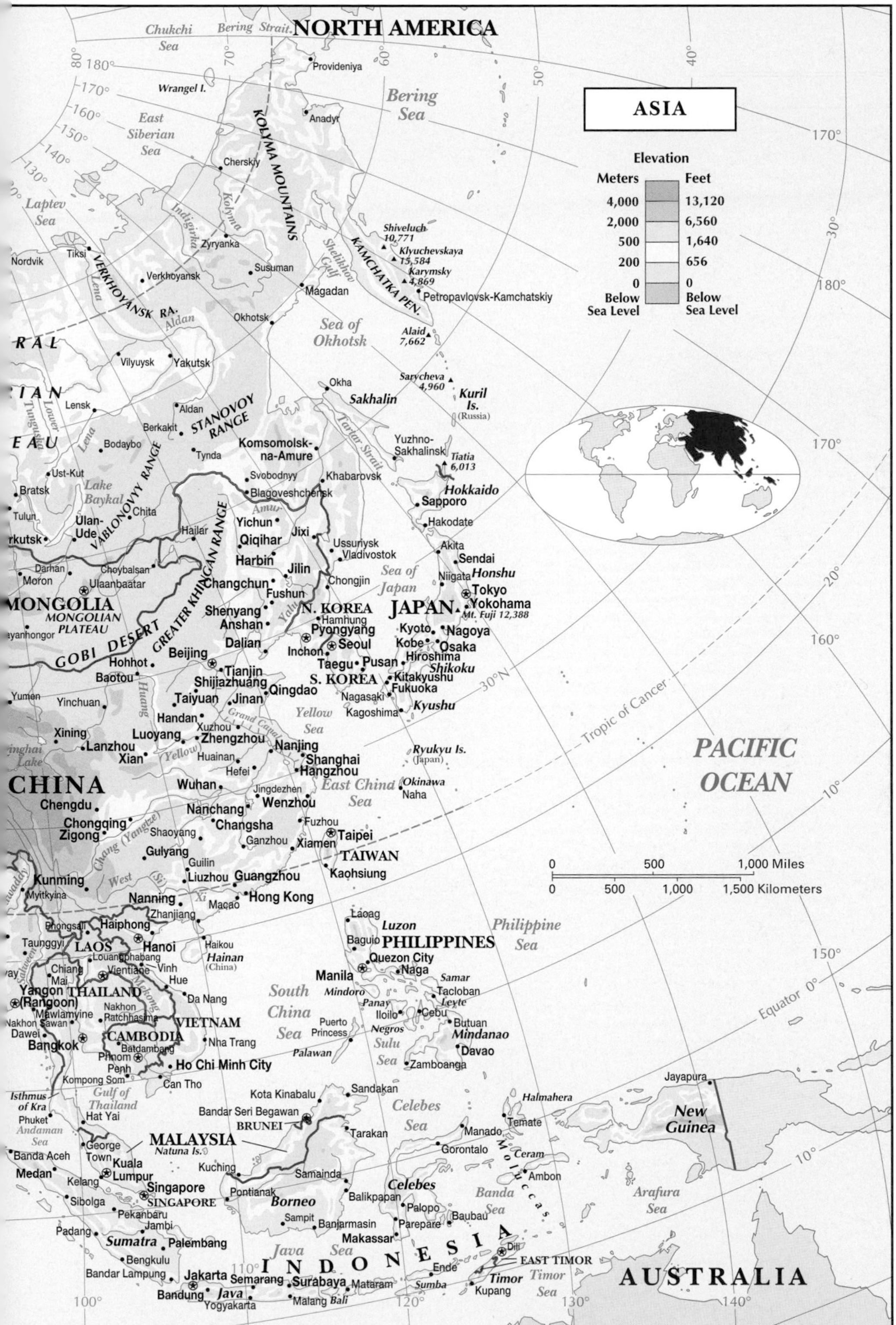
ASIA
Elevation
Meters
Feet
4,000
13,120
2,000
6,560
500
1,640
200
656
0
0
Below Sea Level
Below Sea Level
0
500
1,000 Miles
0
500
1,000
1,500 Kilometers
NORTH AMERICA
Chukchi Sea
Bering Strait
Providenyia
Wrangel I.
Bering Sea
East Siberian Sea
Anadyr
Cherskiy
KOLYMA MOUNTAINS
Kolyma
Indigirka
Laptev Sea
Nordvik
Tiksi
Zyryanka
VERKHOYANSK RA.
Lena
Verkhoyansk
Susuman
Shelikhov Gulf
KAMCHATKA PEN.
Shiveluch 10,771
Klyuchevskaya 15,584
Karymsky 4,869
Petropavlovsk-Kamchatskiy
Magadan
Aldan
Okhotsk
Sea of Okhotsk
Alaid 7,662
Vilyuysk
Yakutsk
Sarycheva 4,960
Kuril Is.
(Russia)
Okha
Sakhalin
Lensk
Lower Tunguska
Aldan
STANOVOY RANGE
Berkakit
Tynda
Bodaybo
Komsomolsk-na-Amure
Tatar Strait
Yuzhno-Sakhalinsk
Tiatia 6,013
Ust-Kut
Bratsk
Lake Baykal
YABLONOVYY RANGE
Svobodnyy
Blagoveshchensk
Khabarovsk
Amur
Hokkaido
Sapporo
Hakodate
Tulun
Chita
Ulan-Ude
Hailar
Yichun
GREATER KHINGAN RANGE
Qiqihar
Jixi
Harbin
Ussuriysk
Vladivostok
Akita
Sendai
Honshu
Darhan
Moron
Choybalsan
Ulaanbaatar
Changchun
Jilin
Chongjin
Sea of Japan
Niigata
Tokyo
Yokohama
Mt. Fuji 12,388
MONGOLIA
MONGOLIAN PLATEAU
Fushun
Shenyang
Anshan
Yalu
N. KOREA
Hamhung
Pyongyang
JAPAN
Kyoto
Nagoya
Kobe
Osaka
GOBI DESERT
Dalian
Seoul
Inchon
Hiroshima
Shikoku
Hohhot
Beijing
Tianjin
Taegu
Pusan
Baotou
Shijiazhuang
Qingdao
S. KOREA
Kitakyushu
Fukuoka
Yumen
Yinchuan
Huang
Taiyuan
Jinan
Nagasaki
Kagoshima
Kyushu
Handan
Yellow Sea
Xining
Lanzhou
Xian
Luoyang
Xuzhou
Zhengzhou
Grand Canal
Nanjing
Ryukyu Is.
(Japan)
Yellow
Huainan
Hefei
Shanghai
Hangzhou
CHINA
Wuhan
Jingdezhen
East China Sea
Okinawa
Naha
Chengdu
Nanchang
Wenzhou
Chongqing
Zigong
Changsha
Chang (Yangtze)
Shaoyang
Fuzhou
Taipei
Ganzhou
Xiamen
Guiyang
TAIWAN
Guilin
Kaohsiung
Kunming
West
Liuzhou
Guangzhou
Myitkyina
Nanning
Xi
Macao
Hong Kong
Zhanjiang
Phongsali
Haiphong
Laoag
Luzon
Philippine Sea
Taunggyi
LAOS
Hanoi
Haikou
Baguio
PHILIPPINES
Louangphabang
Hainan
(China)
Chiang Mai
Vientiane
Vinh
Quezon City
Naga
Manila
Samar
Hue
Mekong
Yangon (Rangoon)
THAILAND
South China Sea
Mindoro
Tacloban
Leyte
Da Nang
Nakhon Ratchasima
Panay
Mawlamyine
Iloilo
Cebu
Nakhon Sawan
VIETNAM
Puerto Princesa
Negros
Butuan
Dawei
Mindanao
Bangkok
CAMBODIA
Nha Trang
Sulu Sea
Davao
Batdambang
Palawan
Phnom Penh
Ho Chi Minh City
Zamboanga
Kompong Som
Can Tho
Sandakan
Jayapura
Isthmus of Kra
Gulf of Thailand
Kota Kinabalu
Celebes Sea
Halmahera
Bandar Seri Begawan
New Guinea
Phuket
Hat Yai
BRUNEI
Ternate
Andaman Sea
MALAYSIA
Natuna Is.
Tarakan
Manado
George Town
Gorontalo
Banda Aceh
Kuala Lumpur
Kuching
Ceram
Moluccas
Medan
Kelang
Samainda
Ambon
Singapore
SINGAPORE
Pontianak
Celebes
Sibolga
Balikpapan
Banda Sea
Arafura Sea
Borneo
Palopo
Pekanbaru
Sampit
Baubau
Jambi
Banjarmasin
Parepare
Padang
Sumatra
Palembang
Makassar
Java Sea
Bengkulu
INDONESIA
Dili
EAST TIMOR
Bandar Lampung
Jakarta
Semarang
Surabaya
Ende
Mataram
Sumba
Timor
Timor Sea
AUSTRALIA
Bandung
Java
Yogyakarta
Malang
Bali
Kupang
PACIFIC OCEAN
Tropic of Cancer
30°N
Equator 0°
180°
170°
160°
150°
140°
130°
120°
110°
100°
80°
70°
60°
50°
40°
30°
20°
10°

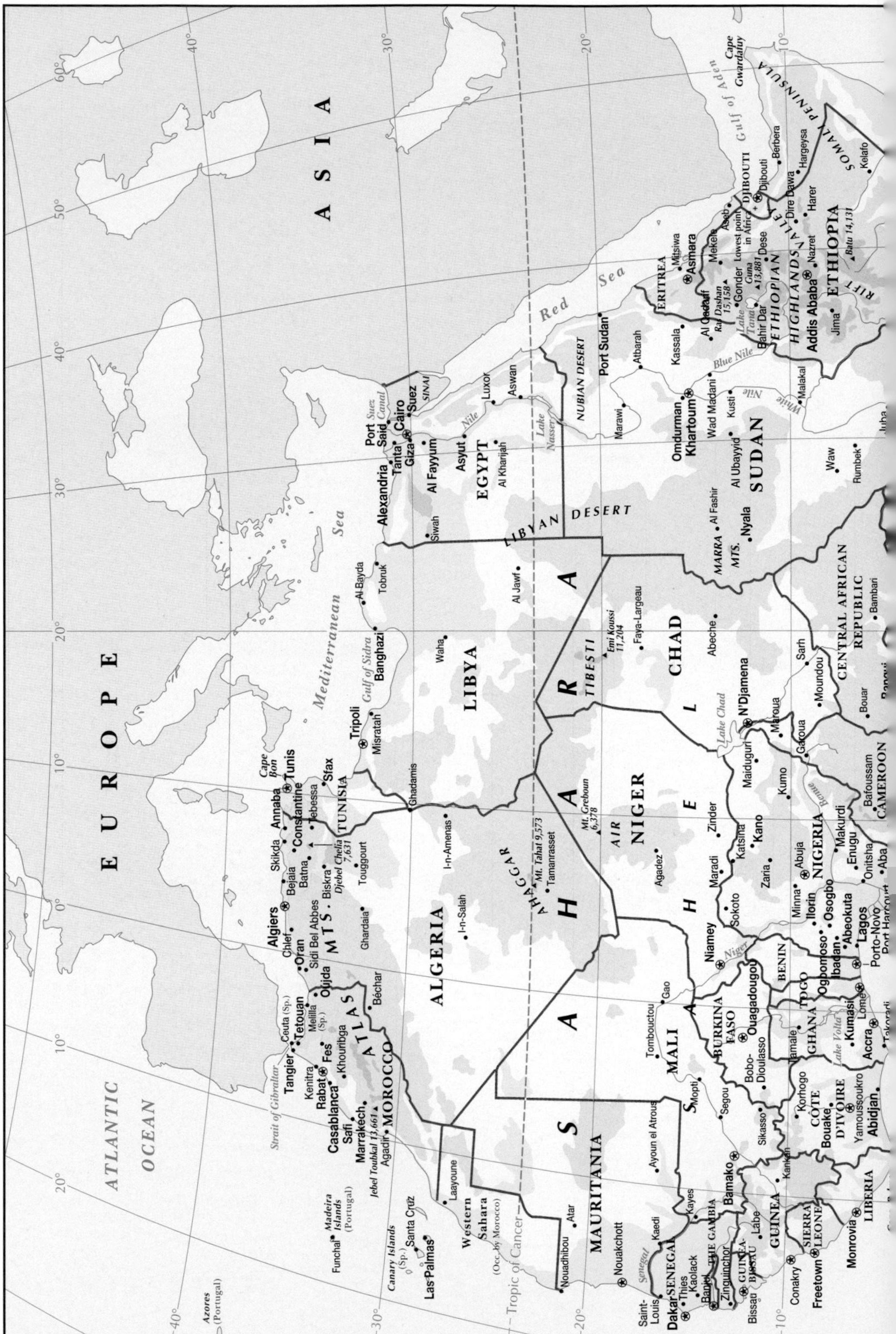
EUROPE
ASIA
ATLANTIC OCEAN
Mediterranean Sea
Red Sea
Gulf of Aden
SAHARA
LIBYAN DESERT
NUBIAN DESERT
MOROCCO
ALGERIA
TUNISIA
LIBYA
EGYPT
SUDAN
ETHIOPIA
ERITREA
DJIBOUTI
MAURITANIA
MALI
NIGER
CHAD
CENTRAL AFRICAN REPUBLIC
CAMEROON
NIGERIA
BENIN
TOGO
GHANA
BURKINA FASO
CÔTE D'IVOIRE
LIBERIA
SIERRA LEONE
GUINEA
GUINEA-BISSAU
THE GAMBIA
SENEGAL
Western Sahara (Occ. by Morocco)
Canary Islands (Sp.)
Madeira Islands (Portugal)
Azores (Portugal)
Tropic of Cancer
Strait of Gibraltar
ATLAS MTS.
AHAGGAR
TIBESTI
AIR
MARRA MTS.
ETHIOPIAN HIGHLANDS
SOMALI PENINSULA
Jebel Toubkal 13,661
Djebel Chelia 7,631
Mt. Tahat 9,573
Mt. Greboun 6,378
Emi Koussi 11,204
Ras Dashan 15,158
Guna 13,881
Batu 14,131
Lake Chad
Lake Tana
Lake Nasser
Lake Volta
Blue Nile
White Nile
Nile
Niger
Benue
Senegal
Suez Canal
SINAI
Gulf of Sidra
Cape Bon
Cape Gwardafuy
Lowest point in Africa
Algiers
Oran
Tunis
Tripoli
Banghazi
Alexandria
Cairo
Giza
Port Said
Suez
Khartoum
Omdurman
Addis Ababa
Asmara
Djibouti
Casablanca
Rabat
Fes
Marrakech
Tangier
Nouakchott
Dakar
Bamako
Niamey
Ouagadougou
N'Djamena
Abuja
Lagos
Accra
Monrovia
Freetown
Conakry
Abidjan
Yamoussoukro
Porto-Novo
Lomé
Las Palmas
Santa Cruz
Funchal
Laayoune
Tamanrasset
Agadez
Timbouctou
Gao
Kano
Port Sudan
Aswan
Luxor
Asyut
Al Fayyum
Siwah
Tobruk
Misratah
Ghadamis
Sfax
Annaba
Constantine
Ghardaia
Béchar
Agadir
Safi
Atar
Nouadhibou
Saint-Louis
Kaolack
Thies
Banjul
Bissau
Kayes
Segou
Mopti
Sikasso
Bobo-Dioulasso
Kumasi
Tamale
Ibadan
Ogbomoso
Abeokuta
Ilorin
Kaduna
Zaria
Sokoto
Maradi
Zinder
Maiduguri
Garoua
Maroua
Moundou
Sarh
Abeche
Faya-Largeau
Bouar
Bambari
Nyala
Al Fashir
Al Ubayyid
Kusti
Wad Madani
Kassala
Atbarah
Malakal
Waw
Rumbek
Gonder
Bahir Dar
Mekele
Dese
Nazret
Dire Dawa
Harer
Hargeysa
Berbera
Jima
Kelafo

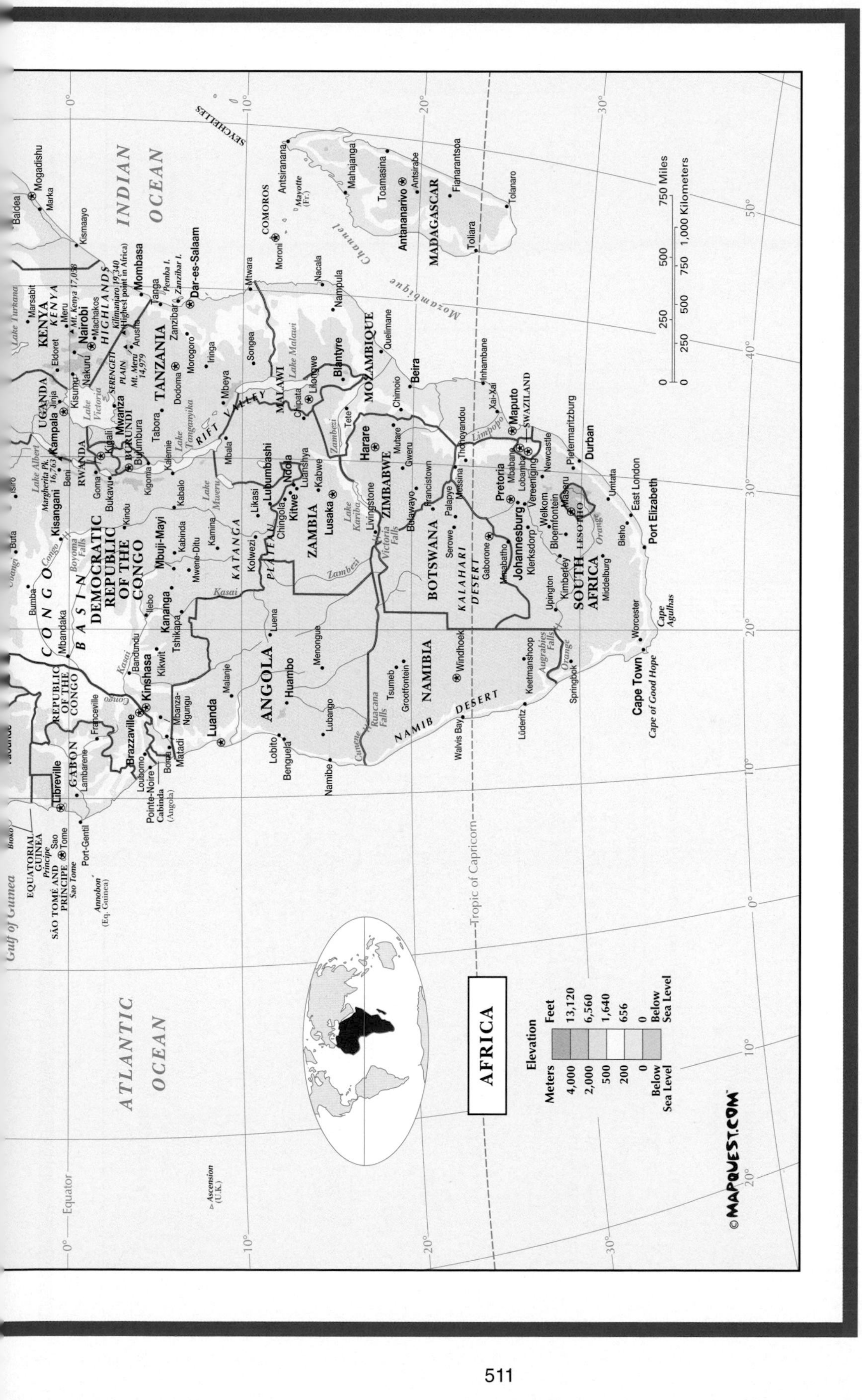

AFRICA
ATLANTIC OCEAN
INDIAN OCEAN
Elevation
Meters
4,000
2,000
500
200
0
Below Sea Level
Feet
13,120
6,560
1,640
656
0
Below Sea Level
0 250 500 750 Miles
0 250 500 750 1,000 Kilometers
Equator
Tropic of Capricorn
Gulf of Guinea
EQUATORIAL GUINEA
SÃO TOMÉ AND PRINCIPE
GABON
REPUBLIC OF THE CONGO
DEMOCRATIC REPUBLIC OF THE CONGO
CONGO BASIN
UGANDA
KENYA
RWANDA
BURUNDI
TANZANIA
ANGOLA
ZAMBIA
MALAWI
MOZAMBIQUE
ZIMBABWE
BOTSWANA
NAMIBIA
SOUTH AFRICA
LESOTHO
SWAZILAND
MADAGASCAR
COMOROS
SEYCHELLES
KALAHARI DESERT
NAMIB DESERT
RIFT VALLEY
KATANGA PLATEAU
SERENGETI PLAIN
KENYA HIGHLANDS
Mozambique Channel
Libreville
Brazzaville
Kinshasa
Luanda
Kampala
Nairobi
Mogadishu
Dar-es-Salaam
Mombasa
Lusaka
Harare
Lilongwe
Blantyre
Maputo
Pretoria
Johannesburg
Durban
Cape Town
Port Elizabeth
East London
Windhoek
Gaborone
Maseru
Mbabane
Antananarivo
Moroni
Beira
Huambo
Lubumbashi
Kisangani
Mbuji-Mayi
Kananga
Kilimanjaro 19,340 (highest point in Africa)
Mt. Kenya 17,058
Cape of Good Hope
Cape Agulhas
Ascension (U.K.)
©MAPQUEST.COM

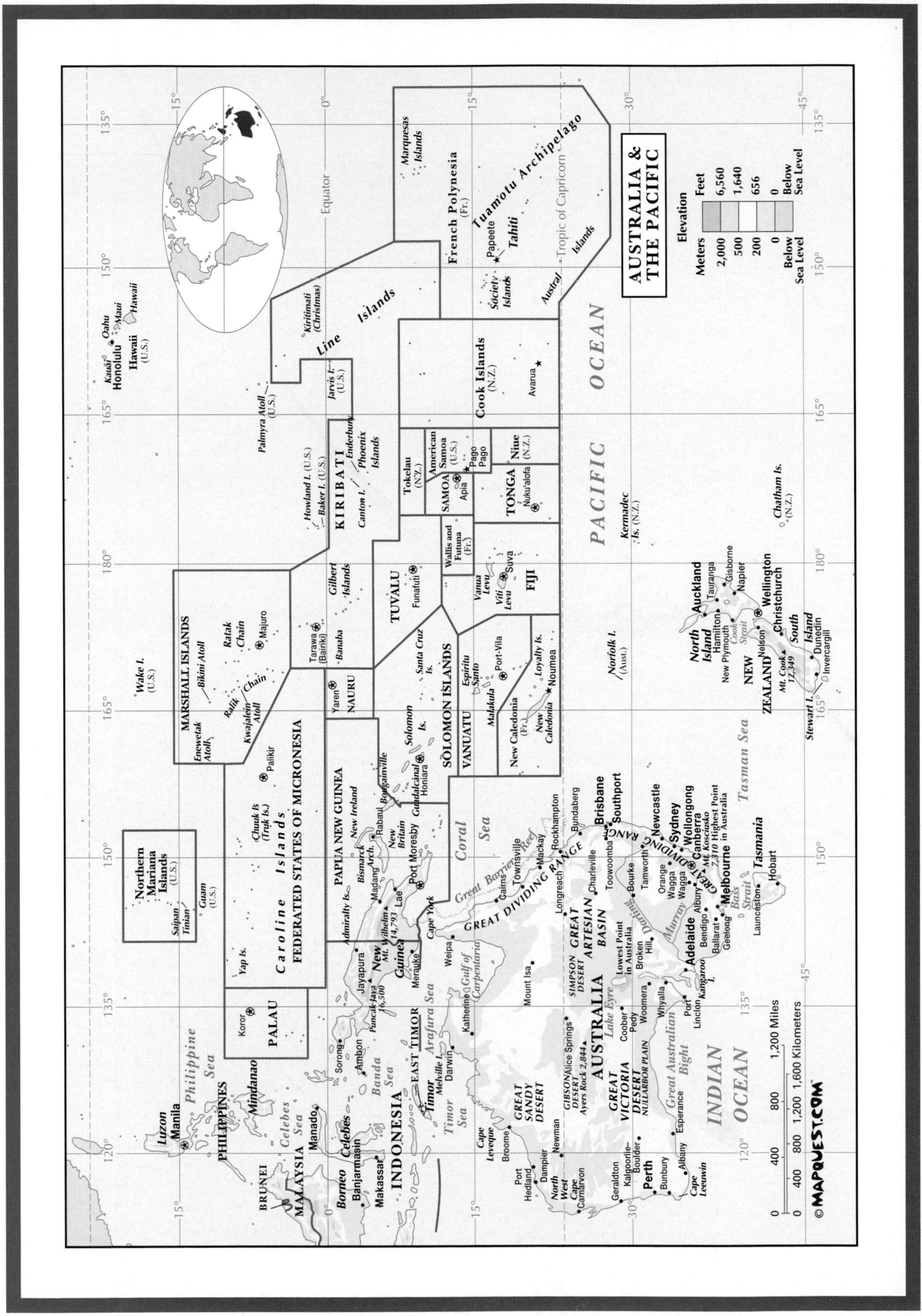
AUSTRALIA & THE PACIFIC
Elevation
Meters Feet
2,000 6,560
500 1,640
200 656
0 0
Below Sea Level Below Sea Level
PACIFIC OCEAN
INDIAN OCEAN
AUSTRALIA
Coral Sea
Tasman Sea
Arafura Sea
Timor Sea
Banda Sea
Celebes Sea
Philippine Sea
Great Australian Bight
Gulf of Carpentaria
Great Barrier Reef
GREAT DIVIDING RANGE
GREAT SANDY DESERT
GREAT VICTORIA DESERT
GIBSON DESERT
SIMPSON DESERT
GREAT ARTESIAN BASIN
NULLARBOR PLAIN
Ayers Rock 2,844
Lake Eyre
Lowest Point in Australia
Mt. Kosciusko 7,310 Highest Point in Australia
Perth
Adelaide
Melbourne
Sydney
Canberra
Brisbane
Darwin
Hobart
Alice Springs
Tasmania
Bass Strait
NEW ZEALAND
North Island
South Island
Auckland
Wellington
Christchurch
Dunedin
Cook Strait
Mt. Cook 12,349
PAPUA NEW GUINEA
Port Moresby
INDONESIA
EAST TIMOR
PHILIPPINES
Manila
MALAYSIA
BRUNEI
PALAU
Koror
FEDERATED STATES OF MICRONESIA
Caroline Islands
Palikir
Northern Mariana Islands (U.S.)
Guam (U.S.)
MARSHALL ISLANDS
Majuro
Wake I. (U.S.)
NAURU
Yaren
KIRIBATI
Tarawa (Bairiki)
Gilbert Islands
Phoenix Islands
Line Islands
TUVALU
Funafuti
SOLOMON ISLANDS
Honiara
VANUATU
Port-Vila
New Caledonia (Fr.)
Nouméa
FIJI
Suva
TONGA
Nuku'alofa
SAMOA
Apia
American Samoa (U.S.)
Pago Pago
Tokelau (N.Z.)
Wallis and Futuna (Fr.)
Niue (N.Z.)
Cook Islands (N.Z.)
Avarua
French Polynesia (Fr.)
Papeete
Tahiti
Society Islands
Tuamotu Archipelago
Marquesas Islands
Austral Islands
Hawaii (U.S.)
Honolulu
Equator
Tropic of Capricorn
0 400 800 1,200 Miles
0 400 800 1,200 1,600 Kilometers
© MAPQUEST.COM

UNITED STATES HISTORY

Chronology of Events

1492

Christopher Columbus and crew sighted land **Oct. 12** in the present-day Bahamas.

1497

John Cabot explored northeast coast to Delaware.

1513

Juan Ponce de León explored Florida coast.

1524

Giovanni da Verrazano led French expedition along coast from Carolina north to Nova Scotia; entered New York harbor.

1539

Hernando de Soto landed in Florida **May 28;** crossed Mississippi River, **1541.**

1540

Francisco Vásquez de Coronado explored Southwest north of Rio Grande. Hernando de Alarcón reached Colorado River; Don Garcia Lopez de Cardenas reached Grand Canyon. Others explored California coast.

1565

St. Augustine, FL, founded **Sept. 8** by Pedro Menéndez. Razed by Francis Drake **1586.**

1579

Francis Drake entered San Francisco Bay and claimed region for Britain.

1607

Capt. John Smith and 105 cavaliers in 3 ships landed on Virginia coast, started first permanent English settlement in New World at **Jamestown** in **May.**

1609

Henry Hudson, English explorer of Northwest Passage, employed by Dutch, sailed into New York harbor in **Sept.,** and up Hudson to Albany. **Samuel de Champlain** explored Lake Champlain, just to the north.

Spaniards settled **Santa Fe, NM.**

1619

House of Burgesses, first representative assembly in New World, elected **July 30** at Jamestown, VA.

First black laborers—indentured servants—in English N. American colonies, landed by Dutch at Jamestown in **Aug.** Chattel slavery legally recognized, **1650.**

1620

Plymouth Pilgrims, Puritan separatists, left Plymouth, England, **Sept. 16** on *Mayflower*. They reached Cape Cod **Nov. 19,** explored coast; 103 passengers landed **Dec. 26** at Plymouth. **Mayflower Compact** was agreement to form a government and abide by its laws. Half of colony died during harsh winter.

1624

Dutch colonies started in Albany and in New York area, where **New Netherland** was established in **May.**

1626

Peter Minuit bought Manhattan for Dutch West India Co. from Man-a-hat-a Indians during summer for goods valued at $24; named island **New Amsterdam.**

1630

Settlement of **Boston** established by Massachusetts colonists led by John Winthrop.

1634

Maryland, founded as a Catholic colony, under a charter granted to Lord Baltimore. Religious toleration granted **1649.**

1636

Roger Williams founded Providence, RI, **June,** as a democratically ruled colony with separation of church and state. Charter granted, **1644.**

Harvard College founded **Oct. 28,** now oldest in U.S.; grammar school, compulsory education established at Boston.

1640

First book was printed in America, the so-called Bay Psalm Book.

1647

Liberal constitution drafted in Rhode Island.

1660

British Parliament passed First **Navigation Act Dec. 1,** regulating colonial commerce to suit English needs.

1664

British troops Sept. 8 seized New Netherland from Dutch. Charles II granted New Netherland and city of New Amsterdam to brother, Duke of York; both renamed **New York.** Dutch recaptured colony **1673,** but ceded it to Britain **Nov. 10, 1674.**

1673

Jacques **Marquette** and Louis **Jolliet** reached the upper **Mississippi** and traveled down it.

1676

Nathaniel Bacon led planters against autocratic British Gov. Sir William Berkeley, burned Jamestown, VA, **Sept. 19.** Rebellion collapsed when Bacon died; 23 followers executed.

Bloody **Indian war** in New England ended **Aug. 12.** King Philip, Wampanoag chief, and Narragansett Indians killed.

1682

Robert Cavelier, Sieur de La Salle, claimed lower Mississippi River country for France, called it Louisiana **Apr. 9.** Had French outposts built in Illinois and Texas, **1684.** Killed during mutiny **Mar. 19, 1687.**

William Penn arrived in **Pennsylvania.**

1683

William Penn signed treaty with Delaware Indians and made payment for Pennsylvania lands.

1692

Witchcraft delusion at Salem, MA; 20 alleged witches executed by special court.

1696

Capt. William Kidd settled in America, was hired by British to fight pirates and take booty, but himself became one. Arrested and sent to England; hanged **1701.**

1699

French settlements made in Mississippi, Louisiana.

1704

Indians attacked Deerfield, MA, **Feb. 28-29;** killed 40, carried off 100.

Boston News Letter, **first regular newspaper,** started by John Campbell, postmaster. (An earlier paper, *Publick Occurences*, was suppressed after one issue **1690.**)

1709

British-Colonial troops captured French fort, Port Royal, Nova Scotia, in **Queen Anne's War 1701-13.** France yielded Nova Scotia by treaty **1713.**

1712

Slaves revolted in New York **Apr. 6.** Six committed suicide; 21 were executed. Second rising, **1741;** 13 slaves hanged, 13 burned, 71 deported.

1716

First theater in colonies opened in Williamsburg, VA.

1726

Poor people **rioted** in Philadelphia.

Great Awakening religious revival began.

1732

Benjamin Franklin published the first *Poor Richard's Almanack;* published annually to **1757.**

Last of the 13 colonies, **Georgia,** chartered.

1735

Editor **John Peter Zenger acquitted** in New York of libeling British governor by criticizing his conduct in office.

1740-41

Capt. Vitus Bering reached Alaska.

1744

King George's War pitted British and colonials vs. French. Colonials captured Louisburg, Cape Breton Is., **June 17, 1745.** Returned to France **1748** by Treaty of Aix-la-Chapelle.

1752

Benjamin Franklin, flying kite in thunderstorm, proved lightning is electricity **June 15;** invented lightning rod.

1754

French and Indian War began when French occupied Ft. Duquesne (Pittsburgh). British moved Acadian French from Nova Scotia to Louisiana **Oct. 8, 1755.** British captured Québec **Sept. 18, 1759,** in battles in which French Gen. Joseph de Montcalm and British Gen. James Wolfe were killed. Peace pact signed **Feb. 10, 1763.** French lost Canada and Midwest.

1764

Sugar Act placed duties on lumber, foodstuffs, molasses, and rum in colonies, to pay French and Indian War debts.

IT'S A FACT: The famous "Battle of Bunker Hill" took place not on Bunker Hill but on nearby Breed's Hill, on June 17, 1775. British forces unexpectedly dug in on a site about 2,000 ft. from Bunker Hill, where they were supposed to engage. Breed's Hill was afterwards renamed Bunker Hill.

1765

Stamp Act, enacted by Parliament **Mar. 22,** required revenue stamps to help fund royal troops. Nine colonies, at **Stamp Act Congress** in New York **Oct. 7-25,** adopted Declaration of Rights. Stamp Act **repealed Mar. 17, 1766.**

1767

Townshend Acts levied taxes on glass, painter's lead, paper, and tea. In **1770** all duties except those on tea were repealed.

1770

British troops fired **Mar. 5** into Boston mob, killed 5 including **Crispus Attucks,** a black man, reportedly leader of group; later called **Boston Massacre.**

1773

East India Co. tea ships turned back at Boston, New York, and Philadelphia in **May.** Cargo ship burned at Annapolis **Oct. 14;** cargo thrown overboard at **Boston Tea Party Dec. 16,** to protest the tea tax.

1774

"Intolerable Acts" of Parliament curtailed Massachusetts self-rule; barred use of Boston harbor till tea was paid for.

First Continental Congress held in Philadelphia **Sept. 5-Oct. 26;** called for civil disobedience against British.

Rhode Island abolished slavery.

1775

Patrick Henry addressed Virginia convention, **Mar. 23,** said "Give me liberty or give me death."

Paul Revere and William Dawes on night of **Apr. 18** rode to alert Patriots that British were on their way to Concord to destroy arms. At Lexington, MA, **Apr. 19,** Minutemen lost 8. On return from Concord, British took 273 casualties.

Col. Ethan Allen (joined by Col. Benedict Arnold) captured **Ft. Ticonderoga, NY, May 10;** also Crown Point. Colonials headed for **Bunker Hill,** fortified Breed's Hill, Charlestown, MA. Repulsed British under Gen. William Howe twice before retreating **June 17;** called Battle of Bunker Hill.

Continental Congress **June 15** named **George Washington** commander in chief.

1776

France and Spain each agreed **May 2** to provide arms.

In Continental Congress **June 7,** Richard Henry Lee (VA) moved "that these united colonies are and of right ought to be free and independent states." Resolution adopted July 2. **Declaration of Independence** approved **July 4.**

Col. William Moultrie's batteries at **Charleston, SC,** repulsed British sea attack **June 28.** Washington lost **Battle of Long Island Aug. 27;** evacuated New York.

Nathan Hale executed as spy by British **Sept. 22.**

Brig. Gen. Arnold's **Lake Champlain** fleet was defeated at Valcour **Oct. 11,** but British returned to Canada. Howe failed to destroy Washington's army at **White Plains Oct. 28.** Hessians captured Ft. Washington, Manhattan, and 3,000 men **Nov. 16;** captured Ft. Lee, NJ, **Nov. 18.**

Washington, in Pennsylvania, recrossed **Delaware River Dec. 25-26,** defeated Hessians at Trenton, NJ, **Dec. 26.**

1777

Washington defeated Lord Cornwallis at **Princeton Jan. 3.** Continental Congress adopted Stars and Stripes.

Maj. Gen. John Burgoyne, force of 8,000 from Canada, captured **Ft. Ticonderoga July 6.** Americans beat back Burgoyne at Bemis Heights **Oct. 7,** cut off British escape route. Burgoyne surrendered 5,000 men at **Saratoga, NY, Oct. 17.**

Articles of Confederation adopted by Continental Congress **Nov. 15.**

1778

France signed treaty of aid with U.S. **Feb. 6.** Sent fleet; British evacuated Philadelphia in consequence **June 18.**

1779

John Paul Jones on the *Bonhomme Richard* defeated *Serapis* in British North Sea waters **Sept. 23.**

1780

Charleston, SC, fell to the British **May 12,** but a British force was defeated near **Kings Mountain, NC, Oct. 7** by militiamen.

Benedict Arnold found to be a traitor **Sept. 23.** Arnold escaped, made brigadier general in British army.

1781

Articles of Confederation took effect **Mar. 1.**

Bank of North America incorporated **May 26.**

Cornwallis retired to **Yorktown, VA.** Adm. Francois Joseph de Grasse landed 3,000 French and stopped British fleet in Hampton Roads. Washington and Jean Baptiste de Rochambeau joined forces, arrived near Williamsburg **Sept. 26.** Siege of Cornwallis began **Oct. 6; Cornwallis surrendered Oct. 19.**

1782

New **British** cabinet agreed **in March** to **recognize U.S.** independence. Preliminary agreement signed in Paris **Nov. 30.**

1783

Massachusetts Supreme Court declared **slavery** illegal in that state.

Britain, U.S. signed Paris **peace treaty Sept. 3** recognizing American independence (Congress ratified it **Jan. 14, 1784).**

Washington ordered army disbanded Nov. 3, bade farewell to his officers at Fraunces Tavern, New York City, **Dec. 4.**

Noah Webster published *American Spelling Book.*

1784

Thomas Jefferson's proposal to **ban slavery** in new territory after 1802 was narrowly defeated **Mar. 1.**

First successful daily newspaper, **Pennsylvania Packet & General Advertiser,** published **Sept. 21.**

1786

Delegates from 5 states at **Annapolis, MD, Sept. 11-14** asked Congress to call a constitutional convention for the 13 states.

1787

Shays's Rebellion of debt-ridden farmers in Massachusetts failed **Jan. 25.**

Northwest Ordinance adopted **July 13** by Continental Congress for Northwest Territory, N of Ohio River, W of New York; made rules for statehood. Guaranteed freedom of religion, support for schools, no slavery.

Constitutional convention opened at Philadelphia **May 25** with Washington presiding. Constitution accepted by delegates **Sept. 17;** ratification by 9th state, New Hampshire, **June 21, 1788,** meant adoption; declared in effect **Mar. 4, 1789.**

1789

George Washington chosen president by all electors voting (73 eligible, 69 voting, 4 absent); John Adams, vice president, got 34 votes. First Congress met at Federal Hall, New York City, **Mar. 4.** Washington inaugurated there **Apr. 30.** Supreme Court created by Federal Judiciary Act **Sept. 24.** Congress submitted Bill of Rights to states **Sept. 25.**

1790

Congress, **Mar. 1,** authorized decennial **U.S. census; Naturalization Act** (2-year residency) passed **Mar. 26.**

Congress met in Philadelphia, new temporary capital, **Dec. 6.**

1791

Bill of Rights went into effect **Dec. 15.**

1792

Coinage Act established **U.S. Mint** in Philadelphia **Apr. 2.**

Gen. **"Mad" Anthony Wayne** made commander in Ohio-Indiana area, trained "American Legion," established string of forts. Routed Indians at Fallen Timbers on Maumee River **Aug. 20, 1794,** checked British at Fort Miami, OH.

White House cornerstone laid **Oct. 13.**

1793

Eli Whitney invented **cotton gin,** reviving Southern slavery.

1794

Whiskey Rebellion, W Pennsylvania farmers protesting liquor tax of **1791,** was suppressed by federal militia **Sept.**

1795

U.S. bought peace from **Algerian pirates** by paying $1 mil ransom for 115 seamen **Sept. 5,** followed by annual tributes.

Gen. Wayne signed peace with Indians at Fort Greenville.

University of North Carolina became first operating state university.

1796

Washington's Farewell Address as president delivered **Sept. 19.** Gave strong warnings against permanent alliances with foreign powers, big public debt, large military establishment, and devices of "small, artful, enterprising minority."

1797

U.S. **frigate United States** launched at Philadelphia **July 10; Constellation** at Baltimore **Sept. 7; Constitution** (Old Ironsides) at Boston **Sept. 20.**

1798

Alien & Sedition Acts passed by Federalists **June-July**; intended to silence political opposition.

War with France threatened over French raids on U.S. shipping and rejection of U.S. diplomats. Navy (45 ships) and 365 privateers captured 84 French ships. USS *Constellation* took French warship *Insurgente* **1799.** Napoleon stopped French raids after becoming First Consul.

1800

Federal government moved to **Washington, DC.**

1801

John Marshall named Supreme Court chief justice, **Jan. 20.**

Tripoli declared war June 10 against U.S., which refused added tribute to commerce-raiding Arab corsairs. Land and naval campaigns forced Tripoli to negotiate **peace June 4, 1805.**

1803

Supreme Court, in **Marbury v Madison** case, for the first time overturned a U.S. law **Feb. 24.**

Napoleon sold all of **Louisiana,** stretching to Canadian border, to U.S., for $11,250,000 in bonds, plus $3,750,000 indemnities to American citizens with claims against France. U.S. took title **Dec. 20.** Purchase doubled U.S. area.

1804

Lewis and Clark expedition ordered by Pres. Thomas Jefferson to explore what is now northwest U.S. Started from St. Louis **May 14; ended Sept. 23, 1806.**

Vice Pres. **Aaron Burr shot Alexander Hamilton** in a duel **July 11** in Weehawken, NJ; Hamilton died the next day.

1807

Robert Fulton made first practical steamboat trip; left New York City **Aug. 17,** reached Albany, 150 mi, in 32 hr.

Embargo Act banned all trade with foreign countries, forbidding ships to set sail for foreign ports **Dec. 22.**

1808

Slave importation outlawed. Some 250,000 slaves were illegally imported **1808-60.**

1811

William Henry Harrison, governor of Indiana, defeated Indians under the Prophet, in battle of **Tippecanoe Nov. 7.**

Cumberland Road begun at Cumberland, MD; became important route to West.

1812

War of 1812 had 3 main causes: Britain seized U.S. ships trading with France; Britain seized 4,000 naturalized U.S. sailors by **1810;** Britain armed Indians who raided western border. U.S. stopped trade with Europe **1807** and **1809.** Trade with Britain only was stopped **1810.**

Unaware that Britain had raised the blockade against France 2 days before, **Congress declared war June 18.**

USS *Essex* captured *Alert* **Aug. 13;** USS *Constitution* destroyed *Guerriere* **Aug. 19;** USS *Wasp* took *Frolic* **Oct. 18;** USS *United States* defeated *Macedonian* off Azores **Oct. 25;** *Constitution* beat *Java* **Dec. 29.** British took Detroit **Aug. 16.**

1813

Oliver H. Perry defeated British fleet at Battle of Lake Erie, **Sept. 10.** U.S. won Battle of the Thames, Ontario, **Oct. 5,** but failed in Canadian invasion attempts. York (Toronto) and Buffalo were burned.

1814

British landed in Maryland in Aug., defeated U.S. force **Aug. 24, burned Capitol and White House.** Maryland militia stopped British advance **Sept. 12.** Bombardment of Ft. McHenry, Baltimore, for 25 hours, **Sept. 13-14,** by British fleet failed; Francis Scott Key wrote words to **"The Star Spangled Banner."**

U.S. won naval Battle of **Lake Champlain Sept. 11.** Peace treaty signed at Ghent **Dec. 24.**

1815

Some 5,300 British, unaware of peace treaty, attacked U.S. entrenchments near **New Orleans, Jan. 8.** British had more than 2,000 casualties; Americans lost 71.

U.S. flotilla finally ended piracy by **Algiers, Tunis, Tripoli** by **Aug. 6.**

1816

Second **Bank of the U.S.** chartered.

1817

Rush-Bagot treaty signed **Apr. 28-29;** limited U.S., British armaments on the Great Lakes.

William Cullen Bryant's poem "Thanatopsis" published.

1819

Spain ceded **Florida** to U.S. **Feb. 22.**

American steamship Savannah made first part-steam-powered, part-sail-powered crossing of Atlantic, Savannah, GA, to Liverpool, England, 29 days.

1820

First organized **immigration of blacks to Africa** from U.S. began with 86 free blacks sailing **Feb.** to Sierra Leone.

Henry Clay's **Missouri Compromise** bill passed by Congress **Mar. 3.** Slavery was allowed in Missouri, but not elsewhere west of the Mississippi River north of 36° 30′ latitude (the southern line of Missouri). Repealed **1854.**

1821

Emma Willard founded Troy Female Seminary, first U.S. women's college.

1823

Monroe Doctrine, opposing European intervention in the Americas, enunciated by Pres. James Monroe **Dec. 2.**

1824

Pawtucket, RI, **weavers strike,** first such action by women.

1825

After a deadlocked election, John Quincy Adams was elected president by the U.S. House, **Feb. 9.**

Erie Canal opened; first boat left Buffalo **Oct. 26,** reached New York City **Nov. 4.**

John Stevens, of Hoboken, NJ, built and operated first experimental **steam locomotive** in U.S.

1826

Thomas Jefferson and John Adams both died **July 4.**

1828

South Carolina **Dec. 19** declared the right of state **nullification of federal laws,** opposing the "Tariff of Abominations."

Noah Webster published his *American Dictionary of the English Language.*

Baltimore & Ohio, the first U.S. passenger railroad, begun **July 4.**

1829

Andrew Jackson inaugurated as president, **Mar. 4.**

1830

Mormon church organized by Joseph Smith in Fayette, NY, **Apr. 6.**

1831

William Lloyd Garrison began abolitionist newspaper *The Liberator,* **Jan. 1.**

Nat Turner, black slave in Virginia, led local slave rebellion, starting **Aug. 21;** 57 whites killed. Troops called in, 100 slaves killed, Turner captured, tried, and hanged **Nov. 11.**

1832

Black Hawk War (IL-WI) **Apr.-Sept.** pushed Sauk and Fox Indians west across Mississippi.

South Carolina convention passed **Ordinance of Nullification Nov. 24** against permanent tariff, threatening to withdraw from Union. Congress **Feb. 1833** passed compromise tariff act, whereupon South Carolina repealed its act.

1833

Oberlin College became first in U.S. to adopt coeducation.

1835

Seminole Indians in Florida under Osceola began attacks **Nov. 1,** protesting forced removal. The unpopular war ended **Aug. 14, 1842;** most of the Indians were sent to Oklahoma.

Texas proclaimed right to secede from Mexico; Sam Houston put in command of Texas army, **Nov. 2-4.**

Gold discovered on **Cherokee land** in Georgia. Indians forced to cede lands **Dec. 20** and to cross Mississippi.

Halley's Comet passed by the Earth.

1836

Texans besieged in Alamo in San Antonio by Mexicans under Santa Anna **Feb. 23-Mar. 6;** entire garrison killed. Texas independence declared, **Mar. 2.** At San Jacinto **Apr. 21,** Sam Houston and Texans defeated Mexicans.

Marcus Whitman, H. H. Spaulding, and wives reached Fort Walla Walla on Columbia River, OR. **First white women to cross plains.**

1838

Cherokee Indians made **"Trail of Tears,"** removed from Georgia to Oklahoma starting **Oct.**

1841

First emigrant **wagon train for California,** 47 persons, left Independence, MO, **May 1,** reached California **Nov. 4.**

Brook Farm commune set up by New England Transcendentalist intellectuals. Lasted to **1846.**

1842

Webster-Ashburton Treaty signed **Aug. 9,** fixing the U.S.-Canada border in Maine and Minnesota.

First use of **anesthetic** (sulfuric ether gas).

Settlement of Oregon began via **Oregon Trail.**

1843

More than 1,000 settlers left Independence, MO, for Oregon **May 22,** arrived **Oct.**

1844

First message over first **telegraph line** sent **May 24** by inventor Samuel F.B. Morse from Washington to Baltimore: "What hath God wrought!"

1845

Texas Congress **voted for annexation** by U.S. **July 4.** U.S. Congress admitted Texas to Union **Dec. 29.**

Edgar Allan Poe's poem "The Raven" published.

1846

Mexican War began after Pres. James K. Polk ordered Gen. Zachary Taylor to seize disputed Texan land settled by Mexicans. After border clash, U.S. declared war **May 13;** Mexico **May 23.**

Bear flag of Republic of California raised by American settlers at Sonoma **June 14.**

About 12,000 U.S. troops took Vera Cruz **Mar. 27, 1847,** and Mexico City **Sept. 14, 1847.** By **treaty,** signed **Feb. 2, 1848,** war was ended, and Mexico ceded claims to Texas, California, and other territory.

Treaty with Britain **June 15** set **boundary in Oregon** territory at 49th parallel (extension of existing line). Expansionists had used slogan "54° 40′ or fight."

Mormons, after violent clashes with settlers over polygamy, left Nauvoo, IL, for West under Brigham Young; settled **July 1847** at **Salt Lake City, UT.**

Elias Howe invented **sewing machine.**

1847

First **adhesive U.S. postage stamps** on sale **July 1;** Benjamin Franklin 5¢, Washington 10¢.

Ralph Waldo Emerson published first book of poems; **Henry Wadsworth Longfellow** published *Evangeline.*

1848

Gold discovered Jan. 24 in California; 80,000 prospectors emigrated in **1849.**

Lucretia Mott and Elizabeth Cady Stanton led **Seneca Falls, NY, Women's Rights Convention July 19-20.**

1850

Sen. Henry Clay's **Compromise of 1850** admitted California as 31st state **Sept. 9,** with slavery forbidden; made Utah and New Mexico territories; made Fugitive Slave Law more harsh; ended District of Columbia slave trade.

Nathaniel Hawthorne's *The Scarlet Letter* published.

1851

Herman Melville's *Moby-Dick* published.

1852

Uncle Tom's Cabin, by **Harriet Beecher Stowe,** published.

1853

Comm. Matthew C. Perry, U.S.N., received by Japan, **July 14;** negotiated **treaty to open Japan** to U.S. ships.

1854

Republican Party formed at Ripon, WI, **Feb. 28.** Opposed Kansas-Nebraska Act (became law **May 30**), which left issue of slavery to vote of settlers.

Henry David Thoreau published *Walden.*

Treaty ratified with Mexico **Apr. 25,** providing for purchase of a strip of land **(Gadsden Purchase).**

1855

Walt Whitman published *Leaves of Grass.*

First railroad train crossed Mississippi on the river's first bridge, Rock Island, IL, Davenport, IA, **Apr. 21.**

1856

Republican Party's first nominee for president, **John C. Fremont,** defeated. Abraham Lincoln made 50 speeches for him.

Lawrence, KS, sacked **May 21** by proslavery group; abolitionist **John Brown** led antislavery men against Missourians at **Osawatomie, KS, Aug. 30.**

1857

Dred Scott decision by Supreme Court **Mar. 6** held that slaves did not become free in a free state, Congress could not bar slavery from a territory, and blacks could not be citizens.

1858

First **Atlantic cable** was completed, by Cyrus W. Field **Aug. 5.**

Lincoln-Douglas debates in Illinois **Aug. 21-Oct. 15.**

1859

First commercially productive **oil well,** drilled near Titusville, PA, by Edwin L. Drake **Aug. 27.**

Abolitionist **John Brown,** with 21 men, seized U.S. Armory at **Harpers Ferry Oct. 16.** U.S. Marines captured raiders, killing several. Brown was hanged for treason **Dec. 2.**

1860

Approximately 20,000 **New England shoe workers** went on strike **Feb. 22** and won higher wages.

Abraham Lincoln, Republican, elected president **Nov. 6** in 4-way race.

First **Pony Express** between Sacramento, CA, and St. Joseph, MO, started **Apr. 3;** service ended **Oct. 24, 1861,** when first transcontinental telegraph line was completed.

1861

Seven southern states set up **Confederate States of America Feb. 8,** with Jefferson Davis as president, captured federal arsenals and forts. **Civil War** began as Confederates fired on **Ft. Sumter** in Charleston, SC, **Apr. 12,** capturing it **Apr. 14.**

Pres. **Lincoln** called for 75,000 volunteers **Apr. 15.** By **May,** 11 states had seceded. Lincoln blockaded Southern ports **Apr. 19,** cutting off vital exports, aid.

Confederates repelled Union forces at first **Battle of Bull Run July 21.**

First **transcontinental telegraph** was put in operation.

1862

Homestead Act approved **May 20;** it granted free family farms to settlers.

Land Grant Act approved **July 7,** providing for public land sale to benefit agricultural education; eventually led to establishment of state university systems.

Union forces were victorious in Western campaigns, took **New Orleans May 1.** Battles in East were inconclusive.

1863

Pres. Lincoln issued **Emancipation Proclamation Jan. 1,** freeing "all slaves in areas still in rebellion."

Entire **Mississippi River** was in Union hands by **July 4.** Union forces won a major victory at **Gettysburg, PA, July 1-3.** Lincoln read his **Gettysburg Address Nov. 19.**

In **draft riots** in New York City about 1,000 were killed or wounded; some blacks were hanged by mobs **July 13-16.**

1864

Gen. William Tecumseh **Sherman marched through Georgia,** taking Atlanta **Sept. 1,** Savannah **Dec. 22.**

Sand Creek massacre of Cheyenne and Arapaho Indians **Nov. 29.** Cavalry attacked Indians awaiting surrender terms.

1865

Gen. Robert E. Lee surrendered 27,800 Confederate troops to Gen. Ulysses S. Grant at Appomattox Court House, VA, **Apr. 9.** J. E. Johnston surrendered 31,200 to Sherman at Durham Station, NC, **Apr. 18.** Last rebel troops surrendered **May 26.**

Pres. Lincoln was shot Apr. 14 by John Wilkes Booth in Ford's Theater, Washington, DC; died the following morning. Vice Pres. **Andrew Johnson** was sworn in as president. Booth was hunted down; fatally wounded, perhaps by his own hand, **Apr. 26.** Four co-conspirators were hanged **July 7.**

13th Amendment, abolishing slavery, ratified **Dec. 6.**

1866

Ku Klux Klan formed secretly in South to terrorize blacks who voted. Disbanded **1869-71.** A 2d Klan organized **1915.**

Congress took control of Southern Reconstruction, backed freedmen's rights.

1867

Alaska sold to U.S. by Russia for $7.2 mil **Mar. 30** through efforts of Sec. of State William H. Seward.

Horatio Alger published first book, *Ragged Dick.*

The **Grange** was organized **Dec. 4,** to protect farmer interests.

1868

The World Almanac, a publication of the *New York World,* appeared for the first time.

Pres. **Johnson** tried to remove Edwin M. Stanton, secretary of war; was impeached by House **Feb. 24** for violation of Tenure of Office Act; acquitted by Senate Mar.-May.

1869

Financial **"Black Friday"** in New York **Sept. 24;** caused by attempt to "corner" gold.

Transcontinental railroad completed; golden spike driven at Promontory, UT, **May 10,** marking the junction of Central Pacific and Union Pacific.

Knights of Labor formed in Philadelphia. By **1886,** this labor union had 700,000 members nationally.

Woman suffrage law passed in Wyoming Territory **Dec. 10.**

1871

Great fire destroyed **Chicago Oct. 8-11.**

1872

Amnesty Act restored civil rights to citizens of the South **May 22** except for 500 Confederate leaders.

Congress established first national park—**Yellowstone.**

1873

First U.S. **postal card** issued **May 1.**

Banks failed, panic began in **Sept.** Depression lasted 5 years.

"Boss" William Tweed of New York City convicted **Nov. 19** of stealing public funds. He died in jail in **1878.**

New York's Bellevue Hospital started **first nursing school.**

1875

Congress passed **Civil Rights Act Mar. 1,** giving equal rights to blacks in public accommodations and jury duty. Act invalidated in **1883** by Supreme Court.

First **Kentucky Derby** held **May 17.**

1876

Samuel J. Tilden, Democrat, received majority of popular votes for president over **Rutherford B. Hayes,** Republican, but 22 electoral votes were in dispute; issue left to Congress. Hayes won the presidency in **Feb. 1877** after Republicans agreed to end Reconstruction of South.

Col. **George A. Custer** and 264 soldiers of the 7th Cavalry killed **June 25** in "last stand," Battle of the Little Big Horn, MT, in Sioux Indian War.

1877

Molly Maguires, Irish terrorist society in Scranton, PA, mining areas, was broken up by the hanging, **June 21,** of 11 leaders for murders of mine officials and police.

Pres. Rutherford B. Hayes sent troops in violent national **railroad strike.**

1878

First commercial **telephone** exchange opened, New Haven, CT, **Jan. 28.**

Thomas A. Edison founded **Edison Electric Light Co.** on **Oct. 15.**

1879

F. W. Woolworth opened his first five-and-ten store, in Utica, NY, **Feb. 22.**

Henry George published *Progress & Poverty,* advocating single tax on land.

1881

Pres. **James A. Garfield shot** in Washington, DC, **July 2;** died **Sept. 19.**

Booker T. Washington founded Tuskegee Institute for blacks.

Helen Hunt Jackson published *A Century of Dishonor,* about mistreatment of Indians.

1883

Pendleton Act passed **Jan. 16,** reformed civil service.

Brooklyn Bridge opened **May 24.**

1884

Mark Twain's masterpiece, *The Adventures of Huckleberry Finn,* appeared.

1886

Haymarket riot and bombing, **May 4,** followed bitter labor battles for 8-hour day in Chicago; 7 police and 4 workers died. Eight anarchists found guilty **Aug. 20,** 4 hanged **Nov. 11.**

Geronimo, Apache Indian, finally surrendered **Sept. 4.**

Statue of Liberty dedicated **Oct. 28.**

American Federation of Labor (AFL) formed **Dec. 8** by 25 craft unions.

1888

Great blizzard struck eastern U.S. **Mar. 11-14,** causing about 400 deaths.

1889

U.S. opened Oklahoma to white settlement **Apr. 22;** within 24 hours **claims for 2 mil acres** were staked by 50,000 settlers.

Johnstown, PA, flood May 31; 2,200 lives lost.

1890

Battle of **Wounded Knee, SD, Dec. 29,** the last major conflict between Indians and U.S. troops. About 200 Indian men, women, and children and 29 soldiers were killed.

Sherman Antitrust Act passed **July 2,** began federal effort to curb monopolies.

Jacob Riis published *How the Other Half Lives,* about city slums.

Poems of **Emily Dickinson** published posthumously.

1891

Forest Reserve Act Mar. 3 let president close public forest land to settlement for establishment of national parks.

1892

Ellis Island, in New York Bay, opened **Jan. 1** to receive immigrants.

Homestead, PA, strike at Carnegie steel mills; 7 guards and 11 strikers and spectators shot to death **July 6;** setback for unions.

1893

Financial panic began, led to 4-year depression.

1894

Thomas A. Edison's kinetoscope (motion pictures) (invented **1887**) given first public showing **Apr. 14.**

The **Pullman strike** began **May 11** at a railroad car plant in Chicago.

Jacob S. Coxey led army of unemployed from the Midwest, reaching Washington, DC, **Apr. 30.** Coxey arrested **May 1** for trespassing on Capitol grounds; his army disbanded.

1896

William Jennings Bryan delivered "Cross of Gold" speech **July 8;** won Democratic Party nomination.

Supreme Court, in **Plessy v. Ferguson,** approved racial segregation under the "separate but equal" doctrine.

1898

U.S. **battleship Maine** blown up **Feb. 15** at Havana; 260 killed.

U.S. blockaded Cuba Apr. 22 in aid of independence forces. U.S. declared war on Spain, **Apr. 24,** destroyed Spanish fleet in Philippines **May 1,** took Guam **June 20.**

Puerto Rico taken by U.S. **July 25-Aug. 12.** Spain agreed **Dec. 10** to cede Philippines, Puerto Rico, and Guam, and approved independence for Cuba.

Annexation of **Hawaii** signed by Pres. William McKinley, **July 7.**

1899

Filipino insurgents, unable to get recognition of independence from U.S., started guerrilla war **Feb. 4.** Their leader, Emilio Aguinaldo, captured **May 23, 1901.** Philippine Insurrection ended **1902.**

U.S. declared **Open Door Policy** to make China an open international market and to preserve its integrity as a nation.

John Dewey published *The School and Society,* advocating "progressive education."

1900

Carry Nation, Kansas antisaloon agitator, began raiding with hatchet.

U.S. helped suppress **"Boxers"** in Beijing.

International Ladies' Garment Workers Union was founded in New York City **June 3.**

1901

Texas had first significant **oil strike, Jan. 10.**

Pres. **McKinley was shot Sept. 6** in Buffalo, NY, by an anarchist, Leon Czolgosz; died **Sept. 14.**

1903

Treaty between U.S. and Colombia to have U.S. dig **Panama Canal** signed **Jan. 22,** rejected by Colombia. Panama declared independence from Colombia with U.S. support **Nov. 3;** recognized by Pres. Theodore Roosevelt **Nov. 6.** U.S., Panama signed canal treaty **Nov. 18.**

Wisconsin set first **direct primary** voting system **May 23.**

First successful flight in heavier-than-air mechanically propelled airplane by **Orville Wright, Dec. 17** near Kitty Hawk, NC, 120 ft in 12 secs. Fourth flight same day by **Wilbur Wright,** 852 ft in 59 secs. Improved plane patented, **1906.**

Great Train Robbery, pioneering film, produced.

1904

Ida Tarbell published muckraking *The History of the Standard Oil Company.*

1905

First **Rotary Club** founded in Chicago.

1906

San Francisco earthquake and fire **Apr. 18-19** left 503 dead, $350 mil damages.

Pure Food and Drug Act and Meat Inspection Act both passed **June 30.**

1907

Financial panic and depression started **Mar. 13.**

First round-world cruise of U.S. **"Great White Fleet";** 16 battleships, 12,000 men.

1908

Henry Ford introduced **Model T** car, priced at $850, **Oct. 1.**

1909

Adm. Robert E. Peary claimed to have reached **North Pole Apr. 6** on 6th attempt, accompanied by Matthew Henson, a black man, and 4 Eskimos; may have fallen short.

National Conference on the Negro convened **May 30,** leading to founding of National Association for the Advancement of Colored People.

1910

Boy Scouts of America founded **Feb. 8.**

1911

Supreme Court dissolved **Standard Oil Co. May 15.**

Building holding New York City's **Triangle Shirtwaist Co.** factory caught fire **Mar. 25;** 146 died.

First **transcontinental airplane flight** (with numerous stops) by C. P. Rodgers, New York to Pasadena, CA, **Sept. 17-Nov. 5;** time in air 82 hr, 4 min.

1912

American Girl Guides founded **Mar. 12;** name changed in **1913** to **Girl Scouts.**

U.S. sent Marines **Aug. 14** to **Nicaragua,** which was in default of loans to U.S. and Europe.

1913

NY Armory Show brought modern art to U.S. **Feb. 17.**

U.S. blockaded Mexico in support of revolutionaries.

Charles Beard published his *Economic Interpretation of the Constitution.*

Federal Reserve System was authorized **Dec. 23,** in a major reform of U.S. banking and finance.

1914

Ford Motor Co. raised basic wage rates from $2.40 for 9-hr day to $5 for 8-hr day **Jan. 5.**

When U.S. sailors were arrested at Tampico, Mexico, **Apr. 9,** Atlantic fleet was sent to **Veracruz,** occupied city.

Pres. Woodrow Wilson proclaimed **U.S. neutrality** in the European war **Aug. 4.**

Panama Canal was officially opened **Aug. 15.**

The **Clayton Antitrust Act** was passed **Oct. 15,** strengthening federal antimonopoly powers.

1915

First transcontinental **telephone call,** New York to San Francisco, completed **Jan. 25,** by Alexander Graham Bell and Thomas A. Watson.

British ship **Lusitania** sunk **May 7** by German submarine; 128 American passengers lost (Germany had warned passengers in advance). As a result of U.S. campaign, Germany issued apology and promise of payments **Oct. 5.** Pres. Wilson asked for a military fund increase **Dec. 7.**

U.S. troops landed in **Haiti July 28.** Haiti became a virtual U.S. protectorate under **Sept. 16** treaty.

1916

Gen. John J. **Pershing entered Mexico** to pursue Francisco (Pancho) Villa, who had raided U.S. border areas. Forces withdrawn **Feb. 5, 1917.**

Rural Credits Act passed **July 17,** followed by Warehouse Act **Aug. 11;** both provided financial aid to farmers.

Bomb exploded during **San Francisco** Preparedness Day parade **July 22,** killed 10. Thomas J. Mooney, labor organizer, and Warren K. Billings, shoe worker, were convicted **1917;** both later pardoned.

U.S. bought **Virgin Islands** from Denmark **Aug. 4.**

Jeannette Rankin (R, MT) elected as **first-ever female** member of U.S. **House.**

U.S. established military government in the **Dominican Republic Nov. 29.**

Trade and loans to **European allies** soared during the year.

1917

Germany, suffering from British blockade, declared almost unrestricted **submarine warfare Jan. 31.** U.S. cut diplomatic ties with Germany **Feb. 3,** and formally declared war **Apr. 6.**

Conscription law was passed **May 18.** First U.S. troops arrived in Europe **June 26.**

18th **(Prohibition)** Amendment to the Constitution was submitted to the states by Congress **Dec. 18.** On **Jan. 16, 1919,** the 36th state (Nevada) ratified it.

1918

Pres. Wilson set out his **14 Points** as basis for peace **Jan. 8.**

More than 1 mil **American troops** were in Europe by **July.** Allied counteroffensive launched at Château-Thierry **July 18.** War ended with signing of armistice **Nov. 11.**

Influenza epidemic killed an estimated 20 mil worldwide, 548,000 in U.S.

1919

First **transatlantic flight,** by U.S. Navy seaplane, left Rockaway, NY, **May 8,** stopped at Newfoundland, Azores, Lisbon **May 27.**

Boston police strike Sept. 9; National Guard breaks strike.

Sherwood Anderson published *Winesburg, Ohio.*

About 250 **alien radicals** were deported **Dec. 22.**

1920

In national **Red Scare,** some 2,700 Communists, anarchists, and other radicals were arrested **Jan.-May.**

Senate refused **Mar. 19** to ratify the **League of Nations Covenant.**

Radicals Nicola **Sacco** and Bartolomeo **Vanzetti** accused of killing 2 men in Massachusetts payroll holdup **Apr. 15.** Found guilty **1921.** A 6-year campaign for their release failed, and both were executed **Aug. 23, 1927.** Controversial verdict repudiated **1977,** by proclamation of Massachusetts Gov. Michael Dukakis.

First regular licensed **radio broadcasting** begun **Aug. 20.**

19th Amendment ratified **Aug. 18,** giving women right to vote.

League of Women Voters founded.

Wall St., New York City, **bomb** explosion killed 30, injured 100, did $2 mil damage **Sept. 16.**

Sinclair Lewis's *Main Street,* **F. Scott Fitzgerald's** *This Side of Paradise* published.

1921

Congress sharply curbed **immigration,** set national quota system **May 19.**

Joint congressional resolution declaring **peace with Germany, Austria,** and **Hungary** signed **July 2** by Pres. Warren G. Harding; treaties were signed in **Aug.**

Limitation of Armaments Conference met in Washington, DC, **Nov. 12-Feb. 6, 1922.** Major powers agreed to curtail naval construction, outlaw poison gas, restrict submarine attacks on merchant vessels, respect integrity of China.

Ku Klux Klan began revival with violence against Catholics in North, South, and Midwest.

1922

Violence during **coal-mine strike** at Herrin, IL, **June 22-23** cost 36 lives, including those of 21 nonunion miners.

Reader's Digest founded.

1923

First **sound-on-film motion picture,** *Phonofilm,* shown at Rivoli Theater, New York City, beginning in **April.**

1924

Law approved by Congress **June 15** making all **Indians citizens.**

Nellie Tayloe Ross elected governor of Wyoming **Nov. 9** as nation's first woman governor. **Miriam (Ma) Ferguson** elected governor of Texas **Nov. 9;** installed **Jan. 20, 1925.**

George Gershwin wrote *Rhapsody in Blue.*

1925

John T. Scopes found guilty of having taught **evolution** in Dayton, TN, high school, fined $100 and costs **July 24.**

1926

Dr. Robert H. Goddard demonstrated practicality of **rockets Mar. 16** at Auburn, MA, with first liquid-fuel rocket; rocket traveled 184 ft in 2.5 sec.

Congress established **Army Air Corps July 2.**

Air Commerce Act passed **Nov. 2,** providing federal aid for airlines and airports.

Ernest Hemingway's *The Sun Also Rises* published.

1927

About 1,000 **marines landed in China Mar. 5** to protect property in civil war.

Capt. **Charles A. Lindbergh** left Roosevelt Field, NY, **May 20** alone in plane *Spirit of St. Louis* on first New York-Paris nonstop flight. Reached Le Bourget airfield **May 21,** 3,610 mi in 33½ hours.

The Jazz Singer, with **Al Jolson,** demonstrated part-talking pictures in New York City **Oct. 6.**

Show Boat opened in New York **Dec. 27.**

O. E. Rolvaag published *Giants in the Earth.*

1928

Herbert Hoover elected president, defeating New York Gov. **Alfred E. Smith,** a Catholic.

Amelia Earhart became first woman to fly the Atlantic, **June 17.**

1929

"St. Valentine's Day massacre" in Chicago **Feb. 14;** gangsters killed 7 rivals.

Farm price stability aided by **Agricultural Marketing Act,** passed **June 15.**

Albert B. Fall, former secretary of the interior, was convicted of accepting bribe of $100,000 in the leasing of the **Elk Hills (Teapot Dome)** naval oil reserve; sentenced **Nov. 1** to a year in prison and fined $100,000.

Stock market crash Oct. 29 marked end of past prosperity as stock prices plummeted. Stock losses for 1929-31 estimated at $50 bil; worst American depression began.

Thomas Wolfe published *Look Homeward, Angel.* **William Faulkner** published *The Sound and the Fury.*

1930

London **Naval Reduction Treaty** signed by U.S., Britain, Italy, France, and Japan **Apr. 22;** in effect **Jan. 1, 1931;** expired **Dec. 31, 1936.**

Hawley-Smoot Tariff signed; rate hikes slash world trade.

1931

Empire State Building opened in New York City **May 1.**

Al Capone was convicted of tax evasion **Oct. 17.**

1932

Reconstruction Finance Corp. established **Jan. 22** to stimulate banking and business. Unemployment at 12 mil.

19-month-old **Charles Lindbergh Jr.** was **kidnapped Mar. 1;** found dead **May 12.** Bruno Hauptmann found guilty in trial **Jan.-Feb. 1935;** executed **Apr. 3, 1936.**

Bonus March on Washington, DC, launched **May 29** by World War I veterans demanding Congress pay their bonus in full.

Franklin D. Roosevelt elected president for the first time.

1933

Pres. Roosevelt named **Frances Perkins** U.S. secretary of labor; first woman in U.S. cabinet.

All **banks in the U.S.** were ordered **closed** by Pres. Roosevelt **Mar. 6.**

In a "100 days" special session, **Mar. 9-June 16,** Congress passed **New Deal** social and economic measures, including measures to regulate banks, distribute funds to the jobless, create jobs, raise agricultural prices, and set wage and production standards for industry.

Tennessee Valley Authority created by act of Congress, **May 18.**

Gold standard dropped by U.S.; announced by Pres. Roosevelt **Apr. 19,** ratified by Congress **June 5.**

Prohibition ended in the U.S. as 36th state ratified 21st Amendment **Dec. 5.**

U.S. foreswore armed intervention in **western hemisphere** nations **Dec. 26.**

1934

U.S. troops pulled out of **Haiti Aug. 6.**

1935

Works Progress Administration **(WPA)** instituted **May 6.** Rural Electrification Administration created **May 11.** National Industrial Recovery Act struck down by Supreme Court **May 27.**

Comedian **Will Rogers** and aviator **Wiley Post killed Aug. 15** in Alaska plane crash.

Social Security Act passed by Congress **Aug. 14.**

Huey Long, senator from Louisiana and national political leader, **assassinated Sept. 8.**

Porgy and Bess opened **Oct. 10** in New York.

Committee for Industrial Organization (CIO; later Congress of Industrial Organizations) formed to expand industrial unionism **Nov. 9.**

1936

Boulder Dam completed.

Margaret Mitchell published *Gone With the Wind.*

1937

Joe Louis knocked out James J. Braddock, became world heavyweight champ **June 22.**

Amelia Earhart, aviator, and copilot Fred Noonan lost **July 2** near Howland Island, in the Pacific.

Pres. Roosevelt asked for 6 additional Supreme Court justices; **"packing" plan** defeated.

1938

Naval Expansion Act passed **May 17.**

National minimum wage enacted **June 25.**

Orson Welles radio dramatization of **Martian invasion,** *War of the Worlds,* caused nationwide scare **Oct. 30.**

1939

Pres. Roosevelt asked for **defense budget hike Jan. 5, 12.**

New York World's Fair opened **Apr. 30,** closed **Oct. 31;** reopened **May 11, 1940,** and finally closed **Oct. 21.**

Albert Einstein alerted Pres. Roosevelt to **A-bomb** opportunity in **Aug. 2** letter.

U.S. declared its neutrality in European war **Sept. 5.**

Roosevelt proclaimed a limited **national emergency Sept. 8,** an unlimited emergency **May 27, 1941.** Both ended by Pres. Harry Truman **Apr. 28, 1952.**

John Steinbeck published *Grapes of Wrath.*

Gone With the Wind and *The Wizard of Oz* appeared on screen.

1940

U.S. okayed sale of **surplus war materiel** to Britain **June 3;** announced transfer of 50 overaged destroyers **Sept. 3.**

First **peacetime draft** approved **Sept. 14.**

Richard Wright published *Native Son.*

1941

Four Freedoms termed essential by Pres. Roosevelt in speech to Congress **Jan. 6:** freedom of speech and religion, freedom from want and fear.

Lend-Lease Act signed **Mar. 11** provided $7 bil in military credits for Britain. Lend-Lease for USSR approved in **Nov.**

U.S. occupied **Iceland July 7.**

The **Atlantic Charter,** 8-point declaration of principles, issued by Roosevelt and British Prime Min. Winston Churchill **Aug. 14.**

Japan attacked **Pearl Harbor,** Hawaii, 7:55 AM Hawaiian time, **Dec. 7;** 19 ships sunk or damaged, 2,300 dead. U.S. declared war on Japan **Dec. 8,** on Germany and Italy **Dec. 11.**

1942

Japanese troops took Bataan peninsula **Apr. 8,** Corregidor **May 6.**

Federal government forcibly moved 110,000 **Japanese-Americans** from West Coast to detention camps. Exclusion lasted 3 years.

Battle of **Midway June 4-7** was Japan's first major defeat.

Marines landed on **Guadalcanal Aug. 7;** last Japanese not expelled until **Feb. 9, 1943.**

U.S., Britain invaded North Africa **Nov. 8.**

IT'S A FACT: The use of war as an instrument of foreign policy was renounced, under the Kellogg-Briand Pact, signed initially by 15 countries, including Germany, Italy, and Japan, in 1928. It failed to halt invasions by Italy and Japan in the 1930s, not to mention World War II.

1943

Oklahoma! opened **Mar. 31** on Broadway.

War contractors barred from **racial discrimination, May 27.**

Pres. Roosevelt signed **June 10** pay-as-you-go income tax bill. Starting **July 1** wage and salary earners were subject to a **paycheck withholding** tax.

Pearl Buck published *The Good Earth.*

Auto, steel labor unions won first big contracts.

First **nuclear chain reaction** (fission of uranium isotope U-235) produced at University of Chicago, under physicists Arthur Compton, Enrico Fermi, others **Dec. 2.**

Race riot in Detroit June 21; 34 dead, 700 injured. Riot in Harlem section of New York City; 6 killed.

U.S., Britain invaded **Sicily July 9,** Italian **mainland Sept. 3.**

Marines recaptured the Gilbert Islands, captured by Japan in 1941 and 1942, in Nov.

1944

U.S., Allied forces invaded Europe at **Normandy June 6** in greatest amphibious landing in history.

GI Bill of Rights signed **June 22,** providing benefits for veterans.

U.S. forces landed on **Leyte,** Philippines, **Oct. 20.**

1945

Yalta Conference met in the Crimea, USSR, **Feb. 4-11.** Roosevelt, Churchill, and Soviet leader Joseph Stalin agreed that their 3 countries, plus France, would occupy Germany and that the Soviet Union would enter war against Japan.

Marines landed on **Iwo Jima Feb. 19,** won control of Iwo Jima **Mar. 16** after heavy casualties. U.S. forces invaded **Okinawa Apr. 1,** captured Okinawa **June 21.**

Pres. Roosevelt, 63, died in Warm Springs, GA, **Apr. 12;** Vice Pres. **Harry S. Truman** became president.

Germany surrendered May 7; May 8 proclaimed V-E Day.

First **atomic bomb,** produced at Los Alamos, NM, exploded at Alamogordo, NM, **July 16.** Bomb dropped on **Hiroshima Aug. 6,** with about 75,000 people killed; bomb dropped on **Nagasaki Aug. 9,** killing about 40,000. Japan agreed to surrender, **Aug. 14;** formally surrendered **Sept. 2.**

At **Potsdam Conference, July 17-Aug. 2,** leaders of U.S., USSR, and Britain agreed on disarmament of Germany, occupation zones, war crimes trials.

U.S. forces entered **Korea** south of 38th parallel to displace Japanese **Sept. 8.**

Gen. Douglas MacArthur took over supervision of Japan **Sept. 9.**

1946

Strike by 400,000 **mine workers** began **Apr. 1;** other industries followed.

Philippines given independence by U.S. **July 4.**

1947

Pres. Truman asked Congress to aid Greece and Turkey to combat Communist terrorism (**Truman Doctrine), Mar. 12.** Approved **May 15.**

UN Security Council voted **Apr. 2** to place under **U.S. trusteeship** the Pacific islands formerly mandated to Japan.

Jackie Robinson joined the Brooklyn Dodgers **Apr. 11,** breaking the color barrier in major league baseball.

Taft-Hartley Labor Act curbing strikes was vetoed by Truman **June 20;** Congress overrode the veto.

The **Marshall Plan,** for U.S. aid to European countries, was proposed by Sec. of State George C. Marshall **June 5.** Congress authorized some $12 bil in next 4 years.

1948

USSR halted all surface traffic into W. Berlin, **June 23;** in response, U.S. and British troops launched an airlift. Soviet blockade halted **May 12, 1949;** airlift ended **Sept. 30.**

Organization of American States founded **Apr. 30.**

Alger Hiss indicted **Dec. 15** for perjury, after denying he had passed secret documents to Whittaker Chambers for transmission to a Communist spy ring. Convicted **Jan. 21, 1950.**

Pres. Truman elected Nov. 2, defeating Gov. Thomas E. Dewey in a historic upset.

Kinsey Report on sexuality in the human male published.

1949

NATO established **Aug. 24** by U.S., Canada, and 10 Western European nations, agreeing that an armed attack against one or more would be considered an attack against all.

Mrs. I. Toguri D'Aquino (**Tokyo Rose** of Japanese wartime broadcasts) was sentenced **Oct. 7** to 10 years in prison for treason. Paroled **1956,** pardoned **1977.**

Eleven leaders of **U.S. Communist Party** convicted **Oct. 14** of advocating violent overthrow of U.S. government; sentenced to prison. Supreme Court upheld convictions **1951.**

1950

Masked bandits robbed **Brink's, Inc.,** Boston express office, **Jan. 17** of $2.8 mil. Case solved **1956;** 8 sentenced to life.

Pres. Truman authorized production of the **H-bomb Jan. 31.**

North Korea forces invaded **South Korea June 25.** UN asked for troops to restore peace.

Truman ordered Air Force and Navy to Korea **June 27.** Truman approved ground forces, air strikes against North Korea **June 30.**

U.S. sent 35 military advisers to **South Vietnam June 27,** and agreed to aid anti-Communist government.

Army seized all railroads Aug. 27 on Truman's order to prevent a general strike; returned to owners in **1952.**

U.S. forces landed at Inchon Sept. 15; UN force took Pyongyang **Oct. 20,** reached China border **Nov. 20;** China sent troops across border **Nov. 26.**

Two members of **Puerto Rican nationalist** movement tried to kill Pres. Truman **Nov. 1.**

U.S. **Dec. 8** banned shipments to **Communist China** and to Asiatic ports trading with it.

1951

Sen. Estes Kefauver led Senate probe into organized crime.

Julius Rosenberg, his wife, **Ethel,** and Morton Sobell found guilty **Mar. 29** of conspiracy to commit wartime **espionage.** Rosenbergs executed **June 19, 1953.** Sobell sentenced to 30 years; released **1969.**

Gen. Douglas MacArthur removed from Korea command **Apr. 11** by Pres. Truman, for unauthorized policy statements.

Korea cease-fire talks began in July; lasted 2 years. **Fighting ended July 27, 1953.**

Tariff concessions by the U.S. to the Soviet Union, Communist China, and all Communist-dominated lands were suspended **Aug. 1.**

The **U.S., Australia,** and **New Zealand** signed a mutual security pact **Sept. 1.**

Transcontinental TV begun **Sept. 4** with Pres. Truman's address at Japanese Peace Treaty Conference in San Francisco.

Japanese peace treaty signed in San Francisco **Sept. 8** by U.S., Japan, and 47 other nations.

J. D. Salinger published *Catcher in the Rye.*

1952

U.S. **seizure of nation's steel mills** was ordered by Pres. Truman **Apr. 8** to avert a strike. Ruled illegal by Supreme Court **June 2.**

Peace contract between West Germany, U.S., Great Britain, and France was signed **May 26.**

The last racial and ethnic barriers to naturalization removed, **June 26-27,** with passage of **Immigration and Naturalization Act of 1952.**

First **hydrogen device** explosion **Nov. 1** in Pacific.

1953

Pres. Dwight D. Eisenhower announced **May 8** that U.S. had given France $60 mil for **Indochina War.** More aid was announced in **Sept.**

Korean War armistice signed **July 27.**

1954

Nautilus, first atomic-powered submarine, was launched at Groton, CT, **Jan. 21.**

Five members of Congress were **wounded** in the House **Mar. 1** by 4 **Puerto Rican independence supporters** who fired at random from a spectators' gallery.

Sen. Joseph McCarthy (R, WI) led televised hearings **Apr. 22-June 17** into alleged Communist influence in the Army.

Racial segregation in public schools unanimously ruled unconstitutional by Supreme Court **May 17,** in *Brown* v. *Board of Education of Topeka.*

Southeast Asia Treaty Organization (**SEATO**) formed by defense pact signed in Manila **Sept. 8** by U.S., Britain, France, Australia, New Zealand, Philippines, Pakistan, and Thailand.

Condemnation of **Sen. McCarthy** voted by Senate, 67-22, **Dec. 2** for contempt of Senate subcommittee, abuse of its members, insults to Senate during Army investigation hearings.

1955

U.S. agreed **Feb. 12** to help train **South Vietnamese** army.

Supreme Court ordered **"all deliberate speed"** in integration of public schools **May 31.**

A **summit meeting** of leaders of U.S., Britain, France, and USSR took place **July 18-23** in Geneva, Switzerland.

Rosa Parks refused **Dec. 1** to give her seat to a white man on a **bus in Montgomery, AL.** Bus segregation ordinance declared unconstitutional by a federal court following boycott and NAACP protest.

America's 2 largest labor organizations merged **Dec. 5,** creating the **AFL-CIO**.

1956

Massive resistance to Supreme Court desegregation rulings was called for **Mar. 12** by 101 Southern congressmen.

Federal-Aid **Highway Act** signed **June 29,** inaugurating interstate highway system.

First transatlantic **telephone cable** activated **Sept. 25.**

1957

Congress approved first **civil rights bill** for blacks since Reconstruction, **Apr. 29,** to protect voting rights.

National Guardsmen, called out by Arkansas Gov. Orval Faubus **Sept. 4,** barred 9 black students from entering all-white high school in **Little Rock.** Faubus complied **Sept. 21** with federal court order to remove Guardsmen, but the blacks were ordered to withdraw by local authorities. Pres. Eisenhower sent federal troops **Sept. 24** to enforce court order.

Jack Kerouac published *On the Road.*

1958

First U.S. **earth satellite** to go into orbit, **Explorer I,** launched by Army **Jan. 31** at Cape Canaveral, FL; discovered Van Allen radiation belt.

U.S. Marines sent to **Lebanon** to protect elected government from threatened overthrow **July-Oct.**

First domestic **jet airline** passenger service in U.S. opened by National Airlines **Dec. 10** between New York and Miami.

1959

Alaska admitted as 49th state **Jan. 3; Hawaii** admitted as 50th **Aug. 21.**

St. Lawrence Seaway opened **Apr. 25.**

Soviet **Premier Nikita Khrushchev** paid unprecedented visit to U.S. **Sept. 15-27;** made transcontinental tour.

1960

Sit-ins began **Feb. 1** when 4 black college students in Greensboro, NC, refused to move from a Woolworth lunch counter when denied service. By **Sept. 1961** more than 70,000 students, whites and blacks, had participated in sit-ins.

Congress approved a strong **voting rights act Apr. 21.**

A U.S. **U-2 reconnaissance plane** was shot down in the Soviet Union **May 1**; pilot Gary Powers captured. The incident led to cancellation of an imminent Paris summit conference.

Vice Pres. Richard Nixon and Sen. John F. Kennedy faced each other, **Sept. 26,** in the first in a series of televised **debates. Kennedy defeated Nixon** to win presidency, **Nov. 8.**

U.S. announced **Dec. 15** it backed rightist group in **Laos,** which took power the next day.

1961

U.S. severed diplomatic and consular relations with **Cuba Jan. 3,** after disputes over nationalizations of U.S. firms, U.S. military presence at Guantanamo base.

Invasion of Cuba's **"Bay of Pigs" Apr. 17** by Cuban exiles trained, armed, and directed by U.S., attempted to overthrow the regime of Premier Fidel Castro, unsuccessfully.

Peace Corps created by executive order, **Mar. 1.**

Commander Alan B. Shepard Jr. was rocketed from Cape Canaveral, FL, 116.5 mi above the earth in a Mercury capsule **May 5,** in first U.S.-crewed suborbital space flight.

"Freedom Rides" from Washington, DC, across deep South were launched **May 20** to **protest segregation** in interstate transportation.

1962

Lt. Col. John H. Glenn Jr. became first American in orbit **Feb. 20** when he circled the earth 3 times in the Mercury capsule **Friendship 7**.

Pres. John F. Kennedy said **Feb. 14** U.S. military advisers in Vietnam would fire if fired upon.

Supreme Court **Mar. 26** backed **"one-man one-vote"** apportionment of seats in state legislatures.

James Meredith became first black student at University of Mississippi **Oct. 1** after 3,000 troops put down riots.

A Soviet **offensive missile buildup in Cuba** was revealed **Oct. 22** by Pres. Kennedy, who ordered a naval and air quarantine on shipment of offensive military equipment to the island. He and Soviet Premier Khrushchev agreed **Oct. 28** on a formula to end the crisis. Kennedy announced **Nov. 2** that Soviet missile bases in Cuba were being dismantled.

1963

Supreme Court ruled **Mar. 18** that all **criminal defendants** must have counsel and that illegally acquired evidence was inadmissible in state as well as federal courts.

University of Alabama **desegregated** after Gov. **George Wallace** stepped aside when confronted by federally deployed National Guard troops, **June 11.**

Civil rights leader **Medgar Evers** was assassinated **June 12.**

Supreme Court ruled, 8-1, **June 17** that laws requiring **recitation of the Lord's Prayer** or Bible verses in public schools were unconstitutional.

A limited **nuclear test-ban treaty** was agreed upon **July 25** by the U.S., the Soviet Union, and Britain.

March on Washington by 200,000 persons **Aug. 28** in support of **black demands** for equal rights. Highlight was "I have a dream" speech by **Dr. Martin Luther King Jr.**

Baptist church in Birmingham, AL, bombed **Sept. 15** in racial violence; 4 black girls killed.

South Vietnam Pres. **Ngo Dinh Diem assassinated Nov. 2;** U.S. had earlier withdrawn support.

Pres. Kennedy shot and fatally wounded Nov. 22 as he rode in a motorcade through downtown Dallas, TX. Vice Pres. **Lyndon B. Johnson sworn in** as president. **Lee Harvey Oswald arrested** and charged with the murder; he was shot and fatally wounded **Nov. 24. Jack Ruby,** a nightclub owner, was convicted of Oswald's murder; he died in **1967,** while awaiting retrial following reversal of his conviction.

Betty Friedan's *Feminine Mystique* was published.

1964

Panama suspended relations with U.S. **Jan. 9** after riots. U.S. offered **Dec. 18** to negotiate a new canal treaty.

Supreme Court ordered **Feb. 17** that **congressional districts** have equal populations.

U.S. reported **May 27** it was sending military planes to **Laos.**

Omnibus **civil rights bill** cleared by Congress **July 2,** signed same day by Pres. Johnson, banning discrimination in voting, jobs, public accommodations.

Three **civil rights workers** were reported missing in Mississippi **June 22;** found buried **Aug. 4.** Twenty-one white men were arrested. On **Oct. 20, 1967,** an all-white federal jury convicted 7 of conspiracy in the slayings.

Bill establishing **Medicare,** government health insurance program for persons over 65, signed **July 30.**

U.S. Congress **Aug. 7** the passed **Tonkin Gulf Resolution,** authorizing presidential action in Vietnam, after N Vietnamese boats reportedly attacked 2 U.S. destroyers **Aug. 2.**

Congress approved **War on Poverty** bill **Aug. 11,** providing for a domestic Peace Corps **(VISTA)**, a **Job Corps,** and anti-poverty funding.

The **Warren Commission** released **Sept. 27** a report concluding that Lee Harvey Oswald was solely responsible for the Kennedy assassination.

Pres. Johnson was elected to a full term, **Nov. 3,** defeating Republican **Sen. Barry Goldwater** (AZ) in a landslide.

1965

Pres. Johnson in **Feb.** ordered continuous **bombing of North Vietnam** below 20th parallel.

Malcolm X assassinated **Feb. 21** at New York City rally.

Some 14,000 U.S. troops sent to **Dominican Republic** during civil war **Apr. 28.** All troops withdrawn by next year.

March from Selma to Montgomery, AL, **begun Mar. 21** by Rev. Martin Luther King Jr. to demand federal protection of **blacks' voting rights.** New **Voting Rights Act** signed **Aug. 6.**

Los Angeles riot by blacks living in **Watts** area resulted in 34 deaths and $200 mil in property damage **Aug. 11-16.**

National **immigration** quota system abolished **Oct. 3.**

Electric power failure blacked out most of northeastern U.S., parts of 2 Canadian provinces the night of **Nov. 9-10.**

1966

U.S. forces began firing into **Cambodia May 1.**

Bombing of Hanoi area of N Vietnam by U.S. planes began **June 29.** By **Dec. 31,** 385,300 U.S. troops were stationed in S Vietnam, plus 60,000 offshore and 33,000 in Thailand.

Medicare began **July 1.**

Edward Brooke (R, MA) elected **Nov. 8** as first black U.S. senator in 85 years.

1967

Black U.S. Rep. **Adam Clayton Powell** (D, NY) was denied **Mar. 1** his seat because of charges he misused government funds. Reelected in **1968,** he was seated, but fined $25,000 and stripped of his seniority.

Rachel Carson's *Silent Spring* launched environmentalist movement.

Pres. Johnson and Soviet Premier Aleksei Kosygin met **June 23 and 25** at **Glassboro State College** in NJ; agreed not to let any crisis push them into war.

The **25th Amendment,** providing for **presidential succession,** was ratified **Feb. 10.**

USS *Liberty,* an intelligence ship, was torpedoed by Israel in the Mediterranean, apparently by accident **June 8;** 34 killed.

Riots by blacks in **Newark, NJ, July 12-17** killed 26, injured 1,500; more than 1,000 arrested. In **Detroit, MI, July 23-30,** more than 40 died; 2,000 injured, 5,000 left homeless by rioting, looting, burning in city's black ghetto.

Thurgood Marshall was sworn in **Oct. 2** as first black U.S. Supreme Court Justice. **Carl B. Stokes** (D, Cleveland) and **Richard G. Hatcher** (D, Gary, IN) were elected first black mayors of major U.S. cities **Nov. 7.**

1968

USS *Pueblo* and 83-man crew seized in Sea of Japan **Jan. 23** by North Koreans; 82 men released **Dec. 22.**

"Tet offensive": Communist troops attacked Saigon, 30 province capitals **Jan. 30,** suffered heavy casualties.

Pres. Johnson **curbed bombing** of North Vietnam **Mar. 31.** Peace talks began in Paris **May 10.** All bombing of North halted **Oct. 31.**

Martin Luther King Jr., 39, assassinated Apr. 4 in Memphis, TN. **James Earl Ray,** an escaped convict, pleaded guilty to the slaying, was sentenced to 99 years.

Sen. Robert F. Kennedy (D, NY), 42, **shot June 5** in Hotel Ambassador, Los Angeles, after celebrating presidential primary victories. Died **June 6.** Sirhan Bishara Sirhan, convicted of murder, **1969;** death sentence commuted to life in prison, **1972.**

Vice Pres. **Hubert Humphrey nominated** for president by Democrats **at national convention in Chicago,** marked by clash between police and **antiwar protesters, Aug. 26-29.**

The Republican nominee, **Richard Nixon, won** the **presidency,** defeating Hubert Humphrey in a close race **Nov. 5.**

Rep. Shirley Chisholm (D, NY) became the first black woman elected to Congress.

1969

Expanded 4-party **Vietnam peace talks** began **Jan. 18.** U.S. force peaked at 543,400 in April. Withdrawal started **July 8.** Pres. Nixon set Vietnamization policy **Nov. 3.**

U.S. astronaut **Neil Armstrong,** commander of the *Apollo 11* mission, became the first person to **set foot on the moon, July 20;** followed by astronaut **Edwin Aldrin;** astronaut **Michael Collins** remained aboard command module.

Woodstock music festival near Bethel, NY, drew 300,000-500,000 people, **Aug. 15-18.**

Anti-Vietnam War **demonstrations peaked** in U.S.; some 250,000 marched in Washington, DC, **Nov. 15.**

Massacre of hundreds of civilians at **Mylai, South Vietnam,** in 1968 incident reported **Nov. 16.**

1970

United Mine Workers official **Joseph A. Yablonski,** his wife, and their daughter found shot to death **Jan. 5;** UMW chief W. A. (Tony) Boyle later convicted of the killing.

A federal jury **Feb. 18** found the **"Chicago 7"** antiwar activists innocent of conspiring to incite riots during the 1968 **Democratic National Convention.** However, 5 were convicted of crossing state lines with intent to incite riots.

Millions of Americans participated in antipollution demonstrations **Apr. 22** to mark the **first Earth Day.**

U.S. and South Vietnamese forces crossed **Cambodian** borders **Apr. 30** to get at enemy bases. Four students were killed **May 4** at **Kent State** University in Ohio by National Guardsmen during a protest against the war.

Two **women generals,** the first in U.S. history, were named by Pres. Nixon **May 15.**

A **postal reform** measure was signed **Aug. 12,** creating an independent U.S. Postal Service.

1971

Charles Manson and 3 of his cult followers were found guilty **Jan. 25** of first-degree murder in **1969** slaying of actress Sharon Tate and 6 others.

The 26th Amendment, lowering the **voting age to 18** in all elections, was ratified **June 30.**

A court-martial jury **Mar. 29** convicted **Lt. William L. Calley Jr.** of premeditated murder of 22 South Vietnamese at My Lai on **Mar. 16, 1968.** He was sentenced to life imprisonment **Mar. 31.** Sentence was reduced to 20 years **Aug. 20.**

Publication of classified **Pentagon papers** on U.S. involvement in Vietnam was begun **June 13** by the *New York Times.* In a 6-3 vote, U.S. Supreme Court **June 30** upheld the right of the *Times* and the *Washington Post* to publish the documents.

U.S. bombers struck massively in North Vietnam for 5 days starting **Dec. 26** in retaliation for alleged violations of agreements reached prior to the 1968 bombing halt.

1972

Pres. Nixon arrived in **Beijing Feb. 21** for an 8-day visit to China, in what he called a "journey for peace."

By a vote of 84 to 8, the Senate, **Mar. 22,** approved banning **discrimination** on the basis of sex, and sent the measure to the states for ratification.

North Vietnamese forces launched the biggest attacks in 4 years across the demilitarized zone **Mar. 30.** The U.S. responded **Apr. 15** by resumption of bombing of Hanoi and Haiphong after a 4-year lull.

Pres. Nixon announced **May 8** the mining of **North Vietnam ports.** Last U.S. combat troops left **Aug. 11.**

Gov. George C. Wallace (AL), campaigning for the presidency at a Laurel, MD, shopping center **May 15, was shot** and seriously wounded. Arthur H. Bremer **convicted Aug. 4,** sentenced to 63 years for shooting Wallace and 3 bystanders.

In **first visit of a U.S. president to Moscow,** Pres. Nixon arrived **May 22** for a week of summit talks with Kremlin leaders that culminated in a landmark **strategic arms pact.**

Five men were arrested **June 17** for breaking into the offices of the Democratic National Committee in the **Watergate** office complex in Washington, DC.

Pres. **Nixon** was **reelected Nov. 7** in a landslide, carrying 49 states to defeat Democratic Sen. George McGovern (SD); he won 61% of the popular vote.

The **Dow Jones** industrial average closed above 1,000 for the first time, **Nov. 14.**

Full-scale **bombing of North Vietnam** resumed after Paris peace negotiations reached an impasse **Dec. 18.**

1973

Five of 7 defendants in **Watergate** break-in trial pleaded guilty **Jan. 11 and 15;** the other 2 were convicted **Jan. 30.**

In **Roe v. Wade,** Supreme Court ruled, 7-2, **Jan. 22,** that states may not ban **abortions** during **first 3 months of pregnancy** and may regulate, but may not ban, abortions during 2d trimester.

Four-party **Vietnam peace pacts** were signed in Paris **Jan. 27,** and North Vietnam released some 590 U.S. prisoners by **Apr. 1.** Last U.S. troops left **Mar. 29.**

End of the military **draft** announced **Jan. 27.**

Top **Nixon aides** H. R. Haldeman, John D. Ehrlichman, and John Dean and Attorney Gen. Richard Kleindienst **resigned Apr. 30,** amid charges of White House efforts to obstruct justice in the Watergate case.

John Dean, former Nixon counsel, told Senate hearings **June 25** that Nixon, his staff and campaign aides, and the Justice Department had conspired to cover up Watergate facts.

The U.S. officially ceased bombing in **Cambodia** at midnight **Aug. 14** in accord with a June congressional action.

Vice Pres. Spiro T. Agnew Oct. 10 resigned and pleaded no contest to a charge of tax evasion on payments made to him by contractors when he was governor of Maryland. **Gerald R. Ford Oct. 12** became **first appointed vice president** under the 25th Amendment; sworn in **Dec. 6.**

A total ban on **oil exports** to the U.S. was imposed by Arab oil-producing nations **Oct. 19-21** after the outbreak of an Arab-Israeli war. The ban was lifted **Mar. 18, 1974.**

Attorney Gen. Elliot Richardson resigned, and his deputy William D. Ruckelshaus and **Watergate Special Prosecutor Archibald Cox** were **fired** by Pres. Nixon **Oct. 20,** when Cox threatened to secure a judicial ruling that Nixon was violating a court order to give tapes to Judge John Sirica. **Leon Jaworski** named **Nov. 1** by the Nixon administration to succeed Cox.

Congress overrode **Nov. 7** Pres. Nixon's veto of the **war powers** bill, which curbed president's power to commit armed forces to hostilities abroad without congressional approval.

1974

Impeachment hearings opened **May 9** against Pres. Nixon by the House Judiciary Committee.

John D. Ehrlichman and 3 **White House "plumbers"** found guilty **July 12** of conspiring to violate the civil rights of Pentagon Papers leaker Daniel Ellsberg's psychiatrist by breaking into his office.

U.S. Supreme Court ruled, 8-0, **July 24** that Nixon had to turn over **64 tapes** of White House conversations.

House Judiciary Committee, in televised hearings **July 24-30,** recommended 3 **articles of impeachment** against Pres. Nixon. The first, voted 27-11 **July 27,** charged conspiracy to obstruct justice in the Watergate cover-up. The 2d, voted 28-10 **July 29,** charged abuses of power. The 3d, voted 21-17 **July 30,** charged defiance of committee subpoenas. The House voted **Aug. 20,** 412-3, to accept the committee report, which included the impeachment articles.

Pres. Nixon announced his resignation, **Aug. 8,** and **resigned Aug. 9;** his support in Congress had begun to collapse **Aug. 5,** after release of tapes implicating him in Watergate cover-up. **Vice Pres. Gerald R. Ford** was **sworn in Aug. 9** as 38th U.S. president.

A **pardon** to ex-Pres. Nixon for any federal crimes he committed while president issued by Pres. Ford **Sept. 8.**

1975

Found guilty of Watergate cover-up charges Jan. 1 were ex-Atty. Gen. John Mitchell and ex-presidential advisers H. R. Haldeman and John Ehrlichman.

U.S. launched **evacuation** of American and some South Vietnamese **from Saigon Apr. 29** as Communist forces completed takeover of South Vietnam; **South Vietnamese** government officially **surrendered Apr. 30.**

U.S. merchant ship ***Mayaguez*** and its crew of 39 were seized by Cambodian forces in Gulf of Siam **May 12.** In rescue operation, U.S. Marines attacked Tang Island, planes bombed air base; Cambodia surrendered ship and crew.

Congress voted $405 mil for **South Vietnam refugees May 16;** 140,000 were flown to the U.S.

Illegal CIA operations described by panel headed by Vice Pres. **Nelson Rockefeller June 10.**

Publishing heiress **Patricia (Patty) Hearst,** kidnapped **Feb. 5, 1974,** by "Symbionese Liberation Army" militants, was captured, in San Francisco **Sept. 18** with others. She was convicted **Mar. 20, 1976,** of bank robbery.

1976

U.S. celebrated **200th anniversary of independence July 4,** with festivals, parades, and New York City's Operation Sail, a gathering of tall ships from around the world.

"Legionnaire's disease" killed 29 persons who attended an American Legion convention **July 21-24** in Philadelphia.

Viking II set down on **Mars'** Utopia Plains **Sept. 3,** following the successful landing by *Viking I* **July 20.**

1977

Pres. Jimmy Carter **Jan. 21** pardoned most Vietnam War **draft evaders.**

Convicted murderer **Gary Gilmore executed** by a Utah firing squad **Jan. 17,** in the first exercise of capital punishment in the U.S. since **1967.**

Pres. Carter signed an act **Aug. 4** creating a new cabinet-level **Energy Department.**

1978

U.S. Senate voted **Apr. 18** to turn over **Panama Canal** to Panama Dec. 31, 1999; **Mar. 16** vote had given approval to a treaty guaranteeing the area's neutrality after the year 2000.

Californians, **June 6,** approved **Proposition 13,** a state constitutional amendment slashing property taxes.

U.S. Supreme Court, **June 28,** ruled against **racial quotas** in *Bakke* v. *University of California.*

1979

Partial meltdown released radioactive material **Mar. 28,** at nuclear reactor on **Three Mile Island** near Middletown, PA.

Federal government announced, **Nov. 1,** a $1.5 bil loan-guarantee plan to aid the ailing **Chrysler Corp.**

Some 90 people, including 63 Americans, **taken hostage, Nov. 4,** at **American embassy in Tehran,** Iran, by militant followers of **Ayatollah Khomeini.** He demanded return of former Shah Muhammad Reza Pahlavi, who was undergoing medical treatment in New York City.

1980

Pres. Carter announced, **Jan. 4, economic sanctions against the USSR,** in retaliation for Soviet invasion of Afghanistan. At Carter's request, **U.S. Olympic Committee** voted, **Apr. 12,** against U.S. participation in Moscow Summer Olympics.

Eight Americans killed and 5 wounded, **Apr. 24, in ill-fated** attempt to **rescue hostages** held by Iranian militants.

Mt. St. Helens, in Washington state, **erupted, May 18.** The blast, with others **May 25** and **June 12,** left 57 dead.

In a sweeping victory, **Nov. 4, Ronald Reagan** (R) was elected 40th president, defeating incumbent Pres. Carter. Republicans gained control of the Senate.

Former Beatle **John Lennon** was shot and **killed, Dec. 8,** in New York City.

1981

Minutes after Reagan's inauguration **Jan. 20,** the **52 Americans** held **hostage in Iran** for 444 days were **freed.**

Pres. Reagan was **shot and seriously wounded, Mar. 30,** in Washington, DC; also seriously wounded were a Secret Service agent, a policeman, and Press Sec. **James Brady. John W. Hinckley Jr.** arrested, found not guilty by reason of insanity in **1982,** and committed to mental institution.

World's first reusable spacecraft, the **space shuttle *Columbia*,** was sent into space, **Apr. 12.**

Congress, **July 29,** passed Pres. Reagan's **tax-cut legislation,** expected to save taxpayers $750 bil over 5 years.

Federal air traffic controllers, Aug. 3, began an illegal **nationwide strike.** Most defied a back-to-work order and were dismissed by Pres. Reagan **Aug. 5.**

In a 99-0 vote, the Senate confirmed, **Sept. 21,** appointment of **Sandra Day O'Connor** as **first woman justice** of U.S. Supreme Court.

1982

The 13-year-old lawsuit against **AT&T** by the **Justice Dept.** was settled **Jan. 8.** AT&T agreed to give up the 22 Bell System companies and was allowed to expand.

The Equal Rights Amendment was **defeated** after a 10-year struggle for ratification.

A retired dentist, **Dr. Barney B. Clark,** 61, became first recipient of a **permanent artificial heart, Dec. 2.**

1983

On **Apr. 20, Pres. Reagan** signed a compromise bipartisan bill designed to save **Social Security** from bankruptcy.

Sally Ride became the first American **woman** to travel in **space, June 18,** when the **space shuttle *Challenger*** was launched from Cape Canaveral, FL.

On **Sept. 1, a South Korean passenger jet** infringing on Soviet air space was **shot down;** 269 people were killed.

On **Oct. 23,** 241 **U.S. Marines and sailors** were killed in Lebanon when a TNT-laden suicide bomb blew up Marine headquarters at **Beirut** International Airport.

U.S. troops, with a small force from 6 **Caribbean** nations, invaded **Grenada Oct. 25.** After a few days, Grenadian militia and Cuban "construction workers" were overcome, U.S. citizens evacuated, and the **Marxist regime deposed.**

1984

The space shuttle *Challenger* was launched on its 4th trip into space, **Feb. 3.** On **Feb. 7,** Navy Capt. Bruce McCandless, followed by Army Lt. Colonel Robert Stewart, became **first humans to fly free of a spacecraft.**

On **May 7,** American **Vietnam war** veterans reached an out-of-court **settlement with 7 chemical companies** in a class-action suit over the herbicide **Agent Orange.**

Former Vice Pres. **Walter Mondale** won the **Democratic presidential nomination, June 6;** he chose **Rep. Geraldine Ferraro** (D, NY), as candidate for **vice president.**

Pres. **Reagan** was **reelected Nov. 6** in a Republican **landslide,** carrying 49 states for a record 525 electoral votes.

1985

"Live Aid," a rock concert broadcast around the world **July 13,** raised $70 mil for starving peoples of Africa.

On **June 14** a **TWA jet was seized** by terrorists after takeoff from Athens; 153 passengers and crew held hostage for 17 days; 1 U.S. serviceman killed.

On **Oct. 7, 4 Palestinian hijackers seized** Italian cruise ship **Achille Lauro** in the Mediterranean and held it hostage for 2 days; one American, Leon Klinghoffer, was killed.

1986

On **Jan. 20,** for the first time, the U.S. officially observed **Martin Luther King Jr. Day.**

Moments after liftoff, **Jan. 28,** the space shuttle **Challenger exploded, killing 6 astronauts and Christa McAuliffe,** a New Hampshire teacher, on board.

Congress, overriding Pres. Reagan's veto in **Sept.,** imposed **economic sanctions on South Africa.**

U.S. Senate confirmed, **Sept. 17,** Reagan's nomination of **William Rehnquist** as chief justice and **Antonin Scalia** as associate justice of the Supreme Court.

Press reports in early **Nov.** broke first news of the **Iran-contra scandal,** involving secret U.S. sale of arms to Iran.

Ivan Boesky, accused of insider trading, agreed, **Nov. 14,** to plead guilty to an unspecified criminal count.

1987

Pres. Reagan produced the nation's first **trillion-dollar budget, Jan. 5.**

Dow Jones closed above 2,000 for first time, **Jan. 8.**

An **Iraqi missile killed 37 sailors** on the frigate USS *Stark* in the Persian Gulf, **May 17.** Iraq called it an accident.

Public hearings by Senate and House committees investigating the **Iran-contra affair** were held **May-Aug.** Lt. Col. **Oliver North** said he had believed all his activities were authorized by his superiors. Pres. Reagan, **Aug. 12,** denied knowing of a diversion of funds to the contras.

Wall Street crashed, Oct. 19, with the Dow Jones plummeting a record 508 points.

Pres. Reagan and Soviet leader **Mikhail Gorbachev, Dec. 8,** signed a **pact to dismantle** all 1,752 **U.S. and** 859 **Soviet missiles** with a 300- to 3,400-mi. range.

1988

Nearly **1.4 mil illegal aliens** met **May 4** deadline for applying for **amnesty** under a new federal policy.

A missile, fired from **U.S. Navy warship *Vincennes,*** in the Persian Gulf, mistakenly struck and **destroyed** a commercial **Iranian airliner, July 3,** killing all 290.

George H. W. Bush, vice president under Reagan, was **elected** 41st U.S. **president, Nov. 8.** Bush decisively defeated the Democratic nominee, Gov. **Michael Dukakis** (MA).

Pan Am Flight 103 exploded and crashed into the town of **Lockerbie, Scotland, Dec. 21,** killing all 259 people aboard, as well as 11 people on the ground. British investigators said, **Dec. 28,** a powerful plastic explosive had destroyed the plane.

Drexel Burnham Lambert agreed, **Dec. 21, to plead guilty** to 6 violations of federal law, including insider trading, and **pay penalties of $650 mil,** the largest such settlement ever.

1989

Major oil spill occurred when the **Exxon Valdez** struck Bligh Reef in Alaska's Prince William Sound, **Mar. 24.**

Former National Security Council staff member **Oliver North** was convicted, **May 4,** on charges related to **Iran-contra** scandal. Conviction thrown out on appeal in **1991.**

A measure to **rescue the savings and loan industry** was signed into law, **Aug. 9,** by Pres. Bush.

Army Gen. Colin Powell was nominated **Aug. 10** by Pres. Bush, as **chairman of the Joint Chiefs of Staff;** he became the first black to hold the post.

Just before a World Series game, **Oct. 17,** an **earthquake struck the San Francisco Bay area,** causing 62 deaths.

L. Douglas Wilder (D) elected governor of Virginia, the **first U.S. black governor** since Reconstruction.

U.S. troops invaded Panama, Dec. 20, overthrowing the government of **Manuel Noriega.** Noriega, wanted by U.S. authorities on drug charges, surrendered **Jan. 3, 1990.**

1990

Pres. Bush signed **Americans With Disabilities Act** on **July 26,** barring discrimination against handicapped.

Justice William Brennan announced, **July 20,** his resignation from the U.S. Supreme Court; his replacement, **Judge David Souter,** was confirmed **Sept. 27.**

Operation Desert Shield forces left for **Saudi Arabia, Aug. 7,** to defend that country following the **invasion** of its neighbor **Kuwait by Iraq, Aug. 2.**

Pres. Bush signed, **Nov. 5,** a bill to **reduce budget deficits** $500 bil over 5 years, by spending curbs and tax hikes.

1991

The **U.S. and its allies defeated Iraq** in the **Persian Gulf War** and liberated Kuwait, which Iraq had overrun in Aug. **1990.** On **Jan. 17,** the allies launched a devastating **attack on Iraq from the air.** In a **ground war** starting **Feb. 24,** which lasted just 100 hours, the U.S.-led forces killed or captured many thousands of Iraqi soldiers and sent the rest into retreat before Pres. Bush ordered a cease-fire **Feb. 27.**

U.S. **House bank** ordered closed **Oct. 3** after revelations House members had written 8,331 bad checks.

The **Senate approved, Oct. 15, nomination of Clarence Thomas** to the Supreme Court, despite allegations of sexual harassment against him by **Anita Hill,** a former aide. He became the 2d African-American to serve on the Court, replacing retiring Justice **Thurgood Marshall,** the 1st black.

Charles Keating convicted of securities fraud **Dec. 4.**

1992

Riots swept South-Central **Los Angeles Apr. 29,** after **jury acquitted 4 white policemen** on all but one count in videotaped 1991 beating of black motorist **Rodney King.** Death toll in the L.A. violence was put at 52.

Bill Clinton (D) was **elected** 42d president, **Nov. 3,** defeating **Pres. Bush** (R) and independent **Ross Perot.**

A UN-sanctioned military force, led by U.S. troops, arrived in **Somalia Dec. 9.**

1993

A bomb exploded in a parking garage beneath the **World Trade Center** in New York City, **Feb. 26,** killing 6 people. Two Islamic militants were convicted in the bombing, **Nov. 12, 1997.** Four men were found guilty, **Mar. 4, 1994.**

Janet Reno became the first woman U.S. attorney general **Mar. 12.**

Four federal agents were killed, Feb. 28, during an unsuccessful raid on the **Branch Davidian compound near Waco, TX.** A 51-day siege of the compound by federal agents ended **Apr. 19,** when the compound **burned down,** leaving more than 70 cult members dead. 11 **cult** members were acquitted **Feb. 26, 1994** of charges in the deaths of the federal agents.

A federal jury, **Apr. 17,** found **2 Los Angeles police officers guilty** and 2 not guilty of violating the civil rights of motorist **Rodney King** in 1991 beating incident.

"The Great Flood of 1993" inundated 8 mil acres in 9 Midwestern states in summer, leaving 50 dead.

Pres. Clinton, **July 19,** announced a "don't ask, don't tell, don't pursue" policy for **homosexuals** in the U.S. military.

Vincent Foster, deputy White House counsel, found shot to death **July 20** in a N Virginia park, an apparent suicide.

Judge Ruth Bader Ginsburg was sworn in, **Aug. 10,** as 107th justice of the Supreme Court.

Pres. Clinton, **Aug. 10,** signed a measure designed to **cut federal budget deficits** $496 bil over 5 years, through spending cuts and new taxes.

The **"Brady Bill,"** a major gun-control measure, was signed into law by Pres. Clinton **Nov. 30.**

1994

North American Free Trade Agreement took effect **Jan. 1.**

Attorney Gen. Janet Reno **Jan. 20** appointed Robert Fiske independent counsel to probe **Whitewater affair;** under a court ruling he was replaced **Aug. 5** by **Kenneth Starr**. Congressional committees, **late July,** began Whitewater hearings.

Byron De La Beckwith convicted Feb. 5 of the 1963 murder of civil rights leader **Medgar Evers.**

Longtime CIA officer **Aldrich Ames** and his wife were **charged Feb. 21 with spying.** Under a plea bargain, he received life in prison, while she drew 63 months.

Major league **baseball players went on strike,** following **Aug. 11** games; strike ended **Apr. 25, 1995.**

Senate Majority Leader George Mitchell (D, ME), **Sept. 26,** dropped efforts to pass Clinton's **health-care reform** package.

1995

When the 104th Congress opened, **Jan. 4, Sen. Bob Dole** (R, KS) became **Senate majority leader** and **Rep. Newt Gingrich** (R, GA) was elected **House Speaker.** A bill to end Congress's exemption from federal labor laws, first in a series of measures in Republicans' **"Contract With America,"** cleared Congress **Jan. 17;** signed into law **Jan. 23.**

Clinton invoked emergency powers, **Jan. 31,** to extend a **$20 bil loan to** help **Mexico** avert financial collapse.

The last UN peacekeeping troops withdrew from **Somalia Feb. 28-Mar. 3,** with the aid of U.S. Marines. In **Haiti,** peacekeeping responsibilities were transferred from U.S. to UN forces **Mar. 31,** with the U.S. providing 2,400 soldiers.

A truck **bomb** exploded outside **a federal office building in Oklahoma City Apr. 19, killing 168** people in all.

The U.S. space shuttle *Atlantis* made the first in a series of planned **dockings with** the Russian space station **Mir, June 29-July 4.**

A U.S. **F-16 fighter jet** piloted by Air Force Capt. **Scott O'Grady** was **shot down over Bosnia and Herzegovina June 2**; O'Grady was **rescued** by U.S. Marines 6 days later.

The U.S. announced on **July 11** that it was reestablishing **diplomatic relations with Vietnam.**

Former football star **O. J. Simpson** was found **not guilty Oct. 3** of the **June 1994** murders of his former wife, Nicole Brown Simpson, and her friend Ronald Goldman.

Hundreds of thousands of African-American men participated in **"Million Man March"** and rally in Washington, DC, **Oct. 16,** organized by Rev. Louis Farrakhan.

The federal **55-mile-per-hour speed limit** was **repealed** by a measure signed **Nov. 28.**

After talks outside Dayton, OH, **warring parties in Bosnia and Herzegovina reached agreement Nov. 21** to end their conflict; treaty was signed **Dec. 14,** after which first of some 20,000 **U.S. peacekeeping troops** arrived in Bosnia.

Five Americans were among 7 **killed, Nov. 13,** when **2 bombs exploded** at a military post in **Riyadh, Saudi Arabia.**

A budget impasse between Congress and Pres. Clinton led to a partial **government shutdown** beginning **Nov. 14.** Operations resumed Nov. 20 under continuing resolutions.

1996

Long-sought records released by White House **Jan. 5** showed **Hillary Rodham Clinton** did 60 hours of work for an S&L linked to **Whitewater** scandal. Responding to a subpoena, she testified **Jan. 26** before a grand jury.

Senate, **Jan. 26,** approved, 87–4, the Second Strategic Arms Reduction Treaty.

On **Feb. 24 Cuban jets shot down** 2 unarmed planes owned by a Cuban exile organization; all 4 persons on the planes were presumed killed.

John Salvi found guilty, **Mar. 18,** in the **1994 murder** of receptionists **at 2 abortion clinics** in Brookline, MA.

Congress, in late **Mar.,** approved a **"line item veto"** bill; struck down by the Supreme Court, **June 25, 1998.**

U.S. Commerce Sec. **Ron Brown** was killed **Apr. 3** in a plane crash in Croatia.

On **Apr. 10,** Pres. Clinton vetoed a bill that would have banned so-called **partial-birth abortions.**

James and Susan McDougal were convicted **May 28** of fraud and conspiracy. Arkansas Gov. **Jim Guy Tucker** was convicted of similar charges by the same jury.

The antitax **Freemen** surrendered to federal authorities **June 13** after an 81-day standoff at a ranch near Jordan, MT. Four of the group's leaders were convicted, **July 8, 1998,** of conspiring to defraud four banks.

Republicans **June 12** chose Sen. **Trent Lott** (MS) as new majority leader to replace Sen. **Robert Dole,** who resigned, **June 11,** to focus on his presidential campaign.

A **bomb** exploded at a military complex near Dhahran, **Saudi Arabia, June 25,** killing 19 American servicemen.

On **July 27 a bomb exploded** in Atlanta, GA, near the **Olympics;** one person was directly killed.

A wide-ranging **welfare reform bill** was signed into law **Aug. 22.**

Shannon Lucid, Sept. 26, completed a space voyage of 188 days, a record for women and for U.S. astronauts.

Pres. Clinton was reelected to 2d term, **Nov. 5.**

1997

Bombs were detonated at **2 abortion clinics** in Tulsa, OK, **Jan. 1,** in Atlanta on **Jan. 16,** and again at the first site in Tulsa on **Jan. 19.** Six people were injured.

Newt Gingrich (R, GA) was reelected Speaker of the U.S. House, **Jan. 7;** he was fined and reprimanded by colleagues for alleged misuse of tax-exempt donations.

Madeleine Albright was sworn in as secretary of state **Jan. 23,** becoming the first woman to head State Dept.

Harold Nicholson, a former CIA official, pleaded guilty, **Mar. 3,** to spying for Russia.

Thirty-nine members of the **Heaven's Gate religious cult** were found dead in a large house in Rancho Santa Fe, CA, **Mar. 26,** in an apparent mass suicide.

James McDougal, former partner with then-Gov. Bill Clinton in the Whitewater Development Corp., was sentenced **Apr. 14** to 3 years in prison for seeking to enrich himself with fraudulent loans. He died in prison, **Mar. 8, 1998.**

Timothy McVeigh was convicted of conspiracy and murder, **June 2,** in 1995 Oklahoma City bombing.

On **Oct. 27,** the **Dow Jones** fell 554.26 points, the largest 1-day point decline yet. On **Oct. 28,** the Dow rebounded, surging 337.17 points, the largest-yet single-day point advance.

Islamic militants **Ramzi Ahmed Yousef** and **Eyad Ismoil Yousef** were convicted, **Nov. 12,** in the 1993 bombing of the World Trade Center in New York City.

On **Nov. 19, Bobbi McCaughey,** 29, in Des Moines, IA, delivered the first set of live septuplets (4 boys, 3 girls) to survive more than a month.

Terry Nichols was convicted **Dec. 23** on some charges related to the 1995 **Oklahoma City bombing**.

1998

It was reported **Jan. 21** that Kenneth Starr, the independent counsel investigating the **Whitewater** scandal, had evidence of a sexual relationship between Pres. Clinton and onetime White House intern Monica Lewinsky. Clinton denied any affair.

Theodore Kaczynski, the so-called **Unabomber**, pleaded guilty **Jan. 22** in connection with California and New Jersey bombings that killed 3 people and injured 2.

The state of Texas, **Feb. 3,** executed its first female convict in 135 years—**Karla Faye Tucker**.

Mitchell Johnson, 13, and **Andrew Golden, 11,** were arrested, **Mar. 24,** for allegedly killing 4 schoolgirls and a teacher outside a **Jonesboro, AR,** middle school. They were later committed to a juvenile detention center.

A federal judge, **Apr. 1,** dismissed the sexual harassment suit brought against **Pres. Clinton** by **Paula Corbin Jones.**

On **Apr. 25,** First Lady **Hillary Rodham Clinton** provided videotaped testimony at the White House for the Little Rock, AR, grand jury in the **Whitewater** case.

The TV show **Seinfeld** aired its last episode **May 14.**

Kipland Kinkel, 15, was arrested in Springfield, OR, **May 21,** and charged with the shotgun **murder** of his parents and 2 students at his high school. He was sentenced **Nov. 10** to life in prison without parole.

Monica Lewinsky, July 28, agreed to testify before a Whitewater grand jury in return for immunity. On **Aug. 6,** she testified to having had a sexual relationship with **Pres. Clinton,** but said she was never asked to lie. In testimony provided to a grand jury, and in an address to the nation on **Aug. 17, Pres. Clinton** acknowledged having had an inappropriate relationship with Lewinsky. On **Sept. 9,** independent counsel **Kenneth Starr** sent to the House what he called "substantial and credible information that may constitute grounds" for impeaching Clinton.

Mark McGwire, **Sept. 8,** hit his 62d **home run** of the season, breaking **Roger Maris**'s season record.

On **Sept. 30,** Pres. Clinton announced a **budget surplus** of $70 billion for fiscal year 1998, the first since 1969.

The House Judiciary Committee, **Oct. 5,** voted 21-16 along party lines to recommend that the Clinton **impeachment** investigation proceed. The House concurred **Oct. 8,** voting 258-176; 31 Democrats voted yes.

Dr. Barnett Slepian, an obstetrician who performed abortions, was shot to death in his home near Buffalo, NY, **Oct. 23** by a sniper.

John Glenn, the first U.S. astronaut to orbit Earth, returned to space **Oct. 29-Nov. 7,** aboard the shuttle *Discovery.*

Pres. Clinton, Nov. 13, settled a suit by agreeing to pay $850,000 to **Paula Corbin Jones.** She alleged that he had made an unwanted sexual advance to her in 1991.

The country's 4 largest **tobacco** companies, in a settlement, **Nov. 23,** with 46 states, the District of Columbia, and 4 territories, agreed to pay $206 billion over 25 years to cover public health costs related to smoking.

The U.S. House of Representatives gave its approval, **Dec. 19,** to 2 articles of **impeachment** charging **Pres. Clinton** with grand jury perjury (228-206) and obstruction of justice (221-212) in connection with a coverup of his sexual relationship with former White House intern Monica Lewinsky. Two other impeachment articles failed. Clinton became only the 2d president in U.S. history to be impeached.

1999

J. Dennis Hastert (IL) was elected Speaker of the House for the 106th Congress, **Jan. 6.**

Pres. Clinton's impeachment trial—the 2d such trial in U.S. history—began in the GOP-controlled Senate **Jan. 7**. 13 House managers presented the case against him, aided by 3 videotaped depositions, with no live witnesses called. He was acquitted, **Feb. 12.** The grand jury perjury article failed, with 45 votes; the obstruction of justice article drew a 50-50 vote, with a two-thirds vote needed for conviction.

Amadou Diallo, 22, an unarmed African immigrant, was killed **Feb. 4** when struck by 19 of 41 shots fired at him by 4 New York City police officers.

John William King was convicted **Feb. 23,** and sentenced to death for the brutal **dragging death** of a black man, James Byrd Jr., in June 1998. A 2d white man was convicted, **Sept. 20**, and sentenced to death; a 3d was convicted and sentenced to life in prison **Nov. 18.**

Dr. Jack Kevorkian, who claimed he had helped 130 people take their own lives, was convicted of 2d-degree murder **Mar. 26** in one of the deaths. On **Apr. 13,** he was sentenced to 10 to 25 years in prison.

Russell Henderson, 21, pleaded guilty **Apr. 5** in the 1998 beating death of **Matthew Shepard,** an openly homosexual student at the Univ. of Wyoming. Aaron McKinney was convicted **Nov. 3** of 2d-degree murder, robbery, and kidnapping. Both were sentenced to life.

U.S. District Judge Susan Webber Wright, in Little Rock, AR, **Apr. 12,** held **Pres. Clinton** in civil **contempt of court** for testifying falsely about his relationship with Monica Lewinsky in a deposition.

Eric Harris, 18, and Dylan Klebold, 17, killed 12 fellow students and a teacher **Apr. 20** at **Columbine** High School in Littleton, CO, then shot themselves fatally. More than 30 people were wounded, some critically.

One NYC police officer pleaded guilty on 6 charges, **May 25,** and another was convicted on an assault charge, **June 8,** in connection with the 1997 torture and sodomizing of Haitian immigrant **Abner Louima** in a police station.

TX Gov. **George W. Bush, June 12,** announced his candidacy for the GOP presidential nomination; Vice Pres. **Al Gore, June 16,** announced his candidacy for the Democratic nomination.

John F. Kennedy Jr., son of the former president, died in a plane crash **July 16** along with his wife, Carolyn Bessette **Kennedy,** and his sister-in-law, Lauren Bessette.

On **July 23,** with the launch of the space shuttle *Columbia*, Air Force Col. **Eileen M. Collins** became the first woman to command a shuttle flight.

Mark Barton, a securities trader who had suffered heavy losses, shot 9 people to death and wounded 13 others in Atlanta, **July 29,** before taking his own life.

Former Sen. **Bill Bradley, Sept. 8,** declared his candidacy for the Democratic nomination for president. Sen. **John McCain** (R, AZ) declared his candidacy for the GOP nomination **Sept. 27**.

Hurricane Floyd, **Sept. 14**, caused some 3 million Americans to evacuate their homes. More than 40 people died in North Carolina, and at least 28 in other states.

Elizabeth Dole withdrew from the 2000 campaign for the White House, **Oct. 20.** On **Oct. 25, Pat Buchanan** resigned from the GOP and announced he would seek the Reform Party nomination for president.

First Lady Hillary Rodham Clinton announced, **Nov. 23**, that she would run for a U.S. Senate seat from New York. She won it a year later.

The **Dow Jones** Industrial Average closed the year at a **record** level of 11497.12 — 25.2% above the 1998 close.

2000

Across the U.S., midnight **celebrations** marked the changeover to the year 2000 on **Jan. 1**.

America Online Inc. announced **Jan. 10** that it would buy **Time Warner Inc.**, in the largest merger to date. The Federal Trade Commission approved the merger **Dec. 14**.

In the New Hampshire primary, **Feb. 1**, Sen. **McCain** received 48% of the Republican vote, upsetting the favorite, Gov. **Bush** (30%). Among the Democrats, **Vice Pres. Gore** defeated former Sen. **Bradley**, 50% to 46%.

Pres. Clinton submitted his last federal budget—for the 2001 fiscal year—on **Feb. 7**, projecting a surplus of $184 billion, the 3d consecutive surplus and the highest ever.

Four white New York City police officers were acquitted on all charges, **Feb. 25**, in the 1999 shooting death of **Amadou Diallo**.

Vice Pres. **Gore** and Gov. **Bush** emerged as decisive winners of the presidential nominations, **Mar. 7**, when 15 states held primaries or caucuses. Former Sen. **Bradley** withdrew **Mar. 9** and Sen. **McCain** suspended his campaign.

A U.S. District Court, **Apr. 3**, found **Microsoft** Corp. guilty of antitrust volations, and ordered, **June 7**, that it split into 2 separate companies.

19 Marines died when an Osprey aircraft crashed at Marana, AZ, **Apr. 8**. 4 more were killed in another Osprey crash, **Dec. 2**.

The **Vermont** legislature, **Apr. 19-25,** approved a bill granting homosexual couples, under civil union status, the same legal rights married heterosexuals have.

A **forest fire** started, **May 4**, by the National Park Service as a preventative measure against wildfires caused widespread damage in New Mexico and nearly reached a facility at Los Alamos National Laboratory that stored plutonium.

Hundreds of thousands protested gun violence in Washington, DC, on Mother's Day, **May 14**, at **"The Million Mom March.**"

The **Federal Reserve Board, May 16**, raised interest rates for the 6th time since June 1999.

The U.S. House, **May 24**, voted, 237-197, to grant permanent "normal trade relations" with **China**. The Senate consented **Sept. 19** in a vote of 83-15.

Gasoline prices rose across the U.S.; in the Midwest the price of a gallon reached $2.13 by early June.

On **June 25**, longtime consumer advocate **Ralph Nader** received the Green Party nomination for president.

A team of scientists from the U.S. and another from Great Britain announced jointly, **June 26** that they had determined the structure of the **human genome.**

Following a bitter legal controversy, 6-year-old Cuban **Elián González** returned to Cuba **June 28**, 7 months after he was rescued from a boatwreck off the coast of Florida. .

The Justice Dept., **July 21**, cleared U.S. agents of any wrongdoing in a 1993 assault on the compound of the Branch Davidian religious sect in **Waco,** TX.

The Episcopal Church and the Evangelical Lutheran Church in America agreed **July 8** to recognize each others' sacraments and share clergy and resources.

On **Aug. 2** the Republican Party, at its convention in Philadelphia, nominated Gov. **George W. Bush** (TX) for president and former U.S. Rep. and Defense Sec. **Richard Cheney** (WY) for vice president.

Bridgestone/Firestone, Inc. announced **Aug. 9** it was recalling 6.5 million tires after widespread reports of blowouts.

At a convention in Long Beach, CA, **Aug. 10-12**, the **Reform Party** split into 2 factions that nominated different candidates. The Federal Election Commission voted 5-1, **Sept. 12**, that **Pat Buchanan** was the legitimate nominee.

Delegates to the Democratic National Convention in Los Angeles, **Aug. 14-17,** nominated Vice Pres. **Al Gore** and Sen. **Joseph Lieberman** for president and vice president.

Gov. Marc Racicot (R) declared all of Montana a disaster area, **Aug. 16**. Fires had burned more than a million acres in 13 Western states since late July.

The government dropped 58 of 59 felony counts against imprisoned scientist **Wen Ho Lee**, **Sept. 10**. He had been suspected of removing classified data from the Los Alamos National Laboratory.

Ending a 6-year investigation, independent counsel Robert Ray said **Sept. 20** that there was **insufficient evidence** to establish criminal wrongdoing by Pres. Clinton or First Lady Hillary Rodham Clinton.

The U.S. Food and Drug Administration announced, **Sept. 28**, approval of **RU-486,** a pill that induces abortions.

On **Election night, Nov. 7**, the winner of Florida's 25 deciding electoral votes remained uncertain. The Florida Supreme Court, **Dec. 8**, ordered a manual recount of all ballots that did not have a vote for president recorded by machine. On **Dec. 12**, the U.S. Supreme Court reversed that decision. Vice Pres. Gore conceded the presidential election to Gov. **George W. Bush** (TX) in a televised address, **Dec. 13**.

The NASDAQ closed the year, **Dec. 29**, at 2,470.52, 39.3% below its value one year earlier. The Dow had also declined 6.2%.

The Mayflower Compact

The threat of James I to "harry them out of the land" sent a band of religious dissenters from England to Holland in 1608. They were known as Separatists because they wished to cut all ties with the established church. In 1620, some of them, known now as the Pilgrims, joined with a larger group in England to set sail on the *Mayflower* for the New World. A joint stock company financed their venture.

In November, they sighted Cape Cod and decided to land an exploring party at Plymouth Harbor. A rebellious group picked up at Southampton and London troubled the Pilgrim leaders, however, and to control their actions 41 Pilgrims drew up the Mayflower Compact and signed it before going ashore. The voluntary agreement to govern themselves was America's first written constitution. It reads as follows:

In the name of God, Amen. We, whose names are underwritten, the Loyal Subjects of our dread Sovereign Lord, King *James,* by the Grace of God, of *Great Britain, France and Ireland,* King, *Defender of the Faith,* etc.

Having undertaken for the Glory of God, and Advancement of the Christian Faith, and the Honour of our King and Country, a voyage to plant the first colony in the northern Parts of Virginia; do by these Presents, solemnly and mutually in the Presence of God and one of another, covenant and combine ourselves together into a civil Body Politick, for our better Ordering and Preservation, and Furtherance of the Ends aforesaid; And by Virtue hereof to enact, constitute, and frame, such just and equal Laws, Ordinances, Acts, Constitutions and Offices, from time to time, as shall be thought most meet and convenient for the General good of the Colony; unto which we promise all due Submission and Obedience.

In Witness whereof we have hereunto subscribed our names at *Cape Cod* the eleventh of *November,* in the Reign of our Sovereign Lord, King *James* of *England, France* and *Ireland,* the eighteenth, and of *Scotland* the fifty-fourth. *Anno Domini, 1620.*

The Continental Congress: Meetings, Presidents

Meeting places	Dates of meetings	Congress presidents	Date elected
Philadelphia, PA	Sept. 5 to Oct. 26, 1774	Peyton Randolph, VA (1)	Sept. 5, 1774
"	"	Henry Middleton, SC	Oct. 22, 1774
Philadelphia, PA	May 10, 1775 to Dec. 12, 1776	Peyton Randolph, VA	May 10, 1775
"	"	John Hancock, MA	May 24, 1775
Baltimore, MD	Dec. 20, 1776 to Mar. 4, 1777	"	
Philadelphia, PA	Mar. 5 to Sept. 18, 1777	"	
Lancaster, PA	Sept. 27, 1777 (one day)	"	
York, PA	Sept. 30, 1777 to June 27, 1778	Henry Laurens, SC	Nov. 1, 1777 (4)
Philadelphia, PA	July 2, 1778 to June 21, 1783	John Jay, NY	Dec. 10, 1778
"	"	Samuel Huntington, CT	Sept. 28, 1779
"	"	Thomas McKean, DE	July 10, 1781
"	"	John Hanson, MD (2)	Nov. 5, 1781
		Elias Boudinot, NJ	Nov. 4, 1782
Princeton, NJ	June 30 to Nov. 4, 1783	Thomas Mifflin, PA	Nov. 3, 1783
Annapolis, MD	Nov. 26, 1783 to June 3, 1784	"	
Trenton, NJ	Nov. 1 to Dec. 24, 1784	Richard Henry Lee, VA	Nov. 30, 1784
New York City, NY	Jan. 11 to Nov. 4, 1785	"	
"	Nov. 7, 1785 to Nov. 3, 1786	John Hancock, MA (3)	Nov. 23, 1785
"	"	Nathaniel Gorham, MA	June 6, 1786
"	Nov. 6, 1786 to Oct. 30, 1787	Arthur St. Clair, PA	Feb. 2, 1787
"	Nov. 5, 1787 to Oct. 21, 1788	Cyrus Griffin, VA	Jan. 22, 1788
"	Nov. 3, 1788 to Mar. 2, 1789	"	

(1) Resigned Oct. 22, 1774. (2) Titled "President of the United States in Congress Assembled," John Hanson is considered by some the first U.S. president because he was the first to serve under the Articles of Confederation. He was, however, little more than presiding officer of the Congress, which retained full executive power. He could be considered the head of government, but not head of state. (3) Elected Nov. 1785, meetings held Nov. 1785-Nov. 1786; resigned May 29, 1786, without having served, because of illness. (4) Articles of Confederation agreed upon, Nov. 15, 1777; last ratification from Maryland, Mar. 1, 1781.

Patrick Henry's Speech to the Virginia Convention

The following is an excerpt from Patrick Henry's speech to the Virginia Convention on Mar. 23, 1775:

Gentlemen may cry, peace, peace—but there is no peace. The war is actually begun! The next gale that sweeps from the north will bring to our ears the clash of resounding arms! Our brethren are already in the field! Why stand we here idle? What is it that gentlemen wish? What would they have? Is life so dear, or peace so sweet, as to be purchased at the price of chains and slavery? Forbid it, Almighty God! I know not what course others may take; but as for me, give me liberty, or give me death!

How the Declaration of Independence Was Adopted

On June 7, 1776, Richard Henry Lee, who had issued the first call for a congress of the colonies, introduced in the Continental Congress at Philadelphia a resolution declaring "that these United Colonies are, and of right ought to be, free and independent states, that they are absolved from all allegiance to the British Crown, and that all political connection between them and the state of Great Britain is, and ought to be, totally dissolved."

The resolution, seconded by John Adams on behalf of the Massachusetts delegation, came up again on June 10 when a committee of 5, headed by Thomas Jefferson, was appointed to express the purpose of the resolution in a declaration of independence. The others on the committee were John Adams, Benjamin Franklin, Robert R. Livingston, and Roger Sherman.

Drafting the Declaration was assigned to Jefferson, who worked on a portable desk of his own construction in a room at Market and 7th Sts. The committee reported the result on June 28, 1776. The members of the Congress suggested a number of changes, which Jefferson called "deplorable." They didn't approve Jefferson's arraignment of the British people and King George III for encouraging and fostering the slave trade, which Jefferson called "an execrable commerce." They made 86 changes, eliminating 480 words and leaving 1,337. In the final form, capitalization was erratic. Jefferson had written that men were endowed with "inalienable" rights; in the final copy it came out as "unalienable" and has been thus ever since.

The Lee-Adams resolution of independence was adopted by 12 yeas on July 2—the actual date of the act of independence. The Declaration, which explains the act, was adopted July 4, in the evening.

After the Declaration was adopted, July 4, 1776, it was turned over to John Dunlap, printer, to be printed on broadsides. The original copy was lost and one of his broadsides was attached to a page in the journal of the Congress. It was read aloud July 8 in Philadelphia, PA, Easton, PA, and Trenton, NJ. On July 9 at 6 PM it was read by order of Gen. George Washington to the troops assembled on the Common in New York City (City Hall Park).

The Continental Congress of July 19, 1776, adopted the following resolution:

"Resolved, That the Declaration passed on the 4th, be fairly engrossed on parchment with the title and stile of 'The Unanimous Declaration of the thirteen United States of America' and that the same, when engrossed, be signed by every member of Congress."

Not all delegates who signed the engrossed Declaration were present on July 4. Robert Morris (PA), William Williams (CT), and Samuel Chase (MD) signed on Aug. 2; Oliver Wolcott (CT), George Wythe (VA), Richard Henry Lee (VA), and Elbridge Gerry (MA) signed in August and September; Matthew Thornton (NH) joined the Congress Nov. 4 and signed later. Thomas McKean (DE) rejoined Washington's army before signing and said later that he signed in 1781.

Charles Carroll of Carrollton was appointed a delegate by Maryland on July 4, 1776, presented his credentials July 18, and signed the engrossed Declaration on Aug. 2. Born Sept. 19, 1737, he was 95 years old and the last surviving signer when he died on Nov. 14, 1832.

Two Pennsylvania delegates who did not support the Declaration on July 4 were replaced.

The 4 New York delegates did not have authority from their state to vote on July 4. On July 9, the New York state convention authorized its delegates to approve the Declaration, and the Congress was so notified on July 15, 1776. The 4 signed the Declaration on Aug. 2.

The original engrossed Declaration is preserved in the National Archives Building in Washington.

Declaration of Independence

The Declaration of Independence was adopted by the Continental Congress in Philadelphia on July 4, 1776. John Hancock was president of the Congress, and Charles Thomson was secretary. A copy of the Declaration, engrossed on parchment, was signed by members of Congress on and after Aug. 2, 1776. On Jan. 18, 1777, Congress ordered that "an authenticated copy, with the names of the members of Congress subscribing the same, be sent to each of the United States, and that they be desired to have the same put upon record." Authenticated copies were printed in broadside form in Baltimore, where the Continental Congress was then in session. The following text is that of the original printed by John Dunlap at Philadelphia for the Continental Congress. The original is on display at the National Archives in Washington, DC.

IN CONGRESS, July 4, 1776.

A DECLARATION

By the REPRESENTATIVES of the

UNITED STATES OF AMERICA,

In GENERAL CONGRESS assembled

When in the Course of human Events, it becomes necessary for one People to dissolve the Political Bands which have connected them with another, and to assume among the Powers of the Earth, the separate and equal Station to which the Laws of Nature and of Nature's God entitle them, a decent Respect to the Opinions of Mankind requires that they should declare the causes which impel them to the Separation.

We hold these Truths to be self-evident, that all Men are created equal, that they are endowed by their Creator with certain unalienable Rights, that among these are Life, Liberty, and the Pursuit of Happiness—That to secure these Rights, Governments are instituted among Men, deriving their just Powers from the Consent of the Governed, that whenever any Form of Government becomes destructive of these Ends, it is the Right of the People to alter or to abolish it, and to institute new Government, laying its Foundation on such Principles, and organizing its Powers in such Form, as to them shall seem most likely to effect their Safety and Happiness. Prudence, indeed, will dictate that Governments long established should not be changed for light and transient Causes; and accordingly all Experience hath shewn, that Mankind are more disposed to suffer, while Evils are sufferable, than to right themselves by abolishing the Forms to which they are accustomed. But when a long Train of Abuses and Usurpations, pursuing invariably the same Object, evinces a Design to reduce them under absolute Despotism, it is their Right, it is their Duty, to throw off such Government, and to provide new Guards for their future Security. Such has been the patient Sufferance of these Colonies; and such is now the Necessity which constrains them to alter their former Systems of Government. The History of the present King of Great-Britain is a History of repeated Injuries and Usurpations, all having in direct Object the Establishment of an absolute Tyranny over these States. To prove this, let Facts be submitted to a candid World.

He has refused his Assent to Laws, the most wholesome and necessary for the public Good.

He has forbidden his Governors to pass Laws of immediate and pressing Importance, unless suspended in their Operation till his Assent should be obtained; and when so suspended, he has utterly neglected to attend to them.

He has refused to pass other Laws for the Accommodation of large Districts of People, unless those People would relinquish the Right of Representation in the Legislature, a Right inestimable to them, and formidable to Tyrants only.

He has called together Legislative Bodies at Places unusual, uncomfortable, and distant from the Depository of their Public Records, for the sole Purpose of fatiguing them into Compliance with his Measures.

He has dissolved Representative Houses repeatedly, for opposing with manly Firmness his Invasions on the Rights of the People.

He has refused for a long Time, after such Dissolutions, to cause others to be elected; whereby the Legislative Powers, incapable of Annihilation, have returned to the People at large for their exercise; the State remaining in the mean time exposed to all the Dangers of Invasion from without, and Convulsions within.

He has endeavoured to prevent the Population of these States; for that Purpose obstructing the Laws for Naturalization of Foreigners; refusing to pass others to encourage their Migrations hither, and raising the Conditions of new Appropriations of Lands.

He has obstructed the Administration of Justice, by refusing his Assent to Laws for establishing Judiciary Powers.

He has made Judges dependent on his Will alone, for the Tenure of their Offices, and the Amount and payment of their Salaries.

He has erected a Multitude of new Offices, and sent hither Swarms of Officers to harrass our People, and eat out their Substance.

He has kept among us, in Times of Peace, Standing Armies, without the consent of our Legislatures.

He has affected to render the Military independent of, and superior to the Civil Power.

He has combined with others to subject us to a Jurisdiction foreign to our Constitution, and unacknowledged by our Laws; giving his Assent to their Acts of pretended Legislation:

For quartering large Bodies of Armed Troops among us:

For protecting them, by a mock Trial, from Punishment for any Murders which they should commit on the Inhabitants of these States:

For cutting off our Trade with all Parts of the World:

For imposing Taxes on us without our Consent:

For depriving us, in many Cases, of the Benefits of Trial by Jury:

For transporting us beyond Seas to be tried for pretended Offences:

For abolishing the free System of English Laws in a neighbouring Province, establishing therein an arbitrary Government, and enlarging its Boundaries, so as to render it at once an Example and fit Instrument for introducing the same absolute Rule into these Colonies:

For taking away our Charters, abolishing our most valuable Laws, and altering fundamentally the Forms of our Governments:

For suspending our own Legislatures, and declaring themselves invested with Power to legislate for us in all Cases whatsoever.

He has abdicated Government here, by declaring us out of his Protection and waging War against us.

He has plundered our Seas, ravaged our Coasts, burnt our towns, and destroyed the Lives of our People.

He is, at this Time, transporting large Armies of foreign Mercenaries to complete the works of Death, Desolation, and Tyranny, already begun with circumstances of Cruelty and Perfidy, scarcely paralleled in the most barbarous Ages, and totally unworthy the Head of a civilized Nation.

He has constrained our fellow Citizens taken Captive on the high Seas to bear Arms against their Country, to become the Executioners of their Friends and Brethren, or to fall themselves by their Hands.

He has excited domestic Insurrections amongst us, and has endeavoured to bring on the Inhabitants of our Frontiers, the merciless Indian Savages, whose known Rule of Warfare, is an undistinguished Destruction, of all Ages, Sexes and Conditions.

In every stage of these Oppressions we have Petitioned for Redress in the most humble Terms: Our repeated Petitions have been answered only by repeated Injury. A Prince, whose Character is thus marked by every act which may define a Tyrant, is unfit to be the Ruler of a free People.

Nor have we been wanting in Attentions to our British Brethren. We have warned them from Time to Time of Attempts by their Legislature to extend an unwarrantable Jurisdiction over us. We have reminded them of the Circumstances of our Emigration and Settlement here. We have appealed to their native Justice and Magnanimity, and we have conjured them by the Ties of our common Kindred to disavow these Usurpations, which, would inevitably interrupt our Connections and Correspondence. They too have been deaf to the Voice of Justice and of Consanguinity. We must, therefore, acquiesce in the Necessity, which denounces our Separation, and hold them, as we hold the rest of Mankind, Enemies in War, in Peace, Friends.

We, therefore, the Representatives of the UNITED STATES OF AMERICA, in General Congress, Assembled, appealing to the Supreme Judge of the World for the Rectitude of our Intentions, do, in the Name, and by Authority of the good People of these Colonies, solemnly Publish and Declare, That these United Colonies are, and of Right ought to be, Free and Independent States; that they are absolved from all Allegiance to the British Crown, and that all political Connection between them and the State of Great-Britain, is and ought to be totally dissolved; and that as Free and Independent States, they have full Power to levy War, conclude Peace, contract Alliances, establish Commerce, and to do all other Acts and Things which Independent States may of right do. And for the support of this declaration, with a firm Reliance on the Protection of Divine Providence, we mutually pledge to each other our lives, our Fortunes, and our sacred Honor.

JOHN HANCOCK, President

Attest.
CHARLES THOMSON, Secretary.

Signers of the Declaration of Independence

Delegate (state)	Occupation	Birthplace	Born	Died
Adams, John (MA)	Lawyer	Braintree (Quincy), MA	Oct. 30, 1735	July 4, 1826
Adams, Samuel (MA)	Political leader	Boston, MA	Sept. 27, 1722	Oct. 2, 1803
Bartlett, Josiah (NH)	Physician, judge	Amesbury, MA	Nov. 21, 1729	May 19, 1795
Braxton, Carter (VA)	Farmer	Newington Plantation, VA	Sept. 10, 1736	Oct. 10, 1797
Carroll, Chas. of Carrollton (MD)	Lawyer	Annapolis, MD	Sept. 19, 1737	Nov. 14, 1832
Chase, Samuel (MD)	Judge	Princess Anne, MD	Apr. 17, 1741	June 19, 1811
Clark, Abraham (NJ)	Surveyor	Roselle, NJ	Feb. 15, 1726	Sept. 15, 1794
Clymer, George (PA)	Merchant	Philadelphia, PA	Mar. 16, 1739	Jan. 23, 1813
Ellery, William (RI)	Lawyer	Newport, RI	Dec. 22, 1727	Feb. 15, 1820
Floyd, William (NY)	Soldier	Brookhaven, NY	Dec. 17, 1734	Aug. 4, 1821
Franklin, Benjamin (PA)	Printer, publisher	Boston, MA	Jan. 17, 1706	Apr. 17, 1790
Gerry, Elbridge (MA)	Merchant	Marblehead, MA	July 17, 1744	Nov. 23, 1814
Gwinnett, Button (GA)	Merchant	Down Hatherly, England	c. 1735	May 19, 1777
Hall, Lyman (GA)	Physician	Wallingford, CT	Apr. 12, 1724	Oct. 19, 1790
Hancock, John (MA)	Merchant	Braintree (Quincy), MA	Jan. 12, 1737	Oct. 8, 1793
Harrison, Benjamin (VA)	Farmer	Berkeley, VA	Apr. 5, 1726	Apr. 24, 1791
Hart, John (NJ)	Farmer	Stonington, CT	c. 1711	May 11, 1779
Hewes, Joseph (NC)	Merchant	Princeton, NJ	Jan. 23, 1730	Nov. 10, 1779
Heyward, Thos. Jr. (SC)	Lawyer, farmer	St. Luke's Parish, SC	July 28, 1746	Mar. 6, 1809
Hooper, William (NC)	Lawyer	Boston, MA	June 28, 1742	Oct. 14, 1790
Hopkins, Stephen (RI)	Judge, educator	Providence, RI	Mar. 7, 1707	July 13, 1785
Hopkinson, Francis (NJ)	Judge, author	Philadelphia, PA	Sept. 21, 1737	May 9, 1791
Huntington, Samuel (CT)	Judge	Windham County, CT	July 3, 1731	Jan. 5, 1796
Jefferson, Thomas (VA)	Lawyer	Shadwell, VA	Apr. 13, 1743	July 4, 1826
Lee, Francis Lightfoot (VA)	Farmer	Westmoreland County, VA	Oct. 14, 1734	Jan. 11, 1797
Lee, Richard Henry (VA)	Farmer	Westmoreland County, VA	Jan. 20, 1732	June 19, 1794
Lewis, Francis (NY)	Merchant	Llandaff, Wales	Mar., 1713	Dec. 31, 1802
Livingston, Philip (NY)	Merchant	Albany, NY	Jan. 15, 1716	June 12, 1778
Lynch, Thomas Jr. (SC)	Farmer	Winyah, SC	Aug. 5, 1749	(at sea) 1779
McKean, Thomas (DE)	Lawyer	New London, PA	Mar. 19, 1734	June 24, 1817
Middleton, Arthur (SC)	Farmer	Charleston, SC	June 26, 1742	Jan. 1, 1787
Morris, Lewis (NY)	Farmer	Morrisania (Bronx County), NY	Apr. 8, 1726	Jan. 22, 1798
Morris, Robert (PA)	Merchant	Liverpool, England	Jan. 20, 1734	May 9, 1806
Morton, John (PA)	Judge	Ridley, PA	1724	Apr., 1777
Nelson, Thos. Jr. (VA)	Farmer	Yorktown, VA	Dec. 26, 1738	Jan. 4, 1789
Paca, William (MD)	Judge	Abingdon, MD	Oct. 31, 1740	Oct. 23, 1799
Paine, Robert Treat (MA)	Judge	Boston, MA	Mar. 11, 1731	May 12, 1814
Penn, John (NC)	Lawyer	Near Port Royal, VA	May 17, 1741	Sept. 14, 1788
Read, George (DE)	Judge	Near North East, MD	Sept. 18, 1733	Sept. 21, 1798
Rodney, Caesar (DE)	Judge	Dover, DE	Oct. 7, 1728	June 29, 1784
Ross, George (PA)	Judge	New Castle, DE	May 10, 1730	July 14, 1779
Rush, Benjamin (PA)	Physician	Byberry, PA (Philadelphia)	Dec. 24, 1745	Apr. 19, 1813
Rutledge, Edward (SC)	Lawyer	Charleston, SC	Nov. 23, 1749	Jan. 23, 1800
Sherman, Roger (CT)	Lawyer	Newton, MA	Apr. 19, 1721	July 23, 1793
Smith, James (PA)	Lawyer	Dublin, Ireland	c. 1719	July 11, 1806
Stockton, Richard (NJ)	Lawyer	Near Princeton, NJ	Oct. 1, 1730	Feb. 28, 1781
Stone, Thomas (MD)	Lawyer	Charles County, MD	1743	Oct. 5, 1787
Taylor, George (PA)	Ironmaster	Ireland	1716	Feb. 23, 1781
Thornton, Matthew (NH)	Physician	Ireland	1714	June 24, 1803
Walton, George (GA)	Judge	Prince Edward County, VA	1741	Feb. 2, 1804
Whipple, William (NH)	Merchant, judge	Kittery, ME	Jan. 14, 1730	Nov. 28, 1785
Williams, William (CT)	Merchant	Lebanon, CT	Apr. 23, 1731	Aug. 2, 1811
Wilson, James (PA)	Judge	Carskerdo, Scotland	Sept. 14, 1742	Aug. 28, 1798
Witherspoon, John (NJ)	Clergyman, educator	Gifford, Scotland	Feb. 5, 1723	Nov. 15, 1794
Wolcott, Oliver (CT)	Judge	Windsor, CT	Dec. 1, 1726	Dec. 1, 1797
Wythe, George (VA)	Lawyer	Elizabeth City Co. (Hampton), VA	1726	June 8, 1806

Origin of the Constitution

The War of Independence was conducted by delegates from the original 13 states, called the Congress of the United States of America and known as the Continental Congress. In 1777 the Congress submitted to the legislatures of the states the Articles of Confederation and Perpetual Union, which were ratified by New Hampshire, Massachusetts, Rhode Island, Connecticut, New York, New Jersey, Pennsylvania, Delaware, Virginia, North Carolina, South Carolina, and Georgia and finally, in 1781, by Maryland.

The first article read: "The stile of this confederacy shall be the United States of America." This did not signify a sovereign nation, because the states delegated only those powers they could not handle individually, such as to wage war, make treaties, and contract debts for general expenses (e.g. paying the army). Taxes for payment of such debts were levied by the individual states. The president signed himself "President of the United States in Congress assembled," but here the United States were considered in the plural, a cooperating group.

When the war was won, it became evident that a stronger federal union was needed. The Congress left the initiative to the legislatures. Virginia in Jan. 1786 appointed commissioners to meet with representatives of other states; delegates from Virginia, Delaware, New York, New Jersey, and Pennsylvania met at Annapolis. Alexander Hamilton prepared their call asking delegates from all states to meet in Philadelphia in May 1787 "to render the Constitution of the federal government adequate to the exigencies of the union." Congress endorsed the plan on Feb. 21, 1787. Delegates were appointed by all states except Rhode Island.

The convention met on May 14, 1787. George Washington was chosen president (presiding officer). The states certified 65 delegates, but 10 did not attend. The work was done by 55, not all of whom were present at all sessions. Of the 55 attending delegates, 16 failed to sign, and 39 actually signed Sept. 17, 1787, some with reservations. Some historians have said 74 delegates (9 more than the 65 actually certified) were named and 19 failed to attend. These 9 additional persons refused the appointment, were never delegates, and were never counted as absentees. Washington sent the Constitution to Congress, and that body, Sept. 28, 1787, ordered it sent to the legislatures, "in order to be submitted to a convention of delegates chosen in each state by the people thereof."

The Constitution was ratified by votes of state conventions as follows: Delaware, Dec. 7, 1787, unanimous; Pennsylvania, Dec. 12, 1787, 43 to 23; New Jersey, Dec. 18, 1787, unanimous; Georgia, Jan. 2, 1788, unanimous; Connecticut, Jan. 9, 1788, 128 to 40; Massachusetts, Feb. 6, 1788, 187 to 168; Maryland, Apr. 28, 1788, 63 to 11; South Carolina, May 23, 1788, 149 to 73; New Hampshire, June 21, 1788, 57 to 46; Virginia, June 25, 1788, 89 to 79; New York, July 26, 1788, 30 to 27. Nine states were needed to establish the operation of the Constitution "between the states so ratifying the same," and New Hampshire was the 9th state. The government did not declare the Constitution in effect until the first Wednesday in Mar. 1789, which was Mar. 4. After that, North Carolina ratified it on Nov. 21, 1789, 194 to 77; and Rhode Island, May 29, 1790, 34 to 32. Vermont in convention ratified it on Jan. 10, 1791, and by act of Congress approved on Feb. 18, 1791, was admitted into the Union as the 14th state, Mar. 4, 1791.

Constitution of the United States

The Original 7 Articles

The text of the Constitution given here (exception for Amendment XXVII) is taken from the pocket-size edition of the Constitution published by the U.S. Government Printing Office as a result of a U.S. House and Senate resolution to print the Constitution in its original form as amended through July 5, 1971. *Text in brackets* indicates that an item has been superseded or amended, or provides background information. **Text preceding** each article, section, or amendment is a brief summary, added by The World Almanac.

PREAMBLE

We, the People of the United States, in Order to form a more perfect Union, establish Justice, insure domestic Tranquility, provide for the common defence, promote the general Welfare, and secure the Blessings of Liberty to ourselves and our Posterity, do ordain and establish this Constitution for the United States of America.

ARTICLE I.

Section 1—Legislative powers; in whom vested:

All legislative Powers herein granted shall be vested in a Congress of the United States, which shall consist of a Senate and House of Representatives.

Section 2—House of Representatives, how and by whom chosen. Qualifications of a Representative. Representatives and direct taxes, how apportioned. Enumeration. Vacancies to be filled. Power of choosing officers, and of impeachment.

The House of Representatives shall be composed of Members chosen every second Year by the People of the several States, and the Electors in each State shall have the Qualifications requisite for Electors of the most numerous Branch of the State Legislature.

No person shall be a Representative who shall not have attained to the Age of twenty-five Years, and been seven Years a Citizen of the United States, and who shall not, when elected, be an Inhabitant of that State in which he shall be chosen.

[Representatives and direct taxes shall be apportioned among the several States which may be included within this Union, according to their respective Numbers, which shall be determined by adding to the whole Number of free Persons, including those bound to Service for a Term of Years, and excluding Indians not taxed, three-fifths of all other persons.] [The previous sentence was superseded by Amendment XIV, section 2.] The actual Enumeration shall be made within three Years after the first Meeting of the Congress of the United States, and within every subsequent Term of ten Years, in such Manner as they shall by Law direct. The Number of Representatives shall not exceed one for every thirty Thousand, but each State shall have at Least one Representative; and until such enumeration shall be made, the State of New Hampshire shall be entitled to chuse three, Massachusetts eight, Rhode-Island and Providence Plantations one, Connecticut five, New-York six, New Jersey four, Pennsylvania eight, Delaware one, Maryland six, Virginia ten, North Carolina five, South Carolina five, and Georgia three.

When vacancies happen in the Representation from any State, the Executive Authority thereof shall issue Writs of Election to fill such Vacancies.

The House of Representatives shall chuse their Speaker and other Officers; and shall have the sole Power of Impeachment.

Section 3—Senators, how and by whom chosen. How classified. Qualifications of a Senator. President of the Senate, his right to vote. President pro tem., and other officers of the Senate, how chosen. Power to try impeachments. When President is tried, Chief Justice to preside. Sentence.

The Senate of the United States shall be composed of two Senators from each State, *[chosen by the Legislature thereof] [the preceding five words were superseded by Amendment XVII, section 1]* for six Years; and each Senator shall have one Vote.

Immediately after they shall be assembled in Consequence of the first Election, they shall be divided as equally as may be into three Classes. The Seats of the Senators of the first Class shall be vacated at the Expiration of the second Year, of the second Class at the Expiration of the fourth Year, and of the third Class at the Expiration of the Sixth year, so that one-third may be chosen every second Year; *[and if Vacancies happen by Resignation, or otherwise, during the Recess of the Legislature of any State, the Executive thereof may make temporary Appointments until the next Meeting of the Legislature, which shall then fill such Vacancies.] [The words in parentheses were superseded by Amendment XVII, section 2.]*

No person shall be a Senator who shall not have attained to the Age of thirty Years, and been nine Years a Citizen of the United States, and who shall not, when elected, be an Inhabitant of that State for which he shall be chosen.

The Vice President of the United States shall be President of the Senate, but shall have no Vote, unless they be equally divided.

The Senate shall chuse their other Officers, and also a President pro tempore, in the absence of the Vice President, or when he shall exercise the Office of President of the United States.

The Senate shall have the sole Power to try all Impeachments. When sitting for that Purpose, they shall be on Oath or Affirmation. When the President of the United States is tried, the Chief Justice shall preside: And no Person shall be convicted without the Concurrence of two thirds of the Members present.

Judgment in Cases of Impeachment shall not extend further than to removal from Office, and disqualification to hold and enjoy any Office of honor, Trust or Profit under the United States: but the Party convicted shall nevertheless be liable and subject to Indictment, Trial, Judgment and Punishment, according to Law.

Section 4—Times, etc., of holding elections, how prescribed. One session each year.

The Times, Places and Manner of holding Elections for Senators and Representatives, shall be prescribed in each State by the Legislature thereof; but the Congress may at any time by Law make or alter such Regulations, except as to the Place of Chusing Senators.

The Congress shall assemble at least once in every Year, and such Meeting shall *[be on the first Monday in December,] [The words in parentheses were superseded by Amendment XX, section 2.]* unless they shall by Law appoint a different Day.

Section 5—Membership, quorum, adjournments, rules. Power to punish or expel. Journal. Time of adjournments, how limited, etc.

Each House shall be the Judge of the Elections, Returns and Qualifications of its own Members, and a Majority of each shall constitute a Quorum to do Business; but a smaller number may adjourn from day to day, and may be authorized to compel the Attendance of absent Members, in such manner, and under such Penalties as each House may provide.

Each House may determine the Rules of its Proceedings, punish its members for disorderly Behavior, and, with the Concurrence of two thirds, expel a Member.

Each House shall keep a Journal of its Proceedings, and from time to time publish the same, excepting such Parts as may in their Judgment require Secrecy; and the Yeas and Nays of the Members of either House on any question shall, at the Desire of one fifth of those Present, be entered on the Journal.

Neither House, during the Session of Congress, shall, without the Consent of the other, adjourn for more than three days, nor to any other Place than that in which the two Houses shall be sitting.

Section 6—Compensation, privileges, disqualifications in certain cases.

The Senators and Representatives shall receive a Compensation for their Services, to be ascertained by Law, and paid out of the Treasury of the United States. They shall in all Cases, except Treason, Felony and Breach of the Peace, be privileged from Arrest during their Attendance at the Session of their respective Houses, and in going to and returning from the same; and for any Speech or Debate in either House, they shall not be questioned in any other Place.

No Senator or Representative shall, during the Time for which he was elected, be appointed to any civil Office under the Authority of the United States, which shall have been created, or the Emoluments whereof shall have been encreased during such time; and no Person holding any Office under the United States, shall be a Member of either House during his Continuance in Office.

Section 7—House to originate all revenue bills. Veto. Bill may be passed by two-thirds of each House, notwithstanding, etc. Bill, not returned in ten days, to become a law. Provisions as to orders, concurrent resolutions, etc.

All bills for raising Revenue shall originate in the House of Representatives; but the Senate may propose or concur with Amendments as on other Bills.

Every Bill which shall have passed the House of Representatives and the Senate, shall, before it become a Law, be presented to the President of the United States; If he approve he shall sign it, but if not he shall return it, with his Objections to that House in which it shall have originated, who shall enter the Objections at large on their Journal, and proceed to reconsider it. If after such Reconsideration two thirds of that House shall agree to pass the Bill, it shall be sent, together with the Objections, to the other House, by which it shall likewise be reconsidered, and if approved by two thirds of that House, it shall become a Law. But in all such Cases the Votes of both Houses shall be determined by Yeas and Nays, and the Names of the Persons voting for and against the Bill shall be entered on the Journal of each House respectively. If any Bill shall not be returned by the President within ten Days (Sundays excepted) after it shall have been presented to him, the Same shall be a Law, in like Manner as if he had signed it, unless the Congress by their Adjournment prevent its Return, in which Case it shall not be a Law.

Every order, Resolution, or Vote to which the Concurrence of the Senate and House of Representatives may be necessary (except on a question of Adjournment) shall be presented to the President of the United States; and before the Same shall take Effect, shall be approved by him, or being disapproved by him, shall be repassed by two thirds of the Senate and House of Representatives, according to the Rules and Limitations prescribed in the Case of a Bill.

Section 8—Powers of Congress.

The Congress shall have Power To lay and collect Taxes, Duties, Imposts and Excises, to pay the Debts and provide for the common Defence and general Welfare of the United States; but all Duties, Imposts and Excises shall be uniform throughout the United States;

To borrow money on the credit of the United States;

To regulate Commerce with foreign Nations, and among the several States, and with the Indian Tribes;

To establish an uniform Rule of Naturalization, and uniform Laws on the subject of Bankruptcies throughout the United States;

To coin Money, regulate the Value thereof, and of foreign Coin, and fix the Standard of Weights and Measures;

To provide for the Punishment of counterfeiting the Securities and current Coin of the United States;

To establish Post Offices and post Roads;

To promote the Progress of Science and useful Arts, by securing for limited Times to Authors and Inventors the exclusive Right to their respective Writings and Discoveries;

To constitute Tribunals inferior to the supreme Court;

To define and punish Piracies and Felonies committed on the high Seas, and Offenses against the Law of Nations;

To declare War, grant Letters of Marque and Reprisal, and make Rules concerning Captures on Land and Water;

To raise and support Armies, but no Appropriation of Money to that Use shall be for a longer Term than two Years;

To provide and maintain a Navy;

To make Rules for the Government and Regulation of the land and naval Forces;

To provide for calling forth the Militia to execute the Laws of the Union, suppress Insurrections and repel Invasions;

To provide for organizing, arming, and disciplining the Militia, and for governing such Part of them as may be employed in the Service of the United States, reserving to the States respectively, the Appointment of the Officers, and the Authority of training the Militia according to the discipline prescribed by Congress;

To exercise exclusive Legislation in all Cases whatsoever, over such District (not exceeding ten Miles square) as may, by Cession of particular States, and the acceptance of Congress, become the Seat of the Government of the United States, and to exercise like Authority over all Places purchased by the Consent of the Legislature of the State in which the Same shall be, for the Erection of Forts, Magazines, Arsenals, dock-Yards, and other needful Buildings;—And

To make all Laws which shall be necessary and proper for carrying into Execution the foregoing Powers, and all other Powers vested by this Constitution in the Government of the United States, or in any Department or Officer thereof.

Section 9—Provision as to migration or importation of certain persons. Habeas corpus, bills of attainder, etc. Taxes, how apportioned. No export duty. No commercial preference. Money, how drawn from Treasury, etc. No titular nobility. Officers not to receive presents, etc.

The Migration or Importation of such Persons as any of the States now existing shall think proper to admit, shall not be prohibited by the Congress prior to the Year one thousand eight hundred and eight, but a tax or duty may be imposed on such Importation, not exceeding ten dollars for each Person.

The privilege of the Writ of Habeas Corpus shall not be suspended, unless when in Cases of Rebellion or Invasion the public Safety may require it.

No Bill of Attainder or ex post facto Law shall be passed.

No capitation, or other direct, Tax shall be laid, unless in Proportion to the Census or Enumeration herein before directed to be taken. *[Modified by Amendment XVI.]*

No Tax or Duty shall be laid on Articles exported from any State.

No Preference shall be given by any Regulation of Commerce or Revenue to the Ports of one State over those of another: nor shall Vessels bound to, or from, one State, be obliged to enter, clear, or pay Duties in another.

No Money shall be drawn from the Treasury, but in Consequence of Appropriations made by Law; and a regular Statement and Account of the Receipts and Expenditures of all public Money shall be published from time to time.

No Title of Nobility shall be granted by the United States: and no Person holding any Office of Profit or Trust under them, shall, without the Consent of the Congress, accept of any present, Emolument, Office, or Title, of any kind whatever, from any King, Prince, or foreign State.

Section 10—States prohibited from the exercise of certain powers.

No State shall enter into any Treaty, Alliance, or Confederation; grant Letters of Marque and Reprisal; coin Money; emit Bills of Credit; make any Thing but gold and silver Coin a Tender in Payment of Debts; pass any Bill of Attainder, ex post facto Law, or Law impairing the Obligation of Contracts, or grant any Title of Nobility.

No State shall, without the Consent of the Congress, lay any Imposts or Duties on Imports or Exports, except what may be absolutely necessary for executing its inspection Laws: and the net Produce of all Duties and Imposts, laid by any State on Imports or Exports, shall be for the Use of the Treasury of the United States; and all such Laws shall be subject to the Revision and Control of the Congress.

No State shall, without the Consent of Congress, lay any duty of Tonnage, keep Troops, or Ships of War in time of Peace, enter into any Agreement or Compact with another State, or with a foreign Power, or engage in War, unless actually invaded, or in such imminent Danger as will not admit of delay.

ARTICLE II.

Section 1—President: his term of office. Electors of President; number and how appointed. Electors to vote on same day. Qualification of President. On whom his duties devolve in case of his removal, death, etc. President's compensation. His oath of office.

The executive Power shall be vested in a President of the United States of America. He shall hold his Office during the Term of four Years, and, together with the Vice President, chosen for the same Term, be elected, as follows.

Each State shall appoint, in such Manner as the Legislature thereof may direct, a Number of Electors, equal to the whole Number of Senators and Representatives to which the State may be entitled in the Congress: but no Senator or Representative, or Person holding an Office of Trust or Profit under the United States, shall be appointed an Elector.

[The Electors shall meet in their respective States, and vote by Ballot for two persons, of whom one at least shall not be an Inhabitant of the same State with themselves. And they shall make a List of all the Persons voted for, and of the Number of Votes for each; which List they shall sign and certify, and transmit sealed to the Seat of the Government of the United States, directed to the President of the Senate. The President of the Senate shall, in the Presence of the Senate and House of Representatives, open all the Certificates, and the Votes shall then be counted. The Person having the greatest Number of Votes shall be the President, if such Number be a Majority of the whole Number of Electors appointed; and if there be more than one who have such Majority, and have an equal Number of Votes, then the House of Representatives shall immediately chuse by Ballot one of them for President; and if no Person have a Majority, then from the five highest on the List the said House shall in like Manner chuse the President. But in chusing the President, the Votes shall be taken by States, the Representation from each State having one Vote; a quorum for this Purpose shall consist of a Member or Members from two thirds of the States, and a Majority of all the States shall be necessary to a Choice. In every Case, after the Choice of the President, the Person having the greatest Number of Votes of the Electors shall be the Vice President. But if there should remain two or more who have equal Votes, the Senate shall chuse from them by Ballot the Vice-President.]

[This clause was superseded by Amendment XII.]

The Congress may detemine the Time of chusing the Electors, and the Day on which they shall give their Votes; which Day shall be the same throughout the United States.

No person except a natural born Citizen, or a Citizen of the United States, at the time of the Adoption of this Constitution, shall be eligible to the Office of President; neither shall any Person be eligible to that Office who shall not have attained to the Age of thirty-five Years, and been fourteen Years a Resident within the United States.

[For qualification of the Vice President, see Amendment XII.]

In Case of the Removal of the President from Office, or of his Death, Resignation, or Inability to discharge the Powers and Duties of the said Office, the same shall devolve on the Vice President, and the Congress may by Law, provide for the Case of Removal, Death, Resignation or Inability, both of the President and Vice President, declaring what Officer shall then act as President, and such Officer shall act accordingly, until the Disability be removed, or a President shall be elected.

[This clause has been modified by Amendments XX and XXV.]

The President shall, at stated Times, receive for his Services, a Compensation, which shall neither be encreased nor diminished during the Period for which he shall have been elected, and he shall not receive within that Period any other Emolument from the United States, or any of them.

Before he enter on the Execution of his Office, he shall take the following Oath or Affirmation:–"I do solemnly swear (or affirm) that I will faithfully execute the Office of President of the United States, and will to the best of my Ability, preserve, protect and defend the Constitution of the United States."

Section 2—President to be Commander-in-Chief. He may require opinions of cabinet officers, etc., may pardon. Treaty-making power. Nomination of certain officers. When President may fill vacancies.

The President shall be Commander in Chief of the Army and Navy of the United States, and of the Militia of the several States, when called into the actual Service of the United States; he may require the Opinion in writing, of the principal Officer in each of the executive Departments, upon any subject relating to the Duties of their respective Offices, and he shall have Power to Grant Reprieves and Pardons for Offenses against the United States, except in Cases of Impeachment.

He shall have Power, by and with the Advice and Consent of the Senate, to make Treaties, provided two-thirds of the Senators present concur; and he shall nominate, and by and with the Advice and Consent of the Senate, shall appoint Ambassadors, other public Ministers and Consuls, Judges of the supreme Court, and all other Officers of the United States, whose Appointments are not herein otherwise provided for, and which shall be established by Law: but the Congress may by Law vest the Appointment of such inferior Officers, as they think proper, in the President alone, in the Courts of Law, or in the Heads of Departments.

The President shall have Power to fill up all Vacancies that may happen during the Recess of the Senate, by granting Commissions which shall expire at the End of their next Session.

Section 3—President shall communicate to Congress. He may convene and adjourn Congress, in case of disagreement, etc. Shall receive ambassadors, execute laws, and commission officers.

He shall from time to time give to the Congress Information of the State of the Union, and recommend to their Consideration such Measures as he shall judge necessary and expedient; he may, on extraordinary Occasions, convene both Houses, or either of them, and in Case of Disagreement between them, with Respect to the Time of Adjournment, he may adjourn them to such Time as he shall think proper; he shall receive Ambassadors and other public Ministers; he shall take Care that the Laws be faithfully executed, and shall Commission all the Officers of the United States.

Section 4—All civil offices forfeited for certain crimes.

The President, Vice President and all civil Officers of the United States, shall be removed from Office on Impeachment for, and Conviction of, Treason, Bribery, or other high Crimes and Misdemeanors.

ARTICLE III.

Section 1—Judicial powers, Tenure. Compensation.

The judicial Power of the United States, shall be vested in one supreme Court, and in such inferior Courts as the Congress may from time to time ordain and establish. The Judges, both of the supreme and inferior Courts, shall hold their Offices during good Behaviour, and shall, at stated Times, receive for their Services, a Compensation, which shall not be diminished during their Continuance in Office.

Section 2—Judicial power; to what cases it extends. Original jurisdiction of Supreme Court; appellate jurisdiction. Trial by jury, etc. Trial, where.

The judicial Power shall extend to all Cases, in Law and Equity, arising under this Constitution, the Laws of the United States, and Treaties made, or which shall be made, under their Authority;–to all Cases affecting Ambassadors, other public Ministers and Consuls;–to all Cases of admiralty and maritime Jurisdiction;–to Controversies to which the United States shall be a Party;–to Controversies between two or more States;–between a State and Citizens of another State;–between Citizens of different States;–between Citizens of the same State claiming Lands under Grants of different States, and between a State, or the Citizens thereof, and foreign States, Citizens or Subjects.

[This section is modified by Amendment XI.]

In all Cases affecting Ambassadors, other public Ministers and Consuls, and those in which a State shall be Party, the supreme Court shall have original Jurisdiction. In all the other Cases before mentioned, the supreme Court shall have appellate Jurisdiction, both as to Law and Fact, with such Exceptions, and under such Regulations as the Congress shall make.

The trial of all Crimes, except in Cases of Impeachment, shall be by Jury; and such Trial shall be held in the State where the said Crimes shall have been committed; but when not committed within any State, the Trial shall be at such Place or Places as the Congress may by Law have directed.

Section 3—Treason Defined, Proof of, Punishment of.

Treason against the United States, shall consist only in levying War against them, or in adhering to their Enemies, giving them Aid and Comfort. No Person shall be convicted of Treason unless on the Testimony of two Witnesses to the same overt Act, or on Confession in open Court.

The Congress shall have Power to declare the Punishment of Treason, but no Attainder of Treason shall work Corruption of Blood, or Forfeiture except during the Life of the Person attainted.

ARTICLE IV.

Section 1—Each State to give credit to the public acts, etc., of every other State.

Full Faith and Credit shall be given in each State to the public Acts, Records, and judicial Proceedings of every other State. And the Congress may by general Laws prescribe the Manner in which such Acts, Records and Proceedings shall be proved, and the Effect thereof.

Section 2—Privileges of citizens of each State. Fugitives from justice to be delivered up. Persons held to service having escaped, to be delivered up.

The Citizens of each State shall be entitled to all Privileges and Immunities of Citizens in the several States.

A Person charged in any State with Treason, Felony, or other Crime, who shall flee from Justice, and be found in another State, shall on demand of the executive Authority of the State from which he fled, be delivered up, to be removed to the State having Jurisdiction of the Crime.

[No Person held to Service or Labour in one State, under the Laws thereof, escaping into another, shall, in Consequence of any Law or Regulation therein, be discharged from such Service or Labour, but shall be delivered up on Claim of the Party to whom such Service or Labour may be due.] [This clause was superseded by Amendment XIII.]

Section 3—Admission of new States. Power of Congress over territory and other property.

New States may be admitted by the Congress into this Union; but no new State shall be formed or erected within the Jurisdiction of any other State; nor any State be formed by the Junction of two or more States, or parts of States, without the Consent of the Legislatures of the States concerned as well as of the Congress.

The Congress shall have Power to dispose of and make all needful Rules and Regulations respecting the Territory or other Property belonging to the United States; and nothing in this Constitution shall be so construed as to Prejudice any Claims of the United States, or of any particular State.

Section 4—Republican form of government guaranteed. Each state to be protected.

The United States shall guarantee to every State in this Union a Republican Form of Government, and shall protect each of them against Invasion; and on Application of the Legislature, or of the Executive (when the Legislature cannot be convened) against domestic Violence.

ARTICLE V.

Constitution: how amended; proviso.

The Congress, whenever two-thirds of both Houses shall deem it necessary, shall propose Amendments to this Constitution, or, on the Application of the Legislatures of two-thirds of the several States, shall call a Convention for proposing Amendments, which, in either Case, shall be valid to all Intents and Purposes, as part of this Constitution, when ratified by the Legislatures of three-fourths of the several States, or by Conventions in three-fourths thereof, as the one or the other Mode of Ratification may be proposed by the Congress: Provided that no Amendment which may be made prior to the Year One thousand eight hundred and eight shall in any Manner affect the first and fourth Clauses in the Ninth Section of the first Article; and that no State, without its Consent, shall be deprived of its equal Suffrage in the Senate.

ARTICLE VI.

Certain debts, etc., declared valid. Supremacy of Constitution, treaties, and laws of the United States. Oath to support Constitution, by whom taken. No religious test.

All Debts contracted and Engagements entered into, before the Adoption of this Constitution, shall be as valid against the United States under this Constitution, as under the Confederation.

This Constitution, and the Laws of the United States which shall be made in Pursuance thereof; and all Treaties made, or which shall be made, under the Authority of the United States, shall be the supreme Law of the Land; and the Judges in every State shall be bound thereby, any Thing in the Constitution or Laws of any State to the Contrary notwithstanding.

The Senators and Representatives before mentioned, and the Members of the several State Legislatures, and all executive and judicial Officers, both of the United States and of the several States, shall be bound by Oath or Affirmation, to support this Constitution; but no religious Test shall ever be required as a Qualification to any Office or public Trust under the United States.

ARTICLE VII.

What ratification shall establish Constitution.

The Ratification of the Conventions of nine States shall be sufficient for the Establishment of this Constitution between the States so ratifying the Same.

Done in Convention by the Unanimous Consent of the States present the Seventeenth Day of September in the Year of our Lord one thousand seven hundred and Eighty seven and of the Independence of the United States of America the Twelfth.

In Witness whereof We have hereunto subscribed our Names.

Go WASHINGTON, Presidt and deputy from Virginia
New Hampshire—John Langdon, Nicholas Gilman
Massachusetts—Nathaniel Gorham, Rufus King
Connecticut—Wm. Saml. Johnson, Roger Sherman
New York—Alexander Hamilton
New Jersey—Wil: Livingston, David Brearley, Wm. Paterson, Jona: Dayton
Pennsylvania—B Franklin, Thomas Mifflin, Robt Morris, Geo. Clymer, Thos. FitzSimons, Jared Ingersoll, James Wilson, Gouv Morris
Delaware—Geo: Read, Gunning Bedford jun, John Dickinson, Richard Bassett, Jaco: Broom
Maryland—James McHenry, Dan of St Thos. Jenifer, Danl Carroll
Virginia—John Blair, James Madison Jr.
North Carolina—Wm. Blount, Rich'd Dobbs Spaight, Hu Williamson
South Carolina—J. Rutledge, Charles Cotesworth Pinckney, Charles Pinckney, Pierce Butler
Georgia—William Few, Abr Baldwin
Attest: William Jackson, Secretary.

Ten Original Amendments: The Bill of Rights

In force Dec. 15, 1791

[The First Congress, at its first session in the City of New York, Sept. 25, 1789, submitted to the states 12 amendments to clarify certain individual and state rights not named in the Constitution. They are generally called the Bill of Rights.

Influential in framing these amendments was the Declaration of Rights of Virginia, written by George Mason (1725-1792) in 1776. Mason, a Virginia delegate to the Constitutional Convention, did not sign the Constitution and opposed its ratification on the ground that it did not sufficiently oppose slavery or safeguard individual rights.

In the preamble to the resolution offering the proposed amendments, Congress said: "The conventions of a number of the States having at the time of their adopting the Constitution, expressed a desire, in order to prevent misconstruction or abuse of its powers, that further declaratory and restrictive clauses should be added, and as extending the ground of public confidence in the government will best insure the beneficent ends of its institution, be it resolved," etc.

Ten of these amendments, now commonly known as one to 10 inclusive, but originally 3 to 12 inclusive, were ratified by the states as follows: New Jersey, Nov. 20, 1789; Maryland, Dec. 19, 1789; North Carolina, Dec. 22, 1789; South Carolina, Jan. 19, 1790; New Hampshire, Jan. 25, 1790; Delaware, Jan. 28, 1790; New York, Feb. 27, 1790; Pennsylvania, Mar. 10, 1790; Rhode Island, June 7, 1790; Vermont, Nov. 3, 1791; Virginia, Dec. 15, 1791; Massachusetts, Mar. 2, 1939; Georgia, Mar. 18, 1939; Connecticut, Apr. 19, 1939. These original 10 ratified amendments follow as Amendments I to X inclusive.

Of the two original proposed amendments that were not ratified promptly by the necessary number of states, the first related to apportionment of Representatives; the second, relating to compensation of members of Congress, was ratified in 1992 and became Amendment 27.]

AMENDMENT I.

Religious establishment prohibited. Freedom of speech, of press, right to assemble and to petition.

Congress shall make no law respecting an establishment of religion, or prohibiting the free exercise thereof; or abridging the freedom of speech, or of the press; or the right of the people peaceably to assemble, and to petition the Government for a redress of grievances.

AMENDMENT II.

Right to keep and bear arms.

A well regulated Militia, being necessary to the security of a free State, the right of the people to keep and bear Arms, shall not be infringed.

AMENDMENT III.

Conditions for quarters for soldiers.

No Soldier shall, in time of peace be quartered in any house, without the consent of the Owner, nor in time of war, but in a manner to be prescribed by law.

AMENDMENT IV.

Protection from unreasonable search and seizure.

The right of the people to be secure in their persons, houses, papers, and effects, against unreasonable searches and seizures, shall not be violated, and no Warrants shall issue, but upon probable cause, supported by Oath or affirmation, and particularly describing the place to be searched, and the persons or things to be seized.

AMENDMENT V.

Provisions concerning prosecution and due process of law. Double jeopardy restriction. Private property not to be taken without compensation.

No person shall be held to answer for a capital, or otherwise infamous crime, unless on a presentment or indictment of a Grand Jury, except in cases arising in the land or naval forces, or in the Militia, when in actual service in time of War or public danger; nor shall any person be subject for the same offence to be twice put in jeopardy of life or limb; nor shall be compelled in any criminal case to be a witness against himself, nor be deprived of life, liberty, or property, without due process of law; nor shall private property be taken for public use, without just compensation.

AMENDMENT VI.

Right to speedy trial, witnesses, etc.

In all criminal prosecutions, the accused shall enjoy the right to a speedy and public trial, by an impartial jury of the State and district wherein the crime shall have been committed, which district shall have been previously ascertained by law, and to be informed of the nature and cause of the accusation; to be confronted with the witnesses against him; to have compulsory process for obtaining witnesses in his favor, and to have the Assistance of Counsel for his defence.

AMENDMENT VII.

Right of trial by jury.

In suits at common law, where the value in controversy shall exceed twenty dollars, the right of trial by jury shall be preserved, and no fact tried by a jury, shall be otherwise reexamined in any Court of the United States, than according to the rules of the common law.

AMENDMENT VIII.

Excessive bail or fines; cruel and unusual punishment.

Excessive bail shall not be required, nor excessive fines imposed, nor cruel and unusual punishments inflicted.

AMENDMENT IX.

Rule of construction of Constitution.

The enumeration in the Constitution, of certain rights, shall not be construed to deny or disparage others retained by the people.

AMENDMENT X.

Rights of States under Constitution.

The powers not delegated to the United States by the Constitution, nor prohibited by it to the States, are reserved to the States respectively, or to the people.

Amendments Since the Bill of Rights

AMENDMENT XI.

Judicial powers construed.

The Judicial power of the United States shall not be construed to extend to any suit in law or equity, commenced or prosecuted against one of the United States by Citizens of another State, or by Citizens or Subjects of any Foreign State.

[This amendment was proposed to the Legislatures of the several States by the Third Congress on March. 4, 1794, and was declared to have been ratified in a message from the President to Congress, dated Jan. 8, 1798.

[It was on Jan. 5, 1798, that Secretary of State Pickering received from 12 of the States authenticated ratifications, and informed President John Adams of that fact.

[As a result of later research in the Department of State, it is now established that Amendment XI became part of the Constitution on Feb. 7, 1795, for on that date it had been ratified by 12 States as follows:

[1. New York, Mar. 27, 1794. 2. Rhode Island, Mar. 31, 1794. 3. Connecticut, May 8, 1794. 4. New Hampshire, June 16, 1794. 5. Massachusetts, June 26, 1794. 6. Vermont, between Oct. 9, 1794, and Nov. 9, 1794. 7. Virginia, Nov. 18, 1794. 8. Georgia, Nov. 29, 1794. 9. Kentucky, Dec. 7, 1794. 10. Maryland, Dec. 26, 1794. 11. Delaware, Jan. 23, 1795. 12. North Carolina, Feb. 7, 1795.

[On June 1, 1796, more than a year after Amendment XI had become a part of the Constitution—but before anyone was officially aware of this—Tennessee had been admitted as a State; but not until Oct. 16, 1797, was a certified copy of the resolution of Congress proposing the amendment sent to the Governor of Tennessee, John Sevier, by Secretary of State Pickering, whose office was then at Trenton, New Jersey, because of the epidemic of yellow fever at Philadelphia; it seems, however, that the Legislature of Tennessee took no action on Amendment XI, owing doubtless to the fact that public announcement of its adoption was made soon thereafter.

[Besides the necessary 12 States, one other, South Carolina, ratified Amendment XI, but this action was not taken until Dec. 4, 1797; the two remaining States, New Jersey and Pennsylvania, failed to ratify.]

AMENDMENT XII.

Manner of choosing President and Vice-President.

[Proposed by Congress Dec. 9, 1803; ratified June 15, 1804.]

The Electors shall meet in their respective states and vote by ballot for President and Vice-President, one of whom, at least, shall not be an inhabitant of the same state with themselves; they shall name in their ballots the person voted for as President, and in distinct ballots the person voted for as Vice-President, and they shall make distinct lists of all persons voted for as President, and of all persons voted for as Vice-President, and of the number of votes for each, which lists they shall sign and certify, and transmit sealed to the seat of the government of the United States, directed to the President of the Senate;–The President of the Senate shall, in presence of the Senate and House of Representatives, open all the certificates and the votes shall then be counted;—The person having the greatest number of votes for President, shall be the President, if such number be a majority of the whole number of Electors appointed; and if no person have such majority, then from the persons having the highest numbers not exceeding three on the list of those voted for as President, the House of Representatives shall choose immediately, by ballot, the President. But in choosing the President, the votes shall be taken by states, the representation from each state having one vote; a quorum for this purpose shall consist of a member or members from two-thirds of the states, and a majority of all the states shall be necessary to a choice. *[And if the House of Representatives shall not choose a President whenever the right of choice shall devolve upon them, before the fourth day of March next following, then the Vice-President shall act as President, as in the case of the death or other constitutional disability of the President.] [The words in parentheses were superseded by Amendment XX, section 3.]* The person having the greatest number of votes as Vice-President, shall be the Vice-President, if such number be a majority of the whole number of Electors appointed, and if no person have a majority, then from the two highest numbers on the list, the Senate shall choose the Vice-President; a quorum for the purpose shall consist of two-thirds of the whole number of Senators, and a majority of the whole number shall be necessary to a choice. But no person constitutionally ineligible to the office of President shall be eligible to that of Vice-President of the United States.

THE RECONSTRUCTION AMENDMENTS

[Amendments XIII, XIV, and XV are commonly known as the Reconstruction Amendments, inasmuch as they followed the Civil War, and were drafted by Republicans who were bent on imposing their own policy of reconstruction on the South. Post-bellum legislatures there—Mississippi, South Carolina, Georgia, for example—had set up laws which, it was charged, were contrived to perpetuate Negro slavery under other names.]

AMENDMENT XIII.

Slavery abolished.

[Proposed by Congress Jan. 31, 1865; ratified Dec. 6, 1865. The amendment, when first proposed by a resolution in Congress, was passed by the Senate, 38 to 6, on Apr. 8, 1864, but was defeated in the House, 95 to 66 on June 15, 1864. On reconsideration by the House, on Jan. 31, 1865, the resolution passed, 119 to 56. It was approved by President Lincoln on Feb. 1, 1865, although the Supreme Court had decided in 1798 that the President has nothing to do with the proposing of amendments to the Constitution, or their adoption.]

1. Neither slavery nor involuntary servitude, except as a punishment for crime whereof the party shall have been duly convicted, shall exist within the United States, or any place subject to their jurisdiction.

2. Congress shall have power to enforce this article by appropriate legislation.

AMENDMENT XIV.

Citizenship rights not to be abridged.

[The following amendment was proposed to the Legislatures of the several states by the 39th Congress, June 13, 1866, ratified July 9, 1868, and declared to have been ratified in a proclamation by the Secretary of State, July 28, 1868.

[The 14th amendment was adopted only by virtue of ratification subsequent to earlier rejections. Newly constituted legislatures in both North Carolina and South Carolina (respectively July 4 and 9, 1868), ratified the proposed amendment, although earlier legislatures had rejected the proposal. The Secretary of State issued a proclamation, which, though doubtful as to the effect of attempted withdrawals by Ohio and New Jersey, entertained no doubt as to the validity of the ratification by North and South Carolina. The following day (July 21, 1868), Congress passed a resolution which declared the 14th Amendment to be a part of the Constitution and directed the Secretary of State so to promulgate it. The Secretary waited, however, until the newly constituted Legislature of Georgia had ratified the amendment, subsequent to an earlier rejection, before the promulgation of the ratification of the new amendment.]

1. All persons born or naturalized in the United States, and subject to the jurisdiction thereof, are citizens of the United States and of the State wherein they reside. No State shall make or enforce any law which shall abridge the privileges or immunities of citizens of the United States; nor shall any State deprive any person of life, liberty, or property, without due process of law; nor deny to any person within its jurisdiction the equal protection of the laws.

2. Representatives shall be apportioned among the several States according to their respective numbers, counting the whole number of persons in each State, excluding Indians not taxed. But when the right to vote at any election for the choice of electors for President and Vice-President of the United States, Representatives in Congress, the Executive and Judicial officers of a State, or the members of the Legislature thereof, is denied to any of the male inhabitants of such State, being twenty-one years of age, and citizens of the United States, or in any way abridged, except for participation in rebellion, or other crime, the basis of representation therein shall be reduced in the proportion which the number of such

male citizens shall bear to the whole number of male citizens twenty-one years of age in such State.

3. No person shall be a Senator or Representative in Congress, or elector of President and Vice-President, or hold any office, civil or military, under the United States, or under any State, who, having previously taken an oath, as a member of Congress, or as an officer of the United States, or as a member of any State legislature, or as an executive or judicial officer of any State, to support the Constitution of the United States, shall have engaged in insurrection or rebellion against the same, or given aid or comfort to the enemies thereof. But Congress may by a vote of two-thirds of each House, remove such disability.

4. The validity of the public debt of the United States, authorized by law, including debts incurred for payment of pensions and bounties for services in suppressing insurrection or rebellion, shall not be questioned. But neither the United States nor any State shall assume or pay any debt or obligation incurred in aid of insurrection or rebellion against the United States, or any claim for the loss or emancipation of any slave; but all such debts, obligations and claims shall be held illegal and void.

The Congress shall have power to enforce, by appropriate legislation, the provisions of this article.

AMENDMENT XV.

Race no bar to voting rights.

[The following amendment was proposed to the legislatures of the several States by the 40th Congress, Feb. 26, 1869, and ratified Feb. 8, 1870.]

1. The right of citizens of the United States to vote shall not be denied or abridged by the United States or by any State on account of race, color, or previous condition of servitude–

2. The Congress shall have power to enforce this article by appropriate legislation.

AMENDMENT XVI.

Income taxes authorized.

[Proposed by Congress July 12, 1909; ratified Feb. 3, 1913.]

The Congress shall have power to lay and collect taxes on incomes, from whatever source derived, without apportionment among the several States, and without regard to any census or enumeration.

AMENDMENT XVII.

United States Senators to be elected by direct popular vote.

[Proposed by Congress May 13, 1912; ratified Apr. 8, 1913.]

The Senate of the United States shall be composed of two Senators from each State, elected by the people thereof, for six years; and each Senator shall have one vote. The electors in each State shall have the qualifications requisite for electors of the most numerous branch of the State legislatures.

When vacancies happen in the representation of any State in the Senate, the executive authority of such State shall issue writs of election to fill such vacancies: *Provided,* That the legislature of any State may empower the executive thereof to make temporary appointments until the people fill the vacancies by election as the legislature may direct.

This amendment shall not be so construed as to affect the election or term of any Senator chosen before it becomes valid as part of the Constitution.

AMENDMENT XVIII.

Liquor prohibition amendment.

[Proposed by Congress Dec. 18, 1917; ratified Jan. 16, 1919. Repealed by Amendment XXI, effective Dec. 5, 1933.]

1. After one year from the ratification of this article the manufacture, sale, or transportation of intoxicating liquors within, the importation thereof into, or the exportation thereof from the United States and all territory subject to the jurisdiction thereof for beverage purposes is hereby prohibited.

2. The Congress and the several States shall have concurrent power to enforce this article by appropriate legislation.

3. This article shall be inoperative unless it shall have been ratified as an amendment to the Constitution by the legislatures of the several States as provided in the Constitution, within seven years from the date of the submission hereof to the States by the Congress.

[The total vote in the Senates of the various States was 1,310 for, 237 against—84.6% dry. In the lower houses of the States the vote was 3,782 for, 1,035 against—78.5% dry.

[The amendment ultimately was adopted by all the States except Connecticut and Rhode Island.]

AMENDMENT XIX.

Giving nationwide suffrage to women.

[Proposed by Congress June 4, 1919; ratified Aug. 18, 1920.]

The right of citizens of the United States to vote shall not be denied or abridged by the United States or by any State on account of sex.

Congress shall have power to enforce this Article by appropriate legislation.

AMENDMENT XX.

Terms of President and Vice President to begin on Jan. 20; those of Senators, Representatives, Jan. 3.

[Proposed by Congress Mar. 2, 1932; ratified Jan. 23, 1933.]

1. The terms of the President and Vice President shall end at noon on the 20th day of January, and the terms of Senators and Representatives at noon on the 3d day of January, of the years in which such terms would have ended if this article had not been ratified; and the terms of their successors shall then begin.

2. The Congress shall assemble at least once in every year, and such meeting shall begin at noon on the 3d day of January, unless they shall by law appoint a different day.

3. If, at the time fixed for the beginning of the term of the President, the President elect shall have died, the Vice President elect shall become President. If a President shall not have been chosen before the time fixed for the beginning of his term, or if the President elect shall have failed to qualify, then the Vice President elect shall act as President until a President shall have qualified; and the Congress may by law provide for the case wherein neither a President elect nor a Vice President elect shall have qualified, declaring who shall then act as President, or the manner in which one who is to act shall be selected, and such person shall act accordingly until a President or Vice President shall have qualified.

4. The Congress may by law provide for the case of the death of any of the persons from whom the House of Representatives may choose a President whenever the right of choice shall have devolved upon them, and for the case of the death of any of the persons from whom the Senate may choose a Vice President whenever the right of choice shall have devolved upon them.

5. Sections 1 and 2 shall take effect on the 15th day of October following the ratification of this article (Oct. 1933).

6. This article shall be inoperative unless it shall have been ratified as an amendment to the Constitution by the legislatures of three-fourths of the several States within seven years from the date of its submission.

AMENDMENT XXI.

Repeal of Amendment XVIII.

[Proposed by Congress Feb. 20, 1933; ratified Dec. 5, 1933.]

1. The eighteenth article of amendment to the Constitution of the United States is hereby repealed.

2. The transportation or importation into any State, Territory, or possession of the United States for delivery or use therein of intoxicating liquors, in violation of the laws thereof, is hereby prohibited.

3. This article shall be inoperative unless it shall have been ratified as an amendment to the Constitution by conventions in the several States, as provided in the Constitution, within seven years from the date of the submission hereof to the States by the Congress.

AMENDMENT XXII.

Limiting Presidential terms of office.

[Proposed by Congress Mar. 24, 1947; ratified Feb. 27, 1951.]

1. No person shall be elected to the office of the President more than twice, and no person who has held the office of

President, or acted as President, for more than two years of a term to which some other person was elected President shall be elected to the office of the President more than once. But this Article shall not apply to any person holding the office of President when this Article was proposed by the Congress, and shall not prevent any person who may be holding the office of President, or acting as President, during the term within which this Article becomes operative from holding the office of President or acting as President during the remainder of such term.

2. This article shall be inoperative unless it shall have been ratified as an amendment to the Constitution by the legislatures of three-fourths of the several States within seven years from the date of its submission to the States by the Congress.

AMENDMENT XXIII.

Presidential vote for District of Columbia.

[Proposed by Congress June 16, 1960; ratified Mar. 29, 1961.]

1. The District constituting the seat of Government of the United States shall appoint in such manner as the Congress may direct:

A number of electors of President and Vice President equal to the whole number of Senators and Representatives in Congress to which the District would be entitled if it were a State, but in no event more than the least populous State; they shall be in addition to those appointed by the States, but they shall be considered, for the purposes of the election of President and Vice President, to be electors appointed by a State; and they shall meet in the District and perform such duties as provided by the twelfth article of amendment.

2. The Congress shall have power to enforce this article by appropriate legislation.

AMENDMENT XXIV.

Barring poll tax in federal elections.

[Proposed by Congress Aug. 27, 1962; ratified Jan. 23, 1964.]

1. The right of citizens of the United States to vote in any primary or other election for President or Vice President, for electors for President or Vice President, or for Senator or Representative in Congress, shall not be denied or abridged by the United States or any State by reason of failure to pay any poll tax or other tax.

2. The Congress shall have power to enforce this article by appropriate legislation.

AMENDMENT XXV.

Presidential disability and succession.

[Proposed by Congress July 6, 1965; ratified Feb. 10, 1967.]

1. In case of the removal of the President from office or of his death or resignation, the Vice President shall become President.

2. Whenever there is a vacancy in the office of the Vice President, the President shall nominate a Vice President who shall take office upon confirmation by a majority vote of both houses of Congress.

3. Whenever the President transmits to the President pro tempore of the Senate and the Speaker of the House of Representatives his written declaration that he is unable to discharge the powers and duties of his office, and until he transmits to them a written declaration to the contrary, such powers and duties shall be discharged by the Vice President as Acting President.

4. Whenever the Vice President and a majority of either the principal officers of the executive departments or of such other body as Congress may by law provide, transmit to the President pro tempore of the Senate and the Speaker of the House of Representatives their written declaration that the President is unable to discharge the powers and duties of his office, the Vice President shall immediately assume the powers and duties of the office as Acting President.

Thereafter, when the President transmits to the President pro tempore of the Senate and the Speaker of the House of Representatives his written declaration that no inability exists, he shall resume the powers and duties of his office unless the Vice President and a majority of either the principal officers of the executive department or of such other body as Congress may by law provide, transmit within four days to the President pro tempore of the Senate and the Speaker of the House of Representatives their written declaration that the President is unable to discharge the powers and duties of his office. Thereupon Congress shall decide the issue, assembling within forty-eight hours for that purpose if not in session. If the Congress, within twenty-one days after receipt of the latter written declaration, or, if Congress is not in session, within twenty-one days after Congress is required to assemble, determines by two-thirds vote of both Houses that the President is unable to discharge the powers and duties of his office, the Vice President shall continue to discharge the same as Acting President; otherwise, the President shall resume the powers and duties of his office.

AMENDMENT XXVI.

Lowering voting age to 18 years.

[Proposed by Congress Mar. 23, 1971; ratified June 30, 1971.]

1. The right of citizens of the United States, who are eighteen years of age or older, to vote shall not be denied or abridged by the United States or by any State on account of age.

2. The Congress shall have the power to enforce this article by appropriate legislation.

AMENDMENT XXVII.

Congressional pay.

[Proposed by Congress Sept. 25, 1789; ratified May 7, 1992.]

No law, varying the compensation for the services of the Senators and Representatives, shall take effect, until an election of Representatives shall have intervened.

How a Bill Becomes a Law

A senator or representative introduces a bill in Congress by sending it to the clerk of the House or the Senate, who assigns it a number and title. This procedure is termed the first reading. The clerk then refers the bill to the appropriate committee of the Senate or House.

If the committee opposes the bill, it will table, or kill, it. Otherwise, the committee holds hearings to listen to opinions and facts offered by members and other interested people. The committee then debates the bill and possibly offers amendments. A vote is taken, and if favorable, the bill is sent back to the clerk of the House or Senate.

The clerk reads the bill to the house—the second reading. Members may then debate the bill and suggest amendments.

After debate and possibly amendment, the bill is given a third reading, simply of the title, and put to a voice or roll-call vote.

If passed, the bill goes to the other house, where it may be defeated or passed, with or without amendments. If defeated, the bill dies. If passed with amendments, a conference committee made up of members of both houses works out the differences and arrives at a compromise.

After passage of the final version by both houses, the bill is sent to the president. If the president signs it, the bill becomes a law. The president may, however, veto the bill by refusing to sign it and sending it back to the house where it originated, with reasons for the veto.

The president's objections are then read and debated, and a roll-call vote is taken. If the bill receives less than a two-thirds majority, it is defeated. If it receives at least two-thirds, it is sent to the other house. If that house also passes it by at least a two-thirds majority, the veto is overridden, and the bill becomes a law.

If the president neither signs nor vetoes the bill within 10 days—not including Sundays—it automatically becomes a law even without the president's signature. However, if Congress has adjourned within those 10 days, the bill is automatically killed; this indirect rejection is termed a pocket veto.

Note: Under "line-item veto" legislation effective Jan. 1, 1997, the president was authorized, under certain circumstances, to veto a bill in part, but the legislation was found unconstitutional by the Supreme Court, June 25, 1998.

Confederate States and Secession

The American Civil War (1861-65) grew out of sectional disputes over the continued existence of slavery in the South and the contention of Southern legislators that the states retained many rights, including the right to secede.

The war was not fought by state against state but by one federal regime against another, the Confederate government in Richmond assuming control over the economic, political, and military life of the South, under protest from Georgia and South Carolina.

South Carolina voted an ordinance of secession from the Union, repealing its 1788 ratification of the U.S. Constitution on Dec. 20, 1860, to take effect on Dec. 24. Other states seceded in 1861. Their votes in conventions were: Mississippi, Jan. 9, 84-15; Florida, Jan. 10, 62-7; Alabama, Jan. 11, 61-39; Georgia, Jan. 19, 208-89; Louisiana, Jan. 26, 113-17; Texas, Feb. 1, 166-7, ratified by popular vote on Feb. 23 (for 34,794, against 11,325); Virginia, Apr. 17, 88-55, ratified by popular vote on May 23 (for 128,884; against 32,134); Arkansas, May 6, 69-1; Tennessee, May 7, ratified by popular vote on June 8 (for 104,019, against 47,238); North Carolina, May 21.

Missouri Unionists stopped secession in conventions Feb. 28 and Mar. 9. The legislature condemned secession Mar. 7. Under the protection of Confederate troops, secessionist members of the legislature adopted a resolution of secession at Neosho, Oct. 31. The Confederate Congress seated the secessionists' representatives.

Kentucky did not secede, and its government remained Unionist. In a part of the state occupied by Confederate troops, Kentuckians approved secession, and the Confederate Congress admitted their representatives.

The Maryland legislature voted against secession Apr. 27, 53-13. Delaware did not secede. Western Virginia held conventions at Wheeling, named a pro-Union governor on June 11, 1861, and was admitted to the Union as West Virginia on June 20, 1863. Its constitution provided for gradual abolition of slavery.

Confederate Government

Forty-two delegates from South Carolina, Georgia, Alabama, Mississippi, Louisiana, and Florida met in convention at Montgomery, AL, on Feb. 4, 1861. They adopted a provisional constitution of the Confederate States of America and elected Jefferson Davis (MS) as provisional president and Alexander H. Stephens (GA) as provisional vice president.

A permanent constitution was adopted Mar. 11. It abolished the African slave trade, but it did not bar interstate commerce in slaves. On July 20 the Congress moved to Richmond, VA. Davis was elected president in October and was inaugurated on Feb. 22, 1862.

The Congress adopted a flag, consisting of a red field with a white stripe, and a blue jack with a circle of white stars. Later the more popular flag was the red field with blue diagonal crossbars that held 13 white stars, for the 11 states in the Confederacy plus Kentucky and Missouri.

Lincoln's Address at Gettysburg, 1863

Fourscore and seven years ago our fathers brought forth on this continent a new nation, conceived in liberty and dedicated to the proposition that all men are created equal.

Now we are engaged in a great civil war, testing whether that nation or any nation so conceived and so dedicated can long endure. We are met on a great battle field of that war. We have come to dedicate a portion of that field, as a final resting-place for those who here gave their lives that that nation might live. It is altogether fitting and proper that we should do this.

But, in a larger sense, we can not dedicate—we can not consecrate—we can not hallow—this ground. The brave men, living and dead, who struggled here, have consecrated it, far above our poor power to add or detract. The world will little note, nor long remember, what we say here, but it can never forget what they did here. It is for us the living, rather, to be dedicated here to the unfinished work which they who fought here have thus far so nobly advanced. It is rather for us to be here dedicated to the great task remaining before us—that from these honored dead we take increased devotion to that cause for which they gave the last full measure of devotion—that we here highly resolve that these dead shall not have died in vain—that this nation, under God, shall have a new birth of freedom—and that government of the people, by the people, for the people, shall not perish from the earth.

Selected Landmark Decisions of the U.S. Supreme Court

1803: Marbury v. Madison. The Court ruled that Congress exceeded its power in the Judiciary Act of 1789; the Court thus established its power to review acts of Congress and declare invalid those it found in conflict with the Constitution.

1819: McCulloch v. Maryland. The Court ruled that Congress had the authority to charter a national bank, under the Constitution's granting of the power to enact all laws "necessary and proper" to responsibilities of government.

1819: Trustees of Dartmouth College v. Woodward. The Court ruled that a state could not arbitrarily alter the terms of a college's contract. (The Court later used a similar principle to limit the states' ability to interfere with business contracts.)

1857: Dred Scott v. Sanford. The Court declared unconstitutional the already-repealed Missouri Compromise of 1820 because it deprived a person of his or her property—a slave—without due process of law. The Court also ruled that slaves were not citizens of any state nor of the U.S. (The latter part of the decision was overturned by ratification of the 14th Amendment in 1868.)

1896: Plessy v. Ferguson. The Court ruled that a state law requiring federal railroad trains to provide separate but equal facilities for black and white passengers neither infringed upon federal authority to regulate interstate commerce nor violated the 13th and 14th Amendments. (The "separate but equal" doctrine remained effective until the 1954 **Brown v. Board of Education** decision.)

1904: Northern Securities Co. v. U.S. The Court ruled that a holding company formed solely to eliminate competition between two railroad lines was a combination in restraint of trade, violating the federal antitrust act.

1908: Muller v. Oregon. The Court upheld a state law limiting the working hours of women. (Louis D. Brandeis, counsel for the state, cited evidence from social workers, physicians, and factory inspectors that the number of hours women worked affected their health and morals.)

1911: Standard Oil Co. of New Jersey et al. v. U.S. The Court ruled that the Standard Oil Trust must be dissolved because of its unreasonable restraint of trade.

1919: Schenck v. U.S. The Court sustained the Espionage Act of 1917, maintaining that freedom of speech and press could be constrained if "the words used . . . create a clear and present danger. . ."

1925: Gitlow v. New York. The Court ruled that the First Amendment prohibition against government abridgment of the freedom of speech applied to the states as well as to the federal government. The decision was the first of a number of rulings holding that the 14th Amendment extended the guarantees of the Bill of Rights to state action.

1935: Schechter Poultry Corp. v. U.S. The Court ruled that Congress exceeded its authority to delegate legislative powers and to regulate interstate commerce when it enacted the National Industrial Recovery Act, which afforded the U.S. president too much discretionary power.

1951: Dennis et al. v. U.S. The Court upheld convictions under the Smith Act of 1940 for invoking Communist theory that advocated the forcible overthrow of the government. (In the **1957 Yates v. U.S.** decision, the Court moderated this ruling by allowing such advocacy in the abstract, if not connected to action to achieve the goal.)

IT'S A FACT: The first session of the Supreme Court, held in February 1790 at the Royal Exchange Bldg. in New York City, adjourned after ten days of minor administrative business, because there were no cases to hear.

1954: Brown v. Board of Education of Topeka. The Court ruled that separate public schools for black and white students were inherently unequal, so that state-sanctioned segregation in public schools violated the equal protection guarantee of the 14th Amendment. And in **Bolling v. Sharpe** the Court ruled that the congressionally mandated segregated public school system in the District of Columbia violated the 5th Amendment's due process guarantee of personal liberty. (The Brown ruling also led to abolition of state-sponsored segregation in other public facilities.)

1957: Roth v. U.S., Alberts v. California. The Court ruled obscene material was not protected by First Amendment guarantees of freedom of speech and press, defining obscene as "utterly without redeeming social value" and appealing to "prurient interests" in the view of the average person. This definition was modified in later decisions, and the "average person" standard was replaced by the "local community" standard in **Miller v. California (1973).**

1961: Mapp v. Ohio. The Court ruled that evidence obtained in violation of the 4th Amendment guarantee against unreasonable search and seizure must be excluded from use at state as well as federal trials.

1962: Engel v. Vitale. The Court held that public school officials could not require pupils to recite a state-composed prayer, even if it was nondenominational and voluntary, because this would be an unconstitutional attempt to establish religion.

1962: Baker v. Carr. The Court held that the constitutional challenges to the unequal distribution of voters among legislative districts could be resolved by federal courts.

1963: Gideon v. Wainwright. The Court ruled that state and federal defendants who are charged with serious crimes must have access to an attorney, at state expense if necessary.

1964: New York Times Co. v. Sullivan. The Court ruled that the First Amendment protected the press from libel suits for defamatory reports about public officials unless an injured party could prove that a defamatory report was made out of malice or "reckless disregard" for the truth.

1965: Griswold v. Conn. The Court ruled that a state unconstitutionally interfered with personal privacy in the marriage relationship when it prohibited anyone, including married couples, from using contraceptives.

1966: Miranda v. Arizona. The Court ruled that, under the guarantee of due process, suspects in custody, before being questioned, must be informed that they have the right to remain silent, that anything they say may be used against them, and that they have the right to counsel.

1973: Roe v. Wade, Doe v. Bolton. The Court ruled that the fetus was not a "person" with constitutional rights and that a right to privacy inherent in the 14th Amendment's due process guarantee of personal liberty protected a woman's decision to have an abortion. During the first trimester of pregnancy, the Court maintained, the decision should be left entirely to a woman and her physician. Some regulation of abortion procedures was allowed in the 2d trimester, and some restriction of abortion in the 3d.

1974: U.S. v. Nixon. The Court ruled that neither the separation of powers nor the need to preserve the confidentiality of presidential communications could alone justify an absolute executive privilege of immunity from judicial demands for evidence to be used in a criminal trial.

1976: Gregg v. Georgia, Profitt v. Fla., Jurek v. Texas. The Court held that death, as a punishment for persons convicted of first degree murder, was not in and of itself cruel and unusual punishment in violation of the 8th Amendment. But the Court ruled that the sentencing judge and jury must consider the individual character of the offender and the circumstances of the particular crime.

1978: Regents of Univ. of Calif. v. Bakke. The Court ruled that a special admissions program for a state medical school, under which a set number of places were reserved for minorities, violated the 1964 Civil Rights Act, which forbids excluding anyone, because of race, from a federally funded program. However, the Court ruled that race could be considered as one of a complex of factors.

1986: Bowers v. Hardwick. The Court refused to extend any constitutional right of privacy to homosexual activity, upholding a Georgia law that in effect made such activity a crime. (Although the Georgia law made no distinction between heterosexual or homosexual sodomy, enforcement had been confined to homosexuals; the statute was invalidated by the state supreme court in 1998.) In **Romer v. Evans (1996),** the Court struck down a Colorado constitutional provision that barred legislation protecting homosexuals from discrimination.

1990: Cruzan v. Missouri. The Court ruled that a person had the right to refuse life-sustaining medical treatment. However, the Court also ruled that, before treatment could be withheld from a comatose patient, a state could require "clear and convincing evidence" that the patient would not have wanted to live. And in 2 **1997** rulings, **Washington v. Glucksberg** and **Vacco v. Quill,** the Court ruled that states could ban doctor-assisted suicide.

1995: Adarand Constructors v. Peña. The Court held that federal programs that classify people by race, unless "narrowly tailored" to accomplish a "compelling governmental interest," may deny individuals the right to equal protection.

1995: U.S. Term Limits Inc. v. Thornton. The Court ruled that neither states nor Congress could limit terms of members of Congress, since the Constitution reserves to the people the right to choose federal lawmakers.

1997: Clinton v. Jones. Rejecting an appeal by Pres. Clinton in a sexual harassment suit, the Court ruled that a sitting president did not have temporary immunity from a lawsuit for actions outside the realm of official duties.

1997: City of Boerne v. Flores. The Court overturned a 1993 law that banned enforcement of laws that "substantially burden" religious practice unless there is a "compelling need" to do so. The Court held that the act was an unwarranted intrusion by Congress on states' prerogatives and an infringement of the judiciary's role.

1997: Reno v. ACLU. Citing the right to free expression, the Court overturned a provision making it a crime to display or distribute "indecent" or "patently offensive" material on the Internet. In **1998**, however, the Court ruled in **NEA v. Finley** that "general standards of decency" may be used as a criterion in federal arts funding.

1998: Clinton v. City of New York. The Court struck down the Line-Item Veto Act (1996), holding that it unconstitutionally gave the president "the unilateral power to change the text of duly enacted statutes."

1998: Faragher v. City of Boca Raton, Burlington Industries, Inc. v. Ellerth. The Court issued new guidelines for workplace sexual harassment suits, holding employers responsible for misconduct by supervisory employees. And in **Oncale v. Sundowner Offshore Services,** the Court ruled that the law against sexual harassment applies regardless of whether harasser and victim are the same sex.

1999: Dept. of Commerce v. U.S. House. Upholding a challenge to plans for the 2000 census, the Court required an actual head count for apportioning the U.S. House of Representatives, but allowed the use of statistical sampling methods for other purposes, such as the allocation of federal funds.

1999: Alden v. Maine, Florida Prepaid v. College Savings Bank, College Savings Bank v. Florida. In a series of rulings, the Court applied the principle of "sovereign immunity" to shield states in large part from being sued under federal law.

2000: Troxel v. Granville. The justices found that a Washington state law allowing grandparents visitation rights, as broadly applied, interfered with parents' right to determine the best care for their children.

2000: Boy Scouts of America v. Dale. The Court ruled that the Boy Scouts of America could dismiss a troop leader after learning he was gay, holding that the right to freedom of association outweighed a New Jersey anti-discrimination statute.

2000: Stenberg v. Carhart. The Court struck down a Nebraska law that banned so-called partial-birth abortion. It argued that the law could be interpreted as banning other abortion procedures and that it should have made exception for reasons of health. (See 1973: *Roe* v. *Wade*.)

2000: Bush v. Gore. The Court ruled that manual recounts of presidential ballots in the Nov. 2000 election could not proceed because inconsistent evaluation standards in different counties violated the equal protection clause. In effect, the ruling meant Bush would win the election.

IT'S A FACT: Vice-Pres. Lyndon B. Johnson took the oath of office as 36th president in the private cabin of *Air Force One* on a runway on Dallas's Love Field, following the assassination of Pres. John F. Kennedy in a Dallas motorcade, Nov. 22, 1963. *The World Almanac* was used as the source for the oath.

Presidential Oath of Office

The Constitution (Article II) directs that the president-elect shall take the following oath or affirmation to be inaugurated as president: "I do solemnly swear [affirm] that I will faithfully execute the office of President of the United States, and will, to the best of my ability, preserve, protect, and defend the Constitution of the United States." (Custom decrees the addition of the words "So help me God" at the end of the oath when taken by the president-elect, with the left hand on the Bible for the duration of the oath, and the right hand slightly raised.)

Law on Succession to the Presidency

If by reason of death, resignation, removal from office, inability, or failure to qualify there is neither a president nor vice president to discharge the powers and duties of the office of president, then the speaker of the House of Representatives shall upon his resignation as speaker and as representative, act as president. The same rule shall apply in the case of the death, resignation, removal from office, or inability of an individual acting as president.

If at the time when a speaker is to begin the discharge of the powers and duties of the office of president there is no speaker, or the speaker fails to qualify as acting president, then the president pro tempore of the Senate, upon his resignation as president pro tempore and as senator, shall act as president.

An individual acting as president shall continue to act until the expiration of the then current presidential term, except that (1) if his discharge of the powers and duties of the office is founded in whole or in part in the failure of both the president-elect and the vice president-elect to qualify, then he shall act only until a president or vice president qualifies, and (2) if his discharge of the powers and duties of the office is founded in whole or in part on the inability of the president or vice president, then he shall act only until the removal of the disability of one of such individuals.

If, by reason of death, resignation, removal from office, or failure to qualify, there is no president pro tempore to act as president, then the officer of the United States who is highest on the following list, and who is not under any disability to discharge the powers and duties of president shall act as president; the secretaries of state, treasury, defense, attorney general; secretaries of interior, agriculture, commerce, labor, health and human services, housing and urban development, transportation, energy, education, veterans affairs.

(Legislation approved July 18, 1947; amended Sept. 9, 1965, Oct. 15, 1966, Aug. 4, 1977, and Sept. 27, 1979. See also Constitutional Amendment XXV.)

Origin of the United States National Motto

In God We Trust, designated as the U.S. National Motto by Congress in 1956, originated during the Civil War as an inscription for U. S. coins, although it was used by Francis Scott Key in a slightly different form when he wrote "The Star-Spangled Banner" in 1814. On Nov. 13, 1861, when Union morale had been shaken by battlefield defeats, the Rev. M. R. Watkinson, of Ridleyville, PA, wrote to Secy. of the Treasury Salmon P. Chase. "From my heart I have felt our national shame in disowning God as not the least of our present national disasters," the minister wrote, suggesting "recognition of the Almighty God in some form on our coins." Secy. Chase ordered designs prepared with the inscription *In God We Trust* and backed coinage legislation that authorized use of this slogan. The motto first appeared on some U.S. coins in 1864, and disappeared and reappeared on various coins until 1955, when Congress ordered it placed on all paper money and all coins.

The Great Seal of the U.S.

On July 4, 1776, the Continental Congress appointed a committee consisting of Benjamin Franklin, John Adams, and Thomas Jefferson "to bring in a device for a seal of the United States of America." The designs submitted by this and a subsequent committee were considered unacceptable. After many delays, a third committee, appointed early in 1782, presented a design prepared by William Barton. Charles Thomson, the secretary of Congress, suggested certain changes, and Congress finally approved the design on June 20, 1782. The obverse side of the seal shows an American bald eagle. In its mouth is a ribbon bearing the motto *e pluribus unum* (one out of many). In the eagle's talons are the arrows of war and an olive branch of peace. The reverse side shows an unfinished pyramid with an eye (the eye of Providence) above it.

The American's Creed

William Tyler Page, Clerk of the U.S. House of Representatives, wrote "The American's Creed" in 1917. It was accepted by the House on behalf of the American people on April 3, 1918.

"I believe in the United States of America as a government of the people, by the people, for the people; whose just powers are derived from the consent of the governed; a democracy in a republic; a sovereign Nation of many sovereign States; a perfect union, one and inseparable; established upon those principles of freedom, equality, justice, and humanity for which American patriots sacrificed their lives and fortunes.

"I therefore believe it is my duty to my country to love it, to support its Constitution, to obey its laws, to respect its flag, and to defend it against all enemies."

The Flag of the U.S.—The Stars and Stripes

The 50-star flag of the United States was raised for the first time officially at 12:01 AM on July 4, 1960, at Fort McHenry National Monument in Baltimore, MD. The 50th star had been added for Hawaii; a year earlier the 49th, for Alaska. Before that, no star had been added since 1912, when New Mexico and Arizona were admitted to the Union.

The true history of the Stars and Stripes has become so cluttered by myth and tradition that the facts are difficult, and in some cases impossible, to establish. For example, it is not certain who designed the Stars and Stripes, who made the first such flag, or even whether it ever flew in any sea fight or land battle of the American Revolution.

All agree, however, that the Stars and Stripes originated as the result of a resolution offered by the Marine Committee of the Second Continental Congress at Philadelphia and adopted on June 14, 1777. It read:

Resolved: that the flag of the United States be thirteen stripes, alternate red and white; that the union be thirteen stars, white in a blue field, representing a new constellation.

Congress gave no hint as to the designer of the flag, no instructions as to the arrangement of the stars, and no information on its appropriate uses. Historians have been unable to find the original flag law.

The resolution establishing the flag was not even published until Sept. 2, 1777. Despite repeated requests, Washington did not get the flags until 1783, after the American Revolution was over. And there is no certainty that they were the Stars and Stripes.

Early Flags

Many historians consider the first flag of the U.S. to have been the Grand Union (sometimes called Great Union) flag, although the Continental Congress never officially adopted it. This flag was a modification of the British Meteor flag, which had the red cross of St. George and the white cross of St. Andrew combined in the blue canton. For the Grand Union flag, 6 horizontal stripes were imposed on the red field, dividing it into 13 alternating red and white stripes. On Jan. 1, 1776, when the Continental Army came into formal existence, this flag was unfurled on Prospect Hill, Somerville, MA. Washington wrote that "we hoisted the Union Flag in compliment to the United Colonies."

One of several flags about which controversy has raged for years is at Easton, PA. Containing the devices of the national flag in reversed order, this flag has been in the public library at Easton for more than 150 years. Some contend that this flag was actually the first Stars and Stripes, first displayed on July 8, 1776. This flag has 13 red and white stripes in the canton, 13 white stars centered in a blue field.

A flag was hastily improvised from garments by the defenders of Fort Schuyler at Rome, NY, Aug. 3-22, 1777. Historians believe it was the Grand Union Flag.

The Sons of Liberty had a flag of 9 red and white stripes, to signify 9 colonies, when they met in New York in 1765 to oppose the Stamp Tax. By 1775, the flag had grown to 13 red and white stripes, with a rattlesnake on it.

At Concord, Apr. 19, 1775, the minutemen from Bedford, MA, are said to have carried a flag having a silver arm with sword on a red field. At Cambridge, MA, the Sons of Liberty used a plain red flag with a green pine tree on it.

In June 1775, Washington went from Philadelphia to Boston to take command of the army, escorted to New York by the Philadelphia Light Horse Troop. It carried a yellow flag that had an elaborate coat of arms—the shield charged with 13 knots, the motto "For These We Strive"—and a canton of 13 blue and silver stripes.

In Feb. 1776, Col. Christopher Gadsden, a member of the Continental Congress, gave the South Carolina Provincial Congress a flag "such as is to be used by the commander-in-chief of the American Navy." It had a yellow field, with a rattlesnake about to strike and the words "Don't Tread on Me."

At the Battle of Bennington, Aug. 16, 1777, patriots used a flag of 7 white and 6 red stripes with a blue canton extending down 9 stripes and showing an arch of 11 white stars over the figure 76 and a star in each of the upper corners. The stars are 7-pointed. This flag is preserved in the Historical Museum at Bennington, VT.

At the Battle of Cowpens, Jan. 17, 1781, the 3d Maryland Regiment is said to have carried a flag of 13 red and white stripes, with a blue canton containing 12 stars in a circle around one star.

Who Designed the Flag? No one knows for certain. Francis Hopkinson, designer of a naval flag, declared he also had designed the flag and in 1781 asked Congress to reimburse him for his services. Congress did not do so. Dumas Malone of Columbia University wrote: "This talented man . . . designed the American flag."

Who Called the Flag "Old Glory"? The flag is said to have been named Old Glory by William Driver, a sea captain of Salem, MA. One legend has it that when he raised the flag on his brig, the *Charles Doggett*, in 1824, he said: "I name thee Old Glory." But his daughter, who presented the flag to the Smithsonian Institution, said he named it at his 21st birthday celebration on Mar. 17, 1824, when his mother presented the homemade flag to him.

The Betsy Ross Legend. The widely publicized legend that Mrs. Betsy Ross made the first Stars and Stripes in June 1776, at the request of a committee composed of George Washington, Robert Morris, and George Ross, an uncle, was first made public in 1870, by a grandson of Mrs. Ross. Historians have been unable to find a historical record of such a meeting or committee.

Adding New Stars

The flag of 1777 was used until 1795. Then, on the admission of Vermont and Kentucky to the Union, Congress passed and Pres. Washington signed an act that after May 1, 1795, the flag should have 15 stripes, alternating red and white, and 15 white stars on a blue field.

When new states were admitted, it became evident that the flag would become burdened with stripes. Congress thereupon ordered that after July 4, 1818, the flag should have 13 stripes, symbolizing the 13 original states; that the union have 20 stars, and that whenever a new state was admitted a new star should be added on the July 4 following admission. No law designates the permanent arrangement of the stars. However, since 1912, when a new state has been admitted, the new design has been announced by executive order. No star is specifically identified with any state.

Code of Etiquette for Display and Use of the U.S. Flag

Reviewed by National Flag Foundation

Although the Stars and Stripes originated in 1777, it was not until 146 years later that there was a serious attempt to establish a uniform code of etiquette for the U.S. flag. On Feb. 15, 1923, the War Department issued a circular on the rules of flag usage. These rules were adopted almost in their entirety June 14, 1923, by a conference of 68 patriotic organizations in Washington, D.C. Finally, on June 22, 1942, a joint resolution of Congress, amended by Public Law 94-344, July 7, 1976, codified "existing rules and customs pertaining to the display and use of the flag . . ."

When to Display the Flag—The flag should be displayed on all days, especially on legal holidays and other special occasions, on official buildings when in use, in or near polling places on election days, and in or near schools when in session. Citizens may fly the flag at any time. It is customary to display it only from sunrise to sunset on buildings and on stationary flagstaffs in the open. It may be displayed at night, however, on special occasions, preferably lighted. The flag now flies over the White House both day and night. It flies over the Senate wing of the Capitol when the Senate is in session and over the House wing when that body is in session. It flies day and night over the east and west fronts of the Capitol, without floodlights at night but receiving illumination from the Capitol Dome. It flies 24 hours a day at several other places, including the Fort McHenry National Monument in Baltimore, where it inspired Francis Scott Key to write "The Star Spangled Banner." The flag also flies 24 hours a day, properly illuminated, at U.S. Customs ports of entry.

Flying the Flag at Half-Staff—Flying the flag at half-staff, that is, halfway up the staff, is a signal of mourning. The flag should be hoisted to the top of the staff for an instant before being lowered to half-staff. It should be hoisted to the peak again before being lowered for the day or night.

As provided by presidential proclamation, the flag should fly at half-staff for 30 days from the day of death of a president or former president; for 10 days from the day of death of a vice president, chief justice or retired chief justice of the U.S., or speaker of the House of Representatives; from day of death until burial of an associate justice of the Supreme Court, cabinet member, former vice president, Senate president pro tempore, or majority or minority Senate or House leader; for a U.S. senator, representative, territorial delegate, or the resident commissioner of Puerto Rico, on day of death and the following day within the metropolitan area of the District of Columbia and from day of death until burial within the decedent's state, congressional district, territory or commonwealth; and for the death of the governor of a state, territory, or possession of the U.S., from day of death until burial.

On Memorial Day, the flag should fly at half-staff until noon and then be raised to the peak. The flag should also fly at half-staff on Korean War Veterans Armistice Day (July 27), National Pearl Harbor Remembrance Day (Dec. 7), and Peace Officers Memorial Day (May 15).

How to Fly the Flag—The flag should be hoisted briskly and lowered ceremoniously and should never be allowed to touch the ground or the floor. When the flag is hung over a sidewalk from a rope extending from a building to a pole, the union should be away from the building. When the flag is hung over the center of a street the union should be to the north in an east-west street and to the east in a north-south street. No other flag may be flown above or, if on the same level, to the right of the U.S. flag, except that at the United Nations Headquarters the UN flag may be placed above flags of all member nations and other national flags may be flown with equal prominence or honor with the flag of the U.S. At services by Navy chaplains at sea, the church pennant may be flown above the flag.

When 2 flags are placed against a wall with crossed staffs, the U.S. flag should be at right—its own right, and its staff should be in front of the staff of the other flag; when a number of flags are grouped and displayed from staffs, it should be at the center and highest point of the group.

Church and Platform Use—In an auditorium, the flag may be displayed flat, above and behind the speaker. When displayed from a staff in a church or in a public auditorium, the flag should hold the position of superior prominence, in advance of the audience, and in the position of honor at the speaker's right as she or he faces the audience. Any other flag so displayed should be placed on the left of the speaker or to the right of the audience.

When the flag is displayed horizontally or vertically against a wall, the stars should be uppermost and at the observer's left.

When used to cover a casket, the flag should be placed so that the union is at the head and over the left shoulder. It should not be lowered into the grave nor touch the ground.

How to Dispose of Worn Flags—When the flag is in such condition that it is no longer a fitting emblem for display, it should be destroyed in a dignified way, preferably by burning.

When to Salute the Flag—All persons present should face the flag, stand at attention, and salute on the following occasions: (1) when the flag is passing in a parade or in a review, (2) during the ceremony of hoisting or lowering, (3) when the national anthem is played, and (4) during the Pledge of Allegiance. Those present in uniform should render the military salute. Those not in uniform should place the right hand over the heart. A man wearing a hat should remove it with his right hand and hold it to his left shoulder during the salute.

Prohibited Uses of the Flag—The flag should not be dipped to any person or thing. (An exception—customarily, ships salute by dipping their colors.) It should never be displayed with the union down save as a distress signal. It should never be carried flat or horizontally, but always aloft and free.

It should not be displayed on a float, an automobile, or a boat except from a staff. It should never be used as a covering for a ceiling, nor have placed on it any word, design, or drawing. It should never be used as a receptacle for carrying anything. It should not be used to cover a statue or a monument.

The flag should never be used for advertising purposes, nor be embroidered on such articles as cushions or handkerchiefs, printed or otherwise impressed on boxes or anything that is designed for temporary use and discard; or used as a costume or athletic uniform. Advertising signs should not be fastened to its staff or halyard.

The flag should never be used as drapery of any sort, never festooned, drawn back, nor up, in folds, but always allowed to fall free. Bunting of blue, white, and red, always arranged with the blue above and the white in the middle, should be used for covering a speaker's desk, draping the front of a platform, and for decoration in general.

An act of Congress approved on Feb. 8, 1917, provided certain penalties for the desecration, mutilation, or improper use of the flag within the District of Columbia. A 1968 federal law provided penalties of as much as a year's imprisonment or a $1,000 fine or both for publicly burning or otherwise desecrating any U.S. flag. In addition, many states have laws against flag desecration. In 1989, the Supreme Court ruled that no laws could prohibit political protesters from burning the flag. The decision had the effect of declaring unconstitutional the flag desecration laws of 48 states, as well as a similar federal statute, in cases of peaceful political expression.

The Supreme Court, in June 1990, declared that a new federal law making it a crime to burn or deface the American flag violated the free-speech guarantee of the First Amendment. The 5-4 decision led to renewed calls in Congress for a constitutional amendment to make it possible to prosecute flag burners.

Pledge of Allegiance to the Flag

I pledge allegiance to the flag of the United States of America and to the republic for which it stands, one nation under God, indivisible, with liberty and justice for all.

This, the current official version of the Pledge of Allegiance, has developed from the original pledge, which was first published in the Sept. 8, 1892, issue of *Youth's Companion*, a weekly magazine then published in Boston. The original pledge contained the phrase "my flag," which was changed more than 30 years later to "flag of the United States of America." A 1954 act of Congress added the words "under God."

The authorship of the pledge had been in dispute for many years. The *Youth's Companion* stated in 1917 that the original draft was written by James B. Upham, an executive of the magazine who died in 1910. A leaflet circulated by the magazine later named Upham as the originator of the draft "afterwards condensed and perfected by him and his associates of the Companion force."

Francis Bellamy, a former member of *Youth's Companion* editorial staff, publicly claimed authorship of the pledge in 1923. In 1939, the United States Flag Association, acting on the advice of a committee named to study the controversy, upheld the claim of Bellamy, who had died 8 years earlier. In 1957 the Library of Congress issued a report attributing the authorship to Bellamy.

The History of the National Anthem

"The Star-Spangled Banner" was ordered played by the military and naval services by Pres. Woodrow Wilson in 1916. It was designated the national anthem by Act of Congress, Mar. 3, 1931. The words were written by Francis Scott Key, of Georgetown, MD, during the bombardment of Fort McHenry, Baltimore, Sept. 13-14, 1814. Key was a lawyer, a graduate of St. John's College, Annapolis, and a volunteer in a light artillery company. When a friend, Dr. Beanes, a Maryland physician, was taken aboard Admiral Cockburn's British squadron for interfering with ground troops, Key and J. S. Skinner, carrying a note from Pres. Madison, went to the fleet under a flag of truce on a cartel ship to ask Beanes's release. Cockburn consented, but as the fleet was about to sail up the Patapsco to bombard Fort McHenry, he detained them, first on HMS *Surprise* and then on a supply ship.

Key witnessed the bombardment from his own vessel. It began at 7 AM, Sept. 13, 1814, and lasted, with intermissions, for 25 hr. The British fired more than 1,500 shells, each weighing as much as 220 lb. They were unable to approach closely because the U.S. had sunk 22 vessels. Only 4 Americans were killed and 24 wounded. A British bomb-ship was disabled.

During the event, Key wrote a stanza on the back of an envelope. Next day at Indian Queen Inn, Baltimore, he wrote out the poem and gave it to his brother-in-law, Judge J. H. Nicholson. Nicholson suggested use of the tune, "Anacreon in Heaven" (attributed to a British composer named John Stafford Smith), and had the poem printed on broadsides, of which 2 survive. On Sept. 20 it appeared in the *Baltimore American*. Later Key made 3 copies; one is in the Library of Congress, and one in the Pennsylvania Historical Society. The copy Key wrote on Sept. 14 remained in the Nicholson family for 93 years. In 1907 it was sold to Henry Walters of Baltimore. In 1934 it was bought at auction by the Walters Art Gallery, Baltimore, for $26,400. In 1953 it was sold to the Maryland Historical Society for the same price.

The flag that Key saw during the bombardment is preserved in the Smithsonian Institution, Washington, DC. It is 30 by 42 ft and has 15 alternating red and white stripes and 15 stars, for the original 13 states plus Kentucky and Vermont. It was made by Mary Young Pickersgill. The Baltimore Flag House, a museum, occupies her premises, which were restored in 1953.

The Star-Spangled Banner

I

Oh, say can you see by the dawn's early light
What so proudly we hailed at the twilight's last gleaming?
Whose broad stripes and bright stars thru the perilous fight,
O'er the ramparts we watched were so gallantly streaming?
And the rocket's red glare, the bombs bursting in air,
Gave proof through the night that our flag was still there.
Oh, say does that star-spangled banner yet wave
O'er the land of the free and the home of the brave?

II

On the shore, dimly seen through the mists of the deep,
Where the foe's haughty host in dread silence reposes,
What is that which the breeze, o'er the towering steep,
As it fitfully blows, half conceals, half discloses?
Now it catches the gleam of the morning's first beam,
In full glory reflected now shines in the stream:
'Tis the star-spangled banner! Oh long may it wave
O'er the land of the free and the home of the brave!

III

And where is that band who so vauntingly swore
That the havoc of war and the battle's confusion,
A home and a country should leave us no more!
Their blood has washed out their foul footsteps' pollution.
No refuge could save the hireling and slave
From the terror of flight, or the gloom of the grave:
And the star-spangled banner in triumph doth wave
O'er the land of the free and the home of the brave!

IV

Oh! thus be it ever, when freemen shall stand
Between their loved home and the war's desolation!
Blest with victory and peace, may the heav'n rescued land
Praise the Power that hath made and preserved us a nation.
Then conquer we must, when our cause it is just,
And this be our motto: "In God is our trust."
And the star-spangled banner in triumph shall wave
O'er the land of the free and the home of the brave!

America (My Country 'Tis of Thee)

First sung in public on July 4, 1831, at a service in the Park Street Church, Boston, the words were written by Rev. Samuel Francis Smith, a Baptist clergyman, who set them to a melody he found in a German songbook, unaware that it was the tune for the British anthem, "God Save the King/Queen."

My country, 'tis of thee,
Sweet land of liberty,
Of thee I sing.
Land where my fathers died!
Land of the Pilgrims' pride!
From ev'ry mountainside,
Let freedom ring!

My native country, thee,
Land of the noble free,
Thy name I love.
I love thy rocks and rills,
Thy woods and templed hills;
My heart with rapture thrills
Like that above.

Let music swell the breeze,
And ring from all the trees
Sweet freedom's song.
Let mortal tongues awake;
Let all that breathe partake;
Let rocks their silence break,
The sound prolong.

Our fathers' God, to Thee,
Author of liberty,
To Thee we sing.
Long may our land be bright
With freedom's holy light;
Protect us by Thy might,
Great God, our King!

America, the Beautiful

Words composed by Katharine Lee Bates, a Massachusetts educator and author, in 1893, inspired by the view she experienced atop Pikes Peak. The final form was established in 1911, and it is set to the music of Samuel A. Ward's "Materna."

O beautiful for spacious skies,
For amber waves of grain,
For purple mountain majesties
Above the fruited plain.
America! America!
God shed His grace on thee,
And crown thy good with brotherhood
From sea to shining sea.

O beautiful for pilgrim feet
Whose stern impassion'd stress
A thorough-fare for freedom beat
Across the wilderness.
America! America!
God mend thine ev'ry flaw,
Confirm thy soul in self control,
Thy liberty in law.

O beautiful for heroes prov'd
In liberating strife,
Who more than self their country lov'd
And mercy more than life.
America! America!
May God thy gold refine
Till all success be nobleness,
And ev'ry gain divine.

O beautiful for patriot dream
That sees beyond the years,
Thine alabaster cities gleam,
Undimmed by human tears.
America! America!
God shed His grace on thee,
And crown thy good with brotherhood
From sea to shining sea.

The Liberty Bell: Its History and Significance

The Liberty Bell is housed in Independence National Historical Park, Philadelphia.

The original bell was ordered by Assembly Speaker and Chairman of the State House Superintendents Isaac Norris and was ordered from Thomas Lester, Whitechapel Foundry, London. It reached Philadelphia at the end of August 1752. It bore an inscription from Leviticus 25:10: "PROCLAIM LIBERTY THROUGHOUT ALL THE LAND UNTO ALL THE INHABITANTS THEREOF."

The bell was cracked by a stroke of its clapper in Sept. 1752 while it hung on a truss in the State House yard for testing. Pass & Stow, Philadelphia founders, recast the bell, adding 1½ ounces of copper to a pound of the original "Whitechapel" metal to reduce its high tone and brittleness. It was found that the bell contained too much copper, injuring its tone, so Pass & Stow recast it again, this time successfully.

In June 1753 the bell was hung in the old wooden steeple of the State House. In use while the Continental Congress was in session in the State House, it rang out in defiance of British tax and trade restrictions, and it proclaimed the Boston Tea Party and the first public reading of the Declaration of Independence.

On Sept. 18, 1777, when the British Army was about to occupy Philadelphia, the Liberty Bell was moved in a baggage train of the American Army to Allentown, PA, where it was hidden until June 27, 1778. The bell was moved back to Philadelphia after the British left the city.

In July 1781 the wooden steeple became insecure and had to be taken down. The bell was lowered into the brick section of the tower, where it remained until 1828. Between 1828 and 1844 the old State House bell continued to ring during special occasions. It rang for the last time on Feb. 23, 1846. In 1852 it was placed on exhibition in the Declaration Chamber of Independence Hall.

In 1876, when many thousands of Americans visited Philadelphia for the Centennial Exposition, the bell was placed in its old wooden support in the tower hallway. In 1877 it was hung from the ceiling of the tower by a chain of 13 links. It was returned again to the Declaration Chamber and in 1896 taken back to the tower hall, where it occupied a glass case. In 1915 the case was removed so that the public might touch it. On Jan. 1, 1976, just after midnight to mark the opening of the Bicentennial Year, the bell was moved to a new glass and steel pavilion behind Independence Hall for easier viewing.

The measurements of the bell are: circumference around the lip, 12 ft ½ in; circumference around the crown, 6 ft 11¼ in; lip to the crown, 3 ft; height over the crown, 2 ft 3 in; thickness at lip, 3 in; thickness at crown, 1¼ in; weight, 2,080 lb; length of clapper, 3 ft 2 in.

The specific source of the crack in the bell is unknown.

Statue of Liberty National Monument

Since 1886, the Statue of Liberty Enlightening the World has stood as a symbol of freedom in New York harbor. It also commemorates French-American friendship, for it was given by the people of France and designed by French sculptor Frederic Auguste Bartholdi (1834-1904).

Edouard de Laboulaye, French historian, suggested the French present a monument to the U.S., the latter to provide pedestal and site. Bartholdi visualized a colossal statue at the entrance of New York harbor, welcoming the peoples of the world with the torch of liberty.

On Washington's Birthday, Feb. 22, 1877, Congress approved the use of a site on Bedloe's Island suggested by Bartholdi. This island of 12 acres had been owned in the 17th century by a Walloon named Isaac Bedloe. It was called Bedloe's until Aug. 3, 1956, when Pres. Eisenhower approved a resolution of Congress changing the name to Liberty Island.

The statue was finished on May 21, 1884, and formally presented to the U.S. minister to France, Levi Parsons Morton, July 4, 1884, by Ferdinand de Lesseps, head of the Franco-American Union, promoter of the Panama Canal, and builder of the Suez Canal.

On Aug. 5, 1884, the Americans laid the cornerstone for the pedestal. This was to be built on the foundations of Fort Wood, which had been erected by the government in 1811. The American committee had raised $125,000, but this was found to be inadequate. Joseph Pulitzer, owner of the *New York World,* appealed on Mar. 16, 1885, for general donations. By Aug. 11, 1885, he had raised $100,000.

The statue arrived dismantled, in 214 packing cases, from Rouen, France, in June 1885. The last rivet of the statue was driven on Oct. 28, 1886, when Pres. Grover Cleveland dedicated the monument.

The statue weighs 450,000 lb, or 225 tons. The copper sheeting weighs 200,000 lb. There are 167 steps from the land level to the top of the pedestal, 168 steps inside the statue to the head, and 54 rungs on the ladder leading to the arm that holds the torch.

A $2.5 million building housing the American Museum of Immigration was opened by Pres. Richard Nixon on Sept. 26, 1972, at the base of the statue. It houses a permanent exhibition of photos, posters, and artifacts tracing the history of American immigration. The Statue of Liberty National Monument is administered by the National Park Service.

Dimensions of the Statue	Ft.	In.
Height from base to torch (45.3 meters)	151	1
Foundation of pedestal to torch (91.5 meters)	305	1
Heel to top of head	111	1
Length of hand	16	5
Index finger	8	0
Size of finger nail, 13x10 in.		
Head from chin to cranium	17	3
Head thickness from ear to ear	10	0
Length of nose	4	6
Right arm, length	42	0
Right arm, greatest thickness	12	0
Thickness of waist	35	0
Width of mouth	3	0
Tablet, length	23	7
Tablet, width	13	7
Tablet, thickness	2	0

Four years of restoration work was completed before the statue's centennial celebration on July 4, 1986. Among other repairs, the $87 million dollar project included replacing the 1,600 wrought iron bands that hold the statue's copper skin to its frame, replacing its torch, and installing an elevator.

A 4-day extravaganza of concerts, tall ships, ethnic festivals, and fireworks, July 3-6, 1986, celebrated the 100th anniversary. The festivities included Chief Justice Warren E. Burger's swearing-in of 5,000 new citizens on Ellis Island, while 20,000 others across the country were simultaneously sworn in through a satellite telecast.

The ceremonies were followed by others on Oct. 28, 1986, to mark the statue's exact 100th birthday.

Emma Lazarus's Famous Poem

Engraved on pedestal below the statue.

The New Colossus

Not like the brazen giant of Greek fame,
With conquering limbs astride from land to land;
Here at our sea-washed, sunset gates shall stand
A mighty woman with a torch, whose flame
Is the imprisoned lightning, and her name
Mother of Exiles. From her beacon-hand
Glows world-wide welcome; her mild eyes command
The air-bridged harbor that twin cities frame.
"Keep ancient lands, your storied pomp!" cries she
With silent lips. "Give me your tired, your poor,
Your huddled masses yearning to breathe free,
The wretched refuse of your teeming shore.
Send these, the homeless, tempest-tost to me,
I lift my lamp beside the golden door!"

Ellis Island

Ellis Island was the gateway to America for more than 12 million immigrants between 1892 and 1924. In the late 18th century, Samuel Ellis, a New York City merchant, purchased the island and gave it his name. From Ellis, it passed to New York State, and the U.S. government bought it in 1808. On Jan. 1, 1892 the government opened the first federal immigration center in the U.S. on the island. The $27^1/_2$-acre site eventually supported more than 35 buildings, including the Main Building with its Great Hall, in which as many as 5,000 people a day were processed.

Closed as an immigration station in 1954, Ellis Island was proclaimed part of the Statue of Liberty National Monument in 1965 by Pres. Lyndon B. Johnson. After a 6-year $170 million restoration project funded by The Ellis Island Fdn. Inc., Ellis Island was reopened as a museum in 1990. Artifacts, historic photographs and documents, oral histories, and ethnic music depicting 400 years of American immigration are housed in the museum. The museum also includes The American Immigrant Wall of Honor® (http://www.wallofhonor.com). With its New Edition to open in late 2001, it will hold more than 600,000 names. The American Family Immigration History Center, a genealogical center, opened in April 2001. It contains an electronic database of ship arrivals through New York harbor from 1892 to 1924 with over 22 million names, as well as an interactive database which features a Living Family Archive and multimedia presentations on various immigration groups and patterns (http://www.ellisislandrecords.org). In 1998, the Supreme Court ruled that nearly 90% of the island (the 24.2 acres which are landfill) lies in New Jersey, while the original 3.3 acres are in New York.

PRESIDENTS OF THE UNITED STATES

U.S. Presidents

No.	Name	Politics	Born	in	Inaug.	at age	Died	at age
1.	George Washington	Fed.	1732, Feb. 22	VA	1789	57	1799, Dec. 14	67
2.	John Adams	Fed.	1735, Oct. 30	MA	1797	61	1826, July 4	90
3.	Thomas Jefferson	Dem.-Rep.	1743, Apr. 13	VA	1801	57	1826, July 4	83
4.	James Madison	Dem.-Rep.	1751, Mar. 16	VA	1809	57	1836, June 28	85
5.	James Monroe	Dem.-Rep.	1758, Apr. 28	VA	1817	58	1831, July 4	73
6.	John Quincy Adams	Dem.-Rep.	1767, July 11	MA	1825	57	1848, Feb. 23	80
7.	Andrew Jackson	Dem.	1767, Mar. 15	SC	1829	61	1845, June 8	78
8.	Martin Van Buren	Dem.	1782, Dec. 5	NY	1837	54	1862, July 24	79
9.	William Henry Harrison	Whig	1773, Feb. 9	VA	1841	68	1841, Apr. 4	68
10.	John Tyler	Whig	1790, Mar. 29	VA	1841	51	1862, Jan. 18	71
11.	James Knox Polk	Dem.	1795, Nov. 2	NC	1845	49	1849, June 15	53
12.	Zachary Taylor	Whig	1784, Nov. 24	VA	1849	64	1850, July 9	65
13.	Millard Fillmore	Whig	1800, Jan. 7	NY	1850	50	1874, Mar. 8	74
14.	Franklin Pierce	Dem.	1804, Nov. 23	NH	1853	48	1869, Oct. 8	64
15.	James Buchanan	Dem.	1791, Apr. 23	PA	1857	65	1868, June 1	77
16.	Abraham Lincoln	Rep.	1809, Feb. 12	KY	1861	52	1865, Apr. 15	56
17.	Andrew Johnson	(1)	1808, Dec. 29	NC	1865	56	1875, July 31	66
18.	Ulysses Simpson Grant	Rep.	1822, Apr. 27	OH	1869	46	1885, July 23	63
19.	Rutherford Birchard Hayes	Rep.	1822, Oct. 4	OH	1877	54	1893, Jan. 17	70
20.	James Abram Garfield	Rep.	1831, Nov. 19	OH	1881	49	1881, Sept. 19	49
21.	Chester Alan Arthur	Rep.	1830, Oct. 5	VT	1881	50	1886, Nov. 18	56
22.	Grover Cleveland	Dem.	1837, Mar. 18	NJ	1885	47	1908, June 24	71
23.	Benjamin Harrison	Rep.	1833, Aug. 20	OH	1889	55	1901, Mar. 13	67
24.	Grover Cleveland	Dem.	1837, Mar. 18	NJ	1893	55	1908, June 24	71
25.	William McKinley	Rep.	1843, Jan. 29	OH	1897	54	1901, Sept. 14	58
26.	Theodore Roosevelt	Rep.	1858, Oct. 27	NY	1901	42	1919, Jan. 6	60
27.	William Howard Taft	Rep.	1857, Sept. 15	OH	1909	51	1930, Mar. 8	72
28.	Woodrow Wilson	Dem.	1856, Dec. 28	VA	1913	56	1924, Feb. 3	67
29.	Warren Gamaliel Harding	Rep.	1865, Nov. 2	OH	1921	55	1923, Aug. 2	57
30.	Calvin Coolidge	Rep.	1872, July 4	VT	1923	51	1933, Jan. 5	60
31.	Herbert Clark Hoover	Rep.	1874, Aug. 10	IA	1929	54	1964, Oct. 20	90
32.	Franklin Delano Roosevelt	Dem.	1882, Jan. 30	NY	1933	51	1945, Apr. 12	63
33.	Harry S. Truman	Dem.	1884, May 8	MO	1945	60	1972, Dec. 26	88
34.	Dwight David Eisenhower	Rep.	1890, Oct. 14	TX	1953	62	1969, Mar. 28	78
35.	John Fitzgerald Kennedy	Dem.	1917, May 29	MA	1961	43	1963, Nov. 22	46
36.	Lyndon Baines Johnson	Dem.	1908, Aug. 27	TX	1963	55	1973, Jan. 22	64
37.	Richard Milhous Nixon (2)	Rep.	1913, Jan. 9	CA	1969	56	1994, Apr. 22	81
38.	Gerald Rudolph Ford	Rep.	1913, July 14	NE	1974	61		
39.	Jimmy Carter	Dem.	1924, Oct. 1	GA	1977	52		
40.	Ronald Reagan	Rep.	1911, Feb. 6	IL	1981	69		
41.	George H.W. Bush	Rep.	1924, June 12	MA	1989	64		
42.	Bill Clinton	Dem.	1946, Aug. 19	AR	1993	46		
43.	George W. Bush	Rep.	1946, July 6	CT	2001	54		

(1) Andrew Johnson was a Democrat, nominated vice president by Republicans, and elected with Lincoln on National Union ticket.
(2) Resigned Aug. 9, 1974.

U.S. Presidents, Vice Presidents, Congresses

	President	Service		Vice President	Congresses
1.	George Washington	Apr. 30, 1789—Mar. 3, 1797	1.	John Adams	1, 2, 3, 4
2.	John Adams	Mar. 4, 1797—Mar. 3, 1801	2.	Thomas Jefferson	5, 6
3.	Thomas Jefferson	Mar. 4, 1801—Mar. 3, 1805	3.	Aaron Burr	7, 8
	"	Mar. 4, 1805—Mar. 3, 1809	4.	George Clinton	9, 10
4.	James Madison	Mar. 4, 1809—Mar. 3, 1813		" (1)	11, 12
	"	Mar. 4, 1813—Mar. 3, 1817	5.	Elbridge Gerry (2)	13, 14
5.	James Monroe	Mar. 4, 1817—Mar. 3, 1825	6.	Daniel D. Tompkins	15, 16, 17, 18
6.	John Quincy Adams	Mar. 4, 1825—Mar. 3, 1829	7.	John C. Calhoun	19, 20
7.	Andrew Jackson	Mar. 4, 1829—Mar. 3, 1833		" (3)	21, 22
	"	Mar. 4, 1833—Mar. 3, 1837	8.	Martin Van Buren	23, 24
8.	Martin Van Buren	Mar. 4, 1837—Mar. 3, 1841	9.	Richard M. Johnson	25, 26
9.	William Henry Harrison (4)	Mar. 4, 1841—Apr. 4, 1841	10.	John Tyler	27
10.	John Tyler	Apr. 6, 1841—Mar. 3, 1845			27, 28
11.	James K. Polk	Mar. 4, 1845—Mar. 3, 1849	11.	George M. Dallas	29, 30
12.	Zachary Taylor (4)	Mar. 5, 1849—July 9, 1850	12.	Millard Fillmore	31
13.	Millard Fillmore	July 10, 1850—Mar. 3, 1853			31, 32
14.	Franklin Pierce	Mar. 4, 1853—Mar. 3, 1857	13.	William R. King (5)	33, 34
15.	James Buchanan	Mar. 4, 1857—Mar. 3, 1861	14.	John C. Breckinridge	35, 36
16.	Abraham Lincoln	Mar. 4, 1861—Mar. 3, 1865	15.	Hannibal Hamlin	37, 38
	" (4)	Mar. 4, 1865—Apr. 15, 1865	16.	Andrew Johnson	39
17.	Andrew Johnson	Apr. 15, 1865—Mar. 3, 1869			39, 40
18.	Ulysses S. Grant	Mar. 4, 1869—Mar. 3, 1873	17.	Schuyler Colfax	41, 42
	"	Mar. 4, 1873—Mar. 3, 1877	18.	Henry Wilson (6)	43, 44
19.	Rutherford B. Hayes	Mar. 4, 1877—Mar. 3, 1881	19.	William A. Wheeler	45, 46
20.	James A. Garfield (4)	Mar. 4, 1881—Sept. 19, 1881	20.	Chester A. Arthur	47
21.	Chester A. Arthur	Sept. 20, 1881—Mar. 3, 1885			47, 48
22.	Grover Cleveland (7)	Mar. 4, 1885—Mar. 3, 1889	21.	Thomas A. Hendricks (8)	49, 50
23.	Benjamin Harrison	Mar. 4, 1889—Mar. 3, 1893	22.	Levi P. Morton	51, 52
24.	Grover Cleveland (7)	Mar. 4, 1893—Mar. 3, 1897	23.	Adlai E. Stevenson	53, 54
25.	William McKinley	Mar. 4, 1897—Mar. 3, 1901	24.	Garret A. Hobart (9)	55, 56
	" (4)	Mar. 4, 1901—Sept. 14, 1901	25.	Theodore Roosevelt	57
26.	Theodore Roosevelt	Sept. 14, 1901—Mar. 3, 1905			57, 58
		Mar. 4, 1905—Mar. 3, 1909	26.	Charles W. Fairbanks	59, 60
27.	William H. Taft	Mar. 4, 1909—Mar. 3, 1913	27.	James S. Sherman (10)	61, 62
28.	Woodrow Wilson	Mar. 4, 1913—Mar. 3, 1921	28.	Thomas R. Marshall	63, 64, 65, 66

	President	Service		Vice President	Congresses
29.	Warren G. Harding (4)	Mar. 4, 1921—Aug. 2, 1923	29.	Calvin Coolidge	67
30.	Calvin Coolidge	Aug. 3, 1923—Mar. 3, 1925			68
	"	Mar. 4, 1925—Mar. 3, 1929	30.	Charles G. Dawes	69, 70
31.	Herbert C. Hoover	Mar. 4, 1929—Mar. 3, 1933	31.	Charles Curtis	71, 72
32.	Franklin D. Roosevelt (11)	Mar. 4, 1933—Jan. 20, 1941	32.	John N. Garner	73, 74, 75, 76
	"	Jan. 20, 1941—Jan. 20, 1945	33.	Henry A. Wallace	77, 78
	" (4)	Jan. 20, 1945—Apr. 12, 1945	34.	Harry S. Truman	79
33.	Harry S. Truman	Apr. 12, 1945—Jan. 20, 1949			79, 80
	"	Jan. 20, 1949—Jan. 20, 1953	35.	Alben W. Barkley	81, 82
34.	Dwight D. Eisenhower	Jan. 20, 1953—Jan. 20, 1961	36.	Richard M. Nixon	83, 84, 85, 86
35.	John F. Kennedy (4)	Jan. 20, 1961—Nov. 22, 1963	37.	Lyndon B. Johnson	87, 88
36.	Lyndon B. Johnson	Nov. 22, 1963—Jan. 20, 1965			88
	"	Jan. 20, 1965—Jan. 20, 1969	38.	Hubert H. Humphrey	89, 90
37.	Richard M. Nixon	Jan. 20, 1969—Jan. 20, 1973	39.	Spiro T. Agnew (12)	91, 92, 93
	" (13)	Jan. 20, 1973—Aug. 9, 1974	40.	Gerald R. Ford (14)	93
38.	Gerald R. Ford (15)	Aug. 9, 1974—Jan. 20, 1977	41.	Nelson A. Rockefeller (16)	93, 94
39.	Jimmy (James Earl) Carter	Jan. 20, 1977—Jan. 20, 1981	42.	Walter F. Mondale	95, 96
40.	Ronald Reagan	Jan. 20, 1981—Jan. 20, 1989	43.	George H.W. Bush	97, 98, 99, 100
41.	George H. W. Bush	Jan. 20, 1989—Jan. 20, 1993	44.	Dan Quayle	101, 102
42.	Bill Clinton	Jan. 20, 1993—Jan. 20, 2001	45.	Al Gore	103, 104, 105, 106
43.	George W. Bush	Jan. 20, 2001—	46.	Richard Cheney	107

(1) Died Apr. 20, 1812. (2) Died Nov. 23, 1814. (3) Resigned Dec. 28, 1832, to become U.S. senator. (4) Died in office. (5) Died Apr. 18, 1853. (6) Died Nov. 22, 1875. (7) Terms not consecutive. (8) Died Nov. 25, 1885. (9) Died Nov. 21, 1899. (10) Died Oct. 30, 1912. (11) First president to be inaugurated under 20th Amendment, Jan. 20, 1937. (12) Resigned Oct. 10, 1973. (13) Resigned Aug. 9, 1974. (14) First nonelected vice president, chosen under 25th Amendment procedure. (15) First president never elected president or vice president. (16) Second nonelected vice president, chosen under 25th Amendment.

Vice Presidents of the U.S.

The numerals given vice presidents do not coincide with those given presidents, because some presidents had none and some had more than one.

	Name	Birthplace	Year	Home	Inaug.	Politics	Place of death	Year	Age
1.	John Adams	Quincy, MA	1735	MA	1789	Fed.	Quincy, MA	1826	90
2.	Thomas Jefferson	Shadwell, VA	1743	VA	1797	Dem.-Rep.	Monticello, VA	1826	83
3.	Aaron Burr	Newark, NJ	1756	NY	1801	Dem.-Rep.	Staten Island, NY	1836	80
4.	George Clinton	Ulster Co., NY	1739	NY	1805	Dem.-Rep.	Washington, DC	1812	73
5.	Elbridge Gerry	Marblehead, MA	1744	MA	1813	Dem.-Rep.	Washington, DC	1814	70
6.	Daniel D. Tompkins	Scarsdale, NY	1774	NY	1817	Dem.-Rep.	Staten Island, NY	1825	51
7.	John C. Calhoun (1)	Abbeville, SC	1782	SC	1825	Dem.-Rep.	Washington, DC	1850	68
8.	Martin Van Buren	Kinderhook, NY	1782	NY	1833	Dem.	Kinderhook, NY	1862	79
9.	Richard M. Johnson (2)	Louisville, KY	1780	KY	1837	Dem.	Frankfort, KY	1850	70
10.	John Tyler	Greenway, VA	1790	VA	1841	Whig	Richmond, VA	1862	71
11.	George M. Dallas	Philadelphia, PA	1792	PA	1845	Dem.	Philadelphia, PA	1864	72
12.	Millard Fillmore	Summerhill, NY	1800	NY	1849	Whig	Buffalo, NY	1874	74
13.	William R. King	Sampson Co., NC	1786	AL	1853	Dem.	Dallas Co., AL	1853	67
14.	John C. Breckinridge	Lexington, KY	1821	KY	1857	Dem.	Lexington, KY	1875	54
15.	Hannibal Hamlin	Paris, ME	1809	ME	1861	Rep.	Bangor, ME	1891	81
16.	Andrew Johnson	Raleigh, NC	1808	TN	1865	(3)	Carter Co., TN	1875	66
17.	Schuyler Colfax	New York, NY	1823	IN	1869	Rep.	Mankato, MN	1885	62
18.	Henry Wilson	Farmington, NH	1812	MA	1873	Rep.	Washington, DC	1875	63
19.	William A. Wheeler	Malone, NY	1819	NY	1877	Rep.	Malone, NY	1887	68
20.	Chester A. Arthur	Fairfield, VT	1830	NY	1881	Rep.	New York, NY	1886	57
21.	Thomas A. Hendricks	Muskingum Co., OH	1819	IN	1885	Dem.	Indianapolis, IN	1885	66
22.	Levi P. Morton	Shoreham, VT	1824	NY	1889	Rep.	Rhinebeck, NY	1920	96
23.	Adlai E. Stevenson (4)	Christian Co., KY	1835	IL	1893	Dem.	Chicago, IL	1914	78
24.	Garret A. Hobart	Long Branch, NJ	1844	NJ	1897	Rep.	Paterson, NJ	1899	55
25.	Theodore Roosevelt	New York, NY	1858	NY	1901	Rep.	Oyster Bay, NY	1919	60
26.	Charles W. Fairbanks	Unionville Centre, OH	1852	IN	1905	Rep.	Indianapolis, IN	1918	66
27.	James S. Sherman	Utica, NY	1855	NY	1909	Rep.	Utica, NY	1912	57
28.	Thomas R. Marshall	N. Manchester, IN	1854	IN	1913	Dem.	Washington, DC	1925	71
29.	Calvin Coolidge	Plymouth, VT	1872	MA	1921	Rep.	Northampton, MA	1933	60
30.	Charles G. Dawes	Marietta, OH	1865	IL	1925	Rep.	Evanston, IL	1951	85
31.	Charles Curtis	Topeka, KS	1860	KS	1929	Rep.	Washington, DC	1936	76
32.	John Nance Garner	Red River Co., TX	1868	TX	1933	Dem.	Uvalde, TX	1967	98
33.	Henry Agard Wallace	Adair County, IA	1888	IA	1941	Dem.	Danbury, CT	1965	77
34.	Harry S. Truman	Lamar, MO	1884	MO	1945	Dem.	Kansas City, MO	1972	88
35.	Alben W. Barkley	Graves County, KY	1877	KY	1949	Dem.	Lexington, VA	1956	78
36.	Richard M. Nixon	Yorba Linda, CA	1913	CA	1953	Rep.	New York, NY	1994	81
37.	Lyndon B. Johnson	Johnson City, TX	1908	TX	1961	Dem.	San Antonio, TX	1973	64
38.	Hubert H. Humphrey	Wallace, SD	1911	MN	1965	Dem.	Waverly, MN	1978	66
39.	Spiro T. Agnew (5)	Baltimore, MD	1918	MD	1969	Rep.	Berlin, MD	1996	77
40.	Gerald R. Ford (6)	Omaha, NE	1913	MI	1973	Rep.			
41.	Nelson A. Rockefeller (7)	Bar Harbor, ME	1908	NY	1974	Rep.	New York, NY	1979	70
42.	Walter F. Mondale	Ceylon, MN	1928	MN	1977	Dem.			
43.	George H. W. Bush	Milton, MA	1924	TX	1981	Rep.			
44.	Dan Quayle	Indianapolis, IN	1947	IN	1989	Rep.			
45.	Al Gore	Washington, DC	1948	TN	1993	Dem.			
46.	Richard Cheney	Lincoln, NE	1941	WY	2001	Rep.			

(1) John C. Calhoun resigned Dec. 28, 1832, having been elected to the Senate to fill a vacancy. (2) Richard M. Johnson was the only vice president to be chosen by the Senate because of a tied vote in the Electoral College. (3) Andrew Johnson was a Democrat, nominated vice president by Republicans, and elected with Lincoln on the National Union Ticket. (4) Adlai E. Stevenson, 23d vice president, was grandfather of Democratic candidate for president in 1952 and 1956. (5) Resigned Oct. 10, 1973. (6) First nonelected vice president, chosen under 25th Amendment procedure. (7) Second nonelected vice president, chosen under 25th Amendment procedure.

Biographies of the Presidents

GEORGE WASHINGTON (1789-97), first president, Federalist, was born on Feb. 22, 1732, in Wakefield on Pope's Creek, Westmoreland Co., VA, the son of Augustine and Mary Ball Washington. He spent his early childhood on a farm near Fredericksburg. His father died when George was 11. He studied mathematics and surveying, and at 16, he went to live with his elder half brother, Lawrence, who built and named Mount Vernon. George surveyed the lands of Thomas Fairfax in the Shenandoah Valley, keeping a diary. He accompanied Lawrence to Barbados, West Indies, where he contracted smallpox and was deeply scarred. Lawrence died in 1752, and George inherited his property. He valued land, and when he died, he owned 70,000 acres in Virginia and 40,000 acres in what is now West Virginia.

Washington's military service began in 1753, when Lt. Gov. Robert Dinwiddie of Virginia sent him on missions deep into Ohio country. He clashed with the French and had to surrender Fort Necessity on July 3, 1754. He was an aide to the British general Edward Braddock and was at his side when the army was ambushed and defeated (July 9, 1755) on a march to Fort Duquesne. He helped take Fort Duquesne from the French in 1758.

After Washington's marriage to Martha Dandridge Custis, a widow, in 1759, he managed his family estate at Mount Vernon. Although not at first for independence, he opposed the repressive measures of the British crown and took charge of the Virginia troops before war broke out. He was made commander of the newly created Continental Army by the Continental Congress on June 15, 1775.

The American victory was due largely to Washington's leadership. He was resourceful, a stern disciplinarian, and the one strong, dependable force for unity. Washington favored a federal government. He became chairman of the Constitutional Convention of 1787 and helped get the Constitution ratified. Unanimously elected president by the Electoral College, he was inaugurated Apr. 30, 1789, on the balcony of New York's Federal Hall. He was reelected in 1792. Washington made an effort to avoid partisan politics as president.

Refusing to consider a 3d term, Washington retired to Mount Vernon in March 1797. He suffered acute laryngitis after a ride in snow and rain around his estate, was bled profusely, and died Dec. 14, 1799.

JOHN ADAMS (1797-1801) 2d president, Federalist, was born on Oct. 30, 1735, in Braintree (now Quincy), MA, the son of John and Susanna Boylston Adams. He was a great-grandson of Henry Adams, who came from England in 1636. He graduated from Harvard in 1755 and then taught school and studied law. He married Abigail Smith in 1764. In 1765 he argued against taxation without representation before the royal governor. In 1770 he successfully defended in court the British soldiers who fired on civilians in the Boston Massacre. He was a delegate to the Continental Congress and a signer of the Declaration of Independence. In 1778, Congress sent Adams and John Jay to join Benjamin Franklin as diplomatic representatives in Europe. Because he ran second to Washington in Electoral College balloting in February 1789, Adams became the nation's first vice president, a post he characterized as highly insignificant; he was reelected in 1792.

In 1796 Adams was chosen president by the electors. His administration was marked by growing conflict with fellow Federalist Alexander Hamilton and with others in his own cabinet who supported Hamilton's strongly anti-French position. Adams avoided full-scale war with France, but became unpopular, especially after securing passage of the Alien and Sedition Acts in 1798. His foreign policy contributed significantly to the election of Thomas Jefferson in 1800.

Adams lived for a quarter century after he left office, during which time he wrote extensively. He died July 4, 1826, on the same day as Jefferson (the 50th anniversary of the Declaration of Independence).

THOMAS JEFFERSON (1801-9), 3d president, Democratic-Republican, was born on Apr. 13, 1743, in Shadwell in Goochland (now Albemarle) Co., VA, the son of Peter and Jane Randolph Jefferson. Peter died when Thomas was 14, leaving him 2,750 acres and his slaves. Jefferson attended (1760-62) the College of William and Mary, read Greek and Latin classics, and played the violin. In 1769 he was elected to the Virginia House of Burgesses. In 1770 he began building his home, Monticello, and in 1772 he married Martha Wayles Skelton, a wealthy widow. Jefferson helped establish the Virginia Committee of Correspondence. As a member of the Second Continental Congress he drafted the Declaration of Independence. He also was a member of the Virginia House of Delegates (1776-79) and was elected governor of Virginia in 1779, succeeding Patrick Henry. He was reelected in 1780 but resigned in 1781 after British troops invaded Virginia. During his term he wrote the statute on religious freedom. After his wife's death in 1782, Jefferson again became a delegate to the Congress, and in 1784 he drafted the report that was the basis for the Ordinances of 1784, 1785, and 1787. He was minister to France from 1785 to 1789, when George Washington appointed him secretary of state.

Jefferson's strong faith in the consent of the governed conflicted with the emphasis on executive control, favored by Alexander Hamilton, secretary of the Treasury, and Jefferson resigned on Dec. 31, 1793. In the 1796 election Jefferson was the Democratic-Republican candidate for president; John Adams won the election, and Jefferson became vice president. In 1800, Jefferson and Aaron Burr received equal Electoral College votes; the House of Representatives elected Jefferson president. Jefferson was a strong advocate of westward expansion, major events of his first term were the Louisiana Purchase (1803) and the Lewis and Clark Expedition. An important development during his second term was passage of the Embargo Act, barring U.S. ships from setting sail to foreign ports. Jefferson established the University of Virginia and designed its buildings. He died July 4, 1826, on the same day as John Adams (the 50th anniversary of the Declaration of Independence).

Following analysis of DNA taken from descendants of Jefferson and Sally Hemings, one of his slaves, it has been widely acknowledged that Jefferson fathered at least one, perhaps all, of her six known children.

JAMES MADISON (1809-17), 4th president Democratic-Republican, was born on Mar. 16, 1751, in Port Conway, King George Co., VA, the son of James and Eleanor Rose Conway Madison. Madison graduated from Princeton in 1771. He served in the Virginia Constitutional Convention (1776), and, in 1780, became a delegate to the Second Continental Congress. He was chief recorder at the Constitutional Convention in 1787 and supported ratification in the *Federalist Papers*, written with Alexander Hamilton and John Jay. In 1789, Madison was elected to the House of Representatives, where he helped frame the Bill of Rights and fought against passage of the Alien and Sedition Acts. In the 1790s, he helped found the Democratic-Republican Party, which ultimately became the Democratic Party. He became Jefferson's secretary of state in 1801.

Madison was elected president in 1808. His first term was marked by tensions with Great Britain, and his conduct of foreign policy was criticized by the Federalists and by his own party. Nevertheless, he was reelected in 1812, the year war was declared on Great Britain. The war that many considered a second American revolution ended with a treaty that settled none of the issues. Madison's most important action after the war was demilitarizing the U.S.-Canadian border.

In 1817, Madison retired to his estate, Montpelier, where he served as an elder statesman. He edited his famous papers on the Constitutional Convention and helped found the University of Virginia, of which he became rector in 1826. He died June 28, 1836.

JAMES MONROE (1817-25), 5th president, Democratic-Republican, was born on Apr. 28, 1758, in Westmoreland Co., VA, the son of Spence and Eliza Jones Monroe. He entered the College of William and Mary in 1774 but left to serve in the 3d Virginia Regiment during the American Revolution. After the war, he studied law with Thomas Jefferson. In 1782 he was elected to the Virginia House of Delegates, and he served (1783-86) as a delegate to the Confederation Congress. He opposed ratification of the Constitution because it lacked a bill of rights. Monroe was elected to the U.S. Senate in 1790. In 1794 President George Washington appointed Monroe minister to France. He served twice as governor of Virginia (1799-1802, 1811). President Jefferson also sent him to France as minister (1803), and from 1803 to 1807 he served as minister to Great Britain.

In 1816 Monroe was elected president; he was reelected in 1820 with all but one Electoral College vote. His administration became known as the Era of Good Feeling. He obtained Florida from Spain, settled boundary disputes with Britain over Canada, and eliminated border forts. He supported the antislavery position that led to the Missouri Compromise. His most significant contribution was the Monroe Doctrine, which opposed European intervention in the Western Hemisphere and became a cornerstone of U.S. foreign policy.

Although Monroe retired to Oak Hill, VA, financial problems forced him to sell his property and move to New York City. He died there on July 4, 1831.

JOHN QUINCY ADAMS (1825-29), 6th president, independent Federalist, later Democratic-Republican, was born on July 11, 1767, in Braintree (now Quincy), MA, the son of John and Abigail Adams. His father was the 2d president. He studied abroad and at Harvard University, from which he graduated in 1787. In 1803, he was elected to the U.S. Senate. President Monroe chose him as his secretary of state in 1817. In this capacity he negotiated the cession of the Floridas from Spain, supported exclusion of slavery in the Missouri Compromise, and helped formulate the Monroe Doctrine.

In 1824 Adams was elected president by the House of Representatives after he failed to win an Electoral College majority. His expansion of executive powers was strongly opposed, and in the 1828 election he lost to Andrew Jackson. In 1831 he entered the House of Representatives and served 17 years with distinction. He opposed slavery, the annexation of Texas, and the Mexican War. He helped establish the Smithsonian Institution.

Adams suffered a stroke in the House and died in the Speaker's Room on Feb. 23, 1848.

ANDREW JACKSON (1829-37), 7th president, Democratic-Republican, later a Democrat, was born on Mar. 15, 1767, in the Waxhaw district, on the border of North Carolina and South Carolina, the son of Andrew and Elizabeth Hutchinson Jackson. At the age of 13, he joined the militia to fight in the American Revolution and was captured. Orphaned at the age of 14, Jackson was brought up by a well-to-do uncle. By age 20, he was practicing law, and he later served as prosecuting attorney in Nashville, TN. In 1796 he helped draft the constitution of Tennessee, and for a year he occupied its one seat in the House of Representatives. The next year he served in the U.S. Senate.

In the War of 1812, Jackson crushed (1814) the Creek Indians at Horseshoe Bend, AL, and, with an army consisting chiefly of backwoodsmen, defeated (1815) General Edward Pakenham's British troops at the Battle of New Orleans. In 1818 he briefly invaded Spanish Florida to quell Seminoles and outlaws who harassed frontier settlements. In 1824 he ran for president against John Quincy Adams. Although he won the most popular and electoral votes, he did not have a majority. The House of Representatives decided the election and chose Adams. In the 1828 election, however, Jackson defeated Adams, carrying the West and the South.

As president, Jackson introduced what became known as the spoils system—rewarding party members with government posts. Perhaps his most controversial act, however, was depositing federal funds in so-called pet banks, those directed by Democratic bankers, rather than in the Bank of the United States. "Let the people rule" was his slogan. In 1832, Jackson killed the congressional caucus for nominating presidential candidates and substituted the national convention. When South Carolina refused to collect imports under his protective tariff, he ordered army and naval forces to Charleston. After leaving office in 1837, he retired to the Hermitage, outside Nashville, where he died on June 8, 1845.

MARTIN VAN BUREN (1837-41), 8th president, Democrat, was born on Dec. 5, 1782, in Kinderhook, NY, the son of Abraham and Maria Hoes Van Buren. After attending local schools, he studied law and became a lawyer at the age of 20. A consummate politician, Van Buren began his career in the New York state senate and then served as state attorney general from 1816 to 1819. He was elected to the U.S. Senate in 1821. He helped swing eastern support to Andrew Jackson in the 1828 election and then served as Jackson's secretary of state from 1829 to 1831. In 1832 he was elected vice president. Known as the Little Magician, Van Buren was extremely influential in Jackson's administration.

In 1836, Van Buren defeated William Henry Harrison for president and took office as the financial panic of 1837 initiated a nationwide depression. Although he instituted the independent treasury system, his refusal to spend land revenues led to his defeat by William Henry Harrison in 1840. In 1844 he lost the Democratic nomination to James Knox Polk. In 1848 he again ran for president on the Free Soil ticket but lost. He died in Kinderhook on July 24, 1862.

WILLIAM HENRY HARRISON (1841), 9th president, Whig, who served only 31 days, was born on Feb. 9, 1773, in Berkeley, Charles City Co., VA, the son of Benjamin Harrison, a signer of the Declaration of Independence, and of Elizabeth Bassett Harrison. He attended Hampden-Sydney College. Harrison served as secretary of the Northwest Territory in 1798 and was its delegate to the House of Representatives in 1799. He was the first governor of the Indiana Territory and served as superintendent of Indian affairs. With 900 men he put down a Shawnee uprising at Tippecanoe, IN, on Nov. 7, 1811. A generation later, in 1840, he waged a rousing presidential campaign, using the slogan "Tippecanoe and Tyler too." The Tyler of the slogan was his running mate, John Tyler.

Although born to one of the wealthiest, most prestigious, and most influential families in Virginia, Harrison was elected president with a "log cabin and hard cider" slogan. He caught pneumonia during the inauguration and died Apr. 4, 1841, after only one month in office.

JOHN TYLER (1841-45), 10th president, independent Whig, was born on Mar. 29, 1790, in Greenway, Charles City Co., VA, the son of John and Mary Armistead Tyler. His father was governor of Virginia (1808-11). Tyler graduated from the College of William and Mary in 1807 and in 1811 was elected to the Virginia legislature. In 1816 he was chosen for the U.S. House of Representatives. He served in the Virginia legislature again from 1823 to 1825, when he was elected governor of Virginia. After a stint in the U.S. Senate (1827-36), he was elected vice president (1840).

When William Henry Harrison died only a month after taking office, Tyler succeeded him. Because he was the first person to occupy the presidency without having been elected to that office, he was referred to as "His Accidency." He gained passage of the Preemption Act of 1841, which gave squatters on government land the right to buy 160 acres at the minimum auction price. His last act as president was to sign a resolution annexing Texas. Tyler accepted renomination in 1844 from some Democrats but withdrew in favor of the official party candidate, James K. Polk. He died in Richmond, VA, on Jan. 18, 1862.

IT'S A FACT: On a visit to England in 1855, Millard Fillmore turned down an honorary degree, written in Latin, from Oxford University, saying one shouldn't accept a degree one couldn't read.

JAMES KNOX POLK (1845-49), 11th president, Democrat, was born on Nov. 2, 1795, in Mecklenburg Co., NC, the son of Samuel and Jane Knox Polk. He graduated from the University of North Carolina in 1818 and served in the Tennessee state legislature from 1823 to 1825. He served in the U.S. House of Representatives from 1825 to 1839, the last 4 years as Speaker. He was governor of Tennessee from 1839 to 1841. In 1844, after the Democratic National Convention became deadlocked, it nominated Polk, who thus became the nation's first "dark horse" candidate for president. He was nominated primarily because he was known to favor annexation of Texas.

As president, Polk reestablished the independent treasury system originated by Van Buren. He was so intent on acquiring California from Mexico that he sent troops to the Mexican border and, when Mexicans attacked, declared that a state of war existed. The Mexican War ended with the annexation of California and much of the Southwest as part of America's "manifest destiny." Polk compromised on the Oregon boundary ("54-40 or fight!") by accepting the 49th parallel and yielding Vancouver Island to the British. A few weeks after leaving office, Polk died in Nashville, TN, on June 15, 1849.

ZACHARY TAYLOR (1849-50), 12th president, Whig, who served only 16 months, was born on Nov. 24, 1784, in Orange Co., VA, the son of Richard and Sarah Strother Taylor. He grew up on his father's plantation near Louisville, KY, where he was educated by private tutors. In 1808 Taylor joined the regular army and was commissioned first lieutenant. He fought in the War of 1812, the Black Hawk War (1832), and the second Seminole War (beginning in 1837). He was called "Old Rough and Ready." In 1846 President Polk sent him with an army to the Rio Grande. When the Mexicans attacked him, Polk declared war. Outnumbered 4-1, Taylor defeated (1847) Santa Anna at Buena Vista.

A national hero, Taylor received the Whig nomination in 1848 and was elected president, even though he had never bothered to vote. He resumed the spoils system and, though a slaveholder, worked to admit California as a free state. He fell ill and died in office on July 9, 1850.

MILLARD FILLMORE (1850-53), 13th president, Whig, was born on Jan. 7, 1800, in Cayuga Co., NY, the son of Nathaniel and Phoebe Millard Fillmore. Although he had little schooling, he became a law clerk at the age of 22 and a year later was admitted to the bar. He was elected to the New York state assembly in 1828 and served until 1831. From 1833 until 1835 and again from 1837 to 1843, he represented his district in the U.S. House of Representatives. He opposed the entrance of Texas as a slave territory and voted for a protective tariff. In 1844 he was defeated for governor of New York.

In 1848 he was elected vice president, and he succeeded as president after Taylor's death. Fillmore favored the Compromise of 1850 and signed the Fugitive Slave Law. His policies pleased neither expansionists nor slaveholders, and he was not renominated in 1852. In 1856 he was nominated by the American (Know-Nothing) Party, but despite the support of the Whigs, he was defeated by James Buchanan. He died in Buffalo, NY, on Mar. 8, 1874.

FRANKLIN PIERCE (1853-57), 14th president, Democrat, was born on Nov. 23, 1804, in Hillsboro, NH, the son of Benjamin Pierce, Revolutionary War general and governor of New Hampshire, and Anna Kendrick. He graduated from Bowdoin College in 1824 and was admitted to the bar in 1827. He was elected to the New Hampshire state legislature in 1829 and was chosen Speaker in 1831. He went to the U.S. House in 1833 and was elected a U.S. senator in 1837. He enlisted in the Mexican War and became brigadier general under Gen. Winfield Scott.

In 1852 Pierce was nominated as the Democratic presidential candidate on the 49th ballot. He decisively defeated Gen. Scott, his Whig opponent, in the election. Although against slavery, Pierce was influenced by pro-slavery Southerners. He supported the controversial Kansas-Nebraska Act, which left the question of slavery in the new territories of Kansas and Nebraska to popular vote. Pierce signed a reciprocity treaty with Canada and approved the Gadsden Purchase of a border area on a proposed railroad route, from Mexico. Denied renomination, he spent most of his remaining years in Concord, NH, where he died on Oct. 8, 1869.

JAMES BUCHANAN (1857-61), 15th president, Federalist, later Democrat, was born on Apr. 23, 1791, near Mercersburg, PA, the son of James and Elizabeth Speer Buchanan. He graduated from Dickinson College in 1809 and was admitted to the bar in 1812. He fought in the War of 1812 as a volunteer. He was twice elected to the Pennsylvania general assembly, and in 1821 he entered the U.S. House of Representatives. After briefly serving (1832-33) as minister to Russia, he was elected U.S. senator from Pennsylvania. As Polk's secretary of state (1845-49), he ended the Oregon dispute with Britain and supported the Mexican War and annexation of Texas. As minister to Great Britain, he signed the Ostend Manifesto (1854), declaring a U.S. right to take Cuba by force should efforts to purchase it fail.

Nominated by Democrats, Buchanan was elected president in 1856. On slavery he favored popular sovereignty and choice by state constitutions but did not consistently uphold this position. He denied the right of states to secede but opposed coercion and attempted to keep peace by not provoking secessionists. Buchanan left office having failed to deal decisively with the situation. He died at Wheatland, his estate, near Lancaster, PA, on June 1, 1868.

ABRAHAM LINCOLN (1861-65), 16th president, Republican, was born on Feb. 12, 1809, in a log cabin on a farm then in Hardin Co., KY, now in Larue, the son of Thomas and Nancy Hanks Lincoln. The Lincolns moved to Spencer Co., IN, near Gentryville, when Abe was 7. After Abe's mother died, his father married (1819) Mrs. Sarah Bush Johnston. In 1830 the family moved to Macon Co., IL.

Defeated in 1832 in a race for the state legislature, Lincoln was elected on the Whig ticket 2 years later and served in the lower house from 1834 to 1842. In 1837 Lincoln was admitted to the bar and became partner in a Springfield, IL, law office. He soon won recognition as an effective and resourceful attorney. In 1846, he was elected to the House of Representatives, where he attracted attention during a single term for his opposition to the Mexican War and his position on slavery. In 1856 he campaigned for the newly founded Republican Party, and in 1858 he became its senatorial candidate against Stephen A. Douglas. Although he lost the election, Lincoln gained national recognition from his debates with Douglas.

In 1860, Lincoln was nominated for president by the Republican Party on a platform of restricting slavery. He ran against Douglas, a northern Democrat; John C. Breckinridge, a Southern proslavery Democrat; and John Bell, of the Constitutional Union Party. As a result of Lincoln's winning the election, South Carolina seceded from the Union on Dec. 20, 1860, followed in 1861 by 10 other Southern states.

The Civil War erupted when Fort Sumter, which Lincoln decided to resupply, was attacked by Confederate forces on Apr. 12, 1861. Lincoln called successfully for recruits from the North. On Sept. 22, 1862, 5 days after the Battle of Antietam, Lincoln announced that slaves in territory then in rebellion would be free Jan. 1, 1863, the date of the Emancipation Proclamation. His speeches, including his Gettysburg and Inaugural addresses, are remembered for their eloquence.

Lincoln was reelected, in 1864, over Gen. George B. McClellan, Democrat. Lee surrendered on Apr. 9, 1865. On Apr. 14, Lincoln was shot by actor John Wilkes Booth in Ford's Theater, in Washington, DC. He died the next day.

ANDREW JOHNSON (1865-69), 17th president, Democrat, was born on Dec. 29, 1808, in Raleigh, NC, the son of Jacob and Mary McDonough Johnson. He was apprenticed to a tailor as a youth, but ran away after two years and eventually settled in Greenville, TN. He became popular with the townspeople and in 1829 was elected councilman and later mayor. In 1835 he was sent to the state general assembly. In 1843 he was elected to the U.S. House of Representatives, where he served for 10 years. Johnson was governor of Tennessee from 1853 to 1857, when he was elected to the U.S. Senate. He supported John C. Breckinridge against Lincoln in the 1860 election. Although Johnson had held slaves, he opposed secession and tried to prevent Tennessee from seceding. In Mar. 1862, Lincoln appointed him military governor of occupied Tennessee.

In 1864, in order to balance Lincoln's ticket with a Southern Democrat, the Republicans nominated Johnson for vice president. He was elected vice president with Lincoln and then succeeded to the presidency upon Lincoln's death. Soon afterward, in a controversy with Congress over the president's power over the South, he proclaimed an amnesty to all Confederates, except certain leaders, if they would ratify the 13th Amendment abolishing slavery. States doing so added anti-Negro provisions that enraged Congress, which restored military control over the South. When Johnson removed Edwin M. Stanton, secretary of war, without notifying the Senate, the House, in Feb. 1868, impeached him. Charging him with thereby having violated the Tenure of Office Act, the House was actually responding to his opposition to harsh congressional Reconstruction, expressed in repeated vetoes. He was tried by the Senate, and in May, in two separate votes on different counts, was acquitted, both times by only one vote.

Johnson was denied renomination but remained politically active. He was reelected to the Senate in 1874. Johnson died July 31, 1875, at Carter Station, TN.

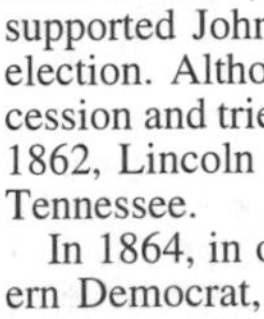

ULYSSES SIMPSON GRANT (1869-77), 18th president, Republican, was born on Apr. 27, 1822, in Point Pleasant, OH, the son of Jesse R. and Hannah Simpson Grant. The next year the family moved to Georgetown, OH. Grant was named Hiram Ulysses, but on entering West Point in 1839, his name was put down as Ulysses Simpson, and he adopted it. He graduated in 1843. During the Mexican War, Grant served under both Gen. Zachary Taylor and Gen. Winfield Scott. In 1854, he resigned his commission because of loneliness and drinking problems, and in the following years he engaged in generally unsuccessful farming and business ventures. With the start of the Civil War, he was named colonel and then brigadier general of the Illinois Volunteers. He took Forts Henry and Donelson and fought at Shiloh. His brilliant campaign against Vicksburg and his victory at Chattanooga made him so prominent that Lincoln placed him in command of all Union armies. Grant accepted Lee's surrender at Appomattox Court House on Apr. 9, 1865. President Johnson appointed Grant secretary of war when he suspended Stanton, but Grant was not confirmed.

Grant was nominated for president by the Republicans in 1868 and elected over Horatio Seymour, Democrat. The 15th Amendment, the amnesty bill, and peaceful settlement of disputes with Great Britain were events of his administration. The Liberal Republicans and Democrats opposed him with Horace Greeley in the 1872 election, but Grant was reelected. His second administration was marked by scandals, including widespread corruption in the Treasury Department and the Indian Service. An attempt by the Stalwarts (Old Guard Republicans) to nominate him in 1880 failed. In 1884 the collapse of an investment firm in which he was a partner left him penniless. He wrote his personal memoirs while ill with cancer and completed them shortly before his death at Mt. McGregor, NY, on July 23, 1885.

RUTHERFORD BIRCHARD HAYES (1877-81), 19th president, Republican, was born on Oct. 4, 1822, in Delaware, OH, the son of Rutherford and Sophia Birchard Hayes. He was reared by his uncle, Sardis Birchard. Hayes graduated from Kenyon College in 1842 and from Harvard Law School in 1845. He practiced law in Lower Sandusky (now Fremont), OH, and was city solicitor of Cincinnati from 1858 to 1861. During the Civil War, he was major of the 23d Ohio Volunteers. He was wounded several times, and by the end of the war he had risen to the rank of brevet major general. While serving (1865-67) in the U.S. House of Representatives, Hayes supported Reconstruction and Johnson's impeachment. He was twice elected governor of Ohio (1867, 1869). After losing a race for the U.S. House in 1872, he was reelected governor of Ohio in 1875.

In 1876, Hayes was nominated for president and believed he had lost the election to Samuel J. Tilden, Democrat. But a few Southern states submitted 2 sets of electoral votes, and the result was in dispute. An electoral commission, consisting of 8 Republicans and 7 Democrats, awarded all disputed votes to Hayes, allowing him to become president by one electoral vote. Hayes, keeping a promise to southerners, withdrew troops from areas still occupied in the South, ending the era of Reconstruction. He proposed civil service reforms, alienating those favoring the spoils system, and advocated repeal of the Tenure of Office Act restricting presidential power to dismiss officials. He supported sound money and specie payments.

Hayes died in Fremont, OH, on Jan. 17, 1893.

JAMES ABRAM GARFIELD (1881), 20th president, Republican, was born on Nov. 19, 1831, in Orange, Cuyahoga Co., OH, the son of Abram and Eliza Ballou Garfield. His father died in 1833, and he was reared in poverty by his mother. He worked as a canal bargeman, a farmer, and a carpenter and managed to secure a college education. He taught at Hiram College and later became principal. In 1859 he was elected to the Ohio legislature. Antislavery and antisecession, he volunteered for military service in the Civil War, becoming colonel of the 42d Ohio Infantry and brigadier in 1862. He fought at Shiloh, was chief of staff for Gen. William Starke Rosecrans, and was made major general for gallantry at Chickamauga. He entered Congress as a radical Republican in 1863, calling for execution or exile of Confederate leaders, but he moderated his views after the Civil War. On the electoral commission in 1877 he voted for Hayes against Tilden on strict party lines.

Garfield was a senator-elect in 1880 when he became the Republican nominee for president. He was chosen as a compromise over Gen. Grant, James G. Blaine, and John Sherman, and won election despite some bitterness among Grant's supporters. Much of his brief tenure as president was concerned with a fight with New York Sen. Roscoe Conkling, who opposed two major appointments made by Garfield. On July 2, 1881, Garfield was shot and seriously wounded by a mentally disturbed office-seeker, Charles J. Guiteau, while entering a railroad station in Washington, DC. He died on Sept. 19, 1881, in Elberon, NJ.

CHESTER ALAN ARTHUR (1881-85), 21st president, Republican, was born on Oct. 5, 1829, in Fairfield, VT, to William and Malvina Stone Arthur. He graduated from Union College in 1848, taught school in Vermont, then studied law and practiced in New York City. In 1853 he argued in a fugitive slave case that slaves transported through New York state were thereby freed. In 1871, he was appointed collector of the Port of New York. President Hayes, an opponent of the spoils system, forced him to resign in 1878. This made the New York machine enemies of Hayes. Arthur and the Stalwarts (Old Guard Republicans) tried to nominate Grant for a 3d term as president in 1880. When Garfield was nominated, Arthur was nominated for vice president in the interests of harmony.

Upon Garfield's assassination, Arthur became president. Despite his past connections, he signed major civil service reform legislation. Arthur tried to dissuade Congress from enacting the high protective tariff of 1883. He was defeated for renomination in 1884 by James G. Blaine. He died in New York City on Nov. 18, 1886.

GROVER CLEVELAND (1885-89; 1893-97) *(According to a ruling of the State Dept., Grover Cleveland should be counted as both the 22d and the 24th president, because his 2 terms were not consecutive.)*

Grover Cleveland, Democrat, was born Stephen Grover Cleveland on Mar. 18, 1837, in Caldwell, NJ, the son of Richard F. and Ann Neal Cleveland. When he was a small boy, his family moved to New York. Prevented by his father's death from attending college, he studied by himself and was admitted to the bar in Buffalo, NY, in 1859. In succession he became assistant district attorney (1863), sheriff (1871), mayor (1881), and governor of New York (1882). He was an independent, honest administrator who hated corruption. Cleveland was nominated for president over Tammany Hall opposition in 1884 and defeated Republican James G. Blaine.

As president, he enlarged the civil service and vetoed many pension raids on the Treasury. In the 1888 election he was defeated by Benjamin Harrison, although his popular vote was larger. Reelected over Harrison in 1892, he faced a money crisis brought about by a lowered gold reserve, circulation of paper, and exorbitant silver purchases under the Sherman Silver Purchase Act. He obtained a repeal of the Sherman Act, but was unable to secure effective tariff reform. A severe economic depression and labor troubles racked his administration, but he refused to interfere in business matters and rejected Jacob Coxey's demand for unemployment relief. In 1894, he broke the Pullman strike. Cleveland was not renominated in 1896. He died in Princeton, NJ, on June 24, 1908.

BENJAMIN HARRISON (1889-93), 23d president, Republican, was born on Aug. 20, 1833, in North Bend, OH, the son of John Scott and Elizabeth Irwin Harrison. His great-grandfather, Benjamin Harrison, was a signer of the Declaration of Independence; his grandfather, William Henry Harrison, was 9th president; his father was a member of Congress. He attended school on his father's farm and graduated from Miami University in Oxford, OH, in 1852. He was admitted to the bar in 1854 and practiced in Indianapolis. During the Civil War, he rose to the rank of brevet brigadier general and fought at Kennesaw Mountain, at Peachtree Creek, at Nashville, and in the Atlanta campaign. He lost the 1876 gubernatorial election in Indiana but succeeded in becoming a U.S. senator in 1881.

In 1888 he defeated Cleveland for president despite receiving fewer popular votes. As president, he expanded the pension list and signed the McKinley high tariff bill, the Sherman Antitrust Act, and the Sherman Silver Purchase Act. During his administration, 6 states were admitted to the Union. He was defeated for reelection in 1892. He died in Indianapolis on Mar. 13, 1901.

WILLIAM MCKINLEY (1897-1901), 25th president, Republican, was born on Jan. 29, 1843, in Niles, OH, the son of William and Nancy Allison McKinley. McKinley briefly attended Allegheny College. When the Civil War broke out in 1861, he enlisted and served for the duration. He rose to captain and in 1865 was made brevet major. After studying law in Albany, NY, he opened (1867) a law office in Canton, OH. He served twice in the U.S. House of Representatives (1877-83; 1885-91) and led the fight there for the McKinley Tariff, which was passed in 1890. However, he was not reelected to the House as a result. He served two terms (1892-96) as governor of Ohio.

In 1896 he was elected president as a proponent of a protective tariff and sound money (gold standard), over William Jennings Bryan, the Democrat and a proponent of free silver. McKinley was reluctant to intervene in Cuba, but the loss of the battleship *Maine* at Havana crystallized opinion. He demanded Spain's withdrawal from Cuba; Spain made some concessions, but Congress announced a state of war as of Apr. 21, 1898. He was reelected in the 1900 campaign, defeating Bryan's anti-imperialist arguments with the promise of a "full dinner pail." McKinley was respected for his conciliatory nature and for his conservative stance on business issues. On Sept. 6, 1901, while welcoming citizens at the Pan-American Exposition, in Buffalo, NY, he was shot by Leon Czolgosz, an anarchist. He died Sept. 14.

WORLD ALMANAC EDITORS' PICKS

The World Almanac staff ranked the following as the most obscure U.S. presidents:

1. Millard Fillmore
2. Franklin Pierce
3. Chester Alan Arthur
4. William Henry Harrison
5. Benjamin Harrison

THEODORE ROOSEVELT (1901-9), 26th president, Republican, was born on Oct. 27, 1858, in New York City, the son of Theodore and Martha Bulloch Roosevelt. He was a 5th cousin of Franklin D. Roosevelt and an uncle of Eleanor Roosevelt. Roosevelt graduated from Harvard University in 1880. He attended Columbia Law School briefly but abandoned the study of law to enter politics. He was elected to the New York state assembly in 1881 and served until 1884. He spent the next 2 years ranching and hunting in the Dakota Territory. Back in politics in 1886, he ran unsuccessfully for mayor of New York City. He was Civil Service commissioner in Washington, DC, from 1889 to 1895. From 1895 to 1897, he served as New York City's police commissioner. He was assistant secretary of the navy under McKinley. The Spanish-American War made Roosevelt a nationally known figure. He organized the 1st U.S. Volunteer Cavalry (Rough Riders) and, as lieutenant colonel, led the charge up Kettle Hill in San Juan. Elected New York governor in 1898, he fought the spoils system and achieved taxation of corporation franchises.

Nominated for vice president in 1900, he became the nation's youngest president when McKinley was assassinated. He was reelected in 1904. As president he fought corruption of politics by big business, dissolved the Northern Securities Co. and others for violating antitrust laws, intervened in the 1902 coal strike on behalf of the public, obtained the Elkins Law (1903) forbidding rebates to favored corporations, and helped pass the Hepburn Railway Rate Act of 1906 (extending Jurisdiction of the Interstate Commerce Commission). He helped obtain passage of the Pure Food and Drug Act (1906), and employers' liability laws. Roosevelt vigorously organized conservation efforts. He mediated (1905) the peace between Japan and Russia, for which he won the Nobel Peace Prize. He abetted the 1903 revolution in Panama that led to U.S. acquisition of territory for the Panama Canal.

In 1908 Roosevelt obtained the nomination of William H. Taft, who was elected. Feeling that Taft had abandoned his policies, Roosevelt unsuccessfully sought the nomination in 1912. He bolted the party and ran on the Progressive "Bull Moose" ticket against Taft and Woodrow Wilson, splitting the Republicans and ensuring Wilson's election. He was shot during the campaign but recovered. In 1916, after unsuccessfully seeking the presidential nomination for himself, Roosevelt supported the Republican candidate, Charles E. Hughes. A strong friend of Britain, he fought for American intervention in World War I. He wrote some 40 books, of which his book *The Winning of the West* is perhaps best known. He died Jan. 6, 1919, at Sagamore Hill, Oyster Bay, NY.

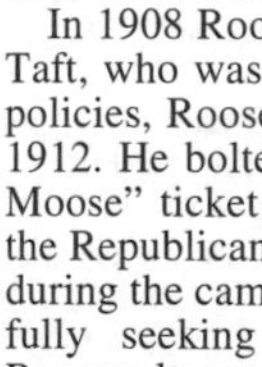

WILLIAM HOWARD TAFT (1909-13), 27th president, Republican, and 10th chief justice of the U.S., was born on Sept. 15, 1857, in Cincinnati, OH, the son of Alphonso and Louisa Maria Torrey Taft. His father was secretary of war and attorney general in Grant's cabinet and minister to Austria and Russia under Arthur. Taft graduated from Yale in 1878 and from Cincinnati Law School in 1880. After working as a law reporter for Cincinnati newspapers, he served as assistant prosecuting attorney (1881-82), assistant county solicitor (1885), judge, superior court (1887), U.S.

solicitor-general (1890), and federal circuit judge (1892). In 1900 he became head of the U.S. Philippines Commission and was the first civil governor of the Philippines (1901-4). In 1904 he served as secretary of war, and in 1906 he was sent to Cuba to help avert a threatened revolution.

Taft was groomed for the presidency by Theodore Roosevelt and elected over William Jennings Bryan in 1908. Taft vigorously continued Roosevelt's trust-busting, instituted the Department of Labor, and drafted the amendments calling for direct election of senators and the income tax. His tariff and conservation policies angered progressives. Although renominated in 1912, he was opposed by Roosevelt, who ran on the Progressive Party ticket; the result was Democrat Woodrow Wilson's election.

Taft, with some reservations, supported the League of Nations. After leaving office, he was professor of constitutional law at Yale (1913-21) and chief justice of the U.S. (1921-30). Taft was the only person in U.S. history to have been both president and chief justice. He died in Washington, DC, on Mar. 8, 1930.

(THOMAS)WOODROW WILSON (1913-21), 28th president, Democrat, was born on Dec. 28, 1856, in Staunton, VA, the son of Joseph Ruggles and Janet (Jessie) Woodrow Wilson. He grew up in Georgia and South Carolina. He attended Davidson College in North Carolina before graduating from Princeton University in 1879. He studied law at the University of Virginia and political science at Johns Hopkins University, where he received his PhD in 1886. He taught at Bryn Mawr (1885-88) and then at Wesleyan (1888-90) before joining the faculty at Princeton. He was president of Princeton from 1902 until 1910, when he was elected governor of New Jersey. In 1912 he was nominated for president with the aid of William Jennings Bryan, who sought to block James "Champ" Clark and Tammany Hall. Wilson won because the Republican vote for Taft was split by the Progressives.

As president, Wilson protected American interests in revolutionary Mexico and fought for American rights on the high seas. He oversaw the creation of the Federal Reserve system, cut the tariff, and developed a reputation as a reformer. His sharp warnings to Germany led to the resignation of his secretary of state, Bryan, a pacifist. In 1916 he was reelected by a slim margin with the slogan, "He kept us out of war," although his attempts to mediate in the war failed. After several American ships had been sunk by the Germans, he secured a declaration of war against Germany on Apr. 6, 1917.

Wilson outlined his peace program on Jan. 8, 1918, in the Fourteen Points, a state paper that had worldwide influence. He enunciated a doctrine of self-determination for the settlement of territorial disputes. The Germans accepted his terms and an armistice on Nov. 11, 1918.

Wilson went to Paris to help negotiate the peace treaty, the crux of which he considered the League of Nations. The Senate demanded reservations that would not make the U.S. subordinate to the votes of other nations in case of war. Wilson refused to consider any reservations and toured the country to get support. He suffered a stroke in Oct. 1919. An invalid for months, he clung to his executive powers while his wife and doctors effectively functioned as president.

Wilson was awarded the 1919 Nobel Peace Prize, but the treaty embodying the League of Nations was ultimately rejected by the Senate in 1920. He left the White House in Mar. 1921. He died in Washington, DC, on Feb. 3, 1924.

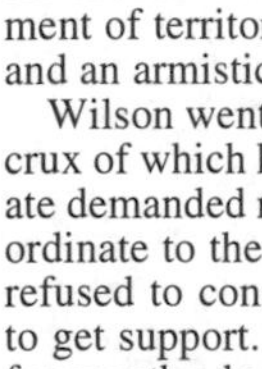

WARREN GAMALIEL HARDING (1921-23), 29th president, Republican, was born on Nov. 2, 1865, near Corsica (now Blooming Grove), OH, the son of George Tyron and Phoebe Elizabeth Dickerson Harding. He attended Ohio Central College, studied law, and became editor and publisher of a county newspaper. He entered the political arena as state senator (1901-4) and then served as lieutenant governor (1904-6). In 1910 he ran unsuccessfully for governor of Ohio; then in 1914 he was elected to the U.S. Senate. In the Senate he voted for antistrike legislation, woman suffrage, and the Volstead Prohibition Enforcement Act over President Wilson's veto. He opposed the League of Nations. In 1920 he was nominated for president and defeated James M. Cox in the election. The Republicans capitalized on war weariness and fear that Wilson's League of Nations would curtail U.S. sovereignty.

Harding stressed a return to "normalcy" and worked for tariff revision and the repeal of excess profits law and high income taxes. His secretary of interior, Albert B. Fall, became involved in the Teapot Dome scandal. As rumors began to circulate about the corruption in his administration, Harding became ill while returning from a trip to Alaska, and he died in San Francisco on Aug. 2, 1923.

(JOHN) CALVIN COOLIDGE (1923-29), 30th president, Republican, was born on July 4, 1872, in Plymouth, VT, the son of John Calvin and Victoria J. Moor Coolidge. Coolidge graduated from Amherst College in 1895. He entered Republican state politics and served as mayor of Northampton, MA, as state senator, as lieutenant governor, and, in 1919, as governor. In Sept. 1919, Coolidge attained national prominence by calling out the state guard in the Boston police strike. He declared: "There is no right to strike against the public safety by anybody, anywhere, anytime." This brought his name before the Republican convention of 1920, where he was nominated for vice president.

Coolige succeeded to the presidency on Harding's death. As president, he opposed the League of Nations and the soldiers' bonus bill, which was passed over his veto. In 1924 he was elected to the presidency by a huge majority. He substantially reduced the national debt. He twice vetoed the McNary-Haugen farm bill, which would have provided relief to financially hard-pressed farmers.

With Republicans eager to renominate him, Coolidge simply announced, Aug. 2, 1927: "I do not choose to run for president in 1928." He died in Northampton, MA, on Jan. 5, 1933.

HERBERT CLARK HOOVER (1929-33), 31st president, Republican, was born on Aug. 10, 1874, in West Branch, IA, the son of Jesse Clark and Hulda Randall Minthorn Hoover. Hoover grew up in Indian Territory (now Oklahoma) and Oregon and graduated from Stanford University with a degree in engineering in 1895. He worked briefly with the U.S. Geological Survey and then managed mines in Australia, Asia, Europe, and Africa. While chief engineer of imperial mines in China, he directed food relief for victims of the Boxer Rebellion. He gained a reputation not only as an engineer but as a humanitarian as he directed the American Relief Committee, London (1914-15) and the U.S. Commission for Relief in Belgium (1915-19). He was U.S. Food Administrator (1917-19), American Relief Administrator (1918-23), and in charge of Russian Relief (1918-23). He served as secretary of commerce under both Harding and Coolidge. Some historians believe that he was the most effective secretary of commerce ever to hold that office.

In 1928 Hoover was elected president over Alfred E. Smith. In 1929 the stock market crashed, and the economy collapsed. During the Great Depression, Hoover inaugurated some government assistance programs, but he was opposed to administration of aid through a federal bureaucracy. As the effects of the depression continued, he was defeated in the 1932 election by Franklin D. Roosevelt. Hoover remained active after leaving office. President Truman named him coordinator of the European Food Program (1946) and chairman of the Commission on Organization of the Executive Branch (1947-49; 1953-55). Hoover died in New York City on Oct. 20, 1964.

IT'S A FACT: The first president born west of the Mississippi river was Herbert Hoover, who was born in West Branch, IA in 1874, and elected president in 1928.

FRANKLIN DELANO ROOSEVELT (1933-45), 32d president, Democrat, was born on Jan. 30, 1882, near Hyde Park, NY, the son of James and Sara Delano Roosevelt. He graduated from Harvard University in 1904. He attended Columbia University Law School without taking a degree and was admitted to the New York state bar in 1907. His political career began when he was elected to the New York state senate in 1910. In 1913 President Wilson appointed him assistant secretary of the navy, a post he held during World War I.

In 1920 Roosevelt ran for vice president with James Cox and was defeated. From 1921 to 1928 he worked in his New York law office and was also vice president of a bank. In Aug. 1921, he was stricken with poliomyelitis, which left his legs paralyzed. As a result of therapy he was able to stand, or walk a few steps, with the aid of leg braces.

Roosevelt served 2 terms as governor of New York (1929-33). In 1932, W. G. McAdoo, pledged to John N. Garner, threw his votes to Roosevelt, who was nominated for president. The Depression and the promise to repeal Prohibition ensured his election. He asked for emergency powers, proclaimed the New Deal, and put into effect a vast number of administrative changes. Foremost was the use of public funds for relief and public works, resulting in deficit financing. He greatly expanded the federal government's regulation of business and by an excess profits tax and progressive income taxes produced a redistribution of earnings on an unprecedented scale. He also promoted legislation establishing the Social Security system. He was the last president inaugurated on Mar. 4 (1933) and the first inaugurated on Jan. 20 (1937).

Roosevelt was the first president to use radio for "fireside chats." When the Supreme Court nullified some New Deal laws, he sought power to "pack" the Court with additional justices, but Congress refused to give him the authority. He was the first president to break the "no 3d term" tradition (1940) and was elected to a 4th term in 1944, despite failing health. Roosevelt was openly hostile to fascist governments before World War II and launched a lend-lease program on behalf of the Allies. With British Prime Min. Winston Churchill he wrote a declaration of principles to be followed after Nazi defeat (the Atlantic Charter of Aug. 14, 1941) and urged the Four Freedoms (freedom of speech, of worship, from want, from fear) Jan. 6, 1941. When Japan attacked Pearl Harbor on Dec. 7, 1941, the U.S. entered the war. Roosevelt conferred with allied heads of state at Casablanca (Jan. 1943), Quebec (Aug. 1943), Tehran (Nov.-Dec. 1943), Cairo (Nov. and Dec. 1943), and Yalta (Feb. 1945).

Roosevelt did not, however, live to see the end of the war. He died of a cerebral hemorrhage in Warm Springs, GA, on Apr. 12, 1945.

HARRY S. TRUMAN (1945-53), 33d president, Democrat, was born on May 8, 1884, in Lamar, MO, the son of John Anderson and Martha Ellen Young Truman. A family disagreement on whether his middle name should be Shippe or Solomon, after names of 2 grandfathers, resulted in his using only the middle initial S. After graduating from high school in Independence, MO, he worked (1901) for the *Kansas City Star,* as a railroad timekeeper, and as a clerk in Kansas City banks until about 1905. He ran his family's farm from 1906 to 1917. He served in France during World War I. After the war he opened a haberdashery shop, was a judge on the Jackson Co. Court (1922-24), and attended Kansas City School of Law (1923-25).

Truman was elected to the U.S. Senate in 1934 and reelected in 1940. In 1944, with Roosevelt's backing, he was nominated for vice president and elected. On Roosevelt's death in 1945, Truman became president. In 1948, in a famous upset victory, he defeated Republican Thomas E. Dewey to win election to a new term.

Truman authorized the first uses of the atomic bomb (Hiroshima and Nagasaki, Aug. 6 and 9, 1945), bringing World War II to a rapid end. He was responsible for what came to be called the Truman Doctrine (to aid nations such as Greece and Turkey, threatened by Communist takeover), and his strong commitment to NATO and to the Marshall Plan helped bring them about. In 1948-49, he broke a Soviet blockade of West Berlin with a massive airlift. When Communist North Korea invaded South Korea (June 1950), he won UN approval for a "police action" and sent in forces under Gen. Douglas MacArthur. When MacArthur opposed his policy of limited objectives, Truman removed him.

He died in Kansas City, MO, on Dec. 26, 1972.

DWIGHT DAVID EISENHOWER (1953-61), 34th president, Republican, was born on Oct. 14, 1890, in Denison, TX, the son of David Jacob and Ida Elizabeth Stover Eisenhower. He grew up on a small farm in Abilene, KS, and graduated from West Point in 1915. He was on the staff of Gen. Douglas MacArthur in the Philippines from 1935 to 1939. In 1942, he was made commander of Allied forces landing in North Africa; the next year he was made full general. He became supreme Allied commander in Europe that same year and as such led the Normandy invasion (June 6, 1944). He was given the rank of general of the army on Dec. 20, 1944, which was made permanent in 1946. On May 7, 1945, Eisenhower received the surrender of Germany at Rheims. He returned to the U.S. to serve as chief of staff (1945-48). His war memoir, *Crusade in Europe* (1948), was a best-seller. In 1948 he became president of Columbia University; in 1950 he became Commander of NATO forces.

Eisenhower resigned from the army and was nominated for president by the Republicans in 1952. He defeated Adlai E. Stevenson in the 1952 election and again in 1956. Eisenhower called himself a moderate, favored the "free market system" vs. government price and wage controls, kept government out of labor disputes, reorganized the defense establishment, and promoted missile programs. He continued foreign aid, sped the end of the Korean War, endorsed Taiwan and SE Asia defense treaties, backed the UN in condemning the Anglo-French raid on Egypt, and advocated the "open skies" policy of mutual inspection with the USSR. He sent U.S. troops into Little Rock, AR, in Sept. 1957, during the segregation crisis.

Eisenhower died on Mar. 28, 1969, in Washington, DC.

JOHN FITZGERALD KENNEDY (1961-63), 35th president, Democrat, was born on May 29, 1917, in Brookline, MA, the son of Joseph P. and Rose Fitzgerald Kennedy. He graduated from Harvard University in 1940. While serving in the navy (1941-45), he commanded a PT boat in the Solomons and won the Navy and Marine Corps Medal. In 1956, while recovering from spinal surgery, he wrote *Profiles in Courage,* which won a Pulitzer Prize in 1957. He served in the House of Representatives from 1947 to 1953 and was elected to the Senate in 1952 and again in 1958. In 1960, Kennedy won the Democratic nomination for president and narrowly defeated Republican Vice Pres. Richard M. Nixon. Kennedy was the youngest president ever elected to the office and the first Roman Catholic.

Kennedy also defied Soviet attempts to force the Allies out of Berlin. He started the Peace Corps, and he backed civil rights and expanded medical care for the aged. Space exploration was greatly developed during his administration.

In Apr. 1961, the new Kennedy administration suffered a severe setback when an invasion force of anti-Castro Cubans, trained and directed by the CIA, failed to establish a beachhead at the Bay of Pigs in Cuba. By the same token, one of Kennedy's most important acts as president was his successful demand on Oct. 22, 1962, that the Soviet Union dismantle its missile bases in Cuba.

On Nov. 22, 1963, Kennedy was assassinated while riding in a motorcade in Dallas, TX.

IT'S A FACT: Zachary Taylor, Ulysses S. Grant, Dwight D. Eisenhower, and Herbert Hoover were all elected to the presidency without having been elected before to any public office. The first three were distinguished generals; Hoover was an engineer and former commerce secretary.

LYNDON BAINES JOHNSON (1963-69), 36th president, Democrat, was born on Aug. 27, 1908, near Stonewall, TX, the son of Sam Ealy and Rebekah Baines Johnson. He graduated from Southwest Texas State Teachers College in 1930 and attended Georgetown University Law School. He taught public speaking in Houston (1930-31) and then served as secretary to Rep. R. M. Kleberg (1931-35). In 1937 Johnson won an election to fill the vacancy caused by the death of a U.S. representative and in 1938 was elected to the full term, after which he returned for 4 terms. During 1941 and 1942 he also served in the Navy in the Pacific, earning a Silver Star for bravery. He was elected U.S. senator in 1948 and reelected in 1954. He became Democratic leader of the Senate in 1953. Johnson had strong support for the Democratic presidential nomination at the 1960 convention, where the nominee, John F. Kennedy, asked him to run for vice president. His campaigning helped overcome religious bias against Kennedy in the South.

Johnson became president when Kennedy was assassinated. He was elected to a full term in 1964. Johnson's domestic program was of considerable importance. He won passage of major civil rights, anti-poverty, aid to education, and health-care (Medicare, Medicaid) legislation—the "Great Society" program. However, his escalation of the war in Vietnam came to overshadow the achievements of his administration. In the face of increasing division in the nation and in his own party over his handling of the war, Johnson declined to seek another term.

Johnson died on Jan. 22, 1973, in San Antonio, TX.

RICHARD MILHOUS NIXON (1969-74), 37th president, Republican, was born on Jan. 9, 1913, in Yorba Linda, CA, the son of Francis Anthony and Hannah Milhous Nixon. He graduated from Whittier College in 1934 and from Duke University Law School in 1937. After practicing law in Whittier and serving briefly in the Office of Price Administration in 1942, he entered the Navy and served in the South Pacific. Nixon was elected to the House of Representatives in 1946 and 1948. He achieved prominence as the House Un-American Activities Committee member who forced the showdown leading to the Alger Hiss perjury conviction. In 1950 he was elected to the Senate.

Nixon was elected vice president in the Eisenhower landslides of 1952 and 1956. He won the Republican nomination for president in 1960 but was narrowly defeated by John F. Kennedy. He ran unsuccessfully for governor of California in 1962. In 1968 he again won the GOP presidential nomination, then defeated Hubert Humphrey for the presidency.

Nixon's 2d term was cut short by scandal, after disclosures relating to a June 1972 burglary of Democratic Party headquarters in the Watergate office complex. The courts and Congress sought tapes of Nixon's office conversations and calls for criminal proceedings against former White House aides and for a House inquiry into possible impeachment. Nixon claimed executive privilege, but the Supreme Court ruled against him. In July the House Judiciary Committee recommended adoption of 3 impeachment articles charging him with obstruction of justice, abuse of power, and contempt of Congress. On Aug. 5, he released transcripts of conversations that linked him to cover-up activities. He resigned on Aug. 9, becoming the first president ever to do so. In later years, Nixon emerged as an elder statesman.

As president, Nixon appointed 4 Supreme Court justices, including the chief justice, moving the court to the right, and as a "new federalist" sought to shift responsibility to state and local governments. He dramatically altered relations with China, which he visited in 1972—the first president to do so. With foreign affairs adviser Henry Kissinger he pursued détente with the Soviet Union. He began a gradual withdrawal from Vietnam, but U.S. troops remained there through his first term. He ordered an incursion into Cambodia (1970) and the bombing of Hanoi and mining of Haiphong Harbor (1972). Reelected by a large majority in Nov. 1972, he secured a Vietnam cease-fire in Jan. 1973.

Nixon died Apr. 22, 1994, in New York City.

GERALD RUDOLPH FORD (1974-77), 38th president, Republican, was born on July 14, 1913, in Omaha, NE, the son of Leslie and Dorothy Gardner King, and was named Leslie Jr. When he was 2, his parents were divorced, and his mother moved with the boy to Grand Rapids, MI. There she met and married Gerald R. Ford, who formally adopted him and gave him his own name. Ford graduated from the University of Michigan in 1935 and from Yale Law School in 1941. He began practicing law in Grand Rapids, but in 1942 joined the navy and served in the Pacific, leaving the service in 1946 as a lieutenant commander. He entered the House of Representatives in 1949 and spent 25 years in the House, 8 of them as Republican leader.

On Oct. 12, 1973, after Vice President Spiro T. Agnew resigned, Ford was nominated by President Nixon to replace him. It was the first use of the procedures set out in the 25th Amendment. When Nixon resigned, Aug. 9, 1974, because of the Watergate scandal, Ford became president; he was the only president who was never elected either to the presidency or to the vice presidency. On Sept. 8, in a controversial move, he pardoned Nixon for any federal crimes he might have committed as president. Ford vetoed 48 bills in his first 21 months in office, mostly in the interest of fighting high inflation; he was less successful in curbing high unemployment. In foreign policy, Ford continued to pursue détente.

Ford was narrowly defeated in the 1976 election.

JIMMY (JAMES EARL) CARTER (1977-81), 39th president, Democrat, was the first president from the Deep South since before the Civil War. He was born on Oct. 1, 1924, in Plains, GA, the son of James and Lillian Gordy Carter.

Carter graduated from the U.S. Naval Academy in 1946 and in 1952 entered the navy's nuclear submarine program as an aide to Capt. (later Adm.) Hyman Rickover. He studied nuclear physics at Union College. Carter's father died in 1953, and he left the navy to take over the family peanut farming businesses. He served in the Georgia state senate (1963-67) and as governor of Georgia (1971-75). In 1976, Carter won the Democratic nomination and defeated President Gerald R. Ford.

On his first full day in office, Carter pardoned all Vietnam draft evaders. He played a major role in the negotiations leading to the 1979 peace treaty between Israel and Egypt, and he won passage of new treaties with Panama providing for U.S. control of the Panama Canal to end at the start of the year 2000. However, Carter was widely criticized for the poor state of the economy and was viewed by some as weak in his handling of foreign policy. In Nov. 1979, Iranian student militants attacked the U.S. embassy in Tehran and held members of the embassy staff hostage. Efforts to obtain release of the hostages were a major preoccupation during the rest of his term. He reacted to the Soviet invasion of Afghanistan by imposing a grain embargo and boycotting the Moscow Olympic Games.

Carter was defeated by Ronald Reagan in the 1980 election. Carter administration efforts finally resulted in the release of the American hostages, but not until Inauguration Day, 1981, just after Reagan officially became president. After leaving office, Carter was hailed for his humanitarian efforts and took a prominent role in mediating international disputes.

RONALD WILSON REAGAN (1981-89), 40th president, Republican, was born on Feb. 6, 1911, in Tampico, IL, the son of John Edward and Nellie Wilson Reagan. Reagan graduated from Eureka College in 1932, after which he worked as a sports announcer in Des Moines, IA. He began a successful career as an actor in 1937, starring in numerous movies, and later in television, until the 1960s. He served as president of the Screen Actors Guild from 1947 to 1952 and in 1959-60. Reagan was elected governor of California in 1966 and reelected in 1970.

In 1980, Reagan gained the Republican presidential nomination and won a landslide victory over Jimmy Carter. He was easily reelected in 1984. Reagan successfully forged a bipartisan coalition in Congress, which led to enactment of his program of large-scale tax cuts, cutbacks in many government programs, and a major defense buildup. He signed a Social Security reform bill designed to provide for the long-term solvency of the system. In 1986, he signed into law a major tax-reform bill. He was shot and wounded in an assassination attempt in 1981.

In 1982, the U.S. joined France and Italy in maintaining a peacekeeping force in Beirut, Lebanon, and the next year Reagan sent a task force to invade the island of Grenada after 2 Marxist coups there. Reagan's opposition to international terrorism led to the U.S. bombing of Libyan military installations in 1986. He strongly supported El Salvador, the Nicaraguan contras, and other anti-communist governments and forces throughout the world. He also held 4 summit meetings with Soviet leader Mikhail Gorbachev. At the 1987 meeting in Washington, DC, a historic treaty eliminating short- and medium-range missiles from Europe was signed.

Reagan faced a crisis in 1986-87, when it was revealed that the U.S. had sold weapons through Israeli brokers to Iran in exchange for release of U.S. hostages being held in Lebanon and that subsequently some of the money was diverted to the Nicaraguan contras (Congress had barred U.S. aid to the contras). The scandal led to the resignation of leading White House aides. As Reagan left office in Jan. 1989, the nation was experiencing its 6th consecutive year of economic prosperity. Over the same period, however, the federal government consistently recorded large budget deficits.

In 1994, in a letter to the American people, Reagan revealed that he was suffering from Alzheimer's disease.

GEORGE HERBERT WALKER BUSH (1989-93), 41st president, Republican, was born on June 12, 1924, in Milton, MA, the son of Prescott and Dorothy Walker Bush. He served as a U.S. Navy pilot in World War II. After graduating from Yale University in 1948, he settled in Texas, where, in 1953, he helped found an oil company. After losing a bid for a U.S. Senate seat in Texas in 1964, he was elected to the House of Representatives in 1966 and 1968. He lost a 2d U.S. Senate race in 1970. Subsequently he served as U.S. ambassador to the United Nations (1971-73), headed the U.S. Liaison Office in Beijing (1974-75), and was director of central intelligence (1976-77).

Following an unsuccessful bid for the 1980 Republican presidential nomination, Bush was chosen by Ronald Reagan as his vice presidential running mate. He served as U.S. vice president from 1981 to 1989.

In 1988, Bush gained the Republican presidential nomination and defeated Democrat Michael Dukakis in the November election. Bush took office faced with the ongoing U.S. budget and trade deficits as well as the rescue of insolvent U.S. savings and loan institutions. He faced a severe budget deficit annually, struggled with military cutbacks in light of reduced cold war tensions, and vetoed abortion-rights legislation. In 1990 he agreed to a budget deficit-reduction plan that included tax hikes.

Bush supported Soviet reforms and Eastern Europe democratization. He was criticized by some for keeping U.S. policy tied closely to Mikhail Gorbachev as the Soviet leader lost power and for underreaction to China's violent repression of pro-democracy demonstrators in 1989. In Dec. 1989, Bush sent troops to Panama; they overthrew the government and captured strongman Gen. Manuel Noriega.

Bush reacted to Iraq's Aug. 1990 invasion of Kuwait by sending U.S. forces to the Persian Gulf area and assembling a UN-backed coalition, including NATO and Arab League members. After a month-long air war, in Feb. 1991, Allied forces retook Kuwait in a 4-day ground assault. The quick victory, with extremely light casualties on the U.S. side, gave Bush at the time one of the highest presidential approval ratings in history. His popularity plummeted by the end of 1991, however, as the economy slipped into recession. He was defeated by his Democratic opponent, Bill Clinton, in the 1992 election.

BILL (WILLIAM JEFFERSON) CLINTON (1993-2001), 42d president, Democrat, was born on Aug. 19, 1946, in Hope, AR, son of William Blythe and Virginia Cassidy Blythe, and was named William Jefferson Blythe IV. Blythe died in an automobile accident before his son was born. His widow married Roger Clinton, and at the age of 16, William Jefferson Blythe IV changed his last name to Clinton. Clinton graduated from Georgetown University in 1968, attended Oxford University as a Rhodes scholar, and earned a degree from Yale Law School in 1973.

Clinton worked on George McGovern's 1972 presidential campaign. He taught at the University of Arkansas from 1973 to 1976, when he was elected state attorney general. In 1978, he was elected governor, becoming the nation's youngest. Defeated for reelection in 1980, he was returned to office in 1982, 1984, 1986, 1988, and 1990. He married Hillary Rodham in 1975.

Despite questions about his character, Clinton won most of the 1992 presidential primaries, moving his party toward the center as he tried to broaden his appeal; as the party's presidential nominee he defeated Pres. George Bush and Reform Party candidate Ross Perot in the November election. In 1993, Clinton won passage of a measure to reduce the federal budget deficit and won congressional approval of the North American Free Trade Agreement. His administration's plan for major health-care reform legislation died in Congress.

After 1994 midterm elections, Clinton faced Republican majorities in both houses of Congress. He followed a centrist course at home, sent troops to Bosnia to help implement a peace settlement, and cultivated relations with Russia and China.

Though accused of improprieties in his involvement in an Arkansas real estate venture (Whitewater) and in other matters, Clinton easily won reelection in 1996. In 1997 he reached agreement with Congress on legislation to balance the federal budget by 2002. In 1998, Clinton became only the 2d U.S. president ever to be impeached by the House of Representatives. Charged with perjury and obstruction of justice in connection with an attempted cover-up of a sexual relationship with a former White House intern, he was acquitted by the Senate in 1999. Despite the scandal he retained wide popularity, aided by a strong economy.

In 1999, the United States, under Clinton, joined other NATO nations in an aerial bombing campaign that ultimately induced Serbia to withdraw troops from the Kosovo region, where they had been terrorizing and driving out ethnic Albanians. During Clinton's last years in office he continued to pursue improved relations with Russia and China and promoted Arab–Israeli talks aimed at a Middle East peace settlement.

After leaving office, Clinton remained very much in the public eye, especially because of controversy over a series of pardons he had issued in the waning days of his presidency.

GEORGE WALKER BUSH (2001-), 43d president, Republican, was born on July 6, 1946, in New Haven, CT. He was the first of six children born to George Herbert Walker Bush and his wife, the former Barbara Pierce. (His brother Jeb won the Florida governorship in 1998.) Bush was the first son of a former president to win the White House since John Quincy Adams in 1824.

Fun-loving, athletic, and popular, the young George Bush grew up in Midland and Houston, TX. In 1961 he was sent to the Phillips Academy in Andover, MA, the same prep school his father had attended. In 1964 he entered Yale University, his father's alma mater, where he majored in history. Eligible for the draft upon graduation from Yale in 1968, at the height of the Vietnam War, he signed on with the Texas Air National Guard.

After earning a master's degree from the Harvard Business School, he returned to Midland in 1975 and went into the oil business. Two years later he was married to Laura Welch, a schoolteacher and librarian; and in 1981 she gave birth to twin daughters.

Bush, who had lost a race for Congress in 1978, returned to the oil business, but success proved elusive, and he barely averted financial disaster in 1986. Realizing by this time that his drinking had become excessive (he had been arrested in 1976 for driving under the influence), he swore off alcohol and renewed his commitment to Christian faith. After assisting in his father's successful 1988 presidential campaign, he put together a group of investors to buy the Texas Rangers baseball club and took a hands-on role as managing partner. Bush ran for governor in 1994, defeating a popular incumbent, Ann Richards. He won reelection by a landslide four years later.

As governor, he concentrated on building personal bonds with Democratic leaders. He backed education reforms, won passage of measures designed to curb so-called "junk lawsuits," and cut property taxes for homeowners. Aspects of the Texas record that were controversial during his 2000 presidential campaign included his steadfast support for the death penalty and the state's record in pollution control.

After defeating Sen. John McCain of Arizona and several other rivals in the Republican party primaries, Bush chose Dick Cheney, a former U.S. representative and defense secretary, as his running mate. Campaign issues included education reform, missile defense, and the size of a proposed tax cut. The November 2000 presidential election was one of the closest in history. Bush came out behind in the popular vote, by about 540,000 out of more than 100 million cast. He was in effect declared the winner by a Supreme Court decision that ended a lengthy and inconclusive attempt to accurately recount the vote in Florida, whose 25 electoral votes decided the election.

On Sept, 11, 2001, Bush faced what appeared to be the key crisis for his presidency when a terrorist attack on the World Trade Center in New York City and the Pentagon near Washington, DC, killed more than 5,500 people. (*See also the front-of-the-book feature article.*)

Wives and Children of the Presidents

Name (Born–died; married)	State	Sons/ Daughters
Martha Dandridge Custis Washington (1731-1802; 1759)	VA	None
Abigail Smith Adams (1744-1818; 1764)	MA	3/2
Martha Wayles Skelton Jefferson (1748-82; 1772)	VA	1/5
Dorothea "Dolley" Payne Todd Madison (1768-1849; 1794)	NC	None
Elizabeth Kortright Monroe (1768-1830; 1786)	NY	0/2(A)
Louisa Catherine Johnson Adams (1775-1852; 1797)	MD(B)	3/1
Rachel Donelson Robards Jackson (1767-1828; 1791)	VA	None
Hannah Hoes Van Buren (1783-1819; 1807)	NY	4/0
Anna Tuthill Symmes Harrison (1775-1864; 1795)	NJ	6/4
Letitia Christian Tyler (1790-1842; 1813)	VA	3/4(A)
Julia Gardiner Tyler (1820-89; 1844)	NY	5/2
Sarah Childress Polk (1803-91; 1824)	TN	None
Margaret Mackall Smith Taylor (1788-1852; 1810)	MD	1/5
Abigail Powers Fillmore (1798-1853; 1826)	NY	1/1
Caroline Carmichael McIntosh Fillmore (1813-81; 1858)	NJ	None
Jane Means Appleton Pierce (1806-63; 1834)	NH	3/0
Mary Todd Lincoln (1818-82; 1842)	KY	4/0
Eliza McCardle Johnson (1810-76; 1827)	TN	3/2
Julia Boggs Dent Grant (1826-1902; 1848)	MO	3/1
Lucy Ware Webb Hayes (1831-89; 1852)	OH	7/1
Lucretia Rudolph Garfield (1832-1918; 1858)	OH	4/1
Ellen Lewis Herndon Arthur (1837-80; 1859)	VA	2/1
Frances Folsom Cleveland (1864-1947; 1886)	NY	2/3
Caroline Lavinia Scott Harrison 1832-92; 1853)	OH	1/1
Mary Scott Lord Dimmick Harrison (1858-1948; 1896)	PA	0/1
Ida Saxton McKinley (1847-1907; 1871)	OH	0/2
Alice Hathaway Lee Roosevelt (1861-84; 1880)	MA	0/1
Edith Kermit Carow Roosevelt (1861-1948; 1886)	CT	4/1
Helen Herron Taft (1861-1943; 1886)	OH	2/1
Ellen Louise Axson Wilson (1860-1914; 1885)	GA	0/3
Edith Bolling Galt Wilson (1872-1961; 1915)	VA	None
Florence Kling De Wolfe Harding (1860-1924; 1891)	OH	None
Grace Anna Goodhue Coolidge (1879-1957; 1905)	VT	2/0
Lou Henry Hoover (1875-1944; 1899)	IA	2/0
Anna Eleanor Roosevelt Roosevelt (1884-1962; 1905)	NY	4/1(A)
Elizabeth Virginia "Bess" Wallace Truman (1885-1982; 1919)	MO	0/1
Mamie Geneva Doud Eisenhower (1896-1979; 1916)	IA	1/0(A)
Jacqueline Lee Bouvier Kennedy (1929-94; 1953)	NY	1/1(A)
Claudia "Lady Bird" Alta Taylor Johnson (1912; 1934)	TX	0/2
Thelma Catherine Patricia Ryan Nixon (1912-1993; 1940)	NV	0/2
Elizabeth Bloomer Warren Ford (1918; 1948)	IL	3/1
Rosalynn Smith Carter (1927; 1946)	GA	3/1
Anne Frances "Nancy" Robbins Davis Reagan (1921; 1952)	NY	1/1(C)
Barbara Pierce Bush (1925; 1945)	NY	4/2
Hillary Rodham Clinton (1947; 1975)	IL	0/1
Laura Welch Bush (1946; 1977)	TX	0/2

NOTE: James Buchanan, 15th president, was unmarried. (A) plus one infant, deceased. (B) Born in London, father a MD citizen. (C) Pres. Reagan married and divorced Jane Wyman; they had a daughter who died in infancy, a son and daughter who lived past infancy.

IT'S A FACT: Lucy Hayes (1831-89) was the first First Lady to have graduated from college; she held a degree from Wesleyan Female College in Cincinnati.

First Lady Laura Welch Bush

Laura Welch Bush was born in Midland, TX, Nov. 4, 1946. She graduated from Southern Methodist University, earned a master's degree in library science at the University of Texas at Austin, and worked as a librarian and teacher in Texas public schools. She and George W. Bush were married in 1977; in 1981, their twin daughters, Jenna and Barbara, were born.

As First Lady of Texas from 1995 to 2001, Mrs. Bush stressed advocacy of educational, reform and literacy programs. In 1998, she launched an early childhood development initiative that included a family literacy project for Texas. She also worked to promote breast cancer awareness and other women's health issues.

Her first solo appearance as U.S. First Lady came at the launch of D.C. Teaching Fellows, a program encouraging professionals to become teachers.

Burial Places of the Presidents

President	Burial Place
Washington	Mt. Vernon, VA
J. Adams	Quincy, MA
Jefferson	Charlottesville, VA
Madison	Montpelier Station, VA
Monroe	Richmond, VA
J. Q. Adams	Quincy, MA
Jackson	Nashville, TN
Van Buren	Kinderhook, NY
W. H. Harrison	North Bend, OH
Tyler	Richmond, VA
Polk	Nashville, TN
Taylor	Louisville, KY
Fillmore	Buffalo, NY
Pierce	Concord, NH
Buchanan	Lancaster, PA
Lincoln	Springfield, IL
A. Johnson	Greeneville, TN
Grant	New York, NY
Hayes	Fremont, OH
Garfield	Cleveland, OH
Arthur	Albany, NY
Cleveland	Princeton, NJ
B. Harrison	Indianapolis, IN
McKinley	Canton, OH
T. Roosevelt	Oyster Bay, NY
Taft	Arlington Natl. Cemetery
Wilson	Wash. Natl. Cathedral
Harding	Marion, OH
Coolidge	Plymouth, VT
Hoover	West Branch, IA
F. Roosevelt	Hyde Park, NY
Truman	Independence, MO
Eisenhower	Abilene, KS
Kennedy	Arlington Natl. Cemetery
L. B. Johnson	Johnson City, TX
Nixon	Yorba Linda, CA

Presidential Facts

- Youngest president: Theodore Roosevelt, who was 42 when sworn in after McKinley's death
- Youngest person elected president: John F. Kennedy, who was 43 when elected
- Oldest president: Ronald Reagan, who was 77 when he left office
- First president to live in the White House: John Adams
- First president to be inaugurated in Washington, D.C.: Thomas Jefferson
- First president born a U.S. citizen: Martin Van Buren
- President who gave the longest inaugural address: William Henry Harrison, who spoke for an hour and 45 minutes, and died of pneumonia only 31 days after being inaugurated
- Only president to also serve as chief justice of the U.S.: William Howard Taft
- Only president to resign: Richard Nixon, after a House committee recommended impeachment for Watergate scandal
- Only president who was never married: James Buchanan. His niece acted as White House hostess
- Only president to serve without having been elected vice president or president in a national election: Gerald Ford
- State where the greatest number of presidents were born: Virginia (8)
- Presidents who died on July 4: John Adams, Thomas Jefferson, and James Monroe
- Presidents who died in office: Eight presidents have died in office. Four of them were assassinated: Abraham Lincoln, James Garfield, William McKinley, and John F. Kennedy. The other four were William Henry Harrison, Zachary Taylor, Warren G. Harding, and Franklin D. Roosevelt

Presidential Libraries

The libraries listed here, except for that of Richard Nixon (which is private), are coordinated by the National Archives and Records Administration (Website: http://www.nara.gov/nara/president/overview.html). NARA also has custody of the Nixon presidential historical materials and those of Bill Clinton. The William J. Clinton Library was under construction in 2001. NARA will release Clinton presidential records to the public at the Clinton Library beginning Jan. 20, 2006. Materials for presidents before Herbert Hoover are held by private institutions.

Herbert Hoover Library
211 Parkside Dr.,
PO Box 488
West Branch, IA 52358-0488
PHONE: 319-643-5301
FAX: 319-643-5825
E-MAIL: hoover.library@nara.gov

Franklin D. Roosevelt Library
511 Albany Post Rd.
Hyde Park, NY 12538-1999
PHONE: 914-229-8114
FAX: 914-229-0872
E-MAIL: roosevelt.library@nara.gov

Harry S. Truman Library
500 West U.S. Hwy. 24
Independence, MO 64050-1798
PHONE: 816-833-1400
FAX: 816-833-4368
E-MAIL: truman.library@nara.gov

Dwight D. Eisenhower Library
200 S.E. 4th St.
Abilene, KS 67410-2900
PHONE: 785-263-4751
FAX: 785-263-4218
E-MAIL: eisenhower.library@nara.gov

John Fitzgerald Kennedy Library
Columbia Pt.
Boston, MA 02125-3398
PHONE: 617-929-4500
FAX: 617-929-4538
E-MAIL: kennedy.library@nara.gov

Lyndon Baines Johnson Library
2313 Red River St.
Austin, TX 78705-5702
PHONE: 512-916-5137
FAX: 512-478-9104
E-MAIL: johnson.library@nara.gov

Richard Nixon Library & Birthplace
18001 Yorba Linda Blvd.
Yorba Linda, CA 92886-3949
PHONE: 714-993-3393
FAX: 714-528-0544
WEBSITE: http://www.nixonfoundation.org
E-MAIL: stedman@chapman.edu

Gerald R. Ford Library
1000 Beal Ave.
Ann Arbor, MI 48109-2114
PHONE: 734-741-2218
FAX: 734-741-2341
E-MAIL: ford.library@nara.gov

Jimmy Carter Library
441 Freedom Pkwy.
Atlanta, GA 30307-1406
PHONE: 404-331-3942
FAX: 404-730-2215
E-MAIL: carter.library@nara.gov

Ronald Reagan Library
40 Presidential Dr.
Simi Valley, CA 93065-0666
PHONE: 805-522-8444
FAX: 805-522-9621
E-MAIL: reagan.library@nara.gov

George Bush Library
1000 George Bush Dr.,
West College Station, TX 77482-0410
PHONE: 409-260-9552
FAX: 409-260-9557
E-MAIL: bush.library@nara.gov

William J. Clinton Library
1000 La Harpe Blvd.
Little Rock, AR 72201
PHONE: 501-254-6866
FAX: 501-244-9764
E-MAIL: clinton.library@nara.gov

Impeachment in U.S. History

The U.S. Constitution provides for impeachment and, upon conviction, removal from office of federal officials on grounds of "Treason, Bribery, or other high Crimes and Misdemeanors" (Article II, Sect. 4). Impeachment is the bringing of charges by the House of Representatives. It is followed by a Senate trial; a two-thirds vote in the Senate is needed for conviction and removal from office, which does not preclude criminal indictment and trial.

In 1868, Andrew Johnson became the first president impeached by the House; he was tried but not convicted by the Senate. In 1974, impeachment articles against Pres. Richard Nixon, in connection with the Watergate scandal, were voted by the House Judiciary Committee; he resigned Aug. 9, before the full House could vote on impeaching him. In 1998, Pres. Bill Clinton was impeached by the House in connection with covering up a relationship with a former White House intern; he was tried in the Senate in 1999 and acquitted.

PRESIDENTIAL ELECTIONS

Popular and Electoral Vote, 1996 and 2000

Source: Voter News Service; Federal Election Commission; totals are official.

	2000								1996					
	Electoral Vote				Democrat	Republican	Green[1]	Reform[2]	Electoral Vote			Democrat	Republican	Reform[2]
State	Gore	Bush	Nader	Buchanan	Gore	Bush	Nader	Buchanan	Clinton	Dole	Perot	Clinton	Dole	Perot
AL	0	9	0	0	692,611	941,173	18,323	6,351	0	9	0	662,165	769,044	92,149
AK	0	3	0	0	79,004	167,398	28,747	5,192	0	3	0	80,380	122,746	26,333
AZ	0	8	0	0	685,341	781,652	45,645	12,373	8	0	0	653,288	622,073	112,072
AR	0	6	0	0	422,768	472,940	13,421	7,358	6	0	0	475,171	325,416	69,884
CA	54	0	0	0	5,861,203	4,567,429	418,707	44,987	54	0	0	5,119,835	3,828,380	697,847
CO	0	8	0	0	738,227	883,748	91,434	10,465	0	8	0	671,152	691,848	99,629
CT	8	0	0	0	816,015	561,094	64,452	4,731	8	0	0	735,740	483,109	139,523
DE	3	0	0	0	180,068	137,288	8,307	777	3	0	0	140,355	99,062	28,719
DC	2[3]	0	0	—	171,923	18,073	10,576	—	3	0	0	158,220	17,339	3,611
FL	0	25	0	0	2,912,253	2,912,790	97,488	17,484	25	0	0	2,545,968	2,243,324	483,776
GA	0	13	—	0	1,116,230	1,419,720	—	10,926	0	13	0	1,053,849	1,080,843	146,337
HI	4	0	0	0	205,286	137,845	21,623	1,071	4	0	0	205,012	113,943	27,358
ID	0	4	—	0	138,637	336,937	—	7,615	0	4	0	165,443	256,595	62,518
IL	22	0	0	0	2,589,026	2,019,421	103,759	16,106	22	0	0	2,341,744	1,587,021	346,408
IN	0	12	0	0	901,980	1,245,836	—	16,959	0	12	0	887,424	1,006,693	224,299
IA	7	0	0	0	638,517	634,373	29,374	5,731	7	0	0	620,258	492,644	105,159
KS	0	6	0	0	399,276	622,332	36,086	7,370	0	6	0	387,659	583,245	92,639
KY	0	8	0	0	638,923	872,520	23,118	4,152	8	0	0	636,614	623,283	120,396
LA	0	9	0	0	792,344	927,871	20,473	14,356	9	0	0	927,837	712,586	123,293
ME	4	0	0	0	319,951	286,616	37,127	4,443	4	0	0	312,788	186,378	85,970
MD	10	0	0	0	1,144,008	813,827	53,768	4,248	10	0	0	966,207	681,530	115,812
MA	12	0	0	0	1,616,487	878,502	173,564	11,149	12	0	0	1,571,509	718,058	227,206
MI	18	0	0	—	2,170,418	1,953,139	84,165	—	18	0	0	1,989,653	1,481,212	336,670
MN	10	0	0	0	1,168,266	1,109,659	126,696	22,166	10	0	0	1,120,438	766,476	257,704
MS	0	7	0	0	404,614	572,844	8,122	2,265	0	7	0	394,022	439,838	52,222
MO	0	11	0	0	1,111,138	1,189,924	38,515	9,818	11	0	0	1,025,935	890,016	217,188
MT	0	3	0	0	137,126	240,178	24,437	5,697	0	3	0	167,922	179,652	55,229
NE	0	5	0	0	231,780	433,862	24,540	3,646	0	5	0	236,761	363,467	71,278
NV	0	4	0	0	279,978	301,575	15,008	4,747	4	0	0	203,974	199,244	43,986
NH	0	4	0	0	266,348	273,559	22,198	2,615	4	0	0	246,166	196,486	48,387
NJ	15	0	0	0	1,788,850	1,284,173	94,554	6,989	15	0	0	1,652,361	1,103,099	262,134
NM	5	0	0	0	286,783	286,417	21,251	1,392	5	0	0	273,495	232,751	32,257
NY	33	0	0	0	4,112,965	2,405,570	244,360	31,554	33	0	0	3,756,177	1,933,492	503,458
NC	0	14	—	0	1,257,692	1,631,163	—	8,874	0	14	0	1,107,849	1,225,938	168,059
ND	0	3	0	0	95,284	174,852	9,486	7,288	0	3	0	106,905	125,050	32,515
OH	0	21	0	0	2,186,190	2,351,209	117,857	26,724	21	0	0	2,148,222	1,859,883	483,207
OK	0	8	—	0	474,276	744,337	—	9,014	0	8	0	488,105	582,315	130,788
OR	7	0	0	0	720,342	713,577	77,357	7,063	7	0	0	649,641	538,152	121,221
PA	23	0	0	0	2,485,967	2,281,127	103,392	16,023	23	0	0	2,215,819	1,801,169	430,984
RI	4	0	0	0	249,508	130,555	25,052	2,273	4	0	0	233,050	104,683	43,723
SC	0	8	0	0	566,039	786,892	20,279	3,309	0	8	0	506,283	573,458	64,386
SD	0	3	—	0	118,804	190,700	—	3,322	0	3	0	139,333	150,543	31,250
TN	0	11	0	0	981,720	1,061,949	19,781	4,250	11	0	0	909,146	863,530	105,918
TX	0	32	0	0	2,433,746	3,799,639	137,994	12,394	0	32	0	2,459,683	2,736,167	378,537
UT	0	5	0	0	203,053	515,096	35,850	9,319	0	5	0	221,633	361,911	66,461
VT	3	0	0	0	149,022	119,775	20,374	2,192	3	0	0	137,894	80,352	31,024
VA	0	13	0	0	1,217,290	1,437,490	59,398	5,455	0	13	0	1,091,060	1,138,350	159,861
WA	11	0	0	0	1,247,652	1,108,864	103,002	7,171	11	0	0	1,123,323	840,712	201,003
WV	0	5	0	0	295,497	336,475	10,680	3,169	5	0	0	327,812	233,946	71,639
WI	11	0	0	0	1,242,987	1,237,279	94,070	11,446	11	0	0	1,071,971	845,029	227,339
WY	0	3	—	0	60,481	147,947	—	2,724	0	3	0	77,934	105,388	25,928
Total	**266[3]**	**271**	**0**	**0**	**51,003,894**	**50,459,211**	**2,834,410**	**446,743**	**379**	**159**	**0**	**47,401,185**	**39,197,469**	**8,085,294**

(—) = Not listed on state's ballot. (1) Listed on the ballot in some states as party other than Green. (2) Listed on the ballot in some states as party other than Reform. (3) One Washington, DC, elector abstained.

2000 Official Presidential General Election Results

Source: Voter News Service; Federal Election Commission

Candidate (Party)	Popular Vote	Percent of Popular Vote
Al Gore (Democrat)	51,003,894	48.41
George W. Bush (Republican)	50,459,211	47.89
Ralph Nader (Green)	2,834,410	2.69
Patrick J. Buchanan (Reform)	446,743	0.42
Harry Browne (Libertarian)	386,041	0.37
Howard Phillips (Constitution)	96,919	0.09
John S. Hagelin (Natural Law)	83,117	0.08
James E. Harris Jr. (Socialist Workers)	7,354	0.01
L. Neil Smith (Libertarian)	5,775	0.01
Monica Moorehead (Workers World)	4,795	0.00
David McReynolds (Socialist)	4,194	0.00
Cathy Gordon Brown (Independent)	1,606	0.00
Denny Lane (Vermont Grassroots)	1,044	0.00
Randall Venson (Independent)	535	0.00
Earl F. Dodge (Prohibition)	208	0.00
Louie G. Youngkeit (Unaffiliated)	161	0.00
Write-in	20,938	0.02
None of These Candidates (Nevada)	3,315	0.00
Total	**105,360,260**	**100.00**

Note: Party designations may vary from one state to another

PRESIDENTIAL ELECTION RETURNS BY COUNTIES

All results official. Results for New England states are for selected cities or towns. All totals statewide. D-Democrat; R-Republican; RF-Reform; I-Independent. (In 1996, Ross Perot was listed on the ballot in some states as "Independent.")

Source: Voter News Service; Federal Election Commission; Alaska Division of Elections

Alabama

	2000		1996		
County	Gore (D)	Bush (R)	Clinton (D)	Dole (R)	Perot (RF)
Autauga	4,942	11,993	5,015	9,509	813
Baldwin	13,997	40,872	12,776	29,487	4,520
Barbour	2,197	1,860	4,787	3,627	515
Bibb	2,710	4,273	2,775	3,037	455
Blount	4,977	12,667	5,061	9,056	985
Bullock	3,395	1,433	3,078	1,154	111
Butler	3,606	4,127	3,828	3,352	538
Calhoun	15,781	22,306	15,725	18,088	2,613
Chambers	5,616	6,037	5,515	4,707	812
Cherokee	3,497	4,154	4,399	3,048	899
Chilton	4,806	10,066	5,354	7,910	929
Choctaw	3,707	3,600	4,074	2,623	413
Clarke	4,679	5,988	4,831	4,785	478
Clay	2,045	3,719	2,306	2,694	538
Cleburne	1,664	3,333	1,737	2,063	385
Coffee	5,220	9,938	5,168	7,805	1,042
Colbert	10,543	10,518	10,226	8,305	1,696
Conecuh	2,783	2,699	2,903	2,093	445
Coosa	2,104	2,382	2,121	1,721	262
Covington	4,440	8,961	4,543	6,035	1,098
Crenshaw	1,934	2,793	2,172	1,939	317
Cullman	9,758	19,157	9,544	14,308	2,440
Dale	4,906	10,593	4,732	8,288	1,216
Dallas	10,967	7,360	10,507	6,612	477
DeKalb	7,056	12,827	6,544	9,823	1,609
Elmore	6,652	16,777	6,530	12,937	1,368
Escambia	4,523	6,975	4,651	5,214	867
Etowah	17,433	21,087	17,976	16,835	2,529
Fayette	3,064	4,582	3,381	3,191	590
Franklin	4,793	6,119	5,028	4,449	966
Geneva	2,769	6,588	3,174	4,725	857
Greene	3,504	850	3,526	796	55
Hale	4,652	2,984	3,372	1,893	190
Henry	2,782	4,054	3,019	3,082	515
Houston	9,412	22,150	8,791	17,476	1,653
Jackson	9,066	8,475	8,204	5,650	1,573
Jefferson	129,889	138,491	120,208	130,980	7,997
Lamar	2,653	4,470	2,843	2,955	597
Lauderdale	13,875	17,478	13,619	14,058	2,574
Lawrence	6,296	5,671	5,254	3,893	964
Lee	14,574	22,433	12,919	17,985	1,949
Limestone	8,992	14,204	8,045	10,862	1,659
Lowndes	4,557	1,638	3,970	1,369	72
Macon	7,665	1,091	7,018	987	150
Madison	48,199	62,151	42,259	50,390	7,437
Marengo	4,841	4,690	4,899	4,013	337
Marion	4,600	6,910	5,049	4,742	979
Marshall	10,381	17,084	8,722	12,323	2,150
Mobile	58,640	78,162	54,749	66,775	7,555
Monroe	3,741	5,153	3,815	4,382	486
Montgomery	40,371	38,827	38,382	37,784	2,036
Morgan	16,060	25,774	14,616	21,765	3,348
Perry	4,020	1,732	4,053	1,703	119
Pickens	4,143	4,306	4,018	3,322	403
Pike	4,357	6,058	4,514	5,281	503
Randolph	3,094	4,666	3,023	3,304	603
Russell	8,396	6,198	7,834	5,025	792
St. Clair	6,485	17,117	6,187	12,762	1,417
Shelby	13,183	47,651	11,280	37,090	2,035
Sumter	4,415	1,629	4,706	1,561	172
Talladega	11,264	13,807	10,385	10,931	1,335
Tallapoosa	6,183	9,805	6,071	7,627	1,038
Tuscaloosa	24,614	34,003	23,067	27,939	3,048
Walker	11,621	13,486	12,929	9,837	2,012
Washington	3,386	4,117	3,935	2,900	819
Wilcox	3,444	1,661	3,303	1,454	71
Winston	2,692	6,413	3,120	4,728	723
Totals	**692,611**	**941,173**	**662,165**	**769,044**	**92,149**

Alabama Vote Since 1952

1952, Eisenhower, Rep., 149,231; Stevenson, Dem., 275,075; Hamblen, Proh., 1,814.

1956, Stevenson, Dem., 290,844; Eisenhower, Rep., 195,694; Independent electors, 20,323.

1960, Kennedy, Dem., 324,050; Nixon, Rep., 237,981; Faubus, States' Rights, 4,367; Decker, Proh., 2,106; King, Afro-Americans, 1,485; scattering, 236.

1964, Dem. (electors unpledged), 209,848; Goldwater, Rep., 479,085; scattering, 105.

1968, Nixon, Rep., 146,923; Humphrey, Dem., 196,579; Wallace, 3d Party, 691,425; Munn, Proh., 4,022.

1972, Nixon, Rep., 728,701; McGovern, Dem., 219,108 plus 37,815 Natl. Dem. Party of Alabama; Schmitz, Conservative, 11,918; Munn., Proh., 8,551.

1976, Carter, Dem., 659,170; Ford, Rep., 504,070; Maddox, Amer. Ind., 9,198; Bubar, Proh., 6,669; Hall, Com., 1,954; MacBride, Libertarian, 1,481.

1980, Reagan, Rep., 654,192; Carter, Dem., 636,730; Anderson, Independent, 16,481; Rarick, Amer. Ind., 15,010; Clark, Libertarian, 13,318; Bubar, Statesman, 1,743; Hall, Com., 1,629; DeBerry, Soc. Workers, 1,303; McReynolds, Socialist, 1,006; Commoner, Citizens, 517.

1984, Reagan, Rep., 872,849; Mondale, Dem., 551,899; Bergland, Libertarian, 9,504.

1988, Bush, Rep., 815,576; Dukakis, Dem., 549,506; Paul, Lib., 8,460; Fulani, Ind., 3,311.

1992, Bush, Rep., 804,283; Clinton, Dem., 690,080; Perot, Ind., 183,109; Marrou, Libertarian, 5,737; Fulani, New Alliance, 2,161.

1996, Dole, Rep., 769,044; Clinton, Dem., 662,165; Perot, Ind. (Ref.), 92,149; Browne, Libertarian, 5,290; Phillips, Ind., 2,365; Hagelin, Natural Law, 1,697; Harris, Ind., 516.

2000, Bush, Rep., 941,173; Gore, Dem., 692,611; Nader, Ind., 18,323; Buchanan, Ind., 6,351; Browne, Libertarian, 5,893; Phillips, Ind., 775 Hagelin, Ind., 447.

Alaska

	2000		1996		
Election District	Gore (D)	Bush (R)	Clinton (D)	Dole (R)	Perot (RF)
No. 1	1,284	4,681	1,480	4,209	696
No. 2	2,081	4,235	2,563	3,247	912
No. 3	3,693	3,135	3,724	2,671	654
No. 4	2,715	4,127	3,037	3,336	694
No. 5	1,931	3,545	2,148	2,564	826
No. 6	1,542	3,862	1,576	2,707	557
No. 7	1,893	4,868	2,177	3,517	907
No. 8	1,498	5,371	1,643	3,624	826
No. 9	1,203	4,789	1,334	3,459	727
No. 10	2,194	5,673	2,203	4,184	642
No. 11	2,043	3,960	1,946	3,073	603
No. 12	2,051	4,626	1,825	3,568	543
No. 13	2,661	3,853	2,780	3,270	608
No. 14	1,626	3,750	1,471	3,005	458
No. 15	2,106	2,453	2,178	1,974	552
No. 16	1,969	1,980	1,629	1,328	414
No. 17	2,230	4,564	1,868	3,284	633
No. 18	2,739	5,421	2,708	4,245	694
No. 19	2,350	4,619	2,014	3,159	636
No. 20	2,259	3,648	2,144	3,025	545
No. 21	2,309	3,263	2,228	2,553	557
No. 22	2,656	4,910	2,511	3,887	624
No. 23	1,282	2,961	1,071	2,127	388
No. 24	1,985	5,063	1,914	3,653	548
No. 25	1,697	5,489	1,629	4,099	691
No. 26	1,608	5,869	1,519	3,913	883
No. 27	2,199	6,714	1,887	4,384	1,122
No. 28	2,116	7,113	1,645	4,202	1,333
No. 29	2,806	4,054	3,023	3,012	658
No. 30	1,698	3,622	1,794	2,785	601
No. 31	1,831	3,326	1,903	2,721	684
No. 32	1,389	4,178	1,275	2,736	675
No. 33	1,765	5,804	1,852	4,089	759
No. 34	1,300	5,243	1,388	3,677	734
No. 35	1,208	4,278	1,447	3,016	875
No. 36	1,945	3,007	2,321	1,992	453
No. 37	1,821	2,725	2,134	1,835	456
No. 38	2.015	2,467	2,436	1,716	393
No. 39	2,282	2,321	2,692	1,618	404
No. 40	1,024	1,831	1,260	1,280	368
Totals	**79,004**	**167,398**	**80,377**	**122,744**	**26,333**

Alaska Vote Since 1960

1960, Kennedy, Dem., 29,809; Nixon, Rep., 30,953.

1964, Johnson, Dem., 44,329; Goldwater, Rep., 22,930.

1968, Nixon, Rep., 37,600; Humphrey, Dem., 35,411; Wallace, 3d Party, 10,024.

1972, Nixon, Rep., 55,349; McGovern, Dem., 32,967; Schmitz, Amer., 6,903.

1976, Carter, Dem., 44,058; Ford, Rep., 71,555; MacBride, Libertarian, 6,785.

1980, Reagan, Rep., 86,112; Carter, Dem., 41,842; Clark, Libertarian, 18,479; Anderson, Ind., 11,155; write-in, 857.
1984, Reagan, Rep., 138,377; Mondale, Dem., 62,007; Bergland, Libertarian, 6,378.
1988, Bush, Rep., 119,251; Dukakis, Dem., 72,584; Paul, Lib., 5,484; Fulani, New Alliance, 1,024.
1992, Bush, Rep., 102,000; Clinton, Dem., 78,294; Perot, Ind., 73,481; Gritz, Populist/America First, 1,379; Marrou, Libertarian, 1,378.
1996, Dole, Rep., 122,746; Clinton, Dem., 80,380; Perot, Ref., 26,333; Nader, Green, 7,597; Browne, Libertarian, 2,276; Phillips, Taxpayers, 925; Hagelin, Natural Law, 729.
2000, Bush, Rep., 167,398; Gore, Dem., 79,004; Nader, Green, 28,747; Buchanan, Reform, 5,192; Browne, Libertarian, 2,636; Hagelin, Natural Law, 919; Phillips, Constitution, 596.

Arizona

	2000		1996		
County	Gore (D)	Bush (R)	Clinton (D)	Dole (R)	Perot (RF)
Apache	13,025	5,947	12,394	4,761	1,296
Cochise	13,360	18,180	13,782	14,365	3,346
Coconino	20,280	17,562	20,475	13,638	3,666
Gila	7,700	9,158	8,577	6,407	2,211
Graham	3,355	6,007	3,938	4,222	1,034
Greenlee	1,216	1,619	1,755	1,159	426
La Paz	1,769	2,543	1,964	1,902	597
Maricopa	386,683	479,967	363,991	386,015	58,479
Mohave	17,470	24,386	16,629	17,997	6,369
Navajo	11,794	12,386	12,912	9,262	2,461
Pima	147,688	124,579	137,983	104,121	18,809
Pinal	19,650	20,122	19,579	13,034	3,972
Santa Cruz	5,233	3,344	5,241	2,256	600
Yavapai	24,063	40,144	21,801	29,921	6,649
Yuma	12,055	15,708	12,267	13,013	2,157
Totals	**685,341**	**781,652**	**653,288**	**622,073**	**112,072**

Arizona Vote Since 1952

1952, Eisenhower, Rep., 152,042; Stevenson, Dem., 108,528.
1956, Eisenhower, Rep., 176,990; Stevenson, Dem., 112,880; Andrews, Ind. 303.
1960, Kennedy, Dem., 176,781; Nixon, Rep., 221,241; Hass, Soc. Labor, 469.
1964, Johnson, Dem., 237,753; Goldwater, Rep., 242,535; Hass, Soc. Labor, 482.
1968, Nixon, Rep., 266,721; Humphrey, Dem., 170,514; Wallace, 3d Party, 46,573; McCarthy, New Party, 2,751; Halstead, Soc. Workers, 85; Cleaver, Peace and Freedom, 217; Blomen, Soc. Labor, 75.
1972, Nixon, Rep., 402,812; McGovern, Dem., 198,540; Schmitz, Amer., 21,208; Soc. Workers, 30,945. Because of ballot peculiarities in 3 counties (particularly Pima), thousands of voters cast ballots for the Soc. Workers Party *and* one of the major candidates. Court ordered both votes counted as official.
1976, Carter, Dem., 295,602; Ford, Rep., 418,642; McCarthy, Ind., 19,229; MacBride, Libertarian, 7,647; Camejo, Soc. Workers, 928; Anderson, Amer., 564; Maddox, Amer. Ind., 85.
1980, Reagan, Rep., 529,688; Carter, Dem., 246,843; Anderson, Ind., 76,952; Clark, Libertarian, 18,784; De Berry, Soc. Workers, 1,100; Commoner, Citizens, 551; Hall, Com., 25; Griswold, Workers World, 2.
1984, Reagan, Rep., 681,416; Mondale, Dem., 333,854; Bergland, Libertarian, 10,585.
1988, Bush, Rep., 702,541; Dukakis, Dem., 454,029; Paul, Lib., 13,351; Fulani, New Alliance, 1,662.
1992, Bush, Rep., 572,086; Clinton, Dem., 543,050; Perot, Ind., 353,741; Gritz, Populist/America First, 8,141; Marrou, Libertarian, 6,759; Hagelin, Natural Law, 2,267.
1996, Clinton, Dem., 653,288; Dole, Rep., 622,073; Perot, Ref., 112,072; Browne, Libertarian, 14,358.
2000, Gore, Bush, Rep., 781,652; Dem., 685,341; Nader, Green, 45,645; Buchanan, Rep., 12,373; Smith, Libertarian, 5,775; Hagelin, Natural Law, 1,120.

Arkansas

	2000		1996		
County	Gore (D)	Bush (R)	Clinton (D)	Dole (R)	Perot (RF)
Arkansas	2,877	3,353	4,220	1,910	463
Ashley	4,253	3,876	5,011	2,428	704
Baxter	6,516	9,538	6,703	6,877	1,572
Benton	17,277	34,838	17,205	23,748	4,147
Boone	4,493	8,569	5,745	6,093	1,132
Bradley	2,122	1,793	2,566	1,146	221
Calhoun	1,017	1,128	1,306	727	237
Carroll	3,595	5,556	3,689	3,957	986
Chicot	2,820	1,564	3,090	1,056	233
Clark	4,661	3,776	5,281	2,112	567
Clay	3,527	2,254	3,848	1,512	464
Cleburne	4,120	5,730	4,475	3,807	1,021
Cleveland	1,414	1,678	1,741	990	268
Columbia	4,003	5,018	4,730	3,376	678
Conway	3,496	3,545	4,055	2,307	746
Craighead	12,376	12,158	13,284	9,210	1,778
Crawford	6,288	10,804	6,749	7,182	1,683
Crittenden	7,224	5,857	8,415	4,673	554
Cross	3,096	3,033	3,631	2,000	466
Dallas	1,710	1,571	2,118	1,041	236
Desha	2,776	1,603	3,230	978	247
Drew	3,060	2,756	3,570	1,657	395
Faulkner	11,950	16,055	12,032	10,178	1,528
Franklin	2,674	3,277	3,269	2,246	626
Fulton	1,976	2,036	2,361	1,351	455
Garland	15,840	19,098	19,211	13,662	2,769
Grant	2,535	3,285	2,948	1,925	557
Greene	6,319	5,831	6,622	3,757	1,014
Hempstead	3,937	3,257	4,983	2,021	501
Hot Spring	5,527	5,042	6,002	2,864	1,123
Howard	2,063	2,326	2,741	1,478	369
Independence	5,146	6,145	6,240	4,021	1,126
Izard	2,587	2,301	2,818	1,678	541
Jackson	3,651	2,280	4,304	1,525	611
Jefferson	17,716	8,765	19,701	6,330	1,284
Johnson	3,270	3,657	3,585	2,367	757
Lafayette	1,806	1,538	2,466	971	374
Lawrence	3,255	2,626	3,652	1,823	609
Lee	2,727	1,351	3,267	1,013	257
Lincoln	1,957	1,526	2,517	907	221
Little River	2,883	2,283	3,183	1,409	480
Logan	3,283	4,487	3,832	2,966	1,048
Lonoke	6,851	10,606	8,049	6,414	1,369
Madison	2,055	3,387	2,504	2,303	461
Marion	2,233	3,402	2,735	2,312	764
Miller	6,278	7,276	6,469	4,874	1,043
Mississippi	7,107	5,199	8,301	3,919	1,016
Monroe	1,910	1,329	2,247	973	202
Montgomery	1,438	2,128	1,830	1,137	427
Nevada	1,867	1,796	2,279	976	345
Newton	1,205	2,529	1,631	1,927	498
Ouachita	5,464	4,739	6,635	3,136	733
Perry	1,648	2,114	1,873	1,143	395
Phillips	6,018	3,154	5,715	2,205	461
Pike	1,604	2,275	2,362	1,401	441
Poinsett	4,102	2,988	4,686	2,034	647
Polk	2,315	4,600	2,824	2,852	876
Pope	6,669	11,244	8,433	8,243	1,891
Prairie	1,563	1,862	2,211	1,025	305
Pulaski	68,320	55,866	75,084	44,780	6,014
Randolph	3,019	2,673	3,213	1,789	561
St. Francis	4,986	3,414	5,562	2,523	506
Saline	12,700	18,617	14,027	11,695	2,612
Scott	1,444	2,399	2,259	1,426	513
Searcy	1,229	2,610	1,669	1,786	381
Sebastian	15,555	23,483	15,514	16,482	2,899
Sevier	2,095	2,111	2,553	1,379	446
Sharp	3,236	3,698	3,573	2,635	687
Stone	2,043	2,623	2,227	1,526	579
Union	6,261	8,647	8,373	6,053	1,073
Van Buren	3,202	3,485	3,521	2,345	830
Washington	21,425	28,231	20,419	19,476	3,133
White	8,342	13,170	10,204	8,659	1,828
Woodruff	1,699	898	2,044	598	186
Yell	3,062	3,223	3,749	2,111	714
Totals	**422,768**	**472,940**	**475,171**	**325,416**	**69,884**

Arkansas Vote Since 1952

1952, Eisenhower, Rep., 177,155; Stevenson, Dem., 226,300; Hamblen, Proh., 886; MacArthur, Christian Nationalist, 458; Hass, Soc. Labor, 1.
1956, Stevenson, Dem., 213,277; Eisenhower, Rep., 186,287; Andrews, Ind., 7,008.
1960, Kennedy, Dem., 215,049; Nixon, Rep., 184,508; Natl. States' Rights, 28,952.
1964, Johnson, Dem., 314,197; Goldwater, Rep., 243,264; Kasper, Natl. States' Rights, 2,965.
1968, Nixon, Rep., 189,062; Humphrey, Dem., 184,901; Wallace, 3d Party, 235,627.
1972, Nixon, Rep., 445,751; McGovern, Dem., 198,899; Schmitz, Amer., 3,016.
1976, Carter, Dem., 498,604; Ford, Rep., 267,903; McCarthy, Ind., 639; Anderson, Amer., 389.
1980, Reagan, Rep., 403,164; Carter, Dem., 398,041; Anderson, Ind., 22,468; Clark, Libertarian, 8,970; Commoner, Citizens, 2,345; Bubar, Statesman, 1,350; Hall, Com., 1,244.
1984, Reagan, Rep., 534,774; Mondale, Dem., 338,646; Bergland, Libertarian, 2,220.
1988, Bush, Rep., 466,578; Dukakis, Dem., 349,237; Duke, Chr. Pop., 5,146; Paul, Lib., 3,297.

1992, Clinton, Dem., 505,823; Bush, Rep., 337,324; Perot, Ind., 99,132; Phillips, U.S. Taxpayers, 1,437; Marrou, Libertarian, 1,261; Fulani, New Alliance, 1,022.
1996, Clinton, Dem., 475,171; Dole, Rep., 325,416; Perot, Ref., 69,884; Nader, Ind., 3,649; Browne, Ind., 3,076; Phillips, Ind., 2,065; Forbes, Ind., 932; Collins, Ind., 823; Masters, Ind., 749; Hagelin, Ind., 729; Moorehead, Ind., 747; Hollis, Ind., 538; Dodge, Ind., 483.
2000, Bush, Rep., 472,940; Gore, Dem., 422,768; Nader, Green, 13,421; Buchanan, Reform, 7,358; Browne, Libertarian, 2,781; Phillips, Constitution, 1,415; Hagelin, Natural Law, 1,098.

California

County	2000 Gore (D)	2000 Bush (R)	1996 Clinton (D)	1996 Dole (R)	1996 Perot (RF)
Alameda	342,889	119,279	303,903	106,581	24,270
Alpine	265	281	258	264	63
Amador	5,906	8,766	5,868	6,870	1,267
Butte	31,338	45,584	30,651	38,961	6,393
Calaveras	7,093	10,599	6,646	8,279	1,612
Colusa	1,745	3,629	2,054	3,047	404
Contra Costa	224,338	141,373	196,512	123,954	20,416
Del Norte	3,117	4,526	3,652	3,670	1,225
El Dorado	26,220	42,045	22,957	32,759	5,077
Fresno	95,059	117,342	94,448	98,813	10,962
Glenn	2,498	5,795	2,841	5,041	788
Humboldt	24,851	23,219	24,628	19,803	5,811
Imperial	15,489	12,524	14,591	9,705	1,778
Inyo	2,652	4,713	2,601	3,924	811
Kern	66,003	110,663	62,658	92,151	13,452
Kings	11,041	16,377	11,254	12,368	1,745
Lake	10,717	8,699	10,432	7,458	2,539
Lassen	2,982	7,080	3,318	5,194	1,080
Los Angeles	1,710,505	871,930	1,430,629	746,544	157,752
Madera	11,650	20,283	11,254	16,510	2,192
Marin	79,135	34,872	67,406	32,714	6,559
Mariposa	2,816	4,727	2,920	3,976	729
Mendocino	16,634	12,272	14,952	9,765	3,685
Merced	22,726	26,102	21,786	20,847	3,427
Modoc	945	2,969	1,368	2,285	528
Mono	1,788	2,296	1,580	1,882	447
Monterey	67,618	43,761	57,700	39,794	7,240
Napa	28,097	20,633	24,588	17,439	4,254
Nevada	17,670	25,998	15,369	21,784	3,330
Orange	391,819	541,299	327,485	446,717	66,195
Placer	42,449	69,835	34,981	49,808	6,542
Plumas	3,458	6,343	3,540	4,905	919
Riverside	202,576	231,955	168,579	178,611	35,481
Sacramento	212,792	195,619	203,019	166,049	23,856
San Benito	9,131	7,015	7,030	5,384	1,044
San Bernardino	214,749	221,757	183,372	180,135	39,330
San Diego	437,666	475,736	389,964	402,876	63,037
San Francisco	241,578	51,496	209,777	45,479	9,659
San Joaquin	79,776	81,773	67,253	65,131	9,692
San Luis Obispo	44,526	56,859	40,395	46,733	8,204
San Mateo	166,757	80,296	152,304	73,508	15,047
Santa Barbara	73,411	71,493	70,650	63,915	9,457
Santa Clara	332,490	188,750	297,639	168,291	34,908
Santa Cruz	66,618	29,627	58,250	27,766	6,555
Shasta	20,127	43,278	20,848	34,736	5,875
Sierra	540	1,172	573	877	170
Siskiyou	6,323	12,198	7,022	8,653	1,879
Solano	75,116	51,604	64,644	40,742	8,682
Sonoma	117,295	63,529	100,738	53,555	13,862
Stanislaus	56,448	67,188	53,738	52,403	8,360
Sutter	8,416	17,350	8,504	14,264	1,533
Tehama	6,507	13,270	7,290	10,292	2,325
Trinity	1,932	3,340	2,203	2,530	856
Tulare	33,006	54,070	32,669	46,272	5,106
Tuolumne	9,359	13,172	8,950	10,386	1,925
Ventura	133,258	136,173	110,772	109,202	23,054
Yolo	33,747	23,057	33,033	18,807	3,150
Yuba	5,546	9,838	5,789	7,971	1,308
Totals	**5,861,203**	**4,567,429**	**5,119,835**	**3,828,380**	**697,847**

California Vote Since 1952

1952, Eisenhower, Rep., 2,897,310; Stevenson, Dem., 2,197,548; Hallinan, Prog., 24,106; Hamblen, Proh., 15,653; MacArthur, (Tenny Ticket), 3,326; (Kellems Ticket) 178; Hass, Soc. Labor, 273; Hoopes, Soc., 206; scattered, 3,249.
1956, Eisenhower, Rep., 3,027,668; Stevenson, Dem., 2,420,136; Holtwick, Proh., 11,119; Andrews, Constitution, 6,087; Hass, Soc. Labor, 300; Hoopes, Soc., 123; Dobbs, Soc. Workers, 96; Smith, Christian Natl., 8.
1960, Kennedy, Dem., 3,224,099; Nixon, Rep., 3,259,722; Decker, Proh., 21,706; Hass, Soc. Labor, 1,051.
1964, Johnson, Dem., 4,171,877; Goldwater, Rep., 2,879,108; Hass, Soc. Labor, 489; DeBerry, Soc. Workers, 378; Munn, Proh., 305; Hensley, Universal, 19.
1968, Nixon, Rep., 3,467,664; Humphrey, Dem., 3,244,318; Wallace, 3d Party, 487,270; Peace and Freedom, 27,707; McCarthy, Alternative, 20,721; Gregory, write-in, 3,230; Mitchell, Com., 260; Munn, Proh., 59; Blomen, Soc. Labor, 341; Soeters, Defense, 17.
1972, Nixon, Rep., 4,602,096; McGovern, Dem., 3,475,847; Schmitz, Amer., 232,554; Spock, Peace and Freedom, 55,167; Hall, Com., 373; Hospers, Libertarian, 980; Munn, Proh., 53; Fisher, Soc. Labor, 197; Jenness, Soc. Workers, 574; Green, Universal, 21.
1976, Carter, Dem., 3,742,284; Ford, Rep., 3,882,244; MacBride, Libertarian, 56,388; Maddox, Amer. Ind., 51,098; Wright, People's, 41,731; Camejo, Soc. Workers, 17,259; Hall, Com., 12,766; write-in, McCarthy, 58,412; other write-in, 4,935.
1980, Reagan, Rep. 4,524,858; Carter, Dem., 3,083,661; Anderson, Ind., 739,833; Clark, Libertarian, 148,434; Commoner, Ind., 61,063; Smith, Peace and Freedom, 18,116; Rarick, Amer. Ind., 9,856.
1984, Reagan, Rep. 5,305,410; Mondale, Dem., 3,815,947; Bergland, Libertarian, 48,400.
1988, Bush, Rep., 5,054,917; Dukakis, Dem., 4,702,233; Paul, Lib., 70,105; Fulani, Ind., 31,181.
1992, Clinton, Dem., 5,121,325; Bush, Rep., 3,630,575; Perot, Ind., 2,296,006; Marrou, Libertarian, 48,139; Daniels, Ind., 18,597; Phillips, U.S. Taxpayers, 12,711.
1996, Clinton, Dem., 5,119,835; Dole, Rep., 3,828,380; Perot, Ref., 697,847; Nader, Green, 237,016; Browne, Libertarian, 73,600; Feinland, Peace & Freedom, 25,332; Phillips, Amer. Ind., 21,202; Hagelin, Natural Law, 15,403.
2000, Gore, Dem., 5,861,203; Bush, Rep., 4,567,429; Nader, Green, 418,707; Browne, Libertarian, 45,520; Buchanan, Reform, 44,987; Phillips, Amer. Ind., 17,042; Hagelin, Natural Law, 10,934.

Colorado

County	2000 Gore (D)	2000 Bush (R)	1996 Clinton (D)	1996 Dole (R)	1996 Perot (RF)
Adams	54,132	47,561	48,314	36,666	7,206
Alamosa	2,455	2,857	2,330	2,038	437
Arapahoe	82,614	97,768	68,306	82,778	8,476
Archuleta	1,432	2,988	997	1,963	360
Baca	531	1,663	659	1,321	203
Bent	783	1,096	1,046	917	209
Boulder	69,983	50,873	63,316	41,922	6,840
Chaffee	2,768	4,300	2,768	3,052	538
Cheyenne	209	957	328	739	91
Clear Creek	2,188	2,247	1,863	1,746	365
Conejos	1,749	1,772	1,726	1,149	245
Costilla	1,054	504	1,168	333	112
Crowley	511	855	559	680	114
Custer	507	1,451	412	920	164
Delta	3,264	8,372	3,584	6,047	1,060
Denver	122,693	61,224	120,312	58,529	8,777
Dolores	293	741	276	417	95
Douglas	27,076	56,007	16,232	32,120	2,662
Eagle	6,772	7,165	5,094	4,637	1,193
Elbert	2,326	6,151	1,894	4,125	507
El Paso	61,799	128,294	55,822	102,403	11,175
Fremont	5,293	9,914	5,344	7,437	1,438
Garfield	6,087	9,103	5,722	6,281	1,562
Gilpin	1,099	1,006	799	682	184
Grand	2,308	3,570	2,012	2,264	473
Gunnison	3,059	3,128	2,812	2,230	570
Hinsdale	188	316	185	289	56
Huerfano	1,495	1,466	1,483	996	210
Jackson	173	682	222	486	107
Jefferson	100,970	120,138	89,494	101,517	12,967
Kiowa	211	728	246	549	74
Kit Carson	809	2,542	1,073	2,068	235
Lake	1,296	1,056	1,338	728	274
La Plata	7,864	9,993	6,509	8,057	1,403
Larimer	46,055	62,429	40,965	45,935	6,823
Las Animas	3,243	2,569	3,611	1,905	427
Lincoln	510	1,630	729	1,272	164
Logan	2,296	5,531	2,765	4,032	609
Mesa	15,465	32,396	17,114	24,761	3,707
Mineral	168	294	192	179	69
Moffat	1,223	3,840	1,635	2,466	649
Montezuma	2,556	6,158	2,578	4,175	827
Montrose	4,041	9,266	4,019	6,730	1,187
Morgan	2,885	5,722	3,347	4,557	687
Otero	2,963	4,082	3,386	3,356	581
Ouray	705	1,279	569	984	167

County	2000 Gore (D)	Bush (R)	1996 Clinton (D)	Dole (R)	Perot (RF)
Park	2,393	3,677	1,844	2,661	534
Philips	564	1,576	706	1,284	156
Pitkin	4,137	2,565	3,949	1,969	535
Prowers	1,361	3,026	1,745	2,504	342
Pueblo	28,888	22,827	28,791	17,402	3,374
Rio Blanco	543	2,185	731	1,697	243
Rio Grande	1,707	3,111	1,720	2,129	379
Routt	4,208	4,472	3,660	3,019	859
Saguache	1,145	1,078	969	712	160
San Juan	149	210	133	153	50
San Miguel	1,598	1,043	1,535	773	231
Sedgwick	384	877	519	715	101
Summit	5,304	4,497	3,970	3,261	823
Teller	2,750	6,477	2,312	4,458	707
Washington	477	1,878	649	1,566	190
Weld	23,436	37,409	21,325	26,518	4,347
Yuma	1,082	3,156	1,439	2,589	319
Totals	**738,227**	**883,748**	**671,152**	**691,848**	**99,629**

Colorado Vote Since 1952

1952, Eisenhower, Rep., 379,782; Stevenson, Dem., 245,504; MacArthur, Constitution, 2,181; Hallinan, Prog., 1,919; Hoopes, Soc., 365; Hass, Soc. Labor, 352.

1956, Eisenhower, Rep., 394,479; Stevenson, Dem., 263,997; Hass, Soc. Lab., 3,308; Andrews, Ind., 759; Hoopes, Soc., 531.

1960, Kennedy, Dem., 330,629; Nixon, Rep., 402,242; Hass, Soc. Labor, 2,803; Dobbs, Soc. Workers, 572.

1964, Johnson, Dem., 476,024; Goldwater, Rep., 296,767; Hass, Soc. Labor, 302; DeBerry, Soc. Workers, 2,537; Munn, Proh., 1,356.

1968, Nixon, Rep., 409,345; Humphrey, Dem., 335,174; Wallace, 3d Party, 60,813; Blomen, Soc. Labor, 3,016; Gregory, New-party, 1,393; Munn, Proh., 275; Halstead, Soc. Workers, 235.

1972, Nixon, Rep., 597,189; McGovern, Dem., 329,980; Fisher, Soc. Labor, 4,361; Hospers, Libertarian, 1,111; Hall, Com., 432; Jenness, Soc. Workers, 555; Munn, Proh., 467; Schmitz, Amer., 17,269; Spock, Peoples, 2,403.

1976, Carter, Dem., 460,353; Ford, Rep., 584,367; McCarthy, Ind., 26,107; MacBride, Libertarian, 5,330; Bubar, Proh., 2,882.

1980, Reagan, Rep., 652,264; Carter, Dem., 367,973; Anderson, Ind., 130,633; Clark, Libertarian, 25,744; Commoner, Citizens, 5,614; Bubar, Statesman, 1,180; Pulley, Socialist, 520; Hall, Com., 487.

1984, Reagan, Rep., 821,817; Mondale, Dem., 454,975; Bergland, Libertarian, 11,257.

1988, Bush, Rep., 728,177; Dukakis, Dem., 621,453; Paul, Lib., 15,482; Dodge, Proh., 4,604.

1992, Clinton, Dem., 629,681; Bush, Rep., 562,850; Perot, Ind., 366,010; Marrou, Libertarian, 8,669; Fulani, New Alliance, 1,608.

1996, Dole, Rep., 691,848; Clinton, Dem., 671,152; Perot, Ref., 99,629; Nader, Green, 25,070; Browne, Libertarian, 12,392; Collins, Ind., 2,809; Phillips, Amer. Constitution, 2,813; Hagelin, Natural Law, 2,547; Hollis, Soc., 669; Moorehead, Workers World, 599; Templin, Amer., 557; Dodge, Proh., 375; Harris, Soc. Workers, 244.

2000, Bush, Rep., 883,748; Gore, Dem, 738,227; Nader, Green, 91,434; Browne, Libertarian, 12,799; Buchanan, Reform, 10,465; Hagelin, Reform, 2,240; Phillips, Amer. Constitution, 1,319; McReynolds, Soc., 712; Harris, Soc. Workers, 216; Dodge, Proh., 208.

Connecticut

City	2000 Gore (D)	Bush (R)	1996 Clinton (D)	Dole (R)	Perot (RF)
Bridgeport	24,303	7,406	22,883	6,785	2,367
Bristol	14,665	7,948	13,616	6,560	3,049
Danbury	12,987	9,371	12,102	7,965	2,158
Fairfield	14,210	13,042	12,639	12,314	2,092
Greenwich	12,780	14,905	11,622	14,308	1,437
Hartford	21,445	3,095	22,929	3,082	1,010
New Britain	13,913	5,059	14,322	4,911	1,717
New Haven	28,145	5,160	26,161	4,822	1,555
Norwalk	19,293	11,519	17,354	10,800	2,237
Stamford	27,430	15,159	25,005	14,696	2,595
Waterbury	18,069	12,415	18,901	12,075	3,169
West Hartford	21,069	10,447	19,037	10,781	1,890
Other	587,706	445,568	519,169	374,010	114,247
Totals	**816,015**	**561,094**	**735,740**	**483,109**	**139,523**

Connecticut Vote Since 1952

1952, Eisenhower, Rep., 611,012; Stevenson, Dem., 481,649; Hoopes, Soc., 2,244; Hallinan, Peoples, 1,466; Hass, Soc. Labor, 535; write-in, 5.

1956, Eisenhower, Rep., 711,837; Stevenson, Dem., 405,079; scattered, 205.

1960, Kennedy, Dem., 657,055; Nixon, Rep., 565,813.

1964, Johnson, Dem., 826,269; Goldwater, Rep., 390,996; scattered, 1,313.

1968, Nixon, Rep., 556,721; Humphrey, Dem., 621,561; Wallace, 3d Party, 76,650; scattered, 1,300.

1972, Nixon, Rep., 810,763; McGovern, Dem., 555,498; Schmitz, Amer., 17,239; scattered, 777.

1976, Carter, Dem., 647,895; Ford, Rep., 719,261; Maddox, George Wallace Party, 7,101; LaRouche, U.S. Labor, 1,789.

1980, Reagan, Rep., 677,210; Carter, Dem., 541,732; Anderson, Ind., 171,807; Clark, Libertarian, 8,570; Commoner, Citizens, 6,130; scattered, 836.

1984, Reagan, Rep., 890,877; Mondale, Dem., 569,597.

1988, Bush, Rep., 750,241; Dukakis, Dem., 676,584; Paul, Lib., 14,071; Fulani, New Alliance, 2,491.

1992, Clinton, Dem., 682,318; Bush, Rep., 578,313; Perot, Ind., 348,771; Marrou, Libertarian, 5,391; Fulani, New Alliance, 1,363.

1996, Clinton, Dem., 735,740; Dole, Rep., 483,109; Perot, Ref., 139,523; Nader, Green, 24,321; Browne, Libertarian, 5,788; Phillips, Concerned Citizens, 2,425; Hagelin, Natural Law, 1,703.

2000, Gore, Dem., 816,015; Bush, Rep., 561,094; Nader, Green, 64,452; Phillips, Concerned Citizens, 9,695; Buchanan, Reform, 4,731; Browne, Libertarian, 3,484.

Delaware

County	2000 Gore Gore (D)	Bush (R)	1996 Clinton (D)	Dole (R)	Perot (RF)
Kent	22,790	24,081	18,327	15,932	4,705
New Castle	127,539	78,587	98,837	60,943	17,748
Sussex	29,739	34,620	23,191	22,187	6,266
Totals	**180,068**	**137,288**	**140,355**	**99,062**	**28,719**

Delaware Vote Since 1952

1952, Eisenhower, Rep., 90,059; Stevenson, Dem., 83,315; Hass, Soc. Labor, 242; Hamblen, Proh., 234; Hallinan, Prog., 155; Hoopes, Soc., 20.

1956, Eisenhower, Rep., 98,057; Stevenson, Dem., 79,421; Oltwick, Proh., 400; Hass, Soc. Labor, 110.

1960, Kennedy, Dem., 99,590; Nixon, Rep., 96,373; Faubus, States' Rights, 354; Decker, Proh., 284; Hass, Soc. Labor, 82.

1964, Johnson, Dem., 122,704; Goldwater, Rep., 78,078; Hass, Soc. Labor, 113; Munn, Proh., 425.

1968, Nixon, Rep., 96,714; Humphrey, Dem., 89,194; Wallace, 3d Party, 28,459.

1972, Nixon, Rep., 140,357; McGovern, Dem., 92,283; Schmitz, Amer., 2,638; Munn, Proh., 238.

1976, Carter, Dem., 122,596; Ford, Rep., 109,831; McCarthy, non-partisan, 2,437; Anderson, Amer., 645; LaRouche, U.S. Labor, 136; Bubar, Proh., 103; Levin, Soc. Labor, 86.

1980, Reagan, Rep., 111,252; Carter, Dem., 105,754; Anderson, Ind., 16,288; Clark, Libertarian, 1,974; Greaves, Amer., 400.

1984, Reagan, Rep., 152,190; Mondale, Dem., 101,656; Bergland, Libertarian, 268.

1988, Bush, Rep., 139,639; Dukakis, Dem., 108,647; Paul, Lib., 1,162; Fulani, New Alliance, 443.

1992, Clinton, Dem., 126,054; Bush, Rep., 102,313; Perot, Ind., 59,213; Fulani, New Alliance, 1,105.

1996, Clinton, Dem., 140,355; Dole, Rep., 99,062; Perot, Ind. (Ref.), 28,719; Browne, Libertarian, 2,052; Phillips, Taxpayers, 348; Hagelin, Natural Law, 274.

2000, Gore, Dem., 180,068; Bush, Rep., 137,288; Nader, Green, 8,307; Buchanan, Reform, 777; Browne, Libertarian, 774; Phillips, Constitution, 208; Hagelin, Natural Law, 107.

District of Columbia

	2000 Gore (D)	Bush (R)	1996 Clinton (D)	Dole (R)	Perot (RF)
Totals	171,923	18,073	158,220	17,339	3,611

District of Columbia Vote Since 1964

1964, Johnson, Dem., 169,796; Goldwater, Rep., 28,801.

1968, Nixon, Rep., 31,012; Humphrey, Dem., 139, 566.

1972, Nixon, Rep., 35,226; McGovern, Dem., 127,627; Reed, Soc. Workers, 316; Hall, Com., 252.

1976, Carter, Dem., 137,818; Ford, Rep., 27,873; Camejo, Soc. Workers, 545; MacBride, Libertarian, 274; Hall, Com., 219; LaRouche, U.S. Labor, 157.

1980, Reagan, Rep., 23,313; Carter, Dem., 130,231; Anderson, Ind., 16,131; Commoner, Citizens, 1,826; Clark, Libertarian, 1,104; Hall, Com., 369; DeBerry, Soc. Workers, 173; Griswold, Workers World, 52; write-ins, 690.

1984, Mondale, Dem., 180,408; Reagan, Rep., 29,009; Bergland, Libertarian, 279.

1988, Bush, Rep., 27,590; Dukakis, Dem., 159,407; Fulani, New Alliance, 2,901; Paul, Lib., 554.
1992, Clinton, Dem., 192,619; Bush, Rep., 20,698; Perot, Ind., 9,681; Fulani, New Alliance, 1,459; Daniels, Ind., 1,186.
1996, Clinton, Dem., 158,220; Dole, Rep., 17,339; Perot, Ref., 3,611; Nader, Green, 4,780; Browne, Libertarian, 588; Hagelin, Natural Law, 283; Harris, Soc. Workers, 257.
2000, Gore, Dem., 171,923; Bush, Rep., 18,073; Nader, Green, 10,576; Browne, Libertarian, 669; Harris, Soc. Workers, 114.

Florida

	2000		1996		
County	Gore (D)	Bush (R)	Clinton (D)	Dole (R)	Perot (RF)
Alachua	47,380	34,135	40,144	25,303	8,072
Baker	2,392	5,611	2,273	3,684	667
Bay	18,873	38,682	17,020	28,290	5,922
Bradford	3,075	5,416	3,356	4,038	819
Brevard	97,341	115,253	80,416	87,980	25,249
Broward	387,760	177,939	320,736	142,834	38,964
Calhoun	2,156	2,873	1,794	1,717	630
Charlotte	29,646	35,428	27,121	27,836	7,783
Citrus	25,531	29,801	22,042	20,114	7,244
Clay	14,668	41,903	13,246	30,332	3,281
Collier	29,939	60,467	23,182	42,590	6,320
Columbia	7,049	10,968	6,691	7,588	1,970
Dade[1]	328,867	289,574	317,378	209,634	24,722
De Soto	3,321	4,256	3,219	3,272	965
Dixie	1,827	2,697	1,731	1,398	652
Duval	108,039	152,460	112,258	126,857	13,844
Escambia	40,990	73,171	37,768	60,839	8,587
Flagler	13,897	12,618	9,583	8,232	2,185
Franklin	2,047	2,454	2,095	1,563	878
Gadsden	9,736	4,770	9,405	3,813	938
Gilchrist	1,910	3,300	1,985	1,939	841
Glades	1,442	1,841	1,530	1,361	521
Gulf	2,398	3,553	2,480	2,424	1,054
Hamilton	1,723	2,147	1,734	1,518	406
Hardee	2,342	3,765	2,417	2,926	851
Hendry	3,240	4,747	3,882	3,855	1,135
Hernando	32,648	30,658	28,520	22,039	7,272
Highlands	14,169	20,207	14,244	15,608	3,739
Hillsborough	169,576	180,794	144,223	136,621	25,154
Holmes	2,177	5,012	2,310	3,248	1,208
Indian River	19,769	28,639	16,373	22,709	4,635
Jackson	6,870	9,139	6,665	7,187	1,602
Jefferson	3,041	2,478	2,543	1,851	393
Lafayette	789	1,670	829	1,166	316
Lake	36,571	50,010	29,750	35,089	8,813
Lee	73,571	106,151	65,692	80,882	18,389
Leon	61,444	39,073	50,058	33,914	6,672
Levy	5,398	6,863	4,938	4,299	1,774
Liberty	1,017	1,317	868	913	376
Madison	3,015	3,038	2,791	2,195	578
Manatee	49,226	58,023	41,835	44,059	10,360
Marion	44,674	55,146	37,033	41,397	11,340
Martin	26,621	33,972	20,851	28,516	5,005
Monroe	16,487	16,063	15,219	12,021	4,817
Nassau	6,955	16,408	7,276	12,134	1,657
Okaloosa	16,989	52,186	16,434	40,631	5,432
Okeechobee	4,589	5,057	4,824	3,415	1,666
Orange	140,236	134,531	105,513	106,026	18,191
Osceola	28,187	26,237	21,870	18,335	6,091
Palm Beach	269,754	152,964	230,621	133,762	30,739
Pasco	69,576	68,607	66,472	48,346	18,011
Pinellas	200,657	184,849	184,728	152,125	36,990
Polk	75,207	90,310	66,735	67,943	14,991
Putnam	12,107	13,457	12,008	9,781	3,272
St. Johns	19,509	39,564	16,713	27,311	4,205
St. Lucie	41,560	34,705	36,168	28,892	8,482
Santa Rosa	12,818	36,339	10,923	26,244	4,957
Sarasota	72,869	83,117	63,648	69,198	14,939
Seminole	59,227	75,790	45,051	59,778	9,357
Sumter	9,637	12,127	7,014	5,960	2,375
Suwannee	4,076	8,009	4,479	5,742	1,874
Taylor	2,649	4,058	3,583	3,188	1,140
Union	1,407	2,332	1,388	1,636	425
Volusia	97,313	82,368	78,905	63,067	17,319
Wakulla	3,838	4,512	3,054	2,931	1,091
Walton	5,643	12,186	5,341	7,706	2,342
Washington	2,798	4,995	2,992	3,522	1,287
Totals	**2,912,253**	**2,912,790**	**2,545,968**	**2,243,324**	**483,776**

(1) In 1997, Dade County changed its name to Miami-Dade County.

Florida Vote Since 1952

1952, Eisenhower, Rep., 544,036; Stevenson, Dem., 444,950; scattered, 351.
1956, Eisenhower, Rep., 643,849; Stevenson, Dem., 480,371.
1960, Kennedy, Dem., 748,700; Nixon, Rep., 795,476.
1964, Johnson, Dem., 948,540; Goldwater, Rep., 905,941.
1968, Nixon, Rep., 886,804; Humphrey, Dem., 676,794; Wallace, 3d Party, 624,207.
1972, Nixon, Rep., 1,857,759; McGovern, Dem., 718,117; scattered, 7,407.
1976, Carter, Dem., 1,636,000; Ford, Rep., 1,469,531; McCarthy, Ind., 23,643; Anderson, Amer., 21,325.
1980, Reagan, Rep., 2,046,951; Carter, Dem., 1,419,475; Anderson, Ind., 189,692; Clark, Libertarian, 30,524; write-ins, 285.
1984, Reagan, Rep., 2,728,775; Mondale, Dem., 1,448,344.
1988, Bush, Rep., 2,616,597; Dukakis, Dem., 1,655,851; Paul, Lib., 19,796, Fulani, New Alliance, 6,655.
1992, Bush, Rep., 2,171,781; Clinton, Dem., 2,071,651; Perot, Ind., 1,052,481; Marrou, Libertarian, 15,068.
1996, Clinton, Dem., 2,545,968; Dole, Rep., 2,243,324; Perot, Ref., 483,776; Browne, Libertarian, 23,312.
2000, Bush, Rep., 2,912,790; Gore, Dem., 2,912,253; Nader, Green, 97,488; Buchanan, Reform, 17,484; Browne, Libertarian, 16,415; Hagelin, Natural Law, 2,281; Moorehead, Workers World, 1,804; Phillips, Constitution, 1,371; McReynolds, Soc., 622; Harris, Soc. Workers, 562.

Georgia

	2000		1996		
County	Gore (D)	Bush (R)	Clinton (D)	Dole (R)	Perot (RF)
Appling	2,093	3,940	2,070	2,572	446
Atkinson	821	1,228	823	784	215
Bacon	956	2,010	1,360	1,580	402
Baker	893	615	955	408	105
Baldwin	5,893	6,041	5,740	4,570	849
Banks	1,220	3,202	1,536	1,925	595
Barrow	3,657	7,925	3,928	5,342	942
Bartow	7,508	14,720	6,853	9,250	1,770
Ben Hill	2,234	2,381	2,198	1,516	358
Berrien	1,640	2,718	2,066	1,950	525
Bibb	24,996	24,071	26,727	20,778	2,268
Bleckley	1,273	2,436	1,365	1,632	300
Brantley	1,372	3,118	1,494	1,738	386
Brooks	2,096	2,406	1,977	1,738	314
Bryan	2,172	4,835	2,152	3,577	513
Bulloch	5,561	8,990	5,396	6,646	939
Burke	3,720	3,381	3,915	2,590	389
Butts	2,281	3,198	2,271	2,027	416
Calhoun	1,107	768	1,217	541	106
Camden	3,636	6,371	3,644	4,222	572
Candler	1,053	1,643	1,097	1,131	264
Carroll	8,752	16,326	8,438	11,157	2,002
Catoosa	5,470	12,033	5,185	8,237	1,257
Charlton	1,015	1,770	1,368	1,374	280
Chatham	37,590	37,847	35,781	31,987	3,028
Chattahoochee	600	590	565	398	115
Chattooga	2,729	3,640	3,003	2,513	796
Cherokee	12,295	38,033	10,802	24,527	2,872
Clarke	15,167	11,850	15,206	10,504	1,201
Clay	821	448	787	293	62
Clayton	40,042	19,966	30,687	20,625	3,494
Clinch	816	1,091	973	789	182
Cobb	86,676	140,494	73,750	114,188	10,438
Coffee	3,593	5,756	3,407	3,934	711
Colquitt	3,297	6,589	4,135	4,847	977
Columbia	8,969	26,660	8,601	21,291	1,709
Cook	1,639	2,279	1,780	1,354	267
Coweta	9,056	21,327	7,794	13,058	1,949
Crawford	1,513	1,987	1,534	1,290	270
Crisp	2,268	3,285	2,504	2,321	445
Dade	1,628	3,333	1,737	2,295	618
Dawson	1,458	4,210	1,434	2,343	473
Decatur	3,398	4,187	3,245	3,035	497
DeKalb	154,509	58,807	137,903	60,255	6,742
Dodge	2,326	3,472	2,696	2,478	587
Dooly	1,901	1,588	1,951	990	207
Dougherty	16,650	12,248	15,600	11,144	1,072
Douglas	11,162	18,893	9,631	14,495	2,109
Early	1,622	1,938	1,648	1,374	246
Echols	272	614	308	335	97
Effingham	3,232	7,326	3,031	5,022	769
Elbert	2,527	3,262	2,900	2,393	552
Emanuel	2,835	3,343	2,947	2,451	450
Evans	1,217	1,841	1,117	1,206	204
Fannin	2,736	5,463	2,741	3,373	782
Fayette	11,912	29,338	9,875	21,005	2,016
Floyd	10,282	16,194	10,464	12,426	2,345
Forsyth	6,694	27,769	5,957	15,013	1,889
Franklin	2,040	3,659	2,338	2,364	665
Fulton	152,039	104,870	143,306	89,809	7,720
Gilmer	2,230	4,941	2,464	3,121	725
Glascock	249	763	348	532	128
Glynn	7,778	14,346	8,058	12,305	1,137
Gordon	4,032	7,944	4,239	5,232	1,284
Grady	2,721	3,894	2,862	2,674	633

	2000		1996		
	Gore	Bush	Clinton	Dole	Perot
County	(D)	(R)	(D)	(R)	(RF)
Greene . . .	2,137	2,980	2,115	1,702	173
Gwinnett . .	61,434	121,756	53,819	96,610	10,236
Habersham	2,530	6,964	3,170	4,730	1,149
Hall	10,259	26,841	10,362	19,280	2,321
Hancock . .	2,414	662	2,135	438	71
Haralson . .	2,869	5,153	2,850	3,260	808
Harris	2,912	5,554	2,779	3,829	489
Hart	3,192	4,242	3,486	2,884	767
Heard	1,178	1,947	1,248	1,170	406
Henry	11,971	25,815	9,498	16,968	2,320
Houston. . .	13,301	23,174	12,760	17,050	2,730
Irwin.	1,105	1,720	1,225	1,085	224
Jackson . . .	3,420	7,878	3,746	4,782	899
Jasper	1,558	2,298	1,553	1,423	243
Jeff Davis .	1,379	2,797	1,576	1,796	428
Jefferson . .	2,973	2,559	3,404	2,077	298
Jenkins . . .	1,250	1,317	1,336	955	166
Johnson. . .	1,065	1,797	1,194	815	242
Jones.	3,102	4,850	3,195	3,272	497
Lamar	2,194	2,912	2,125	1,988	409
Lanier	832	1,048	818	519	160
Laurens . . .	5,724	8,133	5,792	6,118	818
Lee	1,936	5,872	2,005	3,983	506
Liberty	5,347	4,455	4,462	3,042	580
Lincoln. . . .	1,275	1,807	1,334	1,391	208
Long	975	1,320	936	791	236
Lowndes . .	10,616	14,462	9,470	10,578	1,518
Lumpkin. . .	2,121	4,427	1,949	2,576	588
McDuffie . .	2,580	3,926	2,725	3,254	395
McIntosh . .	2,047	1,766	1,927	1,219	293
Macon	2,757	1,566	2,618	1,006	159
Madison. . .	2,285	5,529	2,571	3,992	868
Marion	982	1,187	977	678	159
Meriwether	3,441	3,162	3,492	2,259	480
Miller	783	1,349	909	847	235
Mitchell . . .	2,971	2,790	3,165	2,033	372
Monroe . . .	2,839	4,561	2,768	3,054	488
Montgomery	1,013	1,465	1,233	1,163	284
Morgan . . .	2,238	3,524	2,111	2,118	364
Murray	2,684	5,539	2,861	3,289	938
Muscogee .	28,193	23,479	24,867	19,360	1,891
Newton . . .	6,703	11,127	6,759	7,274	1,258
Oconee . . .	3,184	7,611	2,992	5,116	615
Oglethorpe	1,519	2,706	1,570	1,826	369
Paulding . .	6,743	16,881	5,699	10,152	1,603
Peach	3,540	3,525	3,582	2,676	471
Pickens . . .	2,489	5,488	2,693	3,041	783
Pierce	1,300	3,348	1,420	2,319	333
Pike	1,413	3,358	1,474	2,054	357
Polk	4,112	5,841	4,298	4,130	1,076
Pulaski. . . .	1,390	1,922	1,554	1,196	268
Putnam . . .	2,612	3,596	2,340	2,306	474
Quitman. . .	542	348	514	224	59
Rabun	1,776	3,451	1,943	2,213	585
Randolph. .	1,381	1,174	1,438	816	126
Richmond .	31,413	25,485	30,738	23,670	2,310
Rockdale . .	8,295	15,440	7,656	13,006	1,750
Schley	460	706	576	470	123
Screven . . .	2,233	2,461	2,087	1,862	263
Seminole . .	1,313	1,537	1,265	1,003	250
Spalding . .	5,831	9,271	6,017	7,376	1,059
Stephens . .	2,869	5,370	3,072	3,890	979
Stewart . . .	1,267	675	1,537	525	152
Sumter. . . .	4,748	4,847	4,239	3,358	451
Talbot.	1,662	844	1,579	652	111
Taliaferro . .	556	271	615	235	36
Tattnall. . . .	1,963	3,597	2,369	2,518	541
Taylor.	1,340	1,412	1,450	1,002	195
Telfair.	1,777	1,693	1,856	1,143	322
Terrell	1,584	1,504	1,509	1,111	129
Thomas . . .	4,862	7,093	5,183	5,649	667
Tift	3,547	6,678	4,198	5,613	728
Toombs . . .	2,643	4,487	2,763	3,646	602
Towns	1,495	2,902	1,664	2,030	459
Treutlen . . .	879	1,062	912	723	122
Troup	6,379	11,198	5,940	8,716	1,090
Turner	1,169	1,258	1,272	924	246
Twiggs	1,977	1,570	1,927	958	210
Union.	2,230	4,567	2,175	2,685	622
Upson	3,158	5,019	3,491	3,783	731
Walker	6,341	12,326	6,743	8,817	1,969
Walton	5,484	12,966	5,618	7,934	1,323
Ware	3,480	6,099	4,171	4,746	636
Warren . . .	1,196	933	1,230	735	83
Washington	3,476	3,162	4,057	2,348	488
Wayne	2,736	5,219	2,734	3,709	665
Webster. . .	541	359	529	235	59
Wheeler. . .	752	813	751	460	141
White.	2,014	4,857	1,864	2,959	556
Whitfield . .	7,034	15,852	7,720	12,368	1,637
Wilcox.	962	1,381	1,067	882	171
Wilkes.	1,940	2,044	1,971	1,417	184
Wilkinson . .	1,884	1,800	2,278	1,332	287
Worth	2,214	3,792	2,300	2,752	521
Totals.	**1,116,230**	**1,419,720**	**1,053,849**	**1,080,843**	**146,337**

Georgia Vote Since 1952

1952, Eisenhower, Rep., 198,979; Stevenson, Dem., 456,823; Liberty Party, 1.
1956, Stevenson, Dem., 444,388; Eisenhower, Rep., 222,778; Andrews, Ind., write-in, 1,754.
1960, Kennedy, Dem., 458,638; Nixon, Rep., 274,472; write-in, 239.
1964, Johnson, Dem., 522,557; Goldwater, Rep., 616,600.
1968, Nixon, Rep., 380,111; Humphrey, Dem., 334,440; Wallace, 3d Party, 535,550; write-in, 162.
1972, Nixon, Rep., 881,496; McGovern, Dem., 289,529; scattered, 2,935; Schmitz, Amer., 812.
1976, Carter, Dem., 979,409; Ford, Rep., 483,743; write-in, 4,306.
1980, Reagan, Rep., 654,168; Carter, Dem., 890,955; Anderson, Ind., 36,055; Clark, Libertarian, 15,627.
1984, Reagan, Rep., 1,068,722; Mondale, Dem., 706,628.
1988, Bush, Rep., 1,081,331; Dukakis, Dem., 714,792; Paul, Lib., 8,435; Fulani, New Alliance, 5,099.
1992, Clinton, Dem., 1,008,966; Bush, Rep., 995,252; Perot, Ind., 309,657; Marrou, Libertarian, 7,110.
1996, Dole, Rep., 1,080,843; Clinton, Dem., 1,053,849; Perot, Ref., 146,337; Browne, Libertarian, 17,870.
2000, Bush, Rep., 1,419,720; Gore, Dem., 1,116,230; Browne, Libertarian, 36,332; Buchanan, Independent, 10,926.

Hawaii

	2000		1996		
	Gore	Bush	Clinton	Dole	Perot
County	(D)	(R)	(D)	(R)	(RF)
Hawaii	28,670	17,050	27,262	13,516	5,137
Honolulu.	139,662	101,336	143,793	85,779	17,389
Kauai	13,470	6,583	13,357	5,325	1,568
Maui	23,484	12,876	20,600	9,323	3,264
Totals.	**205,286**	**137,845**	**205,012**	**113,943**	**27,358**

Hawaii Vote Since 1960

1960, Kennedy, Dem., 92,410; Nixon, Rep., 92,295.
1964, Johnson, Dem., 163,249; Goldwater, Rep., 44,022.
1968, Nixon, Rep., 91,425; Humphrey, Dem., 141,324; Wallace, 3d Party, 3,469.
1972, Nixon, Rep., 168,865; McGovern, Dem., 101,409.
1976, Carter, Dem., 147,375; Ford, Rep., 140,003; MacBride, Libertarian, 3,923.
1980, Reagan, Rep., 130,112; Carter, Dem., 135,879; Anderson, Ind., 32,021; Clark, Libertarian, 3,269; Commoner, Citizens, 1,548; Hall, Com., 458.
1984, Reagan, Rep., 184,934; Mondale, Dem., 147,098; Bergland, Libertarian, 2,167.
1988, Bush, Rep., 158,625; Dukakis, Dem., 192,364; Paul, Lib., 1,999; Fulani, New Alliance, 1,003.
1992, Clinton, Dem., 179,310; Bush, Rep., 136,822; Perot, Ind., 53,003; Gritz, Populist/America First, 1,452; Marrou, Libertarian, 1,119.
1996, Clinton, Dem., 205,012; Dole, Rep., 113,943; Perot, Ref., 27,358; Nader, Green, 10,386; Browne, Libertarian, 2,493; Hagelin, Natural Law, 570; Phillips, Taxpayers, 358.
2000, Gore, Dem., 205,286; Bush, Rep., 137,845; Nader, Green, 21,623; Browne, Libertarian, 1,477; Buchanan, Reform, 1,071; Phillips, Constitution, 343; Hagelin, Natural Law, 306.

Idaho

	2000		1996		
	Gore	Bush	Clinton	Dole	Perot
County	(D)	(R)	(D)	(R)	(RF)
Ada.	40,650	75,050	43,040	61,811	11,171
Adams	336	1,476	537	1,053	311
Bannock	10,892	18,223	12,806	14,058	4,158
Bear Lake.	517	2,296	805	1,583	396
Benewah	895	2,606	1,488	1,667	701
Bingham	3,310	10,628	4,304	8,391	2,021
Blaine	3,748	3,528	3,840	3,003	1,193
Boise	745	2,019	879	1,576	440
Bonner	4,318	8,945	5,294	6,207	2,669
Bonneville.	7,235	24,988	9,013	19,977	3,921
Boundary	832	2,797	1,194	1,937	626
Butte.	354	1,054	507	741	233
Camas	113	359	156	283	95
Canyon.	10,588	30,560	11,800	23,988	3,956
Caribou	475	2,601	841	1,740	501
Cassia	1,087	5,983	1,596	4,663	976
Clark.	63	311	117	266	45
Clearwater	841	2,885	1,507	1,658	650

	2000		1996		
County	Gore (D)	Bush (R)	Clinton (D)	Dole (R)	Perot (RF)
Custer	416	1,794	635	1,249	400
Elmore.	1,840	4,891	2,324	3,668	845
Franklin	513	3,594	807	2,435	589
Fremont.	699	4,242	1,114	3,042	630
Gem	1,346	4,376	1,968	3,362	833
Gooding.	1,282	3,502	1,503	2,637	980
Idaho.	1,187	5,806	1,979	3,871	1,083
Jefferson	1,100	6,480	1,427	4,925	994
Jerome	1,360	4,418	1,679	3,358	1,014
Kootenai	13,488	28,162	13,627	18,740	6,083
Latah	5,661	8,161	7,741	6,311	1,828
Lemhi	660	2,859	1,015	2,334	461
Lewis.	335	1,295	674	861	316
Lincoln.	437	1,049	478	744	319
Madison.	816	7,941	1,216	5,706	744
Minidoka	1,344	4,907	1,977	4,008	977
Nez Perce	4,995	10,577	7,491	6,675	2,385
Oneida.	307	1,426	429	993	285
Owyhee.	623	2,450	895	2,033	354
Payette	1,643	4,961	2,119	3,901	906
Power	755	1,872	1,070	1,501	344
Shoshone	2,225	2,879	2,981	1,588	1,283
Teton	720	1,745	866	1,251	326
Twin Falls.	5,777	15,794	6,826	12,393	3,383
Valley.	1,129	2,548	1,564	2,089	568
Washington	980	2,899	1,314	2,318	525
Totals	**138,637**	**336,937**	**164,443**	**256,595**	**62,518**

Idaho Vote Since 1952

1952, Eisenhower, Rep., 180,707; Stevenson, Dem., 95,081; Hallinan, Prog., 443; write-in, 23.

1956, Eisenhower, Rep., 166,979; Stevenson, Dem., 105,868; Andrews, Ind., 126; write-in, 16.

1960, Kennedy, Dem., 138,853; Nixon, Rep., 161,597.

1964, Johnson, Dem., 148,920; Goldwater, Rep., 143,557.

1968, Nixon, Rep., 165,369; Humphrey, Dem., 89,273; Wallace, 3d Party, 36,541.

1972, Nixon, Rep., 199,384; McGovern, Dem., 80,826; Schmitz, Amer., 28,869; Spock, Peoples, 903.

1976, Carter, Dem., 126,549; Ford, Rep., 204,151; Maddox, Amer., 5,935; MacBride, Libertarian, 3,558; LaRouche, U.S. Labor, 739.

1980, Reagan, Rep., 290,699; Carter, Dem., 110,192; Anderson, Ind., 27,058; Clark, Libertarian, 8,425; Rarick, Amer., 1,057.

1984, Reagan, Rep., 297,523; Mondale, Dem., 108,510; Bergland, Libertarian, 2,823.

1988, Bush, Rep., 253,881; Dukakis, Dem., 147,272; Paul, Lib., 5,313; Fulani, Ind., 2,502.

1992, Clinton, Dem., 137,013; Bush, Rep., 202,645; Perot, Ind., 130,395; Gritz, Populist/America First, 10,281; Marrou, Libertarian, 1,167.

1996, Dole, Rep., 256,595; Clinton, Dem., 165,443; Perot, Ref., 62,518; Browne, Libertarian, 3,325; Phillips, Taxpayers, 2,230; Hagelin, Natural Law, 1,600.

2000, Bush, Rep., 336,937; Gore, Dem., 138,637; Buchanan, Reform, 7,615; Browne, Libertarian, 3,488; Phillips, Constitution, 1,469; Hagelin, Natural Law, 1,177.

Illinois

	2000		1996		
County	Gore (D)	Bush (R)	Clinton (D)	Dole (R)	Perot (RF)
Adams.	12,197	17,331	11,336	13,836	3,069
Alexander . .	2,357	1,588	2,753	1,212	321
Bond	3,060	3,804	3,213	3,018	685
Boone	6,481	8,617	5,345	6,181	1,377
Brown	1,077	1,529	997	1,053	237
Bureau.	7,754	8,526	7,651	6,528	1,798
Calhoun. . . .	1,310	1,229	1,676	941	363
Carroll	3,113	3,835	2,926	3,029	792
Cass	2,789	2,968	2,834	2,214	589
Champaign .	35,515	34,645	32,454	28,232	4,806
Christian . . .	6,799	7,537	7,431	5,563	1,727
Clark	2,932	4,398	2,995	3,409	781
Clay	2,212	3,789	2,750	2,703	719
Clinton	6,436	8,588	6,104	6,065	1,580
Coles.	8,904	10,495	8,950	8,038	2,137
Cook	1,280,547	534,542	1,153,289	461,557	96,633
Crawford . . .	3,333	4,974	3,627	3,965	1,057
Cumberland	1,870	2,964	1,776	2,002	657
DeKalb	14,798	17,139	12,715	12,380	3,009
DeWitt	2,870	3,968	2,878	2,978	694
Douglas. . . .	3,215	4,734	2,955	3,272	740
DuPage	152,550	201,037	129,709	164,630	27,419
Edgar.	3,216	4,833	3,552	3,746	935
Edwards . . .	978	2,212	1,089	1,613	384
Effingham . .	4,225	9,855	4,825	7,696	1,555
Fayette	3,886	5,200	3,887	3,881	964
Ford	2,090	3,889	2,065	3,077	590
Franklin . . .	10,201	8,490	9,814	5,354	2,096
Fulton	8,940	6,936	8,857	5,155	1,610
Gallatin. . . .	1,878	1,591	2,113	856	527
Greene	2,490	3,129	2,734	2,245	903
Grundy	7,516	8,709	6,759	6,177	1,860
Hamilton. . .	1,943	2,519	2,242	1,677	560
Hancock . . .	4,256	5,134	4,001	3,961	1,148
Hardin	1,184	1,366	1,323	790	485
Henderson .	2,030	1,708	1,953	1,233	408
Henry	11,921	10,896	11,201	8,393	2,194
Iroquois . . .	4,397	8,685	4,559	6,564	1,522
Jackson . . .	11,773	9,823	12,214	7,422	2,082
Jasper	1,815	3,119	2,038	2,234	641
Jefferson . .	6,685	8,362	7,263	5,937	1,647
Jersey.	4,355	4,699	4,275	3,211	1,186
Jo Daviess .	4,585	5,304	4,171	3,915	1,131
Johnson . . .	1,928	3,285	2,009	2,241	640
Kane.	60,127	76,996	47,902	54,375	11,270
Kankakee . .	19,180	20,049	16,820	14,595	3,574
Kendall	8,444	13,688	6,499	8,958	2,055
Knox.	12,572	9,912	12,487	7,822	2,096
Lake	115,058	120,988	93,315	93,149	16,640
LaSalle	23,355	21,276	21,643	15,299	5,259
Lawrence . .	2,822	3,594	2,871	2,568	916
Lee.	6,111	8,069	5,895	6,677	1,520
Livingston. .	5,829	9,187	5,641	7,653	1,409
Logan	4,600	8,141	4,618	6,518	1,141
McDonough	6,080	6,465	5,632	5,049	1,217
McHenry. . .	40,698	62,112	31,240	41,136	10,082
McLean . . .	24,936	34,008	22,708	26,428	3,816
Macon	24,262	23,830	24,256	18,161	4,540
Macoupin . .	11,015	9,749	11,107	7,235	2,532
Madison . . .	59,077	48,821	53,568	35,758	10,121
Marion	8,068	8,240	7,792	5,999	1,825
Marshall . . .	2,570	3,145	2,640	2,453	586
Mason	3,192	3,411	3,385	2,430	600
Massac. . . .	2,912	3,676	2,841	2,507	675
Menard. . . .	2,164	3,862	2,204	3,106	534
Mercer	4,400	3,688	4,278	2,688	889
Monroe. . . .	5,797	7,632	4,798	5,350	1,276
Montgomery	6,542	6,226	6,338	4,770	1,436
Morgan. . . .	5,899	8,058	6,150	6,352	1,633
Moultrie . . .	2,529	3,058	2,629	2,199	596
Ogle	7,673	12,325	6,765	9,558	1,876
Peoria.	38,604	36,398	37,383	30,990	5,220
Perry.	4,862	4,802	5,347	3,237	1,262
Piatt	3,488	4,619	3,274	3,265	818
Pike	3,198	4,706	3,604	3,225	1,039
Pope.	927	1,346	915	850	277
Pulaski	1,518	1,430	1,524	1,036	235
Putnam. . . .	1,657	1,437	1,425	987	322
Randolph . .	6,794	7,127	7,419	5,422	1,698
Richland. . .	2,491	4,718	2,679	3,137	927
Rock Island	37,957	25,194	34,822	20,626	5,135
St. Clair . . .	55,961	42,299	53,405	33,066	7,027
Saline	5,427	5,933	6,156	3,693	1,752
Sangamon .	38,414	50,374	38,902	42,174	6,446
Schuyler. . .	1,587	2,077	1,636	1,597	483
Scott.	954	1,458	1,012	1,112	396
Shelby	4,018	5,851	4,249	4,215	1,262
Stark.	1,211	1,694	1,262	1,278	312
Stephenson	8,062	10,715	7,145	8,871	1,940
Tazewell . . .	25,379	31,537	24,139	24,395	4,814
Union	3,982	4,397	4,252	3,147	832
Vermilion . .	15,406	15,783	15,525	12,015	3,577
Wabash . . .	1,987	3,406	2,177	2,381	683
Warren	3,524	3,899	3,500	2,974	742
Washington	2,638	4,353	2,744	3,339	790
Wayne	2,209	5,347	3,054	4,029	999
White	2,958	4,521	3,553	2,878	888
Whiteside . .	12,886	11,252	11,913	8,859	2,436
Will	90,902	95,828	69,354	62,506	15,485
Williamson .	12,192	14,012	12,510	9,734	2,877
Winnebago.	51,981	53,816	46,264	44,479	8,192
Woodford . .	5,529	10,905	5,270	8,527	1,170
Totals.	**2,589,026**	**2,019,421**	**2,341,744**	**1,587,021**	**346,408**

Illinois Vote Since 1952

1952, Eisenhower, Rep., 2,457,327; Stevenson, Dem., 2,013,920; Hass, Soc. Labor, 9,363; write-in, 448.

1956, Eisenhower, Rep., 2,623,327; Stevenson, Dem., 1,775,682; Hass, Soc. Labor, 8,342; write-in, 56.

1960, Kennedy, Dem., 2,377,846; Nixon, Rep., 2,368,988; Hass, Soc. Labor, 10,560; write-in, 15.

1964, Johnson, Dem., 2,796,833; Goldwater, Rep., 1,905,946; write-in, 62.

1968, Nixon, Rep., 2,174,774; Humphrey, Dem., 2,039,814; Wallace, 3d Party, 390,958; Blomen, Soc. Labor, 13,878; write-in, 325.

1972, Nixon, Rep. 2,788,179; McGovern, Dem., 1,913,472; Fisher, Soc. Labor, 12,344; Schmitz, Amer., 2,471; Hall, Com., 4,541; others, 2,229.
1976, Carter, Dem., 2,271,295; Ford, Rep., 2,364,269; McCarthy, Ind., 55,939; Hall, Com., 9,250; MacBride, Libertarian, 8,057; Camejo, Soc. Workers, 3,615; Levin, Soc. Labor, 2,422; LaRouche, U.S. Labor, 2,018; write-in, 1,968.
1980, Reagan, Rep., 2,358,049; Carter, Dem., 1,981,413; Anderson, Ind., 346,754; Clark, Libertarian, 38,939; Commoner, Citizens, 10,692; Hall, Com., 9,711; Griswold, Workers World, 2,257; DeBerry, Soc. Workers, 1,302; write-ins, 604.
1984, Reagan, Rep., 2,707,103; Mondale, Dem., 2,086,499; Bergland, Libertarian, 10,086.
1988, Bush, Rep., 2,310,939; Dukakis, Dem., 2,215,940; Paul, Lib., 14,944; Fulani, Solid., 10,276.
1992, Clinton, Dem., 2,453,350; Bush, Rep., 1,734,096; Perot, Ind., 840,515; Marrou, Libertarian, 9,218; Fulani, New Alliance, 5,267; Gritz, Populist/America First, 3,577; Hagelin, Natural Law, 2,751; Warren, Soc. Workers, 1,361.
1996, Clinton, Dem., 2,341,744; Dole, Rep., 1,587,021; Perot, Ref., 346,408; Browne, Libertarian, 22,548; Phillips, Taxpayers, 7,606; Hagelin, Natural Law, 4,606.
2000, Gore, Dem., 2,589,026; Bush, Rep., 2,019,421; Nader, Green, 103,759; Buchanan, Ind., 16,106; Browne, Libertarian, 11,623; Hagelin, Reform, 2,127.

Indiana

	2000		1996		
County	Gore (D)	Bush (R)	Clinton (D)	Dole (R)	Perot (RF)
Adams	3,775	8,555	4,247	6,960	1,346
Allen	41,636	70,426	41,450	59,255	8,808
Bartholomew	9,015	16,200	9,301	13,188	2,815
Benton	1,328	2,441	1,311	1,947	609
Blackford	2,103	2,699	2,335	2,070	681
Boone	4,763	13,161	4,625	11,338	1,498
Brown	2,608	3,871	2,413	2,988	802
Carroll	2,965	5,102	2,747	4,062	1,171
Cass	5,412	9,305	5,419	8,020	2,029
Clark	17,360	19,417	17,799	14,396	3,578
Clay	3,605	6,393	3,605	4,858	1,406
Clinton	3,643	7,141	3,949	6,156	1,355
Crawford	1,817	2,327	2,324	1,759	700
Daviess	2,697	6,872	3,230	5,531	994
Dearborn	6,020	11,452	6,269	8,318	1,731
Decatur	2,889	6,115	3,190	4,782	1,389
Dekalb	4,776	8,701	4,840	6,851	1,534
Delaware	20,876	22,105	20,385	18,126	6,042
Dubois	5,090	10,134	6,499	6,840	1,777
Elkhart	16,402	36,756	16,598	28,770	5,133
Fayette	3,415	5,060	3,822	4,091	1,137
Floyd	13,209	16,486	13,814	12,473	2,609
Fountain	2,717	4,408	2,327	3,984	1,033
Franklin	2,591	5,587	2,808	4,167	943
Fulton	2,960	5,218	2,956	3,934	1,143
Gibson	5,802	7,734	6,488	5,392	1,585
Grant	9,712	16,153	9,818	13,443	3,008
Greene	4,898	7,452	5,277	5,746	1,690
Hamilton	18,002	56,372	14,153	42,792	4,234
Hancock	6,503	15,943	6,123	12,907	2,258
Harrison	5,870	8,711	5,900	6,073	1,839
Hendricks	10,786	28,651	9,392	22,293	3,405
Henry	7,647	10,321	7,667	8,537	2,381
Howard	12,899	20,331	11,999	16,771	4,172
Huntington	4,119	10,113	4,287	8,275	1,400
Jackson	5,330	9,054	5,150	5,883	1,590
Jasper	3,744	7,212	3,554	5,173	1,271
Jay	3,167	4,687	3,356	3,584	1,022
Jefferson	5,117	6,582	5,441	4,827	1,438
Jennings	3,549	5,732	4,223	4,461	1,629
Johnson	11,952	29,404	11,278	23,733	3,975
Knox	6,300	8,485	7,003	6,395	2,022
Kosciusko	5,785	19,040	6,166	15,084	2,531
LaGrange	2,733	5,437	2,704	4,033	949
Lake	109,078	63,389	100,198	47,873	15,051
LaPorte	19,736	18,994	19,879	14,106	5,133
Lawrence	5,071	10,677	5,703	8,107	2,063
Madison	23,403	27,956	23,772	23,151	6,447
Marion	134,189	137,810	124,448	133,329	21,358
Marshall	5,541	10,266	5,486	8,158	1,698
Martin	1,518	3,008	1,848	2,281	485
Miami	4,155	8,401	4,260	6,719	1,657
Monroe	17,523	19,147	18,531	16,744	3,179
Montgomery	3,899	8,891	3,825	7,705	1,766
Morgan	6,228	15,286	5,812	12,872	2,755
Newton	2,101	3,250	1,897	2,075	801
Noble	4,822	9,103	5,101	6,782	1,521
Ohio	951	1,515	1,083	1,098	281
Orange	2,601	4,687	3,016	3,355	938
Owen	2,253	4,019	2,244	3,056	874
Parke	2,481	3,841	2,453	3,151	981
Perry	3,823	3,461	4,427	2,554	913
Pike	2,605	3,566	2,780	2,174	884
Porter	26,790	31,157	24,044	22,931	7,169
Posey	4,430	6,498	4,965	4,638	1,304
Pulaski	1,919	3,497	2,010	2,693	634
Putnam	4,123	7,352	3,962	5,958	1,619
Randolph	3,906	6,020	4,087	4,708	1,557
Ripley	3,498	6,988	4,097	5,303	1,216
Rush	2,370	4,749	2,578	3,827	973
St. Joseph	47,703	47,581	45,704	38,281	8,379
Scott	3,915	3,761	3,798	2,620	760
Shelby	5,374	9,590	5,374	7,778	1,874
Spencer	3,752	5,096	4,058	3,770	739
Starke	4,136	4,349	3,854	3,108	1,096
Steuben	4,103	6,953	4,124	5,513	1,390
Sullivan	3,833	4,319	4,076	3,207	1,178
Switzerland	1,336	1,831	1,496	1,266	403
Tippecanoe	18,220	26,106	17,232	22,556	5,394
Tipton	2,392	4,784	2,478	3,980	861
Union	927	1,838	1,019	1,334	364
Vanderburgh	29,222	35,846	30,934	28,509	6,132
Vermillion	3,370	3,130	3,251	2,334	1,029
Vigo	17,570	18,021	17,974	15,751	4,508
Wabash	4,277	8,321	4,577	6,990	1,294
Warren	1,471	2,218	1,394	1,678	560
Warrick	8,749	13,205	9,285	9,221	2,471
Washington	3,675	5,868	3,819	4,066	1,264
Wayne	10,273	14,273	10,905	12,188	2,525
Wells	3,319	7,755	3,752	6,322	1,157
White	3,655	6,037	3,396	4,642	1,610
Whitley	4,107	8,080	4,176	5,965	1,392
Totals	**901,980**	1,245,836	**887,424**	**1,006,693**	**224,299**

Indiana Vote Since 1952

1952, Eisenhower, Rep., 1,136,259; Stevenson, Dem., 801,530; Hamblen, Proh., 15,335; Hallinan, Prog., 1,222; Hass, Soc. Labor, 979.
1956, Eisenhower, Rep., 1,182,811; Stevenson, Dem., 783,908; Holtwick, Proh., 6,554; Hass, Soc. Labor, 1,334.
1960, Kennedy, Dem., 952,358; Nixon, Rep., 1,175,120; Decker, Proh., 6,746; Hass, Soc. Labor, 1,136.
1964, Johnson, Dem., 1,170,848; Goldwater, Rep., 911,118; Munn, Proh., 8,266; Hass, Soc. Labor, 1,374.
1968, Nixon, Rep., 1,067,885; Humphrey, Dem., 806,659; Wallace, 3d Party, 243,108; Munn, Proh., 4,616; Halstead, Soc. Workers, 1,293; Gregory, write-in, 36.
1972, Nixon, Rep., 1,405,154; McGovern, Dem., 708,568; Reed, Soc. Workers, 5,575; Fisher, Soc. Labor, 1,688; Spock, Peace and Freedom, 4,544.
1976, Carter, Dem., 1,014,714; Ford, Rep., 1,185,958; Anderson, Amer., 14,048; Camejo, Soc. Workers, 5,695; LaRouche, U.S. Labor, 1,947.
1980, Reagan, Rep., 1,255,656; Carter, Dem., 844,197; Anderson, Ind., 111,639; Clark, Libertarian, 19,627; Commoner, Citizens, 4,852; Greaves, Amer., 4,750; Hall, Com., 702; DeBerry, Soc., 610.
1984, Reagan, Rep., 1,377,230; Mondale, Dem., 841,481; Bergland, Libertarian, 6,741.
1988, Bush, Rep., 1,297,763; Dukakis, Dem., 860,643; Fulani, New Alliance, 10,215.
1992, Bush, Rep., 989,375; Clinton, Dem., 848,420; Perot, Ind., 455,934; Marrou, Libertarian, 7,936; Fulani, New Alliance, 2,583.
1996, Dole, Rep., 1,006,693; Clinton, Dem., 887,424; Perot, Ref., 224,299; Browne, Libertarian, 15,632.
2000, Bush, Rep., 1,245,836; Gore, Dem., 901,980; Buchanan, Ind., 16,959; Browne, Libertarian, 15,530.

Iowa

	2000		1996		
County	Gore (D)	Bush (R)	Clinton (D)	Dole (R)	Perot (RF)
Adair	1,753	2,275	1,802	1,655	458
Adams	897	1,170	1,070	920	320
Allamakee	2,883	3,277	2,551	2,457	680
Appanoose	2,560	2,992	2,747	2,233	554
Audubon	1,780	1,909	1,827	1,314	314
Benton	5,915	5,468	5,546	3,835	846
Black Hawk	30,112	23,468	29,651	19,322	3,623
Boone	6,270	5,625	6,446	4,293	987
Bremer	5,169	5,675	5,023	4,213	862
Buchanan	5,045	4,092	4,997	3,043	836
Buena Vista	3,297	4,354	3,420	3,636	831
Butler	2,735	3,837	3,061	3,036	489
Calhoun	2,132	2,776	2,193	2,077	462
Carroll	4,463	4,879	4,333	3,392	998
Cass	2,481	4,206	2,616	3,384	809

County	2000 Gore (D)	2000 Bush (R)	1996 Clinton (D)	1996 Dole (R)	1996 Perot (RF)
Cedar	4,033	4,031	3,856	2,966	756
Cerro Gordo	12,185	9,397	11,943	7,427	1,689
Cherokee	2,845	3,463	2,853	2,629	834
Chickasaw	3,435	2,936	3,355	2,191	759
Clarke	2,081	1,984	2,053	1,401	440
Clay	3,294	3,992	3,659	3,129	802
Clayton	4,238	4,034	4,284	2,944	912
Clinton	12,276	9,229	11,481	7,624	2,300
Crawford	2,838	3,482	3,140	2,686	847
Dallas	8,561	10,306	8,017	6,647	1,198
Davis	1,691	1,956	1,894	1,445	382
Decatur	1,674	1,903	1,846	1,287	452
Delaware	3,808	4,273	3,704	3,065	679
Des Moines	11,351	7,385	10,761	5,778	1,792
Dickinson	3,660	4,225	3,562	3,129	901
Dubuque	22,341	16,462	20,839	13,391	3,304
Emmet	2,165	2,331	2,270	1,641	470
Fayette	4,640	4,747	4,832	3,848	890
Floyd	3,830	3,191	3,769	2,379	689
Franklin	2,122	2,657	2,232	2,054	417
Fremont	1,459	2,069	1,481	1,576	480
Greene	2,301	2,282	2,519	1,861	396
Grundy	2,139	3,851	2,322	2,928	401
Guthrie	2,493	2,840	2,552	2,034	515
Hamilton	3,407	3,968	3,455	3,109	661
Hancock	2,281	2,988	2,399	2,353	529
Hardin	3,734	4,486	4,053	3,505	713
Harrison	2,551	3,802	2,576	3,070	820
Henry	3,907	4,476	3,798	3,478	914
Howard	2,426	1,922	2,303	1,528	555
Humboldt	1,949	2,846	2,080	2,236	590
Ida	1,411	1,968	1,589	1,684	436
Iowa	3,230	3,894	3,354	3,042	575
Jackson	4,945	3,769	4,609	2,827	936
Jasper	8,699	8,729	8,776	6,414	1,263
Jefferson	2,863	3,248	2,597	2,541	571
Johnson	31,174	17,899	27,888	13,402	2,313
Jones	4,690	4,201	4,668	3,083	765
Keokuk	2,181	2,571	2,545	2,080	432
Kossuth	3,960	4,612	4,031	3,477	932
Lee	9,632	6,339	8,831	4,932	1,734
Linn	48,897	40,417	45,497	30,958	5,607
Louisa	2,294	2,207	2,081	1,565	590
Lucas	1,934	2,262	2,168	1,586	433
Lyon	1,313	3,918	1,489	3,396	422
Madison	3,093	3,662	3,070	2,550	654
Mahaska	3,370	5,971	3,737	4,473	656
Marion	5,741	8,358	5,978	6,100	871
Marshall	8,322	8,785	8,669	7,017	1,455
Mills	2,039	3,684	2,068	2,958	683
Mitchell	2,650	2,388	2,596	1,877	563
Monona	2,086	2,304	1,952	1,674	580
Monroe	1,699	1,858	1,884	1,272	329
Montgomery	1,838	3,417	1,912	2,583	663
Muscatine	8,058	7,483	7,674	5,858	1,705
O'Brien	2,170	4,674	2,236	3,877	578
Osceola	913	2,064	1,010	1,736	274
Page	2,293	4,588	2,220	4,032	753
Palo Alto	2,326	2,341	2,371	1,817	477
Plymouth	3,499	6,189	3,745	5,117	997
Pocahontas	1,736	2,242	1,981	1,707	478
Polk	89,715	79,927	83,877	60,884	9,516
Pottawattamie	14,726	18,783	3,276	15,648	3,534
Poweshiek	4,222	4,396	4,183	3,221	681
Ringgold	1,246	1,369	1,439	967	310
Sac	2,099	2,776	2,170	2,209	579
Scott	35,857	32,801	32,694	26,751	4,991
Shelby	2,179	3,655	2,176	3,056	652
Sioux	2,148	12,241	2,392	10,864	718
Story	17,478	16,228	17,234	12,468	2,091
Tama	4,045	4,034	3,994	2,986	713
Taylor	1,247	1,770	1,458	1,419	379
Union	2,540	3,003	2,787	2,156	660
Van Buren	1,440	2,016	1,536	1,460	347
Wapello	8,355	6,313	8,437	4,828	1,376
Warren	9,521	9,621	9,120	6,905	1,267
Washington	3,932	4,827	3,828	3,600	636
Wayne	1,300	1,666	1,650	1,295	310
Webster	8,479	8,172	8,380	6,275	1,580
Winnebago	2,691	2,662	2,679	2,211	590
Winneshiek	4,339	4,647	4,122	3,532	973
Woodbury	17,691	18,864	17,224	16,368	3,436
Worth	2,208	1,659	2,293	1,284	403
Wright	2,796	3,384	2,912	2,473	536
Totals	**638,517**	**634,373**	**620,258**	**492,644**	**105,159**

Iowa Vote Since 1952

1952, Eisenhower, Rep., 808,906; Stevenson, Dem., 451,513; Hallinan, Prog., 5,085; Hamblen, Proh., 2,882; Hoopes, Soc., 219; Hass, Soc. Labor, 139; scattering, 29.

1956, Eisenhower, Rep., 729,187; Stevenson, Dem., 501,858; Andrews (A.C.P. of Iowa), 3,202; Hoopes, Soc., 192; Hass, Soc. Labor, 125.

1960, Kennedy, Dem., 550,565; Nixon, Rep., 722,381; Hass, Soc. Labor, 230; write-in, 634.

1964, Johnson, Dem., 733,030; Goldwater, Rep., 449,148; Hass, Soc. Labor, 182; DeBerry, Soc. Workers, 159; Munn, Proh., 1,902.

1968, Nixon, Rep., 619,106; Humphrey, Dem., 476,699; Wallace, 3d Party, 66,422; Munn, Proh., 362; Halstead, Soc. Workers, 3,377; Cleaver, Peace and Freedom, 1,332; Blomen, Soc. Labor, 241.

1972, Nixon, Rep., 706,207; McGovern, Dem., 496,206; Schmitz, Amer., 22,056; Jenness, Soc. Workers, 488; Fisher, Soc. Labor, 195; Hall, Com., 272; Green, Universal, 199; scattered, 321.

1976, Carter, Dem., 619,931; Ford, Rep., 632,863; McCarthy, Ind., 20,051; Anderson, Amer., 3,040; MacBride, Libertarian, 1,452.

1980, Reagan, Rep., 676,026; Carter, Dem., 508,672; Anderson, Ind., 115,633; Clark, Libertarian, 13,123; Commoner, Citizens, 2,273; McReynolds, Socialist, 534; Hall, Com., 298; DeBerry, Soc. Workers, 244; Greaves, Amer., 189; Bubar, Statesman, 150; scattering, 519.

1984, Reagan, Rep., 703,088; Mondale, Dem., 605,620; Bergland, Libertarian, 1,844.

1988, Bush, Rep., 545,355; Dukakis, Dem., 670,557; LaRouche, Ind., 3,526; Paul, Lib., 2,494.

1992, Clinton, Dem., 586,353; Bush, Rep., 504,891; Perot, Ind., 253,468; Hagelin, Natural Law, 3,079; Gritz, Populist/America First, 1,177; Marrou, Libertarian, 1,076.

1996, Clinton, Dem., 620,258; Dole, Rep., 492,644; Perot, Ref., 105,159; Nader, Green, 6,550; Hagelin, Natural Law, 3,349; Browne, Libertarian, 2,315; Phillips, Taxpayers, 2,229; Harris, Soc. Workers, 331.

2000, Gore, Dem., 638,517; Bush, Rep., 634,373; Nader, Green, 29,374; Buchanan, Reform, 5,731; Browne, Libertarian, 3,209; Hagelin, Ind., 2,281; Phillips, Constitution, 613; Harris, Soc. Workers, 190; McReynolds, Soc., 107.

Kansas

County	2000 Gore (D)	2000 Bush (R)	1996 Clinton (D)	1996 Dole (R)	1996 Perot (RF)
Allen	2,132	3,379	2,299	2,797	793
Anderson	1,327	1,984	1,367	1,636	449
Atchison	3,171	3,378	2,926	2,828	727
Barber	637	1,755	730	1,696	279
Barton	3,238	7,302	3,121	7,855	1,004
Bourbon	2,211	3,852	2,491	3,318	760
Brown	1,512	2,985	1,529	2,688	497
Butler	6,755	13,377	7,294	13,979	2,274
Chase	391	848	496	778	259
Chautauqua	443	1,347	568	1,142	222
Cherokee	3,783	5,014	3,771	4,138	1,072
Cheyenne	350	1,312	422	1,211	174
Clark	292	926	334	855	109
Clay	951	2,998	963	2,793	389
Cloud	1,314	2,918	1,615	2,743	609
Coffey	1,196	2,700	1,118	2,369	572
Comanche	211	760	298	691	133
Cowley	5,535	8,080	5,588	7,872	1,904
Crawford	7,076	7,160	7,504	6,447	1,785
Decatur	424	1,255	417	1,255	156
Dickinson	2,413	5,243	2,423	5,174	888
Doniphan	1,134	2,350	1,050	1,962	0
Douglas	18,249	17,062	18,116	16,116	2,630
Edwards	447	1,062	539	1,088	180
Elk	402	1,080	488	933	206
Ellis	3,926	6,516	4,142	6,809	894
Ellsworth	825	1,845	899	2,078	245
Finney	2,431	6,442	2,420	6,188	805
Ford	2,566	6,050	2,628	5,681	914
Franklin	3,321	5,925	3,552	5,007	1,184
Geary	2,660	3,977	2,444	3,686	618
Gove	296	1,122	351	1,123	141
Graham	346	1,058	432	1,031	152
Grant	683	2,126	633	1,772	250
Gray	482	1,631	404	1,457	164
Greeley	143	628	161	567	47
Greenwood	1,027	2,392	1,108	1,932	552
Hamilton	264	901	342	811	84
Harper	869	2,076	836	1,941	355
Harvey	4,591	8,271	4,918	8,382	1,023
Haskell	263	1,323	304	1,143	96
Hodgeman	217	835	251	808	99
Jackson	1,990	3,001	1,983	2,682	735

County	2000 Gore (D)	2000 Bush (R)	1996 Clinton (D)	1996 Dole (R)	1996 Perot (RF)
Jefferson	3,000	4,423	2,757	3,781	1,030
Jewell	380	1,400	417	1,374	188
Johnson	79,118	129,965	68,129	110,368	10,425
Kearny	320	1,084	335	1,041	106
Kingman	991	2,672	1,006	2,659	409
Kiowa	294	1,262	331	1,264	170
Labette	3,745	4,475	3,931	4,283	1,091
Lane	252	846	271	865	86
Leavenworth	9,733	12,583	9,098	10,778	2,419
Lincoln	469	1,295	528	1,372	212
Linn	1,587	2,513	1,590	2,077	535
Logan	231	1,088	296	1,155	112
Lyon	5,190	6,652	4,884	6,612	1,584
McPherson	3,272	8,501	3,536	8,142	1,115
Marion	1,475	4,156	1,673	4,173	492
Marshall	1,831	3,066	1,932	2,811	713
Meade	400	1,604	426	1,443	173
Miami	4,554	6,611	4,237	5,256	1,339
Mitchell	751	2,350	833	2,435	246
Montgomery	4,770	8,496	5,269	7,428	1,528
Morris	882	1,599	965	1,553	451
Morton	321	1,203	376	1,073	124
Nemaha	1,494	3,578	1,648	3,014	676
Neosho	2,588	4,014	2,527	3,409	907
Ness	383	1,420	428	1,336	186
Norton	598	1,744	640	1,814	265
Osage	2,530	3,770	2,502	3,487	1,101
Osborne	484	1,432	608	1,582	191
Ottawa	631	1,977	752	1,846	261
Pawnee	968	1,850	932	1,927	275
Phillips	611	2,057	758	2,005	242
Pottawatomie	2,037	4,985	1,997	4,504	1,035
Pratt	1,314	2,885	1,367	2,591	408
Rawlins	306	1,349	335	1,393	146
Reno	9,025	15,179	9,108	14,275	2,661
Republic	604	2,239	688	2,283	268
Rice	1,422	2,903	1,434	2,842	482
Riley	6,188	10,672	6,746	11,113	1,478
Rooks	597	2,016	650	1,864	251
Rush	505	1,235	547	1,239	185
Russell	886	2,434	705	3,347	164
Saline	7,487	12,412	7,728	12,475	2,192
Scott	418	1,811	458	1,750	160
Sedgwick	62,561	93,724	59,643	93,397	11,875
Seward	1,126	3,869	1,309	3,812	396
Shawnee	34,818	35,894	32,803	34,845	7,304
Sheridan	281	1,132	264	1,053	95
Sherman	681	1,894	736	2,110	220
Smith	534	1,534	638	1,628	213
Stafford	567	1,546	651	1,604	276
Stanton	215	785	189	628	60
Stevens	345	1,714	405	1,548	213
Sumner	3,549	6,176	3,638	5,952	1,260
Thomas	807	2,822	866	2,725	295
Trego	516	1,220	548	1,205	209
Wabaunsee	1,025	2,182	966	1,884	479
Wallace	103	737	160	738	65
Washington	687	2,446	804	2,397	326
Wichita	207	859	239	796	80
Wilson	1,186	2,748	1,297	2,458	562
Woodson	521	974	598	953	269
Wyandotte	32,411	14,024	31,252	14,011	3,931
Totals	**399,276**	**622,332**	**387,659**	**583,245**	**92,639**

Kansas Vote Since 1952

1952, Eisenhower, Rep., 616,302; Stevenson, Dem., 273,296; Hamblen, Proh., 6,038; Hoopes, Soc., 530.

1956, Eisenhower, Rep., 566,878; Stevenson, Dem., 296,317; Holtwick, Proh., 3,048.

1960, Kennedy, Dem., 363,213; Nixon, Rep., 561,474; Decker, Proh., 4,138.

1964, Johnson, Dem., 464,028; Goldwater, Rep., 386,579; Munn, Proh., 5,393; Hass, Soc. Labor, 1,901.

1968, Nixon, Rep., 478,674; Humphrey, Dem., 302,996; Wallace, 3d Party, 88,921; Munn, Proh., 2,192.

1972, Nixon, Rep., 619,812; McGovern, Dem., 270,287; Schmitz, Conservative, 21,808; Munn, Proh., 4,188.

1976, Carter, Dem., 430,421; Ford, Rep., 502,752; McCarthy, Ind., 13,185; Anderson, Amer., 4,724; MacBride, Libertarian, 3,242; Maddox, Conservative, 2,118; Bubar, Proh., 1,403.

1980, Reagan, Rep., 566,812; Carter, Dem., 326,150; Anderson, Ind., 68,231; Clark, Libertarian, 14,470; Shelton, Amer., 1,555; Hall, Com., 967; Bubar, Statesman, 821; Rarick, Conservative, 789.

1984, Reagan, Rep., 674,646; Mondale, Dem., 332,471; Bergland, Libertarian, 3,585.

1988, Bush, Rep., 554,049; Dukakis, Dem., 422,636; Paul, Ind., 12,553; Fulani, Ind., 3,806.

1992, Clinton, Dem., 390,434; Bush, Rep., 449,951; Perot, Ind., 312,358; Marrou, Libertarian, 4,314.

1996, Dole, Rep., 583,245; Clinton, Dem., 387,659; Perot, Ref., 92,639; Browne, Libertarian, 4,557; Phillips, Ind., 3,519; Hagelin, Ind., 1,655.

2000, Bush, Rep., 622,332; Gore, Dem., 399,276; Nader, Ind., 36,086; Buchanan, Reform, 7,370; Browne, Libertarian, 4,525; Hagelin, Ind., 1,373; Phillips, Constitution, 1,254.

Kentucky

County	2000 Gore (D)	2000 Bush (R)	1996 Clinton (D)	1996 Dole (R)	1996 Perot (RF)
Adair	1,779	5,460	1,821	3,876	790
Allen	1,950	4,415	1,781	3,032	393
Anderson	2,902	4,909	2,898	2,972	751
Ballard	1,880	1,824	2,255	1,064	411
Barren	4,930	8,741	5,044	5,700	1,065
Bath	2,087	2,303	1,886	1,229	428
Bell	4,787	5,585	5,058	3,917	940
Boone	9,248	22,016	8,379	15,085	1,900
Bourbon	3,048	3,881	3,030	2,592	603
Boyd	9,541	9,247	9,668	7,054	2,070
Boyle	3,963	6,126	3,877	4,157	709
Bracken	888	2,065	1,055	1,371	271
Breathitt	2,902	2,084	3,106	1,058	397
Breckinridge	2,595	4,763	2,956	3,151	670
Bullitt	8,195	14,054	7,651	8,697	1,973
Butler	1,299	3,654	1,260	2,531	348
Caldwell	2,223	3,161	2,434	2,067	637
Calloway	5,635	7,705	5,281	4,989	1,223
Campbell	12,040	20,789	11,957	16,640	2,312
Carlisle	1,149	1,405	1,355	816	245
Carroll	1,601	1,818	1,689	1,170	351
Carter	4,182	4,617	3,728	3,240	781
Casey	1,122	4,284	1,106	3,187	525
Christian	6,778	10,787	6,843	8,285	1,064
Clark	4,918	7,297	4,987	4,739	1,095
Clay	1,723	4,926	2,135	3,716	478
Clinton	1,032	3,224	1,072	2,521	350
Crittenden	1,610	2,469	1,480	1,509	400
Cumberland	736	2,220	753	1,654	227
Daviess	14,126	21,361	15,366	15,844	3,344
Edmonson	1,710	3,250	1,595	2,619	298
Elliott	1,525	827	1,298	421	284
Estill	1,591	3,033	1,724	2,220	479
Fayette	47,277	54,495	43,632	42,930	5,345
Fleming	1,813	3,282	1,913	2,313	522
Floyd	10,088	5,068	9,655	3,139	1,518
Franklin	10,853	10,209	11,251	7,132	1,873
Fulton	1,452	1,293	1,614	863	223
Gallatin	1,049	1,345	1,189	838	299
Garrard	1,713	4,043	1,486	2,540	337
Grant	2,568	4,405	2,541	2,697	661
Graves	6,097	7,849	6,991	5,130	1,596
Grayson	2,604	5,843	2,716	4,249	677
Green	1,085	3,615	1,285	2,763	475
Greenup	7,164	7,233	6,883	5,370	1,627
Hancock	1,508	2,032	1,547	1,356	418
Hardin	11,095	18,964	11,031	12,642	2,815
Harlan	5,365	4,980	5,874	3,337	884
Harrison	2,658	3,793	2,934	2,433	801
Hart	2,201	3,725	2,527	2,701	501
Henderson	8,054	7,698	8,051	5,092	1,556
Henry	2,117	3,244	2,324	2,110	564
Hickman	940	1,151	1,220	695	247
Hopkins	6,734	9,490	7,239	6,363	1,512
Jackson	701	4,079	960	3,045	299
Jefferson	149,901	145,052	144,207	114,860	19,413
Jessamine	4,633	10,074	4,428	6,686	1,040
Johnson	3,276	4,811	3,348	3,262	1,010
Kenton	19,100	35,363	19,407	28,579	3,680
Knott	4,349	2,029	4,842	1,201	517
Knox	3,690	6,058	3,736	4,502	811
Larue	1,727	3,384	2,040	2,140	469
Laurel	4,856	13,029	4,306	9,454	1,211
Lawrence	2,258	2,969	2,195	1,812	481
Lee	836	1,893	1,023	1,302	181
Leslie	1,210	3,159	1,466	2,296	304
Letcher	4,698	4,092	4,160	2,222	782
Lewis	1,293	3,217	1,415	2,365	561
Lincoln	2,678	4,795	2,550	3,006	526
Livingston	2,022	2,118	2,228	1,258	449
Logan	3,885	5,344	4,181	3,888	704
Lyon	1,680	1,688	1,641	999	284
McCracken	11,412	14,745	12,670	10,221	2,268
McCreary	1,418	3,321	1,710	2,527	488
McLean	1,747	2,219	1,834	1,368	385
Madison	9,309	13,682	8,142	9,212	1,613
Magoffin	2,603	2,785	2,249	1,434	337

County	2000 Gore (D)	2000 Bush (R)	1996 Clinton (D)	1996 Dole (R)	1996 Perot (RF)
Marion	2,778	3,259	2,922	2,013	757
Marshall	6,203	7,294	6,054	4,579	1,391
Martin	1,714	2,667	1,807	1,612	401
Mason	2,178	3,572	2,444	2,588	484
Meade	3,596	5,319	3,653	2,855	912
Menifee	1,038	1,170	979	608	179
Mercer	3,092	5,362	3,179	3,264	738
Metcalfe	1,318	2,476	1,349	1,651	355
Monroe	1,158	4,377	1,114	3,300	415
Montgomery	3,833	4,534	3,372	2,681	705
Morgan	1,875	2,295	1,843	1,439	380
Muhlenberg	6,295	5,518	6,564	3,569	1,218
Nelson	5,481	7,714	5,392	4,645	1,067
Nicholas	994	1,613	1,092	950	265
Ohio	3,303	5,413	3,487	3,475	1,076
Oldham	6,236	13,580	6,202	10,477	1,521
Owen	1,394	2,582	1,603	1,709	454
Owsley	339	1,466	647	920	153
Pendleton	1,670	3,044	1,926	2,177	462
Perry	5,514	5,300	6,015	3,382	894
Pike	13,611	11,005	14,126	7,160	2,148
Powell	2,008	2,258	2,156	1,526	523
Pulaski	5,415	15,845	5,340	11,945	1,420
Robertson	341	630	360	368	117
Rockcastle	1,174	3,992	1,160	3,106	338
Rowan	3,505	3,546	3,215	2,309	724
Russell	1,710	5,268	1,582	4,017	837
Scott	5,472	7,952	4,258	4,349	977
Shelby	4,435	8,068	4,629	5,307	780
Simpson	2,583	3,169	2,749	2,186	401
Spencer	1,554	3,150	1,404	1,614	341
Taylor	2,790	6,151	2,897	4,573	829
Todd	1,496	2,646	1,744	1,912	424
Trigg	2,110	3,130	2,087	1,975	394
Trimble	1,181	1,837	1,245	999	308
Union	2,547	2,749	2,913	1,554	598
Warren	12,180	20,235	11,642	15,784	1,835
Washington	1,458	3,044	1,639	2,116	383
Wayne	2,312	4,069	2,422	3,122	481
Webster	2,388	2,599	2,852	1,568	660
Whitley	4,101	7,502	4,174	5,402	1,027
Wolfe	1,136	1,267	1,297	772	202
Woodford	3,995	5,890	3,910	4,270	746
Totals	**638,923**	**872,520**	**636,614**	**623,283**	**120,396**

Kentucky Vote Since 1952

1952, Eisenhower, Rep., 495,029; Stevenson, Dem., 495,729; Hamblen, Proh., 1,161; Hass, Soc. Labor, 893; Hallinan, Proh., 336.

1956, Eisenhower, Rep., 572,192; Stevenson, Dem., 476,453; Byrd, States' Rights, 2,657; Holtwick, Proh., 2,145; Hass, Soc. Labor, 358.

1960, Kennedy, Dem., 521,855; Nixon, Rep., 602,607.

1964, Johnson, Dem., 669,659; Goldwater, Rep., 372,977; Kasper, Natl. States Rights, 3,469.

1968, Nixon, Rep., 462,411; Humphrey, Dem., 397,547; Wallace, 3d Party, 193,098; Halstead, Soc. Workers, 2,843.

1972, Nixon, Rep., 676,446; McGovern, Dem., 371,159; Schmitz, Amer., 17,627; Spock, Peoples, 1,118; Jenness, Soc. Workers, 685; Hall, Com., 464.

1976, Carter, Dem., 615,717; Ford, Rep., 531,852; Anderson, Amer., 8,308; McCarthy, Ind., 6,837; Maddox, Amer. Ind., 2,328; MacBride, Libertarian, 814.

1980, Reagan, Rep., 635,274; Carter, Dem., 616,417; Anderson, Ind., 31,127; Clark, Libertarian, 5,531; McCormack, Respect For Life, 4,233; Commoner, Citizens, 1,304; Pulley, Socialist, 393; Hall, Com., 348.

1984, Reagan, Rep., 815,345; Mondale, Dem., 536,756.

1988, Bush, Rep., 734,281; Dukakis, Dem., 580,368; Duke, Pop., 4,494; Paul, Lib., 2,118.

1992, Clinton, Dem., 665,104; Bush, Rep., 617,178; Perot, Ind., 203,944; Marrou, Libertarian, 4,513.

1996, Clinton, Dem., 636,614; Dole, Rep., 623,283; Perot, Ref., 120,396; Browne, Libertarian, 4,009; Phillips, Taxpayers, 2,204; Hagelin, Natural Law, 1,493.

2000, Bush, Rep., 872,520; Gore, Dem., 638,923; Nader, Green, 23,118; Buchanan, Reform, 4,152; Browne, Libertarian, 2,885; Hagelin, Natural Law, 1,513; Phillips, Constitution, 915.

Louisiana

Parish	2000 Gore (D)	2000 Bush (R)	1996 Clinton (D)	1996 Dole (R)	1996 Perot (RF)
Acadia	8,892	13,814	12,300	9,246	2,234
Allen	3,914	4,035	4,930	2,589	1,187
Ascension	13,385	16,818	15,263	10,885	3,027
Assumption	5,222	4,388	6,416	2,698	904
Avoyelles	6,701	7,329	9,689	4,433	1,937
Beauregard	3,958	7,862	4,925	5,526	1,834
Bienville	3,413	3,269	4,335	2,402	457
Bossier	11,933	23,224	15,504	16,852	2,660
Caddo	47,530	46,807	55,543	38,445	4,821
Calcasieu	33,919	38,086	38,238	26,494	8,281
Caldwell	1,359	2,817	2,117	1,842	514
Cameron	1,435	2,593	2,103	1,365	594
Catahoula	1,718	2,912	2,692	1,770	615
Claiborne	2,721	3,384	3,609	2,500	530
Concordia	3,569	4,627	4,565	3,134	855
DeSoto	5,036	5,260	6,221	3,526	646
E. Baton Rouge	76,516	89,128	83,493	77,811	7,990
East Carroll	1,876	1,280	2,149	1,008	186
East Feliciana	3,870	4,051	4,714	2,949	660
Evangeline	5,763	7,290	7,847	5,278	1,447
Franklin	2,792	5,363	4,076	3,961	814
Grant	2,099	4,784	2,980	3,117	1,055
Iberia	11,762	17,236	15,087	12,014	2,448
Iberville	8,355	5,573	9,553	4,031	1,076
Jackson	2,582	4,347	3,368	3,030	571
Jefferson	70,411	105,003	80,407	92,820	9,667
Jefferson Davis	5,162	6,945	6,897	4,311	1,543
Lafayette	27,190	48,491	32,504	36,419	4,631
Lafourche	14,627	18,575	18,810	12,105	2,984
LaSalle	1,397	4,564	2,543	2,925	947
Lincoln	6,851	9,246	7,903	6,973	761
Livingston	11,008	24,889	13,276	16,159	4,150
Madison	2,489	2,127	3,085	1,591	315
Morehouse	5,289	6,641	6,160	5,193	963
Natchitoches	6,924	7,332	8,296	5,471	1,053
Orleans	137,630	39,404	144,720	39,576	3,805
Ouachita	21,457	35,107	24,525	28,559	3,586
Plaquemines	4,425	6,302	5,348	4,493	856
Pointe Coupee	5,813	4,710	6,835	3,545	845
Rapides	18,898	28,831	23,004	21,548	4,670
Red River	2,177	2,200	2,641	1,344	268
Richland	3,282	4,895	4,143	3,765	645
Sabine	2,846	5,754	4,263	3,543	1,043
St. Bernard	11,682	16,255	14,312	13,549	2,664
St. Charles	8,918	11,981	10,612	9,316	1,307
St. Helena	3,059	1,965	3,692	1,455	417
St. James	6,523	3,813	7,247	2,832	608
St. John the Baptist	9,745	7,423	9,937	6,025	966
St. Landry	18,067	15,449	20,636	12,273	2,311
St. Martin	9,853	9,961	12,492	6,296	1,607
St. Mary	9,851	11,325	12,402	8,018	1,850
St. Tammany	22,722	59,193	24,281	44,761	4,741
Tangipahoa	15,843	20,421	18,617	15,517	3,144
Tensas	1,580	1,330	1,882	1,000	176
Terrebonne	14,414	21,314	18,550	13,944	3,359
Union	3,205	5,772	4,260	4,418	696
Vermilion	8,704	12,495	12,609	7,653	1,954
Vernon	4,655	8,794	6,195	5,449	2,068
Washington	7,399	8,983	9,603	6,642	1,643
Webster	7,197	9,420	9,688	6,153	1,324
W. Baton Rouge	5,058	4,924	5,697	3,254	799
West Carroll	1,319	3,220	1,853	2,366	461
W. Feliciana	2,187	2,512	2,416	1,616	388
Winn	2,167	4,028	3,779	2,803	735
Totals	**792,344**	**927,871**	**927,837**	**712,586**	**123,293**

Louisiana Vote Since 1952

1952, Eisenhower, Rep., 306,925; Stevenson, Dem., 345,027.

1956, Eisenhower, Rep., 329,047; Stevenson, Dem., 243,977; Andrews, States' Rights, 44,520.

1960, Kennedy, Dem., 407,339; Nixon, Rep., 230,890; States' Rights (unpledged), 169,572.

1964, Johnson, Dem., 387,068; Goldwater, Rep., 509,225.

1968, Nixon, Rep., 257,535; Humphrey, Dem., 309,615; Wallace, 3d Party, 530,300.

1972, Nixon, Rep., 686,852; McGovern, Dem., 298,142; Schmitz, Amer., 52,099; Jenness, Soc. Workers, 14,398.

1976, Carter, Dem., 661,365; Ford, Rep., 587,446; Maddox, Amer., 10,058; Hall, Com., 7,417; McCarthy, Ind., 6,588; MacBride, Libertarian, 3,325.

1980, Reagan, Rep., 792,853; Carter, Dem., 708,453; Anderson, Ind., 26,345; Rarick, Amer. Ind., 10,333; Clark, Libertarian, 8,240; Commoner, Citizens, 1,584; DeBerry, Soc. Work., 783.

1984, Reagan, Rep., 1,037,299; Mondale, Dem., 651,586; Bergland, Libertarian, 1,876.

1988, Bush, Rep., 883,702; Dukakis, Dem., 717,460; Duke, Pop., 18,612; Paul, Lib., 4,115.

1992, Clinton, Dem., 815,971; Bush, Rep., 733,386; Perot, Ind., 211,478; Gritz, Populist/America First, 18,545; Marrou, Libertarian, 3,155; Daniels, Ind., 1,663; Phillips, U.S. Taxpayers, 1,552; Fulani, New Alliance, 1,434; LaRouche, Ind., 1,136.

1996, Clinton, Dem., 927,837; Dole, Rep., 712,586; Perot, Ref., 123,293; Browne, Libertarian, 7,499; Nader, Liberty, Ecology, Community, 4,719; Phillips, Taxpayers, 3,366; Hagelin, Natural Law, 2,981; Moorehead, Workers World, 1,678.

2000, Bush, Rep., 927,871; Gore, Dem., 792,344; Nader, Green, 20,473; Buchanan, Reform, 14,356; Phillips, Constitution, 5,483; Browne, Libertarian, 2,951; Harris, Soc. Workers, 1,103; Hagelin, Natural Law, 1,075.

Maine

	2000		1996		
City	Gore (D)	Bush (R)	Clinton (D)	Dole (R)	Perot (RF)
Auburn	6,014	4,568	5,750	3,060	1,484
Augusta	5,116	3,344	5,307	2,353	1,100
Bangor	7,311	6,131	7,609	4,476	1,399
Biddeford	5,383	3,126	5,653	1,768	1,019
Brunswick	5,547	3,767	5,258	2,850	841
Gorham	3,394	3,353	2,990	2,269	710
Lewiston	9,663	5,255	10,275	3,182	2,113
Orono	2,701	1,506	2,748	1,106	369
Portland	20,506	8,838	19,755	7,178	2,255
Presque Isle	2,004	2,231	2,015	1,491	594
Saco	4,783	3,402	4,506	2,140	834
Sanford	4,653	3,871	4,368	2,239	1,524
Scarborough	4,278	4,964	3,906	3,214	805
S. Portland	7,267	4,390	6,777	3,241	906
Waterville	4,279	2,115	4,219	1,478	750
Westbrook	4,316	3,258	4,373	2,186	864
Windham	3,550	3,754	3,251	2,396	898
York	3,708	3,462	2,970	2,525	649
Other	215,478	215,281	211,058	137,226	66,856
Totals	**319,951**	**286,616**	**312,788**	**186,378**	**85,970**

Maine Vote Since 1952

1952, Eisenhower, Rep., 232,353; Stevenson, Dem., 118,806; Hallinan, Prog., 332; Hass, Soc. Labor, 156; Hoopes, Soc., 138; scattered, 1.

1956, Eisenhower, Rep., 249,238; Stevenson, Dem., 102,468.

1960, Kennedy, Dem., 181,159; Nixon, Rep., 240,608.

1964, Johnson, Dem., 262,264; Goldwater, Rep., 118,701.

1968, Nixon, Rep., 169,254; Humphrey, Dem., 217,312; Wallace, 3d Party, 6,370.

1972, Nixon, Rep., 256,458; McGovern, Dem., 160,584; scattered, 229.

1976, Carter, Dem., 232,279; Ford, Rep., 236,320; McCarthy, Ind., 10,874; Bubar, Proh., 3,495.

1980, Reagan, Rep., 238,522; Carter, Dem., 220,974; Anderson, Ind., 53,327; Clark, Libertarian, 5,119; Commoner, Citizens, 4,394; Hall, Com., 591; write-ins, 84.

1984, Reagan, Rep., 336,500; Mondale, Dem., 214,515.

1988, Bush, Rep., 307,131; Dukakis, Dem., 243,569; Paul, Lib., 2,700; Fulani, New Alliance, 1,405.

1992, Clinton, Dem., 263,420; Perot, Ind., 206,820; Bush, Rep., 206,504; Marrou, Libertarian, 1,681.

1996, Clinton, Dem., 312,788; Dole, Rep., 186,378; Perot, Ref., 85,970; Nader, Green, 15,279; Browne, Libertarian, 2,996; Phillips, Taxpayers, 1,517; Hagelin, Natural Law, 825.

2000, Gore, Dem., 319,951; Bush, Rep., 286,616; Nader, Green, 37,127; Buchanan, Reform, 4,443; Browne, Libertarian, 3,074; Phillips, Constitution, 579.

Maryland

	2000		1996		
County	Gore (D)	Bush (R)	Clinton (D)	Dole (R)	Perot (RF)
Allegany	10,894	14,656	11,025	12,136	2,652
Anne Arundel	89,624	104,209	72,147	83,574	14,287
Baltimore	160,635	133,033	132,599	114,449	20,393
Calvert	12,986	16,004	10,008	11,509	1,932
Caroline	3,396	5,300	3,251	3,874	947
Carroll	20,146	41,742	17,122	30,316	4,873
Cecil	12,327	15,494	10,144	10,885	3,124
Charles	21,873	21,768	15,890	17,432	2,333
Dorchester	5,232	5,847	4,613	4,337	1,008
Frederick	30,725	45,350	25,081	34,494	4,989
Garrett	2,872	7,514	3,121	5,400	1,200
Harford	35,665	52,862	29,779	39,686	7,939
Howard	58,556	49,809	47,569	40,849	6,011
Kent	3,627	4,155	3,207	3,055	676
Montgomery	232,453	124,580	198,807	117,730	14,450
Prince George's	214,345	50,017	176,612	52,697	9,153
Queen Anne's	6,257	9,970	5,054	7,147	1,312
St. Mary's	11,912	16,856	9,988	11,835	1,827
Somerset	3,785	3,609	3,557	2,919	613
Talbot	5,854	8,874	4,821	6,997	914
Washington	18,221	27,948	16,481	21,434	3,934
Wicomico	14,469	16,338	12,303	12,687	2,160
Worcester	9,389	10,742	7,587	7,621	1,612
City					
Baltimore	158,765	27,150	145,441	28,467	7,473
Totals	**1,144,008**	**813,827**	**966,207**	**681,530**	**115,812**

Maryland Vote Since 1952

1952, Eisenhower, Rep., 499,424; Stevenson, Dem., 395,337; Hallinan, Prog., 7,313.

1956, Eisenhower, Rep., 559,738; Stevenson, Dem., 372,613.

1960, Kennedy, Dem., 565,800; Nixon, Rep., 489,538.

1964, Johnson, Dem., 730,912; Goldwater, Rep., 385,495; write-in, 50.

1968, Nixon, Rep., 517,995; Humphrey, Dem., 538,310; Wallace, 3d Party, 178,734.

1972, Nixon, Rep., 829,305; McGovern, Dem., 505,781; Schmitz, Amer., 18,726.

1976, Carter, Dem., 759,612; Ford, Rep., 672,661.

1980, Reagan, Rep., 680,606; Carter, Dem., 726,161; Anderson, Ind., 119,537; Clark, Libertarian, 14,192.

1984, Reagan, Rep., 879,918; Mondale, Dem., 787,935; Bergland, Libertarian, 5,721.

1988, Bush, Rep., 876,167; Dukakis, Dem., 826,304; Paul, Lib., 6,748; Fulani, New Alliance, 5,115.

1992, Clinton, Dem., 988,571; Bush, Rep., 707,094; Perot, Ind., 281,414; Marrou, Libertarian, 4,715; Fulani, New Alliance, 2,786.

1996, Clinton, Dem., 966,207; Dole, Rep., 681,530; Perot, Ref., 115,812; Browne, Libertarian, 8,765; Phillips, Taxpayers, 3,402; Hagelin, Natural Law, 2,517.

2000, Gore, Dem., 1,144,008; Bush, Rep., 813,827; Nader, Green, 53,768; Browne, Libertarian, 5,310; Buchanan, Reform., 4,248; Phillips, Constitution, 918.

Massachusetts

	2000		1996		
City	Gore (D)	Bush (R)	Clinton (D)	Dole (R)	Perot (RF)
Boston	132,393	36,389	125,529	33,366	8,428
Brockton	18,563	8,288	16,361	6,972	2,738
Brookline	19,384	4,350	18,812	4,579	799
Cambridge	28,846	5,166	29,913	4,976	1,415
Chicopee	13,236	6,512	14,203	5,188	2,495
Fall River	22,051	5,621	22,796	4,287	2,612
Framingham	17,308	7,347	16,836	6,669	1,700
Lawrence	10,048	3,700	8,615	2,804	1,096
Lowell	17,554	7,790	16,912	5,896	2,911
Lynn	18,836	6,776	18,370	5,634	2,726
Medford	16,776	6,353	16,639	5,844	1,741
New Bedford	23,880	5,473	23,620	4,151	2,547
Newton	29,918	8,132	30,005	8,499	1,674
Quincy	23,117	11,282	23,182	9,824	3,066
Somerville	19,984	4,468	20,206	3,983	1,455
Springfield	29,728	10,288	31,266	9,110	3,407
Waltham	13,736	6,700	13,607	5,830	1,663
Weymouth	15,570	8,884	13,536	6,904	2,181
Worcester	35,231	14,402	35,607	12,879	3,925
Other	1,110,328	710,581	1,075,494	570,663	178,627
Totals	**1,616,487**	**878,502**	**1,571,509**	**718,058**	**227,206**

Massachusetts Vote Since 1952

1952, Eisenhower, Rep., 1,292,325; Stevenson, Dem., 1,083,525; Hallinan, Prog., 4,636; Hass, Soc. Labor, 1,957; Hamblen, Proh., 886; scattered, 69; blanks, 41,150.

1956, Eisenhower, Rep., 1,393,197; Stevenson, Dem., 948,190; Hass, Soc. Labor, 5,573; Holtwick, Proh., 1,205; others, 341.

1960, Kennedy, Dem., 1,487,174; Nixon, Rep., 976,750; Hass, Soc. Labor, 3,892; Decker, Proh., 1,633; others, 31; blank and void, 26,024.

1964, Johnson, Dem., 1,786,422; Goldwater, Rep., 549,727; Hass, Soc. Labor, 4,755; Munn, Proh., 3,735; scattered, 159; blank, 48,104.

1968, Nixon, Rep., 766,844; Humphrey, Dem., 1,469,218; Wallace, 3d Party, 87,088; Blomen, Soc. Labor, 6,180; Munn, Proh., 2,369; scattered, 53; blanks, 25,394.

1972, Nixon, Rep., 1,112,078; McGovern, Dem., 1,332,540; Jenness, Soc. Workers, 10,600; Fisher, Soc. Labor, 129; Schmitz, Amer., 2,877; Spock, Peoples, 101; Hall, Com., 46; Hospers, Libertarian, 43; scattered, 342.

1976, Carter, Dem., 1,429,475; Ford, Rep., 1,030,276; McCarthy, Ind., 65,637; Camejo, Soc. Workers, 8,138; Anderson, Amer., 7,555; La Rouche, U.S. Labor, 4,922; MacBride, Libertarian, 135.

1980, Reagan, Rep., 1,057,631; Carter, Dem., 1,053,802; Anderson, Ind., 382,539; Clark, Libertarian, 22,038; DeBerry, Soc. Workers, 3,735; Commoner, Citizens, 2,056; McReynolds, Soc., 62; Bubar, Statesman, 34; Griswold, Workers World, 19; scattered, 2,382.

1984, Reagan, Rep., 1,310,936; Mondale, Dem., 1,239,606.

1988, Bush, Rep., 1,194,635; Dukakis, Dem., 1,401,415; Paul, Lib., 24,251; Fulani, New Alliance, 9,561.

1992, Clinton, Dem., 1,318,639; Bush, Rep., 805,039; Perot, Ind., 630,731; Marrou, Libertarian, 9,021; Fulani, New Alliance, 3,172; Phillips, U.S. Taxpayers, 2,218; Hagelin, Natural Law, 1,812; LaRouche, Ind., 1,027.

1996, Clinton, Dem., 1,571,509; Dole, Rep., 718,058; Perot, Ref., 227,206; Browne, Libertarian, 20,424; Hagelin, Natural Law, 5,183; Moorehead, Workers World, 3,276.

2000, Gore, Dem., 1,616,487; Bush, Rep., 878,502; Nader, Green, 173,564; Browne, Libertarian, 16,366; Buchanan, Reform, 11,149; Hagelin, Natural Law, 2,884.

Michigan

	2000		1996		
County	Gore (D)	Bush (R)	Clinton (D)	Dole (R)	Perot (RF)
Alcona	2,696	3,152	2,619	2,227	669
Alger	2,071	2,142	2,229	1,429	537
Allegan	15,495	28,197	14,361	20,859	3,269
Alpena	7,053	6,769	7,114	4,525	1,730
Antrim	4,329	6,780	4,226	4,630	1,129
Arenac	3,685	3,421	3,472	2,247	844
Baraga	1,400	1,836	1,601	1,209	460
Barry	9,769	15,716	9,467	11,139	2,282
Bay	28,251	22,150	27,835	16,038	5,410
Benzie	3,546	4,172	3,081	2,856	763
Berrien	28,152	35,689	24,614	28,254	5,958
Branch	6,691	8,743	6,567	6,321	1,779
Calhoun	27,312	26,291	26,287	20,953	4,765
Cass	8,808	10,545	8,207	7,373	2,241
Charlevoix	4,958	7,018	4,689	4,864	1,303
Cheboygan	5,484	6,815	5,018	4,244	1,462
Chippewa	6,370	7,526	6,532	5,137	1,453
Clare	6,287	5,937	6,311	3,742	1,531
Clinton	13,394	18,054	11,945	13,694	2,698
Crawford	2,790	3,345	2,666	2,157	840
Delta	7,970	8,871	8,561	5,925	1,543
Dickinson	5,533	6,932	5,614	4,408	1,478
Eaton	23,211	24,803	19,781	20,092	4,378
Emmet	5,451	8,602	4,892	6,002	1,512
Genesee	119,833	66,641	106,065	49,332	17,671
Gladwin	5,573	5,743	5,494	3,670	1,466
Gogebic	4,066	3,929	4,436	2,769	917
Grand Traverse	14,371	22,358	12,987	16,355	3,527
Gratiot	6,538	8,312	6,793	6,214	1,762
Hillsdale	6,495	10,483	5,955	7,947	2,262
Houghton	5,688	7,895	5,957	5,941	1,584
Huron	6,899	8,911	6,827	6,126	1,811
Ingham	69,231	47,314	63,584	43,096	8,640
Ionia	9,481	13,915	9,261	9,574	2,354
Iosco	6,505	6,345	6,240	4,410	1,710
Iron	3,014	2,967	3,232	2,014	755
Isabella	10,228	10,053	9,635	7,460	2,069
Jackson	28,160	32,066	24,633	24,987	5,968
Kalamazoo	48,807	48,254	45,644	40,703	5,867
Kalkaska	2,774	3,842	2,666	2,455	922
Kent	95,442	148,602	85,912	121,335	14,120
Keweenaw	540	740	572	491	169
Lake	2,584	1,961	2,606	1,213	552
Lapeer	15,749	20,351	14,308	13,369	4,793
Leelanau	4,635	6,840	4,019	5,155	924
Lenawee	18,365	20,681	16,924	14,168	4,167
Livingston	28,780	44,637	22,517	30,598	6,337
Luce	956	1,480	1,107	964	366
Mackinac	2,533	3,272	2,700	2,281	742
Macomb	172,625	164,265	151,430	120,616	29,859
Manistee	5,639	5,401	5,383	3,807	1,230
Marquette	15,503	12,577	15,168	8,805	2,492
Mason	5,579	7,066	5,597	5,066	1,525
Mecosta	6,300	8,072	6,370	5,289	1,373
Menominee	4,597	5,529	4,880	4,038	1,205
Midland	15,959	21,887	15,177	16,547	3,964
Missaukee	2,062	4,274	2,256	3,012	719
Monroe	31,555	28,940	26,072	19,678	6,315
Montcalm	9,627	12,696	10,053	8,679	2,530
Montmorency	2,139	2,750	2,120	1,760	682
Muskegon	37,865	30,028	35,328	21,873	5,794
Newaygo	7,677	11,399	7,614	7,868	2,047
Oakland	281,201	274,319	241,884	219,855	36,709
Oceana	4,597	5,913	4,419	3,947	1,286
Ogemaw	4,896	4,706	4,725	2,904	1,369
Ontonagon	1,514	2,472	2,080	1,523	604
Osceola	4,006	5,680	4,085	3,855	1,068
Oscoda	1,677	2,207	1,652	1,545	503
Otsego	4,034	6,108	3,351	3,638	1,280
Ottawa	29,600	78,703	27,024	61,436	6,275
Presque Isle	3,242	3,660	3,449	2,463	932
Roscommon	6,433	6,190	6,092	4,135	1,539
Saginaw	50,825	41,152	47,579	31,577	8,081
St. Clair	33,002	33,571	28,881	22,495	8,134
St. Joseph	8,574	12,906	8,529	9,764	2,319
Sanilac	7,153	10,966	7,092	7,821	2,265
Schoolcraft	2,036	2,088	2,187	1,200	460
Shiawassee	15,520	15,816	14,662	11,714	3,703
Tuscola	10,845	13,213	10,314	9,154	3,013
Van Buren	13,796	14,792	13,355	11,347	2,946
Washtenaw	86,647	52,459	73,106	40,097	8,020
Wayne	530,414	223,021	504,466	175,886	43,554
Wexford	5,326	7,215	5,510	4,866	1,386
Totals	**2,170,418**	**1,953,139**	**1,989,653**	**1,481,212**	**336,670**

Michigan Vote Since 1952

1952, Eisenhower, Rep., 1,551,529; Stevenson, Dem., 1,230,657; Hamblen, Proh., 10,331; Hallinan, Prog., 3,922; Hass, Soc. Labor, 1,495; Dobbs, Soc. Workers, 655; scattered, 3.

1956, Eisenhower, Rep., 1,713,647; Stevenson, Dem., 1,359,898; Holtwick, Proh., 6,923.

1960, Kennedy, Dem., 1,687,269; Nixon, Rep., 1,620,428; Dobbs, Soc. Workers, 4,347; Decker, Proh., 2,029; Daly, Tax Cut, 1,767; Hass, Soc. Labor, 1,718; Ind. Amer., 539.

1964, Johnson, Dem., 2,136,615; Goldwater, Rep., 1,060,152; DeBerry, Soc. Workers, 3,817; Hass, Soc. Labor, 1,704; Proh. (no candidate listed), 699; scattering, 145.

1968, Nixon, Rep., 1,370,665; Humphrey, Dem., 1,593,082; Wallace, 3d Party, 331,968; Halstead, Soc. Workers, 4,099; Blomen, Soc. Labor, 1,762; Cleaver, New Politics, 4,585; Munn, Proh., 60; scattering, 29.

1972, Nixon, Rep., 1,961,721; McGovern, Dem., 1,459,435; Schmitz, Amer., 63,321; Fisher, Soc. Labor, 2,437; Jenness, Soc. Workers, 1,603; Hall, Com., 1,210.

1976, Carter, Dem., 1,696,714; Ford, Rep., 1,893,742; McCarthy, Ind., 47,905; MacBride, Libertarian, 5,406; Wright, People's, 3,504; Camejo, Soc. Workers, 1,804; LaRouche, U.S. Labor, 1,366; Levin, Soc. Labor, 1,148; scattering, 2,160.

1980, Reagan, Rep., 1,915,225; Carter, Dem., 1,661,532; Anderson, Ind., 275,223; Clark, Libertarian, 41,597; Commoner, Citizens, 11,930; Hall, Com., 3,262; Griswold, Workers World, 30; Greaves, Amer., 21; Bubar, Statesman, 9.

1984, Reagan, Rep., 2,251,571; Mondale, Dem., 1,529,638; Bergland, Libertarian, 10,055.

1988, Bush, Rep., 1,965,486; Dukakis, Dem., 1,675,783; Paul, Lib., 18,336; Fulani, Ind., 2,513.

1992, Clinton, Dem., 1,871,182; Bush, Rep., 1,554,940; Perot, Ind., 824,813; Marrou, Libertarian, 10,175; Phillips, U.S. Taxpayers, 8,263; Hagelin, Natural Law, 2,954.

1996, Clinton, Dem., 1,989,653; Dole, Rep., 1,481,212; Perot, Ref., 336,670; Browne, Libertarian, 27,670; Hagelin, Natural Law, 4,254; Moorehead, Workers World, 3,153; White, Soc. Equality, 1,554.

2000, Gore, Dem., 2,170,418; Bush, Rep., 1,953,139; Nader, Green, 84,165; Browne, Libertarian, 16,711; Phillips, U.S. Taxpayers, 3,791; Hagelin, Natural Law, 2,426.

Minnesota

	2000		1996		
County	Gore (D)	Bush (R)	Clinton (D)	Dole (R)	Perot (RF)
Aitkin	3,830	3,755	3,810	2,327	1,155
Anoka	68,008	69,256	63,756	41,745	16,448
Becker	5,253	8,152	5,911	5,461	1,813
Beltrami	7,301	8,346	8,006	5,806	1,635
Benton	6,009	7,663	6,006	4,835	2,133
Big Stone	1,430	1,370	1,619	990	368
Blue Earth	12,329	12,942	12,420	9,082	3,324
Brown	4,650	7,370	4,864	5,580	1,786
Carlton	8,620	5,578	8,052	4,034	1,591
Carver	12,462	20,790	11,554	12,380	3,781
Cass	5,534	7,134	5,437	4,791	1,620
Chippewa	2,952	2,977	3,178	2,119	782
Chisago	9,593	10,937	8,611	5,984	2,812
Clay	10,128	11,712	10,476	8,764	1,733
Clearwater	1,466	2,137	1,578	1,423	471
Cook	1,171	1,295	1,169	1,010	246
Cottonwood	2,503	3,369	2,737	2,633	741
Crow Wing	11,255	15,035	11,156	10,095	3,423
Dakota	85,446	87,250	77,297	57,244	17,095
Dodge	3,370	4,213	3,233	2,888	1,223
Douglas	6,352	9,811	6,450	6,747	2,093
Faribault	3,624	4,336	3,817	3,272	1,103
Fillmore	5,020	4,646	4,732	3,466	1,575
Freeborn	8,514	6,843	8,458	5,166	2,226
Goodhue	9,981	10,852	9,931	7,293	2,806
Grant	1,507	1,804	1,806	1,284	434
Hennepin	307,599	225,657	285,126	173,887	47,663
Houston	4,502	5,077	4,153	3,674	1,439
Hubbard	3,632	5,307	3,802	3,593	1,141
Isanti	6,247	7,668	6,041	4,450	2,242
Itasca	10,583	9,545	10,706	6,506	2,889
Jackson	2,364	2,773	2,727	2,153	908
Kanabec	2,831	3,480	2,927	1,924	996
Kandiyohi	8,220	10,026	9,009	7,119	2,229
Kittson	1,107	1,353	1,394	1,055	270
Koochiching	2,903	3,523	3,472	2,080	1,098
LacQuiParle	2,244	1,941	2,420	1,447	561
Lake	3,579	2,465	3,388	1,684	752

County	2000 Gore (D)	2000 Bush (R)	1996 Clinton (D)	1996 Dole (R)	1996 Perot (RF)
Lake of the Woods. . .	848	1,216	888	814	287
Le Sueur . . .	5,361	6,138	5,457	3,902	1,699
Lincoln.	1,590	1,513	1,641	1,199	504
Lyon.	4,737	6,087	5,062	4,932	1,351
McLeod	5,609	8,782	6,027	5,474	2,402
Mahnomen .	921	1,122	1,026	877	270
Marshall. . . .	1,910	2,912	2,333	2,068	710
Martin	4,166	5,686	4,718	4,303	1,405
Meeker	4,402	5,520	4,531	3,428	1,571
Mille Lacs . .	4,376	5,223	4,336	2,948	1,467
Morrison	5,274	8,197	5,728	5,054	2,310
Mower	10,693	6,873	10,413	4,994	2,464
Murray	2,093	2,407	2,173	1,907	753
Nicollet	7,041	7,221	6,772	5,057	1,737
Nobles	3,760	4,766	4,106	3,769	1,132
Norman	1,575	1,808	1,875	1,392	425
Olmsted. . . .	25,822	30,641	22,857	22,860	5,640
Otter Tail . . .	9,844	16,963	10,519	11,808	3,191
Pennington .	2,458	3,380	2,814	2,129	910
Pine.	6,148	5,854	5,432	3,080	1,597
Pipestone . .	1,970	2,693	1,999	2,096	599
Polk	5,764	7,609	6,369	5,563	1,502
Pope	2,771	2,808	2,803	1,992	665
Ramsey	138,470	87,669	133,878	66,954	20,351
Red Lake. . .	830	1,090	1,053	695	334
Redwood . . .	2,681	4,589	2,997	3,700	1,053
Renville	3,533	4,036	3,956	2,887	1,311
Rice.	13,140	10,876	12,821	7,016	2,872
Rock	2,081	2,772	2,142	2,169	554
Roseau	2,128	4,695	2,759	2,988	1,081
St. Louis . . .	64,237	35,420	60,736	25,553	11,308
Scott	17,503	23,954	14,657	12,734	4,886
Sherburne . .	12,109	16,813	10,551	8,699	3,665
Sibley	2,687	4,087	2,769	2,590	1,226
Stearns	24,800	32,402	24,238	21,474	8,150
Steele	6,900	8,223	6,974	5,617	2,197
Stevens	2,434	2,831	2,741	2,141	467
Swift	2,698	2,376	3,054	1,541	690
Todd	4,132	6,031	4,520	4,078	1,958
Traverse. . . .	884	1,074	1,135	775	295
Wabasha . . .	4,522	5,245	4,523	3,452	1,474
Wadena	2,251	3,733	2,480	2,696	801
Waseca	3,694	4,608	3,819	3,171	1,385
Washington .	49,637	51,502	45,119	31,219	10,106
Watonwan . .	2,258	2,562	2,534	1,997	711
Wilkin.	1,046	2,032	1,319	1,508	358
Winona	11,069	10,773	10,272	7,955	2,907
Wright	16,762	23,861	15,542	13,224	5,550
Yellow Medicine .	2,528	2,598	2,741	2,006	818
Totals	**1,168,266**	**1,109,659**	**1,120,438**	**766,476**	**257,704**

Minnesota Vote Since 1952

1952, Eisenhower, Rep., 763,211; Stevenson, Dem., 608,458; Hallinan, Prog., 2,666; Hass, Soc. Labor, 2,383; Hamblen, Proh., 2,147; Dobbs, Soc. Workers, 618.

1956, Eisenhower, Rep., 719,302; Stevenson, Dem., 617,525; Hass, Soc. Labor (Ind. Gov.), 2,080; Dobbs, Soc. Workers, 1,098.

1960, Kennedy, Dem., 779,933; Nixon, Rep., 757,915; Dobbs, Soc. Workers, 3,077; Industrial Gov., 962.

1964, Johnson, Dem., 991,117; Goldwater, Rep., 559,624; DeBerry, Soc. Workers, 1,177; Hass, Industrial Gov., 2,544.

1968, Nixon, Rep., 658,643; Humphrey, Dem., 857,738; Wallace, 3d Party, 68,931; scattered, 2,443; Halstead, Soc. Workers, 808; Blomen, Ind. Gov't., 285; Mitchell, Com., 415; Cleaver, Peace, 935; McCarthy, write-in, 585; scattered, 170.

1972, Nixon, Rep., 898,269; McGovern, Dem., 802,346; Schmitz, Amer., 31,407; Spock, Peoples, 2,805; Fisher, Soc. Labor, 4,261; Jenness, Soc. Workers, 940; Hall, Com., 662; scattered, 962.

1976, Carter, Dem., 1,070,440; Ford, Rep., 819,395; McCarthy, Ind., 35,490; Anderson, Amer., 13,592; Camejo, Soc. Workers, 4,149; MacBride, Libertarian, 3,529; Hall, Com., 1,092.

1980, Reagan, Rep., 873,268; Carter, Dem., 954,173; Anderson, Ind., 174,997; Clark, Libertarian, 31,593; Commoner, Citizens, 8,406; Hall, Com., 1,117; DeBerry, Soc. Workers, 711; Griswold, Workers World, 698; McReynolds, Soc., 536; write-ins, 281.

1984, Reagan, Rep., 1,032,603; Mondale, Dem., 1,036,364; Bergland, Libertarian, 2,996.

1988, Bush, Rep., 962,337; Dukakis, Dem., 1,109,471; McCarthy, Minn. Prog., 5,403; Paul, Lib., 5,109.

1992, Clinton, Dem., 1,020,997; Bush, Rep., 747,841; Perot, Ind., 562,506; Marrou, Libertarian, 3,373; Gritz, Populist/America First, 3,363; Hagelin, Natural Law, 1,406.

1996, Clinton, Dem., 1,120,438; Dole, Rep., 766,476; Perot, Ref., 257,704; Nader, Green, 24,908; Browne, Libertarian, 8,271; Peron, Grass Roots, 4,898; Phillips, Taxpayers, 3,416; Hagelin, Natural Law, 1,808; Birrenbach, Ind. Grass Roots, 787; Harris, Soc. Workers, 684; White, Soc. Equality, 347.

2000, Gore, Dem., 1,168,266; Bush, Rep., 1,109,659; Nader, Green, 126,696; Buchanan, Reform Minnesota, 22,166; Browne, Libertarian, 5,282; Phillips, Constitution, 3,272; Hagelin, Reform, 2,294; Harris, Soc. Workers, 1,022.

Mississippi

County	2000 Gore (D)	2000 Bush (R)	1996 Clinton (D)	1996 Dole (R)	1996 Perot (RF)
Adams	8,065	6,691	8,218	5,378	779
Alcorn.	5,059	7,254	4,964	4,960	929
Amite	2,673	3,677	2,824	2,521	351
Attala	2,922	4,206	3,092	3,130	383
Benton	1,886	1,561	1,944	993	209
Bolivar	8,436	4,847	8,670	4,027	320
Calhoun	2,251	3,448	2,178	2,470	351
Carroll.	1,726	3,165	2,041	2,629	245
Chickasaw	3,519	3,549	2,971	2,535	401
Choctaw	1,278	2,398	1,247	1,715	247
Claiborne	3,670	883	3,739	784	103
Clarke.	2,368	4,503	2,337	3,470	366
Clay	4,515	3,570	4,267	2,948	337
Coahoma	5,662	3,695	5,776	3,441	256
Copiah	4,845	5,643	4,415	4,138	375
Covington	2,623	4,180	2,628	3,219	417
DeSoto	9,586	24,879	10,282	18,135	2,399
Forrest	8,500	13,281	7,965	11,278	1,094
Franklin	1,486	2,427	1,381	1,586	329
George	1,977	5,143	1,888	3,311	710
Greene	1,317	3,082	1,347	1,947	322
Grenada	3,813	4,743	4,402	4,527	470
Hancock	4,801	9,326	4,303	5,820	1,143
Harrison	19,142	32,256	18,775	25,486	3,726
Hinds	46,789	37,753	45,410	35,653	2,929
Holmes.	5,447	1,937	4,720	1,536	140
Humphreys.	2,288	1,628	2,305	1,382	110
Issaquena	555	366	546	269	42
Itawamba	2,994	5,424	2,987	3,490	732
Jackson	14,193	30,068	13,598	24,918	2,947
Jasper	3,104	3,294	3,170	2,615	353
Jefferson	2,786	600	2,531	489	89
Jefferson Davis .	2,835	2,437	2,663	1,890	264
Jones	7,713	16,341	7,360	13,020	1,362
Kemper.	2,311	1,915	2,048	1,439	188
Lafayette.	5,139	7,081	4,646	4,753	580
Lamar.	3,478	12,795	3,169	8,609	925
Lauderdale	8,412	17,315	8,668	15,055	1,036
Lawrence	2,841	3,674	2,481	2,392	471
Leake	2,793	4,114	2,902	3,017	406
Lee.	9,142	15,551	8,438	11,815	1,361
Leflore	6,401	4,626	6,853	4,456	240
Lincoln	4,358	8,540	4,294	5,960	778
Lowndes.	7,537	11,404	6,220	9,169	750
Madison	10,416	19,109	9,354	14,467	759
Marion	4,114	6,796	4,334	5,023	585
Marshall	7,735	4,723	7,521	3,272	482
Monroe.	5,783	7,397	5,184	5,206	889
Montgomery	2,187	2,630	1,970	1,943	197
Neshoba.	2,563	6,409	2,646	4,545	560
Newton	2,147	5,540	2,163	4,223	464
Noxubee.	3,383	1,530	2,801	1,287	119
Oktibbeha.	6,443	7,959	5,923	6,142	395
Panola	5,880	5,424	5,408	3,701	513
Pearl River	4,611	11,575	4,892	8,212	1,190
Perry.	1,285	3,026	1,413	2,178	450
Pike	6,544	7,464	6,302	5,403	683
Pontotoc	2,771	6,601	2,597	4,289	774
Prentiss	3,287	5,101	3,053	3,473	574
Quitman	2,103	1,280	2,186	1,121	126
Rankin	8,050	32,983	8,614	24,585	2,093
Scott.	3,548	5,601	3,163	4,018	466
Sharkey	1,706	1,074	1,566	906	70
Simpson.	3,227	6,254	2,851	4,455	525
Smith	1,620	4,838	1,858	3,371	522
Stone	1,677	3,702	1,551	2,288	417
Sunflower	4,981	3,369	4,960	2,926	290
Tallahatchie	3,041	2,428	2,990	1,676	251
Tate	3,441	5,148	3,195	3,694	406
Tippah	2,908	5,381	2,992	3,249	661
Tishomingo.	2,747	4,122	2,709	2,766	609
Tunica.	1,539	792	1,263	557	55
Union	3,094	6,087	3,316	4,375	788
Walthall	2,356	3,476	2,240	2,239	444
Warren	7,485	10,892	8,774	9,261	1,259
Washington	10,405	7,367	10,053	6,762	437
Wayne	2,981	4,635	2,652	3,219	595
Webster	1,426	3,069	1,379	2,254	255
Wilkinson	2,551	1,423	2,807	1,016	226
Winston	3,672	4,645	3,488	3,498	434
Yalobusha.	2,674	2,470	2,437	1,711	332
Yazoo	4,997	5,254	4,754	4,152	362
Totals.	**404,614**	**572,844**	**394,022**	**439,838**	**52,222**

Mississippi Vote Since 1952

1952, Eisenhower, Ind. vote pledged to Rep. candidate, 112,966; Stevenson, Dem., 172,566.
1956, Eisenhower, Rep., 56,372; Stevenson, Dem., 144,498; Black and Tan Grand Old Party, 4,313; total, 60,685; Byrd, Ind., 42,966.
1960, Kennedy, Dem., 108,362; Democratic unpledged electors, 116,248; Nixon, Rep., 73,561. Mississippi's victorious slate of 8 unpledged Democratic electors cast their votes for Sen. Harry F. Byrd (D, VA).
1964, Johnson, Dem., 52,618; Goldwater, Rep., 356,528.
1968, Nixon, Rep., 88,516; Humphrey, Dem., 150,644; Wallace, 3d Party, 415,349.
1972, Nixon, Rep., 505,125; McGovern, Dem., 126,782; Schmitz, Amer., 11,598; Jenness, Soc. Workers, 2,458.
1976, Carter, Dem., 381,309; Ford, Rep., 366,846; Anderson, Amer., 6,678; McCarthy, Ind., 4,074; Maddox, Ind., 4,049; Camejo, Soc. Workers, 2,805; MacBride, Libertarian, 2,609.
1980, Reagan, Rep., 441,089; Carter, Dem., 429,281; Anderson, Ind., 12,036; Clark, Libertarian, 5,465; Griswold, Workers World, 2,402; Pulley, Soc. Workers, 2,347.
1984, Reagan, Rep., 582,377; Mondale, Dem., 352,192; Bergland, Libertarian, 2,336.
1988, Bush, Rep., 557,890; Dukakis, Dem., 363,921; Duke, Ind., 4,232; Paul, Lib., 3,329.
1992, Bush, Rep., 487,793; Clinton, Dem., 400,258; Perot, Ind., 85,626; Fulani, New Alliance, 2,625; Marrou, Libertarian, 2,154; Phillips, U.S. Taxpayers, 1,652; Hagelin, Natural Law, 1,140.
1996, Dole, Rep., 439,838; Clinton, Dem., 394,022; Perot, Ind. (Ref.), 52,222; Browne, Libertarian, 2,809; Phillips, Taxpayers, 2,314; Hagelin, Natural Law, 1,447; Collins, Ind., 1,205.
2000, Bush, Rep., 572,844; Gore, Dem., 404,614; Nader, Ind., 8,122; Phillips, Constitution, 3,267; Buchanan, Reform, 2,265; Browne, Libertarian, 2,009; Harris, Ind., 613; Hagelin, Natural Law, 450.

Missouri

County	2000 Gore (D)	2000 Bush (R)	1996 Clinton (D)	1996 Dole (R)	1996 Perot (RF)
Adair	4,101	6,050	4,441	4,656	1,170
Andrew	2,795	4,257	2,807	3,281	964
Atchison	1,013	1,798	1,266	1,327	367
Audrain	4,551	5,256	4,690	3,955	1,046
Barry	4,135	7,885	4,352	5,855	1,494
Barton	1,424	3,836	1,625	2,812	563
Bates	3,386	4,245	3,224	2,904	949
Benton	3,150	4,218	2,996	2,895	764
Bollinger	1,692	3,487	2,044	2,420	506
Boone	28,811	28,426	24,984	22,047	4,083
Buchanan	17,085	16,423	15,848	12,610	4,248
Butler	4,996	9,111	5,780	6,996	1,414
Caldwell	1,488	2,220	1,487	1,464	468
Callaway	6,708	8,238	5,880	5,567	1,530
Camden	6,323	10,358	5,566	7,190	1,809
Cape Girardeau	9,334	19,832	9,957	15,557	1,861
Carroll	1,620	2,880	2,080	1,839	580
Carter	997	1,730	1,172	1,180	301
Cass	14,921	20,113	11,743	13,495	3,474
Cedar	1,979	3,530	2,027	2,484	658
Chariton	1,792	2,300	2,072	1,508	423
Christian	7,896	14,824	6,627	9,477	2,301
Clark	1,812	1,899	1,749	1,081	458
Clay	39,084	39,083	32,603	28,935	7,048
Clinton	3,994	4,323	3,445	2,780	848
Cole	12,056	20,167	10,857	16,140	2,121
Cooper	2,567	4,072	2,753	2,900	891
Crawford	3,350	4,754	3,349	2,990	1,223
Dade	1,193	2,468	1,243	1,822	447
Dallas	2,311	3,723	2,277	2,554	787
Daviess	1,367	2,011	1,534	1,321	466
DeKalb	1,562	2,363	1,679	1,627	492
Dent	1,839	3,996	2,234	2,542	693
Douglas	1,546	3,599	1,744	2,601	775
Dunklin	4,947	5,426	5,428	3,766	934
Franklin	16,172	21,863	13,908	13,715	5,517
Gasconade	2,257	4,190	2,104	2,997	820
Gentry	1,271	1,771	1,493	1,361	416
Greene	41,091	59,178	39,300	48,193	8,569
Grundy	1,563	2,976	2,073	1,883	631
Harrison	1,328	2,552	1,628	1,737	484
Henry	4,459	5,120	4,579	3,260	1,231
Hickory	1,961	2,172	1,858	1,491	531
Holt	871	1,738	1,144	1,323	314
Howard	1,944	2,414	2,014	1,545	568
Howell	4,641	9,018	5,261	5,991	2,066
Iron	2,044	2,237	2,221	1,328	568
Jackson	160,419	104,418	140,317	85,534	21,047
Jasper	11,737	24,899	11,462	18,361	3,545
Jefferson	38,616	36,766	32,073	23,877	8,893
Johnson	6,926	9,339	6,220	6,276	1,911
Knox	787	1,226	891	862	254
Laclede	4,183	8,556	4,047	5,887	1,459
Lafayette	6,343	7,849	6,118	5,489	1,516
Lawrence	4,235	8,305	4,465	6,099	1,613
Lewis	2,023	2,388	2,050	1,453	644
Lincoln	6,961	8,549	5,644	4,897	1,881
Linn	2,646	3,246	2,967	2,097	781
Livingston	2,425	3,709	2,913	2,384	777
McDonald	1,866	4,460	1,980	3,008	923
Macon	2,817	4,232	2,937	2,634	848
Madison	1,828	2,460	2,351	1,595	625
Maries	1,554	2,216	1,540	1,560	516
Marion	4,993	6,550	4,924	4,653	1,082
Mercer	555	1,250	700	660	208
Miller	3,217	5,945	3,110	4,387	1,185
Mississippi	2,756	2,395	3,235	1,595	380
Moniteau	2,176	3,764	2,129	2,603	693
Monroe	1,860	2,175	1,938	1,333	532
Montgomery	2,092	3,106	2,277	2,124	772
Morgan	3,235	4,460	3,006	3,059	1,006
New Madrid	3,738	3,416	4,451	2,417	663
Newton	6,447	14,232	5,840	10,067	1,995
Nodaway	3,553	5,161	3,966	3,362	1,043
Oregon	1,568	2,521	1,795	1,502	475
Osage	1,938	4,154	2,045	2,890	608
Ozark	1,432	2,663	1,445	1,882	595
Pemiscot	3,245	2,750	3,371	1,820	458
Perry	2,085	4,667	2,517	3,427	777
Pettis	5,855	9,533	6,057	7,336	1,716
Phelps	6,262	9,444	6,405	6,990	1,703
Pike	3,557	3,648	3,495	2,209	916
Platte	15,325	17,785	12,705	13,332	3,035
Polk	3,606	6,430	3,307	4,521	1,169
Pulaski	3,800	6,531	3,783	4,089	1,141
Putnam	708	1,593	857	1,091	276
Ralls	2,033	2,446	1,998	1,513	520
Randolph	4,116	4,844	4,502	3,274	1,130
Ray	4,970	4,517	4,714	2,884	1,113
Reynolds	1,298	1,762	1,631	903	386
Ripley	1,820	3,121	2,081	1,988	530
St. Charles	53,806	72,114	41,369	47,705	11,591
St. Clair	1,866	2,731	1,974	1,815	650
St. Francois	9,075	9,327	9,034	6,200	2,266
St. Louis	250,631	224,689	225,524	196,096	34,850
Ste. Genevieve	3,600	3,505	3,597	2,078	942
Saline	4,585	4,572	4,765	2,931	1,090
Schuyler	808	1,159	857	777	287
Scotland	790	1,335	990	773	326
Scott	6,452	8,999	7,011	6,641	1,483
Shannon	1,430	2,245	1,882	1,339	524
Shelby	1,262	1,936	1,410	1,213	413
Stoddard	4,476	7,727	4,883	5,020	1,185
Stone	4,055	7,793	3,497	5,223	1,353
Sullivan	1,127	1,877	1,402	1,275	340
Taney	5,092	9,647	4,623	6,844	1,580
Texas	3,486	6,136	3,897	4,065	1,335
Vernon	3,156	4,985	3,363	3,123	1,135
Warren	4,524	5,979	3,443	3,768	1,254
Washington	4,047	4,020	4,315	2,259	1,169
Wayne	2,387	3,346	2,754	2,172	674
Webster	4,174	7,350	3,855	4,958	1,214
Worth	469	651	572	540	150
Wright	2,250	5,391	2,280	3,754	890
City					
St. Louis	96,557	24,799	91,233	22,121	7,276
Totals	**1,111,138**	**1,189,924**	**1,025,935**	**890,016**	**217,188**

Missouri Vote Since 1952

1952, Eisenhower, Rep., 959,429; Stevenson, Dem., 929,830; Hallinan, Prog., 987; Hamblen, Proh., 885; MacArthur, Christian Nationalist, 302; America First, 233; Hoopes, Soc., 227; Hass, Soc. Labor, 169.
1956, Stevenson, Dem., 918,273; Eisenhower, Rep., 914,299.
1960, Kennedy, Dem., 972,201; Nixon, Rep., 962,221.
1964, Johnson, Dem., 1,164,344; Goldwater, Rep., 653,535.
1968, Nixon, Rep., 811,932; Humphrey, Dem., 791,444; Wallace, 3d Party, 206,126.
1972, Nixon, Rep., 1,154,058; McGovern, Dem., 698,531.
1976, Carter, Dem., 999,163; Ford, Rep., 928,808; McCarthy, Ind., 24,329.
1980, Reagan, Rep., 1,074,181; Carter, Dem., 931,182; Anderson, Ind., 77,920; Clark, Libertarian, 14,422; DeBerry, Soc. Workers, 1,515; Commoner, Citizens, 573; write-ins, 31.
1984, Reagan, Rep., 1,274,188; Mondale, Dem., 848,583.
1988, Bush, Rep., 1,084,953; Dukakis, Dem., 1,001,619; Fulani, New Alliance, 6,656; Paul, write-in, 434.
1992, Clinton, Dem., 1,053,873; Bush, Rep., 811,159; Perot, Ind., 518,741; Marrou, Libertarian, 7,497.
1996, Clinton, Dem., 1,025,935; Dole, Rep., 890,016; Perot, Ref., 217,188; Phillips, Taxpayers, 11,521; Browne, Libertarian, 10,522; Hagelin, Natural Law, 2,287.
2000, Bush, Rep., 1,189,924; Gore, Dem., 1,111,138; Nader, Green, 38,515; Buchanan, Reform, 9,818; Browne, Libertarian, 7,436; Phillips, Constitution, 1,957; Hagelin, Natural Law, 1,104.

Montana

County	2000 Gore (D)	2000 Bush (R)	1996 Clinton (D)	1996 Dole (R)	1996 Perot (RF)
Beaverhead	799	3,113	1,164	2,414	412
Big Horn	2,345	1,651	2,453	1,336	424
Blaine	1,246	1,410	1,316	1,127	435
Broadwater	462	1,488	603	1,029	318
Carbon	1,434	3,008	1,854	2,147	713
Carter	53	573	150	522	89
Cascade	13,137	18,164	15,707	14,291	4,749
Chouteau	686	2,039	1,039	1,660	434
Custer	1,501	3,156	2,115	2,467	695
Daniels	303	750	510	558	240
Dawson	1,364	2,723	1,903	1,890	842
Deer Lodge	2,672	1,493	3,331	883	772
Fallon	256	1,061	452	778	276
Fergus	1,352	4,353	1,866	3,671	605
Flathead	8,329	22,519	10,452	16,542	4,786
Gallatin	10,009	18,833	10,972	14,559	3,146
Garfield	61	651	107	562	69
Glacier	2,211	1,709	2,292	1,270	491
Golden Valley	88	405	128	284	73
Granite	295	1,181	429	733	228
Hill	2,760	3,392	3,517	2,601	950
Jefferson	1,513	3,308	1,775	2,248	729
Judith Basin	278	1,057	452	753	126
Lake	3,884	6,441	4,195	4,723	1,804
Lewis & Clark	9,982	15,091	11,535	11,665	3,140
Liberty	243	752	379	634	144
Lincoln	1,629	5,578	2,705	3,552	1,425
McCone	267	827	390	615	244
Madison	758	2,656	955	1,984	516
Meagher	176	698	281	505	142
Mineral	382	1,078	658	549	383
Missoula	17,241	21,474	21,874	16,034	5,586
Musselshell	512	1,582	652	1,121	291
Park	2,154	4,523	2,564	3,837	959
Petroleum	36	254	62	186	36
Phillips	423	1,727	705	1,392	401
Pondera	792	1,948	1,123	1,438	383
Powder River	115	860	236	663	137
Powell	638	1,971	952	1,274	531
Prairie	164	541	259	417	99
Ravalli	4,451	11,241	5,200	8,138	2,731
Richland	1,018	2,858	1,614	2,021	906
Roosevelt	2,059	1,605	2,118	1,209	645
Rosebud	1,394	1,826	1,681	1,413	547
Sanders	1,165	3,144	1,573	2,043	990
Sheridan	702	1,176	1,187	832	408
Silver Bow	8,967	6,299	11,199	3,909	2,447
Stillwater	925	2,765	1,282	1,871	618
Sweet Grass	305	1,450	469	1,109	186
Teton	847	2,294	1,188	1,701	416
Toole	630	1,639	874	1,203	386
Treasure	106	344	171	237	87
Valley	1,273	2,500	1,674	1,838	645
Wheatland	243	708	391	563	127
Wibaux	121	369	197	284	128
Yellowstone	20,370	33,922	22,992	26,367	6,139
Totals	**137,126**	**240,178**	**167,922**	**179,652**	**55,229**

Montana Vote Since 1952

1952, Eisenhower, Rep., 157,394; Stevenson, Dem., 106,213; Hallinan, Prog., 723; Hamblen, Proh., 548; Hoopes, Soc., 159.

1956, Eisenhower, Rep., 154,933; Stevenson, Dem., 116,238.

1960, Kennedy, Dem., 134,891; Nixon, Rep., 141,841; Decker, Proh., 456; Dobbs, Soc. Workers, 391.

1964, Johnson, Dem., 164,246; Goldwater, Rep., 113,032; Kasper, Natl. States' Rights, 519; Munn, Proh., 499; DeBerry, Soc. Workers, 332.

1968, Nixon, Rep., 138,835; Humphrey, Dem., 114,117; Wallace, 3d Party, 20,015; Halstead, Soc. Workers, 457; Munn, Proh., 510; Caton, New Reform, 470.

1972, Nixon, Rep., 183,976; McGovern, Dem., 120,197; Schmitz, Amer., 13,430.

1976, Carter, Dem., 149,259; Ford, Rep., 173,703; Anderson, Amer., 5,772.

1980, Reagan, Rep., 206,814; Carter, Dem., 118,032; Anderson, Ind., 29,281; Clark, Libertarian, 9,825.

1984, Reagan, Rep., 232,450; Mondale, Dem., 146,742; Bergland, Libertarian, 5,185.

1988, Bush, Rep., 190,412; Dukakis, Dem., 168,936; Paul, Lib., 5,047; Fulani, New Alliance, 1,279.

1992, Clinton, Dem., 154,507; Bush, Rep., 144,207; Perot, Ind., 107,225; Gritz, Populist/America First, 3,658.

1996, Dole, Rep., 179,652; Clinton, Dem., 167,922; Perot, Ref., 55,229; Browne, Libertarian, 2,526; Hagelin, Natural Law, 1,754.

2000, Bush, Rep, 240,178; Gore, Dem., 137,126; Nader, Green, 24,437; Buchanan, Reform, 5,697; Browne, Libertarian, 1,718; Phillips, Constitution, 1,155; Hagelin, Natural Law, 675.

Nebraska

County	2000 Gore (D)	2000 Bush (R)	1996 Clinton (D)	1996 Dole (R)	1996 Perot (RF)
Adams	3,686	8,162	3,935	6,924	1,513
Antelope	678	2,562	884	2,005	457
Arthur	26	235	25	187	46
Banner	65	390	62	309	30
Blaine	43	299	53	284	39
Boone	575	2,196	806	1,695	424
Box Butte	1,614	3,208	1,782	2,458	695
Boyd	265	931	372	778	181
Brown	250	1,375	359	1,105	289
Buffalo	3,927	11,931	4,277	10,004	1,484
Burt	1,223	2,056	1,237	1,707	497
Butler	1,028	2,638	1,099	2,042	512
Cass	3,656	6,144	3,477	4,878	1,239
Cedar	1,062	2,989	1,218	2,171	739
Chase	306	1,505	365	1,277	197
Cherry	446	2,322	551	1,905	332
Cheyenne	844	3,207	1,059	2,571	287
Clay	774	2,326	880	1,982	425
Colfax	863	2,338	1,065	1,954	492
Cuming	857	3,232	1,033	2,520	503
Custer	976	4,245	1,293	3,453	615
Dakota	2,695	3,119	2,632	2,592	721
Dawes	823	2,549	1,108	1,991	442
Dawson	1,740	5,511	2,180	4,794	1,044
Deuel	213	783	245	629	111
Dixon	820	1,834	931	1,478	414
Dodge	5,021	8,871	5,181	7,484	1,894
Douglas	73,347	101,025	70,708	92,334	14,863
Dundy	179	801	224	752	112
Fillmore	848	2,024	1,058	1,696	321
Franklin	420	1,196	483	1,013	215
Frontier	244	1,102	310	901	169
Furnas	534	1,849	663	1,475	207
Gage	3,516	5,538	4,008	4,413	1,346
Garden	203	963	279	851	155
Garfield	202	718	249	625	111
Gosper	228	757	275	609	150
Grant	49	324	84	258	55
Greeley	416	839	472	642	155
Hall	5,952	11,803	6,708	10,183	2,403
Hamilton	1,066	3,251	1,172	2,623	457
Harlan	438	1,358	520	1,120	203
Hayes	66	486	87	439	39
Hitchcock	312	1,126	409	977	173
Holt	846	3,954	1,107	3,436	677
Hooker	74	317	115	308	83
Howard	955	1,760	853	1,294	417
Jefferson	1,361	2,351	1,520	1,979	495
Johnson	794	1,210	770	1,009	309
Kearney	680	2,333	782	1,953	296
Keith	778	2,953	830	2,504	460
Keya Paha	78	422	94	385	47
Kimball	379	1,379	527	1,011	212
Knox	1,037	2,784	1,266	2,123	531
Lancaster	44,650	55,514	43,339	44,812	8,595
Lincoln	5,205	9,220	5,165	7,482	2,043
Logan	60	336	79	294	72
Loup	84	284	74	229	28
McPherson	48	244	50	233	33
Madison	2,772	9,636	3,047	7,965	1,554
Merrick	848	2,380	997	2,084	449
Morrill	460	1,597	620	1,296	262
Nance	497	1,105	585	892	238
Nemaha	1,063	2,177	1,232	1,888	485
Nuckolls	644	1,701	757	1,383	306
Otoe	2,208	4,178	2,279	3,290	877
Pawnee	522	937	580	766	207
Perkins	243	1,170	352	1,018	163
Phelps	934	3,575	1,071	3,015	465
Pierce	570	2,534	697	1,923	446
Platte	2,612	9,861	3,010	7,948	1,353
Polk	610	1,925	750	1,504	268
Red Willow	1,188	3,680	1,365	3,112	499
Richardson	1,382	2,623	1,517	2,089	633
Rock	141	725	180	564	135
Saline	2,321	2,581	2,523	1,945	689
Sarpy	14,637	28,979	12,806	23,023	3,722
Saunders	2,852	5,688	2,777	4,514	1,223
Scotts Bluff	3,937	9,397	4,547	7,641	1,251
Seward	2,250	4,457	2,432	3,479	745
Sheridan	392	2,105	573	1,834	289
Sherman	564	1,072	567	822	266
Sioux	98	629	138	551	75
Stanton	500	1,895	577	1,457	386
Thayer	821	2,096	933	1,698	334
Thomas	55	329	64	303	62
Thurston	924	1,040	962	835	293
Valley	583	1,610	758	1,346	274
Washington	2,550	5,758	2,248	4,391	971
Wayne	1,001	2,774	1,048	2,150	440

	2000		1996		
	Gore	Bush	Clinton	Dole	Perot
County	(D)	(R)	(D)	(R)	(RF)
Webster.	584	1,302	621	1,094	236
Wheeler.	85	351	106	241	69
York	1,407	4,816	1,653	4,266	559
Totals	**231,780**	**433,862**	**236,761**	**363,467**	**71,278**

Nebraska Vote Since 1952

1952, Eisenhower, Rep., 421,603; Stevenson, Dem., 188,057.
1956, Eisenhower, Rep., 378,108; Stevenson, Dem., 199,029.
1960, Kennedy, Dem., 232,542; Nixon, Rep., 380,553.
1964, Johnson, Dem., 307,307; Goldwater, Rep., 276,847.
1968, Nixon, Rep., 321,163; Humphrey, Dem., 170,784; Wallace, 3d Party, 44,904.
1972, Nixon, Rep., 406,298; McGovern, Dem., 169,991; scattered, 817.
1976, Carter, Dem., 233,287; Ford, Rep., 359,219; McCarthy, Ind., 9,383; Maddox, Amer. Ind., 3,378; MacBride, Libertarian, 1,476.
1980, Reagan, Rep., 419,214; Carter, Dem., 166,424; Anderson, Ind., 44,854; Clark, Libertarian, 9,041.
1984, Reagan, Rep., 459,135; Mondale, Dem., 187,475; Bergland, Libertarian, 2,075.
1988, Bush, Rep., 397,956; Dukakis, Dem., 259,235; Paul, Lib., 2,534; Fulani, New Alliance, 1,740.
1992, Bush, Rep., 343,678; Clinton, Dem., 216,864; Perot, Ind., 174,104; Marrou, Libertarian, 1,340.
1996, Dole, Rep., 363,467; Clinton, Dem., 236,761; Perot, Ref., 71,278; Browne, Libertarian, 2,792; Phillips, Ind., 1,928; Hagelin, Natural Law, 1,189.
2000, Bush, Rep., 433,862; Gore, Dem., 231,780; Nader, Green, 24,540; Buchanan, Ind., 3,646; Browne, Libertarian, 2,245; Hagelin, Natural Law, 478; Phillips, Ind., 468.

Nevada

	2000		1996		
	Gore	Bush	Clinton	Dole	Perot
County	(D)	(R)	(D)	(R)	(RF)
Churchill . . .	2,191	6,237	2,282	4,369	821
Clark	196,100	170,932	127,963	103,431	23,177
Douglas. . . .	5,837	11,193	5,109	8,828	1,486
Elko	2,542	11,025	3,149	6,512	1,539
Esmeralda . .	116	333	140	277	91
Eureka.	150	632	158	412	90
Humboldt. . .	1,128	3,638	1,467	2,334	603
Lander.	395	1,619	660	1,107	361
Lincoln.	461	1,372	499	936	255
Lyon.	3,955	7,270	3,419	4,753	1,104
Mineral	916	1,227	1,068	814	361
Nye	4,525	6,904	3,300	3,979	1,544
Pershing . . .	476	1,221	565	743	203
Storey	666	1,014	614	705	244
Washoe	52,097	63,640	44,915	49,477	9,970
White Pine. .	1,069	2,234	1,397	1,399	546
City					
Carson City .	7,354	11,084	7,269	9,168	1,591
Totals	**279,978**	**301,575**	**203,974**	**199,244**	**43,986**

Nevada Vote Since 1952

1952, Eisenhower, Rep., 50,502; Stevenson, Dem., 31,688.
1956, Eisenhower, Rep., 56,049; Stevenson, Dem., 40,640.
1960, Kennedy, Dem., 54,880; Nixon, Rep., 52,387.
1964, Johnson, Dem., 79,339; Goldwater, Rep., 56,094.
1968, Nixon, Rep., 73,188; Humphrey, Dem., 60,598; Wallace, 3d Party, 20,432.
1972, Nixon, Rep., 115,750; McGovern, Dem., 66,016.
1976, Carter, Dem., 92,479; Ford, Rep., 101,273; MacBride, Libertarian, 1,519; Maddox, Amer. Ind., 1,497; scattered, 5,108.
1980, Reagan, Rep., 155,017; Carter, Dem., 66,666; Anderson, Ind., 17,651; Clark, Libertarian, 4,358.
1984, Reagan, Rep., 188,770; Mondale, Dem., 91,655; Bergland, Libertarian, 2,292.
1988, Bush, Rep., 206,040; Dukakis, Dem., 132,738; Paul, Lib., 3,520; Fulani, New Alliance, 835.
1992, Clinton, Dem., 189,148; Bush, Rep., 175,828; Perot, Ind., 132,580; Gritz, Populist/America First, 2,892; Marrou, Libertarian, 1,835.
1996, Clinton, Dem., 203,974; Dole, Rep., 199,244; Perot, Ref., 43,986; "None of These Candidates," 5,608; Nader, Green, 4,730; Browne, Libertarian, 4,460; Phillips, Ind. Amer., 1,732; Hagelin, Natural Law, 545.
2000, Bush, Rep., 301,575; Gore, Dem., 279,978; Nader, Green, 15,008; Buchanan, Citizens First, 4,747; "None of these candidates," 3,315; Browne, Libertarian, 3,311; Phillips, Ind. Amer., 621; Hagelin, Natural Law, 415.

New Hampshire

	2000		1996		
	Gore	Bush	Clinton	Dole	Perot
County	(D)	(R)	(D)	(R)	(RF)
Concord.	10,025	6,981	9,719	5,082	1,164
Derry	5,530	6,093	4,814	4,503	1,083
Dover	6,812	5,008	6,332	3,752	930
Hudson.	4,573	4,527	3,841	3,167	976
Keene.	5,856	3,704	5,401	2,910	621
Laconia	3,015	3,814	2,865	2,842	508
Londonderry. . .	4,348	5,463	3,666	4,076	838
Manchester . . .	19,991	19,152	20,185	14,704	3,053
Merrimack	5,571	6,239	4,934	4,499	949
Nashua.	18,398	14,803	16,584	11,479	2,858
Portsmouth. . . .	6,862	3,896	6,343	3,014	661
Rochester.	5,401	5,522	5,489	3,650	1,108
Salem.	5,711	5,713	5,164	4,257	1,241
Other	164,255	182,644	150,829	128,551	32,397
Totals.	**266,318**	**273,559**	**246,166**	**196,486**	**48,387**

New Hampshire Vote Since 1952

1952, Eisenhower, Rep., 166,287; Stevenson, Dem., 106,663.
1956, Eisenhower, Rep., 176,519; Stevenson, Dem., 90,364; Andrews, Const., 111.
1960, Kennedy, Dem., 137,772; Nixon, Rep., 157,989.
1964, Johnson, Dem., 182,065; Goldwater, Rep., 104,029.
1968, Nixon, Rep., 154,903; Humphrey, Dem., 130,589; Wallace, 3d Party, 11,173; New Party, 421; Halstead, Soc. Workers, 104.
1972, Nixon, Rep., 213,724; McGovern, Dem., 116,435; Schmitz, Amer., 3,386; Jenness, Soc. Workers, 368; scattered, 142.
1976, Carter, Dem., 147,645; Ford, Rep., 185,935; McCarthy, Ind., 4,095; MacBride, Libertarian, 936; Reagan, write-in, 388; La Rouche, U.S. Labor, 186; Camejo, Soc. Workers, 161; Levin, Soc. Labor, 66; scattered, 215.
1980, Reagan, Rep., 221,705; Carter, Dem., 108,864; Anderson, Ind., 49,693; Clark, Libertarian, 2,067; Commoner, Citizens, 1,325; Hall, Com., 129; Griswold, Workers World, 76; DeBerry, Soc. Workers, 72; scattered, 68.
1984, Reagan, Rep., 267,051; Mondale, Dem., 120,377; Bergland, Libertarian, 735.
1988, Bush, Rep., 281,537; Dukakis, Dem., 163,696; Paul, Lib., 4,502; Fulani, New Alliance, 790.
1992, Clinton, Dem., 209,040; Bush, Rep., 202,484; Perot, Ind., 121,337; Marrou, Libertarian, 3,548.
1996, Clinton, Dem., 246,166; Dole, Rep., 196,486; Perot, Ref., 48,387; Browne, Libertarian, 4,214; Phillips, Taxpayers, 1,344.
2000, Bush, Rep., 273,559; Gore, Dem., 266,348; Nader, Green, 22,198; Browne, Libertarian, 2,757; Buchanan, Independence, 2,615; Phillips, Constitution, 328.

New Jersey

	2000		1996		
	Gore	Bush	Clinton	Dole	Perot
County	(D)	(R)	(D)	(R)	(RF)
Atlantic	52,880	35,593	44,434	29,538	8,261
Bergen	202,682	152,731	191,085	141,164	25,512
Burlington. .	99,506	72,254	85,086	57,337	18,407
Camden . . .	127,166	62,464	114,962	52,791	17,433
Cape May. .	22,189	23,794	19,849	19,357	4,978
Cumberland	28,188	18,882	25,444	14,744	5,348
Essex	185,505	66,842	175,387	65,172	9,513
Gloucester .	61,095	42,315	51,928	32,138	14,361
Hudson. . . .	118,206	43,804	116,121	38,288	8,965
Hunterdon .	21,387	32,210	18,446	26,379	5,686
Mercer	83,256	46,670	77,641	40,559	10,536
Middlesex. .	154,998	93,545	145,201	82,433	24,643
Monmouth .	131,476	119,291	120,414	99,975	22,754
Morris	88,039	111,066	81,092	95,830	15,299
Ocean	102,104	105,684	94,243	82,830	22,864
Passaic. . . .	90,324	61,043	85,879	53,584	10,944
Salem.	13,718	12,257	12,044	9,294	4,124
Somerset . .	56,232	59,725	50,673	51,868	8,377
Sussex	21,353	33,277	19,525	26,746	6,705
Union	112,003	68,554	108,102	65,912	12,432
Warren	16,543	22,172	14,805	17,160	4,992
Totals.	**1,788,850**	**1,284,173**	**1,652,361**	**1,103,099**	**262,134**

New Jersey Vote Since 1952

1952, Eisenhower, Rep., 1,373,613; Stevenson, Dem., 1,015,902; Hoopes, Soc., 8,593; Hass, Soc. Labor, 5,815; Hallinan, Prog., 5,589; Krajewski, Poor Man's, 4,203; Dobbs, Soc. Workers, 3,850; Hamblen, Proh., 989.
1956, Eisenhower, Rep., 1,606,942; Stevenson Dem., 850,337; Holtwick, Proh., 9,147; Hass, Soc. Labor, 6,736; Andrews, Cons., 5,317; Dobbs, Soc. Workers, 4,004; Krajewski, Amer. Third Party, 1,829.
1960, Kennedy, Dem., 1,385,415; Nixon, Rep., 1,363,324; Dobbs, Soc. Workers, 11,402; Lee, Cons., 8,708; Hass, Soc. Labor, 4,262.
1964, Johnson, Dem., 1,867,671; Goldwater, Rep., 963,843; DeBerry, Soc. Workers, 8,181; Hass, Soc. Labor, 7,075.
1968, Nixon, Rep., 1,325,467; Humphrey, Dem., 1,264,206; Wallace, 3d Party, 262,187; Halstead, Soc. Workers, 8,667; Gregory, Peace and Freedom, 8,084; Blomen, Soc. Labor, 6,784.

1972, Nixon, Rep., 1,845,502; McGovern, Dem., 1,102,211; Schmitz, Amer., 34,378; Spock, Peoples, 5,355; Fisher, Soc. Labor, 4,544; Jenness, Soc. Workers, 2,233; Mahalchik, Amer. First, 1,743; Hall, Com., 1,263.

1976, Carter, Dem., 1,444,653; Ford, Rep., 1,509,688; McCarthy, Ind., 32,717; MacBride, Libertarian, 9,449; Maddox, Amer., 7,716; Levin, Soc. Labor, 3,686; Hall, Com., 1,662; LaRouche, U.S. Labor, 1,650; Camejo, Soc. Workers, 1,184; Wright, People's, 1,044; Bubar, Proh., 554; Zeidler, Soc., 469.

1980, Reagan, Rep., 1,546,557; Carter, Dem., 1,147,364; Anderson, Ind., 234,632; Clark, Libertarian, 20,652; Commoner, Citizens, 8,203; McCormack, Right to Life, 3,927; Lynen, Middle Class, 3,694; Hall, Com., 2,555; Pulley, Soc. Workers, 2,198; McReynolds, Soc., 1,973; Gahres, Down With Lawyers, 1,718; Griswold, Workers World, 1,288; Wendelken, Ind., 923.

1984, Reagan, Rep., 1,933,630; Mondale, Dem., 1,261,323; Bergland, Libertarian, 6,416.

1988, Bush, Rep., 1,740,604; Dukakis, Dem., 1,317,541; Lewin, Peace and Freedom, 9,953; Paul, Lib., 8,421.

1992, Clinton, Dem., 1,436,206; Bush, Rep., 1,356,865; Perot, Ind., 521,829; Marrou, Libertarian, 6,822; Fulani, New Alliance, 3,513; Phillips, U.S. Taxpayers, 2,670; LaRouche, Ind., 2,095; Warren, Soc. Workers, 2,011; Daniels, Ind., 1,996; Gritz, Populist/America First, 1,867; Hagelin, Natural Law, 1,353.

1996, Clinton, Dem., 1,652,361; Dole, Rep., 1,103,099; Perot, Ref., 262,134; Nader, Green, 32,465; Browne, Libertarian, 14,763; Hagelin, Natural Law, 3,887; Phillips, Taxpayers, 3,440; Harris, Soc. Workers, 1,837; Moorehead, Workers World, 1,337; White, Soc. Equality, 537.

2000, Gore, Dem., 1,788,850; Bush, Rep., 1,284,173; Nader, Ind., 94,554; Buchanan, Ind., 6,989; Browne, Ind., 6,312; Hagelin, Ind., 2,215; McReynolds, Ind., 1,880; Phillips, Ind., 1,409; Harris, Ind., 844.

New Mexico

	2000		1996		
County	Gore (D)	Bush (R)	Clinton (D)	Dole (R)	Perot (RF)
Bernalillo	99,461	95,249	88,140	78,832	8,708
Catron	353	1,273	423	923	114
Chaves	6,340	11,378	7,014	9,991	1,271
Cibola	4,127	2,752	4,030	2,245	488
Colfax	2,653	2,600	2,659	1,975	411
Curry	3,471	8,301	4,116	7,378	842
De Baca	349	612	509	489	86
Dona Ana	23,912	21,263	22,766	17,541	2,269
Eddy	7,108	10,335	8,959	8,534	1,297
Grant	5,673	4,961	5,860	3,993	778
Guadalupe	1,076	548	1,208	436	79
Harding	214	366	264	321	28
Hidalgo	839	954	943	789	209
Lea	3,855	10,157	5,393	7,661	1,465
Lincoln	2,027	4,458	2,209	3,396	666
Los Alamos	4,149	5,623	3,983	4,999	560
Luna	2,975	3,395	3,001	2,616	598
McKinley	10,281	5,070	10,124	4,470	650
Mora	1,456	668	1,646	561	131
Otero	5,465	10,258	5,938	9,065	1,096
Quay	1,471	2,292	1,830	1,943	377
Rio Arriba	8,169	3,495	7,965	2,551	469
Roosevelt	1,762	3,762	2,097	3,245	467
Sandoval	14,899	15,423	13,081	11,015	1,482
San Juan	11,980	21,434	12,070	17,478	2,355
San Miguel	6,540	2,215	6,995	1,938	405
Santa Fe	32,017	13,974	26,349	10,857	1,846
Sierra	1,689	2,721	2,154	2,140	431
Socorro	3,294	3,173	3,374	2,315	455
Taos	7,039	2,744	6,635	2,126	545
Torrance	1,868	2,891	2,072	2,154	332
Union	452	1,269	519	995	125
Valencia	9,819	10,803	9,169	7,779	1,222
Totals	**286,783**	**286,417**	**273,495**	**232,751**	**32,257**

New Mexico Vote Since 1952

1952, Eisenhower, Rep., 132,170; Stevenson, Dem., 105,661; Hamblen, Proh., 297; Hallinan, Ind. Prog., 225; MacArthur, Christian National, 220; Hass, Soc. Labor, 35.

1956, Eisenhower, Rep., 146,788; Stevenson, Dem., 106,098; Holtwick, Proh., 607; Andrews, Ind., 364; Hass, Soc. Labor, 69.

1960, Kennedy, Dem., 156,027; Nixon, Rep., 153,733; Decker, Proh., 777; Hass, Soc. Labor, 570.

1964, Johnson, Dem., 194,017; Goldwater, Rep., 131,838; Hass, Soc. Labor, 1,217; Munn, Proh., 543.

1968, Nixon, Rep., 169,692; Humphrey, Dem., 130,081; Wallace, 3d Party, 25,737; Chavez, 1,519; Halstead, Soc. Workers, 252.

1972, Nixon, Rep., 235,606; McGovern, Dem., 141,084; Schmitz, Amer., 8,767; Jenness, Soc. Workers, 474.

1976, Carter, Dem., 201,148; Ford, Rep., 211,419; Camejo, Soc. Workers, 2,462; MacBride, Libertarian, 1,110; Zeidler, Soc., 240; Bubar, Proh., 211.

1980, Reagan, Rep., 250,779; Carter, Dem., 167,826; Anderson, Ind., 29,459; Clark, Libertarian, 4,365; Commoner, Citizens, 2,202; Bubar, Statesman, 1,281; Pulley, Soc. Workers, 325.

1984, Reagan, Rep., 307,101; Mondale, Dem., 201,769; Bergland, Libertarian, 4,459.

1988, Bush, Rep., 270,341; Dukakis, Dem., 244,497; Paul, Lib., 3,268; Fulani, New Alliance, 2,237.

1992, Clinton, Dem., 261,617; Bush, Rep., 212,824; Perot, Ind., 91,895; Marrou, Libertarian, 1,615.

1996, Clinton, Dem., 273,495; Dole, Rep., 232,751; Perot, Ref., 32,257; Nader, Green, 13,218; Browne, Libertarian, 2,996; Phillips, Taxpayers, 713; Hagelin, Natural Law, 644.

2000, Gore, Dem., 286,783; Bush, Rep., 286,417; Nader, Green, 21,251; Browne, Libertarian, 2,058; Buchanan, Reform, 1,392; Hagelin, Natural Law, 361; Phillips, Constitution, 343.

New York

	2000		1996		
County	Gore (D)	Bush (R)	Clinton (D)	Dole (R)	Perot (RF)
Albany	85,617	47,624	85,993	39,785	11,957
Allegany	6,336	11,436	6,621	8,107	2,730
Bronx	265,801	36,245	248,276	30,435	7,186
Broome	45,381	36,946	44,407	31,327	9,114
Cattaraugus	13,697	18,382	13,029	12,971	5,151
Cayuga	17,031	14,988	15,879	11,093	4,420
Chautauqua	27,016	29,064	26,831	21,261	7,484
Chemung	17,424	18,779	16,977	14,287	3,967
Chenango	9,112	10,033	8,797	7,319	2,822
Clinton	15,542	13,274	15,386	9,759	3,488
Columbia	13,489	13,153	12,910	10,324	3,466
Cortland	9,691	9,857	9,130	7,606	2,398
Delaware	8,450	10,662	8,724	7,684	2,601
Dutchess	52,390	52,669	47,339	41,929	12,294
Erie	240,174	160,176	224,554	132,343	45,679
Essex	7,927	8,822	7,893	6,379	2,363
Franklin	8,870	7,643	8,494	5,072	2,499
Fulton	9,314	11,434	9,779	7,881	3,214
Genesee	10,191	14,459	10,074	10,821	2,996
Greene	8,480	11,332	8,251	8,712	2,790
Hamilton	1,114	2,388	1,228	1,841	492
Herkimer	12,224	14,147	11,910	10,085	4,235
Jefferson	16,799	18,192	16,783	12,362	4,561
Kings	497,513	96,609	432,232	81,406	15,031
Lewis	4,333	6,103	4,402	3,965	1,669
Livingston	10,476	15,244	10,868	10,981	2,889
Madison	12,017	14,879	11,832	11,324	3,379
Monroe	161,743	141,266	164,858	115,694	23,936
Montgomery	10,249	9,765	10,485	7,172	3,253
Nassau	341,610	226,954	303,587	196,820	36,122
New York	454,523	82,113	394,131	67,839	11,144
Niagara	47,781	40,952	44,203	31,438	12,564
Oneida	43,933	47,603	44,399	37,996	11,296
Onondaga	109,896	83,678	100,190	73,771	17,602
Ontario	19,761	23,885	19,156	17,237	4,391
Orange	58,170	62,852	54,995	45,956	11,778
Orleans	5,991	9,202	6,233	6,865	1,986
Oswego	22,857	23,249	20,440	17,159	7,499
Otsego	11,460	12,219	11,470	8,774	3,217
Putnam	18,525	21,853	16,173	17,452	4,032
Queens	416,967	122,052	372,925	107,650	22,288
Rensselaer	34,808	29,562	34,273	23,482	8,405
Richmond	73,828	63,903	64,684	52,207	8,968
Rockland	69,530	48,441	63,127	40,395	6,798
St. Lawrence	21,386	16,449	21,798	10,827	5,309
Saratoga	43,359	46,623	39,832	34,337	10,141
Schenectady	35,534	27,961	35,404	22,106	7,865
Schoharie	5,390	7,459	5,902	5,353	1,796
Schuyler	3,301	4,381	3,303	3,134	1,037
Seneca	6,841	6,734	6,825	5,004	1,889
Steuben	14,600	24,200	14,481	17,710	5,496
Suffolk	306,306	240,992	261,828	182,510	52,209
Sullivan	14,348	12,703	15,052	9,321	3,453
Tioga	9,170	12,239	8,769	9,416	2,721
Tompkins	21,807	13,351	20,772	11,532	2,623
Ulster	38,162	33,447	35,852	26,212	9,246
Warren	12,193	14,993	11,603	11,152	3,623
Washington	9,641	12,596	9,572	8,954	3,648
Wayne	14,977	21,701	15,145	15,837	4,619
Westchester	218,010	139,278	196,310	123,719	18,028
Wyoming	5,935	10,809	5,735	7,477	2,411
Yates	3,962	5,565	4,066	3,925	1,190
Totals	**4,112,963**	**2,405,570**	**3,756,177**	**1,933,492**	**503,458**

New York Vote Since 1952

1952, Eisenhower, Rep., 3,952,815; Stevenson, Dem., 2,687,890; Liberal, 416,711; total, 3,104,601; Hallinan, Amer. Lab., 64,211; Hoopes, Soc., 2,664; Dobbs, Soc. Workers, 2,212; Hass, Ind. Gov't., 1,560; scattering, 178; blank and void, 87,813.

1956, Eisenhower, Rep., 4,340,340; Stevenson, Dem., 2,458,212; Liberal, 292,557; total, 2,750,769; write-in votes for Andrews, 1,027; Werdel, 492; Hass, 150; Hoopes, 82; others, 476.

1960, Kennedy, Dem., 3,423,909; Liberal, 406,176; total, 3,830,085; Nixon, Rep., 3,446,419; Dobbs, Soc. Workers, 14,319; scattering, 256; blank and void, 88,896.
1964, Johnson, Dem., 4,913,156; Goldwater, Rep., 2,243,559; Hass, Soc. Labor, 6,085; DeBerry, Soc. Workers, 3,215; scattering, 188; blank and void, 151,383.
1968, Nixon, Rep., 3,007,932; Humphrey, Dem., 3,378,470; Wallace, 3d Party, 358,864; Blomen, Soc. Labor, 8,432; Halstead, Soc. Workers, 11,851; Gregory, Freedom and Peace, 24,517; blank, void, and scattering, 171,624.
1972, Nixon, Rep., 3,824,642; Cons., 368,136; McGovern, Dem., 2,767,956; Liberal, 183,128; Reed, Soc. Workers, 7,797; Fisher, Soc. Labor, 4,530; Hall, Com., 5,641; blank, void, or scattered, 161,641.
1976, Carter, Dem., 3,389,558; Ford, Rep., 3,100,791; MacBride, Libertarian, 12,197; Hall, Com., 10,270; Camejo, Soc. Workers, 6,996; LaRouche, U.S. Labor, 5,413; blank, void, or scattered, 143,037.
1980, Reagan, Rep., 2,893,831; Carter, Dem., 2,728,372; Anderson, Ind., 467,801; Clark, Libertarian, 52,648; McCormack, Right To Life, 24,159; Commoner, Citizens, 23,186; Hall, Com., 7,414; DeBerry, Soc. Workers, 2,068; Griswold, Workers World, 1,416; scattering, 1,064.
1984, Reagan, Rep., 3,664,763; Mondale, Dem., 3,119,609; Bergland, Libertarian, 11,949.
1988, Bush, Rep., 3,081,871; Dukakis, Dem., 3,347,882; Marra, Right to Life, 20,497; Fulani, New Alliance, 15,845.
1992, Clinton, Dem., 3,444,450; Bush, Rep., 2,346,649; Perot, Ind., 1,090,721; Warren, Soc. Workers, 15,472; Marrou, Libertarian, 13,451; Fulani, New Alliance, 11,318; Hagelin, Natural Law, 4,420.
1996, Clinton, Dem., 3,756,177; Dole, Rep., 1,933,492; Perot, Ind. (Ref.), 503,458; Nader, Green, 75,956; Phillips, Right to Life, 23,580; Browne, Libertarian, 12,220; Hagelin, Natural Law, 5,011; Harris, Soc. Workers, 2,762; Moorehead, Workers World, 3,473.
2000, Gore, Dem., 4,112,965; Bush, Rep., 2,405,570; Nader, Green, 244,360; Buchanan, Reform, 31,554; Hagelin, Independence, 24,369; Browne, Libertarian, 7,664; Harris, Soc. Workers, 1,790; Phillips, Constitution, 1,503.

North Carolina

	2000		1996		
County	Gore (D)	Bush (R)	Clinton (D)	Dole (R)	Perot (RF)
Alamance	17,459	29,305	15,814	22,461	3,395
Alexander	4,166	9,242	3,955	6,748	1,004
Alleghany	1,715	2,531	1,801	1,936	458
Anson	4,792	3,161	4,890	2,193	512
Ashe	4,011	6,226	3,825	5,203	865
Avery	1,686	4,956	1,586	3,870	655
Beaufort	6,634	10,531	6,172	8,154	834
Bertie	4,660	2,488	4,202	1,745	316
Bladen	5,889	4,977	4,952	3,335	655
Brunswick	13,118	15,427	10,041	10,065	1,815
Buncombe	38,545	46,101	31,658	30,518	6,254
Burke	11,924	18,466	11,678	13,853	2,654
Cabarrus	16,284	32,704	14,447	23,035	3,626
Caldwell	8,588	17,337	8,050	12,653	2,099
Camden	1,187	1,628	1,186	1,074	293
Carteret	8,839	17,381	7,566	11,721	1,467
Caswell	4,091	4,270	4,312	3,310	510
Catawba	16,246	34,244	15,601	26,898	3,629
Chatham	10,461	10,248	9,353	7,731	1,113
Cherokee	3,239	6,305	3,129	3,883	785
Chowan	2,430	2,415	2,239	1,659	359
Clay	1,361	2,416	1,462	1,769	387
Cleveland	13,455	19,064	12,728	13,474	1,931
Columbus	9,986	8,342	9,019	6,017	1,170
Craven	12,213	19,494	10,317	13,264	1,528
Cumberland	38,626	38,129	32,739	29,804	3,776
Currituck	2,595	4,095	2,277	2,569	770
Dare	5,589	7,301	4,522	4,977	1,258
Davidson	16,199	35,387	13,593	24,797	3,698
Davie	3,651	10,184	3,525	8,141	915
Duplin	6,475	7,840	6,179	5,432	766
Durham	53,907	30,150	49,186	27,825	3,122
Edgecombe	11,315	6,836	10,568	6,010	660
Forsyth	52,457	67,700	46,543	59,160	5,747
Franklin	7,454	8,501	6,448	5,648	891
Gaston	19,281	39,453	19,458	33,149	3,921
Gates	1,944	1,480	2,155	1,072	307
Graham	1,006	2,304	1,210	1,801	270
Granville	7,733	7,364	6,747	5,498	432
Greene	2,478	3,353	2,224	2,689	280
Guilford	80,787	84,394	69,208	67,727	9,739
Halifax	10,222	6,698	9,551	5,700	816
Harnett	9,155	14,762	8,767	11,596	1,287
Haywood	9,793	12,118	9,350	7,995	2,594
Henderson	12,562	25,688	10,626	19,182	2,679
Hertford	5,484	2,382	4,856	1,823	356
Hoke	5,017	3,439	3,510	1,914	481
Hyde	1,088	1,132	1,109	782	143
Iredell	15,434	29,853	13,102	21,163	2,970
Jackson	5,722	6,237	5,211	4,244	970
Johnston	13,704	27,212	11,175	18,704	2,163
Jones	1,822	2,114	1,829	1,682	197
Lee	6,785	9,406	6,290	7,321	980
Lenoir	9,527	11,512	8,635	9,433	822
Lincoln	8,412	15,951	7,721	11,439	1,619
McDowell	4,747	9,109	4,553	6,407	1,275
Macon	4,683	8,406	4,209	5,267	1,121
Madison	3,505	4,676	3,333	3,110	538
Martin	4,929	4,420	4,500	3,590	445
Mecklenburg	126,911	134,068	103,429	97,719	10,473
Mitchell	1,535	4,984	1,496	3,874	549
Montgomery	3,979	4,946	3,856	3,379	587
Moore	11,232	19,882	9,847	14,760	1,761
Nash	12,376	17,995	11,142	15,309	1,751
New Hanover	29,292	36,503	22,839	27,889	3,615
Northampton	5,513	2,667	5,207	1,881	402
Onslow	10,269	19,657	8,685	13,396	1,857
Orange	30,921	17,930	28,674	15,053	1,534
Pamlico	2,188	2,999	2,204	2,270	297
Pasquotank	5,874	4,943	4,233	2,999	565
Pender	6,415	7,661	5,409	5,538	945
Perquimans	2,033	2,230	2,069	1,561	369
Person	5,042	6,722	4,540	4,883	591
Pitt	19,685	23,192	17,555	18,227	2,037
Polk	3,114	5,074	2,704	3,516	493
Randolph	11,366	30,959	10,783	23,030	3,593
Richmond	7,935	6,263	7,564	3,973	1,230
Robeson	17,834	11,721	17,361	8,146	2,105
Rockingham	13,260	18,979	12,096	14,255	2,528
Rowan	14,891	28,922	13,461	22,754	2,902
Rutherford	7,697	13,755	7,162	9,792	1,585
Sampson	8,768	10,410	8,150	8,241	825
Scotland	5,627	3,740	4,870	2,858	548
Stanly	7,066	15,548	7,131	11,446	1,690
Stokes	5,030	12,028	4,769	9,471	1,025
Surry	7,757	15,401	7,303	11,117	1,538
Swain	2,097	2,224	1,869	1,444	401
Transylvania	5,044	9,011	4,842	6,734	1,183
Tyrrell	849	706	908	488	112
Union	14,890	31,876	11,525	18,802	2,477
Vance	7,092	5,564	6,385	4,651	575
Wake	123,466	142,494	103,574	108,780	11,811
Warren	4,576	2,202	4,141	1,861	319
Washington	2,704	2,169	2,790	1,562	171
Watauga	7,959	10,438	7,349	8,146	1,415
Wayne	13,005	20,758	11,580	16,588	1,178
Wilkes	7,226	16,826	6,793	12,395	1,967
Wilson	11,266	13,466	9,779	10,518	1,100
Yadkin	3,127	10,435	2,927	8,439	913
Yancey	3,714	4,970	3,956	3,973	720
Totals	**1,257,692**	**1,631,163**	**1,107,849**	**1,225,938**	**168,059**

North Carolina Vote Since 1952

1952, Eisenhower, Rep., 558,107; Stevenson, Dem., 652,803.
1956, Eisenhower, Rep., 575,062; Stevenson, Dem., 590,530.
1960, Kennedy, Dem., 713,136; Nixon, Rep., 655,420.
1964, Johnson, Dem., 800,139; Goldwater, Rep., 624,844.
1968, Nixon, Rep., 627,192; Humphrey, Dem., 464,113; Wallace, 3d Party, 496,188.
1972, Nixon, Rep., 1,054,889; McGovern, Dem., 438,705; Schmitz, Amer., 25,018.
1976, Carter, Dem., 927,365; Ford, Rep., 741,960; Anderson, Amer., 5,607; MacBride, Libertarian, 2,219; LaRouche, U.S. Labor, 755.
1980, Reagan, Rep., 915,018; Carter, Dem., 875,635; Anderson, Ind., 52,800; Clark, Libertarian, 9,677; Commoner, Citizens, 2,287; DeBerry, Soc. Workers, 416.
1984, Reagan, Rep., 1,346,481; Mondale, Dem., 824,287; Bergland, Libertarian, 3,794.
1988, Bush, Rep., 1,237,258; Dukakis, Dem., 890,167; Fulani, New Alliance, 5,682; Paul, write-in, 1,263.
1992, Clinton, Dem., 1,114,042; Bush, Rep., 1,134,661; Perot, Ind., 357,864; Marrou, Libertarian, 5,171.
1996, Dole, Rep., 1,225,938; Clinton, Dem., 1,107,849; Perot, Ref., 168,059; Browne, Libertarian, 8,740; Hagelin, Natural Law, 2,771.
2000, Bush, Rep., 1,631,163; Gore, Dem., 1,257,692; Browne, Libertarian, 13,891; Buchanan, Reform, 8,874.

North Dakota

	2000		1996		
County	Gore (D)	Bush (R)	Clinton (D)	Dole (R)	Perot (RF)
Adams	286	826	366	575	200
Barnes	1,933	3,452	2,317	2,449	666
Benson	952	1,055	1,059	850	252
Billings	82	394	116	281	107

County	2000 Gore (D)	2000 Bush (R)	1996 Clinton (D)	1996 Dole (R)	1996 Perot (RF)
Bottineau	1,173	2,349	1,280	1,682	536
Bowman	330	1,080	489	710	261
Burke	296	698	416	483	176
Burleigh	9,842	22,467	10,679	15,464	3,535
Cass	21,451	33,536	21,693	24,238	4,116
Cavalier	618	1,513	941	1,188	326
Dickey	806	1,853	953	1,418	276
Divide	306	443	637	488	209
Dunn	474	1,124	587	830	304
Eddy	458	703	553	517	201
Emmons	405	1,430	544	1,148	441
Foster	474	1,172	664	801	265
Golden Valley	156	611	235	520	163
Grand Forks	10,593	15,875	11,376	11,606	2,663
Grant	235	1,077	300	760	295
Griggs	484	920	670	731	162
Hettinger	353	1,057	418	765	238
Kidder	283	837	434	691	242
La Moure	689	1,590	880	1,220	276
Logan	223	812	360	705	254
McHenry	888	1,682	1,096	1,187	453
McIntosh	350	1,178	470	1,005	295
McKenzie	653	1,634	928	1,338	428
McLean	1,465	2,891	1,759	1,988	618
Mercer	1,011	2,984	1,300	1,953	764
Morton	3,439	6,993	3,745	4,699	1,566
Mountrail	1,256	1,466	1,277	965	360
Nelson	687	1,031	827	745	206
Oliver	244	709	333	499	183
Pembina	1,093	2,430	1,191	1,678	400
Pierce	500	1,348	671	1,017	270
Ramsey	1,658	3,005	2,123	2,077	549
Ransom	1,080	1,488	1,199	920	303
Renville	443	820	562	576	210
Richland	2,490	4,999	2,890	3,345	782
Rolette	2,681	1,416	2,299	823	448
Sargent	959	1,103	1,003	814	241
Sheridan	161	707	252	566	121
Sioux	724	269	393	207	82
Slope	85	316	123	260	60
Stark	2,784	6,387	3,095	4,086	1,456
Steele	475	655	620	486	115
Stutsman	3,067	5,488	3,589	3,784	1,141
Towner	410	694	649	542	187
Traill	1,512	2,392	1,822	1,820	380
Walsh	1,743	3,099	2,082	2,222	599
Ward	7,533	13,997	8,660	10,546	2,587
Wells	661	1,610	962	1,192	373
Williams	2,330	5,187	3,018	3,590	1,174
Totals	**95,284**	**174,852**	**106,905**	**125,050**	**32,515**

North Dakota Vote Since 1952

1952, Eisenhower, Rep., 191,712; Stevenson, Dem., 76,694; MacArthur, Christian Nationalist, 1,075; Hallinan, Prog., 344; Hamblen, Proh., 302.

1956, Eisenhower, Rep., 156,766; Stevenson, Dem., 96,742; Andrews, Amer., 483.

1960, Kennedy, Dem., 123,963; Nixon, Rep., 154,310; Dobbs, Soc. Workers, 158.

1964, Johnson, Dem., 149,784; Goldwater, Rep., 108,207; DeBerry, Soc. Workers, 224; Munn, Proh., 174.

1968, Nixon, Rep., 138,669; Humphrey, Dem., 94,769; Wallace, 3d Party, 14,244; Halstead, Soc. Workers, 128; Munn, Prohibition, 38; Troxell, Ind., 34.

1972, Nixon, Rep., 174,109; McGovern, Dem., 100,384; Jenness, Soc. Workers, 288; Hall, Com., 87; Schmitz, Amer., 5,646.

1976, Carter, Dem., 136,078; Ford, Rep., 153,470; Anderson, Amer., 3,698; McCarthy, Ind., 2,952; Maddox, Amer. Ind., 269; MacBride, Libertarian, 256; scattering, 371.

1980, Reagan, Rep., 193,695; Carter, Dem., 79,189; Anderson, Ind., 23,640; Clark, Libertarian, 3,743; Commoner, Libertarian, 429; McLain, Natl. People's League, 296; Greaves, Amer., 235; Hall, Com., 93; DeBerry, Soc. Workers, 89; McReynolds, Soc., 82; Bubar, Statesman, 54.

1984, Reagan, Rep., 200,336; Mondale, Dem., 104,429; Bergland, Libertarian, 703.

1988, Bush, Rep., 166,559; Dukakis, Dem., 127,739; Paul, Lib., 1,315; LaRouche, Natl. Econ. Recovery, 905.

1992, Clinton, Dem., 99,168; Bush, Rep., 136,244; Perot, Ind., 71,084.

1996, Dole, Rep., 125,050; Clinton, Dem., 106,905; Perot, Ref., 32,515; Browne, Libertarian, 847; Phillips, Ind., 745; Hagelin, Natural Law, 349.

2000, Bush, Rep., 174,852; Gore, Dem., 95,284; Nader, Ind., 9,486; Buchanan, Reform, 7,288; Browne, Ind., 660; Phillips, Constitution, 373; Hagelin, Ind., 313.

Ohio

County	2000 Gore (D)	2000 Bush (R)	1996 Clinton (D)	1996 Dole (R)	1996 Perot (RF)
Adams	3,581	6,380	4,317	4,763	1,223
Allen	13,996	28,647	15,529	24,325	3,799
Ashland	6,685	13,533	6,573	10,402	2,630
Ashtabula	19,831	17,940	19,341	13,287	5,700
Athens	13,158	9,703	13,418	7,154	2,777
Auglaize	5,564	13,770	6,652	10,169	2,641
Belmont	15,980	12,625	17,705	8,213	4,452
Brown	5,972	10,027	6,318	6,970	1,941
Butler	46,390	86,587	43,690	67,023	10,540
Carroll	4,960	6,732	4,792	4,449	2,445
Champaign	5,955	9,220	5,990	6,568	2,219
Clark	27,984	27,660	27,890	22,297	7,083
Clermont	20,927	47,129	21,329	36,457	5,795
Clinton	4,791	9,824	5,303	7,504	1,588
Columbiana	20,657	21,804	20,716	15,386	7,127
Coshocton	5,594	8,243	6,005	6,018	2,183
Crawford	6,721	11,666	7,449	8,730	3,072
Cuyahoga	359,913	192,099	341,357	163,770	50,691
Darke	7,741	14,817	8,871	10,798	3,168
Defiance	6,175	9,540	6,343	7,469	1,929
Delaware	17,134	36,639	13,463	24,123	3,471
Erie	17,732	16,105	16,730	12,204	4,225
Fairfield	19,065	33,523	18,821	26,850	4,660
Fayette	3,363	5,685	3,665	4,831	1,047
Franklin	202,018	197,862	192,795	178,412	25,400
Fulton	6,805	11,546	6,662	8,703	2,412
Gallia	4,872	7,511	5,386	5,135	1,839
Geauga	15,327	25,417	14,143	19,662	4,848
Greene	25,059	37,946	25,082	30,677	5,246
Guernsey	6,643	8,181	6,731	5,970	2,251
Hamilton	161,578	204,175	160,458	186,493	21,335
Hancock	8,798	20,985	9,334	17,252	2,904
Hardin	4,557	7,124	4,930	5,506	1,365
Harrison	3,351	3,417	3,721	2,310	1,302
Henry	4,367	8,530	4,762	6,385	1,550
Highland	5,328	9,728	5,837	7,102	1,629
Hocking	4,474	5,702	4,646	4,017	1,564
Holmes	2,066	6,754	2,531	5,213	1,276
Huron	8,183	12,286	8,858	8,750	3,338
Jackson	5,131	6,958	5,538	4,922	1,529
Jefferson	17,488	15,038	19,402	10,212	4,748
Knox	7,133	13,393	7,562	10,159	2,138
Lake	46,497	51,747	43,186	40,974	12,507
Lawrence	11,307	12,531	11,595	8,832	3,232
Licking	23,196	37,180	22,624	28,276	6,516
Logan	5,945	11,849	6,397	8,325	2,264
Lorain	59,809	47,957	55,744	34,937	14,889
Lucas	108,344	73,342	104,911	58,120	17,282
Madison	5,287	8,892	5,072	6,871	1,386
Mahoning	69,212	40,460	72,716	31,397	13,213
Marion	10,370	13,617	10,482	11,112	2,897
Medina	26,635	37,349	23,727	26,120	8,700
Meigs	3,674	5,750	4,275	3,622	1,453
Mercer	5,212	12,485	6,300	8,832	2,361
Miami	15,584	26,037	15,540	19,509	4,599
Monroe	3,605	3,145	3,914	1,856	1,128
Montgomery	114,597	109,792	115,416	95,391	18,298
Morgan	2,261	3,451	2,385	2,566	922
Morrow	4,529	7,842	4,627	5,655	1,745
Muskingum	13,415	17,995	13,813	13,861	4,880
Noble	2,296	3,435	2,366	2,183	899
Ottawa	9,485	9,917	9,321	6,991	2,438
Paulding	3,384	5,210	3,449	3,760	1,292
Perry	5,895	6,440	5,819	4,606	1,854
Pickaway	6,598	10,717	7,042	8,666	1,702
Pike	4,923	5,333	5,542	3,759	1,402
Portage	31,446	28,271	29,441	18,939	9,178
Preble	6,375	11,176	6,611	8,139	2,235
Putnam	4,063	12,837	4,972	9,294	1,767
Richland	20,572	30,138	20,832	23,697	6,613
Ross	11,662	13,706	12,649	10,286	2,648
Sandusky	11,146	13,699	11,547	10,033	3,617
Scioto	13,997	15,022	15,041	11,679	4,418
Seneca	9,512	13,863	10,044	9,713	3,498
Shelby	6,593	12,476	6,729	8,773	2,686
Stark	75,308	78,153	73,437	60,212	23,004
Summit	119,759	96,721	112,050	73,555	27,723
Trumbull	57,643	34,654	55,604	24,811	13,563
Tuscarawas	15,879	19,549	15,244	13,388	5,682
Union	5,040	11,502	4,989	8,290	1,596
Van Wert	4,209	8,679	4,453	6,999	1,487
Vinton	2,037	2,720	2,350	1,673	728
Warren	19,142	48,318	17,089	33,210	4,689
Washington	10,383	15,342	10,945	11,965	2,832
Wayne	14,779	25,901	14,850	19,628	5,771
Williams	5,454	9,941	5,524	7,747	2,121
Wood	22,687	27,504	23,183	20,518	5,065
Wyandot	3,397	6,113	3,677	4,473	1,347
Totals	**2,186,190**	**2,351,209**	**2,148,222**	**1,859,883**	**483,207**

Ohio Vote Since 1952

1952, Eisenhower, Rep., 2,100,391; Stevenson, Dem., 1,600,367.
1956, Eisenhower, Rep., 2,262,610; Stevenson, Dem., 1,439,655.
1960, Kennedy, Dem., 1,944,248; Nixon, Rep., 2,217,611.
1964, Johnson, Dem., 2,498,331; Goldwater, Rep., 1,470,865.
1968, Nixon, Rep., 1,791,014; Humphrey, Dem., 1,700,586; Wallace, 3d Party, 467,495; Gregory, 372; Munn, Proh., 19; Blomen, Soc. Labor, 120; Halstead, Soc. Workers, 69; Mitchell, Com., 23.
1972, Nixon, Rep., 2,441,827; McGovern, Dem., 1,558,889; Fisher, Soc. Labor, 7,107; Hall, Com., 6,437; Schmitz, Amer., 80,067; Wallace, Ind., 460.
1976, Carter, Dem., 2,011,621; Ford, Rep., 2,000,505; McCarthy, Ind., 58,258; Maddox, Amer. Ind., 15,529; MacBride, Libertarian, 8,961; Hall, Com., 7,817; Camejo, Soc. Workers, 4,717; LaRouche, U.S. Labor, 4,335; scattered, 130.
1980, Reagan, Rep., 2,206,545; Carter, Dem., 1,752,414; Anderson, Ind., 254,472; Clark, Libertarian, 49,033; Commoner, Citizens, 8,564; Hall, Com., 4,729; Congress, Ind., 4,029; Griswold, Workers World, 3,790; Bubar, Statesman, 27.
1984, Reagan, Rep., 2,678,559; Mondale, Dem., 1,825,440; Bergland, Libertarian, 5,886.
1988, Bush, Rep., 2,416,549; Dukakis, Dem., 1,939,629; Fulani, Ind., 12,017; Paul, Ind., 11,926.
1992, Clinton, Dem., 1,984,942; Bush, Rep., 1,894,310; Perot, Ind., 1,036,426; Marrou, Libertarian, 7,252; Fulani, New Alliance, 6,413; Gritz, Populist/America First, 4,699; Hagelin, Natural Law, 3,437; LaRouche, Ind., 2,446.
1996, Clinton, Dem., 2,148,222; Dole, Rep., 1,859,883; Perot, Ref., 483,207; Browne, Ind., 12,851; Moorehead, Ind., 10,813; Hagelin, Natural Law, 9,120; Phillips, Ind., 7,361.
2000, Bush, Rep., 2,351,209; Gore, Dem., 2,186,190; Nader, Ind., 117,857; Buchanan, Ind., 26,724; Browne, Libertarian, 13,475; Hagelin, Natural Law, 6,169; Phillips, Ind., 3,823.

Oklahoma

County	2000 Gore (D)	2000 Bush (R)	1996 Clinton (D)	1996 Dole (R)	1996 Perot (RF)
Adair	2,361	3,503	2,792	2,956	751
Alfalfa	583	1,886	796	1,504	348
Atoka	1,906	2,375	2,281	1,542	532
Beaver	339	2,092	515	1,893	199
Beckham	2,408	4,067	2,797	2,912	817
Blaine	1,402	2,633	1,832	2,127	563
Bryan	5,554	6,084	5,962	3,943	1,396
Caddo	4,272	4,835	4,844	3,422	1,358
Canadian	8,367	22,679	8,977	18,139	3,297
Carter	6,659	9,667	6,979	6,769	1,997
Cherokee	7,256	6,918	6,817	5,046	1,777
Choctaw	2,799	2,461	3,198	1,580	589
Cimarron	227	1,230	361	986	102
Cleveland	27,792	47,393	26,038	36,457	6,785
Coal	1,148	1,196	1,205	734	323
Comanche	11,971	17,103	12,841	14,461	2,819
Cotton	1,068	1,388	1,258	1,042	381
Craig	2,568	2,815	2,649	2,058	758
Creek	9,753	13,580	9,674	9,861	2,837
Custer	3,115	6,527	4,027	4,723	1,101
Delaware	5,514	7,618	5,094	5,230	1,573
Dewey	599	1,607	816	1,179	292
Ellis	468	1,513	619	1,090	279
Garfield	6,543	14,902	7,504	11,712	2,523
Garvin	4,189	5,536	4,639	3,745	1,345
Grady	6,037	10,040	6,256	7,228	2,048
Grant	709	1,762	867	1,382	384
Greer	839	1,287	1,240	905	361
Harmon	507	692	729	448	143
Harper	374	1,296	511	1,036	219
Haskell	2,510	2,039	2,762	1,442	590
Hughes	2,334	2,196	2,748	1,510	730
Jackson	2,515	5,591	3,245	4,422	892
Jefferson	1,245	1,320	1,430	865	337
Johnston	1,809	2,072	1,998	1,229	532
Kay	6,122	11,768	6,882	9,741	2,785
Kingfisher	1,304	4,693	1,626	3,423	621
Kiowa	1,544	2,173	1,973	1,638	510
Latimer	1,865	1,739	2,222	1,189	578
Le Flore	6,536	8,215	6,831	5,689	1,721
Lincoln	4,140	7,387	4,332	5,243	1,500
Logan	4,510	8,187	4,854	5,949	1,410
Love	1,530	1,807	1,675	1,224	385
McClain	3,679	6,750	3,753	4,363	1,289
McCurtain	3,752	6,601	4,350	3,892	1,483
McIntosh	4,206	3,444	4,219	2,400	1,044
Major	635	2,672	900	2,188	410
Marshall	2,210	2,641	2,624	1,605	663
Mayes	6,618	7,132	6,377	5,268	1,617
Murray	2,263	2,609	2,620	1,712	723
Muskogee	12,520	11,820	12,963	8,974	3,163
Noble	1,416	3,230	1,756	2,318	694
Nowata	1,703	2,069	1,788	1,457	586
Okfuskee	1,814	1,910	2,074	1,380	536
Oklahoma	81,590	139,078	80,438	120,429	18,411
Okmulgee	7,186	5,797	7,555	4,246	1,487
Osage	7,540	8,138	7,342	5,827	1,938
Ottawa	5,647	5,625	5,844	4,127	1,496
Pawnee	2,435	3,386	2,663	2,560	756
Payne	9,319	15,256	9,985	11,686	2,472
Pittsburg	7,627	8,514	8,475	5,966	2,217
Pontotoc	5,387	7,299	6,470	5,366	1,712
Pottawatomie	8,763	13,235	9,141	9,802	2,724
Pushmataha	1,969	2,331	2,270	1,458	588
Roger Mills	441	1,234	733	959	233
Rogers	10,813	17,713	9,544	12,883	3,022
Seminole	3,783	4,011	4,225	2,935	1,041
Sequoyah	5,425	6,614	5,665	4,733	1,673
Stephens	6,467	10,860	7,248	8,144	2,312
Texas	1,084	4,964	1,408	4,139	518
Tillman	1,400	1,920	1,827	1,346	471
Tulsa	81,656	134,152	76,924	111,243	18,201
Wagoner	8,244	12,981	7,749	9,392	2,357
Washington	6,644	13,788	6,732	11,605	2,255
Washita	1,564	2,850	1,913	1,994	748
Woods	1,235	2,774	1,431	2,151	497
Woodward	1,950	5,067	2,403	4,093	963
Totals	**474,276**	**744,337**	**488,105**	**582,315**	**130,788**

Oklahoma Vote Since 1952

1952, Eisenhower, Rep., 518,045; Stevenson, Dem., 430,939.
1956, Eisenhower, Rep., 473,769; Stevenson, Dem., 385,581.
1960, Kennedy, Dem., 370,111; Nixon, Rep., 533,039.
1964, Johnson, Dem., 519,834; Goldwater, Rep., 412,665.
1968, Nixon, Rep., 449,697; Humphrey, Dem., 301,658; Wallace, 3d Party, 191,731.
1972, Nixon, Rep., 759,025; McGovern, Dem., 247,147; Schmitz, Amer., 23,728.
1976, Carter, Dem., 532,442; Ford, Rep., 545,708; McCarthy, Ind., 14,101.
1980, Reagan, Rep., 695,570; Carter, Dem., 402,026; Anderson, Ind., 38,284; Clark, Libertarian, 13,828.
1984, Reagan, Rep., 861,530; Mondale, Dem., 385,080; Bergland, Libertarian, 9,066.
1988, Bush, Rep., 678,367; Dukakis, Dem., 483,423; Paul, Lib., 6,261; Fulani, New Alliance, 2,985.
1992, Clinton, Dem., 473,066; Bush, Rep., 592,929; Perot, Ind., 319,878; Marrou, Libertarian, 4,486.
1996, Dole, Rep., 582,315; Clinton, Dem., 488,105; Perot, Ref., 130,788; Browne, Libertarian, 5,505.
2000, Bush, Rep., 744,337; Gore, Dem., 474,276; Buchanan, Reform, 9,014; Browne, Libertarian, 6,602.

Oregon

County	2000 Gore (D)	2000 Bush (R)	1996 Clinton (D)	1996 Dole (R)	1996 Perot (RF)
Baker	2,195	5,618	2,547	3,975	900
Benton	19,444	15,825	17,211	12,450	2,445
Clackamas	76,421	77,539	67,709	59,443	12,304
Clatsop	8,296	6,950	7,732	5,334	1,582
Columbia	10,331	9,369	9,275	6,205	2,330
Coos	11,610	15,626	12,171	10,886	3,460
Crook	2,474	5,363	2,607	3,250	948
Curry	4,090	6,551	4,202	4,790	1,560
Deschutes	22,061	32,132	17,151	21,135	5,306
Douglas	14,193	30,294	15,250	21,855	4,465
Gilliam	359	679	485	398	143
Grant	589	3,078	1,180	2,110	432
Harney	766	2,799	980	1,948	506
Hood River	4,072	3,721	3,654	2,794	721
Jackson	33,153	46,052	29,230	33,771	7,470
Jefferson	2,681	3,838	2,555	2,634	813
Josephine	11,864	22,186	11,113	16,048	3,546
Klamath	7,541	18,855	7,207	12,116	2,538
Lake	707	2,830	962	2,239	385
Lane	78,583	61,578	69,461	48,253	11,498
Lincoln	10,861	8,446	10,552	6,717	2,269
Linn	16,682	25,359	17,041	18,331	4,773
Malheur	2,336	7,624	2,827	6,045	844
Marion	49,430	57,443	48,637	46,415	8,802
Morrow	1,197	2,224	1,426	1,381	455
Multnomah	188,441	83,677	159,878	71,094	17,536
Polk	11,921	14,988	10,942	11,478	2,093
Sherman	326	679	444	476	126
Tillamook	5,762	5,775	5,775	3,884	1,263
Umatilla	7,809	14,140	8,774	9,703	2,500
Union	3,577	7,836	4,379	5,414	1,241
Wallowa	836	3,279	1,321	2,379	483
Wasco	4,616	5,356	4,967	3,662	1,004
Washington	90,662	86,091	76,619	65,221	11,446
Wheeler	202	584	299	418	121
Yamhill	14,254	19,193	13,078	13,900	2,913
Totals	**720,342**	**713,577**	**649,641**	**538,152**	**121,221**

Oregon Vote Since 1952

1952, Eisenhower, Rep., 420,815; Stevenson, Dem., 270,579; Hallinan, Ind., 3,665.
1956, Eisenhower, Rep., 406,393; Stevenson, Dem., 329,204.
1960, Kennedy, Dem., 367,402; Nixon, Rep., 408,060.
1964, Johnson, Dem., 501,017; Goldwater, Rep., 282,779; write-in, 2,509.
1968, Nixon, Rep., 408,433; Humphrey, Dem., 358,866; Wallace, 3d Party, 49,683; write-in, McCarthy, 1,496; N. Rockefeller, 69; others, 1,075.
1972, Nixon, Rep., 486,686; McGovern, Dem., 392,760; Schmitz, Amer., 46,211; write-in, 2,289.
1976, Carter, Dem., 490,407; Ford, Rep., 492,120; McCarthy, Ind., 40,207; write-in, 7,142.
1980, Reagan, Rep., 571,044; Carter, Dem., 456,890; Anderson, Ind., 112,389; Clark, Libertarian, 25,838; Commoner, Citizens, 13,642; scattered, 1,713.
1984, Reagan, Rep., 658,700; Mondale, Dem., 536,479.
1988, Bush, Rep., 560,126; Dukakis, Dem., 616,206; Paul, Lib., 14,811; Fulani, Ind., 6,487.
1992, Clinton, Dem., 621,314; Bush, Rep., 475,757; Perot, Ind., 354,091; Marrou, Libertarian, 4,277; Fulani, New Alliance, 3,030.
1996, Clinton, Dem., 649,641; Dole, Rep., 538,152; Perot, Ref., 121,221; Nader, Pacific, 49,415; Browne, Libertarian, 8,903; Phillips, Taxpayers, 3,379; Hagelin, Natural Law, 2,798; Hollis, Soc., 1,922.
2000, Gore, Dem., 720,342; Bush, Rep., 713,577; Nader, Green, 77,357; Browne, Libertarian, 7,447; Buchanan, Ind., 7,063; Hagelin, Reform, 2,574; Phillips, Constitution, 2,189.

Pennsylvania

	2000		1996		
County	Gore (D)	Bush (R)	Clinton (D)	Dole (R)	Perot (RF)
Adams	11,682	20,848	10,774	15,338	3,186
Allegheny	329,963	235,361	284,480	204,067	42,309
Armstrong	11,127	15,508	11,130	11,052	3,452
Beaver	38,925	32,491	39,578	26,048	8,276
Bedford	5,474	13,598	5,954	10,064	2,041
Berks	59,150	71,273	49,887	56,289	13,788
Blair	15,774	28,376	15,036	21,282	4,014
Bradford	7,911	14,660	7,736	10,393	2,712
Bucks	132,914	121,927	103,313	94,899	24,544
Butler	25,037	44,009	21,990	32,038	6,145
Cambria	30,308	28,001	30,391	20,341	7,837
Cameron	779	1,383	822	1,113	283
Carbon	10,668	9,717	9,457	7,193	2,992
Centre	21,409	26,172	21,145	20,935	4,173
Chester	82,047	100,080	64,783	77,029	14,067
Clarion	5,605	9,796	5,954	6,916	2,064
Clearfield	11,718	18,019	11,991	12,987	3,758
Clinton	5,521	6,064	5,658	4,293	1,424
Columbia	8,975	12,095	8,379	8,234	3,654
Crawford	13,250	18,858	12,943	14,659	3,519
Cumberland	31,053	54,802	28,749	43,943	5,669
Dauphin	44,390	53,631	40,936	44,417	6,967
Delaware	134,861	105,836	115,946	92,628	21,883
Elk	5,754	7,347	5,749	4,889	2,293
Erie	59,399	49,027	57,508	39,884	10,386
Fayette	28,152	20,013	26,359	14,019	5,722
Forest	843	1,371	964	902	325
Franklin	14,922	33,042	14,980	25,392	4,127
Fulton	1,425	3,753	1,620	2,665	554
Greene	7,230	5,890	7,620	4,002	2,052
Huntingdon	5,073	10,408	5,285	7,324	1,813
Indiana	13,667	16,799	13,868	12,874	3,674
Jefferson	5,566	11,473	5,846	8,156	2,322
Juniata	2,656	5,795	2,896	4,128	911
Lackawanna	57,471	35,096	46,377	26,930	8,189
Lancaster	54,968	115,900	49,120	92,875	11,601
Lawrence	20,593	18,060	18,993	13,088	4,002
Lebanon	16,093	28,534	14,187	21,885	4,235
Lehigh	56,667	55,492	48,568	45,103	10,947
Luzerne	62,199	52,328	60,174	43,577	12,424
Lycoming	14,663	27,137	13,516	21,535	3,855
McKean	5,510	9,661	5,509	6,838	2,350
Mercer	23,817	23,132	23,003	17,213	5,108
Mifflin	4,835	9,400	5,327	6,888	1,392
Monroe	21,939	23,265	16,547	17,326	4,650
Montgomery	177,990	145,623	143,664	121,047	24,392
Montour	2,356	3,960	2,183	2,785	784
Northampton	53,097	47,396	43,959	35,726	9,848
Northumberland	13,670	18,142	13,418	13,551	5,173
Perry	4,459	11,184	4,611	8,156	1,609
Philadelphia	449,182	100,959	412,988	85,345	29,329
Pike	7,330	9,339	5,509	6,697	1,873
Potter	2,037	4,858	2,146	3,714	925
Schuylkill	26,215	29,841	24,860	22,920	8,471
Snyder	3,536	8,963	3,405	6,742	1,451
Somerset	12,028	20,218	12,719	14,735	3,968
Sullivan	1,066	1,928	1,071	1,352	418
Susquehanna	6,481	10,226	5,912	7,354	2,266
Tioga	4,617	9,635	4,961	7,382	1,993
Union	4,209	8,523	3,658	6,570	1,431
Venango	8,196	11,642	8,205	8,398	2,777
Warren	7,537	9,290	7,291	7,056	2,504
Washington	44,961	37,339	40,952	27,777	8,661
Wayne	6,904	11,201	5,928	8,077	2,126
Westmoreland	71,792	80,858	63,686	62,058	16,230
Wyoming	4,363	6,922	4,049	4,888	1,414
York	51,958	87,652	49,596	65,188	11,652
Totals	**2,485,967**	**2,281,127**	**2,215,819**	**1,801,169**	**430,984**

Pennsylvania Vote Since 1952

1952, Eisenhower, Rep., 2,415,789; Stevenson, Dem., 2,146,269; Hamblen, Proh., 8,771; Hallinan, Prog., 4,200; Hoopes, Soc., 2,684; Dobbs, Militant Workers, 1,502; Hass, Ind. Gov., 1,347; scattered, 155.
1956, Eisenhower, Rep., 2,585,252; Stevenson, Dem., 1,981,769; Hass, Soc. Labor, 7,447; Dobbs, Militant Workers, 2,035.
1960, Kennedy, Dem., 2,556,282; Nixon, Rep., 2,439,956; Hass, Soc. Labor, 7,185; Dobbs, Soc. Workers, 2,678; scattering, 440.
1964, Johnson, Dem., 3,130,954; Goldwater, Rep., 1,673,657; DeBerry, Soc. Workers, 10,456; Hass, Soc. Labor, 5,092; scattering, 2,531.
1968, Nixon, Rep., 2,090,017; Humphrey, Dem., 2,259,405; Wallace, 3d Party, 378,582; Blomen, Soc. Labor, 4,977; Halstead, Soc. Workers, 4,862; Gregory, Peace and Freedom, 7,821; others, 2,264.
1972, Nixon, Rep., 2,714,521; McGovern, Dem., 1,796,951; Schmitz, Amer., 70,593; Jenness, Soc. Workers, 4,639; Hall, Com., 2,686; others, 2,715.
1976, Carter, Dem., 2,328,677; Ford, Rep., 2,205,604; McCarthy, Ind., 50,584; Maddox, Constitution, 25,344; Camejo, Soc. Workers, 3,009; LaRouche, U.S. Labor, 2,744; Hall, Com., 1,891; others, 2,934.
1980, Reagan, Rep., 2,261,872; Carter, Dem., 1,937,540; Anderson, Ind., 292,921; Clark, Libertarian, 33,263; DeBerry, Soc. Workers, 20,291; Commoner, Consumer, 10,430; Hall, Com., 5,184.
1984, Reagan, Rep., 2,584,323; Mondale, Dem., 2,228,131; Bergland, Libertarian, 6,982.
1988, Bush, Rep., 2,300,087; Dukakis, Dem., 2,194,944; McCarthy, Consumer, 19,158; Paul, Lib., 12,051.
1992, Clinton, Dem., 2,239,164; Bush, Rep., 1,791,841; Perot, Ind., 902,667; Marrou, Libertarian, 21,477; Fulani, New Alliance, 4,661.
1996, Clinton, Dem., 2,215,819; Dole, Rep., 1,801,169; Perot, Ref., 430,984; Browne, Libertarian, 28,000; Phillips, Constitutional, 19,552; Hagelin, Natural Law, 5,783.
2000, Gore, Dem., 2,485,967; Bush, Rep., 2,281,127; Nader, Green, 103,392; Buchanan, Reform, 16,023; Phillips, Constitution, 14,428; Browne, Libertarian, 11,248.

Rhode Island

	2000		1996		
City	Gore (D)	Bush (R)	Clinton (D)	Dole (R)	Perot (RF)
Cranston	21,204	10,420	20,901	9,098	3,457
East Providence	13,033	5,072	12,846	4,199	1,971
Pawtucket	15,429	4,598	14,719	3,877	2,508
Providence	31,979	7,669	29,450	7,068	2,733
Warwick	23,948	12,741	23,152	10,414	4,541
Other	143,915	90,055	131,982	70,027	28,513
Totals	**249,508**	**130,555**	**233,050**	**104,683**	**43,723**

Rhode Island Vote Since 1952

1952, Eisenhower, Rep., 210,935; Stevenson, Dem., 203,293; Hallinan, Prog., 187; Hass, Soc. Labor, 83.
1956, Eisenhower, Rep., 225,819; Stevenson, Dem., 161,790.
1960, Kennedy, Dem., 258,032; Nixon, Rep., 147,502.
1964, Johnson, Dem., 315,463; Goldwater, Rep., 74,615.
1968, Nixon, Rep., 122,359; Humphrey, Dem., 246,518; Wallace, 3d Party, 15,678; Halstead, Soc. Workers, 383.
1972, Nixon, Rep., 220,383; McGovern, Dem., 194,645; Jenness, Soc. Workers, 729.
1976, Carter, Dem., 227,636; Ford, Rep., 181,249; MacBride, Libertarian, 715; Camejo, Soc. Workers, 462; Hall, Com., 334; Levin, Soc. Labor, 188.
1980, Reagan, Rep., 154,793; Carter, Dem., 198,342; Anderson, Ind., 59,819; Clark, Libertarian, 2,458; Hall, Com., 218; McReynolds, Soc., 170; DeBerry, Soc. Workers, 90; Griswold, Workers World, 77.
1984, Reagan, Rep., 212,080; Mondale, Dem., 197,106; Bergland, Libertarian, 277.
1988, Bush, Rep., 177,761; Dukakis, Dem., 225,123; Paul, Lib., 825; Fulani, New Alliance, 280.
1992, Clinton, Dem., 213,299; Bush, Rep., 131,601; Perot, Ind., 105,045; Fulani, New Alliance, 1,878.

1996, Clinton, Dem., 233,050; Dole, Rep., 104,683; Perot, Ref., 43,723; Nader, Green, 6,040; Browne, Libertarian, 1,109; Phillips, Taxpayers, 1,021; Hagelin, Natural Law, 435; Moorehead, Workers World, 186.
2000, Gore, Dem., 249,508; Bush, Rep., 130,555; Nader, Ind., 25,052; Buchanan, Reform, 2,273; Browne, Ind., 742; Hagelin, Ind., 271; Moorehead, Ind., 199; Phillips, Ind., 97; McReynolds, Ind., 52; Harris, Ind., 34.

South Carolina

County	2000 Gore (D)	2000 Bush (R)	1996 Clinton (D)	1996 Dole (R)	1996 Perot (RF)
Abbeville	3,766	4,450	3,493	3,054	537
Aiken	16,409	33,203	14,314	26,539	1,984
Allendale	2,338	967	2,222	941	87
Anderson	19,606	35,827	17,460	24,137	3,896
Bamberg	3,451	2,047	3,380	1,715	192
Barnwell	3,661	4,521	3,620	3,808	310
Beaufort	17,487	25,561	15,764	17,575	1,838
Berkeley	17,707	24,796	13,358	17,691	1,922
Calhoun	3,063	3,216	2,716	2,520	316
Charleston	49,520	58,229	43,571	48,675	3,514
Cherokee	6,138	9,900	5,821	6,689	1,064
Chester	5,242	4,986	5,108	3,157	758
Chesterfield	6,111	6,266	5,734	4,028	768
Clarendon	5,999	5,186	5,930	3,841	395
Colleton	6,449	6,767	5,329	4,462	550
Darlington	10,253	11,290	8,943	8,220	898
Dillon	4,930	3,975	3,992	2,774	275
Dorchester	12,168	20,734	9,931	15,283	1,591
Edgefield	3,950	4,760	3,576	3,640	244
Fairfield	5,263	3,011	4,719	2,414	284
Florence	17,157	23,678	15,804	18,490	1,563
Georgetown	9,445	10,535	8,298	7,023	950
Greenville	43,810	92,714	41,605	71,210	6,761
Greenwood	8,139	12,193	8,193	8,865	985
Hampton	4,896	2,798	4,828	2,111	344
Horry	29,113	40,300	23,722	26,159	4,446
Jasper	3,646	2,414	4,053	2,024	348
Kershaw	7,428	11,911	6,764	8,513	996
Lancaster	8,782	11,676	8,752	7,544	1,598
Laurens	7,920	12,102	7,055	8,057	1,341
Lee	3,899	2,675	3,588	1,973	320
Lexington	22,830	58,095	18,907	39,658	3,703
McCormick	1,896	1,704	1,858	1,104	148
Marion	7,358	4,687	6,359	3,595	356
Marlboro	5,060	2,699	5,348	2,148	494
Newberry	4,428	7,492	4,804	5,670	682
Oconee	7,571	15,364	7,398	10,503	1,961
Orangeburg	19,802	12,657	18,610	10,494	1,112
Pickens	8,927	24,681	8,369	17,151	2,211
Richland	63,179	50,164	52,222	39,092	3,158
Saluda	2,682	4,098	2,486	2,825	371
Spartanburg	29,559	52,114	26,814	35,972	3,885
Sumter	14,365	15,915	12,198	12,080	933
Union	4,662	6,234	5,407	3,855	749
Williamsburg	6,723	4,524	6,987	3,957	375
York	19,251	33,776	16,873	22,222	3,173
Totals	**566,039**	**786,892**	**506,283**	**573,458**	**64,386**

South Carolina Vote Since 1952

1952, Eisenhower ran on two tickets. Under state law votes cast for two Eisenhower slates of electors could not be combined. Eisenhower, Ind., 158,289; Rep., 9,793; total, 168,082; Stevenson, Dem., 173,004; Hamblen, Proh., 1.
1956, Eisenhower, Rep., 75,700; Stevenson, Dem., 136,372; Byrd, Ind., 88,509; Andrews, Ind., 2.
1960, Kennedy, Dem., 198,129; Nixon, Rep., 188,558; write-in, 1.
1964, Johnson, Dem., 215,700; Goldwater, Rep., 309,048; write-ins: Nixon, 1, Wallace, 5; Powell, 1; Thurmond, 1.
1968, Nixon, Rep., 254,062; Humphrey, Dem., 197,486; Wallace, 3d Party, 215,430.
1972, Nixon, Rep., 477,044; McGovern, Dem., 184,559; United Citizens, 2,265; Schmitz, Amer., 10,075; write-in, 17.
1976, Carter, Dem., 450,807; Ford, Rep., 346,149; Anderson, Amer., 2,996; Maddox, Amer. Ind., 1,950; write-in, 681.
1980, Reagan, Rep., 439,277; Carter, Dem., 428,220; Anderson, Ind., 13,868; Clark, Libertarian, 4,807; Rarick, Amer. Ind., 2,086.
1984, Reagan, Rep., 615,539; Mondale, Dem., 344,459; Bergland, Libertarian, 4,359.
1988, Bush, Rep., 606,443; Dukakis, Dem., 370,554; Paul, Lib., 4,935; Fulani, United Citizens, 4,077.
1992, Clinton, Dem., 479,514; Bush, Rep., 577,507; Perot, Ind., 138,872; Marrou, Libertarian, 2,719; Phillips, U.S. Taxpayers, 2,680; Fulani, New Alliance, 1,235.
1996, Dole, Rep., 573,458; Clinton, Dem., 506,283; Perot, Ref./Patriot, 64,386; Browne, Libertarian, 4,271; Phillips, Taxpayers, 2,043; Hagelin, Natural Law, 1,248.
2000, Bush, Rep., 786,892; Gore, Dem., 566,039; Nader, United Citizens, 20,279; Browne, Libertarian, 4,898; Buchanan, Reform, 3,309; Phillips, Constitution, 1,682; Hagelin, Natural Law, 943.

South Dakota

County	2000 Gore (D)	2000 Bush (R)	1996 Clinton (D)	1996 Dole (R)	1996 Perot (RF)
Aurora	513	847	664	709	199
Beadle	3,216	4,347	3,984	3,670	842
Bennett	377	712	507	539	93
Bon Homme	1,162	1,901	1,569	1,428	391
Brookings	4,546	6,212	5,105	5,112	979
Brown	7,173	9,060	7,913	6,801	1,622
Brule	818	1,268	1,091	981	281
Buffalo	256	140	465	134	35
Butte	840	2,760	1,132	1,947	541
Campbell	147	739	202	623	140
Charles Mix	1,300	2,205	1,913	1,711	390
Clark	791	1,272	956	998	272
Clay	2,638	2,363	2,980	2,008	505
Codington	4,192	6,718	4,722	4,995	1,239
Corson	549	629	539	533	216
Custer	955	2,495	1,122	1,740	418
Davison	2,936	4,445	3,364	3,371	737
Day	1,492	1,623	1,840	1,282	395
Deuel	926	1,245	1,090	955	275
Dewey	880	761	1,114	657	195
Douglas	363	1,311	524	1,210	161
Edmunds	676	1,257	973	1,055	263
Fall River	1,133	2,185	1,357	1,636	417
Faulk	388	904	493	726	165
Grant	1,475	2,235	1,805	1,782	471
Gregory	718	1,487	923	1,208	286
Haakon	164	938	284	887	110
Hamlin	923	1,731	1,101	1,352	285
Hand	565	1,419	803	1,187	250
Hanson	457	944	541	801	170
Harding	64	650	151	537	90
Hughes	2,212	5,188	2,788	4,469	531
Hutchinson	1,052	2,497	1,285	2,177	409
Hyde	218	592	309	493	95
Jackson/ Washabaugh	319	687	423	646	88
Jerauld	468	624	656	530	151
Jones	137	509	184	463	75
Kingsbury	1,049	1,612	1,357	1,297	320
Lake	2,331	2,724	2,526	1,966	593
Lawrence	2,797	6,327	3,568	4,430	1,308
Lincoln	3,844	6,546	3,643	4,201	682
Lyman	482	875	646	726	130
McCook	965	1,610	1,166	1,292	245
McPherson	295	1,073	463	1,080	182
Marshall	939	1,097	1,185	861	189
Meade	2,267	6,870	2,960	4,984	1,133
Mellette	222	495	302	417	67
Miner	523	724	739	571	170
Minnehaha	27,042	33,428	29,790	27,432	4,425
Moody	1,318	1,361	1,443	1,024	284
Pennington	11,123	24,696	12,784	19,293	3,149
Perkins	297	1,237	460	983	225
Potter	356	1,112	534	979	181
Roberts	1,700	2,237	2,186	1,646	474
Sanborn	468	767	647	630	151
Shannon	1,667	252	1,926	253	87
Spink	1,274	1,957	1,636	1,651	360
Stanley	402	955	454	795	121
Sully	209	633	321	592	106
Todd	993	478	1,380	482	108
Tripp	799	1,909	1,088	1,680	337
Turner	1,414	2,514	1,682	1,970	385
Union	2,358	3,265	2,378	2,234	555
Walworth	721	1,758	939	1,461	366
Yankton	3,596	4,904	3,775	3,885	1,073
Ziebach	314	384	483	375	62
Totals	**118,804**	**190,700**	**139,333**	**150,543**	**31,250**

South Dakota Vote Since 1952

1952, Eisenhower, Rep., 203,857; Stevenson, Dem., 90,426.
1956, Eisenhower, Rep., 171,569; Stevenson, Dem., 122,288.
1960, Kennedy, Dem., 128,070; Nixon, Rep., 178,417.
1964, Johnson, Dem., 163,010; Goldwater, Rep., 130,108.
1968, Nixon, Rep., 149,841; Humphrey, Dem., 118,023; Wallace, 3d Party, 13,400.
1972, Nixon, Rep., 166,476; McGovern, Dem., 139,945; Jenness, Soc. Workers, 994.
1976, Carter, Dem., 147,068; Ford, Rep., 151,505; MacBride, Libertarian, 1,619; Hall, Com., 318; Camejo, Soc. Workers, 168.
1980, Reagan, Rep., 198,343; Carter, Dem., 103,855; Anderson, Ind., 21,431; Clark, Libertarian, 3,824; Pulley, Soc. Workers, 250.
1984, Reagan, Rep., 200,267; Mondale, Dem., 116,113.
1988, Bush, Rep., 165,415; Dukakis, Dem., 145,560; Paul, Lib., 1,060; Fulani, New Alliance, 730.
1992, Clinton, Dem., 124,888; Bush, Rep., 136,718; Perot, Ind., 73,295.

1996, Dole, Rep., 150,543; Clinton, Dem., 139,333; Perot, Ref., 31,250; Browne, Libertarian, 1,472; Phillips, Taxpayers, 912; Hagelin, Natural Law, 316.
2000, Bush, Rep., 190,700; Gore, Dem., 118,804; Buchanan, Reform, 3,322; Phillips, Ind., 1,781; Browne, Libertarian, 1,662.

Tennessee

	2000		1996		
County	Gore (D)	Bush (R)	Clinton (D)	Dole (R)	Perot (RF)
Anderson	13,556	14,688	13,457	11,943	1,817
Bedford	6,136	5,911	5,735	4,634	823
Benton	3,700	2,484	4,341	2,395	663
Bledsoe	1,756	2,380	1,621	1,626	251
Blount	14,688	25,273	14,687	19,310	2,556
Bradley	8,768	20,167	9,095	15,478	1,856
Campbell	6,492	5,784	6,122	4,393	785
Cannon	2,697	1,924	2,318	1,468	361
Carroll	5,239	5,465	4,912	4,206	697
Carter	6,724	12,111	6,218	10,540	1,383
Cheatham	6,062	6,356	4,883	4,283	705
Chester	2,192	3,487	1,922	2,746	203
Claiborne	3,841	5,023	3,861	4,023	727
Clay	1,931	1,468	1,559	1,108	316
Cocke	3,872	6,185	3,326	4,481	798
Coffee	8,741	8,788	7,951	7,038	1,205
Crockett	2,705	2,676	2,256	1,872	201
Cumberland	7,644	10,994	6,676	8,096	1,399
Davidson	120,508	84,117	110,805	78,453	9,018
Decatur	2,278	2,046	2,262	1,712	229
De Kalb	3,765	2,411	3,213	1,696	342
Dickson	8,332	7,016	7,458	5,283	996
Dyer	5,425	6,282	5,602	5,059	676
Fayette	5,037	6,402	4,655	4,406	416
Fentress	2,529	3,417	2,332	2,307	386
Franklin	7,828	6,560	6,929	5,296	1,057
Gibson	8,663	8,286	8,851	6,614	891
Giles	5,527	4,377	4,948	3,269	733
Grainger	2,361	3,746	2,162	2,875	382
Greene	7,909	12,540	6,885	9,779	1,604
Grundy	2,970	1,553	2,596	1,094	326
Hamblen	7,564	11,824	7,006	9,797	1,106
Hamilton	51,708	66,605	48,008	55,205	6,699
Hancock	690	1,343	760	1,259	116
Hardeman	4,953	3,729	4,859	2,961	346
Hardin	3,735	4,951	3,508	3,980	594
Hawkins	6,753	10,071	6,367	8,164	1,282
Haywood	3,887	2,554	3,565	2,293	154
Henderson	3,166	5,153	2,841	4,002	408
Henry	6,093	5,944	6,153	4,272	992
Hickman	4,239	2,914	3,917	2,002	460
Houston	2,081	993	1,868	742	182
Humphreys	4,205	2,387	3,675	1,892	423
Jackson	3,304	1,384	2,889	944	289
Jefferson	5,226	8,657	4,688	6,446	882
Johnson	1,813	3,740	1,698	3,137	489
Knox	60,969	86,851	61,158	70,761	6,402
Lake	1,419	781	1,273	589	110
Lauderdale	4,224	3,329	4,349	2,481	308
Lawrence	6,643	7,613	6,188	6,115	973
Lewis	2,281	2,037	1,971	1,298	316
Lincoln	5,060	5,435	4,361	4,551	761
Loudon	5,905	10,266	5,552	7,097	889
McMinn	6,142	10,155	5,987	7,655	1,033
McNairy	4,003	4,897	4,050	3,960	519
Macon	3,059	3,366	2,240	2,481	421
Madison	15,781	17,862	13,577	14,908	968
Marion	5,441	4,651	5,194	3,166	768
Marshall	5,107	4,105	4,447	2,781	603
Maury	11,127	11,930	10,367	8,737	1,366
Meigs	1,555	1,797	1,476	1,228	245
Monroe	5,327	7,514	4,872	5,257	713
Montgomery	18,818	19,644	16,498	15,133	1,781
Moore	1,107	1,145	935	846	177
Morgan	2,921	3,144	2,767	2,070	446
Obion	6,056	6,168	6,226	4,310	932
Overton	4,507	2,875	3,800	1,756	431
Perry	1,650	1,165	1,444	747	178
Pickett	939	1,281	901	1,046	116
Polk	2,574	2,907	2,450	1,910	377
Putnam	10,785	11,248	10,047	9,093	1,487
Rhea	3,722	5,900	3,969	4,476	694
Roane	9,575	11,345	9,744	9,044	1,438
Robertson	10,249	9,675	8,465	6,685	993
Rutherford	27,360	33,445	22,815	24,565	3,787
Scott	2,967	3,579	2,506	2,646	431
Sequatchie	1,648	2,169	1,598	1,391	288
Sevier	8,208	16,734	7,136	11,847	1,650
Shelby	190,404	141,756	179,663	136,315	8,307
Smith	4,884	2,384	3,812	1,857	346
Stewart	2,870	1,826	2,962	1,306	386
Sullivan	21,354	33,482	20,571	29,296	3,555
Sumner	22,118	27,601	19,205	20,863	2,783
Tipton	6,300	10,070	6,596	7,585	799
Trousdale	1,966	950	1,615	683	190
Unicoi	2,566	3,780	2,131	3,122	447
Union	2,564	3,199	2,421	2,253	385
Van Buren	1,255	845	1,010	504	128
Warren	7,378	5,552	6,389	4,226	917
Washington	14,769	22,579	13,259	18,960	2,237
Wayne	1,859	3,370	1,574	2,715	323
Weakley	5,570	6,106	5,657	4,622	873
White	4,135	3,525	3,592	2,498	505
Williamson	18,745	38,901	15,231	27,699	2,071
Wilson	16,561	18,844	13,655	13,817	1,841
Totals	**981,720**	**1,061,949**	**909,146**	**863,530**	**105,918**

Tennessee Vote Since 1952

1952, Eisenhower, Rep., 446,147; Stevenson, Dem., 443,710; Hamblen, Proh., 1,432; Hallinan, Prog., 885; MacArthur, Christian Nationalist, 379.
1956, Eisenhower, Rep., 462,288; Stevenson, Dem., 456,507; Andrews, Ind., 19,820; Holtwick, Proh., 789.
1960, Kennedy, Dem., 481,453; Nixon, Rep., 556,577; Faubus, States' Rights, 11,304; Decker, Proh., 2,458.
1964, Johnson, Dem., 635,047; Goldwater, Rep., 508,965; write-in, 34.
1968, Nixon, Rep., 472,592; Humphrey, Dem., 351,233; Wallace, 3d Party, 424,792.
1972, Nixon, Rep., 813,147; McGovern, Dem., 357,293; Schmitz, Amer., 30,373; write-in, 369.
1976, Carter, Dem., 825,879; Ford, Rep., 633,969; Anderson, Amer., 5,769; McCarthy, Ind., 5,004; Maddox, Amer. Ind., 2,303; MacBride, Libertarian, 1,375; Hall, Com., 547; LaRouche, U.S. Labor, 512; Bubar, Proh., 442; Miller, Ind., 316; write-in, 230.
1980, Reagan, Rep., 787,761; Carter, Dem., 783,051; Anderson, Ind., 35,991; Clark, Libertarian, 7,116; Commoner, Citizens, 1,112; Bubar, Statesman, 521; McReynolds, Soc., 519; Hall, Com., 503; DeBerry, Soc. Workers, 490; Griswold, Workers World, 400; write-ins, 152.
1984, Reagan, Rep., 990,212; Mondale, Dem., 711,714; Bergland, Libertarian, 3,072.
1988, Bush, Rep., 947,233; Dukakis, Dem., 679,794; Paul, Ind., 2,041; Duke, Ind., 1,807.
1992, Clinton, Dem., 933,521; Bush, Rep., 841,300; Perot, Ind., 199,968; Marrou, Libertarian, 1,847.
1996, Clinton, Dem., 909,146; Dole, Rep., 863,530; Perot, Ind. (Ref.), 105,918; Nader, Ind., 6,427; Browne, Ind., 5,020; Phillips, Ind., 1,818; Collins, Ind., 688; Hagelin, Ind., 636; Michael, Ind., 408; Dodge, Ind., 324.
2000, Bush, Rep., 1,061,949; Gore, Dem., 981,720; Nader, Green, 19,781; Browne, Libertarian, 4,284; Buchanan, Reform, 4,250; Brown, Ind., 1,606; Phillips, Ind., 1,015; Hagelin, Reform, 613; Venson, Ind., 535.

Texas

	2000		1996		
County	Gore (D)	Bush (R)	Clinton (D)	Dole (R)	Perot (RF)
Anderson	5,041	9,835	5,693	6,458	1,170
Andrews	876	3,091	1,181	2,360	431
Angelina	9,957	16,648	11,346	11,789	2,160
Aransas	2,637	5,390	2,964	3,769	655
Archer	993	2,951	1,235	1,974	437
Armstrong	150	772	272	582	75
Atascosa	4,322	6,231	4,259	4,102	813
Austin	2,407	6,661	2,719	4,669	577
Bailey	488	1,589	706	1,246	109
Bandera	1,426	5,613	1,383	3,700	520
Bastrop	6,973	10,310	6,773	6,323	1,342
Baylor	663	1,285	955	860	262
Bee	3,795	4,429	4,561	3,611	539
Bell	21,011	41,208	22,638	30,348	3,666
Bexar	185,158	215,613	180,308	161,619	17,822
Blanco	811	2,777	1,028	1,919	330
Borden	62	283	93	194	45
Bosque	1,930	4,745	2,427	2,840	739
Bowie	11,662	18,325	13,657	12,750	2,760
Brazoria	24,883	53,445	22,959	36,392	5,869
Brazos	12,359	32,864	13,968	22,082	2,215
Brewster	1,349	1,867	1,643	1,438	299
Briscoe	224	544	408	416	65
Brooks	1,854	556	2,945	413	108
Brown	3,138	9,609	4,138	6,524	1,081
Burleson	2,235	3,542	2,419	2,174	347
Burnet	3,557	9,286	4,123	5,744	1,108
Caldwell	3,872	5,216	3,961	3,239	545
Calhoun	2,766	3,724	2,753	2,832	507
Callahan	1,174	3,656	1,666	2,480	534
Cameron	33,214	27,800	34,891	18,434	2,760
Camp	1,625	2,121	1,912	1,488	252
Carson	480	2,216	742	1,742	227
Cass	4,618	6,295	5,691	4,066	1,038
Castro	727	1,607	1,107	1,231	144

County	2000 Gore (D)	2000 Bush (R)	1996 Clinton (D)	1996 Dole (R)	1996 Perot (RF)
Chambers	2,888	6,769	2,876	4,101	818
Cherokee	4,755	9,599	5,185	6,483	971
Childress	602	1,506	719	1,072	165
Clay	1,460	3,112	1,690	1,997	465
Cochran	344	807	541	667	127
Coke	355	1,137	595	790	157
Coleman	853	2,687	1,488	1,793	349
Collin	42,884	128,179	37,854	83,750	10,443
Collingsworth	429	974	581	729	118
Colorado	2,229	4,913	2,795	3,381	574
Comal	7,131	24,599	7,132	16,763	1,903
Comanche	1,636	3,334	2,138	2,123	511
Concho	268	818	434	488	107
Cooke	3,153	10,128	3,782	7,320	1,150
Coryell	4,493	10,321	5,300	7,143	1,443
Cottle	241	502	404	331	77
Crane	387	1,246	616	984	201
Crockett	467	924	684	714	147
Crosby	705	1,270	1,122	968	189
Culberson	577	413	804	329	99
Dallam	341	1,385	483	970	170
Dallas	275,308	322,345	255,766	260,058	36,759
Dawson	1,463	3,337	1,612	2,319	232
Deaf Smith	1,240	3,687	1,655	3,051	310
Delta	726	1,143	849	744	146
Denton	40,144	102,171	36,138	65,313	9,294
DeWitt	1,570	4,541	2,074	3,577	483
Dickens	284	589	509	421	117
Dimmit	2,678	1,032	2,242	604	128
Donley	360	1,333	495	988	97
Duval	3,990	1,010	3,958	543	136
Eastland	1,774	4,531	2,594	3,272	705
Ector	9,425	22,893	12,017	17,746	2,511
Edwards	261	663	437	511	60
Ellis	10,629	26,091	10,832	16,046	2,750
El Paso	83,848	57,574	83,964	43,255	6,300
Erath	2,804	8,126	3,664	4,750	1,134
Falls	2,417	3,239	3,256	2,260	479
Fannin	4,102	6,074	4,276	3,495	980
Fayette	2,542	6,658	3,119	4,195	708
Fisher	884	968	1,142	537	170
Floyd	580	1,830	986	1,530	126
Foard	263	286	355	166	52
Fort Bend	47,569	73,567	38,163	49,945	4,363
Franklin	1,018	2,420	1,484	1,575	386
Freestone	2,316	4,247	2,630	2,888	568
Frio	2,317	1,774	2,593	1,225	253
Gaines	723	2,691	1,012	1,812	353
Galveston	40,020	50,397	38,458	35,251	5,897
Garza	454	1,302	703	946	103
Gillespie	1,511	8,096	1,655	5,867	542
Glasscock	39	528	70	382	30
Goliad	1,233	2,108	1,135	1,335	148
Gonzales	1,877	4,092	2,110	2,687	354
Gray	1,376	6,732	2,114	6,102	568
Grayson	13,647	25,596	14,338	17,169	3,745
Gregg	11,244	26,739	13,659	21,611	2,079
Grimes	2,450	4,197	2,584	2,564	538
Guadalupe	8,311	21,499	8,079	14,254	1,811
Hale	2,158	6,868	3,204	5,905	605
Hall	472	966	750	626	94
Hamilton	878	2,447	1,200	1,493	323
Hansford	198	1,874	343	1,493	105
Hardeman	566	976	750	610	168
Hardin	5,595	11,962	7,179	8,529	2,112
Harris	418,267	529,159	386,726	421,462	42,364
Harrison	8,878	13,834	10,307	9,835	1,427
Hartley	359	1,645	463	1,242	101
Haskell	1,401	1,488	1,374	966	225
Hays	11,387	20,170	11,580	12,865	1,990
Hemphill	251	1,203	344	986	104
Henderson	8,704	16,607	10,085	10,345	2,274
Hidalgo	61,390	38,301	56,335	24,437	3,536
Hill	3,524	7,054	3,988	4,401	1,052
Hockley	1,419	5,250	2,170	4,230	519
Hood	4,704	12,429	5,459	7,575	1,445
Hopkins	3,692	7,076	4,522	4,341	1,034
Houston	2,833	5,308	3,383	3,443	585
Howard	2,744	6,668	3,732	5,007	1,037
Hudspeth	380	514	427	367	92
Hunt	7,857	16,177	8,801	10,746	2,225
Hutchinson	1,796	7,443	2,553	6,350	864
Irion	162	624	213	386	86
Jack	822	2,107	1,019	1,162	301
Jackson	1,446	3,365	1,785	2,533	309
Jasper	4,533	7,071	5,039	4,523	1,041
Jeff Davis	283	708	370	482	99
Jefferson	45,409	40,320	45,854	32,821	5,314
Jim Hogg	1,512	623	1,437	307	64
Jim Wells	7,418	4,498	7,116	2,989	430
Johnson	11,778	26,202	12,817	16,246	3,250
Jones	1,899	4,080	2,422	2,351	614

County	2000 Gore (D)	2000 Bush (R)	1996 Clinton (D)	1996 Dole (R)	1996 Perot (RF)
Karnes	1,617	2,638	2,154	1,869	291
Kaufman	7,455	15,290	7,383	8,697	1,831
Kendall	1,901	8,788	2,092	5,940	620
Kenedy	119	106	133	71	4
Kent	185	346	260	187	67
Kerr	4,002	14,637	4,192	11,173	1,236
Kimble	328	1,313	521	898	131
King	14	120	46	97	29
Kinney	486	932	503	650	97
Kleberg	4,481	4,526	5,136	3,391	431
Knox	617	947	785	599	149
Lamar	5,553	9,775	6,075	6,393	1,198
Lamb	1,114	3,451	1,683	2,593	283
Lampasas	1,569	4,526	1,819	3,008	509
LaSalle	1,266	731	1,522	570	85
Lavaca	2,171	5,288	2,575	3,697	551
Lee	1,733	3,699	2,008	2,354	421
Leon	1,893	4,362	2,217	2,839	499
Liberty	7,311	12,458	6,877	7,784	2,011
Limestone	2,768	4,212	3,236	2,691	693
Lipscomb	206	1,072	357	869	115
Live Oak	1,114	2,828	1,372	1,929	292
Llano	2,143	6,295	2,633	4,290	762
Loving	29	124	14	48	15
Lubbock	18,469	56,054	22,786	47,304	3,996
Lynn	562	1,507	903	1,151	136
McCulloch	794	2,084	1,231	1,465	296
McLennan	23,462	43,955	27,050	30,666	5,131
McMullen	77	358	117	274	35
Madison	1,241	2,333	1,470	1,576	293
Marion	1,852	2,039	2,028	1,260	353
Martin	415	1,520	643	973	140
Mason	417	1,352	618	949	151
Matagorda	4,696	7,584	5,374	5,876	1,190
Maverick	5,995	3,143	5,307	1,050	202
Medina	4,025	8,590	3,880	5,710	715
Menard	334	642	490	443	102
Midland	7,534	31,514	9,513	25,382	2,079
Milam	3,429	4,706	3,869	3,019	657
Mills	548	1,738	748	1,044	230
Mitchell	837	1,708	1,213	949	232
Montague	2,256	4,951	2,718	3,029	842
Montgomery	23,286	80,600	20,722	51,011	6,065
Moore	1,040	4,201	1,358	3,353	359
Morris	2,455	2,381	2,973	1,449	402
Motley	118	514	164	380	56
Nacogdoches	6,204	13,145	7,641	10,361	1,352
Navarro	5,366	8,358	6,078	5,236	1,140
Newton	2,503	2,423	2,554	1,409	474
Nolan	1,874	3,337	2,582	2,166	613
Nueces	45,349	49,906	50,009	37,470	5,103
Ochiltree	251	2,687	467	2,448	167
Oldham	108	659	213	583	77
Orange	11,887	17,325	13,741	12,560	2,836
Palo Pinto	3,263	5,690	3,938	3,666	1,011
Panola	3,011	5,975	4,168	4,008	777
Parker	8,878	23,651	9,447	14,580	2,703
Parmer	447	2,274	676	2,042	160
Pecos	1,539	2,700	1,816	1,730	369
Polk	6,877	11,746	6,360	6,473	1,347
Potter	7,242	17,629	9,273	14,995	1,799
Presidio	1,064	618	1,205	383	111
Rains	1,225	2,049	1,265	1,123	335
Randall	7,209	33,921	9,177	28,266	1,985
Reagan	282	959	407	645	101
Real	316	1,146	414	845	178
Red River	2,219	2,941	2,339	1,783	433
Reeves	1,872	1,273	2,279	1,007	245
Refugio	1,172	1,721	1,635	1,376	222
Roberts	72	472	122	421	40
Robertson	3,283	3,007	2,912	1,944	315
Rockwall	3,642	13,666	3,289	8,319	1,121
Runnels	969	3,020	1,417	1,941	396
Rusk	4,841	11,611	5,988	8,423	1,072
Sabine	1,753	2,764	1,913	1,660	334
San Augustine	1,636	2,116	1,924	1,296	324
San Jacinto	2,946	4,623	2,771	2,878	810
San Patricio	7,840	10,599	8,132	7,678	1,085
San Saba	618	1,691	726	991	194
Schleicher	338	826	505	587	111
Scurry	1,193	4,060	2,099	2,929	813
Shackelford	264	1,066	502	792	169
Shelby	3,227	5,692	3,720	3,482	815
Sherman	144	998	243	809	89
Smith	16,470	43,320	18,265	32,171	2,933
Somervell	752	2,120	993	1,099	273
Starr	6,505	1,911	6,312	756	157
Stephens	811	2,425	1,218	1,714	336
Sterling	132	520	186	394	86
Stonewall	294	496	487	323	105
Sutton	468	1,063	508	688	102

County	2000 Gore (D)	2000 Bush (R)	1996 Clinton (D)	1996 Dole (R)	1996 Perot (RF)
Swisher	856	1,612	1,224	1,159	195
Tarrant	173,758	286,921	170,431	208,312	28,715
Taylor	10,504	31,701	13,213	23,682	2,912
Terrell	219	243	278	185	47
Terry	1,108	2,910	1,272	2,013	269
Throckmorton	228	608	285	360	90
Titus	3,008	4,995	3,725	3,438	744
Tom Green	9,288	24,733	11,782	18,112	2,757
Travis	125,526	141,235	128,970	98,454	14,008
Trinity	2,142	3,093	2,774	2,058	460
Tyler	2,775	4,236	3,340	2,804	645
Upshur	4,180	8,448	5,032	5,174	1,086
Upton	266	982	424	685	88
Uvalde	3,436	4,855	3,397	3,494	403
Val Verde	5,056	6,223	5,623	4,357	548
Van Zandt	5,245	12,383	5,752	7,453	1,756
Victoria	8,176	18,787	8,238	14,457	1,197
Walker	4,943	9,076	6,088	7,177	1,186
Waller	5,046	5,686	4,535	3,559	499
Ward	1,256	2,534	1,644	1,620	446
Washington	2,996	8,645	3,460	6,319	601
Webb	18,120	13,076	18,997	4,712	936
Wharton	4,838	8,455	5,176	6,163	871
Wheeler	579	1,787	750	1,355	174
Wichita	14,108	27,802	15,775	20,495	3,371
Wilbarger	1,356	3,138	1,730	2,037	465
Willacy	3,218	1,789	3,789	1,332	241
Williamson	26,591	65,041	24,175	36,836	4,931
Wilson	3,997	7,509	3,713	4,530	760
Winkler	556	1,468	872	1,009	218
Wise	4,830	11,234	5,056	6,330	1,516
Wood	3,893	9,810	4,711	6,228	1,184
Yoakum	531	1,911	738	1,485	218
Young	1,843	5,022	2,394	3,647	639
Zapata	1,638	953	1,786	521	131
Zavala	2,616	751	2,629	463	91
Totals	**2,433,746**	**3,799,639**	**2,459,683**	**2,736,167**	**378,537**

Texas Vote Since 1952

1952, Eisenhower, Rep., 1,102,878; Stevenson, Dem., 969,228; Hamblen, Proh., 1,983; MacArthur, Christian Nationalist, 833; MacArthur, Constitution, 730; Hallinan, Prog., 294.

1956, Eisenhower, Rep., 1,080,619; Stevenson, Dem., 859,958; Andrews, Ind., 14,591.

1960, Kennedy, Dem., 1,167,932; Nixon, Rep., 1,121,699; Sullivan, Constitution, 18,169; Decker, Proh., 3,870; write-in, 15.

1964, Johnson, Dem., 1,663,185; Goldwater, Rep., 958,566; Lightburn, Constitution, 5,060.

1968, Nixon, Rep., 1,227,844; Humphrey, Dem., 1,266,804; Wallace, 3d Party, 584,269; write-in, 489.

1972, Nixon, Rep., 2,298,896; McGovern, Dem., 1,154,289; Schmitz, Amer., 6,039; Jenness, Soc. Workers, 8,664; others, 3,393.

1976, Carter, Dem., 2,082,319; Ford, Rep., 1,953,300; McCarthy, Ind., 20,118; Anderson, Amer., 11,442; Camejo, Soc. Workers, 1,723; write-in, 2,982.

1980, Reagan, Rep., 2,510,705; Carter, Dem., 1,881,147; Anderson, Ind., 111,613; Clark, Libertarian, 37,643; write-in, 528.

1984, Reagan, Rep., 3,433,428; Mondale, Dem., 1,949,276.

1988, Bush, Rep., 3,036,829; Dukakis, Dem., 2,352,748; Paul, Lib., 30,355; Fulani, New Alliance, 7,208.

1992, Clinton, Dem., 2,281,815; Bush, Rep., 2,496,071; Perot, Ind., 1,354,781; Marrou, Libertarian, 19,699.

1996, Dole, Rep., 2,736,167; Clinton, Dem., 2,459,683; Perot, Ind. (Ref.), 378,537; Browne, Libertarian, 20,256; Phillips, Taxpayers, 7,472; Hagelin, Natural Law, 4,422.

2000, Bush, Rep., 3,799,639; Gore, Dem., 2,433,746; Nader, Green, 137,994; Browne, Libertarian, 23,160; Buchanan, Ind., 12,394.

Utah

County	2000 Gore (D)	2000 Bush (R)	1996 Clinton (D)	1996 Dole (R)	1996 Perot (RF)
Beaver	541	1,653	687	1,164	217
Box Elder	2,555	12,288	3,170	8,373	1,578
Cache	5,170	25,920	6,595	16,832	2,399
Carbon	3,298	3,758	4,172	2,343	952
Daggett	104	317	131	237	55
Davis	18,845	64,375	19,301	42,768	7,495
Duchesne	779	3,622	892	2,648	566
Emery	958	3,243	1,371	2,033	663
Garfield	178	1,719	283	1,330	222
Grand	1,158	1,822	1,199	1,384	432
Iron	1,789	10,106	1,887	6,550	716
Juab	619	2,023	928	1,290	353
Kane	387	2,254	304	1,682	290
Millard	696	3,850	945	2,681	505
Morgan	553	2,464	859	1,659	337
Piute	133	626	176	475	59
Rich	152	736	179	523	88
Salt Lake	107,576	171,585	117,951	127,951	27,620
San Juan	1,838	2,721	1,675	2,139	271
Sanpete	1,211	5,781	1,568	3,631	801
Sevier	1,046	5,763	1,327	4,031	670
Summit	4,601	6,168	4,177	3,867	971
Tooele	4,001	7,807	3,992	3,881	1,244
Uintah	1,387	6,733	1,714	4,743	899
Utah	16,445	98,255	18,291	69,653	8,106
Wasatch	1,476	3,819	1,374	2,222	558
Washington	5,465	25,481	4,816	17,637	2,069
Wayne	202	953	265	741	121
Weber	19,890	39,254	21,404	27,443	6,204
Totals	**203,053**	**515,096**	**221,633**	**361,911**	**66,461**

Utah Vote Since 1952

1952, Eisenhower, Rep., 194,190; Stevenson, Dem., 135,364.

1956, Eisenhower, Rep., 215,631; Stevenson, Dem., 118,364.

1960, Kennedy, Dem., 169,248; Nixon, Rep., 205,361; Dobbs, Soc. Workers, 100.

1964, Johnson, Dem., 219,628; Goldwater, Rep., 181,785.

1968, Nixon, Rep., 238,728; Humphrey, Dem., 156,665; Wallace, 3d Party, 26,906; Halstead, Soc. Workers, 89; Peace and Freedom, 180.

1972, Nixon, Rep., 323,643; McGovern, Dem., 126,284; Schmitz, Amer., 28,549.

1976, Carter, Dem., 182,110; Ford, Rep., 337,908; Anderson, Amer., 13,304; McCarthy, Ind., 3,907; MacBride, Libertarian, 2,438; Maddox, Amer. Ind., 1,162; Camejo, Soc. Workers, 268; Hall, Com., 121.

1980, Reagan, Rep., 439,687; Carter, Dem., 124,266; Anderson, Ind., 30,284; Clark, Libertarian, 7,226; Commoner, Citizens, 1,009; Greaves, Amer., 965; Rarick, Amer. Ind., 522; Hall, Com., 139; DeBerry, Soc. Workers, 124.

1984, Reagan, Rep., 469,105; Mondale, Dem., 155,369; Bergland, Libertarian, 2,447.

1988, Bush, Rep., 428,442; Dukakis, Dem., 207,352; Paul, Lib., 7,473; Dennis, Amer., 2,158.

1992, Clinton, Dem., 183,429; Bush, Rep., 322,632; Perot, Ind., 203,400; Gritz, Populist/America First, 28,602; Marrou, Libertarian, 1,900; Hagelin, Natural Law, 1,319; LaRouche, Ind., 1,089.

1996, Dole, Rep., 361,911; Clinton, Dem., 221,633; Perot, Ref., 66,461; Nader, Green, 4,615; Browne, Libertarian, 4,129; Phillips, Taxpayers, 2,601; Templin, Ind. Amer., 1,290; Crane, Ind., 1,101; Hagelin, Natural Law, 1,085; Moorehead, Workers World, 298; Harris, Soc. Workers, 235; Dodge, Proh., 111.

2000, Bush, Rep., 515,096; Gore, Dem., 203,053; Nader, Green, 35,850; Buchanan, Reform, 9,319; Browne, Libertarian, 3,616; Phillips, Ind. Amer., 2,709; Hagelin, Natural Law, 763; Harris, Soc. Workers, 186; Youngkeit, Ind., 161.

Vermont

City	2000 Gore (D)	2000 Bush (R)	1996 Clinton (D)	1996 Dole (R)	1996 Perot (RF)
Barre City	1,895	1,676	1,890	1,107	376
Bennington	3,745	2,384	3,454	1,654	960
Brattleboro	3,128	1,486	3,016	1,195	395
Burlington	10,961	4,273	11,600	3,762	1,309
Colchester	3,876	2,989	3,314	2,035	769
Essex	4,632	4,344	4,063	2,944	796
Hartford	2,462	1,957	2,106	1,290	400
Montpelier	2,576	1,265	2,458	1,118	269
Rutland City	3,916	3,003	3,817	2,320	741
S. Burlington	4,393	2,995	3,929	2,274	548
Springfield	2,386	1,720	2,267	1,189	561
Other	105,052	191,683	95,980	59,464	23,900
Totals	**149,022**	**119,775**	**137,894**	**80,352**	**31,024**

Vermont Vote Since 1952

1952, Eisenhower, Rep., 109,717; Stevenson, Dem., 43,355; Hallinan, Prog., 282; Hoopes, Soc., 185.

1956, Eisenhower, Rep., 110,390; Stevenson, Dem., 42,549; scattered, 39.

1960, Kennedy, Dem., 69,186; Nixon, Rep., 98,131.

1964, Johnson, Dem., 107,674; Goldwater, Rep., 54,868.

1968, Nixon, Rep., 85,142; Humphrey, Dem., 70,255; Wallace, 3d Party, 5,104; Halstead, Soc. Workers, 295; Gregory, New Party, 579.

1972, Nixon, Rep., 117,149; McGovern, Dem., 68,174; Spock, Liberty Union, 1,010; Jenness, Soc. Workers, 296; scattered, 318.

1976, Carter, Dem., 77,798; Carter, Ind. Vermonter, 991; Ford, Rep., 100,387; McCarthy, Ind., 4,001; Camejo, Soc. Workers, 430; LaRouche, U.S. Labor, 196; scattered, 99.

1980, Reagan, Rep., 94,598; Carter, Dem., 81,891; Anderson, Ind., 31,760; Commoner, Citizens, 2,316; Clark, Libertarian, 1,900; McReynolds, Liberty Union, 136; Hall, Com., 118; DeBerry, Soc. Workers, 75; scattering, 413.

1984, Reagan, Rep., 135,865; Mondale, Dem., 95,730; Bergland, Libertarian, 1,002.

1988, Bush, Rep., 124,331; Dukakis, Dem., 115,775; Paul, Lib., 1,000; LaRouche, Ind., 275.

1992, Clinton, Dem., 133,590; Bush, Rep., 88,122; Perot, Ind., 65,985.
1996, Clinton, Dem., 137,894; Dole, Rep., 80,352; Perot, Ref., 31,024; Nader, Green, 5,585; Browne, Libertarian, 1,183; Hagelin, Natural Law, 498; Peron, Grass Roots, 480; Phillips, Taxpayers, 382; Hollis, Liberty Union, 292; Harris, Soc. Workers, 199.
2000, Gore, Dem., 149,022; Bush, Rep., 119,775; Nader, Green, 20,374; Buchanan, Reform, 2,192; Lane, Grass Roots, 1,044; Browne, Libertarian, 784; Hagelin, Natural Law, 219; McReynolds, Liberty Union, 161; Phillips, Constitution, 153; Harris, Soc. Workers, 70.

Virginia

County	2000 Gore (D)	2000 Bush (R)	1996 Clinton (D)	1996 Dole (R)	1996 Perot (RF)
Accomack	5,092	6,352	5,220	5,013	1,218
Albemarle	16,255	18,291	14,089	15,243	1,533
Alleghany	2,214	2,808	2,398	2,015	607
Amelia	1,754	2,947	1,625	2,119	323
Amherst	4,812	6,660	4,864	5,094	835
Appomattox	2,132	3,654	2,239	2,625	510
Arlington	50,260	28,555	45,573	26,106	2,782
Augusta	6,643	17,744	5,965	13,458	1,916
Bath	822	1,311	922	847	247
Bedford	8,160	17,224	7,786	11,955	1,976
Bland	851	1,759	939	1,167	385
Botetourt	4,627	8,867	4,576	6,404	1,138
Brunswick	3,387	2,561	3,442	2,059	340
Buchanan	5,745	3,867	6,551	2,785	858
Buckingham	2,561	2,738	2,374	1,974	392
Campbell	6,659	13,162	6,788	10,273	1,505
Caroline	4,314	3,873	3,897	2,816	521
Carroll	3,638	7,142	3,611	5,088	1,158
Charles City	1,981	1,023	1,842	729	178
Charlotte	2,017	2,855	2,007	2,103	431
Chesterfield	38,638	69,924	30,220	56,650	6,004
Clarke	2,166	2,883	1,906	2,201	379
Craig	851	1,580	895	979	262
Culpeper	4,364	7,440	3,907	5,688	787
Cumberland	1,405	1,974	1,303	1,544	275
Dickenson	3,951	3,122	3,913	2,229	660
Dinwiddie	4,001	4,959	3,871	3,503	666
Essex	1,750	1,995	1,668	1,627	188
Fairfax	196,501	202,181	170,150	176,033	16,134
Fauquier	8,296	14,456	6,759	11,063	1,287
Floyd	1,957	3,423	1,909	2,374	545
Fluvanna	3,431	4,962	2,676	3,442	457
Franklin	7,145	11,225	7,300	7,382	2,015
Frederick	7,158	14,574	5,976	10,608	1,599
Giles	3,004	3,574	3,196	2,566	841
Gloucester	4,553	8,718	4,710	6,447	1,266
Goochland	3,197	5,378	2,784	4,119	424
Grayson	2,467	4,236	2,661	3,004	675
Greene	1,774	3,375	1,440	2,351	346
Greensville	2,314	1,565	2,381	1,176	263
Halifax	5,963	7,732	5,599	6,490	876
Hanover	12,044	28,614	9,880	22,086	2,447
Henrico	48,645	62,887	41,121	54,430	5,920
Henry	8,898	11,870	9,061	9,110	2,370
Highland	453	942	446	631	134
Isle of Wight	5,162	7,587	4,952	5,416	893
James City	9,090	14,628	7,247	10,120	1,116
King and Queen	1,387	1,423	1,393	1,073	213
King George	2,070	3,590	1,875	2,597	341
King William	2,125	3,547	1,765	2,346	339
Lancaster	1,937	3,411	1,844	2,709	324
Lee	4,031	4,551	4,444	3,225	822
Loudoun	30,938	42,453	19,942	25,715	3,082
Louisa	4,309	5,461	3,761	3,768	693
Lunenburg	2,026	2,510	1,995	2,063	299
Madison	1,844	2,940	1,734	2,296	360
Mathews	1,499	2,951	1,602	2,206	403
Mecklenburg	4,797	6,600	4,408	4,933	789
Middlesex	1,671	2,844	1,704	2,141	350
Montgomery	11,720	13,991	10,867	10,517	2,594
Nelson	2,907	2,913	2,782	1,988	411
New Kent	2,055	3,934	1,859	2,852	520
Northampton	2,340	2,299	2,569	1,763	522
Northumberland	2,118	3,362	1,957	2,605	375
Nottoway	2,460	2,870	2,327	2,416	346
Orange	4,126	5,991	3,590	4,435	750
Page	2,726	5,089	2,868	3,876	640
Patrick	2,254	4,901	2,301	3,547	719
Pittsylvania	7,834	15,760	7,681	12,127	1,469
Powhatan	2,708	6,820	2,254	4,679	626
Prince Edward	2,922	3,214	2,678	2,530	403
Prince George	4,182	6,579	3,498	5,216	698
Prince William	44,745	52,788	33,462	39,292	4,881
Pulaski	5,255	7,089	5,333	5,387	1,399
Rappahannock	1,462	1,850	1,405	1,505	213
Richmond	1,076	1,784	1,101	1,424	201
Roanoke	16,141	25,740	15,387	20,700	2,934
Rockbridge	2,953	4,522	3,116	3,274	760
Rockingham	5,834	17,482	5,867	14,035	1,318
Russell	5,442	5,065	5,437	3,706	862
Scott	3,552	5,535	3,449	4,086	798
Shenandoah	4,420	9,636	4,224	7,440	1,353
Smyth	4,836	6,580	4,990	4,966	1,407
Southampton	3,359	3,293	3,454	2,275	564
Spotsylvania	13,455	20,739	10,342	13,786	1,860
Stafford	12,596	20,731	9,902	14,098	1,856
Surry	1,845	1,313	1,753	944	181
Sussex	2,006	1,745	2,089	1,378	256
Tazewell	7,227	8,655	7,500	6,131	1,554
Warren	4,313	6,335	3,814	4,657	904
Washington	7,549	12,064	6,939	9,098	1,654
Wesmoreland	2,922	2,932	2,949	2,333	427
Wise	6,412	6,504	6,712	4,660	1,478
Wythe	3,462	6,539	3,275	4,274	955
York	8,622	15,312	7,731	11,396	1,469
Cities					
Alexandria	33,633	19,043	27,968	15,554	1,472
Bedford	1,078	1,269	1,065	990	212
Bristol	2,646	3,495	2,586	2,983	429
Buena Vista	941	980	1,090	713	216
Charlottesville	7,762	4,034	7,916	4,091	565
Chesapeake	33,578	39,684	28,713	29,251	4,456
Clifton Forge	868	613	974	486	147
Colonial Heights	2,100	5,519	1,782	4,632	518
Covington	1,168	966	1,394	763	255
Danville	8,221	9,427	8,168	9,254	762
Emporia	1,116	938	1,103	835	98
Fairfax	4,361	4,762	3,909	4,319	422
Falls Church	3,109	2,131	2,375	1,644	202
Franklin	1,763	1,393	1,962	1,200	201
Fredericksburg	3,360	2,935	3,215	2,579	300
Galax	996	1,160	1,033	910	221
Hampton	27,490	19,561	24,493	16,596	2,783
Harrisonburg	3,482	5,741	3,346	4,945	434
Hopewell	3,024	3,749	2,868	3,493	550
Lexington	1,048	957	1,059	850	112
Lynchburg	10,374	12,518	10,281	11,441	1,155
Manassas	5,262	6,752	4,378	5,799	670
Manassas Park	1,048	1,460	748	916	151
Martinsville	3,048	2,560	2,941	2,446	387
Newport News	29,779	27,006	27,678	23,072	3,090
Norfolk	38,221	21,920	37,655	18,693	3,435
Norton	867	639	802	416	138
Petersburg	8,751	2,109	8,105	2,261	423
Poquoson	1,448	4,271	1,409	3,422	400
Portsmouth	22,286	12,628	22,150	10,686	2,238
Radford	2,063	2,190	2,113	1,742	381
Richmond	42,717	20,265	42,273	20,993	2,762
Roanoke	17,920	14,630	17,282	12,283	2,169
Salem	4,348	6,188	4,282	4,936	796
Staunton	3,324	4,878	3,162	4,526	605
Suffolk	12,471	11,836	10,827	8,572	1,266
Virginia Beach	62,268	83,674	52,142	63,741	9,328
Waynesboro	2,737	4,084	2,398	3,466	462
Williamsburg	1,724	1,777	1,820	1,560	162
Winchester	3,318	4,314	3,027	3,681	434
Totals	**1,217,290**	**1,437,490**	**1,091,060**	**1,138,350**	**159,861**

Virginia Vote Since 1952

1952, Eisenhower, Rep., 349,037; Stevenson, Dem., 268,677; Hass, Soc. Labor, 1,160; Hoopes, Soc. Dem., 504; Hallinan, Prog., 311.
1956, Eisenhower, Rep., 386,459; Stevenson, Dem., 267,760; Andrews, States' Rights, 42,964; Hoopes, Soc. Dem., 444; Hass, Soc. Labor, 351.
1960, Kennedy, Dem., 362,327; Nixon, Rep., 404,521; Coiner, Cons., 4,204; Hass, Soc. Labor, 397.
1964, Johnson, Dem., 558,038; Goldwater, Rep., 481,334; Hass, Soc. Labor, 2,895.
1968, Nixon, Rep., 590,319; Humphrey, Dem., 442,387; Wallace, 3d Party, *320,272; Blomen, Soc. Labor, 4,671; Munn, Proh., 601; Gregory, Peace and Freedom, 1,680.
*10,561 votes for Wallace were omitted in the count.
1972, Nixon, Rep., 988,493; McGovern, Dem., 438,887; Schmitz, Amer., 19,721; Fisher, Soc. Labor, 9,918.
1976, Carter, Dem., 813,896; Ford, Rep., 836,554; Camejo, Soc. Workers, 17,802; Anderson, Amer., 16,686; LaRouche, U.S. Labor, 7,508; MacBride, Libertarian, 4,648.
1980, Reagan, Rep., 989,609; Carter, Dem., 752,174; Anderson, Ind., 95,418; Commoner, Citizens, 14,024; Clark, Libertarian, 12,821; DeBerry, Soc. Workers, 1,986.
1984, Reagan, Rep., 1,337,078; Mondale, Dem., 796,250.
1988, Bush, Rep., 1,309,162; Dukakis, Dem., 859,799; Fulani, Ind., 14,312; Paul, Lib., 8,336.

1992, Clinton, Dem., 1,038,650; Bush, Rep., 1,150,517; Perot, Ind., 348,639; LaRouche, Ind., 11,937; Marrou, Libertarian, 5,730; Fulani, New Alliance, 3,192.
1996, Dole, Rep., 1,138,350; Clinton, Dem., 1,091,060; Perot, Ref., 159,861; Phillips, Taxpayers, 13,687; Browne, Libertarian, 9,174; Hagelin, Natural Law, 4,510.
2000, Bush, Rep., 1,437,490; Gore, Dem., 1,217,290; Nader, Green, 59,398; Browne, Libertarian, 15,198; Buchanan, Reform, 5,455; Phillips, Constitution, 1,809.

Washington

	2000		1996		
County	Gore (D)	Bush (R)	Clinton (D)	Dole (R)	Perot (RF)
Adams	1,406	3,440	1,740	2,356	448
Asotin	2,736	4,909	3,349	2,860	936
Benton	19,512	38,367	20,783	26,664	5,311
Chelan	8,412	16,980	8,595	12,363	2,332
Clallam	13,779	16,251	12,585	12,432	3,187
Clark	61,767	67,219	52,254	46,794	9,663
Columbia	515	1,523	743	948	228
Cowlitz	18,233	16,873	18,054	11,221	3,441
Douglas	3,822	8,512	3,913	5,682	1,132
Ferry	932	1,896	1,197	1,091	408
Franklin	4,653	8,594	4,961	5,946	992
Garfield	300	982	497	623	117
Grant	7,073	15,830	8,065	10,895	2,496
Grays Harbor	13,304	11,225	14,082	7,635	3,757
Island	14,778	16,408	12,157	12,387	2,787
Jefferson	8,281	6,095	7,145	4,607	1,385
King	476,700	273,171	417,846	232,811	51,309
Kitsap	50,302	46,427	44,167	35,304	8,769
Kittitas	5,516	7,727	5,707	5,224	1,214
Klickitat	3,062	4,557	3,214	2,662	875
Lewis	9,891	18,565	10,331	13,238	3,373
Lincoln	1,417	3,546	1,806	2,587	518
Mason	10,876	10,257	10,088	7,149	2,816
Okanogan	4,335	9,384	4,810	5,890	1,797
Pacific	4,895	4,042	5,095	2,598	1,131
Pend Oreille	1,973	3,076	2,126	2,012	709
Pierce	138,249	118,431	120,893	89,295	22,051
San Juan	4,426	3,005	3,663	2,523	508
Skagit	20,432	22,163	18,295	16,397	4,818
Skamania	1,753	2,151	1,724	1,387	450
Snohomish	129,612	109,615	109,624	81,885	22,731
Spokane	74,604	89,299	71,727	66,628	16,532
Stevens	5,560	11,299	5,591	7,524	2,158
Thurston	50,467	39,924	45,522	29,835	7,622
Wahkiakum	803	1,033	924	619	215
Walla Walla	7,188	13,304	8,038	9,085	1,894
Whatcom	34,033	34,287	29,074	27,153	4,854
Whitman	6,509	9,003	7,262	6,734	1,315
Yakima	25,546	39,494	25,676	27,668	4,724
Totals	**1,247,652**	**1,108,864**	**1,123,323**	**840,712**	**201,003**

Washington Vote Since 1952

1952, Eisenhower, Rep., 599,107; Stevenson, Dem., 492,845; MacArthur, Christian Nationalist, 7,290; Hallinan, Prog., 2,460; Hass, Soc. Labor, 633; Hoopes, Soc., 254; Dobbs, Soc. Workers, 119.
1956, Eisenhower, Rep., 620,430; Stevenson, Dem., 523,002; Hass, Soc. Labor, 7,457.
1960, Kennedy, Dem., 599,298; Nixon, Rep., 629,273; Hass, Soc. Labor, 10,895; Curtis, Constitution, 1,401; Dobbs, Soc. Workers, 705.
1964, Johnson, Dem., 779,699; Goldwater, Rep., 470,366; Hass, Soc. Labor, 7,772; DeBerry, Freedom Soc., 537.
1968, Nixon, Rep., 588,510; Humphrey, Dem., 616,037; Wallace, 3d Party, 96,990; Blomen, Soc. Labor, 488; Cleaver, Peace and Freedom, 1,609; Halstead, Soc. Workers, 270; Mitchell, Free Ballot, 377.
1972, Nixon, Rep., 837,135; McGovern, Dem., 568,334; Schmitz, Amer., 58,906; Spock, Ind., 2,644; Fisher, Soc. Labor, 1,102; Jenness, Soc. Workers, 623; Hall, Com., 566; Hospers, Libertarian, 1,537.
1976, Carter, Dem., 717,323; Ford, Rep., 777,732; McCarthy, Ind., 36,986; Maddox, Amer. Ind., 8,585; Anderson, Amer., 5,046; MacBride, Libertarian, 5,042; Wright, People's, 1,124; Camejo, Soc. Workers, 905; LaRouche, U.S. Labor, 903; Hall, Com., 817; Levin, Soc. Labor, 713; Zeidler, Soc., 358.
1980, Reagan, Rep., 865,244; Carter, Dem., 650,193; Anderson, Ind., 185,073; Clark, Libertarian, 29,213; Commoner, Citizens, 9,403; DeBerry, Soc. Workers, 1,137; McReynolds, Soc., 956; Hall, Com., 834; Griswold, Workers World, 341.
1984, Reagan, Rep., 1,051,670; Mondale, Dem., 798,352; Bergland, Libertarian, 8,844.
1988, Bush, Rep., 903,835; Dukakis, Dem., 933,516; Paul, Lib., 17,240; LaRouche, Ind., 4,412.
1992, Clinton, Dem., 993,037; Bush, Rep., 731,234; Perot, Ind., 541,780; Marrou, Libertarian, 7,533; Gritz, Populist/America First, 4,854; Hagelin, Natural Law, 2,456; Phillips, U.S. Taxpayers, 2,354; Fulani, New Alliance, 1,776; Daniels, Ind., 1,171.
1996, Clinton, Dem., 1,123,323; Dole, Rep., 840,712; Perot, Ref., 201,003; Nader, Ind., 60,322; Browne, Libertarian, 12,522; Hagelin, Natural Law, 6,076; Phillips, Taxpayers, 4,578; Collins, Ind., 2,374; Moorehead, Workers World, 2,189; Harris, Soc. Workers, 738.
2000, Gore, Dem., 1,247,652; Bush, Rep., 1,108,864; Nader, Green, 103,002; Browne, Libertarian, 13,135; Buchanan, Freedom, 7,171; Hagelin, Natural Law, 2,927; ; Phillips, Constitution, 1,989; Moorehead, Workers World, 1,729; McReynolds, Soc., 660; Harris, Soc. Workers, 304.

West Virginia

	2000		1996		
County	Gore (D)	Bush (R)	Clinton (D)	Dole (R)	Perot (RF)
Barbour	2,503	3,411	3,076	2,155	784
Berkeley	8,797	13,619	8,321	9,859	2,291
Boone	5,656	3,353	6,048	1,917	927
Braxton	2,719	2,529	3,001	1,441	527
Brooke	4,678	4,195	5,338	2,741	1,375
Cabell	14,896	16,440	16,277	13,179	2,968
Calhoun	1,112	1,425	1,402	1,000	307
Clay	1,617	1,887	2,074	1,137	355
Doddridge	773	1,955	865	1,335	382
Fayette	8,371	5,897	9,471	3,669	1,552
Gilmer	1,092	1,560	1,390	933	316
Grant	891	3,571	1,206	2,599	481
Greenbrier	5,627	6,866	6,286	4,434	1,418
Hampshire	2,069	3,879	2,335	2,814	605
Hancock	6,249	6,458	7,521	4,268	2,158
Hardy	1,621	2,816	1,911	1,895	438
Harrison	13,009	12,948	14,746	8,857	3,135
Jackson	4,937	6,341	4,882	4,235	1,295
Jefferson	6,860	7,045	6,361	5,287	1,307
Kanawha	38,524	36,809	40,357	29,311	6,412
Lewis	2,355	3,606	2,868	2,285	974
Lincoln	3,939	3,389	4,994	2,530	696
Logan	8,927	5,334	10,840	2,627	1,532
McDowell	4,845	2,348	5,989	1,550	655
Marion	12,315	9,972	12,994	6,160	2,881
Marshall	6,000	6,859	7,045	4,460	2,202
Mason	4,963	5,972	5,284	3,581	1,533
Mercer	8,347	10,206	8,721	7,768	2,141
Mineral	3,341	6,180	3,487	4,380	1,170
Mingo	6,049	3,866	7,584	2,229	1,020
Monongalia	12,603	13,595	13,406	10,189	3,040
Monroe	2,094	2,940	2,382	2,131	559
Morgan	1,939	3,639	1,929	2,599	513
Nicholas	4,059	4,359	4,769	2,649	1,071
Ohio	7,653	9,607	8,781	7,267	2,065
Pendleton	1,172	1,996	1,591	1,431	276
Pleasants	1,267	1,884	1,478	1,265	416
Pocahontas	1,392	1,970	1,796	1,242	426
Preston	3,515	6,607	4,237	4,257	1,760
Putnam	7,891	12,173	8,029	8,803	1,901
Raleigh	11,047	12,587	12,547	8,628	2,355
Randolph	4,028	5,248	5,469	3,348	1,184
Ritchie	1,024	2,717	1,385	1,906	522
Roane	2,332	3,172	2,572	2,069	622
Summers	2,299	2,304	2,397	1,505	438
Taylor	2,473	3,124	2,692	1,977	844
Tucker	1,319	1,935	1,649	1,217	424
Tyler	1,214	2,582	1,459	734	563
Upshur	2,770	5,165	3,052	3,325	1,031
Wayne	7,940	7,993	8,300	5,492	1,633
Webster	1,764	1,484	2,292	654	369
Wetzel	2,849	3,239	3,209	2,037	1,004
Wirt	818	1,518	906	928	280
Wood	12,664	20,428	13,261	15,502	3,694
Wyoming	4,289	3,473	5,550	2,155	812
Totals	**295,497**	**336,475**	**327,812**	**233,946**	**71,639**

West Virginia Vote Since 1952

1952, Eisenhower, Rep., 419,970; Stevenson, Dem., 453,578.
1956, Eisenhower, Rep., 449,297; Stevenson, Dem., 381,534.
1960, Kennedy, Dem., 441,786; Nixon, Rep., 395,995.
1964, Johnson, Dem., 538,087; Goldwater, Rep., 253,953.
1968, Nixon, Rep., 307,555; Humphrey, Dem., 374,091; Wallace, 3d Party, 72,560.
1972, Nixon, Rep., 484,964; McGovern, Dem., 277,435.
1976, Carter, Dem., 435,864; Ford, Rep., 314,726.
1980, Reagan, Rep., 334,206; Carter, Dem., 367,462; Anderson, Ind., 31,691; Clark, Libertarian, 4,356.
1984, Reagan, Rep., 405,483; Mondale, Dem., 328,125.
1988, Bush, Rep., 310,065; Dukakis, Dem., 341,016; Fulani, New Alliance, 2,230.
1992, Clinton, Dem., 331,001; Bush, Rep., 241,974; Perot, Ind., 108,829; Marrou, Libertarian, 1,873.

1996, Clinton, Dem., 327,812; Dole, Rep., 233,946; Perot, Ref., 71,639; Browne, Libertarian, 3,062.
2000, Bush, Rep., 336,475; Gore, Dem., 295,497; Nader, Green, 10,680; Buchanan, Reform, 3,169; Browne, Libertarian, 1,912; Hagelin, Natural Law, 367.

Wisconsin

	2000		1996		
County	Gore (D)	Bush (R)	Clinton (D)	Dole (R)	Perot (RF)
Adams	4,826	3,920	4,119	2,450	1,122
Ashland	4,356	3,038	3,808	1,863	861
Barron	8,928	9,848	8,025	6,158	2,692
Bayfield	4,427	3,266	3,895	2,250	899
Brown	49,096	54,258	42,823	38,563	8,036
Buffalo	3,237	3,038	2,681	1,800	972
Burnett	3,626	3,967	3,625	2,452	962
Calumet	8,202	10,837	6,940	7,049	2,112
Chippewa	12,102	12,835	9,647	7,520	3,567
Clark	5,931	7,461	5,540	4,622	2,486
Columbia	12,636	11,987	10,336	8,377	2,377
Crawford	4,005	3,024	3,658	2,149	1,060
Dane	142,317	75,790	109,347	59,487	12,436
Dodge	14,580	21,684	12,625	12,890	3,322
Door	6,560	7,810	5,590	4,948	1,475
Douglas	13,593	6,930	10,976	5,167	2,001
Dunn	9,172	8,911	7,536	4,917	2,555
Eau Claire	24,078	20,921	20,298	13,900	5,160
Florence	816	1,528	869	927	316
Fond du Lac	18,181	26,548	15,542	16,488	4,204
Forest	2,158	2,404	2,092	1,166	678
Grant	10,691	10,240	9,203	7,021	2,648
Green	7,863	6,790	6,136	4,697	1,534
Green Lake	3,301	5,451	3,152	3,565	1,025
Iowa	5,842	4,221	4,690	2,866	1,071
Iron	1,620	1,734	1,725	1,260	469
Jackson	4,380	3,670	3,705	2,262	1,163
Jefferson	15,203	19,204	13,188	12,681	3,177
Juneau	4,813	4,910	4,331	3,226	1,393
Kenosha	32,429	28,891	27,964	18,296	6,507
Kewaunee	4,670	4,883	4,311	3,431	1,161
La Crosse	28,455	24,327	23,647	16,482	4,844
La Fayette	3,710	3,336	3,261	2,172	944
Langlade	4,199	5,125	4,074	3,206	1,249
Lincoln	6,664	6,727	6,166	4,076	1,800
Manitowoc	17,667	19,358	16,750	13,239	3,941
Marathon	26,546	28,883	24,012	19,874	6,749
Marinette	8,676	10,535	8,413	7,231	2,367
Marquette	3,437	3,522	2,859	2,208	915
Menominee	949	225	992	230	107
Milwaukee	252,329	163,491	216,620	119,407	26,027
Monroe	7,460	8,217	6,924	5,299	2,081
Oconto	7,260	8,706	6,723	5,389	1,655
Oneida	8,339	9,512	7,619	6,339	2,604
Outagamie	32,735	39,460	28,815	27,758	7,235
Ozaukee	15,030	31,155	13,269	22,078	2,774
Pepin	1,854	1,631	1,585	1,007	456
Pierce	8,559	8,169	7,970	4,599	2,074
Polk	8,961	9,557	8,334	5,387	2,369
Portage	17,942	13,214	15,901	9,631	3,410
Price	3,413	4,136	3,523	2,545	1,218
Racine	41,563	44,014	38,567	30,107	7,611
Richland	3,837	3,994	3,502	2,642	901
Rock	40,472	27,467	32,450	20,096	6,800
Rusk	3,161	3,758	2,941	2,219	1,331
St. Croix	13,077	15,240	11,384	8,253	3,180
Sauk	13,035	11,586	9,889	7,448	2,448
Sawyer	3,333	3,972	2,773	2,603	962
Shawano	7,335	9,548	6,850	6,396	2,071
Sheboygan	23,569	29,648	22,022	20,067	4,157
Taylor	3,254	5,278	3,253	3,108	1,457
Trempealeau	6,678	5,002	5,848	3,035	1,688
Vernon	6,577	5,684	5,572	3,796	1,523
Vilas	4,706	6,958	4,226	4,496	1,548
Walworth	15,492	22,982	13,283	15,099	3,729
Washburn	3,695	3,912	3,231	2,703	920
Washington	18,115	41,162	17,154	25,829	4,786
Waukesha	64,319	133,105	57,354	91,729	13,109
Waupaca	8,787	12,980	7,800	8,679	2,464
Waushara	4,239	5,571	3,824	3,573	1,264
Winnebago	33,983	38,330	29,564	27,880	6,531
Wood	15,936	17,803	14,650	12,666	4,599
Totals	**1,242,987**	**1,237,279**	**1,071,971**	**845,029**	**227,339**

Wisconsin Vote Since 1952

1952, Eisenhower, Rep., 979,744; Stevenson, Dem., 622,175; Hallinan, Ind., 2,174; Dobbs, Ind., 1,350; Hoopes, Ind., 1,157; Hass, Ind., 770.
1956, Eisenhower, Rep., 954,844; Stevenson, Dem., 586,768; Andrews, Ind., 6,918; Hoopes, Soc., 754; Hass, Soc. Labor, 710; Dobbs, Soc. Workers, 564.
1960, Kennedy, Dem., 830,805; Nixon, Rep., 895,175; Dobbs, Soc. Workers, 1,792; Hass, Soc. Labor, 1,310.
1964, Johnson, Dem., 1,050,424; Goldwater, Rep., 638,495; DeBerry, Soc. Workers, 1,692; Hass, Soc. Labor, 1,204.
1968, Nixon, Rep., 809,997; Humphrey, Dem., 748,804; Wallace, 3d Party, 127,835; Blomen, Soc. Labor, 1,338; Halstead, Soc. Workers, 1,222; scattered, 2,342.
1972 Nixon, Rep., 989,430; McGovern, Dem., 810,174; Schmitz, Amer., 47,525; Spock, Ind., 2,701; Fisher, Soc. Labor, 998; Hall, Com., 663; Reed, Ind., 506; scattered, 893.
1976, Carter, Dem., 1,040,232; Ford, Rep., 1,004,987; McCarthy, Ind., 34,943; Maddox, Amer. Ind., 8,552; Zeidler, Soc., 4,298; MacBride, Libertarian, 3,814; Camejo, Soc. Workers, 1,691; Wright, People's, 943; Hall, Com., 749; LaRouche, U.S. Lab., 738; Levin, Soc. Labor, 389; scattered, 2,839.
1980, Reagan, Rep., 1,088,845; Carter, Dem., 981,584; Anderson, Ind., 160,657; Clark, Libertarian, 29,135; Commoner, Citizens, 7,767; Rarick, Constitution, 1,519; McReynolds, Soc., 808; Hall, Com., 772; Griswold, Workers World, 414; DeBerry, Soc. Workers, 383; scattering, 1,337.
1984, Reagan, Rep., 1,198,584; Mondale, Dem., 995,740; Bergland, Libertarian, 4,883.
1988, Bush, Rep., 1,047,499; Dukakis, Dem., 1,126,794; Paul, Lib., 5,157; Duke, Pop., 3,056.
1992, Clinton, Dem., 1,041,066; Bush, Rep., 930,855; Perot, Ind., 544,479; Marrou, Libertarian, 2,877; Gritz, Populist/America First, 2,311; Daniels, Ind., 1,883; Phillips, U.S. Taxpayers, 1,772; Hagelin, Natural Law, 1,070.
1996, Clinton, Dem., 1,071,971; Dole, Rep., 845,029; Perot, Ref., 227,339; Nader, Green, 28,723; Phillips, Taxpayers, 8,811; Browne, Libertarian, 7,929; Hagelin, Natural Law, 1,379; Moorehead, Workers World, 1,333; Hollis, Soc., 848; Harris, Soc. Workers, 483.
2000, Gore, Dem., 1,242,987; Bush, Rep., 1,237,279; Nader, Green, 94,070; Buchanan, Reform, 11,446; Browne, Libertarian, 6,640; Phillips, Constitution, 2,042; Moorehead, Workers World, 1,063; Hagelin, Reform, 878; Harris, Soc. Workers, 306.

Wyoming

	2000		1996		
County	Gore (D)	Bush (R)	Clinton (D)	Dole (R)	Perot (RF)
Albany	5,069	7,814	6,399	5,967	1,333
Big Horn	1,004	3,720	1,438	2,821	545
Campbell	1,967	10,203	3,468	6,382	1,954
Carbon	2,206	4,498	2,690	2,930	855
Converse	1,076	3,919	1,520	2,702	639
Crook	361	2,289	651	1,698	394
Fremont	4,172	10,560	5,445	7,554	1,840
Goshen	1,439	3,922	1,923	2,989	547
Hot Springs	544	1,733	779	1,348	287
Johnson	555	2,886	815	2,071	378
Laramie	12,162	21,797	13,676	16,924	2,958
Lincoln	1,184	5,415	1,803	3,764	906
Natrona	8,646	18,439	11,240	13,182	3,524
Niobrara	190	888	325	757	209
Park	2,424	9,884	3,240	7,430	1,318
Platte	1,249	2,925	1,631	2,155	579
Sheridan	3,330	8,424	4,594	5,892	1,414
Sublette	458	2,624	677	1,829	401
Sweetwater	5,521	9,425	7,088	5,591	2,792
Teton	4,019	5,454	4,042	3,918	839
Uinta	1,650	5,469	2,414	3,471	1,242
Washakie	806	3,138	1,205	2,250	470
Weston	449	2,521	871	1,763	504
Totals	**60,481**	**147,947**	**77,934**	**105,388**	**25,928**

Wyoming Vote Since 1952

1952, Eisenhower, Rep., 81,047; Stevenson, Dem., 47,934; Hamblen, Proh., 194; Hoopes, Soc., 40; Haas, Soc. Labor, 36.
1956, Eisenhower, Rep., 74,573; Stevenson, Dem., 49,554.
1960, Kennedy, Dem., 63,331; Nixon, Rep., 77,451.
1964, Johnson, Dem., 80,718; Goldwater, Rep., 61,998.
1968, Nixon, Rep., 70,927; Humphrey, Dem., 45,173; Wallace, 3d Party, 11,105.
1972, Nixon, Rep., 100,464; McGovern, Dem., 44,358; Schmitz, Amer., 748.
1976, Carter, Dem., 62,239; Ford, Rep., 92,717; McCarthy, Ind., 624; Reagan, Ind., 307; Anderson, Amer., 290; MacBride, Libertarian, 89; Brown, Ind., 47; Maddox, Amer. Ind., 30.
1980, Reagan, Rep., 110,700; Carter, Dem., 49,427; Anderson, Ind., 12,072; Clark, Libertarian, 4,514.
1984, Reagan, Rep., 133,241; Mondale, Dem., 53,370; Bergland, Libertarian, 2,357.
1988, Bush, Rep., 106,867; Dukakis, Dem., 67,113; Paul, Lib., 2,026; Fulani, New Alliance, 545.
1992, Clinton, Dem., 68,160; Bush, Rep., 79,347; Perot, Ind., 51,263.
1996, Dole, Rep., 105,388; Clinton, Dem., 77,934; Perot, Ind. (Ref.), 25,928; Browne, Libertarian, 1,739; Hagelin, Natural Law, 582.
2000, Bush, Rep., 147,947; Gore, Dem., 60,481; Buchanan, Reform, 2,724; Browne, Libertarian, 1,443; Phillips, Ind., 720; Hagelin, Natural Law, 411.

Voter Turnout in Presidential Elections, 1932-2000

Source: Federal Election Commission; Commission for Study of American Electorate; *Congressional Quarterly*; Facts On File World News Digest

	Candidates	Voter Participation (% of voting-age population)		Candidates	Voter Participation (% of voting-age population)
1932	Roosevelt-Hoover	52.4	1968	Nixon-Humphrey	60.9
1936	Roosevelt-Landon	56.0	1972	Nixon-McGovern	55.2[1]
1940	Roosevelt-Willkie	58.9	1976	Carter-Ford	53.5
1944	Roosevelt-Dewey	56.0	1980	Reagan-Carter	54.0
1948	Truman-Dewey	51.1	1984	Reagan-Mondale	53.1
1952	Eisenhower-Stevenson	61.6	1988	Bush-Dukakis	50.2
1956	Eisenhower-Stevenson	59.3	1992	Clinton-Bush-Perot	55.9
1960	Kennedy-Nixon	62.8	1996	Clinton-Dole-Perot	49.0
1964	Johnson-Goldwater	61.9	2000	Bush-Gore	50.7[2]

(1) The sharp drop in 1972 followed the expansion of eligibility with the enfranchisement of 18- to 20-year-olds. (2) Preliminary.

Electoral Votes for President

(Figures in **boldface**, based on 1990 census, were in force for 1992, 1996, and 2000 elections; where figures for 2004 election will be different, based on 2000 census, these are given in parentheses.)

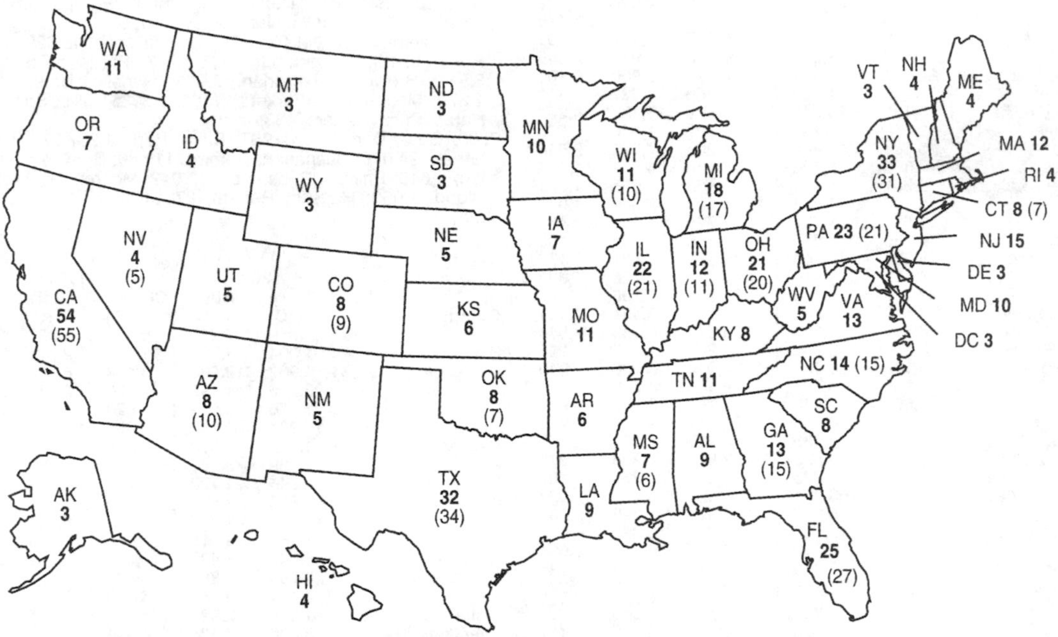

The Electoral College

The president and the vice president are the only elective federal officials not chosen by direct vote of the people. They are elected by the members of the Electoral College, an institution provided for in the U.S. Constitution.

On presidential election day, the first Tuesday after the first Monday in Nov. of every 4th year, each state chooses as many electors as it has senators and representatives in Congress. In 1964, for the first time, as provided by the 23d Amendment to the Constitution, the District of Columbia voted for 3 electors. Thus, with 100 senators and 435 representatives, there are 538 members of the Electoral College, with a majority of 270 electoral votes needed to elect the president and vice president.

Although political parties were not part of the original plan created by the Founding Fathers, today political parties customarily nominate their lists of electors at their respective state conventions. Some states print names of the candidates for president and vice president at the top of the Nov. ballot; others list only the electors' names. In either case, the electors of the party receiving the highest vote are elected. Two states, Maine and Nebraska, allow for proportional allocation.

The electors meet on the first Monday after the 2d Wednesday in Dec. in their respective state capitals or in some other place prescribed by state legislatures. By long-established custom, they vote for their party nominees, although this is not required by federal law; some states do require it.

The Constitution requires electors to cast a ballot for at least one person who is not an inhabitant of that elector's home state. This ensures that presidential and vice presidential candidates from the same party will not be from the same state. (In 2000, Republican vice presidential nominee Dick Cheney changed his voter registration to Wyoming from Gov. George W. Bush's home state of Texas.) Also, an elector cannot be a member of Congress or hold federal office.

Certified and sealed lists of the votes of the electors in each state are sent to the president of the U.S. Senate, who then opens them in the presence of the members of the Senate and House of Representatives in a joint session held in early Jan., and the electoral votes of all the states are then officially counted.

If no candidate for president has a majority, the House of Representatives chooses a president from the top 3 candidates, with all representatives from each state combining to cast one vote for that state. The House decided the outcome of the 1800 and 1824 presidential elections. If no candidate for vice president has a majority, the Senate chooses from the top 2, with the senators voting as individuals. The Senate chose the vice president following the 1836 election.

Under the electoral college system, a candidate who fails to be the top vote getter in the popular vote still may win a majority of electoral votes. This happened in the elections of 1876, 1888, and 2000.

Third-Party and Independent Presidential Candidates

Although many "third party" candidates or independents have pursued the presidency, only 10 of these have polled more than a million votes. In most elections since 1860, fewer than one vote in 20 has been cast for a third-party candidate. In only 5 presidential elections since then have all non-major-party candidates combined polled more than 10% of the vote. The major vote getters in those elections were James B. Weaver (People's Party), 1892; former President Theodore Roosevelt (Progressive Party), 1912; Robert M. La Follette (Progressive Party), 1924; George C. Wallace (American Independent Party), 1968; and H. Ross Perot, as an independent in 1992 and with the Reform Party in 1996.

Roosevelt outpolled the Republican candidate, William Howard Taft, in 1912, capturing 28% of the popular vote and 88 electoral votes. In 1948, Strom Thurmond was able to capture 39 electoral votes (from 5 Southern states); however, all third parties received only 5.75% of the popular vote in the election. Twenty years later, George Wallace's popularity in the same region allowed him to get 46 electoral votes and 13.5% of the popular vote.

In 1992 Perot captured 19% of the popular vote; however, he did not win a single state. In 1996, Perot won 8% of the popular vote; all third-party candidates combined won about 10%. Ralph Nader won about 3% of the vote in 2000.

Despite the difficulty in winning the presidency, independent and third-party candidates sometimes succeed in winning other offices and often bring to the attention of the nation their most prominent issues.

Notable Third Party and Independent Campaigns by Year

Party	Presidential nominee	Year	Issues	Strength in . . .
Anti-Masonic	William Wirt	1832	Against secret societies and oaths	PA, VT
Liberty	James G. Birney	1844	Anti-slavery	North
Free Soil	Martin Van Buren	1848	Anti-slavery	NY, OH
American (Know-Nothing)	Millard Fillmore	1856	Anti-immigrant	Northeast, South
Greenback	Peter Cooper	1876	For "cheap money," labor rights	National
Greenback	James B. Weaver	1880	For "cheap money," labor rights	National
Prohibition	John P. St. John	1884	Anti-liquor	National
People's (Populists)	James B. Weaver	1892	For "cheap money," end of national banks	South, West
Socialist	Eugene V. Debs	1900-12; 1920	For public ownership	National
Progressive (Bull Moose)	Theodore Roosevelt	1912	Against high tariffs	Midwest, West
Progressive	Robert M. La Follette	1924	Farmer and labor rights	Midwest, West
Socialist	Norman Thomas	1928-48	Liberal reforms	National
Union	William Lemke	1936	Anti-New Deal	National
States' Rights (Dixiecrats)	Strom Thurmond	1948	For states' rights	South
Progressive	Henry A. Wallace	1948	Anti-cold war	NY, CA
American Independent	George C. Wallace	1968	For states' rights	South
American	John G. Schmitz	1972	For "law and order"	Far West, OH, LA
None (Independent)	John B. Anderson	1980	A 3d choice	National
None (Independent)	H. Ross Perot	1992	Federal budget deficit	National
Reform	H. Ross Perot	1996	Deficit; campaign finance	National
Green	Ralph Nader	2000	Corporate power	National

Major-Party Nominees for President and Vice President

Asterisk (*) denotes winning ticket

	Democratic		Republican	
Year	President	Vice President	President	Vice President
1856	James Buchanan*	John Breckinridge	John Frémont	William Dayton
1860	Stephen A. Douglas (1)	Herschel V. Johnson	Abraham Lincoln*	Hannibal Hamlin
1864	George McClellan	G.H. Pendleton	Abraham Lincoln*	Andrew Johnson
1868	Horatio Seymour	Francis Blair	Ulysses S. Grant*	Schuyler Colfax
1872	Horace Greeley	B. Gratz Brown	Ulysses S. Grant*	Henry Wilson
1876	Samuel J. Tilden	Thomas Hendricks	Rutherford B. Hayes*	William Wheeler
1880	Winfield Hancock	William English	James A. Garfield*	Chester A. Arthur
1884	Grover Cleveland*	Thomas Hendricks	James Blaine	John Logan
1888	Grover Cleveland	A.G. Thurman	Benjamin Harrison*	Levi Morton
1892	Grover Cleveland*	Adlai Stevenson	Benjamin Harrison	Whitelaw Reid
1896	William J. Bryan	Arthur Sewall	William McKinley*	Garret Hobart
1900	William J. Bryan	Adlai Stevenson	William McKinley*	Theodore Roosevelt
1904	Alton Parker	Henry Davis	Theodore Roosevelt*	Charles Fairbanks
1908	William J. Bryan	John Kern	William H. Taft*	James Sherman
1912	Woodrow Wilson*	Thomas Marshall	William H. Taft	James Sherman (2)
1916	Woodrow Wilson*	Thomas Marshall	Charles Hughes	Charles Fairbanks
1920	James M. Cox	Franklin D. Roosevelt	Warren G. Harding*	Calvin Coolidge
1924	John W. Davis	Charles W. Bryan	Calvin Coolidge*	Charles G. Dawes
1928	Alfred E. Smith	Joseph T. Robinson	Herbert Hoover*	Charles Curtis
1932	Franklin D. Roosevelt*	John N. Garner	Herbert Hoover	Charles Curtis
1936	Franklin D. Roosevelt*	John N. Garner	Alfred M. Landon	Frank Knox
1940	Franklin D. Roosevelt*	Henry A. Wallace	Wendell L. Willkie	Charles McNary
1944	Franklin D. Roosevelt*	Harry S. Truman	Thomas E. Dewey	John W. Bricker
1948	Harry S. Truman*	Alben W. Barkley	Thomas E. Dewey	Earl Warren
1952	Adlai E. Stevenson	John J. Sparkman	Dwight D. Eisenhower*	Richard M. Nixon
1956	Adlai E. Stevenson	Estes Kefauver	Dwight D. Eisenhower*	Richard M. Nixon
1960	John F. Kennedy*	Lyndon B. Johnson	Richard M. Nixon	Henry Cabot Lodge
1964	Lyndon B. Johnson*	Hubert H. Humphrey	Barry M. Goldwater	William E. Miller
1968	Hubert H. Humphrey	Edmund S. Muskie	Richard M. Nixon*	Spiro T. Agnew
1972	George S. McGovern	R. Sargent Shriver Jr.	Richard M. Nixon*	Spiro T. Agnew
1976	Jimmy Carter*	Walter F. Mondale	Gerald R. Ford	Bob Dole
1980	Jimmy Carter	Walter F. Mondale	Ronald Reagan*	George H. W. Bush
1984	Walter F. Mondale	Geraldine Ferraro	Ronald Reagan*	George H. W. Bush
1988	Michael S. Dukakis	Lloyd Bentsen	George H.W. Bush*	Dan Quayle
1992	Bill Clinton*	Al Gore	George H.W. Bush	Dan Quayle
1996	Bill Clinton*	Al Gore	Bob Dole	Jack Kemp
2000	Al Gore	Joseph Lieberman	George W. Bush*	Richard Cheney

(1) Douglas and Johnson were nominated at the Baltimore convention. An earlier convention in Charleston, SC, failed to reach a consensus and resulted in a split in the party. The Southern faction of the Democrats nominated John Breckinridge for president and Joseph Lane for vice president. (2) Died Oct. 30; replaced on ballot by Nicholas Butler.

Popular and Electoral Vote for President, 1789-2000

(D) Democrat; (DR) Democratic Republican; (F) Federalist; (LR) Liberal Republican; (NR) National Republican; (P) People's; (PR) Progressive; (R) Republican; (RF) Reform; (SR) States' Rights; (W) Whig; Asterisk (*)–See notes.

Year	President elected	Popular	Elec.	Major losing candidate(s)	Popular	Elec.
1789	George Washington (F)	Unknown	69	No opposition	—	—
1792	George Washington (F)	Unknown	132	No opposition	—	—
1796	John Adams (F)	Unknown	71	Thomas Jefferson (DR)	Unknown	68
1800*	Thomas Jefferson (DR)	Unknown	73	Aaron Burr (DR)	Unknown	73
1804	Thomas Jefferson (DR)	Unknown	162	Charles Pinckney (F)	Unknown	14
1808	James Madison (DR)	Unknown	122	Charles Pinckney (F)	Unknown	47
1812	James Madison (DR)	Unknown	128	DeWitt Clinton (F)	Unknown	89
1816	James Monroe (DR)	Unknown	183	Rufus King (F)	Unknown	34
1820	James Monroe (DR)	Unknown	231	John Quincy Adams (DR)	Unknown	1
1824*	John Quincy Adams (DR)	105,321	84	Andrew Jackson (DR)	155,872	99
				Henry Clay (DR)	46,587	37
				William H. Crawford (DR)	44,282	41
1828	Andrew Jackson (D)	647,231	178	John Quincy Adams (NR)	509,097	83
1832	Andrew Jackson (D)	687,502	219	Henry Clay (NR)	530,189	49
1836	Martin Van Buren (D)	762,678	170	William H. Harrison (W)	548,007	73
1840	William H. Harrison (W)	1,275,017	234	Martin Van Buren (D)	1,128,702	60
1844	James K. Polk (D)	1,337,243	170	Henry Clay (W)	1,299,068	105
1848	Zachary Taylor (W)	1,360,101	163	Lewis Cass (D)	1,220,544	127
				Martin Van Buren (Free Soil)	291,501	—
1852	Franklin Pierce (D)	1,601,474	254	Winfield Scott (W)	1,386,578	42
1856	James Buchanan (D)	1,927,995	174	John C. Fremont (R)	1,391,555	114
				Millard Fillmore (American)	873,053	8
1860	Abraham Lincoln (R)	1,866,352	180	Stephen A. Douglas (D)	1,375,157	12
				John C. Breckinridge (D)	845,763	72
				John Bell (Const. Union)	589,581	39
1864	Abraham Lincoln (R)	2,216,067	212	George McClellan (D)	1,808,725	21
1868	Ulysses S. Grant (R)	3,015,071	214	Horatio Seymour (D)	2,709,615	80
1872*	Ulysses S. Grant (R)	3,597,070	286	Horace Greeley (D-LR)*	2,834,079	—
1876*	Rutherford B. Hayes (R)	4,033,950	185	Samuel J. Tilden (D)	4,284,757	184
1880	James A. Garfield (R)	4,449,053	214	Winfield S. Hancock (D)	4,442,030	155
1884	Grover Cleveland (D)	4,911,017	219	James G. Blaine (R)	4,848,334	182
1888*	Benjamin Harrison (R)	5,444,337	233	Grover Cleveland (D)	5,540,050	168
1892	Grover Cleveland (D)	5,554,414	277	Benjamin Harrison (R)	5,190,802	145
				James Weaver (P)	1,027,329	22
1896	William McKinley (R)	7,035,638	271	William J. Bryan (D-P)	6,467,946	176
1900	William McKinley (R)	7,219,530	292	William J. Bryan (D)	6,358,071	155
1904	Theodore Roosevelt (R)	7,628,834	336	Alton B. Parker (D)	5,084,491	140
1908	William H. Taft (R)	7,679,006	321	William J. Bryan (D)	6,409,106	162
1912	Woodrow Wilson (D)	6,286,214	435	Theodore Roosevelt (PR)	4,216,020	88
				William H. Taft (R)	3,483,922	8
1916	Woodrow Wilson (D)	9,129,606	277	Charles E. Hughes (R)	8,538,221	254
1920	Warren G. Harding (R)	16,152,200	404	James M. Cox (D)	9,147,353	127
1924	Calvin Coolidge (R)	15,725,016	382	John W. Davis (D)	8,385,586	136
				Robert M. La Follette (PR)	4,822,856	13
1928	Herbert Hoover (R)	21,392,190	444	Alfred E. Smith (D)	15,016,443	87
1932	Franklin D. Roosevelt (D)	22,821,857	472	Herbert Hoover (R)	15,761,841	59
1936	Franklin D. Roosevelt (D)	27,751,597	523	Alfred Landon (R)	16,679,583	8
1940	Franklin D. Roosevelt (D)	27,243,466	449	Wendell Willkie (R)	22,304,755	82
1944	Franklin D. Roosevelt (D)	25,602,505	432	Thomas E. Dewey (R)	22,006,278	99
1948	Harry S. Truman (D)	24,105,812	303	Thomas E. Dewey (R)	21,970,065	189
				Strom Thurmond (SR)	1,169,021	39
				Henry A. Wallace (PR)	1,157,172	—
1952	Dwight D. Eisenhower (R)	33,936,252	442	Adlai E. Stevenson (D)	27,314,992	89
1956*	Dwight D. Eisenhower (R)	35,585,316	457	Adlai E. Stevenson (D)	26,031,322	73
1960*	John F. Kennedy (D)	34,227,096	303	Richard M. Nixon (R)	34,108,546	219
1964	Lyndon B. Johnson (D)	43,126,506	486	Barry M. Goldwater (R)	27,176,799	52
1968	Richard M. Nixon (R)	31,785,480	301	Hubert H. Humphrey (D)	31,275,166	191
				George C. Wallace (3d party)	9,906,473	46
1972*	Richard M. Nixon (R)	47,165,234	520	George S. McGovern (D)	29,170,774	17
1976*	Jimmy Carter (D)	40,828,929	297	Gerald R. Ford (R)	39,148,940	240
1980	Ronald Reagan (R)	43,899,248	489	Jimmy Carter (D)	35,481,435	49
				John B. Anderson (independent)	5,719,437	—
1984	Ronald Reagan (R)	54,281,858	525	Walter F. Mondale (D)	37,457,215	13
1988*	George H. W. Bush (R)	48,881,221	426	Michael S. Dukakis (D)	41,805,422	111
1992	Bill Clinton (D)	44,908,254	370	George H. W. Bush (R)	39,102,343	168
				H. Ross Perot (independent)	19,741,065	—
1996	Bill Clinton (D)	47,401,185	379	Bob Dole (R)	39,197,469	159
				H. Ross Perot (RF)	8,085,294	—
2000*	George W. Bush (R)	50,459,211	271	Al Gore (D)	51,003,894	266
				Ralph Nader (3d party)	2,834,410	—

***1800**—Elected by House of Representatives because of tied electoral vote. **1824**—Elected by House of Representatives because no candidate had polled a majority. By 1824, the Democratic Republicans had become a loose coalition of competing political groups. By 1828, the supporters of Jackson were known as Democrats, and the John Q. Adams and Henry Clay supporters as National Republicans. **1872**—Greeley died Nov. 29, 1872. His electoral votes were split among 4 individuals. **1876**—FL, LA, OR, and SC election returns were disputed. Congress in joint session (Mar. 2, 1877) declared Hayes and Wheeler elected president and vice president. **1888**—Cleveland had more popular votes than Harrison, but since Harrison won 233 electoral votes against 168 for Cleveland, Harrison won the presidency. **1956**—Democrats elected 74 electors, but one from Alabama refused to vote for Stevenson. **1960**—Sen. Harry F. Byrd (D, VA) received 15 electoral votes. **1972**—John Hospers of California received one vote from an elector of Virginia. **1976**—Ronald Reagan of CA received one vote from an elector of Washington. **1988**—Sen. Lloyd Bentsen (D, TX) received 1 vote from an elector of West Virginia. **2000**—One Gore elector from Washington, DC, abstained.

UNITED STATES FACTS

Superlative U.S. Statistics[1]

Source: U.S. Geological Survey, Dept. of the Interior; U.S. Bureau of the Census, Dept. of Commerce; World Almanac research

Area for 50 states and Washington, DC	TOTAL	
	Land, 3,537,440 sq mi; Water, 256,648 sq mi	3,794,085 sq mi[4]
Largest state	Alaska	663,267 sq mi[1]
Smallest state	Rhode Island	1,545 sq mi
Largest county (excluding Alaska)	San Bernardino County, CA	20,105 sq mi
Smallest county	Arlington, VA	26 sq mi
Largest incorporated city	Sitka, AK	4,812 sq mi
Northernmost city	Barrow, AK	71° 17′ N
Northernmost point	Point Barrow, AK	71° 23′ N
Southernmost city	Hilo, HI	19° 44′ N
Southernmost settlement	Naalehu, HI	19° 03′ N
Southernmost point	Ka Lae (South Cape), Island of Hawaii	18° 55′ N (155°41′ W)
Easternmost city	Eastport, ME	66° 59′05″ W
Easternmost settlement[2]	Amchitka Isl., AK	179° 15′ E
Easternmost point[2]	Pochnoi Point, on Semisopochnoi Isl., AK	179° 46′ E
Westernmost city	Atka, AK	174° 12′ W
Westernmost settlement	Adak Station, AK	176° 39′ W
Westernmost point	Amatignak Isl., AK	179° 06′ W
Highest settlement	Climax, CO	11,360 ft
Lowest settlement	Calipatria, CA	–184 ft
Highest point on Atlantic coast	Cadillac Mountain, Mount Desert Isl., ME	1,530 ft
Oldest national park	Yellowstone National Park (1872), WY, MT, ID	2,219,791 acres
Largest national park	Wrangell-St. Elias, AK	8,323,618 acres
Highest waterfall	Yosemite Falls—Total in 3 sections	2,425 ft
	Upper Yosemite Fall	1,430 ft
	Cascades in middle section	675 ft
	Lower Yosemite Fall	320 ft
Longest river system	Mississippi-Missouri-Red Rock	3,710 mi
Highest mountain	Mount McKinley, AK	20,320 ft
Lowest point	Death Valley, CA	–282 ft
Deepest lake	Crater Lake, OR	1,932 ft
Rainiest spot	Mount Waialeale, HI	Annual avg rainfall 460 in
Largest gorge	Grand Canyon, Colorado River, AZ	277 mi long, 600 ft to 18 mi wide, 1 mi deep
Deepest gorge	Hells Canyon, Snake River, OR-ID	7,900 ft
Strongest surface wind	Mount Washington, NH, recorded 1934	231 mph
Largest dam	New Cornelia Tailings, Ten Mile Wash, AZ[3]	274,026,000 cu yds material used
Tallest building	Sears Tower, Chicago, IL	1,450 ft
Largest building	Boeing 747 Manufacturing Plant, Everett, WA	472,000,000 cu ft; covers 98 acres
Tallest structure	TV tower, Blanchard, ND	2,063 ft
Longest bridge span	Verrazano-Narrows, NY	4,260 ft
Highest bridge	Royal Gorge, CO	1,053 ft above water
Deepest well	Gas well, Washita County, OK	31,441 ft

The 48 Contiguous States

Area for 48 states and Washington, DC	TOTAL	3,119,887 sq mi[4]
	Land, 2,959,066 sq mi; Water, 160,824 sq mi	
Largest state	Texas	268,581 sq mi
Northernmost city	Bellingham, WA	48°46′ N
Northernmost settlement	Angle Inlet, MN	49°21′ N
Northernmost point	Northwest Angle, MN	49°23′ N
Southernmost city	Key West, FL	24°33′ N
Southernmost mainland city	Florida City, FL	25°27′ N
Southernmost point	Key West, FL	24°33′ N
Easternmost settlement	Lubec, ME	66°58′49 W
Easternmost point	West Quoddy Head, ME	66°57′W
Westernmost town	La Push, WA	124°38′ W
Westernmost point	Cape Alava, WA	124°44′ W
Highest mountain	Mount Whitney, CA	14,494 ft

(1) All areas are total area, including water, unless otherwise noted. (2) Alaska's Aleutian Islands extend into the eastern hemisphere and thus technically contain the easternmost point and settlement in the U.S. (3) The New Cornelia Tailings Dam is a privately owned industrial dam composed of tailings, remnants of a mining process. (4) Does not add, because of rounding.

Geodetic Datum of North America

In July 1986, the National Oceanic and Atmospheric Administration's National Geodetic Survey (NGS), in cooperation with Canada and Mexico, completed readjustment and redefinition of the system of latitudes and longitudes. The resulting North American Datum of 1983 (NAD 83) replaces the North American Datum of 1927, as well as local reference systems for Hawaii and for Puerto Rico and the Virgin Islands. The change was prompted by Hawaii's increased need for accurate coordinate information. To facilitate use of satellite surveying and navigation systems, such as the Global Positioning System (GPS), the new datum was redefined using the Geodetic Reference System 1980 as the reference ellipsoid because this model more closely approximates the true size and shape of the earth. In addition, the origin of the coordinate system is referenced to the mass center of the earth to coincide with the orbital orientation of the GPS satellites. Positional changes resulting from the datum redefinition can reach 330 ft in the continental U.S., Canada, and Mexico. Changes that exceed 660 ft can be expected in Alaska, Puerto Rico, and the Virgin Islands. Hawaii's coordinates changed about 1,300 ft.

Additional Statistical Information About the U.S.

The annual *Statistical Abstract of the United States,* published by U.S. Dept. of Commerce, contains additional social, political, and economic data about the U.S. For information on this and other printed publications, write to: Superintendent of Documents, Government Printing Office, PO Box 371954, Pittsburgh, PA 15250-7954, or call (202) 512-1800. For information on electronic products, write to: U.S. Dept. of Commerce, U.S. Census Bureau, MS1921, PO Box 277943, Atlanta, GA 30384-7943, or call (301) 457-4100. Parts of *The Statistical Abstract* can be viewed on the Internet at http://www.census.gov/statab/www

Highest and Lowest Altitudes in U.S. States and Territories

Source: U.S. Geological Survey, Dept. of the Interior

(Minus sign means below sea level.)

	HIGHEST POINT			LOWEST POINT		
	Name	County	Elev. (ft)	Name	County	Elev. (ft)
Alabama	Cheaha Mountain	Cleburne	2,405	Gulf of Mexico		Sea level
Alaska	Mount McKinley	Denali	20,320	Pacific Ocean		Sea level
Arizona	Humphreys Peak	Coconino	12,633	Colorado R	Yuma	70
Arkansas	Magazine Mountain	Logan	2,753	Ouachita R	Ashley-Union	55
California	Mount Whitney	Inyo-Tulare	14,494	Death Valley	Inyo	−282
Colorado	Mount Elbert	Lake	14,433	Arkansas R	Prowers	3,350
Connecticut	Mount Frissell	Litchfield	2,380	Long Island Sound		Sea level
Delaware	On Ebright Road	New Castle	448	Atlantic Ocean		Sea level
Dist. of Columbia	Tenleytown	N W part	410	Potomac R		1
Florida	Sec. 30, T6N, R20W[1]	Walton	345	Atlantic Ocean		Sea level
Georgia	Brasstown Bald	Towns-Union	4,784	Atlantic Ocean		Sea level
Guam	Mount Lamlam	Agat District	1,332	Pacific Ocean		Sea level
Hawaii	Mauna Kea	Hawaii	13,796	Pacific Ocean		Sea level
Idaho	Borah Peak	Custer	12,662	Snake R	Nez Perce	710
Illinois	Charles Mound	Jo Daviess	1,235	Mississippi R	Alexander	279
Indiana	Franklin Township	Wayne	1,257	Ohio R	Posey	320
Iowa	Sec. 29, T100N, R41W[1]	Osceola	1,670	Mississippi R	Lee	480
Kansas	Mount Sunflower	Wallace	4,039	Verdigris R	Montgomery	679
Kentucky	Black Mountain	Harlan	4,145	Mississippi R	Fulton	257
Louisiana	Driskill Mountain	Bienville	535	New Orleans	Orleans	−8
Maine	Mount Katahdin	Piscataquis	5,267	Atlantic Ocean		Sea level
Maryland	Backbone Mountain	Garrett	3,360	Atlantic Ocean		Sea level
Massachusetts	Mount Greylock	Berkshire	3,487	Atlantic Ocean		Sea level
Michigan	Mount Arvon	Baraga	1,979	Lake Erie	Monroe	571
Minnesota	Eagle Mountain	Cook	2,301	Lake Superior		600
Mississippi	Woodall Mountain	Tishomingo	806	Gulf of Mexico		Sea level
Missouri	Taum Sauk Mt.	Iron	1,772	St. Francis R	Dunklin	230
Montana	Granite Peak	Park	12,799	Kootenai R	Lincoln	1,800
Nebraska	Johnson Township	Kimball	5,424	Missouri R	Richardson	840
Nevada	Boundary Peak	Esmeralda	13,140	Colorado R	Clark	479
New Hampshire	Mt. Washington	Coos	6,288	Atlantic Ocean		Sea level
New Jersey	High Point	Sussex	1,803	Atlantic Ocean		Sea level
New Mexico	Wheeler Peak	Taos	13,161	Red Bluff Res.	Eddy	2,842
New York	Mount Marcy	Essex	5,344	Atlantic Ocean		Sea level
North Carolina	Mount Mitchell	Yancey	6,684	Atlantic Ocean		Sea level
North Dakota	White Butte	Slope	3,506	Red R	Pembina	750
Ohio	Campbell Hill	Logan	1,549	Ohio R	Hamilton	455
Oklahoma	Black Mesa	Cimarron	4,973	Little R	McCurtain	289
Oregon	Mount Hood	Clackamas-Hood R.	11,239	Pacific Ocean		Sea level
Pennsylvania	Mt. Davis	Somerset	3,213	Delaware R	Delaware	Sea level
Puerto Rico	Cerro de Punta	Ponce District	4,390	Atlantic Ocean		Sea level
Rhode Island	Jerimoth Hill	Providence	812	Atlantic Ocean		Sea level
Samoa	Lata Mountain	Tau Island	3,160	Pacific Ocean		Sea level
South Carolina	Sassafras Mountain	Pickens	3,560	Atlantic Ocean		Sea level
South Dakota	Harney Peak	Pennington	7,242	Big Stone Lake	Roberts	966
Tennessee	Clingmans Dome	Sevier	6,643	Mississippi R	Shelby	178
Texas	Guadalupe Peak	Culberson	8,749	Gulf of Mexico		Sea level
Utah	Kings Peak	Duchesne	13,528	Beaverdam Wash	Washington	2,000
Vermont	Mount Mansfield	Lamoille	4,393	Lake Champlain		95
Virginia	Mount Rogers	Grayson-Smyth	5,729	Atlantic Ocean		Sea level
Virgin Islands	Crown Mountain	St. Thomas Island	1,556	Atlantic Ocean		Sea level
Washington	Mount Rainier West	Pierce	14,410	Pacific Ocean		Sea level
West Virginia	Spruce Knob	Pendleton	4,861	Potomac R	Jefferson	240
Wisconsin	Timms Hill	Price	1,951	Lake Michigan		579
Wyoming	Gannett Peak	Fremont	13,804	Belle Fourche R	Crook	3,099

(1) Sec.=section; T=township; R=range; N=north; W=west.

U.S. Coastline by States

Source: National Oceanic and Atmospheric Administration, U.S. Dept. of Commerce

(in statute miles)

	Coastline[1]	Shoreline[2]
ATLANTIC COAST	**2,069**	**28,673**
Connecticut	0	618
Delaware	28	381
Florida	580	3,331
Georgia	100	2,344
Maine	228	3,478
Maryland	31	3,190
Massachusetts	192	1,519
New Hampshire	13	131
New Jersey	130	1,792
New York	127	1,850
North Carolina	301	3,375
Pennsylvania	0	89
Rhode Island	40	384
South Carolina	187	2,876
Virginia	112	3,315

	Coastline[1]	Shoreline[2]
GULF COAST	**1,631**	**17,141**
Alabama	53	607
Florida	770	5,095
Louisiana	397	7,721
Mississippi	44	359
Texas	367	3,359
PACIFIC COAST	**7,623**	**40,298**
Alaska	5,580	31,383
California	840	3,427
Hawaii	750	1,052
Oregon	296	1,410
Washington	157	3,026
ARCTIC COAST	**1,060**	**2,521**
UNITED STATES	**12,383**	**88,633**

(1) Figures are lengths of general outline of seacoast. Measurements were made with a unit measure of 30 minutes of latitude on charts as near the scale of 1:1,200,000 as possible. Coastline of sounds and bays is included to a point where they narrow to width of unit measure, and includes the distance across at such point. (2) Figures obtained in 1939-40 with a recording instrument on the largest-scale charts and maps then available. Shoreline of outer coast, offshore islands, sounds, bays, rivers, and creeks is included to the head of tidewater or to a point where tidal waters narrow to a width of 100 ft.

Key Data for the 50 States

The 13 colonies that declared independence from Great Britain and fought the War of Independence (American Revolution) became the 13 original states. They were (in the order in which they ratified the Constitution): Delaware, Pennsylvania, New Jersey, Georgia, Connecticut, Massachusetts, Maryland, South Carolina, New Hampshire, Virginia, New York, North Carolina, and Rhode Island.

			Entered Union		Extent in miles		Area in sq mi			Rank
State	Settled[1]	Capital	Date	Order	Long (approx.)	Wide (mean)	Land	Water	Total	in area[2]
AL	1702	Montgomery	Dec. 14, 1819	22	330	190	50,744	1,675	52,419	30
AK	1784	Juneau	Jan. 3, 1959	49	1,480[3]	810	571,951	91,316	663,267	1
AZ	1776	Phoenix	Feb. 14, 1912	48	400	310	113,635	364	113,998	6
AR	1686	Little Rock	June 15, 1836	25	260	240	52,068	1,110	53,179	29
CA	1769	Sacramento	Sept. 9, 1850	31	770	250	155,959	7,736	163,696	3
CO	1858	Denver	Aug. 1, 1876	38	380	280	103,718	376	104,094	8
CT	1634	Hartford	Jan. 9, 1788	5	110	70	4,845	699	5,543	48
DE	1638	Dover	Dec. 7, 1787	1	100	30	1,954	536	2,489	49
DC	NA	NA	NA	NA	. . .	. . .	61	7	68	51
FL	1565	Tallahassee	Mar. 3, 1845	27	500	160	53,927	11,828	65,755	22
GA	1733	Atlanta	Jan. 2, 1788	4	300	230	57,906	1,519	59,425	24
HI	1820	Honolulu	Aug. 21, 1959	50	. . .	. . .	6,423	4,508	10,931	43
ID	1842	Boise	July 3, 1890	43	570	300	82,747	823	83,570	14
IL	1720	Springfield	Dec. 3, 1818	21	390	210	55,584	2,331	57,914	25
IN	1733	Indianapolis	Dec. 11, 1816	19	270	140	35,867	551	36,418	38
IA	1788	Des Moines	Dec. 28, 1846	29	310	200	55,869	402	56,272	26
KS	1727	Topeka	Jan. 29, 1861	34	400	210	81,815	462	82,277	15
KY	1774	Frankfort	June 1, 1792	15	380	140	39,728	681	40,409	37
LA	1699	Baton Rouge	Apr. 30, 1812	18	380	130	43,562	8,278	51,840	31
ME	1624	Augusta	Mar. 15, 1820	23	320	190	30,862	4,523	35,385	39
MD	1634	Annapolis	Apr. 28, 1788	7	250	90	9,774	2,633	12,407	42
MA	1620	Boston	Feb. 6, 1788	6	190	50	7,840	2,715	10,555	44
MI	1668	Lansing	Jan. 26, 1837	26	490	240	56,804	39,912	96,716	11
MN	1805	St. Paul	May 11, 1858	32	400	250	79,610	7,329	86,939	12
MS	1699	Jackson	Dec. 10, 1817	20	340	170	46,907	1,523	48,430	32
MO	1735	Jefferson City	Aug. 10, 1821	24	300	240	68,886	818	69,704	21
MT	1809	Helena	Nov. 8, 1889	41	630	280	145,552	1,490	147,042	4
NE	1823	Lincoln	Mar. 1, 1867	37	430	210	76,872	481	77,354	16
NV	1849	Carson City	Oct. 31, 1864	36	490	320	109,826	735	110,561	7
NH	1623	Concord	June 21, 1788	9	190	70	8,968	382	9,350	46
NJ	1660	Trenton	Dec. 18, 1787	3	150	70	7,417	1,304	8,721	47
NM	1610	Santa Fe	Jan. 6, 1912	47	370	343	121,356	234	121,589	5
NY	1614	Albany	July 26, 1788	11	330	283	47,214	7,342	54,556	27
NC	1660	Raleigh	Nov. 21, 1789	12	500	150	48,711	5,108	53,819	28
ND	1812	Bismarck	Nov. 2, 1889	39	340	211	68,976	1,724	70,700	19
OH	1788	Columbus	Mar. 1, 1803	17	220	220	40,948	3,877	44,825	34
OK	1889	Oklahoma City	Nov. 16, 1907	46	400	220	68,667	1,231	69,898	20
OR	1811	Salem	Feb. 14, 1859	33	360	261	95,997	2,384	98,381	9
PA	1682	Harrisburg	Dec. 12, 1787	2	283	160	44,817	1,239	46,055	33
RI	1636	Providence	May 29, 1790	13	40	30	1,045	500	1,545	50
SC	1670	Columbia	May 23, 1788	8	260	200	30,109	911	32,020	40
SD	1859	Pierre	Nov. 2, 1889	40	380	210	75,885	1,232	77,116	17
TN	1769	Nashville	June 1, 1796	16	440	120	41,217	926	42,143	36
TX	1682	Austin	Dec. 29, 1845	28	790	660	261,797	6,784	268,581	2
UT	1847	Salt Lake City	Jan. 4, 1896	45	350	270	82,144	2,755	84,899	13
VT	1724	Montpelier	Mar. 4, 1791	14	160	80	9,250	365	9,614	45
VA	1607	Richmond	June 25, 1788	10	430	200	39,594	3,180	42,774	35
WA	1811	Olympia	Nov. 11, 1889	42	360	240	66,544	4,756	71,300	18
WV	1727	Charleston	June 20, 1863	35	240	130	24,078	152	24,230	41
WI	1766	Madison	May 29, 1848	30	310	260	54,310	11,188	65,498	23
WY	1834	Cheyenne	July 10, 1890	44	360	280	97,100	713	97,814	10

Note: Land and water areas may not add to totals because of rounding. NA=Not applicable. (1) First permanent settlement by Europeans. (2) Rank is based on total area, including inland and coastal waters. (3) Aleutian Islands and Alexander Archipelago are not considered in these measurements.

IT'S A FACT: At nearly 7,000 feet above sea level, Sante Fe, New Mexico, is the highest capital city in the U.S. The Palace of Governors, built there in 1610, is the oldest government building in the U.S.

The Continental Divide of the U.S.

The Continental Divide of the U.S., also known as the Great Divide, is located at the watershed created by the mountain ranges, or tablelands, of the Rocky Mountains. This watershed separates the waters that drain easterly into the Atlantic Ocean and its marginal seas, such as the Gulf of Mexico, from those waters that drain westerly into the Pacific Ocean. The majority of easterly flowing water in the U.S. drains into the Gulf of Mexico before reaching the Atlantic Ocean. The majority of westerly flowing water, before reaching the Pacific Ocean, drains either through the Columbia River or through the Colorado River, which flows into the Gulf of California before reaching the Pacific Ocean.

The location and route of the Continental Divide across the U.S. can briefly be described as follows:

Beginning at point of crossing the U.S.-Mexican boundary, near long. 108° 45´ W, the Divide, in a northerly direction, crosses New Mexico along the W edge of the Rio Grande drainage basin, entering Colorado near long. 106° 41´ W.

From there by a very irregular route north across Colorado along the W summits of the Rio Grande and of the Arkansas, the South Platte, and the North Platte river basins, and across Rocky Mountain National Park, entering Wyoming near long. 106° 52´ W.

From there in a northwesterly direction, forming the W rims of the North Platte, the Big Horn, and the Yellowstone river basins, crossing the SW portion of Yellowstone National Park.

From there in a westerly and then a northerly direction forming the common boundary of Idaho and Montana, to a point on said boundary near long. 114° 00´ W.

From there northeasterly and northwesterly through Montana and the Glacier National Park, entering Canada near long. 114° 04´ W.

Chronological List of Territories, With State Admissions to Union

Source: National Archives and Records Service

Name of territory	Date of act creating territory	When act took effect	Admission as state	Yrs. terr.
Northwest Territory[1]	July 13, 1787	No fixed date	Mar. 1, 1803[2]	16
Territory southwest of River Ohio	May 26, 1790	No fixed date	June 1, 1796[3]	6
Mississippi	Apr. 7, 1798	When president acted	Dec. 10, 1817	19
Indiana	May 7, 1800	July 4, 1800	Dec. 11, 1816	16
Orleans	Mar. 26, 1804	Oct. 1, 1804	Apr. 30, 1812[4]	7
Michigan	Jan. 11, 1805	June 30, 1805	Jan. 26, 1837	31
Louisiana-Missouri[5]	Mar. 3, 1805	July 4, 1805	Aug. 10, 1821	16
Illinois	Feb. 3, 1809	Mar. 1, 1809	Dec. 3, 1818	9
Alabama	Mar. 3, 1817	When MS became a state	Dec. 14, 1819	2
Arkansas	Mar. 2, 1819	July 4, 1819	June 15, 1836	17
Florida	Mar. 30, 1822	No fixed date	Mar. 3, 1845	23
Wisconsin	Apr. 20, 1836	July 3, 1836	May 29, 1848	12
Iowa	June 12, 1838	July 3, 1838	Dec. 28, 1846	8
Oregon	Aug. 14, 1848	Date of act	Feb. 14, 1859	10
Minnesota	Mar. 3, 1849	Date of act	May 11, 1858	9
New Mexico	Sept. 9, 1850	On president's proclamation	Jan. 6, 1912	61
Utah	Sept. 9, 1850	Date of act	Jan. 4, 1896	46
Washington	Mar. 2, 1853	Date of act	Nov. 11, 1889	36
Nebraska	May 30, 1854	Date of act	Mar. 1, 1867	12
Kansas	May 30, 1854	Date of act	Jan. 29, 1861	6
Colorado	Feb. 28, 1861	Date of act	Aug. 1, 1876	15
Nevada	Mar. 2, 1861	Date of act	Oct. 31, 1864	3
Dakota	Mar. 2, 1861	Date of act	Nov. 2, 1889	28
Arizona	Feb. 24, 1863	Date of act	Feb. 14, 1912	49
Idaho	Mar. 3, 1863	Date of act	July 3, 1890	27
Montana	May 26, 1864	Date of act	Nov. 8, 1889	25
Wyoming	July 25, 1868	When officers were qualified	July 10, 1890	22
Alaska[6]	May 17, 1884	No fixed date	Jan. 3, 1959	75
Oklahoma	May 2, 1890	Date of act	Nov. 16, 1907	17
Hawaii	Apr. 30, 1900	June 14, 1900	Aug. 21, 1959	59

(1) Included what is now Ohio, Indiana, Illinois, Michigan, Wisconsin, E Minnesota. (2) Whole territory admitted as the state of Ohio. (3) Admitted as the state of Tennessee. (4) Admitted as the state of Louisiana. (5) The act creating Missouri Territory (June 4, 1812) became effective Dec. 7, 1812. (6) Although the May 17, 1884, act actually constituted Alaska as a district, it was often referred to as a territory, and administered as such. The Territory of Alaska was formally organized by an act of Aug. 24, 1912.

Geographic Centers, U.S. and Each State

Source: U.S. Geological Survey, Dept. of the Interior

There is no generally accepted definition of geographic center and no uniform method for determining it. Following the U.S. Geological Survey, the geographic center of an area is defined here as the center of gravity of the surface, or that point on which the surface would balance if it were a plane of uniform thickness. All locations in the following list are approximate.

No marked or monumented point has been officially established by any government agency as the geographic center of the 50 states, the conterminous U.S. (48 states), or the North American continent. A group of private citizens erected a monument in Lebanon, KS, marking it as geographic center of the conterminous U.S., and a cairn erected in Rugby, ND, asserts that location as the center of the North American continent.

Geographic centers as reported by the U.S. Geological Survey are indicated below:

United States, including Alaska and Hawaii—W of Castle Rock, Butte County, SD; lat. 44° 58′ N, long. 103° 46′ W
Conterminous U.S. (48 states)—Near Lebanon, Smith Co., Kansas, lat. 39° 50′ N, long. 98° 35′ W
North American continent—6 mi W of Balta, Pierce County, North Dakota; lat. 48° 10′ N, long. 100° 10′ W

Alabama—Chilton, 12 mi SW of Clanton
Alaska—lat. 63° 50′ N, long. 152° W; approx. 60 mi NW of Mt. McKinley
Arizona—Yavapai, 55 mi E-SE of Prescott
Arkansas—Pulaski, 12 mi NW of Little Rock
California—Madera, 38 mi E of Madera
Colorado—Park, 30 mi NW of Pikes Peak
Connecticut—Hartford, at East Berlin
Delaware—Kent, 11 mi S of Dover
District of Columbia—Near 4th and L Sts. NW
Florida—Hernando, 12 mi N-NW of Brooksville
Georgia—Twiggs, 18 mi SE of Macon
Hawaii—Hawaii, lat. 20° 15′ N, long. 156° 20′ W, off Maui Is.
Idaho—Custer, SW of Challis
Illinois—Logan, 28 mi NE of Springfield
Indiana—Boone, 14 mi N-NW of Indianapolis
Iowa—Story, 5 mi NE of Ames
Kansas—Barton, 15 mi NE of Great Bend
Kentucky—Marion, 3 mi N-NW of Lebanon
Louisiana—Avoyelles, 3 mi SE of Marksville
Maine—Piscataquis, 18 mi N of Dover
Maryland—Prince George's, 4.5 mi NW of Davidsonville
Massachusetts—Worcester, N part of city
Michigan—Wexford, 5 mi N-NW of Cadillac
Minnesota—Crow Wing, 10 mi SW of Brainerd
Mississippi—Leake, 9 mi W-NW of Carthage
Missouri—Miller, 20 mi SW of Jefferson City
Montana—Fergus, 11 mi W of Lewistown
Nebraska—Custer, 10 mi NW of Broken Bow
Nevada—Lander, 26 mi SE of Austin
New Hampshire—Belknap, 3 mi E of Ashland
New Jersey—Mercer, 5 mi SE of Trenton
New Mexico—Torrance, 12 mi S-SW of Willard
New York—Madison, 12 mi S of Oneida and 26 mi SW of Utica
North Carolina—Chatham, 10 mi NW of Sanford
North Dakota—Sheridan, 5 mi SW of McClusky
Ohio—Delaware, 25 mi N-NE of Columbus
Oklahoma—Oklahoma, 8 mi N of Oklahoma City
Oregon—Crook, 25 mi S-SE of Prineville
Pennsylvania—Centre, 2.5 mi SW of Bellefonte
Rhode Island—Kent, 1 mi S-SW of Crompton
South Carolina—Richland, 13 mi SE of Columbia
South Dakota—Hughes, 8 mi NE of Pierre
Tennessee—Rutherford, 5 mi NE of Murfreesboro
Texas—McCulloch, 15 mi NE of Brady
Utah—Sanpete, 3 mi N of Manti
Vermont—Washington, 3 mi E of Roxbury
Virginia—Buckingham, 5 mi SW of Buckingham
Washington—Chelan, 10 mi W-SW of Wenatchee
West Virginia—Braxton, 4 mi E of Sutton
Wisconsin—Wood, 9 mi SE of Marshfield
Wyoming—Fremont, 58 mi E-NE of Lander

International Boundary Lines of the U.S.

The length of the N boundary of the conterminous U.S.—the U.S.-Canadian border, excluding Alaska—is 3,987 mi according to the U.S. Geological Survey, Dept. of the Interior. The length of the Alaskan-Canadian border is 1,538 mi. The length of the U.S.-Mexican border, from the Gulf of Mexico to the Pacific Ocean, is approximately 1,933 mi (1963 boundary agreement).

Origins of the Names of U.S. States

Source: State officials, Smithsonian Institution, and Topographic Division, U.S. Geological Survey, Dept. of the Interior

Alabama—Indian for tribal town, later a tribe (Alabamas or Alibamons) of the Creek confederacy.

Alaska—Russian version of Aleutian (Eskimo) word, *alakshak*, for "peninsula," "great lands," or "land that is not an island."

Arizona—Spanish version of Pima Indian word for "little spring place," or Aztec *arizuma*, meaning "silver-bearing."

Arkansas—Algonquin name for the Quapaw Indians, meaning "south wind."

California—Bestowed by the Spanish conquistadors (possibly by Cortez). It was the name of an imaginary island, an earthly paradise, in *Las Serges de Esplandian*, a Spanish romance written by Montalvo in 1510. *Baja California* (Lower California, in Mexico) was first visited by Spanish in 1533. The present U.S. state was called *Alta* (Upper) *California*.

Colorado—From Spanish for "red," first applied to Colorado River.

Connecticut—From Mohican and other Algonquin words meaning "long river place."

Delaware—Named for Lord De La Warr, early governor of Virginia; first applied to river, then to Indian tribe (Lenni-Lenape), and the state.

District of Columbia—For Christopher Columbus, 1791.

Florida—Named by Ponce de Leon *Pascua Florida*, "Flowery Easter," on Easter Sunday, 1513.

Georgia—For King George II of England, by James Oglethorpe, colonial administrator, 1732.

Hawaii—Possibly derived from native word for homeland, *Hawaiki* or *Owhyhee*.

Idaho—Said to be a coined name with an invented meaning: "gem of the mountains"; originally suggested for the Pikes Peak mining territory (Colorado), then applied to the new mining territory of the Pacific Northwest. Another theory suggests *Idaho* may be a Kiowa Apache term for the Comanche.

Illinois—French for *Illini* or "land of *Illini*," Algonquin word meaning "men" or "warriors."

Indiana—Means "land of the Indians."

Iowa—Indian word variously translated as "here I rest" or "beautiful land." Named for the Iowa R., which was named for the Iowa Indians.

Kansas—Sioux word for "south wind people."

Kentucky—Indian word that is variously translated as "dark and bloody ground," "meadowland," and "land of tomorrow."

Louisiana—Part of territory called Louisiana by Sieur de La Salle for French King Louis XIV.

Maine—From Maine, ancient French province. Also: descriptive, referring to the mainland as distinct from the many coastal islands.

Maryland—For Queen Henrietta Maria, wife of Charles I of England.

Massachusetts—From Indian tribe named after "large hill place" identified by Capt. John Smith as being near Milton, MA.

Michigan—From Chippewa words, *mici gama*, meaning "great water," after the lake of the same name.

Minnesota—From Dakota Sioux word meaning "cloudy water" or "sky-tinted water" of the Minnesota River.

Mississippi—Probably Chippewa; *mici zibi*, "great river" or "gathering-in of all the waters." Also: Algonquin word, *messipi*.

Missouri—An Algonquin Indian term meaning "river of the big canoes."

Montana—Latin or Spanish for "mountainous."

Nebraska—From Omaha or Otos Indian word meaning "broad water" or "flat river," describing the Platte River.

Nevada—Spanish, meaning "snow-clad."

New Hampshire—Named, 1629, by Capt. John Mason of Plymouth Council for his home county in England.

New Jersey—The Duke of York, 1664, gave a patent to John Berkeley and Sir George Carteret to be called Nova Caesaria, or New Jersey, after England's Isle of Jersey.

New Mexico—Spaniards in Mexico applied term to land north and west of Rio Grande in the 16th century.

New York—For Duke of York and Albany, who received patent to New Netherland from his brother Charles II and sent an expedition to capture it, 1664.

North Carolina—In 1619 Charles I gave a large patent to Sir Robert Heath to be called Province of Carolana, from *Carolus*, Latin name for Charles. A new patent was granted by Charles II to Earl of Clarendon and others. Divided into North and South Carolina, 1710.

North Dakota—*Dakota* is Sioux for "friend" or "ally."

Ohio—Iroquois word for "fine or good river."

Oklahoma—Choctaw word meaning "red man," proposed by Rev. Allen Wright, Choctaw-speaking Indian.

Oregon—Origin unknown. One theory holds that the name may have been derived from that of the Wisconsin River, shown on a 1715 French map as "Ouaricon-sint."

Pennsylvania—William Penn, the Quaker who was made full proprietor of this area by King Charles II in 1681, suggested "Sylvania," or "woodland," for his tract. The king's government owed Penn's father, Admiral William Penn, 16,000 pounds, and the land was granted as partial settlement. Charles II added the "Penn" to Sylvania, against the desires of the modest proprietor, in honor of the admiral.

Puerto Rico—Spanish for "rich port."

Rhode Island—Exact origin is unknown. One theory notes that Giovanni de Verrazano recorded an island about the size of Rhodes in the Mediterranean in 1524, but others believe the state was named *Roode Eylandt* by Adriaen Block, Dutch explorer, because of its red clay.

South Carolina—See North Carolina.

South Dakota—See North Dakota.

Tennessee—*Tanasi* was the name of Cherokee villages on the Little Tennessee River. From 1784 to 1788 this was the State of Franklin, or Frankland.

Texas—Variant of word used by Caddo and other Indians meaning "friends" or "allies," and applied to them by the Spanish in eastern Texas. Also written *Texias*, *Tejas*, *Teysas*.

Utah—From a Navajo word meaning "upper," or "higher up," as applied to a Shoshone tribe called Ute. Spanish form is *Yutta*. The English is *Uta* or *Utah*. Proposed name *Deseret*, "land of honeybees," from Book of Mormon, was rejected by Congress.

Vermont—From French words *vert* (green) and *mont* (mountain). The Green Mountains were said to have been named by Samuel de Champlain. When the state was formed, 1777, Dr. Thomas Young suggested combining *vert* and *mont* into Vermont.

Virginia—Named by Sir Walter Raleigh, who fitted out the expedition of 1584, in honor of Queen Elizabeth, the Virgin Queen of England.

Washington—Named after George Washington. When the bill creating the Territory of Columbia was introduced in the 32d Congress, the name was changed to Washington because of the existence of the District of Columbia.

West Virginia—So named when western counties of Virginia refused to secede from the U.S. in 1863.

Wisconsin—An Indian name, spelled *Ouisconsin* and *Mesconsing* by early chroniclers. Believed to mean "grassy place" in Chippewa. Congress made it *Wisconsin*.

Wyoming—From the Algonquin words for "large prairie place," "at the big plains," or "on the great plain."

Territorial Sea of the U.S.

According to a Dec. 27, 1988, proclamation by Pres. Ronald Reagan: "The territorial sea of the United States henceforth extends to 12 nautical miles from the baselines of the United States determined in accordance with international law. In accordance with international law, as reflected in the applicable provisions of the 1982 United Nations Convention on the Law of the Sea, within the territorial sea of the United States, the ships of all countries enjoy the right of innocent passage and the ships and aircraft of all countries enjoy the right of transit passage through international straits."

Major Accessions of Territory by the U.S.

Source: U.S. Dept. of the Interior; Bureau of the Census, U.S. Dept. of Commerce

Not including territories such as Panama Canal Zone and the Philippines which are no longer under U.S. jurisdiction; area figures may differ from figures for current areas given elsewhere.

	Acquisition date	Gross area (sq mi)		Acquisition date	Gross area (sq mi)		Acquisition date	Gross area (sq mi)
Territory in 1790[1]	NA	888,685	Texas	1845	390,143	Puerto Rico[2]	1899	3,435
Louisiana Purchase	1803	827,192	Oregon Territory	1846	285,580	Guam[3]	1899	212
Purchase of Florida	1819	58,560	Mexican Cession	1848	529,017	American Samoa[4]	1900	76
Other areas from Spain	1819	13,443	Gadsden Purchase	1853	29,640	U.S. Virgin Islands	1917	133
			Alaska	1867	586,412	Northern Mariana Islands[5]	1986	179
			Hawaii	1898	6,450			

NA = not applicable. (1) Includes that part of a drainage basin of Red River of the North, S of 49th parallel, sometimes considered part of Louisiana Purchase. (2) Ceded by Spain in 1898, ratified in 1899, and became the Commonwealth of Puerto Rico by Act of Congress on July 25, 1952. (3) Acquired in 1898; ratified 1899. (4) Acquired in 1899; ratified 1900. (5) Formerly a part of the U.S. administered Trust Territory of the Pacific Islands; became a U.S. commonwealth, Nov. 3, 1986.

Federally Owned Land, by State, 2000

Source: Office of Governmentwide Policy, General Services Administration; as of Sept. 30, 2000

State	Federal acreage[1]	Total acreage of state[2]	Percentage of federally owned acreage[1]
AL	1,325,698.1	32,678,400	4.057
AK	220,851,882.2	365,481,600	60.428
AZ	32,379,448.9	72,688,000	44.546
AR	3,409,646.8	33,599,360	10.148
CA	47,889,218.7	100,206,720	47.790
CO	24,108,409.8	66,485,760	36.261
CT	14,150.3	3,135,360	0.451
DE	15,557.3	1,265,920	1.229
DC	9,045.0	39,040	23.169
FL	4,599,237.0	34,721,280	13.246
GA	2,027,086.9	37,295,360	5.435
HI	638,621.8	4,105,600	15.555
ID	33,106,382.4	52,933,120	62.544
IL	590,477.8	35,795,200	1.650
IN	510,316.3	23,158,400	2.204
IA	229,867.4	35,860,480	0.641
KS	674,083.8	52,510,720	1.284
KY	1,446,750.4	25,512,320	5.671
LA	1,198,765.1	28,867,840	4.153
ME	173,153.8	19,847,680	0.872
MD	166,213.0	6,319,360	2.630
MA	71,170.3	5,034,880	1.414
MI	4,075,640.0	36,492,160	11.169
MN	4,216,552.0	51,205,760	8.235
MS	1,671,595.8	30,222,720	5.531
MO	4,798,315.7	44,248,320	10.844
MT	27,427,898.6	93,271,040	29.407
NE	650,856.7	49,031,680	1.327
NV	58,319,331.6	70,264,320	83.000
NH	758,632.1	5,768,960	13.150
NJ	123,791.0	4,813,440	2.572
NM	26,572,245.8	77,766,400	34.169
NY	221,538.5	30,680,960	0.722
NC	1,988,594.7	31,402,880	6.333
ND	2,315,952.0	44,452,480	5.210
OH	441,303.3	26,222,080	1.683
OK	1,665,802.5	44,087,680	3.778
OR	32,356,070.1	61,598,720	52.527
PA	716,527.5	28,804,480	2.488
RI	3,607.2	677,120	0.533
SC	1,109,545.1	19,374,080	5.727
SD	3,120,203.3	48,881,920	6.383
TN	2,115,019.2	26,727,680	7.913
TX	2,307,171.0	168,217,600	1.372
UT	34,001,393.8	52,696,960	64.522
VT	374,667.9	5,936,640	6.311
VA	2,279,561.3	25,496,320	8.941
WA	12,176,082.0	42,693,760	28.520
WV	1,222,434.3	15,410,560	7.932
WI	1,819,189.2	35,011,200	5.196
WY	31,069,995.2	62,343,040	49.837
TOTAL	**635,354,700.5**	**2,271,343,360**	**27.973**

Note: Totals do not include inland water. (1) Excludes trust properties. (2) Bureau of the Census, U.S. Dept. of Commerce figures.

Special Recreation Areas Administered by the U.S. Forest Service, 2000

Source: U.S. Forest Service, Dept. of Agriculture

NHS=National Historic Scenic Area; NM=National Monument; NRA=National Recreation Area; NSA=National Scenic Area; NVM=National Volcanic Monument; SRA=Scenic Recreation Area

Area name	Location	Estab.	Acres
Admiralty Island NM	AK	1980	978,881
Allegheny NRA	PA	1984	23,063
Arapaho NRA	CO	1978	30,690
Beech Creek NS & Botanic Area	OK	1988	7,500
Cascade Head NS(-Research)A	OR	1974	6,630
Columbia River Gorge NSA	OR-WA	1986	63,150
Coosa Bald NSA	GA	1991	7,100
Ed Jenkins NRA	GA	1991	23,166
Flaming Gorge NRA	WY-UT	1968	189,825
Giant Sequoia NM	CA	2000	327,769
Grand Island NRA	MI	1990	12,961
Hells Canyon NRA	ID-OR	1975	536,648
Indian Nations NS & Wildlife Area	OK	1988	40,051
Jemez NRA	NM	1993	57,000
Land Between the Lakes NRA	KY-TN	1998	170,000
Misty Fiords NM	AK	1980	2,293,428
Mono Basin NSA	CA	1984	115,600
Mount Baker NRA	WA	1984	8,473
Mount Pleasant NSA	VA	1994	7,580
Mount Rogers NRA	VA	1966	114,520
Mount St. Helens NVM	WA	1989	112,593
Newberry NVM	OR	1990	54,822
North Cascades NSA	WA	1984	87,600
Opal Creek SRA	OR	1996	13,000
Oregon Dunes NRA	OR	1972	27,212
Pine Ridge NRA	NE	1986	6,600
Rattlesnake NRA	MT	1980	59,119
Santa Rosa and San Jacinto Mts. NM	CA	2000	272,000
Sawtooth NRA	ID	1972	729,322
Smith River NRA	CA	1990	305,169
Spring Mt. NRA	NV	1993	316,000
Spruce Knob-Seneca Rocks NRA	WV	1965	57,237
Valles Caldera National Preserve	NM	2000	88,900
Whiskeytown-Shasta-Trinity NRA	CA	1965	176,367
White Rocks NRA	VT	1984	36,400
Winding Stair Mt. NRA	OK	1988	25,890

National Parks, Other Areas Administered by National Park Service

Dates when sites were authorized for initial protection by Congress or by presidential proclamation are given in parentheses. If different, the date the area got its current designation, or was transferred to the National Park Service, follows. Gross area in acres, as of Dec. 31, 2000, follows date(s). Over 84 mil acres of federal land are now administered by the National Park Service.

NATIONAL PARKS

Acadia, ME (1916/1929) 47,657. Includes Mount Desert Isl., half of Isle au Haut, Schoodic Peninsula on mainland. Highest elevation on Eastern seaboard.

American Samoa, AS (1988) 9,000. Features a paleotropical rain forest and a coral reef. No federal facilities.

Arches, UT (1929/1971) 76,519. Contains giant red sandstone arches and other products of erosion.

Badlands, SD (1929/1978) 242,756. Prairie with bison, bighorn, and antelope. Contains animal fossils from 26 to 37 mil years ago.

Big Bend, TX (1935) 801,163. Rio Grande, Chisos Mts.

Biscayne, FL (1968/1980) 172,924. Aquatic park encompassing chain of islands south of Miami.

Black Canyon of the Gunnison, CO (1933/1999) 27,705. Features a canyon 2,900 ft deep and 40 ft wide at its narrowest part.

Bryce Canyon, UT (1923/1928) 35,835. Spectacularly colorful and unusual display of erosion effects.

Canyonlands, UT (1964) 337,598. At junction of Colorado and Green rivers; extensive evidence of prehistoric Indians.

Capitol Reef, UT (1937/1971) 241,904. A 70-mi uplift of sandstone cliffs dissected by high-walled gorges.

Carlsbad Caverns, NM (1923/1930) 46,766. Largest known caverns; not yet fully explored.

Channel Islands, CA (1938/1980) 249,561. Sea lion breeding place, nesting sea birds, unique plants.

Crater Lake, OR (1902) 183,224. Extraordinary blue lake in the crater of Mt. Mazama, a volcano that erupted about 7,700 years ago; deepest U.S. lake.

Death Valley, CA-NV (1933/1994) 3,286,241. Large desert area. Includes the lowest point in the Western Hemisphere; also includes Scottys Castle.

Denali, AK (1917/1980) 4,740,912. Name changed from Mt. McKinley NP. Contains highest mountain in U.S.; wildlife.

Dry Tortugas, FL (1935/1992) 64,701. Formerly Ft. Jefferson National Monument.

Everglades, FL (1934) 1,508,571. Largest remaining subtropical wilderness in continental U.S.

Gates of the Arctic, AK (1978/1984) 7,523,898. Vast wilderness in north central region. Limited federal facilities.

Glacier, MT (1910) 1,013,572. Superb Rocky Mt. scenery, numerous glaciers and glacial lakes. Part of Waterton-Glacier Intl. Peace Park established by U.S. and Canada in 1932.

Glacier Bay, AK (1925/1986) 3,224,840. Great tidewater glaciers that move down mountainsides and break up into the sea; much wildlife.

Grand Canyon, AZ (1893/1919) 1,217,403. Most spectacular part of Colorado River's greatest canyon.

Grand Teton, WY (1929) 309,994. Most impressive part of the Teton Mts., winter feeding ground of largest American elk herd.

Great Basin, NV (1922/1986) 77,180. Includes Wheeler Pk., Lexington Arch, and Lehman Caves.

Great Smoky Mountains, NC-TN (1926/1934) 521,490. Largest Eastern mountain range, magnificent forests.

Guadalupe Mountains, TX (1966) 86,416. Extensive Permian limestone fossil reef; tremendous earth fault.

Haleakala, HI (1916/1960) 29,830. Dormant volcano on Maui with large colorful craters.

Hawaii Volcanoes, HI (1916/1961) 209,695. Contains Kilauea and Mauna Loa, active volcanoes.

Hot Springs, AR (1832/1921) 5,549. Bathhouses are furnished with thermal waters from the park's 47 hot springs; these waters are used for bathing and drinking.

Isle Royale, MI (1931) 571,790. Largest island in Lake Superior, noted for its wilderness area and wildlife.

Joshua Tree, CA (1936/1994) 1,018,198. Desert region includes Joshua trees, other plant and animal life.

Katmai, AK (1918/1980) 3,674,530. "Valley of Ten Thousand Smokes," scene of 1912 volcanic eruption.

Kenai Fjords, AK (1978/1980) 669,983. Abundant marine mammals, birdlife; the Harding Icefield, one of the 4 major icecaps in U.S.

Kings Canyon, CA (1890/1940) 461,901. Mountain wilderness, dominated by Kings River Canyons and High Sierra; contains giant sequoias.

Kobuk Valley, AK (1978/1980) 1,750,737. Contains geological and recreational sites. Limited federal facilities.

Lake Clark, AK (1978/1980) 2,619,733. Across Cook Inlet from Anchorage. A scenic wilderness rich in fish and wildlife. Limited federal facilities.

Lassen Volcanic, CA (1907/1916) 106,372. Contains Lassen Peak, recently active volcano, and other volcanic phenomena.

Mammoth Cave, KY (1926/1941) 52,830. 144 mi of surveyed underground passages, beautiful natural formations, river 300 ft below surface.

Mesa Verde, CO (1906) 52,122. Most notable and best preserved prehistoric cliff dwellings in the U.S.

Mount Rainier, WA (1899) 235,625. Greatest single-peak glacial system in the U.S.

North Cascades, WA (1968) 504,781. Spectacular mountainous region with many glaciers, lakes.

Olympic, WA (1909/1938) 922,651. Mountain wilderness containing finest remnant of Pacific Northwest rain forest, active glaciers, Pacific shoreline, rare elk.

Petrified Forest, AZ (1906/1962) 93,533. Extensive petrified wood and Indian artifacts. Contains part of Painted Desert.

Redwood, CA (1968) 112,613. 40 mi of Pacific coastline, groves of ancient redwoods and world's tallest trees.

Rocky Mountain, CO (1915) 265,769. On the Continental Divide; includes peaks over 14,000 ft.

Saguaro, AZ (1933/1994) 91,446. Part of the Sonoran Desert; includes the giant saguaro cacti, unique to the region.

Sequoia, CA (1890) 402,510. Groves of giant sequoias, highest mountain in conterminous U.S.—Mt. Whitney (14,494 ft). World's largest tree.

Shenandoah, VA (1926) 198,979. Portion of the Blue Ridge Mts.; overlooks Shenandoah Valley; Skyline Drive.

Theodore Roosevelt, ND (1947/1978) 70,447. Contains part of T.R.'s ranch and scenic badlands.

Virgin Islands, VI (1956) 14,689. Authorized to cover 75% of St. John Isl. and Hassel Isl.; lush growth, lovely beaches, Carib Indian petroglyphs, evidence of colonial Danes.

Voyageurs, MN (1971) 218,200. Abundant lakes, forests, wildlife, canoeing, boating.

Wind Cave, SD (1903) 28,295. Limestone caverns in Black Hills. Extensive wildlife includes a herd of bison.

Wrangell-St. Elias, AK (1978/1980) 8,323,618. Largest area in park system, most peaks over 16,000 ft, abundant wildlife; day's drive east of Anchorage. Limited federal facilities.

Yellowstone, ID-MT-WY (1872) 2,219,791. World's first national park. World's greatest geyser area has about 10,000 geysers and hot springs; spectacular falls and impressive canyons of the Yellowstone River; grizzly bear, moose, and bison.

Yosemite, CA (1890) 761,266. Yosemite Valley, the nation's highest waterfall, grove of sequoias, and mountains.

Zion, UT (1909/1919) 146,592. Unusual shapes and landscapes have resulted from erosion and faulting; evidence of past volcanic activity; Zion Canyon, with sheer walls ranging up to 2,640 ft, is readily accessible.

NATIONAL HISTORICAL PARKS

Appomattox Court House, VA (1930/1954) 1,775. Where Lee surrendered to Grant.

Boston, MA (1974) 43. Includes Faneuil Hall, Old North Church, Bunker Hill, Paul Revere House.

Cane River Creole (and heritage area), LA (1994) 207. Preserves the Creole culture as it developed along the Cane R.

Chaco Culture, NM (1907/1980) 33,974. Ruins of pueblos built by prehistoric Indians.

Chesapeake and Ohio Canal, MD-DC-WV (1938/1971) 19,593. 184-mi historic canal; DC to Cumberland, MD.

Colonial, VA (1930/1936) 9,349. Includes most of Jamestown Isl., site of first successful English colony; Yorktown, site of Cornwallis's surrender to George Washington; and the Colonial Parkway.

Cumberland Gap, KY-TN-VA (1940) 20,454. Mountain pass of the Wilderness Road, which carried the first great migration of pioneers into America's interior.

Dayton Aviation Heritage, OH (1992) 86. Commemorates the area's aviation heritage.

George Rogers Clark, Vincennes, IN (1966) 26. Commemorates American defeat of British in West during Revolution.

Harpers Ferry, MD-VA-WV (1944/1963) 2,343. At the confluence of the Shenandoah and Potomac rivers, the site of John Brown's 1859 raid on the Army arsenal.

Hopewell Culture, OH (1923/1992) 1,170. Formerly Mound City Group National Monument.

Independence, PA (1948) 45. Contains several properties associated with the American Revolution and the founding of the U.S. Includes Independence Hall.

Jean Laffite (and preserve), LA (1907/1978) 20,020. Includes Chalmette, site of 1815 Battle of New Orleans; French Quarter.

Kalaupapa, HI (1980) 10,779. Molokai's former leper colony site and other historic areas.

Kaloko-Honokohau, HI (1978) 1,161. Preserves the native culture of Hawaii. No federal facilities.

Keweenaw, MI (1992) 1,937. Site of first significant copper mine in U.S. Federal facilities are under development.

Klondike Gold Rush, AK-WA (1976) 13,191. Alaskan Trails in 1898 Gold Rush. Museum in Seattle.

Lowell, MA (1978) 141. Textile mills, canal, 19th-cent. structures; park shows planned city of Industrial Revolution.

Lyndon B. Johnson, TX (1969/1980) 1,570. President's birthplace, boyhood home, ranch.

Marsh-Billings-Rockefeller, VT (1992) 643. Boyhood home of pioneer conservationist George Perkins Marsh. No federal facilities.

Minute Man, MA (1959) 965. Where the colonial Minute Men battled the British, Apr. 19, 1775. Also contains Nathaniel Hawthorne's home.

Morristown, NJ (1933) 1,703. Sites of important military encampments during the American Revolution; Washington's headquarters, 1777, 1779-80.

Natchez, MS (1988) 108. Mansions, townhouses, and villas related to history of Natchez.

New Bedford Whaling, MA (1996) 34. Preserves structures and relics associated with the city's 19th-cent. whaling industry.

New Orleans Jazz, LA (1994) 5. Preserves, educates, and interprets jazz as it has evolved in New Orleans.

Nez Perce, ID (1965) 2,134. Illustrates the history and culture of the Nez Perce Indian country (38 separate sites).

Pecos, NM (1965/1990) 6,667. Ruins of ancient Pueblo of Pecos, archaeological sites, and 2 associated Spanish colonial missions from the 17th and 18th centuries.

Pu'uhonua o Honaunau, HI (1955/1978) 182. Until 1819, a sanctuary for Hawaiians vanquished in battle and for those guilty of crimes or breaking taboos.

Rosie the Riveter WWII Home Front, CA (2000) 145. Built on the site that was a shipyard employing thousands of women during WWII; commemorates the women who worked in war-time industries.

Salt River Bay (and ecological preserve), St. Croix, VI (1992) 948. The only site known where, 500 years ago, members of a Columbus party landed on what is now territory of the U.S.

San Antonio Missions, TX (1978) 816. Four of finest Spanish missions in U.S., 18th-cent. irrigation system.

San Francisco Maritime, CA (1988) 50. Artifacts, photographs, and historic vessels related to the development of the Pacific Coast.

San Juan Island, WA (1966) 1,752. Commemorates peaceful relations between the U.S., Canada, and Great Britain since the 1872 boundary disputes.

Saratoga, NY (1938) 3,392. Scene of a major 1777 battle that became a turning point in the American Revolution.

Sitka, AK (1910/1972) 107. Scene of last major resistance of the Tlingit Indians to the Russians, 1804.

Tumacacori, AZ (1908/1990) 46. Historic Spanish Catholic mission building stands near the site first visited by Jesuit Father Kino in 1691.

Valley Forge, PA (1976) 3,466. Continental Army campsite in 1777-78 winter.

War in the Pacific, GU (1978) 2,031. Seven distinct units illustrating the Pacific theater of WWII. Limited federal facilities.

Women's Rights, NY (1980) 7. Seneca Falls site where Lucretia Mott, Elizabeth Cady Stanton began rights movement in 1848.

NATIONAL BATTLEFIELDS

Antietam, MD (1890/1978) 3,365. Battle here ended first Confederate invasion of North, Sept. 17, 1862.

Big Hole, MT (1910/1963) 656. Site of major battle with Nez Perce Indians.

Cowpens, SC (1929/1972) 842. American Revolution battlefield.

Fort Donelson, TN-KY (1928/1985) 552. Site of first major Union victory.

Fort Necessity, PA (1931/1961) 903. Site of first battle of French and Indian War.

Monocacy, MD (1934/1976) 1,647. Civil War battle in defense of Washington, DC, fought here, July 9, 1864.

Moores Creek, NC (1926/1980) 88. 1776 battle between Patriots and Loyalists commemorated here.

Petersburg, VA (1926/1962) 2,659. Scene of 10-month Union campaigns, 1864-65.

Stones River, TN (1927/1960) 714. Scene of battle that began federal offensive to trisect the Confederacy.

Tupelo, MS (1929/1961) 1. Site of crucial battle over Sherman's supply line, 1865.

Wilson's Creek, MO (1960/1970) 1,750. Scene of Civil War battle for control of Missouri.

NATIONAL BATTLEFIELD PARKS

Kennesaw Mountain, GA (1917/1935) 2,884. Site of two major battles of Atlanta campaign in Civil War.

Manassas, VA (1940) 5,072. Scene of two battles in Civil War, 1861 and 1862.

Richmond, VA (1936) 1,718. Site of battles defending Confederate capital.

NATIONAL BATTLEFIELD SITE

Brices Cross Roads, MS (1929) 1. Civil War battlefield.

NATIONAL MILITARY PARKS

Chickamauga and Chattanooga, GA-TN (1890) 8,190. Site of major Confederate victory, 1863.

Fredericksburg and Spotsylvania County, VA (1927/1933) 8,383. Sites of several major Civil War battles and campaigns.

Gettysburg, PA (1895/1933) 5,992. Site of decisive Confederate defeat in North and of Gettysburg Address.

Guilford Courthouse, NC (1917/1933) 221. American Revolution battle site.

Horseshoe Bend, AL (1956) 2,040. On Tallapoosa River, where Gen. Andrew Jackson's forces broke the power of the Upper Creek Indian Confederacy.

Kings Mountain, SC (1931/1933) 3,945. Site of American Revolution battle.

Pea Ridge, AR (1956) 4,300. Scene of Civil War battle.

Shiloh, TN (1894/1933) 3,989. Major Civil War battlesite; includes some well-preserved Indian burial mounds.

Vicksburg, MS (1899/1933) 1,738. Union victory gave North control of the Mississippi and split the Confederate forces.

NATIONAL MEMORIALS

Arkansas Post, AR (1960) 747. First permanent French settlement in the lower Mississippi River valley.

Arlington House, the Robert E. Lee Memorial, VA (1925/1972) 28. Lee's home overlooking the Potomac.

Chamizal, El Paso, TX (1966/1974) 55. Commemorates 1963 settlement of 99-year border dispute with Mexico.

Coronado, AZ (1941/1952) 4,750. Commemorates first European exploration of the Southwest.

DeSoto, FL (1948) 27. Commemorates 16th-cent. Spanish explorations.

Federal Hall, NY (1939/1955) 0.45. First seat of U.S. government under the Constitution.

Fort Caroline, FL (1950) 138. On St. Johns River, overlooks site of a French Huguenot colony.

Fort Clatsop, OR (1958) 125. Lewis and Clark encampment, 1805-6.

Franklin Delano Roosevelt, DC (1982) 8. Statues of Pres. Roosevelt and Eleanor Roosevelt, as well as waterfalls and gardens.

General Grant, NY (1958) 0.76. Tomb of Grant and wife.

Hamilton Grange, NY (1962) 1. Home of Alexander Hamilton.

Jefferson National Expansion Memorial, St. Louis, MO (1935) 193. Commemorates westward expansion.

Johnstown Flood, PA (1964) 164. Commemorates tragic flood of 1889.

Korean War Veterans, DC (1986) 2. Dedicated in 1995; honors those who served in the Korean War.

Lincoln Boyhood, IN (1962) 200. Lincoln grew up here.

Lincoln Memorial, DC (1911/1933) 107. Marble statue of the 16th U.S. president.

Lyndon B. Johnson Memorial Grove on the Potomac, DC (1973) 17. Overlooks the Potomac R.; vista of the Capitol.

Mount Rushmore, SD (1925) 1,278. World-famous sculpture of 4 presidents.

Oklahoma City, OK (1997) 6. Commemorates site of April 19, 1995, bombing which killed 168.

Perry's Victory and International Peace Memorial, Put-in-Bay, OH (1936/1972) 25. The world's most massive Doric column, constructed 1912-15, promotes pursuit of peace through arbitration and disarmament.

Roger Williams, Providence, RI (1965) 5. Memorial to founder of Rhode Island.

Thaddeus Kosciuszko, PA (1972) 0.02. Memorial to Polish hero of American Revolution.

Theodore Roosevelt Island, DC (1932/1933) 89. Statue of Roosevelt in wooded island sanctuary.

Thomas Jefferson Memorial, DC (1934) 18. Statue of Jefferson in an inscribed circular, colonnaded structure.

USS Arizona, HI (1980) 11. Memorializes American losses at Pearl Harbor.

Vietnam Veterans, DC (1980) 2. Black granite wall inscribed with names of those missing or killed in action in the Vietnam War.

Washington Monument, DC (1848/1933) 106. Obelisk honoring the first U.S. president.

Wright Brothers, NC (1927/1953) 428. Site of first powered flight.

NATIONAL HISTORIC SITES

Abraham Lincoln Birthplace, Hodgenville, KY (1916/1959) 117. Early 17th-cent. cabin.

Adams, Quincy, MA (1946/1952) 14. Home of Pres. John Adams, John Quincy Adams, and celebrated descendants.

Allegheny Portage Railroad, PA (1964) 1,249. Linked the Pennsylvania Canal system and the West.

Andersonville, Andersonville, GA (1970) 495. Noted Civil War prisoner-of-war camp.

Andrew Johnson, Greeneville, TN (1935/1963) 17. Two homes and the tailor shop of the 17th U.S. president.

Bent's Old Fort, CO (1960) 799. Reconstruction of S Plains outpost.

Boston African-American, MA (1980) 0.59. Pre-Civil War black history structures.

Brown v. Board of Education, KS (1992) 2. Commemorates the landmark 1954 U.S. Supreme Court decision.

Carl Sandburg Home, Flat Rock, NC (1968) 264. Poet's home.

Charles Pinckney, SC (1988) 28. Statesman's farm.

Christiansted, St. Croix, VI (1952/1961) 27. Commemorates Danish colony.

Clara Barton, MD (1974) 9. Home of founder of American Red Cross.

Edgar Allan Poe, PA (1978/1980) 0.52. Writer's home.

Edison, West Orange, NJ (1955/1962) 21. Inventor's home and laboratory.

Eisenhower, Gettysburg, PA (1967) 690. Home of 34th president.

Eleanor Roosevelt, Hyde Park, NY (1977) 181. The former first lady's personal retreat.

Eugene O'Neill, Danville, CA (1976) 13. Playwright's home.

First Ladies, Canton, OH (2000) 0.33. Library devoted to America's first ladies.

Ford's Theatre, DC (1866/1970) 0.29. Includes theater, now restored, where Lincoln was assassinated, house where he died, and Lincoln Museum.

Fort Bowie, AZ (1964) 1,000. Focal point of operations against Geronimo and the Apaches.

Fort Davis, TX (1961) 474. Key frontier outpost in West Texas.

Fort Laramie, WY (1938/1960) 833. Military post on Oregon Trail.

Fort Larned, KS (1964/1966) 718. Military post on Santa Fe Trail.

Fort Point, San Francisco, CA (1970) 29. West Coast fortification.

Fort Raleigh, NC (1941) 513. First attempted English settlement in North America.

Fort Scott, KS (1965/1978) 17. Commemorates U.S. frontier of 1840s and '50s.

Fort Smith, AR-OK (1961) 75. Active post during 1817-90.

Fort Union Trading Post, MT-ND (1966) 444. Principal fur-trading post on upper Missouri, 1829-67.

Fort Vancouver, WA (1948/1961) 209. Headquarters for Hudson's Bay Company in 1825. Early political seat.

Frederick Douglass, DC (1962/1988) 9. Home of famous black abolitionist, writer, and orator.

Frederick Law Olmsted, MA (1979) 7. Home of famous city planner.

Friendship Hill, PA (1978) 675. Home of Albert Gallatin, Jefferson's and Madison's secretary of treasury.

Golden Spike, UT (1957) 2,735. Commemorates completion of first transcontinental railroad in 1869.

Grant-Kohrs Ranch, MT (1972) 1,618. Ranch house and part of 19th-cent. ranch.

Hampton, MD (1948) 62. 18th-cent. Georgian mansion.

Harry S. Truman, MO (1983) 7. Home of Pres. Truman after 1919.

Herbert Hoover, West Branch, IA (1965) 187. Birthplace and boyhood home of 31st president.

Home of Franklin D. Roosevelt, Hyde Park, NY (1944) 800. FDR's birthplace, home, and "summer White House."

Hopewell Furnace, PA (1938/1985) 848. 19th-cent. iron-making village.

Hubbell Trading Post, AZ (1965) 160. Still active today.

James A. Garfield, Mentor, OH (1980) 8. Home of 20th president.

Jimmy Carter, GA (1987) 71. Birthplace and home of 39th president.

John Fitzgerald Kennedy, Brookline, MA (1967) 0.09. Birthplace and childhood home of 35th president.

John Muir, Martinez, CA (1964) 345. Home of early conservationist and writer.

Knife River Indian Villages, ND (1974) 1,758. Remnants of villages last occupied by Hidatsa and Mandan Indians.

Lincoln Home, Springfield, IL (1971) 12. Lincoln's residence at the time he was elected 16th president, 1860.

Little Rock Central High School, AR (1998) 18. Commemorates 1957 desegregation during which federal troops had to be called in to protect 9 black students.

Longfellow, Cambridge, MA (1972) 2. Longfellow's home, 1837-82, and Washington's headquarters during Boston siege, 1775-76.

Maggie L. Walker, VA (1978) 1. Richmond home of black leader and bank president, daughter of an ex-slave.

Manzanar, Lone Pine, CA (1992) 814. Commemorates Manzanar War Relocation Ctr., a Japanese-American internment camp during WWII. No federal facilities.

Martin Luther King Jr., Atlanta, GA (1980) 39. Birthplace, grave, church of the civil rights leader. Limited federal facilities.

Martin Van Buren, NY (1974) 40. Lindenwald, home of 8th president, near Kinderhook.

Mary McLeod Bethune Council House, DC (1982/1991) 0.07. Commemorates Bethune's leadership in the black women's movement.

Minuteman Missile, SD (1999) 15. Missile launch facilities dating back to the Cold War era.

Nicodemus, KS (1996) 161. Only remaining western town established by African-Americans during Reconstruction.

Ninety Six, SC (1976) 989. Colonial trading village.

Palo Alto Battlefield, TX (1978) 3,357. Scene of first battle of the Mexican War.

Pennsylvania Avenue, DC (1965) Acreage undetermined. Also includes area adjacent to the road between Capitol and White House, encompassing Ford's Theatre and a number of other federal structures.

Puukohola Heiau, HI (1972) 86. Ruins of temple built by King Kamehameha.

Sagamore Hill, Oyster Bay, NY (1962) 83. Home of Pres. Theodore Roosevelt from 1885 until his death in 1919.

Saint-Gaudens, Cornish, NH (1964) 148. Home, studio, and gardens of American sculptor Augustus Saint-Gaudens.

Saint Paul's Church, NY, NY (1943) 6. Site associated with John Peter Zenger's "freedom of press" trial.

Salem Maritime, MA (1938) 9. Only port never seized from the patriots by the British. Major fishing and whaling port.

Sand Creek Massacre, Sand Creek, CO (2000) 12,583. Site where over 100 Cheyenne and Arapaho Indians were killed by U.S. soldiers in 1864.

San Juan, PR (1949) 75. 16th-cent. Span. fortifications.

Saugus Iron Works, MA (1974) 9. Reconstructed 17th-cent. colonial ironworks.

Springfield Armory, MA (1974) 55. Small-arms manufacturing center for nearly 200 years.

Steamtown, PA (1986) 62. Railyard, roadhouse, repair shops of former Delaware, Lackawanna & Western Railroad.

Theodore Roosevelt Birthplace, New York, NY (1962) 0.11. Reconstructed brownstone.

Theodore Roosevelt Inaugural, Buffalo, NY (1966) 1. Wilcox House where he took oath of office, 1901.

Thomas Stone, MD (1978) 328. Home of signer of Declaration of Independence, built in 1771.

Tuskegee Airmen, AL (1998) 90. Airfield where pilots of all-black air corps unit of WWII received flight training.

Tuskegee Institute, AL (1974) 58. College founded by Booker T. Washington in 1881 for blacks.

Ulysses S. Grant, St. Louis Co., MO (1989) 10. Home of Grant during pre-Civil War years.

Vanderbilt Mansion, Hyde Park, NY (1940) 212. Mansion of 19th-cent. financier.

Washita Battlefield, OK (1996) 315. Scene of Nov. 27, 1868, battle between Plains tribes and the U.S. army.

Weir Farm, Wilton, CT (1990) 74. Home and studio of American impressionist painter J. Alden Weir.

Whitman Mission, WA (1936/1963) 98. Site where Dr. and Mrs. Marcus Whitman ministered to the Indians until slain by them in 1847.

William Howard Taft, Cincinnati, OH (1969) 3. Birthplace and early home of the 27th president.

NATIONAL MONUMENTS

Name	State	Year[1]	Acreage
Agate Fossil Beds	NE	1965	3,055
Alibates Flint Quarries	TX	1965	1,371
Aniakchak[2]	AK	1978	137,176
Aztec Ruins	NM	1923	318
Bandelier	NM	1916	33,677
Booker T. Washington	VA	1956	224
Buck Island Reef	VI	1961	880
Cabrillo	CA	1913	160
Canyon de Chelly	AZ	1931	83,840
Cape Krusenstern[3]	AK	1978	649,182
Capulin Volcano	NM	1916	793
Casa Grande Ruins	AZ	1889	473
Castillo de San Marcos	FL	1924	20
Castle Clinton	NY	1946	1
Cedar Breaks	UT	1933	6,155

Name	State	Year[1]	Acreage
Chiricahua	AZ	1924	11,985
Colorado	CO	1911	20,534
Congaree Swamp	SC	1976	21,888
Craters of the Moon	ID	1924	714,440
Devils Postpile	CA	1911	798
Devils Tower	WY	1906	1,347
Dinosaur	CO-UT	1915	210,278
Effigy Mounds	IA	1949	2,526
El Malpais	NM	1987	114,277
El Morro	NM	1906	1,279
Florissant Fossil Beds	CO	1969	5,998
Fort Frederica	GA	1936	241
Fort Matanzas	FL	1924	300
Fort McHenry National Monument and Historic Shrine	MD	1925	43
Fort Pulaski	GA	1924	5,623
Fort Stanwix	NY	1935	16
Fort Sumter	SC	1948	200
Fort Union	NM	1954	721
Fossil Butte	WY	1972	8,198
George Washington Birthplace	VA	1930	550
George Washington Carver	MO	1943	210
Gila Cliff Dwellings	NM	1907	533
Grand Portage	MN	1951	710
Great Sand Dunes National Monument and Preserve	CO	2000	80,348
Hagerman Fossil Beds[3]	ID	1988	4,351
Hohokam Pima[4]	AZ	1972	1,690
Homestead National Monument of America	NE	1936	195
Hovenweep	CO-UT	1923	785
Jewel Cave	SD	1908	1,274
John Day Fossil Beds	OR	1974	14,057
Lava Beds	CA	1925	46,560
Little Big Horn Battlefield	MT	1879	765
Montezuma Castle	AZ	1906	858
Muir Woods	CA	1908	554
Natural Bridges	UT	1908	7,636
Navajo	AZ	1909	360
Ocmulgee	GA	1934	702
Oregon Caves	OR	1909	488
Organ Pipe Cactus	AZ	1937	330,689
Petroglyph	NM	1990	7,232
Pinnacles	CA	1908	16,265
Pipe Spring	AZ	1923	40
Pipestone	MN	1937	282
Poverty Point[2]	LA	1988	911
Rainbow Bridge[3]	UT	1910	160
Russell Cave	AL	1961	310
Salinas Pueblo Missions	NM	1909	1,071
Scotts Bluff	NE	1919	3,003
Statue of Liberty	NJ-NY	1924	58
Sunset Crater Volcano	AZ	1930	3,040
Timpanogos Cave	UT	1922	250
Tonto	AZ	1907	1,120
Tuzigoot	AZ	1939	801
Walnut Canyon	AZ	1915	3,579
White Sands	NM	1933	143,733
Wupatki	AZ	1924	35,422
Yucca House[4]	CO	1919	34

NATIONAL PRESERVES

Name	State	Year[1]	Acreage
Aniakchak	AK	1978	465,603
Bering Land Bridge[3]	AK	1978	2,697,406
Big Cypress	FL	1974	720,571
Big Thicket	TX	1974	97,191
Denali	AK	1917	1,334,118
Gates of the Arctic	AK	1978	948,629
Glacier Bay	AK	1925	58,406
Katmai	AK	1918	418,699
Lake Clark	AK	1978	1,410,292
Little River Canyon[2]	AL	1992	13,633
Mojave	CA	1994	1,497,438
Noatak[3]	AK	1978	6,569,904
Tallgrass Prairie	KS	1996	10,894
Timucuan Ecological & Historic Preserve[3]	FL	1988	46,019
Wrangell-St. Elias	AK	1978	4,852,753
Yukon-Charley Rivers[3]	AK	1978	2,526,512

NATIONAL SEASHORES

Name	State	Year[1]	Acreage
Assateague Island	MD-VA	1965	39,733
Canaveral	FL	1975	57,662
Cape Cod	MA	1961	43,604
Cape Hatteras	NC	1937	30,321
Cape Lookout	NC	1966	28,243
Cumberland Island	GA	1972	36,415
Fire Island	NY	1964	19,580
Gulf Islands	FL-MS	1971	137,458
Padre Island	TX	1962	130,434
Point Reyes	CA	1962	71,068

NATIONAL PARKWAYS

Name	State	Year[1]	Acreage
Blue Ridge	NC-VA	1933	90,860
George Washington Memorial	VA-MD-DC	1930	7,248
John D. Rockefeller Jr. Mem.	WY	1972	23,777
Natchez Trace	MS-AL-TN	1938	51,980

NATIONAL LAKESHORES

Name	State	Year[1]	Acreage
Apostle Islands	WI	1970	69,372
Indiana Dunes	IN	1966	15,062
Pictured Rocks	MI	1966	73,228
Sleeping Bear Dunes	MI	1970	71,195

NATIONAL RESERVES

Name	State	Year[1]	Acreage
City of Rocks[3]	ID	1988	14,107
Ebey's Landing[3]	WA	1978	19,019

NATIONAL RIVERS

Name	State	Year[1]	Acreage
Big South Fork Natl. R and Recreation Area	KY-TN	1976	125,310
Buffalo	AR	1972	94,294
Mississippi Natl. R and Recreation Area	MN	1988	53,775
New River Gorge	WV	1978	69,832
Niobrara	NE-SD	1991	5,962
Ozark	MO	1964	80,785

NATIONAL WILD AND SCENIC RIVERS

Name	State	Year[1]	Acreage
Alagnak	AK	1980	30,665
Bluestone[2]	WV	1978	4,310
Delaware	NY-NJ-PA	1978	1,973
Great Egg Harbor	NJ	1992	43,311
Missouri	NE-SD	1991	45,350
Obed	TN	1976	5,174
Rio Grande[2]	TX	1978	9,600
Saint Croix	MN-WI	1968	92,742
Upper Delaware	NY-PA	1978	75,005

NATIONAL RECREATION AREAS

Name	State	Year[1]	Acreage
Amistad	TX	1965	58,500
Bighorn Canyon	MT-WY	1966	120,296
Boston Harbor Islands	MA	1996	1,482
Chattahoochee R.	GA	1978	9,330
Chickasaw	OK	1902	9,889
Curecanti	CO	1965	41,972
Cuyahoga Valley	OH	1974	32,864
Delaware Water Gap	NJ-PA	1965	66,749
Gateway	NJ-NY	1972	26,610
Gauley R.[3]	WV	1988	11,506
Glen Canyon	AZ-UT	1958	1,254,306
Golden Gate	CA	1972	74,825
Lake Chelan	WA	1968	61,958
Lake Mead	AZ-NV	1936	1,495,666
Lake Meredith	TX	1965	44,978
Lake Roosevelt[5]	WA	1946	100,390
Ross Lake	WA	1968	117,575
Santa Monica Mts.[3]	CA	1978	153,687
Whiskeytown-Shasta-Trinity	CA	1965	42,503

NATIONAL SCENIC TRAIL

Name	State	Year[1]	Acreage
Appalachian	ME to GA	1968	214,362
Natchez Trace	MS-TN	1983	10,995
Potomac Heritage	MD-DC-VA-PA	1983	NA

PARKS (no other classification)

Name	State	Year[1]	Acreage
Catoctin Mountain	MD	1954	5,810
Constitution Gardens	DC	1974	52
Fort Washington	MD	1930	341
Greenbelt	MD	1950	1,176
National Capital	DC	1933	6,605
National Mall	DC	1933	146
Piscataway	MD	1961	4,627
Prince William Forest	VA	1948	18,714
Rock Creek	DC	1890	1,755
White House	DC	1933	18
Wolf Trap Farm Park for the Performing Arts	VA	1966	130

INTERNATIONAL HISTORIC SITE

Name	State	Year[1]	Acreage
Saint Croix Island[3]	ME	1949	45

NA=Not available. (1) Year first designated. (2) No federal facilities. (3) Limited federal facilities. (4) Not open to the public. (5) Formerly Coulee Dam National Recreation Area.

20 Most-Visited Sites in the National Park System, 2000

Source: National Park Service, Dept. of the Interior

Attendance at all areas administered by the National Park Service in 2000 totaled 285,891,275 recreation visits.

Site (location)	Recreation visits	Site (location)	Recreation visits
Blue Ridge Parkway	19,153,081	Castle Clinton NM	4,588,273
Golden Gate NRA	14,486,065	Cape Cod NS	4,581,169
Great Smoky Mountains NP	10,175,812	Grand Canyon NP	4,460,228
Lake Mead NRA	8,755,005	Lincoln Memorial	4,009,145
Gateway NRA	7,927,567	Vietnam Veterans Memorial	3,782,445
George Washington Memorial Pkwy	7,897,161	Jefferson National Expansion Memo.	3,458,956
Natchez Trace Parkway	5,737,183	San Francisco Maritime NHP	3,433,066
Statue of Liberty NM	5,509,706	Yosemite NP	3,400,903
Delaware Water Gap NRA	4,900,745	Olympic NP	3,327,722
Gulf Islands NS	4,590,595	Cuyahoga Valley NRA	3,324,918

IT'S A FACT: The Native American population, just counting those who classified themselves only under the one racial category "American Indian and Alaska Native," went up 26% between the 1990 and 2000 censuses. This population was relatively young (34% under 18, compared to 26% for the whole U.S. population) and fast growing (the total U.S. population grew 13%), though Asians grew faster (48%) and the total Hispanic population increased even more rapidly (58%).

U.S. States Ranked by American Indian and Alaska Native Population, 2000

Source: Bureau of the Census, U.S. Dept. of Commerce

Rank	State	One race only[1]	More than one race[2]	Rank	State	One race only[1]	More than one race[2]
1	California	333,346	294,216	27	Georgia	21,737	31,460
2	Oklahoma	273,230	118,719	28	Virginia	21,172	31,692
3	Arizona	255,879	36,673	29	New Jersey	19,492	29,612
4	New Mexico	173,483	17,992	30	Pennsylvania	18,348	34,302
5	Texas	118,362	97,237	31	Arkansas	17,808	19,194
6	North Carolina	99,551	32,185	32	Idaho	17,645	9,592
7	Alaska	98,043	21,198	33	Indiana	15,815	23,448
8	Washington	93,301	65,639	34	Maryland	15,423	24,014
9	New York	82,461	89,120	35	Tennessee	15,152	24, 036
10	South Dakota	62,283	5,998	36	Massachusetts	15,015	23,035
11	Michigan	58,479	65,933	37	Nebraska	14,896	7,308
12	Montana	56,068	10,252	38	South Carolina	13,718	13,738
13	Minnesota	54,967	26,107	39	Mississippi	11,652	7,903
14	Florida	53,541	64,339	40	Wyoming	11,133	3,879
15	Wisconsin	47,228	22,158	41	Connecticut	9,639	14,849
16	Oregon	45,211	40,456	42	Iowa	8,989	9,257
17	Colorado	44,241	35,448	43	Kentucky	8,616	15,936
18	North Dakota	31,329	3,899	44	Maine	7,098	6,058
19	Illinois	31,006	42,155	45	Rhode Island	5,121	5,604
20	Utah	29,684	10,761	46	West Virginia	3,606	7,038
21	Nevada	26,420	15,802	47	Hawaii	3,535	21,347
22	Louisiana	25,477	17,401	48	New Hampshire	2,964	4,921
23	Missouri	25,076	35,023	49	Delaware	2,731	3,338
24	Kansas	24,936	22,247	50	Vermont	2,420	3,976
25	Ohio	24,486	51,589	51	Washington, DC	1,713	3,062
26	Alabama	22,430	22,019		**UNITED STATES**	**2,475,956**	**1,643,345**

(1) Respondents classified themselves only under the category "American Indian and Alaska Native" on Census 2000 questionnaires. (2) Respondents classified themselves as "American Indian and Alaska Native" in combination with one or more other races.

Largest American Indian Tribes

Source: Bureau of the Census, U.S. Dept. of Commerce, as of 1990 census

Tribe	Number	Percent	Tribe	Number	Percent
ALL AMERICAN INDIANS	**1,937,391**	**100.0**	Chickasaw	21,522	1.1
Cherokee	369,035	19.0	Tohono O'Odham	16,876	0.9
Navajo	225,298	11.6	Potawatomi	16,719	0.9
Sioux[1]	107,321	5.5	Seminole	15,564	0.8
Chippewa	105,988	5.5	Pima	15,074	0.8
Choctaw	86,231	4.5	Tlingit	14,417	0.7
Pueblo	55,330	2.9	Alaskan Athabaskans	14,198	0.7
Apache	53,330	2.8	Cheyenne	11,809	0.6
Iroquois[2]	52,557	2.7	Comanche	11,437	0.6
Lumbee	50,888	2.6	Paiute	11,369	0.6
Creek	45,872	2.4	Osage	10,430	0.5
Blackfoot	37,992	2.0	Puget Sound Salish	10,384	0.5
Canadian and Latin American	27,179	1.4	Yaqui	9,838	0.5

(1) Any entry from NC with the spelling "Siouan" in the 1990 census was miscoded to count as Sioux. (2) Reporting and/or processing problems in the 1990 census have affected accuracy of the data for this tribe.

BUILDINGS, BRIDGES, AND TUNNELS

50 Tallest Buildings in the World

Source: Council on Tall Buildings and Urban Habitat, Lehigh Univ.; Jeff Herzer and Marshall Gerometta, WTB/World's Tallest Buildings; Rick Bronson, http://www.worldstallest.com; http://www.skyscrapers.com

Included in this list are the twin towers of the World Trade Center, destroyed in a Sept. 11, 2001, terrorist attack.
Structures under construction are denoted by asterisk *. Year is date of completion or projected completion. Height is in feet.

Name, Year, City, Country	Height	Stories
Petronas Tower I, 1998, Kuala Lumpur, Malaysia	1,483	88
Petronas Tower II, 1998, Kuala Lumpur, Malaysia	1,483	88
Sears Tower, 1974, Chicago, IL, U.S.	1,450	110
Jin Mao Bldg., 1998, Shanghai, China	1,381	88
[World Trade Center One, 1972, New York, U.S.]	*1,368*	*110*
[World Trade Center Two, 1973, New York, U.S.]	*1,362*	*110*
CITIC Plaza, 1997, Guangzhou, China	1,283	80
Shun Hing Square, 1996, Shenzhen, China	1,260	69
Empire State Building, 1931, New York, U.S.	1,250	102
Central Plaza, 1992, Hong Kong, China	1,227	78
Bank of China, 1989, Hong Kong, China	1,209	70
Emirates Towers One, 2000, Dubai, U.A.E.	1,165	55
The Centre, 1998, Hong Kong, China	1,148	79
Tuntex & Chein-Tai Tower, 1998, Kaohsiung, Taiwan	1,140	85
Aon Center, 1973, Chicago, IL, U.S.	1,136	80
*Kingdom Centre, 2001, Riyadh, Saudi Arabia	1,132	30
John Hancock Center, 1969, Chicago, IL, U.S.	1,127	100
Burj al Arab Hotel, 1999, Dubai, U.A.E.	1,053	60
Baiyoke Tower II, 1998, Bangkok, Thailand	1,050	90
Chrysler Bldg., 1930, New York, U.S.	1,046	77
Bank of America Plaza, 1993, Atlanta, GA, U.S.	1,023	55
Library Tower, 1990, Los Angeles, CA, U.S.	1,018	73
Telekom Malaysia Headquarters, 1999, Kuala Lumpur, Malaysia	1,017	55
*Emirates Towers Two, 2000, Dubai, U.A.E.	1,014	54
AT&T Corporate Center, 1989, Chicago, IL, U.S.	1,007	60
Chase Tower, 1982, Houston, TX, U.S.	1,000	75
Two Prudential Plaza, 1990, Chicago, IL, U.S.	995	64
Ryugyong Hotel, 1995, Pyongyang, North Korea	984	105
Commerzbank Tower, 1997, Frankfurt, Germany	981	63
Wells Fargo Plaza, 1983, Houston, TX, U.S.	972	71
Landmark Tower, 1993, Yokohama, Japan	971	70
311 S. Wacker Drive, 1990, Chicago, IL, U.S.	961	65
American International Bldg., 1932, New York, U.S.	952	67
Cheung Kong Centre, 1999, Hong Kong, China	951	70
First Canadian Place, 1975, Toronto, Canada	951	72
Key Tower, 1991, Cleveland, OH, U.S.	950	57
One Liberty Place, 1987, Philadelphia, PA, U.S.	945	61
Plaza 66/Nanjing Xi Lu, 2000, Shanghai, China	945	66
Columbia Seafirst Center, 1984, Seattle, WA, U.S.	943	76
Sunjoy Tomorrow Square, 1999, Shanghai, China	934	59
The Trump Bldg., 1930, New York, U.S.	927	72
Nations Bank Plaza, 1985, Dallas, TX, U.S.	921	72
Overseas Union Bank Centre, 1986, Singapore	919	66
United Overseas Bank Plaza One, 1992, Singapore	919	66
Republic Plaza, 1995, Singapore	919	66
Citicorp Center, 1977, New York, U.S.	915	59
Scotia Plaza, 1988, Toronto, Canada	902	68
Williams Tower, 1983, Houston, TX, U.S.	901	64
Faisaliah Complex, 2000, Riyadh, Saudi Arabia	899	30
Renaissance Tower, 1974, Dallas, TX, U.S.	886	56

IT'S A FACT: The World Trade Center twin towers were the tallest buildings in the world when completed in 1972-73; each of these giant steel tubes weighed 250,000 tons and housed about 22,000 office workers; an estimated 5,000 to 10,000 were in each tower at the time of the terrorist attacks of Sept. 11, 2001.

World's 10 Tallest Free-Standing Towers

Name	City	Country	Height (ft.)	Year
*Indosat Telkom Tower	Jakarta	Indonesia	1,831	2001
CN Tower	Toronto	Canada	1,815	1976
Ostankino Tower	Moscow	Russia	1,772	1967
Oriental Pear Television Tower	Shanghai	China	1,535	1995
*Tehran Telecommunications Tower	Tehran	Iran	1,411	2001
Menara Kuala Lumpur	Kuala Lumpur	Malaysia	1,379	1996
Beijing Radio & T.V. Tower	Beijing	China	1,369	1992
Tianjin Radio & T.V. Tower	Tianjin	China	1,362	1991
Tashkent Tower	Tashkent	Uzbekistan	1,230	1985
Alma-Ata Tower	Alma-Ata	Kazakhstan	1,214	1982

*Under construction.

Tall Buildings in Selected North American Cities

Source: Council on Tall Buildings and Urban Habitat, Lehigh Univ.; Jeff Herzer and Marshall Gerometta, WTB/World's Tallest Buildings; http://www.worldstallest.com; Rick Bronson, http://www.skyscrapers.com

Lists include freestanding towers and other structures that do not have stories and are not technically considered "buildings." Also included are some structures still under construction (denoted by asterisk *). Year in parentheses is date of completion or projected completion. Height is generally measured from sidewalk to roof, including penthouse and tower if enclosed as integral part of structure; stories generally counted from street level. NA = not available or not applicable.

Atlanta, GA

Building	Ht. (ft.)	Stories
Bank of America Plaza, 600 Peachtree (1992)	1,023	55
SunTrust Bank Tower, 303 Peachtree (1992)	871	60
One Atlantic Center, 1201 Peachtree W (1987)	820	50
191 Peachtree Tower (1991)	770	50
Westin Peachtree Plaza, 210 Peachtree NW (1973)	723	73
Georgia Pacific Tower, 133 Peachtree NE (1981)	697	51
Promenade II/A.T.& T. (1989)	691	40
Bellsouth, 675 Peachtree NE (1980)	677	47
GLG Grand/Occidental Hotel, 75 14th St. (1992)	609	53
Wachovia Bank of Georgia Tower, 2 Peachtree NW (1967)	556	44
Marriott Marquis, 265 Peachtree NE (1985)	554	52
Equitable Bldg., 100 Peachtree (1967)	453	34
101 Marietta Tower (1976)	446	36
One Park Tower, 34 Peachtree (1961)	439	32
A.T.& T. Long Line Bldg. (1975)	433	NA
Bell South Enterprises, 1100 Peachtree (1990)	428	28
Atlanta Plaza I, 950 Paces Ferry Rd. E. (1986)	425	32
Park Place, 2660 Peachtree (1986)	420	40
*West Peachtree Villas (2002)	420	37
Oakwood Apts. (1989)	410	38
Peachtree Summit No. 1, 401 Peachtree NE (1975)	406	31
Coca Cola Headquarters Bldg., 310 North Ave. (1979)	403	26
Tower Place, 3361 Piedmont Rd. (1974)	401	29

Baltimore, MD

Building	Ht. (ft.)	Stories
Legg Mason Building (1973)	529	40
Bank of America Bldg. (1929)	509	34
William Donald Schaefer Tower, 6 St. Paul Pl. (1992)	493	29
Commerce Place (1992)	454	31
Marriott Baltimore Inner Harbor East (2000)	430	32
World Trade Center (1977)	405	32

Birmingham, AL

Building	Ht. (ft.)	Stories
Southtrust Tower (1986)	454	34
AmSouth/Harbert Plaza (1989)	437	32

Boston, MA

Building	Ht. (ft.)	Stories
John Hancock Tower, 200 Clarendon St. (1976)	790	60
Prudential Tower, 800 Boylston St. (1964)	750	52
Federal Reserve Bldg., 600 Atlantic Ave. (1983)	604	32
Boston Company Bldg., 1 Boston Place (1970)	601	41
One International Place, 100 Oliver St. (1987)	600	46
First National Bank of Boston, 100 Federal St. (1971)	591	37
One Financial Center (1984)	590	46
*111 Huntingdon Ave, (2001)	564	36
Two International Place (1993)	538	35
One Post Office Square (1981)	525	40
1 Federal St. (1975)	520	38
Exchange Place, 53 State St. (1984)	510	39

Building	Ht. (ft.)	Stories
Sixty State St. (1977)	509	38
1 Beacon St. (1972)	507	36
*1 Lincoln Place (2003)	503	36
28 State Street(1969)	500	40
Mariott's Custom House (1915)	496	32
John Hancock Bldg. (1949)	495	26
*33 Arch St. (2003)	489	31
State St. Bank (1966)	477	34
*Millennium Place 1 (2001)	475	38
125 High St. (1991)	452	30
100 Summer St. (1975)	450	33
*Millennium Place 2 (2001)	445	36
McCormack Bldg.	401	22
Harbor Towers I, 85 E. India (1971)	400	40

Calgary, Alberta

Building	Ht. (ft.)	Stories
Petro Canada Tower (1984)	689	52
Bankers Hall East Tower (1989)	645	50
Bankers Hall West Tower (2000)	645	50
Calgary Tower (1967)	626	NA
TCPL Tower (2000)	608	37
Canterra Tower (1988)	580	46
First Canadian Centre (1983)	530	43
Canada Trust, Calgary Eatons Centre (1991)	530	40
Scotia Square (1975)	525	42
Western Canadian Place–N. Tower (1983)	507	41
Nova Bldg., 801 7th Ave. SW (1982)	500	37
Petro-Canada Tower, E. Tower (1983)	469	33
Two Bow Valley Square (1974)	468	39
Dome Tower (1976)	463	34
5th & 5th Bldg. (1980)	460	35
Shell Tower (1977)	460	34
T.D. Square (1976)	449	33
Four Bow Valley Square (1982)	441	37
Fifth Avenue Place (1981)	435	34
Esso Plaza II (1981)	435	34
Cascade 300	432	31
Western Canadian Place–S. Tower (1983)	420	32
Family Life Bldg.	410	33
Pan Canadian Bldg., 150 9th Ave. SW (1982)	410	28
Serval Tower (1976)	408	33
Alberta Stock Exchange (1979)	407	33

Charlotte, NC

Building	Ht. (ft.)	Stories
Bank of America Corporate Center (1992)	871	60
*Hearst Tower, 214 N. Tyron (2002)	658	50
One First Union Center, 301 S. College St. (1988)	588	42
Bank of America, 101 S. Tryon (1974)	503	40
Interstate Tower, 121 W. Trade St. (1990)	462	32
IJL Financial Center, 201 N.Tyron St. (1997)	447	30
Three First Union Center (1999)	440	32
Two First Union Plaza, 301 S. Tyron St. (1971)	433	32
Wachovia Center, 400 S. Tryon (1974)	420	32

Chicago, IL

Building	Ht. (ft.)	Stories
Sears Tower, 233 S. Wacker Dr. (1974)	1,450	110
Aon Center, 200 E. Randolph (1973)	1,136	83
John Hancock Center, 875 N. Michigan Ave. (1969)	1,127	100
AT&T Corporate Center, 227 W. Monroe (1989)	1,007	61
2 Prudential Plaza, 180 N. Stetson Ave. (1990)	995	64
311 S. Wacker Drive (1990)	961	65
900 N. Michigan Ave. (1989)	871	66
Water Tower Place, 845 N. Michigan Ave.(1976)	859	74
Bank One Plaza (1969)	850	60
Park Tower, 800 N. Michigan Ave. (2000)	844	67
Chicago Title & Trust Center, 161 N. Clark St. (1991)	756	50
3 First National Plaza, 70 W. Madison (1981)	753	57
Olympia Centre, 737 N. Michigan Ave. (1986)	725	63
IBM Bldg., 330 N. Wabash Ave. (1973)	695	52
*River East Center 1, 350 E. Illinois St. (2001)	681	58
Paine Webber Bldg., 181 W. Madison (1990)	680	50
One Magnificent Mile, 980 N. Michigan Ave. (1983)	673	58
R.R. Donnelley Center, 77 W. Wacker Dr. (1992)	668	50
*UBS Tower, 1 N. Wacker (2001)	652	50
*55 E. Erie (2003)	650	59
Daley Center, 55 W. Washington St. (1965)	648	31
Lake Point Tower, 505 N. Lake Shore Dr. (1968)	645	70
*Grand Plaza 1	641	57
Leo Burnett, 35 W. Wacker Dr. (1989)	635	50
NBC Tower, 445 N. Cityfront Plaza Dr. (1989)	627	34
Chicago Place, 700 N. Michigan Ave. (1991)	608	49
Board of Trade (incl. statue), 141 W. Jackson (1930)	605	44
Prudential Bldg., 130 E. Randolph (1955)	601	41
Heller International Tower, 500 W. Monroe (1992)	600	45
CNA Plaza, 325 S. Wabash (1972)	598	45
One Madison Plaza, 200 W. Madison (1982)	597	45
*The Residences at Millennium Centre (2002)	595	60
1000 Lake Shore Plaza Apts. (1964)	590	55
Marina City Apts. 1, 300 N. State (1964)	588	61
Marina City Apts. 2, 300 N. State (1964)	588	61
Citicorp Center, 500 W. Madison (1985)	588	41
Mid Continental Plaza, 55 E. Monroe (1972)	582	50
North Pier Apt. Tower, 474 N. Lake Shore Dr. (1990)	581	61
*Dearborn Center, 131 S. Dearborn (2002)	580	37
Smurfit-Stone , 150 N. Michigan Ave. (1983)	575	41
*The Fordham, 25 E. Superior St. (2002)	573	50
190 S. LaSalle St. (1986)	573	42
Onterie Center, 446 E. Ontario St. (1985)	570	57
Chicago Temple, 77 W. Washington (1923)	558	21
919 N. Michigan Ave. (incl. beacon) (1929)	565	37
Huron Plaza Apts., 30 E. Huron St. (1983)	560	56
Morton Intl. Tower, 100 N. Riverside Plaza (1990)	560	36
Pittsfield, 55 E. Washington (1927)	557	38
The Parkshore, 195 N. Harbor Dr. (1991)	556	56
North Harbor Tower, 175 N. Harbor Dr. (1991)	556	55
Civic Opera Bldg., 20 N. Wacker Dr. (1929)	555	45
Newberry Plaza, State & Oak (1974)	553	53
Boulevard Towers South, 205 N. Michigan Ave. (1985)	553	44
30 N. LaSalle St. (1975)	553	43
Harbor Point, 155 N. Harbor Dr. (1975)	550	54
One S. Wacker Dr. (1983)	550	42
*Park Millennium, 222 Columbus Dr. (2002)	535	53
LaSalle National Bank, 135 S. LaSalle St. (1934)	535	44
Park Place Tower, 655 W. Irving Park Rd. (1973)	531	56
One LaSalle St. (1930)	530	49
The Elysees, 111 E. Chestnut St. (1972)	529	56
River Plaza, Rush & Hubbard (1977)	524	56
35 E. Wacker Dr. (1926)	523	40
Unitrin, 1 E. Wacker Dr. (1962)	522	41
Chicago Mercantile Exchange, 10 S. Wacker Dr. (1987)	520	40
Chicago Merc. Exchange, 30 S. Wacker Dr. (1983)	520	40
Kluczynski Federal Bldg., 230 S. Dearborn (1976)	520	40
*191 N. Wacker (2002)	516	37
One Financial Place, 440 S. LaSalle St. (1985)	515	40
LaSalle-Wacker, 221 N. LaSalle St. (1930)	512	41
Quaker Tower, 321 N. Clark (1987)	510	35
Harris Bank III, 115 S. LaSalle St. (1977)	510	35
Carbide & Carbon, 230 N. Michigan Ave. (1929)	503	37
1 Superior Place (1999)	501	52
Savings of America Tower, 120 N. LaSalle St. (1991)	501	41
Chase Plaza, 10 S. LaSalle St. (1986)	501	37
401 E. Ontario (1990)	500	51
200 S. Wacker Dr. (1981)	500	38
USG Building, 125 S. Franklin (1992)	497	35
Xerox Centre, 55 W. Monroe (1980)	496	40
1 N. Franklin St. (1991)	493	38
Ontario Place, 10 E. Ontario St. (1983)	488	51
333 Wacker Dr. (1983)	487	36
Lincoln Tower, 75 E. Wacker Dr. (1928)	484	38
American National Bank, 33 N. LaSalle St. (1930)	479	40
Park Tower Condos, 5415 N. Sheridan Rd. (1974)	476	54
Bankers, 105 W. Adams St. (1927)	476	41
Britannica Center, 310 S. Michigan Ave. (1924)	475	37
Brunswick Bldg., 69 W. Washington (1965)	475	37
American Furniture Mart, 680 N. Lake Shore Dr. (1926)	474	30
Intercontinental Hotel, 505 N. Michigan (1929)	471	42
City Place, 676 N. Michigan (1990)	470	40
Columbus Plaza, 233 E. Wacker Dr. (1980)	468	49
The Sterling, 345 N. LaSalle St. (2001)	466	50
188 Randolph Tower (1925)	465	45
The Bristol, 57 E. Delaware Pl. (2000)	465	42
200 N. Dearborn (1989)	463	47
Tribune Tower, 435 N. Michigan Ave. (1925)	462	36
The New York, 3660 N. Lake Shore Dr. (1986)	461	50
Presidential Towers, 555 W. Madison St. (1985)	461	49
Presidential Towers, 575 W. Madison St. (1985)	461	49
Presidential Towers, 605 W. Madison St. (1985)	461	49
Presidential Towers, 625 W. Madison St. (1985)	461	49
Chicago Marriott, 540 N. Michigan Ave. (1978)	460	45
Swissotel, 323 E. Wacker Dr. (1989)	457	43
Equitable Life, 401 N. Michigan Ave. (1964)	457	35
Roanoke, 11 S. LaSalle St. (1925)	452	37
*The Residences at River Bend (2001)	451	37
Eugenie Terrace on the Park, 1730 N. Clark St. (1987)	450	35
Gateway Center III, 222 S. Riverside Plaza (1972)	450	35

IT'S A FACT: Egypt's Great Pyramid of Giza ranked as the tallest human-built structure on earth for more than 43 centuries. Its original height was 481 ft.

Cincinnati, OH

Building	Ht. (ft.)	Stories
Carew Tower, 441 Vine St. (1930)	574	48
PNC Tower , 1 W. 4th St. (1913)	495	28
Scripps Center, 312 Walnut St. (1990)	468	36
Atrium Two, 221 E. 4th St. (1984)	428	30
Fifth Third Center, 511 Walnut St. (1969)	423	32
Chemed Center, 255 5th St. (1990)	410	32
Cincinnati Commerce Center, 600 Vine St. (1984)	402	29

Cleveland, OH

Building	Ht. (ft.)	Stories
Key Tower, 127 Public Square (1991)	950	57
Terminal Tower, 50 Public Square (1930)	708	52
BP America, 200 Public Square (1985)	658	46
100 Erieview, 1801 E. 9th St. (1964)	529	40
One Cleveland Center, 1375 E. 9th St. (1983)	450	31
Bank One Center (1991)	446	38
*Federal Courthouse, 801 W. Superior Ave. (2001)	430	24
Justice Center, 1250 Ontario (1976)	420	26
Federal Building (1967)	419	32
National City Center, 1900 E. 9th St. (1980)	410	35

Columbus, OH

Building	Ht. (ft.)	Stories
James A. Rhodes State Office Tower, 30 E. Broad (1973)	624	41
Leveque-Lincoln Tower, 50 W. Broad (1927)	555	47
William Green Building (1990)	530	33
Huntington Center, 41 S. High St. (1983)	512	37
Vern Riffe State Office Tower, 77 S. High St. (1988)	503	33
One Nationwide Plaza (1976)	485	40
Franklin County Courthouse (1991)	464	27
AEP Building, One Riverside Plaza (1983)	456	31
Borden Bldg., 180 E. Broad (1974)	438	34
Three Nationwide Plaza (1989)	408	29

Dallas, TX

Building	Ht. (ft.)	Stories
Bank of America Plaza, 901 Main St. (1985)	921	72
Renaissance Tower, 1201 Elm St. (1974)	886	56
Bank One Center, 1717 Main St. (1987)	787	60
Chase Texas Plaza, 2200 Ross Ave. (1987)	738	55
Fountain Place, 1445 Ross Ave. (1986)	720	58
Trammel Crow Tower, 2001 Ross Ave. (1984)	686	50
1700 Pacific Ave. (1983)	655	50
Thanksgiving Tower, 1600 Pacific Ave. (1982)	645	50
Energy Plaza, 1601 Bryan St. (1983)	629	49
Elm Place, 1401 Elm St. (1965)	625	52
Republic Center Tower II, 325 N. St. Paul (1964)	598	50
One Bell Plaza, 208 S. Akard St. (1984)	580	37
One Lincoln Plaza, 500 Akard St. (1984)	579	45
Cityplace Center East (1989)	560	42
Reunion Tower (1976)	560	NA
Southland Center, 400 Olive St. (1959)	550	42
2001 Bryan St.(1973)	512	40
Harwood Center, 1999 Bryan St. (1982)	483	36
Maxus Energy, 717 N. Harwood St. (1980)	481	34
San Jacinto Tower, 2121 San Jacinto St. (1982)	456	33
Republic Center Tower I, 300 N. Ervay (1954)	452	36
Renaissance Hotel (1983)	451	29
Adam's Mark Hotel North Tower (1980)	448	31
One Dallas Centre, 350 N. Paul St. (1979)	448	30
One Main Place, 1201 Main St. (1968)	445	34
1600 Pacific Bldg. (1964)	434	31
Mercantile National Bank Bldg. (1937)	430	31
Magnolia Bldg., 108 Akard St. (1923)	430	27
Fidelity Union Tower (1959)	400	33
Mart Hotel	400	29

Denver, CO

Building	Ht. (ft.)	Stories
Republic Plaza, 330 17th St. (1984)	714	56
US West Tower, 1801 California (1982)	709	54
Wells Fargo Center (1983)	698	52
1999 Broadway (1985)	544	43
MCI Tower, 707 17th St. (1981)	522	42
Qwest Tower, 555 17th St. (1978)	507	40
Amoco Bldg., 1670 Broadway (1980)	448	36
17th St. Plaza, 1225 17th St. (1982)	438	32
First Interstate Tower North, 633 17th St. (1974)	434	32
Brooks Towers, 1020 15th St. (1968)	420	42
Two Denver Place, 999 18th St. (1981)	416	34
One Tabor Center, 1200 17th St. (1984)	408	32
Manville Plaza, 717 17th St. (1989)	404	29

Des Moines, IA

Building	Ht. (ft.)	Stories
Principal Financial Group, 801 Grand (1990)	630	44
Ruan Center (1974)	457	36

Detroit, MI

Building	Ht. (ft.)	Stories
Marriott Hotel, Renaissance Center I (1977)	725	73
Comercia Tower, 500 Woodward (1991)	619	45
Penobscot Bldg., 633 Griswold (1928)	557	47
Renaissance Center II (1976)	534	39
Renaissance Center III (1976)	534	39
Renaissance Center IV (1976)	534	39
Renaissance Center V (1976)	534	39
Guardian, 500 Griswold (1928)	485	40
Book Tower, 1265 Washington (1925)	472	35
Madden Bldg., 150 W. Jefferson (1988)	470	29
Cadillac Tower, 65 Cadillac Sq. (1928)	437	40
David Stott Bldg., 1150 Griswold (1928)	436	38
ANR Bldg., 1 Wood Ward (1962)	430	30
Fisher Bldg., 311 W. Grand Blvd. (1928)	420	28

Dunwoody, GA

Building	Ht. (ft.)	Stories
Concourse Tower #5 (1988)	570	32
Concourse Tower #6 (1991)	553	32
Ravinia #3 (1991)	444	34

Edmonton, Alberta

Building	Ht. (ft.)	Stories
Manulife Place, 10170-101 St. (1983)	479	39
AGT Tower, 10020-100 St. (1971)	441	34
Canada Trust Tower (1982)	440	31
Commerce Place (1990)	409	30

Fort Worth, TX

Building	Ht. (ft.)	Stories
Burnett Plaza, 801 Cherry St. (1983)	567	40
Center Tower II, 301 Commerce St. (1984)	547	38
UPR Plaza, 777 Main St. (1982)	525	40
Landmark Tower, 200 W. 7th (1957)	481	32
Chase Texas Tower, 201 Main St. (1982)	477	33
Bank One Tower, 400 Throckmorton (1974)	454	36

Hartford, CT

Building	Ht. (ft.)	Stories
City Place (1980)	535	38
CitiGroup (1919)	527	34
Goodwin Square, 255 Asylum St. (1990)	522	30
Hartford Plaza (1967)	420	22

Honolulu, HI

Building	Ht. (ft.)	Stories
First Hawaiian Bank Bldg. (1996)	435	30
Nauru Tower (1991)	418	45
Hawaiki Tower (1999)	400	45
Waterfront Towers (1990)	400	46
Imperial Plaza (1992)	400	40

Houston, TX

Building	Ht. (ft.)	Stories
Chase Tower, 600 Travis (1982)	1,002	75
Wells Fargo Plaza, 1000 Louisiana (1983)	972	71
Williams Tower, 2800 Post Oak Blvd. (1983)	901	64
Bank of America Center, 700 Louisiana (1983)	780	56
Texaco Heritage Plaza, 1111 Bagby (1987)	762	53
Southwest Bank of Texas, 1100 Louisiana (1980)	748	55
Houston Industries Plaza, 1111 Louisiana (1974)	741	53
1600 Smith St. (1984)	732	55
Chevron Tower, 1301 McKinney (1982)	725	52
One Shell Plaza, 900 Louisiana (1970)	714	50
Enron Bldg., 1400 Smith St. (1983)	691	50
Capital National Bank Plaza (1980)	685	50
One Houston Center, 1221 McKinney (1978)	678	47
First City Tower (1984)	662	47
San Felipe Plaza (1984)	625	45
Exxon Bldg., 800 Bell Ave. (1962)	606	44
*Enron Tower (2001)	600	40
America Tower, 2929 Allen Parkway (1983)	590	42
Two Houston Center, 909 Fannin (1974)	579	40
San Jacinto Column (monument) (1983)	570	NA
Marathon Oil Tower, 5555 San Felipe (1983)	562	41
Wedge International Building, 1415 Louisiana (1983)	550	44
Kellogg Tower, 601 Jefferson (1973)	550	40
Pennzoil Bldg. 1, 700 Milam St. (1975)	523	36
Pennzoil Bldg. 2, 700 Milam St. (1975)	523	36
Two Allen Center, 1200 Smith St. (1978)	521	36
1201 Louisiana Bldg. (1971)	518	35
The Huntington (1982)	503	34
El Paso Energy Bldg. (1962)	502	33
Greenway Plaza (1973)	465	32
One Allen Center, 500 Dallas (1974)	452	34
Summit Tower East (1978)	441	31
Summit Tower West (1979)	441	31
Four Leafs Towers I, 5100 San Felipe Blvd. (1982)	439	40
Four Leafs Towers II (1982)	439	40

Building	Ht. (ft.)	Stories
Phoenix Tower, 3200 Southwest Fwy. (1984)	434	34
Chase Bank Bldg., 712 Main St. (1929)	428	37
The Spires (1984)	426	41
Central Tower, 4 Oaks Place (1983)	420	30
One City Center (1960)	410	32
Bob Lanier Public Works Bldg., 611 Walker Ave. (1968)	410	27
Neils Esperson Bldg., 802 Travis St. (1927)	409	31
Hyatt Regency (1972)	401	34

Indianapolis, IN

Building	Ht. (ft.)	Stories
Bank One Tower, 11 Monument Circle (1990)	811	51
American United Life Ins. (1981)	533	37
One Indiana Square (1970)	504	36
Market Tower, 10 W. Market St. (1988)	450	32
300 N. Meridian Bldg. (1988)	408	28

Jacksonville, FL

Building	Ht. (ft.)	Stories
Bank of America Tower (1990)	617	42
Modis Tower (1975)	535	37
BellSouth Tower (1983)	435	27
Riverplace Tower (1967)	433	28

Jersey City, NJ

Building	Ht. (ft.)	Stories
Merrill Lynch Building, 101 Hudson St. (1992)	548	42
Newport Tower, 525 Washington Blvd. (1992)	531	36
Exchange Place Centre, 10 Exchange Place (1989)	490	30
*77 Hudson St. (2002)	491	32
*Harborside Financial Plaza V (2002)	480	34
Liberty Center	400	29

Kansas City, MO

Building	Ht. (ft.)	Stories
One Kansas City Place (1988)	632	42
Transamerica Tower(1986)	591	38
Hyatt Regency (1980)	504	45
Power & Light Bldg. (1931)	476	32
City Hall, 414 E. 12th St. (1937)	443	29
Fidelity Bank and Trust Bldg. (1931)	433	35
Oak Tower, 324 E. 11th St.	430	28
1201 Walnut (1991)	427	30
Federal Office Bldg. (1962)	413	35
Commerce Tower (1965)	407	32
City Center Square (1977)	404	30

Las Vegas, NV

Building	Ht. (ft.)	Stories
Stratosphere Tower (1996)	1,149	NA
Eiffel Tower, Paris Hotel and Casino (1998)	560	NA
New York, New York Hotel and Casino (1997)	525	48
Bellagio Hotel and Casino (1998)	508	36
Eiffel Tower, Paris Hotel and Casino (1998)	492	NA
Rio Masquerade Tower (1996)	483	40
Mandalay Resort-Bay Hotel and Casino (1999)	480	43
Venetian Resort-Hotel and Casino (1999)	480	35
Caesars Palace Hotel Tower (1998)	470	29
Paris Hotel and Casino (1999)	440	34
Aladdin Resort and Casino (2000)	424	38
Harrahs Hotel and Casino	400	35
Fitzgeralds Hotel	400	33

Little Rock, AR

Building	Ht. (ft.)	Stories
TCBY Tower (1986)	546	40
Regions Center (1975)	454	30

Los Angeles, CA

Building	Ht. (ft.)	Stories
Library Tower, 633 W. 5th St. (1990)	1,018	73
First Interstate Tower, 707 Wilshire Blvd. (1974)	858	62
Two California Plaza, 350 S. Grand Ave. (1992)	750	52
Gas Company Tower, 555 W. 5th St. (1991)	749	52
BP Plaza, 333 South Hope (1975)	735	55
777 Tower, 777 S. Figueroa St. (1990)	725	53
Wells Fargo Tower, 333 S. Grand Ave. (1983)	723	54
United California Bank Plaza, 601 S. Figueroa St. (1989)	717	52
Atlantic Richfield Tower, 515 S. Flower St. (1971)	699	52
Bank of America Tower, 555 S. Flower St. (1971)	699	52
Citibank Square, 444 S. Flower St. (1979)	625	48
611 W. 6th St. (1969)	620	42
One California Plaza, 300 S. Grand Ave. (1985)	578	42
Century Plaza Tower 1, 2029 Cent. Park E. (1973)	571	44
Century Plaza Tower 2, 2049 Cent. Park E. (1973)	571	44
KPMG Tower, 355 S. Grand Ave. (1983)	560	44
Ernst & Young, LLP Plaza, 725 S. Figueroa St. (1988)	534	41
SunAmerica Tower, 1999 Ave. of the Stars (1989)	533	39
Manulife Tower, 865 S. Figueroa St. (1990)	517	37
Union Bank Square, 445 S. Figueroa St. (1968)	516	40
10 Universal City Plaza (1984)	506	36
1100 Wilshire (1987)	496	36
Fox Plaza, 2121 Ave. of Stars (1987)	492	34
*MGM Tower , Century City (2003)	488	34
1055 W. 7th St. (1985)	462	33
Equitable Life, 3435 Wilshire Blvd. (1969)	454	34
City Hall, 200 N. Spring St. (1927)	454	28
Transamerica Center, 1150 Olive St. (1965)	452	32
Madison Complex/Pacific Bell Switching Station (1961)	448	17
Peoples Savings Bank (1970)	435	31
550 South Hope (1991)	423	28
Warner Center Plaza III, 21650 Oxnard St., Woodland Hills	415	25
MCI Plaza, 700 S. Flower St. (1973)	414	33

Louisville, KY

Building	Ht. (ft.)	Stories
AEGON Center, 400 W. Market St. (1992)	549	35
National CityTower, 101 S. 5th St. (1972)	512	40
PNC Bank Bldg., 5th & Jefferson (1971)	420	30
Humana Center, 5th & Main (1985)	417	28

Mexico City, Mexico

Building	Ht. (ft.)	Stories
*Torre Mayor (2002)	738	55
Torre de Pemex (1984)	702	52
Torre Altus (1999)	640	42
Torre Latino Americana (1956)	597	45
World Trade Center (1994)	534	50
Los Arcos Bosques I (1997)	529	34
*Los Arcos Bosques II (2001)	529	34
*World Trade Center Hotel (2003)	459	38
Torre Las Lomas (1993)	453	36
Hotel Nikko Mexico	446	38
Torre Mural, Insurgentes Sur 1605 (1995)	440	33
Stoufers President Hotel	427	42
Torre del Caballito	417	37
Nonoalco Tlatelolco Tower (1962)	417	25
Torre Reforma Andres Bello 45	410	28
JW Marriott Hotel, Andres Bello 29	400	27

Miami, FL

Building	Ht. (ft.)	Stories
*Four Seasons Hotel and Tower (2002)	794	64
First Union Financial Center, 200 S. Biscayne Blvd. (1983)	764	55
Bank of America Tower, 100 S. E. 2nd St. (1987)	625	47
Santa Maria, 1643 Brickell Ave. (1997)	520	51
Stephen P. Clark Center (1985)	510	30
*Espirito Santo Plaza, 1301 Brickell Ave. (2002)	487	36
Citicorp Tower, 201 S. Biscayne Blvd. (1986)	484	35
*Three Tequesta Point (2001)	480	46
One Biscayne Tower, 2 S. Biscayne Blvd. (1974)	456	30
701 Brickell Ave. (1986)	450	33
Barclay's Financial Center (2000)	425	30
Mark on Bricknell (2001)	420	36
Courthouse Center (1986)	405	30
The Palace (1982)	400	42
Two Tequesta Point, 808 Brickell Key Dr. (1999)	400	39

Miami Beach, FL

Building	Ht. (ft.)	Stories
Blue Diamond Tower (2000)	565	45
Green Diamond Tower (2000)	565	45
PortofinoTower, 100 S. Pointe Dr. (1997)	484	44
*The Continuum on South Beach, South Tower (2001)	474	43
*Murano at Portofino (2002)	402	38

Milwaukee, WI

Building	Ht. (ft.)	Stories
Firstar Center (1973)	601	42
100 E. Wisconsin Ave (1989)	549	37
Milwaukee Center, 111 E. Kilbourn Ave. (1987)	426	29
411 Bldg., 411 E. Wisconsin Ave. (1983)	408	30

Minneapolis, MN

Building	Ht. (ft.)	Stories
U.S. Bank Place, 601 2nd Ave. (1992)	776	53
IDS Center (1973)	775	57
Wells Fargo Center, 90 S. 7th St. (1988)	774	57
Multifoods Tower, 33 S. 6th St. (1983)	669	51
Piper Jaffray Tower, 222 S. 9th St. (1984)	579	42
Pillsbury Center, 200 S. 6th St. (1981)	561	40
Dain Rauscher Plaza, 60 S. 6th St. (1994)	539	40
Fifth Street Towers II, 150 S. 5th St. (1987)	503	36
American Express Tower, 707 2nd Ave. S. (2000)	498	30

Building	Ht. (ft.)	Stories
*Target Plaza South, 1020 Nicolet Mall (2001)	492	33
Plaza VII, 45 S. 7th St. (1987)	475	36
US Bankcorp Center, 800 Nicolet Mall (2000)	468	30
AT&T Tower, 901 Marquette Ave. (1991)	464	34
Accenture Tower, 333 S. 7th St. (1987)	455	33
Foshay Tower, 821 Marquette Ave. (1929)	447	32
Qwest, 224 S. 5th St. (1931)	416	26
Hennepin Co. Government Center (1973)	403	24
*Dorsey & Whitney Tower, 50 S. Sixth St. (2001)	401	29

Montreal, Quebec

Building	Ht. (ft.)	Stories
1000 Rue de la Gauchetière (1991)	669	51
Marathon (IBM), 1250 Blvd. René Lévesque (1989)	640	47
Tour de la Bourse, 800 Place Victoria (1963)	624	47
Place Villa Marie (1962)	616	42
Canadian Imperial Bank, 1155 Blvd. René Lévesque (1962)	604	43
Montreal Tower (1987)	574	NA
Place Montreal Trust (1988)	519	30
Tour McGill College (1992)	519	38
Le Complexe Desjardins Sud (1975)	498	40
Les Cooperants, 600 Maisonneuve (1987)	479	34
Le Centre Sheraton (1980)	449	38
Place Montreal Trust (1988)	440	30
Maison Royal Trust, 630 Blvd. Réné Lévesque (1962)	429	32
Le Complexe Desjardins Est (1975)	428	32
La Tour Laurier	425	36
Port Royal Apts. (1964)	424	33
Chateau Champlain Hotel, 1 Place du Canada (1967)	420	38
Centre Mount Royal (1976)	420	28
Tour Terminal (1966)	400	30

Nashville, TN

Building	Ht. (ft.)	Stories
BellSouth Tower (1994)	617	33
Sun Trust Bank (1986)	490	31
American General Center (1970)	452	31
Nashville Life & Casualty	409	30
City Center (1987)	402	27

Newark, NJ

Building	Ht. (ft.)	Stories
Midatlantic National Bank, 744 Broad St. (1930)	465	36
Raymond-Commerce, 1180 Raymond Blvd.	448	36

New Orleans, LA

Building	Ht. (ft.)	Stories
One Shell Square (1972)	697	51
Place St. Charles (1985)	645	53
Plaza Tower (1969)	531	45
Energy Centre (1984)	530	39
LL&E Tower, 901 Poydras (1987)	481	36
Sheraton Hotel (1985)	478	47
Marriott Hotel (1972)	450	42
Texaco Bldg. (1983)	442	33
Canal Place One (1979)	439	32
Bank of New Orleans, 1010 Common (1971)	438	31
World Trade Center (1965)	407	33

New York, NY

(C) = collapsed as a result of Sept. 11, 2001, terrorist attack.

Building	Ht. (ft.)	Stories
World Trade Center One (1972) **(C)**	1,368	110
World Trade Center Two (1973) **(C)**	1,362	110
Empire State Bldg., 350 5th Ave. (1931)	1,250	102
Chrysler Bldg., 405 Lexington Ave. (1930)	1,046	77
American International Bldg., 70 Pine St. (1932)	952	67
The Trump Bldg., 40 Wall St. (1930)	927	71
Citigroup Center, 153 E. 53rd St. (1977)	915	59
Trump World Tower, 845 UN Plaza (2001)	881	72
G. E. Bldg., 30 Rockefeller Center (1933)	850	70
Cityspire, 150 W. 56th St. (1989)	814	72
One Chase Manhattan Plaza (1960)	813	60
Condé Nast Bldg., 4 Times Square (1999)	809	48
MetLife Bldg., 200 Park Ave. (1963)	808	59
Woolworth Bldg., 233 Broadway (1913)	792	57
1 Worldwide Plaza, 935 8th Ave. (1989)	778	47
Carnegie Hall Tower, 152 W. 57th St. (1991)	757	60
Bear Stearns World Headquarters, 383 Madison Ave. (2001)	757	47
Equitable Center West, 787 7th Ave. (1985)	752	51
One Penn Plaza, 250 W. 34th St. (1972)	750	57
*Time Warner Center South Tower (2003)	750	55
*Time Warner Center North Tower (2003)	750	55
1251 Ave. of Americas (1971)	750	54
J.P. Morgan Headquarters, 60 Wall St. (1989)	745	50
1 Liberty Plaza, 165 Broadway (1973)	743	54
20 Exchange Place (1931)	741	57
American Express Bldg., Three World Financial Center, 200 Vesey St. (1986)	739	51
One Astor Plaza, 1515 Broadway (1969)	730	54
Metropolitan Tower, 142 W. 57th St. (1985)	716	68
Chase World Headquarters, 270 Park Ave. (1960)	707	52
General Motors, 767 5th Ave. (1968)	705	50
Metropolitan Life Tower, 1 Madison Ave. (1909)	700	50
500 5th Ave. (1931)	697	60
Americas Tower, 1177 Ave. of the Amer. (1992)	692	48
Solow Bldg., 9 W. 57th St. (1974)	689	50
Marine Midland Bank, 140 Broadway (1966)	688	52
55 Water St. (1972)	687	53
Davidson, Lufkin, and Jenrette Building, 277 Park Ave. (1963)	687	50
Solomon Equities Bldg.,1585 Broadway (1989)	685	42
*Random House Tower, 1739 Broadway (2002)	684	52
Four Seasons Hotel, 57 E. 57th St. (1993)	682	52
Trump Intl. Hotel & Tower, 15 Columbus Circle (1970)	679	52
Bertelsmann Building, 1540 Broadway (1990)	676	42
McGraw Hill Bldg., 1221 Ave. of Amer. (1972)	674	51
Lincoln, 60 E. 42nd St. (1930)	673	53
Paramount Plaza, 1633 Broadway (1970)	670	48
Trump Tower, 725 5th Ave. (1982)	664	58
Citicorp, Queens (1990)	658	50
Irving Trust, 1 Wall St. (1932)	654	50
599 Lexington Ave. (1986)	653	51
712 5th Ave. (1990)	650	53
Chanin Bldg., 122 E. 42nd St. (1929)	649	56
245 Park Ave. (1967)	648	47
Sony Bldg., 550 Madison Ave. (1983)	648	37
Merrill Lynch, Two World Financial Center, 225 Liberty St. (1986)	645	44
RCA Victor Bldg., 570 Lexington Ave. (1930)	642	50
345 Park Ave. (1968)	634	44
Grace Plaza, 1114 Ave. of the Amer. (1972)	630	50
One New York Plaza (1968)	630	50
Home Insurance Co., 59 Maiden Lane (1966)	630	44
N.Y. Telephone, 1095 Ave. of the Amer. (1970)	630	40
Central Park Place, 301 W. 57th St. (1988)	628	56
1 Dag Hammarskjold Plaza, 885 2nd Ave. (1972)	628	49
888 7th Ave. (1971)	628	45
Alliance Capital Bldg., 1345 Ave. of the Amer. (1969)	625	50
Waldorf Astoria, 301 Park Ave. (1931)	625	47
Trump Palace, 200 E. 69th St. (1991)	623	55
Olympic Tower, 645 5th Ave. (1976)	620	51
10 E. 40th St. (1929)	620	48
*425 Fifth Avenue (2003)	618	55
101 Park Ave. (1982)	618	50
750 7th Ave. (1989)	615	35
New York Life, 51 Madison Ave. (1928)	615	33
Tower 49, 12 E. 49th St. (1985)	614	44
Credit Lyonnais Bldg., 1301 Ave. of the Amer. (1964)	609	46
Museum Tower Apts., 21 W. 53rd St. (1985)	605	58
IBM, 590 Madison Ave. (1983)	603	41
3 Lincoln Center, 160 W. 66th St. (1993)	595	60
Celanese Bldg., 1211 Ave. of the Amer. (1973)	592	45
Rihga Royal Hotel, 151 W. 54th St. (1990)	590	54
U.S. Court House, 505 Pearl St. (1927)	590	37
Millennium Hilton Hotel, 55 Church St. (1992)	588	58
Time & Life, 1271 Ave. of the Amer. (1959)	587	48
Jacob K. Javits Federal Bldg., 26 Federal Plaza (1967)	587	41
W-Hotel, 1567 Broadway (2000)	584	53
Stevens Tower, 1185 Ave. of Amer. (1971)	580	42
Municipal Bldg., 1 Centre St. (1914)	580	34
Trump International Hotel & Tower (1970)	579	44
520 Madison Ave. (1981)	577	43
Oppenheimer & Co., 1 World Financial Ctr. (1986)	577	37
Merchandise Mart, 41 Madison Ave. (1973)	576	42
Park Ave. Plaza, 55 E. 52nd St. (1981)	575	44
*Ernst & Young Tower, 5 Times Square (2002)	575	40
*Morgan Stanley Dean Whitter Plaza (2002)	575	38
One Financial Square, 33 Old Slip (1987)	575	37
Marriott Marquis Times Square, (1985)	574	50
Westavco Bldg., 299 Park Ave. (1967)	574	42
1166 Ave. of the Americas (1974)	572	44
Socony Mobil, 150 E. 42nd Street (1956)	572	42
Wang Bldg.,780 3rd Ave. (1983)	570	49
7 World Trade Center (1987) **(C)**	570	47
AXA Finance Center, 1290 Ave. of the Amer. (1963)	570	43
600 3rd Ave. (1971)	570	42
450 Lexington Ave. (1991)	568	38
Paramount Tower, 240 E. 39th St. (1998)	567	51
Deutsche Bank, 130 Liberty St. (1974)	565	40
Helmsley Bldg., 230 Park Ave. (1928)	565	35
New York Palace Hotel, 455 Madison Ave. (1980)	563	51

Building	Ht. (ft.)	Stories
30 Broad St. (1932)	562	48
Park Ave. Tower, 65 E. 55th St. (1986)	561	36
Sherry-Netherland, 781 5th Ave. (1927)	560	40
Swiss Bank Tower, 10 E. 50th St. (1990)	560	36
100 UN Plaza (1986)	557	52
Continental Can, 633 3rd Ave. (1962)	557	39
3 Park Ave. (1975)	556	42
Continental Corp., 180 Maiden Lane (1983)	555	41
Sperry & Hutchinson, 330 Madison Ave. (1964)	555	41
Reuters Bldg., 3 Times Square (2001)	555	30
Madison Belvedere, 14 E. 29th St. (1999)	554	48
Interchem, 1133 Ave. of the Amer. (1970)	552	45
Equitable Trust Co. Bldg. (1927)	551	42
Burroughs Bldg., 605 3rd Ave. (1963)	550	44
Bell Atlantic, 33 Thomas St. (1974)	550	29
2 Grand Central Tower, 140 E. 45th St. (1982)	550	44

Oklahoma City, OK

Building	Ht. (ft.)	Stories
Liberty Tower (1971)	500	36
First National Center, 120 N. Robinson St. (1931)	493	33
Apco Tower, 204 N. Robinson St. (1931)	440	32
First Oklahoma Tower (1982)	434	31

Omaha, NE

Building	Ht. (ft.)	Stories
*One First National Center (2002)	634	45
Woodmen Tower (1969)	478	30

Orlando, FL

Building	Ht. (ft.)	Stories
Sun Bank Center Tower (1988)	441	31
Orange County Courthouse (1997)	416	24
Bank of America Center (1988)	409	28

Philadelphia, PA

Building	Ht. (ft.)	Stories
One Liberty Place, 1650 Market St. (1987)	945	61
Two Liberty Place, 1601 Chestnut St. (1989)	848	58
Mellon Bank Center, 1735 Market St. (1990)	792	54
Verizon Tower, 18th & Arch Sts. (1991)	739	53
Blue Cross Tower, 1901 Market St. (1990)	700	50
Commerce Square #1, 2005 Market St. (1990)	572	40
Commerce Square #2, 2001 Market St. (1992)	572	40
City Hall (incl. statue) (1901)	548	9
1818 Market St. (1974)	500	40
Lowe's Philadelphia Hotel , 12 S. 12th St. (1932)	492	39
PNC, 1600 Market St. (1983)	491	40
First Union, 1542 Market St. (1973)	490	38
5 Penn Center (1970)	488	36
1700 Market St. (1969)	482	32
Philadelphia National Bank, 1 S. Broad St. (1930)	475	25
Two Logan Square, 100 N. 18th St. (1988)	435	34
2000 Market St. (1973)	435	29
11 Penn Center, 1835 Market St. (1985)	420	29
Aramark Tower, 1101 Market St. (1984)	417	31
Centre Square, 1500 Market St. (1973)	416	32
First Union Bank, 123 S. Broad St. (1927)	405	30
Lewis Tower, 1419 Locust St. (1929)	400	33
One Logan Square, 130 N. 18th St. (1982)	400	32

Phoenix, AZ

Building	Ht. (ft.)	Stories
Bank One Center, 201 N. Central (1972)	486	40
101 N. Second Ave. (1976)	407	31

Pittsburgh, PA

Building	Ht. (ft.)	Stories
USX Tower, 600 Grant St. (1970)	841	64
One Mellon Bank Center, 500 Grant St. (1983)	725	54
One PPG Place (1984)	635	40
Fifth Ave. Place (1987)	616	32
One Oxford Centre, 301 Grant St. (1982)	615	46
Gulf Tower, 707 Grant St. (1932)	582	44
Univ. of Pittsburgh Cath. of Learning (1936)	535	42
3 Mellon Bank Center, 525 Wm. Penn Way (1951)	520	41
Freemarket Center, 1 Oliver Center (1968)	511	40
Grant Bldg., 330 Grant St. (1928)	485	40
Koppers, 436 7th Ave. (1929)	475	34
2 Oliver Center (1975)	469	32
Equibank (1975)	445	34
CNG Tower (1987)	430	32
Pittsburgh National Bank (1972)	424	30
Alcoa Bldg., 425 6th Ave. (1953)	410	30

Portland, OR

Building	Ht. (ft.)	Stories
Wells Fargo Tower (1973)	546	40
U.S. Bancorp Tower (1983)	536	43
Koin Tower Plaza (1984)	509	31
Pacwest Center (1984)	418	30

Providence, RI

Building	Ht. (ft.)	Stories
Fleet Bank Bldg. (1927)	428	26
FleetBoston Tower (1973)	410	28

Richmond, VA

Building	Ht. (ft.)	Stories
James Monroe Bldg. (1981)	450	29
SunTrust Plaza, 919 E. Main St. (1984)	400	24

St. Louis, MO

Building	Ht. (ft.)	Stories
Gateway Arch (1965)	630	NA
Metropolitan Square Tower (1988)	593	42
One Bell Center, 900 Pine St. (1984)	588	44
Thos. F. Eagleton Federal Courthouse (2000)	557	29
Firstar Center (1976)	485	35
Laclede Gas Bldg., 8th & Olive (1969)	400	31

St. Paul, MN

Building	Ht. (ft.)	Stories
Minnesota World Trade Center (1987)	471	36
Galtier Plaza Jackson Tower (1986)	453	46
First National Bank (1930)	417	32

Salt Lake City, UT

Building	Ht. (ft.)	Stories
American Stores Center (1998)	422	24
L.D.S. Church Office Bldg. (1972)	420	30

San Antonio, TX

Building	Ht. (ft.)	Stories
Tower of the Americas (1968)	622	NA
Marriott Rivercenter, 101 Bowie St. (1988)	546	38
Weston Centre, 112 Pecan St. (1988)	444	32
Tower Life, 310 S. St. Mary's (1929)	404	30

San Diego, CA

Building	Ht. (ft.)	Stories
One American Plaza, 600 W. Broadway (1991)	500	34
Symphony Tower, 759 B St. (1989)	499	34
Hyatt Regency (1992)	497	40
Emerald-Shapery Center, 400 W. Broadway (1991)	450	30
One Harbor Drive (2 bldgs.), 100 Harbor Dr. (1992)	424	41

San Francisco, CA

Building	Ht. (ft.)	Stories
Transamerica Pyramid, 600 Montgomery St. (1972)	853	48
Bank of America, 555 California St. (1969)	779	52
345 California Center (1986)	695	48
101 California St. (1986)	600	48
50 Fremont Center (1983)	600	43
Embarcadero Center No. 4 (1984)	570	45
Embarcadero Center No. 1 (1970)	569	45
Spear Tower, 1 Market St. (1976)	565	43
Wells Fargo, 44 Montgomery St. (1966)	561	43
575 Market St. (1975)	551	39
One Sansome-Citicorp (1984)	550	39
Shaklee Terrace Bldg., 444 Market St. (1979)	537	38
One Post Plaza, 1 Post St.	529	38
525 Market St. (1973)	529	38
One Metro Plaza, 425 Market St. (1973)	524	38
Pacific Telesis Center (1982)	500	38
333 Bush St. (1986)	495	36
Hilton Hotel, 201 Mason St. (1971)	493	46
Pacific Gas & Electric, 77 Beale St. (1970)	492	34
50 California St. (1972)	490	37
*St. Regis Museum Tower (2003)	480	41
100 Pine Center (1972)	476	34
Bechtel Bldg., 45 Fremont St. (1979)	475	34
333 Market Bldg. (1979)	474	33
Hartford Bldg., 650 California St. (1965)	465	33
Four Seasons Hotel & Tower (2001)	443	40
1 California St. (1969)	438	32
Marriott Hotel (1989)	436	39
Russ Bldg., 235 Montgomery St. (1928)	435	31
Pacific Bell Headquarters, 140 Montgomery St. (1925)	435	26
*Chase H&Q Bldg. (2002)	421	31
Pacific Gateway, 201 Mission St. (1983)	416	30
Embarcadero Center No. 2 (1974)	412	31
Embarcadero Center No. 3 (1976)	412	31
595 Market (1977)	410	31
123 Mission Bldg.	406	28
Embarcadero Center West, 275 Battery St. (1988)	405	33
101 Montgomery St. (1983)	405	29

Seattle, WA

Building	Ht. (ft.)	Stories
Bank of America Center, 701 5th Ave. (1985)	954	76
Two Union Square, 600 Union St. (1989)	740	56
Washington Mutual Tower, 1201 3rd Ave. (1988)	730	55
Key Tower, 700 5th Ave. (1990)	722	62
1001 Fourth Avenue Plaza (1969)	609	50
Space Needle, 203 6th Ave. (1962)	605	NA
U.S. Bank Center, 1420 5th Ave. (1989)	580	44
Wells Fargo Center, 999 3rd Ave. (1983)	574	47
800 Fifth Avenue Plaza (1981)	543	42
Security Pacific Bank, 900 4th Ave. (1973)	536	41
Rainier Tower, 1301 5th Ave. (1977)	514	31
*IDX Tower, 915 4th Ave. (2003)	512	40
Civic Center Plaza, 1000 2nd Ave. (1986)	493	40
Henry M. Jackson Bldg. (1974)	487	37
US West Plaza, 1600 7th Ave. (1976)	466	33
Smith Tower, 506 2nd Ave. (1914)	465	38
One Union Square, 600 University Ave. (1981)	456	36
1111 3rd Ave. (1980)	454	34
Westin Hotel NorthTower, 1900 5th Ave. (1982)	448	44
Westin Bldg., 2001 6th Ave. (1981)	409	34

Southfield, MI

Building	Ht. (ft.)	Stories
Prudential, 3000 Town Center (1975)	448	32
1000 Town Center (1988)	405	32

Sunny Isles, FL

Building	Ht. (ft.)	Stories
The Pinnacle, 17555 Collins Ave. (1999)	476	40
*The Residences at Ocean Grande (2003)	443	38
*Ocean Two (2001)	426	40

Tampa, FL

Building	Ht. (ft.)	Stories
100 N. Tampa (1992)	579	42
Bank of America Plaza (1986)	577	36
Tampa City Center (1981)	537	39
Suntrust Financial Center (1992)	525	36
First Financial Tower (1973)	458	36
400 N. Ashley (1988)	454	33

Toledo, OH

Building	Ht. (ft.)	Stories
One Seagate (1962)	411	32
HyTower (1970)	400	30

Toronto, Ontario

Building	Ht. (ft.)	Stories
CN Tower, 301 Front St. W (1976)	1,815	NA
First Canadian Place (1975)	978	72
Scotia Plaza (1988)	902	68
BCE Place, Canada Trust Tower (1990)	863	51
Commerce Court West (1973)	784	57
Toronto Dominion Centre–Toronto Dominion Bank Tower (1967)	730	56
BCE Place, Bay-Wellington Tower (1991)	679	43
Toronto Dominion Centre–Royal Trust Tower (1969)	600	43
Royal Bank Plaza–South Tower (1976)	567	41
Manulife Centre (1975)	545	51
Toronto Dominion Centre–Aetna Tower (1985)	504	37
The 250, 250 Yonge St. (1991)	494	35
Two Bloor West (1974)	486	34
Simcoe Place (1995)	486	33
Exchange Tower (1981)	480	36
CIBC-Commerce Court North, 25 King St. (1931)	477	34
Simpson Tower, 401 Bay Street (1968)	473	33
Cadillac-Fairview (1982)	465	36
One Financial Center, 1 Adelaide Pl. (2002)	458	45
One Palace Pier Court, Etobicoke (1991)	455	46
Three Palace Pier Court, Etobicoke (1978)	453	46
Continental Bank, 130 Adelaide St. W (1980)	450	35
Pantages Tower (2002)	450	32
Sheraton Centre, 123 Queen St. W (1972)	443	43
Two Bloor East (1974)	440	35
Royal York Hotel (1929)	439	26
Ernst & Young Tower (1990)	438	31
Old Toronto Exchange Bldg. (1990)	436	31
Leaside Towers (2 bldgs.) (1970)	423	44
Canadian Pacific Tower (1974)	420	32
Metro Hall, 55 John St. (1991)	420	27
Maple Leaf Mills (1977)	419	30
Marriott Hotel/Plaza 2 Apts., 90 Bloor St. E (1973)	415	41
Sun Life Financial Center East Tower, 150 King St. W (1981)	410	27
Young-Eglington Centre–Triathlon Tower (1975)	408	30

Tulsa, OK

Building	Ht. (ft.)	Stories
Williams Center (1975)	667	52
Cityplex Central Tower (1981)	648	60
First National Bank (1973)	516	41
Mid-Continent Tower (1984)	513	36
Fourth National Bank (1966)	412	33
National Bank of Tulsa, 320 South Boston (1918)	400	24

Vancouver, British Columbia

Building	Ht. (ft.)	Stories
One Wall Centre, 1000 Burrard St. (2001)	491	45
200 Granville Square (1973)	466	32
Scotia Tower, 650 W. Georgia St. (1977)	452	36
Royal Bank Tower, 1055 W. Georgia St. (1973)	461	37
Park Place, 666 Burrard St. (1984)	459	35
Bentall IV Canada Trust, 1055 Dunsmir (1981)	454	36
Scotia Tower, 650 W. Georgia St. (1977)	452	36
Harbour Centre, 555 W. Hastings (1977)	426	28
Toronto Dominion Bank Tower, 700 W. Georgia (1970)	417	30
Bentall III, Bank of Montreal, 595 Burrard St. (1974)	400	31

Winnipeg, Manitoba

Building	Ht. (ft.)	Stories
Toronto Dominion Centre, 201 Portage Ave. (1990)	420	33
Richardson Bldg., 1 Lombard Place (1969)	406	34

Winston-Salem, NC

Building	Ht. (ft.)	Stories
Wachovia Center, 100 N. Main St. (1995)	460	34
301 N. Main St. (1965)	410	30

Other Tall Buildings in North American Cities

Building	City	Ht. (ft.)	Stories
Erastus Corning II Tower (1973)	Albany, NY	589	44
San Jacinto Monument (1936)	La Porte, TX	570	NA
Washington Monument (1884)	Washington, DC	555	NA
Dataflux Tower (2000)	Monterrey, Mexico	549	NA
Marine Midland Ctr. (1970)	Buffalo, NY	529	40
Vehicle Assembly Bldg (1965)	Cape Canaveral, FL.	525	40
Skylon (Tower) (1965)	Niagara Falls, Ont.	520	NA
Fenwick Place (1970)	Halifax, Nova Scotia	486	36
State Capitol (1932)	Baton Rouge, LA	460	34
Burbank Tower, 2900 W. Burbank (1988)	Burbank, CA.	460	32
Hotel Habana	Havana, Cuba	450	30
*The Diplomat (2001)	Hollywood, FL	444	39
Xerox Tower (1965)	Rochester, NY	443	30
One Summit Square (1981)	Fort Wayne, IN	442	27
Two Hanover Square (1991)	Raleigh, NC	431	29
Union Planters Bank, 100 N. Main (1965)	Memphis, TN	430	37
Taj Mahal, 1000 Boardwalk (1990)	Atlantic City, NJ	429	43
AmSouth Bank Bldg.	Mobile, AL	424	33
*Museum Tower, Lincoln Ctr. (2002)	Bellevue, WA	421	40
State Capitol (1932)	Lincoln, NE	419	22
Century 21	Hamilton, Ont.	418	43
Oakbrook Terrace Tower (1985)	Oakbrook, IL	418	31
Hidden Bay 1 (2000)	Aventura, FL	417	40
Galaxie Apts. (3 bldgs.) (1976)	Guttenberg, NJ	415	44
Complexe G (1972)	Quebec City, Que.	415	33
AmSouth Bank Bldg. (1996)	Montgomery, AL.	415	25
One Shoreline Plaza, South Tower (1988)	Corpus Christi, TX.	411	28
Silver Legacy Hotel & Casino, 407 N. Virginia St. (1995)	Reno, NV	410	38
AutoNation Tower (1988)	Ft. Lauderdale, FL.	410	30
Financial Center (1986)	Lexington, KY	410	30
Clark Tower (1972)	Memphis, TN	410	33
*Clayton Plaza (2002)	Clayton, MO	409	30
Kettering Tower (1970)	Dayton, OH	405	30
Wells Fargo Center (1991)	Sacramento, CA	404	30
Ordway Bldg. (1985)	Oakland, CA	404	28
Three Lakeway Center (1987)	Metairie, LA	403	34
Edificio Fosca (1957)	Havana, Cuba	402	37
United American Bank (1977)	Knoxville, TN	400	30
Monarch Place (1987)	Springfield, MA	400	26
Bank of America (1987)	St. Petersburg, FL	400	26

Notable Bridges in North America

Source: Federal Highway Administration, Bridge Division, U.S. Dept. of Transportation; World Almanac research

Asterisk (*) designates railroad bridge. Year is date of completion. Span of a bridge is the distance between its supports.

Suspension

Year	Bridge	Location	Main span (ft.)
1964	Verrazano-Narrows	New York, NY	4,260
1937	Golden Gate	San Fran. Bay, CA	4,200
1957	Mackinac Straits	Sts. of Mackinac, MI	3,800
1931	Geo. Washington	Hudson R., NY–NJ	3,500
1950	Tacoma Narrows	Tacoma, WA	2,800
1950	Tacoma Narrows II	Tacoma, WA	2,800
1936	San. Fran.-Oakland Bay[1]	San Fran. Bay, CA	2,310
1939	Bronx-Whitestone	East R., NY	2,300
1970	Pierre Laporte	Quebec, Canada	2,190
1951	Del. Memorial	Wilmington, DE	2,150
1957	Walt Whitman	Philadelphia, PA	2,000
1929	Ambassador	Detroit, MI–Can.	1,850
1961	Throgs Neck	Long Is. Sound, NY	1,800
1926	Benjamin Franklin	Philadelphia, PA	1,750
1924	Bear Mt.	Hudson R., NY	1,632
1903	Williamsburg	East R., NY	1,600
1952	Wm. Preston La. Mem.[2]	Sandy Point, MD	1,600
1969	Newport	Narragansett Bay, RI	1,600
1883	Brooklyn	East R., NY	1,595
1939	Lion's Gate	Burrard Inlet, BC	1,550
1930	Mid-Hudson	Poughkeepsie, NY	1,500
1963	Vincent Thomas	L. A. Harbor, CA	1,500
1909	Manhattan	East R., NY	1,470
1955	MacDonald Bridge	Halifax, Nova Scotia	1,447
1970	A. Murray Mackay	Halifax, Nova Scotia	1,400
1936	Triborough	East R., NY	1,380
1931	St. Johns	Portland, OR	1,207
1929	Mount Hope	RI	1,200
1960	Ogdensburg	St. Lawrence R., NY	1,150
1965	Bidwell Bar Bridge	Oroville, CA	1,108
1964	Middle Fork Feather	CA	1,105
1939	Deer Isle	ME	1,080
1931	Simon Kenton Memorial	Ohio R., Maysville, KY	1,060
1936	Ile d'Orleans	St. Lawrence R., Quebec	1,059
1867	John A. Roebling	Ohio R., KY	1,057
1971	Dent	Clearwater Co., ID	1,050
1900	Miampimi	Mexico	1,030
1849	Wheeling	Ohio R., WV	1,010

Cantilever

Year	Bridge	Location	Main span (ft.)
1917	Québec Bridge	St. Lawrence R., Quebec	1,800
1988	Greater New Orleans Bridge	Mississippi R., New Orleans, LA	1,575
1995	Gramercy Bridge	Mississippi R., Gramercy, LA	1,460
1936	Transbay	San Fran. Bay, CA	1,400
1968	Baton Rouge Bridge	Mississippi R., Baton Rouge, LA	1,235
1955	Tappan Zee	Hudson R., NY	1,212
1930	Lewis and Clark	Longview, WA–OR	1,200
1976	Patapsco River	Baltimore , MD	1,200
1909	Queensboro	East R., NY	1,182
1927	Carquinez Strait	CA	1,100
1958	Parallel Span	CA	1,100
1930	Jacques Cartier	Montreal, Quebec	1,097
1968	Isaiah D. Hart	Jacksonville, FL	1,088
1956	Richmond[3]	San Fran. Bay, CA	1,070
1929	Grace Memorial	Charleston, SC	1,050
1980	Newburgh-Beacon	Hudson R., NY	1,000
1949	Martin Luther King	St. Louis, MO	963
1975	Caruthersville	Mississippi R., MO–TN	920
1969	Silver Memorial	Pt. Pleasant, WV–OH	900
1977	Saint Marys	Saint Marys, WV–OH	900
1981	Ravenswood	WV	900
1987	Carl Perkins	Ohio R., KY	900
1988	Mississippi R.	Natchez, MS	875
1938	Blue Water	Pt. Huron, MI	871
1972	Mississippi R.	Vicksburg, MS	870
1972	N. Fork American R.	Auburn, CA	862
1940	*Baton Rouge	Mississippi R., LA	848
1899	*Cornwall	St. Lawrence R.	843
1940	Rte. 82	Mississippi R., AR.	840
1961	Mississippi R.	Greenville, MS	840
1961	Rte. 49	Mississippi R., AR	840
1963	Brent Spence	KY–OH	830
1940	Mississippi R.	Vicksburg, MS	825
1963	Mississippi R.	Donaldsonville, LA	825
1929	Clark Memorial	Ohio R., KY	820
1961	Campbellton-Cross Pt.	New Brunswick, Can.	815
1932	Washington Mem.	Seattle, WA	800
1935	Rip Van Winkle	Catskill, NY	800
1938	Cairo	Ohio R., IL–KY	800
1936	McCullough	Coos Bay, OR	793
1892	Memphis	Mississippi R., TN	790
1935	Huey P. Long[4]	New Orleans, LA	790
1949	Rte. 55	Mississippi R., AR–TN	790
1910	*P&LE RR Bridge	Ohio R., PA	750
1932	Bi-State Vietnam Gold Star	Henderson, KY	720
1904	*Norfolk Southern RR	Ohio R., OH	700
1943	*Pit River	Redding, CA	620
1941	Columbia R.	Kettle Falls, WA	600
1954	Columbia R.	Umatilla, OR	600
1954	Columbia R.	The Dalles, OR	576
1968	W. 17th St.	Huntington, WV	562

Simple Truss

Year	Bridge	Location	Main span (ft.)
1976	Chester	Chester, WV	745
1929	Irvin S. Cobb	Ohio R., IL–KY	716
1922	*Tanana R.	Nenana, AK	700
1967	I-77, Ohio R.	Williamstown, WV	650
1917	MacArthur[4]	St. Louis, IL–MO	647
1992	St. Charles	Missouri R, MO	625
1933	Atchafalaya	Morgan City, LA	608
1924	*Castleton	Hudson R., NY	598
1937	Delaware R.	Easton, PA	550
1930	Swindell Bridge	Pittsburgh, PA	545
1952	Allegheny R. Tpk.	Pittsburgh, PA	534
1930	*Martinez	Martinez, CA	528
1951	Rankin	Pittsburgh, PA	525
1914	Old Brownsville	Brownsville, PA	520
1906	Donora-Webster	Donora-Webster, PA	515
1909	Hulton	Pittsburgh, PA	505
1967	Tanana R.	AK	500

Steel Truss

Year	Bridge	Location	Main span (ft.)
1988	Glade Creek	Raleigh Co., WV	784
1973	Atchafalaya R.	Krotz Springs, LA	780
1972	Piscataqua R.	NH–ME	756
1972	Atchafalaya R.	Simmesport, LA	720
1957	SR-3, Rappahannock R.	Middlesex Co., VA.	648
1978	Atchafalaya R.	Morgan City, LA	607
1959	Summit	Summit, DE.	600
1969	Reedy Point	Delaware City, DE	600
1938	US-22	Delaware R., NJ	540
1955	Interstate (I-5)	Columbia R., OR–WA	531
1910	McKinley, St. Louis[4]	Mississippi R., MO	517
1972	Mississippi R.	Muscatine, IA	512
1896	Newport	Ohio R., KY	511
1989	US 190, Atchafalaya R.	Krotz Springs, LA	506
1931	Lucy Jefferson Lewis	Cumberland R., KY	500
1958	Lake Oahe	Gettysburg, SD	500
1958	Lake Oahe	Mobridge, SD	500
1970	Lake Koocanusa	Lincoln Co., MT	500

Continuous Truss

Year	Bridge	Location	Main span (ft.)
1966	Columbia R. (Astoria)	OR–WA	1,232
1977	Francis Scott Key	Baltimore, MD	1,200
1981	Ravenswood/Ohio R.	Ravenswood, WV	902
1995	Central	Ohio R., KY–OH	850
1943	Dubuque	Mississippi R., IA	845
1966	Charles Braga	Fall River, MA	840
1956	Earl C. Clements5	Ohio R., IL–KY	825
1929	U.S. 31	Ohio R., IN–KY	820
1953	John E. Mathews	Jacksonville, FL	810
1950	Maurice J. Tobin	Boston, MA.	801
1940	Gov. Nice Memorial	Potomac River, MD	800
1957	Kingston-Rhinecliff	Hudson R., NY	800
1992	Mark Clark Expy. I-526	Cooper R., Charleston, SC	800
1986	Rochester-Monaca	Rochester-Monaca, PA	780
1940	U.S. 231	Ohio R., IN	750
1974	Carroll L. Cropper	Ohio R., IN–KY	750
1981	Sewickley	Sewickley, PA	750
1984	13th St. Bridge, Ohio R.	Ashland, KY	740
1959	Monaca-E. Rochester	Monaca-E. Rochester, PA.	730
1976	Betsy Ross	Philadelphia, PA	729
1929	U.S. 421	Ohio R., IN–KY	727
1967	Matthew E. Welsh6	Mauckport, IN	725
1962	U.S. 41	Ohio R., IN–KY	720
1994	6th St.	Huntington, WV	720
1970	Vanport	Vanport, PA	715
1962	Champlain	Montreal, Que.	707
1962	John F. Kennedy7	Ohio R., IN–KY	701
1973	Girard Point	Philadelphia, PA	700
1954	PA Tpk., Delaware R.	Philadelphia, PA	682
1938	Port Arthur-Orange	TX	680
1949	George Platt	Philadelphia, PA	680

Year	Bridge	Location	Main span (ft.)
1926	Cape Girardeau	Mississippi R., MO	677
1929	*Cincinnati	Ohio R., OH.	675
1946	Chester	Mississippi R, IL	670
1970	Gulfgate	Port Arthur, TX	664
1994	Williamstown-Marietta	Williamstown, WV	650
1955	Jefferson City	Missouri R., MO	640
1930	Quincy	Mississippi R., IL	628
1959	US 181, over harbor	Corpus Christi, TX	620
1961	Shippingport	Shippingport, PA	620
1935	Bourne-Sagamore	Cape Cod Canal, MA	616
1965	Clarion R. (I-80)	Clarion, PA	612
1975	Donora-Monessen	Donora-Monessen, PA	608
1957	Blatnik	Duluth, MN	600
1965	Rio Grande Gorge	Taos, NM	600
1991	Hoffstadt Creek	Mt. St. Helens, WA	600
1991	Jefferson City	Missouri R., MO	596
1962	W. Branch Feather R.	Oroville, CA	576
1967	Glenwood	Pittsburgh, PA	567
1936	Mark Twain Mem.	Hannibal, MO	562
1957	Mackinac	Mackinac Straits, MI	560
1932	Pulaski Skyway	Passaic R.-Hackensack R., NJ	550
1966	Emlenton	Emlenton, PA.	540
1973	Gold Star Memorial	New London, CT	540
1936	Homestead High Level	Pittsburgh, PA	534
1959	Martinez	Benicia-Martinez, CA	528
1960	Brownsville High Level	Brownsville, PA	518
1971	Grandad	Elk River, ID.	504
1945	Mansfield-Dravosburg	Pittsburgh, PA	500

Continuous Box and Plate Girder

Year	Bridge	Location	Main span (ft.)
1967	San Mateo-Hayward #2	San Fran. Bay, CA	750
1976	Intracoastal Canal	Forked Is., LA	750
1977	Intracoastal Canal	Gibbstown, LA	750
1982	Houston Ship Chan	Houston, TX	750
1969	San Diego-Coronado[8]	San Diego Bay, CA	660
1987	Umatilla, Columbia R.	OR–WA	660
1994	Acosta	Jacksonville, FL	630
1981	Douglas	Juneau, AK	620
1976	Wax L. Outlet	Calumet, LA	618
1963	Poplar St.	St. Louis, MO.	600
1981	Glenn Jackson (I-205)	Columbia R., OR–WA	600
1976	Stanislaus River	Sonora, CA	580
1982	Illinois R.	Pekin, IL	550
1982	I-440	Arkansas R., AR	540
1980	US-64, Tennessee R.	Savannah, TN	525
1965	McDonald-Cartier	Ottawa, Ont.	520
1988	Mon City	Monongahela, PA	520
1984	Columbia R.	Richland, WA.	450
1986	Veterans	Pittsburgh, PA	440
1987	SR 76, Cumberland R.	Dover, TN	440
1987	SR 20, Tennessee R.	Perryville, TN.	440
1970	Willamette R., I-205	West Linn, OR	430
1974	I-430	Arkansas R., AR	430
1965	I-24, Tennessee R.	Marion Co., TN	420
1974	Dunbar-S. Charleston	S. Charleston, WV.	420
1975	36th St.	Charleston, WV	420
1978	Snake R.	Clarkston, WA	420
1984	FAU 3456, TN R.	Chattanooga, TN.	420

Continuous Plate

Year	Bridge	Location	Main span (ft.)
1973	Ship Channel (I-610)	Houston, TX	630
1971	W. Atchafalaya	Henderson, LA	573
1992	State Route 76	Paris, TN	525
1981	Illinois 23	Illinois R., IL.	510
1968	Trinity R.	Dallas, TX	480
1978	San Joaquin R.	Antioch, CA	460
1977	Thomas Johnson Mem.	Solomons, MD	451
1967	Mississippi R.	La Crosse, WI	450
1975	I-129	Missouri R., IA–NE	450
1979	Lewis	St. Louis, MO.	450
1992	Cuba Landing Bridge	Tennessee R., TN	450
1966	I-480	Missouri R., IA–NE	425
1972	Whiskey Bay Pilot	Ramah, LA	425
1972	I-80	Missouri R., IA–NE	425
1972	I-635, Kansas City	Missouri R., KS–MO	425
1983	US-36	Missouri R., KS–MO	425
1987	I-435	Missouri R., KS–MO	425
1978	I-24	Cumberland R., KY	420
1993	Bob Michel Bridge	Peoria, IL.	360

Cable-Stayed

Year	Bridge	Location	Main span (ft.)
1986	Annacis (Alex Fraser)	Vancouver, BC.	1,526
1993	Quetzalapa Bridge	Quetzalapa, Mexico	1,391
1988	Dames Point	Jacksonville, FL	1,300
1995	Houston Ship Channel	Baytown, TX	1,250
1983	Hale Boggs Memorial	Luling, LA	1,222
1987	Sunshine Skyway	Tampa Bay, FL.	1,200
1988	Tampico/Panuco R.	Mexico	1,181

Year	Bridge	Location	Main span (ft.)
1988	ALRT Fraser River Bridge.	Vancouver, BC	1,115
1990	Talmadge Mem.	Savannah, GA.	1,100
1993	Mezcala	Mex. City/Acapulco Hwy.	1,024
1978	Pasco-Kennewick	Columbia R., WA	981
1984	Coatzacoalcos R.	Mexico.	919
1985	E. Huntington	E. Huntington, WV	900
1987	Bayview Bridge	Quincy, IL	900
1970	Burton Bridge	New Brunswick, Canada	850
1990	Weirton-Steubenville	WV–OH.	820
1969	Papineau-Leblanc	Montreal, Que.	790
1991	Cochrane	Mobile, AL.	780
1994	Clark Bridge	Alton, IL.	756
1995	Chesapeake & Delaware Canal Bridge	Dover-Wilmington, DE	750
1966	Longs Creek	New Brunswick, Canada	713
1967	Hawkshaw	New Brunswick, Canada	713
1993	Quetzalapa Bridge	Quetzalapa, Mexico	699
1993	Burlington Bridge	Burlington, IA	660
1991	Neches R.	Port Arthur-Orange, TX	640
1989	James River Bridge.	Richmond, VA.	630

I-Beam Girder

Year	Bridge	Location	Main span (ft.)
1980	Interstate 20.	Shreveport, LA	438
1988	Route 18	Weston's Mill Pond, NJ	276

Steel Arch

Year	Bridge	Location	Main span (ft.)
1977	New River Gorge	Fayetteville, WV	1,700
1931	Bayonne (Kill Van Kull)	Bayonne, NJ	1,652
1973	Fremont	Portland, OR	1,255
1964	Port Mann	Vancouver, BC	1,200
1967	Lavioleete	Three Rivers, Canada	1,100
1967	Trois-Rivieres	St. Lawrence R., Que.	1,100
1992	Roosevelt Lake	Roosevelt Lake, AZ	1,080
1959	Glen Canyon	Page, AZ	1,028
1962	Lewiston-Queenston	Niagara R., Ont.	1,000
1976	Perrine	Twin Falls, ID.	993
1941	Rainbow Bridge	Niagara Falls, NY	984
1917	*Hell Gate	East R., N.Y	977
1977	Moundsville	Ohio R., WV	912
1992	I-255, Miss. R.	St. Louis, MO	909
1972	I-40, Miss. R.[9]	AR–TN	900
1936	Henry Hudson	Harlem R., NY.	840
1967	Lincoln Trail Bridge	Ohio R., IN–KY	825
1978	I-57, Miss. R.	Cairo , IL.	821
1961	I-64, Ohio R.	IN	800
1980	I-65, Mobile R.	Mobile, AL.	800
1930	West End	Pittsburgh, PA	780
1978	I-470 Bridge, Ohio R.	Wheeling, WV.	780
1996	Navajo Bridge	Glen Canyon, AZ	726

Concrete Arch

Year	Bridge	Location	Main span (ft.)
1993	Natchez Trace Pkwy.	Franklin, TN	582
1993	Lake Street Bridge.	St. Paul, MN	556
1971	Selah Creek (twin)	Selah, WA.	549
1968	Cowlitz R.	Mossyrock, WA.	520
1931	Westinghouse	Pittsburgh, PA	460
1923	Cappelen	Minneapolis, MN	435
1930	Jack's Run	Pittsburgh, PA	400

Segmental Concrete

Year	Bridge	Location	Main span (ft.)
1997	Confederation Bridge	Prince Edward Isl., NB	820
1978	Shubenacadie River	S. Maitland, Nova Scotia	790
1982	Jesse H, Jones Memorial	Houston, TX	750
1992	Narragansett Bay Crossing	Jamestown, RI	674
1986	WB I-82 (Columbia R.)	Umatilla, OR	660
1976	Stanislaus River	Parrets Ferry. CA	640
1992	Jamestown-Verrazano	Jamestown, RI	636
1981	Gastineau Channel Br.	Juneau, AK	620
1991	Veterans Memorial Centennial Bridge	Coeur d'Alene, ID	520
1974	Pine Valley Creek	Pine Valley, CA	450
1988	Zilwaukee Bridge (twin)	Zilwaukee, MI	392
1985	Red River Bridge	Boyce, LA	370

Twin Concrete Trestle[10]

Year	Bridge	Location	Main span (ft.)
1979	I-55/I-10	Manchac, LA	181,157
1969	L. Pontchartrain Cswy.	Mandeville, LA	126,720
1972	Atchafalaya Flwy.	Baton Rouge, LA.	93,984
1963	L. Pontchartrain	Slidell, LA	28,547
1983	*Interstate 310	Kenner, LA	25,925

Concrete Slab Dam[10]

Year	Bridge	Location	Main span (ft.)
1927	Conowingo Dam	MD	4,611
1952	SR-4, Roanoke R.	Mecklenburg Co., VA	2,785
1936	Hoover Dam.	Lake Mead, NV	1,324

Drawbridges

Vertical Lift

Year	Bridge	Location	Main span (ft.)
1959	*Arthur Kill	NY–NJ	558
1965	Pennsylvania Railroad	Kirkwood-Mt. Pleas., DE	548
1935	*Cape Cod Canal	Cape Cod, MA	544
1961	*Delair	Delaware R., NJ	542
1931	Burlington-Bristol	Delaware R., NJ–PA	540
1937	Marine Parkway	Jamaica Bay, NY	540
1908	*Willamette R.	Portland, OR	521
1968	Second Narrows	Vancouver, B.C.	493
1912	*A-S-B Fratt	Kansas City, MO	428
1945	*Harry S Truman	Kansas City, MO	427
1955	Roosevelt Island	East R., NY	418
1980	US-17, James R.	Isle of Wight, Co., VA	415
1932	*M-K-T R.R.	Missouri R., MO	414
1969	Cape Fear Mem.	Wilmington, NC	408
1930	Aerial	Duluth, MN	386
1962	Burlington	Ontario, Can.	370
1922	*Cincinnati	Ohio R., OH	365
1941	Main Street	Jacksonville, FL	365
1967	SR-156, James R.	Prince George Co., VA	364
1950	Red R.	Moncla, LA	360
1957	Industrial Canal	New Orleans, LA	360
1936	Tribo	Harlem R., NY	344
1961	Corpus Christi Harbor[4]	Corpus Christi, TX	344
1939	U.S. 1&9, Passaic R.	Newark, NJ	333
1930	*Martinez	Martinez, CA	328
1960	St. Andrews Bay	Panama City, FL	327
1929	*Penn-Lehigh	Newark Bay, PA	322
1987	Industrial Canal	New Orleans, LA	320
1920	*Chattanooga	Tennessee R., TN	310

Bascule

Year	Bridge	Location	Main span (ft.)
1940	Lorain	Black R., OH	333
1917	SR-8, Tennessee R.	Chattanooga, TN	306
1956	Duwamish R.	Seattle, WA	300
1955	Chehalis R.	Aberdeen, WA	288
1968	Elizabeth R.	Chesapeake, VA	280
1913	Broadway	Portland, OR	278
1954	Fuller Warren	Jacksonville, FL	267

Swing Bridges

Year	Bridge	Location	Main span (ft.)
1927	Fort Madison[4]	Mississippi R., IA	545
1991	SW. Spokane St.	Seattle, WA	480
1930	Rigolets Pass	New Orleans, LA	400
1950	Douglass Memorial	Washington, DC	386
1945	Lord Delaware	Mattaponi R., VA	252

Swing Span

Year	Bridge	Location	Main span (ft.)
1952	US-17	York R., VA	500
1897	*Duluth	St. Louis Bay, MN	486
1899	*C.M.&N.R.R.	Chicago, IL	474
1913	Rt. 82, Conn-R.	E. Haddam, CT	465
1914	*Coos Bay	OR	458

Floating Pontoon

Year	Bridge	Location	Main span (ft.)
1963	Evergreen Pt.	Seattle, WA	7,578
1961	Hood Canal	Pt. Gamble, WA	6,521
1993	Lacey V. Murrow[11]	Seattle, WA	6,620
1989	Third Lake Washington	Seattle, WA	5,811

(1) Swing span bridge with 2 spans of 2,310 ft. each. (2) A second bridge in parallel was completed in 1978. (3) The Richmond Bridge has twin spans 1,070 ft. each. (4) Railroad and vehicular bridge. (5) Two spans each 825 ft. (6) Two spans each 707 ft. (7) Two spans each 700 ft. (8) Two spans each 660 ft. (9) Two spans each 900 ft. (10) Length listed is total length of bridge. (11) Replaces the original Lacey V. Murrow bridge, which opened in 1940 and sank in 1990.

Oldest U.S. Bridges in Continuous Use

Built in 1697, the stone-arch Frankford Ave. Bridge crosses Pennypack Creek in Philadelphia, PA. A 3-span bridge with a total length of 75 ft., it was constructed as part of the King's Road, which eventually connected Philadelphia to New York.

The oldest covered bridge, completed in 1827, is the double-span, 278-ft. Haverhill Bath Bridge, which spans the Ammonoosuc River, between the towns of Bath and Haverhill, NH.

Some Notable International Bridges

Span of bridge is the distance between its supports.

Asterisk (*) designates under construction.

Suspension

Year	Bridge	Location	Main span (ft.)
1998	Akashi Kaikyo	Japan	6,570
2003	*Izmit Bay	Turkey	5,538
1998	Storebælt (East Bridge)	Denmark	5,328
1981	Humber	England	4,626
1999	Jiangyin Yangtze	China	4,544
1997	Tsing Ma[1]	China	4,518
1997	Hoga Kusten	Sweden	3,970
1988	Minami Bisan-Seto	Japan	3,609
1988	Bosphorus II	Turkey	3,576
1973	Bosphorus I	Turkey	3,524
1999	Kurushima III	Japan	3,379
1999	Kurushima II	Japan	3,346
1966	Tagus River[2]	Portugal	3,323
1964	Forth Road	Scotland	3,300
1988	Kita Bisan-Seto	Japan	3,248
1966	Severn	England	3,241
1988	Shimotsui Strait	Japan	3,084

Cantilever

Year	Bridge	Location	Main span (ft.)
1890	Forth[3] (rail)	Scotland	1,710
1974	Nanko	Japan	1,673

Steel Arch

Year	Bridge	Location	Main span (ft.)
1932	Sydney Harbour	Australia	1,650
1967	Zdakov	Czech Republic	1,244
1962	Thatcher	Panama Canal Zone	1,128
1961	Runcorn-Widnes	England	1,082
1935	Birchenough	Zimbabwe	1,080

Concrete Arch

Year	Bridge	Location	Main span (ft.)
1980	Krk I	Croatia	1,280
1964	Gladesville	Australia	1,000
1964	Amizade	Brazil	951
1963	Arrabida	Portugal	886
1943	Sando	Sweden	866

Steel Plate and Box Girder

Year	Bridge	Location	Main span (ft.)
1974	President Costa e Silva	Brazil	984
1956	Sava I	Yugoslavia	856
1966	Zoobrüke	Germany	850

Cable-Stayed

Year	Bridge	Location	Main span (ft.)
1999	Tatara	Japan	2,920
1995	Pont de Normandie	France	2,808
1996	Quingzhou Minjang	China	1,985
1993	Yangpu	China	1,975
1997	Xupu	China	1,936
1998	Meiko Chuo	Japan	1,936
1991	Skarnsundet	Norway	1,739
1999	Queshi	China	1,700
1995	Tsurumi Tsubasa	Japan	1,673
2000	Oresund	Denmark/Sweden	1,614
1991	Ikuchi	Japan	1,608
1994	Higashi Kobe	Japan	1,591
1998	Zhanjiang	China	1,575
1997	Ting Kau	China	1,558
1999	Seo Hae Grand	South Korea	1,542
1989	Yokohama Bay	Japan	1,509
1993	Second Hooghly River	India	1,499
1995	Second Severn Crossing	England/Wales	1,496

(1) Double-decked road and rail bridge. (2) Railroad and highway bridge. (3) Two spans of 1,710 ft. each.

Underwater Vehicular Tunnels in North America

(more than 5,000 ft. in length; year in parentheses is year of completion)

Name	Location	Waterway	Feet
Brooklyn-Battery (1950) (twin)	New York, NY	East River	9,117
Holland Tunnel (1927) (twin)	New York, NY	Hudson River	8,557
Ted Williams Tunnel (1995)	Boston, MA	Boston Harbor	8,448
Lincoln Tunnel (1937, 1945, 1957) (3 tubes)	New York, NY	Hudson River	8,216
Thimble Shoal Channel (1964)	Northampton Co., VA	Chesapeake Bay	8,187
Chesapeake Channel (1964)	Northampton Co., VA	Chesapeake Bay	7,941
Fort McHenry Tunnel (1985) (twin)	Baltimore, MD	Baltimore Harbor	7,920
Hampton Roads (1957) (twin)	Hampton, VA	Hampton Roads	7,479
Baltimore Harbor Tunnel (1957) (twin)	Baltimore, MD	Patapsco River	7,392
Queens Midtown (1940) (twin)	New York, NY	East River	6,414
Sumner Tunnel (1934)	Boston, MA	Boston Harbor	5,653
Louis-Hippolyte Lafontaine Tunnel	Montreal, Que.	St. Lawrence River	5,280
Detroit-Windsor (1930)	Detroit, MI	Detroit River	5,160
Callahan Tunnel (1961)	Boston, MA	Boston Harbor	5,070

Land Vehicular Tunnels in the U.S.

Source: Federal Highway Administration

(more than 3,000 ft. in length)

Name	Location	Feet
Anton Anderson Mem. Tunnel[1]	Whittier, AK	13,300
E. Johnson Memorial	I-70, CO	8,959
Eisenhower Memorial	I-70, CO	8,941
Allegheny (twin)	PA Turnpike	6,072
Liberty Tubes	Pittsburgh, PA	5,920
Zion Natl. Park	Rte. 9, UT	5,766
East River Mt.	Mercer Co., VA	5,654
East River Mt. (twin)	VA–WV	5,412
Tuscarora (twin)	PA Turnpike	5,400
Tetsuo Harano (twin)	H-3, HI	5,165
Kittatinny (twin)	PA Turnpike	4,660
Cumberland Gap (twin)	KY–TN	4,600
Blue Mountain (twin)	PA Turnpike	4,435
Lehigh (twin)	PA Turnpike	4,379
Wawona	Yosemite Natl. Pk., CA	4,233
Big Walker Mt. (twin)	Bland Co., VA	4,229
Squirrel Hill	Pittsburgh, PA	4,225
Hanging Lake (twin)	Glenwood Canyon, CO	4,000
Caldecott (3 tubes)	Oakland, CA	3,616
Fort Pitt (twin)	Pittsburgh, PA	3,560
Mount Baker Ridge	Seattle, WA	3,456
Devil's Side Tunnel	U.S. 101 CA	3,400
Dingess Tunnel	Mingo Co., WV	3,400
Mall Tunnel	Dist. of Columbia	3,400
Cody No. 1	U.S. 14, 16, 20, WY	3,202

(1) Tunnel is used for vehicular and railroad traffic.

World's Longest Railway Tunnels

Source: Railway Directory & Year Book

Tunnel	Date	Miles	Operating railway	Country
Seikan	1985	33.50	Japanese Railway	Japan
English Channel Tunnel	1994	31.04	Eurotunnel	United Kingdom-France
Dai-shimizu	1979	14.00	Japanese Railway	Japan
Simplon No. 1 and 2	1906, 1922	12.00	Swiss Fed. & Italian St.	Switzerland-Italy
Kanmon	1975	12.00	Japanese Railway	Japan
Apennine	1934	11.00	Italian State	Italy
Rokko	1972	10.00	Japanese Railway	Japan
Mt. MacDonald	1989	9.10	Canadian Pacific	Canada
Gotthard	1882	9.00	Swiss Federal	Switzerland
Lotschberg	1913	9.00	Bern-Lotschberg-Simplon	Switzerland
Hokuriku	1962	9.00	Japanese Railway	Japan
Mont Cenis (Frejus)	1871	8.00	Italian State	France-Italy
Cascade	1929	8.00	Burlington Northern	United States
Shin-Shimizu	1961	8.00	Japanese Railway	Japan
Flathead	1970	8.00	Burlington Northern	United States
Aki	1975	8.00	Japanese Railway	Japan

World's Largest-Capacity Hydro Plants

Source: U.S. Committee on Large Dams of the Intl. Commission on Large Dams

Rank order#	Name	Country	Rated capacity now (MW)	Rated capacity planned (MW)
1.	Turukhansk (Lower Tungu-ska)*	Russia	—	20,000
2.	Three Gorges Dam*	China	—	18,200
3.	Itaipu	Brazil/ Paraguay	7,400	13,320
4.	Grand Coulee	U.S.	6,495	10,830
5.	Guri (Raúl Leoni)	Venezuela	10,300	10,300
6.	Tucuruí	Brazil	2,640	7,260
7.	Sayano-Shushensk*	Russia	—	6,400
8.	Corpus Posadas	Argentina/ Paraguay	4,700	6,000
8.	Krasnoyarsk	Russia	6,000	6,000
10.	La Grande 2	Canada	5,328	5,328
11.	Churchill Falls	Canada	5,225	5,225
12.	Xingo	Brazil	3,012	5,020
13.	Tarbela	Pakistan	1,750	4,678
14.	Bratsk	Russia	4,500	4,500
14.	Ust-Ilim	Russia	3,675	4,500
16.	Cabora Bassa	Mozambique	2,425	4,150
17.	Boguchany*	Russia	—	4,000
18.	Oak Creek	U.S.	3,600	3,600
19.	Paulo Afonso I	Brazil	1,524	3,409
20.	Pati*	Argentina	—	3,300
21.	Ilha Solteira	Brazil	3,200	3,200
22.	Chapetón*	Argentina	—	3,000
23.	Gezhouba	China	2,715	2,715

#Ranked by rated capacity planned. *Planned or under construction.

Major Dams of the World

Source: U.S. Committee on Large Dams of the Intl. Commission on Large Dams

World's Highest Dams

Rank order	Name	Country	Height above lowest formation (m)
1.	Nurek	Tajikistan	300
2.	Grand Dixence	Switzerland	285
3.	Inguri	Georgia	272
4.	Vajont	Italy	262
5.	Manuel M. Torres	Mexico	261
6.	Alvaro Obregon	Mexico	260
7.	Mauvoisin	Switzerland	250
8.	Mica	Canada	243
9.	Alberto Lleras C.	Colombia	243
10.	Sayano-Shushensk	Russia	242
11.	Ertan	China	240
12.	La Esmeralda	Colombia	237
13.	Oroville	U.S.	235
14.	El Cajón	Honduras	234
15.	Chirkey	Russia	233
16.	Bhakra	India	226
17.	Luzzone	Switzerland	225
18.	Hoover	U.S.	223
19.	Contra	Switzerland	220
20.	Mratinje	Yugoslavia	220

World's Largest-Volume Embankment Dams

Rank order	Name	Country	Volume cubic meters × 1000
1.	Tarbela	Pakistan	148,500
2.	Fort Peck	U.S.	96,050
3.	Tucurui	Brazil	85,200
4.	Ataturk*	Turkey	85,000
5.	Yacireta*	Argentina	81,000
6.	Rogun*	Tajikistan	75,500
7.	Oahe	U.S.	70,339
8.	Guri	Venezuela	70,000
9.	Parambikulam	India	69,165
10.	High Island West	China	67,000
11.	Gardiner	Canada	65,000
12.	Afsluitdijk	Netherlands	63,400
13.	Mangla	Pakistan	63,379
14.	Oroville	U.S.	59,635
15.	San Luis	U.S.	59,559
16.	Nurek	Tajikistan	58,000
17.	Tanda	Pakistan	57,250
18.	Garrison	U.S.	50,843
19.	Cochiti	U.S.	50,228
20.	Oosterschelde	Netherlands	50,000

*Under construction.

World's Largest-Capacity Reservoirs

Source: U.S. Committee on Large Dams of the Intl. Commission on Large Dams, 2001

Rank order	Name	Country	Capacity cubic meters × 1,000,000
1.	Kariba	Zimbabwe/ Zambia	180,600
2.	Bratsk	Russia	169,000
3.	High Aswan	Egypt	162,000
4.	Akosombo	Ghana	147,960
5.	Daniel Johnson	Canada	141,851
6.	Xinfeng	China	138,960
7.	Guri	Venezuela	135,000
8.	W A C Bennett	Canada	74,300
9.	Krasnoyarsk	Russia	73,300
10.	Zeya	Russia	68,400
11.	La Grande 2	Canada	61,715
12.	La Grande 3	Canada	60,020
13.	Ust-Ilim	Russia	59,300
14.	Kuibyshev	Russia	58,000
15.	Serra da Mesa	Brazil	54,400

Major U.S. Dams and Reservoirs

Source: Committee on Register of Dams, Corps of Engineers, U.S. Army, Sept. 2001

Highest U.S. Dams

Rank Order	Dam name	River	State	Type	Height Feet	Height Meters	Year completed
1.	Oroville	Feather	California	E	754	230	1968
2.	Hoover	Colorado	Nevada-Arizona	A	725	221	1936
3.	Dworshak	N. Fork Clearwater	Idaho	G	718	219	1973
4.	Glen Canyon	Colorado	Arizona	A	708	216	1966
5.	New Bullards Bar	North Yuba	California	A	636	194	1970
6.	Seven Oaks	Santa Ana	California	E	632	193	1999
7.	New Melones	Stanislaus	California	R	626	191	1979
8.	Swift	Lewis	Washington	E	610	186	1958
9.	Mossyrock	Cowlitz	Washington	A	607	185	1968
10.	Shasta	Sacramento	California	G	600	183	1945

E = Embankment, Earthfill; R = Embankment, Rockfill; G = Gravity; A = Arch.

Largest U.S. Embankment Dams

Rank Order	Dam name	River	State	Type	Volume Cubic yards × 1000	Volume Cubic meters × 1000	Year completed
1.	Fort Peck	Missouri	Montana	E	125,624	96,050	1937
2.	Oahe	Missouri	South Dakota	E	91,996	70,339	1958
3.	Oroville	Feather	California	E	77,997	59,635	1968
4.	San Luis	San Luis Creek	California	E	77,897	59,559	1967
5.	Garrison	Missouri	North Dakota	E	66,498	50,843	1953
6.	Cochiti	Rio Grande	New Mexico	E	65,693	50,228	1975
7.	Fort Randall	Missouri	South Dakota	E	49,962	38,200	1952
8.	Castaic	Castaic Creek	California	E	43,998	33,640	1973
9.	Ludington P/S	Lake Michigan	Michigan	E	37,699	28,824	1973
10.	Kingsley	N. Platte	Nebraska	E	31,999	24,466	1941

E = Embankment, Earthfill.

Largest U.S. Reservoirs

Rank Order	Dam name, location	Reservoir name	Location	Reservoir capacity Acre-Feet	Reservoir capacity Cubic meters × 1000	Year completed
1.	Hoover, NV/AZ	Lake Mead	AZ/NV	28,255,000	34,850,000	1936
2.	Glen Canyon, AZ	Lake Powell	AZ/UT	27,000,000	33,300,000	1966
3.	Oahe, SD	Lake Oahe	ND/SD	19,300,000	27,430,000	1958
4.	Garrison, ND	Lake Sakakawea	ND	18,500,000	27,920,000	1953
5.	Fort Peck, MT	Fort Peck Lake	MT	15,400,000	22,120,000	1937
6.	Grand Coulee, WA	F. D. Roosevelt Lake	WA	9,562,000	11,790,000	1942
7.	Libby, MT	Lake Koocanusa	MT/B.C.	5,809,000	7,170,000	1973
8.	Shasta, CA	Lake Shasta	CA	4,552,000	5,610,000	1945
9.	Toledo Bend, LA	Toledo Bend Lake	LA/TX	4,477,000	5,520,000	1968
10.	Fort Randall, SD	Lake Francis Case	SD	3,800,000	5,700,000	1952

1 acre-foot = 1 acre of water, 1 foot deep

SCIENCE AND TECHNOLOGY

Inventions

Invention	Date	Inventor	Nationality
Adding machine	**1642**	Pascal	French
Adding machine	**1885**	Burroughs	U.S.
Aerosol spray	**1926**	Rotheim	Norwegian
Airbag	**1952**	Hetrick	U.S.
Air brake	**1868**	Westinghouse	U.S.
Air conditioning	**1902**	Carrier	U.S.
Air pump	**1654**	Guericke	German
Airplane, automatic pilot	**1912**	Sperry	U.S.
Airplane, experimental	**1896**	Langley	U.S.
Airplane, hydro	**1911**	Curtiss	U.S.
Airplane jet engine	**1939**	Ohain	German
Airplane with motor	**1903**	Wright Bros.	U.S.
Airship	**1852**	Giffard	French
Airship, rigid dirigible	**1900**	Zeppelin	German
Arc welder	**1919**	Thomson	U.S.
Aspartame	**1965**	Schlatter	U.S.
Autogyro	**1920**	de la Cierva	Spanish
Automobile, differential gear	**1885**	Benz	German
Automobile, electric	**1892**	Morrison	U.S.
Automobile, exp'mtl	**1864**	Marcus	Austrian
Automobile, gasoline	**1889**	Daimler	German
Automobile, gasoline	**1892**	Duryea	U.S.
Automobile magneto	**1897**	Bosch	German
Automobile muffler	**1904**	Pope	U.S.
Automobile self-starter	**1911**	Kettering	U.S.
Bakelite	**1907**	Baekeland	Belgium, U.S.
Balloon	**1783**	Montgolfier	French
Barometer	**1643**	Torricelli	Italian
Bicycle, modern	**1885**	Starley	English
Bifocal lens	**1780**	Franklin	U.S.
Block signals, railway	**1867**	Hall	U.S.
Bomb, depth	**1916**	Tait	U.S.
Bottle machine	**1895**	Owens	U.S.
Braille printing	**1829**	Braille	French
Bubble gum	**1928**	Diemer	U.S.
Burner, gas	**1855**	Bunsen	German
Calculating machine	**1833**	Babbage	English
Calculator, electronic pocket	**1972**	Merryman, Van Tassel	U.S.
Camera, Kodak	**1888**	Eastman, Walker	U.S.
Camera, Polaroid Land	**1948**	Land	U.S.
Car coupler	**1873**	Janney	U.S.
Carburetor, gasoline	**1893**	Maybach	German
Card time recorder	**1894**	Cooper	U.S.
Carding machine	**1797**	Whittemore	U.S.
Carpet sweeper	**1876**	Bissell	U.S.
Cash register	**1879**	Ritty	U.S.
Cassette, audio	**1963**	Philips Co.	Dutch
Cassette, videotape	**1969**	Sony	Japanese
Cathode-ray tube	**1897**	Braun	German
CAT, or CT, scan	**1973**	Hounsfield	English
Cellophane	**1908**	Brandenberger	Swiss
Celluloid	**1870**	Hyatt	U.S.
Cement, Portland	**1824**	Aspdin	English
Chronometer	**1735**	Harrison	English
Circuit breaker	**1925**	Hilliard	U.S.
Circuit, integrated	**1959**	Kilby, Noyce, Texas Instr.	U.S.
Clock, pendulum	**1657**	Huygens	Dutch
Coaxial cable system	**1929**	Affel, Espensched	U.S.
Coke oven	**1893**	Hoffman	Austrian
Compressed air rock drill	**1871**	Ingersoll	U.S.
Comptometer	**1887**	Felt	U.S.
Computer, automatic sequence	**1944**	Aiken, et al.	U.S.
Computer, electronic	**1942**	Atanasoff, Berry	U.S.
Computer, laptop	**1987**	Sinclair	English
Computer, mini	**1960**	Digital Corp	U.S.
Condenser microphone (telephone)	**1916**	Wente	U.S.
Contact lens, corneal	**1948**	Tuohy	U.S.
Contraceptive, oral	**1954**	Pincus, Rock	U.S.
Corn, hybrid	**1917**	Jones	U.S.
Correction fluid	**1951**	Nesmith	U.S.
Cotton gin	**1793**	Whitney	U.S.
Cream separator	**1878**	DeLaval	Swedish
Cultivator, disc	**1878**	Mallon	U.S.
Cystoscope	**1878**	Nitze	German
Diesel engine	**1895**	Diesel	German
Disc, compact	**1972**	RCA	U.S.
Disc player, compact	**1979**	Sony, Philips Co.	Japan, Dutch
Disk, floppy	**1970**	IBM	U.S.
Disk, video	**1972**	Philips Co.	Dutch
Dynamite	**1866**	Nobel	Swedish
Dynamo, continuous current	**1871**	Gramme	Belgian
Dynamo, hydrogen cooled	**1915**	Schuler	U.S.
Electric battery	**1800**	Volta	Italian
Electric fan	**1882**	Wheeler	U.S.
Electrocardiograph	**1903**	Einthoven	Dutch
Electroencephalograph	**1929**	Berger	German
Electromagnet	**1824**	Sturgeon	English
Electron spectrometer	**1944**	Deutsch, Elliott, Evans	U.S.
Electron tube multigrid	**1913**	Langmuir	U.S.
Electroplating	**1805**	Brugnatelli	Italian
Electrostatic generator	**1929**	Van de Graaff	U.S.
Elevator brake	**1852**	Otis	U.S.
Elevator, push button	**1922**	Larson	U.S.
Engine, automatic transmission	**1910**	Fottinger	German
Engine, coal-gas 4-cycle	**1876**	Otto	German
Engine, compression ignition	**1883**	Daimler	German
Engine, electric ignition	**1883**	Benz	German
Engine, gas, compound	**1926**	Eickemeyer	U.S.
Engine, gasoline	**1872**	Brayton, Geo.	U.S.
Engine, gasoline	**1889**	Daimler	German
Engine, jet	**1930**	Whittle	English
Engine, steam, piston	**1705**	Newcomen	English
Engine, steam, piston	**1769**	Watt	Scottish
Engraving, half-tone	**1852**	Talbot	U.S.
Fiberglass	**1938**	Owens-Corning	U.S.
Fiber optics	**1955**	Kapany	English
Filament, tungsten	**1913**	Coolidge	U.S.
Flanged rail	**1831**	Stevens	U.S.
Flatiron, electric	**1882**	Seely	U.S.
Food, frozen	**1923**	Birdseye	U.S.
Freon	**1930**	Midgley, et al.	U.S.
Furnace (for steel)	**1858**	Siemens	German
Galvanometer	**1820**	Sweigger	German
Gas discharge tube	**1922**	Hull	U.S.
Gas lighting	**1792**	Murdoch	Scottish
Gas mantle	**1885**	Welsbach	Austrian
Gasoline (lead ethyl)	**1922**	Midgley	U.S.
Gasoline, cracked	**1913**	Burton	U.S.
Gasoline, high octane	**1930**	Ipatieff	Russian
Geiger counter	**1913**	Geiger	German
Glass, laminated safety	**1909**	Benedictus	French
Glider	**1853**	Cayley	English
Gun, breechloader	**1811**	Thornton	U.S.
Gun, Browning	**1897**	Browning	U.S.
Gun, magazine	**1875**	Hotchkiss	U.S.
Gun, silencer	**1908**	Maxim, H.P.	U.S.
Guncotton	**1847**	Schoenbein	German
Gyrocompass	**1911**	Sperry	U.S.
Gyroscope	**1852**	Foucault	French
Harvester-thresher	**1818**	Lane	U.S.
Heart, artificial	**1982**	Jarvik	U.S.
Helicopter	**1939**	Sikorsky	U.S.
Hydrometer	**1768**	Baume	French
Iron lung	**1928**	Drinker, Slaw	U.S.
Kaleidoscope	**1817**	Brewster	Scottish
Kinetoscope	**1889**	Edison	U.S.
Lacquer, nitrocellulose	**1921**	Flaherty	U.S.
Lamp, arc	**1847**	Staite	English
Lamp, fluorescent	**1938**	General Electric, Westinghouse	U.S.
Lamp, incandescent	**1879**	Edison	U.S.
Lamp, incand., frosted	**1924**	Pipkin	U.S.
Lamp, incand., gas	**1913**	Langmuir	U.S.
Lamp, klieg	**1911**	Kliegl, A. & J.	U.S.
Lamp, mercury vapor	**1912**	Hewitt	U.S.
Lamp, miner's safety	**1816**	Davy	English
Lamp, neon	**1909**	Claude	French
Lathe, turret	**1845**	Fitch	U.S.
Launderette	**1934**	Cantrell	U.S.
Lens, achromatic	**1758**	Dollond	English
Lens, fused bifocal	**1908**	Borsch	U.S.
Leyden jar (condenser)	**1745**	von Kleist	German
Lightning rod	**1752**	Franklin	U.S.
Linoleum	**1860**	Walton	English
Linotype	**1884**	Mergenthaler	U.S.
Lock, cylinder	**1851**	Yale	U.S.
Locomotive, electric	**1851**	Vail	U.S.
Locomotive, exp'mtl	**1802**	Trevithick	English
Locomotive, exp'mtl	**1812**	Fenton, et al.	English
Locomotive, exp'mtl	**1813**	Hedley	English
Locomotive, exp'mtl	**1814**	Stephenson	English
Locomotive, practical	**1829**	Stephenson	English
Locomotive, 1st U.S.	**1830**	Cooper, P.	U.S.
Loom, power	**1785**	Cartwright	English
Loudspeaker, dynamic	**1924**	Rice, Kellogg	U.S.
Machine gun	**1862**	Gatling	U.S.
Machine gun, improved	**1872**	Hotchkiss	U.S.
Machine gun (Maxim)	**1883**	Maxim, H.S.	U.S., Eng.
Magnet, electro	**1828**	Henry	U.S.
Mantle, gas	**1885**	Welsbach	Austrian
Mason jar	**1858**	Mason, J.	U.S.
Match, friction	**1827**	Walker, J.	English
Mercerized textiles	**1843**	Mercer, J.	English
Meter, induction	**1888**	Shallenberger	U.S.
Metronome	**1816**	Malezel	German
Microcomputer	**1973**	Truong, et al.	French
Micrometer	**1636**	Gascoigne	English
Microphone	**1877**	Berliner	U.S.
Microprocessor	**1971**	Intel Corp.	U.S.

Invention	Date	Inventor	Nationality
Microscope, compound	**1590**	Janssen	Dutch
Microscope, electronic	**1931**	Knoll, Ruska	German
Microscope, field ion	**1951**	Mueller	German
Microwave oven	**1947**	Spencer	U.S.
Minivan	**1983**	Chrysler	U.S.
Monitor, warship	**1861**	Ericsson	U.S.
Monotype	**1887**	Lanston	U.S.
Motor, AC	**1892**	Tesla	U.S.
Motor, DC	**1837**	Davenport	U.S.
Motor, induction	**1887**	Tesla	U.S.
Motorcycle	**1885**	Daimler	German
Movie machine	**1894**	Jenkins	U.S.
Movie, panoramic	**1952**	Waller	U.S.
Movie, talking	**1927**	Warner Bros.	U.S.
Mower, lawn	**1831**	Budding, Ferrabee	English
Mowing machine	**1822**	Bailey	U.S.
Neoprene	**1930**	Carothers	U.S.
Nylon	**1937**	Du Pont lab	U.S.
Nylon synthetic	**1930**	Carothers	U.S.
Oil cracking furnace	**1891**	Gavrilov	Russian
Oil filled power cable	**1921**	Emanueli	Italian
Oleomargarine	**1869**	Mege-Mouries	French
Ophthalmoscope	**1851**	Helmholtz	German
Pacemaker	**1952**	Zoll	U.S.
Paper	**105**	Ts'ai	Chinese
Paper clip	**1900**	Waaler	Norwegian
Paper machine	**1809**	Dickinson	U.S.
Parachute	**1785**	Blanchard	French
Pen, ballpoint	**1888**	Loud	U.S.
Pen, fountain	**1884**	Waterman	U.S.
Pen, steel	**1780**	Harrison	English
Pendulum	**1583**	Galileo	Italian
Percussion cap	**1807**	Forsythe	Scottish
Phonograph	**1877**	Edison	U.S.
Photo, color	**1892**	Ives	U.S.
Photo film, celluloid	**1893**	Reichenbach	U.S.
Photo film, transparent	**1884**	Eastman, Goodwin	U.S.
Photoelectric cell	**1895**	Elster	German
Photocopier	**1938**	Carlson	U.S.
Photographic paper	**1835**	Talbot	English
Photography	**1816**	Niepce	French
Photography	**1835**	Talbot	English
Photography	**1835**	Daguerre	French
Photophone	**1880**	Bell	U.S.-Scot.
Phototelegraphy	**1925**	Bell Labs	U.S.
Piano	**1709**	Cristofori	Italian
Piano, player	**1863**	Fourneaux	French
Pin, safety	**1849**	Hunt	U.S.
Pistol (revolver)	**1836**	Colt	U.S.
Plow, cast iron	**1785**	Ransome	English
Plow, disc	**1896**	Hardy	U.S.
Pneumatic hammer	**1890**	King	U.S.
Post-it note	**1980**	3M	U.S.
Powder, smokeless	**1884**	Vieille	French
Printing press, rotary	**1845**	Hoe	U.S.
Printing press, web	**1865**	Bullock	U.S.
Propeller, screw	**1804**	Stevens	U.S.
Propeller, screw	**1837**	Ericsson	Swedish
Pulsars	**1967**	Bell	English
Punch card accounting	**1889**	Hollerith	U.S.
Quasars	**1963**	Schmidt	U.S.
Radar	**1940**	Watson-Watt	Scottish
Radio, magnetic detector	**1902**	Marconi	Italian
Radio, signals	**1895**	Marconi	Italian
Radio amplifier	**1906**	De Forest	U.S.
Radio beacon	**1928**	Donovan	U.S.
Radio crystal oscillator	**1918**	Nicolson	U.S.
Radio receiver, cascade tuning	**1913**	Alexanderson	U.S.
Radio receiver, heterodyne	**1913**	Fessenden	U.S.
Radio transmitter triode modulation	**1914**	Alexanderson	U.S.
Radio tube diode	**1905**	Fleming	English
Radio tube oscillator	**1915**	De Forest	U.S.
Radio tube triode	**1906**	De Forest	U.S.
Radio FM, 2-path	**1933**	Armstrong	U.S.
Rayon (acetate)	**1895**	Cross	English
Rayon (cuprammonium)	**1890**	Despeissis	French
Rayon (nitrocellulose)	**1884**	Chardonnet	French
Razor, electric	**1917**	Schick	U.S.
Razor, safety	**1895**	Gillette	U.S.
Reaper	**1834**	McCormick	U.S.
Record, cylinder	**1887**	Bell, Tainter	U.S.
Record, disc	**1887**	Berliner	U.S.
Record, long playing	**1947**	Goldmark	U.S.
Record, wax cylinder	**1888**	Edison	U.S.
Refrigerator car	**1868**	David	U.S.
Resin, synthetic	**1931**	Hill	English
Richter scale	**1935**	Richter	U.S.
Rifle, repeating	**1860**	Henry	U.S.
Rocket engine	**1926**	Goddard	U.S.
Rollerblades	**1980**	Olson	U.S.
Rubber, vulcanized	**1839**	Goodyear	U.S.
Saccharin	**1879**	Remsen, Fahlberg	U.S.
Saw, band	**1808**	Newberry	English
Saw, circular	**1777**	Miller	English
Scotch tape	**1930**	Drew	U.S.
Seat belt	**1959**	Volvo	Swedish
Sewing machine	**1846**	Howe	U.S.
Shoe-lasting machine	**1883**	Matzeliger	U.S.
Shoe-sewing machine	**1860**	McKay	U.S.
Shrapnel shell	**1784**	Shrapnel	English
Shuttle, flying	**1733**	Kay	English
Sleeping-car	**1865**	Pullman	U.S.
Slide rule	**1620**	Oughtred	English
Soap, hardwater	**1928**	Bertsch	German
Spectroscope	**1859**	Kirchoff, Bunsen	German
Spectroscope (mass)	**1918**	Dempster	U.S.
Spinning jenny	**c.1764**	Hargreaves	English
Spinning mule	**1779**	Crompton	English
Steamboat, exp'mtl	**1778**	Jouffroy	French
Steamboat, exp'mtl	**1785**	Fitch	U.S.
Steamboat, exp'mtl	**1787**	Rumsey	U.S.
Steamboat, exp'mtl	**1788**	Miller	Scottish
Steamboat, exp'mtl	**1803**	Fulton	U.S.
Steamboat, exp'mtl	**1804**	Stevens	U.S.
Steamboat, practical	**1802**	Symington	Scottish
Steamboat, practical	**1807**	Fulton	U.S.
Steam car	**1770**	Cugnot	French
Steam turbine	**1884**	Parsons	English
Steel (converter)	**1856**	Bessemer	English
Steel alloy	**1891**	Harvey	U.S.
Steel alloy, high-speed	**1901**	Taylor, White	U.S.
Steel, manganese	**1884**	Hadfield	English
Steel, stainless	**1916**	Brearley	English
Stereoscope	**1838**	Wheatstone	English
Stethoscope	**1819**	Laennec	French
Stethoscope, binaural	**1840**	Cammann	U.S.
Stock ticker	**1870**	Edison	U.S.
Storage battery, rechargeable	**1859**	Plante	French
Stove, electric	**1896**	Hadaway	U.S.
Submarine	**1891**	Holland	U.S.
Submarine, even keel	**1894**	Lake	U.S.
Submarine, torpedo	**1776**	Bushnell	U.S.
Superconductivity	**1957**	Bardeen, Cooper, Schreiffer	U.S.
Synthesizer	**1964**	Moog	U.S.
Tank, military	**1914**	Swinton	English
Tape recorder, magnetic	**1899**	Poulsen	Danish
Teflon	**1938**	Du Pont	U.S.
Telegraph, magnetic	**1837**	Morse	U.S.
Telegraph, quadruplex	**1864**	Edison	U.S.
Telegraph, railroad	**1887**	Woods	U.S.
Telegraph, wireless high frequency	**1895**	Marconi	Italian
Telephone	**1876**	Bell	U.S.-Scot.
Telephone, automatic	**1891**	Strowger	U.S.
Telephone, cellular	**1947**	Bell Labs	U.S.
Telephone, radio	**1900**	Poulsen, Fessenden	Danish
Telephone, radio	**1906**	De Forest	U.S.
Telephone, radio, long dist.	**1915**	AT&T	U.S.
Telephone, recording	**1898**	Poulsen	Danish
Telephone, wireless	**1899**	Collins	U.S.
Telephone amplifier	**1912**	De Forest	U.S.
Telescope	**1608**	Lippershey	Neth.
Telescope	**1609**	Galileo	Italian
Telescope, astronomical	**1611**	Kepler	German
Teletype	**1928**	Morkrum, Kleinschmidt	U.S.
Television, color	**1928**	Baird	Scottish
Television, electronic	**1927**	Farnsworth	U.S.
Television, iconoscope	**1923**	Zworykin	U.S.
Television, mech. scanner	**1923**	Baird	Scottish
Thermometer	**1593**	Galileo	Italian
Thermometer	**1730**	Reaumur	French
Thermometer, mercury	**1714**	Fahrenheit	German
Time, self-regulator	**1918**	Bryce	U.S.
Time recorder	**1890**	Bundy	U.S.
Tire, double-tube	**1845**	Thomson	Scottish
Tire, pneumatic	**1888**	Dunlop	Scottish
Toaster, automatic	**1918**	Strite	U.S.
Toilet, flush	**1589**	Harington	English
Tool, pneumatic	**1865**	Law	English
Torpedo, marine	**1804**	Fulton	U.S.
Tractor, crawler	**1904**	Holt	U.S.
Transformer, AC	**1885**	Stanley	U.S.
Transistor	**1947**	Shockley, Brattain, Bardeen	U.S.
Trolley car, electric	**1884-87**	Van DePoele, Sprague	U.S.
Tungsten, ductile	**1912**	Coolidge	U.S.
Tupperware	**1945**	Tupper	U.S.
Turbine, gas	**1849**	Bourdin	French
Turbine, hydraulic	**1849**	Francis	U.S.
Turbine, steam	**1884**	Parsons	English
Type, movable	**1447**	Gutenberg	German
Typewriter	**1867**	Sholes, Soule, Glidden	U.S.

Invention	Date	Inventor	Nationality
Vacuum cleaner, electric	1907	Spangler	U.S.
Vacuum evaporating pan	1846	Rillieux	U.S.
Velcro	1948	de Mestral	Swiss
Video game ("Pong")	1972	Bushnell	U.S.
Video home system (VHS)	1975	Matsushita, JVC	Japan
Washer, electric	1901	Fisher	U.S.
Welding, atomic hydrogen	1924	Langmuir, Palmer	U.S.
Welding, electric	1877	Thomson	U.S.
Windshield wiper	1903	Anderson	U.S.
Wind tunnel	1912	Eiffel	French
Wire, barbed	1874	Glidden	U.S.
Wire, barbed	1875	Haish	U.S.
Wrench, double-acting	1913	Owen	U.S.
X-ray tube	1913	Coolidge	U.S.
Zeppelin	1900	Zeppelin	German

IT'S A FACT: In 1948, Swiss engineer George de Mestral returned from a hike and noticed burrs clinging to his pants. He saw that burrs had small hooks that attached to the clothing, and came up with the idea forVelcro as a strong fastener for fabrics and other materials.

Discoveries and Innovations: Chemistry, Physics, Biology, Medicine

	Date	Discoverer	Nationality
Acetylene gas	1862	Berthelot	French
ACTH	1927	Evans, Long	U.S.
Adrenalin	1901	Takamine	Japan
Aluminum, electrolytic process	1886	Hall	U.S.
Aluminum, isolated	1825	Oersted	Danish
Anesthesia, ether	1842	Long	U.S.
Anesthesia, local	1885	Koller	Austrian
Anesthesia, spinal	1898	Bier	German
Aniline dye	1856	Perkin	English
Anti-rabies	1885	Pasteur	French
Antiseptic surgery	1867	Lister	English
Antitoxin, diphtheria	1891	Von Behring	German
Argyrol	1897	Bayer	German
Arsphenamine	1910	Ehrlich	German
Aspirin	1853	Gerhardt	French
Atabrine	1932	Mietzsch, et al.	German
Atomic numbers	1913	Moseley	English
Atomic theory	1803	Dalton	English
Atomic time clock	1948	Lyons	U.S.
Atomic time clock, cesium beam	1948	Essen	English
Atom-smashing theory	1919	Rutherford	English
Bacitracin	1943	Johnson, Meleneyl	U.S.
Bacteria, description	1676	Leeuwenhoek	Dutch
Barbital	1903	Fischer	German
Bleaching powder	1798	Tennant	English
Blood, circulation	1628	Harvey	English
Blood plasma storage (blood banks)	1940	Drew	U.S.
Bordeaux mixture	1885	Millardet	French
Bromine from the sea	1826	Balard	French
Calcium carbide	1888	Wilson	U.S.
Calculus	1670	Newton	English
Camphor synthetic	1896	Haller	French
Canning (food)	1804	Appert	French
Carbomycin	1952	Tanner	U.S.
Carbon oxides	1925	Fisher	German
Chemotherapy	1909	Ehrlich	German
Chloamphenicol	1947	Burkholder	U.S.
Chlorine	1774	Scheele	Swedish
Chloroform	1831	Guthrie, S.	U.S.
Chlortetracycline	1948	Duggen	U.S.
Classification of plants and animals	1735	Linnaeus	Swedish
Cloning, mammal	1996	Wilmut, et al.	Scottish
Cocaine	1860	Niermann	German
Combustion explained	1777	Lavoisier	French
Conditioned reflex	1914	Pavlov	Russian
Cortisone	1936	Kendall	U.S.
Cortisone, synthesis	1946	Sarett	U.S.
Cosmic rays	1910	Gockel	Swiss
Cyanamide	1905	Frank, Caro	German
Cyclotron	1930	Lawrence	U.S.
DDT (not applied as insecticide until 1939)	1874	Zeidler	German
Deuterium	1932	Urey, Brickwedde, Murphy	U.S.
DNA (structure)	1951	Crick Watson Wilkins	English U.S. English
Electric resistance, law of	1827	Ohm	German
Electric waves	1888	Hertz	German
Electrolysis	1852	Faraday	English
Electromagnetism	1819	Oersted	Danish
Electron	1897	Thomson, J.	English
Electron diffraction	1936	Thomson, G.Davisson	English U.S.
Electroshock treatment	1938	Cerletti, Bini	Italian
Erythromycin	1952	McGuire	U.S.
Evolution, natural selection	1858	Darwin	English
Falling bodies, law of	1590	Galileo	Italian
Gases, law of combining volumes	1808	Gay-Lussac	French
Geometry, analytic	1619	Descartes	French
Gold, cyanide process for extraction	1887	MacArthur, Forest	British
Gravitation, law	1687	Newton	English
Holograph	1948	Gabor	British
Human heart transplant	1967	Barnard	S. African
Human immunodeficiency virus identified	1984	Mortagnier Gallo	French U.S.
Indigo, synthesis of	1880	Baeyer	German
Induction, electric	1830	Henry	U.S.
Insulin	1922	Banting, Best, Macleod	Canadian, Scottish
Intelligence testing	1905	Binet, Simon	French
In vitro fertilization	1978	Steptoe, Edwards	English
Isoniazid	1952	Hoffmann-LaRoche Domagk	U.S. German
Isotopes, theory	1912	Soddy	English
Laser (light amplification by stimulated emission of radiation)	1957	Gould	U.S.
Light, velocity	1675	Roemer	Danish
Light, wave theory	1690	Huygens	Dutch
Lithography	1796	Senefelder	Bohemian
Lobotomy	1935	Egas Moniz	Portuguese
Logarithms	1614	Napier	Scottish
LSD-25	1943	Hoffman	Swiss
Mendelian laws	1866	Mendel	Austrian
Mercator projection (map)	1568	Mercator (Kremer)	Flemish
Methanol	1661	Boyle	Irish
Milk condensation	1853	Borden	U.S.
Molecular hypothesis	1811	Avogadro	Italian
Motion, laws of	1687	Newton	English
Neomycin	1949	Waksman, Lechevalier	U.S.
Neutron	1932	Chadwick	English
Nitric acid	1648	Glauber	German
Nitric oxide	1772	Priestley	English
Nitroglycerin	1846	Sobrero	Italian
Oil cracking process	1891	Dewar	U.S.
Oxygen	1774	Priestley	English
Oxytetracycline	1950	Finlay, et al.	U.S.
Ozone	1840	Schonbein	German
Paper, sulfite process	1867	Tilghman	U.S.
Paper, wood pulp, sulfate process	1884	Dahl	German
Penicillin	1928	Fleming	Scottish
practical use	1941	Florey, Chain	English
Periodic law and table of elements	1869	Mendeleyev	Russian
Physostigmine synthesis	1935	Julian	U.S.
Pill, birth-control	1954	Pincus, Rock	U.S.
Planetary motion, laws	1609	Kepler	German
Plutonium fission	1940	Kennedy, Wahl, Seaborg, Segre	U.S.
Polymyxin	1947	Ainsworth	English
Positron	1932	Anderson	U.S.
Proton	1919	Rutherford	N. Zealand
Psychoanalysis	1900	Freud	Austrian
Quantum theory	1900	Planck	German
Quasars	1963	Matthews, Sandage	U.S.
Quinine synthetic	1946	Woodward, Doering	U.S.
Radioactivity	1896	Becquerel	French
Radiocarbon dating	1947	Libby	U.S.
Radium	1898	Curie, Pierre Curie, Marie	French Pol.-Fr.
Relativity theory	1905	Einstein	German
Reserpine	1949	Jal Vaikl	Indian
Schick test	1913	Schick	U.S.
Silicon	1823	Berzelius	Swedish
Smallpox eradication	1979	World Health Org.	UN
Streptomycin	1944	Waksman, et al	U.S.
Sulfanilamide	1935	Bovet, Trefouel	French
Sulfanilamide theory	1908	Gelmo	German
Sulfapyridine	1938	Ewins, Phelps	English
Sulfathiazole	1939	Fosbinder, Walter	U.S.
Sulfuric acid	1831	Phillips	English
Sulfuric acid, lead	1746	Roebuck	English
Syphilis test	1906	Wassermann	German
Thiacetazone	1950	Belmisch, Mietzsch, Domagk	German

	Date	Discoverer	Nationality
Tuberculin	1890	Koch	German
Uranium fission theory	1939	Hahn, Meitner, Strassmann	German
		Bohr	Danish
		Fermi	Italian
		Einstein, Pegram, Wheeler	U.S.
Uranium fission, atomic reactor	1942	Fermi, Szilard	U.S.
Vaccine, measles	1963	Enders	U.S.
Vaccine, meningitis (first conjugate)	1987	Gordon, et al., Connaught Lab.	U.S.
Vaccine, polio	1954	Salk	U.S.
Vaccine, polio, oral	1960	Sabin	U.S.
Vaccine, rabies	1885	Pasteur	French
Vaccine, smallpox	1796	Jenner	English
Vaccine, typhus	1909	Nicolle	French
Vaccine, varicella	1974	Takahashi	Japan
Van Allen belts, radiation	1958	Van Allen	U.S.
Vitamin A	1913	McCollum, Davis	U.S.
Vitamin B	1916	McCollum	U.S.
Vitamin C	1928	Szent-Gyorgyi, King	U.S.
Vitamin D	1922	McCollum	U.S.
Vitamin K	1935	Dam, Doisy	U.S.
Xerography	1938	Carlson	U.S.
X ray	1895	Roentgen	German

Top 20 Corporations Receiving U.S. Patents in 2000

Source: *Technology Assessment and Forecast Report,* U.S. Patent and Trademark Office, U.S. Department of Commerce

Rank	Company	Number of patents
1.	International Business Machines Corp.	2,886
2.	NEC Corp.	2,021
3.	Canon K. K.	1,890
4.	Samsung Electronics Co., Ltd.	1,441
5.	Lucent Technologies Inc.	1,411
6.	Sony Corp.	1,385
7.	Micron Technology, Inc.	1,304
8.	Toshiba Corp.	1,232
9.	Motorola, Inc.	1,196
10.	Fujitsu Ltd.	1,147
11.	Matsushita Electric Industrial Co., Ltd.	1,137
12.	Advanced Micro Devices, Inc.	1,053
13.	Hitachi, Ltd.	1,036
14.	Mitsubishi Denki K. K.	1,010
15.	Siemens Aktiengesellschaft	912
16.	Hewlett-Packard Company	901
17.	Eastman Kodak Company	875
18.	Intel Corp.	795
19.	General Electric Company	787
20.	U.S. Philips Corp.	693

Breaking the Sound Barrier; Speed of Sound

The prefix **Mach** is used to describe supersonic speed. It was named for Ernst Mach (1838-1916), a Czech-born Austrian physicist, who contributed to the study of sound. When a plane moves at the speed of sound, it is Mach 1. When the plane is moving at twice the speed of sound, it is Mach 2. When it is moving below the speed of sound, the speed can be designated accordingly—for example, Mach 0.90. Mach may be defined as the ratio of the velocity of a rocket or a jet to the velocity of sound in the medium being considered.

When a plane passes the sound barrier—flying faster than sound travels—listeners in the area hear thunderclaps, but the pilot of the plane does not hear them.

Sound is produced by vibrations of an object and is transmitted by alternate increase and decrease in pressures that radiate outward through a material media of molecules—somewhat like waves spreading out on a pond after a rock has been tossed into it.

The **frequency of sound** is determined by the number of times the vibrating waves undulate per second and is measured in cycles per second. The slower the cycle of waves, the lower the frequency. As frequencies increase, the sound is higher in pitch.

Sound is audible to human beings only if the frequency falls within a certain range. The human ear is usually not sensitive to frequencies of fewer than 20 vibrations per second or greater than about 20,000 vibrations per second—although this range varies among individuals. Any sound at a pitch higher than the human ear can hear is termed ultrasonic.

Intensity, or loudness, is the strength of the pressure of these radiating waves and is measured in decibels. The human ear responds to intensity in a range from zero to 120 decibels. Any sound with a pressure of more than 120 decibels is painful to the human ear.

The **speed of sound** is generally defined as 1,088 feet per second at sea level at 32° F. It varies in other temperatures and in different media. Sound travels faster in water than in air, and even faster in iron and steel. It takes about 5 seconds to travel a mile in air, 1 second to move a mile under water, and $^1/_3$ second to move a mile in iron. Sound travels through ice-cold vapor at approximately 4,708 feet per second; for other media, speeds are: ice-cold water, 4,938; granite, 12,960; hardwood, 12,620; brick, 11,960; glass, 16,410 to 19,690; silver, 8,658; gold, 5,717.

Light; Colors of the Spectrum

Light, a form of electromagnetic radiation similar to radiant heat, radio waves, and X rays, is emitted from a source in straight lines and spreads out over a larger and larger area as it travels; the light per unit area diminishes as the square of the distance.

The English mathematician and physicist Sir Isaac Newton (1642-1727) described light as an **emission of particles**; the Dutch astronomer, mathematician, and physicist Christiaan Huygens (1629-95) developed the theory that light travels by a **wave motion**. It is now believed that these 2 theories are essentially complementary, and the development of quantum theory has led to results where light acts like a series of particles in some experiments and like a wave in others.

The **speed of light** was first measured in a laboratory experiment by the French physicist Armand Hippolyte Louis Fizeau (1819-96). Today the speed of light is known very precisely as 299,792.458 km per sec (or 186,282.396 mi per sec) in a vacuum. (Scientists reported in July 2000, however, that they were able to make a pulse of light exceed this speed by shooting light through a tube containing atoms specially prepared with lasers.) The velocity of light in air varies slightly with color, averaging about 3% less than in a vacuum; the speed in water is about 25% less, and in glass, 33% less.

Color sensations are produced through the excitation of the retina of the eye by light vibrating at different frequencies. The different colors of the spectrum may be produced by viewing a light beam that is refracted by passage through a prism, which breaks the light into its wavelengths.

Customarily, the **primary colors** of the spectrum are taken to be the 6 monochromatic colors that occupy relatively large areas of the spectrum: red, orange, yellow, green, blue, and violet. However, Newton named a 7th color, indigo, situated between blue and violet on the spectrum. Aubert estimated (1865) the solar spectrum to contain approximately 1,000 distinguishable hues; of the hues, according to Rood (1881), 2 million tints and shades can be distinguished. Luckiesh stated (1915) that 55 distinctly different hues have been seen in a single spectrum.

Many physicists recognize only 3 primary colors: red, yellow, and blue (Mayer, 1775); red, green, and violet (Thomas Young, 1801); or red, green, and blue (Clerk Maxwell, 1860).

The color sensation of **black** is due to complete lack of stimulation of the retina, that of **white** to complete stimulation. The **infrared and ultraviolet rays**, below the red (long) end of the spectrum and above the violet (short) end respectively, are invisible to the naked eye. Heat is the principal effect of the infrared rays, and chemical action that of the ultraviolet rays.

Weight or Mass of Water

Weight, at 20° C

1	cubic inch	0.0360 pound
12	cubic inches	0.433 pound
1	cubic foot	62.4 pounds
1	cubic foot	7.48052 U.S. gal
1.8	cubic feet	112.0 pounds (1 gross)
35.96	cubic feet	2240.0 pounds (20 gross)
1	U.S. gallon	8.33 pounds

Weight, at 20° C

13.45	U.S. gallons	112.0 pounds
269.0	U.S. gallons	2240.0 pounds

Mass, at 4° C (Maximum Density)

1	cubic centimeter	1 gram
1	liter	1 kilogram
1	cubic meter	1 metric ton

Density of Gases and Vapors

at 0° C and 760 mmHg; kilograms per cubic meter

Gas	Mass	Gas	Mass	Gas	Mass
Acetylene	1.171	Ethylene	1.260	Methyl fluoride	1.545
Air	1.293	Fluorine	1.696	Mono methylamine	1.38
Ammonia	0.759	Helium	0.178	Neon	0.900
Argon	1.784	Hydrogen	0.090	Nitric oxide	1.341
Arsine	3.48	Hydrogen bromide	3.50	Nitrogen	1.250
Butane-iso	2.60	Hydrogen chloride	1.639	Nitrosyl chloride	2.99
Butane-n	2.519	Hydrogen iodide	5.724	Nitrous oxide	1.997
Carbon dioxide	1.977	Hydrogen selenide	3.66	Oxygen	1.429
Carbon monoxide	1.250	Hydrogen sulfide	1.539	Phosphine	1.48
Carbon oxysulfide	2.72	Krypton	3.745	Propane	2.020
Chlorine	3.214	Methane	0.717	Silicon tetrafluoride	4.67
Chlorine monoxide	3.89	Methyl chloride	2.25	Sulfur dioxide	2.927
Ethane	1.356	Methyl ether	2.091	Xenon	5.897

Chemical Elements, Atomic Weights, Discoverers

Source: Darleane C. Hoffman, Ph.D., Lawrence Berkeley National Laboratory and Department of Chemistry, University of California, Berkeley

Atomic weights, based on the exact number 12 as the assigned atomic mass of the principal isotope of carbon, carbon 12, are provided through the courtesy of the International Union of Pure and Applied Chemistry (IUPAC) and Butterworth Scientific Publications. For the radioactive elements, with the exception of uranium and thorium, the mass number listed is that of either the isotope of longest half-life (*) or the better known isotope (**).

Chemical element	Symbol	Atomic number	Atomic weight	Year discov.	Discoverer
Actinium	Ac	89	227.03	1899	Debierne
Aluminum	Al	13	26.9815	1825	Oersted
Americium	Am	95	243*	1944	Seaborg, et al.
Antimony	Sb	51	121.75	1450	Valentine
Argon	Ar	18	39.948	1894	Rayleigh, Ramsay
Arsenic	As	33	74.9216	13th c.	Albertus Magnus
Astatine	At	85	210*	1940	Corson, et al.
Barium	Ba	56	137.33	1808	Davy
Berkelium	Bk	97	247*	1949	Thompson, Ghiorso, Seaborg
Beryllium	Be	4	9.0122	1798	Vauquelin
Bismuth	Bi	83	208.980	15th c.	Valentine
Bohrium	Bh	107	264*	1981	Münzenberg, et al.
Boron	B	5	10.811[a]	1808	Gay-Lussac, Thenard
Bromine	Br	35	79.904[b]	1826	Balard
Cadmium	Cd	48	112.41	1817	Stromeyer
Calcium	Ca	20	40.08	1808	Davy
Californium	Cf	98	251*	1950	Thompson, et al.
Carbon	C	6	12.01115[a]	BC	unknown
Cerium	Ce	58	140.12	1803	Klaproth
Cesium	Cs	55	132.905	1860	Bunsen, Kirchhoff
Chlorine	Cl	17	35.453[b]	1774	Scheele
Chromium	Cr	24	51.996[b]	1797	Vauquelin
Cobalt	Co	27	58.9332	1735	Brandt
Copper	Cu	29	3.546[b]	BC	unknown
Curium	Cm	96	247*	1944	Seaborg, James, Ghiorso
Dubnium (Hahnium)[1]	Db	105	262*	1970	Ghiorso, et al.
Dysprosium	Dy	66	162.50*	1886	Boisbaudran
Einsteinium	Es	99	252*	1952	Ghiorso, et al.
Erbium	Er	68	167.26	1843	Mosander
Europium	Eu	63	151.96	1901	Demarcay
Fermium	Fm	100	257*	1953	Ghiorso, et al.
Fluorine	F	9	18.9984	1771	Scheele
Francium	Fr	87	223*	1939	Perey
Gadolinium	Gd	64	157.25	1886	Marignac
Gallium	Ga	31	69.72	1875	Boisbaudran
Germanium	Ge	32	72.59	1886	Winkler
Gold	Au	79	196.967	BC	unknown
Hafnium	Hf	72	178.49	1923	Coster, Hevesy
Hahnium[1]	Ha	105	262*	1970	Ghiorso, et al.
Hassium	Hs	108	269*	1984	Münzenberg, et al.
Helium	He	2	4.0026	1868	Janssen, Lockyer
Holmium	Ho	67	164.930	1878	Soret, Delafontaine
Hydrogen	H	1	1.00797[a]	1766	Cavendish
Indium	In	49	114.82	1863	Reich, Richter
Iodine	I	53	126.9044	1811	Courtois
Iridium	Ir	77	192.22	1804	Tennant

Chemical element	Symbol	Atomic number	Atomic weight	Year discov.	Discoverer
Iron	Fe	26	55.847[b]	BC	unknown
Krypton	Kr	36	83.80	1898	Ramsay, Travers
Lanthanum	La	57	138.91	1839	Mosander
Lawrencium	Lr	103	262*	1961	Ghiorso, et al.
Lead	Pb	82	207.19	BC	unknown
Lithium	Li	3	6.939	1817	Arfvedson
Lutetium	Lu	71	174.97	1907	Welsbach, Urbain
Magnesium	Mg	12	24.312	1829	Bussy
Manganese	Mn	25	54.9380	1774	Gahn
Meitnerium	Mt	109	268*	1982	Münzenberg, et al.
Mendelevium	Md	101	258*	1955	Ghiorso, et al.
Mercury	Hg	80	200.59	BC	unknown
Molybdenum	Mo	42	95.94	1782	Hjelm
Neodymium	Nd	60	144.24	1885	Welsbach
Neon	Ne	10	20.183	1898	Ramsay, Travers
Neptunium	Np	93	237.05*	1940	McMillan, Abelson
Nickel	Ni	28	58.70	1751	Cronstedt
Niobium[2]	Nb	41	92.906	1801	Hatchett
Nitrogen	N	7	14.0067	1772	Rutherford
Nobelium	No	102	259*	1958	Ghiorso, et al.
Osmium	Os	76	190.2	1804	Tennant
Oxygen	O	8	15.9994[a]	1774	Priestley, Scheele
Palladium	Pd	46	106.4	1803	Wollaston
Phosphorus	P	15	30.9738	1669	Brand
Platinum	Pt	78	195.09	1735	Ulloa
Plutonium	Pu	94	244*	1941	Seaborg, et al.
Polonium	Po	84	210**	1898	P. and M. Curie
Potassium	K	19	39.102	1807	Davy
Praseodymium	Pr	59	140.907	1885	Welsbach
Promethium	Pm	61	147**	1945	Glendenin, Marinsky, Coryell
Protactinium	Pa	91	231.04*	1917	Hahn, Meitner
Radium	Ra	88	226.03*	1898	P. and M. Curie, Bemont
Radon	Rn	86	222*	1900	Dorn
Rhenium	Re	75	186.21	1925	Noddack, Tacke, Berg
Rhodium	Rh	45	102.905	1803	Wollaston
Rubidium	Rb	37	85.47	1861	Bunsen, Kirchhoff
Ruthenium	Ru	44	101.07	1845	Klaus
Rutherfordium	Rf	104	261*	1969	Ghiorso, et al.
Samarium	Sm	62	150.35	1879	Boisbaudran
Scandium	Sc	21	44.956	1879	Nilson
Seaborgium	Sg	106	266*	1974	Ghiorso, et al.
Selenium	Se	34	78.96	1817	Berzelius
Silicon	Si	14	28.086[a]	1823	Berzelius
Silver	Ag	47	107.868[b]	BC	unknown
Sodium	Na	11	22.9898	1807	Davy
Strontium	Sr	38	87.62	1790	Crawford
Sulfur	S	16	32.064[a]	BC	unknown
Tantalum	Ta	73	180.948	1802	Ekeberg
Technetium	Tc	43	99**	1937	Perrier, Segre
Tellurium	Te	52	127.60	1782	Von Reichenstein
Terbium	Tb	65	158.9324	1843	Mosander
Thallium	Tl	81	204.37	1861	Crookes
Thorium	Th	90	232.038	1828	Berzelius
Thulium	Tm	69	168.934	1879	Cleve
Tin	Sn	50	118.69	BC	unknown
Titanium	Ti	22	47.90	1791	Gregor
Tungsten (Wolfram)	W	74	183.85	1783	d'Elhujar
Uranium	U	92	238.03	1789	Klaproth
Vanadium	V	23	50.942	1830	Sefstrom
Xenon	Xe	54	131.30	1898	Ramsay, Travers
Ytterbium	Yb	70	173.04	1878	Marignac
Yttrium	Y	39	88.905	1794	Gadolin
Zinc	Zn	30	65.37	BC	unknown
Zirconium	Zr	40	91.22	1789	Klaproth

Note: 109 elements are listed here. In addition, discovery of elements 110-112 has been reported. Discovery of element 110 was reported by 3 different groups between 1994 and 1996, but since each reported evidence for different isotopes, none can be considered confirmation of the others. A. Ghiorso, et al. at the Lawrence Berkeley National Laboratory (LBNL) in Berkeley, CA, reported evidence for element 110 with mass number 267; S. Hofmann, et al. at the Gesellschaft für Schwerionenforschung (GSI) at Darmstadt, Germany, reported element 110 with mass numbers 269, 270, and 271; Yu. Lazarev, et al. at the Flerov Laboratory for Nuclear Reactions, Dubna, Russia, reported element 110 with mass 273. The group of S. Hofmann has the most convincing data for the discovery of 110, but the half-lives and cross sections of all groups appear reasonable. In 1995-96, Hofmann, et al. also reported discovery of elements 111 and 112 at GSI with mass numbers of 272 and 277, respectively. These elements have not yet been named. In July 1999, a multinational group working at Dubna, Russia, published evidence for observation of element 114 with mass number 287. A Dubna/Lawrence Livermore National Laboratory group published their evidence in Oct. 1999 for element 114 with mass number 289 (first announced in Jan. 1999) and reported observation of element 114 with mass number 288 in Sept. 2000 and element 116 with mass number 292 in Dec. 2000. These reports await confirmation by other groups. (1) The name Dubnium (Db) has been approved by IUPAC for element 105, but the name Hahnium (Ha) is used for element 105 in most of the scientific literature before 1998 and is still sometimes used in the U.S. (2) Formerly Columbium. (a) Atomic weights so designated are known to be variable because of natural variations in isotopic composition. The observed ranges are: hydrogen 0.0001; boron 0.003; carbon 0.005; oxygen 0.0001; silicon 0.001; sulfur 0.003. (b) Atomic weights so designated are believed to have the following experimental uncertainties: chlorine 0.001; chromium 0.001; iron 0.003; copper 0.001; bromine 0.001; silver 0.001.

Periodic Table of the Elements

Source: © 1996 Lawrence Berkeley National Laboratory

Parentheses indicate undiscovered elements.

Key: atomic number (14), atomic weight (28.09), symbol (Si), name (Silicon)

alkali metals	alkaline earth metals																noble gases
1 1.01 **H** Hydrogen																	2 4.003 **He** Helium
												nonmetals					
3 6.94 **Li** Lithium	4 9.01 **Be** Beryllium											5 10.81 **B** Boron	6 12.01 **C** Carbon	7 14.01 **N** Nitrogen	8 15.999 **O** Oxygen	9 18.998 **F** Fluorine	10 20.18 **Ne** Neon
11 22.99 **Na** Sodium	12 24.31 **Mg** Magnesium											13 26.98 **Al** Aluminum	14 28.09 **Si** Silicon	15 30.97 **P** Phosphorus	16 32.06 **S** Sulfur	17 35.45 **Cl** Chlorine	18 39.95 **Ar** Argon
		transitional metals															
19 39.10 **K** Potassium	20 40.08 **Ca** Calcium	21 44.96 **Sc** Scandium	22 47.90 **Ti** Titanium	23 50.94 **V** Vanadium	24 51.996 **Cr** Chromium	25 54.94 **Mn** Manganese	26 55.85 **Fe** Iron	27 58.93 **Co** Cobalt	28 58.70 **Ni** Nickel	29 63.55 **Cu** Copper	30 65.37 **Zn** Zinc	31 69.72 **Ga** Gallium	32 72.59 **Ge** Germanium	33 74.92 **As** Arsenic	34 78.96 **Se** Selenium	35 79.90 **Br** Bromine	36 83.80 **Kr** Krypton
37 85.47 **Rb** Rubidium	38 87.62 **Sr** Strontium	39 88.91 **Y** Yttrium	40 91.22 **Zr** Zirconium	41 92.91 **Nb** Niobium	42 95.94 **Mo** Molybdenum	43 98 **Tc** Technetium	44 101.07 **Ru** Ruthenium	45 102.91 **Rh** Rhodium	46 106.40 **Pd** Palladium	47 107.87 **Ag** Silver	48 112.41 **Cd** Cadmium	49 114.82 **In** Indium	50 118.69 **Sn** Tin	51 121.75 **Sb** Antimony	52 127.60 **Te** Tellurium	53 126.90 **I** Iodine	54 131.30 **Xe** Xenon
55 132.91 **Cs** Cesium	56 137.33 **Ba** Barium	57 138.91 **La** Lanthanum	72 178.49 **Hf** Hafnium	73 180.95 **Ta** Tantalum	74 183.85 **W** Tungsten	75 186.21 **Re** Rhenium	76 190.20 **Os** Osmium	77 192.22 **Ir** Iridium	78 195.09 **Pt** Platinum	79 196.97 **Au** Gold	80 200.59 **Hg** Mercury	81 204.37 **Tl** Thallium	82 207.19 **Pb** Lead	83 208.98 **Bi** Bismuth	84 209 **Po** Polonium	85 210 **At** Astatine	86 222 **Rn** Radon
87 223 **Fr** Francium	88 226.03 **Ra** Radium	89 227.03 **Ac** Actinium	104 261 **Rf** Rutherfordium	105 262 **Db/Ha** Dubnium (Hahnium)	106 266 **Sg** Seaborgium	107 264 **Bh** Bohrium	108 269 **Hs** Hassium	109 268 **Mt** Meitnerium	110 273	111 272	112 277	**(113)**	**(114)**	**(115)**	**(116)**	**(117)**	**(118)**
												other metals					

Lanthanide series	58 140.12 **Ce** Cerium	59 140.91 **Pr** Praseodymium	60 144.24 **Nd** Neodymium	61 145 **Pm** Promethium	62 150.35 **Sm** Samarium	63 151.96 **Eu** Europium	64 157.25 **Gd** Gadolinium	65 158.93 **Tb** Terbium	66 162.50 **Dy** Dysprosium	67 164.93 **Ho** Holmium	68 167.26 **Er** Erbium	69 168.93 **Tm** Thulium	70 173.04 **Yb** Ytterbium	71 174.97 **Lu** Lutetium
Actinide series	90 232.04 **Th** Thorium	91 231.04 **Pa** Protactinium	92 238.03 **U** Uranium	93 237.05 **Np** Neptunium	94 244 **Pu** Plutonium	95 243 **Am** Americium	96 247 **Cm** Curium	97 247 **Bk** Berkelium	98 251 **Cf** Californium	99 252 **Es** Einsteinium	100 257 **Fm** Fermium	101 258 **Md** Mendelevium	102 259 **No** Nobelium	103 262 **Lr** Lawrencium

WEIGHTS AND MEASURES

Source: National Institute of Standards and Technology, U.S. Dept. of Commerce

The International System of Units (SI)

Two systems of weights and measures coexist in the U.S. today: the U.S. Customary System and the International System of Units (SI, after the initials of Systême International). SI, commonly identified with the metric system, is actually a more complete, coherent version of it. Throughout U.S. history, the Customary System (inherited from, but now different from, the British Imperial System) has been generally used; federal and state legislation has given it, through implication, standing as the primary weights and measures system. The metric system, however, is the only system that Congress has ever specifically sanctioned. An 1866 law reads:

It shall be lawful throughout the United States of America to employ the weights and measures of the metric system; and no contract or dealing, or pleading in any court, shall be deemed invalid or liable to objection because the weights or measures expressed or referred to therein are weights or measures of the metric system.

Since that time, use of the metric system in the U.S. has slowly and steadily increased, particularly in the scientific community, in the pharmaceutical industry, and in the manufacturing sector—the last motivated by the practice in international commerce, in which the metric system is now predominantly used.

On Feb. 10, 1964, the National Bureau of Standards (now known as the National Institute of Standards and Technology) issued the following statement:

Henceforth it shall be the policy of the National Bureau of Standards to use the units of the International System (SI), as adopted by the 11th General Conference on Weights and Measures (October 1960), except when the use of these units would obviously impair communication or reduce the usefulness of a report.

On Dec. 23, 1975, Pres. Gerald R. Ford signed the Metric Conversion Act of 1975. It defines the metric system as being the International System of Units as interpreted in the U.S. by the secretary of commerce. The Trade Act of 1988 and other legislation declare the metric system the preferred system of weights and measures for U.S. trade and commerce, call for the federal government to adopt metric specifications, and mandate the Commerce Dept. to oversee the program. However, the metric system has still not become the system of choice for most Americans' daily use.

The following 7 units serve as the base units for the International System: **length**—meter; **mass**—kilogram; **time**—second; **electric current**—ampere; **thermodynamic temperature**—kelvin; **amount of substance**—mole; and **luminous intensity**—candela.

Prefixes

The following prefixes, in combination with the basic unit names, provide the multiples and submultiples in the International System. For example, the unit name *meter*, with the prefix *kilo* added, produces *kilometer*, meaning "1,000 meters."

Prefix	Symbol	Multiples	Equivalent
yotta	Y	10^{24}	septillionfold
zetta	Z	10^{21}	sextillionfold
exa	E	10^{18}	quintillionfold
peta	P	10^{15}	quadrillionfold
tera	T	10^{12}	trillionfold
giga	G	10^{9}	billionfold
mega	M	10^{6}	millionfold
kilo	k	10^{3}	thousandfold
hecto	h	10^{2}	hundredfold
deka	da	10	tenfold

Prefix	Symbol	Multiples	Equivalent
deci	d	10^{-1}	tenth part
centi	c	10^{-2}	hundredth part
milli	m	10^{-3}	thousandth part
micro	μ	10^{-6}	millionth part
nano	n	10^{-9}	billionth part
pico	p	10^{-12}	trillionth part
femto	f	10^{-15}	quadrillionth part
atto	a	10^{-18}	quintillionth part
zepto	z	10^{-21}	sextillionth part
yocto	y	10^{-24}	septillionth part

Tables of Metric Weights and Measures

(**Note:** The SI generally uses the term *mass* instead of *weight*. Mass is a measure of an object's inertial property, or the amount of matter it contains. Weight is a measure of the force exerted on an object by gravity or the force needed to support it. Also, the SI does not make a distinction between "dry volume" and "liquid volume.")

Length

10 millimeters (mm)	= 1 centimeter (cm)
10 centimeters	= 1 decimeter (dm)
	= 100 millimeters
10 decimeters	= 1 meter (m)
	= 1,000 millimeters
10 meters	= 1 dekameter (dam)
10 dekameters	= 1 hectometer (hm)
	= 100 meters
10 hectometers	= 1 kilometer (km)
	= 1,000 meters

Area

100 square millimeters (mm^2)	= 1 square centimeter (cm^2)
10,000 square centimeters	= 1 square meter (m^2)
	= 1,000,000 square millimeters
100 square meters	= 1 are (a)
100 ares	= 1 hectare (ha)
	= 10,000 square meters
100 hectares	= 1 square kilometer (km^2)
	= 1,000,000 square meters

Volume

10 milliliters (mL)	= 1 centiliter (cL)
10 centiliters	= 1 deciliter (dL)
	= 100 milliliters
10 deciliters	= 1 liter (L)
	= 1,000 milliliters
10 liters	= 1 dekaliter (daL)
10 dekaliters	= 1 hectoliter (hL)
	= 100 liters
10 hectoliters	= 1 kiloliter (kL)
	= 1,000 liters

Volume (Cubic Measure)

1,000 cubic millimeters (mm^3)	= 1 cubic centimeter (cm^3)
1,000 cubic centimeters	= 1 cubic decimeter (dm^3)
	= 1,000,000 cubic millimeters
1,000 cubic decimeters	= 1 cubic meter (m^3)
	= 1 stere
	= 1,000,000 cubic centimeters
	= 1,000,000,000 cubic millimeters

Weight (Mass)

10 milligrams (mg)	= 1 centigram (cg)
10 centigrams	= 1 decigram (dg)
	= 100 milligrams
10 decigrams	= 1 gram (g)
	= 1,000 milligrams
10 grams	= 1 dekagram (dag)
10 dekagrams	= 1 hectogram (hg)
	= 100 grams
10 hectograms	= 1 kilogram (kg)
	= 1,000 grams
1,000 kilograms	= 1 metric ton (t)

Table of U.S. Customary Weights and Measures

Length

12 inches (in)	= 1 foot (ft)
3 feet	= 1 yard (yd)
5½ yards	= 1 rod (rd), pole, or perch (16½ feet)
40 rods	= 1 furlong (fur)
	= 220 yards
	= 660 feet
8 furlongs	= 1 statute mile (mi)
	= 1,760 yards
	= 5,280 feet
3 miles	= 1 league
	= 5,280 yards
	= 15,840 feet
6076.11549 feet	= 1 international nautical mile

Volume (Liquid Measure)

When necessary to distinguish the liquid pint or quart from the dry pint or quart, the word *liquid* or the abbreviation *liq* is used in combination with the name or abbreviation of the liquid unit.

4 gills (gi)	= 1 pint (pt)
	= 28.875 cubic inches
2 pints	= 1 quart (qt)
	= 57.75 cubic inches
4 quarts	= 1 gallon (gal)
	= 231 cubic inches
	= 8 pints
	= 32 gills

Volume (Dry Measure)

When necessary to distinguish the dry pint or quart from the liquid pint or quart, the word *dry* is used in combination with the name or abbreviation of the dry unit.

2 pints (pt)	= 1 quart (qt)
	= 67.2006 cubic inches
8 quarts	= 1 peck (pk)
	= 537.605 cubic inches
	= 16 pints
4 pecks	= 1 bushel (bu)
	= 2,150.42 cubic inches
	= 32 quarts

Area

Squares and cubes of units are sometimes abbreviated by using superscripts. For example, ft^2 means square foot, and ft^3 means cubic foot.

144 square inches	= 1 square foot (ft^2)
9 square feet	= 1 square yard (yd^2)
	= 1,296 square inches
30 ¼ square yards	= 1 square rod (rd^2)
	= 272¼ square feet
160 square rods	= 1 acre
	= 4,840 square yards
	= 43,560 square feet
640 acres	= 1 square mile (mi^2)
1 mile square	= 1 section (of land)
6 miles square	= 1 township
	= 36 sections
	= 36 square miles

Cubic Measure

1 cubic foot (ft^3)	= 1,728 cubic inches (in^3)
27 cubic feet	= 1 cubic yard (yd^3)

Gunter's, or Surveyor's, Chain Measure

7.92 inches (in)	= 1 link
100 links	= 1 chain (ch)
	= 4 rods
	= 66 feet
80 chains	= 1 statute mile (mi)
	= 320 rods
	= 5,280 feet

Avoirdupois Weight

When necessary to distinguish the avoirdupois ounce or pound from the troy ounce or pound, the word *avoirdupois* or the abbreviation *avdp* is used in combination with the name or abbreviation of the avoirdupois unit. The *grain* is the same in avoirdupois and troy weight.

27 11/32 grains	= 1 dram (dr)
16 drams	= 1 ounce (oz)
	= 437 ½ grains
16 ounces	= 1 pound (lb)
	= 256 drams
	= 7,000 grains
100 pounds	= 1 hundredweight (cwt)*
20 hundredweights	= 1 ton
	= 2,000 pounds*

In *gross* or *long* measure, the following values are recognized.

112 pounds	= 1 gross or long hundredweight*
20 gross or long hundredweights	= 1 gross or long ton
	= 2,240 pounds*

*When the terms *hundredweight* and *ton* are used unmodified, they are commonly understood to mean the 100-pound hundredweight and the 2,000-pound ton, respectively; these units may be designated *net* or *short* when necessary to distinguish them from the corresponding units in gross or long measure.

Troy Weight

24 grains	= 1 pennyweight (dwt)
20 pennyweights	= 1 ounce troy (oz t)
	= 480 grains
12 ounces troy	= 1 pound troy (lb t)
	= 240 pennyweights
	= 5,760 grains

Tables of Equivalents

In this table it is necessary to distinguish between the *international* and the *survey* foot. The international foot, defined in 1959 as exactly equal to 0.3048 meter, is shorter than the old survey foot by exactly 2 parts in 1 million. The survey foot is still used in data expressed in feet in geodetic surveys within the U.S. In this table the survey foot is indicated with capital letters.

When the name of a unit is enclosed in brackets, e.g., [1 hand], either (1) the unit is not in general current use in the U.S. or (2) the unit is believed to be based on custom and usage rather than on formal definition.

Equivalents involving decimals are, in most instances, rounded to the 3d decimal place; exact equivalents are so designated.

Lengths

1 angstrom (Å)	= 0.1 nanometer (exactly)
	= 0.000 1 micrometer (exactly)
	= 0.000 000 1 millimeter (exactly)
	= 0.000 000 004 inch
1 cable's length	= 120 fathoms (exactly)
	= 720 FEET (exactly)
	= 219 meters
1 centimeter (cm)	= 0.3937 inch
1 chain (ch) (Gunter's or surveyor's)	= 66 FEET (exactly)
	= 20.1168 meters
	= 100 feet
1 chain (engineer's)	= 30.48 meters (exactly)
1 decimeter (dm)	= 3.937 inches
1 degree (geographical)	= 364,566.929 feet
	= 69.047 miles (avg.)
	= 111.123 kilometers (avg.)
of latitude	= 68.708 miles at equator
	= 69.403 miles at poles
of latitude	= 69.171 miles at equator
1 dekameter (dam)	= 32.808 feet
1 fathom	= 6 FEET (exactly)
	= 1.8288 meters
1 foot (ft)	= 0.3048 meters (exactly)
	= 10 chains (surveyors) (exactly)
1 furlong (fur)	= 660 FEET (exactly)
	= 1/8 statute mile (exactly)
	= 201.168 meters
[1 hand] (height measure for horses from ground to top of shoulders)	= 4 inches
1 inch (in)	= 2.54 centimeters (exactly)
1 kilometer (km)	= 0.621371 mile
	= 3,280.8 feet
1 league (land)	= 3 statute miles (exactly)
	= 4.828 kilometers
1 link (Gunter's or surveyor's)	= 7.92 inches (exactly)
	= 0.201 meter
1 link (engineer's)	= 1 foot
	= 0.305 meter
1 meter (m)	= 39.37 inches
	= 1.09361 yards
1 micrometer (µm)	= 0.001 millimeter (exactly)
	= 0.00003937 inch
1 mil	= 0.001 inch (exactly)
	= 0.0254 millimeter (exactly)
1 mile (mi) (statute or land)	= 5,280 FEET (exactly)
	= 1.609344 kilometers (exactly)
1 international nautical mile (nmi)	= 1.852 kilometers (exactly)
	= 1.150779 statute miles
	= 6,076.11549 feet
1 millimeter (mm)	= 0.03937 inch
1 nanometer (nm)	= 0.001 micrometer (exactly)
	= 0.00000003937 inch
1 pica (typography)	= 12 points
1 point (typography)	= 0.013 837 inch (exactly)
	= 0.351 millimeter
1 rod (rd), pole, or perch	= 16½ FEET (exactly)
	= 5.029 meters
1 yard (yd)	= 0.9144 meter (exactly)

Areas or Surfaces

1 acre	= 43,560 square FEET (exactly)
	= 4,840 square yards
	= 0.405 hectare
1 are (a)	= 119.599 square yards
	= 0.025 acre

1 bolt (cloth measure):
length. = 100 yards (on modern looms)
width . . . . = 45 or 60 inches
1 hectare (ha) . . . . = 2.471 acres
[1 square (building)] . . . . = 100 square feet
1 square centimeter (cm^2) . . . . = 0.155 square inch
1 square decimeter (dm^2). . . . . = 15.500 square inches
1 square foot (ft^2) . . . . = 929.030 square centimeters
1 square inch (in^2) . . . . = 6.4516 square centimeters (exactly)
1 square kilometer (km^2) . . . . . = 247.104 acres
= 0.386102 square mile
1 square meter (m^2) . . . . = 1.196 square yards
= 10.764 square feet
1 square mile (mi^2) . . . . = 258.999 hectares
1 square millimeter (mm^2) . . . . = 0.002 square inch
1 square rod (rd^2), sq. pole, or sq. perch . . . . =25.293 square meters
1 square yard (yd^2) . . . . = 0.836127 square meter

Capacities or Volumes

1 barrel (bbl), liquid . . . . = 31 to 42 gallons*

*There are a variety of "barrels" established by law or usage. For example: federal taxes on fermented liquors are based on a barrel of 31 gallons; many state laws fix the "barrel for liquids" as 31½ gallons; one state fixes a 36-gallon barrel for cistern measurement; federal law recognizes a 40-gallon barrel for "proof spirits"; by custom, 42 gallons constitute a barrel of crude oil or petroleum products for statistical purposes, and this equivalent is recognized "for liquids" by 4 states.

1 barrel (bbl), standard for fruits, vegetables, and other dry commodities except dry cranberries . . . . = 7,056 cubic inches
= 1 barrel (bbl), standard for fruits,
1 barrel (bbl), standard, cranberry . . . . = $86\frac{45}{64}$ dry quarts
= 2.709 bushels, struck measure
= 5,826 cubic inches
1 board foot (lumber measure) = a foot-square board 1 inch thick
1 bushel (bu) (U.S.) (struck measure) . . . . = 2,150.42 cubic inches (exactly)
= 35.239 liters
[1 bushel, heaped (U.S.)] . . . . = 2,747.715 cubic inches
= 1.278 bushels, struck measure*

*Frequently recognized as 1¼ bushels, struck measure.

[1 bushel (bu) (British Imperial) (struck measure)] . . . . = 1.032 U.S. bushels, struck measure
= 2,219.36 cubic inches
1 cord (cd) firewood . . . . = 128 cubic feet (exactly)
1 cubic centimeter (cm^3) . . . . . = 0.061 cubic inch
1 cubic decimeter (dm^3) . . . . . = 61.024 cubic inches
1 cubic inch (in^3) . . . . = 0.554 fluid ounce
= 4.433 fluid drams
= 16.387 cubic centimeters
1 cubic foot (ft^3) . . . . = 7.481 gallons
= 28.317 cubic decimeters
1 cubic meter (m^3) . . . . = 1.308 cubic yards
1 cubic yard (yd^3) . . . . = 0.765 cubic meter
1 cup, measuring . . . . = 8 fluid ounces (exactly)
= ½ liquid pint (exactly)
[1 dram, fluid (fl dr) (British)]. . . = 0.961 U.S. fluid dram
= 0.217 cubic inch
= 3.552 milliliters
1 dekaliter (daL) . . . . = 2.642 gallons
= 1.135 pecks
1 gallon (gal) (U.S.) . . . . = 231 cubic inches (exactly)
= 3.785 liters
= 0.833 British gallon
= 128 U.S. fluid ounces (exactly)
[1 gallon (gal) British Imperial] . = 277.42 cubic inches
= 1.201 U.S. gallons
= 4.546 liters
= 160 British fluid ounces (exactly)
1 gill (gi) . . . . = 7.219 cubic inches
= 4 fluid ounces (exactly)
= 0.118 liter
1 hectoliter (hL) . . . . = 26.418 gallons
= 2.838 bushels
1 liter (L) (1 cubic decimeter exactly) . . . . = 1.057 liquid quarts
= 0.908 dry quart
= 61.024 cubic inches
1 milliliter (mL) (1 cu cm exactly)= 0.271 fluid dram
= 16.231 minims
= 0.061 cubic inch
1 ounce, liquid (U.S.) . . . . = 1.805 cubic inches
= 29.574 milliliters
= 1.041 British fluid ounces
[1 ounce, fluid (fl oz) (British)] . .= 0.961 U.S. fluid ounce
= 1.734 cubic inches
= 28.412 milliliters
1 peck (pk) . . . . = 8.810 liters
1 pint (pt), dry . . . . = 33.600 cubic inches
= 0.551 liter
1 pint (pt), liquid . . . . = 28.875 cubic inches (exactly)
= 0.473 liter
1 quart (qt), dry (U.S.) . . . . = 67.201 cubic inches
= 1.101 liters
= 0.969 British quart
1 quart (qt), liquid (U.S.) . . . . . . = 57.75 cubic in (exactly)
= 0.946 liter
= 0.833 British quart
[1 quart (qt) (British)] . . . . = 69.354 cubic inches
= 1.032 U.S. dry quarts
= 1.201 U.S. liquid quarts
1 tablespoon . . . . = 3 teaspoons*(exactly)
= 4 fluid drams
= ½ fluid ounce (exactly)
1 teaspoon . . . . = $^1/_3$ tablespoon*(exactly)
= $1^1/_3$ fluid drams*

*The equivalent "1 teaspoon = $1^1/_3$ fluid drams" has been found to correspond more closely with the actual capacities of teaspoons in use than the equivalent "1 teaspoon = 1 fluid dram" which is given by many dictionaries.

Weights or Masses

1 assay ton** (AT) . . . . = 29.167 grams

** Used in assaying. The assay ton bears the same relation to the milligram that a ton of 2,000 pounds avoirdupois bears to the ounce troy; hence, the weight in milligrams of precious metal obtained from one assay ton of ore gives directly the number of troy ounces to the net ton.

1 bale (cotton measure) . . . . . . = 500 pounds in U.S.
= 750 pounds in Egypt
1 carat (c) . . . . = 200 milligrams (exactly)
= 3.086 grains
1 dram avoirdupois (dr avdp) . . = $27^{11}/_{32}$ (= 27.344) grains
= 1.772 grams
1 gamma (g) . . . . = 1 microgram (exactly), see below
1 grain . . . . = 64.7989 milligrams
1 gram . . . . = 15.432 grains
= 0.035 ounce, avoirdupois
1 hundredweight, gross or long*** (gross cwt) . . . . = 112 pounds (exactly)
= 50.802 kilograms
1 hundredweight, net or short (cwt or net cwt) . . . . = 100 pounds (exactly)
= 45.359 kilograms
1 kilogram (kg) . . . . = 2.20462 pounds
1 microgram (μg) . . . . = 0.000001 gram (exactly)
1 milligram (mg) . . . . = 0.015 grain
1 ounce, avoirdupois (oz avdp) . = 437.5 grains (exactly)
= 0.911 troy ounce
= 28.3495 grams
1 ounce, troy (oz t) . . . . = 480 grains (exactly)
= 1.097 avoirdupois ounces
= 31.103 grams
1 pennyweight (dwt) . . . . = 1.555 grams
1 pound, avoirdupois (lb avdp) . = 7,000 grains (exactly)
= 1.215 troy pounds
= 453.59237 grams (exactly)
1 pound, troy (lb t) . . . . = 5,760 grains (exactly)
= 0.823 pound, avoirdupois
= 373.242 grams
1 ton, gross or long*** (gross ton)= 2,240 pounds (exactly)
= 1.12 net tons (exactly)
= 1.016 metric tons

***The gross or long ton and hundredweight are used commercially in the U.S. to only a limited extent, usually in restricted industrial fields. These units are the same as the British ton and hundredweight.

1 ton, metric (t) . . . . = 2,204.623 pounds
= 0.984 gross ton
= 1.102 net tons
1 ton, net or short (sh ton) . . . . = 2,000 pounds (exactly)
= 0.893 gross ton
= 0.907 metric ton

Tables of Interrelation of Units of Measurement

Units of length and area of the international and survey measures are included in the following tables.
1 international foot = 0.999998 survey foot or 2 x 0.0254 meter (exactly)
1 survey foot = 1200/3937 meter
BOLD indicates exact values.

Units of Length

Units	Inches	Links	Feet	Yards	Rods	Chains	Miles	Cm	Meters
1 inch=	—	0.126263	0.083333	0.027778	0.005051	0.001263	0.000016	**2.54**	**0.0254**
1 link=	**7.92**	—	**0.66***	**0.22***	**0.04***	0.01	0.000125	20.117	0.201168
1 foot=	**12**	1.515152	—	0.333333	0.060606	0.015152	0.000189	**30.48**	**0.3048**
1 yard=	**36**	4.54545	**3**	—	0.181818	0.045455	0.000568	**91.44**	**0.9144**
1 rod=	**198**	**25**	**16.5**	**5.5**	—	**0.25**	**0.003125**	502.92	5.0292
1 chain=	**792**	**100**	**66**	**22**	**4**	—	**0.0125**	2,011.68	20.1168
1 mile=	**63,360**	**8,000**	**5,280**	**1,760**	**320**	**80**	—	**160,934.4**	**1,609.344**
1 cm=	0.3937	0.049710	0.032808	0.010936	0.001988	0.000497	0.000006	—	**0.01**
1 meter=	39.37	4.970960	3.280840	1.093613	0.198838	0.049710	0.000621	**100**	—

Units of Area

Units	Sq. inches	Sq. links	Sq. feet	Sq. yards	Sq. rods	Sq. chains
1 sq. inch=	—	0.0159423	0.006944	0.000771605	0.0000255	0.000001594
1 sq. link=	62.7264	—	**0.4356**	**0.0484**	**0.0016**	0.0001
1 sq. foot=	**144**	2.295684	—	0.1111111	0.00367309	0.000229568
1 sq. yard=	**1,296**	20.66116	**9**	—	0.03305785	0.00206612
1 sq. rod=	**39,204**	**625**	**272.25**	**30.25**	—	0.0625
1 sq. chain=	**627,264**	**10,000**	**4,356**	**484**	**16**	—
1 acre=	**6,272,640**	**100,000**	**43,560**	**4,840**	**160**	**10**
1 sq. mile=	**4,014,489,600**	**64,000,000**	**27,878,400**	**3,097,600**	**102,400**	**6,400**
1 sq. cm=	0.1550003	0.00247105	0.001076	0.000119599	0.000003954	0.000000247
1 sq. meter=	1,550.003	24.71044	10.76391	1.195990	0.03953670	0.002471044
1 hectare=	15,500,031	247,104	107,639.1	11,959.90	395.3670	**24.71044**

Units	Acres	Sq. miles	Sq. cm	Sq. meters	Hectares
1 sq. inch=	0.000000159423	0.00000000024910	**6.4516**	**0.00064516**	0.000000065
1 sq. link=	**0.00001**	**0.000000015625**	404.68564224	0.04046856	0.000004047
1 sq. foot=	0.00002295684	0.0000000358706	929.0341	0.09290341	0.000009290
1 sq. yard=	0.0002066116	0.0000003228306	**8,361.2736**	0.83612736	0.000083613
1 sq. rod=	**0.00625**	**0.000009765625**	252,929.5	25.29295	0.002529295
1 sq. chain=	**0.1**	**0.00015625**	4,046,873	404.6873	0.04046873
1 acre=	—	**0.0015625**	40,468,730	4,046.873	0.4046873
1 sq. mile=	**640**	—	25,899,881,103	2,589,988.11	258.998811034
1 sq. cm=	0.000000024711	0.000000000038610	—	**0.0001**	**0.00000001**
1 sq. meter=	0.0002471044	0.0000003861022	**10,000**	—	**0.0001**
1 hectare=	2.471044	0.003861006	**100,000,000**	**10,000**	—

Units of Weight or Mass Not Greater Than Pounds and Kilograms

Units	Grains	Pennyweights	Avdp drams	Avdp ounces
1 grain=	—	0.04166667	0.03657143	0.00228571
1 pennyweight=	**24**	—	0.8777143	0.05485714
1 dram avdp=	**27.34375**	1.139323	—	**0.0625**
1 ounce avdp=	**437.5**	18.22917	**16**	—
1 ounce troy=	**480**	**20**	17.55429	1.097143
1 pound troy=	**5,760**	**240**	210.6514	13.16571
1 pound avdp=	**7,000**	291.6667	**256**	**16**
1 milligram=	0.015432	0.000643015	0.000564383	0.000035274
1 gram=	15.43236	0.6430149	0.5643834	0.03527396
1 kilogram=	15,432.36	643.0149	564.3834	35.27396

Units	Troy ounces	Troy pounds	Avdp pounds	Milligrams	Grams	Kilograms
1 grain=	0.00208333	0.000173611	0.000142857	**64.79891**	**0.06479891**	0.000064799
1 pennywt.=	**0.05**	0.004166667	0.003428571	**1,555.17384**	**1.55517384**	0.001555174
1 dram avdp=	0.05696615	0.004747179	0.00390625	1,771.845195	1.771845195	0.001771845
1 oz avdp=	0.9114583	0.07595486	0.0625	**28,349.523125**	**28.349523125**	0.02834952
1 oz troy=	—	0.083333333	0.06857143	**31,103.4768**	**31.1034768**	0.03110348
1 lb troy=	**12**	—	0.8228571	**373,241.7216**	**373.2417216**	0.373241722
1 lb avdp=	14.58333	1.215278	—	**453,592.37**	**453.59237**	**0.45359237**
1 milligram=	0.000032151	0.000002679	0.000002205	—	**0.001**	**0.000001**
1 gram=	0.03215075	0.002679229	0.002204623	**1000**	—	**0.001**
1 kilogram=	32.15075	2.679229	2.204623	**1,000,000**	**1000**	—

Units of Weight or Mass Not Less Than Avoirdupois Ounces

Units	Avdp oz	Avdp lb	Short cwt	Short tons	Long tons	Kilograms	Metric tons
1 oz avdp=	—	**0.0625**	**0.000625**	**0.00003125**	0.000027902	0.028349523	0.000028350
1 lb avdp=	**16**	—	**0.01**	**0.0005**	0.000446429	**0.45359237**	0.000453592
1 sh cwt=	**1,600**	**100**	—	**0.05**	0.04464286	**45.359237**	**0.045359237**
1 sh ton=	**32,000**	**2,000**	**20**	—	0.8928571	**907.18474**	**0.90718474**
1 long ton=	**35,840**	**2,240**	**22.4**	**1.12**	—	**1,016.0469088**	1.016046909
1 kg=	35.27396	2.204623	0.02204623	0.001102311	0.000984207	—	**0.001**
1 metric ton=	35,273.96	2,204.623	22.04623	1.102311	0.9842065	**1,000**	—

Units of Volume

Units	Cubic inches	Cubic feet	Cubic yards	Cubic cm	Cubic dm	Cubic meters
1 cubic inch=	—	0.000578704	0.000021433	**16.387064**	0.016387	0.000016387
1 cubic foot=	**1,728**	—	0.03703704	**28,316.846592**	28.316847	0.028316847
1 cubic yard=	**46,656**	**27**	—	**764,554.857984**	764.554858	0.764554858
1 cubic cm=	0.06102374	0.000035315	0.000001308	—	**0.001**	**0.000001**
1 cubic dm=	61.02374	0.03531467	0.001307951	**1,000**	—	**0.001**
1 cubic meter=	61,023.74	35.31467	1.307951	**1,000,000**	**1,000**	—

Units of Capacity (Liquid Measure)

Units	Minims	Fluid drams	Fluid ounces	Gills	Liquid pint
1 minim=	—	0.0166667	0.00208333	0.000520833	0.000130208
1 fluid dram=	**60**	—	**0.125**	**0.03125**	**0.0078125**
1 fluid ounce=	**480**	**8**	—	**0.25**	**0.0625**
1 gill=	**1,920**	**32**	**4**	—	**0.25**
1 liquid pint=	**7,680**	**128**	**16**	**4**	—
1 liquid quart=	**15,360**	**256**	**32**	**8**	**2**
1 gallon=	**61,440**	**1,024**	**128**	**32**	**8**
1 cubic inch=	265.974	4.4329	0.5541126	0.1385281	0.03463203
1 cubic foot=	459,603.1	7,660.052	957.5065	239.3766	59.84416
1 liter=	16,230.73	270.51218	33.81402	8.453506	2.113376

Units	Liquid quarts	Gallons	Cubic inches	Cubic feet	Liters
1 minim=	0.00006510417	0.00001627604	0.003759766	0.000002175790	0.00006161152
1 flu. dram=	**0.00390625**	**0.0009765625**	0.2255859	0.0001305474	0.003696691
1 fluid oz=	**0.03125**	**0.0078125**	**1.8046875**	0.001044379	0.02957353
1 gill=	**0.125**	**0.03125**	**7.21875**	0.004177517	0.118294118
1 liquid pt=	**0.5**	**0.125**	**28.875**	0.01671007	0.473176473
1 liquid qt=	—	**0.25**	**57.75**	0.03342014	0.946352946
1 gallon=	**4**	—	**231**	0.1336806	3.785411784
1 cubic inch=	0.01731602	0.004329004	—	0.0005787037	0.016387064
1 cubic foot=	29.92208	7.480519	**1,728**	—	**28.316846592**
1 liter=	1.056688	0.26417205	61.02374	0.03531467	—

Units of Capacity (Dry Measure)

Units	Dry pints	Dry quarts	Pecks	Bushels	Cubic in.	Liters
1 dry pint=	—	**0.5**	**0.0625**	**0.015625**	**33.6003125**	0.55061047
1 dry quart=	**2**	—	**0.125**	**0.03125**	**67.200625**	1.1012209
1 peck=	**16**	**8**	—	**0.25**	**537.605**	8.8097675
1 bushel=	**64**	**32**	**4**	—	**2,150.42**	35.23907
1 cubic inch=	0.0297616	0.0148808	0.00186010	0.000465025	—	0.01638706
1 liter=	1.816166	0.908083	0.11351037	0.02837759	61.02374	—

IT'S A FACT: The largest reported snowflake was discovered by a ranch owner at Fort Keogh, MT, on Jan. 28, 1887. It measured 38 cm (15 in) wide and 20 cm (8 in) thick, or, as its discoverer put it, was "larger than milkpans."

Miscellaneous Measures

Caliber—the diameter of a gun bore. In the U.S., caliber is traditionally expressed in hundredths of inches, e.g., .22. In Britain, caliber is often expressed in thousandths of inches, e.g., .270. Now it is commonly expressed in millimeters, e.g., the 5.56 mm M16 rifle. Heavier weapons' caliber has long been expressed in millimeters, e.g., the 155 mm howitzer. Naval guns' caliber refers to the barrel length as a multiple of the bore diameter. A 5-inch, 50-caliber naval gun has a 5-inch bore and a barrel length of 250 inches.

Decibel (dB)—a measure of the relative loudness or intensity of sound. A 20-decibel sound is 10 times louder than a 10-decibel sound; 30 decibels is 100 times louder; 40 decibels is 1,000 times louder, etc.

One decibel is the smallest difference between sounds detectable by the human ear. A 120-decibel sound is painful.

10 decibels – a light whisper
20 – quiet conversation
30 – normal conversation
40 – light traffic
50 – typewriter, loud conversation
60 – noisy office
70 – normal traffic, quiet train
80 – rock music, subway
90 – heavy traffic, thunder
100 – jet plane at takeoff

Em—a printer's measure designating the square width of any given type size. Thus, an em of 10-point type is 10 points. An en is half an em.

Gauge—a measure of shotgun bore diameter. Gauge numbers originally referred to the number of lead balls just fitting the gun barrel diameter required to make a pound. Thus, a 16-gauge shotgun's bore was smaller than a 12-gauge shotgun's. Today, an international agreement assigns millimeter measures to each gauge, e.g.:

Gauge	Bore diameter (in mm)
6	23.34
10	19.67
12	18.52
14	17.60
16	16.81
20	15.90

Horsepower—the power needed to lift 550 pounds 1 foot in 1 second or to lift 33,000 pounds 1 foot in 1 minute. Equivalent to 746 watts or 2,546.0756 Btu/h.

Karat or carat—a measure of fineness for gold equal to $^{1}/_{24}$ part of pure gold in an alloy. Thus 24-karat gold is pure; 18-karat gold is ¼ alloy. The *carat* is also used as a unit of weight for precious stones; it is equal to 200 milligrams or 3.086 grains troy.

Knot—a measure of the speed of ships. A knot equals 1 nautical mile per hour.

Quire—25 sheets of paper

Ream—500 sheets of paper

Electrical Units

The **watt** is the unit of power (electrical, mechanical, thermal, etc.). Electrical power is given by the product of the voltage and the current.

Energy is sold by the **joule,** but in common practice the billing of electrical energy is expressed in terms of the **kilowatt-hour,** which is 3,600,000 joules or 3.6 megajoules.

The **horsepower** is a nonmetric unit sometimes used in mechanics. It is equal to 746 watts.

The **ohm** is the unit of electrical resistance and represents the physical property of a conductor that offers a resistance to the flow of electricity, permitting just 1 ampere to flow at 1 volt of pressure.

Spirits Measures

Pony	= 0.5 jigger
Shot	= 0.666 jigger
	= 1.0 ounce
Jigger	= 1.5 shots
Pint	= 16 shots
	= 0.625 fifth
Fifth	= 25.6 shots
	= 1.6 pints
	= 0.8 quart
	= 0.75706 liter
Quart	= 32 shots
	= 1.25 fifths
Magnum	= 2 quarts
	= 2.49797 bottles (wine)

For champagne and brandy only:

Jeroboam	= 6.4 pints
	= 1.6 magnum
	= 0.8 gallon

For champagne only:

Rehoboam	= 3 magnums
Methuselah	= 4 magnums
Salmanazar	= 6 magnums
Balthazar	= 8 magnums
Nebuchadnezzar	= 10 magnums

Wine bottle (standard)	= 0.800633 quart
	= 0.7576778 liter

Temperature Conversion Table

The numbers in the **center column** refer to the temperatures in either degrees Celsius or degrees Fahrenheit that are to be converted. If converting from degrees Fahrenheit to Celsius, refer to the column on the left; if converting from degrees Celsius to Fahrenheit, consult the column on the right.

For temperatures not shown: To convert Fahrenheit to Celsius by formula, subtract 32 degrees and divide by 1.8; to convert Celsius to Fahrenheit, multiply by 1.8 and add 32 degrees.

Note: Although the term *centigrade* is still frequently used, the International Committee on Weights and Measures and the National Institute of Standards and Technology have recommended since 1948 that this scale be called Celsius.

Celsius		Fahrenheit	Celsius		Fahrenheit	Celsius		Fahrenheit
−273.2	**−459.7**	......	−17.8	**0**	32	35.0	**95**	203
−184	**−300**	......	−12.2	**10**	50	37	**98.6**	209.5
−169	**273**	−459.4	−6.67	**20**	68	37.8	**100**	212
−157	**−250**	−418	−1.11	**30**	86	43	**110**	230
−129	**−200**	−328	4.44	**40**	104	49	**120**	248
−101	**−150**	−238	10.0	**50**	122	54	**130**	266
−73.3	**−100**	−148	15.6	**60**	140	60	**140**	284
−45.6	**−50**	−58	21.1	**70**	158	66	**150**	302
−40.0	**−40**	−40	23.9	**75**	167	93	**200**	392
−34.4	**−30**	−22	26.7	**80**	176	121	**250**	482
−28.9	**−20**	−4	29.4	**85**	185	149	**300**	572
−23.3	**−10**	14	32.2	**90**	194			

Boiling and Freezing Points

Water boils at 212° F (100° C) at sea level. For every 550 feet above sea level, boiling point of water is lower by about 1° F. Methyl alcohol boils at 148° F. Average human oral temperature, 98.6° F. Water freezes at 32° F (0° C).

Prime Numbers

A prime number is an integer divisible only by 1 and itself.

Prime Numbers between 1 and 1,000

	2	3	5	7	11	13	17	19	23
29	31	37	41	43	47	53	59	61	67
71	73	79	83	89	97	101	103	107	109
113	127	131	137	139	149	151	157	163	167
173	179	181	191	193	197	199	211	223	227
229	233	239	241	251	257	263	269	271	277
281	283	293	307	311	313	317	331	337	347
349	353	359	367	373	379	383	389	397	401
409	419	421	431	433	439	443	449	457	461
463	467	479	487	491	499	503	509	521	523
541	547	557	563	569	571	577	587	593	599
601	607	613	617	619	631	641	643	647	653
659	661	673	677	683	691	701	709	719	727
733	739	743	751	757	761	769	773	787	797
809	811	821	823	827	829	839	853	857	859
863	877	881	883	887	907	911	919	929	937
941	947	953	967	971	977	983	991	997	(1,009)

Common Fractions Reduced to Decimals

8ths	16ths	32ds	64ths	
			1	= 0.015625
		1	2	= 0.03125
			3	= 0.046875
	1	2	4	= 0.0625
			5	= 0.078125
		3	6	= 0.09375
			7	= 0.109375
1	2	4	8	= 0.125
			9	= 0.140625
		5	10	= 0.15625
			11	= 0.171875
	3	6	12	= 0.1875
			13	= 0.203125
		7	14	= 0.21875
			15	= 0.234375
2	4	8	16	= 0.25
			17	= 0.265625
		9	18	= 0.28125
			19	= 0.296875
	5	10	20	= 0.3125
			21	= 0.328125

8ths	16ths	32ds	64ths	
		11	22	= 0.34375
			23	= 0.359375
3	6	12	24	= 0.375
			25	= 0.390625
		13	26	= 0.40625
			27	= 0.421875
	7	14	28	= 0.4375
			29	= 0.453125
		15	30	= 0.46875
			31	= 0.484375
4	8	16	32	= 0.5
			33	= 0.515625
		17	34	= 0.53125
			35	= 0.546875
		18	36	= 0.5625
			37	= 0.578125
		19	38	= 0.59375
			39	= 0.609375
5	10	20	40	= 0.625
			41	= 0.640625
		21	42	= 0.65625
			43	= 0.671875

8ths	16ths	32ds	64ths	
	11	22	44	= 0.6875
			45	= 0.703125
		23	46	= 0.71875
			47	= 0.734375
6	12	24	48	= 0.75
			49	= 0.765625
		25	50	= 0.78125
			51	= 0.796875
	13	26	52	= 0.8125
			53	= 0.828125
		27	54	= 0.84375
			55	= 0.859375
7	14	28	56	= 0.875
			57	= 0.890625
		29	58	= 0.90625
			59	= 0.921875
	15	30	60	= 0.9375
			61	= 0.953125
		31	62	= 0.96875
			63	= 0.984375
8	16	32	64	= 1.0

Measures of Force and Pressure

Dyne = force necessary to accelerate a 1-gram mass 1 centimeter per second squared = 0.000072 poundal
Poundal = force necessary to accelerate a 1-pound mass 1 foot per second squared = 13,825.5 dynes = 0.138255 newtons
Newton = force needed to accelerate a 1-kilogram mass 1 meter per second squared
Pascal (pressure) = 1 newton per square meter = 0.020885 pound per square foot
Atmosphere (air pressure at sea level) = 2,116.102 pounds per square foot = 14.6952 pounds per square inch = 1.0332 kilograms per square centimeter = 101,323 newtons per square meter

Mathematical Formulas

Note: The value of π (the Greek letter pi) is approximately 3.14159265 (equal to the ratio of the circumference of a circle to the diameter). The equivalence is typically rounded further to 3.1416 or 3.14.

To find the CIRCUMFERENCE of a:

Circle — Multiply the diameter by π.

To find the AREA of a:

Circle — Multiply the square of the radius (equal to $1/2$ the diameter) by π.
Rectangle — Multiply the length of the base by the height.
Sphere (surface) — Multiply the square of the radius by π and multiply by 4.
Square — Square the length of one side.
Trapezoid — Add the 2 parallel sides, multiply by the height, and divide by 2.
Triangle — Multiply the base by the height, divide by 2.

To find the VOLUME of a:

Cone — Multiply the square of the radius of the base by π, multiply by the height, and divide by 3.
Cube — Cube the length of one edge.
Cylinder — Multiply the square of the radius of the base by π and multiply by the height.
Pyramid — Multiply the area of the base by the height and divide by 3.
Rectangular Prism — Multiply the length by the width by the height.
Sphere — Multiply the cube of the radius by π, multiply by 4, and divide by 3.

Playing Cards and Dice Chances

5-Card Poker Hands

Hand	Number possible	Odds against
Royal flush	4	649,739 to 1
Other straight flush	36	72,192 to 1
Four of a kind	624	4,164 to 1
Full house	3,744	693 to 1
Flush	5,108	508 to 1
Straight	10,200	254 to 1
Three of a kind	54,912	46 to 1
Two pairs	123,552	20 to 1
One pair	1,098,240	4 to 3 (1.37 to 1)
Nothing	1,302,540	1 to 1
TOTAL	**2,598,960**	

Note: Although there are only 13 4-of-a-kind combinations, the above numbers take into account the total possibilities when a 5th card is figured in to make a 5-card hand.

Bridge

The odds—against suit distribution in a hand of 4-4-3-2 are about 4 to 1, against 5-4-2-2 about 8 to 1, against 6-4-2-1 about 20 to 1, against 7-4-1-1 about 254 to 1, against 8-4-1-0 about 2,211 to 1, and against 13-0-0-0 about 158,753,389,899 to 1.

Dice

(probabilities of consecutive winning plays)

No. consecutive wins	By 7, 11, or point
1	244 in 495
2	6 in 25
3	3 in 25
4	1 in 17
5	1 in 34
6	1 in 70
7	1 in 141
8	1 in 287
9	1 in 582

Dice

(probabilities on 2 dice)

Total	Odds against (single toss)
2	35 to 1
3	17 to 1
4	11 to 1
5	8 to 1
6	31 to 5
7	5 to 1
8	31 to 5
9	8 to 1
10	11 to 1
11	17 to 1
12	35 to 1

Large Numbers

U.S.	Number of zeros	British[1], French, German
million	6	million
billion	9	milliard
trillion	12	billion
quadrillion	15	1,000 billion
quintillion	18	trillion
sextillion	21	1,000 trillion
septillion	24	quadrillion
octillion	27	1,000 quadrillion
nonillion	30	quintillion
decillion	33	1,000 quintillion
undecillion	36	sextillion
duodecillion	39	1,000 sextillion
tredecillion	42	septillion
quattuordecillion	45	1,000 septillion
quindecillion	48	octillion
sexdecillion	51	1,000 octillion
septendecillion	54	nonillion
octodecillion	57	1,000 nonillion
novemdecillion	60	decillion
vigintillion	63	1,000 decillion
googol	100	googol
centillion	303	—
—	600	centillion
googolplex	googol	googolplex

(1) In recent years, it has become more common in Britain to use American terminology for large numbers.

Roman Numerals

I — 1	V — 5	IX — 9	XX — 20	LX — 60	CD — 400
II — 2	VI — 6	X — 10	XXX— 30	XC — 90	D — 500
III — 3	VII — 7	XI — 11	XL — 40	C — 100	CM — 900
IV — 4	VIII — 8	XIX — 19	L — 50	CC — 200	M — 1,000

Note: The numerals V, X, L, C, D, or M shown with a horizontal line on top denote 1,000 times the original value.

Ancient Measures

Biblical		**Greek**		**Roman**	
Cubit	= 21.8 inches	Cubit	= 18.3 inches	Cubit	= 17.5 inches
Omer	= 0.45 peck	Stadion	= 607.2 or 622 feet	Stadium	= 202 yards
	= 3.964 liters	Obolos	= 715.38 milligrams	As, libra, pondus	= 325.971 grams
Ephah	= 10 omers	Drachma	= 4.2923 grams		= 0.71864 pound
Shekel	= 0.497 ounce	Mina	= 0.9463 pound		
	= 14.1 grams	Talent	= 60 mina		

COMPUTERS AND THE INTERNET

Computer Milestones

More than 625 million **computers** worldwide—among them 182 million in the U.S.—were expected to be in use by the end of 2001, according to projections by Computer Industry Almanac Inc. The overwhelming majority of these machines are desktop or laptop personal computers (PCs). Devices for performing calculations are nothing new—the abacus, a frame with wires on which beads are moved back and forth (still used today in some parts of the world)—traces its origins back to ancient times. But modern PCs—marvels of electronic miniaturization that run under the supervision of an "operating system" such as Microsoft's Windows— are a relatively recent development. They are the direct descendents of vacuum-tube devices introduced in the early 20th century; even their mechanical gear-based forerunners were invented just a few centuries ago. Among the landmark events in computer history are the following:

- In 1623 the first mechanical calculator, capable of adding, subtracting, multiplying, and dividing, was developed by the German mathematician Wilhelm Schikard; the only two models Schikard made, however, were destroyed in a fire.
- In 1642 French mathematician Blaise Pascal built the first of more than four dozen copies of an adding and subtracting machine that he invented.
- In 1790, French inventor Joseph Marie Jacquard devised a new control system for looms. He "programmed" the loom, communicating desired weaving operations to the machine via patterns of holes in paper cards.
- The British mathematician and scientist Charles Babbage used the Jacquard punch-card system in his design for a sophisticated, programmable "Analytical Engine" that contained some of the basic features of today's computers. Babbage's conception was beyond the capabilities of the technology of his time, and the machine remained unfinished at his death in 1871.
- The 1890 U.S. census was expedited by the rapid processing of huge amounts of data with an electrical punch-card tabulating machine developed by American inventor Herman Hollerith, whose company in 1924 became International Business Machines (IBM).

On the eve of World War II researchers experimented with ways to speed up computation, since calculators using solely mechanical components were too slow. One approach was to use electromechanical relays, which basically are electrically controlled switches.

- In 1940, Bell Laboratories mathematician George Stibitz built completed the first electromechanical relay-based calculator. In the same year Stibitz provided the first demonstration of remote operation of a computer, using a teletype to transmit problems to his machine and to receive the results.
- In 1941, German engineer Konrad Zuse completed the Z3, the first fully functional digital computer to be controlled by a program. In 1944, the first large-scale automatic digital computer, the Mark I, built by IBM and Harvard Professor Howard Aiken, went into operation; it was 55 feet long and 8 feet high.

Efforts were also under way to develop fully electronic machines, using vacuum tubes, which can operate much more quickly than relays.

- Between 1937 and 1942 the first rudimentary vacuum-tube calculator was built by the physicist John Vincent Atanasoff and his assistant Clifford Berry at Iowa State College (now University).
- More substantial electronic machines were the Colossus, developed by the British in 1943 to break German codes, and the Eniac (for Electronic Numerical Integrator and Computer), a 30-ton room-sized computer with over 18,000 vacuum tubes, built by physicist John Mauchly and engineer J. Presper Eckert at the University of Pennsylvania for the U.S. Army and completed in 1946. The Colossus was a special-purpose machine; its capabilities were powerful (for its time) but limited. Eniac was a general-purpose machine and could be programmed to do different tasks, although programming could take a couple of days, since cables had to be plugged in and switches set by hand.
- In 1951, Eckert and Mauchly's Univac ("Universal Automatic Computer") became the first computer commercially available in the U.S.; the first customer: the Census Bureau. CBS-TV used a Univac in 1952 to predict the results of the presidential election.

As a result of the invention of the transistor in 1947 and the integrated circuit in 1958, followed by the development of the microprocessor (an entire computer processing unit on a chip)—the first commercial example of which was the Intel 4004 in 1971—computers gradually grew smaller, speedier, more reliable, and more powerful.

- In 1975 the first widely marketed personal computer, the MITS Altair 8800, was introduced in kit form, with no keyboard and no video display, at a price of under $400.
- In 1977 the Apple II was introduced. Capable of displaying text and graphics in color, it enjoyed phenomenal success.
- In 1981, IBM unveiled its "Personal Computer," which used Microsoft's DOS (disk operating system).
- In 1984, Apple Computer introduced the first Macintosh. The easy-to-use Macintosh came with a proprietary operating system and was the first popular computer to have a GUI (graphical user interface) and a mouse—features originally developed by the Xerox Corporation.
- In 1990, Microsoft released Windows 3.0, the first workable version of its own GUI.
- In 1991, Linux, based on the Unix operating system used in high-power computers, was invented for the PC by Helsinki University student Linus Torvalds and made available for free.
- In 1996 the Palm Pilot, the first widely successful handheld computer and personal information manager, was introduced.
- In 1997 the IBM computer Deep Blue beat world chess champion Garry Kasparov in a 6-game match, 3.5-2.5.

About the Internet

The **Internet** is a vast computer network of computer networks. In 1994, 3 million people (most of them in the U.S.) made use of it. By the end of 2000, according to a study by Computer Industry Almanac Inc., more than 400 million people around the world were using the Internet, including over 130 million in the U.S. It was estimated that by 2005, 1 billion people could be connected.

- A Computer Industry Almanac Inc. study in late 2000 found that the top ten countries in Internet users per capita were: Canada, Sweden, Finland, U.S., Iceland, Denmark, Norway, Australia, Singapore, and New Zealand.
- Nielsen//NetRatings reported in mid-2001 that the U.S. and Canada were the only Internet markets where females online outnumbered males.
- Recent research estimates that there are more than 2 billion pages on the World Wide Web readily accessible by general search engines. The total size of the Web, according to search engine developer BrightPlanet, lies in the hundreds of billions of pages.
- By December 1996, about 627,000 Internet domain names had been registered. By mid-2001, about 36 million had been registered.

The Internet is not owned or funded by any one institution, organization, or government. It has no CEO and is not a commercial service. Its development is guided by the Internet Society (ISOC), composed of volunteers. The ISOC appoints the Internet Architecture Board (IAB), which works out issues of standards, network resources, etc. Another volunteer group, the Internet Engineering Task Force (IETF), handles day-to-day issues.

Practically speaking, the Internet is composed of **people**, **hardware**, and **software**. With the proper equipment on both ends, you can sit at your computer and communicate with someone anyplace in the world. You can also use the

Internet to access vast amounts of information, including text, graphics, sound, and video. From your computer you can send e-mail, listen to music, "chat" with people on another continent, do banking, buy stocks, books, flowers, or cars, work with others on an electronic whiteboard, and, with the appropriate equipment, video-conference.

If your computer has a modem connected to a phone line, you can gain access to the Internet, to e-mail, and other special content through an **Internet service provider** (ISP), such as America Online (AOL), CompuServe (owned by AOL), or MSN (The Microsoft Network). Many national companies, such as AT&T, Earthlink, and Prodigy, and local companies also provide access. ISPs generally charge a monthly fee.

Internet Milestones

The Internet grew out of a series of developments in the academic, governmental, and information technology communities. Here are some major historical highlights:

- In 1969, ARPANET, an experimental 4-computer network, was established by the Advanced Research Projects Agency (ARPA) of the U.S. Defense Dept. so that research scientists could communicate.
- By 1971, ARPANET linked about 2 dozen computers ("hosts") at 15 sites, including MIT and Harvard. By 1981, there were over 200 hosts.
- During the 1980s, more and more computers using different operating systems were connected. In 1983, the military portion of ARPANET was moved onto the MILNET, and ARPANET was disbanded in 1990.
- In the late 1980s, the National Science Foundation's NSFNET began its own network and allowed everyone to access it. It was, however, mainly the domain of "techies," computer-science graduates, and professors.
- Legislation in the early 1990s expanded NSFNET, renamed it NREN (National Research and Education Network), and encouraged development of commercial transmission and network services. The mass commercialization of today's Internet is largely a result of such legislation.
- 1991 saw release of the first **browser**, or software for accessing what became known as the **World Wide Web**. In 1993, the National Center for Supercomputing Applications released versions of Mosaic, the first graphical Web browser, for Microsoft Windows, Unix systems running the X Window GUI, and the Apple Macintosh.
- In 1994, Netscape Communications released the Netscape Navigator browser. Microsoft released its Internet Explorer browser the following year but initially failed to make a significant dent in Netscape's dominance of the browser market.
- In 1998, Netscape Navigator's share of the browser market fell below 50 percent, while Internet Explorer's exceeded 25 percent. Also in 1998, the U.S. Justice Dept. and attorneys general from several states filed suit against Microsoft, claiming that the inclusion of Internet Explorer in Windows 98 violated antitrust guidelines.
- In April 2000 a federal district judge found Microsoft guilty of antitrust violations; two months later he ordered the company split into two parts, but implementation of the penalty was stayed while Microsoft appealed. Market researchers reported that Internet Explorer was now being used by over four-fifths of Internet users.
- In June 2001 a federal appeals court issued an opinion skeptical of the breakup remedy, and in Sept. the Bush administration decided not to pursue a breakup; the government continued to seek means to halt alleged anti-competitive practices by Microsoft.

Internet Resources

E-mail. Electronic mail is the most widely used resource on the Internet. An e-mail address consists of a **username**, a **service**, and a **domain**. In Walmanac@waegroup.com (*The World Almanac*'s e-mail address), Walmanac is the username, waegroup the service, and com the domain (in this case, a company).

Domains. Domains are identified in the Domain Name System. For years the registration of the most popular Internet addresses (with such domains as com, net, and org) was administered by Network Solutions Inc., under contract to the U.S. government. A nonprofit corporation, the Internet Corporation for Assigned Names and Numbers (ICANN), was set up in 1998 to oversee the system, with administration of the registration process to be opened up to more companies. Dozens of companies now offer domain registration services; examples include Network Solutions, Register.com, and BulkRegister.com.

Among the most familiar top-level domains are:

Domain	What It Is
.com	generally a commercial organization, business, or company
.edu	a 4-year higher-educational institution
.gov	a nonmilitary U.S. federal government entity
.int	an international organization
.mil	a U.S. military organization
.net	suggested for a network administration
.org	suggested for a nonprofit organization

In late 2000, ICANN authorized seven new domains, which were expected to gradually be phased into use beginning in 2001, although lawsuits threatened to delay full implementation.

Domain	What It Is
.aero	an organization in the air-transport industry
.biz	a business
.coop	a nonprofit business cooperative, such as a rural electric coop
.info	an informational site for an individual or organization, without restriction
.museum	a museum
.name	an individual
.pro	a professional, such as an accountant, lawyer, or physician

Outside the U.S., the final part of a domain name represents the country where the site is located—for example, jp in Japan, uk in the United Kingdom, and ru in Russia.

FTP. File Transfer Protocol is a method of transferring files on the Internet. Using FTP, you log on to a remote site, view files, and copy them to your computer. Sites that offer FTP capability can be accessed with special programs and also with most browsers. The address for such a site when accessed through a browser typically begins with ftp://.

Newsgroups. Newsgroups, a classic institution of the Internet, are found on the part of the Internet called Usenet. In a newsgroup, messages concerning a particular topic are posted in a public forum. You can simply read the postings, or you can post something yourself.

World Wide Web. The World Wide Web was developed in the early 1990s at the European Center for Nuclear Research as an environment in which scientists in Geneva, Switzerland, could share information. It has evolved into a medium with text, graphics, audio, animation, and video. A **website** address begins with http:// (or https:// for "secure" sites that protect the confidentiality of information you may transmit over the Web). The Web is a graphical environment that can be navigated through **hyperlinks**. From one site you click on hyperlinks to go to related sites.

FAQs. Frequently Asked Questions documents contain answers to common questions. A huge collection of FAQs can be found at the site http://www.faqs.org/faqs

How the Internet Works

The Internet involves 3 basic elements: server, client, and network. A **server** is a computer program that makes data available to other programs on the same or other computers—it "serves" them. A **client** is a computer that requests data from a server. A **network** is an interconnected system in which multiple computers can communicate, via copper wire, coaxial cable, fiber-optic cable, satellite transmission, etc. When you use a browser to go to a site on the World Wide Web, you access the site's files. To locate sites of interest you may want to use an online finding service called a **search engine**, such as Google.

Here are the steps in opening and accessing a file:

- In the browser, specify address, or **URL,** of the website.
- The browser sends your request to the Internet service provider's server.
- That server sends the request to the server at the URL.
- The file is sent to the ISP's server, which sends the file back to the browser, which displays the file.

Doing Business Online

Communication via e-mail or online chat and the storing and distribution of information and images are not the only uses to which the Internet is being put. You can also take courses, subscribe to news services, listen to radio, and do business over the Web. **E-commerce**—the conducting of transactions over the Internet—is becoming increasingly common around the globe. Transactions between companies constitute a significant part of the Internet's growing impact in commerce, since more and more firms are turning to the Web to procure goods and services. More and more companies, notably utilities and broadcasting and technology firms, are offering consumers online billing via the Web.

Consumers are also **shopping** more online. Although Americans' online consumer spending in 2001 showed signs of flagging in the short term as a result of a slow economy, the Internet analysis firm Jupiter Media Metrix projected in midyear that online retail sales would grow from an estimated $34 billion in 2001 to $130 billion by 2006. Consumers can also sell things over the Internet; hence the surging popularity of online auctions. According to a study by Nielsen//NetRatings and Harris Interactive, Americans spent $556 million at online auction sites in May 2001, up from $223 million in May 2000. Users making a purchase at online auctions in May 2001 totaled 6.2 million, compared to 1.1 million the year before. Some goods can be received online—e.g., computer programs, newspapers, and music recordings. Music of near-CD quality can be obtained (either for free or for a fee) in the popular format known as **MP3**, which compresses the electronic audio file somewhat in order to reduce download time.

Figures from the Travel Industry Association of America indicate that more than 59 million Americans used the Internet to help plan **travel** in 2000, 4 times as many as 1997. Some 25 million bought travel products or services online, up 384 percent growth from 1997.

If your system has a microphone, speakers, and the right software, you can also use the Internet as a transmission channel for voice phone calls. Worldwide Internet **telephone** usage for the so-called voice-over Internet protocol jumped from 200 million minutes in 1998 to 11 billion in 2000.

Online banking by consumers is a growing segment of the Internet, particularly in Europe and Japan. Not only is it convenient for consumers, who can check their balances, transfer funds, and pay bills while seated at their home computer, but online transactions cost the bank much less than face-to-face interactions with a bank teller. Datamonitor projected that active online bank accounts in Europe, numbering 26.1 million in 2000, would exceed 66 million by 2003.

The **online brokerage** market, has been a major player in commerce over the Internet. The number of American households trading stocks and shares via the Internet reached 3.5 million in January 2000, up from 2.7 million in May 1999.

According to Datamonitor projections, the number of people **gambling** via the Internet in the U.S. and Europe will rise from 2.9 million in 2001 to over 7 million by 2005. Online gambling revenues are expected to grow over the same period from $6.6 billion over $13 billion in the U.S. and from $130 million to about $7 billion in Europe.

Safety and Security on the Internet

Common sense dictates some basic security rules.

- Do not give out your phone number, address, or other personal information, unless needed for a transaction at a site you trust. Be careful about giving out credit card numbers.
- If you feel someone is being threatening or dangerous, inform your Internet service provider.

Viruses and Worms. There is always a risk of acquiring a computer **virus**. Some viruses may merely display a whimsical message on your screen. Some may wreak havoc in your system. Your system can pick up a virus from a program downloaded from the Internet or elsewhere via modem (or received on a floppy disk); a virus can also be communicated via e-mail, as was the case with the Melissa virus in 1999, or the ILOVEYOU virus in 2000.

You should have antivirus software installed on your computer, keep it up to date, and try to keep abreast of reports of new viruses. Be careful about opening e-mail from unknown correspondents, and if you have programs with a macro capability (macros are bits of auxiliary coding that are meant to play a helpful role but can be taken advantage of by some viruses, such as Melissa), make sure the programs' macro virus protection (if any) is turned on. Keep macros disabled if you do not know what you might want to use them for. If you have a high-speed Internet connection that is always on, consider buying protective "firewall" software to guard your system against attacks by hackers.

Worms are another type of mischievous cyber creature. Strictly speaking, while viruses propagate by infecting other programs, worms propagate without the help of a carrier program. A **Trojan horse** is malicious computer code concealed within harmless code or data, but at some point capable of taking control and causing damage.

Filtering. The two major browsers, Netscape Navigator and Internet Explorer, and some search engines contain features that let you filter the content that can be viewed on your computer. Special filtering software is also available, and some ISPs, such as AOL and MSN, make it possible for you to restrict the type of content seen on screen.

Parents can find more information on protecting their children while online at the websites of several U.S. government agencies, such as the FBI (http://www.fbi.gov/publications/pguide/pguide.htm). Another helpful site is http://www.safekids.com

Where to Start on the Web

Many people have a favorite site that they go to first when logging on to the World Wide Web. A convenient choice for such a site is a **portal,** a gateway site typically offering a search engine but also a variety of other services, such as free e-mail (sometimes free voice mail as well), chat, instant messaging, news services, stock updates, weather reports, real estate listings, yellow pages, people finders, TV and movie listings, shopping, and even tools to create and post your own Web page. Many portals permit you to customize the opening screen. Another common feature is a personal calendar to help you schedule activities.

Leading portals include:

AltaVista . . . . http://www.altavista.com
AOL . . . . http://www.aol.com
Excite . . . . http://www.excite.com
iWon . . . . http://www.iwon.com
Lycos. . . . . http://www.lycos.com
MSN . . . . http://www.msn.com
Netscape Netcenter . . . . http://www.netscape.com
Yahoo! . . . . http://www.yahoo.com

By using the portal's **search engine** you can locate information and, in some cases, images on sites throughout a large part of the Internet. No search engine covers the entire Web completely, and some portals offer a list of search engines to choose from. Search engines typically allow you to find occurrences of a particular key word or words. Search engines use different methods for finding, indexing, and retrieving information. Some store only the title and URL of sites; others index every word of a site's content. Some give extra weight to words in titles or other key positions, or to sites for which more hyperlinks exist on the Web. Many search engines work with the help of a program called a "spider," "crawler," or "bot." This visits sites across the Web and extracts information that can be used to create the search engine's index. In addition to a search engine that requires you to submit key words, some portals also offer a subject guide—a menu-like "directory," generally compiled by humans. You drill down through the directory to find a subcategory with websites of interest. Yahoo! is a popular example.

Among other search engines and directories:

- **Ask Jeeves** (http://www.askjeeves.com) provides a directory but also responds to questions entered "in plain English."
- **FAST Search and Transfer** (http://www.alltheweb.com) is very up-to-date; it claims to completely refresh its index every 9 to 12 days.
- **Google** (http://www.google.com) is also available in the Netscape Netcenter and Yahoo! portals. It relies largely on link popularity in ranking the sites it retrieves and claims to cover over 1.3 billion pages.
- **HotBot** (http://hotbot.lycos.com), owned by Lycos, offers useful advanced options.
- **Northern Light** (http://www.northernlight.com) organizes retrieved documents according to topic.
- **Open Directory** (http://dmoz.org) aims to cope with the vast size of the Web and produce the most comprehensive directory by using volunteer editors. Its information is used by such services as AOL, Netscape, Lycos, and HotBot.
- Teoma (http://www.teoma.com) groups search results into topics and also supplies links to related resources.
- A **meta-search engine** submits your request to several different search engines at the same time. However, meta-search engines typically do not exhaust each of the search engines' databases, and they may be unable to transmit complicated search requests. Among the better-known meta-search engines are **Dogpile** (http://www.dogpile.com); **Ixquick** (http://www.ixquick.com); **Queryserver** (http://www.queryserver.com); and **Vivísimo** (http://www.vivisimo.com).
- Large segments of the Web are not readily searchable by general-purpose search engines. Special search tools include those available via the Direct Search site (http://gwis2.circ.gwu.edu/~gprice/direct.htm). Another helpful site is Invisibleweb.com (http://www.invisibleweb.com).

If you would like more information about search engines, including links to specialized search tools, go to **Search Engine Watch,** at http://www.searchenginewatch.com.

Internet Lingo

The following abbreviations are sometimes used on the Internet documents and in e-mail.

BTW	By the way	**GOK**	God only knows	**OTOH**	On the other hand
F2F	Face to face; a personal meeting	**HHOK**	Ha, ha—only kidding	**PLS**	Please
FCOL	For crying out loud	**IMHO**	In my humble opinion	**ROTFL**	Rolling on the floor laughing
FWIW	For what it's worth	**IMO**	In my opinion	**TAFN**	That's all for now
GG	Got to go	**LOL**	Laughing out loud	**TTFN**	Ta-ta for now

Emoticons, or **smileys**, are a series of typed characters that, when turned sideways, resemble a face and express an emotion. Here are some smileys often encountered on the Internet.

:-)	Smile	:-D	Laugh	:-(	Unhappy	:-b..	Drooling
;-)	Wink	:-*	Kiss	:-o	Shouting	{*}	A hug and a kiss

WORLD ALMANAC EDITORS' PICKS

Most Useful Websites

The World Almanac staff ranked the following as the most useful websites:

1. **Google search engine**—http://www.google.com
2. **New York Times**—http://www.nytimes.com
3. **Yahoo search engine**—http://www.yahoo.com
4. **U.S. Census Bureau**—http://www.census.gov
5. **ESPN.com**—http://espn.go.com/main.html

Internet Directory to Selected Sites

The Websites listed are but a sampling of what is available. For some others, see the following *World Almanac* features: the Where to Get Help directory (Health), the Business Directory (Consumer Information), the Sports Directory, Travel and Tourism, Associations and Societies, Cities of the U.S., States of the U.S., U.S. Government, and Nations of the World. (The addresses are subject to change, and sites or products are not endorsed by *The World Almanac*.)

You must type an address exactly as written. You may be unable to connect to a site because (1) you have mistyped the address, (2) the site is busy, or (3) it has moved or no longer exists.

Online Service Providers

America Online
http://www.aol.com
AT&T WorldNet Service
http://www.att.net
CompuServe
http://www.compuserve.com
EarthLink
http://www.earthlink.net
Erols
http://www.erols.com
Microsoft Network
http://www.msn.com
Juno
http://www.juno.com
Prodigy
http://www.prodigy.com
WebTV Networks
http://www.webtv.com

Security and Screening

The National Fraud Information Center
http://www.fraud.org
SET Secure Electronic Transaction
http://www.setco.org

Directories

Bigfoot (e-mail addresses and white page listings)
http://www.bigfoot.com
InfoSpace, the Ultimate Directory
http://www.infospace.com
People Search
http://people.yahoo.com
Switchboard, the People and Business Directory
http://www.switchboard.com
WhoWhere?
http://www.whowhere.lycos.com

What's New

Internet Scout Project (latest resources for researchers)
http://scout.cs.wisc.edu/index.html
Nerd World: Media (what's new in computer world)
http://www.nerdworld.com/whatsnew.html
Netscape's What's New
http://home.netscape.com/netcenter/new.html
Yahoo! What's New (listing of every new site each day; sometimes thousands)
http://www.yahoo.com/new

News

The Associated Press
http://www.ap.org
BBC Online
http://www.bbc.co.uk/home/today
Cable News Network
http://www.cnn.com
The New York Times on the Web
http://www.nytimes.com
Reuters
http://www.reuters.com

Weather

National Weather Service Home Page
http://www.nws.noaa.gov
National Center for Environmental Prediction (includes links to Storm Prediction Center and other sites)
http://www.ncep.noaa.gov
Weather Channel
http://www.weather.com

Audio/Video

LiveUpdate
http://www.liveupdate.com
MP3.com
http://www.mp3.com
Real Networks
http://www.real.com

Bookstores

Amazon.com Inc.
http://www.amazon.com
Barnes and Noble
http://www.barnesandnoble.com
Borders.Com
http://www.borders.com
The Complete Guide to Online Bookstores
http://www.bookarea.com

Economic Data

Bureau of Economic Analysis
http://www.bea.doc.gov
Bureau of Labor Statistics
http://www.bls.gov

Economics Statistics Briefing Room
http://www.whitehouse.gov/fsbr/esbr.html
Economy at a Glance
http://stats.bls.gov/eag/eag.map.htm
Government Information Sharing Project
http://govinfo.kerr.orst.edu
Office of Management and Budget
http://www.access.gpo.gov/usbudget
Statistical Abstract of the United States (a sampling)
http://www.census.gov/statab/www
STAT-USA/Internet (a subscription-based government service)
http://www.stat-usa.gov/stat-usa.html

Your Money

Internal Revenue Service
http://www.irs.gov
Wall Street Journal
http://www.wsj.com
American Stock Exchange
http://www.amex.com
E*TRADE
http://www.etrade.com
MarketWatch
http://www.marketwatch.com
NASDAQ
http://www.nasdaq.com
New York Stock Exchange
http://www.nyse.com
Priceline (for offering a price for goods or services)
http://www.priceline.com
Mortgage Calculator
http://www.weichert.com
Retirement Calculator
http://www.worldi.com/index.htm

Job Search Sites

CareerBuilder
http://www.careerbuilder.com
Headhunter.net
http://www.Headhunter.net
Monster.com
http://www.monster.com

Auctions

eBay
http://www.ebay.com
Onsale
http://www.onsale.com
uBid Online Auction
http://www.ubid.com
Yahoo! Auctions
http://auctions.yahoo.com

Health

CenterWatch Clinical Trials Listing Service
http://www.centerwatch.com
drkoop.com
http://www.drkoop.com
Drugstore.com
http://www.drugstore.com
Healthfinder
http://www.healthfinder.gov
Mayo Clinic Health Oasis
http://www.mayohealth.org
Medscape
http://www.medscape.com
The Merck Manual
http://www.merck.com
National Institutes of Health (Health Information)
http://www.nih.gov/health
PlanetRx
http://www.planetrx.com
U.S. National Library of Medicine
http://www.nlm.nih.gov
WebMD
http://www.webmd.com

Electronic Greeting Cards

Blue Mountain Arts
http://www.bluemountain.com
Egreetings Network
http://www.egreetings.com
Electronic Postcards
http://www.electronicpostcards.net
Micro-Images Multimedia Greeting Cards
http://www.microimg.com/postcards
1001 Postcards
http://www.postcards.org
123 Greetings
http://www.123greetings.com

Chat Sites

America Online
http://www.aol.com/community/chat/allchats.html
Excite
http://www.excite.com/communities
The Globe
http://www.theglobe.com
IVILLAGE: The Women's Network
http://www.ivillage.com
Lycos
http://chat.lycos.com
Yahoo
http://chat.yahoo.com

Sites for Kids

Children's Television Workshop
http://www.sesameworkshop.org
Judy Blume's Home Base
http://www.judyblume.com/index.html
The Newbery Medal
http://www.ala.org/alsc/newbery.html
Peace Corps Kids World
http://www.peacecorps.gov/kids
Rock and Roll Hall of Fame and Museum
http://www.rockhall.com
Seussville
http://www.randomhouse.com/seussville
SuperSite for Kids
http://www.bonus.com
Weekly Reader
http://www.weeklyreader.com
White House for Kids
http://www.whitehouse.gov/kids
Yahooligans (for homework help sites)
http://www.yahooligans.com

Sports

Major League Baseball
http://www.majorleaguebaseball.com
Major League Soccer
http://www.mlsnet.com
National Basketball Association
http://www.nba.com
Women's National Basketball Association
http://www.wnba.com
National Football League
http://www.nfl.com
National Hockey League
http://www.nhl.com
Special Olympics
http://www.specialolympics.org

Genealogy

FamilySearch
http://www.familysearch.org
Genealogy.com
http://www.genealogy.com
National Archives and Records Administration
http://www.nara.gov
RootsWeb
http://www.rootsweb.com
USGenWeb Project
http://www.usgenweb.org

Resources for Families

Babies Online
http://www.babiesonline.com
BabyCenter
http://www.babycenter.com
Family.Com
http://family.go.com
Family Internet
http://www.familyinternet.com
KidsHealth.org
http://www.kidshealth.org
KidSource Online
http://www.kidsource.com
ParenthoodWeb
http://www.parenthoodweb.com
Parent Soup
http://www.parentsoup.com
ParentsPlace.com
http://www.parentsplace.com
Screen It! Entertainment Reviews for Parents
http://www.screenit.com
Zero to Three
http://www.zerotothree.org

Reference

About.com
http://www.about.com
BookWire
http://www.bookwire.com
CIA Publications and Reports
http://www.odci.gov/cia/publications/pubs.html
Explore the Internet; The Library of Congress
http://lcweb.loc.gov/global
Libweb—Library Servers via WWW
http://sunsite.berkeley.edu/Libweb
Miriam-Webster Network Editions
http://www.m-w.com
yourDictionary.com
http://www.yourdictionary.com
Refdesk
http://www.refdesk.com
Roget's Thesaurus
http://www.thesaurus.com

Most-Visited Websites, July 2001

Source: Media Metrix, Inc.

Rank	Website	Visitors[1]
1.	http://www.yahoo.com*	62,179,720
2.	http://www.msn.com	53,212,260
3.	http://www.aol.com	42,779,390
4.	http://www.x10.com	39,566,600
5.	http://www.microsoft.com	36,801,950
6.	http://www.passport.com	31,854,120
7.	http://www.hotmail.com*	28,655,850
8.	http://www.lycos.com*	27,358,680
9.	http://www.amazon.com	20,480,760
10.	http://www.go.com	19,553,300
11.	http://www.ebay.com	19,095,770
12.	http://www.netscape.com	17,538,400
13.	http://www.google.com	15,153,820
14.	http://www.fastclick.net	15,115,310
15.	http://www.excite.com*	14,635,500
16.	http://www.about.com	14,517,150
17.	http://www.real.com	13,413,350
18.	http://www.mapquest.com	13,116,110
19.	http://www.tripod.com	13,057,410
20.	http://www.cnet.com*	12,056,650

*Represents an aggregation of commonly owned domain names. (1) Number of visitors who visited website at least once in July 2001, according to a Media Metrix sample.

Percent of U.S. Households With Internet Access, by Selected Characteristics, 1998-2000

Source: National Telecommunications and Information Administration, U.S. Dept. of Commerce

	2000	1998	'98-'00 increase
TOTAL	**41.5**	**26.2**	**15.3**
Race			
White, not Hispanic	46.1	29.8	16.3
Black, not Hispanic	23.5	11.2	12.3
Hispanic	23.6	12.6	11.0
Asian, Pacific Islander	56.8	36.0	20.8
Age group			
3-5 years	15.3	11.0	4.3
9-17 years	53.4	43.0	10.3
18-24 years	56.8	44.3	12.5
25-49 years	55.4	40.9	14.5
50+ years	29.6	19.3	10.3
Educational attainment			
Some high school	11.7	5.0	6.7
High school diploma or GED	29.9	16.3	13.6
Some college	49.0	30.2	18.8
Bachelor's degree or more	64.0	46.8	17.2

	2000	1998	'98-'00 increase
Household type			
Married couple with children under 18	60.6	39.3	21.3
Male householder with children under 18	35.7	19.5	16.2
Female householder w. children under 18	30.0	15.0	15.0
Family households without children	43.2	27.2	16.0
Nonfamily households	28.1	17.5	10.6
Location			
Urban	42.3	27.5	14.8
Central city	37.7	24.5	13.2
Rural	38.9	22.2	16.7
Annual income			
$15,000 or less	12.7	7.1	5.6
$15,000-24,999	21.3	11.0	10.3
$25,000-34,999	34.0	19.1	14.9
$35,000-49,999	46.2	29.5	16.7
$50,000-74,999	60.9	43.9	17.0
$75,000+	77.7	60.3	17.4

Percent of U.S. Households With a Computer, by Selected Characteristics, 1998-2000

Source: National Telecommunications and Information Administration, U.S. Dept. of Commerce

	2000	1998	'98-'00 increase
TOTAL	**51.0**	**42.1**	**8.9**
Race			
White, not Hispanic	55.7	46.6	9.1
Black, not Hispanic	32.6	23.2	9.4
Hispanic	33.7	25.5	8.2
Asian, Pacific Islander	65.6	55.0	10.6
Educational attainment			
Less than high school	18.2	12.5	5.7
High school diploma or GED	39.6	31.2	8.4
Some college	60.3	49.3	11.3
College graduate	74.0	66.9	7.1

	2000	1998	'98-'00 increase
Annual income			
$15,000 or less	19.2	14.5	4.7
$15,000-24,999	30.1	23.7	6.4
$25,000-34,999	44.6	35.8	8.8
$35,000-49,999	58.6	50.2	8.4
$50,000-74,999	73.2	66.3	6.9
$75,000+	86.3	79.9	6.4
Location			
Urban	51.5	42.9	8.6
Central city	53.7	38.5	15.2
Rural	50.4	39.9	10.5

IT'S A FACT: E-mail is the Internet's most widespread application—about 80% of Internet users use e-mail. The fastest growing Internet activities in 2000 were shopping and bill-paying.

Top-Selling Software, 2001

Source: PC Data, Reston, VA

(based on average U.S. sales, Jan.-June 2001)

All Software

1. TurboTax Deluxe, Intuit
2. TurboTax, Intuit
3. Taxcut 2000 Deluxe, Block Financial
4. Norton Antivirus 7.0, Symantec
5. TurboTax 45 State, Intuit
6. Taxcut 2000, Block Financial
7. The Sims, Electronic Arts
8. VirusScan 5.0, Network Associates
9. The Sims Livin Large Expansion Pack, Electronic Arts
10. Quicken, Intuit
11. The Sims: House Party Expansion Pack, Electronic Arts
12. MP Roller Coaster Tycoon, Infogrames Entertainment
13. Norton System Works 2001 4.0, Symantec
14. Quicken Deluxe, Intuit
15. Black & White, Electronic Arts
16. TurboTax State CA, Intuit
17. Taxcut 2000 All State, Block Financial
18. Symantec Antivirus Solution 7.5 Mnt PVP Lic, Symantec
19. Frogger JC, Infogrames Entertainment
20. QuickBooks 2001 Pro, Intuit

Games

1. The Sims, Electronic Arts
2. The Sims Livin' Large Expansion Pack, Electronic Arts
3. The Sims: House Party Expansion Pack, Electronic Arts
4. MP Roller Coaster Tycoon, Infogrames Entertainment
5. Black & White, Electronic Arts
6. Frogger JC, Infogrames Entertainment
7. Diablo 2, Vivendi Universal Publishing
8. Sim Theme Park, Electronic Arts
9. MS Age Of Empires II: Age of Kings, Microsoft
10. Diablo 2 Expansion Set: Lord of Destruction; Vivendi

Reference Software

1. MS Encarta Reference Suite, Microsoft
2. MS Encarta Encyclopedia 2001 Deluxe, Microsoft
3. Grolier Multimedia Encyclopedia 2001 Deluxe, Grolier
4. MS Encarta Encyclopedia 2001, Microsoft
5. Webster's 2001 Power Pack, Countertop/Topics Entertainment

Home Education Software

1. Instant Immersion Spanish, Countertop/Topics Entertainment
2. Blue's ABC Time Activities JC, Humongous (Infogrames)
3. Jumpstart Learning Games Tray Pack, Vivendi Universal Publishing
4. Clifford's Thinking Adventure, Scholastic
5. Mickey's Preschool Ages 2-4, Disney
6. Adventure Workshop 1st-3rd Grade, The Learning Company
7. Clifford's Reading Adventure, Scholastic
8. Mavis Beacon Teaches Typing 11.0, The Learning Company
9. Mickey's Kindergarten Ages 4-6, Disney
10. Mickey's Toddler, Disney

Personal Productivity Software

1. MS Expedia Streets/Trip Planner, Microsoft
2. Easy CD Creator 5.0 Platinum, Roxio
3. MS Picture It Publishing Platinum, Microsoft
4. Print Perfect Gold JC, Cosmi
5. Print Shop Deluxe, The Learning Company
6. Photosuite 4.0, MGI Software
7. Hallmark Card Studio 2, Vivendi Universal Publishing
8. Hallmark Card Studio 2 Deluxe, Vivendi Universal Publ.
9. Bible Library JC, Valusoft
10. MS Works Suite, Microsoft

Business Software

1. Norton Antivirus 7.0, Symantec
2. VirusScan 5.0, Network Associates
3. Norton System Works 2001 4.0, Symantec
4. Symantec Antivirus Solution 7.5 Mnt PVP Lic, Symantec
5. MS Windows 2000 Svr Clnt Acc OPEN Lic, Microsoft
6. MS Windows ME Upgr, Microsoft
7. MS Windows 98 2nd Ed Upgr, Microsoft
8. Norton Antivirus 7.5 Corporate Ed PVP Lic, Symantec
9. MS Exchange Clnt Acc OPEN Lic, Microsoft
10. Symantec Antivirus Solution 7.5 PVP Lic, Symantec

Glossary of Computer and Internet Terms

Source: *Microsoft Press® Computer Dictionary, Third Edition* with updates. Copyright 1997, 1998, 1999, 2000, 2001 by Microsoft Press. Reproduced by permission of Microsoft Press. All rights reserved.

application A program designed to assist in the performance of a specific task, such as word processing, accounting, or inventory management.

artificial intelligence (AI) The branch of computer science concerned with enabling computers to simulate such aspects of human intelligence as speech recognition, deduction, inference, creative response, and the ability to learn from experience.

ASCII Pronounced "askee." An acronym for American Standard Code for Information Interchange, a coding scheme using 7 or 8 bits that assigns numeric values to up to 256 characters, including letters, numerals, punctuation marks, control characters, and other symbols.

backup (noun); back up (verb) As a noun, a duplicate copy of a program, a disk, or data. As a verb, to make a duplicate copy of a program, a disk, or data.

bandwidth Data transfer capacity of a digital communications system.

baud rate Speed at which a modem can transmit data.

BBS An abbreviation for bulletin board system, a computer system equipped with one or more modems or other means of network access that serves as an information and message-passing center for remote users.

binary The binary number system has 2 as its base, so values are expressed as combinations of two digits, 0 and 1. These two digits can represent the logical values true and false as well as numerals, and they can be represented in an electronic device by the two states on and off, recognized as two voltage levels. Therefore, the binary number system is at the heart of digital computing.

bit Short for binary digit; the smallest unit of information handled by a computer. One bit expresses a 1 or a 0 in a binary numeral, or a true or a false logical condition, and is represented physically by an element such as a high or low voltage at one point in a circuit or a small spot on a disk magnetized one way or the other.

boot The process of starting or resetting a computer.

browser *See* **Web browser.**

bug An error in coding or logic that causes a program to malfunction or to produce incorrect results. Also, a recurring physical problem that prevents a system or set of components from working together properly.

bulletin board system *See* **BBS.**

byte A unit of data, today almost always consisting of 8 bits. A byte can represent a single character, such as a letter, a digit, or a punctuation mark.

CD-ROM Acronym for compact disc read-only memory, a form of storage characterized by high capacity (roughly 650 megabytes) and the use of laser optics rather than magnetic means for reading data.

central processing unit (CPU) The computational and control unit of a computer; the device that interprets and executes instructions.

certificate authority An issuer of digital certificates, the cyberspace equivalent of ID cards.

chat room The informal term for a data communication channel that links computers and permits users to "converse", often about a particular subject that interests them, by sending text messages to one another in real time.

chip *See* **integrated circuit.**

client On a local area network, a computer that accesses shared network resources provided by another computer (called a server). *See also* **server.**

computer Any machine that does three things: accepts structured input, processes it according to prescribed rules, and produces the results as output.

cookie A block of data that a Web server stores on a client system. When a user returns to the same Web site, the browser sends a copy of the cookie back to the server. Cookies are used to identify users, to instruct the server to send a customized version of the requested Web page, to submit account information for the user, and for other administrative purposes.

CPU *See* **central processing unit.**

crash The failure of either a program or a disk drive. A program crash results in the loss of all unsaved data and can leave the operating system unstable enough to require restarting the computer.

cursor A special on-screen indicator, such as a blinking underline or rectangle, that marks the place of which keystrokes will appear when typed.

cyberspace The universe of environments, such as the Internet, in which persons interact by means of connected computers.

cyberspeak Terminology and language (often jargon, slang, and acronyms) relating to the Internet—computer-connected—environment, that is, cyberspace. Most words prefixed by *cyber-* have the same meaning as their "real-world" counterparts, but specifically indicate their use in the online culture of the Internet and the World Wide Web. Examples: cybercafé, cybercash.

database A file composed of records, each of which contains fields, together with a set of operations for searching, sorting, recombining, and other functions.

data compression A means of reducing the space or bandwidth needed to store or transmit a block of data.

debug To detect, locate, and correct logical or syntactical errors in a program or malfunctions in hardware.

defragger A software utility for reuniting parts of a file that have become fragmented through rewriting and updating.

desktop publishing The use of a computer and specialized software to combine text and graphics to create a document that can be printed on either a laser printer or a typesetting machine.

dial-up access Connection to a data communications network through the public switched telecommunication network.

digerati Cyberspace populace that can be roughly compared to *literati.* Digerati are renowned as or claiming to be knowledgeable about topics and issues related to the digital revolution; more specifically, they are "in the know" about the Internet and online activities.

digital certificate 1. An assurance that software downloaded from the Internet comes from a reputable source. 2. A user identity card or "driver's license" for cyberspace. Issued by a certificate authority.

digital video disc The next generation of optical disc storage technology. With digital video disc technology video, audio, and computer data can be encoded onto a compact disc (CD). A digital video disc can store greater amounts of data than traditional CDs. A standard single-layered, single-sided digital video disc can store 4.7 GB of data; a two-layered standard enhances the single-sided layer to 8.5GB. Digital video discs can be double-sided with a maximum storage of 17 GB per disc. A digital video disc player is needed to read digital video discs. This player is equipped to read older optical storage technologies. Advocates of digital video disc technology intend to replace current digital storage formats, such as laserdisc, CD-ROM, and audio CDs, with the single digital format of digital video disc.

directory service A service on a network that returns mail addresses of other users or enables a user to locate hosts and services.

disk A round, flat piece of flexible plastic (floppy disk) or inflexible metal (hard disk) coated with a magnetic material that can be electrically influenced to hold information recorded in digital (binary) format.

disk drive An electromechanical device that reads from and writes to disks.

disk operating system Abbreviated DOS. A generic term describing any operating system that is loaded from disk devices when the system is started or rebooted.

distance learning Broadly, any educational or learning process or system in which the teacher/instructor is separated geographically or in time from his or her students; or in which students are separated from other students or educational resources.

DOS *See* **disk operating system.**

download In communications, to transfer a copy of a file from a remote computer to the requesting computer by means of a modem or network. *See also* **upload.**

DVD *See* **digital video disc.**

dynamic HTML A technology designed to add richness, interactivity, and graphical interest to Web pages by providing those pages with the ability to change and update themselves in response to user actions, without the need for repeated downloads from a server.

encryption The process of encoding data to prevent unauthorized access, especially during transmission. The U.S. National Bureau of Standards created a complex encryption standard, DES (Data Encryption Standard), that provides almost unlimited ways to encrypt documents.

FAQ An abbreviation for Frequently Asked Questions, a document listing common questions and answers on a particular subject. FAQs are often posted on Internet newsgroups where new participants ask the same questions that regular readers have answered many times.

field A location in a record in which a particular type of data is stored.

file A complete, named collection of information, such as a program, a set of data used by a program, or a user-created document.

firewall A security system intended to protect an organization's network against external threats, such as hackers, from another network. *See also* **proxy server.**

flame An abusive or personally insulting e-mail message or newsgroup posting.

format In general, the structure or appearance of a unit of data. As a verb, to change the appearance of selected text or the contents of a selected cell in a spreadsheet.

forum A medium provided by an online service or BBS for users to carry on written discussions of a topic by posting messages and replying to them.

FTP An abbreviation for File Transfer Protocol, the protocol used for copying files to and from remote computer systems on a network using TCP/IP such as the Internet.

gigabyte Abbreviated GB; 1024 megabytes. *See* **megabyte.**

graphical user interface Abbreviated GUI (pronounced "gooey"). A type of environment that represents programs, files, and options by means of icons, menus, and dialog boxes on the screen. The user can select and activate these options by pointing and clicking with a mouse or, often, with the keyboard. *See also* **icon.**

hacker A computerphile—a person who is engrossed in computer technology and programming or who likes to examine the code of operating systems and other programs to see how they work. Also, a person who uses computer expertise for illicit ends, such as for gaining access to computer systems without permission and tampering with programs and data.

hard copy Printed output on paper, film, or other permanent medium.

hit Retrieval of a document, such as a home page, from a website.

home page A document intended to serve as a starting point in a hypertext system, especially the World Wide Web. Also, an entry page for a set of Web pages and other files in a website.

host The main computer in a system of computers or terminals connected by communications links.

HTML An abbreviation for HyperText Markup Language, the markup language used for documents on the World Wide Web.

HTTP An abbreviation for HyperText Transfer Protocol, the client/server protocol used to access information on the Web.

hyperlink A connection between an element in a hypertext document, such as a word, phrase, symbol, or image, and a different element in the document, another hypertext document, a file, or a script. The user activates the link by clicking on the linked element, which is usually highlighted in some way.

hypermedia The integration of any combination of text, graphics, sound, and video into a primarily associative system of information storage and retrieval in which users jump from subject to related subject.

hypertext Text linked together in a complex, nonsequential web of associations in which the user can browse through related topics.

icon A small image displayed on the screen to represent an object that can be manipulated by the user.

import To bring information from one system or program into another.

instant messaging A service that alerts users when friends or colleagues are on line and allows them to communicate with each other in real time through private online chat areas.

integrated circuit Also called a chip. A device consisting of a number of connected circuit elements, such as transistors and resistors, fabricated on a single chip of silicon crystal or other semiconductor material.

interactive Characterized by conversational exchange of input and output, as when a user enters a question or command the system immediately responds.

Internet The worldwide collection of networks and gateways that use the TCP/IP suite of protocols to communicate with each other. At the heart of the Internet is a backbone of high-speed data communication lines between major nodes or host computers, consisting of thousands of commercial, government, educational, and other computer systems, that route data and messages.

intranet A TCP/IP network designed for information processing within a company or organization. It usually employs Web pages for information dissemination and Internet applications, such as Web browsers.

IP address Short for Internet Protocol address, a 32-bit (4-byte) binary number that uniquely identifies a host (computer) connected to the Internet to other Internet hosts, for communication through the transfer of packets.

Java A programming language, developed by Sun Microsystems, Inc., that can be run on any platform.

kilobyte Abbreviated K, KB, or Kbyte; 1,024 bytes.

LAN Rhymes with "can." Acronym for local area network, a group of computers and other devices dispersed over a limited area and connected by a link that enables any device to interact with any other on the network.

manipulated by either optical character recognition software

laptop A small, portable computer that runs on either batteries or AC power, designed for use during travel. Laptops have flat screens and small keyboards. Some weigh as little as 5 pounds.

legacy system A computer, software program, network, or other computer equipment that remains in use after a business or organization installs new systems.

link *See* **hyperlink.**

Linux A version of the UNIX System V Release 3.0 kernel developed for PCs with 80386 and higher microprocessors. Developed by Linus Torvalds of Sweden (for whom it is named) along with numerous collaborators worldwide, Linux is distributed free with source code through BBSs and the Internet, although some companies distribute it as part of a commercial package with Linux-compatible utilites. The Linux kernel works with the GNU utilities developed by the Free Software Foundation, which did not produce a kernel.

local area network *See* **LAN.**

logon The process of identifying oneself to a computer after connecting to it over a communications line. Also called *login.*

lurk To receive and read articles or messages in a newsgroup or other online conference without contributing anything to the ongoing conversation.

mailing list A list of names and e-mail addresses that are grouped under a single name. When a user places the name of the mailing list in a mail client's To: field, the client automatically sends the same message to the machine where the mailing list resides, and that machine sends the message to all the addresses on the list.

mainframe computer A high-level computer designed for the most intensive computational tasks.

markup language A set of codes in a text file that instruct a printer or video display how to format, index, and link the contents of the file. Examples of markup languages are HyperText Markup Language (HTML), which is used in Web pages, and Standard Generalized Markup Language (SGML), which is used for typesetting and desktop publishing purposes and in electronic documents.

megabyte Abbreviated MB. Usually 1,048,576 bytes (2^{20}); sometimes interpreted as 1 million bytes.

meltdown The complete collapse of a computer network caused by a high level of traffic.

memory Circuitry that allows information to be stored and retrieved. In common usage it refers to the fast semiconductor storage (RAM) directly connected to the processor. *See also* **RAM.**

menu A list of options from which a program user can make a selection in order to perform a desired action, such as choosing a command or applying a format.

microcomputer A computer built around a single-chip microprocessor.

microprocessor A central processing unit (CPU) on a single chip. *See also* **integrated circuit.**

minicomputer A mid-level computer built to perform complex computations while dealing efficiently with input and output from users connected via terminals.

modem A communications device that enables a computer to transmit information over a standard telephone line.

monitor The device on which images generated by the computer's video adapter are displayed.

motherboard The main circuit board containing the primary components of a computer system.

mouse A common pointing device. It has a flat-bottomed casing designed to be gripped by one hand.

multimedia The combination of sound, graphics, animation, and video. In the world of computers, multimedia is a subset of hypermedia, which combines the aforementioned elements with hypertext.

multitasking A mode of operation offered by an operating system in which a computer works on more than one task at a time.

Net Short for Internet.

netiquette Short for network etiquette.

netizen A person who participates in online communication through the Internet and other networks, especially conference and chat services.

network A group of computers and associated devices that are connected by communications facilities.

newsgroup A forum on the Internet for threaded discussions on a specified range of subjects. A newsgroup consists of articles and follow-up posts. *See* **post, thread.**

online Activated and ready for operating; capable of communicating with or being controlled by a computer.

operating system The software that controls the allocation and usage of hardware resources such as memory, CPU time, disk space, and peripheral devices.

optical scanner An input device that uses light-sensing equipment to scan paper or another medium, translating the pattern of light and dark or color into a digital signal that can be or graphics software.

packet A unit of information transmitted as a whole from one device to another on a network.

palmtop A portable personal computer whose size enables it to be held in one hand while it is operated with the other hand. A major difference between palmtop computers and laptop computers is that palmtops are usually powered by off-the-shelf batteries such as AA cells. Palmtop computers typically do not have disk drives; rather their programs are stored in ROM and are loaded into RAM when they are switched on. More recent palmtop computers are equipped with PCMCIA slots to provide wider flexibility and greater capability.

password A unique string of characters that a user types in as an identification code.

PC Abbreviation for personal computer, a microcomputer that conforms to the standard developed by IBM for personal computers, which uses an Intel microprocessor (or one that is compatible); also used as a general term for any microcomputer.

PDA Acronym for Personal Digital Assistant. A lightweight palmtop computer designed to provide specific functions like personal organization (calendar, note taking, database, calculator, and so on) as well as communications. More advanced models also offer multimedia features. Many PDA devices rely on a pen or other pointing device for input instead of a keyboard or mouse, although some offer a keyboard too small for touch typing to use in conjunction with pen or pointing device. For data storage, a PDA relies on flash memory instead of power-hungry disk drives.

peripheral A device, such as a disk drive, printer, modem, or joystick, that is connected to a computer and is controlled by the computer's microprocessor.

personal computer *See* **PC**.

pixel Short for picture element; also called *pel.* One spot in a rectilinear grid of thousands of such spots that are individually "painted" to form an image produced on the screen by a computer or on paper by a printer.

portal A website that serves as a gateway to the Internet. A portal is a collection of links, content, and services designed to guide users to information they are likely to find interesting—news, weather, entertainment, commerce sites, chat rooms, and so on.

post To submit an article in a newsgroup or other online conference. *See* **thread.**

program A sequence of instructions that can be executed by a computer.

protocol A set of rules or standards designed to enable computers to communicate with one another and to exchange information with as little error as possible.

proxy server A firewall component that manages Internet traffic to and from a local area network and can provide other features, e.g., document caching and access control.

RAM Pronounced "ram." An acronym for random access memory. Semiconductor-based memory that can be read and written by the CPU or other hardware devices.

ROM 1. Acronym for read-only-memory. A semiconductor circuit into which code or data is permanently installed by the manufacturing process. 2. Any semiconductor circuit serving as a memory that contains instructions or data that can be read but not modified.

routing table In data communications, a table of information that provides network hardware (bridges and routers) with the directions needed to forward packets of data to locations on other networks.

RTF An acronym for rich text format. RTF is used for transferring formatted documents between applications, even those applications running on different platforms, such as between IBM and compatibles and Apple Macintoshes.

search engine On the Internet, a program that searches for keywords in files and documents.

server On a local area network (LAN), a computer running software that controls access to the network and its resources, such as printers and disk drives. On the Internet or other network, a computer or program that responds to commands from a client. *See* **client, LAN.**

sleep mode A power management mode that shuts down all unnecessary computer operations to save energy; also known as suspend mode.

snail mail A phrase popular on the Internet for referring to mail services provided by the United States Postal Service and similar agencies in other countries.

software Computer programs; instructions that make hardware work.

spam An unsolicited e-mail message sent to many recipients at one time, or a news article posted simultaneously to many newsgroups. Electronic junk mail.

spreadsheet program An application commonly used for budgets, forecasting, and other finance-related tasks that organizes data values using cells, where the relationships between cells are defined by formulas.

supercomputer A large, extremely fast, and expensive computer used for complex or sophisticated calculations.

surf To browse among collections of information on the Internet, in newsgroups, and especially the World Wide Web.

system administrator The person responsible for administering use of a multiuser computer system, communications system, or both.

TCP/IP An abbreviation for Transmission Control Protocol/Internet Protocol, a protocol developed by the Department of Defense for communications between computers. It has become the de facto standard for data transmission over networks, including the Internet.

technophobe A person who is afraid of or dislikes technological advances, especially computers.

telecommute To work in one location (often, at home) and communicate with a main office at a different location through a personal computer.

teleconferencing The use of audio, video, or computer equipment linked through a communications system to enable geographically separated individuals to participate in a meeting or discussion.

teleworker A businessperson who substitutes information technologies for work-related travel. Teleworkers include home-based and small business workers who use computer and communications technologies to interact with customers and/or colleagues.

thread In electronic mail and Internet newsgroups, a series of messages and replies related to a specific topic.

upload In communications, the process of transferring a copy of a file from a local computer to a remote computer by means of a modem or network.

URL An abbreviation for Uniform Resource Locator, an address for a resource on the Internet.

Usenet A worldwide network of Unix systems that has a decentralized administration and is used as a bulletin board system by special-interest discussion groups.

user interface The portion of a program with which a user interacts.

user-friendly Easy to learn and easy to use.

virus An intrusive program that infects computer files by inserting in those files copies of itself.

voice recognition The capability of a computer to understand the spoken word for the purpose of receiving commands and data input from the speaker.

WAN *See* **wide area network**

Web *See* **World Wide Web.**

Web browser A client application that enables a user to view HTML documents, follow the hyperlinks among them, transfer files, and execute some programs.

webcasting Popular term for broadcasting information via the World Wide Web, using push and pull technologies to move selected information from a server to a client.

webmaster The person or persons responsible for creating and maintaining a site on the World Wide Web.

website A group of related HTML documents and associated files, scripts, and databases that is served up by an HTTP server on the World Wide Web.

WebTV® Trademark name for technology from Microsoft and WebTV Networks that provide consumers with the ability to access the Internet on a television by means of a set-top box equipped with a modem.

wide area network (WAN) A communications network that connects geographically separated areas.

window In applications and graphical interfaces, a portion of the screen that can contain its own document or message.

word processor A program for manipulating text-based documents; the electronic equivalent of paper, pen, typewriter, eraser, and, most likely, dictionary and thesaurus.

workstation A combination of input, output, and computing hardware used for work by an individual.

World Wide Web (WWW) The total set of interlinked hypertext documents residing on Web, or HTTP, servers all around the world.

WYSIWYG Pronounced "wizzywig." An acronym for "What you see is what you get." A display method that shows documents and graphics characters on the screen as they will appear when printed.

Zip drive A disk drive developed by Iomega that uses 3.5-inch removable disks (Zip disks) capable of storing 100 megabytes of data apiece. *See also* **disk drive.**

TELECOMMUNICATIONS

International Telecommunications Market Data

Source: International Telecommunication Union

	Total market revenue (billions of U.S. $)[1]	International phone traffic (billions of minutes)[2]	Main telephone lines (millions)	Mobile cellular subscribers (millions)		Total market revenue (billions of U.S. $)[1]	International phone traffic (billions of minutes)[2]	Main telephone lines (millions)	Mobile cellular subscribers (millions)
1990	$508	33	520	11	1996	885	71	740	145
1991	539	38	546	16	1997	939	80	794	214
1992	593	43	574	23	1998	1,004	90	848	319
1993	630	48	606	34	1999	1,082	100	906	471
1994	691	56	645	55	2000	1,160	110	970	650
1995	797	62	692	91	2002[3]	1,300	130	1,115	1,000

(1) Revenue from installation, subscription, and local, trunk, and international call charges. (2) From 1994 including traffic between countries of the former Soviet Union. (3) Projected.

Worldwide Use of Cellular Telephones, 2000

Source: International Telecommunication Union; estimated

Country/Region	Number of subscribers (millions)	% of pop.	Country/Region	Number of subscribers (millions)	% of pop.	Country/Region	Number of subscribers (millions)	% of pop.
Luxembourg	0.4	87.2	Ireland	2.5	66.8	Japan	66.8	52.6
Taiwan	17.9	80.3	Portugal	6.7	66.5	France	29.1	49.4
Hong Kong	5.5	80.2	Switzerland	4.6	64.5	Australia	8.6	44.6
Austria	6.5	78.6	Spain	24.7	61.0	Czech Republic	4.3	42.4
Italy	42.2	73.7	Denmark	3.3	61.0	New Zealand	1.5	40.3
Finland	3.8	72.6	Germany	48.1	58.6	Estonia	0.6	38.7
Sweden	6.3	71.4	United Arab Emirates	1.4	58.5	**U.S.**	**100.3**	**36.5**
Israel	4.4	70.2	South Korea	26.8	56.7	Cyprus	0.2	32.1
Singapore	2.8	68.4	Greece	6.0	55.9	Andorra	0.02	30.2
Netherlands	10.7	67.1	Belgium	5.6	54.9	Bahrain	2.1	30.1
United Kingdom	40.0	67.0	Slovenia	1.1	54.7	Seychelles	0.02	30.1
Iceland	0.2	67.0						

IT'S A FACT: An estimated 12 out of every 100 people in the world (36 out of every 100 people in Europe, 21 out of every 100 in the Americas) are cell phone subscribers.

U.S. Cellular Telephone Subscribership, 1985–2000

Source: The CTIA Semi-Annual Wireless Survey. Used with permission of CTI; in thousands of subscribers in December[1]

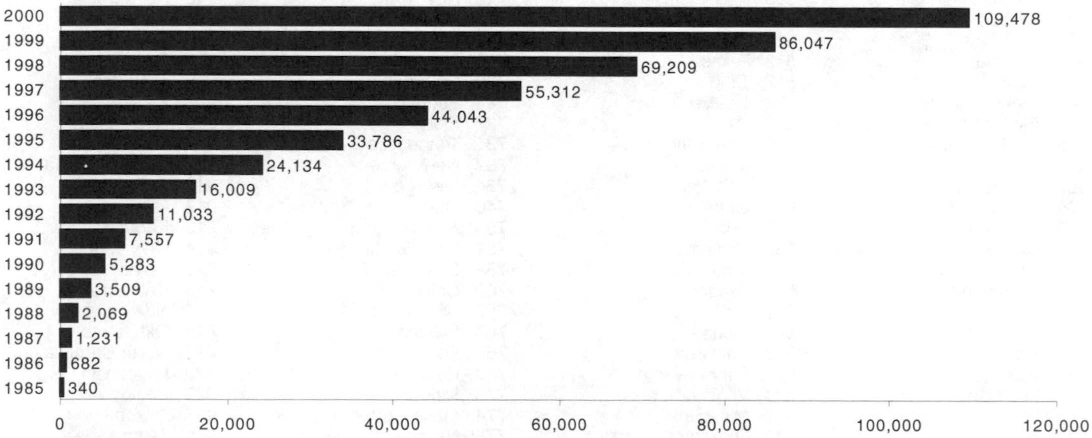

(1) Data may differ slightly from other sources.

U.S. Sales and Household Penetration, Selected Telecommunication Products[1]

Source: eBrain Market Research

	2000		1995		1990		1985	
	Sales[2]	% of all households	Sales	% of all households	Sales	% of all households	Sales	% of all households
Cordless telephones	$1,307	80	$1,141	55	$842	28	$280	11
Pagers	750	40	300	11	118	1	—	—
Modems/fax modems	1,564	55	770	16	191	2.7	10	0
Telephone answering devices	984	75	1,077	57	827	35	325	7
Cellular phones	3,228	60	2,574	29	1,098	5	116	0.1

(1) Data may differ slightly from other sources. (2) In millions of dollars.

Telephone Area Codes, by Number

As of Sept. 2001. For area codes listed by place, see pages 389–422.

Area Code	Location or Service
201	New Jersey
202	District of Columbia
203	Connecticut
204	Manitoba
205	Alabama
206	Washington
207	Maine
208	Idaho
209	California
210	Texas
212	New York
213	California
214	Texas
215	Pennsylvania
216	Ohio
217	Illinois
218	Minnesota
219	Indiana
225	Louisiana
228	Mississippi
229	Georgia
231	Michigan
234	Ohio
240	Maryland
242	Bahamas
246	Barbados
248	Michigan
250	British Columbia
251	Alabama
252	North Carolina
253	Washington
254	Texas
256	Alabama
262	Wisconsin
264	Anguilla
267	Pennsylvania
268	Antigua/Barbuda
270	Kentucky
281	Texas
284	British Virgin Islands
289	Ontario
301	Maryland
302	Delaware
303	Colorado
304	West Virginia
305	Florida
306	Saskatchewan
307	Wyoming
308	Nebraska
309	Illinois
310	California
311	Non-Emergency Access
312	Illinois
313	Michigan
314	Missouri
315	New York
316	Kansas
317	Indiana
318	Louisiana
319	Iowa
320	Minnesota
321	Florida
323	California
330	Ohio
334	Alabama
336	North Carolina
337	Louisiana
339	Massachusetts
340	U.S. Virgin Islands
345	Cayman Islands
347	New York
351	Massachusetts
352	Florida
360	Washington
361	Texas
386	Florida
401	Rhode Island
402	Nebraska
403	Alberta
404	Georgia
405	Oklahoma
406	Montana
407	Florida
408	California
409	Texas
410	Maryland
411	Directory Assistance
412	Pennsylvania
413	Massachusetts
414	Wisconsin
415	California
416	Ontario
417	Missouri
418	Quebec
419	Ohio
423	Tennessee
425	Washington
434	Virginia
435	Utah
440	Ohio
441	Bermuda
443	Maryland
450	Quebec
456	Inbound International
469	Texas
473	Grenada
478	Georgia
480	Arizona
484	Pennyslvania
500	Personal Comm. Serv.
501	Arkansas
502	Kentucky
503	Oregon
504	Louisiana
505	New Mexico
506	New Brunswick
507	Minnesota
508	Massachusetts
509	Washington
510	California
512	Texas
513	Ohio
514	Quebec
515	Iowa
516	New York
517	Michigan
518	New York
519	Ontario
520	Arizona
530	California
540	Virginia
541	Oregon
559	California
561	Florida
562	California
570	Pennsylvania
571	Virginia
573	Missouri
580	Oklahoma
600	(Canadian Services)
601	Mississippi
602	Arizona
603	New Hampshire
604	British Columbia
605	South Dakota
606	Kentucky
607	New York
608	Wisconsin
609	New Jersey
610	Pennsylvania
611	Repair Service
612	Minnesota
613	Ontario
614	Ohio
615	Tennessee
616	Michigan
617	Massachusetts
618	Illinois
619	California
620	Kansas
623	Arizona
626	California
630	Illinois
631	New York
636	Missouri
641	Iowa
646	New York
647	Ontario
649	Turks & Caicos Islands
650	California
651	Minnesota
660	Missouri
661	California
662	Mississippi
664	Montserrat
670	N. Mariana Islands
671	Guam
678	Georgia
682	Texas
700	IC Services
701	North Dakota
702	Nevada
703	Virginia
704	North Carolina
705	Ontario
706	Georgia
707	California
708	Illinois
709	Newfoundland
710	U.S. Government
711	TRS Access
712	Iowa
713	Texas
714	California
715	Wisconsin
716	New York
717	Pennsylvania
718	New York
719	Colorado
720	Colorado
724	Pennsylvania
727	Florida
731	Tennessee
732	New Jersey
734	Michigan
740	Ohio
754	Florida
757	Virginia
758	St. Lucia
760	California
763	Minnesota
765	Indiana
767	Dominica
770	Georgia
773	Illinois
774	Massachusetts
775	Nevada
780	Alberta
781	Massachusetts
784	St. Vincent & Gren.
785	Kansas
786	Florida
787	Puerto Rico
800	Toll-Free Service
801	Utah
802	Vermont
803	South Carolina
804	Virginia
805	California
806	Texas
807	Ontario
808	Hawaii
809	Dominican Republic
810	Michigan
811	Business Office
812	Indiana
813	Florida
814	Pennsylvania
815	Illinois
816	Missouri
817	Texas
818	California
819	Quebec
828	North Carolina
830	Texas
831	California
832	Texas
843	South Carolina
845	New York
847	Illinois
850	Florida
855	Toll-Free Service
856	New Jersey
857	Massachusetts
858	California
859	Kentucky
860	Connecticut
863	Florida
864	South Carolina
865	Tennessee
866	Toll-Free Service
867	Yukon & NW Terr.
868	Trinidad & Tobago
869	St. Kitts & Nevis
870	Arkansas
876	Jamaica
877	Toll-Free Service
880	Toll-Free Service
881	Toll-Free Service
882	Toll-Free Service
888	Toll-Free Service
900	Premium Service
901	Tennessee
902	Nova Scotia
903	Texas
904	Florida
905	Ontario
906	Michigan
907	Alaska
908	New Jersey
909	California
910	North Carolina
911	Emergency
912	Georgia
913	Kansas
914	New York
915	Texas
916	California
917	New York
918	Oklahoma
919	North Carolina
920	Wisconsin
925	California
928	Arizona
931	Tennessee
936	Texas
937	Ohio
940	Texas
941	Florida
949	California
952	Minnesota
954	Florida
956	Texas
970	Colorado
971	Oregon
972	Texas
973	New Jersey
978	Massachusetts
979	Texas
980	North Carolina
985	Louisiana
989	Michigan

Codes for International Direct Dial Calling From the U.S.

Basic station-to-station calls: 011 + country code (as shown) + city code (if required) + local number.

Person-to-person, operator-assisted, collect, credit card calls; calls billed to another number: 01 + country code (below) + city code (if required) + local number.

Selected city codes given below. For countries or territories not listed, contact your long distance company.

Country/Territory	Code
Afghanistan	93
Albania	355
Algeria	213
American Samoa	684
Andorra	376
Angola	244
Anguilla	264*
Antarctica (Scott Base and Casey Base)	672
Antigua & Barbuda	268*
Argentina	54
Armenia	374
Aruba	297
Ascension Island	247
Australia	61
Austria	43
Azerbaijan	994
Bahamas	242*
Bahrain	973
Bangladesh	880
Barbados	246*
Belarus	375
Belgium	32
Belize	501
Benin	229
Bermuda	441*
Bhutan	975
Bolivia	591
Bosnia & Herzegovina	387
Botswana	267
Brazil	55
Brunei	673
Bulgaria	359
Burkina Faso	226
Burundi	257
Cambodia	855
Cameroon	237
Canada	
Alberta	403*/780*
British Columbia	250*
Vancouver	604*
Manitoba	204*
New Brunswick	506*
Newfoundland	709*
NW Territories	604*
Nova Scotia	902*
Nunavut	867*
Ontario	
London	519*
North Bay	705*
Ottawa	613*
Thunder Bay	807*
Toronto Metro	416*
Toronto Vicinity	905*
Prince Edward Isl.	902*
Quebec	
Montreal	514*/450*
Quebec City	418*
Sherbrooke	819*
Saskatchewan	306*
Yukon Territory	403*
Cape Verde	238
Cayman Islands	345*
Central African Rep.	236
Chad Republic	235
Chile	56
China	86
Colombia	57
Comoros	269
Congo Republic	242
Costa Rica	506
Côte d'Ivoire	225
Croatia	385
Cuba	53
Cyprus	357
Czech Republic	420
Denmark	45
Djibouti	253
Dominica	767*
Dominican Republic	809
East Timor	670
Ecuador	593
Egypt	20
El Salvador	503
Equatorial Guinea	240
Estonia	372
Falkland Islands	500
Faroe Islands	298
Fiji	679
Finland	358
France	33
French Antilles	596
French Guiana	594
French Polynesia	689
Gabon	241
Gambia, The	220
Georgia	995
Germany	49
Ghana	233
Gibraltar	350
Greece	30
Greenland	299
Grenada	473*
Guadeloupe	590
Guam	671*
Guantanamo Bay	53
Guatemala	502
Guinea	224
Guinea-Bissau	245
Guyana	592
Haiti	509
Honduras	504
Hong Kong	852
Hungary	36
Iceland	354
India	91
Indonesia	62
Iran	98
Iraq	964
Ireland	353
Israel	972
Italy	39
Jamaica	876*
Japan	81
Jordan	962
Kazakhstan	7
Kenya	254
Kiribati	686
Korea, North	850
Korea, South	82
Kuwait	965
Kyrgyzstan	7
Laos	856
Latvia	371
Lebanon	961
Lesotho	266
Liberia	231
Libya	218
Liechtenstein	423
Lithuania	370
Luxembourg	352
Macao	853
Macedonia	389
Madagascar	261
Malawi	265
Malaysia	60
Maldives	90
Mali	223
Malta	356
Marshall Islands	692
Mauritania	222
Mauritius	230
Mexico	52
Micronesia	691
Moldova	373
Monaco	33
Mongolia	976
Montserrat	664*
Morocco	212
Mozambique	258
Myanmar	95
Namibia	264
Nauru	674
Nepal	977
Netherlands	31
New Caledonia	687
New Zealand	64
Nicaragua	505
Niger	227
Nigeria	234
N. Mariana Isls.	670
Norway	47
Oman	968
Pakistan	92
Palau	680
Panama	507
Papua New Guinea	675
Paraguay	595
Peru	51
Philippines	63
Poland	48
Portugal	351
Puerto Rico	787
Qatar	974
Romania	40
Russia	7
Rwanda	250
St. Kitts & Nevis	869*
St. Lucia	758*
St. Vincent & the Grenadines	809*
Samoa (formerly Western Samoa)	685
San Marino	378
São Tomé & Príncipe	239
Saudi Arabia	966
Senegal	221
Seychelles	248
Sierra Leone	232
Singapore	65
Slovakia	421
Slovenia	386
Solomon Islands	677
Somalia	252
South Africa	27
Spain	34
Sri Lanka	94
Sudan	249
Suriname	597
Swaziland	268
Sweden	46
Switzerland	41
Syria	963
Taiwan	886
Tajikistan	992
Tanzania	255
Thailand	66
Togo	228
Tonga	676
Trinidad & Tobago	868*
Tunisia	216
Turkey	90
Turkmenistan	7
Turks & Caicos Isls.	649*
Tuvalu	688
Uganda	256
Ukraine	380
United Arab Emirates	971
United Kingdom	44
Uruguay	598
Uzbekistan	998
Vanuatu	678
Vatican City	39
Venezuela	58
Vietnam	84
Virgin Islands, British	284*
Virgin Islands, U.S.	340*
Yemen	967
Yugoslavia	381
Zambia	260
Zimbabwe	263

* These numbers are area codes. Follow Domestic Dialing instructions: dial "1" + area code + number you are calling.

Selected city codes: Beijing, 10; Brasilia, 61; Buenos Aires, 11; Dhaka, 2; Dublin, 1; Islamabad, 51; Jakarta, 21; Jerusalem, 2; Lagos, 1; London, 20; Madrid, 91; Mexico City, 5; New Delhi, 11; Paris, 1; Rome, 06; Tokyo, 3.

ASTRONOMY AND CALENDAR

Edited by Dr. Lee T. Shapiro, Planetarium Director, Morehead Planetarium, University of North Carolina at Chapel Hill

Celestial Events Summary, 2002

There will be 5 eclipses in 2002, 2 solar and 3 lunar. The center path of the annular solar eclipse in June crosses the Pacific Ocean from Indonesia to Mexico and will be seen as partial over much of N America. The path of the total solar eclipse in December starts in the lower S Atlantic Ocean, crosses lower Africa and the Indian Ocean, and ends in Australia. All 3 lunar eclipses are penumbral and thus not very noticeable. The most likely viewing successes for meteor showers will be the Lyrids in April, the Perseids in August, the Leonids in November (possible storm), and the Geminids in December.

At the start of the year, Jupiter, Saturn, and Mars are all prominent in the early evening sky, with Jupiter up the whole night. In March, Venus also joins the evening grouping. After mid-April, Mercury also appears, placing all 5 planets visible with the naked eye in the western part of the sky in the early evening. At mid-May, the grouping is even closer and the Moon also appears within the group. At the end of the month, Saturn and Mercury are lost in the glare of sunset. By early June, only Mars, Jupiter, and Venus are in the early evening sky, with Jupiter and Venus very close early in the month. In July, only Venus is left in the early evening sky, where it stays until October. Saturn enters the early morning sky in July, joined by Jupiter in August, Mars in late September, and Venus in November. Finally, in December, Saturn starts the return of the bright planets to the evening sky.

In the middle of April, May, June, July, and August there are pretty views of Venus with the waxing crescent Moon in the evening, and in December the waning crescent Moon pairs with Venus in the morning sky. Other intriguing groupings include Saturn, Jupiter, and the Moon on January 25, February 22, March 21, April 17, and on May 15 as part of the multiple planet grouping mentioned earlier. In early June, Venus and Jupiter pair together, and the Moon makes it a threesome 10 days later.

Astronomical Positions Defined

Two celestial bodies are in **conjunction** when they are due N and S of each other, either in **right ascension** (with respect to the N celestial pole) or in **celestial longitude** (with respect to the N ecliptic pole). If the bodies are seen near each other, they will rise and set at nearly the same time. For the inner planets—Mercury and Venus—**inferior conjunction** occurs when either planet passes between Earth and the Sun, while **superior conjunction** occurs when either Mercury or Venus is on the far side of the Sun. Celestial bodies are in **opposition** when their Right Ascensions differ by exactly 12 hours, or when their Celestial Longitudes differ by 180° . One of the 2 objects in opposition will rise while the other is setting. **Quadrature** refers to the arrangement where the coordinates of 2 bodies differ by exactly 90° . These terms may refer to the relative positions of any 2 bodies as seen from Earth, but one of the bodies is so frequently the Sun that mention of the Sun is omitted in that case; otherwise, both bodies are named. When objects are in conjunction, the alignment is not perfect, and one is usually passing above or below the other. The geocentric angular separation between the Sun and an object is termed **elongation**. Elongation is limited only for Mercury and Venus; the greatest elongation for each of these bodies is noted in the appropriate table and is approximately the time for longest observation. **Perihelion** is the point in an orbit that is nearest to the Sun, and **aphelion,** the point farthest from the Sun. **Perigee** is the point in an orbit that is nearest Earth, **apogee** the point that is farthest from Earth. An **occultation** of a planet or a star is an **eclipse** of it by some other body, usually the Moon.

Astronomical Constants; Speed of Light

The following were adopted as part of the International Astronomical Union System of Astronomical Constants (1976): **Speed of light,** 299,792.458 km per sec., or about 186,282 statute mi per sec.; **solar parallax,** 8".794148; **Astronomical Unit,** 149,597,870 km, or 92,955,807 mi; **constant of nutation,** 9".2025; and **constant of aberration,** 20".49552.

Celestial Events Highlights, 2002

(Coordinated Universal Time, or UTC—the standard time of the prime meridian)

January

Mercury may be seen low in the SW in the evening sky during the first three weeks of the month.

Venus hides in the glare of the Sun during this month.

Mars is prominent in the SW after sunset.

Jupiter, rising in the E after sunset, is up nearly the whole night.

Saturn is noticeable in the E after sunset, higher than Jupiter.

Moon passes Mercury and Uranus on the 15th, Mars on the 18th, occults Saturn on the 24th, and occults Jupiter on the 26th. Watch for the waxing gibbous Moon between Jupiter and Saturn on the 25th.

Jan. 1—Jupiter at opposition. Pluto in Ophiuchus, stays there all year. Neptune in Capricornus, stays there all year. Uranus in Capricornus. Saturn in Taurus. Jupiter in Gemini. Mars in Aquarius. Venus and Sun in Sagittarius.

Jan. 2—Earth at perihelion, closest approach to Sun.

Jan. 9—Mercury passes 1.3° S of Neptune. Mars enters Pisces.

Jan. 11—Mercury at greatest eastern elongation of 19°.

Jan. 14—Venus at superior conjunction, passing behind the Sun.

Jan. 15—Moon passes 4° S of Mercury and 4° S of Uranus.

Jan. 18—Mercury stationary, begins retrograde motion. Moon passes 5° S of Mars. Venus enters Capricornus.

Jan. 19—Mercury at perihelion, closest approach to Sun. Sun enters Capricornus.

Jan. 24—Moon passes 0.08° N of Saturn, occults Saturn.

Jan. 25—Venus at aphelion, farthest from the Sun.

Jan. 26—Moon passes 0.9° N of Jupiter, occults Jupiter.

Jan. 27—Mercury at inferior conjunction passing between Earth and Sun.

Jan. 28—Neptune at conjunction. Mars barely touches constellation of Cetus.

February

Mercury is low in the SE before sunrise about mid-month.

Venus begins to appear low in the west after sunset by the end of the month.

Mars is prominent in the WSW after sunset.

Jupiter is rising higher in the E after sunset.

Saturn is in the S after sunset.

Moon passes Mercury on the 10th, Mars on the 16th, occults Saturn on the 21st (visible in the U.S.) and occults Jupiter on the 23d. Watch for the waxing gibbous Moon between Jupiter and Saturn on the 22d.

Feb. 8—Mercury stationary, resumes direct motion.

Feb. 9—Saturn stationary, resumes direct motion. Venus enters Aquarius.

Feb. 10—Moon passes 5° S of Mercury.

Feb. 13—Uranus at conjunction.

Feb. 16—Moon passes 5° S of Mars. Sun enters Aquarius.

Feb. 21—Moon passes 0.2° N of Saturn, occults Saturn. Mercury at greatest western elongation of 27°.

Feb. 23 —Moon passes 0.9° N of Jupiter, occults Jupiter.

Feb. 24—Mercury passes 0.5° S of Neptune.

Feb. 28—Mars enters Aries.

March

Mercury is low in the ESE early in the month, then disappearing into the glare of the sunrise.

Venus low in the W after sunset.

Mars in the W after sunset.

Jupiter is high in the S after sunset.

Saturn still prominent in the W after sunset, is near the star Aldebaran during the month.

Moon passes Uranus on the 11th, Mercury on the 12th, Mars on the 18th, occults Saturn on the 20th and occults Jupiter on the 22d. Watch for the waxing crescent Moon between Jupiter and Saturn on the 21st. See if you can spot the very thin crescent Moon with Venus below on the 15th.

Mar. 1—Jupiter stationary, resumes direct motion.

Mar. 3—Venus enters Pisces.

Mar. 4—Mercury at aphelion, farthest from the Sun.

Mar. 9—Mercury passes 1.2° S of Uranus.

Mar. 10—Moon passes 4° S of Neptune.

Mar. 11—Moon passes 4° S of Uranus.

Mar. 12—Moon passes 3° S of Mercury. Sun enters Pisces.

Mar. 13—Venus enters Cetus.

Mar. 16—Venus enters Pisces.

Mar. 18—Moon passes 4° S of Mars.

Mar. 20—Moon passes 0.5° N of Saturn, occults Saturn. Vernal Equinox at 2:16 PM EST (19:16 UTC), spring begins in the northern hemisphere, autumn in the southern hemisphere.

Mar. 21—Pluto stationary, begins retrograde motion.

Mar. 22—Moon passes 1.1° N of Jupiter, occults Jupiter.

Mar. 28—Sun barely touches constellation of Cetus.

Mar. 31—Saturn passes 4° N of Aldebaran. Venus enters Aries.

April

Mercury returns to the evening sky late in the month low in the WNW, making 5 planets visible with the naked eye in the early evening during the remainder of the month. During the last week of the month watch for Mercury lined up with three other planets—Venus, Mars, and Saturn.

Venus slowly getting higher in the E after sunset.

Moon passes Uranus on the 8th, Venus on the 14th, Mars on the 15th, occults Saturn on the 16th, and passes Jupiter on the 18th. Watch the waxing crescent Moon pass four planets during five days, the 14th through the 18th.

Mars continues low in the W after sunset, passes Aldebaran at the end of the month.

Jupiter is prominent in the W after sunset.

Saturn is getting lower in the W after sunset.

Apr. 1—Uranus enters Aquarius

Apr. 4—Mars enters Taurus.

Apr. 6—Moon passes 4° S of Neptune.

Apr. 7—Mercury at superior conjunction, passing behind the Sun.

Apr. 8—Moon passes 4° S of Uranus.

Apr. 14—Moon passes 3° S of Venus.

Apr. 15—Moon passes 2° S of Mars.

Apr. 16—Moon passes 0.8° N of Saturn, occults Saturn.

Apr. 17—Mercury at perihelion.

Apr. 18—Moon passes 1.6° N of Jupiter.

Apr. 19—Sun enters Aries.

Apr. 20—Venus enters Taurus.

Apr. 22—Lyrid meteor shower early in the morning before sunrise.

Apr. 29—Mars passes 6° N of Aldebaran.

May

Mercury low in the WNW after sunset during the first half of the month, continuing 5 naked-eye planets in the early evening sky.

Venus low in the WNW, passes Aldebaran during the first week of the month.

Mars low in the WNW, passes Saturn during the first week of the month.

Jupiter gets lower in the W after sunset.

Saturn very low in the WNW after sunset, disappears into the glare of sunset after the middle of the month.

Moon passes Uranus on the 5th, Mercury on the 13th, occults Saturn, Mars, and Venus on the 14th, and passes Jupiter on the 16th. Watch the spectacular groupings of the waxing crescent Moon with planets on the 13th, 14th, and 15th.

May 4—Moon passes 4° S of Neptune. Mercury at greatest eastern elongation of 21°. Venus passes 6° N of Aldebaran. Mars passes 2° N of Saturn.

May 5—Moon passes 4° S of Uranus.

May 7—Venus passes 2° N of Saturn.

May 13—Neptune stationary, begins retrograde motion. Moon passes 3° S of Mercury.

May 14—Moon passes 1.1° N of Saturn, 0.6° S of Mars, and 0.8° S of Venus, occults all three planets. Sun enters Taurus.

May 16—Mercury stationary, begins retrograde motion. Moon passes 2° N of Jupiter.

May 17—Venus at perihelion, closest to the Sun.

May 21—Venus enters Gemini.

May 26—Penumbral lunar eclipse; see details under Eclipses.

May 27—Mercury at inferior conjunction.

May 28—Mars enters Gemini.

May 31—Moon passes 4° S of Neptune. Mercury at aphelion.

June

Mercury lost in the glare of the Sun, reappears in the morning sky in the NE, low in the sky about mid-month.

Venus is bright in the W after sunset throughout the month, passes Jupiter on the 3d and Pollux on the 9th.

Mars, very low in the WNW, is lost in the glare of the Sun by the end of the month.

Jupiter, also very low in the WNW, is also lost in the glare of sunset by the end of the month.

Saturn, hidden during most of the month, reemerges from the glare of the Sun in the ENE just before sunrise.

Moon passes Uranus on the 1st and 29th, Mercury on the 9th, occults Mars on the 12th, and passes Jupiter and Venus on the 13th. Watch for the triple grouping of the three brightest night-time objects, Moon, Venus, and Jupiter, on the 13th.

June 1—Moon passes 4° S of Uranus.

June 3—Uranus stationary, begins retrograde motion. Venus passes 1.6° N of Jupiter.

June 7—Pluto at opposition.

June 8—Mercury stationary, resumes direct motion.

June 9—Saturn at conjunction. Moon passes 3° N of Mercury. Venus passes 5° S of Pollux.

June 10—Annular solar eclipse; see details under Eclipses.

June 12—Moon passes 0.9° N of Mars, occults Mars.

June 13—Moon passes 2° N of Jupiter and 1.5° N of Venus. Venus enters Cancer.

June 21—Northern solstice at 9:24 AM EDT (13:24 UTC), summer begins in the northern hemisphere, winter in the southern hemisphere. Mercury at greatest western elongation of 23°. Sun enters Gemini.

June 24—Mercury passes 2° N of Aldebaran. Penumbral lunar eclipse; see details under Eclipses.

June 27—Moon passes 4° S of Neptune.

June 29—Moon passes 4° S of Uranus. Venus enters Leo.

July

Mercury is visible low in the NE before sunrise at the start of the month, passes Saturn on the 2d.

Venus remains bright in the W after sunset, passing Regulus on the 10th.

Mars is absent from dark skies during the month.

Jupiter, like Mars, is too close to the Sun to be seen during the month.

Saturn, appearing low in the ENE in the morning, gets higher during the month.

Moon passes Jupiter on the 3d, Saturn on the 8th, Venus on the 10th, and Uranus on the 26th.

July 2—Mercury passes 0.2° N of Saturn.

July 3—Mars passes 0.8° N of Jupiter.

July 4—Mars passes 6° S of Pollux.

July 6—Earth at aphelion.

July 8—Moon passes 1.7° N of Saturn.

July 10—Venus passes 1.1° N of Regulus. Mars enters Cancer

July 13—Moon passes 4° N of Venus.

July 14—Mercury at perihelion.

July 20—Jupiter at conjunction. Sun enters Cancer.

July 21—Mercury at superior conjunction.

July 23—Jupiter enters Cancer.

July 24—Moon passes 4° S of Neptune.

July 26—Moon passes 4° S of Uranus.

August

Mercury is very low in the W after sunset.

Venus is bright in the ESE in the early evening.

Mars still too close to the Sun to be seen.

Jupiter is now low in the ENE before sunrise.

Saturn is higher in the E in the morning.

Moon passes Saturn on the 5th, Mercury on the 10th, Venus on the 16th, and Uranus on the 22d.

Aug. 1—Venus enters Virgo.

Aug. 2—Neptune at opposition.

Aug. 5—Moon passes 2° N of Saturn.

Aug. 6—Mercury passes 0.9° N of Regulus.

Aug. 10—Moon passes 4° N of Mercury. Mars at conjunction. Mars and Sun enter Leo.

Aug. 12—Perseid meteor shower early in the morning before sunrise.

Aug. 16—Moon passes 6° N of Venus.

Aug. 20—Uranus at opposition.

Aug. 21—Moon passes 4° S of Neptune.

Aug. 22—Venus at greatest eastern elongation of 46°. Moon passes 4° S of Uranus.

Aug. 27—Pluto stationary, resumes direct motion. Mercury at aphelion.

Aug. 30—Saturn enters Orion.

September

Mercury is very low in the WSW before disappearing into the Sun's glare after mid-month.

Venus, very low in the SW after sunset, passes Spica on the 1st.

Mars reappears very low in the W before sunrise.

Jupiter is now high in the E before sunrise.

Saturn, nearly overhead before sunrise, rises about midnight.

Moon passes Saturn on the 1st and the 29th, Jupiter on the 4th, Mercury on the 8th, Venus on the 10th, and Uranus on the 18th.

Sept. 1—Venus passes 0.9° S of Spica. Mercury at greatest eastern elongation of 27°. Moon passes 2° N of Saturn.

Sept. 4—Moon passes 4.0° N of Jupiter.

Sept. 7—Venus at aphelion.

Sept. 8—Moon passes 9° N of Mercury.

Sept. 10—Moon passes 8° N of Venus.

Sept. 14—Mercury stationary, begins retrograde motion.

Sept. 16—Sun enters Virgo.

Sept. 17—Moon passes 4° S of Neptune.

Sept. 18—Moon passes 4° S of Uranus.

Sept. 21—Mars at aphelion, furthest from the Sun.

Sept. 22—Venus enters Libra.

Sept. 23—Autumnal Equinox at 12:55 AM EDT (4:55 UTC), autumn begins in the northern hemisphere, spring begins in the southern hemisphere

Sept. 27—Mercury at inferior conjunction.

Sept. 29—Moon passes 3° N of Saturn.

October

Mercury, very low in the E before sunrise in the middle of the month, passes Spica on the 27th.

Venus, very low in the SE after sunset, disappears into the glare of the Sun.

Mars is low in the ESE before sunrise.

Jupiter, high in the E before sunrise, rises a little after midnight.

Saturn, high in the W before sunrise, rises before the middle of the night.

Moon passes Jupiter on the 2d and 29th, Mars on the 5th, Venus on the 8th, Uranus on the 15th, and Saturn on the 26th.

Oct. 2—Moon passes 4° N of Jupiter.

Oct. 5—Moon passes 4° N of Mars. Mars enters Virgo.

Oct. 6—Mercury stationary, resumes direct motion.

Oct. 8—Moon passes 10° N of Venus.

Oct. 10—Venus stationary, begins retrograde motion. Mercury at perihelion.

Oct. 11—Saturn stationary, begins retrograde motion.

Oct. 13—Mercury at greatest western elongation of 18°.

Oct. 14—Moon passes 5° S of Neptune.

Oct. 15—Moon passes 4° S of Uranus.

Oct. 20—Neptune stationary, resumes direct motion.

Oct. 22—Venus enters Virgo.

Oct. 26—Moon passes 3° N of Saturn.

Oct. 27—Mercury passes 4° N of Spica.

Oct. 29—Moon passes 4° N of Jupiter.

Oct. 31—Venus at inferior conjunction, passing between Earth and Sun. Sun enters Libra.

November

Mercury stays hidden from view during the month.

Venus reappears in the morning sky in the SE before sunrise, approaching Mars at the month's end.

Mars, in the SE before sunrise, passes Spica on the 20th.

Jupiter, high in the S before sunrise, rises about the middle of the night.

Saturn, in the W at sunrise, rises a couple of hours after sunset.

Moon passes Mars on the 2d, Uranus on the 12th, Saturn on the 22d, and Jupiter on the 26th.

Nov. 2—Moon passes 4° N of Mars.

Nov. 4—Uranus stationary, resumes direct motion.

Nov. 10—Moon passes 5° S of Neptune.

Nov. 12—Moon passes 5° S of Uranus.

Nov. 14—Mercury at superior conjunction.

Nov. 17—Leonid meteor shower (possible storm) after the middle of the night.

Nov. 19—Venus stationary, resumes direct motion. Penumbral lunar eclipse; see details under Eclipses.

Nov. 20—Mars passes 3° N of Spica.

Nov. 22—Moon passes 3° N of Saturn. Saturn enters Taurus. Jupiter enters Leo. Sun enters Scorpius.

Nov. 23—Mercury at aphelion.

Nov. 26—Moon passes 4° N of Jupiter.

Nov. 30—Sun enters Ophiuchus.

December

Mercury appears very low in the SW after sunset, higher during the 2d half of the month.

Venus is low in the SE before sunrise, near Mars for most of the month.

Mars is in the SE before sunrise.

Jupiter, high in the west at sunrise, rises before the middle of the night.

Saturn, rising in the ENE around sunset, is up most of the night.

Moon passes Mars and Venus on the 1st and the 30th, Uranus on the 9th, Saturn on the 19th, and Jupiter on the 23d. Watch for two triple groupings of the waning crescent Moon with Venus and Mars on the mornings of the 1st and the 30th.

Dec. 1—Moon passes 3° N of Mars and 2° N of Venus.

Dec. 4—Penumbral lunar eclipse; see details under Eclipses. Jupiter stationary, begins retrograde motion.

Dec. 8—Moon passes 5° S of Neptune.

Dec. 9— Moon passes 5° S of Uranus. Pluto at conjunction.

Dec. 11—Venus enters Libra.

Dec. 14—Geminid meteor shower after the middle of the night.

Dec. 17—Saturn at opposition. Jupiter enters Cancer.

Dec. 18—Sun enters Sagittarius.

Dec. 19—Moon passes 3° N of Saturn.

Dec. 21—Southern solstice at 8:14 PM EST (1:14 UTC on Dec. 22), winter begins in the northern hemisphere, summer begins in the southern hemisphere.

Dec. 23—Moon passes 4° N of Jupiter.

Dec. 26—Mercury at greatest eastern elongation of 20°.

Dec. 30—Moon passes 1.2° N of Mars and 2° S of Venus.

Meteorites and Meteor Showers

When a chunk of material, ice or rock, plunges into Earth's atmosphere and burns up in a fiery display, the event is a **meteor**. While the chunk of material is still in space, it is a **meteoroid**. If a portion of the material survives passage through the atmosphere and reaches the ground, the remnant on the ground is a **meteorite**.

Meteorites found on Earth are classified into types, depending on their composition: **irons**, those composed chiefly of iron, a small percentage of nickel, and traces of other metals such as cobalt; **stones**, stony meteors consisting of silicates; and **stony irons**, containing varying proportions of both iron and stone.

The serious study of meteorites as non-earth objects began in the 20th century. Scientists now use sophisticated chemical analysis, X rays, and mass spectrography in determining their origin and composition. In 1996, the results of a study of a Mars rock recovered 12 years earlier from the Allan Hills region of Antarctica suggested that life once existed on that planet. Although most meteorites are now believed to be fragments of asteroids or comets, geochemical studies have shown that a few Antarctic stones came from the moon or from Mars, from which they presumably were ejected by the explosive impact of asteroids.

The **largest known meteorite**, estimated to weigh about 55 metric tons, is situated at Hoba West near Grootfontein, Namibia. The Manicouagan impact crater in Quebec, Canada, is one of the largest crater structures still visible on the surface of the Earth. Other large impact craters include the Chicxulub crater off the coast of Mexico, estimated at 185 mi across, the Sudbury crater in Ontario, Canada, estimated at 125 mi across, and the Acraman crater in Australia, estimated at 100 mi across.

Sporadic meteors, which enter the atmosphere throughout the year, seem to originate from the asteroid belt. Other meteors that come in groups and tend to occur at the same time each year create what are called **meteor showers**; these are the meteors associated with comets. As a comet orbits the Sun, the Sun slowly boils away some of the comet's material, and the comet leaves a trail of tiny particles which are dispersed along the comet's path. If Earth's orbit and this path intersect, then once a year, as Earth reaches that particular point in its orbit, there will be a meteor shower.

Meteor showers vary in strength, but usually the 3 best meteor showers of the year are the **Perseids**, which occur around Aug. 12, the **Orionids**, which occur around Oct. 21, and the **Geminids**, which occur around Dec. 13. These showers feature meteors at the rate of about 60 per hour. Best observing conditions occur with the absence of moonlight, usually when the Moon's phase is between waning crescent Moon and waxing quarter Moon. Meteor showers are also usually seen better after the middle of the night.

For most meteor showers the cometary debris is relatively uniformly scattered along the comet's orbit. However, in the case of the Leonid meteor shower, which occurs every year around Nov. 17-18, the cometary debris, from Comet Temple-Tuttle, seems to be bunched up in one stretch. That means that most years when Earth crosses the orbit of this comet, the meteor shower produced is relatively weak. However, approximately every 33 years Earth encounters the bunched-up debris. Sometimes the storm is a disappointment, as it was in 1899 and 1933; at other times it is a roaring success, as in 1833 and 1866.

In 1966 observers on the west coast of the United States were treated to an awesome display of meteors in the early morning as the rate peaked at 150,000 meteors per hour. In 1999, the predictions were very accurate, and the measured peak rate in the Middle East was about 4,000 meteors per hour. In 2000, not expected to be as good, the reported rates were 200 meteors per hour in parts of Europe, Africa, and the Americas. There were expectations for better displays in 2001 and for a strong storm in 2002, perhaps along the east coast of the United States.

IT'S A FACT: The biggest known crater produced on Earth by a crashing meteorite is a pit in northwest Quebec that is 2.5 miles in diameter, containing a lake and surrounded by piles of shattered granite.

Rising and Setting of Planets, 2002

Coordinated Universal Time (0 in the *h* col. designates midnight)

Venus, 2002

Date	20° N Latitude		30° N Latitude		40° N Latitude		50° N Latitude		60° N Latitude	
	Rise	Set	Rise	Set	Rise	Set	Rise	Set	Rise	Set
	h m	h m	h m	h m	h m	h m	h m	h m	h m	h m
Jan. 1	6 24	17 17	6 45	16 55	7 13	16 28	7 51	15 49	9 00	14 41
11	6 37	17 33	6 58	17 13	7 24	16 47	8 00	16 11	9 04	15 07
21	6 48	17 51	7 06	17 33	7 29	17 10	8 01	16 38	8 55	15 44
31	6 55	18 08	7 10	17 53	7 29	17 34	7 55	17 08	8 38	16 26
Feb. 10	6 59	18 24	7 11	18 13	7 25	17 59	7 44	17 39	8 15	17 09
20	7 00	18 39	7 08	18 32	7 17	18 23	7 29	18 11	7 49	17 52
Mar. 2	7 00	18 53	7 03	18 50	7 07	18 46	7 12	18 41	7 20	18 34
12	6 58	19 07	6 57	19 08	6 56	19 10	6 54	19 12	6 51	19 16
22	6 57	19 20	6 51	19 26	6 45	19 33	6 35	19 43	6 21	19 58
Apr. 1	6 56	19 34	6 46	19 44	6 34	19 56	6 18	20 13	5 51	20 41
11	6 57	19 49	6 43	20 03	6 26	20 20	6 02	20 45	5 23	21 25
21	7 00	20 04	6 42	20 22	6 20	20 44	5 49	21 16	4 57	22 09
May 1	7 06	20 21	6 45	20 41	6 19	21 08	5 42	21 45	4 36	22 53
11	7 15	20 37	6 52	20 59	6 23	21 29	5 42	22 11	4 23	23 30
21	7 27	20 51	7 03	21 15	6 33	21 45	5 49	22 30	4 25	23 54
31	7 41	21 04	7 18	21 27	6 48	21 57	6 05	22 40	4 43	f 0 01
June 10	7 56	21 13	7 34	21 34	7 06	22 02	6 27	22 41	5 15	23 51
20	8 10	21 18	7 51	21 37	7 27	22 00	6 53	22 34	5 55	23 31
30	8 23	21 20	8 07	21 35	7 48	21 54	7 21	22 21	6 36	23 04
July 10	8 34	21 18	8 22	21 29	8 08	21 43	7 48	22 03	7 16	22 34
20	8 43	21 13	8 36	21 20	8 26	21 29	8 13	21 42	7 53	22 01
30	8 50	21 05	8 47	21 08	8 43	21 12	8 37	21 18	8 28	21 26
Aug. 9	8 56	20 56	8 56	20 55	8 57	20 54	8 59	20 52	9 00	20 50
19	8 59	20 45	9 04	20 40	9 10	20 33	9 18	20 25	9 31	20 12
29	9 00	20 32	9 09	20 23	9 20	20 12	9 35	19 56	9 59	19 32
Sept. 8	8 58	20 17	9 11	20 04	9 27	19 48	9 49	19 26	10 24	18 50
18	8 51	19 59	9 08	19 42	9 28	19 21	9 56	18 53	10 43	18 06
28	8 36	19 34	8 56	19 15	9 20	18 51	9 54	18 17	10 51	17 19
Oct. 8	8 09	19 01	8 30	18 40	8 57	18 14	9 34	17 37	10 39	16 31
18	7 25	18 17	7 46	17 56	8 12	17 30	8 49	16 54	9 54	15 49
28	6 25	17 24	6 43	17 06	7 06	16 43	7 38	16 11	8 32	15 17
Nov. 7	5 19	16 31	5 33	16 17	5 51	15 59	6 16	15 34	6 56	14 55
17	4 24	15 48	4 35	15 37	4 49	15 23	5 07	15 04	5 37	14 35
27	3 46	15 16	3 55	15 07	4 07	14 55	4 23	14 39	4 47	14 15
Dec. 7	3 23	14 54	3 33	14 45	3 44	14 33	4 00	14 18	4 24	13 53
17	3 12	14 40	3 22	14 29	3 35	14 16	3 52	13 59	4 20	13 31
27	3 08	14 30	3 20	14 18	3 35	14 03	3 56	13 42	4 28	13 10

Mars, 2002

Date	20° N Latitude		30° N Latitude		40° N Latitude		50° N Latitude		60° N Latitude	
	Rise	Set	Rise	Set	Rise	Set	Rise	Set	Rise	Set
	h m	h m	h m	h m	h m	h m	h m	h m	h m	h m
Jan. 1	10 36	22 24	10 41	22 20	10 47	22 14	10 54	22 07	11 05	21 56
11	10 19	22 16	10 21	22 14	10 24	22 12	10 27	22 09	10 31	22 04
21	10 02	22 07	10 01	22 08	10 00	22 09	9 59	22 10	9 57	22 13
31	9 44	21 59	9 41	22 02	9 37	22 06	9 32	22 12	9 23	22 21
Feb. 10	9 27	21 50	9 21	21 56	9 14	22 03	9 05	22 13	8 49	22 29
20	9 10	21 42	9 02	21 50	8 52	22 00	8 38	22 14	8 16	22 37
Mar. 2	8 53	21 33	8 43	21 44	8 30	21 57	8 11	22 15	7 43	22 45
12	8 37	21 25	8 24	21 38	8 08	21 54	7 46	22 16	7 10	22 52
22	8 22	21 16	8 07	21 31	7 48	21 50	7 22	22 17	6 39	23 00
Apr. 1	8 07	21 08	7 50	21 25	7 28	21 47	6 58	22 17	6 08	23 08
11	7 53	21 00	7 34	21 19	7 10	21 42	6 37	22 16	5 39	23 14
21	7 39	20 51	7 19	21 12	6 53	21 37	6 17	22 14	5 12	23 19
May 1	7 27	20 43	7 05	21 04	6 38	21 32	5 59	22 11	4 48	23 22
11	7 15	20 34	6 52	20 56	6 24	21 25	5 43	22 06	4 27	23 22
21	7 04	20 24	6 41	20 47	6 12	21 16	5 29	21 58	4 10	23 18
31	6 53	20 14	6 30	20 37	6 01	21 06	5 18	21 49	3 58	23 08
June 10	6 43	20 03	6 20	20 25	5 51	20 54	5 09	21 36	3 51	22 54
20	6 33	19 51	6 10	20 13	5 42	20 41	5 02	21 21	3 47	22 36
30	6 23	19 38	6 01	19 59	5 34	20 26	4 56	21 04	3 46	22 13
July 10	6 13	19 24	5 53	19 44	5 27	20 09	4 51	20 45	3 48	21 48
20	6 03	19 09	5 44	19 27	5 20	19 51	4 47	20 23	3 50	21 20
30	5 52	18 53	5 35	19 10	5 14	19 31	4 44	20 01	3 54	20 50
Aug. 9	5 41	18 36	5 26	18 51	5 07	19 10	4 41	19 36	3 57	20 19
19	5 30	18 19	5 17	18 32	5 00	18 48	4 38	19 11	4 01	19 47
29	5 19	18 00	5 08	18 11	4 54	18 25	4 34	18 44	4 04	19 15
Sept. 8	5 07	17 42	4 58	17 51	4 47	18 02	4 31	18 17	4 06	18 41
18	4 55	17 22	4 48	17 29	4 39	17 38	4 28	17 50	4 09	18 08
28	4 43	17 03	4 38	17 08	4 32	17 14	4 24	17 22	4 11	17 34
Oct. 8	4 31	16 43	4 28	16 46	4 25	16 49	4 20	16 54	4 13	17 01
18	4 19	16 23	4 18	16 24	4 17	16 24	4 16	16 25	4 14	16 27
28	4 06	16 04	4 08	16 02	4 10	16 00	4 12	15 57	4 16	15 53
Nov. 7	3 54	15 44	3 58	15 40	4 03	15 35	4 09	15 29	4 18	15 20
17	3 42	15 25	3 48	15 19	3 55	15 11	4 05	15 02	4 20	14 46
27	3 30	15 06	3 38	14 58	3 48	14 48	4 01	14 34	4 22	14 14
Dec. 7	3 19	14 48	3 29	14 37	3 41	14 25	3 58	14 08	4 24	13 42
17	3 07	14 30	3 20	14 18	3 34	14 03	3 55	13 42	4 27	13 10
27	2 57	14 13	3 10	13 59	3 28	13 41	3 51	13 18	4 29	12 40

Jupiter, 2002

Date	20° N Latitude Rise h m	Set h m	30° N Latitude Rise h m	Set h m	40° N Latitude Rise h m	Set h m	50° N Latitude Rise h m	Set h m	60° N Latitude Rise h m	Set h m
Jan. 1	17 23	6 42	17 02	7 03	16 34	7 30	15 55	8 09	14 45	9 20
11	16 38	5 57	16 16	6 18	15 49	6 46	15 09	7 25	13 58	8 36
21	15 53	5 12	15 31	5 34	15 04	6 01	14 24	6 41	13 12	7 53
31	15 09	4 29	14 47	4 50	14 20	5 18	13 40	5 58	12 27	7 10
Feb. 10	14 26	3 46	14 04	4 08	13 37	4 35	12 57	5 15	11 44	6 28
20	13 45	3 05	13 23	3 27	12 55	3 54	12 15	4 34	11 02	5 47
Mar. 2	13 05	2 25	12 43	2 47	12 15	3 14	11 35	3 55	10 22	5 08
12	12 26	1 46	12 05	2 08	11 37	2 36	10 57	3 16	9 43	4 29
22	11 49	1 09	11 27	1 31	11 00	1 59	10 20	2 39	9 06	3 52
Apr. 1	11 14	0 33	10 52	0 55	10 24	1 23	9 44	2 03	8 31	3 16
11	10 39	23 55	10 17	0 20	9 49	0 48	9 09	1 28	7 57	2 40
21	10 05	23 21	9 44	23 43	9 16	0 14	8 36	0 53	7 24	2 06
May 1	9 33	22 48	9 11	23 10	8 44	23 37	8 04	0 20	6 52	1 32
11	9 01	22 16	8 39	22 37	8 12	23 05	7 33	23 44	6 22	0 58
21	8 30	21 44	8 08	22 05	7 41	22 32	7 03	23 11	5 53	0 24
31	7 59	21 13	7 38	21 34	7 11	22 01	6 33	22 39	5 24	23 47
June 10	7 29	20 42	7 08	21 03	6 42	21 29	6 04	22 07	4 57	23 14
20	6 59	20 11	6 39	20 32	6 13	20 57	5 36	21 34	4 30	22 40
30	6 30	19 40	6 10	20 01	5 44	20 26	5 08	21 02	4 04	22 06
July 10	6 01	19 10	5 41	19 30	5 16	19 54	4 40	20 30	3 38	21 32
20	5 31	18 39	5 12	18 58	4 47	19 23	4 13	19 57	3 13	20 57
30	5 02	18 08	4 43	18 27	4 19	18 51	3 46	19 25	2 47	20 23
Aug. 9	4 33	17 37	4 14	17 56	3 51	18 19	3 18	18 52	2 22	19 48
19	4 03	17 06	3 45	17 24	3 23	17 47	2 51	18 18	1 57	19 12
29	3 33	16 35	3 16	16 52	2 54	17 14	2 23	17 45	1 31	18 37
Sept. 8	3 03	16 03	2 46	16 20	2 25	16 41	1 55	17 11	1 05	18 01
18	2 32	15 31	2 16	15 47	1 55	16 08	1 26	16 37	0 38	17 25
28	2 01	14 58	1 45	15 14	1 25	15 34	0 57	16 02	0 10	16 48
Oct. 8	1 29	14 24	1 13	14 40	0 54	14 59	0 26	15 26	23 38	16 11
18	0 56	13 50	0 41	14 05	0 22	14 24	23 52	14 51	23 08	15 34
28	0 22	13 15	0 07	13 30	23 45	13 49	23 19	14 14	22 37	14 57
Nov. 7	23 44	12 39	23 29	12 54	23 11	13 12	22 45	13 38	22 04	14 19
17	23 07	12 02	22 53	12 17	22 35	12 35	22 09	13 00	21 28	13 41
27	22 30	11 24	22 15	11 39	21 57	11 57	21 32	12 22	20 51	13 03
Dec. 7	21 50	10 45	21 36	11 00	21 18	11 18	20 53	11 43	20 12	12 24
17	21 10	10 05	20 55	10 20	20 37	10 38	20 12	11 03	19 31	11 44
27	20 28	9 24	20 13	9 39	19 55	9 57	19 29	10 23	18 47	11 05

Saturn, 2002

Date	20° N Latitude Rise h m	Set h m	30° N Latitude Rise h m	Set h m	40° N Latitude Rise h m	Set h m	50° N Latitude Rise h m	Set h m	60° N Latitude Rise h m	Set h m
Jan. 1	15 14	4 22	14 56	4 41	14 33	5 04	14 00	5 36	13 04	6 32
11	14 32	3 40	14 14	3 59	13 51	4 22	13 19	4 54	12 23	5 50
21	13 51	2 59	13 33	3 18	13 10	3 41	12 38	4 13	11 42	5 08
31	13 11	2 19	12 53	2 37	12 30	3 00	11 57	3 33	11 02	4 28
Feb. 10	12 31	1 39	12 13	1 58	11 50	2 21	11 18	2 53	10 22	3 49
20	11 53	1 01	11 34	1 19	11 11	1 42	10 38	2 15	9 43	3 11
Mar. 2	11 14	0 23	10 56	0 41	10 33	1 05	10 00	1 37	9 04	2 34
12	10 37	23 42	10 18	0 04	9 55	0 28	9 22	1 01	8 25	1 58
22	10 00	23 06	9 41	23 25	9 18	23 48	8 45	0 25	7 47	1 22
Apr. 1	9 24	22 30	9 05	22 49	8 41	23 13	8 08	23 47	7 10	0 48
11	8 48	21 55	8 29	22 14	8 05	22 38	7 31	23 12	6 32	0 14
21	8 13	21 20	7 54	21 40	7 30	22 04	6 55	22 38	5 56	23 38
May 1	7 38	20 46	7 19	21 05	6 54	21 30	6 20	22 05	5 19	23 05
11	7 04	20 12	6 44	20 32	6 19	20 56	5 44	21 31	4 43	22 33
21	6 29	19 38	6 10	19 58	5 45	20 23	5 09	20 58	4 07	22 00
31	5 55	19 05	5 35	19 25	5 10	19 50	4 35	20 25	3 32	21 28
June 10	5 21	18 31	5 01	18 51	4 36	19 16	4 00	19 52	2 57	20 56
20	4 47	17 58	4 27	18 18	4 02	18 43	3 25	19 19	2 22	20 23
30	4 13	17 24	3 53	17 44	3 27	18 10	2 51	18 46	1 47	19 50
July 10	3 39	16 50	3 19	17 10	2 53	17 36	2 16	18 13	1 12	19 17
20	3 05	16 16	2 44	16 36	2 19	17 02	1 42	17 39	0 37	18 44
30	2 30	15 41	2 10	16 02	1 44	16 28	1 07	17 05	0 02	18 10
Aug. 9	1 55	15 07	1 35	15 27	1 09	15 53	0 32	16 30	23 23	17 35
19	1 20	14 31	0 59	14 52	0 33	15 18	23 53	15 55	22 47	17 00
29	0 44	13 55	0 23	14 16	23 54	14 42	23 17	15 19	22 11	16 24
Sept. 8	0 07	13 19	23 43	13 39	23 17	14 05	22 40	14 42	21 35	15 48
18	23 27	12 42	23 06	13 02	22 40	13 28	22 03	14 05	20 57	15 11
28	22 49	12 04	22 28	12 24	22 02	12 50	21 25	13 27	20 20	14 33
Oct. 8	22 10	11 25	21 49	11 45	21 24	12 11	20 47	12 48	19 41	13 54
18	21 30	10 46	21 10	11 06	20 44	11 32	20 07	12 09	19 02	13 14
28	20 50	10 05	20 30	10 26	20 04	10 52	19 27	11 29	18 21	12 34
Nov. 7	20 09	9 24	19 49	9 45	19 23	10 11	18 46	10 47	17 40	11 53
17	19 27	8 43	19 07	9 03	18 41	9 29	18 04	10 06	16 59	11 11
27	18 45	8 00	18 25	8 21	17 59	8 47	17 22	9 23	16 17	10 29
Dec. 7	18 03	7 18	17 42	7 38	17 16	8 04	16 40	8 41	15 34	9 46
17	17 20	6 35	16 59	6 55	16 34	7 21	15 57	7 58	14 52	9 03
27	16 37	5 52	16 17	6 12	15 51	6 38	15 14	7 15	14 09	8 20

Brightest Stars

This table lists stars of greatest visual magnitude as seen in the night sky (the lower the number, the brighter the star). The common name of the star is in parentheses. Stars of variable magnitude are designated by v. Coordinates are for mid-2002. Greek letters in the star names indicate perceived degree of brightness within the constellation, alpha being the brightest.

To find the time when the star is on the meridian, subtract Right Ascension of Mean Sun (see the table Greenwich Sidereal Time for 0^h UTC) from the star's Right Ascension, first adding 24h to the latter if necessary. Mark this result PM if less than 12h; if greater than 12, subtract 12h and mark the remainder AM.

Star	Magnitude	Parallax "	Light-yrs	Right ascen. h	Right ascen. m	Declination ° '
α Canis Majoris (Sirius)	−1.44v	0.379	9	6	45.2	−16 43
α Carinae (Canopus)	−0.62v	0.010	313	6	24.0	−52 42
α Bootis (Arcturus)	−0.05v	0.089	37	14	15.8	+19 10
α Centauri (Rigel Kentaurus)	−0.01	00.74	4.4	14	39.8	−60 51
α Lyrae (Vega)	0.03	0.129	25.3	18	37.0	+38 47
α Aurigae (Capella)	0.08v	0.077	42	5	16.8	+46 00
β Orionis (Rigel)	0.18v	0.004	773	5	14.6	− 8 12
α Canis Minoris (Procyon)	0.40	0.286	11	7	39.4	+ 5 13
α Eridani (Achernar)	0.45v	0.023	144	1	37.8	−57 13
α Orionis (Betelgeuse)	0.45v	0.008	427	5	55.3	+ 7 24
β Centauri (Hadar)	0.61v	0.006	525	14	04.0	−60 23
α Aquilae (Altair)	0.76v	0.194	16.8	19	50.9	+ 8 52
α Crucis (Acrux)	0.77	0.010	320	12	26.7	−63 07
α Tauri (Aldebaran)	0.87v	0.050	65	4	36.0	+16 31
α Virginis (Spica)	0.98v	0.012	262	13	25.3	−11 10
α Scorpii (Antares)	1.06v	0.005	604	16	29.6	−26 26
β Geminorum (Pollux)	1.16v	0.097	34	7	45.4	+28 01
α Piscis Austrinis (Fomalhaut)	1.17	0.130	25.1	22	57.8	−29 36
β Crucis (Becrux)	1.25v	0.009	352	12	47.8	−59 42
α Cygni (Deneb)	1.25v	0.001	3228	20	41.5	+45 17
α Leonis (Regulus)	1.36	0.042	77	10	08.5	+11 57
ε Canis Majoris (Adhara)	1.50v	0.008	431	6	58.7	−28 58
α Geminorum (Castor)	1.58	0.063	52	7	34.7	+31 53
γ Crucis (Gacrux)	1.59v	0.037	88	12	31.3	−57 08
λ Scorpii (Shaula)	1.62v	0.005	703	17	33.8	−37 06
γ Orionis (Bellatrix)	1.64v	0.013	243	5	25.2	+ 6 21
β Tauri (Elnath)	1.65	0.025	131	5	26.4	+28 37
β Carinae (Miaplacidus)	1.67v	0.029	111	9	13.2	−69 44
ε Orionis (Alnilam)	1.69v	0.002	1341	5	36.3	− 1 12
α Gruis (Al Nair)	1.73v	0.032	101	22	08.4	−46 57
ζ Orionis (Alnitak)	1.74	0.004	836	5	40.9	− 1 57
γ Velorum (Al Suhail)	1.75v	0.004	840	8	09.6	−47 21
ε Ursae Majoris (Alioth)	1.76v	0.040	81	12	54.1	+55 57
ε Sagittarii (Kaus Australis)	1.79	0.023	145	18	24.3	−34 23
α Persei (Mirfak)	1.79v	0.006	592	3	24.5	+49 52
α Ursae Majoris (Dubhe)	1.81	0.026	124	11	03.8	+61 45
δ Canis Majoris (Wezen)	1.83v	0.002	1791	7	08.5	−26 24
η Ursae Majoris (Alkaid)	1.85v	0.032	101	13	47.6	+49 18
θ Scorpii	1.86	0.012	272	17	37.5	−43 00
ε Carinae (Avior)	1.86v	0.005	632	8	22.5	−59 31
β Aurigae (Menkalinan)	1.90v	0.040	82	5	59.7	+44 57
α Trianguli Australis (Atria)	1.91v	0.008	415	16	49.0	−69 02
γ Geminorum (Alhena)	1.93	0.031	105	6	37.8	+16 24
δ Velorum	1.93	0.041	80	8	44.6	−54 43
α Pavonis (Peacock)	1.94v	0.018	183	20	25.9	−56 44
α Ursae Minoris (Polaris)	1.97v	0.008	431	2	33.3	+89 16
β Canis Majoris (Mirzam)	1.98v	0.007	499	6	22.8	−17 57
α Hydrae (Alphard)	1.99v	0.018	177	9	27.7	− 8 40
α Arietis (Hamal)	2.01	0.049	66	2	07.3	+23 28
γ Leonis (Algieba)	2.01v	0.026	126	10	20.1	+19 50
β Ceti (Deneb Kaitos)	2.04v	0.034	96	0	43.7	−17 58
σ Sagittarii (Nunki)	2.05v	0.015	224	18	55.4	−26 18
θ Centauri (Menkent)	2.06	0.054	61	14	06.8	−36 23
α Andromedae (Alpheratz)	2.07v	0.034	97	0	08.5	+29 06
β Andromedae (Mirach)	2.07v	0.016	199	1	09.8	+35 38
β Gruis	2.07v	0.019	170	22	42.8	−46 52
κ Orionis (Saiph)	2.07v	0.005	721	5	47.8	− 9 40
β Ursae Minoris (Kochab)	2.07v	0.026	126	14	50.8	+74 09
α Ophiuchi (Rasalhague)	2.08	0.070	47	17	35.1	+12 34
β Persei (Algol)	2.09v	0.035	93	3	08.3	+40 58
γ Andromedae (Almaak)	2.10	0.009	355	2	04.0	+42 20
β Leonis (Denebola)	2.14	0.090	36	11	49.2	+14 34
γ Cassiopeiae	2.15v	0.005	613	0	56.8	+60 43
γ Centauri	2.20	0.025	130	12	41.6	48 59
ι Carinae (Tureis)	2.21	0.005	692	9	17.1	−59 17
ζ Puppis (Naos)	2.21v	0.002	1399	8	03.6	−40 01
α Coronae Borealis (Alphecca)	2.22v	0.044	75	15	34.8	+26 43
ζ Ursae Majoris (Mizar)	2.23	0.042	78	13	24.0	+54 55
γ Cygni (Sadr)	2.23v	0.002	1523	20	22.3	+40 16
λ Velorum (Suhail)	2.23v	0.006	573	9	08.1	−43 27
γ Draconis (Eltanin)	2.24v	0.022	148	17	56.7	+51 29
δ Orionis (Mintaka)	2.25v	0.004	916	5	32.1	− 0 18
β Cassiopeiae (Caph)	2.28v	0.060	54	0	09.3	+59 09
ε Scorpii	2.29	0.050	65	16	50.3	−34 18
ε Centauri	2.29v	0.009	376	13	40.0	−53 29
δ Scorpii (Dschubba)	2.29v	0.008	401	16	00.5	−22 38
α Lupi	2.30v	0.006	548	14	42.1	−47 24
η Centauri	2.33v	0.011	308	14	35.7	−42 10
β Ursae Majoris (Merak)	2.34	0.041	79	11	02.0	+56 23
ε Boo (Izar)	2.35	0.016	210	14	44.9	+27 04
κ Scorpii	2.39v	0.007	464	17	42.7	−39 02

IT'S A FACT: The star Sirius—also known as the Dog Star—is visible low on the horizon in August. The ancients believed this bright star, in conjunction with the Sun, caused the temperature to rise in late summer. Hence, the expression "the Dog Days of Summer."

Morning and Evening Stars, 2002

(Coordinated Universal Time)

	Morning	Evening
Jan.	Mercury from Jan. 27 Venus to Jan. 14 Jupiter to Jan. 1 Neptune from Jan. 28 Pluto	Mercury to Jan. 27 Venus from Jan. 14 Mars Jupiter from Jan. 1 Saturn Uranus Neptune to Jan. 28
Feb.	Mercury Uranus from Feb. 13 Neptune Pluto	Venus Mars Jupiter Saturn Uranus to Feb. 13
Mar.	Mercury Uranus Neptune Pluto	Venus Mars Jupiter Saturn
Apr.	Mercury to Apr. 7 Uranus Neptune Pluto	Mercury from Apr. 7 Venus Mars Jupiter Saturn

	Morning	Evening
May	Mercury from May 27 Uranus Neptune Pluto	Mercury to May 27 Venus Mars Jupiter Saturn
June	Mercury Saturn from June 9 Uranus Neptune Pluto to June 7	Venus Mars Jupiter Saturn to June 9 Pluto from June 7
July	Mercury to July 21 Jupiter from July 20 Saturn Uranus Neptune	Mercury from July 21 Venus Mars Jupiter to July 20 Pluto
Aug.	Mars from Aug. 10 Jupiter Saturn Uranus to Aug. 20 Neptune to Aug. 2	Mercury Venus Mars to Aug. 10 Uranus from Aug. 20 Neptune from Aug. 2 Pluto
Sept.	Mercury from Sept. 27 Mars Jupiter Saturn Pluto	Mercury to Sept. 27 Venus Uranus Neptune
Oct.	Mercury Venus from Oct. 31 Mars Jupiter Saturn	Venus to Oct. 31 Uranus Neptune Pluto
Nov.	Mercury to Nov. 14 Venus Mars Jupiter Saturn	Mercury from Nov. 14 Uranus Neptune Pluto
Dec.	Venus Mars Jupiter Saturn to Dec. 17 Pluto from Dec. 9	Mercury Saturn from Dec. 17 Uranus Neptune Pluto to Dec. 9

Greenwich Sidereal Time for 0^h UTC, 2002

(Add 12 hours to obtain Right Ascension of Mean Sun)

Date	d	h	m	Date	d	h	m	Date	d	h	m	Date	d	h	m
Jan.	1	6	41.9	**Apr.**	1	12	36.7	**July**	10	19	11.0	**Oct.**	8	1	05.8
	11	7	21.3		11	13	16.2		20	19	50.4		18	1	45.3
	21	8	00.8		21	13	55.6		30	20	29.8		28	2	24.7
	31	8	40.2	**May**	1	14	35.0	**Aug.**	9	21	09.3	**Nov.**	7	3	04.1
Feb.	10	9	19.6		11	15	14.4		19	21	48.7		17	3	43.5
	20	9	59.0		21	15	53.9		29	22	28.1		27	4	23.0
Mar.	2	10	38.5		31	16	33.3	**Sept.**	8	23	07.6	**Dec.**	7	5	02.4
	12	11	17.9	**June**	10	17	12.7		18	23	47.0		17	5	41.8
	22	11	57.3		20	17	52.1		28	0	26.4		27	6	21.2
					30	18	31.6								

The Zodiac

The Sun's apparent yearly path among the stars is known as the **ecliptic**. The zone, 18° wide, 9° on each side of the ecliptic, is known as the **zodiac**. Inside this zone are the apparent paths of the Sun, Moon, Earth, and the other planets. Only Pluto regularly strays outside this band on the celestial sphere. The zodiac is used both astrologically and astronomically. Though the two had a common beginning, they are no longer the same.

Beginning at the point on the ecliptic that marks the position of the Sun at the vernal equinox and proceeding eastward, the **astrological zodiac** is divided into 12 signs of approximately 30° each. These signs are named from the 12 constellations of the zodiac with which the signs coincided in the time of the astronomer Hipparchus, about 2,000 years ago.

Owing to the precession of the equinoxes, that is to say, to the retrograde motion of the equinoxes along the ecliptic, each sign in the zodiac has, in the course of 2,000 years, moved backward about 30° into the constellation W of it; the sign Aries is now in the constellation Pisces, for example, and so on. The vernal equinox will move from Pisces into Aquarius about the middle of the 26th century.

The **astronomical** constellations of the **zodiac,** unlike the astrological signs, are not equal in size. The ecliptic actually moves through parts of 13, not 12, astronomical constellations, the 13th being Ophiuchus. Also, the constellation of the scorpion is called Scorpius, while the sign is called Scorpio. In actuality, the planets (other than Pluto) may appear in parts of 21 different constellations

The traditional **signs of the zodiac,** with their Latin and English names are given in the next column.

Spring	1.	♈	Aries	The Ram
	2.	♉	Taurus	The Bull
	3.	♊	Gemini	The Twins
Summer	4.	♋	Cancer	The Crab
	5.	♌	Leo	The Lion
	6.	♍	Virgo	The Virgin
Autumn	7.	♎	Libra	The Balance
	8.	♏	Scorpio	The Scorpion
	9.	♐	Sagittarius	The Archer
Winter	10.	♑	Capricorn	The Goat
	11.	♒	Aquarius	The Water Bearer
	12.	♓	Pisces	The Fishes

On Mar. 28, 2002, the disk of the Sun clips a corner of the constellation of Cetus. The constellations of the **astronomical** (not astrological) **zodiac,** with the approximate dates that the Sun is in each constellation in 2002, are as follows:

Jan. 1	- Jan. 19	Sagittarius
Jan. 19	- Feb. 16	Capricornus
Feb. 16	- Mar. 12	Aquarius
Mar. 12	- Apr. 19	Pisces
Apr. 19	- May 14	Aries
May 14	- June 21	Taurus
June 21	- July 20	Gemini
July 20	- Aug. 10	Cancer
Aug. 10	- Sept. 16	Leo
Sept. 16	- Oct. 31	Virgo
Oct. 31	- Nov. 22	Libra
Nov. 22	- Nov. 30	Scorpius
Nov. 30	- Dec. 18	Ophiuchus
Dec. 18	- Dec. 31	Sagittarius

Constellations

Culturally, constellations are imagined patterns among the stars that, in some cases, have been recognized through millennia. Knowledge of constellations was once necessary in order to function as an astronomer. For today's astronomers, constellations are simply areas on the entire sky in which interesting objects await observation and interpretation.

Because Western culture has prevailed in establishing modern science, equally viable and interesting constellations and celestial traditions of other cultures are not well known outside their regions of origin. Even the patterns with which we are most familiar today have undergone considerable change over the centuries.

Today, 88 constellations are officially recognized. Although many have ancient origins, some are "modern," devised out of unclaimed stars by astronomers a few centuries ago. Unclaimed stars were those too faint or inconveniently placed to be included in the more prominent constellations. Stars in a constellation are not necessarily near each other; they are just located in the same direction on the celestial sphere.

When astronomers began to travel to S Africa in the 16th and 17th centuries, they found an unfamiliar sky that showed numerous brilliant stars. Thus, we find constellations in the southern hemisphere that depict technological marvels of the time, as well as some arguably traditional forms, such as the "fly."

Many of the commonly recognized constellations had their origins in ancient Asia Minor. These were adopted by the Greeks and Romans, who translated their names and stories into their own languages, modifying some details in the process. After the declines of these cultures, most such knowledge entered oral tradition or remained hidden in monastic libraries. From the 8th century, the Muslim explosion spread through the Mediterranean world. Wherever possible, everything was translated into Arabic to be taught in the universities the Muslims established all over their new-found world.

In the 13th century, Alfonso X of Castile, an avid student of astronomy, had Ptolemy's *Almagest* translated into Latin. It thus became widely available to European scholars. In the process, the constellation names were translated, but the star names were retained in their Arabic forms. Transliterating Arabic into the Roman alphabet has never been an exact art, so many of the star names we use today seem Arabic only to those who are not scholars.

Until the 1920s, astronomers used curved boundaries for the constellation areas. As these were rather arbitrary at best, the International Astronomical Union adopted new constellation boundaries that ran due north-south and east-west, filling the sky much as the contiguous states fill up the area of the "lower 48" United States.

Common names of stars often indicated what parts of the traditional figures they represented: Deneb, the tail of the swan; Betelgeuse, the armpit of the giant. Avoiding traditional names, astronomers may label stars by using Greek letters, generally to denote order of brightness. Thus, the "alpha star" would generally be the brightest star of that constellation. The "of" implies possession, so the genitive (possessive) form of the constellation name is used, as in Alpha Orionis, the first star of Orion (Betelgeuse). Astronomers usually use a 3-letter abbreviation for the constellation name, as indicated here.

Within these boundaries, and occasionally crossing them, popular "asterisms" are recognized: the so-called Big Dipper is a small part of the constellation Ursa Major, the big bear; the Sickle is the traditional head and mane of Leo, the lion; one of the horn tips of Taurus, the bull, properly belongs to Auriga, the charioteer; the northeast star of the Great Square of Pegasus is Alpha Andromedae.

Name	Genitive Case	Abbr.	Meaning
Andromeda	Andromedae	And	Chained Maiden
Antlia	Antliae	Ant	Air Pump
Apus	Apodis	Aps	Bird of Paradise
Aquarius	Aquarii	Aqr	Water Bearer
Aquila	Aquilae	Aql	Eagle
Ara	Arae	Ara	Altar
Aries	Arietis	Ari	Ram
Auriga	Aurigae	Aur	Charioteer
Boötes	Boötis	Boo	Herdsmen
Caelum	Caeli	Cae	Chisel
Camelopardalus	Camelopardalis	Cam	Giraffe
Cancer	Cancri	Cnc	Crab
Canes Venatici	Canum Venaticorum	CVn	Hunting Dogs
Canis Major	Canis Majoris	CMa	Greater Dog
Canis Minor	Canis Minoris	CMi	Littler Dog
Capricornus	Capricorni	Cap	Sea-goat
Carina	Carinae	Car	Keel
Cassiopeia	Cassiopeiae	Cas	Queen
Centaurus	Centauri	Cen	Centaur
Cepheus	Cephei	Cep	King
Cetus	Ceti	Cet	Whale
Chamaeleon	Chamaeleontis	Cha	Chameleon
Circinus	Circini	Cir	Compasses (art)
Columba	Columbae	Col	Dove
Coma Berenices	Comae Berenices	Com	Berenice's Hair
Corona Australis	Coronae Australis	CrA	Southern Crown
Corona Borealis	Coronae Borealis	CrB	Northern Crown
Corvus	Corvi	Crv	Crow
Crater	Crateris	Crt	Cup
Crux	Crucis	Cru	Cross (southern)
Cygnus	Cygni	Cyg	Swan
Delphinus	Delphini	Del	Dolphin
Dorado	Doradus	Dor	Goldfish
Draco	Draconis	Dra	Dragon
Equuleus	Equulei	Equ	Little Horse
Eridanus	Eridani	Eri	River
Fornax	Fornacis	For	Furnace
Gemini	Geminorum	Gem	Twins
Grus	Gruis	Gru	Crane (bird)
Hercules	Herculis	Her	Hercules
Horologium	Horologii	Hor	Clock
Hydra	Hydrae	Hya	Water Snake (female)
Hydrus	Hydri	Hyi	Water Snake (male)
Indus	Indi	Ind	Indian
Lacerta	Lacertae	Lac	Lizard
Leo	Leonis	Leo	Lion
Leo Minor	Leonis Minoris	LMi	Littler Lion
Lepus	Leporis	Lep	Hare
Libra	Librae	Lib	Balance
Lupus	Lupi	Lup	Wolf
Lynx	Lyncis	Lyn	Lynx
Lyra	Lyrae	Lyr	Lyre
Mensa	Mensae	Men	Table Mountain
Microscopium	Microscopii	Mic	Microscope
Monoceros	Monocerotis	Mon	Unicorn
Musca	Muscae	Mus	Fly
Norma	Normae	Nor	Square (rule)
Octans	Octantis	Oct	Octant
Ophiuchus	Ophiuchi	Oph	Serpent Bearer
Orion	Orionis	Ori	Hunter
Pavo	Pavonis	Pav	Peacock
Pegasus	Pegasi	Peg	Flying Horse
Perseus	Persei	Per	Hero
Phoenix	Phoenicis	Phe	Phoenix
Pictor	Pictoris	Pic	Painter
Pisces	Piscium	Psc	Fishes
Piscis Austrinius	Piscis Austrini	PsA	Southern Fish
Puppis	Puppis	Pup	Stern (deck)
Pyxis	Pyxidis	Pyx	Compass (sea)
Reticulum	Reticuli	Ret	Reticle
Sagitta	Sagittae	Sge	Arrow
Sagittarius	Sagittarii	Sgr	Archer
Scorpius	Scorpii	Sco	Scorpion
Sculptor	Sculptoris	Scl	Sculptor
Scutum	Scuti	Sct	Shield
Serpens	Serpentis	Ser	Serpent
Sextans	Sextantis	Sex	Sextant
Taurus	Tauri	Tau	Bull
Telescopium	Telescopii	Tel	Telescope
Triangulum	Trianguli	Tri	Triangle
Triangulum Australe	Trianguli Australis	TrA	Southern Triangle
Tucana	Tucanae	Tuc	Toucan
Ursa Major	Ursae Majoris	UMa	Greater Bear
Ursa Minor	Ursae Minoris	UMi	Littler Bear
Vela	Velorum	Vel	Sail
Virgo	Virginis	Vir	Maiden
Volans	Volantis	Vol	Flying Fish
Vulpecula	Vulpeculae	Vul	Fox

Aurora Borealis and Aurora Australis

The **Aurora Borealis,** also called the **Northern Lights,** is a broad display of rather faint light in the northern skies at night. The **Aurora Australis,** a similar phenomenon, appears at the same time in southern skies. The aurora appears in a wide variety of forms. Sometimes it is seen as a quiet glow, almost foglike in character; sometimes as vertical streamers in which there may be considerable motion; sometimes as a series of luminous expanding arcs. There are many colors, with white, yellow, and red predominating.

The auroras are most vivid and most frequently seen at about 20° from the magnetic poles, along the northern coast of the N American continent and the eastern part of the northern coast of Europe. The Aurora Borealis has been seen as far S as Key West, and the Aurora Australis has been seen as far N as Australia and New Zealand. Such occurrences are rare, however.

The Sun produces a stream of charged particles, called the solar wind. These particles, mainly electrons and protons, approach Earth at speeds on the order of 300 mi per second. Coronal mass ejections are large-scale, high-speed releases of as much as 10 billion tons of coronal material. Some of these particles are trapped by Earth's magnetic field, forming the Van Allen belts—2 donut-shaped radiation bands around Earth. Excess amounts of these charged particles, often produced by solar flares, follow Earth's magnetic lines of force toward Earth's magnetic poles. High in the atmosphere, collisions between solar and terrestrial atoms result in the glow in the upper atmosphere called the aurora. The glow may be vivid where the lines of magnetic force converge near the magnetic poles.

The auroral displays appear at heights ranging from 50 to about 600 mi and have given us a means of estimating the extent of Earth's atmosphere.

The auroras are often accompanied by magnetic storms whose forces, also guided by the lines of force of Earth's magnetic field, disrupt electrical communication. On Apr. 6, 2000, an unusually intense geomagnetic storm triggered auroral displays as far south as Florida and Texas. In February 2001, the Sun's magnetic field reversed, a strong indication that the Sun had reached its peak of the current solar cycle (#23). The higher level of sunspots is expected to have a continued effect on both aurora and electrical communication.

Eclipses, 2002

(in Coordinated Universal Time, standard time of the prime meridian)

There are five eclipses in 2002, a total eclipse of the Sun, an annular eclipse of the Sun, and three penumbral eclipses of the Moon.

I. Penumbral eclipse of the Moon, May 26

Penumbral eclipses of the Moon are not very noticeable, since direct sunlight still reaches all portions of the daytime side of the Moon. Unlike partial or total lunar eclipses, there is no distinct shadow (the umbra) observable on the Moon. The beginning of the eclipse will be visible in most of N America, Central America, western S America, eastern Asia, Australia, most of Antarctica, the Pacific Ocean, and the SE Indian Ocean. The end of the eclipse will be visible in SW Alaska, most of Asia, Australia, the E Indian Ocean, and most of the Pacific Ocean.

Circumstances of the Eclipse

Event	Date		h	m
Eclipse begins	May	26	10	12.7
Middle of eclipse		26	12	3.3
Eclipse ends		26	13	53.8

II. Annular eclipse of the Sun, June 10-11

The path of annularity starts in Indonesia, crosses the Pacific, and ends at the W coast of Mexico.

Circumstances of the Eclipse

Event	Date		h	m
Partial eclipse begins	June	10	20	51.8
Annular eclipse begins		10	21	54.2
Central eclipse at midday		10	23	48.2
Annular eclipse ends		11	1	34.2
Partial eclipse ends		11	2	36.6

III. Penumbral eclipse of the Moon, June 24

As described above, penumbral eclipses of the Moon are not very noticeable. The beginning of the eclipse will be visible in Australia, Indonesia, S and W Asia, most of Europe, Africa, Antarctica, the Indian Ocean, the E and S Atlantic Ocean, and the SW Pacific Ocean. The end of the eclipse will be visible in Africa, most of Europe, most of S America, Antarctica, W Australia, SW Asia, the Indian Ocean, the E and S Atlantic Ocean, and the SE Pacific Ocean.

Circumstances of the Eclipse

Event	Date		h	m
Eclipse begins	June	24	20	18.4
Middle of eclipse		24	21	27.1
Eclipse ends		24	22	35.4

IV. Penumbral eclipse of the Moon, Nov. 19-20

As described above, penumbral eclipses of the Moon are not very noticeable. The beginning of the eclipse will be visible in Africa, Europe, Greenland, most of N America, the Arctic region, Central America, most of S America, the Atlantic Ocean, and the W Indian Ocean. The end of the eclipse will be visible in N America, the Arctic region, Central America, S America, Greenland, Europe, NW Russia, the W Middle East, W Africa, the Antarctica Peninsula, the Atlantic Ocean, and the E Pacific Ocean.

Circumstances of the Eclipse

Event	Date		h	m
Eclipse begins	Nov.	19	23	32.0
Middle of eclipse		20	1	46.5
Eclipse ends		20	4	1.1

V. Total eclipse of the Sun, Dec. 4

The path of totality begins in the S Atlantic, crosses S Africa, crosses the S Indian Ocean, to end in Australia.

Circumstances of the Eclipse

Event	Date		h	m
Partial eclipse begins	Dec.	4	4	51.3
Total eclipse begins		4	5	50.4
Central eclipse at midday		4	7	38.7
Total eclipse ends		4	9	11.8
Partial eclipse ends		4	10	11.0

Eclipses in the U.S. in the 21st Century

During the 21st century Halley's Comet will return (2061-62), and there will be 8 total solar eclipses that are visible somewhere in the continental United States. The first comes after a long gap; the last one to be seen there was on Feb. 26, 1979, in the northwestern U.S.

Date	Path of Totality
Aug. 21, 2017	Oregon to South Carolina
Apr. 8, 2024	Mexico to Texas and up through Maine
Aug. 23, 2044	Montana to North Dakota
Aug. 12, 2045	N California to Florida
Mar. 30, 2052	Florida to Georgia
May 11, 2078	Louisiana to North Carolina
May 1, 2079	New Jersey to the lower edge of New England
Sept. 14, 2099	North Dakota to Virginia

Total Solar Eclipses, 1961-2025

Total solar eclipses actually take place nearly as often as total lunar eclipses; they occur at a rate of about 3 every 4 years, while total lunar eclipses come at a rate of about 5 every 6 years. However, total lunar eclipses are visible over at least half of the Earth, while total solar eclipses can be seen only along a very narrow path up to a few hundred miles wide and a few thousand miles long. Observing a total solar eclipse is thus a rarity for most people. Unlike lunar eclipses, solar eclipses can be dangerous to observe. This is not because the Sun emits more potent rays during a solar eclipse, but because the Sun is always dangerous to observe directly and people are particularly likely to stare at it during a solar eclipse.

Date	Duration[1] m	s	Width (mi)	Path of Totality
1961, Feb. 15	2	45	160	Europe, Soviet Union
1962, Feb. 5	4	8	91	Borneo, New Guinea, Pacific Ocean
1963, July 20	1	39	63	Pacific Ocean, Alaska, Canada, Maine
1965, May 30	5	15	123	New Zealand, Pacific Ocean
1966, Nov. 12	1	57	52	Pacific Ocean, S America, Atlantic Ocean
1968, Sept. 22	0	39	64	Soviet Union, China
1970, Mar. 7	3	27	95	Pacific Ocean, Mexico, Eastern U.S., Canada
1972, July 10	2	35	109	Siberia, Alaska, Canada
1973, June 30	7	3	159	Atlantic Ocean, Central Africa, Indian Ocean
1974, June 20	5	8	214	Indian Ocean, Australia
1976, Oct. 23	4	46	123	Africa, Indian Ocean, Australia
1977, Oct. 12	2	37	61	Pacific Ocean, Colombia, Venezuela
1979, Feb. 26	2	49	185	NW U.S., Canada, Greenland
1980, Feb. 16	4	8	92	Africa, Indian Ocean, India, Burma, China
1981, July 31	2	2	67	Soviet Union, Pacific Ocean
1983, June 11	5	10	123	Indian Ocean, Indonesia, New Guinea
1984, Nov. 22	1	59	53	New Guinea, Pacific Ocean
1985, Nov. 12	1	58	430	Antarctica
1986, Oct. 3h	0	1	1	N Atlantic Ocean
1987, Mar. 29h	0	7	3	S Atlantic Ocean, Africa
1988, Mar. 18	3	46	104	Sumatra, Borneo, Philippines, Pacific Ocean
1990, July 22	2	32	125	Finland, Soviet Union, Aleutian Islands
1991, July 11	6	53	160	Hawaii, Mexico, Central America, Colombia, Brazil
1992, June 30	5	20	182	S Atlantic Ocean
1994, Nov. 3	4	23	117	Peru, Bolivia, Paraguay, Brazil
1995, Oct. 24	2	9	48	Iran, India, SE Asia
1997, Mar. 9	2	50	221	Mongolia, Siberia
1998, Feb. 26	4	8	94	Galapagos Islands, Panama, Colombia, Venezuela
1999, Aug. 11	2	22	69	Europe, Middle East, India
2001, June 21	4	56	125	Atlantic Ocean, Africa, Madagascar
2002, Dec. 4	2	4	54	S Africa, Indian Ocean, Australia
2003, Nov. 23	1	57	338	Antarctica
2005, Apr. 8h	0	42	17	Pacific Ocean, NW S America
2006, Mar. 29	4	7	118	Atlantic Ocean, Africa, Asia
2008, Aug. 1	2	27	157	Arctic Ocean, Asia
2009, July 22	6	39	160	Asia, Pacific Ocean
2010, July 11	5	20	164	Pacific Ocean, southern S America
2012, Nov. 13	4	2	112	N Australia, Pacific Ocean
2013, Nov. 3h	1	40	36	Atlantic Ocean, Africa
2015, Mar. 20	2	47	304	N Atlantic Ocean, Arctic Ocean
2016, Mar. 9	4	10	96	Indonesia, Pacific Ocean
2017, Aug. 21	2	40	71	Pacific Ocean, U.S., Atlantic Ocean
2019, July 2	4	33	125	S Pacific Ocean, S America
2020, Dec. 14	2	10	56	S Pacific Ocean, S America, S Atlantic Ocean
2021, Dec. 4	1	55	282	Antarctica, S Atlantic Ocean
2023, Apr. 20h	1	16	31	Indian Ocean, New Guinea, Pacific Ocean
2024, Apr. 8	4	28	127	Pacific Ocean, Mexico, N America, Atlantic Ocean

h = indicates annular-total hybrid eclipse. (1) Duration refers to length of time at optimal viewing area.

Beginnings of the Universe

One of the dominating astronomical discoveries of the 20th century was the realization that the galaxies of the universe all seem to be moving away from us. It turned out that they are moving away not just from us but from one another—that is, the universe seems to be expanding. Hence, scientists conclude that the universe must once, very long ago, have been extremely compact and dense. Although there are alternatives to this theory, much of the observational evidence currently available supports the idea that the universe we know began its existence between 8 and 20 bil years ago as an explosion of a super-dense, super-small concentration of matter.

This explosion of matter giving birth to the universe is called the **Big Bang.** On the subatomic level, according to this theory, there were vast changes of energy and matter and the way physical laws operated during the first 5 minutes. After those minutes the percentages of the basic matter of the universe—hydrogen, helium, and lithium—were set. Everything was so compact and so hot that radiation dominated the early universe and there were no stable, un-ionized atoms. At first, the universe was opaque, in the sense that any energy emitted was quickly absorbed and then re-emitted by free electrons. As the universe expanded, the density and the temperature continued to drop. A few hundred thousand years after the initial Big Bang, the temperature had dropped far enough that electrons and nuclei could combine to form stable atoms as the universe became transparent. Once that had occurred, the radiation which had been trapped was free to escape.

In the 1940s, George Gamov and others predicted that astronomers should be able to see remnants of this escaped radiation. Astronomers continued to refine the theories and were preparing to build equipment to search for this background radiation when physicists Arno Penzias and Robert Wilson of the Bell Telephone Laboratories inadvertently beat them to the punch (the 2 were later awarded a Nobel Prize). Despite the Big Bang's success at predicting the existence of **cosmic background radiation,** there are still many unresolved questions, and astronomers are still working on modifications of the theory.

A possibly related mystery is the evidence available on the scale of galaxies which suggests that there is hidden matter that cannot directly be seen. This **dark matter** may be composed of gas, large numbers of cool, small objects, or even sub-atomic particles. The visible matter we see seems to constitute only about 10% of the total mass of the universe. In June 2001, NASA launched the MAP (Microwave Anisotropy Probe) to study the cosmic microwave background radiation in greater detail than previously and attempt to better understand the early history of the universe.

The Solar System

The planets of the solar system, in order of mean distance from the Sun, are Mercury, Venus, Earth, Mars, Jupiter, Saturn, Uranus, Neptune, and Pluto (Pluto sometimes nearer than Neptune). Both Uranus and Neptune are visible through good binoculars, but Pluto is so distant and so small that only large telescopes or long-exposure photographs can make it visible. All the planets orbit or revolve counterclockwise around the Sun.

Because Mercury and Venus are nearer to the Sun than is Earth, their motions about the Sun are seen from Earth as wide swings first to one side of the Sun then to the other, though both planets move continuously around the Sun in almost circular orbits. When their passage takes them either between Earth and the Sun or beyond the Sun as seen from Earth, they are invisible to us. Because of geometry of the planetary orbits, Mercury and Venus require much less time to pass between Earth and the Sun than around the far side of the Sun; so their periods of visibility and invisibility are unequal.

The planets that lie farther from the Sun than does Earth may be seen for longer periods and are invisible only when so located in our sky that they rise and set at about the same time as the Sun—and thus become overwhelmed by the Sun's great brilliance. Although several of the giant planets emit their own energy, they are observed from Earth as a result of sunlight reflecting from their surfaces or cloud layers. Mercury and Venus, because they are between Earth and the Sun, show phases very much as the Moon does. The planets farther from the Sun are always seen as full, although Mars does occasionally present a slightly gibbous phase—like the Moon when not quite full.

The planets appear to move rapidly among the stars because of being closer. The stars are also in motion, some at tremendous speeds, but they are so far away that their motion does not change their apparent positions in the heavens sufficiently to be perceived. The nearest star is about 7,000 times farther away than the most distant planet in our solar system.

Planets and the Sun, by Selected Characteristics

Sun and Planets	Semi-Diameter: at unit distance "	at mean least distance "	in mi mean s.d.	Volume[1]	Mass[1]	Density[1]	Sidereal period d	h	m	s	Gravity at surface[1]	Reflecting power Pct°	Daytime surface temp. ° F
Sun	959.5	976	432,474	1,304,000	332,950	0.26	25	9	7	12	28.0		+9,941
Mercury	3.36	6.3	1,516	0.056	0.0553	0.98	58	15	36		0.38	0.11	846
Venus	8.34	33	3,760	0.857	0.815	0.95	243	12	R		0.91	0.65	867
Earth	8.8		3,959	1.000	1.000	1.00		23	56	4.2	1.00	0.37	68
Moon	2.40	986.4	1,080	0.0203	0.0123	0.60	27	7	43	41	0.17	0.12	260
Mars	4.67	12.85	2,106	0.151	0.107	0.71		24	37	22	0.38	0.15	–24
Jupiter	96.39	24.5	43,441	1,321	317.83	0.24		9	55	30	2.36	0.52	–162
Saturn	80.29	10.05	36,184	764	95.16	0.12		10	39	22	0.92	0.47	–218
Uranus	34.97	2.05	15,759	63	14.54	0.23		17	14	24R	0.89	0.51	–323
Neptune	33.95	1.2	15,301	58	17.15	0.30		16	6	36	1.12	0.41	–330
Pluto	1.65	0.06	743	0.007	0.0021	0.32	6	9	17	34R	0.06	0.3	–369

(1) Earth = 1. R= Retrograde rotation.

Planet Superlatives

Largest, most massive, planet	Jupiter	Smallest, least massive planet	Pluto
Fastest orbiting planet	Mercury	Slowest orbiting planet	Pluto
Most eccentric orbit	Pluto	Most circular orbit	Venus
Longest (synodic) day	Mercury	Shortest (synodic) day	Jupiter
Coldest planet	Pluto	Hottest planet	Venus
Most moons	Saturn (30)	No moons	Mercury, Venus
Planet with largest moon	Jupiter	Planet with moon with most eccentric orbit	Neptune
Greatest average density	Earth	Lowest average density	Saturn
Tallest mountain	Mars	Deepest oceans	Jupiter

Largest Telescopes

Astronomers indicate the size of telescopes not by length or magnification, but by the diameter of the primary light- gathering component of the system—such as the lens or mirror. This measurement is a direct indication of the telescope's light-gathering power. The bigger the diameter, the fainter the objects you are enabled to see. For larger telescopes, the Earth's atmosphere limits the resolution of what you see. That is why the Hubble Space Telescope, which is outside the atmosphere, can have better resolution than larger telescopes on the Earth. Large mirror telescopes can be made less expensively than large lens telescopes, so all modern large optical telescopes are made with mirrors rather than lenses. Radio telescopes view at wavelengths not visible to optical telescopes, which are limited to the wavelengths detectable by the human eye. Radio telescopes have to be made larger than optical telescopes because resolving power requires larger diameters at longer wavelengths such as radio wavelengths.

Largest Refracting (lens) Optical Telescope:
Yerkes Observatory—1 m (40 in), at Williams Bay, WI

Largest Reflecting (mirror) Optical Telescope:
Keck—10 m (394 in), on Mauna Kea in Hawaii (segmented mirror; 2 equal-size telescopes)

Largest Space Telescope:
Hubble Space Telescope—2.4 m (94 in), in orbit around the Earth

Largest Single Radio Dish:
Arecibo Observatory—305 m (1,000 ft), in Puerto Rico

Largest Radio Interferometer:
10 telescopes of the Very Long Baseline Array (VLBA), scattered from Hawaii to the Virgin Islands with a resolution equal to a radio dish of 6,000 km (3,700 mi)

The Planets: Motion, Distance, and Brightness

Planet	Mean daily motion[1]	Orbital velocity mi per sec.[2]	Sidereal revolution days[3]	Synodic revolution days[4]	Distance from Sun in millions of mi Max.	Distance from Sun in millions of mi Min.	Distance from Earth in millions of mi Max.	Distance from Earth in millions of mi Min.	Light at[5] perihelion	Light at[5] aphelion
Mercury . . .	14,727	29.74	88.0	115.9	43.4	28.6	138	48	10.56	4.59
Venus	5,768	21.76	224.7	583.9	67.7	66.8	162	24	1.94	1.89
Earth	3,548	18.50	365.3	—	94.5	91.4	—		1.03	0.97
Mars	1,886	14.99	687.0	779.9	154.9	128.4	249	34	0.52	0.36
Jupiter	299	8.12	4,332.6	398.9	507.4	460.1	602	366	0.041	0.034
Saturn	120	6.02	10,759.2	378.1	941.1	840.4	1,031	743	0.012	0.0098
Uranus	42	4.23	30,685.4	369.7	1,866.4	1,703.4	1,962	1,604	0.0030	0.0025
Neptune . . .	22	3.37	60,189.0	367.5	2,824.5	2,761.6	2,913	2,676	0.0011	0.0011
Pluto	14	2.93	90,465.0	366.7	4,538.7	2,755.8	4,681	2,668	0.0011	0.00041

(1) Average angular motion measured in seconds of arc per day. (2) Speed of revolution around Sun. (3) Number of Earth days to orbit Sun with respect to background stars. (4) Number of Earth days to get back to the same position in its orbit around Sun, relative to Earth. (5) Light at perihelion and aphelion is solar illumination measured in units of mean illumination at Earth.

Planets of the Solar System

Note: AU = astronomical unit (92.96 mil mi, mean distance of Earth from the Sun); d = 1 Earth synodic (solar) day (24 hrs); synodic day = rotation period of a planet measured with respect to the Sun (the "true" day, i.e. the time from midday to midday, or from sunrise to sunrise); sidereal day = the rotation period of a planet with respect to the stars

Mercury

Distance from Sun	
Perihelion	28.6 mil mi
Semi-major axis	0.387 AU
Aphelion	43.4 mil mi
Period of revolution around Sun	87.97 d
Orbital eccentricity	0.2056
Orbital inclination	7.00°
Synodic day (midday to midday)	175.94 d
Sidereal day	58.65 d
Rotational inclination	0.01°
Mass (Earth = 1)	0.0553
Mean radius	1,516 mi
Mean density (Earth = 1)	0.984
Natural satellites	0
Average surface temperature	333° F

Mercury, the nearest planet to the Sun, is the 2d-smallest of the 9 known planets. Its diameter is 3,032 mi; its mean distance from the Sun is 35,980,000 mi.

Mercury moves with great speed around the Sun, averaging about 30 mi per second to complete its circuit in about 88 Earth days. Mercury rotates upon its axis over a period of nearly 59 days, thus exposing all its surface periodically to the Sun. Because its orbital period is only about 50% longer than its sidereal rotation, the solar (synodic) day on Mercury, or the time from one sunrise to the next, is about 176 days, twice as long as a Mercurian year. It is believed that the surface passing before the Sun may reach a temperature of about 845° F, while the temperature on the nighttime side may fall as low as –300° F.

Uncertainty about conditions on Mercury and its motion arises from its short angular distance from the Sun as seen from Earth. Mercury is too much in line with the Sun to be observed against a dark sky, but is always seen during either morning or evening twilight.

Mariner 10 passed Mercury 3 times in 1974 and 1975. Less than half of the surface was photographed, revealing a degree of cratering similar to that of the Moon. The most imposing feature on Mercury, the Caloris Basin, is a huge impact crater more than 800 mi in diameter. Mercury also has a higher percentage of iron than any other planet. A very thin atmosphere of hydrogen and helium may be made up of gases of the solar wind temporarily concentrated by the presence of Mercury. The discovery of a weak but permanent magnetic field was a surprise to scientists. It has been held that both a fluid core and rapid rotation are necessary for the generation of a planetary magnetic field. Mercury may demonstrate the contrary; the field may reveal something about the history of Mercury. In 1992, radar mapping of Mercury with radio telescopes on Earth revealed evidence of possible water ice near its north and south poles. In 2004, NASA plans to launch its *MESSENGER* spacecraft to orbit Mercury and provide further studies.

Venus

Distance from Sun	
Perihelion	66.8 mil mi
Semi-major axis	0.723 AU
Aphelion	67.7 mil mi
Period of revolution around Sun	224.70 d
Orbital eccentricity	0.0067
Orbital inclination	3.39°
Synodic day (midday to midday)	116.75 d (retrograde)
Sidereal day	243.02 d (retrograde)
Rotational inclination	177.4°
Mass (Earth = 1)	0.815
Mean radius	3,760 mi
Mean density (Earth = 1)	0.951
Natural satellites	0
Average surface temperature	867° F

Venus, slightly smaller than Earth, moves about the Sun at a mean distance of 67,240,000 mi in 225 Earth days. Its synodical revolution—its return to the same relationship with Earth and the Sun, which is a result of the combination of its own motion with that of Earth—is 584 days. As a result, every 19 months Venus is nearer to Earth than any other planet. Venus is covered with a dense, white, cloudy atmosphere that conceals whatever is below it. This same cloud reflects sunlight efficiently so that Venus is the 3d-brightest object in the sky, exceeded only by the Sun and the Moon.

Spectral analysis of sunlight reflected from Venus's cloud tops has shown features that can best be explained by identifying material of the clouds as sulfuric acid. In 1956, radio astronomers at the Naval Research Laboratories in Washington, DC, found a temperature for Venus of about 600° F. Subsequent data from the *Mariner 2* space probe in 1962 confirmed a high temperature. *Mariner 2* was unable to detect the existence of a magnetic field even as weak as 1/100,000 of Earth's magnetic field.

In 1967, a Soviet space probe, *Venera 4*, and the American *Mariner 5* arrived at Venus within a few hours of each other. *Venera 4* was designed to allow an instrument package to land gently on the surface, but it ceased to transmit information when its temperature reading went above 500° F, when it was still about 20 mi above the surface. The orbiting *Mariner 5*'s radio signals passed to Earth through Venus's atmosphere twice (once on the night side and once on the day side). The results were startling. Venus's atmosphere is nearly all carbon dioxide (96.5%), with 3.5% nitrogen and trace amounts of sulfur dioxide, carbon monoxide, argon, water, helium, and neon. It exerts a pressure at the planet's surface more than 90 times Earth's normal sea-level pressure of one atmosphere.

Because Earth and Venus are about the same size and were presumably formed at the same time by the same general process and from the same mixture of chemical elements, one is faced with the question: Why the difference? Recent measurements indicate that Venus has a surface temperature of over 865° F as a result of an extreme greenhouse effect. Because of the thick atmosphere, the temperature is essentially the same both day and night.

Radar astronomers determined the rotation period of Venus to be 243 days clockwise—in other words, contrary to the spin of the other planets and contrary to its own motion around the Sun. If it were exactly 243.16 days, Venus would present the same face toward Earth at every inferior conjunction. This rate and sense of rotation allows a solar day (sunrise to sunrise) on Venus of 116.8 Earth days. Any part of Venus will receive sunlight on its clouds for more than 58 days and then return to darkness for 58 days.

Mariner 10 passed Venus before traveling on to Mercury in 1974. The carbon dioxide found in abundance in the atmosphere is rather opaque to certain ultraviolet wavelengths, enabling sensitive cameras to photograph the cloud cover. Soviet spacecraft discovered that the clouds are confined in a 12-mi layer 30 to 42 mi above the surface.

In 1978, two U.S. *Pioneer* probes confirmed expected high surface temperatures and high winds aloft. Winds of about 200 mi per hour there may account for the transfer of heat into the night side despite the low rotation speed of the planet. However, at the surface, the winds are very slow. Soviet scientists obtained, in 1975 and later in 1982, 4 photos of surface rocks. Sulfur seems to play a large role in the chemistry of Venus, and reactions involving sulfur may be responsible for the glow. The *Pioneer* orbiter confirmed the cloud pattern and its circulation shown by *Mariner 10*. Radar produced maps of the entire planet showing large craters, continent-size highlands, and extensive dry lowlands.

The Venus orbiter *Magellan* launched in 1989 used sophisticated radar techniques to observe Venus and map more than 99% of the surface. The spacecraft observed over 1,600 volcanoes and volcanic features, enabling creation of a 3-dimensional map of the Venusian surface. *Magellan* has shown that more than 85% of the surface is covered by volcanic flows. Additionally, there are highly deformed mountain belts.

Craters more than 20 mi wide are believed to have been caused by impacting bodies. Theia Mons, a huge shield volcano, has a diameter of over 600 mi and a height of over 3.5 mi. (Compare this to the largest Hawaiian volcano, which is only about 125 mi in diameter, but with a height of nearly 5.5 mi from the ocean floor.)

Erosion is a very slow process on Venus due to the extreme lack of water, and features persist for long periods of time. There are indications of only restricted wind movement of dust and sand.

Tectonic actions on Venus are distinctly different from such actions on Earth. No activity on Venus has been found to be similar to Earth's moving tectonic plates, but a system of global rift zones and numerous broad, low dome-like structures, which are called coronae, may be produced by the upwelling and subsidence of magma from the mantle. Volcanic surface features, such as vast lava plains, fields of small lava domes, and large shield volcanoes, are common. The few impact craters on Venus suggest that the surface is generally geologically young—less than 800 million years old. A channel about 4,200 mi long, due to lava flows, has been mapped.

The orbit of *Magellan* was adjusted to a nearly circular shape about 300 mi from the planet's surface in 1993. In this mode, variation in *Magellan*'s orbital speed revealed information on irregularities in the gravitational field, presumably due to details in the internal structure of the planet. Although *Magellan* ceased operating in 1994, its data about the topography of Venus's surface kept teams of analysts and theoreticians busy in subsequent years. A number of spacecraft missions to other planets have flown by Venus en route to their final destinations, including *Galileo* to Jupiter in 1989 and *Cassini* to Saturn in 1997. *MESSENGER* to Mercury is scheduled to fly by in 2005.

Mars

Distance from Sun	
Perihelion	128.4 mil mi
Semi-major axis	1.524 AU
Aphelion	154.9 mil mi
Period of revolution around Sun	686.98 d (1.88 y)
Orbital eccentricity	0.0935
Orbital inclination	1.85°
Synodic day (midday to midday)	24h 39m 35s
Sidereal day	24h 37m 22s
Rotational inclination	25.19°
Mass (Earth = 1)	0.107
Mean radius	2,106 mi
Mean density (Earth = 1)	0.713
Natural satellites	2
Average surface temperature	–81° F

Mars is the first planet beyond Earth, away from the Sun. Mars's diameter is about 4,213 mi. Although Mars's orbit is nearly circular, it is somewhat more eccentric than the orbits of many of the other planets, and Mars is more than 26 mil mi farther from the Sun in some parts of its year than it is in others. Mars takes 687 Earth days to make one circuit of the Sun, traveling at about 15 mi a second. The planet rotates upon its axis in almost the same period of time as Earth—24 hours and 37 minutes. Mars's mean distance from the Sun is 142 mil mi, so its temperature would be lower than that on Earth even if its atmosphere were not so thin. In 1965, *Mariner 4* became the first spacecraft to fly by Mars, reporting that atmospheric pressure on Mars is between 1% and 2% of Earth's atmospheric pressure. As is the case with Venus, the atmosphere is composed largely of carbon dioxide. The planet is exposed to an influx of cosmic radiation about 100 times as intense as that on Earth.

Mars's position in its orbit and its speed around that orbit in relation to Earth's position and speed bring the planet fairly close to Earth on occasions about 2 years apart, then move Mars and Earth too far apart for favorable observation. Every 15-17 years the close approaches are especially favorable for observation.

Although early Earth telescopic observations led some to believe the colors they saw were indications of some sort of vegetation, this would only be possible if Mars had water and oxygen at that time.

Mars's axis of rotation is inclined from a vertical to the plane of its orbit about the Sun by about 25°, and therefore Mars has seasons as does Earth. White caps form about the poles of Mars, growing in the winter and shrinking in the summer. These polar caps are now believed to be both water ice and carbon dioxide ice. It is the carbon dioxide that is seen to come and go with the seasons. The water ice is apparently in many layers with dust between them, indicating climatic cycles.

Mariners 6 and *7* in 1969 sent back many photographs of higher quality showing cratering similar to the earlier views, but also other types of terrain. Some regions seemed featureless over large areas; others were chaotic, showing high relief without apparent organization into mountain chains or craters. *Mariner 9,* the first spacecraft to orbit Mars (1971), transmitted photos and other data showing that Mars resembles no other planet we know, yet there were features clearly of volcanic origin. One of these is Olympus Mons, a shield volcano whose caldera is more than 40 mi wide and whose outer slopes are 300 mi in diameter; it stands 15 mi above the surrounding plain—the tallest known mountain in the solar system. Some features may have been produced by cracking (faulting) and stretching of the surface. Valles Marineris, extending nearly 2,500 mi, is an example on a colossal scale. Many craters seem to have been produced by impacting bodies that may have come from the nearby asteroid belt. Features near the S pole may have been produced by glaciers no longer present.

In 1976, the U.S. landed 2 *Viking* spacecraft on the Martian surface. The landers had devices aboard to perform chemical analyses of the soil in search of evidence of life; results were inconclusive. The 2 *Viking* orbiters returned pictures of Martian topographic features that scientists believe can be explained only if Mars once had large quantities of flowing water.

Two U.S. spacecraft—the *Mars Pathfinder* and the *Mars Global Surveyor*—were launched toward Mars in 1996. On July 4, 1997, using a unique array of balloons, *Pathfinder*, with its small movable robot named Sojourner, bounced to a safe landing on Mars. It actually bounded about 40 feet high after striking the ground at 40 mph and bounced 15 more times before coming to a halt. Sojourner spent 3 months examining rocks near *Pathfinder*. Geological results from the *Pathfinder* indicate that in its beginning stages Mars melted to a sufficient extent to separate into dense and lighter layers. It also appears that there was an era when the planet had large amounts of flooding waters on its surface.

The *Surveyor* did extensive mapping of the planet and reported the presence of a very weak magnetic field that may have been stronger in the distant past. *Surveyor* results support a view of the southern hemisphere of Mars covered with ancient craters like Earth's Moon. Interestingly, there is a significant difference in the northern hemisphere, which consists mainly of plains that are much younger and lower in elevation. The *Global Surveyor* has produced a dramatic 3-D map that clearly shows this dramatic contrast.

Mars has 2 satellites, discovered in 1877 by Asaph Hall. The outer satellite, Deimos, revolves around the planet in about 31 hours. The inner satellite, Phobos, whips around Mars in a little more than 7 hours, making 3 trips around the planet each Martian day. Since it orbits Mars faster than the planet rotates, Phobos rises in the W and sets in the E, opposite to what other bodies appear to do in the Martian sky. *Mariner* and *Viking* photos show these satellites to be irregularly shaped and pitted with numerous craters. Phobos also exhibits a system of linear grooves, each about 1/3 mi across and roughly parallel. Phobos measures about 8 by 12 mi and Deimos about 5 by 7.5 mi.

Of the tens of thousands of meteorites found on Earth, approximately a dozen may have originated on Mars. In 1996, a NASA research team concluded that a meteorite found in 1984 on an Antarctic ice field not only might be a rock blasted from the surface of Mars but also might contain evidence that life existed on Mars more than 3.5 bil years ago. The meteorite has been age-dated to about 4.5 bil years. The scientists theorize that 3.5 bil years ago, Mars may have been warmer and wetter, and microscopic life may have formed and left evidence in the rock, including possible fossilized microscopic organisms. Then, 16 mil years ago, it is believed that a huge asteroid or comet struck Mars, blasting material, including this rock, into space. The rock may have entered Earth's atmosphere about 13,000 years ago, landing in Antarctica. The evidence is intriguing, but not conclusive, in suggesting that Mars may have had microscopic life, at least far in the past.

In 2000, pictures from the *Mars Global Surveyor* showed evidence for the presence of liquid water on Mars in recent times. Accessible water supplies would make future human exploration and settlement of Mars much easier. It was hoped that the unmanned Mars Odyssey spacecraft, launched in 2001 and scheduled to begin orbital studies the same year, would provide further data relevant to this issue.

Jupiter

Jupiter, largest of the planets, has an equatorial diameter of nearly 89,000 mi, 11 times the diameter of Earth. Its polar diameter is almost 5,800 mi shorter. This noticeable oblateness is a result of the liquidity of the planet and its extremely rapid rate of rotation; a day is less than 10 Earth hours long. For a planet this size, this rotational speed is amazing. A point on Jupiter's equator moves at a speed of 22,000 mph, as compared with 1,000 mph for a point on Earth's equator. Jupiter is at an average distance of 484 mil mi from the Sun and takes almost 12 Earth years to make one complete circuit of the Sun.

The major chemical constituents of Jupiter's atmosphere are molecular hydrogen (H_2—90%) and helium (He—10%). Minor constituents include methane (CH_4), ammonia (NH_3), hydrogen deuteride (HD), ethane (C_2H_6), and water (H_2O).

Distance from Sun	
Perihelion	460.1 mil mi
Semi-major axis	5.204 AU
Aphelion	507.4 mil mi
Period of revolution around Sun	11.86 y
Orbital eccentricity	0.0489
Orbital inclination	1.304°
Synodic day (midday to midday)	9h 55m 33s
Sidereal day	9h 55m 30s
Rotational inclination	3.13°
Mass (Earth = 1)	317.8
Mean radius	43,441 mi
Mean density (Earth = 1)	0.24
Natural satellites	28
Average temperature*	–162° F

*i.e., temperature where atmosphere pressure equals 1 Earth atmosphere.

The temperature at the tops of clouds may be about –280° F. The gases become denser with depth, until they may turn into a slush or slurry. There is no sharp interface between the gaseous atmosphere and the hydrogen ocean that accounts for most of Jupiter's volume. *Pioneer 10* and *11*, passing Jupiter in 1973 and 1974, provided evidence for considering Jupiter almost entirely liquid hydrogen. Jupiter apparently has a liquid hydrogen ocean more than 35,000 mi deep. It likely has a rocky core about the size of Earth, but 13 times more massive.

Jupiter's magnetic field is by far the strongest of any planet. Electrical activity caused by this field is so strong that it discharges billions of watts into Earth's magnetic field daily. At lower layers, under enormous pressure, the liquid hydrogen takes on the properties of a metal. It is likely that this liquid metallic hydrogen is the source for both Jupiter's persistent radio noise and its improbably strong magnetic field.

Fourteen of Jupiter's 28 known satellites were found through Earth-based observations. Four of the moons, Io, Europa, Ganymede, and Callisto—all discovered by Galileo in 1610—are large and bright, rivaling Earth's Moon and Mercury in diameter, and may be seen through binoculars. They move rapidly around Jupiter, and it is easy to observe their change of position from night to night. The other satellites are much smaller, in all but one instance much farther from Jupiter, and cannot be seen except through powerful telescopes. Eleven additional moons were reported in 2000, 9 of which are in retrograde orbits. The 14 outermost satellites revolve around Jupiter clockwise as seen from the north, contrary to the motions of most satellites in the solar system and to the direction of revolution of planets around the Sun. These moons may be captured asteroids. Jupiter's mass is more than twice the mass of all the other planets, moons, and asteroids put together.

Photographs from *Pioneer 10* and *11* were far surpassed by those of *Voyager 1* and *2*, both of which rendezvoused with Jupiter in 1979. The Great Red Spot exhibited internal counterclockwise rotation. Much turbulence was seen in adjacent material passing N or S of it. The satellites Amalthea, Io, Europa, Ganymede, and Callisto were photographed, some in great detail. Io has active volcanoes that probably have ejected material into a doughnut-shaped ring enveloping its orbit about Jupiter. This is not to be confused with the thin, flat disklike ring closer to Jupiter's surface.

In 1994, 21 large fragments of Comet Shoemaker-Levy 9 collided with Jupiter in a dramatic barrage. Moving at 134,000 mph, stretched out like a 21-car freight train, the fragments impacted one after another against Jupiter. Massive plumes of gas erupted from the impact sites, forming brilliant fireballs and leaving dark blotches and smears behind. One of the largest chunks, labeled the G fragment, impacted with the force of 6 mil megatons of TNT, 100,000 times the power of the largest nuclear bomb ever detonated. It produced a plume 1,200-1,600 mi high and 5,000 mi wide and left a dark discoloration larger than Earth.

The *Galileo* spacecraft went into orbit around Jupiter and released an atmospheric probe into the Jovian atmosphere in

Dec. 1995. The probe, traveling at a speed of over 100,000 mph, plunged into Jupiter's atmosphere relaying information about it for 57.6 minutes. The probe revealed a relatively dry atmosphere, with the upper part warmer and denser than expected. It also gave evidence of wind speeds of more than 400 mph and a relative absence of lightning. The probe found the atmosphere to be quite turbulent, driven by Jupiter's own internal heat. *Galileo* continued an extended mission to study the 4 large moons. *Galileo* observations show extensive ongoing volcanic eruptions on Io. Europa may have a 30-mi-deep liquid ocean beneath its icy crust, perhaps a small metallic core, and a very tenuous atmosphere. Ganymede, with a magnetosphere and a thin oxygen atmosphere, seems to be differentiated into 3 levels—a small metallic core and a rocky silicate mantle topped by an icy shell. Callisto has the oldest, most heavily cratered surface in the solar system, a very thin atmophere of carbon dioxide, and possibly also a subsurface liquid ocean.

Saturn

Distance from Sun	
Perihelion	840.4 mil mi
Semi-major axis	9.582 AU
Aphelion	941.1 mil mi
Period of revolution around Sun	29.46 y
Orbital eccentricity	0.0565
Orbital inclination	2.485°
Synodic day (midday to midday)	10h 39m 23s
Sidereal day	10h 39m 22s
Rotational Inclination	26.73°
Mass (Earth = 1)	95.16
Mean radius	36,184 mi
Mean density (Earth = 1)	0.125
Natural satellites	30
Average temperature*	–218° F

*i.e., temperature where atmosphere pressure equals 1 Earth atmosphere.

Saturn, last of the planets visible to the unaided eye, is almost twice as far from the Sun as Jupiter. It is 2d in size to Jupiter, but its mass is much smaller. Saturn's specific gravity is less than that of water. Its diameter is almost 74,900 mi at the equator while its polar diameter is almost 7,300 mi shorter—even more extreme than Jupiter. This noticeable oblateness is a result of the liquidity of the planet and its extremely rapid rate of rotation; a day is little more than 10 Earth hours long. Saturn's atmosphere is much like that of Jupiter, except that the temperature at the top of its cloud layer is at least 50° F colder. At about 300° F below zero, the ammonia would be frozen out of Saturn's clouds. The theoretical construction of Saturn resembles that of Jupiter; it likely has a small dense center surrounded by a layer of liquid and a deep atmosphere.

Until *Pioneer 11* passed Saturn in 1979, only 10 satellites of the planet were known from ground-based observations. *Pioneer 11* discovered 2 more, and the other 6 were found in the *Voyager 1* and *2* flybys, which also yielded more information about Saturn's icy satellites. 12 more moons were reported in 2000. Like Jupiter, Saturn is composed of about 75% hydrogen, 25% helium, and traces of water, ammonia, methane, and rock.

Saturn's ring system begins about 4,000 mi above the visible disk of Saturn, lying above its equator and extending about 260,000 mi into space. The diameter of the ring system visible from Earth is about 170,000 mi; the rings are estimated to be about 700 feet thick. In 1973, radar observation showed the ring particles to be large chunks of material averaging a meter on a side.

Voyager 1 and *2* observations showed the rings to be considerably more complex than had been believed. To the untrained eye, the *Voyager* photographs could be mistaken for pictures of a colorful phonograph record. Launched in Oct. 1997, the *Cassini* spacecraft was scheduled to reach Saturn in the summer of 2004 to study the planet, its rings, and its satellites, dropping a probe into the atmosphere of its largest moon, Titan.

Uranus

Distance from Sun	
Perihelion	1,703 mil mi
Semi-major axis	19.201 AU
Aphelion	1,866 mil mi
Period of revolution around Sun	84.01 y
Orbital eccentricity	0.0457
Orbital inclination	0.772°
Synodic day (midday to midday)	17h 14m 23s (retrograde)
Sidereal day	17h 14m 24s (retrograde)
Rotational inclination	97.77°
Mass (Earth = 1)	14.54
Mean radius	15,759 mi
Mean density (Earth = 1)	0.230
Natural satellites	21
Average temperature*	–323° F

*i.e., temperature where atmosphere pressure equals 1 Earth atmosphere.

Voyager 2, after passing Saturn in 1981, headed for a rendezvous with Uranus, culminating in a flyby in 1986.

Uranus, discovered by Sir William Herschel on Mar. 13, 1781, lies 1.8 bil mi from the Sun, taking 84 years to make its circuit around our star. Uranus has a diameter of over 31,000 mi and spins once in some 17.4 hours, according to flyby magnetic data.

One of the most fascinating features of Uranus is how far over it is tipped. Its N pole lies 98° from being directly up and down to its orbit plane. Thus, its seasons are extreme. When the Sun rises at the N pole, it stays up for 42 Earth years; then it sets, and the N pole is in darkness (and winter) for 42 Earth years.

Uranus has 21 known moons, which have orbits lying in the plane of the planet's equator. 5 moons are relatively large, while 16 are very small and more recently discovered. In that plane there is also a complex of 10 rings, 9 of which were discovered in 1978. Invisible from Earth, the 9 original rings were found by observers watching Uranus pass before a star. As they waited, they saw their photoelectric equipment register several short eclipses of the star; then the planet occulted the star as expected. After the star came out from behind Uranus, the star winked out several more times. Subsequent observations and analyses indicated the 9 narrow, nearly opaque rings circling Uranus. Evidence from the *Voyager 2* flyby showed the ring particles to be predominantly a yard or so in diameter.

In addition to photos of the 11 new, very small satellites, *Voyager 2* returned detailed photos of the 5 large satellites. As in the case of other satellites newly observed in the *Voyager* program, these bodies proved to be quite different from one another and from any others. Miranda has grooved markings, reminiscent of Jupiter's Ganymede, but often arranged in a chevron pattern. Ariel shows rifts and channels. Umbriel is extremely dark, prompting some observers to regard its surface as among the oldest in the system. Titania has rifts and fractures, but not the evidence of flow found on Ariel. Oberon's main feature is its surface saturated with craters, unrelieved by other formations.

Uranus likely does not have a rocky core, but rather a mixture of rocks and assorted ices with less than 20% hydrogen and little helium. The atmosphere is about 83% hydrogen, 15% helium, and 2% methane. In addition to its rotational tilt, Uranus's magnetic field axis is tipped an incredible 58.6° from its rotational axis and is displaced about 30% of its radius away from the planet's center.

Neptune

Neptune lies at an average distance of 2.8 bil mi. It was the last planet visited in *Voyager 2*'s epic 12-year trek (1977-89) from Earth.

As with other giant planets, Neptune may have no solid surface, or exact diameter. However, a mean value of 30,600 mi may be assigned to a diameter between atmosphere levels where the pressure is about the same as sea level on Earth. Without a solid surface to view, it is challenging to determine a "true" rotation rate for a giant planet.

Distance from Sun	
Perihelion	2,762 mil mi
Semi-major axis	30.05 AU
Aphelion	2,824 mil mi
Period of revolution around Sun	164.79 y
Orbital eccentricity	0.0113
Orbital inclination	1.769°
Synodic day (midday to midday)	16h 6m 37s
Sidereal day	16h 6m 36s
Rotational inclination	28.32°
Mass (Earth = 1)	17.15
Mean radius	15,301 mi
Mean density (Earth = 1)	0.297
Natural satellites	8
Average temperature*	–330° F

*i.e., temperature where atmosphere pressure equals 1 Earth atmosphere.

Astronomers use a determination of the rotation rate of the planet's magnetic field to indicate the internal rotation rate, which in the case of Neptune is 16.1 hours. Neptune orbits the Sun in 164.8 years in a nearly circular orbit. Neptune was discovered in 1846; not until 2010 will it have completed one full trip around the Sun since its discovery.

Voyager 2, which passed 3,000 mi from Neptune's N pole, found a magnetic field that is considerably asymmetric to the planet's structure, similar to, but not so extreme as, that found at Uranus. Neptune's magnetic field axis is tipped 46.9° from its rotational axis and is displaced more than 55% of its radius away from the planet's center.

Neptune's atmosphere was seen to be quite blue, with quickly changing white clouds often suspended high above an apparent surface. There is a Great Dark Spot, reminiscent of the Great Red Spot of Jupiter. Observations with the Hubble Space Telescope have shown that the Great Dark Spot originally seen by *Voyager* has apparently dissipated, but a new dark spot has since appeared.

Neptune's atmosphere is about 80% hydrogen, 19% helium, and 1% methane. Although lightning and auroras have been found on other giant planets, only the aurora phenomenon has been seen on Neptune.

Six new satellites were definitively discerned around Neptune by *Voyager 2*. Five of them orbit Neptune in a half day or less. Of the 8 satellites of Neptune in all, the largest, Triton, is in a retrograde orbit, suggesting that it was captured rather than being coeval with Neptune. Triton's large size, sufficient to raise significant tides on the planet, may one day, billions of years from now, cause Triton to come close enough to Neptune for it to be torn apart. Nereid was found in 1949 and has the highest orbital eccentricity (0.75) of any moon. Its long looping orbit suggests that it, too, was captured.

Each of the satellites that has been photographed by the 2 *Voyagers* in the planetary encounters has been different from any of the other satellites, and certainly different from any of the planets. Only about half of Triton has been observed, but its terrain shows cratering and a strange regional feature described as resembling the skin of a cantaloupe. Triton has a tenuous atmosphere of nitrogen with a trace of hydrocarbons and evidence of active geysers injecting material into it. At –390° F, the wintertime parts of Triton are the coldest regions yet found in the solar system.

Voyager 2 also confirmed the existence of 6 rings composed of very fine particles. There may be some clumpiness in the rings' structure. It is not known whether Neptune's satellites influence the formation or maintenance of the rings.

As with the other giant planets, Neptune is emitting more energy than it receives from the Sun. *Voyager* found the excess to be 2.7 times the solar contribution. Cooling from internal heat sources and from the heat of formation of the planets is thought to be responsible.

Pluto

Although Pluto on the average stays about 3.6 bil mi from the Sun, its orbit is so eccentric that its minimum dis-tance of 2.76 bil mi is less than Neptune's distance from the Sun.

Distance from Sun	
Perihelion	2,756 mil mi
Semi-major axis	39.24 AU
Aphelion	4,539 mil mi
Period of revolution around Sun	247.68 y
Orbital eccentricity	0.2444
Orbital inclination	17.16°
Synodic day (midday to midday)	6d 9h 17m (retrograde)
Sidereal day	6d 9h 18m (retrograde)
Rotational inclination	122.53°
Mass (Earth = 1)	0.0021
Mean radius	743 mi
Mean density (Earth = 1)	0.317
Natural satellites	1
Average surface temperature	–369° F

Pluto is currently the most distant planet, but for about 20 years of its orbit, Pluto is closer to the Sun than Neptune. Pluto takes 247.7 years to circumnavigate the Sun, a 3/2 resonance with Neptune.

About a century ago, a hypothetical planet was believed to lie beyond Neptune and Uranus because neither planet followed paths predicted by astronomers when all known gravitational influences were considered. In little more than a guess, a mass of 1 Earth was assigned to the mysterious body, and mathematical searches were begun. Amid some controversy about the validity of the predictive process, Pluto was discovered nearly where it had been predicted to lie, by Clyde Tombaugh at the Lowell Observatory in Flagstaff, AZ, in 1930.

At the U.S. Naval Observatory in Flagstaff, in 1978, James Christy obtained a photograph of Pluto that was distinctly elongated. Repeated observations of this shape and its variation were convincing evidence of the discovery of a satellite of Pluto, now named Charon. Later observations showed it to be 743 mi across, over 12,000 mi from Pluto, and taking 6.4 days to move around Pluto. In this same length of time, Pluto and Charon both rotate once around their axes. The Pluto-Charon system thus appears to rotate as virtually a rigid body. Gravitational laws allow these interactions to give the mass of Pluto as 0.0021 of Earth. This mass, together with a new diameter for Pluto of 1,413 mi, make the density about twice that of water. Theorists predict that Pluto has a rocky core, surrounded by a thick mantle of ice.

It is now clear that Pluto could not have influenced Neptune and Uranus to go astray. Besides being the smallest planet, Pluto is actually smaller than 7 of the solar system's moons. Although a 10th planet might be out there somewhere, theorists no longer believe there are unexplained perturbations in the orbit of Uranus or Neptune that might be caused by it. Astronomers have found nearly 300 asteroid-size objects, somewhat beyond Pluto, in a region called the Kuiper Belt, where some comets are believed to originate.

Because the rotational axis of the system is tipped more than 120°, there is only a few-years interval every 125 years when Pluto and Charon alternately eclipse each other. Both worlds are roughly spherical and have comparable densities. Large regions on Pluto are dark, others light; Pluto has spots and perhaps polar caps. Although extremely cold, Pluto appears to have a thin nitrogen–carbon dioxide–methane atmosphere, at least while it is closer to the Sun. When Pluto occulted a star, the star's light faded in such a way as to have passed through a haze layer lying above the planet's surface, indicating an inversion of temperatures, suggesting Pluto has primitive weather. A recent controversy raised the issue of Pluto's planet status. Pluto is clearly different from both the rocky terrestrial planets and the giant planets. Although some astronomers think Pluto most closely resembles the Kuiper Belt Objects and should be grouped with them, most still classify Pluto as a planet. By way of comparison, Mercury, the 2d-smallest planet is about 2 times the radius of Pluto, while Pluto is about 2.65 times the radius of the largest known Kuiper Belt Object, Varuna. Closest in size is Neptune's largest moon, Triton, which is only 1.13 times the radius of Pluto.

The Sun

The Sun, the controlling body of Earth's solar system, is a star often described as average. Yet, the Sun's mass and luminosity are greater than that of 80% of the stars in our Milky Way galaxy. On the other hand, most of the stars that can be easily seen on any clear night are bigger and brighter than the Sun. It is the Sun's proximity to Earth that makes it appear tremendously large and bright. The Sun is 400,000 times as bright as the full moon and gives Earth 6 mil times as much light as do all the other stars put together. A series of nuclear fusion reactions where hydrogen nuclei are converted to helium nuclei produces the heat and light that make life possible on Earth.

The Sun has a diameter of 865,000 mi and, on average, is 92,956,000 mi from Earth. It is 1.41 times as dense as water. The light of the Sun reaches Earth in 499 seconds, or in slightly more than 8 minutes. The average solar surface temperature has been measured at a value of 5,778 K, or about 9,941° F. The interior temperature of the Sun is theorized to be about 28,000,000° F.

When sunlight is analyzed with a spectroscope, it is found to consist of a continuous spectrum composed of all the colors of the rainbow in order, crossed by many dark lines. The dark "absorption lines" are produced by gaseous materials in the outer layers of the Sun. More than 60 of the natural terrestrial elements have been identified in the Sun, all in gaseous form because of the Sun's intense heat.

Spheres and Corona

The radiating surface of the Sun is called the **photosphere;** just above it is the **chromosphere.** The chromosphere is visible to the naked eye only at total solar eclipses, appearing then to be a pinkish-violet layer with occasional great prominences projecting above its general level. With proper instruments, the chromosphere can be seen or photographed whenever the Sun is visible without waiting for a total eclipse. Above the chromosphere is the **corona,** also visible to the naked eye only at times of total eclipse. Instruments also permit the brighter portions of the corona to be studied whenever conditions are favorable. The pearly light of the corona surges mil of mi from the Sun. Iron, nickel, and calcium are believed to be principal contributors to the composition of the corona, all in a state of extreme attenuation and high ionization that indicates temperatures nearly 2 mil degrees Fahrenheit.

Sunspots

There is an intimate connection between sunspots and the corona. At times of low sunspot activity, the fine streamers of the corona are longer above the Sun's equator than over the polar regions of the Sun; during periods of high sunspot activity, the corona extends fairly evenly outward from all regions of the Sun, but to a much greater distance in space. Sunspots are dark, irregularly shaped regions whose diameters may reach tens of thousands of mi. The average life of a sunspot group is 2 months, but some have lasted for more than a year by being carried repeatedly around as the Sun rotated upon its axis.

Sunspots reach a low point, on average, every 11.3 years, with a peak of activity occurring irregularly between 2 successive minima. Launched in December 1995, the SOHO spacecraft was designed to provide several years of study of the Sun from an orbit around the Sun. We are just past the maximum of the current sunspot cycle, for which SOHO has provided extraordinary views of the Sun's activity. Observations from SOHO show that magnetic arches, called prominences, extending tens of thousands of mi into the corona may release enormous amounts of energy heating the corona. SOHO has also highlighted enormous releases of solar energy called coronal mass ejections. Coronal holes are regions where the corona appears dark, and it is in these regions where the high-speed solar wind originates.

IT'S A FACT: When the Sun, later in its evolution, becomes a red giant star—scientists predict this will happen in about 5 billion years—it will grow to 70 times its current size, placing its surface near Earth's orbit.

The Moon

Distance from Earth	
Perigee	225,744 mi
Semi-major axis	238,855 mi
Apogee	251,966 mi
Period of revolution	27.322 d
Synodic orbital period (period of phases)	29.53 d
Orbital eccentricity	0.0549
Orbital inclination	5.145°
Sidereal day (rotation period	27.322 d
Rotational inclination	6.68°
Mass (Earth = 1)	0.0123
Mean radius	1,080 mi
Mean density (Earth = 1)	0.605
Average surface temperature	–10° F

The Moon completes a circuit around Earth in a period whose mean or average duration is 27 days, 7 hours, 43.2 minutes. This is the Moon's **sidereal period**. Because of the motion of the Moon in common with Earth around the Sun, the mean duration of the lunar month—the period from one New Moon to the next New Moon—is 29 days, 12 hours, 44.05 minutes. This is the Moon's **synodic period.**

The mean distance of the Moon from Earth is 238,855 mi. Because the orbit of the Moon about Earth is not circular but elliptical, however, the actual distance varies considerably. The maximum distance from Earth that the Moon may reach is 251,966 mi and the least distance is 225,744 mi. (All distances given here are from the center of one body to the center of the other.)

The Moon rotates on its axis in a period of time that is exactly equal to its sidereal revolution about Earth: 27.322 days. Thus the backside or farside of the Moon always faces away from Earth. This does not mean that the backside is always dark, since the Sun is the main source of light in the Solar System. The farside of the Moon gets just as much direct sunlight as the nearside. At New Moon phase, the farside of the Moon is fully lit. With its long day and night, the daytime temperature can reach 260° F, while the coldest nighttime temperature may reach –280° F. This day-to-night contrast is exceeded only by that on Mercury.

The Moon's revolution about Earth is irregular because of its elliptical orbit. The Moon's rotation, however, is regular, and this, together with the irregular revolution, produces what is called "libration in longitude," which permits the observer on Earth to see first farther around the E side and then farther around the W side of the Moon. The Moon's variation N or S of the ecliptic permits one to see farther over first one pole and then the other of the Moon; this is called "libration in latitude." These two libration effects permit observers on Earth to see a total of about 60% of the Moon's surface over a period of time.

The hidden side of the Moon was first photographed in 1959 by the Soviet space vehicle *Lunik III*. The moon's farside does appear noticeably different from the nearside, in that the farside has practically none of the large lava plains, called maria, so prominent on the nearside.

From 1969 through 1972, 6 American spacecraft brought 12 astronauts to walk on the surface of the Moon. In 1998 NASA's *Lunar Prospector* spacecraft provided evidence for the presence of 300 million metric tons of water ice at the lunar poles. *Lunar Prospector* results also indicate that the Moon has a small core, supporting the idea that most of the mass of the Moon was ripped away from the early Earth when a Mars-size object collided with Earth.

Tides on Earth are caused mainly by the Moon, because of its proximity to Earth. The ratio of the tide-raising power of the Moon to that of the Sun is 11 to 5.

Harvest Moon and Hunter's Moon

The Harvest Moon, the full Moon nearest the autumnal equinox, ushers in a period of several successive days when the Moon rises soon after sunset. This phenomenon gives farmers in temperate latitudes extra hours of light in which to harvest their crops before frost and winter. The 2002 Harvest Moon falls on Sept. 21 UTC. Harvest Moon in the southern hemisphere temperate latitudes falls on Mar. 28.

The next full Moon after Harvest Moon is called the Hunter's Moon; it is accompanied by a similar but less marked phenomenon. In 2002, the Hunter's Moon occurs on Oct. 21, in the northern hemisphere and on Apr. 27 in the southern hemisphere.

Moon's Perigee and Apogee, 2002

(Coordinated Universal Time, standard time of the prime meridian)

Perigee				Apogee			
Date	Hour	Date	Hour	Date	Hour	Date	Hour
Jan. 2	7	July 14	13	Jan. 18	9	July 30	2
Jan. 30	9	Aug. 10	23	Feb. 14	22	Aug. 26	18
Feb. 27	20	Sept. 8	3	Mar. 14	1	Sept. 23	3
Mar. 28	8	Oct. 6	13	Apr. 10	5	Oct. 20	5
Apr. 25	16	Nov. 4	1	May 7	19	Nov. 16	11
May 23	16	Dec. 2	9	June 4	13	Dec. 14	4
June 19	7	Dec. 30	1	July 2	8		

Moon Phases, 2002

(Coordinated Universal Time, standard time of the prime meridian)

New Moon				Waxing Quarter				Full Moon				Waning Quarter			
Month	d	h	m	Month	d	h	m	Month	d	h	m	Month	d	h	m
												Jan.	6	3	55
Jan.	13	13	29	Jan.	21	17	46	Jan.	28	22	50	Feb.	4	13	33
Feb.	12	7	41	Feb.	20	12	2	Feb.	27	9	17	Mar.	6	1	24
Mar.	14	2	2	Mar.	22	2	28	Mar.	28	18	25	Apr.	4	15	29
Apr.	12	19	21	Apr.	20	12	48	Apr.	27	3	00	May	4	7	16
May	12	10	45	May	19	19	42	May	26	11	51	June	3	00	5
June	10	23	46	June	18	00	29	June	24	21	42	July	2	17	19
July	10	10	26	July	17	4	47	July	24	9	7	Aug.	1	10	22
Aug.	8	19	15	Aug.	15	10	12	Aug.	22	22	29	Aug.	31	2	31
Sept.	7	3	10	Sept.	13	18	8	Sept.	21	13	59	Sept.	29	17	3
Oct.	6	11	18	Oct.	13	5	33	Oct.	21	7	20	Oct.	29	5	28
Nov.	4	20	34	Nov.	11	20	52	Nov.	20	1	34	Nov.	27	15	46
Dec.	4	7	34	Dec.	11	15	49	Dec.	19	19	10	Dec.	27	00	31

Searching for Planets

People have known of the existence of the planets in the Solar System that are closest to the Sun (Mercury, Venus, Mars, Jupiter, and Saturn) since ancient times because they could be seen with the naked eye. However, the 3 farthest (Uranus, Neptune, and Pluto) were discovered only since the invention of the telescope. The first, Uranus, was discovered in 1781 by the English astronomer William Herschel. Next, Neptune's existence and location were predicted through its action upon Uranus, by both John Couch Adams of England and Urbain Jean Joseph Le Verrier of France in 1845, leading to its discovery the following year. Finally, Pluto was discovered in 1930 by the American astronomer Clyde Tombaugh.

During the last 10 years of the 20th century astronomers began to detect the presence of planets orbiting stars other than the Sun. As of yet, they are not actually seeing those objects, but merely inferring their existence by their effect on their parent star. Although the Sun is not an average star, it is typical. With over 200 bil stars in the Milky Way galaxy, it seems plausible that other stars might have planets.

Using the Doppler Effect to detect radial velocity changes in the motions of individual stars, astronomers are more likely to find high-mass planets in close and eccentric orbits around stars, because that situation produces larger and more noticeable changes. More than 3 dozen star systems have been found that appear to have at least one planet less than 13 times the mass of Jupiter. Another dozen systems may have more massive planets. In 2 cases planets may have been detected in orbit around pulsars.

The star Upsilon Andromedae seems to have 3 planets, with masses 0.71, 2.11, and 4.61 times the mass of Jupiter, yet 2 of the planets are closer to their star than Earth is to the Sun. Astronomers are puzzled as to how planets the size of Jupiter or larger can exist so close to a star. In November 2000 a Saturn-size planet and another half the mass of Saturn (the smallest extra-solar planet yet detected) were discovered orbiting a star in the constellation Vela. In August 2001 astronomers found evidence of a 2d large planet in nearly circular orbit around a star in Ursa Major that is similar to our Sun. The recent discoveries raised hopes for finding Earth-like planets outside our own solar system. The total of star systems with confirmed planets is now more than 60.

Earth: Size, Computation of Time, Seasons

Distance from the Sun	
Perihelion	91.4 mil mi
Semi-major axis	1.0000 AU
Aphelion	94.5 mil mi
Period of revolution	365.256 d
Orbital eccentricity	0.0167
Orbital inclination	0.0°
Sidereal day (Rotation period)	23h 56m 4.1s
Synodic day (midday to midday	24h 0m 0s
Rotational inclination	23.45°
Mass (Earth = 1)	1.00
Mean radius	3,959 mi
Mean density (Earth = 1)	1.00
Natural satellites	1
Average surface temperature	59° F

Earth is the 5th-largest planet and the 3d from the Sun. Its mass is 6,580,000,000,000,000,000,000 tons. Earth's equatorial diameter is 7,926 mi while its polar diameter is only 7,900 mi.

Size and Dimensions

Earth is considered a solid mass, yet it has a large, liquid iron, magnetic core with a radius of about 2,165 mi. Surprisingly, it has a solid inner core that may be a large iron crystal, with a radius of 800 mi. Around the core is a thick shell or mantle of dense rock. This mantle is composed of materials rich in iron and magnesium. It is somewhat plastic-like and under slow steady pressure can flow like a liquid. The mantle, in turn, is covered by a thin crust forming the solid granite and basalt base of the continents and ocean basins.

Over broad areas of Earth's surface, the crust has a thin cover of sedimentary rock such as sandstone, shale, and limestone formed by weathering of Earth's surface and deposits of sands, clays, and plant and animal remains.

The temperature inside the Earth increases about 1° F with every 100 to 200 feet in depth, in the upper 100 km of Earth, and reaches nearly 8,500° F at the center. The heat is believed to be derived from radioactivity in the rocks, pressures developed within Earth, and the original heat of formation.

Atmosphere of Earth

Earth's atmosphere is a blanket composed of nitrogen, oxygen, and argon, in amounts of about 78%, 21%, and 1% by volume. Also present in minute quantities are carbon dioxide, hydrogen, neon, helium, krypton, and xenon. Water vapor displaces other gases and varies from nearly zero to about 4% by volume. The atmosphere rests on Earth's surface with the weight equivalent to a layer of water 34 ft deep. For about 300,000 ft upward, the gases remain in the proportions stated. Gravity holds the gases to Earth. The weight of the air compresses it at the bottom so that the greatest density is at Earth's surface. Pressure and density decrease as height increases because the weight pressing upon any layer is always less than that pressing upon the layers below.

The lowest layer of the atmosphere extending up about 7.5 mi is the **troposphere**, which contains 90% of the air and the tallest mountains. This is also where most weather phenomena occur. The temperature drops with increasing height throughout this layer. The atmosphere for about 23 mi above the troposphere is the **stratosphere,** where the temperature generally increases with height. The stratosphere contains ozone, which prevents ultraviolet rays from reaching Earth's surface. Since there is very little convection in the stratosphere, jets regularly cruise in the lower parts to provide a smoother ride for passengers.

Above the stratosphere is the **mesosphere,** where the temperature again decreases with height for another 19 mi. Extending above the mesosphere to the outer fringes of the atmosphere is the **thermosphere,** a region where temperature once more increases with height to a value measured in thousands of degrees Fahrenheit. The lower portion of this region, extending from 50 to about 400 mi in altitude, is characterized by a high ion density and is thus called the **ionosphere.** Most meteors are in the lower thermosphere or the mesosphere at the time they are observed.

Longitude, Latitude

Position on the globe is measured by meridians and parallels. Meridians, which are imaginary lines drawn around Earth through the poles, determine **longitude**. The meridian running through Greenwich, England, is the **prime meridian** of longitude, and all others are either E or W. Parallels, which are imaginary circles parallel with the equator, determine **latitude**. The length of a degree of longitude varies as the cosine of the latitude. At the equator a degree of longitude is 69.171 statute mi; this is gradually reduced toward the poles. Value of a longitude degree at the poles is zero.

Latitude is reckoned by the number of degrees N or S of the **equator,** an imaginary circle on Earth's surface everywhere equidistant between the two poles. According to the International Astronomical Union ellipsoid of 1964, the length of a degree of latitude is 68.708 statute mi at the equator and varies slightly N and S because of the oblate form of the globe; at the poles it is 69.403 statute mi.

Definitions of Time

Earth rotates on its axis and follows an elliptical orbit around the Sun. The rotation makes the Sun appear to move across the sky from E to W. This rotation determines day and night, and the complete rotation, in relation to the Sun, is called the **apparent** or **true solar day**. A sundial thus measures **apparent solar time**. This length of time varies, but an average determines the mean solar day of 24 hours.

The mean solar day and **mean solar time** are in universal use for civil purposes. Mean solar time may be obtained from apparent solar time by correcting observations of the Sun for the **equation of time**. Mean solar time may be as much as 16 minutes behind or 14 minutes ahead of apparent solar time.

Sidereal time is the measure of time defined by the diurnal motion of the vernal equinox and is determined from observation of the meridian transits of stars. One complete rotation of Earth relative to the equinox is called the **sidereal day**. The **mean sidereal day** is 23 hours, 56 minutes, 4.091 seconds of mean solar time.

The interval required for Earth to make one absolute revolution around the Sun is a **sidereal year;** it consisted of 365 days, 6 hours, 9 minutes, and 9.5 seconds of mean solar time (approximately 24 hours per day) in 1900 and has been increasing at the rate of 0.0001 second annually.

The **tropical year,** upon which our calendar is based, is the interval between 2 consecutive returns of the Sun to the vernal equinox. The tropical year consisted of 365 days, 5 hours, 48 minutes, and 46 seconds in 1900. It has been decreasing at the rate of 0.530 second per century. The **calendar year** begins at 12 o'clock midnight precisely, local clock time, on the night of Dec. 31-Jan. 1. The day and the calendar month also begin at midnight by the clock.

On Jan. 1, 1972, the Bureau International des Poids et Mesures in Paris introduced International Atomic Time (TAI) as the most precisely determined time scale for astronomical usage. The fundamental unit of TAI in the international system of units is the **second,** defined as the duration of 9,192,631,770 periods of the radiation corresponding to the transition between 2 hyperfine levels of the ground state of the cesium 133 atom. **Coordinated Universal Time** (UTC), which serves as the basis for civil timekeeping and is the standard time of the prime meridian, is officially defined by a formula which relates UTC to mean sidereal time in Greenwich, England. (UTC has replaced GMT as the basis for standard time for the world.)

The Zones and Seasons

The 5 zones of Earth's surface are the Torrid, lying between the Tropics of Cancer and Capricorn; the N Temperate, between Cancer and the Arctic Circle; the S Temperate, between Capricorn and the Antarctic Circle; and the 2 Frigid Zones, between the Polar Circles and the Poles.

The inclination or tilt of Earth's axis, 23° 27′ away from a perpendicular to the Earth's orbit of the Sun, determines the seasons. These are commonly marked in the N Temperate Zone, where spring begins at the vernal equinox, summer at the summer solstice, autumn at the autumnal equinox, and winter at the winter solstice.

In the S Temperate Zone, the seasons are reversed. Spring begins at the autumnal equinox, summer at the winter solstice, etc.

The points at which the Sun crosses the equator are the equinoxes, when day and night are most nearly equal. The points at which the Sun is at a maximum distance from the equator are the solstices. Days and nights are then most unequal. However, at the equator, day and night are equal throughout the year.

In June, the North Pole is tilted 23° 27′ toward the Sun, and the days in the northern hemisphere are longer than the nights, while the days in the southern hemisphere are shorter than the nights. In Dec., the North Pole is tilted 23° 27′ away from the Sun, and the situation is reversed.

The Seasons in 2002

In 2002 the 4 seasons begin in the northern hemisphere as shown. (Add one hour to Eastern Standard Time for Atlantic Time; subtract one hour for Central, 2 for Mountain, 3 for Pacific, 4 for Alaska, 5 for Hawaii-Aleutian. Also shown is Coordinated Universal Time.)

Seasons	Date	EST	UTC
Vernal Equinox (spring)	Mar. 20	14:16	19:16
Northern Solstice (summer)	June 21	8:24	13:24
Autumnal Equinox (autumn)	Sept. 23	23:55*	4:55
Southern Solstice (winter)	Dec. 22	20:14*	1:14

* previous day

Poles of Earth

The geographic (rotation) poles, or points where Earth's axis of rotation cuts the surface, are not absolutely fixed in the body of Earth. The pole of rotation describes an irregular curve about its mean position.

Two periods have been detected in this motion: (1) an annual period due to seasonal changes in barometric pressure, to load of ice and snow on the surface, and to other phenomena of seasonal character; (2) a period of about 14 months due to the shape and constitution of Earth.

In addition, there are small but as yet unpredictable irregularities. The whole motion is so small that the actual pole at any time remains within a circle of 30 or 40 feet in radius centered at the mean position of the pole.

The pole of rotation for the time being is of course the pole having a latitude of 90° and an indeterminate longitude.

Magnetic Poles

Although Earth's magnetic field resembles that of an ordinary bar magnet, this magnetic field is probably produced by electric currents in the liquid currents of the Earth's outer core. The **north magnetic pole** of Earth is that region where the magnetic force is vertically downward, and the **south magnetic pole** is that region where the magnetic force is vertically upward. A compass placed at the magnetic poles experiences no directive force in azimuth.

There are slow changes in the distribution of Earth's magnetic field. This slow temporal change is referred to as the Secular change of the main magnetic field and the magnetic poles shift due this. The location of the N magnetic pole was first measured in 1831 at Cape Adelaide on the west coast of Boothia Peninsula in Canada's Northwest Territories (about latitude 70° N and longitude 96° W). Since then it has moved over 500 miles. In 1994, the N magnetic pole was located at Ellef Ringnes Island in northern Canada and it is now northwest of that island. Measurement for the past several decades by Canadian scientists indicate the NW motion of the pole continues, averaging about 9 mi per year.

The direction of the horizontal components of the magnetic field at any point is known as magnetic N at that point, and the angle by which it deviates E or W of true N is known as the magnetic declination.

A compass without error points in the direction of magnetic north. (In general, this is not the direction of the magnetic north pole.) If one follows the direction indicated by the N end of the compass, he or she will travel along a rather irregular curve that eventually reaches the north magnetic pole (though not usually by a great-circle route). However, the action of the compass should not be thought of as due to any influence of the distant pole, but simply as an indication of the distribution of Earth's magnetism at the place of observation.

Rotation of Earth

The speed of rotation of Earth about its axis has been found to be slightly variable. The variations may be classified as:

(A) **Secular.** Tidal friction acts as a brake on the rotation and causes a slow secular increase in the length of the day, about 1 millisecond per century.

(B) **Irregular.** The speed of rotation may increase for a number of years, about 5 to 10, and then start decreasing. The maximum difference from the mean in the length of the day during a century is about 5 milliseconds. The accumulated difference in time has amounted to approximately 44 seconds since 1900. The cause is probably motion in the interior of Earth.

(C) **Periodic.** Seasonal variations exist with periods of 1 year and 6 months. The cumulative effect is such that each year, Earth is late about 30 milliseconds near June 1 and is ahead about 30 milliseconds near Oct. 1. The maximum seasonal variation in the length of the day is about 0.5 millisecond. It is believed that the principal cause of the annual variation is the seasonal change in the wind patterns of the northern and southern hemispheres. The semiannual variation is due chiefly to tidal action of the Sun, which distorts the shape of Earth slightly.

The secular and irregular variations were discovered by comparing time based on the rotation of Earth with time based on the orbital motion of the Moon about Earth and of the planets about the Sun. The periodic variation was determined largely with the aid of quartz-crystal clocks. The introduction of the cesium-beam atomic clock in 1955 made it possible to determine in greater detail than before the nature of the irregular and periodic variations.

Chronological Eras

Era	Year	Begins in 2002
Byzantine	7511	Sept. 14
Jewish	5763	Sept. 6[1]
Roman (Ab Urbe Condita)	2755	Jan. 14
Nabonassar (Babylonian)	2751	Apr. 23
Japanese	2662	Jan. 1
Grecian (Seleucidae)	2314	Sept. 14 or Oct. 14
Diocletian	1719	Sept. 11
Indian (Saka)	1924	Mar. 22
Islamic/Muslim (Hijra)	1423	Mar. 14[1]

(1) Year begins at sunset.

Chronological Cycles, 2002

Dominical Letter	F	Roman Indiction	10	Solar Cycle	23
Golden Number (Lunar Cycle)	VIII	Epact	16	Julian Period (year of)	6715

Twilight

Twilight is that evening period of waning light from the time of sunset to dark, often termed dusk. Morning twilight, a time of increasing light, is called **dawn**. The source of this light is the Sun shining on the atmosphere above the observer. Twilight is a time of very slowly changing sky illumination with no abrupt variations. Nevertheless, there are 3 commonly accepted divisions in this smooth continuum defined by the distance the Sun lies below the astronomical horizon: civil twilight, nautical twilight, and astronomical twilight.

The **astronomical horizon** is that great circle lying 90° from the zenith, the point directly over the observer's head. Twilight ends in the evening or begins in the morning at a particular time. Nominally, evening events are repeated in reverse order in the morning.

Civil twilight is the time from the moment of sunset, when the Sun's apparent upper edge is just at the horizon, until the center of the Sun is 6° directly below the horizon. In many states, this is the time in the evening when automobile headlights must be turned on, not to see better, but to be seen by other drivers. After this time, a newspaper becomes increasingly difficult to read in the absence of artificial light. **Nautical twilight** ends when the Sun's center is 12° below the horizon. By this time in the evening, the bright stars used by navigators have appeared, and the horizon may still be seen. After this time, the horizon is more difficult to perceive, preventing navigators from sighting stars. **Astronomical twilight** ends in the evening when the Sun is 18° below the horizon and the sky is dark enough, at least away from the Sun's location, to allow astronomical work to proceed. Sunlight, however, is still shining on the higher levels of the atmosphere from the observer's zenith to the horizon toward the Sun. Although not named as a period of twilight, when the Sun is 24° below the horizon, no part of the observer's atmosphere, even toward the Sun, receives any sunlight.

In the tropics, the Sun moves nearly vertically, accomplishing its 6°, 12°, or 18° depression very quickly. In the polar regions, the Sun's diurnal motion may actually be nearly along the horizon, prolonging the twilight period or even not permitting darkness to fall at all. In mid-latitudes, civil twilight may last about a half hour; nautical twilight, an hour; and astronomers can go to work after approximately 90 minutes. The twilight tables given in *The World Almanac* are for the beginning of morning twilight and the end of evening astronomical twilight. Although the instant of the Sun's horizontal depression may be calculated precisely, the phenomena associated with the event are sufficiently imprecise that the table need not be recalculated each year.

Astronomical Twilight—Meridian of Greenwich

Date	20°		30°		40°		50°		60°	
	Morn.	Eve.	Morn.	Eve.	Morn.	Eve.	Morn.	Eve.	Morn.	Eve.
	h m	h m	h m	h m	h m	h m	h m	h m	h m	h m
Jan. 1	5 16	18 50	5 30	18 36	5 45	18 22	6 00	18 07	6 18	17 49
11	5 20	18 56	5 33	18 43	5 45	18 30	5 59	18 17	6 15	18 01
21	5 21	19 02	5 32	18 50	5 43	18 40	5 53	18 29	6 05	18 18
Feb. 1	5 20	19 07	5 29	18 59	5 36	18 51	5 43	18 45	5 49	18 39
11	5 17	19 12	5 23	19 06	5 27	19 02	5 30	19 00	5 29	19 01
21	5 12	19 15	5 15	19 13	5 16	19 12	5 13	19 15	5 04	19 25
Mar. 1	5 06	19 18	5 06	19 19	5 03	19 22	4 56	19 30	4 39	19 47
11	4 59	19 22	4 55	19 25	4 48	19 33	4 34	19 47	4 07	20 15
21	4 50	19 25	4 43	19 32	4 31	19 44	4 10	20 05	3 32	20 45
Apr. 1	4 40	19 28	4 28	19 40	4 11	19 58	3 42	20 27	2 46	21 25
11	4 30	19 32	4 15	19 48	3 53	20 10	3 15	20 48	1 55	22 12
21	4 21	19 37	4 02	19 56	3 34	20 24	2 47	21 13	0 18	
May 1	4 13	19 42	3 50	20 05	3 16	20 39	2 17	21 40		
11	4 06	19 47	3 40	20 14	3 00	20 54	1 45	22 11		
21	4 01	19 53	3 31	20 23	2 46	21 08	1 09	22 48		
June 1	3 57	19 59	3 25	20 31	2 34	21 22				
11	3 56	20 03	3 22	20 37	2 28	21 31				
21	3 57	20 06	3 23	20 41	2 28	21 36				
July 1	4 00	20 07	3 26	20 41	2 33	21 35				
11	4 05	20 06	3 32	20 38	2 42	21 28		23 54		
21	4 10	20 02	3 40	20 32	2 54	21 17	1 14	22 55		
Aug. 1	4 16	19 56	3 49	20 22	3 10	21 02	1 54	22 16		
11	4 21	19 49	3 58	20 11	3 24	20 45	2 24	21 44		
21	4 25	19 40	4 06	19 59	3 38	20 27	2 50	21 14		23 33
Sept. 1	4 30	19 30	4 15	19 44	3 52	20 06	3 16	20 42	1 56	21 59
11	4 33	19 20	4 22	19 31	4 04	19 48	3 36	20 15	2 40	21 10
21	4 35	19 10	4 28	19 17	4 16	19 29	3 55	19 50	3 15	20 29
30	4 38	19 02	4 33	19 06	4 25	19 14	4 10	19 28	3 42	19 56
Oct. 1	4 38	19 01	4 34	19 05	4 26	19 12	4 12	19 26	3 44	19 53
11	4 40	18 53	4 40	18 53	4 36	18 56	4 28	19 04	4 11	19 21
21	4 43	18 46	4 46	18 43	4 47	18 42	4 44	18 44	4 35	18 53
Nov. 1	4 47	18 40	4 53	18 34	4 58	18 29	5 00	18 26	5 00	18 26
11	4 51	18 37	5 00	18 28	5 08	18 20	5 15	18 13	5 21	18 06
21	4 56	18 36	5 07	18 25	5 17	18 14	5 28	18 03	5 39	17 52
Dec. 1	5 01	18 37	5 14	18 24	5 27	18 12	5 40	17 58	5 56	17 42
11	5 07	18 40	5 21	18 26	5 35	18 12	5 50	17 57	6 08	17 38
21	5 12	18 45	5 26	18 30	5 41	18 16	5 56	18 00	6 16	17 41
31	5 16	18 50	5 30	18 36	5 44	18 22	6 00	18 07	6 18	17 48

Calculation of Rise Times

The Daily Calendar on pages 663-674 contain rise and set times for the Sun and Moon for the Greenwich Meridian at N latitudes 20°, 30°, 40°, 50°, and 60°. From day to day, the values for the Sun at any particular latitude do not change very much. This means that whatever time the Sun rises or sets at the 0° meridian, it will rise or set at the same time at the Standard Time meridian of your time zone. Standard Time meridians occur every 15° of longitude (15° E and W, 30° E and W, etc.). The corrections necessary to observe that event from your location will be to account for your distance from the Standard Time meridian and for your latitude. Thus, if your latitude is about 45°, sunrise on Jan. 1, 2000, is roughly halfway between 7:22 and 7:59 AM on the Standard Time meridian for your time zone. If you are 7.5° west of your Standard Time meridian, sunrise will be about ½ hour later than this; if 7.5° east, about ½ hour earlier.

The Moon, however, moves its own diameter, about one-half degree, in an hour, or about 13.2° in one complete turn of Earth—one day. Most of this is eastward against the background stars of the sky, but some is also N or S movement. All this motion considerably affects the times of rise or set, as you can see from the adjacent entries in the table. Thus, it is necessary to take your longitude into account in addition to your latitude. If you have no need for total accuracy, simply note that the time will be between the 4 values (see example below) you find surrounding your location and the dates of interest.

The process of finding more accurate corrections is called interpolation. In the example, linear interpolation involving simple differences is used. In extreme cases, higher order interpolation should be used. If such cases are important to you, it is suggested that you plot the times, draw smooth curves through the plots, and interpolate by eye between the relevant curves. Some people find this exercise fun.

Let's find the times of the moonrise for the May Waxing Quarter Moon and sunset the same day at Salem, OR.

First, where is Salem, OR? Find Salem's latitude and longitude in the "Latitude, Longitude, and Altitude of U.S. and Canadian Cities" table found in the World Exploration and Geography section of *The World Almanac*. You must also know the time zone in which the city is located, which you can estimate from the "International Time Zones" map in the map section of *The World Almanac*.

I. Salem, OR: 44° 56′ 35″ N
123° 2′ 2″ W

IA. Convert these values to decimals:
35/60 = 0.58
56 + 0.58 = 56.58
56.58/60 = 0.94
44 + 0.94 = 44.94 N
2/60 = 0.03
2 + 0.03 = 2.03
2.03/60 = 0.03
123 + 0.03 = 123.03 W

IB. Fraction Salem lies between 40° and 50°:
44.94 – 40 = 4.94; 4.94/10 = 0.494

IC. Fraction world must turn between Greenwich and Salem:
123.03/360 = 0.342

ID. Salem is in the Pacific Standard Time zone and the PST meridian is 120°; thus 123.03 is 123.03 – 120 = 3.03° W of the Pacific Standard Meridian. In 24 hours, there are 24 x 60 = 1,440 minutes; 1,440/360 = 4 minutes for every degree around Earth. So events happen 4 x 3.03 = 12.1 minutes later in Salem than at the 120° meridian. (If the location is E of the Standard Meridian, events happen earlier.)

IE. The values IB and IC are interpolates for Salem; ID is the time correction from local to Standard time for Salem. These values need never be calculated again for Salem.

IIA. To find the time of moonrise we start from the table of Moon Phases, 2002. We see that May's Waxing Quarter Moon occurs on May 19. We need the Greenwich times for moonrise at latitudes 40° and 50°, and for May 19 and 20, the day of the Waxing Quarter Moon and the next day. These values are found in the Astronomy Daily Calendar 2002; we then compute the difference between the two latitudes.

	40°	Diff.	50°
May 19	10:59	0:29	10:30
May 20	12:11	0:20	11:51

IIB. We want IB and the May 19 time difference:
0.494 x –29 = –14.3
Add this to the May 19, 40° rise time:
10:59 + (–14.3) = 10:44.7
And for May 20:
0.494 x –20 = –9.9
Add this to the May 20, 40° rise time:
12:11 + (–9.9) = 12:01.1
These 2 times are for the latitude of Salem, but for the Greenwich meridian.

IIC. To get the time for Salem meridian, take the difference between these 2 times just determined,
12:01.1 – 10:44.7 = 76.4 minutes,
and calculate what fraction of this 24–hour change took place while Earth turned between Greenwich and Salem, (See IC).
76.4 x 0.342 = 26.1 minutes after 10:44.7
Thus 10:44.7 + 26.1 = 11:10.8 is the time the Waxing Quarter Moon will rise in the local time of Salem.

IID. But this happens 12.1 minutes (See ID) later by PST clock time at Salem, thus
11:10.8 + 12.1 = 11:22.9 PST
But this is early spring, and daylight time is in effect;
11:23 + 1:00 = 12:23 PDT is the rise time for the Waxing Quarter Moon at Salem the morning of May 19, 2002.

IIIA. To find the time of sunset we need the Greenwich times for sunset at latitudes 40° and 50°. These values are found in the Astronomy Daily Calendar 2002; we then compute the difference between the two latitudes.

	40°	Diff.	50°
May 19	19:12	0:33	19:45

IIIB. We want IB and the May 19 time difference:
0.494 x 33 = 16.3
Add this to the May 19, 40° set time:
19:12 + 16.3 = 19:28.3
This is the local time for the latitude of Salem.

IIIC. But this happens 12.1 minutes (See ID) later by PST clock time at Salem, thus
19:28.3 + 12.1 = 19:40.4
But daylight time is in effect;
19:40 + 1:00 = 20:40 PDT is sunset at Salem on May 19, 2002.

JANUARY 2002

1st Month **31 days**

Coordinated Universal Time (Greenwich Mean Time)

NOTE: For each day, numbers on first line indicate Sun. Numbers on second line indicate Moon.

Degrees are North Latitude.

Moon Phases: FM = Full Moon: LQ = Last (Waning) Quarter: NM = New Moon, FQ = First (Waxing) Quarter

Sun's distance is in Astronomical Units

CAUTION: Must be converted to local time. For instructions see "Calculation of Rise Times."

Day of month, of week, of year	Sun on Meridian Moon Phase h m s	Sun's Declination ° ′ Distance	20° Rise Sun Moon h m	20° Set Sun Moon h m	30° Rise Sun Moon h m	30° Set Sun Moon h m	40° Rise Sun Moon h m	40° Set Sun Moon h m	50° Rise Sun Moon h m	50° Set Sun Moon h m	60° Rise Sun Moon h m	60° Set Sun Moon h m
1 TU	12 03 32	- 23 02	6 35	17 32	6 56	17 11	7 22	16 45	7 59	16 09	9 02	15 05
1		.9833	19 51	8 28	19 32	8 48	19 08	9 14	18 33	9 50	17 34	10 52
2 WE	12 04 00	- 22 57	6 35	17 33	6 56	17 12	7 22	16 46	7 58	16 10	9 02	15 06
2		.9833	20 56	9 20	20 41	9 37	20 22	9 58	19 56	10 27	19 13	11 14
3 TH	12 04 27	- 22 52	6 36	17 33	6 56	17 13	7 22	16 47	7 58	16 11	9 01	15 08
3		.9833	21 58	10 08	21 48	10 21	21 36	10 36	21 19	10 56	20 51	11 28
4 FR	12 04 54	- 22 46	6 36	17 34	6 57	17 13	7 22	16 48	7 58	16 12	9 01	15 10
4		.9833	22 58	10 52	22 54	11 00	22 48	11 08	22 40	11 20	22 28	11 38
5 SA	12 05 22	- 22 40	6 36	17 35	6 57	17 14	7 22	16 49	7 58	16 13	9 00	15 11
5		.9833	23 57	11 34	23 57	11 36	23 58	11 38	23 59	11 41	none	11 46
6 SU	12 05 48	- 22 33	6 37	17 35	6 57	17 15	7 22	16 50	7 58	16 14	8 59	15 13
6	03 55 LQ	.9833	none	12 14	none	12 10	none	12 07	none	12 01	0 01	11 53
7 MO	12 06 14	- 22 25	6 37	17 36	6 57	17 16	7 22	16 51	7 57	16 15	8 58	15 15
7		.9833	0 54	12 54	1 00	12 45	1 07	12 35	1 17	12 22	1 33	12 01
8 TU	12 06 40	- 22 18	6 37	17 36	6 57	17 17	7 22	16 52	7 57	16 17	8 57	15 17
8		.9834	1 52	13 35	2 02	13 22	2 16	13 06	2 35	12 45	3 05	12 11
9 WE	12 07 05	- 22 10	6 37	17 37	6 57	17 17	7 22	16 53	7 56	16 18	8 56	15 19
9		.9834	2 49	14 19	3 05	14 02	3 24	13 40	3 51	13 11	4 36	12 23
10 TH	12 07 29	- 22 01	6 37	17 38	6 57	17 18	7 22	16 54	7 56	16 19	8 55	15 21
10		.9834	3 47	15 06	4 06	14 45	4 31	14 19	5 05	13 43	6 05	12 42
11 FR	12 07 53	- 21 52	6 37	17 38	6 57	17 19	7 21	16 55	7 55	16 21	8 54	15 23
11		.9835	4 45	15 56	5 07	15 33	5 35	15 04	6 15	14 23	7 27	13 10
12 SA	12 08 17	- 21 43	6 38	17 39	6 57	17 20	7 21	16 56	7 55	16 22	8 52	15 25
12		.9835	5 41	16 48	6 04	16 25	6 34	15 55	7 16	15 12	8 35	13 54
13 SU	12 08 40	- 21 33	6 38	17 40	6 57	17 21	7 21	16 57	7 54	16 23	8 51	15 27
13	13 29 NM	.9836	6 34	17 42	6 57	17 20	7 26	16 51	8 08	16 10	9 24	14 55
14 MO	12 09 02	- 21 22	6 38	17 40	6 57	17 21	7 20	16 58	7 54	16 25	8 49	15 29
14		.9836	7 23	18 36	7 44	18 16	8 11	17 50	8 49	17 13	9 55	16 08
15 TU	12 09 24	- 21 12	6 38	17 41	6 57	17 22	7 20	16 59	7 53	16 26	8 48	15 31
15		.9837	8 07	19 29	8 26	19 12	8 49	18 50	9 22	18 20	10 16	17 28
16 WE	12 09 44	- 21 01	6 38	17 42	6 57	17 23	7 20	17 00	7 52	16 28	8 46	15 34
16		.9837	8 48	20 21	9 03	20 07	9 22	19 50	9 47	19 27	10 29	18 48
17 TH	12 10 04	- 20 49	6 38	17 42	6 56	17 24	7 19	17 01	7 51	16 29	8 45	15 36
17		.9838	9 25	21 10	9 36	21 01	9 50	20 49	10 09	20 33	10 38	20 08
18 FR	12 10 24	- 20 37	6 38	17 43	6 56	17 25	7 19	17 02	7 50	16 31	8 43	15 38
18		.9839	10 00	21 59	10 07	21 54	10 15	21 47	10 27	21 39	10 45	21 25
19 SA	12 10 43	- 20 25	6 38	17 44	6 56	17 26	7 18	17 03	7 49	16 32	8 41	15 41
19		.9839	10 33	22 47	10 36	22 46	10 39	22 45	10 44	22 44	10 51	22 42
20 SU	12 11 00	- 20 12	6 38	17 44	6 56	17 27	7 18	17 05	7 48	16 34	8 40	15 43
20		.9840	11 06	23 35	11 05	23 39	11 03	23 43	11 00	23 50	10 57	23 59
21 MO	12 11 18	- 19 59	6 38	17 45	6 55	17 27	7 17	17 06	7 47	16 36	8 38	15 46
21	17 46 FQ	.9841	11 40	none	11 34	none	11 27	none	11 17	none	11 03	none
22 TU	12 11 34	- 19 46	6 38	17 46	6 55	17 28	7 17	17 07	7 46	16 37	8 36	15 48
22		.9842	12 16	0 24	12 05	0 33	11 53	0 43	11 36	0 57	11 10	1 19
23 WE	12 11 50	- 19 32	6 38	17 46	6 55	17 29	7 16	17 08	7 45	16 39	8 34	15 51
23		.9843	12 55	1 16	12 40	1 29	12 22	1 45	11 58	2 06	11 20	2 42
24 TH	12 12 04	- 19 18	6 37	17 47	6 54	17 30	7 15	17 09	7 44	16 40	8 32	15 53
24		.9844	13 38	2 11	13 20	2 28	12 57	2 49	12 26	3 18	11 34	4 08
25 FR	12 12 18	- 19 04	6 37	17 48	6 54	17 31	7 15	17 10	7 43	16 42	8 30	15 56
25		.9845	14 28	3 09	14 06	3 29	13 39	3 55	13 01	4 31	11 56	5 35
26 SA	12 12 32	- 18 49	6 37	17 48	6 54	17 32	7 14	17 12	7 42	16 44	8 28	15 58
26		.9846	15 23	4 09	15 00	4 32	14 31	5 01	13 49	5 42	12 33	6 57
27 SU	12 12 44	- 18 34	6 37	17 49	6 53	17 33	7 13	17 13	7 41	16 45	8 26	16 01
27		.9847	16 25	5 11	16 02	5 34	15 33	6 04	14 51	6 46	13 34	8 03
28 MO	12 12 56	- 18 18	6 37	17 49	6 53	17 34	7 12	17 14	7 39	16 47	8 23	16 03
28	22 50 FM	.9848	17 31	6 11	17 10	6 33	16 43	7 01	16 05	7 40	14 59	8 48
29 TU	12 13 06	- 18 02	6 36	17 50	6 52	17 34	7 11	17 15	7 38	16 49	8 21	16 06
29		.9849	18 38	7 07	18 20	7 26	17 59	7 49	17 29	8 22	16 38	9 16
30 WE	12 13 16	- 17 46	6 36	17 51	6 52	17 35	7 11	17 16	7 37	16 50	8 19	16 08
30		.9850	19 43	7 59	19 31	8 13	19 16	8 31	18 55	8 55	18 21	9 33
31 TH	12 13 25	- 17 30	6 36	17 51	6 51	17 36	7 10	17 18	7 35	16 52	8 17	16 11
31		.9852	20 47	8 46	20 40	8 55	20 32	9 07	20 20	9 22	20 02	9 45

FEBRUARY 2002

2d Month **28 days**

Coordinated Universal Time (Greenwich Mean Time)

NOTE: For each day, numbers on first line indicate Sun. Numbers on second line indicate Moon.

Degrees are North Latitude.

Moon Phases: FM = Full Moon: LQ = Last (Waning) Quarter: NM = New Moon, FQ = First (Waxing) Quarter

Sun's distance is in Astronomical Units

CAUTION: Must be converted to local time. For instructions see "Calculation of Rise Times."

Day of month, of week, of year	Sun on Meridian / Moon Phase h m s	Sun's Declination ° ′ / Distance	20° Rise Sun / Moon h m	20° Set Sun / Moon h m	30° Rise Sun / Moon h m	30° Set Sun / Moon h m	40° Rise Sun / Moon h m	40° Set Sun / Moon h m	50° Rise Sun / Moon h m	50° Set Sun / Moon h m	60° Rise Sun / Moon h m	60° Set Sun / Moon h m
1 FR	12 13 34	- 17 13	6 36	17 52	6 50	17 37	7 09	17 19	7 34	16 54	8 14	16 14
32		.9853	21 48	9 30	21 47	9 34	21 46	9 39	21 44	9 45	21 41	9 54
2 SA	12 13 41	- 16 56	6 35	17 52	6 50	17 38	7 08	17 20	7 32	16 56	8 12	16 16
33		.9855	22 48	10 12	22 52	10 10	22 57	10 08	23 05	10 06	23 16	10 02
3 SU	12 13 48	- 16 38	6 35	17 53	6 49	17 39	7 07	17 21	7 31	16 57	8 10	16 19
34		.9856	23 46	10 53	23 56	10 46	none	10 38	none	10 27	none	10 10
4 MO	12 13 54	- 16 21	6 35	17 53	6 49	17 40	7 06	17 22	7 29	16 59	8 07	16 22
35	13 33 LQ	.9858	none	11 34	none	11 23	0 08	11 08	0 24	10 49	0 50	10 19
5 TU	12 13 59	- 16 03	6 34	17 54	6 48	17 40	7 05	17 24	7 28	17 01	8 05	16 24
36		.9859	0 45	12 18	0 59	12 02	1 17	11 42	1 42	11 14	2 22	10 30
6 WE	12 14 03	- 15 45	6 34	17 55	6 47	17 41	7 04	17 25	7 26	17 02	8 02	16 27
37		.9861	1 42	13 03	2 01	12 44	2 24	12 19	2 57	11 45	3 53	10 46
7 TH	12 14 07	- 15 26	6 33	17 55	6 47	17 42	7 03	17 26	7 25	17 04	8 00	16 29
38		.9863	2 40	13 52	3 01	13 30	3 28	13 02	4 07	12 22	5 17	11 11
8 FR	12 14 10	- 15 07	6 33	17 56	6 46	17 43	7 02	17 27	7 23	17 06	7 57	16 32
39		.9864	3 36	14 43	3 59	14 20	4 28	13 50	5 11	13 07	6 28	11 49
9 SA	12 14 12	- 14 48	6 32	17 56	6 45	17 44	7 00	17 28	7 21	17 08	7 55	16 35
40		.9866	4 29	15 36	4 52	15 13	5 22	14 44	6 05	14 02	7 23	12 44
10 SU	12 14 13	- 14 29	6 32	17 57	6 44	17 45	6 59	17 30	7 20	17 09	7 52	16 37
41		.9868	5 19	16 30	5 41	16 09	6 09	15 42	6 48	15 03	7 59	13 54
11 MO	12 14 14	- 14 09	6 31	17 57	6 43	17 45	6 58	17 31	7 18	17 11	7 49	16 40
42		.9870	6 04	17 23	6 24	17 04	6 49	16 41	7 23	16 08	8 22	15 12
12 TU	12 14 13	- 13 50	6 31	17 58	6 43	17 46	6 57	17 32	7 16	17 13	7 47	16 43
43	07 41 NM	.9872	6 46	18 15	7 02	18 00	7 23	17 41	7 51	17 15	8 36	16 32
13 WE	12 14 12	- 13 30	6 30	17 58	6 42	17 47	6 56	17 33	7 15	17 15	7 44	16 45
44		.9874	7 24	19 05	7 37	18 54	7 52	18 41	8 13	18 22	8 47	17 52
14 TH	12 14 10	- 13 10	6 30	17 59	6 41	17 48	6 55	17 34	7 13	17 16	7 41	16 48
45		.9875	7 59	19 54	8 08	19 47	8 18	19 39	8 32	19 28	8 54	19 11
15 FR	12 14 08	- 12 49	6 29	17 59	6 40	17 49	6 53	17 36	7 11	17 18	7 39	16 51
46		.9877	8 33	20 42	8 37	20 40	8 42	20 37	8 49	20 33	9 00	20 28
16 SA	12 14 04	- 12 28	6 29	18 00	6 39	17 49	6 52	17 37	7 09	17 20	7 36	16 53
47		.9879	9 06	21 30	9 06	21 32	9 06	21 35	9 06	21 39	9 06	21 45
17 SU	12 14 00	- 12 08	6 28	18 00	6 38	17 50	6 51	17 38	7 07	17 21	7 33	16 56
48		.9881	9 39	22 18	9 35	22 25	9 29	22 33	9 22	22 45	9 11	23 02
18 MO	12 13 56	- 11 47	6 28	18 01	6 37	17 51	6 49	17 39	7 05	17 23	7 30	16 58
49		.9883	10 13	23 08	10 05	23 19	9 54	23 33	9 39	23 52	9 17	none
19 TU	12 13 50	- 11 25	6 27	18 01	6 36	17 52	6 48	17 40	7 04	17 25	7 28	17 01
50		.9885	10 50	none	10 37	none	10 21	none	9 59	none	9 25	0 23
20 WE	12 13 44	- 11 04	6 26	18 01	6 36	17 52	6 47	17 41	7 02	17 27	7 25	17 04
51	12 02 FQ	.9887	11 30	0 00	11 14	0 16	10 53	0 35	10 24	1 02	9 37	1 46
21 TH	12 13 38	- 10 42	6 26	18 02	6 35	17 53	6 45	17 43	7 00	17 28	7 22	17 06
52		.9889	12 16	0 55	11 55	1 14	11 30	1 38	10 54	2 12	9 54	3 11
22 FR	12 13 30	- 10 21	6 25	18 02	6 34	17 54	6 44	17 44	6 58	17 30	7 19	17 09
53		.9892	13 07	1 53	12 44	2 15	12 16	2 43	11 35	3 22	10 22	4 34
23 SA	12 13 22	-9 59	6 24	18 03	6 33	17 55	6 43	17 45	6 56	17 32	7 16	17 11
54		.9894	14 04	2 52	13 41	3 16	13 11	3 45	12 28	4 28	11 09	5 47
24 SU	12 13 13	-9 37	6 24	18 03	6 32	17 55	6 41	17 46	6 54	17 33	7 14	17 14
55		.9896	15 07	3 52	14 44	4 15	14 16	4 44	13 35	5 26	12 21	6 41
25 MO	12 13 04	-9 15	6 23	18 03	6 31	17 56	6 40	17 47	6 52	17 35	7 11	17 17
56		.9898	16 13	4 49	15 53	5 10	15 29	5 36	14 54	6 13	13 54	7 15
26 TU	12 12 54	-8 52	6 22	18 04	6 29	17 57	6 38	17 48	6 50	17 37	7 08	17 19
57		.9900	17 20	5 43	17 05	6 00	16 46	6 21	16 20	6 50	15 37	7 36
27 WE	12 12 44	-8 30	6 21	18 04	6 28	17 57	6 37	17 49	6 48	17 38	7 05	17 22
58	09 17 FM	.9903	18 26	6 33	18 16	6 45	18 04	7 00	17 48	7 20	17 22	7 50
28 TH	12 12 33	-8 07	6 21	18 05	6 27	17 58	6 35	17 50	6 46	17 40	7 02	17 24
59		.9905	19 30	7 20	19 26	7 26	19 22	7 34	19 15	7 45	19 05	8 00

MARCH 2002

3d Month **31 days**

Coordinated Universal Time (Greenwich Mean Time)

NOTE: For each day, numbers on first line indicate Sun. Numbers on second line indicate Moon.

Degrees are North Latitude.

Moon Phases: FM = Full Moon: LQ = Last (Waning) Quarter: NM = New Moon, FQ = First (Waxing) Quarter

Sun's distance is in Astronomical Units

CAUTION: Must be converted to local time. For instructions see "Calculation of Rise Times."

Day of month, of week, of year	Sun on Meridian / Moon Phase (h m s)	Sun's Declination (° ′) / Distance	20° Rise Sun / Moon (h m)	20° Set Sun / Moon (h m)	30° Rise Sun / Moon (h m)	30° Set Sun / Moon (h m)	40° Rise Sun / Moon (h m)	40° Set Sun / Moon (h m)	50° Rise Sun / Moon (h m)	50° Set Sun / Moon (h m)	60° Rise Sun / Moon (h m)	60° Set Sun / Moon (h m)
1 FR	12 12 21	-7 45	6 20	18 05	6 26	17 59	6 34	17 52	6 44	17 42	6 59	17 27
60		.9908	20 33	8 04	20 35	8 04	20 37	8 06	20 41	8 07	20 46	8 09
2 SA	12 12 09	-7 22	6 19	18 05	6 25	18 00	6 32	17 53	6 42	17 43	6 56	17 29
61		.9910	21 34	8 46	21 42	8 42	21 51	8 36	22 04	8 28	22 25	8 17
3 SU	12 11 57	-6 59	6 18	18 06	6 24	18 00	6 31	17 54	6 40	17 45	6 53	17 32
62		.9913	22 35	9 29	22 48	9 19	23 04	9 07	23 26	8 51	none	8 25
4 MO	12 11 44	-6 36	6 18	18 06	6 23	18 01	6 29	17 55	6 38	17 47	6 50	17 34
63		.9915	23 35	10 13	23 53	9 59	none	9 40	none	9 15	0 02	8 35
5 TU	12 11 30	-6 13	6 17	18 06	6 22	18 02	6 28	17 56	6 36	17 48	6 47	17 37
64		.9918	none	11 00	none	10 41	0 14	10 17	0 45	9 45	1 37	8 50
6 WE	12 11 17	-5 50	6 16	18 07	6 21	18 02	6 26	17 57	6 33	17 50	6 44	17 39
65	01 24 LQ	.9920	0 34	11 48	0 55	11 27	1 22	10 59	1 59	10 20	3 06	9 12
7 TH	12 11 02	-5 26	6 15	18 07	6 20	18 03	6 25	17 58	6 31	17 52	6 41	17 42
66		.9923	1 32	12 39	1 55	12 16	2 24	11 46	3 06	11 04	4 24	9 45
8 FR	12 10 48	-5 03	6 15	18 07	6 18	18 04	6 23	17 59	6 29	17 53	6 39	17 44
67		.9926	2 26	13 32	2 50	13 09	3 20	12 39	4 03	11 56	5 24	10 36
9 SA	12 10 33	-4 40	6 14	18 08	6 17	18 04	6 22	18 00	6 27	17 55	6 36	17 47
68		.9928	3 17	14 26	3 40	14 03	4 08	13 35	4 50	12 55	6 04	11 42
10 SU	12 10 18	-4 16	6 13	18 08	6 16	18 05	6 20	18 01	6 25	17 56	6 33	17 49
69		.9931	4 03	15 19	4 24	14 59	4 50	14 34	5 27	13 59	6 29	12 58
11 MO	12 10 02	-3 53	6 12	18 08	6 15	18 06	6 18	18 02	6 23	17 58	6 30	17 52
70		.9934	4 46	16 11	5 03	15 54	5 25	15 34	5 56	15 06	6 46	14 18
12 TU	12 09 46	-3 29	6 11	18 09	6 14	18 06	6 17	18 03	6 21	18 00	6 27	17 54
71		.9936	5 24	17 01	5 38	16 49	5 56	16 34	6 19	16 13	6 56	15 39
13 WE	12 09 30	-3 05	6 10	18 09	6 13	18 07	6 15	18 04	6 19	18 01	6 24	17 57
72		.9939	6 00	17 50	6 10	17 42	6 22	17 32	6 39	17 19	7 04	16 58
14 TH	12 09 14	-2 42	6 10	18 09	6 11	18 07	6 14	18 05	6 16	18 03	6 21	17 59
73	02 02 NM	.9942	6 34	18 38	6 40	18 35	6 47	18 31	6 56	18 25	7 10	18 15
15 FR	12 08 57	-2 18	6 09	18 10	6 10	18 08	6 12	18 07	6 14	18 05	6 18	18 02
74		.9944	7 07	19 26	7 09	19 27	7 10	19 29	7 12	19 30	7 15	19 33
16 SA	12 08 40	-1 54	6 08	18 10	6 09	18 09	6 10	18 08	6 12	18 06	6 15	18 04
75		.9947	7 40	20 15	7 37	20 20	7 33	20 27	7 28	20 36	7 20	20 50
17 SU	12 08 23	-1 31	6 07	18 10	6 08	18 09	6 09	18 09	6 10	18 08	6 12	18 07
76		.9950	8 14	21 04	8 06	21 14	7 57	21 26	7 45	21 43	7 26	22 10
18 MO	12 08 06	-1 07	6 06	18 10	6 07	18 10	6 07	18 10	6 08	18 09	6 08	18 09
77		.9952	8 49	21 55	8 38	22 09	8 23	22 27	8 03	22 52	7 33	23 32
19 TU	12 07 48	-0 43	6 05	18 11	6 05	18 11	6 06	18 11	6 06	18 11	6 05	18 12
78		.9955	9 28	22 48	9 12	23 07	8 52	23 29	8 26	none	7 42	none
20 WE	12 07 31	-0 20	6 04	18 11	6 04	18 11	6 04	18 12	6 03	18 13	6 02	18 14
79		.9958	10 11	23 44	9 51	none	9 27	none	8 53	0 01	7 56	0 56
21 TH	12 07 13	+0 04	6 03	18 11	6 03	18 12	6 02	18 13	6 01	18 14	5 59	18 16
80		.9961	10 58	none	10 36	0 05	10 08	0 32	9 28	1 11	8 19	2 19
22 FR	12 06 55	+0 28	6 03	18 12	6 02	18 12	6 01	18 14	5 59	18 16	5 56	18 19
81	02 28 FQ	.9963	11 51	0 41	11 28	1 04	10 58	1 34	10 15	2 17	8 56	3 35
23 SA	12 06 37	+0 52	6 02	18 12	6 01	18 13	5 59	18 15	5 57	18 17	5 53	18 21
82		.9966	12 50	1 39	12 26	2 02	11 57	2 33	11 14	3 16	9 55	4 35
24 SU	12 06 19	+1 15	6 01	18 12	5 59	18 14	5 57	18 16	5 55	18 19	5 50	18 24
83		.9969	13 52	2 35	13 31	2 57	13 04	3 25	12 25	4 05	11 17	5 16
25 MO	12 06 00	+1 39	6 00	18 12	5 58	18 14	5 56	18 17	5 53	18 20	5 47	18 26
84		.9972	14 57	3 29	14 39	3 48	14 17	4 12	13 46	4 45	12 54	5 41
26 TU	12 05 42	+2 02	5 59	18 13	5 57	18 15	5 54	18 18	5 50	18 22	5 44	18 29
85		.9974	16 02	4 19	15 50	4 34	15 34	4 52	15 12	5 17	14 37	5 56
27 WE	12 05 24	+2 26	5 58	18 13	5 56	18 15	5 53	18 19	5 48	18 24	5 41	18 31
86		.9977	17 07	5 07	17 00	5 16	16 51	5 28	16 39	5 43	16 21	6 07
28 TH	12 05 06	+2 49	5 57	18 13	5 55	18 16	5 51	18 20	5 46	18 25	5 38	18 33
87	18 25 FM	.9980	18 11	5 51	18 10	5 55	18 08	6 00	18 07	6 06	18 04	6 16
29 FR	12 04 48	+3 13	5 56	18 13	5 53	18 17	5 49	18 21	5 44	18 27	5 35	18 36
88		.9983	19 14	6 35	19 19	6 33	19 25	6 31	19 33	6 28	19 46	6 23
30 SA	12 04 29	+3 36	5 56	18 14	5 52	18 17	5 48	18 22	5 42	18 28	5 32	18 38
89		.9986	20 17	7 19	20 28	7 11	20 41	7 02	20 59	6 50	21 28	6 31
31 SU	12 04 11	+3 59	5 55	18 14	5 51	18 18	5 46	18 23	5 40	18 30	5 29	18 41
90		.9989	21 20	8 03	21 36	7 51	21 55	7 35	22 23	7 14	23 08	6 40

APRIL 2002

4th Month **30 days**

Coordinated Universal Time (Greenwich Mean Time)

NOTE: For each day, numbers on first line indicate Sun. Numbers on second line indicate Moon.

Degrees are North Latitude.

Moon Phases: FM = Full Moon: LQ = Last (Waning) Quarter: NM = New Moon, FQ = First (Waxing) Quarter

Sun's distance is in Astronomical Units

CAUTION: Must be converted to local time. For instructions see "Calculation of Rise Times."

Day of month, of week, of year	Sun on Meridian / Moon Phase (h m s)	Sun's Declination (° ′) / Distance	20° Rise Sun/Moon (h m)	20° Set Sun/Moon (h m)	30° Rise Sun/Moon (h m)	30° Set Sun/Moon (h m)	40° Rise Sun/Moon (h m)	40° Set Sun/Moon (h m)	50° Rise Sun/Moon (h m)	50° Set Sun/Moon (h m)	60° Rise Sun/Moon (h m)	60° Set Sun/Moon (h m)
1 MO	12 03 53	+4 23	5 54	18 14	5 50	18 19	5 45	18 24	5 37	18 31	5 26	18 43
91		.9992	22 22	8 50	22 42	8 33	23 07	8 11	23 43	7 41	none	6 52
2 TU	12 03 36	+4 46	5 53	18 14	5 49	18 19	5 43	18 25	5 35	18 33	5 23	18 46
92		.9995	23 22	9 40	23 45	9 19	none	8 52	none	8 15	0 45	7 10
3 WE	12 03 18	+5 09	5 52	18 15	5 47	18 20	5 41	18 26	5 33	18 35	5 20	18 48
93		.9997	none	10 32	none	10 08	0 14	9 39	0 55	8 56	2 12	7 39
4 TH	12 03 00	+5 32	5 51	18 15	5 46	18 20	5 40	18 27	5 31	18 36	5 17	18 51
94	15 29 LQ	1.0000	0 20	11 26	0 44	11 02	1 14	10 31	1 58	9 47	3 21	8 24
5 FR	12 02 43	+5 55	5 50	18 15	5 45	18 21	5 38	18 28	5 29	18 38	5 14	18 53
95		1.0003	1 13	12 20	1 37	11 57	2 07	11 28	2 50	10 45	4 08	9 27
6 SA	12 02 26	+6 17	5 50	18 16	5 44	18 22	5 37	18 29	5 27	18 39	5 11	18 55
96		1.0006	2 02	13 14	2 23	12 53	2 51	12 27	3 30	11 49	4 38	10 43
7 SU	12 02 09	+6 40	5 49	18 16	5 43	18 22	5 35	18 30	5 24	18 41	5 08	18 58
97		1.0009	2 45	14 06	3 04	13 49	3 28	13 27	4 01	12 56	4 56	12 03
8 MO	12 01 52	+7 03	5 48	18 16	5 41	18 23	5 33	18 31	5 22	18 42	5 05	19 00
98		1.0012	3 25	14 57	3 41	14 44	4 00	14 27	4 26	14 03	5 07	13 24
9 TU	12 01 36	+7 25	5 47	18 16	5 40	18 23	5 32	18 32	5 20	18 44	5 02	19 03
99		1.0015	4 02	15 47	4 13	15 37	4 27	15 26	4 46	15 09	5 15	14 44
10 WE	12 01 20	+7 47	5 46	18 17	5 39	18 24	5 30	18 33	5 18	18 46	4 59	19 05
100		1.0018	4 36	16 35	4 43	16 30	4 52	16 24	5 04	16 15	5 21	16 02
11 TH	12 01 04	+8 10	5 45	18 17	5 38	18 25	5 29	18 34	5 16	18 47	4 56	19 08
101		1.0021	5 09	17 23	5 12	17 23	5 15	17 22	5 20	17 21	5 26	17 20
12 FR	12 00 48	+8 32	5 45	18 17	5 37	18 25	5 27	18 35	5 14	18 49	4 53	19 10
102	19 21 NM	1.0024	5 42	18 11	5 40	18 16	5 38	18 21	5 35	18 27	5 31	18 38
13 SA	12 00 33	+8 54	5 44	18 18	5 36	18 26	5 26	18 36	5 12	18 50	4 50	19 13
103		1.0026	6 15	19 01	6 09	19 09	6 01	19 20	5 51	19 35	5 36	19 58
14 SU	12 00 18	+9 15	5 43	18 18	5 35	18 26	5 24	18 37	5 10	18 52	4 47	19 15
104		1.0029	6 50	19 52	6 40	20 05	6 27	20 21	6 09	20 43	5 42	21 20
15 MO	12 00 03	+9 37	5 42	18 18	5 34	18 27	5 23	18 38	5 08	18 53	4 44	19 18
105		1.0032	7 28	20 44	7 13	21 02	6 55	21 23	6 30	21 53	5 50	22 44
16 TU	11 59 49	+9 58	5 41	18 18	5 32	18 28	5 21	18 39	5 06	18 55	4 41	19 20
106		1.0035	8 09	21 39	7 50	22 00	7 27	22 26	6 55	23 03	6 02	none
17 WE	11 59 35	+ 10 20	5 41	18 19	5 31	18 28	5 20	18 40	5 04	18 57	4 38	19 22
107		1.0038	8 55	22 36	8 33	22 59	8 06	23 28	7 27	none	6 20	0 08
18 TH	11 59 21	+ 10 41	5 40	18 19	5 30	18 29	5 18	18 41	5 02	18 58	4 35	19 25
108		1.0040	9 46	23 33	9 22	23 57	8 52	none	8 09	0 10	6 51	1 28
19 FR	11 59 08	+ 11 02	5 39	18 19	5 29	18 30	5 17	18 42	5 00	19 00	4 32	19 27
109		1.0043	10 41	none	10 17	none	9 47	0 27	9 03	1 11	7 41	2 33
20 SA	11 58 55	+ 11 22	5 38	18 20	5 28	18 30	5 15	18 43	4 58	19 01	4 30	19 30
110	12 48 FQ	1.0046	11 41	0 28	11 18	0 52	10 50	1 21	10 09	2 03	8 54	3 19
21 SU	11 58 43	+ 11 43	5 38	18 20	5 27	18 31	5 14	18 44	4 56	19 03	4 27	19 32
111		1.0048	12 43	1 21	12 23	1 42	11 59	2 08	11 24	2 45	10 24	3 47
22 MO	11 58 31	+ 12 03	5 37	18 20	5 26	18 31	5 12	18 45	4 54	19 04	4 24	19 35
112		1.0051	13 46	2 11	13 30	2 28	13 12	2 49	12 45	3 18	12 02	4 05
23 TU	11 58 19	+ 12 23	5 36	18 21	5 25	18 32	5 11	18 46	4 52	19 06	4 21	19 37
113		1.0054	14 48	2 58	14 38	3 10	14 26	3 25	14 09	3 45	13 43	4 16
24 WE	11 58 08	+ 12 43	5 36	18 21	5 24	18 33	5 10	18 47	4 50	19 07	4 18	19 40
114		1.0056	15 51	3 42	15 46	3 49	15 41	3 57	15 34	4 08	15 24	4 25
25 TH	11 57 58	+ 13 03	5 35	18 21	5 23	18 33	5 08	18 48	4 48	19 09	4 15	19 42
115		1.0059	16 53	4 25	16 55	4 26	16 57	4 27	17 00	4 29	17 05	4 32
26 FR	11 57 47	+ 13 23	5 34	18 22	5 22	18 34	5 07	18 49	4 46	19 11	4 12	19 45
116		1.0062	17 56	5 07	18 03	5 03	18 13	4 57	18 26	4 50	18 46	4 39
27 SA	11 57 38	+ 13 42	5 34	18 22	5 21	18 35	5 06	18 50	4 44	19 12	4 10	19 47
117	03 00 FM	1.0064	18 59	5 51	19 12	5 41	19 28	5 29	19 51	5 12	20 28	4 46
28 SU	11 57 28	+ 14 01	5 33	18 22	5 20	18 35	5 04	18 51	4 42	19 14	4 07	19 50
118		1.0067	20 03	6 37	20 21	6 22	20 43	6 03	21 15	5 37	22 10	4 56
29 MO	11 57 20	+ 14 20	5 32	18 23	5 19	18 36	5 03	18 52	4 40	19 15	4 04	19 52
119		1.0069	21 06	7 26	21 28	7 07	21 55	6 42	22 35	6 08	23 46	5 11
30 TU	11 57 12	+ 14 39	5 32	18 23	5 18	18 36	5 02	18 53	4 39	19 17	4 01	19 55
120		1.0072	22 07	8 19	22 31	7 56	23 01	7 27	23 45	6 47	none	5 34

MAY 2002

5th Month **31 days**

Coordinated Universal Time (Greenwich Mean Time)

NOTE: For each day, numbers on first line indicate Sun. Numbers on second line indicate Moon.

Degrees are North Latitude.

Moon Phases: FM = Full Moon: LQ = Last (Waning) Quarter: NM = New Moon, FQ = First (Waxing) Quarter

Sun's distance is in Astronomical Units

CAUTION: Must be converted to local time. For instructions see "Calculation of Rise Times."

Day of month, of week, of year	Sun on Meridian / Moon Phase h m s	Sun's Decli-nation ° ′ / Distance	20° Rise Sun / Moon h m	20° Set Sun / Moon h m	30° Rise Sun / Moon h m	30° Set Sun / Moon h m	40° Rise Sun / Moon h m	40° Set Sun / Moon h m	50° Rise Sun / Moon h m	50° Set Sun / Moon h m	60° Rise Sun / Moon h m	60° Set Sun / Moon h m
1 WE	11 57 04	+ 14 57	5 31	18 23	5 17	18 37	5 00	18 54	4 37	19 18	3 59	19 57
121		1.0075	23 04	9 14	23 28	8 49	23 59	8 19	none	7 34	1 08	6 11
2 TH	11 56 57	+ 15 15	5 30	18 24	5 17	18 38	4 59	18 55	4 35	19 20	3 56	20 00
122		1.0077	23 56	10 10	none	9 46	none	9 15	0 43	8 31	2 07	7 09
3 FR	11 56 51	+ 15 33	5 30	18 24	5 16	18 38	4 58	18 56	4 33	19 21	3 53	20 02
123		1.0080	none	11 05	0 19	10 43	0 48	10 15	1 29	9 35	2 43	8 22
4 SA	11 56 45	+ 15 51	5 29	18 24	5 15	18 39	4 57	18 57	4 32	19 23	3 51	20 05
124	07 16 LQ	1.0082	0 42	12 00	1 03	11 40	1 28	11 16	2 04	10 42	3 05	9 44
5 SU	11 56 40	+ 16 08	5 29	18 25	5 14	18 40	4 56	18 58	4 30	19 24	3 48	20 07
125		1.0085	1 24	12 52	1 41	12 36	2 02	12 17	2 31	11 50	3 18	11 06
6 MO	11 56 35	+ 16 25	5 28	18 25	5 13	18 40	4 54	18 59	4 28	19 26	3 45	20 10
126		1.0087	2 02	13 42	2 15	13 31	2 31	13 17	2 53	12 58	3 27	12 27
7 TU	11 56 31	+ 16 42	5 28	18 26	5 12	18 41	4 53	19 00	4 27	19 27	3 43	20 12
127		1.0090	2 37	14 31	2 46	14 24	2 56	14 16	3 11	14 04	3 33	13 46
8 WE	11 56 27	+ 16 59	5 27	18 26	5 12	18 42	4 52	19 01	4 25	19 29	3 40	20 15
128		1.0092	3 10	15 19	3 15	15 17	3 20	15 14	3 27	15 10	3 38	15 04
9 TH	11 56 24	+ 17 15	5 27	18 26	5 11	18 42	4 51	19 02	4 23	19 30	3 38	20 17
129		1.0095	3 43	16 07	3 43	16 09	3 43	16 12	3 43	16 16	3 43	16 22
10 FR	11 56 22	+ 17 31	5 26	18 27	5 10	18 43	4 50	19 03	4 22	19 32	3 35	20 19
130		1.0097	4 16	16 56	4 11	17 03	4 06	17 11	3 58	17 23	3 47	17 42
11 SA	11 56 20	+ 17 46	5 26	18 27	5 09	18 44	4 49	19 04	4 20	19 33	3 33	20 22
131		1.0099	4 50	17 46	4 41	17 58	4 30	18 12	4 15	18 32	3 52	19 04
12 SU	11 56 19	+ 18 02	5 25	18 28	5 09	18 44	4 48	19 05	4 19	19 35	3 30	20 24
132	10 45 NM	1.0102	5 27	18 39	5 14	18 55	4 57	19 15	4 35	19 43	3 59	20 28
13 MO	11 56 18	+ 18 17	5 25	18 28	5 08	18 45	4 47	19 06	4 17	19 36	3 28	20 27
133		1.0104	6 07	19 34	5 50	19 54	5 28	20 19	4 58	20 54	4 10	21 55
14 TU	11 56 18	+ 18 32	5 24	18 28	5 07	18 46	4 46	19 07	4 16	19 38	3 25	20 29
134		1.0106	6 52	20 31	6 31	20 54	6 05	21 22	5 28	22 03	4 25	23 18
15 WE	11 56 18	+ 18 46	5 24	18 29	5 07	18 46	4 45	19 08	4 14	19 39	3 23	20 31
135		1.0108	7 42	21 28	7 19	21 52	6 49	22 23	6 07	23 07	4 51	none
16 TH	11 56 19	+ 19 00	5 24	18 29	5 06	18 47	4 44	19 09	4 13	19 40	3 21	20 34
136		1.0110	8 37	22 25	8 12	22 49	7 42	23 19	6 57	none	5 34	0 30
17 FR	11 56 21	+ 19 14	5 23	18 30	5 06	18 47	4 43	19 10	4 12	19 42	3 18	20 36
137		1.0112	9 35	23 18	9 12	23 40	8 42	none	7 59	0 02	6 41	1 22
18 SA	11 56 23	+ 19 27	5 23	18 30	5 05	18 48	4 42	19 11	4 10	19 43	3 16	20 38
138		1.0114	10 36	none	10 15	none	9 49	0 08	9 11	0 47	8 06	1 54
19 SU	11 56 26	+ 19 41	5 23	18 30	5 04	18 49	4 42	19 12	4 09	19 45	3 14	20 41
139	19 42 FQ	1.0116	11 37	0 08	11 20	0 27	10 59	0 50	10 30	1 22	9 41	2 14
20 MO	11 56 29	+ 19 53	5 22	18 31	5 04	18 49	4 41	19 13	4 08	19 46	3 12	20 43
140		1.0118	12 38	0 55	12 26	1 09	12 11	1 26	11 51	1 50	11 18	2 27
21 TU	11 56 32	+ 20 06	5 22	18 31	5 03	18 50	4 40	19 14	4 07	19 47	3 10	20 45
141		1.0120	13 38	1 38	13 32	1 47	13 24	1 58	13 13	2 13	12 56	2 35
22 WE	11 56 37	+ 20 18	5 22	18 32	5 03	18 51	4 39	19 14	4 06	19 48	3 08	20 47
142		1.0122	14 39	2 20	14 38	2 23	14 37	2 28	14 35	2 34	14 33	2 42
23 TH	11 56 41	+ 20 30	5 21	18 32	5 02	18 51	4 38	19 15	4 04	19 50	3 05	20 49
143		1.0124	15 39	3 01	15 44	2 59	15 50	2 57	15 58	2 53	16 11	2 48
24 FR	11 56 46	+ 20 41	5 21	18 32	5 02	18 52	4 38	19 16	4 03	19 51	3 04	20 52
144		1.0126	16 40	3 43	16 51	3 35	17 04	3 26	17 22	3 14	17 51	2 55
25 SA	11 56 52	+ 20 52	5 21	18 33	5 02	18 52	4 37	19 17	4 02	19 52	3 02	20 54
145		1.0128	17 43	4 26	17 59	4 14	18 18	3 58	18 46	3 37	19 32	3 03
26 SU	11 56 58	+ 21 03	5 21	18 33	5 01	18 53	4 36	19 18	4 01	19 53	3 00	20 56
146	11 51 FM	1.0129	18 46	5 13	19 07	4 56	19 32	4 34	20 08	4 04	21 12	3 14
27 MO	11 57 05	+ 21 13	5 21	18 34	5 01	18 54	4 36	19 19	4 00	19 55	2 58	20 58
147		1.0131	19 49	6 04	20 12	5 43	20 42	5 16	21 24	4 38	22 43	3 32
28 TU	11 57 12	+ 21 23	5 20	18 34	5 01	18 54	4 35	19 19	3 59	19 56	2 56	21 00
148		1.0133	20 49	6 59	21 14	6 35	21 45	6 05	22 30	5 21	23 55	4 02
29 WE	11 57 20	+ 21 33	5 20	18 34	5 00	18 55	4 35	19 20	3 58	19 57	2 54	21 02
149		1.0135	21 45	7 55	22 09	7 31	22 39	7 00	23 22	6 15	none	4 50
30 TH	11 57 28	+ 21 42	5 20	18 35	5 00	18 55	4 34	19 21	3 58	19 58	2 53	21 03
150		1.0136	22 35	8 53	22 57	8 29	23 24	8 00	none	7 17	0 42	5 58
31 FR	11 57 36	+ 21 51	5 20	18 35	5 00	18 56	4 34	19 22	3 57	19 59	2 51	21 05
151		1.0138	23 20	9 49	23 38	9 28	none	9 02	0 03	8 25	1 10	7 19

JUNE 2002

6th Month **30 days**

Coordinated Universal Time (Greenwich Mean Time)

NOTE: For each day, numbers on first line indicate Sun. Numbers on second line indicate Moon.

Degrees are North Latitude.

Moon Phases: FM = Full Moon: LQ = Last (Waning) Quarter: NM = New Moon, FQ = First (Waxing) Quarter

Sun's distance is in Astronomical Units

CAUTION: Must be converted to local time. For instructions see "Calculation of Rise Times."

Day of month, of week, of year	Sun on Meridian / Moon Phase	Sun's Declination ° ′ / Distance	20° Rise Sun / Moon	20° Set Sun / Moon	30° Rise Sun / Moon	30° Set Sun / Moon	40° Rise Sun / Moon	40° Set Sun / Moon	50° Rise Sun / Moon	50° Set Sun / Moon	60° Rise Sun / Moon	60° Set Sun / Moon
	h m s		h m	h m	h m	h m	h m	h m	h m	h m	h m	h m
1 SA	11 57 45	+ 22 00	5 20	18 36	4 59	18 56	4 33	19 22	3 56	20 00	2 50	21 07
152		1.0140	24 00	10 43	none	10 26	0 01	10 04	0 33	9 34	1 27	8 44
2 SU	11 57 54	+ 22 08	5 20	18 36	4 59	18 57	4 33	19 23	3 55	20 01	2 48	21 09
153		1.0141	none	11 34	0 14	11 22	0 33	11 05	0 57	10 43	1 37	10 07
3 MO	11 58 04	+ 22 16	5 20	18 36	4 59	18 57	4 33	19 24	3 55	20 02	2 47	21 10
154	00 05 LQ	1.0143	0 36	12 24	0 47	12 16	1 00	12 05	1 17	11 51	1 44	11 28
4 TU	11 58 14	+ 22 23	5 20	18 37	4 59	18 58	4 32	19 25	3 54	20 03	2 46	21 12
155		1.0144	1 10	13 13	1 16	13 09	1 24	13 04	1 34	12 57	1 49	12 46
5 WE	11 58 25	+ 22 30	5 20	18 37	4 59	18 58	4 32	19 25	3 53	20 04	2 44	21 14
156		1.0146	1 43	14 00	1 45	14 01	1 47	14 02	1 49	14 03	1 54	14 04
6 TH	11 58 36	+ 22 36	5 20	18 38	4 59	18 59	4 32	19 26	3 53	20 05	2 43	21 15
157		1.0147	2 15	14 49	2 13	14 54	2 09	15 00	2 05	15 09	1 58	15 23
7 FR	11 58 47	+ 22 43	5 20	18 38	4 58	18 59	4 31	19 26	3 52	20 06	2 42	21 16
158		1.0148	2 49	15 39	2 42	15 48	2 33	16 01	2 21	16 17	2 03	16 43
8 SA	11 58 58	+ 22 48	5 20	18 38	4 58	19 00	4 31	19 27	3 52	20 06	2 41	21 18
159		1.0150	3 25	16 31	3 13	16 45	2 59	17 03	2 39	17 27	2 09	18 07
9 SU	11 59 10	+ 22 54	5 20	18 39	4 58	19 00	4 31	19 28	3 52	20 07	2 40	21 19
160		1.0151	4 04	17 25	3 48	17 44	3 28	18 07	3 01	18 39	2 18	19 34
10 MO	11 59 22	+ 22 59	5 20	18 39	4 58	19 01	4 31	19 28	3 51	20 08	2 39	21 20
161	23 46 NM	1.0152	4 47	18 22	4 27	18 44	4 03	19 11	3 29	19 51	2 31	21 01
11 TU	11 59 34	+ 23 03	5 20	18 39	4 58	19 01	4 31	19 29	3 51	20 08	2 38	21 21
162		1.0153	5 36	19 21	5 13	19 44	4 45	20 15	4 04	20 58	2 52	22 19
12 WE	11 59 47	+ 23 07	5 20	18 40	4 58	19 01	4 31	19 29	3 51	20 09	2 38	21 22
163		1.0154	6 30	20 19	6 06	20 43	5 35	21 14	4 51	21 58	3 29	23 20
13 TH	11 59 59	+ 23 11	5 20	18 40	4 58	19 02	4 31	19 30	3 51	20 10	2 37	21 23
164		1.0155	7 28	21 14	7 04	21 37	6 34	22 06	5 50	22 46	4 29	23 59
14 FR	12 00 12	+ 23 14	5 20	18 40	4 58	19 02	4 31	19 30	3 50	20 10	2 37	21 24
165		1.0156	8 29	22 06	8 08	22 26	7 40	22 51	7 01	23 25	5 50	none
15 SA	12 00 25	+ 23 17	5 20	18 40	4 58	19 02	4 31	19 30	3 50	20 11	2 36	21 25
166		1.0157	9 31	22 54	9 13	23 10	8 50	23 29	8 18	23 55	7 24	0 22
16 SU	12 00 38	+ 23 20	5 21	18 41	4 59	19 03	4 31	19 31	3 50	20 11	2 36	21 26
167		1.0158	10 33	23 38	10 19	23 49	10 02	none	9 39	none	9 01	0 36
17 MO	12 00 51	+ 23 22	5 21	18 41	4 59	19 03	4 31	19 31	3 50	20 12	2 36	21 26
168		1.0159	11 33	none	11 24	none	11 14	0 02	11 00	0 19	10 38	0 46
18 TU	12 01 04	+ 23 24	5 21	18 41	4 59	19 03	4 31	19 31	3 50	20 12	2 36	21 27
169	00 29 FQ	1.0160	12 32	0 20	12 29	0 25	12 26	0 31	12 21	0 40	12 14	0 53
19 WE	12 01 17	+ 23 25	5 21	18 42	4 59	19 04	4 31	19 32	3 50	20 12	2 36	21 27
170		1.0160	13 30	1 00	13 33	1 00	13 37	1 00	13 42	0 59	13 49	0 59
20 TH	12 01 30	+ 23 26	5 21	18 42	4 59	19 04	4 31	19 32	3 50	20 13	2 36	21 27
171		1.0161	14 29	1 40	14 38	1 34	14 48	1 28	15 03	1 19	15 25	1 05
21 FR	12 01 43	+ 23 26	5 21	18 42	4 59	19 04	4 31	19 32	3 51	20 13	2 36	21 28
172		1.0162	15 30	2 21	15 43	2 11	16 01	1 58	16 24	1 40	17 03	1 12
22 SA	12 01 56	+ 23 26	5 22	18 42	5 00	19 04	4 31	19 32	3 51	20 13	2 36	21 28
173		1.0162	16 31	3 06	16 50	2 50	17 13	2 31	17 45	2 04	18 41	1 22
23 SU	12 02 09	+ 23 26	5 22	18 42	5 00	19 04	4 32	19 33	3 51	20 13	2 36	21 28
174		1.0163	17 34	3 54	17 55	3 34	18 23	3 09	19 03	2 34	20 16	1 36
24 MO	12 02 22	+ 23 25	5 22	18 43	5 00	19 05	4 32	19 33	3 51	20 13	2 37	21 28
175	21 42 FM	1.0163	18 34	4 46	18 58	4 23	19 29	3 54	20 13	3 13	21 37	1 59
25 TU	12 02 34	+ 23 24	5 22	18 43	5 00	19 05	4 32	19 33	3 52	20 13	2 37	21 28
176		1.0164	19 32	5 41	19 57	5 17	20 27	4 46	21 12	4 01	22 35	2 37
26 WE	12 02 47	+ 23 22	5 23	18 43	5 01	19 05	4 33	19 33	3 52	20 13	2 38	21 27
177		1.0164	20 25	6 39	20 48	6 15	21 17	5 44	21 58	5 00	23 11	3 37
27 TH	12 02 59	+ 23 20	5 23	18 43	5 01	19 05	4 33	19 33	3 53	20 13	2 38	21 27
178		1.0165	21 13	7 36	21 33	7 14	21 58	6 46	22 33	6 06	23 32	4 54
28 FR	12 03 12	+ 23 18	5 23	18 43	5 01	19 05	4 33	19 33	3 53	20 13	2 39	21 27
179		1.0165	21 55	8 32	22 12	8 13	22 32	7 49	23 00	7 16	23 45	6 19
29 SA	12 03 24	+ 23 15	5 23	18 43	5 02	19 05	4 34	19 33	3 54	20 13	2 40	21 26
180		1.0166	22 33	9 25	22 46	9 10	23 01	8 52	23 21	8 26	23 53	7 44
30 SU	12 03 36	+ 23 12	5 24	18 43	5 02	19 05	4 34	19 33	3 54	20 13	2 41	21 26
181		1.0166	23 09	10 16	23 16	10 06	23 26	9 53	23 39	9 35	23 59	9 07

JULY 2002

7th Month **31 days**

Coordinated Universal Time (Greenwich Mean Time)

NOTE: For each day, numbers on first line indicate Sun. Numbers on second line indicate Moon.

Degrees are North Latitude.

Moon Phases: FM = Full Moon: LQ = Last (Waning) Quarter: NM = New Moon, FQ = First (Waxing) Quarter

Sun's distance is in Astronomical Units

CAUTION: Must be converted to local time. For instructions see "Calculation of Rise Times."

Day of month, of week, of year	Sun on Meridian / Moon Phase h m s	Sun's Declination ° ′ / Distance	20° Rise Sun / Moon h m	20° Set Sun / Moon h m	30° Rise Sun / Moon h m	30° Set Sun / Moon h m	40° Rise Sun / Moon h m	40° Set Sun / Moon h m	50° Rise Sun / Moon h m	50° Set Sun / Moon h m	60° Rise Sun / Moon h m	60° Set Sun / Moon h m
1 MO	12 03 47	+ 23 08	5 24	18 43	5 02	19 05	4 35	19 33	3 55	20 12	2 42	21 25
182		1.0166	23 42	11 05	23 45	10 59	23 49	10 52	23 55	10 42	none	10 27
2 TU	12 03 59	+ 23 04	5 24	18 43	5 03	19 05	4 35	19 33	3 55	20 12	2 43	21 24
183	17 19 LQ	1.0166	none	11 53	none	11 52	none	11 50	none	11 48	0 04	11 45
3 WE	12 04 10	+ 23 00	5 25	18 44	5 03	19 05	4 36	19 32	3 56	20 12	2 44	21 23
184		1.0167	0 14	12 41	0 13	12 44	0 12	12 49	0 10	12 54	0 08	13 03
4 TH	12 04 21	+ 22 55	5 25	18 44	5 04	19 05	4 36	19 32	3 57	20 11	2 45	21 22
185		1.0167	0 47	13 30	0 42	13 38	0 35	13 48	0 26	14 01	0 13	14 22
5 FR	12 04 32	+ 22 49	5 25	18 44	5 04	19 05	4 37	19 32	3 58	20 11	2 47	21 21
186		1.0167	1 22	14 21	1 12	14 33	1 00	14 48	0 43	15 10	0 18	15 44
6 SA	12 04 42	+ 22 44	5 26	18 44	5 04	19 05	4 37	19 32	3 58	20 10	2 48	21 20
187		1.0167	1 59	15 14	1 45	15 30	1 27	15 51	1 03	16 20	0 25	17 09
7 SU	12 04 52	+ 22 38	5 26	18 44	5 05	19 05	4 38	19 31	3 59	20 10	2 49	21 19
188		1.0167	2 40	16 10	2 22	16 30	1 59	16 56	1 28	17 32	0 36	18 36
8 MO	12 05 01	+ 22 31	5 26	18 43	5 05	19 04	4 39	19 31	4 00	20 09	2 51	21 18
189		1.0167	3 26	17 08	3 05	17 31	2 38	18 00	2 00	18 42	0 54	20 00
9 TU	12 05 10	+ 22 24	5 27	18 43	5 06	19 04	4 39	19 31	4 01	20 09	2 53	21 17
190		1.0167	4 19	18 07	3 55	18 31	3 25	19 02	2 42	19 46	1 23	21 10
10 WE	12 05 19	+ 22 17	5 27	18 43	5 06	19 04	4 40	19 30	4 02	20 08	2 54	21 15
191	10 26 NM	1.0166	5 16	19 05	4 52	19 28	4 21	19 58	3 37	20 41	2 14	21 58
11 TH	12 05 28	+ 22 10	5 28	18 43	5 07	19 04	4 41	19 30	4 03	20 07	2 56	21 14
192		1.0166	6 18	20 00	5 55	20 21	5 26	20 47	4 45	21 24	3 29	22 26
12 FR	12 05 36	+ 22 02	5 28	18 43	5 07	19 04	4 41	19 29	4 04	20 07	2 58	21 12
193		1.0166	7 21	20 50	7 02	21 07	6 37	21 28	6 02	21 57	5 02	22 44
13 SA	12 05 43	+ 21 53	5 28	18 43	5 08	19 03	4 42	19 29	4 05	20 06	2 59	21 11
194		1.0165	8 25	21 36	8 09	21 49	7 51	22 03	7 24	22 24	6 41	22 55
14 SU	12 05 50	+ 21 44	5 29	18 43	5 08	19 03	4 43	19 28	4 06	20 05	3 01	21 09
195		1.0165	9 26	22 19	9 16	22 26	9 04	22 35	8 47	22 46	8 21	23 03
15 MO	12 05 56	+ 21 35	5 29	18 43	5 09	19 03	4 44	19 28	4 07	20 04	3 03	21 07
196		1.0165	10 26	23 00	10 22	23 02	10 17	23 03	10 10	23 06	9 58	23 09
16 TU	12 06 02	+ 21 26	5 29	18 43	5 10	19 02	4 44	19 27	4 08	20 03	3 05	21 06
197		1.0164	11 25	23 40	11 27	23 36	11 28	23 31	11 30	23 25	11 34	23 15
17 WE	12 06 08	+ 21 16	5 30	18 42	5 10	19 02	4 45	19 27	4 09	20 02	3 07	21 04
198	04 47 FQ	1.0163	12 24	none	12 31	none	12 39	none	12 51	23 45	13 09	23 22
18 TH	12 06 12	+ 21 06	5 30	18 42	5 11	19 02	4 46	19 26	4 11	20 01	3 09	21 02
199		1.0163	13 23	0 21	13 35	0 12	13 50	0 00	14 11	none	14 45	23 30
19 FR	12 06 16	+ 20 55	5 31	18 42	5 11	19 01	4 47	19 25	4 12	20 00	3 11	21 00
200		1.0162	14 23	1 03	14 40	0 49	15 01	0 32	15 31	0 08	16 21	23 42
20 SA	12 06 20	+ 20 44	5 31	18 42	5 12	19 01	4 48	19 25	4 13	19 59	3 13	20 58
201		1.0161	15 24	1 49	15 45	1 31	16 11	1 07	16 49	0 35	17 56	none
21 SU	12 06 23	+ 20 33	5 31	18 41	5 12	19 00	4 48	19 24	4 14	19 58	3 15	20 56
202		1.0161	16 24	2 39	16 48	2 17	17 17	1 49	18 01	1 10	19 21	0 01
22 MO	12 06 26	+ 20 21	5 32	18 41	5 13	19 00	4 49	19 23	4 16	19 57	3 17	20 54
203		1.0160	17 22	3 32	17 47	3 08	18 18	2 37	19 03	1 53	20 27	0 32
23 TU	12 06 28	+ 20 09	5 32	18 41	5 13	18 59	4 50	19 22	4 17	19 55	3 20	20 52
204		1.0159	18 16	4 28	18 40	4 03	19 10	3 32	19 53	2 48	21 11	1 23
24 WE	12 06 29	+ 19 57	5 32	18 40	5 14	18 59	4 51	19 22	4 18	19 54	3 22	20 50
205	09 07 FM	1.0158	19 06	5 25	19 27	5 02	19 54	4 33	20 31	3 51	21 37	2 34
25 TH	12 06 30	+ 19 45	5 33	18 40	5 15	18 58	4 52	19 21	4 19	19 53	3 24	20 47
206		1.0157	19 50	6 21	20 08	6 01	20 30	5 36	21 01	5 00	21 52	3 57
26 FR	12 06 30	+ 19 32	5 33	18 40	5 15	18 57	4 53	19 20	4 21	19 51	3 26	20 45
207		1.0156	20 30	7 16	20 44	6 59	21 01	6 39	21 24	6 10	22 02	5 22
27 SA	12 06 30	+ 19 18	5 34	18 39	5 16	18 57	4 54	19 19	4 22	19 50	3 29	20 43
208		1.0155	21 07	8 08	21 16	7 56	21 28	7 41	21 44	7 20	22 08	6 46
28 SU	12 06 29	+ 19 05	5 34	18 39	5 16	18 56	4 54	19 18	4 23	19 49	3 31	20 40
209		1.0154	21 40	8 58	21 46	8 50	21 52	8 41	22 00	8 28	22 13	8 08
29 MO	12 06 27	+ 18 51	5 34	18 38	5 17	18 56	4 55	19 17	4 25	19 47	3 33	20 38
210		1.0153	22 13	9 46	22 14	9 43	22 15	9 40	22 16	9 35	22 17	9 27
30 TU	12 06 25	+ 18 37	5 35	18 38	5 18	18 55	4 56	19 16	4 26	19 46	3 35	20 36
211		1.0152	22 46	10 34	22 42	10 36	22 37	10 38	22 31	10 40	22 22	10 44
31 WE	12 06 22	+ 18 22	5 35	18 38	5 18	18 54	4 57	19 15	4 28	19 44	3 38	20 33
212		1.0151	23 19	11 22	23 11	11 28	23 01	11 36	22 47	11 46	22 27	12 02

AUGUST 2002

8th Month **31 days**

Coordinated Universal Time (Greenwich Mean Time)

NOTE: For each day, numbers on first line indicate Sun. Numbers on second line indicate Moon.

Degrees are North Latitude.

Moon Phases: FM = Full Moon: LQ = Last (Waning) Quarter: NM = New Moon, FQ = First (Waxing) Quarter

Sun's distance is in Astronomical Units

CAUTION: Must be converted to local time. For instructions see "Calculation of Rise Times."

Day of month, of week, of year	Sun on Meridian / Moon Phase h m s	Sun's Declination ° ´ / Distance	20° Rise Sun / Moon h m	20° Set Sun / Moon h m	30° Rise Sun / Moon h m	30° Set Sun / Moon h m	40° Rise Sun / Moon h m	40° Set Sun / Moon h m	50° Rise Sun / Moon h m	50° Set Sun / Moon h m	60° Rise Sun / Moon h m	60° Set Sun / Moon h m
1 TH	12 06 19	+ 18 07	5 35	18 37	5 19	18 53	4 58	19 14	4 29	19 43	3 40	20 31
213	10 22 LQ	1.0150	23 54	12 11	23 42	12 22	23 26	12 35	23 05	12 53	22 33	13 22
2 FR	12 06 15	+ 17 52	5 36	18 37	5 19	18 53	4 59	19 13	4 30	19 41	3 42	20 28
214		1.0149	none	13 03	none	13 18	23 56	13 36	23 27	14 02	22 42	14 45
3 SA	12 06 10	+ 17 37	5 36	18 36	5 20	18 52	5 00	19 12	4 32	19 40	3 45	20 26
215		1.0147	0 33	13 57	0 16	14 15	none	14 39	23 55	15 13	22 55	16 11
4 SU	12 06 05	+ 17 21	5 36	18 36	5 21	18 51	5 01	19 11	4 33	19 38	3 47	20 23
216		1.0146	1 16	14 53	0 56	15 15	0 31	15 43	none	16 23	23 18	17 36
5 MO	12 05 59	+ 17 05	5 37	18 35	5 21	18 50	5 02	19 10	4 35	19 36	3 50	20 21
217		1.0145	2 05	15 51	1 42	16 15	1 13	16 46	0 32	17 30	23 58	18 52
6 TU	12 05 53	+ 16 49	5 37	18 34	5 22	18 50	5 03	19 08	4 36	19 35	3 52	20 18
218		1.0143	3 00	16 50	2 36	17 14	2 05	17 45	1 21	18 29	none	19 50
7 WE	12 05 46	+ 16 32	5 37	18 34	5 22	18 49	5 04	19 07	4 37	19 33	3 54	20 15
219		1.0142	4 01	17 47	3 37	18 09	3 07	18 37	2 23	19 17	1 03	20 27
8 TH	12 05 39	+ 16 15	5 38	18 33	5 23	18 48	5 05	19 06	4 39	19 31	3 57	20 13
220	19 15 NM	1.0140	5 04	18 40	4 43	18 59	4 16	19 22	3 38	19 55	2 30	20 49
9 FR	12 05 30	+ 15 58	5 38	18 33	5 24	18 47	5 06	19 05	4 40	19 30	3 59	20 10
221		1.0139	6 09	19 29	5 52	19 43	5 31	20 01	5 01	20 25	4 10	21 02
10 SA	12 05 22	+ 15 41	5 38	18 32	5 24	18 46	5 07	19 04	4 42	19 28	4 02	20 07
222		1.0137	7 14	20 15	7 02	20 24	6 47	20 34	6 26	20 49	5 53	21 11
11 SU	12 05 13	+ 15 24	5 39	18 31	5 25	18 45	5 07	19 02	4 43	19 26	4 04	20 05
223		1.0136	8 16	20 57	8 10	21 01	8 02	21 05	7 51	21 10	7 35	21 18
12 MO	12 05 03	+ 15 06	5 39	18 31	5 25	18 44	5 08	19 01	4 45	19 24	4 06	20 02
224		1.0134	9 17	21 39	9 17	21 36	9 16	21 34	9 15	21 30	9 14	21 24
13 TU	12 04 52	+ 14 48	5 39	18 30	5 26	18 43	5 09	19 00	4 46	19 22	4 09	19 59
225		1.0132	10 18	22 20	10 23	22 12	10 29	22 03	10 38	21 50	10 52	21 30
14 WE	12 04 42	+ 14 29	5 40	18 29	5 27	18 42	5 10	18 58	4 48	19 21	4 11	19 56
226		1.0130	11 18	23 02	11 28	22 49	11 42	22 34	12 00	22 12	12 30	21 38
15 TH	12 04 30	+ 14 11	5 40	18 29	5 27	18 41	5 11	18 57	4 49	19 19	4 14	19 54
227	10 12 FQ	1.0128	12 18	23 47	12 34	23 30	12 53	23 08	13 21	22 38	14 07	21 48
16 FR	12 04 18	+ 13 52	5 40	18 28	5 28	18 40	5 12	18 56	4 51	19 17	4 16	19 51
228		1.0127	13 18	none	13 38	none	14 04	23 47	14 40	23 10	15 42	22 05
17 SA	12 04 05	+ 13 33	5 41	18 27	5 28	18 39	5 13	18 54	4 52	19 15	4 18	19 48
229		1.0125	14 18	0 35	14 41	0 14	15 11	none	15 53	23 50	17 11	22 31
18 SU	12 03 52	+ 13 14	5 41	18 27	5 29	18 38	5 14	18 53	4 54	19 13	4 21	19 45
230		1.0123	15 17	1 27	15 41	1 03	16 12	0 33	16 57	none	18 23	23 15
19 MO	12 03 38	+ 12 55	5 41	18 26	5 30	18 37	5 15	18 51	4 55	19 11	4 23	19 42
231		1.0121	16 11	2 21	16 35	1 57	17 06	1 26	17 50	0 41	19 12	none
20 TU	12 03 24	+ 12 35	5 41	18 25	5 30	18 36	5 16	18 50	4 57	19 09	4 26	19 39
232		1.0119	17 02	3 17	17 24	2 54	17 52	2 24	18 32	1 40	19 42	0 19
21 WE	12 03 10	+ 12 15	5 42	18 24	5 31	18 35	5 17	18 49	4 58	19 07	4 28	19 37
233		1.0117	17 47	4 14	18 06	3 52	18 30	3 25	19 04	2 47	19 59	1 39
22 TH	12 02 55	+ 11 55	5 42	18 24	5 31	18 34	5 18	18 47	5 00	19 05	4 30	19 34
234	22 29 FM	1.0115	18 28	5 08	18 43	4 51	19 02	4 28	19 28	3 57	20 10	3 04
23 FR	12 02 39	+ 11 35	5 42	18 23	5 32	18 33	5 19	18 46	5 01	19 03	4 33	19 31
235		1.0112	19 05	6 01	19 16	5 47	19 30	5 30	19 49	5 07	20 17	4 28
24 SA	12 02 23	+ 11 15	5 42	18 22	5 32	18 32	5 20	18 44	5 03	19 01	4 35	19 28
236		1.0110	19 40	6 52	19 47	6 42	19 55	6 31	20 06	6 15	20 23	5 51
25 SU	12 02 07	+ 10 54	5 43	18 21	5 33	18 31	5 21	18 43	5 04	18 59	4 38	19 25
237		1.0108	20 13	7 41	20 15	7 36	20 18	7 30	20 21	7 23	20 27	7 11
26 MO	12 01 50	+ 10 33	5 43	18 20	5 33	18 30	5 22	18 41	5 06	18 57	4 40	19 22
238		1.0106	20 45	8 29	20 43	8 29	20 40	8 29	20 36	8 29	20 31	8 29
27 TU	12 01 33	+ 10 13	5 43	18 20	5 34	18 29	5 23	18 40	5 07	18 55	4 42	19 19
239		1.0104	21 18	9 16	21 11	9 21	21 03	9 27	20 52	9 35	20 35	9 47
28 WE	12 01 15	+9 52	5 43	18 19	5 35	18 27	5 24	18 38	5 09	18 53	4 45	19 16
240		1.0102	21 52	10 05	21 41	10 14	21 27	10 25	21 09	10 41	20 40	11 06
29 TH	12 00 57	+9 30	5 44	18 18	5 35	18 26	5 25	18 37	5 10	18 51	4 47	19 13
241		1.0100	22 29	10 55	22 13	11 08	21 55	11 25	21 29	11 49	20 48	12 27
30 FR	12 00 39	+9 09	5 44	18 17	5 36	18 25	5 26	18 35	5 12	18 49	4 50	19 10
242		1.0097	23 09	11 47	22 50	12 04	22 26	12 26	21 53	12 58	20 58	13 50
31 SA	12 00 21	+8 48	5 44	18 16	5 36	18 24	5 26	18 34	5 13	18 47	4 52	19 07
243	02 31 LQ	1.0095	23 55	12 41	23 33	13 02	23 05	13 29	22 25	14 07	21 16	15 14

SEPTEMBER 2002

9th Month **30 days**

Coordinated Universal Time (Greenwich Mean Time)

NOTE: For each day, numbers on first line indicate Sun. Numbers on second line indicate Moon.

Degrees are North Latitude.

Moon Phases: FM = Full Moon: LQ = Last (Waning) Quarter: NM = New Moon, FQ = First (Waxing) Quarter

Sun's distance is in Astronomical Units

CAUTION: Must be converted to local time. For instructions see "Calculation of Rise Times."

Day of month, of week, of year	Sun on Meridian / Moon Phase h m s	Sun's Declination ° ′ / Distance	20° Rise Sun / Moon h m	20° Set Sun / Moon h m	30° Rise Sun / Moon h m	30° Set Sun / Moon h m	40° Rise Sun / Moon h m	40° Set Sun / Moon h m	50° Rise Sun / Moon h m	50° Set Sun / Moon h m	60° Rise Sun / Moon h m	60° Set Sun / Moon h m
1 SU	12 00 02	+8 26	5 44	18 15	5 37	18 23	5 27	18 32	5 15	18 44	4 54	19 04
244		1.0093	none	13 37	none	14 01	23 51	14 31	23 07	15 14	21 46	16 34
2 MO	11 59 43	+8 04	5 45	18 15	5 37	18 22	5 28	18 30	5 16	18 42	4 57	19 01
245		1.0091	0 46	14 35	0 22	14 59	none	15 30	none	16 15	22 38	17 40
3 TU	11 59 23	+7 42	5 45	18 14	5 38	18 20	5 29	18 29	5 18	18 40	4 59	18 58
246		1.0088	1 43	15 31	1 18	15 55	0 47	16 25	0 02	17 08	23 55	18 25
4 WE	11 59 04	+7 20	5 45	18 13	5 38	18 19	5 30	18 27	5 19	18 38	5 01	18 55
247		1.0086	2 44	16 26	2 21	16 47	1 52	17 13	1 11	17 49	none	18 52
5 TH	11 58 44	+6 58	5 45	18 12	5 39	18 18	5 31	18 26	5 21	18 36	5 04	18 52
248		1.0083	3 48	17 17	3 29	17 34	3 04	17 54	2 30	18 22	1 30	19 08
6 FR	11 58 24	+6 36	5 45	18 11	5 40	18 17	5 32	18 24	5 22	18 34	5 06	18 49
249		1.0081	4 53	18 04	4 39	18 16	4 21	18 30	3 55	18 49	3 14	19 18
7 SA	11 58 04	+6 13	5 46	18 10	5 40	18 16	5 33	18 22	5 23	18 32	5 08	18 46
250	03 10 NM	1.0078	5 58	18 49	5 49	18 55	5 38	19 02	5 22	19 12	4 58	19 26
8 SU	11 57 43	+5 51	5 46	18 09	5 41	18 14	5 34	18 21	5 25	18 29	5 11	18 43
251		1.0076	7 02	19 32	6 58	19 32	6 55	19 32	6 49	19 32	6 42	19 32
9 MO	11 57 22	+5 28	5 46	18 08	5 41	18 13	5 35	18 19	5 26	18 27	5 13	18 40
252		1.0073	8 04	20 14	8 07	20 09	8 11	20 02	8 16	19 52	8 24	19 38
10 TU	11 57 02	+5 06	5 46	18 07	5 42	18 12	5 36	18 17	5 28	18 25	5 15	18 37
253		1.0071	9 06	20 58	9 15	20 47	9 26	20 33	9 41	20 14	10 05	19 45
11 WE	11 56 41	+4 43	5 46	18 07	5 42	18 11	5 37	18 16	5 29	18 23	5 18	18 34
254		1.0068	10 09	21 43	10 23	21 27	10 41	21 07	11 06	20 39	11 47	19 54
12 TH	11 56 20	+4 20	5 47	18 06	5 43	18 09	5 38	18 14	5 31	18 21	5 20	18 31
255		1.0065	11 11	22 31	11 30	22 11	11 54	21 45	12 28	21 09	13 26	20 08
13 FR	11 55 58	+3 57	5 47	18 05	5 43	18 08	5 39	18 13	5 32	18 19	5 23	18 28
256	18 08 FQ	1.0063	12 13	23 23	12 35	22 59	13 04	22 30	13 45	21 47	15 00	20 31
14 SA	11 55 37	+3 34	5 47	18 04	5 44	18 07	5 40	18 11	5 34	18 16	5 25	18 25
257		1.0060	13 12	none	13 37	23 52	14 08	23 21	14 53	22 35	16 19	21 09
15 SU	11 55 16	+3 11	5 47	18 03	5 44	18 06	5 41	18 09	5 35	18 14	5 27	18 22
258		1.0057	14 08	0 17	14 33	none	15 04	none	15 50	23 33	17 15	22 08
16 MO	11 54 54	+2 48	5 48	18 02	5 45	18 04	5 42	18 08	5 37	18 12	5 30	18 19
259		1.0055	15 00	1 13	15 23	0 48	15 52	0 18	16 34	none	17 49	23 24
17 TU	11 54 33	+2 25	5 48	18 01	5 45	18 03	5 43	18 06	5 38	18 10	5 32	18 16
260		1.0052	15 46	2 09	16 07	1 46	16 32	1 18	17 08	0 38	18 09	none
18 WE	11 54 11	+2 02	5 48	18 00	5 46	18 02	5 43	18 04	5 40	18 08	5 34	18 13
261		1.0049	16 28	3 03	16 45	2 44	17 05	2 20	17 34	1 46	18 21	0 48
19 TH	11 53 50	+1 39	5 48	17 59	5 47	18 01	5 44	18 03	5 41	18 05	5 37	18 10
262		1.0046	17 06	3 56	17 19	3 41	17 34	3 23	17 55	2 56	18 28	2 13
20 FR	11 53 28	+1 15	5 48	17 58	5 47	17 59	5 45	18 01	5 43	18 03	5 39	18 07
263		1.0043	17 41	4 47	17 49	4 37	17 59	4 23	18 13	4 05	18 33	3 36
21 SA	11 53 07	+0 52	5 49	17 57	5 48	17 58	5 46	17 59	5 44	18 01	5 41	18 04
264	13 59 FM	1.0041	18 14	5 37	18 18	5 31	18 22	5 23	18 28	5 13	18 38	4 56
22 SU	11 52 46	+0 29	5 49	17 56	5 48	17 57	5 47	17 58	5 46	17 59	5 44	18 01
265		1.0038	18 46	6 25	18 45	6 23	18 44	6 22	18 43	6 19	18 41	6 15
23 MO	11 52 25	+0 05	5 49	17 56	5 49	17 56	5 48	17 56	5 47	17 56	5 46	17 57
266		1.0035	19 19	7 13	19 13	7 16	19 07	7 20	18 58	7 25	18 45	7 33
24 TU	11 52 04	-0 18	5 49	17 55	5 49	17 54	5 49	17 54	5 49	17 54	5 48	17 54
267		1.0032	19 52	8 01	19 42	8 09	19 30	8 18	19 14	8 31	18 49	8 52
25 WE	11 51 43	-0 41	5 49	17 54	5 50	17 53	5 50	17 53	5 50	17 52	5 51	17 51
268		1.0030	20 28	8 50	20 13	9 02	19 56	9 18	19 33	9 39	18 55	10 12
26 TH	11 51 22	-1 05	5 50	17 53	5 50	17 52	5 51	17 51	5 52	17 50	5 53	17 48
269		1.0027	21 06	9 41	20 48	9 57	20 26	10 18	19 55	10 47	19 04	11 35
27 FR	11 51 02	-1 28	5 50	17 52	5 51	17 51	5 52	17 49	5 53	17 48	5 55	17 45
270		1.0024	21 49	10 34	21 28	10 54	21 01	11 19	20 23	11 56	19 17	12 59
28 SA	11 50 42	-1 51	5 50	17 51	5 51	17 49	5 53	17 48	5 55	17 46	5 58	17 42
271		1.0021	22 37	11 28	22 13	11 51	21 43	12 21	21 00	13 03	19 41	14 21
29 SU	11 50 21	-2 15	5 50	17 50	5 52	17 48	5 54	17 46	5 56	17 43	6 00	17 39
272	17 03 LQ	1.0018	23 30	12 24	23 05	12 49	22 34	13 20	21 48	14 06	20 21	15 32
30 MO	11 50 02	-2 38	5 51	17 49	5 53	17 47	5 55	17 44	5 58	17 41	6 02	17 36
273		1.0016	none	13 20	none	13 44	23 33	14 15	22 49	15 00	21 26	16 24

OCTOBER 2002

10th Month **31 days**

Coordinated Universal Time (Greenwich Mean Time)

NOTE: For each day, numbers on first line indicate Sun. Numbers on second line indicate Moon.

Degrees are North Latitude.

Moon Phases: FM = Full Moon: LQ = Last (Waning) Quarter: NM = New Moon, FQ = First (Waxing) Quarter

Sun's distance is in Astronomical Units

CAUTION: Must be converted to local time. For instructions see "Calculation of Rise Times."

Day of month, of week, of year	Sun on Meridian / Moon Phase h m s	Sun's Declination ° ′ / Distance	20° Rise Sun / Moon h m	20° Set Sun / Moon h m	30° Rise Sun / Moon h m	30° Set Sun / Moon h m	40° Rise Sun / Moon h m	40° Set Sun / Moon h m	50° Rise Sun / Moon h m	50° Set Sun / Moon h m	60° Rise Sun / Moon h m	60° Set Sun / Moon h m
1 TU	11 49 42	-3 01	5 51	17 48	5 53	17 46	5 56	17 43	6 00	17 39	6 05	17 33
274		1.0013	0 28	14 13	0 04	14 36	none	15 04	none	15 45	22 53	16 56
2 WE	11 49 23	-3 25	5 51	17 47	5 54	17 45	5 57	17 41	6 01	17 37	6 07	17 30
275		1.0010	1 29	15 04	1 08	15 24	0 40	15 47	0 02	16 20	none	17 15
3 TH	11 49 04	-3 48	5 51	17 47	5 54	17 43	5 58	17 40	6 03	17 35	6 10	17 27
276		1.0007	2 32	15 52	2 15	16 07	1 53	16 25	1 23	16 49	0 31	17 26
4 FR	11 48 46	-4 11	5 52	17 46	5 55	17 42	5 59	17 38	6 04	17 32	6 12	17 24
277		1.0004	3 36	16 38	3 24	16 47	3 09	16 58	2 48	17 12	2 15	17 34
5 SA	11 48 28	-4 34	5 52	17 45	5 56	17 41	6 00	17 36	6 06	17 30	6 14	17 21
278		1.0002	4 40	17 21	4 33	17 24	4 26	17 28	4 15	17 33	3 59	17 41
6 SU	11 48 10	-4 57	5 52	17 44	5 56	17 40	6 01	17 35	6 07	17 28	6 17	17 18
279	11 18 NM	.9999	5 43	18 04	5 43	18 01	5 43	17 58	5 43	17 53	5 42	17 46
7 MO	11 47 52	-5 20	5 52	17 43	5 57	17 39	6 02	17 33	6 09	17 26	6 19	17 15
280		.9996	6 47	18 48	6 53	18 39	7 00	18 28	7 10	18 14	7 27	17 52
8 TU	11 47 36	-5 43	5 53	17 42	5 57	17 37	6 03	17 32	6 10	17 24	6 22	17 12
281		.9993	7 51	19 33	8 03	19 19	8 18	19 02	8 38	18 38	9 12	18 00
9 WE	11 47 19	-6 06	5 53	17 41	5 58	17 36	6 04	17 30	6 12	17 22	6 24	17 09
282		.9990	8 56	20 22	9 13	20 03	9 35	19 39	10 05	19 06	10 57	18 11
10 TH	11 47 03	-6 29	5 53	17 41	5 59	17 35	6 05	17 29	6 14	17 20	6 27	17 06
283		.9987	10 00	21 14	10 22	20 51	10 49	20 23	11 28	19 42	12 39	18 29
11 FR	11 46 47	-6 52	5 53	17 40	5 59	17 34	6 06	17 27	6 15	17 18	6 29	17 03
284		.9984	11 03	22 09	11 27	21 44	11 58	21 13	12 43	20 27	14 09	19 01
12 SA	11 46 32	-7 14	5 54	17 39	6 00	17 33	6 07	17 25	6 17	17 16	6 31	17 01
285		.9981	12 02	23 06	12 28	22 41	12 59	22 09	13 46	21 23	15 15	19 54
13 SU	11 46 17	-7 37	5 54	17 38	6 00	17 32	6 08	17 24	6 18	17 13	6 34	16 58
286	05 33 FQ	.9978	12 57	none	13 21	23 40	13 51	23 10	14 35	22 27	15 56	21 07
14 MO	11 46 03	-7 59	5 54	17 37	6 01	17 31	6 09	17 22	6 20	17 11	6 36	16 55
287		.9976	13 45	0 03	14 07	none	14 34	none	15 12	23 36	16 19	22 31
15 TU	11 45 49	-8 22	5 55	17 37	6 02	17 30	6 10	17 21	6 21	17 09	6 39	16 52
288		.9973	14 28	0 59	14 47	0 38	15 09	0 13	15 40	none	16 32	23 57
16 WE	11 45 36	-8 44	5 55	17 36	6 02	17 28	6 11	17 19	6 23	17 07	6 41	16 49
289		.9970	15 07	1 52	15 21	1 36	15 39	1 15	16 02	0 46	16 40	none
17 TH	11 45 24	-9 06	5 55	17 35	6 03	17 27	6 12	17 18	6 25	17 05	6 44	16 46
290		.9967	15 43	2 44	15 53	2 32	16 04	2 16	16 20	1 55	16 45	1 21
18 FR	11 45 12	-9 28	5 56	17 34	6 04	17 26	6 13	17 16	6 26	17 03	6 46	16 43
291		.9964	16 16	3 33	16 21	3 26	16 28	3 16	16 36	3 03	16 49	2 43
19 SA	11 45 00	-9 49	5 56	17 34	6 04	17 25	6 14	17 15	6 28	17 01	6 49	16 40
292		.9961	16 48	4 22	16 49	4 19	16 50	4 15	16 51	4 10	16 53	4 02
20 SU	11 44 50	- 10 11	5 56	17 33	6 05	17 24	6 15	17 14	6 29	16 59	6 51	16 37
293		.9958	17 20	5 09	17 17	5 11	17 12	5 13	17 05	5 16	16 56	5 20
21 MO	11 44 40	- 10 33	5 57	17 32	6 06	17 23	6 17	17 12	6 31	16 57	6 54	16 35
294	07 20 FM	.9956	17 53	5 57	17 45	6 04	17 35	6 12	17 21	6 22	17 00	6 39
22 TU	11 44 30	- 10 54	5 57	17 32	6 06	17 22	6 18	17 11	6 33	16 55	6 56	16 32
295		.9953	18 28	6 46	18 15	6 57	18 00	7 11	17 38	7 30	17 05	7 59
23 WE	11 44 21	- 11 15	5 58	17 31	6 07	17 21	6 19	17 09	6 34	16 54	6 59	16 29
296		.9950	19 06	7 37	18 49	7 52	18 28	8 11	17 59	8 38	17 12	9 22
24 TH	11 44 13	- 11 36	5 58	17 30	6 08	17 20	6 20	17 08	6 36	16 52	7 01	16 26
297		.9947	19 47	8 29	19 27	8 49	19 01	9 13	18 25	9 47	17 23	10 47
25 FR	11 44 06	- 11 57	5 58	17 30	6 08	17 19	6 21	17 07	6 38	16 50	7 04	16 23
298		.9945	20 33	9 24	20 10	9 46	19 40	10 15	18 58	10 55	17 42	12 10
26 SA	11 43 59	- 12 18	5 59	17 29	6 09	17 18	6 22	17 05	6 39	16 48	7 06	16 21
299		.9942	21 24	10 19	20 59	10 43	20 27	11 14	19 41	12 00	18 15	13 26
27 SU	11 43 53	- 12 38	5 59	17 28	6 10	17 17	6 23	17 04	6 41	16 46	7 09	16 18
300		.9939	22 19	11 13	21 54	11 38	21 22	12 10	20 37	12 56	19 09	14 24
28 MO	11 43 48	- 12 58	6 00	17 28	6 11	17 17	6 24	17 03	6 43	16 44	7 11	16 15
301		.9937	23 17	12 07	22 54	12 30	22 25	13 00	21 44	13 43	20 27	15 01
29 TU	11 43 44	- 13 18	6 00	17 27	6 11	17 16	6 25	17 02	6 44	16 43	7 14	16 13
302	05 28 LQ	.9934	none	12 57	23 58	13 18	23 34	13 44	22 59	14 21	22 00	15 23
30 WE	11 43 40	- 13 38	6 00	17 27	6 12	17 15	6 26	17 00	6 46	16 41	7 16	16 10
303		.9932	0 18	13 44	none	14 01	none	14 22	none	14 50	23 38	15 36
31 TH	11 43 38	- 13 58	6 01	17 26	6 13	17 14	6 28	16 59	6 48	16 39	7 19	16 07
304		.9929	1 19	14 29	1 04	14 41	0 46	14 55	0 20	15 14	none	15 44

NOVEMBER 2002

11th Month **30 days**

Coordinated Universal Time (Greenwich Mean Time)

NOTE: For each day, numbers on first line indicate Sun. Numbers on second line indicate Moon.

Degrees are North Latitude.

Moon Phases: FM = Full Moon: LQ = Last (Waning) Quarter: NM = New Moon, FQ = First (Waxing) Quarter

Sun's distance is in Astronomical Units

CAUTION: Must be converted to local time. For instructions see "Calculation of Rise Times."

Day of month, of week, of year	Sun on Meridian / Moon Phase h m s	Sun's Declination ° ´ / Distance	20° Rise Sun / Moon h m	20° Set Sun / Moon h m	30° Rise Sun / Moon h m	30° Set Sun / Moon h m	40° Rise Sun / Moon h m	40° Set Sun / Moon h m	50° Rise Sun / Moon h m	50° Set Sun / Moon h m	60° Rise Sun / Moon h m	60° Set Sun / Moon h m
1 FR	11 43 36	- 14 17	6 01	17 26	6 14	17 13	6 29	16 58	6 49	16 37	7 22	16 05
305		.9926	2 20	15 11	2 11	15 18	1 59	15 25	1 44	15 35	1 19	15 50
2 SA	11 43 34	- 14 37	6 02	17 25	6 14	17 12	6 30	16 57	6 51	16 36	7 24	16 02
306		.9924	3 22	15 53	3 19	15 54	3 14	15 54	3 09	15 55	3 00	15 56
3 SU	11 43 34	- 14 56	6 02	17 25	6 15	17 12	6 31	16 56	6 53	16 34	7 27	15 59
307		.9921	4 24	16 36	4 27	16 30	4 30	16 23	4 35	16 14	4 42	16 01
4 MO	11 43 35	- 15 14	6 03	17 24	6 16	17 11	6 32	16 55	6 54	16 32	7 29	15 57
308	20 34 NM	.9919	5 28	17 20	5 37	17 09	5 47	16 55	6 02	16 36	6 26	16 07
5 TU	11 43 36	- 15 33	6 03	17 24	6 17	17 10	6 33	16 53	6 56	16 31	7 32	15 54
309		.9916	6 33	18 08	6 48	17 51	7 06	17 30	7 31	17 02	8 13	16 15
6 WE	11 43 38	- 15 51	6 04	17 23	6 18	17 09	6 34	16 52	6 58	16 29	7 34	15 52
310		.9914	7 39	18 59	7 59	18 38	8 24	18 11	8 59	17 34	10 01	16 29
7 TH	11 43 41	- 16 09	6 04	17 23	6 18	17 09	6 36	16 51	6 59	16 28	7 37	15 49
311		.9911	8 45	19 55	9 09	19 30	9 38	19 00	10 22	18 15	11 42	16 54
8 FR	11 43 45	- 16 27	6 05	17 22	6 19	17 08	6 37	16 50	7 01	16 26	7 40	15 47
312		.9909	9 49	20 53	10 14	20 28	10 46	19 55	11 33	19 08	13 04	17 37
9 SA	11 43 50	- 16 44	6 05	17 22	6 20	17 07	6 38	16 49	7 03	16 24	7 42	15 45
313		.9906	10 47	21 52	11 12	21 28	11 44	20 57	12 30	20 12	13 57	18 45
10 SU	11 43 55	- 17 01	6 06	17 22	6 21	17 07	6 39	16 48	7 04	16 23	7 45	15 42
314		.9904	11 40	22 50	12 03	22 28	12 32	22 01	13 13	21 21	14 26	20 09
11 MO	11 44 01	- 17 18	6 06	17 21	6 22	17 06	6 40	16 47	7 06	16 22	7 47	15 40
315	20 52 FQ	.9902	12 26	23 46	12 46	23 28	13 10	23 05	13 45	22 33	14 42	21 38
12 TU	11 44 08	- 17 35	6 07	17 21	6 22	17 06	6 41	16 46	7 07	16 20	7 50	15 38
316		.9899	13 07	none	13 23	none	13 42	none	14 09	23 43	14 52	23 04
13 WE	11 44 16	- 17 51	6 08	17 21	6 23	17 05	6 43	16 46	7 09	16 19	7 52	15 35
317		.9897	13 44	0 39	13 55	0 25	14 09	0 08	14 28	none	14 57	none
14 TH	11 44 25	- 18 07	6 08	17 21	6 24	17 05	6 44	16 45	7 11	16 18	7 55	15 33
318		.9894	14 18	1 29	14 25	1 20	14 33	1 08	14 44	0 52	15 02	0 27
15 FR	11 44 35	- 18 22	6 09	17 20	6 25	17 04	6 45	16 44	7 12	16 16	7 57	15 31
319		.9892	14 50	2 18	14 52	2 13	14 55	2 07	14 59	1 59	15 05	1 47
16 SA	11 44 45	- 18 38	6 09	17 20	6 26	17 04	6 46	16 43	7 14	16 15	8 00	15 29
320		.9890	15 22	3 06	15 20	3 06	15 17	3 06	15 13	3 06	15 08	3 05
17 SU	11 44 56	- 18 53	6 10	17 20	6 26	17 03	6 47	16 42	7 16	16 14	8 02	15 27
321		.9888	15 55	3 53	15 48	3 58	15 39	4 04	15 28	4 12	15 11	4 24
18 MO	11 45 09	- 19 07	6 10	17 20	6 27	17 03	6 48	16 42	7 17	16 13	8 05	15 25
322		.9886	16 29	4 42	16 17	4 51	16 03	5 03	15 45	5 19	15 15	5 44
19 TU	11 45 22	- 19 21	6 11	17 20	6 28	17 02	6 49	16 41	7 19	16 11	8 07	15 23
323		.9883	17 05	5 32	16 50	5 46	16 30	6 03	16 04	6 27	15 22	7 06
20 WE	11 45 35	- 19 35	6 12	17 19	6 29	17 02	6 50	16 40	7 20	16 10	8 10	15 21
324	01 34 FM	.9881	17 46	6 25	17 26	6 43	17 02	7 05	16 28	7 37	15 31	8 31
21 TH	11 45 50	- 19 49	6 12	17 19	6 30	17 02	6 52	16 40	7 22	16 09	8 12	15 19
325		.9879	18 30	7 19	18 08	7 40	17 39	8 08	16 59	8 47	15 47	9 57
22 FR	11 46 05	- 20 02	6 13	17 19	6 31	17 01	6 53	16 39	7 23	16 08	8 15	15 17
326		.9877	19 20	8 14	18 55	8 38	18 24	9 09	17 39	9 53	16 14	11 17
23 SA	11 46 22	- 20 15	6 13	17 19	6 31	17 01	6 54	16 39	7 25	16 07	8 17	15 15
327		.9875	20 14	9 10	19 49	9 35	19 17	10 07	18 31	10 53	17 02	12 22
24 SU	11 46 38	- 20 28	6 14	17 19	6 32	17 01	6 55	16 38	7 26	16 06	8 19	15 13
328		.9874	21 11	10 03	20 48	10 28	20 17	10 59	19 34	11 43	18 13	13 06
25 MO	11 46 56	- 20 40	6 15	17 19	6 33	17 01	6 56	16 38	7 28	16 06	8 21	15 12
329		.9872	22 10	10 54	21 50	11 16	21 24	11 44	20 46	12 23	19 41	13 31
26 TU	11 47 15	- 20 51	6 15	17 19	6 34	17 00	6 57	16 37	7 29	16 05	8 24	15 10
330		.9870	23 10	11 42	22 54	12 00	22 33	12 23	22 04	12 54	21 16	13 46
27 WE	11 47 34	- 21 03	6 16	17 19	6 35	17 00	6 58	16 37	7 31	16 04	8 26	15 09
331	15 46 LQ	.9868	none	12 26	23 58	12 40	23 44	12 56	23 24	13 19	22 53	13 55
28 TH	11 47 54	- 21 14	6 17	17 19	6 36	17 00	6 59	16 36	7 32	16 03	8 28	15 07
332		.9866	0 10	13 07	none	13 16	none	13 26	none	13 40	none	14 01
29 FR	11 48 15	- 21 24	6 17	17 19	6 36	17 00	7 00	16 36	7 34	16 02	8 30	15 06
333		.9865	1 09	13 48	1 03	13 51	0 55	13 54	0 45	13 59	0 30	14 06
30 SA	11 48 36	- 21 34	6 18	17 19	6 37	17 00	7 01	16 36	7 35	16 02	8 32	15 04
334		.9863	2 08	14 28	2 08	14 25	2 08	14 22	2 08	14 17	2 07	14 11

DECEMBER 2002

12th Month **31 days**

Coordinated Universal Time (Greenwich Mean Time)

NOTE: For each day, numbers on first line indicate Sun. Numbers on second line indicate Moon.

Degrees are North Latitude.

Moon Phases: FM = Full Moon: LQ = Last (Waning) Quarter: NM = New Moon, FQ = First (Waxing) Quarter

Sun's distance is in Astronomical Units

CAUTION: Must be converted to local time. For instructions see "Calculation of Rise Times."

Day of month, of week, of year	Sun on Meridian / Moon Phase (h m s)	Sun's Declination (° ′) / Distance	20° Rise Sun / Moon (h m)	20° Set Sun / Moon (h m)	30° Rise Sun / Moon (h m)	30° Set Sun / Moon (h m)	40° Rise Sun / Moon (h m)	40° Set Sun / Moon (h m)	50° Rise Sun / Moon (h m)	50° Set Sun / Moon (h m)	60° Rise Sun / Moon (h m)	60° Set Sun / Moon (h m)
1 SU	11 48 58	- 21 44	6 18	17 19	6 38	17 00	7 02	16 35	7 36	16 01	8 34	15 03
335		.9862	3 09	15 09	3 14	15 01	3 22	14 51	3 31	14 37	3 47	14 16
2 MO	11 49 21	- 21 53	6 19	17 19	6 39	17 00	7 03	16 35	7 38	16 01	8 36	15 02
336		.9860	4 11	15 54	4 23	15 40	4 37	15 23	4 57	15 00	5 30	14 22
3 TU	11 49 45	- 22 02	6 20	17 20	6 40	17 00	7 04	16 35	7 39	16 00	8 38	15 01
337		.9858	5 16	16 43	5 33	16 24	5 54	16 00	6 25	15 27	7 16	14 33
4 WE	11 50 09	- 22 11	6 20	17 20	6 40	17 00	7 05	16 35	7 40	16 00	8 40	15 00
338	07 34 NM	.9857	6 22	17 37	6 44	17 14	7 11	16 45	7 51	16 04	9 01	14 50
5 TH	11 50 33	- 22 19	6 21	17 20	6 41	17 00	7 06	16 35	7 41	15 59	8 42	14 59
339		.9856	7 28	18 34	7 52	18 09	8 24	17 37	9 09	16 51	10 36	15 23
6 FR	11 50 58	- 22 26	6 22	17 20	6 42	17 00	7 07	16 35	7 43	15 59	8 44	14 58
340		.9854	8 30	19 35	8 56	19 10	9 28	18 38	10 15	17 51	11 46	16 20
7 SA	11 51 24	- 22 33	6 22	17 20	6 43	17 00	7 08	16 35	7 44	15 59	8 46	14 57
341		.9853	9 28	20 35	9 52	20 12	10 22	19 43	11 06	19 00	12 27	17 40
8 SU	11 51 50	- 22 40	6 23	17 21	6 43	17 00	7 09	16 35	7 45	15 58	8 47	14 56
342		.9851	10 18	21 34	10 39	21 14	11 06	20 49	11 44	20 13	12 49	19 10
9 MO	11 52 17	- 22 46	6 23	17 21	6 44	17 00	7 10	16 35	7 46	15 58	8 49	14 55
343		.9850	11 03	22 29	11 20	22 14	11 42	21 54	12 12	21 26	13 01	20 40
10 TU	11 52 44	- 22 52	6 24	17 21	6 45	17 01	7 11	16 35	7 47	15 58	8 50	14 55
344		.9849	11 42	23 22	11 55	23 11	12 11	22 57	12 33	22 37	13 08	22 07
11 WE	11 53 11	- 22 58	6 25	17 22	6 45	17 01	7 11	16 35	7 48	15 58	8 52	14 54
345	15 49 FQ	.9847	12 17	none	12 26	none	12 36	23 57	12 51	23 46	13 13	23 29
12 TH	11 53 39	- 23 03	6 25	17 22	6 46	17 01	7 12	16 35	7 49	15 58	8 53	14 54
346		.9846	12 50	0 12	12 54	0 05	12 59	none	13 06	none	13 16	none
13 FR	11 54 07	- 23 07	6 26	17 22	6 47	17 01	7 13	16 35	7 50	15 58	8 55	14 53
347		.9845	13 22	1 00	13 22	0 58	13 21	0 56	13 20	0 53	13 19	0 48
14 SA	11 54 36	- 23 11	6 26	17 23	6 47	17 02	7 14	16 35	7 51	15 58	8 56	14 53
348		.9844	13 54	1 48	13 49	1 50	13 43	1 54	13 35	1 59	13 22	2 06
15 SU	11 55 04	- 23 15	6 27	17 23	6 48	17 02	7 14	16 36	7 52	15 58	8 57	14 53
349		.9843	14 28	2 36	14 18	2 43	14 06	2 53	13 50	3 05	13 26	3 26
16 MO	11 55 33	- 23 18	6 28	17 23	6 49	17 02	7 15	16 36	7 53	15 58	8 58	14 53
350		.9842	15 03	3 25	14 49	3 37	14 32	3 52	14 08	4 13	13 31	4 47
17 TU	11 56 02	- 23 20	6 28	17 24	6 49	17 03	7 16	16 36	7 53	15 59	8 59	14 53
351		.9842	15 42	4 17	15 24	4 33	15 02	4 54	14 31	5 23	13 39	6 11
18 WE	11 56 32	- 23 22	6 29	17 24	6 50	17 03	7 16	16 37	7 54	15 59	9 00	14 53
352		.9840	16 26	5 11	16 04	5 31	15 37	5 57	14 59	6 33	13 52	7 37
19 TH	11 57 01	- 23 24	6 29	17 25	6 50	17 04	7 17	16 37	7 55	15 59	9 01	14 53
353	19 10 FM	.9839	17 14	6 06	16 50	6 30	16 20	6 59	15 36	7 42	14 15	9 01
20 FR	11 57 31	- 23 25	6 30	17 25	6 51	17 04	7 18	16 37	7 55	16 00	9 01	14 54
354		.9838	18 08	7 03	17 43	7 28	17 11	7 59	16 24	8 46	14 55	10 14
21 SA	11 58 01	- 23 26	6 30	17 26	6 52	17 04	7 18	16 38	7 56	16 00	9 02	14 54
355		.9838	19 05	7 58	18 41	8 23	18 10	8 55	17 25	9 40	16 00	11 06
22 SU	11 58 30	- 23 26	6 31	17 26	6 52	17 05	7 19	16 38	7 56	16 01	9 02	14 55
356		.9837	20 05	8 51	19 43	9 14	19 15	9 43	18 36	10 24	17 25	11 37
23 MO	11 59 00	- 23 26	6 31	17 27	6 53	17 05	7 19	16 39	7 57	16 01	9 03	14 55
357		.9836	21 05	9 40	20 47	10 00	20 25	10 24	19 53	10 58	19 00	11 54
24 TU	11 59 30	- 23 25	6 32	17 27	6 53	17 06	7 20	16 39	7 57	16 02	9 03	14 56
358		.9836	22 04	10 25	21 51	10 41	21 35	10 59	21 13	11 24	20 37	12 05
25 WE	12 00 00	- 23 24	6 32	17 28	6 53	17 07	7 20	16 40	7 58	16 03	9 03	14 57
359		.9835	23 03	11 07	22 55	11 17	22 46	11 30	22 33	11 46	22 13	12 12
26 TH	12 00 30	- 23 23	6 33	17 28	6 54	17 07	7 20	16 41	7 58	16 03	9 04	14 58
360		.9835	none	11 47	23 59	11 52	23 56	11 58	23 53	12 05	23 48	12 17
27 FR	12 00 59	- 23 21	6 33	17 29	6 54	17 08	7 21	16 41	7 58	16 04	9 04	14 59
361	00 31 LQ	.9834	0 01	12 26	none	12 25	none	12 25	none	12 23	none	12 21
28 SA	12 01 29	- 23 18	6 34	17 29	6 55	17 08	7 21	16 42	7 58	16 05	9 04	15 00
362		.9834	0 59	13 06	1 03	12 59	1 07	12 52	1 13	12 41	1 23	12 26
29 SU	12 01 58	- 23 15	6 34	17 30	6 55	17 09	7 21	16 43	7 58	16 06	9 03	15 01
363		.9834	1 59	13 47	2 08	13 36	2 19	13 21	2 35	13 02	3 01	12 31
30 MO	12 02 27	- 23 12	6 34	17 31	6 55	17 10	7 21	16 44	7 59	16 07	9 03	15 02
364		.9834	3 00	14 33	3 15	14 16	3 33	13 55	3 59	13 26	4 42	12 39
31 TU	12 02 56	- 23 08	6 35	17 31	6 56	17 10	7 22	16 44	7 59	16 07	9 03	15 03
365		.9834	4 04	15 23	4 23	15 01	4 48	14 35	5 23	13 57	6 25	12 53

Julian and Gregorian Calendars; Leap Year; Century

Calendars based on the movements of the sun and moon have been used since ancient times, but none has been perfect. The **Julian calendar**, under which Western nations measured time until AD 1582, was authorized by Julius Caesar in 46 BC, the year 709 of Rome. His expert was a Greek, Sosigenes. The Julian calendar, on the assumption that the length of the true year was 365 ¼ days, gave every 4th year 366 days. St. Bede the Venerable, an Anglo-Saxon monk, announced in AD 730 that the 365 ¼-day Julian year was 11 min, 14 sec too long, a cumulative error of about a day every 128 years, but nothing was done about this for more than 800 years.

By 1582 the accumulated error was estimated to amount to 10 days. In that year Pope Gregory XIII decreed that the day following Oct. 4, 1582, should be called Oct. 15, thus dropping 10 days and initiating what became known as the **Gregorian calendar**.

However, with common years 365 days and a 366-day leap year every 4th year, the error in the length of the year would have recurred at the rate of a little more than 3 days every 400 years. Therefore, 3 of every 4 centesimal years (years ending in 00) were made common years, not leap years. Under this plan, 1600 and 2000 are leap years; 1700, 1800, and 1900 are not. **Leap years** are those years divisible by 4, except centesimal years, which are common unless divisible by 400.

The Gregorian calendar was adopted at once by France, Italy, Spain, Portugal, and Luxembourg. Within 2 years most German Catholic states, Belgium, and parts of Switzerland and the Netherlands were brought under the new calendar, and Hungary followed in 1587. The rest of the Netherlands, along with Denmark and the German Protestant states, made the change in 1699-1700. (German Protestants retained the Julian calendar's reckoning of the movable feast of Easter until 1776.)

The British government imposed the Gregorian calendar on all its possessions, including the American colonies, in 1752, decreeing that the day following Sept. 2, 1752, should be called Sept. 14, a loss of 11 days. All dates preceding were marked OS, for Old Style. In addition, New Year's Day was moved to Jan. 1 from Mar. 25 (under the old reckoning, for example, Mar. 24, 1700, had been followed by Mar. 25, 1701). Thus George Washington's birthdate, which was Feb. 11, 1731, OS, became Feb. 22, 1732, NS (New Style). In 1753 Sweden also went Gregorian, although it retained the Julian calendar's rules for Easter until 1844.

In 1793 the French revolutionary government adopted a calendar of 12 months of 30 days with 5 extra days in September of each common year and a 6th every 4th year. Napoleon reinstated the Gregorian calendar in 1806.

The Gregorian system later spread to non-European regions, first in the European colonies and then in independent countries, replacing traditional calendars at least for official purposes. Japan in 1873, Egypt in 1875, China in 1912, and Turkey in 1925 made the change, usually in conjunction with political upheaval. In China, the republican government began reckoning years from its 1911 founding. After 1949, the Communists adopted the Common, or Christian Era, year count, even for the traditional lunar calendar.

In 1918 the Soviet Union decreed that the day after Jan. 31, 1918, OS, would be Feb. 14, 1918, NS. Greece changed over in 1923. For the first time in history, all major cultures now have one calendar. (The Russian Orthodox Church, however, has retained the Julian calendar, as have various Middle Eastern Christian sects.)

To convert from the Julian to the Gregorian calendar, add 10 days to dates Oct. 5, 1582, through Feb. 28, 1700; after that date add 11 days through Feb. 28, 1800; 12 days through Feb. 28, 1900; and 13 days through Feb. 28, 2100.

A **century** consists of 100 consecutive years. The 1st century AD may be said to have run from the years 1 through 100. The 20th century by this reckoning consisted of the years 1901 through 2000 and technically ended Dec. 31, 2000, as did the 2d millennium AD. The 21st century thus technically began Jan. 1, 2001.

Julian Calendar

To find which of the 14 calendars of the Perpetual Calendar applies to any year, starting Jan. 1, under the Julian system, find the century for the desired year in the 3 leftmost columns below. Read across and find the year in the 4 top rows. Then read down. The number in the intersection is the calendar designation for that year.

Year (last 2 figures of desired year)

				01	**02**	**03**	**04**	**05**	**06**	**07**	**08**	**09**	**10**	**11**	**12**	**13**	**14**	**15**	**16**	**17**	**18**	**19**	**20**	**21**	**22**	**23**	**24**	**25**	**26**	**27**	**28**
				29	**30**	**31**	**32**	**33**	**34**	**35**	**36**	**37**	**38**	**39**	**40**	**41**	**42**	**43**	**44**	**45**	**46**	**47**	**48**	**49**	**50**	**51**	**52**	**53**	**54**	**55**	**56**
				57	**58**	**59**	**60**	**61**	**62**	**63**	**64**	**65**	**66**	**67**	**68**	**69**	**70**	**71**	**72**	**73**	**74**	**75**	**76**	**77**	**78**	**79**	**80**	**81**	**82**	**83**	**84**
Century			**00**	**85**	**86**	**87**	**88**	**89**	**90**	**91**	**92**	**93**	**94**	**95**	**96**	**97**	**98**	**99**													
0	**700**	**1400**	12	7	1	2	10	5	6	7	8	3	4	5	13	1	2	3	11	6	7	1	9	4	5	6	14	2	3	4	12
100	**800**	**1500**	11	6	7	1	9	4	5	6	14	2	3	4	12	7	1	2	10	5	6	7	8	3	4	5	13	1	2	3	11
200	**900**	**1600**	10	5	6	7	8	3	4	5	13	1	2	3	11	6	7	1	9	4	5	6	14	2	3	4	12	7	1	2	10
300	**1000**	**1700**	9	4	5	6	14	2	3	4	12	7	1	2	10	5	6	7	8	3	4	5	13	1	2	3	11	6	7	1	9
400	**1100**	**1800**	8	3	4	5	13	1	2	3	11	6	7	1	9	4	5	6	14	2	3	4	12	7	1	2	10	5	6	7	8
500	**1200**	**1900**	14	2	3	4	12	7	1	2	10	5	6	7	8	3	4	5	13	1	2	3	11	6	7	1	9	4	5	6	14
600	**1300**	**2000**	13	1	2	3	11	6	7	1	9	4	5	6	14	2	3	4	12	7	1	2	10	5	6	7	8	3	4	5	13

Gregorian Calendar

Choose the desired year from the table below or from the Perpetual Calendar (for years 1803 to 2080). The number after each year designates which calendar to use for that year, as shown in the Perpetual Calendar. (The Gregorian calendar was inaugurated Oct. 15, 1582. From that date to Dec. 31, 1582, use calendar 6.)

1583-1802

1583	7	1603	4	1623	1	1643	5	1663	2	1683	6	1703	2	1723	6	1743	3	1763	7	1783	4
1584	8	1604	12	1624	9	1644	13	1664	10	1684	14	1704	10	1724	14	1744	11	1764	8	1784	12
1585	3	1605	7	1625	4	1645	1	1665	5	1685	2	1705	5	1725	2	1745	6	1765	3	1785	7
1586	4	1606	1	1626	5	1646	2	1666	6	1686	3	1706	6	1726	3	1746	7	1766	4	1786	1
1587	5	1607	2	1627	6	1647	3	1667	7	1687	4	1707	7	1727	4	1747	1	1767	5	1787	2
1588	13	1608	10	1628	14	1648	11	1668	8	1688	12	1708	8	1728	12	1748	9	1768	13	1788	10
1589	1	1609	5	1629	2	1649	6	1669	3	1689	7	1709	3	1729	7	1749	4	1769	1	1789	5
1590	2	1610	6	1630	3	1650	7	1670	4	1690	1	1710	4	1730	1	1750	5	1770	2	1790	6
1591	3	1611	7	1631	4	1651	1	1671	5	1691	2	1711	5	1731	2	1751	6	1771	3	1791	7
1592	11	1612	8	1632	12	1652	9	1672	13	1692	10	1712	13	1732	10	1752	14	1772	11	1792	8
1593	6	1613	3	1633	7	1653	4	1673	1	1693	5	1713	1	1733	5	1753	2	1773	6	1793	3
1594	7	1614	4	1634	1	1654	5	1674	2	1694	6	1714	2	1734	6	1754	3	1774	7	1794	4
1595	1	1615	5	1635	2	1655	6	1675	3	1695	7	1715	3	1735	7	1755	4	1775	1	1795	5
1596	9	1616	13	1636	10	1656	14	1676	11	1696	8	1716	11	1736	8	1756	12	1776	9	1796	13
1597	4	1617	1	1637	5	1657	2	1677	6	1697	3	1717	6	1737	3	1757	7	1777	4	1797	1
1598	5	1618	2	1638	6	1658	3	1678	7	1698	4	1718	7	1738	4	1758	1	1778	5	1798	2
1599	6	1619	3	1639	7	1659	4	1679	1	1699	5	1719	1	1739	5	1759	2	1779	6	1799	3
1600	14	1620	11	1640	8	1660	12	1680	9	1700	6	1720	9	1740	13	1760	10	1780	14	1800	4
1601	2	1621	6	1641	3	1661	7	1681	4	1701	7	1721	4	1741	1	1761	5	1781	2	1801	5
1602	3	1622	7	1642	4	1662	1	1682	5	1702	1	1722	5	1742	2	1762	6	1782	3	1802	6

Perpetual Calendar

The number shown for each year indicates which Gregorian calendar to use. For 1583-1802, see "Gregorian Calendar" on page 675. For 1803-20, use numbers for 1983-2000, respectively. For Julian Calendar, see "Julian Calendar" on page 675.

1821 2	1847 6	1873 4	1899 1	1925 5	1951 2	1977 7	2003 4	2029 2	2055 6
1822 3	1848 . . .14	1874 5	1900 2	1926 6	1952 . . .10	1978 1	2004 . . .12	2030 3	2056 . . .14
1823 4	1849 2	1875 6	1901 3	1927 7	1953 5	1979 2	2005 7	2031 4	2057 2
1824 . . .12	1850 3	1876 . . .14	1902 4	1928 8	1954 6	1980 . . .10	2006 1	2032 . . .12	2058 3
1825 7	1851 4	1877 2	1903 5	1929 3	1955 7	1981 5	2007 2	2033 7	2059 4
1826 1	1852 . . .12	1878 3	1904 . . .13	1930 4	1956 8	1982 6	2008 . . .10	2034 1	2060 . . .12
1827 2	1853 7	1879 4	1905 1	1931 5	1957 3	1983 7	2009 5	2035 2	2061 7
1828 . . .10	1854 1	1880 . . .12	1906 2	1932 . . .13	1958 4	1984 8	2010 6	2036 . . .10	2062 1
1829 5	1855 2	1881 7	1907 3	1933 1	1959 5	1985 3	2011 7	2037 5	2063 2
1830 6	1856 . . .10	1882 1	1908 . . .11	1934 2	1960 . . .13	1986 4	2012 8	2038 6	2064 . . .10
1831 7	1857 5	1883 2	1909 6	1935 3	1961 1	1987 5	2013 3	2039 7	2065 5
1832 8	1858 6	1884 . . .10	1910 7	1936 . . .11	1962 2	1988 . . .13	2014 4	2040 8	2066 6
1833 3	1859 7	1885 5	1911 1	1937 6	1963 3	1989 1	2015 5	2041 3	2067 7
1834 4	1860 8	1886 6	1912 9	1938 7	1964 . . .11	1990 2	2016 . . .13	2042 4	2068 8
1835 5	1861 3	1887 7	1913 4	1939 1	1965 6	1991 3	2017 1	2043 5	2069 3
1836 . . .13	1862 4	1888 8	1914 5	1940 9	1966 7	1992 . . .11	2018 2	2044 . . .13	2070 4
1837 1	1863 5	1889 3	1915 6	1941 4	1967 1	1993 6	2019 3	2045 1	2071 5
1838 2	1864 . . .13	1890 4	1916 . . .14	1942 5	1968 9	1994 7	2020 . . .11	2046 2	2072 . . .13
1839 3	1865 1	1891 5	1917 2	1943 6	1969 4	1995 1	2021 6	2047 3	2073 1
1840 . . .11	1866 2	1892 . . .13	1918 3	1944 . . .14	1970 5	1996 9	2022 7	2048 . . .11	2074 2
1841 6	1867 3	1893 1	1919 4	1945 2	1971 6	1997 4	2023 1	2049 6	2075 3
1842 7	1868 . . .11	1894 2	1920 . . .12	1946 3	1972 . . .14	1998 5	2024 9	2050 7	2076 . . .11
1843 1	1869 6	1895 3	1921 7	1947 4	1973 2	1999 6	2025 4	2051 1	2077 6
1844 9	1870 7	1896 . . .11	1922 1	1948 . . .12	1974 3	2000 . . .14	2026 5	2052 9	2078 7
1845 4	1871 1	1897 6	1923 2	1949 7	1975 4	2001 2	2027 6	2053 4	2079 1
1846 5	1872 9	1898 7	1924 . . .10	1950 1	1976 . . .12	2002 3	2028 . . .14	2054 5	2080 9

1

JANUARY

S	M	T	W	T	F	S
1	2	3	4	5	6	7
8	9	10	11	12	13	14
15	16	17	18	19	20	21
22	23	24	25	26	27	28
29	30	31				

FEBRUARY

S	M	T	W	T	F	S
			1	2	3	4
5	6	7	8	9	10	11
12	13	14	15	16	17	18
19	20	21	22	23	24	25
26	27	28				

MARCH

S	M	T	W	T	F	S
			1	2	3	4
5	6	7	8	9	10	11
12	13	14	15	16	17	18
19	20	21	22	23	24	25
26	27	28	29	30	31	

APRIL

S	M	T	W	T	F	S
						1
2	3	4	5	6	7	8
9	10	11	12	13	14	15
16	17	18	19	20	21	22
23	24	25	26	27	28	29
30						

MAY

S	M	T	W	T	F	S
	1	2	3	4	5	6
7	8	9	10	11	12	13
14	15	16	17	18	19	20
21	22	23	24	25	26	27
28	29	30	31			

JUNE

S	M	T	W	T	F	S
				1	2	3
4	5	6	7	8	9	10
11	12	13	14	15	16	17
18	19	20	21	22	23	24
25	26	27	28	29	30	

JULY

S	M	T	W	T	F	S
						1
2	3	4	5	6	7	8
9	10	11	12	13	14	15
16	17	18	19	20	21	22
23	24	25	26	27	28	29
30	31					

AUGUST

S	M	T	W	T	F	S
		1	2	3	4	5
6	7	8	9	10	11	12
13	14	15	16	17	18	19
20	21	22	23	24	25	26
27	28	29	30	31		

SEPTEMBER

S	M	T	W	T	F	S
					1	2
3	4	5	6	7	8	9
10	11	12	13	14	15	16
17	18	19	20	21	22	23
24	25	26	27	28	29	30

OCTOBER

S	M	T	W	T	F	S
1	2	3	4	5	6	7
8	9	10	11	12	13	14
15	16	17	18	19	20	21
22	23	24	25	26	27	28
29	30	31				

NOVEMBER

S	M	T	W	T	F	S
			1	2	3	4
5	6	7	8	9	10	11
12	13	14	15	16	17	18
19	20	21	22	23	24	25
26	27	28	29	30		

DECEMBER

S	M	T	W	T	F	S
					1	2
3	4	5	6	7	8	9
10	11	12	13	14	15	16
17	18	19	20	21	22	23
24	25	26	27	28	29	30
31						

2 2001

JANUARY

S	M	T	W	T	F	S
	1	2	3	4	5	6
7	8	9	10	11	12	13
14	15	16	17	18	19	20
21	22	23	24	25	26	27
28	29	30	31			

FEBRUARY

S	M	T	W	T	F	S
				1	2	3
4	5	6	7	8	9	10
11	12	13	14	15	16	17
18	19	20	21	22	23	24
25	26	27	28			

MARCH

S	M	T	W	T	F	S
				1	2	3
4	5	6	7	8	9	10
11	12	13	14	15	16	17
18	19	20	21	22	23	24
25	26	27	28	29	30	31

APRIL

S	M	T	W	T	F	S
1	2	3	4	5	6	7
8	9	10	11	12	13	14
15	16	17	18	19	20	21
22	23	24	25	26	27	28
29	30					

MAY

S	M	T	W	T	F	S
		1	2	3	4	5
6	7	8	9	10	11	12
13	14	15	16	17	18	19
20	21	22	23	24	25	26
27	28	29	30	31		

JUNE

S	M	T	W	T	F	S
					1	2
3	4	5	6	7	8	9
10	11	12	13	14	15	16
17	18	19	20	21	22	23
24	25	26	27	28	29	30

JULY

S	M	T	W	T	F	S
1	2	3	4	5	6	7
8	9	10	11	12	13	14
15	16	17	18	19	20	21
22	23	24	25	26	27	28
29	30	31				

AUGUST

S	M	T	W	T	F	S
			1	2	3	4
5	6	7	8	9	10	11
12	13	14	15	16	17	18
19	20	21	22	23	24	25
26	27	28	29	30	31	

SEPTEMBER

S	M	T	W	T	F	S
						1
2	3	4	5	6	7	8
9	10	11	12	13	14	15
16	17	18	19	20	21	22
23	24	25	26	27	28	29
30						

OCTOBER

S	M	T	W	T	F	S
	1	2	3	4	5	6
7	8	9	10	11	12	13
14	15	16	17	18	19	20
21	22	23	24	25	26	27
28	29	30	31			

NOVEMBER

S	M	T	W	T	F	S
				1	2	3
4	5	6	7	8	9	10
11	12	13	14	15	16	17
18	19	20	21	22	23	24
25	26	27	28	29	30	

DECEMBER

S	M	T	W	T	F	S
						1
2	3	4	5	6	7	8
9	10	11	12	13	14	15
16	17	18	19	20	21	22
23	24	25	26	27	28	29
30	31					

3 2002

JANUARY

S	M	T	W	T	F	S
		1	2	3	4	5
6	7	8	9	10	11	12
13	14	15	16	17	18	19
20	21	22	23	24	25	26
27	28	29	30	31		

FEBRUARY

S	M	T	W	T	F	S
					1	2
3	4	5	6	7	8	9
10	11	12	13	14	15	16
17	18	19	20	21	22	23
24	25	26	27	28		

MARCH

S	M	T	W	T	F	S
					1	2
3	4	5	6	7	8	9
10	11	12	13	14	15	16
17	18	19	20	21	22	23
24	25	26	27	28	29	30
31						

APRIL

S	M	T	W	T	F	S
	1	2	3	4	5	6
7	8	9	10	11	12	13
14	15	16	17	18	19	20
21	22	23	24	25	26	27
28	29	30				

MAY

S	M	T	W	T	F	S
			1	2	3	4
5	6	7	8	9	10	11
12	13	14	15	16	17	18
19	20	21	22	23	24	25
26	27	28	29	30	31	

JUNE

S	M	T	W	T	F	S
						1
2	3	4	5	6	7	8
9	10	11	12	13	14	15
16	17	18	19	20	21	22
23	24	25	26	27	28	29
30						

JULY

S	M	T	W	T	F	S
	1	2	3	4	5	6
7	8	9	10	11	12	13
14	15	16	17	18	19	20
21	22	23	24	25	26	27
28	29	30	31			

AUGUST

S	M	T	W	T	F	S
				1	2	3
4	5	6	7	8	9	10
11	12	13	14	15	16	17
18	19	20	21	22	23	24
25	26	27	28	29	30	31

SEPTEMBER

S	M	T	W	T	F	S
1	2	3	4	5	6	7
8	9	10	11	12	13	14
15	16	17	18	19	20	21
22	23	24	25	26	27	28
29	30					

OCTOBER

S	M	T	W	T	F	S
		1	2	3	4	5
6	7	8	9	10	11	12
13	14	15	16	17	18	19
20	21	22	23	24	25	26
27	28	29	30	31		

NOVEMBER

S	M	T	W	T	F	S
					1	2
3	4	5	6	7	8	9
10	11	12	13	14	15	16
17	18	19	20	21	22	23
24	25	26	27	28	29	30

DECEMBER

S	M	T	W	T	F	S
1	2	3	4	5	6	7
8	9	10	11	12	13	14
15	16	17	18	19	20	21
22	23	24	25	26	27	28
29	30	31				

4 2003

JANUARY

S	M	T	W	T	F	S
			1	2	3	4
5	6	7	8	9	10	11
12	13	14	15	16	17	18
19	20	21	22	23	24	25
26	27	28	29	30	31	

FEBRUARY

S	M	T	W	T	F	S
						1
2	3	4	5	6	7	8
9	10	11	12	13	14	15
16	17	18	19	20	21	22
23	24	25	26	27	28	

MARCH

S	M	T	W	T	F	S
						1
2	3	4	5	6	7	8
9	10	11	12	13	14	15
16	17	18	19	20	21	22
23	24	25	26	27	28	29
30	31					

APRIL

S	M	T	W	T	F	S
		1	2	3	4	5
6	7	8	9	10	11	12
13	14	15	16	17	18	19
20	21	22	23	24	25	26
27	28	29	30			

MAY

S	M	T	W	T	F	S
				1	2	3
4	5	6	7	8	9	10
11	12	13	14	15	16	17
18	19	20	21	22	23	24
25	26	27	28	29	30	31

JUNE

S	M	T	W	T	F	S
1	2	3	4	5	6	7
8	9	10	11	12	13	14
15	16	17	18	19	20	21
22	23	24	25	26	27	28
29	30					

JULY

S	M	T	W	T	F	S
		1	2	3	4	5
6	7	8	9	10	11	12
13	14	15	16	17	18	19
20	21	22	23	24	25	26
27	28	29	30	31		

AUGUST

S	M	T	W	T	F	S
					1	2
3	4	5	6	7	8	9
10	11	12	13	14	15	16
17	18	19	20	21	22	23
24	25	26	27	28	29	30
31						

SEPTEMBER

S	M	T	W	T	F	S
	1	2	3	4	5	6
7	8	9	10	11	12	13
14	15	16	17	18	19	20
21	22	23	24	25	26	27
28	29	30				

OCTOBER

S	M	T	W	T	F	S
			1	2	3	4
5	6	7	8	9	10	11
12	13	14	15	16	17	18
19	20	21	22	23	24	25
26	27	28	29	30	31	

NOVEMBER

S	M	T	W	T	F	S
						1
2	3	4	5	6	7	8
9	10	11	12	13	14	15
16	17	18	19	20	21	22
23	24	25	26	27	28	29
30						

DECEMBER

S	M	T	W	T	F	S
	1	2	3	4	5	6
7	8	9	10	11	12	13
14	15	16	17	18	19	20
21	22	23	24	25	26	27
28	29	30	31			

5

JANUARY

S	M	T	W	T	F	S
				1	2	3
4	5	6	7	8	9	10
11	12	13	14	15	16	17
18	19	20	21	22	23	24
25	26	27	28	29	30	31

FEBRUARY

S	M	T	W	T	F	S
1	2	3	4	5	6	7
8	9	10	11	12	13	14
15	16	17	18	19	20	21
22	23	24	25	26	27	28

MARCH

S	M	T	W	T	F	S
1	2	3	4	5	6	7
8	9	10	11	12	13	14
15	16	17	18	19	20	21
22	23	24	25	26	27	28
29	30	31				

APRIL

S	M	T	W	T	F	S
			1	2	3	4
5	6	7	8	9	10	11
12	13	14	15	16	17	18
19	20	21	22	23	24	25
26	27	28	29	30		

MAY

S	M	T	W	T	F	S
					1	2
3	4	5	6	7	8	9
10	11	12	13	14	15	16
17	18	19	20	21	22	23
24	25	26	27	28	29	30
31						

JUNE

S	M	T	W	T	F	S
	1	2	3	4	5	6
7	8	9	10	11	12	13
14	15	16	17	18	19	20
21	22	23	24	25	26	27
28	29	30				

JULY

S	M	T	W	T	F	S
			1	2	3	4
5	6	7	8	9	10	11
12	13	14	15	16	17	18
19	20	21	22	23	24	25
26	27	28	29	30	31	

AUGUST

S	M	T	W	T	F	S
						1
2	3	4	5	6	7	8
9	10	11	12	13	14	15
16	17	18	19	20	21	22
23	24	25	26	27	28	29
30	31					

SEPTEMBER

S	M	T	W	T	F	S
		1	2	3	4	5
6	7	8	9	10	11	12
13	14	15	16	17	18	19
20	21	22	23	24	25	26
27	28	29	30			

OCTOBER

S	M	T	W	T	F	S
				1	2	3
4	5	6	7	8	9	10
11	12	13	14	15	16	17
18	19	20	21	22	23	24
25	26	27	28	29	30	31

NOVEMBER

S	M	T	W	T	F	S
1	2	3	4	5	6	7
8	9	10	11	12	13	14
15	16	17	18	19	20	21
22	23	24	25	26	27	28
29	30					

DECEMBER

S	M	T	W	T	F	S
		1	2	3	4	5
6	7	8	9	10	11	12
13	14	15	16	17	18	19
20	21	22	23	24	25	26
27	28	29	30	31		

6

JANUARY

S	M	T	W	T	F	S
					1	2
3	4	5	6	7	8	9
10	11	12	13	14	15	16
17	18	19	20	21	22	23
24	25	26	27	28	29	30
31						

FEBRUARY

S	M	T	W	T	F	S
	1	2	3	4	5	6
7	8	9	10	11	12	13
14	15	16	17	18	19	20
21	22	23	24	25	26	27
28						

MARCH

S	M	T	W	T	F	S
	1	2	3	4	5	6
7	8	9	10	11	12	13
14	15	16	17	18	19	20
21	22	23	24	25	26	27
28	29	30	31			

APRIL

S	M	T	W	T	F	S
				1	2	3
4	5	6	7	8	9	10
11	12	13	14	15	16	17
18	19	20	21	22	23	24
25	26	27	28	29	30	

MAY

S	M	T	W	T	F	S
						1
2	3	4	5	6	7	8
9	10	11	12	13	14	15
16	17	18	19	20	21	22
23	24	25	26	27	28	29
30	31					

JUNE

S	M	T	W	T	F	S
		1	2	3	4	5
6	7	8	9	10	11	12
13	14	15	16	17	18	19
20	21	22	23	24	25	26
27	28	29	30			

JULY

S	M	T	W	T	F	S
				1	2	3
4	5	6	7	8	9	10
11	12	13	14	15	16	17
18	19	20	21	22	23	24
25	26	27	28	29	30	31

AUGUST

S	M	T	W	T	F	S
1	2	3	4	5	6	7
8	9	10	11	12	13	14
15	16	17	18	19	20	21
22	23	24	25	26	27	28
29	30	31				

SEPTEMBER

S	M	T	W	T	F	S
			1	2	3	4
5	6	7	8	9	10	11
12	13	14	15	16	17	18
19	20	21	22	23	24	25
26	27	28	29	30		

OCTOBER

S	M	T	W	T	F	S
					1	2
3	4	5	6	7	8	9
10	11	12	13	14	15	16
17	18	19	20	21	22	23
24	25	26	27	28	29	30
31						

NOVEMBER

S	M	T	W	T	F	S
	1	2	3	4	5	6
7	8	9	10	11	12	13
14	15	16	17	18	19	20
21	22	23	24	25	26	27
28	29	30				

DECEMBER

S	M	T	W	T	F	S
			1	2	3	4
5	6	7	8	9	10	11
12	13	14	15	16	17	18
19	20	21	22	23	24	25
26	27	28	29	30	31	

7

JANUARY

S	M	T	W	T	F	S
						1
2	3	4	5	6	7	8
9	10	11	12	13	14	15
16	17	18	19	20	21	22
23	24	25	26	27	28	29
30	31					

FEBRUARY

S	M	T	W	T	F	S
		1	2	3	4	5
6	7	8	9	10	11	12
13	14	15	16	17	18	19
20	21	22	23	24	25	26
27	28					

MARCH

S	M	T	W	T	F	S
		1	2	3	4	5
6	7	8	9	10	11	12
13	14	15	16	17	18	19
20	21	22	23	24	25	26
27	28	29	30	31		

APRIL

S	M	T	W	T	F	S
					1	2
3	4	5	6	7	8	9
10	11	12	13	14	15	16
17	18	19	20	21	22	23
24	25	26	27	28	29	30

MAY

S	M	T	W	T	F	S
1	2	3	4	5	6	7
8	9	10	11	12	13	14
15	16	17	18	19	20	21
22	23	24	25	26	27	28
29	30	31				

JUNE

S	M	T	W	T	F	S
			1	2	3	4
5	6	7	8	9	10	11
12	13	14	15	16	17	18
19	20	21	22	23	24	25
26	27	28	29	30		

JULY

S	M	T	W	T	F	S
					1	2
3	4	5	6	7	8	9
10	11	12	13	14	15	16
17	18	19	20	21	22	23
24	25	26	27	28	29	30
31						

AUGUST

S	M	T	W	T	F	S
	1	2	3	4	5	6
7	8	9	10	11	12	13
14	15	16	17	18	19	20
21	22	23	24	25	26	27
28	29	30	31			

SEPTEMBER

S	M	T	W	T	F	S
				1	2	3
4	5	6	7	8	9	10
11	12	13	14	15	16	17
18	19	20	21	22	23	24
25	26	27	28	29	30	

OCTOBER

S	M	T	W	T	F	S
						1
2	3	4	5	6	7	8
9	10	11	12	13	14	15
16	17	18	19	20	21	22
23	24	25	26	27	28	29
30	31					

NOVEMBER

S	M	T	W	T	F	S
		1	2	3	4	5
6	7	8	9	10	11	12
13	14	15	16	17	18	19
20	21	22	23	24	25	26
27	28	29	30			

DECEMBER

S	M	T	W	T	F	S
				1	2	3
4	5	6	7	8	9	10
11	12	13	14	15	16	17
18	19	20	21	22	23	24
25	26	27	28	29	30	31

8

JANUARY

S	M	T	W	T	F	S
1	2	3	4	5	6	7
8	9	10	11	12	13	14
15	16	17	18	19	20	21
22	23	24	25	26	27	28
29	30	31				

FEBRUARY

S	M	T	W	T	F	S
			1	2	3	4
5	6	7	8	9	10	11
12	13	14	15	16	17	18
19	20	21	22	23	24	25
26	27	28	29			

MARCH

S	M	T	W	T	F	S
				1	2	3
4	5	6	7	8	9	10
11	12	13	14	15	16	17
18	19	20	21	22	23	24
25	26	27	28	29	30	31

APRIL

S	M	T	W	T	F	S
1	2	3	4	5	6	7
8	9	10	11	12	13	14
15	16	17	18	19	20	21
22	23	24	25	26	27	28
29	30					

MAY

S	M	T	W	T	F	S
		1	2	3	4	5
6	7	8	9	10	11	12
13	14	15	16	17	18	19
20	21	22	23	24	25	26
27	28	29	30	31		

JUNE

S	M	T	W	T	F	S
					1	2
3	4	5	6	7	8	9
10	11	12	13	14	15	16
17	18	19	20	21	22	23
24	25	26	27	28	29	30

JULY

S	M	T	W	T	F	S
1	2	3	4	5	6	7
8	9	10	11	12	13	14
15	16	17	18	19	20	21
22	23	24	25	26	27	28
29	30	31				

AUGUST

S	M	T	W	T	F	S
			1	2	3	4
5	6	7	8	9	10	11
12	13	14	15	16	17	18
19	20	21	22	23	24	25
26	27	28	29	30	31	

SEPTEMBER

S	M	T	W	T	F	S
						1
2	3	4	5	6	7	8
9	10	11	12	13	14	15
16	17	18	19	20	21	22
23	24	25	26	27	28	29
30						

OCTOBER

S	M	T	W	T	F	S
	1	2	3	4	5	6
7	8	9	10	11	12	13
14	15	16	17	18	19	20
21	22	23	24	25	26	27
28	29	30	31			

NOVEMBER

S	M	T	W	T	F	S
				1	2	3
4	5	6	7	8	9	10
11	12	13	14	15	16	17
18	19	20	21	22	23	24
25	26	27	28	29	30	

DECEMBER

S	M	T	W	T	F	S
						1
2	3	4	5	6	7	8
9	10	11	12	13	14	15
16	17	18	19	20	21	22
23	24	25	26	27	28	29
30	31					

9

JANUARY

S	M	T	W	T	F	S
	1	2	3	4	5	6
7	8	9	10	11	12	13
14	15	16	17	18	19	20
21	22	23	24	25	26	27
28	29	30	31			

FEBRUARY

S	M	T	W	T	F	S
				1	2	3
4	5	6	7	8	9	10
11	12	13	14	15	16	17
18	19	20	21	22	23	24
25	26	27	28	29		

MARCH

S	M	T	W	T	F	S
					1	2
3	4	5	6	7	8	9
10	11	12	13	14	15	16
17	18	19	20	21	22	23
24	25	26	27	28	29	30
31						

APRIL

S	M	T	W	T	F	S
	1	2	3	4	5	6
7	8	9	10	11	12	13
14	15	16	17	18	19	20
21	22	23	24	25	26	27
28	29	30				

MAY

S	M	T	W	T	F	S
			1	2	3	4
5	6	7	8	9	10	11
12	13	14	15	16	17	18
19	20	21	22	23	24	25
26	27	28	29	30	31	

JUNE

S	M	T	W	T	F	S
						1
2	3	4	5	6	7	8
9	10	11	12	13	14	15
16	17	18	19	20	21	22
23	24	25	26	27	28	29
30						

JULY

S	M	T	W	T	F	S
	1	2	3	4	5	6
7	8	9	10	11	12	13
14	15	16	17	18	19	20
21	22	23	24	25	26	27
28	29	30	31			

AUGUST

S	M	T	W	T	F	S
				1	2	3
4	5	6	7	8	9	10
11	12	13	14	15	16	17
18	19	20	21	22	23	24
25	26	27	28	29	30	31

SEPTEMBER

S	M	T	W	T	F	S
1	2	3	4	5	6	7
8	9	10	11	12	13	14
15	16	17	18	19	20	21
22	23	24	25	26	27	28
29	30					

OCTOBER

S	M	T	W	T	F	S
		1	2	3	4	5
6	7	8	9	10	11	12
13	14	15	16	17	18	19
20	21	22	23	24	25	26
27	28	29	30	31		

NOVEMBER

S	M	T	W	T	F	S
					1	2
3	4	5	6	7	8	9
10	11	12	13	14	15	16
17	18	19	20	21	22	23
24	25	26	27	28	29	30

DECEMBER

S	M	T	W	T	F	S
1	2	3	4	5	6	7
8	9	10	11	12	13	14
15	16	17	18	19	20	21
22	23	24	25	26	27	28
29	30	31				

10

JANUARY

S	M	T	W	T	F	S
		1	2	3	4	5
6	7	8	9	10	11	12
13	14	15	16	17	18	19
20	21	22	23	24	25	26
27	28	29	30	31		

FEBRUARY

S	M	T	W	T	F	S
					1	2
3	4	5	6	7	8	9
10	11	12	13	14	15	16
17	18	19	20	21	22	23
24	25	26	27	28	29	

MARCH

S	M	T	W	T	F	S
						1
2	3	4	5	6	7	8
9	10	11	12	13	14	15
16	17	18	19	20	21	22
23	24	25	26	27	28	29
30	31					

APRIL

S	M	T	W	T	F	S
		1	2	3	4	5
6	7	8	9	10	11	12
13	14	15	16	17	18	19
20	21	22	23	24	25	26
27	28	29	30			

MAY

S	M	T	W	T	F	S
				1	2	3
4	5	6	7	8	9	10
11	12	13	14	15	16	17
18	19	20	21	22	23	24
25	26	27	28	29	30	31

JUNE

S	M	T	W	T	F	S
1	2	3	4	5	6	7
8	9	10	11	12	13	14
15	16	17	18	19	20	21
22	23	24	25	26	27	28
29	30					

JULY

S	M	T	W	T	F	S
		1	2	3	4	5
6	7	8	9	10	11	12
13	14	15	16	17	18	19
20	21	22	23	24	25	26
27	28	29	30	31		

AUGUST

S	M	T	W	T	F	S
					1	2
3	4	5	6	7	8	9
10	11	12	13	14	15	16
17	18	19	20	21	22	23
24	25	26	27	28	29	30
31						

SEPTEMBER

S	M	T	W	T	F	S
	1	2	3	4	5	6
7	8	9	10	11	12	13
14	15	16	17	18	19	20
21	22	23	24	25	26	27
28	29	30				

OCTOBER

S	M	T	W	T	F	S
			1	2	3	4
5	6	7	8	9	10	11
12	13	14	15	16	17	18
19	20	21	22	23	24	25
26	27	28	29	30	31	

NOVEMBER

S	M	T	W	T	F	S
						1
2	3	4	5	6	7	8
9	10	11	12	13	14	15
16	17	18	19	20	21	22
23	24	25	26	27	28	29
30						

DECEMBER

S	M	T	W	T	F	S
	1	2	3	4	5	6
7	8	9	10	11	12	13
14	15	16	17	18	19	20
21	22	23	24	25	26	27
28	29	30	31			

11

JANUARY

S	M	T	W	T	F	S
			1	2	3	4
5	6	7	8	9	10	11
12	13	14	15	16	17	18
19	20	21	22	23	24	25
26	27	28	29	30	31	

FEBRUARY

S	M	T	W	T	F	S
						1
2	3	4	5	6	7	8
9	10	11	12	13	14	15
16	17	18	19	20	21	22
23	24	25	26	27	28	29

MARCH

S	M	T	W	T	F	S
1	2	3	4	5	6	7
8	9	10	11	12	13	14
15	16	17	18	19	20	21
22	23	24	25	26	27	28
29	30	31				

APRIL

S	M	T	W	T	F	S
			1	2	3	4
5	6	7	8	9	10	11
12	13	14	15	16	17	18
19	20	21	22	23	24	25
26	27	28	29	30		

MAY

S	M	T	W	T	F	S
					1	2
3	4	5	6	7	8	9
10	11	12	13	14	15	16
17	18	19	20	21	22	23
24	25	26	27	28	29	30
31						

JUNE

S	M	T	W	T	F	S
	1	2	3	4	5	6
7	8	9	10	11	12	13
14	15	16	17	18	19	20
21	22	23	24	25	26	27
28	29	30				

JULY

S	M	T	W	T	F	S
			1	2	3	4
5	6	7	8	9	10	11
12	13	14	15	16	17	18
19	20	21	22	23	24	25
26	27	28	29	30	31	

AUGUST

S	M	T	W	T	F	S
						1
2	3	4	5	6	7	8
9	10	11	12	13	14	15
16	17	18	19	20	21	22
23	24	25	26	27	28	29
30	31					

SEPTEMBER

S	M	T	W	T	F	S
		1	2	3	4	5
6	7	8	9	10	11	12
13	14	15	16	17	18	19
20	21	22	23	24	25	26
27	28	29	30			

OCTOBER

S	M	T	W	T	F	S
				1	2	3
4	5	6	7	8	9	10
11	12	13	14	15	16	17
18	19	20	21	22	23	24
25	26	27	28	29	30	31

NOVEMBER

S	M	T	W	T	F	S
1	2	3	4	5	6	7
8	9	10	11	12	13	14
15	16	17	18	19	20	21
22	23	24	25	26	27	28
29	30					

DECEMBER

S	M	T	W	T	F	S
		1	2	3	4	5
6	7	8	9	10	11	12
13	14	15	16	17	18	19
20	21	22	23	24	25	26
27	28	29	30	31		

12 2004

JANUARY

S	M	T	W	T	F	S
				1	2	3
4	5	6	7	8	9	10
11	12	13	14	15	16	17
18	19	20	21	22	23	24
25	26	27	28	29	30	31

FEBRUARY

S	M	T	W	T	F	S
1	2	3	4	5	6	7
8	9	10	11	12	13	14
15	16	17	18	19	20	21
22	23	24	25	26	27	28
29						

MARCH

S	M	T	W	T	F	S
	1	2	3	4	5	6
7	8	9	10	11	12	13
14	15	16	17	18	19	20
21	22	23	24	25	26	27
28	29	30	31			

APRIL

S	M	T	W	T	F	S
				1	2	3
4	5	6	7	8	9	10
11	12	13	14	15	16	17
18	19	20	21	22	23	24
25	26	27	28	29	30	

MAY

S	M	T	W	T	F	S
						1
2	3	4	5	6	7	8
9	10	11	12	13	14	15
16	17	18	19	20	21	22
23	24	25	26	27	28	29
30	31					

JUNE

S	M	T	W	T	F	S
		1	2	3	4	5
6	7	8	9	10	11	12
13	14	15	16	17	18	19
20	21	22	23	24	25	26
27	28	29	30			

JULY

S	M	T	W	T	F	S
				1	2	3
4	5	6	7	8	9	10
11	12	13	14	15	16	17
18	19	20	21	22	23	24
25	26	27	28	29	30	31

AUGUST

S	M	T	W	T	F	S
1	2	3	4	5	6	7
8	9	10	11	12	13	14
15	16	17	18	19	20	21
22	23	24	25	26	27	28
29	30	31				

SEPTEMBER

S	M	T	W	T	F	S
			1	2	3	4
5	6	7	8	9	10	11
12	13	14	15	16	17	18
19	20	21	22	23	24	25
26	27	28	29	30		

OCTOBER

S	M	T	W	T	F	S
					1	2
3	4	5	6	7	8	9
10	11	12	13	14	15	16
17	18	19	20	21	22	23
24	25	26	27	28	29	30
31						

NOVEMBER

S	M	T	W	T	F	S
	1	2	3	4	5	6
7	8	9	10	11	12	13
14	15	16	17	18	19	20
21	22	23	24	25	26	27
28	29	30				

DECEMBER

S	M	T	W	T	F	S
			1	2	3	4
5	6	7	8	9	10	11
12	13	14	15	16	17	18
19	20	21	22	23	24	25
26	27	28	29	30	31	

13

JANUARY

S	M	T	W	T	F	S
					1	2
3	4	5	6	7	8	9
10	11	12	13	14	15	16
17	18	19	20	21	22	23
24	25	26	27	28	29	30
31						

FEBRUARY

S	M	T	W	T	F	S
	1	2	3	4	5	6
7	8	9	10	11	12	13
14	15	16	17	18	19	20
21	22	23	24	25	26	27
28	29					

MARCH

S	M	T	W	T	F	S
		1	2	3	4	5
6	7	8	9	10	11	12
13	14	15	16	17	18	19
20	21	22	23	24	25	26
27	28	29	30	31		

APRIL

S	M	T	W	T	F	S
					1	2
3	4	5	6	7	8	9
10	11	12	13	14	15	16
17	18	19	20	21	22	23
24	25	26	27	28	29	30

MAY

S	M	T	W	T	F	S
1	2	3	4	5	6	7
8	9	10	11	12	13	14
15	16	17	18	19	20	21
22	23	24	25	26	27	28
29	30	31				

JUNE

S	M	T	W	T	F	S
			1	2	3	4
5	6	7	8	9	10	11
12	13	14	15	16	17	18
19	20	21	22	23	24	25
26	27	28	29	30		

JULY

S	M	T	W	T	F	S
					1	2
3	4	5	6	7	8	9
10	11	12	13	14	15	16
17	18	19	20	21	22	23
24	25	26	27	28	29	30
31						

AUGUST

S	M	T	W	T	F	S
	1	2	3	4	5	6
7	8	9	10	11	12	13
14	15	16	17	18	19	20
21	22	23	24	25	26	27
28	29	30	31			

SEPTEMBER

S	M	T	W	T	F	S
				1	2	3
4	5	6	7	8	9	10
11	12	13	14	15	16	17
18	19	20	21	22	23	24
25	26	27	28	29	30	

OCTOBER

S	M	T	W	T	F	S
						1
2	3	4	5	6	7	8
9	10	11	12	13	14	15
16	17	18	19	20	21	22
23	24	25	26	27	28	29
30	31					

NOVEMBER

S	M	T	W	T	F	S
		1	2	3	4	5
6	7	8	9	10	11	12
13	14	15	16	17	18	19
20	21	22	23	24	25	26
27	28	29	30			

DECEMBER

S	M	T	W	T	F	S
				1	2	3
4	5	6	7	8	9	10
11	12	13	14	15	16	17
18	19	20	21	22	23	24
25	26	27	28	29	30	31

14 2000

JANUARY

S	M	T	W	T	F	S
						1
2	3	4	5	6	7	8
9	10	11	12	13	14	15
16	17	18	19	20	21	22
23	24	25	26	27	28	29
30	31					

FEBRUARY

S	M	T	W	T	F	S
		1	2	3	4	5
6	7	8	9	10	11	12
13	14	15	16	17	18	19
20	21	22	23	24	25	26
27	28	29				

MARCH

S	M	T	W	T	F	S
			1	2	3	4
5	6	7	8	9	10	11
12	13	14	15	16	17	18
19	20	21	22	23	24	25
26	27	28	29	30	31	

APRIL

S	M	T	W	T	F	S
						1
2	3	4	5	6	7	8
9	10	11	12	13	14	15
16	17	18	19	20	21	22
23	24	25	26	27	28	29
30						

MAY

S	M	T	W	T	F	S
	1	2	3	4	5	6
7	8	9	10	11	12	13
14	15	16	17	18	19	20
21	22	23	24	25	26	27
28	29	30	31			

JUNE

S	M	T	W	T	F	S
				1	2	3
4	5	6	7	8	9	10
11	12	13	14	15	16	17
18	19	20	21	22	23	24
25	26	27	28	29	30	

JULY

S	M	T	W	T	F	S
						1
2	3	4	5	6	7	8
9	10	11	12	13	14	15
16	17	18	19	20	21	22
23	24	25	26	27	28	29
30	31					

AUGUST

S	M	T	W	T	F	S
		1	2	3	4	5
6	7	8	9	10	11	12
13	14	15	16	17	18	19
20	21	22	23	24	25	26
27	28	29	30	31		

SEPTEMBER

S	M	T	W	T	F	S
					1	2
3	4	5	6	7	8	9
10	11	12	13	14	15	16
17	18	19	20	21	22	23
24	25	26	27	28	29	30

OCTOBER

S	M	T	W	T	F	S
1	2	3	4	5	6	7
8	9	10	11	12	13	14
15	16	17	18	19	20	21
22	23	24	25	26	27	28
29	30	31				

NOVEMBER

S	M	T	W	T	F	S
			1	2	3	4
5	6	7	8	9	10	11
12	13	14	15	16	17	18
19	20	21	22	23	24	25
26	27	28	29	30		

DECEMBER

S	M	T	W	T	F	S
					1	2
3	4	5	6	7	8	9
10	11	12	13	14	15	16
17	18	19	20	21	22	23
24	25	26	27	28	29	30
31						

The Julian Period

How many days have you lived? To determine this, multiply your age by 365, add the number of days since your last birthday, and account for all leap years. Chances are your calculations will go wrong somewhere. Astronomers, however, find it convenient to express dates and time intervals in days rather than in years, months, and days. This is done by placing events within the Julian period.

The Julian period was devised in 1582 by the French classical scholar Joseph Scaliger (1540-1609), and it was named after his father, Julius Caesar Scaliger, not after the Julian calendar as might be supposed.

Scaliger began Julian Day (JD) #1 at noon, Jan. 1, 4713 BC, the most recent time that 3 major chronological cycles began on the same day: (1) the 28-year solar cycle, after which dates in the Julian calendar (e.g., Feb. 11) return to the same days of the week (e.g., Monday); (2) the 19-year lunar cycle, after which the phases of the moon return to the same dates of the year; and (3) the 15-year indiction cycle, used in ancient Rome to regulate taxes. It will take 7,980 years to complete the period, the product of 28, 19, and 15.

Noon of Dec. 31, 2001, marks the beginning of JD 2,452,275; that many days will have passed since the start of the Julian period. The JD at noon of any date in 2001 may be found by adding to this figure the day of the year for that date, which can be obtained from the left half of the "How Far Apart Are Two Dates?" chart.

How Far Apart Are Two Dates?

This table covers a period of 2 years. To use, find the **number** for each date and subtract the smaller from the larger. Example—for days from Feb. 10, 2002, to Dec. 15, 2003, subtract 41 from 714; the result is 673. For leap years, such as 2000, one day must be added; thus Feb. 10, 2000, and Dec. 15, 2001, are 674 days apart.

First Year

Date	Jan.	Feb.	Mar.	April	May	June	July	Aug.	Sept.	Oct.	Nov.	Dec.
1	**1**	**32**	**60**	**91**	**121**	**152**	**182**	**213**	**244**	**274**	**305**	**335**
2	**2**	**33**	**61**	**92**	**122**	**153**	**183**	**214**	**245**	**275**	**306**	**336**
3	**3**	**34**	**62**	**93**	**123**	**154**	**184**	**215**	**246**	**276**	**307**	**337**
4	**4**	**35**	**63**	**94**	**124**	**155**	**185**	**216**	**247**	**277**	**308**	**338**
5	**5**	**36**	**64**	**95**	**125**	**156**	**186**	**217**	**248**	**278**	**309**	**339**
6	**6**	**37**	**65**	**96**	**126**	**157**	**187**	**218**	**249**	**279**	**310**	**340**
7	**7**	**38**	**66**	**97**	**127**	**158**	**188**	**219**	**250**	**280**	**311**	**341**
8	**8**	**39**	**67**	**98**	**128**	**159**	**189**	**220**	**251**	**281**	**312**	**342**
9	**9**	**40**	**68**	**99**	**129**	**160**	**190**	**221**	**252**	**282**	**313**	**343**
10	**10**	**41**	**69**	**100**	**130**	**161**	**191**	**222**	**253**	**283**	**314**	**344**
11	**11**	**42**	**70**	**101**	**131**	**162**	**192**	**223**	**254**	**284**	**315**	**345**
12	**12**	**43**	**71**	**102**	**132**	**163**	**193**	**224**	**255**	**285**	**316**	**346**
13	**13**	**44**	**72**	**103**	**133**	**164**	**194**	**225**	**256**	**286**	**317**	**347**
14	**14**	**45**	**73**	**104**	**134**	**165**	**195**	**226**	**257**	**287**	**318**	**348**
15	**15**	**46**	**74**	**105**	**135**	**166**	**196**	**227**	**258**	**288**	**319**	**349**
16	**16**	**47**	**75**	**106**	**136**	**167**	**197**	**228**	**259**	**289**	**320**	**350**
17	**17**	**48**	**76**	**107**	**137**	**168**	**198**	**229**	**260**	**290**	**321**	**351**
18	**18**	**49**	**77**	**108**	**138**	**169**	**199**	**230**	**261**	**291**	**322**	**352**
19	**19**	**50**	**78**	**109**	**139**	**170**	**200**	**231**	**262**	**292**	**323**	**353**
20	**20**	**51**	**79**	**110**	**140**	**171**	**201**	**232**	**263**	**293**	**324**	**354**
21	**21**	**52**	**80**	**111**	**141**	**172**	**202**	**233**	**264**	**294**	**325**	**355**
22	**22**	**53**	**81**	**112**	**142**	**173**	**203**	**234**	**265**	**295**	**326**	**356**
23	**23**	**54**	**82**	**113**	**143**	**174**	**204**	**235**	**266**	**296**	**327**	**357**
24	**24**	**55**	**83**	**114**	**144**	**175**	**205**	**236**	**267**	**297**	**328**	**358**
25	**25**	**56**	**84**	**115**	**145**	**176**	**206**	**237**	**268**	**298**	**329**	**359**
26	**26**	**57**	**85**	**116**	**146**	**177**	**207**	**238**	**269**	**299**	**330**	**360**
27	**27**	**58**	**86**	**117**	**147**	**178**	**208**	**239**	**270**	**300**	**331**	**361**
28	**28**	**59**	**87**	**118**	**148**	**179**	**209**	**240**	**271**	**301**	**332**	**362**
29	**29**	—	**88**	**119**	**149**	**180**	**210**	**241**	**272**	**302**	**333**	**363**
30	**30**	—	**89**	**120**	**150**	**181**	**211**	**242**	**273**	**303**	**334**	**364**
31	**31**	—	**90**	—	**151**	—	**212**	**243**	—	**304**	—	**365**

Second Year

Date	Jan.	Feb.	Mar.	April	May	June	July	Aug.	Sept.	Oct.	Nov.	Dec.
1	**366**	**397**	**425**	**456**	**486**	**517**	**547**	**578**	**609**	**639**	**670**	**700**
2	**367**	**398**	**426**	**457**	**487**	**518**	**548**	**579**	**610**	**640**	**671**	**701**
3	**368**	**399**	**427**	**458**	**488**	**519**	**549**	**580**	**611**	**641**	**672**	**702**
4	**369**	**400**	**428**	**459**	**489**	**520**	**550**	**581**	**612**	**642**	**673**	**703**
5	**370**	**401**	**429**	**460**	**490**	**521**	**551**	**582**	**613**	**643**	**674**	**704**
6	**371**	**402**	**430**	**461**	**491**	**522**	**552**	**583**	**614**	**644**	**675**	**705**
7	**372**	**403**	**431**	**462**	**492**	**523**	**553**	**584**	**615**	**645**	**676**	**706**
8	**373**	**404**	**432**	**463**	**493**	**524**	**554**	**585**	**616**	**646**	**677**	**707**
9	**374**	**405**	**433**	**464**	**494**	**525**	**555**	**586**	**617**	**647**	**678**	**708**
10	**375**	**406**	**434**	**465**	**495**	**526**	**556**	**587**	**618**	**648**	**679**	**709**
11	**376**	**407**	**435**	**466**	**496**	**527**	**557**	**588**	**619**	**649**	**680**	**710**
12	**377**	**408**	**436**	**467**	**497**	**528**	**558**	**589**	**620**	**650**	**681**	**711**
13	**378**	**409**	**437**	**468**	**498**	**529**	**559**	**590**	**621**	**651**	**682**	**712**
14	**379**	**410**	**438**	**469**	**499**	**530**	**560**	**591**	**622**	**652**	**683**	**713**
15	**380**	**411**	**439**	**470**	**500**	**531**	**561**	**592**	**623**	**653**	**684**	**714**
16	**381**	**412**	**440**	**471**	**501**	**532**	**562**	**593**	**624**	**654**	**685**	**715**
17	**382**	**413**	**441**	**472**	**502**	**533**	**563**	**594**	**625**	**655**	**686**	**716**
18	**383**	**414**	**442**	**473**	**503**	**534**	**564**	**595**	**626**	**656**	**687**	**717**
19	**384**	**415**	**443**	**474**	**504**	**535**	**565**	**596**	**627**	**657**	**688**	**718**
20	**385**	**416**	**444**	**475**	**505**	**536**	**566**	**597**	**628**	**658**	**689**	**719**
21	**386**	**417**	**445**	**476**	**506**	**537**	**567**	**598**	**629**	**659**	**690**	**720**
22	**387**	**418**	**446**	**477**	**507**	**538**	**568**	**599**	**630**	**660**	**691**	**721**
23	**388**	**419**	**447**	**478**	**508**	**539**	**569**	**600**	**631**	**661**	**692**	**722**
24	**389**	**420**	**448**	**479**	**509**	**540**	**570**	**601**	**632**	**662**	**693**	**723**
25	**390**	**421**	**449**	**480**	**510**	**541**	**571**	**602**	**633**	**663**	**694**	**724**
26	**391**	**422**	**450**	**481**	**511**	**542**	**572**	**603**	**634**	**664**	**695**	**725**
27	**392**	**423**	**451**	**482**	**512**	**543**	**573**	**604**	**635**	**665**	**696**	**726**
28	**393**	**424**	**452**	**483**	**513**	**544**	**574**	**605**	**636**	**666**	**697**	**727**
29	**394**	—	**453**	**484**	**514**	**545**	**575**	**606**	**637**	**667**	**698**	**728**
30	**395**	—	**454**	**485**	**515**	**546**	**576**	**607**	**638**	**668**	**699**	**729**
31	**396**	—	**455**	—	**516**	—	**577**	**608**	—	**669**	—	**730**

Chinese Calendar, Asian Festivals

Source: Chinese Information and Culture Center, New York, NY

The Chinese calendar (like the Islamic calendar; see Religious Information section) is a lunar calendar. It is divided into 12 months of 29 or 30 days (compensating for the lunar month's mean duration of 29 days, 12 hr, 44.05 min). This calendar is synchronized with the solar year by the addition of extra months at fixed intervals.

The Chinese calendar runs on a 60-year cycle. The cycles 1876-1935 and 1936-95, with the years grouped under their 12 animal designations, are printed below, along with the first 24 years of the current cycle. It began in 1996 and will last until 2055. The year 2002 (Lunar Year 4700) is found in the 7th column, under Horse, and is known as a Year of the Horse. Readers can find the animal name for the year of their birth in the same chart. (Note: The first 3-7 weeks of each Western year belong to the previous Chinese year and animal designation.)

Both the Western (Gregorian) and traditional lunar calendars are used publicly in China and in North and South Korea, and 2 New Year's celebrations are held. In Taiwan, in overseas Chinese communities, and in Vietnam, the lunar calendar is used only to set the dates for traditional festivals, with the Gregorian system in general use.

The 4-day Chinese New Year, Hsin Nien, the 3-day Vietnamese New Year festival, Tet, and the 3-to-4-day Korean festival, Suhl, begin at the 2d new moon after the winter solstice. The new moon in the Far East, which is west of the International Date Line,may be one day later than the new moon in the U.S. The festivals may start, therefore, anywhere between Jan. 21 and Feb. 19 of the Gregorian calendar. Feb. 12 marks the start of the new Chinese year in 2002.

Rat	Ox	Tiger	Hare (Rabbit)	Dragon	Snake	Horse	Sheep (Goat)	Monkey	Rooster	Dog	Pig
1876	1877	1878	1879	1880	1881	1882	1883	1884	1885	1886	1887
1888	1889	1890	1891	1892	1893	1894	1895	1896	1897	1898	1899
1900	1901	1902	1903	1904	1905	1906	1907	1908	1909	1910	1911
1912	1913	1914	1915	1916	1917	1918	1919	1920	1921	1922	1923
1924	1925	1926	1927	1928	1929	1930	1931	1932	1933	1934	1935
1936	1937	1938	1939	1940	1941	1942	1943	1944	1945	1946	1947
1948	1949	1950	1951	1952	1953	1954	1955	1956	1957	1958	1959
1960	1961	1962	1963	1964	1965	1966	1967	1968	1969	1970	1971
1972	1973	1974	1975	1976	1977	1978	1979	1980	1981	1982	1983
1984	1985	1986	1987	1988	1989	1990	1991	1992	1993	1994	1995
1996	1997	1998	1999	2000	2001	2002	2003	2004	2005	2006	2007
2008	2009	2010	2011	2012	2013	2014	2015	2016	2017	2018	2019

Standard Time, Daylight Saving Time, and Others

Source: National Imagery and Mapping Agency; U.S. Dept. of Transportation

Standard Time

Standard Time is reckoned from the Prime Meridian of Longitude in Greenwich, England. The world is divided into 24 zones, each 15 deg of arc, or one hour in time apart. The Greenwich meridian (0 deg) extends through the center of the initial zone, and the zones to the east are numbered from 1 to 12, with the prefix "minus" indicating the number of hours to be subtracted to obtain Greenwich Time. Each zone extends 7.5 deg on either side of its central meridian.

Westward zones are similarly numbered, but prefixed "plus," showing the number of hours that must be added to get Greenwich Time. Although these zones apply generally to sea areas, the Standard Time maintained in many countries does not coincide with zone time. A graphical representation of the zones is shown on the Standard Time Zone Chart of the World (WOBZC76) published by the National Imagery and Mapping Agency. This chart is available from the National Ocean Service (NOS), 6501 Lafayette Avenue, Riverdale, MD 20737-1199; telephone: (800) 638-8972.

The U.S. and possessions are divided into 10 Standard Time zones. Each zone is approximately 15 deg of longitude in width. All places in each zone use, instead of their own local time, the time counted from the transit of the "mean sun" across the Standard Time meridian that passes near the middle of that zone. These time zones are designated as Atlantic, Eastern, Central, Mountain, Pacific, Alaska, Hawaii-Aleutian, Samoa, Wake Island, and Guam; the time in these zones is reckoned from the 60th, 75th, 90th, 105th, 120th, 135th, 150th, and 165th meridians west of Greenwich and the 165th and 150th meridians east of Greenwich. The time zone line wanders to conform to local geographical regions. The time in the various zones in the U.S. and U.S. territories west of Greenwich is earlier than Greenwich Time by 4, 5, 6, 7, 8, 9, 10, and 11 hours, respectively. However, Wake Island and Guam cross the International Date Line and are 12 and 10 hours later than Greenwich Time, respectively.

24-Hour Time

Twenty-four-hour time is widely used in scientific work throughout the world. In the U.S. it is also used in operations of the armed forces. In Europe it is frequently used by the transportation networks in preference to the 12-hour AM and PM system. With the 24-hour system the day begins at midnight, and times are designated 00:00 through 23:59.

International Date Line

The Date Line, approximately coinciding with the 180th meridian, separates the calendar dates. The date must be advanced one day when crossing in a westerly direction and set back one day when crossing in an easterly direction. The Date Line frequently deviates from the 180th meridian because of decisions made by individual nations affected. The line is deflected eastward through the Bering Strait and westward of the Aleutians to prevent separating these areas by date. The line is deflected eastward of the Tonga and New Zealand Islands in the South Pacific for the same reason. More recently it was deflected much farther eastward to include all of Kiribati. The line is established by international custom; there is no international authority prescribing its exact course.

Daylight Saving Time

Daylight Saving Time is achieved by advancing the clock one hour. Daylight Saving Time in the U.S. begins each year at 2 AM on the first Sunday in Apr. and ends at 2 AM on the last Sunday in Oct.

Daylight Saving Time was first observed in the U.S. during World War I, and then again during World War II. In the intervening years, some states and communities observed Daylight Saving Time, using whatever beginning and ending dates they chose. In 1966, Congress passed the Uniform Time Act, which provided that any state or territory that chooses to observe Daylight Saving Time must begin and end on the federal dates. Any state could, by law, exempt itself; a 1972 amendment to the act authorized states split by time zones to observe Daylight Saving Time in one time zone and standard time in the other time zone. Currently, Arizona, Hawaii, the eastern time zone portion of Indiana, Puerto Rico, the U.S. Virgin Islands, and American Samoa do not observe Daylight Saving Time.

Congress and the secretary of transportation both have authority to change time zone boundaries. Since 1966 there have been a number of changes to U.S. time zone boundaries. In addition, efforts to conserve energy have prompted various changes in the times that Daylight Saving Time is observed.

International Usage

Adjusting clock time so as to gain the added daylight on summer evenings is common throughout the world.

Canada, which extends over 6 time zones, generally observes Daylight Saving Time from the first Sunday of Apr. until the last Sunday of Oct. Saskatchewan remains on standard time all year. Communities elsewhere in Canada also may exempt themselves from Daylight Saving Time. Mexico, which occupies 3 time zones, observes Daylight Saving Time during the same period as most of Canada.

Member nations of the European Union (EU) observe a "summer-time period," the EU's version of Daylight Saving Time, from the last Sunday of Mar. until the last Sunday in Oct.

Russia, which extends over 11 time zones, maintains its Standard Time 1 hour fast for its zone designation. Additionally, it proclaims Daylight Saving Time from the last Sunday in Mar. until the 4th Sunday in Oct.

China, which extends across 5 time zones, has decreed that the entire country be placed on Greenwich Time plus 8 hours. Daylight Saving Time is not observed. Japan, which lies within one time zone, also does not modify its legal time during the summer months.

Many countries in the Southern Hemisphere maintain Daylight Saving Time, generally from Oct. to Mar.; however, most countries near the equator do not deviate from Standard Time.

Standard Time Differences—World Cities

The time indicated in the table is fixed by law and is called the legal time or, more generally, Standard Time. Use of Daylight Saving Time varies widely. * Indicates morning of the following day. At 12:00 noon, Eastern Standard Time, the Standard Time (in 24-hour time) in selected cities is as follows:

Addis Ababa	20 00	Casablanca	17 00	Madrid	18 00	Sarajevo	18 00
Amsterdam	18 00	Copenhagen	18 00	Manila	1 00*	Seoul	2 00*
Athens	19 00	Dhaka	23 00	Mecca	20 00	Shanghai	1 00*
Auckland	5 00*	Dublin	17 00	Melbourne	3 00*	Singapore	1 00*
Baghdad	20 00	Geneva	18 00	Montevideo	14 00	Stockholm	18 00
Bangkok	0 00*	Helsinki	19 00	Moscow	20 00	Sydney	3 00*
Beijing	1 00*	Ho Chi Minh City	0 00*	Munich	18 00	Taipei	1 00*
Belfast	17 00	Hong Kong	1 00*	Nagasaki	2 00*	Tashkent	22 00
Berlin	18 00	Istanbul	19 00	Nairobi	20 00	Tehran	20 30
Bogotá	12 00	Jakarta	0 00*	New Delhi	22 30	Tel Aviv	19 00
Bombay (Mumbai)	22 30	Jerusalem	19 00	Oslo	18 00	Tokyo	2 00*
Brussels	18 00	Johannesburg	19 00	Paris	18 00	Vladivostok	3 00*
Bucharest	19 00	Karachi	22 00	Prague	18 00	Vienna	18 00
Budapest	18 00	Kathmandu	22 45	Quito	12 00	Warsaw	18 00
Buenos Aires	14 00	Kiev	19 00	Rio de Janeiro	14 00	Wellington	5 00*
Cairo	19 00	Lagos	18 00	Rome	18 00	Yangon (Rangoon)	23 30
Calcutta	22 30	Lima	12 00	St. Petersburg	20 00	Yokohama	2 00*
Cape Town	19 00	Lisbon	17 00	Santiago	13 00	Zurich	18 00
Caracas	13 00	London	17 00				

Standard Time Differences—North American Cities

At 12:00 noon, Eastern Standard Time, the Standard Time in selected North American cities is as follows:

Akron, OH	12 00 Noon
Albuquerque, NM	10 00 AM
Atlanta, GA	12 00 Noon
Austin, TX	11 00 AM
Baltimore, MD	12 00 Noon
Birmingham, AL	11 00 AM
Bismarck, ND	11 00 AM
Boise, ID	10 00 AM
Boston, MA	12 00 Noon
Buffalo, NY	12 00 Noon
Butte, MT	10 00 AM
Calgary, Alta.	10 00 AM
Charleston, SC	12 00 Noon
Charleston, WV	12 00 Noon
Charlotte, NC	12 00 Noon
Charlottetown, PEI	1 00 PM
Chattanooga, TN	12 00 Noon
Cheyenne, WY	10 00 AM
Chicago, IL	11 00 AM
Cleveland, OH	12 00 Noon
Colorado Spr., CO	10 00 AM
Columbus, OH	12 00 Noon
Dallas, TX	11 00 AM
*Dawson, Yuk.	9 00 AM
Dayton, OH	12 00 Noon
Denver, CO	10 00 AM
Des Moines, IA	11 00 AM
Detroit, MI	12 00 Noon
Duluth, MN	11 00 AM
Edmonton, Alta.	10 00 AM
El Paso, TX	10 00 AM
Erie, PA	12 00 Noon
Evansville, IN	11 00 AM
Fairbanks, AK	8 00 AM
Flint, MI	12 00 Noon
*Fort Wayne, IN	12 00 Noon
Fort Worth, TX	11 00 AM
Frankfort, KY	12 00 Noon
Galveston, TX	11 00 AM
Grand Rapids, MI	12 00 Noon
Halifax, NS	1 00 PM
Hartford, CT	12 00 Noon
Havana, Cuba	12 00 Noon
Helena, MT	10 00 AM
*Honolulu, HI	7 00 AM
Houston, TX	11 00 AM
*Indianapolis, IN	12 00 Noon
Jacksonville, FL	12 00 Noon
Juneau, AK	8 00 AM
Kansas City, MO	11 00 AM
*Kingston, Jamaica	12 00 Noon
Knoxville, TN	12 00 Noon
Lexington, KY	12 00 Noon
Lincoln, NE	11 00 AM
Little Rock, AR	11 00 AM
Los Angeles, CA	9 00 AM
Louisville, KY	12 00 Noon
Mexico City, Mexico	11 00 AM
Memphis, TN	11 00 AM
Miami, FL	12 00 Noon
Milwaukee, WI	11 00 AM
Minneapolis, MN	11 00 AM
Mobile, AL	11 00 AM
Montreal, Que	12 00 Noon
Nashville, TN	11 00 AM
Nassau, Bahamas	12 00 Noon
New Haven, CT	12 00 Noon
New Orleans, LA	11 00 AM
New York, NY	12 00 Noon
Nome, AK	8 00 AM
Norfolk, VA	12 00 Noon
Oklahoma City, OK	11 00 AM
Omaha, NE	11 00 AM
Ottawa, Ont	12 00 Noon
*Panama City, Panama	12 00 Noon
Peoria, IL	11 00 AM
Philadelphia, PA	12 00 Noon
*Phoenix, AZ	10 00 AM
Pierre, SD	11 00 AM
Pittsburgh, PA	12 00 Noon
Portland, ME	12 00 Noon
Portland, OR	9 00 AM
Providence, RI	12 00 Noon
Quebec, Que.	12 00 Noon
*Regina, Sask.	11 00 AM
Reno, NV	9 00 AM
Richmond, VA	12 00 Noon
Rochester, NY	12 00 Noon
Sacramento, CA	9 00 AM
St. John's, Nfld.	1 30 PM
St. Louis, MO	11 00 AM
St. Paul, MN	11 00 AM
Salt Lake City, UT	10 00 AM
San Antonio, TX	11 00 AM
San Diego, CA	9 00 AM
San Francisco, CA	9 00 AM
*San Juan, PR	1 00 PM
Santa Fe, NM	10 00 AM
Savannah, GA	12 00 Noon
Seattle, WA	9 00 AM
Shreveport, LA	11 00 AM
Sioux Falls, SD	11 00 AM
Spokane, WA	9 00 AM
Tampa, FL	12 00 Noon
Toledo, OH	12 00 Noon
Topeka, KS	11 00 AM
Toronto, Ont	12 00 Noon
*Tucson, AZ	10 00 AM
Tulsa, OK	11 00 AM
Vancouver, BC	9 00 AM
Washington, DC	12 00 Noon
Wichita, KS	11 00 AM
Wilmington, DE	12 00 Noon
Winnipeg, Man.	11 00 AM

Note: This same table can be used for Daylight Saving Time when it is in effect, but allowance must be made for cities that do not observe it; they are marked with an asterisk (*). Daylight Saving Time is one hour later than Standard Time.

U.S. Legal or Public Holidays, 2002

Technically, the U.S. observes no national holidays; each state has jurisdiction over its holidays, which are designated by legislative enactment or executive proclamation. The president and the U.S. Congress can legally designate holidays only for the District of Columbia and for federal employees. In practice, however, most states observe the federal legal public holidays. Federal legal public holidays are New Year's Day, Martin Luther King Jr.'s Birthday, Washington's Birthday (often called Presidents' Day), Memorial Day, Independence Day, Labor Day, Columbus Day, Veterans Day, Thanksgiving, and Christmas.

Chief Legal or Public Holidays, 2002

When a holiday falls on Saturday or Sunday, it is usually observed on the preceding Friday or the next Monday. For some holidays, government and business closing practices vary. In most states, the secretary of state's office can provide details.

The following will be legal or public holidays in most states in 2002:

Jan. 1 (Tues.) — New Year's Day
Jan. 21 (3d Mon. in Jan.) — Martin Luther King Jr.'s Birthday
Feb. 12 (Tues.) — Lincoln's Birthday
Feb. 18 (3d Mon. in Feb.) — Washington's Birthday, or Presidents' Day, or Washington-Lincoln Day
May 27 (last Mon. in May) — Memorial Day, or Decoration Day
July 4 (Thurs.) — Independence Day
Sept. 2 (1st Mon. in Sept.) — Labor Day
Nov. 11 (Mon.) — Veterans Day
Nov. 28 (4th Thurs. in Nov.) — Thanksgiving
Dec. 25 (Wed.) — Christmas Day

In some states these also will be holidays in 2002:

Mar. 29 (Fri.) — Good Friday (In some states, observed for half or part of day.)
Oct. 14 (2d Mon. in Oct.) — Columbus Day, or Discoverers' Day, or Pioneers' Day
Nov. 5 (1st Tues. after 1st Mon. in Nov.) — Election Day

IT'S A FACT: Juneteenth, a day celebrated in Texas and some other states, especially by African-Americans, is short for "June 19." That was the date in 1865, two years after the Emancipation Proclamation, when a Union general, arriving in Galveston, TX, with his troops, announced that slaves in Texas should be freed.

Selected International Holidays, 2002

Jan. 26 — Australia Day, Australia
Feb. 5 — Constitution Day, Mexico
Feb. 9–12 — Carnival, Brazil
Feb. 12— Chinese New Year
Mar. 11 — Commonwealth Day, Canada, Great Britain
Mar. 17 — St. Patrick's Day, Ireland
Mar. 21 — Benito Juarez's Birthday, Mexico
Apr. 8 — Buddha's Birthday, Korea, Japan
Apr. 23 — National Sovereignty Day, Turkey
May 5 — Cinco de Mayo (Battle of Puebla Day), Mexico
May 17 — Constitution Day, Norway
May 20 — Victoria Day, Canada
June 15 — Dragon Boat Festival, China
June 23 — Midsummer Eve, Baltics, Scandinavia
July 1 — Canada Day, Canada
July 14 — Bastille Day, France
Aug. 30 — St. Rose of Lima, Peru
Sept. 2 — Labor Day, Canada
Sept. 16 — Independence Day, Mexico
Sept. 19 — St. Gennaro, Italy
Oct. 3 — German Unification Day, Germany
Oct. 12 — Día de la Raza, Mexico
Oct. 14 — Thanksgiving Day, Canada
Nov. 2 — Day of the Dead, Mexico
Nov. 5 — Guy Fawkes Day, Great Britain
Nov. 11 — Remembrance Day, Canada, Great Britain
Dec. 12 — Jamhuri Day, Kenya; Guadalupe Day, Mexico
Dec. 26 — Boxing Day, Australia, Canada, Great Britain, New Zealand

RELIGIOUS INFORMATION

Membership of Religious Groups in the U.S.

Sources: *2001 Yearbook of American & Canadian Churches*, © National Council of the Churches of Christ in the USA; *World Almanac* research

These membership figures generally are based on reports made by officials of each group, and not on any religious census. Figures from other sources may vary. Many groups keep careful records; others only estimate. Not all groups report annually. Church membership figures reported in this table are generally inclusive and do not refer simply to full communicants or confirmed members. Specific definitions of "member" vary from one denomination to another.

The number of houses of worship appears in parentheses. * Indicates that the group declines to make membership figures public. Groups reporting fewer than 5,000 members are not included; where membership numbers are not available, only those groups with 50 or more houses of worship are listed.

Religious Group	Members
Adventist churches:	
Advent Christian Ch. (302)	25,702
Seventh-day Adventist Ch. (4,421)	861,860
American Catholic Church (100)	**25,000**
Apostolic Christian Churches of America (91)	**12,800**
Bahá'í Faith (7,159 centers)	**133,709**[1]
Baptist churches:	
American Baptist Assn. (1,760)	275,000
American Baptist Chs. in the U.S.A. (5,755)	1,454,388
Baptist Bible Fellowship Intl. (4,500)	1,200,000
Baptist General Conference (880)	142,871
Baptist Missionary Assn. of America (1,334)	234,732
Conservative Baptist Assn. of America (1,200)	200,000
Free Will Baptists, Natl. Assn. of (2,476)	216,711
General Assn. of General Baptists (719)	67,314
General Assn. of Regular Baptist Chs. (1,398)	92,129
Natl. Baptist Convention, U.S.A., Inc. (2,500)	3,500,000
Natl. Missionary Baptist Convention of America	4,500,000
North American Baptist Conference (271)	45,738
Progressive National Baptist Convention (2,000)	2,500,000
Separate Baptists in Christ (100)	8,000
Southern Baptist Convention (41,099)	15,851,756
Brethren in Christ (252)	**20,010**
Brethren (German Baptists):	
Brethren Ch. (Ashland, OH) (118)	13,227
Church of the Brethren (1,095)	141,400
Grace Brethren Chs., Fellowship of (260)	30,371
Old German Baptist Brethren (55)	6,050
Buddhist Churches of America (60)	**15,750**[1]
Christian Brethren (Plymouth Brethren) (1,150)	**100,000**
Christian Church (Disciples of Christ) (3,765)	**831,125**
Christian Ch. of N.A., Gen. Council (96)	**7,200**
Christian Congregation, Inc. (1,438)	**118,209**
Christian and Missionary Alliance (1,973)	**347,973**
Christian Union, Churches of Christ in (216)	**10,104**
Church of Christ (Holiness) U.S.A. (167)	**10,383**
Church of Christ, Scientist (2,200)	*
Church of the United Brethren in Christ (228)	**23,585**
Churches of Christ (15,000)	**1,500,000**
Churches of God:	
Chs. of God, General Conference (340)	32,045
Ch. of God (Anderson, IN) (2,353)	234,311
Ch. of God (Seventh Day), Denver, CO (185)	11,000
Ch. of God by Faith, Inc. (145)	8,235
Ch. of God, Mountain Assembly (118)	6,140
Church of the Nazarene (5,101)	**627,054**
Community Churches, Intl. Council of (180)	**200,000**
Congreg. Christian Chs., Nat'l Assoc. of (426)	**65,502**
Conservative Congregational Christian Conference (242)	**40,414**
Eastern Orthodox churches:	
American Carpatho-Russian Orthodox Greek Catholic Ch. (80)	13,120
Antiochian Orthodox Christian Archdiocese of N.A. (227)	65,000
Apostolic Catholic Assyrian Ch. of the East, N.A. Dioceses (22)	120,000
Armenian Apostolic Ch. of America (36)	360,000
Dioceses of America, Armenian Apostolic Church (72)	414,000
Coptic Orthodox Ch. (85)	180,000
Greek Orthodox Archdiocese of America (523)	1,954,500
Mar Thoma Syrian Chief of India (65)	30,000
Orthodox Ch. in America (710)	1,000,000
Patriarchal Parishes of the Russian Orthodox Ch. in the USA (38)	9,780
Romanian Orthodox Episcopate of N. America (56)	25,000
Russian Orthodox Church Outside of Russia (177)	*
Syrian Orthodox Ch. of Antioch (22)	32,500
Episcopal Church (7,390)	**2,317,794**
Apostolic Episcopal Church (300)	14,000
Evangelical Church (132)	**12,369**
Evangelical Congregational Church (143)	**22,349**
Evangelical Covenant Church (636)	**98,526**
Evangelical Free Church of America (1,224)	**242,619**
Friends:	
Evangelical Friends Intl.-N.A. Region (288)	36,760
Friends General Conference (620)	32,000
Friends United Meeting (479)	33,908
Religious Society of Friends (Conservative) (1,200)	104,000
Full Gospel Assemblies Intl. (286)	52,500
Full Gospel Fellowship of Churches and Ministers Intl. (896)	**325,000**
General Church of the New Jerusalem (34)	**5,553**
Grace Gospel Fellowship (128)	**60,000**
Hindu	**1,285,000**[1]
Independent Fundamental Churches of America (659)	**61,655**
Islam	**5,780,000**[1]
Jehovah's Witnesses (11,257)	**990,340**
Jewish organizations:	
Union of American Hebrew Congregations (Reform) (896)	1,500,000
Union of Orthodox Jewish Congregations of America (800)	1,075,000
United Synagogue of Conservative Judaism, The (762)	1,500,000
Jewish Reconstructionist Federation (100)	65,000
Latter-day Saints:	
Ch. of Jesus Christ of Latter-day Saints (Mormon) (11,315)	5,113,409
Reorganized Ch. of Jesus Christ of Latter-day Saints (1,236)	137,038
Liberal Catholic Church—Province of the U.S.A. (27)	**6,500**
Lutheran churches:	
Apostolic Lutheran Ch. of America (58)	*
Ch. of the Lutheran Brethren of America (115)	13,920
Ch. of the Lutheran Confession (72)	8,631
Evangelical Lutheran Ch. in America (10,851)	5,149,668
Evangelical Lutheran Synod (139)	22,264
Free Lutheran Congregations, Assn. of (241)	32,984
Latvian Evangelical Lutheran Church in America (71)	15,012
Lutheran Ch.—Missouri Synod (6,220)	2,582,440
Lutheran Chs., American Assn. of (101)	18,252
Wisconsin Evangelical Lutheran Synod (1,239)	722,754
Mennonite churches:	
Beachy Amish Mennonite Chs. (114)	7,853
Church of God in Christ (Mennonite) (102)	12,144
Hutterian Brethren (428)	42,800
Mennonite Brethren Chs., Gen. Conf. (368)	82,130
Mennonite Church (935)	92,002
Mennonite Ch., General Conference of (295)	35,759
Old Order Amish Ch. (898)	80,820
Methodist churches:	
African Methodist Episcopal Ch. (6,200)	2,500,000
African Methodist Episcopal Zion Ch. (3,125)	1,276,662
Evangelical Methodist Ch. (123)	8,615
Free Methodist Ch. of North America (971)	70,556
Primitive Methodist Ch. in the U.S.A. (77)	6,031
Southern Methodist Ch. (117)	7,686
United Methodist Ch. (35,609)	8,377,662
The Wesleyan Church (1,594)	112,615
Metropolitan Community Churches, Universal Fellowship of (300)	**44,000**
Missionary Church (354)	**46,015**

Religious Group	Members
Moravian Ch. in America, Northern Province (93)	**26,130**
Natl. Organization of the New Apostolic Ch. of North America (385)	**36,254**
Pentecostal churches:	
Apostolic Faith Mission Ch. of God (19)	10,651
Apostolic Overcoming Holy Catholic Church of God Inc.(129)	10,714
Assemblies of God (12,055)	2,574,531
Bible Church of Christ (6)	6,850
Bible Fellowship Church (56)	7,169
Church of God (Cleveland, TN) (6,328)	870,039
Church of God in Christ (15,300)	5,499,875
Church of God of Prophecy (1,862)	75,112
Elim Fellowship (98)	*
Intl. Ch. of the Foursquare Gospel (1,836)	233,412
Intl. Pentecostal Church of Christ (69)	5,572
Intl. Pentecostal Holiness Church (1,771)	185,431
Open Bible Standard Chs. (357)	35,700
Pentecostal Assemblies of the World Inc. (1,750)	1,500,000
Pentecostal Church of God (1,237)	105,200
Pentecostal Free Will Baptist Ch. (150)	28,000
United Pentecostal Ch. Intl. (3,790)	*
Presbyterian churches:	
Associated Reformed Presbyterian Ch. (General Synod) (244)	40,113
Cumberland Presbyterian Ch. (775)	86,049
Cumberland Presbyterian Ch. in America (152)	15,142
Evangelical Presbyterian Ch. (197)	63,447
Genl. Assembly of the Korean Presbyterian Church in America (310)	50,221
Orthodox Presbyterian Ch. (204)	25,302
Presbyterian Ch. in America (1,206)	299,055
Presbyterian Ch. (U.S.A.) (11,216)	3,561,184
Reformed Presbyterian Ch. of N. America (86)	6,105
Reformed churches:	
Christian Reformed Ch. in N. America (732)	198,400
Hungarian Reformed Ch. in America (27)	60,000
Netherlands Reformed Congregations (24)	9,047
Protestant Reformed Churches in America (27)	6,730
Reformed Ch. in America (901)	293,147
United Church of Christ (5,961)	1,401,682
Reformed Episcopal Church (125)	**6,400**
Roman Catholic Church (19,627)	**62,391,484**
Salvation Army (1,410)	**472,871**
Unitarian Universalist Assn. of Congregations (1,050)	**216,931**

(1) Estimate; figures from other sources may vary.

Headquarters of Selected Religious Groups in the U.S.

Sources: *2001 Yearbook of American & Canadian Churches*, © National Council of the Churches of Christ in the USA; *World Almanac* research

(Year organized in parentheses)

African Methodist Episcopal Church (1787), 1134 11th St. NW, Washington, DC 20001; Senior Bishop, Bishop John Hurst Adams

African Methodist Episcopal Zion Church (1796), PO Box 32843, Charlotte, NC 28232; Pres. Samuel Chuka Ekemam, Sr. (Note: Presidency rotates every 6 mos. according to seniority.)

American Baptist Churches in the U.S.A. (1907), PO Box 851, Valley Forge, PA 19482; http://www.abc-usa.org; Pres., Trinette V. McCray

American Hebrew Congregations, Union of, 633 3rd Ave., New York, NY 10017; http://www.uahc.org; Pres., Rabbi Eric Yoffie

American Rescue Workers (1890), 25 Ross St., Williamsport, PA 17701; http://www.arwus.com; Commander-in-Chief & Pres., Gen. Claude S. Astin Jr., Rev.

Antiochian Orthodox Christian Archdiocese of North America (1895), 358 Mountain Rd., Englewood, NJ 07631; http://www.archdiocese@antiochian.org; Primate, Metropolitan Philip Saliba

Armenian Apostolic Church of America (1887), **Eastern Prelacy**: 138 E. 39th St., New York, NY 10016; http://www.armprelacy.org; Prelate, Bishop Oshagan Choloyan; **Western Prelacy**: 4401 Russel Ave., Los Angeles, CA 90027; Prelate, Bishop Moushegh Mardirossian

Assemblies of God (1914), 1445 Boonville Ave., Springfield, MO 65802; http://www.agifellowship.org; Gen. Supt., Thomas E. Trask

Bahá'í Faith, National Spiritual Assembly of the Bahá'í's of the U.S., 536 Sheridan Rd., Wilmette, IL 60091; http://www.bahai.org; Secy. Gen., Dr. Robert Henderson

Baptist Bible Fellowship Intl. (1950), Baptist Bible Fellowship Missions Bldg., 720 E. Kearney St., Springfield, MO 65803; Pres., Ken Gillming Sr.

Baptist Convention, Southern (1845), 901 Commerce St., Ste. 750, Nashville, TN 37203; http://www.sbcnet.org; Pres., Paige Patterson

Baptist Convention, U.S.A., Inc., National 1700 Baptist World Center Dr., Nashville, TN 37207; Pres. Dr. William J. Shaw

Baptist Convention of America, Inc., National (1880), 777 S. R.L. Thornton Freeway, Ste. 205, Dallas, TX 75203; http://www.greatertempleofgod.com; Pres., Dr. E. Edward Jones

Baptist Convention of America, Natl. Missionary (1988), 1404 E. Firestone, Los Angeles, CA 90001; Pres., Dr. W. T. Snead Sr.

Baptist General Conference (1852), 2002 S. Arlington Heights Rd., Arlington Heights, IL 60005; http://www.bgc.bethel.edu; Pres., Dr. Robert S. Ricker

Brethren in Christ Church (1778), PO Box A, Grantham, PA 17027; Moderator, Dr. Warren L. Hoffman

Buddhist Churches of America (1899), 1710 Octavia St., San Francisco, CA 94109; Presiding Bishop, Hakubun Watanabe

Christian and Missionary Alliance (1897), PO Box 35000, Colorado Springs, CO 80935; http://www.cmalliance.org; Pres., Rev. Peter N. Nanfelt, D.D.

Christian Church (Disciples of Christ) (1832), 130 E. Washington St., PO Box 1986, Indianapolis, IN 46206; http://www.disciples.org; Gen. Minister and Pres., Richard L. Hamm

Christian Churches and Churches of Christ, 4210 Bridgetown Rd., Box 11326, Cincinnati, OH 45211; http://www.nacc-online.org

Christian Congregation, Inc., The (1887), 804 W. Hemlock St., LaFollette, TN 37766; Gen. Supt., Rev. Ora W. Eads, D.D.

Christian Methodist Episcopal Church (1870), 4466 Elvis Presley Blvd., Memphis, TN 38116; Executive Secretary, Attorney Juanita Bryant

Christian Reformed Church in North America (1857), 2850 Kalamazoo Ave. SE, Grand Rapids, MI 49560; http://www.crcna.org; Gen. Secy., Dr. David H. Engelhard

Church of the Brethren (1708), 1451 Dundee Ave., Elgin, IL 60120; Moderator, Phill Carlos Archbold

Church of Christ (1830), PO Box 472, Independence, MO 64051; Council of Apostles, Secy., Apostle Smith N. Brickhouse

Church of God (Anderson, IN) (1881), Box 2420, Anderson, IN 46018; http://www.chog.org; Gen. Dir., Robert W. Pearson

Church of God (Cleveland, TN) (1886), PO Box 2430, Cleveland, TN 37320; Gen. Overseer, R. Lamar Vest

Church of God in Christ (1907), Mason Temple, 939 Mason St., Memphis, TN 38126; Presiding Bishop, Bishop Chandler D. Owens

Church of Jesus Christ (Bickertonites) (1862), 6th & Lincoln Sts., Monongahela, PA 15063; Pres., Dominic Thomas

Church of Jesus Christ of Latter-day Saints (Mormon), The (1830), 47 E. South Temple St., Salt Lake City, UT 84150; http://www.lds.org; Pres., Gordon B. Hinckley

Church of the Nazarene (1907), 6401 The Paseo, Kansas City, MO 64131; Gen. Secy., Jack Stone

Community Churches, International Council of (1950), 21116 Washington Pkwy., Frankfort, IL 60423; Pres., Rev. Judson Souers

Conservative Judaism, United Synagogue of, 155 5th Ave., New York, NY 10010; http://www.uscj.org; Pres., Stephen Wolnek

Coptic Orthodox Church, 427 West Side Ave., Jersey City, NJ 07304

Cumberland Presbyterian Church (1810), 1978 Union Ave., Memphis, TN 38104; http://www.cumberland.org; Moderator, Bob G. Roberts

Episcopal Church (1789), 815 Second Ave., New York, NY 10017; http://www.ecusa.anglican.org; Presiding Bishop and Primate, Most Rev. Frank Tracy Griswold III

Evangelical Free Church of America (1884), 901 E. 78th St., Minneapolis, MN 55420; Acting Pres., Rev. William Hamel

Evangelical Lutheran Church in America (1987), 8765 W. Higgins Rd., Chicago, IL 60631; http://www.elca.org; Presiding Bishop, Rev. Dr. H. George Anderson

Fellowship of Grace Brethren Churches (1882), PO Box 386, Winona Lake, IN 46590; http://www.fgbc.org; Moderator, Dr. Galen Wiley

First Church of Christ, Scientist, The (1879), 175 Huntington Ave., Boston, MA 02115; http://www.tfccs.com; Pres., Thomas J. Black

Free Methodist Church of North America (1860), World Ministries Center, 770 N. High School Rd., Indianapolis, IN 46214

Friends General Conference (1900), 1216 Arch St. 2B, Philadelphia, PA 19107; Gen. Secy., Bruce Birchard

Greek Orthodox Archdiocese of America (1922), 8-10 E. 79th St., New York, NY 10021; http://www.goarch.org; Primate of Greek Orthodox Church in America, Archbishop Demitrios

International Church of the Foursquare Gospel (1927), 1910 W. Sunset Blvd., Ste. 200, PO Box 26902, Los Angeles, CA 90026; http://www.foursquare.org; Pres., Dr. Paul C. Risser

Islamic Society of North America, P.O. Box 38, Plainfield, IN 46168; http://www.isna.net; Genl. Secy., Dr. Sayyid M. Syeed

Jehovah's Witnesses, 25 Columbia Heights, Brooklyn, NY 11201; Pres., Milton G. Henschel

Jewish Reconstructionist Federation (1922), Beit Devora, 7804 Montgomery Ave., Suite 9, Elkins Park, PA 19027; http://www.jrf.org; Exec. Vice Pres. Mark Seal

Lutheran Church—Missouri Synod (1847), 1333 S. Kirkwood Rd., St. Louis, MO 63122; http://www.lcms.org; Pres., Dr. A. L. Barry

Mennonite Brethren Churches, General Conference of (1860), 4812 E. Butler Ave., Fresno CA 93727; Moderator, Ed Boschman

Mennonite Church (1893), 421 S. Second St., Ste. 600, Elkhart, IN 46516; http://www.mennonites.org; Moderator, Ervin Stutzman

Mennonite Church, The General Conference (1860), 722 Main, P.O. Box 347, Newton, KS 67114; http://www2.southwind.net/~gcmc; Moderator, Lee Snyder

Moravian Church in America (1735), **Northern Prov.:** 1021 Center St., PO Box 1245, Bethlehem, PA 18016; http://www.moravian.org; Pres., Rev. R. Burke Johnson; **Southern Prov.:** 459 S. Church St., Winston-Salem, NC 27101; Pres., Rev. Dr. Robert E. Sawyer; **Alaska Prov.**: PO Box 545, Bethel, AK 99559; Pres., Rev. Frank Chingliak

National Baptist Convention, Inc. Progressive (1961), 601 50th St., NE, Washington, DC 20019; http://www.pribc.org; Pres., Dr. Bennett W. Smith Sr.

Orthodox Church in America (1794), PO Box 675, Syosset, NY 11791; http://www.oca.org; Primate, Most Blessed Theodosius

Orthodox Jewish Congregations in America, Union of, 333 17th Ave., New York, NY 10001; http://www.ou.org; Pres., Mandell I. Ganchrow, M.D.

Pentecostal Assemblies of the World, Inc., 3939 Meadows Dr., Indianapolis, IN 46205; Presiding Bishop, Norman L. Wagner

Presbyterian Church (U.S.A.), (1983), 100 Witherspoon St., Louisville, KY 40202; http://www.pcusa.org; Moderator, Syngman Rhee

Presbyterian Church in America (1973), 1852 Century Pl., Atlanta, GA 30345; http://www.pcanet.org; Moderator, Rev. Kennedy Smartt

Reformed Church in America (1628), 475 Riverside Dr., New York, NY 10115; http://www.rca.org; Pres., Carol Mutch

Reorganized Church of Jesus Christ of Latter-day Saints (1830), PO Box 1059, Independence, MO 64051; Pres. W. Grant McMurray

Roman Catholic Church (1634), National Conference of Catholic Bishops, 3211 Fourth St., Washington, DC 20017; Pres., Bishop Joseph A. Fiorenza

Romanian Orthodox Episcopate of America (1929), PO Box 309, Grass Lake, MI 49240; http://www.roea.org; Ruling Bishop, His Grace Bishop Nathaniel Popp

Salvation Army (1865), 615 Slaters Lane, Alexandria, VA 22313; National Comdr., Commissioner John A. Busby

Seventh-Day Adventist Church (1863), 12501 Old Columbia Pike, Silver Spring, MD 20904; Pres., Jan Paulsen

Swedenborgian Church (1792), 11 Highland Ave., Newtonville, MA 02460; http://www.swedenborg.org; Pres., Rev. Ronald P. Brugler

Unitarian Universalist Association of Congregations (1961), 25 Beacon St., Boston, MA 02108; http://www.uua.org; Pres., The Rev. Dr. John A. Buehrens

United Church of Christ (1957), 700 Prospect Ave., Cleveland, OH 44115; http://www.ucc.org; Pres., Rev. John H. Thomas

United Methodist Church (1968), 1204 Freedom Rd., Cranberry Twp., PA 16066; http://www.umc.org; Pres. Council of Bishops, Bishop William B. Oden

United Pentecostal Church Intl. (1925), 8855 Dunn Rd., Hazelwood, MO 63042; http://www.upcimain@aol.com; Gen. Superintendent, Rev. Nathaniel A. Urshan

Volunteers of America (1896), 110 S. Union St., Alexandria, VA 22314; Chairperson, Walt Patterson

Wesleyan Church (1968), PO Box 50434, Indianapolis, IN 46250; http://www.wesleyan.org; Gen. Supts., Dr. Earle L. Wilson, Dr. David H. Holdren, Dr. Thomas E. Armiger

Membership of Religious Groups in Canada

Sources: *2001 Yearbook of American & Canadian Churches*, © National Council of the Churches of Christ in the USA; *World Almanac* research

Figures are generally based on reports by officials of each group. The numbers are generally inclusive and not restricted to full communicants or the like. Specific definitions of "member" may vary, however. Some groups keep careful records; others only estimate. Not all groups report annually. The number of houses of worship appears in parentheses. *Indicates membership figures were not reported. Groups reporting fewer than 5,000 members are not included. Where membership numbers are not available, only groups with 50 or more houses of worship are listed.

Religious Group	Members
Anglican Church of Canada (2,957)	739,699
Antiochian Orthodox Christian Archdiocese of North America (215)	350,000
Apostolic Church of Pentecost of Canada, Inc. (153)	16,000
Armenian Holy Apostolic Church (Canadian Diocese) (15)	80,000
Associated Gospel Churches (not reported)	10,076
Bahá'í Faith (1,480)	28,500
Baptist Conference, North American (124)	17,531
Baptist Convention of Ontario and Quebec (386)	57,800
Baptist Ministries, Canadian (1,133)	129,055
Baptist Union of Western Canada (155)	20,427
Christian and Missionary Alliance in Canada (376)	87,197
Christian Brethren (also known as Plymouth Brethren) (600)	50,000
Christian Reformed Church in North America (237)	80,557
Church of God (Cleveland, TN) (130)	10,914
Church of Jesus Christ of Latter-day Saints in Canada (433)	151,000
Church of the Nazarene Canada (166)	12,042
Churches of Christ in Canada (140)	8,000
Estonian Evangelical Lutheran Church (11)	5,089
Evangelical Baptist Churches in Canada, Fellowship of (506)	*
Evangelical Christian Churches, Canadian (30)	10,000
Evangelical Free Church of Canada (147)	24,054
Evangelical Lutheran Church in Canada (631)	193,915
Evangelical Mennonite Conference of Canada (53)	7,000
Evangelical Missionary Church of Canada (145)	12,217
Free Methodist Church in Canada (128)	11,396
Greek Orthodox Metropolis of Toronto (Canada) (76)	350,000
Hindu	90,000[1]
Independent Assemblies of God Intl. (Canada) (214)	*
Islam	150,000[1]
Jehovah's Witnesses (1,383)	184,787
Jewish congregations (270+)	70,0001
Lutheran Church–Canada (329)	79,844
Mennonite Brethren Churches, Canadian Conference of (214)	33,214
Mennonite Church (Canada) (117)	8,172
Mennonites in Canada, Conference of (223)	35,995
Open Bible Faith Fellowship of Canada (68)	*
Orthodox Church in America (Canada Section) (606)	1,000,000
Pentecostal Assemblies of Canada (1,100)	218,782
Pentecostal Assemblies of Newfoundland (140)	29,361
Presbyterian Church in Canada (1,100)	200,738
Reformed Church in Canada (44)	6,609
Reformed Churches, Canadian and American (49)	15,213
Reorganized Church of Jesus Christ of Latter Day Saints (75)	11,264
Roman Catholic Church in Canada (5,716)	12,498,605
Salvation Army in Canada (376)	80,180
Serbian Orthodox Church in the U.S.A. and Canada, Diocese of Canada (23)	230,000
Seventh-Day Adventist Church in Canada (324)	48,900
Southern Baptists, Canadian Convention of (152)	9,626
United Baptist Convention of the Atlantic Provinces (553)	62,784
United Church of Canada (3,764)	1,589,886
United Pentecostal Church in Canada (199)	*
Wesleyan Church of Canada (75)	5,992

(1) Estimate; figures from other sources may vary.

Headquarters of Selected Religious Groups in Canada

Sources: *2001 Yearbook of American & Canadian Churches*, © National Council of the Churches of Christ in the USA; *World Almanac* research

(Year organized in parentheses)

Anglican Church of Canada (1700), Church House, 600 Jarvis St., Toronto, ON M4Y 2J6; http://www.anglican.ca; Primate, Most Rev. Michael G. Peers

Bahá'í National Centre of Canada, 7200 Leslie St., Thornhill, ON L3T 6L8; Gen'l.-Secy., Judy Filson

Baptist Ministries, Canadian, 7185 Millcreek Dr., Mississauga, ON L5N 5R4; http://www.cbmin.org; Pres., Doug Coomas

Christian and Missionary Alliance in Canada (1887), Box 7900, Str. B, Willowdale, ON M2K 2R6; http://www.cmacan.org; Pres., Dr. Arnold Cook

Church of Jesus Christ of Latter-day Saints (Mormon), The (1830), 50 E. North Temple St., Salt Lake City, UT 84150

Church of the Nazarene in Canada (1902), 20 Regan Rd. Unit 9, Brampton, ON L7A 1C3; http://web.1-888.com.nazarene/national; Natl. Dir., Dr. William E. Stewart

Evangelical Baptist Churches in Canada, Fellowship of (1953), 679 Southgate Dr., Guelph, ON N1G 4S2; Pres., Rev. Terry D. Cuthbert

Evangelical Lutheran Church in Canada (1985), 302-393 Portage Ave., Winnipeg, MB R3B 3H6; http://www.elcic.ca; Bishop, Rev. Telmor G. Sartison

Evangelical Missionary Church of Canada (1993), 4031 Brentwood Rd., NW, Calgary, AB T2L 1L1; Pres., Rev. Mark Bolender

Greek Orthodox Metropolis of Toronto, 86 Overlea Blvd., Toronto, ON M4H 1C6; http://www.gocanada.org; His Eminence Metropolitan Archbishop Sotirios

Jehovah's Witnesses (1879), Canadian office: Box 4100, Halton Hills, ON L7G 4Y4; Pres., Milton G. Henschel

Jewish Congress, Canadian (1919), 100 Sparks St., Ste. 650, OHawa, Ont. K1P 5B7; http://www.cjc.ca; Pres., Moshe Ronen (Nonreligious umbrella organization of Jewish groups)

Lutheran Church—Canada (1959), 3074 Portage Ave., Winnipeg, MB R3K OY2; Pres., Rev. Ralph Mayan

Mennonite Church (1898), 421 S. Second St., Ste. 600, Elkhart, IN 46516; Mod., Dwight McFadden Jr.

Muslim Communities in Canada, Council of, 1250 Ramsey View Court, Suite. 504, Sudbury, ON P3E 2E7; Director, Mir Iqbal Ali

North American Shi'a Muslim Communities Organization (NASIMCO), 300 John St., PO Box 87629, Dawnhill, ON L3T 7R3; Pres. Ghulamabbas Sajan

Pentecostal Assemblies of Canada (1919), 6745 Century Ave., Mississauga, ON L5N 6P7; http://www.paoc.org; Gen. Supt., Rev. William D. Morrow

Presbyterian Church in Canada (1925), 50 Wynford Dr., North York, ON M3C 1J7; http://www.presbycan.cal; Principal Clerk: Rev. Stephen Kendall

Roman Catholic Church (1618), Canadian Conference of Catholic Bishops, 90 Parent Ave., Ottawa, ON K1N 7B1; http://www.cccb.ca; Pres., Jean-Claude Cardinal Turcotte

Salvation Army (1909), 2 Overlea Blvd., Toronto, ON M4H 1P4; http://www.sallynet.org; Territorial Cmdr., Commissioner Norman Howe

Seventh-Day Adventist Church (1901), 1148 King St. E., Oshawa, ON L1H 1H8; Pres., Orville Parchment

Ukrainian Orthodox Church (1918), Office of the Consistory, 9 St. John's Ave., Winnipeg, MB R2W 1G8; http://www.uocc.ca; Primate, Most Rev. Metropolitan Wasyly Fedak

United Brethren Church (1767) 302 Lake St., Huntington, IN 46750; Pres., Rev. Brian Magnus

United Church of Canada (1925), The United Church House, 3250 Bloor St. W., Ste. 300, Etobicoke, ON M8X 2Y4; http://www.uccan.org; Mod., Marion Pardy

Wesleyan Church (1968), The Wesleyan Church Intl. Center, PO Box 50434, Indianapolis, IN 46250; Dist. Supt., Rev. Donald E. Hodgins

Adherents of All Religions by Six Continental Areas, Mid-2000

Source: *2001 Encyclopædia Britannica Book of the Year*

	Africa	Asia	Europe	Latin America	Northern America	Oceania	World
Baha'is	1,733,000	3,475,000	130,000	873,000	786,000	110,000	7,107,000
Buddhists	134,000	354,651,000	1,547,000	647,000	2,701,000	301,000	359,981,000
Chinese folk religionists	32,000	383,408,000	255,000	194,000	854,000	64,000	384,807,000
Christians	360,232,000	312,849,000	559,643,000	481,102,000	260,624,000	25,110,000	1,999,560,000
Anglicans	42,542,000	727,000	26,637,000	1,090,000	3,244,000	5,409,000	79,649,000
Orthodox	35,304,000	14,113,000	158,105,000	558,000	6,342,000	706,000	215,128,000
Protestants	89,001,000	49,967,000	77,529,000	48,132,000	69,978,000	7,392,000	341,999,000
Roman Catholics	120,386,000	110,480,000	285,978,000	461,220,000	71,035,000	8,228,000	1,057,327,000
Confucianists	0	6,264,000	11,000	0	0	24,000	6,299,000
Ethnic religionists	96,805,000	128,298,000	1,263,000	1,288,000	444,000	268,000	228,366,000
Hindus	2,351,000	805,120,000	1,416,000	768,000	1,327,000	355,000	811,337,000
Jains	66,000	4,145,000	0	0	7,000	0	4,218,000
Jews	214,000	4,429,000	2,527,000	1,142,000	6,024,000	97,000	14,433,000
Mandeans	0	39,000	0	0	0	0	39,000
Muslims	317,374,000	832,879,000	31,566,000	1,672,000	4,450,000	301,000	1,188,242,000
New-Religionists	28,000	100,639,000	158,000	623,000	842,000	66,000	102,356,000
Shintoists	0	2,699,000	0	7,000	56,000	0	2,762,000
Sikhs	53,000	22,421,000	239,000	0	528,000	18,000	23,259,000
Spiritists	3,000	2,000	133,000	12,039,000	151,000	7,000	12,335,000
Zoroastrians	1,000	2,463,000	1,000	0	78,000	1,000	2,544,000
Other religionists	5,092,000	611,262,000	107,076,000	16,026,000	29,079,000	3,306,000	771,841,000
Nonreligious	5,024,000	608,594,000	106,841,000	15,928,000	28,473,000	3,298,000	768,158,000
Atheists	420,000	121,945,000	23,922,000	2,757,000	1,680,000	365,000	150,089,000

Adherents. As defined in the 1948 Universal Declaration of Human Rights, a person's religion is what he or she says it is. Totals are enumerated following the methodology of the *World Christian Encyclopedia*, 2nd ed. (2000), using recent censuses, polls, literature, and other data. As a result of the varieties of sources used, totals may differ from standard estimates for total populations.

Buddhists. 56% Mahayana, 38% Theravada (Hinayana), 6% Tantrayana (Lamaism).

Chinese folk religionists. Followers of traditional Chinese religion (local deities, ancestor veneration, Confucian ethics, Taoism, universism, divination, some Buddhist elements).

Christians. Total Christians include those affiliated with churches not shown, plus other persons professing in censuses or polls to be Christians but not affiliated with any church. Figures for the subgroups of Christians do not add up to the totals because all subgroups are not shown and some Christians adhere to more than one denomination.

Confucians. Non-Chinese followers of Confucius and Confucianism, mostly Koreans in Korea.

Ethnic religionists. Followers of local, tribal, animistic, or shamanistic religions.

Hindus. 70% Vaishnavites, 25% Shaivites, 2% neo-Hindus and reform Hindus.

Jews. Adherents of Judaism.

Muslims. 83% Sunni Muslims, 16% Shia Muslims (Shi'ites), 1% other schools.

New-Religionists. Followers of Asian 20th-cent. New Religions, New Religious movements, radical new crisis religions, and non-Christian syncretistic mass religions, all founded since 1800 and most since 1945.

Other religionists. Including 70 minor world religions and more than 10,000 national or local religions and a large number of spiritist religions, New Age religions, quasi-religions, pseudo religions, parareligions, religious or mystic systems, and religious and semireligious brotherhoods of numerous varieties.

Nonreligious. Persons professing no religion, nonbelievers, agnostics, freethinkers, dereligionized secularists indifferent to all religion.

Atheists. Persons professing atheism, skepticism, disbelief, or irreligion, including antireligious (opposed to all religion).

Episcopal Church Liturgical Colors and Calendar

Source: Church Publishing Incorporated, New York

The liturgical colors in the Episcopal Church are as follows: **White**—from Christmas Day through the First Sunday after Epiphany; Maundy Thursday (as an alternative to crimson at the Eucharist); from the Vigil of Easter to the Day of Pentecost (Whitsunday); Trinity Sunday; Feasts of the Lord (except Holy Cross Day); the Confession of St. Peter; the Conversion of St. Paul; St. Joseph; St. Mary Magdalene; St. Mary the Virgin; St. Michael and All Angels; All Saints' Day; St. John the Evangelist; memorials of other saints who were not martyred; Independence Day and Thanksgiving Day; weddings and funerals. **Red**—the Day of Pentecost; Holy Cross Day; feasts of apostles and evangelists (except those listed above); feasts and memorials of martyrs (including Holy Innocents' Day). **Violet**—Advent and Lent. **Crimson** or oxblood (dark red)—Holy Week. **Green**—the seasons after Epiphany and after Pentecost. **Black**—optional alternative for funerals and Good Friday. Alternative colors used in some churches: **Blue**—Advent; **Lenten White (unbleached linen)**—Ash Wednesday to Palm Sunday.

In the Episcopal Church the days of fasting are Ash Wednesday and Good Friday. Other days of special devotion (penitence) are the 40 days of Lent and all Fridays of the year, except those in Christmas and Easter seasons and any Feasts of the Lord that occur on a Friday or during Lent. Ember Days (optional) are days of prayer for the church's ministry. They fall on the Wednesday, Friday, and Saturday after the first Sunday in Lent, the Day of Pentecost, Holy Cross Day, and the Third Sunday of Advent. Rogation Days (also optional), the 3 days before Ascension Day, are days of prayer for God's blessing on the crops, on commerce and industry, and for conservation of the earth's resources.

Days, etc.	2001	2002	2003	2004	2005
Golden Number	7	8	9	10	11
Sunday Letter	G	F	E	D & C	B
Sundays after Epiphany	8	5	8	7	5
Ash Wednesday	Feb. 28	Feb. 13	Mar. 5	Feb. 25	Feb. 9
First Sunday in Lent	Mar. 4	Feb. 17	Mar. 9	Feb. 29	Feb. 13
Passion/Palm Sunday	Apr. 8	Mar. 24	Apr. 13	Apr. 4	Mar. 20
Good Friday	Apr. 13	Mar. 29	Apr. 18	Apr. 9	Mar. 25
Easter Day	Apr. 15	Mar. 31	Apr. 20	Apr. 11	Mar. 27
Ascension Day	May 24	May 9	May 29	May 20	May 5
The Day of Pentecost	June 3	May 19	June 8	May 30	May 15
Trinity Sunday	June 10	May 26	June 15	June 6	May 22
Numbered Proper of 2 Pentecost	#6	#4	#7	#6	#4
First Sunday of Advent	Dec. 2	Dec. 1	Nov. 30	Nov. 28	Nov. 27

Greek Orthodox Movable Ecclesiastical Dates, 2001-2005

This 5-year chart has the dates of feast days and fasting days, which are determined annually on the basis of the date of Holy Pascha (Easter). This ecclesiastical cycle begins with the first day of the Triodion and ends with the Sunday of All Saints, a total of 18 weeks.

	2001	2002	2003	2004	2005
Triodion begins	Feb. 4	Feb. 24	Feb. 16	Feb. 1	Feb. 21
Sat. of Souls	Feb. 17	Mar. 9	Mar. 1	Feb. 14	Mar. 5
Meat Fare	Feb. 18	Mar. 10	Mar. 2	Feb. 15	Mar. 6
2d Sat. of Souls	Feb. 24	Mar. 16	Mar. 8	Feb. 21	Mar. 12
Lent Begins	Feb. 26	Mar. 18	Mar. 10	Feb. 23	Mar. 14
St. Theodore—3d Sat. of Souls	Mar. 3	Mar. 23	Mar. 15	Feb. 28	Mar. 19
Sunday of Orthodoxy	Mar. 4	Mar. 24	Mar. 16	Feb. 29	Mar. 20
Sat. of Lazarus	Apr. 7	Apr. 27	Apr. 19	Apr. 3	Apr. 23
Palm Sunday	Apr. 8	Apr. 28	Apr. 20	Apr. 4	Apr. 24
Holy (Good) Friday	Apr. 13	May 3	Apr. 25	Apr. 9	Apr. 29
Western Easter	Apr. 15	Mar. 31	Apr. 20	Apr. 11	May 1
Orthodox Easter	Apr. 15	May 5	Apr. 27	Apr. 11	Mar. 27
Ascension	May 24	June 13	June 5	May 20	June 9
Sat. of Souls	June 2	June 22	June 14	May 29	June 18
Pentecost	June 3	June 23	June 15	May 30	June 19
All Saints	June 10	June 30	June 22	June 7	June 26

Important Islamic Dates, 1422-26 (2001-2006)

Source: Imad-ad-Dean, Inc., Bethesda, MD 20814

The Islamic calendar is a strict lunar calendar reckoned from the year of the Hijra (Muhammad's flight from Mecca to Medina). Each year consists of 12 lunar months of 29 or 30 days beginning and ending with each new moon's visible crescent. Common years have 354 days; leap years have 355 days. Some Muslim countries employ a conventionalized calendar with the leap day added to the last month, Dhûl Hijah, but for religious purposes the leap date is taken into account by tracking each new moon sighting. The dates given below are based on the convention that the first new moon must be seen before the following dawn on the East Coast of the Americas. Actual (local) Western Hemisphere sightings may occur a day later, but never a day earlier, than these dates reflect.

	(1422) 2001-02	(1423) 2002-03	(1424) 2003-04	(1425) 2004-05	(1426) 2005-06
New Year's Day (Muharram1)	Mar. 26, 2001	Mar. 15, 2002	Mar. 4, 2003	Feb. 21, 2004	Feb. 10, 2005
Ashura (Muharram 10)	Apr. 4, 2001	Mar. 24, 2002	Mar. 13, 2003	Mar. 1, 2004	Feb. 19, 2005
Mawlid (Rabi'l 12)	June 4, 2001	May 24, 2002	May 13, 2003	May 1, 2004	April 21, 2005
Ramadan 1	Nov. 16, 2001	Nov. 6, 2002	Oct. 26, 2003	Oct. 15, 2004	Oct. 4, 2005
Eid al-Fitr (Shawwal)	Dec. 16, 2001	Dec. 5, 2002	Nov. 25, 2003	Nov. 13, 2004	Nov. 3, 2005
Eid al-Adha (Dhûl-Hijjah 10)	Feb. 22, 2002	Feb. 11, 2003	Feb. 1, 2004	Jan. 20, 2005	Jan. 10, 2006

IT'S A FACT: The Hindu holy day begins at sunrise, the Jewish holy day at sunset, and the Christian holy day at midnight.

Jewish Holy Days, Festivals, and Fasts 5761-5766 (2001-2005)

	(5761-62) 2001		(5762-63) 2002		(5763-64) 2003		(5764-65) 2004		(5765-66) 2005	
Tu B'Shvat	Feb. 8	Thu.	Jan. 28	Mon.	Jan. 18	Sat.	Feb. 7	Sat.	Jan. 25	Tue.
Ta'anis Esther (Fast of Esther)	Mar. 8	Thu.	Feb. 25	Mon.	Mar. 17	Mon.	Mar. 4	Thu.*	Mar. 24	Thu.
Purim	Mar. 9	Fri.	Feb. 26	Tue.	Mar. 18	Tue.	Mar. 7	Sun.	Mar. 25	Fri.
Pesach (Passover)	Apr. 8	Sun.	Mar. 28	Thu.	Apr. 17	Thu.	Apr. 6	Tue.	Apr. 24	Sun.
	Apr. 15	Sun.	Apr. 4	Thu.	Apr. 24	Thu.	Apr. 13	Tue.	May 1	Sun.
Lag B'Omer	May 11	Fri.	Apr. 30	Tue.	May. 20	Tue.	May 9	Sun.	May 27	Fri.
Shavuot (Pentecost)	May 28	Mon.	May 17	Fri.	June 6	Fri.	May 26	Wed.	June 13	Mon.
	May 29	Tue.	May 18	Sat.	June 7	Sat.	May 27	Thu.	June 14	Tue.
Fast of the 17th Day of Tammuz	July 8	Sun.	June 27	Thu.	July 17	Thu.	July 6	Tue.	July 24	Sun.
Fast of the 9th Day of Av	July 29	Sun.	July 18	Thu.	Aug. 7	Thu.	July 27	Tue.	Aug. 14	Sun.
Rosh Hashanah (Jewish New Year)	Sept. 18	Tue.	Sept. 7	Sat.	Sept. 27	Sat.	Sept. 16	Thu.	Oct. 4	Tue.
	Sept. 19	Wed.	Sept. 8	Sun.	Sept. 28	Sun.	Sept. 17	Fri.	Oct. 5	Wed.
Fast of Gedalya	Sept. 20	Thu.	Sept. 9	Mon.	Sept. 29	Mon.	Sept. 19	Sun.*	Oct. 6	Thu.
Yom Kippur (Day of Atonement)	Sept. 27	Thu.	Sept. 16	Mon.	Oct. 6	Mon.	Sept. 25	Sat.	Oct. 13	Thu.
Sukkot	Oct. 2	Tue.	Sept. 21	Sat.	Oct. 11	Sat.	Sept. 30	Thu.	Oct. 18	Tue.
	Oct. 8	Mon.	Sept. 27	Fri.	Oct. 17	Fri.	Oct. 1	Wed.	Oct. 24	Mon.
Shmini Atzeret	Oct. 9	Tue.	Sept. 28	Sat.	Oct. 18	Sat.	Oct. 7	Thu.	Oct. 25	Tue.
	Oct. 10	Wed.	Sept. 29	Sun.	Oct. 19	Sun.	Oct. 8	Fri.	Oct. 26	Wed.
Hanukkah	Dec. 10	Mon.	Nov. 30	Sat.	Dec. 20	Sat.	Dec. 8	Wed.	Dec. 26	Mon.
	Dec. 17	Mon.	Dec. 7	Sat.	Dec. 27	Sat.	Dec. 15	Wed.	Jan. 2, 2006	Mon.
Fast of the 10th of Tevet	Dec. 25	Tue.	Jan. 15, 2003	Sun.	Jan. 4, 2004	Sun.	Dec. 22	Wed.	Jan. 10, 2006	Tue.

The months of the Jewish year are: 1) Tishri; 2) Cheshvan (also Marcheshvan); 3) Kislev; 4) Tevet (also Tebeth); 5) Shebat (also Shebhat); 6) Adar; 6a) Adar Sheni (II) added in leap years; 7) Nisan; 8) Iyar; 9) Sivan; 10) Tammuz; 11) Av (also Abh); 12) Elul. All Jewish holy days, etc., begin at sunset on the previous day. *Date changed to avoid Sabbath.

Ash Wednesday and Easter Sunday (Western churches), 1901-2100

Year	Ash Wed.	Easter Sunday	Year	Ash Wed.	Easter Sunday	Year	Ash Wed.	Easter Sunday	Year	Ash Wed.	Easter Sunday
1901	Feb. 20	Apr. 7	1952	Feb. 27	Apr. 13	2003	Mar. 5	Apr. 20	2053	Feb. 19	Apr. 6
1902	Feb. 12	Mar. 30	1953	Feb. 18	Apr. 5	2004	Feb. 25	Apr. 11	2054	Feb. 11	Mar. 29
1903	Feb. 25	Apr. 12	1954	Mar. 3	Apr. 18	2005	Feb. 9	Mar. 27	2055	Mar. 3	Apr. 18
1904	Feb. 17	Apr. 3	1955	Feb. 23	Apr. 10	2006	Mar. 1	Apr. 16	2056	Feb. 16	Apr. 2
1905	Mar. 8	Apr. 23	1956	Feb. 15	Apr. 1	2007	Feb. 21	Apr. 8	2057	Mar. 7	Apr. 22
1906	Feb. 28	Apr. 15	1957	Mar. 6	Apr. 21	2008	Feb. 6	Mar. 23	2058	Feb. 27	Apr. 14
1907	Feb. 13	Mar. 31	1958	Feb. 19	Apr. 6	2009	Feb. 25	Apr. 12	2059	Feb. 12	Mar. 30
1908	Mar. 4	Apr. 19	1959	Feb. 11	Mar. 29	2010	Feb. 17	Apr. 4	2060	Mar. 3	Apr. 18
1909	Feb. 24	Apr. 11	1960	Mar. 2	Apr. 17	2011	Mar. 9	Apr. 24	2061	Feb. 23	Apr. 10
1910	Feb. 9	Mar. 27	1961	Feb. 15	Apr. 2	2012	Feb. 22	Apr. 8	2062	Feb. 8	Mar. 26
1911	Mar. 1	Apr. 16	1962	Mar. 7	Apr. 22	2013	Feb. 13	Mar. 31	2063	Feb. 28	Apr. 15
1912	Feb. 21	Apr. 7	1963	Feb. 27	Apr. 14	2014	Mar. 5	Apr. 20	2064	Feb. 20	Apr. 6
1913	Feb. 5	Mar. 23	1964	Feb. 12	Mar. 29	2015	Feb. 18	Apr. 5	2065	Feb. 11	Mar. 29
1914	Feb. 25	Apr. 12	1965	Mar. 3	Apr. 18	2016	Feb. 10	Mar. 27	2066	Feb. 24	Apr. 11
1915	Feb. 17	Apr. 4	1966	Feb. 23	Apr. 10	2017	Mar. 1	Apr. 16	2067	Feb. 16	Apr. 3
1916	Mar. 8	Apr. 23	1967	Feb. 8	Mar. 26	2018	Feb. 14	Apr. 1	2068	Mar. 7	Apr. 22
1917	Feb. 21	Apr. 8	1968	Feb. 28	Apr. 14	2019	Mar. 6	Apr. 21	2069	Feb. 27	Apr. 14
1918	Feb. 13	Mar. 31	1969	Feb. 19	Apr. 6	2020	Feb. 26	Apr. 12	2070	Feb. 12	Mar. 30
1919	Mar. 5	Apr. 20	1970	Feb. 11	Mar. 29	2021	Feb. 17	Apr. 4	2071	Mar. 4	Apr. 19
1920	Feb. 18	Apr. 4	1971	Feb. 24	Apr. 11	2022	Mar. 2	Apr. 17	2072	Feb. 24	Apr. 10
1921	Feb. 9	Mar. 27	1972	Feb. 16	Apr. 2	2023	Feb. 22	Apr. 9	2073	Feb. 8	Mar. 26
1922	Mar. 1	Apr. 16	1973	Mar. 7	Apr. 22	2024	Feb. 14	Mar. 31	2074	Feb. 28	Apr. 15
1923	Feb. 14	Apr. 1	1974	Feb. 27	Apr. 14	2025	Mar. 5	Apr. 20	2075	Feb. 20	Apr. 7
1924	Mar. 5	Apr. 20	1975	Feb. 12	Mar. 30	2026	Feb. 18	Apr. 5	2076	Mar. 4	Apr. 19
1925	Feb. 25	Apr. 12	1976	Mar. 3	Apr. 18	2027	Feb. 10	Mar. 28	2077	Feb. 24	Apr. 11
1926	Feb. 17	Apr. 4	1977	Feb. 23	Apr. 10	2028	Mar. 1	Apr. 16	2078	Feb. 16	Apr. 3
1927	Mar. 2	Apr. 17	1978	Feb. 8	Mar. 26	2028	Mar. 1	Apr. 16	2079	Mar. 8	Apr. 23
1928	Feb. 22	Apr. 8	1979	Feb. 28	Apr. 15	2029	Feb. 14	Apr. 1	2080	Feb. 21	Apr. 7
1929	Feb. 13	Mar. 31	1980	Feb. 20	Apr. 6	2030	Mar. 6	Apr. 21	2081	Feb. 12	Mar. 30
1930	Mar. 5	Apr. 20	1981	Mar. 4	Apr. 19	2031	Feb. 26	Apr. 13	2082	Mar. 4	Apr. 19
1931	Feb. 18	Apr. 5	1982	Feb. 24	Apr. 11	2032	Feb. 11	Mar. 28	2083	Feb. 17	Apr. 4
1932	Feb. 10	Mar. 27	1983	Feb. 16	Apr. 3	2033	Mar. 2	Apr. 17	2084	Feb. 9	Mar. 26
1933	Mar. 1	Apr. 16	1984	Mar. 7	Apr. 22	2034	Feb. 22	Apr. 9	2085	Feb. 28	Apr. 15
1934	Feb. 14	Apr. 1	1985	Feb. 20	Apr. 7	2035	Feb. 7	Mar. 25	2086	Feb. 13	Mar. 31
1935	Mar. 6	Apr. 21	1986	Feb. 12	Mar. 30	2036	Feb. 27	Apr. 13	2087	Mar. 5	Apr. 20
1936	Feb. 26	Apr. 12	1987	Mar. 4	Apr. 19	2037	Feb. 18	Apr. 5	2088	Feb. 25	Apr. 11
1937	Feb. 10	Mar. 28	1988	Feb. 17	Apr. 3	2038	Mar. 10	Apr. 25	2089	Feb. 16	Apr. 3
1938	Mar. 2	Apr. 17	1989	Feb. 8	Mar. 26	2039	Feb. 23	Apr. 10	2090	Mar. 1	Apr. 16
1939	Feb. 22	Apr. 9	1990	Feb. 28	Apr. 15	2040	Feb. 15	Apr. 1	2091	Feb. 21	Apr. 8
1940	Feb. 7	Mar. 24	1991	Feb. 13	Mar. 31	2041	Mar. 6	Apr. 21	2092	Feb. 13	Mar. 30
1941	Feb. 26	Apr. 13	1992	Mar. 4	Apr. 19	2042	Feb. 19	Apr. 6	2093	Feb. 25	Apr. 12
1942	Feb. 18	Apr. 5	1993	Feb. 24	Apr. 11	2043	Feb. 11	Mar. 29	2094	Feb. 17	Apr. 4
1943	Mar. 10	Apr. 25	1994	Feb. 16	Apr. 3	2044	Mar. 2	Apr. 17	2095	Mar. 9	Apr. 24
1944	Feb. 23	Apr. 9	1995	Mar. 1	Apr. 16	2045	Feb. 22	Apr. 9	2096	Feb. 29	Apr. 15
1945	Feb. 14	Apr. 1	1996	Feb. 21	Apr. 7	2046	Feb. 7	Mar. 25	2097	Feb. 13	Mar. 31
1946	Mar. 6	Apr. 21	1997	Feb. 12	Mar. 30	2047	Feb. 27	Apr. 14	2098	Mar. 5	Apr. 20
1947	Feb. 19	Apr. 6	1998	Feb. 25	Apr. 12	2048	Feb. 19	Apr. 5	2099	Feb. 25	Apr. 12
1948	Feb. 11	Mar. 28	1999	Feb. 17	Apr. 4	2049	Mar. 3	Apr. 18	2100	Feb. 10	Mar. 28
1949	Mar. 2	Apr. 17	2000	Mar. 8	Apr. 23	2050	Feb. 23	Apr. 10			
1950	Feb. 22	Apr. 9	2001	Feb. 28	Apr. 15	2051	Feb. 15	Apr. 2			
1951	Feb. 7	Mar. 25	2002	Feb. 13	Mar. 31	2052	Mar. 6	Apr. 21			

The Ten Commandments

According to Judeo-Christian tradition, as related in the Bible, the Ten Commandments were revealed by God to Moses and form the basic moral component of God's covenant with Israel. The Ten Commandments appear in 2 places in the Old Testament—Exodus 20:1-17 and Deuteronomy 5:6-21.

Following is the text of the Ten Commandments as it appears in Exodus 20:1-17, in the King James version of the Bible.

I. I am the LORD thy God, which have brought thee out of the land of Egypt, out of the house of bondage. Thou shalt have no other gods before me.

II. Thou shalt not make unto thee any graven image, or any likeness of any thing that is in heaven above, or that is in the earth beneath, or that is in the water under the earth. Thou shalt not bow down thyself to them, nor serve them: for I the LORD thy God am a jealous God, visiting the iniquity of the fathers upon the children unto the third and fourth generation of them that hate me.

III. Thou shalt not take the name of the LORD thy God in vain; for the LORD will not hold him guiltless that taketh his name in vain.

IV. Remember the sabbath day, to keep it holy.

V. Honour thy father and thy mother: that thy days may be long upon the land which the LORD thy God giveth thee.

VI. Thou shalt not kill.

VII. Thou shalt not commit adultery.

VIII. Thou shalt not steal.

IX. Thou shalt not bear false witness against thy neighbour.

X. Thou shalt not covet thy neighbour's house, thou shalt not covet thy neighbour's wife, nor his manservant, nor his maidservant, nor his ox, nor his ass, nor any thing that is thy neighbour's.

Most Protestant, Anglican, and Orthodox Christians follow Jewish tradition, which considers the introduction ("I am the Lord . . .") the first commandment and makes the prohibition against idolatry the second. Roman Catholic and Lutheran traditions combine I and II and split the last commandment into 2 that separately prohibit coveting of a neighbor's wife and a neighbor's goods. This arrangement alters the numbering of the other commandments by one.

Books of the Bible

Old Testament—Standard Protestant List

Genesis	I Kings	Ecclesiastes	Obadiah
Exodus	II Kings	Song of Solomon	Jonah
Leviticus	I Chronicles	Isaiah	Micah
Numbers	II Chronicles	Jeremiah	Nahum
Deuteronomy	Ezra	Lamentations	Habakkuk
Joshua	Nehemiah	Ezekiel	Zephaniah
Judges	Esther	Daniel	Haggai
Ruth	Job	Hosea	Zechariah
I Samuel	Psalms	Joel	Malachi
II Samuel	Proverbs	Amos	

New Testament List

Matthew	Ephesians	Hebrews
Mark	Phillippians	James
Luke	Colossians	I Peter
John	I Thessalonians	II Peter
Acts	II Thessalonians	I John
Romans	I Timothy	II John
I Corinthians	II Timothy	III John
II Corinthians	Titus	Jude
Galatians	Philemon	Revelation

The standard Protestant Old Testament consists of the same 39 books as in the Bible of Judaism, but the latter is organized differently. The Old Testament used by Roman Catholics has 7 additional "deuterocanonical" books, plus some additional parts of books. The 7 are: **Tobit, Judith, Wisdom, Sirach (Ecclesiasticus), Baruch, I Maccabees,** and **II Maccabees.** Both Catholic and Protestant versions of the New Testament have 27 books, with the same names.

Roman Catholic Hierarchy

Source: U.S. Catholic Conference

Supreme Pontiff

At the head of the Roman Catholic Church is the supreme pontiff, Pope John Paul II, Karol Wojtyla, born at Wadowice (Kraków), Poland, May 18, 1920; ordained priest Nov. 1, 1946; appointed bishop July 4, 1958; named archbishop of Kraków Jan. 13, 1964; proclaimed cardinal June 26, 1967; elected pope Oct. 16, 1978; installed Oct. 22, 1978.

College of Cardinals

Members of the Sacred College of Cardinals are chosen by the pope to be his chief assistants and advisers in the administration of the church. Among their duties is the election of the pope when the Holy See becomes vacant.

In its present form, the College of Cardinals dates from the 12th century. The first cardinals, from about the 6th century, were deacons and priests of the leading churches of Rome and were bishops of neighboring dioceses. The title of cardinal was limited to members of the college in 1567. The number of cardinals was set at 70 in 1586 by Pope Sixtus V. From 1959 Pope John XXIII began to increase the number; however, the number eligible to participate in papal elections was limited to 120. In Feb. 2001, Pope John Paul II waived the limit on number of electors when he created 44 new cardinals. As of Aug. 2001, there were 179 members of the College, of whom 131 were electors. In 1918 the Code of Canon Law specified that all cardinals must be priests. Pope John XXIII in 1962 established that all cardinals must be bishops. In 1971 Pope Paul VI decreed that at age 80 cardinals must retire from curial departments and offices and from participation in papal elections.

North American Cardinals

Name	Office	Born	Named Cardinal
Aloysius M. Ambrozic	Archbishop of Toronto	1930	1998
William W. Baum	Major Penitentiary of Apostolic Penitentiary, the Vatican	1926	1976
Anthony J. Bevilacqua	Archbishop of Philadelphia	1923	1991
G. Emmett Carter[1]	Archbishop emeritus of Toronto	1912	1979
Ernesto Corripio Ahumada[1]	Archbishop emeritus of Mexico	1919	1979
Avery Robert Dulles[1]	Professor, Fordham University	1918	2001
Edward M. Egan	Archbishop of New York	1932	2001
Edouard Gagnon[1]	Pres. of Pontifical Commission of Intl. Eucharistic Congresses	1918	1985
Francis E. George	Archbishop of Chicago	1937	1998
James A. Hickey[1]	Archbishop emeritus of Washington, DC	1920	1988
William Henry Keeler	Archbishop of Baltimore	1931	1994
Bernard F. Law	Archbishop of Boston	1931	1985
Roger Mahony	Archbishop of Los Angeles	1936	1991
Adam Joseph Maida	Archbishop of Detroit	1930	1994
Theodore E. McCarrick	Archbishop of Washington, DC	1930	2001
Norberto Rivera Carrera	Archbishop of Mexico City	1942	1998
Juan Sandoval Iniguez	Archbishop of Guadalajara	1933	1994
James F. Stafford	President of the Pontifical Council for the Laity	1932	1998
Adolfo Antonio Suarez Rivera	Archbishop of Monterrey	1927	1994
Edmund C. Szoka	Pres. of Prefecture of Economic Affairs of Holy See, the Vatican	1927	1988
Jean-Claude Turcotte	Archbishop of Montreal	1936	1994
Louis-Albert Vachon[1]	Archbishop emeritus of Quebec	1912	1985

(1) Ineligible to take part in papal elections (as of Sept. 2001).

Chronological List of Popes

Source: Annuario Pontificio. Table lists year of accession of each pope.

The Roman Catholic Church named the Apostle Peter as founder of the church in Rome and the first pope. He arrived there c 42, was martyred there c 67, and was ultimately canonized as a saint. **The pope's temporal title is:** Sovereign of the State of Vatican City. **The pope's spiritual titles are:** Bishop of Rome, Vicar of Jesus Christ, Successor of St. Peter, Prince of the Apostles, Supreme Pontiff of the Universal Church, Patriarch of the West, Primate of Italy, Archbishop and Metropolitan of the Roman Province.

The names of antipopes are *in italics* and followed by an *. Antipopes were illegitimate claimants to the papal throne.

Year	Pope
	St. Peter
67	St. Linus
76	St. Anacletus or Cletus
88	St. Clement I
97	St. Evaristus
105	St. Alexander I
115	St. Sixtus I
125	St. Telesphorus
136	St. Hyginus
140	St. Pius I
155	St. Anicetus
166	St. Soter
175	St. Eleutherius
189	St. Victor I
199	St. Zephyrinus
217	St. Callistus I
217	_St. Hippolytus_*
222	St. Urban I
230	St. Pontian
235	St. Anterus
236	St. Fabian
251	St. Cornelius
251	_Novatian_*
253	St. Lucius I
254	St. Stephen I
257	St. Sixtus II
259	St. Dionysius
269	St. Felix I
275	St. Eutychian
283	St. Caius
296	St. Marcellinus
308	St. Marcellus I
309	St. Eusebius
311	St. Melchiades
314	St. Sylvester I
336	St. Marcus
337	St. Julius I
352	Liberius
355	_Felix II_*
366	St. Damasus I
366	_Ursinus_*
384	St. Siricius
399	St. Anastasius I
401	St. Innocent I
417	St. Zosimus
418	St. Boniface I
418	_Eulabus_*
422	St. Celestine I
432	St. Sixtus III
440	St. Leo I
461	St. Hilary
468	St. Simplicius
483	St. Felix III (II)
492	St. Gelasius I
496	Anastasius II
498	St. Symmachus
498	_Lawrence_* (501-505)
514	St. Hormisdas
523	St. John I, Martyr
526	St. Felix IV (III)
530	Boniface II
530	_Dioscorus_*
533	John II
535	St. Agapitus I
536	St. Silverius, Martyr
537	Vigilius
556	Pelagius I
561	John III
575	Benedict I
579	Pelagius II
590	St. Gregory I
604	Sabinian
607	Boniface III
608	St. Boniface IV
615	St. Deusdedit or Adeodatus
619	Boniface V
625	Honorius I
640	Severinus
640	John IV
642	Theodore I
649	St. Martin I, Martyr
654	St. Eugene I
657	St. Vitalian
672	Adeodatus II
676	Donus
678	St. Agatho
682	St. Leo II
684	St. Benedict II
685	John V
686	Conon
687	_Theodore_*
687	_Paschal_*
687	St. Sergius I
701	John VI
705	John VII
708	Sisinnius
708	Constantine
715	St. Gregory II
731	St. Gregory III
741	St. Zachary
752	Stephen II (III)
757	St. Paul I
767	_Constantine_*
768	_Philip_*
768	Stephen III (IV)
772	Adrian I
795	St. Leo III
816	Stephen IV (V)
817	St. Paschal I
824	Eugene II
827	Valentine
827	Gregory IV
844	_John_*
844	Sergius II
847	St. Leo IV
855	Benedict III
855	_Anastasius_*
858	St. Nicholas I
867	Adrian II
872	John VIII
882	Marinus I
884	St. Adrian III
885	Stephen V (VI)
891	Formosus
896	Boniface VI
896	Stephen VI (VII)
897	Romanus
897	Theodore II
898	John IX
900	Benedict IV
903	Leo V
903	_Christopher_*
904	Sergius III
911	Anastasius III
913	Landus
914	John X
928	Leo VI
928	Stephen VII(VIII)
931	John XI
936	Leo VII
939	Stephen VIII(IX)
942	Marinus II
946	Agapitus II
955	John XII
963	Leo VIII
964	Benedict V
965	John XIII
973	Benedict VI
974	_Boniface VII_*
974	Benedict VII
983	John XIV
985	John XV
996	Gregory V
997	_John XVI_*
999	Sylvester II
1003	John VII
1004	John XVIII
1009	Sergius IV
1012	Benedict VIII
1012	_Gregory_*
1024	John XIX
1032	Benedict IX
1045	Sylvester III
1045	Benedict IX
1045	Gregory VI
1046	Clement II
1047	Benedict IX
1048	Damasus II
1049	St. Leo IX
1055	Victor II
1057	Stephen IX (X)
1058	_Benedict X_*
1059	Nicholas II
1061	Alexander II
1061	_Honorius II_*
1073	St. Gregory VII
1080	_Clement III_*
1086	Bl. Victor III
1088	Bl. Urban II
1099	Paschal II
1100	_Theodoric_*
1102	_Albert_*
1105	_Sylvester IV_*
1118	Gelasius II
1118	_Gregory VIII_*
1119	Callistus II
1124	Honorius II
1124	_Celestine II_*
1130	Innocent II
1130	_Anacletus II_*
1138	_Victor IV_*
1143	Celestine II
1144	Lucius II
1145	Bl. Eugene III
1153	Anastasius IV
1154	Adrian IV
1159	Alexander III
_1159_	_Victor IV_*
1164	_Paschal III_*
1168	_Callistus III_*
1179	_Innocent III_*
1181	Lucius III
1185	Urban III
1187	Clement III
1187	Gregory VIII
1191	Celestine III
1198	Innocent III
1216	Honorius III
1227	Gregory IX
1241	Celestine IV
1243	Innocent IV
1254	Alexander IV
1261	Urban IV
1265	Clement IV
1271	Bl. Gregory X
1276	Bl. Innocent V
1276	Adrian V
1276	John XXI
1277	Nicholas III
1281	Martin IV
1285	Honorius IV
1288	Nicholas IV
1294	St. Celestine V
1294	Boniface VIII
1303	Bl. Benedict XI
1305	Clement V
1316	John XXII
1328	_Nicholas V_*
1334	Benedict XII
1342	Clement VI
1352	Innocent VI
1362	Bl. Urban V
1370	Gregory XI
1378	Urban VI
1378	_Clement VII_*
1389	Boniface IX
1394	_Benedict XIII_*
1404	Innocent VII
1406	Gregory XII
1409	_Alexander V_*
1410	_John XXIII_*
1417	Martin V
1431	Eugene IV
1439	_Felix V_*
1447	Nicholas V
1455	Callistus III
1458	Pius II
1464	Paul II
1471	Sixtus IV
1484	Innocent VIII
1492	Alexander VI
1503	Pius III
1503	Julius II
1513	Leo X
1522	Adrian VI
1523	Clement VII
1534	Paul III
1550	Julius III
1555	Marcellus II
1555	Paul IV
1559	Pius IV
1566	St. Pius V
1572	Gregory XIII
1585	Sixtus V
1590	Urban VII
1590	Gregory XIV
1591	Innocent IX
1592	Clement VIII
1605	Leo XI
1605	Paul V
1621	Gregory XV
1623	Urban VIII
1644	Innocent X
1655	Alexander VII
1667	Clement IX
1670	Clement X
1676	Bl. Innocent XI
1689	Alexander VIII
1691	Innocent XII
1700	Clement XI
1721	Innocent XIII
1724	Benedict XIII
1730	Clement XII
1740	Benedict XIV
1758	Clement XIII
1769	Clement XIV
1775	Pius VI
1800	Pius VII
1823	Leo XII
1829	Pius VIII
1831	Gregory XVI
1846	Pius IX
1878	Leo XIII
1903	St. Pius X
1914	Benedict XV
1922	Pius XI
1939	Pius XII
1958	John XXIII
1963	Paul VI
1978	John Paul I
1978	John Paul II

Major Non-Christian World Religions

Sources: Reviewed by Anthony Padovano, PhD, STD, prof. of literature & relig. studies, Ramapo College, NJ, adj. prof. of theol., Fordham U., NYC; Islam reviewed by Abdulaziz Sachedina, PhD, prof. of Islamic studies, Univ. of Virginia

Buddhism

Founded: About 525 BC, reportedly near Benares, India.

Founder: Gautama Siddhartha (c 563-483 BC), the Buddha, who achieved enlightenment through intense meditation.

Sacred Texts: The *Tripitaka*, a collection of the Buddha's teachings, rules of monastic life, and philosophical commentaries on the teachings; also a vast body of Buddhist teachings and commentaries, many of which are called *sutras*.

Organization: The basic institution is the *sangha*, or monastic order, through which the traditions are passed to from generation to generation. Monastic life tends to be democratic and anti-authoritarian. Large lay organizations have developed in some sects.

Practice: Varies widely according to the sect, and ranges from austere meditation to magical chanting and elaborate temple rites. Many practices, such as exorcism of devils, reflect pre-Buddhist beliefs.

Divisions: A variety of sects grouped into 3 primary branches: Theravada (sole survivor of the ancient Hinayana schools), which emphasizes the importance of pure thought and deed; Mahayana (includes Zen and Soka-gakkai), which ranges from philosophical schools to belief in the saving grace of higher beings or ritual practices and to practical meditative disciplines; and Tantrism, a combination of belief in ritual magic and sophisticated philosophy.

Location: Throughout Asia, from Sri Lanka to Japan. Zen and Soka-gakkai have some 15,000 adherents in the U.S.

Beliefs: Life is misery and decay, and there is no ultimate reality in it or behind it. The cycle of endless birth and rebirth continues because of desire and attachment to the unreal "self." Right meditation and deeds will end the cycle and achieve Nirvana, the Void, nothingness.

Hinduism

Founded: About 500 BC by Aryans who migrated to India, where their Vedic religion intermixed with the practices and beliefs of the natives.

Sacred texts: The *Veda,* including the *Upanishads,* a collection of rituals and mythological and philosophical commentaries; a vast number of epic stories about gods, heroes, and saints, including the *Bhagavadgita,* a part of the *Mahabharata,* and the *Ramayana;* and a great variety of other literature.

Organization: None, strictly speaking. Generally, rituals should be performed or assisted by Brahmins, the priestly caste, but in practice, simpler rituals can be performed by anyone. Brahmins are the final judges of ritual purity, the vital element in Hindu life. Temples and religious organizations are usually presided over by Brahmins.

Practice: A variety of private rituals, primarily passage rites (e.g., initiation, marriage, death, etc.) and daily devotions, and a similar variety of public rites in temples. Of the public rites, the *puja,* a ceremonial dinner for a god, is the most common.

Divisions: There is no concept of orthodoxy in Hinduism, which presents a variety of sects, most of them devoted to the worship of one of the many gods. The 3 major living traditions are those devoted to the gods Vishnu and Shiva and to the goddess Shakti; each is divided into further subsects. Numerous folk beliefs and practices, often in amalgamation with the above groups, exist side by side with sophisticated philosophical schools and exotic cults.

Location: Mainly India, Nepal, Malaysia, Guyana, Suriname, and Sri Lanka.

Beliefs: There is only one divine principle; the many gods are only aspects of that unity. Life in all its forms is an aspect of the divine, but it appears as a separation from the divine, a meaningless cycle of birth and rebirth (*samsara*) determined by the purity or impurity of past deeds (*karma*). To improve one's *karma* or escape *samsara* by pure acts, thought, and/or devotion is the aim of every Hindu.

Islam

Founded: About AD 622 in Mecca, Arabian Peninsula.

Founder: Muhammad (c 570-632), the Prophet.

Sacred texts: The *Koran* (al-Qur'an), the Word of God; *Sunna*, collections of *adth*, describing what Muhammad said or did.

Organization: Since the founder was both a prophet and a statesman, Muslim leadership has combined the civil and moral function of a state. Within the larger community, there are cultural and national groups, held together by a common religious law, the *Shari'a*, enforced uniformly in matters of religion only. In social transactions the community has often departed from traditional formulations. Although Islam is basically egalitarian and suspicious of authoritarianism, Muslim culture tends to be dominated by the conservative spirit of its religious establishment, the *ulema*.

Practice: Besides the general moral guidance that determines everyday life, there are "Five Pillars of Islam": profession of faith (oneness of God and prophethood of Muhammad); prayer 5 times a day; alms *(zakat)* from one's savings and estate; dawn-to-dusk fasting in the month of Ramadan; and once in a lifetime, pilgrimage to Mecca, if possible.

Divisions: There are 2 major groups: the majority known as Sunni and the minority Shiites. Shiites believe in Twelve Imams (perfect teachers) after the Prophet, of whom the last Imam has lived an invisible existence since 874, continuing to guide his community. Sunni Muslims believe in God's overpowering will over their affairs and tend to be predestinarian; Shiites believe in free will and give a substantial role to human reason in daily life. Sufism (mystical dimension of Islam) is prevalent among both Sunni and Shiites. Sufis emphasize personal relation to God and obedience informed by love of God.

Location: W Africa to Philippines, across band including E Africa, Central Asia and W China, India, Malaysia, Indonesia. Islam has several million adherents in North America.

Beliefs: Strictly monotheistic. God is creator of the universe, omnipotent, omniscient, just, forgiving, and merciful. The human is God's highest creation, but weak and egocentric, prone to forget the goal of life, constantly tempted by the Satan, an evil being. God revealed the Koran to Muhammad to guide humanity to truth and justice. Those who repent and sincerely "submit" (literal meaning of "islam") to God attain salvation. The forgiven enter the Paradise, and the wicked burn in Hell.

Judaism

Founded: About 1300 BC.

Founder: Abraham is regarded as the founding patriarch, but the Torah of Moses is the basic source of the teachings.

Sacred Texts: The 5 books of Moses constitute the written Torah. Special sanctity is also assigned other writings of the Hebrew Bible—the teachings of oral Torah are recorded in the Talmud, in the Midrash, and in various commentaries.

Organization: Originally theocratic, Judaism has evolved a congregational polity. The basic institution is the local synagogue, operated by the congregation and led by a rabbi of their choice. Chief rabbis in France and Great Britain have authority only over those who accept it; in Israel, the 2 chief rabbis have civil authority in family law.

Practice: Among traditional practicioners, almost all areas of life are governed by strict religious discipline. Sabbath and holidays are marked by special observances, and attendance at public worship is considered especially important then. Chief annual observances are Passover, celebrating liberation of the Israelites from Egypt and marked by the Seder meal in homes, and the 10 days from Rosh Hashana (New Year) to Yom Kippur (Day of Atonement), a period of fasting and penitence.

Divisions: Judaism is an unbroken spectrum from ultraconservative to ultraliberal, largely reflecting different points of view regarding the binding character of the prohibitions and duties—particularly the dietary and Sabbath observations—traditionally prescribed for the daily life of the Jew.

Location: Almost worldwide, with concentrations in Israel and the U.S.

Beliefs: Strictly monotheistic. God is the creator and absolute ruler of the universe. Men and women are free to choose to rebel against God's rule. God established a particular relationship with the Hebrew people: by obeying a divine law God gave them, they would be a special witness to God's mercy and justice. Judaism stresses ethical behavior (and, among the traditional, careful ritual obedience) as true worship of God.

Major Christian Denominations:

Brackets indicate some features that tend to

Denom-ination	Origins	Organization	Authority	Special rites
Baptists	In radical Reformation, objections to infant baptism, demands for church and state separation; John Smyth, English Separatist, in 1609; Roger Williams, 1638, Providence, RI.	Congregational; each local church is autonomous.	Scripture; some Baptists, particularly in the South, interpret the Bible literally.	*[Baptism, usually early teen years and after, by total immersion;]* Lord's Supper.
Church of Christ (Disciples)	Among evangelical Presbyterians in KY (1804) and PA (1809), in distress over Protestant factionalism and decline of fervor; organized in 1832.	Congregational.	*["Where the Scriptures speak, we speak; where the Scriptures are silent, we are silent."]*	Adult baptism; Lord's Supper (weekly).
Episco-palians	Henry VIII separated English Catholic Church from Rome, 1534, for political reasons; Protestant Episcopal Church in U.S. founded in 1789.	*[Diocesan bishops, in apostolic succession, are elected by parish representatives; the national Church is headed by General Convention and Presiding Bishop; part of the Anglican Communion.]*	Scripture as interpreted by tradition, especially *39 Articles* (1563); tri-annual convention of bishops, priests, and lay people.	Infant baptism, Eucharist, and other sacraments; sacrament taken to be symbolic, but as having real spiritual effect.
Jehovah's Witnesses	Founded in 1870 in PA by Charles Taze Russell; i ncorporated as Watch Tower Bible and Tract Society of PA, 1884; name Jehovah's Witnesses adopted in 1931.	A governing body located in NY coordinates worldwide activities; each congregation cared for by a body of elders; each Witness considered a minister.	The Bible.	Baptism by immersion; annual Lord's Meal ceremony.
Latter-day Saints (Mormons)	In a vision of the Father and the Son reported by Joseph Smith (1820s) in NY. Smith also reported receiving new scripture on golden tablets: The Book of Mormon.	Theocratic; 1st Presidency (church president, 2 counselors), 12 Apostles preside over international church. Local congregations headed by lay priesthood leaders.	Revelation to living prophet (church president). The Bible, Book of Mormon, and other revelations to Smith and his successors.	Baptism, at age 8; laying on of hands (which confers the gift of the Holy Ghost); Lord's Supper; temple rites: baptism for the dead, marriage for eternity, others.
Lutherans	Begun by Martin Luther in Wittenberg, Germany, in 1517; objection to Catholic doctrine of salvation and sale of indulgences; break complete, 1519.	Varies from congregational to episcopal; in U.S., a combination of regional synods and congregational polities is most common.	Scripture alone. *The Book of Concord* (1580), which includes the three Ecumenical Creeds, is subscribed to as a correct exposition of Scripture.	Infant baptism; Lord's Supper; Christ's true body and blood present "in, with, and under the bread and wine."
Methodists	Rev. John Wesley began movement in 1738, within Church of England; first U.S. denomination, Baltimore (1784).	Conference and superintendent system; *[in United Methodist Church, general superintendents are bishops—not a priestly order, only an office—who are elected for life.]*	Scripture as interpreted by tradition, reason, and experience.	Baptism of infants or adults; Lord's Supper commanded; other rites include marriage, ordination, solemnization of personal commitments.
Orthodox	Developed in original Christian proselytizing; broke with Rome in 1054, after centuries of doctrinal disputes and diverging traditions	Synods of bishops in autonomous, usually national, churches elect a patriarch, archbishop, or metropolitan; these men, as a group, are the heads of the church.	Scripture, tradition, and the first 7 church councils up to Nicaea II in 787; bishops in council have authority in doctrine and policy.	Seven sacraments: infant baptism and anointing, Eucharist, ordination, penance, marriage, and anointing of the sick.
Pente-costal	In Topeka, KS (1901) and Los Angeles (1906), in reaction to perceived loss of evangelical fervor among Methodists and others.	Originally a movement, not a formal organization, Pentecostalism now has a variety of organized forms and continues also as a movement.	Scripture; individual charismatic leaders, the teachings of the Holy Spirit.	*[Spirit baptism, especially as shown in "speaking in tongues"; healing and sometimes exorcism;]* adult baptism; Lord's Supper.
Presby-terians	In 16th-cent. Calvinist eformation; differed with Lutherans over sacraments, church government; John Knox founded Scotch Presbyterian church about 1560.	*[Highly structured representational system of ministers and lay persons (presbyters) in local, regional, and national bodies (synods).]*	Scripture.	Infant baptism; Lord's Supper; bread and wine symbolize Christ's spiritual presence.
Roman Catholics	Traditionally, founded by Jesus who named St. Peter the 1st vicar; developed in early Christian proselytizing, especially after the conversion of imperial Rome in the 4th cent.	*[Hierarchy with supreme power vested in pope elected by cardinals;]* councils of bishops advise on matters of doctrine and policy.	*[The pope, when speaking for the whole church in matters of faith and morals; and tradition (which is expressed in church councils and in part contained in Scripture).]*	Mass; 7 sacraments: baptism, reconciliation, Eucharist, confirmation, marriage, ordination, and anointing of the sick (unction).
United Church of Christ	*[By ecumenical union, in 1957, of Congregationalists and Evangelical & Reformed, representing both Calvinist and Lutheran traditions.]*	Congregational; a General Synod, representative of all congregations, sets general policy.	Scripture.	Infant baptism; Lord's Supper.

How Do They Differ?

distinguish a denomination sharply from others.

Practice	Ethics	Doctrine	Other	Denomination
Worship style varies from staid to evangelistic; extensive missionary activity.	Usually opposed to alcohol and tobacco; some tendency toward a perfectionist ethical tandard.	*[No creed; true church is of believers only, who are all equal.]*	Believing no authority can stand between the believer and God, the Baptists are strong supporters of church and state separation.	**Baptists**
Tries to avoid any rite not considered part of the 1st-century church; some congregations may reject instrumental music.	Some tendency toward perfectionism; increasing interest in social action programs.	Simple New Testament faith; avoids any elaboration not firmly based on Scripture.	Highly tolerant in doctrinal and religious matters; strongly supportive of scholarly education.	**Church of Christ (Disciples)**
Formal, based on "Book of Common Prayer," updated 1979; services range from austerely simple to highly liturgical.	Tolerant, sometimes permissive; some social action programs.	Scripture; the "historic creeds," which include the Apostles, Nicene, and Athanasian, and the "Book of Common Prayer"; ranges from Anglo-Catholic to low church, with Calvinist influences.	Strongly ecumenical, holding talks with many branches of Christendom.	**Episcopalians**
Meetings are held in Kingdom Halls and members' homes for study and worship; *[extensive door-to-door visitations.]*	High moral code; stress on marital fidelity and family values; avoidance of tobacco and blood transfusions.	*[God, by his first creation, Christ, will soon destroy all wickedness; 144,000 faithful ones will rule in heaven with Christ over others on a paradise earth.]*	Total allegiance proclaimed only to God's kingdom or heavenly government by Christ; main periodical, *The Watchtower,* is printed in 115 languages.	**Jehovah's Witnesses**
Simple service with prayers, hymns, sermon; private temple ceremonies may be more elaborate.	Temperance; strict moral code; *[tithing];* a strong work ethic with communal self-reliance; *[strong missionary activity];* family emphasis.	Jesus Christ is the Son of God, the Eternal Father. Jesus' atonement saves all humans; those who are obedient to God's laws may become joint-heirs with Christ in God's kingdom.	Mormons believe theirs is the true church of Jesus Christ, restored by God through Joseph Smith. Official name: The Church of Jesus Christ of Latter-day Saints.	**Latter-day Saints (Mormons)**
Relatively simple, formal liturgy with emphasis on the sermon.	Generally conservative in personal and social ethics; doctrine of "2 kingdoms" (worldly and holy) supports conservatism in secular affairs.	Salvation by grace alone through faith; Lutheranism has made major contributions to Protestant theology.	Though still somewhat divided along ethnic lines (German, Swedish, etc.), main divisions are between fundamentalists and liberals.	**Lutherans**
Worship style varies widely by denomination, local church, geography.	Originally pietist and perfectionist; always strong social activist elements.	No distinctive theological development; 25 Articles abridged from Church of England's 39, not binding.	In 1968, The United Methodist Church was formed by the union of The Methodist Church and The Evangelical United Brethren Church.	**Methodists**
[Elaborate liturgy, usually in the vernacular, though extremely traditional; the liturgy is the essence of Orthodoxy; veneration of icons.]	Tolerant; little stress on social action; divorce, remarriage permitted in some cases; bishops are celibate; priests need not be.	Emphasis on Christ's resurrection, rather than crucifixion; the Holy Spirit proceeds from God the Father only.	Orthodox Church in America originally under Patriarch of Moscow, was granted autonomy in 1970; Greek Orthodox do not recognize this autonomy.	**Orthodox**
Loosely structured service with rousing hymns and sermons, culminating in spirit baptism.	Usually, emphasis on perfectionism, with varying degrees of tolerance.	Simple traditional beliefs, usually Protestant, with emphasis on the immediate presence of God in the Holy Spirit.	Once confined to lower-class "holy rollers," Pentecostalism now appears in mainline churches and has established middle-class congregations.	**Pentecostal**
A simple, sober service in which the sermon is central.	Traditionally, a tendency toward strictness, with firm church- and self-discipline; otherwise tolerant.	Emphasizes the sovereignty and justice of God; no longer dogmatic.	Although traces of belief in predestination (that God has foreordained salvation for the "elect") remain, this idea is no longer a central element in Presbyterianism.	**Presbyterians**
Relatively elaborate ritual centered on the Mass; also rosary recitation, novenas, etc.	Traditionally strict, but increasingly tolerant in practice; divorce and remarriage not accepted, but annulments sometimes granted; celibate clergy, except in Eastern rite.	Highly elaborated; salvation by merit gained through grace; dogmatic; special veneration of Mary, the mother of Jesus.	Relatively rapid change followed Vatican Council II; Mass now in vernacular; more stress on social action, tolerance, ecumenism.	**Roman Catholics**
Usually simple services with emphasis on the sermon.	Tolerant; some social action emphasis.	Standard Protestant; "Statement of Faith" (1959) is not binding.	The 2 main churches in the 1957 union represented earlier unions with small groups of almost every Protestant denomination.	**United Church of Christ**

AWARDS — MEDALS — PRIZES

The Alfred B. Nobel Prize Winners, 1901-2000

Alfred B. Nobel (1833-96), inventor of dynamite, bequeathed $9 mil, the interest to be distributed yearly to those judged to have had most benefited humankind in physics, chemistry, medicine-physiology, literature, and the promotion of peace. These prizes were first awarded in 1901. The first Nobel Memorial Prize in Economic Science was awarded in 1969, funded by the central bank of Sweden. If the year is omitted, no award was given.

Each prize is now worth more than $900,000.

Physics

2000 Jack S. Kilby, U.S.; Zhores I. Alferov, Russ.
1999 Gerardus 't Hooft and Martinus J. G. Veltman, Netherlands
1998 Robert B. Laughlin, Horst L. Störmer, Daniel C. Tsui, U.S.
1997 Steven Chu, William D. Phillips, U.S.; Claude Cohen-Tannoudji, Fr.
1996 David M. Lee, Douglas D. Osheroff, Robert C. Richardson, U.S.
1995 Martin Perl, Frederick Reines, U.S.
1994 Bertram N. Brockhouse, Can.; Clifford G. Shull, U.S.
1993 Joseph H. Taylor, Russell A. Hulse, U.S.
1992 Georges Charpak, Pol.-Fr.
1991 Pierre-Giles de Gennes, Fr.
1990 Richard E. Taylor, Can.; Jerome I. Friedman, Henry W. Kendall, U.S.
1989 Norman F. Ramsey, U.S.; Hans G. Dehmelt, Ger.-U.S.; Wolfgang Paul, Ger.
1988 Leon M. Lederman, Melvin Schwartz, Jack Steinberger, U.S.
1987 K. Alex Müller, Swiss; J. Georg Bednorz, Ger.
1986 Ernest Ruska, Ger.; Gerd Binnig, Ger.; Heinrich Rohrer, Swiss
1985 Klaus von Klitzing, Ger.
1984 Carlo Rubbia, It.; Simon van der Meer, Dutch
1983 Subrahmanyan Chandrasekhar, William A. Fowler, U.S.
1982 Kenneth G. Wilson, U.S.
1981 Nicolaas Bloembergen, Arthur Schaalow, U.S.; Kai M. Siegbahn, Swed.
1980 James W. Cronin, Val L. Fitch, U.S.
1979 Steven Weinberg, Sheldon L. Glashow, U.S.; Abdus Salam, Pakistani
1978 Pyotr Kapitsa, USSR; Arno Penzias, Robert Wilson, U.S.
1977 John H. Van Vleck, Philip W. Anderson, U.S.; Nevill F. Mott, Br.
1976 Burton Richter, Samuel C.C. Ting, U.S.
1975 James Rainwater, U.S.; Ben Mottelson, U.S.-Dan.; Aage Bohr, Dan.
1974 Martin Ryle, Antony Hewish, Br.
1973 Ivar Giaever, U.S.; Leo Esaki, Jpn.; Brian D. Josephson, Br.
1972 John Bardeen, Leon N. Cooper, John R. Schrieffer, U.S.
1971 Dennis Gabor, Br.
1970 Louis Neel, Fr.; Hannes Alfven, Swed.
1969 Murray Gell-Mann, U.S.
1968 Luis W. Alvarez, U.S.
1967 Hans A. Bethe, U.S.
1966 Alfred Kastler, Fr.
1965 Richard P. Feynman, Julian S. Schwinger, U.S.; Shinichiro Tomonaga, Jpn.
1964 Nikolai G. Basov, Aleksander M. Prochorov, USSR; Charles H. Townes, U.S.
1963 Maria Goeppert-Mayer, Eugene P. Wigner, U.S.; J. Hans D. Jensen, Ger.
1962 Lev. D. Landau, USSR
1961 Robert Hofstadter, U.S.; Rudolf L. Mossbauer, Ger.
1960 Donald A. Glaser, U.S.
1959 Owen Chamberlain, Emilio G. Segre, U.S.
1958 Pavel Cherenkov, Ilya Frank, Igor Y. Tamm, USSR
1957 Tsung-dao Lee, Chen Ning Yang, U.S.
1956 John Bardeen, Walter H. Brattain, William Shockley, U.S.
1955 Polykarp Kusch, Willis E. Lamb, U.S.
1954 Max Born, Br.; Walter Bothe, Ger.
1953 Frits Zernike, Dutch
1952 Felix Bloch, Edward M. Purcell, U.S.
1951 Sir John D. Cockroft, Br.; Ernest T. S. Walton, Ir.
1950 Cecil F. Powell, Br.
1949 Hideki Yukawa, Jpn.
1948 Patrick M. S. Blackett, Br.
1947 Sir Edward V. Appleton, Br.
1946 Percy W. Bridgman, U.S.
1945 Wolfgang Pauli, U.S.
1944 Isidor Isaac Rabi, U.S.
1943 Otto Stern, U.S.
1939 Ernest O. Lawrence, U.S.
1938 Enrico Fermi, It.-U.S.
1937 Clinton J. Davisson, U.S.; Sir George P. Thomson, Br.
1936 Carl D. Anderson, U.S.; Victor F. Hess, Aus.
1935 Sir James Chadwick, Br.
1933 Paul A. M. Dirac, Br.; Erwin Schrodinger, Austria
1932 Werner Heisenberg, Ger.
1930 Sir Chandrasekhara V. Raman, Indian
1929 Prince Louis-Victor de Broglie, Fr.
1928 Owen W. Richardson, Br.
1927 Arthur H. Compton, U.S.; Charles T. R. Wilson, Br.
1926 Jean B. Perrin, Fr.
1925 James Franck, Gustav Hertz, Ger.
1924 Karl M. G. Siegbahn, Swed.
1923 Robert A. Millikan, U.S.
1922 Niels Bohr, Dan.
1921 Albert Einstein, Ger.-U.S.
1920 Charles E. Guillaume, Fr.
1919 Johannes Stark, Ger.
1918 Max K. E. L. Planck, Ger.
1917 Charles G. Barkla, Br.
1915 Sir William H. Bragg, Sir William L. Bragg, Br.
1914 Max von Laue, Ger.
1913 Heike Kamerlingh-Onnes, Dutch
1912 Nils G. Dalen, Swed.
1911 Wilhelm Wien, Ger.
1910 Johannes D. van der Waals, Dutch
1909 Carl F. Braun, Ger.; Guglielmo Marconi, It.
1908 Gabriel Lippmann, Fr.
1907 Albert A. Michelson, U.S.
1906 Sir Joseph J. Thomson, Br.
1905 Philipp E. A. von Lenard, Ger.
1904 John W. Strutt, Lord Rayleigh, Br.
1903 Antoine Henri Becquerel, Pierre Curie, Fr.; Marie Curie, Pol.-Fr.
1902 Hendrik A. Lorentz, Pieter Zeeman, Dutch
1901 Wilhelm C. Roentgen, Ger.

Chemistry

2000 Alan J. Heeger, U.S.; Alan G. MacDiarmid, NZ-U.S.; Hideki Shirakawa, Japan
1999 Ahmed H. Zewail, U.S.
1998 Walter Kohn, U.S.; John A. Pople, Br.
1997 Paul D. Boyer, U.S., & John E. Walker, Br.; Jens C. Skou, Dan.
1996 Harold W. Kroto, Br.; Robert F. Curl Jr., Richard E. Smalley, U.S.
1995 Paul Crutzen, Dutch; Mario Molina, Mex.-U.S.; Sherwood Rowland, U.S.
1994 George A. Olah, U.S.
1993 Kary B. Mullis, U.S.; Michael Smith, Br.-Can.
1992 Rudolph A. Marcus, Can.-U.S.
1991 Richard R. Ernst, Swiss
1990 Elias James Corey, U.S.
1989 Thomas R. Cech, Sidney Altman, U.S.
1988 Johann Deisenhofer, Robert Huber, Hartmut Michel, Ger.
1987 Donald J. Cram, Charles J. Pedersen, U.S.; Jean-Marie Lehn, Fr.
1986 Dudley Herschbach, Yuan T. Lee, U.S.; John C. Polanyi, Can.
1985 Herbert A. Hauptman, Jerome Karle, U.S.
1984 Bruce Merrifield, U.S.
1983 Henry Taube, Can.
1982 Aaron Klug, S. Afr.
1981 Kenichi Fukui, Jpn.; Roald Hoffmann, U.S.
1980 Paul Berg, Walter Gilbert, U.S.; Frederick Sanger, Br.
1979 Herbert C. Brown, U.S.; George Wittig, Ger.
1978 Peter Mitchell, Br.
1977 Ilya Prigogine, Belg.
1976 William N. Lipscomb, U.S.
1975 John Cornforth, Austral.-Br.; Vladimir Prelog, Yugo.-Swiss
1974 Paul J. Flory, U.S.
1973 Ernst Otto Fischer, Ger.; Geoffrey Wilkinson, Br.
1972 Christian B. Anfinsen, Stanford Moore, William H. Stein, U.S.
1971 Gerhard Herzberg, Canadian
1970 Luis F. Leloir, Arg.
1969 Derek H. R. Barton, Br.; Odd Hassel, Nor.
1968 Lars Onsager, U.S.
1967 Manfred Eigen, Ger.; Ronald G. W. Norrish, George Porter, Br.
1966 Robert S. Mulliken, U.S.
1965 Robert B. Woodward, U.S.
1964 Dorothy C. Hodgkin, Br.
1963 Giulio Natta, It.; Karl Ziegler, Ger.
1962 John C. Kendrew, Max F. Perutz, Br.
1961 Melvin Calvin, U.S.
1960 Willard F. Libby, U.S.
1959 Jaroslav Heyrovsky, Czech.
1958 Frederick Sanger, Br.
1957 Sir Alexander R. Todd, Br.
1956 Sir Cyril N. Hinshelwood, Br.; Nikolai N. Semenov, USSR
1955 Vincent du Vigneaud, U.S.
1954 Linus C. Pauling, U.S.
1953 Hermann Staudinger, Ger.
1952 Archer J. P. Martin, Richard L. M. Synge, Br.
1951 Edwin M. McMillan, Glenn T. Seaborg, U.S.
1950 Kurt Alder, Otto P. H. Diels, Ger.
1949 William F. Giauque, U.S.
1948 Arne W. K. Tiselius, Swed.
1947 Sir Robert Robinson, Br.
1946 James B. Sumner, John H. Northrop, Wendell M. Stanley, U.S.
1945 Artturi I. Virtanen, Fin.
1944 Otto Hahn, Ger.
1943 Georg de Hevesy, Hung.
1939 Adolf F. J. Butenandt, Ger.; Leopold Ruzicka, Swiss
1938 Richard Kuhn, Ger.
1937 Walter N. Haworth, Br.; Paul Karrer, Swiss
1936 Peter J. W. Debye, Dutch
1935 Frederic & Irene Joliot-Curie, Fr.
1934 Harold C. Urey, U.S.
1932 Irving Langmuir, U.S.
1931 Friedrich Bergius, Karl Bosch, Ger.

1930 Hans Fischer, Ger.
1929 Sir Arthur Harden, Br.; Hans von Euler-Chelpin, Swed.
1928 Adolf O. R. Windaus, Ger.
1927 Heinrich O. Wieland, Ger.
1926 Theodor Svedberg, Swed.
1925 Richard A. Zsigmondy, Ger.
1923 Fritz Pregl, Austrian
1922 Francis W. Aston, Br.
1921 Frederick Soddy, Br.
1920 Walther H. Nernst, Ger.
1918 Fritz Haber, Ger.
1915 Richard M. Willstatter, Ger.
1914 Theodore W. Richards, U.S.
1913 Alfred Werner, Swiss
1912 Victor Grignard, Paul Sabatier, Fr.
1911 Marie Curie, Pol.-Fr.
1910 Otto Wallach, Ger.
1909 Wilhelm Ostwald, Ger.
1908 Ernest Rutherford, Br.
1907 Eduard Buchner, Ger.
1906 Henri Moissan, Fr.
1905 Adolf von Baeyer, Ger.
1904 Sir William Ramsay, Br.
1903 Svante A. Arrhenius, Swed.
1902 Emil Fischer, Ger.
1901 Jacobus H. van't Hoff, Dutch

Physiology or Medicine

2000 Arvid Carlsson, Swed.; Paul Greengard, U.S.; Eric R. Kandel, Aus-U.S.
1999 Günter Blobel, U.S.
1998 Robert F. Furchgott, Louis J. Ignarro, Ferid Murad, U.S.
1997 Stanley B. Prusiner, U.S.
1996 Peter C. Doherty, Austral.; Rolf M. Zinkernagel, Swiss
1995 Edward B. Lewis, Eric F. Wieschaus, U.S.; Christiane Nuesslein-Volhard, Ger.
1994 Alfred G. Gilman, Martin Rodbell, U.S.
1993 Phillip A. Sharp, U.S.; Richard J. Roberts, Br.
1992 Edmond H. Fisher, Edwin G. Krebs, U.S.
1991 Edwin Neher, Bert Sakmann, Ger.
1990 Joseph E. Murray, E. Donnall Thomas, U.S.
1989 J. Michael Bishop, Harold E. Varmus, U.S.
1988 Gertrude B. Elion, George H. Hitchings, U.S; Sir James Black, Br.
1987 Susumu Tonegawa, Jpn.
1986 Rita Levi-Montalcini, It.-U.S., Stanley Cohen, U.S.
1985 Michael S. Brown, Joseph L. Goldstein, U.S.
1984 Cesar Milstein, Br.-Arg.; Georges J. F. Koehler, Ger.; Niels K. Jerne, Br.-Dan.
1983 Barbara McClintock, U.S.
1982 Sune Bergstrom, Bengt Samuelsson, Swed.; John R. Vane, Br.
1981 Roger W. Sperry, David H. Hubel, Torsten N. Wiesel, U.S.
1980 Baruj Benacerraf, George Snell, U.S.; Jean Dausset, Fr.
1979 Allan M. Cormack, U.S.; Godfrey N. Hounsfield, Br.
1978 Daniel Nathans, Hamilton O. Smith, U.S.; Werner Arber, Swiss
1977 Rosalyn S. Yalow, Roger C.L. Guillemin, Andrew V. Schally, U.S.
1976 Baruch S. Blumberg, Daniel Carleton Gajdusek, U.S.
1975 David Baltimore, Howard Temin, U.S.; Renato Dulbecco, It.-U.S.
1974 Albert Claude, Lux.-U.S.; George Emil Palade, Rom.-U.S.; Christian Rene de Duve, Belg.
1973 Karl von Frisch, Ger.; Konrad Lorenz, Ger.-Aus.; Nikolaas Tinbergen, Br.
1972 Gerald M. Edelman, U.S.; Rodney R. Porter, Br.
1971 Earl W. Sutherland Jr., U.S.
1970 Julius Axelrod, U.S.; Sir Bernard Katz, Br.; Ulf von Euler, Swed.
1969 Max Delbrück, Alfred D. Hershey, Salvador Luria, U.S.
1968 Robert W. Holley, H. Gobind Khorana, Marshall W. Nirenberg, U.S.
1967 Ragnar Granit, Swed.; Haldan Keffer Hartline, George Wald, U.S.
1966 Charles B. Huggins, Francis Peyton Rous, U.S.
1965 François Jacob, Andre Lwoff, Jacques Monod, Fr.
1964 Konrad E. Bloch, U.S.; Feodor Lynen, Ger.
1963 Sir John C. Eccles, Austral.; Alan L. Hodgkin, Andrew F. Huxley, Br.
1962 Francis H. C. Crick, Maurice H. F. Wilkins, Br.; James D. Watson, U.S.
1961 Georg von Bekesy, U.S.
1960 Sir F. MacFarlane Burnet,Austral.; Peter B. Medawar, Br.
1959 Arthur Kornberg, Severo Ochoa, U.S.
1958 George W. Beadle, Edward L. Tatum, Joshua Lederberg, U.S.
1957 Daniel Bovet, It.
1956 Andre F. Cournand, Dickinson W. Richards Jr., U.S.; Werner Forssmann, Ger.
1955 Alex H. T. Theorell, Swed.
1954 John F. Enders, Frederick C. Robbins, Thomas H. Weller, U.S.
1953 Hans A. Krebs, Br.; Fritz A. Lipmann, U.S.
1952 Selman A. Waksman, U.S.
1951 Max Theiler, U.S.
1950 Philip S. Hench, Edward C. Kendall, U.S.; Tadeus Reichstein, Swiss
1949 Walter R. Hess, Swiss; Antonio Moniz, Port.
1948 Paul H. Müller, Swiss
1947 Carl F. Cori, Gerty T. Cori, U.S.; Bernardo A. Houssay, Arg.
1946 Hermann J. Muller, U.S.
1945 Ernst B. Chain, Sir Alexander Fleming, Sir Howard W. Florey, Br.
1944 Joseph Erlanger, Herbert S. Gasser, U.S.
1943 Henrik C. P. Dam, Dan.; Edward A. Doisy, U.S.
1939 Gerhard Domagk, Ger.
1938 Corneille J. F. Heymans, Belg.
1937 Albert Szent-Gyorgyi, Hung.-U.S.
1936 Sir Henry H. Dale, Br.; Otto Loewi, U.S.
1935 Hans Spemann, Ger.
1934 George R. Minot, William P. Murphy, G. H. Whipple, U.S.
1933 Thomas H. Morgan, U.S.
1932 Edgar D. Adrian, Sir Charles S. Sherrington, Br.
1931 Otto H. Warburg, Ger.
1930 Karl Landsteiner, U.S.
1929 Christiaan Eijkman, Dutch; Sir Frederick G. Hopkins, Br.
1928 Charles J. H. Nicolle, Fr.
1927 Julius Wagner-Jauregg, Austrian
1926 Johannes A. G. Fibiger, Dan.
1924 Willem Einthoven, Dutch
1923 Frederick G. Banting, Can.; John J. R. Macleod, Scot.
1922 Archibald V. Hill, Br.; Otto F. Meyerhof, Ger.
1920 Schack A. S. Krogh, Dan.
1919 Jules Bordet, Belg.
1914 Robert Barany, Aus.
1913 Charles R. Richet, Fr.
1912 Alexis Carrel, Fr.
1911 Allvar Gullstrand, Swed.
1910 Albrecht Kossel, Ger.
1909 Emil T. Kocher, Swiss
1908 Paul Ehrlich, Ger.; Elie Metchnikoff, Fr.
1907 Charles L. A. Laveran, Fr.
1906 Camillo Golgi, It.; Santiago Ramon y Cajal, Span.
1905 Robert Koch, Ger.
1904 Ivan P. Pavlov, Russ.
1903 Niels R. Finsen, Dan.
1902 Sir Ronald Ross, Br.
1901 Emil A. von Behring, Ger.

Literature

2000 Gao Xingjian, Chin.
1999 Günter Grass, Ger.
1998 José Saramago, Por.
1997 Dario Fo, It.
1996 Wislawa Szymborska, Pol.
1995 Seamus Heaney, Ir.
1994 Kenzaburo Oe, Jpn.
1993 Toni Morrison, U.S.
1992 Derek Walcott, W. Ind.
1991 Nadine Gordimer, S. Afr.
1990 Octavio Paz, Mex.
1989 Camilo José Cela, Span.
1988 Naguib Mahfouz, Egy.
1987 Joseph Brodsky, USSR-U.S.
1986 Wole Soyinka, Nig.
1985 Claude Simon, Fr.
1984 Jaroslav Siefert, Czech.
1983 William Golding, Br.
1982 Gabriel Garcia Marquez, Colombian-Mex.
1981 Elias Canetti, Bulg.-Br.
1980 Czeslaw Milosz, Pol.-U.S.
1979 Odysseus Elytis, Gk.
1978 Isaac Bashevis Singer, U.S.
1977 Vicente Aleixandre, Span.
1976 Saul Bellow, U.S.
1975 Eugenio Montale, It.
1974 Eyvind Johnson, Harry Edmund Martinson, Swed.
1973 Patrick White, Austral.
1972 Heinrich Böll, Ger.
1971 Pablo Neruda, Chil.
1970 Aleksandr I. Solzhenitsyn, USSR
1969 Samuel Beckett, Ir.
1968 Yasunari Kawabata, Jpn.
1967 Miguel Angel Asturias, Guat.
1966 Samuel Joseph Agnon, Isr.; Nelly Sachs, Swed.
1965 Mikhail Sholokhov, USSR
1964 Jean Paul Sartre, Fr. (declined)
1963 Giorgos Seferis, Gk.
1962 John Steinbeck, U.S.
1961 Ivo Andric, Yugo.
1960 Saint-John Perse, Fr.
1959 Salvatore Quasimodo, It.
1958 Boris L. Pasternak, USSR (declined)
1957 Albert Camus, Fr.
1956 Juan Ramon Jimenez, Span.
1955 Halldor K. Laxness, Ice.
1954 Ernest Hemingway, U.S.
1953 Sir Winston Churchill, Br.
1952 Francois Mauriac, Fr.
1951 Par F. Lagerkvist, Swed.
1950 Bertrand Russell, Br.
1949 William Faulkner, U.S.
1948 T.S. Eliot, Br.
1947 Andre Gide, Fr.
1946 Hermann Hesse, Ger.-Swiss
1945 Gabriela Mistral, Chil.
1944 Johannes V. Jensen, Dan.
1939 Frans E. Sillanpaa, Fin.
1938 Pearl S. Buck, U.S.
1937 Roger Martin du Gard, Fr.
1936 Eugene O'Neill, U.S.
1934 Luigi Pirandello, It.
1933 Ivan A. Bunin, USSR
1932 John Galsworthy, Br.
1931 Erik A. Karlfeldt, Swed.
1930 Sinclair Lewis, U.S.
1929 Thomas Mann, Ger.
1928 Sigrid Undset, Nor.
1927 Henri Bergson, Fr.
1926 Grazia Deledda, It.
1925 George Bernard Shaw, Ir.-Br.
1924 Wladyslaw S. Reymont, Pol.
1923 William Butler Yeats, Ir.
1922 Jacinto Benavente, Span.
1921 Anatole France, Fr.
1920 Knut Hamsun, Nor.
1919 Carl F. G. Spitteler, Swiss

1917 Karl A. Gjellerup, Henrik Pontoppidan, Dan.
1916 Verner von Heidenstam, Swed.
1915 Romain Rolland, Fr.
1913 Rabindranath Tagore, Indian
1912 Gerhart Hauptmann, Ger.
1911 Maurice Maeterlinck, Belg.
1910 Paul J. L. Heyse, Ger.
1909 Selma Lagerlof, Swed.
1908 Rudolf C. Eucken, Ger.
1907 Rudyard Kipling, Br.
1906 Giosue Carducci, It.
1905 Henryk Sienkiewicz, Pol.
1904 Frederic Mistral, Fr.; Jose Echegaray, Span.
1903 Bjornsterne Bjornson, Nor.
1902 Theodor Mommsen, Ger.
1901 Rene F. A. Sully Prudhomme, Fr.

Peace

2000 Kim Dae Jung, S. Kor.
1999 Doctors Without Borders (Médecins Sans Frontières), Fr.
1998 John Hume, David Trimble, N. Ir.
1997 Jody Williams, U.S.; International Campaign to Ban Landmines
1996 Bishop Carlos Ximenes Belo, José Ramos-Horta, Timorese
1995 Joseph Rotblat, Pol.-Br.; Pugwash Conference
1994 Yasir Arafat, Pal.; Shimon Peres, Yitzhak Rabin, Isr.
1993 Frederik W. de Klerk, Nelson Mandela, S. Afr.
1992 Rigoberta Menchú, Guat.
1991 Aung San Suu Kyi, Myanmarese
1990 Mikhail S. Gorbachev, USSR
1989 Dalai Lama, Tibet
1988 UN Peacekeeping Forces
1987 Oscar Arias Sanchez, Costa Rican
1986 Elie Wiesel, Rom.-U.S.
1985 Intl. Physicians for the Prevention of Nuclear War, U.S.
1984 Bishop Desmond Tutu, S. Afr.
1983 Lech Walesa, Pol.
1982 Alva Myrdal, Swed.; Alfonso Garcia Robles, Mex.
1981 Office of UN High Com. for Refugees
1980 Adolfo Perez Esquivel, Arg.
1979 Mother Teresa of Calcutta, Alb.-Ind.
1978 Anwar Sadat, Egy.; Menachem Begin, Isr.
1977 Amnesty International
1976 Mairead Corrigan, Betty Williams, N. Ir.
1975 Andrei Sakharov, USSR
1974 Eisaku Sato, Jpn.; Sean MacBride, Ir.
1973 Henry Kissinger, U.S.; Le Duc Tho, N. Viet. (Tho declined)
1971 Willy Brandt, Ger.
1970 Norman E. Borlaug, U.S.
1969 Intl. Labor Organization
1968 Rene Cassin, Fr.
1965 UN Children's Fund (UNICEF)
1964 Martin Luther King Jr., U.S.
1963 International Red Cross, League of Red Cross Societies
1962 Linus C. Pauling, U.S.
1961 Dag Hammarskjold, Swed.
1960 Albert J. Luthuli, S. Afr.
1959 Philip J. Noel-Baker, Br.
1958 Georges Pire, Belg.
1957 Lester B. Pearson, Can.
1954 Office of UN High Com. for Refugees
1953 George C. Marshall, U.S.
1952 Albert Schweitzer, Fr.
1951 Leon Jouhaux, Fr.
1950 Ralph J. Bunche, U.S.
1949 Lord John Boyd Orr of Brechin Mearns, Br.
1947 Friends Service Council, Br.; Amer. Friends Service Committee, U.S.
1946 Emily G. Balch, John R. Mott, U.S.
1945 Cordell Hull, U.S.
1944 International Red Cross
1938 Nansen International Office for Refugees
1937 Viscount Cecil of Chelwood, Br.
1936 Carlos de Saavedra Lamas, Arg.
1935 Carl von Ossietzky, Ger.
1934 Arthur Henderson, Br.
1933 Sir Norman Angell, Br.
1931 Jane Addams, Nicholas Murray Butler, U.S.
1930 Nathan Soderblom, Swed.
1929 Frank B. Kellogg, U.S.
1927 Ferdinand E. Buisson, Fr.; Ludwig Quidde, Ger.
1926 Aristide Briand, Fr.; Gustav Stresemann, Ger.
1925 Sir J. Austen Chamberlain, Br.; Charles G. Dawes, U.S.
1922 Fridtjof Nansen, Nor.
1921 Karl H. Branting, Swed.; Christian L. Lange, Nor.
1920 Leon V.A. Bourgeois, Fr.
1919 Woodrow Wilson, U.S.
1917 International Red Cross
1913 Henri La Fontaine, Belg.
1912 Elihu Root, U.S.
1911 Tobias M.C. Asser, Dutch; Alfred H. Fried, Austrian
1910 Permanent Intl. Peace Bureau
1909 Auguste M. F. Beernaert, Belg.; Paul H. B. B. d'Estournelles de Constant, Fr.
1908 Klas P. Arnoldson, Swed.; Fredrik Bajer, Dan.
1907 Ernesto T. Moneta, It.; Louis Renault, Fr.
1906 Theodore Roosevelt, U.S.
1905 Baroness Bertha von Suttner, Austrian
1904 Institute of International Law
1903 Sir William R. Cremer, Br.
1902 Elie Ducommun, Charles A. Gobat, Swiss
1901 Jean H. Dunant, Swiss; Frederic Passy, Fr.

Nobel Memorial Prize in Economic Science

2000 James J. Heckman, Daniel L. McFadden, U.S.
1999 Robert A. Mundell, Can.
1998 Amartya Sen, Indian
1997 Robert C. Merton, U.S.; Myron S. Scholes, Can.-U.S.
1996 James A. Mirrlees, Br.; William Vickrey, Can.-U.S.
1995 Robert E. Lucas Jr., U.S.
1994 John C. Harsanyi, John F. Nash, U.S.; Reinhard Selten, Ger.
1993 Robert W. Fogel, Douglass C. North, U.S.
1992 Gary S. Becker, U.S.
1991 Ronald H. Coase, Br.-U.S.
1990 Harry M. Markowitz, William F. Sharpe, Merton H. Miller, U.S.
1989 Trygve Haavelmo, Nor.
1988 Maurice Allais, Fr.
1987 Robert M. Solow, U.S.
1986 James M. Buchanan, U.S.
1985 Franco Modigliani, It.-U.S.
1984 Richard Stone, Br.
1983 Gerard Debreu, Fr.-U.S.
1982 George J. Stigler, U.S.
1981 James Tobin, U.S.
1980 Lawrence R. Klein, U.S.
1979 Theodore W. Schultz, U.S.; Sir Arthur Lewis, Br.
1978 Herbert A. Simon, U.S.
1977 Bertil Ohlin, Swed.; James E. Meade, Br.
1976 Milton Friedman, U.S.
1975 Tjalling Koopmans, Dutch-U.S.; Leonid Kantorovich, USSR
1974 Gunnar Myrdal, Swed.; Friedrich A. von Hayek, Austrian
1973 Wassily Leontief, U.S.
1972 Kenneth J. Arrow, U.S.; John R. Hicks, Br.
1971 Simon Kuznets, U.S.
1970 Paul A. Samuelson, U.S.
1969 Ragnar Frisch, Nor.; Jan Tinbergen, Dutch

Pulitzer Prizes in Journalism, Letters, and Music

The Pulitzer Prizes were endowed by Joseph Pulitzer (1847-1911), publisher of the *New York World*, in a bequest to Columbia Univ. and have been awarded annually, in years shown, for work the previous year. Prizes are now $5,000 in each category, except Public Service (in Journalism), for which a medal is given. For letters and music, prizes in past years are listed; if a year is omitted, no award was given that year.

Journalism, 2001

Public Service: *Oregonian, Portland, OR; for series examining problems and abuse within the U.S. Immigration and Naturalization Service.*

Breaking News Reporting: *Miami Herald* staff; for first-hand coverage of the raid that took Elian Gonzalez from his Miami relatives and returned him to his Cuban father.

Investigative Reporting: David Willman, *Los Angeles Times; for exposé on unsafe prescription drugs.*

Explanatory Reporting: *Chicago Tribune* staff; "Gateway to Gridlock," a profile of the American air traffic system.

Beat Reporting: David Cay Johnston, *The New York Times;* for reporting on loopholes in the U.S. tax code.

National Reporting: *The New York Times* staff; for series on racial experiences and attitudes in the U.S.

International Reporting: Ian Johnson, *Wall Street Journal; for series on the Chinese government's suppression of the Falun Gong movement.* Paul Salopek, *Chicago Tribune*; reporting on political strife and disease epidemics in Africa.

Feature Writing: Tom Hallman, Jr.,*Oregonian*, Portland, OR; for profile of disfigured 14-year-old boy who elects to have life-threatening surgery to improve his appearance.

Commentary: Dorothy Rabinowitz, *Wall Street Journal;* for articles on American society and culture.

Criticism: Gail Caldwell, *Boston Globe;* for observations on contemporary life and literature.

Editorial Writing: David Moats, *Rutland Herald,* VT; for editorials on the issue of civil unions for same-sex couples.

Editorial Cartooning: Ann Telnaes, *Los Angeles Times Syndicate*

Breaking News Photography: Alan Diaz, *Associated Press*; for photos of armed federal agents seizing Elian Gonzalez in Miami.

Feature Photography: Matt Rainey of the Newark, NJ, *Star-Ledger;* for photographs on the recovery of 2 students critically burned in a dormitory fire at Seton Hall Univ.

Letters

Fiction

1918—Ernest Poole, *His Family*
1919—Booth Tarkington, *The Magnificent Ambersons*
1921—Edith Wharton, *The Age of Innocence*
1922—Booth Tarkington, *Alice Adams*
1923—Willa Cather, *One of Ours*
1924—Margaret Wilson, *The Able McLaughlins*
1925—Edna Ferber, *So Big*
1926—Sinclair Lewis, *Arrowsmith* (refused prize)
1927—Louis Bromfield, *Early Autumn*
1928—Thornton Wilder, *Bridge of San Luis Rey*
1929—Julia M. Peterkin, *Scarlet Sister Mary*
1930—Oliver LaFarge, *Laughing Boy*
1931—Margaret Ayer Barnes, *Years of Grace*
1932—Pearl S. Buck, *The Good Earth*
1933—T. S. Stribling, *The Store*
1934—Caroline Miller, *Lamb in His Bosom*
1935—Josephine W. Johnson, *Now in November*
1936—Harold L. Davis, *Honey in the Horn*
1937—Margaret Mitchell, *Gone With the Wind*
1938—John P. Marquand, *The Late George Apley*
1939—Marjorie Kinnan Rawlings, *The Yearling*
1940—John Steinbeck, *The Grapes of Wrath*
1942—Ellen Glasgow, *In This Our Life*
1943—Upton Sinclair, *Dragon's Teeth*
1944—Martin Flavin, *Journey in the Dark*
1945—John Hersey, *A Bell for Adano*
1947—Robert Penn Warren, *All the King's Men*
1948—James A. Michener, *Tales of the South Pacific*
1949—James Gould Cozzens, *Guard of Honor*
1950—A. B. Guthrie Jr., *The Way West*
1951—Conrad Richter, *The Town*
1952—Herman Wouk, *The Caine Mutiny*
1953—Ernest Hemingway, *The Old Man and the* Sea
1955—William Faulkner, *A Fable*
1956—MacKinlay Kantor, *Andersonville*
1958—James Agee, *A Death in the Family*
1959—Robert Lewis Taylor, *The Travels of Jaimie McPheeters*
1960—Allen Drury, *Advise and Consent*
1961—Harper Lee, *To Kill a Mockingbird*
1962—Edwin O'Connor, *The Edge of Sadness*
1963—William Faulkner, *The Reivers*
1965—Shirley Ann Grau, *The Keepers of the House*
1966—Katherine Anne Porter, *Collected Stories*
1967—Bernard Malamud, *The Fixer*
1968—William Styron, *The Confessions of Nat Turner*
1969—N. Scott Momaday, *House Made of Dawn*
1970—Jean Stafford, *Collected Stories*
1972—Wallace Stegner, *Angle of Repose*
1973—Eudora Welty, *The Optimist's Daughter*
1975—Michael Shaara, *The Killer Angels*
1976—Saul Bellow, *Humboldt's Gift*
1978—James Alan McPherson, *Elbow Room*
1979—John Cheever, *The Stories of John Cheever*
1980—Norman Mailer, *The Executioner's Song*
1981—John Kennedy Toole, *A Confederacy of Dunces*
1982—John Updike, *Rabbit Is Rich*
1983—Alice Walker, *The Color Purple*
1984—William Kennedy, *Ironweed*
1985—Alison Lurie, *Foreign Affairs*
1986—Larry McMurtry, *Lonesome Dove*
1987—Peter Taylor, *A Summons to Memphis*
1988—Toni Morrison, *Beloved*
1989—Anne Tyler, *Breathing Lessons*
1990—Oscar Hijuelos, *The Mambo Kings Play Songs of Love*
1991—John Updike, *Rabbit at Rest*
1992—Jane Smiley, *A Thousand Acres*
1993—Robert Olen Butler, *A Good Scent From a Strange Mountain*
1994—E. Annie Proulx, *The Shipping News*
1995—Carol Shields, *The Stone Diaries*
1996—Richard Ford, *Independence Day*
1997—Steven Millhauser, *Martin Dressler: The Tale of an American Dreamer*
1998—Philip Roth, *American Pastoral*
1999—Michael Cunningham, *The Hours*
2000—Jhumpa Lahiri, *Interpreter of Maladies*
2001—Michael Chabon, *The Amazing Adventures of Kavalier & Clay*

Drama

1918—Jesse Lynch Williams, *Why Marry?*
1920—Eugene O'Neill, *Beyond the Horizon*
1921—Zona Gale, *Miss Lulu Bett*
1922—Eugene O'Neill, *Anna Christie*
1923—Owen Davis, *Icebound*
1924—Hatcher Hughes, *Hell-Bent for Heaven*
1925—Sidney Howard, *They Knew What They Wanted*
1926—George Kelly, *Craig's Wife*
1927—Paul Green, *In Abraham's Bosom*
1928—Eugene O'Neill, *Strange Interlude*
1929—Elmer Rice, *Street Scene*
1930—Marc Connelly, *The Green Pastures*
1931—Susan Glaspell, *Alison's House*
1932—George S. Kaufman, Morrie Ryskind, and Ira Gershwin, *Of Thee I Sing*
1933—Maxwell Anderson, *Both Your Houses*
1934—Sidney Kingsley, *Men in White*
1935—Zoe Akins, *The Old Maid*
1936—Robert E. Sherwood, *Idiot's Delight*
1937—George S. Kaufman and Moss Hart, *You Can't Take It With You*
1938—Thornton Wilder, *Our Town*
1939—Robert E. Sherwood, *Abe Lincoln in Illinois*
1940—William Saroyan, *The Time of Your Life*
1941—Robert E. Sherwood, *There Shall Be No Night*
1943—Thornton Wilder, *The Skin of Our Teeth*
1945—Mary Chase, *Harvey*
1946—Russel Crouse and Howard Lindsay, *State of the Union*
1948—Tennessee Williams, *A Streetcar Named Desire*
1949—Arthur Miller, *Death of a Salesman*
1950—Richard Rodgers, Oscar Hammerstein 2d and Joshua Logan, *South Pacific*
1952—Joseph Kramm, *The Shrike*
1953—William Inge, *Picnic*
1954—John Patrick, *Teahouse of the August Moon*
1955—Tennessee Williams, *Cat on a Hot Tin Roof*
1956—Frances Goodrich and Albert Hackett, *The Diary of Anne Frank*
1957—Eugene O'Neill, *Long Day's Journey Into Night*
1958—Ketti Frings, *Look Homeward, Angel*
1959—Archibald MacLeish, *J. B.*
1960—George Abbott, Jerome Weidman, Sheldon Harnick, and Jerry Bock, *Fiorello!*
1961—Tad Mosel, *All the Way Home*
1962—Frank Loesser and Abe Burrows, *How to Succeed in Business Without Really Trying*
1965—Frank D. Gilroy, *The Subject Was Roses*
1967—Edward Albee, *A Delicate Balance*
1969—Howard Sackler, *The Great White Hope*
1970—Charles Gordone, *No Place to Be Somebody*
1971—Paul Zindel, *The Effect of Gamma Rays on Man-in-the-Moon Marigolds*
1973—Jason Miller, *That Championship Season*
1975—Edward Albee, *Seascape*
1976—Michael Bennett, James Kirkwood, Nicholas Dante, Marvin Hamlisch, and Edward Kleban, *A Chorus Line*
1977—Michael Cristofer, *The Shadow Box*
1978—Donald L. Coburn, *The Gin Game*
1979—Sam Shepard, *Buried Child*
1980—Lanford Wilson, *Talley's Folly*
1981—Beth Henley, *Crimes of the Heart*
1982—Charles Fuller, *A Soldier's Play*
1983—Marsha Norman, *'night, Mother*
1984—David Mamet, *Glengarry Glen Ross*
1985—Stephen Sondheim and James Lapine, *Sunday in the Park With George*
1987—August Wilson, *Fences*
1988—Alfred Uhry, *Driving Miss Daisy*
1989—Wendy Wasserstein, *The Heidi Chronicles*
1990—August Wilson, *The Piano Lesson*
1991—Neil Simon, *Lost in Yonkers*
1992—Robert Schenkkan, *The Kentucky Cycle*
1993—Tony Kushner, *Angels in America: Millennium Approaches*
1994—Edward Albee, *Three Tall Women*
1995—Horton Foote, *The Young Man From Atlanta*
1996—Jonathan Larson, *Rent*
1998—Paula Vogel, *How I Learned to Drive*
1999—Margaret Edson, *Wit*
2000—Donald Margulies, *Dinner With Friends*
2001—David Auburn, *Proof*

History (U.S.)

1917—J. J. Jusserand, *With Americans of Past and Present Days*
1918—James Ford Rhodes, *History of the Civil War*
1920—Justin H. Smith, *The War With Mexico*
1921—William Sowden Sims, *The Victory at Sea*
1922—James Truslow Adams, *The Founding of New England*
1923—Charles Warren, *The Supreme Court in United States History*
1924—Charles Howard McIlwain, *The American Revolution: A Constitutional Interpretation*
1925—Frederick L. Paxton, *A History of the American Frontier*
1926—Edward Channing, *A History of the U.S.*
1927—Samuel Flagg Bemis, *Pinckney's Treaty*
1928—V. L Parrington, *Main Currents in American Thought*

1929—Fred A. Shannon, *The Organization and Administration of the Union Army, 1861-65*
1930—Claude H. Van Tyne, *The War of Independence*
1931—Bernadotte E. Schmitt, *The Coming of the War, 1914*
1932—Gen. John J. Pershing, *My Experiences in the World War*
1933—Frederick J. Turner, *The Significance of Sections in American History*
1934—Herbert Agar, *The People's Choice*
1935—Charles McLean Andrews, *The Colonial Period of American History*
1936—Andrew C. McLaughlin, *The Constitutional History of the United States*
1937—Van Wyck Brooks, *The Flowering of New England*
1938—Paul Herman Buck, *The Road to Reunion, 1865-1900*
1939—Frank Luther Mott, *A History of American Magazines*
1940—Carl Sandburg, *Abraham Lincoln: The War Years*
1941—Marcus Lee Hansen, *The Atlantic Migration, 1607-1860*
1942—Margaret Leech, *Reveille in Washington*
1943—Esther Forbes, *Paul Revere and the World He Lived In*
1944—Merle Curti, *The Growth of American Thought*
1945—Stephen Bonsal, *Unfinished Business*
1946—Arthur M. Schlesinger Jr., *The Age of Jackson*
1947—James Phinney Baxter 3d, *Scientists Against Time*
1948—Bernard De Voto, *Across the Wide Missouri*
1949—Roy F. Nichols, *The Disruption of American Democracy*
1950—O. W. Larkin, *Art and Life in America*
1951—R. Carlyle Buley, *The Old Northwest: Pioneer Period 1815-1840*
1952—Oscar Handlin, *The Uprooted*
1953—George Dangerfield, *The Era of Good Feelings*
1954—Bruce Catton, *A Stillness at Appomattox*
1955—Paul Horgan, *Great River: The Rio Grande in North American History*
1956—Richard Hofstadter, *The Age of Reform*
1957—George F. Kennan, *Russia Leaves the War*
1958—Bray Hammond, *Banks and Politics in America—From the Revolution to the Civil War*
1959—Leonard D. White and Jean Schneider, *The Republican Era; 1869-1901*
1960—Margaret Leech, *In the Days of McKinley*
1961—Herbert Feis, *Between War and Peace: The Potsdam Conference*
1962—Lawrence H. Gibson, *The Triumphant Empire: Thunderclouds Gather in the West*
1963—Constance McLaughlin Green, *Washington: Village and Capital, 1800-1878*
1964—Sumner Chilton Powell, *Puritan Village: The Formation of a New England Town*
1965—Irwin Unger, *The Greenback Era*
1966—Perry Miller, *Life of the Mind in America*
1967—William H. Goetzmann, *Exploration and Empire: The Explorer and Scientist in the Winning of the American West*
1968—Bernard Bailyn, *The Ideological Origins of the American Revolution*
1969—Leonard W. Levy, *Origin of the Fifth Amendment*
1970—Dean Acheson, *Present at the Creation: My Years in the State Department*
1971—James McGregor Burns, *Roosevelt: The Soldier of Freedom*
1972—Carl N. Degler, *Neither Black nor White*
1973—Michael Kammen, *People of Paradox: An Inquiry Concerning the Origins of American Civilization*
1974—Daniel J. Boorstin, *The Americans: The Democratic Experience*
1975—Dumas Malone, *Jefferson and His Time*
1976—Paul Horgan, *Lamy of Santa Fe*
1977—David M. Potter, *The Impending Crisis*
1978—Alfred D. Chandler Jr., *The Visible Hand: The Managerial Revolution in American Business*
1979—Don E. Fehrenbacher, *The Dred Scott Case: Its Significance in American Law and Politics*
1980—Leon F. Litwack, *Been in the Storm So Long*
1981—Lawrence A. Cremin, *American Education: The National Experience, 1783-1876*
1982—C. Vann Woodward, ed., *Mary Chesnut's Civil War*
1983—Rhys L. Issac, *The Transformation of Virginia, 1740-1790*
1985—Thomas K. McCraw, *Prophets of Regulation*
1986—Walter A. McDougall, *The Heavens and the Earth*
1987—Bernard Bailyn, *Voyagers to the West*
1988—Robert V. Bruce, *The Launching of Modern American Science, 1846-1876*
1989—Taylor Branch, *Parting the Waters: America in the King Years, 1954-63*; and James M. McPherson, *Battle Cry of Freedom: The Civil War Era*
1990—Stanley Karnow, *In Our Image: America's Empire in the Philippines*
1991—Laurel Thatcher Ulrich, *A Midwife's Tale: The Life of Martha Ballard,* based on her diary, 1785-1812
1992—Mark E. Neely Jr., *The Fate of Liberty: Abraham Lincoln and Civil Liberties*
1993—Gordon S. Wood, *The Radicalism of the American Revolution*
1995—Doris Kearns Goodwin, *No Ordinary Time: Franklin and Eleanor Roosevelt: The Home Front in World War II*
1996—Alan Taylor, *William Cooper's Town: Power and Persuasion on the Frontier of the Early American Republic*
1997—Jack N. Rakove, *Original Meanings: Politics and Ideas in the Making of the Constitution*
1998—Edward J. Larson, *Summer for the Gods: The Scopes Trial and America's Continuing Debate Over Science and Religion*
1999—Edwin G. Burrows and Mike Wallace, *Gotham: A History of New York City to 1898*
2000—David M. Kennedy, *Freedom From Fear: The American People in Depression and War, 1929-1945*
2001—Joseph J. Ellis, *Founding Brothers: The Revolutionary Generation*

Biography or Autobiography

1917—Laura E. Richards and Maude Howe Elliott, assisted by Florence Howe Hall, *Julia Ward Howe*
1918—William Cabell Bruce, *Benjamin Franklin, Self-Revealed*
1919—Henry Adams, *The Education of Henry Adams*
1920—Albert J. Beveridge, *The Life of John Marshall*
1921—Edward Bok, *The Americanization of Edward Bok*
1922—Hamlin Garland, *A Daughter of the Middle Border*
1923—Burton J. Hendrick, *The Life and Letters of Walter H. Page*
1924—Michael Pupin, *From Immigrant to Inventor*
1925—M. A. DeWolfe Howe, *Barrett Wendell and His Letters*
1926—Harvey Cushing, *Life of Sir William Osler*
1927—Emory Holloway, *Whitman: An Interpretation in Narrative*
1928—Charles Edward Russell, *The American Orchestra and Theodore Thomas*
1929—Burton J. Hendrick, *The Training of an American: The Earlier Life and Letters of Walter H. Page*
1930—Marquis James, *The Raven* (Sam Houston)
1931—Henry James, *Charles W. Eliot*
1932—Henry F. Pringle, *Theodore Roosevelt*
1933—Allan Nevins, *Grover Cleveland*
1934—Tyler Dennett, *John Hay*
1935—Douglas Southall Freeman, *R. E. Lee*
1936—Ralph Barton Perry, *The Thought and Character of William James*
1937—Allan Nevins, *Hamilton Fish: The Inner History of the Grant Administration*
1938—Divided between Odell Shepard, *Pedlar's Progress* (Bronson Alcott) and Marquis James, *Andrew Jackson*
1939—Carl Van Doren, *Benjamin Franklin*
1940—Ray Stannard Baker, *Woodrow Wilson, Life and Letters*
1941—Ola Elizabeth Winslow, *Jonathan Edwards*
1942—Forrest Wilson, *Crusader in Crinoline* (Harriet Beecher Stowe)
1943—Samuel Eliot Morison, *Admiral of the Ocean Sea* (Christopher Columbus)
1944—Carleton Mabee, *The American Leonardo: The Life of Samuel F. B. Morse*
1945—Russell Blaine Nye, *George Bancroft: Brahmin Rebel.*
1946—Linny Marsh Wolfe, *Son of the Wilderness* (John Muir)
1947—William Allen White, *Autobiography of William Allen White*
1948—Margaret Clapp, *Forgotten First Citizen: John Bigelow*
1949—Robert E. Sherwood, *Roosevelt and Hopkins*
1950—Samuel Flagg Bemis, *John Quincy Adams and the Foundations of American Foreign Policy*
1951—Margaret Louise Coit, *John C. Calhoun: American Portrait*
1952—Merlo J. Pusey, *Charles Evans Hughes*
1953—David J. Mays, *Edmund Pendleton, 1721-1803*
1954—Charles A. Lindbergh, *The Spirit of St. Louis*
1955—William S. White, *The Taft Story*
1956—Talbot F. Hamlin, *Benjamin Henry Latrobe*
1957—John F. Kennedy, *Profiles in Courage*
1958—Douglas Southall Freeman (Vols. I-VI) and John Alexander Carroll and Mary Wells Ashworth (Vol. VII), *George Washington*
1959—Arthur Walworth, *Woodrow Wilson: American Prophet*
1960—Samuel Eliot Morison, *John Paul Jones*
1961—David Donald, *Charles Sumner and the Coming of the Civil War*
1963—Leon Edel, *Henry James: Vols. 2-3*
1964—Walter Jackson Bate, *John Keats*
1965—Ernest Samuels, *Henry Adams*
1966—Arthur M. Schlesinger Jr., *A Thousand Days*
1967—Justin Kaplan, *Mr. Clemens and Mark Twain*
1968—George F. Kennan, *Memoirs (1925-1950)*
1969—B. L. Reid, *The Man From New York: John Quinn and His Friends*
1970—T. Harry Williams, *Huey Long*
1971—Lawrence Thompson, *Robert Frost: The Years of Triumph, 1915-1938*
1972—Joseph P. Lash, *Eleanor and Franklin*

1973—W. A. Swanberg, *Luce and His Empire*
1974—Louis Sheaffer, *O'Neill, Son and Artist*
1975—Robert A. Caro, *The Power Broker: Robert Moses and the Fall of New York*
1976—R.W.B. Lewis, *Edith Wharton: A Biography*
1977—John E. Mack, *A Prince of Our Disorder: The Life of T. E. Lawrence*
1978—Walter Jackson Bate, *Samuel Johnson*
1979—Leonard Baker, *Days of Sorrow and Pain: Leo Baeck and the Berlin Jews*
1980—Edmund Morris, *The Rise of Theodore Roosevelt*
1981—Robert K. Massie, *Peter the Great: His Life and World*
1982—William S. McFeely, *Grant: A Biography*
1983—Russell Baker, *Growing Up*
1984—Louis R. Harlan, *Booker T. Washington*
1985—Kenneth Silverman, *The Life and Times of Cotton Mather*
1986—Elizabeth Frank, *Louise Bogan: A Portrait*
1987—David J. Garrow, *Bearing the Cross: Martin Luther King Jr. and the Southern Christian Leadership Conference*
1988—David Herbert Donald, *Look Homeward: A Life of Thomas Wolfe*
1989—Richard Ellmann, *Oscar Wilde*
1990—Sebastian de Grazia, *Machiavelli in Hell*
1991—Steven Naifeh and Gregory White Smith, *Jackson Pollock: An American Saga*
1992—Lewis B. Puller Jr., *Fortunate Son: The Healing of a Vietnam Vet*
1993—David McCullough, *Truman*
1994—David Levering Lewis, *W.E.B. DuBois: Biography of a Race, 1868-1919*
1995—Joan D. Hedrick, *Harriet Beecher Stowe: A Life*
1996—Jack Miles, *God: A Biography*
1997—Frank McCourt, *Angela's Ashes: A Memoir*
1998—Katharine Graham, *Personal History*
1999—A. Scott Berg, *Lindbergh*
2000—Stacy Schiff, *Véra (Mrs. Vladimir Nabokov)*
2001—David Levering Lewis, *W.E.B. Du Bois: The Fight for Equality and the American Century, 1919-1963*

American Poetry

Before 1922, awards were funded by the Poetry Society: **1918**—*Love Songs*, by Sara Teasdale; **1919**—*Old Road to Paradise*, by Margaret Widdemer; *Corn Huskers*, by Carl Sandburg.

1922—Edwin Arlington Robinson, *Collected Poems*
1923—Edna St. Vincent Millay, *The Ballad of the Harp-Weaver; A Few Figs From Thistles; other works*
1924—Robert Frost, *New Hampshire: A Poem With Notes and Grace Notes*
1925—Edwin Arlington Robinson, *The Man Who Died Twice*
1926—Amy Lowell, *What's O'Clock*
1927—Leonora Speyer, *Fiddler's Farewell*
1928—Edwin Arlington Robinson, *Tristram*
1929—Stephen Vincent Benet, *John Brown's Body*
1930—Conrad Aiken, *Selected Poems*
1931—Robert Frost, *Collected Poems*
1932—George Dillon, *The Flowering Stone*
1933—Archibald MacLeish, *Conquistador*
1934—Robert Hillyer, *Collected Verse*
1935—Audrey Wurdemann, *Bright Ambush*
1936—Robert P. Tristram Coffin, *Strange Holiness*
1937—Robert Frost, *A Further Range*
1938—Marya Zaturenska, *Cold Morning Sky*
1939—John Gould Fletcher, *Selected Poems*
1940—Mark Van Doren, *Collected Poems*
1941—Leonard Bacon, *Sunderland Capture*
1942—William Rose Benet, *The Dust Which Is God*
1943—Robert Frost, *A Witness Tree*
1944—Stephen Vincent Benet, *Western Star*
1945—Karl Shapiro, *V-Letter and Other Poems*
1947—Robert Lowell, *Lord Weary's Castle*
1948—W. H. Auden, *The Age of Anxiety*
1949—Peter Viereck, *Terror and Decorum*
1950—Gwendolyn Brooks, *Annie Allen*
1951—Carl Sandburg, *Complete Poems*
1952—Marianne Moore, *Collected Poems*
1953—Archibald MacLeish, *Collected Poems*
1954—Theodore Roethke, *The Waking*
1955—Wallace Stevens, *Collected Poems*
1956—Elizabeth Bishop, *Poems, North and South*
1957—Richard Wilbur, *Things of This World*
1958—Robert Penn Warren, *Promises: Poems 1954-1956*
1959—Stanley Kunitz, *Selected Poems 1928-1958*
1960—W. D. Snodgrass, *Heart's Needle*
1961—Phyllis McGinley, *Times Three: Selected Verse From Three Decades*
1962—Alan Dugan, *Poems*
1963—William Carlos Williams, *Pictures From Breughel*
1964—Louis Simpson, *At the End of the Open Road*
1965—John Berryman, *77 Dream Songs*
1966—Richard Eberhart, *Selected Poems*
1967—Anne Sexton, *Live or Die*
1968—Anthony Hecht, *The Hard Hours*
1969—George Oppen, *Of Being Numerous*
1970—Richard Howard, *Untitled Subjects*
1971—William S. Merwin, *The Carrier of Ladders*
1972—James Wright, *Collected Poems*
1973—Maxine Winokur Kumin, *Up Country*
1974—Robert Lowell, *The Dolphin*
1975—Gary Snyder, *Turtle Island*
1976—John Ashbery, *Self-Portrait in a Convex Mirror*
1977—James Merrill, *Divine Comedies*
1978—Howard Nemerov, *Collected Poems*
1979—Robert Penn Warren, *Now and Then: Poems 1976-1978*
1980—Donald Justice, *Selected Poems*
1981—James Schuyler, *The Morning of the Poem*
1982—Sylvia Plath, *The Collected Poems*
1983—Galway Kinnell, *Selected Poems*
1984—Mary Oliver, *American Primitive*
1985—Carolyn Kizer, *Yin*
1986—Henry Taylor, *The Flying Change*
1987—Rita Dove, *Thomas and Beulah*
1988—William Meredith, *Partial Accounts: New and Selected Poems*
1989—Richard Wilbur, *New and Collected Poems*
1990—Charles Simic, *The World Doesn't End*
1991—Mona Van Duyn, *Near Changes*
1992—James Tate, *Selected Poems*
1993—Louise Glück, *The Wild Iris*
1994—Yusef Komunyakaa, *Neon Vernacular*
1995—Philip Levine, *The Simple Truth*
1996—Jorie Graham, *The Dream of the Unified Field*
1997—Lisel Mueller, *Alive Together: New and Selected Poems*
1998—Charles Wright, *Black Zodiac*
1999—Mark Strand, *Blizzard of One*
2000—C. K. Williams, *Repair*
2001—Stephen Dunn, *Different Hours*

General Nonfiction

1962—Theodore H. White, *The Making of the President 1960*
1963—Barbara W. Tuchman, *The Guns of August*
1964—Richard Hofstadter, *Anti-Intellectualism in American Life*
1965—Howard Mumford Jones, *O Strange New World*
1966—Edwin Way Teale, *Wandering Through Winter*
1967—David Brion Davis, *The Problem of Slavery in Western Culture*
1968—Will and Ariel Durant, *Rousseau and Revolution*
1969—Norman Mailer, *The Armies of the Night;* Rene Jules Dubos, *So Human an Animal: How We Are Shaped by Surroundings and Events*
1970—Eric H. Erikson, *Gandhi's Truth*
1971—John Toland, *The Rising Sun*
1972—Barbara W. Tuchman, *Stilwell and the American Experience in China, 1911-1945*
1973—Frances FitzGerald, *Fire in the Lake: The Vietnamese and the Americans in Vietnam*; Robert Coles, *Children of Crisis,* Volumes II & III
1974—Ernest Becker, *The Denial of Death*
1975—Annie Dillard, *Pilgrim at Tinker Creek*
1976—Robert N. Butler, *Why Survive? Being Old in America*
1977—William W. Warner, *Beautiful Swimmers*
1978—Carl Sagan, *The Dragons of Eden*
1979—Edward O. Wilson, *On Human Nature*
1980—Douglas R. Hofstadter, *Gödel, Escher, Bach: An Eternal Golden Braid*
1981—Carl E. Schorske, *Fin-de-Siecle Vienna: Politics and Culture*
1982—Tracy Kidder, *The Soul of a New Machine*
1983—Susan Sheehan, *Is There No Place on Earth for Me?*
1984—Paul Starr, *Social Transformation of American Medicine*
1985—Studs Terkel, *The Good War*
1986—Joseph Lelyveld, *Move Your Shadow;* J. Anthony Lukas, *Common Ground*
1987—David K. Shipler, *Arab and Jew*
1988—Richard Rhodes, *The Making of the Atomic Bomb*
1989—Neil Sheehan, *A Bright Shining Lie: John Paul Vann and America in Vietnam*
1990—Dale Maharidge and Michael Williamson, *And Their Children After Them*
1991—Bert Holldobler and Edward O. Wilson, *The Ants*
1992—Daniel Yergin, *The Prize: The Epic Quest for Oil*
1993—Garry Wills, *Lincoln at Gettysburg*
1994—David Remnick, *Lenin's Tomb: The Last Days of the Soviet Empire*
1995—Jonathan Weiner, *The Beak of the Finch: A Story of Evolution in Our Time*
1996—Tina Rosenberg, *The Haunted Land: Facing Europe's Ghosts After Communism*
1997—Richard Kluger, *Ashes to Ashes: America's Hundred-Year Cigarette War, the Public Health, and the Unabashed Triumph of Philip Morris*
1998—Jared Diamond, *Guns, Germs, and Steel: The Fates of Human Societies*

1999—John McPhee, *Annals of the Former World*
2000—John W. Dower, *Embracing Defeat: Japan in the Wake of World War II*
2001—Herbert P. Bix, *Hirohito and the Making of Modern Japan*

Special Citation in Letters

1944—Richard Rodgers and Oscar Hammerstein II, for *Oklahoma!*
1957—Kenneth Roberts, for his historical novels
1960—*The Armada*, by Garrett Mattingly
1961—*American Heritage Picture History of the Civil War*
1973—*George Washington, Vols. I-IV*, by James Thomas Flexner
1977—Alex Haley, for *Roots*
1978—E.B. White
1984—Theodore Seuss Geisel (Dr. Seuss)
1992—Art Spiegleman, for *Maus*

Music

1943—William Schuman, *Secular Cantata No. 2, A Free Song*
1944—Howard Hanson, *Symphony No. 4, Op. 34*
1945—Aaron Copland, *Appalachian Spring*
1946—Leo Sowerby, *The Canticle of the Sun*
1947—Charles E. Ives, *Symphony No. 3*
1948—Walter Piston, *Symphony No. 3*
1949—Virgil Thomson, *Louisiana Story*
1950—Gian-Carlo Menotti, *The Consul*
1951—Douglas Moore, *Giants in the Earth*
1952—Gail Kubik, *Symphony Concertante*
1954—Quincy Porter, *Concerto for Two Pianos and Orchestra*
1955—Gian-Carlo Menotti, *The Saint of Bleecker Street*
1956—Ernest Toch, *Symphony No. 3*
1957—Norman Dello Joio, *Meditations on Ecclesiastes*
1958—Samuel Barber, *Vanessa*
1959—John La Montaine, *Concerto for Piano and Orchestra*
1960—Elliott Carter, *Second String Quartet*
1961—Walter Piston, *Symphony No. 7*
1962—Robert Ward, *The Crucible*
1963—Samuel Barber, *Piano Concerto No. 1*
1966—Leslie Bassett, *Variations for Orchestra*
1967—Leon Kirchner, *Quartet No. 3*
1968—George Crumb, *Echoes of Time and The River*
1969—Karel Husa, *String Quartet No. 3*
1970—Charles W. Wuorinen, *Time's Encomium*
1971—Mario Davidovsky, *Synchronisms No. 6*
1972—Jacob Druckman, *Windows*
1973—Elliott Carter, *String Quartet No. 3*
1974—Donald Martino, *Notturno*
1975—Dominick Argento, *From the Diary of Virginia Woolf*
1976—Ned Rorem, *Air Music*
1977—Richard Wernick, *Visions of Terror and Wonder*
1978—Michael Colgrass, *Deja Vu for Percussion and Orchestra*
1979—Joseph Schwantner, *Aftertones of Infinity*
1980—David Del Tredici, *In Memory of a Summer Day*
1982—Roger Sessions, *Concerto for Orchestra*
1983—Ellen T. Zwilich, *Three Movements for Orchestra*
1984—Bernard Rands, *Canti del Sole*
1985—Stephen Albert, *Symphony, RiverRun*
1986—George Perle, *Wind Quintet IV*
1987—John Harbison, *The Flight Into Egypt*
1988—William Bolcom, *12 New Etudes for Piano*
1989—Roger Reynolds, *Whispers Out of Time*
1990—Mel Powell, *Duplicates: A Concerto for Two Pianos and Orchestra*
1991—Shulamit Ran, *Symphony*
1992—Wayne Peterson, *The Face of the Night, The Heart of the Dark*
1993—Christopher Rouse, *Trombone Concerto*
1994—Gunther Schuller, *Of Reminiscences and Reflections*
1995—Morton Gould, *Stringmusic*
1996—George Walker, *Lilacs*
1997—Wynton Marsalis, *Blood on the Fields*
1998—Aaron Jay Kernis, *String Quartet No. 2*
1999—Melinda Wagner, *Concerto for Flute, Strings and Percussion*
2000—Lewis Spratlan, *Life is a Dream, Opera in Three Acts: Act II, Concert Version*
2001—John Corigliano, *Symphony No. 2 for String Orchestra*

Special Citation in Music

1974—Roger Sessions
1976—Scott Joplin
1982—Milton Babbitt
1985—William Schuman
1998—George Gershwin
1999—Edward Kennedy "Duke" Ellington

Miscellaneous Book Awards

Year in parentheses is year awarded

Academy of American Poets Awards (2000). Lenore Marshall Poetry Prize, $10,000: David Ferry, *Of No Country I Know: New and Selected Poems and Translations*; Raiziss/de Palchi Translation Award (fellowship) $20,000: Emanuel di Pasquale, *Sharing a Trip: Selected Poems* by Silvio Ramat; Academy Fellowship, $20,000 stipend: Lyn Hejinian; Wallace Stevens Award, for mastery in the art of poetry, $150,000: Frank Bidart. (2001) Landon Translation Award, $1,000: (co-winners) Clayton Eshleman, *Trilce* by César Vallejo; Edward Snow, Duino Elegies by Rainer Maria Rilke; Walt Whitman Award, $5,000: John Canaday, *The Invisible World*

American Academy of Arts and Letters (2001). Gold Medal for Fiction: Philip Roth; Award of Merit for the Short Story: Frederick Busch; Academy Awards in Literature ($7,500 each): Guy Davenport, David Ferry, Charles Mee, Alice Notley, Carl Phillips, Frederic Tuten, Tobias Wolff, James Wood. Witter Bynner Prize for Poetry, $2,500: Rachel Wetzsteon; E. M. Forster Award, $15,000: Marina Carr; Sue Kaufman Prize for First Fiction, $2,500: Akhil Sharma, *An Obedient Father;* Addison Metcalf Award, $5,000: Dave Eggers, *A Heartbreaking Work of Staggering Genius*; Richard and Hinda Rosenthal Foundation Award, $5,000: David Ebershoff, *The Danish Girl*; Harold D. Vursell Memorial Award in Literature, $5,000: Anne Hollander; Morton Dauwen Zabel Award for Fiction, $5,000: Paul Violi; Rome Fellowship in Literature, one-year residence at the American Academy in Rome (2001-2002): Mark Halliday, Vincent Katz

Booker Prize (2000) by Booker PLC for best novel written in English by a UK, Commonwealth, or South African author. Margaret Atwood, *The Blind Assassin*

Golden Kite Awards (2001), by Society of Children's Book Writers and Illustrators. Fiction: Kathleen Karr, *The Boxer;* nonfiction: Ellen Levine, *Darkness Over Denmark;* picture-illustration: David Shannon, *The Rain Came Down*; picture book text: Jane Kurtz, *River Friendly, River Wild*

Hugo Awards (2001), by the World Science Fiction Convention. Novel: *Harry Potter and the Goblet of Fire*, J.K. Rowling; novella: *The Ultimate Earth*, Jack Williamson; novelette: *Millennium Babies*, Kristine Kathryn Rusch; short story: "Different Kinds of Darkness," David Langford

Nebula Awards (2001), by the Science Fiction Writers of America. Grand Master award: Philip José Farmer; Novel: *Darwin's Radio*, Greg Bear; novella: *Goddesses*, Linda Nagata; novelette: *Daddy's World*, Walter Jon Williams; short story: "macs," Terry Bisson

Coretta Scott King Award (2001), by the American Library Assn. for African American authors and illustrators of outstanding books for children and young adults. Author: Jacqueline Woodson, *Miracle's Boys*; Illustrator: Bryan Collier, *Uptown*

Lincoln Prize (2001), by Lincoln and Soldiers Institute at Gettysburg College, for contribution to Civil War studies, $35,000 and a bronze bust of Lincoln. Russell F. Weigley, *A Great Civil War: A Military and Political History, 1861-1865*; e-Lincoln Prize ($50,000): Edward L. Ayers, Anne S. Rubin, and William G. Thomas, for CD-ROM, web site, and book, *Valley of the Shadow: The Eve of War*

National Book Awards (2000), by National Book Foundation, $10,000 each. Fiction: Susan Sontag, *In America;* nonfiction: Nathaniel Philbrick, *In the Heart of the Sea: The Tragedy of the Whaleship Essex*; poetry: Lucille Clifton, *Blessing the Boats: New & Selected Poems 1988-2000*; young people's literature: Gloria Whelan, *Homeless Bird*; Medal for Distinguished Contr. to American Letters: Ray Bradbury

National Book Critics Circle Awards (2001). Fiction: Jim Crace, *Being Dead;* nonfiction: Ted Conover, *Newjack: Guarding Sing Sing;* criticism: Cynthia Ozick, *Quarrel & Quandry;* biography and autobiography: Herbert P. Bix, *Hirohito and the Making of Modern Japan;* poetry: Judy Jordan, *Carolina Ghost Woods;* Nona Balakian Citation for Excellence in Reviewing: Daniel Mendelsohn; Ivan Sandrof Lifetime Achievement Award: Barney Rosset

PEN/Faulkner Award (2001), for fiction, $15,000. Philip Roth, *The Human Stain*

Edgar Awards (2001), by the Mystery Writers of America: Grand Master award: Edward D. Hoch; best novel: *The Bottoms*, Joe R. Lansdale; best short story: "Missing in Action," Peter Robinson

Newbery Medal Books

The Newbery Medal is awarded annually by the Association for Library Service to Children, a division of the American Library Association, to the author of the most distinguished contribution to American literature for children.

Year Given	Book, Author
1922	*The Story of Mankind,* Hendrik Willem van Loon
1923	*The Voyages of Dr. Dolittle,* Hugh Lofting
1924	*The Dark Frigate,* Charles Boardman Hawes
1925	*Tales From Silver Lands,* Charles Joseph Finger
1926	*Shen of the Sea,* Arthur Bowie Chrisman
1927	*Smoky, the Cowhorse,* Will James
1928	*Gay-Neck,* Dhan Gopal Mukerji
1929	*The Trumpeter of Krakow,* Eric P. Kelly
1930	*Hitty, Her First Hundred Years,* Rachel Field
1931	*The Cat Who Went to Heaven,* Elizabeth Coatsworth
1932	*Waterless Mountain,* Laura Adams Armer
1933	*Young Fu of the Upper Yangtze,* Elizabeth Foreman Lewis
1934	*Invincible Louisa,* Cornelia Lynde Meigs
1935	*Dobry,* Monica Shannon
1936	*Caddie Woodlawn,* Carol Ryrie Brink
1937	*Roller Skates,* Ruth Sawyer
1938	*The White Stag,* Kate Seredy
1939	*Thimble Summer,* Elizabeth Enright
1940	*Daniel Boone,* James Daugherty
1941	*Call It Courage,* Armstrong Sperry
1942	*The Matchlock Gun,* Walter D. Edmonds
1943	*Adam of the Road,* Elizabeth Janet Gray
1944	*Johnny Tremain,* Esther Forbes
1945	*Rabbit Hill,* Robert Lawson
1946	*Strawberry Girl,* Lois Lenski
1947	*Miss Hickory,* Carolyn S. Bailey
1948	*Twenty-One Balloons,* William Pène Du Bois
1949	*King of the Wind,* Marguerite Henry
1950	*The Door in the Wall,* Marguerite de Angeli
1951	*Amos Fortune, Free Man,* Elizabeth Yates
1952	*Ginger Pye,* Eleanor Estes
1953	*Secret of the Andes,* Ann Nolan Clark
1954	*. . . And Now Miguel,* Joseph Krumgold
1955	*The Wheel on the School,* Meindert DeJong
1956	*Carry On, Mr. Bowditch,* Jean Lee Latham
1957	*Miracles on Maple Hill,* Virginia Sorensen
1958	*Rifles for Watie,* Harold Keith
1959	*The Witch of Blackbird Pond,* Elizabeth George Speare
1960	*Onion John,* Joseph Krumgold
1961	*Island of the Blue Dolphins,* Scott O'Dell
1962	*The Bronze Bow,* Elizabeth George Speare
1963	*A Wrinkle in Time,* Madeleine L'Engle
1964	*It's Like This, Cat,* Emily Cheney Neville
1965	*Shadow of a Bull,* Maja Wojciechowska
1966	*I, Juan de Pareja,* Elizabeth Borton de Trevino
1967	*Up a Road Slowly,* Irene Hunt
1968	*From the Mixed-Up Files of Mrs. Basil E. Frankweiler,* E. L. Konigsburg
1969	*The High King,* Lloyd Alexander
1970	*Sounder,* William H. Armstrong
1971	*The Summer of the Swans,* Betsy Byars
1972	*Mrs. Frisby and the Rats of NIMH,* Robert C. O'Brien
1973	*Julie of the Wolves,* Jean George
1974	*The Slave Dancer,* Paula Fox
1975	*M. C. Higgins the Great,* Virginia Hamilton
1976	*Grey King,* Susan Cooper
1977	*Roll of Thunder, Hear My Cry,* Mildred D. Taylor
1978	*Bridge to Terabithia,* Katherine Paterson
1979	*The Westing Game,* Ellen Raskin
1980	*A Gathering of Days,* Joan Blos
1981	*Jacob Have I Loved,* Katherine Paterson
1982	*A Visit to William Blake's Inn: Poems for Innocent and Experienced Travelers,* Nancy Willard
1983	*Dicey's Song,* Cynthia Voigt
1984	*Dear Mr. Henshaw,* Beverly Cleary
1985	*The Hero and the Crown,* Robin McKinley
1986	*Sarah, Plain and Tall,* Patricia MacLachlan
1987	*The Whipping Boy,* Sid Fleischman
1988	*Lincoln: A Photobiography,* Russell Freedman
1989	*Joyful Noise: Poems for Two Voices,* Paul Fleischman
1990	*Number the Stars,* Lois Lowry
1991	*Maniac Magee,* Jerry Spinelli
1992	*Shiloh,* Phyllis Reynolds Naylor
1993	*Missing May,* Cynthia Rylant
1994	*The Giver,* Lois Lowry
1995	*Walk Two Moons,* Sharon Creech
1996	*The Midwife's Apprentice,* Karen Cushman
1997	*The View From Saturday,* E. L. Konigsburg
1998	*Out of the Dust,* Karen Hesse
1999	*Holes,* Louis Sachar
2000	*Bud, Not Buddy,* Christopher Paul Curtis
2001	*A Year Down Yonder,* Richard Peck

Caldecott Medal Books

The Caldecott Medal is awarded annually by the Association for Library Service to Children, a division of the American Library Association, to the illustrator of the most distinguished American picture book for children.

Year Given	Book, Illustrator
1938	*Animals of the Bible,* Dorothy P. Lathrop
1939	*Mei Li,* Thomas Handforth
1940	*Abraham Lincoln,* Ingri & Edgar Parin d'Aulaire
1941	*They Were Strong and Good,* Robert Lawson
1942	*Make Way for Ducklings,* Robert McCloskey
1943	*The Little House,* Virginia Lee Burton
1944	*Many Moons,* Louis Slobodkin
1945	*Prayer for a Child,* Elizabeth Orton Jones
1946	*The Rooster Crows,* Maude & Miska Petersham
1947	*The Little Island,* Leonard Weisgard
1948	*White Snow, Bright Snow,* Roger Duvoisin
1949	*The Big Snow,* Berta & Elmer Hader
1950	*Song of the Swallows,* Leo Politi
1951	*The Egg Tree,* Karherine Milhous
1952	*Finders Keepers,* Nicolas, pseud. (Nicholas Mordvinoff)
1953	*The Biggest Bear,* Lynd Ward
1954	*Madeline's Rescue,* Ludwig Bemelmans
1955	*Cinderella, or the Little Glass Slipper,* Marcia Brown
1956	*Frog Went A-Courtin',* Feodor Rojankovsky
1957	*A Tree Is Nice,* Marc Simont
1958	*Time of Wonder,* Robert McCloskey
1959	*Chanticleer and the Fox,* Barbara Cooney
1960	*Nine Days to Christmas,* Marie Hall Ets
1961	*Baboushka and the Three Kings,* Nicolas Sidjakov
1962	*Once a Mouse,* Marcia Brown
1963	*The Snowy Day,* Ezra Jack Keats
1964	*Where the Wild Things Are,* Maurice Sendak
1965	*May I Bring a Friend?,* Beni Montressor
1966	*Always Room for One More,* Nonny Hogrogian
1967	*Sam, Bang, and Moonshine,* Evaline Ness
1968	*Drummer Hoff,* Ed Emberley
1969	*The Fool of the World and the Flying Ship,* Uri Shulevitz
1970	*Sylvester and the Magic Pebble,* William Steig
1971	*A Story A Story,* Gail E. Haley
1972	*One Fine Day,* Nonny Hogrogian
1973	*The Funny Little Woman,* Blair Lent
1974	*Duffy and the Devil,* Margot Zemach
1975	*Arrow to the Sun,* Gerald McDermott
1976	*Why Mosquitoes Buzz in People's Ears,* Leo & Diane Dillon
1977	*Ashanti to Zulu: African Traditions,* Leo & Diane Dillon
1978	*Noah's Ark,* Peter Spier
1979	*The Girl Who Loved Wild Horses,* Paul Goble
1980	*Ox-Cart Man,* Barbara Cooney
1981	*Fables,* Arnold Lobel
1982	*Jumanji,* Chris Van Allsburg
1983	*Shadow,* Marcia Brown
1984	*The Glorious Flight: Across the Channel with Louis Bleriot,* Alice and Martin Provensen
1985	*Saint George and the Dragon,* Trina Schart Hyman
1986	*The Polar Express,* Chris Van Allsburg
1987	*Hey, Al,* Richard Egielski
1988	*Owl Moon,* John Schoenherr
1989	*Song and Dance Man,* Stephen Grammell
1990	*Lon Po Po: A Red-Riding Hood Story From China,* Ed Young
1991	*Black and White,* David Macaulay
1992	*Tuesday,* David Wiesner
1993	*Mirette on the High Wire,* Emily Arnold McCully
1994	*Grandfather's Journey,* Allen Say
1995	*Smoky Night,* David Diaz
1996	*Officer Buckle and Gloria,* Peggy Rathmann
1997	*Golem,* David Wisniewski
1998	*Rapunzel,* Paul O. Zelinsky
1999	*Snowflake Bentley,* Mary Azarian
2000	*Joseph Had a Little Overcoat,* Simms Taback
2001	*So You Want to be President?,* David Small

Journalism

Year in parentheses is year awarded

National Journalism Awards (2001), by Scripps Howard Foundation, $2,500 each. Editorial writing: Debra Decker, *Dallas Morning News*; human interest writing: Tom Hallman, Jr., *The Oregonian* (Portland); environmental reporting (over 100,000 circ.): Michael Grunwald, *Washington Post*; environmental reporting (under 100,000 circ.): Sherry Devlin, *Missoulian (MT)*; public service reporting (over 100,000 circ.): *Detroit News*, (Melvin Claxton, Charles Hurt); public service reporting (under 100,000 circ.): *Chronicle-Tribune* (Marion, IN); commentary: Dennis Roddy, *Pittsburgh Post-Gazette*; photojournalism: Marc Andrew Piscotty, *Rocky Mountain News* (Denver); editorial cartooning: James Casciari, *Press Journal (Vero Beach, FL)*; college cartooning: Barry Deutsch, *Portland State Univ. Vanguard;* distinguished service to literacy: *Bakersfield Californian*; distinguished service to First Amendment: *Des Moines Register* (IA); business/economics reporting: Doris Hajewski, *Milwaukee Journal Sentinel*; web reporting: MTVi News (NY); Electronic journalism—Small market radio: KCSD-FM, South Dakota Public Radio, Sioux Falls, SD; large market radio: Minnesota Public Radio (St. Paul); small market TV/cable: WSET-TV, Lynchburg, VA; large market TV/cable: KHOU-TV, Houston, TX

National Magazine Awards (2001), by American Society of Magazine Editors and Columbia Univ. Graduate School of Journalism. Gen. excel., circ. over 1 mil: *Teen People*; 400,000 to 1 mil: *The New Yorker*, 100,000-400,000: *Mother Jones*; under 100,000: *The American Scholar*; personal service: *National Geographic Adventure*; spec. interests: *The New Yorker;* feature writing: *Rolling Stone*; fiction: *Zeotrope: All-Story;* design: *Nest;* photography: *National Geographic;* reporting: *Esquire*; public interest: *Time;* profiles: *The New Yorker;* essays: *The New Yorker;* criticism/reviews: *The New Yorker;* gen. excellence online: *U.S. News Online*; best interactive design: *SmartMoney.com*

George Foster Peabody Awards (2001) by the Univ. of Georgia. *60 Minutes 11: Death by Denial*, CBS News, NY; *Dateline NBC: The Paper Chase*, NBC, NY; *48 Hours: Heroes Under Fire*, CBS News, NY; *Treading on Danger?*, KHOU-TV, Houston, TX; *An Eighty Four Year Old Youngful Man Lives in the Cabin*, KBS, Seoul, Korea; *Behind Closed Doors*, WJXT-TV, Jacksonville, FL; *Regret to Inform*, P.O.V., NY, presented on PBS; *King Gimp*, HBO (L.A.), Whiteford-Hadary, Univ. of Maryland, and Tapestry International; *1900 House*, Wall to Wall Productions for Channel 4 in assoc. with Thirteen/WNET, on PBS; *Cancer: Evolution to Revolution*, HBO and Lovett Productions; *Frontline: Drug Wars*, WGBH/Frontline (Boston), on PBS; *Ali-Frazier 1: One Nation . . . Divisible*, HBO Sports, NY; *CNN Perspectives: Cry Freetown*, CNN Productions (Atlanta), Insight News TV and Channel 4 Internat.; *Napoleon*, David Grubin Productions, Inc. (NY), Devillier Donegan Enterprises, on PBS; *Walking with Dinosaurs,* BBC Discovery Channel/TV Asahi co-prod. in assoc. with ProSieben and France 3; *Marketplace*, Minnesota Public Radio, on Public Radio Internat.; *Witness to an Execution*, Sound Portraits Productions (NY), on NPR; *The NPR 100*, NPR, Washington, DC; *Slavery*, True Vision Productions Ltd. for Channel 4 and HBO; *Katie Couric: Confronting Colon Cancer*, NBC News, NY; *School Sleuth: The Case of an Excellent School*, Learning Matters, Inc., and The Herrow Report (NY), on PBS; *Arthur*, WGBH-TV (Boston) and Cinar Corp., on PBS; *Hearts and Minds: Teens and Mental Illness*, Idaho Public TV (Boise) and Idaho Dept. of Health and Welfare; *Building Big*, WGBH-TV (Boston) and Production Group, Inc., on PBS; *The West Wing*, NBC, John Wells Productions in assoc. with Warner Bros. (L.A.); *The Sopranos,* HBO (NY), Chase Films, Brad Grey TV; *Sharing the Secret*, Robert Greenwald Productions and Pearson TV, Internat. (Santa Monica, CA), on CBS; *Howard Goodall's Big Bangs*, Tiger Aspect Production for Channel 4, London; *ExxonMobil Masterpiece Theater: David Copperfield*, BBC America and WGBH-TV (Boston) co-prod., on PBS; *The Crossing*, A&E TV Networks (NY) and Columbia TriStar TV Productions in assoc. with Chris/Rose Productions, Inc.; *The Corner*, HBO (L.A.); *Malcolm in the Middle*, Fox and Regency TV; *The Daily Show with Jon Stewart: Indecision 2000*, Mad Cow Productions, Inc. in assoc. with Comedy Central, NY; H. Martin "Marty" Haag, personal award

Reuben Award, by National Cartoonists Society. Best cartoonist of 2000: Jack Davis

IT'S A FACT: In 1960, William Hanna and Joseph Barbera's *The Huckleberry Hound Show* became the first TV cartoon series to win an Emmy Award. Among other characters, the show featured Yogi Bear (who got his own show the next year) as well as Pixie and Dixie.

The Spingarn Medal

The Spingarn Medal has been awarded annually since 1915 (except in 1938) by the National Assoc. for the Advancement of Colored People for the highest achievement by a black American in the previous year.

1915 Ernest E. Just
1916 Charles Young
1917 Harry T. Burleigh
1918 William S. Braithwaite
1919 Archibald H. Grimké
1920 W. E. B. Du Bois
1921 Charles S. Gilpin
1922 Mary B. Talbert
1923 George W.Carver
1924 Roland Hayes
1925 James W. Johnson
1926 Carter G. Woodson
1927 Anthony Overton
1928 Charles W. Chesnutt
1929 Mordecai W. Johnson
1930 Henry A. Hunt
1931 Richard B. Harrison
1932 Robert R. Moton
1933 Max Yergan
1934 William T. B. Williams
1935 Mary McLeod Bethune
1936 John Hope
1937 Walter White
1939 Marian Anderson
1940 Louis T. Wright
1941 Richard Wright
1942 A. Philip Randolph
1943 William H. Hastie
1944 Charles Drew
1945 Paul Robeson
1946 Thurgood Marshall
1947 Dr. Percy L. Julian
1948 Channing H. Tobias
1949 Ralph J. Bunche
1950 Charles H. Houston
1951 Mabel K. Staupers
1952 Harry T. Moore
1953 Paul R. Williams
1954 Theodore K. Lawless
1955 Carl Murphy
1956 Jack R. Robinson
1957 Martin Luther King Jr.
1958 Daisy Bates and the Little Rock Nine
1959 Edward Kennedy (Duke) Ellington
1960 Langston Hughes
1961 Kenneth B. Clark
1962 Robert C. Weaver
1963 Medgar W. Evers
1964 Roy Wilkins
1965 Leontyne Price
1966 John H. Johnson
1967 Edward W. Brooke
1968 Sammy Davis Jr.
1969 Clarence M. Mitchell Jr.
1970 Jacob Lawrence
1971 Leon H. Sullivan
1972 Gordon Parks
1973 Wilson C. Riles
1974 Damon Keith
1975 Henry (Hank) Aaron
1976 Alvin Ailey
1977 Alex Haley
1978 Andrew Young
1979 Rosa L. Parks
1980 Dr. Rayford W. Logan
1981 Coleman Young
1982 Dr. Benjamin E. Mays
1983 Lena Horne
1984 Thomas Bradley
1985 Bill Cosby
1986 Dr. Benjamin L. Hooks
1987 Percy E. Sutton
1988 Frederick D. Patterson
1989 Jesse Jackson
1990 L. Douglas Wilder
1991 Gen. Colin L. Powell
1992 Barbara Jordan
1993 Dorothy I. Height
1994 Maya Angelou
1995 John Hope Franklin
1996 A. Leon Higginbotham
1997 Carl T. Rowan
1998 Myrlie Evers-Williams
1999 Earl G. Graves Sr.
2000 Oprah Winfrey
2001 Vernon E. Jordan, Jr.

Miscellaneous Awards

Year in parentheses is year awarded.

American Academy of Arts and Letters (2001). Gold Medal for Sculpture: Richard Serra; Award for Distinguished Service to the Arts: Elizabeth Barlow Rogers; Charles Ives Living, $225,000: Chen Yi; Arnold W. Brunner Memorial Prize in Architecture, $5,000: Henry Smith-Miller and Laurie Hawkinson; Academy Awards, $7,500 each, in Architecture: Vincent James, Sharples Holden Pasquarelli; in Art: Chakaia Booker, Lucky DeBellevue, Martha Diamond, Jacqueline Humphries, Justen Ladda; in Music: Gerald Plain, Allen Shawn, Bright Sheng, Augusta Read Thomas; Jimmy Ernst Award in Art, $5,000: Bernard Chaet; Walter Hinrichsen Award: Kurt Rohde; Charles Ives Fellowships in Music, $15,000: Sally Lamb, Russell Platt, Erik Santos; Charles Ives Scholarships in Music, $7,500 each: James Barry, Michael Djupstrom, Gabriela Frank, Hubert Ho, Jonathan Newman, Tom Swafford; Vladimir and Rhoda Lakond Award in Music, $5,000: Braxton Blake; Goddard Lieberson Fellowships in Music, $12,500 each: Louis Karchin, Mark Kilstofte; Willard L. Metcalf Award in Art, $5,000: Laura Owens; Richard Rodgers Awards for the Musical Theater, $100,000: production: *The Spitfire Grill*, Fred Alley and James Valcq; development: *Heading East*, Leon Ko and Robert Lee; Richard and Hinda Rosenthal Foundation Award in Art, $5,000: Blake Rayne

National Humanities Medal (formerly Charles Frankel Prize), by National Endowment for the Humanities. $5,000 each (2000): Robert N. Bellah, Will D. Campbell, Judy Crichton, David C. Driskell, Ernest J. Gaines, Herman T. Guerrero, Quincy Jones, Barbara Kingsolver, Edmund S. Morgan, Toni Morrison, Earl Shorris, Virginia Driving Hawk Sneve

Intel Science Talent Search (formerly given by Westinghouse) (2001). First ($100,000 schol.): Mariangela Lisanti, Westport, CT;

Second ($75,000 schol.): Nathaniel Jay Craig, Sacramento, CA; Third ($50,000 schol.): Gabriel Drew Carroll, Oakland, CA

National Inventor of the Year Awards (2001), by Intellectual Property Owners. Craig A. Smith, Raymond G. Goodwin, and M. Patricia Beckmann, Immunex Corp.

John F. Kennedy Center for the Performing Arts Awards (2001). Julie Andrews, Van Cliburn, Quincy Jones, Jack Nicholson, Luciano Pavarotti

Library of the Year Award (2001), by Gale Research, Inc., and *Library Journal.* The Richland County Public Library, Columbia, SC

Congressional Gold Medal (2001), by Congress. The 29 "Navajo Code Talkers" of World War II, including John Brown, Jr., Allen Dale June, Chester Nez, Lloyd Oliver, Joe Palmer and 24 members honored posthumously.

National Medal of the Arts (2000), by the National Endowment for the Arts and the White House. Maya Angelou, Eddy Arnold, Mikhail Baryshnikov, Benny Carter, Chuck Close, Horton Foote, Lewis Manilow, NPR, Claes Oldenburg, Itzhak Perlman, Harold Prince, Barbra Streisand

Pritzker Architecture Prize (2001), by the Hyatt Foundation, $100,000: Jacques Herzog and Pierre de Meuron, Switzerland

Teacher of the Year (2001), by Council of Chief State School Officers and Scholastic, Inc. Michele Forman, Middlebury, VT

Templeton Prize for Progress in Religion (2001), by Templeton Foundation, £700,000 (about $1 million): Arthur Peacocke

Miss America Winners, for 1921-2002

1921	Margaret Gorman, Washington, DC
1922-23	Mary Campbell, Columbus, Ohio
1924	Ruth Malcolmson, Philadelphia, Pennsylvania
1925	Fay Lamphier, Oakland, California
1926	Norma Smallwood, Tulsa, Oklahoma
1927	Lois Delander, Joliet, Illinois
1933	Marion Bergeron, West Haven, Connecticut
1935	Henrietta Leaver, Pittsburgh, Pennsylvania
1936	Rose Coyle, Philadelphia, Pennsylvania
1937	Bette Cooper, Bertrand Island, New Jersey
1938	Marilyn Meseke, Marion, Ohio
1939	Patricia Donnelly, Detroit, Michigan
1940	Frances Marie Burke, Philadelphia, Pennsylvania
1941	Rosemary LaPlanche, Los Angeles, California
1942	Jo-Caroll Dennison, Tyler, Texas
1943	Jean Bartel, Los Angeles, California
1944	Venus Ramey, Washington, D.C.
1945	Bess Myerson, New York City, New York
1946	Marilyn Buferd, Los Angeles, California
1947	Barbara Walker, Memphis, Tennessee
1948	BeBe Shopp, Hopkins, Minnesota
1949	Jacque Mercer, Litchfield, Arizona
1951	Yolande Betbeze, Mobile, Alabama
1952	Coleen Kay Hutchins, Salt Lake City, Utah
1953	Neva Jane Langley, Macon, Georgia
1954	Evelyn Margaret Ay, Ephrata, Pennsylvania
1955	Lee Meriwether, San Francisco, California
1956	Sharon Ritchie, Denver, Colorado
1957	Marian McKnight, Manning, South Carolina
1958	Marilyn Van Derbur, Denver, Colorado
1959	Mary Ann Mobley, Brandon, Mississippi
1960	Lynda Lee Mead, Natchez, Mississippi
1961	Nancy Fleming, Montague, Michigan
1962	Maria Fletcher, Asheville, North Carolina
1963	Jacquelyn Mayer, Sandusky, Ohio
1964	Donna Axum, El Dorado, Arkansas
1965	Vonda Kay Van Dyke, Phoenix, Arizona
1966	Deborah Irene Bryant, Overland Park, Kansas
1967	Jane Anne Jayroe, Laverne, Oklahoma
1968	Debra Dene Barnes, Moran, Kansas
1969	Judith Anne Ford, Belvidere, Illinois
1970	Pamela Anne Eldred, Birmingham, Michigan
1971	Phyllis Ann George, Denton, Texas
1972	Laurie Lea Schaefer, Columbus, Ohio
1973	Terry Anne Meeuwsen, DePere, Wisconsin
1974	Rebecca Ann King, Denver, Colorado
1975	Shirley Cothran, Fort Worth, Texas
1976	Tawney Elaine Godin, Yonkers, New York
1977	Dorothy Kathleen Benham, Edina, Minnesota
1978	Susan Perkins, Columbus, Ohio
1979	Kylene Barker, Galax, Virginia
1980	Cheryl Prewitt, Ackerman, Mississippi
1981	Susan Powell, Elk City, Oklahoma
1982	Elizabeth Ward, Russellville, Arkansas
1983	Debra Maffett, Anaheim, California
1984	Vanessa Williams*, Milwood, New York
	Suzette Charles, Mays Landing, New Jersey
1985	Sharlene Wells, Salt Lake City, Utah
1986	Susan Akin, Meridian, Mississippi
1987	Kellye Cash, Memphis, Tennessee
1988	Kaye Lani Rae Rafko, Monroe, Michigan
1989	Gretchen Carlson, Anoka, Minnesota
1990	Debbye Turner, Columbia, Missouri
1991	Marjorie Vincent, Oak Park, Illinois
1992	Carolyn Suzanne Sapp, Honolulu, Hawaii
1993	Leanza Cornett, Jacksonville, Florida
1994	Kimberly Aiken, Columbia, South Carolina
1995	Heather Whitestone, Birmingham, Alabama
1996	Shawntel Smith, Muldrow, Oklahoma
1997	Tara Dawn Holland, Overland Park, Kansas
1998	Kate Shindle, Evanston, Illinois
1999	Nicole Johnson, Roanoke, Virginia
2000	Heather Renee French, Maysville, Kentucky
2001	Angela Perez Baraquio, Honolulu, Hawaii
2002	Katie Harman, Gresham, Oregon

* Resigned July 23, 1984.

Entertainment Awards

Tony Awards, 1948-2001

Year	Play	Musical
1948	Mister Roberts	No Award
1949	Death of a Salesman	Kiss Me Kate
1950	The Cocktail Party	South Pacific
1951	The Rose Tattoo	Guys and Dolls
1952	The Fourposter	The King and I
1953	The Crucible	Wonderful Town
1954	The Teahouse of the August Moon	Kismet
1955	The Desperate Hours	The Pajama Game
1956	The Diary of Anne Frank	Damn Yankees
1957	Long Day's Journey Into Night	My Fair Lady
1958	Sunrise at Campobello	The Music Man
1959	J.B.	Redhead
1960	The Miracle Worker	(tie) Fiorello!, The Sound of Music
1961	Becket	Bye, Bye Birdie
1962	A Man for All Seasons	How to Succeed in Business Without Really Trying
1963	Who's Afraid of Virginia Woolf?	A Funny Thing Happened on the Way to the Forum
1964	Luther	Hello, Dolly!
1965	The Subject Was Roses	Fiddler on the Roof
1966	Marat/Sade	Man of La Mancha
1967	The Homecoming	Cabaret
1968	Rosencrantz and Guildenstern Are Dead	Hallelujah, Baby!
1969	The Great White Hope	1776
1970	Borstal Boy	Applause
1971	Sleuth	Company
1972	Sticks and Bones	Two Gentleman of Verona
1973	That Championship Season	A Little Night Music
1974	The River Niger	Raisin
1975	Equus	The Wiz
1976	Travesties	A Chorus Line
1977	The Shadow Box	Annie
1978	Da	Ain't Misbehavin'
1979	The Elephant Man	Sweeney Todd
1980	Children of a Lesser God	Evita
1981	Amadeus	42nd Street
1982	The Life and Adventures of Nicholas Nickelby	Nine
1983	Torch Song Trilogy	Cats
1985	Biloxi Blues	Big River
1986	I'm Not Rappaport	The Mystery of Edwin Drood
1987	Fences	Les Miserables
1988	M. Butterfly	Phantom of the Opera
1989	The Heidi Chronicles	Jerome Robbins' Broadway
1990	The Grapes of Wrath	City of Angels
1991	Lost in Yonkers	The Will Rogers Follies
1992	Dancing at Lughnasa	Crazy for You
1993	Angels in America: Millennium Approaches	Kiss of the Spider Woman
1994	Angels in America: Perestroika	Passion
1995	Love! Valour! Compassion!	Sunset Boulevard
1996	Master Class	Rent
1997	The Last Night of Ballyhoo	Titanic
1998	Art	The Lion King
1999	Side Man	Fosse
2000	Copenhagen	Contact
2001	Proof	The Producers

Tony (Antoinette Perry) Awards (for Broadway Theater)

Tony Awards Given in 2001

Play: *Proof*
Musical: *The Producers*
Book of a musical: Mel Brooks and Thomas Meehan, *The Producers*
Actor, play: Richard Easton, *The Invention of Love*
Actress, play: Mary-Louise Parker, *Proof*
Actor, musical: Nathan Lane, *The Producers*
Actress, musical: Christine Ebersole, *42nd Street*
Musical score: Mel Brooks, *The Producers*
Director, play: Daniel Sullivan, *Proof*
Director, musical: Susan Stroman, *The Producers*
Play revival: *One Flew Over the Cuckoo's Nest*
Musical revival: *42nd Street*
Featured actor, play: Robert Sean Leonard, *The Invention of Love*
Featured actress, play: Viola Davis, *King Hedley II*
Featured actor, musical: Gary Beach, *The Producers*
Featured actress, musical: Cady Huffman, *The Producers*
Choreography: Susan Stroman, *The Producers*
Costume design: William Ivey Long, *The Producers*
Scenic design: Robin Wagner, *The Producers*
Lighting design: Peter Kaczorowski, *The Producers*
Orchestrations: Douglas Besterman, *The Producers*
Lifetime achievement: Paul Gemignani, musical director
Special Theatrical Event: *Blast!*
Regional Theater: Victory Gardens Theatre of Chicago, IL

Emmy Awards, 1952-2000

The National Academy of Television Arts and Science, presented the first Emmy Awards in 1949. Through the years, the number and names of award categories have changed, but since 1952, the Academy has recognized an outstanding comedy and drama each year.

Year Given	Comedy	Drama
1952	*Red Skelton Show*, NBC	*Studio One*, CBS
1953	*I Love Lucy*, CBS	*Robert Montgomery Presents*, NBC
1954	*I Love Lucy*, CBS	*The U.S. Steel Hour*, ABC
1955	*Make Room for Daddy*, ABC	*The U.S. Steel Hour*, ABC
1956	*Phil Silvers Show*, CBS	*Producer's Showcase*, NBC
1957	*Phil Silvers Show*, CBS	*Requiem for a Heavyweight*, CBS[1]
1958	*Phil Silvers Show*, CBS	*Gunsmoke*, CBS
1959[2]	*Jack Benny Show*, CBS	(3)
1960	*Art Carney Special*, NBC	*Playhouse 90*, CBS
1961	*Jack Benny Show*, CBS	*Hallmark Hall of Fame*: *Macbeth*, NBC
1962	*Bob Newhart Show*, CBS	*The Defenders*, CBS
1963	*Dick Van Dyke Show*, CBS	*The Defenders*, CBS
1964	*Dick Van Dyke Show*, CBS	*The Defenders*, CBS
1965	*Dick Van Dyke Show*, CBS	*Hallmark Hall of Fame*: *The Magnificent Yankee*, NBC
1966	*Dick Van Dyke Show*, CBS	*The Fugitive*, ABC
1967	*The Monkees*, NBC	*Mission: Impossible*, CBS
1968	*Get Smart*, NBC	*Mission: Impossible*, CBS
1969	*Get Smart*, NBC	*NET Playhouse*, NET
1970	*My World and Welcome to It*, NBC	*Marcus Welby, M.D.*, ABC
1971	*All in the Family*, CBS	*The Bold Ones: "The Senator,"* NBC
1972	*All in the Family*, CBS	*Masterpiece Theatre: Elizabeth R*, PBS
1973	*All in the Family*, CBS	*The Waltons*, CBS
1974	*M*A*S*H*, CBS	*Masterpiece Theatre: Upstairs, Downstairs;* PBS
1975	*Mary Tyler Moore Show*, CBS	*Masterpiece Theatre: Upstairs, Downstairs;* PBS
1976	*Mary Tyler Moore Show*, CBS	*Police Story*, NBC
1977	*Mary Tyler Moore Show*, CBS	*Masterpiece Theatre: Upstairs, Downstairs;* PBS
1978	*All in the Family*, CBS	*The Rockford Files*, NBC
1979	*Taxi*, ABC	*Lou Grant*, CBS
1980	*Taxi*, ABC	*Lou Grant*, CBS
1981	*Taxi*, ABC	*HIll Street Blues*, NBC
1982	*Barney Miller*, ABC	*HIll Street Blues*, NBC
1983	*Cheers*, NBC	*HIll Street Blues*, NBC
1984	*Cheers*, NBC	*HIll Street Blues*, NBC
1985	*The Cosby Show*, NBC	*Cagney & Lacey*, CBS
1986	*Golden Girls*, NBC	*Cagney & Lacey*, CBS
1987	*Golden Girls*, NBC	*L.A. Law*, NBC
1988	*The Wonder Years*, ABC	*thirtysomething*, ABC
1989	*Cheers*, NBC	*L.A. Law*, NBC
1990	*Murphy Brown*, CBS	*L.A. Law*, NBC
1991	*Cheers*, NBC	*L.A. Law*, NBC
1992	*Murphy Brown*, CBS	*Northern Exposure*, CBS
1993	*Seinfeld*, NBC	*Picket Fences*, CBS
1994	*Frasier*, NBC	*Picket Fences*, CBS
1995	*Frasier*, NBC	*NYPD Blue*, ABC
1996	*Frasier*, NBC	*ER*, NBC
1997	*Frasier*, NBC	*Law & Order*, NBC
1998	*Frasier*, NBC	*The Practice*, ABC
1999	*Ally McBeal*, Fox	*The Practice*, ABC
2000	*Will & Grace*, NBC	*The West Wing*, NBC

(1) "Best Single Program of the Year," shown on *Playhouse 90*, which was named "Best New Series." (2) Beginning in 1959, Emmys awarded for work in the season encompassing the previous and current year. (3) *Playhouse 90* (CBS) was best drama of one hour or longer; *Alcoa-Goodyear Theatre* (NBC) was best drama of less than one hour.

2001 Golden Globe Awards

(Awarded for work in 2000)

Film

Drama: *Gladiator*
Musical/comedy: *Almost Famous*
Actress, drama: Julia Roberts, *Erin Brockovich*
Actor, drama: Tom Hanks, *Cast Away*
Actress, musical/comedy: Renee Zellweger, *Nurse Betty*
Actor, musical/comedy: George Clooney, *O Brother, Where Art Thou?*
Sup. actress, drama: Kate Hudson, *Almost Famous*
Sup. actor, drama: Benicio Del Toro, *Traffic*
Director: Ang Lee, *Crouching Tiger, Hidden Dragon*
Screenplay: Stephen Gaghan, *Traffic*
Foreign-language film: *Crouching Tiger, Hidden Dragon* (Taiwan)
Original score: Hans Zimmer and Lisa Gerrard, *Gladiator*
Original song: "Things Have Changed," from *Wonder Boys*, Bob Dylan
Cecil B. De Mille award for lifetime achievement: Al Pacino

Television

Series, drama: *The West Wing*, NBC
Actress, drama: Sela Ward, *Once and Again*, ABC
Actor, drama: Martin Sheen, *The West Wing*
Series, musical/comedy: *Sex and the City*, HBO
Actress, musical/comedy: Sarah Jessica Parker, *Sex and the City*
Actor, musical/comedy: Kelsey Grammer, *Frasier*, NBC
Miniseries, movie made for TV: *Dirty Pictures*, Showtime
Actress, miniseries/movie: Judi Dench, *Last of the Blonde Bombshells*, HBO
Actor, miniseries/movie: Brian Dennehy, *Arthur Miller's Death of a Salesman*, Showtime
Sup. actress, miniseries/movie: Vanessa Redgrave, *If These Walls Could Talk 2*, HBO
Sup. actor, miniseries/movie: Robert Downey Jr., *Ally McBeal*, Fox

Academy Awards (Oscars) for 1927-2000

1927-28

Picture: *Wings*
Actor: Emil Jannings, *The Way of All Flesh*
Actress: Janet Gaynor, *Seventh Heaven*
Director: Frank Borzage, *Seventh Heaven;* Lewis Milestone, *Two Arabian Knights*

1928-29

Picture: *Broadway Melody*
Actor: Warner Baxter, *In Old Arizona*
Actress: Mary Pickford, *Coquette*
Director: Frank Lloyd, *The Divine Lady*

1929-30

Picture: *All Quiet on the Western Front*
Actor: George Arliss, *Disraeli*
Actress: Norma Shearer, *The Divorcee*
Director: Lewis Milestone, *All Quiet on the Western Front*

1930-31

Picture: *Cimarron*
Actor: Lionel Barrymore, *Free Soul*
Actress: Marie Dressler, *Min and Bill*
Director: Norman Taurog, *Skippy*

1931-32

Picture: *Grand Hotel*
Actor: Fredric March, *Dr. Jekyll and Mr. Hyde;* Wallace Beery, *The Champ* (tie)
Actress: Helen Hayes, *The Sin of Madelon Claudet*
Director: Frank Borzage, *Bad Girl*
Special: Walt Disney, *Mickey Mouse*

1932-33

Picture: *Cavalcade*
Actor: Charles Laughton, *The Private Life of Henry VIII*

Actress: Katharine Hepburn, *Morning Glory*
Director: Frank Lloyd, *Cavalcade*

1934
Picture: *It Happened One Night*
Actor: Clark Gable, *It Happened One Night*
Actress: Claudette Colbert, *It Happened One Night*
Director: Frank Capra, *It Happened One Night*

1935
Picture: *Mutiny on the Bounty*
Actor: Victor McLaglen, *The Informer*
Actress: Bette Davis, *Dangerous*
Director: John Ford, *The Informer*

1936
Picture: *The Great Ziegfeld*
Actor: Paul Muni, *Story of Louis Pasteur*
Actress: Luise Rainer, *The Great Ziegfeld*
Sup. Actor: Walter Brennan, *Come and Get It*
Sup. Actress: Gale Sondergaard, *Anthony Adverse*
Director: Frank Capra, *Mr. Deeds Goes to Town*

1937
Picture: *Life of Emile Zola*
Actor: Spencer Tracy, *Captains Courageous*
Actress: Luise Rainer, *The Good Earth*
Sup. Actor: Joseph Schildkraut, *Life of Emile Zola*
Sup. Actress: Alice Brady, *In Old Chicago*
Director: Leo McCarey, *The Awful Truth*

1938
Picture: *You Can't Take It With You*
Actor: Spencer Tracy, *Boys Town*
Actress: Bette Davis, *Jezebel*
Sup. Actor: Walter Brennan, *Kentucky*
Sup. Actress: Fay Bainter, *Jezebel*
Director: Frank Capra, *You Can't Take It With You*

1939
Picture: *Gone With the Wind*
Actor: Robert Donat, *Goodbye, Mr. Chips*
Actress: Vivien Leigh, *Gone With the Wind*
Sup. Actor: Thomas Mitchell, *Stage Coach*
Sup. Actress: Hattie McDaniel, *Gone With the Wind*
Director: Victor Fleming, *Gone With the Wind*

1940
Picture: *Rebecca*
Actor: James Stewart, *The Philadelphia Story*
Actress: Ginger Rogers, *Kitty Foyle*
Sup. Actor: Walter Brennan, *The Westerner*
Sup. Actress: Jane Darwell, *The Grapes of Wrath*
Director: John Ford, *The Grapes of Wrath*

1941
Picture: *How Green Was My Valley*
Actor: Gary Cooper, *Sergeant York*
Actress: Joan Fontaine, *Suspicion*
Sup. Actor: Donald Crisp, *How Green Was My Valley*
Sup. Actress: Mary Astor, *The Great Lie*
Director: John Ford, *How Green Was My Valley*

1942
Picture: *Mrs. Miniver*
Actor: James Cagney, *Yankee Doodle Dandy*
Actress: Greer Garson, *Mrs. Miniver*
Sup. Actor: Van Heflin, *Johnny Eager*
Sup. Actress: Teresa Wright, *Mrs. Miniver*
Director: William Wyler, *Mrs. Miniver*

1943
Picture: *Casablanca*
Actor: Paul Lukas, *Watch on the Rhine*
Actress: Jennifer Jones, *The Song of Bernadette*
Sup. Actor: Charles Coburn, *The More the Merrier*
Sup. Actress: Katina Paxinou, *For Whom the Bell Tolls*
Director: Michael Curtiz, *Casablanca*

1944
Picture: *Going My Way*
Actor: Bing Crosby, *Going My Way*
Actress: Ingrid Bergman, *Gaslight*
Sup. Actor: Barry Fitzgerald, *Going My Way*
Sup. Actress: Ethel Barrymore, *None But the Lonely Heart*
Director: Leo McCarey, *Going My Way*

1945
Picture: *The Lost Weekend*
Actor: Ray Milland, *The Lost Weekend*
Actress: Joan Crawford, *Mildred Pierce*
Sup. Actor: James Dunn, *A Tree Grows in Brooklyn*
Sup. Actress: Anne Revere, *National Velvet*
Director: Billy Wilder, *The Lost Weekend*

1946
Picture: *The Best Years of Our Lives*
Actor: Fredric March, *The Best Years of Our Lives*
Actress: Olivia de Havilland, *To Each His Own*
Sup. Actor: Harold Russell, *The Best Years of Our Lives*
Sup. Actress: Anne Baxter, *The Razor's Edge*
Director: William Wyler, *The Best Years of Our Lives*

1947
Picture: *Gentleman's Agreement*
Actor: Ronald Colman, *A Double Life*
Actress: Loretta Young, *The Farmer's Daughter*
Sup. Actor: Edmund Gwenn, *Miracle on 34th Street*
Sup. Actress: Celeste Holm, *Gentleman's Agreement*
Director: Elia Kazan, *Gentleman's Agreement*

1948
Picture: *Hamlet*
Actor: Laurence Olivier, *Hamlet*
Actress: Jane Wyman, *Johnny Belinda*
Sup. Actor: Walter Huston, *Treasure of Sierra Madre*
Sup. Actress: Claire Trevor, *Key Largo*
Director: John Huston, *Treasure of Sierra Madre*

1949
Picture: *All the King's Men*
Actor: Broderick Crawford, *All the King's Men*
Actress: Olivia de Havilland, *The Heiress*
Sup. Actor: Dean Jagger, *Twelve O'Clock High*
Sup. Actress: Mercedes McCambridge, *All the King's Men*
Director: Joseph L. Mankiewicz, *Letter to Three Wives*

1950
Picture: *All About Eve*
Actor: Jose Ferrer, *Cyrano de Bergerac*
Actress: Judy Holliday, *Born Yesterday*
Sup. Actor: George Sanders, *All About Eve*
Sup. Actress: Josephine Hull, *Harvey*
Director: Joseph L. Mankiewicz, *All About Eve*

1951
Picture: *An American in Paris*
Actor: Humphrey Bogart, *The African Queen*
Actress: Vivien Leigh, *A Streetcar Named Desire*
Sup. Actor: Karl Malden, *A Streetcar Named Desire*
Sup. Actress: Kim Hunter, *A Streetcar Named Desire*
Director: George Stevens, *A Place in the Sun*

1952
Picture: *The Greatest Show on Earth*
Actor: Gary Cooper, *High Noon*
Actress: Shirley Booth, *Come Back, Little Sheba*
Sup. Actor: Anthony Quinn, *Viva Zapata!*
Sup. Actress: Gloria Grahame, *The Bad and the Beautiful*
Director: John Ford, *The Quiet Man*

1953
Picture: *From Here to Eternity*
Actor: William Holden, *Stalag 17*
Actress: Audrey Hepburn, *Roman Holiday*
Sup. Actor: Frank Sinatra, *From Here to Eternity*
Sup. Actress: Donna Reed, *From Here to Eternity*
Director: Fred Zinnemann, *From Here to Eternity*

1954
Picture: *On the Waterfront*
Actor: Marlon Brando, *On the Waterfront*
Actress: Grace Kelly, *The Country Girl*
Sup. Actor: Edmond O'Brien, *The Barefoot Contessa*
Sup. Actress: Eva Marie Saint, *On the Waterfront*
Director: Elia Kazan, *On the Waterfront*

1955
Picture: *Marty*
Actor: Ernest Borgnine, *Marty*
Actress: Anna Magnani, *The Rose Tattoo*
Sup. Actor: Jack Lemmon, *Mister Roberts*
Sup. Actress: Jo Van Fleet, *East of Eden*
Director: Delbert Mann, *Marty*

1956
Picture: *Around the World in 80 Days*
Actor: Yul Brynner, *The King and I*
Actress: Ingrid Bergman, *Anastasia*
Sup. Actor: Anthony Quinn, *Lust for Life*
Sup. Actress: Dorothy Malone, *Written on the Wind*
Director: George Stevens, *Giant*

1957
Picture: *The Bridge on the River Kwai*
Actor: Alec Guinness, *The Bridge on the River Kwai*
Actress: Joanne Woodward, *The Three Faces of Eve*
Sup. Actor: Red Buttons, *Sayonara*
Sup. Actress: Miyoshi Umeki, *Sayonara*
Director: David Lean, *The Bridge on the River Kwai*

1958
Picture: *Gigi*
Actor: David Niven, *Separate Tables*
Actress: Susan Hayward, *I Want to Live*
Sup. Actor: Burl Ives, *The Big Country*
Sup. Actress: Wendy Hiller, *Separate Tables*
Director: Vincente Minnelli, *Gigi*

1959
Picture: *Ben-Hur*
Actor: Charlton Heston, *Ben-Hur*
Actress: Simone Signoret, *Room at the Top*
Sup. Actor: Hugh Griffith, *Ben-Hur*
Sup. Actress: Shelley Winters, *Diary of Anne Frank*
Director: William Wyler, *Ben-Hur*

1960
Picture: *The Apartment*
Actor: Burt Lancaster, *Elmer Gantry*
Actress: Elizabeth Taylor, *Butterfield 8*
Sup. Actor: Peter Ustinov, *Spartacus*
Sup. Actress: Shirley Jones, *Elmer Gantry*
Director: Billy Wilder, *The Apartment*

1961
Picture: *West Side Story*
Actor: Maximilian Schell, *Judgment at Nuremberg*
Actress: Sophia Loren, *Two Women*
Sup. Actor: George Chakiris, *West Side Story*
Sup. Actress: Rita Moreno, *West Side Story*
Director: Jerome Robbins, Robert Wise, *West Side Story*

1962
Picture: *Lawrence of Arabia*
Actor: Gregory Peck, *To Kill a Mockingbird*
Actress: Anne Bancroft, *The Miracle Worker*
Sup. Actor: Ed Begley, *Sweet Bird of Youth*
Sup. Actress: Patty Duke, *The Miracle Worker*
Director: David Lean, *Lawrence of Arabia*

1963
Picture: *Tom Jones*
Actor: Sidney Poitier, *Lilies of the Field*
Actress: Patricia Neal, *Hud*
Sup. Actor: Melvyn Douglas, *Hud*
Sup. Actress: Margaret Rutherford, *The V.I.P.s*
Director: Tony Richardson, *Tom Jones*

1964
Picture: *My Fair Lady*
Actor: Rex Harrison, *My Fair Lady*
Actress: Julie Andrews, *Mary Poppins*
Sup. Actor: Peter Ustinov, *Topkapi*
Sup. Actress: Lila Kedrova, *Zorba the Greek*
Director: George Cukor, *My Fair Lady*

1965
Picture: *The Sound of Music*
Actor: Lee Marvin, *Cat Ballou*
Actress: Julie Christie, *Darling*
Sup. Actor: Martin Balsam, *A Thousand Clowns*
Sup. Actress: Shelley Winters, *A Patch of Blue*
Director: Robert Wise, *The Sound of Music*

1966
Picture: *A Man for All Seasons*
Actor: Paul Scofield, *A Man for All Seasons*
Actress: Elizabeth Taylor, *Who's Afraid of Virginia Woolf?*
Sup. Actor: Walter Matthau, *The Fortune Cookie*
Sup. Actress: Sandy Dennis, *Who's Afraid of Virginia Woolf?*
Director: Fred Zinnemann, *A Man for All Seasons*

1967
Picture: *In the Heat of the Night*
Actor: Rod Steiger, *In the Heat of the Night*
Actress: Katharine Hepburn, *Guess Who's Coming to Dinner*
Sup. Actor: George Kennedy, *Cool Hand Luke*
Sup. Actress: Estelle Parsons, *Bonnie and Clyde*
Director: Mike Nichols, *The Graduate*

1968
Picture: *Oliver!*
Actor: Cliff Robertson, *Charly*
Actress: Katharine Hepburn, *The Lion in Winter;* Barbra Streisand, *Funny Girl* (tie)
Sup. Actor: Jack Albertson, *The Subject Was Roses*
Sup. Actress: Ruth Gordon, *Rosemary's Baby*
Director: Sir Carol Reed, *Oliver!*

1969
Picture: *Midnight Cowboy*
Actor: John Wayne, *True Grit*
Actress: Maggie Smith, *The Prime of Miss Jean Brodie*
Sup. Actor: Gig Young, *They Shoot Horses, Don't They?*
Sup. Actress: Goldie Hawn, *Cactus Flower*
Director: John Schlesinger, *Midnight Cowboy*

1970
Picture: *Patton*
Actor: George C. Scott, *Patton* (refused)
Actress: Glenda Jackson, *Women in Love*
Sup. Actor: John Mills, *Ryan's Daughter*
Sup. Actress: Helen Hayes, *Airport*
Director: Franklin Schaffner, *Patton*

1971
Picture: *The French Connection*
Actor: Gene Hackman, *The French Connection*
Actress: Jane Fonda, *Klute*
Sup. Actor: Ben Johnson, *The Last Picture Show*
Sup. Actress: Cloris Leachman, *The Last Picture Show*
Director: William Friedkin, *The French Connection*

1972
Picture: *The Godfather*
Actor: Marlon Brando, *The Godfather* (refused)
Actress: Liza Minnelli, *Cabaret*
Sup. Actor: Joel Grey, *Cabaret*
Sup. Actress: Eileen Heckart, *Butterflies Are Free*
Director: Bob Fosse, *Cabaret*

1973
Picture: *The Sting*
Actor: Jack Lemmon, *Save the Tiger*
Actress: Glenda Jackson, *A Touch of Class*
Sup. Actor: John Houseman, *The Paper Chase*
Sup. Actress: Tatum O'Neal, *Paper Moon*
Director: George Roy Hill, *The Sting*

1974
Picture: *The Godfather, Part II*
Actor: Art Carney, *Harry and Tonto*
Actress: Ellen Burstyn, *Alice Doesn't Live Here Anymore*
Sup. Actor: Robert DeNiro, *The Godfather, Part II*
Sup. Actress: Ingrid Bergman, *Murder on the Orient Express*
Director: Francis Ford Coppola, *The Godfather, Part II*

1975
Picture: *One Flew Over the Cuckoo's Nest*
Actor: Jack Nicholson, *One Flew Over the Cuckoo's Nest*
Actress: Louise Fletcher, *One Flew Over the Cuckoo's Nest*
Sup. Actor: George Burns, *The Sunshine Boys*
Sup. Actress: Lee Grant, *Shampoo*
Director: Milos Forman, *One Flew Over the Cuckoo's Nest*

1976
Picture: *Rocky*
Actor: Peter Finch, *Network*
Actress: Faye Dunaway, *Network*
Sup. Actor: Jason Robards, *All the President's Men*
Sup. Actress: Beatrice Straight, *Network*
Director: John G. Avildsen, *Rocky*

1977
Picture: *Annie Hall*
Actor: Richard Dreyfuss, *The Goodbye Girl*
Actress: Diane Keaton, *Annie Hall*
Sup. Actor: Jason Robards, *Julia*
Sup. Actress: Vanessa Redgrave, *Julia*
Director: Woody Allen, *Annie Hall*

1978
Picture: *The Deer Hunter*
Actor: Jon Voight, *Coming Home*
Actress: Jane Fonda, *Coming Home*
Sup. Actor: Christopher Walken, *The Deer Hunter*
Sup. Actress: Maggie Smith, *California Suite*
Director: Michael Cimino, *The Deer Hunter*

1979
Picture: *Kramer vs. Kramer*
Actor: Dustin Hoffman, *Kramer vs. Kramer*
Actress: Sally Field, *Norma Rae*
Sup. Actor: Melvyn Douglas, *Being There*
Sup. Actress: Meryl Streep, *Kramer vs. Kramer*
Director: Robert Benton, *Kramer vs. Kramer*

1980
Picture: *Ordinary People*
Actor: Robert DeNiro, *Raging Bull*
Actress: Sissy Spacek, *Coal Miner's Daughter*
Sup. Actor: Timothy Hutton, *Ordinary People*
Sup. Actress: Mary Steenburgen, *Melvin & Howard*
Director: Robert Redford, *Ordinary People*

1981
Picture: *Chariots of Fire*
Actor: Henry Fonda, *On Golden Pond*
Actress: Katharine Hepburn, *On Golden Pond*
Sup. Actor: John Gielgud, *Arthur*
Sup. Actress: Maureen Stapleton, *Reds*
Director: Warren Beatty, *Reds*

1982
Picture: *Gandhi*
Actor: Ben Kingsley, *Gandhi*
Actress: Meryl Streep, *Sophie's Choice*
Sup. Actor: Louis Gossett Jr., *An Officer and a Gentleman*
Sup. Actress: Jessica Lange, *Tootsie*
Director: Richard Attenborough, *Gandhi*

1983
Picture: *Terms of Endearment*
Actor: Robert Duvall, *Tender Mercies*
Actress: Shirley MacLaine, *Terms of Endearment*
Sup. Actor: Jack Nicholson, *Terms of Endearment*
Sup. Actress: Linda Hunt, *The Year of Living Dangerously*
Director: James L. Brooks, *Terms of Endearment*

1984
Picture: *Amadeus*
Actor: F. Murray Abraham, *Amadeus*
Actress: Sally Field, *Places in the Heart*
Sup. Actor: Haing S. Ngor, *The Killing Fields*
Sup. Actress: Peggy Ashcroft, *A Passage to India*
Director: Milos Forman, *Amadeus*

1985
Picture: *Out of Africa*
Actor: William Hurt, *Kiss of the Spider Woman*
Actress: Geraldine Page, *The Trip to Bountiful*
Sup. Actor: Don Ameche, *Cocoon*
Sup. Actress: Anjelica Huston, *Prizzi's Honor*
Director: Sydney Pollack, *Out of Africa*

1986
Picture: *Platoon*
Actor: Paul Newman, *The Color of Money*
Actress: Marlee Matlin, *Children of a Lesser God*
Sup. Actor: Michael Caine, *Hannah and Her Sisters*
Sup. Actress: Dianne Wiest, *Hannah and Her Sisters*
Director: Oliver Stone, *Platoon*

1987
Picture: *The Last Emperor*
Actor: Michael Douglas, *Wall Street*
Actress: Cher, *Moonstruck*
Sup. Actor: Sean Connery, *The Untouchables*
Sup. Actress: Olympia Dukakis, *Moonstruck*
Director: Bernardo Bertolucci, *The Last Emperor*

1988
Picture: *Rain Man*
Actor: Dustin Hoffman, *Rain Man*
Actress: Jodie Foster, *The Accused*
Sup. Actor: Kevin Kline, *A Fish Called Wanda*
Sup. Actress: Geena Davis, *The Accidental Tourist*
Director: Barry Levinson, *Rain Man*

1989
Picture: *Driving Miss Daisy*
Actor: Daniel Day-Lewis, *My Left Foot*
Actress: Jessica Tandy, *Driving Miss Daisy*
Sup. Actor: Denzel Washington, *Glory*
Sup. Actress: Brenda Fricker, *My Left Foot*
Director: Oliver Stone, *Born on the Fourth of July*

1990
Picture: *Dances With Wolves*
Actor: Jeremy Irons, *Reversal of Fortune*
Actress: Kathy Bates, *Misery*
Sup. Actor: Joe Pesci, *Goodfellas*
Sup. Actress: Whoopi Goldberg, *Ghost*
Director: Kevin Costner, *Dances With Wolves*

1991
Picture: *The Silence of the Lambs*
Actor: Anthony Hopkins, *The Silence of the Lambs*
Actress: Jodie Foster, *The Silence of the Lambs*
Sup. Actor: Jack Palance, *City Slickers*
Sup. Actress: Mercedes Ruehl, *The Fisher King*
Director: Jonathan Demme, *The Silence of the Lambs*

1992
Picture: *Unforgiven*
Actor: Al Pacino, *Scent of a Woman*
Actress: Emma Thompson, *Howards End*
Sup. Actor: Gene Hackman, *Unforgiven*
Sup. Actress: Marisa Tomei, *My Cousin Vinny*
Director: Clint Eastwood, *Unforgiven*

1993
Picture: *Schindler's List*
Actor: Tom Hanks, *Philadelphia*
Actress: Holly Hunter, *The Piano*
Sup. Actor: Tommy Lee Jones, *The Fugitive*
Sup. Actress: Anna Paquin, *The Piano*
Director: Steven Spielberg, *Schindler's List*

1994
Picture: *Forrest Gump*
Actor: Tom Hanks, *Forrest Gump*
Actress: Jessica Lange, *Blue Sky*
Sup. Actor: Martin Landau, *Ed Wood*
Sup. Actress: Dianne Wiest, *Bullets Over Broadway*
Director: Robert Zemeckis, *Forrest Gump*

1995
Picture: *Braveheart*
Actor: Nicolas Cage, *Leaving Las Vegas*
Actress: Susan Sarandon, *Dead Man Walking*
Sup. Actor: Kevin Spacey, *The Usual Suspects*
Sup. Actress: Mira Sorvino, *Mighty Aphrodite*
Director: Mel Gibson, *Braveheart*

1996
Picture: *The English Patient*
Actor: Geoffrey Rush, *Shine*
Actress: Frances McDormand, *Fargo*
Sup. Actor: Cuba Gooding Jr., *Jerry Maguire*
Sup. Actress: Juliette Binoche, *The English Patient*
Director: Anthony Minghella, *The English Patient*

1997
Picture: *Titanic*
Actor: Jack Nicholson, *As Good As It Gets*
Actress: Helen Hunt, *As Good As It Gets*
Sup. Actor: Robin Williams, *Good Will Hunting*
Sup. Actress: Kim Basinger, *L.A. Confidential*
Director: James Cameron, *Titanic*

1998
Picture: *Shakespeare in Love*
Actor: Roberto Benigni, *Life Is Beautiful*
Actress: Gwyneth Paltrow, *Shakespeare in Love*
Sup. Actor: James Coburn, *Affliction*
Sup. Actress: Judi Dench, *Shakespeare in Love*
Director: Steven Spielberg, *Saving Private Ryan*

1999
Picture: *American Beauty*
Actor: Kevin Spacey, *American Beauty*
Actress: Hilary Swank, *Boys Don't Cry*
Sup. Actor: Michael Caine, *The Cider House Rules*
Sup. Actress: Angelina Jolie, *Girl, Interrupted*
Director: Sam Mendes, *American Beauty*

2000
Picture: *Gladiator*
Actor: Russell Crowe, *Gladiator*
Actress: Julia Roberts, *Erin Brockovich*
Sup. Actor: Benicio Del Toro, *Traffic*
Sup. Actress: Marcia Gay Harden, *Pollock*
Director: Steven Soderbergh, *Traffic*
Foreign Film: *Crouching Tiger, Hidden Dragon,* Taiwan
Original Screenplay: Cameron Crow, *Almost Famous*
Adapted Screenplay: Stephen Gaghan, *Traffic*
Cinematography: Peter Pau, *Crouching Tiger, Hidden Dragon*
Art Direction: Tim Yip, *Crouching Tiger, Hidden Dragon*
Film Editing: Stephen Mirrione, *Traffic*
Original Song: "Things Have Changed," *Wonder Boys,* Bob Dylan
Original Score: Tan Dun, *Crouching Tiger, Hidden Dragon*
Costume Design: Janty Yates, *Gladiator*
Makeup: Rick Baker and Gail Ryan, *Dr. Seuss' How the Grinch Stole Christmas*
Sound: Scott Millan, Bob Beemer, and Ken Weston, *Gladiator*
Documentary Feature: Mark Jonathan Harris and Deborah Oppenheimer, *Into the Arms of Strangers: Stories of the Kindertransport*
Documentary Short Subject: Tracy Seretean, *Big Mama*
Short Film, Live: Florian Gallenberger, *Quiero Ser (I Want to Be)*
Short Film, Animated: Michael Dudok de Wit, *Father and Daughter*
Visual Effects: John Nelson, Neil Corbould, Tim Burke, and Rob Harvey, *Gladiator*
Sound Effects Editing: Jon Johnson, *U-571*
Gordon E. Sawyer Award: Irwin W. Young
Irving G. Thalberg Memorial Award: Dino De Laurentiis
Honorary Oscar: Jack Cardiff, Ernest Lehman

Other Film Awards

Year in parentheses is year awarded

Cannes Film Festival Awards (2001), Feature Films—Palme d'Or (Golden Palm): *The Son's Room,* Nanni Moretti (Italy); Grand Prize: *The Piano Teacher*, Michael Haneke (Austria); best actress: Isabelle Huppert (France), *The Piano Teacher*, best actor: Benoit Magimel (France), *The Piano Teacher*, best director: David Lynch, *Mulholland Drive* and Joel Coen, *The Man Who Wasn't There*, co-winners (both U.S.); best screenplay: *No Man's Land*, Danis Tanovic (Bosnia). Technical Jury Prize (sound): Tu Duu-Chih (Taiwan), *Millennium Mambo* and *What Time Is It There?;* Camera d'Or (Golden Camera, first-time director): Zacharias Kunuk (Canada), *Atanarjuat* (*The Fast Runner*). Short Films—Palme d'Or: *Beancake*, David Greenspan (U.S.); Jury Prize for Fiction: *Daddy's Girl*, Irvine Allan (Scotland); Jury Prize for Animation: *Pizza Passionata*, Kari Juusonen (Finland). Lifetime Achievement Award: Melanie Griffith.

Directors Guild of America Awards (2001), Feature film: Ang Lee, *Crouching Tiger, Hidden Dragon*; documentary: Chuck Braverman, *High School Boot Camp*

Sundance Film Festival Awards (2001), Grand Jury Prize: (drama) *The Believer,* Henry Bean; (docu.) *Southern Comfort, Kate Davis.* Directing Award: (drama) John Cameron Mitchell, *Hedwig and the Angry Inch*; (docu.) Stacy Peralta, *Dogtown and Z-Boys.* Waldo Salt Screenwriting Award: Christopher Nolan, *Memento.* Freedom of Expression Award (docu.): *Scout's Honor, Tom Shepard.* Audience Award: (drama) *Hedwig and the Angry Inch*; (docu.) *Scout's Honor and Dogtown and Z-Boys (split)*; (world) *The Road Home*, Zhang Yimou. Cinematography Award: (drama) Giles Nuttgens, *The Deep End*; (docu.) Albert Maysles, *Lalee's Kin: The Legacy of Cotton.* Special Jury Awards: (drama) *Tom Wilkinson and Sissy Spacek, In the Bedroom*; (docu.) *Children Underground,* Edet Belzberg. Latin Amer. Cinema Award: (Jury Prize) *Possible Loves, Sandra Werneck and Without a Trace, Maria Navaro (split).* Short Filmmaking: (Jury Prize) *Gina, an Actress, Age 29*, Paul Harrill.

2001 Academy of Country Music Awards

Entertainer of the Year: Dixie Chicks
Single of the Year: "I Hope You Dance," Lee Ann Womack
Album of the Year: *How Do You LIke Me Now?!,*Toby Keith; James Stroud/Toby Keith, producers; DreamWorks Records
Single of the Year: "I Hope You Dance," Lee Ann Womack/Sons of the Desert; Mark Wright, producer; MCA Nashville
Top Female Vocalist: Faith Hill
Top Male Vocalist: Toby Keith
Top Vocal Duo: Brooks & Dunn
Vocal Group of the Year: Dixie Chicks
Top New Female Vocalist: Jamie O'Neal
Top New Male Vocalist: Keith Urban
Top New Vocal Duo or Group: Rascal Flatts
Video of the Year: "Goodbye Earl," Dixie Chicks; Keely Gould, producer; Evan Bernard, director
Song of the Year: "I Hope You Dance," Lee Ann Womack/Sons of the Desert; Mark D. Sanders/Tia Sellers, composers
Vocal Event of the Year: "I Hope You Dance," Lee Ann Womack/Sons of the Desert; Mark Wright, producer
Pioneer Award: Barbara Mandrell
Career Achievement Award: Kenny Rogers

2001 MTV Video Music Awards

Video of the Year: Christina Aguilera, Pink, Lil' Kim and Mya; featuring Missy "Misdemeanor" Elliott, "Lady Marmalade"
Best Male Video: Moby featuring Gwen Stefani, "South Side"
Best Female Video: Eve featuring Gwen Stefani, "Let Me Blow Ya Mind"
Best Group Video: 'N Sync, "Pop"
Best Rap Video: Nelly, "Ride Wit Me"
Best Dance Video: 'N Sync, "Pop"
Best Pop Video: 'N Sync, "Pop"
Best Rock Video: Limp Bizkit, "Rollin'"
Best Hip Hop Video: OutKast, "Ms. Jackson"
Best New Artist: Alicia Keys, "Fallin'"
Breakthrough Video: Fatboy Slim, "Weapon of Choice"
Best R&B Video: Destiny's Child, "Survivor"
Best Video From a Film: Christina Aguilera, Pink, Lil' Kim, Mya, featuring Missy "Misdemeanor" Elliott, "Lady Marmalade" (*Moulin Rouge*)
Best MTV2 Video: Mudvayne, "Dig"
Best Direction: Spike Jonze for Fatboy Slim's, "Weapon of Choice"
Best Choreography: Fatboy Slim, "Weapon of Choice"
Best Art Direction: Fatboy Slim, "Weapon of Choice"
Best Editing: Fatboy Slim, "Weapon of Choice"
Best Cinematography: Fatboy Slim, "Weapon of Choice"
Viewers' Choice: 'N Sync, "Pop"

Grammy Awards

Source: National Academy of Recording Arts & Sciences

Selected Grammy Awards for 2000

Record of the Year (single): "Beautiful Day," U2
Album of the Year: *Two Against Nature,* Steely Dan
Song of the Year: "Beautiful Day," U2, songwriters (U2)
New artist: Shelby Lynne
Pop vocal perf., female : "I Try," Macy Gray
Pop vocal perf., male: "She Walks This Earth (Soberana Rosa)," Sting
Pop vocal perf., duo/group: "Cousin Dupree," Steely Dan
Pop vocal album, traditional: *Both Sides Now*, Joni Mitchell
Pop instrumental album: *Symphony No. 1*, Joe Jackson
Pop album: *Two Against Nature,* Steely Dan
Dance recording: "Who Let the Dogs Out," Baha Men
Rock vocal perf., female: "There Goes the Neighborhood," Sheryl Crow
Rock vocal perf., male: "Again," Lenny Kravitz
Rock vocal perf., duo/group: "Beautiful Day," U2
Hard rock perf.: "Guerrilla Radio," Rage Against The Machine
Metal perf.: "Elite," Deftones
Rock song: "With Arms Wide Open," Scott Stapp and Mark Tremonti, songwriters (Creed)
Rock album: *There Is Nothing Left to Lose,* Foo Fighters
R&B vocal perf., female: "He Wasn't Man Enough," Toni Braxton
R&B vocal perf., male: "Untitled (How Does It Feel)," D'Angelo
R&B vocal perf., duo/group: "Say My Name," Destiny's Child
R&B song: "Say My Name," LaShawn Daniels, Fred Jerkins III, Rodney Jerkins, Beyoncé Knowles, LeToya Luckett, LaTavia Roberson, and Kelendria Rowland, songwriters (Destiny's Child)
R&B vocal album, traditional: *Ear-Resistible,* The Temptations
R&B album: *Voodoo,* D'Angelo
Rap solo perf.: "The Real Slim Shady," Eminem
Rap vocal perf., duo/group: "Forgot About Dre," Dr. Dre featuring Eminem
Rap album: *The Marshall Mathers LP,* Eminem
Country vocal perf., female: "Breathe," Faith Hill
Country vocal perf., male: "Solitary Man," Johnny Cash
Country perf. with vocal, duo/group: "Cherokee Maiden," Asleep at the Wheel
Country song: "I Hope You Dance," Mark D. Sanders and Tia Sillers, songwriters (Lee Ann Womack)
Country album: *Breathe,* Faith Hill
Bluegrass album: *The Grass Is Blue,* Dolly Parton
Jazz album, vocal: *In The Moment–Live In Concert,* Dianne Reeves
Jazz album, instr.: *Contemporary Jazz,* Branford Marsalis
Jazz album, contemporary: *Outbound,* Béla Fleck & The Flecktones
Blues album, contemporary: *Shoutin' In Key,* Taj Mahal & The Phantom Blues Band
Blues album, traditional: *Riding With The King,* B.B. King & Eric Clapton
Folk album, contemporary: *Red Dirt Girl,* Emmylou Harris
Folk album, traditional: *Public Domain–Songs From the Wild Land,* Dave Alvin
Reggae album: *Art and Life,* Beenie Man
Latin pop album: *Shakira–MTV Unplugged,* Shakira
Producer: non-classical, Dr. Dre; classical, Steven Epstein
Opera recording: *Busoni: Doktor Faust;* Kent Nagano, Kim Begley, Dietrich Fischer-Dieskau, Dietrich Henschel, Markus Hollop, Eva Jenis
Classical vocal perf.: *The Vivaldi Album* (Dell'aura al sussurrar, Alma oppressa, Etc.); Cecilia Bartoli (mezzo soprano)
Classical album: *Shostakovich: The String Quartets*; Emerson String Quartet; Da-Hong Seetoo & Max Wilcox, producers

Grammy Awards for 1958-2000

Record of the Year (single)	Year	Album of the Year
Domenico Modugno, "Nel Blu Dipinto Di Blu (Volare)"	1958	Henry Mancini, *The Music From Peter Gunn*
Bobby Darin, "Mack the Knife"	1959	Frank Sinatra, *Come Dance With Me*
Percy Faith, "Theme From a Summer Place"	1960	Bob Newhart, *Button Down Mind*
Henry Mancini, "Moon River"	1961	Judy Garland, *Judy at Carnegie Hall*
Tony Bennett, "I Left My Heart in San Francisco"	1962	Vaughn Meader, *The First Family*
Henry Mancini, "The Days of Wine and Roses"	1963	Barbra Streisand, *The Barbra Streisand Album*
Stan Getz, Astrud Gilberto, "The Girl From Ipanema"	1964	Stan Getz, Astrud Gilberto, *Getz/Gilberto*
Herb Alpert, "A Taste of Honey"	1965	Frank Sinatra, *September of My Years*
Frank Sinatra, "Strangers in the Night"	1966	Frank Sinatra, *A Man and His Music*
5th Dimension, "Up, Up and Away"	1967	The Beatles, *Sgt. Pepper's Lonely Hearts Club Band*
Simon & Garfunkel, "Mrs. Robinson"	1968	Glen Campbell, *By the Time I Get to Phoenix*
5th Dimension, "Aquarius/Let the Sunshine In"	1969	Blood Sweat and Tears, *Blood, Sweat and Tears*
Simon & Garfunkel, "Bridge Over Troubled Water"	1970	Simon & Garfunkel, *Bridge Over Troubled Water*
Carole King, "It's Too Late"	1971	Carole King, *Tapestry*
Roberta Flack, "The First Time Ever I Saw Your Face"	1972	George Harrison and friends, *The Concert for Bangla Desh*
Roberta Flack, "Killing Me Softly With His Song"	1973	Stevie Wonder, *Innervisions*
Olivia Newton-John, "I Honestly Love You"	1974	Stevie Wonder, *Fulfillingness' First Finale*
Captain & Tennille, "Love Will Keep Us Together"	1975	Paul Simon, *Still Crazy After All These Years*
George Benson, "This Masquerade"	1976	Stevie Wonder, *Songs in the Key of Life*
Eagles, "Hotel California"	1977	Fleetwood Mac, *Rumours*
Billy Joel, "Just the WayYou Are"	1978	Bee Gees, *Saturday Night Fever*
The Doobie Brothers, "What a Fool Believes"	1979	Billy Joel, *52nd Street*
Christopher Cross, "Sailing"	1980	Christopher Cross, *Christopher Cross*
Kim Carnes, "Bette Davis Eyes"	1981	John Lennon, Yoko Ono, *Double Fantasy*
Toto, "Rosanna"	1982	Toto, *Toto IV*
Michael Jackson, "Beat It"	1983	Michael Jackson, *Thriller*
Tina Turner, "What's Love Got to Do With It"	1984	Lionel Richie, *Can't Slow Down*
USA for Africa, "We Are the World"	1985	Phil Collins, *No Jacket Required*
Steve Winwood, "Higher Love"	1986	Paul Simon, *Graceland*
Paul Simon, "Graceland"	1987	U2, *The Joshua Tree*
Bobby McFerrin, "Don't Worry, Be Happy"	1988	George Michael, *Faith*
Bette Midler, "Wind Beneath My Wings"	1989	Bonnie Raitt, *Nick of Time*
Phil Collins, "Another Day in Paradise"	1990	Quincy Jones, *Back on the Block*
Natalie Cole, with Nat "King" Cole, "Unforgettable"	1991	Natalie Cole, with Nat "King" Cole, *Unforgettable*
Eric Clapton, "Tears in Heaven"	1992	Eric Clapton, *Unplugged*
Whitney Houston, "I Will Always Love You"	1993	Whitney Houston, *The Bodyguard*

Record of the Year (single)	Year	Album of the Year
Sheryl Crow, "All I Wanna Do"	**1994**	Tony Bennett, *MTV Unplugged*
Seal, "Kiss From a Rose"	**1995**	Alanis Morissette, *Jagged Little Pill*
Eric Clapton, "Change the World"	**1996**	Celine Dion, *Falling Into You*
Shawn Colvin, "Sunny Came Home"	**1997**	Bob Dylan, *Time Out of Mind*
Celine Dion, "My Heart Will Go On"	**1998**	Lauryn Hill, *The Miseducation of Lauryn Hill*
Santana featuring Rob Thomas, "Smooth"	**1999**	*Santana, Supernatural*
U2, "Beautiful Day"	**2000**	Steely Dan, *Two Against Nature*

National Book Awards, 1950-2000

The National Book Awards (known as the American Book Awards from 1980 to 1986) are administered by the National Book Foundation and have been given annually since 1950. The prizes, each valued at $10,000, are awarded to U.S. citizens for works published in the U.S. in the 12 months prior to the nominations. In some years, multiple awards were given for nonfiction in various categories; in such cases, the history and biography (if any) or biography winner is listed. Selected additional awards in nonfiction are given in footnotes. Nonfiction winners in certain separate categories may not be shown.

Fiction

Year	Author, Title
1950	Nelson Algren, *The Man With the Golden Arm*
1951	William Faulkner, *The Collected Stories*
1952	James Jones, *From Here to Eternity*
1953	Ralph Ellison, *Invisible Man*
1954	Saul Bellow, *The Adventures of Augie March*
1955	William Faulkner, *A Fable*
1956	John O'Hara, *Ten North Frederick*
1957	Wright Morris, *The Field of Vision*
1958	John Cheever, *The Wapshot Chronicle*
1959	Bernard Malamud, *The Magic Barrel*
1960	Philip Roth, *Goodbye, Columbus*
1961	Conrad Richter, *The Waters of Kronos*
1962	Walker Percy, *The Moviegoer*
1963	J.F. Powers, *Morte d'Urban*
1964	John Updike, *The Centaur*
1965	Saul Bellow, *Herzog*
1966	Katherine Anne Porter, *The Collected Stories*
1967	Bernard Malamud, *The Fixer*
1968	Thornton Wilder, *The Eighth Day*
1969	Jerzy Kosinski, *Steps*
1970	Joyce Carol Oates, *Them*
1971	Saul Bellow, *Mr. Sammler's Planet*
1972	Flannery O'Connor, *The Complete Stories*
1973	John Barth, *Chimera*
1974	Thomas Pynchon, *Gravity's Rainbow*
1974	Isaac Bashevis Singer, *A Crown of Feathers*

Year	Author, Title
1975	Robert Stone, *Dog Soldiers*
1976	William Gaddis, *JR*
1977	Wallace Stegner, *The Spectator Bird*
1978	Mary Lee Settle, *Blood Ties*
1979	Tim O'Brien, *Going After Cacciato*
1980	William Styron, *Sophie's Choice*
1981	Wright Morris, *Plains Song*
1982	John Updike, *Rabbit Is Rich*
1983	Alice Walker, *The Color Purple*
1984	Ellen Gilchrist, *Victory Over Japan*
1985	Don DeLillo, *White Noise*
1986	E.L. Doctorow, *World's Fair*
1987	Larry Heinemann, *Paco's Story*
1988	Pete Dexter, *Paris Trout*
1989	John Casey, *Spartina*
1990	Charles Johnson, *Middle Passage*
1991	Norman Rush, *Mating*
1992	Cormac McCarthy, *All the Pretty Horses*
1993	E. Annie Proulx, *The Shipping News*
1994	William Gaddis, *A Frolic of His Own*
1995	Philip Roth, *Sabbath's Theater*
1996	Andrea Barrett, *Ship Fever and Other Stories*
1997	Charles Frazier, *Cold Mounatin*
1998	Alice McDermott, *Charming Billy*
1999	Ha Jin, *Waiting*
2000	Susan Sontag, *In America*

Nonfiction

Year	Author, Title
1950	Ralph L. Rusk, *Ralph Waldo Emerson*
1951	Newton Arvin, *Herman Melville*
1952	Rachel Carson, *The Sea Around Us*
1953	Bernard A. De Voto, *The Course of an Empire*
1954	Bruce Catton, *A Stillness at Appomattox*
1955	Joseph Wood Krutch, *The Measure of Man*
1956	Herbert Kubly, *An American in Italy*
1957	George F. Kennan, *Russia Leaves the War*
1958	Catherine Drinker Bowen, *The Lion and the Throne*
1959	J. Christopher Herold, *Mistress to an Age: A Life of Madame De Stael*
1960	Richard Ellman, *James Joyce*
1961	William L. Shirer, *The Rise and Fall of the Third Reich*
1962	Lewis Mumford, *The City in History: Its Origins, Its Transformations, and Its Prospects*
1963	Leon Edel, *Henry James: Vol. II: The Conquest of London; Vol. III: The Middle Years*
1964	WIlliam H. McNeill, *The Rise of the West: A History of the Human Community*
1965	Louis Fisher, *The Life of Lenin*
1966	Arthur M. Schlesinger, Jr., *A Thousand Days: John F. Kennedy in the White House*
1967	Peter Gay, *The Enlightenment, An Interpretation Vol I: The Rise of Modern Paganism*
1968	George F. Kennan, *Memoirs: 1925–1950*[1]
1969	Winthrop D. Jordan, *White Over Black: American Attitudes Toward the Negro, 1550-1812*[2]
1970	T. Harry Williams, *Huey Long*[3]
1971	James MacGregor Burns, *Roosevelt: The Soldier of Freedom*
1972	Joseph P. Lash, *Eleanor and Franklin: The Story of Their Relationship, Based on Eleanor Roosevelt's Private Papers*
1973	James Thomas Flexner, *George Washington, Vol. IV: Anguish and Farewell, 1793-1799*[4]
1974	John Clive, *Macaulay, The Shaping of the Historian*; Douglas Day, *Malcolm Lowry: A Biography*[5]
1975	Richard B. Sewall, *The Life of Emily Dickinson*[6]

Year	Author, Title
1976	David Brion Davis, *The Problem of Slavery in the Age of Revolution, 1770-1823*
1977	W.A. Swanberg, *Norman Thomas: The Last Idealist*[7]
1978	W. Jackson Bate, *Samuel Johnson*
1979	Arthur M. Schlesinger, Jr., *Robert Kennedy and His Times*
1980	Tom Wolfe, *The Right Stuff*
1981	Maxine Hong Kingston, *China Men*
1982	Tracy Kidder, *The Soul of a New Machine*
1983	Fox Butterfield, *China: Alive in the Bitter Sea*
1984	Robert V. Remini, *Andrew Jackson and the Course of American Democracy, 1833-1845*
1985	J. Anthony Lukas, *Common Ground: A Turbulent Decade in the Lives of Three American Families*
1986	Barry Lopez, *Arctic Dreams*
1987	Richard Rhodes, *The Making of the Atom Bomb*
1988	Neil Sheehan, *A Bright Shining Lie: John Paul Vann and America in Vietnam*
1989	Thomas L. Friedman, *From Beirut to Jerusalem*
1990	Ron Chernow, *The House of Morgan: An American Banking Dynasty and the Rise of Modern Finance*
1991	Orlando Patterson, *Freedom*
1992	Paul Monette, *Becoming a Man: Half a Life Story*
1993	Gore Vidal, *United States: Essays 1952-1992*
1994	Sherwin B. Nuland, *How We Die: Reflections on Life's Final Chapter*
1995	Tina Rosenberg, *The Haunted Land: Facing Europe's Ghosts After Communism*
1996	James Carroll, *An American Requiem: God, My Father, and the War That Came Between Us*
1997	Joseph J. Ellis, *American Sphinx: The Character of Thomas Jefferson*
1998	Edward Ball, *Slaves in the Family*
1999	John W. Dower, *Embracing Defeat: Japan in the Wake of World War II*
2000	Nathaniel Philbrick, *In the Heart of the Sea: The Tragedy of the Whaleship Essex*

(1) Science, Philosophy, and Religion: Jonathan Kozol, *Death at an Early Age.* (2) Arts and Letters: Norman Mailer, *The Armies of the Night: History as a Novel, The Novel as History.* (3) Arts and Letters: Lillian Hellman, *An Unfinished Woman: A Memoir.* (4) Contemporary Affairs: Frances FitzGerald, *Fire in the Lake: The Vietnamese and the Americans in Vietnam.* (5) Arts and Letters: Pauline Kael, *Deeper Into the Movies.* (6) Arts and Letters: Roger Shattuck, *Marcel Proust;* Lewis Thomas, *The Lives of a Cell: Notes of a Biology Watcher.* (7) Contemporary Thought: Bruno Bettelheim, *The Uses of Enchantment: The Meaning and Importance of Fairy Tales.*

CONSUMER INFORMATION

Business Directory

Listed below are major U.S. corporations offering products and services to consumers. Alphabetization is by first key word. Listings generally include examples of products offered. **Note:** Information does not include changes resulting from the Sept. 11, 2001, attack that destroyed the World Trade Center and damaged other buildings in New York City.

Company Name; Address; Telephone Number; Website; Top Executive; Business, Products, or Services.

Abbott Laboratories; One Abbott Park Rd., North Chicago, IL 60064; (847) 937-6100; Website: http://www.abbott.com; Miles D. White; health care prods. (Murine, Selsun Blue).

Aetna, Inc.; 151 Farmington Ave., Hartford, CT 06156; (203) 273-0123; Website: http://www.aetna.com; William H. Donaldson; health insurance, financial services.

Alberto-Culver; 2525 Armitage Ave., Melrose Park, IL 60160; (708) 450-3000; Website: http://www.alberto.com; Leonard H. Lavin; hair care (VO5), consumer prods. (Mrs. Dash, Sugar Twin), personal care prods. (St. Ives), Sally Beauty Supply stores.

Albertson's, Inc.; 250 Parkcenter Blvd., Boise, ID 83726; (208) 395-6200; Website: http://www.albertsons.com; Gary Michael; supermarkets. (Co. merged with American Stores Co. 6/24/99, making Albertson's, Inc., the largest retail food and drug co. in the U.S.)

Allegheny Technologies, Inc.; 1000 Six PPG Place, Pittsburgh, PA 15222-5479; (412) 394-2800; Website: http://www.alleghenytechnologies.com; James L. Murdy; electronics, aerospace, industrial, consumer prods. (Water Pik); specialty metals.

Allstate Corp.; Allstate Plaza, Northbrook, IL 60062; (847) 402-5000; Website: http://www.allstate.com; Edward Liddy; property/casualty, life insurance.

Aluminum Co. of America (Alcoa); 201 Isabella St., Pittsburgh, PA 15212; (412) 553-4545; Website: http://www.shareholder.com/Alcoa; Alain J.P. Belda; world's largest aluminum producer.

Amazon.com Inc.; 1200 12th Ave. S., Suite 1200 Seattle, WA 98144; (206) 622-2335; Website: http://www.amazon.com; Jeff Bezos; on-line bookseller.

Amerada Hess Corp.; 1185 Ave. of the Americas, NY, NY 10036; (212) 997-8500; Website: http://www.hess.com; J. B. Hess; integrated international oil co.

AOL Time Warner Inc.; 75 Rockefeller Plaza, New York, NY 10009; (212) 484-8000; Website: http://www.aoltimewarner.com; Stephen M. Case; world's largest Internet online service; magazine publishing (*Time, Sports Illustrated, Fortune, Money, People,* DC Comics), TV and CATV (WB Network, HBO, Cinemax, CNN, TBS, TNT), book publishing (Little, Brown; Warner Books), motion pictures (Warner Bros., New Line Cinema), recordings, sports teams (Atlanta Braves, Atlanta Hawks), retailing (Warner Bros. stores). (America Online and Time Warner completed the largest corporate merger in history 1/11/01, becoming the largest media company in the U.S.)

American Express Co.; 200 Vesey St., NY, NY 10285; (212) 640-2000; Website: http://www.americanexpress.com; Kenneth I. Chenault; travel, financial, and information services.

American Greetings Corp.; 1 American Rd., Cleveland, OH 44144; (216) 252-7300; Website: http://www.americangreetings.com; Morry Weiss; greeting cards, stationery, party goods, gift items.

American Home Prods. Corp.; 5 Giralda Farms, Madison, NJ 07940; (973) 660-5000; Website: http://www.ahp.com; John R. Stafford; prescription and over-the-counter drugs (Advil, Anacin, Dristan, Robitussin).

American Intl. Group; 70 Pine St., NY, NY 10270; (212) 770-7000; Website: http://www.aig.com; Maurice R. Greenberg; insurance, financial services.

AMR Corp.; PO Box 619616, Dallas/Ft. Worth Airport, TX 75261; (817) 963-1234; Website: http://www.amrcorp.com; Donald J. Carty; air transportation (American Airlines, American Eagle). (American Airlines announced agreement, 1/11/01, to acquire Trans World Airlines, or TWA, for $500 million.)

Anheuser-Busch Cos., Inc.; 1 Busch Pl., St. Louis, MO 63118; (314) 577-2000; Website: http://www.anheuser-busch. com; August A. Busch 3d; world's largest brewer (Budweiser, Michelob, BudLight, Natural Light, Busch, O'Doul's), aluminum can manuf. and recycling, theme parks.

Apple Computer, Inc.; 1 Infinite Loop, Cupertino, CA 95014-2084; (408) 996-1010; Website: http://www.apple.com; Steve Jobs; manuf. of personal computers, software, peripherals.

Aramark Corp.; Aramark Tower, 1101 Market St., Philadelphia, PA 19107; (215) 238-3000; Website: http://www.aramark.com; Joseph Neubauer; food and support services, uniforms and career apparel, child care and early education.

Archer Daniels Midland Co.; 4666 Faries Pkwy., Decatur, IL 62526; (217) 424-5200; Website: http://www.admworld.com; G. Allen Andreas; agricultural commodities and prods.

Armstrong World Industries, Inc.; 2500 Columbia Ave., PA 17604; (717) 397-0611; Website: http://www.armstrong.com; Michael D. Lockhart; interior furnishings, specialty prods.

Arvinmeritor Industries, Inc.; 2135 West Maple Road, Troy, MI 48084; (248) 435-1000; Website: http://www.arvinmeritor.com; V. William Hunt; auto emission and ride control systems.

Ashland Inc.; 50 E. River Center, PO Box 391, Covington, KY 41012; (606) 815-3333; Website: http://www.ashland.com; Paul W. Chellgren; petroleum producer and refiner (Valvoline), chemicals, road construction.

AT&T Corp.; 32 Ave. of the Americas, NY, NY 10013-2412; (212) 387-5400; Website: http://www.att.com; C. Michael Armstrong; communications, global information management.

Avon Prods., Inc.; 1345 Ave. of Americas, NY, NY 10105; (212) 282-5000; Website: http://www.avon.com; Stanley Gault; cosmetics, fragrances, toiletries, fashion jewelry, gift items, casual apparel, lingerie.

Bank of America Corp.; Bank of America Corporate Center, Charlotte, NC 28255; (704) 386-5000; Website: http://www.bankofamerica.com; Hugh L. McColl Jr; largest U.S. bank.

Bausch & Lomb Inc.; One Bausch & Lomb Place, Rochester, NY 14604; (716) 338-6000; Website: http://www.bausch.com; William M. Carpenter; vision and health-care prods., accessories.

Baxter International Inc.; 1 Baxter Pkwy., Deerfield, IL 60015; (847) 948-2000; Website: http://www.baxter.com; H. M. Kraemer Jr; health care prods. & services.

Bear Stearns Cos. Inc.; 245 Park Ave., NY, NY 10167; (212) 272-2000; Website: http://www.bearstearns.com; Alan C. Greenberg; investment banking, securities trading, brokerage.

Becton, Dickinson & Co.; 1 Becton Dr., Franklin Lakes, NJ 07417; (201) 847-6800; Website: http://www.bd.com; C. Castellini; medical, laboratory, diagnostic prods.

BellSouth Corp.; 1155 Peachtree St. NE, Atlanta, GA 30309; (404) 249-2000; Website: http://www.bellsouth.com; John L. Clendenin; telephone service in southern U.S.

Best Buy Co., Inc.; 7075 Flying Cloud Dr., Eden Prairie, MN 55344; (612) 947-2000; Website: http://www.bestbuy.com; R. M. Schulze; retailer of software, appliances, electronics, cameras, home office equipment.

Bethlehem Steel Corp.; 1170 8th Ave., Bethlehem, PA 18016; (610) 694-2424; Website: http://www.bethsteel.com; Duane R. Dunham; steel & steel prods.

Black & Decker Corp.; 701 E. Joppa Rd., Towson, MD 21204; (410) 716-3900; Website: http://www.blackanddecker.com; Nolan D. Archibald; manuf. power tools (DeWalt), household prods. (Kwikset, Price Pfister), small appliances (Black & Decker).

H & R Block, Inc.; 4410 Main St., Kansas City, MO 64111; (816) 753-6900; Website: http://www.hrblock.com; Henry Bloch; tax return preparation.

Boeing Co.; 7755 E. Marginal Way, Seattle, WA 98108; (206) 655-2121; Website: http://www.boeing.com; Philip M. Condit; leading manufacturer of commercial, jet aircraft.

Boise Cascade Corp.; 1111 W. Jefferson St., Boise, ID 83728; (208) 384-6161; Website: http://www.bc.com; George J. Harad; distributor of office products & building materials; paper, wood prods.

Borden, Inc.; 180 E. Broad St., Columbus OH 43215-3707; (614) 225-4000; Website: http://www.bordenfamily.com; C. Robert Kidder; snacks (Wise, Cheez Doodles), adhesives (Elmer's, Krazy Glue), pasta (Prince, Creamette, Goodman's), pasta sauce (Aunt Millie's, Classico), Wyler's bouillon, Soup Starter, Corning Consumer Prods. (Corningware, Corelle, Pyrex, Revere).

Bristol-Myers Squibb Co.; 345 Park Ave., NY, NY 10022; (212) 546-4000; Website: http://www.bms.com; Charles A. Heimhold; toiletries (Ban antiperspirant), haircare (Clairol), drugs (Bufferin, Comtrex, Excedrin, Pravachol), infant formula (Enfamil).

Brown-Forman Corp.; PO Box 1080, Louisville, KY 40201-1080; (502) 585-1100; Website: http://www.brown-forman.com; Owsley Brown 2d; distilled spirits (Jack Daniel's, Southern Comfort), wines (Bolla, Fetzer, Korbel), china and crystal (Dansk, Lenox), Gorham, Kirk Steiff silver prods., Hartmann luggage.

Brown Shoe Co., Inc.; 8300 Maryland Ave., P.O. Box 29, St. Louis, MO 63166; (314) 854-4000; Website: http://www.brownshoe.com; Ronald A. Fromm; manuf. and retailer (Famous Footwear) of women's, men's, and children's shoes (Buster Brown, Naturalizer).

Brunswick Corp.; 1 N. Field Ct., Lake Forest, IL 60045-4811; (847) 735-4700; Website: http://www.brunswickcorp.com; George W. Buckley; largest U.S. maker of leisure and recreation prods., marine, camping, fitness and fishing equip., bowling centers and equip.

Burlington Northern Santa Fe Inc.; 2650 Lou Menk Dr., Ft. Worth, TX 76131-2830; (817) 333-2000; Website: http://www.bnsf.com; Robert Krebs; one of the largest U.S. rail transportation cos.

Campbell Soup Co.; Campbell Pl., Camden, NJ 08103; (609) 342-4800; Website: http://www.campbellsoup.com; Douglas R. Conant; soups, Franco-American spaghetti, V-8 vegetable juice, Godiva chocolates, Swanson frozen dinners, Prego spaghetti sauce, Pepperidge Farm.

Carter-Wallace, Inc.; Half Acre Road, P.O. Box 1001, Cranbury, NJ, 08512; (609) 655-6000; Website: http//www.carter-wallace.com; H. H. Hoyt; personal care (Arrid, Rise, Pearl Drops, Nair, Trojan condoms).

Caterpillar Inc.; 100 N.E. Adams St., Peoria, IL 61629; (309) 675-1000; Website: http://www.cat.com; Glen A. Barton; world's largest producer of earth moving equip.

Chevron Corp.; 575 Market St., San Francisco, CA 94105; (415) 894-7700; Website: http://www.chevron.com; Kenneth T. Derr; integrated oil co.(Chevron and Texaco announced, 9/7/01, that their proposed merger had received FTC approval; the merger would create the 2d largest U.S.-based oil co.)

Chiquita Brands International, Inc.; 250 E. 5th St., Cincinnati, OH 45202; (513) 784-8000; Website: http://www.chiquita.com; Carl H. Lindner; bananas, fruits, vegetables.

Church & Dwight Co., Inc.; 469 N. Harrison St., Princeton, NJ 08543; (609) 683-5900; Website: http://www.armhammer.com; R.A. Davies; world's largest producer of sodium bicarbonate (Arm & Hammer); Brillo.

CIGNA Corp.; 1 Liberty Pl., Philadelphia, PA 19103; (215) 761-1000; Website: http://www.cigna.com; H. Edward Hanway; insurance holding co.

Circuit City Stores, Inc.; 9950 Mayland Dr., Richmond, VA 23233-1464; (804) 527-4000; Website: http://www.circuitcity.com; Allan McCollough; retailer of electronic, audio/video equip., consumer appliances; new and used-car stores (CarMax).

Citigroup; 153 E. 53rd St., NY, NY 10043; (212) 559-1000; Website: http://www.citigroup.com; Michael Carpenter; diversified financial services.

Clorox Co.; 1221 Broadway, Oakland, CA 94612; (510) 271-7000; Website: http://www.clorox.com; G. Craig Sullivan; retail consumer prods. (Clorox, Formula 409, Pine-Sol, S.O.S., Soft Scrub cleansers; Armor All, STP, Rain Dance automotive prods.; Jonny Cat, Fresh Step cat litters; Kingsford charcoal briquets; StarterLogg; Combat and Black Flag insecticides; Hidden Valley dressing; K.C. Masterpiece barbecue sauce; Brita water systems).

Liz Claiborne, Inc.; 1441 Broadway, New York, NY 10018; (212) 354-4900; Website: http://www.lizclaiborne.com; P. Charron; apparel, accessories.

Coca-Cola Co.; 1 Coca-Cola Plaza, Atlanta, GA 30313; (404) 676-2121; Website: http://www.cocacola.com; Douglas N. Daft; world's largest soft drink co. (Coca-Cola, Sprite, Nestea), world's largest dist. of juice prods. (Minute Maid, Five Alive, Hi-C, Fruitopia).

Colgate-Palmolive Co.; 300 Park Ave., NY, NY 10022; (212) 310-2000; Website: http://www.colgate.com; Reuben Mark; soap (Palmolive, Irish Spring), detergent (Fab, Ajax, Fresh Start), toothpaste (Colgate, Ultra Brite), Hill's pet food.

Compaq Computer Corp.; 20555 SH 249, Houston, TX 77070; (281) 370-0670; Website: http://www.compaq.com; Michael D. Capellas; laptop and desktop computers. (Hewlett-Packard agreed, 9/3/01, to acquire Compaq Computer Corp. in a $25 billion stock deal.)

CompUSA Inc.; 14951 N. Dallas Pkwy., Dallas, TX 75254; (972) 982-4000; Website: http://www.compusa.com; Hal Compton; largest U.S. superstore retailer of microcomputers and peripherals.

Computer Sciences Corp.; 2100 E. Grand Ave., El Segundo, CA 90245; (310) 615-0311; Website: http://www.csc.com; Van B. Honeycutt; technology services.

ConAgra; 1 ConAgra Dr., Omaha, NE 68102; (402) 595-4000; Website: http://www.conagra.com; Bruce Rohde; 2d largest U.S. food processor.

Continental Airlines, Inc.; 1600 Smith St. HQ511, Houston, TX 77002; (713) 324-5242; Website: http://www.continental.com; Gordon M. Bethune; air transportation.

Adolph Coors Co.; Golden, CO 80401; (303) 279-6565; Website: http://www.coorsinvestor.com; Peter Coors; brewer (Coors, Killian's, Zima).

Corning Inc.; 1 Riverfront Plaza, Corning, NY 14831; (607) 974-9000; Website: http://www.corning.com; John W. Loose; telecommunications, specialty materials, optical fiber and cable.

Costco Wholesale Corp.; 999 Lake Dr., Issaquah, WA 98027; (425) 313-8100; Website: http://www.costco.com; James D. Sinegal; wholesale-membership warehouses.

Crane Co.; 100 First Stamford Place, Stamford, CT 06902; (203) 363-7300; Website: http://shareholder.com/crane; R. S. Evans; manuf. fluid control devices, vending machines, fiberglass panels, aircraft brakes.

A. T. Cross Co.; 1 Albion Rd., Lincoln, RI 02865; (401) 333-1200; Website: http://www.cross.com; David Whalen; writing instruments.

Crown Cork & Seal Co.; 1 Crown Way, Philadelphia, PA 19154-4599; (215) 698-5100; Website: http://www.crowncork.com; William J. Avery; world's leading supplier of packaging prods.

CSX Corp.; 901 E. Cary St., Richmond, VA 23219; (804) 782-1400; Website: http://www.csx.com; John W. Snow; rail, ocean, barge freight transport.

CVS Corp.; 1 CVS Dr., Woonsocket, RI 02895; (401) 765-1500; Website: http://www.CVS.com; Thomas M. Ryan; drugstore chain.

Dana Corp.; 4500 Dorr St., Toledo, OH 43615; (419) 535-4500; Website: http://www.dana.com; Joseph M. Magliochetti; truck and auto parts, supplies.

Deere & Co.; John Deere Rd., Moline, IL 61265; (309) 765-8000; Website: http://www.deere.com; Robert W. Lane; world's largest manuf. of farm equip.; industrial equip.; lawn and garden tractors.

Dell Computer Corp.; 1 Dell Way, Round Rock, TX 78682; (512) 338-4400; Website: http://www.dell.com; Michael S. Dell; laptop and desktop computers.

Delta Air Lines, Inc.; Hartsfield Atlanta Intl. Airport, Atlanta, GA 30320; (404) 715-2600; Website: http://www.delta-air.com; Leo F. Mullin; air transportation.

Dial Corp.; 15501 N. Dial Blvd., Scottsdale, AZ 85260-1619; (602) 754-3425; Website: http://www.dialcorp.com; Herbert Baum; consumer prods. (Dial, Tone soap, Breck shampoo, Armour Star meats, Renuzit air fresheners).

Diebold, Inc.; PO Box 8230, Canton, OH 44711; (330) 490-4000; Website: http://www.diebold.com; Walden W. O'Dell; manuf. ATMs, security systems and prods.

Dillard's; 1600 Cantrell Rd., Little Rock, AR 72201; (501) 376-5200; Website: http://www.dillards.com; William Dillard, Sr.; 2d largest dept. store chain in U.S.

Walt Disney Co.; 500 S. Buena Vista St., Burbank, CA 91521-7320; (818) 560-1000; Website: http://www.disney.com; Michael D. Eisner; motion pictures, television (ESPN, ABC, A&E, Lifetime), radio stations, theme parks (Walt Disney World, Disneyland) and resorts, publishing, recordings, retailing (Disney Stores).

Dole Food Co., Inc.; One Dole Drive, Westlake Village, CA 91362; (818) 879-6600; Website: http://www.dole.com; David H. Murdock; food prods., fresh fruits and vegetables.

R. R. Donnelley & Sons Co.; 77 W. Wacker Dr., Chicago, IL 60601-1696; (312) 326-8000; Website: http://www.rrdonnelley.com; William L. Davis; commercial printer.

Dow Chemical Co.; 2030 Dow Center, Midland, MI 48674; (517) 636-1000; Website: http://www.dow.com; William S. Stavropoulos; chemicals, plastics (Dow Chemical and Union Carbide merged 2/7/01 to become the world's 2nd largest chemical company.)

Dow Jones & Co., Inc.; 200 Liberty St., NY, NY 10281; (212) 416-2000; Website: http://www.dowjones.com; Peter R. Kann; financial news service, publishing (*Wall Street Journal*, *Barron's*, Ottaway Newspapers).

Dun & Bradstreet Corp.; 1 Diamond Hill Rd., Murray Hill, NJ 07974; (908) 665-5000; Website: http://www.dnbcorp.com; Allen Z. Loren.; business information, publishing (Moody's, "Yellow Pages" phone books).

E. I. du Pont de Nemours & Co. (Dupont); 1007 Market St., Wilmington, DE 19898; (302) 774-1000; Website: http://www.dupont.com; Charles Holliday; largest U.S. chemical co.; petroleum, consumer prods.

Eastman Kodak Co.; 343 State St., Rochester, NY 14650-0205; (716) 724-5492; Website: http://www.kodak.com; D. Carp; world's largest producer of photographic prods.

Eaton Corp.; 1111 Superior Ave., Cleveland, OH 44114; (216) 523-5000; Website: http://www.eaton.com; Alexander M. Cutler; manuf. of vehicle powertrain components, controls.

El Paso Corp.; 1001 Louisiana Street, Houston, TX 77002; (713) 420-2600; Website: http://www.elpaso.com; William Wise; diversified energy company primarily engaged in interstate transmission of natural gas. (El Paso Corp. merged with Coastal Corp. 1/29/01 to become the 4th largest energy company in the U.S.)

Emerson Electric Co.; 8000 West Florissant Avenue, St. Louis, MO 63136; (314) 553-2000; Website: http://www.gotoemerson.com; C. F. Knight; electrical, electronics prods. & systems.

Exxon Mobil Corp.; 5959 Las Colinas Blvd., Irving, TX 75039-2298; (972) 444-1000; Website: http://www.exxonmobil.com; Lee R. Raymond; world's largest publicly owned integrated oil co.; Exxon merged with Mobil 12/1/99.

Fedders Corp.; 505 Martinsville Road, P.O. Box 813, Liberty Corner, NJ 07938; (908) 604-8686; Website: http://www.fedders.com; Salvatore Giordano Jr; manuf. of room air conditioners (Fedders, Airtemp), dehumidifiers.

FedEx Corp.; Box 727, Memphis, TN 38194; (901) 369-3600; Website: http://www.fedex.com; F. W. Smith; express delivery service.

Freddie Mac; 8200 Jones Branch Dr., McLean, VA 22102; (703) 903-2000; Website: http://www.freddiemac.com; Leland C. Brendsel; residential mortgage provider.

Fannie Mae; 3900 Wisconsin Ave. NW, Washington, DC 20016; (202) 752-7000; Website: http://www.fanniemae.com; Franklin Raines; largest U.S. provider of residential mortgage funds.

Federated Dept. Stores; 7 W. 7th St., Cincinnati, OH 45202; (513) 579-7000; Website: http://www.Federated-fds.com; James Zimmerman; full-line dept. stores Macy's, Bloomingdale's, Stern's.

First Data Corp.; 5660 New Northside Dr., Atlanta, GA 30328; (770) 857-0001; Website: http://www.firstdatacorp.com; Henry C. Duques; info. retrieval, data processing.

Fleetwood Enterprises, Inc.; 3125 Myers St., Riverside, CA 92503; (909) 351-3500; Website: http://www.fleetwood.com; Glenn F. Kummer; manufactured homes, recreational vehicles.

Fleming Cos. Inc.; P.O. Box 299013, Lewisville, TX 75057; (972) 906-8000; Website: http://www.fleming.com; Mark S. Hansen; one of largest U.S. wholesale food distrib.

Fluor Corp.; One Enterprise Dr., Aliso Viejo, CA 92698; (949) 349-2000; Website: http://www.fluor.com; Philip J. Carroll; largest international engineering and construction co. in U.S.

Ford Motor Co.; American Rd., Dearborn, MI 48121; (313) 845-8540; Website: http://www.ford.com; Jacques Nassar; 2nd largest auto manufacturer, motor vehicle sales (Ford, Lincoln-Mercury, Volvo), rentals (Hertz).

Fortune Brands, Inc.; 300 Tower Parkway, Lincolnshire, IL 60069; (847) 484-4400; Website: http://www.fortunebrands.com; Norman H. Wesley; whiskey (Jim Beam), hardware, office prods. (Swingline), golf and leisure prods. (Titleist, Cobra, Foot-Joy).

Fruit of the Loom, Inc.; 1 Fruit of the Loom Dr., Bowling Green, KY; (270) 781-6400; Website: http://www.fruit.com; Dennis Bookshester; manuf. of underwear, activewear.

Gannett Co., Inc.; 1100 Wilson Blvd., Arlington, VA 22234; (703) 284-6000; Website: http://www.gannett.com; D.H. McCorkindale; newspaper publishing (*USA Today*), network and cable TV.

The Gap, Inc.; 2 Folsom St., San Francisco, CA 94105; (415) 952-4400; Website: http://www.gap.com; Donald G. Fisher; casual and activewear retailer (Gap, Banana Republic, Old Navy).

General Dynamics; 3190 Fairview Park Drive, Falls Church, VA 22042-4523; (703) 876-3000; Website: http://www.generaldynamics.com; Nicholas D. Chabraja; nuclear submarines (Trident, Seawolf), armored vehicles, combat systems, computing devices, defense systems.

General Electric Co.; 3135 Easton Tpke., Fairfield, CT 06431; (203) 373-2211; Website: http://www.ge.com; Jeffrey Immelt; electrical, electronic equip., radio and television broadcasting (NBC), aircraft engines, power generation, appliances.

General Mills, Inc.; PO Box 1113, Minneapolis, MN 55440; (763) 764-2311; Website: http://www.generalmills.com; S. W. Sanger; foods (Total, Wheaties, Cheerios, Chex, Hamburger Helper, Betty Crocker, Bisquick).

General Motors; 100 Renaissance Center, Detroit, MI 48243; (313) 556-5000; Website: http://www.gm.com; John F. Smith Jr; world's largest auto manuf. (Chevrolet, Pontiac, Cadillac, Buick).

Genuine Parts Co.; 2999 Circle 75 Pkwy., Atlanta, GA 30339; (404) 953-1700; Website: http://www.genpt.com; Larry L. Prince; distributes auto replacement parts (NAPA).

Georgia-Pacific Corp.; 133 Peachtree St. NE, Atlanta, GA 30303; (404) 521-5210; Website: http://www.gp.com; A. D. Correll; manuf. of paper and wood prods.

Gillette; Prudential Tower Bldg., Boston, MA 02199; (617) 463-3000; Website: http://www.gillette.com; James M. Kitts; stationery prods. (PaperMate, Parker, Waterman pens), personal care prods. (Sensor, Atra razors, Right Guard, Soft and Dri), appliances (Braun), batteries (Duracell).

The Goodyear Tire & Rubber Co.; 1144 E. Market St., Akron, OH 44316; (330) 796-2121; Website: http://www.goodyear.com; Samir F. Gibara; world's largest rubber manuf.; tires and other auto prods.

W. R. Grace & Co.; 7500 Grace Dr., Columbia, MD 21044; (410) 531-4000; Website: http://www.grace.com; Paul J. Norris; chemicals, construction prods.

Great Atlantic & Pacific Tea Co. (A&P); 2 Paragon Dr., Montvale, NJ 07645; (201) 573-9700; Website: http://www.aptea.com; Christian Haub; supermarkets (A&P, Waldbaum's, Kohl's, Dominion).

Halliburton Co.; 500 N. Akard St., Dallas, TX 75201; (214) 978-2600; Website: http://www.halliburton.com; Dave Lesar; energy, engineering, and construction services.

Harley-Davidson, Inc.; 3700 West Juneau Avenue, Milwaukee, WI 53208; (414) 343-4680; Website: http://www.harley-davidson.com; Jeffrey Bleustein; manuf. of motorcycles, parts and accessories.

Harrah's Entertainment, Inc.; One Harrah's Court, Las Vegas, NV 89119; (702) 407-6000; Website: http://www.harrahs.com; Philip G. Satre; casino-hotels and riverboats.

Hartford Life, Inc.; Hartford Plaza, 690 Asylum Ave., Hartford, CT 06115; (860) 547-5000; Website: http://www.thehartford.com; Ramani Ayer; insurance, finl. svces.

Hartmarx; 101 N. Wacker Dr., Chicago, IL 60606; (312) 372-6300; Website: http://www.hartmarx.com; Elbert O. Hand; apparel manuf. (Hart Schaffner & Marx, Hickey Freeman, Claiborne, Tommy Hilfiger, Pierre Cardin, Perry Ellis).

Hasbro, Inc.; 1027 Newport Ave., Pawtucket, RI 02862; (401) 431-8697; Website: http://www.hasbro.com; Alan G. Hassenfeld; toy and game manuf. (Milton Bradley, Playskool, G. I. Joe, Parker Bros., Tiger Electronics, Play-Doh).

HCA-The Healthcare Co.; 1 Park Plaza, Nashville, TN 37203; (615) 344-9551; Website: http://www.columbia.net; T.F. Frist Jr; largest hospital mgmt. co. in the U.S.

H. J. Heinz Co.; PO Box 57, Pittsburgh, PA 15230; (412) 456-6014; Website: http://www.heinz.com; William R. Johnson; foods (Star-Kist, Ore-Ida, 57 Varieties), pet food (Ken-L Ration, 9 Lives), Weight Watchers.

Hershey Foods Corp.; 100 Crystal A Dr., Hershey, PA 17033; (717) 534-6799; Website: http://www.hersheys.com; Kenneth L. Wolfe; largest U.S. producer of chocolate and confectionery prods. (Reese's, Kit Kat, Mounds, Almond Joy, Cadbury, Jolly Rancher, Twizzler, Milk Duds, Good 'n' Plenty), pasta (San Giorgio, Ronzoni).

Hewlett-Packard Co.; 3000 Hanover St., Palo Alto, CA 94304; (650) 857-1501; Website: http://www.hp.com; Carly Fiorina; manuf. computers, electronic prods. and systems. (Hewlett-Packard agreed to acquire Compaq Computer Company, 9/3/01, in a $25 billion stock deal.)

Hillenbrand Industries, Inc.; 700 State Rte. 46, Batesville, IN 47006; (812) 934-8400; Website: http://www.Hillenbrand.com; R.J. Hillenbrand; manuf. caskets, adjustable hospital beds, locks (Medeco).

Hilton Hotels Corp.; 9336 Civic Center Dr., Beverly Hills, CA 90210; (310) 278-4321; Website: http://www.hilton.com; Stephen F. Bollenbach; hotels, casinos.

Home Depot, Inc.; 2455 Paces Ferry Rd. NW, Atlanta, GA 30339; (770) 433-8211; Website: http://www.homedepot.com; Bernard Marcus and Arthur M. Block; retail building supply, home improvement warehouse stores.

Honeywell Inc.; 101 Columbia Road, Morristown, NJ 07962; (973) 455-2000; Website: http://www.honeywell.com; Lawrence A. Bossidy; merger in 12/99 with AlliedSignal Corp. industrial and home control systems, aerospace guidance systems.

Hormel Foods Corp.; 1 Hormel Pl., Austin, MN 55912-3680; (507) 437-5611; Website: http://www.hormel.com; Joel W. Johnson; meat processor, pork and beef prods. (SPAM, Dinty Moore, Little Sizzlers).

Houghton Mifflin Co.; 222 Berkeley St., Boston, MA 02116; (617) 351-5000; Website: http://www.hmco.com; Nader F. Darehshori; publisher of textbooks, reference, general interest books.

Huffy Corp.; 225 Byers Rd., Miamisburg, OH 45342; (937) 866-6251; Website: http://www.huffy.com; Don R. Graber; largest U.S. bicycle manuf., sports and hardware equip.

Humana, Inc.; 500 W. Main Street, P.O. Box 1438, Louisville, KY 40201-1438; (502) 580-1000; Website: http://www.humana.com; David A. Jones; managed healthcare service provider, financial services.

IBP, Inc.; 800 Stevens Port Dr., Dakota Dunes, SD 57049; (605) 235-2061; Website: http://www.ibpinc.com; Robert L. Peterson; world's largest processor of fresh beef and pork.

Illinois Toolworks; 3600 Westlake Ave., Glenview, IL 60025; (847) 724-7500; Website: http://www.itwinc.com; W. James Farrell; food equip. (Hobart), home appliances and cookware (West Bend).

Ingersoll-Rand; Woodcliff Lake, NJ 07675; (201) 573-0123; Website: http://www.ingersoll-rand.com; H. Henkel; industrial machinery.

Intel Corp.; 2200 Mission College Blvd., Santa Clara, CA 95052-8119; (408) 765-8080; Website: http://www.intc.com; A. S. Grove; manuf. integrated circuits (Pentium).

International Business Machines Corp. (IBM); New Orchard Rd., Armonk, NY 10504; (914) 499-1900; Website: http://www.ibm.com; Louis V. Gerstner Jr; world's largest supplier of advanced information processing technology equip., services.
International Paper Co.; 2 Manhattanville Road, Purchase, NY 10577; (914) 397-1500; Website: http://www.internationalpaper.com; John T. Dillon; world's largest paper/forest prods. co., chemicals, minerals.
Interstate Bakeries Corp.; 12 E. Armour Blvd., Kansas City, MO 64111; (816) 502-4000; Website: http://www.irin.com/ibc; Charles A. Sullivan; baked goods wholesaler, distributor (Wonder, Hostess, Dolly Madison, Beefsteak, Home Pride).
Jo-Ann Stores, Inc.; 5555 Darrow Rd., Hudson, OH 44236; (330) 656-2600; Website: http://www.joann.com; Alan Rosskamm; nation's largest specialty fabric and craft stores (Jo-Ann Fabric and Crafts, Jo-Ann etc.).
S.C. Johnson & Son, Inc.; 1525 Howe St., Racine, WI 53403; (800) 494-4855; Website: http://www.scjohnson.com; William Perez; cleaning and other household prods. (Johnson's Wax, Windex, Pledge, fantastik, Raid, Off!, Shout, Glade, Scrubbing Bubbles, Ziploc bags).
Johnson & Johnson; 1 Johnson & Johnson Plaza, New Brunswick, NJ 08933; (732) 524-0400; Website: http://www.jnj.com; Dr. H. Fisk-Johnson; surgical dressings (Band-Aid), pharmaceuticals (Tylenol), toiletries (Neutrogena).
Johnson Controls; 5757 North Green Bay Avenue, Milwaukee, WI 53201; (414) 228-1200; Website: http://www.johnsoncontrols.com; James H. Keyes; fire protection services, auto seats and batteries.
Jostens Inc.; 5501 Norman Center Dr., Minneapolis, MN 55437; (612) 830-3300; Website: http://www.jostens.com; Bob Buhrmaster; school rings, yearbooks, plaques.
JPMorgan Chase & Co. Inc.; 270 Park Ave., NY, NY 10017; (212) 270-6000; Website: http://www.jpmorganchase.com; Douglas A. Warner III; global financial firm; (JPMorgan & Co. and Chase Manhattan Corp. completed a merger 12/31/00).
Kellogg Co.; 1 Kellogg Sq., Battle Creek, MI 49016; (616) 961-2000; Website: http://www.kelloggs.com; Carlos Gutierrez; world's largest mfgr. of ready-to-eat cereals, other food prods. (Frosted Flakes, Rice Krispies, Froot Loops, Pop-Tarts, Nutri-Grain, Eggo).
Kimberly-Clark Corp.; PO Box 619100, Dallas, TX 75261-9100; (972) 281-1200; Website: http://www.kimberly-clark.com; Wayne R. Sanders; personal care prods. (Kleenex, Scott, Cottonelle, Huggies, Viva, Kotex).
King World Productions, Inc.; 1700 Broadway, 33rd Floor, NY, NY 10019; (212) 315-4000; Website: http://www.kingworld.com; Roger King; distributor of TV programs (*Oprah Winfrey Show, Wheel of Fortune, Jeopardy!, Inside Edition*).
Kmart Corp.; 3100 W. Big Beaver Rd., Troy, MI 48084; (248) 643-1000; Website: http://www.bluelight.com; Charles C. Conaway; discount stores, home improvement centers (Builders Square).
Knight Ridder, Inc.; 50 West San Fernando Street, San Jose, CA 95113-2413; (408) 938-7700; Website: http://www.Knightridder.com; P. A. Ridder; newspaper publishing.
Kroger Co.; 1014 Vine St., Cincinnati, OH 45202; (513) 762-4000; Website: http://www.kroger.com; Joseph A. Pichler; largest U.S. retail grocery chain.
(Estee) Lauder Cos.; 767 5th Ave., NY, NY 10153; (212) 572-4200; Leonard A. Lauder; cosmetics (Clinique), fragrance prods. (Aramis, Aveda, Tommy Hilfiger).
La-Z-Boy Inc.; 1284 N. Telegraph Rd., Monroe, MI 48161; (734) 242-1444; Website: http://www.lazboy.com; Patrick H. Norton; reclining chairs, other furniture.
Leggett & Platt, Inc.; No. 1 Leggett Rd., Carthage, MO 64836; (417) 358-8131; Website: http://www.leggett.com; Harry M. Cornell Jr; furniture and furniture components.
Lehman Bros. Holdings, Inc.; 3 World Financial Ctr., NY, NY 10285; (212) 526-7000; Website: http://www.lehman.com; Richard S. Fuld Jr; investment bank.
Levi Strauss & Co.; 1155 Battery St., San Francisco, CA 94111; (415) 501-6000; Website: http://www.levistrauss.com; Robert D. Haas; blue jeans, casual sportswear.
Eli Lilly and Company; Lilly Corporate Center, Indianapolis, IN 46285; (317) 276-2000; Website: http://www.lilly.com; Sidney Tauvel; pharmaceuticals (Axid, Ceclor, Prozac) and animal health prods.
The Limited, Inc.; 3 Limited Pkwy., P.O. Box 16000, Columbus, OH 43216; (614) 479-7000; Website: http://www.limited.com; Leslie H. Wexner; women's apparel stores (Lane Bryant, Lerner, Limited, Express, Structure, Victoria's Secret).
Lockheed Martin Corp.; 6801 Rockledge Dr., Bethesda, MD 20817; (301) 897-6000; Website: http://www.lockheedmartin.com; Vance Coffman; commercial and military aircraft, electronics, missiles.
Loews Corp.; 635 Madison Ave., NY, NY 10021; (212) 521-2000; Website: http://www.loews.com; James S. Tisch; tobacco prods. (Kent, True, Newport), watches (Bulova), hotels, insurance (CNA Fin'l.), offshore drilling.
Longs Drug Stores, Inc.; 141 N. Civic Dr., P.O. Box 5222, Walnut Creek, CA 94596; (925) 937-1170; Website: http://www.longs.com; Robert M. Long; drug store chain.
Lowe's Cos., Inc.; Box 1111, N. Wilkesboro, NC 28656; (336) 658-4000; Website: http://www.lowes.com; Robert L. Tillman; building materials and home improvement superstores.
Luby's, Inc.; 2211 NE Loop 410, P.O. Box 33069, San Antonio, TX 78265; (210) 654-9000; Website: http://www.lubys.com; Harris Pappas; operates cafeterias in S and SW.
Lucent Technologies, Inc.; 600 Mountain Ave., Murray Hill, NJ 07974; (888)4LU-CENT; Website: http://www.lucent.com; Henry Schacht; leading developer, designer, and manuf. of telecommunications systems, software, and prods.
Mandalay Resort Group; 3950 Las Vegas Boulevard South, Las Vegas, NV 89109; (702) 632-6700; Website: http://www.mandalayresortgroup.org; Michael Ensign; casino-resort operator (Excalibur, Luxor).
Manpower Inc.; 5301 N. Ironwood Rd., Milwaukee, WI 53201; (414) 961-1000; Website: http://www.manpower.com; Jeffrey A. Joerres; second largest non-gov't. employment services co. in the world.
Marriott International, Inc.; Marriot Drive, Washington, D.C. 20058; (301) 380-3000; Website: http://www.marriott.com; John Willard Marriott Jr; hotels, retirement communities, food service dist.
Masco Corp.; 21001 Van Born Rd., Taylor, MI 48180; (313) 274-7400; Website: http://www.masco.com; Richard A. Manoogian; manuf. kitchen, bathroom prods. (Delta, Peerless faucets; Fieldstone, Merillat cabinets).
Mattel, Inc.; 333 Continental Blvd., El Segundo, CA 90245; (310) 252-2000; Website: http://www.mattel.com; Robert A. Eckert; largest U.S. toymaker (Barbie, Fisher-Price, Hot Wheels, Matchbox, American Girls, Reader Rabbit).
May Department Stores Co.; 611 Olive St., St. Louis, MO 63101; (314) 342-6300; Website: http://www.maycompany.com; Judith K. Hofer; department stores (Hecht's, Lord & Taylor, Filene's, Foley's).
Maytag Corp.; 403 W. Fourth St. N., Newton, IA 50208; (515) 792-8000; Website: http://www.maytagcorp.com; Leonard A. Hadley; major appliance mfgr. (Magic Chef, Admiral, Jenn-Air), Hoover vacuum cleaners, floor care systems.
McDonald's Corp.; 1 McDonald's Plaza, Oak Brook, IL 60523; (630) 623-3000; Website: http://www.mcdonalds.com; Jack Greenberg; fast-food restaurants.
McGraw-Hill Cos.; 1221 Ave. of the Americas, NY, NY 10020; (212) 512-2000; Website: http://www.mcgraw-hill.com; Harold (Terry) McGraw III; book, textbooks, magazine publishing (*Business Week*), information and financial services (Standard and Poor's), TV stations.
McKesson HBOC Corp.; 1 Post St., San Francisco, CA 94104; (415) 983-8300; http://www.mckhboc.com; John Hammergen; distributor of drugs and toiletries and provides software and services in U.S.; bottled water.
Mead Corporation; Courthouse Plaza NE, Dayton, OH 45463; (937) 495-6323; Website: http://www.mead.com; Jerome F. Tatar; printing and writing paper, paperboard, packaging, shipping containers.
Medtronic, Inc.; 7000 Central Ave. NE, Minneapolis, MN 55432; (612) 514-4000; Website: http://www.medtronic.com; W. W. George; world's largest manuf. of implantable biomedical devices.
Merck & Co., Inc.; PO Box 100, Whitehouse Station, NJ 08889-0100; (908) 423-1000; Raymond V. Gilmartin; pharmaceuticals (Pepcid, Zocor), animal health care prods.
Meredith Corp.; 1716 Locust St., Des Moines, IA 50309; (515) 284-3000; Website: http://www.meredith.com; William T. Kerr; magazine publishing (*Better Homes and Gardens, Ladies Home Journal*), book publishing, broadcasting.
Merrill Lynch & Co., Inc.; World Financial Ctr., North Tower, NY, NY 10281-1332; (212) 449-1000; Website: http://www.ml.com; David H. Komansky; securities broker, financial services.
Metropolitan Life Ins. Co.; 1 Madison Ave., NY, NY 10010; (212) 578-2211; Website: http://www.metlife.com; Robert H. Benmosche; insurance, financial services.
MGM Mirage Resorts, Inc.; 300 Las Vegas Blvd. S, Las Vegas, NV 89109; (702) 693-7111; Website: http://www.mirage.com; Terrence Lanni; hotel-casino operator (Mirage, Treasure Island, Golden Nugget).
Microsoft Corp.; 1 Microsoft Way, Redmond, WA 98052-6399; (425) 882-8080; Website: http://www.microsoft.com; William H. Gates; largest independent software maker (Windows, Word, Excel).
Minnesota Mining & Manuf. Co.; 3M Center, St. Paul, MN 55144-1000; (651) 733-6501; Website: http://www.mmm.com; Linda G. Alvarado; abrasives, adhesives, electrical, health care, cleaning (Scotch-Brite, O-Cel-O sponges), printing, consumer prods. (Scotch Tape, Post-It).

Morgan Stanley Dean Witter & Co.; 1585 Broadway, NY, NY 10036; (212) 761-4000; Website: http://www.msdw.com; Phillip J. Purcell; diversified financial services, major U.S. credit-card issuer.

Motorola, Inc.; 1303 E. Algonquin Rd., Schaumburg, IL 60196; (847) 576-5000; Website: http://www.motorola.com; G. L. Tooker; electronic equipment and components.

National Semiconductor Corp.; 2900 Semiconductor Dr., P.O. Box 58090; Santa Clara, CA 95052-8090; (408) 721-5000; Website: http://www.national.com; B. Halla; manuf. of semiconductors, integrated circuits.

Navistar Intl. Corp.; 455 N. Cityfront Plaza Dr., Chicago, IL 60611; (312) 836-2000; Website: http://www.navistar.com; John R. Horne; manuf. heavy-duty trucks, parts, school buses.

New York Times Co.; 229 W. 43d St., NY, NY 10036; (212) 556-1234; Website: http://www.nytco.com; A. O. Sulzberger Jr; newspapers (*Boston Globe*), radio and TV stations, magazines (*Golf Digest*).

Newell Rubbermaid Inc.; Newell Center, 29 E. Stephenson Street, Freeport, IL 61032; (815) 235-4171; Website: http://www.newellco.com; Joseph Galli; cookware (Calphalon, WearEver); hair accessories (Goody, Ace); glassware (Anchor Hocking); kitchen products (Rubbermaid); window treatments (Levelor, Kirsch, Newell); home storage (Lee Ravan); writing instruments (Eberhard Faber, Sanford); address card files (Rolodex); infant and juvenile prods (Little Tykes, Graco, Century).

Nike, Inc.; 1 Bowerman Dr., Beaverton, OR 97005; (503) 671-6453; Website: http://www.NikeBiz.com; Philip H. Knight; athletic and leisure footware, apparel.

Nordstrom, Inc.; 1617 6th Ave., Seattle, WA 98101; (206) 628-2111; Website: http://www.nordstrom.com; Blake W. Nordstrom; upscale dept. store chain.

Norfolk Southern Corp.; 3 Commercial Pl., Box 227, Norfolk, VA 23510; (757) 629-2600; Website: http://www.nscorp.com; David R. Goode; operates railway, freight carrier.

Northrop Grumman Corp.; 1840 Century Park East, Los Angeles, CA 90067; (310) 553-6262; Website: http://www.northgrum.com; Kent Kresa; aircraft, electronics, data systems, information systems, missiles. (Northrop Grumman Corp. acquired Litton Industries, Inc. 6/12/01 for $5.1 billion.)

Northwest Airlines Corp.; 2700 Lone Oak Pkwy., Eagan, MN 55121; (612) 726-2111; Website: http://www.nwa.com; John H. Dasburg; air transportation.

Occidental Petroleum Corp.; 10889 Wilshire Blvd., Los Angeles, CA 90024; (310) 208-8800; Website: http://www.oxy.com; Ray R. Irani; oil, natural gas, chemicals, plastics, fertilizers.

Office Depot.; 2200 Old Germantown Rd., Delray Beach, FL 33445; (561) 278-4800; Website: http://www.officedepot.com; David I. Fuente; retail office supply stores.

Owens Corning; 1 Owens Corning Parkway, Toledo, OH 43659; (419) 248-8000; Website: http://www.owenscorning.com; Glen H. Hiner; world leader in advanced glass, composite materials.

Owens-Illinois; 1 SeaGate, Toledo, OH 43666; (419) 247-5000; J. H. Lemieux; Website: http://www.o-i.com world's largest producer of glass bottles.

Pacific Gas & Electric Corp. (PG&E); 77 Beale St., San Francisco, CA 94105; (415) 973-7000; Website: http://www.pgecorp.com; Robert D. Glynn Jr.; energy supplier.

PaineWebber Group, Inc.; 1285 Ave. of the Americas, NY, NY 10019; (212) 713-2000; Website: http://www.painewebber.com; Donald B. Marron; controls full-service securities firm.

J.C. Penney Co.; 6501 Legacy Dr., Plano, TX 75024; (972) 431-1000; Website: http://www.jcpenney.com; Allen Duestrom; dept. stores, catalog sales, drug stores (Eckerd, Fay's), insurance.

Pennzoil-Quaker State Co.; Pennzoil Pl., P.O. Box 2967, Houston, TX 77252; (713) 546-4000; Website: http://www.pennzoil-quakerstate.com; James J. Postl; automotive consumer products co., franchises Jiffy Lube and Q-Lube service centers.

PepsiCo, Inc.; 700 Anderson Hill Rd., Purchase, NY 10577; (914) 253-2000; Website: http://www.pepsico.com; Steven S. Reinemund; soft drinks (Pepsi-Cola, Mountain Dew), fruit juice (Tropicana), snacks (Ruffles, Lay's, Fritos, Doritos, Rold Gold).

Pfizer, Inc.; 235 E. 42d St., NY, NY 10017; (212) 573-2323; Website: http://www.pfizer.com; Hank McKinnell; pharmaceuticals (Celebrex, Diflucan, Viagra, Zithromax), hospital, agricultural, chemical prods., consumer prods. (Visine, Barbasol, Halls, Desitin, Benadryl, Listerine, Lubriderm, Schick, Zantac 75, Ben-Gay). (Co. merged with Warner-Lambert 6/19/2001, making it the largest pharmaceutical co. in the world in sales, and the 5th largest co. in the world.)

Pharmacia Corp.; 100 Route 206 N., Peapack, NJ 07977; (908) 901-8853; Website: http://www.pharmacia.com; Fred Hassan; pharmaceuticals (Motrin, Rogaine, Halcion, Xanax), chemicals, agricultural, health-care prods., consumer prods. (Equal, NutraSweet). (Created from merger of Pharmacia & Upjohn and Monsanto, March 2000.)

Philip Morris Cos. Inc.; 120 Park Ave., NY, NY 10017; (212) 880-5000; Website: http://www.philipmorris.com; http://www.kraft.com. Geoffrey C. Bible; cigarettes (largest U.S. tobacco company; Marlboro, Merit, Virginia Slims), beer (Miller, Molson, Red Dog), Kraft Foods products (Jell-O, Maxwell House coffee, Kool-Aid, Oscar Mayer, Tang, Cheez Whiz and Velveeta cheese prods., Post cereals, Lender's Bagels, Tombstone Pizza, and Toblerone chocolate); Nabisco products (Oreo cookies, Ritz, Triscuit crackers, Mallomars).

Phillips Petroleum Co.; 4th and Keeler Ave., Bartlesville, OK 74004; (918) 661-6600; Website: http://www.phillips66.com; J. J. Mulva; integrated oil and petrochemical co.

Pillowtex Corp.; 1 Lake Circle Dr., Kannopolis, NC 28081; (704) 939-2000; Website: http://www.pillowtex.com; Tony Williams; household textile prods.

Pitney Bowes, Inc.; Walter H. Wheeler Jr. Dr., Stamford, CT 06926; (203) 356-5000; Website: http://www.pitneybowes.com; Michael J. Critelli; world's largest mfgr. of postage meters, mailing equip.

Polaroid Corp.; 784 Memorial Drive, Cambridge, MA 02139; (617) 386-2000; Website: http://www.polaroid.com; Gary T. DiCamillo; photographic equip. and supplies, optical goods.

PPG Industries, Inc.; 1 PPG Place, Pittsburgh, PA 15272; (412) 434-3131; Website: http://www.ppg.com; Raymond W. Le Boeuf; glass prods., fiberglass, chemicals; world's leading supplier of automobile/industrial coatings.

PRIMEDIA Inc.; 745 Fifth Avenue, New York, NY 10151; (212) 745-0100; Website: http://www.primedia.com; Tom Rogers; consumer magazines (*New York, Seventeen, Modern Bride, American Baby, Soap Opera Digest)*, professional magazines, classroom learning, business directories.

Procter & Gamble Co.; 1 Procter & Gamble Plaza, Cincinnati, OH 45202; (513) 983-1100; Website: http://www.pg.com; Alan Lafley; soaps and detergents (Ivory, Cheer, Tide, Mr. Clean, Comet, Spic and Span, Zest); toiletries (Crest, Scope, Prell, Head and Shoulders, Noxzema, Oil of Olay, Old Spice); pharmaceuticals (Pepto-Bismol, Vicks cough medicines); foods (Folger's coffee, Pringles); paper prods. (Charmin toilet tissues, Bounty towels, Tampax tampons, Pampers & Luvs disposable diapers); Cover Girl and Max Factor cosmetics, Crisco shortening.

Prudential Ins. Co. of America; 751 Broad St., Newark, NJ 0710; (973) 802-6000; Website: http://www.prudential.com; Arthur F. Ryan; insurance, financial services.

Quaker Oats Co.; 321 N. Clark St., Chicago, IL 60610; (312) 222-7111; Website: http://www.quakeroats.com; David R. Banks; cereal (Life, Cap'n Crunch), foods (Aunt Jemima, Rice-A-Roni), beverages (Gatorade).

Ralcorp Holdings, Inc.; 800 Market St., St. Louis, MO 63101; (314) 877-7000; Website: http://www.ralcorp.com; David R. Banks; private-label breakfast cereals, snack foods, baby food (Beech-Nut).

Ralston Purina Group; Checkerboard Sq., St. Louis, MO 63164; (314) 982-2161; Website: http://www.ralston.com; W. P. Stiritz; world's largest producer of dog and cat food (Purina), and dry-cell batteries (Eveready, Energizer).

Raytheon Co.; 141 Spring St., Lexington, MA 02173; (781) 862-6600; Website: http://www.raytheon.com; Daniel P. Burnham; defense systems, electronics.

Reader's Digest Assn., Inc.; Reader's Digest Road, Pleasantville, NY 10570; (914) 238-1000; Website: http://www.readersdigest.com; Thomas Ryder; direct-mail marketer of magazines, books, music and video prods.

Reebok Intl., Ltd.; 1895 J.W. Foster Blvd., Canton, MA 02021; (781) 401-5000; Website: http://www.reebok.com; Paul Fireman; athletic and leisure footwear, apparel.

Revlon, Inc.; 625 Madison Ave., NY, NY 10022; (212) 527-4000; Website: http://www.revlon.com; Ronald O. Perelman; cosmetics, beauty aids, skin care.

Reynolds Metals Co.; P.O. Box 27003, Richmond, VA 23261; (804) 281-3939; Jeremiah J. Sheehan; Website: http://www.rmc.com; aluminum prods.

Rite Aid Corp.; 30 Hunter Lane, Camp Hill, PA 17011-2404; (717) 761-2633; Website: http://www.riteaid.com; Robert G. Miller; discount drug stores.

RJ Reynolds Tobacco; 401 N. Main St., Winston-Salem, NC 27102; (336) 741-5000; Website: http://www.rjrt.com; Andrew J. Schindler; 2d-largest U.S. producer of cigarettes (Winston, Salem, Camel).

Rockwell Intl. Corp.; 77 E. Wisconsin Ave., Suite 1400, Milwaukee, WI 53202; (414) 212-5200; Website: http://www.rockwell.com; Donald H. Davis; diversified high-tech. co.

Ryder System, Inc.; 3600 NW 82d Ave., Miami, FL 33166; (305) 593-3726; Website: http://www.ryder.com; Greg Swienton; truck-leasing service.

Safeway Inc.; 5918 Stoneridge Mall Rd., Pleasanton, CA 94588-3229; (925) 467-3000; Website: http://www.safeway.com; Steven A. Burd; supermarkets.

Sara Lee Corp.; 3 First National Plaza, Chicago, IL 60602; (312) 726-2600; Website: http://www.saralee.com; John H. Bryan Jr; baked goods, fresh and processed meats (Ball Park, Jimmy Dean, Hillshire Farms, Kahn's), hosiery, intimate apparel and knitwear (Hanes, L'eggs, Playtex, Champion), Coach leather goods.

SBC Communications, Inc.; 175 E. Houston, San Antonio, TX 78205; (210) 821-4105; Website: http://www.sbc.com; Edward Whitacre Jr; telephone services (Ameritech, Southwestern Bell, Pacific Bell).

Schering-Plough Corp.; 1 Giralda Farms, Madison, NJ 07940; (973) 822-7000; Website: http://www.sch-plough.com; R.J. Kogan; pharmaceuticals (Claritin, Proventil), consumer prods. (Afrin, Coppertone), animal health prods.

Seagate Technology; 920 Disc Dr., Scotts Valley, CA 95066; (831) 438-6550; Website: http://www.seagate.com; Stephen J. Luczo; manuf. disk drives.

Sears, Roebuck and Co.; 3333 Beverly Rd., Hoffman Estates, IL 60179; (847) 286-7385; Website: http://www.sears.com; Alan J. Lacey; 2nd largest U.S. retailer, department, specialty stores.

Service Merchandise Co., Inc.; 7100 Service Merchandise Rd., Nashville, TN 37202-4600; (615) 660-6000; Website: http://www.servicemerchandise.com; Sam Cusano; discount merchandiser and leading jewelry retailer.

Shaw Industries, Inc.; P.O. Box 2128, Dalton, GA 30720; (706) 278-3812; Website: http://www.shawinc.com; Robert E. Shaw; world's largest carpet mfgr. (Armstrong, Magee, Cabin Craft).

Sherwin-Williams Co.; 101 Prospect Ave. NW, Cleveland, OH 44115; (216) 566-2000; Website: http://www.sherwin.com; Christopher M. Connor; largest North American paint and varnish producer (Dutch Boy, Pratt & Lambert, Minwax).

J. M. Smucker Co.; Strawberry Lane, Orrville, OH 44667; (216) 682-3000; Website: http://www.smucker.com; Timothy P. Smucker; preserves, jams, jellies (Dickinson's), toppings (Magic Shell), syrups, juices.

Smurfit-Stone Container Corp.; 150 N. Michigan Ave., Chicago, IL 60601; (312) 346-6600; Website: http://www.smurfit-stone.net; Ray M. Curran; industry leader for corrugated containers, paper bags and sacks.

Sprint Corp.; PO Box 11315, Kansas City, MO 64112; (913) 624-3000; Website: http://www.sprint.com; William T. Esrey; long-distance and local telecommunications.

Staples, Inc.; 500 Staples Dr., Framingham, MA 01702; (508) 253-5000; Website: http://www.staples.com; Thomas Stemberg; office-supply superstores.

Starwood Hotels and Resorts Worldwide; 777 Westchester Ave., White Plains, NY 10604; (914) 640-8100; Website: http://www.starwoodlodging.com; Barry S. Sternlicht; hotels and leisure company.

State Farm Mutual Automobile Ins. Co.; 1 State Farm Plaza, Bloomington, IL 61701; (309) 766-2311; Website: http://www.statefarm.com; Edward B. Rust Jr; major insurance co.

Stride Rite Corp.; 191 Spring St., P.O. Box 9191, Lexington, MA 02420; (617) 824-6000; Website: http://www.striderite.com; David Chamberlain; high-quality adult's and children's footwear (Keds, Sperry Top-Sider).

Sun Microsystems, Inc.; 2550 Garcia Ave., Mountain View, CA 94043; (650) 960-1300; Website: http://www.sun.com; Scott G. McNealy; supplier of network-based distributed computer systems (Java programming language).

Sunoco, Inc.; 1801 Market St., Philadelphia, PA 19103-1699; (215) 977-3000; Website: http://www.sunocoinc.com; J.D. Drosdick; energy resources co., markets Sunoco gasoline.

SUPERVALU Inc.; PO Box 990, Minneapolis, MN 55440; (952) 828-4000; Website: http://www.supervalu.com; Michael W. Wright; food wholesaler, retailer.

Sysco Corp.; 1390 Enclave Pkwy, Houston, TX 77077-2099; (281) 584-1390; Website: http://www.sysco.com; Charles H. Cotros; leading U.S. food distributor.

Tandy Corp.; 100 Throckmorton St., Suite 1800, Fort Worth, TX 76102; (817) 415-3700; Website: http://www.tandy.com; Leonard H. Roberts; consumer electronics retailer (Computer City, Radio Shack).

Target Corp.; 777 Nicollet Mall, Minneapolis, MN 55402; (612) 370-6948; Website: http://www.targetcorp.com; Robert J. Ulrich; department, specialty stores (Target, Marshall Field's, Dayton's, Hudson's, Mervyn's California).

Tenneco Automotive, Inc.; 500 North Field Drive, Lake Forest, IL 60045; (847) 482-5000; Website: http://www.tenneco-automotive.com; Mark P. Frissora; automotive parts (Monroe, Walker).

Texaco Inc.; 2000 Westchester Ave., White Plains, NY 10650; (914) 253-4000; Website: http://www.texaco.com; Peter I. Bijur; integrated international oil co. (Texaco and Chevron announced, 9/7/01, that their proposed merger had received FTC approval; the merger would create the 2d-largest U.S. oil company.)

Texas Instruments Inc.; 12500 TI Blvd., P.O. Box 660199, Dallas, TX 75266-0199; (972) 995-3773; Website: http://www.ti.com; T. J. Engibous; electronics.

Textron, Inc.; 40 Westminster St., Providence, RI 02903; (401) 421-2800; Website: http://www.textron.com; Lewis B. Campbell; aerospace, industrial, automotive prods., financial services.

Times Mirror Publishing Co.; 202 West 1st St., Los Angeles, CA 90012; (213) 237-3700; Website: http://www.tm.com; John Madigan; newspapers, magazines (*Field & Stream*, *Popular Science*), professional books (Matthew Bender).

The TJX Cos., Inc.; 770 Cochituate Rd., Framingham, MA 01701; (508) 390-1000; Website: http://www.tjx.com; Edmond English; world's largest off-price apparel retailer (T.J. Maxx, Marshalls).

Tootsie Roll Industries, Inc.; 7401 S. Cicero Ave., Chicago, IL 60629; (773) 838-3400; Website: http://www.tootsie.com; M. J. Gordon; candy (Tootsie Roll, Mason Dots, Charms, Sugar Daddy, Charleston Chew, Junior Mints).

Toro Co.; 8111 Lyndale Ave. S, Bloomington, MN 55420; (612) 888-8801; Website: http://www.toro.com; Kendrick B. Melrose; lawn and turf maintenance (Lawn-Boy), snow removal equipment, lighting and irrigation systems.

Toys "R" Us; 461 From Rd., Paramus, NJ 07652; (201) 262-7800; Website: http://www.toysrus.com; John Eyler; world's largest children's specialty retailer (Toys "R" Us, Kids "R" Us, Babies "R" Us).

Transamerica Corp.; 1150 South Olive St. Los Angeles, CA 90015; (213) 742-2111; Website: http://www.transamerica.com; Ron F. Wagley; insurance, financial services.

Triarc Cos., Inc.; 280 Park Ave., NY, NY 10017; (212) 451-3000; Website: http://www.triarc.com; Nelson Peltz; fast-food restaurants (Arby's), beverages (Royal Crown, Mystic, Nehi, Snapple, Stewart's).

Tribune Co.; 435 N. Michigan Ave., Chicago, IL 60611; (312) 222-9100; Website: http://www.tribune.com; J. W. Madigan; newspaper and book publishing, broadcasting, Chicago Cubs baseball team.

TRICON Global Restaurants, Inc.; 1441 Gardiner Lane, Louisville, KY 40213; (502) 874-8300; Website: http://www.triconglobal.com; David C. Novack; fast food (Pizza Hut, KFC, Taco Bell).

Trinity Industries, Inc.; P.O. Box 568887, 2525 Stemmons Freeway, Dallas, TX 75207; (214) 631-4420; Website: http://www.trin.net; Timothy R. Wallace; manufactures metal prods., rail and freight prods.

TRW Inc.; 1900 Richmond Rd., Cleveland, OH 44124; (216) 291-7000; Website: http://www.trw.com; David M. Cote; car and truck operations, electronics, space and defense systems.

Tyco Intl., Ltd.; 1 Tyco Pk., Exeter, NH 03833; (603) 778-9700; Website: http://www.tycoint.com; L. D. Kozlowski; fire protection systems, pipes, power cables, medical supplies, packaging.

Tyson Foods, Inc.; P.O. Box 2020, Springdale, AR 72765; (501) 290-4000; Website: http://www.tyson.com; John Tyson; fresh and processed poultry and seafood prods. (Holly Farms, Weaver, Louis Kemp).

UAL Corp.; 1200 E. Algonquin Rd., Elk Grove Twp., IL 60007; (847) 700-4000; Website: http://www.ual.com; James E. Goodwin; air transportation (United Airlines).

Unilever Bestfoods; 700 Sylvan Ave., Englewood Cliffs, NJ 07632; (201) 894-4000; Website: http://www.bestfoods.com; Neil Beckerman; food (Hellmann's, Best Foods mayonnaise, Knorr soups, NutraBlend soy products, Arnold breads, Mazola oils and margarine, Entenmann's cakes, Lipton Tea, Skippy Peanut Butter). (Unilever and Best Foods merged 10/4/00.)

Union Pacific Corp.; 1717 Main St., Suite 5900, Dallas, TX 75201; (214) 743-5600; Website: http://www.up.com; Richard Davidson; largest railroad, trucking co. in U.S.

Unisys Corp.; Unisys Way, Blue Bell, PA 19424-0001; (215) 986-6999; Website: http://www.unisys.com; Lawrence A. Weinbach; designs, manuf. computer information systems and related prods..

UnitedHealth Group Corp.; 300 Opus Center, 9900 Bren Rd. East, Minnetonka, MN 55343; (612) 936-1300; Website: http://www.unitedhealthgroup.com; William W. McGuire; owns, manages health maintenance organizations.

United Parcel Service of America, Inc.; 55 Glenlake Pkwy. NE, Atlanta, GA 30328; (404) 828-6000; Website: http://www.ups.com; James Kelly; courier services, truck rentals.

United Technologies Corp.; 1 Financial Plaza, Hartford, CT 06101; (860) 728-7000; Website: http://www.utc.com; George David; aerospace, industrial prods. and services (Otis Elevator, Pratt & Whitney, Sikorsky Aircraft).

Unocal Corp.; 2141 Rosecrans Ave., Ste. 4000, El Segundo, CA 90245; (310) 726-7667; Website: http://www.unocal.com; Roger Beach; integrated oil co.

US Airways Group, Inc.; 2345 Crystal Dr., Arlington, VA 22202; (703) 872-5306; Website: http://www.usairways.com; Stephen M. Wolf; air transportation.

UST Inc.; 100 W. Putnam Ave., Greenwich, CT 06830; (203) 661-1100; Website: http://www.ustshareholder.com; Vincent A. Gierer Jr.; smokeless tobacco (Copenhagen, Skoal), pipe tobacco, wine (Chateau St. Michelle, Conn Creek, Columbia Crest).

USX-Marathon Group; 600 Grant St., Pittsburgh, PA 15230; (412) 433-1121; Website: http://www.marathon.com; Thomas J. Usher; integrated oil co.

Venator Group; 112 West 34th St., NY, NY 10120; (212) 720-3700; Website: http://www.venatorgroup.com; Matthew D. Serra; operates retail stores: shoes (Kinney), apparel (Northern group), athletic footwear (Foot Locker), athletic merchandise (Champs), San Francisco Music Box Company.

Verizon Communications; 1095 Avenue of the Americas, New York, NY 10036; (212) 395-2121; Website: http://www.verizon.com; Charles R. Lee, Ivan Seidenberg; largest U.S. wireline and wireless provider; world's lgst. provider of print and on-line directory info. (co. formed from merger of Bell Atlantic and GTE, 6/30/00.)

V.F. Corp.; 628 Green Valley Rd., Suite 500, Greensboro, NC 27408; (336) 547-6000; Website: http://www.vfc.com; M. McDonald; apparel (Lee, Wrangler jeans, Vanity Fair, Healthtex, Jantzen).

Viacom, Inc.; 1515 Broadway, NY, NY 10036; (212) 258-6000; Website: http://www.viacom.com; Mel Karmazin; TV broadcast stations and cable systems, channels (CBS, UPN, TNN, Showtime, MTV, VH-1, Nickelodeon); book publishing (Simon & Schuster, Macmillan); produces, distributes movies, TV shows (Paramount); video stores (Blockbuster), theme parks.

Walgreen Co.; 200 Wilmot Rd., Deerfield, IL 60015; (847) 940-2500; Website: http://www.walgreens.com; L. Daniel Jorndt; nation's largest drugstore chain.

Wal-Mart Stores, Inc.; Box 116, Bentonville, AR 72716; (501) 273-4000; Website: http://www.walmartstores.com; S. Robson Walton; world's largest retailer; discount stores, wholesale clubs.

Washington Post Co.; 1150 15th St. NW, Washington, DC 20071; (202) 334-6000; Website: http://www.washpost.com; D. E. Graham; newspapers, *Newsweek* magazine, TV and CATV stations, Stanley H. Kaplan Educational Centers.

Waste Management; 1001 Fannin, Suite 4000, Houston, TX 77002; (713) 512-6548; Website: http://www.wm.com; Maurice Myers; N. America's largest solid waste collection and disposal co.

Wells Fargo & Co.; 420 Montgomery St., San Francisco, CA 94163; (800) 411-4932; Website: http://www.wellsfargo.com; Dick Kovacevich; bank holding co.

Wendy's Intl., Inc.; 4288 W. Dublin-Granville Rd., Dublin, OH 43017; (614) 764-3100; Website: http://www.wendys.com; John T. Schuessler; quick-service restaurants.

Weyerhaeuser Co.; Tacoma, WA 98477; (253) 924-2345; Website: http://www.weyerhaeuser.com; George H. Weyerhaeuser; world's largest private owner of softwood timber, distrib. paper and wood prods.

Whirlpool Corp.; 2000 N.M. 63, Benton Harbor, MI 49022; (616) 923-5000; Website: http://www.whirlpoolcorp.com; David Whitwam; world's largest manuf. of major home appliances (KitchenAid, Kenmore, Roper).

Whitman Corp.; 3501 Algonquin Road, Rolling Meadows, IL 60008; (708) 818-5000; Website: http://www.whitmancorp.com; Larry D. Young; beverage bottler and distributor (Pepsi-Cola).

Winn-Dixie Stores, Inc.; 5050 Edgewood Ct., Jacksonville, FL 32254; (904) 783-5000; Website: http://www.winn-dixie.com; Al Rowland; supermarkets.

Winnebago Industries, Inc.; PO Box 152, Forest City, IA 50436; (641) 585-3535; Website: http://www.winnebagoind.com; Bruce D. Hertzke; manuf. and financing of motor homes, recreational vehicles.

Wm. Wrigley Jr. Co.; 410 N. Michigan Ave., Chicago, IL 60611; (312) 644-2121; Website: http://www.wrigley.com; William Wrigley; world's largest mfgr. of chewing gum.

WorldCom, Inc.; 500 Clinton Ctr. Dr., Clinton, MS 39056; (877) 624-9266; Website: http://www.wcom.com; Bernard Ebbers; long-distance telephone service.

WRC Media Inc.; 512 Seventh Ave., New York, NY 10018; (212) 768-0455; Website: http://www.wrcmedia.com; Martin E. Kenney Jr.; publisher of educational and reference media; World Almanac Education Group (*World Almanac and Book of Facts*, Funk and Wagnalls New Encyclopedia), FACTS.com, CompassLearning, Weekly Reader, American Guidance, ChildU.

Xerox Corp.; PO Box 1600, Stamford, CT 06904; (203) 968-3000; Website: http://www.xerox.com; Paul Allaire; copiers, printers, document publishing equip.

Yahoo! Inc.; 701 First Ave. Sunnyvale, CA 94089; (408) 349-3300; Website: http://www.yahoo.com; Terry Semel; global internet media company.

IT'S A FACT: Walmart, the world's largest retailer, is bigger than Sears, Kmart, and J.C. Penney combined and has over 4,150 stores worldwide. Heirs of founder Sam Walton own about 38% of the company.

Who Owns What: Familiar Consumer Products

Listed here are consumer brands and their parent companies. For company address and website, see Business Directory.

A-1 steak sauce: Philip Morris
ABC broadcasting: Walt Disney
Admiral appliances: Maytag
Advil: American Home Products
Ajax cleanser: Colgate-Palmolive
Almond Joy candy bar: Hershey
American Girl: Mattel
Anacin: American Home Products
Arm & Hammer: Church & Dwight
Arrid antiperspirant: Carter-Wallace
Aunt Jemima Pancake mix: Quaker Oats
Aunt Millie's pasta sauce: Borden
Baggies: Tenneco
Ban antiperspirant: Bristol-Myers Squibb
Banana Republic stores: The Gap
Band-Aids: Johnson & Johnson
Barbie dolls: Mattel
Beech-Nut baby food: Ralcorp
Ben-Gay: Pfizer
Betty Crocker prods.: General Mills
Black Flag insecticides: Clorox
Blockbuster video stores: Viacom
Bounty paper towels: Procter & Gamble
Breck shampoo: Dial
Brillo soap pads: Church & Dwight
Brita water systems: Clorox
Budweiser beer: Anheuser-Busch
Bufferin: Bristol-Myers Squibb
Bulova watches: Loews
Business Week magazine: McGraw-Hill
Buster Brown shoes: Brown Group
Cadbury: Hershey
Cap'n Crunch cereal: Quaker Oats
Calphalon cookware: Newell Rubbermaid
CBS Broadcasting: Viacom
Charmin toilet tissue: Procter & Gamble
Cheer detergent: Procter & Gamble
Cheerios cereal: General Mills
Cheez Whiz: Philip Morris
Cinemax: AOL Time Warner
Clairol hair prods.: Bristol-Myers Squibb
CNN: AOL Time Warner
Coach leather goods: Sara Lee
Combat insecticides: Clorox
Comet cleanser: Procter & Gamble
Coppertone sun care prods.: Schering-Plough
Crest toothpaste: Procter & Gamble
Crisco shortening: Procter & Gamble
Desitin Ointment: Pfizer
Doritos chips: PepsiCo
Dristan: American Home Prods.
Duracell batteries: Gillette
Dutch Boy paints: Sherwin-Williams
Efferdent dental cleanser: Warner-Lambert
Elmer's glue: Borden
ESPN: Walt Disney
Eveready batteries: Ralston Purina
Excedrin: Bristol-Myers Squibb
Fab detergent: Colgate-Palmolive
Fantastik: S.C. Johnson
Fisher Price Toys: Mattel
Foamy shaving cream: Gillette
Folger's coffee: Procter & Gamble
Formula 409 spray cleaner: Clorox
Franco-American spaghetti: Campbell Soup
Frito-Lays snacks: PepsiCo
Fruitopia drinks: Coca-Cola
Gatorade: Quaker Oats
Godiva chocolate: Campbell Soup
Halcion: Pharmacia
Halls coughdrops: Pfizer
Hamburger Helper: General Mills
Hanes hosiery: Sara Lee
Hawaiian Punch: Procter & Gamble

HBO: AOL Time Warner
Head and Shoulders shampoo: Procter & Gamble
Healthtex: V.F. Corp.
Hellmann's mayonnaise: Unilever Bestfoods
Hi-C fruit drinks: Coca-Cola
Hidden Valley prods.: Clorox
Hillshire Farms meats: Sara Lee
Holly Farms: Tyson
Hostess cakes: Interstate Bakeries
Huggies diapers: Kimberly-Clark
Ivory soap: Procter & Gamble
Jack Daniel's Whiskey: Brown-Forman
Java programming language: Sun Microsystems
Jell-O: Philip Morris
Jenn-Air stoves: Maytag
Jif peanut butter: Procter & Gamble
Jim Beam bourbon: Fortune Brands
Keds footwear: Stride Rite
Ken-L-Ration pet foods: H. J. Heinz
Kent cigarettes: Loews
KFC restaurants: TRICON
Kinney shoe stores: Venator Group
KitchenAid appliances: Whirlpool
Kit Kat candy: Hershey's
Kleenex: Kimberly-Clark
Knorr soups: Unilever Bestfoods
Kool-Aid: Philip Morris
Krazy Glue: Borden
Kwikset doorknobs: Black & Decker
Ladies Home Journal magazine: Meredith
Lee jeans: V.F. Corp.
L'eggs hosiery: Sara Lee
Lender's bagels: Philip Morris
Lenox china: Brown-Forman
Lerner stores: The Limited
Life Savers candy: Philip Morris
Lipton tea: Unilever Bestfoods
Listerine mouthwash: Pfizer
Lord & Taylor: May Dept. Stores
Marlboro cigarettes: Philip Morris
Max Factor beauty products: Procter & Gamble
Maxwell House coffee: Philip Morris
Mazola oils and margarine: Unilever Bestfoods
Metamucil: Procter & Gamble
Michelob beer: Anheuser-Busch
Miller beer: Philip Morris
Milton Bradley games: Hasbro
Minute Maid juices: Coca-Cola
Monroe automotive parts: Tenneco Automotive
MTV: Viacom
Nature Valley granola bars: General Mills
NBC broadcasting: General Electric
Neutrogena soap: Johnson & Johnson
Newsweek magazine: Washington Post
9 Lives cat food: H.J. Heinz
Oil of Olay: Procter & Gamble
Old Navy Clothing: The Gap
Oreo cookies: Philip Morris
Oscar Mayer meats: Philip Morris
Pampers: Procter & Gamble
PaperMate pens: Gillette
People magazine: AOL Time Warner
Pepperidge Farm prods.: Campbell Soup
Pepto-Bismol: Procter & Gamble
Philadelphia Cream Cheese: Philip Morris
Pine-Sol cleaner: Clorox
Pizza Hut restaurants: TRICON
Planters nuts: Philip Morris
Playskool toys: Hasbro
Playtex apparel: Sara Lee
Post cereals: Philip Morris
Post-It stickers: Minn. Mining & Manuf.
Prego pasta sauce: Campbell Soup
Prell shampoo: Procter & Gamble
Prentice Hall publishing: Viacom
Prozac: Eli Lilly
Radio Shack retail outlets: Tandy
Red Dog beer: Philip Morris
Reese's candy: Hershey
Rice-A-Roni: Quaker Oats
Rice Krispies: Kellogg
Right Guard deodorant: Gillette
Ritz crackers: Philip Morris
Robitussin: American Home Products
Rogaine hair growth aide: Pharmacia
Ronzoni pasta: Hershey
Ruffles chips: PepsiCo
San Francisco Music Box Co.: Venator Group
San Giorgio pasta: Hershey
Schick razors: Warner-Lambert
Scope mouthwash: Procter & Gamble
Scotch tape: Minn. Mining & Manuf.
Seventeen magazine: PRIMEDIA
Simon & Schuster publishing: Viacom
Skippy peanut butter: Unilever Bestfoods
SnackWell's cookies: Philip Morris
Snapple beverages: Triarc
S.O.S. cleanser: Clorox
Southern Comfort liquor: Brown-Forman
SPAM meat: Hormel
Spic and Span: Proctor & Gamble
Sports Illustrated magazine: AOL Time Warner
Sprite soda: Coca-Cola
Star-Kist tuna: H.J. Heinz
Sugar Twin: Alberto Culver
Swanson frozen dinners: Campbell Soup
Taco Bell restaurants: TRICON
Tampax tampons: Procter & Gamble
Thomas' English muffins: Unilever Bestfoods
Tide detergent: Procter & Gamble
Titleist: Fortune Brands
Tombstone pizza: Philip Morris
Triscuits: Philip Morris
Trojan condoms: Carter-Wallace
Tylenol: Johnson & Johnson
Ultra Brite toothpaste: Colgate-Palmolive
USA Today newspaper: Gannett
V-8 vegetable juice: Campbell Soup
Vanity Fair apparel: V.F. Corp.
Velveeta cheese prods.: Philip Morris
Viagra: Pfizer
Vicks cough medicines: Procter & Gamble
Victoria's Secret stores: The Limited
Visine eye drops: Pfizer
Wall Street Journal: Dow Jones
Waterman pens: Gillette
Weight Watchers: H.J. Heinz
Wheaties cereal: General Mills
Windex: S.C. Johnson
Windows software applications: Microsoft
Wise snacks: Borden
Wonder bread: Interstate Bakeries
The World Almanac: WRC Media
Zest soap: Procter & Gamble
Ziploc storage bags: S.C. Johnson

Top Brands in Selected Categories

Source: Information Resources, Inc., a Chicago-based marketing research company; figures are for 12-month period ending 7/15/01.

Ready-to-Eat Cold Cereals

	Sales	Market Share (%)
Private Label	$557,846,080	7.4
General Mills Cheerios	368,852,832	4.9
Kelloggs Frosted Flakes	339,588,640	4.5
General Mills Honey Nut Cheerios	230,870,224	3.1
Kelloggs Frosted Mini Wheats	205,670,528	2.7

Toothpaste

	Sales	Market Share (%)
Crest	$329,677,760	19.4
Colgate	293,230,592	17.2
Aquafresh	135,791,040	8.0
Colgate Total	112,349,296	6.6
Crest Multicare	107,167,736	6.3

Ground Coffee (excluding Decaf)

	Sales	Market Share (%)
Folgers	$505,476,256	25.1
Maxwell House	326,058,144	16.2
Folgers Coffee House	172,561,568	8.6
Maxwell House Master Blend	169,179,360	8.4
Private Label	152,643,888	7.6

Cookies

	Sales	Market Share (%)
Nabisco Oreos	$605,509,120	13.2
Private Label	429,969,536	9.4
Nabisco Chips Ahoy	403,303,648	8.8
Keebler Chips Deluxe	175,569,488	3.8
Nabisco Newtons	157,576,848	3.4

At-Home Shopping—Consumer Tips and Rights

Source: Federal Trade Commission, Consumer Information Center; American Express

TIPS

- Deal only with reliable firms. Check with your local consumer protection agency or the Better Business Bureau (BBB) nearest the business.
- Review the advertising offer carefully.
- Inquire about warranty, refund, and exchange policies.
- Never send cash. Pay by money order, check, charge, or credit card so that you have a record of your purchase.
- Keep the ad you responded to and a copy of the order form. If there is no order form, record the company's name, address, phone number, date, the item you purchased, amount paid, and the promised delivery date.
- Be careful about giving out your credit, debit, charge card, or bank account number.

RIGHTS

Late deliveries. By federal law, a company must ship your order within 30 days, unless the advertisement promises a different shipping time. If the company cannot ship in time, it must give you an "Option Notice." You can either wait longer or cancel and get a prompt refund. If you cancel and your order was charged, the seller has one billing cycle to tell the card issuer to credit your account. *Exceptions:*

(1) If a company does not promise a shipping time and if you are applying for credit to pay for your purchase, the company has 50 days after receiving your order to ship.

(2) Other exceptions include spaced deliveries such as magazine subscriptions (except for 1st shipment), items that continue until you cancel (e.g., book or record clubs), COD orders, services, and seeds or growing plants.

Unordered merchandise. If you are shipped a product that you did not order, it's yours. It is illegal for a company to pressure you to pay for it or to return it.

Damaged or spoiled items. If damage is obvious, and if you decide not to accept the package, write "REFUSED" on the wrapper and return it unopened to the seller. No new postage is needed, unless you signed for it.

Disputes or billing errors. If there is a problem with your order these steps are suggested:

(1) Write immediately to the company, explaining the problem and asking for a specific resolution. Include all relevent information, and a copy of the canceled check.

(2) If you charged your purchase or arranged for payment to be withdrawn from a bank account, send a copy of your letter to the card issuer or bank.

You usually have 60 days to dispute charges.

Postal rules allow you to write a check, payable to the sender, on COD orders. If, after examining the merchandise, you believe that there has been misrepresentation or fraud, you can stop payment on the check and file a complaint with the U.S. Postal Inspector's Office.

ON THE INTERNET

When shopping on the Internet:

- Consider using a secured browser, which will encrypt or scramble purchase information that can be intercepted.
- If you are unfamiliar with a company, ask for a paper brochure or catalog in the mail.
- Be cautious about giving out personal information. It is rarely necessary to give your Social Security number. Never give out your password for your Internet service provider.
- Print out a copy of your order and confirmation number.

For further questions, contact: The Federal Trade Commission, Public Reference, CRC-240, Washington, DC 20580; 877-FTC-HELP; or website at http://www.ftc.gov

Top 10 Shopping Websites

Source: Media Metrix, Inc.

Rank	Website address	Visitors[1]
1.	http://www.x10.com	39,566,600
2.	http://www.amazon.com	20,480,760
3.	http://www.mypoints.com	7,632,928
4.	http://www.bizrate.com	7,307,290
5.	http://www.americangreetings.com	7,097,776
6.	http://www.half.com	6,963,683
7.	http://www.bmgmusicservice.com	6,873,555
8.	http://www.columbiahouse.com	6,289,343
9.	http://ticketmaster.com	5,351,590
10.	http://www.barnesandnoble.com	5,274,415

(1) Number of visitors who visited website at least once in July 2001, according to a Media Metrix sample.

Your Credit File

Any individual can investigate the contents of his or her credit file by contacting one or more of the approximately 2,000 credit bureaus, or consumer credit clearinghouses, in the U.S. The nearest ones can be found in the telephone Yellow Pages.

Although the Fair Credit Reporting Act requires that a bureau provide no more than an oral or written credit history review, many bureaus will furnish the same computer-generated compilation of facts they give the banks, retailers, and other companies that subscribe to their service. Anyone denied credit on the basis of negative information from a credit bureau can obtain a review free of charge within 30 days of the denial.

A consumer can question any item in the file. The credit bureau must investigate and remove any item that cannot be substantiated. When a bureau affirms a questionable item, the consumer can present a 100-word explanation that must be placed in the credit file. Whenever an adverse item is deleted or an explanation added, a consumer may request that the credit bureau inform every credit grantor who received a report within the last 6 months.

The Cost of Raising a Child Born in 2000

Source: Center for Nutrition Policy and Promotion, U.S. Dept. of Agriculture

Estimated annual expenditures in 2000 dollars for a child born in 2000, by income group. Estimates are for the younger child in a 2-parent family with 2 children, for the overall U.S.

		Income group[1]		
Year	Age	Lowest	Middle	Highest
2000	under 1	$6,280	$8,740	$13,000
2001	1	6,520	9,070	13,490
2002	2	6,770	9,420	14,010
2003	3	7,180	10,040	14,850
2004	4	7,450	10,420	15,420
2005	5	7,740	10,820	16,000
2006	6	8,160	11,240	16,460
2007	7	8,470	11,670	17,090
2008	8	8,790	12,120	17,740
2009	9	9,130	12,520	18,210
2010	10	$9,480	$13,000	$18,910
2011	11	9,840	13,490	19,620
2012	12	11,550	15,160	21,700
2013	13	11,980	15,740	22,520
2014	14	12,440	16,330	23,380
2015	15	12,740	17,250	24,950
2016	16	13,220	17,910	25,900
2017	17	13,720	18,590	26,880
TOTAL		$171,460	$233,530	$340,130

(1) In 2000, low annual income is less than $38,000 (average in this range = $23,800); middle income is $38,000-$63,999 (average = $50,600); high income is $64,000 or more (average = $95,800). Projected annual inflation rate is 3.8%.

Wedding Anniversaries

The traditional names for wedding anniversaries go back many years in social usage and have been used to suggest types of appropriate anniversary gifts. Traditional products for gifts are listed here in capital letters, with a few allowable revisions in parentheses, followed by common modern gifts in each category.

1st	PAPER, clocks
2d	COTTON, china
3d	LEATHER, crystal, glass
4th	LINEN (SILK), appliances
5th	WOOD, silverware
6th	IRON, wood objects
7th	WOOL (COPPER), desk sets
8th	BRONZE, linens, lace
9th	POTTERY (CHINA), leather goods
10th	TIN, ALUMINUM, diamond
11th	STEEL, fashion jewelry
12th	SILK, pearls, colored gems
13th	LACE, textiles, furs
14th	IVORY, gold jewelry
15th	CRYSTAL, watches
20th	CHINA, platinum
25th	SILVER, sterling silver
30th	PEARL, diamond
35th	CORAL (JADE), jade
40th	RUBY, ruby
45th	SAPPHIRE, sapphire
50th	GOLD, gold
55th	EMERALD, emerald
60th	DIAMOND, diamond

Birthstones

Source: Jewelry Industry Council

MONTH	Ancient	Modern
January	Garnet	Garnet
February	Amethyst	Amethyst
March	Jasper	Bloodstone or Aquamarine
April	Sapphire	Diamond
May	Agate	Emerald
June	Emerald	Pearl, Moonstone, or Alexandrite
July	Onyx	Ruby
August	Carnelian	Sardonyx or Peridot
September	Chrysolite	Sapphire
October	Aquamarine	Opal or Tourmaline
November	Topaz	Topaz
December	Ruby	Turquoise or Zircon

Median Price of Existing Single-Family Homes

Source: National Association of REALTORS®

Metropolitan Area	1999	2000	Second Quarter 2001
Akron, OH	$104,900	$110,100	$113,400
Albany/Schenectady/Troy, NY	106,100	111,100	116,500
Albuquerque, NM	130,300	130,400	132,300
Amarillo, TX	81,400	86,300	90,300
Anaheim/Santa Ana, CA	280,900	318,000	352,700
Appleton/Oshkosh/Neenah, WI	93,300	100,500	107,100
Atlanta, GA	123,700	131,200	139,700
Atlantic City, NJ	117,000	121,500	128,200
Aurora/Elgin, IL	151,900	163,000	178,900
Austin/San Marcos, TX	128,600	142,800	155,000
Baltimore, MD	127,400	153,000	155,100
Baton Rouge, LA	103,600	109,100	113,500
Beaumont/Port Arthur, TX	76,300	80,800	82,500
Biloxi/Gulfport, MS	92,200	NA	102,600
Birmingham, AL	127,100	125,500	134,200
Boise City, ID	123,900	126,000	127,300
Boston, MA	290,000	314,200	356,200
Bradenton, FL	117,200	127,300	134,500
Buffalo/Niagara Falls, NY	81,400	79,800	87,400
Canton, OH	105,000	NA	110,100
Cedar Rapids, IA	105,800	112,900	114,900
Champaign/Urbana/Rantoul, IL	90,600	98,800	99,300
Charleston, SC	131,700	137,900	150,200
Charleston, WV	NA	99,400	110,100
Charlotte/Gast./Rock Hill, NC/SC	138,200	140,300	145,900
Chattanooga, TN/GA	99,100	101,100	110,100
Chicago, IL	171,200	171,800	198,000
Cincinnati, OH/KY/IN	119,900	126,700	132,700
Cleveland, OH	125,100	NA	135,800
Colorado Springs, CO	144,900	154,100	NA
Columbia, SC	109,500	112,800	116,800
Columbus, OH	125,000	129,100	137,400
Corpus Christi, TX	85,000	87,900	91,500
Dallas, TX	115,700	122,500	133,600
Davenport/Moline/Rock Isl., IA/IL	82,800	86,300	86,500
Dayton/Springfield, OH	104,100	105,100	112,600
Daytona Beach, FL	84,500	85,300	92,600
Denver, CO	171,300	196,800	219,000
Des Moines, IA	110,500	116,400	124,900
Detroit, MI	140,000	NA	NA
El Paso, TX	78,100	80,200	86,000
Eugene/Springfield, OR	129,500	132,800	NA
Fargo/Moorhead, ND/MN	93,400	97,100	99,500
Ft. Lauderdale/Hollywood/ Pompano Beach, FL	136,100	148,700	160,900
Ft. Myers/Cape Coral, FL	94,400	97,600	117,000
Ft. Wayne, IN	92,200	91,600	93,700
Ft. Worth/Arlington, TX	NA	NA	NA
Gainesville, FL	108,000	113,100	119,700
Gary/Hammond, IN	107,100	107,000	114,900
Grand Rapids, MI	106,700	114,900	122,400
Green Bay, WI	107,600	118,100	124,600
Greensboro/Winston-Salem/ High Point, NC	124,800	129,300	134,800
Greenville/Spartanburg, SC	113,800	118,100	128,400
Hartford, CT	150,700	159,900	164,700
Honolulu, HI	290,000	295,000	295,000
Houston, TX	105,300	116,100	125,800
Indianapolis, IN	110,900	112,300	118,100
Jackson, MS	$95,100	$99,500	NA
Jacksonville, FL	95,200	100,000	$111,900
Kalamazoo, MI	110,900	109,900	119,600
Kansas City, MO/KS	120,700	127,400	138,500
Knoxville, TN	108,300	110,800	117,500
Lake County, IL	164,000	169,400	182,800
Lansing/East Lansing, MI	105,200	111,200	124,700
Las Vegas, NV	130,800	137,400	147,300
Lexington/Fayette, KY	111,900	118,200	121,200
Lincoln, NE	101,000	109,300	116,500
Little Rock-N. Little Rock, AR	91,200	87,800	98,200
Los Angeles Area, CA	199,000	215,900	234,400
Louisville, KY/IN	109,700	116,700	NA
Madison, WI	136,500	153,600	161,100
Melbourne/Titusville/Palm Bay, FL	90,300	96,900	97,400
Memphis, TN/AR/MS	111,300	115,600	123,800
Miami/Hialeah, FL	134,600	144,600	159,700
Milwaukee, WI	135,300	140,700	156,000
Minneapolis/St. Paul, MN/WI	138,700	151,400	170,300
Mobile, AL	93,300	97,600	113,100
Montgomery, AL	99,100	NA	NA
Nashville, TN	116,400	147,500	137,300
New Haven/Meriden, CT	145,700	151,600	164,400
New Orleans, LA	109,100	112,000	117,400
New York metropolitan area, NY/NJ/CT	203,200	230,200	248,100
Norfolk/Virginia Bch/Newport News, VA	113,500	112,300	115,400
Ocala, FL	70,600	70,900	NA
Oklahoma City, OK	84,200	85,400	95,000
Omaha, NE/IA	109,400	116,900	118,400
Orlando, FL	105,300	111,200	125,000
Pensacola, FL	98,900	101,100	106,400
Peoria, IL	86,200	87,200	89,000
Philadelphia, PA/NJ	124,800	125,200	127,000
Phoenix, AZ	126,400	134,400	141,200
Pittsburgh, PA	89,900	93,600	99,700
Portland, ME	NA	131,100	152,100
Portland, OR	165,000	170,100	171,800
Providence, RI	128,800	137,800	153,900
Raleigh/Durham, NC	165,000	158,400	167,000
Reno, NV	150,600	157,300	160,100
Richland/Kennewick/Pasco, WA	109,100	119,600	NA
Richmond/Petersburg, VA	128,500	129,800	135,000
Riverside/San Bernardino, CA	128,700	138,600	157,500
Rochester, NY	87,700	87,600	92,900
Rockford, IL	94,600	95,900	102,100
Sacramento, CA	131,500	146,500	176,000
Saginaw/Bay City/Midland, MI	81,900	80,200	85,400
Saint Louis, MO/IL	102,900	108,400	122,900
Salt Lake City/Ogden, UT	137,900	141,500	146,500
San Antonio, TX	91,100	96,000	104,800
San Diego, CA	231,600	269,400	298,300
San Francisco Bay Area, CA	340,800	454,600	483,600
Sarasota, FL	134,800	132,000	NA
Seattle, WA	NA	230,100	244,200
Shreveport, LA	83,200	83,800	87,100
Sioux Falls, SD	NA	106,500	116,800
South Bend/Mishawaka, IN	86,700	82,200	97,700

Metropolitan Area	1999	2000	Second Quarter 2001
Spokane, WA	$106,800	$104,200	$106,800
Springfield, IL	86,100	85,000	87,700
Springfield, MA	114,600	120,400	127,400
Springfield, MO	85,800	86,000	NA
Syracuse, NY	82,100	81,000	85,100
Tacoma, WA	NA	151,100	158,300
Tallahassee, FL	117,800	122,500	127,200
Tampa/St. Pete./Clearwater, FL	94,000	110,800	121,200
Toledo, OH	98,100	104,000	113,500
Topeka, KS	80,200	80,600	91,000
Trenton, NJ	144,200	150,900	164,200
Tucson, AZ	$117,700	$120,500	$129,800
Tulsa, OK	92,800	100,000	108,000
Washington, DC/MD/VA	176,500	182,600	206,700
Waterloo/Cedar Falls, IA	74,800	80,200	86,100
W. Palm Beach/Boca Raton/ Delray Beach, FL	131,000	138,400	147,600
Wichita, KS	91,500	90,800	95,800
Wilmington, DE/NJ/MD	120,600	127,600	130,600
Worcester, MA	117,000	131,800	147,900
Youngstown/Warren, OH	76,100	74,100	NA
UNITED STATES	**$133,300**	**$139,000**	**$146,900**

NA = not available.

Housing Affordability, 1990-2001

Source: National Association of REALTORS®

Year	Median priced existing home	Average mortgage rate[1]	Monthly principal & interest payment	Payment as percentage of median income
1990	$ 92,000	10.04%	$648	22.0%
1991	97,100	9.30	642	21.4
1992	99,700	8.11	591	19.3
1993	103,100	7.16	558	18.1
1994	107,200	7.47	598	18.5
1995	110,500	7.85	639	18.9
1996	$115,800	7.71%	$661	18.8%
1997	121,800	7.68	693	18.7
1998	128,400	7.10	690	17.4
1999	133,300	7.33	733	18.0
2000	139,000	8.03	818	19.0
2001[2]	152,600	7.18	827	18.8

(1) Based on effective rate on loans closed on existing homes monitored by the FHA. (2) Preliminary figures for June 2001.

U.S. Home Ownership Rates, by Selected Characteristics, 1997, 2001[1]

	1997	2001
TOTAL, U.S.	**65.7%**	**67.7%**
Region		
Northeast	62.4	63.2
Midwest	70.3	72.7
South	68.1	69.7
West	59.9	62.9

Age	1997	2001
Under 35	38.6%	40.8%
35-44	66.3	68.1
45-54	75.6	77.2
55-64	80.3	81.5
65+	79.1	79.7

Race/Ethnicity	1997	2001
White, non-Hispanic	72.1%	74.1%
Black	44.4	47.9
Hispanic	43.3	46.1
Other	52.7	55.2

Income	1997	2001
Median family income or more	80.8%	82.0%
Below median family income	50.0	51.7

(1) For second quarter of the year.

Mortgage Loan Calculator

Source: Joyce E. Boulanger, Mortgage Access Corp.

To determine monthly payments, divide loan amount by 1,000 and then multiply the resulting figure by the appropriate factor from this table. To find the appropriate factor use the mortgage term in years and the interest rate percentage. More information on calculating mortgages can be found at http://www.interest.com/calculators

EXAMPLE: For a 30-year mortgage at 7.25%, the factor would be 6.82. If the mortgage amount is $220,000, divide by 1,000, which comes to 220. 220 x 6.82 (factor) = $1,500.40 monthly mortgage payment of principal and interest only (there will also be property taxes, home insurance, and other possible costs).

INTEREST RATE	MORTGAGE TERM IN YEARS 5	10	15	20	25	30	35	40
5.00	18.88	10.61	7.91	6.60	5.85	5.37	5.05	4.83
5.25	18.99	10.73	8.04	6.74	6.00	5.53	5.21	4.99
5.50	19.11	10.86	8.18	6.88	6.15	5.68	5.38	5.16
5.75	19.22	10.98	8.31	7.03	6.30	5.84	5.54	5.33
6.00	19.33	11.10	8.44	7.16	6.44	6.00	5.70	5.50
6.25	19.45	11.23	8.57	7.31	6.60	6.16	5.87	5.68
6.50	19.57	11.35	8.71	7.46	6.75	6.32	6.04	5.85
6.75	19.68	11.48	8.85	7.60	6.91	6.49	6.21	6.03
7.00	19.80	11.61	8.99	7.75	7.07	6.65	6.39	6.21
7.25	19.92	11.74	9.13	7.90	7.23	6.82	6.56	6.40
7.50	20.04	11.87	9.27	8.06	7.39	6.99	6.74	6.58
7.75	20.16	12.00	9.41	8.21	7.55	7.16	6.92	6.77
8.00	20.28	12.13	9.56	8.36	7.72	7.34	7.10	6.95
8.25	20.40	12.27	9.70	8.52	7.88	7.51	7.28	7.14
8.50	20.52	12.40	9.85	8.68	8.06	7.69	7.47	7.34
8.75	20.64	12.54	10.00	8.84	8.23	7.87	7.66	7.53
9.00	20.76	12.67	10.15	9.00	8.40	8.05	7.84	7.72
9.25	20.88	12.81	10.30	9.16	8.57	8.23	8.03	7.91
9.50	21.01	12.94	10.45	9.33	8.74	8.41	8.22	8.11
9.75	21.13	13.08	10.60	9.49	8.92	8.60	8.41	8.30
10.00	21.25	13.22	10.75	9.66	9.09	8.78	8.60	8.50
10.25	21.38	13.36	10.90	9.82	9.27	8.97	8.79	8.69
10.50	21.50	13.50	11.06	9.99	9.45	9.15	8.99	8.89
10.75	21.62	13.64	11.21	10.16	9.63	9.34	9.18	9.09
11.00	21.75	13.78	11.37	10.33	9.81	9.53	9.37	9.29
11.25	21.87	13.92	11.53	10.50	9.99	9.72	9.57	9.49
11.50	22.00	14.06	11.69	10.67	10.17	9.91	9.77	9.69
11.75	22.12	14.21	11.85	10.84	10.35	10.10	9.96	9.89
12.00	22.25	14.35	12.01	11.02	10.54	10.29	10.16	10.09
12.25	22.38	14.50	12.17	11.19	10.72	10.48	10.36	10.29
12.50	22.50	14.64	12.33	11.37	10.91	10.68	10.56	10.49
12.75	22.63	14.79	12.49	11.54	11.10	10.87	10.76	10.70
13.00	22.76	14.94	12.66	11.72	11.28	11.07	10.96	10.90
13.25	22.89	15.08	12.82	11.90	11.47	11.26	11.16	11.10
13.50	23.01	15.23	12.99	12.08	11.66	11.46	11.36	11.31
13.75	23.14	15.38	13.15	12.26	11.85	11.66	11.56	11.51
14.00	23.27	15.53	13.32	12.44	12.04	11.85	11.76	11.72

Tracing Your Roots

By Richard Hantula

Richard Hantula is a freelance editor and writer whose interests include genealogy.

Family history is one of Americans' favorite pastimes, and its popularity is growing. The fruits of their labors run the gamut from a simple listing of one's lineage (or "pedigree"); to a more wide-ranging annotated family tree that records names, marriages, births, deaths, dates, places, and the like; to an elaborate narrative that delves as far back as is possible and runs up to the present day and that may be illustrated, at least for the most recent period, with photos and even audio and video clips.

Getting Started

Genealogy is detective work. In some cases you may need to apply a deft combination of ingenuity and patience to get the information you want. And sometimes you may find that the answers you are looking for simply cannot be obtained, because records are inaccessible or do not exist. Getting started, however, is easy. To cut the job down to a manageable size, you will probably want to focus your research, at least in the beginning, on a particular branch of your family. It also makes things easier to proceed step-by-step and to search for data on one person at a time. A few pointers:

Set up a system for organizing the information you will accumulate. Note cards and loose-leaf binders are common tools of the trade. A computer can be a big labor saver.

Start your data collection by writing down what you already know about your family.

Gather together the documents in your home that may contain relevant data, such as birth, marriage, and death certificates; land deeds; school records; medical records; and military papers. Helpful information may be found written in old letters and diaries or inscribed in Bibles or on dishware. Old photos can prove invaluable as well.

Ask your relatives to write down what they know of the family history. You may want to interview them, especially those who knew relatives who are now dead.

Record names, dates, and places accurately. Where feasible, make photocopies of papers that document such facts. Consider videotaping or tape-recording your interviews with relatives.

Be aware that your family may be documented under different names at different times and in different places. Adoptions, divorces, and illegitimate births all complicate the issue, as does the fact that unfamiliar-sounding names are sometimes transcribed incorrectly in official records. The names of many immigrants to the U.S. were arbitrarily anglicized by officials at ports of entry, and some immigrants altered their names themselves, hoping to improve their chances of assimilating to American life.

You can get detailed advice on how to go about tracing your roots from numerous books and videos that can be found in bookstores and libraries, as well as from a Public Broadcasting Service (PBS) television series, *Ancestors*, available to local stations from mid-2000 to mid-2004. Other places where you can learn about genealogy include websites on the Internet, local genealogical clubs, and genealogy courses offered by schools and universities (some of them available online via the Internet).

Filling in the Holes

To flesh out your family history, you may want to do a little travel. Possible sources of data in the U.S. and abroad include churches, cemeteries, archives of old newspapers, public libraries, historical societies, and national, regional, and local government depositories of records. The U.S. National Archives and Records Administration, for example, contains a wealth of material such as passport applications, ships' passenger lists, Bureau of Indian Affairs documents, and census, federal court, and military records. The Family History Library of the Church of Jesus Christ of Latter Day Saints (Mormons) in Salt Lake City, UT, houses a huge collection of information on billions of individuals, some of it copied from archives in other countries (more than 3,500 branches of the library, called Family History Centers, are located around the globe, and much of the data can be accessed via the Internet). The availability of records outside the U.S. varies from country to country. In China, for example, genealogies were recorded for some families for hundreds or even thousands of years, but many documents were destroyed in the Cultural Revolution of the mid-1960s to mid-1970s.

Computer Software

Scores of genealogical computer programs are available, some for free. The typical family tree program organizes the data you collect (and, if you have access to the Internet, even helps you collect it), prints it out in attractive chart and report formats, and saves it in standard GEDCOM (for "Genealogical Data Communications") genealogy files. Some programs store electronic versions of photos, videos, and sound clips. Commercial family tree software often comes packaged with batches of CD-ROMs containing genealogical resource data that may be helpful to some researchers, although more and more data are becoming available on the Internet for free or for a small fee.

If you are considering buying a program, check the computer requirements listed on the package. Beginners will probably want to choose from programs boasting more elaborate tutorials and easier-to-use interfaces. Among the most popular commercial programs for PCs using Microsoft Windows are Family Tree Maker (by Broderbund) and Generations (Sierra). Also easy to use is Family Origins (FormalSoft), which lacks some of the add-ons that the first two programs offer. Master Genealogist (Wholly Genes Software) enjoys a wide following but may be daunting for beginners. The leading software for Macintosh computers is Reunion (Leister Productions). The most popular free software for PCs with Windows is the Mormons' Personal Ancestral File, or PAF, which can be downloaded from the Internet at no charge (a Macintosh version can be purchased on disks). PAF lacks some of the bells and whistles of the major commercial products, but is logically constructed and easy to use.

The Burgeoning Internet

Sites on the World Wide Web have traditionally been valuable sources of indexes and other references to records located in "offline" archives. More and more genealogical resource data, however, is becoming directly available on the Web. Information pertaining to the millions of people who entered the U.S. through Ellis Island and the Port of New York in 1892-1924, for instance, is available from the American Family Immigration History Center (http://www.ellisislandrecords.org). Also, websites and e-mail offer splendid ways of exchanging information with other family history researchers and making contact with relatives.

Some of the most popular genealogy programs are associated with websites that provide how-to advice, along with access to search engines and resource databases. The Mormons' FamilySearch site (http://www.familysearch.org) performs this role for PAF, and the program can be downloaded from it. General-purpose genealogy websites of note include Cyndi's List (http://www.cyndislist.com) and RootsWeb (http://www.rootsweb.com), which claims to be the oldest and largest free genealogy site. Also helpful are About.com (http://genealogy.about.com) and the website associated with the PBS *Ancestors* show (http://www.ancestors.com). Wide-ranging sites like Ancestry.com (http://www.ancestry.com) and Genealogy.com (http://www.genealogy.com) offer access to numerous databases of potential interest to genealogists, some of them on a subscription basis. There are also myriad sites focusing on individual ethnic groups, among them JewishGen (http://www.jewishgen.org) for researchers of Jewish genealogy and AfriGeneas (http://www.afrigeneas.com) and Christine's Genealogy Website (http://www.ccharity.com) for African-Americans.

How to Obtain Birth, Marriage, Death Records

The pamphlet "Where to Write for Vital Records: Births, Deaths, Marriages, and Divorces" (Stock # 017-022-01196-4) is available from the Superintendent of Documents, PO Box 371954, Pittsburgh, PA 15250-7954; advance payment of $3.00 is required. Orders can also be placed by calling (866) 512-1800, via fax at (202) 512-2250, or on the website http://bookstore.gpo.gov

POSTAL INFORMATION

Basic U.S. Postal Service

The Postal Reorganization Act, creating a government-owned postal service under the executive branch and replacing the old Post Office Department, was signed into law by Pres. Richard Nixon, Aug. 12, 1970. The service officially came into being on July 1, 1971.

The U.S. Postal Service is governed by an 11-person Board of Governors. Nine of the members are appointed by the president with Senate approval. These 9 choose a postmaster general. The board and the postmaster general choose the 11th member, who serves as deputy postmaster general. An independent Postal Rate Commission of 5 members, appointed by the president, reviews and rules on proposed postal rate increases submitted by the Board of Governors.

> ***IT'S A FACT:*** The first official U.S. postage stamps were a 5-cent stamp, picturing Benjamin Franklin, the nation's first U.S. postmaster general, and a 10-cent stamp picturing George Washington. The were both issued July 1, 1847.

U.S. Domestic Rates

(Domestic rates apply to the U.S., to its territories and possessions, and to APOs and FPOs. Many changes for domestic postal rates, fees, services, and terminology took effect Jan. 7, 2001.)

First Class

First Class includes written matter such as letters, postal cards, and postcards (private mailing cards), plus all other matter wholly or partly in writing, whether sealed or unsealed, except book manuscripts, periodical articles and music, manuscript copy accompanying proofsheets or corrected proofsheets of the same, and the writing authorized by law on matter of other classes. Also included: matter sealed or closed against inspection, bills, and statements of accounts.

Written letters and matter sealed against inspection cost 34¢ for first ounce or fraction, 23¢ for each additional ounce or fraction up to and including 13 ounces. U.S. Postal Service cards and private postcards alike cost 21¢. Presort and automation-compatible mail can qualify for lower rates if certain piece minimums, mailing permits, and other requirements are met.

Express Mail

Express mail provides guaranteed expedited service for any mailable article (up to 70 lbs and not over 108 in. in combined length and girth). Offers next day delivery by noon to most destinations; no extra charge for Saturday, Sunday, or holiday delivery. Second-day service is available to locations not on the Next Day Delivery Network. The basic rate for Express Mail weighing up to 8 oz is $12.45. All rates include insurance up to $500, shipment receipt, and record of delivery at the destination post office. Express Mail tracking is available on the USPS Web site (http://www.usps.com).

Express Mail Flat Rate: $16.25, regardless of weight, if matter fits into a special Postal Service flat-rate envelope.

Pickup service is available for $10.25 per stop, regardless of the number of pieces or service used (e.g., Express Mail, Priority Mail, or Parcel Post can be picked up together).

Contact your local post office for further information.

Standard Mail

Standard Mail is limited to 16 ounces and bulk mailings (at least 200 pieces or 50 lbs) of such items as solicitations, newsletters, and advertising materials.

For mailing Standard Mail in bulk (at least 200 pieces or 50 lb of such items as solicitations, newsletters, advertising materials, books, and cassettes, each item of which individually weighs less than 1 lb), the minimum rate per piece, basic, non-letter, is $0.322 for pieces weighing 3.3 oz or less. For pieces weighing more than 3.3 oz, the rate is $0.184 per piece plus $0.668 per pound. Contact your post office for the discounts offered for presorted, letter-shaped, destination entry, and automation-compatible mail.

Separate rates are available for some nonprofit organizations provided with a permit. Permit requires a one-time $125 imprint fee plus an annual (calendar year) fee of $125.

Priority Mail

The most expeditious handling and transportation available will be used for fast delivery by "Priority Mail." Priority Mail may include packages up to 70 lbs and not over 108 in. in length and girth combined, whether sealed or unsealed, including written and other First Class material.

Priority Mail Flat Rate: $3.95, regardless of weight, if matter fits into a special Postal Service flat-rate envelope. Other rates are as follows (fractions of a pound are rounded up to the next full pound):

Up to 1lb	2 lb	3 lb	4 lb	5 lb
$3.50	$3.95	$5.20	$6.45	$7.70

Zoned rates apply for parcels over 5 lb. Mileage between specific geographic locations of 3-digit ZIP codes determines the zone number used. The mileage range by zone number is: Zone 1—up to 50 mi; 2—51 to 150 mi; 3—151 to 300 mi; 4—301 to 600 mi; 5—601 to 1,000 mi; 6—1,001 to 1,400 mi; 7—1,401 to 1,800 mi; 8—over 1,800 mi.

Parcels weighing less than 15 lb and measuring over 84 in., but less than 108 in., in length and girth combined, cost the same as a 15-lb parcel mailed to the same zone.

Pickup service costs an additional $10.25 per stop.

Periodicals

Periodicals include newspapers and magazines.

For the general public, the applicable Standard Mail or First Class postage is paid for periodicals.

For publishers, rates vary according to (1) whether item is sent to same county, (2) percentage of reading and advertising matter, (3) weight, (4) distance, (5) level of presort, (6) automation compatibility.

Package Services

Package Services, formerly "Standard Mail (B)," is any mailable matter that is not included in First Class or Periodicals (unless permitted or required by regulations). There are currently four subclasses of Package Services: Parcel Post, Bound Printed Matter, Media Mail (formerly "Special Standard Mail"), and Library Mail. The post office determines charges for Package Services according to the weight of the package in pounds and the zone distance shipped (Media Mail and Library Mail rates are determined weight alone). There is no minimum weight; see separate headings for maximum weight. Presort and automation-compatible mail for all Package Services can qualify for lower rates if certain piece minimums, mailing permits, and other requirements are met. Contact your local post office for further information. Package Services is not sealed against postal inspection.

Parcel Post

Parcel Post is any Package Services not mailed as Bound Print Matter, Media Mail, or Library Mail. Any Package Services matter may be mailed at the Parcel Post rates, subject to these basic standards: not to exceed 70 lbs or 108 in. in combined length and girth (packages over 108 in., but not more than 130 in. in combined length and girth are subject to oversize rates). All fractions of a pound are counted as a full pound.

Parcel Post Basic Rate Schedule

(Inter BMC/ASF ZIP codes only, machinable[1] parcels, no discount, no surcharge)

Weight up to but not exceeding—(pounds)	ZONES 1 & 2	3	4	5	6	7	8
2	$3.42	$3.45	$3.45	$3.45	$3.45	$3.45	$3.45
3	3.90	4.23	4.66	4.71	4.76	4.81	4.86
4	4.05	4.51	5.33	5.80	5.95	6.00	6.05
5	4.19	4.76	5.78	7.00	7.15	7.20	7.25
6	4.33	5.01	6.20	7.70	8.03	8.25	8.84
7	4.46	5.23	6.59	8.38	8.90	9.49	10.69
8	4.60	5.44	6.92	8.96	9.60	10.74	12.53
9	4.70	5.63	7.28	9.50	10.30	11.99	14.20
10	4.83	5.82	7.58	10.01	11.00	13.24	15.26
11	4.93	6.00	7.89	10.48	11.70	14.20	16.14
12	5.03	6.16	8.17	10.92	12.40	15.15	16.98
13	5.13	6.30	8.43	11.33	13.10	16.10	17.79
14	5.23	6.48	8.69	11.72	13.80	17.05	18.57
15	5.32	6.62	8.94	12.08	14.44	17.66	19.33
16	5.40	6.76	9.17	12.42	14.86	18.20	20.05
17	5.50	6.88	9.40	12.74	15.28	18.72	20.76
18	5.58	7.01	9.60	13.04	15.65	19.19	21.44
19	5.67	7.14	9.81	13.33	16.01	19.66	22.10
20[2]	5.74	7.25	9.98	13.61	16.35	20.09	22.74

(1) Machinable parcels must be: not less than 6 in. long, 3 in. high, and .25 in. thick or more than 34 in. long, 17 in. high, and 17 in. thick; at least 6 oz. but not more than 35 lbs. (2) Consult postmaster for pieces greater than 20 lbs.

Bound Printed Matter

(minimum weight: none; maximum weight: 15 lbs)

Applies to advertising, promotional, directory, or editorial material that is bound by permanent fastening and consists of sheets of which at least 90% are imprinted by any process other than handwriting or typewriting. Does not include stationery (or pads of blank forms) or personal correspondence. Packages may not exceed 108 in. in combined length and girth, marked "Bound Printed Matter" or "BPM."

Bound Printed Matter Rates

(single-piece zone rate)

Weight (lbs)	ZONES 1&2	3	4	5	6	7	8
1.5	$1.80	$1.83	$1.87	$1.93	$1.99	$2.06	$2.21
2	1.84	1.88	1.94	2.02	2.10	2.19	2.38
2.5	1.90	1.95	2.00	2.11	2.21	2.33	2.57
3	1.94	2.00	2.08	2.20	2.32	2.46	2.75
3.5	1.99	2.06	2.15	2.29	2.43	2.60	2.93
4	2.03	2.11	2.21	2.37	2.55	2.72	3.11
4.5	2.07	2.17	2.29	2.47	2.65	2.87	3.30
5	2.13	2.23	2.36	2.55	2.77	3.00	3.47
6	2.22	2.35	2.49	2.74	2.99	3.26	3.83
7	2.31	2.46	2.63	2.92	3.21	3.53	4.19
8	2.40	2.57	2.78	3.10	3.44	3.81	4.55
9	2.50	2.68	2.91	3.27	3.66	4.07	4.92
10	2.60	2.80	3.05	3.45	3.87	4.34	5.27
11	2.68	2.91	3.19	3.63	4.09	4.61	5.64
12	2.78	3.03	3.33	3.81	4.32	4.88	6.00
13	2.87	3.14	3.47	3.99	4.54	5.15	6.36
14	2.97	3.26	3.61	4.17	4.76	5.42	6.73
15	3.06	3.37	3.75	4.35	4.98	5.69	7.09

Media Mail

(minimum weight: none; maximum weight: 70 lbs)

Formerly "Special Standard Mail." Applies to books of at least 8 printed pages; 16-mm or narrower-width films; printed music; printed test materials; sound recordings, playscripts, and manuscripts for books; printed educational charts; loose-leaf pages and binders consisting of medical information; computer-readable media. Advertising restrictions apply. Packages must be marked "Media Mail" and may not exceed 108 in. in combined length and girth. Contact your local post office for further information.

Rates are calculated by weight only. Single-piece rates are: $1.33, up to 1 lb; 45¢ for each additional pound or fraction, to 7 lbs; additional pounds thereafter, 30¢ each.

Library Mail

(minimum weight: none; maximum weight: 70 lbs)

Applies to books, printed music, bound academic theses, periodicals, sound recordings, museum materials, and other library materials mailed between schools, colleges, universities, public libraries, museums, veteran and fraternal organizations, and nonprofit religious, educational, scientific, and labor organizations or associations. Advertising restrictions apply. All packages must be marked "Library Mail," and may not exceed 108 in. in combined length and girth. Contact your local post office for further information.

Rates are calculated by weight only. Single-piece rates are: $1.26, up to 1 lb; 43¢ for each additional pound or fraction, to 7 lbs; additional pounds thereafter, 29¢ each.

Domestic Mail Special Services

Special Handling

Provides preferential handling, but not preferential delivery, to the extent practicable in dispatch and transportation. Available for First-Class Mail, Priority Mail, and Package Services for the following surcharge: up to 10 lb, $5.40; over 10 lb, $7.50. Pieces must be marked "Special Handling."

Registered Mail

Provides sender with mailing receipt, and a delivery record is maintained. Only matter prepaid with postage at First Class postage rates may be registered. Stamps or meter stamps must be attached. The face of the article must be at least 5″ long, 3½″ high. The mailer is required to declare the value of mail presented for registration.

Declared Value	Registration Fee[1]
$0.00	$7.25
$0.01 to $100	7.50
$100.01 to $500	8.25
$500.01 to $1,000	9.00
$1,000.01 to $2,000	9.75
$2,000.01 to $3,000	10.50
$3,000.01 to $4,000	11.25
$4,000.01 to $5,000	12.00
$5,000.01 to $6,000	12.75
$6,000.01 to $7,000	13.50
$7,000.01 to $8,000	14.25
$8,000.01 to $9,000	15.00
$9,000.01 to $10,000	15.75

(1) Fee for articles with declared value over $0.00 includes insurance; fee is in addition to postage.

C.O.D.: Unregistered: Applicable to First Class, Priority Mail, Express Mail, and Package Services. Items must be sent as bona fide orders or be in conformity with agreements between senders and addressees. Maximum amount collectible is $1,000. **Registered:** For details, consult postmaster.

Certified mail: Available for any matter having no intrinsic value on which First Class or Priority Mail postage is paid. A receipt is furnished at the time of mailing, and evidence of delivery is obtained. Basic fee is $2.10 in addition to regular postage. Return receipt and restricted delivery available upon payment of additional fees. No indemnity.

Insured Mail

Applicable to Standard Mail, Package Services, and First-Class or Priority Mail items eligible to be mailed as Package Services. Matter for sale addressed to prospective purchasers who have not ordered it or authorized its sending cannot be insured. Note: for Express Mail, insurance is included up to $500. Add $1.00 per $100 or fraction thereof over $500 up to $5,000.

Declared Value	Insured Mail Fee[1]
$0.01 to $50	$1.10
$50.01 to $100	2.00
$100.01 to $200	3.00
$200.01 to $300	4.00
$300.01 to $400	5.00
$400.01 to $500	6.00
$500.01 to $600	7.00
$600.01 to $5,000	7.00 plus $1.00 for each $100 or fraction over $600 in value.

(1) In addition to postage. (Maximum liability is $5,000.) See postmaster for further details on bulk discounts.

Delivery Confirmation

Applies to Priority Mail and Package Services. Available for purchase at the time of mailing only. Provides mailer with the date and time an article was delivered and, if delivery was attempted but not successful, the date and time of the attempt. Electronic confirmation is available for bar-coded matter. Manual confirmation is available for retail purchasers on the Internet (http://www.usps.com) or toll-free by phone (800-222-1811).

Priority Mail fee: manual, 40¢; electronic, free. Package Services fee: 50¢ and 12¢.

Forwarding Addresses

To obtain a forwarding address, the mailer must write on the envelope or cover the words "Address Correction Requested." The destination post office then will check for a forwarding address on file and provide it for 60¢ per manual correction, 20¢ per automated correction.

International Mail Special Services

Registration: Available to practically all countries for letter-post items only. Fee $7.25. The maximum indemnity payable—generally only in case of complete loss (of both contents and wrapper)—is $40.45. To Canada only, the fee is $7.50, providing indemnity for loss up to $100, $8.25 for loss up to $500, and $9.00 for loss up to $1,000. Contact your post office for more details.

Return Receipt: Shows to whom and when delivered; Fee: $1.50 (must be purchased at time of mailing).

Special Delivery: Not available as of June 8, 1997.

Air Mail: Available daily to practically all countries.

Aerogrammes — Aerogrammes are letter sheets that can be folded into the form of an envelope and sealed. Intended for personal communication only and may not include enclosures. Fee: 70¢ from U.S. to all countries.

Air mail postcards (single) — 50¢ to Canada and Mexico; 70¢ to all other countries.

International Reply Coupons (IRC): Provide foreign addressees with a prepaid means of responding to communications initiated by a U.S. sender. Each IRC is equivalent to the destination country's minimum postage rate for an unregistered airmail letter. Fee: $1.75 per coupon.

Restricted Delivery: Available to many countries for registered mail; some limitations. Fee: $3.20.

Insurance: Available to many countries for loss of or damage to items paid at parcel post rate. Consult postmaster for indemnity limits for individual countries.

	Fees	
Limit of indemnity Not over	Canada[1]	All other countries[1]
$ 50	$1.10	$1.85
100	2.00	2.60
200	3.00	3.60
300	4.00	4.60
400	5.00	5.60
500	6.00	6.60
600	7.00	7.60
700	8.00	8.60
800		9.60
900		10.60
1,000[2]		11.60

(1) Not all countries insure items up to the amounts listed in the table. Canada does not insure items for more than $675. (2) For amounts more than $1,000, add $1.00 for each $100 or fraction.

Post Office-Authorized 2-Letter State Abbreviations

The abbreviations below are approved by the U.S. Postal Service for use in addresses.

Alabama	AL	Kentucky	KY	Ohio	OH
Alaska	AK	Louisiana	LA	Oklahoma	OK
American Samoa	AS	Maine	ME	Oregon	OR
Arizona	AZ	Marshall Islands[1]	MH	Palau[1]	PW
Arkansas	AR	Maryland	MD	Pennsylvania	PA
California	CA	Massachusetts	MA	Puerto Rico	PR
Colorado	CO	Michigan	MI	Rhode Island	RI
Connecticut	CT	Minnesota	MN	South Carolina	SC
Delaware	DE	Mississippi	MS	South Dakota	SD
Dist. of Col.	DC	Missouri	MO	Tennessee	TN
Federated States of Micronesia[1]	FM	Montana	MT	Texas	TX
Florida	FL	Nebraska	NE	Utah	UT
Georgia	GA	Nevada	NV	Vermont	VT
Guam	GU	New Hampshire	NH	Virgin Islands	VI
Hawaii	HI	New Jersey	NJ	Virginia	VA
Idaho	ID	New Mexico	NM	Washington	WA
Illinois	IL	New York	NY	West Virginia	WV
Indiana	IN	North Carolina	NC	Wisconsin	WI
Iowa	IA	North Dakota	ND	Wyoming	WY
Kansas	KS	Northern Mariana Is.	MP		

(1) Although an independent nation, this country is currently subject to domestic rates and fees.

Canadian Province and Territory Postal Abbreviations

Source: Canada Post

Alberta	AB	Newfoundland and Labrador	NF	Prince Edward Island	PE
British Columbia	BC	Northwest Territories/Nunavut	NT	Quebec	QC[1]
Manitoba	MB	Nova Scotia	NS	Saskatchewan	SK
New Brunswick	NB	Ontario	ON	Yukon Territory	YT

(1) PQ is also acceptable.

International Postal Rates

Letter-post—Encompasses all classes of international mail formerly categorized as LC (Letters and Cards) and AO (Other Articles), including letters, letter packages, postcards and postal cards, small packets, etc. Airmail and Economy (surface) rates are available. Maximum weight: 4 lbs (64 oz). Minimum dimensions for envelopes and packages is 5.5 in. x 3.5 in. (0.007 in. thick) and a max. of 36 in. combined length and girth (max. length: 24 in.). Consult your local post office regarding rolls (tubes), printed matter, and nonstandard sized items.

Parcel Post—Airmail or Economy (surface) rates available. Weight limits range from 22 lbs to 70 lbs, see Country Rate Group table for specific standards. Most countries require: min. length and width, 5.5 in. x 3.5 in.; max. length, 42 in.; max. length and girth combined, 79 in. Exceptions: Belgium, Canada, Germany, Great Britain, Hong Kong, Ireland, Japan, Liechtenstein, Macao, Sweden, and Switzerland. Consult your local post office for countries with exceptional size limits.

Global Priority Mail (GPM) Flat-Rate: If matter weighs 4 lbs or less, and fits into a special Postal Service flat-rate envelope, the rates are as follows: small envelope (6 in. x 10 in.), $4.00 to Canada or Mexico and $5.00 to all other countries; large envelope (9.5 in. x 12.5 in.), $7.00 to Canada or Mexico and $9.00 to all other countries. Variable-weight option available.

The U.S. Postal Service also offers expedited delivery services (**Global Express Guaranteed** and **Global Express Mail**) to many international destinations. Consult your local post office for details.

Letter-Post Rates

Not over	AIR MAIL RATE GROUPS[1]					ECONOMY (SURFACE) RATE GROUPS[1]				
(oz.)	1	2	3	4	5	1	2	3	4	5
1	$0.60	$0.60	$0.80	$0.80	$0.80	$2.70	$4.35	$3.80	$4.05	$4.95
2	0.85	0.85	1.60	1.70	1.55	—	—	—	—	—
3	1.10	1.25	2.40	2.60	2.30	—	—	—	—	—
4	1.35	1.65	3.20	3.50	3.05	—	—	—	—	—
5	1.60	2.05	4.00	4.40	3.80	—	—	—	—	—
6	1.85	2.45	4.80	5.30	4.55	—	—	—	—	—
7	2.10	2.85	5.60	6.20	5.30	—	—	—	—	—
8	2.35	3.25	6.40	7.10	6.05	—	—	—	—	—
12	3.10	4.00	7.55	8.40	7.65	—	—	—	—	—
16	3.75	5.15	8.70	9.70	9.25	2.70	4.35	3.80	4.05	4.95
20	4.40	6.30	9.85	11.00	10.85	4.05	5.15	4.45	4.70	5.70
24	5.05	7.45	11.00	12.30	12.45	4.55	5.95	5.10	5.35	6.50
28	5.70	8.60	12.15	13.60	14.05	5.05	6.70	5.70	6.00	7.30
32	6.35	9.75	13.30	14.90	15.65	5.60	7.50	6.30	6.65	8.10
36	7.00	10.95	14.50	16.25	17.35	6.00	8.15	6.90	7.25	8.75
40	7.65	12.15	15.70	17.60	19.05	6.40	8.80	7.50	7.85	9.40
44	8.30	13.35	16.90	18.95	20.75	6.80	9.45	8.10	8.45	10.05
48	8.95	14.55	18.10	20.30	22.45	7.20	10.10	8.70	9.05	10.70
52	9.65	15.80	19.35	21.70	24.20	7.60	10.75	9.30	9.65	11.35
56	10.35	17.05	20.60	23.10	25.95	8.00	11.40	9.90	10.25	12.00
60	11.05	18.30	21.85	24.50	27.70	8.40	12.05	10.50	10.85	12.65
64	11.75	19.55	23.10	25.90	29.45	8.80	12.70	11.10	11.45	13.30

(1) Rate Groups: Canada-1, Mexico-2, Australia, Japan, New Zealand-4; for other countries, see "Letter-post rate group" in Country Rate Groups table, page 724.

Air Mail Parcel Post Rates

Not Over	RATE GROUPS[1]												
	1[2]	2	3	4	5	6	7	8	9	10	11	12	13
1 lb.	$13.25	$13.00	$16.00	$16.25	$15.25	$14.00	$16.50	$12.50	$14.50	$16.00	$18.00	$14.00	$17.00
2	13.25	15.50	20.00	20.50	19.75	15.50	19.00	19.00	18.75	18.50	22.00	15.50	19.00
3	14.25	17.75	24.00	24.50	24.50	17.50	21.75	20.00	23.25	21.50	26.00	17.25	22.00
4	15.50	20.25	28.00	29.00	29.75	20.25	24.50	24.25	26.75	24.00	30.00	19.25	25.00
5	16.75	23.00	32.00	33.50	35.00	22.75	27.25	28.75	32.75	26.50	34.00	21.25	28.00
6	17.85	25.00	35.00	36.80	39.25	25.65	30.25	32.65	36.50	29.50	37.50	23.75	31.25
7	18.95	27.00	38.00	40.10	43.50	28.55	33.25	36.55	40.40	32.50	41.00	26.25	34.50
8	20.05	29.00	41.00	43.40	47.75	31.45	36.25	40.45	44.30	35.50	44.50	28.75	37.75
9	21.15	31.00	44.00	46.70	52.00	34.35	39.25	44.35	48.20	38.50	48.00	34.25	41.00
10	22.25	33.00	47.00	50.00	56.25	37.25	42.25	48.25	52.10	41.50	51.50	33.75	44.25
Add'l[3]	1.10	2.00	3.00	2.30	4.25	2.90	3.00	3.90	3.90	3.00	3.50	2.50	3.25

(1) Rate Groups: Canada-1, Mexico-2, Great Britain, Northern Ireland-3, Japan-4, China-5; for other countries, see "Parcel Post rate group" in Country Rate Groups table page 724. (2) Canada: minimum 1 lb; maximum 66 lbs, 22 lbs limit to members of the Canadian Armed Forces based outside Canada (CFPOs). (3) Price of each additional pound or fraction.

Economy (Surface) Parcel Post Rates

Not Over	RATE GROUPS[1]											
	1[2]	2	3	4	5	6	7	8	9	10	11	12
5 lbs[3]	$15.25	$19.50	$23.00	$23.25	$21.25	$18.25	$22.00	$21.50	$28.75	$21.75	$26.25	$20.25
6	15.75	20.75	25.00	25.00	22.75	19.35	24.00	22.80	30.95	23.50	28.75	22.00
7	16.50	22.00	27.00	26.25	24.25	20.45	26.00	24.10	33.15	25.00	31.00	23.75
8	17.25	23.00	29.00	27.75	25.75	21.55	28.00	25.40	35.35	26.75	33.25	25.50
9	17.75	24.00	31.00	29.00	27.25	22.65	30.00	26.70	37.55	29.00	35.50	27.25
10	18.25	24.75	32.75	30.25	28.75	23.75	32.00	28.10	39.75	32.00	37.75	28.90
11	18.70	25.50	34.45	31.30	30.00	24.70	33.60	29.40	41.65	33.40	39.80	30.55
12	19.15	26.25	36.15	32.35	31.25	25.65	35.20	30.70	43.55	34.80	41.85	32.20
13	19.60	27.00	37.85	33.40	32.50	26.60	36.80	32.00	45.45	36.20	43.90	33.85
14	20.05	27.75	39.55	34.45	33.75	27.55	38.40	33.30	47.35	37.60	45.95	35.50
15	20.50	28.50	41.25	35.50	35.00	28.50	40.00	34.60	49.25	39.00	48.00	37.15
16	20.95	29.25	42.95	36.55	36.25	29.45	41.60	35.90	51.15	40.40	50.05	38.80
17	21.40	30.00	44.65	37.60	37.50	30.40	43.20	37.20	53.05	41.80	52.10	40.45
18	21.85	30.75	46.35	38.65	38.75	31.35	44.80	38.50	54.95	43.20	54.15	42.10
19	22.30	31.50	48.05	39.70	40.00	32.30	46.40	39.80	56.85	44.60	56.20	43.75
20[4]	22.75	32.25	49.75	40.75	41.25	33.25	48.00	41.10	58.75	46.00	58.25	45.40

(1) Rate Groups: Canada-1, Mexico-2, Great Britain, Northern Ireland-3, Japan-4, China-5; for other countries, see "Parcel Post rate group" in Country Rate Groups table page 724. (2) Canada: minimum 1 lb; maximum 66 lbs, 22 lbs limit to members of the Canadian Armed Forces based outside Canada (CFPOs). (3) 5-lb rate is the minimum, even if parcel weighs less. (4) Consult postmaster for pieces greater than 20 lbs.

Country Rate Groups

(For further information, consult your local post office.)

Country or territory	Letter-post	Parcel Post: Airmail	Parcel Post: Surface	Parcel Post: Max. Wt. (lbs)
Afghanistan[1]	5	7	7	44
Albania	5	7	7	44
Algeria	5	10	11	44
Andorra	3	7	6	44
Angola	5	10	11	22
Anguilla	5	12	12	22
Antigua & Barbuda	5	12	12	22
Argentina	5	13	12	44
Armenia	5	7	7	44
Aruba	5	12	12	44
Ascension	5	—	11	44s
Australia	4	9	8	44
Austria	5	7	6	70
Azerbaijan	5	7	7	70
Azores	5	7	7	66
Bahamas	5	12	12	44
Bahrain	5	10	10	44
Bangladesh	5	8	8	66
Barbados	5	12	12	44
Belarus	5	6	7	70
Belgium	3	6	6	70
Belize	5	12	12	44
Benin	5	10	10	66
Bermuda	5	13	12	44
Bhutan	5	9	9	22
Bolivia	5	13	12	70
Bosnia & Herzegovina	5	6	6	44
Botswana	5	11	11	70
Brazil	5	13	12	66
British Virgin Isl.	5	12	12	44
Brunei	5	8	8	44
Bulgaria	5	6	7	70
Burkina Faso	5	10	11	66
Burma	see Myanmar			
Burundi	5	11	11	66
Cambodia	5	8	—	66a
Cameroon	5	11	11	66
Cape Verde	5	10	11	66
Cayman Islands	5	12	12	44
Central African Republic	5	11	11	44
Chad	5	10	—	44a
Chile	5	13	12	44
China	5	5	5	70
Colombia	5	12	12	44
Comoros	5	10	10	44
Congo, Dem. Rep. of the	5	11	11	66
Congo, Rep. of the	5	10	10	44
Costa Rica	5	12	12	66
Côte d'Ivoire	5	11	11	70
Croatia	5	6	6	70
Cuba[1]	5	no parcel post		
Cyprus	5	6	6	70
Czech Republic	5	6	7	66
Denmark	3	6	6	70
Djibouti	5	10	10	44
Dominica	5	12	12	44
Dominican Republic	5	12	12	44
East Timor	5	8	8	44
Ecuador	5	13	12	70
Egypt	5	11	11	66
El Salvador	5	12	12	44
Equatorial Guinea	5	10	10	**
Eritrea	5	11	11	66
Estonia	5	7	7	66
Ethiopia	5	10	10	66
Falkland Islands	5	—	12	66s
Faroe Islands	3	6	6	70
Fiji	5	8	8	44
Finland	3	6	6	70
France[2]	3	6	6	66
French Guiana	5	13	12	66
French Polynesia	5	9	9	66
Gabon	5	10	11	44
Gambia, The	5	11	11	22
Georgia, Republic of	5	7	7	44
Germany	3	6	6	70
Ghana	5	11	11	70
Gibraltar	3	6	6	44
Great Britain & N. Ireland	3	3	3	66
Greece	3	6	6	44
Greenland	3	6	6	66
Grenada	5	12	12	44
Guadeloupe	5	13	12	66
Guatemala	5	12	12	44
Guinea	5	10	10	70
Guinea-Bissau[1]	5	11	11	22
Guyana	5	12	12	44
Haiti	5	12	12	55
Honduras	5	13	12	44
Hong Kong, China	5	9	8	44
Hungary	5	6	6	66
Iceland	3	6	6	70
India	5	9	8	44
Indonesia[3]	5	8	8	44
Iran	5	11	11	44
Iraq[1]	5	11	11	44
Ireland	3	6	6	66
Israel[4]	3	10	10	44
Italy	3	6	6	44
Ivory Coast	see Côte d'Ivoire			
Jamaica	5	12	12	22
Japan	4	4	4	44
Jordan	5	10	10	70
Kazakhstan	5	6	7	44
Kenya	5	10	10	70
Kiribati	5	8	8	44
Korea, Dem. People's Rep. of (North)[1]	no parcel post			
Korea, Republic of (South)	5	9	8	44
Kuwait	5	10	10	66
Kyrgyzstan	5	6	7	70
Laos	5	9	9	44
Latvia	5	6	6	70
Lebanon[1]	5	10	—	22a
Lesotho	5	11	11	44
Liberia[1]	5	10	10	44
Libya	5	7	7	44
Liechtenstein	3	6	6	66
Lithuania	5	6	7	70
Luxembourg	3	6	6	70
Macau	5	9	9	44
Macedonia	5	6	7	44
Madagascar	5	11	11	66
Madeira Islands	3	7	7	66
Malawi	5	11	11	44
Malaysia	5	8	8	44
Maldives	5	9	9	70
Mali	5	10	11	44
Malta	5	7	7	44
Martinique	5	13	12	66
Mauritania	5	10	11	44
Mauritius	5	10	10	44
Moldova	5	7	7	66
Mongolia	5	9	9	**
Montserrat	5	8	8	44
Morocco	5	10	11	70
Mozambique	5	11	11	44
Myanmar	5	6	6	44
Namibia	5	11	11	44
Nauru	5	8	8	44
Nepal	5	9	9	44
Netherlands	3	6	6	44
Netherlands Antilles	5	12	12	44
New Caledonia	5	9	9	66
New Zealand	4	8	8	66
Nicaragua	5	12	12	44
Niger	5	10	10	66
Nigeria	5	10	10	66
Norway	3	6	6	55
Oman	5	10	10	44
Pakistan	5	9	8	66
Panama	5	12	12	70
Papua New Guinea	5	9	9	44
Paraguay	5	13	12	70
Peru	5	13	12	70
Philippines	5	9	8	44
Pitcairn Island	5	8	8	22
Poland	5	6	6	44
Portugal	3	7	7	66
Qatar	5	10	10	70
Reunion	5	13	12	66
Romania	5	7	7	70
Russia	5	7	7	44
Rwanda	5	10	11	66
Saint Helena	5	11	11	44
Saint Kitts & Nevis	5	12	12	44
Saint Lucia	5	12	12	44
Saint Pierre & Miquelon	5	6	6	66
Saint Vincent & Grenadines	5	13	12	22
San Marino	3	9	8	44
São Tomé & Príncipe	5	10	10	44
Saudi Arabia	5	10	10	44
Senegal	5	10	10	44
Serbia-Montenegro[1]	5	7	7	33
Seychelles	5	10	11	70
Sierra Leone	5	10	10	44
Singapore	5	8	8	66
Slovakia	5	6	6	66
Slovenia	5	6	7	33
Solomon Islands	5	8	8	44
Somalia[1]	5	10	10	44
South Africa	5	11	10	66
Spain	3	7	6	44
Sri Lanka	5	9	8	66
Sudan	5	11	11	44
Suriname	5	12	12	44
Swaziland	5	10	10	44
Sweden	3	7	7	44
Switzerland	3	6	6	66
Syria	5	10	10	66
Taiwan	5	9	8	44
Tajikistan	5	6	6	66
Tanzania	5	10	10	44
Thailand	5	8	8	44
Togo	5	10	10	44
Tonga	5	8	8	44
Trinidad & Tobago	5	12	12	22
Tristan da Cuñha	5	10	11	22
Tunisia	5	10	10	44
Turkey	5	10	10	70
Turkmenistan	5	12	12	22
Turks & Caicos Isl.	5	12	12	22
Tuvalu	5	8	8	55
Uganda	5	10	11	44
Ukraine	5	7	7	22
United Arab Emirates	5	10	10	70
United Kingdom	5	7	7	66
Uruguay	5	13	12	44
Uzbekistan	5	7	7	44
Vanuatu	5	8	8	44
Vatican City	3	6	6	44
Venezuela	5	12	12	44
Vietnam	5	9	8	44
Wallis & Futuna Isl.	5	9	9	66
Western Samoa	5	8	8	44
Yemen	5	10	11	44
Yugoslavia[1]	5	7	7	33
Zambia	5	10	11	66
Zimbabwe	5	11	11	44

(a) Air only. (s) Surface only. (1) Mailing restrictions currently apply. Consult local post office for details. (2) Includes Monaco and Corsica. (3) Includes East Timor. (4) West Bank and Gaza Strip are same rate group as Israel. **Surface, 44 lbs; air, 22 lbs.

ASSOCIATIONS AND SOCIETIES

Source: World Almanac questionnaire; World Almanac research

Selected list, by first distinctive key word in each title. (Listed by acronym when that is the official name.) Founding year in parentheses; last figure after ZIP code = membership as reported. Information, especially website addresses, subject to change. For other organizations, see Directory of Sports Organizations; Where to Get Help directory in Health chapter; Labor Union Directory in Employment chapter; Membership of Religious Groups in the U.S.; Major International Organizations in Nations chapter.

Note: Information does not include changes resulting from the Sept. 11, 2001, attack that destroyed the World Trade Center and damaged other buildings in New York City.

AACSB-The Intl. Assoc. for Management Education (1916), 600 Emerson Rd., Ste. 300, St. Louis, MO 63141; 850 institutions; http://www.aacsb.edu

Abortion Federation, National (1977), 1755 Massachusetts Ave. NW, Wash., DC 20036; 400 institutions; http://www.prochoice.org

Academic Assistance Program, Intl. (1994), 5904 Snaffle Bit Place, Ste. 17-057, Bonita, CA 91902; http://www.iaap.org.mx

Academies, Natl. (1863), 2101 Constitution Ave. NW, Wash., DC 20418; 2,276; http://www.nationalacademies.org

Accountants, American Institute of Certified Public (1887), 1211 Ave. of the Americas, New York, NY 10036; 330,000+; http://www.aicpa.org

Acoustical Society of America (1929), Ste. 1NO1, 2 Huntington Quad., Melville, NY 11747; 7,000; http://asa.aip.org

Actuaries, Society of (1949), 475 N. Martingale Rd., Ste. 800, Schaumburg, IL 60173; 16,500; http://www.soa.org

Administrative Professionals, Intl. Assn. of (1942), 10502 NW Ambassador Dr., Kansas City, MO 64195-0404; 40,000; http://www.iaap-hq.org

Advancement and Support of Education, Council for (1974); 1307 New York Ave. NW, Ste. 1000, Wash., DC 20005; 3000 member schools; http://www.case.org

Advertisers, Assn. of Natl. (1910), 708 Third Ave., New York, NY 10017; 312 cos.; http://www.ana.net

Aeronautic Assn., Natl. (1905), 1815 N. Fort Myer Dr., Ste. 500, Arlington, VA 22209; 6,000; http://www.naa-usa.org

Aerospace Industries Assn. of America Inc. (1919), 1250 Eye St. NW, Wash., DC 20005-3924; 63 cos.; http://www.aia-aerospace.org

Aerospace Medical Assn. (1929), 320 S. Henry St., Alexandria, VA 22314; 3,400; http://www.asma.org

AFCEA (Armed Forces Communications and Electronics Assn.) (1946), 4400 Fair Lakes Ct., Fairfax, VA 22033; 27,000 indiv., 12,000 corp.; http://www.afcea.org

African-American Life and History, Assn. for the Study of (1915), 7961 Eastern Ave., Ste. 301, Silver Spring, MD 20910; 1,400; http://www.artnoir.com/asalh

African Violet Soc. of America Inc. (1946), 2375 North St., Beaumont, TX 77702; 9,000; http://www.avsa.org

AFS International Programs USA (1947), 198 Madison Ave., 8th Fl., New York, NY 10016; http://www.afs.org/usa

Agricultural Economics Assn., American (1914), 415 S. Duff Ave., Ste. C, Ames, IA 50010; 3,300; http://www.aaea.org

Agricultural Engineers, American Soc. of (ASAE) (1907), 2950 Niles Road, St. Joseph, MI 49085; 9,000; http://www.asae.org

Agronomy, American Society of (1907), 677 S. Segoe Rd., Madison, WI 53711; 11,500; http://www.agronomy.org

Air & Waste Management Assn. (1907), One Gateway Center, 3d Fl., Pittsburgh, PA 15222; 12,000.

Aircraft Owners and Pilots Assn. (1939), 421 Aviation Way, Frederick, MD 21701; 360,000+; http://www.aopa.org

Air Force Assn. (1946), 1501 Lee Hwy., Arlington, VA 22209; 150,000; http://www.afa.org

Al-Anon Family Group Headquarters, Inc. (1951), 1600 Corporate Landing Pkwy., Virginia Beach, VA 23454; 350,000+ worldwide; http://www.al-anon.alateen.org

Alcoholics Anonymous (1935), 475 Riverside Dr., New York, NY 10115; 2,160,013; http://www.aa.org

Alcoholism and Drug Dependence, Inc., Natl. Council on (1944), 20 Exchange Pl., Suite 2902, New York, NY 10005; 100 affil.; http://www.ncadd.org

Alexander Graham Bell Assn. for the Deaf (1890), 3417 Volta Pl. NW, Wash., DC 20007; 5,000; http://www.agbell.org

Allergy, Asthma, and Immunology, American Academy of (1943), 611 E. Wells St., Milwaukee, WI 53202; 6,000+; http://www.aaaai.org

Alpha Delta Kappa Sorority Inc. (1947), 1615 West 92d St., Kansas City, MO 64114; 51,000; http://www.alphadeltakappa.org

Alpha Lambda Delta, Natl. (1924), P.O. Box 4403, Macon, GA 31208-4403; 600,000; http://www.mercer.edu/ald

Alpine Club, American (1902), 710 Tenth St., Ste. 100, Golden, CO 80401; 6,000+; http://www.americanalpineclub.org

Alzheimer's Assn. (1980), 919 N. Michigan Ave., Ste. 1100, Chicago, IL 60611; http://www.alz.org

Amateur Chamber Music Players, Inc. (1969), 1123 Broadway, Rm. 304, New York, NY 10010-2007; 5,100; http://www.acmp.net

Amateur Radio Union, Intl. (IARU) (1925), P.O. Box 310905, Newington, CT 06131; 150 org.; http://www.iaru.org

Amateur Racquetball Assn., U.S. (1969), 1685 W. Vintah, Colorado Springs, CO 80904-2906; 40,000; http://www.usra.org

Amateur Speedskating Union of the U.S. (1927), O S 651 Forest, Winfield, IL 60190; 2,000; http://www.speedskating.org

AMBUCS, Inc., Natl. (1922), P.O. Box 5127, High Point, NC 27262; 5,400; http://www.ambucs.com

American Indians, Natl. Congress of (1944), 1301 Connecticut Ave. NW, Ste. 200, NW, Wash., DC 20036; 250+ member tribes; http://www.ncai.org

American-Islamic Relations, Council on, 453 New Jersey Ave. SE, Wash., DC 20003; http://www.cair-net.org

American Legion (1919), P.O. Box 1055, 700 N. Pennsylvania St., Indianapolis, IN 46206; 3 mil.+; http://www.legion.org

American Legion Auxiliary (1919), 777 N. Meridian St., 3d Fl., Indianapolis, IN 46204; 900,000+; http://www.legion-aux.org

Americares Foundation (1982), 161 Cherry St., New Canaan, CT 06840; http://www.americares.org

AMIDEAST (formerly American Mideast Educational & Training Services) (1951), 1730 M St. NW, Ste. 1100, Wash., DC 20036; http://www.amideast.org

Amnesty Intl. USA (1961), 322 8th Ave., New York, NY 10001; 1,000,000+; http://rights.amnesty.org

Amputation Foundation, Inc., Natl. (1919), 38-40 Church St., Malverne, NY 11565; 2,500; http://www.nationalamputation.org

AMVETS (American Veterans) (1943); **AMVETS Natl. Auxiliary** (1946), 4647 Forbes Blvd., Lanham, MD 20706; 250,000; http://www.amvets.org

Amusement Parks and Attractions, Intl. Assn. of (IAAPA) (1918), 1448 Duke St., Alexandria, VA 22314; 5,000; http://www.iaapa.org

Animals, American Society for Prevention of Cruelty to (ASPCA) (1866), 424 E. 92d St., New York, NY 10128; 475,000+; http://www.aspca.org

Animal Protection Institute (1968), 1122 S St., Sacramento, CA 95820; 85,000; http://www.api4animals.org

Animal Welfare Institute (1951), P.O. Box 3650, Wash., DC 20007; approx. 25,000; http://www.awionline.org

Anthropological Assn., American (1902), 4350 N. Fairfax Dr., Ste. 640, Arlington, VA 22203; 11,000; http://www.aaanet.org

Anti-Vivisection Society, American (AAVS), (1883), 801 Old York Road, #204, Jenkintown, PA 19046; 10,000; http://www.aavs.org

Antiquarian Society, American (1812), 185 Salisbury St., Worcester, MA 01609; 675; http://www.americanantiquarian.org

APICS (1957), 5301 Shawnee Rd., Alexandria, VA 22312- 2317; 70,000; http://www.apics.org

Appalachian Mountain Club (1876), 5 Joy St., Boston, MA 02108; 90,000+; http://www.outdoors.org

Appalachian Trail Conference (1925), 799 Washington St., P.O. Box 807, Harpers Ferry, WV 25425; 23,000; http://www.atconf.org

Appraisers, American Society of (1936), 555 Herndon Parkway, Ste. 125, Herndon, VA 20170; 6,500; http://www.appraisers.org

Arbitration Assn., American (1926), 335 Madison Ave., Fl. 10, New York, NY 10017; 7,000; http://www.adr.org

Arc of the United States, The (1950), 1010 Wayne Avenue, Suite 650, Silver Spring, MD 20910; 140,000+; http://www.thearc.org

Archaeological Institute of America (1879), 656 Beacon St., 4th Fl., Boston, MA 02215; 11,000+; http://www.archaeological.org

Archery Assn. of the United States, Natl. (1879), One Olympic Plaza, Colorado Springs, CO 80909; 6,000; http://www.USArchery.org

Architects, American Institute of (1857), 1735 New York Ave. NW, Wash., DC 20006; 63,000; http://www.aiaonline.com

Architectural Historians, Society of (1940), 1365 N. Astor St., Chicago, IL 60610; 3500; http://www.sah.org

ARMA Intl. (formerly Assn. of Records Managers & Administrators) (1955), 4200 Somerset Dr., Ste. 215, Prairie Village, KS 66208; 10,000; http://www.arma.org

Army, Assn. of the United States (1950), 2425 Wilson Blvd., Arlington, VA 22201; 117,000; http://www.ausa.org

Arthritis Foundation (1948), 1330 W. Peachtree St., Atlanta, GA 30309; http://www.arthritis.org

Arts, American Federation of (1909), 41 E. 65th St., New York, NY 10021; 520+ museums/institutions; http://www.afaweb.org

Arts, Americans for the (1996), 1000 Vermont Ave. NW, Ste. 1200, Wash., D.C. 20005; http://www.artsusa.org

Arts and Letters, American Academy and Institute of (1898), 633 W. 155 St., New York, NY 10032; 250; http://www.nyc-arts.org/nyc-arts/name/name-by-borough/manhattan/upper/artsltrs.html

Arts and Letters, Natl. Society of (1944), 4227 46th St. NW, Wash., DC 20016; 1,450+; http://www.arts-nsal.org

Arts and Sciences, American Academy of (1780), Norton's Woods, 136 Irving St., Cambridge, MA 02138; 4,300 fellows; http://www.amacad.org

ASPRS, The Imaging and Geospatial Information Society (1934), 5410 Grosvenor Ln., Ste. 210, Bethesda, MD 20814; 7,000; http://www.asprs.org

Associated Press (1848), 50 Rockefeller Plaza, New York, NY 10020; 1,700 newspapers, 5,000 U.S. broadcast stations, 8,500 intl. subscribers; http://www.ap.org

Association Executives, American Society of (1920), 1575 I St. NW, Wash., DC 20005; 25,000; http://www.asaenet.org

Association Managers Inc., Intl. (1974), 1224 N. Nokomis NE, Alexandria, MN 56308; 20,000; http://www.iami.org

Astrologers, Inc., American Federation of (AFA, Inc.) (1938), St. 6535 S. Rural Rd., Tempe, AZ 85283; 5,000+; http://www.astrologers.com

Astronautical Society, American (1954), 6352 Rolling Mill Place, Springfield,VA 22152-2354; 1,500; http://www.astronautical.org

Astronomical Society, American (1899), 2000 Florida Ave. NW, #400, Wash., DC 20009; 6,500; http://www.aas.org

Ataxia Foundation, Natl. (1957), 2600 Fernbrook Ln., Ste. 119, Minneapolis, MN 55447-4752; 9,000; http://www.ataxia.org

Atheists, American (1967), P.O. Box 5733, Parsippany, NJ 07054; 1,500; http://www.atheists.org

Auctioneers Assn., Natl. (1948), 8880 Ballentine St., Overland Park, KS 66214; approx. 6,000; http://www.auctioneers.org

Audubon Soc., Natl. (1905), 700 Broadway, New York, NY 10003; 600,000; http://www.audubon.org

Authors Guild, The (1912), 31 E. 28th St., New York, NY 10016; 8,200; http://www.authorsguild.org

Authors Registry, The (1995), 31 E. 28th St., New York, NY 10016; 30,000; http://www.authorsregistry.org

Autism Soc. of America (1965), 7910 Woodmont Ave., Ste. 300, Bethesda, MD 20814; 22,000; http://www.autism-society.org

Autograph Collectors Club, Universal (1965), P.O. Box 6181, Wash., DC 20044-6181; 1,535; http://www.uacc.org

Automobile Assn., American (AAA) (1902), 1000 AAA Dr., Heathrow, FL 32746; 44 mil.; http://www.aaa.com

Automobile Club of America, Antique (1935), 501 W. Governor Road, P.O. Box 417, Hershey, PA 17033; 60,000; http://www.aaca.org

Automobile Dealers Assn., Natl. (1917), 8400 Westpark Dr., McLean, VA 22102; 19,600; http://www.nada.org

Automobile License Plate Collectors Assn. (1954), 7365 Main. St., #214, Stratford, CT 06614; 3,000; http://www.alpca.org

Automotive Hall of Fame (1939), 21400 Oakwood Blvd., Dearborn, MI, 48124; http://www.automotivehalloffame.org

Badminton, USA (1938), One Olympic Plaza, Colorado Springs, CO 80909; 4,000; http://www.usabadminton.org

Bald-Headed Men of America (1973), 102 Bald Dr., Morehead City, NC 28557; approx. 30,000.

Bankers Assn., American (1875), 1120 Connecticut Ave. NW, Wash., DC 20036; http://www.aba.com

Bar Assn., American (1878), 541 N. Fairbanks Ct., Chicago, IL 60611; 400,000+; http://www.abanet.org

Bar Assn., Federal (1920), 2215 M Street NW, Wash., DC 20037; 15,002; http://www.fedbar.org

Barber Shop Quartet Singing in America, Inc., Soc. for the Preservation & Encouragement of (1938), 6315 Harmony Lane, Kenosha, WI 53143; 33,000+; http://www.spebsqsa.org

Baseball Congress, American Amateur (1935), 118-119 Redfield Plaza, P.O. Box 467, Marshall, MI 49068; 14,500 teams; http://www.voyager.net/aabc

Baseball Congress, Natl. (1931), 300 S. Sycamore, P.O. Box 1420, Wichita, KS 67201; 7,500; http://www.nationalbaseballcongress.com

Baseball Players of America, Assn. of Prof. (1924), 1820 West Orangewood Ave., Ste. 206, Orange, CA 92868; 16,000+.

Baseball Research, Inc., Society for American (1971), 812 Huron Road E #719, Cleveland, OH 44115; 7,000+; http://www.sabr.org

Battleship Assn., American (1964), P.O. Box 711247, San Diego, CA 92171; 1,200.

Beer Can Collectors of America (1970), 747 Merus Ct., Fenton, MO 63026; 4,200; http://www.bcca.com

Beta Gamma Sigma, Inc. (1913), 11701 Borman Dr., Ste. 295, St. Louis, MO 63146-4199; 430,000; http://www.betagammasigma.org

Beta Sigma Phi (1931), 1800 W. 91st Pl., Kansas City, MO 64114; 200,000; http://www.betasigmaphi.org

Better Business Bureaus, Council of (1970), 4200 Wilson Blvd., Suite 800, Arlington, VA 22203; 150 bureaus; http://www.bbb.org

Bible Society, American (1816), 1865 Broadway, New York, NY 10023; 650,000; http://www.americanbible.org

Biblical Literature, Society of (1880), 825 Houston Mill Rd., Ste. 350, Atlanta, GA 30329; 8,000; http://www.sbl-site.org

Bibliographical Society of America (1904), P.O. Box 1537, Lenox Hill Station, New York, NY 10021; 1,200; http://www.bibsocamer.org

Big Brothers/Big Sisters of America (1904), 230 N. 13th St., Philadelphia, PA 19107; 494 agencies; http://bbbsa.org

Biochemistry and Molecular Biology, American Society for (1906), 9650 Rockville Pike, Bethesda, MD 20814; 10,240; http://www.faseb.org/asbmb

Biological Sciences, American Institute of (1947), 1444 I St. NW, Ste. 200, Wash., DC 20005; 6,000; http://www.aibs.org

Blind, American Council of the (1961), 1155 15th St. NW, Ste. 1004, Wash., DC 20005; 25,000; http://www.acb.org

Blind, Natl. Federation of the (1940), 1800 Johnson St., Baltimore, MD 21230; 50,000; http://www.nfb.org

Blinded Veterans Assn. (1945), 477 H St. NW, Wash., DC 20001; 10,000+.; http://www.bva.org

Blindness America, Prevent (1908), 500 E. Remington Rd., Schaumburg, IL 60173; http://www.preventblindness.org

Blueberry Council, North American (1965) P.O. Box 1736, Folsom, CA 95763; http://www.blueberry.org

B'nai B'rith Intl. (1843), 1640 Rhode Island Ave. NW, Wash., DC 20036; 250,000; http://www.bbinet.org

Boat Owners Assn. of the U.S. (1966), 880 S. Pickett St., Alexandria, VA 22304; 500,000+; http://www.boatus.com

Bookplate Collectors and Designers, American Soc. of (1922), P.O. Box 380340, Cambridge, MA 02238-0340; http://bookplate.org

Booksellers Assn., American (1900), 828 S. Broadway, Tarrytown, NY 10591; 8,000; http://www.bookweb.org/aba

Boy Scouts of America (1910), 1325 Walnut Hill Lane, Irving, TX 75015; 4.7 mil; http://www.bsa.scouting.org

Boys & Girls Clubs of America (1906), 1230 W. Peachtree St. NW, Atlanta, GA 30309; 3.3 mil; http://www.bgca.org

Bread for the World (1974), 50 F St. NW, Ste. 500, Washington, DC 2001; 44,000; http://www.bread.org

Brewing Chemists, American Society for (1934), 3340 Pilot Knob Road, St. Paul, MN 55121-2097; approx. 1,000; http://www.asbcnet.org

Broadcasters, Natl. Assn. of (1923), 1771 N St. NW, Wash., DC 20036; http://www.nab.org

Burroughs Bibliophiles, The (1960), 454 Elaine Dr., Pittsburgh, PA 15236-2417; 887.

Business Communicators, Intl. Assn. of (1970), 5 3rd St. #724, San Francisco, CA 94103; 13,700; http://www.iabc.com/homepage.htm

Business Education Assn., Natl. (1946), 1914 Association Drive, Reston, VA 20191; 12,000; http://www.nbea.org

Business Women's Assn., American (1949), 9100 Ward Pkwy., P.O. Box 8728, Kansas City, MO 64114; 70,000; http://www.abwahq.org

Button Society, Natl. (1938), c/o Lois Pool, 2733 Juno Pl., Akron, OH 44333-4137; 4,500.

Camp Fire USA (formerly Camp Fire Boys & Girls) (1910), 4601 Madison Ave., Kansas City, MO 64112; 650,000; http://www.campfireusa.org

Camping Assn., American (1951), 5000 State Rd. 67 N., Martinsville, IN 46151; 6,000+; http://www.acacamps.org

Cancer Society, American (1913), 2200 Century Pkwy., Ste. 950, Atlanta, GA 30345; 2 mil.; http://www.cancer.org

Cartoonists Society, Natl. (1946), PO Box 713, Suffield, CT 06078; 500+; http://www.reuben.org

Cat Fanciers' Assn., The (1906), 1805 Atlantic Ave., P.O. Box 1005, Manasquan, NJ 08736-0805; http://www.cfainc.org

Catholic Bishops, Natl. Conference of (1634), 3211 4th St. NE, Wash., DC 20017; 402 members, 350 staff; http://www.nccbuscc.org

Catholic Church Extension Society of the USA (1905), 150 S. Wacker Dr., Chicago, IL 60606; 54 staff; http://www.catholic-extension.org

Catholic Daughters of the Americas (1903), 10 West 71st Street, New York, NY 10023; 110,806; http://www.catholicdaughters.org

Catholic Educational Assn., Natl. (1904), 1077 30th St. NW, Ste. 100, Wash., DC 20007; 26,000; http://www.ncea.org

Catholic Historical Soc., American (1886), 263 S. Fourth St., Philadelphia, PA 19106-3819; 500; http://www.AMCHS.org

Catholic Library Association (1921), 100 North St., Ste. 224, Pittsfield, MA 01201-5109; 1,100; http://www.cathla.org

Catholic War Veterans, USA Inc. (1935), 441 N. Lee St., Alexandria, VA 22314-2301; 25,000; http://cwv.org

Cemetery and Funeral Assn., Intl. (1887), 1895 Preston White Dr., #220, Reston, VA 22091; 6,000; http://www.icfa.org

Ceramic Society, The American (1898), 735 Ceramic Pl., Westerville, OH 43081; 10,200; http://www.acers.org

Cereal Chemists, American Society of (1915), 3340 Pilot Knob Road, St. Paul, MN 55121-2097; 3,500; http://www.scisoc.org/aacc

Cerebral Palsy Assns., Inc., United (1949), 1660 L St. NW, Ste. 700, Wash., DC 20036; 150; http://www.ucpa.org

Certification of Computing Professionals, Institute for (1973), 2350 E. Devon Ave., Ste. 115, Des Plaines, IL 60018-4610; http://www.Iccp.org

Chamber of Commerce of the U.S.A. (1912), 1615 H St. NW, Wash., DC 20062; 215,000; http://uschamber.com

Chamber Music Players, Inc., Amateur (1947), 1123 Broadway, New York, NY 10010; 4,200; http://www.acmp.net
Checker Federation, American (1949), P.O.Box 241, Petal, MS 39465; 1,000.
Chemical Engineers, American Inst. of (1908), 3 Park Ave., New York, NY 10016; 50,000+; http://www.aiche.org
Chemical Manufacturers Assn. (1872), 1300 Wilson Blvd., Arlington, VA 22209; 191 cos.; http://www.cmahq.com
Chemical Society, American (1876), 1155 16th St. NW, Wash., DC 20036; 163,000; http://www.acs.org
Chess Federation, U.S. (1939), 3054 NYS Rt. 9W, New Windsor, NY 12553; 88,000+; http://www.uschess.org
Chess League of America, Correspondence (1897), P.O. Box 59625, Schaumburg, IL 60159; 1,000; http://www.chessbymail.com
Chiefs of Police, Intl. Assn. of (1893), 515 N. Washington St., Alexandria, VA 22314; 19,000; http://www.theiacp.org
Childhood Education Intl., Assn. for (1892), 17904 Georgia Ave., Ste. 215, Olney, MD 20832; 12,000; http://www.udel.edu/bateman/acei
Children's Aid Society (1912), 181 West Valley Ave., Ste. 300, Homewood, AL 35209; http://www.childrensaid.org
Children's Book Council, The (1945), 12 W. 37th St., 2nd Floor, New York, NY 10018; 85 publishers; http://www.cbcbooks.org.
Child Welfare League of America (1920), 440 First St. NW, Wash., DC 20001; 1,100 agencies; http://www.cwla.org
Chiropractic Assn., American (1963), 1701 Clarendon Blvd., Arlington, VA 22209; 19,000; http://www.amerchiro.org
Chris-Craft Antique Boat Club (1973), 217 S. Adams St., Tallahassee, FL 32301-1708; 2,800; http://www.chris-craft.org
Christian Children's Fund (1938), 2821 Emerywood Pkwy., Richmond, VA 23294-3725;160; http://www.christianchildrensfund.org
Christian Endeavor Union, World's (1895), 3575 Valley Road, P.O. Box 326, Liberty Corner, NJ 07938-0326; http://www.christianendeavorworldwide.org
Cities, Natl. League of (1924), 1301 Pennsylvania Ave. NW, Ste. 550, Washington, DC 20004; 1,780; http://www.nlc.org
Civil Air Patrol (1941), 105 S. Hansell St., Maxwell AFB, AL 36112; 60,000; http://www.capnhq.gov
Civil Engineers, American Society of (1852), 1801 Alexander Bell Dr., Reston, VA 20191; 123,000+; http://www.asce.org
Civil Liberties Union, American (ACLU) (1920), 125 Broad St., 18 Fl., New York, NY 10004; 275,000; http://www.aclu.org
Civitan International, Inc. (1917), P.O. Box 130744, Birmingham, AL 35213-0744; 30,000; http://www.civitan.org
Clean Energy Research Inst. (1974), Univ. of Miami, Coral Gables, FL, 33124; 500.
Clinical Pathologists, American Society of (1922), 2100 W. Harrison St., Chicago, IL 60612; 79,000; http://www.ascp.org
Coaster Enthusiasts, American (1978), 5800 Foxridge Dr., Ste. 115, Mission, KS 26202-2333; 8,420; http://www.aceonline.org
Coast Guard Combat Veterans Assn. (1985), 295 Shalimar Dr., Shalimar, FL 32579; 1,800; http://www.aug.edu/~libwrw/cgcva/cgcva.htm
Co-dependents Anonymous (1986), PO Box 33577; Phoenix, AZ 85067; http://www.codependents.org
College Admission Counseling, Natl. Assn. for (1937), 1631 Prince Street, Alexandria, VA 22314; 7,100; http://www.nacac.com
College Board, The (1900), 45 Columbus Ave., New York, NY 10023; 2,900 institutions; http://www.collegeboard.org
College English Assn. (1939), English Dept., Winthrop Univ., Rock Hill, SC 29733; 1,000; http://www.winthrop.edu/cea
College Music Society, The (1958), 202 W. Spruce St., Missoula, MT 59802; 8,000; http://www.music.org
Colleges and Employers, Natl. Assn. of (1956), 62 Highland Ave., Bethlehem, PA 18017; 3,700; http://www.jobweb.org
Colleges and Universities, Assn. of American (1915), 1818 R St. NW, Wash., DC 20009; 700+ institutions; http://www.aacu-edu.org
Collegiate Schools of Business, Assn. to Advance (1916), 600 Emerson Rd., Ste. 300, St. Louis, MO 63141; approx. 900 org; http://www.aacsb.edu
Colonial Dames XVII Century, Natl. Soc. (1915), 1300 New Hampshire Ave. NW, Wash., DC 20036; 13,240; http://www.execpc.com/~sril/ilcd17.html
Commercial Collectors, Inc., Int'l. Assn. of (1970), 4040 W. 70th Street, Minneapolis, MN 55435; 370; http://www.commercialcollector.com
Commercial Law League of America (1895), 150 N. Michigan Avenue, # 600, Chicago, IL 60601; approx. 5,000; http://www.clla.org
Common Cause (1970), 1250 Connecticut Ave. NW, Ste. 600, Wash., DC 20036; 215,000; http://www.commoncause.org
Communication Assn., Natl. (1914), 1765 N St. NW, Wash., DC, 20036; 7,000; http://www.natcom.org
Community and Justice, National Conference for (1927), 475 Park Ave. S.; New York, NY 10016; 62 chapters; http://www.nccj.org
Community Colleges, American Assn. of (1920), One Dupont Circle NW, Ste. 410, Wash., DC 20036; 1,113 inst; http://www.aacc.nche.edu
Composers, Authors & Publishers, American Soc. of (ASCAP) (1914), One Lincoln Plaza, New York, NY 10023; 120,000+; http://www.ascap.com
Composers/USA, Natl. Assn. of (1932), Box 49256, Barrington Station, Los Angeles, CA 90049; 600; http://www.music-usa.org/nacusa
Computing Machinery, Assn. for (1947), 1515 Broadway, 17th Fl., New York, NY 10036; 80,000+; http://www.acm.org
Concerned Women for America (1979), 1015 Fifteenth St. NW, Suite 1100, Wash., DC 20005; 500,000; http://www.cwfa.org
Concrete Institute, American (1904), P.O. Box 9094, Farmington Hills, MI 48333; 17,000; http://www.aci-int.org
Congress of Racial Equality (CORE) (1942), 817 Broadway, 3d Floor, New York, NY 10003; 100,000; http://www.core-online.org
Conscientious Objectors, Central Committee for (1948), 630 20th St., #302, Oakland, CA 94612; 4,500; http://www.objector.org
Constantian Society, The (1970), 840 Old Washington Rd., Macmurray, PA 15317-5374; 750; http://members.tripod.com/~constantian/index.html
Construction Industry Manufacturers Assn. (1911), 111 E. Wisconsin Ave., Milwaukee, WI 53202; 550+ cos.; http://www.cimanet.com
Construction Inspectors, Assn. of (1981), 1224 N. Nokomis NE, Alexandria, MN 56308; 5,000; http://www.iami.org
Construction Specifications Institute (1948); 99 Canal Center Plaza, Ste. 300, Alexandria, VA 22301; 18,000; http://www. csi-net.org
Consumer Federation of America (1968), 1424 16th St. NW, Ste. 604, Wash., DC 20036; 250 organizations; http://www.stateandlocal.org
Consumer Information Center, Federal (1970), Pueblo, CO 81009; http://www.pueblo.gsa.gov
Consumer Interests, American Council on (ACCI) (1953), 240 Stanley Hall, Univ. of Missouri, Columbia, MO 65211-0001; 1,100; http://www.consumerinterests.org
Consumers Union of the U.S. (1936), 101 Truman Ave., Yonkers, NY 10703; 405,990; http://www.consumersunion.org
Contract Bridge League, American (1938), 2990 Airways Blvd., Memphis, TN 38116; 170,000; http://www.acbl.org
Co-op America (1982), 1612 K St. NW, Ste. 600, Wash., DC 20006; 50,000 individuals, 2,000 businesses; http://www.coopamerica.org
Correctional Assn., American (1870), 4380 Forbes Blvd., Lanham, MD 20706; 22,000; http://www.aca.org
Cosmetology Assn., Nat. (1921); 30,000; http://www.salon professionals.org
Cotton Council of America, Natl. (1938), 1918 N. Pkwy., Memphis, TN 38112; http://www.cotton.org
Counseling Assn., American (1952), 5999 Stevenson Ave., Alexandria, VA 22304; 51,521; http://www.counseling.org
Country Music Assn. (1958), One Music Circle S, Nashville, TN 37203; 6,700; http://www.CMAworld.com
Crafts & Creative Industries, Assn. of (ACCI) (1976), 1100-H Brandywine Blvd., P.O. Box 3388, Zanesville, OH 43702; 6,327; http://www.creative-industries.com
Credit Union Natl. Assn. & Affiliates (1934), P.O. Box 431, Madison, WI 53701; 51 credit union leagues; http://www.cuna.org
Crime and Delinquency, Natl. Council on (1907), 1970 Broadway, Ste. 500, Oakland, CA 94612; 300+; http://www.nccd-crc.org
Criminology, American Society of (1941), 1314 Kinnear Rd., Ste. 212, Columbus, OH 43212; 2,600; http://www.asc41.com
Crop Protection Assn., American (1933), 1156 15th St. NW, Ste. 400, Wash., DC 20005; 80 cos.; http://www.acpa.org
Crop Science Society of America (1955), 677 S. Segoe Rd., Madison, WI 53711; 4,700; http://www.crops.org
Cryogenic Soc. of America, Inc. (1964), 1033 South Blvd., Ste. 13, Oak Park, IL 60302; 500; http://www.cryogenicsociety.org
Customs Brokers and Forwarders Assn. of America, Inc., Natl. (1897), 1200 18th St. NW, Ste. 901, Wash., DC 20036; 750; http://www.ncbfaa.org
Cystic Fibrosis Foundation (1955), 6931 Arlington Rd., Bethesda, MD 20814; 30,000; http://www.cff.org

Dairy Management Inc. (1995), 10255 W. Higgins Rd., Ste. 900, Rosemont, IL 60018-5616; http://www.dairyinfo.com
Dark-Sky Association, Intl. (1988), 3225 N. First Ave., Tucson, AZ 85719-2103; 4,500; http://www.darksky.org
Daughters of the American Revolution, Natl. Society (1890), 1776 D Street NW, Wash., DC 20006; 172,000; http://www.dar.org
Daughters of the British Empire, Natl. Society (1909), P.O. Box 872, Ambler, PA 19002; 5,000; http://www.mindspring.com/~dbesociety
Daughters of the Confederacy, United (1894), 328 North Blvd., Richmond, VA 23220; 25,000; http;//www.hqudc.org
Deaf, Natl. Assn. of the (1880), 814 Thayer Ave., Ste. 250, Silver Spring, MD 20910; 5,500; http://www.nad.org
Defenders of Wildlife (1947), 1101 14th St. NW, Ste. 1400, Wash., DC 20005; 480,000; http://www.defenders.org

Delta Kappa Gamma Society Intl. (1929), P.O. Box 1589., Austin, TX 78767; 165,000; http://deltakappagamma.org/international/portal.html

Delta Mu Delta Honor Soc. (1913), P.O. Box 46935, St. Louis, MO 63146-6935; 100,000; http://www.deltamudelta.org

Democratic Natl. Committee (1848), 430 S. Capitol Street, SE, Wash., DC 20003; 432 elected members; http://www.democrats.org/index.html

DeMolay International (1919), 10200 N. Ambassador Dr., Kansas City, MO 64153; 30,000; http://www.demolay.org

Dental Assn., American (1859), 211 E. Chicago Ave., Chicago, IL 60611; 141,000; http://www.ada.org

Diabetes Assn., American (1940), 1701 North Beauregard St., Alexandria, VA 22311; 380,000+; http://www.diabetes.org

Dialect Society, American (1889), c/o Allan Metcalf, English Dept., MacMurray College, Jacksonville, IL 62650; 500; http://www.americandialect.org

Digital Printing & Imaging Assn. (1992), 10015 Main St., Fairfax, VA 22031; 900 firms; http://www.dpia.org

Directors Guild of America (1936), 7920 Sunset Blvd., Los Angeles, CA 90046; 12,000+; http://dga.org

Disabled American Veterans (1920), P.O. Box 14301, Cincinnati, OH 45250; 1,050,000; http://www.dav.org

Disabled Sports USA (1967), 451 Hungerford Dr., Ste. 100, Rockville, MD 20850; 60,000+; http://www.dsusa.org/~dsusa/dsusa.html

Dogs on Stamps Study Unit (1979), 202A Newport Rd., Monroe Twp., NJ 08531-3920; 400; http://www.dossu.org

Down Syndrome Society, Natl. (1979), 666 Broadway, New York, NY 10012; 50,000; http://www.ndss.org

Dozenal Society of America (1944), Math Dept., Nassau Community College, Garden City, NY 11530-6793; 144; http://www.polar.sunynassau.edu/~dozenal/

Ducks Unlimited (1937), One Waterfowl Way, Memphis, TN 38120; 620,000; http://www.ducks.org

Eagles, Fraternal Order of (1898), P.O. Box 250972, Milwaukee, WI 53225; 1.1 mil; http://www.foe.org

Easter Seals (1919), 230 W. Monroe St., Ste. 1800, Chicago, IL 60606; http://www.easter-seals.org

Eastern Star, General Grand Chapter, Order of the (1876), 1618 New Hampshire Ave. NW, Wash., DC 20009; 1.5 mil.; http://www.easternstar.org

Edsel Club (1967), 19296 Tuckaway Ct., N. Fort Myers, FL 33903; 300; http://www.edselworld.com

Education, American Council on (1918), One Dupont Circle NW, Wash., DC 20036; 1,700 org; http://www.acenet.edu

Education, Council for Advancement & Support of (1974), 1307 New York Ave. NW, Wash., D.C. 20005; 2,950 schools; http://www.case.org

Education of Young Children, Natl. Assn. for the (1926), 1509 16th St. NW, Wash., DC 20036-1426; 103,000; http://www.naeyc.org

Educators for World Peace, Intl. Assn. of (1969), P.O. Box 3282, Mastin Lake Station, Huntsville, AL 35810-0282; 35,000; http://www.earthportals.com/portal_messenger/mercieca.html

Egalitarian Communities, Federation of (1976), FEC East Wind, MO 65760; 250; http://www.thefec.org

8th Air Force Historical Society (1975), P.O. Box 7215, St. Paul, MN 55107; 18,000; http://www.visi.com/~mbacklund/8thaf.htm

88th Infantry Division Assn. (1947), c/o Frederick L. Lincoln, Membership Chairman, 11 Lovett Ave., Brockton, MA 02301-1750; 4,100.

84th Infantry Div. Railsplitters Soc., The, (1945), P.O. Box 827, Sioux Falls, SD 57101-0827; 2,700.

82d Airborne Division Assn., Inc. (1946), P.O. Box 9308, Fayetteville, NC 28311-9308; 27,000+; http://www.fayettevillenc.com/airborne82dassn

Electrical and Electronics Engineers, Institute of (1963), 445 Hoes Lane, Piscataway, N.J. 08854; 356,000; http://www.ieee.org

Electrical Manufacturers Assn., Natl. (1926), 1300 N. 17th St., Ste. 1847, Rosslyn, VA 22209; 560 cos.; http://www.nema.org

Electrochemical Society, Inc.,The (ECS, Inc.) (1902), 65 South Main St., Bldg. D, Pennington, NJ 08534-2839; 8,000+; http://www.electrochem.org

Electronic Industries Assn. (1924), 2500 Wislon Blvd., Arlington, VA 22201; 1,058 cos.; http://www.eia.org

Electronics Technicians, Intl. Society of Certified (1970), 3608 Pershing Ave., Ft. Worth, TX 76107; 1,800; http://www.iscet.org

Elks of the U.S.A., Benevolent and Protective Order of (1868), 2750 N. Lakeview Ave., Chicago, IL 60614; 1.2 mil.; http://www.elks.org

Energy Engineers, Assn. of (1977), 4025 Pleasantdale Rd., Ste. 420, Atlanta, GA 30340; 8,000; http://www.aeecenter.org

Engineers, Natl. Society of Professional (1934), 1420 King St., Alexandria, VA 22314; 54,000; http://www.nspe.org

English, U.S. (1985), 1747 Pennsylvania Ave. NW, Ste. 1100, Wash., DC 20006; 1.4 mil; http://www.us-english.org

English-Speaking Union of the U.S. (1920), 25 West 45th Street, Ste. 1303; New York, NY 10036; 18,000; http://www.english-speakingunion.org

Entomological Society of America (1889), 9301 Annapolis Rd., Lanham, MD 20706-3115; 6,700; http://www.entsoc.org

Environmental Health Assn., Natl. (1938), 720 S. Colorado Blvd., South Tower #970, Denver, CO 80246-1925; 5,200; http://www.neha.org

Environmental Medicine, American Academy of (1965), 7701 E. Kellog, Ste. 625, Witchita, KS 67207; 380; http://www.aaem.com

Esperanto League for North America Inc. (1953), 5712 Hollis St., Emeryville, CA 94608; 800; http://www.esperanto-usa.org

Evangelism Crusades, Inc., Intl. (1959) 14617 Victory Blvd., Van Nuys, CA 91411

Exchange Club, Natl. (1911), 3050 Central Ave., Toledo, OH 43606; 33,000; http://www.nationalexchangeclub.com

Experimental Aircraft Assn. (1953), 3000 Poberezeny, Rd., Oshkosh, WI 54902; 170,000; http://www.eaa.org

Exploration Geophysicists, Society of (1930), 8801 South Yale, Tulsa, OK 74137; 16,500; http://www.seg.org

Ex-Prisoners of War, American (1942), 3201 E. Pioneer Pkwy., Arlington, TX 76010; 30,000; http://www.axpow.org

Fairs & Expositions, Intl. Assn. of (1919), P.O. Box 985, Springfield, MO 65809; 2,600. http://www.iafenet.org

Family, Career and Community Leaders of America (1945), 1910 Association Dr., Reston, VA 20191; 222,000; http://www.fhahero.org

Family Physicians, American Academy of (1947), 11400 Tomahawk Creek Parkway, Leawood, KS 66211; 93,100; http://www.aafp.org

Family Relations, Natl. Council on (1938), 3989 Central Avenue NE, Suite 550, Minneapolis, MN 55421; 4,000; http://www.ncfr.org

Farm Bureau Federation, American (1919), 225 Touhy Ave., Park Ridge, IL 60068; 5 mil+ families; http://www.fb.com

Farmers of America Org., Natl. Future (1928), P.O. Box 68960, 6060 FFA Drive, Indianapolis, IN 4626; 452,000; http://www.ffa.org

Farmers Union, Natl. (1902), 11900 E. Cornell Ave., Denver, CO 80014; 300,000; http://www.nfu.org

Fat Acceptance, Inc., Natl. Assn. to Advance (NAAFA) (1969), P.O. Box 188620, Sacramento, CA 95818; 5,000; http://www.naafa.org

Fellowship of Reconciliation, The (1914), P.O. Box 271, Nyack, NY 10960; 26,000; http://www.forusa.org/

Feminists for Life of America (1972), 733 15th St. NW, Ste. 1100, Wash., DC 20005; approx. 5,000 http://www.feministsforlife.org

Financial Executives Institute (1938), 10 Madison Ave., Morristown, NJ 5,000; 07962; 15,000; http://www.fei.org

Financial Professionals, Assn. for (formerly Treasury Management Assn.) (1979), 7315 Wisconsin Ave., Ste. 600W, Bethesda, MD 20814; 15,000; http://www.AFPonline.org

Financial Service Professionals, Soc. of (formerly American Society of CLU & ChFC) (1928), 270 S. Bryn Mawr Ave., Bryn Mawr, PA 19010; 32,000; http://www.financialpro.org

Financial Women Intl. (1921 as the National Assoc. of Bank Women), 200 N. Glebe Rd., Ste. 820, Arlington, VA 22203; 3,500; http://www.fwi.org

Financiers, Inc., Intl. Society of (1979), P.O. Box 398, Naples, NC 28760; http://insofin.com

Fire Chiefs, Intl. Assn. of (1873), 4025 Fair Ridge Dr., Ste. 300, Fairfax, VA 22033; 12,000; http://www.iafc.org

Fire Protection Assn., Natl.(NFPA) (1896), 1 Batterymarch Park, Quincy, MA 02269-9101;75,000; http://www.nfpa.org

Fire Protection Engineers, Soc. of (1950), 7315 Wisconsin Avenue, Ste. 1225W, Bethesda, MD 20814; 3,500; http://www.sfpe.org

First Amendment Studies, Inc., Institute for (1984), P.O. Box 589, Great Barrington, MA 01230; 10,000; http://www.ifas.org

Fisheries Soc., American (1870), 5410 Grosvenor Ln., Bethesda, MD 20814; 10,000; http://www.fisheries.org

Food Industry Suppliers (1911), 1451 Dolley Madison Blvd., McLean, VA 22101; 700 cos.; http://www.iafis.org

Food Technologists, Institute of (1939), 221 N. LaSalle, Ste. 300, Chicago, IL 60601; 28,000; http://www.ift.org

Foreign Study, American Institute for, The (1964), River Plaza, 9 W. Broad St., Stamford, CT 06902; 1 mil+; http://www.aifs.com

Foreign Trade Council, Inc., Natl. (1914), 1625 K St. NW, Wash., DC 20006; 570 cos.; http://www.usaengage.org

Forensic Sciences, American Academy of (1948), P.O. Box 669, Colorado Springs, CO 80901; 5,100; http://www.aafs.org

Foresters, Society of American (1900), 5400 Grosvenor La., Bethesda, MD 20814; 17,500; http://www.safnet.org

Forest History Society (1946), 701 Wm. Vickers Ave., Durham, NC 27701-3162; 1200; http://www.lib.duke.edu/forest

Forest & Paper Assn., American (1993), 1111 19th St. NW, Wash., DC 20036; 400 cos.; http://www.afandpa.org

Forests, American (1875), 910 17th St. NW, Ste. 600, Wash., DC 20001; 160,000; http:www.americanforests.org

Fortean Org., Intl. (1965), P.O. Box N, Dept. W, College Park, MD 20740; 1,000; http://www.research.umbc.edu/~frizzell/info

Foundrymen's Society, American (1896), 505 State St., Des Plaines, IL 60016; 13,000; http://www.afsinc.org
4-H Clubs (1914), 1400 Independence Ave., U.S. Dept of Agriculture, Wash., DC 20250; 6.5 mil; http://www.4h-usa.org
Frederick A. Cook Society, (1940), Sullivan County Museum, Hurleyville, NY 12747; 234; http://www.cookpolar.org
Freedom From Religion Foundation (1978), P.O. Box 750, Madison, WI 53701; 4,450; http://www.ffrf.org
Freedoms Foundation at Valley Forge (1949), 1601 Valley Forge Rd., Valley Forge, PA 19482; http://www.ffvf.org
Freedom of Information Center (1958), School of Journalism, Univ. of Missouri, Columbia, MO 65211-0012; http://www.missouri.edu/~foiwww
Freemasonry, Supreme Council Ancient and Accepted Scottish Rite of, Northern Masonic Jurisdiction (1872), P.O. Box 519, Lexington, MA 02420; 300,000; http://supremecouncil.org
Free Men, Natl. Coalition of (1977), P.O. Box 129, Manhasset, NY 11030; 2,000; http://www.ncfm.org
French Institute/Alliance Française (1972), 22 E. 60th St., New York, NY 10022; 6,500; http://www.fiaf.org
Frozen Food Institute, American (1942), 2000 Corporate Ridge, Suite 1000, McLean, VA 22102; 550; http://www.affi.com
Funeral Consumers Alliance (FAMSA) (1964), P.O. Box 10, Hinesburg, VT 05461; 500,000; http://www.funerals.org/famsa
Future Business Leaders of America/Phi Beta Lambda, Inc. (1942), 1912 Association Drive, Reston, VA 20191; 240,000; http://www.fbla-pbl.org

Gamblers Anonymous (1957), 3255 Wilshire Blvd., Los Angeles, CA 90010; approx. 30,0000; http://www.gamblersanonymous.org
Garden Club of America (1913), 14 E. 60th St., 3rd Floor, New York, NY 10022; 15,000; http://www.gcamerica.org
Garden Clubs, Inc., National Council of State (1929), 4401 Magnolia Ave., St. Louis, MO 63110; 253,316; http://www.gardenclub.org
Gas Assn., American (1918), 400 North Capitol St. NW, Wash., DC 20001; 187 cos.; http://www.aga.org
Gay and Lesbian Task Force, Natl. (1973), 1700 Kalorama Rd. NW, Wash., DC 20009; 30,000; http://www.ngltf.org.
Genealogical Society, Natl. (1903), 4527 17th St. N, Arlington, VA 22207; 18,000; http://www.ngsgenealogy.org
General Contractors of America, The Associated (1918), 333 John Carlyle St., Ste. 200, Alexandria, VA 22314; 36,000+ co.'s; http://www.agc.org
Genetic Association, American (1903), P.O. Box 257, Buckeystown, MD 21704; http://svl.la.asu.edu/agal
Geographers, Assn. of American (1904), 1710 16th St. NW, Wash., DC 20009; 6,500; http://www.aag.org
Geographic Education, Natl. Council for (1915), 16A Leonard Hall, IUP, Indiana, PA 15705; 2,400; http://www.ncge.org
Geographic Society, Natl. (1888), 1145 17th St. NW, Wash., DC 20036; 8.5 mil; http://www.nationalgeographic.com
Geographical Society, The American (1851), 120 Wall St., New York, NY 10005; http://amergeog.org.
Geological Society of America (1888), 3300 Penrose Pl., Boulder, CO 80301; 17,000; http://www.geosociety.org
Geriatrics Society, American (1942), 350 5th Ave., Ste. 801, New York, NY 10118; 6,000; http://www.americangeriatrics.org.
Gideons Intl. (1899), 2900 Lebanon Rd., Nashville, TN 37214; 131,000; http://www.gideons.org
Gifted Children, Natl. Assn. for (1954), 1707 L Street NW, Suite 550, Washington, DC 20036; 8,000; http://www.nagc.org
Girl Scouts of the U.S.A. (1912), 420 5th Ave., New York, NY 10018; 3.8 mil; http://www.girlscouts.org
Gold Star Mothers of America, Inc. (1928), 2128 Leroy Place NW, Wash., DC 20008; 1500; http://www.goldstarmoms.com.
Golden Key National Honor Society (1977), 1189 Ponce de Leon Ave., Atlanta, GA 30306; 1 mil.+; http://gknhs.gsu.edu
Golf Assn., U.S. (1894), Golf House, P.O. Box 708, Far Hills, NJ 07931; 800,000; http://www.usga.org
Gospel Music Assn. (1964), 1205 Division St., Nashville, TN 37203; 5,500; http://www.gospelmusic.org
Government Finance Officers Assn. (1906), 180 N. Michigan Avenue, Suite 800, Chicago, IL 60601; 13,600; http://www.financenet.gov/gfoa.htm
Governors' Assn., Natl. (1908), Hall of the States, 444 N. Capitol, Wash., DC 20001; 55 govs.; http://www.nga.org
Graduate Schools, Council of (1960), One Dupont Circle NW, #430 Wash., DC 20036; 415 instits.; http://www.cgsnet.org
Grange of the Order of Patrons of Husbandry, Natl. (1867), 1616 H Street NW, Wash., DC 20006; 300,000; http://www.nationalgrange.org
Graphic Arts, American Institute of (1914), 164 5th Ave., New York, NY 10010; 15,000; http://www.aiga.org
Gray Panthers (1970), 733 15th St. NW, Ste 437, Wash., DC 20005; approx. 17,000; http://www.graypanthers.org.
Green Mountain Club (1910), 4711 Waterbury-Stowe Rd., Waterbury Ctr., VT 05677; 8,000; http://www.greenmountainclub.org
Green Party (1984), P.O. Box 1134, Lawrence, MA 01842; 1,500+; http://www.greenparty.org
Greenpeace U.S.A. (1971), 702 H St. NW, Wash., DC 20001; 250,000; http://www.greenpeaceusa.org.
Grocery Manufacturers of America (1908), 1010 Wisconsin Avenue., 9th Floor., Wash., DC 20007; 140 cos.; http://www.gmabrands.com
Ground Water Assn., Natl. (1948), 601 Dempsey Rd., Westerville, OH 43081; 16,500; http://www.ngwa.org
Group Against Smokers' Pollution, Inc. (GASP) (1971), P.O. Box 632, College Park, MD 20741; 10,000+.
Guide Dog Foundation for the Blind, Inc. (1946), 371 E. Jericho Turnpike, Smithtown, NY 11787; 162,500; http://www.guidedog.org

Hadassah, the Women's Zionist Organization of America (1912), 50 W. 58th St., New York, NY 10019; 385,000; http://www.hadassah.org
Handball Assn., U.S. (1951), 2333 N. Tucson Blvd., Tucson, AZ 85716; 8,000; http://www.ushandball.org
Health Council, Natl. (1920), 1730 M St. NW, Ste. 500, Wash., DC 20036; http://www.nhcouncil.org
Health Info. Management Assn., American (AHIMA) (1928), 233 N. Michigan Ave., Ste. 2150, Chicago, IL 60601; 39,000; http://www.ahima.org
Hearing Society, Intl. (1951), 16880 Middlebelt Rd., Ste. 4, Livonia, MI 48154; 3,100; http://www.hearingihs.org
Heart Assn., American (1924), 7272 Greenville Ave., Dallas, TX 75231; 31,000; http://www.americanheart.org
Heating, Refrigerating & Air-Conditioning Engineers, Inc., American Soc. of (1894), 1791 Tullie Cir. NE, Atlanta, GA 30329; 55,000; http://www.ashrae.org
Helicopter Society, American (1943), 217 N. Washington St., Alexandria, VA 22314; 6,140; http://www.vtol.org
Hemispheric Affairs, Council on (1975), 1444 I St. NW, Ste. 211, Wash., DC 20005; 1,800; http://www.coha.org
HIAS, Inc. (1881) 333 Seventh Ave., 17th Fl., New York, NY 10001; approx. 9,000; http://www.hias.org
Hibernians in America, Ancient Order of (1836), 1301 S.W. 26th Avenue, Ft. Lauderdale, FL 33312; 200,000; http://www.aoh.com
Highpointers Club (1987), P.O. Box 1496, Golden, CO 80402; 2,500; http://www.highpointers.org
High School Band Directors Hall of Fame, Natl. (1985), 519 N. Halifax Ave., Daytona Beach, FL 32118; http://www.halloffames.homestead.com
Hiking Society, American (1976), 1422 Fenwick Lane, Silver Spring, MD 20910; 7,000; http://www.americanhiking.org
Historians, Organization of American (1907), 112 N. Bryan St., Bloomington, IN 47408; 8,500; http://www.oah.org
Historic Preservation, Natl. Trust for (1949), 1785 Massachusetts Avenue NW, Wash., DC 20036; 250,000; http://www.nationaltrust.org
Historical Assn., American (1884), 400 A St. SE, Wash., DC 20003; 16,000; http://www.theaha.org
Historical Society Doll Collection, United States (1971), 1st and Main Sts., Richmond, VA 23219; 250,000; http://www.ushsdolls.com
Hockey, U.S.A. (1936), 1775 Bob Johnson Dr., Colorado Springs, CO 80906; 575,000; www.usahockey.com.
Home Builders, Natl. Assn. of (1942), 1201 15th St. NW, Wash., DC 20005; 203,000; http://www.nahb.com
Home Energy Research Organization (H.E.R.O) (2001), 10799 Sherman Grove Ave.,#18, Sunland, CA, 91040-2364; 5,600.
Homeless, Natl. Coalition for the (1984), 1012 14th St., Ste. 600, Wash., DC 20005; http://www.nationalhomeless.org
Honor Society, Natl. (1921), 1904 Association Dr., Reston, VA 20191; app. 750,000; http://dsa.principals.org.
Horatio Alger Soc. (1965), P.O. Box 70361, Richmond, VA 23255; 250; http://www.ihot.com/~has
Horse Council, American (1969), 1700 K St. NW, #300, Wash., DC 20006; 195 org.,1,800 ind.; http://www.horsecouncil.org
Hospital Assn., American (1899), 1 N. Franklin, Chicago, IL 60606; 5,100 hospitals; http://www.aha.org
Hostelling Intl.-American Youth Hostels (1934), 733 15th Street NW, Suite 840, Wash., DC 20005; 125,000; http://www.hiayh.org
Hotel & Motel Assn., American (1910), 1201 New York Ave., #600, NW, Wash., DC 20005; 10,000+; http://www.ahma.com
Hot Rod Assn., Natl. (1951), 2035 Financial Way, Glendora, CA 91741; 85,000; http://www.nhra.com
Housing Inspection Foundation (1979), 1224 N. Nokomis NE, Alexandria, MN 56308; 5,000; http:\\www.iami.org
Huguenot Society, Natl. (1951), 9033 Lyndale Ave. S, #108, Bloomington, MN 55420; 5,000.
Humane Society of the U.S. (1954), 2100 L St. NW, Wash., DC 20037; 650,000; http://www.hsus.org
Human Resource Management, Society for (SHRM) (1948), 1800 Duke St., Alexandria, VA 22314; 115,000; http://www.shrm.org
Hydrogen Energy, Intl. Assn. for (1974), P.O. Box 248266, Coral Gables, FL 33124; 2,500; http://www.iahe.org

Identification, Intl. Assn. for (1915), 2535 Pilot Knob Road, Ste. 117, Mondota Heights, MN 55120; 5,000; http://www.theiai.org
Illuminating Engineering Society of N. America (1906), 120 Wall Street, 17th Floor, New York, NY 10005; 8,500; http://www.iesna.org
Illustrators, Inc., Society of (1901), 128 E. 63d St., New York, NY 10021; 1,000; http://www.societyillustrators.org
Independent Community Bankers of America (1930), One Thomas Circle, Ste. 400, Wash., DC 20005; 5,000; http://www.icba.org
Industrial and Applied Mathematics, Society for (1952), 3600 Univ. City Science Ctr., Philadelphia, PA 19104; 9,000; http://www.siam.org
Industrial Designers Society of America (1965), 1142 Walker Rd., Ste. E, Great Falls, VA 22066; 3,200; http:www.idsa.org
Industrial Security, American Soc. for (1955), 1625 Prince St., Alexandria, VA 22313; 30,000; http://www.asisonline.org
Insurance Assn., American (1964), 1130 Connecticut Avenue NW, Suite 1000, Wash., DC 20036; 350; http://www.aiadc.org
Integrative and Comparative Biology, Society for (1890), 1313 Dolley Madison Blvd., Ste. 402, McClean, VA 22101; 2,400; http://www.sicb.org
Intellectual Property Owners Assoc. (1972), 1255 23d St. NW, Ste. 200, Wash., DC 20037; 344; http://www.ipo.org
Intelligence Officers, Assoc. of Former (1976), 6723 Whittier Ave., Ste. 303A, McLean, VA 22101-4533; 3,100+; http://www.afio.com
Intercollegiate Athletics, Natl. Assn. of (1937), 23500 W. 105th St. P.O. Box 1325, Olathe, KS 66051-1325; 331 member colleges/universities; http://www.naia.org
Interior Designers, American Society of (1975), 608 Massachusetts Avenue NE, Wash., DC 20008; 30,000; http://www.asid.org
Intl. Education, Institute of (1919), 809 United Nations Plaza, New York, NY 10017; 650 U.S. colleges and universities; http://www.iie.org
Intl. Educational Exchange, Council on (1947), 205 E. 42d Street, New York, NY 10017; 240 organizations; http://www.ciee.org
Intl. Educators, Assn. of (NAFSA) (1948), 1307 New York Ave., 8th Fl.,Wash., DC 20005; 7,500; http://www.nafsa.org
Intl. Law, American Society of (1906), 2223 Massachusetts Ave. NW, Wash., DC 20008; 4,500; http://www.asil.org
Inventors, American Soc. of (1953), P.O. Box 58426, Philadelphia, PA 19102; 150; http://americaninventor.org
Investigative Pathology, American Soc. for (1900), 9650 Rockville Pike, Bethesda, MD 20814; 1,753; http://www.asip.uthscsa.edu
Investment Management and Research, Assn. for (AIMR) (1990), 560 Ray C. Hunt Dr., Charlottesville, VA 22903-0668; 49,000; http://www.aimr.org
Investors Corp., Natl. Assn. of (1951), 711 W. Thirteen Mile Rd., Madison Heights, MI 48701; 700,000; http://www.better-investing.org
IPC-Association Connecting Electronics Industries (formerly The Institute for Interconnecting & Packaging Electronic Circuits) (1957), 2215 Sanders Rd., Northbrook, IL 60062; 2,800; http://www.ipc.org
Irish American Cultural Inst. (1962), 1 Lackawanna Pl., Morristown, NJ 07960; 4,500; http://www.irishaci.org
Irish Historical Society, American (1897), 991 5th Ave., New York, NY 10028; 850; http://www.aihs.org
Iron and Steel Engineers, Assn. of (1907), Three Gateway Center, Suite 1900, Pittsburgh, PA 15222; 11,650; http://www.aise.org
Islamic Relations, Council on American (1994), 453 New Jersey SE, Wash., DC 20003; http://www.cair-net.org
Italian Historical Society of America (1949), 111 Columbia Heights, Brooklyn, NY 11201; http://www.italianhistorical.org

Jail Assn., American (1981), 2053 Day Rd., Ste. 100, Hagerstown, MD 21740; 4,800+; http://www.corrections.com/aja
Japanese-American Citizens League (1929), 1765 Sutter St., San Francisco, CA 94115; 23,900; http://www.jacl.org
Jewish Committee, American (1906), 165 E. 56th St., New York, NY 10022; 110,000; http://www.ajc.org
Jewish Community Centers Assn. of North America (1917), 15 E. 26th St., New York, NY 10014; 1,000,000+; http://www.jcca.org
Jewish Congress, American (1918), 15 E. 84th St., New York, NY 10028; 50,000; http://www.ajcongress.org
Jewish Historical Society, American (1892), 15 West 16th St. New York, NY 10011; 4,000; http://www.ajhs.org
Jewish War Veterans of the U.S.A. (1896), 1811 R St. NW, Wash., DC 20009; 50,000; http://jwv@jwv.org
Jewish Women, Natl. Council of (1893), 53 W. 23d St., 6th Fl., New York, NY 10010; 90,000; http://www.ncjw.org
John Birch Society (1958), 770 Westhill Blvd, P.O. Box 8040, Appleton, WI 54912; http://www.jbs.org
Joint Action in Community Service (JACS) (1967), 5225 Wisconsin Ave. NW, Ste. 404, Wash., DC 20015; http:www.jacsinc.org
Joseph Diseases Foundation, Inc., Intl. (1977), P.O. Box 2550, Livermore, CA 94551; 1,550; http://www.ijdf.net
Journalists, Society of Professional (1909), 3909 N. Meridian St., Indianapolis, IN 46200-4505; 10,000+; http://spj.org
Journalists and Authors, American Society of (1948), 1501 Broadway, Ste. 302, New York, NY 10036; 1,012; http://www.asja.org
Judicature Society, American (1913), 180 N. Michigan Ave., Ste. 600, Chicago, IL 60601; 8,000; http://www.ajs.org
Jugglers Assn., Intl. (1947), P.O. Box 218, Montague, MA 01351; 2,500; http://www.juggle.org
Junior Achievement, Inc. (1919), One Education Way, Colorado Springs, CO 80906; http://www.ja.org
Junior Auxiliaries, Natl. Assn. of (1941), 845 South Main St., Greenville, MS 38701; 12,876; http://www.najanet.org
Junior Chamber of Commerce, U.S. (1920), P.O. Box 7, 4 W. 21st St., Tulsa, OK 74114; 200,000; http://www.usjaycees.org
Junior College Athletic Assn., Natl. (1938), P.O. Box 7305, Colorado Springs, CO 80933; 981; http://www.njcaa.org
Junior Honor Society, Natl. (1929), 1904 Association Dr., Reston, VA 20191; approx. 250,000; http://dsa.principals.org
Junior Leagues, Assn. of (1901), 132 West 31st St., New York, NY 10016; 193,000; http://www.ajli.org

Kidney Fund, The American (1971), 6110 Executive Blvd., Ste. 1010, Rockville, MD 20852; http://www.akfinc.org
Kiwanis International (1915), 3636 Woodview Trace, Indianapolis, IN 46268; 543,041; http://www.kiwanis.org
Knights of Columbus (1882), One Columbus Plaza, New Haven, CT 06510; 1,632,439; http://www.kofc.org
Knights of Pythias, (1864), 59 Caddington Street, #202, Quincy, MA 02169; approx. 75,000; http://www.pythias.org
Krishna Consciousness, Intl. Soc. for (ISKON, Inc.) (1966), 3764 Watseka Ave., Los Angeles, CA 90034; approx. 20,000; http://www.harekrishna.com

La Leche League Intl. (1956), 1400 N. Meacham Rd., P.O. Box 4079, Schaumburg, IL 60168; http://www.lalecheleague.org
Lady Bird Johnson Wildflower Center (1982), 4801 La Crosse Avenue, Austin, TX 78739; 22,000; http://www.wildflower.org
Landscape Architects, American Society of (1899), 636 I St. NW, Wash., DC 20001-3736; 13,500; http://www.asla.org
Law Libraries, American Assn. of (1906), 53 W. Jackson Blvd., #940, Chicago, IL 60604; 5,150; http://www.aallnet.org
Learned Societies, American Council of (1919), 228 E. 45th St., New York, NY 10017; 64 societies; http://www.acls.org
Lefthanders Intl. (1975), P.O. Box 8249, Topeka, KS 66608; 25,000.
Legal Administrators, Assn. of (1971), 175 E. Hawthorn Parkway, Suite 325, Vernon Hills, IL 60061-1428; 9,000; http://www.alanet.org
Legal Secretaries, Natl. Assn. of (NALS) (1942), 314 E 3rd St., Ste. 210, Tulsa, OK 74120; 8,500; http://www.nals.org
Legion of Valor of the U.S.A., Inc. (1890), c/o Legion of Valor Museum, 2425 Fresno St., Ste. 103, Fresno, CA 93721; 600; http://www.legionofvalor.com
Leprosy Missions, Inc., American (1906), One Alm Way, Greenville, SC 29601; http://www.leprosy.org
Leukemia and Lymphoma Society (1949), 1311 Mamaroneck Ave., White Plains, NY 10605; 58 chapters nationwide; http://www.leukemia-lymphoma.org
Lewis and Clark Trail Heritage Foundation. (1969), 600 Central Plaza, Ste.300, Great Falls, MT 59403; 43,200; http://www.lewisandclark.org
Libertarian Party (1971), 2600 Virginia Ave. NW, Ste. 100, Wash., DC 20037; 33,000+; http://www.lp.org
Liberty Lobby (1955), 300 Independence Ave. SE, Wash., DC 20003; 90,000; http://www.spotlight.org
Libraries Assn., Special (1909), 1700 18th St. NW, Wash., DC 20009; 15,000; http://www.sla.org
Library Assn., American (1879), 50 E. Huron St., Chicago, IL 60611; 56,000; http://www.ala.org
Lighter-Than-Air Society (1952), 1436 Triplett Blvd., Akron, OH 44306; 700+.
Linguistic Society of America (1924), 1325 18th St. NW, Ste. 211, Wash., DC 20036; 4,000 indiv., 2,200 inst.; http://www.lsadc.org
Lions Clubs, Intl., Assn. of (1917), 300 W. 22d St., Oak Brook, IL 60523; 1,400,000; http://www.lionsclubs.org
Literacy Volunteers of America, Inc. (1962), 635 James St., Syracuse, NY 13203; 4,000 indiv., 2,200 Inst.; http://www.literacyvolunteers.org
Little League Baseball, Inc. (1939), P.O. Box 3485, S. Williamsport, PA 17701; approx. 4 mil; http://www.littleleague.org
Little People of America, Inc. (1961), Box 745, Lubbock, TX 79408; 6,000; http://www.lpaonline.org
Logistics, International Society of (SOLE) (1966), 8100 Professional Place, Ste. 211, Hayatsville, MD 20785; 3,500; http://www.sole.org
London Club (1975), 214 North 2100 Rd., Lecompton, KS 66050; 100+.
Lung Assn., American (1904), 1740 Broadway, New York, NY 10019; http://www.lungusa.org

Magazine Publishers of America (1919), 919 Third Ave., New York, NY 10022; 275; http://www.magazine.org
Magicians, Intl. Brotherhood of (1922), 11155 S. Towne Sq., Ste. C, St. Louis, MO 63123-7813; 14,000; http://www.magician.org
Management Accountants, Institute of (1919), 10 Paragon Dr., Montvale, NJ 07645-1760; 75,000; http://www.imanet.org
Management Assn. Intl., American (1923), 1601 Broadway, New York, NY 10019; 70,000+; http://www.manet.org
Management Consulting Firms, Assn. of (1929), 380 Lexington Ave., Ste. 1700, New York, NY 10168; 55 cos; http://www.amcf.org
Manufacturing Engineers, Soc. of (1932), One SME Dr., Dearborn, MI 48121-0930; 55,000+; http://www.sme.org
Manufacturers, Natl. Assn. of (1895), 1331 Pennsylvania Ave. NW, Suite 1500 N. Tower, Wash., DC 20004; 14,000 cos.; http://www.nam.org
March of Dimes Birth Defects Foundation (1938), 1275 Mamaroneck Avenue, White Plains, NY 10605; http://www.modimes.org
Marine Corps League (1937), P.O. Box 3070, Merrifield, VA 22116; 42,000; http://www.mcleague.org
Market Technicians Assn., Inc. (1973), One World Trade Center, Suite 4447, New York, NY 10048; 1,550; http://www.mta.org
Marketing Assn., Am. (1915), 311 S. Wacker Dr., Ste. 5800, Chicago, IL 60606; 45,000; http://www.ama.org
Materials and Process Engineering, Soc. for the Advancement of (1944), 1161 Parkview Drive, Covina, CA 91724-3748; 5,000; http://www.sampe.org
Mathematical Society, American (1888), 201 Charles St., Providence, RI 02904; 30,000; http://www.ams.org
Mayflower Descendants, General Society of (1897), 4 Winslow St., Plymouth, MA 02361; 27,000; http://www.mayflower.org
Mayors, U.S. Conference of (1932), 1620 Eye St. NW, Wash., DC 20006; http://www.usmayors.org
Mechanical Engineers, American Soc. of (1880), 3 Park Ave., New York, NY 10016; 125,000; http://www.asme.org
Medical Assn., American (1847), 515 N. State St., Chicago, IL 60610; 300,000; http://www.ama-assn.org
Medical Corps, International (1984), 11500 W. Olympic Blvd., Ste. 506, Los Angeles, CA 90064; www.imc-la.org
Medical Library Assn. (1898), 65 E. Wacker Pl., Ste. 1900, Chicago, IL 60602; 5,000; http://www.mlanet.org
Medieval Academy of America (1925), 1430 Massachusetts Avenue, Suite 313, Cambridge, MA 02138; 4,500; http://www.medievalacademy.org
Meeting Planners, Intl. Society of (1981) 1224 N. Nokomis NE, Alexandria, MN 56308; 6,000; http://www.iami.org
MENC: The Natl. Assn. for Music Education (formerly Music Educators Natl. Conference) (1907), 1806 Robert Fulton Dr., Reston, VA 20191; 85,000; http://www.menc.org
Mended Hearts, Inc. (1951), 7272 Greenville Ave., Dallas, TX 75231; 25,000; http://www.mendedhearts.org
Mensa, Ltd., American (1960), 1229 Corporate Dr. W, Arlington, TX 76006; 47,000+; http://www.us.mensa.org
Mental Health Assn., Natl. (1909), 1021 Prince St., Alexandria, VA 22314; http://www.nmha.org.
Mentally Ill, Natl. Alliance for the (1980), 2107 Wilson Blvd. Ste. 300, Arlington, VA 22201; 170,000.
Merrill's Marauders Assn. (1947), 11244 N. 33rd St., Phoenix, AZ 85028-2723; 1,758; http://www.marauder.org
Meteorological Society, American (1919), 45 Beacon St., Boston, MA 02108; 11,000; http://www.ametsoc.org/AMS
Metric Assn., Inc., U.S. (1916), 10245 Andasol Ave., Northridge, CA 91325; 1,200; http://usmetric.org
Microbiology, American Society for (1899), 1752 N. St. NW, Wash., DC 20036; 42,000; http://www.asmusa.org
Military Order of the Purple Heart of the USA (1958), 5413-B Backlick Road, Springfield, VA 22151; 30,000; http://www.purpleheart.org
Military Order of the World Wars (1919), 435 N. Lee St., Alexandria, VA 22314; 11,250; http://www.militaryorder.org
Military Surgeons of the U.S., Assn. of (1898), 9320 Old Georgetown Road, Bethesda, MD 20814; 11,000; http://www.amsus.org
Mining, Metallurgy and Exploration, Inc., Society for (1871), 8307 Shaffer Pkwy., Littleton, CO 80127; 16,141; http://www.smenet.org
Mining, Metallurgical and Petroleum Engineers, American Institute of (1871), 3 Park Ave., New York, NY 10016; 90,000; http://www.aimeny.org
Missing and Exploited Children, Natl. Center for (1984), The Charles B. Wang International Children's Building, Alexandria, VA 22314; http://www.missingkids.com
Model A Ford Club of America, Inc. (1955), 250 S Cypress St., La Habra, CA 90631; 15,500; http://www.mafca.com
Model Railroad Assn., Natl. (1935), 4121 Cromwell Rd., Chattanooga, TN 37421-2119; 24,000; http://www.nmra.org
Modern Language Assn. of America (1883), 26 Bdwy., 3d Fl., New York, NY 10014; 32,000; http://www.mla.org
Moose Intl., Inc. (1888), Rte. 31, Mooseheart, IL 60539; 1.5 mil; http://www.mooseintl.org
Mothers, Inc.®, American (1935), P.O. Box 400, Pound Ridge, NY 10576; 6,000; http://www.americanmothers.org
Mothers of Twins Clubs, Natl. Organization of (1960), P.O. Box 438, Thompson Station, TN 37179-0438; 23,000; http://www.nomotc.org
Motion Picture Arts & Sciences, Academy of (1927), 8949 Wilshire Blvd., Beverly Hills, CA 90211; 6,300; http://www.oscars.org
Motion Picture & Television Engineers, Soc. of (1916), 595 W. Hartsdale Ave., White Plains, NY 10607; 10,000; http://www.smpte.org
Motorcyclist Assn., American (1924), 13515 Yarmouth Dr., Pickerington, OH 43147; 270,000; http://www.ama-cycle.org
Motorists Association, Natl. (1982), 402 W. 2nd St., Waunakee, WI 53597; 7,500; http://www.motorists.org
Multiple Sclerosis Society, Natl. (1946), 733 Third Ave., New York, NY 10017; 518,567; http://www.nmss.org
Muscular Dystrophy Assn., Inc. (1950), 3300 E. Sunrise Dr., Tucson, AZ 85718; 2 mil. volunteers; http://www.mdausa.org
Museums, American Assn. of (1906), 1575 Eye St. NW, Ste. 400, Wash., DC 20005; 15,800; http://www.aam-us.org
Music Center, American (1939), 30 W. 26th St., #1001, New York, NY 10010; 2,500; http://www.amc.net
Music Teachers Natl. Assn. (1876), 441 Vine St., Ste. 505, Cincinnati, OH 45202; 24,000; http://www.mtna.org
Musicological Society, American (1934), 201 S. 34th St., Philadelphia, PA 19104-6313; 4,600; http://www.ams-net.org
Muzzle Loading Rifle Assn., Natl. (1933), P.O. Box 67, Friendship, IN 47021; 21,000; http://nmlra@nmlra.org
Myasthenia Gravis Foundation of America (1952), 5841 Cedar Lake Rd., Ste. 204, Minneapolis, MN 55416; 55,000; http://www.myasthenia.org
Mystery Writers of America, Inc. (1945), 17 E. 47th St., 6th Fl., New York, NY 10017; 2,235; http://www.mysterywriters.org

NA'AMAT USA (1925), 350 Fifth Ave., Ste. 4700, New York, NY 10118; 50,000, U.S.; 900,000 worldwide; http://www.naamat.org
Name Society, American (1951), Dept. of Modern Languages, Baruch College, 17 Lexington Ave., New York, NY 10010; 750.
Narcotics Anonymous World Services (1953), P.O Box 9999, Van Nuys, CA 94139; 100,000+; http://www.na.org
Natl. Assn. for the Advancement of Colored People (NAACP) (1909), 4805 Mt. Hope Dr., Baltimore, MD 21215; http://www.naacp.org
National Guard Assn. of the U.S. (1878), One Massachusetts Ave. NW, Wash., DC 20001; 56,000; http://www.ngaus.org
National Press Club (1908), 529 14th St., 13th Fl., NW, Wash., DC 20045; 4,350; http://www.press.org
Nature Conservancy, The (1951), 4245 N. Fairfax Drive, Ste. 100, Arlington, VA 22203; 1 mil+; http://www.tnc.org
Naturists, Inc., The (1980), P.O. Box 132, Oshkosh, WI 54902; 27,000; http://www.naturistsociety.com
Naval Engineers, American Society of (1888), 1452 Duke St., Alexandria, VA 22314; 5,500; http://www.navalengineers.org
Naval Institute, U.S. (1873), 291 Wood Rd., Annapolis, MD 21402; 70,000; http://www.usni.org
Naval Reserve Assn. (1954), 1619 King St., Alexandria, VA 22314; 23,000; http://www.navy-reserve.org
Navy Vets, Sampson WW2 (1987) P.O. Box 63, Seneca Falls, NY 13148; 7000.
Navigation, The Institute of (1945), 1800 Diagonal Rd., Ste. 480, Alexandria, VA 22314; 3,200; http://www.ion.org
Navy League of the United States (1902), 2300 Wilson Blvd., Arlington,VA 22201-3308; 74,000; http://www.navyleague.org
Negro College Fund, United (1944), 8260 Willow Oaks Corporate Drive, Fairfax, VA 22031; 39 institutions; http://www.uncf.org
Neurofibromatosis Foundation, Natl. (1978), 95 Pine St., 16th Fl., New York, NY 10005; 30,000; http://www.nf.org
Newspaper Assn. of America (NAA) (1887), 1921 Gallows Rd., Ste. 600, Vienna, VA 22182-3900; 2,000+; http://www.naa.org
NGA, Inc. (1896) 820 Newton Rd., Walminster, PA 18974; www.nga-inc.org
Ninety-Nines (Intl. Organization of Women Pilots) (1929), 7100 Terminal Dr., Oklahoma City, OK 73159; 6,000; http://www.ninety-nines.org
Non-Commissioned Officers Assn. (1960), 10635 IH 35 North, San Antonio, TX 78233; 160,000; http://www.ncoausa.org
Northern Cross Society (1983), 214 N. 2100 Rd., Lecompton, KS 66050; 100+.
NOT-SAFE: Nat'l Organization Taunting Safety and Fairness Everywhere (1984), P.O. Box 5743-WS, Santa Barbara, CA 93150; 8,475; http://www.notsafe.org.
Notaries, American Society of (1965), P.O. Box 5707, Tallahassee, FL 32314; approx. 20,000; http://www.notaries.org
NSAC (Natl. Soc. of Accountants for Cooperatives) (1936), 6320 Augusta Dr., Ste. 800, Springfield, VA 22150; 2,000; http://www.nsacoop.org
Nuclear Society, American (1954), 555 N. Kensington Ave., La Grange Park, IL 60526; 11,000; http://www.ans.org
Nude Recreation Inc., American Assn. for (1931), 1703 N. Main Street, Suite E, Kissimmee, FL 34744; 50,000; http://www.aanr.com

Numismatic Assn., American (1891), 818 N. Cascade Ave., Colorado Springs, CO 80903; 30,000; http://www.money.org
Numismatic Society, The American (1858), Broadway at 155th St., New York, NY 10032; http://www.amnumsoc.org
Nursing, Natl. League for (1952), 61 Broadway, New York, NY 10006; 4,000; http://www.nln.org
Nutritional Sciences, American Society for (1928), 9650 Rockville Pike, Ste. 4500, Bethesda, MD 20814; 3,400; http://www.faseb.org/asns

Ocean Conservancy (1972), 1725 DeSales St. NW, #600, Wash.,DC 20036; 200,000; http://www.oceanconservancy.org
Odd Fellows, Independent Order of(1819), 422 Trade St., Winston-Salem, NC 27101; 295,077; http://www.ioof.org
Old Crows, Assn. of (1964), 1000 N. Payne St., Alexandria, VA 22314; 16,485; http://www.crows.org
Opthalmology, American Academy of (1979), P.O. Box 7424, San Francisco, CA 94120; 21,000; http://www.eyenet.org
Optical Society of America (1932), 2010 Massachusetts Ave. NW, Wash., DC 20036; http://www.osa.org
Optimist Intl. (1919), 4494 Lindell Blvd., St. Louis, MO 63108; 155,000; http://www.optimist.org
Optometric Assn., American (1898), 243 N. Lindbergh Blvd., St. Louis, MO 63141; 32,000; http://www.aoanet.org
Organ Sharing, United Network for (1977), 1100 Boulders Parkway, Ste. 500, P.O. Box 13770, Richmond, VA 23225; 434; http://www.unos.org
Organists, American Guild of (1896), 475 Riverside Dr., Ste. 1260, New York, NY 10115; 20,200; http://www.agohq.org
Oriental Society, American (1842), Univ. of Michigan, Hatcher Graduate Library, 110D, Ann Arbor, MI 48109; 1,350; http://www.umich.edu/~aos
ORT Federation, American (Org. for Rehabilitation Through Training) (1922), 817 Broadway, 10th Fl., New York, NY 10003; 10,000; http://www.aort.org.
Ornithologists' Union, American (1883), c/o Division of Birds, MRC-116, Smithsonian Institution, Wash., DC 20560-0116; 4,200; http://www.aou.org
Osteopathic Assn., American (1897), 142 E. Ontario, Chicago, IL 60611; 28,974; http://www.aoa-net.org
Ostomy Assn., Inc., United (1962), 19772 MacArthur Blvd., Ste. 200, Irvine, CA 92612, 30,000; http://www.uoa.org
Outlaw and Lawman History, Inc., Natl. Assn. for (NOLA) (1974), 1201 Holly Ct., Harker Heights, TX 76548-1538; approx. 450; http://www.outlawlawman.com
Overeaters Anonymous (1960) 6075 Zenith Court NE, Rio Rancho, NM 87124-4020; http://www.overeatersanonymous.org
Oxfam America (1970) 26 West St., Boston, MA 02111; 100,000; http://www.oxfamamerica.org

Paralyzed Veterans of America (1947), 801 18th St. NW, Wash., DC 20006; 18,000; http://www.pva.org
Parapsychology Institute of America (1971), P.O. Box 5442, Babylon, NY, 11707; 1,420.
Parents Without Partners, Inc. (1958), 1650 S. Dixie Highway, Suite 510, Boca Raton, FL 33432; 35,000; http://www.parentswithoutpartners.org
Parkinson's Disease Foundation, Inc. (1957), William Black Medical Bldg., Columbia-Presbyterian Medical Center, 710 W. 168th St., New York, NY 10032; 95,000; http://www.pdf.org
Parliamentarians, Natl. Assn. of (1930), 213 S. Main St., Independence, MO 64050; 4,000; http://www.parliamentarians.org
Patton, George S. Jr. Society (1970), 3116 Thorn St., San Diego, CA 92104; 200; http://www.geocities.com/pattonhq/homeghq.html
PBY Catalina International Association (1988), 1510 Kabel Dr., New Orleans, LA, 70131; 850; http://www.pbycia.org
Peace Corps (1961), 111 20th St., NW, Wash., DC 20526; 8,500; http://www.peacecorps.gov
Pearl Harbor History Associates, Inc. (1985), P.O. Box 1007, Stratford, CT 06615; approx. 275; http://www.ibilio.org/phha.
PEN American Center, Inc. (1922), 568 Broadway, Rm. 401, New York, NY 10012; 2,300; http://www.pen.org
Pen Friends, Intl. (1967), 758 Kapahulu Ave. #101, Honolulu, HI 96816; 300,000; http://www.pen-pals.net
Pension Plan, Committee for a Natl. (1979), P.O. Box 27851, Las Vegas, NV 89126; 325.
Pen Women, Natl. League of American (1897), 1300 17th St. NW, Wash., DC 20036; approx. 4,000.
People for the Ethical Treatment of Animals (PETA) (1980), 501 Front St., Norfolk, VA 23510; 600,000; http://www.peta-online.org
Performance Improvement, Intl. Society for (1962), 1400 Spring St., Ste. 260, Silver Spring, MD 20910; 6,100; http://www.ispi.org
Petroleum Institute, American (1919), 1220 L St. NW, Wash., DC 20005; 400 companies; http://www.api.org
Pharmaceutical Assn., American (1852), 2215 Constitution Ave. NW, Wash., DC 20037; 50,000; http://www.aphanet.org
Phi Beta Kappa Society (1776), 1785 Massachusetts Ave., N.W., 4th Fl., Wash., D.C. 20036; approx. 600,000; http://www.pbk.org
Phi Delta Kappa Intl., Inc. (1906), 408 N. Union, Bloomington, IN 47408; 95,223; http://www.pdkintl.org
Phi Kappa Phi (1897), P.O. Box 16000-Louisiana State University, Baton Rouge, LA 70893; 900,000+; http://www.phikappaphi.org
Phi Theta Kappa Int'l. Honor Society (1918), 1625 Eastover Drive, Jackson, MS 39211; 800,000; http://www.pk.org
Philatelic Society, American (1886), 100 Oakwood Ave., State College, PA 16803; 50,200; http://www.stamps.org
Philatelic Golf Society, Intl. (1987), P.O. Box 2183, Norfolk, VA 23501-2183; 250; http:www.ipgsonline.org
Philological Association, American (1869), 291 Logan Hall, Univ. of Penn., 249 S. 36th St.. Philadelphia, PA, 19104-6304; 3000; http://www.apaclassics.org
Philosophical Assn., American (1901), Univ. of Delaware, Newark, DE 19716; 10,800; http://www.udel.edu/apa
Photographers of America, Inc., Professional (1880), 229 Peachtree Street, NE, Atlanta, GA 30303; 14,000; http://www.ppa.org
Photographic Society of America, Inc. (1934), 3000 United Founders Blvd., Ste. 103, Oklahoma City, OK 73112; 6,200; http://www.psa-photo.org
Physical Therapy Assn., American (1930), 1111 N. Fairfax St., Alexandria, VA 22314; 68,000; http://www.apta.org
Physically Handicapped, Inc., Natl. Assn. of the (1958), Scarlet Oaks, 440 Lafayette Ave., #GA4, Cincinnati, OH 45220-1022; approx. 400; http://www.naph.net.
Physics, American Inst. of (1931), One Physics Ellipse, College Park, MD 20740; 123,500; http://www.aip.org
Physiological Society, American (1887), 9650 Rockville Pike, Bethesda, MD 20814-3991; 10,668; http://www.the-aps.org
Phytopathological Society, American (1908), 3340 Pilot Knob Rd., St. Paul, MN 55121; 5,000; http://www.apsnet.org
Pilgrims Natl. Soc., Sons and Daughters of (1909), 3917 Heritage Dr., #104, Bloomington, MN 55437-2633; 2,000.
Pilot Intl. & Pilot Intl. Foundation (1921), 244 College St., Macon, GA 31201; 14,557; http://www.pilotinternational.org
Planetary Society (1980), 65 N. Catalina Ave., Pasadena, CA 91106; approx. 100,000; http://www.planetary.org
Planned Parenthood Federation of America, Inc. (1916), 810 Seventh Avenue, New York, NY 10019; http://www.plannedparenthood.org
Plastic Modelers Society, Intl. (1964), P.O. Box 2475, North Canton, OH 44720; 5,000; http://www.ipmsusa.org
Plastics Engineers, Society of (1942), 14 Fairfield Dr., P.O. Box 403, Brookfield, CT 06804; 33,000; http://www.4spe.org
Plastics Industry, Inc., Society of the (1937), 1801 K Street, NW, Ste. 600K, Wash., DC 20006; 2,000+ companies; http://www.socplas.org
Platform Association, Intl. (1831), 101 N. Center St., Westminster, MD, 21157; 2,000; http://www.internationalplatform.com
Poetry Society of America (1910), 15 Gramercy Park, New York, NY 10003; approx. 3,000; http://www.poetrysociety.org
Poets, The Academy of American (1934), 588 Broadway, Ste. 1203, New York, NY 10012; 8,000; http://www.poets.org
Police Assn., Intl. (1950 in UK, 1961 in U.S.), 100 Chase Ave., Yonkers, NY 10703; 291,000+; http://www.ipa-usa.org
Polish Army Veterans Assn. of America, Inc. (1921), 119 E. 15th St., Ste. 1, New York, NY 10003; 3,000.
Political Items Collectors, American (1945), P.O. Box 1149, Cibolo, TX 78108; 3,200; http://apic.ws
Political Science Assn., American (1903), 1527 New Hampshire Ave. NW, Wash., DC 20036; 16,200; http://www.apsanet.org
Political Science Assn., Southern (1928), Dept. of Political Science, 210 Deupree Hall, P.O. Box 1848, University of Mississippi, University, MS 38677; 1,800; http://www.olemiss.edu/orgs/spsa
Political Science, Academy of (1880), 475 Riverside Drive, Ste. 1274, New York, NY 10115; 6,000; http://www.psqonline.org
Political & Social Science, American Academy of (1891), 3937 Chestnut St., Philadelphia, PA 19104; 5,000; http:www.asc.upenn.edu/aapss.
Polo Assn., U.S. (1890), 771 Corporate Dr., Ste. 505, Lexington, KY 40503; 3,545; http://www.uspolo.org
Population Assn. of America (1931), 8630 Fenton St., Ste. 722, Silver Spring, MD 20910; 3,000; http://www.popssoc.org
Portuguese-American Federation, Inc., (1974), P.O. Box 694, Bristol, RI 02809; 250; http://www.apol.net/dightonrock/portuguese_american_federation1.htm.
Portuguese Continental Union of the U.S.A. (1925), 30 Cummings Park, Woburn, MA 01801; 5,488; http://members.aol.com/upceua
Postal Stationery Society, United (1995) P.O. Box 1792, Norfolk, VA 23501-1792; 1,200.
Postcard Dealers, Inc., International Federation of (1979), P.O. Box 1765, Manassas, VA 20108; 303.
Postmasters of the U.S., Natl. Assn. of (1898), 8 Herbert St., Arlington, VA 22305; 43,000, http://www.napus.org.
Postmasters of the U.S., Natl. League of (1887), 1023 N. Royal St., Alexandria, VA 22314; 25,000; http://www.postmasters.org
Powder Metallurgy Institute, American (formerly APMI International) (1959), 105 College Rd. E, Princeton, NJ 08540; approx. 2,700; http://www.mpif.org

Power Boat Assn., American (1903), 17640 E. Nine Mile Rd., Eastpointe, MI 48021; 6,500.
Printing Industries of America, Inc. (1887), 100 Dangerfield Rd., Alexandria, VA 22314; 14,000; http://www.gain.org
Procrastinators Club of America (1956), P.O. Box 712, Bryn Athyn, PA 19006; 14,500.
Professional Ball Players of America, Assn. of (1924), 1820 W. Orangewood Ave., Orange, CA 92868; http://www.apbpa.org
Protection of Old Fishes, Soc. for the (1967), NOAA HAZMAT, 7600 Sand Point Way, N.E., Seattle, WA 98115; 150.
Psi Chi (1929), 825 Vine St., P.O. Box 709, Chattanooga, TN 37401; http://www.psichi.org
Psoriasis Foundation, Natl. (1968), 6600 SW 92d Ave., Ste. 300, Portland, OR 97223; 40,000; http://www.psoriasis.org
Psychiatric Assn., American (1844), 1400 K St. NW, Wash., DC 20005; 40,453; http://www.psych.org
Psychical Research, American Society for (1885), 5 W. 73d St., New York, NY 10023; http://www.aspr.com
Psychoanalytic Assn., American (1911), 309 E. 49th St., New York, NY 10017; 3,500; http://apsa.org
Psychological Assn., American (1892), 750 1st St. NE, Wash., DC 20002; 159,000; http://www.apa.org
Psychological Assn. for Psychoanalysis, Inc., Natl. (1948), 150 W. 13th Street, New York, NY 10011; 365; http://www.npap.org
PTA, Natl. (1897), 330 N. Wabash Ave., Ste. 2100, Chicago, IL 60611; app 6.5 mil; http://www.pta.org
Public Administration, American Soc. for (1939), 1120 G St. NW, Wash., DC 20005; 11,000+; http://www.aspanet.org
Public Health Assn., American (1872), 800 I St. NW, Wash., DC 20001; http://www.apha.org
Public Relations Soc. of America, Inc. (1947), 33 Irving Pl., 3d Fl., New York, NY 10003; 17,383; http://www.prsa.org
Publishers, Assn. of American (1970), 71 5th Ave., New York, NY 10003; 300 cos; http://www.publishers.org
Pulp and Paper Industries, Technical Assn. of the (TAPPI) (1915), 15 Technology Pkwy. S, Norcross, GA 30092; 34,000; http://www.tappi.org

Quill and Scroll Society (1926), School of Journalism, The University of Iowa, Iowa City, IA, 52242; http://www.uiowa.edu/~quill.sc
Quota International, Inc. (1919), 1420 21st St. NW, Wash., DC 20036; 11,000+; http://www.quota.org

Rabbis, Central Conference of American (1889), 355 Lexington Ave., New York, NY 10017; 1,800; http://ccarnet.org
Racquetball Assn., U.S. (1968), 1685 W. Uintah, Colorado Springs, CO 80904; 20,000; http://www.usra.org
Radio Relay League, American (1914), 225 Main St., Newington, CT 06111; 172,000; http://www.arrl.org
Radio and Television Society Foundation, Intl. (1939), 420 Lexington Ave., Ste. 1714, New York, NY 10170; 1,787; http://www.irts.org
Railway Historical Society, Natl. (1936), P.O. Box 58547, Philadelphia, PA 19102; app. 15,000; http://www.nhrs.com
Railway Progress Institute (1908), 700 N. Fairfax St., #601, Alexandria, VA 22314-2098; 98 cos.; http://www.rpi.org
Range Management, Society for (1948), 445 Union Blvd., Ste. 230, Lakewood, CO 80228; 3,700; http://www.srm.org
Reading Assn., Intl. (1956), 800 Barksdale Rd., P.O. Box 8139, Newark, DE 19714; 90,000; http://www.reading.org
Real Estate Institute, Intl. (1968), 1224 N. Nokomis, Alexandria, MN 56308; 5,000; http://www.iami.org
Real Estate Appraisers, Natl. Assn. of (1966) 1224 N. Nokomis NE, Alexandria, MN 56308; 15,000; http://www.iami.org
Rebekah Assemblies, Intl. Assn. of (1922), 422 Trade St., Winston-Salem, NC 27101; 97,052.
Recreation and Park Assn., Natl. (1965), 22377 Belmont Ridge Road, Ashburn, VA 20148; 23,425; http://www.activeparks.org
Recycling Coalition, Natl. (1979), 1727 King St., Ste. 105, Alexandria, VA 22514; 3,500; http://www.nrc-recycle.org
Red Cross, American (1881), 431 18th St., NW, Wash., DC 20006; 1.3 mil volunteers; http://www.redcross.org
Reform Party (1992), 3281 N. Meadow Mine Place, Tucson, AZ 85745; http://www.reformparty.org
Refugee Committee, American (1978), 430 Oak St., Ste. 204., Ste. 350, Minneapolis, MN 55403; http://www.archq.org
Rehabilitation Assn., Natl. (1927), 633 South Washington Street, Alexandria, VA 22314; approx. 11,000; http://www.nationalrehab.org
Religion, American Academy of (1964), 825 Houston Mill Rd., NE, Atlanta, GA 30329; 10,000; http://www.aarweb.org
Renaissance Society of America (1954), 24 W. 12th St., New York, NY 10011; 2,600; http://www.r-s-a.org
Republican National Committee (1856), 310 1st St. SE, Wash., DC 20003; http://www.rnc.org
Review Appraisers/Mortage Underwriters, Natl. Assn. of (1975) 1224 N. Nokomis NE, Alexandria, MN 56308; 14,000; http://www.iami.org
Reserve Officers Assn. of the U.S. (1922), One Constitution Ave. NE, Wash., DC 20002; 95,000; http://www.roa.org
Restaurant Assn., Natl. (1919), 1200 17th St. NW, Wash., DC 20036; 33,000; http://www.restaurant.org
Retail Federation, Natl. (1908), 325 7th St. NW, Ste. 1100, Wash., DC 20004; 50,000; http://www.nrf.com
Retired Federal Employees, Natl. Assn. of (1921), 606 N. Washington St., Alexandria, VA 22314; 422,000 http://www.narfe.org
Retired Officers Assn. (1940), 201 N. Washington St., Alexandria, VA 22314; 391,000; http://www.troa.org
Retired Persons, American Assn. of (1958), 601 E St. NW, Wash., DC 20049; 32 mil.; http://www.aarp.org
Reye's Syndrome Foundation, Natl. (1974), 426 N. Lewis, Bryan, OH 43506; 4,760; http://www.reyessyndrome.org
Richard III Society, Inc. (1953),1421 Wisteria, Metairie, LA 70005; 800; http://www.r3.org
Rifle Assn., Natl. (1871), 11250 Waples Mill Rd., Fairfax, VA 22030; approx 3 mil; http://www.nra.org
Road & Transportation Builders Assn., American (1902), The ARTBA Building, 1010 Massachusetts Ave. NW, Wash., DC 20001; 5,000; http://www.artba.org
Roller Sports, U.S.A. (1937), 4730 South St., Lincoln, NE 68506; 30,000; http://www.usacrs.com
Rose Society, American (1892), 8877 Jefferson Page Rd, Shreveport, LA 71119; 22,000; http://www.ars.org
Rotary Intl. (1905), 1560 Sherman Ave., Evanston, IL 60201; 1,203,726; http://www.rotary.org
Running and Fitness Assn., American (1968), 4405 East West Highway, Ste. 405, Bethesda, MD 20814; approx. 16,500; http://www.americanrunning.org
Ruritan Natl., Inc. (1928), P.O. Box 487, Dublin, VA 24084; 33,447.

Safety Council, Natl. (1913), 1121 Spring Lake Dr., Itasca, IL 60143; 16,000; http://www.nsc.org
Safety Engineers, American Soc. of (1911), 1800 E. Oakton St., Des Plaines, IL 60018; 32,000; http://www.asse.org
Salt Institute (1914), 700 N. Fairfax St., Ste. 600, Alexandria, VA, 22314; 7 U.S., 37 Int'l.; http://www.saltinstitute.org
Sand Castle Builders, Intl. Assn. of (1988), 172 N. Pershing Ave., Akron, OH 44313; 200.
Save-the-Redwoods League (1918), 114 Sansome St., Ste. 1200, San Francisco, CA 94104; 65,000; http://www.savetheredwoods.org
School Administrators, American Assn. of (1865), 1801 N. Moore St., Arlington, VA 22209; 14,000+; http://www.aasa.org
School Boards Assn., Natl. (1940), 1680 Duke St., Alexandria, VA 22314; http://www.nsba.org
School Counselor Assn., American (1952), 801 N. Fairfax Street, Suite 310, Alexandria, VA 22314; 12,000; http://www.schoolcounselor.org
Science, American Assn. for the Advancement of (1848), 1200 New York Ave. NW, Wash., DC 20005; 138,000+; http://www.aaas.org
Science Fiction Society, World (1939), P.O. Box 8442, Van Nuys, CA 91409; 10,000.
Science Service Inc. (1921), 1719 N St. NW, Wash., DC 20036; http://www.sciserv.org
Sciences, Natl. Academy of (1863), 2101 Constitution Ave. NW, Wash., DC 20418; 4,000+; http://www.nas.edu
Science Teachers Assn., Natl. (1944), 1840 Wilson Blvd., Arlington, VA 22201; 53,000; http://www.nsta.org
Science Writers, Natl. Assn. of (1934), P.O. Box 294, Greenlawn, NY 11740; 2,380; http://www.nasw.org
Scrabble® Assn., Natl. (1980), P.O. Box 700, 403 Front St., Greenport, NY 11946; 10,000+; http://www.scrabble-assoc.com
Screen Actors Guild (1933), 5757 Wilshire Blvd., Los Angeles, CA 90036; 90,000; http://www.sag.com
Screenprinting & Graphic Imaging Assn., Intl. (1948), 10015 Main St., Fairfax, VA 22031; 4,000 ; http://www.sgia.org
2d Air Division Assn. of the 8th Air Force (1948), P.O. Box 484, Elkhorn, WI 53121; 6,000+.
Secondary School Principals, Natl. Assn. of (1916), 1904 Association Drive, Reston, VA 20191; 41,000; http://www. nassp.org
Secular Humanism, Council for (1980), P.O. Box 664, Amherst, NY 14226; 24,000; http://www.secularhumanism.org
Securities Industry Assn. (1972), 120 Broadway, 35th Fl., New York, NY 10271; 740+ firms; http://www.sia.com
Separation of Church & State, Americans United for (1947), 518 C St. NE, Wash., DC 20002; 60,000; http://www.au.org
Sertoma International (1912), 1912 E. Meyer Blvd., Kansas City, MO 64132; 24,992; http://www.sertoma.org
Sexuality Information & Education Council of the U.S. (SIECUS) (1964), 130 W. 42d St., Ste. 350, New York, NY 10036-7802; http://www.siecus.org
Sharkhunters Intl. (1983), P.O. Box 1539, Hernando, FL 34442; 6,400+; http://www.sharkhunters.com
Shipbuilders Council of America (1921), 1600 Wilson Blvd., Ste. 1000, Arlington, VA 22209; 37 member cos; http://www.shipbuilders.org
Ships in Bottles Assn. of America (1983), P.O. Box 180550, Coronado, CA 92178; 250.

Shrine of North America, The (1872), 2900 N. Rocky Point Dr., Tampa, FL 33607; approx 600,000+; http://shrinershq.org
Sierra Club (1892), 85 2d St., 2d Fl., San Francisco, CA 94105; 600,000+; http://www.sierraclub.org
Sigma Beta Delta (1994), Univ. of MO-St. Louis, 801 Natural Bridge Rd., 346 Woods Hall, St. Louis, MO 63121-0570; 17,000; http://www.sigmabetadelta.org
Skeet Shooting Assn., Natl. (1946), 5931 Roft Rd., San Antonio, TX 78253; 11,875; http://nssa-nsca.com
Small Business United, Natl. (1937), 1156 15th St. NW, Ste. 1100, Wash., DC 20005; 65,000+; http://www.nsbu.org
Social Work Education, Council on (1952), 1725 Duke St., Ste. 500, Alexandria, VA 22314; 3,500; http://www.cswe.org
Sociological Assn., American (1905), 1307 New York Avenue NW, Suite 700, Wash., DC 20005; 13,000; http://www.asanet.org
Softball Assn./USA Softball, Amateur (1933), 2801 Northeast 50th St., Oklahoma City, OK 73111; 240,000+ teams; http://www.softball.org
Software and Information Industry Assn. (formerly Information Industry Assn.) (1999), 1090 Vermont Ave. NW, Wash. DC 20005; 1,400; http://www.siia.net
Soil Science Society of America (1936), 677 S. Segoe Rd., Madison, WI 53711; 5,714; http://www.soils.org
Soldiers', Sailors', Marines' and Airmen's Club (1919), 283 Lexington Avenue, New York, NY 10016; 190; http://www.ssmaclub.org
Songwriters Guild of America (1931), 1500 Harbor Blvd., Weehawken, NJ 07087; 5,000+; http://www.songwriters.org
Sons of the American Colonists, Natl. Society of (1970) 9033 Lyndale Ave. S., Ste. 108, Bloomington, MD 55420-3535; 500.
Sons of the American Legion (1932), Box 1055, Indianapolis, IN 46206; 212,000; http://www.sal.legion.org
Sons of the American Revolution, Natl. Society of (1889), 1000 S. Fourth St., Louisville, KY 40203; 26,000; http://www.sar.org
Sons of Confederate Veterans (1896), P.O. Box 59, Columbia, TN 38402; 26,000; http://www.scv.org
Sons of the Desert Laurel & Hardy Appreciation Society (1965), P.O. Box 8341, Universal City, CA 91608; 2,500; http://www.wayoutwest.org.
Sons of Italy in America, Order (1905), 219 E St. NE, Wash., DC 20002; 500,000; http://www.osia.org
Sons of Norway (1895), 1455 W. Lake St., Minneapolis, MN 55408; 65,000; http://www.sofn.com
Soroptimist Intl. of the Americas (1921), Two Penn Center Plaza, Ste. 1000, Philadelphia, PA 19102; 47,000; http://www.soroptimist.org
Southern Christian Leadership Conference (1957), 334 Auburn Ave. NE, Atlanta, GA 30303; 1 mil.
Space Society, Natl. (1974), 600 Pennsylvania Ave SE, Ste. 201, Wash., DC 20003; 22,000; http://www.nss.org
Speech-Language-Hearing Assn., American (1925), 10801 Rockville Pike, Rockville, MD 20852; 99,000+; http://www.asha.org
Speleological Society, Natl. (1941), 2813 Cave Ave., Huntsville, AL 35810; 12,000; http://www.caves.org
Sports Car Club of America (1944), 9033 E. Eastern Pl., Englewood, CO 80112; 50,000+; http://www.scca.org
Sportscasters Assn., The American (1980), 225 Broadway, Ste. 2030, New York, NY 10007; 500+.
State & Local History, American Assn. for (1944), 1717 Church St., Nashville, TN 37203; 5,600; http://www.aaslh.org
State Governments, Council of (1933), 2760 Research Park Drive, P.O. Box 11910, Lexington, KY 40517; 50 states, 4 territories; http://www.csg.org
Statistical Assn., American (1839), 1429 Duke St., Alexandria, VA 22314; 18,000; http://www.amstat.org
Steamship Historical Society of America, Inc. (1935), 300 Ray Dr., Ste. 4, Providence, RI 02906; 3,500; http://www.sshsa.org
Stock Exchange, American (1911), 86 Trinity Pl., New York, NY 10006; 864; http://www.amex.com
Stock Exchange, New York (1792), 11 Wall St., New York, NY 10005; http://www.nyse.com
Stock Exchange, Philadelphia (1790), 1900 Market St., Philadelphia, PA 19103; 504; http://www.phlx.com
Student Councils, Natl. Society of (1931) 1904 Association Dr., Reston, VA 20191; approx. 1.5 mil; http://dsa.principals.org
Stuttering Project, Natl. (1977), 5100 E. LaPalma Ave., #208, Anaheim Hills, CA 92807; 2,800; http://www.nspstutter.org
Sudden Infant Death Syndrome Alliance (1987), 1314 Bedford Avenue, Suite 210, Baltimore, MD 21208; http://www.sidsalliance.org
Supreme Council, 33°, Scottish Rite of Freemasonry, Southern Jurisdiction (1801), 1733 16th St. NW, Wash., DC 20009-3103; 413,793; http://www.srmason-sj.org
Surgeons, American College of (1913), 633 N. Saint Clair St., Chicago, IL 60611; 60,000; http://www.facs.org
Symphony Orchestra League, American (1942), 910 17th St. NW, Wash., DC 20006; 850; http://www.symphony.org
Table Tennis Assn., U.S. (1933), One Olympic Plaza, Colorado Springs, CO 80909; 8,500; http://www.usatt.org
Tailhook Assn. (1956), 9696 Business Park Ave., San Diego, CA 92131; 11,800; http//www.tailhook.org
Tall Buildings and Urban Habitat, Council on (1969), Lehigh Univ., 117 ATLSS Dr., Bethlehem, PA 18015; 800; http://www.lehigh.edu/~inctbuh/inctbuh.html
Tau Beta Pi Association (1885), 508 Daugherty, Engineering Bldg., Univ of Tenn., Knoxville, TN 37901-2697; 400,000; http://www.tbp.org
Tax Administrators, Federation of (1932), 444 N. Capitol St. NW, Ste. 348, Wash., DC 20001; http://www.taxadmin.org
Tax Foundation (1937), 1250 H St. NW, Ste. 750, Wash., DC 20005; 50 U.S. states; http://www.taxfoundation.org
Taxpayers Union, Natl. (1969), 108 N. Alfred St., Alexandria, VA 22314; 335,000; http://www.ntu.org
Tea Assn. of the U.S.A., Inc. (1899), 420 Lexington Ave., New York, NY 10170; 150 corps; http://www.teausa.com
Teachers of English, Natl. Council of (1911), 1111 W. Kenyon Rd., Urbana, IL 61801; 77,000; http://www.ncte.org
Teachers of English to Speakers of Other Languages (1966), 700 Washington St., Ste. 200, Alexandria, VA 22314; 16,000; http://www.tesol.edu
Teachers of French, American Assn. of (1927), Mailcode 4510, Dept. of Foriegn Languages, Southern Illinois University, Carbondale, IL 62901-4501; 10,000; http://aatf.utsa.edu
Teachers of German, Inc., American Assn. of (AATG) (1926), 112 Haddontowne Ct. #104, Cherry Hill, NJ 08034-3668; 6,500; http://www.aatg.org
Teachers of Mathematics, Natl. Council of (1920), 1906 Association Drive, Reston, VA 20191-9988; 100,000; http://www.nctm.org
Teachers of Singing, Natl. Assn. of (1944), 6406 Merrill Road, Ste. B, Jacksonville, FL 32277; 5,443; http://www.nats.org
Teachers of Spanish & Portuguese, American Assn. of (1917), Univ. of Northern Colorado, 210 Butler-Hancock, Greeley, CO 80639; 12,000; http://www.aatsp.org
Telecommunications Pioneer Assn., Independent (1920), 1401 H St. NW, Ste. 600, Wash., DC 20005; 26,000+; http://www.telecom-pioneers.com
Television Arts & Sciences, Natl. Academy of (1955), 111 W. 57th St., Ste. 1020, New York, NY 10019; 11,000; http://www.emmyonline.org
Testing & Materials, American Society for (1898), 100 Barr Harbor Dr., P.O. Box C700, West Conshohocken, PA 19428; 32,000; http://www.astm.org
Theodore Roosevelt Assn. (1920), P.O. Box 719, Oyster Bay, NY 11771-0719; 2,100; www.theodoreroosevelt.org.
Theological Library Assn., American (1946), 250 S. Wacker Dr., Ste. 1600, Chicago, IL 60606; 808; http://www.atla.com
Theological Schools in the U.S. and Canada, The Assn. of (1918), 10 Summit Park Dr., Pittsburgh, PA 15275-1103; 243; http://www.ats.edu
Theological Seminary of California, Intl.(1976) 14617 Victory Blvd., Van Nuys, CA 94411; 2,000.
Theosophical Society in America (1875), 1926 N. Main St., Wheaton, IL 60187; 4,671; http://www.theosophical.org
Therapy Dogs Intl., Inc (1976), 88 Bartley Rd., Flanders, NJ 07836; approx. 7,000; http://www.tdi-dog.org
Thoreau Society (1941), 44 Baker Farm, Lincoln, MA 01773; 1,700+; http://www.walden.org
Thoroughbred Racing Assns. (1942), 420 Fair Hill Dr., Ste. 1, Elkton, MD 21921; 49 racing associations; http://www.tra-online.com
318th Service Group Assn. 9th AF (1991) 2114 West 29th St., Erie, PA 16508-1066; 310.
Tin Can Sailors (1976), P.O. Box 100, Somerset, MA 02726; 23,000; http://www.destroyers.org
Titanic Historical Society, Inc. (1963), 208 Main St., P.O. Box 51053, Indian Orchard, MA 01151-0053; 7,000; http://www.titanic1.org
Toastmasters Intl. (1924), P.O. Box 9052, Mission Viejo, CA 92690; 180,000+; http://www.toastmasters.org
Topical Assn., American (1949), 301 Embank St., Albuquerque, NM 87123-0820; 6,000; http://home.prcn.org/~pauld/ata
Totally Useless Skills, Institute of (1987), P.O. Box 181, Temple, NH 03084; 387; http://www.jlc.net/~useless
Toy Industry Assn., Inc. (1916), 1115 Broadway, Suite 400, New York, NY 10010; 300+ cos; http://www.toy-tma.com
Translators Assn., American (1959), 225 Reinekers Lane, Ste. 590, Alexandria, VA 22314; 7,200; http://www.atanet.org
Transportation Alternatives (1973), 115 W. 30th St., #1203, New York, NY 10001; 3,500; http://www.transalt.org
Transportation Assn., American Public (1974), 1666 K Street NW Suite 1100, Wash., DC 20006; 1,100 organizations; http://www.apta.com
Transportation Engineers, Inst. of (1930), 1099 14th St. NW, Suite 300 West, Wash., DC 20005-3438; 15,000; http://www.ite.org
Trapshooting Assn. of America, Amateur (1923), 601 W. National Road, Vandalia, OH 45377; 54,000; http://www.shootata.com
Travel Agents, American Soc. of (1931), 1101 King St., Ste. 200, Alexandria, VA 22314; 26,000; http://www.astanet. com

Travelers Protective Assn. of America (1890), 3755 Lindell Blvd., St. Louis, MO 63108; 109,641.
Trilateral Commission (1973), 1156 15th St., NW, Wash., DC 20005; 365; http://www.trilateral.org
Truck Historical Soc., American (1971), P.O. Box 531168, Birmingham, AL 35253; 21,950; http://www.aths.org
Trucking Assns., American (1933), 2200 Mill Rd., Alexandria, VA 22314; 4,000 cos.; http://www.truckline.com
Tuberous Sclerosis Assn., Natl. (1974), 8000 Corporate Dr., Ste. 120, Landover, MD 20785; 9,000+; http://www.goals.com/transrow/ntsa.htm

U.F.O. Society of America (1997), 10799 Sherman Grove Ave, #18, Sunland, CA 91040; 560.
UFOs, Natl. Investigations Committee on (1967) P.O. Box 53, Van Nuys, CA 91411; 2,000;
UNICEF, U.S. Fund for (1947), 333 E. 38th St., New York, NY 10016; http://www.unicefusa.org
Underwriters, Natl. Assn. of Life (1890), 1922 F St. NW, Wash., DC 20006; 143,000.
Underwriters (CPCU), Soc. of Chartered Property and Casualty (1944), 720 Providence Rd., P.O. Box 3009, Malvern, PA 19355; 28,000; http://www.cpusociety.org
Uniformed Services, Natl. Assn. for (1968), 5535 Hempstead Way, Springfield, VA 22151; 160,000+; http://www.naus.org
United Nations Assn. of the U.S.A. (1943), 801 2nd Ave., New York, NY 10017; 23,000; http://www.unausa.org
United Order True Sisters, Inc. (1846), 100 State St., Ste. 1020, Albany, NY 12207; approx. 2,000; http://uots.org
United Press Intl. (1907), 1510 H St. NW, Wash., DC 20005; http://www.upi.com
United Service Organizations (USO) (1941), Washington Navy Yard, 1008 Eberle Place SE, Ste. 301, Wash., DC 20374; 12,000+, http://www.uso.org
United Way of America (1918), 701 N. Fairfax St., Alexandria, VA 22314; 1,353; http://www.unitedway.org
Universities, Assn. of American (1900), 1200 New York Ave., NW, Ste. 550, Wash., DC 20005; 61 institutions; http://www.aau.edu
University Continuing Education Assn. (1915), One Dupont Circle NW, Ste. 615, Wash., DC 20036; 441 institutions; http://www.nucea.edu
University Women, American Assn. of (1881), 1111 16th St. NW, Wash., DC 20036; 150,000; http://www.aauw.org
Urban League, Natl. (1910), 120 Wall St., New York, NY 10005; 50,000; http://www.nul.org
USENIX Association (1975), 2560 Ninth Street, Ste. 215, Berkeley, CA 94710; 28,000; http://www.usenix.org
U.S. Term Limits (1992), 10 G St.; Ste. 410, Wash., DC 20002; http://www.termlimits.org
USO World Headquarters (1941), 1008 Eberle Place SE, Ste. 301, Wash., DC 20374-5096; http://www.uso.org
USS *Forrestal* CVA/CV/AVT-59 Assn., Inc. (1991), 300 Cassady Ave., Virginia Beach, VA 23452; 2,001; http://www.uss-forrestal.com
USS *Idaho* Assn. (1957), P.O. Box 711247, San Diego, CA 92171; 410.
USS *Los Angeles CA-135* (1978) 5933 Holgate Ave., San Jose, CA 95123; 600; http://www.uss-la-ca135.org
USS *North Carolina BB-55* Battleship Association (1962), P.O. Box 480, Wilmington, NC 28402; 791.
Utility Commissioners, Natl. Assn. of Regulatory (1898), 1101 Vermont Ave., NW, Ste. 200, Wash., DC 20005; 425; http://www.naruc.org

Vampire Research Center (1972), P.O. Box 5442, Babylon, NY 11707; 210.
Ventriloquists, North American Assn. of (1944), P.O. Box 420, Littleton, CO 80160; 1,450.
Veterans of Foreign Wars of the U.S. (1899), 406 W. 34th St., Kansas City, MO 64111; 1.9 mil.; http://www.vfw.org
Veterans of Foreign Wars of the U.S., Ladies Auxiliary to the (1914), 406 W. 34th St., Kansas City, MO 64111; 713,038; http://www.ladiesauxvfw.com
Veterans of the Vietnam War, Inc. (1980), 805 S. Township Blvd., Pittston, PA 18640-3327; 15,000; http://www.vvnw.org
Veterinary Medical Assn., American (1863), 1931 N. Meacham Rd., Ste. 100, Schaumburg, IL 60173; 64,000; http://www.avma.org
Victorian Society in America (1966), 219 S. Sixth St., Philadelphia, PA 19106; 1,600; http://www.victoriansociety.org
Volleyball, USA (1928), 715 S. Circle Dr., Colorado Springs, CO 80910; 110,000; http://www.usavolleyball.org
Volunteers of America (1896), 1660 Duke St., Alexandria, VA 22314-3421; 11,000 staff; http://www.voa.org

War Mothers, American (1917), 5415 Connecticut Ave., NW, Ste. L-30, Wash., DC 20015; under 800.
Watch & Clock Collectors, Inc., Natl. Assn. of (NAWCC) (1943), 514 Poplar St., Columbia, PA 17512; 35,000; http://www.nawcc.org
Watercolor Society, American (1866), 47 5th Ave., New York, NY 10003; 500+; http://www.watercolor-online.com/aws
Water Environment Federation (1928), 601 Wythe St., Alexandria, VA 22314; 40,000; http://www.wef.org
Water Works Assn., American (1881), 6666 W. Quincy Ave., Denver, CO 80235; 55,000; http://www.awwa.org
Welding Society, American (1919), 550 NW LeJeune Rd., Miami, FL 33126; 50,400; http://www.aws.org
Wheelchair Sports, USA (1957), 3595 E. Fountain Blvd., Ste. L-1, Colorado Springs, CO 80910; 4,000; http://www.wsusa.org
Wildlife Federation, Natl. (1936),11100 Wildlife Center Dr., Reston, VA, 20190; 4 mil.;http://www.nwf.org
Wildlife Management Institute (1911), 1101 14th St. NW, Ste. 801, Wash., DC 20005; 250; http://www.wildlifemgt.org/wmi
Wireless Pioneers Inc., The Society of (1967), P.O. Box 86, Geyserville, CA 95441; 1,500; http://www.sowp.org
Wizard of Oz Club, Intl. (1957), 1407 A St., Ste. D, Antioch, CA 94509; app.1,300; http://www.ozclub.org
Women, Natl. Organization for (NOW) (1966), 733 15th St. NW, 2nd Fl., Wash., DC 20005; 500,000; http://www.now.org
Women and Families, Natl. Partnership for (1971), 1875 Connecticut Ave. NW, Ste. 710, Wash., DC 20009; 2,500; http://www.nationalpartnership.org
Women Artists, Inc., Natl. Assn. of (1889), 41 Union Sq. W, #906, New York, NY 10003; 800; http://www.nwanet.org
Women in Communications, The Association for (1909 as Theta Sigma Phi), 780 Ritchie Hwy., Ste. 5-28, Severna Park, MD 21146; 7,500; http://www.womcom.org
Women in Radio and Television Inc., Amer. (1951), 1595 Spring Hill Rd., Vienna, VA 22182; http://www.awrt.org
Women Engineers, Society of (1950), 120 Wall St., 11th Fl., New York, NY 10005; 16,500; http://www.swe.org
Women Voters of the U.S., League of (1920), 1730 M St. NW, #1000, Wash., DC 20036; 130,000; http://www.lwv.org
Women's Army Corps Veterans Assn. (1946), P.O. Box 5577, Ft. McClellan, AL 36205; 3,200; http://www.armywomen.org
Women's Christian Temperance Union, Natl. (1874), 1730 Chicago Ave., Evanston, IL 60201-4585; http://www.wctu.org
Women's Clubs, General Federation of (1890), 1734 N St. NW, Wash., DC, 20036; 300,000 U.S; http://www.gfwc.org
Woodmen of America, Modern (1883), 1701 1st Ave., Rock Island, IL 61204; 750,000; http://www.modern-woodmen.org
Workmen's Circle (1900), 45 E. 33d St., New York, NY 10016; 35,000; http://www.circle.org
World Council of Churches, U.S. Office (1948), 475 Riverside Drive, Rm. 915, New York, NY 10115; 330+ denominations.
World Federalist Assn. (1947), 420 7th St. SE, Wash., DC 20003; 11,000; http://www.wfa.org
World Future Society (1966), 7910 Woodmont Ave., Ste. 450, Bethesda, MD 20814; 30,000; http://www.wfs.org
World Learning (1932), Kipling Rd., P.O. Box 676, Brattleboro, VT 05302-0676; 100,000; http://www.worldlearning.org
World Wildlife Fund (1961), 1250 24th St. NW, P.O. Box 97180, Wash., DC 20037; 1 mil+; http://www.worldwildlife.org
World's Fair Collectors Soc., Inc. (1968), P.O. Box 20806, Sarasota, FL 34216-3806; 400; http://members.aol.com/bbqprod/wfcs.html
Writers Guild of America, West (1954), 7000 W. Third St., Los Angeles, CA 90048; 8,300; http://www.wga.org

Yachting Assn., Southern California (1921), 5855 Naples Plaza, Ste. 211, Long Beach, CA 90803; 90 clubs & orgs., 21,500 families; http://www.scya.org
YMCA (Young Men's Christian Assns.) of the U.S.A. (1851) 101 N. Wacker Dr., Chicago, IL 60606; 17.5 mil.; http://www.ymca.net
Young Women's Christian Assn. of the U.S.A. (1907), Empire State Bldg., 350 Fifth Ave., Ste. 301, New York, NY 10118; approx. 2 mil; http://www.ywca.org

Zero Population Growth (1968), 1400 16th St. NW, Ste. 320, Wash., DC 20036; 55,000+; http://www.zpg.org
Zionist Organization of America (1897), 4 E. 34th St., New York, NY 10016; 50,000+; http://www.zoa.org
Zoo and Aquarium Assn., American (1924), 8403 Colesville Road, Suite 710, Silver Spring, MD 20910; 196 zoos + aquariums; http://www.aza.org

IT'S A FACT: The YMCA, which celebrated its 150th anniversary in 2001, can take credit for at least three modern sports that were invented under its auspices: basketball (by James Naismith, at a YMCA training school in Springfield, MA, in 1891), volleyball (by William Morgan at a Holyoke, MA, Y in 1895), and racquetball (by Y member Joe Sobek at a Greenwich, CT, YMCA facility in 1950).

SOCIAL SECURITY

Social Security Programs

Source: Social Security Administration; World Almanac research; data as of Sept. 2001.

Old-Age, Survivors, and Disability Insurance; Medicare; Supplemental Security Income

Social Security Benefits

Social Security benefits are based on a worker's primary insurance amount (PIA), which is related by law to the average indexed monthly earnings (AIME) on which Social Security contributions have been paid. The full PIA is payable to a retired worker who becomes entitled to benefits at age 65 and to an entitled disabled worker at any age. Spouses and children of retired or disabled workers and survivors of deceased workers receive set proportions of the PIA subject to a family maximum amount. The PIA is calculated by applying varying percentages to succeeding parts of the AIME. The formula is adjusted annually to reflect changes in average annual wages.

Automatic increases in Social Security benefits are initiated for December of each year, assuming the Consumer Price Index (CPI) for the 3d calendar quarter of the year increased relative to the base quarter, which is either the 3d calendar quarter of the preceding year or the quarter in which an increase legislated by Congress became effective. The size of the benefit increase is determined by the percentage rise of the CPI between the quarters measured.

The average monthly benefit payable to all retired workers amounted to $845 in Dec. 1999. The average benefit for disabled workers in that month amounted to $786.

Minimum and maximum monthly retired-worker benefits payable to individuals who retired at age 65[1]

Year of attainment of age 65	Minimum benefit[2]: Payable at retirement	Minimum benefit[2]: Payable effective Dec. 1999	Maximum benefit[2]: Payable at retirement, Men	Maximum benefit[2]: Payable at retirement, Women[3]	Maximum benefit[2]: Payable effective Dec. 1999, Men	Maximum benefit[2]: Payable effective Dec. 1999, Women[3]
1970	$64.00	$307.30	$189.80	$196.40	$899.90	$944.10
1980	133.90	307.30	572.00	—	1,297.90	—
1990	(4)	(4)	975.00	—	1,277.40	—
1993	(4)	(4)	1,128.80	—	1,313.80	—
1994	(4)	(4)	1,147.50	—	1,301.80	—
1995	(4)	(4)	1,199.10	—	1,323.30	—
1996	(4)	(4)	1,248.90	—	1,343.40	—
1997	(4)	(4)	1,326.60	—	1,386.20	—
1998	(4)	(4)	1,342.80	—	1,375.00	—
1999	(4)	(4)	1,373.10	—	1,405.00	—

(1) Assumes retirement at beginning of year. (2) The final benefit amount payable is rounded to next lower $1 (if not already a multiple of $1). (3) Benefits for women are the same as for men except where shown. (4) Minimum eliminated for workers who reached age 62 after 1981.

Amount of Work Required

To qualify for benefits, the worker generally must have worked a certain length of time in covered employment. Just how long depends on when the worker reaches age 62 or, if earlier, when he or she dies or becomes disabled.

A person is fully insured who has 1 quarter of coverage for every year after 1950 (or year age 21 is reached, if later) up to but not including the year the worker reaches 62, dies, or becomes disabled. In 2001, a person earns 1 quarter of coverage for each $830 of annual earnings in covered employment, up to 4 quarters per year.

The law permits special monthly payments under the Social Security program to certain very old persons who are not eligible for regular benefits since they had little or no opportunity to earn work credits during their working lifetime (so-called special age-72 beneficiaries).

To receive disability benefits, the worker, in addition to being fully insured, must generally have credit for 20 quarters of coverage out of the 40 calendar quarters before he or she became disabled. A disabled blind worker need meet only the fully insured requirement. Persons disabled before age 31 can qualify with a briefer period of coverage. Certain survivor benefits are payable if the deceased worker had 6 quarters of coverage in the 13 quarters preceding death.

Work credit for fully insured status for benefits

Born after 1929; die, become disabled, or reach age 62 in	Years needed	Born after 1929; die, become disabled, or reach age 62 in	Years needed
1983	8	1987	9
1984	8½	1988	9¼
1985	8½	1989	9½
1986	8¾	1990	9¾
		1991 and after	10

Contribution and benefit base

Calendar year	OASDI[1]	HI[2]
1990	$51,300	$51,300
1992	55,500	130,200
1993	57,600	135,000
1994	60,600	no limit
1995	61,200	no limit
1996	62,700	no limit
1997	65,400	no limit
1998	68,400	no limit
1999	72,600	no limit
2000	76,200	no limit
2001	80,400	no limit
2002	84,900 (est.)	no limit

(1) Old-Age, Survivors, and Disability Insurance. (2) Hospital Insurance.

Tax-rate schedule

(percentage of covered earnings)

Year	Total	OASDI	HI
	(for employees and employers, each)		
1979-80	6.13	5.08	1.05
1981	6.65	5.35	1.30
1982-83	6.70	5.40	1.30
1984	7.00	5.70	1.30
1985	7.05	5.70	1.35
1986-87	7.15	5.70	1.45
1988-89	7.51	6.06	1.45
1990 and after	7.65	6.20	1.45
	For self-employed		
1979-80	8.10	7.05	1.05
1981	9.30	8.00	1.30
1982-83	9.35	8.05	1.30
1984	14.00	11.40	2.60
1985	14.10	11.40	2.70
1986-87	14.30	11.40	2.90
1988-89	15.02	12.12	2.90
1990 and after	15.30	12.40	2.90

What Aged Workers Receive

When a person has enough work in covered employment and reaches retirement age (currently age 65 for full benefit, age 62 for reduced benefit), he or she may retire and receive monthly old-age benefits. The age when unreduced benefits become payable will increase gradually from 65 to 67 over a 21-year period beginning with workers age 62 in the year 2000 (reduced benefits will still be available as early as age 62, but with a larger reduction at that age).

Beginning with year 2000, the retirement earnings test has been eliminated beginning with the month in which the beneficiary reaches full-benefit retirement age (FRA). A person at and above FRA will not have Social Security benefits reduced because of earnings. In the calendar year in which a beneficiary reaches FRA, benefits are reduced $1 for every $3 of earnings above the limit allowed by law ($17,000 in 2000, $25,000 in 2001), but this reduction is only to months prior to attainment of FRA. For years before the year when the beneficiary attains FRA, the reduction in benefits is $1 for every $2 of earnings over the annual exempt amount $10,680 for year 2001).

For workers who reached age 65 between 1982 and 1989, Social Security benefits are raised by 3% for each year for which the worker between ages 65 and 70 (72 before 1984) failed to receive benefits, whether because of earnings from work or because the worker had not applied for benefits. The delayed retirement credit is 1% per year for workers who reached age 65 before 1982. The delayed retirement credit will gradually rise to 8% per year by 2008. The rate for workers who reached age 65 in 1998-99 is 5.5%; 2000-2001 will be 6.0%; 2002-2003, 6.5%.

Effective Dec. 1999, the special benefit for persons aged 72 or over who do not meet the regular coverage requirements became $210.60 a month. Like other monthly benefits, these payments are subject to cost-of-living increases. They are not made to persons on the public assistance or supplemental security income rolls.

For workers retiring before age 65, benefits are permanently reduced 5/9 of 1% for each month before FRA, up to 36 months. If the number of months exceeds 36, then the benefit is further reduced 5/12 of 1% per month. For example, when FRA reaches 67, for workers who retire at exactly age 62, there are a total of 60 months of reduction. The reduction for the first 36 months is 5/9 of 36%, or 20%. The reduction for the remaining 24 months is 5/12 of 24%, or 10%. Thus, when the FRA reaches 67, the amount of reduction at age 62 will be 30%. The nearer to age 65 the worker is when he or she begins collecting a benefit, the larger the benefit will be. The nearer to the FRA the worker is when he or she begins collecting a benefit, the larger the benefit will be.

Benefits for Worker's Spouse

The spouse of a worker who is getting Social Security retirement or disability payments may become entitled to an insurance benefit of one-half of the worker's PIA, when he or she reaches 65. Reduced spouse's benefits are available at age 62 and are permanently reduced 25/36 of 1% for each month before FRA, up to 36 months. If the number of months exceeds 36, then the benefit is further reduced 5/12 of 1% per month. Benefits are also payable to the aged divorced spouse of an insured worker if he or she was married to the worker for at least 10 years.

Benefits for Children of Workers

If a retired or disabled worker has a child under age 18, the child will get a benefit equal to half of the worker's unreduced benefit. So will the worker's spouse, even if under age 62, if he or she is caring for an entitled child of the worker who is under 16 or became disabled before age 22. Total benefits paid on a worker's earnings record are subject to a maximum; if the total that would be paid to a family exceeds that maximum, the dependents' benefits are adjusted downward. (Total monthly benefits paid to the family of a worker who retired in Jan. 2000 at age 65 and always had the maximum earnings creditable under Social Security cannot exceed $2,509.80.)

When entitled children reach age 18, their benefits generally stop, but a child disabled before age 22 may get a benefit as long as the disability meets the definition in the law. Benefits will be paid until age 19 to a child attending elementary or secondary school full-time.

Benefits may also be paid to a grandchild or step-grandchild of a worker or of his or her spouse, in special circumstances.

OASDI	June 2001	May 2000	May 1999	May 1998
Monthly beneficiaries, total (in thousands)[1]	45,669	**45,132**	**44,353**	**44,080**
Aged 65 and over, total	32,777	32,434	31,880	31,806
Retired workers	25,670	25,644	25,050	24,873
Survivors and dependents	6,948	6,790	6,830	6,933
Under age 65, total	12,978	12,697	12,473	12,274
Retired workers	2,638	2,564	2,505	2,458
Disabled workers	5,139	4,944	4,769	4,581
Survivors and dependents	5,201	5,189	5,199	5,235
Total monthly benefits (in millions)	**$35,200**	**$33,212**	**$31,449**	**$30,603**

(1) Totals may not add because of rounding.

What Disabled Workers Receive

A worker who becomes so disabled as to be unable to work may be eligible for a monthly disability benefit. Benefits continue until it is determined that the individual is no longer disabled. When a disabled-worker beneficiary reaches age 65, the disability benefit becomes a retired-worker benefit.

Benefits generally like those for dependents of retired-worker beneficiaries may be paid to dependents of disabled beneficiaries. However, the maximum family benefit in disability cases is generally lower than in retirement cases.

Survivor Benefits

If an insured worker should die, one or more types of benefits may be payable to survivors, again subject to a maximum family benefit as described above.

1. If claiming benefits at age 65, the surviving spouse will receive a benefit equal to 100% of the deceased worker's PIA. Benefits claimed before FRA are reduced for age with a maximum reduction of 28.5 percent at age 60. However, for those whose spouses claimed their benefits before age 65, these are limited to the reduced amount the worker would be getting if alive, but not less than 82% of the worker's PIA. Remarriage after the worker's death ends the surviving spouse's benefit rights. However, if the widow(er) marries and the marriage is ended, he or she regains benefit rights. (A marriage after age 60, age 50 if disabled, is deemed not to have occurred for benefit purposes.) Survivor benefits may also be paid to a divorced spouse if the marriage lasted for at least 10 years.

Disabled widows and widowers may under certain circumstances qualify for benefits after attaining age 50 at the rate of 71.5% of the deceased worker's PIA. The widow or widower must have become totally disabled before or within 7 years after the spouse's death or the last month in which he or she received mother's or father's insurance benefits.

2. There is a benefit for each child until the child reaches age 18. The monthly benefit for each child of a deceased worker is three-quarters of the amount the worker would have received if he or she had lived and drawn full retirement benefits. A child with a disability that began before age 22 may also receive benefits. Also, a child may receive benefits until reaching age 19 if he or she is in full-time attendance at an elementary or secondary school.

3. There is a mother's or father's benefit for the widow(er) if children of the worker under age 16 are in his or her care. The benefit is 75% of the PIA, and it continues until the youngest child reaches age 16, at which time payments stop even if the child's benefit continues. However, if the widow(er) has a disabled child beneficiary age 16 or over in care, benefits may continue.

4. Dependent parents may be eligible for benefits if they have been receiving at least half their support from the worker before his or her death, have reached age 62, and (except in certain circumstances) have not remarried since the worker's death. Each parent gets 75% of the worker's PIA; if only one parent survives, the benefit is 82 %.

5. A lump sum cash payment of $255 is made when there is a spouse who was living with the worker or a spouse or child who is eligible for immediate monthly survivor benefits.

Self-Employed Workers

A self-employed person who has net earnings of $400 or more in a year must report such earnings for Social Security tax and credit purposes. The person reports net returns from the business. Income from real estate, savings, dividends, loans, pensions, or insurance policies are not included unless it is part of the business.

A self-employed person receives 1 quarter of coverage for each $830 (for 2001), up to a maximum of 4 quarters.

The nonfarm self-employed have the option of reporting their earnings as $^2/_3$ of their gross income from self-employment, but not more than $1,600 a year and not less than their actual net earnings. This option can be used only if actual net earnings from self-employment income are less than $1,600, and may be used only 5 times. Also, the self-employed person must have actual net earnings of $400 or more in 2 of the 3 taxable years immediately preceding the year in which he or she uses the option.

When a person has both taxable wages and earnings from self-employment, wages are credited for Social Security purposes first; only as much self-employment income as brings total earnings up to the current taxable maximum becomes subject to the self-employment tax.

Farm Owners and Workers

Self-employed farmers whose gross annual earnings from farming are $2,400 or less may report 2/3 of their gross earnings instead of net earnings for Social Security purposes. Farmers whose gross income is over $2,400 and whose net earnings are less than $1,600 can report $1,600. Cash or crop shares received from a tenant or share farmer count if the owner participated materially in production or management. The self-employed farmer pays contributions at the same rate as other self-employed persons.

Agricultural employees. A worker's earnings from farm work count toward benefits (1) if the employer pays the worker $150 or more in cash during the year; or (2) if the employer spends $2,500 or more in the year for agricultural labor. Under these rules a person gets credit for 1 calendar quarter for each $830 in cash pay in 2001 up to 4 quarters.

Foreign farm workers admitted to the U.S. on a temporary basis are not covered.

Household Workers

Anyone 18 or older employed as maid, cook, laundry worker, nurse, babysitter, chauffeur, gardener, or other worker in the house of another is covered by Social Security if paid $1,300 or more in cash in calendar year 2001 by any one employer. Room and board do not count, but transportation costs count if paid in cash. The job need not be regular or full-time. The employee should get a Social Security card at the Social Security office and show it to the employer.

The employer deducts the amount of the employee's Social Security tax from the worker's pay, adds an identical amount as the employer's Social Security tax, and sends the total amount to the federal government.

Medicare Coverage

The Medicare health insurance program provides acute-care coverage for Social Security and Railroad Retirement beneficiaries age 65 and over, for persons entitled for 24 months to receive Social Security or Railroad Retirement disability benefits, and for certain persons with end-stage kidney disease. What follows is a basic description and may not cover all circumstances.

The basic Medicare plan, available nationwide, is a fee-for-service arrangement, where the beneficiary may use any provider accepting Medicare; some services are not covered and there are some out-of-pocket costs.

Under "Medicare + Choice," persons eligible for Medicare may have the option of getting services through a health maintenance organization (HMO) or other managed care plan. Any such plan must provide at least the same benefits, except for hospice services, and may provide added benefits—such as lower or no deductibles and coverage for some prescription drugs—but is usually subject to restrictions in choice of health care providers. In some plans services by outside providers are still covered for an extra out-of-pocket cost. Also available as options in some areas are Medicare-approved private fee-for-service plans and Medicare medical savings accounts.

Hospital insurance (Part A). The basic hospital insurance program pays covered services for hospital and posthospital care including the following:

- All necessary inpatient hospital care for the first 60 days of each benefit period, except for a deductible ($792 in 2001). For days 61-90, Medicare pays for services over and above a coinsurance amount ($198 per day in 2001). After 90 days, the beneficiary has 60 reserve days for which Medicare helps pay. The coinsurance amount for reserve days was $396 in 2001.
- Up to 100 days' care in a skilled-nursing facility in each benefit period. Hospital insurance pays for all covered services for the first 20 days; for the 21-100th day, the beneficiary pays coinsurance ($99 a day in 2001).
- Part-time home health care provided by nurses or other health workers.
- Limited coverage of hospice care for individuals certified to be terminally ill.

There is a premium for this insurance in certain cases.

Medical insurance (Part B). Elderly persons can receive benefits under this supplementary program only if they sign up for them and agree to a monthly premium ($50 if you sign up upon being eligible in 2000). The federal government pays the rest of the cost. The medical insurance program usually pays 80% of the approved amount (after the first $100 in each calendar year) for the following services:

- Covered services received from a doctor in his or her office, in a hospital, in a skilled-nursing facility, at home, or in other locations.
- Medical and surgical services, including anesthesia.
- Diagnostic tests and procedures that are part of the patient's treatment.
- Radiology and pathology services by doctors while the individual is a hospital inpatient or outpatient.
- Other services such as X-rays, services of a doctor's office nurse, drugs and biologicals that cannot be self-administered, transfusions of blood and blood components, medical supplies, physical/occupational therapy and speech pathology services.

In addition to the above, certain other tests or preventive measures are now covered without an additional premium. These include mammograms, bone mass measurement, colorectal cancer screening, and flu shots. Outpatient prescription drugs are generally not covered under the basic plan, nor are routine physical exams, dental care, hearing aids, or routine eye care. There is limited coverage for nonhospital treatment of mental illness.

To get medical insurance protection, persons approaching age 65 may enroll in the 7-month period that includes 3 months before the 65th birthday, the month of the birthday, and 3 months after the birthday, but if they wish coverage to begin in the month they reach age 65, they must enroll in the 3 months before their birthday. Persons not enrolling within their first enrollment period may enroll later, during the first 3 months of each year (coverage begins July 1), but their premium may be 10% higher for each 12-month period elapsed since they first could have enrolled.

The monthly premium is deducted from the cash benefit for persons receiving Social Security, Railroad Retirement, or Civil Service retirement benefits. Income from the medical premiums and the federal matching payments are put in a Supplementary Medical Insurance Trust Fund, from which benefits and administrative expenses are paid.

Further details are available on the Internet at http://www.medicare.gov or by calling 1-800-638-6833.

Medicare card. Persons qualifying for hospital insurance under Social Security receive a health insurance card similar to cards now used by Blue Cross and other health insurers. The card indicates whether the individual has taken out medical insurance protection. It is to be shown to the hospital, skilled-nursing facility, home health agency, doctor, or whoever provides the covered services.

Payments are generally made only in the 50 states, Puerto Rico, Virgin Islands, Guam, and American Samoa.

Social Security Financing

Social Security is paid for by a tax on certain earnings (for 2001, on earnings up to $80,400) for Old Age, Survivors, and Disability Insurance and on all earnings (no upper limit) for Hospital Insurance with the Medicare Program; the taxable earnings base for OASDI has been adjusted annually to reflect increases in average wages. The employed worker and his or her employer share Social Security taxes equally.

Employers remit amounts withheld from employee wages for Social Security and income taxes to the Internal Revenue Service; employer Social Security taxes are also payable at the same time. (Self-employed workers pay Social Security taxes when filing their regular income tax forms.) The Social Security taxes (along with revenues arising from partial taxation of the Social Security benefits of certain high-income people) are transferred to the Social Security Trust Funds—the Federal Old-Age and Survivors Insurance (OASI) Trust Fund, the Federal Disability Insurance (DI) Trust Fund, and the Federal Hospital Insurance (HI) Trust Fund; they can be used only to pay benefits, the cost of rehabilitation services, and administrative expenses. Money not immediately needed for these purposes is by law invested in obligations of the federal government, which must pay interest on the money borrowed and must repay the principal when the obligations are redeemed or mature.

Supplemental Security Income

On Jan. 1, 1974, the Supplemental Security Income (SSI) program established by the 1972 Social Security Act amendments replaced the former federal grants to states for aid to the needy aged, blind, and disabled in the 50 states and the District of Columbia. The program provides both for federal payments, based on uniform national standards and eligibility requirements, and for state supplementary payments varying from state to state. The Social Security Administration administers the federal payments financed from general funds of the Treasury—and the state supplements as well, if the state elects to have its supplementary program federally administered. States may supplement the federal payment for all recipients and must supplement it for persons otherwise adversely affected by the transition from the former public assistance programs. In May 2000, the number of persons receiving federally administered payments was 6,633,706 and the payments totaled $2.7 billion.

The maximum monthly federal SSI payment for individuals with no other countable income, living in their own household, was $531 in 2001. For couples it was $796.

Social Security Statement

On Oct. 1, 1999, the Social Security Administration initiated the mailing of an annual *Social Security Statement* to all workers age 25 and older not already receiving benefits. Workers will automatically receive statements about 3 months before their birth month. The statement provides estimates of potential monthly Social Security retirement, disability, and survivor benefits as well as a record of lifetime earnings. The statement also provides workers an easy way to determine whether their earnings are accurately posted in Social Security records. For further information contact the Social Security Administration toll-free at 1-800-772-1213 or visit its website at http://www.ssa.gov

Examples of Monthly Benefits Available

Description of benefit or beneficiary	For low earnings ($14,258 in 2000)[1]	For avg. earnings ($31,685[2] in 2000)	For max. earnings ($76,200 in 2000)
Primary insurance amount (worker retiring at 65)	$597.50	$986.50	$1,433.90
Maximum family benefit (worker retiring at 65)	896.30	1,797.60	2,509.80
Maximum family disability benefit (worker disabled at 55; in 2000)*	919.70	1,616.20	2,450.40
Disabled worker (worker disabled at 55)			
Worker alone	654.20	1,077.50	1,633.60
Worker, spouse, and 1 child	918.00	1,615.00	2,449.00
Retired worker claiming benefits at age 62:			
Worker alone[3]	518.00	853.00	1,241.00
Worker with spouse claiming benefits at—			
Age 65 or over	845.00	1,392.00	2,025.00
Age 62[3]	763.00	1,257.00	1,829.00
Widow or widower claiming benefits at—			
Age 65 or over[4]	597.00	986.00	1,433.00
Age 60 (spouse died at 65 without receiving reduced benefits)	427.00	705.00	1,025.00
Disabled widow or widower claiming benefits at age 50-59[5]	427.00	705.00	1,025.00
1 surviving child	448.00	739.00	1,075.00
Widow or widower age 65 or over and 1 child[6]	1,045.00	1,725.00	2,509.00
Widowed mother or father and 1 child[6]	896.00	1,478.00	2,150.00
Widowed mother or father and 2 children[6]	894.00	1,797.00	2,508.00

Effective Jan. 1999, for beneficiaries with first entitlement in 1998. *Assumes work beginning at age 22. (1) 45% of average. (2) Estimate. (3) Assumes maximum reduction. (4) A widow(er)'s benefit amount is limited to the amount the spouse would have been receiving if still living, but not less than 82.5% of the PIA. (5) Effective Jan. 1984, disabled widow(er)s claiming a benefit at ages 50-59 receive a benefit equal to 71.5% of the PIA. (6) Based on worker dying at age 65.

IT'S A FACT: The Social Security Act was signed into law by Pres. Franklin D. Roosevelt on Aug. 14, 1935, and the first Social Security cards were issued in 1936. Since then, approximately 415 million Social Security cards have been issued, with 217 million remaining active as of July 2001.

Social Security Trust Funds

Old-Age and Survivors Insurance Trust Fund, 1940-2000

(in millions)

	INCOME					DISBURSEMENTS					
Fiscal year[1]	Total	Net contributions[2]	Income from taxing benefits	Payments from the Treasury fund[3]	Net interest[4]	Total	Benefit payments[5]	Administrative expenses	Transfers to Railroad Retirement program	Net increase in fund	Fund at end of period
1940	$368	$325	—	—	$43	$62	$35	$26	—	$306	$2,031
1950	2,928	2,667	—	$4	257	1,022	961	61	—	1,905	13,721
1960	11,382	10,866	—	—	516	11,198	10,677	203	$318	184	20,324
1970	32,220	30,256	—	449	1,515	29,848	28,798	471	579	2,371	32,454
1980	105,841	103,456	—	540	1,845	107,678	105,083	1,154	1,442	–1,837	22,823
1990	286,653	267,530	$4,848	–2,089	16,363	227,519	222,987	1,563	2,969	59,134	214,197
1996	363,741	321,557	6,471	7	35,706	308,217	302,861	1,802	3,554	575,096	589,121
1997	397,169	349,946	7,426	2	39,795	322,073	316,257	2,128	3,688	67,916	567,395
1998	424,848	371,207	9,149	1	44,491	332,324	326,762	1,899	3,662	92,524	681,645
1999	457,040	396,352	10,899	—	49,788	339,874	334,383	1,809	3,681	117,167	798,812
2000	490,513	421,391	11,594	—	57,529	358,339	352,652	2,149	3,538	132,174	930,986

(1) Fiscal years 1980 and later consist of the 12 months ending on Sept. 30 of each year. Fiscal years prior to 1977 consisted of the 12 months ending on June 30 of each year. (2) Beginning in 1983, includes transfers from general fund of Treasury representing contributions that would have been paid on deemed wage credits for military service in 1957 and later, if such credits were considered covered wages. (3) Includes payments (a) in 1947-52 and in 1967 and later, for costs of noncontributory wage credits for military service performed before 1957; (b) in 1972-83, for costs of deemed wage credits for military service performed after 1956; and (c) in 1969 and later, for costs of benefits to certain uninsured persons who attained age 72 before 1968. (4) Net interest includes net profits or losses on marketable investments. Beginning in 1967, administrative expenses were charged currently to the trust fund on an estimated basis, with a final adjustment, including interest, made in the next fiscal year. The amounts of these interest adjustments are included in net interest. For years prior to 1967, the method of accounting for administrative expenses is described in the 1970 Annual Report. Beginning in Oct. 1973, the figures shown include relatively small amounts of gifts to the fund. During 1983-91, interest paid from the trust fund to the general fund on advance tax transfers is reflected. (5) Beginning in 1967, includes payments for vocational rehabilitation services furnished to disabled persons receiving benefits because of their disabilities. Beginning in 1983, amounts are reduced by amount of reimbursement for unnegotiated benefit checks.

Disability Insurance Trust Fund, 1970-2000

(in millions)

Fiscal year[1]	INCOME: Total	Net contributions[2]	Income from taxation of benefits	Payments from the Treasury fund[3]	Net interest[4]	DISBURSEMENTS: Total	Benefit payments[5]	Administrative expenses	Transfers to Railroad Retirement program	Net increase in fund	Fund at end of period
1970	$4,774	$4,481	—	$16	$277	$3,259	$3,085	$164	$10	$1,514	$5,614
1980	13,871	13,255	—	130	485	15,872	15,515	368	–12	2,001	3,629
1990	28,791	28,539	$144	–775	883	25,616	24,829	707	80	3,174	11,079
1996	60,710	57,325	373	—	3,012	45,351	44,189	1,160	2	15,359	52,924
1997	60,499	56,037	470	—	3,992	47,034	45,695	1,280	59	13,465	66,389
1998	64,357	58,966	558	—	4,832	49,931	48,207	1,567	157	14,425	80,815
1999	69,541	63,203	661	—	5,677	53,035	51,381	1,519	135	16,507	97,321
2000	77,920	71,093	721	–836	6,942	56,782	54,983	1,639	159	21,138	118,459

(1) Fiscal years 1977 and later consist of the 12 months ending Sept. 30 of each year. Fiscal years prior to 1977 consisted of the 12 months ending June 30 of each year. (2) Beginning in 1983, includes transfers from general fund of Treasury representing contributions that would have been paid on deemed wage credits for military service in 1957 and later, if such credits were considered to be covered wages. (3) Includes payments (a) for costs of noncontributory wage credits for military service performed before 1957; and (b) in 1972-83, for costs of deemed wage credits for military service performed after 1956. (4) Net interest includes net profits or losses on marketable investments. Administrative expenses are charged currently to the trust fund on an estimated basis, with a final adjustment, including interest, made in the following fiscal year. Figures shown include relatively small amounts of gifts to the fund. During the years 1983-91, interest paid from the trust fund to the general fund on advance tax transfers is reflected. (5) Includes payments for vocational rehabilitation services. Beginning in 1983, amounts are reduced by amount of reimbursement for unnegotiated benefit checks. **NOTE:** Totals may not add because of rounding.

Supplementary Medical Insurance Trust Fund (Medicare), 1975-2000

(in millions)

Fiscal year[1]	INCOME: Premium from participants	Government contributions[2]	Interest and other income[3]	Total Income	DISBURSEMENTS: Benefit payments	Administrative expenses	Total disbursements	Balance in fund at end of year[4]
1975	$1,887	$2,330	$105	$4,322	$3,765	$405	$4,170	$1,424
1980	2,928	6,932	415	10,275	10,144	593	10,737	4,532
1990	11,494[5]	33,210	1,434[5]	46,138[5]	41,498	1,524[5]	43,022[5]	14,527[5]
1995	19,244	36,988	1,937	58,169	63,491	1,722	65,213	13,874
1996	18,931	61,702	1,392	82,025	67,176	1,771	68,946	26,953
1997	19,141	59,471	2,193	80,806	71,133	1,420	72,553	35,206
1998	19,427	59,919	2,608	81,955	74,837[6]	1,435	76,272	40,889
1999	20,160	62,185	2,933	85,278	79,008[6]	1,510	80,518	45,649
2000	20,515	65,561	3,164	89,239	87,212[6]	1,780	88,992	45,896

(1) Fiscal year 1975 consists of the 12 months ending on June 30, 1975; fiscal years 1980 and later consist of the 12 months ending on September 30 of each year. (2) General fund matching payments, plus certain interest-adjustment items. (3) Other income includes recoveries of amounts reimbursed from the trust fund that are not obligations of the trust fund and other miscellaneous income. (4) The financial status of the program depends on both the assets and the liabilities of the program. (5) Includes the impact of the Medicare Catastrophic Coverage Act of 1988 (PL 100-360). (6) Benefit payments less monies transferred from the HI trust fund for home health agency costs, as provided for by PL 105-33. **NOTE:** Totals do not necessarily equal the sums of rounded components.

Hospital Insurance Trust Fund (Medicare), 1975-2000

(in millions)

Fiscal year[1]	INCOME: Payroll taxes	Income from taxation of benefits	Transfers from railroad retirement acct.	Reimbursement for uninsured persons	Premiums from voluntary enrollees	Pymts. for military wage credits	Interest on investments and other income[2]	Total income	DISBURSEMENTS: Benefit pymts.[3]	Administrative expense[4]	Total disbursements	Net increase in fund	Fund at end of year
1975	$11,291	—	$132	$481	$6	$48	$609	$12,568	$10,353	$259	$10,612	$1,956	$9,870
1980	23,244	—	244	697	17	141	1,072	25,415	23,790	497	24,288	1,127	14,490
1990	70,655	—	367	413	113	107	7,908	79,563	65,912	774	66,687	12,876	95,631
1995	98,053	3,913	396	462	998	61	10,963	114,847	113,583	1,300	114,883	–36	129,520
1996	106,934	4,069	401	419	1,107	–2,293[5]	10,496	121,135	124,088	1,229	125,317	–4,182	125,338
1997	112,725	3,558	419	481	1,279	70	10,017	128,548	136,175	1,661	137,836	–9,287	116,050
1998	121,913	5,067	419	34	1,320	67	9,382	138,203	135,487[6]	1,653	137,140	1,063	117,113
1999	134,385	6,552	430	652	1,401	71	9,523	153,015	129,463[6]	1,978	131,441	21,574	138,687
2000	137,738	8,787	465	470	1,392	2	10,827	159,681	127,934[6]	2,340	130,284	29,397	168,084

(1) Fiscal year 1975 consists of the 12 months ending on June 30, 1975; fiscal years 1980 and later consist of the 12 months ending Sept. 30 of each year. (2) Other income includes recoveries of amounts reimbursed from the trust fund that are not obligations of the trust fund, receipts from the fraud and abuse control program, and a small amount of miscellaneous income.(3)Includes costs of Peer Review Organizations (beginning with the implementation of the Prospective Payment System on Oct. 1, 1983). (4) Includes costs of experiments and demonstration projects. Beginning in 1997, includes fraud and abuse control expenses, as provided for by PL 104-191.(5) Includes the lump-sum general revenue adjustment of $–2,366 mil, as provided for by PL 98-21. (6) Includes monies transferred to the SMI trust fund for home health agency costs, as provided for by PL 105-33. **NOTE:** Totals do not necessarily equal the sums of rounded components.

HEALTH

Basic First Aid

Knowing what to do for an injured victim until a doctor or other trained person gets to the accident scene can save a life, especially in cases of severe bleeding, stoppage of breathing, poisoning, and shock.

People with special medical problems, such as diabetes, cardiovascular disease, epilepsy, or allergy, are urged to wear some sort of emblem identifying the problem, as a safeguard against receiving medication that might be harmful or even fatal. Emblems may be obtained from Medic Alert Foundation, 2323 Colorado Ave., Turlock, CA 95382; 800-344-3226.

It is important to get medical assistance as soon as possible.

Animal bite — Wash wound with soap under running water and apply antibiotic ointment and dressing. When possible, the animal should be caught alive for rabies testing.

Asphyxiation — Start rescue breathing immediately after getting patient to fresh air.

Bleeding — Elevate the wound above the heart if possible. Press hard on wound with sterile compress until bleeding stops. Send for doctor if bleeding is severe.

Burn — If mild, with skin unbroken and no blisters, flush with cool water until pain subsides. Apply a loose sterile dry dressing if necessary. If severe, send for doctor. Apply sterile compresses and keep patient comfortably warm until doctor's arrival. Do not try to clean burn or break blisters.

Chemical in eye — With patient lying down, pour cupfuls of water immediately into corner of eye, letting it run to other side to remove chemicals thoroughly. Cover with sterile compress. Get medical attention immediately. Continue to flush until medical help arrives.

Choking — See **Abdominal Thrust**.

Convulsions — Place person on back on bed or rug. Loosen clothing. Turn head to side. Do not place a blunt object between the patient's teeth. If convulsions do not stop, get medical attention immediately.

Cut (minor) — Apply mild antiseptic and sterile compress after washing with soap under warm running water.

Fainting — If victim feels faint, lower head to knees. Lay patient down on back with head turned to side if he or she becomes unconscious. Elevate the legs 8 to 10 inches. Loosen clothing and open windows. Keep patient lying quietly for at least 15 minutes after he or she regains consciousness. Call doctor if faint lasts for more than a few minutes.

Foreign body in eye — Touch object with moistened corner of handkerchief if it can be seen. If it cannot be seen or does not come out after a few attempts, take patient to doctor. Do not rub the eye.

Frostbite — Handle frostbitten area gently. Do not rub. Soak affected area in water no warmer than 105F. Do not allow frostbitten area to touch the container. Soak until frostbitten part looks red and feels warm. Loosely bandage. If fingers or toes are frostbitten, put gauze between them.

Heat Stroke and Heat Exhaustion — Remove the patient from the heat. Loosen any tight clothing and apply cool, wet cloths to the skin. Give the victim cool water, to drink slowly. Call an ambulance if the victim refuses water, vomits, or experiences changes in consciousness.

Hypothermia — Move victim to a warm place. Remove wet clothing and dry victim, if necessary. Warm patient gradually by wrapping the person in warm blankets or clothing. Apply heat pads or other heat sources if available, but not directly to the body. Give the victim warm liquids. Call an ambulance if breathing is slowed or stopped or if the pulse is slow or irregular.

Loss of Limb — If a limb is severed, it is important to properly protect the limb so that it can possibly be reattached. After the patient is cared for, the limb should be wrapped in a sterile gauze or clean material and placed in a clean plastic bag, garbage can, or other suitable container. Pack ice around the limb on the OUTSIDE of the bag to keep the limb cold. Call ahead to the hospital to alert staff there of the situation.

Poisoning — Call ambulance and Poison Control Center and follow their directions. Use antidote listed on label if container is found. Do not give the victim any food or drink or induce vomiting, unless specified by the Poison Control Center.

Shock (injury-related) — Keep the victim lying down on back; if uncertain as to his or her injuries, keep the patient flat on the back. Otherwise elevate feet and legs 12 inches. Maintain normal body temperature; if the weather is cold or damp, place blankets or extra clothing over and under the victim; if weather is hot, provide shade.

Snakebite — Wash the injury. Keep the area still and at a lower level than the heart. Keep the victim quiet. Use a snakebite kit if available.

Sprains and fractures — Apply ice to reduce swelling and pain. Do not try to straighten or move broken limbs. Apply a splint to immobilize the injured area if the victim must be transported.

Sting from insect — If possible, remove stinger. Wash the area with soap and water; cover it to keep it clean. Apply a cold pack to reduce pain and swelling. Call physician immediately if body swells or patient collapses.

Unconsciousness — Send for doctor and place person on his or her back. Start rescue breathing if victim stops breathing. Never give food or liquids.

Abdominal Thrust (Heimlich Maneuver)

The American Red Cross and the American Heart Association both agree that the recommended first aid for choking victims is the abdominal thrust, also known as the Heimlich maneuver, after its creator, Dr. Henry Heimlich. Slaps on the back are no longer advised and may even prove detrimental to a choking victim.

- Get behind the victim and wrap your arms around him or her about 1-2 inches above the navel.
- Make a fist with one hand and place it, with the thumb knuckle pressing inward at the abdomen.
- Grasp the fist with the other hand and give upward thrusts until object is removed or help arrives.

Rescue Breathing

Stressing that your breath can save a life, the American Red Cross gives the following directions for rescue breathing if the victim is not breathing:

- Determine consciousness by tapping the victim on the shoulder and asking loudly, "Are you okay?"
- Tilt the victim's head back so that the chin is pointing upward. Do not press on the soft tissue under the chin, as this might obstruct the airway. If you suspect that an accident victim might have neck or back injuries, open the airway by placing the tips of your index and middle fingers on the corners of the person's jaw to lift it forward without tilting the head.
- Place your cheek and ear close to the victim's mouth and nose. Look at the chest to see if it rises and falls. Listen and feel for air to be exhaled for about 5 seconds.
- If there is no breathing, pinch the victim's nostrils shut with the thumb and index finger of your hand that is pressing on the victim's forehead. Another way to prevent leakage of air when the lungs are inflated is to press your cheek against the victim's nose.
- Blow air into the mouth by taking a deep breath and then sealing your mouth tightly around the victim's mouth. Initially, give 2, slow (approx. 1.5 seconds each), full breaths.
- Watch the patient's chest to see if it rises.
- Stop when the chest is expanded. Raise your mouth; turn your head to the side and listen for exhalation.
- Watch the chest to see if it falls. Check pulse. If there is a pulse, continue rescue breathing. If there is no pulse, start CPR.
- Repeat giving 1 breath every 5 seconds until the victim starts breathing.

Note: Infants (up to 1 year) and children (1 to 8 years) should be treated as described above, except for the following:

- Do not tilt the head as far back as an adult's head.
- Both the mouth and nose of an infant should be sealed by the mouth.
- Give breaths to a child once every 3 seconds.
- Blow into the infant's mouth and nose once every 3 seconds with less pressure and volume than for a child.

Heart and Blood Vessel Disease

Source: American Heart Association, 7272 Greenville Ave., Dallas, TX 75231-4596; phone: (800) 242-8721

Warning Signs

Of Heart Attack

- Uncomfortable pressure, fullness, squeezing, or pain in the center of the chest lasting 2 minutes or longer
- Pain may radiate to the shoulder, arm, neck, or jaw
- Sweating may accompany pain or discomfort
- Nausea and vomiting also may occur
- Shortness of breath, dizziness, or fainting may accompany other signs

The American Heart Association advises immediate action at the onset of these symptoms. The association points out that more than half of heart attack victims die within 1 hour of the onset of symptoms and before they have reached the hospital.

Of Stroke

- Sudden numbness or weakness of face, arm or leg, especially on one side of the body
- Sudden confusion, trouble speaking or understanding
- Sudden trouble seeing in one or both eyes
- Sudden trouble walking, dizziness, loss of balance or coordination
- Sudden severe headache with no known cause

Some Major Risk Factors

Blood pressure—High blood pressure increases the risk of stroke, heart attack, kidney failure, and congestive heart failure.

Cholesterol—A blood cholesterol level over 240 mg/dl (milligrams of cholesterol per deciliter of blood) approximately doubles the risk of coronary heart disease; over 40 mil have a cholesterol level above 240 mg/dl. Levels between 200 and 240 mg/dl are in a zone of moderate and increasing risk.

Smoking—Cigarette smokers have more than twice the risk of heart attack and 2-4 times the risk of sudden cardiac death as nonsmokers. Young smokers have a higher risk for early death from stroke.

Obesity—Using a body mass index (BMI) of 25 and higher for overweight and 30 and higher for obesity, 106.9 mil Americans age 20 and over are overweight and 43.6 mil are obese.

Understanding Blood Pressure

High blood pressure, or hypertension, affects people of all races, sexes, ethnic origins, and ages. Various causes can trigger this often symptomless disease. Since hypertension can increase one's risk for stroke, heart attack, kidney failure, and congestive heart failure, it is recommended that individuals have a blood pressure reading at least once every 2 years (more often if advised by a physician).

A blood pressure reading is really two measurements in one, with one written over the other, such as 122/78. The **upper number (systolic pressure)** represents the amount of pressure in the blood vessels when the heart contracts (beats) and pushes blood through the circulatory system. The **lower number (diastolic pressure)** represents the pressure in the blood vessels between beats, when the heart is resting. According to National Institutes of Health guide-lines, normal blood pressure is below 130/85 and "high normal" is between 130/85 and 139/89.

High blood pressure is divided into 3 stages, based upon severity:

- **Stage 1** is from 140/90 through 159/99
- **Stage 2** is from 160/100 through 179/109
- **Stage 3** is 180/110 or greater

The diagnosis of hypertension can be based on either the systolic or the diastolic reading.

High blood pressure usually cannot be cured, but it can be controlled in a variety of ways, including lifestyle modifications and medication. Treatment always should be at the direction and under the supervision of a physician.

> ***IT'S A FACT:*** Eight million blood cells are produced in the body every second. They replace eight million blood cells that die every second.

Examples of Moderate[1] Amounts of Exercise

Source: *Physical Activity and Health: A Report of the Surgeon General*, U.S. Dept. of Health and Human Services, 1996

ACTIVITY	DURATION[2] (min)	ACTIVITY	DURATION[2] (min)
Washing and waxing a car	45-60	Raking leaves	30
Washing windows or floors	45-60	Walking 2 mi (15 min/mi)	30
Playing touch football	30-45	Swimming laps	20
Wheeling self in wheelchair	30-40	Basketball (playing a game)	15-20
Walking 1¾ mi (20 min/mi)	35	Bicycling 4 mi	15
Basketball (shooting baskets)	30	Jumping rope	15
Bicycling 5 mi	30	Running 1½ mi (10 min/mi)	15
Dancing fast (social)	30	Shoveling snow	15

Note: The activities are arranged from less vigorous, and using more time, to more vigorous, and using less time. (1) A "moderate" amount of physical activity uses about 150 calories (kcal), or 1,000 if done daily for a week. (2) Activities can be performed at various intensities; the suggested durations are based on the expected intensity of effort.

Finding Your Target Heart Rate

Source: Carole Casten, EdD, *Aerobics Today;* Peg Jordan, RN, Aerobics and Fitness Assoc. of America

The target heart rate is the heartbeat rate a person should have during aerobic exercise (such as running, fast walking, cycling, or cross-country skiing) to get the full benefit of the exercise for cardiovascular conditioning.

First, determine the intensity level at which one would like to exercise. A sedentary person may want to begin an exercise regimen at the 60% level and work up gradually to the 70% level. Athletes and highly fit individuals must work at the 85-95% level to receive benefits.

Second, calculate the target heart rate. One common way of doing this is by using the American College of Sports Medicine Method.

To obtain cardiovascular fitness benefits from aerobic exercise, it is recommended that an individual participate in an aerobic activity at least 3-5 times a week for 20-30 minutes per session, although cardiac patients and very sedentary individuals can obtain benefits with shorter periods (15-20 minutes). Generally, training changes occur in 4-6 weeks, but they can occur in as little as 2 weeks.

The American College of Sports Medicine Method

Using the American College of Sports Medicine Method to calculate one's target heart rate, an individual should subtract his or her age from 220, then multiply by the desired intensity level of the workout. Then divide the answer by 6 for a 10-second pulse count. (The 10-second pulse count is useful for checking whether the target heart rate is being achieved during the workout. One can easily check one's pulse—at the wrist or side of the neck—counting the number of beats in 10 seconds.)

For example, a 20-year-old wishing to exercise at 70% intensity would employ the following steps:

Maximum Heart Rate	220 – 20 = 200
Target Heart Rate	200 × .70 = 140
10-second Pulse Count	140 6 = 23

To work at the desired level of intensity, this 20-year-old would strive for a target heart rate of 140 beats per minute, or a 10-second pulse count of 23.

Cancer Prevention

Source: American Cancer Society, 1599 Clifton Road NE, Atlanta, GA 30329-4251; phone: (800) 227-2345

PRIMARY PREVENTION: Modifiable determinants of cancer risk.

Smoking Lung cancer mortality rates are about 20 times higher for current male smokers, and 12 times higher for current female smokers, than for those who have never smoked. Smoking accounts for about 30% of all cancer deaths in the U.S. Tobacco use is responsible for nearly 1 in 5 deaths in the U.S. Smoking is associated with cancer of the lung, mouth, pharynx, larynx, esophagus, pancreas, uterine cervix, kidney, and bladder.

Nutrition and Diet Risk for colon, rectum, breast (among postmenopausal women), kidney, prostate, and endometrial cancers increases in obese people. A diet high in fat may be a factor in the development of certain cancers, particularly cancer of the colon and rectum, prostate, and endometrium. . Eating 5 or more servings of fruits and vegetables each day, and eating other foods from plant sources (especially grains and beans), may reduce risk for many cancers. Physical activity can help protect against some cancers.

Sunlight Many of the one million skin cancers that are expected to be diagnosed in 2001 could have been prevented by protection from the sun's rays. Epidemiological evidence shows that sun exposure is a major factor in the development of melanoma and that the incidence rates are increasing around the world.

Alcohol Heavy drinking, especially when accompanied by cigarette smoking or smokeless tobacco use, increases risk of cancers of the mouth, larynx, pharynx, esophagus, and liver. Studies have also noted an association between alcohol consumption and an increased risk of breast cancer.

Smokeless Tobacco Use of chewing tobacco or snuff increases risk of cancers of the mouth and pharynx. The excess risk of cancer of the cheek and gum may reach nearly 50-fold among long-term snuff users.

Estrogen Estrogen replacement therapy (ERT) to control menopausal symptoms can increase the risk of endometrial cancer. However, adding progesterone to estrogen (hormone replacement therapy, or HRT) helps to minimize this risk. Most studies suggest that long-term use (5 years or more) of ERT or HRT after menopause may slightly increase the risk of breast cancer. The benefits and risks of the use of estrogen by menopausal women should be discussed carefully by the woman and her doctor. Research in this area continues.

Radiation Excessive exposure to ionizing radiation can increase cancer risk. Medical and dental X rays are adjusted to deliver the lowest dose possible without sacrificing image quality. Excessive radon exposure in the home may increase lung cancer risk, especially in cigarette smokers. If levels are found to be too high, remedial actions should be taken.

Environmental Hazards Exposure to various chemicals (including benzene, asbestos, vinyl chloride, arsenic, and aflatoxin) increases risk of various cancers. Risk of lung cancer from asbestos is greatly increased when combined with smoking. Pesticides, low-frequency radiation, toxic wastes, and proximity to nuclear power plants have not been proven to cause cancer.

Cancer-Detection Guidelines

Source: American Cancer Society, 1599 Clifton Road NE, Atlanta, GA 30329-4251; phone: (800) 227-2345

SECONDARY PREVENTION: Steps to diagnose a cancer or precursor as early as possible after it has developed.

A cancer-related checkup is recommended every 3 years for people aged 20-40 and every year for people 40 years of age and older. This exam should include health counseling and, depending on a person's age, might include examinations for cancers of the thyroid, oral cavity, skin, lymph nodes, testes, and ovaries, as well as for some nonmalignant diseases. Special tests for certain cancer sites are recommended as outlined below:

Breast Cancer Breast self-exam monthly, beginning at age 20.
Breast clinical physical examination for women aged 20-39, every 3 years; 40 and over, every year.
Mammography for women aged 40 and over, every year.

Cervical Cancer Annual Pap test and pelvic exam for women who are or have been sexually active or have reached age 18. After 3 or more consecutive satisfactory normal annual exams, the Pap test may be performed less frequently at the discretion of the physician.

Colorectal Cancer Beginning at age 50, both men and women should follow this testing schedule:
Yearly fecal occult blood test, plus flexible sigmoidoscopy and digital rectal examination every 5 years, or
Colonoscopy and digital rectal examination every 10 years, or
Double-contrast barium enema and digital rectal examination every 5-10 years.

Endometrial Cancer For women with or at high risk of hereditary nonpolyposis colon cancer (HNPCC), annual screening including endometrial biopsy should be obtained beginning at age 35.

Prostate Cancer Both Prostate-Specific Antigen (PSA) and Digital Rectal Examination (DRE) should be offered annually, beginning at age 50, to men who have at least a 10-year life expectancy. Men at high risk, such as African-Americans and men who have a first-degree relative (father, brother, or son) diagnosed with prostate cancer at an early age, should begin testing at age 45. Health care professionals should give men the opportunit to openly discuss the benefits and risks of testing at annual checkups. Men should actively participate in the decision by learning about prostate cancer and the pros and cons of early detection and treatment of prostate cancer.

Skin Cancer Adults should practice skin self-exam regularly. Suspicious lesions should be evaluated promptly by a physician.

Breast Cancer

Source: American Cancer Society, Inc., 1599 Clifton Road NE, Atlanta, GA 30329-4251; phone: (800) 227-2345

It is estimated that, in 2001, about 192,200 women and 1,500 men in the United States will be diagnosed with breast cancer, and about 40,200 women and 400 men will die from it. Breast cancer is the second largest cause of cancer death for women in the U.S. (lung cancer ranks first), but mortality rates have been declining, especially among younger women, probably because of earlier detection and improved treatment.

The risk for breast cancer increases as a woman ages. The risk is also higher for women with a personal or family history; a long menstrual history (menstrual periods that started early and ended late in life); recent use of oral contraceptives (birth control pills) or postmenopausal hormone replacement therapy; and no children or no live birth until age 30 or older. Other risk factors for the disease include alcohol consumption and obesity. Inherited mutations such as in the BRCA1 and BRCA2 genes greatly increase a woman's risk for breast cancer, but these mutations probably account for less than 10% of all breast cancers. By far, the majority of women who develop breast cancer have no family history.

Breast cancer is often manifested first as an abnormality that appears on a mammogram, which is a special type of x-ray. Physical signs and symptoms that show up later, which may be detectable by a woman or her doctor, include a breast lump and, less commonly, breast thickening, swelling, distortion, or tenderness; skin irritation or dimpling; or pain, scaliness, or retraction of the nipple. Breast pain is more commonly associated with benign (noncancerous) conditions.

The risk for breast cancer increases as a woman ages. The risk is also higher for women with a personal or family history of the disease; a long menstrual history(menstrual periods that started early and ended late in life); recent use of oral contraceptives (birth control pills) or post menopausal hormone replacement therapy; and no children or no live birth until age 30 or older. Other risk factors for the disease include alcohol consumption and obesity. Inherited mutations such as in the BRCA1 and BRCA2 genes greatly increase a woman's risk for breast cancer, but these mutations probably account for less than 10% of breast cancers. By far the majority of women who develop breast cancer have no family history.

Studies show that **early detection** increases survival and treatment options. The American Cancer Society (ACS) recommends that women 40 and older should have an annual mammogram, have an annual clinical breast exam by a health care professional, and perform monthly breast self-examinations. The ACS recommends that women ages 20-39 should have a clinical breast exam every 3 years and should also perform monthly breast self-examinations. Although most breast lumps that are detected are noncancerous, any suspicious lump needs to be biopsied.

Treatment for breast cancer may involve lumpectomy (local removal of a tumor), mastectomy (surgical removal of the breast), radiation therapy, chemotherapy, hormone therapy, immunotherapy, or some combination of these. For early-stage breast cancer, long-term survival rates following lumpectomy plus radiation therapy are similar to survival rates after modified radical mastectomy.

Numerous **drugs** that may **prevent** breast cancer or improve its treatment are being studied. One is **tamoxifen**, a synthetic hormone that blocks the action of estrogen in the breast. Already used for treating breast cancer, it has been shown to reduce the likelihood of developing the disease in women considered at higher than average risk, including women age 60 and older. Unfortunately, tamoxifen also has dangerous side effects, such as increased risk of uterine cancer and blood clots in the lungs. Research is also being done on another drug, **Raloxifene**, which is approved for preventing osteoporosis in postmenopausal women. It is now being directly compared to tamoxifen in a large clinical study to evaluate its effect on breast cancer risk.

Trends in Daily Use of Cigarettes, for U.S. 8th, 10th, and 12th Graders

Source: *Monitoring the Future,* Univ. of Michigan Inst. for Social Research and National Inst. on Drug Abuse

(percent who smoked daily in last 30 days; change 99-00 in percentage points)

	8th grade						10th grade						12th grade					
	1996	1997	1998	1999	2000	'99-'00 change	1996	1997	1998	1999	2000	'99-'00 change	1996	1997	1998	1999	2000	'99-'00 change
TOTAL	**10.4**	**9.0**	**8.8**	**8.1**	**7.4**	**-0.7**	**18.3**	**18.0**	**15.8**	**15.9**	**14.0**	**-1.9**	**22.2**	**24.6**	**22.4**	**23.1**	**20.6**	**-2.5**
Sex																		
Male	10.5	9.0	8.1	7.4	7.0	-0.4	18.1	17.2	14.7	15.6	13.7	-1.8	22.2	24.8	22.7	23.6	20.9	-2.8
Female	10.1	8.7	9.0	8.4	7.5	-1.0	18.6	18.5	16.8	15.9	14.1	-1.8	21.8	23.6	21.5	22.2	19.7	-2.5
College plans																		
None or under 4 yrs.	26.0	25.4	25.2	25.2	21.7	-3.5	34.3	35.4	31.7	32.1	28.8	-3.3	33.2	35.6	34.6	34.2	31.7	-2.5
Complete 4 yrs.	8.0	6.9	6.6	5.9	5.6	-0.3	15.5	15.0	12.9	13.2	11.6	-1.6	18.9	20.6	18.4	19.5	16.6	-2.9
Region																		
Northeast	11.0	8.8	6.1	7.2	6.9	-0.3	18.8	18.0	18.7	17.7	14.1	-3.6	27.0	29.4	23.4	23.2	22.8	-0.4
North central	12.4	10.3	11.2	11.5	9.0	-2.6	20.6	19.5	17.3	19.6	16.3	-3.3	26.1	28.0	27.8	25.9	23.6	-2.3
South	10.4	9.5	10.2	8.5	7.8	-0.8	20.5	20.5	17.1	16.3	15.7	-0.6	20.5	22.6	21.8	24.2	19.4	-4.8
West	7.5	6.8	5.8	3.8	4.9	+1.1	10.7	11.1	8.8	9.1	7.8	-1.4	13.8	17.5	15.5	17.3	16.9	-0.4
Race/Ethnicity [1]																		
White	11.7	11.4	10.4	9.7	9.0	-0.8	20.0	21.4	20.3	19.1	17.7	-1.4	25.4	27.8	28.3	26.9	25.7	-1.2
Black	3.2	3.7	3.8	3.8	3.2	0.6	5.1	5.6	5.8	5.3	5.2	-0.1	7.0	7.2	7.4	7.7	8.0	+0.3
Hispanic	8.0	8.1	8.4	8.5	7.1	-1.4	11.6	10.8	9.4	9.1	8.8	-0.3	12.9	14.0	13.6	14.0	15.7	+1.7

(1) For each of these groups, data for the specified year and previous year have been combined to increase sample size and thus provide a more reliable estimate.

Some Benefits of Quitting Smoking

Source: American Cancer Society, Inc., 1599 Clifton Road NE, Atlanta, GA 30329-4251; phone: (800) 227-2345

Within 20 Minutes
- Blood pressure drops to a level close to that before the last cigarette
- Temperature of hands and feet increases to normal

Within 8 Hours
- Carbon monoxide level in the blood drops to normal

Within 24 Hours
- Chance of heart attack decreases

Within 2 Weeks to 3 Months
- Circulation improves
- Lung function increases up to 30%

Within 1 to 9 Months
- Coughing, sinus congestion, fatigue, and shortness of breath decrease
- Cilia regain normal function in the lungs, increasing the ability to handle mucus, clean the lungs, reduce infection

Within 1 Year
- Excess risk of coronary heart disease is half that of a smoker's

Within 5 Years
- Stroke risk is reduced to that of a nonsmoker 5-15 years after quitting

Within 10 Years
- Lung cancer death rate about half that of a continuing smoker's
- Risk of cancer of the mouth, throat, esophagus, bladder, kidney, and pancreas decreases

Within 15 Years
- Risk of coronary heart disease is that of a nonsmoker's

Diabetes

Source: American Diabetes Association, 1701 N Beauregard St., Alexandria, VA 22311; phone: (800) 342-2382

Diabetes is a chronic disease in which the body does not produce or properly use **insulin**, a hormone needed to convert sugar, starches, and other foods into energy necessary for daily life. Both genetics and environment appear to play roles in the onset of diabetes. This disease, which has no cure, is the 6th-leading cause of death by disease in the U.S. According to death certificate data, diabetes contributed to 198,140 deaths in 1996.

In 1997, the American Diabetes Association issued **new guidelines for diagnosing diabetes**. The recommendations include: lowering the acceptable level of blood sugar from 140 mg of glucose/deciliter of blood to 126 mg/deciliter, possibly identifying 2 million more people with the disease; testing all adults 45 years and older, and then every 3 years if normal; and testing at a younger age, or more frequently, in high-risk individuals. The American Diabetes Association believes that detection at an earlier stage will help prevent or delay complications of diabetes.

There are 2 major types of diabetes:

- **Type 1 (formerly known as insulin dependent).** The body produces very little or no insulin; disease most often begins in childhood or early adulthood. People with type 1 diabetes must take daily insulin injections to stay alive.
- **Type 2 (formerly known as non-insulin dependent).** The body does not produce enough or cannot properly use insulin. It is the most common form of the disease (90-95% of cases in people over age 20) and often begins later in life.

Warning Signs of Diabetes

Type 1 Diabetes (usually occurs suddenly):

- frequent urination
- unusual thirst
- extreme hunger
- unusual weight loss
- extreme fatigue
- irritability

Type 2 Diabetes (occurs less suddenly):

- any type 1 symptoms
- frequent infections
- blurred vision
- cuts/bruises slow to heal
- tingling/numbness in hands or feet
- recurring skin, gum, or bladder infections

Complications of Diabetes

More than one-third of all individuals with diabetes do not know that they have the disease until one of its life-threatening complications occurs. Potential complications include:

Blindness. Diabetes is the leading cause of blindness in people ages 20-74. Each year, from 12,000 to 24,000 people lose their sight because of diabetes.

Kidney disease. 10% to 21% of all people with diabetes develop kidney disease. In 1995, more than 27,900 people initiated treatment for end-stage renal disease (kidney failure) because of diabetes.

Amputations. Diabetes is the most frequent cause of non-traumatic lower limb amputations. The risk of a leg amputation is 15 to 40 times greater for a person with diabetes than for the average American. Each year, an estimated 56,000 people lose a foot or leg as a result of complications brought on by diabetes.

Heart disease and stroke. People with diabetes are 2 to 4 times more likely to have heart disease (more than 77,000 deaths due to heart disease annually). And they are 2 to 4 times more likely to suffer a stroke.

Health-care and related costs for the treatment of the disease, added to the cost of lost productivity, total nearly $100 billion annually in the U.S.

Alzheimer's Disease

Source: Alzheimer's Association, 919 N Michigan Ave., Suite 1100, Chicago, IL 60611-1676; phone: (800) 272-3900

Alzheimer's disease, the most common form of dementia, is a progressive, degenerative disease of the brain in which nerve cells deteriorate and die for unknown reasons. Its first symptoms usually involve impaired memory and confusion about recent events. As the disease advances, it results in greater impairment of memory, thinking, behavior, and physical health.

The **rate of progression** of Alzheimer's varies, ranging from 3 to 20 years; the average length of time from onset of symptoms until death is 8 years. Eventually, affected individuals lose their ability to care for themselves and become susceptible to infections of the lungs, urinary tract, or other organs as they grow progressively more debilitated.

Alzheimer's disease affects an estimated 4 million Americans, striking men and women of all ethnic groups. Although most people diagnosed with Alzheimer's are older than age 60, some cases occur in people in their 40s and 50s. By age 65, an estimated 10 percent of the population has Alzheimer's, and the disease affects almost half of those over 85. In the United States, annual costs of diagnosis, treatment, and long-term care are estimated at $100 billion.

Diagnosis involves a comprehensive evaluation that may include a complete health history, a physical examination, neurological and mental status assessments, and other testing as needed. Skilled health care professionals can generally diagnose Alzheimer's with about 90 percent accuracy. Other conditions that can cause similar symptoms include depression, drug interactions, nutritional imbalances, infections such as AIDS, meningitis, and syphilis, and other forms of dementia, such as those associated with stroke, Huntington's disease, Parkinson's disease, frontotemporal dementia, and vascular disease. Absolute confirmation of diagnosis requires a brain biopsy or autopsy.

Treatments for cognitive and behavioral symptoms are available, but no intervention has yet been developed that prevents Alzheimer's or reverses its course. Providing care for people with Alzheimer's is physically and psychologically demanding. Nearly 70 percent of affected individuals live at home, where family or friends care for them. In advanced stages of the disease, many individuals require care in a nursing home. Nearly half of all nursing home residents in the United States have Alzheimer's.

People with Alzheimer's need a safe, stable environment and a regular daily schedule offering appropriate stimulation. Physical exercise and social interaction are important, as is proper nutrition. Security is also a consideration, because many people with Alzheimer's tend to wander. An identification bracelet listing the person's name, address, and condition may help ensure the safe return of an individual who wanders.

Warning Signs of Alzheimer's Disease

- Recent memory loss that affects job performance
- Inability to learn new information
- Difficulty with everyday tasks such as cooking or dressing oneself
- Inability to remember simple words
- Use of inappropriate words when communicating
- Disorientation of time and place
- Poor or decreased judgment
- Problems with abstract thinking
- Putting objects in inappropriate places
- Rapid changes in mood or behavior
- Increased irritability, anxiety, depression, confusion, and restlessness
- Prolonged loss of initiative

Acquired Immune Deficiency Syndrome

Source: Centers for Disease Control and Prevention

AIDS (Acquired Immune Deficiency Syndrome) is caused by the human immunodeficiency virus (**HIV**). HIV kills or disables crucial cells of the immune system, progressively destroying the body's ability to fight disease.

HIV is commonly spread through unprotected sexual contact of some kind with an infected partner. HIV is also spread through contact with infected blood. Where modern screening techniques are used it is rare to contract HIV from transfusion, but it can often be contracted when intravenous drug users share syringes. Though HIV can be spread through semen, vaginal fluids, and breast milk, there is no evidence it can be spread through saliva. About one-quarter to one-third of untreated HIV-positive pregnant women transmit HIV to their fetuses, but with drug treatment and cesarean section delivery that risk can be reduced to 1%. Studies have indicated no evidence of HIV transmission through casual contact such as the sharing of food utensils, towels and bedding, telephones, or toilet seats.

Some people experience flu-like symptoms a short time after infection with HIV, and scientists estimate that about half of those infected by HIV develop serious symptoms within ten years. Even when symptoms are not present, HIV is active in the body, multiplying, infecting, and killing CD4+ T cells, or "T-helper cells," the crucial immune cells that signal other cells in the immune system to perform their functions.

The term **AIDS** applies to the most advanced stages of HIV infection. According to the official definition set by the Centers for Disease Control and Prevention (CDC), an HIV–infected person with fewer than 200 CD4+ T cells can be said to have AIDS. (Healthy adults usually have 1,000 or more). An HIV-infected person, regardless of T cell count, is diagnosed with AIDS if he or she develops one of 26 conditions that typically affect people with advanced HIV. Most of these conditions are "opportunistic infections" that occur when the immune system is so ravaged by HIV that the body cannot fight off certain bacteria, viruses and microbes.

Months or years prior to the onset of AIDS, many people experience such symptoms as swollen glands, lack of energy, fevers and sweats, and skin rashes. People with full-blown AIDS may develop infections of the intestinal tract, lungs, brain, eyes, and other organs, with a variety of symptoms, and may become severely debilitated. They also are prone to developing certain cancers, especially those caused by viruses, such as Kaposi's sarcoma, cervical cancer, and lymphoma. Children with AIDS may have delayed development or failure to thrive.

HIV is primarily **detected** by testing a person's blood for the presence of antibodies (disease-fighting proteins) to HIV. In about 5% of infected individuals, HIV antibodies take as long as six months after exposure to reach detectable levels, but in most cases the virus is detectable in about six weeks. HIV testing may also be performed on saliva and urine samples.

The **U.S. Food and Drug Administration** has approved a number of **drugs** that may slow down the spread of HIV in the body and treat the infections and cancers associated with AIDS. The first group of drugs used to treat HIV, called nucleoside analog reverse transcriptase inhibitors (NRTIs), include the drug zidovudine (commonly known as AZT). Nonnucleoside reverse transcriptase inhibitors (NNRTIs) have also been approved to treat HIV. A third class of drugs, called protease inhibitors, have also been approved for the treatment of HIV. Patients are typically given a combination of different drugs, because HIV can become resistant to one particular drug. While these drugs extend the period between HIV infection and serious illness, they do not prevent the spread of the disease and can have severe side effects.

Since there is no vaccine or cure for AIDS, the only protection is to avoid activities that carry a risk. There is no evidence that spermicides are effective in preventing HIV or AIDS. When it cannot be known with certainty whether a sexual partner has the HIV virus, the CDC recommends abstinence (the only certain protection), mutual monogamy with an uninfected partner, or correct and consistent use of male latex condoms.

Allergies and Asthma

Source: Asthma and Allergy Foundation of America, 1233 20th St., NW, Suite 402, Washington, DC 20036; phone: (800) 7-ASTHMA

One out of five Americans suffers from **allergies**. People with allergies have extra-sensitive immune systems that react to normally harmless substances. Allergens that may produce this reaction include plant pollens, dust mites, or animal dander; plants such as poison ivy; certain drugs, such as penicillin; and certain foods such as eggs, milk, nuts, or seafood.

The tendency to develop allergies is usually inherited, and allergies usually begin to appear in childhood, but they can show up at any age. Common allergies for infants include food allergies and eczema (patches of dry skin). Older children and adults may often develop allergic rhinitis (hay fever), a reaction to an inhaled allergen; common symptoms include nasal congestion, runny nose, and sneezing.

It is best to avoid contact with the allergen, if feasible. In some cases, medications such as antihistamines are used to decrease the reaction, and there are treatments aimed at gradually desensitizing the patient to the allergen. Other effective allergy treatments include decongestants, eye drops, and ointments.

Some people with allergies also have **asthma**, and allergens are a common asthma trigger. Asthma is a disease of chronic inflammation, affecting the passages that carry air into and out of the lungs. It is most often seen in children but can develop at any age.

People with asthma have inflamed, supersensitive airways that tighten and become filled with mucus during an asthma episode. Wheezing, difficulty in breathing, tightening of the chest, and coughing are common symptoms. Asthma can progress through stages to become life-threatening if not controlled. Emergency symptoms of asthma include a bluish cast to the face and lips, severe anxiety, increased pulse rate, and sweating.

Besides common allergens, tobacco smoke, cold air, and pollution can trigger an asthma attack, as can viral infections or physical exercise that taxes the breathing. Of course, an accurate diagnosis by a physician is important. Although there is no cure for asthma or allergies, they can be controlled with medications and lifestyle changes.

IT'S A FACT: Whenever you sneeze, you close your eyes. In fact, every time you sneeze, you use six sets of muscles.

Arthritis

Source: Arthritis Foundation, 1330 West Peachtree Street, Atlanta, GA 30309; phone: (800) 283-7800

The term "arthritis" refers to more than 100 different diseases that cause pain, stiffness, swelling, and restricted movement in joints and connective tissue. The condition is usually chronic. Nearly 43 million people in the U.S. have some form of arthritis—about 23 million are women and almost 300,000 are children. The cause for most types of arthritis is unknown; scientists are studying the roles played by genetics, lifestyle, and the environment.

Symptoms of arthritis may develop either slowly or suddenly. A visit to the doctor is indicated when pain, stiffness, or swelling in a joint or difficulty in moving a joint persists for more than two weeks. The doctor analyzes the patient's symptoms, to see if they are consistent with those of arthritis. The doctor examines joint movement, looks for any swelling, and checks for skin rashes. Finally, the doctor may test the blood, urine, or joint fluid, or take X rays of the joints.

Medications to treat arthritis include drugs that relieve pain and swelling, such as analgesics, anti-inflammatory drugs, biologic response modifiers, glucocorticoids, or disease-modifying antirheumatic drugs, which tend to slow the disease process; and sleep medications, which promote deeper sleep and help relax muscles. Most treatment programs call for exercise; use of heat or cold; and joint-protection techniques (such as avoiding excess stress on joints, using assistive devices, and controlling weight). In some cases, surgery can help when other treatments fail.

Of the three most prevalent forms of arthritis, **osteoarthritis** is the most common, affecting more than 20 million Americans; it usually occurs after age 45. In this type, which is also called degenerative arthritis, the cartilage and bones deteriorate, causing pain and stiffness as bones rub against each other. It usually occurs in the fingers, knees, feet, hips, and back.

Fibromyalgia, another common arthritis condition, affects more than 2 million Americans and affects more women than men. In this form, widespread pain and tenderness occur in muscles and their attachments to the bone. Common symptoms include fatigue, disturbed sleep, stiffness, and psychological distress.

Rheumatoid arthritis, which also affects more than 2 million people in the U.S., is one of the most serious and disabling forms of the disease. In this type, which is also more common in women, the joints become inflamed because of an abnormality in the body's immune system. The chronic inflammation may then damage the cartilage and bone. The areas of the body that can be affected are the hands, wrists, feet, knees, ankles, shoulders neck, jaw, and elbows.

Other forms of arthritis and related conditions include lupus, gout, ankylosing spondylitis, and scleroderma; also related are bursitis and tendinitis, which may result from injuring or overusing a joint.

Alternative Medicine

Source: National Center for Complementary and Alternative Medicine, National Institutes of Health (NIH)

Alternative medicine comprises a wide variety of healing philosophies, approaches, and therapies. It includes treatments and health care practices not widely taught in medical schools, not generally used in hospitals, and not usually reimbursed by health insurance companies. The NIH cautions people not to seek alternative therapies without the consultation of a licensed health care provider.

Some alternative therapies are described as **holistic**—meaning that the practitioner considers the whole person, including physical, mental, emotional, and spiritual aspects. Some therapies are known as **preventive**, meaning that the practitioner stresses preventing health problems before they arise.

People may use an alternative therapy alone, along with other alternative therapies, or in combination with more standard therapies. Worldwide, only about 10-30% of health care is provided by conventional practitioners; the remaining 70-90% involves alternative practices. An estimated 1 in 3 Americans uses some form of alternative medicine.

An advisory panel to the National Center for Complementary and Alternative Medicine (formerly called the Office of Alternative Medicine) at the National Institutes of Health classified 7 general fields of practice:

Alternative systems of medical practice range from self-care based on folk traditions to care given by practitioners according to established procedures. Included are such therapies as acupuncture, Ayurveda (India's traditional system of natural medicine), environmental medicine (treatment of certain illnesses believed to be caused by exposure to particular foods or chemicals), homeopathic medicine (use of remedies made from naturally occurring plant, animal, or mineral substances), Native American practices, naturopathic medicine (integration of traditional, natural therapeutics with modern scientific medicine), and traditional Oriental medicine.

Bioelectromagnetic applications explore how living things interact with electromagnetic fields. Such therapies include blue light treatment and artificial lighting, electroacupuncture, and electrostimulation.

Diet, nutrition, and lifestyle changes are intended to prevent illness, maintain good health, and reverse the effects of chronic disease. Examples include use of macrobiotics, nutritional supplements, and megavitamins.

Herbal medicine employs plants and plant products for pharmacological use. Some common plants used are echinacea, garlic, ginkgo biloba, ginseng, St. John's wort, and saw palmetto.

Manual healing uses touch and manipulation with the hands therapeutically. Some types are acupressure, chiropractic medicine, massage therapy, osteopathy, and reflexology.

Mind/body control explores the mind's ability to affect the body. Therapies include hypnosis, meditation, psychotherapy, support groups, tai chi, and yoga.

Pharmacological and biological treatments involve drugs and vaccines that are not accepted by mainstream medicine. These include anti-oxidizing agents, metabolic therapy, and oxidizing agents.

Alternative Health Services in the U.S., 2000

Source: *Nutrition Business Journal*

HEALTH CARE PRACTICE	Licensed practitioners	Lay or other practitioners	Total revenues[1]
Acupuncture	5,800	3,000	$730
Chiropractic	62,000	4,000	15,620
Homeopathy	1,900	1,000	500
Massage therapy	38,000	150,000	7,540
Naturopathy	21,000	2,000	$3,310
Traditional Oriental medicine	11,000	15,000	3,150
TOTAL	**139,700**	**175,000**	**$30,850**

(1) In millions of dollars.

Top-Selling Medicinal Herbs in the U.S., 1998–2000

Source: *Nutrition Business Journal*

HERB	Sales in 1998[1]	Sales in 1999[1]	Sales in 2000[1]	% change 1998-2000
Echinacea	$208	$214	$210	1
Garlic[2]	198	176	174	-12
Ginkgo biloba	300	298	248	-17
Ginseng	217	192	173	-20
St. John's wort	308	233	170	-45
Saw palmetto	$105	$117	$131	25
Combinations	1,762	1,740	1,821	3
All other	862	1,101	1,204	40
TOTAL	**$3,960**	**$4,070**	**$4,131**	**4**

(1) In millions of dollars. (2) Does not include nonmedicinal use.

Performance of Global Health Systems

Source: *The World Health Report 2000,* World Health Organization

Health Expenditure Per Capita

Top 25	Rank	Bottom 25	Rank
United States	1	Somalia	191
Switzerland	2	Madagascar	190
Germany	3	Ethiopia	189
France	4	Dem. Rep. of the Congo	188
Luxembourg	5	Eritrea	187
Austria	6	Burundi	186
Sweden	7	Niger	185
Denmark	8	Afghanistan	184
Netherlands	9	Sierra Leone	183
Canada	10	Yemen	182
Italy	11	Liberia	181
Monaco	12	Togo	180
Japan	13	Mali	179
Iceland	14	Central African Rep.	178
Belgium	15	Rwanda	177
Norway	16	Nigeria	176
Australia	17	Chad	175
Finland	18	Tanzania	174
Israel	19	Burkina Faso	173
New Zealand	20	Dem. People's Rep. of Korea	172
San Marino	21	Benin	171
Bahamas	22	Nepal	170
Andorra	23	Sudan	169
Spain	24	Uganda	168
Ireland	25	São Tomé and Príncipe	167

Overall Goal Attainment[1]

Top 25	Rank	Bottom 25	Rank
Japan	1	Sierra Leone	191
Switzerland	2	Central African Republic	190
Norway	3	Somalia	189
Sweden	4	Niger	188
Luxembourg	5	Liberia	187
France	6	Ethiopia	186
Canada	7	Mozambique	185
Netherlands	8	Nigeria	184
United Kingdom	9	Afghanistan	183
Austria	10	Malawi	182
Italy	11	Angola	181
Australia	12	Guinea-Bissau	180
Belgium	13	Dem. Rep. of the Congo	179
Germany	14	Mali	178
United States	15	Chad	177
Iceland	16	Eritrea	176
Andorra	17	Myanmar (Burma)	175
Monaco	18	Zambia	174
Spain	19	Lesotho	173
Denmark	20	Guinea	172
San Marino	21	Rwanda	171
Finland	22	Djibouti	170
Greece	23	Mauritania	169
Israel	24	Botswana	168
Ireland	25	Madagascar	167

(1) A composite measure that factors in level of health, distribution of health based on the equality of child survival, responsiveness to the needs of disadvantaged groups, and financial fairness, based on in-depth surveys of health care practitioners.

Food Guide Pyramid

The Food Guide Pyramid was developed by the U.S. Dept. of Agriculture and was revised for the year 2000. The Pyramid is an outline of what to eat each day. It is not meant as a rigid prescription, but as a general guide to help in choosing a healthful diet. It calls for eating a variety of foods to get needed nutrients and at the same time the right amount of calories to maintain or improve your weight. The Pyramid focuses heavily on fat because most Americans' diets are too high in fat, especially saturated fat.

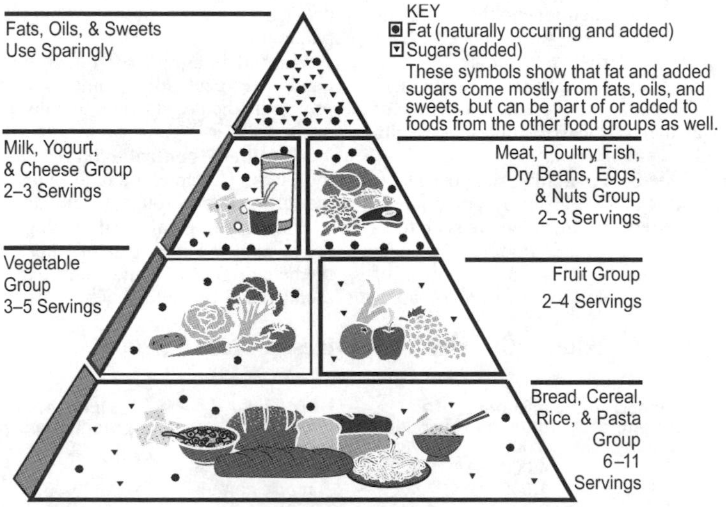

What Counts as a Serving?

Bread, Cereal, Rice, and Pasta
- 1 slice of bread
- 1 ounce of ready to-eat cereal
- 1/2 cup of cooked cereal, rice, or pasta

Vegetable
- 1 cup of raw leafy vegetables
- 1/2 cup of other vegetables, cooked or chopped raw
- 3/4 cup of vegetable juice

Fruit
- 1 medium apple, banana, orange
- 1/2 cup of chopped, cooked, or canned fruit
- 3/4 cup of fruit juice

Milk, Yogurt, and Cheese
- 1 cup of milk or yogurt
- 1-1/2 ounces of natural cheese
- 2 ounces of process cheese

Meat, Poultry, Fish, Dry Beans, Eggs, and Nuts
- 2-3 ounces of cooked lean meat, poultry, or fish
- 1/2 cup of cooked dry beans or 1 egg counts as 1 ounce of lean meat.
- 2 tablespoons of peanut butter or 1/3 cup of nuts count as 1 ounce of meat.

Food and Nutrition

The U.S. Dept. of Health and Human Services and the Dept. of Agriculture reissued guidelines in 1996 that offer dietary and exercise advice for children age 2 and over, as well as for adults. Recommended were: (1) no more than 30 percent of calories from fat, or about 65 grams of fat in a 2,000-calorie daily diet; with no more than 10% of calories, or 20 grams of fat, from saturated fats; (2) maximum alcohol consumption of about 1 drink a day for women, 2 for men; (3) daily consumption of vegetables, 3-5 servings; fruits, 2-4; pastas, cereals, or breads, 6-11; milk, 2-3; meat, poultry, fish, beans, and eggs, 2-3. (For vegetables, 1 serving equals about 1 cup raw leafy greens or one-half cup other kinds; fruit, 1 medium apple, banana, or orange, or cup of fruit juice; grains, 1 slice of bread, cup of pasta, or 1 oz. cereal; milk, 1 cup or 1.5 oz. of cheese; meat and poultry, 2-3 oz. cooked lean beef or chicken without skin; cooked dry beans, cup.)

PROTEIN

Proteins, composed of amino acids, are essential to good nutrition. They build, maintain, and repair the body. Best sources: eggs, milk, fish, meat, poultry, soybeans, nuts. High-quality proteins such as eggs, meat, or fish supply all 8 amino acids needed in the diet. Plant foods can be combined to meet protein needs as well: whole grain breads and cereals, rice, oats, soybeans, other beans, split peas, and nuts.

FATS

Fats provide energy by furnishing calories to the body, and they also carry vitamins A, D, E, and K. They are the most concentrated source of energy in the diet. Best sources of polyunsaturated and monounsaturated fats: margarine, vegetable/plant oils, nuts. Meats, cheeses, butter, cream, egg yolks, lard are concentrated sources of saturated fats.

CARBOHYDRATES

Carbohydrates provide energy for body function and activity by supplying immediate calories. The carbohydrate group includes sugars, starches, fiber, and starchy vegetables. Best sources: grains, legumes, potatoes, vegetables, fruits.

FIBER

The portion of plant foods that our bodies cannot digest is known as fiber. There are 2 basic types: *insoluble* ("roughage") and *soluble*. Insoluble fibers help move food materials through the digestive tract; soluble fibers tend to slow them down. Both types absorb water, thus prevent and treat constipation by softening and increasing the bulk of the undigested food components passing through the digestive tract. Soluble fibers have also been reported to be helpful in reducing blood cholesterol levels. Best sources: beans, bran, fruits, whole grains, vegetables.

WATER

Water dissolves and transports other nutrients throughout the body, aiding the processes of digestion, absorption, circulation, and excretion. It helps regulate body temperature.

VITAMINS

Vitamin A—promotes good eyesight and helps keep the skin and mucous membranes resistant to infection. Best sources: liver, sweet potatoes, carrots, kale, cantaloupe, turnip greens, collard greens, broccoli, fortified milk.

Vitamin B_1 (thiamine)—prevents beriberi. Essential to carbohydrate metabolism and health of nervous system. Best sources: pork, enriched cereals, grains, soybeans, nuts.

Vitamin B_2 (riboflavin)—protects the skin, mouth, eyes, eyelids, and mucous membranes. Essential to protein and energy metabolism. Best sources: milk, meat, poultry, cheese, broccoli, spinach.

Vitamin B_6 (pyridoxine)—important in the regulation of the central nervous system and in protein metabolism. Best sources: whole grains, meats, fish, poultry, nuts, brewers' yeast.

Vitamin B_{12} (cobalamin)—needed to form red blood cells. Best sources: meat, fish, poultry, eggs, dairy products.

Niacin—maintains health of skin, tongue, digestive system. Best sources: poultry, peanuts, fish, enriched flour and bread.

Folic acid (folacin)—required for normal blood cell formation, growth, and reproduction and for important chemical reactions in body cells. Best sources: yeast, orange juice, green leafy vegetables, wheat germ, asparagus, broccoli, nuts.

Other B vitamins—biotin, pantothenic acid.

Vitamin C (ascorbic acid)—maintains collagen, a protein necessary for the formation of skin, ligaments, and bones. It helps heal wounds and mend fractures and aids in resisting some types of viral and bacterial infections. Best sources: citrus fruits and juices, cantaloupe, broccoli, brussels sprouts, potatoes and sweet potatoes, tomatoes, cabbage.

Vitamin D—important for bone development. Best sources: sunlight, fortified milk and milk products, fish-liver oils, egg yolks.

Vitamin E (tocopherol)—helps protect red blood cells. Best sources: vegetable oils, wheat germ, whole grains, eggs, peanuts, margarine, green leafy vegetables.

Vitamin K—necessary for formation of prothrombin, which helps blood to clot. Also made by intestinal bacteria. Best dietary sources: green leafy vegetables, tomatoes.

MINERALS

Calcium—works with phosphorus in building and maintaining bones and teeth. Best sources: milk and milk products, cheese, blackstrap molasses, some types of tofu.

Phosphorus—performs more functions than any other mineral, and plays a part in nearly every chemical reaction in the body. Best sources: cheese, milk, meats, poultry, fish, tofu.

Iron—Necessary for the formation of myoglobin, which is a reservoir of oxygen for muscle tissue, and hemoglobin, which transports oxygen in the blood. Best sources: lean meats, beans, green leafy vegetables, shellfish, enriched breads and cereals, whole grains.

Other minerals—chromium, cobalt, copper, fluorine, iodine, magnesium, manganese, molybdenum, potassium, selenium, sodium, sulfur, and zinc.

Understanding Food Label Claims

Source: Food Labeling Education Information Center, Beltville, MD

The federal Nutrition Labeling and Education Act of 1990 provides that manufacturers can make certain claims on processed food labels only if they meet the definitions specified here:

SUGAR

Sugar free: less than 0.5 g per serving

No added sugar; Without added sugar; No sugar added:

- No sugars added during processing or packing, including ingredients that contain sugars (for example, fruit juices, applesauce, or dried fruit).
- Processing does not increase the sugar content above the amount naturally present in the ingredients. (A functionally insignificant increase in sugars is acceptable from processes used for purposes other than increasing sugar content.)
- The food for which it substitutes normally contains added sugars.

Reduced sugar: at least 25% less sugar than reference food

CALORIES

Low calorie: 40 calories or less per serving; if the serving is 30 g or less or 2 tablespoons or less, 40 calories or less per 50 g of food

Calorie free: under 5 calories per serving

Reduced or Fewer calories: at least 25% fewer calories than reference food

FAT

Fat free: less than 0.5 g of fat per serving

Saturated fat free: less than 0.5 g of saturated fat per serving, and the level of trans fatty acids does not exceed 1% of total fat

Low fat: 3 g or less per serving and, if the serving is 30 g or less or 2 tbs or less, per 50 g of the food

Low saturated fat: 1 g or less per serving and not more than 15% of calories from saturated fatty acids

Reduced or Less fat: at least 25% less per serving than reference food

CHOLESTEROL

Cholesterol free: less than 2 mg of cholesterol and 2 g or less of saturated fat per serving

Low cholesterol: 20 mg or less and 2 g or less of saturated fat per serving and, if the serving is 30 g or less or 2 tbs or less, per 50 g of the food

Reduced or Less cholesterol: at least 25% less than reference food

SODIUM

Sodium free: less than 5 mg per serving

Low sodium: 140 mg or less per serving and, if the serving is 30 g or less or 2 tbs or less, per 50 g of the food

Very low sodium: 35 mg or less per serving and, if the serving is 30 g or less or 2 tbs or less, per 50 g of the food

Reduced or Less sodium: at least 25% less per serving than reference food

FIBER

High fiber: 5 g or more per serving. (Also, must meet low-fat definition, or must state level of total fat.)

Good source of fiber: 2.5 g to 4.9 g per serving

More or Added fiber: at least 2.5 g more per serving than reference food

Nutritive Value of Food (Calories, Proteins, etc.)

Source: *Home and Garden Bulletin No. 72;* U.S. Dept. of Agriculture

FOOD	Measure	Grams	Food Energy (calories)	Protein (grams)	Fat (grams)	Saturated fats (grams)	Carbohydrate (grams)	Calcium (milligrams)	Iron (milligrams)	Sodium (milligrams)	Vitamin A (I.U.)	Ascorbic Acid (milligrams)
DAIRY PRODUCTS												
Cheese, cheddar, cut pieces	1 oz.	28	115	7	9	6.0	T	204	0.2	176	300	0
Cheese, cottage, small curd	1 cup	210	215	26	9	6.0	6	126	0.3	850	340	T
Cheese, cream	1 oz.	28	100	2	10	6.2	1	23	0.3	84	400	0
Cheese, Swiss	1 oz.	28	95	7	7	4.5	1	219	0.2	388	230	0
Half-and-half	1 tbsp.	15	20	T	2	1.1	1	16	T	6	70	T
Cream, sour	1 tbsp.	12	25	T	3	1.6	1	14	T	6	90	T
Milk, whole	1 cup	244	150	8	8	5.1	11	291	0.1	120	310	2
Milk, nonfat (skim)	1 cup	245	85	8	T	0.3	12	302	0.1	126	500	2
Milkshake, chocolate	10 oz.	283	355	9	8	4.8	60	374	0.9	314	240	0
Ice cream, hardened	1 cup	133	270	5	14	8.9	32	176	0.1	116	540	1
Sherbet	1 cup	193	270	2	4	2.4	59	103	0.3	88	190	4
Yogurt, fruit-flavored	8 oz.	227	230	10	2	1.6	43	345	0.2	133	100	1
EGGS												
Fried in margarine	1	46	90	6	7	1.9	1	25	0.7	162	390	0
Hard-cooked	1	50	75	6	5	1.6	1	25	0.6	62	280	0
Scrambled (milk added) in margarine	1	61	100	7	7	2.2	1	44	0.7	171	420	T
FATS & OILS												
Butter, salted	1 tbsp.	14	100	T	11	7.1	T	3	T	116	430	0
Margarine, salted	1 tbsp.	14	100	T	11	2.2	T	4	T	132	460	T
Olive oil	1 tbsp.	14	125	0	14	1.9	0	0	0	0	0	0
Salad dressing, blue cheese	1 tbsp.	15	75	1	8	1.5	1	12	T	164	30	T
Salad dressing, French, regular	1 tbsp.	16	85	T	9	1.4	1	2	T	188	T	T
Salad dressing, French, low calorie	1 tbsp.	16	25	T	2	0.2	2	6	T	306	T	T
Salad dressing, Italian	1 tbsp.	15	80	T	9	1.3	1	1	T	162	30	T
Mayonnaise	1 tbsp.	14	100	T	11	1.7	T	3	0.1	80	40	0
FISH, MEAT, POULTRY												
Clams, raw, meat only	3 oz.	85	65	11	1	0.3	2	59	2.6	102	90	9
Crabmeat, canned	1 cup	135	135	23	3	0.5	1	61	1.1	1,350	50	0
Fish sticks, frozen, reheated	1 fish stick	28	70	6	3	0.8	4	11	0.3	53	20	0
Salmon canned (pink), solids and liquid	3 oz.	85	120	17	5	0.9	0	167	0.7	443	60	0
Sardines, Atlantic, canned in oil, drained solids	3 oz.	85	175	20	9	2.1	0	371	2.6	425	190	0
Shrimp, French fried	3 oz.	85	200	16	10	2.5	11	61	2.0	384	90	0
Trout, broiled, with butter and lemon juice	3 oz.	85	175	21	9	4.1	T	26	1.0	122	230	1
Tuna, canned in oil	3 oz.	85	165	24	7	1.4	0	7	1.6	303	70	0
Bacon, broiled or fried crisp	3 slices	19	110	6	9	3.3	T	2	0.3	303	0	6
Ground beef, broiled, regular	3 oz.	85	245	20	18	6.9	0	9	2.1	70	T	0
Roast beef, relatively lean (lean only)	2.6 oz.	75	135	22	5	1.9	0	3	1.5	46	T	0
Beef steak, lean and fat	3 oz.	85	240	23	15	6.4	0	9	2.6	53	T	0
Beef & vegetable stew	1 cup	245	220	16	11	4.4	15	29	2.9	292	5,690	17
Lamb, chop, broiled loin, lean and fat	2.8 oz.	80	235	22	16	7.3	0	16	1.4	62	T	0
Liver, beef, fried	3 oz.	85	185	23	7	2.5	7	9	5.3	90	30,690	23
Ham, light cure, roasted, lean and fat	3 oz.	85	205	18	14	5.1	0	6	0.7	1,009	0	0
Pork, chop, broiled, lean and fat	3.1 oz.	87	275	24	19	7.0	0	3	0.7	61	10	T
Bologna	2 slices	57	180	7	16	6.1	2	7	0.9	581	0	12
Frankfurter, pork, cooked	1	45	145	5	13	4.8	1	5	0.5	504	0	12
Sausage, pork link, cooked	1 link	13	50	3	4	1.4	T	4	0.2	168	0	T
Veal, cutlet, braised or broiled	3 oz.	85	185	23	9	4.1	0	9	0.8	56	T	0
Chicken, drumstick, fried, bones removed	2.5 oz.	72	195	16	11	3.0	6	12	1.0	194	60	0
Chicken, roasted, half breast, without skin	3 oz.	86	140	27	3	0.9	0	13	0.9	64	20	0
Turkey, roasted, chopped light and dark meat	1 cup	140	240	41	7	2.3	0	35	2.5	98	0	0
Frankfurter, chicken, cooked	1	45	115	6	9	2.5	3	43	0.9	616	60	0
FRUITS & FRUIT PRODUCTS												
Apple, raw, 2-3/4 in. diam.	1	138	80	T	T	0.1	21	10	0.2	T	70	8
Apple juice	1 cup	248	115	T	T	T	29	17	0.9	7	T	2
Apricots, raw	3	106	50	1	T	T	12	15	0.6	1	2,770	11
Banana, raw	1	114	105	1	1	0.2	27	7	0.4	1	90	10
Cherries, sweet, raw	10	68	50	1	1	0.1	11	10	0.3	T	150	5
Cranberry juice cocktail, sweetened	1 cup	253	145	T	T	T	38	8	0.4	10	10	108
Fruit cocktail, canned, in heavy syrup	1 cup	255	185	1	T	T	48	15	0.7	15	520	5
Grapefruit, raw, medium, white	1/2	120	40	1	T	T	10	14	0.1	T	10	41
Grapes, Thompson seedless	10	50	35	T	T	0.1	9	6	0.1	1	40	5
Lemonade, frozen, unsweetened	6 oz.	244	55	1	1	0.1	16	20	0.3	2	30	77
Cantaloupe, 5-in. diam.	1/2	267	95	2	1	0.1	22	29	0.6	24	8,610	113
Orange, 2-5/8 in. diam.	1	131	60	1	T	T	15	52	0.1	T	270	70
Orange juice, frozen, diluted	1 cup	249	110	2	T	T	27	22	0.2	2	190	97
Peach, raw, 2-1/2 in. diam.	1	87	35	1	T	T	10	4	0.1	T	470	6
Raisins, seedless	1 cup	145	435	5	1	0.2	115	71	3.0	17	10	5
Strawberries, whole	1 cup	149	45	1	1	T	10	21	0.6	1	40	84
Tomatoes, raw	1	123	25	1	T	T	5	9	0.6	10	1,390	22
Watermelon, 4 by 8 in. wedge	1 piece	482	155	3	2	0.3	35	39	0.8	10	1,760	46
GRAIN PRODUCTS												
Bagel, plain	1	68	200	7	2	0.3	38	29	1.8	245	0	0
Biscuit, 2 in. diam., from home recipe	1	28	100	2	5	1.2	13	47	0.7	195	10	T
Bread, pita, enriched, white, 6-1/2 in. diam	1 pita	60	165	6	1	0.1	12	15	0.7	124	0	0
Bread, white, enriched	1 slice	25	65	2	1	0.3	12	32	0.7	129	T	T
Bread, whole-wheat	1 slice	28	70	3	1	0.4	13	20	1.0	180	T	T
Oatmeal or rolled oats, without added salt	1 cup	234	145	6	2	0.4	25	19	1.6	2	40	0
Bran flakes (40% bran), added sugar, salt, iron, vitamins	1 oz.	28	90	4	1	0.1	22	14	8.1	264	1,250	0
Corn flakes, added sugar, salt, iron, vitamins	1 oz.	28	110	2	T	T	24	1	1.8	351	1,250	15
Rice, puffed, added iron, thiamine, niacin	1 oz.	28	110	2	T	T	25	4	1.8	340	1,250	15
Wheat, shredded, plain, 1 biscuit or 2/3 cup	1 oz.	28	100	3	1	0.1	23	11	1.2	3	0	0
Bulgur, uncooked	1 cup	170	600	19	3	1.2	129	49	9.5	7	0	0
Cake, angel food, 1/12 of cake	1	53	125	3	T	T	29	44	0.2	269	0	0
Cupcake, 2-1/2 in. diam., with chocolate icing	1	35	120	2	4	1.8	20	21	0.7	92	50	T

FOOD	Measure	Grams	Food Energy (calories)	Protein (grams)	Fat (grams)	Saturated fats (grams)	Carbohydrate (grams)	Calcium (milligrams)	Iron (milligrams)	Sodium (milligrams)	Vitamin A (I.U.)	Ascorbic Acid (milligrams)
Plain sheet cake with white, uncooked frosting, 1/9 of cake	1	121	445	4	14	4.6	77	61	1.2	275	240	T
Fruitcake, dark, 1/32 of loaf	1	43	165	2	7	1.5	25	41	1.2	67	50	16
Cake, pound, 1/17 of loaf	1	29	110	2	5	3.0	15	8	0.5	108	160	0
Cheesecake, 1/12 of 9-in. diam. cake	1	92	280	5	18	9.9	26	52	0.4	204	230	5
Brownies, with nuts, from commercial recipe	1	25	100	1	4	1.6	16	13	0.6	59	70	T
Cookies, chocolate chip, from home recipe	4	40	185	2	11	3.9	26	13	1.0	82	20	0
Crackers, graham, 2-1/2 in. squares	2	14	60	1	1	0.4	11	6	0.4	86	0	0
Crackers, saltines	4	12	50	1	1	0.5	9	3	0.5	165	0	0
Danish pastry, round piece	1	57	220	4	12	3.6	26	60	1.1	218	60	T
Doughnut, cake type	1	50	210	3	12	2.8	24	22	1.0	192	20	T
Macaroni, firm stage (hot)	1 cup	130	190	7	1	0.1	39	14	2.1	1	0	0
Muffin, bran, commercial mix	1	45	140	3	4	1.3	24	27	1.7	385	100	0
Muffin, corn, from home recipe	1	45	145	3	5	1.5	21	66	0.9	169	80	T
Noodles, enriched, cooked	1 cup	160	200	7	2	0.5	37	16	2.6	3	110	0
Pie, apple, 1/6 of pie	1	158	405	3	18	4.6	60	13	1.6	476	50	2
Pie, cherry, 1/6 of pie	1	158	410	4	18	4.7	61	22	1.6	480	700	0
Pie, lemon meringue, 1/6 of pie	1	140	355	5	14	4.3	53	20	1.4	395	240	4
Pie, pecan, 1/6 of pie	1	138	575	7	32	4.7	71	65	4.6	305	220	0
Popcorn, air-popped, plain	1 cup	8	30	1	T	T	6	1	0.2	T	10	0
Pretzels, stick	10	3	10	T	T	T	2	1	0.1	48	0	0
Rolls, enriched, brown & serve	1	28	85	2	2	0.5	14	33	0.8	155	T	T
Rolls, frankfurter & hamburger	1	40	115	3	2	0.5	20	54	1.2	241	T	T
Tortillas, corn	1	30	65	2	1	0.1	13	42	0.6	1	80	0
LEGUMES, NUTS, SEEDS												
Beans, Black	1 cup	171	225	15	1	0.1	41	47	2.9	1	T	0
Beans, Great Northern, cooked	1 cup	180	210	14	1	0.1	38	90	4.9	13	0	0
Peanuts, roasted in oil, salted	1 cup	145	840	39	71	9.9	27	125	2.8	626	0	0
Peanut butter	1 tbsp.	16	95	5	8	1.4	3	5	0.3	75	0	0
Refried beans, canned	1 cup	290	295	18	3	0.4	51	141	5.1	1,228	0	17
Tofu	1 piece	120	85	9	5	0.7	3	108	2.3	8	0	0
Sunflower seeds, hulled	1 oz.	28	160	6	14	1.5	5	33	1.9	1	10	T
MIXED FOODS												
Chop suey with beef and pork, home recipe	1 cup	250	300	26	17	4.3	13	60	4.8	1,053	600	33
Enchilada	1	230	235	20	16	7.7	24	97	3.3	1,332	2,720	T
Pizza, cheese, 1/8 of 15 in.-diam. pie	1	120	290	15	9	4.1	39	220	1.6	699	750	2
Spaghetti with meatballs & tomato sauce	1 cup	248	330	19	12	3.9	39	124	3.7	1,009	1,590	22
SUGARS & SWEETS												
Candy, caramels	1 oz.	28	115	1	3	2.2	22	42	0.4	64	T	T
Candy, milk chocolate	1 oz.	28	145	2	9	5.4	16	50	0.4	23	30	T
Fudge, chocolate	1 oz.	28	115	1	3	2.1	21	22	0.3	54	T	T
Gelatin dessert, from prepared powder	1/2 cup	120	70	2	0	0.0	17	2	T	55	0	0
Candy, hard	1 oz.	28	110	0	0	0.0	28	T	0.1	7	0	0
Honey	1 tbsp.	21	65	T	0	0.0	17	1	0.1	1	0	T
Jams & preserves	1 tbsp.	20	55	T	T	0.0	14	4	0.2	2	T	T
Popsicle, 3 fl. oz.	1	95	70	0	0	0.0	18	0	T	11	0	0
Sugar, white, granulated	1 tbsp.	12	45	0	0	0.0	12	T	T	T	0	0
VEGETABLES												
Asparagus, spears, cooked from raw	4 spears	60	15	2	T	T	3	14	0.4	2	500	16
Beans, green, from frozen, cuts	1 cup	135	35	2	T	T	8	61	1.1	18	710	11
Broccoli, cooked from raw	1 spear	180	50	5	1	0.1	10	82	2.1	20	2,540	113
Cabbage, raw, coarsely shredded or sliced	1 cup	70	15	1	T	T	4	33	0.4	13	90	33
Carrots, raw, 7-1/2 by 1-1/8 in.	1	72	30	1	T	T	7	19	0.4	25	20,250	7
Cauliflower, cooked, drained, from raw	1 cup	125	30	2	T	T	6	34	0.5	8	20	69
Celery, raw	1 stalk	40	5	T	T	T	1	14	0.2	35	50	3
Collards, cooked from raw	1 cup	190	25	2	T	0.1	5	148	0.8	36	4,220	19
Corn, sweet, yellow, cooked from raw	1 ear	77	85	3	1	0.2	19	2	0.5	13	170	5
Eggplant, cooked, steamed	1 cup	96	25	1	T	T	6	6	0.3	3	60	1
Lettuce, iceberg, chopped	1 cup	55	5	1	T	T	1	10	0.3	5	180	2
Lettuce, looseleaf (such as romaine)	1 cup	56	10	1	T	T	2	38	0.8	5	1,060	10
Mushrooms, raw	1 cup	70	20	1	T	T	3	4	0.9	3	0	2
Onions, raw, chopped	1 cup	160	55	2	T	0.1	12	40	0.6	3	0	13
Peas, green, frozen, cooked	1 cup	160	125	8	T	0.1	23	38	2.5	139	1,070	16
Potatoes, baked, peeled	1	156	145	3	T	T	34	8	0.5	8	0	20
Potatoes, frozen, French fried (oven-heated)	10	50	110	2	4	2.1	17	5	0.7	16	0	5
Potatoes, mashed, milk added	1 cup	210	160	4	1	0.7	37	55	0.6	636	40	14
Potato chips	10	20	105	1	7	1.8	10	5	0.2	94	0	8
Potato salad	1 cup	250	360	7	21	3.6	28	48	1.6	1,323	520	25
Spinach, drained, cooked from raw	1 cup	180	40	5	T	0.1	7	245	6.4	126	14,740	18
Sweet potatoes, baked in skin, peeled	1	114	115	2	T	T	28	32	0.5	11	24,880	28
Vegetable juice cocktail, canned	1 cup	242	45	2	T	T	11	27	1.0	883	2,830	67
MISCELLANEOUS												
Beer, regular	12 fl. oz.	360	150	1	0	0.0	13	14	0.1	18	0	0
Gin, rum, vodka, whisky, 86 proof	1-1/2 fl. oz.	42	105	0	0	0.0	T	T	T	T	0	0
Wine, table, white	3-1/2 fl. oz.	102	80	T	0	0.0	3	9	0.3	5	(1)	0
Cola-type beverage	12 fl. oz.	369	160	0	0	0.0	41	11	0.2	18	0	0
Ginger ale	12 fl. oz	366	125	0	0	0.0	32	11	0.1	29	0	0
Coffee, brewed	6 fl. oz.	180	T	T	T	T	T	4	T	2	0	0
Tea, brewed	8 fl. oz.	240	T	T	T	T	T	0	T	1	0	0
Catsup	1 tbsp.	15	15	T	T	T	4	3	0.1	156	210	2
Mustard, prepared, yellow	1 tsp.	5	5	T	T	T	T	4	0.1	63	0	T
Olives, canned, green	4 medium	13	15	T	2	0.2	T	8	0.2	312	40	0
Pickles, dill, whole	1	65	5	T	T	T	1	17	0.7	928	70	4
Relish, finely chopped, sweet	1 tbsp.	15	20	T	T	T	5	3	0.1	107	20	1
Soup, tomato, prepared with milk	1 cup	248	160	6	6	2.9	22	159	1.8	932	850	68
Soup, chicken noodle, prepared with water	1 cup	241	75	4	2	0.7	9	17	0.8	1,106	710	T
Soup, green pea, prepared with water	1 cup	250	165	9	3	1.4	27	28	2.0	988	200	2
Soup, vegetarian, prepared with water	1 cup	241	70	2	2	0.3	12	22	1.1	822	3,010	1

T — Indicates trace. (1) — Value not determined. **NOTE:** Values shown here for these foods may be from several different manufacturers and, therefore, may differ somewhat from the values provided by one source.

Dietary Requirements

In April 1998, the Institute of Medicine of the Food and Nutrition Board, National Academy of Sciences, released a report on Dietary Reference Intakes (DRIs), which updated and expanded dietary requirements previously set for thiamin, riboflavin, niacin, B_6, folate, B_{12}, pantothenic acid, biotin, and choline. A year earlier, the Institute updated the requirements for calcium, phosphorus, magnesium, vitamin D, and fluoride. The new values are based on the latest research relevant to optimizing health at all stages of life, not simply to protecting against nutritional deficiencies. Reports on other nutrients are under development. In the meantime, the previously established Recommended Dietary Allowances (RDAs) for these nutrients apply.

The DRIs include 4 categories for daily consumption: **RDA**—the intake that meets the nutrient requirements of almost all (97-98%) healthy individuals in a specified group; **Estimated Average Requirement (EAR)**—the intake that meets the estimated nutrient need of half the individuals in a specified group; **Adequate Intake (AI)**—the intake specified when sufficient broad scientific evidence is not available to calculate an EAR (for healthy breast-fed infants, the AI is the mean intake; the AI for other life stage groups is believed to cover their needs, but lack of data or uncertainty in the data prevent clear specification of this coverage); and **Tolerable Upper Intake Level (UL)**—the maximum intake that is unlikely to pose risks of adverse health effects in almost all healthy individuals in a specified group. RDAs and AIs may both be used as goals for individual intake. The UL is not recommended as a goal.

Recommended Levels for B Vitamins and Choline

Source: Food and Nutrition Board, National Academy of Sciences—Institute of Medicine, 1998

	Thiamin (mg/d)	Riboflavin (mg/d)	Niacin (mg/d)[1]	B_6 (mg/d)	Folate (mi/d)[2]	B_{12} (mi/d)	Pantothenic Acid (mg/d)	Biotin (mi/d)	Choline[3] (mg/d)
Infants									
0-5 mos	0.2*	0.3*	2*	0.1*	65*	0.4*	1.7*	5*	125*
6-11 mos	0.3*	0.4*	3*	0.3*	80*	0.5*	1.8*	6*	150*
Children									
1-3 yrs	0.5	0.5	6	0.5	150	0.9	2*	8*	200*
4-8 yrs	0.6	0.6	8	0.6	200	1.2	3*	12*	250*
Males									
9-13 yrs	0.9	0.9	12	1.0	300	1.8	4*	20*	375*
14-18 yrs	1.2	1.3	16	1.3	400	2.4	5*	25*	550*
19-30 yrs	1.2	1.3	16	1.3	400	2.4	5*	30*	550*
31-50 yrs	1.2	1.3	16	1.3	400	2.4	5*	30*	550*
51-70 yrs	1.2	1.3	16	1.7	400	2.4[4]	5*	30*	550*
over 70 yrs	1.2	1.3	16	1.7	400	2.4[4]	5*	30*	550*
Females									
9-13 yrs	0.9	0.9	12	1.0	300	1.8	4*	20*	375*
14-18 yrs	1.0	1.0	14	1.2	400[5]	2.4	5*	25*	400*
19-30 yrs	1.1	1.1	14	1.3	400[5]	2.4	5*	30*	425*
31-50 yrs	1.1	1.1	14	1.3	400[5]	2.4	5*	30*	425*
51-70 yrs	1.1	1.1	14	1.5	400[5]	2.4[4]	5*	30*	425*
over 70 yrs	1.1	1.1	14	1.5	400	2.4[4]	5*	30*	425*
Pregnant (all ages)	1.4	1.4	18	1.9	600[6]	2.6	6*	30*	450*
Lactating (all ages)	1.5	1.6	17	2.0	500	2.8	7*	35*	550*

mg/d = milligrams/day. mi = micrograms. mi/d = micrograms/day. **NOTE:** Adequate Intakes are followed by an asterisk. (1) As niacin equivalents. 1 mg of niacin = 60 mg of tryptophan. (2) As dietary folate equivalents (DFE). 1 DFE = 1 mi food folate = 0.6 mi of folic acid (from fortified food or supplement) consumed with food = 0.5 mi of synthetic (supplemental) folic acid taken on an empty stomach. (3) Although AIs have been set for choline, there is little evidence to assess whether a dietary supply of choline is needed at all stages of the life cycle, and it may be that the body can produce the required amount at some of these stages. (4) Since 10-30% of older people may malabsorb food-bound B_{12}, it is advisable for those over 50 yrs to meet this RDA mainly by taking foods fortified with B_{12} or a B_{12}-containing supplement. (5) In view of evidence linking folate intake with neural tube defects in the fetus, it is recommended that all women capable of becoming pregnant consume 400 mi of synthetic folic acid from fortified foods and/or supplements in addition to taking in food folate from a varied diet. (6) It is assumed that women will continue taking 400 mi of folic acid until their pregnancy is confirmed and they enter prenatal care, which ordinarily occurs after the critical time for formation of the neural tube.

Recommended Levels for Calcium, Phosphorus, Magnesium, Vitamin D, and Fluoride

Source: Food and Nutrition Board, National Academy of Sciences—Institute of Medicine, 1997

	Calcium		Phosphorus				Magnesium							Vitamin D		Fluoride		
							EAR[1]		RDA[1]		AI[1]					AI[1]		
	AI[1]	UL[2]	EAR[1]	RDA[1]	AI[1]	UL[2]	m	f	m	f	m	f	UL[1, 3]	AI[4, 5]	UL[4]	m	f	UL[1]
Infants																		
0-6 mos	210	ND	—	—	100	ND	—	—	—	—	30	30	ND	5	25	0.01	0.01	0.7
6-12 mos	270	ND	—	—	275	ND	—	—	—	—	75	75	ND	5	25	0.50	0.50	0.9
Children																		
1-3 yrs	500	2.5	380	460	—	3.0	65	65	80	80	—	—	65	5	50	0.70	0.70	1.3
4-8 yrs	800	2.5	405	500	—	3.0	110	110	130	130	—	—	110	5	50	1.10	1.10	2.2
9-13 yrs	1,300	2.5	1,055	1,250	—	4.0	200	200	240	240	—	—	350	5	50	2.00	2.00	10.0
14-18 yrs	1,300	2.5	1,055	1,250	—	4.0	340	300	410	360	—	—	350	5	50	3.20	2.90	10.0
Adults																		
19-30 yrs	1,000	2.5	580	700	—	3.4	330	255	400	310	—	—	350	5	50	3.80	3.10	10.0
31-50 yrs	1,000	2.5	580	700	—	3.4	350	265	420	320	—	—	350	5	50	3.80	3.10	10.0
51-70 yrs	1,200	2.5	580	700	—	3.4	350	265	420	320	—	—	350	10	50	3.80	3.10	10.0
over 70 yrs	1,200	2.5	580	700	—	3.0	350	265	420	320	—	—	350	15	50	3.80	3.10	10.0
Pregnant																		
18 yrs or less	1,300	2.5	1,055	1,250	—	3.5	—	335	—	400	—	—	350	5	50	—	2.90	10.0
19-30 yrs	1,000	2.5	580	700	—	3.5	—	290	—	350	—	—	350	—	50	—	3.10	10.0
31-50 yrs	1,000	2.5	580	700	—	3.5	—	300	—	360	—	—	350	—	50	—	3.10	10.0
Lactating																		
18 yrs or less	1,300	2.5	1,055	1,250	—	4.0	—	300	—	360	—	—	350	5	50	—	2.90	10.0
19-30 yrs	1,000	2.5	580	700	—	4.0	—	255	—	310	—	—	350	—	50	—	3.10	10.0
31-50 yrs	1,000	2.5	580	700	—	4.0	—	265	—	320	—	—	350	—	50	—	3.10	10.0

m = male. f = female. ND = Not determinable, because of a lack of data on adverse effects in this age group and a concern over body's lack of ability to handle excess amounts. Source of intake in this case should be from food only. (1) mg/day. (2) g/day. (3) The UL for magnesium represents intake from a pharmacological agent only and does not include intake from food and water. (4) microgram/day. (5) In the absence of adequate exposure to sunlight.

Recommended Dietary Allowances (RDAs)

Source: Food and Nutrition Board, National Academy of Sciences—Institute of Medicine, 1989

				Fat soluble vitamins				Minerals			
		Weight (lbs)	Protein (g)	Vitamin A[1]	Vitamin E[2]	Vitamin K (mi)	Vitamin C[3] (mi)	Iron (mg)	Zinc (mg)	Iodine (mi)	Selenium (mi)
Infants	to 6 mos.	13	13	375	3	25	0.3	6	5	40	10
	6 mos. to 1 yr.	20	14	375	4	35	0.5	10	5	50	15
Children	1-3	29	16	400	6	50	0.7	10	10	70	20
	4-6	44	24	500	7	75	1.0	10	10	90	20
	7-10	62	28	700	7	100	1.4	10	10	120	30
Males	11-14	99	45	1,000	10	150	2.0	12	15	150	40
	15-18	145	59	1,000	10	200	2.0	12	15	150	50
	19-24	160	58	1,000	10	200	2.0	10	15	150	70
	25-50	174	63	1,000	10	200	2.0	10	15	150	70
	51+	170	63	1,000	10	200	2.0	10	15	150	70
Females	11-14	101	46	800	8	150	2.0	15	12	150	45
	15-18	120	44	800	8	180	2.0	15	12	150	50
	19-24	128	46	800	8	180	2.0	15	12	150	55
	25-50	138	50	800	8	180	2.0	15	12	150	55
	51+	143	50	800	8	180	2.0	10	12	150	55
Pregnant		—	60	800	10	400	2.2	30	15	175	65
Lactating	1st 6 mos.	—	65	1,300	12	280	2.6	15	19	200	75
	2d 6 mos.	—	62	1,200	11	260	2.6	15	16	200	75

g = grams. mi = micrograms. (1) Retinol equivalents. (2) Milligrams alpha-tocopherol equivalents. (3) Vitamin C is water-soluble.

U.S. Per Capita Consumption of Selected Foods, 1909-99

Source: Economic Research Service, U.S. Dept. of Agriculture

	Whole milk[1]	Low-fat & skim milk[1]	Butter[2]	Margarine[2]	Red meat[2]	Poultry[2]	Fish & shellfish[2]
1909	26.85	7.30	17.9	1.2	101.7	11.2	11.0
1939	29.24	4.79	17.4	2.3	86.9	11.9	10.8
1969	26.60	5.46	5.6	10.7	129.5	32.9	11.2
1999	8.0	13.90	4.8	8.1	117.2	68.3	15.2

(1) Gallons. (2) Pounds.

IT'S A FACT: The average consumption of added fat in the American diet has increased by 32% since 1970, even as consumption of many fatty foods such as butter and lard has actually decreased.

Weight Guidelines for Adults

Source: *Clinical Guidelines on the Identification, Evaluation, and Treatment of Overweight and Obesity in Adults*, National Heart, Lung, and Blood Institute, National Institutes of Health, 1998

Guidelines on identification, evaluation, and treatment of overweight and obesity in adults were released in June 1998 by the National Heart, Lung, and Blood Institute (NHLBI), in cooperation with the National Institute of Diabetes and Digestive and Kidney Diseases. The guidelines, based on research into risk factors in heart disease, stroke, and other conditions, define degrees of overweight and obesity in terms of body mass index (BMI), which is based on weight and height and is strongly correlated with total body fat content. A BMI of 25-29 is said to indicate overweight; a BMI of 30 or above is said to indicate obesity. Weight reduction is advised for persons with a BMI of 25 or higher, about 55% of the adult population. (Previous guidelines have been less stringent.) Factors such as large waist circumference, high blood pressure or cholesterol, and a family history of obesity-related disease may increase risk.

The table given here shows the BMI for certain heights and weights. For weight reduction tips, write to the NHLBI Information Center, PO Box 30105, Bethesda, MD 20824-0105. See also the NHLBI website: http://www.nhlbi.nih.gov/index.htm

Weight (lbs)

Height	HEALTHY						OVERWEIGHT					OBESE								
4'10"	91	96	100	105	110	115	119	124	129	134	138	143	148	153	158	162	167	172	177	181
4'11"	94	99	104	109	114	119	124	128	133	138	143	148	153	158	163	168	173	178	183	188
5'0"	97	102	107	112	118	123	128	133	138	143	148	153	158	163	168	174	179	184	189	194
5'1"	100	106	111	116	122	127	132	137	143	148	153	158	164	169	174	180	185	190	195	201
5'2"	104	109	115	120	126	131	136	142	147	153	158	164	169	175	180	186	191	196	202	207
5'3"	107	113	118	124	130	135	141	146	152	158	163	169	175	180	186	191	197	203	208	214
5'4"	110	116	122	128	134	140	145	151	157	163	169	174	180	186	192	197	204	209	215	221
5'5"	114	120	126	132	138	144	150	156	162	168	174	180	186	192	198	204	210	216	222	228
5'6"	118	124	130	136	142	148	155	161	167	173	179	186	192	198	204	210	216	223	229	235
5'7"	121	127	134	140	146	153	159	166	172	178	185	191	198	204	211	217	223	230	236	242
5'8"	125	131	138	144	151	158	164	171	177	184	190	197	203	210	216	223	230	236	243	249
5'9"	128	135	142	149	155	162	169	176	182	189	195	203	209	216	223	230	236	243	250	257
5'10"	132	139	146	153	160	167	174	181	188	195	202	209	216	222	229	236	243	250	257	264
5'11"	136	143	150	157	165	172	179	186	193	200	208	215	222	229	236	243	250	257	265	272
6'0"	140	147	154	162	169	177	184	191	199	206	213	221	228	235	242	250	258	265	272	279
6'1"	144	151	159	166	174	182	189	197	204	212	219	227	235	242	250	257	265	272	280	288
6'2"	148	155	163	171	179	186	194	202	210	218	225	233	241	249	256	264	272	280	287	295
6'3"	152	160	168	176	184	192	200	208	216	224	232	240	248	256	264	272	279	287	295	303
6'4"	156	164	172	180	189	197	205	213	221	230	238	246	254	263	271	279	287	295	304	312
BMI[1]	**19**	**20**	**21**	**22**	**23**	**24**	**25**	**26**	**27**	**28**	**29**	**30**	**31**	**32**	**33**	**34**	**35**	**36**	**37**	**38**

(1) The BMI numbers apply to both men and women. Some very muscular people may have a high BMI without health risks.

Where to Get Help

Source: Based on Health & Medical Year Book. Copyright © by Collier Newfield, Inc.; additional data, World Almanac research

Listed here are some of the major U.S. and Canadian organizations providing information about good health practices generally, or about specific conditions and how to deal with them. (Canadian sources are identified as such.) Where a toll-free number is not available, an address is given when possible.

Some entries conclude with an e-mail address for the organization and/or an address for its Internet site, where you can also obtain useful information. In addition to these selected sites, there is a vast array of medical information on the Internet; however, it is very important to be certain that the source of information is reliable and accurate. Always check with a physician before embarking on any new health-related venture.

General Sources

Centers for Disease Control and Prevention Voice Information System
888-232-3228
Recorded information about public health topics, such as AIDS and Lyme disease. Also, you can request to talk with a CDC expert or have information faxed to you.
Website: http://www.cdc.gov

National Health Information Center
800-336-4797; in Maryland, 301-565-4167
Phone numbers for more than 1,000 health-related organizations in the United States. Printed materials offered.
E-mail: nhicinfo@health.org

National Institutes of Health
Bethesda, MD 20892
301-496-4000
Free information, including the latest research findings, on many diseases.
E-mail: nininfo@oo.nih.gov
Website: http://www.nih.gov

Tel-Med
Check the phone book for local listings or call Tel-Med at 909-478-0330.
Recorded information on over 600 health topics. Sponsored by local medical societies, health organizations, or hospitals.
E-mail: telmed@ix.netcom.com
Website: http://www.tel-med.com

Aging

National Association of Area Agencies on Aging's Eldercare Locator Line
800-677-1116
Information and assistance on a wide range of services and programs including adult day-care and respite services, consumer fraud, hospital and nursing home information, legal services, elder abuse/protective services, Medicaid/Medigap information, tax assistance, and transportation.
Hours 9 am-8 pm EST M-F.

National Institute on Aging
800-222-2225
Information and publications about disabling conditions, support groups, and community resources.
E-mail: niaic@JBS1.com
Website: http://www.nih.gov/nia

AIDS

AIDS Clinical Trials Information Service
800-874-2572
Information on federally and privately sponsored clinical trials for patients with AIDS or HIV.
E-mail: actis@actis.org
Website: http://www.actis.org

Canadian AIDS Society
613-230-3580.
Written materials and referrals.
E-mail: CASinfo@cdnaids.ca

Centers for Disease Control and Prevention National AIDS/HIV Hotline
800-342-AIDS 24 hours; in Spanish, 800-344-SIDA, everyday, 8 am-2 am; for the hearing impaired, 800-AIDS-TTY, M-F, 10 am-10 pm
Information on the prevention and spread of AIDS, along with referrals.
Website: http://www.ashastd.org

HIV-AIDS Treatment Information Service
800-HIV-0440
Treatment information to people with AIDS, their families, and health care providers.
E-mail: atis@hivatis.org
Website: http://www.hivatis.org

Alcoholism and Drug Abuse

Alcohol and Drug Helpline
800-821-4357, 24 hours
Referrals to local facilities

Alcoholics Anonymous
212-870-3400
Worldwide support groups for alcoholics. Check phone book for local chapters.
Website: http://www.alcoholics-anonymous.org or http://www.AA.org

American Council on Alcoholism
800-527-5344
Treatment referrals and counseling for recovering alcoholics.
Website: http://www.aca-usa.org

National Clearinghouse for Alcohol and Drug Information
800-729-6686
Provides written materials on alcohol and drug-related subjects.
Website: http://www.health.org

National Council on Alcoholism and Drug Dependence Hopeline
800-622-2255
An answering machine for callers to request information.

DrugHelp
800-DRUGHELP
Answers questions on substance abuse and provides referrals to treatment centers. Operates 24 hours.
Website: http://www.drughelp.org

Alzheimer's Disease

Alzheimer's Association
800-272-3900
Gives referrals to local chapters and support groups; offers information on publications available from the association.
E-mail: info@alz.org
Website: http://www.alz.org

Alzheimer's Society of Canada
20 Eglinton Ave., W., Suite 1200
Toronto, ON M4R 1K8
416-488-8772
Gives phone numbers for local support chapters. Publishes support materials.
E-mail: info@alzheimer.ca
Website: http://www.alzheimer.ca

Amyotrophic Lateral Sclerosis

ALS Association
800-782-4747; in the San Fernando Valley, 818-880-9007
Information about ALS (Lou Gehrig's Disease) and referrals to ALS specialists, local chapters and support groups.
Website: http://www.alsa.org

Arthritis

Arthritis Foundation
800-283-7800
Information, publications, and referrals to local groups.
Website: http://www.arthritis.org

Arthritis Society (Canada)
393 University Ave., Suite 1700
Toronto, ON M5G 1E6
416-979-7228; in Ontario only, 800-321-1433
Phone numbers for local chapters.
E-mail: info@arthritis.ca
Website: http://www.arthritis.ca

National Arthritis and Musculoskeletal and Skin Diseases Information Clearinghouse
877-226-4267
Subject searches and resource referrals.
Website: http://www.nih.gov/niams

Asthma and Allergies

See also *Lung Diseases*

Asthma and Allergy Foundation Information Clearinghouse
800-7-ASTHMA
Written information.

American Academy of Allergy, Asthma, and Immunology Referral Line
800-822-ASMA, 24 hours
Written materials on asthma and allergies.
Website: http://www.aaaai.org

Blindness and Eye Care

Canadian National Institute for the Blind
1929 Bayview Avenue
Toronto, ON M4G 3E8
416-480-7520 or contact your local chapter.
National office offers training and library with braille books and audiotapes. Local chapters provide core services: orientation in mobility, sight enhancement, counseling, referrals, career aid, technology services.
Website: http://www.cnib.ca

Foundation Fighting Blindness
888-394-3937; in Maryland, 410-785-1414; for the hearing impaired, 800-683-5551
Answers questions about retinal degenerative diseases; has written materials.
Website: http://www.blindness.org

Library of Congress National Library Service for the Blind and Physically Handicapped
800-424-9100; in Spanish, 800-345-8901; in Washington, DC, 202-707-5100; for the hearing impaired, 202-707-0744
Information on libraries that offer talking books and books in braille.
Website: http://lcweb.loc.gov

National Association for Parents of the Visually Impaired
800-562-6265
Support and information for parents of individuals who are visually impaired.
Website: http://www.napvi.org

Blood Disorders

Cooley's Anemia Foundation
800-522-7222
Information on patient care and support groups; makes referrals to local chapters.
E-mail: scdaa@sicklecelldisease.org
Website: http://www.thalassemia.org

Sickle Cell Disease Association of America
800-421-8453; in California, 310-216-6363
Genetic counseling and information packet.
E-mail: scdaa@sicklecelldisease.org
Website: http://www.sicklecelldisease.org

Burns

Phoenix Society
800-888-2876
Counseling for burn survivors and information on self-help services for burn survivors and their families.
E-mail: info@phoenix-society.org
Website: http://www.phoenix-society.org

Cancer

American Cancer Society
800-ACS-2345
Publications and information about cancer and coping with cancer; makes referrals to local chapters for support services.
Website: http://www.cancer.org

Canadian Cancer Information Service
888-939-3333, in Canada only, 9 am-6pm, Mon.-Fri.
Information on prevention, treatment, drugs, clinical trails, local services.

National Cancer Institute's Cancer Information Service
800-4-CANCER
Information about clinical trials, treatments, symptoms, prevention, referrals to support groups, and screening.
Website: http://www.nci.nih.gov

Y-Me Breast Cancer Support Program
800-221-2141, 24 hours
Information and literature on breast cancer, counseling, and referrals.
Website: http://www.y-me.org

Cerebral Palsy

Ontario Federation for Cerebral Palsy
104-1630 Lawrence Avenue West
Toronto, ON M6L 1C5
877-244-9686; 416-244-9686
Canada does not have a national cerebral palsy organization, but the provincial organizations offer information on housing, services, and coping with life, and each one will provide contact numbers for the others.
E-mail: ofcp@ofcp.on.ca
Website: http://www.ofcp.on.ca

United Cerebral Palsy Associations
877-835-7335; in Washington, DC, 202-776-0406
Written materials.
Website: http://www.ucpa.org

Child Abuse

See *Domestic Violence*

Children

American Academy of Pediatrics
847-228-5005
Child-care publications and materials; referrals to pediatricians.
Website: http://www.aap.org

Childhelp's USA National Child Abuse Hotline
800-4-A-CHILD
Crisis intervention, professional counseling, referrals to local groups and shelters for runaways, and literature. Operates 24 hours.

National Center for Missing and Exploited Children
800-843-5678; for the hearing impaired, 800-826-7653
Hotline for reporting missing children and sightings of missing children.
Website: http://www.missingkids.org

Chronic Fatigue Syndrome

CFIDS Association of America
800-442-3437
Literature and a list of support groups.
E-mail: info@cfids.org
Website: http://www.cfids.org

Crisis

National Runaway Switchboard
800-621-4000
Crisis intervention and referrals for runaways. Runaways can leave messages for parents, and vice versa. Operates 24 hours.
Website: http://nrscrisisline.org

Cystic Fibrosis

Canadian Cystic Fibrosis Foundation
416-485-9149; in Canada only, 800-378-2233
Information and brochures; makes referrals to local chapters.
Website: http://www.ccff.ca

Cystic Fibrosis Foundation
800-FIGHT-CF
Answers questions and offers literature and referrals to local clinics.
Website: http://www.cff.org

Diabetes

American Diabetes Association
800-342-2383; in Virginia and Washington, DC, 703-549-1500
Information about diabetes, nutrition, exercise, and treatment; offers referrals.
Website: http://www.diabetes.org

Canadian Diabetes Association
15 Toronto Street, Suite 800,Toronto, ON M5C 2E3
416-363-3373; in Ontario only, 800-361-1306
Information and publications.
Website: http://www.diabetes.ca

Juvenile Diabetes Foundation Hotline
800-533-2873; in NY: 212-785-9500
Answers questions, provides literature (some in Spanish). Offers referrals to local chapters, physicians, and clinics.
Website: http://www.jdf.org

Digestive Diseases

Crohn's and Colitis Foundation of America
800-932-2423; in New York, 212-685-3440
Educational materials; offers referrals to local chapters, which can provide referrals to support groups and physicians.
Website: http://www.ccfa.org

Crohn's and Colitis Foundation of Canada
301-21 St. Clair Avenue East, Suite 301, Toronto, ON M4T 1L9
416-920-5035; in Canada only, 800-387-1479
Will send out educational materials upon request.
Website: http://www.ccfc.ca

Domestic Violence

National Council on Child Abuse and Family Violence
in Washington, DC, 202-429-6695
A recording provides toll-free numbers to call for information or referrals.
E-mail: nccafv@aol.com
Website: http://www.nccafv.org

Down Syndrome

National Down Syndrome Congress
800-232-6372; in Georgia, 770-604-9500
Answers questions on all aspects of Down syndrome. Provides referrals.
E-mail: ndsccenter@aol.com
Website: http://www.ndsCcenter.org

National Association for Down Syndrome
630-325-9112 (Chicago area only)
Counseling and support; advocacy, referral, and information services.
Website: http://www.nads.org

Drug Abuse

See *Alcoholism and Drug Abuse*

Dyslexia

International Dyslexia Association
800-ABCD-123; in Maryland, 410-296-0232
Information used on testing, tutoring, and computers used to aid people with dyslexia and related disorders.
E-mail: info@interdys.org

Eating Disorders

National Association of Anorexia Nervosa and Associated Disorders
Box 7, Highland Park, IL 60035
847-831-3438
Written materials, referrals to health professionals treating eating disorders, telephone counseling, offers 3 self-help groups and information on how to set up a self-help group.
E-mail: anad20@aol.com
Website: http://www.anad.org

Endometriosis

Endometriosis Association
800-992-ENDO; in Canada, 800-426-2END
An answering machine for callers to request information.

Epilepsy

Epilepsy and Seizure Disorder Service at the Epilepsy Foundation of America
800-332-1000, Mon. through Thurs., 9 am to 5 pm (EST), Fri., 9 am to 3 pm.
Information and referrals to local chapters.
Website: http://www.efa.org

Food Safety and Nutrition

Meat and Poultry Hotline of the U.S. Department of Agriculture's Food, Safety, and Inspection Service
800-535-4555
Information on prevention of food-borne illness and the proper handling, preparation, storage, labeling, and cooking of meat, poultry, and eggs.
Website: http://www.fsis.usda.gov

FDA Center for Food Safety and Applied Nutrition Outreach & Information Center
800-SAFE-FOOD
Information on how to buy and use food products and on their proper handling and storage, women's health, and cosmetics & colors. Callers may speak to food specialists, Mon. through Fri., 10 am to 4 pm (EST).
Website: http://www.cfsan.fda.gov

Headaches

National Headache Foundation
800-843-2256
Literature on headaches and treatment.
Website: http://www.headaches.org

Heart Disease and Stroke

American Heart Association
800-242-8721
Information, publications, and referrals to organizations.
Website: http://www.americanheart.org

National Institute of Neurological Disorders and Stroke
800-352-9424
Literature and information.
Website: http://www.ninds.nih.gov

National Stroke Association
800-787-6537
Information on support networks for stroke victims and their families; referrals to local support groups.
Website: http://www.stroke.org

Hospices

Children's Hospice International
800-242-4453; in Virginia, 703-684-0330
Information, referrals to children's hospices.
E-mail: chiorg@aol.com
Website: http://www.chionline.org

Hospice Education Institute Hospicelink
800-331-1620; in Maine, 207-255-8800
Information, referrals to local programs.
E-mail: hospiceall@aol.com
Website: http://www.hospiceworld.org

Huntington's Disease

Huntington's Disease Society of America
800-345-4372; in New York, 212-242-1968
Information and referrals to physicians and support groups.
Website: http://www.hdsa.org

Huntington Society of Canada
151 Frederick Street, Suite 400, Kitchener, ON N2H 2M2
519-749-7063
Information, including telephone numbers of local services; publications and referrals.
E-mail: info@hsc-ca.org
Website: http://www.hsc-ca.org

Impotence

Impotence Information Center
800-843-4315
Information on treatment of impotence, incontinence, and prostate problems.
Website: http://www.visitams.com

Impotence World Association
800-669-1603
Written materials, physician referrals, and telephone numbers of local Impotents Anonymous chapters.
Website: http://www.impotenceworld.org

Kidney Diseases

Kidney Foundation of Canada
514-369-4806; in Canada only, 800-361-7494
Educational materials and general information.
Website: http://www.kidney.ca

National Kidney and Urologic Diseases Information Clearinghouse
3 Information Way
Bethesda, MD 20892-3580
301-654-4415
Information, referrals to organizations.
Website: http://www.niddk.nih.gov

National Kidney Foundation
800-622-9010
Information and referrals.
Website: http://www.kidney.org

Lead Exposure

National Lead Information Center
800-424-LEAD
Recommendations (in English and Spanish) for reducing a child's exposure to lead. Referrals to state and local agencies.
Website: http://www.epa.gov/lead

Liver Diseases

American Liver Foundation
800-223-0179; in NJ, 973-256-2550
Information on hepatitis, liver disease, and gallbladder disease.
Website: http://www.liverfoundation.org

Lung Diseases
See also *Asthma and Allergies*
American Lung Association
Check the phone book for local listings or call the national office at 800-LUNG-USA for automatic connection to the office nearest you. Answers questions about asthma and lung diseases; publications and referrals.
Website: http://www.lungusa.org
Lung Line Information Service at the National Jewish Medical and Research Center
800-222-LUNG; outside the U.S.: 303-388-4461
Answers questions on asthma, emphysema, allergies, smoking, and other respiratory and immune system disorders.
Website: http://www.njc.org

Lupus
Lupus Foundation of America
800-558-0121; in Colorado, 301-670-9292
Sends information to those who leave name and address on answering machine.
Website: http://www.lupus.org

Lyme Disease
Lyme Disease Foundation
800-886-LYME, 24 hours
Written information; doctor referrals.

Mental Health
National Depressive and Manic Depressive Association
800-826-3632
Support for patients and families, provides publications, and makes referrals to affiliated organizations.
Website: http://www.ndmda.org
National Foundation for Depressive Illness
P.O. Box 2257, NY, NY 10116
800-248-4344, 24 hours
Recorded message describing the symptoms of depression and offering an address for more information and physician referral.
National Institute of Mental Health
6001 Executive Blvd., Room 8184, MSC 9663, Bethesda, MD 20892-9663
301-443-4513
Information on a range of topics, from children's mental disorders to schizophrenia, depression, eating disorders, and others.
Website: http://www.nimh.nih.gov
National Mental Health Association
800-969-6642
Referrals to mental health groups.
Website: http://www.nmha.org

Multiple Sclerosis
Multiple Sclerosis Society of Canada
416-922-6065
Counseling, literature, and referrals to local chapters.
Website: http://www.mssociety.ca
National Multiple Sclerosis Society
800-344-4867
Information about local chapters.
Website: http://www.nmss.org

Muscular Dystrophy
Muscular Dystrophy Association
800-572-1717
Written materials on 40 neuromuscular diseases, including muscular dystrophy. Will give information over the phone about such matters as MDA clinics, support groups, summer camps, and wheelchair purchase assistance.
Website: http://www.mdausa.org

Nutrition
See *Food Safety and Nutrition*

Organ Donation
Living Bank
800-528-2971, 24 hours
A registry and referral service for people wanting to commit organs to transplantation or research.
Website: http://www.livingbank.org

Osteoporosis
National Osteoporosis Foundation
800-223-9994, in Washington, DC, 202-223-2226
Information packet available on request.
Website: http://www.nof.org

Pain
National Chronic Pain Outreach Association
540-862-9437
Information packet available on request.

Parkinson's Disease
National Parkinson Foundation
800-327-4545; in Florida, 800-433-7022; in Miami, 305-547-6666
Answers questions, makes physician referrals, and provides written information in English and Spanish.
E-mail: mailbox@parkinson.org
Website: http://www.parkinson.org
Parkinson Foundation of Canada
800-565-3000, Canada only
Information; referrals to support groups.
Website: http://www.parkinson.ca

Plastic Surgery
Plastic Surgery Information Service
800-635-0635
Referrals to board-certified plastic surgeons in the U.S. and Canada; general information.
Website: http://www.plasticsurgery.org

Polio
International Polio Network
4207 Lindell Blvd., #110
St. Louis, MO 63108-2915
314-534-0475
Information on coping with the late effects of polio; referrals to other organizations.
E-mail: gini_intl@msn.com
Website: http://www.post-polio.org

Prostate Problems
American Foundation for Urologic Disease
800-242-2383
Information and publications.
Website: http://www.afud.org

Rare Disorders
National Organization for Rare Disorders
800-999-6673
Information on diseases and networking programs; referrals to organizations for specific disorders.
Website: http://www.rarediseases.org

Rehabilitation
National Rehabilitation Information Center
800-34-NARIC; in Maryland, 301-562-2400
Research referrals and information on rehabilitation issues.
Website: http://www.naric.com

Scleroderma
United Scleroderma Foundation
800-722-4673
Referrals to local support groups and treatment centers, as well as information on scleroderma and related skin disorders.
E-mail: sfinfo@scleroderma.org
Website: http://www.scleroderma.org

Sexually Transmitted Diseases
See also *AIDS*
National STD Hotline
800-227-8922
Information; confidential referrals.
Website: http://www.ashastd.org

Sjogren's Syndrome
Sjogren's Syndrome Foundation
800-475-6473; in New York, 516-933-6365
Provides an answering machine for callers to request treatment literature.
Website: http://www.sjogrens.org

Skin Problems
National Psoriasis Foundation
800-723-9166
Information and referrals.
Website: http://www.psoriasis.org

Speech and Hearing
American Speech-Language-Hearing Association Action Center
800-638-8255 (also TTY); 888-321-2742, Answerline; in Maryland, 301-897-5700
Materials on speech and language disorders and hearing impairment; referrals.
Website: http://www.asha.org
Canadian Hard of Hearing Association
2435 Holly Lane, Suite 205
Ottawa, ON K1V 7P2
613-526-1584; TTY 613-526-2692
Publications; answers general questions.
E-mail: chhanational@chha.ca
Website: http://www.chha.ca
Dial a Hearing Screening Test
800-222-EARS
Answers questions on hearing problems. Makes referrals to local telephone numbers for a two-minute hearing test. Also to ear, nose, and throat specialists and to organizations that can provide specialized ear and hearing aid information. 9 am-5 pm EST
E-mail: dahst@aol.com
Hearing Aid Helpline
800-521-5247, ext. 333
Information and distributes a directory of hearing aid specialists certified by the International Hearing Society.
Website: http://www.hearingihs.org
National Center for Stuttering
800-221-2483; in New York, 212-532-1460
Information on stuttering in all age groups.
Website: http://www.stuttering.com
Stuttering Foundation of America
800-992-9392
Referrals to speech pathologists; resource lists, publications.
E-mail: stutter@vantek.net
Website: http://www.stutteringhelp.org

Spinal Injuries
National Spinal Cord Injury Association
800-962-9629; in Maryland, 301-588-6959
Peer counseling; referrals to local chapters and other organizations.
Website: http://www.spinalcord.org
National Spinal Cord Injury Hotline
800-526-3456
Written materials on spinal cord injuries; referrals to organizations and support groups.
Website: http://www.paralinks.net/national_sci_injury_hotline.html

Stroke
See *Heart Disease and Stroke*

Sudden Infant Death Syndrome
American Sudden Infant Death Syndrome Institute
800-232-SIDS
Answers questions; literature; referrals to other organizations.
E-mail: prevent@sids.org
Website: http://www.sids.org
National SIDS Foundation
800-221-SIDS; in Maryland, 410-653-8226
Literature on medical information, referrals, and support groups.
Website: http://www.sidsalliance.org

Tourette Syndrome
Tourette Syndrome Association
800-237-0717; in New York, 718-224-2999
Printed information.
E-mail: ts@tsa-usa.org
Website: http://ts-usa.org

Urinary Incontinence
National Association for Continence
800-BLADDER
Information on bladder control, services available for incontinence, and assistive devices.
Website: http://www.nafc.org
Simon Foundation for Continence
800-23-SIMON
Support and literature on incontinence.
Website: http://www.simonfoundation.org

Women's Health
National Women's Health Network
514 10th Street NW
Suite 400, Washington, DC
202-347-1140; 202-628-7814 (clearinghouse)
Information and referrals on more than 70 women's health concerns.
Website: http://www.womenshealthnetwork.org
National Women's Health Resource Center
120 Albany St., Ste. 820
New Brunswick, NJ 08901
877-986-9472
A national clearinghouse for women's health information.
Website: http://www.healthywomen.org

TRAVEL AND TOURISM

Travel in 2000

Source: World Tourism Organization

World tourism in 2000 grew by an estimated 7.4%. That growth, the highest in nearly a decade, was spurred by a strong global economy and by special events to commemorate the new millennium. According to preliminary data, the number of international arrivals for the year was a record 698 million. Receipts from international tourism climbed 4.5% over 1999 levels. Tourism grew in all regions of the globe. East Asia and the Pacific experienced the fastest development, hosting about 14.5% more tourists than in 1999. The Middle East, where arrivals increased by 10.2%, was next. African tourism grew most slowly, at 1.5%.

In U.S. tourism, leisure trips accounted for 75% of domestic travel by U.S. citizens, while business trips made up 22%. Business travel volume fell by 2.4 %; 36% of business trips were for combined purposes of business and pleasure.

The growing popularities of cultural tourism, ecotourism, and adventure travel (which includes activities such as mountain climbing, scuba diving, bird-watching, and sailing), were among industry trends cited by the World Tourism Organization.

World Tourism Receipts, 1990-2000

Source: World Tourism Organization

(in billions; figures rounded)

1990 . . . . . $269	1992 . . . . . $315	1994 . . . . . $354	1996 . . . . $436	1998 . . . . $445	2000 . . . . $476
1991 278	1993 324	1995 405	1997 436	1999 455	

World's Top 10 Tourist Destinations, 2000

Source: World Tourism Organization

(number of arrivals in millions; excluding same-day visitors)

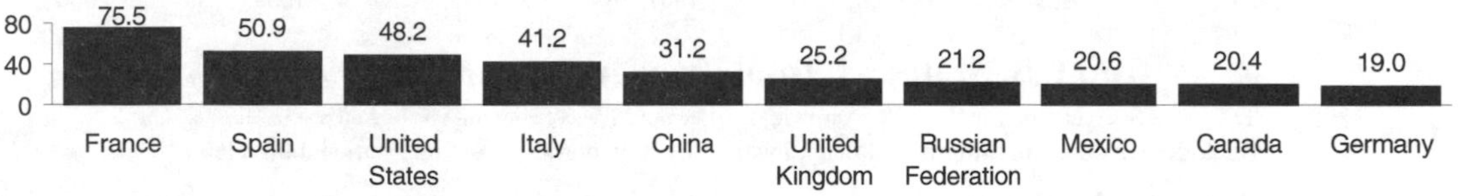

Top Countries in Tourism Earnings, 2000

Source: World Tourism Organization

International tourism receipts (excluding transportation) (in billions of dollars)

Rank 2000	Rank 1990	Country	Receipts 2000
1	1	United States	85.2
2	4	Spain.	31.0
3	2	France.	29.9
4	3	Italy.	27.4
5	5	United Kingdom . .	19.5
6	6	Germany.	17.8
7	25	China	16.2
8	7	Austria	11.4
9	9	Canada.	10.8
10	9	Greece	9.2
11	15	Australia.	8.4
12	10	Mexico	8.3
13	11	Hong Kong, China .	7.9
14	NA	Turkey	7.6

NA = Not Available

Average Number of Vacation Days per Year, Selected Countries

Source: World Tourism Organization

Country	Days
Italy.	42
France	37
Germany	35
Brazil	34
United Kingdom.	28
Canada	26
Korea.	25
Japan	25
United States.	13

IT'S A FACT: Tourism transcended domestic and international categories in 2001 with the visit of American Dennis Tito to the International Space Station. Tito payed a reported $20 million to join the Russian crew that blasted off from Kazakhstan on Apr. 28 and returned safely to earth on May 6. Of his space vacation, Tito said: "It was paradise."

International Travel to the U.S., 1986-2000

Source: Tourism Industries, International Trade Administration, Dept. of Commerce

(Visitors each year are in millions; some figures are revised, may differ from other sources.)

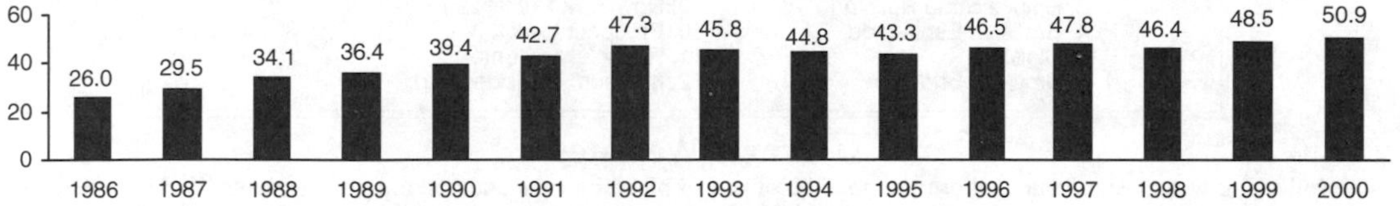

International Visitors to the U.S.[1]

Source: Tourism Industries, International Trade Administration, Dept. of Commerce

Country of origin	Visitors (thousands)[2]	Expenditures (millions)[3]	Expenditures per visitor[3]
Canada.	14,594	$6,206	$440
Mexico	10,322	4,112	415
Japan	5,061	9,711	2,012
United Kingdom .	4,703	8,398	1,975
Germany	1,786	4,398	2,216
France	1,087	2,330	2,200
Brazil	737	$2,034	3,058
South Korea	662	1,251	2,507
Italy	612	1,691	2,701
Venezuela	577	1,697	3,074
All countries. . .	**50,891**	**$74.9 bil[4]**	**$1,472**

(1) Excludes cruise travel. (2) Preliminary 2000 data. (3)Figures for expenditures and expenditures per visitor are for 1999; excludes international passenger fare payments. (4) Does not include international traveler spending on U.S. carriers for transactions made outside the U.S.

Traveler Spending in the U.S., 1987-99

Source: Tourism Industries, International Trade Administration, Dept. of Commerce

(in billions)

Year	Domestic Travelers	International Travelers
1987	$235	$31
1988	258	38
1989	273	47
1990	291	43
1991	296	48
1992	$306	$55
1993	323	58
1994	340	58
1995	360	63
1996	$386	$70
1997	407	73
1998	426	71
1999	446	75

U.S. Domestic Leisure Travel Volume, 1994-2000

Source: "Tourism Works for America," 2001 edition, Travel Industry Assn. of America

(in millions of person-trips of 50 mi or more, one-way)

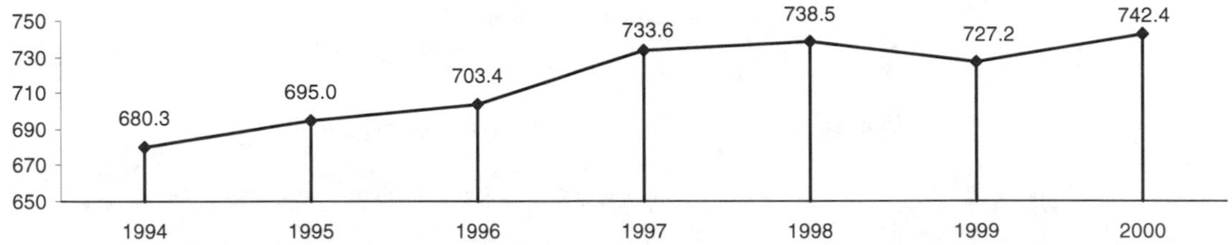

Top U.S. States by Total Traveler Spending, 1999

Source: "Tourism Works for America," 2001 edition, Travel Industry Assn. of America

(includes spending, in billions of dollars, in states by both domestic and international travelers)

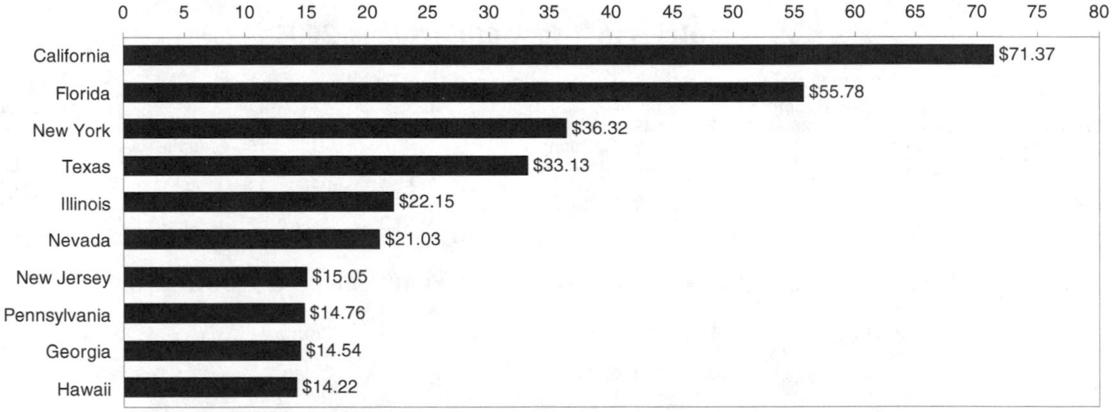

WORLD ALMANAC EDITORS' PICKS

New York CIty remained a major tourist attraction, even as it sought to recover from the terrorist attack on Sept. 11, 2001, that destroyed the World Trade Center's famous twin towers. The World Almanac staff ranked the following as their favorite places/sites to visit in New York City:

1. Central Park
2. Metropolitan Museum of Art
3. Empire State Building
4. Brooklyn Esplanade
5. Cloisters
6. Brooklyn Bridge
7. Ellis Island
8. Coney Island
9. New York Public Library
10. Prospect Park
11. Rockefeller Center
12. Museum of Modern Art

Travel Websites

The following websites are among those that may be of use in planning trips and making arrangements. Websites listed under "Maps" enable the user to plot a route to a destination. Inclusion here does not represent endorsement by *The World Almanac.*

AIRLINES

American Airlines
http://www.aa.com

America West Airlines
http://www.americawest.com

Continental Airlines
http://www.flycontinental.com

Delta Air Lines
http://www.delta.com

Northwest Airlines
http://www.nwa.com

Southwest Airlines
http://www.iflyswa.com

Trans World Airlines
http://www.twa.com

United Airlines
http://www.ual.com

USAirways
http://www.usair.com

BUSES

Gray Line Worldwide
http://www.grayline.com

Greyhound Lines
http://www.greyhound.com

Peter Pan Bus Lines
http://www.peterpanbus.com

TRAINS

Amtrak
http://www.amtrak.com

BC Rail (Canada)
http://www.bcrail.com

Rail Europe
http://www.raileurope.com

CAR RENTALS

Alamo Rent A Ca
http://www.goalamo.com

Avis Rent-A-Car
http://www.avis.com

Budget Rent A Car
http://www.budgetrentacar.com

Dollar Rent A Car
http://www.dollarcar.com

Enterprise Rent-A-Car
http://www.enterprise.com

Hertz
http://www.hertz.com

National Car Rental
http://www.nationalcar.com

Rent-A-Wreck
http://www.rent-a-wreck.com

Thrifty Rent-A-Car
http://www.thrifty.com

CRUISE LINES

Carnival Cruise Lines
http://www.carnival.com

Celebrity Cruises
http://www.celebrity-cruises.com

Costa Cruise Lines
http://www.costacruises.com

Cunard Line
http://www.cunardline.com

Holland America Line
http://www.hollandamerica.com

Norwegian Cruise Line
http://www.ncl.com

Princess Cruises
http://www.princesscruises.com

Renaissance Cruises
http://www.renaissancecruises.com

Royal Caribbean Int'l.
http://www.rccl.com

Windjammer Barefoot Cruises
http://www.windjammer.com

HOTELS/RESORTS

Best Western Int'l.
http://www.bestwestern.com

Choice Hotels Int'l.,
Clarion Hotels & Resorts,
Comfort Inns,
Econo Lodges,
Quality Inns,
Rodeway Inns,
Sleep Inns
http://www.hotelchoice.com

Days Inn of America
http://www.daysinn.com

Doubletree Hotels
http://www.doubletree.com

Embassy Suites
http://www.embassy-suites.com

Four Seasons Hotels
http://www.fshr.com

Hilton Hotels
http://www.hilton.com

Holiday Inn Worldwide
http://www.holiday-inn.com

Hyatt Hotels and Resorts
http://www.hyatt.com

Inter-Continental Hotels
http://www.interconti.com

Loews Hotels
http://www.loewshotels.com

Marriott Int'l.
http://www.marriott.com

Radisson Hotels Int'l.
http://www.radisson.com

Sheraton Hotels & Resorts
http://www.sheraton.com

Westin Hotels & Resorts
http://www.westin.com

Wyndham Hotels & Resorts
http://www.wyndham.com

TRAVEL PLANNING

http://www.travelocity.com
http://www.priceline.com
http://www.Expedia.com
http://www.itn.net
http://www.lowestfare.com
http://www.trip.com

MAPS

http://www.freetrip.com
http://www.mapquest.com
http://www.mapsonus.com

Top 15 Travel Websites

Source: Media Metrix, Inc.

Rank		Visitors[1]
1.	http://www.mapquest.com	13,116,000
2.	http://www.expedia travel*	9,508,000
3.	http://www.travelocity.com	8,754,000
4.	http://www.americanexpress.com	7,954,000
5.	http://www.orbitz.com	5,336,000
6.	http://www.southwest.com	4,402,000
7.	http://www.aa.com	3,994,000
8.	http://www.delta.com	3,505,000
9.	http://www.priceline.com	3,156,000
10.	http://www.united.com	3,093,000
11.	http://www.cheaptickets.com	3,008,000
12.	http://www.nwa.com	2,600,000
13.	http://www.travelersadvantage.com	2,511,000
14.	http://www.hotwire.com	2,495,000
15.	http://www.secureredemption.com	2,349,000

(1) Number who visited at least once in July 2001. *Represents aggregation of commonly owned domain names.

Some Notable Roller Coasters

Source: http://www.rollercoaster.com, http://www.rcdb.com, *World Almanac* research, 2000

Fastest Roller Coasters

Name	Speed	Location
Tower of Terror	100 mph	Dreamworld; Australia
Superman The Escape	100 mph	Six Flags Magic Mountain; Valencia, CA
Steel Dragon 2000	95 mph	Nagashima Spaland; Mie, Japan
Millennium Force	93 mph	Cedar Point; Sandusky, OH
Goliath	85 mph	Six Flags Magic Mountain; Valencia, CA
Titan	85 mph	Six Flags Over Texas; Dallas, TX

Tallest Roller Coasters

Name	Drop	Location
Superman The Escape	415 ft	Six Flags Magic Mountain; Valencia, CA
Tower of Terror	377 ft	Dreamworld; Australia
Steel Dragon 2000	318 ft	Nagashima Spaland; Mie, Japan
Millennium Force	310 ft	Cedar Point; Sandusky, OH
Fujiyama	259 ft	Fujikyu Highland; Yamanashi, Japan

Longest Roller Coasters

Name	Length	Location
Steel Dragon 2000	8133 ft	Nagashima Spaland; Mie, Japan
Daidarasaurus	7677 ft	Expoland, Osaka, Japan
Ultimate	7442 ft	Lightwater Valley; Yorkshire, UK
Beast	7400 ft	Paramount's Kings Island; Kings Island, OH
Son of Beast	7032 ft	Paramount's Kings Island; Kings Island, OH

Roller Coasters With Longest Drop

Name	Drop	Location
Superman The Escape	328 ft	Six Flags Magic Mountain; Valencia, CA
Tower of Terror	328 ft	Dreamworld; Australia
Steel Dragon 2000	307 ft	Nagashima Spaland; Mie, Japan
Millennium Force	300 ft	Cedar Point; Sandusky, OH
Goliath	255 ft	Six Flags Magic Mountain; Valencia, CA
Titan	255 ft	Six Flags Over Texas; Dallas, TX

IT'S A FACT: The first "roller coasters" were ice slides built throughout Russia in the 16th century. These wooden structures, covered in thick sheets of ice, were built up to 80 feet high and hundreds of feet long, with drops of 50 degrees. Ice-sliding is said to have been a favorite pastime of Catherine the Great.

Top 50 Amusement/Theme Parks Worldwide, 2000

Source: Amusement Business, Aug. 2001, ranked by attendance

Rank	Park & Location	Country	Attendance
1.	Tokyo Disneyland	Japan	16,507,000
2.	Magic Kingdom at at Walt Disney World, Lake Buena Vista, FL	United States	15,400,000*
3.	Disneyland, Anaheim, CA	United States	13,900,000*
4.	Disneyland Paris, Marne-La-Vallee	France	12,000,000
5.	Epcot at Walt Disney World, Lake Buena Vista, FL	United States	10,600,000*
6.	Everland, Kyunggi-Do	South Korea	9,153,000
7.	Disney-MGM Studios at Walt Disney World, Lake Buena Vista, FL	United States	8,900,000*
8.	Disney Animal Kingdom at Walt Disney World, Lake Buena Vista, FL	United States	8,300,000*
9.	Universal Studios Florida, Orlando, FL	United States	8,100,000*
10.	Lotte World, Seoul	South Korea	7,200,000
11.	Blackpool (England) Pleasure Beach	United Kingdom	6,800,000*
12.	Universal's Island of Adventures, Orlando, FL	United States	6,000,000*
13.	Yokohama Hakkeijima Sea Paradise	Japan	5,332,000
14.	Universal Studios Hollywood, Universal City, CA	United States	5,200,000*
	Seaworld Florida, Orlando, FL	United States	5,200,000*
16.	Busch Garden Tampa Bay, FL	United States	5,000,000*
17.	Nagashima Spa Land, Kuwana	Japan	4,500,000
18.	Tivoli Gardens, Copenhagen	Denmark	3,900,000
19.	Huis Ten Bosch, Sasebo	Japan	3,861,000
20.	Seaworld California, San Diego, CA	United States	3,600,000*
21.	Six Flags Great Adventure, Jackson, NJ	United States	3,500,000*
22.	Knott's Berry Farm, Buena Park, CA	United States	3,456,000*
23.	Cedar Point, Sandusky, OH	United States	3,432,000*
24.	Morey's Piers, Wildwood, NJ	United States	3,300,000*
	Six Flags Magic Mountain, Valencia, CA	USA	3,300,000
26.	Paramount's Kings Island, Kings Island, OH	United States	3,200,000*
27.	Ocean Park, Hong Kong	China	3,190,000
28.	Seoul Land, Kyunggi-Do	South Korea	3,150,000
	Port Aventura, Salou	Spain	3,150,000*
30.	Suzuka Circuit	Japan	3,042,000
31.	Santa Cruz Beach Boardwalk, CA	USA	3,000,000*
	Liseburg, Gothenburg	Sweden	3,000,000*
	Europa-Park, Rust	Germany	3,000,000
34.	Adventuredome at Circus Circus, Las Vegas, NV	USA	2,977,000*
35.	Paramount Canada's Wonderland, Maples, Ontario	Canada	2,975,000
36.	De Efteling, Kaatsheuvel	The Netherlands	2,900,000
	Gardaland, Castelnuovo Del Garda	Italy	2,900,000*
38.	Six Flags Great America, Gumee, IL	USA	2,875,000*
39.	Six Flags Over Texas, Arlington, TX	USA	2,775,000*
40.	Six Flags Mexico (Reino Adventura), Mexico City	Mexico	2,750,000*
41.	Knott's Camp Snoopy, Bloomington, MN	USA	2,600,000*
42.	Bakken, Klampenborg	Denmark	2,500,000
43.	Alton Towers, North Staffordshire	UK	2,450,000*
	Hersheypark, Hershey, PA	USA	2,450,000
45.	Six Flags Over Georgia, Atlanta, GA	USA	2,400,000*
46.	Kurashiki (JAPAN) Tivoli Park	Japan	2,380,000
47.	Parque Espana, Shima-gun	Japan	2,313,000
48.	Busch Gardens The Old Country, Williamsburg, VA	USA	2,300,000
	Dollywood, Pigeon Forge, TN	USA	2,300,000
50.	Six Flags Fiesta Texas, San Antonio, TX	USA	2,200,000

* Figures are projected from previous year's attendance.

Passports, Health Regulations, and Travel Warnings for Foreign Travel

Source: Bureau of Consular Affairs, U.S. Dept. of State

Passports are issued by the U.S. Department of State to citizens and nationals of the U.S. for the purpose of documenting them for foreign travel and identifying them as U.S. citizens. For U.S. citizens traveling on business or as tourists, especially in Europe, a U.S. passport is often sufficient to gain admission for a limited stay. For many countries, however, a **visa** must also be obtained before entering. It is the responsibility of the traveler to check in advance and obtain any visas where required, from the appropriate embassy or nearest consulate of each country.

Each country has its own specific guidelines concerning length and purpose of visit, etc. Some may require visitors to display proof that they (1) have sufficient funds to stay for the intended time period and (2) have onward/return tickets.

Some countries, including **Canada, Mexico,** and some **Caribbean** islands, do not require a passport or a visa for limited stays. Such countries do require proof of U.S. citizenship, and may have other requirements that must be met. For further information, check with the embassy or nearest consulate of the country you plan to visit.

How to Obtain a Passport

Those who have never been issued a passport in their own name must apply in person before (1) a passport agent; (2) a clerk of any federal court or state court of record or a clerk or judge of a probate court accepting applications; (3) a postal clerk at a post office that is authorized to accept passport applications; or (4) a U.S. diplomatic or consular officer abroad.

A DSP-11 is the correct form to use for those who must apply in person. All persons are required to obtain individual passports in their own name. However, a parent or legal guardian must execute the application for children under 13.

Persons who possess their most recent passport, if it was issued within the last 12 years and after their 18th birthday, may be eligible to apply for a new passport by mail. The form DSP-82, *Application for Passport by Mail*, must be filled out and mailed to the address shown on the form, together with the previous passport, 2 recent identical photographs (see below), and a fee of $40. The DSP-82 may not be used if the most recent passport has been altered or mutilated.

Proof of citizenship—A full validity passport previously issued to the applicant or one in which he or she was included will be accepted as proof of U.S. citizenship. If the applicant has no prior passport and was born in the U.S., a certified copy of the birth certificate generally must be presented. It must generally show the given name and surname, the date and place of birth, and that the birth record was filed shortly after birth. A delayed birth certificate (filed more than 1 year after date of birth) is acceptable if it shows that acceptable secondary evidence was used for creating this record.

If a birth certificate is not obtainable, a notice from a state registrar must be submitted stating that no birth record exists. It must be accompanied by the best obtainable secondary evidence, such as a baptismal certificate or hospital birth record.

A naturalized citizen with no previous passport must present a Certificate of Naturalization. A person born abroad claiming U.S. citizenship through either a native-born or a naturalized citizen parent must normally submit a Certificate of Citizenship issued by the Immigration and Naturalization Service or a Consular Report of Birth or Certification of Birth Abroad issued by the Dept. of State. If such a document has not been obtained, evidence of citizenship of the parent(s) through whom citizenship is claimed and evidence that would establish the parent/child relationship must be submitted. Additionally, if citizenship is derived through birth to citizen parent(s), the applicant must submit parents' marriage certificate plus an affidavit from parent(s) showing periods and places of residence or presence in the U.S. and abroad, and specifying periods spent abroad in the employment of the U.S. government, including the armed forces, or with certain international organizations. If citizenship is derived through naturalization of parents, evidence of admission to the U.S. for permanent residence also is required.

It is important to apply for a passport as far in advance as possible. Passport offices are busiest between March and September. It can take several weeks to receive a passport.

Photographs—Passport applicants must submit 2 identical photographs that are recent (normally not more than 6 months old) and that are a good likeness of and satisfactorily identify the applicant. Photographs should be 2 x 2 in. in size. The image size, from bottom of chin to top of head (including hair), should not be less than 1 inch or more than 1-3/8 in. Photographs should be portrait-type prints. They must be clear, front view, full face, with a plain white or off-white background. Photos that depict the applicant as relaxed and smiling are encouraged.

Identity—Applicants must establish their identity to the satisfaction of the authorities. Generally acceptable documents of identity include a previous U.S. passport, a Certificate of Naturalization, a Certificate of Citizenship, a valid driver's license, or a government identification card. Applicants may not use a Social Security card, learner's or temporary driver's license, credit card, or expired ID card. Extremely old documents cannot be used by themselves.

Applicants unable to establish identity must present some documentation in their own name and be accompanied by a person who has known them at least 2 years and is a U.S. citizen or legal U.S. permanent resident alien. That person must sign an affidavit before the individual who executes the application, and must establish his or her own identity.

Fees—For persons under 16 years of age, the basic passport fee is $40. These passports are valid for 5 years from date of issue. The basic fee is $60 for passports issued to persons 16 and older. These passports are valid for 14 years from date of issuance. To receive a passport within 10 days or less, a $35 expedite fee is required. There is no execution fee when using DSP-82, *Application for Passport by Mail*. This form is available on-line at the Passport Services website. Applicants eligible to use this form pay only a $40 passport fee.

Passport loss—The loss or theft of a valid passport should be reported immediately in writing to Passport Services, Correspondence Branch, 1111 19th St., NW, Suite 510, Washington, DC 20036, telephone: (202) 647-4000, or to the nearest passport agency or nearest U.S. embassy or consulate when abroad. The DSP-64, *Statement Regarding Lost or Stolen Passport*, is available on-line at the address below.

General Information—Visit Passport Services on the internet at http://www.travel.state.gov/passport_services.html

Health Regulations

Under the regulations adopted by the World Health Organization, a country may require International Certificates of Vaccination against yellow fever. A cholera immunization may be required for travelers from infected areas. Check with health care providers or your records to see that other immunizations (e.g., for tetanus and polio) are up-to-date.

Prophylactic medication for malaria and certain other preventive measures are advisable for travel to some countries. No immunizations are needed to return to the U.S. An increasing number of countries have regulations regarding AIDS testing, particularly for longtime visitors. Detailed information is included in *Health Information for International Travel*, available from the U.S. Government Printing Office, Washington, DC 20402, for $24. Information may also be obtained from your local health department or physician, or by calling the Centers for Disease Control and Prevention at 1-877-FYI-TRIP (1-877-394-8747).

General information—The booklets *Passports—Applying for the Easy Way* and *Foreign Entry Requirements* are available for 50¢ each from the Consumer Information Center, Pueblo, CO 81009. For online information, as well as HIV Testing Requirements, go to the Consular Affairs website—http://www.travel.state.gov

Travel Warnings

Travel Warnings are issued when the State Dept. decides, based on relevant information, to recommend that Americans avoid travel to a certain country; these are subject to change. For the latest information, 24 hours a day, dial 202/647-5225 from a touch-tone telephone. As of Sept. 25, 2001, travel warnings were in effect for: Afghanistan, Albania, Algeria, Angola, Bosnia and Herzegovina, Burundi, Central African Republic, Colombia, Congo, Guinea-Bissau, Iran, Iraq, Indonesia, Israel (West Bank and Gaza), Kyrgyzstan, Lebanon, Liberia, Libya, Macedonia, Nigeria, Pakistan, Sierra Leone, Solomon Islands, Somalia, Sri Lanka, Sudan, Tajikistan, Turkmenistan, Yemen, and Yugoslavia.

Customs Exemptions for Travelers

Source: U.S. Dept. of the Treasury, U.S. Customs Service

U.S. residents returning after a stay abroad of at least 48 hours are usually granted customs exemptions of $400 each (this and all exemptions figured according to fair retail value). The duty-free articles must accompany the traveler at the time of return, be for personal or household use, have been acquired as an incident of the trip, and be properly declared to Customs. No more than 1 liter of alcoholic beverages or more than 100 cigars and 200 cigarettes (1 carton) may be included in the $400 exemption. The exemption for alcoholic beverages holds only if the returning resident is at least 21 years old at the time of arrival. Cuban cigars may be included only if purchased in Cuba.

If a U.S. resident arrives directly or indirectly from a U.S. island possession—American Samoa, Guam, or U.S. Virgin Islands—a customs exemption of $1,200 is allowed. Up to 1,000 cigarettes may be included, but only 200 of them may have been purchased elsewhere. If a U.S. resident returns from any one of the following places, the exemption is $600: Antigua and Barbuda, Aruba, Bahamas, Barbados, Belize, British Virgin Islands, Costa Rica, Dominica, Dominican Republic, El Salvador, Grenada, Guatemala, Guyana, Haiti, Honduras, Jamaica, Montserrat, Netherlands Antilles, Nicaragua, Panama, St. Kitts and Nevis, St. Lucia, St. Vincent and the Grenadines, Trinidad and Tobago.

The $400, $600, or $1,200 exemption may be granted only if the exemption has not been used in whole or part within the preceding 30-day period and only if the stay abroad was for at least 48 hours. The 48-hr absence requirement does not apply to travelers returning from Mexico or U.S. Virgin Islands. Travelers who cannot claim the $400, $600, or $1,200 exemption because of the 30-day or 48-hr provisions may bring in free of duty and tax articles acquired abroad for personal or household use up to a value of $25.

There are also allowances for goods when shipped. Goods shipped for personal use may be imported free of duty and tax if the total value is no more than $200. This exemption does not apply to perfume containing alcohol if it is valued at more than $5 retail, to alcoholic beverages, or to cigars and cigarettes. The $200 mail exemption does not apply to merchandise subject to absolute or tariff-rate quotas unless the item is for personal use. Tailor-made suits ordered from Hong Kong, however, are subject to quota/visa requirements even if imported for personal use.

Bona fide gifts of not more than $100 in value, when shipped, can be received in the U.S. free of duty and tax, provided that the same person does not receive more than $100 in gift shipments in one day. The limit is increased to $200 for bona fide gift items shipped from U.S. Virgin Islands, American Samoa, or Guam. (Shipping of alcoholic beverages, including wine and beer, by mail is prohibited by U.S. postal laws.) These gifts are not declared by the traveler upon return to the U.S.

The U.S. Customs Service booklet "Know Before You Go" answers frequently asked customs questions and is available free by writing U.S. Customs Services, KBYG, PO Box 7407, Washington, DC 20044, or by visiting the U.S. Customs website—http://www.customs.ustreas.gov

CRIME

Crime in U.S. Steady in 2000 After 8-Year Decline

Serious crimes reported to law enforcement agencies in the United States remained relatively unchanged from 1999 to 2000, according to preliminary figures from the Federal Bureau of Investigation's Uniform Crime Reporting Program, released May 30, 2001. This stagnation marked an end to the trend of recent years; reported crime in the U.S. had gone down 7.8% in 1999, 5% in 1998, 2% in 1997, 3% in 1996, 1% in both 1994 and 1995, 2% in 1993, and 3% in 1992.

Serious crime is measured by the Crime Index, which includes 4 violent crimes and 3 property crimes. Violent crime increased slightly, by 0.1%, from 1999 to 2000, and property crime showed virtually no change.

In the violent crime category, murder declined 1.1%, robbery by 0.7%; forcible rape increased by 0.7%, and aggravated assault by 0.4%.

Among property crimes, burglary decreased 2.1% from 1999 numbers; motor vehicle theft increased by 2.7%, larceny-theft by 0.1%. Arson, which is not included in the crime index proper, increased by 0.6%.

Declines in the overall Crime Index totals of 2.4% and 1.1%, respectively, were reported in the Northeast and Midwest regions of the country. The total increased 1.1% in the West and 1% in the South.

Cities with populations between 50,000 and 99,999 showed the largest decline in reported crime last year, an average of –1.1%. The decline was 0.5% in cities with over 1 million people. Increases of 0.5% were recorded by cities with populations from 25,000 to 49,999 and from 100,000 to 249,999. The Crime Index increased 0.7% in suburban counties and decreased 0.8% in rural counties.

U.S. Crime Index Trends, 2000

Source: FBI, *Uniform Crime Reports,* 2000, preliminary

(percentage change 2000 over 1999, offenses known to the police)

	No. of agencies[1]	Pop. (thousands)	Crime Index (total)	Modified total[2]	Violent crime[3]	Property crime[4]	Murder	Forcible rape	Robbery	Aggravated assault	Burglary	Larceny/theft	Motor vehicle theft	Arson
Total			**0.0**	**0.0**	**+0.1**	**0.0**	**-1.1**	**+0.7**	**-0.7**	**+0.4**	**-2.1**	**+0.1**	**+2.7**	**-0.6**
Cities:														
Over 1,000,000	10	23,598	-0.5	-0.6	-0.2	-0.6	+2.5	+0.1	-2.9	+1.7	-2.5	-0.4	+0.5	-7.3
500,000 to 999,999	19	12,038	-0.1	-0.1	+0.3	-0.2	-6.2	+3.0	+2.9	-1.4	+0.2	-1.2	+4.2	+5.3
250,000 to 499,999	36	12,582	+0.1	+0.1	-1.7	+0.4	-1.0	-2.2	-2.2	-1.4	-3.3	+0.6	+4.3	+0.2
100,000 to 249,999	163	23,870	+0.5	+0.5	+1.2	+0.5	+1.3	+0.7	+1.3	+1.2	-2.5	+0.6	+4.7	+0.4
50,000 to 99,999	299	20,293	-1.1	-1.1	+0.4	-1.3	-0.7	+1.5	-0.5	+0.7	-1.9	-1.6	+1.8	-6.4
25,000 to 49,999	575	19,922	+0.5	+0.6	+0.3	+0.6	+2.2	+3.9	+0.6	-0.3	-1.5	+1.0	+0.9	+5.0
10,000 to 24,999	1,251	19,648	+0.1	+0.2	+1.2	0.0	-4.7	+1.9	+0.2	+1.5	-1.7	+0.5	-0.2	+3.0
Under 10,000	4,027	14,600	-0.2	-0.2	+0.8	-0.3	+12.6	+3.9	+1.1	+0.3	-1.8	-0.1	+0.4	+2.2
Counties:														
Suburban[5]	876	34,867	+0.7	+0.7	+0.2	+0.8	-7.7	-0.9	+1.7	+0.1	-2.7	+1.2	+6.3	+1.9
Rural[6]	1,667	18,536	-0.8	-0.8	-0.7	-0.8	-7.8	-2.7	-2.5	-0.2	-2.5	+0.1	-0.8	-4.9
Area:														
Suburban[7]	4,523	72,841	+0.7	+0.7	+0.9	+0.7	-3.6	-0.7	+1.0	+1.2	-1.7	+0.9	+3.7	+2.0
Metropolitan	2,206	16,195	-0.6	-0.6	-0.9	-0.6	-1.0	+8.3	+0.7	-2.2	-3.2	+0.2	-2.4	+6.0

Note: All data are preliminary. (1) Law-enforcement agencies. (2) The modified total includes arson, which is not included in the Crime Index Total. (3) Violent crimes are murder, forcible rape, robbery, and aggravated assault. (4) Property crimes are burglary, larceny-theft, and motor vehicle theft. Data for the property crime of arson are not included. (5) Includes crimes reported to sheriffs' department, county police departments, and state police within Metropolitan Statistical Areas. (6) Includes crimes reported to sheriffs' department, county police departments, and state police outside Metropolitan Statistical Areas. (7) Includes crimes reported to city, county, and state law enforcement agencies within Metropolitan Statistical Areas, but outside the central cities.

Crime Index Trends by Geographic Region, 2000

Source: FBI, *Uniform Crime Reports,* 2000, preliminary

(percentage change 2000 over 1999, offenses known to the police)

	Crime index total	Modified Total[1]	Violent crime	Property Crime	Murder	Forcible Rape	Robbery	Aggravated Assault	Burglary	Larceny -Theft	Motor Vehicle Theft	Arson
TOTAL U.S.	**0.0**	**0.0**	**+0.1**	**0.0**	**-1.1**	**+0.7**	**-0.7**	**+0.4**	**-2.1**	**+0.1**	**+2.7**	**-0.6**
Northeast	-2.4	-2.5	-1.7	-2.6	+1.3	-0.2	-5.8	+1.0	-6.7	-1.4	-2.7	-8.1
Midwest	-1.1	-1.1	-0.7	-1.2	-2.3	+1.2	-0.3	-1.2	-2.9	-1.0	+0.8	+2.3
South	+1.0	+1.0	+0.7	+1.0	+0.9	-1.9	+2.5	+0.1	-1.3	+1.6	+2.0	-1.1
West	+1.1	+1.1	+1.2	+1.0	-3.9	+3.5	+0.6	+1.2	-0.1	+0.1	+7.3	+1.7

Note: Data are preliminary. (1) The modified total includes arson, which is not included in the Crime Index total.

Crime Index Trends, 1992-2000

Source: FBI, *Uniform Crime Reports,* 2000, preliminary

(percentage change over previous year, offenses known to police)

Year	Crime Index (total)	Violent crime	Property crime[1]	Murder	Forcible rape	Robbery	Aggravated assault	Burglary	Larceny-theft	Motor vehicle theft
1992	–3	+1	–4	–4	+2	–2	+3	–6	–3	–3
1993	–2	0	–2	+3	–4	–2	+1	–5	–1	–3
1994	–1	–3	–1	–5	–4	–6	–1	–4	+1	–2
1995	–1	–1	–1	–7	–6	–7	–3	–5	+1	–5
1996	–3	–7	–2	–9	–2	–8	–6	–4	–1	–5
1997	–2	–3	–2	–8	0	–7	–1	–2	–2	–3
1998	–5	–6	–5	–7	–3	–10	–5	–5	–5	–8
1999	–7	–7	–7	–9	–4	–8	–6	–10	–6	–8
2000[2,3]	0	0	0	–1	+1	–1	0	–2	0	+3

(1) Data for arson not included. (2) The crime index declined 19.5% from 1992 to 2000. (3) Data are preliminary.

Crime in the U.S., 1979-99

Source: FBI, *Uniform Crime Reports*, 1999; later data available at www.fbi.gov/ucr/ucr.htm

Population[1]	Crime Index (total)[2]	Violent crime	Property crime[3]	Murder and non-negligent manslaughter	Forcible rape	Robbery	Burglary	Larceny-theft
Population by year	**NUMBER OF REPORTED OFFENSES**							
1979–220,099,000	12,249,500	1,208,030	11,041,500	21,460	76,390	480,700	3,327,700	6,601,000
1980–225,349,264	13,408,300	1,344,520	12,063,700	23,040	82,990	565,840	3,795,200	7,136,900
1981–229,146,000	13,423,800	1,361,820	12,061,900	22,520	82,500	592,910	3,779,700	7,194,400
1982–231,534,000	12,974,400	1,322,390	11,652,000	21,010	78,770	553,130	3,447,100	7,142,500
1983–233,981,000	12,108,600	1,258,090	10,850,500	19,310	78,920	506,570	3,129,900	6,712,800
1984–236,158,000	11,881,800	1,273,280	10,608,500	18,690	84,230	485,010	2,984,400	6,591,900
1985–238,740,000	12,431,400	1,328,770	11,102,600	18,980	88,670	497,870	3,073,300	6,926,400
1986–241,077,000	13,211,900	1,489,170	11,722,700	20,610	91,460	542,780	3,241,400	7,257,200
1987–243,400,000	13,508,700	1,484,000	12,024,700	20,100	91,110	517,700	3,236,200	7,499,900
1988–245,807,000	13,923,100	1,566,220	12,356,900	20,680	92,490	542,970	3,218,100	7,705,900
1989–248,239,000	14,251,400	1,646,040	12,605,400	21,500	94,500	578,330	3,168,200	7,872,400
1990–248,709,873	14,475,600	1,820,130	12,655,500	23,440	102,560	639,270	3,073,900	7,945,700
1991–252,177,000	14,872,900	1,911,770	12,961,100	24,700	106,590	687,730	3,157,200	8,142,200
1992–255,082,000	14,438,200	1,932,270	12,505,900	23,760	109,060	672,480	2,979,900	7,915,200
1993–257,908,000	14,144,800	1,926,020	12,218,800	24,530	106,010	659,870	2,834,800	7,820,900
1994–260,341,000	13,989,500	1,857,670	12,131,900	23,330	102,220	618,950	2,712,800	7,879,800
1995–262,755,000	13,862,700	1,798,790	12,063,900	21,610	97,470	580,510	2,593,800	7,997,700
1996–265,284,000	13,493,900	1,688,540	11,805,300	19,650	96,250	535,590	2,506,400	7,904,700
1997–267,637,000	13,194,600	1,636,100	11,558,500	18,210	96,150	498,530	2,460,500	7,743,800
1998–270,296,000	12,485,700	1,533,890	10,951,800	16,970	93,140	447,190	2,332,700	7,376,300
1999–272,691,000	11,635,100	1,430,690	10,204,500	15,530	89,110	409,670	2,099,700	6,957,400
	PERCENT CHANGE: NUMBER OF OFFENSES							
1999/1998	–6.8	–6.7	–6.8	–8.5	–4.3	–8.4	–10.0	–5.7
1999/1995	–16.1	–20.5	–15.4	–28.1	–8.6	–29.4	–19.0	–13.0
1999/1990	–19.6	–21.4	–19.4	–33.7	–13.1	–35.9	–31.7	–12.4
Year	**RATE PER 100,000 INHABITANTS**							
1979	5,565.5	548.9	5,016.6	9.7	34.7	218.4	1,511.9	2,999.1
1980	5,950.0	596.6	5,353.3	10.2	36.8	251.1	1,684.1	3,167.0
1981	5,858.2	594.3	5,263.9	9.8	36.0	258.7	1,649.5	3,139.7
1982	5,603.6	571.1	5,032.5	9.1	34.0	238.9	1,488.8	3,084.8
1983	5,175.0	537.7	4,637.4	8.3	33.7	216.5	1,337.7	2,868.9
1984	5,031.3	539.2	4,492.1	7.9	35.7	205.4	1,263.7	2,791.3
1985	5,207.1	556.6	4,650.5	8.0	37.1	208.5	1,287.3	2,901.2
1986	5,480.4	617.7	4,862.6	8.6	37.9	225.1	1,344.6	3,010.3
1987	5,550.0	609.7	4,940.3	8.3	37.4	212.7	1,329.6	3,081.3
1988	5,664.2	637.2	5,027.1	8.4	37.6	220.9	1,309.2	3,134.9
1989	5,741.0	663.1	5,077.9	8.7	38.1	233.0	1,276.3	3,171.3
1990	5,820.3	731.8	5,088.5	9.4	41.2	257.0	1,235.9	3,194.8
1991	5,897.8	758.1	5,139.7	9.8	42.3	272.7	1,252.0	3,228.8
1992	5,660.2	757.5	4,902.7	9.3	42.8	263.6	1,168.2	3,103.0
1993	5,484.4	746.8	4,737.6	9.5	41.1	255.9	1,099.2	3,032.4
1994	5,373.5	713.6	4,660.0	9.0	39.3	237.7	1,042.0	3,026.7
1995	5,275.9	684.6	4,591.3	8.2	37.1	220.9	987.1	3,043.8
1996	5,086.6	636.5	4,450.1	7.4	36.3	201.9	944.8	2,979.7
1997	4,930.0	611.3	4,318.7	6.8	35.9	186.3	919.4	2,893.4
1998	4,619.3	567.5	4,051.8	6.3	34.5	165.4	863.0	2,729.0
1999	4,266.8	524.7	3,742.1	5.7	32.7	150.2	770.0	2,551.4
	PERCENT CHANGE: RATE PER 100,000 INHABITANTS							
1999/1998	–7.6	–7.5	–7.6	–9.5	–5.2	–9.2	–10.8	–6.5
1999/1995	–19.1	–23.4	–18.5	–30.5	–11.9	–32.0	–22.0	–16.2
1999/1990	–26.7	–28.3	–26.5	–39.4	–20.6	–41.6	–37.7	–20.1

Note: All rates were calculated on the offenses before rounding. (1) Populations are Bureau of the Census provisional estimates as of July 1, except 1980 and 1990, which are the decennial census counts. (2) Because of rounding, violent and property crime may not add to total. Not all categories of violent and property crime appear separately. (3) Data for arson not included.

Law Enforcement Officers, 1999

Source: FBI, *Uniform Crime Reports*, 1999; later data available at www.fbi.gov/ucr/ucr.htm

The U.S. law enforcement community employed an average of 2.5 full-time officers for every 1,000 inhabitants as of Oct. 31, 1999.

Including full-time civilian employees, the overall law enforcement employee rate was 3.6 per 1,000 inhabitants, according to 13,313 city, county, and state police agencies. These agencies collectively offered law enforcement service covering a population of about 253 million, employing 637,551 officers and 261,567 civilians.

The law enforcement employee average for all cities nationwide was 3.2 per 1,000 inhabitants. The highest city law enforcement employee average was 4.4 per 1,000 inhabitants, in cities with populations of 250,000 or more. Averages of 4.3 and 4.1 were recorded in suburban and rural counties, respectively.

Regionally, the law enforcement employee rate was 3.8 in the Northeast, 3.6 in the South, 2.8 in the Midwest, and 2.5 in the West. Nationally, males constituted 89 percent of all sworn employees. Ninety-two percent of the officers in rural counties were males, in suburban counties males accounted for 88 percent.

Civilians made up 29 percent of the total U.S. law enforcement employee force. They represented 23 percent of the police employees in cities and 38 percent in both rural and suburban counties. Females accounted for 64 percent of all civilian employees.

Forty-two law enforcement officers were feloniously slain in the line of duty in 1999, 19 fewer than in 1998. Another 65 officers were killed as a result of accidents occurring while performing official duties, 16 lower than in 1998.

Crime Rates by Region, Geographic Division, and State, 1999

Source: FBI, *Uniform Crime Reports*, 1999; later data available at www.fbi.gov/ucr/ucr.htm

(rate per 100,000 population)

	Total rate	Violent crime[1]	Property crime[2]	Murder	Rape	Robbery	Aggravated assault	Burglary	Larceny-theft	Motor vehicle theft
U.S. TOTAL	**4,266.8**	**524.7**	**3,742.1**	**5.7**	**32.7**	**150.2**	**336.1**	**770.0**	**2,551.4**	**420.7**
Northeast	**3,232.5**	**468.0**	**2,764.5**	**4.1**	**22.5**	**169.9**	**271.5**	**520.2**	**1,901.4**	**343.0**
New England	**3,174.0**	**381.2**	**2,792.8**	**2.4**	**25.4**	**84.0**	**269.4**	**543.6**	**1,916.3**	**332.8**
Connecticut	3,389.3	345.6	3,043.7	3.3	19.9	123.5	198.9	588.0	2,111.5	344.2
Maine	2,815.0	112.2	2,762.8	2.2	19.1	19.4	71.6	601.1	2,026.5	135.2
Massachusetts	3,262.5	551.0	2,711.5	2.0	26.9	96.0	426.0	533.8	1,762.7	415.0
New Hampshire	2,281.9	96.5	2,185.4	1.5	28.7	21.4	44.9	307.9	1,764.8	112.7
Rhode Island	3,581.9	286.6	3,295.4	3.6	39.5	79.5	164.0	639.9	2,248.6	406.9
Vermont	2,817.3	113.8	2,703.5	2.9	22.9	10.9	77.1	595.5	1,954.5	153.5
Middle Atlantic	**3,253.2**	**498.6**	**2,754.6**	**4.6**	**21.5**	**200.2**	**272.2**	**512.0**	**1,896.1**	**346.6**
New Jersey	3,400.1	411.9	2,988.2	3.5	17.3	174.9	216.1	577.2	1,976.9	434.2
New York	3,279.3	588.8	2,690.5	5.0	19.6	240.8	323.5	512.3	1,858.1	320.2
Pennsylvania	3,113.7	420.5	2,693.2	4.9	27.3	155.7	232.5	467.2	1,898.9	327.1
Midwest	**4,040.6**	**448.7**	**3,591.9**	**5.3**	**34.8**	**129.1**	**279.5**	**700.6**	**2,516.6**	**374.7**
East North Central	**4,095.0**	**486.8**	**3,608.2**	**5.8**	**35.5**	**148.7**	**296.8**	**716.1**	**2,482.1**	**410.0**
Illinois	4,506.6	732.5	3,774.1	7.7	34.2	219.4	471.2	712.3	2,632.1	429.7
Indiana	3,765.9	374.6	3,391.3	6.6	27.0	109.3	231.7	714.5	2,335.4	341.4
Michigan	4,324.8	574.9	3,749.9	7.0	49.2	143.0	375.7	777.9	2,396.1	575.8
Ohio	3,996.4	316.4	3,680.1	3.5	36.7	128.0	148.2	773.1	2,558.8	348.2
Wisconsin	3,296.4	245.9	3,050.6	3.4	20.1	84.7	137.6	488.2	2,299.1	263.2
West North Central	**3,912.1**	**358.8**	**3,553.3**	**4.2**	**33.1**	**82.9**	**238.6**	**663.9**	**2,598.0**	**291.4**
Iowa	3,224.0	280.0	2,944.0	1.5	27.2	36.6	214.7	593.0	2,172.0	179.0
Kansas	4,438.7	382.8	4,055.9	6.0	40.1	77.1	259.5	824.2	3,003.8	227.9
Minnesota	3,597.2	274.0	3,323.2	2.8	42.7	82.0	146.5	580.1	2,465.2	278.0
Missouri	4,578.7	500.2	4,078.5	6.6	26.3	130.7	336.6	776.8	2,881.3	420.3
Nebraska	4,108.3	430.2	3,678.1	3.6	24.8	75.9	325.9	609.7	2,741.8	326.5
North Dakota	2,393.1	66.9	2,326.2	1.6	22.4	8.8	34.1	368.6	1,794.2	163.4
South Dakota	2,644.7	167.4	2,477.4	2.5	45.8	14.1	105.0	444.1	1,915.8	117.5
South	**4,932.2**	**600.0**	**4,332.2**	**6.9**	**36.0**	**156.3**	**400.7**	**958.6**	**2,934.8**	**438.8**
South Atlantic	**5,156.0**	**656.6**	**4,499.4**	**7.0**	**35.2**	**179.3**	**435.2**	**982.9**	**3,052.0**	**464.5**
Delaware	4,835.0	734.0	4,101.1	3.2	70.2	197.9	462.7	695.6	3,001.9	403.6
District of Columbia	8,067.1	1,627.7	6,439.3	46.4	47.8	644.3	889.2	976.3	4,181.3	1,281.7
Florida	6,205.5	854.0	5,351.6	5.7	46.3	211.6	590.5	1,200.3	3,534.5	616.7
Georgia	5,148.5	534.0	4,614.6	7.5	29.8	166.4	330.3	917.2	3,182.3	515.2
Maryland	4,919.2	743.4	4,175.8	9.0	30.0	263.7	440.7	835.8	2,848.0	492.0
North Carolina	5,175.4	542.1	4,633.3	7.2	28.2	158.0	348.7	1,286.9	3,012.2	334.0
South Carolina	5,324.4	847.1	4,477.3	6.6	40.8	148.2	651.4	1,019.8	3,085.7	371.7
Virginia	3,373.9	314.7	3,059.2	5.7	25.0	101.1	182.8	471.6	2,326.4	261.2
West Virginia	2,720.6	350.6	2,370.0	4.4	18.6	36.6	291.0	570.2	1,591.6	208.2
East South Central	**4,115.2**	**489.1**	**3,626.1**	**7.0**	**36.9**	**121.5**	**323.6**	**864.2**	**2,396.8**	**365.1**
Alabama	4,412.3	490.2	3,922.2	7.9	34.6	121.2	326.5	884.4	2,737.2	300.5
Kentucky	2,878.1	300.6	2,577.5	5.4	26.3	80.0	189.0	610.9	1,748.7	217.9
Mississippi	4,269.8	349.3	3,920.5	7.7	41.7	111.6	188.2	1,051.2	2,380.6	488.7
Tennessee	4,693.9	694.9	3,998.9	7.1	44.0	156.8	487.0	936.6	2,601.8	460.5
West South Central	**5,013.1**	**568.0**	**4,445.1**	**6.8**	**36.8**	**137.9**	**386.6**	**970.5**	**3,037.6**	**437.0**
Arkansas	4,042.8	425.2	3,617.5	5.6	27.8	79.3	312.5	850.3	2,506.0	261.2
Louisiana	5,746.8	732.7	5,014.2	10.7	33.1	173.6	515.2	1,092.7	3,425.2	496.2
Oklahoma	4,683.9	508.2	4,175.7	6.9	40.9	82.9	377.5	1,026.6	2,787.8	361.3
Texas	5,031.8	560.3	4,471.5	6.1	38.0	146.7	369.5	949.7	3,062.6	459.2
West	**4,327.6**	**532.4**	**3,795.2**	**5.5**	**33.9**	**145.8**	**347.3**	**756.0**	**2,533.4**	**505.8**
Mountain	**4,860.0**	**448.7**	**4,411.3**	**5.9**	**38.3**	**109.8**	**294.7**	**839.9**	**3,069.1**	**502.3**
Arizona	5,896.5	551.2	5,345.4	8.0	28.9	152.5	361.6	1,034.4	3,510.5	800.5
Colorado	4,063.4	340.5	3,722.9	4.6	41.4	75.3	219.2	665.2	2,693.0	364.8
Idaho	3,149.3	244.9	2,904.4	2.0	33.3	17.8	191.8	610.3	2,142.5	151.6
Montana	4,069.9	206.5	3,863.4	2.6	28.3	25.8	149.7	428.5	3,220.2	214.7
Nevada	4,653.7	570.0	4,083.7	9.1	52.1	232.7	276.1	973.6	2,386.2	723.8
New Mexico	5,962.1	834.5	5,127.6	9.8	54.3	148.2	622.2	1,234.5	3,426.0	467.0
Utah	4,976.5	275.5	4,700.9	2.1	37.8	54.4	181.3	685.1	3,669.3	346.6
Wyoming	3,454.8	232.3	3,222.5	2.3	28.5	15.4	186.0	489.4	2,609.0	124.2
Pacific	**4,120.5**	**565.0**	**3,555.5**	**5.4**	**32.2**	**159.7**	**367.7**	**723.3**	**2,325.0**	**507.1**
Alaska	4,363.2	631.5	3,731.7	8.6	83.5	91.4	448.0	611.8	2,690.5	429.4
California	3,805.0	627.2	3,177.8	6.0	28.2	181.1	411.7	675.3	1,994.2	508.3
Hawaii	4,837.5	235.0	4,602.4	3.7	29.9	88.1	113.3	795.0	3,414.2	393.2
Oregon	5,002.0	374.9	4,627.1	2.7	36.8	86.2	249.3	806.7	3,409.3	411.1
Washington	5,255.5	377.3	4,878.3	3.0	47.1	100.9	226.3	949.5	3,341.5	587.3

Note: Offense totals are based on all reporting agencies and estimates for unreported areas. Totals may not add because of rounding. (1) Violent crimes are murder, forcible rape, robbery, and aggravated assault. (2) Property crimes are burglary, larceny-theft, and motor vehicle theft. Data not included for property crime of arson.

State and Federal Prison Population, Death Penalty, 1999-2000[1]

Source: Bureau of Justice Statistics, U.S. Dept. of Justice

The total number of prisoners under the jurisdiction of federal or state adult correctional authorities was 1,381,892 at year-end 2000. Overall, the nation's prison population grew 1.3%, which was less than the average annual growth of 6.0% since 1990. During 2000, the prison population rose at the lowest rate since 1972 and had the smallest absolute increase since 1980. At year-end 2000, state and federal prisons housed nearly two-thirds of the incarcerated population (1,312,354 out of 2,071,686). Jails, which are locally operated and typically hold persons awaiting trial and those with sentences of a year or less, held the remainder. Relative to the number of U.S. residents, the rate of incarceration in prisons was 478 sentenced inmates per 100,000 residents, up from 292 in 1990 (1 in every 109 men and 1 in every 1,695 women were sentenced prisoners under the jurisdicition of state or federal authorities).

	SENTENCED TO MORE THAN 1 YEAR			DEATH PENALTY, 1999		
	Advance[2] 2000	Final[3] 1999	% change 1999-2000	Under sentence of death	Executions	Death penalty
U.S. TOTAL	**1,321,146**	**1,304,074**	**1.3**	**3,527**	**84**	**—**
Federal institutions	**125,044**	**114,275**	**9.4**	**20**	**0**	**Yes**
State institutions	**1,196,102**	**1,189,799**	**0.5**	**3,507**	**84**	**38**
Northeast	**166,631**	**171,237**	**–2.7**	**255**	**1**	**—**
Connecticut	13,155	13,032	0.9	6	0	Yes
Maine	1,635	1,663	–1.7	—	—	No
Massachusetts	9,479	10,282	–7.8	—	—	No
New Hampshire	2,257	2,257	0.0	0	0	Yes
New Jersey	29,784	31,493	–5.4	14	0	Yes
New York	70,198	72,899	–3.7	5	0	Yes
Pennsylvania	36,844	36,525	0.9	230	1	Yes
Rhode Island	1,966	1,908	3.0	—	—	No
Vermont	1,313	1,178	11.5	—	—	No
Midwest	**236,185**	**232,314**	**1.7**	**496**	**12**	**—**
Illinois	45,281	44,660	1.4	156	1	Yes
Indiana	20,081	19,260	4.3	43	1	Yes
Iowa	7,955	7,232	10.0	—	—	No
Kansas	8,344	8,567	–2.6	3	0	Yes
Michigan	47,718	46,617	2.4	—	—	No
Minnesota	6,238	5,955	4.8	—	—	No
Missouri	27,299	26,133	4.5	83	9	Yes
Nebraska	3,816	3,632	5.1	9	0	Yes
North Dakota	994	866	14.8	—	—	No
Ohio	45,833	46,842	–2.2	199	1	Yes
South Dakota	2,613	2,498	4.6	3	0	Yes
Wisconsin	20,013	19,699	1.6	—	—	No
South	**529,025**	**526,764**	**0.4**	**1,916**	**64**	**—**
Alabama	24,123	24,109	—	180	2	Yes
Arkansas	11,851	11,336	4.5	40	4	Yes
Delaware	3,937	3,730	5.5	17	2	Yes
District of Columbia	5,008	6,730	–25.6	—	—	No
Florida	71,318	69,594	2.5	365	1	Yes
Georgia	44,141	42,008	5.1	116	0	Yes
Kentucky	14,919	15,317	–2.6	39	1	Yes
Louisiana	35,047	34,066	2.9	85	1	Yes
Maryland	22,490	22,184	1.4	17	0	Yes
Mississippi	19,239	17,410	10.5	60	0	Yes
North Carolina	27,043	26,672	1.4	202	4	Yes
Oklahoma	23,181	22,393	3.5	139	6	Yes
South Carolina	21,017	21,228	–1.0	65	4	Yes
Tennessee	22,166	22,502	–1.5	100	0	Yes
Texas	150,107	154,865	–3.1	460	35	Yes
Virginia	29,643	29,088	1.9	31	4	Yes
West Virginia	3,795	3,532	7.4	—	—	No
West	**264,252**	**259,484**	**1.8**	**840**	**8**	**—**
Alaska	2,128	2,325	–8.5	—	—	No
Arizona	25,412	23,944	6.1	116	4	Yes
California	160,412	160,517	-0.1	553	2	Yes
Colorado	16,833	15,670	7.4	4	0	Yes
Hawaii	3,553	3,817	–6.9	—	—	No
Idaho	5,526	4,842	14.1	21	0	Yes
Montana	3,105	2,951	5.2	6	0	Yes
Nevada	9,921	9,413	5.4	86	1	Yes
New Mexico	4,887	4,730	3.3	4	0	Yes
Oregon	10,603	9,840	7.8	25	0	Yes
Utah	5,526	5,164	7.0	10	1	Yes
Washington	14,666	14,558	0.7	13	0	Yes
Wyoming	1,680	1,713	–1.9	2	0	Yes

(1) All information applies to Dec. 31 of the year indicated. (2) The advance estimate of prisoners is made in Jan. and may be revised. (3) Revised from previous tabulations.

Sentences vs. Time Served for Selected Crimes

Source: Bureau of Justice Statistics, *Truth in Sentencing in State Prisons,* 1999

The following is a comparison of the average maximum sentence lengths (excluding both life and death sentences) and the actual time served for selected state-court convictions.

Type of offense	Average sentence	Avg. time served[1]
All violent	7 years, 1 month	3 years, 3 months
Homicide	15 years	7 years
Rape	9 years, 8 months	5 years, 1 month
Other sexual assault	6 years, 9 months	3 years, 3 months
Robbery	7 years, 8 months	3 years, 4 months
Negligent manslaughter	8 years, 1 month	3 years, 5 months
Assault	5 years, 1 month	2 years, 4 months
Other	5 years, 7 months	2 years, 5 months

(1) Includes jail credit and prison time

Prison Situation Among the States and in the Federal System, 2000

Source: *Prisoners in 2000,* Bureau of Justice Statistics, U.S. Dept. of Justice; Aug. 2001

10 largest prison populations, 2000	Number of inmates	10 highest incarceration rates, 2000	Prisoners per 100,000 residents[1]	10 largest % increases in prison population: 1998-2000	% increase	Growth since 1990	% increase[2]
California	163,001	Louisiana	801	Idaho	14.1	Idaho	10.9
Texas	157,997	Texas	730	North Dakota	14.1	Texas	10.2
Federal	145,416	Mississippi	688	Mississippi	10.9	Federal	9.5
Florida	71,319	Oklahoma	685	Vermont	10.5	West Virginia	9.3
New York	70,198	Georgia	550	Iowa	10.0	Mississippi	9.1
Michigan	47,718	Alabama	549	Rhode Island	9.4	North Dakota	8.6
Ohio	45,833	South Carolina	532	West Virginia	9.2	Washington	8.4
Illinois	45,281	Nevada	518	Oregon	7.8	Colorado	8.2
Georgia	44,232	Arizona	515	Federal	7.5	Montana	8.1
Pennsylvania	36,847	Delaware	513	Colorado	7.4	Tennessee	7.9

(1) Prisoners with sentences of more than 1 year. The Federal Bureau of Prisons and the District of Columbia are excluded. (2) The average annual percent change from 1990 to 2000.

Executions, by State and Method, 1977-2000

Source: Bureau of Justice Statistics, *Capital Punishment 1999,* Dec. 2000; Death Penalty Information Center, NAACP Legal Defense and Education Fund, *Death Row, U.S.A.*

	No.	Lethal injection	Electro-cution	Lethal gas	Firing squad	Hang-ing		No.	Lethal injection	Electro-cution	Lethal gas	Firing squad	Hang-ing
TOTAL U.S.	**683**	**519**	**149**	**11**	**2**	**3**	Missouri	46	46	0	0	0	0
Alabama	23	0	23	0	0	0	Montana	2	2	0	0	0	0
Arizona	22	20	0	2	0	0	Nebraska	3	0	3	0	0	0
Arkansas	23	22	1	0	0	0	Nevada	8	7	0	1	0	0
California	8	6	0	2	0	0	North Carolina	16	14	0	2	0	0
Colorado	1	1	0	0	0	0	Ohio	1	1	0	0	0	0
Delaware	11	10	0	0	0	1	Oklahoma	30	30	0	0	0	0
Florida	50	6	44	0	0	0	Oregon	2	2	0	0	0	0
Georgia	23	0	23	0	0	0	Pennsylvania	3	3	0	0	0	0
Idaho	1	1	0	0	0	0	South Carolina	25	20	5	0	0	0
Illinois	12	12	0	0	0	0	Tennessee	1	1	0	0	0	0
Indiana	7	4	3	0	0	0	Texas	239	239	0	0	0	0
Kentucky	2	1	1	0	0	0	Utah	6	4	0	0	2	0
Louisiana	26	6	20	0	0	0	Virginia	81	55	26	0	0	0
Maryland	3	3	0	0	0	0	Washington	3	1	0	0	0	2
Mississippi	4	0	0	4	0	0	Wyoming	1	1	0	0	0	0

Note: This table shows methods used since 1977. Lethal injection was used in 76%. 12 states—Arizona, Arkansas, California, Delaware, Indiana, Louisiana, Nevada, North Carolina, South Carolina, Utah, Virginia, and Washington—have employed 2 methods.

IT'S A FACT: The use of imprisonment as a punishment for major crimes developed in the U.S. during the 19th century; before then, people accused of most crimes (other than vagrancy and the like) were jailed prior to trial and sentencing, but punishments were usually corporal, ranging from flogging and branding to execution.

Total Estimated Arrests,[1] 1999

Source: FBI, *Uniform Crime Reports*, 1999

TOTAL[2]	**14,031,070**	Stolen property: buying, receiving, possessing	121,900
Murder and nonnegligent manslaughter	14,790	Vandalism	278,200
Forcible rape	28,830	Weapons: carrying, possessing, etc.	172,400
Robbery	108,850	Prostitution and commercialized vice	92,100
Aggravated assault	483,530	Sex offenses (except forcible rape and prostitution)	92,400
Burglary	296,100	Drug abuse violations	1,532,200
Larceny–theft	1,189,400	Gambling	10,400
Motor vehicle theft	142,200	Offenses against family and children	151,200
Arson	16,800	Driving under the influence	1,511,300
Violent crimes[3]	**635,990**	Liquor laws	657,900
Property crime[4]	**1,644,500**	Drunkenness	656,100
Crime Index total[5]	**2,280,500**	Disorderly conduct	633,100
Other assaults	1,294,400	Vagrancy	30,000
Forgery and counterfeiting	106,900	All other offenses	3,728,100
Fraud	363,800	Curfew and loitering law violations	167,200
Embezzlement	17,100	Runaways	148,300

(1) Arrest totals are based on all reporting agencies and estimates for unreported areas. Arrests on suspicion not included. (2) Because of rounding, figures may not add to totals. (3) Violent crimes are murder, forcible rape, robbery, and aggravated assault. (4) Property crimes are burglary, larceny-theft, motor vehicle theft, and arson. (5) Includes arson.

Federal Bureau of Investigation

The Federal Bureau of Investigation was created July 26, 1908, and was referred to as Office of Chief Examiner. It became the Bureau of Investigation (Mar. 16, 1909), United States Bureau of Investigation (July 1, 1932), Division of Investigation (Aug. 10, 1933), and Federal Bureau of Investigation (July 1, 1935).

Director	Assumed office	Director	Assumed office
Stanley W. Finch	July 26, 1908	Clarence M. Kelley	July 9, 1973
A(lexander) Bruce Bielaski	Apr. 30, 1912	William H. Webster	Feb. 23, 1978
William E. Allen, act.	Feb. 10, 1919	John E. Otto, act.	May 26, 1987
William J. Flynn	July 1, 1919	William S. Sessions	Nov. 2, 1987
William J. Burns	Aug. 22, 1921	Floyd I. Clarke, act.	July 19, 1993
J. Edgar Hoover, act.	May 10, 1924	Louis J. Freeh	Sept. 1, 1993
J. Edgar Hoover	Dec. 10, 1924	Thomas J. Pickard, act.	June 25, 2001
L. Patrick Gray, act.	May 3, 1972	Robert S. Mueller III	Sept. 4, 2001
William D. Ruckelshaus, act.	Apr. 27, 1973		

NATIONS OF THE WORLD

Initials used include: AL (Arab League), APEC (Asia-Pacific Economic Cooperation Group), ASEAN (Association of Southeast Asian Nations), Benelux (Belgium, Netherlands, and Luxembourg), CARICOM (Caribbean Community and Common Market), CIS (Commonwealth of Independent States), EU (European Union), FAO (UN Food & Agriculture Org.), ILO (Intl. Labor Org.), IMF (Intl. Monetary Fund), IMO (Intl. Maritime Org.), NATO (North Atlantic Treaty Org.), OAS (Org. of American States), OAU (Org. of African Unity), OECD (Org. for Economic Cooperation and Development), OECS (Org. of Eastern Caribbean States), OSCE (Org. for Security and Cooperation in Europe), UN (United Nations), WHO (World Health Org.), WTrO (World Trade Org., formerly GATT). FY = fiscal year.

Sources: American Automobile Manufacturers Assn.; (U.S.) Census Bureau: Intl. Data Base; (U.S.) Central Intelligence Agency: *The World Factbook;* (U.S.) Dept. of Commerce; (U.S.) Dept. of Energy; Intl. Institute for Strategic Studies: *The Military Balance;* Intl. Monetary Fund; (U.S.) Dept. of State; UN Demographic Yearbook; UN Food and Agriculture Organization; UN Population Division: *World Urbanization Prospects;* UN Statistical Yearbook; Ward's Communications; World Tourism Organization; Encyclopaedia Britannica Book of the Year; The Europa World Year Book; The Statesman's Yearbook. Telephone data (1999 unless otherwise indicated) supplied by the Intl. Telecommunication Union, from the World Telecommunication Indicators database, copyright ITU.

Note: Because of rounding or incomplete enumeration, some percentages may not add to 100%. **National population and health** figures are mid-2001 estimates, unless otherwise noted. Percentage of urban population is for 1999. **City** populations are the latest available and apply to the city proper, except where an urban agglomeration ("urban agg."), i.e. whole metropolitan area, is indicated. Urban agglomeration figures are estimates for 2000. **Defense** figures are for 2000 unless otherwise noted. **Electricity production** figures are 1999. **GDP** estimates are based on purchasing power parity calculations, which involve use of intl. dollar price weights applied to quantities of goods and services produced. **Tourism** figures and represent receipts from international tourism and are 1999 unless otherwise noted. **Budget** figures are for expenditures, unless otherwise noted. **Motor vehicle** statistics are for 1996 unless otherwise noted; comm. (commercial) vehicles include trucks and buses. Per-person figures in **communications** data are 1995 or later. **Literacy rates** given generally measure the percent of population able to read and write on a lower elementary school level, not the (smaller) percent able to read instructions necessary for a job or license. **Embassy addresses** are Wash., DC, area code (202), unless otherwise noted.

For further details and later information on the response to the Sept. 11, 2001, terrorist attack, see the front-of-the-book feature. For this and other stories in detail, see also Chronology of the Year's Events.

See pages 497-512 for full-color maps and flags of all nations.

Afghanistan

Islamic State of Afghanistan

People: Population: 26,813,057. **Age distrib.** (%): <15: 42.2; 65+: 2.8. **Pop. density:** 107 per sq. mi. **Urban:** 21%. **Ethnic groups:** Pashtun 38%, Tajik 25%, Hazara 19%, Uzbek 6%. **Principal languages:** Pashtu 35%, Afghan Persian (Dari) 50% (both official), Turkic (incl. Uzbek, Turkmen) 11%. **Chief religions:** Sunni Muslim 84%, Shi'a Muslim 15%.

Geography: Area: 249,700 sq. mi. **Location:** In SW Asia, NW of the Indian subcontinent. **Neighbors:** Pakistan on E, S; Iran on W; Turkmenistan, Tajikistan, Uzbekistan on N. The NE tip touches China. **Topography:** The country is landlocked and mountainous, much of it over 4,000 ft. above sea level. The Hindu Kush Mts. tower 16,000 ft. above Kabul and reach a height of 25,000 ft. to the E. Trade with Pakistan flows through the 35-mile-long Khyber Pass. The climate is dry, with extreme temperatures, and there are large desert regions, though mountain rivers produce intermittent fertile valleys. **Capital:** Kabul, (urban agg.) 2,590,000.

Government: Type: In transition. **Local divisions:** 32 provinces. **Defense budget** (1998): $250 mil. **Active troops:** NA.

Economy: Industries: Textiles, soap, furniture, cement. **Chief crops:** Nuts, wheat, fruits. **Minerals:** Gas, oil, copper, coal, zinc, iron. **Other resources:** Wool, karakul pelts, mutton. **Arable land:** 12%. **Livestock** (2000): cattle: 3.48 mil; chickens: 7.2 mil; goats: 7.37 mil; sheep: 18.0 mil. **Electricity prod.:** 420 mil kWh. **Labor force:** 68% agric.; 16% industry.

Finance: Monetary unit: Afghani (Sept. 2001: 4,750.00 = $1 U.S.). **GDP** (1999 est.): $21 bil. **Per capita GDP:** $800. **Imports** (1996 est.): $150 mil; partners: Japan 14%, EU 11%. **Exports** (1996 est.): $80 mil; partners: EU 10%. **Tourism:** $1 mil.

Transport: Railroad: Length: 16 mi. **Motor vehicles:** 35,000 pass. cars, 32,000 comm. vehicles. **Civil aviation:** 98.3 mil pass.-mi.; 3 airports.

Communications: TV sets: 10 per 1,000 pop. **Radios:** 73.7 per 1,000 pop. **Telephones:** 29,000 main lines. **Daily newspaper circ.:** 11 per 1,000 pop.

Health: Life expectancy: 46.97 male; 45.47 female. **Births** (per 1,000 pop): 41.42. **Deaths** (per 1,000 pop.): 17.72. **Natural inc.:** 2.37%. **Infant mortality** (per 1,000 live births): 147.02.

Education: Compulsory: ages 7-13. **Literacy:** 31.5%.

Major Intl. Organizations: UN (FAO, IBRD, ILO, IMF, WHO).

Embassy: 2341 Wyoming Ave. NW 20008; 234-3770.

Website: http://www.afghan-web.com

Afghanistan, occupying a favored invasion route since antiquity, has been variously known as Ariana or Bactria (in ancient times) and Khorasan (in the Middle Ages). Foreign empires alternated rule with local emirs and kings until the 18th century, when a unified kingdom was established. In 1973, a military coup ushered in a republic.

Pro-Soviet leftists took power in a bloody 1978 coup and concluded an economic and military treaty with the USSR. In Dec. 1979 the USSR began a massive airlift into Kabul and backed a new coup, leading to installation of a more pro-Soviet leader. Soviet troops fanned out over Afghanistan and waged a protracted guerrilla war with Muslim rebels, in which some 15,000 Soviet troops reportedly died.

A UN-mediated agreement was signed Apr. 14, 1988, providing for withdrawal of Soviet troops, a neutral Afghan state, and repatriation of refugees. Afghan rebels rejected the pact, vowing to continue fighting while "Soviets and their puppets" remained in Afghanistan. The Soviets completed their troop withdrawal Feb. 15, 1989; fighting between Afghan rebels and government forces ensued.

Communist Pres. Najibullah resigned Apr. 16, 1992, as competing guerrilla forces advanced on Kabul. The rebels achieved power Apr. 28, ending 14 years of Soviet-backed regimes. More than 2 million Afghans had been killed and 6 million had left the country since 1979.

Following the rebel victory there were clashes between moderates and Islamic fundamentalist forces. Burhanuddin Rabbani, a guerrilla leader, became president June 28, 1992, but fierce fighting continued around Kabul and elsewhere. The Taliban, an insurgent Islamic fundamentalist faction, gained increasing control and in Sept. 1996 captured Kabul and set up a government. The Taliban executed former President Najibullah and empowered Islamic religious police to enforce codes of dress and behavior that were especially restrictive to women. Rabbani and other ousted leaders fled to the north.

Victories in the northern cities of Mazar-e Sharif, Aug. 8, 1998, and Taloqan, Aug. 8-11, 1998, gave the Taliban control over more than 90% of the country; the killing of several Iranian diplomats during the Mazar-e Sharif takeover further heightened tensions with Iran. On Aug. 20, 1998, U.S. cruise missiles struck SE of Kabul, hitting facilities the U.S. alleged were terrorist training camps run by a wealthy businessman, Osama bin Laden. The UN imposed sanctions Nov. 14, 1999, when Afghanistan refused to turn over bin Laden to the U.S. for prosecution; a UN ban on all military aid to the Taliban took effect Jan. 19, 2001. By March, aid agencies reported that drought, cold, and continued warfare had put more than 1 million people at risk of famine. Meanwhile, the Taliban launched a campaign to destroy non-Islamic antiquities.

Ahmed Shah Massoud, leader of the anti-Taliban resistance, died Sept. 8, 2001, of wounds sustained in a suicide bombing by assassins posing as journalists. After the Sept. 11 attacks on the World Trade Center and Pentagon, the U.S., blaming bin Laden, demanded that the Taliban surrender him and shut down his terrorist network. When the Taliban refused, the U.S. intensified political, economic, and military pressure on the Afghan rulers.

(For other developments, see also the front-of-the-book feature and the Chronology of the Year's Events.)

Albania

Republic of Albania

People: Population: 3,510,484. **Age distrib.** (%): <15: 29.5; 65+: 7.0. **Pop. density:** 332 per sq. mi. **Urban:** 41%. **Ethnic groups:** Albanians (Gegs in N, Tosks in S) 95%, Greeks 3%. **Principal languages:** Albanian (official; Tosk is the official dialect), Greek. **Chief religions:** Muslim 70%, Albanian Orthodox 20%, Roman Catholic 10%.

Geography: Area: 10,600 sq. mi. **Location:** SE Europe, on SE coast of Adriatic Sea. **Neighbors:** Greece on S, Yugoslavia on N, Macedonia on E. **Topography:** Apart from a narrow coastal plain, Albania consists of hills and mountains covered with scrub forest, cut by small E-W rivers. **Capital:** Tirana, (1999 est.) 279,000.

Government: Type: Republic. **Head of state:** Pres. Rexhep Meidani; b Aug. 17, 1944; in office: July 24, 1997. **Head of gov.:** Prime Min. Ilir Meta; b Mar. 24, 1969; in office: Oct. 29, 1999. **Local divisions:** 36 districts, 1 municipality. **Defense budget (1999):** $43 mil . **Active troops:** NA.

Economy: Industries: Cement, textiles, food processing. **Chief crops:** Corn, wheat, potatoes, watermelon, vegetables. **Minerals:** Chromium, coal, oil, gas. **Crude oil reserves** (2000): 165 mil bbls. **Other resources:** Timber. **Arable land:** 21%. **Livestock** (2000): cattle: 720,000; chickens: 4.0 mil; goats: 1.12 mil; pigs: 81,000; sheep: 1.94 mil. **Electricity prod.:** 5.332 bil kWh. **Labor force:** 49.5% agric, 50.5% industry and services.

Finance: Monetary unit: Lek (Sept. 2001: 145.65 = $1 U.S.). **GDP** (1999 est.): $5.6 bil. **Per capita GDP:** $1,650. **Imports** (1999 est.): $925 mil; partners: Italy 43%, Greece 29%. **Exports** (1999 est.): $242 mil; partners: Italy 63%, Greece 12%. **Tourism** (1998): $54 mil. **Budget** (1997): $996 mil. **Intl. reserves less gold** (June 2000): $373.46 mil. **Gold:** 120,000 oz t. **Consumer prices** (change in 1999): 0.4%.

Transport: Railroad: Length: 419 mi. **Chief ports:** Durres, Sarande, Vlore. **Civil aviation:** 21.9 mil pass.-mi.; 1 airport.

Communications: TV sets: 89 per 1,000 pop. **Radios:** 157 per 1,000 pop. **Telephones:** (1999): 140,400. **Daily newspaper circ.:** 54 per 1,000 pop.

Health: Life expectancy: 69.01 male; 74.87 female. **Births** (per 1,000 pop): 19.01. **Deaths** (per 1,000 pop.): 6.5. **Natural inc.:** 1.25%. **Infant mortality** (per 1,000 live births): 39.99.

Major Intl. Organizations: UN (IBRD, ILO, IMF, IMO, WHO), OSCE.

Education: Free, compulsory: ages 6-14. **Literacy** (1993): 100%.

Embassy: 2100 S St. NW 20008; 223-4942.

Websites: http://www.undp.org.al
http://www.albanian.com

Ancient Illyria was conquered by Romans, Slavs, and Turks (15th century); the latter Islamized the population. Independent Albania was proclaimed in 1912, republic was formed in 1920. King Zog I ruled 1925-39, until Italy invaded.

Communist partisans took over in 1944, allied Albania with USSR, then broke with USSR in 1960 over de-Stalinization. Strong political alliance with China followed, leading to several billion dollars in aid, which was curtailed after 1974. China cut off aid in 1978 when Albania attacked its policies after the death of Chinese ruler Mao Zedong. Large-scale purges of officials occurred during the 1970s.

Enver Hoxha, the nation's ruler for 4 decades, died Apr. 11, 1985. Eventually the new regime introduced some liberalization, including measures in 1990 providing for freedom to travel abroad. Efforts were begun to improve ties with the outside world. Mar. 1991 elections left the former Communists in power, but a general strike and urban opposition led to the formation of a coalition cabinet including non-Communists.

Albania's former Communists were routed in elections Mar. 1992, amid economic collapse and social unrest. Sali Berisha was elected as the first non-Communist president since World War II. Berisha's party claimed a landslide victory in disputed parliamentary elections, May 26 and June 2, 1996. Public protests over the collapse of fraudulent investment schemes in Jan. 1997 led to armed rebellion and anarchy. The UN Security Council, Mar. 28, authorized a 7,000-member force to restore order. Socialists and their allies won parliamentary elections, June 29 and July 6, and international peacekeepers completed their pullout by Aug. 11, 1997. During NATO's air war against Yugoslavia, Mar.-June 1999, Albania hosted some 465,000 Kosovar refugees.

Algeria

Democratic and Popular Republic of Algeria

People: Population: 31,736,053. **Age distrib.** (%): <15: 34.2 65+: 4.1. **Pop. density:** 35 per sq. mi. **Urban:** 60%. **Ethnic groups:** Arab-Berber 99%. **Principal languages:** Arabic (official), French, Berber dialects. **Chief religion:** Sunni Muslim (state religion) 99%.

Geography: Area: 918,500 sq. mi. **Location:** In NW Africa, from Mediterranean Sea into Sahara Desert. **Neighbors:** Morocco on W; Mauritania, Mali, Niger on S; Libya, Tunisia on E. **Topography:** The Tell, located on the coast, comprises fertile plains 50-100 miles wide, with a moderate climate and adequate rain. Two major chains of the Atlas Mts., running roughly E-W and reaching 7,000 ft., enclose a dry plateau region. Below lies the Sahara, mostly desert with major mineral resources. **Capital:** Algiers (El Djazair), (urban agg.) 1,885,000.

Government: Type: Republic. **Head of state:** Pres. Abdelaziz Bouteflika; b Mar. 2, 1937; in office: Apr. 27, 1999. **Head of gov.:** Prime Min. Ali Benflis; b Sept. 8, 1944; in office: Aug. 26, 2000. **Local divisions:** 48 provinces. **Defense budget:** $1.8 bil. **Active troops:** 124,000.

Economy: Industries: Oil, natural gas, light industries, food processing. **Chief crops:** Grains, grapes, citrus, olives. **Minerals:** Iron, oil, gas, phosphates, zinc, lead. **Crude oil reserves** (2000): 9.2 bil bbls. **Arable land:** 3%. **Livestock** (2000): cattle: 1.65 mil; chickens: 110.0 mil; goats: 3.4 mil; pigs: 5,700; sheep: 18.2 mil. **Fish catch** (1999): 99,332 metric tons. **Electricity prod.:** 23.215 bil kWh. **Labor force:** 30% govt.; 27% industry, serv., commerce; 22% agric.

Finance: Monetary unit: Dinar (Sept. 2001: 78.24 = $1 U.S.). **GDP** (1999 est.): $147.6 bil. **Per capita GDP:** $4,700. **Imports** (1999 est.): $9.3 bil; partners: France 29.5%, Italy 9.8%. **Exports** (1999 est.): $13.7 bil; partners: Italy 21.2%, U.S. 15%. **Tourism** (1998): $24 mil. **Budget** (1998 est.): $14.4 bil. **Intl. reserves less gold** (June 2000): $7.7 bil. **Gold:** 5.58 mil oz t. **Consumer prices** (change in 1997): 2.5%.

Transport: Railroad: Length: 2,965 mi. **Motor vehicles:** 500,000 pass. cars, 420,000 comm. vehicles. **Civil aviation:** 1.95 bil pass.-mi.; 28 airports. **Chief ports:** Algiers, Annaba, Oran.

Communications: TV sets: 71 per 1,000 pop. **Radios:** 122 per 1,000 pop. **Telephones:** 1,761,300. **Daily newspaper circ.:** 52 per 1,000 pop.

Health: Life expectancy: 68.6 male; 71.34 female. **Births** (per 1,000 pop): 22.76. **Deaths** (per 1,000 pop.): 5.22. **Natural inc.:** 1.75%. **Infant mortality** (per 1,000 live births): 40.56.

Education: Compulsory: ages 6-15. **Literacy:** 62%.

Major Intl. Organizations: UN (FAO, IBRD, ILO, IMF, IMO, WHO), AL, OAU, OPEC.

Embassy: 2118 Kalorama Rd. NW 20008; 265-2800.

Earliest known inhabitants were ancestors of Berbers, followed by Phoenicians, Romans, Vandals, and, finally, Arabs. Turkey ruled 1518 to 1830, when France took control.

Large-scale European immigration and French cultural inroads did not prevent an Arab nationalist movement from launching guerrilla war. Peace, and French withdrawal, was negotiated with French Pres. Charles de Gaulle. One million Europeans left. Independence came July 5, 1962. Ahmed Ben Bella was the victor of infighting and ruled until 1965, when an army coup installed Col. Houari Boumedienne as leader; Boumedienne led until his death from a blood disease, 1978.

In 1967, Algeria declared war on Israel, broke ties with U.S., and moved toward eventual military and political ties with the USSR. Some 500 died in riots protesting economic hardship in 1988. In 1989, voters approved a new constitution, which cleared the way for a multiparty system.

The government canceled the Jan. 1992 elections that Islamic fundamentalists were expected to win, and banned all nonreligious activities at Algeria's 10,000 mosques. Pres. Mohammed Boudiaf was assassinated June 29, 1992. There were repeated attacks on high-ranking officials, security forces, foreigners, and others by militant Muslim fundamentalists over the next 7 years; pro-government death squads also were active.

Liamine Zeroual won the presidential election of Nov. 16, 1995. A new constitution banning Islamic political parties and increasing the president's powers passed in a referendum on Nov. 28, 1996. Pro-government parties won the parliamentary election of June 6, 1997. Abdelaziz Bouteflika, who became president after a flawed election on Apr. 15, 1999, made peace with rebels and won approval for an amnesty plan in a referendum on Sept. 16; by then, some 100,000 people had died in the civil war. Some 100 people died and thousands were injured in violent protests Apr.-June 2001, chiefly by Algeria's Berber minority.

Andorra

Principality of Andorra

People: Population: 67,627. **Age distrib.** (%): <15: 15.3; 65+: 12.6. **Pop. density:** 389 per sq. mi. **Urban:** 93%. **Ethnic groups:** Spanish 61%, Andorran 30%, French 6%. **Principal languages:** Catalan (official), French, Castilian. **Chief religion:** Predominently Roman Catholic.

Geography: Area: 174 sq. mi. **Location:** SW Europe, in Pyrenees Mts. **Neighbors:** Spain on S, France on N. **Topography:** High mountains and narrow valleys cover the country. **Capital:** Andorra la Vella, (1999 est.) 25,000.

Government: Type: Parliamentary co-principality. **Heads of state:** President of France & Bishop of Urgel (Spain), as co-princes. **Head of gov.:** Marc Forné Molné; b Dec. 30, 1946; in office: Dec. 21, 1994. **Local divisions:** 7 parishes. **Defense budget:** Responsibility of France and Spain.

Economy: Industries: Tourism, sheep, timber, tobacco. **Minerals:** Iron, lead. **Arable land:** 2%. **Labor force:** 72% serv., 21% ind

Finance: Monetary unit: French Franc (Sept. 2001: 7.25 = $1 U.S.). Spanish Peseta (Sept. 2001: 183.93 = $1 U.S.). **GDP** (1996 est.): $1.2 bil. **Per capita GDP:** $18,000. **Imports** (1998): $1.077 bil; partners: Spain 48%, France 35%. **Exports** (1998): $58 mil; partners: France 34%, Spain 58%.

Transport: Motor vehicles: 35,358 pass. cars, 4,238 comm. vehicles.

Communications: TV sets: 315 per 1,000 pop. **Radios:** 156 per 1,000 pop. **Telephones:** 34,200 main lines. **Daily newspaper circ.:** 62 per 1,000 pop.

Health: Life expectancy: 80.57 male; 86.57 female. **Births** (per 1,000 pop): 10.29. **Deaths** (per 1,000 pop.): 5.41. **Natural inc.:** .49%. **Infant mortality** (per 1,000 live births): 4.08.

Education: Free, compulsory: ages 6-16. **Literacy** (1997): 100%.

Major Intl. Organizations: UN.

Embassy: 2 UN Plaza, 25th floor, New York, NY 10017; (212) 750-8064.

Website: http://www.andorra.ad/cniauk.html

Andorra was a co-principality, with joint sovereignty by France and the bishop of Urgel, from 1278 to 1993.

Tourism, especially skiing, is the economic mainstay. A free port, allowing for an active trading center, draws some 13 million tourists annually. Andorran voters chose to end a feudal system that had been in place for 715 years and adopt a parliamentary system of government Mar. 14, 1993.

Angola

Republic of Angola

People: Population: 10,366,031. **Age distrib.** (%): <15: 43.3; 65+: 2.7. **Pop. density:** 22 per sq. mi. **Urban:** 34%. **Ethnic groups:** Ovimbundu 37%, Kimbundu 25%, Bakongo 13%. **Principal languages:** Portuguese (official), various Bantu and other African languages. **Chief religions:** Indigenous beliefs 47%, Roman Catholic 38%, Protestant 15%.

Geography: Area: 480,800 sq. mi. **Location:** In SW Africa on Atlantic coast. **Neighbors:** Namibia on S, Zambia on E, Congo-Kinshasa (formerly Zaire) on N; Cabinda, an enclave separated from rest of country by short Atlantic coast of Congo-Kinshasa, borders Congo-Brazzaville. **Topography:** Most of Angola consists of a plateau elevated 3,000 to 5,000 feet above sea level, rising from a narrow coastal strip. There is also a temperate highland area in the west-central region, a desert in the S, and a tropical rain forest covering Cabinda. **Capital:** Luanda, (urban agg.) 2,677,000.

Government: Type: Republic. **Head of state and gov.:** Pres. José Eduardo dos Santos; b Aug. 28, 1942; in office: Sept. 20, 1979. **Local divisions:** 18 provinces. **Defense budget:** $542 mil. **Active troops:** 107,500.

Economy: Industries: Food processing, textiles, mining, brewing, oil. **Chief crops:** Coffee, sugarcane, bananas. **Minerals:** Iron, diamonds (over 1 mil carats a year), gold, phosphates, oil. **Livestock** (2000): cattle: 4.04 mil; chickens: 6.4 mil; goats: 2.15 mil; pigs: 800,000; sheep: 350,000. **Crude oil reserves** (2000): 5.4 bil bbls. **Arable land:** 2%. **Fish catch** (1999): 72,189 metric tons. **Electricity prod.:** 1.475 bil. kWh. **Labor force:** 85% agric., 15% industry & services.

Finance: Monetary unit: Readjusted Kwanza (Sept. 2001: 22.12 = $1 U.S.). **GDP** (1999 est.): $11.6 bil. **Per capita GDP:** $1,030. **Imports** (1999 est.): $3 bil; partners: Portugal 20%, U.S. 17%, South Africa 10%. **Exports** (1999 est.): $5 bil; partners: U.S. 63%, Benelux 9%.**Tourism:** $13 mil. **Intl. reserves less gold** (Apr. 2000): $342.52 mil. **Consumer prices** (change in 1999): 286.1%.

Transport: Railroad: Length: 1,739 mi. **Motor vehicles:** 197,000 pass. cars, 26,000 comm. vehicles. **Civil aviation:** 385.3 mil pass.-mi.; 17 airports. **Chief ports:** Cabinda, Lobito, Luanda.

Communications: TV sets: 48 per 1,000 pop. **Radios:** 39 per 1,000 pop. **Telephones:** 69,700 main lines. **Daily newspaper circ.:** 11 per 1,000 pop.

Health: Life expectancy: 37.36 male; 39.87 female. **Births** (per 1,000 pop): 46.54. **Deaths** (per 1,000 pop.): 24.68. **Natural inc.:** 2.19%. **Infant mortality** (per 1,000 live births): 193.72.

Education: Free, compulsory: ages 7-15. **Literacy** (1992): 40%.

Major Intl. Organizations: UN (FAO, IBRD, ILO, IMF, IMO, WHO, WTrO), OAU.

Embassy: 1615 M St. NW, Suite 900, 20036; 785-1156.

Website: http://www.angola.org

From the early centuries AD to 1500, Bantu tribes penetrated most of the region. Portuguese came in 1583, allied with the Bakongo kingdom in the north, and developed the slave trade. Large-scale colonization did not begin until the 20th century, when 400,000 Portuguese immigrated.

A guerrilla war begun in 1961 lasted until 1975, when Portugal granted independence. Fighting then erupted between three rival rebel groups—the National Front, based in Zaire (now Congo), the Soviet-backed Popular Movement for the Liberation of Angola (MPLA), and the National Union for the Total Independence of Angola (UNITA), aided by the U.S. and South Africa. Cuban troops and Soviet aid helped the MPLA win control of most of the country by 1976, although fighting continued through the 1980s. A peace accord between the MPLA government and UNITA was signed May 1, 1991.

Elections were held in Sept. 1992, but fighting again broke out, as UNITA rejected the results. UNITA signed a new peace treaty with the government, Nov. 20, 1994, but the rebels were slow to demobilize. The UN Security Council voted, Aug. 28, 1997, to impose sanctions on UNITA. In Aug. 1998, Angola sent thousands of troops into Congo-Kinshasa (formerly Zaire) to support Laurent Kabila's regime. The UN ended its mission in Angola in Mar. 1999, as the civil war continued.

As of 2001, the UN estimated that the war with UNITA had claimed some 1 million lives and left another 2.5 million people homeless. More than 250 died when UNITA rebels ambushed a train Aug. 10.

Antigua and Barbuda

People: Population: 66,970. **Age distrib.** (%): <15: 28; 65+: 4.9. **Pop. density:** 395 per sq. mi. **Urban:** 37%. **Ethnic groups:** Primarily black. **Principal language:** English (official). **Chief religion:** Predominantly Anglican.

Geography: Area: 170 sq. mi. **Location:** Eastern Caribbean. **Neighbors:** St. Kitts & Nevis to W, Guadeloupe (Fr.) to S. **Capital:** Saint John's, (1999) 25,000.

Government: Type: Constitutional monarchy with British-style parliament. **Head of state:** Queen Elizabeth II; represented by Gov.-Gen. James Carlisle; b Aug. 5, 1937; in office: June 10, 1993. **Head of gov.:** Prime Min. Lester Bird; b Feb. 21, 1938; in office: Mar. 9, 1994. **Local divisions:** 6 parishes, 2 dependencies. **Defense budget:** $4 mil. **Active troops:** 150.

Economy: Industries: Tourism, light manufacturing, construction. **Arable land:** 18%. **Livestock** (2000): cattle: 15,700; chickens: 90,000; goats: 11,800; pigs: 2,200; sheep: 12,200. **Electricity prod.:** 95 mil kWh. **Labor force:** 82% comm. and ser., 11% agric.

Finance: Monetary unit: East Caribbean Dollar (Sept. 2001: 2.67 = $1 U.S.). **GDP** (1999 est.): $524 mil. **Per capita GDP:** $8,200. **Imports** (1998): $330 mil; partners: U.S. 27%, U.K. 16%. **Exports** (1998): $38 mil; partners: OECS 26%, Barbados 15%. **Tourism** (1998): $256 mil. **Budget** (1997 est.): $141.2 mil. **Intl. reserves less gold** (Jan. 2000): $64.73 mil.

Transport: Motor vehicles: 13,250 pass. cars, 1,423 comm. vehicles. **Civil aviation:** 155.4 mil pass.-mi.; 2 airports.

Communications: TV sets: 435 per 1,000 pop. **Radios:** 776 per 1,000 pop. **Telephones:** 38,300 main lines.

Health: Life expectancy: 68.45 male; 73.14 female. **Births** (per 1,000 pop): 19.50. **Deaths** (per 1,000 pop.): 5.87. **Natural inc.:** 1.36%. **Infant mortality** (per 1,000 live births): 22.33.

Education: Compulsory: ages 5-16. **Literacy** (1992): 90%.

Major Intl. Organizations: UN (FAO, IBRD, ILO, IMF, IMO, WHO, WTrO), Caricom, the Commonwealth, OAS, OECS.

Embassy: 3216 New Mexico Ave. NW 20016; 362-5211.

Website: http://www.antigua-barbuda.com

Columbus landed on Antigua in 1493. The British colonized it in 1632.

The British associated state of Antigua achieved independence as Antigua and Barbuda on Nov. 1, 1981. The government maintains close relations with the U.S., United Kingdom, and Venezuela. The country was hit hard by Hurricane Luis, Sept. 1995. About 3,000 refugees fleeing a volcanic eruption on Montserrat have settled in Antigua since 1995.

Argentina

Argentine Republic

People: Population: 37,384,816. **Age distrib.** (%): <15: 26.5; 65+: 10.4. **Pop. density:** 35 per sq. mi. **Urban:** 90%. **Ethnic groups:** White 85% (mostly Spanish, Italian); mestizo, Amerindian, other nonwhites 15%. **Principal languages:** Spanish (official), English, Italian. **Chief religion:** Nominally Roman Catholic 90%.

Geography: Area: 1,055,400 sq. mi., second largest country in South America. **Location:** Occupies most of southern South America. **Neighbors:** Chile on W; Bolivia, Paraguay on N; Brazil, Uruguay on NE. **Topography:** Mountains in the W are: the Andean, Central, Misiones, and Southern ranges. Aconcagua is the highest peak in the western hemisphere, alt. 22,834 ft. E of the Andes are heavily wooded plains, called the Gran Chaco in the N, and the fertile, treeless Pampas in the central region. Patagonia, in the S, is bleak and arid. Rio de la Plata, an estuary in the NE, 170 by 140 mi., is mostly fresh water, from 2,485-mi Parana and 1,000-mi Uruguay rivers. **Capital:** Buenos Aires (the Senate has approved moving the capital to the Patagonia Region). **Cities** (urban agg.): Buenos Aires, 12,560,000, (2000 city proper: 2 mil.); Cordoba, 1,434,000; Rosario, 1,278,000.

Government: Type: Republic. **Head of state and gov.:** Pres. Fernando de la Rúa; b Sept. 15, 1937; in office: Dec. 10, 1999. **Local divisions:** 23 provinces, 1 federal district. **Defense budget:** $3.8 bil. **Active troops:** 71,100.

Economy: Industries: Food processing, autos, chemicals, textiles, printing. **Chief crops:** Sunflower seeds, lemons, grapes, peanuts, corn, soybeans. **Minerals:** Oil, lead, zinc, iron, copper, tin, uranium. **Crude oil reserves** (2000): 2.75 bil bbls. **Arable land:** 9%. **Livestock** (2000): cattle: 55.0 mil; chickens: 65.0 mil; goats: 4.5 mil; pigs: 4.2 mil; sheep: 14.5 mil. **Fish catch** (1999): 1.35 mil metric tons. **Electricity prod.:** 77.087 bil kWh.

Finance: Monetary unit: Peso (Sept. 2001: 1.00 = $1 U.S.). **GDP** (1999 est.): $367 bil. **Per capita GDP:** $10,000. **Imports** (1999 est.): $25 bil; partners: EU 28%, Brazil 21%, U.S. 22%. **Exports** (1999 est.): $23 bil; partners: Brazil 24%, EU 21%. **Tourism:** $2.81 bil. **Budget** (1998 est.): $60 bil. **Intl. reserves less gold** (June 2000): $25.68 bil. **Gold:** 333,000 oz t. **Consumer prices** (change in 1999): –1.2%.

Transport: Railroad: Length: 21,015 mi. **Motor vehicles:** 4.78 mil pass. cars, 1.29 mil comm. vehicles. **Civil aviation:** 8.9 bil pass.-mi.; 39 airports. **Chief ports:** Buenos Aires, Bahia Blanca, La Plata.

Communications: TV sets: 289 per 1,000 pop. **Radios:** 595 per 1,000 pop. **Telephones:** 7,894,200 main lines. **Daily newspaper circ.:** 123 per 1,000 pop.

Health: Life expectancy: 71.88 male; 78.82 female. **Births** (per 1,000 pop): 18.41. **Deaths** (per 1,000 pop.): 7.58. **Natural inc.:** 1.08%. **Infant mortality** (per 1,000 live births): 17.75.

Education: Free, compulsory: ages 6-14. **Literacy:** 96%.

Major Intl. Organizations: UN (FAO, IBRD, ILO, IMF, IMO, WHO, WTrO), OAS.

Embassy: 1600 New Hampshire Ave. NW 20009; 238-6400.

Nomadic Indians roamed the Pampas when Spaniards arrived, 1515-16, led by Juan Diaz de Solis. Nearly all the Indians were killed by the late 19th century. The colonists won independence, 1816, and a long period of disorder ended in a strong centralized government.

Large-scale Italian, German, and Spanish immigration in the decades after 1880 spurred modernization. Social reforms were enacted in the 1920s, but military coups prevailed 1930-46, until the election of Gen. Juan Perón as president.

Perón, with his wife, Eva Duarte (d 1952), effected labor reforms, but also suppressed speech and press freedoms, closed religious schools, and ran the country into debt. A 1955 coup exiled Perón, who was followed by a series of military and civilian regimes. Perón returned to in 1973, and was once more elected president. He died 10 months later, succeeded by his wife Isabel, who had been elected vice president, and who became the first woman head of state in the western hemisphere.

A military junta ousted Mrs. Perón in 1976 amid charges of corruption. Under a continuing state of siege, the army battled guerrillas and leftists, killed 5,000 people, and jailed and tortured others. On Dec. 9, 1985, after a trial of 5 months and nearly 1,000 witnesses, 5 former junta members were found guilty of murder and human rights abuses.

Argentine troops seized control of the British-held Falkland Islands on Apr. 2, 1982. Both countries had claimed sovereignty over the islands, located 250 miles off the Argentine coast, since 1833. The British dispatched a task force and declared a total air and sea blockade around the Falklands. Fighting began May 1; several hundred lost their lives as the result of the destruction of a British destroyer and the sinking of an Argentine cruiser.

British troops landed on East Falkland Island May 21 and eventually surrounded Stanley, the capital city and Argentine stronghold. The Argentine troops surrendered, June 14; Argentine Pres. Leopoldo Galtieri resigned June 17.

Democratic rule returned to in 1983 as Raul Alfonsín's Radical Civic Union party gained an absolute majority in the presidential electoral college and Congress. By 1989 the nation was plagued by severe financial and political problems, as hyperinflation sparked looting and rioting in several cities. The government of Perónist Pres. Carlos Saúl Menem, installed 1989, introduced harsh economic measures to curtail inflation, control government spending, and restructure the foreign debt.

About 100 people were killed in the terrorist bombing of a Jewish cultural center in Buenos Aires, July 18, 1994. Following passage of a new constitution in Aug. 1994, Menem was reelected president on May 14, 1995. A pact restoring commercial air links between Argentina and the Falklands was signed July 14, 1999.

Buenos Aires Mayor Fernando de la Rúa won the presidential election Oct. 24, 1999. A prolonged recession and $150 billion foreign debt left Argentina facing an economic crisis in 2001. Menem was indicted July 4 for illegal weapons trading while in office. The IMF announced an $8 billion bailout package Aug. 21.

Armenia

Republic of Armenia

People: Population: 3,336,100. **Age distrib.** (%): <15: 23.2; 65+: 9.7. **Pop. density:** 290 per sq. mi. **Urban:** 70%. **Ethnic groups:** Armenian 93%, Azeri 3%, Russian 2%, Kurd and others 2%. **Principal language:** Armenian (official). **Chief religion:** Armenian Orthodox 94%.

Geography: Area: 11,500 sq. mi. **Location:** SW Asia. **Neighbors:** Georgia on N, Azerbaijan on E, Iran on S, Turkey on W. **Topography:** Mountainous with many peaks above 10,000 ft. **Capital:** Yerevan, (urban agg.) 1,284,000.

Government: Type: Republic. **Head of state:** Pres. Robert Kocharian; b Aug. 31, 1954; in office: Apr. 9, 1998. **Head of gov.:** Prime Min. Andranik Markarian; b June 12, 1951; in office: May 12, 2000. **Local divisions:** 10 provinces, 1 city. **Defense budget:** $96 mil. **Active troops:** NA.

Economy: Industries: Manufacturing, machinery, chemicals. Note: most industry is shut down. **Chief crops:** Vegetables, grapes. **Minerals:** Copper, gold, zinc. **Arable land:** 17%. **Livestock** (2000): cattle: 478,730; chickens: 4.1 mil; goats: 10,580; pigs: 70,556; sheep: 540,000. **Electricity prod.:** 6.668 bil kWh. **Labor force:** 55% agric.; 25% services.

Finance: Monetary unit: Dram (Sept. 2001: 555.45 = $1 U.S.). **GDP** (1999 est.): $9.2 bil. **Per capita GDP:** $2,700. **Imports** (1999 est.): $782 mil; partners: Russia 20%, Turkmenistan 19%. **Exports** (1999 est.): $240 mil; Russia 33%, Turkmenistan 25%. **Tourism:** $27 mil. **Budget** (1998 est.): $424 mil. **Intl. reserves less gold** (June 2000): $316.72 mil. **Gold:** 44,400 oz t. **Consumer prices** (change in 1999): 0.7%.

Transport: Railroad: Length: 515 mi. **Civil aviation:** 476.6 mil pass.-mi.; 1 airport.

Communications: TV sets: 241 per 1,000 pop. **Telephones** (1999): 547,300. **Daily newspaper circ.:** 23 per 1,000 pop.

Health: Life expectancy: 62.12 male; 71.08 female. **Births** (per 1,000 pop): 11.47. **Deaths** (per 1,000 pop.): 9.74. **Natural inc.:** .17%. **Infant mortality** (per 1,000 live births): 41.27.

Education: Compulsory: ages 6-17. **Literacy** (1989): 99%.

Major Intl. Organizations: UN (FAO, IBRD, ILO, IMF, WHO), CIS, OSCE.

Embassy: 2225 R St. NW 20008; 319-1976.

Ancient Armenia extended into parts of what are now Turkey and Iran. Present-day Armenia was set up as a Soviet republic Apr. 2, 1921. It joined Georgian and Azerbaijan SSRs Mar. 12, 1922, to form the Transcaucasian SFSR, which became part of the USSR Dec. 30, 1922. Armenia became a constituent republic of the USSR Dec. 5, 1936. An earthquake struck Armenia Dec. 7, 1988; approximately 55,000 were killed and several cities and towns were left in ruins.

Armenia declared independence Sept. 23, 1991, and became an independent state when the USSR disbanded Dec. 26, 1991.

Fighting between mostly Christian Armenia and mostly Muslim Azerbaijan escalated in 1992 and continued through 1993. Each country claimed Nagorno-Karabakh, an enclave in Azerbaijan that has a majority population of ethnic Armenians. A temporary cease-fire was announced in May 1994, with Armenian forces in control of the enclave. Voters approved, July 5, 1995, a new constitution strengthening presidential powers. Pres. Levon Ter-Petrosian won reelection on Sept. 22, 1996, amid claims of fraud; he resigned Feb. 3, 1998, in a conflict over Nagorno-Karabakh. Robert Kocharian, a nationalist born in the disputed region, won the presidency on Mar. 30, 1998. Gunmen stormed Parliament Oct. 27, 1999, killing Prime Min. Vazgen Sarkissian and 7 others.

Australia

Commonwealth of Australia

People: Population: 19,357,594. **Age distrib.** (%): <15: 20.6; 65+: 12.5. **Pop. density:** 7 per sq. mi. **Urban:** 85%. **Ethnic groups:** Caucasian 92%, Asian 7%, aboriginal and other 1%. **Principal languages:** English (official), aboriginal languages. **Chief religions:** Anglican 26%, Roman Catholic 26%, other Christian 24%.

Geography: Area: 2,937,800 sq. mi. **Location:** SE of Asia, Indian O. is W and S, Pacific O. (Coral, Tasman seas) is E; they meet N of Australia in Timor and Arafura seas. Tasmania lies 150 mi. S of Victoria state, across Bass Strait. **Neighbors:** Nearest are Indonesia, Papua New Guinea on N; Solomons, Fiji, and New Zealand on E. **Topography:** An island continent. The Great Dividing Range along the E coast has Mt. Kosciusko, 7,310 ft. The W plateau rises to 2,000 ft., with arid areas in the Great Sandy and Great Victoria deserts. The NW part of Western Australia and Northern Terr. are arid and hot. The NE has heavy rainfall and Cape York Peninsula has jungles. **Capital:** Canberra. **Cities** (urban agg.): Sydney, 3,664,000; Melbourne, 3,187,000; Brisbane, 1,591,000; Perth, 1,313,000; Adelaide, 1,063,000.

Government: Type: Democratic, federal state system. **Head of state:** Queen Elizabeth II, represented by Gov.-Gen. Peter John Hollingworth; b Apr. 10, 1935; in office: June 29, 2001. **Head of gov.:** Prime Min. John Howard; b July 26, 1939; in office: Mar. 11, 1996. **Local divisions:** 6 states, 2 territories. **Defense budget** (2001): $7.1 bil. **Active troops:** 50,600.

Economy: Industries: Mining, steel, industrial & transportation equip., chemicals, food processing. **Chief crops:** Wheat (a leading export), barley, fruit, sugarcane. **Minerals:** Bauxite, coal, copper, iron, lead, tin, uranium, zinc. **Crude oil reserves** (2000): 2.9 bil bbls. **Other resources:** Wool (world's leading producer), beef. **Arable land:** 6%. **Livestock** (2000): cattle: 26.72 mil; chickens: 96.0 mil; goats: 200,000; pigs: 2.43 mil; sheep: 115.69 mil. **Fish catch** (1999): 214,227 metric tons. **Electricity prod.:** 191.727 bil kWh. **Labor force:** 73% services; 22% industry; 5% agric.

Finance: Monetary unit: Australian Dollar (Sept. 2001: 1.93 = $1 U.S.). **GDP** (1999 est.): $416.2 bil. **Per capita GDP:** $22,200. **Imports** (1999 est.): $67 bil; partners: EU 24%, U.S. 22%. **Exports** (1999 est.): $58 bil; partners: Japan 20%, EU 14%. **Tourism:** $7.53 bil. **Budget** (FY 1998-99 est.): $89.04 bil. **Intl. reserves less gold** (June 2000): $15.99 bil. **Gold:** 2.56 mil oz t. **Consumer prices** (change in 1999): 1.5%.

Transport: Railroad: Length: 20,567 mi. **Motor vehicles:** 8.7 mil pass. cars, 2.05 mil comm. vehicles. **Civil aviation:** 47.2 bil pass.-mi.; 400 airports. **Chief ports:** Sydney, Melbourne, Brisbane, Adelaide, Fremantle, Geelong.

Communications: TV sets: 639 per 1,000 pop. **Radios:** 1,120 per 1,000 pop. **Telephones:** 10,040,000 main lines. **Daily newspaper circ.:** 297 per 1,000 pop.

Health: Life expectancy: 77.02 male; 82.87 female. **Births** (per 1,000 pop): 12.86. **Deaths** (per 1,000 pop.): 7.18. **Natural inc.:** .57%. **Infant mortality** (per 1,000 live births): 4.97.

Education: Free, compulsory: ages 6-15. **Literacy** (1996): 100%.

Major Intl. Organizations: UN and all of its specialized agencies, APEC, the Commonwealth, OECD.

Embassy: 1601 Massachusetts Ave. NW 20036; 797-3000.

Websites: http://www.austemb.org
www.abs.gov.au

Australia harbors many plant and animal species not found elsewhere, including kangaroos, koalas, platypuses, dingos (wild dogs), Tasmanian devils (raccoon-like marsupials), wombats (bear-like marsupials), and barking and frilled lizards.

Capt. James Cook explored the E coast in 1770, when the continent was inhabited by a variety of different tribes. The first settlers, beginning in 1788, were mostly convicts, soldiers, and government officials. By 1830, Britain had claimed the entire continent, and the immigration of free settlers began to accelerate. The Commonwealth was proclaimed Jan. 1, 1901. Northern Terr. was granted limited self-rule July 1, 1978.

State/Territory, Capital	Area (sq. mi.)	Population (1997)
New South Wales, Sydney	309,500	6,274,400
Victoria, Melbourne	87,900	4,605,100
Queensland, Brisbane	666,990	3,401,200
Western Australia, Perth	975,100	1,798,100
South Australia, Adelaide	379,900	1,479,800
Tasmania, Hobart	26,200	473,500
Australian Capital Terr., Canberra	900	309,800
Northern Terr., Darwin	519,800	187,100

Racially discriminatory immigration policies were abandoned in 1973, after 3 million Europeans (half British) had entered since 1945. The 50,000 aborigines and 150,000 part-aborigines are mostly detribalized, but there are several preserves in the Northern Territory. They remain economically disadvantaged.

Australia's agricultural success makes the country among the top exporters of beef, lamb, wool, and wheat. Major mineral deposits have been developed, largely for export. Industrialization has been completed. The nation endured a deep recession 1990-93 but has rebounded strongly.

The Labor Party won a majority in Feb. 1983 general elections and was reelected in 1984, 1987, 1990, and 1993. After an election that focused mainly on economic issues, conservatives swept into power in elections Mar. 2, 1996.

Prime Min. John Howard retained power, but with a reduced majority, in parliamentary elections Oct. 3, 1998. Australia led an international peacekeeping force into East Timor in Sept. 1999. In a referendum Nov. 6, voters rejected a proposal that would have made Australia a republic. Sydney hosted the Summer Olympics Sept. 15-Oct. 1, 2000.

Australian External Territories

Norfolk Isl., area 13.3 sq. mi., pop. (1996 est.) 2,209, was taken over, 1914. The soil is very fertile, suitable for citrus, bananas, and coffee. Many of the inhabitants are descendants of the *Bounty* mutineers, moved to Norfolk 1856 from Pitcairn Isl. Australia offered the island limited home rule in 1978.

Coral Sea Isls. Territory, area 1 sq. mi., is administered from Norfolk Isl.

Territory of Ashmore and Cartier Isls., area 2 sq. mi., in the Indian O., came under Australian authority 1934 and are administered as part of Northern Territory. **Heard Isl. and McDonald Isls.,** area 159 sq. mi., are administered by the Dept. of Science.

Cocos (Keeling) Isls., 27 small coral islands in the Indian O. 1,750 mi. NW of Australia. Pop. (1996 est.) 609; area 5.5 sq. mi. The residents voted to become part of Australia, Apr. 1984.

Christmas Isl., area 52 sq. mi., pop. (1996 est.) 813; 230 mi. S of Java, was transferred by Britain in 1958. It has phosphate deposits.

Australian Antarctic Territory was claimed by Australia in 1933, including 2,362,000 sq. mi. of territory S of 60th parallel S Lat. and between 160th-45th meridians E Long. It does not include Adelie Coast.

Austria

Republic of Austria

People: Population: 8,150,835. **Age distrib.** (%): <15: 16.6; 65+: 15.4. **Pop. density:** 255 per sq. mi. **Urban:** 65%. **Ethnic groups:** German 99%, Croatian, Slovene. **Principal language:** German (official). **Chief religions:** Roman Catholic 78%, Protestant 5%.

Geography: Area: 31,900 sq. mi. **Location:** In S Central Europe. **Neighbors:** Switzerland, Liechtenstein on W; Germany, Czech Rep. on N; Slovakia, Hungary on E; Slovenia, Italy on S. **Topography:** Austria is primarily mountainous, with the Alps and foothills covering the western and southern provinces. The eastern provinces and Vienna are located in the Danube River Basin. **Capital:** Vienna, (urban agg.) 2,070,000.

Government: Type: Parliamentary democracy. **Head of state:** Pres. Thomas Klestil; b Nov. 4, 1932; in office: July 8, 1992. **Head of gov.:** Chancellor Wolfgang Schüssel; b June 7, 1945; in office: Feb. 4, 2000. **Local divisions:** 9 bundeslaender (states), each with a legislature. **Defense budget (2001):** $1.5 bil . **Active troops:** 35,500.

Economy: Industries: Machinery, autos, tourism, paper, chemicals, food. **Chief crops:** Grains, fruits, potatoes, sugar beets. **Minerals:** Iron ore, oil, magnesite. **Crude oil reserves** (2000): 85.68 mil bbls. **Other resources:** Forests, hydropower. **Arable land:** 17%. **Livestock** (2000): cattle: 2.15 mil; chickens: 13.54 mil; goats: 54,200; pigs: 3.79 mil; sheep: 360,812. **Electricity prod.:** 59.283 bil kWh. **Labor force:** 68% services; 29% ind. & crafts.

Finance: Monetary unit: Schilling (Sept. 2001: 15.18 = $1 U.S.). Euro (Sept. 2001: 1.09 = $1 U.S.). **GDP** (1999 est.): $190 bil. **Per capita GDP:** $23,400. **Imports** (1999): $69.9 bil; partners: EU 70%. **Exports** (1999): $62.9 bil; partners: EU 65%. **Tourism:** $11.09 bil. **Budget** (1998 est.): $55.9 bil. **Intl. reserves less gold** (June 2000): $14.50 bil. **Gold:** 13.10 mil oz t. **Consumer prices** (change in 1999): 0.6%.

Transport: Railroad: Length: 3,524 mi. **Motor vehicles in use (1997):** 3.78 mil pass. cars, 324,776 comm. vehicles. **Civil aviation:** 6.3 bil pass.-mi.; 6 airports. **Chief ports:** Linz, Vienna, Enns, Krems.

Communications: TV sets: 496 per 1,000 pop. **Radios:** 744 per 1,000 pop. **Telephones:** 3,889,000 main lines. **Daily newspaper circ.:** 296 per 1,000 pop.

Health: Life expectancy: 74.68 male; 81.15 female. **Births** (per 1,000 pop): 9.74. **Deaths** (per 1,000 pop.): 9.8. **Natural inc.:** -.01%. **Infant mortality** (per 1,000 live births): 4.44.

Education: Free, compulsory: ages 6-15. **Literacy** (1994): 100%.

Major Intl. Organizations: UN and all of its specialized agencies, EU, OECD, OSCE.

Embassy: 3524 International Ct. NW 20008; 895-6700.

Website: http://www.austria.org

Rome conquered Austrian lands from Celtic tribes around 15 BC. In 788 the territory was incorporated into Charlemagne's empire. By 1300, the House of Hapsburg had gained control; they added vast territories in all parts of Europe to their realm in the next few hundred years.

Austrian dominance of Germany was undermined in the 18th century and ended by Prussia by 1866. But the Congress of Vienna, 1815, confirmed Austrian control of a large empire in southeast Europe consisting of Germans, Hungarians, Slavs, Italians, and others. The dual Austro-Hungarian monarchy was established in 1867, giving autonomy to Hungary and almost 50 years of peace.

World War I, started after the June 28, 1914, assassination of Archduke Franz Ferdinand, the Hapsburg heir, by a Serbian nationalist, destroyed the empire. By 1918 Austria was reduced to a small republic, with the borders it has today.

Nazi Germany invaded Austria Mar. 13, 1938. The republic was reestablished in 1945, under Allied occupation. Full independence and neutrality were restored in 1955. Austria joined the European Union Jan. 1, 1995. The rise of the right-wing, anti-immigrant Austrian Freedom Party challenged the dominance of the Austrian Social Democratic Party in the late 1990s. When Freedom Party members joined the cabinet, Feb. 4, 2000, the EU imposed political sanctions on Austria, Feb. 4-Sept. 12, 2000.

Azerbaijan

Azerbaijani Republic

People: Population: 7,771,092. **Age distrib.** (%): <15: 28.9; 65+: 7.1. **Pop. density:** 233 per sq. mi. **Urban:** 57%. **Ethnic groups:** Azeri 90%, Dagestani Peoples 3%, Russian 3%, Armenian 2%. **Principal languages:** Azeri (official) 89%, Russian 3%, Armenian 2%. **Chief religions:** Muslim 93%, Orthodox 5%.

Geography: Area: 33,400 sq. mi. **Location:** SW Asia. **Neighbors:** Russia, Georgia on N; Iran on S; Armenia on W; Caspian Sea on E. **Capital:** Baku, (urban agg.) 1,936,000.

Government: Type: Republic. **Head of state:** Pres. Haydar A. Aliyev; b May 10, 1923; in office: June 30, 1993. **Head of gov.:** Prime Min. Artur Rasizade; b Feb. 26, 1935; in office: Nov. 26, 1996. **Local division:** 59 rayons, 11 cities, 1 autonomous republic. **Defense budget:** $119 mil. **Active troops:** 72,100.

Economy: Industries: Oil refining, chemicals, textiles. **Chief crops:** Grain, rice, cotton, grapes. **Minerals:** Oil, gas, iron. **Crude oil reserves** (2000): 1.2 bil bbls. **Arable land:** 18%. **Livestock** (2000): cattle: 1.95 mil; chickens: 13.9 mil; goats: 400,000; pigs: 19,600; sheep: 5.39 mil.. **Electricity prod.:** 16.378 bil kWh. **Labor force:** 53% services; 32% agric., forestry; 15% ind., const.

Finance: Monetary unit: Manat (Sept. 2001: 4,648 = $1 U.S.). **GDP** (1999 est.): $14 bil. **Per capita GDP:** $1,770. **Imports** (1999 est.): $1.62 bil; partners: Turkey , Russia . **Exports** (1999 est.): $885 mil; partners: Turkey, Russia . **Tourism** (1998): $125 mil. **Budget** (1996 est.): $682 mil. **Intl. reserves less gold** (Feb. 2000): $657.13 mil. **Consumer prices** (change in 1999): –8.6%.

Transport: Railroad: Length: 1,305 mi. **Motor vehicles:** 289,000 pass. cars; 89,000 comm. vehicles. **Civil aviation:** 797.1 mil. pass.-mi.; 3 airports. **Chief port:** Baku.

Communications: TV sets: 212 per 1,000 pop. **Daily newspaper circ.:** 28 per 1,000 pop. **Telephones:** 801,200 main lines.

Health: Life expectancy: 58.65 male; 67.49 female. **Births** (per 1,000 pop): 18.44. **Deaths** (per 1,000 pop.): 9.55. **Natural inc.:** .89%. **Infant mortality** (per 1,000 live births): 83.08.

Education: Compulsory: ages 6-17. **Literacy**: 100%.

Major Intl. Organizations: UN (FAO, IBRD, ILO, IMF, IMO, WHO), CIS, OSCE.

Embassy: 927 15th St. NW 20005; 842-0001.

Website: http://www.president.az

Azerbaijan was the home of Scythian tribes and part of the Roman Empire. Overrun by Turks in the 11th century and conquered by Russia in 1806 and 1813, it joined the USSR Dec. 30, 1922, and became a constituent republic in 1936. Azerbaijan declared independence Aug. 30, 1991, and became an independent state when the Soviet Union disbanded Dec. 26, 1991.

Fighting between mostly Muslim Azerbaijan and mostly Christian Armenia escalated in 1992 and continued in 1993 and 1994. Each country claimed Nagorno-Karabakh, an enclave in Azerbaijan with a majority population of ethnic Armenians. A temporary cease-fire was announced in May 1994, with Armenian forces in control of the enclave.

A National Council ousted Communist Pres. Mutaibov and took power May 19, 1992. Abulfez Elchibey became the nation's first democratically elected president June 7, but was ousted from office by Surat Huseynov, commander of a private militia, June 30, 1993. Huseynov became prime minister, and Haydar Aliyev, a pro-Russian former Communist, became president. Huseynov fled the country after his supporters staged an unsuccessful coup attempt Oct. 1994. Voters approved a new constitution expanding presidential powers, Nov. 12, 1995. Pres. Aliyev was reelected Oct. 11, 1998, but international monitors called the election seriously flawed.

The Bahamas

Commonwealth of The Bahamas

People: Population: 297,852. **Age distrib.** (%): <15: 29.4; 65+: 6.1. **Pop. density:** 77 per sq. mi. **Urban:** 88%. **Ethnic groups:** Black 85%, white 15%. **Principal languages:** English (official), Creole. **Chief religions:** Baptist 32%, Anglican 20%, Roman Catholic 19%, other Christian 24%.

Geography: Area: 3,900 sq. mi. **Location:** In Atlantic O., E of Florida. **Neighbors:** Nearest are U.S. on W, Cuba on S. **Topography:** Nearly 700 islands (29 inhabited) and over 2,000 islets in the W Atlantic O. extend 760 mi. NW to SE. **Capital:** Nassau. **Cities** (1999 est.): Nassau, 214,000; Grand Bahama, 40,898.

Government: Type: Independent commonwealth. **Head of state:** Queen Elizabeth II, represented by Gov.-Gen. Orville A. Turnquest; b July 19, 1929; in office: Jan. 2, 1995. **Head of gov.:** Prime Min. Hubert Ingraham; b Aug. 4, 1947; in office: Aug. 21, 1992. **Local divisions:** 21 districts. **Defense budget:** $26 mil. **Active troops:** 860.

Economy: Industries: Tourism (more than 60% of GDP), rum, cement, banking, pharmaceuticals. **Chief crops:** Citrus, vegetables. **Minerals:** Salt, aragonite. **Other resources:** Lobsters, timber. **Arable land:** 1%. **Livestock** (2000): cattle: 700; chickens: 5.0 mil; goats: 14,500; pigs: 5,900; sheep: 5,700. **Fish Catch:** (1999): 10,440 metric tons. **Electricity prod.:** 1.465 bil kWh. **Labor force:** 40% tourism; 50% serv.; 5% agric.

Finance: Monetary unit: Dollar (Sept. 2001: .99 = $1 U.S.). **GDP** (1998 est.): $5.58 bil. **Per capita GDP:** $20,000. **Imports** (1998): $1.74 bil; partners: U.S. 27.3%, Italy 26.5%, Japan 10%. **Exports** (1998): $362.8 mil; partners: U.S. 22.3%, Switzerland 15.6%, UK 15%. **Tourism** (1998): $1.41 bil. **Budget** (FY 1997-98): $845 mil. **Intl. reserves less gold** (May 2000): $489.6 mil. **Consumer prices** (change in 1999): 1.3%.

Transport: Motor vehicles: 69,000 pass. cars, 14,000 comm. vehicles. **Civil aviation:** 86.9 mil pass.-mi.; 22 airports. **Chief ports:** Nassau, Freeport.

Communications: TV sets: 179 per 1,000 pop. **Radios:** 282 per 1,000 pop. **Telephones**: 114,300 main lines. **Daily newspaper circ.:** 126 per 1,000 pop.

Health: Life expectancy: 67.27 male; 73.71 female. **Births** (per 1,000 pop): 19.10. **Deaths** (per 1,000 pop.): 7.14. **Natural inc.:** 1.2%. **Infant mortality** (per 1,000 live births): 17.03.

Education: Free, compulsory: ages 5-14. **Literacy**: 98%.

Major Intl. Organizations: UN (FAO, IBRD, ILO, IMF, IMO, WHO), Caricom, the Commonwealth, OAS.

Embassy: 2220 Massachusetts Ave. NW 20008; 319-2660.

Website:http://www.bahamas.net

Christopher Columbus first set foot in the New World on San Salvador (Watling Isl.) in 1492, when Arawak Indians inhabited the islands. British settlement began in 1647; the islands became a British colony in 1783. Internal self-government was granted in 1964; full independence within the Commonwealth was attained July 10, 1973.

International banking and investment management have become major industries alongside tourism.

Bahrain

State of Bahrain

People: Population: 645,361. **Age distrib.** (%): <15: 29.6; 65+: 3.0. **Pop. density:** 2,704 per sq. mi. **Urban:** 92%. **Ethnic groups:** Bahraini 63%, Asian 13%, other Arab 10%, Iranian 8%. **Principal languages:** Arabic (official), English, Farsi, Urdu. **Chief religions:** Shi'a Muslim 75%, Sunni Muslim 25%.

Geography: Area: 240 sq. mi. **Location:** SW Asia, in Persian Gulf. **Neighbors:** Nearest are Saudi Arabia on W, Qatar on E. **Topography:** Bahrain Island, and several adjacent, smaller islands, are flat, hot, and humid, with little rain. **Capital:** Manama, (1995 est.) 148,000.

Government: Type: Traditional monarchy. **Head of state:** Emir Hamad bin Isa al-Khalifa; b Jan. 28, 1950; in office: Mar. 6, 1999. **Head of gov.:** Prime Min. Khalifa bin Sulman al-Khalifa; b 1935; in office: Jan. 19, 1970. **Local divisions:** 12 municipalities. **Defense budget:** $306 mil (1999). **Active troops:** 11,000.

Economy: Industries: Oil products, aluminum smelting. **Chief crops:** Fruits, vegetables. **Minerals:** Oil, gas. **Crude oil reserves** (2000): 148.1 mil bbls. **Arable land:** 1%. **Livestock** (2000): cattle: 11,000; chickens: 465,000; goats: 16,300; sheep: 17,500. **Fish Catch:** (1999): 10,050 metric tons. **Electricity prod.:** 6.185 bil kWh. **Labor force:** 79% ind., commerce, services, 20% govt.

Finance: Monetary unit: Dinar (Sept. 2001: 0.37 = $1 U.S.). **GDP** (1999 est.): $8.6 bil. **Per capita GDP:** $13,700. **Imports** (1998): $3.5 bil; partners: Saudi Arabia 45%, U.S. 10%. **Exports** (1998): $3.3 bil; partners: India 18%, Japan 11%. **Tourism** (1998): $366 mil. **Budget** (1999): $1.9 bil. **Intl. reserves less gold** (June 2000): $1.46 bil. **Gold:** 150,000 oz t. **Consumer prices** (change in 1998): –0.4%.

Transport: Motor vehicles: 141,901 pass. cars, 30,243 comm. vehicles. **Civil aviation:** 1.6 bil pass.-mi.; 1 airport. **Chief ports:** Manama, Sitrah.

Communications: TV sets: 442 per 1,000 pop. **Radios:** 555 per 1,000 pop. **Telephones**: 171,000 main lines. **Daily newspaper circ.:** 128 per 1,000 pop.

Health: Life expectancy: 70.81 male; 75.67 female. **Births** (per 1,000 pop): 20.07. **Deaths** (per 1,000 pop.): 3.92. **Natural inc.:** 1.62%. **Infant mortality** (per 1,000 live births): 19.77.

Education: Free, compulsory: ages 6-17. **Literacy:** 85%.

Major Intl. Organizations: UN (FAO, IBRD, ILO, IMF, IMO, WHO, WTrO), AL.

Embassy: 3502 International Dr. NW 20008; 342-0741.

Long ruled by the Khalifa family, Bahrain was a British protectorate from 1861 to Aug. 15, 1971, when it regained independence.

Pearls, shrimp, fruits, and vegetables were the mainstays of the economy until oil was discovered in 1932. By the 1970s, oil reserves were depleted; international banking thrived.

Bahrain took part in the 1973-74 Arab oil embargo against the U.S. and other nations. The government bought controlling interest in the oil industry in 1975. Shiite dissidents have clashed with the Sunni-led government since 1996.

Bangladesh

People's Republic of Bangladesh

People: Population: 131,269,860. **Age distrib.** (%): <15: 35.0; 65+: 3.4 **Pop. density:** 2,542 per sq. mi. **Urban:** 24%. **Ethnic groups:** Bengali 98%, Bihari, tribals. **Principal languages:** Bangla (official), English. **Chief religions:** Muslim 88%, Hindu 11%.

Geography: Area: 51,600 sq. mi. **Location:** In S Asia, on N bend of Bay of Bengal. **Neighbors:** India nearly surrounds country on W, N, E; Myanmar on SE. **Topography:** The country is mostly a low plain cut by the Ganges and Brahmaputra rivers and their delta. The land is alluvial and marshy along the coast, with hills only in the extreme SE and NE. A tropical monsoon climate prevails, among the rainiest in the world. **Capital:** Dhaka. **Cities (urban agg.)**: Dhaka, 12,317,000; Chittagong, 3,581,000; Khulna, 1,426,000.

Government: Type: Parliamentary democracy. **Head of state:** Pres. Shahabuddin Ahmed; b 1930; in office: Oct. 9, 1996. **Head of gov.:** Chief Adviser Latifur Rahman; b Mar. 1, 1936; in office (interim): July 15, 2001. **Local divisions:** 6 divisions. **Defense budget** (1999): $612 mil. **Active troops:** 137,000.

Economy: Industries: Food processing, jute, textiles, fertilizers, steel. **Chief crops:** Jute, rice, tea. **Minerals:** Natural gas. **Crude oil reserves** (2000): 56.9 mil bbls. **Arable land:** 73%. **Livestock**(2000): cattle: 23.65 mil; chickens: 139.3 mil; goats: 33.8 mil; sheep: 1.12 mil.**Fish catch** (1999): 1.34 mil metric tons. **Electricity prod.:** 12.060 bil kWh. **Labor force:** 63% agric.; 26% services; 11% ind. & mining.

Finance: Monetary unit: Taka (Sept. 2001: 56.95 = $1 U.S.). **GDP** (1999 est.): $187 bil. **Per capita GDP:** $1,470. **Imports** (1998): $8.01 bil; partners: India 12%, China 9%. **Exports** (1998): $5.1 bil; partners: U.S. 33%, Germany 10%. **Tourism:** $50 mil. **Budget** (1997): $5.5 bil. **Intl. reserves less gold** (June 2000): $1.57 bil. **Gold:** 107,000 oz t. **Consumer prices** (change in 1999): 6.3%.

Transport: Railroad: Length: 1,681 mi. **Motor vehicles** (1997): 134,073 pass. cars, 92,133 comm. vehicles. **Civil aviation:** 2.0 bil pass.-mi.; 8 airports. **Chief ports:** Chittagong, Dhaka, Mongla Port.

Communications: TV sets: 5 per 1,000 pop. **Radios:** 63 per 1,000 pop. **Telephones**: 471,800 main lines. **Daily newspaper circ.:** 9 per 1,000 pop.

Health: Life expectancy: 60.74 male; 60.33 female. **Births** (per 1,000 pop): 25.30. **Deaths** (per 1,000 pop.): 8.6. **Natural inc.:** 1.67%. **Infant mortality** (per 1,000 live births): 69.85.

Education: Free, compulsory: ages 6-11. **Literacy:** 38%.

Major Intl. Organizations: UN (FAO, IBRD, ILO, IMF, IMO, WHO, WTrO), the Commonwealth.

Embassy: 2201 Wisconsin Ave. NW 20007; 342-8372.

Website: http://www.virtualbangladesh.com

Muslim invaders conquered the formerly Hindu area in the 12th century. British rule lasted from the 18th century to 1947, when East Bengal became part of Pakistan.

Charging West Pakistani domination, the Awami League, based in the East, won National Assembly control in 1971. Assembly sessions were postponed; riots broke out. Pakistani troops attacked Mar. 25; Bangladesh independence was proclaimed the next day. In the ensuing civil war, one million died and 10 million fled to India.

War between India and Pakistan broke out Dec. 3, 1971. Pakistan surrendered in the East on Dec. 16. Mujibur Rahman, known as Sheikh Mujib, became prime minister; he was killed in a coup Aug. 15, 1975. During the 1970s the country moved into the Indian and Soviet orbits in response to U.S. support of Pakistan, and much of the economy was nationalized.

On May 30, 1981, Pres. Ziaur Rahman was killed in an unsuccessful coup attempt by army rivals. Vice Pres. Abdus Sattar assumed the presidency but was ousted in a coup led by army chief of staff Gen. H. M. Ershad, Mar. 1982. Ershad declared Bangladesh an Islamic Republic in 1988; a parliamentary system of government was adopted in 1991.

Bangladesh is subject to devastating storms and floods that kill thousands. A cyclone struck Apr. 1991, killing over 131,000 people and causing $2.7 billion in damages. Chronic destitution in the densely crowded population has been worsened by the decline of jute as a world commodity.

Political turmoil led to the resignation, Mar. 30, 1996, of Prime Minister Khaleda Zia, the widow of Ziaur Rahman. Sheikh Mujib's daughter, Hasina Wazed (known as Sheikh Hasina), led the country after the June 12, 1996 election. Bangladesh and India signed a treaty, Dec. 12, resolving their long-standing dispute over the use of water from the Ganges River. A cyclone in May 1997 left an estimated 800,000 people homeless. Floods in July-Sept. 1998 inundated most of the country, killed over 1,400 people (many through disease), and stranded at least 30 million.

An interim government was installed July 2001 pending national elections, scheduled for October.

Barbados

People: Population: 275,330. **Age distrib.** (%): <15: 21.7; 65+: 8.9. **Pop. density:** 1,619 per sq. mi. **Urban:** 49%. **Ethnic groups:** Black 80%, white 4%, other 16%. **Principal language:** English (official). **Chief religions:** Protestant 67%, Roman Catholic 4%.

Geography: Area: 170 sq. mi. **Location:** In Atlantic O., farthest E of West Indies. **Neighbors:** Nearest are St. Lucia and St. Vincent & the Grenadines to the W. **Topography:** The island lies alone in the Atlantic almost completely surrounded by coral reefs. Highest point is Mt. Hillaby, 1,115 ft. **Capital:** Bridgetown, (1999 est.) 133,000.

Government: Type: Parliamentary democracy. **Head of state:** Queen Elizabeth II, represented by Gov.-Gen. Sir Clifford Husbands; b Aug. 5, 1926; in office: June 1, 1996. **Head of gov.:** Prime Min. Owen Arthur; b Oct. 17, 1949; in office: Sept. 7, 1994. **Local divisions:** 11 parishes and Bridgetown. **Defense budget:** $13 mil. **Active troops:** 610.

Economy: Industries: Sugar, tourism. **Chief crops:** Sugar, vegetables, cotton. **Minerals:** Oil, gas. **Crude oil reserves** (2000): 3.2 mil bbls. **Other resources:** Fish. **Arable land:** 37%. **Livestock** (2000): cattle: 23,000; chickens: 3.6 mil; goats: 4,500; pigs: 33,000; sheep: 41,000.. **Electricity prod.:** 718 mil kWh. **Labor force:** 75% services; 15% ind.; 10% agric.

Finance: Monetary unit: Dollar (Sept. 2001: 1.99 = $1 U.S.). **GDP** (1998 est.): $2.9 bil. **Per capita GDP:** $11,200. **Imports** (1998): $1.01 bil; partners: U.S. 30.7%, Trinidad and Tobago 10.2%. **Exports** (1998): $211.2 mil; partners: UK 14.8%, US 11.6%. **Tourism** (1998): $703 mil. **Budget** (FY 1997-98 est.): $750.6 mil. **Intl. reserves less gold** (June 1999): $368.44 mil. **Consumer prices** (change in 1999): 1.6%.

Transport: Motor vehicles: 45,000 pass. cars; 3,500 comm. vehicles. **Civil aviation:** 204.9 mil pass.-mi.; 1 airport. **Chief port:** Bridgetown.

Communications: TV sets: 287 per 1,000 pop. **Radios:** 1,134 per 1,000 pop. **Telephones** (1999): 115,000 main lines. **Daily newspaper circ.:** 157 per 1,000 pop.

Health: Life expectancy: 70.66 male; 75.86 female. **Births** (per 1,000 pop): 13.47. **Deaths** (per 1,000 pop.): 8.53. **Natural inc.:** .49%. **Infant mortality** (per 1,000 live births): 12.04.

Education: Compulsory: ages 5-16. **Literacy:** 97%.

Major Intl. Organizations: UN (FAO, IBRD, ILO, IMF, IMO, WHO, WTrO), Caricom, the Commonwealth, OAS.

Embassy: 2144 Wyoming Ave. NW 20008; 939-9200.

Barbados was probably named by Portuguese sailors in reference to bearded fig trees. An English ship visited in 1605, and British settlers arrived on the uninhabited island in 1627. Slaves worked the sugar plantations until slavery was abolished in 1834. Self-rule came gradually, with full independence proclaimed Nov. 30, 1966. British traditions have remained.

Belarus

Republic of Belarus

People: Population: 10,350,194. **Age distrib.** (%): <15: 17.9; 65+: 13.9. **Pop. density:** 129 per sq. mi. **Urban:** 71%. **Ethnic groups:** Byelorussian 78%, Russian 13%, Polish 4%. **Principal languages:** Byelorussian (official), Russian. **Chief religions:** Eastern Orthodox 80%, other 20%.

Geography: Area: 80,100 sq. mi. **Location:** E Europe. **Neighbors:** Poland on W; Latvia, Lithuania on N; Russia on E; Ukraine on S. **Capital:** Minsk, (urban agg.) 1,772,000.

Government: Type: Republic. **Head of state:** Pres. Aleksandr Lukashenko; b Aug. 30, 1954; in office: July 20,1994. **Head of gov.:** Prime Min. Vladimir Yermoshin; b Oct. 26, 1942; in office, Feb. 18, 2000. **Local divisions:** 6 oblasts and 1 municipality. **Defense budget:** $75 mil. **Active troops:** 83,100.

Economy: Industries: Manufacturing, chemical fibers, textiles, agricultural and industrial machinery. **Chief crops:** Grain, vegetables, potatoes. **Crude oil reserves** (2000): 198 mil bbls. **Arable land:** 29%. **Livestock** (2000): cattle: 4.33 mil; chickens: 30.0 mil; goats: 58,000; pigs: 3.57 mil; sheep: 91,900.. **Electricity prod.:** 24.911 bil kWh. **Labor force:** 41% services; 40% ind. & const.; 19% agric. & forestry.

Finance: Monetary unit: Ruble (Sept. 2001: 1,475 = $1 U.S.). **GDP** (1999 est.): $55.2 bil. **Per capita GDP:** $5,300. **Imports** (1999): $6.4 bil; partners: Russia 54%. **Exports** (1999): $6 bil; partners: Russia 66%. **Tourism** (1998): $22 mil. **Budget** (1997 est.): $4.1 bil. **Intl. reserves less gold** (May 2000): $301.17 mil. **Consumer prices** (change in 1999): 293.7%.

Transport: Railroad: Length: 3,480 mi. **Motor vehicles:** 842,500 pass. cars, 10,000 comm. vehicles. **Civil aviation:** 247.9 pass.-mi.; 1 airport. **Chief port:** Mazyr.

Communications: TV sets: 265 per 1,000 pop. **Radios:** 311 per 1,000 pop. **Telephones**: 2,751,900 main lines. **Daily newspaper circ.:** 187 per 1,000 pop.

Health: Life expectancy: 62.06 male; 74.52 female. **Births** (per 1,000 pop): 9.57. **Deaths** (per 1,000 pop.): 13.97. **Natural inc.:** -.44%. **Infant mortality** (per 1,000 live births): 14.38.

Education: Compulsory: ages 6-17. **Literacy** (1994): 98%.

Major Intl. Organizations: UN (IBRD, ILO, IMF, WHO), CIS, OSCE.

Embassy: 1619 New Hampshire Ave. NW 20009; 986-1604.

The region was subject to Lithuanians and Poles in medieval times, and was a prize of war between Russia and Poland beginning in 1503. It became part of the USSR in 1922 although the western part of the region was controlled by Poland. Belarus was overrun by German armies in 1941; recovered by Soviet troops in 1944. Following World War II, Belarus increased in area through Soviet annexation of part of NE Poland. Belarus declared independence Aug. 25, 1991. It became an independent state when the Soviet Union disbanded Dec. 26, 1991.

A new constitution was adopted, Mar. 15, 1994, and a new president was chosen in elections concluding July 1. Russia and Belarus signed a pact Apr. 2, 1996, linking their political and economic systems. An authoritarian constitution enacted in Nov. gave Pres. Aleksandr Lukashenko vast new powers. Lukashenko's insistence on tightening ties with Russia resulted in the signing of new accords in 1997 and 1998. Opponents charged harassment and fraud in the presidential election of Sept. 9, 2001, won by Lukashenko.

Belgium

Kingdom of Belgium

People: Population: 10,258,762. **Age distrib.** (%): <15: 17.5; 65+: 16.9. **Pop. density:** 880 per sq. mi. **Urban:** 97%. **Ethnic groups:** Fleming 55%, Walloon 33%. **Principal languages:** Flemish (Dutch) 56%, French 32%, German 1% (all official). **Chief religions:** Roman Catholic 75%; Protestant, other 25%.

Geography: Area: 11,700 sq. mi. **Location:** In W Europe, on North Sea. **Neighbors:** France on W and S, Luxembourg on SE, Germany on E, Netherlands on N. **Topography:** Mostly flat, the country is trisected by the Scheldt and Meuse, major commercial rivers. The land becomes hilly and forested in the SE (Ardennes) region. **Capital:** Brussels, (urban agg.) 1,122,000.

Government: Type: Parliamentary democracy under a constitutional monarch. **Head of state:** King Albert II; b June 6, 1934; in office: Aug. 9, 1993. **Head of gov.:** Premier Guy Verhofstadt; b Apr. 11, 1953; in office: July 12, 1999. **Local divisions:** 10 provinces and Brussels. **Defense budget:** $2.4 bil (2001). **Active troops:** 39,250.

Economy: Industries: Metal products, glassware, autos, textiles, chemicals. **Chief crops:** Grain, fruits, sugar beets, vegetables. **Minerals:** Coal, gas. **Arable land**: 24%. **Livestock** (2000): cattle: 3.09 mil; chickens: 45.0 mil; goats: 12,000; pigs: 7.32 mil; sheep: 152,000. **Fish catch** (1999): 31,346 metric tons. **Electricity prod.:** 79.829 bil kWh. **Labor force:** 73% services; 25% industry; 2% agric.

Finance: Monetary unit: Franc (Sept. 2001: 44.50 = $1 U.S.). Euro (Sept. 2001: 1.09 = $1 U.S.). **GDP** (1999 est.): $243.4 bil. **Per capita GDP** $23,900. *Note:* Import/Export data include Luxembourg. **Imports** (1999 est.): $172.8 bil; partners: EU 71%. **Exports** (1999 est.): $187.3 bil; partners: EU 76%.**Tourism** (1998): $5.44 bil. **Intl. reserves less gold** (June 2000): $9.55 bil. **Gold:** 8.30 mil oz t. **Consumer prices** (change in 1999): 1.1%.

Transport: Railroad: Length: 2,093 mi. **Motor vehicles** (1997): 4.42 mil pass. cars, 541,422 comm. vehicles. **Civil aviation:** 7.0 bil pass.-mi.; 2 airports. **Chief ports:** Antwerp (one of the world's busiest), Zeebrugge, Ghent.

Communications: TV sets: 510 per 1,000 pop. **Radios:** 792 per 1,000 pop. **Telephones:** 5,074,000 main lines. **Daily newspaper circ.:** 161 per 1,000 pop.

Health: Life expectancy: 74.63 male; 81.46 female. **Births** (per 1,000 pop): 10.74. **Deaths** (per 1,000 pop.): 10.1. **Natural inc.:** .06%. **Infant mortality** (per 1,000 live births): 4.7.

Education: Compulsory: ages 6-18. **Literacy:** 99%.

Major Intl. Organizations: UN and all of its specialized agencies, EU, NATO, OECD, OSCE.

Embassy: 3330 Garfield St. NW 20008; 333-6900.

Website: http://www.belgium.fgov.be

Belgium derives its name from the Belgae, the first recorded inhabitants, probably Celts. The land was conquered by Julius Caesar, and was ruled for 1800 years by conquerors, including Rome, the Franks, Burgundy, Spain, Austria, and France. After 1815, Belgium was made a part of the Netherlands, but it became an independent constitutional monarchy in 1830.

Belgian neutrality was violated by Germany in both world wars. King Leopold III surrendered to Germany, May 28, 1940. After the war, he was forced by political pressure to abdicate in favor of his son, King Baudouin. Baudouin was succeeded by his brother, Albert II, Aug. 9, 1993.

The Flemings of northern Belgium speak Dutch, while French is the language of the Walloons in the south. The language difference has been a perennial source of controversy and led to antagonism between the 2 groups. Parliament has passed measures aimed at transferring power from the central government to 3 regions—Wallonia, Flanders, and Brussels. Constitutional changes in 1993 made Belgium a federal state.

Belize

People: Population: 256,062. **Age distrib.** (%): <15: 42.0; 65+: 3.5. **Pop. density:** 29 per sq. mi. **Urban:** 54%. **Ethnic groups:** Mestizo 44%, Creole 30%, Maya 11%, Garifuna 7%. **Principal languages:** English (official), Spanish, Mayan, Garifuna (Carib). **Chief religions:** Roman Catholic 62%, Protestant 30%.

Geography: Area: 8,800 sq. mi. **Location:** Eastern coast of Central America. **Neighbors:** Mexico on N, Guatemala on W and S. **Capital:** Belmopan (1997 est.): 6,785.

Government: Type: Parliamentary democracy. **Head of state:** Queen Elizabeth II, represented by Gov.-Gen. Colville Young; b Nov. 20, 1932; in office: Nov. 17, 1993. **Head of gov.:** Prime Min. Said Musa; b Mar. 19, 1944; in office: Aug. 28, 1998. **Local divisions:** 6 districts. **Defense budget:** $8 mil(2001). **Active troops:** 1,050.

Economy: Industries: Garments, food processing, tourism. **Chief crops:** Sugar (main export), citrus, bananas. **Arable land:** 2%. **Livestock** (2000): cattle: 59,000; chickens: 1.4 mil; goats: 1,300; pigs: 24,000; sheep: 2,800. **Electricity prod.:** 185 mil kWh. **Labor force:** 38% agric.; 32% ind., 30% services.

Finance: Monetary unit: Dollar (Sept. 2001: 1.97= $1 U.S.). **GDP** (1999 est.): $740 mil. **Per capita GDP:** $3,100. **Imports** (1998): $320 mil; partners: U.S. 52%. **Exports** (1998): $150 mil; partners: U.S. 45.5%, UK 30%. **Tourism** (1998): $99 mil. **Budget** (FY1997-98 est.): $142 mil. **Intl. reserves less gold** (June 2000): $97.00 mil. **Consumer prices** (change in 1999): –1.2%.

Transport: Motor vehicles: 2,300 pass. cars, 3,100 comm. vehicles. **Chief ports:** Belize City, Big Creek. **Civil aviation:** 9 airports.

Communications: TV sets: 109 per 1,000 pop. **Radios:** 133 per 1,000 pop. **Telephones:** 36,000 main lines.

Health: Life expectancy: 68.91 male; 73.57 female. **Births** (per 1,000 pop): 31.69. **Deaths** (per 1,000 pop.): 4.7. **Natural inc.:** 2.7%. **Infant mortality** (per 1,000 live births): 25.14.

Education: Compulsory: ages 5-14. **Literacy** (1993): 93%.

Major Intl. Organizations: UN (FAO, IBRD, ILO, IMF, IMO, WHO, WTrO), Caricom, the Commonwealth, OAS.

Embassy: 2535 Massachusetts Ave. NW 20008; 332-9636.

Website: http://www.belizenet.com

Belize (formerly British Honduras) was Britain's last colony on the American mainland; independence was achieved Sept. 21, 1981. Relations with neighboring Guatemala, initially tense, have improved in recent years. Belize has become a center for drug trafficking between Colombia and the U.S.

Benin

Republic of Benin

People: Population: 6,590,782. **Age distrib.** (%): <15: 47.3; 65+: 2.3. **Pop. density:** 154 per sq. mi. **Urban:** 42%. **Ethnic groups:** African (Fon, Adja, Bariba, Yoruba, others) 99%. **Principal languages:** French (official), Fon, Yoruba, various tribal. **Chief religions:** Indigenous beliefs 70%, Muslim 15%, Christian 15%.

Geography: Area: 42,700 sq. mi. **Location:** In W Africa on Gulf of Guinea. **Neighbors:** Togo on W; Burkina Faso, Niger on N; Nigeria on E. **Topography:** Most of Benin is flat and covered with dense vegetation. The coast is hot, humid, and rainy. **Capital:** Porto-Novo. **Cities** (1994 est.): Cotonou 750,000, Porto-Novo 200,000.

Government: Type: Republic. **Head of state and gov.:** Pres. Mathieu Kerekou; b Sept. 2, 1933; in office: Apr. 4, 1996. **Local divisions:** 6 departments. **Defense budget:** $37 mil. **Active troops:** 4,750.

Economy: Industries: textiles, food, beverages, construction materials, petroleum. **Chief crops:** Palm oil, sorghum, cassava, peanuts, cotton, corn, rice. **Minerals:** Oil, limestone, marble. **Crude oil reserves** (2000): 8.2 mil bbls. **Arable land:** 13%. **Livestock** (2000): cattle: 1.44 mil; chickens: 23.0 mil; goats: 1.18 mil; pigs: 470,000; sheep: 644,997. **Fish catch** (1999): 43,771 metric tons. **Electricity prod.:** 226 mil kWh.

Finance: Monetary unit: CFA Franc (Sept. 2001: 724.11 = $1 U.S.). **GDP** (1999 est.): $8.1 bil. **Per capita GDP:** $1,300. **Imports** (1999): $566 mil; partners: France 22%, China 16%. **Exports** (1999): $396 mil; partners: Brazil, Libya, Indonesia.. **Tourism** (1998): $33 mil. **Budget** (1995 est.): $445 mil. **Intl. reserves less gold** (Apr. 2000): $460.1 mil. **Consumer prices** (change in 1999): 0.3%.

Transport: Railroad: Length: 359 mi. **Motor vehicles:** 35,600 pass. cars, 19,300 comm. vehicles. **Civil aviation:** 150.5 mil pass.-mi.; 1 airport. **Chief port:** Cotonou.

Communications: TV sets: 4 per 1,000 pop. **Radios:** 73 per 1,000 pop. **Telephones:** (1998) 38,400 main lines. **Daily newspaper circ.:** 2 per 1,000 pop.

Health: Life expectancy: 49.02 male; 50.88 female. **Births** (per 1,000 pop): 44.23. **Deaths** (per 1,000 pop.): 14.51. **Natural inc.:** 2.97%. **Infant mortality** (per 1,000 live births): 89.68.

Education: Free, compulsory: ages 6-12. **Literacy:** 37%.

Major Intl. Organizations: UN (FAO, IBRD, ILO, IMF, IMO, WHO, WTrO), OAU.

Embassy: 2737 Cathedral Ave. NW 20008; 232-6656.

The Kingdom of Abomey, rising to power in wars with neighboring kingdoms in the 17th century, came under French domination in the late 19th century and was incorporated into French West Africa by 1904.

Under the name Dahomey, the country gained independence Aug. 1, 1960; it became Benin in 1975. In the fifth coup since independence Col. Ahmed Kerekou took power in 1972; two years later he declared a socialist state with a "Marxist-Leninist" philosophy. In Dec. 1989, Kerekou announced Marxism-Leninism would no longer be the state ideology.

In Mar. 1991, Kerekou lost to Nicéphore Soglo in Benin's first free presidential election in 30 years. Kerekou defeated Soglo in Mar. 1996 to reclaim the presidency. He won reelection in a runoff Mar. 22, 2001.

Bhutan

Kingdom of Bhutan

People: Population: 2,049,412. **Age distrib.** (%): <15: 40.0; 65+: 4.0. **Pop. density:** 113 per sq. mi. **Urban:** 7%. **Ethnic groups:** Drukpa 50%, Nepalese 35%. **Principal languages:** Dzongkha (official), Tibetan, Nepalese dialects. **Chief religions:** Lamaistic Buddhist (state religion) 75%, Hindu 25%.

Geography: Area: 18,100 sq. mi. **Location:** S Asia, in eastern Himalayan Mts. **Neighbors:** India on W (Sikkim) and S, China on N. **Topography:** Bhutan is comprised of very high mountains in the N, fertile valleys in the center, and thick forests in the Duar Plain in the S. **Capital:** Thimphu, (1995 est.) 22,000.

Government: Type: Monarchy. **Head of state and gov.:** King Jigme Singye Wangchuk; b Nov. 11, 1955; in office: July 21, 1972. **Head of gov.:** Prime Min. Lyonpo Yeshey Zimba; in office: July 20, 2000. **Local divisions:** 18 districts. **Defense budget:** NA. **Active troops:** NA.

Economy: Industries: Cement, wood products. **Chief crops:** Rice, corn, citrus. **Other resources:** Timber, hydropower. **Livestock** (2000): cattle: 435,000; chickens: 310,000; goats: 42,100; pigs: 74,900; sheep: 58,500. **Arable land:** 2%. **Electricity prod.:** 1.856 bil kWh. **Labor force:** 93% agric.; 5% services.

Finance: Monetary unit: Ngultrum (Sept. 2001: 46.48 = $1 U.S.; Indian Rupee also used). **GDP** (1999 est.): $2.1 bil. **Per capita GDP:** $1,060. **Imports** (1998 est.): $136 mil; partners: India 77%. **Exports** (1998 est.): $111 mil; partners: India 94%. **Tourism:** $9 mil. **Budget** (FY 1995-96): $152 mil. **Intl. reserves less gold** (Mar. 2000): $278.39 mil. **Consumer prices** (change in 1999): 8.5%.

Transport: Civil aviation: 30.1 mil pass.-mi.; 1 airport.

Communications: Radios: 27 per 1,000 pop. **Telephones:** (1999): 11,800 main lines.

Health: Life expectancy: 53.16 male; 52.41 female. **Births** (per 1,000 pop): 35.73. **Deaths** (per 1,000 pop.): 14.03. **Natural inc.:** 2.17%. **Infant mortality** (per 1,000 live births): 108.89.

Education: Not compulsory. **Literacy:** 42%.

Major Intl. Organizations: UN (FAO, IBRD, IMF, WHO).

Website: http://www.kingdomofbhutan.com

The region came under Tibetan rule in the 16th century. British influence grew in the 19th century. A Buddhist monarchy was set up in 1907. According to a 1910 treaty, Britain guided Bhutan's external affairs, while the country remained internally self-governing. Upon independence, India assumed Britain's role in a 1949 revision of the treaty. Isolated for much of its history, Bhutan took tentative steps toward modernization in the 1990s.

Bolivia

Republic of Bolivia

People: Population: 8,300,463. **Age distrib.** (%): <15: 38.5; 65+: 4.5. **Pop. density:** 20 per sq. mi. **Urban:** 62%. **Ethnic groups:** Quechua 30%, mestizo 30%, Aymara 25%, White 15%. **Principal languages:** Spanish, Quechua, Aymara (all official). **Chief religion:** Roman Catholic 95%.

Geography: Area: 418,200 sq. mi. **Location:** In W central South America, in the Andes Mts. (one of 2 landlocked countries in South America). **Neighbors:** Peru and Chile on W, and Paraguay on S, Brazil on E and N. **Topography:** The great central plateau, at an altitude of 12,000 ft., over 500 mi. long, lies between two great cordilleras having 3 of the highest peaks in South America. Lake Titicaca, on Peruvian border, is highest lake in world on which steamboats ply (12,506 ft.). The E central region has semitropical forests; the llanos, or Amazon-Chaco lowlands are in E. **Capitals:** La Paz (administrative), Sucre (judicial). **Cities (urban agg.):** La Paz, 1,480,000; Santa Cruz, 1,065,000.

Government: Type: Republic. **Head of state and gov.:** Pres. Jorge Quiroga Ramirez: b May 5, 1960: in office: Aug. 7, 2001. **Local divisions:** 9 departments. **Defense budget:** $130 mil . **Active troops:** 35,000.

Economy: Industries: Mining, smelting, tobacco, handicrafts, clothing. **Chief crops:** Coffee, sugarcane, potatoes, cotton, corn, coca. **Minerals:** Antimony, tin, tungsten, silver, zinc, oil, gas, iron. **Crude oil reserves** (2000): 132 mil bbls. **Other resources:** Timber. **Arable land:** 2%. **Livestock** (2000): cattle: 6.72 mil; chickens: 73.86 mil; goats: 1.5 mil; pigs: 2.79 mil; sheep: 8.75 mil. **Electricity prod.:** 3.625 bil kWh.

Finance: Monetary unit: Boliviano (Sept. 2001: 6.71 = $1 U.S.). **GDP** (1999 est.): $24.2 bil. **Per capita GDP:** $3,000. **Im-**

ports (1999): $1.6 bil; partners: U.S. 32%, Japan 24%, Brazil 12%. **Exports** (1999): $1.1 bil; partners: U.K. 16%, U.S. 12%, Peru 11%.. **Tourism:** $170 mil. **Budget** (1998): $2.7 bil. **Intl. reserves less gold** (June 2000): $823.2 mil. **Gold:** 940,000 oz t. **Consumer prices** (change in 1999): 2.2%.

Transport: Railroad: Length: 2,295 mi. **Motor vehicles:** 199,309 pass. cars, 230,245 comm. vehicles. **Civil aviation:** 1.3 bil pass.-mi.; 14 airports.

Communications: TV sets: 202 per 1,000 pop. **Radios:** 560 per 1,000 pop. **Telephones**: 502,500 main lines. **Daily newspaper circ.:** 69 per 1,000 pop.

Health: Life expectancy: 61.53 male; 66.72 female. **Births** (per 1,000 pop): 27.27. **Deaths** (per 1,000 pop.): 8.2. **Natural inc.:** 1.91%. **Infant mortality** (per 1,000 live births): 58.98.

Education: Free, compulsory: ages 6-14. **Literacy:** 83%.

Major Intl. Organizations: UN (FAO, IBRD, ILO, IMF, IMO, WHO, WTrO), OAS.

Embassy: 3014 Massachusetts Ave. NW 20008; 483-4410.

The Incas conquered the region from earlier Indian inhabitants in the 13th century. Spanish rule began in the 1530s and lasted until Aug. 6, 1825. The country is named after Simon Bolivar, independence fighter.

In a series of wars, Bolivia lost its Pacific coast to Chile, the oil-bearing Chaco to Paraguay, and rubber-growing areas to Brazil, 1879-1935.

Economic unrest, especially among the militant mine workers, has contributed to continuing political instability. A reformist government under Victor Paz Estenssoro, 1951-64, nationalized tin mines and attempted to improve conditions for the Indian majority but was overthrown by a military junta. A long series of coups and countercoups continued until constitutional government was restored in 1982.

U.S. pressure on the government to reduce the country's coca output, the raw material for cocaine, has led to clashes between police and coca growers and increased anti-U.S. feeling among Bolivians. Gen. Hugo Banzer Suárez, who ruled as a dictator, 1971-78, became president in Aug. 1997. 105 people died in earthquakes near Aiquile May 22, 1998. Stricken with cancer, Banzer resigned and was succeeded Aug. 7, 2001, by Vice-Pres. Jorge Quiroga Ramírez.

Bosnia and Herzegovina

People: Population: 3,922,205. **Age distrib.** (%): <15: 20.1; 65+: 9.1. **Pop. density:** 199 per sq. mi. **Urban:** 43%. **Ethnic groups:** Serb 40%, Muslim 38%, Croat 22%. **Principal language:** Serbo-Croatian (official) 99%. **Chief religions:** Muslim 40%, Orthodox 31%, Catholic 15%.

Geography: Area: 19,700 sq. mi. **Location:** On Balkan Peninsula in SE Europe. **Neighbors:** Yugoslavia on E and SE, Croatia on N and W. **Topography:** Hilly with some mountains. About 36% of the land is forested. **Capital:** Sarajevo, (1999 est.) 415,631.

Government: Type: Republic. **Heads of state: Collective presidency with rotating leadership. Head of gov.: Prime Min. Zlatko Lagumdzija; b Dec. 26, 1955; in office: July 18, 2001. Local divisions:** Muslim-Croat Federation, divided into 10 cantons; Republika Srpska. **Defense budget:** 163 mil. Active troops: NA.

Economy: Industries: Steel, mining, textiles, timber. **Chief crops:** Corn, wheat, fruits, vegetables. **Minerals:** Bauxite, iron, coal. **Arable land:** 14%. **Livestock** (2000): cattle: 461,928; chickens: 3.0 mil; pigs: 150,000; sheep: 661,641.. **Electricity prod.:** 2.585 bil kWh.

Finance: Monetary unit: Conv. Mark (Sept. 2001: 2.16 = $1 U.S.). **GDP** (1999 est.): $6.2 bil. **Per capita GDP:** $1,770. **Imports** (1997): $2.95 bil. **Exports** (1997): $450 mil. **Tourism:** $13 mil.

Transport: Railroad: Length: 634 mi. **Chief port:** Bosanski Brod. **Civil aviation:** 1 airport.

Communications: TV sets: 94 per 1,000 pop. **Telephones:** 407,600 main lines. **Daily newspaper circ.:** 150 per 1,000 pop.

Health: Life expectancy: 69.04 male; 74.65 female. **Births** (per 1,000 pop): 12.86. **Deaths** (per 1,000 pop.): 7.99. **Natural inc.:** .49%. **Infant mortality** (per 1,000 live births): 24.35.

Education: Free, compulsory: ages 7-15. **Literacy** (1991): 86%.

Major Intl. Organizations: UN (FAO, IBRD, ILO, IMF, IMO, WHO), OSCE.

Embassy: 2109 E St. NW, 20037; 337-1500.

Website: http://www.bosnianembassy.org

Bosnia was ruled by Croatian kings c. AD 958, and by Hungary 1000-1200. It became organized c. 1200 and later took control of Herzegovina. The kingdom disintegrated from 1391, with the southern part becoming the independent duchy Herzegovina. It was conquered by Turks in 1463 and made a Turkish province. The area was placed under control of Austria-Hungary in 1878, and made part of the province of **Bosnia and Herzegovina,** which was formally annexed to Austria-Hungary 1908; Bosnia became a province of Yugoslavia in 1918. It was reunited with Herzegovina as a federated republic in the 1946 Yugoslavian constitution.

Bosnia and Herzegovina declared sovereignty Oct. 15, 1991. A referendum for independence was passed Feb. 29, 1992. Ethnic Serbs' opposition to the referendum spurred violent clashes and bombings. The U.S. and EU recognized the republic Apr. 7. Fierce three-way fighting continued between Bosnia's Serbs, Muslims, and Croats. Serb forces massacred thousands of Bosnian Muslims and engaged in "ethnic cleansing" (the expulsion of Muslims and other non-Serbs from areas under Bosnian Serb control). The capital, Sarajevo, was surrounded and besieged by Bosnian Serb forces. Muslims and Croats in Bosnia reached a cease fire Feb. 23, 1994, and signed an accord, Mar. 18, to create a Muslim-Croat confederation in Bosnia. However, by mid-1994, Bosnian Serbs controlled over 70% of the country.

As fighting continued in 1995, the balance of power began to shift toward the Muslim-Croat alliance. Massive NATO air strikes at Bosnian Serb targets beginning Aug. 30 triggered a new round of peace talks, and the siege of Sarajevo was lifted Sept. 15. The new talks produced an agreement in principle to create autonomous regions within Bosnia, with the Serb region (Republika Srpska) constituting 49% of the country. A Croat-Muslim offensive in Sept. recaptured significant territory, leaving Bosnian Serbs in control of approximately half that percentage.

A peace agreement initialed in Dayton, Ohio, Nov. 21, 1995, was signed in Paris, Dec. 14, by leaders of Bosnia, Croatia, and Serbia. Some 60,000 NATO troops (about 20,000 from the U.S.) moved in to police the accord. Meanwhile, a UN tribunal began bringing charges against suspected war criminals. Elections were held Sept. 14, 1996, for a 3-person collective presidency, for seats in a federal parliament, and for regional offices. In Dec. a revamped NATO "stabilization force" (SFOR) of over 30,000 members (more than 8,000 from the U.S.) received an 18-month mandate, which was later extended. By mid-2001, SFOR's troop strength had been reduced to 18,000

In a landmark verdict Aug. 2, 2001, the UN tribunal found Radislav Krstic, a Bosnian Serb general, guilty of genocide for the mass killing of over 7,000 Muslims at Srebrenica in 1995.

Botswana

Republic of Botswana

People: Population: 1,586,119. **Age distrib.** (%): <15: 40.3; 65+: 4.1. **Pop. density:** 7 per sq. mi. **Urban:** 50%. **Ethnic groups:** Batswana 95%, Kalanga, Basarwa, Kgalagadi. **Principal languages:** English (official), Setswana. **Chief religions:** Indigenous beliefs 50%, Christian 50%.

Geography: Area: 225,700 sq. mi. **Location:** In southern Africa. **Neighbors:** Namibia on N and W, South Africa on S, Zimbabwe on NE; Botswana claims border with Zambia on N. **Topography:** The Kalahari Desert, supporting nomadic Bushmen and wildlife, spreads over SW; there are swamplands and farming areas in N, and rolling plains in E where livestock are grazed. **Capital:** Gaborone, (1998 est.) 192,845.

Government: Type: Parliamentary republic. **Head of state and gov.:** Pres. Festus Mogae; b Aug. 21, 1939; in office: Apr. 1, 1998. **Local divisions:** 10 districts, 4 town councils. **Defense budget:** $234 mil. **Active troops:** 9,000.

Economy: Industries: Livestock processing, mining. **Chief crops:** Maize, sorghum, millet, pulses, beans. **Minerals:** Copper, coal, nickel, diamonds, salt, silver. **Arable land:** 1%. **Livestock** (2000): cattle: 2.35 mil; chickens: 3.5 mil; goats: 2.2 mil; pigs: 6,000; sheep: 350,000. **Electricity prod.:** 610 mil kWh.

Finance: Monetary unit: Pula (Sept. 2001: 5.77 = $1 U.S.). **GDP** (1999 est.): $5.7 bil. **Per capita GDP:** $3,900. **Imports** (1999 est.): $2.05 bil.; partners: Southern African Customs Union (SACU) 78% **Exports** (1999 est.): $2.36 bil.; partners: EU 74% **Tourism** (1998): $175 mil. **Budget** (FY 1996-97): $1.8 bil. **Intl. reserves less gold** (May 2000): $6.14 bil. **Consumer prices** (change in 1999): 7.1%.

Transport: Railroad: Length: 603 mi. **Motor vehicles:** 80,000 pass. cars, 19,869 comm. vehicles. **Civil aviation:** 34.2 mil pass.-mi.; 4 airports.

Communications: TV sets: 24 per 1,000 pop. **Radios:** 821 per 1,000 pop. **Telephones** (1999): 123,800 main lines. **Daily newspaper circ.:** 29 per 1,000 pop.

Health: Life expectancy: 36.77 male; 37.51 female. **Births** (per 1,000 pop): 28.85. **Deaths** (per 1,000 pop.): 24.18. **Natural inc.:** .47%. **Infant mortality** (per 1,000 live births): 63.2.

Education: Not compulsory. **Literacy:** 70%.

Major Intl. Organizations: UN (FAO, IBRD, ILO, IMF, WHO, WTrO), the Commonwealth, OAU.

Embassy: 1531-33 New Hampshire Ave. NW, 20036; 244-4990.

First inhabited by bushmen, then Bantus, the region became the British protectorate of Bechuanaland in 1886, halting encroachment by Boers and Germans from the south and southwest. The country became fully independent Sept. 30, 1966, as Botswana. Cattle raising and mining (diamonds, copper, nickel) have contributed to economic growth; economy is closely tied to South Africa. According to UN estimates, more than one-third of the adult population has HIV/AIDS.

Brazil

Federative Republic of Brazil

People: Population: 174,468,575. **Age distrib.** (%): <15: 28.6; 65+: 5.5. **Pop. density:** 53 per sq. mi. **Urban:** 81%. **Ethnic groups:** White (incl. Portuguese, German, Italian, Spanish, Polish) 55%, mixed black and white 38%, black 6%. **Principal languages:** Portuguese (official), Spanish, English, French. **Chief religion:** Roman Catholic 70%.

Geography: Area: 3,261,200 sq. mi., largest country in South America. **Location:** Occupies E half of South America. **Neighbors:** French Guiana, Suriname, Guyana, Venezuela on N; Colombia, Peru, Bolivia, Paraguay, on W; Uruguay on S. **Topography:** Brazil's Atlantic coastline stretches 4,603 miles. In N is the heavily wooded Amazon basin covering half the country. Its network of rivers is navigable for 15,814 mi. The Amazon itself flows 2,093 miles in Brazil, all navigable. The NE region is semiarid scrubland, heavily settled and poor. The S central region, favored by climate and resources, has almost half of the population, produces 75% of farm goods and 80% of industrial output. The narrow coastal belt includes most of the major cities. Almost the entire country has a tropical or semitropical climate. **Capital:** Brasília. **Cities** (urban agg.): São Paulo, 17,755,000, (2001 city est.: 10.4 mil.); Rio de Janeiro, 10,582,000; Belo Horizonte, 4,170,000, Brasília, 1,990,000.

Government: Type: Federal republic. **Head of state and gov.:** Pres. Fernando Henrique Cardoso; b June 18, 1931; in office: Jan. 1, 1995. **Local divisions:** 26 states, 1 federal district (Brasília). **Defense budget:** $9.9 bil. **Active troops:** 287,600.

Economy: Industries: Steel, autos, textiles, shoes, chemicals, machinery. **Chief crops:** Coffee (leading grower), soybeans, sugarcane, cocoa, rice, corn, wheat, citrus. **Minerals:** Iron (largest producer in the world), manganese, phosphates, uranium, gold, nickel, tin, bauxite, oil. **Crude oil reserves** (2000): 7.4 bil bbls. **Arable land:** 5%. **Livestock** (2000): cattle: 167.47 mil; chickens: 1.01 bil; goats: 8.5 mil; pigs: 27.32 mil; sheep: 15.0 mil. **Fish catch** (1999): 820,480 metric tons. **Electricity prod.:** 337.440 bil kWh. **Labor force:** 42% services; 31% agric.; 27% industry.

Finance: Monetary unit: Real (Sept. 2001: 2.58 = $1 U.S.). **GDP** (1999 est.): $1.057 tril. **Per capita GDP:** $6,150. **Imports** (1999): $48.7 bil; partners: U.S. 23%, Argentina 12%. **Exports** (1999): $46.9 bil; partners: U.S. 18%, Argentina 13%. **Tourism:** $3.99 bil. **Budget** (1998): $149 bil. **Intl. reserves less gold** (May 2000): $27.49 bil. **Gold:** 1.72 mil oz t. **Consumer prices** (change in 1999): 4.9%.

Transport: Railroad: Length: 18,578 mi. **Motor vehicles** (1997): 14.00 mil pass. cars, 4.03 mil comm. vehicles. **Civil aviation:** 26.3 bil pass.-mi.; 139 airports. **Chief ports:** Santos, Rio de Janeiro, Vitoria, Salvador, Rio Grande, Recife.

Communications: TV sets: 317 per 1,000 pop. **Radios:** 446 per 1,000 pop. **Telephones:** 30,926,300 main lines. **Daily newspaper circ.:** 41 per 1,000 pop.

Health: Life expectancy: 58.96 male; 67.73 female. **Births** (per 1,000 pop): 18.45. **Deaths** (per 1,000 pop.): 9.34. **Natural inc.:** .91%. **Infant mortality** (per 1,000 live births): 36.96.

Education: Free, compulsory: ages 7-14. **Literacy** (1996): 85%.

Major Intl. Organizations: UN and most of its specialized agencies, OAS.

Embassy: 3006 Massachusetts Ave. NW 20008; 238-2700.

Website: http://www.ibge.gov.br

Pedro Alvares Cabral, a Portuguese navigator, is generally credited as the first European to reach Brazil, in 1500. The country was thinly settled by various Indian tribes. Only a few have survived to the present, mostly in the Amazon basin.

In the next centuries, Portuguese colonists gradually pushed inland, bringing along large numbers of African slaves. (Slavery was not abolished until 1888.)

The King of Portugal, fleeing before Napoleon's army, moved the seat of government to Brazil in 1808. Brazil thereupon became a kingdom under Dom Joao VI. After his return to Portugal, his son Pedro proclaimed the independence of Brazil, Sept. 7, 1822, and was crowned emperor. The second emperor, Dom Pedro II, was deposed in 1889, and a republic proclaimed, called the United States of Brazil. In 1967 the country was renamed the Federative Republic of Brazil.

A military junta took control in 1930; dictatorial power was assumed by Getulio Vargas, until finally forced out by the military in 1945. A democratic regime prevailed 1945-64, during which time the capital was moved from Rio de Janeiro to Brasília. In 1964, Pres. Joao Belchoir Marques Goulart instituted economic policies that aggravated Brazil's inflation; he was overthrown by an army revolt. The next 5 presidents were all military leaders. Censorship was imposed, and much of the opposition was suppressed amid charges of torture.

Since 1930, successive governments have pursued industrial and agricultural growth and interior area development. Exploiting vast natural resources and a huge labor force, Brazil became the leading industrial power of Latin America by the 1970s, while agricultural output soared. By the 1990s, Brazil had one of the world's largest economies; income was poorly distributed, however, and more than one out of four Brazilians continued to survive on less than $1 a day. Despite protective environmental legislation, development has destroyed much of the Amazon ecosystem. Brazil hosted delegates from 178 countries at the Earth Summit, June 3-14, 1992.

Democratic presidential elections were held in 1985 as the nation returned to civilian rule. Fernando Collor de Mello was elected president in Dec. 1989. In Sept. 1992, Collor was impeached for corruption. He resigned on Dec. 29 as his trial was beginning, and Itamar Franco, who had been acting president, was sworn in as president. In elections held on Oct. 3, 1994, Fernando Henrique Cardoso was elected president. Reelected Oct. 4, 1998, he guided Brazil through a series of financial crises.

A new civil code guaranteeing legal equality for women was enacted Aug. 15, 2001.

Brunei

State of Brunei Darussalam

People: Population: 343,653. **Age distrib.** (%): <15: 30.8; 65+: 2.7. **Pop. density:** 169 per sq. mi. **Urban:** 72%. **Ethnic groups:** Malay 64%, Chinese 20%. **Principal languages:** Malay (official), English, Chinese. **Chief religions:** Muslim (official) 63%, Buddhist 14%, Christian 8%.

Geography: Area: 2,000 sq. mi. **Location:** In SE Asia, on the N coast of the island of Borneo; it is surrounded on its landward side by the Malaysian state of Sarawak. **Capital:** Bandar Seri Begawan, (1999 met. area est.) 85,000.

Government: Type: Independent sultanate. **Head of state and gov.:** Sultan Sir Muda Hassanal Bolkiah Mu'izzadin Waddaulah; b July 15, 1946; in office: Jan. 1, 1984. **Local divisions:** 4 districts. **Defense budget:** $281 mil. **Active troops:** 5,000.

Economy: Industries: Oil & gas (over half of GDP is derived from petroleum sector). **Chief crops:** Rice, bananas, cassava. **Minerals:** Oil, gas. **Crude oil reserves** (2000): 1.35 bil bbls. **Arable land:** 1%. **Livestock** (2000): cattle: 2,200; chickens: 5.5 mil; goats: 3,500; pigs: 5,500. **Electricity prod.:** 2.445 bil kWh. **Labor force:** 48% govt.; 42% oil, gas production, etc.

Finance: Monetary unit: Dollar (Sept. 2001: 1.75 = $1 U.S.). **GDP** (1999 est.): $5.6 bil. **Per capita GDP:** $17,400. **Imports** (1998 est.): $1.38 bil; partners: Singapore 32%, UK 17%. **Exports** (1998 est.): $2.04 bil; partners: UK 14%, Japan 51%. **Tourism** (1998): $37 mil.

Transport: Railroad: Length: 12 mi. **Motor vehicles:** 146,000 pass. cars, 17,780 comm. vehicles. **Civil aviation:** 1.8 bil pass.-mi.; 1 airport.

Communications: TV sets: 308 per 1,000 pop. **Radios:** 417 per 1,000 pop. **Telephones** (1999): 79,100 main lines. **Daily newspaper circ.:** 70 per 1,000 pop.

Health: Life expectancy: 71.45 male; 76.31 female. **Births** (per 1,000 pop): 20.45. **Deaths** (per 1,000 pop.): 3.38. **Natural inc.:** 1.7%. **Infant mortality** (per 1,000 live births): 14.40.

Education: Free, compulsory: ages 5-17. **Literacy:** 88%.

Major Intl. Organizations: UN and some of its specialized agencies, APEC, ASEAN, the Commonwealth.

Embassy: 2600 Virginia Ave. NW, 20037; 342-0159.

Website: http://www.brunet.bn

The Sultanate of Brunei was a powerful state in the early 16th century, with authority over all of the island of Borneo as well as parts of the Sulu Islands and the Philippines. In 1888, a treaty placed the state under the protection of Great Britain.

Brunei became a fully sovereign and independent state on Jan. 1, 1984. Much of the country's oil wealth has been squandered in recent years by members of the royal family.

Bulgaria

Republic of Bulgaria

People: Population: 7,707,495. **Age distrib.** (%): <15: 15.1; 65+: 16.7. **Pop. density:** 181 per sq. mi. **Urban:** 69%. **Ethnic groups:** Bulgarian 85%, Turk 9%. **Principal languages:** Bulgarian (official). **Chief religions:** Bulgarian Orthodox 85%, Muslim 13%.

Geography: Area: 42,600 sq. mi. **Location:** SE Europe, in E Balkan Peninsula on Black Sea. **Neighbors:** Romania on N; Yugoslavia, Macedonia on W; Greece, Turkey on S. **Topography:** The Stara Planina (Balkan) Mts. stretch E-W across the center of the country, with the Danubian plain on N, the Rhodope Mts. on SW, and Thracian Plain on SE. **Capital:** Sofia, (1999 est.) 1,192,000.

Government: Type: Republic. **Head of state:** Pres. Petar Stoyanov; b May 25, 1952; in office: Jan. 19, 1997. **Head of gov.:** Prime Min. Simeon Sakskoburggotski (Simeon II); b June 16, 1937; in office: July 24, 2001. **Local divisions:** 9 provinces. **Defense budget:** $389 mil (2001). **Active troops:** 79,760.

Economy: Industries: Chemicals, machinery, metals, textiles, food processing. **Chief crops:** Grain, fruit, oilseed, vegetables, tobacco. **Minerals:** Bauxite, copper, zinc, lead, coal. **Crude oil reserves** (2000): 15 mil bbls. **Arable land:** 37%. **Livestock** (2000): cattle: 682,000; chickens: 13.92 mil; goats: 1.05 mil; pigs: 1.51 mil; sheep: 2.55 mil.**Fish catch** (1999): 16,674 metric tons. **Electricity prod.:** 36.217 bil kWh. **Labor force:** 26% agric., 31% ind., 43% services.

Finance: Monetary unit: Lev (Sept. 2001: 2.15 = $1 U.S.). **GDP** (1999 est.): $34.9 bil. **Per capita GDP:** $4,300. **Imports** (1999 est.): $5.3 bil; partners: Russia 20%, Germany 14%. **Exports** (1999 est.): $3.8 bil; partners: Italy 13%, Germany 10%. **Tourism:** $930 mil. **Budget** (1998 est.): $3.8 bil. **Intl. reserves less gold** (June 2000): $3.01 bil. **Gold:** 1.03 mil oz t. **Consumer prices** (change in 1999): 2.6%.

Transport: Railroad: Length: 4,043 mi. **Motor vehicles:** 1.65 mil pass. cars, 264,196 comm. vehicles. **Civil aviation:** 1.1 bil pass.-mi.; 3 airports. **Chief ports:** Burgas, Varna.

Communications: TV sets: 359 per 1,000 pop. **Telephones:** 2,881,800 main lines. **Daily newspaper circ.:** 141 per 1,000 pop.

Health: Life expectancy: 67.72 male; 74.89 female. **Births** (per 1,000 pop): 8.06. **Deaths** (per 1,000 pop.): 14.53. **Natural inc.:** -.65%. **Infant mortality** (per 1,000 live births): 14.65.

Education: Free, compulsory: ages 7-16. **Literacy:** 98%.

Major Intl. Organizations: UN (FAO, IBRD, ILO, IMF, IMO, WHO, WTrO), OSCE.

Embassy: 1621 22d St. NW 20008; 387-7969.

Bulgaria was settled by Slavs in the 6th century. Turkic Bulgars arrived in the 7th century, merged with the Slavs, became Christians by the 9th century, and set up powerful empires in the 10th and 12th centuries. The Ottomans prevailed in 1396 and remained for 500 years.

A revolt in 1876 led to an independent kingdom in 1908. Bulgaria expanded after the first Balkan War but lost its Aegean coastline in World War I, when it sided with Germany. Bulgaria joined the Axis in World War II but withdrew in 1944. Communists took power with Soviet aid; the monarchy was abolished Sept. 8, 1946.

On Nov. 10, 1989, Communist Party leader and head of state Todor Zhivkov, who had held power for 35 years, resigned. Zhivkov was imprisoned, Jan. 1990, and convicted, Sept. 1992, of corruption and abuse of power. In Jan. 1990, Parliament voted to revoke the constitutionally guaranteed dominant role of the Communist Party. A new constitution took effect July 13, 1991. An economic austerity program was launched in May 1996. Former Prime Min. Andrei Lukanov, a longtime Communist leader, was assassinated Oct. 2 in Sofia. Petar Stoyanov won a presidential runoff election Nov. 3.

Bulgaria's deteriorating economy provoked nationwide strikes and demonstrations in Jan. 1997. The Union of Democratic Forces, an anti-Communist group, won national elections on Apr. 19, 1997. The UDF lost the elections of June 17, 2001, to a party headed by the former king, Simeon II.

Burkina Faso

People: Population: 12,272,289. **Age distrib.** (%): <15: 47.5; 65+ 2.9. **Pop. density:** 116 per sq. mi. **Urban:** 18%. **Ethnic groups:** Mossi (approx. 24%), Gurunsi, Senufo, Lobi, Bobo, Mande, Fulani. **Principal languages:** French (official), Sudanic tribal languages. **Chief religions:** Muslim 50%, indigenous beliefs 40%, Christian (mostly Roman Catholic) 10%.

Geography: Area: 105,600 sq. mi. **Location:** In W Africa, S of the Sahara. **Neighbors:** Mali on NW; Niger on NE; Benin, Togo, Ghana, Côte on S. **Topography:** Landlocked Burkina Faso is in the savanna region of W Africa. The N is arid, hot, and thinly populated. **Capital:** Ouagadougou, (urban agg.) 1,130,000.

Government: Type: Republic. **Head of state:** Pres. Blaise Compaoré; b 1951; in office: Oct. 15, 1987. **Head of gov.:** Prime Min. Paramanga Ernest Yonli; b 1956; in office: Nov. 7, 2000. **Local divisions:** 45 provinces. **Defense budget:** $69 mil. **Active troops:** 10,000.

Economy: Industries: Agricultural processing, beverages, soap, textiles. **Chief crops:** Millet, sorghum, rice, peanuts, cotton. **Minerals:** Manganese, limestone, marble. **Arable land:** 13%. **Livestock** (2000): cattle: 4.7 mil; chickens: 21.77 mil; goats: 8.4 mil; pigs: 610,287; sheep: 6.58 mil. **Electricity prod.:** 285 mil kWh.

Finance: Monetary unit: CFA Franc (Sept. 2001: 724.11= $1 U.S.). **GDP** (1999 est.): $12.4 bil. **Per capita GDP:** $1,100. **Imports** (1998 est.): $572 mil; partners: , Côte d'Ivoire ., France, Senegal. **Exports** (1998 est.): $311 mil; partners: Côte d' Ivoire, Taiwan, France. **Tourism** (1998): $42 mil. **Budget** (1995 est.): $492 mil. **Intl. reserves less gold** (Apr. 2000): $255.6 mil. **Consumer prices** (change in 1999): –1.1%.

Transport: Railroad: Length: 386 mi. **Motor vehicles:** 35,460 pass. cars, 19,473 comm. vehicles. **Civil aviation:** 154.2 mil pass.-mi.; 2 airports.

Communications: TV sets: 4.4 per 1,000 pop. **Radios:** 48.3 per 1,000 pop. **Telephones:** 53,800 main lines.

Health: Life expectancy: 45.86 male; 46.98 female. **Births** (per 1,000 pop): 44.79. **Deaths** (per 1,000 pop.): 17.05. **Natural inc.:** 2.77%. **Infant mortality** (per 1,000 live births): 106.92.

Education: Free, compulsory: ages 7-14. **Literacy:** 19%.

Major Intl. Organizations: UN and many of its specialized agencies, OAU.

Embassy: 2340 Massachusetts Ave. NW 20008; 332-5577.

The Mossi tribe entered the area in the 11th to 13th centuries. Their kingdoms ruled until they were defeated by the Mali and Songhai empires.

French control came by 1896, but Upper Volta (renamed Burkina Faso on Aug. 4, 1984) was not established as a separate territory until 1947. Full independence came Aug. 5, 1960, and a pro-French government was elected. The military seized power in 1980. A 1987 coup established the current regime, which instituted a multiparty democracy in the early 1990s.

Several hundred thousand farm workers migrate each year to Côte d'Ivoire and Ghana. Burkina Faso is heavily dependent on foreign aid.

Burma

(See Myanmar)

Burundi

Republic of Burundi

People: Population: 6,223,897. **Age distrib.** (%): <15: 46.8; 65+: 2.8. **Pop. density:** 629 per sq. mi. **Urban:** 9%. **Ethnic groups:** Hutu (Bantu) 85%, Tutsi 14%, Twa (Pygmy) 1%. **Principal languages:** Kirundi, French (both official), Swahili. **Chief religions:** Roman Catholic 62%, indigenous beliefs 32%, Protestant 5%.

Geography: Area: 9,900 sq. mi. **Location:** In central Africa. **Neighbors:** Rwanda on N, Dem. Rep. of the Congo (formerly Zaire) on W, Tanzania on E and S. **Topography:** Much of the country is grassy highland, with mountains reaching 8,900 ft. The southernmost source of the White Nile is located in Burundi. Lake Tanganyika is the second deepest lake in the world. **Capital:** Bujumbura, (1999 est.) 321,000.

Government: Type: In transition. **Head of state and gov.:** Pres. Pierre Buyoya; b Nov. 14, 1949; in office: July 25, 1996. **Local divisions:** 15 provinces. **Defense**: $62 mil. **Active troops:** 45,500.

Economy: Industries: Light consumer goods, food processing. **Chief crops:** Coffee, cotton, tea. **Minerals:** Nickel, uranium. **Arable land:** 44%. **Livestock** (2000): cattle: 320,000; chickens: 4.0 mil; goats: 550,000; pigs: 50,000; sheep: 120,000.**Fish catch:** (1999): 20,306 metric tons. **Electricity prod.:** 141 mil kWh. **Labor force:** 93% agric.

Finance: Monetary unit: Franc (Sept. 2001: 844.8 = $1 U.S.). **GDP** (1999 est.): $4.2 bil. **Per capita GDP:** $730. **Imports** (1999 est.): $108 mil; partners: Benelux, France, Zambia. **Exports** (1999 est.): $56 mil; partners: U.K., Germany, Benelux. **Tourism** (1998): $1 mil. **Budget** (1998 est.): $165 mil. **Intl. reserves less gold** (June 2000): $42.95 mil. **Gold:** 17,000 oz t. **Consumer prices** (change in 1999): 3.4%.

Transport: Motor vehicles: 8,200 pass. cars, 11,800 comm. vehicles. **Civil aviation:** 5.2 mil pass.-mi.; 1 airport. **Chief port:** Bujumbura.

Communications: TV sets: 7 per 1,000 pop. **Radios:** 50 per 1,000 pop. **Telephones** (1999): 19,000 main lines.

Health: Life expectancy: 45.15 male; 46.99 female. **Births** (per 1,000 pop): 40.13. **Deaths** (per 1,000 pop.): 16.36. **Natural inc.:** 2.38%. **Infant mortality** (per 1,000 live births): 70.74.

Education: Free, compulsory: ages 7-13. **Literacy:** 35%.

Major Intl. Organizations: UN (FAO, IBRD, ILO, IMF, WHO, WTrO), OAU.

Embassy: 2233 Wisconsin Ave. NW 20007; 342-2574.

The pygmy Twa were the first inhabitants, followed by Bantu Hutus, who were conquered in the 16th century by the Tutsi (Watusi), probably from Ethiopia. Under German control in 1899, the area fell to Belgium in 1916, which exercised successively a League of Nations mandate and UN trusteeship over Ruanda-Urundi (now the two countries of Rwanda and Burundi). Burundi became independent July 1, 1962.

An unsuccessful Hutu rebellion in 1972-73 left 10,000 Tutsi and 150,000 Hutu dead. Over 100,000 Hutu fled to Tanzania and Zaire (now Congo). In the 1980s, Burundi's Tutsi-dominated regime pledged itself to ethnic reconciliation and democratic reform. In the nation's first democratic presidential election, in June 1993, a Hutu, Melchior Ndadaye, was elected. He was killed in an attempted coup, Oct. 21, 1993. At least 150,000 Burundians died as a result of ethnic conflict during the next three years. Pres. Cyprien Ntaryamira, elected Jan. 1994, was killed with the president of Rwanda in a mysterious plane crash, Apr. 6. The incident sparked massive carnage in Rwanda; violence in Burundi, initially far more limited, intensified in 1995. Ethnic strife continued after a military coup, July 25, 1996. Former South African Pres. Nelson Mandela mediated peace talks from Dec. 1999; most warring groups signed a draft peace treaty in Arusha, Tanzania, Aug. 28, 2000, witnessed by Mandela and U.S. Pres. Bill Clinton. Coup attempts were suppressed Apr. 18 and July 23, 2001.

Cambodia

Kingdom of Cambodia

People: Population: 12,491,501. **Age distrib.** (%): <15: 41.3; 65+: 3.5. **Pop. density:** 184 per sq. mi. **Urban:** 16%. **Ethnic groups:** Khmer 90%, Vietnamese 5%, Chinese 1%. **Principal languages:** Khmer (official), French. **Chief religion:** Theravada Buddhism 95%.

Geography: Area: 68,100 sq. mi. **Location:** SE Asia, on Indochina Peninsula. **Neighbors:** Thailand on W and N, Laos on NE, Vietnam on E. **Topography:** The central area, formed by the Mekong R. basin and Tonle Sap lake, is level. Hills and mountains are in SE, a long escarpment separates the country from Thailand on NW. 76% of the area is forested. **Capital:** Phnom Penh, (1999 est.) 938,000.

Government: Type: Constitutional monarchy. **Head of state:** King Norodom Sihanouk; b Oct. 31, 1922; in office: Sept. 24, 1993. **Head of gov.:** Prime Min. Hun Sen; b Apr. 4, 1952; in office: Nov. 30, 1998. **Local divisions:** 20 provinces and 3 municipalities. **Defense budget:** $120 mil. **Active troops:** 140,000.

Economy: Industries: Rice milling, wood & wood products, fishing. **Chief crops:** Rice, corn, rubber, vegetables. **Minerals:** Gemstones, phosphates, manganese. **Other resources:** Timber. **Arable land:** 13%. **Livestock** (2000): cattle: 3.0 mil; chickens: 13.2 mil; pigs: 2.6 mil. **Fish catch** (1999): 114,600 metric tons. **Electricity prod.:** 147 mil kWh. **Labor force:** 80% agric.

Finance: Monetary unit: Riel (Sept. 2001: 3,835.00 = $1 U.S.). **GDP** (1999 est.): $8.2 bil. **Per capita GDP:** $710. **Imports** (1999 est.): $1.2 bil; partners: SingaporeVietnam, Japan. **Exports** (1999 est.): $821 mil; partners: U.S., Singapore, Japan. **Tourism:** $190 mil. **Budget** (1995 est.): $496 mil. **Intl. reserves less gold** (June 2000): $490.88 mil. **Consumer prices** (change in 1999): 4.0%.

Transport: Railroad: Length: 380 mi. **Motor vehicles:** 15,000 pass. cars, 15,000 comm. vehicles. **Civil aviation:** 8 airports. **Chief port:** Kampong Saom (Sihanoukville).

Communications: TV sets: 8 per 1,000 pop. **Radios:** 124 per 1,000 pop. **Telephones:** 33,900 main lines.

Health: Life expectancy: 54.62 male; 59.12 female. **Births** (per 1,000 pop): 33.16. **Deaths** (per 1,000 pop.): 10.65. **Natural inc.:** 2.25%. **Infant mortality** (per 1,000 live births): 65.41.

Education: Compulsory: ages 6-12. **Literacy** (1993): 65%.

Major Intl. Organizations: UN (FAO, IBRD, ILO, IMF, IMO, WHO), ASEAN.

Embassy: 4500 16th St. NW 20011; 726-7742.

Website: http://www.cambodia.org

Early kingdoms dating from that of Funan in the 1st century AD culminated in the great Khmer empire that flourished from the 9th century to the 13th, encompassing present-day Thailand, Cambodia, Laos, and southern Vietnam. The peripheral areas were lost to invading Siamese and Vietnamese, and France established a protectorate in 1863. Independence came in 1953.

Prince Norodom Sihanouk, king 1941-1955 and head of state from 1960, tried to maintain neutrality. Relations with the U.S. were broken in 1965, after South Vietnam planes attacked Vietcong forces within Cambodia. Relations were restored in 1969, after Sihanouk charged Viet Communists with arming Cambodian insurgents.

In 1970, pro-U.S. Prem. Lon Nol seized power, demanding removal of 40,000 North Viet troops; the monarchy was abolished. Sihanouk formed a government-in-exile in Beijing, and open war began between the government and Communist Khmer Rouge guerrillas. The U.S. provided heavy military and economic aid.

Khmer Rouge forces captured Phnom Penh Apr. 17, 1975. The new government evacuated all cities and towns, and shuffled the rural population, sending virtually the entire population to clear jungle, forest, and scrub. Over one million people were killed in executions and enforced hardships.

Severe border fighting broke out with Vietnam in 1978 and developed into a full-fledged Vietnamese invasion. Formation of a Vietnamese-backed government was announced, Jan. 8, 1979, one day after the Vietnamese capture of Phnom Penh. Thousands of refugees flowed into Thailand, and widespread starvation was reported.

On Jan. 10, 1983, Vietnam launched an offensive against rebel forces in the west. They overran a refugee camp, Jan. 31, driving 30,000 residents into Thailand. In March, Vietnam launched a major offensive against camps on the Cambodian-Thailand border, engaged Khmer Rouge guerrillas, and crossed the border, instigating clashes with Thai troops. Vietnam withdrew nearly all its troops by Sept. 1989.

Following UN-sponsored elections in Cambodia that ended May 28, 1993, the 2 leading parties agreed to share power in an interim government until a new constitution was adopted. On Sept. 21, a constitution reestablishing a monarchy was adopted by the National Assembly. It took effect Sept. 24, with Sihanouk as king. The Khmer Rouge, which had boycotted the elections, opposed the new government, and armed violence continued in the mid-1990s. Ieng Sary, a Khmer Rouge leader, broke with the guerrillas, formed a rival group, and announced his support for the monarchy in Aug. 1996, as Khmer Rouge strength rapidly diminished.

Co-Prime Min. Hun Sen staged a coup July 5, 1997, ousting his rival, Prince Norodom Ranariddh. Pol Pot, the Khmer Rouge leader who held power during the late 1970s, was denounced by his former comrades at a show trial, July 25, and sentenced to house arrest; he died Apr. 15, 1998. Hun Sen's party won parliamentary elections on July 26. Cambodia was formally admitted to ASEAN on Apr. 30, 1999.

Cameroon

Republic of Cameroon

People: Population: 15,803,220. **Age distrib.** (%): <15: 42.4; 65+: 3.4. **Pop. density:** 87 per sq. mi. **Urban:** 48%. **Ethnic groups:** Cameroon Highlander 31%, Equatorial Bantu 19%, Kirdi 11%, Fulani 10%, NW Bantu 8%. **Principal languages:** English, French (both official), 24 African groups. **Chief religions:** Indigenous beliefs 51%, Christian 33%, Muslim 16%.

Geography: Area: 181,000 sq. mi. **Location:** Between W and central Africa. **Neighbors:** Nigeria on NW; Chad, Central African Republic on E; Congo, Gabon, Equatorial Guinea on S. **Topography:** A low coastal plain with rain forests is in S; plateaus in center lead to forested mountains in W, including Mt. Cameroon, 13,350 ft.; grasslands in N lead to marshes around Lake Chad. **Capital:** Yaoundé. **Cities** (urban agg.): Douala, 1,670,000; Yaoundé, 1,444,000.

Government: Type: Republic. **Head of state:** Pres. Paul Biya; b Feb. 13, 1933; in office: Nov. 6, 1982. **Head of gov.:** Prime Min. Peter Mafani Musonge; b Dec. 3, 1942; in office: Sept. 19, 1996. **Local divisions:** 10 provinces. **Defense budget:** $155 mil. **Active troops:** 22,100.

Economy: Industries: Oil production and refining, food processing, light consumer goods. **Chief crops:** Cocoa, coffee, cotton. **Crude oil reserves** (2000): 400 mil bbls. **Minerals:** Oil, bauxite, iron ore. **Other resources:** Timber. **Arable land:** 13%. **Livestock**(2000): cattle: 5.9 mil; chickens: 30.0 mil; goats: 3.85 mil; pigs: 1.43 mil; sheep: 3.88 mil.**Fish catch** (1999): 89,055 metric tons. **Electricity prod.:** 3.470 bil kWh. **Labor force:** 70% agric., 13% ind. and commerce

Finance: Monetary unit: CFA Franc (Sept. 2001: 724.11 = $1 U.S.). **GDP** (1999 est.): $31.5 bil. **Per capita GDP:** $2,000. **Imports** (1999 est.): $1.5 bil; partners: France 25%, Nigeria 8%, US 8%. **Exports** (1999 est.): $1.5 bil; partners: Italy 25%, Spain 20%, France, 16%. **Tourism** (1998): $40 mil. **Budget** (FY 1996-97 est.): $2.23 bil. **Intl. reserves less gold** (Apr. 2000): $2.35 mil. **Consumer prices** (change in 1998): 0.1%.

Transport: Railroad: Length: 625 mi. **Motor vehicles:** 92,200 pass. cars, 60,800 comm. vehicles. **Civil aviation:** 339.9 mil pass.-mi.; 5 airports. **Chief ports:** Douala, Kribi.

Communications: TV sets: 72 per 1,000 pop. **Radios:** 325 per 1,000 pop. **Telephones** (1999): 94,600 main lines.

Health: Life expectancy: 53.76 male; 55.44 female. **Births** (per 1,000 pop): 36.12. **Deaths** (per 1,000 pop.): 11.99. **Natural inc.:** 2.41%. **Infant mortality** (per 1,000 live births): 69.83.

Education: Free, compulsory: ages 6-12. **Literacy:** 63%.

Major Intl. Organizations: UN (FAO, IBRD, ILO, IMF, IMO, WHO, WTrO), the Commonwealth, OAU.

Embassy: 2349 Massachusetts Ave. NW 20008; 265-8790.

Website: http://www.ecameroun.com

Portuguese sailors were the first Europeans to reach Cameroon, in the 15th century. The European and American slave trade was very active in the area. German control lasted from 1884 to 1916, when France and Britain divided the territory, later receiving League of Nations mandates and UN trusteeships. French Cameroon became independent Jan. 1, 1960; one part of British Cameroon joined Nigeria in 1961, the other part joined Cameroon. Stability has allowed for development of roads, railways, agriculture, and petroleum production.

Pres. Paul Biya retained his office in Oct. 1992 elections, but the results were widely disputed. A new constitution won legislative approval in Dec. 1995. Fraud charges accompanied legislative elections, May 17, 1997, which Biya's party won.

Canada

People: Population: 31,592,805. **Age distrib.** (%): <15: 18.9; 65+: 12.8. **Pop. density:** 9 per sq. mi. **Urban:** 77%. **Ethnic groups:** British Isles 40%, French 27%, other European 20%, Amerindian 1.5%, other (mostly Asian) 11.5%. **Principal languages:** English, French (both official). **Chief religions:** Roman Catholic 45%, United Church 12%, Anglican 8%.

Geography: Area: 3,556,000 sq. mi., the largest country in land size in the western hemisphere. **Topography:** Canada stretches 3,426 miles from east to west and extends southward from the North Pole to the U.S. border. Its seacoast includes 36,356 miles of mainland and 115,133 miles of islands, including the Arctic islands almost from Greenland to near the Alaskan border. **Climate:** While generally temperate, varies from freezing winter cold to blistering summer heat. **Capital:** Ottawa. **Cities (urban agg.)**: Toronto, 4.7 mil; Montreal, 3.4 mil; Vancouver, 2.0 mil; Ottawa-Hull, 1.1 mil; Edmonton, 908,000; Calgary, 899,000.

Government: Type: Confederation with parliamentary democracy. **Head of state:** Queen Elizabeth II, represented by Gov.-Gen. Adrienne Clarkson; b Feb. 10, 1939; in office: Oct. 7, 1999. **Head of gov.:** Prime Min. Jean Chrétien; b Jan. 11, 1934; in office: Nov. 4, 1993. **Local divisions:** 10 provinces, 3 territories. **Defense budget:** $7.6 bil. **Active troops:** 59,100.

Economy: Industries: Mining, wood and food prods., transport equip., chemicals, oil, gas. **Chief crops:** Grains, oilseed, tobacco, fruit, vegetables. **Minerals:** Nickel, zinc, copper, gold, lead, molybdenum, potash, silver. **Crude oil reserves** (2000): 4.93 bil bbls. **Arable land:** 5%. **Livestock** (2000): cattle: 12.79 mil; chickens: 158.0 mil; goats: 30,000; pigs: 12.24 mil; sheep: 694,800 **Fish catch** (1999): 1.03 mil metric tons. **Electricity prod.:** 567.193 bil kWh. **Labor force:** 75% services, 16% manuf., 3% agric.

Finance: Monetary unit: Dollar (Sept. 2001: 1.57 = $1 U.S.). **GDP** (1999 est.): $722.3 bil. **Per capita GDP:** $23,300. **Imports** (1999 est.): $259.3 bil; partners: U.S. 77%. **Exports** (1999 est.): $277 bil; partners: U.S. 84%. **Tourism:** $10.03 bil. **Budget** (1998): $112.6 bil. **Intl. reserves less gold** (June 2000): $29.80 bil. **Gold:** 1.24 mil oz t. **Consumer prices** (change in 1999): 1.7%.

Transport: Railroad: Length: 44,182 mi. **Motor vehicles:** 13.3 mil pass. cars, 3.52 mil comm. vehicles. **Civil aviation:** 38.4 bil pass.-mi.; 269 airports. **Chief ports:** Halifax, Montreal, Quebec, Saint John, Toronto, Vancouver.

Communications: TV sets: 708 per 1,000 pop. **Radios:** 1,078 per 1,000 pop. **Telephones**: 20,802,900 main lines. **Daily newspaper circ.:** 157 per 1,000 pop.

Health: Life expectancy: 76.16 male; 83.13 female. **Births** (per 1,000 pop): 11.21. **Deaths** (per 1,000 pop.): 7.47. **Natural inc.:** .37%. **Infant mortality** (per 1,000 live births): 5.02.

Education: Compulsory primary education. **Literacy** (1994): 97%.

Major Intl. Organizations: UN and all of its specialized agencies, APEC, the Commonwealth, NATO, OAS, OECD, OSCE.

Embassy: 501 Pennsylvania Ave. NW 20001; 682-1740.

Websites: http://www.statcan.ca
http://canada.gc.ca/main_e.html

Provinces/Territories	Area (sq. mi.)	Population (1996 cen.)
Alberta	255,287	2,696,826
British Columbia	365,948	3,724,500
Manitoba	250,947	1,113,898
New Brunswick	28,355	738,133
Newfoundland	156,649	551,792
Nova Scotia	21,425	909,282
Ontario	412,581	10,753,573
Prince Edward Island	2,185	134,557
Quebec	594,860	7,138,795
Saskatchewan	251,866	990,237
Northwest Territories	503,951	39,672
Yukon Territory	186,661	30,766
Nunavut	818,959	24,730

French explorer Jacques Cartier, who reached the Gulf of St. Lawrence in 1534, is generally regarded as Canada's founder. But English seaman John Cabot sighted Newfoundland in 1497, and Vikings are believed to have reached the Atlantic coast centuries before either explorer.

Canadian settlement was pioneered by the French who established Quebec City (1608) and Montreal (1642) and declared New France a colony in 1663.

Britain acquired Acadia (later Nova Scotia) in 1717 and, through military victory over French forces in Canada, captured Quebec (1759) and obtained control of the rest of New France in 1763. The French, through the Quebec Act of 1774, retained the rights to their own language, religion, and civil law. The British presence in Canada increased during the American Revolution when many colonials, proudly calling themselves United Empire Loyalists, moved north to Canada. Fur traders and explorers led Canadians westward across the continent. Sir Alexander Mackenzie reached the Pacific in 1793 and scrawled on a rock by the ocean, "from Canada by land."

In Upper and Lower Canada (later called Ontario and Quebec) and in the Maritimes, legislative assemblies appeared in the 18th century and reformers called for responsible government. But the War of 1812 intervened. The war, a conflict between Great Britain and the United States fought mainly in Upper Canada, ended in a stalemate in 1814.

In 1837 political agitation for more democratic government culminated in rebellions in Upper and Lower Canada. Britain sent Lord Durham to investigate; in a famous report (1839), he recommended union of the 2 parts into one colony called Canada. The union lasted until Confederation, July 1, 1867, when proclamation of the British North America (BNA) Act (now known as the Constitution Act, 1867) launched the Dominion of Canada, consisting of Ontario, Quebec, and the former colonies of Nova Scotia and New Brunswick.

Since 1840 the Canadian colonies had held the right to internal self-government. The BNA Act, which was the basis for the country's written constitution, established a federal system of government on the model of a British parliament and cabinet structure under the crown. Canada was proclaimed a self-governing Dominion within the British Empire in 1931. With the ratification of the Constitution Act, 1982, Canada severed its last formal legislative link with Britain by obtaining the right to amend its constitution.

The so-called Meech Lake Agreement was signed (subject to provincial ratification) June 3, 1987. The accord would have assured constitutional protection for Quebec's efforts to preserve its French language and culture. Critics charged it did not make any provision for other minority groups and it gave Quebec too much power, which might enable Quebec to override the nation's 1982 Charter of Rights and Freedoms (an integral part of the constitution). The accord died June 22, 1990.

Its failure sparked a separatist revival in Quebec, which culminated in Aug. 1992 in the Charlottetown agreement. This called for changes to the constitution, such as recognition of Quebec as a "distinct society" within the Canadian confederation. It was defeated in a national referendum Oct. 26, 1992.

Canada became the first nation to ratify the North American Free Trade Agreement between Canada, Mexico, and the U.S. June 23, 1993. It went into effect Jan. 1, 1994.

On Feb. 24, 1993, Brian Mulroney resigned as prime minister after more than 8 years in office; he was succeeded by Kim Campbell. In elections Oct. 25, 1993, the ruling Conservatives were defeated in a landslide that left them only 2 of the 295 seats in the House of Commons. Jean Chrétien became prime minister. In a Quebec referendum held Oct. 30, 1995, proponents of secession lost by a razor-thin margin. The elections of June 2, 1997, left the Liberals with a slim majority.

On Jan. 7, 1998, the government apologized to native peoples for 150 years of mistreatment and pledged to set up a "healing fund." Canada's highest court ruled, Aug. 20, that Quebec cannot secede unilaterally, even if a majority of the province approves. Nunavut ("Our Land"), carved from Northwest Territories as a homeland for the Inuit, was established Apr. 1, 1999.

Victory by the Liberals in national elections Nov. 27, 2000, made Chrétien the 1st Canadian prime minister in over 50 years to head a 3d successive majority government.

Prime Ministers of Canada

Canada is a constitutional monarchy with a parliamentary system of government. It is also a federal state. Canada's official head of state, Queen Elizabeth II, is represented by a resident Governor-General. However, in practice the nation is governed by the Prime Minister, leader of the party that commands the support of a majority of the House of Commons, dominant chamber of Canada's bicameral Parliament.

Name	Party	Term
Sir John A. MacDonald	Conservative	1867-1873
Alexander Mackenzie	Liberal	1873-1878
Sir John A. MacDonald	Conservative	1878-1891
Sir John J. C. Abbott	Conservative	1891-1892
Sir John S. D. Thompson	Conservative	1892-1894
Sir Mackenzie Bowell	Conservative	1894-1896
Sir Charles Tupper	Conservative	1896[1]
Sir Wilfrid Laurier	Liberal	1896-1911
Sir Robert Laird Borden	Cons./Union.[2]	1911-1920
Arthur Meighen	Unionist	1920-1921
W. L. Mackenzie King	Liberal	1921-1926
Arthur Meighen	Conservative	1926[3]
W. L. Mackenzie King	Liberal	1926-1930
Richard Bedford Bennett	Conservative	1930-1935
W. L. Mackenzie King	Liberal	1935-1948
Louis St. Laurent	Liberal	1948-1957
John G. Diefenbaker	Prog. Cons.	1957-1963
Lester Bowles Pearson	Liberal	1963-1968
Pierre Elliott Trudeau	Liberal	1968-1979
Joe Clark	Prog. Cons.	1979-1980
Pierre Elliott Trudeau	Liberal	1980-1984
John Napier Turner	Liberal	1984[4]
Brian Mulroney	Prog. Cons.	1984-1993
Kim Campbell	Prog. Cons.	1993[5]
Jean Chrétien	Liberal	1993-

(1) May-July. (2) Conservative 1911-1917, Unionist 1917-1920. (3) June-Sept. (4) June-Sept. (5) June-Oct.

Cape Verde

Republic of Cape Verde

People: Population: 405,163. **Age distrib.** (%): <15: 42.8; 65+: 6.5. **Pop. density:** 261 per sq. mi. **Urban:** 61%. **Ethnic groups:** Creole (mulatto) 71%, African 28%, **Principal languages:** Portuguese (official), Crioulo. **Chief religion:** Roman Catholic.

Geography: Area: 1,600 sq. mi. **Location:** In Atlantic O., off W tip of Africa. **Neighbors:** Nearest are Mauritania, Senegal to E. **Topography:** Cape Verde Islands are 15 in number, volcanic in origin (active crater on Fogo). The landscape is eroded and stark, with vegetation mostly in interior valleys. **Capital:** Praia, (1999 est.) 76,000.

Government: Type: Republic. **Head of state:** Pres. Pedro Pires; b Apr. 29, 1934; in office: Mar. 22, 2001. **Head of gov.:** Prime Min. José Maria Neves; b 1959; in office: Feb. 1, 2001. **Local divisions:** 16 districts. **Defense budget:** $8 mil. **Active troops:** 1,150.

Economy: Industries: Food and beverages, fish processing, shoes and garments. **Chief crops:** Bananas, coffee, sweet potatoes, corn, beans. **Minerals:** Salt. **Other resources:** Fish. **Arable land:** 11%. **Livestock** (2000): cattle: 21,500; chickens: 480,000; goats: 110,000; pigs: 640,000; sheep: 8,450. **Fish catch:** (1999): 10,039 metric tons. **Electricity prod.:** 40 mil kWh.

Finance: Monetary unit: Escudo (Sept. 2001 119.75 = $1 U.S.). **GDP** (1999 est.): $618 mil. **Per capita GDP:** $1,500. **Imports** (1999 est.): $225 mil; partners: Portugal, Netherlands, France. **Exports** (1999 est.): $38 mil; partners: Portugal, Germany, Spain. **Tourism** (1998): $20 mil. **Budget** (1996): $228 mil. **Intl. reserves less gold** (May 2000): $36.66 mil. **Consumer prices** (change in 1999): 4.4%.

Transport: Motor vehicles: 11,000 pass. cars, 7,000 comm. vehicles. **Civil aviation:** 166.5 mil pass.-mi.; 9 airports. **Chief ports:** Mindelo, Praia.

Communications: TV sets: 2.6 per 1,000 pop. **Radios:** 146 per 1,000 pop. **Telephones:** 52,900 main lines.

Health: Life expectancy: 65.93 male; 72.6 female. **Births** (per 1,000 pop): 28.71. **Deaths** (per 1,000 pop.): 7.19. **Natural inc.:** 2.15%. **Infant mortality** (per 1,000 live births): 53.22.

Education: Compulsory: ages 7-11. **Literacy:** 72%.

Major Intl. Organizations: UN (FAO, IBRD, ILO, IMF, IMO, WHO), OAU.

Embassy: 3415 Massachusetts Ave. NW 20007; 965-6820.

The uninhabited Cape Verdes were discovered by the Portuguese in 1456 or 1460. The first Portuguese colonists landed in 1462; African slaves were brought soon after, and most Cape Verdeans descend from both groups. Cape Verde independence came July 5, 1975. Antonio Mascarenhas Monteiro won the nation's first free presidential election Feb. 17, 1991; he was reelected without opposition five years later. Pedro Pires won a presidential runoff election Feb. 25, 2001.

Central African Republic

People: Population: 3,576,884. **Age distrib.** (%): <15: 43.2; 65+: 3.8. **Pop. density:** 15 per sq. mi. **Urban:** 41%. **Ethnic groups:** Baya 34%, Banda 27%, Mandjia 21%, Sara 10%. **Principal languages:** French (official), Sangho (national), Arabic, Hunsa, Swahili. **Chief religions:** Protestant 25%, Roman Catholic 25%, indigenous beliefs 24%, Muslim 15%.

Geography: Area: 240,200 sq. mi. **Location:** In central Africa. **Neighbors:** Chad on N, Cameroon on W, Congo-Brazzaville and Congo-Kinshasa (formerly Zaire) on S, Sudan on E. **Topography:** Mostly rolling plateau, average altitude 2,000 ft., with rivers draining S to the Congo and N to Lake Chad. Open, well-watered savanna covers most of the area, with an arid area in NE, and tropical rain forest in SW. **Capital:** Bangui, (1999 est.) 622,000.

Government: Type: Republic. **Head of state:** Pres. Ange-Félix Patassé; b Jan. 25, 1937; in office: Oct. 22, 1993. **Head of gov.:** Prime Min. Martin Ziguéle; b Feb. 12, 1957; in office: Apr. 1, 2001. **Local divisions:** 14 prefectures, 2 economic prefectures, 1 commune. **Defense budget:** $44 mil. **Active troops:** 4,150.

Economy: Industries: Textiles, breweries, sawmills, diamond mining. **Chief crops:** Cotton, coffee, corn, tobacco, yams. **Minerals:** Diamonds, uranium. **Other resources:** Timber. **Arable land:** 3%. **Livestock** ((2000): cattle: 2.95 mil; chickens: 4.1 mil; goats: 2.6 mil; pigs: 650,000; sheep: 210,000. **Fish catch:** (1999): 12,860 metric tons. **Electricity prod.:** 102 mil kWh.

Finance: Monetary unit: CFA Franc (Sept. 2001: 724.11 = $1 U.S.). **GDP** (1999 est.): $5.8 bil. **Per capita GDP:** $1,700. **Imports** (1999 est.): $170 mil; partners: France 30%; Cote d'Ivoire 18%. **Exports** (1999 est.): $195 mil; partners: Benelux. 36%, Cote d'Ivoire 5%. **Tourism** (1998): $6 mil. **Budget** (1994 est.): $1.9 bil. **Intl. reserves less gold** (Apr. 2000): $130.30 mil. **Consumer prices** (change in 1999): –1.9%.

Transport: Motor vehicles: 11,000 pass. cars, 9,000 comm. vehicles. **Civil aviation:** 150.5 mil pass.-mi.; 1 airport. **Chief port:** Bangui.

Communications: TV sets: 5 per 1,000 pop. **Radios:** 75 per 1,000 pop. **Telephones:** 9,500 main lines.

Health: Life expectancy: 42.17 male; 45.48 female. **Births** (per 1,000 pop): 37.05. **Deaths** (per 1,000 pop.): 18.53. **Natural inc.:** 1.85%. **Infant mortality** (per 1,000 live births): 105.25.

Education: Compulsory: ages 6-14. **Literacy:** 60%.

Major Intl. Organizations: UN (FAO, IBRD, ILO, IMF, WHO, WTrO), OAU.

Embassy: 1618 22d St. NW 20008; 483-7800.

Various Bantu tribes migrated through the region for centuries before French control was asserted in the late 19th century, when the region was named Ubangi-Shari. Complete independence was attained Aug. 13, 1960.

All political parties were dissolved in 1960, and the country became a center for Chinese political influence in Africa. Relations with China were severed after 1965. Pres. Jean-Bedel Bokassa, who seized power in a 1965 military coup, proclaimed himself constitutional emperor of the renamed Central African Empire Dec. 1976.

Bokassa's rule was characterized by ruthless authoritarianism and human rights violations. He was ousted in a bloodless coup aided by the French government, Sept. 20, 1979. In 1981, Gen. André Kolingba became head of state in another bloodless coup. Multiparty legislative and presidential elections were held in Oct. 1992 but were canceled by the government when Kolingba was losing. New elections, held in Aug. and Sept. 1993, led to the replacement of Kolingba with a civilian government under Pres. Ange-Félix Patassé. France sent in troops to suppress army mutinies in 1996 and 1997. Patassé loyalists won a narrow majority in legislative elections on Nov. 22 and Dec. 13, 1998, and he was reelected to a 2d 6-year term on Sept. 19, 1999. A coup attempt launched May 28, 2001, was suppressed.

Chad

Republic of Chad

People: Population: 8,707,078. **Age distrib.** (%): <15: 47.7; 65+: 2.8. **Pop. density:** 18 per sq. mi. **Urban:** 23%. **Ethnic groups:** Sara 28%, Sudanic Arab 12%, many others. **Principal languages:** French, Arabic (both official), Sara, Sango, more than 100 other languages. **Chief religions:** Muslim 50%, Christian 25%, indigenous beliefs 25%.

Geography: Area: 485,600 sq. mi. **Location:** In central N Africa. **Neighbors:** Libya on N; Niger, Nigeria, Cameroon on W; Central African Republic on S; Sudan on E. **Topography:** Wooded savanna, steppe, and desert in the S; part of the Sahara in the N. Southern rivers flow N to Lake Chad, surrounded by marshland. **Capital:** N'Djamena, (urban agg.) 1,043,000.

Government: Type: Republic. **Head of state:** Pres. Idriss Déby; b 1952; in office: Dec. 4, 1990. **Head of gov.:** Prime Min. Nagoum Yamassoum; in office: Dec. 13, 1999. **Local divisions:** 14 prefectures. **Defense budget:** $48 mil. **Active troops:** 30,350.

Economy: Industries: Cotton textiles, meat packing, beer brewing, soap. **Chief crops:** Cotton, sorghum, millet. **Minerals:** Uranium. **Arable land:** 3%. **Livestock** (2000): cattle: 5.6 mil; chickens: 4.9 mil; goats: 5.05 mil; pigs: 21,000; sheep: 2.5 mil. **Fish catch** (1999): 85,000 metric tons. **Electricity prod.:** 90 mil kWh. **Labor force:** 85% agric.

Finance: Monetary unit: CFA Franc (Sept. 2001: 724.11 = $1 U.S.). **GDP** (1999 est.): $7.6 bil. **Per capita GDP:** $1,000. **Imports** (1999 est.): $359 mil; partners: France 41%, Nigeria 10%. **Exports** (1999 est.): $288 mil; partners: Portugal 30%, Germany 14%. **Tourism** (1998): $10 mil. **Budget** (1998 est.): $218 mil. **Intl. reserves less gold** (Apr. 2000): $104.00 mil. **Consumer prices** (change in 1997): –6.8%.

Transport: Motor vehicles: 9,630 pass. cars, 14,360 comm. vehicles. **Civil aviation:** 153.3 mil pass.-mi.; 1 airport.

Communications: TV sets: 8 per 1,000 pop. **Radios:** 206 per 1,000 pop. **Telephones** (1999): 9,700 main lines.

Health: Life expectancy: 48.86 male; 52.98 female. **Births** (per 1,000 pop): 48.28. **Deaths** (per 1,000 pop.): 15.4. **Natural inc.:** 3.29%. **Infant mortality** (per 1,000 live births): 95.06.

Education: Compulsory: ages 6-14. **Literacy:** 48%.

Major Intl. Organizations: UN (FAO, IBRD, ILO, IMF, WHO, WTrO), OAU.

Embassy: 2002 R St. NW 20009; 462-4009.

Chad was the site of paleolithic and neolithic cultures before the Sahara Desert formed. A succession of kingdoms and Arab slave traders dominated Chad until France took control around 1900. Independence came Aug. 11, 1960.

Northern Muslim rebels have fought animist and Christian southern government and French troops from 1966, despite numerous cease-fires and peace pacts.

Libyan troops entered the country at the request of a pro-Libyan Chad government, Dec. 1980. The troops were withdrawn from Chad in Nov. 1981. Rebel forces, led by Hissène Habré, captured the capital and forced Pres. Goukouni Oueddei to flee the country in June 1982.

In 1983, France sent some 3,000 troops to Chad to assist Pres. Habré in opposing Libyan-backed rebels. France and Libya agreed to a simultaneous withdrawal of troops from Chad in Sept. 1984, but Libyan forces remained in the north until Mar. 1987, when Chad forces drove them from their last major stronghold. In Dec. 1990, Habré was overthrown by a Libyan-supported insurgent group, the Patriotic Salvation Movement.

On Feb. 3, 1994, the World Court dismissed a long-standing territorial claim by Libya to the mineral-rich Aozou Strip, on the Libyan border. Libyan troops reportedly withdrew at the end of May. Following approval of a new constitution in March 1996, Chad's first multiparty presidential election was held in June and July. The U.S. Peace Corps withdrew from Chad in Apr. 1998 because of clashes between rebels and Chad government forces.

Pres. Idriss Déby won reelection May 20, 2001, to another 5-year term.

Chile

Republic of Chile

People: Population: 15,328,467. **Age distrib.** (%): <15: 27.3; 65+: 7.4. **Pop. density:** 53 per sq. mi. **Urban:** 85%. **Ethnic groups:** White and White-Amerindian 95%, Amerindian 3%. **Principal language:** Spanish (official). **Chief religions:** Roman Catholic 89%, Protestant 11%.

Geography: Area: 288,800 sq. mi. **Location:** Occupies western coast of S South America. **Neighbors:** Peru on N, Bolivia on NE, on E. **Topography:** Andes Mts. on E border incl. some of the world's highest peaks; on W is 2,650-mile Pacific coast. Width varies between 100 and 250 miles. In N is Atacama Desert, in center are agricultural regions, in S, forests and grazing lands. **Capital:** Santiago, (urban agg.) 5,538,000.

Government: Type: Republic. **Head of state and gov.:** Pres. Ricardo Lagos Escobar; b Mar. 2, 1938; in office: Mar. 11, 2000. **Local divisions:** 13 regions. **Defense budget:** $2.1 bil. **Active troops:** 87,000.

Economy: Industries: Fish processing, wood products, iron, steel. **Chief crops:** Grain, grapes, fruits, beans, potatoes, sugar beets. **Minerals:** Copper (world's largest producer and exporter), molybdenum, nitrates, iron. **Crude oil reserves** (2000): 150 mil bbls. **Other resources:** Timber. **Arable land:** 5%. **Livestock** (2000): cattle: 4.07 mil; chickens: 70.0 mil; goats: 740,000; pigs: 2.47 mil; sheep: 4.14 mil.. **Fish catch** (1999): 6.08 mil metric tons. **Electricity prod.:** 38.092 bil kWh. **Labor force:** 59% serv.; 27% ind.; 14% agric..

Finance: Monetary unit: Peso (Sept. 2001: 537.86 = $1 U.S.). **GDP** (1999 est.): $185.1 bil. **Per capita GDP:** $12,400. **Imports** (1999): $13.9 bil; partners: U.S. 24%, EU 23%. **Exports** (1999): $15.6 bil; partners: EU 27%, U.S. 16%, Japan 14%. **Tourism** (1998): $1.06 bil. **Budget** (1996 est.): $17 bil. **Intl. reserves less gold** (June 2000): $14.59 bil. **Gold:** 74,000 oz. t. **Consumer prices** (change in 1999): 3.3%.

Transport: Railroad: Length: 4,084 mi. **Motor vehicles:** 900,000 pass. cars, 475,000 comm. vehicles. **Civil aviation:** 5.3 bil pass.-mi.; 23 airports. **Chief ports:** Valparaiso, Arica, Antofagasta.

Communications: TV sets: 280 per 1,000 pop. **Radios:** 305 per 1,000 pop. **Telephones**: 3,365,000 main lines. **Daily newspaper circ.:** 101 per 1,000 pop.

Health: Life expectancy: 72.63 male; 79.42 female. **Births** (per 1,000 pop): 16.80. **Deaths** (per 1,000 pop.): 5.55. **Natural inc.:** 1.13%. **Infant mortality** (per 1,000 live births): 9.36.

Education: Free and compulsory, from age 6 or 7, for 8 years. **Literacy:** 95%.

Major Intl. Organizations: UN and all of its specialized agencies, APEC, OAS.

Embassy: 1732 Massachusetts Ave. NW 20036; 785-1746.

Website: http://www.segegob.cl/seg-ingl/index2i.html

Northern Chile was under Inca rule before the Spanish conquest, 1536-40. The southern Araucanian Indians resisted until the late 19th century. Independence was gained 1810-18, under José de San Martin and Bernardo O'Higgins; the latter, as supreme director 1817-23, sought social and economic reforms until deposed. Chile defeated Peru and Bolivia in 1836-39 and 1879-84, gaining mineral-rich northern land.

In 1970, Salvador Allende Gossens, a Marxist, became president with a third of the national vote. His government improved conditions for the poor, but illegal and violent actions by extremist supporters of the government, the regime's failure to attain majority support, and poorly planned socialist economic programs led to political and financial chaos.

A military junta seized power Sept. 11, 1973, and said Allende had killed himself. The junta, headed by Gen. Augusto Pinochet Ugarte, named a mostly military cabinet and announced plans to "exterminate Marxism." Repression continued during the 1980s with little sign of any political liberalization.

In a plebiscite held Oct. 5, 1988, voters rejected the incumbent president, Pinochet. He agreed to presidential elections. In Dec. 1989 voters elected a civilian president, although Pinochet continued to head the army until Mar. 10, 1998. In Mar. 1994 a Chilean human rights group estimated that human rights violations had claimed more than 3,100 lives during Pinochet's rule. Attempts to prosecute him failed when he was declared mentally unfit to stand trial by courts in Britain and Chile. Ricardo Lagos Escobar, Chile's 1st Socialist president since the 1973 coup, took office Mar. 11, 2000.

Tierra del Fuego is the largest (18,800 sq. mi.) island in the archipelago of the same name at the southern tip of South America, an area of majestic mountains, tortuous channels, and high winds. It was visited 1520 by Magellan and named Land of Fire because of its many Indian bonfires. Part of the island is in Chile, part in Argentina. Punta Arenas, on a mainland peninsula, is a center of sheep raising and the world's southernmost city (pop. about 70,000); Puerto Williams is the southernmost settlement.

China

People's Republic of China

(Statistical data do not include Hong Kong or Macao.)

People: Population: 1,273,111,290. **Age distrib.** (%): <15: 25; 65+: 7.1. **Pop. density:** 354 per sq. mi. **Urban:** 32%. **Ethnic groups:** Han Chinese 91.9%, Tibetan, Mongol, Korean, Manchu, others. **Principal languages:** Mandarin (official), Yue, Wu, Hakka, Xiang, Gan, Minbei, Minnan, others. **Chief religions:** Officially atheist; Buddhism, Taoism; some Muslims, Christians.

Geography: Area: 3,596,600 sq. mi. **Location:** Occupies most of the habitable mainland of E Asia. **Neighbors:** Mongolia on N; Russia on NE and NW; Afghanistan, Pakistan, Tajikistan, Kyrgystan, Kazakhstan on W; India, Nepal, Bhutan, Myanmar, Laos, Vietnam on S; North Korea on NE. **Topography:** Two-thirds of the vast territory is mountainous or desert; only one-tenth is cultivated. Rolling topography rises to high elevations in the N in the Daxinganlingshanmai separating Manchuria and Mongolia; the Tien Shan in Xinjiang; the Himalayan and Kunlunshanmai in the SW and in Tibet. Length is 1,860 mi. from N to S, width E to W is more than 2,000 mi. The eastern half of China is one of the world's best-watered lands. Three great river systems, the Chang (Yangtze), Huang (Yellow), and Xi, provide water for vast farmlands. **Capital:** Beijing. **Cities (urban agg.):** Shanghai 12,887,000; Beijing 10,839,000; Tianjin 9,156,000; Chongqing 5,312,000; Shenyang 4,828,000; Guangzhou 3,893,000.

Government: Type: Communist Party-led state. **Head of state:** Pres. Jiang Zemin; b Aug. 17, 1926; in office: Mar. 27, 1993. **Head of gov.:** Premier Zhu Rongji; b Oct. 1, 1928; in office: Mar. 17, 1998. **Local divisions:** 22 provinces (not including Taiwan), 5 autonomous regions, and 4 municipalities, plus the special administrative regions of Hong Kong (as of July 1, 1997) and Macao (as of Dec. 20, 1999). **Defense budget:** $14.5 bil. **Active troops:** 2,470 mil.

Economy: Industries: Iron and steel, textiles and apparel, machine building, armaments, cement (world's leading producer of cotton cloth, cement, steel). **Chief crops:** Grain, rice, cotton, potatoes, tea. **Minerals:** Tungsten, antimony, coal, oil, mercury, iron, lead, manganese, molybdenum, tin. **Crude oil reserves** (2000): 24 bil bbls. **Other resources:** Hydropower. **Arable land:** 10%. **Livestock** (2000): cattle: 104.58 mil; chickens: 3.63 bil; goats: 148.4 mil; pigs: 437.55 mil; sheep: 131.1 mil. **Fish catch** (1999): 36.33 mil metric tons. **Electricity prod.:** 1.17 tril kWh. **Labor force:** 50% agric., 24% ind.

Finance: Monetary unit: Renminbi (Yuan) (Sept. 2001: 8.28 = $1 U.S.). **GDP** (1999 est.): $4.8 tril. **Per capita GDP:** $3,800. **Imports** (1999): $165.8 bil; partners: Japan 20%, U.S. 12%, Taiwan 12%. **Exports** (1999): $194.9 bil; partners: U.S. 22%, Hong Kong 19%, Japan 17% . **Tourism:** $14.10 bil. **Intl. reserves less gold** (May 2000): $160.70 bil. **Gold:** 12.7 mil oz t. **Consumer prices** (change in 1999): –1.4%.

Transport: Railroad: Length: 47,672 mi. **Motor vehicles:** 4.7 mil pass. cars, 6.75 mil comm. vehicles. **Civil aviation:** 45.3 bil pass.-mi.; 113 airports. **Chief ports:** Shanghai, Qinhuangdao, Dalian, Guangzhou (Canton).

Communications: TV sets: 319 per 1,000 pop. **Radios:** 195 per 1,000 pop. **Telephones:** 144,000,000 main lines. **Daily newspaper circ.:** 23 per 1,000 pop.

Health: Life expectancy: 69.81 male; 73.59 female. **Births** (per 1,000 pop): 15.95. **Deaths** (per 1,000 pop.): 6.74. **Natural inc.:** .92%. **Infant mortality** (per 1,000 live births): 28.08.

Education: Compulsory 7-17. **Literacy** (1996): 82%.

Major Intl. Organizations: UN (FAO, IBRD, ILO, IMF, IMO, WHO), APEC.

Embassy: 2300 Conn. Ave. NW 20008; 328-2500.

Website: http://www.china-embassy.org

Remains of various humanlike creatures who lived as early as several hundred thousand years ago have been found in many parts of China. Neolithic agricultural settlements dotted the Huang (Yellow) R. basin from about 5000 BC. Their language, religion, and art were the sources of later Chinese civilization.

Bronze metallurgy reached a peak and Chinese pictographic writing, similar to today's, was in use in the more developed culture of the Shang Dynasty (c. 1500 BC-c. 1000 BC), which ruled much of North China.

A succession of dynasties and interdynastic warring kingdoms ruled China for the next 3,000 years. They expanded Chinese political and cultural domination to the south and west, and developed a brilliant technologically and a culturally advanced society. Rule by foreigners (Mongols in the Yuan Dynasty, 1271-1368, and Manchus in the Ch'ing Dynasty, 1644-1911) did not alter the underlying culture.

A period of relative stagnation left China vulnerable to internal and external pressures in the 19th century. Rebellions left tens of millions dead, and Russia, Japan, Britain, and other powers exercised political and economic control in large parts of the country. China became a republic Jan. 1, 1912, following the Wuchang Uprising inspired by Dr. Sun Yat-sen, founder of the Kuomintang (Nationalist) party. By 1928, the Kuomintang, led by Chiang Kai-shek, succeeded in nominal reunification of China. About the same time, a bloody purge of Communists from the ranks of the Kuomintang fomented hostilities between the two groups that would continue for decades.

For over 50 years, 1894-1945, China was involved in conflicts with Japan. In 1895, China ceded Korea, Taiwan, and other areas. On Sept. 18, 1931, Japan seized the Northeastern Provinces (Manchuria) and set up a puppet state called Manchukuo. The border province of Jehol was cut off as a buffer state in 1933. Taking advantage of Chinese dissension, Japan invaded China proper July 7, 1937. On Nov. 20 the retreating Nationalist government moved its capital to Chongqing (Chungking) from Nanking (Nanjing), which Japanese troops then ravaged Dec. 13.

From 1939 the Sino-Japanese War (1937-45) became part of the broader world conflict. After its defeat in World War II, Japan gave up all seized land, and internal conflicts involving the Kuomintang, Communists, and other factions resumed. China came under the domination of Communist armies, 1949-1950. The Kuomintang government moved to Taiwan, Dec. 8, 1949.

The Chinese People's Political Consultative Conference convened Sept. 21, 1949; The People's Republic of China was proclaimed in Beijing (Peking) Oct. 1, 1949, under Mao Zedong. China and the USSR signed a 30-year treaty of "friendship, alliance and mutual assistance," Feb. 15, 1950. The U.S. refused recognition of the new regime. On Nov. 26, 1950, the People's Republic sent armies into Korea against U.S. troops and forced a stalemate in the Korean War.

After an initial period of consolidation, 1949-52, industry, agriculture, and social and economic institutions were forcibly molded according to Maoist ideals. However, frequent drastic changes in policy and violent factionalism interfered with economic development. In 1957, Mao admitted an estimated 800,000 people had been executed 1949-54; opponents claimed much higher figures.

The Great Leap Forward, 1958-60, tried to force the pace of economic development through intensive labor on huge new rural communes, and through emphasis on ideological purity. The program caused resistance and was largely abandoned.

By the 1960s, relations with the USSR deteriorated, with disagreements on borders, ideology, and leadership of world Communism. The USSR canceled aid accords, and China, with Albania, launched anti-Soviet propaganda drives.

The Great Proletarian Cultural Revolution, 1965, was an attempt to oppose pragmatism and bureaucratic power and instruct a new generation in revolutionary principles. Massive purges took place. A program of forcibly relocating millions of urban teenagers into the countryside was launched. By 1968 the movement had run its course; many purged officials returned to office in subsequent years, and reforms that had placed ideology above expertise were gradually weakened.

On Oct. 25, 1971, the UN General Assembly ousted the Taiwan government from the UN and seated the People's Republic in its place. The U.S. had supported the mainland's admission but opposed Taiwan's expulsion.

U.S. Pres. Richard Nixon visited China Feb. 21-28, 1972, on invitation from Premier Zhou Enlai, ending years of antipathy between the 2 nations. China and the U.S. opened liaison offices in each other's capitals, May-June 1973. The U.S., Dec. 15, 1978, formally recognized the People's Republic of China as the sole legal government of China; diplomatic relations between the 2 nations were established, Jan. 1, 1979.

Mao died Sept. 9, 1976. By 1978, Vice Premier Deng Xiaoping had consolidated his power, succeeding Mao as "paramount leader" of China. The new ruling group modified Maoist policies in education, culture, and industry, and sought better ties with non-Communist countries. During this "reassessment" of Mao's policies his widow, Jiang Qing, and other "Gang of Four" leftists were convicted of "committing crimes during the 'Cultural Revolution,' " Jan. 25, 1981.

By the mid-1980s, China had enacted far-reaching economic reforms, deemphasizing centralized planning and incorporating market-oriented incentives. Some 100,000 students and workers staged a march in Beijing to demand political reforms, May 4, 1989. The demonstrations continued during a visit to Beijing by Soviet leader Mikhail Gorbachev May 15-18; it was the first Sino-Soviet summit since 1959. As the unrest spread, martial law was imposed, May 20. Troops entered Beijing, June 3-4, and crushed the pro-democracy protests, as tanks and armored personnel carriers rolled through Tiananmen Square. It is estimated that 5,000 died, 10,000 were injured, and hundreds of students and workers were arrested.

China had one of the world's fastest-growing economies in the 1990s. Although human rights violations have persisted, the U.S. has continued to renew China's most-favored-nation trading status. Deng died Feb. 19, 1997, leaving his chosen successor, Jiang Zemin, in firm control as president. Pres. Jiang paid a state visit to the U.S., Oct. 26-Nov. 3, and U.S. Pres. Clinton visited China, June 25-July 3, 1998. Floods in July and Aug. killed at least 3,000 people, left millions homeless, and caused an estimated $20 billion in property damage.

NATO bombs hit the Chinese embassy in Belgrade, Yugoslavia, on May 7, 1999, killing 3 people and wounding 27; the U.S. agreed on July 30 to pay $4.5 million to compensate victims and their families, and on Dec. 16 to pay $28 million for damage to the embassy. The government banned a popular religious sect, the Falun Gong, July 22, after it staged the largest unauthorized demonstrations in Beijing since 1989. The U.S. and China signed a comprehensive trade agreement Nov. 15; normalization of China trade won U.S. congressional approval Sept. 19, 2000.

After a midair collision Apr. 1, 2001, a Chinese jet fighter crashed into the S China Sea and a U.S. Navy surveillance plane made an emergency landing on Hainan Is.; the 24-member U.S. crew was freed Apr. 12. Beijing was chosen, July 13, to host the 2008 Summer Olympic Games. China and Russia signed a 20-year friendship and cooperation treaty July 16.

By agreement with Great Britain, Hong Kong reverted to Chinese sovereignty July 1, 1997. Portugal returned Macao to China Dec. 20, 1999.

Manchuria. Home of the Manchus, rulers of China 1644-1911, Manchuria has accommodated millions of Chinese settlers in the 20th century. Under Japanese rule 1931-45, the area became industrialized. The region is divided into the 3 NE provinces of Heilongjiang, Jilin, and Liaoning.

Guangxi is in SE China, bounded on N by Guizhou and Hunan provinces, E and S by Guangdong, on SW by Vietnam, and on W by Yunnan. It produces rice in the river valleys and has valuable forest products.

Inner Mongolia was organized by the People's Republic in 1947. Its boundaries have undergone frequent changes, reaching its greatest extent in 1956 (and restored in 1979), with an area of 454,600 sq. mi., allegedly in order to dilute the minority Mongol population. Chinese settlers outnumber the Mongols more than 10 to 1. Pop. (1996 est.): 23.07 mil. Capital: Hohhot.

Xinjiang, in Central Asia, is 635,900 sq. mi., pop. (1996 est.): 16.89 mil (75% Uygurs, a Turkic Muslim group, with a heavy Chinese increase in recent years). Capital: Urumqi. It is China's richest region in strategic minerals.

Tibet, 471,700 sq. mi., is a thinly populated region of high plateaus and massive mountains, the Himalayas on the S, the Kunluns on the N. High passes connect with India and Nepal; roads lead into China proper. Capital: Lhasa. Average altitude is 15,000 ft. Jiachan, 15,870 ft., is believed to be the highest inhabited town on earth. Agriculture is primitive. Pop. (1996 est.): 2.44 mil (of whom about 500,000 are Chinese). Another 4 million Tibetans form the majority of the population of vast adjacent areas that have long been incorporated into China.

China ruled all of Tibet from the 18th century, but independence came in 1911. China reasserted control in 1951, and a Communist government was installed in 1953, revising the theocratic Lamaist Buddhist rule. Serfdom was abolished, but all land remained collectivized.

A Tibetan uprising within China in 1956 spread to Tibet in 1959. The rebellion was crushed with Chinese troops, and Buddhism was almost totally suppressed. The Dalai Lama and 100,000 Tibetans fled to India.

Hong Kong

Hong Kong (Xianggang), located at the mouth of the Zhu Jiang (Pearl R.) in SE China, 90 mi. S of Canton (Guangzhou), was a British dependency from 1842 until July 1, 1997, when it became a Special Administrative Region of China. Its nucleus is Hong Kong Isl., 31 sq. mi., occupied by the British in 1841 and formally ceded to them in 1842, on which is located the seat of government. Opposite is Kowloon Peninsula, 3 sq. mi., and Stonecutters Isl., added to the territory in 1860. An additional 355 sq. mi. known as the New Territories, a mainland area and islands, were leased from China, 1898, for 99 years. Total area 422 sq. mi.; pop. (2001 est.) 7,210,505, including fewer than 20,000 British.

Hong Kong is a major center for trade and banking. Per capita GDP, $23,100 (1999 est.), is among the highest in the world. Principal industries are textiles and apparel; also tourism ($7.21 bil expenditures in 1999), electronics, shipbuilding, iron and steel, fishing, cement, and small manufactures. Hong Kong's spinning mills are among the best in the world.

Hong Kong harbor was long an important British naval station and one of the world's great transshipment ports. The colony was often a place of refuge for exiles from mainland China. It was occupied by Japan during World War II.

From 1949 to 1962 Hong Kong absorbed more than a million refugees fleeing Communist China. Starting in the 1950s, cheap labor led to a boom in light manufacturing, while liberal tax policies attracted foreign investment; Hong Kong became one of the wealthiest, most productive areas in the Far East. Poor living and working conditions and low wages for many led to political unrest in the 1960s, but legislation and public works programs raised the standard of living by the 1970s.

With the end of the 99-year lease on the New Territories drawing near, Britain and China signed an agreement, Dec. 19, 1984, under which all of Hong Kong was to be returned to China in 1997; under this agreement Hong Kong was to be allowed to keep its capitalist system for 50 years. In Dec. 1996, an electoral college appointed by China chose a shipping magnate, Tung Chee-hwa, to be Hong Kong's chief executive when it reverted to Chinese control.

The July 1 transfer of government was marked by an elaborate ceremony. In the immediate wake of the changeover, Hong Kong retained its street names and its currency, the Hong Kong dollar (but without the queen's picture). Official languages remained Chinese (Cantonese dialect) and English. The Legislative Council was disbanded, and an appointed Provisional Legislature installed in its place. The new legislature imposed limits on opposition activities and sharply cut back the number of people eligible to vote in legislative elections; despite the restrictions, pro-democracy candidates did well in May 24, 1998, balloting.

Macao

Macao, area of 6 sq. mi., is an enclave, a peninsula and 2 small islands, at the mouth of the Xi (Pearl) R. in China. It was established as a Portuguese trading colony in 1557. In 1849, Portugal claimed sovereignty over the territory; this claim was accepted by China in an 1887 treaty. Portugal granted broad autonomy in 1976. Under a 1987 agreement, Macao reverted to China Dec. 20, 1999. As in the case of Hong Kong, the Chinese government guaranteed Macao it would not interfere in its way of life and capitalist system for a period of 50 years. Pop. (2001 est.): 453,733.

Colombia

Republic of Colombia

People: Population: 40,349,388. **Age distrib.** (%): <15: 31.9; 65+: 4.8. **Pop. density:** 101 per sq. mi. **Urban:** 74%. **Ethnic groups:** Mestizo 58%, white 20%, mulatto 14%, black 4%. **Principal language:** Spanish (official). **Chief religion:** Roman Catholic 95%.

Geography: Area: 400,600 sq. mi. **Location:** At the NW corner of South America. **Neighbors:** Panama on NW, Ecuador and Peru on S, Brazil and Venezuela on E. **Topography:** Three ranges of Andes—Western, Central, and Eastern Cordilleras—run through the country from N to S. The eastern range consists mostly of high tablelands, densely populated. The Magdalena R. rises in the Andes, flows N to Caribbean, through a rich alluvial plain. Sparsely settled plains in E are drained by Orinoco and Amazon systems. **Capital:** Bogotá. (Full name: Santa Fe de Bogotá.) **Cities** (urban agg.): Bogotá 6,288,000; Medellin 2,951,000; Cali 2,710,000; Barranquilla 1,736,000.

Government: Type: Republic. **Head of state and gov.:** Pres. Andrés Pastrana Arango; b Aug. 17, 1954; in office: Aug. 7, 1998. **Local divisions:** 32 departments, capital district of Bogota. **Defense budget:** $2 bil. **Active troops:** 153,000.

Economy: Industries: Textiles, food processing, clothing, cement, chemicals. **Chief crops:** Coffee, rice, bananas, oilseed, corn, sugar, tobacco, cocoa. **Minerals:** Oil, gas, emeralds, gold, copper, coal, iron, nickel. **Crude oil reserves** (2000): 2.58 bil bbls. **Other resources:** Forest products, cut flowers. **Arable land:** 4%. **Livestock** (2000): cattle: 26.0 mil; chickens: 100.0 mil; goats: 1.12 mil; pigs: 2.8 mil; sheep: 2.2 mil. **Fish catch** (1999): 199,227 metric tons. **Electricity prod.:** 43.574 bil kWh. **Labor force:** 46% serv.; 30% agric.; 24% ind.

Finance: Monetary unit: Peso (Sept. 2001: 2,320.5 = $1 U.S.). **GDP** (1999 est.): $245.1 bil. **Per capita GDP:** $6,200. **Imports** (1999): $10 bil; partners: U.S. 35%, EU 20%. **Exports** (1999): $11.5 bil; partners: U.S. 39%, EU 24%. **Tourism** (1998): $939 mil. **Budget** (1996 est.): $30 bil. **Intl. reserves less gold** (June 2000): $8.36 bil. **Gold:** 328,000 oz t. **Consumer prices** (change in 1999): 11.2%.

Transport: Railroad: Length: 2,007 mi. **Motor vehicles:** 1.15 mil pass. cars, 550,000 comm. vehicles. **Civil aviation:** 4.3 bil pass.-mi.; 43 airports. **Chief ports:** Buenaventura, Barranquilla, Cartagena.

Communications: TV sets: 188 per 1,000 pop. **Radios:** 151 per 1,000 pop. **Telephones:** 7,158,600 main lines. **Daily newspaper circ.:** 55 per 1,000 pop.

Health: Life expectancy: 66.71 male; 74.55 female. **Births** (per 1,000 pop): 22.41. **Deaths** (per 1,000 pop.): 5.69. **Natural inc.:** 1.67%. **Infant mortality** (per 1,000 live births): 23.96.

Education: Free and compulsory for 5 years between ages 6-12. **Literacy:** 91%.

Major Intl. Organizations: UN (FAO, IBRD, ILO, IMF, IMO, WHO, WTrO), OAS.

Embassy: 2118 Leroy Pl. NW 20008; 387-8338.

Spain subdued the local Indian kingdoms (Funza, Tunja) by the 1530s and ruled Colombia and neighboring areas as New Granada for 300 years. Independence was won by 1819. Venezuela and Ecuador broke away in 1829-30, and Panama withdrew in 1903.

Colombia is plagued by rural and urban violence. "La Violencia" of 1948-58 claimed 200,000 lives; since 1989, political violence has resulted in more than 35,000 deaths. Attempts at land and social reform and progress in industrialization have not reduced massive social problems.

The government's increased activity against local drug traffickers sparked a series of retaliation killings. On Aug. 18, 1989, Luis Carlos Galán, the ruling party's presidential hopeful for the 1990 election, was assassinated. In 1990, 2 other presidential candidates were assassinated, as drug traffickers carried on a campaign of intimidation.

Charges that Ernesto Samper Pizano's 1994 campaign received money from the Cali drug cartel engulfed his administration in scandal, although the legislature voted, June 12, 1996, not to impeach him. Andrés Pastrana Arango, son of former Pres. Misael Pastrana Borrero (in office 1970-74), won a presidential runoff election, June 21, 1998. An earthquake Jan. 25, 1999, in western Colombia killed at least 1,185 people and left 250,000 homeless. At least 5 million people in more than 700 cities took part in protests Oct. 24 against continuing violence and human rights abuses.

The U.S. authorized $1.3 billion in antidrug aid to Colombia Aug. 22, 2000. Right-wing paramilitaries launched a violent campaign Dec. 22 against suspected left-wing guerrillas. Legislation expanding the powers of the military was signed Aug. 13, 2001.

Comoros

Federal Islamic Republic of the Comoros

People: Population: 596,202. **Age. distrib.** (%): <15: 42.8; 65+: 2.9. **Pop. density:** 745 per sq. mi. **Urban:** 33%. **Ethnic groups:** Antalote, Cafre, Makoa, Oimatsaha, Sakalava. **Principal languages:** Arabic, French, Comorian (all official). **Chief religions:** Sunni Muslim 86%, Roman Catholic 14%.

Geography: Area: 800 sq. mi. **Location:** 3 islands—Grande Comore (Njazidja), Anjouan (Nzwani), and Moheli (Mwali)—in the Mozambique Channel between NW Madagascar and SE Africa. **Neighbors:** Nearest are Mozambique on W, Madagascar on E. **Topography:** The islands are of volcanic origin, with an active volcano on Grande Comore. **Capital:** Moroni (1999 met. est.) 44,000.

Government: Type: In transition. **Head of state:** Pres. Azali Assoumani; b 1951; in office: May 6, 1999. **Head of gov.:** Prime Min. Hamada Madi Bolero; b 1965; in office: Nov. 29, 2000. **Local divisions:** 3 main islands with 4 municipalities.

Economy: Industries: Perfume, textiles. **Chief crops:** Vanilla, copra, perfume essences, cloves. **Arable land:** 35%. **Livestock** (2000): cattle: 52,000; chickens: 490,000; goats: 140,000; sheep: 20,000. **Fish catch:** (1999): 12,500 metric tons. **Electricity prod.:** 17 mil kWh. **Labor force:** 80% agric.

Finance: Monetary unit: Franc (Sept. 2001: 550.4 = $1 U.S.). **GDP** (1998 est.): $410 mil. **Per capita GDP:** $725. **Imports** (1998 est.): $49.5 mil; partners: France 59%, South Africa 15%. **Exports** (1998 est.): $9.3 mil; partners: France 43%, U.S. 43%, Germany 7%. **Tourism:** $17 mil. **Budget** (1997 est.): $53 mil. **Intl. reserves less gold** (Mar. 2000): $36.08 mil.

Transport: Civil aviation: 2.1 mil pass.-mi.; 2 airports. **Chief ports:** Fomboni, Moroni, Moutsamoudou.

Communications: Radios: 122 per 1,000 pop. **Telephones**: 7,000.

Health: Life expectancy: 58.2 male; 62.68 female. **Births** (per 1,000 pop): 39.52. **Deaths** (per 1,000 pop.): 9.35. **Natural inc.:** 3.02%. **Infant mortality** (per 1,000 live births): 84.07.

Education: Compulsory: ages 7-16. **Literacy:** 57%.

Major Intl. Organizations: UN (FAO, IBRD, ILO, IMF, WHO), AL, OAU.

Embassy: 336 E. 45th St., 2d Fl., New York, NY 10017; (212) 349-2030.

Website: http://www.ksu.edu/sasw/comoros/comoros.html

The islands were controlled by Muslim sultans until the French acquired them 1841-1909. They became a French overseas territory in 1947. A 1974 referendum favored independence, with only the Christian island of Mayotte preferring association with France. The French National Assembly decided to allow each of the islands to decide its own fate. The Comore Chamber of Deputies declared independence July 6, 1975, with Ahmed Abdallah as president. In a referendum in 1976, Mayotte voted to remain French.

A leftist regime that seized power from Abdallah in 1975 was deposed in a pro-French 1978 coup in which he regained the presidency. In Nov. 1989, Pres. Abdallah was assassinated; soon after, a multiparty system was instituted. A Sept. 1995 military coup, assisted by French mercenaries, ousted Pres. Said Mohamed Djohar. French troops invaded, Oct. 4, and forced coup leaders to surrender. Djohar returned from exile in Jan. 1996, and in Mar. a new presidential election was held. A hijacked Ethiopian Airlines Boeing 767 crashed offshore on Nov. 23, killing 123 of the 175 people on board.

Seeking to resume ties with France, Anjouan seceded from the Comoros, Aug. 3, 1997. Comorian troops were unable to put down the rebellion, which was joined by Moheli. Unrest on Grande Comore culminated in a military coup, Apr. 30, 1999. Anjouans endorsed secession in a disputed vote Jan. 23, 2000. A military junta took power on Anjouan Aug. 9, 2001.

Congo (formerly Zaire)

Democratic Republic of the Congo

(Congo, officially Democratic Republic of the Congo, is also known as Congo-Kinshasa. It should not be confused with Republic of the Congo, commonly called Congo Republic, and also known as Congo-Brazzaville.)

People: Population: 53,624,718. **Age distrib.** (%): <15: 48.2; 65+: 2.5. **Pop. density:** 61 per sq. mi. **Urban:** 30%. **Ethnic groups:** More than 200 tribes, mostly Bantu. **Principal languages:** French (official). **Chief religions:** Roman Catholic 50%, Protestant 20%, Muslim 10%, Kimbanguist 10%.

Geography: Area: 874,500 sq. mi. **Location:** In central Africa. **Neighbors:** Congo-Brazzaville on W; Central African Republic, Sudan on N; Uganda, Rwanda, Burundi, Tanzania on E; Zambia, Angola on S. **Topography:** Congo includes the bulk of the Congo R. basin. The vast central region is a low-lying plateau covered by rain forest. Mountainous terraces in the W, savannas in the S and SE, grasslands toward the N, and the high Ruwenzori Mts. on the E surround the central region. A short strip of territory borders the Atlantic O. The Congo R. is 2,718 mi. long. **Capital:** Kinshasa. **Cities (urban agg.):** Kinshasa 5,064,000; Lubumbashi 967,000.

Government: Type: Republic with strong presidential authority (in transition). **Head of state and gov.:** Pres. Joseph Kabila; b June 24, 1971; in office: Jan. 26, 2001. **Local divisions:** 10 provinces, 1 city. **Defense budget:** $400 mil. **Active troops:** 55,900.

Economy: Industries: Mining, consumer prods., food processing. **Chief crops:** Coffee, sugar, palm oil, rubber, tea. **Minerals:** Cobalt, copper, cadmium, oil, diamonds, gold, silver, tin, germanium, zinc, iron, manganese, uranium, radium. **Crude oil reserves** (2000): 187 mil bbls. **Other resources:** Timber. **Arable land:** 3%. **Livestock** (2000): cattle: 822,355; chickens: 21.56 mil; goats: 4.13 mil; pigs: 1.05 mil; sheep: 924,924.. **Fish catch** (1999): 162,961 metric tons. **Electricity prod.:** 5.268 bil kWh. **Labor force:** 65% agric., 16% ind., 19% serv.

Finance: Monetary unit: Congolese Franc (Oct. Sept. 2001: 280 = $1 U.S.). **GDP** (1999 est.): $35.7 bil. **Per capita GDP:** $710. **Imports** (1998 est.): $460 mil; partners: South Africa 25%, Belg.-Lux. 14%. **Exports** (1998 est.): $530 mil; partners: Belg.-Lux. 52%, U.S. 14%. **Tourism** (1998): $2 mil. **Budget** (1996 est.): $244 mil. **Consumer prices** (change in 1997): 176%.

Transport: Railroad: Length: 3,162 mi. **Motor vehicles:** 330,000 pass. cars, 200,000 comm. vehicles. **Civil aviation:** 189.7 mil pass.-mi.; 22 airports. **Chief ports:** Matadi, Boma, Kinshasa.

Communications: Radios: 79 per 1,000 pop. **Telephones** (1999): 20,000. **Daily newspaper circ.:** 3 per 1,000 pop.

Health: Life expectancy: 46.96 male; 50.98 female. **Births** (per 1,000 pop): 46.02. **Deaths** (per 1,000 pop.): 15.15. **Natural inc.:** 3.09%. **Infant mortality** (per 1,000 live births): 99.88.

Education: Compulsory: ages 6-12. **Literacy** 77%.

Major Intl. Organizations: UN and most of its specialized agencies, OAU.

Embassy: 1800 New Hampshire Ave. NW 20009; 234-7690.

The earliest inhabitants of Congo may have been the pygmies, followed by Bantus from the E and Nilotic tribes from the N. The large Bantu Bakongo kingdom ruled much of Congo and Angola when Portuguese explorers visited in the 15th century.

Leopold II, king of the Belgians, formed an international group to exploit the Congo region in 1876. In 1877 Henry M. Stanley explored the Congo, and in 1878 the king's group sent him back to organize the region and win over the native chiefs. The Conference of Berlin, 1884-85, organized the Congo Free State with Leopold as king and chief owner. Exploitation of native laborers on the rubber plantations caused international criticism and led to granting of a colonial charter, 1908; the colony became known as the Belgian Congo. Millions of Congolese are believed to have died between 1880 and 1920 as a result of slave labor and other causes under European rule.

Belgian and Congolese leaders agreed Jan. 27, 1960, the Congo would become independent in June. In the first general elections, May 31, the National Congolese movement of Patrice Lumumba won 35 of 137 seats in the National Assembly. He was appointed premier June 21, and formed a coalition cabinet. The Republic of the Congo was proclaimed June 30.

Widespread violence caused Europeans and others to flee. The UN Security Council, Aug. 9, 1960, called on Belgium to withdraw its troops and sent a UN contingent. Pres. Joseph Kasavubu removed Lumumba as premier in Sept.; Lumumba was murdered Jan. 17, 1961.

The last UN troops left the Congo June 30, 1964, and Moise Tshombe became president.

On Sept. 7, 1964, leftist rebels set up a "People's Republic" in Stanleyville (now Kisangani). Tshombe hired foreign mercenaries and sought to rebuild the Congolese Army. In Nov. and Dec. 1964 rebels killed scores of white hostages and thousands of Congolese; Belgian paratroopers, dropped from U.S. transport planes, rescued hundreds. By July 1965 the rebels had lost their effectiveness.

In late 1965 Gen. Joseph D. Mobutu was named president. He later changed his name to Mobutu Sese Seko. The country became the Democratic Republic of the Congo (1966) and the Republic of Zaire (1971).

Economic decline and government corruption plagued Zaire in the 1980s and worsened in the 1990s. In 1990, Pres. Mobutu announced an end to a 20-year ban on multiparty politics. He sought to retain power despite mounting international pressure and internal opposition.

During 1994, Zaire was inundated with refugees from the massive ethnic bloodshed in Rwanda. Ethnic violence spread to E Zaire in 1996. In Oct. militant Hutus, who dominated in the refugee camps, fought against rebels (mostly Tutsis) in Zaire, precipitating intervention by government troops. As a result of the fighting, Rwandan refugees abandoned the camps; hundreds of thousands returned to Rwanda, while hundreds of thousands more were dispersed throughout E Zaire. The rebels, led by Gen. Laurent Kabila—a former Marxist and longtime opponent of Mobutu—gained momentum and began to move W across Zaire. As turmoil engulfed his nation, Mobutu stayed in W Europe for most of the last 4 months of 1996, receiving treatment for prostate cancer.

With Mobutu out of the country, the Zairean army put up little resistance; rebels were aided by several of Mobutu's enemies, notably Rwanda and Uganda. Mobutu returned to Zaire in March 1997, but attempts to negotiate with Kabila were ineffectual. On May 17, Kabila's troops entered Kinshasa and Mobutu went into exile. The country again assumed the name Democratic Republic of the Congo. Mobutu died Sept. 7 in Rabat, Morocco.

Kabila, who ruled by decree, alienated UN officials, international aid donors, and former allies. Rebels assisted by Rwanda and Uganda threatened Kinshasa in Aug. 1998, but the assault was turned back with help from Angola, Namibia, and Zimbabwe. Rebel groups agreed to a cease-fire on Aug. 31, 1999, but the truce was widely violated. Kabila was assassinated Jan. 16, 2001, apparently by one of his bodyguards, and was succeeded by his son Joseph.

According to UN estimates, more than 1 million adult Congolese have HIV/AIDS.

Congo Republic

Republic of the Congo

(Congo Republic, officially Republic of the Congo, is also known as Congo-Brazzaville. It should not be confused with Democratic Republic of the Congo [formerly Zaire], now commonly called Congo, and also known as Congo-Kinshasa.)

People: Population: 2,894,336. **Age distrib.** (%): <15: 42.4; 65+: 3.3. **Pop. density:** 22 per sq. mi. **Urban:** 62%. **Ethnic groups:** Kongo 48%, Sangha 20%, Teke 17%, M'Bochi 12%. **Principal languages:** French (official); Lingala, Kikongo, other African languages. **Chief religions:** Christian 50%, animist 48%, Muslim 2%.

Geography: Area: 131,700 sq. mi. **Location:** In W central Africa. **Neighbors:** Gabon and Cameroon on W, Central African Republic on N, Congo-Kinshasa (formerly Zaire) on E, Angola on SW. **Topography:** Much of the Congo is covered by thick forests. A coastal plain leads to the fertile Niari Valley. The center is a plateau; the Congo R. basin consists of flood plains in the lower and savanna in the upper portion. **Capital:** Brazzaville, (urban agg.): 1,234,000.

Government: Type: Republic. **Head of state and gov.:** Pres. Denis Sassou-Nguesso; b 1943; in office: Oct. 25, 1997. **Local divisions:** 10 regions, 6 communes. **Defense budget:** $73 mil. **Active troops:** 10,000.

Economy: Industries: Oil, wood products, brewing, cement. **Chief crops:** Cassava, rice, corn, sugar, cocoa, coffee. **Minerals:** Oil, potash, lead, copper, zinc. **Crude oil reserves** (2000): 1.5 bil bbls. **Livestock** (2000): cattle: 77,000; chickens: 1.9 mil; goats: 285,000; pigs: 46,000; sheep: 116,000. **Fish catch** (1999): 38,181 metric tons. **Electricity prod.:** 302 mil kWh.

Finance: Monetary unit: CFA Franc (Sept. 2001: 724.11 = $1 U.S.). **GDP** (1999 est.): $4.15 bil. **Per capita GDP:** $1,530. **Imports** (1999): $770 mil; partners: France 23%. **Exports** (1999): $1.7 bil; partners: U.S. 23%, Belneux 14%, . **Tourism** (1998): $10 mil. **Budget** (1997 est.): $970 mil. **Intl. reserves less gold** (Mar. 1999): $66.29 mil.

Transport: Railroad: Length: 494 mi. **Motor vehicles:** 26,000 pass. cars, 21,100 comm. vehicles. **Civil aviation:** 10 airports. **Chief ports:** Pointe-Noire, Brazzaville.

Communications: TV sets: 17 per 1,000 pop. **Radios:** 312 per 1,000 pop. **Telephones** (1999): 22,000 main lines. **Daily newspaper circ.:** 8 per 1,000 pop.

Health: Life expectancy: 44.38 male; 50.85 female. **Births** (per 1,000 pop): 38.24. **Deaths** (per 1,000 pop.): 16.22. **Natural inc.:** 2.2%. **Infant mortality** (per 1,000 live births): 99.73.

Education: Compulsory: ages 6-16. **Literacy:** 75%.

Major Intl. Organizations: UN (FAO, IBRD, ILO, IMF, IMO, WHO), OAU.

Embassy: 4891 Colorado Ave. NW 20011; 726-5500.

Website: http://www.gksoft.com/govt/en/cg.html

The Loango Kingdom flourished in the 15th century, as did the Anzico Kingdom of the Batekes; by the late 17th century they had become weakened. By 1885, France established control of the region, then called the Middle Congo. Republic of the Congo gained independence Aug. 15, 1960.

After a 1963 coup sparked by trade unions, the country adopted a Marxist-Leninist stance, with the USSR and China vying for influence. France remained a dominant trade partner and source of technical assistance, however, and French-owned private enterprise retained a major economic role. In 1970, the country was renamed People's Republic of the Congo.

In 1990, Marxism was renounced and opposition parties legalized. In 1991 the country's name was changed back to Republic of the Congo, and a new constitution was approved. A democratically elected government came into office in 1992; one of its key problems was a resurgence of ethnic and regional hostilities. Factional fighting broke out in Brazzaville, June 5, 1997, and intensified during the summer, devastating the capital and forcing international aid workers to flee. Troops loyal to former Marxist dictator Denis Sassou-Nguesso took control of the city Oct. 15, 1997.

Costa Rica

Republic of Costa Rica

People: Population: 3,773,057. **Age distrib.** (%): <15: 31.4; 65+: 5.3. **Pop. density:** 193 per sq. mi. **Urban:** 48%. **Ethnic groups:** White and mestizo 96%. **Principal language:** Spanish (official). **Chief religion:** Roman Catholic 95%.

Geography: Area: 19,500 sq. mi. **Location:** In Central America. **Neighbors:** Nicaragua on N, Panama on S. **Topography:** Lowlands by the Caribbean are tropical. The interior plateau, with an altitude of about 4,000 ft., is temperate. **Capital:** San José, (urban agg.): 988,000.

Government: Type: Republic. **Head of state and gov.:** Pres. Miguel Angel Rodríguez Echeverrîa; b Jan. 9, 1940; in office: May 8, 1998. **Local divisions:** 7 provinces. **Defense budget:** N/A. **Active troops:** N/A.

Economy: Industries: Food processing, textiles, construction materials, fertilizer, plastics. **Chief crops:** Coffee, bananas, sugar, rice, potatoes. **Other resources:** Fish, forests, hydropower. **Arable land:** 6%. **Livestock** (2000): cattle: 1.72 mil; chickens: 17.14 mil; goats: 1,650; pigs: 390,000; sheep: 2,500. **Fish catch** (1999): 33,613 metric tons. **Electricity prod.:** 5.805 bil kWh. **Labor force:** 58% serv.; 20% agric., industry and commerce 22%.

Finance: Monetary unit: Colon (Sept. 2001: 332.87 = $1 U.S.). **GDP** (1999 est.): $26 bil. **Per capita GDP:** $7,100. **Imports** (1999 est.): $6.5 bil; partners: U.S. 41%. **Exports** (1999 est.): $6.4 bil; partners: U.S. 49%. **Tourism:** $1.00 bil. **Intl. reserves less gold** (June 2000): $1.28 bil. **Gold:** 2,000 oz t. **Consumer prices** (change in 1999): 10.0%.

Transport: Railroad: Length: 590 mi. **Motor vehicles:** 48,684 pass. cars, 70,308 comm. vehicles. **Civil aviation:** 1.2 bil pass.-mi.; 14 airports. **Chief ports:** Limon, Puntarenas, Golfito.

Communications: TV sets: 102 per 1,000 pop. **Radios:** 224 per 1,000 pop. **Telephones:** 1,003,400 main lines. **Daily newspaper circ.:** 102 per 1,000 pop.

Health: Life expectancy: 73.49 male; 78.68 female. **Births** (per 1,000 pop): 20.27. **Deaths** (per 1,000 pop.): 4.3. **Natural inc.:** 1.6%. **Infant mortality** (per 1,000 live births): 11.18.

Education: Free, compulsory: ages 6-15. **Literacy:** 95%.

Major Intl. Organizations: UN (FAO, IBRD, ILO, IMF, IMO, WHO, WTrO), OAS.

Embassy: 2114 S St. NW 20008; 234-2945.

Guaymi Indians inhabited the area when Spaniards arrived, 1502. Independence came in 1821. Costa Rica seceded from the Central American Federation in 1838. Since the civil war of

1948-49, there has been little violent social conflict, and free political institutions have been preserved. During 1993 there was an unusual wave of kidnappings and hostage-taking, some of it related to the international cocaine trade.

Costa Rica, though still a largely agricultural country, has achieved a relatively high standard of living, and land ownership is widespread. Tourism is growing rapidly.

Côte d'Ivoire

Republic of Ivory Coast

People: Population: 16,393,221. **Age distrib.** (%): <15: 46.2; 65+: 2.2. **Pop. density:** 134 per sq. mi. **Urban:** 46%. **Ethnic groups:** Baoule 23%, Bete 18%, Senoufou 15%, Malinke 11%, Agni, foreign Africans. **Principal languages:** French (official), Dioula and other native dialects. **Chief religions:** Muslim 60%, indigenous beliefs 18%, Christian 22%.

Geography: Area: 122,600 sq. mi. **Location:** On S coast of W Africa. **Neighbors:** Liberia, Guinea on W; Mali, Burkina Faso on N; Ghana on E. **Topography:** Forests cover the W half of the country, and range from a coastal strip to halfway to the N on the E. A sparse inland plain leads to low mountains in NW. **Capital:** Yamoussoukro (official); Abidjan (de facto). **Cities (urban agg.):** Abidjan 3,305,000.

Government: Type: Republic. **Head of state:** Pres. Laurent Gbagbo; b May 31, 1945; in office: Oct. 26, 2000. **Head of gov.**: Prime Min. Affi N'Guessan; b 1953; in office: Oct. 27, 2000. **Local divisions:** 45 provinces. **Defense budget:** $134 mil. **Active troops:** 13,900.

Economy: Industries: Food processing, wood products, vehicles, textiles. **Chief crops:** Coffee, cocoa, rubber, palm kernels. **Minerals:** Oil, diamonds, manganese. **Crude oil reserves** (2000): 100 mil bbls. **Other resources:** Timber. **Arable land:** 8%. **Livestock** (2000): cattle: 1.35 mil; chickens: 30.0 mil; goats: 1.09 mil; pigs: 280,000; sheep: 1.39 mil. **Fish catch** (1999): 67,617 metric tons. **Electricity prod.:** 4.060 bil kWh. **Labor force:** 51% agric.; 12% manuf. & mining.

Finance: Monetary unit: CFA Franc (Sept. 2001: 724.11 = $1 U.S.). **GDP** (1999 est.): $25.7 bil. **Per capita GDP:** $1,600. **Imports** (1999 est.): $2.6 bil; partners: France 29%, U.S. 5%. **Exports** (1999 est.): $3.9 bil; partners: France 17%. Netherlands 12%. **Tourism** (1998): $108 mil. **Budget** (1997 est.): $2.6 bil. **Intl. reserves less gold** (Apr. 2000): $779.2 mil. **Gold:** 45,000 oz t. **Consumer prices** (change in 1999): 0.8%.

Transport: Railroad: Length: 405 mi. **Motor vehicles:** 160,000 pass. cars, 95,000 comm. vehicles. **Civil aviation:** 187.7 mil pass.-mi.; 5 airports. **Chief ports:** Abidjan, Dabou, San-Pédro.

Communications: TV sets: 57 per 1,000 pop. **Radios:** 112 per 1,000 pop. **Telephones:** (1998): 267,000 main lines. **Daily newspaper circ.**: 14 per 1,000 pop.

Health: Life expectancy: 43.58 male; 46.33 female. **Births** (per 1,000 pop): 40.38. **Deaths** (per 1,000 pop.): 16.65. **Natural inc.:** 2.37%. **Infant mortality** (per 1,000 live births): 93.65.

Education: Free, compulsory: ages 7-13. **Literacy:** 40%.

Major Intl. Organizations: UN and all of its specialized agencies, OAU.

Embassy: 2424 Massachusetts Ave. NW 20008; 797-0300.

Website: http://lcweb2.loc.gov/frd/cs/citoc.html

A French protectorate from 1842, Côte d'Ivoire became independent in 1960. It is the most prosperous of all the tropical African nations, as a result of diversification of agriculture for export, close ties to France, and encouragement of foreign investment. About 20% of the population are workers from neighboring countries. Côte d'Ivoire officially changed its name from Ivory Coast in Oct. 1985.

Students and workers protested, Feb. 1990, demanding the ouster of longtime Pres. Félix Houphouët-Boigny. Côte d'Ivoire held its first multiparty presidential election Oct. 1990, and Houphouët-Boigny retained his office. He died Dec. 7, 1993. The National Assembly named a successor, Henri Konan Bédié, who was reelected Oct. 22, 1995; he was ousted in a military coup Dec. 24, 1999. The coup leader, Robert Guéi, apparently lost a presidential vote Oct. 22, 2000, but claimed victory anyway. After mass protests, he fled, and Laurent Gbagbo became president.

Croatia

Republic of Croatia

People: Population: 4,334,142. **Age distrib.** (%): <15: 18.2; 65+: 15.2. **Pop. density:** 199 per sq. mi. **Urban:** 57%. **Ethnic groups:** Croat 78%, Serb 12%. **Principal language:** Serbo-Croatian (official) 96%. **Chief religions:** Catholic 77%, Orthodox 11%.

Geography: Area: 21,800 sq. mi. **Location:** SE Europe, on the Balkan Peninsula. **Neighbors:** Slovenia, Hungary on N; Bosnia and Herzegovina, Yugoslavia on E. **Topography:** Flat plains in NE; highlands, low mtns. along Adriatic coast. **Capital:** Zagreb, (urban agg.): 1,060,000.

Government: Type: Parliamentary democracy. **Head of state:** Pres. Stipe Mesic; b Dec. 24, 1934; in office: Feb. 18, 2000. **Head of gov.:** Prime Min. Ivica Racan; b Feb. 24, 1944; in office: Jan. 27, 2000. **Local divisions:** 21 counties. **Defense budget:** $590 mil. **Active troops:** 61,000.

Economy: Industries: Chemicals, plastics, machine tools, aluminum, steel, paper. **Chief crops:** Olives, wheat, corn, sugar beets, fruits. **Minerals:** Oil, bauxite, iron, coal. **Crude oil reserves** (2000): 92 mil bbls. **Arable land:** 21%. **Livestock** (2000): cattle: 427,000; chickens: 11.26 mil; goats: 80,000; pigs: 1.23 mil; sheep: 528,000.**Fish catch:** (1999): 19,885 metric tons. **Electricity prod.:** 10.960 bil kWh. **Labor force:** 31.1% industry & mining.

Finance: Monetary unit: Kuna (Sept. 2001: 8.17 = $1 U.S.). **GDP** (1999 est.): $23.9 bil. **Per capita GDP:** $5,100. **Imports** (1998): $8.4 bil; partners: Germany 20%, Italy 19%. **Exports** (1998): $4.5 bil; partners: Germany 18%, Italy 21%, Slovenia 12%. **Tourism:** $2.50 bil. **Budget** (1997 est.): $6.3 bil. **Intl. reserves less gold** (June 2000): $3.24 bil. **Consumer prices** (change in 1999): 3.7%.

Transport: Railroad: Length: 1,676 mi. **Motor vehicles:** 698,000 pass. cars, 54,000 comm. vehicles. **Civil aviation:** 291.4 mil pass.-mi.; 4 airports. **Chief ports:** Rijeka, Split, Dubrovnik.

Communications: TV sets: 230 per 1,000 pop. **Radios:** 230 per 1,000 pop. **Telephones** (1999): 1,633,600 main lines. **Daily newspaper circ.:** 575 per 1,000 pop.

Health: Life expectancy: 70.28 male; 77.73 female. **Births** (per 1,000 pop): 12.82. **Deaths** (per 1,000 pop.): 11.41. **Natural inc.:** .14%. **Infant mortality** (per 1,000 live births): 7.21.

Education: Free, compulsory: ages 7-15. **Literacy** (1993): 97%.

Major Intl. Organizations: UN (FAO, IBRD, ILO, IMF, IMO, WHO), OSCE.

Embassy: 2343 Massachusetts Ave. NW 20008; 588-5899.

From the 7th century the area was inhabited by Croats, a south Slavic people. It was formed into a kingdom under Tomislav in 924, and joined with Hungary in 1102. The Croats became westernized and separated from Slavs under Austro-Hungarian influence. The Croats retained autonomy under the Hungarian crown. Slavonia was taken by Turks in the 16th century; the northern part was restored by the Treaty of Karlowitz in 1699. Croatia helped Austria put down the Hungarian revolution 1848-49 and as a result was set up with Slavonia as the separate Austrian crownland of Croatia and Slavonia, which was reunited to Hungary as part of Ausgleich in 1867. It united with other Yugoslav areas to proclaim the Kingdom of Serbs, Croats, and Slovenes in 1918. At the reorganization of Yugoslavia in 1929, Croatia and Slavonia became Savska county, which in 1939 was united with Primorje county to form the county of Croatia. A nominally independent state between 1941 and 1945, it became a constituent republic in the 1946 constitution.

On June 25, 1991, Croatia declared independence from Yugoslavia. Fighting began between ethnic Serbs and Croats, with the former gaining control of about 30% of Croatian territory. A cease-fire was declared in Jan. 1992, but new hostilities broke out in 1993. A cease-fire with Serb rebels forming a self-declared republic of Krajina was agreed to Mar. 30, 1994. Croatian government troops recaptured most of the Serb-held territory Aug. 1995. Pres. Franjo Tudjman signed a peace accord with leaders of Bosnia and Serbia in Paris, Dec. 14. Tudjman won reelection June 15, 1997; international monitors called the vote "free but not fair." The last Serb-held enclave, E Slavonia, returned to Croatian control Jan. 15, 1998.

Tudjman died Dec. 10, 1999. Stipe Mesic, a moderate, won a presidential runoff election Feb. 7, 2000.

Cuba

Republic of Cuba

People: Population: 11,184,023. **Age distrib.** (%): <15: 21; 65+: 9.9. **Pop. density:** 262 per sq. mi. **Urban:** 75%. **Ethnic groups:** Mulatto 51%, white 37%, black 11%. **Principal language:** Spanish (official). **Chief religion:** Roman Catholic 85% prior to Castro.

Geography: Area: 42,800 sq. mi. **Location:** In the Caribbean, westernmost of West Indies. **Neighbors:** Bahamas and U.S. to N, Mexico to W, Jamaica to S, Haiti to E. **Topography:** The coastline is about 2,500 miles. The N coast is steep and rocky, the S coast low and marshy. Low hills and fertile valleys cover more than half the country. Sierra Maestra, in the E, is the highest of 3 mountain ranges. **Capital:** Havana, (urban agg.): 2,256,000.

Government: Type: Communist state. **Head of state and gov.:** Pres. Fidel Castro Ruz; b Aug. 13, 1926; in office: Dec. 3,

1976 (formerly prime min. since Feb. 16, 1959). **Local divisions:** 14 provinces, 1 special municipality. **Defense budget:** $31 mil. **Active troops:** 58,000.

Economy: Industries: Oil, food, tobacco, sugar. **Chief crops:** Sugarcane, tobacco, rice, coffee, citrus. **Minerals:** Cobalt, nickel, iron, copper, manganese, salt. **Crude oil reserves** (2000): 283.5 mil bbls. **Other resources:** Timber. **Arable land:** 24%. **Livestock** (2000): cattle: 4.7 mil; chickens: 15.0 mil; goats: 140,000; pigs: 2.8 mil; sheep: 310,000. **Fish catch** (1999): 122,823 metric tons. **Electricity prod.:** 14.358 bil kWh. **Labor force:** 23% agric., 53% services, 24% ind.

Finance: Monetary unit: Peso (Sept. 2001: 1.00 = $1 U.S.). **GDP** (1999 est.): $18.6 bil. **Per capita GDP:** $1,700. **Imports** (1999 est.): $3.2 bil; partners: Spain 16%, Venezuala 15%. **Exports** (1999 est.): $1.4 bil; partners: Russia 25%, Netherlands 23%. **Tourism:** $1.71 bil.

Transport: Railroad: Length: 2,987 mi. **Motor vehicles:** 16,500 pass. cars, 30,000 comm. vehicles. **Civil aviation:** 2.2 bil pass.-mi.; 14 airports. **Chief ports:** Havana, Matanzas, Cienfuegos, Santiago de Cuba.

Communications: TV sets: 200 per 1,000 pop. **Radios:** 327 per 1,000 pop. **Telephones**: 488,600 main lines. **Daily newspaper circ.:** 122 per 1,000 pop.

Health: Life expectancy: 74.02 male; 78.94 female. **Births** (per 1,000 pop): 12.36. **Deaths** (per 1,000 pop.): 7.33. **Natural inc.:** .5%. **Infant mortality** (per 1,000 live births): 7.39.

Education: Free, compulsory: ages 6-11. **Literacy:** 96%.

Major Intl. Organizations: UN (FAO, ILO, IMO, WHO, WTrO).

Some 50,000 Indians lived in Cuba when it was reached by Columbus in 1492. Its name derives from the Indian Cubanacan. Except for British occupation of Havana, 1762-63, Cuba remained Spanish until 1898. A slave-based sugar plantation economy developed from the 18th century, aided by early mechanization of milling. Sugar remains the chief product and chief export despite government attempts to diversify.

A ten-year uprising ended in 1878 with guarantees of rights by Spain, which Spain failed to carry out. A full-scale movement under Jose Marti began Feb. 24, 1895.

The U.S. declared war on Spain in Apr. 1898, after the sinking of the USS *Maine* in Havana harbor, and defeated it in the Spanish-American War. Spain gave up all claims to Cuba. U.S. troops withdrew in 1902, but under 1903 and 1934 agreements, the U.S. leases a site at Guantánamo Bay in the SE as a naval base. U.S. and other foreign investments acquired a dominant role in the economy. In 1952, former Pres. Fulgencio Batista seized control and established a dictatorship, which grew increasingly harsh and corrupt. Fidel Castro assembled a rebel band in 1956; guerrilla fighting intensified in 1958. Batista fled Jan. 1, 1959, and in the resulting political vacuum Castro took power, becoming premier Feb. 16.

The government began a program of sweeping economic and social changes, without restoring promised liberties. Opponents were imprisoned, and some were executed. Some 700,000 Cubans emigrated in the first years after the Castro takeover, mostly to the U.S.

Cattle and tobacco lands were nationalized, while a system of cooperatives was instituted. By 1960 all banks and industrial companies had been nationalized, including over $1 billion worth of U.S.-owned properties, mostly without compensation.

Poor sugar crops resulted in farm collectivization, tight labor controls, and rationing, despite continued aid from the USSR and other Communist nations. A U.S.-imposed export embargo in 1962 severely damaged the economy.

In 1961, some 1,400 Cubans, trained and backed by the U.S. Central Intelligence Agency, unsuccessfully tried to invade and overthrow the regime. In the fall of 1962, the U.S. learned the USSR had brought nuclear missiles to Cuba. After an Oct. 22 warning from Pres. John F. Kennedy, the missiles were removed.

In 1977, Cuba and the U.S. signed agreements to exchange diplomats, without restoring full ties, and to regulate offshore fishing. In 1978 and 1980, the U.S. agreed to accept political prisoners released by Cuba, some of whom were criminals and mental patients. A 1987 agreement provided for 20,000 Cubans to emigrate to the U.S. each year; Cuba agreed to take back some 2,500 jailed in the U.S. since 1980.

In 1975-78, Cuba sent troops to aid one faction in the Angola civil war; the last Cuban troops were withdrawn by May 1991. Cuba's involvement in Central America, Africa, and the Caribbean contributed to poor relations with the U.S.

Cuba's economy, dependent on aid from other Communist countries, was severely shaken by the collapse of the Communist bloc in the late 1980s. Stiffer trade sanctions enacted by the U.S. in 1992 made things worse. Antigovernment demonstrations in Aug. 1994 prompted Castro to loosen emigration restrictions. A new U.S.-Cuba accord in Sept. ended the exodus of "boat people" after more than 30,000 had left Cuba. In another policy shift, the U.S. announced May 2, 1995, it would admit 20,000 Cuban refugees held at the Guantánamo base but would send further boat people back to Cuba.

The U.S. imposed additional sanctions after Cuba, Feb. 24, 1996, shot down 2 aircraft operated by an anti-Castro exile group based in Miami. Cuba blamed exile groups for bombings at Havana tourist hotels, July-Sept. 1997. Pope John Paul II visited Cuba, Jan. 21-25, 1998; he called for an end to U.S. trade sanctions, while pressing Castro to release political prisoners and allow political and religious freedom. U.S. restrictions on contact with Cuba were eased in 1999. On June 28, 2000, Elián González was returned to Cuba to live with his father, ending a 7-month legal battle that began when the boy was rescued off Florida from a shipwreck in which his mother was killed; the boy's Miami relatives had sought to keep him in the U.S.

Cyprus

Republic of Cyprus

(Figures below marked with a # do not include Turkish-held area—Turkish Republic of Northern Cyprus.)

People: Population: 762,887. **Age distrib.** (%): <15: 22.9; 65+: 10.8. **Pop. density:** 214 per sq. mi. **Urban:** 56%. **Ethnic groups:** Greek 78%, Turkish 18%. **Principal languages:** Greek, Turkish, English. **Chief religions:** Greek Orthodox 78%, Muslim 18%.

Geography: Area: 3,600 sq. mi. **Location:** In eastern Mediterranean Sea, off Turkish coast. **Neighbors:** Nearest are Turkey on N, Syria and Lebanon on E. **Topography:** Two mountain ranges run E-W, separated by a wide, fertile plain. **Capital:** Nicosia, (1999 est.) 197,800.

Government: Type: Republic. **Head of state and gov.:** Pres. Glafcos Clerides; b Apr. 24, 1919; in office: Mar. 1, 1993. **Local divisions:** 6 districts. **Defense budget:** $573 mil. **Active troops#:** 10,000.

Economy: Industries: Food, beverages, textiles. **Chief crops:** Barley, grapes, vegetables, citrus, potatoes, olives. **Minerals:** Copper, pyrites, asbestos. **Arable land:** 12%. **Livestock** (2000): cattle: 55,841; chickens: 3.8 mil; goats: 322,000; pigs: 455,000; sheep: 240,000.**Electricity prod.:** 2.951 bil. kWh. **Labor force:** 66.6% serv.; 23.2% ind.; 10.2% agric.

Finance: Monetary unit: Pound (Sept. 2001: 0.63 = $0.66 U.S.). **GDP** (1999 est.): $9 bil. **Per capita GDP**: $15,400. **Imports** (1998 est.): $3.5 bil; partners: U.S. 12.5%, UK 11.3%, Italy 9.4%. **Exports** (1998 est.): $1.1 bil; partners: Russia 14.5%, U.K. 14.5%. **Tourism:** $1.89 bil. **Budget** (1997 est.): $3.4 bil. **Intl. reserves less gold** (May 2000): $1.88 bil. **Gold:** 464,000 oz t. **Consumer prices** (change in 1999): 1.6%.

Transport: Motor vehicles (1997): 234,976 pass. cars, 108,452 comm. vehicles. **Civil aviation**: 1.7 bil pass.-mi.; 2 airports. **Chief ports:** Famagusta, Limassol.

Communications: Television sets: 160 per 1000 pop. **Radios:** 287 per 1,000 pop. **Telephones**: 440,100 main lines. **Daily newspaper circ.**: 135 per 1,000 pop.

Health: Life expectancy: 74.6 male; 79.3 female. **Births** (per 1,000 pop): 13.08. **Deaths** (per 1,000 pop.): 7.65. **Natural inc.:** .54%. **Infant mortality** (per 1,000 live births): 7.89.

Education: Free, compulsory: ages 5½-15. **Literacy** (1994): 95%.

Major Intl. Organizations: UN (FAO, IBRD, ILO, IMF, IMO, WHO, WTrO), the Commonwealth, OSCE.

Embassy: 2211 R St. NW 20008; 462-5772.

Agitation for enosis (union) with Greece increased after World War II, with the Turkish minority opposed, and broke into violence in 1955-56. In 1959, Britain, Greece, Turkey, and Cypriot leaders approved a plan for an independent republic, with constitutional guarantees for the Turkish minority and permanent division of offices on an ethnic basis. Greek and Turkish Communal Chambers dealt with religion, education, and other matters.

Archbishop Makarios III, formerly the leader of the enosis movement, was elected president, and full independence became final Aug. 16, 1960. Further communal strife led the United Nations to send a peacekeeping force in 1964; its mandate has been repeatedly renewed.

The Cypriot National Guard, led by officers from the army of Greece, seized the government July 15, 1974. On July 20, Turkey invaded the island; Greece mobilized its forces but did not intervene. A cease-fire was arranged but collapsed. By Aug. 16, Turkish forces had occupied the NE 40% of the island, despite the presence of UN peacekeeping forces.

Turkish Cypriots voted overwhelmingly, June 8, 1975, to form a separate Turkish Cypriot federated state. A president and assembly were elected in 1976. Some 200,000 Greeks have been expelled from the Turkish-controlled area, replaced by thousands of Turks, some from the mainland.

Turkish Republic of Northern Cyprus

A declaration of independence was announced by Turkish-Cypriot leader Rauf Denktash, Nov. 15, 1983. The state is not internationally recognized, although it does have trade relations with some countries. Area of TRNC: 1,295 sq mi.; pop. (1995 est.): 134,000, 99% Turkish; capital: Lefkosa (Nicosia).

Czech Republic

People: Population: 10,264,212. **Age distrib.** (%): <15: 16.1; 65+: 13.9. **Pop. density:** 338 per sq. mi. **Urban:** 75%. **Ethnic groups:** Czech 94.4%, Slovak 3%. **Principal languages:** Czech, Slovak. **Chief religions:** Atheist 39.8%, Roman Catholic 39.2%, Protestant 4.6%, Orthodox 3%.

Geography: Area: 30,300 sq. mi. **Location:** In E central Europe. **Neighbors:** Poland on N, Germany on N and W, Austria on S, Slovakia on E and SE. **Topography:** Bohemia, in W, is a plateau surrounded by mountains; Moravia is hilly. **Capital:** Prague, (urban agg.): 1,226,000.

Government: Type: Republic. **Head of state:** Vaclav Havel; b Oct. 5, 1936; in office: Feb. 15, 1993. **Head of gov.:** Prime Min. Milos Zeman; b Sept. 28, 1944; in office: Jul. 17, 1998. **Local divisions:** 73 districts, 4 municipalities. **Defense budget:** $1.15 bil. **Active troops:** 57,700.

Economy: Industries: Machinery, fuels, glass, motor vehicles. **Chief crops:** Wheat, sugar beets, potatoes, hops, fruit. **Minerals:** Coal, kaolin. **Arable land:** 41%. **Crude oil reserves** (2000): 15 mil bbls. **Livestock** (2000): cattle: 1.57 mil; chickens: 29.5 mil; goats: 31,988; pigs: 3.69 mil; sheep: 84,108. . **Fish catch:** (1999): 20,881 metric tons. **Electricity prod.:** 60.671 bil kWh. **Labor force:** 32% ind.; 8.7% constr.; 46.8% serv.

Finance: Monetary unit: Koruna (Sept. 2001: 37.70 = $1 U.S.). **GDP** (1999 est.): $120.8 bil. **Per capita GDP:** $11,700. **Imports** (1999): $29 bil; partners: Germany 34%. Slovakia 6%. **Exports** (1999): $26.9 bil; partners: Germany 42%. Slovakia 8%. **Tourism:** $3.04 bil. **Budget** (1997 est.): $16.6 bil. **Intl. reserves less gold** (May 2000): $12.82 bil. **Gold:** 446,000 mil oz t. **Consumer prices** (change in 1999): 2.1%.

Transport: Railroad: Length: 5,860 mi. **Motor vehicles:** 4.41 mil pass. cars, 514,589 comm. vehicles. **Civil aviation:** 1.5 bil pass.-mi.; 2 airports. **Chief ports:** Decin, Prague, Usti nad Labem.

Communications: TV sets: 446 per 1,000 pop. **Telephones:** 3,871,700 main lines. **Daily newspaper circ.:** 254 per 1,000 pop.

Health: Life expectancy: 71.23 male; 78.43 female. **Births** (per 1,000 pop): 9.11. **Deaths** (per 1,000 pop.): 10.81. **Natural inc.:** -.17%. **Infant mortality** (per 1,000 live births): 5.55.

Education: Compulsory: ages 6-15. **Literacy** (1998 est.): 99%.

Major Intl. Organizations: UN (FAO, IBRD, ILO, IMF, IMO, WHO, WTrO), NATO, OECD, OSCE.

Embassy: 3900 Spring of Freedom St. NW 20008; 363-6315.

Bohemia and Moravia were part of the Great Moravian Empire in the 9th century and later became part of the Holy Roman Empire. Under the kings of Bohemia, Prague in the 14th century was the cultural center of Central Europe. Bohemia and Hungary became part of Austria-Hungary.

In 1914-18 Thomas G. Masaryk and Eduard Benes formed a provisional government with the support of Slovak leaders including Milan Stefanik. They proclaimed the Republic of Czechoslovakia Oct. 28, 1918.

Czechoslovakia

By 1938 Nazi Germany had worked up disaffection among German-speaking citizens in Sudetenland and demanded its cession. British Prime Min. Neville Chamberlain, with the acquiescence of France, signed with Hitler at Munich, Sept. 30, 1938, an agreement to the cession, with a guarantee of peace by Hitler and Mussolini. Germany occupied Sudetenland Oct. 1-2.

Hitler on Mar. 15, 1939, dissolved Czechoslovakia, made protectorates of Bohemia and Moravia, and supported the autonomy of Slovakia, proclaimed independent Mar. 14, 1939.

Soviet troops with some Czechoslovak contingents entered eastern Czechoslovakia in 1944 and reached Prague in May 1945; Benes returned as president. In May 1946 elections, the Communist Party won 38% of the votes, and Benes accepted Klement Gottwald, a Communist, as prime minister.

In Feb. 1948, the Communists seized power in advance of scheduled elections. In May 1948 a new constitution was approved. Benes refused to sign it. On May 30 the voters were offered a one-slate ballot and the Communists won full control. Benes resigned June 7 and Gottwald became president. The country was renamed the Czechoslovak Socialist Republic. A harsh Stalinist period followed, with complete and violent suppression of all opposition.

In Jan. 1968 a liberalization movement spread nations explosively through Czechoslovakia. Antonin Novotny, long the Stalinist ruler, was deposed as party leader and succeeded by Alexander Dubcek, a Slovak, who supported democratic reforms. On Mar. 22 Novotny resigned as president and was succeeded by Gen. Ludvik Svoboda. On Apr. 6, Prem. Joseph Lenart resigned and was succeeded by Oldrich Cernik, a reformer.

In July 1968 the USSR and 4 Warsaw Pact nations demanded an end to liberalization. On Aug. 20, the Soviet, Polish, East German, Hungarian, and Bulgarian armies invaded Czechoslovakia. Despite demonstrations and riots by students and workers, press censorship was imposed, liberal leaders were ousted from office and promises of loyalty to Soviet policies were made by some old-line Communist Party leaders.

On Apr. 17, 1969, Dubcek resigned as leader of the Communist Party and was succeeded by Gustav Husak. In Jan. 1970, Cernik was ousted. Censorship was tightened, and the Communist Party expelled a third of its members. In 1973, amnesty was offered to some of the 40,000 who fled the country after the 1968 invasion, but repressive policies continued.

More than 700 leading Czechoslovak intellectuals and former party leaders signed a human rights manifesto in 1977, called Charter 77, prompting a renewed crackdown by the regime.

The police crushed the largest antigovernment protests since 1968, when tens of thousands of demonstrators took to the streets of Prague, Nov. 17, 1989. As protesters demanded free elections, the Communist Party leadership resigned Nov. 24; millions went on strike Nov. 27.

On Dec. 10, 1989, the first cabinet in 41 years without a Communist majority took power; Vaclav Havel, playwright and human rights campaigner, was chosen president, Dec. 29. In Mar. 1990 the country was officially renamed the Czech and Slovak Federal Republic. Havel failed to win reelection July 3, 1992; his bid was blocked by a Slovak-led coalition.

Slovakia declared sovereignty, July 17. Czech and Slovak leaders agreed, July 23, on a basic plan for a peaceful division of Czechoslovakia into 2 independent states.

Czech Republic

Czechoslovakia split into 2 separate states—the Czech Republic and Slovakia—on Jan. 1, 1993. Havel was elected president of the Czech Republic on Jan. 26. Record floods in July 1997 caused more than $1.7 billion in damage. The country became a full member of NATO on Mar. 12, 1999.

Denmark

Kingdom of Denmark

People: Population: 5,352,815. **Age distrib.** (%): <15: 18.6; 65+: 14.9. **Pop. density:** 328 per sq. mi. **Urban:** 85%. **Ethnic groups:** Scandinavian, Eskimo, Faroese, German. **Principal languages:** Danish, Faroese. **Chief religion:** Evangelical Lutheran 91%.

Geography: Area: 16,300 sq. mi. **Location:** In N Europe, separating the North and Baltic seas. **Neighbors:** Germany on S, Norway on NW, Sweden on NE. **Topography:** Denmark consists of the Jutland Peninsula and about 500 islands, 100 inhabited. The land is flat or gently rolling and is almost all in productive use. **Capital:** Copenhagen, (urban agg.): 1,388,000.

Government: Type: Constitutional monarchy. **Head of state:** Queen Margrethe II; b Apr. 16, 1940; in office: Jan. 14, 1972. **Head of gov.:** Prime Min. Poul Nyrup Rasmussen; b June 15, 1943; in office: Jan. 25, 1993. **Local divisions:** 14 counties, 2 kommunes. **Defense budget:** $2.3 bil (2001). **Active troops:** 21,810.

Economy: Industries: Food processing, machinery, textiles, furniture, electronics. **Chief crops:** Grains, potatoes, sugar beets. **Minerals:** Oil, gas, salt. **Crude oil reserves** (2000): 1.07 bil bbls. **Arable land:** 60%. **Livestock** (2000): cattle: 1.85 mil; chickens: 19.97 mil; pigs: 11.55 mil; sheep: 142,900. **Fish catch** (1999): 1.87 mil metric tons. **Electricity prod.:** 37.885 bil kWh. **Labor force:** 71% serv.; 25% ind., 4% agric.

Finance: Monetary unit: Danish Krone (Sept. 2001: 8.20 = $1 U.S.). **GDP** (1999 est.): $127.7 bil. **Per capita GDP:** $23,800. **Imports** (1999): $43.9 bil; partners: EU 72.5%. **Exports** (1999): $49.5 bil; partners: EU 66.6%. **Tourism:** $3.68 bil. **Budget** (1996 est.): $66.4 bil. **Intl. reserves less gold** (May 2000): $16.07 bil. **Gold:** 2.0 mil oz t. **Consumer prices** (change in 1999): 2.5%.

Transport: Railroad: Length: 1,780 mi. **Motor vehicles** (1997): 1.79 mil pass. cars, 306,403 comm. vehicles. **Civil aviation:** 3.5 bil pass.-mi.; 13 airports. **Chief ports:** Copenhagen, Alborg, Arhus, Odense.

Communications: TV sets: 569 per 1,000 pop. **Radios:** 1,145 per 1,000 pop. **Telephones** : 4,011,100 main lines. **Daily newspaper circ.:** 309 per 1,000 pop.

Health: Life expectancy: 74.12 male; 79.47 female. **Births** (per 1,000 pop): 11.96. **Deaths** (per 1,000 pop.): 10.9. **Natural inc.:** .11%. **Infant mortality** (per 1,000 live births): 5.04.

Education: Compulsory: ages 7-15. **Literacy** (1998): 100%.

Major Intl. Organizations: UN and all of its specialized agencies, EU, NATO, OECD, OSCE.

Embassy: 3200 Whitehaven St. NW 20008; 234-4300.

Website: http://www.denmark.org

The origin of Copenhagen dates back to ancient times, when the fishing and trading place named Havn (port) grew up on a cluster of islets, but Bishop Absalon (1128-1201) is regarded as the actual founder of the city.

Danes formed a large component of the Viking raiders in the early Middle Ages. The Danish kingdom was a major power until the 17th century, when it lost its land in southern Sweden. Norway was separated in 1815, and Schleswig-Holstein in 1864. Northern Schleswig was returned in 1920.

Voters ratified the Maastricht Treaty, the basic document of the European Union, in May 1993, after rejecting it in 1992. On Sept. 28, 2000, Danes voted not to join the euro currency zone.

The **Faroe Islands** in the North Atlantic, about 300 mi. NW of the Shetlands, and 850 mi. from Denmark proper, 18 inhabited, have an area of 540 sq. mi. and pop. (2001 est.) of 45,661. They are an administrative division of Denmark, self-governing in most matters. Torshavn is the capital. Fish is a primary export (345, 415 metric tons in 1999).

Greenland (Kalaallit Nunaat)

Greenland, a huge island between the North Atlantic and the Polar Sea, is separated from the North American continent by Davis Strait and Baffin Bay. Its total area is 840,000 sq. mi., 84% of which is ice-capped. Most of the island is a lofty plateau 9,000 to 10,000 ft. in altitude. The average thickness of the cap is 1,000 ft. The population (2000 est.) is 56,309. Under the 1953 Danish constitution the colony became an integral part of the realm with representatives in the Folketing (Danish legislature). The Danish parliament, 1978, approved home rule for Greenland, effective May 1, 1979. With home rule, Greenlandic place names came into official use. The technically correct name for Greenland is now Kalaallit Nunaat; the official name for its capital is Nuuk, rather than Godthab. Fish is the principal export (120,596 metric tons in 1999).

Djibouti

Republic of Djibouti

People: Population: 460,700. **Age distrib.** (%): <15: 42.6; 65+: 2.8. **Pop. density:** 54 per sq. mi. **Urban:** 83%. **Ethnic groups:** Somali 60%, Afar 35%. **Principal languages:** French, Arabic (both official); Afar, Somali. **Chief religions:** Muslim 94%, Christian 6%.

Geography: Area: 8,500 sq. mi. **Location:** On E coast of Africa, separated from Arabian Peninsula by the strategically vital strait of Bab el-Mandeb. **Neighbors:** Ethiopia on W and SW, Eritrea on NW, Somalia on SE. **Topography:** The territory, divided into a low coastal plain, mountains behind, and an interior plateau, is arid, sandy, and desolate. The climate is generally hot and dry. **Capital:** Djibouti (1995): 383,000.

Government: Type: Republic. **Head of state:** Pres. Ismail Omar Guelleh; b 1947; in office: May 8, 1999. **Head of gov.:** Prime Min. Dileita Mohamed Dileita; b Mar. 12, 1958; in office: Mar. 7, 2001. **Local divisions:** 5 districts. **Defense budget:** $23 mil. **Active troops:** 9,600.

Economy: Based on service activities. **Livestock** (2000): cattle: 269,000; goats: 513,000; sheep: 465,000. **Electricity prod.:** 180 mil kWh. **Labor force:** 75% agric., 11% ind., 14% serv.

Finance: Monetary unit: Djibouti Franc (Sept. 2001: 172 = $1 U.S.). **GDP** (1999 est.): $550 mil. **Per capita GDP:** $1,200. **Imports** (1999 est.): $440 mil; partners: France 13%, Ethiopia 12%. **Exports** (1999 est.): $260 mil; partners: Somalia 53%, Yemen 23%. **Tourism** (1998): $4 mil. **Budget** (1997 est.): $175 mil. **Intl. reserves less gold** (June 2000): $62.51 mil.

Transport: Railroad: Length: 66 mi. **Motor vehicles:** 13,000 pass. cars, 3,000 comm. vehicles. **Civil aviation:** 1 airport. **Chief port:** Djibouti.

Communications: TV sets: 43 per 1,000 pop. **Radios:** 80 per 1,000 pop. **Telephones** (1999): 8,800 main lines. **Daily newspaper circ.:** 8 per 1,000 pop.

Health: Life expectancy: 49.37 male; 53.1 female. **Births** (per 1,000 pop): 40.66. **Deaths** (per 1,000 pop.): 14.66. **Natural inc.:** 2.6%. **Infant mortality** (per 1,000 live births): 101.51.

Education: Literacy: 46%.

Major Intl. Organizations: UN (FAO, IBRD, ILO, IMF, IMO, WHO, WTrO), AL, OAU.

Embassy: Suite 515, 1156 15th St. NW 20005; 331-0270.

France gained control of the territory in stages between 1862 and 1900. As French Somaliland it became an overseas territory of France in 1945; in 1967 it was renamed the French Territory of the Afars and the Issas.

Ethiopia and Somalia have renounced their claims to the area, but each has accused the other of trying to gain control. There were clashes between Afars (ethnically related to Ethiopians) and Issas (related to Somalis) in 1976. Immigrants from both countries continued to enter the country up to independence, which came June 27, 1977.

French aid is the mainstay of the economy, as well as assistance from Arab countries. A peace accord Dec. 1994 ended a 3-year-long uprising by Afar rebels.

Dominica

Commonwealth of Dominica

People: Population: 70,786. **Age distrib.:** (%): <15: 28.7; 65+: 7.8. **Pop. density:** 244 per sq. mi. **Urban:** 71%. **Ethnic groups:** Black, Carib Amerindian. **Principal languages:** English (official), French patois. **Chief religions:** Roman Catholic 77%, Protestant 15%.

Geography: Area: 300 sq. mi. **Location:** In Eastern Caribbean, most northerly Windward Isl. **Neighbors:** Guadeloupe to N, Martinique to S. **Topography:** Mountainous, a central ridge running from N to S, terminating in cliffs; volcanic in origin, with numerous thermal springs; rich deep topsoil on leeward side, red tropical clay on windward coast. **Capital:** Roseau (1995 est.): 21,000.

Government: Type: Parliamentary democracy. **Head of state:** Pres. Vernon Lorden Shaw; b May 13, 1930; in office: Oct. 6, 1998. **Head of gov.:** Prime Min. Pierre Charles; b 1954; in office: Oct. 3, 2000. **Local divisions:** 10 parishes.

Economy: Industries: Soap, tourism. **Chief crops:** Bananas, citrus, mangoes, coconuts. **Other resources:** Forests. **Arable land:** 9%. **Livestock** (2000): cattle: 13,400; chickens: 190,000; goats: 9,700; pigs: 5,000; sheep: 7,600. **Electricity prod.:** 62 mil kWh. **Labor force:** 40% agric.; 32% ind. & commerce; 28% services.

Finance: Monetary unit: East Caribbean Dollar (Sept. 2001: 2.67 = $1 U.S.). **GDP** (1998 est.): $225 mil. **Per capita GDP:** $3,400. **Imports** (1998): $120.4 mil; partners: U.S. 41%. **Exports** (1998): $60.8 mil; partners: UK 36%. **Tourism** (1998): $38 mil. **Budget** (FY 1995-96): $78 mil. **Intl. reserves less gold** (Dec. 1999): $31.57 mil. **Consumer prices** (change in 1999): 1.2%.

Transport: Motor vehicles (1997): 7,560 pass. cars, 3,673 comm. vehicles. **Civil aviation:** 2 airports. **Chief port:** Roseau.

Communications: TV sets: 70 per 1,000 pop. **Radios:** 875 per 1,000 pop. **Telephones** (1999): 21,300 main lines.

Health: Life expectancy: 70.74 male; 76.61 female. **Births** (per 1,000 pop): 17.81. **Deaths** (per 1,000 pop.): 7.19. **Natural inc.:** 1.06%. **Infant mortality** (per 1,000 live births): 16.54.

Education: Free, compulsory: ages 5-15. **Literacy** (1993): 90%.

Major Intl. Organizations: UN (FAO, IBRD, ILO, IMF, IMO, WHO, WTrO), Caricom, the Commonwealth, OAS, OECS.

Embassy: 3216 New Mexico Ave. NW 20016; 364-6781.

A British colony since 1805, Dominica was granted self-government in 1967. Independence was achieved Nov. 3, 1978.

Hurricane David struck, Aug. 30, 1979, devastating the island and destroying the banana plantations, Dominica's economic mainstay. Coups were attempted in 1980 and 1981.

Dominica participated in the 1983 U.S.-led invasion of nearby Grenada.

Dominican Republic

People: Population: 8,581,477. **Age distrib.** (%): <15: 34.1; 65+: 4.9. **Pop. density:** 460 per sq. mi. **Urban:** 64%. **Ethnic groups:** Mixed 73%, white 16%, black 11%. **Principal language:** Spanish (official). **Chief religion:** Roman Catholic 95%.

Geography: Area: 18,700 sq. mi. **Location:** In West Indies, sharing isl. of Hispaniola with Haiti. **Neighbors:** Haiti on W, Puerto Rico (U.S.) to E. **Topography:** The Cordillera Central range crosses the center of the country, rising to over 10,000 ft., highest in the Caribbean. The Cibao Valley to the N is major agricultural area. **Capital:** Santo Domingo. **Cities** (urban agg.): Santo Domingo 3,599,000; Santiago de los Caballeros 1,539,000.

Government: Type: Republic. **Head of state and gov.:** Pres. Hipólito Mejía; b Feb. 22, 1941; in office: Aug. 16, 2000. **Local divisions:** 29 provinces and national district. **Defense budget:** $2.4 bil. **Active troops:** 24,500.

Economy: Industries: Sugar processing, cement, tourism. **Chief crops:** Sugar, cocoa, coffee, cotton, rice. **Minerals:**

Nickel, bauxite, gold, silver. **Arable land:** 21%. **Livestock** (2000): cattle: 1.9 mil; chickens: 46.0 mil; goats: 170,000; pigs: 538,599; sheep: 105,454.. **Fish catch:** (1999): 15,276 metric tons. **Electricity prod.:** 7.290 bil kWh. **Labor force:** 58.7% serv. and govt., 24.3% ind., agric. 17%.

Finance: Monetary unit: Peso (Sept. 2001: 16.40 = $1 U.S.). **GDP** (1999 est.): $43.7 bil. **Per capita GDP:** $5,400. **Imports** (1999): $8.2 bil; partners: U.S. 56%, Venezuela 23%. **Exports** (1999): $5.1 bil; partners: U.S. 61.6%, Belgium 11.1%. **Tourism:** $2.52 bil. **Budget** (1999 est.): $2.9 bil. **Intl. reserves less gold** (June 2000): $494.3 mil. **Gold:** 18,000 oz t. **Consumer prices** (change in 1997): 6.5%.

Transport: Railroad: Length: 1,083 mi. **Motor vehicles:** 113,835 pass. cars, 92,198 comm. vehicles. **Civil aviation:** 9.8 mil pass.-mi.; 7 airports. **Chief ports:** Santo Domingo, San Pedro de Macoris, Puerto Plata.

Communications: TV sets: 97 per 1,000 pop. **Radios:** 154 per 1,000 pop. **Telephones** (1999): 820,900 main lines. **Daily newspaper circ.:** 35 per 1,000 pop.

Health: Life expectancy: 71.34 male; 75.64 female. **Births** (per 1,000 pop): 24.77. **Deaths** (per 1,000 pop.): 4.7. **Natural inc.:** 2.01%. **Infant mortality** (per 1,000 live births): 34.67.

Education: Compulsory: ages 6-14. **Literacy**: 82%.

Major Intl. Organizations: UN (FAO, IBRD, ILO, IMF, IMO, WHO, WTrO), OAS.

Embassy: 1715 22d St. NW 20008; 332-6280.

Carib and Arawak Indians inhabited the island of Hispaniola when Columbus landed in 1492. The city of Santo Domingo, founded 1496, is the oldest settlement by Europeans in the hemisphere and has the supposed ashes of Columbus in an elaborate tomb in its ancient cathedral.

The western third of the island was ceded to France in 1697. Santo Domingo itself was ceded to France in 1795. Haitian leader Toussaint L'Ouverture seized it, 1801. Spain returned intermittently 1803-21, as several native republics came and went. Haiti ruled again, 1822-44; Spanish occupation occurred 1861-63.

The country was occupied by U.S. Marines from 1916 to 1924, when a constitutionally elected government was installed.

In 1930, Gen. Rafael Leonidas Trujillo Molina was elected president. Trujillo ruled brutally until his assassination in 1961. Pres. Joaquín Balaguer, appointed by Trujillo in 1960, resigned under pressure in 1962.

Juan Bosch, elected president in the first free elections in 38 years, was overthrown in 1963. On Apr. 24, 1965, a revolt was launched by followers of Bosch and others, including a few Communists. Four days later U.S. Marines intervened against pro-Bosch forces. Token units were later sent by 5 South American countries as a peacekeeping force. A provisional government supervised a June 1966 election, in which Balaguer defeated Bosch. Balaguer remained in office for most of the next 28 years, but his May 1994 reelection was widely denounced as fraudulent. He cut short his term and on June 30, 1996, Leonel Fernández Reyna was elected.

Hurricane Georges struck Sept. 22, 1998, causing extensive property damage and claiming more than 200 lives. The leftist candidate, Hipólito Mejía, won a presidential vote May 16, 2000.

Ecuador

Republic of Ecuador

People: Population: 13,183,978. **Age distrib.** (%): <15: 35.8; 65+: 4.4. **Pop. density:** 123 per sq. mi. **Urban:** 64%. **Ethnic groups:** Mestizo 55%, Amerindian 25%, Spanish 10%, black 10%. **Principal languages:** Spanish (official), Quechua, other Amerindian. **Chief religion:** Roman Catholic 95%.

Geography: Area: 106,800 sq. mi. **Location:** In NW South America, on Pacific coast, astride the Equator. **Neighbors:** Colombia on N, Peru on E and S. **Topography:** Two ranges of Andes run N and S, splitting the country into 3 zones: hot, humid lowlands on the coast; temperate highlands between the ranges; and rainy, tropical lowlands to the E. **Capital:** Quito. **Cities (urban agg.):** Guayaquil: 2,293,000; Quito: 1,754,000.

Government: Type: Republic. **Head of state and gov.:** Pres. Gustavo Noboa Bejarano; b Aug. 21, 1937; in office: Jan. 22, 2000. **Local divisions:** 21 provinces. **Defense budget:** $400 mil. **Active troops:** 57,500.

Economy: Industries: Oil, food processing, metalwork, textiles. **Chief crops:** Bananas, cocoa, coffee, rice, sugar, potatoes, plantains. **Minerals:** Oil. **Crude oil reserves** (2000): 2.1 bil bbls. **Other resources:** Forests (leading balsawood producer), seafood (world's 2d largest shrimp producer.). **Arable land:** 6%. **Livestock** (2000): cattle: 5.11 mil; chickens: 130.0 mil; goats: 284,393; pigs: 2.87 mil; sheep: 2.13 mil. **Fish catch** (1999): 688,297 metric tons. **Electricity prod.:** 10.065 bil kWh. **Labor force:** 30% agric., 25% ind., 45% serv.

Finance: Monetary unit: U.S. dollar and sucre (Sept. 2001: 1 = $1 U.S.). **GDP** (1999 est.): $54.5 bil. **Per capita GDP:** $4,300. **Imports** (1999): $2.8 bil; partners: , U.S. 39%, Colombia 11%. **Exports** (1999): $4.1 bil; partners: U.S. 39%, Colombia 7%. **Tourism:** $343 mil. **Budget** (1999): $5.1 bil. **Intl. reserves less gold** (June 2000): $1.11 bil. **Gold:** 415,000 oz t. **Consumer prices** (change in 1999): 52.2%.

Transport: Railroad: Length: 600 mi. **Motor vehicles:** 255,640 pass. cars, 424,120 comm. vehicles. **Civil aviation:** 1.3 bil pass.-mi.; 14 airports. **Chief ports:** Guayaquil, Manta, Esmeraldas, Puerto Bolivar.

Communications: TV sets: 79 per 1,000 pop. **Radios:** 277 per 1,000 pop. **Telephones:** 1,265,200 main lines. **Daily newspaper circ.:** 72 per 1,000 pop.

Health: Life expectancy: 68.52 male; 74.28 female. **Births** (per 1,000 pop): 25.99. **Deaths** (per 1,000 pop.): 5.44. **Natural inc.:** 2.06%. **Infant mortality** (per 1,000 live births): 34.08.

Education: Free and compulsory for 6 years between ages 6-14. **Literacy**: 90%.

Major Intl. Organizations: UN (FAO, IBRD, ILO, IMF, IMO, WHO, WTrO), OAS.

Embassy: 2535 15th St. NW 20009; 234-7200.

The region, which was the northern Inca empire, was conquered by Spain in 1533. Liberation forces defeated the Spanish May 24, 1822, near Quito. Ecuador became part of the Great Colombia Republic but seceded, May 13, 1830.

Since 1972, the economy has revolved around petroleum exports; oil revenues have declined since 1982, causing severe economic problems. Ecuador suspended interest payments for 1987 on its estimated $8.2 billion foreign debt following a Mar. 5-6 earthquake that left 20,000 homeless and destroyed a stretch of the country's main oil pipeline.

Ecuadoran Indians staged protests in the 1990s to demand greater rights. A border war with Peru flared from Jan. 26, 1995, until a truce took effect Mar. 1. Vice-Pres. Alberto Dahik resigned and fled Ecuador, Oct. 11, 1995, to avoid arrest on corruption charges. Elected president in a runoff, July 7, 1996, Abdalá Bucaram—a populist known as El Loco, or "The Crazy One"—imposed stiff price increases and other austerity measures. His rising unpopularity and erratic behavior led the National Congress, Feb. 6, 1997, to dismiss him for "mental incapacity." Bucaram went into exile, and Congress, on Feb. 11, confirmed its leader, Fabián Alarcón, as president for 18 months. Voters endorsed the actions in a referendum May 25.

Jamil Mahuad Witt, mayor of Quito, won a presidential runoff election July 12, 1998. In Sept. 1998 and Mar. 1999 he imposed emergency measures to cope with a continuing economic crisis. Opposed by Indian groups and military leaders, he was ousted Jan. 21, 2000, and succeeded by Vice-Pres. Gustavo Noboa Bejarano. Noboa went ahead with a plan introduced by Mahuad to replace the sucre with the U.S. dollar as Ecuador's currency.

The **Galápagos Islands,** pop. (2001 est.) 16,000, about 600 mi. to the W, are the home of huge tortoises and other unusual animals. The oil tanker *Jessica* ran aground Jan. 16, 2001, off San Cristóbal Is., spilling some 185,000 gallons of fuel.

Egypt

Arab Republic of Egypt

People: Population: 69,536,644. **Age distrib** (%) <15: 34.6; 65+: 3.8. **Pop. density:** 181 per sq. mi. **Urban:** 45%. **Ethnic groups:** Eastern Hamitic stock (Egyptian, Bedouin, Berber) 99%. **Principal languages:** Arabic (official), English, French. **Chief religions:** Muslim (mostly Sunni) 94%, Coptic Christian and other 6%.

Geography: Area: 383,900 sq. mi. **Location:** Northeast corner of Africa. **Neighbors:** Libya on W, Sudan on S, Israel and Gaza Strip on E. **Topography:** Almost entirely desolate and barren, with hills and mountains in E and along Nile. The Nile Valley, where most of the people live, stretches 550 miles. **Capital:** Cairo. **Cities (urban agg.)**: Cairo 10,552,000; Alexandria 4,113,000.

Government: Type: Republic. **Head of state:** Pres. Hosni Mubarak; b May 4, 1928; in office: Oct. 14, 1981. **Head of gov.:** Prime Min. Atef Obeid; b Apr. 14, 1932; in office: Oct. 5, 1999. **Local divisions:** 26 governorates. **Defense budget:** $2.5 bil (1999). **Active troops:** 448,500.

Economy: Industries: Textiles, tourism, chemicals, oil, food processing, cement. **Chief crops:** Cotton, rice, beans, fruits, wheat, vegetables, corn. **Minerals:** Oil, gas, phosphates, gypsum, iron, manganese, limestone. **Crude oil reserves** (2000): 2.9 bil bbls. **Arable land:** 2%. **Livestock** (2000): cattle: 3.18 mil; chickens: 88.0 mil; goats: 3.3 mil; pigs: 29,500; sheep: 4.45

mil. **Fish catch** (1999): 418,694 metric tons. **Electricity prod.:** 64.685 bil kWh. **Labor force:** 38% serv. & gov't; 40% agric.; 22% ind.

Finance: Monetary unit: Pound (Sept. 2001: 4.26 = $1 U.S.). **GDP** (1999 est.): $200 bil. **Per capita GDP:** $3,000. **Imports** (1999 est.): $15.8 bil; partners: U.S. 16%, EU 42%. **Exports** (1999 est.): $4.6 bil; partners: EU 47%, U.S. 14%. **Tourism:** $3.90 bil. **Budget** (FY 1997-98): $20.8 bil. **Intl. reserves less gold** (Apr. 2000): $13.83 bil. **Gold:** 2.43 mil oz t. **Consumer prices** (change in 1999): 3.1%.

Transport: Railroad: Length: 2,989 mi. **Motor vehicles:** 1.28 mil pass. cars, 423,300 comm. vehicles. **Civil aviation:** 5.6 bil pass.-mi.; 11 airports. **Chief ports:** Alexandria, Port Said, Suez, Damietta.

Communications: TV sets: 127 per 1,000 pop. **Radios:** 312 per 1,000 pop. **Telephones**: 5,483,600 main lines. **Daily newspaper circ.:** 38 per 1,000 pop.

Health: Life expectancy: 61.62 male; 65.85 female. **Births** (per 1,000 pop): 24.89. **Deaths** (per 1,000 pop.): 7.7. **Natural inc.:** 1.72%. **Infant mortality** (per 1,000 live births): 60.46.

Education: Compulsory for 5 years between ages 6-13. **Literacy**: 51%.

Major Intl. Organizations: UN (FAO, IBRD, ILO, IMF, IMO, WHO, WTrO), AL, OAU.

Embassy: 3521 International Ct. NW 20008; 895-5400.

Website: http://www.idsc.gov.eg

Archaeological records of ancient Egyptian civilization date back to 4000 BC. A unified kingdom arose around 3200 BC and extended its way south into Nubia and as far north as Syria. A high culture of rulers and priests was built on an economic base of serfdom, fertile soil, and annual flooding of the Nile.

Imperial decline facilitated conquest by Asian invaders (Hyksos, Assyrians). The last native dynasty fell in 341 BC to the Persians, who were in turn replaced by Greeks (Alexander and the Ptolemies), Romans, Byzantines, and Arabs, who introduced Islam and the Arabic language. The ancient Egyptian language is preserved only in Coptic Christian liturgy.

Egypt was ruled as part of larger Islamic empires for several centuries. The Mamluks, a military caste of Caucasian origin, ruled Egypt from 1250 until defeat by the Ottoman Turks in 1517. Under Turkish sultans the khedive as hereditary viceroy had wide authority. Britain intervened in 1882 and took control of administration, though nominal allegiance to the Ottoman Empire continued until 1914.

The country was a British protectorate from 1914 to 1922. A 1936 treaty strengthened Egyptian autonomy, but Britain retained bases in Egypt and a condominium over the Sudan. Britain fought German and Italian armies from Egypt, 1940-42. In 1951 Egypt abrogated the 1936 treaty; the Sudan became independent in 1956.

The uprising of July 23, 1952 was led by the Society of Free Officers, who named Maj. Gen. Mohammed Naguib commander in chief and forced King Farouk to abdicate. When the republic was proclaimed June 18, 1953, Naguib became its first president and premier. Lt. Col. Gamal Abdel Nasser removed Naguib and became premier in 1954. In 1956, he was voted president. Nasser died in 1970 and was replaced by Vice Pres. Anwar Sadat.

The Aswan High Dam, completed 1971, provides irrigation for more than a million acres of land. Artesian wells, drilled in the Western Desert, reclaimed 43,000 acres, 1960-66.

When the state of Israel was proclaimed in 1948, Egypt joined other Arab nations invading Israel and was defeated.

After terrorist raids across its border, Israel invaded Egypt's Sinai Peninsula, Oct. 29, 1956. Egypt rejected a cease-fire demand by Britain and France; on Oct. 31 the 2 nations dropped bombs and on Nov. 5-6 landed forces. Egypt and Israel accepted a UN cease-fire; fighting ended Nov. 7.

A UN Emergency Force guarded the 117-mile-long border between Egypt and Israel until May 19, 1967, when it was withdrawn at Nasser's demand. Egyptian troops entered the Gaza Strip and the heights of Sharm el Sheikh and 3 days later closed the Strait of Tiran to all Israeli shipping. Full-scale war broke out June 5; before it ended under a UN cease-fire June 10, Israel had captured Gaza and the Sinai Peninsula, controlled the east bank of the Suez Canal, and reopened the gulf. After sporadic fighting, Israel and Egypt agreed, Aug. 7, 1970, to a new cease-fire.

In a surprise attack Oct. 6, 1973, Egyptian forces crossed the Suez Canal into the Sinai. (At the same time, Syrian forces attacked Israelis on the Golan Heights.) Egypt was supplied by a USSR military airlift; the U.S. responded with an airlift to Israel. Israel counterattacked, crossed the canal, surrounded Suez City. A UN cease-fire took effect Oct. 24.

Under an agreement signed Jan. 18, 1974, Israeli forces withdrew from the canal's W bank; limited numbers of Egyptian forces occupied a strip along the E bank. A second accord was signed in 1975, with Israel yielding Sinai oil fields. Pres. Sadat's surprise visit to Jerusalem, Nov. 1977, opened the prospect of peace with Israel. On Mar. 26, 1979, Egypt and Israel signed a formal peace treaty, ending 30 years of war, and establishing diplomatic relations. Israel returned control of the Sinai to Egypt in Apr. 1982.

Tension between Muslim fundamentalists and Christians in 1981 caused street riots and culminated in a nationwide security crackdown in Sept. Pres. Sadat was assassinated on Oct. 6; he was succeeded by Hosni Mubarak.

Egypt was a political and military supporter of the Allied forces in their defeat of Iraq in the Persian Gulf War, 1991.

Egypt saw a rising tide of Islamic fundamentalist violence in the 1990s. Egyptian security forces conducted raids against Islamic militants, some of whom were executed for terrorism. Naguib Mahfouz, winner of the 1988 Nobel Prize for Literature, was stabbed by Islamic militants Oct. 14, 1994. Pres. Mubarak escaped assassination in Ethiopia, June 26, 1995; Egypt blamed Sudan for the attack. On Nov. 17, 1997, near Luxor, Muslim extremists killed 58 foreign tourists and 4 Egyptians.

Mubarak, who was grazed by a knife-wielding assailant Sept. 6, 1999, was confirmed by popular vote Sept. 26 for a 4th presidential term. An EgyptAir jetliner bound from New York to Cairo plunged into the Atlantic near Nantucket Is., Oct. 31, 1999, killing all 217 people on board.

The **Suez Canal,** 103 mi. long, links the Mediterranean and Red seas. It was built by a French corporation 1859-69, but Britain obtained controlling interest in 1875. The last British troops were removed June 13, 1956. On July 26, Egypt nationalized the canal.

El Salvador

Republic of El Salvador

People: Population: 6,237,662. **Age distrib.** (%): <15: 37.7; 65+: 5.1. **Pop. density:** 781 per sq. mi. **Urban:** 46%. **Ethnic groups:** Mestizo 94%, Amerindian 5%. **Principal language:** Spanish (official). **Chief religions:** Roman Catholic 75%, many Protestant groups.

Geography: Area: 8,000 sq. mi. **Location:** In Central America. **Neighbors:** Guatemala on W, Honduras on N. **Topography:** A hot Pacific coastal plain in the south rises to a cooler plateau and valley region, densely populated. The N is mountainous, including many volcanoes. **Capital:** San Salvador, (urban agg.): 1,408,000.

Government: Type: Republic. **Head of state and gov.:** Pres. Francisco Flores; b Oct. 17, 1959; in office: June 1, 1999. **Local divisions:** 14 departments. **Defense budget:** $112 mil. **Active troops:** 16,800.

Economy: Industries: Food and beverages, oil products, chemicals. **Chief crops:** Coffee, corn, sugar, rice. **Other resources:** Hydropower. **Arable land:** 27%. **Livestock** (2000): cattle: 1.21 mil; chickens: 8.1 mil; goats: 15,300; pigs: 300,000; sheep: 5,200. **Fish catch:** (1999): 10,987 metric tons. **Electricity prod.:** 3.641 bil kWh. **Labor force:** 55% serv; 30% agric.; ; 15% ind.

Finance: Monetary unit: Colon (Sept. 2001: 8.75 = $1 U.S.). **GDP** (1999 est.): $18.1 bil. **Per capita GDP:** $3,100. **Imports** (1999): $4.15 bil; partners: U.S. 51%, Guatemala 9%. **Exports** (1999): $2.5 bil; partners: U.S. 59%, Guatemala 12%. **Tourism:** $211 mil. **Budget** (1997 est.): $1.82 bil. **Intl. reserves less gold** (June 2000): $1.94 bil. **Gold:** 469,000 oz t. **Consumer prices** (change in 1999): 0.5%.

Transport: Railroad: Length: 349 mi. **Motor vehicles:** 35,300 pass. cars, 44,800 comm. vehicles. **Civil aviation:** 1.3 bil pass.-mi.; 1 airport. **Chief ports:** La Union, Acajutla, La Libertad.

Communications: TV sets: 91 per 1,000 pop. **Radios:** 373 per 1,000 pop. **Telephones:** 570,000 main lines. **Daily newspaper circ.:** 53 per 1,000 pop.

Health: Life expectancy: 66.43 male; 73.81 female. **Births** (per 1,000 pop): 28.67. **Deaths** (per 1,000 pop.): 6.18. **Natural inc.:** 2.25%. **Infant mortality** (per 1,000 live births): 28.4.

Education: Free, compulsory: ages 7-16. **Literacy**: 71%.

Major Intl. Organizations: UN (FAO, IBRD, ILO, IMF, IMO, WHO, WTrO), OAS.

Embassy: 2308 California St. NW 20008; 265-9671.

El Salvador became independent of Spain in 1821, and of the Central American Federation in 1839.

A fight with Honduras in 1969 over the presence of 300,000 Salvadoran workers left 2,000 dead.

A military coup overthrew the government of Pres. Carlos Humberto Romero in 1979, but the ruling military-civilian junta failed to quell a rebellion by leftist insurgents, armed by Cuba and Nicaragua. Extreme right-wing death squads organized to eliminate suspected leftists were blamed for thousands of deaths in the 1980s. The Reagan administration staunchly supported the government with military aid. The 12-year civil war ended Jan. 16, 1992, as the government and leftist rebels signed a formal peace treaty. The civil war had taken the lives of some 75,000 people. The treaty provided for military and political reforms.

Nine soldiers, including 3 officers, were indicted Jan. 1990 in the Nov. 1989 slaying of 6 Jesuit priests in San Salvador. Two of the officers received maximum 30-year jail sentences. They were released Mar. 20, 1993, when the National Assembly passed a sweeping amnesty.

Francisco Flores, candidate of the right-wing ARENA party, won the presidential election of Mar. 7, 1999. Earthquakes Jan. 13 and Feb. 13, 2001, left more than 1,150 people dead.

Equatorial Guinea

Republic of Equatorial Guinea

People: Population: 486,060. **Age distrib.** (%): <15: 42.6; 65+: 3.8. **Pop. density:** 45 per sq. mi. **Urban:** 47%. **Ethnic groups:** Fang 83%, Bubi 10%. **Principal languages:** Spanish, French (both official), Fang, Bubi. **Chief religion:** Predominantly Roman Catholic.

Geography: Area: 10,800 sq. mi. **Location:** Bioko Isl. off W Africa coast in Gulf of Guinea, and Rio Muni, mainland enclave. **Neighbors:** Gabon on S, Cameroon on E and N. **Topography:** Bioko Isl. consists of 2 volcanic mountains and a connecting valley. Rio Muni, with over 90% of the area, has a coastal plain and low hills beyond. **Capital:** Malabo (1995 est.): 30,000.

Government: Type: Republic. **Head of state:** Pres. Teodoro Obiang Nguema Mbasogo; b June 5, 1942; in office: Oct. 10, 1979. **Head of gov.:** Prime Min. Cándido Muatetema Rivas; b 1961; in office: Mar. 4, 2001. **Local divisions:** 7 provinces. **Defense budget:** $11 mil. **Active troops:** 1,320.

Economy: Industries: Oil (68% of export earnings), fishing, sawmilling. **Chief crops:** Cocoa, coffee, rice, bananas, yams cassava. **Minerals:** Oil. **Other resources:** Timber. **Crude oil reserves** (2000): 12 mil bbls. **Arable land:** 5%. **Livestock** (2000): cattle: 4,800; chickens: 245,000; goats: 8,100; pigs: 5,300; sheep: 36,000. **Electricity prod.:** 21 mil kWh.

Finance: Monetary unit: CFA Franc (Sept. 2001: 724.11 = $1 U.S.). **GDP** (1999 est.): $960 mil. **Per capita GDP:** $2,000. **Imports** (1999): $300 mil; partners: U.S. 35%, Spain 10%, France 15%. **Exports** (1999): $555 mil; partners: U.S. 62%, Spain 17%. **Tourism** (1998): $2 mil. **Budget** (1996 est.): $43 mil. **Intl. reserves less gold** (Apr. 2000): $7.84 mil.

Transport: Motor vehicles: 4,000 pass. cars, 3,600 comm. vehicles. **Civil aviation:** 2.8 mil pass.-mi.; 1 airport. **Chief ports:** Malabo, Bata.

Communications: TV sets: 88 per 1,000 pop. **Radios:** 464 per 1,000 pop. **Telephones** (1998): 5,600 main lines.

Health: Life expectancy: 51.89 male; 56.07 female. **Births** (per 1,000 pop): 37.72. **Deaths** (per 1,000 pop.): 13.11. **Natural inc.:** 2.46%. **Infant mortality** (per 1,000 live births): 92.9.

Education: Free, compulsory: ages 6-11. **Literacy**: 78%.

Major Intl. Organizations: UN (FAO, IBRD, ILO, IMF, IMO, WHO), OAU.

Embassy: 1712 I St. NW, Suite 410, 20005; 393-0525.

Fernando Po (now Bioko) Island was reached by Portugal in the late 15th century and ceded to Spain in 1778. Independence came Oct. 12, 1968. Riots occurred in 1969 over disputes between the island and the more backward Rio Muni province on the mainland. Masie Nguema Biyogo, a mainlander, became president for life in 1972.

Masie's reign was one of the most brutal in Africa, resulting in a bankrupted nation. Most of the nation's 7,000 Europeans emigrated. He was ousted in a military coup, Aug. 1979, and Teodoro Mbasogo, leader of the coup, became president. His regime eventually agreed to elections, held Nov. 21, 1993. These were nominally won by the ruling party, but boycotted by opposition parties that maintained the rules were rigged. Elections for president, Feb. 25, 1996, and for the legislature, Mar. 6, 1999, were similarly condemned.

Eritrea

State of Eritrea

People: Population: 4,298,269. **Age distrib.** (%): <15: 42.8; 65+: 3.3. **Pop. density:** 92 per sq. mi. **Urban:** 18%. **Ethnic groups:** Tigrinya 50%, Tigre and Kunama 40%, Afar 4%. **Principal languages:** Tigrinya, Tigre and Kunama, Afar, Amhanc, Arabic. **Chief religions:** Muslim, Coptic Christian, Roman Catholic, Protestant.

Geography: Area: 46,800 sq. mi. **Location:** In E Africa, on SW coast of Red Sea. **Neighbors:** Ethiopia on S, Djibouti on SE, Sudan on W. **Topography:** Includes many islands of the Dahlak Archipelago, low coastal plains in S, mountain range with peaks to 9,000 ft. in N. **Capital:** Asmara (1999 est.): 514,000.

Government: Type: In transition. **Head of state and gov.:** Isaias Afwerki; b Feb. 2, 1946; in office: May 24, 1993. **Local divisions:** 8 provinces. **Defense budget:** $263 mil. **Active troops:** 200,000-250,000 (est.).

Economy: Industries: Food processing, textiles, beverages. **Chief crops:** Cotton, coffee, vegetables, maize, tobacco, lentils, sorghum. **Minerals:** Gold, potash, zinc, copper. **Arable land:** 12%. **Livestock** (2000): cattle: 1.8 mil; chickens: 1.0 mil; goats: 1.5 mil; sheep: 1.54 mil. **Electricity prod.:** 165 mil kWh. **Labor force:** 80% agric., 20% ind. and comm.

Finance: Monetary unit: Birr (Sept. 2000: 9.5 = $1 U.S.); the changeover to a new currency, the nakfa, began Nov. 1997. **GDP** (1999 est.): $2.9 bil. **Per capita GDP:** $750. **Imports** (1997): $489.4 mil; partners: Saudi Arabia 16%, Italy 14%. **Exports** (1997): $52.9 mil; partners: Ethiopia 64%, Sudan 17%. **Tourism:** $28 mil. **Budget** (1996 est.): $453 mil.

Transport: Civil aviation: 2 airports. **Chief ports:** Mitsiwa, Aseb.

Communications: TV sets: 6 per 1,000 pop. **Telephones:** 30,600 main lines.

Health: Life expectancy: 53.73 male; 58.71 female. **Births** (per 1,000 pop): 42.52. **Deaths** (per 1,000 pop.): 12.07. **Natural inc.:** 3.05%. **Infant mortality** (per 1,000 live births): 75.14.

Education: Free, compulsory: ages 7-13. **Literacy** (1994): 20%.

Major Intl. Organizations: UN (FAO, IBRD, ILO, IMF, IMO, WHO), OAU.

Embassy: 1708 New Hampshire Ave. NW 20009; 319-1991.

Website: http://www.eritrea.org

Eritrea was part of the Ethiopian kingdom of Aksum. It was an Italian colony from 1890 to 1941, when it was captured by the British. Following a period of British and UN supervision, Eritrea was awarded to Ethiopia as part of a federation in 1952. Ethiopia annexed Eritrea as a province in 1962. This led to a 31-year struggle for independence, which ended when Eritrea formally declared itself an independent nation May 24, 1993. A border war with Ethiopia which erupted in June 1998 intensified in May 2000, as Ethiopian troops plunged into W Eritrea; a cease-fire signed June 18 provided for UN peacekeepers to patrol a buffer zone on Eritrean territory. A peace treaty was signed Dec. 12, 2000.

Estonia

Republic of Estonia

People: Population: 1,423,316. **Age distrib.** (%): <15: 17.1; 65+: 14.8. **Pop. density:** 82 per sq. mi. **Urban:** 69%. **Ethnic groups:** Estonian 65%, Russian 28%. **Principal languages:** Estonian (official), Russian. **Chief religion:** Evangelical Lutheran, Russian Orthodox.

Geography: Area: 17,400 sq. mi. **Location:** E Europe, bordering the Baltic Sea and Gulf of Finland. **Neighbors:** Russia on E, Latvia on S. **Capital:** Tallinn (2000 est.): 408,329.

Government: Type: Republic. **Head of state:** Pres. Lennart Meri; b Mar. 29, 1929; in office: Oct. 5, 1992. **Head of gov.:** Prime Min. Mart Laar; b Apr. 22, 1960; in office: Mar. 25, 1999. **Local divisions:** 15 counties. **Defense budget:** $88 mil (2001). **Active troops:** 4,800.

Economy: Industries: Shipbuilding, electric motors, cement. **Chief crops:** Potatoes, fruits, vegetables. **Minerals:** Shale oil, peat, phosphorite. **Other resources:** Dairy prods. **Arable land:** 25%. **Livestock** (2000): cattle: 285,600; chickens: 2.46 mil; pigs: 281,200; sheep: 29,400.. **Fish catch** (1999): 123,873 metric tons. **Electricity prod.** 7.782 bil kWh. **Labor force:** 69% services; 20% industry; 11% agric., forestry.

Finance: Monetary unit: Kroon (Sept. 2001: 17.27 = $1 U.S.). **GDP** (1999 est.): $7.9 bil. **Per capita GDP:** $5,600. **Imports** (1999): $3.4 bil; partners: Finland 23%, Russia 13.2%. **Exports** (1999): $2.5 bil; partners: Sweden 19.3%, Finland 18.8%, Russia 8.8%. **Tourism:** $560 mil. **Budget** (1997 est.): $1.37 bil. **Intl. reserves less gold** (June 2000): $810.75 mil. **Gold:** 8,000 oz t. **Consumer prices** (change in 1999): 3.3%.

Transport: Railroad: Length: 636 mi. **Motor vehicles:** 338,000 pass. cars, 60,000 comm. vehicles. **Civil aviation:** 83.6 mil pass.-mi.; 1 airport. **Chief port:** Tallinn.

Communications: TV sets: 411 per 1,000 pop. **Telephones:** 522,800 main lines. **Daily newspaper circ.:** 242 per 1,000 pop.

Health: Life expectancy: 63.72 male; 76.05 female. **Births** (per 1,000 pop): 8.70. **Deaths** (per 1,000 pop.): 13.48. **Natural inc.:** -.48%. **Infant mortality** (per 1,000 live births): 12.62.

Education: Compulsory: ages 7-16. **Literacy** (1994): 100%.

Major Intl. Organizations: UN (FAO, IBRD, ILO, IMF, IMO, WHO), OSCE.

Embassy: 2131 Massachusetts Ave. NW 20008; 588-0101.

Website: http://www.ciesin.ee/ESTCG

Estonia was a province of imperial Russia before World War I, was independent between World Wars I and II. It was conquered by the USSR in 1940 and incorporated as the Estonian SSR. Estonia declared itself an "occupied territory," and proclaimed itself a free nation Mar. 1990. During an abortive Soviet coup, Estonia declared immediate full independence, Aug. 20, 1991; the Soviet Union recognized its independence in Sept. 1991. The first free elections in over 50 years were held Sept. 20, 1992. The last occupying Russian troops were withdrawn by Aug. 31, 1994. Center-right parties won the legislative election of Mar. 7, 1999.

Ethiopia

Federal Democratic Republic of Ethiopia

People: Population: 65,891,874. **Age distrib.** (%): <15: 47.2; 65+: 2.8. **Pop. density:** 153 per sq. mi. **Urban:** 17%. **Ethnic groups:** Oromo 40%, Amhara and Tigrean 32%, Sidamo 9%. **Principal languages:** Amharic (official), Tigrinya, Orominga. **Chief religions:** Muslim 45-50%, Ethiopian Orthodox 35-40%, animist 12%.

Geography: Area: 431,800 sq. mi. **Location:** In East Africa. **Neighbors:** Sudan on W, Kenya on S, Somalia and Djibouti on E, Eritrea on N. **Topography:** A high central plateau, between 6,000 and 10,000 ft. high, rises to higher mountains near the Great Rift Valley, cutting in from the SW. The Blue Nile and other rivers cross the plateau, which descends to plains on both W and SE. **Capital:** Addis Ababa, (urban agg.): 2,639,000.

Government: Type: Federal republic. **Head of state:** Pres. Negasso Gidada; b Sept. 8, 1943; in office: Aug. 22, 1995. **Head of gov.:** Prime Min. Meles Zenawi; b May 8, 1955; in office: Aug. 23, 1995. **Local divisions:** 9 states, 2 charted cities. **Defense budget:** $457 mil. **Active troops:** 352,500.

Economy: Industries: Food processing, chemicals, textiles. **Chief crops:** Coffee (60% of export earnings), cereals, sugarcane, pulses, oilseed. **Minerals:** Platinum, gold, copper. **Arable land:** 12%. **Crude oil reserves:** 428,000 bbls. **Livestock** (2000): cattle: 35.0 mil; chickens: 55.6 mil; goats: 16.8 mil; pigs: 24,000; sheep: 21.0 mil. **Fish catch:** (1999): 10,414 metric tons. **Electricity prod.:** 1.625 bil kWh. **Labor force:** 80% agric.

Finance: Monetary unit: Birr (Sept. 2001: 8.40 = $1 U.S.). **GDP** (1999 est.): $33.3 bil. **Per capita GDP:** $560. **Imports** (1998 est.): $1.25 bil; partners: Italy 10%, U.S. 9%. **Exports** (1998 est.): $420 mil; partners: Germany 22%, Japan 12%, Italy 9%. **Tourism** (1998): $11 mil. **Budget** (FY 1996-97): $1.48 bil. **Intl. reserves less gold** (Mar. 2000): $384.1 mil. **Gold:** 2,000 oz t. **Consumer prices** (change in 1997): –3.7%.

Transport: Railroad: Length: 486 mi. **Motor vehicles:** 45,559 pass. cars, 20,462 comm. vehicles. **Civil aviation:** 1.2 bil pass.-mi.; 31 airports.

Communications: TV sets: 4 per 1,000 pop. **Radios:** 153 per 1,000 pop. **Telephones**: 231,900 main lines.

Health: Life expectancy: 43.88 male; 45.51 female. **Births** (per 1,000 pop): 44.68. **Deaths** (per 1,000 pop.): 17.84. **Natural inc.:** 2.68%. **Infant mortality** (per 1,000 live births): 99.96.

Education: Free, compulsory: ages 7-13. **Literacy**: 35%.

Major Intl. Organizations: UN (FAO, IBRD, ILO, IMF, IMO, WHO), OAU.

Embassy: 2134 Kalorama Rd. NW 20008; 234-2281.

Ethiopian culture was influenced by Egypt and Greece. The ancient monarchy was invaded by Italy in 1880 but maintained its independence until another Italian invasion in 1936. British forces freed the country in 1941.

The last emperor, Haile Selassie I, established a parliament and judiciary system in 1931 but barred all political parties.

A series of droughts in the 1970s killed hundreds of thousands. An army mutiny, strikes, and student demonstrations led to the dethronement of Selassie in 1974; he died Aug. 1975, while being held by the ruling junta. The junta pledged to form a one-party socialist state and instituted a successful land reform; opposition was violently suppressed. The influence of the Coptic Church, embraced in AD 330, was curbed, and the monarchy was abolished in 1975.

The regime, torn by bloody coups, faced uprisings by tribal and political groups in part aided by Sudan and Somalia. Ties with the U.S., once a major ally, deteriorated, while cooperation accords were signed with the USSR in 1977. In 1978, Soviet advisers and Cuban troops helped defeat Somalian forces. Ethiopia and Somalia signed a peace agreement in 1988.

A worldwide relief effort began in 1984, as an extended drought threatened the country with famine; up to a million people may have died as a result of starvation and disease.

The Ethiopian People's Revolutionary Democratic Front (EPRDF), an umbrella group of 6 rebel armies, launched a major push against government forces, Feb. 1991. In May, Pres. Mengistu Haile Mariam resigned and left the country. The EPRDF took over and set up a transitional government. Ethiopia's first multiparty general elections were held in 1995.

Eritrea, a province on the Red Sea, declared its independence May 24, 1993. Fighting along the border with Eritrea, which erupted in June 1998, intensified in May 2000, as Ethiopian forces plunged into Eritrean territory; a cease-fire was signed June 18 and a peace treaty Dec. 12. The war displaced 350,000 Ethiopians and is estimated to have cost the country nearly $3 billion.

Fiji

Republic of the Fiji Islands

People: Population: 844,330. **Age distrib.** (%): <15: 32.9; 65+: 3.6. **Pop. density:** 120 per sq. mi. **Urban:** 49%. **Ethnic groups:** Fijian 51%, Indian 44%. **Principal languages:** English (official), Fijian, Hindustani. **Chief religions:** Christian 52%, Hindu 38%, Muslim 8%.

Geography: Area: 7,000 sq. mi. **Location:** In western South Pacific O. **Neighbors:** Nearest are Vanuatu to W, Tonga to E. **Topography:** 322 islands (106 inhabited), many mountainous, with tropical forests and large fertile areas. Viti Levu, the largest island, has over half the total land area. **Capital:** Suva (1999): 196,000.

Government: Type: In transition. **Head of state:** Pres. Ratu Josefa Iloilo; b Dec. 29, 1920; in office: July 18, 2000. **Head of gov.:** Laisenia Qarase; b Feb. 4, 1941; in office: Mar. 16, 2001. **Local divisions:** 4 divisions comprising 14 provinces and 1 dependency. **Defense budget:** 27 mil. **Active troops:** 3,500.

Economy: Industries: Sugar refining, light industry, tourism. **Chief crops:** Sugarcane, cassava, coconuts. **Minerals:** Gold, copper. **Other resources:** Timber, fish. **Arable land:** 10%. **Livestock** (2000): cattle: 350,000; chickens: 4.0 mil; goats: 235,000; pigs: 115,000; sheep: 7,000. **Fish catch: (1999):** 36,374 metric tons. **Electricity prod.:** 510 mil kWh. **Labor force:** 67% agric.

Finance: Monetary unit: Dollar (Sept. 2001: 2.24 = $1.00 U.S.). **GDP** (1999 est.): $5.9 bil. **Per capita GDP:** $7,300. **Imports** (1998): $612 mil; partners: Australia 45%, N.Z. 15%. **Exports** (1998): $393 mil; partners: Australia 34%, UK 18%. **Tourism** (1998): $266 mil. **Budget** (1997 est.): $742.65 mil. **Intl. reserves less gold** (May 2000): $382.01 mil. **Gold:** 1,000 oz t. **Consumer prices** (change in 1999): 2.0%.

Transport: Railroad: Length: 370 mi. **Motor vehicles:** 30,000 pass. cars, 29,000 comm. vehicles. **Civil aviation:** 1.2 bil pass.-mi.; 13 airports. **Chief ports:** Suva, Lautoka.

Communications: TV sets: 89 per 1,000 pop. **Radios:** 561 per 1,000 pop. **Telephones** (1999): 81,500 main lines. **Daily newspaper circ.:** 68 per 1,000 pop.

Health: Life expectancy: 65.83 male; 70.78 female. **Births** (per 1,000 pop): 23.33. **Deaths** (per 1,000 pop.): 5.75. **Natural inc.:** 1.76%. **Infant mortality** (per 1,000 live births): 14.08.

Education: Free: ages 6-14. **Literacy** (1996): 91%.

Major Intl. Organizations: UN (FAO, IBRD, ILO, IMF, IMO, WHO, WTrO), the Commonwealth.

Embassy: 2233 Wisconsin Ave. NW 20007; 337-8320.

A British colony since 1874, Fiji became an independent parliamentary democracy Oct. 10, 1970. Cultural differences between the Indian community (descendants of contract laborers brought to the islands in the 19th century) and indigenous Fijians have led to political polarization.

In 1987, a military coup ousted the government; order was restored May 21 under a compromise granting Lt. Col. Sitiveni Rabuka, the coup's leader, increased power. Rabuka staged a second coup Sept. 25 and declared Fiji a republic. Civilian government was restored in Dec. A new constitution favoring indigenous Fijians was issued July 25, 1990; amendments enacted in July 1997 made the constitution more equitable.

Fiji's 1st Indian prime minister, Mahendra Chaudhry, took office May 19, 1999. He and other government officials were taken captive May 19, 2000, by indigenous Fijian gunmen led by George Speight. The hostage crisis led to a military takeover, May 29. Release of the last remaining hostages in July coincided with the installation of an interim military-backed government. Speight was charged with treason, Aug. 4. Awaiting trial in jail, he was elected to parliament in voting ended Sept. 1, 2001, which returned caretaker Prime Min. Laisenia Qarase to office. The government was reconstituted in Mar. 2001 after an appellate court ruled it illegal.

Finland

Republic of Finland

People: Population: 5,175,783. **Age distrib.** (%): <15: 18.0; 65+: 15.0. **Pop. density:** 44 per sq. mi. **Urban:** 67%. **Ethnic groups:** Finn 93%, Swede 6%. **Principal languages:** Finnish, Swedish (both official). **Chief religion:** Evangelical Lutheran 89%.

Geography: Area: 117,800 sq. mi. **Location:** In northern Europe. **Neighbors:** Norway on N, Sweden on W, Russia on E. **Topography:** South and central Finland are generally flat areas with low hills and many lakes. The N has mountainous areas, 3,000-4,000 ft. above sea level. **Capital:** Helsinki, (urban agg.): 1,167,000.

Government: Type: Constitutional republic. **Head of state:** Pres. Tarja Halonen; b Dec. 24, 1943; in office: Mar. 1, 2000. **Head of gov.:** Prime Min. Paavo Lipponen; b Apr. 23, 1941; in office: Apr. 13, 1995. **Local divisions:** 6 laanit (provinces). **Defense budget:** $1.6 bil (2001). **Active troops:** 31,700.

Economy: Industries: Metal prods., shipbuilding, wood processing, chemicals, textiles. **Chief crops:** Grains, sugar beets, potatoes. **Minerals:** Copper, iron, silver, zinc. **Other resources:** Timber, dairy prods. **Arable land:** 8%. **Livestock** (2000): cattle: 1.09 mil; chickens: 6.0 mil; goats: 7,900; pigs: 1.35 mil; sheep: 106,600. **Fish catch** (1999): 196,513 metric tons. **Electricity prod.:** 75.792 bil kWh. **Labor force:** 46% ind., commerce & finance; 32% public serv.; 8% agric. and forestry.

Finance: Monetary unit: Markka (Sept. 2001: 6.56 = $1 U.S.). Euro (Sept. 2001: 1.09 = $1 U.S.). **GDP** (1999 est.): $108.6 bil. **Per capita GDP:** $21,000. **Imports** (1998): $30.7 bil; partners: EU 60%, US 8%, Russia 7%. **Exports** (1998): $43 bil; partners: EU 56%, US 7%, Russia 6%. **Tourism:** $1.46 bil. **Budget** (1996 est.): $40 bil. **Intl. reserves less gold** (June 2000): $7.75 bil. **Gold:** 1.58 mil oz t. **Consumer prices** (change in 1999): 1.2%.

Transport: Railroad: Length: 3,641 mi. **Motor vehicles** (1997): 1.94 mil pass. cars, 291,235 comm. vehicles. **Civil aviation:** 5.9 bil pass.-mi.; 24 airports. **Chief ports:** Helsinki, Turku, Rauma, Kotka.

Communications: TV sets: 535 per 1,000 pop. **Radios:** 1,385 per 1,000 pop. **Telephones:** 2,831,000 main lines. **Daily newspaper circ.:** 455 per 1,000 pop.

Health: Life expectancy: 73.92 male; 81.36 female. **Births** (per 1,000 pop): 10.69. **Deaths** (per 1,000 pop.): 9.75. **Natural inc.:** .09%. **Infant mortality** (per 1,000 live births): 3.79.

Education: Free, compulsory: ages 7-16. **Literacy** (1997): 100%.

Major Intl. Organizations: UN (FAO, IBRD, ILO, IMF, IMO, WHO, WTrO), EU, OECD, OSCE.

Embassy: 3301 Massachusetts Ave. NW 20008; 298-5800.

Website: http://www.finland.org

The early Finns probably migrated from the Ural area at about the beginning of the Christian era. Swedish settlers brought the country into Sweden, 1154 to 1809, when Finland became an autonomous grand duchy of the Russian Empire. Russian exactions created a strong national spirit; on Dec. 6, 1917, Finland declared its independence and in 1919 became a republic.

On Nov. 30, 1939, the Soviet Union invaded, and the Finns were forced to cede 16,173 sq. mi. of territory. After World War II, further cessions were exacted. In 1948, Finland signed a treaty of mutual assistance with the USSR; Finland and Russia nullified this treaty with a new pact in Jan. 1992.

Following approval by Finnish voters in an advisory referendum Oct. 16, 1994, Finland joined the European Union effective Jan. 1, 1995.

Aland or **Ahvenanmaa,** constituting an autonomous province, is a group of small islands, 590 sq. mi., in the Gulf of Bothnia, 25 mi. from Sweden, 15 mi. from Finland. Mariehamn is the principal port.

France

French Republic

People: Population: 59,551,227. **Age distrib.** (%): <15: 18.7; 65+: 16.1. **Pop. density:** 283 per sq. mi. **Urban:** 75%. **Ethnic groups:** Celtic and Latin; Teutonic, Slavic, North African, Indochinese, Basque minorities. **Principal language:** French (official). **Chief religion:** Roman Catholic 90%.

Geography: Area: 210,400 sq. mi. **Location:** In western Europe, between Atlantic O. and Mediterranean Sea. **Neighbors:** Spain on S; Italy, Switzerland, Germany on E; Luxembourg, Belgium on N. **Topography:** A wide plain covers more than half of the country, in N and W, drained to W by Seine, Loire, Garonne rivers. The Massif Central is a mountainous plateau in center. In E are Alps (Mt. Blanc is tallest in W Europe, 15,771 ft.), the lower Jura range, and the forested Vosges. The Rhone flows from Lake Geneva to Mediterranean. Pyrenees are in SW, on border with Spain. **Capital:** Paris. **Cities** (urban agg.): Paris 9,624,000, (1997 city est.: 2,152,000); Lyon 1,318,000; Marseilles 1,241,000; Lille 991,000.

Government: Type: Republic. **Head of state:** Pres. Jacques Chirac; b Nov. 29, 1932; in office: May 17, 1995. **Head of gov.:** Prime Min. Lionel Jospin; b July 12, 1937; in office: June 3, 1997. **Local divisions:** 22 administrative regions containing 96 departments. **Defense budget:** $27.0 bil. **Active troops:** 294,430.

Economy: Industries: Steel, chemicals, textiles, tourism, wine, perfume, aircraft, machinery, electronics. **Chief crops:** Grains, sugar beets, winegrapes, fruits, potatoes, vegetables. France is largest food producer, exporter, in W Europe. **Minerals:** Bauxite, iron, coal. **Crude oil reserves** (2000): 107 mil bbls. **Other resources:** Timber, dairy. **Arable land:** 33%. **Livestock** (2000): cattle: 20.53 mil; chickens: 232.97 mil; goats: 1.19 mil; pigs: 14.64 mil; sheep: 10.0 mil. **Fish catch** (1999): 829,914 metric tons. **Electricity prod.:** 497.260 bil kWh. **Labor force:** 69% services; 26% ind.; 5% agric.

Finance: Monetary unit: Franc (Sept. 2001: 7.24 = $1 U.S.). Euro (Sept. 2001: 1.09 = $1 U.S.). **GDP** (1999 est.): $1.373 tril. **Per capita GDP:** $23,300. **Imports** (1999): $280.8 bil; partners: Germany 17%, Italy 10%, U.S. 9%. **Exports** (1999): $304.7 bil; partners: Germany 16%, UK 10%, Italy 9%. **Tourism:** $31.70 bil. **Budget** (1998 est.): $265 bil. **Intl. reserves less gold** (June 2000): $40.88 bil. **Gold:** 97.25 mil oz t. **Consumer prices** (change in 1999): 0.5%.

Transport: Railroad: Length: 19,847 mi. **Motor vehicles:** in use: 25.50 mil pass. cars, 5.26 mil comm. vehicles. **Civil aviation:** 52.6 bil pass.-mi.; 61 airports. **Chief ports:** Marseille, Le Havre, Bordeaux, Rouen.

Communications: TV sets: 606 per 1,000 pop. **Radios:** 943 per 1,000 pop. **Telephones:** 34,114,000 main lines. **Daily newspaper circ.:** 218 per 1,000 pop.

Health: Life expectancy: 75.01 male; 83.01 female. **Births** (per 1,000 pop): 12.10. **Deaths** (per 1,000 pop.): 9.09. **Natural inc.:** .3%. **Infant mortality** (per 1,000 live births): 4.46.

Education: Free, compulsory: ages 6-16. **Literacy** (1994): 99%.

Major Intl. Organizations: UN and most of its specialized agencies, EU, NATO, OECD, OSCE.

Embassy: 4101 Reservoir Rd. NW 20007; 944-6000.

Website: http://www.info-france-usa.org

Celtic Gaul was conquered by Julius Caesar 58-51 BC; Romans ruled for 500 years. Under Charlemagne, Frankish rule extended over much of Europe. After his death France emerged as one of the successor kingdoms.

The monarchy was overthrown by the French Revolution (1789-93) and succeeded by the First Republic; followed by the First Empire under Napoleon (1804-15), a monarchy (1814-48), the Second Republic (1848-52), the Second Empire (1852-70), the Third Republic (1871-1946), the Fourth Republic (1946-58), and the Fifth Republic (1958 to present).

France suffered severe losses in manpower and wealth in the First World War, when it was invaded by Germany. By the Treaty of Versailles, France exacted return of Alsace and Lorraine, provinces seized by Germany in 1871. Germany invaded France again in May 1940, and signed an armistice with a government based in Vichy. After France was liberated by the Allies in Sept. 1944, Gen. Charles de Gaulle became head of the provisional government, serving until 1946.

De Gaulle again became premier in 1958, during a crisis over Algeria, and obtained voter approval for a new constitution, ushering in the Fifth Republic. He became president Jan. 1959. Using strong executive powers, he promoted French economic and technological advances in the context of the European Economic Community and guarded French foreign policy independence.

France had withdrawn from Indochina in 1954, and from Morocco and Tunisia in 1956. Most of its remaining African territories were freed 1958-62. In 1966, France withdrew all its troops from the integrated military command of NATO, though 60,000 remained stationed in Germany.

In May 1968 rebellious students in Paris and other centers rioted, battled police, and were joined by workers who launched nationwide strikes. The government awarded pay increases to the strikers May 26. De Gaulle resigned from office in Apr. 1969, after losing a nationwide referendum on constitutional reform. Georges Pompidou, who was elected to succeed him, continued De Gaulle's emphasis on French independence from the two superpowers. After Pompidou's death, in 1974, Valery Giscard d'Estaing was elected president; he continued the basically conservative policies of his predecessors.

On May 10, 1981, France elected François Mitterrand, a Socialist, president. Under Mitterrand the government nationalized 5 major industries and most private banks. After 1986, however, when rightists won a narrow victory in the National Assembly, Mitterrand chose conservative Jacques Chirac as premier. A 2-year period of "cohabitation" ensued, and France began to pursue a privatization program in which many state-owned companies were sold. After Mitterrand was elected to a 2d 7-year term in 1988, he appointed a Socialist as premier. The center-right won a large majority in 1993 legislative elections, ushering in another period of "cohabitation" with a conservative premier.

In 1993, France set tighter rules for entry into the country and made it easier for the government to expel foreigners. In 1994, France sent troops to Rwanda in an effort to help protect civilians there from ongoing massacres. The international terrorist known as Carlos the Jackal (Ilich Ramirez Sánchez) was arrested in Sudan in Aug. 1994 and extradited to France, where he had been sentenced in absentia to life imprisonment.

Former conservative Prime Min. Jacques Chirac won the presidency in a runoff May 7, 1995. A series of terrorist bombings and bombing attempts began in summer 1995; Islamic extremists, opposed to France's support of the Algerian government and its struggle with Islamic fundamentalists, were believed responsible. In Sept. 1995, France stirred widespread protests by resuming nuclear tests in the South Pacific, after a 3-year moratorium; the tests ended Jan. 1996.

Chirac cut government spending to help the French economy meet the budgetary goals set for the introduction of a common European currency. With unemployment at nearly 13%, legislative elections completed June 1, 1997, produced a decisive victory for the leftist parties. The result was a new period of "cohabitation," this time between a conservative president and a Socialist prime minister, Lionel Jospin. France contributed 7,000 troops to the NATO-led security force (KFOR) that entered Kosovo in June 1999.

After a sensational trial, Roland Dumas, a former foreign minister, was convicted May 30, 2001, on corruption charges; also found guilty were 4 co-defendants, including his former mistress.

The island of **Corsica,** in the Mediterranean W of Italy and N of Sardinia, is a territorial collectivity and region of France comprising 2 departments. It elects a total of 2 senators and 3 deputies to the French Parliament. Area: 3,369 sq. mi.; pop. (1996 est.): 258,000. The capital is Ajaccio, birthplace of Napoleon I. Violence by Corsican separatist groups has hurt tourism, a leading industry on the island.

Overseas Departments

French Guiana is on the NE coast of South America with Suriname on the W and Brazil on the E and S. Its area is 34,400 sq. mi.; pop. (2001 est.): 177,562. Guiana sends one senator and 2 deputies to the French Parliament. Guiana is administered by a prefect and has a Council General of 16 elected members; capital is Cayenne.

The famous penal colony, Devil's Island, was phased out between 1938 and 1951. The European Space Agency maintains a satellite-launching center (established by France in 1964) in the city of Kourou.

Immense forests of rich timber cover 88% of the land. Fishing (especially shrimp), forestry, and gold mining are the most important industries.

Guadeloupe, in the West Indies' Leeward Islands, consists of 2 large islands, Basse-Terre and Grande-Terre, separated by the Salt River, plus Marie Galante and the Saintes group to the S and, to the N, Desirade, St. Barthelemy, and over half of St. Martin (the Netherlands' portion is called St. Maarten). A French possession since 1635, the department is represented in the French Parliament by 2 senators and 4 deputies; administration consists of a prefect (governor) as well as an elected general and regional councils.

Area of the islands is 687 sq. mi.; pop. (2001 est.) 431,170, mainly descendants of slaves; capital is Basse-Terre on Basse-Terre Island. The land is fertile; sugar, rum, and bananas are exported. Tourism is an important industry.

Martinique, the northernmost of the Windward Islands, in the West Indies, has been a possession since 1635, and a department since Mar. 1946. It is represented in the French Parliament by 2 senators and 4 deputies. The island was the birthplace of Napoleon's Empress Josephine.

It has an area of 436 sq. mi.; pop. (2001 est.) 418,454, mostly descendants of slaves. The capital is Fort-de-France (pop. 1991: 101,000). It is a popular tourist stop. The chief exports are rum, bananas, and petroleum products.

Réunion is a volcanic island in the Indian O. about 420 mi. E of Madagascar, and has belonged to France since 1665. Area, 970 sq. mi.; pop. (2001 est.) 732,570, 30% of French extraction. Capital: Saint-Denis. The chief export is sugar. It elects 5 deputies, 3 senators to the French Parliament.

Overseas Territorial Collectivities

Mayotte, claimed by Comoros and administered by France, voted in 1976 to become a territorial collectivity of France. An island NW of Madagascar, area is 144 sq. mi., pop. (2001 est.) 163,366. The capital is Mamoutzou.

St. Pierre and Miquelon, formerly an overseas territory (1816-1976) and department (1976-85), made the transition to territorial collectivity in 1985. It consists of 2 groups of rocky islands near the SW coast of Newfoundland, inhabited by fishermen. The exports are chiefly fish products. The St. Pierre group has an area of 10 sq. mi.; Miquelon, 83 sq. mi. Total pop. (2001 est.), 6,928. The capital is St. Pierre.

Both Mayotte and St. Pierre and Miquelon elect a deputy and a senator to the French Parliament.

Overseas Territories

Territory of **French Polynesia** comprises 130 islands widely scattered among 5 archipelagos in the South Pacific; administered by a Council of Ministers (headed by a president). Territorial Assembly and the Council have headquarters at Papeete, on Tahiti, one of the **Society Islands** (which include the **Windward** and **Leeward** islands). Two deputies and a senator are elected to the French Parliament.

Other groups are the **Marquesas Islands,** the **Tuamotu Archipelago,** including the **Gambier Islands,** and the **Austral Islands.**

Total area of the islands administered from Tahiti is 1,400 sq. mi.; pop. (2001 est.), 253,506, more than half on Tahiti. Tahiti is picturesque and mountainous with a productive coastline bearing coconuts, citrus, pineapples, and vanilla. Cultured pearls are also produced.

Tahiti was visited by Capt. James Cook in 1769 and by Capt. Bligh in the *Bounty*, 1788-89. Its beauty impressed Herman Melville, Paul Gauguin, and Charles Darwin. Tahitians angered by French nuclear testing rioted Sept. 1995.

Territory of the **French Southern and Antarctic Lands** comprises **Adelie Land,** on Antarctica, and 4 island groups in the Indian O. Adelie, reached 1840, has a research station, a coastline of 185 mi., and tapers 1,240 mi. inland to the South Pole. The U.S. does not recognize national claims in Antarctica. There are 2 huge glaciers, Ninnis, 22 mi. wide, 99 mi. long, and Mentz, 11 mi. wide, 140 mi. long. The Indian O. groups are:

Kerguelen Archipelago, visited 1772, consists of one large and 300 small islands. The chief is 87 mi. long, 74 mi. wide, and has Mt. Ross, 6,429 ft. tall. Principal research station is Port-aux-Français. Seals often weigh 2 tons; there are blue whales, coal, peat, semiprecious stones. **Crozet Archipelago,** reached 1772, covers 195 sq. mi. Eastern Island rises to 6,560 ft. **Saint Paul,** in southern Indian O., has warm springs with earth at places heating to 120× to 390× F. **Amsterdam** is nearby; both produce cod and rock lobster.

Territory of **New Caledonia** and Dependencies is a group of islands in the Pacific O. about 1,115 mi. E of Australia and approx. the same distance NW of New Zealand. Dependencies are the **Loyalty Islands, Isle of Pines, Belep Archipelago,** and **Huon Islands.**

The largest island, New Caledonia, is 6,530 sq. mi. Total area of the territory is 8,548 sq. mi.; population (2001 est.) 204,863. The group was acquired by France in 1853.

The territory is administered by a High Commissioner. There is a popularly elected Territorial Congress. Two deputies and a senator are elected to the French Parliament. Capital: Noumea.

Mining is the chief industry. New Caledonia is one of the world's largest nickel producers. Other minerals found are chrome, iron, cobalt, manganese, silver, gold, lead, and copper. Agricultural products include yams, sweet potatoes, potatoes, manioc (cassava), corn, and coconuts.

In 1987, New Caledonian voters chose by referendum to remain within the French Republic. There were clashes between French and Melanesians (Kanaks) in 1988. An agreement Apr. 21, 1998, between France and rival New Caledonian factions specified a 15- to 20-year period of "shared sovereignty." The French constitution was amended, July 6, to allow the territory a gradual increase in autonomy, and New Caledonian voters approved the plan Nov. 8, 1998, by a 72% majority.

Territory of the **Wallis and Futuna Islands** comprises 2 island groups in the SW Pacific S of Tuvalu, N of Fiji, and W of Western Samoa; became an overseas territory July 29, 1961. The islands have a total area of 106 sq. mi. and population (2001 est.) of 15,435. **Alofi,** attached to Futuna, is uninhabited. Capital: Mata-Utu. Chief products are copra, yams, taro roots, bananas, and coconuts. A senator and a deputy are elected to the French Parliament.

Gabon

Gabonese Republic

People: Population: 1,221,175. **Age distrib.** (%): <15: 33.3; 65+: 5.9. **Pop. density:** 12 per sq. mi. **Urban:** 81%. **Ethnic groups:** Fang, Eshira, Bapounou, Bateke, other Bantu, other Africans, Europeans. **Principal languages:** French (official), Bantu dialects. **Chief religions:** Christian 55%-75%.

Geography: Area: 99,400 sq. mi. **Location:** On Atlantic coast of W central Africa. **Neighbors:** Equatorial Guinea and Cameroon on N, Congo on E and S. **Topography:** Heavily forested, the country consists of coastal lowlands; plateaus in N, E, and S; mountains in N, SE, and center. The Ogooue R. system covers most of Gabon. **Capital:** Libreville (1999): 523,000.

Government: Type: Republic. **Head of state:** Pres. Omar Bongo; b Dec. 30, 1935; in office: Dec. 2, 1967. **Head of gov.:** Prime Min. Jean-François Ntoutoume-Emane; b Oct. 6, 1939; in office: Jan. 23, 1999. **Local divisions:** 9 provinces. **Defense budget:** $126 mil. **Active troops:** 4,700.

Economy: Industries: Oil products, textiles, food and beverages, wood products. **Chief crops:** Cocoa, coffee, palm products. **Minerals:** Oil, manganese, uranium, iron, gold. **Crude oil reserves** (2000): 2.5 bil bbls. **Other resources:** Timber. **Arable land:** 1%. **Livestock** (2000): cattle: 36,000; chickens: 3.2 mil; goats: 91,000; pigs: 213,000; sheep: 198,000. **Fish catch:** (1999): 44,772 metric tons. **Electricity prod.:** 1.020 bil kWh. **Labor force:** 60% agric.; 25% serv. and gov.; 15% ind. & commerce.

Finance: Monetary unit: CFA Franc (Sept. 2001: 724.11 = $1 U.S.). **GDP** (1999 est.): $7.9 bil. **Per capita GDP:** $6,500. **Imports** (1999 est.): $1.2 bil; partners: France 39%. **Exports** (1999 est.): $2.4 bil; partners: U.S. 68%. **Tourism:** $11 mil. **Budget** (1996 est.): $1.3 bil. **Intl. reserves less gold** (Apr. 2000): $129.67 mil. **Gold:** 13,000 oz t. **Consumer prices** (change in 1997): 4.0%.

Transport: Railroad: Length: 415 mi. **Motor vehicles:** 23,800 pass. cars, 15,700 comm. vehicles. **Civil aviation:** 513.4 mil pass.-mi.; 17 airports. **Chief ports:** Port-Gentil, Owendo, Libreville.

Communications: TV sets: 35 per 1,000 pop. **Radios:** 173 per 1,000 pop. **Telephones**: 39,000 main lines. **Daily newspaper circ.:** 34 per 1,000 pop.

Health: Life expectancy: 48.47 male; 50.75 female. **Births** (per 1,000 pop): 27.42. **Deaths** (per 1,000 pop.): 17.22. **Natural inc.:** 1.02%. **Infant mortality** (per 1,000 live births): 94.91.

Education: Compulsory: ages 6-16. **Literacy**: 63%.

Major Intl. Organizations: UN (FAO, IBRD, ILO, IMF, IMO, WHO, WTrO), OAU.

Embassy: Suite 200, 2034 20th St. NW 20009; 797-1000.

France established control over the region in the second half of the 19th century. Gabon became independent Aug. 17, 1960. A multiparty political system was introduced in 1990, and a new constitution was enacted Mar. 14, 1991. However, the reelection of longtime Pres. Omar Bongo, on Dec. 5, 1993, prompted rioting and charges of vote fraud; another Bongo victory on Dec. 6, 1998, was likewise allegedly marred by irregularities.

Gabon is one of the most prosperous black African countries, thanks to abundant natural resources, foreign private investment, and government development programs.

The Gambia

Republic of The Gambia

People: Population: 1,411,205. **Age distrib.** (%): <15: 45.2; 65+: 2.7. **Pop. density:** 366 per sq. mi. **Urban:** 32%. **Ethnic groups:** Mandinka 42%, Fula 18%, Wolof 16%, other African. **Principal languages:** English (official), Mandinka, Wolof, Fula. **Chief religions:** Muslim 90%, Christian 9%.

Geography: Area: 3,900 sq. mi. **Location:** On Atlantic coast near W tip of Africa. **Neighbors:** Surrounded on 3 sides by Senegal. **Topography:** A narrow strip of land on each side of the lower Gambia R. **Capital:** Banjul (1999): 229,000.

Government: Type: Republic. **Head of state and gov.:** Yahya Jammeh; b May 25, 1965; in office: July 23, 1994. **Local divisions:** 5 divisions, 1 city. **Defense budget:** $15 mil. **Active troops:** 800.

Economy: Industries: Tourism, peanut processing. **Chief crops:** Peanuts (main export), rice. **Arable land:** 18%. **Livestock** (2000): cattle: 370,000; chickens: 730,000; goats: 270,000; pigs: 14,000; sheep: 195,000. **Fish catch** (1999): 32,258 metric tons. **Electricity prod.:** 75 mil kWh. **Labor force:** 75% agric.; 19% ind., comm., serv.

Finance: Monetary unit: Dalasi (Oct. 2000: 17.03 = $1.00 U.S.). **GDP** (1999 est.): $1.4 bil. **Per capita GDP:** $1,030. **Imports** (1998): $201 mil; partners: Hong Kong, UK, Netherlands. **Exports** (1998): $132 mil; partners: Benelux 78%, Japan. **Tourism** (1998): $33 mil. **Budget** (FY 1996-97 est.): $98.2 mil. **Intl. reserves less gold** (Mar. 2000): $118.96 mil. **Consumer prices** (change in 1999): 3.8%.

Transport: Motor vehicles: 8,000 pass. cars, 1,000 comm. vehicles. **Civil aviation:** 31.1 mil pass.-mi.; 1 airport. **Chief port:** Banjul.

Communications: Radios: 126 per 1,000 pop. **Telephones** (1999): 29,200 main lines.

Health: Life expectancy: 51.65 male; 55.58 female. **Births** (per 1,000 pop): 41.76. **Deaths** (per 1,000 pop.): 12.92. **Natural inc.:** 2.88%. **Infant mortality** (per 1,000 live births): 77.84.

Education: Free: ages 7-13. **Literacy**: 39%.

Major Intl. Organizations: UN (FAO, IBRD, ILO, IMF, IMO, WHO, WTrO), the Commonwealth, OAU.

Embassy: Suite 1000, 1155 15th St. NW 20005; 785-1399.

Website: http://www.Gambia.com

The tribes of Gambia were at one time associated with the West African empires of Ghana, Mali, and Songhay. The area became Britain's first African possession in 1588.

Independence came Feb. 18, 1965; republic status within the Commonwealth was achieved in 1970. The country suffered from severe famine in the 1970s. After a coup attempt in 1981, The Gambia formed the confederation of Senegambia with Senegal that lasted until 1989.

On July 23, 1994, after 24 years in power, Pres. Dawda K. Jawara was deposed in a bloodless coup by a military officer, Yahya Jammeh. Jammeh barred political activity, detained potential opponents, and governed by decree. A new constitution was approved by referendum, Aug. 8, 1996. On Sept. 27 Jammeh won the presidential election. Parliamentary balloting on Jan. 2, 1997, completed the nominal return to civilian rule, but Jammeh retained a firm grip on power.

Georgia

People: Population: 4,989,285. **Age distrib.** (%): <15: 19.6; 65+: 12.5. **Pop. density:** 186 per sq. mi. **Urban:** 60%. **Ethnic groups:** Georgian 70%, Armenian 8%, Russian 6%. **Principal languages:** Georgian (official), Russian. **Chief religions:** Georgian Orthodox 65%, Muslim 11%, Russian Orthodox 10%.

Geography: Area: 26,900 sq. mi. **Location:** SW Asia, on E coast of Black Sea. **Neighbors:** Russia on N and NE, Turkey and Armenia on S, Azerbaijan on SE. **Topography:** Separated from Russia on NE by main range of the Caucasus Mts. **Capital:** Tbilisi, (urban agg.): 1,310,000.

Government: Type: Republic. **Head of state and gov.:** Pres. Eduard A. Shevardnadze; b Jan. 25, 1928; in office: Mar. 10, 1992. **Local divisions:** 53 rayons, 9 cities, and 2 autonomous republics. **Defense budget:** $22 mil. **Active troops:** 26,900.

Economy: Industries: Steel, machinery, trucks, textiles. **Chief crops:** Citrus, potatoes, vegetables, grapes, tea. **Minerals:** Manganese, iron, copper, coal. **Other resources:** Forests. **Crude oil reserves** (2000): 35 mil bbls. **Arable land:** 9%. **Livestock** (2000): cattle: 1.12 mil; chickens: 8.47 mil; goats: 73,400; pigs: 411,100; sheep: 560,000. **Electricity prod.:** 7.975 bil kWh. **Labor force:** 20% ind. and const., 40% agric. and forestry, 40% services.

Finance: Monetary unit: Lavi (Oct. 2000: 1.96 = $1 U.S.). **GDP** (1999 est.): $11.7 bil. **Per capita GDP:** $2,300. **Imports** (1999 est.): $840 mil; partners: EU 22%, Russia 15%, Azerbaijan 12%. **Exports** (1999 est.): $330 mil; partners: Russia 27%, Turkey 20%. **Tourism:** $400 mil. **Intl. reserves less gold** (June 2000): $109.97 mil. **Gold:** 2,100 oz t. **Consumer prices** (change in 1999): 19.1%.

Transport: Railroad: Length: 983 mi. **Motor vehicles:** 442,000 pass. cars, 50,000 comm. vehicles. **Civil aviation:** 128.1 mil pass.-mi.; 1 airport. **Chief ports:** Batumi, Sukhumi.

Communications: TV sets: 220 per 1,000 pop. **Telephones** (1999): 671,500 main lines.

Health: Life expectancy: 61.04 male; 68.28 female. **Births** (per 1,000 pop): 11.18. **Deaths** (per 1,000 pop.): 14.58. **Natural inc.:** -.34%. **Infant mortality** (per 1,000 live births): 52.37.

Education: Compulsory: ages 6-14. **Literacy**: 99%.

Major Intl. Organizations: UN (FAO, IBRD, ILO, IMF, IMO, WHO), CIS, OSCE.

Embassy: Suite 300, 1615 New Hampshire Ave. NW 20009; 393-5959.

Website: http://www.parliament.ge

The region, which contained the ancient kingdoms of Colchis and Iberia, was Christianized in the 4th century and conquered by Arabs in the 8th century. It expanded to include an area from the Black Sea to the Caspian and parts of Armenia and Persia before its disintegration under the impact of Mongol and Turkish invasions. Annexation by Russia in 1801 led to the Russian war with Persia, 1804-1813. Georgia entered the USSR in 1922 and became a constituent republic in 1936.

Georgia declared independence Apr. 9, 1991. It became an independent state when the Soviet Union disbanded Dec. 26, 1991. There was fighting during 1991 between rebel forces and loyalists of Pres. Zviad Gamsakhurdia, who fled the capital Jan. 6, 1992. The ruling Military Council picked former Soviet Foreign Minister Eduard A. Shevardnadze to chair a newly created State Council. An attempted coup by forces loyal to Gamsakhurdia was crushed June 24, 1992. Shevardnadze was later elected president. Gamsakhurdia died Jan. 1994, reportedly by suicide.

In Abkhazia, an autonomous republic within Georgia, ethnic Abkhazis, reportedly aided by Russia, launched a bloody military campaign and, by late 1993, had gained control of much of the region. A cease-fire providing for Russian peacekeepers was signed in Moscow May 14, 1994. Intermittent clashes continued into the late 1990s.

On Feb. 3, 1994, Georgia signed agreements with Russia for economic and military cooperation. On Mar. 1, Georgia's Supreme Council ratified membership by Georgia in the Commonwealth of Independent States.

Shevardnadze was wounded by a car bomb Aug. 29, 1995, while on his way to Parliament to sign a new constitution. He was reelected president Nov. 5. Shevardnadze escaped another assassination attempt, Feb. 9, 1998, when gunmen ambushed his motorcade. A mutiny by more than 200 soldiers opposed to Shevardnadze was crushed Oct. 19. He won another 5-year presidential term Apr. 9, 2000.

Germany

Federal Republic of Germany

People: Population: 83,029,536. **Age distrib.** (%): <15: 15.6; 65+: 16.6. **Pop. density:** 615 per sq. mi. **Urban:** 87%. **Ethnic groups:** German 92%, Turkish 2%. **Principal language:** German (official). **Chief religions:** Protestant 38%, Roman Catholic 34%.

Geography: Area: 135,100 sq. mi. **Location:** In central Europe. **Neighbors:** Denmark on N; Netherlands, Belgium, Luxembourg, France on W; Switzerland, Austria on S; Czech Rep., Poland on E. **Topography:** Germany is flat in N, hilly in center and W, and mountainous in Bavaria in the S. Chief rivers are Elbe, Weser, Ems, Rhine, and Main, all flowing toward North Sea, and Danube, flowing toward Black Sea. **Capital:** Berlin. **Cities** (2001 est.): Berlin 3.4 mil; Munich 1.2 mil; Hamburg 1.7 mil; Cologne 964,000; Frankfurt 644,000; Essen 603,000.

Government: Type: Federal republic. **Head of state:** Pres. Johannes Rau; b Jan. 16, 1931; in office: July 1, 1999. **Head of gov.:** Chan. Gerhard Schröder; b Apr. 7, 1944; in office: Oct. 27, 1998. **Local divisions:** 16 laender (states). **Defense budget:** $23.1 bil (2001). **Active troops:** 321,000.

Economy: Industries: Steel, ships, vehicles, machinery, electronics, coal, chemicals, iron, cement, food and beverages. **Chief crops:** Grains, potatoes, sugar beets. **Minerals:** Coal, potash, lignite, iron, uranium. **Crude oil reserves** (2000): 356.97 mil bbls. **Other resources:** Timber. **Arable land:** 33%. **Livestock** (2000): cattle: 14.66 mil; chickens: 107.66 mil; goats: 135,000; pigs: 27.05 mil; sheep: 2.1 mil. **Fish catch** (1999): 318,785 metric tons. **Electricity prod.:** 531.377 bil kWh. **Labor force:** 63.6% services, 33.7% ind.; 2.7% agric.

Finance: Monetary unit: Mark (Sept. 2001: 2.16 = $1 U.S.). Euro (Sept. 2001: 1.09 = $1 U.S.). **GDP** (1999 est.): $1.864 tril. **Per capita GDP:** $22,700. **Imports** (1999 est.): $587 bil; partners: EU 53.7%. **Exports** (1999 est.): $610 bil; partners: EU 56.4%. **Tourism:** $16.83 bil. **Budget** (1998 est.): $1.02 tril. **Intl. reserves less gold** (June 2000): $59.52 bil. **Gold:** 111.52 mil oz t. **Consumer prices** (change in 1999): 0.6%.

Transport: Railroad: Length: 54,994 mi. **Motor vehicles** (1997): 41.33 mil pass. cars, 3.17 mil comm. vehicles. **Civil aviation:** 53.6 bil pass.-mi.; 35 airports. **Chief ports:** Hamburg, Bremen, Bremerhaven, Lubeck, Rostock.

Communications: TV sets: 571 per 1,000 pop. **Radios:** 946 per 1,000 pop. **Telephones:** 49,400,000 main lines. **Daily newspaper circ.:** 311 per 1,000 pop.

Health: Life expectancy: 74.47 male; 80.92 female. **Births** (per 1,000 pop): 9.16. **Deaths** (per 1,000 pop.): 10.42. **Natural inc.:** -.13%. **Infant mortality** (per 1,000 live births): 4.71.

Education: Compulsory: ages 6-15. **Literacy** (1993): 100%.

Major Intl. Organizations: UN and all of its specialized agencies, EU, NATO, OECD, OSCE.

Embassy: 4645 Reservoir Rd. NW 20007; 298-4000.

Websites: http://www.undp.org/missions/germany
http://eng.bundesregierung.de/frameset/index.jsp

Germany is a central European nation originally composed of numerous states, with a common language and traditions, that were united in one country in 1871; Germany was split into 2 countries from the end of World War II until 1990, when it was reunified.

History and government. Germanic tribes were defeated by Julius Caesar, 55 and 53 BC, but Roman expansion N of the Rhine was stopped in AD 9. Charlemagne, ruler of the Franks, consolidated Saxon, Bavarian, Rhenish, Frankish, and other lands; after him the eastern part became the German Empire. The Thirty Years' War, 1618-1648, split Germany into small principalities and kingdoms. After Napoleon, Austria contended with Prussia for dominance, but lost the Seven Weeks' War to Prussia, 1866. Otto von Bismarck, Prussian chancellor, formed the North German Confederation, 1867.

In 1870 Bismarck maneuvered Napoleon III into declaring war. After the quick defeat of France, Bismarck formed the **German Empire** and on Jan. 18, 1871, in Versailles, proclaimed King Wilhelm I of Prussia German emperor (Deutscher kaiser).

The German Empire reached its peak before World War I in 1914, with 208,780 sq. mi., plus a colonial empire. After that war Germany ceded Alsace-Lorraine to France; West Prussia and Posen (Poznan) province to Poland; part of Schleswig to Denmark; lost all colonies and ports of Memel and Danzig.

Republic of Germany, 1919-1933, adopted the Weimar constitution; met reparation payments and elected Friedrich Ebert and Gen. Paul von Hindenburg presidents.

Third Reich, 1933-1945, Adolf Hitler led the National Socialist German Workers' (Nazi) party after World War I. In 1923 he attempted to unseat the Bavarian government and was imprisoned. Pres. von Hindenburg named Hitler chancellor Jan. 30, 1933; on Aug. 3, 1934, the day after Hindenburg's death, the cabinet joined the offices of president and chancellor and made Hitler fuehrer (leader). Hitler abolished freedom of speech and assembly, and began a long series of persecutions climaxed by the murder of millions of Jews and others.

He repudiated the Versailles treaty and reparations agreements, remilitarized the Rhineland (1936), and annexed Austria (Anschluss, 1938). At Munich he made an agreement with Neville Chamberlain, British prime minister, which permitted Germany to annex part of Czechoslovakia. He signed a nonaggression treaty with the USSR, 1939 and declared war on Poland Sept. 1, 1939, precipitating World War II. With total defeat near, Hitler committed suicide in Berlin Apr. 1945. The victorious Allies voided all acts and annexations of Hitler's Reich.

Division of Germany. Germany was sectioned into 4 zones of occupation, administered by the Allied Powers (U.S., USSR, U.K., and France). The USSR took control of many E German states. The territory E of the so-called Oder-Neisse line was assigned to, and later annexed by, Poland. Northern East Prussia (now Kaliningrad) was annexed by the USSR. Administration of the remaining regions, in the W and S (which make up about two-thirds of present-day Germany), was split among the Western Allies.

There was also created the area of Greater Berlin, within but not part of the Soviet zone, administered by the 4 occupying powers under the Allied Command. In 1948 the USSR withdrew, established its single command in East Berlin, and cut off supplies. The Western Allies utilized a gigantic airlift to bring food to West Berlin, 1948-49.

In 1949, 2 separate German states were established; in May the zones administered by the Western Allies became West Germany, capital: Bonn; in Oct. the Soviet sector became East Germany, capital: East Berlin. West Berlin was considered an enclave of West Germany, although its status was disputed by the Soviet bloc.

East Germany. The German Democratic Republic (East Germany) was proclaimed in the Soviet sector of Berlin Oct. 7, 1949. It was proclaimed fully sovereign in 1954, but Soviet troops remained on grounds of security and the 4-power Potsdam agreement.

Coincident with the entrance of West Germany into the European defense community in 1952, the East German government decreed a prohibited zone 3 miles deep along its 600-mile border with West Germany and cut Berlin's telephone system in two. Berlin was further divided by erection of a fortified wall in 1961, after over 3 million East Germans had emigrated West; an exodus of refugees to the West continued, though on a smaller scale.

East Germany suffered severe economic problems at least until the mid-1960s. Then a "new economic system" was introduced, easing central planning controls and allowing factories to make profits provided they were reinvested in operations or redistributed to workers as bonuses. By the early 1970s, the economy was highly industrialized, and the nation was credited with the highest standard of living among Warsaw Pact countries. But growth slowed in the late 1970s, because of shortages of natural resources and labor, and a huge debt to lenders in the West. Comparison with the lifestyle in the West caused many young people to leave the country.

The government firmly resisted following the USSR's policy of *glasnost,* but by Oct. 1989, was faced with nationwide demonstrations demanding reform. Pres. Erich Honecker, in office since 1976, was forced to resign, Oct. 18. On Nov. 4, the border with Czechoslovakia was opened and permission granted for refugees to travel to the West. On Nov. 9, the East German government announced its decision to open the border with the West, signaling the end of the "Berlin Wall," which was the supreme emblem of the cold war. On Aug. 23, 1990, the East German parliament agreed to formal unification with West Germany; this occurred Oct. 3.

West Germany. The Federal Republic of Germany (West Germany) was proclaimed May 23, 1949, in Bonn, after a constitution had been drawn up by a consultative assembly formed by representatives of the 11 laender (states) in the French, British, and American zones. Later reorganized into 9 units, the laender numbered 10 with the addition of the Saar, 1957. Berlin also was granted land (state) status, but the 1945 occupation agreements placed restrictions on it.

The occupying powers, the U.S., Britain, and France, restored civil status, Sept. 21, 1949. The Western Allies ended the state of war with Germany in 1951 (the U.S. resumed diplomatic relations July 2), while the USSR did so in 1955. The powers lifted controls and the republic became fully independent May 5, 1955.

Dr. Konrad Adenauer, Christian Democrat, was made chancellor Sept. 15, 1949, reelected 1953, 1957, 1961. Willy Brandt, heading a coalition of Social Democrats and Free Democrats, became chancellor Oct. 21, 1969. (He resigned May 1974 because of a spy scandal.)

In 1970 Brandt signed friendship treaties with the USSR and Poland. In 1971, the U.S., Britain, France, and the USSR signed an agreement on Western access to West Berlin. In 1972 East and West Germany signed their first formal treaty, implementing the agreement easing access to West Berlin. In 1973 a West Germany-Czechoslovakia pact normalized relations and nullified the 1938 "Munich Agreement."

West Germany experienced strong economic growth starting in the 1950s. The country led Europe in provisions for worker participation in the management of industry.

In 1989 the changes in the East German government and opening of the Berlin Wall sparked talk of reunification of the 2 Germanys. In 1990, under Chancellor Kohl's leadership, West Germany moved rapidly to reunite with East Germany.

A New Era. As Communism was being rejected in East Germany, talks began concerning German reunification. At a meeting in Ottawa, Feb. 1990, the foreign ministers of the World War II "Big Four" Allied nations and of East Germany and West Germany reached agreement on a format for high-level talks on German reunification.

In May, NATO ministers adopted a package of proposals on reunification, including the inclusion of the united Germany as a full member of NATO and the barring of the new Germany from having its own nuclear, chemical, or biological weapons. In July, the USSR agreed to conditions that would allow Germany to become a member of NATO.

The 2 nations agreed to monetary unification under the West German mark beginning in July. The merger of the 2 Germanys took place Oct. 3, and the first all-German elections since 1932 were held Dec. 2. Eastern Germany received more than $1 trillion in public and private funds from western Germany between 1990 and 1995.

In 1991, Berlin again became the capital of Germany; the legislature, most administrative offices, and most foreign embassies had shifted from Bonn to Berlin by late 1999.

Germany's highest court ruled, July 12, 1994, that German troops could participate in international military missions abroad, when approved by Parliament. Ceremonies were held marking the final withdrawal of Russian troops from Germany, Aug. 31. Ceremonies were held the following week marking the final withdrawal of American, British, and French troops from Berlin. General elections Oct. 16 left Chancellor Helmut Kohl's governing coalition with a slim parliamentary majority. On Oct. 31, 1996, after more than 14 years in office, Kohl surpassed Adenauer as Germany's longest-serving chancellor in the 20th century.

Unemployment hit a postwar high of 12.6% in Jan. 1998. The Kohl era ended with the defeat of the Christian Democrats in parliamentary elections Sept. 27; Gerhard Schröder, of the Social Democratic Party, became chancellor. Germany contributed 8,500 troops to the NATO-led security force (KFOR) that entered Kosovo in June 1999. Kohl resigned as honorary party chairman Jan. 18, 2000, amid allegations of illegal fund-raising. Kohl reached an agreement with prosecutors Feb. 8, 2001, in which he acknowledged committing a "breach of trust" and agreed to pay a fine, but did not plead guilty to any criminal charges.

Helgoland, an island of 130 acres in the North Sea, was taken from Denmark by a British Naval Force in 1807 and later ceded to Germany to become part of Schleswig-Holstein province in return for rights in East Africa. The heavily fortified island was surrendered to UK, May 23, 1945, demilitarized in 1947, and returned to West Germany, Mar. 1, 1952. It is a free port.

Ghana

Republic of Ghana

People: Population: 19,894,014. **Age distrib.** (%): <15: 41.2; 65+: 3.5. **Pop. density:** 224 per sq. mi. **Urban:** 38%. **Ethnic groups:** Akan 44%, Moshi-Dagomba 16%, Ewe 13%, Ga 8%. **Principal languages:** English (official), Akan, Moshi-Dagomba, Ewe, Ga. **Chief religions:** Indigenous beliefs 38%, Muslim 30%, Christian 24%.

Geography: Area: 88,700 sq. mi. **Location:** On southern coast of W Africa. **Neighbors:** Côte d'Ivoire on W, Burkina Faso on N, Togo on E. **Topography:** Most of Ghana consists of low fertile plains and scrubland, cut by rivers and by the artificial Lake Volta. **Capital:** Accra, (urban agg.): 1,976,000.

Government: Type: Republic. **Head of state and gov.:** Pres. John Agyekum Kufuor; b Dec. 8, 1938; in office: Jan. 7, 2001. **Local divisions:** 10 regions. **Defense budget:** $45 mil.. **Active troops:** 7,000.

Economy: Industries: Aluminum, light manufacturing, mining, lumbering, food processing. **Chief crops:** Cocoa, coffee, rice, cassava, peanuts, corn. **Minerals:** Gold, manganese, industrial diamonds, bauxite. **Crude oil reserves** (2000): 16.5 mil bbls. **Other resources:** Timber, rubber. **Arable land:** 12%. **Livestock** (2000): cattle: 1.29 mil; chickens: 18.0 mil; goats: 2.8 mil; pigs: 350,000; sheep: 2.57 mil. **Fish catch** (1999): 446,883 metric tons. **Electricity prod.:** 5.466 bil kWh. **Labor force:** 60% agric.; 25% services, 15% ind.

Finance: Monetary unit: Cedi (Sept. 2001: 7,090 = $1 U.S.). **GDP** (1999 est.): $35.5 bil. **Per capita GDP:** $1,900. **Imports** (1999): $2.5 bil; partners: Germany, UK, U.S. . **Exports** (1999): $1.7 bil; partners: Togo, UK, Italy. **Tourism** (1998): $274 mil. **Budget** (1996 est.): $1.47 bil. **Intl. reserves less gold** (June 2000): $396.3 mil. **Gold:** 280,000 oz t. **Consumer prices** (change in 1999): 12.4%.

Transport: Railroad: Length: 592 mi. **Motor vehicles:** 90,000 pass. cars, 45,000 comm. vehicles. **Civil aviation:** 436.4 mil pass.-mi.; 1 airport. **Chief ports:** Tema, Takoradi.

Communications: TV sets: 15 per 1,000 pop. **Radios:** 249 per 1,000 pop. **Telephones:** 237,200 main lines. **Daily newspaper circ.:** 64 per 1,000 pop.

Health: Life expectancy: 55.86 male; 58.66 female. **Births** (per 1,000 pop): 28.95. **Deaths** (per 1,000 pop.): 10.26. **Natural inc.:** 1.87%. **Infant mortality** (per 1,000 live births): 56.54.

Education: Compulsory: ages 6-16. **Literacy**: 64%.

Major Intl. Organizations: UN and all of its specialized agencies, the Commonwealth, OAU.

Embassy: 3512 International Dr. NW 20008; 686-4520.

Named for an African empire along the Niger River, AD 400-1240, Ghana was ruled by Britain for 113 years as the Gold Coast. The UN in 1956 approved merger with the British Togoland trust territory. Independence came Mar. 6, 1957, and republic status within the Commonwealth in 1960.

Pres. Kwame Nkrumah built hospitals and schools, promoted development projects like the Volta R. hydroelectric and aluminum plants but ran the country into debt, jailed opponents, and was accused of corruption. A 1964 referendum gave Nkrumah dictatorial powers and set up a one-party socialist state. Nkrumah was overthrown in 1966 by a police-army coup, which expelled Chinese and East German teachers and technicians. Elections were held in 1969, but 4 further coups occurred in 1972, 1978, 1979, and 1981. The 1979 and 1981 coups, led by Flight Lieut. Jerry Rawlings, were followed by suspension of the constitution and banning of political parties. A new constitution, allowing multiparty politics, was approved in April 1992.

In Feb. 1993 more than 1,000 people were killed in ethnic clashes in northern Ghana. Rawlings won the presidential election of Dec. 7, 1996. Kofi Annan, a career UN diplomat from Ghana, became UN secretary general on Jan. 1, 1997. Opposition leader John Agyekum Kufuor won a runoff vote Dec. 28, 2000, and was sworn in Jan. 7, 2001, marking Ghana's 1st peaceful transfer of power from one elected president to another.

Greece

Hellenic Republic

People: Population: 10,623,835. **Age distrib.** (%): <15: 15.0; 65+: 17.7. **Pop. density:** 211 per sq. mi. **Urban:** 60%. **Ethnic groups:** Greek 98%. (**Note:** Greek govt. states there are no ethnic divisions in Greece.) **Principal languages:** Greek (official), English, French. **Chief religion:** Greek Orthodox 98% (official).

Geography: Area: 50,400 sq. mi. **Location:** Occupies southern end of Balkan Peninsula in SE Europe. **Neighbors:** Albania, Macedonia, Bulgaria on N; Turkey on E. **Topography:** About three-quarters of Greece is nonarable, with mountains in all areas. Pindus Mts. run through the country N to S. The heavily indented coastline is 9,385 mi. long. Of over 2,000 islands, only 169 are inhabited, among them Crete, Rhodes, Milos, Kerkira (Corfu), Chios, Lesbos, Samos, Euboea, Delos, Mykonos. **Capital:** Athens, (urban agg.): 3,116,000, (1999 city proper: 748,110).

Government: Type: Parliamentary republic. **Head of state:** Pres. Konstantinos Stephanopoulos; b Aug. 15, 1926; in office: Mar. 8, 1995. **Head of gov.:** Prime Min. Costas Simitis; b June 23, 1936; in office: Jan. 18, 1996. **Local divisions:** 13 regions comprising 51 prefectures. **Defense budget:** $3.3 bil. **Active troops:** 159,170.

Economy: Industries: Tourism, textiles, chemicals, metals, wine, food processing. **Chief crops:** Grains, corn, sugar beets, cotton, tobacco, olives, grapes, citrus and other fruits, tomatoes. **Minerals:** Bauxite, lignite, magnesite, marble, oil. **Crude oil reserves** (2000): 10 mil bbls. **Arable land:** 19%. **Livestock** (2000): cattle: 590,000; chickens: 28.0 mil; goats: 5.29 mil; pigs: 906,000; sheep: 9.04 mil. **Fish catch** (1999): 214,228 metric tons. **Electricity prod.:** 46.432 bil kWh. **Labor force:** 59% services; 21% ind.; 20% agric.

Finance: Monetary unit: Drachma (Sept. 2001: 375.92 = $1 U.S.). **GDP** (1999 est.): $149.2 bil. **Per capita GDP:** $13,900. **Imports** (1998): $27.7 bil; partners: Italy 16%, Germany 16%. **Exports** (1998 est.): $12.4 bil; partners: Germany 25%, Italy 11%. **Tourism:** $8.77 bil. **Budget** (1998 est.): $47.6 bil. **Intl. reserves less gold** (June 2000): $14.89 bil. **Gold:** 4.25 mil oz t. **Consumer prices** (change in 1999): 2.6%.

Transport: Railroad: Length: 1,537 mi. **Motor vehicles:** 2.34 mil pass. cars, 939,923 comm. vehicles. **Civil aviation:** 5.8 bil pass.-mi.; 36 airports. **Chief ports:** Piraeus, Thessaloníki, Patrai.

Communications: TV sets: 466 per 1,000 pop. **Radios:** 477 per 1,000 pop. **Telephones:** 5,659,300 main lines. **Daily newspaper circ.:** 153 per 1,000 pop.

Health: Life expectancy: 76.03 male; 81.32 female. **Births** (per 1,000 pop): 9.83. **Deaths** (per 1,000 pop.): 9.73. **Natural inc.:** .01%. **Infant mortality** (per 1,000 live births): 6.38.

Education: Free, compulsory: ages 6-15. **Literacy** (1993): 95%.

Major Intl. Organizations: UN (FAO, IBRD, ILO, IMF, IMO, WHO, WTrO), EU, NATO, OECD, OSCE.

Embassy: 2221 Massachusetts Ave. NW 20008; 939-5800.

Website: http://www.gr-indexer.gr

The achievements of ancient Greece in art, architecture, science, mathematics, philosophy, drama, literature, and democracy became legacies for succeeding ages. Greece reached the height of its glory and power, particularly in the Athenian city-state, in the 5th century BC. Greece fell under Roman rule in the 2d and 1st centuries BC. In the 4th century AD it became part of the Byzantine Empire and, after the fall of Constantinople to the Turks in 1453, part of the Ottoman Empire.

Greece won its war of independence from Turkey 1821-1829, and became a kingdom. A republic was established 1924; the monarchy was restored, 1935, and George II, King of the Hellenes, resumed the throne. In Oct. 1940, Greece rejected an ultimatum from Italy. Nazi support resulted in its defeat and occupation by Germans, Italians, and Bulgarians. By the end of 1944 the invaders withdrew. Communist resistance forces were defeated by Royalist and British troops. A plebiscite again restored the monarchy.

Communists waged guerrilla war 1947-49 against the government but were defeated with the aid of the U.S. A period of reconstruction and rapid development followed, mainly with conservative governments under Premier Constantine Karamanlis. The Center Union, led by George Papandreou, won elections in 1963 and 1964, but King Constantine, who acceded in 1964, forced Papandreou to resign. A period of political maneuvers ended in the military takeover of April 21, 1967, by Col. George Papadopoulos. King Constantine tried to reverse the consolidation of the harsh dictatorship Dec. 13, 1967, but failed and fled to Italy. Papadopoulos was ousted Nov. 25, 1973.

Greek army officers serving in the National Guard of Cyprus staged a coup on the island July 15, 1974. Turkey invaded Cyprus a week later, precipitating the collapse of the Greek junta, which was implicated in the Cyprus coup. Democratic government returned (and in 1975 the monarchy was abolished).

The 1981 electoral victory of the Panhellenic Socialist Movement (Pasok) of Andreas Papandreou brought substantial changes in Greece's internal and external policies. A scandal centered on George Kostokas, a banker and publisher, led to the arrest or investigation of leading Socialists, implicated Papandreou, and contributed to the defeat of the Socialists at the polls in 1989. However, Papandreou, who was narrowly acquitted Jan. 1992 of corruption charges, led the Socialists to a comeback victory in general elections Oct. 10, 1993.

Tensions between Greece and the Former Yugoslav Republic of Macedonia eased when the 2 countries agreed to normalize relations Sept. 13, 1995. The ailing Papandreou was replaced as prime minister by Costas Simitis, Jan. 18, 1996. Simitis led the Socialists to victory in the election of Sept. 22. The International Olympic Committee, Sept. 5, 1997, chose Athens to host the Summer Games in 2004. An earthquake that shook Athens Sept. 7, 1999, killed at least 143 people and left over 60,000 homeless. The Socialists retained power by a narrow margin in the elections of Apr. 9, 2000.

Grenada

People: Population: 89,227. **Age distrib.** (%): <15: 37.0; 65+: 3.9. **Pop. density:** 686 per sq. mi. **Urban:** 37%. **Ethnic groups:** Mostly black African. **Principal languages:** English (official), French patois. **Chief religions:** Roman Catholic 53%, Protestant 33%.

Geography: Area: 130 sq. mi. **Location:** In Caribbean, 90 mi. N of Venezuela. **Neighbors:** Venezuela, Trinidid & Tobago to S; St. Vincent & the Grenadines to N. **Topography:** Main island is mountainous; country includes Carriacou and Petit Martinique islands. **Capital:** Saint George's, (1997): 35,742.

Government: Type: Parliamentary democracy. **Head of state:** Queen Elizabeth II, represented by Gov.-Gen. Daniel Williams; b Nov. 4, 1935; in office: Aug. 8, 1996. **Head of gov.:** Prime Min. Keith Mitchell; b Nov. 12, 1946; in office: June 22, 1995. **Local divisions:** 6 parishes, 1 dependency.

Economy: Industries: Tourism, textiles, food and beverages. **Chief crops:** Nutmeg, bananas, cocoa, mace. **Resources:** Timber. **Arable land:** 15%. **Livestock** (2000): cattle: 4,380; chickens: 220,000; goats: 7,000; pigs: 5,300; sheep: 13,000.. **Electricity prod.:** 120 mil kWh. **Labor force:** 62% services; 24% agric.; 14% const.

Finance: Monetary unit: East Caribbean Dollar (Sept. 2001: 2.67 = $1 U.S.). **GDP** (1999 est.): $360 mil. **Per capita GDP:** $3,700. **Imports** (1998): $200 mil; partners: U.S. 31%, Caricom 24%, UK 14%. **Exports** (1998): $26.8 mil; partners: Caricom 32%; UK 20%. **Tourism:** $63 mil. **Budget** (1997): $102.1 mil. **Intl. reserves less gold** (Jan. 2000): $49.80 mil. **Consumer prices** (change in 1997): 0.2%.

Transport: Civil aviation: 2 airports. **Chief ports:** Saint George's, Grenville.

Communications: TV sets: 154 per 1,000 pop. **Radios:** 460 per 1,000 pop. **Telephones:** 31,400 main lines.

Health: Life expectancy: 62.74 male; 66.31 female. **Births** (per 1,000 pop): 23.12. **Deaths** (per 1,000 pop.): 7.82. **Natural inc.:** 1.53%. **Infant mortality** (per 1,000 live births): 14.63.

Education: Free, compulsory: ages 5-16. **Literacy** (1994): 85%.

Major Intl. Organizations: UN (FAO, IBRD, ILO, IMF, WHO, WTrO), Caricom, the Commonwealth, OAS, OECS.

Embassy: 1701 New Hampshire Ave. NW 20009; 265-2561.

Website: http://www.grenada.org

Columbus sighted Grenada in 1498. First European settlers were French, 1650. The island was held alternately by France and England until final British occupation, 1784. Grenada became fully independent Feb. 7, 1974, during a general strike. It is the smallest independent nation in the western hemisphere.

On Oct. 14, 1983, a military coup ousted Prime Minister Maurice Bishop, who was put under house arrest, later freed by supporters, rearrested, and, finally, on Oct. 19, executed. U.S. forces, with a token force from 6 area nations, invaded Grenada, Oct. 25. Resistance from the Grenadian army and Cuban advisors was quickly overcome as most people welcomed the invading forces. U.S. troops left Grenada in June 1985. Cuban Pres. Castro received an enthusiastic greeting when visiting Grenada Aug. 2-3, 1998.

Guatemala

Republic of Guatemala

People: Population: 12,974,361. **Age distrib.** (%): <15: 42.1; 65+: 3.6. **Pop. density:** 310 per sq. mi. **Urban:** 39%. **Ethnic groups:** Mestizo 56%, Amerindian 44%. **Principal languages:** Spanish (official), Mayan languages. **Religion:** Mostly Roman Catholic, some Protestant, traditional Mayan.

Geography: Area: 41,800 sq. mi. **Location:** In Central America. **Neighbors:** Mexico on N and W, El Salvador on S, Honduras and Belize on E. **Topography:** The central highland and mountain areas are bordered by the narrow Pacific coast and the lowlands and fertile river valleys on the Caribbean. There are numerous volcanoes in S, more than half a dozen over 11,000 ft. **Capital:** Guatemala City, (urban agg.): 3,242,000.

Government: Type: Republic. **Head of state and gov.:** Pres. Alfonso Portillo Cabrera; b Sept. 24, 1951; in office: Jan. 14, 2000. **Local divisions:** 22 departments. **Defense budget:** **Active troops:** 31,400.

Economy: Industries: Furniture, rubber, sugar, chemicals, textiles. **Chief crops:** Coffee, sugar, bananas, corn, cardamom. **Minerals:** Oil, nickel. **Crude oil reserves** (2000): 526 mil bbls. **Other resources:** Rare woods, fish, chicle. **Arable land:** 12%. **Livestock** (2000): cattle: 2.3 mil; chickens: 24.0 mil; goats: 110,000; pigs: 825,000; sheep: 551,000. **Fish catch:** (1999): 11,303 metric tons. **Electricity prod.:** 3.785 bil kWh. **Labor force:** 50% agric.; 35% serv.; 15% ind.

Finance: Monetary unit: Quetzal (Sept. 2001: 7.92 = $1 U.S.). **GDP** (1999 est.): $47.9 bil. **Per capita GDP:** $3,900. **Imports** (1999): $4.5 bil; partners: U.S. 46%. **Exports** (1999): $2.4 bil; partners: U.S. 48%. **Tourism:** $570 mil. **Budget** (1996 est.): $1.88 bil. **Intl. reserves less gold** (June 2000): $1.65 bil. **Gold:** 217,000 oz t. **Consumer prices** (change in 1999): 4.9%.

Transport: Railroad: Length: 549 mi. **Motor vehicles:** 102,000 pass. cars, 97,000 comm. vehicles. **Civil aviation:** 228.9 mil pass.-mi.; 2 airports. **Chief ports:** Puerto Barrios, San Jose.

Communications: TV sets: 45 per 1,000 pop. **Radios:** 52 per 1,000 pop. **Telephones:** 650,000 main lines. **Daily newspaper circ.:** 29 per 1,000 pop.

Health: Life expectancy: 63.85 male; 69.31 female. **Births** (per 1,000 pop): 34.61. **Deaths** (per 1,000 pop.): 6.79. **Natural inc.:** 2.78%. **Infant mortality** (per 1,000 live births): 45.79.

Education: Free, compulsory: ages 7-14. **Literacy:** 56%.

Major Intl. Organizations: UN (FAO, IBRD, ILO, IMF, IMO, WHO, WTrO), OAS.

Embassy: 2220 R St. NW 20008; 745-4952.

The old Mayan Indian empire flourished in what is today Guatemala for over 1,000 years before the Spanish.

Guatemala was a Spanish colony 1524-1821; briefly a part of Mexico and then of the U.S. of Central America, the republic was established in 1839.

Since 1945 when a liberal government was elected to replace the long-term dictatorship of Jorge Ubico, the country has seen a variety of military and civilian governments and periods of civil war. Dissident army officers seized power Mar. 23, 1982, denouncing a presidential election as fraudulent and pledging to restore "authentic democracy" to the nation. Political violence caused large numbers of Guatemalans to seek refuge in Mexico. Another military coup occurred Oct. 8, 1983. The nation returned to civilian rule in 1986.

The crisis-ridden government of Pres. Jorge Serrano Elías was ousted by the military June 1, 1993. Ramiro de León Carpio was elected president by Congress June 6. A conservative businessman, Alvaro Arzú Irigoyen, won the presidency, Jan. 7, 1996. On Sept. 19 the Guatemalan government and leftist rebels approved a peace accord; the final agreement was signed Dec. 29. During more than 35 years of armed conflict, some 200,000 people were killed or "disappeared" (and are presumed dead); most of these casualties were attributed to the government and its paramilitary allies.

Violent episodes in 1998 included the daylight ambush of a busload of U.S. college students, Jan. 16, resulting in the rape of five young women, and the murder of Bishop Juan José Girardi, a human rights activist, Apr. 26. U.S. Pres. Bill Clinton, on a visit to Guatemala Mar. 10, 1999, apologized for aid the U.S. had given to forces which he said "engaged in violence and widespread repression." Candidates of the right-wing populist Guatemalan Republican Front won control of Congress, Nov. 7, 1999, and the presidency, Dec. 26.

Guinea

Republic of Guinea

People: Population: 7,613,870. **Age distrib.** (%): <15: 43.1; 65+: 2.7. **Pop. density:** 80 per sq. mi. **Urban:** 32%. **Ethnic groups:** Peuhl 40%, Malinke 30%, Soussou 20%, smaller tribes 10%. **Principal languages:** French (official), tribal languages. **Chief religions:** Muslim 85%, Christian 8%.

Geography: Area: 94,800 sq. mi. **Location:** On Atlantic coast of W Africa. **Neighbors:** Guinea-Bissau, Senegal, Mali on N; Côte d'Ivoire on E; Liberia on S. **Topography:** A narrow coastal belt leads to the mountainous middle region, the source of the Gambia, Senegal, and Niger rivers. Upper Guinea, farther inland, is a cooler upland. The SE is forested. **Capital:** Conakry, (urban agg.): 1,824,000.

Government: Type: Republic. **Head of state:** Pres. Gen. Lansana Conté; b 1934; in office: Apr. 5, 1984. **Head of gov.:** rem. Lamine Sidimé; b 1944; in office: Mar. 8, 1999. **Local divisions:** 4 administrative regions, 1 special zone. **Defense budget:**$55 mil . **Active troops:** 9,700.

Economy: Industries: Mining, light manufacturing, agricultural processing. **Chief crops:** Bananas, pineapples, rice, palm kernels, coffee, cassava. **Minerals:** Bauxite, iron, diamonds, gold. **Arable land:** 2%. **Livestock** (2000): cattle: 2.37 mil; chickens: 8.9 mil; goats: 864,000; pigs: 54,000; sheep: 687,000. **Fish catch** (1999): 102,589 metric tons. **Electricity prod.:** 750 mil kWh. Labor force: 80% agric., 11% ind. and comm.

Finance: Monetary unit: Franc (Sept. 2001: 1,960.00 = $1 U.S.). **GDP** (1999 est.): $9.2 bil. **Per capita GDP:** $1,200. **Imports** (1998 est.): $560 mil; partners: France , Côte d'Ivoire. **Exports** (1998 est.): $695 mil; partners: Benelux., U.S., Russia. **Tourism:** $7 mil. **Budget** (1995 est.): $652 mil. **Intl. reserves less gold** (June 1998): $202.16 mil.

Transport: Railroad: Length: 411 mi. **Motor vehicles:** 13,700 pass. cars, 19,300 comm. vehicles. **Civil aviation:** 33.9 mil pass.-mi.; 1 airport. **Chief port:** Conakry.

Communications: TV sets: 10 per 1,000 pop. **Radios:** 34 per 1,000 pop. **Telephones:** 62,400 main lines.

Health: Life expectancy: 43.49 male; 48.42 female. **Births** (per 1,000 pop): 39.78. **Deaths** (per 1,000 pop.): 17.53. **Natural inc.:** 2.23%. **Infant mortality** (per 1,000 live births): 129.03.

Education: Free, compulsory: ages 7-13. **Literacy:** 36%.

Major Intl. Organizations: UN and most of its specialized agencies, OAU.

Embassy: 2112 Leroy Pl. NW 20008; 483-9420.

Part of the ancient West African empires, Guinea fell under French control 1849-98. Under Sékou Touré, it opted for full independence in 1958, and France withdrew all aid.

Touré turned to Communist nations for support and set up a militant one-party state. Thousands of opponents were jailed in the 1970s, in the aftermath of an unsuccessful Portuguese invasion. Many were tortured and killed.

The military took control in a bloodless coup after the March 1984 death of Touré. A new constitution was approved in 1991, but movement toward democracy was slow. When presidential elections were finally held, in Dec. 1993, the incumbent, Gen. Lansana Conté, was the official winner; outside monitors called the elections flawed. Parliamentary elections June 11, 1995, raised similar complaints. Conté suppressed an army mutiny in Conakry, Feb. 2-3, 1996, and won reelection in Dec. 1998.

Fighting in early 2001 along the border with Liberia and Sierra Leone created what UN officials described as a refugee crisis.

Guinea-Bissau

Republic of Guinea-Bissau

People: Population: 1,315,822. **Age distrib.** (%): <15: 42.1; 65+: 2.9. **Pop. density:** 122 per sq. mi. **Urban:** 23%. **Ethnic groups:** Balanta 30%, Fula 20%, Manjaca 14%, Mandinga 13%. **Principal languages:** Portuguese (official), Crioulo, tribal languages. **Chief religions:** Indigenous beliefs 50%, Muslim 45%, Christian 5%.

Geography: Area: 10,800 sq. mi. **Location:** On Atlantic coast of W Africa. **Neighbors:** Senegal on N, Guinea on E and S. **Topography:** A swampy coastal plain covers most of the country; to the east is a low savanna region. **Capital:** Bissau (1999 est.): 274,000.

Government: Type: Republic. **Head of state:** Pres. Kumba Yala; in office: Feb. 17, 2000. **Head of gov.:** Prime Min. Faustino Imbali; in office: Mar. 21, 2001. **Local divisions:** 9 regions. **Defense budget:** $3 mil. **Active troops:** 9,250.

Economy: Chief crops: Peanuts, cashews, corn, beans, cotton, rice. **Minerals:** Bauxite, phosphates. **Arable land:** 11%. **Livestock** (2000): cattle: 530,000; chickens: 850,000; goats: 325,000; pigs: 345,000; sheep: 285,000.. **Electricity prod.:** 55 mil kWh. **Labor force:** 78% agric..

Finance: Monetary unit: CFA Franc (Sept. 2001: 724.11 = $1 U.S.). **GDP** (1999 est.): $1.1 bil. **Per capita GDP:** $900. **Imports** (1998): $22.9 mil; partners: France 8%, Portugal 26%. **Exports** (1998): $26.8 mil; partners: Singapore 12%, India 59%. **Intl. reserves less gold** (Apr. 2000): $38.36 mil. **Consumer prices** (change in 1997): –0.7%.

Transport: Motor vehicles: 3,500 pass. cars, 2,500 comm. vehicles. **Civil aviation:** 6.2 mil pass.-mi.; 2 airports. **Chief port:** Bissau.

Communications: Radios: 42 per 1,000 pop. **Telephones** (1998): 8,100 main lines.

Health: Life expectancy: 47.12 male; 51.78 female. **Births** (per 1,000 pop): 39.29. **Deaths** (per 1,000 pop.): 15.33. **Natural inc.:** 2.4%. **Infant mortality** (per 1,000 live births): 110.4.

Education: Compulsory: ages 7-13. **Literacy:** 55%.

Major Intl. Organizations: UN (FAO, IBRD, ILO, IMF, IMO, WHO, WTrO), OAU.

Embassy: 1511 K St. NW 20005; 347-3950.

Portuguese mariners explored the area in the mid-15th century; the slave trade flourished in the 17th and 18th centuries, and colonization began in the 19th.

Beginning in the 1960s, an independence movement waged a guerrilla war and formed a government in the interior that had international support. Independence came Sept. 10, 1974, after the Portuguese regime was overthrown.

The November 1980 coup gave Vieira absolute power. Vieira eventually initiated political liberalization; multiparty elections were held July 3, 1994. An army uprising June 7, 1998, triggered a civil war, with Senegal and Guinea aiding the Vieira regime. After a peace accord signed on Nov. 2 broke down, rebel troops ousted Vieira on May 7, 1999. Elections Nov. 28-29, 1999, and Jan. 16, 2000, brought a return of civilian rule.

Guyana

Co-operative Republic of Guyana

People: Population: 697,181. **Age distrib.** (%): <15: 28.2; 65+: 4.9. **Pop. density:** 9 per sq. mi. **Urban:** 38%. **Ethnic groups:** East Indian 49%, black 32%, mixed 12%, Amerindian 6%. **Principal languages:** English (official), Amerindian dialects. **Chief religions:** Christian 57%, Hindu 33%, Muslim 9%.

Geography: Area: 75,900 sq. mi. **Location:** On N coast of South America. **Neighbors:** Venezuela on W, Brazil on S, Suriname on E. **Topography:** Dense tropical forests cover much of the land, although a flat coastal area up to 40 mi. wide, where 90% of the population lives, provides rich alluvial soil for agriculture. A grassy savanna divides the 2 zones. **Capital:** Georgetown (1999 est.): 275,000.

Government: Type: Republic. **Head of state:** Pres. Bharrat Jagdeo; b Jan. 23, 1964; in office: Aug. 11, 1999. **Head of gov.:** Prime Min. Samuel Hinds; b Dec. 27, 1943; in office: Dec. 22, 1997. **Local divisions:** 10 regions. **Defense budget:** $5 mil. **Active troops:** 1,600.

Economy: Industries: Mining, textiles. **Chief crops:** Sugar, rice, wheat. **Minerals:** Bauxite, gold, diamonds. **Other resources:** Timber, shrimp, dairy prods. **Arable land:** 2%. **Livestock** (2000): cattle: 220,000; chickens: 12.5 mil; goats: 79,000; pigs: 20,000; sheep: 130,000. **Fish catch:** (1999): 57,409 metric tons. **Electricity prod.:** 455 mil kWh. **Labor force:** 39% agric., forestry, fishing; 24% mining, manuf., const.

Finance: Monetary unit: Dollar (Sept. 2001: 180.50 = $1 U.S.). **GDP** (1999 est.): $1.86 bil. **Per capita GDP:** $2,500. **Imports** (1999 est.): $620 mil; partners: U.S. 28%, Trin. & Tob. 21%. **Exports** (1999 est.): $574 mil; partners: Canada 24%, U.S. 25%. **Tourism** (1998): $52 mil. **Budget** (1996 est): $299 mil. **Intl. reserves less gold** (Apr. 2000): $288.82 mil. **Consumer prices** (change in 1999): 7.5%.

Transport: Motor vehicles: 24,000 pass. cars, 9,000 comm. vehicles. **Civil aviation:** 154.1 mil pass.-mi.; 1 airport. **Chief port:** Georgetown.

Communications: TV sets: 197 per 1,000 pop. **Radios:** 454 per 1,000 pop. **Telephones** (1999): 64,000 main lines. **Daily newspaper circ.:** 585 per 1,000 pop.

Health: Life expectancy: 60.52 male; 66.24 female. **Births** (per 1,000 pop): 17.92. **Deaths** (per 1,000 pop.): 8.87. **Natural inc.:** .91%. **Infant mortality** (per 1,000 live births): 38.72.

Education: Free, compulsory: ages 6-14. **Literacy**: 98%.

Major Intl. Organizations: UN (FAO, IBRD, ILO, IMF, IMO, WHO, WTrO), Caricom, the Commonwealth, OAS.

Embassy: 2490 Tracy Pl. NW 20008; 265-6900.

Guyana became a Dutch possession in the 17th century, but sovereignty passed to Britain in 1815. Indentured servants from India soon outnumbered African slaves. Ethnic tension has affected political life.

Guyana became independent May 26, 1966. A Venezuelan claim to the western half of Guyana was suspended in 1970 but renewed in 1982; an agreement was reached in 1989. The Suriname border is disputed. The government has nationalized most of the economy, which has remained severely depressed.

The Port Kaituma ambush of U.S. Rep. Leo J. Ryan and others investigating mistreatment of American followers of the Rev. Jim Jones's People's Temple cult triggered a mass suicide-execution of 911 cultists at Jonestown in the jungle, Nov. 18, 1978.

The People's National Congress, the party in power since Guyana became independent, was voted out of office with the election of Cheddi Jagan in Oct. 1992. When Pres. Jagan died Mar. 6, 1997, Prime Min. Samuel Hinds succeeded him; his widow, Janet Jagan, became prime min. Mar. 17. She won the presidency in a disputed election Dec. 15. She resigned because of ill health Aug. 11, 1999, and was succeeded by Bharrat Jagdeo, then 35, who became the youngest head of state in the Americas. He was reelected Mar. 19, 2001.

Haiti

Republic of Haiti

People: Population: 6,964,549. **Age distrib.** (%): <15: 40.3; 65+: 4.2. **Pop. density:** 655 per sq. mi. **Urban:** 35%. **Ethnic groups:** Black 95%. **Principal languages:** Haitian Creole, French (both official). **Chief religions:** Roman Catholic 80%, Protestant 16%; Voodoo widely practiced.

Geography: Area: 10,600 sq. mi. **Location:** In Caribbean, occupies western third of Isl. of Hispaniola. **Neighbors:** Dominican Republic on E, Cuba to W. **Topography:** About two-thirds of Haiti is mountainous. Much of the rest is semiarid. Coastal areas are warm and moist. **Capital:** Port-au-Prince, (urban agg.): 1,769,000.

Government: Type: Republic. **Head of state:** Pres. Jean-Baptiste Aristide; b July 15, 1953; in office Feb. 7, 2001. **Head of gov.:** Jean-Marie Chérestal; b June 18,1947; in office: Mar. 2, 2001. **Local divisions:** 9 departments. **Defense budget:** Nil. **Active troops:** NA.

Economy: Industries: Sugar refining, textiles. **Chief crops:** Coffee, sugar, mangoes, corn, rice. **Arable land:** 20%. **Livestock** (2000): cattle: 1.43 mil; chickens: 5.5 mil; goats: 1.94 mil; pigs: 1.0 mil; sheep: 151,800.**Electricity prod.:** 672 mil kWh. **Labor force:** 66% agric.; 25% services; 9% ind.

Finance: Monetary unit: Gourde (Sept. 2001: 23.75 = $1 U.S.). **GDP** (1999 est.): $9.2 bil. **Per capita GDP:** $1,340. **Imports** (1999): $762 mil; partners: U.S. 60%. **Exports** (1999): $322 mil; partners: U.S. 86%. **Tourism** (1998): $57 mil. **Budget** (FY 1997-98 est.): $363 mil. **Intl. reserves less gold** (June 1998): $61.7 mil. **Gold:** 1,000 oz t. **Consumer prices** (change in 1999): 8.7%.

Transport: Motor vehicles: 32,000 pass. cars, 21,000 comm. vehicles. **Civil aviation:** 2 airports. **Chief ports:** Port-au-Prince, Les Cayes, Cap-Haitien.

Communications: TV sets: 4 per 1,000 pop. **Radios:** 41 per 1,000 pop. **Telephones**: 72,500 main lines. **Daily newspaper circ.:** 7 per 1,000 pop.

Health: Life expectancy: 47.67 male; 51.17 female. **Births** (per 1,000 pop): 31.68. **Deaths** (per 1,000 pop.): 15.. **Natural inc.:** 1.67%. **Infant mortality** (per 1,000 live births): 95.23.

Education: Compulsory: ages 6-12. **Literacy**: 45%.

Major Intl. Organizations: UN and most of its specialized agencies, OAS.

Embassy: 2311 Massachusetts Ave. NW 20008; 332-4090.

Website: http://www.haiti.org

Haiti, visited by Columbus, 1492, and a French colony from 1697, attained its independence, 1804, following the rebellion led by former slave Toussaint L'Ouverture. Following a period of political violence, the U.S. occupied the country 1915-34.

Francois Duvalier was elected president in Sept. 1957; in 1964 he was named president for life. Upon his death in 1971, he was succeeded by his son, Jean Claude. Drought in 1975-77 brought famine, and Hurricane Allen in 1980 destroyed most of the rice, bean, and coffee crops. Following several weeks of unrest, President Jean Claude Duvalier fled Haiti aboard a U.S. Air Force jet Feb. 7, 1986, ending the 28-year dictatorship by the Duvalier family.

A military-civilian council headed by Gen. Henri Namphy assumed control. In 1987, voters approved a new constitution, but the Jan. 1988 elections were marred by violence and boycotted by the opposition. Gen. Namphy seized control, June 20, but was ousted by a military coup in Sept.

Father Jean-Bertrand Aristide was elected president Dec. 1990. In Sept. 1991, Aristide was arrested by the military and expelled from the country. Some 35,000 Haitian refugees were intercepted by the U.S. Coast Guard as they tried to enter the U.S., 1991-92. Most were returned to Haiti. There was a new upsurge of refugees starting in late 1993.

The UN imposed a worldwide oil, arms, and financial embargo on Haiti June 23, 1993. The embargo was suspended when the military agreed to Aristide's return to power on Oct. 30, but the military effectively blocked his return. After renewed sanctions, the UN Security Council authorized, July 31, 1994, an invasion of Haiti by a multinational force. With U.S. troops already en route, an invasion was averted, Sept. 18, by a new agreement for military leaders to step down and Aristide to resume office. As part of the agreement, thousands of U.S. troops began arriving in Haiti, Sept. 19. Aristide returned to Haiti and was restored in office Oct. 15. A UN peacekeeping force exercised responsibility in Haiti from Mar. 31, 1995 to Nov. 30, 1997.

Aristide transferred power to his elected successor, René Préval, on Feb. 7, 1996. Prime Min. Rosny Smarth announced his resignation June 9, 1997, and quit running the government Oct. 20, but Préval and Parliament deadlocked for another 17 months until a successor was appointed by presidential decree. At least 140 people died and more than 160,000 became homeless when Hurricane Georges struck Haiti Sept. 22, 1998.

Aristide's Lavalas Family party swept parliamentary and local elections, May 21 and June 9, 2000. Aristide won the presidency Nov. 26, 2000, in an election boycotted by opposition groups.

Honduras

Republic of Honduras

People: Population: 6,406,052. **Age distrib.** (%): <15: 42.2; 65+: 3.6. **Pop. density:** 148 per sq. mi. **Urban:** 52%. **Ethnic groups:** Mestizo 90%, Amerindian 7%. **Principal language:** Spanish (official). **Chief religion:** Roman Catholic 97%.

Geography: Area: 43,100 sq. mi. **Location:** In Central America. **Neighbors:** Guatemala on W, El Salvador and Nicaragua on S. **Topography:** The Caribbean coast is 500 mi. long. Pacific coast, on Gulf of Fonseca, is 40 mi. long. Honduras is mountainous, with wide fertile valleys and rich forests. **Capital:** Tegucigalpa, (urban agg.): 950,000.

Government: Type: Republic. **Head of state:** Pres. Carlos Roberto Flores Facussé; b Mar. 1, 1950; in office: Jan. 27, 1998. **Local divisions:** 18 departments. **Defense budget:**$35 mil. **Active troops:** 8,300.

Economy: Industries: Textiles, wood prods. **Chief crops:** Bananas, coffee, citrus. **Minerals:** Gold, silver, copper, lead, zinc, iron, antimony, coal. **Other resources:** Timber, fish. **Arable land:** 15%. **Livestock** (2000): cattle: 1.95 mil; chickens: 18.0 mil; goats: 30,000; pigs: 800,000; sheep: 13,700. **Fish catch:** (1999): 23,585 metric tons. **Electricity prod.:** 3.319 bil kWh. **Labor force:** 60% services; 29% agric.; 21% industry.

Finance: Monetary unit: Lempira (Sept. 2001: 15.57 = $1 U.S.). **GDP** (1999 est.): $14.1 bil. **Per capita GDP:** $2,050. **Imports** (1999 est.): $2.7 bil; partners: U.S. 60%. **Exports** (1999 est.): $1.6 bil; partners: U.S. 73%. **Tourism:** $165 mil. **Budget** (1997 est.): $850 mil. **Intl. reserves less gold** (May 2000): $1.36 bil. **Gold:** 21,000 oz t. **Consumer prices** (change in 1999): 11.7%.

Transport: Railroad: Length: 614 mi. **Motor vehicles:** 80,000 pass. cars, 105,000 comm. vehicles. **Civil aviation:** 189.5 mil pass.-mi.; 8 airports. **Chief ports:** Puerto Cortes, La Ceiba.

Communications: TV sets: 29 per 1,000 pop. **Radios:** 337 per 1,000 pop. **Telephones**: 298,700 main lines. **Daily newspaper circ.:** 45 per 1,000 pop.

Health: Life expectancy: 67.51 male; 71.28 female. **Births** (per 1,000 pop): 31.94. **Deaths** (per 1,000 pop.): 5.52. **Natural inc.:** 2.64%. **Infant mortality** (per 1,000 live births): 30.88.

Education: Free, compulsory: ages 7-13. **Literacy:** 73%.

Major Intl. Organizations: UN, (FAO, IBRD, ILO, IMF, IMO, WHO, WTrO), OAS.

Embassy: 3007 Tilden St. NW 20008; 966-7702.

Website: http://www.honduras.com

Mayan civilization flourished in Honduras in the 1st millennium AD. Columbus arrived in 1502. Honduras became independent after freeing itself from Spain, 1821, and from the Fed. of Central America, 1838.

Gen. Oswaldo Lopez Arellano, president for most of the period 1963-75 by virtue of one election and 2 coups, was ousted by the army in 1975 over charges of pervasive bribery by United Brands Co. of the U.S. An elected civilian government took power in 1982. Some 3,200 U.S. troops were sent to Honduras after the Honduran border was violated by Nicaraguan forces, Mar. 1988.

Already one of the poorest countries in the western hemisphere, Honduras was devastated in late Oct. 1998 by Hurricane Mitch, which killed at least 5,600 people and caused more than $850 million in damage to crops and livestock.

Hungary

Republic of Hungary

People: Population: 10,106,017. **Age distrib.** (%): <15: 16.6; 65+: 14.7. **Pop. density:** 284 per sq. mi. **Urban:** 64%. **Ethnic groups:** Hungarian 90%, Gypsy 4%, German 3%. **Principal language:** Hungarian (Magyar; official). **Chief religions:** Roman Catholic 68%, Calvinist 20%, Lutheran 5%.

Geography: Area: 35,600 sq. mi. **Location:** In E central Europe. **Neighbors:** Slovakia, Ukraine on N; Austria on W; Slovenia, Yugoslavia, Croatia on S; Romania on E. **Topography:** The Danube R. forms the Slovak border in the NW, then swings S to bisect the country. The eastern half of Hungary is mainly a great fertile plain, the Alfold; the W and N are hilly. **Capital:** Budapest, (urban agg.): 1,825,000.

Government: Type: Parliamentary democracy. **Head of state:** Pres. Ferenc Mádl; b Jan. 29, 1931; in office: Aug. 4, 2000. **Head of gov.:** Prime Min. Viktor Orbán; b May 31, 1963; in office: July 8, 1998. **Local divisions:** 19 counties, 20 urban counties, 1 capital. **Defense budget:** $791 mil. **Active troops:** 43,790.

Economy: Industries: Mining, metallurgy, construction materials, processed foods, pharmaceuticals, vehicles. **Chief crops:** Wheat, corn, sunflowers, potatoes, sugar beets. **Minerals:** Bauxite, coal, gas. **Crude oil reserves (2000): 109.7 mil bbls. Arable land:** 51%. **Livestock** (2000): cattle: 857,000; chickens: 25.89 mil; goats: 160,000; pigs: 5.34 mil; sheep: 934,000. **Fish catch:** (1999): 21,916 metric tons. **Electricity prod.:** 34.868 bil kWh. **Labor force:** 65% services; 27% ind.; 8% agric.

Finance: Monetary unit: Forint (Sept. 2001: 279.22 = $1 U.S.). **GDP** (1999 est.): $79.4 bil. **Per capita GDP:** $7,800. **Imports** (1999): $25.1 bil; partners: Germany 28%, Austria 10%, Italy 8%. **Exports** (1999): $22.6 bil; partners: Germany 37%, Austria 11%. **Budget** (1998 est.): $13.2 bil. **Tourism:** $3.39 bil. **Intl. reserves less gold** (June 2000): $10.50 bil. **Gold:** 101,000 oz t. **Consumer prices** (change in 1999): 10.3%.

Transport: Railroad: Length: 8,190 mi. **Motor vehicles:** 2.28 mil pass. cars, 319,424 comm. vehicles. **Civil aviation:** 1.5 bil. pass.-mi.; 1 airport.

Communications: TV sets: 438 per 1,000 pop. **Radios:** 689 per 1,000 pop. **Telephones** (1999): 3,725,800 main lines. **Daily newspaper circ.:** 186 per 1,000 pop.

Health: Life expectancy: 67.28 male; 76.3 female. **Births** (per 1,000 pop): 9.32. **Deaths** (per 1,000 pop.): 13.21. **Natural inc.:** -.39%. **Infant mortality** (per 1,000 live births): 8.96.

Education: Compulsory: ages 6-16. **Literacy** (1993): 99%.

Major Intl. Organizations: UN (FAO, IBRD, ILO, IMF, IMO, WHO, WTrO), NATO, OECD, OSCE.

Embassy: 3910 Shoemaker St. NW 20008; 966-7726.

Website: http://www.hungaryemb.org

Earliest settlers, chiefly Slav and Germanic, were overrun by Magyars from the E. Stephen I (997-1038) was made king by Pope Sylvester II in AD 1000. The country suffered repeated Turkish invasions in the 15th-17th centuries. After the defeats of the Turks, 1686-1697, Austria dominated, but Hungary obtained concessions until it regained internal independence in 1867, with the emperor of Austria as king of Hungary in a dual monarchy with a single diplomatic service. Defeated with the Central Powers in 1918, Hungary lost Transylvania to Romania, Croatia and Bacska to Yugoslavia, Slovakia and Carpatho-Ruthenia to Czechoslovakia, all of which had large Hungarian minorities. A republic under Michael Karolyi and a bolshevist revolt under Bela Kun were followed by a vote for a monarchy in 1920 with Admiral Nicholas Horthy as regent.

Hungary joined Germany in World War II, and was allowed to annex most of its lost territories. Russian troops captured the country, 1944-1945. By terms of an armistice with the Allied powers Hungary agreed to give up territory acquired by the 1938 dismemberment of Czechoslovakia and to return to its borders of 1937.

A republic was declared Feb. 1, 1946; Zoltan Tildy was elected president. In 1947 the Communists forced Tildy out. Premier Imre Nagy, who had been in office since mid-1953, was ousted for his moderate policy of favoring agriculture and consumer production, April 18, 1955.

In 1956, popular demands to oust Erno Gero, Communist Party secretary, and for formation of a government by Nagy, resulted in the latter's appointment Oct. 23; demonstrations against Communist rule developed into open revolt. On Nov. 4 Soviet forces launched a massive attack against Budapest with 200,000 troops, 2,500 tanks and armored cars.

About 200,000 persons fled the country. Thousands were arrested and executed, including Nagy in June 1958. In spring 1963 the regime freed many captives from the 1956 revolt.

Hungarian troops participated in the 1968 Warsaw Pact invasion of Czechoslovakia. Major economic reforms were launched early in 1968, switching from a central planning system to one based on market forces and profit.

In 1989 Parliament passed legislation legalizing freedom of assembly and association as Hungary shifted away from communism. In Oct. the Communist Party was formally dissolved. The last Soviet troops left Hungary June 19, 1991. Hungary became a full member of NATO on Mar. 12, 1999.

Iceland

Republic of Iceland

People: Population: 277,906. **Age distrib.** (%): <15: 23.2; 65+: 11.8. **Pop. density:** 7 per sq. mi. **Urban:** 92%. **Ethnic groups:** Homogeneous descendants of Norwegians, Celts. **Principal language:** Icelandic (Islenska; official). **Chief religion:** Evangelical Lutheran 96%.

Geography: Area: 38,700 sq. mi. **Location:** Isl. at N end of Atlantic O. **Neighbors:** Nearest is Greenland (Den.), to W. **Topography:** Recent volcanic origin. Three-quarters of the surface is wasteland: glaciers, lakes, a lava desert. There are geysers and hot springs, and the climate is moderated by the Gulf Stream. **Capital:** Reykyavík (1999 est.): 109,763.

Government: Type: Constitutional republic. **Head of state:** Pres. Olafur Ragnar Grímsson; b May 14, 1943; in office: Aug.

1, 1996. **Head of gov.:** Prime Min. David Oddsson; Jan. 17, 1948; in office: Apr. 30, 1991. **Local divisions:** 23 counties, 14 independent towns. **Defense budget:** Icelandic Defense Force provided by the U.S.

Economy: Industries: Fish products (75% of exports), aluminum smelting. **Chief crops:** Potatoes, turnips. **Livestock** (2000): cattle: 72,100; chickens: 160,000; goats: 426; pigs: 43,000; sheep: 465,000. **Fish catch** (1999): (1999): 2.21 mil metric tons. **Electricity prod.:** 7.069 bil kWh. **Labor force:** 60% commerce & services; 13% manuf.; 12% fishing.

Finance: Monetary unit: Krona (Sept. 2001: 99.19 = $1 U.S.). Euro (Sept 2001: 1.09 = $1 US). **GDP** (1999 est.): $6.42 bil. **Per capita GDP:** $23,500. **Imports** (1998): $2.4 bil; partners: Germany 12%, Norway 9%, UK 10%. **Exports** (1998): $1.9 bil; partners: UK 19%, Germany 15%. **Tourism** (1998): $207 mil. **Budget** (1996 est.): $2.1 bil. **Intl. reserves less gold** (May 2000): $412.2 mil. **Gold:** 59,000 oz t. **Consumer prices** (change in 1999): 3.2%.

Transport: Motor vehicles (1997): 132,468 pass. cars, 17,511 comm. vehicles. **Civil aviation:** 2.0 bil. pass.-mi.; 24 airports. **Chief port:** Reykjavík.

Communications: TV sets: 285 per 1,000 pop. **Radios:** 733 per 1,000 pop. **Telephones** (1999): 188,800 main lines. **Daily newspaper circ.:** 515 per 1,000 pop.

Health: Life expectancy: 77.31 male; 81.92 female. **Births** (per 1,000 pop): 14.62. **Deaths** (per 1,000 pop.): 6.89. **Natural inc.:** .77%. **Infant mortality** (per 1,000 live births): 3.56.

Education: Free, compulsory: ages 7-15. **Literacy** (1997): 100%.

Major Intl. Organizations: UN (FAO, IBRD, ILO, IMF, IMO, WHO, WTrO), EFTA, NATO, OECD, OSCE.

Embassy: Suite 1200, 1156 15th St. NW 20005; 265-6653.

Website: http://www.iceland.org

Iceland was an independent republic from 930 to 1262, when it joined with Norway. Its language has maintained its purity for 1,000 years. Danish rule lasted from 1380-1918; the last ties with the Danish crown were severed in 1941. The Althing, or assembly, is the world's oldest surviving parliament.

India

Republic of India

People: Population: 1,029,991,145. **Age distrib.** (%): <15: 33.1; 65+: 4.7. **Pop. density:** 898 per sq. mi. **Urban:** 28%. **Ethnic groups:** Indo-Aryan 72%, Dravidian 25%. **Principal languages:** Hindi (official), English (associate official), 14 regional official languages, others. **Chief religions:** Hindu 80%, Muslim 14%.

Geography: Area: 1,146,600 sq. mi. **Location:** Occupies most of the Indian subcontinent in S Asia. **Neighbors:** Pakistan on W; China, Nepal, Bhutan on N; Myanmar, Bangladesh on E. **Topography:** The Himalaya Mts., highest in world, stretch across India's northern borders. Below, the Ganges Plain is wide, fertile, and among the most densely populated regions of the world. The area below includes the Deccan Peninsula. Close to one quarter of the area is forested. The climate varies from tropical heat in S to near-Arctic cold in N. Rajasthan Desert is in NW; NE Assam Hills get 400 in. of rain a year. **Capital:** New Delhi. **Cities (urban agg.):** Mumbai (Bombay) 18,066,000, (2000 city est.: 15 mil); Kolkata (Calcutta) 12,918,000; Delhi 11,695,000; Hyderabad 6,842,000; Chennai (Madras) 6,648,000; Bangalore 5,561,000.

Government: Type: Federal republic. **Head of state:** Pres. Kocheril Raman Narayanan; b Oct. 17, 1920; in office: July 25, 1997. **Head of gov.:** Prime Min. Atal Bihari Vajpayee; b Dec. 25, 1924; in office Mar. 19, 1998. **Local divisions:** 28 states, 6 union territories, 1 national capital territory. **Defense budget:** $15.9 bil. **Active troops:** 1,303,000

Economy: Industries: Textiles, steel, processed foods, cement, machinery, chemicals, mining. **Chief crops:** Rice, grains, sugar, spices, tea, cashews, cotton, potatoes, jute, oilseed. **Minerals:** Coal (4th largest reserves in the world), iron, manganese, mica, bauxite, titanium, chromite, diamonds, gas, oil. **Crude oil reserves** (2000): 4.84 bil bbls. **Other resources:** Timber. **Arable land:** 56%. **Livestock** (2000): cattle: 218.8 mil; chickens: 402.0 mil; goats: 123.0 mil; pigs: 16.5 mil; sheep: 57.9 mil.**Fish catch** (1999): 5.38 mil metric tons. **Electricity prod.:** 454.561 bil kWh. **Labor force:** 67% agric.; 18% services; 15% industry.

Finance: Monetary unit: Rupee (Sept. 2001: 47.22 = $1 U.S.). **GDP** (1999 est.): $1.805 tril. **Per capita GDP:** $1,800. **Imports** (1999 est.): $50.2 bil; partners: U.S. 10%, Belgium 7%, UK 6%. **Exports** (1999 est.): $36.3 bil; partners: U.S. 21%, Japan 5%, UK 6%. **Tourism:** $3.04 bil. **Budget** (FY 1998-99): $63.79 bil. **Intl. reserves less gold** (May 2000): $35.04 bil. **Gold:** 11.50 mil oz t. **Consumer prices** (change in 1999): 4.7%.

Transport: Railroad: Length: 38,935 mi. **Motor vehicles:** 4.25 mil pass. cars, 2.51 mil comm. vehicles. **Civil aviation:** 15.0 bil pass.-mi.; 66 airports. **Chief ports:** Kolkata (Calcutta), Mumbai (Bombay), Chennai (Madras), Vishakhapatnam, Kandla.

Communications: TV sets: 68 per 1,000 pop. **Radios:** 117 per 1,000 pop. **Telephones**: 32,436,100 main lines. **Daily newspaper circ.:** 21 per 1,000 pop.

Health: Life expectancy: 62.22 male; 63.53 female. **Births** (per 1,000 pop): 24.28. **Deaths** (per 1,000 pop.): 8.74. **Natural inc.:** 1.55%. **Infant mortality** (per 1,000 live births): 63.19.

Education: Theoretically compulsory in 23 states to age 14. **Literacy**: 52%.

Major Intl. Organizations: UN (FAO, IBRD, ILO, IMF, IMO, WHO, WTrO), the Commonwealth.

Embassy: 2107 Massachusetts Ave. NW 20008; 939-7000.

Websites: http://www.nic.in
http://www.indianembassy.org
http://www.tourindia.com

India has one of the oldest civilizations in the world. Excavations trace the Indus Valley civilization back for at least 5,000 years. Paintings in the mountain caves of Ajanta, richly carved temples, the Taj Mahal in Agra, and the Kutab Minar in Delhi are among relics of the past.

Aryan tribes, speaking Sanskrit, invaded from the NW around 1500 BC, and merged with the earlier inhabitants to create classical Indian civilization.

Asoka ruled most of the Indian subcontinent in the 3d century BC, and established Buddhism. But Hinduism revived and eventually predominated. During the Gupta kingdom, 4th-6th century AD, science, literature, and the arts enjoyed a "golden age."

Arab invaders established a Muslim foothold in the W in the 8th century, and Turkish Muslims gained control of North India by 1200. The Mogul emperors ruled 1526-1857.

Vasco da Gama established Portuguese trading posts 1498-1503. The Dutch followed. The British East India Co. sent Capt. William Hawkins, 1609, to get concessions from the Mogul emperor for spices and textiles. Operating as the East India Co. the British gained control of most of India. The British parliament assumed political direction; under Lord Bentinck, 1828-35, rule by rajahs was curbed. After the Sepoy troops mutinied, 1857-58, the British supported the native rulers.

Nationalism grew rapidly after World War I. The Indian National Congress and the Muslim League demanded constitutional reform. A leader emerged in Mohandas K. Gandhi (called Mahatma, or Great Soul), born Oct. 2, 1869, assassinated Jan. 30, 1948. He advocated self-rule, nonviolence, and removal of the caste system of untouchability. In 1930 he launched a program of civil disobedience, including a boycott of British goods and rejection of taxes without representation.

In 1935 Britain gave India a constitution providing a bicameral federal congress. Muhammad Ali Jinnah, head of the Muslim League, sought creation of a Muslim nation, Pakistan.

The British government partitioned British India into the dominions of India and Pakistan. India became a member of the UN in 1945, a self-governing member of the Commonwealth in 1947, and a democratic republic, Jan. 26, 1950. More than 12 million Hindu and Muslim refugees crossed the India-Pakistan borders in a mass transferral of some of the 2 peoples during 1947; about 200,000 were killed in communal fighting.

After Pakistan troops began attacks on Bengali separatists in East Pakistan, Mar. 25, 1971, some 10 million refugees fled into India. India and Pakistan went to war Dec. 3, 1971, on both the East and West fronts. Pakistan troops in the east surrendered Dec. 16; Pakistan agreed to a cease-fire in the west Dec. 17.

Indira Gandhi, India's prime minister since Jan. 1966, invoked emergency powers in June 1975. Thousands of opponents were arrested and press censorship imposed. These and other actions, including enforcement of coercive birth control measures in some areas, were widely resented. Opposition parties, united in the Janata coalition, turned Gandhi's New Congress Party from power in federal and state parliamentary elections in 1977.

Gandhi became prime minister for the second time, Jan. 14, 1980. She was assassinated by 2 of her Sikh bodyguards Oct. 31, 1984, in response to the government suppression of a Sikh uprising in Punjab in June 1984, which included an assault on the Golden Temple at Amritsar, the holiest Sikh shrine. Widespread rioting followed the assassination. Thousands of Sikhs were killed and some 50,000 left homeless.

Rajiv, Indira Gandhi's son, replaced her as prime minister. He was swept from office in 1989 amid charges of incompetence and corruption, and assassinated May 21, 1991, while campaigning to recapture the prime ministership.

Sikhs ignited several violent clashes during the 1980s. The government's May 1987 decision to bring the state of Punjab under rule of the central government led to violence. Many died during a government siege of the Golden Temple, May 1988. Another trouble spot was Assam in NW India, where thousands were killed in ethnic violence in Feb. 1993; a renewed outburst in July 1994 led to more than 60 deaths.

Nationwide riots followed the destruction of a 16th-century mosque by Hindu militants in Dec. 1992. In the biggest wave of criminal violence in Indian history, a series of bombs jolted Bombay and Calcutta, Mar. 12-19, 1993, killing over 300.

Corruption scandals dominated Indian politics in the mid-1990s. After an inconclusive election, a Hindu nationalist party was unable to form a government, and a center-left coalition took office June 1, 1996. An aircraft collision in midair near New Delhi killed 349 passengers and crew on Nov. 12.

India's 1st lowest-caste pres., K. R. Narayanan, took office July 25, 1997. Mother Teresa of Calcutta, renowned for her work among the poor, died Sept. 5. Parliamentary elections in Feb. 1998 resulted in a Hindu nationalist victory, and Atal Bihari Vajpayee was sworn in as prime minister Mar. 19. India conducted a series of nuclear tests in mid-May, drawing worldwide condemnation and raising tensions with Pakistan.

An alliance led by Vajpayee won a majority in legislative elections, Sept. 5-Oct. 3, 1999. A cyclone that hit the state of Orissa, E India, on Oct. 29, 1999, left some 10,000 people dead. A powerful earthquake in Gujarat state on Jan. 26, 2001, claimed more than 20,000 lives and left more than 166,00 people injured.

Sikkim, bordered by Tibet, Bhutan, and Nepal, formerly British protected, became a protectorate of India in 1950. Area, 2,740 sq. mi; pop., 1994 est., 444,000; capital: Gangtok. In Sept. 1974, India's parliament voted to make Sikkim an associate Indian state, absorbing it into India.

Kashmir, a predominantly Muslim region in the NW, has been in dispute between India and Pakistan since 1947. A cease-fire was negotiated by the UN Jan. 1, 1949; it gave Pakistan control of one-third of the area, in the west and northwest, and India the remaining two-thirds, the Indian state of **Jammu and Kashmir**, which enjoys internal autonomy.

In the 1990s there were repeated clashes between Indian army troops and pro-independence demonstrators triggered by India's decision to impose central government rule. The clashes strained relations between India and Pakistan, which India charged was aiding the Muslim separatists; the heaviest fighting in more than 2 decades took place during May-June 1999.

France, 1952-54, peacefully yielded to India its 5 colonies, former French India, comprising Pondicherry, Karikal, Mahe, Yanaon (which became **Pondicherry Union Territory**, area 190 sq. mi; pop., 1994 est., 894,000) and Chandernagor (which was incorporated into the state of **West Bengal**).

Indonesia

Republic of Indonesia

People: Population: 228,437,870. **Age distrib.** (%): <15: 30.3; 65+: 4.6. **Pop. density:** 324 per sq. mi. **Urban:** 40%. **Ethnic groups:** Javanese 45%, Sundanese 14%, Madurese 8%, Malay 8%. **Principal languages:** Bahasa Indonesian (official), English, Dutch, Javanese. **Chief religions:** Muslim 87%, Protestant 6%.

Geography: Area: 704,400 sq. mi. **Location:** Archipelago SE of Asian mainland along the Equator. **Neighbors:** Malaysia on N, Papua New Guinea on E. **Topography:** Indonesia comprises over 13,500 islands (6,000 inhabited), including Java (one of the most densely populated areas in the world with over 2,000 persons per sq. mi.), Sumatra, Kalimantan (most of Borneo), Sulawesi (Celebes), and West Irian (Irian Jaya, the W half of New Guinea). Also: Bangka, Billiton, Madura, Bali, Timor. The mountains and plateaus on the major islands have a cooler climate than the tropical lowlands. **Capital:** Jakarta. **Cities (urban agg.)**: Jakarta 11,018,000, (2000 city proper: 8.8 mil); Bandung 3,409,000; Surabaja 2,461,000.

Government: Type: Republic. **Head of state and gov.:** Megawati Sukarnoputri; b Jan. 23, 1947; in office: July 23, 2001. **Local divisions:** 24 provinces, 2 special regions, 1 capital district. **Defense budget:** $2,271 bil. **Active troops:** 297,000.

Economy: Industries: Oil, gas, food processing, textiles, cement, mining. **Chief crops:** Rice, cocoa, peanuts, rubber. **Minerals:** Nickel, tin, oil, bauxite, copper, gas. **Crude oil reserves** (2000): 4.98 bil bbls. **Other resources:** Timber. **Arable land:** 10%. **Livestock** (2000): cattle: 12.1 mil; chickens: 800.0 mil; goats: 14.12 mil; pigs: 9.35 mil; sheep: 7.5 mil.**Fish catch** (1999): 4.40 mil metric tons. **Electricity prod.:** 78.674 bil kWh. **Labor force:** 45% agric.; trd., rest., and hotel 19%; 11% manuf.

Finance: Monetary unit: Rupiah (Sept. 2001: 9,150 = $1 U.S.). **GDP** (1999 est.): $610 bil. **Per capita GDP:** $2,800. **Imports** (1999 est.): $24 bil; partners: Japan 17%, U.S. 13%. **Exports** (1999 est.): $48 bil; partners: Japan 18%, EU 15%, U.S. 14%. **Tourism** (1998): $4.05 bil. **Budget** (FY 1998-99 est.): $35 bil. **Intl. reserves less gold** (Apr. 2000): $28.09 bil. **Gold:** 3.10 mil oz t. **Consumer prices** (change in 1999): 20.5%.

Transport: Railroad: Length: 4,090 mi. **Motor vehicles** (1997): 2.64 mil pass. cars, 2.16 mil comm. vehicles. **Civil aviation:** 14.6 bil pass.-mi.; 81 airports. **Chief ports:** Jakarta, Surabaya, Palembang, Semarang, Ujungpandang.

Communications: TV sets: 134 per 1,000 pop. **Radios:** 128 per 1,000 pop. **Telephones**: 6,662,600 main lines. **Daily newspaper circ.:** 23 per 1,000 pop.

Health: Life expectancy: 65.9 male; 70.75 female. **Births** (per 1,000 pop): 22.26. **Deaths** (per 1,000 pop.): 6.3. **Natural inc.:** 1.6%. **Infant mortality** (per 1,000 live births): 40.91.

Education: Compulsory: ages 7-16. **Literacy**: 84%.

Major Intl. Organizations: UN and all of its specialized agencies, APEC, ASEAN, OPEC.

Embassy: 2020 Massachusetts Ave. NW 20036; 775-5200.

Hindu and Buddhist civilization from India reached Indonesia nearly 2,000 years ago, taking root especially in Java. Islam spread along the maritime trade routes in the 15th century, and became predominant by the 16th century. The Dutch replaced the Portuguese as the area's most important European trade power in the 17th century, securing territorial control over Java by 1750. The outer islands were not finally subdued until the early 20th century, when the full area of present-day Indonesia was united under one rule for the first time.

Following Japanese occupation, 1942-45, nationalists led by Sukarno and Hatta declared independence. The Netherlands ceded sovereignty Dec. 27, 1949, after 4 years of fighting. A republic was declared, Aug. 17, 1950, with Sukarno as president. West Irian, on New Guinea, remained under Dutch control. After the Dutch in 1957 rejected proposals for new negotiations over West Irian, Indonesia stepped up the seizure of Dutch property. In 1963 the UN turned the area over to Indonesia, which promised a plebiscite. In 1969, voting by tribal chiefs favored staying with Indonesia, despite an uprising and widespread opposition.

Sukarno suspended Parliament in 1960, and was named president for life in 1963. He made close alliances with Communist governments. Russian-armed Indonesian troops staged raids in 1964 and 1965 into Malaysia, whose formation Sukarno had opposed. (In 1966 Indonesia and Malaysia signed an agreement ending hostility.)

In 1965 an attempted coup in which several military officers were murdered was successfully put down. The regime blamed the coup on the Communist Party, some of whose members were known to have been involved. In its wake more than 300,000 alleged Communists were killed in army-initiated massacres.

Gen. Suharto, head of the army, was named president in 1968. With military backing he developed a strong government party, restricted the opposition, and allied the country with the West; meanwhile, oil exports spurred economic growth. During Aug.-Nov. 1997, haze from forest fires in Indonesia blanketed large areas of SE Asia. A plane crash near Medan airport, Sept. 26, 1997, killed 234 persons.

Parliament reelected Suharto to a 7th consecutive 5-year term Mar. 10, 1998, as a severe economic downturn focused public anger on nepotism, cronyism, and corruption in the Suharto regime. Price increases in May sparked mass protests and then mob violence in Jakarta and other cities, claiming some 500 lives. Suharto resigned May 21 and was succeeded by his vice-president, Bacharuddin Jusuf Habibie. Abdurrahman Wahid, leader of Indonesia's largest Muslim organization, was elected president Oct. 20, 1999. In Aug. 2000, under pressure from the legislature, he agreed to share power with Vice-Pres. Megawati Sukarnoputri, the daughter of the late Pres. Sukarno.

Clashes between Muslims and Christians in the Maluku (Molucca) Is. have claimed more than 2,500 lives since Jan. 1999; in addition, some 550 people, many refugees from the fighting, died when their ferry sank June 29, 2000. Ethnic violence in Kalimantan, Borneo, killed more than 400 in Feb. 2001.

In Dec. 1975, Indonesia invaded **East Timor** as Portuguese rule collapsed. Indonesia annexed it in 1976, despite international condemnation; an estimated 200,000 Timorese died as a result of famine, oppression and fighting. In a referendum held Aug. 30, 1999, under UN auspices, Timorese voted overwhelmingly for independence. Pro-Indonesian militias then went on a rampage, terrorizing the population. Under pressure, the government allowed entrance of an international peacekeeping force, which began arriving in Sept.; a UN interim administration formally took command Oct. 26, 1999. Pro-independence forces won elections for a constituent assembly Aug. 30, 2001.

Iran

Islamic Republic of Iran

People: Population: 66,128,965. **Age distrib.** (%): <15: 33.0; 65+: 4.6. **Pop. density:** 105 per sq. mi. **Urban:** 61%. **Ethnic groups:** Persian 51%, Azerbaijani 24%, Kurd 7%. **Principal languages:** Persian (Farsi; official), Turkic, Kurdish, Luri. **Chief religions:** Shi'a Muslim 89%, Sunni Muslim 10%.

Geography: Area: 630,900 sq. mi. **Location:** Between the Middle East and S Asia. **Neighbors:** Turkey, Iraq on W; Armenia, Azerbaijan, Turkmenistan on N; Afghanistan, Pakistan on E. **Topography:** Interior highlands and plains surrounded by high mountains, up to 18,000 ft. Large salt deserts cover much of area, but there are many oases and forest areas. Most of the population inhabits the N and NW. **Capital:** Tehran. **Cities** (urban agg.): Tehran 7,225,000; Esfahan 2,589,000; Mashhad 2,329,000.

Government: Type: Islamic republic. **Religious head:** Ayatollah Sayyed Ali Khamenei; b 1939; in office: June 4, 1989. **Head of state and gov.:** Pres. Mohammad Khatami; b 1943; in office: Aug. 3, 1997. **Local divisions:** 25 provinces. **Defense budget:** $7.5 bil. **Active troops:** 513,000.

Economy: Industries: Oil, petrochemicals, cement, sugar refining, carpets. **Chief crops:** Grains, rice, fruits, nuts, sugar beets, cotton. **Minerals:** Chromium, coal, oil, gas. **Crude oil reserves** (2000): 89.7 bil bbls. **Arable land:** 10%. **Livestock** (2000): cattle: 8.1 mil; chickens: 250.0 mil; goats: 26.0 mil; sheep: 55.0 mil.. **Fish catch** (1999): 380,200 metric tons. **Electricity prod.:** 103.054 bil kWh. **Labor force:** 42% services; 25% ind.; 33% agric.

Finance: Monetary unit: Rial (Sept. 2001: 1,750 = $1 U.S.). **GDP** (1999 est.): $347.6 bil. **Per capita GDP:** $5,300. **Imports** (1998 est.): $13.8 bil; partners: Germany, Italy. **Exports** (1998 est.): $12.2 bil; partners: Japan, Italy, Greece. **Tourism:** $662 mil. **Budget** (FY 1996-97): $34.9 bil. **Consumer prices** (change in 1999): 21.0%.

Transport: Railroad: Length: 4,527 mi. **Motor vehicles:** 1.63 mil pass. cars, 609,000 comm. vehicles. **Civil aviation:** 5.5 bil pass.-mi.; 19 airports. **Chief port:** Bandar-e Abbas.

Communications: TV sets: 148 per 1,000 pop. **Radios:** 273 per 1,000 pop. **Telephones**: 9,486,300 main lines. **Daily newspaper circ.:** 28 per 1,000 pop.

Health: Life expectancy: 68.61 male; 71.37 female. **Births** (per 1,000 pop): 17.10. **Deaths** (per 1,000 pop.): 5.41. **Natural inc.:** 1.17%. **Infant mortality** (per 1,000 live births): 29.04.

Education: Free, compulsory: ages 6-10. **Literacy** (1997): 79%.

Major Intl. Organizations: UN (FAO, IBRD, ILO, IMF, IMO, WHO), OPEC.

Iran was once called Persia. The Iranians, who supplanted an earlier agricultural civilization, came from the E during the 2d millennium BC; they were an Indo-European group related to the Aryans of India.

In 549 BC Cyrus the Great united the Medes and Persians in the Persian Empire, conquered Babylonia in 538 BC, and restored Jerusalem to the Jews. Alexander the Great conquered Persia in 333 BC, but Persians regained independence in the next century under the Parthians, themselves succeeded by Sassanian Persians in AD 226. Arabs brought Islam to Persia in the 7th century, replacing the indigenous Zoroastrian faith. After Persian political and cultural autonomy was reasserted in the 9th century, arts and sciences flourished.

Turks and Mongols ruled Persia in turn from the 11th century to 1502, when a native dynasty reasserted full independence. The British and Russian empires vied for influence in the 19th century; Afghanistan was severed from Iran by Britain in 1857.

Reza Khan abdicated as shah, 1941; succeeded by his son, Mohammad Reza Pahlavi. He brought economic and social change to Iran, but political opposition was not tolerated.

Conservative Muslim protests led to 1978 violence. Martial law was declared in 12 cities Sept. 8. A military government was appointed Nov. 6 to deal with striking oil workers. The shah, who left Iran Jan. 16, 1979, appointed Prime Min. Shahpur Bakhtiar to head a regency council in his absence.

Exiled religious leader Ayatollah Ruhollah Khomeini named a provisional government council in preparation for his return to Tehran, Feb. 1. Clashes between Khomeini's supporters and government troops culminated in a rout of Iran's elite Imperial Guard Feb. 11, leading to the fall of Bakhtiar's government.

The Iranian revolution was marked by revolts among ethnic minorities and by a continuing struggle between the clerical forces and westernized intellectuals and liberals. The Islamic Constitution established final authority to be vested in a Faghi, the Ayatollah Khomeini.

Iranian militants seized the U.S. embassy, Nov. 4, 1979, and took hostages including 62 Americans. Despite international condemnations and U.S. efforts, including an abortive Apr. 1980 rescue attempt, the crisis continued. The U.S. broke diplomatic relations with Iran, Apr. 7. The shah died in Egypt, July 27. The hostage drama ended Jan. 20, 1981, when an accord, involving the release of frozen Iranian assets, was reached.

A dispute over the Shatt al-Arab waterway that divides the two countries brought Iran and Iraq, Sept. 22, 1980, into open warfare. Iraqi troops occupied Iranian territory, including the port city of Khorramshahr in October. Iranian troops recaptured the city and drove Iraqi troops back across the border, May 1982. Iraq, and later Iran, attacked several oil tankers in the Persian Gulf during 1984.

In Nov. 1986 it became known that senior U.S. officials had secretly visited Iran and that the U.S. had provided arms in exchange for Iran's help in obtaining the release of U.S. hostages held by terrorists in Lebanon. The revelation sparked a major scandal in the Reagan administration.

A U.S. Navy warship shot down an Iranian commercial airliner, July 3, 1988, after mistaking it for an F-14 fighter jet; all 290 aboard the plane died. In Aug. 1988, Iran agreed to accept a UN resolution calling for a cease-fire with Iraq.

An earthquake struck northern Iran June 21, 1990, killing more than 45,000, injuring 100,000, and leaving 400,000 homeless. Some one million Kurdish refugees fled from Iraq to Iran following the Persian Gulf War. To curb Iran's alleged support for international terrorism, the U.S. in 1996 authorized sanctions on foreign companies that invest there.

Mohammad Khatami, a moderate Shiite Muslim cleric, was elected president on May 23, 1997, winning nearly 70% of the vote. During the next 3 years, hardline Islamists clashed repeatedly and sometimes violently with reformers, who won a majority in parliamentary elections Feb. 18 and May 5, 2000. Inviting rapprochement with Iran, the U.S. eased some sanctions Mar. 18. Reelected June 8, 2001, with a 77% majority, Khatami continued to face resistance from religious conservatives.

Iraq

Republic of Iraq

People: Population: 23,331,985. **Age distrib.** (%): <15: 41.6; 65+: 3.1. **Pop. density:** 139 per sq. mi. **Urban:** 76%. **Ethnic groups:** Arab 75-80%, Kurd 15-20%, Turkoman. **Principal languages:** Arabic (official), Kurdish. **Chief religions:** Muslim 97% (Shi'a 60-65%, Sunni 32-37%).

Geography: Area: 167,400 sq. mi. **Location:** In the Middle East, occupying most of historic Mesopotamia. **Neighbors:** Jordan and Syria on W, Turkey on N, Iran on E, Kuwait and Saudi Arabia on S. **Topography:** Mostly an alluvial plain, including the Tigris and Euphrates rivers, descending from mountains in N to desert in SW. Persian Gulf region is marshland. **Capital:** Baghdad. **Cities (urban agg.):** Baghdad 4,797,000; Arbil 2,369,000; Mosul 1,034,000.

Government: Type: Republic. **Head of state and gov.:** Pres. Saddam Hussein; b. Apr. 28, 1937; in office: July 16, 1979; also assumed post of prime minister, May 29, 1994. **Local divisions:** 18 governorates (3 in Kurdish Autonomous Region). **Defense budget:** $1.4 bil (1999) **Active troops:** 429,000.

Economy: Industries: Textiles, chemicals, oil refining, cement. **Chief crops:** Grains, dates, cotton. **Minerals:** Oil, gas. **Arable land:** 12%. **Crude oil reserves** (2000): 112.5 bil bbls. **Other resources:** Wool, hides. **Livestock** (2000): cattle: 1.35 mil; chickens: 23.0 mil; goats: 1.6 mil; sheep: 6.78 mil.**Fish catch:** (1999): 34,702 metric tons. **Electricity prod.:** 29.420 bil kWh.

Finance: Monetary unit: Dinar (Sept. 2001: 0.31 = $1 U.S.). **GDP** (1999 est.): $59.9 bil. **Per capita GDP:** $2,700. **Imports** (1999 est.): $8.9 bil; partners: Russia, France, Egypt, . **Exports** (1999 est.): $12.7 bil; partners: Russia, France, China. **Tourism** (1998): $13 mil.

Transport: Railroad: Length: 1,263 mi. **Motor vehicles:** 672,000 pass. cars, 368,000 comm. vehicles. **Civil aviation:** 12.4 mil pass.-mi. **Chief port:** Basra.

Communications: TV sets: 48 per 1,000 pop. **Radios:** 167 per 1,000 pop. **Telephones**: 675,000 main lines. **Daily newspaper circ.:** 27 per 1,000 pop.

Health: Life expectancy: 65.92 male; 68.03 female. **Births** (per 1,000 pop): 34.64. **Deaths** (per 1,000 pop.): 6.21. **Natural inc.:** 2.84%. **Infant mortality** (per 1,000 live births): 60.05.

Education: Free, compulsory: ages 6-12. **Literacy**: 58%.

Major Intl. Organizations: UN (FAO, IBRD, ILO, IMF, IMO, WHO), AL, OPEC.

Website: http://www.Iraqi-mission.org

The Tigris-Euphrates valley, formerly called Mesopotamia, was the site of one of the earliest civilizations in the world. The Sumerian city-states of 3,000 BC originated the culture later developed by the Semitic Akkadians, Babylonians, and Assyrians.

Mesopotamia ceased to be a separate entity after the Persian, Greek, and Arab conquests. The latter founded Baghdad, from where the caliph ruled a vast empire in the 8th and 9th centuries. Mongol and Turkish conquests led to a decline in population, economy, cultural life, and the irrigation system.

Britain secured a League of Nations mandate over Iraq after World War I. Independence under a king came in 1932. A leftist, pan-Arab revolution established a republic in 1958, which oriented foreign policy toward the USSR. Most industry has been nationalized, and large land holdings broken up.

A local faction of the international Baath Arab Socialist party has ruled by decree since 1968. The USSR and Iraq signed an aid pact in 1972, and arms were sent along with several thousand advisers. The 1978 execution of 21 Communists and a shift of trade to the West signalled a more neutral policy, straining relations with the USSR. In the 1973 Arab-Israeli war Iraq sent forces to aid Syria. Within a month of assuming power, Saddam Hussein instituted a bloody purge in the wake of a reported coup attempt against the new regime.

Years of battling with the Kurdish minority resulted in total defeat for the Kurds in 1975, when Iran withdrew support. The fighting led to Iraqi bombing of Kurdish villages in Iran, causing relations with Iran to deteriorate.

After skirmishing intermittently for 10 months over the sovereignty of the disputed Shatt al-Arab waterway that divides the two countries, Iraq and Iran entered into open warfare on Sept. 22, 1980. In the following days, there was heavy ground fighting around Abadan and the port of Khorramshahr, as Iraq launched an attack on Iran's oil-rich province of Khuzistan.

Israeli planes destroyed a nuclear reactor near Baghdad June 7, 1981, claiming it could be used to produce nuclear weapons.

Iraq and Iran expanded their war to the Persian Gulf in Apr. 1984. There were several attacks on oil tankers. An Iraqi warplane launched a missile attack on the USS *Stark,* a U.S. Navy frigate on patrol in the Persian Gulf, May 17, 1987; 37 U.S. sailors died. Iraq apologized for the attack, claiming it was inadvertent. The fierce war ended Aug. 1988, when Iraq accepted a UN resolution for a cease-fire.

Iraq attacked and overran Kuwait Aug. 2, 1990, sparking an international crisis. The UN, Aug. 6, imposed a ban on all trade with Iraq and called on member countries to protect the assets of the legitimate government of Kuwait. Iraq declared Kuwait its 19th province, Aug. 28.

A U.S.-led coalition launched air and missile attacks on Iraq, Jan. 16, 1991, after the expiration of a UN Security Council deadline for Iraq to withdraw from Kuwait. Iraq retaliated by firing scud missiles at Saudi Arabia and Israel. The coalition began a ground attack to retake Kuwait Feb. 23. Iraqi forces showed little resistance and were soundly defeated in 4 days. Some 175,000 Iraqis were taken prisoner, and casualties were estimated at over 85,000. As part of the cease-fire agreement, Iraq agreed to scrap all poison gas and germ weapons and allow UN observers to inspect the sites. UN trade sanctions would remain in effect until Iraq complied with all terms.

In the aftermath of the war, there were revolts against Pres. Saddam Hussein throughout Iraq. In Feb., Iraqi troops drove Kurdish insurgents and civilians to the borders of Iran and Turkey, causing a refugee crisis. The U.S. and allies established havens inside Iraq for the Kurds. Iraqi cooperation with UN weapons inspection teams was intermittent.

The U.S. launched a missile attack aimed at Iraq's intelligence headquarters in Baghdad June 26, 1993. The U.S. justified the attack by citing evidence that Iraq had sponsored a plot to kill former Pres. George Bush during his visit to Kuwait in Apr. 1993. In Aug. 1995, two of Saddam Hussein's sons-in-law, who held high positions in the Iraqi military, defected to Jordan; both were killed after returning to Iraq in Feb. 1996. After fighting between two Kurdish factions (one allied with Iraq, the other with Iran) erupted in the protected zone of northern Iraq, the Baghdad government intervened in the conflict by sending troops into Arbil, Aug. 31, 1996. The U.S. retaliated with missile strikes against air defense sites in the south. On Dec. 9 the UN allowed Baghdad to begin selling limited amounts of oil for food and medicine. Saddam Hussein's son Odai was seriously wounded in an assassination attempt in Baghdad Dec. 12.

Iraqi resistance to unrestricted UN access to suspected weapons sites led to diplomatic crises in Nov. 1997, Feb. 1998, and Oct.-Dec. 1998. Threatened with imminent air strikes by the U.S., Iraq on Feb. 22, 1998, embraced peace proposals brought to Baghdad by UN Secretary General Kofi Annan. Renewed disputes over inspections culminated in intensive U.S. and British aerial bombardment of Iraqi military targets, Dec. 16-19, 1998. After 2 years of intermittent activity, U.S. and British warplanes struck harder at sites near Baghdad on Feb. 16, 2001.

Ireland

People: Population: 3,840,838. **Age distrib.** (%): <15: 21.6; 65+: 11.4. **Pop. density:** 145 per sq. mi. **Urban:** 59%. **Ethnic groups:** Principally Celtic, English minority. **Principal languages:** English predominates, Irish (Gaelic) spoken by minority (both official). **Chief religions:** Roman Catholic 93%, Anglican 3%.

Geography: Area: 26,600 sq. mi. **Location:** In the Atlantic O. just W of Great Britain. **Neighbors:** United Kingdom (Northern Ireland) on E. **Topography:** Ireland consists of a central plateau surrounded by isolated groups of hills and mountains. The coastline is heavily indented by the Atlantic O. **Capital:** Dublin, (urban agg.): 985,000.

Government: Type: Parliamentary republic. **Head of state:** Pres. Mary McAleese; b June 27, 1951; in office: Nov. 11, 1997. **Head of gov.:** Prime Min. Bertie Ahern; b Sept. 12, 1951; in office: June 26, 1997. **Local divisions:** 26 counties. **Defense budget:** $725 mil. **Active troops:** 11,460.

Economy: Industries: Food processing, textiles, chemicals, brewing, machinery, crystal. **Chief crops:** Potatoes, grains, sugar beets, turnips. **Minerals:** Zinc, lead, gas, barite, copper, gypsum. **Arable land:** 13%. **Livestock** (2000): cattle: 6.71 mil; chickens: 11.5 mil; pigs: 1.76 mil; sheep: 5.39 mil. **Fish catch** (1999): 329,496 metric tons. **Electricity prod.:** 19.542 bil kWh. **Labor force:** 63% services; 28% ind.; 9% agric., forestry & fish.

Finance: Monetary unit: Punt (Sept. 2001: 0.87 = $1 U.S.). Euro (Sept. 2001: 1.09 = $1 U.S.). **GDP** (1999 est.): $73.7 bil. **Per capita GDP:** $20,300. **Imports** (1999 est.): $44 bil; partners: UK 31%, U.S. 16%. **Exports** (1999 est.): $66 bil; partners: UK 22%, Germany 15%. **Tourism:** $3.31 bil. **Budget** (1998): $20.6 bil. **Intl. reserves less gold** (June 2000): $5.25 bil. **Gold:** 176,000 oz t. **Consumer prices** (change in 1999): 1.6%.

Transport: Railroad: Length: 1,210 mi. **Motor vehicles:** 1.06 mil pass. cars, 161,355 comm. vehicles. **Civil aviation:** 4.5 bil pass.-mi.; 9 airports. **Chief ports:** Dublin, Cork.

Communications: TV sets: 457 per 1,000 pop. **Radios:** 580 per 1,000 pop. **Telephones:** 1,590,000 main lines. **Daily newspaper circ.:** 150 per 1,000 pop.

Health: Life expectancy: 74.23 male; 79.93 female. **Births** (per 1,000 pop): 14.57. **Deaths** (per 1,000 pop.): 8.07. **Natural inc.:** .65%. **Infant mortality** (per 1,000 live births): 5.53.

Education: Compulsory: ages 6-15. **Literacy** (1993): 100%.

Major Intl. Organizations: UN (FAO, IBRD, ILO, IMF, IMO, WHO, WTrO), EU, OECD, OSCE.

Embassy: 2234 Massachusetts Ave. NW 20008; 462-3939.

Websites: http://www.cso.ie/index.html
http://www.genuki.org.uk

Celtic tribes invaded the islands about the 4th century BC; their Gaelic culture and literature flourished and spread to Scotland and elsewhere in the 5th century AD, the same century in which St. Patrick converted the Irish to Christianity. Invasions by Norsemen began in the 8th century, ended with defeat of the Danes by the Irish King Brian Boru in 1014. English invasions started in the 12th century; for over 700 years the Anglo-Irish struggle continued with bitter rebellions and savage repressions.

The Easter Monday Rebellion in 1916 failed but was followed by guerrilla warfare and harsh reprisals by British troops called the "Black and Tans." The Dail Eireann (Irish parliament) reaffirmed independence in Jan. 1919. The British offered dominion status to Ulster (6 counties) and southern Ireland (26 counties) Dec. 1921. The constitution of the Irish Free State, a British dominion, was adopted Dec. 11, 1922. Northern Ireland remained part of the United Kingdom.

A new constitution adopted by plebiscite came into operation Dec. 29, 1937. It declared the name of the state Eire in the Irish language (Ireland in the English) and declared it a sovereign democratic state.On Dec. 21, 1948, an Irish law declared the country a republic rather than a dominion and withdrew it from the Commonwealth. The British Parliament recognized both actions, 1949, but reasserted its claim to incorporate the 6 northeastern counties in the United Kingdom. This claim has not been recognized by Ireland *(see United Kingdom—Northern Ireland).*

Irish governments have favored peaceful unification of all Ireland and cooperated with Britain against terrorist groups. On Dec. 15, 1993, Irish and British governments agreed on outlines of a peace plan to resolve the Northern Ireland issue. On Aug. 31, 1994, the Irish Republican Army announced a cease-fire; when peace talks lagged, however, the IRA returned to its terror campaign on Feb. 9, 1996. The IRA proclaimed a new cease-fire as of July 20, 1997, and peace talks resumed Sept. 15.

Ireland's first woman president, Mary Robinson, resigned Sept. 12 to become UN high commissioner for human rights. She was succeeded by Mary McAleese, a law professor from Northern Ireland and the first northerner to hold the office. After negotiators in Northern Ireland approved a peace settlement on Good Friday, April 10, 1998, voters in the Irish Republic endorsed the accord on May 22.

Israel

State of Israel

People: Population: 5,938,093. **Age distrib.** (%): <15: 27.4; 65+: 9.9. **Pop. density:** 757 per sq. mi. **Urban:** 91%. **Ethnic groups:** Jewish 80%, non-Jewish (mostly Arab) 20%. **Principal languages:** Hebrew (official), Arabic (used officially for Arab minority), English. **Chief religions:** Judaism 80%, Muslim (mostly Sunni) 15%.

Geography: Area: 7,800 sq. mi. **Location:** Middle East, on E end of Mediterranean Sea. **Neighbors:** Lebanon on N; Syria, West Bank, and Jordan on E; Gaza Strip and Egypt on W. **Topography:** The Mediterranean coastal plain is fertile and well-watered. In the center is the Judean Plateau. A triangular-shaped semi-desert region, the Negev, extends from south of Beersheba to an apex at the head of the Gulf of Aqaba. The E border drops sharply into the Jordan Rift Valley, including Lake Tiberias (Sea of Galilee) and the Dead Sea, which is 1,312 ft. below sea level, lowest point on the earth's surface. **Capital:** Jerusalem (most countries maintain their embassy in Tel Aviv). **Cities:** Jerusalem (1997 est.) 591,400; Tel Aviv-Yafo 2,181,000; Haifa (1997 est.) 255,300.

Government: Type: Republic. **Head of state:** Pres. Moshe Katsav; b 1945; in office: Aug. 1, 2000. **Head of gov.:** Prime Min. Ariel Sharon; b 1928; in office: Mar. 7, 2001. **Local divisions:** 6 districts. **Defense budget:** $7 bil. **Active troops:** 172,500.

Economy: Industries: Diamond cutting, textiles, electronics, food processing. **Chief crops:** Citrus, fruit, vegetables, cotton. **Minerals:** Copper, phosphates, bromide, potash, clay. **Crude oil reserves** (2000): 3.9 mil bbls. **Arable land:** 17%. **Livestock** (2000): cattle: 388,000; chickens: 28.0 mil; goats: 70,000; pigs: 163,000; sheep: 350,000.. **Fish catch:** (1999): 23,274 metric tons. **Electricity prod.:** 35.437 bil kWh. **Labor force:** 31% public services; 20% mfg.; 13% commerce.

Finance: Monetary unit: New Shekel (Sept. 2001: 4.30 = $1 U.S.). **GDP** (1999 est.): $105.4 bil. **Per capita GDP:** $18,300. **Imports** (1999): $30.6 bil; partners: U.S. 19%, Benelux 12%. **Exports** (1999): $23.5 bil; partners: U.S. 32%. **Tourism:** $3.10 bil. **Budget** (1998 est.): $58 bil. **Intl. reserves less gold** (June 2000): $22.27 bil. **Consumer prices** (change in 1999): 5.2%.

Transport: Railroad: Length: 379 mi. **Motor vehicles** (1997): 1.24 mil pass. cars, 304,033 comm. vehicles. **Civil aviation:** 7.3 bil pass.-mi.; 7 airports. **Chief ports:** Haifa, Ashdod, Elat.

Communications: TV sets: 335 per 1,000 pop. **Radios:** 530 per 1,000 pop. **Telephones:** 2,900,000 main lines. **Daily newspaper circ.:** 291 per 1,000 pop.

Health: Life expectancy: 76.69 male; 80.84 female. **Births** (per 1,000 pop): 19.12. **Deaths** (per 1,000 pop.): 6.22. **Natural inc.:** 1.29%. **Infant mortality** (per 1,000 live births): 7.72.

Education: Free, compulsory: ages 5-15. **Literacy**: 96%.

Major Intl. Organizations: UN (FAO, IBRD, ILO, IMF, IMO, WHO, WTrO).

Embassy: 3514 International Dr. NW 20008; 364-5500.

Website: http://www.israel.org

Occupying the SW corner of the ancient Fertile Crescent, Israel contains some of the oldest known evidence of agriculture and of primitive town life. A more advanced civilization emerged in the 3d millennium BC. The Hebrews probably arrived early in the 2d millennium BC. Under King David and his successors (c.1000 BC-597 BC), Judaism was developed and secured. After conquest by Babylonians, Persians, and Greeks, an independent Jewish kingdom was revived, 168 BC, but Rome took effective control in the next century, suppressed Jewish revolts in AD 70 and AD 135, and renamed Judea Palestine, after the earlier coastal inhabitants, the Philistines.

Arab invaders conquered Palestine in 636. The Arabic language and Islam prevailed within a few centuries, but a Jewish minority remained. The land was ruled from the 11th century as a part of non-Arab empires by Seljuks, Mamluks, and Ottomans (with a crusader interval, 1098-1291).

After 4 centuries of Ottoman rule, during which the population declined to a low of 350,000 (1785), the land was taken in 1917 by Britain, which pledged in the Balfour Declaration to support a Jewish national homeland there. In 1920 a British Palestine Mandate was recognized; in 1922 the land east of the Jordan was detached.

Jewish immigration, begun in the late 19th century, swelled in the 1930s with refugees from the Nazis; heavy Arab immigration from Syria and Lebanon also occurred. Arab opposition to Jewish immigration turned violent in 1920, 1921, 1929, and 1936. The UN General Assembly voted in 1947 to partition Palestine into an Arab and a Jewish state. Britain withdrew in May 1948.

Israel was declared an independent state May 14, 1948; the Arabs rejected partition. Egypt, Jordan, Syria, Lebanon, Iraq, and Saudi Arabia invaded, but failed to destroy the Jewish state, which gained territory. Separate armistices with the Arab nations were signed in 1949; Jordan occupied the West Bank, Egypt occupied Gaza; neither granted Palestinian autonomy.

After persistent terrorist raids, Israel invaded Egypt's Sinai, Oct. 29, 1956, aided briefly by British and French forces. A UN cease-fire was arranged Nov. 6.

An uneasy truce between Israel and the Arab countries, supervised by a UN Emergency Force, prevailed until May 19, 1967, when the UN force withdrew at Egypt's demand. Egyptian forces reoccupied the Gaza Strip and closed the Gulf of Aqaba to Israeli shipping. In a 6-day war that started June 5, the Israelis took the Gaza Strip, occupied the Sinai Peninsula to the Suez Canal, and captured East Jerusalem, Syria's Golan Heights, and Jordan's West Bank. The fighting was halted June 10 by UN-arranged cease-fire agreements.

Egypt and Syria attacked Israel, Oct. 6, 1973 (on Yom Kippur, the most solemn day on the Jewish calendar). Israel counter-attacked, driving the Syrians back, and crossed the Suez Canal. A cease-fire took effect Oct. 24 and a UN peacekeeping force went to the area. Under a disengagement agreement signed Jan. 18, 1974, Israel withdrew from the canal's west bank.

Israeli forces raided Entebbe, Uganda, July 3, 1976, and rescued 103 hostages who had been seized by Arab and German terrorists.

In 1977, the conservative opposition, led by Menachem Begin, was voted into office for the first time. Egypt's Pres. Anwar al-Sadat visited Jerusalem Nov. 1977, and on Mar. 26, 1979, Egypt and Israel signed a formal peace treaty, ending 30 years of war and establishing diplomatic relations. Israel returned the Sinai to Egypt in 1982.

Israel invaded S Lebanon, Mar. 1978, following a Lebanon-based terrorist attack in Israel. Israel withdrew in favor of a 6,000-man UN force, but continued to aid Lebanese Christian militiamen. Israel affirmed the whole of Jerusalem as its capital, July 1980, encompassing the annexed East Jerusalem.

On June 7, 1981, Israeli jets destroyed an Iraqi atomic reactor near Baghdad that, Israel claimed, would have enabled Iraq to manufacture nuclear weapons. Israeli forces invaded Lebanon, June 6, 1982, to destroy PLO strongholds there. After massive Israeli bombing of West Beirut, the PLO agreed to evacuate the city. Israeli troops entered West Beirut after newly elected Lebanese Pres. Bashir Gemayel was assassinated on Sept. 14. Israel drew widespread condemnation when Lebanese Christian forces, Sept. 16, entered two West Beirut refugee camps and slaughtered hundreds of Palestinian refugees.

In 1989, violence escalated over the Israeli military occupation of the West Bank and Gaza Strip. In a series of uprisings known as the intifada, Palestinian protesters defied Israeli troops, who forcibly retaliated. Israeli police and stone-throwing Palestinians clashed, Oct. 8, 1990, around the al-Aqsa mosque on the Temple Mount in Jerusalem; some 20 Palestinians died.

During the Persian Gulf War in early 1991, Iraq fired a series of Scud missiles at Israel. The Labor Party of Yitzhak Rabin won a clear victory in elections held June 23, 1992.

Ongoing peace talks led to historic agreements between Israel and the PLO, Sept. 1993. The PLO recognized Israel's right to exist; Israel recognized the PLO as the Palestinians' representative; the two sides then signed, Sept. 13, an agreement for limited Palestinian self-rule and the West Bank and Gaza.

Israel and Jordan signed, July 25, 1994, in Washington, DC, a declaration ending their 46-year state of war. A formal peace treaty was signed Oct. 26.

Arab and Jewish extremists repeatedly challenged the peace process. A Jewish gunman opened fire on Arab worshippers at a mosque in Hebron, Feb. 25, 1994, killing at least 29 before he himself was killed. On Nov. 4, 1995, an Orthodox Jewish Israeli assassinated Rabin as he left a peace rally in Tel Aviv.

Support for Rabin's successor, Shimon Peres, was shaken by a series of suicide bombings and rocket attacks against Israel by Islamic militants. In Apr. 1996, Israel attacked suspected guerrilla bases in southern Lebanon. Emphasizing security issues, the candidate of the conservative Likud bloc, Benjamin Netanyahu, was elected prime minister on May 29.

On Sept. 24, 1996, Israel opened a tunnel entrance near a sacred Muslim site in Jerusalem, setting off several days of violence between Israeli soldiers and Palestinian demonstrators and police. Pres. Clinton hosted a summit meeting between Netanyahu and PLO leader Yasir Arafat soon after, on Oct. 1-2, and peace talks were resumed.

Two suicide bombings in a Jerusalem market July 30, 1997, left 15 people dead and more than 170 wounded. The parliament (Knesset) reelected Ezer Weizman as president Mar. 4, 1998, despite opposition from Netanyahu.

CYCLE OF VIOLENCE ►

On Aug. 9 a Palestinian suicide bomber struck a crowded Jerusalem restaurant (right), killing 15 people and himself. Israel retaliated with raids the next day, including missile attacks on the West Bank city of Ramallah, where police headquarters (below) were reduced to rubble. In 12 months Israeli-Palestinian violence had claimed more than 700 lives, an estimated three-quarters of them Palestinian.

AP/WIDE WORLD PHOTOS

AP/WIDE WORLD PHOTOS

© REUTERS NEWMEDIA INC./CORBIS

SWEEPING VICTORY ►

Ariel Sharon, of the conservative Likud party, was elected prime minister of Israel, Feb. 6, by the largest winning margin in Israeli history; on taking office Mar. 7, he cited "bolstering Israel's security" as his government's main mission.

AP/WIDE WORLD PHOTOS

AP/WIDE WORLD PHOTOS

▲ NEPAL MASSACRE

Bodies of members of Nepal's royal family are cremated in Kathmandu, following a palace massacre June 1. An official report found that Crown Prince Dipendra, standing (center) in the undated family portrait above, shot and killed King Birendra and Queen Aiswarya (sitting), Princess Shruti (standing, left), Prince Nirajan (right), and 6 others, before fatally wounding himself.

AP/WIDE WORLD PHOTOS

◄ KOFI ANNAN

UN Sec. Gen. Kofi Annan addresses the summit of the Organization of African Unity, July 9 in Lusaka, Zambia. Annan was unanimously reelected by the UN General Assembly, June 29, to a 2d 5-year term as UN chief.

AP/WIDE WORLD PHOTOS

AP/WIDE WORLD PHOTOS

STATUE DESTROYED ►

A Taliban soldier in the mountains of Bamiyan, Afghanistan, views remains of a 175-foot, 2000-year-old Buddha statue (once the world's tallest; see inset). It was blasted away, Mar. 26, by order of the ruling Taliban, in a campaign to destroy statues thought insulting to Islam.

FOOT AND MOUTH SLAUGHTER ►

Cattle are burned on a pyre in Ecclefechan, Scotland, Apr. 17. By the end of August, nearly 3.8 million cattle and other animals had been culled in Britain to curb the spread of foot and mouth disease; the epidemic cost the British economy an estimated $3.9 billion or more.

AP/WIDE WORLD PHOTOS

AP/WIDE WORLD PHOTOS

◄ LABOUR REPEATS

Britain's Tony Blair became the first prime minister in the history of the Labour Party to win a second consecutive full term in office, following elections June 7.

AP/WIDE WORLD PHOTOS

MILOSEVIC IN CAPTIVITY ►

Extradited former Yugoslav Pres. Slobodan Milosevic, under guard, appears July 2 at a tribunal in The Hague, where he was arraigned for alleged war crimes against ethnic Albanians during a 1999 conflict in the Serbian province of Kosovo.

AP/WIDE WORLD PHOTOS

◄ SPY PLANE

In a photo released by China, a U.S. Navy EP-3E surveillance plane, crippled in an Apr. 1 collision with a Chinese fighter plane, sits at an air base on China's Hainan island, where the crew had safely landed and were being held. After 11 days of tense talks, the crew was allowed to go home; in July the plane was sent back in pieces.

2002 ARTS AND ENTERTAINMENT

AP/WIDE WORLD PHOTOS

AP/WIDE WORLD PHOTOS

◄ TONY PRODUCER
The cast of *The Producers*, starring Matthew Broderick and Nathan Lane (center), sing at the Tony awards, June 3; the smash hit musical, based on a Mel Brooks movie, with new lyrics and music by Brooks (inset), won a record 12 Tonys.

***SHREK* ►**
Shrek, an animated film featuring voices by Mike Myers (in the title role), Eddie Murphy, Cameron Diaz, and John Lithgow, delighted critics and ranked number one in summer domestic box office receipts, raking in more than $260 million by the end of August.

AFP

©REUTERS NEWMEDIA INC./CORBIS

◄ SUPERSTAR
Julia Roberts gleefully accepts her Best Actress Oscar at the 73rd annual Academy Awards in Los Angeles, Mar. 25. She won for her role in the film *Erin Brockovich*.

AP/WIDE WORLD PHOTOS

◄ BOZO BOWS OUT

Joey D'Auria, the most recent Bozo the Clown, checks his make-up on one of his last days on the job at WGN-TV, Chicago; the character, created in 1946, appeared continuously from 1961 until the show's final taping in July 2001.

AP/WIDE WORLD PHOTOS

MISTER ROGERS RETIRES—IN PART ►

After more than 30 seasons, the last new episode of *Mister Rogers' Neighborhood* was broadcast on TV in August 2001. But Mr. (Fred) Rogers (shown here in 1996 on the show's Pittsburgh set) will still be around, as the treasury of back episodes is replayed for new generations of children.

◄ JOHN ADAMS REVISITED

Founding father and second president John Adams was in the public eye in 2001, as a biography by David McCullough hit the best-seller list and refurbished his image. (This portrait was painted by Gilbert Stuart around 1821.)

AP/WIDE WORLD PHOTOS

NUMBER ONE AGAIN ►

The Beatles arrive in New York, Feb. 7, 1964, for their first U.S. appearance. In early 2001 their album *1*, with 27 number-one hits from 1962 through 1970, topped charts in 34 countries.

AP/WIDE WORLD PHOTOS

ICHIRO ►
A Tokyo salesclerk watches TV news, June 22, reporting that Seattle Mariner Ichiro Suzuki had become batting leader in the American League. The popular "Ichiro" was a 7-time batting champ of Japan's Pacific League.

AP/WIDE WORLD PHOTOS

◄ IRON MAN RETIRES
Baltimore's Cal Ripken Jr. homers in the 72d All-Star Game, July 10, to earn his 2d All-Star MVP award. Ripken, who played in a record 2,632 consecutive games, had announced in June that the 2001 season would be his last.

AP/WIDE WORLD PHOTOS

DOMINANT DUO ►
The Los Angeles Lakers—led by All-Stars Kobe Bryant (left) and Shaquille O'Neal—won their 2d NBA Championship in a row in 2001, defeating the Philadelphia 76ers, 108-96, in Game Five of the NBA Finals, June 15 in Philadelphia. O'Neal won his 2d NBA Finals MVP Award.

AP/WIDE WORLD PHOTOS

◄ TIGER SLAM
Golfer Tiger Woods won his 2d Masters championship at Augusta National, Apr. 8, and earned an unprecedented 4th consecutive major golf championship. After finishing 5th at the Masters in 2000, Woods went on to win the 2000 U.S. Open, British Open, and PGA Championship before his 2001 Masters triumph.

59! ►
Golfer Annika Sorenstam, right, takes a celebratory leap into caddie Terry McNamara's arms after shooting an LPGA-record 59 in the 2d round of the Standard Register Ping tournament, Mar. 16, in Phoenix.

AP/WIDE WORLD PHOTOS

AP/WIDE WORLD PHOTOS

◄ SUPER BOWL MVP
The NFL's Defensive Player of the Year, Baltimore Ravens linebacker Ray Lewis, smiles on the sidelines during the 4th quarter of Super Bowl XXXV, Jan. 28 in Tampa, FL. Lewis was named Super Bowl MVP, after the Ravens swept away the NY Giants, 34-7. In 2000, Lewis had accepted a plea bargain and probation, after having been charged with murder in connection with a nightclub brawl.

VENUS WILLIAMS ►

Venus Williams (right), Sept. 8, won the U.S. Open title in New York City, defeating her sister, Serena. Venus had won the Wimbledon crown July 8.

AP/WIDE WORLD PHOTOS

AP/WIDE WORLD PHOTOS

▲ CAPRIATI COMEBACK

Jennifer Capriati returned to the top ranks of the tennis world, winning both the Australian Open (Jan. 27) and the French Open, above (June 9).

AP/WIDE WORLD PHOTOS

▲ TOUR DE LANCE

Texan Lance Armstrong passes the Arc de Triomphe in Paris, July 29, after winning his 3d consecutive Tour de France, the world's premier cycling event.

AP/WIDE WORLD PHOTOS

▲ SO YOUNG, SO FAST

Alan Webb, 18, reacts after shattering the U.S. high school record in the men's mile, at the international Prefontaine Classic, May 27 in Eugene, OR. Webb's time of 3:53.43 broke Jim Ryun's U.S. record of 3:55.3 that had stood since 1965.

Under an interim accord brokered by Clinton and signed by Netanyahu and Arafat at the White House, Oct. 23, 1998, Israel yielded more West Bank territory to the Palestinians, in exchange for new security guarantees. Negotiations bogged down, however, and full implementation did not begin until Sept. 1999. In the interim, Netanyahu lost by a landslide to the Labor party candidate, Ehud Barak, in the general election of May 17.

Israel pulled virtually all its troops out of S Lebanon by May 24, 2000. Marathon summit talks in the U.S. between Barak and Arafat, July 11-25, failed to reach agreement. A new wave of violence began in late Sept. in Israel and the Palestinian territories. Barak called new elections for prime minister but lost Feb. 6, 2001, to Ariel Sharon, a hardliner; he took office Mar. 7, heading a national unity government with Peres as foreign minister. The bloodshed intensified during the summer, as Palestinian suicide bombers hit a Tel Aviv nightclub June 1, killing 20 young Israelis, and a Jerusalem pizzeria Aug. 9, killing 15 patrons. Israel launched offensives against Palestinian-controlled territory and carried out an assassination campaign against dozens of suspected terrorists. As of late Sept., at least 560 Palestinians and over 170 Israelis had been killed during the previous 12 months.

Gaza Strip

The Gaza Strip, also known as Gaza, extends NE from the Sinai Peninsula for 40 km (25 mi), with the Mediterranean Sea to the W and Israel to the E. The Palestinian Authority is responsible for civil government, but Israel retains control over security. Nearly all the inhabitants are Palestinian Arabs, more than 35% of whom live in refugee camps. Population (2001 est.): 1,178,119. Area: 140 sq. mi.

Israel captured Gaza from Egypt in the 1967 war. It remained under Israeli occupation until May 1994, when the Israel Defense Forces withdrew. Agreements between Israel and the PLO in 1993 and 1994 provided for interim self-rule in Gaza, pending the completion of final status negotiations.

West Bank

Located W of the Jordan R. and Dead Sea, the West Bank is bounded by Jordan on the E and by Israel on the N, W, and S. The Palestinian Authority administers several major cities, but Israel retains control over much land, including Jewish settlements. Population (2001 est.): 2,090,713. Area: 2,200 sq. mi.

Israel captured the West Bank from Jordan in the 1967 war. A 1974 Arab summit conference designated the PLO as sole representative of West Bank Arabs. In 1988 Jordan cut legal and administrative ties with the territory. Jericho was returned to Palestinian control in May 1994. An accord between Israel and the PLO expanding Palestinian self-rule in the West Bank was signed Sept. 28, 1995. Later agreements gave Palestinians full or shared control of 40% of West Bank territory.

Italy

Italian Republic

People: Population: 57,679,825. **Age distrib.** (%): <15: 14.2; 65+: 18.3. **Pop. density:** 509 per sq. mi. **Urban:** 67%. **Ethnic groups:** Italian, small minorities of German, French, Slovene, Albanian. **Principal languages:** Italian (official), German, French, Slovene. **Chief religion:** Roman Catholic 98%.

Geography: Area: 113,400 sq. mi. **Location:** In S Europe, jutting into Mediterranean Sea. **Neighbors:** France on W, Switzerland and Austria on N, Slovenia on E. **Topography:** Occupies a long boot-shaped peninsula, extending SE from the Alps into the Mediterranean, with the islands of Sicily and Sardinia offshore. The alluvial Po Valley drains most of N. The rest of the country is rugged and mountainous, except for intermittent coastal plains, like the Campania, S of Rome. Apennine Mts. run down through center of peninsula. **Capital:** Rome. **Cities** (1998 city proper): Milan 1,308,000; Naples 1,020,000; Rome 2,646,000; Turin 910,000.

Government: Type: Republic. **Head of state:** Pres. Carlo Azeglio Ciampi; b Dec. 9, 1920; in office: May 18, 1999. **Head of gov.:** Prime Min. Silvic Berlusconi; b Sept. 29, 1936; in office: June 11, 2001. **Local divisions:** 20 regions divided into 94 provinces. **Defense:** 2.0% of GDP. **Active troops:** 298,400

Economy: Industries: Tourism, steel, machinery, autos, textiles, shoes, clothing, chemicals. **Chief crops:** Grapes, olives, fruits, vegetables, grain. **Minerals:** Mercury, potash, marble, sulphur. **Crude oil reserves** (2000): 621.8 mil bbls. **Arable land:** 31%. **Livestock** (2000): cattle: 7.18 mil; chickens: 100.0 mil; goats: 1.36 mil; pigs: 8.4 mil; sheep: 10.97 mil. **Fish catch** (1999): 562,196 metric tons. **Electricity prod.:** 247.679 bil kWh. **Labor force:** 61% services; 32% ind.; 7% agric.

Finance: Monetary unit: Lira (Sept. 2001: 2,136.10 = $1 U.S.). Euro (Sept. 2001: 1.09 = $1 U.S.). **GDP** (1999 est.): $1.212 tril. **Per capita GDP:** $21,400. **Imports** (1998): $206.9 bil; partners: Germany 18.8%, France 13.12%, UK 6.47%. **Exports** (1998): $242.6 bil; partners: Germany 16.5%, France 12.7%, U.S. 8.5%. **Tourism:** $28.36 bil. **Budget** (1998 est.): $589 bil. **Intl. reserves less gold** (June 2000): $24.55 bil. **Gold:** 78.83 mil oz t. **Consumer prices** (change in 1999): 1.7%.

Transport: Railroad: Length: 9,944 mi. **Motor vehicles** (1997): 31.00 mil pass. cars, 2.99 mil comm. vehicles. **Civil aviation:** 23.6 bil pass.-mi.; 34 airports. **Chief ports:** Genoa, Venice, Trieste, Palermo, Naples, La Spezia.

Communications: TV sets: 483 per 1,000 pop. **Radios:** 874 per 1,000 pop. **Telephones:** 27,153,000 main lines. **Daily newspaper circ.:** 104 per 1,000 pop.

Health: Life expectancy: 75.97 male; 82.52 female. **Births** (per 1,000 pop): 9.05. **Deaths** (per 1,000 pop.): 10.07. **Natural inc.:** -.1%. **Infant mortality** (per 1,000 live births): 5.84.

Education: Free, compulsory: ages 6-13. **Literacy** (1994): 97%.

Major Intl. Organizations: UN and all of its specialized agencies, EU, NATO, OECD, OSCE.

Embassy: 1601 Fuller St. NW 20009; 328-5500.

Website: http://www.istat.it

Rome emerged as the major power in Italy after 500 BC, dominating the Etruscans to the N and Greeks to the S. Under the Empire, which lasted until the 5th century AD, Rome ruled most of Western Europe, the Balkans, the Middle East, and N Africa. In 1988, archaeologists unearthed evidence showing Rome as a dynamic society in the 6th and 7th centuries BC.

After the Germanic invasions, lasting several centuries, a high civilization arose in the city-states of the N, culminating in the Renaissance. But German, French, Spanish, and Austrian intervention prevented the unification of the country. In 1859 Lombardy came under the crown of King Victor Emmanuel II of Sardinia. By plebiscite in 1860, Parma, Modena, Romagna, and Tuscany joined, followed by Sicily and Naples, and by the Marches and Umbria. The first Italian Parliament declared Victor Emmanuel king of Italy Mar. 17, 1861. Mantua and Venetia were added in 1866 as an outcome of the Austro-Prussian war. The Papal States were taken by Italian troops Sept. 20, 1870, on the withdrawal of the French garrison. The states were annexed to the kingdom by plebiscite. Italy recognized Vatican City as independent Feb. 11, 1929.

Fascism appeared in Italy Mar. 23, 1919, led by Benito Mussolini, who took over the government at the invitation of the king Oct. 28, 1922. Mussolini acquired dictatorial powers. He made war on Ethiopia and proclaimed Victor Emmanuel III emperor, defied the sanctions of the League of Nations, sent troops to fight for Franco against the Republic of Spain, and joined Germany in World War II.

After Fascism was overthrown in 1943, Italy declared war on Germany and Japan and contributed to the Allied victory. It surrendered conquered lands and lost its colonies. Mussolini was killed by partisans Apr. 28, 1945. Victor Emmanuel III abdicated May 9, 1946; his son Humbert II was king until June 10, when Italy became a republic after a referendum, June 2-3.

Since World War II, Italy has enjoyed growth in industrial output and living standards, in part a result of membership in the European Community (now European Union). Political stability has not kept pace with economic prosperity, and organized crime and corruption have been persistent problems.

Christian Democratic leader and former Prime Min. Aldo Moro was abducted and murdered in 1978 by Red Brigade terrorists. The wave of left-wing political violence, including other kidnappings and assassinations, continued into the 1980s.

In the early 1990s, scandals implicated some of Italy's most prominent politicians. In Mar. 1994 voting, under reformed election rules, right-wing parties won a majority, dislodging Italy's long-powerful Christian Democratic Party. After a series of short-lived governments, a coalition of center-left parties won the election of Apr. 21, 1996. Italy led a 7,000-member international peacekeeping force in Albania, Apr.-Aug. 1997. Two earthquakes in central Italy Sept. 26 killed 11 people, left about 12,000 homeless, and damaged priceless frescoes in Assisi.

On Feb. 3, 1998, a low-flying U.S. military aircraft severed a gondola cable at a ski resort in N Italy, killing 20 people. Implementation of a deficit reduction plan enabled Italy to qualify in May to adopt the euro, a common European currency. Italy contributed 2,000 troops to the NATO-led security force (KFOR) that entered Kosovo in June 1999. Turin was chosen June 19 to host the Winter Olympics in 2006.

Supporters of Silvio Berlusconi, a multibillionaire media magnate, won the parliamentary elections of May 13, 2001.

Sicily, 9,926 sq. mi., pop. (1994 est.) 5,025,000, is an island 180 by 120 mi., seat of a region that embraces the island of **Pantelleria,** 32 sq. mi., and the **Lipari** group, 44 sq. mi., including 2 active volcanoes: **Vulcano,** 1,637 ft., and **Stromboli,** 3,038 ft. From prehistoric times Sicily has been settled by various peoples; a Greek state had its capital at Syracuse. Rome took Sicily from Carthage 215 BC. **Mt. Etna,** an 11,053-ft. active volcano, is its tallest peak.

Sardinia, 9,301 sq. mi., pop. (1994 est.) 1,657,000, lies in the Mediterranean, 115 mi. W of Italy and $7^{1}/_{2}$ mi. S of Corsica. It is 160 mi. long, 68 mi. wide, and mountainous, with mining of coal, zinc, lead, copper. In 1720 Sardinia was added to the possessions of the Dukes of Savoy in Piedmont and Savoy to form the Kingdom of Sardinia. Giuseppe Garibaldi is buried on the nearby isle of Caprera. **Elba,** 86 sq. mi., lies 6 mi. W of Tuscany. Napoleon I lived in exile on Elba 1814-1815.

Jamaica

People: Population: 2,665,636. **Age distrib.** (%): <15: 29.7; 65+: 6.8. **Pop. density:** 638 per sq. mi. **Urban:** 56%. **Ethnic groups:** Black 90%. **Principal languages:** English (official), Jamaican Creole. **Chief religions:** Protestant 61%, Roman Catholic 4%, spiritual cults and other 35%.

Geography: Area: 4,200 sq. mi. **Location:** In West Indies. **Neighbors:** Nearest are Cuba to N, Haiti to E. **Topography:** Four-fifths of Jamaica is covered by mountains. **Capital:** Kingston (1999 urban agg.): 655,000.

Government: Type: Parliamentary democracy. **Head of state:** Queen Elizabeth II, represented by Gov.-Gen. Sir Howard Cooke; b Nov. 13, 1915; in office: Aug. 1, 1991. **Head of gov.:** Prime Min. Percival J. Patterson; b Apr. 10, 1935; in office: Mar. 30, 1992. **Local divisions:** 14 parishes. **Defense budget:** $50 mil . **Active troops:** 2,830.

Economy: Industries: Bauxite mining, tourism. **Chief crops:** Sugar, coffee, bananas, potatoes, citrus. **Minerals:** Bauxite, limestone, gypsum. **Arable land:** 14%. **Livestock**(2000): cattle: 400,000; chickens: 11.0 mil; goats: 440,000; pigs: 180,000; sheep: 1,400.. **Fish catch** (1999): 11,458 metric tons. **Electricity prod.:** 6.530 bil kWh. **Labor force:** 60% services; 21% agric.; 19% industry.

Finance: Monetary unit: Dollar (Sept. 2001: 45.55 = $1 U.S.). **GDP** (1999 est.): $8.8 bil. **Per capita GDP:** $3,350. **Imports** (1999 est.): $2.7 bil; partners: U.S. 50.9%. **Exports** (1999 est.): $1.4 bil; partners: U.S. 39.5%. **Tourism:** $1.23 bil. **Budget** (FY 1998-99 est.): $1.27 bil. **Intl. reserves less gold** (Mar. 2000): $803.3 mil. **Consumer prices** (change in 1999): 6.0%.

Transport: Railroad: Length: 129 mi. **Motor vehicles:** 43,500 pass. cars, 15,400 comm. vehicles. **Civil aviation:** 1.7 bil pass.-mi.; 4 airports. **Chief ports:** Kingston, Montego Bay.

Communications: TV sets: 306 per 1,000 pop. **Radios:** 739 per 1,000 pop. **Telephones**: 511,700 main lines. **Daily newspaper circ.:** 65 per 1,000 pop.

Health: Life expectancy: 73.45 male; 77.49 female. **Births** (per 1,000 pop): 18.12. **Deaths** (per 1,000 pop.): 5.48. **Natural inc.:** 1.26%. **Infant mortality** (per 1,000 live births): 14.16.

Education: Free, compulsory: ages 6-12. **Literacy**: 85%.

Major Intl. Organizations: UN (FAO, IBRD, ILO, IMF, IMO, WHO, WTrO), Caricom, the Commonwealth, OAS.

Embassy: 1520 New Hampshire Ave. NW 20036; 452-0660.

Website: http://www.jamaica.com/islands/jamaica/default.html

Jamaica was visited by Columbus, 1494, and ruled by Spain (under whom Arawak Indians died out) until seized by Britain, 1655. Jamaica won independence Aug. 6, 1962.

In 1974 Jamaica sought an increase in taxes paid by U.S. and Canadian bauxite mines. The socialist government acquired 50% ownership of the companies' Jamaican interests in 1976, and was reelected that year. Rudimentary welfare state measures were passed. Relations with the U.S. improved in the 1980s when Jamaican politics entered a more conservative phase. Violent clashes between government forces and West Kingston slum residents claimed at least 20 lives July 7-10, 2001.

Japan

People: Population: 126,771,662. **Age distrib.** (%): <15: 14.6; 65+: 17.5. **Pop. density:** 867 per sq. mi. **Urban:** 79%. **Ethnic groups:** Japanese 99.4%. **Principal language:** Japanese (official). **Chief religions:** Buddhism, Shintoism shared by 84%.

Geography: Area: 152,200 sq. mi. **Location:** Archipelago off E coast of Asia. **Neighbors:** Russia to N, South Korea to W. **Topography:** Japan consists of 4 main islands: Honshu ("mainland"), 87,805 sq. mi.; Hokkaido, 30,144 sq. mi.; Kyushu, 14,114 sq. mi.; and Shikoku, 7,049 sq. mi. The coast, deeply indented, measures 16,654 mi. The northern islands are a continuation of the Sakhalin Mts. The Kunlun range of China continues into southern islands, the ranges meeting in the Japanese Alps. In a vast transverse fissure crossing Honshu E-W rises a group of volcanoes, mostly extinct or inactive, including 12,388 ft. Mt. Fuji (Fujiyama) near Tokyo. **Capital:** Tokyo. **Cities** (urban agg.): Tokyo 26,444,000 (1998 city proper: 7,854,000); Osaka 11,013,000, (1998 city proper: 2,599,642); Nagoya 3,157,000; Sapporo 1,984,000; Kyoto 1,849,000.

Government: Type: Parliamentary democracy. **Head of state:** Emp. Akihito; b Dec. 23, 1933; in office: Jan. 7, 1989. **Head of gov.:** Prime Min. Junichiro Koizumi; b Jan. 8, 1942; in office: Apr. 26, 2001. **Local divisions:** 47 prefectures. **Defense budget:** $45.6 bil. **Active troops:** 236,700.

Economy: Industries: Electrical & electronic equip., vehicles, machinery, steel, metallurgy, chemicals, fishing. **Chief crops:** Rice, sugar beets, vegetables, fruits. **Crude oil reserves** (2000): 58.6 mil bbls. **Arable land:** 11%. **Livestock** (2000): cattle: 4.59 mil; chickens: 298.0 mil; goats: 31,000; pigs: 9.88 mil; sheep: 11,000. **Fish catch** (1999): 6.69 mil metric tons. **Electricity prod.:** 1.02 tril. kWh. **Labor force:** 65% services & trade; 30 % ind.; ; 5% agric., forestry, & fish.

Finance: Monetary unit: Yen (Sept. 2001: 120.18 = $1 U.S.). **GDP** (1999 est.): $2.95 tril. **Per capita GDP:** $23,400. **Imports** (1999 est.): $306 bil; partners: U.S. 22%, China 14%. **Exports** (1999 est.): $413 bil; partners: U.S. 31%, Taiwan 7%. **Tourism:** $3.43 bil. **Budget** (FY 1999-2000 est.): $711 bil. **Intl. reserves less gold** (Mar. 2000): $337.87 bil. **Gold:** 24.23 mil oz t. **Consumer prices** (change in 1999): –0.3%.

Transport: Railroad: Length: 12,511 mi. **Motor vehicles** (1997): 46.64 mil pass. cars, 21.39 mil comm. vehicles. **Civil aviation:** 93.9 bil pass.-mi.; 73 airports. **Chief ports:** Tokyo, Kobe, Osaka, Nagoya, Chiba, Kawasaki, Hakodate.

Communications: TV sets: 708 per 1,000 pop. **Radios:** 957 per 1,000 pop. **Telephones**: 82,916,000 main lines. **Daily newspaper circ.:** 578 per 1,000 pop.

Health: Life expectancy: 77.62 male; 84.15 female. **Births** (per 1,000 pop): 10.04. **Deaths** (per 1,000 pop.): 8.34. **Natural inc.:** .17%. **Infant mortality** (per 1,000 live births): 3.88.

Education: Compulsory: ages 6-15. **Literacy:** 100%.

Major Intl. Organizations: UN and all its specialized agencies, APEC, OECD.

Embassy: 2520 Massachusetts Ave. NW 20008; 238-6700.

Website: http://www.mofa.go.jp

According to Japanese legend, the empire was founded by Emperor Jimmu, 660 BC, but earliest records of a unified Japan date from 1,000 years later. Chinese influence was strong in the formation of Japanese civilization. Buddhism was introduced before the 6th century AD.

A feudal system, with locally powerful noble families and their samurai warrior retainers, dominated from 1192. Central power was held by successive families of shoguns (military dictators), 1192-1867, until recovered by Emperor Meiji, 1868. The Portuguese and Dutch had minor trade with Japan in the 16th and 17th centuries; U.S. Commodore Matthew C. Perry opened the country to U.S. trade in a treaty ratified 1854. Industrialization was begun in the late 19th century. Japan fought China, 1894-95, gaining Taiwan. After war with Russia, 1904-5, Russia ceded S half of Sakhalin and gave concessions in China. Japan annexed Korea 1910.

In World War I Japan ousted Germany from Shandong in China and took over German Pacific islands. Japan took Manchuria in 1931and launched full-scale war in China in 1937. Japan launched war against the U.S. by attacking Pearl Harbor Dec. 7, 1941. The U.S. dropped atomic bombs on Hiroshima, Aug. 6, and Nagasaki, Aug. 9, 1945. Japan surrendered Aug. 14, 1945.

In a new constitution adopted May 3, 1947, Japan renounced the right to wage war; the emperor gave up claims to divinity; the Diet became the sole law-making authority. The U.S. and 48 other non-Communist nations signed a peace treaty and the U.S. a bilateral defense agreement with Japan, in San Francisco Sept. 8, 1951, restoring Japan's sovereignty as of April 28, 1952.

Rebuilding after World War II, Japan emerged as one of the most powerful economies in the world, and as a leader in technology.The U.S. and Western Europe criticized Japan for its restrictive policy on imports, which eventually allowed Japan to accumulate huge trade surpluses.

On June 26, 1968, the U.S. returned to Japanese control the Bonin Isls., Volcano Isls. (including Iwo Jima), and Marcus Isls. On May 15, 1972, Okinawa, the other Ryukyu Isls., and the Daito Isls. were returned by the U.S.; it was agreed the U.S. would continue to maintain military bases on Okinawa.

The Recruit scandal, the nation's worst political scandal since World War II, which involved illegal political donations and stock trading, led to the resignation of Premier Noboru Takeshita in May 1989. Following new political and economic scandals, the ruling Liberal Democratic Party (LDP) was denied a majority in general elections July 18, 1993. On June 29, 1994, Tomiichi Murayama became Japan's first Socialist premier since 1947-48.

An earthquake in the Kobe area in Jan. 1995 claimed more than 5,000 lives, injured nearly 35,000, and caused over $90 billion in property damage. On Mar. 20, a nerve gas attack in the Tokyo subway (blamed on a religious cult) killed 12 and injured thousands. Public anger at the rape of a 12-year-old Okinawa schoolgirl by 3 U.S. servicemen, Sept. 4, led the U.S. to begin reducing its military presence there.

Murayama resigned as prime minister, Jan. 5, 1996, and was replaced by Ryutaro Hashimoto of the LDP. Hashimoto signed a joint security declaration with U.S. Pres. Bill Clinton in Tokyo, Apr. 17, 1996. Nagano hosted the Winter Olympics, Feb. 7-22, 1998.

With Japan mired in a lengthy recession, the LDP suffered a sharp rebuke in elections for parliament's upper house, July 12, 1998. Hashimoto resigned, and on July 24, the LDP chose Keizo Obuchi as prime minister. After Obuchi had a stroke Apr. 3, 2000, an LDP stalwart, Yoshiro Mori, succeeded him on Apr. 5. Obuchi died May 14. Parliamentary elections June 25 left the LDP and its allies with a reduced majority in the lower house. The unpopular Mori was replaced as LDP leader and prime minister in Apr. 2001 by Junichiro Koizumi, a populist reformer.

Jordan

Hashemite Kingdom of Jordan

People: Population: 5,153,378. **Age distrib.** (%): <15: 37.2; 65+: 3.3. **Pop. density:** 146 per sq. mi. **Urban:** 74%. **Ethnic groups:** Arab 98%. **Principal language:** Arabic (official), English. **Chief religions:** Sunni Muslim 96%, Christian 4%.

Geography: Area: 35,300 sq. mi. **Location:** In Middle East. **Neighbors:** Israel and West Bank on W, Saudi Arabia on S, Iraq on E, Syria on N. **Topography:** About 88% of Jordan is arid. Fertile areas are in W. Only port is on short Aqaba Gulf coast. Country shares Dead Sea (1,312 ft. below sea level) with Israel. **Capital:** Amman (urban agg.): 1,430,000.

Government: Type: Constitutional monarchy. **Head of state:** King Abdullah II; b Jan. 30, 1962; in office: Feb. 7, 1999. **Head of gov.:** Prime Min. Ali Abu al-Ragheb; b 1946; in office: June 19, 2000. **Local divisions:** 12 governorates. **Defense budget:** $488 mil. **Active troops:** 103,880.

Economy: Industries: Oil refining, cement, light manufacturing. **Chief crops:** Grains, olives, fruits. **Minerals:** Phosphates, potash. **Arable land:** 4%. **Livestock** (2000): cattle: 57,000; chickens: 25.0 mil; goats: 630,000; sheep: 1.6 mil. **Electricity prod.:** 6.657 bil kWh. **Crude oil reserves** (2000): 900,000 bbls. **Labor force:** 11.4% ind., 10% const., 52% other serv.

Finance: Monetary unit: Dinar (Sept. 2001: 0.71 = $1 U.S.). **GDP** (1999 est.): $16 bil. **Per capita GDP:** $3,500. **Imports** (1999 est.): $3.3 bil; partners: Germany, Iraq, U.S. **Exports** (1999 est.): $1.8 bil; partners: Saudi Arabia, Iraq ., India. **Tourism:** $795 mil. **Budget** (1999 est.): $3 bil. **Intl. reserves less gold** (June 2000): $2.29 bil. **Gold:** 399,000 oz t. **Consumer prices** (change in 1999): 0.6%.

Transport: Railroad: Length: 421 mi. **Motor vehicles:** 175,000 pass. cars, 90,000 comm. vehicles. **Civil aviation:** 3.0 bil pass.-mi.; 2 airports. **Chief port:** Al Aqabah.

Communications: TV sets: 176 per 1,000 pop. **Radios:** 224 per 1,000 pop. **Telephones**: 620,000 main lines. **Daily newspaper circ.:** 62 per 1,000 pop.

Health: Life expectancy: 75.1 male; 80.12 female. **Births** (per 1,000 pop): 25.44. **Deaths** (per 1,000 pop.): 2.62. **Natural inc.:** 2.28%. **Infant mortality** (per 1,000 live births): 20.36.

Education: Free, compulsory: ages 6-16. **Literacy**: 87%.

Major Intl. Organizations: UN (FAO, IBRD, ILO, IMF, IMO, WHO), AL.

Embassy: 3504 International Dr. NW 20008; 966-2664.

Website: http://www.nic.gov.jo

From ancient times to 1922 the lands to the E of the Jordan River were culturally and politically united with the lands to the W. Arabs conquered the area in the 7th century; the Ottomans took control in the 16th. Britain's 1920 Palestine Mandate covered both sides of the Jordan. In 1921, Abdullah, son of the ruler of Hejaz in Arabia, was installed by Britain as emir of an autonomous Transjordan, covering two-thirds of Palestine. An independent kingdom was proclaimed, 1946.

During the 1948 Arab-Israeli war the West Bank and East Jerusalem were added to the kingdom, which changed its name to Jordan. All these territories were lost to Israel in the 1967 war, which swelled the number of Arab refugees on the East Bank.

Some 700,000 refugees entered Jordan following Iraq's invasion of Kuwait, Aug. 1990. Jordan was viewed as supporting Iraq during the 1990-1991 Persian Gulf crisis.

Jordan and Israel officially agreed, July 25, 1994, to end their state of war; a formal peace treaty was signed Oct. 26. Following a prolonged bout with cancer, King Hussein died Feb. 7, 1999; his eldest son and designated successor immediately assumed the throne as Abdullah II.

Kazakhstan

Republic of Kazakhstan

People: Population: 16,731,303. **Age distrib.** (%): <15: 26.7; 65+: 7.2. **Pop. density:** 16 per sq. mi. **Urban:** 56%. **Ethnic groups:** Kazakh 46%, Russian 35%, Ukrainian 5%. **Principal languages:** Kazakh, Russian (both official). **Chief religions:** Muslim 47%, Russian Orthodox 44%.

Geography: Area: 1,047,900 sq. mi. **Location:** In Central Asia. **Neighbors:** Russia on N; China on E; Kyrgyzstan, Uzbekistan, Turkmenistan on S; Caspian Sea on W. **Topography:** Extends from the lower reaches of Volga in Europe to the Altay Mts. on the Chinese border. **Capital:** Astana. **Cities** (1997 est.): Alma-Ata 1,248,000; (1999 est.): Astana 303,000.

Government: Type: Republic. **Head of state:** Pres. Nursultan A. Nazarbayev; b July 6, 1940; in office: Apr. 1990. **Head of gov.:** Prime Min. Kasymzhomart Tokayev; b May 17, 1953; in office: Oct. 12, 1999. **Local divisions:** 14 oblystar, 1 city. **Defense budget:** $115 mil. **Active troops:** 64,000.

Economy: Industries: Oil, steel, mining, agricultural machinery. **Chief crops:** Grain, cotton. **Minerals:** Oil, gas, coal, iron, manganese, chrome ore, copper. **Crude oil reserves** (2000): 5.42 bil bbls. **Arable land:** 12%. **Livestock** (2000): cattle: 4.0 mil; chickens: 17.9 mil; goats: 705,400; pigs: 1.03 mil; sheep: 9.78 mil.**Fish catch:** (1999): 41,367 metric tons. **Electricity prod.:** 44.360 bil kWh. **Labor force:** 27% industry; 23% agric., forestry.

Finance: Monetary unit: Tenge (Sept. 2001: 147.76 = $1 U.S.). **GDP** (1999 est.): $54.5 bil. **Per capita GDP:** $3,200. **Imports** (1999 est.): $4.8 bil; partners: Russia 39%. **Exports** (1999 est.): $5.2 bil; partners: EU 32 %, China 29%, Russia 29%. **Tourism** (1998): $289 mil. **Budget** (1998 est.): $4.2 bil. **Intl. reserves less gold** (May 2000): $1.32 bil. **Gold:** 1.82 mil oz t. **Consumer prices** (change in 1999): 8.2%.

Transport: Railroad: Length: 13,422 mi. **Motor vehicles:** 1.0 mil pass. cars, 515,000 comm. vehicles. **Civil aviation:** 826.5 mil pass.-mi.; 20 airports. **Chief ports:** Aqtau, Atyrau.

Communications: TV sets: 275 per 1,000 pop. **Telephones** (1999): 1,759,800 main lines.

Health: Life expectancy: 57.87 male; 68.97 female. **Births** (per 1,000 pop): 17.30. **Deaths** (per 1,000 pop.): 10.61. **Natural inc.:** .67%. **Infant mortality** (per 1,000 live births): 59.17.

Education: Free, compulsory: ages 7-18. **Literacy:** 98%.

Major Intl. Organizations: UN (IBRD, ILO, IMF, IMO, WHO), CIS, OSCE.

Embassy: 1401 16th St. NW 20036; 232-5488.

Website: http://www.un.int/kazakhstan

The region came under the Mongols' rule in the 13th century and gradually came under Russian rule, 1730-1853. It was admitted to the USSR as a constituent republic 1936. Kazakhstan declared independence Dec. 16, 1991. It became an independent state when the Soviet Union dissolved Dec. 26, 1991. The party chief, Nursultan Nazarbayev, was elected president unopposed. In legislative elections Mar. 7, 1994, criticized by international monitors, his party won a sweeping victory. Kazakhstan agreed, Feb. 14, to dismantle nuclear missiles and adhere to the 1968 Nuclear Nonproliferation Treaty; the U.S. pledged increased aid. Private land ownership was legalized Dec. 26, 1995.

Astana (formerly Akmola) was dedicated as the nation's new capital on June 9, 1998. Pres. Nazarbayev won reelection to a 7-year term Jan. 10, 1999, after his leading opponent, former Prime Min. Akezhan Kazhegeldin, was barred on a technicality.

Kenya

Republic of Kenya

People: Population: 30,765,916. **Age distrib.** (%): <15: 41.9; 65+: 2.8. **Pop. density:** 140 per sq. mi. **Urban:** 32%. **Ethnic groups:** Kikuyu 22%, Luhya 14%, Luo 13%, Kalenjin 12%, Kamba 11%, others including Asian, Arab, European. **Principal languages:** Swahili, English (both official), numerous indigenous languages. **Chief religions:** Protestant 38%, Roman Catholic 28%, indigenous beliefs 26%.

Geography: Area: 219,500 sq. mi. **Location:** E Africa, on coast of Indian O. **Neighbors:** Uganda on W, Tanzania on S, Somalia on E, Ethiopia on N, Sudan on NW. **Topography:** The northern three-fifths of Kenya is arid. To the S, a low coastal area and a plateau varying from 3,000 to 10,000 ft. The Great Rift Valley enters the country N-S, flanked by high mountains. **Capital:** Nairobi. **Cities** (urban agg.): Nairobi 2,310,000; Mombasa (1991 est.) 600,000.

Government: Type: Republic. **Head of state and gov.:** Pres. Daniel arap Moi; b Sept. 2, 1924; in office: Aug. 22, 1978. **Local divisions:** Nairobi and 7 provinces. **Defense budget:** $235 mil. **Active troops:** 22,200.

Economy: Industries: Tourism, light industry, agricultural processing, oil refining. **Chief crops:** Coffee, corn, tea. **Minerals:** Gold, limestone, salt, rubies, fluorspar, garnets. **Other resources:** Hides, dairy products, cut flowers (world's 4th lgst. exporter). **Arable land:** 7%. **Livestock** (2000): cattle: 13.79 mil; chickens: 27.0 mil; goats: 9.6 mil; pigs: 170,000; sheep: 7.0 mil. **Fish catch** (1999): 161,183 metric tons. **Electricity prod.:** 4.225 bil kWh. **Labor force:** 75-80% agric.

Finance: Monetary unit: Shilling (Sept. 2001: 79.03 = $1 U.S.). **GDP** (1999 est.): $45.1 bil. **Per capita GDP:** $1,600. **Imports** (1999): $3.3 bil; partners: UK 13%, UAE 9%, U.S. 8%. **Exports** (1999): $2.2 bil; partners: Uganda 16%, UK 13%, Tanzania 13%. **Tourism:** $256 mil. **Budget** (1997 est.): $2.7 bil. **Intl. reserves less gold** (June 2000): $808.6 mil. **Consumer prices** (change in 1999): 2.6%.

Transport: Railroad: Length: 1,885 mi. **Motor vehicles:** 271,000 pass. cars, 75,900 comm. vehicles. **Civil aviation:** 1.1 bil pass.-mi.; 11 airports. **Chief ports:** Mombasa, Kisumu, Lamu.

Communications: TV sets: 18 per 1,000 pop. **Radios:** 103 per 1,000 pop. **Telephones**: 310,000 main lines.

Health: Life expectancy: 46.57 male; 48.44 female. **Births** (per 1,000 pop): 28.50. **Deaths** (per 1,000 pop.): 14.35. **Natural inc.:** 1.42%. **Infant mortality** (per 1,000 live births): 67.99.

Education: Free, compulsory: ages 6-14. **Literacy:** 78%.

Major Intl. Organizations: UN and all of its specialized agencies, the Commonwealth, OAU.

Embassy: 2249 R St. NW 20008; 387-6101.

Arab colonies exported spices and slaves from the Kenya coast as early as the 8th century. Britain obtained control in the 19th century. Kenya won independence Dec. 12, 1963, 4 years after the end of the violent Mau Mau uprising.

Kenya had steady growth in industry and agriculture under a modified private enterprise system, and enjoyed a relatively free political life. But stability was shaken in 1974-75, with opposition charges of corruption and oppression. Jomo Kenyatta, the country's leader since independence, died Aug. 22, 1978. He was succeeded by his vice president, Daniel arap Moi.

During the first half of the 1990s, Kenya suffered widespread unemployment and high inflation. Tribal clashes in the western provinces claimed thousands of lives and left tens of thousands homeless. Pres. Moi won a third term in Dec. 1992 elections, which were marred by violence and fraud. Clashes in the Mombasa region, Aug. 1997, left more than 40 people dead. Pres. Moi was reelected Dec. 29, in an election again plagued by irregularities.

A truck bomb explosion at the U.S. embassy in Nairobi, Aug. 7, 1998, killed more than 200 people and injured about 5,000. The U.S. blamed the attack and a near-simultaneous embassy bombing in Tanzania on Islamic terrorists associated with wealthy Saudi-born Osama bin Laden, believed to be sheltered in Afghanistan. After a trial in New York City, 4 conspirators were convicted May 29, 2001.

Kiribati

Republic of Kiribati

People: Population: 94,149. **Age distrib. (%): <15: 40.5; 65+: 3.2. Pop. density:** 340 per sq. mi. **Urban:** 39%. **Ethnic groups:** Micronesian. **Principal languages:** English (official), Gilbertese. **Chief religions:** Roman Catholic 53%, Protestant 41%.

Geography: Area: 277 sq. mi. **Location:** 33 Micronesian islands (the Gilbert, Line, and Phoenix groups) in the mid-Pacific scattered in a 2-mil sq. mi. chain around the point where the International Date Line formerly cut the Equator. In 1997 the Date Line was moved to follow Kiribati's E border. **Neighbors:** Nearest are Nauru to SW, Tuvalu and Tokelau Isls. to S. **Topography:** Except Banaba (Ocean) Isl., all are low-lying, with soil of coral sand and rock fragments, subject to erratic rainfall. **Capital:** Tarawa (1990 census): 25,000.

Government: Type: Republic. **Head of state and gov.:** Pres. Teburoro Tito; b Aug. 25, 1953; in office: Oct. 1, 1994. **Local divisions:** 3 units, 6 districts.

Economy: Industries: Fishing, handicrafts. **Chief crops:** Copra, taro, breadfruit, sweet potatoes, vegetables. **Livestock** (2000): chickens: 365,000; pigs: 12,000. **Fish catch:** (1999): 23,052 metric tons. **Electricity prod.:** 7 mil kWh.

Finance: Monetary unit: Australian Dollar (Sept. 2001: 1.92 = $1 U.S.). **GDP** (1999 est.): $74 mil. **Per capita GDP:** $860. **Imports** (1998 est.): $37 mil; partners: Australia 46%. **Exports** (1998 est.): $6 mil; partners: U.S., Australia, New Zealand. **Tourism** (1998): $1 mil. **Budget** (1996 est.): $47.7 mil.

Transport: Chief port: Tarawa. **Civil aviation:** 7.0 mil pass.-mi.; 17 airports.

Communications: Radios: 75 per 1,000 pop. **Telephones** (1999): 3,500.

Health: Life expectancy: 57.25 male; 63.22 female. **Births** (per 1,000 pop): 31.98. **Deaths** (per 1,000 pop.): 8.88. **Natural inc.:** 2.31%. **Infant mortality** (per 1,000 live births): 54.

Education: Free, compulsory: ages 6-14. **Literacy:** 90%.

Major Intl. Organizations: UN (IBRD, IMF, WHO), the Commonwealth.

A British protectorate since 1892, the Gilbert and Ellice Islands colony was completed with the inclusion of the Phoenix Islands, 1937. Tarawa Atoll was the scene of some of the bloodiest fighting in the Pacific during World War II.

Self-rule was granted 1971; the Ellice Islands separated from the colony 1975 and became independent Tuvalu, 1978. Kiribati (pronounced *Kiribass)* independence was attained July 12, 1979. Under a treaty of friendship the U.S. relinquished its claims to several Line and Phoenix islands, including Christmas (Kiritimati), Canton, and Enderbury. Kiribati was admitted to the UN Sept. 14, 1999.

Korea, North

Democratic People's Republic of Korea

People: Population: 21,968,228. **Age distrib.** (%): <15: 25.5; 65+: 6.8. **Pop. density:** 473 per sq. mi. **Urban:** 60%. **Ethnic group:** Korean. **Principal language:** Korean (official). **Chief religions:** Activities almost nonexistent; traditionally Buddhism, Confucianism, Chondogyo.

Geography: Area: 46,400 sq. mi. **Location:** In northern E Asia. **Neighbors:** China and Russia on N, South Korea on S. **Topography:** Mountains and hills cover nearly all the country, with narrow valleys and small plains in between. The N and the E coasts are the most rugged areas. **Capital:** Pyongyang. **Cities (urban agg.):** Pyongyang 3,197,000; Nampo 1,046,000.

Government: Type: Communist state. **Leader:** Kim Jong Il; b Feb. 16, 1942; officially assumed post Oct. 8, 1997. **Local divisions:** 9 provinces, 3 special cities. **Defense**: $1.3 bil. **Active troops:** 1,082,000.

Economy: Industries: Textiles, chemicals, machinery, food processing. **Chief crops:** Corn, potatoes, soybeans, rice. **Minerals:** Coal, lead, tungsten, zinc, graphite, magnesite, iron, copper, gold, salt. **Arable land:** 14%. **Livestock** (2000): cattle: 600,000; chickens: 10.37 mil; goats: 2.1 mil; pigs: 2.97 mil; sheep: 190,000. **Fish catch** (1999): 306,636 metric tons. **Electricity prod.:** 28.600 bil kWh. **Labor force:** 36% agric.

Finance: Monetary unit: Won (Sept. 2001: 2.20 = $1 U.S.). **GDP** (1999 est.): $22.6 bil. **Per capita GDP:** $1,000. **Imports** (1998 est.): $954 mil; partners: China 33%, Japan 17%. **Exports** (1998 est.): $680 mil; partners: Japan 28%, South Korea 21%.

Transport: Railroad: Length: 5,302 mi. **Civil aviation:** 177.5 mil pass.-mi.; 1 airport. **Chief ports:** Chongjin, Hamhung, Nampo.

Communications: TV sets: 85 per 1,000 pop. **Radios:** 200 per 1,000 pop. **Telephones** (1998): 1,100,000 main lines. **Daily newspaper circ.:** 213 per 1,000 pop.

Health: Life expectancy: 68.04 male; 74.15 female. **Births** (per 1,000 pop): 19.10. **Deaths** (per 1,000 pop.): 6.92. **Natural inc.:** 1.22%. **Infant mortality** (per 1,000 live births): 23.55.

Education: Free, compulsory: ages 6-17. **Literacy** (1992): 95%.

Major Intl. Organizations: UN (FAO, IMO, WHO).

The Democratic People's Republic of Korea was founded May 1, 1948, in the zone occupied by Russian troops after World War II. Its armies tried to conquer the south, 1950. After 3 years of fighting, with Chinese and U.S. intervention, a cease-fire was proclaimed. For the next four decades, a hardline Communist regime headed by Kim Il Sung kept tight control over the nation's political, economic, and cultural life. The nation used its abundant mineral and hydroelectric resources to develop its military strength and heavy industry.

In Mar. 1993, North Korea became the first nation to formally withdraw from the Nuclear Nonproliferation Treaty, the international pact designed to limit the spread of nuclear weapons. The nation suspended its withdrawal in June in reaction to threats of UN economic sanctions, but was widely believed to be developing nuclear weapons. The U.S. and North Korea reached an interim agreement, Aug. 13, 1994, intended to resolve the nuclear issue, and further negotiations followed.

Kim Il Sung died July 8, 1994. He was succeeded by his son, Kim Jong Il. North Korea suffered from defections by high officials, a deteriorating economy, and severe food shortages in the late 1990s. On Sept. 17, 1999, the U.S. eased travel and trade restrictions on North Korea after Pyongyang agreed to suspend long-range missile testing. A first-ever summit conference in Pyongyang between North and South Korean leaders, June 13-15, 2000, marked an unexpected improvement in relations between the 2 Koreas, and brought an end to many U.S. sanctions.

Korea, South

Republic of Korea

People: Population: 47,904,370. **Age distrib.** (%): <15: 21.6; 65+: 7.3. **Pop. density:** 1,265 per sq. mi. **Urban:** 81%. **Ethnic group:** Korean. **Principal language:** Korean (official). **Chief religions:** Christianity 49%, Buddhism 47%.

Geography: Area: 37,900 sq. mi. **Location:** In northern E Asia. **Neighbors:** North Korea on N. **Topography:** The country is mountainous, with a rugged east coast. The western and southern coasts are deeply indented, with many islands and harbors. **Capital:** Seoul. **Cities (urban agg.)**: Seoul 9,888,000; Pusan 3,830,000; Inch'on 2,884,000: Taegu 2,675,000.

Government: Type: Republic, with power centralized in a strong executive. **Head of state:** Pres. Kim Dae Jung; b Dec. 3, 1925; in office: Feb. 25, 1998. **Head of gov.:** Prime Min. Lee Han Dong; b 1934; in office: May 22, 2000. **Local divisions:** 9 provinces, 6 special cities. **Defense budget:** $12.8 bil. **Active troops:** 683,000

Economy: Industries: Electronics, autos, chemicals, ships, textiles, clothing. **Chief crops:** Rice, barley, vegetables. **Minerals:** Tungsten, coal, graphite. **Arable land:** 19%. **Livestock** (2000): cattle: 2.49 mil; chickens: 97.0 mil; goats: 505,050; pigs: 7.86 mil; sheep: 1,355. **Fish catch** (1999): 2.60 mil metric tons. **Electricity prod.:** 250.287 bil kWh. **Labor force:** 68% services & other; 20% manuf. & mining; 12% agric.

Finance: Monetary unit: Won (Sept. 2001: 1,285 = $1 U.S.). **GDP** (1999 est.): $625.7 bil. **Per capita GDP:** $13,300. **Imports** (1999): $116 bil; partners: U.S. 22%, Japan 18%, China 7%. **Exports** (1999): $144 bil; partners: U.S. 17%, Japan 9%, China 9%. **Tourism**: $5.62 bil. **Budget** (1997 est.): $100.5 bil. **Intl. reserves less gold** (June 2000): $90.11 bil. **Gold:** 438,000 oz t. **Consumer prices** (change in 1999): 0.8%.

Transport: Railroad: Length: 4,072 mi. **Motor vehicles** (1997): 7.59 mil pass. cars, 2.83 mil comm. vehicles. **Civil aviation:** 34.62 bil pass.-mi.; 14 airports. **Chief ports:** Pusan, Inchon.

Communications: TV sets: 337 per 1,000 pop. **Radios:** 1,037 per 1,000 pop. **Telephones**: 21,931,700 main lines. **Daily newspaper circ.:** 394 per 1,000 pop.

Health: Life expectancy: 70.97 male; 78.74 female. **Births** (per 1,000 pop): 14.85. **Deaths** (per 1,000 pop.): 5.93. **Natural inc.:** .89%. **Infant mortality** (per 1,000 live births): 7.71.

Education: Free, compulsory: ages 6-12. **Literacy**: 98%.

Major Intl. Organizations: UN (FAO, IBRD, ILO, IMF, IMO, WHO, WTrO), APEC, OECD.

Embassy: 2450 Massachusetts Ave. NW 20008; 939-5600.

Korea, once called the Hermit Kingdom, has a recorded history since the 1st century BC. It was united in a kingdom under the Silla Dynasty, AD 668. It was at times associated with the Chinese empire; the treaty that concluded the Sino-Japanese war of 1894-95 recognized Korea's complete independence. In 1910 Japan forcibly annexed Korea as Chosun.

At the Potsdam conference, July 1945, the 38th parallel was designated as the line dividing the Soviet and the American occupation. Russian troops entered Korea Aug. 10, 1945; U.S. troops entered Sept. 8, 1945. The Soviet military organized socialists and Communists and blocked efforts to let the Koreans unite their country.

The South Koreans formed the Republic of Korea in May 1948 with Seoul as the capital. Dr. Syngman Rhee was chosen president. A separate, Communist regime was formed in the N; its army attacked the S in June 1950, initiating the Korean War. UN troops, under U.S. command, supported the S in the war, which ended in an armistice (July 1953) leaving Korea divided by a "no-man's land" along the 38th parallel.

Rhee's authoritarian rule became increasingly unpopular, and a movement spearheaded by college students forced his resignation Apr. 26, 1960. In an army coup May 16, 1961, Gen. Park Chung Hee became chairman of a ruling junta. He was elected president, 1963; a 1972 referendum allowed him to be reelected for an unlimited series of 6-year terms. Park was assassinated by the chief of the Korean CIA, Oct. 26, 1979. In May 1980, Gen. Chun Doo Hwan, head of military intelligence, reinstated full martial law and ordered the brutal suppression of pro-democracy demonstrations in Kwangju.

In July 1972 South and North Korea agreed on a common goal of reunifying the 2 nations by peaceful means. But there was no sign of a thaw in relations between the two regimes until 1985, when they agreed to discuss economic issues.

On June 10, 1987, middle-class office workers, shopkeepers, and business executives joined with students in antigovernment protests in Seoul calling for democratic reforms. Following weeks of rioting and violence, Chun, July 1, agreed to permit election of the next president by direct popular vote and other reforms. In Dec., Roh Tae Woo was elected president. In 1990, the nation's 3 largest political parties merged; some 100,000 students protested the merger as undemocratic.

Kim Young Sam took office in 1993 as the first civilian president since 1961. Convicted of mutiny, treason, and corruption, Chun was sentenced to death by a Seoul court, Aug. 26, 1996, for his role in the 1979 coup and 1980 Kwangju massacre; Roh received a 22-1/2 year prison sentence. On Dec. 16, Chun's term was reduced to life in prison, and Roh's to 17 years.

The collapse in Jan. 1997 of the Hanbo steel firm triggered a new round of corruption scandals. With currency and stock values plummeting, the nation averted default by agreeing, Dec. 4, on a $57 billion bailout from the IMF. Kim Dae Jung, a longtime dissident, won the presidential election Dec. 18. Chun and Roh were released and pardoned Dec. 22, 1997.

At an unprecedented summit meeting in Pyongyang, June 13-15, 2000, Pres. Kim Dae Jung and North Korean leader Kim Jong Il agreed to work for reconciliation and eventual reunification of their 2 countries. On Oct. 13, 2000, Kim Dae Jung was named the winner of the 2000 Nobel Peace Prize.

Kuwait

State of Kuwait

People: Population: 2,041,961. **Age distrib.** (%): <15: 28.8; 65+: 2.4. **Pop. density:** 297 per sq. mi. **Urban:** 98%. **Ethnic groups:** Kuwaiti 45%, other Arab 35%. **Principal languages:** Arabic (official), English. **Chief religion:** Muslim 85%.

Geography: Area: 6,900 sq. mi. **Location:** In Middle East, at N end of Persian Gulf. **Neighbors:** Iraq on N, Saudi Arabia on S. **Topography:** The country is flat, very dry, and extremely hot. **Capital:** Kuwait City (1995): 28,559.

Government: Type: Constitutional monarchy. **Head of state:** Emir Sheikh Jabir al-Ahmad al-Jabir as-Sabah; b 1928; in office: Jan. 1, 1978. **Head of gov.:** Prime Min. Sheikh Saad Abdulla as-Salim as-Sabah; b 1930; in office: Feb. 8, 1978. **Local divisions:** 5 governorates. **Defense budget:** $2.6 bil. **Active troops:** 15,300.

Economy: Industries: Oil products. **Minerals:** Oil, gas. **Crude oil reserves** (2000): 94 bil bbls. **Livestock** (2000): cattle: 20,400; chickens: 33.0 mil; goats: 150,000; sheep: 450,000. **Electricity prod.:** 31.567 bil kWh. **Labor force:** 50% gov't. and social services; 40% services; 10% industry and agric.

Finance: Monetary unit: Dinar (Sept. 2001: 0.31 = $1 U.S.). **GDP** (1999 est.): $44.8 bil. **Per capita GDP:** $22,500. **Imports** (1999): $8.1 bil; partners: U.S. 22%, Japan 15%, UK 13%. **Exports** (1999): $13.5 bil; partners: Japan 24%, India 16%, U.S. 13%. **Tourism** (1998): $207 mil. **Budget** (FY 1998-99 est.): $14.5 bil. **Intl. reserves less gold** (June 2000): $5.71 bil. **Gold:** 2.54 mil oz t. **Consumer prices** (change in 1999): 3.0%.

Transport: Motor vehicles: 538,000 pass. cars, 155,000 comm. vehicles. **Civil aviation:** 3.7 bil pass.-mi.; 1 airport. **Chief port:** Mina al-Ahmadi.

Communications: TV sets: 505 per 1,000 pop. **Radios:** 678 per 1,000 pop. **Telephones**: 467,100 main lines. **Daily newspaper circ.:** 377 per 1,000 pop.

Health: Life expectancy: 75.42 male; 77.15 female. **Births** (per 1,000 pop): 21.91. **Deaths** (per 1,000 pop.): 2.45. **Natural inc.:** 1.95%. **Infant mortality** (per 1,000 live births): 11.18.

Education: Free, compulsory: ages 6-14. **Literacy**: 79%.

Major Intl. Organizations: UN (FAO, IBRD, ILO, IMF, IMO, WHO, WTrO), AL, OPEC.

Embassy: 2940 Tilden St. NW 20008; 966-0702.

Kuwait is ruled by the Al-Sabah dynasty, founded 1759. Britain ran foreign relations and defense from 1899 until independence in 1961. The majority of the population is non-Kuwaiti, with many Palestinians, and cannot vote.

Oil is the fiscal mainstay, providing most of Kuwait's income. Oil pays for free medical care, education, and social security. There are no taxes, except customs duties.

Kuwaiti oil tankers came under frequent attack by Iran because of Kuwait's support of Iraq in the Iran-Iraq War. In July 1987, U.S. Navy warships began escorting Kuwaiti tankers in the Persian Gulf.

Kuwait was attacked and overrun by Iraqi forces Aug. 2, 1990. The emir and senior members of the ruling family fled to Saudi Arabia to establish a government in exile. On Aug. 28, Iraq announced that Kuwait was its 19th province. Following several weeks of aerial attacks on Iraq and Iraqi forces in Kuwait, a U.S.-led coalition began a ground attack Feb. 23, 1991. By Feb. 27, Iraqi forces were routed and Kuwait liberated. Following liberation, there were reports of abuse of Palestinians and others suspected of collaborating with Iraqi occupiers.

Former U.S. Pres. George Bush visited Kuwait, Apr. 14-16, 1993, and was honored as the leader of the Persian Gulf War alliance that expelled Iraqi troops. Kuwaiti authorities arrested 14 Iraqis and Kuwaitis for allegedly plotting to assassinate Bush during his visit; 13 were convicted and sentenced to prison or death, June 4, 1994. The UN Security Council ruled, Sept. 27, 2000, that Iraq had to pay the Kuwait Petroleum Corp. $15.9 billion for damage to Kuwaiti oil fields during the Persian Gulf War.

Kyrgyzstan

Kyrgyz Republic

People: Population: 4,753,003. **Age distrib. (%):** <15: 35.0; 65+: 6.1. **Pop density:** 62 per sq. mi. **Urban:** 33%. **Ethnic groups:** Kyrgyz 52%, Russian 18%, Uzbek 13%. **Principal languages:** Kyrgyz, Russian (both official). **Chief religions:** Muslim 75%, Russian Orthodox 20%.

Geography: Area: 76,600 sq. mi. **Location:** In Central Asia. **Neighbors:** Kazakhstan on N, China on E, Uzbekistan on W, Tajikistan on S. **Capital:** Bishkek (1999 est.): 619,000.

Government: Type: Republic. **Head of state:** Pres. Askar Akayev; b Nov. 10, 1944; in office: Oct. 28, 1990. **Head of gov.:** Prime Min. Kurmanbek Saliyevich Bakiyev; b Aug. 1, 1949; in office: Dec. 21, 2000. **Local divisions:** 6 oblasts, 1 city. **Defense budget:** $29 mil. **Active troops:** 9,000.

Economy: Industries: Textiles, mining, food processing, cement, small machinery. **Chief crops:** Tobacco, cotton, fruits. **Minerals:** Gold, coal, oil. **Crude oil reserves** (2000): 40 mil bbls. **Arable land:** 7%. **Livestock** (2000): cattle: 932,273; chickens: 2.3 mil; goats: 234,000; pigs: 104,830; sheep: 3.26 mil. **Electricity prod.:** 12.981 bil kWh. **Labor force:** 55% agric. & forestry; 30% services; 15% ind. & const.

Finance: Monetary unit: Som (Oct. 2000: 46.98 = $1 U.S.). **GDP** (1999 est.): $10.3 bil. **Per capita GDP:** $2,300. **Imports** (1999 est.): $590 mil; partners: Russia 24%, Uzbekistan 14%, Kazakhstan 9%. **Exports** (1999 est.): $515 mil; partners: Germany 37%, Kazakhstan 17%, Russia 16%. **Tourism** (1998): $7 mil. **Budget** (1996 est.): $308 mil. **Intl. reserves less gold** (June 2000): $257.3 mil. **Gold:** 83,100 oz t. **Consumer prices** (change in 1999): 35.9%.

Transport: Railroad: Length: 249 mi. **Motor vehicles:** 164,000 pass. cars. **Civil aviation:** 280.7 mil pass.-mi.; 2 airports. **Chief port:** Ysyk-Kol.

Communications: TV sets: 238 per 1,000 pop. **Telephones:** 376,100 main lines. **Daily newspaper circ.:** 11 per 1,000 pop.

Health: Life expectancy: 59.2 male; 67.94 female. **Births** (per 1,000 pop): 26.18. **Deaths** (per 1,000 pop.): 9.13. **Natural inc.:** 1.71%. **Infant mortality** (per 1,000 live births): 76.5.

Education: Compulsory: ages 6-15. **Literacy** (1993): 97%.

Major Intl. Organizations: UN (FAO, IBRD, ILO, IMF, WHO), CIS, OSCE.

Embassy: 1732 Wisconsin Ave. NW, 20007; 338-5141.

The region was inhabited around the 13th century by the Kyrgyz. It was annexed to Russia 1864. After 1917, it was nominally a Kara-Kyrgyz autonomous area, which was reorganized 1926, and made a constituent republic of the USSR in 1936. Kyrgyzstan declared independence Aug. 31, 1991. It became an independent state when the USSR disbanded Dec. 26, 1991. A constitution was adopted May 5, 1993.

Reelected Dec. 24, 1995, Pres. Askar Akayev gained approval by referendum of a constitutional amendment expanding his presidential powers, Feb. 10, 1996. Amendments restricting the powers of parliament and allowing private ownership of land were ratified by referendum Oct. 17, 1998. Akayev won a 3d 5-year term in the Oct. 29, 2000, election.

Laos

Lao People's Democratic Republic

People: Population: 5,635,967. **Age distrib.** (%): <15: 42.7; 65+: 3.3. **Pop. density:** 63 per sq. mi. **Urban:** 23%. **Ethnic groups:** Lao Loum 68%, Lao Theung 22%, Lao Soung (includes Hmong and Yao) 9%. **Principal languages:** Lao (official), French, English. **Chief religions:** Buddhism 60%, animist and other 40%.

Geography: Area: 89,000 sq. mi. **Location:** In Indochina Peninsula in SE Asia. **Neighbors:** Myanmar and China on N, Vietnam on E, Cambodia on S, Thailand on W. **Topography:** Landlocked, dominated by jungle. High mountains along eastern border are the source of the E-W rivers slicing across the country to the Mekong R., which defines most of the western border. **Capital:** Vientiane (1999 urban agg.): 640,000.

Government: Type: Communist. **Head of state:** Pres. Khamtai Siphandon; b Feb. 8, 1924; in office: Feb. 24, 1998. **Head of gov.:** Prime Min. Boungnang Vorachith; in office: Mar. 27, 2001. **Local divisions:** 16 provinces, 1 municipality, 1 special zone. **Defense budget:** $15 mil (1999). **Active troops:** 29,100.

Economy: Industries: Wood products, mining. **Chief crops:** Sweet potatoes, corn, cotton, vegetables, coffee. **Minerals:** Gypsum, tin, gold. **Arable land:** 3%. **Livestock** (2000): cattle: 986,600; chickens: 12.03 mil; goats: 100,300; pigs: 1.1 mil. **Fish catch:** (1999): 40,000 metric tons. **Electricity prod.:** 792 mil kWh. **Labor force:** 80% agric.

Finance: Monetary unit: Kip (Sept. 2001: 7,900.00 = $1 U.S.). **GDP** (1999 est.): $7 bil. **Per capita GDP:** $1,300. **Imports** (1999): $497 mil; partners: Thailand, Japan, Vietnam. **Exports** (1999): $271 mil; partners: Vietnam, Thailand. **Tourism:** $103 mil. **Budget** (1996): $365.9 mil. **Intl. reserves less gold** (Dec. 1999): $101.19 mil. **Gold:** 17,100 oz t. **Consumer prices** (change in 1999): 128.4%.

Transport: Motor vehicles: 9,000 pass. cars, 9,000 comm. vehicles. **Civil aviation:** 29.9 mil pass.-mi.; 11 airports.

Communications: TV sets: 17 per 1,000 pop. **Radios:** 116 per 1,000 pop. **Telephones:** 40,900 main lines.

Health: Life expectancy: 51.58 male; 55.44 female. **Births** (per 1,000 pop): 37.84. **Deaths** (per 1,000 pop.): 13.02. **Natural inc.:** 2.48%. **Infant mortality** (per 1,000 live births): 92.89.

Education: Compulsory for 5 years between ages 6-15. **Literacy:** 57%.

Major Intl. Organizations: UN (FAO, IBRD, ILO, IMF, WHO), ASEAN.

Embassy: 2222 S St. NW 20008; 332-6416.

Website: http://www.laoembassy.com/discover/index.htm

Laos became a French protectorate in 1893, but regained independence as a constitutional monarchy July 19, 1949.

Conflicts among neutralist, Communist, and conservative factions created a chaotic political situation. Armed conflict increased after 1960.

The 3 factions formed a coalition government in June 1962, with neutralist Prince Souvanna Phouma as premier. A 14-nation conference in Geneva signed agreements, 1962, guaranteeing neutrality and independence. By 1964 the Pathet Lao had withdrawn from the coalition, and, with aid from North Vietnamese troops, renewed sporadic attacks. U.S. planes bombed the Ho Chi Minh trail, supply line from North Vietnam to Communist forces in Laos and South Vietnam.

In 1970 the U.S. stepped up air support and military aid. After Pathet Lao military gains, Souvanna Phouma in May 1975 ordered government troops to cease fighting; the Pathet Lao took control. The Lao People's Democratic Republic was proclaimed Dec. 3, 1975.

From the mid-1970s through the 1980s, the Laotian government relied on Vietnam for military and financial aid. Since easing its foreign investment laws in 1988, Laos has attracted more than $5 billion from Thailand, the U.S., and other nations. Laos was admitted to ASEAN on July 23, 1997.

Latvia

Republic of Latvia

People: Population: 2,385,231. **Age distrib.** (%): <15: 16.6 65+: 15.3. **Pop density:** 96 per sq. mi. **Urban:** 69%. **Ethnic groups:** Latvian 57%, Russian 30%. **Principal languages:** Lettish (official), Lithuanian, Russian. **Chief religions:** Lutheran, Roman Catholic, Russian Orthodox.

Geography: Area: 24,900 sq. mi. **Location:** E Europe, on the Baltic Sea. **Neighbors:** Estonia on N, Lithuania and Belarus on S, Russia on E. **Capital:** Riga (urban agg.): 775,000.

Government: Type: Republic. **Head of state:** Pres. Vaira Vike-Freiberga; b Dec. 1, 1937; in office: July 8, 1999. **Head of gov.:** Prime Min. Andris Berzins; b Aug. 4, 1951; in office; May 5, 2000. **Local divisions:** 26 counties, 7 municipalities. **Defense budget:** $72 mil (2001). **Active troops:** 5,050.

Economy: Industries: Machinery, vehicles, railway cars. **Chief crops:** Grains, sugar beets, potatoes. **Minerals:** Amber, peat. **Arable land:** 27%. **Livestock** (2000): cattle: 378,400; chickens: 2.8 mil; goats: 8,100; pigs: 404,900; sheep: 27,000. **Fish catch** (1999): 106,027 metric tons. **Electricity prod.:** 3.996 bil kWh. **Labor force:** 43% services; 41% ind.; 16% agric. & forestry.

Finance: Monetary unit: Lat (Sept. 2001: 0.62 = $1 U.S.). **GDP** (1999 est.): $9.8 bil. **Per capita GDP:** $4,200. **Imports** (1998): $2.8 bil; partners: Germany 17%, Russia 12%, Finland 10%. **Exports** (1999): $1.9 bil; partners: Germany 16%, UK 14%, Russia 12%. **Tourism:** $111 mil. **Budget** (1998 est.): $1.27 bil. **Intl. reserves less gold** (June 2000): $831.97 mil. **Gold:** 248,700 oz t. **Consumer prices** (change in 1999): 2.4%.

Transport: Railroad: Length: 1,499 mi. **Motor vehicles:** 252,000 pass. cars, 74,000 comm. vehicles. **Civil aviation:** 135.0 mil pass.-mi.; 1 airport. **Chief port:** Riga.

Communications: TV sets: 452 per 1,000 pop. **Radios:** 560 per 1,000 pop. **Telephones** (1999): 731,500 main lines. **Daily newspaper circ.:** 235 per 1,000 pop.

Health: Life expectancy: 62.8 male; 74.9 female. **Births** (per 1,000 pop): 8.03. **Deaths** (per 1,000 pop.): 14.8. **Natural inc.:** –.68%. **Infant mortality** (per 1,000 live births): 15.34.

Education: Compulsory: ages 7-16. **Literacy** (1989): 100%.

Major Intl. Organizations: UN (FAO, IBRD, ILO, IMF, IMO, WHO), OSCE.

Embassy: 4325 17th St. NW 20011; 726-8213.

Websites: http://www.latvia-usa.org
http://www.csb.lv

Prior to 1918, Latvia was occupied by the Russians and Germans. It was an independent republic, 1918-39. The Aug. 1939 Soviet-German agreement assigned Latvia to the Soviet sphere of influence. It was officially accepted as part of the USSR on Aug. 5, 1940. It was overrun by the German army in 1941, but retaken in 1945.

During an abortive Soviet coup, Latvia declared independence, Aug. 21, 1991. The Soviet Union recognized Latvia's independence in Sept. 1991. The last Russian troops in Latvia withdrew by Aug. 31, 1994. Responding to international pressure, Latvian voters on Oct. 3, 1998, eased citizenship laws that had discriminated against some 500,000 ethnic Russians. On June 17, 1999, the legislature elected Vaira Vike-Freiberga as Latvia's 1st woman president.

Lebanon

Republic of Lebanon

People: Population: 3,627,774. **Age distrib.** (%): <15: 27.6; 65+: 6.7. **Pop. density:** 920 per sq. mi. **Urban:** 89%. **Ethnic groups:** Arab 95%, Armenian 4%. **Principal languages:** Arabic (official), French, English, Armenian. **Chief religions:** Islam 70%, Christian 30%.

Geography: Area: 3,900 sq. mi. **Location:** In Middle East, on E end of Mediterranean Sea. **Neighbors:** Syria on E, Israel on S. **Topography:** There is a narrow coastal strip, and 2 mountain ranges running N-S enclosing the fertile Beqaa Valley. The Litani R. runs S through the valley, turning W to empty into the Mediterranean. **Capital:** Beirut (urban agg.): 2,055,000.

Government: Type: Republic. **Head of state:** Pres. Emile Lahoud; b 1936; in office: Nov. 24, 1998. **Head of gov.:** Prime Min. Rafiq al-Hariri; b 1944; in office: Oct. 23, 2000. **Local divisions:** 5 governorates. **Defense budget:** $560 mil (1999). **Active troops:** 63,570.

Economy: Industries: Banking, food products, textiles, cement, oil refining. **Chief crops:** Citrus, olives, tobacco, potatoes, vegetables. **Minerals:** Limestone, iron. **Arable land:** 21%. **Livestock** (2000): cattle: 77,000; chickens: 32.0 mil; goats: 445,000; pigs: 63,500; sheep: 380,000. **Electricity prod.:** 7.748 bil kWh. **Labor force:** 62% services; 31% industry; 7% agric.

Finance: Monetary unit: Pound (Sept. 2001: 1,514 = $1 U.S.). **GDP** (1999 est.): $16.2 bil. **Per capita GDP:** $4,500. **Imports** (1999): $5.7 bil; partners: Italy 12%, France 10%, U.S. 9%. **Exports** (1999): $866 mil; partners: Saudi Arabia 12%, UAE 10%, France 9%. **Tourism:** $807 mil. **Budget** (1998 est): $7.9 bil. **Intl. reserves less gold** (June 2000): $6.87 bil. **Gold:** 9.22 mil oz t.

Transport: Railroad: Length: 138 mi. **Motor vehicles:** 1.1 mil pass. cars, 83,000 comm. vehicles. **Civil aviation:** 1.3 bil pass.-mi.; 1 airport. **Chief ports:** Beirut, Tripoli, Sidon.

Communications: TV sets: 291 per 1,000 pop. **Radios:** 608 per 1,000 pop. **Telephones:** 698,000 main lines. **Newspaper circ.:** 172 per 1,000 pop.

Health: Life expectancy: 69.13 male; 74.03 female. **Births** (per 1,000 pop): 20.16. **Deaths** (per 1,000 pop.): 6.39. **Natural inc.:** 1.38%. **Infant mortality** (per 1,000 live births): 28.35.

Education: Literacy: 92%.

Major Intl. Organizations: UN (FAO, IBRD, ILO, IMF, IMO, WHO), AL.

Embassy: 2560 28th St. NW 20008; 939-6300.

Website: http://www.erols.com/lebanon/stat.htm

Formed from 5 former Turkish Empire districts, Lebanon became an independent state Sept. 1, 1920, administered under French mandate 1920-41. French troops withdrew in 1946.

Under the 1943 National Covenant, all public positions were divided among the various religious communities, with Christians in the majority. By the 1970s, Muslims became the majority and demanded a larger political and economic role.

U.S. Marines intervened, May-Oct. 1958, during a Syrian-aided revolt. Continued raids against Israeli civilians, 1970-75, brought Israeli attacks against guerrilla camps and villages. Israeli troops occupied S Lebanon, Mar. 1978, and again in Apr. 1980.

An estimated 60,000 were killed and billions of dollars in damage inflicted in a 1975-76 civil war. Palestinian units and leftist Muslims fought against the Maronite militia, the Phalange, and other Christians. Several Arab countries provided political and arms support to the various factions, while Israel aided Christian forces. Up to 15,000 Syrian troops intervened in 1976 to fight Palestinian groups. A cease-fire was mainly policed by Syria.

New clashes between Syrian troops and Christian forces erupted, Apr. 1, 1981. By Apr. 22, fighting had also broken out between two Muslim factions. In July, Israeli air raids on Beirut killed or wounded some 800 persons.

Israeli forces invaded Lebanon June 6, 1982, in a coordinated land, sea, and air attack aimed at crushing strongholds of the Palestine Liberation Organization (PLO). Israeli and Syrian forces engaged in the Bekaa Valley. By June 14, Israeli troops had encircled Beirut. On Aug. 21, the PLO evacuated west Beirut after massive Israeli bombings there. Israeli troops entered west Beirut following the Sept. 14 assassination of newly elected Lebanese Pres. Bashir Gemayel. On Sept. 16, Lebanese Christian troops entered 2 refugee camps and massacred hundreds of Palestinian refugees. An agreement May 17, 1983, between Lebanon, Israel, and the U.S. (but not Syria) provided for the withdrawal of Israeli troops; at least 30,000 Syrian troops remained in Lebanon, and Israeli forces continued to occupy a "security zone" in the south.

In 1983, terrorist bombings became a way of life in Beirut as some 50 people were killed in an explosion at the U.S. Embassy, Apr. 18; 241 U.S. servicemen and 58 French soldiers died in separate Muslim suicide attacks, Oct. 23.

Kidnapping of foreign nationals by Islamic militants became common in the 1980s. U.S., British, French, and Soviet citizens were victims. All were released by 1992.

A treaty signed May 22, 1991, between Lebanon and Syria recognized Lebanon as a separate state for the first time since the 2 countries gained independence in 1943.

Israeli forces conducted air raids and artillery strikes against guerrilla bases and villages in S Lebanon, causing over 200,000 to flee their homes July 25-29, 1993. Some 500,000 civilians fled their homes in Apr. 1996 when Israel again struck suspected guerrilla bases in the south. Pope John Paul II visited Lebanon May 10-11, 1997. During May-June 1998 the nation held its 1st municipal elections in 35 years. With Syria's approval, the legislature unanimously elected Lebanese armed forces chief Emile Lahoud as president Oct. 15.

Israel withdrew virtually all its troops from S Lebanon by May 24, 2000, leaving Hezbollah, an Iranian-backed guerrilla group, in control of much of the region.

Lesotho

Kingdom of Lesotho

People: Population: 2,177,062. **Age distrib.** (%): <15: 39.3; 65+: 4.7. **Pop. density:** 186 per sq. mi. **Urban:** 27%. **Ethnic groups:** Sotho 99.7%. **Principal languages:** English, Sesotho (both official). **Chief religions:** Christian 80%, indigenous beliefs 20%.

Geography: Area: 11,700 sq. mi. **Location:** In southern Africa. **Neighbors:** Completely surrounded by Republic of South Africa. **Topography:** Landlocked and mountainous, altitudes from 5,000 to 11,000 ft. **Capital:** Maseru (1999 est.): 373,000.

Government: Type: Modified constitutional monarchy. **Head of state:** King Letsie III; b July 17, 1963; in office: Feb. 7, 1996. **Head of gov.:** Pakalitha Mosisili; b Mar. 14, 1945; in office: May 29, 1998. **Local divisions:** 10 districts. **Defense budget:** $26 mil. **Active troops:** 2,000.

Economy: Industries: Food, textiles. **Chief crops:** Corn, grains, pulses, sorghum. **Other resources:** Diamonds. **Arable land:** 11%. **Livestock** (2000): cattle: 520,000; chickens: 1.8 mil; goats: 580,000; pigs: 65,000; sheep: 750,000. **Labor force:** 86% subsistence agric.

Finance: Monetary unit: Maloti (Sept. 2001: 8.45 = $1 U.S.). **GDP:** (1998 est.): $4.7 bil. **Per capita GDP:** $2,240. **Imports** (1998 est.): $700 mil; partners: South Africa 90%. **Exports** (1998 est.): $235 mil; partners: South Africa 65%. **Tourism:** $19 mil. **Budget** (FY 1996-97): $487 mil. **Intl. reserves less gold** (May 2000): $432.48 mil. **Consumer prices** (change in 1999): 7.3%.

Transport: Motor vehicles: 5,000 pass. cars, 18,000 comm. vehicles. **Civil aviation:** 5.7 mil pass.-mi.

Communications: TV sets: 7 per 1,000 pop. **Radios:** 558 per 1,000 pop. **Telephones:** 23,700 main lines. **Daily newspaper circ.:** 7 per 1,000 pop.

Health: Life expectancy: 47.97 male; 49.74 female. **Births** (per 1,000 pop): 31.24. **Deaths** (per 1,000 pop.): 15.7. **Natural inc.:** 1.55%. **Infant mortality** (per 1,000 live births): 82.77.

Education: Free, compulsory: ages 6-13. **Literacy:** 71%.

Major Intl. Organizations: UN (FOA, IBRD, ILO, IMF, WHO, WTrO), the Commonwealth, OAU.

Embassy: 2511 Massachusetts Ave. NW 20008; 797-5533.

Lesotho (once called Basutoland) became a British protectorate in 1868 when Chief Moshesh sought protection against the Boers. Independence came Oct. 4, 1966. Elections were suspended in 1970. Most of Lesotho's GNP is provided by citizens working in South Africa. Livestock raising is the chief industry; diamonds are the chief export.

South Africa imposed a blockade, Jan. 1, 1986, because Lesotho had given sanctuary to anti-apartheid groups. The blockade sparked a Jan. 20 military coup, and was lifted, Jan. 25, when the new leaders agreed to expel the rebels.

In Mar. 1990, King Moshoeshoe was exiled by the military government. Letsie III became king Nov. 12. In Mar. 1993, Ntsu Mokhehle, a civilian, was elected prime minister, ending 23 years of military rule. After a series of violent disturbances, the king dismissed the Mokhele government Aug. 17, 1994; constitutional rule was restored Sept. 14. Letsie abdicated and Moshoeshoe was reinstated Jan. 25, 1995.

Moshoeshoe died in an automobile accident, Jan. 15, 1996. Letsie was reinstated Feb. 7; his formal coronation was Oct. 31, 1997. South Africa and Botswana sent troops Sept. 22, 1998, to help suppress violent antigovernment protests.

According to UN estimates, nearly one-fourth of the adult population has HIV/AIDS.

Liberia

Republic of Liberia

People: Population: 3,225,837. **Age distrib.** (%): <15: 43.2; 65+: 3.5. **Pop. density:** 87 per sq. mi. **Urban:** 44%. **Ethnic groups:** Indigenous tribes 95%, Americo-Liberians 2.5%. **Principal languages:** English (official), tribal languages. **Chief religions:** Traditional beliefs 70%, Muslim 20%, Christian 10%.

Geography: Area: 37,100 sq. mi. **Location:** On SW coast of W Africa. **Neighbors:** Sierra Leone on W, Guinea on N, Côte d'Ivoire on E. **Topography:** Marshy Atlantic coastline rises to low mountains and plateaus in the forested interior; 6 major rivers flow in parallel courses to the ocean. **Capital**: Monrovia (urban agg.): 962,000.

Government: Type: Republic. **Head of state and gov.:** Pres. Charles Taylor; b Jan. 29, 1948; in office: Aug. 2, 1997. **Local divisions:** 13 counties. **Defense budget:** $15 mil. **Active troops:** 11-15,000.

Economy: Industries: Rubber/palm oil processing, diamond mining. **Chief crops:** Rice, cassava, coffee, cocoa, sugar. **Minerals:** Iron, diamonds, gold. **Other resources:** Rubber, timber. **Arable land:** 1%. **Livestock** (2000): cattle: 36,000; chickens: 3.5 mil; goats: 220,000; pigs: 120,000; sheep: 210,000. **Electricity prod.:** 432 mil kWh. **Labor force:** 70% agric.; 22% services.

Finance: Monetary unit: Dollar (Sept. 2001: 1.00 = $1 U.S.). **GDP:** (1999 est.): $2.85 bil. **Per capita GDP:** $1,000. **Imports** (1998): $142 mil; partners: S. Korea 38%, Japan 14%, Italy 11%. **Exports** (1998): $39 mil; partners: Benelux 36%, Norway 18%, Ukraine 15%.

Transport: Motor vehicles: 17,400 pass. cars, 10,700 comm. vehicles. **Civil aviation:** 4.3 mil pass.-mi; 1 airport. **Chief ports:** Monrovia, Buchanan, Greenville, Harper.

Communications: TV sets: 20 per 1,000 pop. **Radios:** 263 per 1,000 pop. **Telephones:** 6,700 main lines. **Daily newspaper circ.:** 15 per 1,000 pop.

Health: Life expectancy: 49.96 male; 52.91 female. **Births** (per 1,000 pop): 46.55. **Deaths** (per 1,000 pop.): 16.36. **Natural inc.:** 3.02%. **Infant mortality** (per 1,000 live births): 132.42.

Education: Free, compulsory: ages 7-16. **Literacy**: 38%.

Major Intl. Organizations: UN and most of its specialized agencies, OAU.

Embassy: 5201 16th St. NW 20011; 723-0437.

Liberia was founded in 1822 by U.S. black freedmen who settled at Monrovia with the aid of colonization societies. It became a republic July 26, 1847, with a constitution modeled on that of the U.S. Descendants of freedmen dominated politics.

Charging rampant corruption, an Army Redemption Council of enlisted men staged a bloody predawn coup, April 12, 1980, in which Pres. Tolbert was killed and replaced as head of state by Sgt. Samuel Doe. Doe was chosen president in a disputed election, and survived a subsequent coup, in 1985.

A civil war began Dec. 1989. Rebel forces seeking to depose Pres. Doe made major territorial gains and advanced on the capital, June 1990. In Sept., Doe was captured and put to death. Despite the introduction of peacekeeping forces from several countries, factional fighting intensified, and a series of cease-fires failed. A transitional Council of State was instituted Sept. 1, 1995. Factional fighting flared up again in Apr. 1996, devastating Monrovia.

On Sept. 3, 1996, Ruth Perry became modern Africa's first female head of state, leading another transitional government. By then, the civil war had claimed more than 150,000 lives and uprooted over half the population.

Former rebel leader Charles Taylor was elected president July 19, 1997, in Liberia's 1st national election in 12 years. The UN imposed sanctions May 4, 2001, to punish Liberia for aiding the Revolutionary United Front (RUF) insurgency in Sierra Leone.

Libya

Socialist People's Libyan Arab Jamahiriya

People: Population: 5,240,599. **Age distrib.** (%): <15: 35.4; 65+: 3.9. **Pop. density:** 8 per sq. mi. **Urban:** 87%. **Ethnic groups:** Arab-Berber 97%. **Principal language:** Arabic (official), Italian, English. **Chief religion:** Sunni Muslim 97%.

Geography: Area: 678,600 sq. mi. **Location:** On Mediterranean coast of N Africa. **Neighbors:** Tunisia, Algeria on W; Niger, Chad on S; Sudan, Egypt on E. **Topography:** Desert and semidesert regions cover 92% of the land, with low mountains in N, higher mountains in S, and a narrow coastal zone. **Capital**: Tripoli (urban agg.): 1,822,000.

Government: Type: Islamic Arabic Socialist "Mass-State." **Head of state and gov.:** Col. Muammar al-Qaddafi; b Sept. 1942; in power: Sept. 1969. **Local divisions:** 25 municipalities. **Defense budget:** $1.2 bil. **Active troops:** 76,000.

Economy: Industries: Oil, food processing, textiles. **Chief crops:** Dates, olives, citrus, barley, wheat. **Minerals:** Gypsum, oil, gas. **Crude oil reserves** (2000): 29.5 bil bbls. **Arable land:** 1%. **Livestock** (2000): cattle: 143,000; chickens: 24.8 mil; goats: 1.9 mil; sheep: 5.1 mil. **Fish catch:** (1999): 32,849 metric tons. **Electricity prod.:** 18.900 bil kWh. **Labor force:** 54% services & govt.; 29% ind.; 17% agric.

Finance: Monetary unit: Dinar (Sept. 2001: 0.64 = $1 U.S.). **GDP:** (1999 est.): $39.3 bil. **Per capita GDP:** $7,900. **Imports** (1998): $7 bil; partners: Italy 23%, Germany 12%, UK 9%. **Exports** (1998): $6.6 bil; partners: Italy 40%, Germany 17%, Spain 12%. **Tourism:** $28 mil. **Budget** (1998): $5.1 bil. **Intl. reserves less gold** (Apr. 2000): $6.65 bil.

Transport: Motor vehicles: 592,000 pass. cars, 312,000 comm. vehicles. **Civil aviation:** 234.1 mil pass.-mi. **Chief ports:** Tripoli, Banghazi.

Communications: TV sets: 105 per 1,000 pop. **Radios:** 191 per 1,000 pop. **Telephones**: 610,000 main lines. **Daily newspaper circ.:** 15 per 1,000 pop.

Health: Life expectancy: 73.53 male; 77.88 female. **Births** (per 1,000 pop): 27.67. **Deaths** (per 1,000 pop.): 3.51. **Natural inc.:** 2.42%. **Infant mortality** (per 1,000 live births): 28.99.

Education: Compulsory: ages 6-15. **Literacy**: 76%.

Major Intl. Organizations: UN (FAO, IBRD, ILO, IMF, IMO, WHO), AL, OAU, OPEC.

First settled by Berbers, Libya was ruled in succession by Carthage, Rome, the Vandals, and the Ottomans. Italy ruled from 1912, and Britain and France after WW II. Libya became an independent constitutional monarchy Jan. 2, 1952. In 1969 a junta led by Col. Muammar al-Qaddafi seized power.

Libya and Egypt fought several air and land battles along their border in July 1977. Chad charged Libya with military occupation of its uranium-rich northern region in 1977. Libyan troops were driven from their last major stronghold by Chad forces in 1987, leaving over $1 billion in military equipment behind.

Libya reportedly helped arm violent revolutionary groups in Egypt and Sudan and aided terrorists of various nationalities.

On Jan. 7, 1986, the U.S. imposed economic sanctions against Libya, ordered all Americans to leave that country, and froze all Libyan assets in the U.S. The U.S. commenced flight operations over the Gulf of Sidra, Jan. 27, and a U.S. Navy task force began conducting exercises in the Gulf, Mar. 23. When Libya fired antiaircraft missiles at American warplanes, the U.S. responded by sinking 2 Libyan ships and bombing a missile site in Libya. The U.S. withdrew from the Gulf, Mar. 27.

The U.S. accused Qaddafi of ordering the Apr. 5, 1986, bombing of a West Berlin discotheque, which killed 3, including a U.S. serviceman. In response, the U.S. sent warplanes to attack terrorist-related targets in Tripoli and Banghazi, Libya, Apr. 14.

The UN imposed limited sanctions, Apr. 15, 1992, for Libya's failure to extradite 2 agents linked to the 1988 bombing of Pan American World Airways Flight 103 over Lockerbie, Scotland, and 4 others linked to an airplane bombing over Niger. Sanctions were tightened Dec. 1, 1993. The international embargo ended, although U.S. sanctions remained, after Libya, Apr. 5, 1999, handed over two Lockerbie suspects for trial in the Netherlands under Scottish law. One of the two defendants, a Libyan intelligence official, Abdel Basset Ali al-Meghri, was convicted of murder Jan. 31, 2001.

Liechtenstein

Principality of Liechtenstein

People: Population: 32,528. **Age distrib.** (%): <15: 18.4; 65+: 11.0. **Pop. density:** 542 per sq. mi. **Urban:** 22%. **Ethnic groups:** Alemannic 88%. **Principal languages:** German (official), Alemannic dialect. **Chief religions:** Roman Catholic 80%, Protestant 7.4%.

Geography: Area: 60 sq. mi. **Location:** Central Europe, in the Alps. **Neighbors:** Switzerland on W, Austria on E. **Topog-**

raphy: The Rhine Valley occupies one-third of the country, the Alps cover the rest. **Capital:** Vaduz (1997 est.): 5,017.

Government: Type: Hereditary constitutional monarchy. **Head of state:** Prince Hans-Adam II; b Feb. 14, 1945; in office: Nov. 13, 1989. **Head of gov.:** Otmar Hasler; b Sept. 28, 1953; in office: Apr. 5, 2001. **Local divisions:** 11 communes.

Economy: Industries: Precision instruments, electronics, textiles, ceramics. **Chief crops:** Grain, corn, potatoes. **Arable land:** 24%. **Livestock: (2000): cattle: 6,000; goats: 280; pigs: 3,000; sheep: 2,900. Labor force:** 53% services; 45% industry, trade, constr.

Finance: Monetary unit: Franc (Sept. 2001: 1.67 = $1 U.S.). **GDP:** (1998 est.): $730 mil. **Per capita GDP:** $23,000. **Imports** (1996): $917.3 mil. **Exports** (1996): $2.47 bil; partners: Switzerland 15.7%. **Budget** (1996 est.): $435 mil.

Transport: Railroad: Length: 12 mi.

Communications: TV sets: 371 per 1,000 pop. **Radios:** 384 per 1,000 pop. **Daily newspaper circ.:** 564 per 1,000 pop.

Health: Life expectancy: 75.32 male; 82.6 female. **Births** (per 1,000 pop): 11.53. **Deaths** (per 1,000 pop.): 6.7. **Natural inc.:** .48%. **Infant mortality** (per 1,000 live births): 4.99.

Education: Compulsory: ages 7-16. **Literacy** (1997): 100%.

Major Intl. Organizations: UN (WTrO), EFTA, OSCE.

Liechtenstein became sovereign in 1806. Austria administered Liechtenstein's ports up to 1920; Switzerland has administered its postal services since 1921. Liechtenstein is united with Switzerland by a customs and monetary union. Taxes are low; many international corporations have headquarters there. Foreign workers comprise 60% of the labor force.

Lithuania

Republic of Lithuania

People: Population: 3,610,535. **Age distrib.** (%): <15: 18.7; 65+: 13.6. **Pop. density:** 144 per sq. mi. **Urban:** 68%. **Ethnic groups:** Lithuanian 80.6%, Russian 8.7%, Polish 7%. **Principal languages:** Lithuanian (official), Polish, Russian. **Chief religions:** Primarily Roman Catholic.

Geography: Area: 25,100 sq. mi. **Location:** In E Europe, on SE coast of Baltic. **Neighbors:** Latvia on N, Belarus on E, S, Poland and Russia on W. **Capital:** Vilnius. **Cities** (2000 est.): Vilnius 578,000; Kaunas 412,639.

Government: Type: Republic. **Head of state:** Pres. Valdas Adamkus; b Nov. 3, 1926; in office: Feb. 26, 1998. **Head of gov.:** Prime Min. Algirdas Brazauskas; b Sept. 22, 1932; in office: July 3, 2001. **Local divisions:** 10 provinces. **Defense budget** (2001): $210 mil. **Active troops:** 12,700.

Economy: Industries: Machinery, shipbuilding, textiles. **Chief crops:** Sugar beets, grain, potatoes, vegetables. **Crude oil reserves** (2000): 12 mil bbls. **Arable land:** 35%. **Livestock** (2000): cattle: 897,800; chickens: 6.0 mil; goats: 24,700; pigs: 936,100; sheep: 13,800. **Fish catch:** (1999): 19,837 metric tons. **Electricity prod.:** 13.567 bil kWh. **Labor force:** 50% services; 30% industry; 20% agric.

Finance: Monetary unit: Litas (Sept. 2001: 4.00 = $1 U.S.). **GDP:** (1999 est.): $17.3 bil. **Per capita GDP:** $4,800. **Imports** (1999): $4.5 bil; partners: Russia 20.4%, Germany 16.5%. **Exports** (1999): $3.3 bil; partners: Russia 17.4%, Germany 15.8%, Latvia 12.7%. **Tourism:** $550 mil. **Budget** (1997 est.): $1.7 bil. **Intl. reserves less gold** (June 2000): $1.38 bil. **Gold:** 186,300 oz t. **Consumer prices** (change in 1999): 0.8%.

Transport: Railroad: Length: 1,802 mi. **Motor vehicles:** 653,000 pass. cars, 111,000 comm. vehicles. **Civil aviation:** 187.2 mil pass.-mi; 3 airports. **Chief port:** Klaipeda.

Communications: TV sets: 364 per 1,000 pop. **Radios:** 404 per 1,000 pop. **Telephones:** 1,187,700 main lines. **Daily newspaper circ.:** 136 per 1,000 pop.

Health: Life expectancy: 63.3 male; 75.5 female. **Births** (per 1,000 pop): 10.00. **Deaths** (per 1,000 pop.): 12.86. **Natural inc.:** -.29%. **Infant mortality** (per 1,000 live births): 14.5.

Education: Free, compulsory: ages 7-16. Literacy (1989): 98%.

Major Intl. Organizations: UN (FAO, IBRD, ILO, IMF, IMO, WHO), OSCE.

Embassy: 2622 16th St. NW 20009; 234-5860.

Website: http://www.std.lt

Lithuania was occupied by the German army, 1914-18. It was annexed by the Soviet Russian army, but the Soviets were overthrown, 1919. Lithuania was a democratic republic until 1926, when the regime was ousted by a coup. In 1939 the Soviet-German treaty assigned most of Lithuania to the Soviet sphere of influence. Lithuania was annexed by the USSR Aug. 3, 1940.

Lithuania formally declared its independence from the Soviet Union Mar. 11, 1990. During an abortive Soviet coup in Aug., the Western nations recognized Lithuania's independence, which was ratified by the Soviet Union in Sept. 1991.

The last Russian troops withdrew on Aug. 31, 1993. Lithuania applied to join the European Union, Dec. 8, 1995. The conservative Homeland Union defeated the former Communists in parliamentary elections Oct. 20 and Nov. 10, 1996. A Lithuanian-American, Valdas Adamkus, won the presidency in a runoff election Jan. 4, 1998. Parliamentary elections Oct. 8, 2000, dealt conservatives a major setback.

Luxembourg

Grand Duchy of Luxembourg

People: Population: 442,972. **Age distrib.** (%): <15: 18.9; 65+: 14.1. **Pop. density:** 444 per sq. mi. **Urban:** 91%. **Ethnic groups:** Mixture of French and Germans predominates. **Principal languages:** French, German, Luxembourgian, English. **Chief religion:** Roman Catholic 97%.

Geography: Area: 1,000 sq. mi. **Location:** In W Europe. **Neighbors:** Belgium on W, France on S, Germany on E. **Topography:** Heavy forests (Ardennes) cover N, S is a low, open plateau. **Capital:** Luxembourg (2000 est.): 80,700.

Government: Type: Constitutional monarchy. **Head of state:** Grand Duke Henri; b Apr. 16, 1955; in office: Oct. 7, 2000. **Head of gov.:** Prime Min. Jean-Claude Juncker; b Dec. 9, 1954; in office: Jan. 19, 1995. **Local divisions:** 3 districts. **Defense budget:** $100 mil. **Active troops:** 899.

Economy: Industries: Steel, chemicals, food processing, tires, banking, engineering, metal products. **Chief crops:** Grains, potatoes, wine grapes. **Arable land:** 24%. **Electricity prod.:** 648 mil kWh. **Labor force:** 83.2% services; 14.3% ind.

Finance: Monetary unit: Lux. Franc (Sept. 2001: 44.52 = $1 U.S.). Euro (Sept. 2001: 1.09 = $1 U.S.). **GDP:** (1999 est.): $14.7 bil. **Per capita GDP:** $34,200. **Imports** (1998) $9.6 bil; partners: Belgium 36%, Germany 27%, France 12%. **Exports** (1998): $7.5 bil; partners: Germany 33%, France 20%, Belgium 12%. **Tourism** (1998): $309 mil. **Budget** (1997 est.): $5.46 bil. **Intl. reserves less gold** (June 2000): $77.4 mil. **Gold:** 76,000 oz t. **Consumer prices** (change in 1999): 1.0%.

Transport: Railroad: Length: 170 mi. **Motor vehicles:** 231,666 pass. cars, 16,665 comm. vehicles. **Civil aviation:** 174.8 mil pass.-mi; 1 airport. **Chief port:** Mertert.

Communications: TV sets: 916 per 1,000 pop. **Radios:** 586 per 1,000 pop. **Telephones:** 331,000 main lines. **Daily newspaper circ.:** 381 per 1,000 pop.

Health: Life expectancy: 74.02 male; 80.8 female. **Births** (per 1,000 pop): 12.25. **Deaths** (per 1,000 pop.): 8.88. **Natural inc.:** .34%. **Infant mortality (per 1,000 live births):** 4.77.

Education: Compulsory: ages 6-15. **Literacy:** 100%.

Major Intl. Organizations: UN (FAO, IBRD, ILO, IMF, IMO, WHO, WTrO), EU, NATO, OECD, OSCE.

Embassy: 2200 Massachusetts Ave. NW 20008; 265-4171.

Luxembourg, founded about 963, was ruled by Burgundy, Spain, Austria, and France from 1448 to 1815. It left the Germanic Confederation in 1866. Overrun by Germany in 2 world wars, Luxembourg ended its neutrality in 1948, when a customs union with Belgium and Netherlands was adopted.

Macedonia

Former Yugoslav Republic of Macedonia

People: Population: 2,046,209. **Age distrib.** (%): <15: 22.9; 65+: 10.1. **Pop. density:** 206 per sq. mi. **Urban:** 62%. **Ethnic groups:** Macedonian 66%, Albanian 23%. **Principal languages:** Macedonian (official), Albanian, Serbo-Croatian. **Chief religions:** Eastern Orthodox 67%, Muslim 30%.

Geography: Area: 9,900 sq. mi. **Location:** In SE Europe. **Neighbors:** Bulgaria on E, Greece on S, Albania on W, Serbia on N. **Capital:** Skopje (1994 census): 429,964.

Government: Type: Republic. **Head of state:** Pres. Boris Trajkovski; b June 25, 1956; in office: Dec. 15, 1999. **Head of gov.:** Prime Min. Ljupco Georgievski; b Jan. 17, 1966; in office: Nov. 30, 1998. **Local divisions:** 123 municipalities. **Defense budget:** $77 mil. **Active troops:** 16,000.

Economy: Industries: Mining, textiles. **Chief crops:** Wheat, rice, cotton, tobacco. **Minerals:** Chromium, lead, zinc. **Arable land:** 24%. **Livestock** (2000): cattle: 290,000; chickens: 3.35 mil; pigs: 197,000; sheep: 1.6 mil. **Electricity prod.:** 6.395 bil kWh.

Finance: Monetary unit: Denar (Oct. 2000: 65.40 = $1 U.S.). **GDP:** (1999 est.): $7.6 bil. **Per capita GDP:** $3,800. **Imports** (1999): $1.56 bil; partners: Germany 13%, Serbia & Montenegro 13%, Slovenia 8%. **Exports** (1999): $1.2 bil.; partners: Germany 21%, Serbia & Montnegro 18%, U.S. 13%. **Tourism** (1998): $15 mil. **Budget** (1996 est.): $1.0 bil. **Intl. reserves less gold** (June 2000): $513.37 mil. **Gold:** 108,000 oz t. **Consumer prices** (change in 1999): –1.3%.

Transport: Railroad: Length: 573 mi. **Motor vehicles:** 263,000 pass. cars, 23,000 comm. vehicles. **Civil aviation:** 161.1 mil pass.-mi; 2 airports.

Communications: TV sets: 179 per 1,000 pop. **Radios:** 179 per 1,000 pop. **Telephones** (1999): 471,000 main lines. **Daily newspaper circ.:** 21 per 1,000 pop.

Health: Life expectancy: 71.79 male; 76.43 female. **Births** (per 1,000 pop): 13.50. **Deaths** (per 1,000 pop.): 7.7. **Natural inc.:** .58%. **Infant mortality** (per 1,000 live births): 12.95.

Education: Free, compulsory: ages 7-15. **Literacy** (1996): 89%.

Major Intl. Organizations: UN (FAO, IBRD, ILO, IMF, IMO, WHO).

Embassy: 3050 K St. NW 20007; 337-3063.

Macedonia, as part of a larger region also called Macedonia, was ruled by Muslim Turks from 1389 to 1912, when native Greeks, Bulgarians, and Slavs won independence. Serbia received the largest part of the territory, with the rest going to Greece and Bulgaria. In 1913, the area was incorporated into Serbia, which in 1918 became part of the Kingdom of Serbs, Croats, and Slovenes (later Yugoslavia). In 1946, Macedonia became a constituent republic of Yugoslavia.

Macedonia declared its independence Sept. 8, 1991, and was admitted to the UN under a provisional name in 1993. A UN force, which included several hundred U.S. troops, was deployed there to deter the warring factions in Bosnia from carrying their dispute into other areas of the Balkans.

In Feb. 1994 both Russia and the U.S. recognized Macedonia. Greece, which objected to Macedonia's use of what it considered a Hellenic name and symbols, imposed a trade blockade on the landlocked nation; the 2 countries agreed to normalize relations Sept. 13, 1995. A car bombing, Oct. 3, seriously injured Pres. Kiro Gligorov. Macedonia and Yugoslavia signed a treaty normalizing relations Apr. 8, 1996.

By the end of NATO's air war against Yugoslavia, Mar.-June 1999, Macedonia had a Kosovar refugee population of more than 250,000; over 90% had been repatriated by Sept. 1. Boris Trajkovski, candidate of the ruling center-right coalition, won a presidential runoff vote Nov. 14.

Ethnic Albanian guerrillas launched an offensive Mar. 2001 in NW Macedonia. A peace accord signed Aug. 13 paved the way for the introduction of a NATO peacekeeping force.

WORLD ALMANAC EDITORS' PICKS

The World Almanac staff ranked the following as their favorite Flags for nations other than the U.S. (see pages 497-500 for all the national flags):

1. Bhutan	6. Canada
2. Macedonia	7. Japan
3. Kiribati	8. Swaziland
4. Sri Lanka	9. Seychelles
5. Brazil	10. France

Madagascar

Republic of Madagascar

People: Population: 15,982,563. **Age distrib.** (%): <15: 45.0; 65+: 3.2. **Pop. density:** 71 per sq. mi. **Urban:** 29%. **Ethnic groups:** Malayo-Indonesian, Cotiers, French, Indian, Creole, Comoran. **Principal languages:** Malagasy, French (both official). **Chief religions:** Indigenous beliefs 52%, Christian 41%, Muslim 7%.

Geography: Area: 224,300 sq. mi. **Location:** In the Indian O., off the SE coast of Africa. **Neighbors:** Comoro Isls. to NW, Mozambique to W. **Topography:** Humid coastal strip in the E, fertile valleys in the mountainous center plateau region, and a wider coastal strip on the W. **Capital:** Antananarivo (urban agg.): 1,507,000.

Government: Type: Republic. **Head of state:** Pres. Didier Ratsiraka; b Nov. 4, 1936; in office: Jan. 31, 1997. **Head of gov.:** Tantely Andrianarivo; b May 25, 1954; in office: July 23, 1998. **Local divisions:** 6 provinces. **Defense budget:** $42 mil. **Active troops:** 21,000.

Economy: Industries: Meat processing, textiles. **Chief crops:** Coffee, cloves, vanilla beans, rice, sugar, cassava, peanuts. **Minerals:** Chromite, graphite, coal, bauxite. **Arable land:** 4%. **Livestock** (2000): cattle: 10.36 mil; chickens: 20.0 mil; goats: 1.37 mil; pigs: 900,000; sheep: 800,000. **Fish catch** (1999): 124,973 metric tons. **Electricity prod:** 810 mil kWh.

Finance: Monetary unit: Malagasy franc (Sept. 2001: 6,365.00 = $1 U.S.). **GDP:** (1999 est.): $11.5 bil. **Per capita GDP:** $780. **Imports** (1998 est.): $881 mil; partners: France 39%. **Exports** (1998 est.): $600 mil; partners: France 40%, U.S. 9%, Germany 8%. **Tourism:** $100 mil. **Budget** (1996 est.): $706 mil. **Intl. reserves less gold** (Apr. 2000): $212.4 mil. **Consumer prices** (change in 1999): 9.9%.

Transport: Railroad: Length: 640 mi. **Motor vehicles:** 58,100 pass. cars, 15,860 comm. vehicles. **Civil aviation:** 471.0 mil pass.-mi; 44 airports. **Chief ports:** Toamasina, Antsiranana, Mahajanga, Toliara, Antsohimbondrona.

Communications: TV sets: 20 per 1,000 pop. **Radios:** 193 per 1,000 pop. **Telephones:** 56,700 main lines. **Daily newspaper circ.:** 4 per 1,000 pop.

Health: Life expectancy: 53.08 male; 57.68 female. **Births** (per 1,000 pop): 42.66. **Deaths** (per 1,000 pop.): 12.42. **Natural inc.:** 3.02%. **Infant mortality** (per 1,000 live births): 83.58.

Health: Life expectancy: 52.31 male; 54.92 female. **Births** (per 1,000 pop.): 42.92. **Deaths** (per 1,000 pop.): 12.69. **Natural inc.:** 3.023%. **Infant mortality** (per 1,000 live births): 87.62.

Education: Compulsory for 5 years between ages 6 and 13. **Literacy:** 46%.

Major Intl. Organizations: UN (FAO, IBRD, ILO, IMF, IMO, WHO, WTrO), OAU.

Embassy: 2374 Massachusetts Ave. NW 20008; 265-5525.

Madagascar was settled 2,000 years ago by Malayan-Indonesian people, whose descendants still predominate. A unified kingdom ruled the 18th and 19th centuries. The island became a French protectorate, 1885, and a colony 1896. Independence came June 26, 1960.

Discontent with inflation and French domination led to a coup in 1972. The new regime nationalized French-owned financial interests, closed French bases and a U.S. space-tracking station, and obtained Chinese aid. The government conducted a program of arrests, expulsion of foreigners, and repression of strikes, 1979.

In 1990, Madagascar ended a ban on multiparty politics that had been in place since 1975. Albert Zafy was elected president in 1993, ending the 17-year rule of Adm. Didier Ratsiraka. After Zafy was impeached by the legislature, Madagascar's constitutional court removed him from office, Sept. 5, 1996. Prime Min. Norbert Ratsirahonana then became interim president pending national elections, Nov. 3 and Dec. 29, in which Ratsiraka edged Zafy. A cholera epidemic, exacerbated by cyclones in Feb. and Apr. 2000, claimed at least 1,600 lives.

Malawi

Republic of Malawi

People: Population: 10,548,250. **Age distrib.** (%): <15: 44.4; 65+: 2.8. **Pop. density:** 291 per sq. mi. **Urban:** 24%. **Ethnic groups:** Chewa, Nyanja, Lomwe, other Bantu tribes. **Principal languages:** English, Chichewa (both official). **Chief religions:** Protestant 55%, Muslim 20%, Roman Catholic 20%.

Geography: Area: 36,300 sq. mi. **Location:** In SE Africa. **Neighbors:** Zambia on W, Mozambique on S and E, Tanzania on N. **Topography:** Malawi stretches 560 mi. N-S along Lake Malawi (Lake Nyasa), most of which belongs to Malawi. High plateaus and mountains line the Rift Valley the length of the nation. **Capital:** Lilongwe. **Cities** (1998 est.): Blantyre 2,000,000; Lilongwe 1,000,000.

Government: Type: Multiparty democracy. **Head of state and gov.:** Pres. Bakili Muluzi; b Mar. 17, 1943; in office: May 21, 1994. **Local divisions:** 3 regions, 26 districts. **Defense budget:** $26 mil. **Active troops:** 5,000.

Economy: Industries: Agricultural processing, cement. **Chief crops:** Tea, tobacco, sugar, cotton, corn, potatoes. **Arable land:** 18%. **Livestock** (2000): cattle: 760,000; chickens: 15.0 mil; goats: 1.27 mil; pigs: 240,000; sheep: 115,000. **Fish catch** (1999): 56,564 metric tons. **Electricity prod.:** 1.025 bil kWh. **Labor force** (1990): 86% agric.

Finance: Monetary unit: Kwacha (Sept. 2001: 62.99 = $1 U.S.). **GDP:** (1999 est.): $9.4 bil. **Per capita GDP:** $940. **Imports** (1999): $512 mil; partners: South Africa 38%, Zimbabwe 18%, Zambia 8%. **Exports** (1999): $510 mil; partners: South Africa 15%, U.S.9%, Germany 9%. **Tourism:** $20 mil. **Intl. reserves less gold** (June 2000): $217.94 mil. **Gold:** 10,000 oz t. **Consumer prices** (change in 1999): 44.9%.

Transport: Railroad: Length: 490 mi. **Motor vehicles:** 25,400 pass. cars, 28,900 comm. vehicles. **Civil aviation:** 208.6 mil pass.-mi; 5 airports.

Communications: Radios: 112 per 1,000 pop. **Telephones:** 45,000 main lines.

Health: Life expectancy: 36.61 male; 37.55 female. **Births (per 1,000 pop):** 37.80. **Deaths (per 1,000 pop.): 22.81. Natural inc.:** 1.5%. **Infant mortality (per 1,000 live births):** 121.12.

Health: Life expectancy: 36.34 male; 35.7 female. **Births** (per 1,000 pop.): 38.49. **Deaths** (per 1,000 pop.): 22.44. **Natural inc.:** 1.605%. **Infant mortality** (per 1,000 live births): 130.52.

Education: Compulsory: ages 6-14. **Literacy:** 56%.

Major Intl. Organizations: UN (FAO, IBRD, ILO, IMF, IMO, WHO, WTrO), the Commonwealth, OAU.

Embassy: 2408 Massachusetts Ave. NW 20008; 797-1007.

Bantus came to the land in the 16th century, Arab slavers in the 19th. The area became the British protectorate Nyasaland in 1891. It became independent July 6, 1964, and a republic in 1966. After 3 decades as a one-party state under Pres. Hastings Kamuzu Banda, Malawi adopted a new constitution and, in multiparty elections held May 17, 1994, chose a new leader, Bakili Muluzi. Banda was acquitted, Dec. 23, 1995, of complicity in the deaths of 4 political opponents in 1983; he died Nov. 25, 1997.

According to UN estimates, more than 15% of the adult population has HIV/AIDS.

Malaysia

People: Population: 22,229,040. **Age distrib.** (%): <15: 34.5; 65+: 4.2. **Pop. density:** 175 per sq. mi. **Urban:** 57%. **Ethnic groups:** Malay and other indigenous 58%, Chinese 26%, Indian 7%. **Principal languages:** Malay (official), English, Chinese dialects. **Chief religions:** Muslim, Hindu, Buddhist, Christian.

Geography: Area: 126,700 sq. mi. **Location:** On the SE tip of Asia, plus the N coast of the island of Borneo. **Neighbors:** Thailand on N, Indonesia on S. **Topography:** Most of W Malaysia is covered by tropical jungle, including the central mountain range that runs N-S through the peninsula. The western coast is marshy, the eastern, sandy. E Malaysia has a wide, swampy coastal plain, with interior jungles and mountains. **Capital:** Kuala Lumpur (urban agg.): 1,378,000.

Government: Type: Federal parliamentary democracy with a constitutional monarch. **Head of state:** Paramount Ruler Sultan Salahuddin Abdul Aziz Shah Alhaj; b Mar. 8, 1926; in office: Apr. 26, 1999. **Head of gov.:** Prime Min. Datuk Seri Mahathir bin Mohamad; b Dec. 20, 1925; in office: July 16, 1981. **Local divisions:** 13 states, 2 federal territories. **Defense budget:** $1.6 bil. **Active troops:** 96,000.

Economy: Industries: Rubber goods, logging, electronics, petroleum production. **Chief crops:** Palm oil (world's leading producer), rice. **Minerals:** Tin (a leading producer), oil, gas, bauxite, copper, iron. **Crude oil reserves** (2000): 3.9 bil bbls. **Other resources:** Rubber, timber. **Arable land:** 3%. **Livestock** (2000): cattle: 723,346; chickens: 120.0 mil; goats: 231,834; pigs: 1.83 mil; sheep: 174,661. **Fish catch** (1999): 1.28 mil metric tons. **Electricity prod.:** 59.044 bil kWh. **Labor force:** 27% manuf.; 16% agric., forestry & fisheries; 17% local trade & tourism.

Finance: Monetary unit: Ringgit (Sept. 2001: 3.80 = $1 U.S.). **GDP:** (1999 est.): $229.1 bil. **Per capita GDP:** $10,700. **Imports** (1999): $61.5 bil; partners: Japan 21%, U.S. 18%, Singapore 14%. **Exports** (1999): $83.5 bil; partners: U.S. 23%, Singapore 16%, Japan 11%. **Tourism:** $2.82 bil. **Budget** (1996 est.): $22 bil. **Intl. reserves less gold** (June 2000): $33.67 bil. **Gold:** 1.17 mil oz t. **Consumer prices** (change in 1999): 2.7%.

Transport: Railroad: Length: 1,113 mi. **Motor vehicles** (1997): 3.33 mil pass. cars, 618,066 comm. vehicles. **Civil aviation:** 17.8 bil pass.-mi; 39 airports. **Chief ports:** Kuantan, Kelang, Kota Kinabalu, Kuching.

Communications: TV sets: 424 per 1,000 pop. **Radios:** 442 per 1,000 pop. **Telephones**: 4,637,000 main lines. **Daily newspaper circ.:** 163 per 1,000 pop.

Health: Life expectancy: 68.48 male; 73.92 female. **Births** (per 1,000 pop): 24.75. **Deaths** (per 1,000 pop.): 5.2. **Natural inc.:** 1.96%. **Infant mortality** (per 1,000 live births): 20.31.

Education: Free, compulsory: ages 6-16. **Literacy**: 83%.

Major Intl. Organizations: UN (FAO, IBRD, ILO, IMF, IMO, WHO, WTrO), APEC, ASEAN, the Commonwealth.

Embassy: 2401 Massachusetts Ave. NW 20008; 328-2700.

European traders appeared in the 16th century; Britain established control in 1867. Malaysia was created Sept. 16, 1963. It included Malaya (which had become independent in 1957 after the suppression of Communist rebels), plus the formerly British Singapore, Sabah (N Borneo), and Sarawak (NW Borneo). Singapore was separated in 1965, in order to end tensions between Chinese, the majority in Singapore, and Malays in control of the Malaysian government.

A monarch is elected by a council of hereditary rulers of the Malayan states every 5 years.

Abundant natural resources have bolstered prosperity, and foreign investment has aided industrialization. Work on a new federal capital at Putrajaya, south of Kuala Lumpur, began in 1995. However, sagging stock and currency prices forced the postponement of major development projects in Sept. 1997.

As the recession deepened and political unrest grew, Prime Min. Mahathir bin Mohamad imposed new currency controls and fired his popular deputy prime minister, Anwar bin Ibrahim, Sept. 2, 1998. Anwar, who then called for Mahathir's resignation, was arrested Sept. 20; he was convicted of corruption, Apr. 14, 1999, and sentenced to 6 years in prison. Another conviction, Aug. 8, 2000, for sodomy, resulted in an additional 9-year sentence.

Maldives

Republic of Maldives

People: Population: 310,764. **Age distrib.** (%): <15: 45.6; 65+: 3.0. **Pop. density:** 3,108 per sq. mi. **Urban:** 26%. **Ethnic groups:** Sinhalese, Dravidian, Arab, African. **Principal languages:** Maldivian Divehi (Sinhalese dialect; official), English. **Chief religion:** Sunni Muslim.

Geography: Area: 100 sq. mi. **Location:** In the Indian O., SW of India. **Neighbors:** Nearest is India on N. **Topography:** 19 atolls with 1,190 islands, 198 inhabited. None of the islands are over 5 sq. mi. in area, and all are nearly flat. **Capital:** Male (1999 est.): 72,000.

Government: Type: Republic. **Head of state and gov.:** Pres. Maumoon Abdul Gayoom; b Dec. 29, 1937; in office: Nov. 11, 1978. **Local divisions:** 19 atolls and Male. **Defense budget:** NA. **Active troops:** NA.

Economy: Industries: Fish processing, tourism. **Chief crops:** Coconuts, corn, sweet potatoes. **Arable land:** 10%. **Fish catch** (1999): 107,676 metric tons. **Electricity prod.:** 101 mil kWh. **Labor force** (1995): 60% services, 22% agric.; 18% ind.

Finance: Monetary unit: Rufiyaa (Sept. 2001: 11.71 = $1 U.S.). **GDP:** (1999 est.): $540 mil. **Per capita GDP:** $1,800. **Imports** (1998): $312 mil; partners: Singapore, India, Sri Lanka. **Exports** (1998): $98 mil; partners: UK, Sri Lanka, Japan. **Tourism:** $334 mil. **Budget** (1995 est.): $141 mil. **Intl. reserves less gold** (June 2000): $126.98 mil. **Consumer prices** (change in 1999): 3.0%.

Transport: Civil aviation: 181.2 mil pass.-mi; 5 airports. **Chief ports:** Male, Gan.

Communications: TV sets: 19 per 1,000 pop. **Radios:** 96 per 1,000 pop. **Telephones**: 24,400 main lines. **Daily newspaper circ.:** 12 per 1,000 pop.

Health: Life expectancy: 61.39 male; 63.8 female. **Births** (per 1,000 pop): 38.15. **Deaths** (per 1,000 pop.): 8.09. **Natural inc.:** 3.01%. **Infant mortality** (per 1,000 live births): 63.72.

Education: Literacy: 93%.

Major Intl. Organizations: UN (FAO, IBRD, IMF, IMO, WHO, WTrO), the Commonwealth.

Websites: http://www.maldives-info.com
http://www.undp.org/missions/maldives

The islands had been a British protectorate since 1887. The country became independent July 26, 1965. Long a sultanate, the Maldives became a republic in 1968. Natural resources and tourism are being developed; however, the Maldives remains one of the world's poorest countries.

Mali

Republic of Mali

People: Population: 11,008,518. **Age distrib.** (%): <15: 47.2; 65+: 3.1. **Pop. density:** 23 per sq. mi. **Urban:** 29%. **Ethnic groups:** Mande (Bambara, Malinke, Sarakole) 50%, Peul 17%, Voltaic 12%, Tuareg and Moor 10%, Songhai 6%. **Principal languages:** French (official), Bambara, numerous African languages. **Chief religions:** Muslim 90%, indigenous beliefs 9%.

Geography: Area: 470,500 sq. mi. **Location:** In the interior of W Africa. **Neighbors:** Mauritania, Senegal on W; Guinea, Côte d'Ivoire, Burkina Faso on S; Niger on E; Algeria on N. **Topography:** A landlocked grassy plain in the upper basins of the Senegal and Niger rivers, extending N into the Sahara. **Capital** (urban agg.): Bamako: 1,131,000.

Government: Type: Republic. **Head of state:** Pres. Alpha Oumar Konare; b Feb. 2, 1946; in office: June 8, 1992. **Head of gov.:** Prime Min. Mande Sidibe; b 1940; in office: Feb. 15, 2000. **Local divisions:** 8 regions, 1 capital district. **Defense budget:** $30 mil. **Active troops:** 7,350.

Economy: Industries: construction, mining. **Chief crops:** Millet, rice, peanuts, corn, vegetables, cotton. **Minerals:** Gold, phosphates, kaolin. **Arable land:** 2%. **Livestock** (2000): cattle: 6.2 mil; chickens: 25.0 mil; goats: 8.55 mil; pigs: 65,000; sheep: 6.0 mil. **Fish catch** (1999): 99,610 metric tons. **Electricity prod.:** 445 mil kWh. **Labor force** (1998): 80% agric. and fishing.

Finance: Monetary unit: CFA Franc (Sept. 2001: 724.11 = $1 U.S.). **GDP:** (1999 est.): $8.5 bil. **Per capita GDP:** $820. **Imports** (1999 est.): $650 mil; partners: Côte d'Ivoire 19%, France 17%. **Exports** (1999 est.): $640 mil; partners: Thailand 20%, Italy 20%, China 9%. **Tourism** (1998): $50 mil. **Budget** (1997 est.): $770 mil. **Intl. reserves less gold** (Apr. 2000): $308.2 mil. **Gold:** 19,000 oz t. **Consumer prices** (change in 1999): –1.2%.

Transport: Railroad: Length: 398 mi. **Motor vehicles:** 24,700 pass. cars, 17,100 comm. vehicles. **Civil aviation:** 150.5 mil pass.-mi; 9 airports. **Chief port:** Koulikoro.

Communications: TV sets: 12 per 1,000 pop. **Radios:** 168 per 1,000 pop. **Telephones** (1998): 26,800 main lines.

Health: Life expectancy: 45.84 male; 48.24 female. **Births** (per 1,000 pop): 48.79. **Deaths** (per 1,000 pop.): 18.71. **Natural inc.:** 3.01%. **Infant mortality** (per 1,000 live births): 121.44.

Education: Free, compulsory: ages 7-16. **Literacy:** 31%.

Major Intl. Organizations: UN and most of its specialized agencies, OAU.

Embassy: 2130 R St. NW 20008; 332-2249.

Until the 15th century the area was part of the great Mali Empire. Timbuktu (Tombouctou) was a center of Islamic study. French rule was secured, 1898. The Sudanese Rep. and Senegal became independent as the Mali Federation June 20, 1960, but Senegal withdrew, and the Sudanese Rep. was renamed Mali.

Mali signed economic agreements with France and, in 1963, with Senegal. In 1968, a coup ended the socialist regime. Famine struck in 1973-74, killing as many as 100,000 people. Drought conditions returned in the 1980s.

The military, Mar. 26, 1991, overthrew the government of Pres. Moussa Traoré, who had been in power since 1968. Oumar Konare, a coup leader, was elected president, Apr. 26, 1992. A peace accord between the government and a Tuareg rebel group was signed in June 1994. Konare and his party won a series of flawed elections, Apr.-Aug. 1997. Twice condemned to death for crimes committed in office, Traoré had his sentences commuted to life imprisonment in Dec. 1997 and Sept. 1999.

Malta

Republic of Malta

People: Population: 394,583. **Age distrib.** (%): <15: 20.0; 65+: 12.5. **Pop. density:** 3,288 per sq. mi. **Urban:** 90%. **Ethnic group:** Maltese. **Principal languages:** Maltese, English (both official). **Chief religion:** Roman Catholic 98%.

Geography: Area: 120 sq. mi. **Location:** In center of Mediterranean Sea. **Neighbors:** Nearest is Italy on N. **Topography:** Island of Malta is 95 sq. mi.; other islands in the group: Gozo, 26 sq. mi.; Comino, 1 sq. mi. The coastline is heavily indented. Low hills cover the interior. **Capital:** Valletta (1999 est.): 7,073.

Government: Type: Parliamentary democracy. **Head of state:** Pres. Guido de Marco; b July 22, 1931; in office: Apr. 4, 1999. **Head of gov.:** Prime Min. Edward Fenech-Adami; b Feb. 7, 1934; in office: Sept. 6, 1998. **Local divisions:** 3 regions comprising 67 local councils. **Defense budget:** $27.6 mil. **Active troops:** 2,140.

Economy: Industries: Tourism, electronics, construction, textiles, food & beverages. **Chief crops:** Potatoes, cauliflower, tomatoes. **Minerals:** Salt, limestone. **Arable land:** 38%. **Livestock** (2000): cattle: 19,200; chickens: 820,000; goats: 9,000; pigs: 80,074; sheep: 16,000. **Electricity prod.:** 1.650 bil kWh. **Labor force:** 71% services; 24% ind.

Finance: Monetary unit: Lira (Sept. 2001: .44 = $1 U.S.). **GDP:** (1999 est.): $5.3 bil. **Per capita GDP:** $13,800. **Imports** (1998): $2.7 bil; partners: Italy 19.3%, France 17.8%, UK 12.4%. **Exports** (1998): $1.8 bil; partners: France 20.7%, U.S. 18.1%, Germany 12.6%. **Tourism:** $675 mil. **Budget** (1997 est.): $1.76 bil. **Intl. reserves less gold** (May 2000): $1.71 bil. **Gold:** 6,000 oz t. **Consumer prices** (change in 1999): 2.1%.

Transport: Motor vehicles: 122,100 pass. cars, 19,100 comm. vehicles. **Civil aviation:** 1.0 bil pass.-mi; 1 airport. **Chief ports:** Valletta, Marsaxlokk.

Communications: TV sets: 739 per 1,000 pop. **Radios:** 525 per 1,000 pop. **Telephones:** 204,200 main lines. **Daily newspaper circ.:** 145 per 1,000 pop.

Health: Life expectancy: 75.64 male; 80.79 female. **Births** (per 1,000 pop): 12.75. **Deaths** (per 1,000 pop.): 7.74. **Natural inc.:** .5%. **Infant mortality** (per 1,000 live births): 5.83.

Education: Free, compulsory: ages 5-16. **Literacy:** 91%.

Major Intl. Organizations: UN (FAO, IBRD, ILO, IMF, IMO, WHO, WTrO), the Commonwealth, OSCE.

Embassy: 2017 Connecticut Ave. NW 20008; 462-3611.

Website: http://www.magnet.mt/home/cos

Malta was ruled by Phoenicians, Romans, Arabs, Normans, the Knights of Malta, France, and Britain (since 1814). It became independent Sept. 21, 1964. Malta became a republic in 1974. The withdrawal of the last British sailors, Apr. 1, 1979, ended 179 years of British military presence on the island. From 1971 to 1987 and again from 1996 to 1998, Malta was governed by the socialist Labour Party. The Nationalist Party, which held office 1987-96 and favors Malta's entry into the EU, returned to power after elections Sept. 5, 1998.

Marshall Islands

Republic of the Marshall Islands

People: Population: 70,822. **Age. distrib.** (%): <15: 49.3; 65+: 2.1. **Pop. density:** 1,012 per sq. mi. **Urban:** 70%. **Ethnic groups:** Micronesian. **Principal languages:** English (official), Marshallese, Japanese. **Chief religion:** Protestant 63%.

Geography: Area: 70 sq. mi. **Location:** In N Pacific Ocean; composed of two 800-mi-long parallel chains of coral atolls. **Neighbors:** Nearest are Micronesia to W, Nauru and Kiribati to S. **Capital:** Majuro (1999 est.): 33,000.

Government: Type: Republic. **Head of state and gov.:** Pres. Kessai Note; in office: Jan. 10, 2000. **Local divisions:** 33 municipalities.

Economy: Agriculture and tourism are mainstays.

Finance: Monetary unit: U.S. Dollar. **GDP:** (1998 est.): $105 mil. **Per capita GDP:** $1,670. **Imports** (1997 est.): $58 mil; partners: U.S., Japan, Australia. **Exports** (1997 est.): $28 mil; partners: U.S., Japan, Australia. **Tourism:** $4 mil. **National budget** (FY 1995-96 est.): $77.4 mil.

Transport: Civil aviation: 16.1 mil pass.-mi; 25 airports. **Chief port:** Majuro.

Communications: Telephones (1999): 4,000 main lines.

Health: Life expectancy: 64.04 male; 67.73 female. **Births** (per 1,000 pop): 45.07. **Deaths** (per 1,000 pop.): 6.23. **Natural inc.:** 3.88%. **Infant mortality** (per 1,000 live births): 39.82.

Education: Compulsory: ages 6-14. **Literacy** (1994): 93%.

Major Intl. Organizations: UN (IBRD, IMF, WHO).

Embassy: 2433 Massachusetts Ave. NW 20008; 234-5414.

The Marshall Islands were a German possession until World War I and were administered by Japan between the World Wars. After WW II, they were administered as part of the UN Trust Territory of the Pacific Islands by the U.S.

The Marshall Islands secured international recognition as an independent nation on Sept. 17, 1991. Amata Kabua, the islands' first and only president since 1979, died Dec. 19, 1996. His cousin Imata Kabua, elected president Jan. 13, 1997, was succeeded by Kessai Note on Jan. 10, 2000.

Mauritania

Islamic Republic of Mauritania

People: Population: 2,747,312. **Age distrib.** (%): <15: 46.1; 65+: 2.3. **Pop. density:** 7 per sq. mi. **Urban:** 56%. **Ethnic groups:** Mixed Maur/black 40%, Maur 30%, black 30%. **Principal languages:** Hasaniya Arabic, Wolof (both official), Pular, Soninke. **Chief religion:** Muslim 100%.

Geography: Area: 397,400 sq. mi. **Location:** In NW Africa. **Neighbors:** Morocco on N, Algeria and Mali on E, Senegal on S. **Topography:** The fertile Senegal R. valley in the S gives way to a wide central region of sandy plains and scrub trees. The N is arid and extends into the Sahara. **Capital:** Nouakchott (1999 est.): 881,000.

Government: Type: Islamic republic. **Head of state:** Pres. Maaouya Ould Sidi Ahmed Taya; b 1943; in office: Apr. 18, 1992. **Head of gov.:** Prime Min. Cheikh El Afia Ould Mohamed Khouna; b 1956; in office: Nov. 16, 1998. **Local divisions:** 12 regions, 1 capital district. **Defense budget:** $23.6 mil. **Active troops:** 15,650.

Economy: Industries: Fish processing, iron mining. **Chief crops:** Dates, millet. **Minerals:** Iron ore, gypsum. **Livestock** (2000): cattle: 1.44 mil; chickens: 4.1 mil; goats: 4.14 mil; sheep: 6.2 mil. **Fish catch** (1999): 82,000 metric tons. **Electricity prod.:** 151 mil kWh. **Labor force:** 47% agric.; 39% services; 14% ind.

Finance: Monetary unit: Ouguiya (Sept. 2001: 257.61 = $1 U.S.). **GDP:** (1999 est.): $4.9 bil. **Per capita GDP:** $1,910. **Imports** (1997): $444 mil; partners: France 26%, Spain 8%. **Exports** (1997): $425 mil; partners: Japan 24%, Italy 17%, France 14%. **Tourism** (1998): $21 mil. **Budget** (1996 est.): $265 mil. **Intl. reserves less gold** (May 2000): $193.9 mil. **Gold:** 12,000 oz t. **Consumer prices** (change in 1999): 4.1%.

Transport: Railroad: Length: 437 mi. **Motor vehicles:** 17,300 pass. cars, 9,210 comm. vehicles. **Civil aviation:** 201.1 mil pass.-mi; 9 airports. **Chief ports:** Nouakchott, Nouadhibou.

Communications: Radios: 428 per 1,000 pop. **Telephones:** 19,000 main lines.

Health: Life expectancy: 49.06 male; 53.29 female. **Births** (per 1,000 pop): 42.95. **Deaths** (per 1,000 pop.): 13.65. **Natural inc.:** 2.93%. **Infant mortality** (per 1,000 live births): 76.7.

Education: Compulsory: ages 6-12. **Literacy:** 38%.

Major Intl. Organizations: UN (FAO, IBRD, ILO, IMF, IMO, WHO, WTrO), AL, OAU.

Embassy: 2129 Leroy Pl. NW 20008; 232-5700.

Website: http://www.embassy.org/mauritania

Mauritania was a French protectorate from 1903. It became independent Nov. 28, 1960 and annexed the south of former Spanish Sahara (now Western Sahara) in 1976. Saharan guerrillas of the Polisario Front stepped up attacks in 1977; 8,000 Moroccan troops and French bomber raids aided the government. Mauritania signed a peace treaty with the Polisario Front, 1979, resumed diplomatic relations with Algeria while breaking a defense treaty with Morocco, and renounced sovereignty over its share of Western Sahara. Opposition parties were legalized and a new constitution approved in 1991.

Although slavery has been repeatedly abolished, most recently in 1980, thousands of Mauritanians continued to live under conditions of servitude.

Mauritius

Republic of Mauritius

People: Population: 1,189,825. **Age distrib.** (%): <15: 25.5; 65+: 6.2. **Pop. density:** 1,669 per sq. mi. **Urban:** 41%. **Ethnic groups:** Indo-Mauritian 68%, Creole 27%. **Principal languages:** English (official), French, Creole, Hindi, Bojpoori. **Chief religions:** Hindu 52%, Christian 28.3%, Muslim 16.6%.

Geography: Area: 700 sq. mi. **Location:** In the Indian O., 500 mi. E of Madagascar. **Neighbors:** Nearest is Madagascar to W. **Topography:** A volcanic island nearly surrounded by coral reefs. A central plateau is encircled by mountain peaks. **Capital:** Port Louis (1999 est.): 172,000.

Government: Type: Republic. **Head of state:** Pres. Cassam Uteem; b Mar. 22, 1941; in office: June 30, 1992. **Head of gov.:** Prime Min. Anerood Jugnauth; b Mar. 29, 1930; in office: Sept. 17, 2000. **Local divisions:** 9 districts, 3 dependencies. **Defense budget:** $9 mil.

Economy: Industries: Tourism, textiles, food processing. **Chief crops:** Sugarcane, corn, potatoes, tea. **Arable land:** 49%. **Livestock** (2000): cattle: 29,400; chickens: 4.4 mil; goats: 94,000; pigs: 20,000; sheep: 7,300. **Fish catch:** (1999): 13,852 metric tons. **Electricity prod.:** 1.260 bil kWh. **Labor force:** 36% const. & ind.; 24% services; 16% trade, restaurants & hotels.

Finance: Monetary unit: Rupee (Sept. 2001: 29.60 = $1 U.S.). **GDP:** (1999 est.): $12.3 bil. **Per capita GDP:** $10,400. **Imports** (1998 est.): $2.3 bil; partners: France 19%. **Exports** (1999 est.): $1.6 bil; partners: UK 32%, France 19%, U.S. 14%. **Tourism:** $545 mil. **Budget** (FY 1995-96 est.): $1 bil. **Intl. reserves less gold** (June 2000): $676.0 mil. **Gold:** 62,000 oz t. **Consumer prices** (change in 1999): 6.9%.

Transport: Motor vehicles: 69,945 pass. cars, 12,328 comm. vehicles. **Civil aviation:** 2.4 bil pass.-mi; 1 airport. **Chief port:** Port Louis.

Communications: TV sets: 150 per 1,000 pop. **Radios:** 353 per 1,000 pop. **Telephones**: 280,900 main lines. **Daily newspaper circ.:** 49 per 1,000 pop.

Health: Life expectancy: 67.26 male; 75.31 female. **Births** (per 1,000 pop): 16.50. **Deaths** (per 1,000 pop.): 6.82. **Natural inc.:** .97%. **Infant mortality** (per 1,000 live births): 17.19.

Education: Compulsory: ages 5-12. **Literacy:** 83%.

Major Intl. Organizations: UN and all of its specialized agencies, the Commonwealth, OAU.

Embassy: 4301 Connecticut Ave. NW, Suite 441, 20008; 244-1491.

Mauritius was uninhabited when settled in 1638 by the Dutch, who introduced sugarcane. France took over in 1721, bringing African slaves. Britain ruled from 1810 to Mar. 12, 1968, bringing Indian workers for the sugar plantations.

Mauritius formally severed its association with the British crown Mar. 12, 1992.

Mexico

United Mexican States

People: Population: 101,879,171. **Age distrib.** (%): <15: 33.3; 65+: 4.4. **Pop. density:** 137 per sq. mi. **Urban:** 74%. **Ethnic groups:** Mestizo 60%, Amerindian 30%, Caucasian 9%. **Principal languages:** Spanish (official), Mayan dialects. **Chief religions:** Roman Catholic 89%, Protestant 6%.

Geography: Area: 741,600 sq. mi. **Location:** In southern North America. **Neighbors:** U.S. on N, Guatemala and Belize on S. **Topography:** The Sierra Madre Occidental Mts. run NW-SE near the west coast; the Sierra Madre Oriental Mts. run near the Gulf of Mexico. They join S of Mexico City. Between the 2 ranges lies the dry central plateau, 5,000 to 8,000 ft. alt., rising toward the S, with temperate vegetation. Coastal lowlands are tropical. About 45% of land is arid. **Capital:** Mexico City. **Cities** (1995 est.): Mexico City 8,489,007; Guadalajara 1,633,216; Puebla 1,222,569.

Government: Type: Federal republic. **Head of state and gov.:** Pres. Vicente Fox Quesada; b July 2, 1942; in office: Dec. 1, 2000. **Local divisions:** 31 states, 1 federal district. **Defense budget:** $3 bil. **Active troops:**192,770.

Economy: Industries: Steel, food & beverages, chemicals, consumer durables, textiles, tourism. **Chief crops:** Cotton, coffee, wheat, rice, beans, soybeans, corn. **Minerals:** Silver, lead, zinc, gold, oil, gas, copper. **Crude oil reserves** (2000): 28.40 bil bbls. **Arable land:** 12%. **Livestock** (2000): cattle: 30.29 mil; chickens: 476.0 mil; goats: 9.6 mil; pigs: 13.69 mil; sheep: 5.9 mil. **Fish catch** (1999): 1.53 mil metric tons. **Electricity prod.:** 182.492 bil kWh. **Labor force** (1997): 55% services; 24% agric.; 21% ind.

Finance: Monetary unit: New Peso (Sept. 2001: 9.36 = $1 U.S.). **GDP:** (1999 est.): $865.5 bil. **Per capita GDP:** $8,500. **Imports** (1999 est.): $142.1 bil; partners: U.S. 74.8%. **Exports** (1999 est.): $136.8 bil; partners: U.S. 89.3%. **Tourism:** $7.59 bil. **Budget** (1998 est.): $123 bil. **Intl. reserves less gold** (June 2000): $32.97 bil. **Gold:** 296,000 oz t. **Consumer prices** (change in 1997): 16.6%.

Transport: Railroad: Length: 16,543 mi. **Motor vehicles:** 8.2 mil pass. cars, 4.03 mil comm. vehicles. **Civil aviation:** 14.7 bil pass.-mi; 83 airports. **Chief ports:** Coatzacoalcos, Mazatlan, Tampico, Veracruz.

Communications: TV sets: 257 per 1,000 pop. **Radios:** 329 per 1,000 pop. **Telephones**: 12,332,600 main lines. **Daily newspaper circ.:** 97 per 1,000 pop.

Health: Life expectancy: 68.73 male; 74.93 female. **Births** (per 1,000 pop): 22.77. **Deaths** (per 1,000 pop.): 5.02. **Natural inc.:** 1.78%. **Infant mortality** (per 1,000 live births): 25.36.

Education: Free, compulsory: ages 6-12. **Literacy**: 90%.

Major Intl. Organizations: UN (FAO, IBRD, ILO, IMF, IMO, WHO, WTrO), APEC, OAS, OECD.

Embassy: 1911 Pennsylvania Ave. NW 20006; 728-1600.

Website: http://www.inegi.gob.mx

Mexico was the site of advanced Indian civilizations. The Mayas, an agricultural people, moved up from Yucatan, built immense stone pyramids, invented a calendar. The Toltecs were overcome by the Aztecs, who founded Tenochtitlan AD 1325, now Mexico City. Hernando Cortes, Spanish conquistador, destroyed the Aztec empire, 1519-21.

After 3 centuries of Spanish rule the people rose, under Fr. Miguel Hidalgo y Costilla, 1810, Fr. Morelos y Payon, 1812, and Gen. Agustin Iturbide, who made himself emperor as Agustin I, 1821. A republic was declared in 1823.

Mexican territory extended into the present American Southwest and California until Texas revolted and established a republic in 1836; the Mexican legislature refused recognition but was unable to enforce its authority there. After numerous clashes, the U.S.-Mexican War, 1846-48, resulted in the loss by Mexico of the lands north of the Rio Grande.

French arms supported an Austrian archduke on the throne of Mexico as Maximilian I, 1864-67, but pressure from the U.S. forced France to withdraw. Dictatorial rule by Porfirio Diaz, president 1877-80, 1884-1911, led to a period of rebellion and factional fighting. A new constitution, Feb. 5, 1917, brought social reform.

The Institutional Revolutionary Party (PRI) dominated politics from 1929 until the late 1990s. Radical opposition, including some guerrilla activity, was contained by strong measures. Some gains in agriculture, industry, and social services were achieved, but much of the work force remained jobless or underemployed. Although prospects brightened with the discovery of vast oil reserves, inflation and a drop in world oil prices aggravated Mexico's economic problems in the 1980s.

Mexico reached agreement with the U.S. and Canada on the North American Free Trade Agreement (NAFTA) Aug. 12, 1992; it took effect Jan. 1, 1994.

Guerrillas of the Zapatista National Liberation Army (EZLN) launched an uprising, Jan. 1, 1994, in southern Mexico. A tentative peace accord was reached Mar. 2. The presidential candidate of the governing PRI, Luis Donaldo Colosio Murrieta, was assassinated at a political rally in Tijuana, Mar. 23. The new PRI candidate, Ernesto Zedillo Ponce de León, won election Aug. 21 and was inaugurated Dec. 1, 1994.

An austerity plan and pledges of aid from the U.S. saved Mexico's currency from collapse in early 1995. Popular Revolutionary Army guerrillas launched coordinated attacks on government targets in Aug. 1996. In elections July 6, 1997, the PRI failed to win a congressional majority for the first time since 1929. An armed gang massacred 45 peasants in Chiapas on Dec. 22, 1997. In the presidential election of July 2, 2000, the PRI lost for the 1st time in over 7 decades; the winner, opposition candidate Vicente Fox Quesada, took office Dec. 1, 2000.

Micronesia

Federated States of Micronesia

People: Population: 134,597. **Pop. density:** 497 per sq. mi. **Urban:** 28%. **Ethnic groups:** 9 ethnic Micronesian and Polynesian groups. **Principal languages:** English (official), Trukese, Pohnpeian, Yapese. **Chief religions:** Roman Catholic 50%, Protestant 47%.

Geography: Area: 271 sq. mi. **Location:** Consists of 607 islands in the W Pacific Ocean. **Capital:** Palikir, on Pohnpei; (1994 island pop.) 33,372.

Government: Type: Republic. **Head of state and gov.:** Pres. Leo A. Falcam; b Nov. 20, 1935; in office: May 11, 1999. **Local divisions:** 4 states.

Economy: Industries: Tourism, fish processing. **Chief crops:** Tropical fruits, vegetables, black pepper. **Livestock** (2000): cattle: 13,900; chickens: 185,000; goats: 4,000; pigs: 32,000. **Labor force:** 67% govt.

Finance: Monetary unit: U.S. Dollar. **GDP:** (1997 est.): $240 mil. **Per capita GDP:** $2,000. **Imports** (1996 est.): $168 mil; partners: U.S., Japan, Australia. **Exports** (1996 est.): $73 mil; partners: Japan, U.S., Guam. **Budget** (FY 1995-96 est.): $52 mil.

Transport: 4 airports. **Chief ports:** Colonia (Yap), Kolonia (Pohnpei), Lele, Moen.

Communications: TV sets: 19 per 1,000 pop. **Radios:** 664 per 1,000 pop. **Telephones**: 9,800 main lines.

Health (2000): **Life expectancy:** 66.67 male; 70.62 female. **Births** (per 1,000 pop): 27.09. **Deaths** (per 1,000 pop.): 5.95. **Natural inc.:** 2.11%. **Infant mortality** (per 1,000 live births): 33.48.

Education: Compulsory: ages 6-14. **Literacy** (1991): 90%.

Major Intl. Organizations: UN (IBRD, IMF, WHO).

Embassy: 1725 N St. NW 20036; 223-4383.

The Federated States of Micronesia, formerly known as the Caroline Islands, was ruled successively by Spain, Germany, Japan, and the U.S. It was internationally recognized as an independent nation Sept. 17, 1991.

Moldova

Republic of Moldova

People: Population: 4,431,570. **Age distrib.** (%): <15: 22.4; 65+: 9.9. **Pop. density:** 341 per sq. mi. **Urban:** 46%. **Ethnic groups:** Moldovan/Romanian 64.5%, Ukrainian 13.8%, Russian 13%. **Principal languages:** Moldovan (official), Russian. **Chief religion:** Eastern Orthodox 98.5%.

Geography: Area: 13,000 sq. mi. **Location:** In E Europe. **Neighbors:** Romania on W; Ukraine on N, E, and S. **Capital** (1999 est.): Chisinau 655,000.

Government: Type: Republic. **Head of state:** Pres. Vladimir Voronin; b May 25, 1941; in office: Apr. 7, 2001. **Head of gov.:** Prime Min. Vasile Tarlev; b Oct. 9, 1963; in office: Apr. 19, 2001. **Local divisions:** 21 cities and towns, 48 urban settlements, more than 1,600 villages. **Defense budget:** $5.5 mil (2001). **Active troops:** 9,500.

Economy: Industries: Food processing, machinery, textiles. **Chief crops:** Grain, vegetables, fruits, wine. **Minerals:** Lignite, phosphorites, gypsum. **Arable land:** 53%. **Livestock** (2000): cattle: 416,000; chickens: 13.5 mil; goats: 95,000; pigs: 705,000; sheep: 974,000. **Electricity prod.:** 4.155 bil kWh. **Labor force:** 40.2% agric.; 14.3% industry.

Finance: Monetary unit: Leu (Sept. 2001: 12.87 = $1 U.S.). **GDP:** (1999 est.): $9.7 bil. **Per capita GDP:** $2,200. **Imports** (1999): $560 mil; partners: Russia 22%, Ukraine 16%, Romania 12%. **Exports** (1999): $470 mil; partners: Russia 53%, Romania 10%, Ukraine 8%. **Tourism** (1998): $2 mil. **Budget** (1997 est.): $594 mil. **Intl. reserves less gold** (June 1999): $198.77 mil. **Consumer prices** (change in 1999): 45.9%.

Transport: Railroad: Length: 746 mi. **Motor vehicles:** 169,000 pass. cars, 71,000 comm. vehicles. **Civil aviation:** 37.8 mil pass.-mi; 1 airport.

Communications: TV sets: 30 per 1,000 pop. **Radios:** 209 per 1,000 pop. **Telephones**: 583,800 main lines. **Daily newspaper circ.:** 24 per 1,000 pop.

Health: Life expectancy: 60.15 male; 69.26 female. **Births** (per 1,000 pop): 13.35. **Deaths** (per 1,000 pop.): 12.6. **Natural inc.:** .08%. **Infant mortality** (per 1,000 live births): 42.74.

Education: Compulsory: ages 7-16. **Literacy:** 96%.

Major Intl. Organizations: UN (FAO, IBRD, ILO, IMF, WHO, WTrO), CIS, OSCE.

Embassy: 2101 S St. NW 20008; 667-1130.

In 1918, Romania annexed all of Bessarabia that Russia had acquired from Turkey in 1812 by the Treaty of Bucharest. In 1924, the Soviet Union established the Moldavian Autonomous Soviet Socialist Republic on the eastern bank of the Dniester. It was merged with the Romanian-speaking districts of Bessarabia in 1940 to form the Moldavian SSR.

During World War II, Romania, allied with Germany, occupied the area. It was recaptured by the USSR in 1944. Moldova declared independence Aug. 27, 1991. It became an independent state when the USSR disbanded Dec. 26, 1991.

Fighting erupted Mar. 1992 in the Dnestr (Dniester) region between Moldovan security forces and Slavic separatists—ethnic Russians and ethnic Ukrainians—who feared Moldova would merge with neighboring Romania. In a plebiscite on Mar. 6, 1994, voters in Moldova supported independence, without unification with Romania.

Defying the Moldovan government, voters in the breakaway Dnestr region held legislative elections and approved a separatist constitution Dec. 24, 1995. Petru Lucinschi, a former Communist, won a presidential runoff election Dec. 1, 1996. A peace accord with Dnestr separatists was signed in Moscow May 8, 1997. The Communists won the most seats in parliamentary elections Mar. 22, 1998, but a coalition of three center-right parties formed the government. New elections Feb. 25, 2001, brought a decisive Communist victory.

Monaco

Principality of Monaco

People: Population: 31,842. **Age distrib.** (%): <15: 15.3; 65+: 22.5. **Pop. density:** 42,458 per sq. mi. **Urban:** 100%. **Ethnic groups:** French 47%, Italian 16%, Monegasque 16%. **Principal languages:** French (official), English, Italian, Monegasque. **Chief religion:** Roman Catholic 95%.

Geography: Area: 0.75 sq. mi. **Location:** On the NW Mediterranean coast. **Neighbors:** France to W, N, E. **Topography:** Monaco-Ville sits atop a high promontory, the rest of the principality rises from the port up the hillside. **Capital:** Monaco-ville (1990): 1,151.

Government: Type: Constitutional monarchy. **Head of state:** Prince Rainier III; b May 31, 1923; in office: May 9, 1949. **Head of gov.:** Min. of State Patrick Leclercq; b 1938; in office: Jan. 5, 2000. **Local divisions:** 4 quarters.

Economy: Industries: Tourism, gambling, chemicals, precision instruments.

Finance: Monetary unit: French Franc (Sept. 2001: 7.24 = $1 U.S.) or Monegasque Franc. **GDP:** (1999 est.): $870 mil. **Per capita GDP:** $27,000. **Budget** (1995 est.): $531 mil.

Transport: Motor vehicles: 17,000 pass. cars, 4,000 comm. vehicles. **Civil aviation:** 820,000 pass.-mi; 1 airport. **Chief port:** Monaco.

Communications: TV sets: 690 per 1,000 pop. **Radios:** 941 per 1,000 pop.

Health: Life expectancy: 75.04 male; 83.12 female. **Births** (per 1,000 pop): 9.74. **Deaths** (per 1,000 pop.): 13.. **Natural inc.:** -.33%. **Infant mortality** (per 1,000 live births): 5.83.

Education: Compulsory: ages 6-16.

Major Intl. Organizations: UN (IMO, WHO), OSCE.

An independent principality for over 300 years, Monaco has belonged to the House of Grimaldi since 1297, except during the French Revolution. It was placed under the protectorate of Sardinia in 1815, and under France, 1861. The Prince of Monaco was an absolute ruler until the 1911 constitution. Monaco was admitted to the UN on May 28, 1993.

Monaco's fame as a tourist resort is widespread. It is noted for its mild climate, magnificent scenery, and elegant casinos.

Mongolia

People: Population: 2,654,999. **Age distrib.** (%): <15: 33.0; 65+: 3.9. **Pop. density:** 4 per sq. mi. **Urban:** 63%. **Ethnic groups:** Mongol 90%. **Principal language:** Khalkha Mongol (official). **Chief religion:** Mostly Tibetan Buddhist.

Geography: Area: 603,500 sq. mi. **Location:** In E Central Asia. **Neighbors:** Russia on N, China on E, W, and S. **Topography:** Mostly a high plateau with mountains, salt lakes, and vast grasslands. Arid lands in the S are part of the Gobi Desert. **Capital:** Ulaanbaatar (1999 est.): 744,000.

Government: Type: Republic. **Head of state:** Pres. Natsagiyn Bagabandi; b Apr. 22, 1950; in office: June 20, 1997. **Head of gov.:** Nambaryn Enkhbayar; b June 1, 1958; in office: July 26, 2000. **Local divisions:** 18 provinces, 3 municipalities. **Defense budget:** $24.6 mil. **Active troops:** 9,100.

Economy: Industries: Food processing, mining, construction materials. **Chief crops:** Grain, potatoes. **Minerals:** Coal, oil, tungsten, copper, molybdenum, gold, phosphates, tin. **Arable land:** 1%. **Livestock** (2000): cattle: 3.5 mil; chickens: 70,000; goats: 10.0 mil; pigs: 19,000; sheep: 14.0 mil. **Electricity prod.:** 2.671 bil kWh. **Labor force:** primarily agricultural.

Finance: Monetary unit: Tugrik (Sept. 2001: 1,100.00 = $1 U.S.). **GDP:** (1999 est.): $6.1 bil. **Per capita GDP:** $2,320. **Imports** (1998 est.): $472.4 mil; partners: Russia 30.6%, China 13.3%, Japan 11.7%. **Exports** (1998 est.): $316.8 mil; partners: China 30.1%, Switzerland 21.5%, Russia 12.1%. **Tourism:** $28 mil. **Intl. reserves less gold** (May 2000): $148.80 mil. **Gold:** 1,000 oz t. **Consumer prices** (change in 1999): 7.6%.

Transport: Railroad: Length: 1,294 mi. **Motor vehicles:** 21,000 pass. cars, 27,000 comm. vehicles. **Civil aviation:** 121.0 mil pass.-mi; 1 airport.

Communications: TV sets: 60.7 per 1,000 pop. **Radios:** 74 per 1,000 pop. **Telephones**: 132,200 lines. **Daily newspaper circ.:** 92 per 1,000 pop.

Health: Life expectancy: 62.14 male; 66.5 female. **Births** (per 1,000 pop): 21.80. **Deaths** (per 1,000 pop.): 7.1. **Natural inc.:** 1.47%. **Infant mortality** (per 1,000 live births): 53.5.

Education: Compulsory: ages 6-16. **Literacy** (1991): 83%.

Major Intl. Organizations: UN (FAO, IBRD, ILO, IMF, IMO, WHO, WTrO).

Embassy: 2833 M St. NW 20007; 333-7117.

Website: http://www.MongoliaOnline.mn

One of the world's oldest countries, Mongolia reached the zenith of its power in the 13th century when Genghis Khan and his successors conquered all of China and extended their influence as far west as Hungary and Poland. In later centuries, the empire dissolved and Mongolia became a province of China.

With the advent of the 1911 Chinese revolution, Mongolia, with Russian backing, declared its independence. A Communist regime was established July 11, 1921.

In 1990, the Mongolian Communist Party yielded its monopoly on power but won election in July. A new constitution took effect Feb. 12, 1992. A democratic alliance won legislative elections, June 30, 1996. Natsagiyn Bagabandi, a former Communist, won the presidential election of May 18, 1997. A protracted political crisis took a violent turn Oct. 2, 1998, with the murder of Sanjaasuregiyn Zorig, a popular cabinet member seeking to become prime minister. The former Communists won 72 of 76 seats in parliamentary elections, July 2, 2000.

Morocco

Kingdom of Morocco

People: Population: 30,645,305. **Age distrib.** (%): <15: 34.4; 65+: 4.7. **Pop. density:** 178 per sq. mi. **Urban:** 55%. **Ethnic groups:** Arab-Berber 99%. **Principal languages:** Arabic (official), Berber dialects. **Chief religion:** Muslim 98.7%.

Geography: Area: 172,100 sq. mi. **Location:** On NW coast of Africa. **Neighbors:** Western Sahara on S, Algeria on E. **Topography:** Consists of 5 natural regions: mountain ranges (Riff in the N, Middle Atlas, Upper Atlas, and Anti-Atlas); rich plains in the W; alluvial plains in SW; well-cultivated plateaus in the center; a pre-Sahara arid zone extending from SE. **Capital:** Rabat. **Cities:** Casablanca 3,541,000; Rabat 1,496,000.

Government: Type: Constitutional monarchy. **Head of state:** King Mohammed VI; b Aug. 21, 1963; in office: Jul 23, 1999. **Head of gov.:** Prime Min. Abderrahmane El Youssoufi; b Mar. 8, 1924; in office: Feb. 4, 1998. **Local divisions:** 16 regions. **Defense budget** (1999): $1.7 bil. **Active troops:** 198,500.

Economy: Industries: Food processing, textiles, leather goods, mining, tourism. **Chief crops:** Grain, citrus, wine grapes, olives. **Minerals:** Phosphates, iron ore, manganese, lead, zinc. **Crude oil reserves** (2000): 1.9 mil bbls. **Arable land:** 21%. **Livestock** (2000): cattle: 2.67 mil; chickens: 100.0 mil; goats: 5.12 mil; pigs: 8,000; sheep: 17.3 mil. **Fish catch** (1999): 785,843 metric tons. **Electricity prod.:** 13.695 bil kWh. **Labor force:** 50% agric.; 35% services; 15% ind.

Finance: Monetary unit: Dirham (Sept. 2001: 11.36 = $1 U.S.). **GDP:** (1999 est.): $108 bil. **Per capita GDP:** $3,600. **Imports** (1998): $9.5 bil; partners: France 22%, Spain 10%, U.S. 7%. **Exports** (1998): $7.1 bil; partners: France 27%, Spain 11%, India 7%. **Tourism:** $1.96 bil. **Budget** (1996 est.): $10 bil. **Intl. reserves less gold** (May 2000): $4.86 bil. **Gold:** 705,000 oz t. **Consumer prices** (change in 1999): 0.7%.

Transport: Railroad: Length: 1,099 mi. **Motor vehicles** (1997): 1.10 mil pass. cars, 333,152 comm. vehicles. **Civil aviation:** 3.3 bil pass.-mi; 11 airports. **Chief ports:** Tangier, Casablanca, Kenitra.

Communications: TV sets: 92.7 per 1,000 pop. **Radios:** 222 per 1,000 pop. **Telephones**: 1,425,000 main lines. **Daily newspaper circ.:** 14.5 per 1,000 pop.

Health: Life expectancy: 67.2 male; 71.76 female. **Births** (per 1,000 pop): 24.16. **Deaths** (per 1,000 pop.): 5.94. **Natural inc.:** 1.82%. **Infant mortality** (per 1,000 live births): 48.11.

Education: Compulsory: ages 7-13. **Literacy**: 44%.

Major Intl. Organizations: UN (FAO, IBRD, ILO, IMF, IMO, WHO, WTrO), AL.

Embassy: 1601 21st St. NW 20009; 462-7979.

Berbers were the original inhabitants, followed by Carthaginians and Romans. Arabs conquered in 683. In the 11th and 12th centuries, a Berber empire ruled all NW Africa and most of Spain from Morocco.

Part of Morocco came under Spanish rule in the 19th century; France controlled the rest in the early 20th. Tribal uprisings lasted from 1911 to 1933. The country became independent Mar. 2, 1956. Tangier, an internationalized seaport, was turned over to Morocco, 1956. Ifni, a Spanish enclave, was ceded in 1969. Morocco annexed the disputed territory of Western Sahara during the second half of the 1970s.

King Hassan II assumed the throne in 1961, reigning until his death on July 23, 1999; he was immediately succeeded by his eldest son. Political reforms in the 1990s included the establishment of a bicameral legislature in 1997.

Western Sahara

Western Sahara, formerly the protectorate of Spanish Sahara, is bounded the in N by Morocco, the NE by Algeria, the E and S by Mauritania, and on the W by the Atlantic Ocean. Phosphates are the major resource. Population (2001 est.): 250,559; capital: Laayoune (El Aaiun). Area: 102,600 sq mi.

Spain withdrew from its protectorate in Feb. 1976. On Apr. 14, 1976, Morocco annexed over 70,000 sq. mi, with the remainder annexed by Mauritania. A guerrilla movement, the Polisario Front, which had proclaimed the region independent Feb. 27, launched attacks with Algerian support. After Mauritania signed a treaty with Polisario on Aug. 5, 1979, Morocco occupied Mauritania's portion of Western Sahara.

After years of bitter fighting, Morocco controlled the main urban areas, but Polisario guerrillas moved freely in the vast, sparsely populated deserts. The 2 sides implemented a ceasefire in 1991, when a UN peacekeeping force was deployed. A UN-sponsored referendum on self-determination for Western Sahara has been repeatedly postponed.

Mozambique

Republic of Mozambique

People: Population: 19,371,057. **Age distrib.** (%): <15: 42.7; 65+: 2.8. **Pop. density:** 64 per sq. mi. **Urban:** 39%. **Ethnic groups:** Indigenous tribal groups. **Principal languages:** Portuguese (official), indigenous dialects. **Chief religions:** Indigenous beliefs 50%, Christian 30%, Muslim 20%.

Geography: Area: 302,400 sq. mi. **Location:** On SE coast of Africa. **Neighbors:** Tanzania on N; Malawi, Zambia, Zimbabwe on W; South Africa, Swaziland on S. **Topography:** Coastal lowlands comprise nearly half the country with plateaus rising in steps to the mountains along the western border. **Capital:** Maputo (urban agg.): 3,025,000.

Government: Type: Republic. **Head of state:** Pres. Joaquim Chissano; b Oct. 22, 1939; in office: Oct. 19, 1986. **Head of gov.:** Prime Min. Pascoal Mocumbi; b Apr. 10, 1941; in office: Dec. 21, 1994. **Local divisions:** 10 provinces. **Defense budget:** $87 mil **Active troops:** 5,100–6,100.

Economy: Industries: Chemicals, petroleum products, textiles. **Chief crops:** Cashews, cotton, sugar, corn, cassava, tea. **Minerals:** Coal, titanium. **Arable land:** 4%. **Livestock** (2000): cattle: 1.32 mil; chickens: 28.0 mil; goats: 392,000; pigs: 180,000; sheep: 125,000. **Fish catch** (1999): 39,579 metric tons. **Electricity prod.:** 2.300 bil kWh. **Labor force:** 81% agric.; 13% services.

Finance: Monetary unit: Metical (Sept. 2001: 21,700 = $1 U.S.). **GDP:** (1999 est.): $18.7 bil. **Per capita GDP:** $1000. **Imports** (1999 est.): $1.6 bil; partners: South Africa 55%, Zimbabwe 7%. **Exports** (1999 est.): $300 mil; partners: Spain 17%, South Africa 16%, Portugal 12%. **Budget** (1997 est.): $799 mil. **Intl. reserves less gold** (May 2000): $684.20 mil. **Consumer prices** (change in 1999): 2.0%.

Transport: Railroad: Length: 1,940 mi. **Motor vehicles:** 67,600 pass. cars, 21,200 comm. vehicles. **Civil aviation:** 180.6 mil pass.-mi; 7 airports. **Chief ports:** Maputo, Beira, Nacala, Inhambane.

Communications: TV sets: 3.5 per 1,000 pop. **Radios:** 38 per 1,000 pop. **Telephones**: 85,700 main lines. **Daily newspaper circ.:** 8 per 1,000 pop.

Health: Life expectancy: 37.25 male; 35.62 female. **Births** (per 1,000 pop): 37.20. **Deaths** (per 1,000 pop.): 24.21. **Natural inc.:** 1.3%. **Infant mortality** (per 1,000 live births): 139.2.

Education: Compulsory: ages 7-14. **Literacy**: 40%.

Major Intl. Organizations: UN (FAO, IBRD, ILO, IMF, IMO, WHO, WTrO), the Commonwealth, OAU.

Embassy: 1990 M St. NW, Suite 570, 20036; 293-7146.

Website: http://www.mbendi.co.za/cymzcy.htm

The first Portuguese post on the Mozambique coast was established in 1505, on the trade route to the East. Mozambique became independent June 25, 1975, after a ten-year war against Portuguese colonial domination. The 1974 revolution in Portugal had paved the way for the orderly transfer of power to Frelimo (Front for the Liberation of Mozambique). Frelimo took over local administration Sept. 20, 1974, although opposed, in part violently, by some blacks and whites.

The new government, led by Maoist Pres. Samora Machel, provided for a gradual transition to a Communist system. Economic problems included the emigration of most of the country's whites, a politically untenable economic dependence on white-ruled South Africa, and a large external debt.

In the 1980s, severe drought and civil war caused famine and heavy loss of life. Pres. Machel was killed in a plane crash just inside the South African border, Oct. 19, 1986.

The ruling party formally abandoned Marxist-Leninism in 1989, and a new constitution, effective Nov. 30, 1990, provided for multiparty elections and a free-market economy.

On Oct. 4, 1992, a peace agreement was signed aimed at ending hostilities between the government and the rebel Mozambique National Resistance (MNR). Repatriation of 1.7 million Mozambican refugees officially ended June 1995. In Mar. 1999 the heaviest floods in 4 decades left nearly 200,000 people stranded. Even worse flooding in Feb.-Mar. 2000 claimed more than 600 lives, displaced over 1 million people, and devastated the economy.

Myanmar (*formerly* Burma)

Union of Myanmar

People: Population: 41,734,853. **Age distrib.** (%): <15: 29.8; 65+: 4.7. **Pop. density:** 159 per sq. mi. **Urban:** 27%. **Ethnic groups:** Burman 68%, Shan 9%, Karen 7%, Rakhine 4%. **Principal language:** Burmese (official). **Chief religions:** Buddhist 89%, Christian 4%, Muslim 4%.

Geography: Area: 262,000 sq. mi. **Location:** Between S and SE Asia, on Bay of Bengal. **Neighbors:** Bangladesh, India on W; China, Laos, Thailand on E. **Topography:** Mountains surround Myanmar on W, N, and E, and dense forests cover much of the nation. N-S rivers provide habitable valleys and communications, especially the Irrawaddy, navigable for 900 miles. The country has a tropical monsoon climate. **Capital:** Yangon (Rangoon); pop. (urban agg.): 4,196,000.

Government: Type: Military. **Head of state and gov.:** Gen. Than Shwe; b Feb. 2, 1933; in office: Apr. 24, 1992. **Local divisions:** 7 states, 7 divisions. **Defense budget** (1999): $1.7 bil. **Active troops:** 429,000.

Economy: Industries: Textiles, footwear, wood products, agric. processing. **Chief crops:** Rice, sugarcane, corn, pulses. **Minerals:** Oil, lead, copper, tin, tungsten, precious stones. **Crude oil reserves** (2000): 50 mil bbls. **Arable land:** 15%. **Livestock** (2000): cattle: 10.96 mil; chickens: 43.52 mil; goats: 1.39 mil; pigs: 3.91 mil; sheep: 389,781. **Fish catch** (1999): 917,666 metric tons. **Electricity prod.:** 4.813 bil kWh. **Labor force:** 65% agric.; 25% services; 10% industry.

Finance: Monetary unit: Kyat (Sept. 2001: 6.63 = $1 U.S.). **GDP:** (1999 est.): $59.4 bil. **Per capita GDP:** $1,200. **Imports** (1998): $2.5 bil; partners: Singapore 31%, Japan 12%, Thailand 12%. **Exports** (1998): $1.2 bil; partners: India 13%, China 11%, Singapore 10%. **Tourism:** $35 mil. **Budget** (FY 1996-97): $12.2 bil. **Intl. reserves less gold** (May 2000): $286.1 mil. **Gold:** 231,000 oz t. **Consumer prices** (change in 1999): 18.4%.

Transport: Railroad: Length: 3,144 mi. **Motor vehicles:** 35,000 pass. cars, 34,000 comm. vehicles. **Civil aviation:** 91.5 mil pass.-mi; 19 airports. **Chief ports:** Bassein, Moulmein.

Communications: TV sets: 22 per 1,000 pop. **Radios:** 72 per 1,000 pop. **Telephones**: 266,200 main lines. **Daily newspaper circ.:** 23 per 1,000 pop.

Health: Life expectancy: 53.73 male; 56.68 female. **Births** (per 1,000 pop.): 20.13. **Deaths** (per 1,000 pop.): 12.30. **Natural inc.:** 0.783%. **Infant mortality** (per 1,000 live births): 73.71.

Education: Free, compulsory: ages 5-10. **Literacy**: 83%.

Major Intl. Organizations: UN (FAO, IBRD, ILO, IMF, IMO, WHO, WTrO), ASEAN.

Embassy: 2300 S St. NW 20008; 332-9044.

Website: http://www.myanmar.com/e-index.html

The Burmese arrived from Tibet before the 9th century, displacing earlier cultures, and a Buddhist monarchy was established by the 11th. Burma was conquered by the Mongol dynasty of China in 1272, then ruled by Shans as a Chinese tributary, until the 16th century.

Britain subjugated Burma in 3 wars, 1824-84, and ruled the country as part of India until 1937, when it became self-governing. Independence outside the Commonwealth was achieved Jan. 4, 1948.

Gen. Ne Win dominated politics from 1962 to 1988, first as military ruler then as constitutional president. His regime drove Indians from the civil service and Chinese from commerce. Economic socialization was advanced, isolation from foreign countries enforced. In 1987 Burma, once the richest nation in SE Asia, was granted less-developed status by the UN.

Ne Win resigned July 1988, following waves of antigovernment riots. Rioting and street violence continued, and in Sept. the military seized power, under Gen. Saw Maung. In 1989 the country's name was changed to Myanmar.

The first free multiparty elections in 30 years took place May 27, 1990, with the main opposition party winning a decisive victory, but the military refused to hand over power. A key opposition leader, Aung San Suu Kyi, awarded the Nobel Peace Prize in 1991, was held under house arrest from July 20, 1989, to July 10, 1995; after her release, the military government continued to restrict her activities and to harass and imprison her supporters. New U.S. economic sanctions took effect on May 21, 1997. Myanmar was admitted to ASEAN July 23, 1997.

Namibia

Republic of Namibia

People: Population: 1,797,677. **Age distrib.** (%): <15: 42.7; 65+: 3.7. **Pop. density:** 6 per sq. mi. **Urban:** 30%. **Ethnic groups:** Ovambo 50%, Kavangos 9%, Herero 7%, Damara 7%. **Principal languages:** Afrikaans, English (official), German, indigenous languages. **Chief religions:** Lutheran 50%, other Christian 30%.

Geography: Area: 317,500 sq. mi. **Location:** In S Africa on the coast of the Atlantic Ocean. **Neighbors:** Angola on N, Botswana on E, South Africa on S. **Capital:** Windhoek (1999 est.): 202,000.

Government: Type: Republic. **Head of state:** Pres. Sam Nujoma; b May 12, 1929; in office: Mar. 21, 1990. **Head of gov.:** Prime Min. Hage Geingob; b Aug. 3, 1941; in office: Mar. 21, 1990. **Local divisions:** 13 regions. **Defense budget:** $96 mil. **Active troops:** 9,000.

Economy: Mining accounts for 20% of GDP. **Minerals:** Diamonds, copper, gold, tin, lead, uranium. **Arable land:** 1%. **Livestock** (2000): cattle: 2.06 mil; chickens: 2.3 mil; goats: 1.65 mil; pigs: 17,000; sheep: 2.1 mil. **Fish catch** (1997): 291,164 metric tons. **Labor force:** 47% agric.; 28% services; 25% ind.

Finance: Monetary unit: Rand (Sept. 2001: 8.46 = $1 U.S.). **GDP:** (1999 est.): $7.1 bil. **Per capita GDP:** $4,300. **Imports** (1999 est.): $1.5 bil; partners: South Africa 84%. **Exports** (1999 est.): $1.4 bil; partners: UK 43%, South Africa 26%, Spain 14%. **Tourism** (1998): $288 mil. **Budget** (FY 1996-97 est.): $1.2 bil. **Intl. reserves less gold** (May 2000): $300.76 mil. **Consumer prices** (change in 1999): 8.6%.

Transport: Railroad: Length: 1,480 mi. **Motor vehicles:** 62,500 pass. cars, 66,500 comm. vehicles. **Civil aviation:** 563.0 mil pass.-mi; 11 airports. **Chief ports:** Luderitz, Walvis Bay.

Communications: TV sets: 27.6 per 1,000 pop. **Radios:** 152 per 1,000 pop. **Telephones:** 104,400 main lines. **Daily newspaper circ.:** 27.4 per 1,000 pop.

Health: Life expectancy: 42.48 male; 38.71 female. **Births** (per 1,000 pop): 34.71. **Deaths** (per 1,000 pop.): 20.9. **Natural inc.:** 1.38%. **Infant mortality** (per 1,000 live births): 71.66.

Education: Compulsory: ages 6-16. **Literacy** (1993): 76%.

Major Intl. Organizations: UN (FAO, IBRD, ILO, IMF, IMO, WHO, WTrO), the Commonwealth, OAU.

Embassy: 1605 New Hampshire Ave. NW 20009; 986-0540.

Namibia was declared a German protectorate in 1890 and officially called South-West Africa. South Africa seized the territory from Germany in 1915 during World War I; the League of Nations gave South Africa a mandate over the territory in 1920. In 1966, the Marxist South-West Africa People's Organization (SWAPO) launched a guerrilla war for independence. The UN General Assembly named the area Namibia in 1968.

After many years of guerrilla warfare and failed diplomatic efforts, South Africa, Angola, and Cuba signed a U.S.-mediated agreement Dec. 22, 1988, to end South African administration of Namibia and provide for a cease-fire and transition to independence, in accordance with a 1978 UN plan. A separate accord between Cuba and Angola provided for a phased withdrawal of Cuban troops from Namibia. A constitution providing for multiparty government was adopted Feb. 9, 1990, and Namibia gained independence Mar. 21.

Walvis Bay, the principal deepwater port, had been turned over to South African administration in 1922. It remained in South African hands after independence, but South Africa turned control of the port back to Namibia, as of Mar. 1, 1994. Separatist violence flared in the Caprivi Strip in the late 1990s.

According to UN estimates, about one-fifth of the adult population has HIV/AIDS.

Nauru

Republic of Nauru

People: Population: 12,088. **Age distrib. (%): <15: 40.3; 65+: 1.7. Pop. density:** 1,1511 per sq. mi. **Urban:** 100%. **Ethnic groups:** Nauruan 58%, other Pacific Islander 26%, Chinese 8%, European 8%. **Principal languages:** Nauruan (official), English. **Chief religion:** Predominantly Christian.

Geography: Area: 8 sq. mi. **Location:** In W Pacific O. just S of the Equator. **Neighbors:** Nearest is Kiribati to E. **Topography:** Mostly a plateau bearing high-grade phosphate deposits, surrounded by a sandy shore and coral reef in concentric rings. **Capital:** Govt. offices in Yaren district.

Government: Type: Republic. **Head of state and gov.:** Pres. Rene Harris; b 1948; in office: Mar. 30, 2001. **Local divisions:** 14 districts.

Economy: Industries: Phosphate mining. **Minerals:** Phosphates. **Livestock:** (2000): chickens: 5,000; pigs: 2,800. **Electricity prod.:** 30 mil kWh.

Finance: Monetary unit: Australian Dollar (Sept. 2001: 1.92 = $1 U.S.). **GDP:** (1993 est.): $100 mil. **Per capita GDP:** $10,000. **Imports** (1991 est.): $21.1 mil; partners: Australia, UK, New Zealand. **Exports** (1991 est.): $25.3 mil; partners: Australia, New Zealand. **Budget** (FY1995-96): $64.8 mil.

Transport: Civil aviation: 151.0 mil pass.-mi. **Chief port:** Nauru.

Communications: Radios: 385 per 1,000 pop.

Health: Life expectancy: 57.7 male; 64.88 female. **Births** (per 1,000 pop): 27.22. **Deaths** (per 1,000 pop.): 7.2. **Natural inc.:** 2.%. **Infant mortality** (per 1,000 live births): 10.71.

Education: Free, compulsory: ages 6-16. **Literacy**: 99%.

Major Intl. Organizations: UN (WHO), the Commonwealth.

The island was discovered in 1798 by the British but was formally annexed to the German Empire in 1886. After World War I, Nauru became a League of Nations mandate administered by Australia. During World War II the Japanese occupied the island and shipped 1,200 Nauruans to the fortress island of Truk as slave laborers.

In 1947 Nauru was made a UN trust territory, administered by Australia. It became an independent republic Jan. 31, 1968, and was admitted to the UN Sept. 14, 1999.

Phosphate exports have provided Nauru with per capita revenues that are among the highest in the Third World. Phosphate reserves, however, are nearly depleted, and environmental damage from strip-mining has been severe. Lax banking practices have made Nauru a haven for money laundering.

Nepal

Kingdom of Nepal

People: Population: 25,284,463. **Age distrib.** (%): <15: 40.3; 65+: 3.5. **Pop. density:** 479 per sq. mi. **Urban:** 12%. **Ethnic groups:** Newars, Indians, Tibetans, Gunings, Sherpas, others. **Principal languages:** Nepali (official), many dialects. **Chief religions:** Hindu (official) 90%, Buddhist 5%, Muslim 3%.

Geography: Area: 52,800 sq. mi. **Location:** Astride the Himalaya Mts. **Neighbors:** China on N, India on S. **Topography:** The Himalayas stretch across the N, the hill country with its fertile valleys extends across the center, while the S border region is part of the flat, subtropical Ganges Plain. **Capital:** Kathmandu. **Cities** (1995 met. est.): Kathmandu 533,000; Lalitpur 190,000; Biratnagar 132,000.

Government: Type: Constitutional monarchy. **Head of state:** King Gyanendra Bir Bikram Shah Dev; b July 7, 1947; in office: June 4, 2001. **Head of gov.:** Prime Min. Sher Bahadur Deuba; b June 13, 1946; in office: July 22, 2001. **Local divisions:** 5 regions subdivided into 14 zones. **Defense budget:** $29 mil. **Active troops:** 46,000.

Economy: Industries: Sugar and jute mills, tourism. **Chief crops:** Sugar, rice, grain. **Minerals:** Quartz. **Other resources:** Forests. **Arable land:** 17%. **Livestock** (2000): cattle: 7.03 mil; chickens: 17.8 mil; goats: 6.5 mil; pigs: 900,000; sheep: 870,000. **Fish catch:** (1999): 23,206 metric tons. **Electricity prod.:** 1.255 bil kWh. **Labor force:** 81% agric., 16% services.

Finance: Monetary unit: Rupee (Sept. 2001: 75 = $1 U.S.). **GDP:** (1999 est.): $27.4 bil. **Per capita GDP:** $1,100. **Imports** (1998): $1.2 bil; partners: India 31%, China/Hong Kong 16%, Singapore 14%. **Exports** (1998): $485 mil; partners: India 33%, U.S. 26%, Germany 25%. **Tourism:** $168 mil. **Budget** (FY 1996-97 est.): $818 mil. **Intl. reserves less gold** (May 2000): $952.5 mil. **Gold:** 153,000 oz t. **Consumer prices** (change in 1999): 8.0%.

Transport: Railroad: Length: 63 mi. **Civil aviation:** 564.3 mil pass.-mi; 24 airports.

Communications: TV sets: 12 per 1,000 pop. **Radios:** 30 per 1,000 pop. **Telephones**: 266,900 main lines. **Daily newspaper circ.:** 8 per 1,000 pop.

Health: Life expectancy: 58.65 male; 57.77 female. **Births** (per 1,000 pop): 33.40. **Deaths** (per 1,000 pop.): 10.22. **Natural inc.:** 2.32%. **Infant mortality** (per 1,000 live births): 74.14.

Education: Free, compulsory: ages 6-11. **Literacy**: 27%.

Major Intl. Organizations: UN (FAO, IBRD, ILO, IMF, IMO, WHO).

Embassy: 2131 Leroy Pl. NW 20008; 667-4550.

Website: http://www.info-nepal.com

Nepal was originally a group of petty principalities, the inhabitants of one of which, the Gurkhas, became dominant about 1769. In 1951 King Tribhubana Bir Bikram, member of the Shah family, ended the system of rule by hereditary premiers of the Ranas family, who had kept the kings virtual prisoners, and established a cabinet system of government.

Virtually closed to the outside world for centuries, Nepal is now linked to India and Pakistan by roads and air service and to Tibet by road. Polygamy, child marriage, and the caste system were officially abolished in 1963.

The government announced the legalization of political parties in 1990. Elections on Nov. 15, 1994, led to the installation of Nepal's first Communist government, which held power until a no-confidence vote Sept. 10, 1995. An insurgency by Maoist rebels has claimed more than 1,200 lives since 1996.

Nine members of Nepal's royal family, including King Birendra and Queen Aishwarya, died as the result of a massacre on the night of June 1, 2001. An official inquiry blamed the carnage on a 10th member of the family, Crown Prince Dipendra, who reportedly shot himself that night and died 3 days later, allowing Birendra's brother Gyanendra to assume the throne

Netherlands

Kingdom of the Netherlands

People: Population: 15,981,472. **Age distrib.** (%): <15: 18.4; 65+: 13.7. **Pop. density:** 1,221 per sq. mi. **Urban:** 89%. **Ethnic groups:** Dutch 94%. **Principal language:** Dutch (official). **Chief religions:** Roman Catholic 34%, Protestant 25%.

Geography: Area: 13,100 sq. mi. **Location:** In NW Europe on North Sea. **Neighbors:** Germany on E, Belgium on S. **Topography:** The land is flat, an average alt. of 37 ft. above sea level, with much land below sea level reclaimed and protected by some 1,500 miles of dikes. Since 1920 the government has been draining the IJsselmeer, formerly the Zuider Zee. **Capital:** Amsterdam. **Cities** (1998 est.): Amsterdam 718,151; Rotterdam 590,478; The Hague 442,799.

Government: Type: Parliamentary democracy under a constitutional monarch. **Head of state:** Queen Beatrix; b Jan. 31, 1938; in office: Apr. 30, 1980. **Head of gov.:** Prime Min. Wim Kok; b Sept. 29, 1938; in office: Aug. 22, 1994. **Seat of govt.:** The Hague. **Local divisions:** 12 provinces. **Defense budget:** $6.2 bil. **Active troops:** 51,940.

Economy: Industries: Metals, machinery, chemicals, oil, microelectronics. **Chief crops:** Grains, potatoes, sugar beets, vegetables, fruits. **Minerals:** Natural gas, oil. **Crude oil reserves** (2000): 106.9 mil bbls. **Arable land:** 25%. **Livestock** (2000): cattle: 4.2 mil; chickens: 105.55 mil; goats: 153,000; pigs: 13.14 mil; sheep: 1.4 mil. **Fish catch** (1999): 550,009 metric tons. **Electricity prod.:** 85.294 bil kWh. **Labor force:** 73% services; 23% ind.; 4% agric.

Finance: Monetary unit: Guilder (Sept. 2001: 2.43 = $1 U.S.). Euro (Sept. 2001: 1.09 = $1 U.S.). **GDP:** (1999 est.): $365.1 bil. **Per capita GDP:** $23,100. **Imports** (1998): $152 bil; partners: EU 61%. **Exports** (1998): $169 bil; partners: EU 78%. **Tourism:** $7.09 bil. **Budget** (1999 est.): $170 bil. **Intl. reserves less gold** (June 2000): $8.94 bil. **Gold:** 29.32 mil oz t. **Consumer prices** (change in 1999): 2.2%.

Transport: Railroad: Length: 1,702 mi. **Motor vehicles** (1997): 5.81 mil pass. cars, 715,000 comm. vehicles. **Civil aviation:** 41.4 bil pass.-mi; 6 airports. **Chief ports:** Rotterdam, Amsterdam, IJmuiden.

Communications: TV sets: 545 per 1,000 pop. **Radios:** 764 per 1,000 pop. **Telephones** (1999): 9,610,000 main lines. **Daily newspaper circ.:** 306 per 1,000 pop.

Health: Life expectancy: 75.55 male; 81.44 female. **Births** (per 1,000 pop): 11.85. **Deaths** (per 1,000 pop.): 8.69. **Natural inc.:** .32%. **Infant mortality** (per 1,000 live births): 4.37.

Education: Compulsory: ages 5-18. **Literacy**: 100%.

Major Intl. Organizations: UN and all of its specialized agencies, EU, NATO, OECD, OSCE.

Embassy: 4200 Linnean Ave. NW 20008; 244-5300.

Website: http://www.cbs.nl/enindex.htm

Julius Caesar conquered the region in 55 BC, when it was inhabited by Celtic and Germanic tribes.

After the empire of Charlemagne fell apart, the Netherlands (Holland, Belgium, Flanders) split among counts, dukes, and bishops, passed to Burgundy and thence to Charles V of Spain. His son, Philip II, tried to check the Dutch drive toward political freedom and Protestantism (1568-1573). William the Silent, prince of Orange, led a confederation of the northern provinces, called Estates, in the Union of Utrecht, 1579. The Estates retained individual sovereignty, but were represented jointly in the States-General, a body that had control of foreign affairs and defense. In 1581 they repudiated allegiance to Spain. The rise of the Dutch republic to naval, economic, and artistic eminence came in the 17th century.

The United Dutch Republic ended 1795 when the French formed the Batavian Republic. Napoleon made his brother Louis king of Holland, 1806; Louis abdicated 1810 when Napoleon annexed Holland. In 1813 the French were expelled. In 1815 the Congress of Vienna formed a kingdom of the Netherlands, including Belgium, under William I. In 1830, the Belgians seceded and formed a separate kingdom.

The constitution, promulgated 1814, and subsequently revised, provides for a hereditary constitutional monarchy.

The Netherlands maintained its neutrality in World War I, but was invaded and brutally occupied by Germany, 1940-45.

In 1949, after several years of fighting, the Netherlands granted independence to Indonesia. In 1963, West New Guinea (now Irian Jaya) was turned over to Indonesia. Immigration from former Dutch colonies has been substantial.

Although the Netherlands is heavily industrialized, its small farms export large quantities of pork and dairy foods. Rotterdam, located along the principal mouth of the Rhine, is one of the world's leading cargo ports. Canals, extending over 3,400 miles, are important in transportation.

Netherlands Dependencies

The **Netherlands Antilles,** constitutionally on a level of equality with the Netherlands homeland within the kingdom, consist of 2 groups of islands in the West Indies. **Curaçao** and **Bonaire** are near the coast of Venezuela; **St. Eustatius, Saba,** and the southern part of **St. Maarten** are SE of Puerto Rico. The northern two-thirds of St. Maarten belongs to French Guadeloupe; the French call the island St. Martin. Total area of the 2 groups is 309 sq. mi., including Bonaire (111), Curaçao (171), St. Eustatius (8), Saba (5), St. Maarten (Dutch part) (13). St. Maarten suffered extensive damage from Hurricane Luis, Sept. 1995. Total pop. of the Netherlands Antilles (2001 est.) was 212,226. Willemstad, on Curaçao, is the capital. The principal industry is the refining of crude oil from Venezuela. Tourism is also an important industry, as is shipbuilding.

Aruba, about 26 mi. W of Curaçao, was separated from the Netherlands Antilles on Jan. 1, 1986; it is an autonomous member of the Netherlands, the same status as the Netherland Antilles. Area 75 sq. mi.; pop. (2001 est.) 70,007; capital Oranjestad. Chief industries are oil refining and tourism.

New Zealand

People: Population: 3,864,129. **Age distrib.** (%): <15: 22.4; 65+: 11.5. **Pop. density:** 37 per sq. mi. **Urban:** 86%. **Ethnic groups:** New Zealand European 75%, Maori 10%. **Principal languages:** English (official), Maori. **Chief religions:** Anglican 24%, Presbyterian 18%, Roman Catholic 15%.

Geography: Area: 103,600 sq. mi. **Location:** In SW Pacific O. **Neighbors:** Nearest are Australia on W, Fiji and Tonga on N. **Topography:** Each of the 2 main islands (North and South Isls.) is mainly hilly and mountainous. The east coasts consist of fertile plains, especially the broad Canterbury Plains on South Isl. A volcanic plateau is in center of North Isl. South Isl. has glaciers and 15 peaks over 10,000 ft. **Capital:** Wellington. **Cities**: Auckland (urban agg.): 1,102,000; (met. area, 1996 cen.): Wellington 335,468; Christchurch 331,443.

Government: Type: Parliamentary democracy. **Head of state:** Queen Elizabeth II, represented by Gov.-Gen. Dame Silvia Cartwright; b Nov. 7, 1943; in office: Apr. 4, 2001. **Head of gov.:** Prime Min. Helen Clark; b Feb. 26, 1950; in office: Dec. 10, 1999. **Local divisions:** 93 counties, 9 districts, 3 town districts. **Defense budget:** $804 mil. **Active troops:**9,230.

Economy: Industries: Food processing, textiles, machinery. **Chief crops:** Grains, potatoes, fruits. **Minerals:** Gold, gas, iron, coal. **Crude oil reserves** (2000): 127 mil bbls. **Other resources:** Wool, timber. **Arable land:** 9%. **Livestock** (2000): cattle: 9.46 mil; chickens: 13.0 mil; goats: 186,000; pigs: 344,000; sheep: 45.8 mil. **Fish catch** (1999): 669,267 metric tons. **Electricity prod.:** 37.952 bil kWh. **Labor force:** 65% services; 25% ind.; 10% agric.

Finance: Monetary unit: N.Z. Dollar (Sept. 2001: 2.30 = $1 U.S.). **GDP:** (1999 est.): $63.8 bil. **Per capita GDP:** $17,400. **Imports** (1998 est.): $11.2 bil; partners: Australia 22%, U.S. 20%, Japan 11%. **Exports** (1998 est.): $12.2 bil; partners: Australia 21%, Japan 13%, U.S. 13%. **Tourism** (1998): $1.73 bil. **Budget** (FY 1997-98 est.): $23.7 bil. **Intl. reserves less gold** (May 2000): $3.37 bil. **Consumer prices** (change in 1999): –0.1%.

Transport: Railroad: Length: 2,433 mi. **Motor vehicles** (1997): 1.50 mil pass. cars, 322,889 comm. vehicles. **Civil aviation:** 13.0 bil pass.-mi; 36 airports. **Chief ports:** Auckland, Christchurch, Wellington, Dunedin, Tauranga.

Communications: TV sets: 496.5 per 1,000 pop. **Radios:** 1,027 per 1,000 pop. **Telephones** (1999): 1,889,000 main lines. **Daily newspaper circ.:** 223 per 1,000 pop.

Health: Life expectancy: 75.01 male; 81.1 female. **Births** (per 1,000 pop): 14.28. **Deaths** (per 1,000 pop.): 7.56. **Natural inc.:** .67%. **Infant mortality** (per 1,000 live births): 6.28.

Education: Free, compulsory: ages 6-16. **Literacy** (1997): 100%.

Major Intl. Organizations: UN (FAO, IBRD, ILO, IMF, IMO, WHO, WTrO), APEC, the Commonwealth, OECD.

Embassy: 37 Observatory Cir. NW 20008; 328-4800.

Website: http://www.stats.govt.nz/statsweb.nsf

The Maoris, a Polynesian group from the eastern Pacific, reached New Zealand before and during the 14th century. The first European to sight New Zealand was Dutch navigator Abel Janszoon Tasman, but Maoris refused to allow him to land. British Capt. James Cook explored the coasts, 1769-1770.

British sovereignty was proclaimed in 1840, with organized settlement beginning in the same year. Representative institutions were granted in 1853. Maori Wars ended in 1870 with British victory. The colony became a dominion in 1907, and is an independent member of the Commonwealth.

A progressive tradition in politics dates back to the 19th century, when New Zealand was internationally known for social experimentation; much of the nation's economy has been deregulated in recent years. The National Party, led by Jim Bolger, won general elections in 1990 and 1993. After inconclusive elections, Oct. 12, 1996, Bolger remained as prime minister, heading a National/New Zealand First party coalition. When Bolger lost his party's support, Jenny Shipley became the nation's first female prime minister, Dec. 8, 1997. The Labour Party won the general election of Nov. 27, 1999.

The native Maoris number about 550,000. Six of 120 members of the House of Representatives are elected directly by the Maori people.

New Zealand comprises **North Island,** 44,702 sq. mi.; **South Island,** 58,384 sq. mi.; **Stewart Island,** 674 sq. mi.; **Chatham Islands,** 372 sq. mi.; and several groups of smaller islands.

In 1965, the **Cook Islands** (pop., 2001 est., 20,611; area 100 sq. mi.), located halfway between New Zealand and Hawaii, became self-governing although New Zealand retains responsibility for defense and foreign affairs. **Niue** attained the same status in 1974; it lies 400 mi. to W (pop., 1995 est., 1,800; area 100 sq. mi.). **Tokelau** (pop., 1995 est., 1,500; area 4 sq. mi.) comprises 3 atolls 300 mi. N of Samoa.

Ross Dependency, administered by New Zealand since 1923, comprises 160,000 sq. mi. of Antarctic territory.

Nicaragua

Republic of Nicaragua

People: Population: 4,918,393. **Age distrib.** (%): <15: 39.0; 65+: 2.9. **Pop. density:** 106 per sq. mi. **Urban:** 56%. **Ethnic groups:** Mestizo 69%, white 17%, black 9%, Amerindian 5%. **Principal languages:** Spanish (official). **Chief religion:** Roman Catholic 95%.

Geography: Area: 46,400 sq. mi. **Location:** In Central America. **Neighbors:** Honduras on N, Costa Rica on S. **Topography:** Both Caribbean and Pacific coasts are over 200 mi. long. The Cordillera Mts., with many volcanic peaks, run NW-SE through the middle of the country. Between this and a volcanic range to the E lie Lakes Managua and Nicaragua. **Capital:** Managua (urban agg.): 959,000.

Government: Type: Republic. **Head of state and gov.:** Pres. Arnoldo Alemán Lacayo; b Jan. 23, 1946; in office Jan. 10, 1997. **Local divisions:** 15 departments, 2 autonomous regions. **Defense budget:**$26 mil. **Active troops:** 16,000.

Economy: Industries: Oil refining, food processing, chemicals, textiles. **Chief crops:** Bananas, cotton, citrus, coffee, sugar, corn, rice. **Minerals:** Gold, silver, copper, tungsten. **Other resources:** Forests, seafood. **Arable land:** 9%. **Livestock** (2000): cattle: 1.66 mil; chickens: 10.0 mil; goats: 6,500; pigs: 400,000; sheep: 4,000. **Fish catch:** (1999): 16,130 metric tons. **Electricity prod.:** 2.349 bil kWh. **Labor force:** 43% services; 42% agric.; 15%ind.

Finance: Monetary unit: Gold Cordoba (Sept. 2001: 13.62 = $1 U.S.). **GDP:** (1999 est.): $12.5 bil. **Per capita GDP:** $2,650. **Imports** (1999): $1.5 bil partners: U.S. 31%, Costa Rica 11%, Guatemala 8%. **Exports** (1998): $573 mil; partners: U.S. 35%, Germany 13%, El Salvador 10%. **Tourism:** $113 mil. **Budget** (1996 est.): $551 mil. **Intl. reserves less gold** (May 2000): $473.72 mil. **Consumer prices** (change in 1999): 11.2%.

Transport: Motor vehicles: 72,413 pass. cars, 72,227 comm. vehicles. **Civil aviation:** 52.8 mil pass.-mi; 10 airports. **Chief ports:** Corinto, Puerto Sandino, San Juan del Sur.

Communications: TV sets: 48 per 1,000 pop. **Radios:** 206 per 1,000 pop. **Telephones** (1999): 150,300 main lines. **Daily newspaper circ.:** 31 per 1,000 pop.

Health: Life expectancy: 67.1 male; 71.11 female. **Births** (per 1,000 pop): 27.64. **Deaths** (per 1,000 pop.): 4.82. **Natural inc.:** 2.28%. **Infant mortality** (per 1,000 live births): 33.66.

Education: Free, compulsory: ages 7-13. **Literacy**: 66%.

Major Intl. Organizations: UN and most of its specialized agencies, OAS.

Embassy: 1627 New Hampshire Ave. NW 20009; 939-6570.

Nicaragua, inhabited by various Indian tribes, was conquered by Spain in 1552. After gaining independence from Spain, 1821, Nicaragua was united for a short period with Mexico, then with the United Provinces of Central America, finally becoming an independent republic, 1838.

U.S. Marines occupied the country at times in the early 20th century, the last time from 1926 to 1933.

Gen. Anastasio Somoza Debayle was elected president in 1967. He resigned in 1972, but was re-elected president in 1974. Martial law was imposed in Dec. 1974, after officials were kidnapped by the Marxist Sandinista guerrillas. Violent opposition spread to nearly all classes in 1978; nationwide strikes called against the government touched off a civil war, which ended when Somoza fled Nicaragua and the Sandinistas took control of Managua in July 1979. Somoza was assassinated in Paraguay, Sept. 17, 1980.

Relations with the U.S. were strained as a result of Nicaragua's aid to leftist guerrillas in El Salvador and U.S. backing of anti-Sandinista contra guerrilla groups. In 1983 the contras launched a major offensive; the Sandinistas imposed rule by decree. In 1985 the U.S. House rejected Pres. Reagan's request for military aid to the contras. The subsequent diversion of funds to the contras from the proceeds of a secret arms sale to Iran caused a major scandal in the U.S.

In a stunning upset, Violeta Barrios de Chamorro defeated Sandinista leader Daniel Ortega Saavedra in national elections, Feb. 25, 1990. Arnoldo Alemán Lacayo, a conservative former mayor of Managua, defeated Ortega in the presidential election of Oct. 20, 1996. Up to 2,000 people died in W Nicaragua Oct. 30, 1998, in a mudslide caused by rains from Hurricane Mitch. Drought and a drop in coffee prices plunged Nicaragua into an economic crisis in 2001.

Niger

Republic of Niger

People: Population: 10,355,156. **Age distrib.** (%): <15: 48.0; 65+: 2.3. **Pop. density:** 21 per sq. mi. **Urban:** 20%. **Ethnic groups:** Hausa 56%, Djerma 22%, Fula 9%, Tuareg 8%. **Principal languages:** French (official), Hausa, Djerma. **Chief religion:** Muslim 80%.

Geography: Area: 488,500 sq. mi. **Location:** In the interior of N Africa. **Neighbors:** Libya, Algeria on N; Mali, Burkina Faso on W; Benin, Nigeria on S; Chad on E. **Topography:** Mostly arid desert and mountains. A narrow savanna in the S and the Niger R. basin in the SW contain most of the population. **Capital:** Niamey (1999 est.): 731,000.

Government: Type: Republic. **Head of state:** Pres. Tandja Mamadou; b 1938; in office: Dec. 22, 1999. **Head of gov.:** Prime Min. Hama Amadou; b 1950; in office: Jan. 3, 2000. **Local divisions:** 7 departments, 1 capital district. **Defense budget** (1999): $27 mil. **Active troops:** 5,300.

Economy: Chief crops: Peanuts, cowpeas, cotton. **Minerals:** Uranium, coal, iron. **Arable land:** 3%. **Livestock** (2000): cattle: 2.22 mil; chickens: 20.0 mil; goats: 6.6 mil; pigs: 39,000; sheep: 4.3 mil. **Electricity prod.:** 200 mil kWh. **Labor force:** 90% agric.; 6% ind. & commerce; 4% govt.

Finance: Monetary unit: CFA Franc (Oct. 2000: 752.63 = $1 U.S.). **GDP:** (1999 est.): $9.6 bil. **Per capita GDP:** $1,000. **Imports** (1997): $295 mil; partners: France, Cote d'Ivoire, U.S. **Exports** (1997): $269 mil; partners: U.S., Greece, Japan. **Tourism:** $21 mil. **Budget** (1998 est.): $370 mil. **Intl. reserves less gold** (Apr. 2000): $39.2 mil. **Gold:** 11,000 oz t. **Consumer prices** (change in 1999): –2.3%.

Transport: Motor vehicles: 37,500 pass. cars, 14,100 comm. vehicles. **Civil aviation:** 150.5 mil pass.-mi; 6 airports.

Communications: TV sets: 2.8 per 1,000 pop. **Radios:** 48 per 1,000 pop. **Telephones** (1998): 18,100 main lines.

Health: Life expectancy: 41.74 male; 41.44 female. **Births** (per 1,000 pop): 50.68. **Deaths** (per 1,000 pop.): 22.71. **Natural inc.:** 2.8%. **Infant mortality** (per 1,000 live births): 123.57.

Education: Free, compulsory: ages 7-15. **Literacy**: 14%.

Major Intl. Organizations: UN (FAO, IBRD, ILO, IMF, WHO, WTrO), OAU.

Embassy: 2204 R St. NW 20008; 483-4224.

Niger was part of ancient and medieval African empires. European explorers reached the area in the late 18th century. The French colony of Niger was established 1900-22, after the defeat of Tuareg fighters, who had invaded the area from the N a century before. The country became independent Aug. 3, 1960. The next year it signed a bilateral agreement with France.

In 1993, Niger held its first free and open elections since independence; an opposition leader, Mahamane Ousmane, won the presidency. A peace accord Apr. 24, 1995, ended a Tuareg rebellion that began in 1990. A coup, Jan. 27, 1996, followed by a disputed presidential election in July, left the military in control of Niger. On Apr. 9, 1999, Gen. Ibrahim Bare Mainassara, Niger's president since 1996, was assassinated, apparently by members of his security team. Elections were held Oct. 17 and Nov. 24, 1999, under a new constitution, approved by referendum July 18, that provided for a return to civilian rule.

Nigeria

Federal Republic of Nigeria

People: Population: 126,635,626. **Age distrib.** (%): <15: 43.7; 65+: 2.8. **Pop. density:** 361 per sq. mi. **Urban:** 43%. **Ethnic groups:** Hausa, Yoruba, Ibo, Fulani, others. **Principal languages:** English (official), Hausa, Yoruba, Ibo. **Chief religions:** Muslim (in N) 50%, Christian (in S) 40%.

Geography: Area: 351,200 sq. mi. **Location:** On the S coast of W Africa. **Neighbors:** Benin on W, Niger on N, Chad and Cameroon on E. **Topography:** 4 E-W regions divide Nigeria: a coastal mangrove swamp 10-60 mi. wide, a tropical rain forest 50-100 mi. wide, a plateau of savanna and open woodland, and semidesert in the N. **Capital:** Abuja. **Cities** (urban agg.): Lagos 13,427,000; Ibadan 1,731,000.

Government: Type: Republic. **Head of state and gov.:** Pres. Olusegun Obasanjo; b Mar. 6, 1935; in office: May 29, 1999. **Local divisions:** 30 states, 1 capital territory. **Defense budget:** $340 mil. **Active troops:** 76,500.

Economy: Industries: Crude oil, coal, palm oil, cotton, textiles. **Chief crops:** Cocoa (main export crop), palm products, corn, rice, yams, cassava. **Minerals:** Oil, gas, lead, zinc, coal, iron, limestone, columbite, tin. **Crude oil reserves** (2000): 22.5 bil bbls. **Other resources:** Timber, rubber, hides. **Arable land:** 33%. **Livestock** (2000): cattle: 19.83 mil; chickens: 126.0 mil; goats: 24.3 mil; pigs: 4.86 mil; sheep: 20.5 mil. **Fish catch** (1999): 383,417 metric tons. **Electricity prod.:** 18.700 bil kWh. **Labor force:** 54% agric.; 40% services; 6% ind.

Finance: Monetary unit: Naira (Sept. 2001: 112.66 = $1 U.S.). **GDP:** (1999 est.): $110.5 bil. **Per capita GDP:** $970. **Imports** (1999): $10 bil; partners: UK 13%, U.S. 12%, Germany 10%. **Exports** (1999): $13.1 bil; partners: U.S. 35%, Spain 11%, India 9%. **Tourism** (1998): $142 mil. **Budget** (1998 est.): $13.9 bil. **Consumer prices** (change in 1999): 6.6%.

Transport: Railroad: Length: 2,178 mi. **Motor vehicles:** 589,600 pass. cars, 363,900 comm. vehicles. **Civil aviation:** 137.2 mil pass.-mi; 12 airports. **Chief ports:** Port Harcourt, Lagos, Warri, Calabar.

Communications: TV sets: 61 per 1,000 pop. **Radios:** 197 per 1,000 pop. **Telephones**: 492,000 main lines. **Daily newspaper circ.:** 24 per 1,000 pop.

Health: Life expectancy: 51.07 male; 51.07 female. **Births** (per 1,000 pop): 39.69. **Deaths** (per 1,000 pop.): 13.91. **Natural inc.:** 2.58%. **Infant mortality** (per 1,000 live births): 73.34.

Education: Free, compulsory: ages 6-15. **Literacy**: 57%.

Major Intl. Organizations: UN (FAO, IBRD, ILO, IMF, IMO, WHO, WTrO), the Commonwealth, OAU, OPEC.

Embassy: 1333 16th St. NW 20036; 986-8400.

Early cultures in Nigeria date back to at least 700 BC. From the 12th to the 14th centuries, more advanced cultures developed in the Yoruba area, at Ife, and in the north, where Muslim influence prevailed.

Portuguese and British slavers appeared from the 15th-16th centuries. Britain seized Lagos, 1861, and gradually extended control inland until 1900. Nigeria became independent Oct. 1, 1960, and a republic Oct. 1, 1963.

On May 30, 1967, the Eastern Region seceded, proclaiming itself the Republic of Biafra, plunging the country into civil war. Casualties in the war were estimated at over 1 million, including many "Biafrans" (mostly Ibos) who died of starvation despite international efforts to provide relief. The secessionists, after steadily losing ground, capitulated Jan. 12, 1970.

Nigeria emerged as one of the world's leading oil exporters in the 1970s, but much of the revenue has been squandered through corruption and mismanagement.

After 13 years of military rule, the nation made a peaceful return to civilian government, Oct. 1979. Military rule resumed, Dec. 31, 1983; a second coup came in 1985.

Headed by Gen. Ibrahim Babangida, the military regime held elections June 12, 1993, but annulled the vote June 23 when it appeared that Moshood Abiola would win. Riots followed and many were killed. Babangida resigned and appointed a civilian to head an interim government, Aug. 26, but that government was ousted Nov. 17 in a coup led by Gen. Sani Abacha. On June 11, 1994, Abiola declared himself president; he was jailed June 23.

Abacha's brutal rule ended June 8, 1998, when he died of an apparent heart attack. Abiola died in prison July 7, as Abacha's successor, Gen. Abdulsalam Abubakar, was reportedly preparing to free him. Abiola's death (also apparently of natural causes) sparked riots in Lagos and other cities; on July 20, Abubakar promised early elections and a return to civilian rule. Olusegun Obasanjo (a former military ruler) won the presidential vote Feb. 27, 1999, to become the head of Nigeria's 1st civilian government in 15 years.

An oil fire that exploded from a ruptured pipeline in S. Nigeria, Oct. 17, 1998, killed at least 700 people who were scavenging for fuel. The imposition of strict Islamic law in northern states led to clashes, Jan.-Mar. 2000, in which at least 800 people died. U.S. Pres. Bill Clinton visited Nigeria Aug. 26-27, 2000, making the 1st visit there by a U.S. head of state in 22 years.

Norway

Kingdom of Norway

People: Population: 4,503,440. **Age distrib.** (%): <15: 20.0; 65+: 15.1. **Pop. density:** 38 per sq. mi. **Urban:** 75%. **Ethnic groups:** Germanic (Nordic, Alpine, Baltic), Lapps. **Principal languages:** Norwegian (official). **Chief religion:** Evangelical Lutheran 87.8%.

Geography: Area: 118,700 sq. mi. **Location:** W part of Scandinavian peninsula in NW Europe (extends farther north than any European land). **Neighbors:** Sweden, Finland, Russia on E. **Topography:** A highly indented coast is lined with tens of thousands of islands. Mountains and plateaus cover most of the country, which is only 25% forested. **Capital:** Oslo. **Cities**: Oslo (urban agg.): 978,000 (city proper 2000 est., 507,467); Bergen (1996 est.) 223,773.

Government: Type: Hereditary constitutional monarchy. **Head of state:** King Harald V; b Feb. 21, 1937; in office: Jan. 17, 1991. **Head of gov.:** Prime Min. Jens Stoltenberg; b Mar. 16, 1959; in office: Mar. 17, 2000. **Local divisions:** 19 provinces. **Defense budget** (2001): $2.9 bil. **Active troops:** 26,700.

Economy: Industries: Wood & paper prods., shipbuilding, metals, chemicals, food processing, fish, oil, gas. **Chief crops:** Grains, oats. **Minerals:** Oil, gas, copper, pyrites, nickel, iron, zinc, lead. **Crude oil reserves** (2000): 10.79 bil bbls. **Other resources:** Fish, livestock. **Arable land:** 3%. **Livestock** (2000): cattle: 1.04 mil; chickens: 23.37 mil; goats: 82,500; pigs: 689,600; sheep: 2.4 mil. **Fish catch** (1999): 3.22 mil metric tons. **Electricity prod.:** 121.084 bil kWh. **Labor force:** 74% services; 22% industry; 4% agric., forestry & fishing.

Finance: Monetary unit: Krone (Sept. 2001: 8.82 = $1 U.S.). **GDP:** (1999 est.): $111.3 bil. **Per capita GDP:** $25,100. **Imports** (1999): $38.6 bil; partners: EU 69%, U.S. 7%. **Exports** (1999): $47.3 bil; partners: EU 77%, U.S. 7%. **Tourism:** $2.23 bil. **Intl. reserves less gold** (Apr. 2000): $20.24 bil. **Gold:** 1.18 mil oz t. **Consumer prices** (change in 1999): 2.3%.

Transport: Railroad: Length: 2,485 mi. **Motor vehicles** (1997): 1.76 mil pass. cars, 412,183 comm. vehicles. **Civil aviation:** 5.7 bil pass.-mi; 50 airports. **Chief ports:** Bergen, Stavanger, Oslo, Kristiansand.

Communications: TV sets: 579 per 1,000 pop. **Radios:** 913 per 1,000 pop. **Telephones**: 3,270,000 main lines. **Daily newspaper circ.:** 588 per 1,000 pop.

Health: Life expectancy: 75.87 male; 81.92 female. **Births** (per 1,000 pop): 12.60. **Deaths** (per 1,000 pop.): 9.83. **Natural inc.:** .28%. **Infant mortality** (per 1,000 live births): 3.94.

Education: Compulsory: ages 6-16. **Literacy** (1994): 100%.

Major Intl. Organizations: UN and all of its specialized agencies, EFTA, NATO, OECD, OSCE.

Embassy: 2720 34th St. NW 20008; 333-6000.

Website: http://www.ssb.no/www-open/english

The first ruler of Norway was Harald the Fairhaired, who came to power in AD 872. Between 800 and 1000, Norway's Vikings raided and occupied widely dispersed parts of Europe.

The country was united with Denmark 1381-1814, and with Sweden, 1814-1905. In 1905, the country became independent with Prince Charles of Denmark as king.

Norway remained neutral during World War I. Germany attacked Norway Apr. 9, 1940, and held it until liberation May 8, 1945. The country abandoned its neutrality after the war, and joined NATO. In a referendum Nov. 28, 1994, Norwegian voters rejected European Union membership.

Abundant hydroelectric resources provided the base for industrialization, giving Norway one of the highest living standards in the world. The country is a leading producer and exporter of crude oil, with extensive reserves in the North Sea. Norway's merchant marine is one of the world's largest.

Svalbard is a group of mountainous islands in the Arctic O., area 23,957 sq. mi., pop. (1997 est.) 3,231. The largest, Spitsbergen (formerly called West Spitsbergen), 15,060 sq. mi., seat of the governor, is about 370 mi. N of Norway. By a treaty signed in Paris, 1920, major European powers recognized the sovereignty of Norway, which incorporated it in 1925.

Jan Mayen, area 144 sq. mi., is a volcanic island located about 565 mi. WNW of Norway; it was annexed in 1929.

Oman

Sultanate of Oman

People: Population: 2,622,198. **Age distrib.** (%): <15: 41.5; 65+: 2.4. **Pop. density:** 32 per sq. mi. **Urban:** 83%. **Ethnic groups:** Arab, Indian. **Principal languages:** Arabic (official). **Chief religion:** Ibadhi Muslim 75%.

Geography: Area: 81,900 sq. mi. **Location:** On SE coast of Arabian peninsula. **Neighbors:** United Arab Emirates, Saudi Arabia, Yemen on W. **Topography:** Oman has a narrow coastal plain up to 10 mi. wide, a range of barren mountains reaching 9,900 ft., and a wide, stony, mostly waterless plateau, avg. alt. 1,000 ft. Also, an exclave at the tip of the Musandam peninsula controls access to the Persian Gulf. **Capital:** Muscat (1993): 51,969.

Government: Type: Absolute monarchy. **Head of state and gov.:** Sultan Qabus bin Said; b Nov. 18, 1940; in office: July 23, 1970 (also prime min. since Jan. 2, 1972). **Local divisions:** 6 regions and 2 governorates. **Defense budget:** $1.75 bil. **Active troops:** 43,500.

Economy: Industries: Oil, gas, construction. **Chief crops:** Dates, limes, vegetables, alfalfa, bananas. **Minerals:** Oil (75% of exports). **Livestock** (2000): cattle: 149,000; chickens: 3.4 mil; goats: 729,000; sheep: 180,000. **Fish catch:** (1999): 117,049 metric tons. **Crude oil reserves** (2000): 5.28 bil bbls. **Electricity prod.:** 8.630 bil kWh.

Finance: Monetary unit: Rial Omani (Sept. 2001: 0.39 = $1 U.S.). **GDP:** (1999 est.): $19.6 bil. **Per capita GDP:** $8,000. **Imports** (1999 est.): $5.4 bil; partners: UAE 23%, Japan 16%, UK 13%. **Exports** (1999 est.): $7.2 bil; partners: Japan 21%, China 16%, Thailand 16%. **Tourism:** $104 mil. **Budget** (1999 est.): $5.6 bil. **Intl. reserves less gold** (June 2000): $509.3 mil. **Gold:** 291,000 oz t. **Consumer prices** (change in 1999): 0.4%.

Transport: Motor vehicles (1997): 246,097 pass. cars, 101,223 comm. vehicles. **Civil aviation:** 2.0 bil pass.-mi; 6 airports. **Chief ports:** Matrah, Mina' al Fahl.

Communications: TV sets: 711 per 1,000 pop. **Radios:** 426 per 1,000 pop. **Telephones:** 225,400 main lines. **Daily newspaper circ.:** 31 per 1,000 pop.

Health: Life expectancy: 69.9 male; 74.29 female. **Births** (per 1,000 pop): 37.96. **Deaths** (per 1,000 pop.): 4.1. **Natural inc.:** 3.39%. **Infant mortality** (per 1,000 live births): 22.52.

Education: Literacy (1993): 59%.

Major Intl. Organizations: UN (FAO, IBRD, ILO, IMF, IMO, WHO), AL.

Embassy: 2535 Belmont Rd. NW 20008; 387-1980.

Oman was originally called Muscat and Oman. A long history of rule by other lands, including Portugal in the 16th century, ended with the ouster of the Persians in 1744. By the early 19th century, Muscat and Oman was one of the most important countries in the region, controlling much of the Persian and Pakistan coasts, and also ruling far-away Zanzibar, which was separated in 1861 under British mediation.

British influence was confirmed in a 1951 treaty, and Britain helped suppress an uprising by traditionally rebellious interior tribes against control by Muscat in the 1950s.

On July 23, 1970, Sultan Said bin Taimur was overthrown by his son, who changed the nation's name to Sultanate of Oman.

Oil is the major source of income.

Oman opened its air bases to Western forces following the Iraqi invasion of Kuwait on Aug. 2, 1990.

Pakistan

Islamic Republic of Pakistan

People: Population: 144,616,639. **Age distrib.** (%): <15: 40.5; 65+: 4.1. **Pop. density:** 482 per sq. mi. **Urban:** 37%. **Ethnic groups:** Punjabi, Sindhi, Pashtun (Pathan), Baloch. **Principal languages:** Urdu, English (both official), Punjabi, Sindhi, Pashtu. **Chief religions:** Sunni Muslim 77%, Shi'a Muslim 20%.

Geography: Area: 300,300 sq. mi. **Location:** In W part of South Asia. **Neighbors:** Iran on W, Afghanistan and China on N, India on E. **Topography:** The Indus R. rises in the Hindu Kush and Himalaya Mts. in the N (highest is K2, or Godwin Austen, 28,250 ft., 2d highest in world), then flows over 1,000 mi. through fertile valley and empties into Arabian Sea. Thar Desert, Eastern Plains flank Indus Valley. **Capital:** Islamabad. **Cities (urban agg.):** Karachi 11,794,000; Lahore 6,040,000; Faisalabad 2,232,000.

Government: Type: In transition. **Head of state and gov:** Pres. Pervez Musharraf; b Aug. 11, 1943; in office: Oct. 15, 1999 (as pres. from June 20, 2001). **Local divisions:** 4 provinces and 1 capital territory, plus federally administered tribal areas. **Defense budget:** $3.3 bil. **Active troops:** 612,000.

Economy: Industries: Textiles, food processing, beverages. **Chief crops:** Rice, wheat, cotton. **Minerals:** Natural gas. **Crude oil reserves** (2000): 208 mil bbls. **Arable land:** 27%.

Livestock (2000): cattle: 22.0 mil; chickens: 148.0 mil; goats: 47.4 mil; sheep: 24.1 mil. **Fish catch** (1999): 597,201 metric tons. **Electricity prod.:** 62.078 bil kWh. **Labor force:** 44% agric.; 39% services; 17% ind.

Finance: Monetary unit: Rupee (Sept. 2001: 63.85 = $1 U.S.). **GDP:** (1999 est.): $282 bil. **Per capita GDP**: $2,000. **Imports** (1999 est.): $9.8 bil; partners: U.S. 8%, Japan 8%, Malaysia 7%. **Exports** (1999 est.): $8.4 bil; partners: U.S. 22%, Hong Kong 7%, UK 7%. **Tourism** (1998): $98 mil. **Budget** (FY 1996-97): $12 bil. **Intl. reserves less gold** (June 2000): $1.40 bil. **Gold:** 2.09 mil oz t. **Consumer prices** (change in 1999): 4.1%.

Transport: Railroad: Length: 5,453 mi. **Motor vehicles:** 800,000 pass. cars, 300,000 comm. vehicles. **Civil aviation:** 7.2 bil pass.-mi; 35 airports. **Chief port:** Karachi.

Communications: TV sets: 62 per 1,000 pop. **Radios:** 92 per 1,000 pop. **Telephones** (1999): 2,986,100 main lines. **Daily newspaper circ.:** 21 per 1,000 pop.

Health: Life expectancy: 60.61 male; 62.32 female. **Births** (per 1,000 pop): 31.21. **Deaths** (per 1,000 pop.): 9.26. **Natural inc.:** 2.2%. **Infant mortality** (per 1,000 live births): 80.5.

Education: Literacy: 38%.

Major Intl. Organizations: UN (FAO, IBRD, ILO, IMF, IMO, WHO, WTrO).

Embassy: 2315 Massachusetts Ave. NW 20008; 939-6200.

Website: http://www.pak.gov.pk

Present-day Pakistan shares the 5,000-year history of the India-Pakistan subcontinent. At present-day Harappa and Mohenjo Daro, the Indus Valley Civilization, with large cities and elaborate irrigation systems, flourished c. 4,000-2,500 BC.

Aryan invaders from the NW conquered the region around 1,500 BC, forging a Hindu civilization that dominated Pakistan as well as India for 2,000 years.

Beginning with the Persians in the 6th century BC, and continuing with Alexander the Great and with the Sassanians, successive nations to the west ruled or influenced Pakistan. The first Arab invasion, AD 712, introduced Islam. Under the Mogul empire (1526-1857), Muslims ruled most of India, yielding to British encroachment and resurgent Hindus.

After World War I the Muslims of British India began agitation for minority rights in elections. Muhammad Ali Jinnah (1876-1948) was the principal architect of Pakistan. A leader of the Muslim League from 1916, he worked for dominion status for India; from 1940 he advocated a separate Muslim state.

When the British withdrew Aug. 14, 1947, the Islamic majority areas of India acquired self-government as Pakistan, with dominion status in the Commonwealth. Pakistan was divided into 2 sections, West Pakistan and East Pakistan. The 2 areas were nearly 1,000 mi. apart on opposite sides of India. Pakistan became a republic in 1956.

In Oct. 1958, Gen. Mohammad Ayub Khan took power in a coup. He was elected president in 1960, reelected in 1965. He resigned Mar. 25, 1969, after several months of violent rioting and unrest, most of it in East Pakistan, which demanded autonomy. The government was turned over to Gen. Agha Mohammad Yahya Khan and martial law was declared.

The Awami League, which sought regional autonomy for East Pakistan, won a majority in Dec. 1970 elections to a constituent assembly. In March 1971 Yahya postponed the assembly. Rioting and strikes broke out in the East.

On Mar. 25, 1971, government troops launched attacks in the East. The Easterners, aided by India, proclaimed the independent nation of Bangladesh. In months of widespread fighting, countless thousands were killed. Some 10 million Easterners fled into India. Full-scale war between India and Pakistan had spread to both the East and West fronts by Dec. 3. Pakistan troops in the East surrendered Dec. 16; Pakistan agreed to a cease-fire in the West Dec. 17. On July 3, 1972, Pakistan and India signed a pact agreeing to withdraw troops from their borders and seek peaceful solutions to all problems.

Zulfikar Ali Bhutto, leader of the Pakistan People's Party, which had won the most West Pakistan votes in Dec. 1970 elections, became president Dec. 20. Bhutto was overthrown in a military coup July 1977. Convicted of complicity in a 1974 political murder, he was executed Apr. 4, 1979. Over 3 million Afghan refugees flooded into Pakistan after the USSR invaded Afghanistan Dec. 1979; over 1.2 million remained in 1999.

Pres. Mohammad Zia ul-Haq was killed when his plane exploded in Aug. 1988. Following Nov. elections, Benazir Bhutto, daughter of Zulfikar Ali Bhutto, was named prime minister, becoming the first woman leader of a Muslim nation. She was accused of corruption and dismissed by the president, Aug. 1990; her party was soundly defeated in Oct. 1990 elections, and Nawaz Sharif became prime minister. She regained power after elections in Oct. 1993. Opposition to Bhutto centered around Karachi, which was crippled by violent strikes and ethnic clashes during 1995 and 1996. Accusing the Bhutto government of corruption and mismanagement, Pres. Farooq Leghari appointed a caretaker prime minister Nov. 5, 1996. Elections on Feb. 3, 1997, gave Sharif a parliamentary majority.

Responding to nuclear weapons tests by India, Pakistan conducted its own tests, May 28-30, 1998; the U.S. imposed economic sanctions on both countries. Tried in absentia, the exiled Bhutto was convicted Apr. 15, 1999, of receiving kickbacks; the conviction was overturned and a new trial ordered Apr. 6, 2001.

In mid-1999, Muslim infiltrators, apparently including Pakistani troops, seized Indian-held positions in the disputed territory of Kashmir, which witnessed its heaviest fighting in over 2 decades. After meeting with Pres. Bill Clinton on July 4, Sharif agreed to a Pakistani pullback. Growing conflict between Sharif and the military climaxed in his firing on Oct. 12 of army chief Gen. Pervez Musharraf, whose supporters staged a bloodless coup later that day. Martial law was imposed and the constitution suspended Oct. 15. Because of the military takeover, Pakistan was suspended, Oct. 18, from the Commonwealth. Sharif was sentenced to life imprisonment on Apr. 6, 2000; he was released and exiled to Saudi Arabia Dec. 10. Musharraf assumed the presidency June 20, 2001.

Following the Sept. 11, 2001, terrorist attack on the U.S., Pres. Musharraf, Sept. 19, pledged cooperation with the U.S. in any action against accused terrorist Osama bin Laden and the ruling Taliban said to be harboring him in neighboring Afghanistan. Pakistan has been a major ally of the Taliban, whom many Pakistanis support.

For further details and developments see the front-of-the-book feature and the Chronology of the Year's Events.

Palau

Republic of Palau

People: Population: 19,092. **Age distrib.** (%): <15: 26.9; 65+: 4.7. **Pop. density:** 108 per sq. mi. **Urban:** 72%. **Ethnic groups:** Polynesian, Malayan, Melanesian. **Principal languages:** English (official), Palauan, Sonsorolese, Angaur, Japanese, Tobi (all official within certain states). **Chief religions:** Catholic, Modekngei.

Geography: Area: 177 sq. mi. **Location:** Archipelago (26 islands, more than 300 islets) in the W Pacific Ocean, about 530 mi SE of the Philippines. **Neighbors:** Micronesia to E, Indonesia to S. **Capital:** Koror (2000) 30,303. (Note: a new capital is being built in Babelthuap.)

Government: Type: Republic. **Head of state and gov.:** Pres. Tommy Remengesau; b 1956; in office: Jan. 19, 2001. **Local divisions:** 18 states..

Economy: Industries: Tourism, fish. **Chief crops:** Coconuts, copra, cassava, sweet potatoes. **Minerals:** Gold.

Finance: Monetary unit: U.S. Dollar. **GDP:** (1997 est.): $160 mil. **Per capita GDP:** $8,800. **Imports** (1996): $72.4 mil; partners: U.S. **Exports** (1996): $14.3 mil; partners: U.S., Japan. **Budget** (1997 est.): $59.9 mil.

Transport: 1 airport.

Communications: TV sets: 98 per 1,000 pop. **Radios:** 550 per 1,000 pop.

Health: Life expectancy: 65.77 male; 72.19 female. **Births** (per 1,000 pop): 19.64. **Deaths** (per 1,000 pop.): 7.23. **Natural inc.:** 1.24%. **Infant mortality** (per 1,000 live births): 16.67.

Education: Compulsory: ages 6-14. **Literacy** (1990): 98%.

Major Intl. Organizations: UN (WHO).

Embassy: 1150 18th Street NW, Suite 750, 20036; 452-6814.

Spain acquired the Palau Islands in 1886 and sold them to Germany in 1899. Japan seized them in 1914. American forces occupied the islands in 1944; in 1947, they became part of the U.S.-administered UN Trust Territory of the Pacific Islands. In 1981 Palau became an autonomous republic; in 1993 the republic ratified a compact of free association with the U.S., which provides financial aid in return for U.S. use of Palauan military facilities over 15 years. Palau became an independent nation on Oct. 1, 1994. Vice-Pres. Tommy Remengesau won the presidential election held Nov. 7, 2000.

Panama

Republic of Panama

People: Population: 2,845,647 **Age distrib.** (%): <15: 30.1; 65+: 6.0. **Pop. density:** 97 per sq. mi. **Urban:** 56%. **Ethnic groups:** Mestizo 70%, West Indian 14%, white 10%, Amerindian 6%. **Principal languages:** Spanish (official), English. **Chief religions:** Roman Catholic 85%, Protestant 15%.

Geography: Area: 29,300 sq. mi. **Location:** In Central America. **Neighbors:** Costa Rica on W, Colombia on E. **Topography:** 2 mountain ranges run the length of the isthmus.

Tropical rain forests cover the Caribbean coast and eastern Panama. **Capital**: Panama City (urban agg.): 1,173,000.

Government: Type: Constitutional republic. **Head of state and gov.:** Pres. Mireya Elisa Moscoso; b July 1, 1946; in office: Sept. 1, 1999. **Local divisions:** 9 provinces, 3 territories. **Defense budget:** $135 mil. **Active troops**: Nil. (11,800 paramilitary).

Economy: Industries: Oil refining, cement, construction. **Chief crops:** Bananas, rice, corn, coffee, sugar. **Minerals:** Copper. **Other resources:** Forests (mahogany), shrimp. **Arable land:** 7%. **Livestock** (2000): cattle: 1.36 mil; chickens: 11.79 mil; goats: 5,200; pigs: 279,510. **Fish catch:** (1999): 169,718 metric tons. **Electricity prod.:** 4.413 bil kWh. **Labor force:** 64% services; 18% agric.; 18% ind.

Finance: Monetary unit: Balboa (Sept. 2001: 1.00 = $1 U.S.). **GDP:** (1999 est.): $21 bil. **Per capita GDP:** $7,600. **Imports** (1999 est.): $6.4 bil; partners: U.S. 40%. **Exports** (1999 est.): $4.7 bil; partners: U.S. 40%. **Tourism** (1998): $379 mil. **Budget** (1997 est.): $2.4 bil. **Intl. reserves less gold** (May 2000): $797.1 mil. **Consumer prices** (change in 1999): 1.3%.

Transport: Railroad: Length: 220 mi. **Motor vehicles:** 144,000 pass. cars, 82,800 comm. vehicles. **Civil aviation:** 679.9 mil pass.-mi; 10 airports. **Chief ports:** Balboa, Cristobal.

Communications: TV sets: 13 per 1,000 pop. **Radios:** 5.1 per 1,000 pop. **Telephones** (1999): 462,500 main lines. **Daily newspaper circ.:** 62 per 1,000 pop.

Health: Life expectancy: 72.94 male; 78.53 female. **Births** (per 1,000 pop): 19.06. **Deaths** (per 1,000 pop.): 4.95. **Natural inc.:** 1.41%. **Infant mortality** (per 1,000 live births): 20.18.

Education: Free, compulsory for 6 years between ages 6-15. **Literacy**: 91%.

Major Intl. Organizations: UN (FAO, IBRD, ILO, IMF, IMO, WHO), OAS.

Embassy: 2862 McGill Terrace NW 20008; 483-1407.

The coast of Panama was sighted by Rodrigo de Bastidas, sailing with Columbus for Spain in 1501, and was visited by Columbus in 1502. Vasco Nunez de Balboa crossed the isthmus and "discovered" the Pacific Ocean, Sept. 13, 1513. Spanish colonies were ravaged by Francis Drake, 1572-95, and Henry Morgan, 1668-71. Morgan destroyed the old city of Panama which had been founded in 1519. Freed from Spain, Panama joined Colombia in 1821.

Panama declared its independence from Colombia Nov. 3, 1903, with U.S. recognition. In support of Panama, U.S. naval forces deterred action by Colombia. Panama granted use, occupation, and control of the Canal Zone to the U.S. by treaty, ratified Feb. 26, 1904. In 1978, a new treaty provided for a gradual takeover by Panama of the canal, and withdrawal of U.S. troops, to be completed before the end of the century. U.S. payments were substantially increased in the interim.

President Delvalle was ousted by the National Assembly, Feb. 26, 1988, after he tried to fire the head of the Panama Defense Forces, Gen. Manuel Antonio Noriega, who was under U.S. federal indictment on drug charges. U.S. troops invaded Panama Dec. 20, 1989, and Noriega surrendered Jan. 3, 1990.

On Aug. 30, 1998, voters rejected a constitutional change that would have allowed Pres. Ernesto Pérez Balladares to run for reelection in 1999. Mireya Moscoso, widow of former Pres. Arnulfo Arias, was elected president May 2, 1999, becoming Panama's first female head of state. The U.S.handed over control of the Panama Canal to Panama Dec. 31, 1999.

Papua New Guinea

Independent State of Papua New Guinea

People: Population: 5,049,055. **Age distrib.** (%): <15: 38.7; 65+: 3.7. **Pop. density:** 29 per sq. mi. **Urban:** 17%. **Ethnic groups:** Papuan, Melanesian. **Principal languages:** English (official), Motu, 715 indigenous dialects. **Chief religions:** Indigenous beliefs 34%, Roman Catholic 22%, Lutheran 16%.

Geography: Area: 174,200 sq. mi. **Location:** SE Asia, occupying E half of island of New Guinea and about 600 nearby islands. **Neighbors:** Indonesia (West Irian) on W, Australia on S. **Topography:** Thickly forested mts. cover much of the center of the country, with lowlands along the coasts. Included are some islands of Bismarck and Solomon groups, such as the Admiralty Isls., New Ireland, New Britain, and Bougainville. **Capital:** Port Moresby (1995): 193,242.

Government: Type: Parliamentary democracy. **Head of state:** Queen Elizabeth II, represented by Gov-Gen. Silas Atopare; in office: Nov. 1997. **Head of gov.:** Prime Min. Mekere Morauta; b 1946; in office: July 14, 1999. **Local divisions:** 20 provinces. **Defense budget:** $36 mil. **Active troops:** 4,400.

Economy: Industries: Wood products, mining, oil. **Chief crops:** Coffee, coconuts, cocoa. **Minerals:** Gold, copper, silver. **Crude oil reserves** (2000): 333 mil bbls. **Livestock** (2000): cattle: 87,000; chickens: 3.6 mil; goats: 2,200; pigs: 1.55 mil; sheep: 6,200. **Fish catch:** (1999): 45,025 metric tons. **Electricity prod.:** 1.820 bil kWh.

Finance: Monetary unit: Kina (Sept. 2001: 3.39 = $1 U.S.). **GDP:** (1999 est.): $11.6 bil. **Per capita GDP:** $2,500. **Imports** (1999 est.): $1 bil; partners: Australia 51%, Singapore 10%, Japan 8%. **Exports** (1999 est.): $1.9 bil; partners: Australia 20%, Japan 13%, Germany 7%. **Tourism:** $104 mil. **Budget** (1997 est.): $1.35 bil. **Intl. reserves less gold** (Mar. 2000): $175.70 mil. **Gold:** 63,000 oz t. **Consumer prices** (change in 1999): 14.9%.

Transport: Motor vehicles: 21,600 pass. cars, 77,700 comm. vehicles. **Civil aviation:** 456.5 mil pass.-mi; 129 airports. **Chief ports:** Port Moresby, Lae.

Communications: TV sets: 23 per 1,000 pop. **Radios:** 68 per 1,000 pop. **Telephones**: 64,800 main lines. **Daily newspaper circ.:** 15 per 1,000 pop.

Health: Life expectancy: 61.39 male; 65.64 female. **Births** (per 1,000 pop): 32.15. **Deaths** (per 1,000 pop.): 7.88. **Natural inc.:** 2.43%. **Infant mortality** (per 1,000 live births): 58.21.

Education: Literacy: 72%.

Major Intl. Organizations: UN (FAO, IBRD, ILO, IMF, IMO, WHO, WTrO), the Commonwealth, APEC.

Embassy: 1779 Massachusetts Ave NW, 20036; 745-3680.

Website: http://www.pngembassy.org

Human remains have been found in the interior of New Guinea dating back at least 10,000 years and possibly much earlier. Successive waves of peoples probably entered the country from Asia through Indonesia. The indigenous population consists of a huge number of tribes, many living in almost complete isolation with mutually unintelligible languages.

Europeans visited in the 15th century, but actual land claims did not begin until the 19th century, when the Dutch took control of the island's western half. The southern half of eastern New Guinea was first claimed by Britain in 1884, and transferred to Australia in 1905. The northern half was claimed by Germany in 1884, but captured in World War I by Australia, which was first granted a League of Nations mandate and then a UN trusteeship over the area. The 2 territories were administered jointly after 1949, given self-government Dec. 1, 1973, and became independent Sept. 16, 1975.

Secessionist rebels clashed with government forces on Bougainville beginning in 1988; a truce signed Oct. 10, 1997, brought a halt to the fighting, which had claimed an estimated 20,000 lives.The country suffered from a severe drought in 1997. A tsunami killed at least 3,000 people July 17, 1998. Soldiers mutinied Mar. 2001 to protest defense cuts, which were then rescinded. A Bougainville autonomy agreement was signed Aug. 30, 2001.

Paraguay

Republic of Paraguay

People: Population: 5,734,139. **Age distrib.** (%): <15: 38.9; 65+: 4.7. **Pop. density:** 37 per sq. mi. **Urban:** 55%. **Ethnic groups:** Mestizo 95%, white & Amerindian 5%. **Principal languages:** Spanish (official), Guarani. **Chief religion:** Roman Catholic 90%.

Geography: Area: 153,200 sq. mi. **Location:** Landlocked country in central South America. **Neighbors:** Bolivia on N, Argentina on S, Brazil on E. **Topography:** Paraguay R. bisects the country. To E are fertile plains, wooded slopes, grasslands. To W is the Gran Chaco plain, with marshes and scrub trees. Extreme W is arid. **Capital**: Asunción (urban agg.): 1,262,000.

Government: Type: Republic. **Head of state and gov.:** Pres. Luis Angel González Macchi; b Dec. 13, 1947; in office: Mar. 28, 1999. **Local divisions:** 18 departments and capital city. **Defense budget:** $83 mil. **Active troops:** 20,200.

Economy: Industries: Textiles, cement. **Chief crops:** Corn, cotton, soybeans, sugarcane. **Minerals:** Iron, manganese, limestone. **Other resources:** Forests. **Arable land:** 6%. **Livestock** (2000): cattle: 9.91 mil; chickens: 25.0 mil; goats: 137,500; pigs: 2.7 mil; sheep: 412,500. **Fish catch:** (1999): 28,000 metric tons. **Electricity prod.:** 51.554 bil kWh. **Labor force:** 45% agric.

Finance: Monetary unit: Guarani (Sept. 2001: 4,352.00 = $1 U.S.). **GDP:** (1999 est.): $19.9 bil. **Per capita GDP:** $3,650. **Imports** (1999 est.): $3.2 bil; partners: Brazil 34%, U.S., Argentina. **Exports** (1999 est.): $3.1 bil; partners: Brazil, Argnetina, EU. **Tourism** (1998): $595 mil. **Budget** (1995 est.): $1.66 bil. **Intl. reserves less gold** (June 2000): $922.04 mil. **Gold:** 35,000 o zt. **Consumer prices** (change in 1999): 6.8%.

Transport: Railroad: Length: 274 mi. **Motor vehicles:** 71,000 pass. cars, 50,000 comm. vehicles. **Civil aviation:** 133.7 mil pass.-mi; 5 airports. **Chief port:** Asunción.

Communications: TV sets: 144 per 1,000 pop. **Radios:** 141 per 1,000 pop. **Telephones** (1999): 268,100 main lines. **Daily newspaper circ.:** 40 per 1,000 pop.

Health: Life expectancy: 71.44 male; 76.52 female. **Births** (per 1,000 pop): 30.88. **Deaths** (per 1,000 pop.): 4.75. **Natural inc.:** 2.61%. **Infant mortality** (per 1,000 live births): 29.78.

Education: Compulsory: ages 6-12. **Literacy**: 92%.

Major Intl. Organizations: UN (FAO, IBRD, ILO, IMF, IMO, WHO, WTrO), OAS.

Embassy: 2400 Massachusetts Ave. NW, 20008; 483-6960.

The Guarani Indians were settled farmers speaking a common language before the arrival of Europeans.

Visited by Sebastian Cabot in 1527 and settled as a Spanish possession in 1535, Paraguay gained its independence from Spain in 1811. It lost much of its territory to Brazil, Uruguay, and Argentina in the War of the Triple Alliance, 1865-1870. Large areas were won from Bolivia in the Chaco War, 1932-35.

Gen. Alfredo Stroessner, who had ruled since 1954, was ousted in a military coup led by Gen. Andrés Rodríguez on Feb. 3, 1989. Rodríguez was elected president May 1. Juan Carlos Wasmosy was elected president May 9, 1993, becoming the nation's first civilian head of state in many years.

A prolonged power struggle involving a popular military leader, Gen. Lino César Oviedo, who was accused of insubordination, culminated in his surrender Dec. 12, 1997. He was freed Aug. 18, 1998, following the inauguration of Pres. Raúl Cubas Grau, Oviedo's successor as Colorado Party nominee. The assassination of Vice Pres. Luis María Argaña, Mar. 23, 1999, by an unidentified gunman, was widely attributed to Cubas and triggered protests and an impeachment vote; Cubas resigned Mar. 28 and was succeeded by Senate leader Luis Angel González Macchi. An attempted military coup was suppressed May 18, 2000.

Peru

Republic of Peru

People: Population: 27,483,864. **Age distrib.** (%): <15: 34.4; 65+: 4.8. **Pop. density:** 56 per sq. mi. **Urban:** 72%. **Ethnic groups:** Amerindian 45%, mestizo 37%, white 15%. **Principal languages:** Spanish, Quechua (both official), Aymara. **Chief religion:** Roman Catholic.

Geography: Area: 493,600 sq. mi. **Location:** On the Pacific coast of South America. **Neighbors:** Ecuador, Colombia on N; Brazil, Bolivia on E; Chile on S. **Topography:** An arid coastal strip, 10 to 100 mi. wide, supports much of the population thanks to widespread irrigation. The Andes cover 27% of land area. The uplands are well-watered, as are the eastern slopes reaching the Amazon basin, which covers half the country with its forests and jungles. **Capital:** Lima. **Cities**: Lima (urban agg.): 7,443,000; (1998 met. est.): Arequipa 710,103; Callao 424,294.

Government: Type: Republic. **Head of state:** Pres. Alejandro Toledo; b Mar. 28, 1946; in office: July 28, 2001. **Head of gov.:** Prime Min. Roberto Dañino; b Mar. 2,1951; in office: July 28, 2001. **Local divisions:** 12 regions, 24 departments, 1 constitutional province. **Defense:** 2.2% of GDP. **Active troops:** 125,000.

Economy: Industries: Fishing, mining, food processing, textiles. **Chief crops:** Cotton, sugar, coffee, rice. **Minerals:** Copper, silver, gold, iron, oil. **Crude oil reserves** (2000): 355.0 mil bbls. **Other resources:** Wool, fish. **Arable land:** 3%. **Livestock** (2000): cattle: 4.9 mil; chickens: 81.3 mil; goats: 2.07 mil; pigs: 2.79 mil; sheep: 14.4 mil. **Fish catch** (1999): 7.88 mil metric tons. **Electricity prod.:** 18.886 bil kWh.

Finance: Monetary unit: New Sol (Sept. 2001: 3.48 = $1 U.S.). **GDP:** (1999 est.): $116 bil. **Per capita GDP:** $4,400. **Imports** (1999): $8.4 bil; partners: U.S. 19%, Colombia 6%. **Exports** (1999 est.): $5.9 bil; partners: U.S. 25%, China 8%, Japan 7%. **Tourism** (1998): $913 mil. **Budget** (1996 est.): $9.3 bil. **Intl. reserves less gold** (May 2000): $9.18 bil. **Gold:** 1.10 mil oz t. **Consumer prices** (change in 1999): 3.5%.

Transport: Railroad: Length: 1,318 mi. **Motor vehicles:** 500,000 pass. cars, 275,000 comm. vehicles. **Civil aviation:** 1.8 bil pass.-mi; 27 airports. **Chief ports:** Callao, Chimbote, Matarani, Salaverry.

Communications: TV sets: 85 per 1,000 pop. **Radios:** 221 per 1,000 pop. **Telephones**: 1,635,900 main lines. **Daily newspaper circ.:** 87 per 1,000 pop.

Health: Life expectancy: 67.9 male; 72.81 female. **Births** (per 1,000 pop): 23.90. **Deaths** (per 1,000 pop.): 5.78. **Natural inc.:** 1.81%. **Infant mortality** (per 1,000 live births): 39.39.

Education: Free, compulsory: ages 6-11. **Literacy**: 89%.

Major Intl. Organizations: UN and all of its specialized agencies, APEC, OAS.

Embassy: 1700 Massachusetts Ave. NW 20036; 833-9860.

The powerful Inca empire had its seat at Cuzco in the Andes and covered most of Peru, Bolivia, and Ecuador, as well as parts of Colombia, Chile, and Argentina. Building on the achievements of 800 years of Andean civilization, the Incas had a high level of skill in architecture, engineering, textiles, and social organization.

A civil war had weakened the empire when Francisco Pizarro, Spanish conquistador, began raiding Peru for its wealth, 1532. In 1533 he seized the ruling Inca, Atahualpa, filled a room with gold as a ransom, then executed him and enslaved the natives.

Lima was the seat of Spanish viceroys until the Argentine liberator, José de San Martin, captured it in 1821; Spanish forces were ultimately routed by Simón Bolívar, 1824.

On Oct. 3, 1968, a military coup ousted Pres. Fernando Belaunde Terry. In 1968-74, the military government started socialist programs. Food shortages, escalating foreign debt, and strikes led to another coup, Aug. 29, 1976.

After 12 years of military rule, Peru returned to democratic leadership in 1980 but was plagued by economic problems and by leftist Shining Path (Sendero Luminoso) guerrillas.

Elected president in June 1990, Alberto Fujimori, the son of Japanese immigrants, dissolved the National Congress, suspended parts of the constitution, and initiated press censorship, Apr. 5, 1992. The leader of Shining Path was captured Sept. 12.

With the economy booming and signs of significant progress in curtailing guerrilla activity, Fujimori won reelection Apr. 9, 1995. Repressive antiterrorism tactics, however, drew international criticism. On Dec. 17, 1996, leftist Tupac Amaru guerrillas infiltrated a reception at the Japanese ambassador's residence in Lima and took hundreds of hostages, most of whom were later released. Peruvian soldiers stormed the embassy Apr. 22, 1997, rescuing 71 of the remaining hostages; 1 hostage, 2 soldiers, and all 14 guerrillas were killed.

Fujimori's path to a 3d term was cleared when his lone remaining challenger withdrew, charging electoral fraud, 6 days before a runoff vote on May 28, 2000. Scandals involving his top aide and intelligence chief, Vladimiro Montesinos, led Fujimori to resign his office Nov. 20, while on a visit to Japan; instead of accepting his resignation, Congress ousted him as "morally unfit."

Alejandro Toledo won a presidential runoff election June 3, 2001. Montesinos was captured in Venezuela June 23 and extradited to Peru. Charges were filed Sept. 5 against the exiled Fujimori, alleging his complicity in the killings by a paramilitary death squad of at least 25 people during 1991-92.

Philippines

Republic of the Philippines

People: Population: 82,841,518. **Age distrib.** (%): <15: 36.9; 65+: 3.7. **Pop. density:** 720 per sq. mi. **Urban:** 58%. **Ethnic groups:** Christian Malay 92%, Muslim Malay 4%. **Principal languages:** Pilipino, English (both official). **Chief religions:** Roman Catholic 83%, Protestant 9%, Muslim 5%.

Geography: Area: 115,000 sq. mi. **Location:** An archipelago off the SE coast of Asia. **Neighbors:** Nearest are Malaysia and Indonesia on S, Taiwan on N. **Topography:** The country consists of some 7,100 islands stretching 1,100 mi. N-S. About 95% of area and population are on 11 largest islands, which are mountainous, except for the heavily indented coastlines and for the central plain on Luzon. **Capital:** Manila. **Cities** (2000 census): Quezon City 2,160,000.

Government: Type: Republic. **Head of state and gov.:** Pres. Gloria Macapagal Arroyo; b Apr. 5, 1947; in office: Jan. 20, 2001. **Local divisions:** 72 provinces, 61 chartered cities. **Defense budget:** $1.3 bil. **Active troops:** 106,000.

Economy: Industries: Food processing, textiles, chemicals, pharmaceuticals, wood prods. **Chief crops:** Sugar, rice, pineapples, corn, coconuts. **Minerals:** Cobalt, copper, gold, nickel, silver, oil. **Other resources:** Forests (46% of area). **Crude oil reserves** (2000): 289 mil bbls. **Arable land:** 19%. **Livestock** (2000): cattle: 2.55 mil; chickens: 142.0 mil; goats: 6.78 mil; pigs: 10.4 mil; sheep: 30,000. **Fish catch** (1999): 2.14 mil metric tons. **Electricity prod.:** 40.745 bil kWh. **Labor force:** 39.8% agric.; 19.4% govt. & social services; 17.7% services.

Finance: Monetary unit: Peso (Sept. 2001: 51.40 = $1 U.S.). **GDP:** (1999 est.): $282 bil. **Per capita GDP:** $3,600. **Imports** (1999 est.): $30.7 bil; partners: U.S. 22%, Japan 20%, S. Korea 8%. **Exports** (1999 est.): $34.8 bil; partners: U.S. 34%, EU 20%, Japan 14%. **Tourism:** $2.53 bil. **Budget** (1998 est.): $12.6 bil. **Intl. reserves less gold** (May 2000): $13.58 bil. **Gold:** 6.74 mil oz t. **Consumer prices** (change in 1999): 6.7%.

Transport: Railroad: Length: 557 mi. **Motor vehicles:** 702,578 pass. cars, 1.35 mil comm. vehicles. **Civil aviation:** 10.1 bil pass.-mi; 21 airports. **Chief ports:** Cebu, Manila, Iloilo, Davao.

Communications: TV sets: 109 per 1,000 pop. **Radios:** 113 per 1,000 pop. **Telephones**: 3,000,000 main lines. **Daily newspaper circ.:** 82 per 1,000 pop.

Health: Life expectancy: 64.96 male; 70.79 female. **Births** (per 1,000 pop): 27.37. **Deaths** (per 1,000 pop.): 6.04. **Natural inc.:** 2.13%. **Infant mortality** (per 1,000 live births): 28.7.

Education: Free, compulsory: ages 7-12. **Literacy**: 95%.

Major Intl. Organizations: UN (FAO, IBRD, ILO, IMF, IMO, WHO, WTrO), ASEAN.

Embassy: 1600 Massachusetts Ave. NW 20036; 467-9300.

Website: http://www.census.gov.ph

The Malay peoples of the Philippine Islands, whose ancestors probably migrated from Southeast Asia, were mostly hunters, fishers, and unsettled cultivators.

The archipelago was visited by Magellan, 1521. The Spanish founded Manila, 1571. The islands, named for King Philip II of Spain, were ceded by Spain to the U.S. for $20 million, 1898, following the Spanish-American War. U.S. troops suppressed a guerrilla uprising in a brutal 6-year war, 1899-1905.

Japan attacked the Philippines Dec. 8, 1941, and occupied the islands during WW II. On July 4, 1946, independence was proclaimed in accordance with an act passed by the U.S. Congress in 1934. A republic was established.

On Sept. 21, 1972, Pres. Ferdinand Marcos declared martial law. Marcos proclaimed a new constitution, Jan. 17, 1973, with himself as president. His wife, Imelda, received wide powers in 1978 to supervise planning and development. Political corruption was widespread. Martial law was lifted Jan. 17, 1981, but Marcos retained broad emergency powers. He was reelected in June to a new 6-year term as president.

The assassination of prominent opposition leader Benigno S. Aquino Jr., Aug. 21, 1983, sparked demonstrations calling for the resignation of Marcos. After a bitter presidential campaign, amid allegations of widespread election fraud, Marcos was declared the victor Feb. 16, 1986, over Corazon Aquino, widow of the slain opposition leader. With his support collapsing, Marcos fled the country Feb. 25.

Recognized as president by the U.S. and other nations, Aquino was plagued by a weak economy, widespread poverty, Communist and Muslim insurgencies, and lukewarm military support. Rebel troops seized military bases and TV stations and bombed the presidential palace, Dec. 1, 1989. Government forces defeated the attempted coup aided by air cover provided by U.S. F-4s. Aquino endorsed Fidel Ramos in the May 1992 presidential election, which he won.

The U.S. vacated the Subic Bay Naval Station at the end of 1992, ending its long military presence in the Philippines.

The government signed a cease-fire agreement, Jan. 30, 1994, with Muslim separatist guerrillas, but some rebels refused to abide by the accord. A new treaty providing for expansion and development of an autonomous Muslim region on Mindanao was signed Sept. 2, 1996, formally ending a rebellion that had claimed more than 120,000 lives since 1972.

Running as a populist, Joseph (Erap) Estrada, a former movie actor, won the presidential election of May 11, 1998. Charged with bribery and corruption, he was impeached Nov. 13, 2000. When the Supreme Court ruled the presidency vacant Jan. 20, 2001, Vice-Pres. Gloria Macapagal Arroyo became president.

Poland

Republic of Poland

People: Population: 38,633,912. **Age distrib.** (%): <15: 18.4; 65+: 12.4. **Pop. density:** 329 per sq. mi. **Urban:** 65%. **Ethnic groups:** Polish 98%. **Principal language:** Polish (official). **Chief religion:** Roman Catholic 95%.

Geography: Area: 117,400 sq. mi. **Location:** On the Baltic Sea in E central Europe. **Neighbors:** Germany on W; Czech Rep., Slovakia on S; Lithuania, Belarus, Ukraine on E; Russia on N. **Topography:** Mostly lowlands forming part of the Northern European Plain. The Carpathian Mts. along the S border rise to 8,200 ft. **Capital:** Warsaw. **Cities** (1997): Warsaw 1,632,500; Lodz 812,300; Krakow 740,500.

Government: Type: Republic. **Head of state:** Pres. Aleksander Kwasniewski; b Nov. 15, 1954; in office: Dec. 23, 1995. **Head of gov.:** Prime Min. Jerzy Buzek; b July 3, 1940; in office: Oct. 31, 1997. **Local divisions:** 16 provinces. **Defense budget:** $3.4 bil (2001). **Active troops:** 217,290.

Economy: Industries: Shipbuilding, coal mining, chemicals, metals, machinery, food processing. **Chief crops:** Grains, potatoes, fruits, vegetables. **Minerals:** Coal, copper, silver, lead, sulfur, natural gas. **Crude oil reserves** (2000): 114.88 mil bbls. **Arable land:** 47%. **Livestock** (2000): cattle: 6.08 mil; chickens: 49.53 mil; pigs: 17.12 mil; sheep: 361,582. **Fish catch** (1999): 390,586 metric tons. **Electricity prod.:** 134.351 bil kWh. **Labor force:** 50% services; 25% ind.; 25% agric.

Finance: Monetary unit: Zloty (Sept. 2001: 4.22 = $1 U.S.). **GDP:** (1999 est.): $276.5 bil. **Per capita GDP:** $7,200. **Imports** (1999): $40.8 bil; partners: Germany 25.8%, Italy 9.4%, France 6.5%. **Exports** (1999): $27.8 bil; partners: Germany 36%. **Tourism:** $6.10 bil. **Budget** (1997 est.): $38.3 bil. **Intl. reserves less gold** (May 2000): $24.87 bil. **Gold:** 3.31 mil oz t. **Consumer prices** (change in 1999): 7.3%.

Transport: Railroad: Length: 14,904 mi. **Motor vehicles:** 7.52 mil pass. cars, 1.55 mil comm. vehicles. **Civil aviation:** 2.6 bil pass.-mi; 8 airports. **Chief ports:** Gdansk, Gdynia, Ustka, Szczecin.

Communications: TV sets: 414 per 1,000 pop. **Radios:** 522.6 per 1,000 pop. **Telephones**: 10,945,600 main lines. **Daily newspaper circ.:** 113 per 1,000 pop.

Health: Life expectancy: 69.26 male; 77.82 female. **Births** (per 1,000 pop): 10.20. **Deaths** (per 1,000 pop.): 9.98. **Natural inc.:** .02%. **Infant mortality** (per 1,000 live births): 9.39.

Education: Free, compulsory: ages 7-14. **Literacy** (1994): 99%.

Major Intl. Organizations: UN (FAO, IBRD, ILO, IMF, IMO, WHO, WTrO), NATO, OECD, OSCE.

Embassy: 2640 16th St. NW 20009; 234-3800.

Website: http://www.polishworld.com

Slavic tribes in the area were converted to Latin Christianity in the 10th century. Poland was a great power from the 14th to the 17th centuries. In 3 partitions (1772, 1793, 1795) it was apportioned among Prussia, Russia, and Austria. Overrun by the Austro-German armies in World War I, it declared its independence on Nov. 11, 1918, and was recognized as independent by the Treaty of Versailles, June 28, 1919. Large territories to the east were taken in a war with Russia, 1921.

Germany and the USSR invaded Poland Sept. 1-27, 1939, and divided the country. During the war, some 6 million Polish citizens, half of them Jews, were killed by the Nazis. With Germany's defeat, a Polish government-in-exile in London was recognized by the U.S., but the USSR pressed the claims of a rival group. The election of 1947 was completely dominated by the Communists.

In compensation for 69,860 sq. mi. ceded to the USSR, in 1945 Poland received approx. 40,000 sq. mi. of German territory E of the Oder-Neisse line comprising Silesia, Pomerania, West Prussia, and part of East Prussia.

In 12 years of rule by Stalinists, large estates were abolished, industries nationalized, schools secularized, and Roman Catholic prelates jailed. Farm production fell off. Harsh working conditions caused a riot in Poznan, June 28-29, 1956. A new Politburo, committed to a more independent Polish Communism, was named Oct. 1956, with Wladyslaw Gomulka as first secretary of the party. Collectivization of farms was ended. Gomulka agreed to permit religious liberty and religious publications, provided the church kept out of politics.

In Dec. 1970 workers in port cities rioted because of price rises and new incentive wage rules. On Dec. 20 Gomulka resigned as party leader; he was succeeded by Edward Gierek. The rules were dropped and price rises revoked.

After 2 months of labor turmoil had crippled the country, the Polish government, Aug. 30, 1980, met the demands of striking workers at the Lenin Shipyard, Gdansk. Among the 21 concessions granted were the right to form independent trade unions and the right to strike. By 1981, 9.5 mil workers had joined the independent trade union (Solidarity). Solidarity leaders proposed, Dec. 12, a nationwide referendum on establishing a non-Communist government if the government failed to agree to a series of demands.

Spurred by fear of Soviet intervention, the government, Dec. 13, imposed martial law. Lech Walesa and other Solidarity leaders were arrested. The U.S. imposed sanctions, which were lifted when martial law was suspended Dec. 1982

On Apr. 5, 1989, an accord was reached between the government and opposition factions on political and economic reforms, including free elections. Candidates endorsed by Solidarity swept the parliamentary elections, June 4. Lech Walesa became president Dec. 22, 1990.

A radical economic program designed to transform the economy into a free-market system led to inflation and unemployment. In Sept. 1993, former Communists and other leftists won a majority in the lower house of Parliament. Walesa lost to a former Communist, Aleksander Kwasniewski, in a presidential runoff election, Nov. 19, 1995.

A new constitution was approved by referendum May 25, 1997. Flooding in July caused more than $1 billion in property damage. Solidarity won parliamentary elections held Sept. 21. Poland became a full member of NATO on Mar. 12, 1999. Pres. Kwasniewski was reelected Oct. 8, 2000. The former Communists won a plurality in parliamentary voting Sept. 23, 2001.

Portugal

Portuguese Republic

People: Population: 10,066,253. **Age distrib.** (%): <15: 17.0; 65+: 15.6. **Pop. density:** 285 per sq. mi. **Urban:** 63%. **Ethnic groups:** Homogeneous Mediterranean stock, small African minority. **Principal languages:** Portuguese (official). **Chief religion:** Roman Catholic 97%.

Geography: Area: 35,300 sq. mi., incl. the Azores and Madeira Islands. **Location:** At SW extreme of Europe. **Neighbors:** Spain on N, E. **Topography:** Portugal N of Tajus R., which bisects the country NE-SW, is mountainous, cool and rainy. To the S there are drier, rolling plains, and a warm climate. **Capital:** Lisbon. **Cities** (urban agg.): Lisbon 3,826,000; Porto 1,922,000.

Government: Type: Republic. **Head of state:** Pres. Jorge Sampaio; b Sept. 18, 1939; in office: Mar. 9, 1996. **Head of gov.:** Prime Min. António Guterres; b Apr. 30, 1949; in office: Oct. 30, 1995. **Local divisions:** 18 districts, 2 autonomous regions. **Defense budget:** $1.6 bil. **Active troops:** 44,650.

Economy: Industries: Textiles, footwear, cork, wood pulp, chemicals, fish canning, metal working, oil refining, wine, paper. **Chief crops:** Grains, potatoes, grapes, olives. **Minerals:** Tungsten, uranium, iron. **Other resources:** Forests (world leader in cork production). **Arable land:** 26%. **Livestock** (2000): cattle: 1.26 mil; chickens: 28 mil; goats: 800,000; pigs: 2.33 mil; sheep: 5.85 mil. **Fish catch** (1999): 229,108 metric tons. **Electricity prod.:** 41.696 bil kWh. **Labor force:** 60% services; 30% ind.; 10% agric.

Finance: Monetary unit: Escudo (Sept. 2001: 221.17 = $1 U.S.). Euro (Sept. 2001: 1.09 = $1 U.S.). **GDP:** (1999 est.): $151.4 bil. **Per capita GDP:** $15,300. **Imports** (1998): $34.9 bil; partners: EU 77%. **Exports** (1998): $25 bil; partners: EU 82%. **Tourism:** $5.17 bil. **Budget** (1996 est.): $52 bil. **Intl. reserves less gold** (June 2000): $8.30 bil. **Gold:** 19.51 mil oz t. **Consumer prices** (change in 1999): 2.3%.

Transport: Railroad: Length: 1,909 mi. **Motor vehicles** (1997): 2.95 mil pass. cars, 960,300 comm. vehicles. **Civil aviation:** 5.23 bil pass.-mi; 16 airports. **Chief ports:** Lisbon, Setubal, Leixoes.

Communications: TV sets: 523 per 1,000 pop. **Radios:** 306 per 1,000 pop. **Telephones:** 4,229,800 main lines. **Daily newspaper circ.:** 75 per 1,000 pop.

Health: Life expectancy: 72.44 male; 79.68 female. **Births** (per 1,000 pop): 11.51. **Deaths** (per 1,000 pop.): 10.21. **Natural inc.:** .13%. **Infant mortality** (per 1,000 live births): 5.94.

Education: Free, compulsory: ages 6-15. **Literacy:** 90%.

Major Intl. Organizations: UN (FAO, IBRD, ILO, IMF, IMO, WHO, WTrO), EU, NATO, OECD, OSCE.

Embassy: 2125 Kalorama Rd. NW 20008; 328-8610.

Website: http://www.portugal.org

Portugal, an independent state since the 12th century, was a kingdom until a revolution in 1910 drove out King Manoel II and a republic was proclaimed.

From 1932 a strong, repressive government was headed by Premier Antonio de Oliveira Salazar. Illness forced his retirement in Sept. 1968.

On Apr. 25, 1974, the government was seized by a military junta led by Gen. Antonio de Spinola, who became president. The new government reached agreements providing independence for Guinea-Bissau, Mozambique, Cape Verde Islands, Angola, and São Tomé and Príncipe. Banks, insurance companies, and other industries were nationalized.

Parliament approved, June 1, 1989, a package of reforms that did away with the socialist economy and created a "democratic" economy, denationalizing industries. Portugal returned Macao to China on Dec. 20, 1999.

Azores Islands, in the Atlantic, 740 mi. W of Portugal, have an area of 868 sq. mi. and a pop. (1993 est.) of 238,000. A 1951 agreement gave the U.S. rights to use defense facilities in the Azores. The **Madeira Islands,** 350 mi. off the NW coast of Africa, have an area of 306 sq. mi. and a pop. (1993 est.) of 437,312. Both groups were offered partial autonomy in 1976.

Qatar

State of Qatar

People: Population: 769,152. **Age distrib.** (%): <15: 25.8; 65+: 2.5. **Pop. density:** 181 per sq. mi. **Urban:** 92%. **Ethnic groups:** Arab 40%, Pakistani 18%, Indian 18%, Iranian 10%. **Principal languages:** Arabic (official), English. **Chief religion:** Muslim 95%.

Geography: Area: 4,200 sq. mi. **Location:** Middle East, occupying peninsula on W coast of Persian Gulf. **Neighbors:** Saudi Arabia on S. **Topography:** Mostly a flat desert, with some limestone ridges; vegetation of any kind is scarce. **Capital:** Doha (1999 est.): 391,000.

Government: Type: Traditional monarchy. **Head of state:** Emir Hamad bin Khalifa ath-Thani; b 1950; in office: June 27, 1995. **Head of gov.:** Prime Min. Abdullah bin Khalifa ath-Thani; in office: Oct. 29, 1996. **Local divisions:** 9 municipalities. **Defense budget:** $1.3 bil. **Active troops:** 12,330.

Economy: Industries: Oil production and refining, petrochemicals, cement. **Minerals:** Oil, gas. **Crude oil reserves** (2000): 3.7 bil bbls. **Livestock** (2000): cattle: 14,200; chickens: 40 mil; goats: 179,000; sheep: 215,000. **Electricity prod.:** 9 bil kWh. **Arable land:** 10%.

Finance: Monetary unit: Riyal (Sept. 2001: 3.64 = $1 U.S.). **GDP:** (1999 est.): $12.3 bil. **Per capita GDP:** $17,000. **Imports** (1999 est.): $4.2 bil; partners: UK 25%, France 13%, Japan 10%. **Exports** (1999 est.): $6.7 bil; partners: Japan 50%, Singapore 12%, S. Korea 9%. **Budget** (FY 1998-99 est.): $4.3 bil. **Gold:** 19,000 oz t.

Transport: Motor vehicles: 96,800 pass. cars, 85,600 comm. vehicles. **Civil aviation:** 1.6 bil pass.-mi; 1 airport. **Chief ports:** Doha, Umm Sáid.

Communications: TV sets: 451 per 1,000 pop. **Radios:** 322 per 1,000 pop. **Telephones:** 160,200 main lines. **Daily newspaper circ.:** 143 per 1,000 pop.

Health: Life expectancy: 70.16 male; 75.21 female. **Births** (per 1,000 pop): 15.91. **Deaths** (per 1,000 pop.): 4.26. **Natural inc.:** 1.17%. **Infant mortality** (per 1,000 live births): 21.44.

Education: Literacy: 79%.

Major Intl. Organizations: UN (FAO, IBRD, ILO, IMF, IMO, WHO, WTrO), AL, OPEC.

Embassy: 4200 Wisconsin Ave. NW 20016; 274-1600.

Website: http://www.arab.net/qatar/qatar_contents.html

Qatar was under Bahrain's control until the Ottoman Turks took power, 1872 to 1915. In a treaty signed 1916, Qatar gave Great Britain responsibility for its defense and foreign relations. After Britain announced it would remove its military forces from the Persian Gulf area by the end of 1971, Qatar sought a federation with other British-protected states in the area; this failed and Qatar declared itself independent, Sept. 1, 1971. Crown Prince Hamad bin Khalifa ath-Thani ousted his father, Emir Khalifa bin Hamad ath-Thani, June 27, 1995. In municipal elections held Mar. 8, 1999, women participated for the 1st time as candidates and voters.

Oil and natural gas revenues give Qatar a per capita income among the world's highest.

Romania

People: Population: 22,364,022. **Age distrib.** (%): <15: 17.9; 65+: 13.5. **Pop. density:** 252 per sq. mi. **Urban:** 56%. **Ethnic groups:** Romanian 89.1%, Hungarian 8.9%. **Principal languages:** Romanian (official), Hungarian, German. **Chief religions:** Romanian Orthodox 70%, Roman Catholic 6%, Protestant 6%.

Geography: Area: 88,800 sq. mi. **Location:** SE Europe, on the Black Sea. **Neighbors:** Moldova on E, Ukraine on N, Hungary and Serbia and Montenegro on W, Bulgaria on S. **Topography:** The Carpathian Mts. encase the north-central Transylvanian plateau. There are wide plains S and E of the mountains, through which flow the lower reaches of the rivers of the Danube system. **Capital:** Bucharest (urban agg.): 2,054,000.

Government: Type: Republic. **Head of state:** Pres. Ion Iliescu; b Mar. 3, 1930; in office: Dec. 20, 2000. **Head of gov.:** Prime Min. Adrian Nastase; b June 22, 1950; in office: Dec. 28, 2000. **Local divisions:** 40 counties, 1 municipality. **Defense budget:** $607 mil (1999). **Active troops:** 207,000.

Economy: Industries: Mining, timber, construction materials, metals, machinery, oil products, chemicals, food processing. **Chief crops:** Grains, grapes, sunflower seeds, sugar beets, potatoes. **Minerals:** Oil, gas, coal, iron. **Crude oil reserves** (2000): 1.43 bil bbls. **Other resources:** Timber. **Arable land:** 41%. **Livestock** (2000): cattle: 3.15 mil; chickens: 72.0 mil; goats: 554,200; pigs: 5.95 mil; sheep: 7.97 mil. **Fish catch** (1999): 19,322 metric tons. **Electricity prod.:** 49.036 bil kWh. **Labor force (1994):** 36.5% agric; 34.4% ind.; services 29.1%.

Finance: Monetary unit: Leu (Sept. 2001: 30,180 = $1 U.S.). **GDP:** (1999 est.): $87.4 bil. **Per capita GDP:** $3,900. **Imports** (1999 est.): $9.6 bil; partners: Germany 17.5%, Italy 17.4%, France 6.9%. **Exports** (1999 est.): $8.4 bil; partners: Italy 22%; Germany 19.6%; France 5.9%. **Tourism:** $254 mil. **Budget** (1997 est.): $11.7 bil. **Intl. reserves less gold** (May 2000): $2.80 bil. **Gold:** 3.35 mil oz t. **Consumer prices** (change in 1999): 45.8%.

Transport: Railroad: Length: 7,062 mi. **Motor vehicles:** 2.39 mil pass. cars; 513,312 comm. vehicles. **Civil aviation:** 1.1 bil pass.-mi; 8 airports. **Chief ports:** Constanta, Braila.

Communications: TV sets: 201 per 1,000 pop. **Radios:** 198 per 1,000 pop. **Telephones:** 3,899,200 main lines. **Daily newspaper circ.:** 297 per 1,000 pop.

Health: Life expectancy: 66.36 male; 74.19 female. **Births** (per 1,000 pop): 10.80. **Deaths** (per 1,000 pop.): 12.28. **Natural inc.:** -.15%. **Infant mortality** (per 1,000 live births): 19.36.

Education: Compulsory: ages 6-16. **Literacy** (1992): 97%.
Major Intl. Organizations: UN (FAO, IBRD, ILO, IMF, IMO, WHO, WTrO), OSCE.
Embassy: 1607 23d St. NW 20008; 332-4846.
Website: http://www.embassy.org/romania

Romania's earliest known people merged with invading Proto-Thracians, preceding by centuries the Dacians. The Dacian kingdom was occupied by Rome, AD 106-271; people and language were Romanized. The principalities of Wallachia and Moldavia, dominated by Turkey, were united in 1859, became Romania in 1861. In 1877 Romania proclaimed independence from Turkey, and became an independent state by the Treaty of Berlin, 1878; a kingdom under Carol I, 1881; and a constitutional monarchy with a bicameral legislature, 1886.

Romania helped Russia in its war with Turkey, 1877-78. After World War I it acquired Bessarabia, Bukovina, Transylvania, and Banat. In 1940 it ceded Bessarabia and Northern Bukovina to the USSR, part of southern Dobrudja to Bulgaria, and northern Transylvania to Hungary.

In 1941, Prem. Marshal Ion Antonescu led Romania in support of Germany against the USSR. In 1944 he was overthrown by King Michael and Romania joined the Allies.

After occupation by Soviet troops a People's Republic was proclaimed, Dec. 30, 1947; Michael was forced to abdicate.

On Aug. 22, 1965, a new constitution proclaimed Romania a Socialist Republic. Pres. Nicolae Ceausescu maintained an independent course in foreign affairs, but his domestic policies were repressive. All industry was state-owned, and state farms and cooperatives owned almost all arable land.

On Dec. 16, 1989, security forces opened fire on antigovernment demonstrators in Timisoara; hundreds were buried in mass graves. Ceausescu declared a state of emergency as protests spread to other cities. On Dec. 21, in Bucharest, security forces fired on protesters. Army units joined the rebellion, Dec. 22, and a group known as the Council of National Salvation announced that it had overthrown the government. Fierce fighting took place between the army, which backed the new government, and forces loyal to Ceausescu.

Ceausescu and his wife were captured and, following a trial in which they were found guilty of genocide, were executed Dec. 25, 1989. Former Communists dominated the government in succeeding years. A new constitution providing for a multiparty system took effect Dec. 8, 1991. Many of Romania's state-owned companies were privatized in 1996. The former Communists were swept from power in elections Nov. 3 and 17, 1996, but made a comeback in balloting Nov. 26 and Dec. 10, 2000.

Russia

Russian Federation

People: Population: 145,470,197. **Age distrib.** (%): <15: 17.4; 65+: 12.8 . **Pop. density:** 22 per sq. mi. **Urban:** 77%. **Ethnic groups:** Russian 81.5%, Tatar 3.8%. **Principal languages:** Russian (official), many others. **Chief religions:** Russian Orthodox, Muslim, others.

Geography: Area: 6,585,000 sq. mi., more than 76% of total area of the former USSR and the largest country in the world. **Location:** Stretches from E Europe across N Asia to the Pacific O. **Neighbors:** Finland, Norway, Estonia, Latvia, Belarus, Ukraine on W; Georgia, Azerbaijan, Kazakhstan, China, Mongolia, North Korea on S; Kaliningrad exclave bordered by Poland on the S, Lithuania on the N and E. **Topography:** Russia contains every type of climate except the distinctly tropical, and has a varied topography. The European portion is a low plain, grassy in S, wooded in N, with Ural Mts. on the E, and Caucasus Mts. on the S. Urals stretch N-S for 2,500 mi. The Asiatic portion is also a vast plain, with mountains on the S and in the E; tundra covers extreme N, with forest belt below; plains, marshes are in W, desert in SW. **Capital:** Moscow. **Cities** (urban agg.): Moscow 9,321,000; St. Petersburg 5,133,000; Nizhniy Novgorod 1,458,000; Novosibirsk 1,478,000.

Government: Type: Federal republic. **Head of state:** Vladimir Putin; b Oct. 7, 1952; in office: May 7, 2000. **Head of gov.:** Prime Min. Mikhail Kasyanov; b Dec. 8, 1957; in office: May 17, 2000. **Local divisions:** 21 autonomous republics, 68 autonomous territories and regions. **Defense budget:** $29 bil. **Active troops:** 1.004 mil.

Economy: Industries: Steel, machinery, machine tools, vehicles, chemicals, mining, footwear, textiles, appliances, paper. **Chief crops:** Grains, sugar beets, vegetables, sunflowers. **Minerals:** Manganese, mercury, potash, bauxite, cobalt, chromium, copper, coal, gold, lead, molybdenum, nickel, phosphates, silver, tin, tungsten, zinc, oil, gas, iron, potassium. **Crude oil reserves** (2000): 48.57 bil bbls. **Other resources:** Forests. **Arable land:** 8%. **Livestock** (2000): cattle: 27.5 mil; chickens: 340.0 mil; goats: 1.72 mil; pigs: 18.3 mil; sheep: 14.0 mil. **Fish catch** (2000): 4.72 mil metric tons. **Electricity prod.:** 798.065 bil kWh. **Labor force (1999 est.):** services 55%; industry 30%; agric.15%.

Finance: Monetary unit: Ruble (Sept. 2001: 29.44 = $1 U.S. NOTE: On Jan 1, 1998, Russia eliminated 3 digits from the ruble.) **GDP:** (1999 est.): $620.3 bil. **Per capita GDP:** (1999 est.)$4,200. **Imports** (1999 est.): $48.2 bil; partners: Germany, Belarus, Ukraine. **Exports** (1999): $75.4 bil; partners: Ukraine, Germany, U.S. **Tourism:** $7.77 bil. **Budget** (1998 est.): $63 bil. **Intl. reserves less gold** (June 2000): $7.68 bil. **Gold:** 11.04 mil oz t. **Consumer prices** (change in 1999): 85.7%.

Transport: Railroad: Length: 94,400 mi. **Motor vehicles:** 13.71 mil pass. cars, 9.86 mil comm. vehicles. **Civil aviation:** 30.6 bil pass.-mi; 75 airports. **Chief ports:** St. Petersburg, Murmansk, Arkhangelsk.

Communications: TV sets: 389 per 1,000 pop. **Radios:** 417 per 1,000 pop. **Telephones**: 32,070,000 main lines. **Daily newspaper circ.:** 105 per 1,000 pop.

Health: Life expectancy: 62.12 male; 72.83 female. **Births** (per 1,000 pop): 9.35. **Deaths** (per 1,000 pop.): 13.85. **Natural inc.:** -.45%. **Infant mortality** (per 1,000 live births): 20.05.

Education: Free, compulsory: ages 7-17. **Literacy**: 99%.

Major Intl. Organizations: UN (IBRD, ILO, IMF, IMO, WHO), APEC, CIS, OSCE.

Embassy: 2650 Wisconsin Ave. NW 20007; 298-5700.

Website: http://www.un.int/russia/

History. Slavic tribes began migrating into Russia from the W in the 5th century AD. The first Russian state, founded by Scandinavian chieftains, was established in the 9th century, centering in Novgorod and Kiev. In the 13th century the Mongols overran the country. It recovered under the grand dukes and princes of Muscovy, or Moscow, and by 1480 freed itself from the Mongols. Ivan the Terrible was the first to be formally proclaimed Tsar (1547). Peter the Great (1682-1725) extended the domain and, in 1721, founded the Russian Empire.

Western ideas and the beginnings of modernization spread through the huge Russian empire in the 19th and early 20th centuries. But political evolution failed to keep pace.

Military reverses in the 1905 war with Japan and in World War I led to the breakdown of the Tsarist regime. The 1917 Revolution began in March with a series of sporadic strikes for higher wages by factory workers. A provisional democratic government under Prince Georgi Lvov was established but was quickly followed in May by the second provisional government, led by Alexander Kerensky. The Kerensky government and the freely-elected Constituent Assembly were overthrown in a Communist coup led by Vladimir Ilyich Lenin Nov. 7.

Soviet Union

Lenin's death Jan. 21, 1924, resulted in an internal power struggle from which Joseph Stalin eventually emerged on top. Stalin secured his position at first by exiling opponents, but from the 1930s to 1953, he resorted to a series of "purge" trials, mass executions, and mass exiles to work camps. These measures resulted in millions of deaths, according to most estimates.

Germany and the Soviet Union signed a non-aggression pact Aug. 1939; Germany launched a massive invasion of the Soviet Union, June 1941. Notable heroic episode was the "900 days" siege of Leningrad (now St. Petersburg), lasting to Jan. 1944, and causing a million deaths; the city was never taken. Russian winter counterthrusts, 1941-42 and 1942-43, stopped the German advance. Turning point was the failure of German troops to take and hold Stalingrad (now Volgograd), Sept. 1942 to Feb. 1943. With British and U.S. Lend-Lease aid and sustaining great casualties, the Russians drove the German forces from eastern Europe and the Balkans in the next 2 years.

After Stalin died, Mar. 5, 1953, Nikita Khrushchev was elected first secretary of the Central Committee. In 1956 he condemned Stalin and "de-Stalinization" began.

Under Khrushchev the open antagonism of Poles and Hungarians toward domination by Moscow was brutally suppressed in 1956. He advocated peaceful co-existence with the capitalist countries, but continued arming the Soviet Union with nuclear weapons. He aided the Cuban revolution under Fidel Castro but withdrew Soviet missiles from Cuba during confrontation by U.S. Pres. Kennedy, Sept.-Oct. 1962. Khrushchev was suddenly deposed, Oct. 1964, and replaced by Leonid I. Brezhnev.

In Aug. 1968 Russian, Polish, East German, Hungarian, and Bulgarian military forces invaded Czechoslovakia to put a curb on liberalization policies of the Czech government.

Massive Soviet military aid to North Vietnam in the late 1960s and early 1970s helped assure Communist victories throughout Indo-China. Soviet arms aid and advisers were sent to several African countries in the 1970s.

In Dec. 1979, Soviet forces entered Afghanistan to support that government against rebels. In Apr. 1988, the Soviets agreed to withdraw their troops, ending a futile 8-year war.

Mikhail Gorbachev was chosen gen. secy. of the Communist Party, Mar. 1985. He held 4 summit meetings with U.S. Pres. Ronald Reagan. In 1987, in Washington, a treaty was signed eliminating intermediate-range nuclear missiles from Europe.

In 1987, Gorbachev initiated a program of reforms, including expanded freedoms and the democratization of the political process, through openness (*glasnost*) and restructuring (*perestroika*). The reforms were opposed by some Eastern bloc countries and many old-line Communists in the USSR. Gorbachev faced economic problems as well as ethnic and nationalist unrest in the republics.

When an apparent coup against Gorbachev became known on Aug. 19, 1991, the pres. of the Russian Republic, Boris Yeltsin, denounced it and called for a general strike. Some 50,000 demonstrated at the Russian Parliament in support of Yeltsin. By Aug. 21, the coup had failed and Gorbachev was restored as president. On Aug. 24, Gorbachev resigned as leader of the Communist Party. Several republics declared their independence, including Russia, Ukraine, and Kazakhstan. On Aug. 29, the Soviet Parliament voted to suspend all activities of the Communist Party.

The Soviet Union officially broke up Dec. 26, 1991, one day after Gorbachev resigned. The Soviet hammer and sickle flying over the Kremlin was lowered and replaced by the flag of Russia, ending the domination of the Communist Party over all areas of national life since 1917.

Russian Federation

In a first major step in radical economic reform, Russia eliminated state subsidies of most goods and services, Jan. 1992. The effect was to allow prices to soar far beyond the means of ordinary workers. In June, Pres. Yeltsin and U.S. Pres. George Bush agreed to massive arms reductions.

Russia launched a drive to privatize thousands of large and medium-sized state-owned enterprises in 1993. Yeltsin narrowly survived an impeachment vote by the Congress of People's Deputies, Mar. 28. He received strong support from voters in a referendum Apr. 25, but he continued to face a legislature dominated by conservatives and former Communists.

On Sept. 21, 1993, Yeltsin called for early elections and dissolved Parliament, which in turn declared him deposed. Anti-Yeltsin legislators then barricaded themselves in the Parliament building. On Oct. 3, anti-Yeltsin forces attacked some facilities in Moscow and broke into the Parliament building. Yeltsin ordered the army to attack and seize the building. About 140 people were killed in the fighting, according to medical authorities. More than 150 were arrested.

In elections Dec. 12, 1993, a Yeltsin-supported constitution was approved, but ultranationalists and Communist hard-liners made strong showings in legislative contests. In Dec. 1994 the Russian government sent troops into the breakaway republic of Chechnya. Grozny, the Chechen capital, fell in Feb. 1995 after heavy fighting, but Chechen rebels continued to resist.

Communists made further gains in parliamentary elections Dec. 17, 1995. Despite poor health, Yeltsin won a presidential runoff election over a Communist opponent, July 3, 1996. On Aug. 14, after rebels embarrassed the Russian military by retaking Grozny, Yeltsin gave his security chief, Alexander Lebed, broad powers to negotiate an end to the Chechnya war. Lebed and Chechen leaders signed a peace accord Aug. 31. On Oct. 17, Yeltsin dismissed Lebed for insubordination. Yeltsin survived quintuple-bypass heart surgery Nov. 5.

Russian troops remaining in Chechnya were pulled out Jan. 1997. A revitalized Yeltsin revamped his cabinet in Mar. to strengthen the hand of reformers. On May 27, he signed a "founding act" increasing cooperation with NATO and paving the way for NATO to admit Eastern European nations.

Russia's economic crisis deepened throughout 1998; in Aug. the ruble plummeted and the country defaulted on its debt. Yeltsin dismissed Prime Min. Viktor Chernomyrdin on Mar. 23 and Chernomyrdin's successor, Sergei Kiriyenko, on Aug. 23. Each move triggered a confrontation with parliament. Yevgeny Primakov became premier Sept. 11, 1998; in subsequent cabinet upheavals, Sergei Stepashin took over on May 19, 1999, followed by Vladimir Putin on Aug. 16. Disagreements over the war in Kosovo strained relations with the U.S. and NATO from Mar. to July. Russia moved forcibly in Aug. to suppress Islamic rebels in Dagestan; the conflict soon spread to neighboring Chechnya, where Russia launched a full-scale assault.

Yeltsin unexpectedly resigned Dec. 31, 1999, naming Putin as his interim successor. Russian troops took control of Grozny in early Feb. 2000. Putin defeated 10 opponents in a presidential election Mar. 26. The Russian parliament ratified 2 nuclear weapons treaties, the START II arms-reduction accord Apr. 14 and the Comprehensive Test Ban Treaty Apr. 21. A reorganization plan announced May 17 sought to reassert Moscow's control over Russia's regional governments. In a tragedy that raised both political and military issues, the Russian nuclear submarine *Kursk* sank in the Barents Sea Aug. 12, killing 118 sailors.

Russia and China signed a 20-year friendship and cooperation treaty July 16, 2001.

Rwanda

Republic of Rwanda

People: Population: 7,312,756. **Age distrib.** (%): <15: 42.4; 65+: 2.9. **Pop. density:** 760 per sq. mi. **Urban:** 6%. **Ethnic groups:** Hutu 80%, Tutsi 19%, Twa (Pygmoid) 1%. **Principal languages:** French, Kinyarwanda, English (all official). **Chief religions:** Roman Catholic 65%, indigenous beliefs 25%.

Geography: Area: 9,600 sq. mi. **Location:** In E central Africa. **Neighbors:** Uganda on N, Congo (formerly Zaire) on W, Burundi on S, Tanzania on E. **Topography:** Grassy uplands and hills cover most of the country, with a chain of volcanoes in the NW. The source of the Nile R. has been located in the headwaters of the Kagera (Akagera) R., SW of Kigali. **Capital:** Kigali (1999): 369,000.

Government: Type: Republic. **Head of state:** Pres. Paul Kagame; b Oct. 1957; in office: Apr. 22, 2000 (de facto from Mar. 24). **Head of gov.:** Prime Min. Bernard Makuza; b 1961; in office: Mar. 8, 2000. **Local divisions:** 12 prefectures subdivided into 155 communes. **Defense budget:** $125 mil. **Active troops:** 55-70,000.

Economy: Industries: Mining, cement. **Chief crops:** Coffee, tea, pyrethrum, bananas. **Minerals:** Tin, gold, wolframite. **Arable land:** 35%. **Livestock** (2000): cattle: 725,000; chickens: 1.4 mil; goats: 700,000; pigs: 160,000; sheep: 320,000. **Electricity prod.:** 132 mil kWh. **Labor force:** agric 90% .

Finance: Monetary unit: Franc (Sept. 2001: 438 = $1 U.S.). **GDP:** (1999 est.): $5.9 bil. **Per capita GDP:** $720. **Imports** (1999 est.): $242 mil; partners: Kenya, Tanzania, U.S. **Exports** (1999 est.): $70.8 mil; partners: Brazil, Germany, Belgium. **Tourism** (1998): $19 mil. **Budget** (1996 est.): $319 mil. **Intl. reserves less gold** (June 2000): $134.68 mil. **Consumer prices** (change in 1999): –2.4%.

Transport: Motor vehicles: 11,900 pass. cars, 15,900 comm. vehicles. **Civil aviation:** 1.2 mil pass.-mi; 2 airports. **Chief ports:** Gisenyi, Cyangugu.

Communications: Radios: 78.4 per 1,000 pop. **Telephones**: 17,600 main lines.

Health: Life expectancy: 38.35 male; 39.65 female. **Births** (per 1,000 pop): 33.97. **Deaths** (per 1,000 pop.): 21.13. **Natural inc.:** 1.28%. **Infant mortality** (per 1,000 live births): 118.92.

Education: Compulsory: ages 7-14. **Literacy**: 60%.

Major Intl. Organizations: UN (FAO, IBRD, ILO, IMF, WHO, WTrO), OAU.

Embassy: 1714 New Hampshire Ave. NW 20009; 232-2882.

For centuries, the Tutsi (an extremely tall people) dominated the Hutu (90% of the population). A civil war broke out in 1959 and Tutsi power was ended. Many Tutsi went into exile. A referendum in 1961 abolished the monarchic system. Rwanda, which had been part of the Belgian UN trusteeship of Rwanda-Urundi, became independent July 1, 1962.

In 1963 Tutsi exiles invaded in an unsuccessful coup; a large-scale massacre of Tutsi followed. Rivalries among Hutu led to a bloodless coup July 1973 in which Juvénal Habyarimana took power. After an invasion and coup attempt by Tutsi exiles in 1990, a multiparty democracy was established.

Renewed ethnic strife led to an Aug. 1993 peace accord between the government and rebels of the Tutsi-led Rwandan Patriotic Front (RPF). But after Habyarimana and the president of Burundi were killed Apr. 6, 1994, in a suspicious plane crash, massive violence broke out. At least 500,000 died in massacres, mainly of Tutsi by Hutu militias, and in civil warfare as the RPF sought power. About 2 million Tutsi and Hutu fled to camps in Zaire (now Congo) and other countries, where many died of cholera and other natural causes. French troops under a UN mandate moved into SW Rwanda June 23 to establish a so-called safe zone. The RPF claimed victory, installing a government in July led by a moderate Hutu president. French troops pulled out Aug. 22. A UN peacekeeping mission ended Mar. 8, 1996, but the Rwandan government and a UN-sponsored tribunal in Tanzania continued to gather evidence against those responsible for genocide. More than 1 million refugees (mostly Hutu) flooded back to Rwanda from Tanzania and Zaire in Nov. and Dec. 1996.

Firing squads in Rwanda on Apr. 24, 1998, executed 22 people convicted of genocide. Former Prime Min. Jean Kambanda pleaded guilty May 1 before the UN tribunal and received a life sentence Sept. 4, 1998. Maj. Gen. Paul Kagame, leader of the RPF, was sworn in as Rwanda's 1st Tutsi president Apr. 22, 2000. A Belgian court June 8, 2001, convicted 2 Roman Catholic nuns and 2 other Rwandans for their role in the 1994 genocide.

Saint Kitts and Nevis

Federation of Saint Kitts and Nevis

People: Population: 38,756. **Age distrib.** (%):<15: 29.8; 65+: 8.8. **Pop. density:** 373 per sq. mi. **Urban:** 34%. **Ethnic group:** Black. **Principal language:** English (official). **Chief religion:** Protestant.

Geography: Area: 104 sq. mi. **Location:** In the N part of the Leeward group of the Lesser Antilles in the E Caribbean Sea. **Neighbors:** Antigua and Barbuda to E. **Capital:** Basseterre (1994 est.): 12,600.

Government: Type: Constitutional monarchy. **Head of state:** Queen Elizabeth II, represented by Gov-Gen. Sir Cuthbert M. Sebastian; b Oct. 22, 1921; in office: Jan. 1, 1996. **Head of gov.:** Prime Min. Denzil Llewellyn Douglas; b Jan. 14, 1953; in office: July 7, 1995. **Local divisions:** 14 parishes.

Economy: Industries: Sugar (main industry), tourism. **Arable land:** 22%. **Livestock** (2000): cattle: 3,600; chickens: 60,000; goats: 14,500; pigs: 3,000; sheep: 7,400. **Electricity prod.:** 90 mil kWh. **Labor force:** N/A.

Finance: Monetary unit: East Caribbean Dollar (Oct. 2000: 2.70 = $1 U.S.). **GDP:** (1998 est.): $244 mil. **Per capita GDP:** $6,000. **Imports** (1998 est.): $160 mil; partners: U.S. 42.4%, Caricom nations 17.2%, UK 11.3%. **Exports** (1998 est.): $42 mil; partners: U.S. 68.5%, UK 22.3%. **Tourism:** $66 mil. **Budget:** (1997 est.) $73.3 mil. **Intl. reserves less gold** (Jan. 2000): $38.71 mil. **Consumer prices** (change in 1999): 3.9%.

Transport: Civil aviation: 2 airports. **Chief ports:** Basseterre, Charlestown.

Communications: TV sets: 241 per 1,000 pop. **Radios:** 659 per 1,000 pop. **Telephones** (1999): 20,100 main lines.

Health: Life expectancy: 68.22 male; 73.97 female. **Births** (per 1,000 pop): 18.78. **Deaths** (per 1,000 pop.): 9.21. **Natural inc.:** .96%. **Infant mortality** (per 1,000 live births): 16.28.

Education: Compulsory for 12 years between ages 5-17. **Literacy** (1992): 90%.

Major Intl. Organizations: UN (FAO, IBRD, ILO, IMF, WHO, WTrO), Caricom, the Commonwealth, OAS, OECS.

Embassy: 3216 New Mexico Ave., NW 20016; 686-2636.

St. Kitts (formerly St. Christopher; known by the natives as Liamuiga) and Nevis were reached (and named) by Columbus in 1493. They were settled by Britain in 1623, but ownership was disputed with France until 1713. They were part of the Leeward Islands Federation, 1871-1956, and the Federation of the West Indies, 1958-62. The colony achieved self-government as an Associated State of the UK in 1967, and became fully independent Sept. 19, 1983. A secession referendum on Nevis, Aug. 10, 1998, fell short of the two-thirds majority required.

Saint Lucia

People: Population: 158,178. **Age distrib.** (%): <15: 32.1; 65+: 5.3. **Urban:** 38%. **Pop. density:** 659 per sq. mi. **Ethnic groups:** Black 90%. **Principal languages:** English (official), French patois. **Chief religions:** Roman Catholic 90%, Protestant 7%.

Geography: Area: 240 sq. mi. **Location:** In E Caribbean, 2d largest of the Windward Isls. **Neighbors:** Martinique to N, St. Vincent to S. **Topography:** Mountainous, volcanic in origin; Soufriere, a volcanic crater, in the S. Wooded mountains run N-S to Mt. Gimie, 3,145 ft., with streams through fertile valleys. **Capital:** Castries (1999 est.): 57,000.

Government: Type: Parliamentary democracy. **Head of state:** Queen Elizabeth II, represented by Gov.-Gen. Calliopa Pearlette Louisy; b June 8, 1946; in office: Sept. 17, 1997. **Head of gov.:** Prime Min. Kenny Anthony; b Jan. 8, 1951; in office: May 24, 1997. **Local divisions:** 11 quarters.

Economy: Industries: Clothing, beverages, tourism. **Chief crops:** Bananas, coconuts, vegetables, root crops, cocoa, citrus. **Other resources:** Forests. **Arable land:** 8%. **Livestock** (2000): cattle: 12,400; chickens: 210,000; goats: 9,800; pigs: 14,750; sheep: 12,500. **Electricity prod.:** 110 mil kWh. **Labor force (1983 est.):** 43.4% agric.; 38.9% services;17.7% ind.

Finance: Monetary unit: East Caribbean Dollar (Sept. 2001: 2.67 = $1 U.S.). **GDP:** (1998 est.): $656 mil. **Per capita GDP:** $4,300. **Imports** (1998 est.): $290 mil; partners: U.S. 36%, Caricom countries 22%, UK 11%. **Exports** (1998 est.): $75 mil; partners: UK 50%, U.S. 24%, Caricom countries 16%. **Tourism** (1998): $291 mil. **Budget** (FY 1997-98 est.): $146.7 mil. **Intl. reserves less gold** (Jan. 2000): $78.00 mil. **Consumer prices** (change in 1999): 1.0%.

Transport: Motor vehicles: 10,000 pass. cars, 9,100 comm. vehicles. **Civil aviation:** 2 airports. **Chief ports:** Castries, Vieux Fort.

Communications: TV sets: 172 per 1,000 pop. **Radios:** 619 per 1,000 pop. **Telephones** (1999): 44,500 main lines.

Health: Life expectancy: 69. male; 76.39 female. **Births** (per 1,000 pop): 21.80. **Deaths** (per 1,000 pop.): 5.36. **Natural inc.:** 1.64%. **Infant mortality** (per 1,000 live births): 15.22.

Education: Compulsory: ages 5-15. **Literacy** (1993): 80%.

Major Intl. Organizations: UN (FAO, IBRD, ILO, IMF, IMO, WHO, WTrO), Caricom, the Commonwealth, OAS, OECS.

Embassy: 3216 New Mexico Ave. NW 20016; 364-6792.

St. Lucia was ceded to Britain by France at the Treaty of Paris, 1814. Self-government was granted with the West Indies Act, 1967. Independence was attained Feb. 22, 1979.

Saint Vincent and the Grenadines

People: Population: 115,942. **Age distrib.** (%):<15: 29.6; 65+: 6.3. **Pop. density:** 892 per sq. mi. **Urban:** 54%. **Ethnic groups:** Black 82%, mixed 14%. **Principal languages:** English (official), French patois. **Chief religions:** Anglican, Methodist, Roman Catholic.

Geography: Area: 130 sq. mi. **Location:** In the E Caribbean, St. Vincent (133 sq. mi.) and the northern islets of the Grenadines form a part of the Windward chain. **Neighbors:** St. Lucia to N, Barbados to E, Grenada to S. **Topography:** St. Vincent is volcanic, with a ridge of thickly wooded mountains running its length. **Capital:** Kingstown (1999): 28,000.

Government: Constitutional monarchy. **Head of State:** Queen Elizabeth II, represented by Gov.-Gen. Sir Charles James Antrobus; b May 14, 1933; in office: June 1, 1996. **Head of gov.:** Prime Min. Ralph Gonsalves; b Aug.. 8, 1946; in office: Mar. 29, 2001. **Local divisions:** 6 parishes.

Economy: Industries: Food processing, cement, furniture, clothing. **Chief crops:** Bananas, coconuts, sweet potatoes. **Arable land:** 10%. **Livestock** (2000): cattle: 6,200; chickens: 200,000; goats: 6,000; pigs: 9,500; sheep: 13,000. **Electricity prod.:** 82 mil kWh. **Labor force** (1980 est.): 57% services; 26% agric.; 17% ind.

Finance: Monetary unit: East Caribbean Dollar (Sept. 2001: 2.67 = $1 U.S.). **GDP:** (1999 est.): $309 mil. **Per capita GDP:** $2,600. **Imports** (1998 est.): $180 mil; partners: U.S. 36%, Caricom countries 28%, UK 13%. **Exports** (1998 est.): $47.8 mil; partners: Caricom countries 49%, UK 16%, U.S. 10%. **Tourism:** $77 mil. **Budget** (1997 est.): $98.6 mil. **Intl. reserves less gold** (Jan. 2000): $42.82 mil. **Consumer prices** (change in 1999): 1.0%.

Transport: Motor vehicles: 5,000 pass. cars, 3,200 comm. vehicles. **Civil aviation:** 5 airports. **Chief port:** Kingstown.

Communications: TV sets: 161 per 1,000 pop. **Radios:** 591 per 1,000 pop. **Telephones**: 24,900 main lines.

Health: Life expectancy: 70.83 male; 74.34 female. **Births** (per 1,000 pop): 17.91. **Deaths** (per 1,000 pop.): 6.16. **Natural inc.:** 1.18%. **Infant mortality** (per 1,000 live births): 16.61.

Education: Literacy (1994): 82%.

Major Intl. Organizations: UN (FAO, IBRD, ILO, IMF, IMO, WHO, WTrO), Caricom, the Commonwealth, OAS, OECS.

Embassy: 3216 New Mexico Ave. NW 20016; 364-6730.

Website: http://www.stvincentandgrenadines.com

Columbus landed on St. Vincent on Jan. 22, 1498 (St. Vincent's Day). Britain and France both laid claim to the island in the 17th and 18th centuries; the Treaty of Versailles, 1783, finally ceded it to Britain. Associated State status was granted 1969; independence was attained Oct. 27, 1979.

Samoa (*formerly* Western Samoa)

Independent State of Samoa

People: Population: 179,058. **Age distrib.** (%): <15: 31.9; 65+: 5.7. **Pop. density:** 163 per sq. mi. **Urban:** 21%. **Ethnic groups:** Samoan 92.6%, Euronesian (mixed) 7%, **Principal languages:** Samoan, English (both official). **Chief religion:** Christian 99.7%.

Geography: Area: 1,100 sq. mi. **Location:** In the S Pacific O. **Neighbors:** Nearest are Fiji to SW, Tonga to S. **Topography:** Main islands, Savaii (659 sq. mi.) and Upolu (432 sq. mi.), both ruggedly mountainous, and small islands Manono and Apolima. **Capital:** Apia (1999 est.): 38,000.

Government: Type: Constitutional monarchy. **Head of state:** Malietoa Tanumafili II; b Jan. 4, 1913; in office: Jan. 1, 1962. **Head of gov.:** Prime Min. Tuilaepa Sailele Malielegaoi; b Apr. 14, 1945; in office: Nov. 23, 1988. **Local divisions:** 11 districts.

Economy: Industries: Timber, tourism. **Chief crops:** Coconuts, taro, yams, bananas. **Other resources:** Hardwoods, fish. **Arable land:** 19%. **Livestock** (2000): cattle: 26,000; chickens: 350,000; pigs: 178,800. **Electricity prod.:** 100 mil kWh. **Labor force:** 65% agric.; 30% services; 5% industry.

Finance: Monetary unit: Tala (Sept. 2001: 3.43 = $1 U.S.). **GDP:** (1998 est.): $485 mil. **Per capita GDP:** $2,100. **Imports** (1998): $96.6 mil; partners: Australia, New Zealand, Japan. **Ex-**

ports (1998): $20.3 mil; partners: American Samoa, Australia, New Zealand. **Budget** (FY1996-97 est.): $99 mil. **Tourism:** $42 mil. **Intl. reserves less gold** (May 2000): $62.27 mil. **Consumer prices** (change in 1998): 2.2%.

Transport: Motor vehicles (1997): 1,200 pass. cars, 1,400 comm. vehicles. **Civil aviation:** 3 airports. **Chief ports:** Apia, Asau.

Communications: TV sets: 30 per 1,000 pop. **Radios:** 448 per 1,000 pop. **Telephones:** 8,600 main lines.

Health: Life expectancy: 66.77 male; 72.37 female. **Births** (per 1,000 pop): 15.59. **Deaths** (per 1,000 pop.): 6.29. **Natural inc.:** .93%. **Infant mortality** (per 1,000 live births): 31.75.

Education: Free, compulsory: ages 6-16. **Literacy** (1989): 100%.

Major Intl. Organizations: UN (FAO, IBRD, IMF, IMO, WHO), the Commonwealth.

Embassy: 820 2nd Ave., Suite 800D, New York, NY 10017; (212) 599-6196.

Samoa (formerly known as Western Samoa to distinguish it from American Samoa, a small U.S. territory) was a German colony, 1899 to 1914, when New Zealand landed troops and took over. It became a New Zealand mandate under the League of Nations and, in 1945, a New Zealand UN Trusteeship.

An elected local government took office in Oct. 1959, and the country became fully independent Jan. 1, 1962.

San Marino

Most Serene Republic of San Marino

People: Population: 27,336. **Age distrib.** (%): <15: 15.9; 65+: 16.2. **Pop. density:** 1,367 per sq. mi. **Urban:** 89%. **Ethnic groups:** Sammarinese, Italian. **Principal language:** Italian. **Chief religion:** Roman Catholic.

Geography: Area: 20 sq. mi. **Location:** In N central Italy near Adriatic coast. **Neighbors:** Completely surrounded by Italy. **Topography:** The country lies on the slopes of Mt. Titano. **Capital:** San Marino (1999 est.): 4,464.

Government: Type: Republic. **Heads of state and gov.:** Two co-regents appt. every 6 months. **Local divisions:** 9 castelli.

Economy: Industries: Tourism, textiles, electronics, wine, cement, ceramics. **Chief crops:** Wheat, grapes, maize. **Arable land:** 17%. **Labor force:** 60% services, 38% industry.

Finance: Monetary unit: Italian Lira (Sept. 2001: 2137.05 = $1 U.S.). **GDP:** (1997 est.): $500 mil. **Per capita GDP:** $20,000. **Budget** (1995 est.): $320 mil.

Transport: Motor vehicles (1997): 24,825 pass. cars, 4,149 comm. vehicles.

Communications: Radios: 514 per 1,000 pop. **Daily newspaper circ.:** 82 per 1,000 pop.

Health: Life expectancy: 77.68 male; 85.1 female. **Births** (per 1,000 pop): 10.76. **Deaths** (per 1,000 pop.): 7.68. **Natural inc.:** .31%. **Infant mortality** (per 1,000 live births): 6.21.

Education: Compulsory: ages 6-13. **Literacy** (1997): 99%.

Major Intl. Organizations: UN (ILO, IMF, WHO), OSCE.

San Marino claims to be the oldest state in Europe and to have been founded in the 4th century. A Communist-led coalition ruled 1947-57; a similar coalition ruled 1978-86. It has had a treaty of friendship with Italy since 1862.

São Tomé and Príncipe

Democratic Republic of São Tomé and Príncipe

People: Population: 165,034. **Age distrib.** (%): <15: 47.7; 65+: 4.0. **Pop. density:** 413 per sq. mi. **Urban:** 46%. **Ethnic groups:** Mestico (Portuguese-African), African minority (Angola, Mozambique immigrants). **Principal language:** Portuguese (official). **Chief religions:** Roman Catholic, Protestant.

Geography: Area: 400 sq. mi. **Location:** In the Gulf of Guinea about 125 miles off W central Africa. **Neighbors:** Gabon, Equatorial Guinea to E. **Topography:** São Tomé and Príncipe islands, part of an extinct volcano chain, are both covered by lush forests and croplands. **Capital:** São Tomé (1993 est.): 43,000.

Government: Type: Republic. **Head of state:** Pres. Fradique de Menezes; in office: Sept. 3, 2001. **Head of gov.:** Prime Min. Guilherme Posser da Costa; b 1945; in office: Jan. 5, 1999. **Local divisions:** 2 provinces.

Economy: Industries: light construction, textiles, soap, beer. **Chief crops:** Cocoa, coconuts. **Arable land:** 2%. **Livestock** (2000): cattle: 4,050; chickens: 300,000; goats: 4,800; pigs: 2,200; sheep: 2,600. **Electricity prod.:** 17 mil kWh.

Finance: Monetary unit: Dobra (Sept. 2001: 8203.50 = $1 U.S.). **GDP:** (1999 est.): $169 mil. **Per capita GDP:** $1,100. **Imports** (1999 est.): $19.5 mil; partners: Portugal 26%; France 18%. **Exports** (1999 est.): $4.9 mil; partners: Netherlands 51%; Germany 6%; Portugal 6%. **Tourism** (1998): $2 mil. **Intl. reserves less gold** (Dec. 1999): $10.88 mil.

Transport: Civil aviation: 5.8 mil pass.-mi; 2 airports. **Chief ports:** São Tomé, Santo Antonio.

Communications: TV sets: 154 per 1,000 pop. **Radios:** 232 per 1,000 pop. **Telephones:** 4,600 main lines.

Health: Life expectancy: 64.15 male; 67.07 female. **Births** (per 1,000 pop): 42.74. **Deaths** (per 1,000 pop.): 7.54. **Natural inc.:** 3.52%. **Infant mortality** (per 1,000 live births): 48.96.

Education: Compulsory for 4 years between ages 7-14. **Literacy** (1991): 73%.

Major Intl. Organizations: UN (FAO, IBRD, ILO, IMF, IMO, WHO), OAU.

The islands were discovered in 1471 by the Portuguese, who brought the first settlers—convicts and exiled Jews. Sugar planting was replaced by the slave trade as the chief economic activity until coffee and cocoa were introduced in the 19th century.

Portugal agreed, 1974, to turn the colony over to the Gabon-based Movement for the Liberation of São Tomé and Príncipe, which proclaimed as first president its East German-trained leader, Manuel Pinto da Costa. Independence came July 12, 1975. Democratic reforms were instituted in 1987. In 1991 Miguel Trovoada won the first free presidential election following da Costa's withdrawal. A military coup that ousted Trovoada Aug. 15, 1995, was reversed a week later after Angolan mediation. Trovoada defeated da Costa in a presidential runoff election, July 21, 1996.

Fradique de Menezes, a wealthy cocoa exporter, easily beat da Costa in the presidential election of July 29, 2001.

Saudi Arabia

Kingdom of Saudi Arabia

People: Population: 22,757,092. **Age distrib.** (%): <15: 42.5; 65+: 2.7. **Pop. density:** 27 per sq. mi. **Urban:** 85%. **Ethnic groups:** Arab 90%, Afro-Asian 10%. **Principal language:** Arabic (official). **Chief religion:** Muslim 100%.

Geography: Area: 829,000 sq. mi. **Location:** Occupies most of Arabian Peninsula in Mid-East. **Neighbors:** Kuwait, Iraq, Jordan on N; Yemen, Oman on S; United Arab Emirates, Qatar on E. **Topography:** Bordered by Red Sea on the W. The highlands on W, up to 9,000 ft., slope as an arid, barren desert to the Persian Gulf on the E. **Capital:** Riyadh. **Cities** (urban agg.): Riyadh 3,324,000; Jeddah 1,810,000; Mecca 919,000.

Government: Type: Monarchy with council of ministers. **Head of state and gov.:** King Fahd ibn Abdul Aziz; b 1923; in office: June 13, 1982 (prime min. since 1982). **Local divisions:** 13 provinces. **Defense budget:** $18.7 bil. **Active troops:** 126,500.

Economy: Industries: Oil, oil products. **Chief crops:** Dates, wheat, barley, tomatoes, melon, citrus. **Minerals:** Oil, gas, gold, copper, iron. **Crude oil reserves** (2000): 261 bil bbls. **Arable land:** 2%. **Livestock** (2000): cattle: 297,000; chickens: 130.0 mil; goats: 4.31 mil; sheep: 7.58 mil. **Fish catch:** (1999): 54,085 metric tons. **Electricity prod.:** 120 bil kWh. **Labor force:** 63% services; 25% industry & oil; 12% agric.

Finance: Monetary unit: Riyal (Sept. 2001: 3.75 = $1 U.S.). **GDP:** (1999 est.): $191 bil. **Per capita GDP:** $9,000. **Imports** (1999): $28 bil; partners: U.S. 21%, UK 9%, Japan 9%. **Exports** (1999): $48 bil; partners: Japan 17%, U.S. 15%, Korea 11%. **Tourism** (1998): $1.46 bil. **Budget** (1999 est.): $44 bil. **Intl. reserves less gold** (May 2000): $16.84 bil. **Gold:** 4.60 mil oz t. **Consumer prices** (change in 1999): –1.4%.

Transport: Railroad: Length: 864 km. **Motor vehicles:** 1.71 mil pass. cars, 1.17 mil comm. vehicles. **Civil aviation:** 11.8 bil pass.-mi; 25 airports. **Chief ports:** Jiddah, Ad Dammam.

Communications: TV sets: 252 per 1,000 pop. **Radios:** 309 per 1,000 pop. **Telephones:** 2,964,700 main lines. **Daily newspaper circ.:** 59 per 1,000 pop.

Health: Life expectancy: 66.4 male; 69.85 female. **Births** (per 1,000 pop): 37.34. **Deaths** (per 1,000 pop.): 5.94. **Natural inc.:** 3.14%. **Infant mortality** (per 1,000 live births): 51.25.

Education: Literacy: 63%.

Major Intl. Organizations: UN (FAO, IBRD, ILO, IMF, IMO, WHO), AL, OPEC.

Embassy: 601 New Hampshire Ave. NW 20037; 342-3800.

Before Muhammad, Arabia was divided among numerous warring tribes and small kingdoms and was at times dominated by larger Arabian and non-Arabian kingdoms. It was united for the first time by Muhammad, in the early 7th century AD. His successors conquered the entire Near East and North Africa, bringing Islam and the Arabic language. But Arabia itself soon returned to its former status.

Nejd, in central Arabia, long an independent state and center of the Wahhabi sect, fell under Turkish rule in the 18th century. In 1913 Ibn Saud, founder of the Saudi dynasty, overthrew the Turks and captured the Turkish province of Hasa in E Arabia; he took the Hejaz region in W Arabia in 1925 and most of Asir, in SW Arabia, by 1926. The discovery of oil in the 1930s transformed the new country.

Ibn Saud reigned until his death, Nov. 1953. Subsequent kings have been sons of Ibn Saud. The king exercises authority together with a Council of Ministers. The Islamic religious code is the law of the land. Alcohol and public entertainments are restricted, and women have an inferior legal status. There is no constitution and no parliament, although a Consultative Council was established by the king in 1993.

Saudi Arabia has often allied itself with the U.S. and other Western nations, and billions of dollars of advanced arms have been purchased from Britain, France, and the U.S.; however, Western support for Israel has often strained relations. Saudi units fought against Israel in the 1948 and 1973 Arab-Israeli wars. Beginning with the 1967 Arab-Israeli war, Saudi Arabia provided large annual financial gifts to Egypt; aid was later extended to Syria, Jordan, and Palestinian groups, as well as to other Islamic countries.

King Faisal played a leading role in the 1973-74 Arab oil embargo against the U.S. and other nations. Crown Prince Khalid was proclaimed king on Mar. 25, 1975, after the assassination of Faisal. Fahd became king on June 13, 1982, following Khalid's death.

The Hejaz contains the holy cities of Islam—Medina, where the Mosque of the Prophet enshrines the tomb of Muhammad, and Mecca, his birthplace. More than 2 million Muslims make pilgrimage to Mecca annually. In 1987, Iranians making a pilgrimage to Mecca clashed with anti-Iranian pilgrims and Saudi police; more than 400 were killed. Some 1,426 Muslim pilgrims died July 2, 1990, in a stampede in a pedestrian tunnel leading to Mecca. Nearly 300 pilgrims were killed in a stampede in Mecca, May 26, 1994. More than 340 pilgrims died in a tent fire near Mecca, Apr. 15, 1997.

Following Iraq's attack on Kuwait, Aug. 2, 1990, Saudi Arabia accepted the Kuwait royal family and more than 400,000 Kuwaiti refugees. King Fahd invited Western and Arab troops to deploy on its soil in support of Saudi defense forces. During the Persian Gulf War, 28 U.S. soldiers were killed when an Iraqi missile hit their barracks in Dhahran, Feb. 25, 1991. The nation's northern Gulf coastline suffered severe pollution as a result of Iraqi sabotage of Kuwaiti oil fields. Islamic extremists were blamed for truck bombs that killed 7 (5 from the U.S.) at a military training center in Riyadh, Nov. 13, 1995, and 19 Americans at a base in Dhahran, June 25, 1996. U.S. officials repeatedly chided the Saudi government for failing to cooperate fully in the investigation.

With King Fahd ailing, his half-brother, Crown Prince Abdullah, has taken a leading role in recent years.

For further details and developments see the front-of-the-book feature and the Chronology of the Year's Events.

Senegal

Republic of Senegal

People: Population: 10,284,929. **Age distrib.** (%): <15: 44.1; 65+: 3.1. **Pop. density:** 139 per sq. mi. **Urban:** 47%. **Ethnic groups:** Wolof 43.3%, Serer 14.7%, Diola 3.7%. **Principal languages:** French (official), Wolof, Pulaar, Diola, Mandingo. **Chief religions:** Muslim 92%, indigenous beliefs 6%, Christian 2%.

Geography: Area: 74,000 sq. mi. **Location:** At W extreme of Africa. **Neighbors:** Mauritania on N, Mali on E, Guinea and Guinea-Bissau on S; surrounds Gambia on three sides. **Topography:** Low rolling plains cover most of Senegal, rising somewhat in the SE. Swamp and jungles are in SW. **Capital:** Dakar (urban agg.): 2,079,000.

Government: Type: Republic. **Head of state:** Pres. Abdoulaye Wade; b May 29, 1926; in office: Apr. 1, 2000. **Head of gov.:** Prime Min. Mame Madior Boye; b 1940; in office: Mar. 3, 2001. **Local divisions:** 10 regions. **Defense:** 1.7% of GDP. **Active troops:** 11,000

Economy: Industries: Agricultural processing, fishing, phosphate mining. **Chief crops:** Peanuts, millet, corn, sorghum, rice. **Minerals:** Phosphates, iron. **Arable land:** 12%. **Livestock:** (2000): cattle: 2.96 mil; chickens: 45.0 mil; goats: 3.6 mil; pigs: 330,000; sheep: 4.3 mil. **Fish catch** (1999): 507,040 metric tons. **Electricity prod.:** 1.270 bil kWh. **Labor force:** 60% agric.

Finance: Monetary unit: CFA Franc (Sept. 2001: 724.11 = $1 U.S.). **GDP:** (1999 est.): $16.6 bil. **Per capita GDP:** $1,650. **Imports** (1998): $1.2 bil; partners: France 36%. **Exports** (1998): $925 mil; partners: France 22%. **Tourism:** $166 mil. **Budget** (1996 est.): $885 mil. **Intl. reserves less gold** (Apr. 2000): 421.0 mil. **Gold:** 29,000 oz t. **Consumer prices** (change in 1999): 0.8%.

Transport: Railroad: Length: 562 mi. **Motor vehicles:** 110,000 pass. cars, 50,000 comm. vehicles. **Civil aviation:** 164.6 mil pass.-mi; 7 airports. **Chief ports:** Dakar, Saint-Louis.

Communications: TV sets: 6.9 per 1,000 pop. **Radios:** 93 per 1,000 pop. **Telephones:** 205,900 main lines.

Health: Life expectancy: 60.94 male; 64.22 female. **Births** (per 1,000 pop): 37.46. **Deaths** (per 1,000 pop.): 8.35. **Natural inc.:** 2.91%. **Infant mortality** (per 1,000 live births): 56.75.

Education: Compulsory: ages 7-13. **Literacy:** 33%.

Major Intl. Organizations: UN and all of its specialized agencies, OAU.

Embassy: 2112 Wyoming Ave. NW 20008; 234-0540.

Portuguese settlers arrived in the 15th century, but French control grew from the 17th century. The last independent Muslim state was subdued in 1893. Dakar became the capital of French West Africa.

Independence as part, along with the Sudanese Rep., of the Mali Federation, came June 20, 1960. Senegal withdrew Aug. 20. French political and economic influence remained strong.

Senegal, Dec. 17, 1981, signed an agreement with The Gambia for confederation of the 2 countries, without loss of individual sovereignty, under the name of Senegambia. The confederation collapsed in 1989, although in 1991 the 2 nations signed a friendship and cooperation treaty.

Separatists in Casamance Province of S Senegal have clashed with government forces since 1982. Senegal sent troops in June 1998 to help the Guinea-Bissau government suppress an army uprising. Forty years of Socialist Party rule ended when Abdoulaye Wade, leader of the Senegalese Democratic Party, won a presidential runoff election Mar. 19, 2000.

Seychelles

Republic of Seychelles

People: Population: 79,715. **Age distrib.** (%): <15: 28.3; 65+: 6.3. **Pop. density:** 453 per sq. mi. **Urban:** 63%. **Ethnic groups:** Seychellois (mixture of Asians, Africans, Europeans). **Principal languages:** English, French (both official), Creole. **Chief religions:** Roman Catholic 90%, Anglican 8%.

Geography: Area: 176 sq. mi. **Location:** In the Indian O. 700 miles NE of Madagascar. **Neighbors:** Nearest are Madagascar on SW, Somalia on NW. **Topography:** A group of 86 islands, about half of them composed of coral, the other half granite, the latter predominantly mountainous. **Capital:** Victoria (1999 est.): 28,000.

Government: Type: Republic. **Head of state and gov.:** Pres. France-Albert René, b. Nov. 16, 1935; in office: June 5, 1977. **Local divisions:** 23 districts. **Defense budget:** $11 mil. **Active troops:** 450.

Economy: Industries: Tourism, food processing, fishing. **Chief crops:** Coconuts, cinnamon, vanilla. **Arable land:** 2%. **Livestock** (2000): cattle: 1,400; chickens: 550,000; goats: 5,200; pigs: 18,300. **Electricity prod.:** 160 mil kWh. **Labor force:** 57% services; 19% industry; 14% government; 10% agric.

Finance: Monetary unit: Rupee (Sept. 2001: 5.62 = $1 U.S.). **GDP:** (1999 est.): $590 mil. **Per capita GDP:** $7,500. **Imports** (1998): $403 mil; partners: South Africa, UK, China. **Exports** (1998): $91 mil; partners: France, UK, Netherlands. **Tourism** (1998): $111 mil. **Budget** (1994 est.): $241 mil. **Intl. reserves less gold** (May 2000): $27.21 mil. **Consumer prices** (change in 1999): 6.3%.

Transport: Motor vehicles: 6,620 pass. cars, 1,880 comm. vehicles. **Civil aviation:** 526.5 mil pass.-mi; 2 airports. **Chief port:** Victoria.

Communications: TV sets: 173.4 per 1,000 pop. **Radios:** 667 per 1,000 pop. **Telephones**: 20,700 main lines. **Daily newspaper circ.:** 41 per 1,000 pop.

Health: Life expectancy: 65.17 male; 76.37 female. **Births** (per 1,000 pop): 17.66. **Deaths** (per 1,000 pop.): 6.65. **Natural inc.:** 1.1%. **Infant mortality** (per 1,000 live births): 17.3.

Education: Free, compulsory: ages 6-15. **Literacy:** 84%.

Major Intl. Organizations: UN (FAO, IBRD, ILO, IMF, IMO, WHO), the Commonwealth, OAU.

Embassy: 800 2d Ave., Suite 400C, New York, NY 10017; 212-972-1785.

The islands were occupied by France in 1768, and seized by Britain in 1794. Ruled as part of Mauritius from 1814, the Seychelles became a separate colony in 1903. The ruling party had opposed independence as impractical, but pressure from the OAU and the UN became irresistible, and independence was declared June 29, 1976. The first president was ousted in a coup a year later by a socialist leader. A new constitution, approved June 1993, provided for a multiparty state.

Sierra Leone

Republic of Sierra Leone

People: Population: 5,426,618. **Age distrib.** (%): <15: 44.7; 65+: 3.1. **Pop. density:** 196 per sq. mi. **Urban:** 36%. **Ethnic groups:** Temne 30%, Mende 30%, other tribes 30%. **Principal**

languages: English (official), Mende, Temne, Krio. **Chief religions:** Muslim 60%, indigenous beliefs 30%, Christian 10%.

Geography: Area: 27,600 sq. mi. **Location:** On W coast of W Africa. **Neighbors:** Guinea on N and E, Liberia on S. **Topography:** The heavily-indented, 210-mi. coastline has mangrove swamps. Behind are wooded hills, rising to a plateau and mountains in the E. **Capital:** Freetown (1999 est.): 822,000.

Government: Type: Republic. **Head of state and gov.:** Ahmad Tejan Kabbah; b Feb. 16, 1932; in office: Mar. 10, 1998. **Local divisions:** 3 provinces, 1 area. **Defense budget:** $9 mil. **Active troops:** 3,000.

Economy: Industries: Mining, light manufacturing. **Chief crops:** Cocoa, coffee, palm kernels, rice. **Minerals:** Diamonds, titanium, bauxite. **Arable land:** 7%. **Livestock** (2000): cattle: 420,000; chickens: 6.0 mil; goats: 200,000; pigs: 52,000; sheep: 365,000. **Fish catch** (1999): 68,739 metric tons. **Electricity prod.:** 240 mil kWh.

Finance: Monetary unit: Leone (Sept. 2001: 1,900 = $1 U.S.). **GDP:** (1999 est.): $2.5 bil. **Per capita GDP:** $500. **Imports** (1998): $166 mil; partners: U.K. 24%, Coite d'Ivoire 14%, Benelux 10%. **Exports** (1998): $41 mil; partners: Benelux 49%, Spain 10%, U.S. 8%. **Budget** (1996 est.): $150 mil. **Intl. reserves less gold** (June 2000): $38.8 mil. **Consumer prices** (change in 1999): 34.1%.

Transport: Railroad: Length: 52 mi. **Motor vehicles:** 20,860 pass. cars, 21,074 comm. vehicles. **Civil aviation:** 14.9 mil pass.-mi; 1 airport. **Chief ports:** Freetown, Bonthe.

Communications: Radios: 72 per 1,000 pop. **Telephones** (1998): 19,000 main lines.

Health: Life expectancy: 42.69 male; 48.61 female. **Births** (per 1,000 pop): 45.11. **Deaths** (per 1,000 pop.): 19.19. **Natural inc.:** 2.59%. **Infant mortality** (per 1,000 live births): 146.52.

Education: Literacy: 31%.

Major Intl. Organizations: UN (FAO, IBRD, ILO, IMF, IMO, WHO, WTrO), the Commonwealth, OAU.

Embassy: 1701 19th St. NW 20009; 939-9261.

Website: http://www.Sierra-Leone.org

Freetown was founded in 1787 by the British government as a haven for freed slaves. Their descendants, known as Creoles, number more than 60,000.

Successive steps toward independence followed the 1951 constitution. Ten years later, full independence arrived Apr. 27, 1961. Sierra Leone declared itself a republic Apr. 19, 1971. A one-party state approved by referendum in 1978 brought political stability, but mismanagement and corruption plagued the economy.

Mutinous soldiers ousted Pres. Joseph Momoh Apr. 30, 1992. Another coup, Jan. 16, 1996, paved the way for multiparty elections and a return to civilian rule. A peace accord, signed Nov. 30 with the Revolutionary United Front (RUF), brought a temporary halt to a civil war that had claimed over 10,000 lives in 5 years.

A coup on May 25, 1997, was met with widespread international opposition. Armed intervention by Nigeria restored Pres. Ahmad Tejan Kabbah to power on Mar. 10, 1998, but RUF rebels mounted a guerrilla counteroffensive, reportedly killing thousands of civilians and mutilating thousands more. The Kabbah government signed a power-sharing agreement with the RUF on July 7, 1999. The accord collapsed in early May 2000, as RUF guerrillas took more than 500 UN peacekeepers hostage. Rebel leader Foday Sankoh was captured in Freetown May 17. The hostages were freed by the end of May, and 233 more UN personnel behind rebel lines were rescued July 15. The RUF rebellion continued in 2001, fueled by funds from diamond smuggling.

Singapore

Republic of Singapore

People: Population: 4,300,419. **Age distrib.** (%): <15: 17.9; 65+: 7.0. **Pop. density:** 17,202 per sq. mi. **Urban:** 100%. **Ethnic groups:** Chinese 76.4%, Malay 14.9%, Indian 6.4%. **Principal languages:** Chinese, Malay, Tamil, English (all official). **Chief religions:** Buddhist, Taoist, Muslim, Christian, Hindu.

Geography: Area: 250 sq. mi. **Location:** Off tip of Malayan Peninsula in SE Asia. **Neighbors:** Nearest are Malaysia on N, Indonesia on S. **Topography:** Singapore is a flat, formerly swampy island. The nation includes 40 nearby islets. **Capital:** Singapore (urban agg.): 3,567,000.

Government: Type: Republic. **Head of state:** Pres. S. R. Nathan; b July 3, 1924; in office: Sept. 1, 1999. **Head of gov.:** Prime Min. Goh Chok Tong; b May 20, 1941; in office: Nov. 28, 1990. **Defense budget:** $4.4 bil. **Active troops:** 60,500.

Economy: Industries: Oil refining, electronics, banking, food and rubber processing, biotechnology. **Chief crops:** Copra, rubber, fruit, vegetables. **Arable land:** 2%. **Livestock** (2000): cattle: 200; chickens: 2.0 mil; goats: 300; pigs: 190,000. **Fish catch** (1999): 13,338 metric tons. **Electricity prod.:** 27.381 bil kWh. **Labor force:** 38% finance, business, other serv.; 21.6% manuf.; 21.4% commerce.

Finance: Monetary unit: Dollar (Sept. 2001: 1.75 = $1 U.S.). **GDP:** (1999 est.): $98 bil. **Per capita GDP:** $27,800. **Imports** (1999 est.): $111 bil; partners: U.S. 17%, Japan 17%, Malaysia 16%, . **Exports** (1999 est.): $114 bil; partners: U.S. 19%, Malaysia 17%, Hong Kong 8%. **Tourism:** $5.79 bil. **Budget** (FY 1997-98 est.): $13.6 bil. **International reserves** (May 2000): $76.07 bil. **Consumer prices** (change in 1998): –0.3%.

Transport: Railroad: Length: 52 mi. **Motor vehicles** (1997): 379,497 pass. cars, 140,827 comm. vehicles. **Civil aviation:** 34.5 bil pass.-mi; 1 airport. **Chief port:** Singapore.

Communications: TV sets: 223 per 1,000 pop. **Radios:** 260 per 1,000 pop. **Telephones**: 1,946,500 main lines. **Daily newspaper circ.:** 360 per 1,000 pop.

Health: Life expectancy: 77.22 male; 83.35 female. **Births** (per 1,000 pop): 12.80. **Deaths** (per 1,000 pop.): 4.24. **Natural inc.:** .86%. **Infant mortality** (per 1,000 live births): 3.62.

Education: Literacy: 91%.

Major Intl. Organizations: UN (IBRD, ILO, IMF, IMO, WHO, WTrO), the Commonwealth, APEC, ASEAN.

Embassy: 3501 International Pl. NW 20008; 537-3100.

Website: http://www.singstat.gov.sg

Founded in 1819 by Sir Thomas Stamford Raffles, Singapore was a British colony until 1959, when it became autonomous within the Commonwealth. On Sept. 16, 1963, it joined with Malaya, Sarawak, and Sabah to form the Federation of Malaysia. Tensions between Malayans, dominant in the federation, and ethnic Chinese, dominant in Singapore, led to an accord under which Singapore became a separate nation, Aug. 9, 1965.

Singapore is one of the world's largest ports. Standards in health, education, and housing are generally high. International banking has grown rapidly in recent years. The government, dominated by a single party, has taken strong actions to suppress dissent.

Slovakia

Slovak Republic

People: Population: 5,414,937. **Age distrib.** (%): <15: 18.9; 65+: 11.5. **Pop. density:** 288 per sq. mi. **Urban:** 57%. **Ethnic groups:** Slovak 85.7%, Hungarian 10.7%. **Principal languages:** Slovak (official), Hungarian. **Chief religions:** Roman Catholic 60%, Protestant 8%.

Geography: Area: 18,800 sq. mi. **Location:** In E central Europe. **Neighbors:** Poland on N, Hungary on S, Austria and Czech Rep. on W, Ukraine on E. **Topography:** Mountains (Carpathians) in N, fertile Danube plane in S. **Capital:** Bratislava. **Cities** (1999 est.): Bratislava 460,000; (1997 est.): Kosice 242,000.

Government: Type: Republic. **Head of state:** Rudolf Schuster; b Jan. 4, 1934; in office: June 15, 1999. **Head of gov.:** Prime Min. Mikulás Dzurinda; b Feb. 4, 1955; in office: Oct. 30, 1998. **Local divisions:** 8 departments. **Defense budget:** $348 mil. **Active troops:** 38,600.

Economy: Industries: Metal products, food and beverages, oil, chemicals. **Chief crops:** Grains, potatoes, sugar beets, hops, fruit. **Minerals:** Coal, lignite, iron, copper. **Crude oil reserves** (2000): 9.0 mil bbls. **Arable land:** 31%. **Livestock** (2000): cattle: 665,055; chickens: 12.25 mil; goats: 51,075; pigs: 1.56 mil; sheep: 340,346. **Electricity prod.:** 22.582 bil kWh. **Labor force:** 29.3% ind.; 8.9% agric.; 8% constr.

Finance: Monetary unit: Koruna (Sept. 2001: 47.69 = $1 U.S.). **GDP:** (1999 est.): $45.9 bil. **Per capita GDP:** $8,500. **Imports** (1999): $11.2 bil; partners: EU 50%, Czech Rep. 18%. **Exports** (1999): $10.1 bil; partners: EU 56%, Czech Rep. 20%. **Tourism:** $461 mil. **Budget** (1997): $6.5 bil. **Intl. reserves less gold** (May 2000): $4.03 bil. **Gold:** 1.29 mil oz t. **Consumer prices** (change in 1999): 10.6%.

Transport: Railroad: Length: 2,277 mi. **Motor vehicles:** 994,000 pass. cars, 94,000 comm. vehicles. **Civil aviation:** 63.8 mil pass.-mi; 2 airports. **Chief ports:** Bratislava, Komarno.

Communications: TV sets: 216 per 1,000 pop. **Telephones:** 1,698,000 main lines. **Daily newspaper circ.:** 256 per 1,000 pop.

Health: Life expectancy: 69.95 male; 78.2 female. **Births** (per 1,000 pop): 10.05. **Deaths** (per 1,000 pop.): 9.25. **Natural inc.:** .08%. **Infant mortality** (per 1,000 live births): 8.97.

Education: Compulsory: ages 6-14. **Literacy** (1994): 100%.

Major Intl. Organizations: UN (FAO, IBRD, ILO, IMF, IMO, WHO, WTrO), OSCE.

Embassy: 2201 Wisconsin Ave. NW 20007; 965-5161.

Website: http://www.slovakemb.com/index.html

Slovakia was originally settled by Illyrian, Celtic, and Germanic tribes and was incorporated into Great Moravia in the 9th century. It became part of Hungary in the 11th century.

Overrun by Czech Hussites in the 15th century, it was restored to Hungarian rule in 1526. The Slovaks disassociated themselves from Hungary after World War I and joined the Czechs of Bohemia to form the Republic of Czechoslovakia, Oct. 28, 1918.

Germany invaded Czechoslovakia, 1939, and declared Slovakia independent. Slovakia rejoined Czechoslovakia in 1945.

Czechoslovakia split into 2 separate states—the Czech Republic and Slovakia—on Jan. 1, 1993. Slovakia, with its less developed economy, applied to join the European Union in 1995. A prolonged parliamentary standoff left the country without a president for much of 1998.

Prime Min. Vladimir Meciar, a nationalist, suffered a setback in legislative elections Sept. 25-26, 1998, and was defeated in a presidential runoff vote by Rudolf Schuster, May 29, 1999.

Slovenia

Republic of Slovenia

People: Population: 1,930,132. **Age distrib.** (%): <15: 16.1; 65+: 14.3. **Pop. density:** 247 per sq. mi. **Urban:** 50%. **Ethnic groups:** Slovene 91%, Croat 3%. **Principal languages:** Slovenian (official), Serbo-Croatian. **Chief religion:** Roman Catholic 70.8%.

Geography: Area: 7,800 sq. mi. **Location:** In SE Europe. **Neighbors:** Italy on W, Austria on N, Hungary on NE, Croatia on SE, S. **Topography:** Mostly hilly; 42% of the land is forested. **Capital:** Ljubljana (1997 est.): 330,000.

Government: Type: Republic. **Head of state:** Pres. Milan Kucan; b Jan. 14, 1941; in office: Apr. 1990. **Head of gov.:** Prime Min. Janez Drnovsek; b May 17, 1950; in office: Nov. 17, 2000. **Local divisions:** 136 municipalities, 11 urban municipalities. **Defense budget** (2001): $373 mil. **Active troops:** 9,000.

Economy: Industries: Metallurgy, electronics, trucks. **Minerals:** Coal, lead, zinc, mercury. **Chief crops:** Potatoes, hops, wheat. **Arable land:** 12%. **Livestock** (2000): cattle: 471,425; chickens: 7.15 mil; goats: 14,643; pigs: 558,459; sheep: 72,500. **Electricity prod.:** 12.451 bil kWh.

Finance: Monetary unit: Tolar (Sept. 2001: 242.52 = $1 U.S.). **GDP:** (1999 est.): $21.4 bil. **Per capita GDP:** $10,900. **Imports** (1999): $9.7 bil; partners: Germany 21%, Italy 17%. **Exports** (1998): $8.4 bil; partners: Germany 28%, Italy 14%. **Tourism:** $1.01 bil. **Budget** (1996 est.): $8.53 bil. **Intl. reserves less gold** (May 2000): $3.11 bil. **Gold:** 3,000 oz t. **Consumer prices** (change in 1999): 6.6%.

Transport: Railroad: Length: 746 mi. **Motor vehicles:** 657,000 pass. cars, 37,000 comm. vehicles. **Civil aviation:** 233.0 mil pass.-mi; 1 airport. **Chief ports:** Izola, Koper, Piran.

Communications: TV sets: 352 per 1,000 pop. **Radios:** 317 per 1,000 pop. **Telephones** (1999): 751,800 main lines. **Daily newspaper circ.:** 199 per 1,000 pop.

Health: Life expectancy: 71.2 male; 79.17 female. **Births** (per 1,000 pop): 9.32. **Deaths** (per 1,000 pop.): 9.98. **Natural inc.:** -.07%. **Infant mortality** (per 1,000 live births): 4.51.

Education: Free, compulsory: ages 6-15. **Literacy** (1993): 99%.

Major Intl. Organizations: UN (FAO, IBRD, ILO, IMF, IMO, WHO, WTrO), OSCE.

Embassy: 1525 New Hampshire Ave. NW 20036; 667-5363.

The Slovenes settled in their current territory during the period from the 6th to the 8th century. They fell under German domination as early as the 9th century. Modern Slovenian political history began after 1848 when the Slovenes, who were divided among several Austrian provinces, began their struggle for political and national unification. In 1918 a majority of Slovenes became part of the Kingdom of Serbs, Croats, and Slovenes, later renamed Yugoslavia.

Slovenia declared independence June 25, 1991, and joined the UN May 22, 1992. Linked by trade with the European Union, Slovenia applied for full membership June 10, 1996.

Solomon Islands

People: Population: 480,442. **Age distrib.** (%): <15: 43.8; 65+: 3.1. **Pop. density:** 45 per sq. mi. **Urban:** 19%. **Ethnic groups:** Melanesian 93%, Polynesian 4%. **Principal languages:** English (official); Melanesian, Polynesian languages. **Chief religions:** Anglican 34%, Roman Catholic 19%, Baptist 17%, other Christian 26%.

Geography: Area: 10,600 sq. mi. **Location:** Melanesian Archipelago in the W Pacific O. **Neighbors:** Nearest is Papua New Guinea to W. **Topography:** 10 large volcanic and rugged islands and 4 groups of smaller ones. **Capital:** Honiara (1999 est.): 68,000.

Government: Type: Parliamentary democracy within the Commonwealth of Nations. **Head of state:** Queen Elizabeth II, represented by Gov.-Gen. John Lapli; in office: July 7, 1999. **Head of gov.:** Prime Min. Manasseh Sogavare; b 1954; in office: June 30, 2000. **Local divisions:** 9 provinces and Honiara.

Economy: Industries: Copra, tuna. **Chief crops:** Coconuts, rice, cocoa, beans. **Minerals:** Gold, bauxite. **Other resources:** Forests. **Arable land:** 1%. **Livestock** (2000): cattle: 12,000; chickens: 190,000; pigs: 59,000. **Fish catch** (1999): 53,442 metric tons. **Electricity prod.:** 30 mil kWh.

Finance: Monetary unit: Dollar (Sept. 2001: 5.30 = $1 U.S.). **GDP:** (1999 est.): $1.21 bil. **Per capita GDP:** $2,650. **Imports** (1998 est.): $160 mil; partners: Australia 42%, Japan 10%. **Exports** (1998): $142 mil; partners: Japan 50%,Spain 16%. **Tourism** (1998): $13 mil. **Budget** (1997 est.): $168 mil. **Intl. reserves less gold** (Apr. 2000): $47.08 mil. **Consumer prices:** (change in 1999): 8.3%.

Transport: Civil aviation: 46.0 mil pass.-mi; 21 airports. **Chief port:** Honiara.

Communications: TV sets: 16 per 1,000 pop. **Radios:** 96 per 1,000 pop. **Telephones** (1998): 7,900 main lines.

Health: Life expectancy: 69.12 male; 74.1 female. **Births** (per 1,000 pop): 34.05. **Deaths** (per 1,000 pop.): 4.27. **Natural inc.:** 2.98%. **Infant mortality** (per 1,000 live births): 24.47.

Education: Literacy (1994): 54%.

Major Intl. Organizations: UN (FAO, IBRD, ILO, IMF, IMO, WHO, WTrO), the Commonwealth.

Embassy: 800 Second Ave., Suite 400L, New York, NY 10017; (212) 599-6193.

The Solomon Islands were sighted in 1568 by an expedition from Peru. Britain established a protectorate in the 1890s over most of the group, inhabited by Melanesians. The islands saw major World War II battles. Self-government came Jan. 2, 1976, and independence was formally attained July 7, 1978.

A coup attempt June 5, 2000, sparked factional fighting in Honiara.

Somalia

People: Population: 7,488,773. **Age distrib.** (%): <15: 44.5; 65+: 2.8. **Pop. density:** 31 per sq. mi. **Urban:** 27%. **Ethnic groups:** Somali 85%, Bantu, Arab. **Principal languages:** Somali (official), Arabic, Italian, English. **Chief religion:** Sunni Muslim.

Geography: Area: 241,900 sq. mi. **Location:** Occupies the eastern horn of Africa. **Neighbors:** Djibouti, Ethiopia, Kenya on W. **Topography:** The coastline extends for 1,700 mi. Hills cover the N; the center and S are flat. **Capital:** Mogadishu (urban agg.): 1,219,000.

Government: Type: In transition. **Head of state**: Abdiqassim Salad Hassan; in office: Aug. 27, 2000. **Head of gov.**: Prime Min. Ali Khalif Galaid; b Oct. 15, 1941; in office: Oct. 8, 2000. Local divisions: 18 regions. **Defense budget:** $15 mil. **Active troops:** Nil.

Economy: Chief crops: Sugar, bananas, sorghum, corn, mangoes. **Minerals:** Uranium, iron, tin, gypsum, bauxite. **Arable land:** 2%. **Livestock** (2000): cattle: 5.1 mil; chickens: 3.2 mil; goats: 12.3 mil; pigs: 3,900; sheep: 13.1 mil **Fish catch** (1999): 15,700 metric tons. **Electricity prod.:** 260 mil kWh. **Labor force:** 71% nomadic agric.; 29% industry & services.

Finance: Monetary unit: Shilling (Sept. 2001: 2,606.90 = $1 U.S.). **GDP:** (1999 est.): $4.3 bil. **Per capita GDP**: $600. **Imports** (1998 est.): $327 mil; partners: Djibouti 20%; Kenya 11%, Belarus 11%. **Exports** (1998 est.): $187 mil; partners: Saudi Arabia 57%, UAE 15%, Italy 12%.

Transport: Motor vehicles: 10,000 pass. cars, 10,000 comm. vehicles. **Civil aviation:** 86.9 mil pass.-mi; 1 airport. **Chief ports:** Mogadishu, Berbera.

Communications: TV sets: 18 per 1,000 pop. **Radios:** 45 per 1,000 pop. **Telephones:** 15,000 main lines.

Health: Life expectancy: 44.99 male; 48.25 female. **Births** (per 1,000 pop): 47.23. **Deaths** (per 1,000 pop.): 18.35. **Natural inc.:** 2.89%. **Infant mortality** (per 1,000 live births): 123.97.

Education: Free, compulsory: ages 6-14. **Literacy** (1990): 24%.

Major Intl. Organizations: UN (FAO, IBRD, ILO, IMF, IMO, WHO), AL, OAU.

Website: http://gaia.info.usaid.gov/horn/somalia/somalia.html

British Somaliland (present-day N Somalia) was formed in the 19th century, as was Italian Somaliland (now central and S Somalia). Italy lost its African colonies in World War II. In 1949, the UN approved eventual independence for the former Italian colony (designated the UN Trust Territory of Somalia) after a 10-year period under Italian administration.

British Somaliland gained independence, June 26, 1960, and by prearrangement, merged July 1 with the trust territory of Somalia to create the independent Somali Republic (Somalia).

On Oct. 16, 1969, Pres. Abdi Rashid Ali Shirmarke was assassinated. On Oct. 21, a military group led by Maj. Gen. Muhammad Siad Barre seized power. In 1970, Barre declared the country a socialist state—the Somali Democratic Republic.

Somalia has laid claim to Ogaden, the huge eastern region of Ethiopia, peopled mostly by Somalis. Ethiopia battled Somali rebels in 1977. Some 11,000 Cuban troops with Soviet arms defeated Somali army troops and ethnic Somali rebels in Ethiopia, 1978. As many as 1.5 million refugees entered Somalia. Guerrilla fighting in Ogaden continued until 1988, when a peace agreement was reached with Ethiopia.

The civil war intensified again and Barre was forced to flee the capital, Jan. 1991. Fighting between rival factions caused 40,000 casualties in 1991 and 1992, and by mid-1992 the civil war, drought, and banditry combined to produce a famine that threatened some 1.5 million people with starvation.

In Dec. 1992 the UN accepted a U.S. offer of troops to safeguard food delivery to the starving. The UN took control of the multinational relief effort from the U.S. May 4, 1993. While the operation helped alleviate the famine, efforts to reestablish order foundered, and there were significant U.S. and other casualties. The U.S. withdrew its peacekeeping forces Mar. 25, 1994.

When the last UN troops pulled out Mar. 3, 1995, Mogadishu had no functioning central government, and armed factions controlled different regions. By 1999 a joint police force was operating in the capital, but much of the country, especially in S Somalia, faced continued violence and food shortages.

South Africa

Republic of South Africa

People: Population: 43,586,097. **Age distrib.** (%): <15: 32.0; 65+: 4.9. **Pop. density:** 93 per sq. mi. **Urban:** 50%. **Ethnic groups:** Black 75.2%, white 13.6%, colored 8.6%. **Principal languages:** 11 official languages incl. Afrikaans, English, Ndebele, Pedi, Sotho. **Chief religions:** Christian 68%; traditional, animistic 28.5%.

Geography: Area: 470,900 sq. mi. **Location:** At the southern extreme of Africa. **Neighbors:** Namibia, Botswana, Zimbabwe on N; Mozambique, Swaziland on E; surrounds Lesotho. **Topography:** The large interior plateau reaches close to the country's 2,700-mi. coastline. There are few major rivers or lakes; rainfall is sparse in W, more plentiful in E. **Capitals:** Cape Town (legislative), Pretoria (administrative), and Bloemfontein (judicial). **Cities** (urban agg.): Cape Town 2,993,000; Johannesburg 2,335,000; Pretoria 1,508,000.

Government: Type: Republic. **Head of state and gov.:** Pres. Thabo Mvuyelwa Mbeki; b: June 18, 1942; in office: June 16 1999. **Local divisions:** 9 provinces. **Defense budget:** $2.1 bil(2001). **Active troops:** 63,389.

Economy: Industries: Mining, steel, chemicals, vehicles, machinery, textiles. **Chief crops:** Corn, wheat, vegetables, sugar, fruit. **Minerals:** Platinum, chromium, antimony, coal, iron, manganese, nickel, phosphates, tin, uranium, gem diamonds, copper, vanadium; world's largest producer of gold (approx. 30% of total world prod.) **Crude oil reserves** (2000): 29.36 mil bbls. **Other resources:** Wool, dairy products. **Arable land:** 10%. **Livestock** (2000): cattle: 13.7 mil; chickens: 61.0 mil; goats: 6.5 mil; pigs: 1.54 mil; sheep: 28.7 mil. **Fish catch** (1999): 513,586 metric tons. **Electricity prod.:** 186.903 bil kWh. **Labor force:** 45% services; 30% agric.; 25% ind.

Finance: Monetary unit: Rand (Sept. 2001: 8.47 = $1 U.S.). **GDP:** (1999 est.): $296.1 bil. **Per capita GDP** $6,900. **Imports** (1999): $26 bil; partners: Germany, UK, U.S. **Exports** (1999): $28 bil; partners: U.K., Italy, Japan. **Tourism:** $2.74 bil. **Budget** (FY 1994-95 est.): $38 bil. **Intl. reserves less gold** (June 2000): $6.57 bil. **Gold:** 3.98 mil oz t. **Consumer prices** (change in 1999): 5.2%.

Transport: Railroad: Length: 13,418 mi. **Motor vehicles** (1997): 4.35 mil pass. cars, 1.65 mil comm. vehicles. **Civil aviation:** 10.5 bil pass.-mi; 24 airports. **Chief ports:** Durban, Cape Town, East London, Port Elizabeth.

Communications: TV sets: 128 per 1,000 pop. **Radios:** 322 per 1,000 pop. **Telephones**: 4,961,700 main lines. **Daily newspaper circ.:** 31 per 1,000 pop.

Health: Life expectancy: 47.64 male; 48.56 female. **Births** (per 1,000 pop): 21.12. **Deaths** (per 1,000 pop.): 16.77. **Natural inc.:** .44%. **Infant mortality** (per 1,000 live births): 60.33.

Education: Compulsory: ages 7-16. **Literacy**: 82%.

Major Intl. Organizations: UN (FAO, IBRD, ILO, IMF, IMO, WHO, WTrO), the Commonwealth, OAU.

Embassy: 3051 Massachusetts Ave. NW 20008; 232-4400.

Website: http://www.statssa.gov.za

Bushmen and Hottentots were the original inhabitants. Bantus, including Zulu, Xhosa, Swazi, and Sotho, had occupied the area from NE to S South Africa before the 17th century.

The Cape of Good Hope area was settled by Dutch, beginning in the 17th century. Britain seized the Cape in 1806. Many Dutch trekked north and founded 2 republics, Transvaal and Orange Free State. Diamonds were discovered, 1867, and gold, 1886. The Dutch (Boers) resented encroachments by the British and others; the Anglo-Boer War followed, 1899-1902. Britain won and, effective May 31, 1910, created the Union of South Africa, incorporating 2 British colonies (Cape and Natal) with Transvaal and Orange Free State. After a referendum, the Union became the Republic of South Africa, May 31, 1961, and withdrew from the Commonwealth.

With the election victory of Daniel Malan's National Party in 1948, the policy of separate development of the races, or apartheid, already existing unofficially, became official. Under apartheid, blacks were severely restricted to certain occupations, and paid far lower wages than whites for similar work. Only whites could vote or run for public office. Persons of Asian Indian ancestry and those of mixed race (Coloureds) had limited political rights. In 1959 the government passed acts providing for the eventual creation of several Bantu nations, or Bantustans.

Protests against apartheid were brutally suppressed. At Sharpeville on Mar. 21, 1960, 69 black protesters were killed by government troops. At least 600 persons, mostly Bantus, were killed in 1976 riots protesting apartheid. In 1981, South Africa launched military operations in Angola and Mozambique to combat guerrilla groups.

A new constitution was approved by referendum, Nov. 1983, extending the parliamentary franchise to the Coloured and Asian minorities. Laws banning interracial sex and marriage were repealed in 1985.

In 1986, Nobel Peace Prize winner Bishop Desmond Tutu called for Western nations to apply sanctions against South Africa to force an end to apartheid. Pres. P. W. Botha announced in Apr. the end to the nation's system of racial pass laws and offered blacks an advisory role in government. On May 19, South Africa attacked 3 neighboring countries—Zimbabwe, Botswana, Zambia—to strike at guerrilla strongholds of the black nationalist African National Congress (ANC). A nationwide state of emergency was declared June 12, giving almost unlimited power to the security forces.

Some 2 million South African black workers staged a massive strike, June 6-8, 1988. Pres. Botha, head of the government since 1978, resigned Aug. 14, 1989, and was replaced by F. W. de Klerk. In 1990 the government lifted its ban on the ANC. Black nationalist leader Nelson Mandela was freed Feb. 11 after more than 27 years in prison. In Feb. 1991, Pres. de Klerk announced plans to end all apartheid laws.

In 1993 negotiators agreed on basic principles for a new democratic constitution. South Africa's partially self-governing black territories, or "homelands," were dissolved and incorporated into a national system of 9 provinces. In elections Apr. 26-29, 1994, the ANC won 62.7% of the vote, making Mandela president. The National Party won 20.4%. The Inkatha Freedom Party won 10.5% and control of the legislature in a mainly Zulu province. By then, fighting between the ANC and Inkatha (aided, during the apartheid era, by South African defense forces) had killed more than 14,000 people in the Zulu region since the mid-1980s.

In 1995, Mandela appointed a truth commission, led by Desmond Tutu, to document human rights abuses under apartheid. A post-apartheid constitution, modified to meet the objections of the Constitutional Court, became law Dec. 10, 1996, with provisions to take effect over a 3-year period.

The ANC won a landslide victory in elections held June 2, 1999. ANC leader Thabo Mbeki, Mandela's deputy president, thus became South Africa's 2d popularly elected president.

According to UN estimates, more than 4 million South Africans, including 20% of all adults, have HIV/AIDS. South Africa was the site, in July 2000, of an international AIDS conference. Pharmaceutical firms, Apr. 19, 2001, dropped their challenge to a 1997 law that allowed cheaper, generic versions of patented AIDS drugs to be imported.

Spain

Kingdom of Spain

People: Population: 40,037,995. **Age distrib.** (%): <15: 14.6; 65+: 17.2. **Pop. density:** 208 per sq. mi. **Urban:** 77%. **Ethnic groups:** Mix of Mediterranean and Nordic types. **Principal languages:** Castilian Spanish (official), Catalan, Galician, Basque. **Chief religion:** Roman Catholic 99%.

Geography: Area: 192,600 sq. mi. **Location:** In SW Europe. **Neighbors:** Portugal on W, France on N. **Topography:** The interior is a high, arid plateau broken by mountain ranges and river valleys. The NW is heavily watered, the S has lowlands and a Mediterranean climate. **Capital:** Madrid. **Cities** (urban agg.): Madrid 4,072,000, (1998 city proper: 2,881,506); Barcelona 2,819,000; Valencia 754,000.

Government: Type: Constitutional monarchy. **Head of state:** King Juan Carlos I de Borbon y Borbon; b Jan. 5, 1938; in office: Nov. 22, 1975. **Head of gov.:** Prime Min. José María Aznar; b Feb. 25, 1953; in office: May 5, 1996. **Local divisions:** 17 automomous communities. **Defense budget:** $7.0 bil. **Active troops:** 166,050.

Economy: Industries: Machinery, metals, textiles, shoes, vehicles, processed foods, tourism. **Chief crops:** Grains, olives, grapes, citrus, vegetables. **Minerals:** Lignite, uranium, iron, mercury, pyrites, fluorspar, gypsum, zinc, lead, coal. **Crude oil reserves** (2000): 14 mil bbls. **Other resources:** Forests. **Arable land:** 30%. **Livestock** (2000): cattle: 6.2 mil; chickens: 128.0 mil; goats: 2.87 mil; pigs: 23.68 mil; sheep: 23.7 mil. **Fish catch** (1999): 1.34 mil metric tons. **Electricity prod.:** 197.694 bil kWh. **Labor force:** 64% serv.; 28% manuf., mining, const.; 8% agric.

Finance: Monetary unit: Peseta (Sept. 2001: 183.56 = $1 U.S.). **Euro** (Sept. 2001: 1.09 = $1 U.S.) **GDP:** (1999 est.): $677.5 bil. **Per capita GDP:** $17,300. **Imports** (1999): $137.5 bil; partners: EU 67%. **Exports** (1999): $112.3 bil; partners: EU 72%. **Tourism:** $32.91 bil. **Budget** (1995): $139 bil. **Intl. reserves less gold** (June 2000): $32.74 bil. **Gold:** 16.83 mil oz t. **Consumer prices** (change in 1999): 2.3%.

Transport: Railroad: Length: 8,252 mi. **Motor vehicles** (1997): 15.30 mil pass. cars, 3.36 mil comm. vehicles. **Civil aviation:** 23.1 bil pass.-mi; 25 airports. **Chief ports:** Barcelona, Bilbao, Valencia, Cartagena.

Communications: TV sets: 500 per 1,000 pop. **Radios:** 332 per 1,000 pop. **Telephones:** 17,101,700 main lines. **Daily newspaper circ.:** 99 per 1,000 pop.

Health: Life expectancy: 75.47 male; 82.62 female. **Births** (per 1,000 pop): 9.26. **Deaths** (per 1,000 pop.): 9.13. **Natural inc.:** .01%. **Infant mortality** (per 1,000 live births): 4.92.

Education: Free, compulsory: ages 6-16. **Literacy**: 97%.

Major Intl. Organizations: UN and all of its specialized agencies, EU, NATO, OECD, OSCE.

Embassy: 2375 Pennsylvania Ave. NW 20037; 452-0100.

Website: http://www.sispain.org

Initially settled by Iberians, Basques, and Celts, Spain was successively ruled (wholly or in part) by Carthage, Rome, and the Visigoths. Muslims invaded Iberia from North Africa in 711. Reconquest of the peninsula by Christians from the N laid the foundations of modern Spain. In 1469 the kingdoms of Aragon and Castile were united by the marriage of Ferdinand II and Isabella I. Moorish rule ended with the fall of the kingdom of Granada, 1492. Spain's large Jewish community was expelled the same year.

Spain obtained a colonial empire with the "discovery" of America by Columbus, 1492, the conquest of Mexico by Cortes, and Peru by Pizarro. It also controlled the Netherlands and parts of Italy and Germany. Spain lost its American colonies in the early 19th century. It lost Cuba, the Philippines, and Puerto Rico during the Spanish-American War, 1898.

Primo de Rivera became dictator in 1923. King Alfonso XIII revoked the dictatorship, 1930, but was forced to leave the country in 1931. A republic was proclaimed, which disestablished the church, curtailed its privileges, and secularized education. During 1936-39 a Popular Front composed of socialists, Communists, republicans, and anarchists governed Spain.

Army officers under Francisco Franco revolted against the government, 1936. In a destructive 3-year war, in which some one million died, Franco received massive help and troops from Italy and Germany, while the USSR, France, and Mexico supported the republic. The war ended Mar. 28, 1939. Franco was named caudillo, leader of the nation. Spain was officially neutral in World War II, but its cordial relations with fascist countries caused its exclusion from the UN until 1955.

In July 1969, Franco and the Cortes (Parliament) designated Prince Juan Carlos as the future king and chief of state. After Franco's death, Nov. 20, 1975, Juan Carlos was sworn in as king. In free elections June 1977, moderates and democratic socialists emerged as the largest parties.

In 1981 a coup attempt by right-wing military officers was thwarted by the king. The Socialist Workers' Party, under Felipe González Márquez, won 4 consecutive general elections, from 1982 to 1993, but lost to a coalition of conservative and regional parties in the election of Mar. 3, 1996.

Catalonia and the Basque country were granted autonomy, Jan. 1980, following overwhelming approval in home-rule referendums. Basque extremists, however, have pushed for independence. The militant Basque separatist group ETA proclaimed a cease-fire as of Sept. 18, 1998, but announced an end to the truce Nov. 28, 1999. The Popular Party of conservative Prime Min. José María Aznar won a majority in the parliamentary election of Mar. 12, 2000.

The **Balearic Islands** in the W Mediterranean, 1,927 sq. mi., are a province of Spain; they include **Majorca** (Mallorca; capital Palma de Mallorca), **Minorca, Cabrera, Ibiza,** and **Formentera.** The **Canary Islands,** 2,807 sq. mi., in the Atlantic W of Morocco, form 2 provinces, and include the islands of **Tenerife, Palma, Gomera, Hierro, Grand Canary, Fuerteventura,** and **Lanzarote;** Las Palmas and Santa Cruz are thriving ports. **Ceuta** and **Melilla,** small Spanish enclaves on Morocco's Mediterranean coast, gained limited autonomy in Sept. 1994.

Spain has sought the return of Gibraltar, in British hands since 1704.

Sri Lanka

Democratic Socialist Republic of Sri Lanka

People: Population: 19,408,635. **Age distrib.** (%): <15: 26.0; 65+: 6.6. **Pop. density:** 777 per sq. mi. **Urban:** 23%. **Ethnic groups:** Sinhalese 74%, Tamil 18%, Moor 7%. **Principal languages:** Sinhala (official), Tamil, English. **Chief religions:** Buddhist 69%, Hindu 15%, Christian 8%, Muslim 8%.

Geography: Area: 25,000 sq. mi. **Location:** In Indian O. off SE coast of India. **Neighbors:** India on NW. **Topography:** The coastal area and the northern half are flat; the S-central area is hilly and mountainous. **Capital:** Colombo (1999): 690,000.

Government: Type: Republic. **Head of state:** Pres. Chandrika Bandaranaike Kumaratunga; b June 29, 1945; in office: Nov. 12, 1994. **Head of gov.:** Prime Min. Ratnasiri Wickremanayake; b May 5, 1933; in office: Aug. 10, 2000. **Local divisions:** 8 provinces. **Defense budget:** $700 mil. **Active troops:** 110,000–115,000.

Economy: Industries: Clothing, agric. processing, oil refining, textiles. **Chief crops:** Tea, coconuts, rice, sugar. **Minerals:** Graphite, limestone, gems, phosphates. **Other resources:** Forests, rubber. **Arable land:** 14%. **Livestock** (2000): cattle: 1.62 mil; chickens: 10.2 mil; goats: 514,400; pigs: 73,600; sheep: 12,100. **Fish catch** (1999): 247,000 metric tons. **Electricity prod.** 6.026 bil kWh. **Labor force:** 45% services; 38% agric.

Finance: Monetary unit: Rupee (Sept. 2001: 90.15 = $1 U.S.). **GDP:** (1999 est.): $50.5 bil. **Per capita GDP:** $2,600. **Imports** (1998): $5.3 bil; partners: Japan 10%, India 10%. **Exports** (1998): $4.7 bil; partners: U.S. 40%, UK 11%. **Tourism:** $275 mil. **Budget** (1997): $4.2 bil. **Intl. reserves less gold** (May 2000): $1.38 bil. **Gold:** 63,000 oz t. **Consumer prices** (change in 1999): 4.7%.

Transport: Railroad: Length: 928 mi. **Motor vehicles:** 220,000 pass. cars, 248,900 comm. vehicles. **Civil aviation:** 2.6 bil pass.-mi; 1 airport. **Chief ports:** Colombo, Trincomalee, Galle.

Communications: TV sets: 91 per 1,000 pop. **Radios:** 210 per 1,000 pop. **Telephones**: 767,400 main lines. **Daily newspaper circ.:** 29 per 1,000 pop.

Health: Life expectancy: 69.58 male; 74.73 female. **Births** (per 1,000 pop): 16.58. **Deaths** (per 1,000 pop.): 6.43. **Natural inc.:** 1.02%. **Infant mortality** (per 1,000 live births): 16.08.

Education: Free, compulsory: ages 5-12. **Literacy**: 88%.

Major Intl. Organizations: UN (FAO, IBRD, ILO, IMF, IMO, WHO, WTrO), the Commonwealth.

Embassy: 2148 Wyoming Ave. NW 20008; 483-4025.

The island was known to the ancient world as Taprobane (Greek for copper-colored) and later as Serendip (from Arabic). Colonists from N India subdued the indigenous Veddahs about 543 BC; their descendants, the Buddhist Sinhalese, still form most of the population. Hindu descendants of Tamil immigrants from S India account for about one-fifth of the population.

Parts were occupied by the Portuguese in 1505 and the Dutch in 1658. The British seized the island in 1796. As Ceylon it became an independent member of the Commonwealth in 1948, and the Republic of Sri Lanka May 22, 1972.

Prime Min. W. R. D. Bandaranaike was assassinated Sept. 25, 1959. In new elections, the Freedom Party was victorious under Mrs. Sirimavo Bandaranaike, widow of the former prime minister. After May 1970 elections, Mrs. Bandaranaike became prime minister again. In 1971 the nation suffered economic problems and terrorist activities by ultra-leftists, thousands of whom were executed. Massive land reform and nationalization of foreign-owned plantations were undertaken in the mid-1970s. Mrs. Bandaranaike was ousted in 1977 elections. Presidential powers were increased in 1978 in an effort to restore stability.

Tensions between the Sinhalese and Tamil separatists erupted into violence in the early 1980s. More than 60,000 have died in the civil war, which continued through the late 1990s; another 12,000, mostly young Tamils, have "disappeared" after they were taken into custody by government security forces.

Pres. Ranasinghe Premadasa was assassinated May 1, 1993, by a Tamil rebel. Mrs. Bandaranaike's daughter, Chandrika Bandaranaike Kumaratunga, became prime minister after the Aug. 16, 1994, general elections. Elected president Nov. 9, Kumaratunga appointed her mother prime minister. Kumaratunga, who was injured in a suicide bomb attack at a campaign rally Dec. 18, 1999, won a 2d 6-year term 3 days later. In failing health, Mrs. Bandaranaike resigned Aug. 10 and died Oct. 10, 2000. Facing a possible no-confidence motion, Pres. Kumaratunga suspended parliament July 10, 2001.

Sudan

Republic of the Sudan

People: Population: 36,080,373. **Age distrib.** (%): <15: 44.6; 65+: 2.1. **Pop. density:** 39 per sq. mi. **Urban:** 35%. **Ethnic groups:** Black 52%, Arab 39%, Beja 6%. **Principal languages:** Arabic (official), Nubian, Ta Bedawie. **Chief religions:** Sunni Muslim 70%, indigenous beliefs 25%.

Geography: Area: 916,300 sq. mi., the largest country in Africa. **Location:** At the E end of Sahara desert zone. **Neighbors:** Egypt on N; Libya, Chad, Central African Republic on W; Congo (formerly Zaire), Uganda, Kenya on S; Ethiopia, Eritrea on E. **Topography:** The N consists of the Libyan Desert in the W, and the mountainous Nubia Desert in E, with narrow Nile valley between. The center contains large, fertile, rainy areas with fields, pasture, and forest. The S has rich soil, heavy rain. **Capital:** Khartoum. **Cities**: Khartoum (urban agg.): 2,731,000; Omdurman (1993) 1,271,403.

Government: Type: Republic with strong military influence. **Head of state and gov.:** Pres. Gen. Omar Hassan Ahmad Al-Bashir; b Jan. 1, 1944; in office: June 30, 1989. **Local divisions:** 26 states. **Defense budget:** $425 mil. **Active troops:** 104,500.

Economy: Industries: Cotton ginning, textiles, cement. **Chief crops:** Gum arabic, sorghum, cotton (main export), wheat. **Minerals:** Petroleum, iron, chromium, copper. **Crude oil reserves** (2000): 262.1 mil bbls. **Arable land:** 5%. **Livestock** (2000): cattle: 37.09 mil; chickens: 41.5 mil; goats: 37.8 mil; sheep: 42.8 mil. **Fish catch:** (1999): 48,072 metric tons. **Electricity prod.:** 1.760 bil kWh. **Labor force:** 80% agric.; 10% ind. & comm.; 6% govt.

Finance: Monetary unit: Pound (Sept. 2001: 258.70 = $1 U.S.), Dinar (Oct. 2000): 256.00 = $1 U.S.). **GDP:** (1999 est.): $32.6 bil. **Per capita GDP:** $940. **Imports** (1999): $1.4 bil; partners: China 27%, France 14% UK 10%. **Exports** (1999): $580 mil; partners: Saudi Arabia 24%, Italy 10%, Germany 5%. **Tourism** (1998): $8 mil. **Budget** (1996): $1.5 bil. **Intl. reserves less gold** (Apr. 2000): $238.4 mil. **Consumer prices** (change in 1999): 16.0%.

Transport: Railroad: Length: 2,960 mi. **Motor vehicles:** 35,000 pass. cars, 40,000 comm. vehicles. **Civil aviation:** 292.8 mil pass.-mi; 3 airports. **Chief port:** Port Sudan.

Communications: TV sets: 8.2 per 1,000 pop. **Radios:** 182 per 1,000 pop. **Telephones:** 386,800 main lines. **Daily newspaper circ.:** 21 per 1,000 pop.

Health: Life expectancy: 55.85 male; 58.08 female. **Births** (per 1,000 pop): 37.89. **Deaths** (per 1,000 pop.): 10.04. **Natural inc.:** 2.79%. **Infant mortality** (per 1,000 live births): 68.67.

Education: Literacy: 46%.

Major Intl. Organizations: UN (FAO, IBRD, ILO, IMF, IMO, WHO), AL, OAU.

Embassy: 2210 Massachusetts Ave. NW 20008; 338-8565.

Website: http://www.sudan.net

Northern Sudan, ancient Nubia, was settled by Egyptians in antiquity. The population was converted to Coptic Christianity in the 6th century. Arab conquests brought Islam to the area in the 15th century.

In the 1820s Egypt took over Sudan, defeating the last of earlier empires, including the Fung. In the 1880s a revolution was led by Muhammad Ahmad, who called himself the Mahdi (leader of the faithful), and his followers, the dervishes.

In 1898 an Anglo-Egyptian force crushed the Mahdi's successors. In 1951 the Egyptian Parliament abrogated its 1899 and 1936 treaties with Great Britain and amended its constitution to provide for a separate Sudanese constitution. Sudan voted for complete independence as a parliamentary government effective Jan. 1, 1956.

In 1969, a Revolutionary Council took power, but a civilian premier and cabinet were appointed; the government announced it would create a socialist state.

Economic problems plagued the nation in the 1980s and 1990s, aggravated by civil war and influxes of refugees from neighboring countries. After 16 years in power, Pres. Jaafar al-Nimeiry was overthrown in a bloodless military coup, Apr. 6, 1985. Sudan held its first democratic parliamentary elections in 18 years in 1986, but the elected government was overthrown in a bloodless coup June 30, 1989.

In the mid-1980s, rebels in the south (populated largely by black Christians and followers of tribal religions) took up arms against government domination by northern Sudan, mostly Arab-Muslim. War and related famine cost an estimated 2 million lives and displaced millions of southerners by the late 1990s. In 1993, Amnesty International accused Sudan of "ethnic cleansing" against the Nuba people in the South.

Egypt publicly blamed Sudan for an attempted assassination of Egyptian Pres. Hosni Mubarak in Ethiopia, June 26, 1995. Opposition groups boycotted elections Mar. 1996.

A new constitution based on Islamic law took effect June 30, 1998. On Aug. 20, in retaliation for bombings in Kenya and Tanzania, U.S. missiles destroyed a Khartoum pharmaceutical plant the U.S. alleged was associated with terrorist activities; independent inquiries later cast some doubt on the U.S. claim. Embroiled in a power struggle, Pres. Omar Hassan Ahmad Al-Bashir dissolved parliament and declared a state of emergency Dec. 12, 1999. The main opposition parties boycotted presidential and legislative elections Dec. 13-22, 2000, won by Bashir.

Suriname

Republic of Suriname

People: Population: 433,998. **Age distrib.** (%): <15: 31.6; 65+: 5.7. **Pop. density:** 7 per sq. mi. **Urban:** 74%. **Ethnic groups:** Hindustani 37%, Creole 31%, Javanese 15%. **Principal languages:** Dutch (official), Sranang Tongo, English, Hindustani. **Chief religions:** Hindu 27%, Protestant 25%, Roman Catholic 23%, Muslim 20%.

Geography: Area: 62,300 sq. mi. **Location:** On N shore of South America. **Neighbors:** Guyana on W, Brazil on S, French Guiana on E. **Topography:** A flat Atlantic coast, where dikes permit agriculture. Inland is a forest belt; to the S, largely unexplored hills cover 75% of the country. **Capital:** Paramaribo (1997 est.): 289,000.

Government: Type: Republic. **Head of state and gov.:** Pres. Runaldo Ronald Venetiaan; b June 18, 1936; in office: Aug. 12, 2000. **Local divisions:** 10 districts. **Defense budget:** $11 mil. **Active troops:** 2,040.

Economy: Industries: Aluminum, mining, food processing. **Chief crops:** Rice, bananas, palm kernels. **Minerals:** Kaolin, bauxite, gold. **Crude oil reserves** (2000): 74 mil bbls. **Other resources:** Forests, fish, shrimp. **Livestock** (2000): cattle: 106,000; chickens: 2.0 mil; goats: 12,000; pigs: 32,000; sheep: 12,400. **Fish catch:** (1999): 13,001 metric tons. **Electricity prod.:** 1.937 bil kWh.

Finance: Monetary unit: Guilder (Sept. 2001: 981.00 = $1 U.S.). **GDP:** (1999 est.): $1.48 bil. **Per capita GDP:** $3,400. **Imports** (1998 est.): $461.4 mil; partners: U.S. 31.2%, Netherlands 17.3%. **Exports** (1998 est.): $406.1 mil; partners: Norway 24%, Netherlands 23.8%, U.S. 21.7%. **Tourism** (1998): $44 mil. **Budget** (1997 est.): $403 mil. **Consumer prices** (change in 1999): 98.9%.

Transport: Railroad: Length: 187 mi. **Motor vehicles:** 46,408 pass. cars, 19,255 comm. vehicles. **Civil aviation:** 663.5 mil pass.-mi; 3 airports. **Chief ports:** Paramaribo, New Nickerie, Albina.

Communications: TV sets: 146 per 1,000 pop. **Radios:** 719 per 1,000 pop. **Telephones:** 75,300 main lines. **Daily newspaper circ.:** 107 per 1,000 pop.

Health: Life expectancy: 68.97 male; 74.42 female. **Births** (per 1,000 pop): 20.53. **Deaths** (per 1,000 pop.): 5.68. **Natural inc.:** 1.49%. **Infant mortality** (per 1,000 live births): 24.27.

Education: Free, compulsory: ages 6-16. **Literacy**: 93%.

Major Intl. Organizations: UN (FAO, IBRD, ILO, IMF, IMO, WHO, WTrO), Caricom, OAS.

Embassy: 4301 Connecticut Ave. NW 20008; 244-7488.

The Netherlands acquired Suriname in 1667 from Britain, in exchange for New Netherlands (New York). The 1954 Dutch constitution raised the colony to a level of equality with the Netherlands and the Netherlands Antilles. Independence was granted Nov. 25, 1975, despite objections from East Indians. Some 40% of the population (mostly East Indians) immigrated to the Netherlands in the months before independence.

The National Military Council took control of the government, Feb. 1982. Civilian rule was restored in 1987, but political turmoil continued until 1992, disrupting the nation's economy.

Swaziland

Kingdom of Swaziland

People: Population: 1,104,343. **Age distrib.** (%): <15: 45.5; 65+: 2.6. **Pop. density:** 166 per sq. mi. **Urban:** 26%. **Ethnic groups:** African 97%, European 3%. **Principal languages:** siSwati, English (both official). **Chief religions:** Christian 60%, indigenous beliefs 40%.

Geography: Area: 6,600 sq. mi. **Location:** In southern Africa, near Indian O. coast. **Neighbors:** South Africa on N, W, S; Mozambique on E. **Topography:** The country descends from W-E in broad belts, becoming more arid in the low veld region, then rising to a plateau in the E. **Capitals:** Mbabane (administrative), Lobamba (legislative). **Cities:** Mbabane (1999 est.): 73,000.

Government: Type: Constitutional monarchy. **Head of state:** King Mswati III; b Apr. 19, 1968; in office: Apr. 25, 1986. **Head of gov.:** Prime Min. Barnabas Sibusiso Dlamini; b May 15, 1942; in office: July 26, 1996. **Local divisions:** 4 districts.

Economy: Industries: Wood pulp, mining. **Chief crops:** Sugar, corn, cotton, rice, pineapples, tobacco, citrus, peanuts. **Minerals:** Asbestos, clay, coal. **Other resources:** Forests. **Arable land:** 11%. **Livestock** (2000): cattle: 610,000; chickens: 3.0 mil; goats: 440,000; pigs: 33,000; sheep: 30,000.. **Electricity prod.:** 375 mil kWh. **Labor force: private sector: 70%, public sector: 30%.**

Finance: Monetary unit: Lilangeni (Sept. 2001: 8.50 = $1 U.S.). **GDP:** (1999 est.): $4.2 bil. **Per capita GDP:** $4,200. **Imports** (1999): $1.05 bil; partners: South Africa 83%, EU 6% . **Exports** (1999): $825 mil; partners: South Africa 74%, EU 12%, Mozambique 5%. **Tourism** (1998): $37 mil. **Budget** (FY1996-97): $450 mil. **Intl. reserves less gold** (May 2000): $342.70 mil. **Consumer prices** (change in 1999): 6.1%.

Transport: Railroad: Length: 187 mi. **Motor vehicles:** 28,523 pass. cars, 8,232 comm. vehicles. **Civil aviation:** 26.5 mil pass.-mi; 1 airport.

Communications: TV sets: 96 per 1,000 pop. **Radios:** 129 per 1,000 pop. **Telephones** (1999): 30,600 main lines. **Daily newspaper circ.:** 40 per 1,000 pop.

Health: Life expectancy: 37.86 male; 39.4 female. **Births** (per 1,000 pop): 40.12. **Deaths** (per 1,000 pop.): 21.84. **Natural inc.:** 1.83%. **Infant mortality** (per 1,000 live births): 109.19.

Education: Literacy: 77%.

Major Intl. Organizations: UN (FAO, IBRD, ILO, IMF, WHO, WTrO), the Commonwealth, OAU.

Embassy: 3400 International Dr. NW 20008; 362-6683.

Website: http://www.realnet.co.sz

The royal house of Swaziland traces back 400 years, and is one of Africa's last ruling dynasties. The Swazis, a Bantu people, were driven to Swaziland from lands to the N by the Zulus in 1820. Their autonomy was later guaranteed by Britain and Transvaal (later part of South Africa), with Britain assuming control after 1903. Independence came Sept. 6, 1968. In 1973 the king repealed the constitution and assumed full powers.

A new constitution banning political parties took effect Oct. 13, 1978. As Swaziland slowly moved toward political reform, student and labor unrest grew in the 1990s. The UN estimates that about one-fourth of the adult population has HIV/AIDS.

Sweden

Kingdom of Sweden

People: Population: 8,875,053. **Age distrib.** (%): <15: 18.2; 65+: 17.3. **Pop. density:** 56 per sq. mi. **Urban:** 83%. **Ethnic groups:** Swedish 89%, Finnish 2%. **Principal language:** Swedish. **Chief religion:** Evangelical Lutheran 94%.

Geography: Area: 158,700 sq. mi. **Location:** On Scandinavian Peninsula in N Europe. **Neighbors:** Norway on W, Denmark on S (across Kattegat), Finland on E. **Topography:** Mountains along NW border cover 25% of Sweden, flat or rolling terrain covers the central and southern areas, which include several large lakes. **Capital:** Stockholm. **Cities** (urban agg.): Stockholm 1,583,000; Göteborg 766,000.

Government: Type: Constitutional monarchy. **Head of state:** King Carl XVI Gustaf; b Apr. 30, 1946; in office: Sept. 19, 1973. **Head of gov.:** Prime Min. Goran Persson; b June 20, 1949; in office: Mar. 21, 1996. **Local divisions:** 21 counties. **Defense budget** (2001): $4.8 bil. **Active troops:** 52,700.

Economy: Industries: Steel, precision equipment, vehicles, processed foods, paper. **Chief crops:** Grains, potatoes, sugar beets. **Minerals:** Zinc, iron, lead, copper, silver. **Other resources:** Forests (half the country); yield about 17% of exports. **Arable land:** 7%. **Livestock** (2000): cattle: 1.71 mil; chickens: 7.85 mil; pigs: 1.92 mil; sheep: 437,249. **Fish catch** (1999): 364,115 metric tons. **Electricity prod.:** 146.633 bil kWh. **Labor force**: 74% services; industry 24%; agriculture 2%.

Finance: Monetary unit: Krona (Sept. 2001: 10.50 = $1 U.S.). **GDP:** (1999 est.): $184 bil. **Per capita GDP:** $20,700. **Imports** (1999): $67.9 bil; partners: EU 68%, Norway 8%, US 6%. **Exports** (1999): $85.7 bil; partners: EU 57%; Norway 9%; US 9%. **Tourism:** $3.89 bil. **Budget** (FY 1995-96): $146.1 bil. **Intl. reserves less gold** (June 2000): $14.54 bil. **Gold:** 5.93 mil oz t. **Consumer prices** (change in 1999): –0.5%.

Transport: Railroad: Length: 6,756 mi. **Motor vehicles** (1997): 3.70 mil pass. cars, 336,593 comm. vehicles. **Civil aviation:** 5.54 bil pass.-mi; 48 airports. **Chief ports:** Göteborg, Stockholm, Malmö.

Communications: TV sets: 531 per 1,000 pop. **Radios:** 904 per 1,000 pop. **Telephones**: 6,056,800 main lines. **Daily newspaper circ.:** 484 per 1,000 pop.

Health: Life expectancy: 77.07 male; 82.5 female. **Births** (per 1,000 pop): 9.91. **Deaths** (per 1,000 pop.): 10.61. **Natural inc.:** -.07%. **Infant mortality** (per 1,000 live births): 3.47.

Education: Compulsory: ages 6-15. **Literacy**: 100%.

Major Intl. Organizations: UN and all of its specialized agencies, EU, OECD, OSCE.

Embassy: 1501 M St. NW 20005; 467-2600.

Website: http://www.scb.se/eng/index.htm

The Swedes have lived in present-day Sweden for at least 5,000 years, longer than nearly any other European people. Gothic tribes from Sweden played a major role in the disintegration of the Roman Empire. Other Swedes helped create the first Russian state in the 9th century.

The Swedes were Christianized from the 11th century, and a strong centralized monarchy developed. A parliament, the Riksdag, was first called in 1435, the earliest parliament on the European continent, with all classes of society represented.

Swedish independence from rule by Danish kings (dating from 1397) was secured by Gustavus I in a revolt, 1521-23; he built up the government and military and established the Lutheran Church. In the 17th century Sweden was a major European power, gaining most of the Baltic seacoast, but its international position subsequently declined.

The Napoleonic wars, 1799-1815, in which Sweden acquired Norway (it became independent 1905), were the last in which Sweden participated. Armed neutrality was maintained in both world wars.

More than 4 decades of Social Democratic rule ended in the 1976 parliamentary elections; the party returned to power in the 1982 elections. After Prime Min. Olof Palme was shot to death in Stockholm, Feb. 28, 1986, Ingvar Carlsson took office. Carl Bildt, a non-Socialist, became prime minister Oct. 1991, with a mandate to restore Sweden's economic competitiveness. The Social Democrats returned to power following 1994 elections.

Swedish voters approved membership in the European Union Nov. 13, 1994, and Sweden entered the EU as of Jan. 1, 1995. Carlsson retired and was succeeded by Goran Persson in Mar. 1996. Persson forged a coalition with the Left and Green parties after his Social Democrats lost ground in elections Sept. 20, 1998.

Switzerland

Swiss Confederation

People: Population: 7,283,274. **Age distrib.** (%): <15: 17.0; 65+: 15.3. **Pop. density:** 475 per sq. mi. **Urban:** 68%. **Ethnic groups:** German 65%, French 18%, Italian 10%, Romansch 1%. **Principal languages:** German, French, Italian, Romansch (all official). **Chief religions:** Roman Catholic 46.1%, Protestant 40%.

Geography: Area: 15,300 sq. mi. **Location:** In the Alps Mts. in central Europe. **Neighbors:** France on W, Italy on S, Austria on E, Germany on N. **Topography:** The Alps cover 60% of the land area; the Jura, near France, 10%. Running between, from NE to SW, are midlands, 30%. **Capitals:** Bern (administrative), Lausanne (judicial). **Cities**: Zurich (urban agg.): 983,000; (1999 est.): Basel 166,700; Geneva 398,910; Bern 941,144.

Government: Type: Federal republic. **Head of state and gov.:** The president is elected by the Federal Assembly to a non-renewable 1-year term. **Local divisions:** 20 full cantons, 6 half cantons. **Defense budget** (2001): $2.8 bil. **Active troops:** 3,470.

Economy: Industries: Machinery, chemicals, precision instruments, watches, textiles, foodstuffs (cheese, chocolate), banking, tourism. **Chief crops:** Grains, fruits, vegetables. **Minerals:** Salt. **Other resources:** Hydropower potential, timber. **Arable land:** 10%. **Livestock** (2000): cattle: 1.6 mil; chickens: 6.72 mil; goats: 65,000; pigs: 1.45 mil; sheep: 450,000. **Electricity prod.:** 66.768 bil kWh. **Labor force:** 67% serv.; 28% manuf. & const.

Finance: Monetary unit: Franc (Sept. 2001: 1.67 = $1 U.S.). **GDP:** (1999 est.): $197 bil. **Per capita GDP:** $27,100. **Imports** (1999): $99 bil; partners: EU 80%. **Exports** (1999): $98.5 bil; partners: EU 62%. **Tourism:** $7.36 bil. **Budget** (1998): $34.89 bil. **Intl. reserves less gold** (June 2000): $30.26 bil. **Gold:** 81.61 mil oz t. **Consumer prices** (change in 1999): 0.7%.

Transport: Railroad: Length: 3,132 mi. **Motor vehicles (1997):** 3.32 mil pass. cars, 302,707 comm. vehicles. **Civil aviation:** 13.83 bil pass.-mi; 5 airports. **Chief port:** Basel.

Communications: TV sets: 536 per 1,000 pop. **Radios:** 990 per 1,000 pop. **Telephones**: 5,158,000 main lines. **Daily newspaper circ.:** 337 per 1,000 pop.

Health: Life expectancy: 76.85 male; 82.76 female. **Births** (per 1,000 pop): 10.12. **Deaths** (per 1,000 pop.): 8.77. **Natural inc.:** .14%. **Infant mortality** (per 1,000 live births): 4.48.

Education: Compulsory: ages 7-16. **Literacy** (1994): 100%.

Major Intl. Organizations: Many UN specialized agencies (though not a member), EFTA, OECD, OSCE.

Embassy: 2900 Cathedral Ave. NW 20008; 745-7900.

Websites: http://www.swissembassy.org.uk
http://www.admin.ch/bfs/eindex.htm

Switzerland, the former Roman province of Helvetia, traces its modern history to 1291, when 3 cantons created a defensive league. Other cantons were subsequently admitted to the Swiss Confederation, which obtained its independence from the Holy Roman Empire through the Peace of Westphalia (1648). The cantons were joined under a federal constitution in 1848, with large powers of local control retained by each.

Switzerland has maintained an armed neutrality since 1815, and has not been involved in a foreign war since 1515. It is the seat of many UN and other international agencies.

Switzerland is a world banking center. In an effort to crack down on criminal transactions, the nation's strict bank-secrecy rules have been eased since 1990. Stung by charges that assets seized by the Nazis and deposited in Swiss banks in World War II had not been properly returned, the government announced, March 5, 1997, a $4.7 billion fund to compensate victims of the Holocaust and other catastrophies. Swiss banks agreed Aug. 12, 1998, to pay $1.25 billion in reparations.

Syria

Syrian Arab Republic

People: Population: 16,728,808. **Age distrib.** (%): <15: 39.9; 65+: 3.2. **Pop. density:** 236 per sq. mi. **Urban:** 54%. **Ethnic groups:** Arab 90%. **Principal languages:** Arabic (official), Kurdish, Armenian. **Chief religions:** Sunni Muslim 74%, other Muslims 16%, Christian 10%.

Geography: Area: 71,000 sq. mi. **Location:** Middle East, at E end of Mediterranean Sea. **Neighbors:** Lebanon and Israel on W, Jordan on S, Iraq on E, Turkey on N. **Topography:** Syria has a short Mediterranean coastline, then stretches E and S with fertile lowlands and plains, alternating with mountains and large desert areas. **Capital:** Damascus. **Cities** (urban agg.): Damascus 2,335,000; Aleppo 2,173,000.

Government: Type: Republic (under military regime). **Head of state:** Pres. Bashar al-Assad; b Sept. 1965; in office: July 17, 2000. **Head of gov.:** Prime Min. Muhammad Mustafa Mero; b 1941; in office: Mar. 13, 2000. **Local divisions:** 14 provinces. **Defense budget:** $1.8 bil. **Active troops:** 316,000.

Economy: Industries: Oil prods., textiles, food processing, tobacco, phosphate mining. **Chief crops:** Cotton, grains, lentils, chickpeas. **Minerals:** Oil, phosphates, chrome, manganese, asphalt, iron. **Crude oil reserves** (2000): 2.5 bil bbls. **Other resources:** Wool, dairy prods. **Arable land:** 28%. **Livestock** (2000): cattle: 920,000; chickens: 22.0 mil; goats: 1.1 mil; pigs: 770; sheep: 14.5 mil. **Electricity prod.:** 17.940 bil kWh. **Labor force:** 40% agric.; 40% services; 20% ind.

Finance: Monetary unit: Pound (Sept. 2001: 50.20 = $1 U.S.). **GDP:** (1999 est.): $42.2 bil. **Per capita GDP:** $2,500. **Imports** (1999): $3.2 bil.; partners: Ukraine 16%, Italy 6%, Germany 6% **Exports** (1999): $3.3 bil.; partners: Germany 14%, Turkey 13%, Italy 12%. **Tourism:** $1.36 bil. **Budget** (1997 est.): $4.2 bil. **Gold:** 833,000 oz t. **Consumer prices** (change in 1998): –0.5%.

Transport: Railroad: Length: 1,097 mi. **Motor vehicles:** 134,000 pass. cars, 218,900 comm. vehicles. **Civil aviation:** 767.6 mil pass.-mi; 5 airports. **Chief ports:** Latakia, Tartus.

Communications: TV sets: 49 per 1,000 pop. **Radios:** 211 per 1,000 pop. **Telephones**: 1,675,000 main lines. **Daily newspaper circ.:** 19 per 1,000 pop.

Health: Life expectancy: 67.63 male; 69.98 female. **Births** (per 1,000 pop): 30.64**. Deaths** (per 1,000 pop.): 5.21. **Natural inc.:** 2.54%. **Infant mortality** (per 1,000 live births): 33.8.

Education: Compulsory: ages 6-12. **Literacy**: 79%.

Major Intl. Organizations: UN (FAO, IBRD, ILO, IMF, IMO, WHO), AL.

Embassy: 2215 Wyoming Ave. NW 20008; 232-6313.

Syria contains some of the most ancient remains of civilization. It was the center of the Seleucid empire, but later became absorbed in the Roman and Arab empires. Ottoman rule prevailed for 4 centuries, until the end of World War I.

The state of Syria was formed from former Turkish districts, separated by the Treaty of Sevres, 1920, and divided into the states of Syria and Greater Lebanon. Both were administered under a French League of Nations mandate 1920-1941.

Syria was proclaimed a republic by the occupying French Sept. 16, 1941, and exercised full independence Apr. 17, 1946. Syria joined the Arab invasion of Israel in 1948.

Syria joined Egypt Feb. 1958 in the United Arab Republic but seceded Sept. 1961. The Socialist Baath party and military leaders seized power Mar. 1963. The Baath, a pan-Arab organization, became the only legal party. The government has been dominated by the Alawite minority.

In the Arab-Israeli war of June 1967, Israel seized and occupied the Golan Heights, from which Syria had shelled Israeli settlements. On Oct. 6, 1973, Syria joined Egypt in an attack on Israel. Arab oil states agreed in 1974 to give Syria $1 billion a year to aid anti-Israel moves. Some 30,000 Syrian troops entered Lebanon in 1976 to mediate in a civil war. They fought Palestinian guerrillas and, later, Christian militiamen. Syrian troops again battled Christian forces in Lebanon, Apr. 1981.

Following Israel's invasion of Lebanon, June 6, 1982, Israeli planes destroyed 17 Syrian antiaircraft missile batteries in the Bekaa Valley, June 9. Some 25 Syrian planes were downed during the engagement. Israel and Syria agreed to a cease-fire June 11. In 1983, Syria backed the PLO rebels who ousted Yasir Arafat's forces from Tripoli.

Syria's role in promoting international terrorism led to the breaking of diplomatic relations with Great Britain and to limited sanctions by the European Community in 1986.

Syria condemned the Aug. 1990 Iraqi invasion of Kuwait and sent troops to help Allied forces in the Gulf War. In 1991, Syria accepted U.S. proposals for the terms of an Arab-Israeli peace conference. Syria subsequently participated in negotiations with Israel, but progress toward peace was slow. Turkey has accused Syria of aiding Kurdish separatists.

Former Prime Min. Mahmoud Al-Zoubi killed himself May 21, 2000, after being charged with corruption. Hafez al-Assad, president of Syria since 1971, died June 10, 2000, and was succeeded by his son Bashar al-Assad.

Taiwan

Republic of China

People: Population: 22,370,461. **Age distrib.** (%): <15: 21.2; 65+: 8.8. **Pop. density:** 1,798 per sq. mi. **Urban:** 75%. **Ethnic groups:** Taiwanese 84%, mainland Chinese 14%. **Principal languages:** Mandarin Chinese (official), Taiwanese. **Chief religions:** Buddhist, Taoist, and Confucian 93%, Christian 5%.

Geography: Area: 12,400 sq. mi. **Location:** Off SE coast of China, between East and South China seas. **Neighbors:** Nearest is China. **Topography:** A mountain range forms the backbone of the island; the eastern half is very steep and craggy, the western slope is flat, fertile, and well cultivated. **Capital:** Taipei. **Cities** (1997 est.): Taipei 2,595,699; Kaohsiung 1,434,907; Taichung 881,870.

Government: Type: Democracy. **Head of state:** Pres. Chen Shui-bian; b Feb. 18, 1951; in office: May 20, 2000. **Head of gov.:** Prime Min. Chang Chun-hsiung; b Mar. 23, 1938; in office: Oct. 4, 2000. **Local divisions:** 16 counties, 5 municipalities, 2 special municipalities (Taipei, Kaohsiung). **Defense budget:** $12.8 bil. **Active troops:** 370,000.

Economy: Industries: Textiles, clothing, electronics, processed foods, chemicals. **Chief crops:** Vegetables, rice, fruit, tea. **Minerals:** Coal, gas, limestone, marble. **Crude oil reserves** (2000): 4 mil bbls. **Arable land:** 24%. **Fish catch** (1997): 1.04 mil metric tons. **Electricity prod.:** 139.676 bil kWh. **Labor force:** 55% services; 37% ind.; 8% agric.

Finance: Monetary unit: New Taiwan Dollar (Sept. 2001: 34.65 = $1 U.S.). **GDP:** (1999 est.): $357 bil. **Per capita GDP:** $16,100. **Imports** (1999): $101.7 bil; partners: Japan 27%, U.S. 18%, Europe 16%. **Exports** (1999): $122.1 bil; partners: U.S. 26%, Hong Kong 21%, EU 18%. **Tourism:** $3.57 bil. **Budget** (1998 est.): $55 bil.

Transport: Railroad: Length: 2,410 mi. **Motor vehicles (1997):** 4.40 mil pass. cars, 833,545 comm. vehicles. **Civil aviation:** 22.8 bil pass.-mi; 13 airports. **Chief ports:** Kaohsiung, Chilung (Keelung), Hualien, Taichung.

Communications: TV sets: 327 per 1,000 pop. **Radios:** 402 per 1,000 pop. **Telephones:** 12,642,200 main lines. **Daily newspaper circ.:** 20.2 per 1,000 pop.

Health: Life expectancy: 73.81 male; 79.51 female. **Births** (per 1,000 pop): 14.31**. Deaths** (per 1,000 pop.): 6.. **Natural inc.:** .83%. **Infant mortality** (per 1,000 live births): 6.93.

Education: Free, compulsory: ages 6-15. **Literacy**: 94%.

Major Intl. Organizations: APEC.

Website: http://www.gio.gov.tw

Large-scale Chinese immigration began in the 17th century. The island came under mainland control after an interval of Dutch rule, 1620-62. Taiwan (also called Formosa) was ruled by Japan 1895-1945. Two million Kuomintang supporters fled to the island in 1949, establishing Taiwan as the seat of the Republic of China. The U.S., upon recognizing the People's Republic of China, Dec. 15, 1978, severed diplomatic ties with Taiwan. The U.S. and Taiwan maintain contact via quasi-official agencies.

Land reform, government planning, U.S. aid and investment, and free universal education brought huge advances in industry, agriculture, and living standards. In 1987 martial law was lifted after 38 years, and in 1991 the 43-year period of emergency rule ended. Taiwan held its first direct presidential election

Mar. 23, 1996. An earthquake on Sept. 21, 1999, killed more than 2,300 people and injured thousands more. Five decades of Nationalist Party rule ended with the presidential election of Mar. 18, 2000, won by Chen Shui-bian, leader of the pro-independence Democratic Progressive Party.

Both the Taipei and Beijing governments long considered Taiwan an integral part of China, although Taiwanese officials appeared to signal a departure from that policy in July 1999. Taiwan has resisted Beijing's efforts at reunification, including military pressure, but economic ties with the mainland expanded in the 1990s. Taiwan has one of the world's strongest economies and is among the 10 leading capital exporters.

The **Penghu Isls.** (Pescadores), 49 sq. mi., pop. (1996 est.) 90,142, lie between Taiwan and the mainland. **Quemoy** and **Matsu,** pop. (1996 est.) 53,286, lie just off the mainland.

Tajikistan

Republic of Tajikistan

People: Population: 6,578,681. **Age distrib.** (%): <15: 41.2; 65+: 4.6. **Pop. density:** 119 per sq. mi. **Urban:** 32%. **Ethnic groups:** Tajik 65%, Uzbek 25%. **Principal languages:** Tajik (official), Russian. **Chief religion:** Sunni Muslim 80%.

Geography: Area: 55,200 sq. mi. **Location:** Central Asia. **Neighbors:** Uzbekistan on N and W, Kyrgyzstan on N, China on E, Afghanistan on S. **Topography:** Mountainous region that contains the Pamirs, Trans-Alai mountain system. **Capital:** Dushanbe (1999 est.): 523,000.

Government: Type: Republic. **Head of state:** Pres. Imomali Rakhmonov; b Oct. 5, 1952; in office: Nov. 19, 1994. **Head of gov.:** Akil Akilov; b 1944; in office: Dec. 20, 1999. **Local divisions:** 2 viloyats, 1 autonomous viloyat. **Defense budget** (1999): $18 mil. **Active troops:** 6,000.

Economy: Industries: Aluminum, cement. **Chief crops:** Cotton, grains, fruits, vegetables. **Minerals:** Oil, uranium, mercury, coal, lead, zinc. **Crude oil reserves** (2000): 12 mil bbls. **Arable land:** 6%. **Livestock** (2000): cattle: 1.04 mil; chickens: 800,000; goats: 590,000; pigs: 920; sheep: 1.59 mil. **Electricity prod.:** 15.623 bil kWh. **Labor force:** 50% agric. & forestry; 30% serv.; 20% ind.

Finance: Monetary unit: Ruble (Sept. 2000: 2,090.00 = $1 U.S.). **GDP:** (1999 est.): $6.2 bil. **Per capita GDP:** $1020. **Imports** (1999 est.): $770 mil; partners: Netherlands 32%, Uzbekistan 29%, Switzerland 20%. **Exports** (1999 est.): $634 mil; partners: Uzbekistan 37%, Liechtenstein 26% Russia 16%.

Transport: Railroad: Length: 294.5 mi. **Motor vehicles:** 185,000 pass. cars, 3,600 comm. vehicles. **Civil aviation:** 1.1 bil pass.-mi; 1 airport.

Communications: TV sets: 259 per 1,000 pop. **Telephones**: 218,500 main lines. **Daily newspaper circ.:** 13.7 per 1,000 pop.

Health: Life expectancy: 61.09 male; 67.42 female. **Births** (per 1,000 pop): 33.23. **Deaths** (per 1,000 pop.): 8.57. **Natural inc.:** 2.47%. **Infant mortality** (per 1,000 live births): 116.09.

Education: Compulsory for 9 years between ages 7-17. **Literacy**: 100%.

Major International Organizations: UN (FAO, IBRD, ILO, IMF, WHO), CIS, OSCE.

Website: http://www.eurasianet.org/resource/tajikistan/index.shtml

There were settled societies in the region from about 3000 BC. Throughout history, it has undergone invasions by Iranians (Arabs who converted the population to Islam), Mongols, Uzbeks, Afghans, and Russians. The USSR gained control of the region 1918-25. In 1924, the Tajik ASSR was created within the Uzbek SSR. The Tajik SSR was proclaimed in 1929.

Tajikistan declared independence Sept. 9, 1991. It became an independent state when the Soviet Union disbanded Dec. 26, 1991. Conservative Communist Pres. Rakhmon Nabiyev was forced to resign, Sept. 1992, by a coalition of Islamic, nationalist, and Western-oriented parties.

Factional fighting led to the installation of a pro-Communist regime, Jan. 1993. A new constitution establishing a presidential system was approved by referendum Nov. 6, 1994. Clashes between Muslim rebels, reportedly armed by Afghanistan, and troops loyal to the government and supported by Russia, claimed an estimated 55,000 lives by mid-1997, despite a series of peace accords. Constitutional changes including legalization of Islamic political parties were approved by referendum Sept. 26, 1999. Pres. Imomali Rakhmonov won a Nov. 6 election called "a farce" by human-rights observers.

Tanzania

United Republic of Tanzania

People: Population: 36,232,074. **Age distrib.** (%): <15: 44.8; 65+: 2.9. **Pop. density:** 106 per sq. mi. **Urban:** 32%. **Ethnic groups:** African 99%. **Principal languages:** Swahili, English (both official), many others. **Chief religions:** Christian 45%, Muslim 35%, indigenous beliefs 20%; Zanzibar is 99% Muslim.

Geography: Area: 341,700 sq. mi. **Location:** On coast of E Africa. **Neighbors:** Kenya, Uganda on N; Rwanda, Burundi, Congo (formerly Zaire) on W; Zambia, Malawi, Mozambique on S. **Topography:** Hot, arid central plateau, surrounded by the lake region in the W, temperate highlands in N and S, the coastal plains. Mt. Kilimanjaro, 19,340 ft., is highest in Africa. **Capital:** Dar-es-Salaam (capital is being moved to Dodoma). **Cities** (urban agg.): Dar-es-Salaam 2,347,000.

Government: Type: Republic. **Head of state:** Pres. Benjamin William Mkapa; b Nov. 12, 1938; in office: Nov. 23, 1995. **Head of gov.:** Prime Min. Frederick Tluway Sumaye; b May 29, 1950; in office: Nov. 28, 1995. **Local divisions:** 25 regions. **Defense budget:** $144 mil. **Active troops:** 34,000.

Economy: Industries: Agricultural processing, mining, textiles. **Chief crops:** Sisal, cotton, coffee, tea, tobacco, corn, cloves. **Minerals:** Tin, phosphates, iron, coal, gemstones, diamonds, gold. **Other resources:** Pyrethrum (insecticide made from chrysanthemums). **Arable land:** 3%. **Livestock** (2000): cattle: 14.38 mil; chickens: 27.8 mil; goats: 9.95 mil; pigs: 350,000; sheep: 4.2 mil. **Fish catch** (1999): 357,210 metric tons. **Electricity prod.:** 2.248 bil kWh. **Labor force:** 90% agric.; 10% ind. & comm.

Finance: Monetary unit: Shilling (Sept. 2001: 894.00 = $1 U.S.). **GDP:** (1999 est.): $23.3 bil. **Per capita GDP:** $550. **Imports** (1999): $1.44 bil; partners: South Africa 12.9%, Kenya 9.6%, UK 8.7%. **Exports** (1999): $828 mil; partners: India 9.8% Germany 8.9%, Japan 7.8%. **Tourism:** $733 mil. **Budget** (FY 1998-99 est.): $1.1 bil. **Intl. reserves less gold** (May 2000): $745.1 mil. **Consumer prices** (change in 1999): 7.9%.

Transport: Railroad: Length: 2,218 mi. **Motor vehicles:** 55,000 pass. cars; 78,800 comm. vehicles. **Civil aviation:** 143.4 mil pass.-mi; 11 airports. **Chief ports:** Dar-es-Salaam, Mtwara, Tanga.

Communications: TV sets: 2.8 per 1,000 pop. **Radios:** 20 per 1,000 pop. **Telephones**: 173,600 main lines.

Health: Life expectancy: 51.04 male; 52.95 female. **Births** (per 1,000 pop): 39.65. **Deaths** (per 1,000 pop.): 12.95. **Natural inc.:** 2.67%. **Infant mortality** (per 1,000 live births): 79.41.

Education: Free, compulsory: ages 7-14. **Literacy**: 68%.

Major Intl. Organizations: UN and all of its specialized agencies, the Commonwealth, OAU.

Embassy: 2139 R St. NW 20008; 518-6647.

The Republic of Tanganyika in E Africa and the island Republic of Zanzibar, off the coast of Tanganyika, both of which had recently gained independence, joined into a single nation, the United Republic of Tanzania, Apr. 26, 1964. Zanzibar retains internal self-government.

Until resigning as president in 1985, Julius K. Nyerere, a former Tanganyikan independence leader, dominated Tanzania's politics, which emphasized government planning and control of the economy, with single-party rule. In 1992 the constitution was amended to establish a multiparty system. Privatization of the economy was undertaken in the 1990s.

At least 500 people died when an overcrowded Tanzanian ferry sank in Lake Victoria, May 21, 1996. About 460,000 Rwandan refugees, mostly Hutu, returned from Tanzania to Rwanda in Dec. 1996. A bomb at the U.S. embassy in Dar-es-Salaam, Aug. 7, 1998, killed 11 people and injured at least 70 others. The U.S. blamed the attack and a near-simultaneous embassy bombing in Kenya on Islamic terrorists associated with wealthy Saudi-born Osama bin Laden, believed to be sheltered in Afghanistan. After a trial in New York City, 4 conspirators were convicted May 29, 2001.

Former Pres. Nyerere died in London Oct. 14, 1999. President since 1995, Benjamin Mkapa was reelected Oct. 29, 2000.

Tanganyika. Arab colonization and slaving began in the 8th century AD; Portuguese sailors explored the coast by about 1500. Other Europeans followed.

In 1885 Germany established German East Africa of which Tanganyika formed the bulk. It became a League of Nations mandate and, after 1946, a UN trust territory, both under Britain. It became independent Dec. 9, 1961, and a republic within the Commonwealth a year later.

Zanzibar, the Isle of Cloves, lies 23 mi. off mainland Tanzania; area 640 sq. mi. and pop. (1995 est.) 456,934. The island of **Pemba,** 25 mi. to the NE, area 380 sq. mi. and pop. (1995 est.) 322,466, is included in the administration.

Chief industry is cloves and clove oil production, of which Zanzibar and Pemba produce most of the world's supply.

Zanzibar was for centuries the center for Arab slave traders. Portugal ruled the region for 2 centuries until ousted by Arabs around 1700. Zanzibar became a British Protectorate in 1890; independence came Dec. 10, 1963. Revolutionary forces overthrew the Sultan Jan. 12, 1964. The new government ousted Western diplomats and newsmen, slaughtered thousands of Arabs, and nationalized farms. Union with Tanganyika followed.

Thailand

Kingdom of Thailand

People: Population: 61,797,751. **Age distrib.** (%): <15: 23.4; 65+: 6.6. **Pop. density:** 313 per sq. mi. **Urban:** 21%. **Ethnic groups:** Thai 75%, Chinese 14%. **Principal languages:** Thai (official), English. **Chief religions:** Buddhist 95%, Muslim 4%.

Geography: Area: 197,400 sq. mi. **Location:** On Indochinese and Malayan peninsulas in SE Asia. **Neighbors:** Myanmar on W and N, Laos on N, Cambodia on E, Malaysia on S. **Topography:** A plateau dominates the NE third of Thailand, dropping to the fertile alluvial valley of the Chao Phraya R. in the center. Forested mountains are in the N, with narrow fertile valleys. The S peninsula region is covered by rain forests. **Capital:** Bangkok (1999 est.): 7,133,000.

Government: Type: Constitutional monarchy. **Head of state:** King Bhumibol Adulyadej; b Dec. 5, 1927; in office: June 9, 1946. **Head of gov.:** Prime Min. Thaksin Shinawatra; b July 26, 1949; in office: Feb. 18, 2001. **Local divisions:** 76 provinces. **Defense:** 1.5% of GDP. **Active troops:** 306,000.

Economy: Industries: Textiles, agric. processing, tourism. **Chief crops:** Rice (world's largest exporter), corn, cassava, sugarcane. **Minerals:** Tin, tungsten, gas. **Crude oil reserves** (2000): 296.25 mil bbls. **Other resources:** Forests, rubber, seafood (world's largest exporter of farmed shrimp). **Arable land:** 34%. **Livestock** (2000): cattle: 6.1 mil; chickens: 172.0 mil; goats: 130,000; pigs: 7.68 mil; sheep: 42,000. **Fish catch** (1999): 3.49 mil metric tons. **Electricity prod.:** 89.431 bil kWh. **Labor force:** 54% agric.; 31% serv. & govt.; 15% ind.

Finance: Monetary unit: Baht (Sept. 2001: 44.58 = $1 U.S.). **GDP:** (1999 est.): $388.7 bil. **Per capita GDP:** $6,400. **Imports** (1999): $45 bil; partners: Japan 23.6%, U.S. 14%, Singapore 5.5%. **Exports** (1999): $58.5 bil; partners: U.S. 22.3%, Japan 13.7%; Singapore 8.6%. **Tourism:** $7.00 bil. **Budget** (FY 1996-97): $25 bil. **Intl. reserves less gold** (June 2000): $31.43 bil. **Gold:** 2.37 mil oz t. **Consumer prices** (change in 1999): 0.3%.

Transport: Railroad: Length: 2,471 mi. **Motor vehicles:** 1.55 mil pass. cars, 4.15 mil comm. vehicles. **Civil aviation:** 19.2 bil pass.-mi; 25 airports. **Chief ports:** Bangkok, Sattahip.

Communication: TV sets: 54 per 1,000 pop. **Radios:** 163 per 1,000 pop. **Telephones** (1999): 5,215,600 main lines. **Daily newspaper circ.:** 63 per 1,000 pop.

Health: Life expectancy: 65.64 male; 72.24 female. **Births** (per 1,000 pop): 16.63. **Deaths** (per 1,000 pop.): 7.54. **Natural inc.:** .91%. **Infant mortality** (per 1,000 live births): 30.49.

Education: Compulsory: ages 6-15. **Literacy**: 94%.

Major Intl. Organizations: UN (FAO, IBRD, ILO, IMF, IMO, WHO, WTrO), ASEAN, APEC.

Embassy: 1024 Wisconsin Ave. NW 20007; 944-3600.

Websites: http://mahidol.ac.th/Thailand/Thailand-main.html
http://necter.or.th/WWW-VL-Thailand.html

Thais began migrating from southern China during the 11th century. A unified Thai kingdom was established in 1350.

Thailand, known as Siam until 1939, is the only country in SE Asia never taken over by a European power, thanks to King Mongkut and his son King Chulalongkorn. Ruling successively from 1851 to 1910, they modernized the country and signed trade treaties with Britain and France. A bloodless revolution in 1932 limited the monarchy. Thailand was an ally of Japan during World War II and of the U.S. during the postwar period.

The military took over the government in a bloody 1976 coup. Kriangsak Chomanan, prime minister, resigned Feb. 1980 because of soaring inflation, oil price increases, labor unrest, and growing crime. Vietnamese troops crossed the border but were repulsed by Thai forces in the 1980s.

Chatichai Choonhavan was chosen prime minister in a democratic election, Aug. 1988. In Feb. 1991, the military ousted Choonhavan in a bloodless coup. A violent crackdown on street demonstrations in May 1992 led to more than 50 deaths. AIDS reached epidemic proportions in Thailand in the mid-1990s.

A steep downturn in the economy forced Thailand to seek more than $15 billion in emergency international loans in Aug. 1997. A new constitution won legislative approval Sept. 27. As the economic crisis deepened, Chuan Leekpai became prime minister Nov. 9, 1997, and implemented financial reforms.

Following elections in Jan. 2001, Thaksin Shinawatra, a wealthy former telecommunications executive, became prime minister. Thailand's Constitutional Court acquitted him Aug. 3 of corruption while he was deputy prime minister in 1997.

By the end of the 1990s, according to UN estimates, more than 750,000 people in Thailand had HIV/AIDS.

Togo

Togolese Republic

People: Population: 5,153,088. **Age distrib.** (%): <15: 45.6; 65+: 2.4. **Pop. density:** 246 per sq. mi. **Urban:** 33%. **Ethnic groups:** Ewe, Mina, Kabre, 37 other tribes. **Principal languages:** French (official), Ewe, Mina, Dagomba, Kabye. **Chief religions:** Indigenous beliefs 70%, Christian 20%, Muslim 10%.

Geography: Area: 21,000 sq. mi. **Location:** On S coast of W Africa. **Neighbors:** Ghana on W, Burkina Faso on N, Benin on E. **Topography:** A range of hills running SW-NE splits Togo into 2 savanna plains regions. **Capital:** Lomé (1999 met. est.): 790,000.

Government: Type: Republic. **Head of state:** Pres. Gnassingbé Eyadéma; b Dec. 26, 1937; in office: Apr. 14, 1967. **Head of gov.:** Prime Min. Agbeyome Messan Kodjo; b. Oct. 12, 1954; in office: Aug. 29, 2000. **Local divisions:** 5 regions. **Defense budget:** $31 mil. **Active troops:** 6,950.

Economy: Industries: Textiles, handicrafts, agric. processing. **Chief crops:** Coffee, cocoa, yams, cotton, millet, rice. **Minerals:** Phosphates, limestone, marble. **Arable land:** 38%. **Livestock** (2000): cattle: 215,000; chickens: 7.5 mil; goats: 1.11 mil; pigs: 850,000; sheep: 740,000.. **Fish catch:** (1999): 14,310 metric tons. **Electricity prod.:** 92 mil kWh. **Labor force:** 65% agric.; 30% services.

Finance: Monetary unit: CFA Franc (Sept. 2001: 724.11 = $1 U.S.). **GDP:** (1999 est.): $8.6 bil. **Per capita GDP:** $1,700. **Imports** (1999): $450 mil; partners: Ghana, France, Cote d'Ivoire. **Exports** (1999): $400 mil; partners: Canada, Phillipines, Ghana. **Budget** (1997 est.): $252 mil. **Intl. reserves less gold** (Apr. 2000): $142.0 mil. **Gold:** 13,000 oz t. **Consumer prices** (change in 1999): –0.1%.

Transport: Railroad: Length: 245 mi. **Motor vehicles:** 74,662 pass. cars, 34,605 comm. vehicles. **Civil aviation:** 150.5 mil pass.-mi; 2 airports. **Chief port:** Lomé.

Communications: TV sets: 36 per 1,000 pop. **Radios:** 212 per 1,000 pop. **Telephones**: 42,800 main lines.

Health: Life expectancy: 52.38 male; 56.38 female. **Births** (per 1,000 pop): 37.04. **Deaths** (per 1,000 pop.): 11.24. **Natural inc.:** 2.58%. **Infant mortality** (per 1,000 live births): 70.43.

Education: Compulsory: ages 6-12. **Literacy**: 52%.

Major Intl. Organizations: UN (FAO, IBRD, ILO, IMF, IMO, WHO, WTrO), OAU.

Embassy: 2208 Massachusetts Ave. NW 20008; 234-4212.

The Ewe arrived in southern Togo several centuries ago. The country later became a major source of slaves. Germany took control in 1884. France and Britain administered Togoland as UN trusteeships. The French sector became the republic of Togo Apr. 27, 1960.

The population is divided between Bantus in the S and Hamitic tribes in the N. Togo has actively promoted regional integration, as a means of stimulating the economy.

In Jan. 1993 police fired on antigovernment demonstrators, killing at least 22. Some 25,000 people fled to Ghana and Benin as a result of civil unrest. In Jan. 1994 at least 40 people were killed when gunmen reportedly attacked an army base. Further violence marred Togo's 1st multiparty legislative elections, held Feb. 1994. In office since 1967, Pres. Gnassingbé Eyadéma was reelected June 21, 1998, in a vote that was disputed as in previous elections.

Tonga

Kingdom of Tonga

People: Population: 104,227. **Age distrib. (%): <15: 40.9; 65+: 4.1. Pop. density:** 361 per sq. mi. **Urban:** 37%. **Ethnic groups:** Polynesian. **Principal languages:** Tongan, English (both official). **Chief religions:** Free Wesleyan 41%, Roman Catholic 16%, Mormon 14%.

Geography: Area: 289 sq. mi. **Location:** In western South Pacific O. **Neighbors:** Nearest are Fiji to W, Samoa to NE. **Topography:** Tonga comprises 170 volcanic and coral islands, 36 inhabited. **Capital:** Nuku'alofa (1999 est.): 37,000.

Government: Type: Constitutional monarchy. **Head of state:** King Taufa'ahau Tupou IV; b July 4, 1918; in office: Dec. 16, 1965. **Head of gov.:** Prime Min. Prince Ulukalala Lavaka Ata; b July 12, 1959; in office: Jan. 3, 2000. **Local divisions:** 5 divisions, 23 districts.

Economy: Industries: Tourism, fishing. **Chief crops:** Coconuts, copra, bananas, vanilla beans. **Arable land:** 24%. **Livestock** (2000): cattle: 9,318; chickens: 266,000; goats: 13,939; pigs: 80,853.. **Electricity prod.:** 35 mil kWh. **Labor force:** 65% agric.

Finance: Monetary unit: Pa'anga (Sept. 2001: 2.18 = $1 U.S.). **GDP** (1998 est.): $238 mil. **Per capita GDP:** $2,200. **Imports** (1998): $69 mil; partners: N.Z. 30%, Australia 19%. **Exports** (1998): $8 mil; partners: Japan 53%, U.S. 18%, N.Z. 6%. **Tourism** (1998): $15 mil. **Budget** (FY 1996-97 est.): $120 mil. **Intl. reserves less gold** (May 2000): $22.32 mil. **Consumer prices** (change in 1999): 4.5%.

Transport: Motor vehicles: 3,400 pass. cars, 3,900 comm. vehicles. **Civil aviation:** 6.4 mil pass.-mi; 6 airports. **Chief port:** Nuku'alofa.

Communications: TV sets: 20 per 1,000 pop. **Radios:** 397 per 1,000 pop. **Telephones** (1999): 9,100 main lines. **Daily newspaper circ.:** 70 per 1,000 pop.

Health: Life expectancy: 65.83 male; 70.78 female. **Births** (per 1,000 pop): 23.59. **Deaths** (per 1,000 pop.): 5.74. **Natural inc.:** 1.79%. **Infant mortality** (per 1,000 live births): 14.08.

Education: Free, compulsory: ages 5-14. **Literacy** (1992): 93%.

Major Intl. Organizations: UN (FAO, IBRD, IMF, WHO), the Commonwealth.

The islands were first visited by the Dutch in the early 17th century. A series of civil wars ended in 1845 with establishment of the Tupou dynasty. In 1900 Tonga became a British protectorate. On June 4, 1970, Tonga became independent and a member of the Commonwealth. It joined the UN on Sept. 14, 1999.

Trinidad and Tobago

Republic of Trinidad and Tobago

People: Population: 1,169,682. **Age distrib.** (%): <15: 24.1; 65+: 6.7. **Pop. density:** 591 per sq. mi. **Urban:** 74%. **Ethnic groups:** Black 40%, East Indian 40%, mixed 14%. **Principal languages:** English (official), Hindi, French, Spanish. **Chief religions:** Roman Catholic 32%, Protestant 14%, Hindu 24%.

Geography: Area: 2,000 sq. mi. **Location:** In Caribbean, off E coast of Venezuela. **Neighbors:** Nearest is Venezuela to SW. **Topography:** Three low mountain ranges cross Trinidad E-W, with a well-watered plain between N and central ranges. Parts of E and W coasts are swamps. Tobago, 116 sq. mi., lies 20 mi. NE. **Capital:** Port-of-Spain (1999 est.): 53,000.

Government: Type: Parliamentary democracy. **Head of state:** Pres. Arthur N. R. Robinson; b Dec. 16, 1926; in office: Mar. 19, 1997. **Head of gov.:** Prime Min. Basdeo Panday; b May 25, 1933; in office: Nov. 9, 1995. **Local divisions:** 8 counties, 3 municipalities, 1 ward. **Defense budget** (1999): $62 mil. **Active troops:** 2,700.

Economy: Industries: Oil products, chemicals, tourism. **Chief crops:** Sugar, cocoa, coffee, citrus, rice. **Minerals:** Asphalt, oil, gas. **Crude oil reserves** (2000): 605 mil bbls. **Arable land:** 15%. **Livestock** (2000): cattle: 35,000; chickens: 10.0 mil; goats: 59,000; pigs: 41,000; sheep: 12,000. **Fish catch:** (1999): 15,012 metric tons. **Electricity prod.:** 4.900 bil kWh. **Labor force:** 64.1% services; 14% manuf.; 12.4% const.; 9.5% agric.

Finance: Monetary unit: Dollar (Sept. 2001: 6.06 = $1 U.S.). **GDP:** (1999 est.): $9.41 bil. **Per capita GDP:** $8,500. **Imports** (1998): $3 bil; partners: U.S. 44.7%; Latin America 18.9%; EU 13.7%. **Exports** (1998): $2.4 bil; partners: U.S. 36.9%, Caricom countries 29.4%. **Tourism** (1998): $201 mil. **Budget** (1997 est.): $1.54 bil. **Intl. reserves less gold** (Mar. 2000): $889.8 mil. **Gold:** 58,000 o zt. **Consumer prices** (change in 1999): 3.4%.

Transport: Motor vehicles: 128,000 pass. cars, 27,000 comm. vehicles. **Civil aviation:** 1.5 bil pass.-mi; 2 airports. **Chief ports:** Port-of-Spain, Scarborough.

Communications: TV sets: 198 per 1,000 pop. **Radios:** 433 per 1,000 pop. **Telephones:** 299,100 main lines. **Daily newspaper circ.:** 139 per 1,000 pop.

Health: Life expectancy: 65.74 male; 70.92 female. **Births** (per 1,000 pop): 13.73. **Deaths** (per 1,000 pop.): 8.82. **Natural inc.:** .49%. **Infant mortality** (per 1,000 live births): 24.98.

Education: Free, compulsory: ages 5-12. **Literacy**: 98%.

Major Intl. Organizations: UN (FAO, IBRD, ILO, IMF, IMO, WHO, WTrO), Caricom, the Commonwealth, OAS.

Embassy: 1708 Massachusetts Ave. NW 20036; 467-6490.

Columbus sighted Trinidad in 1498. A British possession since 1802, Trinidad and Tobago won independence Aug. 31, 1962. It became a republic in 1976.

The nation is one of the most prosperous in the Caribbean. Oil production has increased with offshore finds. Middle Eastern oil is refined and exported, mostly to the U.S.

In July 1990, some 120 Muslim extremists captured the Parliament building and TV station and took about 50 hostages, including Prime Min. Arthur N. R. Robinson, who was beaten, shot in the legs, and tied to explosives. After a 6-day siege, the rebels surrendered.

Basdeo Panday, the country's first prime minister of East Indian ancestry, took office Nov. 9, 1995. Robinson became president on Mar. 19, 1997.

Tunisia

Republic of Tunisia

People: Population: 9,705,102. **Age distrib.** (%) <15: 28.7; 65+: 6.1. **Pop. density:** 162 per sq. mi. **Urban:** 65%. **Ethnic groups:** Arab 98%. **Principal languages:** Arabic (official), French. **Chief religion:** Muslim 98%.

Geography: Area: 59,900 sq. mi. **Location:** On N coast of Africa. **Neighbors:** Algeria on W, Libya on E. **Topography:** The N is wooded and fertile. The central coastal plains are given to grazing and orchards. The S is arid, approaching Sahara Desert. **Capital:** Tunis (urban agg.): 1,897,000.

Government: Type: Republic. **Head of state:** Pres. Gen. Zine al-Abidine Ben Ali; b Sept. 3, 1936; in office: Nov. 7, 1987. **Head of gov.:** Prime Min. Mohamed Ghannouchi; b Aug. 18, 1941; in office: Nov. 17, 1999. **Local divisions:** 23 governorates. **Defense budget:** $365 mil. **Active troops:** 35,000.

Economy: Industries: Food processing, textiles, oil products, mining, tourism. **Chief crops:** Grains, dates, olives, sugar beets, grapes. **Minerals:** Phosphates, iron, oil, lead, zinc. **Crude oil reserves** (2000): 307.56 mil bbls. **Arable land:** 19%. **Livestock** (2000): cattle: 790,000; chickens: 37.0 mil; goats: 1.4 mil; pigs: 6,000; sheep: 6.6 mil. **Fish catch** (1999): 89,027 metric tons. **Electricity prod.:** 9.173 bil kWh. **Labor force:** 55% services; 23% industry; 22% agric.

Finance: Monetary unit: Dinar (Sept. 2001: 1.44 = $1 U.S.). **GDP:** (1999 est.): $52.6 bil. **Per capita GDP:** $5,500. **Imports** (1999 est.): $8.3 bil; partners: France 27%; Italy 20%; Germany 12%. **Exports** (1999 est.): $5.8 bil; partners: France 27%; Italy 22%; Germany 15%. **Tourism:** $1.61 bil. **Budget** (1998 est.): $6.5 bil. **Intl. reserves less gold** (May 2000): $1.50 bil. **Gold:** 218,000 oz t. **Consumer prices** (change in 1999): 2.7%.

Transport: Railroad: Length: 1,337 mi. **Motor vehicles:** 248,000 pass. cars, 283,000 comm. vehicles. **Civil aviation:** 1.5 bil pass.-mi; 5 airports. **Chief ports:** Tunis, Sfax, Bizerte.

Communications: TV sets: 156 per 1,000 pop. **Radios:** 188 per 1,000 pop. **Telephones** (1999): 850,400 main lines. **Daily newspaper circ.:** 45 per 1,000 pop.

Health: Life expectancy: 72.35 male; 75.62 female. **Births** (per 1,000 pop): 17.11. **Deaths** (per 1,000 pop.): 4.99. **Natural inc.:** 1.21%. **Infant mortality** (per 1,000 live births): 29.04.

Education: Compulsory: ages 6-16. **Literacy:** 67%.

Major Intl. Organizations: UN (FAO, IBRD, ILO, IMF, IMO, WHO, WTrO), AL, OAU.

Embassy: 1515 Massachusetts Ave. NW 20005; 862-1850.

Website: http://www.tunisiaonline.com

Site of ancient Carthage and a former Barbary state under the suzerainty of Turkey, Tunisia became a protectorate of France under a treaty signed May 12, 1881. The nation became independent Mar. 20, 1956, and ended the monarchy the following year. Habib Bourguiba, an independence leader, served as president until 1987, when he was deposed by his prime minister, Zine al-Abidine Ben Ali.

Tunisia has actively repressed Islamic fundamentalism.

Turkey

Republic of Turkey

People: Population: 66,493,970. **Age distrib.** (%): <15: 28.4; 65+: 6.1. **Pop. density:** 224 per sq. mi. **Urban:** 74%. **Ethnic groups:** Turk 80%, Kurd 20%. **Principal languages:** Turkish (official), Kurdish, Arabic. **Chief religion:** Muslim 99.8%.

Geography: Area: 297,200 sq. mi. **Location:** Occupies Asia Minor, stretches into continental Europe; borders on Mediterranean and Black seas. **Neighbors:** Bulgaria, Greece on W; Georgia, Armenia on N; Iran on E; Iraq, Syria on S. **Topography:** Central Turkey has wide plateaus, with hot, dry summers and cold winters. High mountains ring the interior on all but W, with more than 20 peaks over 10,000 ft. Rolling plains are in W; mild, fertile coastal plains are in S, W. **Capital:** Ankara. **Cities (urban agg.)**: Istanbul 9,451,000; Ankara 3,203,000; Izmir 2,409,000.

Government: Type: Republic. **Head of state:** Pres. Ahmet Necdet Sezer; b Sept. 13, 1941; in office: May 16, 2000. **Head of gov.:** Prime Min. Bülent Ecevit; b 1925; in office: Jan. 11, 1999. **Local divisions:** 80 provinces. **Defense budget:** $7.7 bil. **Active troops:** 609.700.

Economy: Industries: Textiles, steel, mining, processed foods. **Chief crops:** Tobacco, grains, cotton, pulses, citrus, olives, sugar beets. **Minerals:** Antimony, chromium, mercury, copper, coal. **Crude oil reserves** (2000): 298.7 mil bbls. **Ara-**

ble land: 32%. **Livestock** (2000): cattle: 11.03 mil; chickens: 237.0 mil; goats: 8.06 mil; pigs: 5,000; sheep: 29.44 mil. **Fish catch** (1999): 500,260 metric tons. **Electricity prod.:** 111.502 bil kWh. **Labor force:** 45.8% agric.; 33.7% serv.; 20.5% ind.

Finance: Monetary unit: Lira (Sept. 2001: 1,407,500.00 = $1 U.S.). **GDP:** (1999 est.): $409.4 bil. **Per capita GDP:** $6,200. **Imports** (1999): $40 bil; partners: Germany 14%, Italy 8%, U.S. 8%. **Exports** (1999): $26 bil; partners: Germany 21%, U.S 9%, U.K. 7%. **Tourism:** $5.20 bil. **Budget** (1998): $58.5 bil. **Intl. reserves less gold** (June 2000): $24.74 bil. **Gold:** 3.74 mil oz t. **Consumer prices** (change in 1999): 64.9%.

Transport: Railroad: Length: 5,348 mi. **Motor vehicles:** 3.27 mil pass. cars, 1.05 mil comm. vehicles. **Civil aviation:** 7.7 bil pass.-mi; 26 airports. **Chief ports:** Istanbul, Izmir, Mersin.

Communications: TV sets: 288 per 1,000 pop. **Radios:** 181 per 1,000 pop. **Telephones:** 18,395,200 main lines. **Daily newspaper circ:** 111 per 1,000 pop.

Health: Life expectancy: 68.89 male; 73.71 female. **Births** (per 1,000 pop): 18.31. **Deaths** (per 1,000 pop.): 5.95. **Natural inc.:** 1.24%. **Infant mortality** (per 1,000 live births): 47.34.

Education: Free, compulsory: ages 6-14. **Literacy:** 82%.

Major Intl. Organizations: UN (FAO, IBRD, ILO, IMF, IMO, WHO, WTrO), NATO, OECD, OSCE.

Embassy: 1714 Massachusetts Ave. NW 20036; 659-8200.

Website: http://www.turkey.org

Ancient inhabitants of Turkey were among the world's first agriculturalists. Such civilizations as the Hittite, Phrygian, and Lydian flourished in Asiatic Turkey (Asia Minor), as did much of Greek civilization. After the fall of Rome in the 5th century, Constantinople (now Istanbul) was the capital of the Byzantine Empire for 1,000 years. It fell in 1453 to Ottoman Turks, who ruled a vast empire for over 400 years.

Just before World War I, Turkey, or the Ottoman Empire, ruled what is now Syria, Lebanon, Iraq, Jordan, Israel, Saudi Arabia, Yemen, and islands in the Aegean Sea.

Turkey joined Germany and Austria in World War I, and its defeat resulted in the loss of much territory and the fall of the sultanate. A republic was declared Oct. 29, 1923, with Mustafa Kemal (later Kemal Ataturk) as its first president. Ataturk led Turkey until his death in 1938. The Caliphate (spiritual leadership of Islam) was renounced in 1924.

Long embroiled with Greece over Cyprus, off Turkey's south coast, Turkey invaded the island July 20, 1974, after Greek officers seized the Cypriot government as a step toward unification with Greece. Turkey sought a new government for Cyprus, with Greek Cypriot and Turkish Cypriot zones. In reaction to Turkey's moves, the U.S. cut off military aid in 1975. Turkey, in turn, suspended the use of most U.S. bases. Aid was restored in 1978. There was a military takeover, Sept. 12, 1980.

Religious and ethnic tensions and active left and right extremists have caused endemic violence. The military formally transferred power to an elected Parliament in 1983. Martial law, imposed in 1978, was lifted in 1984.

Turkey was a member of the Allied forces that ousted Iraq from Kuwait, 1991. In the aftermath of the war, millions of Kurdish refugees fled to Turkey's border to escape Iraqi forces. The Turkish government mounted sporadic offensives against separatist Kurds in this border area and in N Iraq, causing heavy casualties among guerrillas and civilians.

Kurdish militants, demanding an independent state for the Kurds, raided Turkish diplomatic missions in some 25 Western European cities June 24, 1993. Tansu Ciller officially became Turkey's first woman prime minister July 5, 1993. The Welfare Party, an Islamic group, gained strength in the 1990s but was unable to form a government until June 1996, when it came to power in coalition with Ciller's True Path Party.

The pro-Islamic government resigned June 18, 1997, under pressure from the military. The European Union rebuffed Turkey's membership bid Dec. 12, 1997. The military stepped up its campaign against Islamic fundamentalism in 1998.

Kurdish rebel leader Abdullah Öcalan was captured Feb. 15, 1999; convicted of terrorism June 29, he was sentenced to death by a Turkish security court. His organization, the Kurdistan Workers' Party, announced Aug. 5 that it would abandon its 14-year-old armed insurgency. A major earthquake Aug. 17 in NW Turkey killed over 17,000 people and injured thousands more. Another quake in the same region Nov. 12 claimed at least 675 lives.

The IMF announced $7.5 billion in emergency loans Dec. 6, 2000, to help Turkey cope with a severe financial crisis.

Turkmenistan

People: Population: 4,603,244. **Age distrib.** (%): <15: 37.9; 65+: 4.0. **Pop. density:** 24 per sq. mi. **Urban:** 45%. **Ethnic groups:** Turkmen 77%, Uzbek 9%, Russian 7%. **Principal languages:** Turkmen (official), Russian, Uzbek. **Chief religions:** Muslim 89%, Eastern Orthodox 9%.

Geography: Area: 188,200 sq. mi. **Neighbors:** Kazakhstan on N, Uzbekistan on N and E, Afghanistan and Iran on S. **Topography:** The Kara Kum Desert occupies 80% of the area. Bordered on W by Caspian Sea. **Capital:** Ashgabat (1999 est.): 525,000.

Government: Type: Republic. **Head of state and gov.:** Pres. Saparmurad Niyazov; b Feb. 18, 1940; in office: Oct. 27, 1990. **Local divisions:** 5 regions. **Defense budget:** $157 mil. **Active troops:** 17,500.

Economy: Industries: Oil, natural gas, food processing, textiles. **Chief crops:** Grain, cotton. **Minerals:** Coal, sulfur, oils, gas, salt. **Crude oil reserves** (2000): 546 mil bbls. **Arable land:** 3%. **Livestock** (2000): cattle: 850,000; chickens: 4.35 mil; goats: 368,000; pigs: 46,000; sheep: 5.6 mil. **Electricity prod.:** 8.371 bil kWh. **Labor force:** 44% agric. & forestry; 19% ind. & constr.

Finance: Monetary unit: Manat (Sept. 2001: 33.79 = $1 U.S.). **GDP:** (1999 est.): $7.7 bil. **Per capita GDP:** $1,800. **Imports** (1999): $1.25 bil; partners: Ukraine, Turkey, Russia. **Exports** (1999): $1.1 bil; partners: Iran, Turkey, Russia. **Tourism** (1998): $192 mil. **Budget** (1996 est.): $548 mil.

Transport: Railroad: Length: 1,317 mi. **Civil aviation:** 679.2 mil pass.-mi; 1 airport. **Chief port:** Turkmenbashi.

Communications: TV sets: 189 per 1,000 pop. **Radios:** 189 per 1,000 pop. **Telephones** (1999): 358,900 main lines.

Health: Life expectancy: 57.43 male; 64.76 female. **Births** (per 1,000 pop): 28.55. **Deaths** (per 1,000 pop.): 8.98. **Natural inc.:** 1.96%. **Infant mortality** (per 1,000 live births): 73.25.

Education: Literacy: 100%.

Major Intl. Organizations: UN (FAO, IBRD, ILO, IMF, IMO, WHO), CIS, OSCE.

Embassy: 2207 Massachusetts Ave., NW 20008; 588-1500.

Website: http://www.turkmenistan.com

The region has been inhabited by Turkic tribes since the 10th century. It became part of Russian Turkestan in 1881, and a constituent republic of the USSR in 1925. Turkmenistan declared independence Oct. 27, 1991, and became an independent state when the USSR disbanded Dec. 26, 1991.

Extensive oil and gas reserves place Turkmenistan in a more favorable economic position than other former Soviet republics. A new rail line linking Iran and Turkmenistan was inaugurated May 13, 1996. Political power centered around the former Communist Party apparatus, and Pres. Saparmurad Niyazov became the object of a personality cult.

Tuvalu

People: Population: 10,991. **Age distrib.** (%): <15: 33.3; 65+: 5.1. **Pop. density:** 1,099 per sq. mi. **Urban:** 51%. **Ethnic group:** Polynesian 96%. **Principal languages:** Tuvaluan, English. **Chief religion:** Church of Tuvalu (Congregationalist) 97%.

Geography: Area: 10 sq. mi. **Location:** 9 islands forming a NW-SE chain 360 mi. long in the SW Pacific O. **Neighbors:** Nearest are Kiribati to N, Fiji to S. **Topography:** The islands are all low-lying atolls, nowhere rising more than 15 ft. above sea level, composed of coral reefs. **Capital:** Funafuti Atoll (1995 est.): 4,000.

Government: Head of state: Queen Elizabeth II, represented by Gov.-Gen. Tomasi Puapua; b 1938; in office: June 26, 1998. **Head of gov.:** Prime Min. Faimalaga Luka; in office: Feb. 24, 2001.

Economy: Industries: Copra, fishing, tourism. **Chief crops:** Coconuts. **Livestock:** (2000): chickens: 27,000; pigs: 12,600.

Finance: Monetary unit: Australian Dollar (Oct. 2000: 1.88 = $1 U.S.). **GDP:** (1995 est.): $7.8 mil. **Per capita GDP:** $800.

Transport: Civil aviation: 1 airport. **Chief port:** Funafuti.

Communications: Radios: 320 per 1,000 pop.

Health: Life expectancy: 64.52 male; 68.88 female. **Births** (per 1,000 pop): 21.56. **Deaths** (per 1,000 pop.): 7.55. **Natural inc.:** 1.4%. **Infant mortality** (per 1,000 live births): 22.65.

Education: Compulsory: ages 7-15. **Literacy** (1990): 95%.

Major Intl. Organizations: UN, WHO, the Commonwealth.

Website: http://www.emulateme.com/tuvalu.htm

The Ellice Islands separated from the British Gilbert and Ellice Islands Colony in 1975 and became Tuvalu; independence came Oct. 1, 1978. In 2000, Tuvalu joined the United Nations.

Uganda

Republic of Uganda

People: Population: 23,985,712. **Age distrib.** (%): <15: 51.1; 65+: 2.1. **Pop. density:** 311 per sq. mi. **Urban:** 14%. **Ethnic groups:** Baganda 17%, Karamojong 12%, many others. **Principal languages:** English (official), Luganda, Swahili. **Chief religions:** Protestant 33%, Roman Catholic 33%, indigenous beliefs 18%, Muslim 16%.

Geography: Area: 77,000 sq. mi. **Location:** In E Central Africa. **Neighbors:** Sudan on N, Congo (formerly Zaire) on W, Rwanda and Tanzania on S, Kenya on E. **Topography:** Most of Uganda is a high plateau 3,000-6,000 ft. high, with high Ruwenzori range in W (Mt. Margherita 16,750 ft.), volcanoes in SW; NE is arid, W and SW rainy. Lakes Victoria, Edward, Albert form much of borders. **Capital:** Kampala (urban agg.): 1,212,000.

Government: Type: Republic. **Head of state:** Pres. Yoweri Kaguta Museveni; b Mar. 1944; in office: Jan. 29, 1986. **Head of gov.:** Prime Min. Apollo Nsibambi; b Nov. 27, 1938; in office: Apr. 5, 1999. **Local divisions:** 39 districts. **Defense budget:** $132 mil. **Active troops:** 50,000–60,000.

Economy: Industries: Brewing, textiles, cement. **Chief crops:** Coffee, cotton, tea, corn, tobacco. **Minerals:** Copper, cobalt. **Arable land:** 25%. **Livestock** (2000): cattle: 5.97 mil; chickens: 25.0 mil; goats: 3.7 mil; pigs: 970,000; sheep: 1.98 mil. **Fish catch** (1999): 218,236 metric tons. **Electricity prod.:** 1.326 bil kWh. **Labor force:** 82% agriculture.

Finance: Monetary unit: Shilling (Sept. 2001: 1,770.00 = $1 U.S.). **GDP** (1999 est.): $24.2 bil. **Per capita GDP:** $1,060. **Imports** (1999): $1.1 bil; partners: Kenya 12%, UK 6%. **Exports** (1999): $471 mil; partners: EU 51%. **Tourism:** $142 mil. **Budget** (1995-96 est.): $985 mil. **Intl. reserves less gold** (Mar. 2000): $721.2 mil. **Consumer prices** (change in 1999): 6.4%.

Transport: Railroad: Length: 771 mi. **Motor vehicles:** 24,400 pass. cars, 26,600 comm. vehicles. **Civil aviation:** 68.4 mil pass.-mi; 1 airport. **Chief ports:** Entebbe, Jinja.

Communications: TV sets: 27 per 1,000 pop. **Radios:** 485 per 1,000 pop. **Telephones** (1999): 57,200 main lines.

Health: Life expectancy: 42.59 male; 44.17 female. **Births** (per 1,000 pop): 47.52. **Deaths** (per 1,000 pop.): 17.97. **Natural inc.:** 2.96%. **Infant mortality** (per 1,000 live births): 91.3.

Education: Literacy: 62%.

Major Intl. Organizations: UN (FAO, IBRD, ILO, IMF, WHO, WTrO), the Commonwealth, OAU.

Embassy: 5911 16th St. NW 20011; 726-7100.

Website: http://www.government.go.ug

Britain obtained a protectorate over Uganda in 1894. The country became independent Oct. 9, 1962, and a republic within the Commonwealth a year later. In 1967, the traditional kingdoms, including the powerful Buganda state, were abolished and the central government strengthened.

Gen. Idi Amin seized power from Prime Min. Milton Obote in 1971. During his eight years of dictatorial rule, he was responsible for the deaths of up to 300,000 of his opponents. In 1972 he expelled nearly all of Uganda's 45,000 Asians. Amin was named president for life in 1976. Tanzanian troops and Ugandan exiles and rebels ousted Amin, Apr. 11, 1979.

Obote held the presidency from Dec. 1980 until his ouster in a military coup July 27, 1985. Guerrilla war and rampant human rights abuses plagued Uganda under Obote's regime.

Conditions improved after Yoweri Museveni took power in Jan. 1986. In 1993 the government authorized restoration of the Buganda and other monarchies, but only for ceremonial purposes. Under a constitution ratified Oct. 1995, nonparty presidential and legislative elections were held in 1996. Uganda helped Laurent Kabila seize power in the Congo (formerly Zaire) in 1997 but sent troops in 1998 to aid insurgents seeking his ouster. Museveni faced several regional insurgencies in the late 1990s.

At least 330 members of the Movement for the Restoration of the Ten Commandments of God died in a church fire in Kanungu, Mar. 17, 2000; officials later estimated a total of more than 900 recent deaths could be associated with the cult. An ebola virus outbreak Oct.-Dec. 2000 killed more than 150 people. Pres. Museveni won reelection Mar. 12, 2001.

Ukraine

People: Population: 48,760,474. **Age distrib.** (%): <15: 17.3; 65+: 14.1. **Pop. density:** 209 per sq. mi. **Urban:** 68%. **Ethnic groups:** Ukrainian 73%, Russian 22%. **Principal languages:** Ukrainian, Russian. **Chief religions:** Ukrainian Orthodox, Ukrainian Catholic.

Geography: Area: 232,800 sq. mi. **Location:** In E Europe. **Neighbors:** Belarus on N; Russia on NE and E; Moldova and Romania on SW; Hungary, Slovakia, and Poland on W. **Topography:** Part of the E European plain. Mountainous areas include the Carpathians in the SW and Crimean chain in the S. Arable black soil constitutes a large part of the country. **Capital:** Kiev. **Cities** (urban agg.): Kiev (Kyiv) 2,670,000; Kharkov 1,526,000; Dnipropetrovsk 1,129,000.

Government: Type: Constitutional republic. **Head of state:** Pres. Leonid Danylovich Kuchma; b Aug. 9, 1938; in office: July 19, 1994. **Head of gov.:** Prime Min. Anatoly Kinakh; b Aug. 4, 1954; in office: May 29, 2001. **Local divisions:** 24 oblasts, 2 municipalities, 1 autonomous republic. **Defense budget:** $441 mil. **Active troops:** 303,800.

Economy: Industries: Chemicals, machinery, food processing. **Chief crops:** Grains, sugar beets, vegetables. **Minerals:** Iron, manganese, coal, gas, oil, sulfur, salt. **Other resources:** Forests. **Crude oil reserves** (2000): 395 mil bbls. **Arable land:** 58%. **Livestock** (2000): cattle: 10.63 mil; chickens: 93.5 mil; goats: 825,200; pigs: 10.07 mil; sheep: 1.06 mil. **Fish catch** (1999): 403,005 metric tons. **Electricity prod.:** 157.823 bil kWh. **Labor force:** 32% ind. & constr.; 24% agric. & forestry; 17% health, edu., culture.

Finance: Monetary unit: Hryvnya (Sept. 2001: 5.34 = $1 U.S.). **GDP:** (1999 est.): $109.5 bil. **Per capita GDP:** $2,200. **Imports** (1999 est.): $11.8 bil; partners: Russia 48%, EU 23%. **Exports** (1999 est.): $11.6 bil; partners: Russia 20%, EU 17%. **Tourism:** $541 mil. **Budget** (1997 est.): $21 bil. **Intl. reserves less gold** (June 2000): $811.7 mil. **Gold:** 437,800 oz t. **Consumer prices** (change in 1997): 15.9%.

Transport: Railroad: Length: 14,100 mi. **Motor vehicles:** 4.5 mil pass. cars. **Civil aviation:** 1.2 bil pass.-mi; 12 airports. **Chief ports:** Odesa, Kiev, Berdiansk.

Communications: TV sets: 233 per 1,000 pop. **Radios:** 346 per 1,000 pop. **Telephones** (1999): 10,074,000 main lines. **Daily newspaper circ.:** 118 per 1,000 pop.

Health: Life expectancy: 60.62 male; 71.96 female. **Births** (per 1,000 pop): 9.31. **Deaths** (per 1,000 pop.): 16.43. **Natural inc.:** -.71%. **Infant mortality** (per 1,000 live births): 21.4.

Education: Compulsory: ages 7-15. **Literacy**: 99%.

Major Intl. Organizations: UN (IBRD, ILO, IMF, IMO, WHO), CIS, OSCE.

Embassy: 3350 M St. NW 20007; 333-0606.

Website: http://www.ukremb.com

Trypilians flourished along the Dnieper River, Ukraine's main artery, from 6000-1000 BC. Ukrainians' Slavic ancestors inhabited modern Ukrainian territory well before the first century AD.

In the 9th century, the princes of Kiev established a strong state called Kievan Rus, which included much of present-day Ukraine. A strong dynasty was established, with ties to virtually all major European royal families. St. Vladimir the Great, ruler of Kievan Rus, accepted Christianity as the national faith in 988. At the crossroads of European trade routes, Kievan Rus reached its zenith under Yaroslav the Wise (1019-1054). Internal conflicts led to the disintegration of the Ukrainian state by the 13th century. Mongol rule was supplanted by Poland and Lithuania in the 14th and 15th centuries. The N Black Sea coast and Crimea came under the control of the Turks in 1478.

Ukrainian Cossacks, starting in the late 16th century, waged numerous wars of liberation against the occupiers of Ukraine: Russia, Poland, and Turkey. By the late 18th century, Ukrainian independence was lost. Ukraine's neighbors once again divided its territory. At the turn of the 19th century, Ukraine was occupied by Russia and Austria-Hungary.

An independent Ukrainian National Republic was proclaimed on January 22, 1918. In 1921, Ukraine's neighbors occupied and divided Ukrainian territory. In 1922, Ukraine became a constituent republic of the USSR as the Ukrainian SSR. In 1932-33, the Soviet government engineered a man-made famine in eastern Ukraine, resulting in the deaths of 7-10 million Ukrainians.

In March 1939, independent Carpatho-Ukraine was the first European state to wage war against Nazi-led aggression in the region. During World War II the Ukrainian nationalist underground and its Ukrainian Insurgent Army (UPA) fought both Nazi German and Soviet forces. The restoration of Ukrainian independence was declared on June 30, 1941. Over 5 million Ukrainians lost their lives during the war. With the reoccupation of Ukraine by Soviet troops in 1944 came a renewed wave of mass arrests, executions, and deportations of Ukrainians.

The world's worst nuclear power plant disaster occurred in Chernobyl, Ukraine, in April 1986; many thousands were killed or disabled as a result of the radiation leak.The plant was finally shut down Dec. 15, 2000.

Ukrainian independence was restored in Dec. 1991 with the dissolution of the Soviet Union. In the post-Soviet period Ukraine was burdened with a deteriorating economy.

Following a 1994 accord with Russia and the U.S., Ukraine's large nuclear arsenal was transferred to Russia for destruction. A new constitution legalizing private property and establishing Ukrainian as the sole official language was approved by parliament June 29, 1996. In May 1997, Russia and Ukraine resolved disputes over the Black Sea fleet and the future of Sevastopol and signed a long-delayed treaty of friendship. President since 1994, Leonid Kuchma won a 2d 5-year term in a runoff vote Nov. 14, 1999; a referendum expanding his powers passed Apr. 16, 2000.

United Arab Emirates

People: Population: 2,407,460. **Age distrib.** (%): <15: 28.9; 65+: 2.4. **Pop. density:** 75 per sq. mi. **Urban:** 86%. **Ethnic groups:** Arab, Iranian, Pakistani, Indian. **Principal languages:** Arabic (official), Persian, English, Hindi, Urdu. **Chief religions:** Muslim 96%, Christian, Hindu.

Geography: Area: 32,000 sq. mi. **Location:** Middle East, on the S shore of the Persian Gulf. **Neighbors:** Saudi Arabia on W and S, Oman on E. **Topography:** A barren, flat coastal plain gives way to uninhabited sand dunes on the S. Hajar Mts. are on E. **Capital:** Abu Dhabi (urban agg.): 927,000.

Government: Type: Federation of emirates. **Head of state:** Pres. Zaid ibn Sultan an-Nahayan; b 1918; in office: Dec. 2, 1971. **Head of gov.:** Prime Min. Sheik Maktum ibn Rashid al-Maktum; b 1946; in office: Nov. 20, 1990. **Local divisions:** 7 autonomous emirates: Abu Dhabi, Ajman, Dubai, Fujaira, Ras al-Khaimah, Sharjah, Umm al-Qaiwain. **Defense budget:**$3.9 bil. **Active troops:** 65,000.

Economy: Industries: Oil, fishing, petrochemicals. **Chief crops:** Vegetables, dates. **Minerals:** Oil, natural gas. **Crude oil reserves** (2000): 92.2 bil bbls. **Livestock** (2000): cattle: 110,000; chickens: 14.65 mil; goats: 1.2 mil; sheep: 467,281. **Fish catch:** (1999): 114,358 metric tons. **Electricity prod.:** 36.700 bil kWh. **Labor force:** 60% services; 32% ind. and commerce; 8% agric.

Finance: Monetary unit: Dirham (Sept. 2001: 3.67 = $1 U.S.). **GDP:** (1999 est.): $41.5 bil. **Per capita GDP:** $17,700. **Imports** (1999 est.): $27.5 bil; partners: US 10% Japan 9% UK 9%. **Exports** (1999 est.): $34 bil; partners: Japan 30%, South Korea 10%, India 6%. **Budget** (1998 est.): $5.8 bil. **Intl. reserves less gold** (Dec. 1999): $10.68 bil. **Gold:** 397,000 oz t.

Transport: Motor vehicles: 320,000 pass. cars, 80,000 comm. vehicles. **Civil aviation:** 8.4 bil pass.-mi; 6 airports. **Chief ports:** Ajman, Das Island.

Communications: TV sets: 260 per 1,000 pop. **Radios:** 206 per 1,000 pop. **Telephones:** 1,020,100 main lines. **Daily newspaper circ.:** 170 per 1,000 pop.

Health: Life expectancy: 71.84 male; 76.86 female. **Births** (per 1,000 pop): 18.11. **Deaths** (per 1,000 pop.): 3.79. **Natural inc.:** 1.43%. **Infant mortality** (per 1,000 live births): 16.68.

Education: Compulsory: ages 6-12. **Literacy**: 79%.

Major Intl. Organizations: UN (FAO, IBRD, ILO, IMF, IMO, WHO, WTrO), AL, OPEC.

Embassy: Suite 700, 1255 22nd Street NW, 20037, 955-7999.

Websites: http://www.uae.org.ae
http://www.emirates.org

The 7 "Trucial Sheikdoms" gave Britain control of defense and foreign relations in the 19th century. They merged to become an independent state Dec. 2, 1971.

The Abu Dhabi Petroleum Co. was fully nationalized in 1975. Oil revenues have given the UAE one of the highest per capita GDPs in the world. International banking has grown in recent years.

United Kingdom

United Kingdom of Great Britain and Northern Ireland

People: Population: 59,647,790. **Age distrib.** (%): <15: 18.9; 65+: 15.7. **Pop. density:** 640 per sq. mi. **Urban:** 89%. **Ethnic groups:** English 81.5%, Scottish 9.6%, Irish 2.4%, Welsh 1.9%, Ulster 1.8%; West Indian, Indian, Pakistani, others 2.8%. **Principal languages:** English, Welsh, Scottish, Gaelic. **Chief religions:** Anglican, Roman Catholic, other Christian, Muslim.

Geography: Area: 93,200 sq. mi. **Location:** Off the NW coast of Europe, across English Channel, Strait of Dover, and North Sea. **Neighbors:** Ireland to W, France to SE. **Topography:** England is mostly rolling land, rising to Uplands of southern Scotland; Lowlands are in center of Scotland, granite Highlands are in N. Coast is heavily indented, especially on W. British Isles have milder climate than N Europe due to the Gulf Stream and ample rainfall. Severn, 220 mi., and Thames, 215 mi., are longest rivers. **Capital:** London. **Cities** (urban agg.): London 7,640,000; Birmingham 2,272,000; Leeds 1,433,000; Liverpool 914,000; Manchester 2,252,000.

Government: Type: Constitutional monarchy. **Head of state:** Queen Elizabeth II; b Apr. 21, 1926; in office: Feb. 6, 1952. **Head of gov.:** Prime Min. Tony Blair; b May 6, 1953; in office: May 2, 1997. **Local divisions:** 467 local authorities, including England: 387; Wales: 22; Scotland: 32; Northern Ireland: 26. **Defense budget** (2001): $34.8 bil. **Active troops:** 212,450.

Economy: Industries: Metals, vehicles, shipbuilding, textiles, chemicals, electronics, aircraft, machinery. **Chief crops:** Cereals, oilseeds, potatoes, vegetables. **Minerals:** Coal, tin, oil, gas, limestone, iron, salt, clay. **Crude oil reserves** (2000): 5.15 bil bbls. **Arable land:** 25%. **Livestock** (2000): cattle: 11.13 mil; chickens: 157.05 mil; pigs: 6.48 mil; sheep: 42.26 mil. **Fish catch** (1999): 1.03 mil metric tons. **Electricity prod.:** 342.771 bil kWh. **Labor force:** 69% services; 18% manuf. & constr.; 11% govt.

Finance: Monetary unit: Pound (Sept. 2001: .68 = $1 U.S.). **GDP:** (1999 est.): $1.29 tril. **Per capita GDP:** $21,800. **Imports** (1998): $305.9 bil; partners: EU 53%, U.S. 14%. **Exports** (1998): $271 bil; partners: EU 58%, U.S. 13%. **Tourism:** $20.97 bil. **Budget** (1997 est.): $492.6 bil. **Intl. reserves less gold** (Mar. 2000): $28.34 bil. **Gold:** 19.73 mil oz t. **Consumer prices** (change in 1999): 1.6%.

Transport: Railroad: Length: 23,518 mi. **Motor vehicles (1997):** 25.59 mil pass. cars, 3.22 mil comm. vehicles. **Civil aviation:** 98.1 bil pass.-mi; 57 airports. **Chief ports:** London, Liverpool, Cardiff, Belfast.

Communications: TV sets: 641 per 1,000 pop. **Radios:** 1,445 per 1,000 pop. **Telephones** (1999): 33,750,000 main lines. **Daily newspaper circ.:** 332 per 1,000 pop.

Health: Life expectancy: 75.13 male; 80.66 female. **Births** (per 1,000 pop): 11.54. **Deaths** (per 1,000 pop.): 10.35. **Natural inc.:** .12%. **Infant mortality** (per 1,000 live births): 5.54.

Education: Compulsory: ages 5-16. **Literacy** (1993): 100%.

Major Intl. Organizations: UN and all of its specialized agencies, the Commonwealth, EU, NATO, OECD, OSCE.

Embassy: 3100 Massachusetts Ave. NW 20008; 588-6500.

Websites: http://www.statistics.gov.uk/ons_f.htm
http://www.genuki.org.uk

The United Kingdom of Great Britain and Northern Ireland comprises England, Wales, Scotland, and Northern Ireland.

Queen and Royal Family. The ruling sovereign is Elizabeth II of the House of Windsor, b Apr. 21, 1926, elder daughter of King George VI. She succeeded to the throne Feb. 6, 1952, and was crowned June 2, 1953. She was married Nov. 20, 1947, to Lt. Philip Mountbatten, b June 10, 1921, former Prince of Greece. He was created Duke of Edinburgh, and given the title H.R.H., Nov. 19, 1947; he was named Prince of the United Kingdom and Northern Ireland Feb. 22, 1957. Prince Charles Philip Arthur George, b Nov. 14, 1948, is the Prince of Wales and heir apparent. His 1st son, William Philip Arthur Louis, b June 21, 1982, is second in line to the throne.

Parliament is the legislative body for the UK, with certain powers over dependent units. It consists of 2 houses: The **House of Commons** has 659 members, elected by direct ballot and divided as follows: England 529; Wales 40; Scotland 72; Northern Ireland 18. Following a drastic reduction in the number of hereditary peerages, the **House of Lords** (July 2001) comprised 91 hereditary peers, 592 life peers, and 2 archbishops and 24 bishops of the Church of England, for a total of 709.

Resources and Industries. Great Britain's major occupations are manufacturing and trade. Metals and metal-using industries contribute more than 50% of exports. Of about 60 million acres of land in England, Wales, and Scotland, 46 million are farmed, of which 17 million are arable, the rest pastures.

Large oil and gas fields have been found in the North Sea. Commercial oil production began in 1975. There are large deposits of coal.

Britain imports all of its cotton, rubber, sulphur, about 80% of its wool, half of its food and iron ore, also certain amounts of paper, tobacco, chemicals. Manufactured goods made from these basic materials have been exported since the industrial age began. Main exports are machinery, chemicals, woolen and synthetic textiles, clothing, autos and trucks, iron and steel, locomotives, ships, jet aircraft, farm machinery, drugs, radio, TV, radar and navigation equipment, scientific instruments, arms, whisky.

Religion and Education. The Church of England is Protestant Episcopal. The queen is its temporal head, with rights of appointments to archbishoprics, bishoprics, and other offices. There are 2 provinces, Canterbury and York, each headed by an archbishop. The most famous church is Westminster Abbey (1050-1760), site of coronations, tombs of Elizabeth I, Mary, Queen of Scots, kings, poets, and of the Unknown Warrior.

The most celebrated British universities are Oxford and Cambridge, each dating to the 13th century. There are about 70 other universities.

History. Britain was part of the continent of Europe until about 6,000 BC, but migration across the English Channel continued long afterward. Celts arrived 2,500 to 3,000 years ago. Their language survives in Welsh, and Gaelic enclaves.

England was added to the Roman Empire in AD 43. After the withdrawal of Roman legions in 410, waves of Jutes, Angles, and Saxons arrived from German lands. They contended with Danish raiders for control from the 8th through 11th centuries. The last successful invasion was by French speaking Normans in 1066, who united the country with their dominions in France.

Opposition by nobles to royal authority forced King John to sign the Magna Carta in 1215, a guarantee of rights and the rule of law. In the ensuing decades, the foundations of the parliamentary system were laid.

English dynastic claims to large parts of France led to the Hundred Years War, 1338-1453, and the defeat of England. A long civil war, the War of the Roses, lasted 1455-85, and ended with the establishment of the powerful Tudor monarchy. A distinct English civilization flourished. The economy prospered over long periods of domestic peace unmatched in continental Europe. Religious independence was secured when the Church of England was separated from the authority of the pope in 1534.

Under Queen Elizabeth I, England became a major naval power, leading to the founding of colonies in the new world and the expansion of trade with Europe and the Orient. Scotland was united with England when James VI of Scotland was crowned James I of England in 1603.

A struggle between Parliament and the Stuart kings led to a bloody civil war, 1642-49, and the establishment of a republic under the Puritan Oliver Cromwell. The monarchy was restored in 1660, but the "Glorious Revolution" of 1688 confirmed the sovereignty of Parliament: a Bill of Rights was granted 1689.

In the 18th century, parliamentary rule was strengthened. Technological and entrepreneurial innovations led to the Industrial Revolution. The 13 North American colonies were lost, but replaced by growing empires in Canada and India. Britain's role in the defeat of Napoleon, 1815, strengthened its position as the leading world power.

The extension of the franchise in 1832 and 1867, the formation of trade unions, and the development of universal public education were among the drastic social changes that accompanied the spread of industrialization and urbanization in the 19th century. Large parts of Africa and Asia were added to the empire during the reign of Queen Victoria, 1837-1901.

Though victorious in World War I, Britain suffered huge casualties and economic dislocation. Ireland became independent in 1921, and independence movements became active in India and other colonies. The country suffered major bombing damage in World War II, but held out against Germany singlehandedly for a year after France fell in 1940.

Industrial growth continued in the postwar period, but Britain lost its leadership position to other powers. Labor governments passed socialist programs nationalizing some basic industries and expanding social security. Prime Min. Margaret Thatcher's Conservative government, however, tried to increase the role of private enterprise. In 1987, Thatcher became the first British leader in 160 years to be elected to a 3d consecutive term as prime minister. Falling on unpopular times, she resigned as prime minister in Nov. 1990. Her successor, John Major, led Conservatives to an upset victory at the polls, Apr. 9, 1992.

The UK supported the UN resolutions against Iraq and sent military forces to the Persian Gulf War.

The Channel Tunnel linking Britain to the Continent was officially inaugurated May 6, 1994. Britain's relations with the European Union were frayed in 1996 when the EU banned British beef because of the threat of "mad cow" disease.

On May 1, 1997, the Labour Party swept into power in a landslide victory, the largest of any party since 1935. Labour Party leader Tony Blair, 43, became Britain's youngest prime minister since 1812. Diana, Princess of Wales, the divorced wife of Prince Charles and the mother of Prince William, died in a car crash in Paris, Aug. 31. Britain played a leading role in the NATO air war against Yugoslavia, Mar.-June 1999, and contributed 12,000 troops to the multinational security force in Kosovo (KFOR).

Farming and tourism were hit hard in Mar.-Apr. 2001 by an epidemic of foot-and-mouth disease. Blair led Labour to another landslide election victory June 7, 2001.

Wales

The Principality of Wales in western Britain has an area of 8,019 sq. mi. and a population (1997 est.) of 2,927,000. Cardiff is the capital, pop. (1996 est.) 315,040.

Less than 20% of Wales residents speak English and Welsh; about 32,000 speak Welsh solely. A 1979 referendum rejected, 4-1, the creation of an elected Welsh assembly; a similar proposal passed by a thin margin on Sept. 18, 1997. Elections were held May 6, 1999.

Early Anglo-Saxon invaders drove Celtic peoples into the mountains of Wales, terming them Waelise (Welsh, or foreign). There they developed a distinct nationality. Members of the ruling house of Gwynedd in the 13th century fought England but were crushed, 1283. Edward of Caernarvon, son of Edward I of England, was created Prince of Wales, 1301.

Scotland

Scotland, a kingdom now united with England and Wales in Great Britain, occupies the northern 37% of the main British island, and the Hebrides, Orkney, Shetland, and smaller islands. Length 275 mi., breadth approx. 150 mi., area 30,418 sq. mi., population (1992 est.) 5,111,000.

The Lowlands, a belt of land approximately 60 mi. wide from the Firth of Clyde to the Firth of Forth, divide the farming region of the Southern Uplands from the granite Highlands of the North; they contain 75% of the population and most of the industry. The Highlands, famous for hunting and fishing, have been opened to industry by many hydroelectric power stations.

Edinburgh, pop. (1996 est.) 448,850, is the capital. Glasgow, pop. (1996 est.) 616,430, is Britain's greatest industrial center. It is a shipbuilding complex on the Clyde and an ocean port. Aberdeen, pop. (1996 est.) 227,430, NE of Edinburgh, is a major port, center of granite industry, fish-processing, and North Sea oil exploration. Dundee, pop. (1996 est.) 150,250, NE of Edinburgh, is an industrial and fish-processing center. About 90,000 persons speak Gaelic as well as English.

History. Scotland was called Caledonia by the Romans who battled early Celtic tribes and occupied southern areas from the 1st to the 4th centuries. Missionaries from Britain introduced Christianity in the 4th century; St. Columba, an Irish monk, converted most of Scotland in the 6th century.

The Kingdom of Scotland was founded in 1018. William Wallace and Robert Bruce both defeated English armies 1297 and 1314, respectively.

In 1603 James VI of Scotland, son of Mary, Queen of Scots, succeeded to the throne of England as James I, and effected the Union of the Crowns. In 1707 Scotland received representation in the British Parliament, resulting from the union of former separate Parliaments. Its executive in the British cabinet is the Secretary of State for Scotland. The growing Scottish National Party urges independence. A 1979 referendum on the creation of an elected Scottish assembly was defeated, but a proposal to create a regional legislature with limited taxing authority passed by a landslide Sept. 11, 1997. Elections were held May 6, 1999.

Memorials of Robert Burns, Sir Walter Scott, John Knox, and Mary, Queen of Scots, draw many tourists, as do the beauties of the Trossachs, Loch Katrine, Loch Lomond, and abbey ruins.

Industries. Engineering products are the most important industry, with growing emphasis on office machinery, autos, electronics, and other consumer goods. Oil has been discovered offshore in the North Sea, stimulating on-shore support industries.

Scotland produces fine woolens, worsteds, tweeds, silks, fine linens, and jute. It is known for its special breeds of cattle and sheep. Fisheries have large hauls of herring, cod, whiting. Whisky is the biggest export.

The Hebrides are a group of c. 500 islands, 100 inhabited, off the W coast. The Inner Hebrides include **Skye, Mull,** and **Iona,** the last famous for the arrival of St. Columba, AD 563. The Outer Hebrides include **Lewis** and **Harris.** Industries include sheep raising and weaving. The **Orkney Islands,** c. 90, are to the NE. The capital is Kirkwall, on Pomona Isl. Fish curing, sheep raising, and weaving are occupations. NE of the Orkneys are the 200 **Shetland Islands,** 24 inhabited, home of Shetland pony. The Orkneys and Shetlands are centers for the North Sea oil industry.

Northern Ireland

Northern Ireland was constituted in 1920 from 6 of the 9 counties of Ulster, the NE corner of Ireland. Area 5,452 sq. mi., pop. (1996 est.) 1,663,300, capital and chief industrial center, Belfast, pop. (1996 est.) 297,300.

Industries. Shipbuilding, including large tankers, has long been an important industry, centered in Belfast, the largest port. Linen manufacture is also important, along with apparel, rope, and twine. Growing diversification has added engineering products, synthetic fibers, and electronics. There are large numbers of cattle, hogs, and sheep. Potatoes, poultry, and dairy foods are also produced.

Government. An act of the British Parliament, 1920, divided Northern from Southern Ireland, each with a parliament and government. When Ireland became a dominion, 1921, and later a republic, Northern Ireland chose to remain a part of the United Kingdom. It elects 18 members to the British House of Commons.

During 1968-69, large demonstrations were conducted by Roman Catholics who charged they were discriminated against in voting rights, housing, and employment. The Catholics, a minority comprising about a third of the population, demanded abolition of property qualifications for voting in local elections. Violence and terrorism intensified, involving branches of the Irish Republican Army (outlawed in the Irish Republic), Protestant groups, police, and British troops.

A succession of Northern Ireland prime ministers pressed reform programs but failed to satisfy extremists on both sides. Between 1969 and 1994 more than 3,000 were killed in sectarian violence, many in England itself. Britain suspended the Northern Ireland parliament Mar. 30, 1972, and imposed direct British rule. A coalition government was formed in 1973 when moderates won election to a new one-house Assembly. But a Protestant general strike overthrew the government in 1974 and direct rule was resumed.

The agony of Northern Ireland was dramatized in 1981 by the deaths of 10 Irish nationalist hunger strikers in Maze Prison near Belfast. In 1985 the Hillsborough agreement gave the Rep. of Ireland a voice in the governing of Northern Ireland; the accord was strongly opposed by Ulster loyalists. On Dec. 12, 1993, Britain and Ireland announced a declaration of principles to resolve the Northern Ireland conflict.

On Aug. 31, 1994, the IRA announced a cease-fire, saying it would rely on political means to achieve its objectives; the IRA resumed its terrorist tactics on Feb. 9, 1996. Reinstatement of the IRA cease-fire as of July 20, 1997, led to the resumption of peace talks Sept. 15.

A settlement reached on Good Friday, April 10, 1998, provided for restoration of home rule and election of a 108-member assembly with safeguards for minority rights. Both Ireland and Great Britain agreed to give up their constitutional claims on Northern Ireland. The accord was approved May 22 by voters in Northern Ireland and the Irish Republic, and elections to the assembly were held June 25. IRA dissidents seeking to derail the agreement were responsible for a bomb at Omagh Aug. 15 that killed 29 people and injured over 330.

London transferred authority to a Northern Ireland power-sharing government Dec. 2, 1999. Self-rule was suspended Feb. 11-May 29, 2000, because of IRA reluctance to disarm.

Education and Religion. Northern Ireland is about 58% Protestant, 42% Roman Catholic. Education is compulsory between the ages of 5 and 16 years.

Channel Islands

The Channel Islands, area 75 sq. mi., pop. (1997 est.) 152,241, off the NW coast of France, the only parts of the one-time Dukedom of Normandy belonging to England, are Jersey, Guernsey and the dependencies of Guernsey—Alderney, Brechou, Great Sark, Little Sark, Herm, Jethou and Lihou. Jersey and Guernsey have separate legal existences and lieutenant governors named by the Crown. The islands were the only British soil occupied by German troops in World War II.

Isle of Man

The Isle of Man, area 227 sq. mi., pop. (2001 est.) 73,489, is in the Irish Sea, 20 mi. from Scotland, 30 mi. from Cumberland. It is rich in lead and iron. The island has its own laws and a lieutenant governor appointed by the Crown. The Tynwald (legislature) consists of the Legislative Council, partly elected, and House of Keys, elected. Capital: Douglas. Farming, tourism, and fishing (kippers, scallops) are chief occupations. Man is famous for the Manx tailless cat.

Gibraltar

Gibraltar, a dependency on the southern coast of Spain, guards the entrance to the Mediterranean. The Rock of Gibraltar has been in British possession since 1704. The Rock is 2.75 mi. long, 3/4 of a mi. wide and 1,396 ft. in height; a narrow isthmus connects it with the mainland. Pop. (2001 est.) 27,649.

Gibraltar has historically been an object of contention between Britain and Spain. Residents voted with near unanimity to remain under British rule, in a 1967 referendum held in pursuance of a UN resolution on decolonization. A new constitution, May 30, 1969, increased Gibraltarian control of domestic affairs (the UK continues to handle defense and internal security matters). Following a 1984 agreement between Britain and Spain, the border, closed by Spain in 1969, was fully reopened in Feb. 1985. A UN General Assembly resolution requested Britain to end Gibraltar's colonial status by Oct. 1, 1996. No settlement has been reached.

British West Indies

Swinging in a vast arc from the coast of Venezuela NE, then N and NW toward Puerto Rico are the Leeward Islands, forming a coral and volcanic barrier sheltering the Caribbean from the open Atlantic. Many of the islands are self-governing British possessions. Universal suffrage was instituted 1951-54; ministerial systems were set up 1956-1960.

The **Leeward Islands** still associated with the UK are **Montserrat**, area 32 sq. mi., pop. (2001 est.) 7,574, capital Plymouth; the **British Virgin Islands,** 59 sq. mi., pop. (2001 est.) 20,812, capital Road Town; and **Anguilla**, the most northerly of the Leeward Islands, 60 sq. mi., pop. (2001 est.) 12,132, capital The Valley. Montserrat has been devastated by the Soufrière Hills volcano, which began erupting July 18, 1995.

The three **Cayman Islands,** a dependency, lie S of Cuba, NW of Jamaica. Pop. (2001 est.) 35,527, most of it on Grand Cayman. It is a free port; in the 1970s Grand Cayman became a tax-free refuge for foreign funds and branches of many Western banks were opened there. Total area 102 sq. mi., capital Georgetown.

The **Turks and Caicos Islands** are a dependency at the SE end of the Bahama Islands. Of about 30 islands, only 6 are inhabited; area 193 sq. mi., pop. (2001 est.) 18,122; capital Grand Turk. Salt, shellfish, and conch shells are the main exports.

Bermuda

Bermuda is a British dependency governed by a royal governor and an assembly, dating from 1620, the oldest legislative body among British dependencies. Capital is Hamilton.

It is a group of about 150 small islands of coral formation, 20 inhabited, comprising 20.0 sq. mi. in the western Atlantic, 580 mi. E of North Carolina. Pop. (2001 est.) 63,503 (about 61% of African descent). Pop. density is high.

The U.S. maintains a NASA tracking facility; a U.S. naval air base was closed in 1995.

Tourism is the major industry; Bermuda boasts many resort hotels. The government raises most revenue from import duties. Exports: petroleum products, medicine. In a referendum Aug. 15, 1995, voters rejected independence by nearly a 3-to-1 majority.

South Atlantic

The **Falkland Islands,** a dependency, lie 300 mi. E of the Strait of Magellan at the southern end of South America.

The Falklands or Islas Malvinas include 2 large islands and about 200 smaller ones, area 4,700 sq. mi., pop. (1995 est.) 2,317, capital Stanley. The licensing of foreign fishing vessels has become the major source of revenue. Sheep-grazing is a main industry; wool is the principal export. There are indications of large oil and gas deposits. The islands are also claimed by Argentina, though 97% of inhabitants are of British origin. Argentina invaded the islands Apr. 2, 1982. The British responded by sending a task force to the area, landing their main force on the Falklands, May 21, and forcing an Argentine surrender at Port Stanley, June 14. A pact resuming commercial air service with Argentina was signed July 14, 1999.

British Antarctic Territory, south of 60° S lat., formerly a dependency of the Falkland Isls., was made a separate colony in 1962 and includes the **South Shetland Islands,** the **South Orkneys**, and the Antarctic Peninsula. A chain of meteorological stations is maintained.

South Georgia and the South Sandwich Islands, formerly administered by the Falklands Isls., became a separate dependency in 1985. South Georgia, 1,450 sq. mi., with no permanent population, is about 800 mi. SE of the Falklands; the South Sandwich Isls., 130 sq. mi., are uninhabited, about 470 mi. SE of South Georgia.

St. Helena, an island 1,200 mi. off the W. coast of Africa and 1,800 mi. E of South America, 47 sq. mi. and pop. (2001 est.) 7,266. Flax, lace, and rope-making are the chief industries. After Napoleon Bonaparte was defeated at Waterloo the Allies exiled him to St. Helena, where he lived from Oct. 16, 1815, to his death, May 5, 1821. Capital is Jamestown.

Tristan da Cunha is the principal of a group of islands of volcanic origin, total area 40 sq. mi., halfway between the Cape of Good Hope and South America. A volcanic peak 6,760 ft. high erupted in 1961. The 262 inhabitants were removed to England, but most returned in 1963. The islands are dependencies of St. Helena. Pop. (1993) 300.

Ascension is an island of volcanic origin, 34 sq. mi. in area, 700 mi. NW of St. Helena, through which it is administered. It is a communications relay center for Britain, and has a U.S. satellite tracking center. Pop. (1993) was 1,117, half of them communications workers. The island is noted for sea turtles.

Hong Kong

(*See* China/Hong Kong)

British Indian Ocean Territory

Formed Nov. 1965, embracing islands formerly dependencies of Mauritius or Seychelles: the Chagos Archipelago (including Diego Garcia), Aldabra, Farquhar, and Des Roches. The latter 3 were transferred to Seychelles, which became independent in 1976. Area 23 sq. mi. No permanent civilian population remains; the U.K. and the U.S. maintain a military presence.

Pacific Ocean

Pitcairn Island is in the Pacific, halfway between South America and Australia. The island was discovered in 1767 by Philip Carteret but was not inhabited until 23 years later when the mutineers of the *Bounty* landed there. The area is 1.7 sq. mi. and 1995 pop. was 54. It is a British dependency and is administered by a British High Commissioner in New Zealand and a local Council. The uninhabited islands of **Henderson, Ducie,** and **Oeno** are in the Pitcairn group.

United States

United States of America

People: Population: 278,058,881 (incl. 50 states & Dist. of Columbia). (Note: U.S. pop. figures may differ elsewhere in *The World Almanac.*) **Age distrib.** (%): <15: 21.1; 65+: 12.6. **Pop. density:** 79 per sq. mi. **Urban:** 76%.

Geography: Area: 3,535,000 sq. mi. (incl. 50 states and DC). **Topography:** Vast central plain, mountains in west, hills and low mountains in east. **Capital:** Washington, D.C.

Government: Federal republic, strong democratic tradition. **Head of state and gov.:** Pres. George W. Bush; b July 6, 1946; in office: Jan. 20, 2001. **Local divisions:** 50 states and Dist. of Columbia. **Defense budget:** $291.2 bil. **Active troops:** 1,365,800.

Economy: Minerals: Coal, oil, gas, copper, lead, molybdenum, phosphates, uranium, bauxite, gold, iron, mercury, nickel, potash, silver, tungsten, zinc. **Crude oil reserves** (2000): 21.03 bil bbls. **Other resources:** forests. **Arable land:** 19%. **Livestock** (2000): cattle: 98.05 mil; chickens: 1.72 bil; goats: 1.35 mil; pigs: 59.34 mil; sheep: 7.22 mil. **Fish catch** (1999): 5.45 mil metric tons. **Electricity prod.:** 3.67 tril kWh. **Labor force** (1999): 30% managerial; 29.2% technical; 24.5% manuf.

Finance: GDP (1999 est.): $9.255 tril. **Per capita GDP:** $33,900. **Imports** (1998): $912 bil; partners: Canada 19%, Japan 13%; Mexico 10%. **Exports** (1998): $663 bil; partners: Canada 23%, Mexico 12%, Japan 8%. **Tourism:** $74.49 bil. **Budget** (1998): $1.653 tril. **Intl. reserves less gold** (June 2000): $56.91 bil. **Gold:** 261.65 mil oz t. **Consumer prices** (change in 1999): 2.2%.

Transport: Railroad: Length: 137,900 mi. **Motor vehicles:** 129.73 mil pass. cars, 76.64 mil comm. vehicles. **Civil aviation:** 599.4 bil pass.-mi; 834 airports.

Communications: TV sets: 847 per 1,000 pop. **Radios:** 2,115 per 1,000 pop. **Telephones:** 192,518,800 main lines. **Daily newspaper circ.:** 215 per 1,000 pop.

Health: Life expectancy: 74.37 male; 80.05 female. **Births** (per 1,000 pop): 14.20. **Deaths** (per 1,000 pop.): 8.7. **Natural inc.:** .55%. **Infant mortality** (per 1,000 live births): 6.76.

Education: Free, compulsory: ages 7-16. **Literacy** (1994): 97%.

Major Intl. Organizations: UN (FAO, IBRD, ILO, IMF, IMO, WHO, WTrO), APEC, NATO, OAS, OECD, OSCE.

Websites: http://www.census.gov
http://www.whitehouse.gov
http://www.firstgov.gov

***See also* United States History chapter.**

Uruguay

Oriental Republic of Uruguay

People: Population: 3,360,105. **Age distrib.** (%): <15: 24.4; 65+: 13.0. **Pop. density:** 50 per sq. mi. **Urban:** 91%. **Ethnic groups:** White 88%, mestizo 8%, black 4%. **Principal language:** Spanish. **Chief religion:** Roman Catholic 66%.

Geography: Area: 67,000 sq. mi. **Location:** In southern South America, on the Atlantic O. **Neighbors:** Argentina on W, Brazil on N. **Topography:** Uruguay is composed of rolling, grassy plains and hills, well watered by rivers flowing W to Uruguay R. **Capital:** Montevideo (urban agg.): 1,236,000.

Government: Type: Republic. **Head of state and gov.:** Pres. Jorge Batlle Ibáñez; b Oct. 25, 1927; in office: Mar. 1, 2000. **Local divisions:** 19 departments. **Defense budget:** $227 mil. **Active troops:** 23,700.

Economy: Industries: Meat processing, wool and hides, textiles, wine, oil refining. **Chief crops:** Corn, wheat, sorghum, rice. **Arable land:** 7%. **Livestock** (2000): cattle: 10.8 mil; chickens: 13.0 mil; goats: 14,800; pigs: 380,000; sheep: 13.03 mil. **Fish catch** (1999): 136,912 metric tons. **Electricity prod.:** 5.704 bil kWh. **Labor force:** N/A

Finance: Monetary unit: Peso (Oct. 2000: 13.42 = $1 U.S.). **GDP:** (1998 est.): $28 bil. **Per capita GDP:** $8,500. **Imports** (1999): $3.4 bil; partners: MERCOSUR partners 43%, EU 20% US 11%. **Exports** (1999): $2.1 bil; partners: MERCOSUR partners 45%, EU 20%, US 7%. **Tourism:** $653 mil. **Budget** (1997 est.): $4.3 bil. **Intl. reserves less gold** (May 2000): $2.17 bil. **Gold:** 1.82 mil oz t. **Consumer prices** (change in 1999): 5.7%.

Transport: Railroad: Length: 1,288 mi. **Motor vehicles:** 475,000 pass. cars, 50,000 comm. vehicles. **Civil aviation:** 470.0 mil pass.-mi; 1 airport. **Chief port:** Montevideo.

Communications: TV sets: 191 per 1,000 pop. **Radios:** 586 per 1,000 pop. **Telephones:** 929,100 main lines. **Daily newspaper circ.:** 241 per 1,000 pop.

Health: Life expectancy: 72.11 male; 78.96 female. **Births** (per 1,000 pop): 17.36. **Deaths** (per 1,000 pop.): 9.03. **Natural inc.:** .83%. **Infant mortality** (per 1,000 live births): 14.7.

Education: Free, compulsory for 6 years between ages 6-14. **Literacy:** 97%.

Major Intl. Organizations: UN (FAO, IBRD, ILO, IMF, IMO, WHO, WTrO), OAS.

Embassy: 2715 M St. NW 20007; 331-1313.

Website: http://www.embassy.org/uruguay

Spanish settlers began to supplant the indigenous Charrua Indians in 1624. Portuguese from Brazil arrived later, but Uruguay was attached to the Spanish Viceroyalty of Rio de la Plata in the 18th century. Rebels fought against Spain beginning in 1810. An independent republic was declared Aug. 25, 1825.

Terrorist activities led Pres. Juan María Bordaberry to agree to military control of his administration Feb. 1973. In June he abolished Congress and set up a Council of State in its place. Bordaberry was removed by the military in a 1976 coup. Civilian government was restored in 1985.

Socialist measures were adopted in the early 1900s, and the state retains a dominant role in the power, telephone, railroad, cement, oil-refining, and other industries. Uruguay's standard of living remains one of the highest in South America, and political and labor conditions among the freest.

Uzbekistan

Republic of Uzbekistan

People: Population: 25,155,064. **Age distrib.** (%): <15: 36.3; 65+: 4.6. **Pop. density:** 146 per sq. mi. **Urban:** 37%. **Ethnic groups:** Uzbek 80%, Russian 6%, Tajik 5%. **Principal languages:** Uzbek, Russian. **Chief religions:** Muslim (mostly Sunni) 88%, Eastern Orthodox 9%.

Geography: Area: 172,500 sq. mi. **Location:** Central Asia. **Neighbors:** Kazakhstan on N and W, Kyrgyzstan and Tajikistan on E, Afghanistan and Turkmenistan on S. **Topography:** Mostly plains and desert. **Capital:** Tashkent (urban agg.) 2,148,000.

Government: Type: Republic. **Head of state:** Pres. Islam A. Karimov; b Jan. 30, 1938; in office: Mar. 24, 1990. **Head of gov.:** Prime Min. Utkir Sultanov; b July 14, 1939; in office: Dec. 21, 1995. **Local divisions:** 12 regions, 1 autonomous republic, 1 city. **Defense budget** (1999): $285 mil. **Active troops:** 59,100.

Economy: Industries: Machine building, food processing, natural gas, textiles. **Chief crops:** Vegetables, cotton, fruits, grain. **Minerals:** Gas, oil, coal, gold, uranium, silver, copper. **Crude oil reserves** (2000): 594 mil bbls. **Arable land:** 9%. **Livestock** (2000): cattle: 5.27 mil; chickens: 14.41 mil; goats: 638,500; pigs: 80,000; sheep: 8.92 mil. **Fish catch:** (1999): 10,565 metric tons. **Electricity prod.:** 42.876 bil kWh. **Labor force:** 44% agric. & forestry; 20% ind.

Finance: Monetary unit: Som (Oct. 2000: 775.00 = $1 U.S.). **GDP:** (1999 est.): $59.3 bil. **Per capita GDP:** $2,500. **Imports** (1999): $3.1 bil; partners: Russia 16%, S. Korea 11%, Germany. 11%. **Exports** (1999): $2.9 bil; partners: Russia 15%, Switzerland 10%, UK 10%. **Tourism** (1998): $21 mil.

Transport: Railroad: Length: 2,100 mi. **Motor vehicles:** 865,000 pass. cars, 14,500 comm. vehicles. **Civil aviation:** 2.2 bil pass.-mi; 9 airports. **Chief port:** Termiz.

Communications: TV sets: 176 per 1,000 pop. **Telephones**: (1999): 1,599,400 main lines.

Health: Life expectancy: 60.24 male; 67.56 female. **Births** (per 1,000 pop): 26.10. **Deaths** (per 1,000 pop.): 8.. **Natural inc.:** 1.81%. **Infant mortality** (per 1,000 live births): 71.92.

Education: Compulsory: ages 6-14. **Literacy** (1993): 97%.

Major Intl. Organizations: UN (IBRD, ILO, IMF, WHO), CIS, OSCE.

Embassy: 1746 Massachusetts Ave. NW 20036; 887-5300.

Website: http://www.gov.uz

The region was overrun by the Mongols under Genghis Khan in 1220. In the 14th century, Uzbekistan became the center of a native empire—that of the Timurids. In later centuries Muslim feudal states emerged. Russian military conquest began in the 19th century.

The Uzbek SSR became a Soviet Union republic in 1925. Uzbekistan declared independence Aug. 29, 1991. It became an independent republic when the Soviet Union disbanded Dec. 26, 1991. Subsequently, the government of Uzbekistan was dominated by former Communists.

Vanuatu

Republic of Vanuatu

People: Population: 192,910. **Age distrib.** (%): <15: 36.3; 65+: 3.2. **Pop. density:** 34 per sq. mi. **Urban:** 20%. **Ethnic groups:** Melanesian 94%, French 4%. **Principal languages:** French, English, Bislama (all official). **Chief religions:** Presbyterian 37%, Anglican 15%, Catholic 15%, other Christian 10%, indigenous beliefs 8%.

Geography: Area: 5,700 sq. mi. **Location:** SW Pacific, 1,200 mi. NE of Brisbane, Australia. **Neighbors:** Fiji to E, Solomon Isls. to NW. **Topography:** Dense forest with narrow coastal strips of cultivated land. **Capital:** Port-Vila (1999 est.): 26,000.

Government: Type: Republic. **Head of state:** Pres. John Bani; b. July 1, 1941; in office: Mar. 24, 1999. **Head of gov.:** Prime Min. Edward Natapei; b 1955; in office: Apr. 13, 2001. **Local divisions:** 6 provinces

Economy: Industries: Fish-freezing, meat canneries, wood processing. **Chief crops:** Copra, coconuts, cocoa, coffee. **Minerals:** Manganese. **Arable land:** 2%. **Other resources:** Forests, cattle. **Fish catch** (1996): 2,729 metric tons. **Livestock** (2000): cattle: 152,000; chickens: 340,000; goats: 12,000; pigs: 62,000. **Electricity prod.:** 35 mil kWh.**Labor force:** 65% agric.; 32% services.

Finance: Monetary unit: Vatu (Sept. 2001: 144.10 = $1 U.S.). **GDP** (1999 est.): $245 mil. **Per capita GDP:** $1,300. **Imports** (1998): $76.2 mil; partners: Japan 52%, Australia 20%. **Exports** (1998): $33.8 mil; partners: Japan 32%, Germany 14%, Spain 8%. **Tourism:** $56 mil. **Budget** (1996 est.): $99.8 mil. **Intl. reserves less gold** (Mar. 2000): $36.10 mil. **Consumer prices** (change in 1998): 3.3%.

Transport: Motor vehicles: 4,000 pass. cars, 2,500 comm. vehicles. **Civil aviation:** 93.2 mil pass.-mi; 29 airports. **Chief ports:** Forai, Port-Vila.

Communications: Radios: 319 per 1,000 pop. **Telephones**: 6,600 main lines.

Health: Life expectancy: 59.58 male; 62.39 female. **Births** (per 1,000 pop): 25.40. **Deaths** (per 1,000 pop.): 8.38. **Natural inc.:** 1.7%. **Infant mortality** (per 1,000 live births): 61.05.

Education: Literacy (1997): 36%.

Major Intl. Organizations: UN (FAO, IBRD, IMF, IMO, WHO), the Commonwealth.

The Anglo-French condominium of the New Hebrides, administered jointly by France and Great Britain since 1906, became the independent Republic of Vanuatu on July 30, 1980.

Vatican City (The Holy See)

People: Population: 870. **Urban:** 100%. **Ethnic groups:** Italian, Swiss. **Principal languages:** Italian, Latin. **Chief religion:** Roman Catholic.

Geography: Area: 108.7 acres. **Location:** In Rome, Italy. **Neighbors:** Completely surrounded by Italy.

Monetary unit: Vatican Lira, Italian Lira (equal value) (Oct. 2000: 2,221.64 = $1 U.S.).

Apostolic Nunciature in U.S.: 3339 Massachusetts Ave. NW 20008; 333-7121.

Website: http://www.vatican.va

The popes for many centuries, with brief interruptions, held temporal sovereignty over mid-Italy (the so-called Papal States), comprising an area of some 16,000 sq. mi., with a population in the 19th century of more than 3 million. This territory was incorporated in the new Kingdom of Italy (1861), the sovereignty of the pope being confined to the palaces of the Vatican and the Lateran in Rome and the villa of Castel Gandolfo, by an Italian law, May 13, 1871. This law also guaranteed to the pope and his successors a yearly indemnity of over $620,000. The allowance, however, remained unclaimed.

A Treaty of Conciliation, a concordat, and a financial convention were signed Feb. 11, 1929, by Cardinal Gasparri and Premier Mussolini. The documents established the independent state of Vatican City and gave the Roman Catholic church special status in Italy. The treaty (Lateran Agreement) was made part of the Constitution of Italy (Article 7) in 1947. Italy and the Vatican signed an agreement in 1984 on revisions of the concordat; the accord eliminated Roman Catholicism as the state religion and ended required religious education in Italian schools.

Vatican City includes the Basilica of Saint Peter, the Vatican Palace and Museum covering over 13 acres, the Vatican gardens, and neighboring buildings between Viale Vaticano and the church. Thirteen buildings in Rome, outside the boundaries, enjoy extraterritorial rights; these buildings house congregations or officers necessary for the administration of the Holy See.

The legal system is based on the code of canon law, the apostolic constitutions, and laws especially promulgated for the Vatican City by the pope. The Secretariat of State represents the Holy See in its diplomatic relations. By the Treaty of Conciliation the pope is pledged to a perpetual neutrality unless his mediation is specifically requested. This, however, does not prevent the defense of the Church whenever it is persecuted.

The present sovereign of the State of Vatican City is the Supreme Pontiff John Paul II, born Karol Wojtyla in Wadowice, Poland, May 18, 1920, elected Oct. 16, 1978 (the first non-Italian to be elected pope in 456 years).

The U.S. restored formal relations in 1984 after the U.S. Congress repealed an 1867 ban on diplomatic relations with the Vatican. The Vatican and Israel agreed to establish formal relations Dec. 30, 1993.

Venezuela

Bolivarian Republic of Venezuela

People: Population: 23,916,810. **Age distrib.** (%): <15: 32.1; 65+: 4.7. **Pop. density:** 67 per sq. mi. **Urban:** 87%. **Ethnic groups:** Spanish, Portuguese, Italian. **Principal language:** Spanish (official). **Chief religion:** Roman Catholic 96%.

Geography: Area: 340,200 sq. mi. **Location:** On Caribbean coast of South America. **Neighbors:** Colombia on W, Brazil on S, Guyana on E. **Topography:** Flat coastal plain and Orinoco Delta are bordered by Andes Mts. and hills. Plains, called llanos, extend between mountains and Orinoco. Guiana Highlands and plains are S of Orinoco, which stretches 1,600 mi. and drains 80% of Venezuela. **Capital:** Caracas. **Cities** (urban agg.): Caracas 3,153,000; Maracaibo 1,901,000; Valencia 1,893,000.

Government: Type: Federal republic. **Head of state and gov.:** Pres. Hugo Rafael Chávez Frías; b July 28, 1954; in office: Feb. 2, 1999. **Local divisions:** 22 states, 1 federal district (Caracas), 1 federal dependency (72 islands). **Defense budget:** $1,404 mil. **Active troops:** 79,000.

Economy: Industries: Iron mining, steel, oil, textiles. **Chief crops:** Rice, corn, sorghum, bananas, sugar. **Minerals:** Oil, gas, iron, gold. **Crude oil reserves** (2000): 72.6 bil bbls. **Arable land:** 4%. **Livestock** (2000): cattle: 15.8 mil; chickens: 110.0 mil; goats: 3.6 mil; pigs: 4.9 mil; sheep: 781,000. **Fish catch** (1999): 502,728 metric tons. **Electricity prod.:** 81.215 bil kWh. **Labor force:** 64% services; 23% ind.; 13% agric.

Finance: Monetary unit: Bolivar (Sept. 2001: 747.00 = $1 U.S.). **GDP** (1999): $182.8 bil. **Per capita GDP:** $8,000. **Imports** (1999): $11.8 bil; partners: U.S. 53%. **Exports** (1999): $20.9 bil; partners: U.S. & Puerto Rico 57%. **Tourism:** $656 mil. **Budget** (1996 est.): $11.48 bil. **Intl. reserves less gold** (June 2000): $12.15 bil. **Gold:** 10.29 mil oz t. **Consumer prices** (change in 1999): 23.6%.

Transport: Railroad: Length: 390 mi. **Motor vehicles:** 1.50 mil pass. cars, 525,000 comm. vehicles. **Civil aviation:** 2.8 bil pass.-mi; 20 airports. **Chief ports:** Maracaibo, La Guaira, Puerto Cabello.

Communications: TV sets: 183 per 1,000 pop. **Radios:** 372 per 1,000 pop. **Telephones**: 2,605,600 main lines. **Daily newspaper circ.:** 215 per 1,000 pop.

Health: Life expectancy: 70.29 male; 76.56 female. **Births** (per 1,000 pop): 20.65. **Deaths** (per 1,000 pop.): 4.92. **Natural inc.:** 1.57%. **Infant mortality** (per 1,000 live births): 25.37.

Education: Free, compulsory: ages 5-15. **Literacy**: 91%.

Major Intl. Organizations: UN (FAO, IBRD, ILO, IMF, IMO, WHO, WTrO), OAS, OPEC.

Embassy: 1099 30th St. NW 20007; 342-2214.

Website: http://www.embavenez-us.org

Columbus first set foot on the South American continent on the peninsula of Paria, Aug. 1498. Alonso de Ojeda, 1499, was the first European to see Lake Maracaibo. He called the land Venezuela, or Little Venice, because the Indians had houses on stilts. Spain dominated Venezuela until Simón Bolívar's victory near Carabobo in June 1821. The republic was formed after secession from the Colombian Federation in 1830.

Military strongmen ruled Venezuela for most of the 20th century. They promoted the oil industry; some social reforms were implemented. Since 1959, the country has had democratically elected governments.

Venezuela helped found the Organization of Petroleum Exporting Countries (OPEC). The government, Jan. 1, 1976, nationalized the oil industry with compensation. Oil accounts for most of Venezuela's export earnings; the economy suffered a severe cash crisis in the 1980s and 1990s as a result of depressed oil revenues. Government attempts to reduce dependence on oil have met with limited success.

An attempted coup by midlevel military officers was thwarted by loyalist troops Feb. 4, 1992. A second coup attempt was thwarted in Nov. Pres. Carlos Andrés Pérez was removed from office on corruption charges, May 1993; he was convicted, May 1996, of mismanaging a $17 million secret government security fund. Citing an economic crisis, Pres. Rafael Caldera, a populist elected Dec. 5, 1993, suspended many civil liberties June 27, 1994; rights were restored in most regions July 6, 1995.

A 1992 coup leader, Hugo Chávez, who ran as a populist, was elected president Dec. 6, 1998. A constitutional assembly elected July 25, 1999, and controlled by Chávez supporters slashed the powers of Congress and moved to dismiss corrupt judges. Voters on Dec. 15 approved a new constitution greatly increasing the powers of the president. Floods and mudslides in Dec. 1999 killed, by official estimates, at least 30,000 people.

Vietnam

Socialist Republic of Vietnam

People: Population: 79,939,014. **Age distrib.** (%): <15: 32.1; 65+: 5.4. **Pop. density:** 637 per sq. mi. **Urban:** 20%. **Ethnic groups:** Vietnamese 85-90%, Chinese 3%, Muong, Tai, Meo, Khmer, Man, Cham. **Principal languages:** Vietnamese (official), French, Chinese, English. **Chief religions:** Buddhist, Taoist, Roman Catholic, indigenous beliefs.

Geography: Area: 125,500 sq. mi. **Location:** SE Asia, on the E coast of the Indochinese Peninsula. **Neighbors:** China on N, Laos and Cambodia on W. **Topography:** Vietnam is long and narrow, with a 1,400-mi. coast. About 22% of country is readily arable, including the densely settled Red R. valley in the N, narrow coastal plains in center, and the wide, often marshy Mekong R. Delta in the S. The rest consists of semi-arid plateaus and barren mountains, with some stretches of tropical rain forest. **Capital:** Hanoi. **Cities (urban agg.):** Ho Chi Minh City 4,615,000; Hanoi 3,734,000.

Government: Type: Communist. **Head of state:** Pres. Tran Duc Luong; b May 1937; in office: Sept. 24, 1997. **Head of gov.:** Prime Min. Phan Van Khai; b Dec. 1933; in office: Sept. 25, 1997. **Local divisions:** 58 provinces, 3 cities, 1 capital region. **Defense budget:**1.0 bil. **Active troops:** 484,000.

Economy: Industries: Food processing, garments, shoes, chemical fertilizer. **Chief crops:** Rice, potatoes, soybeans, coffee, tea, corn. **Minerals:** Phosphates, coal, gas, manganese, bauxite, chromate, oil. **Crude oil reserves** (2000): 600 mil bbls. **Other resources:** Forests. **Arable land:** 17%. **Livestock** (2000): cattle: 4.14 mil; chickens: 196.19 mil; goats: 543,867; pigs: 20.19 mil. **Fish catch** (1999): 1.55 mil metric tons. **Electricity prod.:** 22.985 bil kWh. **Labor force:** 67% agric; 33% ind.

Finance: Monetary unit: Dong (Sept. 2001: 15,027.00 = $1 U.S.). **GDP:** (1999 est.): $143.1 bil. **Per capita GDP:** $1,850. **Imports** (1999 est.): $11.6 bil; partners: Singapore, S. Korea. **Exports** (1999 est.): $11.5 bil; partners: Japan, Germany, Singapore. **Tourism** (1998): $86 mil. **Budget** (1996 est.): $6 bil.

Transport: Railroad: Length: 1,619 mi. **Motor vehicles:** 79,079 pass. cars, 97,104 comm. vehicles. **Civil aviation:** 2.4 bil pass.-mi; 12 airports. **Chief ports:** Ho Chi Minh City, Haiphong, Da Nang.

Communications: TV sets: 182 per 1,000 pop. **Radios:** 109 per 1,000 pop. **Telephones**: 2,542,700 main lines. **Daily newspaper circ.:** 4.1 per 1,000 pop.

Health: Life expectancy: 67.12 male; 72.19 female. **Births** (per 1,000 pop): 21.23. **Deaths** (per 1,000 pop.): 6.22. **Natural inc.:** 1.5%. **Infant mortality** (per 1,000 live births): 30.24.

Education: Compulsory: ages 6-11. **Literacy**: 94%.

Major Intl. Organizations: UN (FAO, IBRD, ILO, IMF, IMO, WHO), APEC, ASEAN.

Embassy: Suite 400, 1233 20th St. NW 20036; 861-0737.

Website: http://www.batin.com.vn

Vietnam's recorded history began in Tonkin before the Christian era. Settled by Viets from central China, Vietnam was held by China, 111 BC-AD 939, and was a vassal state during subsequent periods. Vietnam defeated the armies of Kublai Khan, 1288. Conquest by France began in 1858 and ended in 1884 with the protectorates of Tonkin and Annam in the N and the colony of Cochin-China in the S.

Japan occupied Vietnam in 1940; nationalist aims gathered force. A number of groups formed the Vietminh (Independence) League, headed by Ho Chi Minh, Communist guerrilla leader. In Aug. 1945 the Vietminh forced out Bao Dai, former emperor of Annam, head of a Japan-sponsored regime. France, seeking to reestablish colonial control, battled Communist and nationalist forces, 1946-1954, and was defeated at Dienbienphu, May 8, 1954. Meanwhile, on July 1, 1949, Bao Dai had formed a State of Vietnam, with himself as chief of state, with French approval. China backed Ho Chi Minh.

A cease-fire signed in Geneva July 21, 1954, provided for a buffer zone, withdrawal of French troops from the North, and elections to determine the country's future. Under the agreement the Communists gained control of territory north of the 17th parallel, with its capital at Hanoi and Ho Chi Minh as president. South Vietnam came to comprise the 39 southern provinces. Some 900,000 North Vietnamese fled to South Vietnam.

On Oct. 26, 1955, Ngo Dinh Diem, premier of the interim government of South Vietnam, proclaimed the Republic of Vietnam and became its first president.

The North adopted a constitution Dec. 31, 1959, based on Communist principles and calling for reunification of all Vietnam. North Vietnam sought to take over South Vietnam beginning in 1954. Fighting persisted from 1956, with the Communist Vietcong, aided by North Vietnam, pressing war in the South. Northern aid to Vietcong guerrillas was intensified in 1959, and large-scale troop infiltration began in 1964, with Soviet and Chinese arms assistance. Large Northern forces were stationed in border areas of Laos and Cambodia.

A serious political conflict arose in the South in 1963 when Buddhists denounced authoritarianism and brutality. This paved the way for a military coup Nov. 1-2, 1963, which overthrew Diem. Several other military coups followed.

In 1964, the U.S. began air strikes against North Vietnam. Beginning in 1965, the raids were stepped up and U.S. troops became combatants. U.S. troop strength in Vietnam, which reached a high of 543,400 in Apr. 1969, was ordered reduced by President Nixon in a series of withdrawals, beginning in June 1969. U.S. bombings were resumed in 1972-73.

A cease-fire agreement was signed in Paris Jan. 27, 1973 by the U.S., North and South Vietnam, and the Vietcong. It was never implemented.

North Vietnamese forces attacked remaining government outposts in the Central Highlands in the first months of 1975. Government retreats turned into a rout, and the Saigon regime surrendered April 30. North Vietnam assumed control, and began transforming society along Communist lines.

The war's toll included—Combat deaths: U.S. 47,369; South Vietnam more than 200,000; other allied forces 5,225. Total U.S. fatalities numbered more than 58,000. Vietnamese civilian casualties were more than a million. Displaced war refugees in South Vietnam totaled more than 6.5 million.

The country was officially reunited July 2, 1976. The Northern capital, flag, anthem, emblem, and currency were applied to the new state. Nearly all major government posts went to officials of the former Northern government.

Heavy fighting with Cambodia took place, 1977-80, amid mutual charges of aggression and atrocities against civilians. Increasing numbers of Vietnamese civilians, ethnic Chinese, escaped the country, via the sea or the overland route across Cambodia. Vietnam launched an offensive against Cambodian refugee strongholds along the Thai-Cambodian border in 1985; they also engaged Thai troops.

Relations with China soured as 140,000 ethnic Chinese left Vietnam charging discrimination; China cut off economic aid. Reacting to Vietnam's invasion of Cambodia, China attacked 4 Vietnamese border provinces, Feb. 1979.

Vietnam announced reforms aimed at reducing central control of the economy in 1987, as many of the old revolutionary followers of Ho Chi Minh were removed from office.

Citing Vietnamese cooperation in returning remains of U.S. soldiers killed in the Vietnam War, the U.S. announced an end, Feb. 3, 1994, to a 19-year-old U.S. embargo on trade with Vietnam. The U.S. extended full diplomatic recognition to Vietnam July 11, 1995. The Communist Party replaced the country's ill and aging leadership in Sept. 1997.

Floods in central Vietnam, Oct.-Nov. 1999, killed some 550 people and left over 600,000 families homeless. The U.S. and Vietnam signed a comprehensive trade deal July 13, 2000. U.S. Pres. Bill Clinton made a historic visit to Vietnam Nov. 17-19. Nong Duc Manh, a moderate, was named to head the Communist Party Apr. 22, 2001.

Western Samoa

See Samoa (*formerly* Western Samoa)

Yemen

Republic of Yemen

People: Population: 18,078,035. **Age distrib.** (%): <15: 47.2; 65+: 3.0. **Pop. density:** 89 per sq. mi. **Urban:** 24%. **Ethnic groups:** Predominantly Arab, Afro-Arab, South Asian. **Principal language:** Arabic. **Chief religions:** Muslim (Sha'fi-Sunni, Zaydi-Shi'a).

Geography: Area: 203,600 sq. mi. **Location:** Middle East, on the S coast of the Arabian Peninsula. **Neighbors:** Saudi Arabia on N, Oman on the E. **Topography:** A sandy coastal strip leads to well-watered fertile mountains in interior. **Capital:** Sanaa. **Cities:** Sanaa (urban agg.): 1,303,000; Aden (1995 est.): 562,000.

Government: Type: Republic. **Head of state:** Pres. Ali Abdullah Saleh; b. 1942; in office: July 17, 1978. **Head of gov.:** Prime Min. Abd-al-Qadir Bajamal; b 1946; in office: Apr. 4, 2001. **Local divisions:** 17 governorates and capital region. **Defense budget** (1999): $374 mil. **Active troops:** 66,300

Economy: Industries: Oil, food processing. **Chief crops:** Grains, fruits, qat, coffee, cotton. **Minerals:** Oil, salt. **Crude oil reserves** (2000): 4 bil bbls. **Arable land:** 3%. **Livestock**(2000): cattle: 1.28 mil; chickens: 28.0 mil; goats: 4.21 mil; sheep: 4.76 mil. **Fish catch** (1999): 115,654 metric tons. **Electricity prod.:** 2.400 bil kWh.

Finance: Monetary unit: Rial (Sept. 2001: 169.52 = $1 U.S.). **GDP:** (1999 est.): $12.7 bil. **Per capita GDP:** $750. **Im-**

ports (1999 est.): $2.3 bil; partners: U.S. 9%, UAE 8%, France 8%. **Exports** (1999 est.): $2 bil; partners: China 31%, S. Korea 25%. **Tourism** (1998): $84 mil. **Budget** (1998 est.): $2.6 bil. **International reserves less gold** (Oct. 1999): $1.31 bil. **Gold:** 50,000 oz t. **Consumer prices** (change in 1998): 7.9%.

Transport: Motor vehicles: 229,084 pass. cars, 282,615 comm. vehicles. **Civil aviation:** 650.0 mil pass.-mi; 11 airports. **Chief ports:** Al Hudaydah, Al Mukalla, Aden.

Communications: TV sets: 6.5 per 1,000 pop. **Radios:** 43 per 1,000 pop. **Telephones:** 417,100 main lines.

Health: Life expectancy: 58.45 male; 62.05 female. **Births** (per 1,000 pop): 43.36. **Deaths** (per 1,000 pop.): 9.58. **Natural inc.:** 3.38%. **Infant mortality** (per 1,000 live births): 68.53.

Education: Compulsory: ages 6-15. **Literacy** (1994): 43%.

Major Intl. Organizations: UN (FAO, IBRD, ILO, IMF, IMO, WHO), AL.

Embassy: Suite 705, 2600 Virginia Ave. NW 20037; 965-4760.

Website: http://www.arab.net/yemen/yemen_contents.html

Yemen's territory once was part of the ancient Kingdom of Sheba, or Saba, a prosperous link in trade between Africa and India. The Bible speaks of its gold, spices, and precious stones as gifts borne by the Queen of Sheba to King Solomon.

Yemen became independent in 1918, after years of Ottoman Turkish rule, but remained politically and economically backward. Imam Ahmed ruled 1948-1962. Army officers headed by Brig. Gen. Abdullah al-Salal declared the country to be the Yemen Arab Republic.

The Imam Ahmed's heir, the Imam Mohamad al-Badr, fled to the mountains where tribesmen joined royalist forces; internal warfare between them and the republican forces continued. About 150,000 people died in the fighting.

There was a bloodless coup Nov. 5, 1967. In April 1970 hostilities ended with an agreement between Yemen and Saudi Arabia. On June 13, 1974, an army group, led by Col. Ibrahim al-Hamidi, seized the government. He was killed in 1977.

Meanwhile, South Yemen won independence from Britain in 1967, formed out of the British colony of Aden and the British protectorate of South Arabia. It became the Arab world's only Marxist state, taking the name People's Democratic Republic of Yemen in 1970 and signing a friendship treaty with the USSR in 1979 that allowed for the stationing of Soviet troops.

More than 300,000 Yemenis fled from the south to the north after independence, contributing to 2 decades of hostility between the 2 states that flared into warfare twice in the 1970s.

An Arab League-sponsored agreement between North and South Yemen on unification of the 2 countries was signed Mar. 29, 1979. An agreement providing for widespread political and economic cooperation was signed in 1988.

The 2 countries were formally united May 21, 1990, but regional clan-based rivalries led to full-scale civil war in 1994. Secessionists declared a breakaway state in S Yemen, May 21, 1994, but northern troops captured the former southern capital of Aden in July.

A new constitution was approved Sept. 28. Parliamentary elections were held Apr. 27, 1997.

A dispute between Yemen and Eritrea over the Hanish Isls. in the Red Sea, which led to armed clashes in 1995, was resolved by arbitration in 1998.

While on a refueling stop in Aden, Oct. 12, 2000, the destroyer U.S.S. *Cole* was bombed, leaving 17 Americans dead and more than 3 dozen injured; the U.S. government blamed the attack on terrorists associated with Osama bin Laden.

Yugoslavia

Federal Republic of Yugoslavia

People: Population: 10,677,290. **Age distrib.** (%): <15: 20.2; 65+: 13.1. **Pop. density:** 271 per sq. mi. **Urban:** 52%. **Ethnic groups:** Serbian 63%, Albanian 14%, Montenegrin 6%. **Principal languages:** Serbo-Croatian (official) 95%, Albanian 5%. **Chief religions:** Orthodox 65%, Muslim 19%, Roman Catholic 4%.

Geography: Area: 39,400 sq. mi. **Location:** On the Balkan Peninsula in SE Europe. Present-day Yugoslavia consists of the republics of Serbia and Montenegro. **Neighbors:** Croatia, Bosnia and Herzegovina on W; Hungary on N; Romania, Bulgaria on E; Albania, Macedonia on S. **Capital:** Belgrade. **Cities** (urban agg.): Belgrade 1,482,000.

Government: Type: Republic. **Head of state:** Pres. Vojislav Kostunica; b Mar. 24, 1944; in office: Oct. 7, 2000. **Head of gov.:** Prime Min. Dragisa Pesic; b 1954; in office: July 24, 2001. **Local divisions:** 2 republics, 2 autonomous provinces. **Defense budget:** $1.3 bil. **Active troops:** 97,700.

Economy: Industries: Steel, machinery, consumer goods, mining, electronics. **Chief crops:** Cereals, fruits, vegetables. **Minerals:** Oil, gas, coal, antimony, lead, nickel, gold, zinc, pyrite, copper, chrome. **Crude oil reserves** (2000): 48.57 bil bbls. **Livestock** (2000): cattle: 1.45 mil; chickens: 21.12 mil; goats: 241,000; pigs: 4.09 mil; sheep: 1.92 mil. **Electricity prod.:** 34.455 bil kWh. **Labor force:** N/A

Finance: Monetary unit: New Dinar (Sept. 2001: 65.23 = $1 U.S.). **GDP** (1999 est.): $20.6 bil. **Per capita GDP:** $1,800. **Imports** (1999 est.): $3.3 bil; partners: Germany, Italy. **Exports** (1999 est.): $1.5 bil; partners: Bosnia and Herzegovina, Italy, Macedonia. **Tourism:** $17 mil.

Transport: Railroad: Length: 2,505 mi. **Motor vehicles:** 1.00 mil pass. cars, 331,000 comm. vehicles. **Civil aviation:** 93 mil pass.-mi; 4 airports. **Chief ports:** Bar, Novi Sad.

Communications: TV sets: 27 per 1,000 pop. **Radios:** 118 per 1,000 pop. **Telephones:** 2,280,700 main lines. **Daily newspaper circ.:** 256 per 1,000 pop.

Health (Serbia only): **Life expectancy:** 70.51 male; 76.49 female. **Births** (per 1,000 pop.): 12.46. **Deaths** (per 1,000 pop.): 10.72. **Natural inc.:** 0.17%. **Hosp. beds** (1995): 1 per 188 persons. **Physicians** (1995): 1 per 495 persons. **Infant mortality** (per 1,000 live births): 17.97.

Education: Free, compulsory: ages 7-15. **Literacy:** 98%.

Major Intl. Organizations: Currently suspended from UN and its agencies.

Embassy: 2410 California St. NW 20008; 462-6566.

Website: http://www.gov.yu

Serbia, which had since 1389 been a vassal principality of Turkey, was established as an independent kingdom by the Treaty of Berlin, 1878. Montenegro, independent since 1389, also obtained international recognition in 1878. After the Balkan wars, Serbia's boundaries were enlarged by the annexation of Old Serbia and Macedonia, 1913.

When the Austro-Hungarian empire collapsed after World War I, the Kingdom of Serbs, Croats, and Slovenes was formed from the former provinces of Croatia, Dalmatia, Bosnia, Herzegovina, Slovenia, Vojvodina, and the independent state of Montenegro. The name became Yugoslavia in 1929.

Nazi Germany invaded in 1941. Many Yugoslav partisan troops continued to operate. Among these were the Chetniks led by Draja Mikhailovich, who fought other partisans led by Josip Broz, known as Marshal Tito. Tito, backed by the USSR and Britain from 1943, was in control by the time the Germans had been driven from Yugoslavia in 1945. Mikhailovich was executed July 17, 1946, by the Tito regime.

A constituent assembly proclaimed Yugoslavia a republic Nov. 29, 1945. It became a federal republic Jan. 31, 1946, with Tito, a Communist, heading the government. Tito rejected Stalin's policy of dictating to all Communist nations, and he accepted economic and military aid from the West.

Pres. Tito died May 4, 1980. After his death, Yugoslavia was governed by a collective presidency, with a rotating succession. On Jan. 22, 1990, the Communist Party renounced its leading role in society.

Croatia and Slovenia formally declared independence June 25, 1991. In Croatia, fighting began between Croats and ethnic Serbs. Serbia sent arms and medical supplies to the Serb rebels in Croatia. Croatian forces clashed with Yugoslav army units and their Serb supporters.

The republics of Serbia and Montenegro proclaimed a new "Federal Republic of Yugoslavia" Apr. 17, 1992. Serbia, under Pres. Slobodan Milosevic, was the main arms supplier to ethnic Serb fighters in Bosnia and Herzegovina. The UN imposed sanctions May 30 on the newly reconstituted Yugoslavia as a means of ending the bloodshed in Bosnia.

A peace agreement initialed in Dayton, Ohio, Nov. 21, 1995, was signed in Paris, Dec. 14, by Milosevic and leaders of Bosnia and Croatia. In May 1996, a UN tribunal in the Netherlands began trying suspected war criminals from the former Yugoslavia. The UN lifted sanctions against Yugoslavia Oct. 1, 1996, after elections were held in Bosnia. Mass protests erupted when Milosevic refused to accept opposition victories in local elections Nov. 17; non-Communist governments took office in Belgrade and other cities in Feb. 1997. Barred from running for a 3d term as Serbian president, Milosevic had himself inaugurated as president of Yugoslavia on July 23, 1997.

Defeated in a presidential election Sept. 24, 2000, by opposition leader Vojislav Kostunica, Milosevic initially refused to accept the result. A rising tide of mass demonstrations forced him to resign Oct. 6, and Kostunica was sworn in the next day. Charged with corruption and abuse of power, Milosevic surrendered to Serbian authorities Apr. 1, 2001. He was extradited June 28, 2001, to The Hague, where a UN tribunal had indicted him for war crimes.

Kosovo: A nominally autonomous province in southern Serbia (4,203 sq. mi.), with a population of about 2,000,000, mostly Albanians. The capital is Pristina. Revoking provincial autonomy, Serbia began ruling Kosovo by force in 1989. Albanian secessionists proclaimed an independent Republic of Kosovo in July

1990. Guerrilla attacks by the Kosovo Liberation Army in 1997 brought a ferocious counteroffensive by Serbian authorities.

Fearful that the Serbs were employing "ethnic cleansing" tactics, as they had in Bosnia, the U.S. and its NATO allies sought to pressure the Yugoslav government. When Milosevic refused to comply, NATO launched an air war against Yugoslavia, Mar.-June 1999; the Serbs retaliated by terrorizing the Kosovars and forcing hundreds of thousands to flee, mostly to Albania and Macedonia. A 50,000-member multinational force (KFOR) entered Kosovo in June, and most of the Kosovar refugees had returned by Sept. 1.

Vojvodina: A nominally autonomous province in northern Serbia (8,304 sq. mi.), with a population of about 2,000,000, mostly Serbian. The capital is Novi Sad.

Zaire

See Congo (*formerly* Zaire)

Zambia

Republic of Zambia

People: Population: 9,770,199. **Age distrib.** (%): <15: 47.4; 65+: 2.5. **Pop. density:** 34 per sq. mi. **Urban:** 40%. **Ethnic groups:** African 98.7%, European 1.1%. **Principal languages:** English (official), indigenous. **Chief religions:** Christian 50-75%, Hindu and Muslim 24-49%.

Geography: Area: 285,700 sq. mi. **Location:** In S central Africa. **Neighbors:** Congo (formerly Zaire) on N; Tanzania, Malawi, Mozambique on E; Zimbabwe, Namibia on S; Angola on W. **Topography:** Zambia is mostly high plateau country covered with thick forests, and drained by several important rivers, including the Zambezi. **Capital:** Lusaka (urban agg.): 1,640,000.

Government: Type: Republic. **Head of state and gov.:** Pres. Frederick Chiluba; b Apr. 30, 1943; in office: Nov. 2, 1991. **Local divisions:** 9 provinces. **Defense budget:** $65 mil. **Active troops:** 21,600.

Economy: Industries: Mining, construction, foodstuffs, chemicals. **Chief crops:** Corn, cassava, sorghum, sugar. **Minerals:** Cobalt, copper, zinc, emeralds, gold, lead, silver, uranium, coal. **Arable land:** 7%. **Livestock** (2000): cattle: 2.37 mil; chickens: 29.0 mil; goats: 1.25 mil; pigs: 330,000; sheep: 140,000. **Fish catch** (1999): 70,702 metric tons. **Electricity prod.:** 7.642 bil kWh. **Labor force:** 85% agric.

Finance: Monetary unit: Kwacha (Sept. 2001: 3,635.00 = $1 U.S.). **GDP** (1999 est.): $8.5 bil. **Per capita GDP:** $880. **Imports** (1999 est.): $1.15 bil; partners: South Africa 48%. **Exports** (1999 est.): $900 mil; partners: Japan, Saudi Arabia. **Tourism:** $85 mil. **Budget** (1995 est.): $835 mil. **Intl. reserves less gold** (Dec. 1999): $45.4 mil. **Consumer prices** (change in 1997): 24.8%.

Transport: Railroad: Length: 791 mi. **Motor vehicles:** 142,000 pass. cars, 73,500 comm. vehicles. **Civil aviation:** 27.7 mil pass.-mi; 4 airports. **Chief port:** Mpulungu.

Communications: TV sets: 32 per 1,000 pop. **Radios:** 99 per 1,000 pop. **Telephones** (1999): 83,100 main lines. **Daily newspaper circ.:** 13 per 1,000 pop.

Health: Life expectancy: 37.06 male; 37.53 female. **Births** (per 1,000 pop): 41.46. **Deaths** (per 1,000 pop.): 21.97. **Natural inc.:** 1.95%. **Infant mortality** (per 1,000 live births): 90.89.

Education: Compulsory: ages 7-14. **Literacy**: 78%.

Major Intl. Organizations: UN (FAO, IBRD, ILO, IMF, WHO, WTrO), the Commonwealth, OAU.

Embassy: 2419 Massachusetts Ave. NW 20008; 265-9717.

Website: http://www.zamnet.zm

As Northern Rhodesia, the country was under the administration of the South Africa Company, 1889 until 1924, when the office of governor was established, and, subsequently, a legislature. The country became an independent republic within the Commonwealth Oct. 24, 1964.

After the white government of Rhodesia (now Zimbabwe) declared its independence from Britain Nov. 11, 1965, relations between Zambia and Rhodesia became strained.

As part of a program of government participation in major industries, a government corporation in 1970 took over 51% of the ownership of 2 foreign-owned copper-mining companies. Privately-held land and other enterprises were nationalized in 1975. In the 1980s and 1990s lowered copper prices hurt the economy and severe drought caused famine.

Food riots erupted in June 1990, as the nation suffered its worst violence since independence. Elections held Oct. 1991 brought an end to one-party rule. The new government sought to sell state enterprises, including the copper industry. Pres. Frederick Chiluba won reelection Nov. 18, 1996, but international observers cited harassment of opposition parties. A coup attempt was suppressed Oct. 28, 1997.

According to UN estimates, the AIDS epidemic had orphaned some 650,000 children in Zambia by the end of the 1990s; at that time, about 20% of the adult population had HIV/AIDS.

Zimbabwe

Republic of Zimbabwe

People: Population: 11,365,366. **Age distrib.** (%): <15: 38.7; 65+: 3.6. **Pop. density:** 76 per sq. mi. **Urban:** 35%. **Ethnic groups:** Shona 71%, Ndebele 16%. **Principal languages:** English (official), Shona, Sindebele. **Chief religions:** Syncretic (Christian-indigenous mix) 50%, Christian 25%, indigenous beliefs 24%.

Geography: Area: 149,100 sq. mi. **Location:** In southern Africa. **Neighbors:** Zambia on N, Botswana on W, South Africa on S, Mozambique on E. **Topography:** Zimbabwe is high plateau country, rising to mountains on eastern border, sloping down on the other borders. **Capital:** Harare (urban agg.): 1,752,000.

Government: Type: Republic. **Head of state and gov.:** Pres. Robert Mugabe; b Feb. 21, 1924; in office: Dec. 31, 1987. **Local divisions:** 8 provinces, 2 cities. **Defense budget:** $235 mil. **Active troops:** 40,000

Economy: Industries: Clothing, mining, steel, chemicals. **Chief crops:** Tobacco, sugar, cotton, wheat, corn. **Minerals:** Chromium, gold, nickel, asbestos, copper, iron, coal. **Arable land:** 7%. **Livestock** (2000): cattle: 5.55 mil; chickens: 16.0 mil; goats: 2.79 mil; pigs: 275,000; sheep: 530,000. **Fish catch** (1999): 18,241 metric tons. **Electricity prod.:** 5.780 bil kWh. **Labor force:** 66% agric., 24% serv., 10% ind.

Finance: Monetary unit: Dollar (Sept. 2001: 55.43 = $1 U.S.). **GDP** (1999 est.): $26.5 bil. **Per capita GDP:** $2,400. **Imports** (1999 est.): $2 bil; partners: South Africa 37%, UK 7%, US 6%. **Exports** (1999 est.): $2 bil; partners: South Africa 12%, UK 11%. **Tourism:** $145 mil. **Budget** (FY 1996-97): $2.9 bil. **Intl. reserves less gold** (May 2000): $206.2 mil. **Gold:** 420,000 oz t. **Consumer prices** (change in 1998): 31.8%.

Transport: Railroad: Length: 1,714 mi. **Motor vehicles:** 250,000 pass. cars, 108,000 comm. vehicles. **Civil aviation:** 582.7 mil pass.-mi; 7 airports. **Chief ports:** Binga, Kariba.

Communications: TV sets: 12 per 1,000 pop. **Radios:** 113 per 1,000 pop. **Telephones** (1999): 239,000 main lines. **Daily newspaper circ.:** 17 per 1,000 pop.

Health: Life expectancy: 38.51 male; 35.7 female. **Births** (per 1,000 pop): 24.68. **Deaths** (per 1,000 pop.): 23.22. **Natural inc.:** .15%. **Infant mortality** (per 1,000 live births): 62.61.

Education: Compulsory: ages 6-13. **Literacy**: 85%.

Major Intl. Organizations: UN (FAO, IBRD, ILO, IMF, WHO, WTrO), the Commonwealth, OAU.

Embassy: 1608 New Hampshire Ave. NW 20009; 332-7100.

Britain took over the area as Southern Rhodesia in 1923 from the British South Africa Co. (which, under Cecil Rhodes, had conquered it by 1897) and granted internal self-government. Under a 1961 constitution, voting was restricted to keep whites in power. On Nov. 11, 1965, Prime Min. Ian D. Smith announced his country's unilateral declaration of independence.

Britain termed the act illegal and demanded that the country (known as Rhodesia until 1980) broaden voting rights to provide for eventual rule by the black African majority. The UN imposed sanctions and, in May 1968, a trade embargo.

Intermittent negotiations between the government and various black nationalist groups failed to prevent increasing guerrilla warfare. An "internal settlement" signed Mar. 1978 in which Smith and 3 popular black leaders would share control of the government until a transfer of power to the black majority was rejected by guerrilla leaders.

In the country's first universal-franchise election, Apr. 21, 1979, Bishop Abel Muzorewa's United African National Council gained a bare majority of the black-dominated Parliament. A cease-fire was accepted by all parties, Dec. 5. Independence as Zimbabwe was finally achieved Apr. 18, 1980.

On Mar. 6, 1992, Pres. Robert Mugabe declared a national disaster because of drought and appealed to foreign donors for food, money, and medicine. An economic adjustment program caused widespread hardship. Mugabe was reelected Mar. 1996 after opposition candidates withdrew. A land redistribution campaign launched by Mugabe triggered violent attacks in Apr. 2000 against some white farmers; whites make up less than 1% of the population but hold 70% of the land. Mugabe's opponents gained in legislative elections June 24-25, 2000.

According to UN estimates, about one-fourth of the adult population has HIV/AIDS.

Area and Population of the World

Source: Bureau of the Census, U.S. Dept. of Commerce; prior to 1950, Rand McNally & Co.

Continent or Region	AREA (1,000 sq km)	% of Earth	2001	% World Total, 2001	POPULATION (est., in thousands) 1650	1750	1850	1900	1950	1980	2000
North America	21,400	14.8	486,000	7.9	5,000	5,000	39,000	106,000	221,000	372,000	481,000
South America	17,500	12.1	351,000	5.7	8,000	7,000	20,000	38,000	111,000	242,000	347,000
Europe	22,800	15.7	729,000	11.8	100,000	140,000	265,000	400,000	392,000	484,000	729,000
Asia	31,000	21.4	3,737,000	60.7	335,000	476,000	754,000	932,000	1,411,000	2,601,000	3,688,000
Africa	29,800	20.5	823,000	13.4	100,000	95,000	95,000	118,000	229,000	470,000	805,000
Oceania, incl. Australia	8,400	5.8	31,000	0.5	2,000	2,000	2,000	6,000	12,000	23,000	31,000
Antarctica	14,000	9.7	No indigenous inhabitants								
WORLD	**145,000**	**—**	**6,157,000**	**—**	**550,000**	**725,000**	**1,175,000**	**1,600,000**	**2,556,000**	**4,458,000**	**6,080,000**

Note: Areas are as defined by the U.S. Bureau of the Census and strictly apply only to 1950 and after; before than, areas may be defined differently. Census Bureau area for Europe includes all of Russia. Figures may not add to totals because of rounding.

IT'S A FACT: The continent of Antarctica is about 1½ times the size of the United States, but 98% of the surface is covered by ice, with rest of it barren rock. It is populated mostly by scientific researchers (an average of about 4,000 live there in the summer, and about 1,000 in the winter) from 29 countries, operating under an international treaty signed in 1959.

National Rankings by Population, Area, Population Density, 2001

Source: Bureau of the Census, U.S. Dept. of Commerce

China had the highest population in the world, with an estimated 1.27 billion inhabitants in mid-2001, one-fifth of the world total. India, the second-largest country in population, passed the 1-billion mark in 1999. The U.S. ranked third, with about 278 million in 2001. Russia is the largest country in land area, followed by Canada and China.

Largest Populations

Rank	Country	Population
1.	China	1,273,111,290
2.	India	1,029,991,145
3.	United States	278,058,881
4.	Indonesia	228,437,870
5.	Brazil	174,468,575
6.	Russia	145,470,197
7.	Pakistan	144,616,639
8.	Bangladesh	131,269,860
9.	Japan	126,771,662
10.	Nigeria	126,635,626
11.	Mexico	101,879,171
12.	Germany	83,029,536
13.	Philippines	82,841,518
14.	Vietnam	79,939,014
15.	Egypt	69,536,644
16.	Turkey	66,493,970
17.	Iran	66,128,965
18.	Ethiopia	65,891,874
19.	Thailand	61,797,751
20.	United Kingdom	59,647,790

Smallest Populations

Rank	Country	Population
1.	Vatican City	880
2.	Tuvalu	10,991
3.	Nauru	12,088
4.	Palau	19,092
5.	San Marino	27,336
6.	Monaco	31,842
7.	Liechtenstein	32,528
8.	Saint Kitts and Nevis	38,756
9.	Antigua and Barbuda	66,970
10.	Andorra	67,627

Largest Land Areas

Rank	Country	Area (sq km)
1.	Russia	17,075,400
2.	China	9,326,411
3.	Canada	9,220,970
4.	United States	9,166,601
5.	Brazil	8,456,511
6.	Australia	7,617,931
7.	India	2,973,190
8.	Argentina	2,736,690
9.	Kazakhstan	2,717,300
10.	Algeria	2,381,741

Smallest Land Areas

Rank	Country	Area (sq km)
1.	Vatican City	0.4
2.	Monaco	2
3.	Nauru	21
4	Tuvalu	26
5.	San Marino	60
6.	Liechtenstein	161
7.	Marshall Islands	181
8.	Maldives	300
9.	Malta	321
10.	Grenada	339

Most Densely Populated

Rank	Country	Persons per sq km
1.	Monaco	15,921.0
2.	Singapore	6,891.7
3.	Vatican City	2,200.0
4.	Malta	1,229.2
5.	Bahrain	1,042.6
6.	Maldives	1,035.9
7.	Bangladesh	980.3
8.	Taiwan	693.4
9.	Mauritius	643.5
10.	Barbados	640.3

Most Sparsely Populated

Rank	Country	Persons per sq km
1.	Mongolia	1.7
2.	Namibia	2.2
3.	Australia	2.5
4	Suriname	2.7
5.	Botswana	2.7
6.	Mauritania	2.7
7.	Iceland	2.8
8.	Libya	3.0
9.	Canada	3.4
10.	Guyana	3.5

Current Population and Projections for All Countries: 2001, 2025, and 2050

Source: Bureau of the Census, U.S. Dept. of Commerce

(midyear figures, in thousands)

COUNTRY	2001	2025	2050
Afghanistan	26,813	48,045	76,231
Albania	3,510	4,306	4,609
Algeria	31,736	47,676	58,880
Andorra	68	88	69
Angola	10,366	21,598	34,465
Antigua and Barbuda	67	65	51
Argentina	37,385	48,351	56,258
Armenia	3,336	3,434	3,428
Australia	19,358	22,191	22,846
Austria	8,151	7,822	6,136
Azerbaijan	7,771	9,429	10,585
Bahamas	298	369	404
Bahrain	645	923	1,098
Bangladesh	131,270	179,129	211,020
Barbados	275	279	266
Belarus	10,350	10,248	9,100
Belgium	10,259	9,533	7,609
Belize	256	383	489
Benin	6,591	13,541	22,171
Bhutan	2,049	3,341	4,935
Bolivia	8,300	12,007	15,240
Bosnia and Herzegovina	3,922	3,471	2,833
Botswana	1,586	1,634	2,146
Brazil	174,469	209,587	228,145
Brunei	344	530	704
Bulgaria	7,707	7,292	5,905
Burkina Faso	12,272	21,360	34,956
Burma (Myanmar)	41,995	68,107	87,778
Burundi	6,224	10,469	17,304
Cambodia	12,492	21,434	35,065
Cameroon	15,803	29,108	48,606
Canada	31,593	37,987	40,491

COUNTRY	2001	2025	2050
Cape Verde	405	532	545
Central African Republic	3,577	5,545	7,915
Chad	8,707	14,360	22,504
Chile	15,328	18,681	19,453
China	1,273,111	1,407,739	1,322,435
Colombia	40,349	58,287	73,349
Comoros	596	1,160	1,953
Congo (Brazzaville)	2,894	4,246	6,081
Congo (Kinshasa)	53,625	105,737	184,456
Costa Rica	3,773	5,327	6,321
Côte d'Ivoire	16,393	27,840	44,509
Croatia	4,334	4,348	3,486
Cuba	11,184	11,722	10,594
Cyprus	763	870	878
Czech Republic	10,264	10,128	8,626
Denmark	5,353	5,334	4,476
Djibouti	461	841	1,329
Dominica	71	67	69
Dominican Republic	8,581	11,781	14,586
Ecuador	13,184	17,800	21,059
Egypt	69,537	97,431	117,121
El Salvador	6,238	8,382	10,814
Equatorial Guinea	486	876	1,394
Eritrea	4,298	8,438	13,736
Estonia	1,423	1,237	1,047
Ethiopia	65,892	98,763	159,170
Fiji	844,330	1,085	1,285
Finland	5,176	5,009	4,170
France	59,551	57,806	48,219
Gabon	1,221	1,800	2,518
Gambia, The	1,411	2,678	4,038
Georgia	4,989	4,718	4,365
Germany	83,030	75,372	57,429
Ghana	19,894	28,191	34,324
Greece	10,624	10,473	8,362
Grenada	89	154	210
Guatemala	12,974	22,344	32,185
Guinea	7,614	13,135	20,034
Guinea-Bissau	1,316	2,102	2,970
Guyana	697	710	726
Haiti	6,965	10,171	12,746
Honduras	6,406	8,612	11,001
Hong Kong S.A.R.	7,211	7,816	6,647
Hungary	10,106	9,374	7,684
Iceland	278	298	279
India	1,029,991	1,415,274	1,706,951
Indonesia	228,438	287,985	330,566
Iran	66,129	91,889	110,326
Iraq	23,332	44,146	65,529
Ireland	3,841	3,913	3,600
Israel	5,938	7,778	8,961
Italy	57,680	50,352	38,290
Jamaica	2,666	3,355	3,712
Japan	126,772	119,865	101,334
Jordan	5,153	8,223	11,303
Kazakhstan	16,731	18,565	20,426
Kenya	30,766	34,774	43,852
Kiribati	94	99	100
Kuwait	2,042	3,559	4,159
Kyrgyzstan	4,753	6,066	7,394
Laos	5,636	9,805	13,844
Latvia	2,385	1,965	1,659
Lebanon	3,628	4,831	5,598
Lesotho	2,177	2,724	3,533
Liberia	3,226	6,524	10,992
Libya	5,241	8,297	10,704
Liechtenstein	33	36	31
Lithuania	3,611	3,417	3,063
Luxembourg	443	447	360
Macau	454	644	762
Macedonia, The Former Yugo. Rep. of	2,046	2,171	1,977
Madagascar	15,983	29,306	48,327
Malawi	10,548	12,475	16,884
Malaysia	22,229	34,248	47,289
Maldives	311	623	949
Mali	11,009	22,647	40,433
Malta	395	391	325
Marshall Islands	71	171	348
Mauritania	2,747	5,446	9,329
Mauritius	1,190	1,488	1,614
Mexico	101,879	141,593	167,479
Micronesia, Federated States of	135	143	143
Moldova	4,432	4,830	4,811
Monaco	32	34	34
Mongolia	2,655	3,555	4,057

NA = Not available.

COUNTRY	2001	2025	2050
Montenegro	674	692	603
Morocco	30,645	43,228	52,069
Mozambique	19,371	33,308	47,805
Namibia	1,798	2,310	3,757
Nauru	12	12	12
Nepal	25,284	42,576	60,661
Netherlands	15,981	15,852	12,974
New Zealand	3,864	4,445	4,561
Nicaragua	4,918	8,112	10,817
Niger	10,355	20,424	33,896
Nigeria	126,636	203,423	337,591
North Korea	21,968	25,485	25,930
Norway	4,503	4,592	4,012
Oman	2,622	5,307	8,453
Pakistan	144,617	211,675	260,247
Palau	19	24	26
Panama	2,846	3,796	4,418
Papua New Guinea	5,049	7,597	10,049
Paraguay	5,734	9,929	15,001
Peru	27,484	39,170	47,899
Philippines	82,842	120,519	150,272
Poland	38,634	40,117	36,465
Portugal	10,066	9,012	7,256
Qatar	769	1,208	1,348
Romania	22,364	21,417	18,483
Russia	145,470	138,842	121,777
Rwanda	7,313	12,159	19,607
Saint Kitts and Nevis	39	60	69
Saint Lucia	158	203	224
Saint Vincent and the Grenadines	116	151	163
Samoa	179	367	471
San Marino	27	27	27
São Tomé and Príncipe	165	331	518
Saudi Arabia	22,757	50,374	97,120
Senegal	10,285	22,456	39,690
Serbia	10,003	10,552	9,195
Seychelles	80	91	95
Sierra Leone	5,427	11,010	18,369
Singapore	4,300	4,231	4,161
Slovakia	5,415	5,718	5,215
Slovenia	1,930	1,864	1,484
Solomon Islands	480	840	1,158
Somalia	7,489	15,192	26,243
South Africa	43,586	49,851	58,972
South Korea	47,904	54,256	52,625
Spain	40,038	36,841	29,405
Sri Lanka	19,409	24,088	26,146
Sudan	36,080	64,757	93,625
Suriname	434	460	380
Swaziland	1,104	1,589	3,059
Sweden	8,875	9,158	8,052
Switzerland	7,283	7,064	5,614
Syria	16,729	31,684	43,463
Taiwan	22,370	25,897	25,189
Tajikistan	6,579	9,634	13,261
Tanzania	36,232	50,661	76,500
Thailand	61,798	70,316	69,741
Togo	5,153	11,712	20,725
Tonga	104	133	156
Trinidad and Tobago	1,170	1,083	1,057
Tunisia	9,705	12,760	14,399
Turkey	66,494	89,736	103,656
Turkmenistan	4,603	6,514	8,422
Tuvalu	11	15	20
Uganda	23,986	49,181	91,398
Ukraine	48,760	45,096	39,096
United Arab Emirates	2,407	3,444	4,057
United Kingdom	59,648	59,985	54,116
United States	278,059	335,360	394,241
Uruguay	3,360	3,916	4,256
Uzbekistan	25,155	34,348	42,762
Vanuatu	193	282	347
Vatican City	880	NA	NA
Venezuela	23,917	32,474	37,773
Vietnam	79,939	103,909	119,464
Yemen	18,078	40,439	76,008
Zambia	9,770	16,156	26,967
Zimbabwe	11,365	12,366	16,064
REGIONS			
Asia	3,736,874	4,765,675	5,368,505
Africa	823,490	1,273,302	1,845,701
Europe	728,977	714,309	642,447
South America	350,819	435,601	480,270
North America	486,033	611,876	722,284
Oceania, incl. Australia	31,207	39,897	44,999
WORLD	**6,157,401**	**7,840,660**	**9,104,206**

Population of the World's Largest Cities

Source: United Nations, Dept. for Economic and Social Information and Policy Analysis

Population figures are revised UN estimates and projections for "urban agglomerations"—that is, contiguous densely populated urban areas, not demarcated by administrative boundaries. Data may differ from figures for cities elsewhere in *The World Almanac*.

Rank	City, Country	Pop. (thousands) 2000	Pop. (thousands, projected) 2015	Annual growth rate (percent) 1995-2000	Percentage increase for: 1975-2000	2000-2015[1]	Pop. of city as percentage of nation's: Total pop.	Urban pop.
1.	Tokyo, Japan	26,444	26,444	0.51	34	0	21	27
2.	Mexico City, Mexico	18,131	19,180	1.81	61	6	18	25
3.	Mumbai (Bombay), India	18,066	26,138	3.54	164	45	2	6
4.	Sao Paulo, Brazil	17,755	20,397	1.43	77	15	10	13
5.	New York City, U.S.	16,640	17,432	0.37	5	5	6	8
6.	Lagos, Nigeria	13,427	23,173	5.33	307	73	12	27
7.	Los Angeles, U.S.	13,140	14,080	1.15	47	7	5	6
8.	Calcutta, India	12,918	17,252	1.60	63	34	1	4
9.	Shanghai, China	12,887	14,575	–0.35	13	13	1	3
10.	Buenos Aires, Argentina	12,560	14,076	1.14	37	12	34	38
11.	Dhaka, Bangladesh	12,317	21,119	5.37	467	71	10	39
12.	Karachi, Pakistan	11,794	19,211	3.84	196	63	8	20
13.	Delhi, India	11,695	16,808	3.24	164	44	1	4
14.	Jakarta, Indonesia	11,018	17,256	3.69	129	57	5	13
15.	Osaka, Japan	11,013	11,013	–0.05	12	0	9	11

(1) Projected.

The World's Refugees, 2000

Source: *World Refugee Survey 2001,* U.S. Committee for Refugees, a nonprofit corp.

These estimates are conservative. The refugees in this table include only those considered in need of protection and/or assistance and generally do not include those who have achieved permanent resettlement.

(as of Dec. 31, 2000; only countries estimated to host 50,000 or more refugees are listed)

Place of asylum	Origin of Most	Number
TOTAL AFRICA		**3,346,000**
Algeria	Western Sahara, Palestinians	85,000*
Central African Republic	Sudan, Democratic Rep. of the Congo, Chad	54,000
Democratic Republic of the Congo	Angola, Sudan, Republic of the Congo, Burundi, Uganda, Rwanda	276,000*
Republic of the Congo	Democratic Rep. of the Congo, Angola, Rwanda	126,000
Côte d'Ivoire	Liberia, Sierra Leone	94,000
Egypt	Palestinians, Somalia, Sudan	57,000*
Guinea	Liberia, Sierra Leone	390,000*
Kenya	Somalia, Sudan, Ethiopia	233,000*
Liberia	Sierra Leone	70,000*
Sudan	Eritrea, Ethiopia, Chad, Uganda	385,000*
Tanzania	Burundi, Democratic Rep. of the Congo, Rwanda, Somalia	543,000*
Uganda	Sudan, Rwanda, Congo-Kinshasa, Somalia	230,000
Zambia	Angola, Democratic Rep. of the Congo	255,000*
TOTAL EUROPE		**1,909,000**
Germany	Serbia and Montenegro, Bosnia and Herzegovina	180,000
Switzerland	Serbia and Montenegro, Other	62,600
United Kingdom		87,800
Serbia and Montenegro	Croatia, Bosnia and Herzegovina	484,200
TOTAL AMERICAS AND THE CARIBBEAN		**562,000**
Canada		54,400
United States	El Salvador, Guatemala, Haiti	481,500**
TOTAL EAST ASIA AND THE PACIFIC		**792,000**
China	Vietnam, North Korea	350,000
Indonesia	East Timor	120,800
Malaysia	Philippines, Indonesia	57,400
Thailand	Burma	217,300
TOTAL MIDDLE EAST		**6,035,000**
Gaza Strip	Palestinians	824,600
Iran	Afghanistan, Iraq	1,895,000*
Iraq	Palestinians, Iran, Turkey, Eritrea, Somalia, Sudan	127,700
Jordan	Palestinians	1,580,000
Kuwait	Palestinians, Iraq, Somalia	52,000
Lebanon	Palestinians	383,200
Saudi Arabia	Palestinians, Iraq, Afghanistan	128,500
Syria	Palestinians	389,000
West Bank	Palestinians	583,000
Yemen	Somalia, Palestinians, Ethiopia, Eritrea	67,600
TOTAL SOUTH AND CENTRAL ASIA		**2,656,000**
Bangladesh	Burma, Other	121,600*
India	China (Tibet), Sri Lanka, Burma, Bhutan, Afghanistan, Other	290,000
Nepal	Bhutan, China (Tibet)	129,000
Pakistan	Afghanistan, India	2,019,000*
TOTAL REFUGEES		**14,544,000**

*Estimates vary widely in number reported. **Includes asylum seekers with cases pending in the United States.

Principal Sources of Refugees, 2000

Sources: *World Refugee Survey 2001,* U.S. Committee for Refugees (as of Dec. 31, 2000)

Palestinians	4,000,000
Afghanistan	3,600,000
Sudan	460,000
Iraq	450,000*
Burundi	420,000
Angola	400,000
Sierra Leone	400,000*
Burma	380,000*
Somalia	370,000*
Democratic Republic of the Congo	350,000
Eritrea	350,000*
Croatia	315,000*
Vietnam	300,000
Bosnia and Herzegovina	250,000*
Azerbaijan	230,000
El Salvador	230,000**
Liberia	200,000*
Yugoslavia	190,000*
Armenia	188,000
China (Tibet)	145,000
Bhutan	144,000*
East Timor	120,000
Sri Lanka	110,000*
Western Sahara	110,000*
Guatemala	100,000**

*Estimates vary widely in number reported. **Includes asylum seekers with cases pending in the United States.

Estimated HIV Infection and Reported AIDS Cases, Year-end 2000

Source: UNAIDS, Joint United Nations Program on HIV/AIDS

Studies, primarily in industrialized nations, have indicated that about 60% of adults infected by the human immunodeficiency virus (HIV) will develop acquired immune deficiency syndrome (AIDS) within 12-13 years of becoming infected; development of the disease might be more rapid in Third World countries. About 75-85% of adult HIV infections worldwide have been transmitted through unprotected sexual intercourse.

The number of people living with HIV/AIDS worldwide as of Dec. 2000 was an estimated 36.1 million, with the largest number in sub-Saharan Africa. The total includes about 34.7 million adults and 1.4 million children (under 15 years old; most children are believed to have acquired their HIV infection from their mother before or at birth, or through breastfeeding). The virus continues to spread with great rapidity, causing almost 15,000 new infections daily. Of those, approximately 1,700 are under the age of 15, or 1 child every minute. UNAIDS estimates that nearly 5.3 million new HIV infections occurred in 2000 (10 men, women, and children per minute) and that 3 million people died that year (more than ever before in a single year), including roughly 500,000 children. Since the start of the global epidemic in the late 1970s, HIV has infected an estimated 58 million people; an estimated 21.8 million people have died of AIDS, including 4.3 million children. In all parts of the world except sub-Saharan Africa, more men than women are infected with HIV and die of AIDS. Recognizing the role of male behavior and influence in spreading or curbing the spread of HIV, the world AIDS Campaign chose "Men Make a Difference" as its theme for 2001.

Estimated Current HIV/AIDS Cases by Region, Year-end 2000

Region	Current cases[1]	Percent[2]
Sub-Saharan Africa	25,300,000	70
South/Southeast Asia	5,800,000	16
Latin America	1,400,000	4
North America	920,000	3
East Asia/Pacific	640,000	2
Western Europe	540,000	1
Eastern Europe/Central Asia	700,000	2
Caribbean	390,000	1
North Africa/Middle East	400,000	1
Australasia	15,000	—
WORLD[3]	**36,100,000**	**100**

(1) Adults and children living with HIV/AIDS. (2) Percentage of total number of people worldwide living with HIV. (3) Details do not add to total because of rounding. (—) Dash means less than 1%.

Naturalization: How to Become an American Citizen

Source: Federal Statutes

A person who wishes to be naturalized as a citizen of the United States may obtain the necessary application form as well as detailed information from the nearest office of the Immigration and Naturalization Service.

An applicant must be at least 18 years old and must have been continuously resident in the U.S. for at least 5 years after admission for permanent residence. For husbands and wives of U.S. citizens the period is 3 years in most instances. Special provisions apply to certain veterans of the armed forces.

An applicant must have been physically present in the country for at least half of the required 5 years before filing an application and must:

(1) have been a person of good moral character, attached to the principles of the Constitution, and well disposed to the good order and happiness of the United States for 5 years just before filing the application or for whatever other period of residence is required in the particular case and continue to be such a person;

(2) demonstrate an understanding of the English language, including an ability to read, write, and speak words in ordinary usage in English (persons who are unable to demonstrate this requirement because of physical or developmental disability or mental impairment, are exempt. Persons who, on the date of filing the application, are over 50 years of age and have lived in the U.S. as lawful permanent residents for at least 20 years, or who are over 55 and have been residents for at least 15 years, are exempt); and

(3) demonstrate a knowledge and understanding of the fundamentals of the history, and the principles and form of government, of the United States. This must be done before an INS officer at the interview. Persons who are unable to demonstrate this requirement because of physical or developmental disability or mental impairment, are exempt.

At the interview the applicant may be represented by a lawyer or other representative. If action is favorable, there is a swearing in ceremony. The following oath of allegiance is given:

I hereby declare, on oath, that I absolutely and entirely renounce and abjure all allegiance and fidelity to any foreign prince, potentate, state or sovereignty, to whom or which I have heretofore been a subject or citizen; that I will support and defend the Constitution and laws of the United States of America against all enemies, foreign and domestic; that I will bear true faith and allegiance to the same; that I will bear arms on behalf of the United States when required by the law; that I will perform noncombatant service in the armed forces of the United States when required by the law; that I will perform work of national importance under civilian direction when required by the law; and that I take this obligation freely without any mental reservation or purpose of evasion; so help me God.

Major International Organizations

Asia-Pacific Economic Cooperation Group (APEC), founded Nov. 1989 as a forum to further cooperation on trade and investment between nations of the region and the rest of the world. Members of APEC in 2001 were Australia, Brunei, Canada, Chile, China, Indonesia, Japan, Malaysia, Mexico, New Zealand, Papua New Guinea, Peru, Philippines, Russia, Singapore, South Korea, Taiwan, Thailand, the United States, and Vietnam. Headquarters: Singapore. Website: http://www.apecsec.org.sg

Association of Southeast Asian Nations (ASEAN), formed Aug. 1967 to promote economic, social, and cultural cooperation and development among states of the Southeast Asian region. Members in 2001 were Brunei, Cambodia, Indonesia, Laos, Malaysia, Myanmar, Philippines, Singapore, Thailand, and Vietnam. Annual ministerial meetings set policy; the organization has a central Secretariat and specialized intergovernmental committees. Headquarters: Jakarta. Website: http://www.asean.or.id

Caribbean Community and Common Market (CARICOM), established July 4, 1973. Its aim is to further cooperation in economics, health, education, culture, science and technology, and tax administration, as well as the coordination of foreign policy. Members in 2001 were Antigua and Barbuda, Bahamas (Community only), Barbados, Belize, Dominica, Grenada, Guyana, Haiti (provisional), Jamaica, Montserrat, Saint Kitts and Nevis, Saint Lucia, Saint Vincent and the Grenadines, Suriname, and Trinidad and Tobago. Headquarters: Georgetown, Guyana. Website: http://www.caricom.org

Commonwealth of Independent States (CIS), created Dec. 1991 upon the disbanding of the Soviet Union. An alliance of independent states, it is made up of former Soviet constituent republics. Members in 2001 were 12 of the 15: Armenia, Azerbaijan, Belarus, Georgia, Kazakhstan, Kyrgyzstan, Moldova, Russia, Tajikistan, Turkmenistan, Ukraine, and Uzbekistan. Policy is set through coordinating bodies such as a Council of Heads of State and Council of Heads of Government. Capital of the commonwealth: Minsk, Belarus.

The Commonwealth, originally called the British Commonwealth of Nations, then the Commonwealth of Nations; an association of nations and dependencies that were once parts of the former British Empire. The British monarch is the symbolic head of the Commonwealth.

There are 53 independent nations in the Commonwealth. As of 2001, regular members included the United Kingdom and 14 other nations recognizing the British monarch, represented by a governor-general, as their head of state: Antigua and Barbuda, Australia, Bahamas, Barbados, Belize, Canada, Grenada, Jamaica, New Zealand, Papua New Guinea, Saint Kitts and Nevis, Saint Lucia, Saint Vincent and the Grenadines, and the Solomon Islands. (In addition, Tuvalu, which also recognizes the queen as head of state, was a special member.) Also members in good standing were 37 countries with their own heads of state: Bangladesh, Botswana, Brunei, Cameroon, Cyprus, Dominica, Fiji, The Gambia, Ghana, Guyana, India, Kenya, Kiribati, Lesotho, Malawi, Malaysia, Maldives, Malta, Mauritius, Mozambique, Namibia, Nauru, Nigeria, Samoa, Seychelles, Sierra Leone, Singapore, South Africa, Sri Lanka, Swaziland, Tanzania, Tonga, Trinidad and Tobago, Uganda, Vanuatu, Zambia, and Zimbabwe. Following a military coup in Oct. 1999, Pakistan was suspended from the councils of the Commonwealth. The Commonwealth facilitates consultation among members through meetings of prime ministers and finance ministers and through a permanent Secretariat. Headquarters: London. Website: http://www.thecommonwealth.org/index1.htm

European Free Trade Association (EFTA), created May 3, 1960, to promote expansion of free trade. By Dec. 31, 1966, tariffs and quotas between member nations had been eliminated. Members entered into free trade agreements with the EU in 1972 and 1973. In 1992 the EFTA and EU agreed to create a single market—with free flow of goods, services, capital, and labor—among nations of the 2 organizations. Members in 2001 were Iceland, Liechtenstein, Norway, and Switzerland. Many former EFTA members are now EU members. Headquarters: Geneva. Website: http://www.efta.int/structure/main/index.html

European Union (EU)—known as the European Community (EC) until 1994—the collective designation of 3 organizations with common membership: the European Economic Community (Common Market), the European Coal and Steel Community, and the European Atomic Energy Community (Euratom). The 15 full members in 2001 were Austria, Belgium, Denmark, Finland, France, Germany, Greece, Ireland, Italy, Luxembourg, Netherlands, Portugal, Spain, Sweden, and United Kingdom. Austria, Finland, and Sweden entered the EU on Jan. 1, 1995. Some 70 nations in Africa, the Caribbean, and the Pacific are affiliated under the Lomé Convention. Website: http://europa.eu.int/index.htm

A merger of the 3 communities' executives went into effect July 1, 1967, though the component organizations date back to 1951 and 1958. The Council of Ministers, European Commission, European Parliament, and European Court of Justice comprise the permanent structure. The EU aims to integrate the economies, coordinate social developments, and bring about political union of the member states. Effective Dec. 31, 1992, there are no restrictions on the movement of goods, services, capital, workers, and tourists within the EU. There are also common agricultural, fisheries, and nuclear research policies.

Leaders of member nations (12 at the time) met Dec. 9-11, 1991, in Maastricht, the Netherlands. Treaties and accompanying protocols agreed upon by the leaders committed the organization to launching a common currency (the euro) by 1999; sought to establish common foreign policies; laid the groundwork for a common defense policy; gave the organization a leading role in social policy (Britain was not included in this plan); pledged increased aid for poorer member nations; and slightly increased the powers of the 567-member European Parliament. The treaties went into effect Nov. 1, 1993, following ratification by all 12 members.

In June 1998 the European Central Bank was established. In Jan. 1999, 11 of the 15 EU countries began using the euro for some purposes: Austria, Belgium, Finland, France, Germany, Ireland, Italy, Luxembourg, Netherlands, Portugal, and Spain. This includes all EU countries that wished to participate, except Greece, which did not meet all criteria for inclusion. On July 1, 2002, the 11 countries will change over completely to the euro; at that time current national currencies will no longer be legal tender.

Group of Eight (G-8), established Sept. 22, 1985; organization of 7 major industrial democracies (Canada, France, Germany, Italy, Japan, United Kingdom, and United States) and (later) Russia, meeting periodically to discuss world economic and other issues. At its annual economic summit in May 1998, the name was changed to G-8 from G-7. The original 7 were still free to meet without Russia on some issues, especially those relating to global finance.

International Criminal Police Organization (Interpol), created June 13, 1956, to promote mutual assistance among all police authorities within the limits of the law existing in the different countries. There were 178 members (independent nations), plus 12 subbureaus (dependencies) in 2001.

League of Arab States (Arab League), created Mar. 22, 1945. The League promotes economic, social, political, and military cooperation, mediates disputes, and represents Arab states in certain international negotiations. Members in 2001 were Algeria, Bahrain, Comoros, Djibouti, Egypt, Iraq, Jordan, Kuwait, Lebanon, Libya, Mauritania, Morocco, Oman, Palestine (considered an independent state by the League), Qatar, Saudi Arabia, Somalia, Sudan, Syria, Tunisia, United Arab Emirates, and Yemen. Headquarters: Cairo.

North Atlantic Treaty Organization (NATO), created by treaty (signed Apr. 4, 1949; in effect Aug. 24, 1949). Members in 2001 were Belgium, Canada, Czech Republic, Denmark, France, Germany, Greece, Hungary, Iceland, Italy, Luxembourg, Netherlands, Norway, Poland, Portugal, Spain, Turkey, United Kingdom, and United States. Members agreed to settle disputes by peaceful means, develop their individual and collective capacity to resist armed attack, to regard an attack on one as an attack on all, and take necessary action to repel an attack under Article 51 of the UN Charter. Website: http://www.nato.int

The NATO structure consists of a Council, the Defense Planning Committee, the Military Committee (consisting of 2 commands: Allied Command Europe, Allied Command Atlantic), Nuclear Planning Group, and Canada-U.S. Regional Planning Group. France detached itself from the military command structure in 1966.

With the dissolution of the Soviet Union and the end of the cold war in the early 1990s, members sought to modify the NATO mission, putting greater stress on political action and creating a rapid deployment force to react to local crises. By the mid-1990s, 27 nations, including Russia and other former Soviet republics, had joined with NATO in the so-called Partnership for Peace (PfP; drafted Dec. 1993), which provided for limited joint military exercises, peace-keeping missions, and information exchange. NATO has proceeded gradually toward extending full membership to former Eastern bloc nations. On Mar. 12, 1999, 3 former Warsaw Pact members, Hungary, Poland, and the Czech Republic, formally became members.

In Dec. 1995, a NATO-led multinational force was deployed to help keep the peace in Bosnia and Herzegovina. Headquarters: Brussels.

In response to the terrorist attack on the U.S., Sept. 11, 2001, the NATO Council agreed, Sept. 12, that each member state would take whatever actions it deemed necessary to restore and maintain the security of the North Atlantic area. This was the first instance of terrorism motivating NATO to invoke Article 5 of the 1949 treaty, which stipulates the conditions for collective defense.

Organization of African Unity (OAU), formed May 25, 1963, by 32 African countries (53 members in 2001) to promote peace and security as well as economic and social development. It holds annual conferences of heads of state. Headquarters: Addis Ababa, Ethiopia. Website: http://www.oau-oua.org

Organization of American States (OAS), formed in Bogotá, Colombia, Apr. 30, 1948. It has a Permanent Council, Inter-American Council for Integral Development, Juridical Committee, and Commission on Human Rights. The Permanent Council can call meetings of foreign ministers to deal with urgent security matters. A General Assembly meets annually.

Members in 2001 were: Antigua and Barbuda, Argentina, Bahamas, Barbados, Belize, Bolivia, Brazil, Canada, Chile, Colombia, Costa Rica, Cuba, Dominica, Dominican Republic, Ecuador, El Salvador, Grenada, Guatemala, Guyana, Haiti, Honduras, Jamaica, Mexico, Nicaragua, Panama, Paraguay, Peru, Saint Kitts and Nevis, Saint Lucia, Saint Vincent and the Grenadines, Suriname, Trinidad and Tobago, United States, Uruguay, and Venezuela. In 1962, the OAS suspended Cuba from participation in OAS activities but not from OAS membership. Headquarters: Washington, DC. Website: http://www.oas.org

Organization for Economic Cooperation and Development (OECD), established Sept. 30, 1961, to promote the economic and social welfare of all its member countries and to stimulate efforts on behalf of developing nations. The OECD also collects and disseminates economic and environmental information.

Members in 2001 were Australia, Austria, Belgium, Canada, Czech Republic, Denmark, Finland, France, Germany, Greece, Hungary, Iceland, Ireland, Italy, Japan, Luxembourg, Mexico, Netherlands, New Zealand, Norway, Poland, Portugal, South Korea, Spain, Sweden, Switzerland, Turkey, United Kingdom, and the United States. Headquarters: Paris. Website: http://www.oecd.org

Organization of Petroleum Exporting Countries (OPEC), created Sept. 14, 1960. The group attempts to set world oil prices by controlling oil production. It also pursues members' interests in trade and development dealings with industrialized oil-consuming nations. Members in 2001 were Algeria, Indonesia, Iran, Iraq, Kuwait, Libya, Nigeria, Qatar, Saudi Arabia, United Arab Emirates, and Venezuela. Headquarters: Vienna. Website: http://www.opec.org

Organization for Security and Cooperation in Europe (OSCE), established in 1972 as the Conference on Security and Cooperation in Europe; current name adopted Jan. 1, 1995. The group, formed by NATO and Warsaw Pact members, is interested in furthering East-West relations through a commitment to nonaggression and human rights as well as cooperation in economics, science and technology, cultural exchange, and environmental protection.

There were 55 member states in 2001. Headquarters: Vienna. Website: http://www.osce.org

United Nations

The opening of the 56th regular session of United Nations General Assembly, scheduled for Sept. 11, 2001, was postponed to the following day because of the terrorist attack in New York City; in its first resolution, Sept. 12, the Assembly unanimously passed a resolution condemning terrorism.

UN headquarters is in New York, NY, between First Ave. and Roosevelt Drive and E. 42d St. and E. 48th St. The General Assembly Bldg., Secretariat, Conference and Library bldgs. are interconnected.

Some 52,100 people work in the UN system, which includes the Secretariat and 29 other organizations.

The UN has a post office originating its own stamps.

Proposals to establish an organization of nations for maintenance of world peace led to convening of the United Nations Conference on International Organization at San Francisco, Apr. 25-June 26, 1945, where the charter of the United Nations was drawn up.

The charter was signed June 26 by 50 nations, and by Poland, one of the original 51 members of the United Nations, on Oct. 15, 1945. The charter came into effect Oct. 24, 1945, upon ratification by the permanent members of the Security Council and a majority of other signatories.

Purposes: To maintain international peace and security; to develop friendly relations among nations; to achieve international cooperation in solving economic, social, cultural, and humanitarian problems and in promoting respect for human rights and fundamental freedoms; to be a center for harmonizing the actions of nations in attaining these common ends.

Visitors to the UN: Headquarters is open to the public every day except Thanksgiving, Christmas, and New Year's Day. Guided tours are given approximately every half hour from 9:30 A.M. to 4:45 P.M. daily, except on weekends in January and February.

Groups of 12 or more should write to the Group Program Unit, Public Services Section, Room GA-63, United Nations, New York, NY 10017, or telephone (212) 963-4440. Children under 5 not permitted on tours.

Roster of the United Nations

The 189 members of the United Nations, with the years in which they became members; as of Sept. 2001.

Member	Year
Afghanistan	1946
Albania	1955
Algeria	1962
Andorra	1993
Angola	1976
Antigua and Barbuda	1981
Argentina	1945
Armenia	1992
Australia	1945
Austria	1955
Azerbaijan	1992
Bahamas	1973
Bahrain	1971
Bangladesh	1974
Barbados	1966
Belarus	1945
Belgium	1945
Belize	1981
Benin	1960
Bhutan	1971
Bolivia	1945
Bosnia and Herzegovina	1992
Botswana	1966
Brazil	1945
Brunei	1984
Bulgaria	1955
Burkina Faso	1960
Burundi	1962
Cambodia	1955
Cameroon	1960
Canada	1945
Cape Verde	1975
Central African Republic	1960
Chad	1960
Chile	1945
China[1]	1945
Colombia	1945
Comoros	1975
Congo, Democratic Republic of the (Zaire)	1960
Congo, Republic of the	1960
Costa Rica	1945
Côte d'Ivoire	1960
Croatia	1992
Cuba	1945
Cyprus	1960
Czech Republic[2]	1993
Denmark	1945
Djibouti	1977
Dominica	1978
Dominican Republic	1945
Ecuador	1945
Egypt[3]	1945
El Salvador	1945
Equatorial Guinea	1968
Eritrea	1993
Estonia	1991
Ethiopia	1945
Fiji	1970
Finland	1955
France	1945
Gabon	1960
Gambia, The	1965
Georgia	1992
Germany	1973
Ghana	1957
Greece	1945
Grenada	1974
Guatemala	1945
Guinea	1958
Guinea-Bissau	1974
Guyana	1966
Haiti	1945
Honduras	1945
Hungary	1955
Iceland	1946
India	1945
Indonesia[4]	1950
Iran	1945
Iraq	1945
Ireland	1955
Israel	1949
Italy	1955
Jamaica	1962
Japan	1956
Jordan	1955
Kazakhstan	1992
Kenya	1963
Kiribati	1999
Korea, North	1991
Korea, South	1991
Kuwait	1963
Kyrgyzstan	1992
Laos	1955
Latvia	1991
Lebanon	1945
Lesotho	1966
Liberia	1945
Libya	1955
Liechtenstein	1990
Lithuania	1991
Luxembourg	1945
Macedonia[5]	1993
Madagascar	1960
Malawi	1964
Malaysia[6]	1957
Maldives	1965
Mali	1960
Malta	1964
Marshall Islands	1991
Mauritania	1961
Mauritius	1968
Mexico	1945
Micronesia	1991
Moldova	1992
Monaco	1993
Mongolia	1961
Morocco	1956
Mozambique	1975
Myanmar (Burma)	1948
Namibia	1990
Nauru	1999
Nepal	1955

Member	Year	Member	Year	Member	Year	Member	Year
Netherlands	1945	Russia[7]	1945	Solomon Islands	1978	Turkey	1945
New Zealand	1945	Rwanda	1962	Somalia	1960	Turkmenistan	1992
Nicaragua	1945	Saint Kitts and Nevis	1983	South Africa[8]	1945	Tuvalu	2000
Niger	1960	Saint Lucia	1979	Spain	1955	Uganda	1962
Nigeria	1960	Saint Vincent and the		Sri Lanka	1955	Ukraine	1945
Norway	1945	Grenadines	1980	Sudan	1956	United Arab	
Oman	1971	Samoa (formerly		Suriname	1975	Emirates	1971
Pakistan	1947	Western Samoa)	1976	Swaziland	1968	United Kingdom	1945
Palau	1994	San Marino	1992	Sweden	1946	United States	1945
Panama	1945	São Tomé and		Syria[3]	1945	Uruguay	1945
Papua New Guinea	1975	Príncipe	1975	Tajikistan	1992	Uzbekistan	1992
Paraguay	1945	Saudi Arabia	1945	Tanzania[9]	1961	Vanuatu	1981
Peru	1945	Senegal	1960	Thailand	1946	Venezuela	1945
Philippines	1945	Seychelles	1976	Togo	1960	Vietnam	1977
Poland	1945	Sierra Leone	1961	Tonga	1999	Yemen[10]	1947
Portugal	1955	Singapore[6]	1965	Trinidad and		Yugoslavia[11]	1945
Qatar	1971	Slovakia[2]	1993	Tobago	1962	Zambia	1964
Romania	1955	Slovenia	1992	Tunisia	1956	Zimbabwe	1980

(1) The General Assembly voted in 1971 to expel the Chinese government on Taiwan and admit the Beijing government in its place. (2) Czechoslovakia, which split into the separate nations of the Czech Republic and Slovakia on Jan. 1, 1993, was a UN member from 1945 to 1992. (3) Egypt and Syria were original members of the UN. In 1958, the United Arab Republic was established by a union of Egypt and Syria and continued as a single member of the UN. In 1961, Syria resumed its separate membership. (4) Indonesia withdrew from the UN in 1965 and rejoined in 1966. (5) Admitted under the provisional name of The Former Yugoslav Republic of Macedonia. (6) Malaya joined the UN in 1957. In 1963, its name was changed to Malaysia following the accession of Singapore, Sabah, and Sarawak. Singapore became an independent UN member in 1965. (7) The Union of Soviet Socialist Republics was an original member of the UN from 1945. After the USSR's dissolution in 1991, Russia informed the UN it would be continuing the USSR's membership in the Security Council and all other UN organs with the support of the Commonwealth of Independent States (comprised of most of the former Soviet republics). (8) In 1994, the General Assembly accepted the credentials of the South African delegation, which had been rejected for 24 years because of the country's former apartheid policies. (9) Tanganyika was a member of the UN from 1961 and Zanzibar was a member from 1963. Following the ratification in 1964 of Articles of Union between Tanganyika and Zanzibar, the United Republic of Tanganyika and Zanzibar continued as a single member of the UN, later changing its name to United Republic of Tanzania. (10) The Yemen Arab Republic was admitted in 1947; the People's Republic of Yemen, in 1967. The two nations merged in 1990. (11) The Socialist Federal Republic of Yugoslavia became a member in 1945. After four of its six republics (Bosnia and Herzegovina, Croatia, Macedonia, and Slovenia) declared independence in 1991-92, the two remaining republics, Montenegro and Serbia, reconstituted themselves as the Federal Republic of Yugoslavia, which assumed Yugoslavia's UN seat Apr. 8, 1992. In Sept. 1992, the General Assembly decided the Federal Republic of Yugoslavia should apply for membership as it could not automatically take the seat of the former Yugoslavia. Membership was granted in Nov. 2000 by a vote of the General Assembly.
NOTE: The following sovereign countries are not members of the UN: China (Taiwan), Switzerland, Vatican City (Holy See). Switzerland and Vatican City are, however, permanent observers.

United Nations Secretaries General

Took Office	Secretary, Nation	Took Office	Secretary, Nation	Took Office	Secretary, Nation
1946	Trygve Lie, Norway	1972	Kurt Waldheim, Austria	1992	Boutros Boutros-Ghali, Egypt
1953	Dag Hammarskjold, Sweden	1982	Javier Perez de Cuellar, Peru	1997	Kofi Annan, Ghana
1961	U Thant, Burma				

U.S. Representatives to the United Nations

The U.S. Representative to the United Nations is the Chief of the U.S. Mission to the United Nations in New York and holds the rank and status of Ambassador Extraordinary and Plenipotentiary (A.E.P.). Year given is the year each took office.

Year	Representative	Year	Representative	Year	Representative
1946	Edward R. Stettinius, Jr.	1968	James Russell Wiggins	1981	Jeane J. Kirkpatrick
1946	Herschel V. Johnson (act.)	1969	Charles W. Yost	1985	Vernon A. Walters
1947	Warren R. Austin	1971	George H. W. Bush	1989	Thomas R. Pickering
1953	Henry Cabot Lodge, Jr.	1973	John A. Scali	1992	Edward J. Perkins
1960	James J. Wadsworth	1975	Daniel P. Moynihan	1993	Madeleine K. Albright
1961	Adlai E. Stevenson	1976	William W. Scranton	1997	Bill Richardson
1965	Arthur J. Goldberg	1977	Andrew Young	1999	Richard C. Holbrooke
1968	George W. Ball	1979	Donald McHenry	2001	John D. Negroponte

Organization of the United Nations

The text of the UN Charter may be obtained from the Public Inquiries Unit, Department of Public Information, United Nations, New York, NY 10017. (212) 963-4475.

General Assembly. The General Assembly is composed of representatives of all the member nations. Each nation is entitled to one vote.

The General Assembly meets in regular annual sessions and in special session when necessary. Special sessions are convoked by the secretary general at the request of the Security Council or of a majority of the members of the UN.

On important questions a two-thirds majority of members present and voting is required; on other questions a simple majority is sufficient.

The General Assembly must approve the UN budget and apportion expenses among members. A member in arrears can lose its vote if the amount of arrears equals or exceeds the amount of the contributions due for the preceding 2 full years.

Security Council. The Security Council consists of 15 members, 5 with permanent seats. The remaining 10 are elected for 2-year terms by the General Assembly; they are not eligible for immediate reelection.

Permanent members of the Council are: China, France, Russia, United Kingdom, and the United States.

Nonpermanent members are: (with terms expiring Dec. 31, 2000) Argentina, Canada, Malaysia, Namibia, and the Netherlands; (with terms expiring Dec. 31, 2001) Bangladesh, Jamaica, Mali, Tunisia, and Ukraine.

The Security Council has the primary responsibility within the UN for maintaining international peace and security. The Council may investigate any dispute that threatens international peace and security.

Any member of the UN at UN headquarters may, if invited by the Council, participate in its discussions and a nation not a member of the UN may appear if it is a party to a dispute.

Decisions on procedural questions are made by an affirmative vote of 9 members. On all other matters the affirmative vote of 9 members must include the concurring votes of all permanent members; it is this clause which gives rise to the so-called veto power of permanent members. A party to a dispute must refrain from voting.

The Security Council directs the various peacekeeping forces deployed throughout the world.

Ongoing UN Peacekeeping Missions, 2001

Source: United Nations Cartographic Section

(Year given is the year each mission began operation)

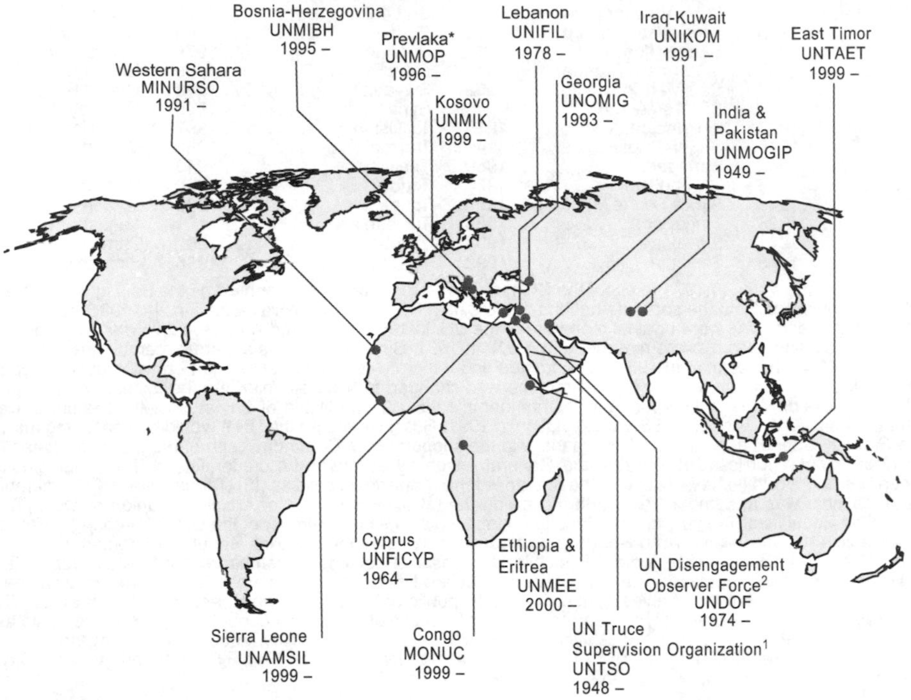

*Prevlaka is on border between Croatia and Montenegro. (1) Functions in 5 Mideast nations. (2) In Golan Heights.

Economic and Social Council. The Economic and Social Council consists of 54 members elected by the General Assembly for 3-year terms. The council is responsible for carrying out UN functions with regard to international economic, social, cultural, educational, health, and related matters. It meets once a year.

Trusteeship Council. The administration of trust territories was under UN supervision; however, all 11 Trust Territories have attained their right to self-determination. The work of the Council has, therefore, been suspended.

Secretariat. The Secretary General is the chief administrative officer of the UN. The Secretary General reports to the General Assembly and may bring to the attention of the Security Council any matter that threatens international peace.

Budget: The General Assembly approved a total budget for the biennium 2000–2001 of $2.53 billion.

International Court of Justice (World Court). The International Court of Justice is the principal judicial organ of the United Nations. All members are *ipso facto* parties to the statute of the Court. Other states may become parties to the Court's statute.

The Court has jurisdiction over cases which the parties submit to it and matters especially provided for in the charter or in treaties. The Court gives advisory opinions and renders judgments. Its decisions are binding only between parties concerned and in respect to a particular dispute. If any party to a case fails to heed a judgment, the other party may have recourse to the Security Council.

The 15 judges are elected for 9-year terms by the General Assembly and the Security Council. Retiring judges are eligible for reelection. The Court remains permanently in session, except during vacations. All questions are decided by majority. The International Court of Justice sits in The Hague, Netherlands.

Selected Specialized and Related Agencies

These agencies are autonomous, with their own memberships and organs, and have a functional relationship or working agreement with the UN (headquarters), except for UNICEF and UNHCR, which report directly to the Economic and Social Council and to the General Assembly.

Food and Agriculture Organization (FAO), aims to increase production from farms, forests, and fisheries; improve food distribution and marketing, nutrition, and the living conditions of rural people. (Viale delle Terme di Caracalla, 00100 Rome, Italy.)

International Atomic Energy Agency (IAEA), aims to promote the safe, peaceful uses of atomic energy. (Vienna International Centre, PO Box 100, A-1400, Vienna, Austria.)

International Bank for Reconstruction and Development (IBRD) (World Bank), provides loans and technical assistance for projects in developing member countries; encourages cofinancing for projects from other public and private sources. The IBRD has 4 affiliates: (1) The **International Development Association (IDA)** provides funds for development projects on concessionary terms to the poorer developing member countries. (2) The **International Finance Corporation (IFC)** promotes the growth of the private sector in developing member countries; encourages the development of local capital markets; stimulates the international flow of private capital. (3) The **Multilateral Investment Guarantee Agency (MIGA)** promotes private investment in developing countries; guarantees investments to protect investors from noncommercial risks, such as nationalization; advises governments on attracting private investment. (4) The **International Center for Settlement of Investment Disputes (ICSID)** provides conciliation and arbitration services for disputes between foreign investors and host governments which arise out of an investment. (1818 H St., NW, Washington, DC 20433.)

International Civil Aviation Org. (ICAO), promotes international civil aviation standards and regulations. (999 University St., Montreal, Quebec, Canada H3C 5H7.)

International Fund for Agricultural Development (IFAD), aims to mobilize funds for agricultural and rural projects in developing countries. (107 Via del Seratico, 00142 Rome, Italy.)

International Labor Org. (ILO), aims to promote employment; improve labor conditions and living standards. (4 route des Morillons, CH-1211 Geneva 22, Switzerland.)

International Maritime Org. (IMO), aims to promote cooperation on technical matters affecting international shipping. (4 Albert Embankment, London SE1 7SR, England.)

International Monetary Fund (IMF), aims to promote international monetary cooperation and currency stabilization and expansion of international trade. (700 19th St., NW, Washington, DC 20431.)

International Telecommunication Union (ITU), establishes international regulations for radio, telegraph, telephone, and space radio-communications, allocates radio frequencies. (Place des Nations, 1211 Geneva 20, Switzerland.)

United Nations Children's Fund (UNICEF), provides financial aid and development assistance to programs for children and mothers in developing countries. (3 UN Plaza, New York, NY 10017.)

United Nations Educational, Scientific, and Cultural Org. (UNESCO), aims to promote collaboration among nations through education, science, and culture. (7 Place de Fontenoy, 75352 Paris 07SP, France.)

United Nations High Commissioner for Refugees (UNHCR), provides essential assistance for refugees. (Place des Nations, 1211 Geneva 10, Switzerland.)

Universal Postal Union (UPU), aims to perfect postal services and promote international collaboration. (Weltpoststrasse 4, 3000 Berne, 15 Switzerland.)

World Health Org. (WHO), aims to aid the attainment of the highest possible level of health. (Avenue Appia 20, CH-1211 Geneva 27, Switzerland.)

World Intellectual Property Org. (WIPO), seeks to protect, through international cooperation, literary, industrial, scientific, and artistic works. (34, Chemin des Colom Bettes, 1211 Geneva, Switzerland.)

World Meteorological Org. (WMO), aims to coordinate and improve world meteorological work. (7 bis Avenue de la Paix, CP2300, 1211 Geneva 2, Switzerland.)

World Trade Org. (WTrO), replacing the General Agreement on Tariffs and Trade (GATT), is the major body overseeing international trade. The WTrO administers trade agreements and treaties, examines the trade regimes of members, keeps track of various trade measures and statistics, and attempts to settle trade disputes. (Centre William Rappard, 154 rue de Lausanne, 1211 Geneva 21, Switzerland.)

IT'S A FACT: The term "United Nations" was coined at the White House on New Year's Day, 1942, when Pres. Franklin D. Roosevelt approached his house guest British Prime Min. Winston Churchill as the latter was taking a bath; FDR suggested it as the name for the wartime alliance being forged against the Axis powers. Later that day, Great Britain and the U.S., with the Soviet Union and China, signed a "United Nations" pact to fight on and avoid agreeing to a separate peace.

Geneva Conventions

The Geneva Conventions are 4 international treaties governing the protection of civilians in time of war, the treatment of prisoners of war, and the care of the wounded and sick in the armed forces. The first convention, covering the sick and wounded, was concluded in Geneva, Switzerland, in 1864; it was amended and expanded in 1906. A third convention, in 1929, covered prisoners of war. Outrage at the treatment of prisoners and civilians during World War II by some belligerents, notably Germany and Japan, prompted the conclusion, in Aug. 1949, of 4 new conventions. Three of these restated and strengthened the previous conventions, and the fourth codified general principles of international law governing the treatment of civilians in wartime.

The 1949 convention for civilians provided for special safeguards for the following categories of people: wounded persons, children under 15 years of age, pregnant women, and the elderly. Discrimination was forbidden on racial, religious, national, or political grounds. Torture, collective punishment, reprisals, the unwarranted destruction of property, and the forced use of civilians for an occupier's armed forces were also prohibited under the 1949 conventions.

Also included in the new 1949 treaties was a pledge to treat prisoners humanely, feed them adequately, and deliver relief supplies to them. They were not to be forced to disclose more than minimal information.

Most countries have formally accepted all or most of the humanitarian conventions as binding. A nation is not free to withdraw its ratification of the conventions during wartime. However, there is no permanent machinery in place to apprehend, try, or punish violators.

Biological and Chemical Weapons

Sources: Center for Civilian Biodefense Studies, Schools of Medicine and Public Health, Johns Hopkins University; Centers for Disease Control and Prevention; *Journal of the American Medical Association;* Federal Emergency Management Agency; *World Almanac* research

The terrorist attacks of Sept. 11, 2001, increased concerns about the possibility of terrorist or state-sponsored attacks using chemical or biological agents. Experts caution that creating or acquiring such weapons of mass destruction would generally require great financial and scientific resources. But even before Sept. 11, Pres. George W. Bush had warned that "the threat of chemical, biological, or nuclear weapons being used against the United States—while not immediate—is very real."

According to the Federal Emergency Management Agency, "Biological agents are infectious microbes or toxins used to produce illness or death in people, animals, or plants. Biological agents can be dispersed as aerosols or airborne particles. Terrorists may use biological agents to contaminate food or water because they are extremely difficult to detect. Chemical agents kill or incapacitate people, destroy livestock, or ravage crops. Some chemical agents are odorless and tasteless and are difficult to detect. They can have an immediate effect (a few seconds to a few minutes) or a delayed effect (several hours to several days)."

Potential Biological Weapons

The Centers for Disease Control and Prevention lists the biological toxins discussed below as among "organisms that pose a risk to national security because they can be easily disseminated or transmitted person-to-person; cause high mortality, with potential for major public health impact; might cause public panic and social disruption; and require special action for public health preparedness."

Anthrax (*Bacillus anthracis*). Anthrax is a deadly bacterial disease that normally afflicts grazing animals, such as sheep or cows. It normally enters the body through skin wounds, inhalation, or ingestion. Human cases rarely occur. In the second half of the 20th century, anthrax was developed as part of a larger biological weapons program by several countries, including the Soviet Union and the U.S. The most significant experience with inhalation anthrax occurred after the accidental release of aerosolized anthrax spores in 1979 at a military biology facility in Sverdlovsk, Russia. Some 79 cases of inhalation anthrax were reported, of which 68 were fatal. In a biological attack, anthrax would likely be distributed by means of aerosol-spray containers in the form of spores. Anthrax spores can survive harsh conditions and cover a great distance before disintegrating, exposing a large number of people to infection. The disease attacks quickly, leading to overwhelming infection within 36 hours. With proper antibiotic treatment, the disease is fatal in fewer than 1% of cases. Yet symptoms, including fever and respiratory failure, often do not develop until the disease is in an advanced stage, at which point medical treatments are rarely effective. An anthrax vaccine exists but is not currently available for civilians.

Botulism (*Clostridium botulinum* toxin). Botulism is a rare but serious paralytic illness. The classic symptoms include double vision, blurred vision, drooping eyelids, slurred speech, difficulty swallowing, dry mouth, and muscle weakness. These are all symptoms of the muscle paralysis caused by the bacterial toxin. If untreated, these symptoms may progress to cause paralysis of the arms, legs, trunk, and respiratory muscles. The respiratory failure and paralysis that occur with severe botulism may require a patient to be on a breathing machine (ventilator) for weeks, needing intensive medical and nursing care. Terrorists have already attempted to use botulinum toxin as a weapon. Aerosols were dispersed at multiple sites in downtown Tokyo, Japan, and at U.S. military installations in Japan on at least 3 occasions between 1990 and 1995 by the Japanese cult Aum Shinrikyo. The attacks failed, apparently because of faulty microbiological technique, deficient aerosol-generating equipment, or internal sabotage. Botulism spread via an aerosol would cause symptoms that would typically present 12 to 72 hours after exposure. If diagnosed early, botulism can be treated with an antitoxin which blocks the action of toxin circulating in the blood.

Plague (*Yersinia pestis*). Plague is the lethal bacterial disease that triggered the "Black Death," which killed approximately 25 million people in Europe in the mid-14th century. It is usually transmitted by fleas that have fed on bacteria-infected rodents. As a weapon, the bacteria would likely be transmitted via aerosol containers, which would lead to a form of the disease called pneumonic plague. Symptoms, which generally appear 2 to 10 days after exposure, include high fever, malaise, headaches, and extremely tender, swollen lymph nodes. Because of the visibility of its symptoms, plague is easier to diagnose than anthrax or smallpox. If diagnosed within 24 hours of the first symptoms, plague is treatable with antibiotics.

Smallpox (*variola major*). Smallpox is a lethal virus that causes severe infection, resulting in the blisters from which the disease gets it name. Smallpox is highly contagious and spreads from person to person through the air, when infected people cough or sneeze. The incubation period—the time before symptoms develop—is 12 days. Within two weeks of developing the first symptom, a red rash, about 30% of victims die. Although there are no known treatments for smallpox, a prototype for a highly effective vaccine, which protects recipients for 10 years, was developed in 1798. Vaccination programs from 1967 to 1980 were highly successful, completely eradicating the disease and leading to the dismantling of production facilities for vaccines. Currently, samples of the virus are known to exist only in two laboratories, one in the U.S. and one in Russia. Few Americans born after 1973, however, have been vaccinated, and even vaccinated people are likely to be vulnerable again. As a result, if the disease were to be released back into civilization, an outbreak would be difficult to contain. In 2000, Centers for Disease Control awarded a contract to Oravax of Cambridge, MA, to produce a smallpox vaccine, but the first full-scale production lots were not expected until 2004.

Tularemia (*Francisella tularensis*). Tularemia is one of the most infectious pathogenic bacteria known, requiring inoculation or inhalation of as few as 10 organisms to cause disease. It is considered to be a dangerous potential biological weapon because of its extreme infectivity, ease of dissemination, and substantial capacity to cause illness and death. Human-to-human transmission has not been documented. During World War II, the potential of *F. tularensis* as a biological weapon was studied by the Japanese as well as by the U.S. and its allies. It was one of several biological weapons stockpiled by the U.S. military in the late 1960s, all of which were destroyed by 1973. The Soviet Union continued weapons production of antibiotic- and vaccine-resistant strains into the early 1990s. Aerosol dissemination of *F. tularensis* in a populated area would be expected to result in the abrupt onset of large numbers of cases of acute, non-specific feverish illness beginning 3 to 5 days later (incubation range, 1-14 days), with inflammation of the lung sacs developing in a significant proportion of cases over the ensuing days and weeks. Without antibiotic treatment, the clinical course could progress to respiratory failure, shock, and death. Several antibiotics have each been used with success in the treatment of tularemia. Vaccination is not recommended for patients already exposed.

Viral hemorrhagic fever (VHF). The term viral hemorrhagic fever (VHF) refers to a group of illnesses that are caused by several distinct families of viruses. While some types of hemorrhagic fever viruses can cause relatively mild illnesses, many of these viruses cause severe, life-threatening diseases. Viruses associated with most VHFs naturally reside in an animal host. However, the hosts of some viruses remain unknown—Ebola and Marburg viruses are well-known examples. Some viruses that cause hemorrhagic fever, for example, Ebola, Marburg, Lassa, and Crimean-Congo, can spread from one person to another. Specific signs and symptoms vary by the type of VHF, but initial signs and symptoms often include marked fever, fatigue, dizziness, muscle aches, loss of strength, and exhaustion. Patients with severe cases of VHF often show signs of bleeding under the skin, in internal organs, or from body orifices like the mouth, eyes, or ears. Patients receive supportive therapy, but generally speaking, there is no other treatment or established cure for VHF. With the exception of yellow fever and Argentine hemorrhagic fever, no vaccines exist that can protect against these diseases. Up to 90% of Ebola cases lead to death within a week.

Potential Chemical Weapons

Chemical weapons could take the form of common substances, like pesticides or mercury, used as poisons. More sophisticated agents for chemical weapons include the following types, which act against different systems in the body.

Blood Agents. Blood agents interfere with the absorption of oxygen into the bloodstream. The chemicals, which included hydrocyanic acid and cyanogen chloride, are stored as a liquid in shells, but convert to a thick gas upon detonation. In high concentrations, cyanide inhalation kills quickly. In two to three minutes after exposure, respiratory failure occurs, followed by heart failure. Six to eight minutes after exposure, victims die. Because the chemicals used in blood agents are highly volatile and are lethal only in large doses, their military usefulness is considered limited. Protective masks and clothing provide a safeguard against attack, and antidotes, which must be intravenously administered immediately after exposure, are highly effective.

Nerve Agents. Nerve agents are the most toxic of all known chemical agents. They include: sarin, tabun, soman, GF, and VX. (Sarin was used in a notorious terrorist attack on the Tokyo subway system by the Aum Shinirkyo cult in March 1995 that killed 12 people and injured several thousand.) When they touch the skin or are inhaled, nerve agents can cause death within minutes. The chemicals disable enzymes needed for the transmission of nerve impulses. Initial symptoms may include runny nose, watery eyes, drooling, excessive sweating, tightness of the chest, and difficulty in breathing. Strong doses may lead to the loss of consciousness, convulsions, and death within 10 minutes. As with most chemical weapons, special protective clothing and masks can prevent exposure to nerve agents. Although antidotes exist, the extremely rapid action of nerve agents often demands immediate treatment. As a result, most American soldiers facing possible chemical attack carry the necessary drugs with them.

Pulmonary (Choking) Agents. Choking agents, such as phosgene and chlorine, are dispersed when a liquid-filled shell explodes, creating a cloud of gas. When the gas is inhaled, the lungs become filled with liquid, making breathing increasingly difficult. Victims often choke to death or die from a respiratory infection. Depending on which chemical agent is used, lethal effects can be felt anywhere from 10 minutes to 24 hours after exposure. There are no treatments or cures for choking agents. Gas masks provide the only defense.

Vesicant (Blister) Agents. Mustard, the best-known vesicant weapon, has been considered a major military threat since its widespread use in gas form during World War I (1914-18). In both liquid and gas forms, mustard causes burning blisters on the skin, eyes, and lungs within 2 to 24 hours of exposure. Symptoms include respiratory failure, pneumonia, and immune system failure, causing death. Other vesicants, such as lewisite, have similar effects and act even more rapidly. Although no specific antidote for vesicants exists, masks and protective clothing can prevent exposure. Often, the effects of the attack can be lessened through the application of ointments and antibiotics on the blisters, or by rinsing exposed flesh with water.

VITAL STATISTICS

Recent Trends in Vital Statistics

Source: National Center for Health Statistics, U.S. Dept. of Health and Human Services; latest years available

Highlights

Provisional data for 2000 reported by the National Center for Health Statistics show that the teen birth rate continued a steady decline that began after 1991 (48.7 births per 1,000 women aged 15-19 years in 2000, compared with 62.1 in 1991). Marriage rates have declined slightly, while divorce rates have held steady. Life expectancy reached an all-time high of 76.7 years in 1998 and remained unchanged for 1999.

Births

An estimated 4,063,000 babies were born in the U.S. in 2000, an increase from 3,959,417 births in 1999. The birth rate increased to 14.8 per 1,000 population in 2000, compared to 14.5 in 1999. The fertility rate (number of live births per 1,000 women aged 15-44 years) for 2000 was estimated at 67.6, higher than the rate for 1999 (65.9).

Deaths

The number of deaths during 2000 was estimated at 2,404,000, slightly higher than during the previous year (2,391,399). Provisional data for 2000 showed a slightly lower death rate (8.7 per 1,000 population) than the previous year (8.8). The provisional infant mortality rate of 6.7 infant deaths per 1,000 live births in 2000 was significantly lower than the 1999 rate of 7.1.

Natural Increase

As a result of natural increase (the excess of births over deaths) by itself, an estimated 1,659,000 persons were added to the population in 2000. The rate of increase (6.1 per 1,000 population) was significantly higher than for 1999 (5.7). The rising rate of natural increase reflected a higher fertility rate and a greater increase in the birth rate than in the death rate, as shown in provisional 2000 data.

Marriages

An estimated 2,329,000 marriages were performed in 2000, about 1% fewer than were performed in 1999 (2,358,000). The marriage rate for 2000 (8.5 per 1,000 population) was down from the 1999 rate of 8.6.

Divorces

About 1,135,000 divorces were granted in the U.S. in 1998, 2% fewer than the number for 1997 (1,163,000), and about 7% fewer than the all-time high of 1,215,000 in 1992. Totals for 1999 and 2000 were not available, but provisional data indicated a divorce rate, for both years, of 4.1 per 1,000 population, the lowest in over 25 years, but somewhat high compared to earlier periods.

Births and Deaths in the U.S.

Source: National Center for Health Statistics, U.S. Dept. of Health and Human Services

	BIRTHS	
Year	Total number	Rate
1960	4,257,850	23.7
1970	3,731,386	18.4
1980	3,612,258	15.9
1990	4,158,212	16.7
1991	4,110,907	16.3
1992	4,065,014	15.9
1993	4,000,240	15.5
1994	3,952,767	15.2
1995	3,899,589	14.8
1996	3,891,494	14.7
1997	3,880,894	14.5
1998	3,941,553	14.6
1999	3,959,417	14.5
2000 (P)	4,063,000	14.5

	DEATHS	
Year	Total number	Rate
1960	1,711,982	9.5
1970	1,921,031	9.5
1980	1,989,841	8.7
1990	2,148,463	8.6
1991	2,169,518	8.6
1992	2,175,613	8.5
1993	2,268,000	8.8
1994	2,278,994	8.8
1995	2,312,132	8.8
1996	2,314,690	8.7
1997	2,314,245	8.6
1998	2,337,258	8.7
1999	2,391,399	8.8
2000 (P)	2,404,000	8.7

(P) = provisional data. **NOTE:** Statistics cover only events occurring within the U.S. and exclude fetal deaths. Rates per 1,000 population; enumerated as of Apr. 1 for 1960 and 1970; estimated as of July 1 for all other years. Beginning 1970 statistics exclude births and deaths occurring to nonresidents of the U.S. Data include revisions.

Marriage and Divorce Rates, 1924-2000

Source: National Center for Health Statistics, U.S. Dept. of Health and Human Services

The U.S. marriage rate dipped during the Depression and peaked just after World War II; by 1998 the rate had fallen to 8.4 per 1,000, with little change in 1999 and 2000. The divorce rate has generally risen since the 1920s; it peaked at 5.3 per 1,000 in 1981, before declining somewhat. The graph below shows marriage and divorce rates per 1,000 population since 1924. Some data provisional.

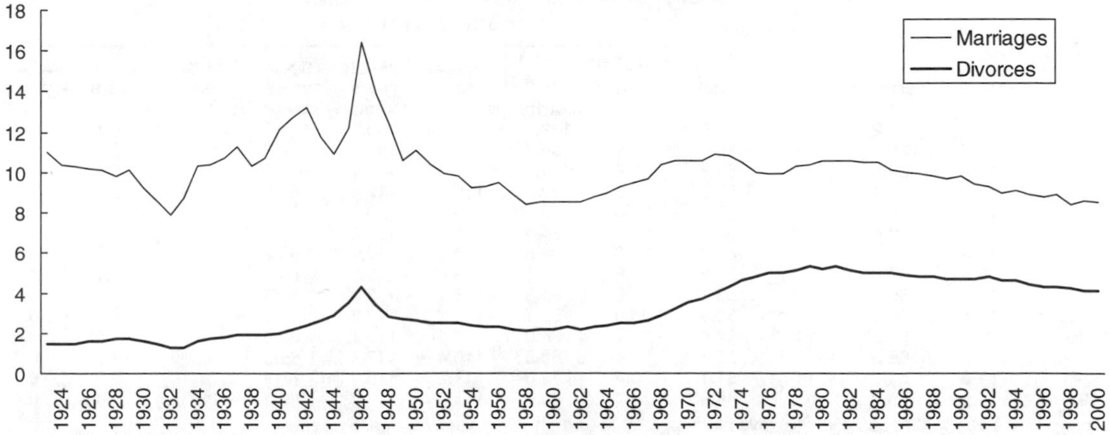

Births and Deaths, by States, 1998-99

Source: National Center for Health Statistics, U.S. Dept. of Health and Human Services

	LIVE BIRTHS				DEATHS			
	1999		1998		1999		1998	
	Number	Rate	Number	Rate	Number	Rate	Number	Rate
Alabama	62,122	14.2	62,074	14.3	44,806	1,025.3	43,950	1,009.9
Alaska	9,950	16.1	9,926	16.2	2,708	437.1	2,571	418.7
Arizona	81,145	17.0	78,243	16.8	40,050	838.2	38,300	820.4
Arkansas	36,729	14.4	36,865	14.5	27,925	1,094.5	27,510	1,083.8
California	518,508	15.6	521,661	16.0	229,380	692.0	226,954	694.8
Colorado	62,167	15.3	59,577	15.0	27,114	668.5	26,640	670.9
Connecticut	43,310	13.2	43,820	13.4	29,446	897.2	29,710	907.4
Delaware	10,676	14.2	10,578	14.2	6,666	884.6	6,578	884.6
District of Columbia	7,522	14.5	7,686	14.7	6,076	1,170.7	6,054	1,157.3
Florida	197,023	13.0	195,637	13.1	163,224	1,080.1	158,167	1,060.4
Georgia	126,717	16.3	122,368	16.0	62,028	796.4	60,428	790.7
Hawaii	17,038	14.4	17,583	14.7	8,270	697.6	8,091	678.2
Idaho	19,872	15.9	19,391	15.8	9,579	765.3	9,155	745.1
Illinois	182,068	15.0	182,588	15.2	108,436	894.1	104,480	867.4
Indiana	86,031	14.5	85,122	14.4	55,303	930.6	53,477	906.5
Iowa	37,558	13.1	37,282	13.0	28,411	990.1	28,362	990.8
Kansas	38,782	14.6	38,422	14.6	24,472	922.1	24,057	915.0
Kentucky	54,403	13.7	54,329	13.8	39,321	992.7	37,832	961.1
Louisiana	67,136	15.4	66,888	15.3	41,238	943.2	40,337	923.3
Maine	13,616	10.9	13,733	11.0	12,261	978.5	12,135	975.3
Maryland	71,967	13.9	71,972	14.0	43,089	833.2	42,059	819.1
Massachusetts	80,939	13.1	81,411	13.2	55,840	904.3	55,237	898.6
Michigan	133,607	13.5	133,666	13.6	87,232	884.4	85,160	867.5
Minnesota	65,990	13.8	65,202	13.8	38,537	807.0	37,195	787.1
Mississippi	42,684	15.4	42,939	15.6	28,185	1,018.0	27,847	1,011.8
Missouri	75,432	13.8	75,358	13.9	55,931	1,022.8	55,070	1,012.6
Montana	10,785	12.2	10,795	12.3	8,128	920.7	7,981	906.5
Nebraska	23,907	14.3	23,534	14.2	15,579	935.1	15,198	914.0
Nevada	29,362	16.2	28,699	16.4	15,082	833.6	14,464	828.0
New Hampshire	14,041	11.7	14,429	12.2	9,537	794.0	9,495	801.2
New Jersey	114,105	14.0	114,550	14.1	73,981	908.5	71,611	882.5
New Mexico	27,191	15.6	27,318	15.7	13,676	786.0	12,907	743.1
New York	255,612	14.0	258,207	14.2	159,927	878.9	156,619	861.7
North Carolina	113,795	14.9	111,688	14.8	69,600	909.7	67,993	901.0
North Dakota	7,639	12.1	7,932	12.4	6,103	963.1	5,920	927.5
Ohio	152,584	13.6	152,794	13.6	108,517	964.0	105,891	944.7
Oklahoma	49,010	14.6	49,461	14.8	34,700	1,033.3	33,929	1,013.8
Oregon	45,204	13.6	45,273	13.8	29,422	887.2	29,383	895.3
Pennsylvania	145,347	12.1	145,899	12.2	130,283	1,086.2	126,700	1,055.7
Rhode Island	12,366	12.5	12,599	12.7	9,708	979.8	9,604	971.6
South Carolina	54,948	14.1	53,877	14.0	36,053	927.8	34,827	907.9
South Dakota	10,524	14.4	10,288	13.9	6,953	948.4	6,867	930.3
Tennessee	77,803	14.2	77,396	14.3	53,765	980.5	53,415	983.6
Texas	349,245	17.4	342,283	17.3	146,858	732.7	142,605	721.7
Utah	46,290	21.7	45,165	21.5	12,058	566.1	11,824	563.1
Vermont	6,567	11.1	6,582	11.1	4,993	840.9	4,948	837.4
Virginia	95,469	13.9	94,351	13.9	55,320	804.9	54,446	801.7
Washington	79,586	13.8	79,663	14.0	43,865	762.0	42,706	750.6
West Virginia	20,728	11.5	20,747	11.5	21,049	1,164.9	20,767	1,146.6
Wisconsin	68,208	13.0	67,450	12.9	46,672	888.9	45,947	879.6
Wyoming	6,129	12.8	6,252	13.0	4,042	842.8	3,853	801.2
United States	**3,959,417**	**14.5**	**3,941,553**	**14.6**	**2,391,399**	**877.0**	**2,337,256**	**864.7**

Note: Birth rates are per 1,000 population. Death rates are per 100,000 population.

Birth Rates; Fertility Rates by Age of Mother, 1950-99

Source: National Center for Health Statistics, U.S. Dept. of Health and Human Services

			AGE OF MOTHER									
			10-14 years	15-19 years			20-24 years	25-29 years	30-34 years	35-39 years	40-44 years	45-49 years
	Birth rate[1]	Fertility rate[2]		Total	15-17	18-19						
			Live births per 1,000 women by age group									
1950	24.1	106.2	1.0	81.6	40.7	132.7	196.6	166.1	103.7	52.9	15.1	1.2
1960	23.7	118.0	0.8	89.1	43.9	166.7	258.1	197.4	112.7	56.2	15.5	0.9
1970	18.4	87.9	1.2	68.3	38.8	114.7	167.8	145.1	73.3	31.7	8.1	0.5
1980	15.9	68.4	1.1	53.0	32.5	82.1	115.1	112.9	61.9	19.8	3.9	0.2
1990	16.7	70.9	1.4	59.9	37.5	88.6	116.5	120.2	80.8	31.7	5.5	0.2
1991	16.3	69.6	1.4	62.1	38.7	94.4	115.7	118.2	79.5	32.0	5.5	0.2
1992	15.9	68.9	1.4	60.7	37.8	94.5	114.6	117.4	80.2	32.5	5.9	0.3
1993	15.5	67.6	1.4	59.6	37.8	92.1	112.6	115.5	80.8	32.9	6.1	0.3
1994	15.2	66.7	1.4	58.9	37.6	91.5	111.1	113.9	81.5	33.7	6.4	0.3
1995	14.8	65.6	1.3	56.8	36.0	89.1	109.8	112.2	82.5	34.3	6.6	0.3
1996	14.7	65.3	1.2	54.4	33.8	86.0	110.4	113.1	83.9	35.3	6.8	0.3
1997	14.5	65.0	1.1	52.3	32.1	83.6	110.4	113.8	85.3	36.1	7.1	0.4
1998	14.6	65.6	1.0	51.1	30.4	82.0	111.2	115.9	87.4	37.4	7.3	0.4
1999	14.5	65.9	0.9	49.6	28.7	80.3	111.0	117.8	89.6	38.3	7.4	0.4

(1) Live births per 1,000 population. (2) Live births per 1,000 women 15-44 years of age.

Nonmarital Childbearing in the U.S., 1970-99

Source: National Center for Health Statistics, U.S. Dept. of Health and Human Services

	1970	1975	1980	1985	1990	1993	1994	1995	1996	1997	1998	1999
Race of Mother	**Percent of live births to unmarried mothers**											
All races	10.7	14.3	18.4	22.0	28.0	31.0	32.6	32.2	32.4	32.4	32.8	33.0
White	5.5	7.1	11.2	14.7	20.4	23.6	25.4	25.3	25.7	25.8	26.3	26.8
Black	37.5	49.5	56.1	61.2	66.5	68.7	70.4	69.9	69.8	69.2	69.1	68.9
American Indian or Alaska Native	22.4	32.7	39.2	46.8	53.6	55.8	57.0	57.2	58.0	58.7	59.3	58.9
Asian or Pacific Islander	—	—	7.3	9.5	13.2	15.7	16.2	16.3	16.7	15.6	15.6	15.4
Hispanic origin (selected states)[1,2]	—	—	23.6	29.5	36.7	40.0	43.1	40.8	40.7	40.9	41.6	42.2
White, non-Hispanic (selected states)[1]	—	—	9.6	12.4	16.9	19.5	20.8	21.2	21.5	21.5	21.9	22.1
Black, non-Hispanic (selected states)[1]	—	—	57.3	62.1	66.7	68.9	70.7	70.0	70.0	69.4	69.3	69.1
Live births to unmarried mothers	399	448	666	828	1,165	1,240	1,290	1,254	1,260	1,257	1,294	1,309
Maternal age	**Percent distribution of live births to unmarried mothers**											
Under 20 years	50.1	52.1	40.8	33.8	30.9	29.7	30.5	30.9	30.4	30.7	30.1	29.3
20–24 years	31.8	29.9	35.6	36.3	34.7	35.4	34.8	34.5	34.2	34.9	35.6	36.4
25 years and over	18.1	18.0	23.5	29.9	34.4	34.9	34.6	34.7	35.3	34.4	34.3	34.3
	Live births per 1,000 unmarried women 15–44 years of age[3]											
All races and origins	26.4	24.5	29.4	32.8	43.8	45.3	46.9	45.1	44.8	44.0	44.3	44.4
White[4]	13.9	12.4	18.1	22.5	32.9	35.9	38.3	37.5	37.6	37.0	37.5	38.1
Black[4]	95.5	84.2	81.1	77.0	90.5	84.0	82.1	75.9	74.4	73.4	73.3	71.5
Hispanic origin (selected states)[1,2]	—	—	—	—	89.6	95.2	101.2	95.0	93.2	91.4	90.1	93.4
White, non-Hispanic	—	—	—	—	—	—	28.5	28.2	28.3	27.0	27.4	27.9

— Data not available. (1) Data for Hispanics and non-Hispanics are affected by expansion of the reporting area for an Hispanic-origin item on the birth certificate and by immigration. These 2 factors affect numbers of events, composition of the Hispanic population, and maternal and infant health characteristics. The states in the reporting area increased from 22 in 1980, to 23 and the District of Columbia in 1983, 48 and DC by 1990, and 50 and DC by 1993 and later years. (2) Includes mothers of all races. (3) Rates computed by relating births to unmarried mothers, regardless of mother's age, to unmarried women 15–44 years of age. (4) For 1970 and 1975, birth rates are by race of child.

Top Countries for U.S. Foreign Adoptions, 1991-2000

Source: Holt International Children's Service

Country	2000	1999	1998	1997	1996	1995	1994	1993	1992	1991
China	5,053	4,101	4,206	3,597	3,333	2,130	787	330	206	61
Russia	4,269	4,348	4,491	3,816	2,454	1,896	1,530	746	324	0
South Korea	1,794	2,008	1,829	1,654	1,516	1,666	1,795	1,775	1,840	1,818
Guatemala	1,511	1,002	911	788	427	449	436	512	418	329
Romania	1,122	895	406	612	555	275	199	97	121	2,594
Vietnam	724	712	603	425	354	318	220	110	22	37
Ukraine	659	323	180	NA	1	4	164	273	55	0
India	503	500	478	349	380	371	412	331	352	445
Cambodia	402	248	351	233	32	10	3	1	15	60
Kazakhstan	399	113	NA	NA	0	0	0	0	0	0
Colombia	246	231	NA	NA	255	350	351	426	404	521
Bulgaria	214	221	151	148	163	110	97	133	91	9
Philippines	173	198	200	163	229	298	314	360	357	393
Haiti	131	96	121	144	68	49	61	51	16	49
Mexico	106	137	168	152	76	83	85	91	91	97
Ethiopia	95	103	96	82	44	63	54	30	37	15
Thailand	88	77	84	NA	55	53	47	69	86	131
Poland	83	NA	NA	NA	62	30	94	70	109	92
Moldova	79	63	NA	NA	40	44	5	1	20	0
Bolivia	60	NA	NA	NA	33	21	37	124	73	46

NA = Not available.

Numbers of Multiple Births in the U.S., 1990-99

Source: National Center for Health Statistics, U.S. Dept. of Health and Human Services

The general upward trend in multiple births reflects greater numbers of births to older women and increased use of fertility drugs.

Year	Twins	Triplets	Quadruplets	Quintuplets and higher	Year	Twins	Triplets	Quadruplets	Quintuplets and higher
1990	93,865	2,830	185	13	1996	100,750	5,298	560	81
1992	95,372	3,547	310	26	1997	104,137	6,148	510	79
1993	96,445	3,834	277	57	1998	110,670	6,919	627	79
1994	97,064	4,233	315	46	1999	114,307	6,742	512	67
1995	96,736	4,551	365	57					

10 Leading Causes of Infant Death in the U.S., 1999

Source: National Center for Health Statistics, U.S. Dept. of Health and Human Services

Cause	Number	Rate[1]	Percent change 1998-99
Congenital malformations, deformations, and chromosomal abnormalities	5,473	138.2	–3.2
Disorders relating to short gestation and low birthweight, not elsewhere classified	4.392	110.9	–3.5
Sudden infant death syndrome	2,648	66.9	–9.9
Newborn affected by maternal complications of pregnancy	1,399	35.3	0.7
Respiratory distress syndrome	1,110	28.0	–16.8
Newborn affected by complications of placenta, cord, and membranes	1,025	25.9	1.5
Accidents (unintentional injuries)	845	21.3	13.5
Bacterial sepsis[2] of newborn	691	17.5	2.1
Diseases of the circulatory system	667	16.8	–6.4
Atelectasis[3]	647	16.3	45.9
All other causes	9,040	228.3	NA
All causes	**27,937**	**705.6**	**–2.0**

NA = Not available. (1) Infant deaths per 100,000 live births. (2) Toxic condition resulting from the spread of bacteria. (3) Defective expansion of the air-containing cells of the lungs.

U.S. Infant Mortality Rates, by Race and Sex, 1960-99

Source: National Center for Health Statistics, U.S. Dept. of Health and Human Services

	ALL RACES			WHITE			BLACK		
Year	Total	Male	Female	Total	Male	Female	Total	Male	Female
1960	26.0	29.3	22.6	22.9	26.0	19.6	44.3	49.1	39.4
1970	20.0	22.4	17.5	17.8	20.0	15.4	32.6	36.2	29.0
1980	12.6	13.9	11.2	11.0	12.3	9.6	21.4	23.3	19.4
1985	10.6	11.9	9.3	9.3	10.6	8.0	18.2	19.9	16.5
1986	10.4	11.5	9.1	8.9	10.0	7.8	18.0	20.0	16.0
1987	10.1	11.2	8.9	8.6	9.6	7.6	17.9	19.6	16.0
1988	10.0	11.0	8.9	8.5	9.5	7.4	17.6	19.0	16.1
1989	9.8	10.8	8.8	8.1	9.0	7.1	18.6	20.0	17.2
1990	9.2	10.3	8.1	7.6	8.5	6.6	18.0	19.6	16.2
1991	8.9	10.0	7.8	7.3	8.3	6.3	17.6	19.4	15.7
1992	8.5	9.4	7.6	6.9	7.7	6.1	16.8	18.4	15.3
1993	8.4	9.3	7.4	6.8	7.6	6.0	16.5	18.3	14.7
1994	8.0	8.8	7.2	6.6	7.2	5.9	15.8	17.5	14.1
1995	7.6	8.3	6.8	6.3	7.0	5.6	15.1	16.3	13.9
1996	7.3	8.0	6.6	6.1	6.7	5.4	14.7	16.0	13.3
1997	7.2	8.0	6.5	6.0	6.7	5.4	14.2	15.5	12.8
1998	7.2	7.8	6.5	6.0	6.5	5.4	14.3	15.7	12.8
1999	7.1	7.7	6.4	5.8	6.4	5.2	14.6	15.9	13.2

Note: Rates per 1,000 live births.

Years of Life Expected at Birth in U.S., 1900-99

Source: National Center for Health Statistics, U.S. Dept. of Health and Human Services

	ALL RACES			WHITE			BLACK		
Year[1]	Total	Male	Female	Total	Male	Female	Total	Male	Female
1900	47.3	46.3	48.3	47.6	46.6	48.7	NA	NA	NA
1910	50.0	48.4	51.8	50.3	48.6	52.0	NA	NA	NA
1920	54.1	53.6	54.6	54.9	54.4	55.6	NA	NA	NA
1930	59.7	58.1	61.6	61.4	59.7	63.5	NA	NA	NA
1940	62.9	60.8	65.2	64.2	62.1	66.6	NA	NA	NA
1950	68.2	65.6	71.1	69.1	66.5	72.2	NA	NA	NA
1960	69.7	66.6	73.1	70.6	67.4	74.1	NA	NA	NA
1970	70.8	67.1	74.7	71.7	68.0	75.6	64.1	60.0	68.3
1975	72.6	68.8	76.6	73.4	69.5	77.3	68.8	62.4	71.3
1980	73.7	70.0	77.5	74.4	70.7	78.1	68.1	63.8	72.5
1981	74.2	70.4	77.8	74.8	71.1	78.4	68.9	64.5	73.2
1982	74.5	70.9	78.1	75.1	71.5	78.7	69.4	65.1	73.6
1983	74.6	71.0	78.1	75.2	71.7	78.7	69.4	65.2	73.5
1984	74.7	71.2	78.2	75.3	71.8	78.7	69.5	65.3	73.6
1985	74.7	71.2	78.2	75.3	71.9	78.7	69.3	65.0	73.4
1986	74.8	71.3	78.3	75.4	72.0	78.8	69.1	64.8	73.4
1987	75.0	71.5	78.4	75.6	72.2	78.9	69.1	64.7	73.4
1988	74.9	71.5	78.3	75.6	72.3	78.9	68.9	64.4	73.2
1989	75.1	71.7	78.5	75.9	72.5	79.2	68.8	64.3	73.3
1990	75.4	71.8	78.8	76.1	72.9	79.4	69.1	64.5	73.6
1991	75.5	72.0	78.9	76.3	72.9	79.2	69.3	64.6	73.8
1992	75.5	72.1	78.9	76.4	73.0	79.5	69.6	65.0	73.9
1993	75.5	72.1	78.9	76.3	73.0	79.5	69.2	64.6	73.7
1994	75.7	72.4	79.0	76.5	73.3	79.6	69.5	64.9	73.9
1995	75.8	72.5	78.9	76.5	73.4	79.6	69.6	65.2	73.9
1996	76.1	73.1	79.1	76.8	73.9	79.7	70.2	66.1	74.2
1997	76.5	73.6	79.4	77.1	74.3	79.9	71.1	67.2	74.7
1998	76.7	73.8	79.5	77.3	74.5	80.0	71.3	67.6	74.8
1999	76.7	73.9	79.4	77.3	74.6	79.9	71.4	67.8	74.7

NA = Not available. (1) Data prior to 1940 for death-registration states only.

U.S. Life Expectancy at Selected Ages, 1999

Source: National Center for Health Statistics, U.S. Dept. of Health and Human Services

	ALL RACES[1]			WHITE			BLACK		
Exact age in years	Both sexes	Male	Female	Both sexes	Male	Female	Both sexes	Male	Female
0	76.7	73.9	79.4	77.3	74.6	79.9	71.4	67.8	74.7
1	76.3	73.5	78.9	76.8	74.1	79.3	71.5	67.9	74.7
5	72.4	69.6	75.0	72.9	70.2	75.4	67.6	64.1	70.9
10	67.4	64.7	70.0	67.9	65.2	70.5	62.7	59.2	66.0
15	62.5	59.8	65.1	63.0	60.3	65.5	57.8	54.3	61.0
20	57.7	55.0	60.2	58.2	55.6	60.6	53.1	49.6	56.2
25	53.0	50.4	55.4	53.4	50.9	55.8	48.5	45.2	51.4
30	48.2	45.7	50.5	48.6	46.2	50.9	43.9	40.7	46.6
35	43.5	41.1	45.7	43.9	41.5	46.1	39.3	36.3	41.9
40	38.8	36.5	41.0	39.2	36.9	41.3	34.8	31.9	37.4
45	34.2	32.0	36.3	34.6	32.4	36.6	30.6	27.8	33.0
50	29.8	27.7	31.7	30.1	28.0	32.0	26.6	24.0	28.7
55	25.5	23.5	27.3	25.7	23.8	27.5	22.8	20.4	24.7
60	21.5	19.6	23.1	21.6	19.8	23.2	19.3	17.2	20.9
65	17.7	16.1	19.1	17.8	16.1	19.2	16.0	14.3	17.3
70	14.3	12.8	15.4	14.4	12.9	15.5	13.0	11.6	14.0
75	11.2	10.0	12.1	11.2	10.0	12.1	10.4	9.2	11.1
80	8.5	7.5	9.1	8.5	7.5	9.0	8.2	7.2	8.6
85	6.3	5.5	6.6	6.2	5.4	6.6	6.2	5.6	6.5
90	4.6	4.1	4.8	4.5	4.0	4.7	4.8	4.4	4.8
95	3.4	3.0	3.5	3.2	2.9	3.3	3.6	3.5	3.6
100	2.6	2.4	2.6	2.3	2.2	2.4	2.8	2.8	2.7

(1) Includes races other than white and black.

The 10 Leading Causes of Death, 1999

Source: National Center for Health Statistics, U.S. Dept. of Health and Human Services

	Number	Death rate[1]	Percentage of total deaths
ALL CAUSES	**2,391,399**	**877.0**	**100.0**
1. Heart disease	725,192	265.9	30.3
2. Cancer	549,838	201.6	23.0
3. Stroke	167,366	61.4	7.0
4. Chronic lower respiratory diseases	124,181	45.5	5.2
5. Accidents (unintentional injuries)	97,860	35.9	4.1
6. Diabetes mellitus	68,399	25.1	2.9
7. Influenza and pneumonia	63,730	23.4	2.7
8. Alzheimer's disease	44,536	16.3	1.9
9. Kidney disease	35,525	13.0	1.5
10. Blood poisoning	30,680	11.3	1.3

(1) Per 100,000 population.

Death Rates[1] for Suicide at Selected Ages, 1970, 1980, 1999

Source: *Health, United States, 2001,* National Center for Health Statistics, U.S. Dept. of Health and Human Services

	1999			1980			1970		
AGE	BOTH SEXES	MALE	FEMALE	BOTH SEXES	MALE	FEMALE	BOTH SEXES	MALE	FEMALE
15-24	10.3	17.1	3.1	12.3	19.9	4.3	13.1	13.5	4.2
25-44	13.9	22.3	5.6	15.6	24.0	7.7	8.8	20.9	10.2
45-64	13.4	21.2	6.0	15.9	23.7	8.9	20.6	30.0	12.0
65 and older	15.8	32.1	4.3	17.6	35.0	6.1	20.8	38.4	8.1
All ages	**10.6**	**18.1**	**4.0**	**12.2**	**19.9**	**5.7**	**13.1**	**19.8**	**7.4**

(1) Per 100,000 population.

U.S. Abortions, by State, 1992-96

Source: Alan Guttmacher Institute, New York, NY; latest year reported

	Number of reported abortions[1]			Rate per 1,000 women[2]			% change
	1992	1995	1996	1992	1995	1996	1992-96
TOTAL U.S.	**1,528,930**	**1,363,690**	**1,365,730**	**25.9**	**22.9**	**22.9**	**–12**
Alabama	17,450	14,580	15,150	18.2	15.0	15.6	–15
Alaska	2,370	1,990	2,040	16.5	14.2	14.6	–11
Arizona	20,600	18,120	19,310	24.1	19.1	19.8	–18
Arkansas	7,130	6,010	6,200	13.5	11.1	11.4	–15
California	304,230	240,240	237,830	42.1	33.4	33.0	–22
Colorado	19,880	15,690	18,310	23.6	18.0	20.9	–12
Connecticut	19,720	16,680	16,230	26.2	23.0	22.5	–14
Delaware	5,730	5,790	4,090	35.2	34.4	24.1	–32
District of Columbia	21,320	21,090	20,790	138.4	151.7	154.5	12
Florida	84,680	87,500	94,050	30.0	30.0	32.0	7
Georgia	39,680	36,940	37,320	24.0	21.2	21.1	–12
Hawaii	12,190	7,510	6,930	46.0	29.3	27.3	–41
Idaho	1,710	1,500	1,600	7.2	5.8	6.1	–15
Illinois	68,420	68,160	69,390	25.4	25.6	26.1	3
Indiana	15,840	14,030	14,850	12.0	10.6	11.2	–7
Iowa	6,970	6,040	5,780	11.4	9.8	9.4	–17
Kansas	12,570	10,310	10,630	22.4	18.3	18.9	–16
Kentucky	10,000	7,770	8,470	11.4	8.8	9.6	–16
Louisiana	13,600	14,820	14,740	13.4	14.7	14.7	10
Maine	4,200	2,690	2,700	14.7	9.6	9.7	–34
Maryland	31,260	30,520	31,310	26.4	25.6	26.3	0
Massachusetts	40,660	41,190	41,160	28.4	29.2	29.3	3
Michigan	55,580	49,370	48,780	25.2	22.6	22.3	–11
Minnesota	16,180	14,910	14,660	15.6	14.2	13.9	–11
Mississippi	7,550	3,420	4,490	12.4	5.5	7.2	–42
Missouri	13,510	10,540	10,810	11.6	8.9	9.1	–21
Montana	3,300	3,010	2,900	18.2	16.2	15.6	–14
Nebraska	5,580	4,360	4,460	15.7	12.1	12.3	–22
Nevada	13,300	15,600	15,450	44.2	46.7	44.6	1
New Hampshire	3,890	3,240	3,470	14.6	12.0	12.7	–13
New Jersey	55,320	61,130	63,100	31.0	34.5	35.8	16
New Mexico	6,410	5,450	5,470	17.7	14.4	14.4	–19
New York	195,390	176,420	167,600	46.2	42.8	41.1	–11
North Carolina	36,180	34,600	33,550	22.4	21.0	20.2	–10
North Dakota	1,490	1,330	1,290	10.7	9.6	9.4	–13
Ohio	49,520	40,940	42,870	19.5	16.2	17.0	–13
Oklahoma	8,940	9,130	8,400	12.5	12.9	11.8	–5
Oregon	16,060	15,590	15,050	23.9	22.6	21.6	–10
Pennsylvania	49,740	40,760	39,520	18.6	15.5	15.2	–18
Rhode Island	6,990	5,720	5,420	30.0	25.5	24.4	–19
South Carolina	12,190	11,020	9,940	14.2	12.9	11.6	–19
South Dakota	1,040	1,040	1,030	6.8	6.6	6.5	–4
Tennessee	19,060	18,240	17,990	16.2	15.2	14.8	–8
Texas	97,400	89,240	91,270	23.1	20.5	20.7	–10
Utah	3,940	3,740	3,700	9.3	8.1	7.8	–16
Vermont	2,900	2,420	2,300	21.2	17.9	17.1	–19
Virginia	35,020	31,480	29,940	22.7	20.0	18.9	–16
Washington	33,190	25,190	26,340	27.7	20.2	20.9	–24
West Virginia	3,140	3,050	2,610	7.7	7.6	6.6	–14
Wisconsin	15,450	13,300	14,160	13.6	11.6	12.3	–9
Wyoming	460	280	280	4.3	2.7	2.7	–37

(1) Rounded to the nearest 10. (2) Only for women aged 15-44 years old.

Contraceptive Use in the U.S., 1995

Source: National Center for Health Statistics, U.S. Dept. of Health and Human Services

Age	15-44	15-19	20-24	25-29	30-34	35-39	40-44
				Percent distribution			
Using contraception	64.2	29.8	63.4	69.3	72.7	72.9	71.5
Female sterilization	17.8	0.1	2.5	11.8	21.4	29.8	35.6
Male sterilization	7.0	—	0.7	3.1	7.6	13.6	14.5
Pill	17.3	13.0	33.1	27.0	20.7	8.1	4.2
Implant	0.9	0.8	2.4	1.4	0.5	0.2	0.1
Injectable	1.9	2.9	3.9	2.9	1.3	0.8	0.2
Intrauterine device (IUD)	0.5	—	0.2	0.5	0.6	0.7	0.9
Diaphragm	1.2	0.0	0.4	0.6	1.7	2.2	1.9
Condom	13.1	10.9	16.7	16.8	13.4	12.3	8.8
Female condom	0.0	—	0.1	—	—	—	—
Periodic abstinence	1.5	0.4	0.6	1.2	2.3	2.1	1.8
Natural family planning	0.2	—	0.1	0.2	0.3	0.4	0.2
Withdrawal	2.0	1.2	2.1	2.6	2.1	2.3	1.4
Other methods[1]	1.0	0.3	0.9	1.2	1.3	0.9	1.8

(1) Includes morning-after pill, foam, cervical cap, Today sponge, suppository, jelly or cream (without diaphragm), and other methods not shown separately.

U.S. Median Age at First Marriage, 1890-2000

Source: Bureau of the Census, U.S. Dept. of Commerce

Year[1]	Men	Women	Year[1]	Men	Women	Year[1]	Men	Women	Year[1]	Men	Women
2000	26.9	25.1	1994	26.7	24.5	1980	24.7	22.0	1940	24.3	21.5
1999	26.8	25.1	1993	26.5	24.5	1975	23.5	21.1	1930	24.3	21.3
1998	26.7	25.0	1992	26.5	24.4	1970	23.2	20.8	1920	24.6	21.2
1997	26.8	25.0	1991	26.3	24.1	1965	22.8	20.6	1910	25.1	21.6
1996	27.1	24.8	1990	26.1	23.9	1960	22.8	20.3	1900	25.9	21.9
1995	26.9	24.5	1985	25.5	23.3	1950	22.8	20.3	1890	26.1	22.0

(1) Figures after 1940 based on Current Population Survey data; figures for 1900-40 based on decennial censuses.

Interracial Married Couples in the U.S., 1960-2000

Source: Bureau of the Census, U.S. Dept. of Commerce; numbers in thousands

		INTERRACIAL MARRIED COUPLES				
			Black/White			
Year[1]	TOTAL MARRIED COUPLES	Total interracial	Black husband/ white wife	White husband/ black wife	White/ Other race[2]	Black/ Other race[2]
2000	56,497	1,464	268	95	1,051	50
1999	55,849	1,481	240	124	1,086	31
1998	55,305	1,348	210	120	975	43
1997	54,666	1,264	201	110	896	57
1996	54,664	1,260	220	117	884	39
1995	54,937	1,392	206	122	988	76
1990	53,256	964	150	61	720	33
1980	49,714	651	122	45	450	34
1970	44,598	310	41	24	233	12
1960	40,491	149	25	26	90	7

(1) Data from Mar. of year, except for 1970 and 1960, which are from decennial census. (2) Any race other than white or black.

Cigarette Use in the U.S., 1985-2000

Source: Substance Abuse and Mental Health Services Administration (SAMHSA), U.S. Dept. of Health and Human Services

(percentage reporting use in the month prior to the survey; figures exclude persons under age 12)

	1985	1998	1999	2000
TOTAL	**38.7**	**27.7**	**25.8**	**24.9**
Sex				
Male	43.4	29.7	28.3	26.9
Female	34.5	25.7	23.4	23.1
Race/Ethnicity				
White	38.9	27.9	27.0	25.9
Black	38.0	29.4	22.5	23.3
Hispanic	40.0	25.8	22.6	20.7
Other	(1)	23.8	(1)	(1)

	1985	1998	1999	2000
Age group				
12-17	29.4	18.2	14.9	13.4
18-25	47.4	41.6	39.7	38.3
26-34	45.7	32.5	24.9[2]	24.2[2]
35 and older	35.5	25.1	NA	NA
Education[3]				
Non-high school graduate	37.3	36.9	39.9	32.4
High school graduate	37.0	34.3	36.4	31.1
Some college	32.6	29.2	32.5	27.7
College graduate	23.0	15.2	18.2	13.9

NA = Not available. (1) No estimate reported. (2) 1999 and 2000 figures are for all persons aged 26 and older. (3) Estimates for Education are for persons aged 18 and older.

Drug Use in the General U.S. Population, 2000

Source: Substance Abuse and Mental Health Services Administration (SAMHSA), U.S. Dept. of Health and Human Services

According to the Substance Abuse and Mental Health Services Administration's 2000 National Household Survey on Drug Abuse, an estimated 87 milllion Americans 12 years of age and older (39%) had used an illicit drug at least once during their lifetimes, 11% had used one during the previous year, and 6.3% had used one in the month before the survey was conducted. As in prior years, men continued to have a higher rate of current illicit drug use than women (7.7% to 5.0%) in 2000. The rate of past-month marijuana use among women aged 12 and older, however, increased from 3.1% in 1999 to 3.5% in 2000.

The Substance Abuse and Mental Health Services Administration's Drug Abuse Warning Network (DAWN) reported 601,776 drug-related episodes in hospital emergency departments nationwide in 2000, up 9% from in 1999. Cocaine was the illicit drug most frequently involved.

Drug Use: America's Middle and High School Students, 2000

Source: *Monitoring the Future,* Univ. of Michigan Inst. for Social Research and National Inst. on Drug Abuse

Use of illicit drugs by American young people held steady in 2000, according to the University of Michigan's 26th annual survey of high school seniors and 10th annual survey of 8th and 10th graders.

While drug use was lower than the recent peaks in 1996 and 1997, not much of the decline occurred in 2000. Drug use for 10th and 12th graders remained steady. Although it was the 4th year of decline for 8th graders, the proportion taking illicit drugs in the 12 months prior to the survey (19.5%) was still nearly double the 1991 level (11%). The proportion of 10th graders using illicit drugs in the prior 12 months has increased by more than two-thirds (from 21% to 36.4%) since 1991, and among 12th graders the proportion has increased by more than 40% (from 29% to 41%).

Marijuana remained the most commonly used illegal drug for all 3 grade levels. In 2000, the proportion of students that reported using marijuana in the past year declined to 15.6% of 8th graders and 36.5% of 12th graders, while increasing slightly to 32.2% of 10th graders. Use of marijuana on a daily basis remained steady among all grades. About 1 in 17 high school seniors (6.0%), 1 in 26 10th graders (3.8%), and 1 in 77 (1.3%) 8th graders were daily users.

Use of LSD and other hallucinogens decreased or remained the same in all 3 grades, except for ecstasy, which increased for all 3 grades. Use of stimulants increased slightly. While the use of inhalants by 8th graders declined, use by 10th and 12th graders increased. Heroin use remained relatively low. Alcohol use decreased for 8th and 12th graders, but increased slightly for 10th graders.

Cigarette smoking decreased for all grades. About 14.6% of 8th graders, 23.9% of 10th graders, and 31.4% of 12th graders reported having smoked daily during the 30 days before they responded to the survey.

In 2000, about 17,300 seniors, 14,600 10th graders, and 13,300 8th graders from 435 public and private secondary schools participated in the survey. It should be noted that the surveys missed the 3-6% of a class group that drops out of school early, and about 9-17% who were absentees. These populations tend to have higher rates of drug use overall.

Drug Use: America's High School Seniors, 1975-2000

Source: *Monitoring the Future,* Univ. of Michigan Inst. for Social Research and National Inst. on Drug Abuse

	PERCENTAGE EVER USED									
	Class of 1975	**Class of 1980**	**Class of 1985**	**Class of 1990**	**Class of 1995**	**Class of 1997**	**Class of 1998**	**Class of 1999**	**Class of 2000**	**'99-'00 change**
Marijuana/hashish	47.3	60.3	54.2	40.7	41.7	49.6	49.1	49.7	48.8	–0.9
Inhalants[1]	NA	17.3	18.1	18.5	17.8	16.9	16.5	16.0	14.2	–1.2
Amyl & butyl nitrites	NA	11.1	7.9	2.1	1.5	2.0	2.7	1.7	0.8	–0.8
Hallucinogens[2]	NA	15.6	12.1	9.7	12.7	15.1	14.1	13.7	13.0	–0.7
LSD	11.3	9.3	7.5	8.7	11.7	13.6	12.6	12.2	11.1	–1.1
PCP	NA	9.6	4.9	2.8	2.7	3.9	3.9	3.4	3.4	–0.1
Ecstasy	NA	NA	NA	NA	NA	6.9	5.8	8.0	11.0	+3.0
Cocaine	9.0	15.7	17.3	9.4	6.0	8.7	9.3	9.8	8.6	–1.2
Crack	NA	NA	NA	3.5	3.0	3.9	4.4	4.6	3.9	–0.7
Heroin[3]	2.2	1.1	1.2	1.3	1.6	2.1	2.0	2.0	2.4	+0.4
Other opiates[4]	9.0	9.8	10.2	8.3	7.2	9.7	9.8	10.2	10.6	+0.4
Stimulants[4,5]	22.3	26.4	26.2	17.5	15.3	16.5	16.4	16.3	15.6	–0.7
Sedatives[4]	18.2	14.9	11.8	7.5	7.6	8.7	9.2	9.5	NA	+0.3
Barbiturates[4]	16.9	11.0	9.2	6.8	7.4	8.1	8.7	8.9	9.2	+0.2
Methaqualone[4]	8.1	9.5	6.7	2.3	1.2	1.7	1.6	1.8	NA	+0.2
Tranquilizers[4]	17.0	15.2	11.9	7.2	7.1	7.8	8.5	9.3	8.9	–0.5
Alcohol[6]	90.4	93.2	92.2	89.5	80.7	81.7	81.4	80.0	80.3	+0.2
Cigarettes	73.6	71.0	68.8	64.4	64.2	65.4	65.3	64.6	62.5	–2.1
Steroids	NA	NA	NA	2.9	2.3	2.4	2.7	2.9	2.5	–0.4

NA = Not available. (1) Adjusted for underreporting of amyl and butyl nitrites. (2) Adjusted for underreporting of PCP. (3) Reflects use with or without injection. (4) Includes only drug use that was not under a doctor's orders. (5) Data for 1990-2000 are not directly comparable to prior years. (6) Data for 1994-2000 are not directly comparable to prior years.

Alcohol Use by 8th and 12th Graders, 1980-2000

Source: *Monitoring the Future,* Univ. of Michigan Inst. for Social Research and National Inst. on Drug Abuse

	1980	**1990**	**1991**	**1992**	**1993**	**1994**	**1995**	**1996**	**1997**	**1998**	**1999**	**2000**
ALCOHOL[1]			**Percent using alcohol in the month before the survey**									
All 12th graders	72.0	57.1	54.0	51.3	48.6	50.1	51.3	50.8	52.7	52.0	51.0	50.0
Male	77.4	61.3	58.4	55.8	54.2	55.5	55.7	54.8	56.2	57.6	55.3	54.0
Female	66.8	52.3	49.0	46.8	43.4	45.2	47.0	46.9	48.9	46.9	46.8	46.1
White	75.8	62.2	57.7	56.0	53.4	54.8	54.8	54.7	57.9	57.6	54.9	55.1
Black	47.7	32.9	34.4	29.5	35.1	33.1	37.4	35.7	33.1	33.6	30.8	30.0
All 8th graders	—	—	25.1	26.1	24.3	25.5	24.6	26.2	24.5	23.0	24.0	22.4
Male	—	—	26.3	26.3	25.3	26.5	25.0	26.6	25.2	24.0	24.8	22.5
Female	—	—	23.8	25.9	28.7	24.7	24.0	25.8	23.9	21.9	23.3	22.0
White	—	—	26.0	27.3	25.1	25.4	25.4	27.7	25.7	24.0	25.6	24.7
Black	—	—	17.8	19.2	17.7	20.2	17.3	19.0	16.9	15.4	16.8	16.0
HEAVY ALCOHOL[2]			**Percent heavily using the 2 weeks before the survey**									
All 12th graders	41.2	32.2	29.8	27.9	27.5	28.2	29.8	30.2	31.3	31.5	30.8	30.0
Male	52.1	39.1	37.8	35.6	34.6	37.0	36.9	37.0	37.9	39.2	38.1	36.7
Female	30.5	24.4	21.2	20.3	20.7	20.2	23.0	23.5	24.4	24.0	23.6	23.5
White	44.6	36.2	32.9	31.3	31.3	31.7	32.9	34.0	36.1	36.6	34.8	34.6
Black	17.0	11.6	11.8	10.8	14.6	14.2	15.5	15.1	12.0	12.7	11.9	11.5
All 8th graders	—	—	12.9	13.4	13.5	14.5	14.5	15.6	14.5	13.7	15.2	14.1
Male	—	—	14.3	13.9	14.8	16.0	15.1	16.5	15.3	14.4	16.4	14.4
Female	—	—	11.4	12.8	12.3	13.0	13.9	14.5	13.5	12.7	13.9	13.6
White	—	—	12.6	12.9	12.4	13.4	14.5	15.7	14.6	13.5	15.2	14.9
Black	—	—	9.9	9.3	11.9	11.8	10.0	10.9	8.8	9.1	10.8	10.0

— Data not available. **Note**: *Monitoring the Future* study excludes high school dropouts (about 3-6% of the class group, according to a 1996 report) and absentees (about 16-17% of 12th graders and about 9-10% of 8th graders). High school dropouts and absentees have higher alcohol usage than those included in the survey. (1) In 1993 the alcohol question was changed to indicate that a "drink" meant "more than a few sips." (2) Five or more drinks in a row at least once in the prior 2-week period.

Principal Types of Accidental Deaths in the U.S., 1970-2000

Source: National Safety Council

Year	Motor vehicle	Falls	Poison (solid, liquid)	Drowning	Fires, burns	Ingestion of food, object	Firearms	Poison (gases)
1970	54,633	16,926	3,679	7,860	6,718	2,753	2,406	1,620
1980	53,172	13,294	3,089	7,257	5,822	3,249	1,955	1,242
1985	45,901	12,001	4,091	5,316	4,938	3,551	1,649	1,079
1990	46,814	12,313	5,055	4,685	4,175	3,303	1,416	748
1991	43,536	12,662	5,698	4,818	4,120	3,240	1,441	736
1992	40,982	12,646	6,449	3,542	3,958	3,182	1,409	633
1993	41,893	13,141	7,877	3,807	3,900	3,160	1,521	660
1994	42,524	13,450	8,309	3,942	3,986	3,065	1,356	685
1995	43,363	13,986	8,461	4,350	3,761	3,185	1,225	611
1996	43,649	14,986	8,872	3,959	3,741	3,206	1,134	638
1997	43,458	15,447	9,587	4,051	3,490	3,275	981	576
1998	43,501	16,274	10,255	4,406	3,255	3,515	866	546
1999[1]	43,000	16,000	11,000	4,000	3,100	3,100	800	400
2000[2]	43,000	16,200	11,700	3,900	3,600	3,400	600	400
			Death rates per 100,000 population					
1970	26.8	8.3	1.8	3.9	3.3	1.4	1.2	0.8
1980	23.4	5.9	1.4	3.2	2.6	1.4	0.9	0.5
1985	19.3	5.0	1.7	2.2	2.1	1.5	0.7	0.5
1990	18.8	4.9	2.0	1.9	1.7	1.3	0.6	0.3
1991	17.3	5.0	2.3	1.8	1.6	1.3	0.6	0.3
1992	16.1	5.0	2.5	1.4	1.6	1.2	0.6	0.2
1993	16.3	5.1	3.1	1.5	1.5	1.2	0.6	0.3
1994	16.3	5.2	3.2	1.5	1.5	1.2	0.5	0.3
1995	16.5	5.3	3.2	1.7	1.4	1.2	0.5	0.2
1996	16.5	5.6	3.3	1.5	1.4	1.2	0.4	0.2
1997	16.2	5.8	3.6	1.5	1.3	1.2	0.4	0.2
1998	16.1	6.0	3.8	1.6	1.2	1.3	0.3	0.2
1999[1]	15.8	5.9	4.0	1.5	1.1	1.1	0.3	0.1
2000[2]	15.6	5.9	4.2	1.4	1.3	1.2	0.2	0.1

Note: There were 14,500 other accidental deaths in 2000; the most frequently occurring types involved medical and surgical complications, machinery, air transport, water transport (except drownings), mechanical suffocation, and excessive cold. (1) Revised figures. (2) Preliminary figures.

U.S. Motor Vehicle Accidents

Source: National Safety Council

Motor vehicle deaths in the U.S. were unchanged from 1999 to 2000, but down 1% from 1998. Among the 189,800,000 licensed drivers in 2000, there were slightly more male drivers than female (95,469,000 male vs 94,331,000 female), but males accounted for an estimated 63% of all miles driven.

Male drivers were involved in more fatal accidents than female drivers in 2000. About 45,600 men and 15,800 women drivers were involved in fatal accidents.

About 15.2 mil male drivers and 9.9 mil female drivers were involved in some type of accident in 2000. However, women had higher accident involvement rates. Accident rates were 90 per 10 million miles driven for men and 100 per 10 million miles driven for women.

About 38% of all traffic fatalities in 1999 involved an intoxicated or alcohol-impaired driver or nonoccupant. In 1987 alcohol-related fatalities accounted for a much larger 51% of all traffic deaths. Of these 15,786 alcohol-related traffic fatalities, an estimated 12,321 occurred in accidents where a driver or pedestrian was intoxicated, and the remainder involved a driver or pedestrian who had been drinking but was not legally intoxicated. Alcohol was a factor in about 7% of all traffic accidents.

	Death total 2000	Percentage change from 1999	Death rate 2000[1]
All motor vehicle accidents	43,000	0	15.6
Collision between motor vehicles	20,600	+3	7.5
Collision with fixed object	11,200	–1	4.1
Pedestrian accidents	5,300	–9	1.9
Noncollision accidents	4,600	–2	1.7
Collision with pedalcycle	800	0	0.3
Collision with railroad train	400	+33	0.1
Other collision (animal, animal-drawn vehicles)	100	0	(2)

(1) Deaths per 100,000 population. (2) Death rate was less than 0.05.

Improper Driving Reported in Accidents, 1999-2000

Source: National Safety Council

Type	Percentage of fatal accidents 1999	Percentage of fatal accidents 2000	Percentage of injury accidents 1999	Percentage of injury accidents 2000	Percentage of all accidents 1999	Percentage of all accidents 2000
Improper driving	**72.6**	**61.6**	**67.2**	**60.3**	**62.2**	**57.8**
Speed too fast or unsafe	23.0	18.6	13.0	16.3	10.6	13.6
Right of way	20.1	10.1	25.8	19.9	22.9	20.1
Failed to yield	10.8	4.6	19.2	15.0	13.8	12.7
Passed stop sign	4.6	3.8	1.7	3.6	3.2	5.3
Disregarded signal	4.7	8.2	4.9	1.3	5.9	2.2
Drove left of center	9.6	0.7	1.7	1.1	1.3	1.0
Improper overtaking	1.1	0.9	0.9	2.0	1.2	2.4
Made improper turn	1.2	0.7	2.4	0.6	3.0	0.9
Followed too closely	0.5	0.9	3.4	4.3	6.3	5.7
Other improper driving	17.1	9.0	20.3	16.1	16.9	14.1
No improper driving stated	**27.4**	**38.4**	**32.8**	**39.7**	**37.8**	**42.2**

Note: Based on reports from 12 state traffic authorities. When a driver was under the influence of alcohol or drugs, the accident was considered a result of the driver's physical condition—not a driving error. For this reason, accidents in which the driver was reported to be under the influence are included under "no improper driving."

Risk Behaviors in High School Students, 1999

Source: CDC, *Youth Risk Behavior Surveillance—United States, 1999*

	Percent rarely or never wore seatbelts[1]			Percent rarely or never wore bicycle helmets[2]			Percent injured while playing sports or being physically active[3]		
	Female	Male	Total	Female	Male	Total	Female	Male	Total
Race									
Non-Hispanic White	11.2	19.6	15.5	82.1	86.0	84.3	33.5	43.6	38.7
Non-Hispanic Black	17.4	27.9	22.5	94.1	90.3	91.9	25.8	39.7	32.6
Hispanic	9.5	19.6	14.4	83.4	88.5	86.3	28.5	40.9	34.6
Grade									
9	14.4	19.8	17.1	77.1	82.8	80.3	35.4	45.8	40.7
10	12.3	17.7	15.0	86.0	87.7	86.9	35.9	41.9	38.9
11	9.8	17.7	13.8	85.4	87.1	86.4	32.1	37.6	34.9
12	10.3	28.1	19.1	89.9	91.4	90.7	26.1	43.9	34.9
Total	**11.9**	**20.8**	**16.4**	**83.6**	**86.7**	**85.3**	**32.7**	**42.5**	**37.7**

(1) When riding in a car or truck driven by someone else. (2) Among the 70.8% of students who rode bicycles during the 12 months preceding the survey. (3) Seriously enough to be treated by a doctor or nurse during the 12 months preceding the survey.

Deaths in the U.S. Involving Firearms, by Age, 1998

Source: National Safety Council

	All ages	Under 5	5-14	15-19	20-24	25-44	45-64	65-74	75 & over
Total firearms deaths[1]	**30,078**	**83**	**529**	**3,180**	**4,240**	**11,999**	**6,137**	**2,081**	**2,459**
Male	26,189	51	371	2,817	3,791	10,012	5,084	1,834	2,229
Female	4,519	32	158	363	449	1,987	1,053	247	230
Unintentional	866	19	102	141	119	280	140	34	31
Male	762	15	81	130	113	242	123	32	26
Female	104	4	21	11	6	38	17	2	5
Suicides	17,424	0	154	1,087	1,423	6,118	4,525	1,831	2,286
Male	15,104	0	118	949	1,270	5,162	3,821	1,660	2,124
Female	2,320	0	36	138	153	956	704	171	162
Homicides	11,798	63	251	1,870	2,614	5,311	1,365	202	122
Male	9,771	36	152	1,662	2,334	4,347	1,049	129	62
Female	2,027	27	99	208	280	964	316	73	60
Undetermined[2]	316	1	54	19	37	111	66	10	18
Male	254	0	49	19	27	83	52	9	15
Female	62	1	5	0	10	28	14	1	3

(1) Figures exclude firearms deaths by legal intervention. These deaths totaled 304 in 1998. (2) "Undetermined" means that the intention involved (whether accident, suicide, or homicide) could not be determined.

Home Accident Deaths in the U.S., 1950-2000

Source: National Safety Council

Year	Total	Falls	Poison (solid, liquid)	Fires, burns[1]	Suffoc.: ingesting object	Suffoc.: mechanical	Firearms	Poison (gases)	All other[3]
1950	29,000	14,800	1,300	5,000	(2)	1,600	950	1,250	4,100
1960	28,000	12,300	1,350	6,350	1,850	1,500	1,200	900	2,550
1970	27,000	9,700	3,000	5,600	1,800[4]	1,100[4]	1,400[4]	1,100	3,300[4]
1980	22,800	7,100	2,500	4,800	2,000	500	1,100	700	4,100[5]
1990	21,500	6,700	4,000	3,400	2,300	600	800	500	3,200
1991	22,100	6,900	4,500	3,400	2,200	700	800	500	3,100
1992	24,000	7,700	4,800	3,700	1,500	700	1,000	400	4,200
1993	26,100	7,900	6,000	3,700	1,700	700	1,100	500	4,500
1994	26,300	8,100	6,300	3,700	1,600	800	900	500	4,400
1995	27,200	8,400	6,600	3,500	1,500	800	900	400	5,100
1996	27,500	9,000	6,800	3,500	1,500	800	800	500	4,500
1997	27,700	9,100	7,400	3,200	1,500	800	700	400	3,500
1998[6]	29,000	9,500	8,000	2,900	1,800	800	600	400	4,000
1999[6]	28,200	8,800	8,700	2,700	1,400	900	500	300	4,100
2000[7]	29,500	9,300	9,400	3,200	1,700	900	400	300	3,400

(1) Includes deaths resulting from conflagration, regardless of nature of injury. (2) Included under "All other" category. (3) Includes drowning; estimated at 900 per yr., 1991-97, 1,000 for 1998, 800 for 1999. Data for this year and later not comparable with earlier data because of classification changes. (4) Data for this year and later not comparable with earlier data because of classification changes. (5) Includes about 1,000 deaths attributed to summer heat wave. (6) Revised figures. (7) Data for 2000 are preliminary.

Worldwide Airline Fatalities, 1980-2000[1]

Source: National Safety Council.

Year	Aircraft accidents[2]	Passenger deaths	Death rate[3]	Year	Aircraft accidents[2]	Passenger deaths	Death rate[3]	Year	Aircraft accidents[2]	Passenger deaths	Death rate[3]
1980	21	734	0.13	1987	23	889	0.10	1994	23	962	0.08
1981	22	365	0.06	1988	26	712	0.08	1995	20	541	0.04
1982	25	762	0.13	1989	29	879	0.09	1996	21	1,125	0.08
1983	21	817	0.13	1990	23	473	0.05	1997	25	867	0.05
1984	16	218	0.03	1991	24	518	0.05	1998	20	904	0.06
1985	25	1,037	0.14	1992	24	978	0.09	1999	19	487	0.03
1986	19	427	0.05	1993	31	806	0.07	2000[4]	18	755	0.04

(1) Some figures for 1999 and earlier have been revised from previous figures. (2) Involving 1 or more fatalities only. (3) Passenger deaths per 100 mil passenger mi. (4) Preliminary.

U.S. Fires, 2000

Source: National Fire Protection Assn.

Fires

- Public fire departments responded to 1,703,000 fires in 2000, a decrease of 6.3% from 1999.
- There were 505,000 structure fires in 2000, a decrease of 3.3% from the 1999 figure.
- 75% of all structure fires, or 379,500 fires, occurred in residential properties.
- There were 348,500 vehicle fires in 2000, a decrease of 5.4% from the previous year.
- There were 854,000 fires in outside properties, a decrease of 8.3% from 1999.
- The South had the highest fire incident rate in the country, with 7.2 fires per 1,000 population.

Civilian deaths

- There were 4,045 civilian fire deaths in 2000, a significant increase of 13.3% from 1999.
- The number of deaths from fire in the home increased by 18.1%, to 3,420.
- About 85% of all fire deaths occurred in the home.
- The South had the highest regional fire death rate, with 17.7 civilian deaths per million population.
- Nationwide, someone died in a fire every 130 minutes.

Civilian injuries

- There were an estimated 22,350 civilian fire injuries in 2000, an increase of 2.2% from 1999. This estimate is traditionally low because of underreporting of civilian fire injuries to the fire service.
- Residential properties were the site of 17,400 civilian fire injuries, or 77.9% of injuries overall; 2,200 injuries, or 9.8%, occurred in nonresidential structure fires.
- The Northeast had the highest regional injury rate in the U.S., with 111.7 civilian injuries per million population. The next highest rate was in the North Central region, with 92.1 injuries per million.
- Nationwide, a civilian was injured in a fire every 23 minutes.

Property damage

- Property damage resulting from fires increased significantly in 2000, by 11.8%, to an estimated $11.2 billion.
- Structure fires resulted in 85% of all property damage, or $9.5 billion.
- 46% of all structure property loss occurred in residential properties, accounting for $5.092 billion.
- The West had the highest property loss rate in the U.S.—about $46.20 per person—in part due to the New Mexico Wildland Fire, which caused an estimated $1.0 billion in losses.

Incendiary and suspicious fires

- About 14.8% of all structure fires, or an estimated 75,000 fires in 2000 were deliberately set or suspected of having been deliberately set. This represents an increase of 4.2% from 1999.
- Incendiary or suspicious structure fires resulted in 505 civilian deaths, an increase of 36.5% from the previous year. Incendiary or suspicious fires caused $1.34 billion in property damage. This represents 15.7% of all property loss from structure fires.
- The number of vehicle fires of an incendiary or suspicious origin in 2000 was 46,500, a 3.3% increase over the previous year. They caused an estimated $186 million in property damage, a decrease of 4.6% from 1999.

Physicians by Age, Sex, and Specialty, 1999

Source: American Medical Assn., as of Dec. 31, 1999

	Total Physicians[1]		Under 35 yrs		35-44 yrs		45-54 yrs		55-64 yrs	
	Male	Female	Male	Female	Male	Female	Male	Female	Male	Female
All Specialties	**611,028**	**186,606**	**82,466**	**53,550**	**148,131**	**65,277**	**152,343**	**40,881**	**99,113**	**14,715**
Aerospace Medicine	455	29	10	1	97	10	142	14	99	3
Allergy & Immunology	3,059	627	101	88	699	326	978	264	717	93
Anesthesiology	27,667	7,080	3,076	1,135	10,679	2,857	7,497	1,840	3,963	900
Cardiovascular Disease	18,688	1,517	1,479	246	5,810	681	6,158	425	3,341	119
Child Psychiatry	3,518	2,353	210	226	913	866	1,124	747	752	322
Colon/Rectal Surgery	1,002	84	50	25	306	43	328	14	188	2
Dermatology	6,411	2,994	634	773	1,381	1,232	1,956	705	1,526	217
Diagnostic Radiology	16,462	3,935	2,698	940	5,019	1,648	4,972	1,046	2,914	244
Emergency Medicine	17,960	4,065	3,547	1,316	5,249	1,478	6,411	982	1,897	223
Family Practice	50,176	18,887	7,773	6,606	14,588	7,202	16,049	3,896	5,779	834
Forensic Pathology	399	161	19	12	91	60	116	51	95	24
Gastroenterology	9,363	864	751	170	3,160	423	3,223	221	1,620	44
General Practice	13,571	2,372	65	24	776	351	2,269	769	3,191	645
General Preventive Med.	2,440	1,056	97	93	475	340	669	315	498	142
General Surgery	35,572	3,739	6,310	1,687	7,695	1,224	8,280	650	7,197	126
Internal Medicine	94,244	34,465	17,799	11,452	26,112	12,870	27,310	7,556	13,158	1,864
Medical Genetics	176	134	20	18	48	44	52	46	40	20
Neurological Surgery	4,678	226	713	72	1,120	94	1,120	48	1,053	10
Neurology	9,389	2,418	887	478	2,591	947	3,225	725	1,787	192
Nuclear Medicine	1,189	257	79	31	231	72	370	92	307	44
Obstetrics/Gynecology	26,158	13,205	2,262	4,341	5,579	4,812	7,741	2,824	6,034	911
Occupational Medicine	2,530	480	7	6	413	173	769	186	487	73
Ophthalmology	15,211	2,442	1,400	547	3,814	1,042	4,203	609	3,626	179
Orthopedic Surgery	20,833	720	3,041	241	5,453	278	5,539	160	4,462	26
Otolaryngology	8,271	770	1,195	248	2,065	325	2,038	160	1,970	24
Pathology-Anat./Clin.	12,848	5,293	1,054	812	2,952	1,906	3,507	1,495	2,874	746
Pediatric Cardiology	1,108	367	110	75	392	169	291	62	192	38
Pediatrics	31,050	28,499	5,551	9,303	8,009	9,770	8,622	6,214	5,173	2,339
Physical Med./Rehab.	4,164	1,980	660	364	1,617	784	990	490	494	229
Plastic Surgery	5,362	548	316	87	1,502	221	1,694	180	1,299	44
Psychiatry	27,790	11,266	1,866	1,638	5,226	3,553	7,496	3,340	6,523	1,603
Pulmonary Diseases	7,211	976	673	197	2,333	483	2,678	203	1,092	57
Radiation Oncology	2,971	816	346	127	989	322	783	240	561	99
Radiology	7,378	1,095	405	104	1,407	381	1,309	325	2,413	208
Thoracic Surgery	220	19	111	9	107	10	1	0	0	0
Urological Surgery	9,700	301	1,050	121	2,270	116	2,538	53	2,500	7
Other	4,968	875	57	9	554	191	1,211	275	1,226	185
Unspecified	6,191	2,451	2,560	1,431	1,627	605	1,061	271	503	86

(1) Includes physicians 65 and older, those living in U.S. possessions, "Not Classified," "Inactive," and "Address Unknown."

U.S. Health Expenditures, 1965-99

Source: *Health, United States, 2001,* National Center for Health Statistics, U.S. Dept. of Health and Human Services

	1960	1965	1970	1975	1980	1985	1990	1995	1997	1998	1999
					Amount in billions						
National health expenditures	**$26.7**	**$41.0**	**$73.1**	**$129.8**	**$245.8**	**$426.5**	**$695.6**	**$987.0**	**$1,093.9**	**$1,146.1**	**$1,210.7**
					Percent distribution						
Health services and supplies	93.6	91.1	92.2	93.2	95.0	95.8	96.2	96.7	96.6	96.7	96.7
Personal health care	87.6	84.7	86.5	87.0	87.3	87.3	87.6	87.7	87.7	87.5	87.4
Hospital care	34.4	33.7	37.8	40.0	41.3	39.1	36.5	34.8	33.6	32.9	32.3
Professional services	31.3	30.3	28.3	27.8	27.4	29.4	31.2	32.1	32.2	32.6	32.8
Physician and clinical services	20.1	20.3	19.1	19.1	19.2	21.1	22.6	22.3	22.0	22.2	22.2
Other professional services	1.5	1.3	1.0	1.1	1.5	2.0	2.6	2.9	3.1	3.1	3.1
Dental services	7.4	6.8	6.4	6.1	5.4	5.1	4.5	4.5	4.6	4.6	4.6
Other personal health care	2.4	1.9	1.7	1.5	1.3	1.2	1.4	2.3	2.5	2.6	2.7
Home health care	0.2	0.2	0.3	0.5	1.0	1.3	1.8	3.1	3.2	2.9	2.7
Nursing home care	3.2	3.6	5.8	6.7	7.2	7.2	7.6	7.6	7.8	7.7	7.4
Retail outlet sales of medical products	18.6	16.9	14.3	12.0	10.5	10.4	10.5	10.2	10.9	11.4	12.2
Government administration and net cost of private health insurance	4.5	4.9	3.8	3.9	4.9	5.8	5.7	5.8	5.6	5.8	5.9
Government public health activities	1.5	1.5	1.9	2.3	2.7	2.7	2.9	3.2	3.3	3.4	3.4
Investment	6.4	8.9	7.8	6.8	5.0	4.2	3.8	3.3	3.4	3.3	3.3
Research	2.6	3.7	2.7	2.6	2.2	1.9	1.8	1.7	1.7	1.8	1.8
Construction	3.8	5.2	5.2	4.3	2.8	2.2	2.0	1.6	1.7	1.5	1.5
					Average annual percent change from previous year shown						
National health expenditures	**NA**	**9.0**	**12.2**	**12.2**	**13.6**	**11.7**	**10.3**	**7.2**	**5.3**	**4.8**	**5.6**
Health services and supplies	**NA**	**8.4**	**12.5**	**12.4**	**14.1**	**11.9**	**10.4**	**7.4**	**5.2**	**4.9**	**5.7**
Personal health care	NA	8.2	12.7	12.3	13.7	11.7	10.4	7.3	5.2	4.5	5.5
Hospital care	NA	8.5	14.9	13.4	14.4	10.4	8.8	6.2	3.4	2.6	3.7
Physician and clinical services	NA	9.2	10.9	12.2	13.7	13.8	11.9	7.0	4.5	5.5	6.0
Other professional services	NA	6.3	6.9	13.2	21.1	18.6	16.4	9.5	8.1	7.5	5.6
Dental services	NA	7.3	10.8	11.2	10.9	10.2	7.8	7.1	6.2	5.8	5.6
Other personal health care	NA	4.2	10.2	9.4	10.5	10.0	12.9	18.9	10.0	8.6	9.9
Home health care	NA	9.6	19.7	23.2	30.7	18.9	17.3	19.4	6.4	–3.0	–1.4
Nursing home care	NA	11.6	23.4	15.5	15.3	11.7	11.4	7.2	6.8	3.5	2.3
Retail outlet sales of medical products	NA	7.0	8.6	8.3	10.5	11.4	10.7	6.5	8.9	9.2	13.0
Government administration and net cost of private health insurance	NA	10.6	6.7	12.8	19.2	15.4	9.7	7.7	3.8	8.7	7.4
Government public health activities	NA	9.6	16.9	16.7	18.1	11.4	11.8	9.2	7.1	7.1	6.6
Investment	NA	16.5	9.5	9.1	6.7	7.6	8.3	4.3	7.2	2.2	4.1
Research	NA	17.1	5.1	11.2	10.4	8.7	8.8	6.2	4.6	9.8	8.2
Construction	NA	16.1	12.2	8.0	4.2	6.7	7.8	2.4	9.9	–5.3	–0.6

Note: NA = Not applicable. Numbers may not add to totals because of rounding. (1) Includes personal care services delivered by government public health agencies.

IT'S A FACT: As a cost-cutting measure, more and more surgical procedures in community hospitals are performed on outpatients—62% in 1999, compared to 51% in 1990 and 16% in 1980.

Ownership of Life Insurance in the U.S. and Assets of U.S. Life Insurance Companies, 1940-99

Source: American Council of Life Insurance

(amounts in millions)

	PURCHASES OF LIFE INSURANCE				INSURANCE IN FORCE					
Year	Ordinary	Group	Industrial	Total	Ordinary	Group	Industrial	Credit	Total	Assets
1940	$6,689	$691	$3,350	$10,730	$79,346	$14,938	$20,866	$380	$115,530	$30,802
1950	17,326	6,068	5,402	28,796	149,116	47,793	33,415	3,844	234,168	64,020
1960	52,883	14,645	6,880	74,408	341,881	175,903	39,563	29,101	586,448	119,576
1970	122,820	63,690[1]	6,612	193,122[1]	734,730	551,357	38,644	77,392	1,402,123	207,254
1975	188,003	95,190[1]	6,729	289,922[1]	1,083,421	904,695	39,423	112,032	2,139,571	289,304
1980	385,575	183,418	3,609	572,602	1,760,474	1,579,355	35,994	165,215	3,541,038	479,210
1985	910,944	319,503[2]	722	1,231,169[2]	3,247,289	2,561,595	28,250	215,973	6,053,107	825,901
1990	1,069,660	459,271	220	1,529,151	5,366,982	3,753,506	24,071	248,038	9,392,597	1,408,208
1991	1,041,508	573,953[1]	198	1,615,659[1]	5,677,777	4,057,606	22,475	228,478	9,986,336	1,551,201
1992	1,048,135	440,143	222	1,488,500	5,941,810	4,240,919	20,973	202,090	10,405,792	1,664,531
1993	1,101,327	576,823	149	1,678,299	6,428,434	4,456,338	20,451	199,518	11,104,741	1,839,127
1994	1,056,976	560,232	257	1,617,465	6,429,811	4,443,179	18,947	189,398	11,081,335	1,942,273
1995	1,039,102	537,828	156	1,577,086	6,872,252	4,604,856	18,134	201,083	11,696,325	2,143,544
1996	1,089,137	614,565	130	1,703,832	7,407,682	5,067,804	18,064	210,746	12,704,296	2,327,924
1997	1,203,552	688,589	128	1,892,269	7,854,570	5,279,042	17,991	212,255	13,363,858	2,579,078
1998	1,324,565	739,508	106	2,064,179	8,505,894	5,735,273	17,365	212,917	14,471,449	2,819,992
1999[3]	1,399,848	966,858	—	2,508,019	9,172,397	6,110,218	—	213,453	15,496,069	—

Note: — = Data not available. Ordinary purchases, ordinary in force, and group in force numbers were revised for 1994-97. (1) Includes Servicemen's Group Life Insurance, which amounted to $17.1 billion in 1970, $1.7 billion in 1975, and $166.7 billion in 1991. (2) Includes Federal Employees' Group Life Insurance of $10.8 billion. (3) For 1999, category of "ordinary" is combined with data for "industrials."

Health Insurance Coverage,[1] by State, 1990, 1999-2000

Source: Bureau of the Census, U.S. Dept. of Commerce

STATE	Not covered 2000[2]	% not covered 2000	% not covered 1999	% not covered 1990	STATE	Not covered 2000[2]	% not covered 2000	% not covered 1999	% not covered 1990
AL.	588	13.6	13.2	17.4	MT	166	18.4	17.8	14.0
AK	118	19.3	18.3	15.6	NE	159	10.0	10.1	8.5
AZ.	805	16.0	20.0	15.5	NV	289	15.7	18.3	16.5
AR	371	13.7	14.5	17.4	NH	76	6.8	9.2	9.9
CA	6,372	17.9	19.0	19.1	NJ	1,066	12.5	11.9	10.0
CO	518	13.3	15.2	14.7	NM.	435	23.8	24.1	22.2
CT	253	8.5	9.0	6.9	NY.	2,842	15.1	15.1	12.1
DE	80	10.6	9.9	13.9	NC.	993	12.9	14.3	13.8
DC	72	14.5	13.7	19.1	ND.	67	11.4	11.5	6.3
FL.	2,666	17.2	18.0	18.0	OH.	1,285	10.8	10.2	10.3
GA	1,149	14.4	15.0	15.3	OK.	634	19.3	16.4	18.6
HI	112	10.2	10.3	7.4	OR.	441	13.8	13.9	12.5
ID	200	15.5	18.1	15.1	PA	920	7.5	8.3	10.1
IL	1,687	13.4	13.0	10.9	RI.	53	6.2	6.1	11.1
IN	646	11.9	9.4	10.7	SC.	436	12.0	15.6	16.2
IA	239	8.8	7.6	8.1	SD.	79	11.7	10.8	11.6
KS	287	11.7	11.7	10.8	TN	575	10.3	10.3	13.7
KY	505	13.0	13.2	13.2	TX	4,501	21.4	22.0	21.1
LA.	821	18.9	21.7	9.7	UT	296	13.4	13.2	9.0
ME	138	11.8	10.9	11.2	VT	62	11.0	11.1	9.6
MD	476	9.9	10.9	12.7	VA	869	12.7	13.0	15.7
MA	607	9.5	9.0	9.1	WA.	743	13.7	13.8	11.4
MI	1,001	9.8	10.1	9.4	WV.	250	14.3	15.6	13.8
MN	406	9.1	7.3	8.9	WI	372	7.4	10.0	6.6
MS	370	12.9	15.6	19.9	WY.	64	14.4	15.1	12.5
MO	561	10.8	6.9	12.7	**TOTAL U.S. .**	**38,729**	**14.0**	**14.3**	**13.9**

(1) For population, all ages, including those 65 or over, an age group largely covered by Medicare. (2) In thousands.

Persons Not Covered by Health Insurance, by Selected Characteristics, 2000

Source: Bureau of the Census, U.S. Dept. of Commerce

	Number[1]	Percent
Total	**38,729**	**14.0**
Sex		
Male	20,177	14.9
Female	18,552	13.1
Race and Ethnicity		
White	29,285	12.9
Non-Hispanic	18,898	9.7
Black	6,629	18.5
Asian and Pacific Islander	2,051	18.0
Hispanic[2]	10,835	32.0
Age		
Under 18 years.	8,451	11.6
18 to 24 years	7,349	27.3
25 to 34 years	7,926	21.2
35 to 44 years	6,939	15.5
45 to 64 years	7,819	12.6
65 years and over.	245	0.7
Nativity		
Native	29,270	11.9
Foreign born.	9,459	31.6
Naturalized citizen	1,807	15.9
Not a citizen.	7,652	41.3
Region		
Northeast.	6,019	11.4
Midwest	6,787	10.8
South.	15,357	15.8
West	10,566	16.7
Household Income		
Less than $25,000.	13,898	22.7
$25,000 to $49,999	12,783	17.0
$50,000 to $74,999	6,496	11.0
$75,000 or more	5,552	6.9
Education (18 years and older)		
Total	**30,278**	**14.8**
No high school diploma.	9,026	26.6
High school graduate only. . . .	10,816	16.4
Some college, no degree	5,369	13.4
Associate degree.	1,619	10.3
Bachelor's degree or higher . .	3,448	7.1

(1) In thousands. (2) Persons of Hispanic origin may be of any race.

Health Coverage for Persons Under 65, by Characteristics, 1984, 1997-99

Source: *Health, United States, 2001,* National Center for Health Statistics, U.S. Dept. of Health and Human Services

	PRIVATE INSURANCE				MEDICAID[1]				NOT COVERED[2]			
	1984	1997[3]	1998	1999	1984	1997[3]	1998	1999	1984	1997[3]	1998	1999
Age	Percent of each population group											
Under 18 years	72.6	66.1	68.4	68.8	11.9	18.4	17.1	18.1	13.9	14.0	12.7	11.9
18-44 years	76.5	69.4	71.1	72.0	5.1	6.6	5.8	5.7	17.1	22.4	21.4	21.0
45-64 years	83.3	79.0	79.0	79.3	3.4	4.6	4.5	4.4	9.6	12.4	12.2	12.2
Race and Hispanic origin[4]												
White, non-Hispanic	82.4	78.0	79.6	80.2	3.7	6.2	5.7	6.0	11.8	13.7	12.5	12.1
Black, non-Hispanic.	59.4	56.3	56.1	58.3	19.1	20.3	19.4	18.6	19.7	20.1	20.7	19.3
All Hispanic	57.1	47.9	49.9	50.3	12.2	16.0	14.1	14.1	29.1	34.3	34.0	33.9
Percent of poverty level[4]												
Below 100%	33.0	23.6	24.1	26.1	30.5	38.8	37.9	36.8	34.7	34.4	34.6	34.4
100-149%	61.8	42.0	43.3	40.1	7.5	17.5	16.0	18.6	27.0	36.1	36.5	35.8
150-199%	77.2	63.6	61.4	59.4	3.1	7.4	7.2	9.8	17.4	25.9	26.7	27.7
200% or more	91.6	87.6	88.3	88.7	0.6	1.7	1.8	2.0	5.8	8.8	8.0	7.7
Geographic region[4]												
Northeast.	80.7	74.3	76.4	77.1	8.5	11.2	9.8	10.1	10.1	13.4	12.3	12.2
Midwest.	80.9	77.3	79.1	80.2	7.2	8.2	7.5	7.3	11.1	13.1	11.9	11.5
South.	74.5	67.5	67.8	68.0	5.0	8.6	8.6	8.9	17.4	20.7	20.0	19.8
West	72.3	65.8	67.8	68.9	6.9	11.4	9.7	10.3	17.8	20.4	19.9	18.6

Note: Data based on household interviews of a sample of the civilian noninstitutionalized population. Percents do not add to 100 because other types of health insurance (e.g., Medicare, military) are not shown and persons with both private insurance and Medicaid appear in both columns. (1) Includes Medicaid or other public assistance. In 1999, the age-adjusted percent of the population under 65 covered by Medicaid was 8.2%; 0.6% were covered by state-sponsored health plans and 0.3% were covered by Child Health Insurance Program (CHIP). (2) Includes persons not covered by private insurance, Medicaid or other public assistance, Medicare, or military plans. (3) The questionnaire changed compared with previous years. (4) Age adjusted.

Enrollment in Health Maintenance Organizations (HMOs), 1976-2000

Source: *Health, United States, 2001,* National Center for Health Statistics, U.S. Dept. of Health and Human Services

	1976	1980	1990	1992	1993	1994	1995	1996	1997	1998	1999	2000
	Number of enrolled in millions											
TOTAL	**6.0**	**9.1**	**33.0**	**36.1**	**38.4**	**45.1**	**50.9**	**59.1**	**66.8**	**76.6**	**81.3**	**80.9**
Model type[1]												
Individual practice assoc.[2]	0.4	1.7	13.7	14.7	15.3	17.8	20.1	26.0	26.7	32.6	32.8	33.4
Group[3]	5.6	7.4	19.3	16.5	15.4	13.9	13.3	14.1	11.0	13.8	15.9	15.2
Mixed	—	—	—	4.9	7.7	13.4	17.6	19.0	29.0	30.1	32.6	32.3
Federal program[4]												
Medicaid[5]	—	0.3	1.2	1.7	1.7	2.6	3.5	4.7	5.6	7.8	10.4	10.8
Medicare	—	0.4	1.8	2.2	2.2	2.5	2.9	3.7	4.8	5.7	6.5	6.6
	Percent of population enrolled in HMOs											
TOTAL	**2.8**	**4.0**	**13.4**	**14.3**	**15.1**	**17.3**	**19.4**	**22.3**	**25.2**	**28.6**	**30.1**	**30.0**
Geographic region												
Northeast	2.0	3.1	14.6	16.1	18.0	20.8	24.4	25.9	32.4	37.8	36.7	36.5
Midwest	1.5	2.8	12.6	12.8	13.2	15.2	16.4	18.8	19.5	22.7	23.3	23.2
South	0.4	0.8	7.1	7.8	8.4	10.2	12.4	15.2	17.9	21.0	23.9	22.6
West	9.7	12.2	23.2	24.7	25.1	27.4	28.6	33.2	36.4	39.1	41.4	41.7

— = Not available. **Note:** Data as of June 30 in 1976-80, Jan. 1 in 1990-2000. Medicaid enrollment in 1990 as of June 30. HMOs in Guam included starting in 1994; Puerto Rico, 1998. Open-ended enrollment in HMO plans, amounting to 11.6 million on Jan. 1, 1998, included from 1994 onwards. (1) In 1976, 11 HMOs with 35,000 enrollment did not report model type. In 1997, 11 HMOs with 153,000 enrollment did not report model type. In 1998, 6 HMOs with 109,000 enrollment did not report model type. In 1999, 3 HMO's with 18,000 enrollment did not report model type. In 2000, 1 HMO did not report model type. (2) This type of HMO contracts with an association of physicians from various settings (a mixture of solo and group practices) to provide health services. (3) Group includes staff, group, and network model types. (4) Enrollment by Medicaid or Medicare beneficiaries, where the Medicaid or Medicare program contracts directly with the HMO to pay the premium. (5) Data for 1990 and later include enrollment in managed-care health insuring organizations.

Health Care Visits, by Selected Characteristics, 1998, 1999

Source: Centers for Disease Control and Prevention, National Center for Health Statistics. National Health Interview Survey, family core and sample adult questionnaires.

	No visits		1-3 visits		4-9 visits		10 or more visits	
	1998	1999	1998	1999	1998	1999	1998	1999
	Percent distribution							
All persons	**16.0**	**17.5**	**46.8**	**45.8**	**23.8**	**23.3**	**13.5**	**13.4**
Age								
Under 6 years	4.9	5.9	46.7	45.9	36.5	36.8	11.8	11.3
6–17 years	15.0	15.5	58.4	58.5	20.2	19.4	6.3	6.7
18–24 years	22.6	24.8	47.7	46.1	18.4	17.8	11.2	11.4
25–44 years	21.3	24.0	47.6	45.7	18.6	17.8	12.5	12.6
45–54 years	17.2	18.4	44.9	43.2	22.5	22.8	15.4	15.7
55–64 years	13.8	14.7	41.6	41.1	27.1	28.4	17.5	15.8
65–74 years	8.4	8.6	36.5	36.9	34.3	33.2	20.8	21.3
75 years and over	6.0	7.2	30.8	31.1	36.5	35.1	26.7	26.6
Sex								
Male	20.7	23.1	47.3	45.5	21.2	20.6	10.8	10.8
Female	11.3	12.0	46.4	46.1	26.3	25.9	16.0	15.9
Race and Hispanic origin								
White non-Hispanic	14.7	15.5	47.1	45.9	24.6	24.5	14.0	14.1
Black, non-Hispanic	16.5	18.3	46.5	46.1	23.2	22.1	13.8	13.5
Hispanic[1]	24.0	28.2	44.9	44.3	19.7	19.2	11.5	10.3
Geographic region								
Northeast	12.1	12.8	47.8	46.4	25.4	25.6	14.7	15.2
Midwest	15.6	16.2	46.9	46.7	24.2	23.8	13.4	13.3
South	17.0	18.9	46.7	45.5	23.1	22.5	13.2	13.2
West	18.3	20.9	46.0	44.8	22.8	21.9	12.9	12.4

(1) Persons of Hispanic origin may be of any race.

Major Reasons Given by Patients for Emergency Room Visits, 1999

Source: National Center for Health Statistics, U.S. Dept. of Health and Human Services

Principal reason for visit	Number of visits (in thousands)	Percent distribution
All visits	**102,765**	**100.0**
Stomach and abdominal pain, cramps, and spasms	6,610	6.4
Chest pain and related symptoms	5,608	5.5
Fever	4,678	4.6
Headache, pain in head	2,809	2.7
Cough	2,596	2.5
Shortness of breath	2,575	2.5
Laceration and cuts—upper extremity	2,462	2.4
Back symptoms	2,445	2.4
Vomiting	2,319	2.3
Pain, site not referable to a specified body system	2,058	2.0

Principal reason for visit	Number of visits (in thousands)	Percent distribution
Symptoms referable to throat	1,957	1.9
Accident, not otherwise specified	1,928	1.9
Earache or ear infection	1,707	1.7
Motor vehicle accident, type of injury unspecified	1,546	1.5
Labored or difficult breathing (dyspnea)	1,527	1.5
Laceration and cuts—facial area	1,328	1.3
Skin rash	1,309	1.3
Vertigo—dizziness	1,306	1.3
Injury, other and unspecified type—hand and finger(s)	1,278	1.2
Injury, other and unspecified type—head, neck, and face	1,277	1.2
ALL OTHER REASONS	**53,443**	**52.0**

Top 20 Reasons Given by Patients for Physicians' Office Visits, 1999

Source: National Center for Health Statistics, U.S. Dept. of Health and Human Services

	No of visits (1,000)	PERCENT DISTRIBUTION Total	Female	Male
ALL VISITS	**756,734**	**100.0**	**100.0**	**100.0**
1. General medical examination	46,039	6.1	5.8	6.4
2. Progress visit, not otherwise specified	33,975	4.5	4.0	5.2
3. Postoperative visit	22,513	3.0	3.1	2.7
4. Cough	20,654	2.7	2.6	2.9
5. Routine prenatal examination	17,899	2.4	4.0	*
6. Symptoms referable to throat	15,315	2.0	1.9	2.2
7. Well-baby examinations	13,111	1.7	1.4	2.2
8. Vision dysfunctions	12,243	1.6	1.6	1.6
9. Knee symptoms	11,778	1.6	1.5	1.6
10. Hypertension	11,130	1.5	1.4	1.5
11. Earache or ear infection	11,047	1.5	1.2	1.8
12. Back symptoms	10,482	1.4	1.4	1.4
13. Skin rash	10,446	1.4	1.2	1.6
14. Stomach pain, cramps, and spasms	10,077	1.3	1.4	1.2
15. Fever	9,963	1.3	1.0	1.7
16. Depression	9,664	1.3	1.5	1.0
17. Medication, other and unspecified kinds	9,284	1.2	1.2	1.2
18. Low back symptoms	9,186	1.2	1.1	1.4
19. Nasal congestion	9,067	1.2	1.1	1.3
20. Headache, pain in the head	8,599	1.1	1.3	1.0
ALL OTHER REASONS	**454,263**	**60.0**	**60.1**	**59.8**

Drugs Most Frequently Prescribed in Physicians' Offices, 1999

Source: National Center for Health Statistics, U.S. Dept. of Health and Human Services; *Physicians' Desk Reference*; in thousands

Rank	Name of drug (principal generic substance)[1]	Times prescribed	Therapeutic use
1.	Claritin (loratadine)	15,978	Antihistamine
2.	Lasix (furosemide)	12,910	Diuretic, antihypertensive
3.	Prednisone	12,705	Steroid replacement therapy, anti-inflammatory agent
4.	Synthroid (levothyroxine)	12,520	Thyroid hormone therapy
5.	Lipitor (atorvastatin calcium)	12,319	Lowers cholesterol
6.	Premarin (estrogens)	11,878	Estrogen replacement therapy
7.	Prilosec (omeprazole)	11,704	For duodenal or gastric ulcer
8.	Tylenol (acetaminophen)	11,366	Analgesic (for pain relief)
9.	Amoxicillin	10,623	Antibiotic
10.	Celebrex (celecoxib)	9,531	Anti-inflammatory agent
11.	Norvasc(amlodipine besylate)	8,769	For high blood pressure
12.	Albuterol Sulfate	8,612	Antiasthmatics/bronchodilators
13.	Zoloft (sertraline hydrochloride)	8,351	Antidepressant
14.	Coumadin (crystalline warfarin sodium)	8,081	Anticoagulant
15.	Prozac (fluoxetine hydrochloride)	7,922	Antidepressant
16.	Paxil (paroxetine hydrochloride)	7,858	Antidepressant
17.	Atenolol	7,750	For high blood pressure
18.	Lanoxin(digoxin)	7,707	For congestive heart failure, irregular heartbeat
19.	Motrin (ibuprofen)	7,417	Anti-inflammatory agent
20.	Keflex (cepha lexin)	7,190	Antibiotic
	ALL OTHER	**935,497**	

(1) The trade or generic name used by the physician on the prescription or other medical records. The use of trade names is for identification only and does not imply endorsement by the Public Health Service or the U.S. Dept. of Health and Human Services.

Hospitals and Nursing Homes in the U.S., 1999

Source: *1999 Hospital Statistics*, Health Forum, L.L.C., An American Hospital Association Company, © 2001; *Health, United States, 2001*

For information on choosing a nursing home, go to the website http://www.medicare.gov/nursing/overview.asp

STATE	Hospitals[1]	% of beds occupied[1]	Nursing homes	% of beds occupied
AL	120	60.5	225	92.5
AK	27	54.1	15	76.4
AZ	82	61.7	162	76.4
AR	93	59.9	263	77.3
CA	461	63.6	1,390	81.3
CO	79	57.1	225	82.8
CT	49	72.1	257	91.5
DE	10	67.5	42	74.0
DC	18	76.1	20	93.3
FL	248	61.7	742	82.6
GA	184	62.2	362	91.1
HI	28	72.4	44	92.2
ID	47	54.4	82	74.7
IL	228	60.1	880	76.3
IN	126	57.0	568	69.5
IA	125	57.8	470	79.3
KS	141	52.7	399	81.4
KY	121	60.4	313	90.0
LA	147	56.7	336	78.7
ME	42	64.8	125	89.7
MD	76	71.2	248	82.1
MA	118	70.7	548	88.6
MI	165	66.0	44.3	84.3
MN	148	67.4	44.4	89.8
MS	106	60.5	198	93.7
MO	141	57.5	559	71.7
MT	59	68.2	105	79.8
NE	93	60.0	237	83.3
NV	32	74.0	49	70.7
NH	33	55.3	84	91.6
NJ	99	68.6	363	89.7
NM	56	59.0	82	88.3
NY	271	76.9	659	94.8
NC	142	67.1	408	90.1
ND	49	60.1	89	92.2
OH	188	59.0	1,007	78.9
OK	132	56.3	409	71.0
OR	65	58.1	157	76.8
PA	263	67.9	786	88.7
RI	15	71.2	101	88.7
SC	80	66.4	176	86.5
SD	58	66.4	114	91.6
TN	139	57.1	352	90.4
TX	467	57.9	1,254	67.8
UT	48	57.3	93	77.2
VT	17	64.8	45	91.8
VA	120	65.5	272	90.0
WA	101	58.7	279	81.3
WV	66	60.5	136	89.8
WI	139	58.6	425	85.5
WY	28	52.4	41	81.9
U.S.	**5,890**	**63.4**	**17,259**	**82.7**

(1) Community hospitals (excludes federal hospitals, hospital units of institutions, facilities for the mentally retarded, and alcoholism and chemical dependency hospitals).

Expected New Cancer Cases and Deaths, by Sex, for Leading Sites, 2001

Source: American Cancer Society

The estimates of expected new cases are offered as a rough guide only. They exclude basal and squamous cell skin cancers and in situ carcinomas, except urinary bladder. Carcinoma in situ of the breast accounts for about 46,400 new cases annually, melanoma carcinoma in situ for about 31,400. More than 1.3 million basal cell and squamous cell skin cancers are expected to be diagnosed in 2001. About 1,900 nonmelanoma skin cancer deaths are included among deaths expected in all sites.

EXPECTED NEW CASES

Both sexes		Women		Men	
Prostate	198,100	Breast	192,200	Prostate	198,100
Breast	193,700	Lung	78, 800	Lung	90,700
Lung	169,500	Colorectal	68,100	Colorectal	67,300
Colorectal	135,400	Endometrium (uterus)	38,300	Urinary bladder	39,200
Non-Hodgkin's lymphoma	56,200	Non-Hodgkin's lymphoma	25,100	Non-Hodgkin's lymphoma	31,100
ALL SITES	**1,268,000**	**ALL SITES**	**625,000**	**ALL SITES**	**643,000**

EXPECTED DEATHS

Both sexes		Women		Men	
Lung	157,400	Lung	67,300	Lung	90,100
Colorectal	56,700	Breast	40,200	Prostate	31,500
Breast	40,600	Colorectal	29,000	Colorectal	27,700
Prostate	31,500	Pancreas	14,800	Non-Hodgkin's lymphoma	13,800
Pancreas	28,900	Ovary	13,900	Pancreas	14,100
ALL SITES	**553,400**	**ALL SITES**	**267,300**	**ALL SITES**	**286,100**

U.S. Cancer Incidence for Top 15 Sites, 1990-98

Source: Surveillance, Epidemiology, and End Results (SEER) Program, National Cancer Institute

	Rate[1]	% change 1997-98		Rate[1]	% change 1997-98
ALL SITES	**397.8**	**−1.1**	Ovary	14.5	−1.3
Prostate	149.7	−5.1	Melanomas of the skin	12.9	+2.8
Breast (female)	111.2	+1.2	Leukemias	10.4	−1.4
Lung	54.4	−1.6	Oral cavity and pharynx	9.9	−1.9
Colon and rectum	43.3	−0.7	Kidney and renal pelvis	9.1	+0.8
Uterus	21.2	+0.4	Pancreas	8.8	−0.6
Urinary bladder	16.3	−0.9	Stomach	7.4	−1.8
Non-Hodgkin's lymphomas	15.7	+0.1	Thyroid	5.3	+2.7

(1) Per 100,000 population; annual average for 8-year period.

U.S. Cancer Mortality forTop 15 Sites, 1990-98

Source: Surveillance, Epidemiology, and End Results (SEER) Program, National Cancer Institute

	Rate[1]	% change 1997-98		Rate[1]	% change 1997-98
ALL SITES	**167.8**	**−1.1**	Leukemias	6.2	−0.8
Lung	49.1	−0.8	Brain and other nervous system	4.2	−0.6
Breast (female)	24.7	−2.4	Stomach	4.1	−2.7
Prostate	24.5	−3.5	Esophagus	3.6	+0.5
Colon and rectum	17.1	−1.8	Liver and intraheptic bile duct	3.5	+2.9
Pancreas	8.4	−0.4	Kidney and renal pelvis	3.5	+0.1
Ovary	7.5	−1.2	Urinary bladder	3.2	−0.4
Non-Hodgkin's lymphomas	6.8	+1.1	Multiple myeloma (bone marrow)	3.1	0.0

(1) Per 100,000 population; annual average for 8-year period.

Cardiovascular Diseases Statistical Summary, 1999

Source: American Heart Association

Numbers preliminary for 1999.

Prevalence — An estimated 61,800,000 Americans had one or more forms of heart and blood vessel disease in 1999.
- hypertension (high blood pressure) — 50,000,000
- coronary heart disease — 12,560,000
- stroke — 4,600,000

Mortality — 959,052 in 1999 (40.1% of all deaths).
- Someone died from cardiovascular disease every 33 seconds in the U.S. in 1998.

Congenital or inborn heart defects —
- Mortality from such heart defects was 4,657 in 1998.

Coronary heart disease (heart attack and angina pectoris) — caused 529,544 deaths in 1999.
- 12,560,000 Americans had a history of heart attack and/or angina pectoris.
- As many as 1,100,000 Americans had coronary attacks in 1999.

Congestive heart failure — 4,790,000; killed 54,920 in 1999.

Stroke — killed 167,340 Americans in 1999.

Rheumatic heart disease — killed 3,676 in 1999.

Transplant Waiting List, Sept. 2001

Source: United Network for Organ Sharing

Type of transplant	Patients waiting
Kidney	50,003
Liver	18,414
Pancreas	1,158
Kidney-pancreas	2,497
Intestine	177
Heart	4,171
Heart-lung	214
Lung	3,741
Total[1]	**78,197**

Transplants Performed, 2000

Source: United Network for Organ Sharing

Type of transplant	Number
Kidney	13,332
Liver	4,950
Pancreas	436
Kidney-pancreas	910
Intestine	79
Heart	2,196
Heart-lung	48
Lung	956
Total	**22,908**

(1) Some patients are waiting for more than one organ; therefore the total number of patients is less than the sum of patients waiting for each organ.

AIDS Deaths and New AIDS Cases in the U.S., 1985-2000

Source: *Health, United States, 2001; HIV/AIDS Surveillance Report,* Vol. 12, No. 2, covering through 2000; National Center for Health Statistics, U.S. Dept. of Health and Human Services

	All Years[2]	1985	1990	1995	1996	1997	1998	1999	1st ½ 2000
TOTAL DEATHS[1]	**427,002**	**6,854**	**31,145**	**50,877**	**37,983**	**22,070**	**18,210**	**16,767**	**NA**
					NEW AIDS CASES				
TOTAL NEW CASES	**729,326**	**8,160**	**41,486**	**70,632**	**66,138**	**57,880**	**46,088**	**44,891**	**20,550**
Male									
All males, 13 years and over	**601,471**	**7,508**	**36,234**	**56,894**	**52,369**	**45,016**	**35,278**	**34,254**	**15,383**
Race									
White, non-Hispanic	295,990	4,752	20,853	26,122	23,096	17,408	13,868	12,775	5,759
Black, non-Hispanic	205,630	1,710	10,256	20,891	20,014	18,709	14,639	14,885	6,667
Hispanic[4]	92,440	989	4,752	9,146	8,544	8,219	6,195	6,085	2,674
American Indian or Alaska Native	1,804	8	80	199	172	170	118	132	64
Asian Pacific Islander	4,792	49	264	491	480	382	323	297	147
Age									
13-19 years	2,161	28	107	223	204	181	141	126	58
20-29 years	95,991	1,503	6,934	8,402	7,060	5,767	4,275	3,973	1,647
30-39 years	273,004	3,588	16,694	25,774	23,755	20,075	15,238	14,508	6,447
40-49 years	163,297	1,633	8,838	16,223	15,416	13,513	10,912	10,953	4,947
50-59 years	49,644	597	2,651	4,709	4,415	4,102	3,514	3,506	1,700
60 years and over	17,374	159	1,010	1,563	1,519	1,378	1,198	1,188	584
Females									
All females, 13 years and over	**119,454**	**522**	**4,528**	**12,992**	**13,115**	**12,422**	**10,442**	**10,382**	**5,074**
Race:									
White, non-Hispanic	27,205	142	1,222	3,050	2,835	2,458	2,006	1,908	940
Black, non-Hispanic	71,656	279	2,544	7,597	8,076	7,793	6,702	6,743	3,278
Hispanic	19,359	98	730	2,227	2,058	2,030	1,607	1,608	763
American Indian or Alaska Native	399	2	9	37	44	37	32	41	41
Asian Pacific Islander	663	1	19	72	79	63	57	62	40
Age:									
13-19 years	1,533	4	66	157	173	173	143	168	82
20-29 years	25,582	177	1,114	2,670	2,666	2,406	1,914	1,884	889
30-39 years	53,562	232	2,077	5,944	5,874	5,448	4,442	4,278	2,028
40-49 years	27,342	45	780	3,069	3,248	3,228	2,869	2,811	1,477
50-59 years	7,661	26	272	818	828	819	788	918	435
60 years and over	3,774	38	219	334	326	348	286	323	163
Children									
All children, under 13 years	**8,401**	**130**	**724**	**746**	**654**	**442**	**368**	**255**	**93**
Race									
White, non-Hispanic	1,535	26	158	117	96	62	59	32	18
Black, non-Hispanic	5,148	86	389	484	431	288	236	170	55
Hispanic	1,623	18	168	135	123	85	70	48	17
American Indian or Alaska Native	31	0	5	2	3	2	0	2	1
Asian Pacific Islander	49	0	4	5	1	3	2	2	1
Age:									
Under 1 year	3,296	63	318	269	222	131	96	88	35
1-12 years	5,105	67	406	477	432	311	272	167	58

NA = Not available. **Note:** The definition of AIDS cases for reporting purposes was expanded in 1985, 1987, and 1993, as more was learned about the spectrum of human immunodeficiency virus-associated diseases. Data exclude residents of U.S. territories. Figures were updated June 30, 2000 to include temporarily delayed case reports and may differ from previous reports of *Health, United States.* (1) Based on preliminary figures and subject to revision. (2) Revised figures; includes cases and deaths prior to 1985 and for years not shown. (3) Includes Aleut and Eskimo. (4) Includes Chinese, Japanese, Filipino, Hawaiian and part-Hawaiian, and other Asian or Pacific Islander.

New AIDS Cases in the U.S., 1985-2000, by Transmission Category

Source: *Health, United States, 2001,* CDC, National Center for HIV, STD, and TB Prevention, Div. of HIV/AIDS Prevention

TRANSMISSION CATEGORY	All Years[1]	1985	1990	1995	1996	1997	1998	1999	1st ½ 2000
All males 13 years and older	**601,471**	**7,508**	**36,234**	**56,894**	**52,369**	**45,016**	**35,278**	**34,254**	**15,383**
Men who have sex with men	344,619	5,353	23,687	30,997	27,687	21,432	16,825	15,444	6,782
Injecting drug use	127,491	1,103	6,938	13,416	11,932	10,245	7,443	6,844	2,902
Men who have sex with men and injecting drug use	45,991	660	2,931	4,093	3,521	2,669	2,118	1,832	770
Hemophilia/coagulation disorder	4,791	68	333	438	322	186	149	139	44
Heterosexual contact[2]	25,609	32	712	2,898	3,241	3,140	2,653	2,805	1,132
Sex with injecting drug user	8,436	25	452	870	830	787	645	625	253
Transfusion[3]	4,831	102	440	321	251	204	146	132	73
Undetermined[4]	48,139	190	1,193	4,731	5,415	7,140	5,944	7,058	3,680
All females 13 years and older	**119,454**	**522**	**4,528**	**12,992**	**13,115**	**12,422**	**10,442**	**10,382**	**5,074**
Injecting drug use	49,539	286	2,333	5,404	4,831	4,315	3,220	2,902	1,299
Hemophilia/coagulation disorder	266	3	16	28	25	34	22	13	NA
Heterosexual contact[2]	47,050	119	1,537	5,515	5,874	5,215	4,261	4,229	1,877
Sex with injecting drug user	18,494	82	1,033	1,915	1,923	1,519	1,221	1,109	467
Transfusion[3]	3,637	63	330	253	259	166	122	127	69
Undetermined[4]	18,962	51	312	1,792	2,126	2,692	2,817	3,111	1,829

NA = Not available. **Note**: The definition of AIDS cases for reporting purposes was expanded in 1985, 1987, and 1993, as more was learned about the spectrum of human immunodeficiency virus-associated diseases. Data exclude residents of U.S. territories. Figures were updated June 30, 2000 to include temporarily delayed case reports and may differ from previous reports of *Health, United States.* (1) Includes cases prior to 1985 and for years not shown. (2) Includes persons who have had heterosexual contact with a person with human immunodeficiency virus (HIV) infection or at risk of HIV infection. (3) Receipt of blood transfusion, blood components, or tissue. (4) Includes persons for whom risk information is incomplete, persons still under investigation, men reported only to have had heterosexual contact with prostitutes, and interviewed persons for whom no specific risk is identified.

SPORTS

SPORTS HIGHLIGHTS OF 2001

San Francisco's Barry Bonds broke the single-season record of 70 home runs (set by Mark McGwire, 1998), hitting his 71st and 72d homers of the year at home off LA pitcher Chan Ho Park, Oct. 5. He hit his 73d homer on the final day of the season. Bonds also broke Babe Ruth's season records for slugging percentage (.847) and walks (170), finishing with .863 and 177. With 107 extra-base hits, he tied Chuck Klein's NL record, 3d all time behind Ruth (119) and Gehrig (117).

On Apr. 8, golfer Tiger Woods won his 2d Masters championship at Augusta National. It was an unprecedented 4th consecutive major championship on the PGA Tour for Woods, who in 2000 also won the U.S. Open, the British Open, and the PGA Championship.

Lance Armstrong, riding for the U.S. Postal Service team, won his 3d straight Tour de France July 29. He is the 1st American, and the 5th rider overall, to accomplish this feat in the world's premier cycling event.

The Los Angeles Lakers won their 2d NBA Championship in a row, defeating the Philadelphia 76ers, 108-96, in Game 5 of the NBA Finals on June 15, in Philadelphia. Lakers center Shaquille O'Neal became only the 3d player to be named Finals MVP in consecutive years. The Lakers made a nearly perfect run to the title, finishing with a playoff record of 15-1, the best ever.

Led by a record-setting defense, the AFC's Baltimore Ravens defeated the NFC's NY Giants, 34-7, in Super Bowl XXXV, Jan. 28, at Raymond James Stadium in Tampa, FL. Baltimore's All-Pro linebacker Ray Lewis was named MVP.

In the first U.S. Open shown in network prime-time, Sept. 8, defending champion Venus Williams, age 21, defeated younger sister Serena Williams (19), the 1999 U.S. Open champ, 6-2, 6-4. It was the 1st time in professional tennis, and the 1st time since the 1884 Wimbledon, that sisters faced each other in the finals of a Grand Slam event. The sisters were also the 1st African Americans to oppose each other in a Grand Slam final.

With a 3-1 victory in Game 7, June 9, the Colorado Avalanche took the Stanley Cup from the defending champion New Jersey Devils. Goalie Patrick Roy won the Conn Smythe Trophy (MVP in the playoffs) a record 3d time. Ray Bourque, traded from Boston to Colorado in 2000 (at his request), finally won the Cup after 22 seasons. The NHL's highest scoring defenseman and 6-time Norris Trophy winner, who skated on a record 19 All-Star teams, retired June 26.

Michael Jordan announced his return to the NBA as a player on Sept. 25. The 5-time MVP, who led the Chicago Bulls to 6 NBA titles, had retired for the 2d time in Jan. 1999. Jordan, 38, who signed with the Washington Wizards, was required under NBA rules to sell his ownership stake in the team and resign from his position as president of basketball operations. He said he would donate his 1st year's salary—$1 million—to victims of the Sept. 11 terrorist attacks.

Duke topped Arizona, 82-72, for the NCAA men's basketball title, Apr. 2. Duke's Shane Battier was named Most Outstanding Player of the tournament. It was the Blue Devils' 3d title under coach Mike Krzyzewski. In the women's championship game on Apr. 1, Notre Dame defeated Purdue, 68-66, for their 1st women's title ever. Ruth Riley, of Notre Dame, was named Most Outstanding Player of the tournament.

On May 27, at the Univ. of Oregon's Hayward Field, 18-year-old high school senior Alan Webb (Reston, VA) became the 19th-fastest U.S. miler ever, at 3 mins., 53.43 sec., smashing Jim Ryun's 1965 schoolboy record of 3:55.3. In Jan., at the New Balance Games in New York, Webb became the 1st U.S. prep runner to break 4 minutes indoors (3:59.86).

WORLD ALMANAC EDITORS' PICKS

The *World Almanac* staff rated the following as the ten most dramatic moments in Olympic sports:

1. Jesse Owens wins 4 track and field golds in Berlin (1936).
2. U.S. ice hockey team upsets the Soviets and goes on to win gold in Lake Placid (1980).
3. Mark Spitz swims to 7 golds (4 indiv., 3 team), all in world-record times, in Munich (1972).
4. 14-year-old Romanian gymnast Nadia Comaneci earns the first perfect 10s on her way to 3 golds in Montreal (1976).
5. Bob Beamon breaks the long-jump record by nearly 2 feet in the thin air of Mexico City (1968).
6. Ethiopian Abebe Bikila wins the marathon running barefoot through Rome, becoming the first black African to win a gold medal (1960).
7. Eric Heiden wins 5 speed-skating golds in Lake Placid (1980).
8. Carl Lewis wins 4 track and field golds in Los Angeles (1984).
9. Billy Mills shocks the field to win the 10,000 meter gold in Tokyo (1964).
10. *(tied)*: Figure-skater Peggy Fleming wins the only gold medal for the U.S. in Grenoble (1968). Marion Jones wins 5 medals (3 gold, 2 bronze) in Sydney, the most ever for a female track athlete (2000).

Winter Olympic Games

Sites of Winter Olympic Games

1924 Chamonix, France
1928 St. Moritz, Switzerland
1932 Lake Placid, New York
1936 Garmisch-Partenkirchen, Germany
1948 St. Moritz, Switzerland
1952 Oslo, Norway
1956 Cortina d'Ampezzo, Italy
1960 Squaw Valley, California
1964 Innsbruck, Austria
1968 Grenoble, France
1972 Sapporo, Japan
1976 Innsbruck, Austria
1980 Lake Placid, New York
1984 Sarajevo, Yugoslavia
1988 Calgary, Alberta
1992 Albertville, France
1994 Lillehammer, Norway
1998 Nagano, Japan
2002 Salt Lake City, Utah
2006 Turin, Italy

The Winter Olympic Games in Salt Lake City

Salt Lake City, Utah, Feb. 8-24, 2002

In Feb. 2002, the world's attention will focus on Utah's capital, on the southern shore of the Great Salt Lake, where 3,500 athletes from 80 countries are expected to take part in the Winter Olympics. Despite concerns following the terrorist attacks of Sept. 11, 2001, Olympic organizers said the Games would be held as scheduled and would be secure. Congressional leaders, Oct. 3, pledged some $40 mil. for increased security measures, bringing total federal security spending on the event to about $240 mil. Some 5,000 law enforcement officers from 60 local, state, and federal agencies were assigned to the Games. The Games will include 78 medal events in 7 sports. Olympic sites in the city and surrounding area include 11 competition venues, athlete residences, a stadium for opening and closing ceremonies, and a medal awards plaza.

Winter Olympic Games in 1998

Nagano, Japan, Feb. 7-22, 1998

More than 2,400 athletes from 72 nations participated in the 1998 Winter Olympics in Nagano, Japan, where snowboarding and curling made Olympic debuts. Germany won the most medals, 29, and the most gold medals, 12. The U.S. finished 6th in medals, with 13 (6 gold), trailing Germany, Norway (25), Russia (18), Austria (17), and Canada (15).

The Japanese ski jumping team won 4 medals, including a gold in the team event. Hermann Maier won the gold for Austria in both the giant and the super giant slalom, after recovering from a crash in the downhill; cross-country skier Bjoern Daehlie won 3 golds and 1 silver for Norway, giving him an unprecedented 12 total and 8 gold Winter Olympic medals for his career. The Czech Republic, led by the NHL's Dominik Hasek, captured the gold in men's hockey; the U.S. women's hockey team upset Team Canada for the gold; U.S. skater Tara Lipinski, 15, became the youngest Olympic figure-skating gold medalist.

Final 1998 Medal Standings

	Gold	Silver	Bronze	Total
Germany	12	9	8	**29**
Norway	10	10	5	**25**
Russia	9	6	3	**18**
Austria	3	5	9	**17**
Canada	6	5	4	**15**
U.S.	6	3	4	**13**
Finland	2	4	6	**12**
Netherlands	5	4	2	**11**
Japan	5	1	4	**10**
Italy	2	6	2	**10**
France	2	1	5	**8**
China	0	6	2	**8**
Switzerland	2	2	3	**7**
Korea	3	1	2	**6**
Czech Republic	1	1	1	**3**
Sweden	0	2	1	**3**
Belarus	0	0	2	**2**
Kazakhstan	0	0	2	**2**
Bulgaria	1	0	0	**1**
Denmark	0	1	0	**1**
Ukraine	0	1	0	**1**
Australia	0	0	1	**1**
Belgium	0	0	1	**1**
Great Britain	0	0	1	**1**
TOTAL	**69**	**68**	**68**	**205**

Winter Olympic Games Champions, 1924-98

In 1992, the Unified Team represented the former Soviet republics of Russia, Ukraine, Belarus, Kazakhstan, and Uzbekistan.

ALPINE SKIING

	Men's Downhill	Time
1948	Henri Oreiller, France	2:55.0
1952	Zeno Colo, Italy	2:30.8
1956	Anton Sailer, Austria	2:52.2
1960	Jean Vuarnet, France	2:06.0
1964	Egon Zimmermann, Austria	2:18.16
1968	Jean-Claude Killy, France	1:59.85
1972	Bernhard Russi, Switzerland	1:51.43
1976	Franz Klammer, Austria	1:45.73
1980	Leonhard Stock, Austria	1:45.50
1984	Bill Johnson, U.S.	1:45.59
1988	Pirmin Zurbriggen, Switzerland	1:59.63
1992	Patrick Ortlieb, Austria	1:50.37
1994	Tommy Moe, U.S.	1:45.75
1998	Jean-Luc Cretier, France	1:50.11

	Men's Super Giant Slalom	Time
1988	Franck Piccard, France	1:39.66
1992	Kjetil-Andre Aamodt, Norway	1:13.04
1994	Markus Wasmeier, Germany	1:32.53
1998	Hermann Maier, Austria	1:34.82

	Men's Giant Slalom	Time
1952	Stein Eriksen, Norway	2:25.0
1956	Anton Sailer, Austria	3:00.1
1960	Roger Staub, Switzerland	1:48.3
1964	Francois Bonlieu, France	1:46.71
1968	Jean-Claude Killy, France	3:29.28
1972	Gustavo Thoeni, Italy	3:09.62
1976	Heini Hemmi, Switzerland	3:26.97
1980	Ingemar Stenmark, Sweden	2:40.74
1984	Max Julen, Switzerland	2:41.18
1988	Alberto Tomba, Italy	2:06.37
1992	Alberto Tomba, Italy	2:06.98
1994	Markus Wasmeier, Germany	2:52.46
1998	Hermann Maier, Austria	2:38.51

	Men's Slalom	Time
1948	Edi Reinalter, Switzerland	2:10.3
1952	Othmar Schneider, Austria	2:00.0
1956	Anton Sailer, Austria	3:14.7
1960	Ernst Hinterseer, Austria	2:08.9
1964	Josef Stiegler, Austria	2:11.13
1968	Jean-Claude Killy, France	1:39.73
1972	Francisco Fernandez Ochoa, Spain	1:49.27
1976	Piero Gros, Italy	2:03.29
1980	Ingemar Stenmark, Sweden	1:44.26
1984	Phil Mahre, U.S.	1:39.41
1988	Alberto Tomba, Italy	1:39.47
1992	Finn Christian Jagge, Norway	1:44.39
1994	Thomas Stangassinger, Austria	2:02.02
1998	Hans-Petter Buraas, Norway	1:49.31

	Men's Combined	Time
1988	Hubert Strolz, Austria	36.55 (pts.)
1992	Josef Polig, Italy	14.58 (pts.)
1994	Lasse Kjus, Norway	3:17.53
1998	Mario Reiter, Austria	3:08.06

	Women's Downhill	Time
1948	Hedi Schlunegger, Switzerland	2:28.3
1952	Trude Jochum-Beiser, Austria	1:47.1
1956	Madeleine Berthod, Switzerland	1:40.7
1960	Heidi Biebl, Germany	1:37.6
1964	Christl Haas, Austria	1:55.39
1968	Olga Pall, Austria	1:40.87
1972	Marie Therese Nadig, Switzerland	1:36.68
1976	Rosi Mittermaier, W. Germany	1:46.16
1980	Annemarie Proell Moser, Austria	1:37.52
1984	Michela Figini, Switzerland	1:13.36
1988	Marina Kiehl, W. Germany	1:25.86
1992	Kerrin Lee-Gartner, Canada	1:52.55
1994	Katja Seizinger, Germany	1:35.93
1998	Katja Seizinger, Germany	1:28.89

	Women's Super Giant Slalom	Time
1988	Sigrid Wolf, Austria	1:19.03
1992	Deborah Compagnoni, Italy	1:21.22
1994	Diann Roffe-Steinrotter, U.S.	1:22.15
1998	Picabo Street, U.S.	1:18.02

	Women's Giant Slalom	Time
1952	Andrea Mead Lawrence, U.S.	2:06.8
1956	Ossi Reichert, Germany	1:56.5
1960	Yvonne Ruegg, Switzerland	1:39.9
1964	Marielle Goitschel, France	1:52.24
1968	Nancy Greene, Canada	1:51.97
1972	Marie Therese Nadig, Switzerland	1:29.90
1976	Kathy Kreiner, Canada	1:29.13
1980	Hanni Wenzel, Liechtenstein (2 runs)	2:41.66
1984	Debbie Armstrong, U.S.	2:20.98
1988	Vreni Schneider, Switzerland	2:06.49
1992	Pernilla Wiberg, Sweden	2:12.74
1994	Deborah Compagnoni, Italy	2:30.97
1998	Deborah Compagnoni, Italy	2:50.59

	Women's Slalom	Time
1948	Gretchen Fraser, U.S.	1:57.2
1952	Andrea Mead Lawrence, U.S.	2:10.6
1956	Renee Colliard, Switzerland	1:52.3
1960	Anne Heggtveigt, Canada	1:49.6
1964	Christine Goitschel, France	1:29.86
1968	Marielle Goitschel, France	1:25.86
1972	Barbara Cochran, U.S.	1:31.24
1976	Rosi Mittermaier, W. Germany	1:30.54
1980	Hanni Wenzel, Liechtenstein	1:25.09
1984	Paoletta Magoni, Italy	1:36.47
1988	Vreni Schneider, Switzerland	1:36.69
1992	Petra Kronberger, Austria	1:32.68
1994	Vreni Schneider, Switzerland	1:56.01
1998	Hilde Gerg, Germany	1:32.40

	Women's Combined	Time
1988	Anita Wachter, Austria	29.25 (pts.)
1992	Petra Kronberger, Austria	2.55 (pts.)
1994	Pernilla Wiberg, Sweden	3:05.16
1998	Katja Seizinger, Germany	2:40.74

BIATHLON

	Men's 10 Kilometers	Time
1980	Frank Ullrich, E. Germany	32:10.69
1984	Eirik Kvalfoss, Norway	30:53.80
1988	Frank-Peter Roetsch, E. Germany	25:08.10
1992	Mark Kirchner, Germany	26:02.30
1994	Serguei Tchepikov, Russia	28:07.00
1998	Ole Einar Bjoerndalen, Norway	27:16.20

	Men's 20 Kilometers	Time
1960	Klas Lestander, Sweden	1:33:21.6
1964	Vladimir Melanin, USSR	1:20:26.8
1968	Magnar Solberg, Norway	1:13:45.9
1972	Magnar Solberg, Norway	1:15:55.50
1976	Nikolai Kruglov, USSR	1:14:12.26
1980	Anatoly Aljabiev, USSR	1:08:16.31
1984	Peter Angerer, W. Germany	1:11:52.7
1988	Frank-Peter Roetsch, E. Germany	0:56:33.33
1992	Yevgeny Redkine, Unified Team	0:57:34.4
1994	Serguei Tarasov, Russia	0:57:25.3
1998	Halvard Hanevold, Norway	0:56:16.4

	Men's 30-Kilometer Relay	Time
1968	USSR, Norway, Sweden (40 km)	2:13:02.4
1972	USSR, Finland, E. Germany (40 km)	1:51:44.92
1976	USSR, Finland, E. Germany (40 km)	1:57:55.64
1980	USSR, E. Germany, W. Germany	1:34:03.27
1984	USSR, Norway, W. Germany	1:38:51.70

	Men's 30-Kilometer Relay	Time
1988	USSR, W. Germany, Italy	1:22:30.00
1992	Germany, Unified Team, Sweden	1:24:43.50
1994	Germany, Russia, France	1:30:22.1
1998	Germany, Norway, Russia	1:19:43.3

	Women's 7.5 Kilometers	Time
1992	Anfissa Restsova, Unified Team	24:29.20
1994	Myriam Bedard, Canada	26:08.8
1998	Galina Koukleva, Russia	23:08.0

	Women's 15 Kilometers	Time
1992	Antje Misersky, Germany	51:47.2
1994	Myriam Bedard, Canada	52:06.6
1998	Ekaterina Dafovska, Bulgaria	54:52.0

	Women's 22.5-Kilometer Relay	Time
1992	France, Germany, Unified Team	1:15:55.6

	Women's 30-Kilometer Relay	Time
1994	Russia, Germany, France	1:47:19.5
1998	Germany, Russia, Norway	1:40:13.6

BOBSLEDDING

(Driver in parentheses)

	4-Man Bob	Time
1924	Switzerland (Eduard Scherrer)	5:45.54
1928	United States (William Fiske) (5-man)	3:20.50
1932	United States (William Fiske)	7:53.68
1936	Switzerland (Pierre Musy)	5:19.85
1948	United States (Francis Tyler)	5:20.10
1952	Germany (Andreas Ostler)	5:07.84
1956	Switzerland (Franz Kapus)	5:10.44
1964	Canada (Victor Emery)	4:14.46
1968	Italy (Eugenio Monti) (2 races)	2:17.39
1972	Switzerland (Jean Wicki)	4:43.07
1976	E. Germany (Meinhard Nehmer)	3:40.43
1980	E. Germany (Meinhard Nehmer)	3:59.92
1984	E. Germany (Wolfgang Hoppe)	3:20.22
1988	Switzerland (Ekkehard Fasser)	3:47.51
1992	Austria (Ingo Appelt)	3:53.90
1994	Germany (Wolfgang Hoppe)	3:27.28
1998	Germany-2 (Christoph Langen)	2:39.41

	2-Man Bob	Time
1932	United States (Hubert Stevens)	8:14.74
1936	United States (Ivan Brown)	5:29.29
1948	Switzerland (F. Endrich)	5:29.20
1952	Germany (Andreas Ostler)	5:24.54
1956	Italy (Dalla Costa)	5:30.14
1964	Great Britain (Anthony Nash)	4:21.90
1968	Italy (Eugenio Monti)	4:41.54
1972	W. Germany (Wolfgang Zimmerer)	4:57.07
1976	E. Germany (Meinhard Nehmer)	3:44.42
1980	Switzerland (Erich Schaerer)	4:09.36
1984	E. Germany (Wolfgang Hoppe)	3:25.56
1988	USSR (Janis Kipours)	3:54.19
1992	Switzerland (Gustav Weber)	4:03.26
1994	Switzerland (Gustav Weber)	3:30.81
1998	Canada (Pierre Lueders), Italy (Guenther Huber) (tie)	3:37.24

CURLING

	Men
1998	Switzerland, Canada, Norway

	Women
1998	Canada, Denmark, Sweden

FIGURE SKATING

(#) Event was held at Summer Olympics.

	Men's Singles
1908#	Ulrich Salchow, Sweden
1920#	Gillis Grafstrom, Sweden
1924	Gillis Grafstrom, Sweden
1928	Gillis Grafstrom, Sweden
1932	Karl Schaefer, Austria
1936	Karl Schaefer, Austria
1948	Richard Button, U.S.
1952	Richard Button, U.S.
1956	Hayes Alan Jenkins, U.S.
1960	David W. Jenkins, U.S.
1964	Manfred Schnelldorfer, Germany
1968	Wolfgang Schwartz, Austria
1972	Ondrej Nepela, Czechoslovakia
1976	John Curry, Great Britain
1980	Robin Cousins, Great Britain
1984	Scott Hamilton, U.S.
1988	Brian Boitano, U.S.
1992	Viktor Petrenko, Unified Team
1994	Aleksei Urmanov, Russia
1998	Ilya Kulik, Russia

	Women's Singles
1908#	Madge Syers, Great Britain
1920#	Magda Julin-Mauroy, Sweden
1924	Herma von Szabo-Planck, Austria
1928	Sonja Henie, Norway
1932	Sonja Henie, Norway
1936	Sonja Henie, Norway
1948	Barbara Ann Scott, Canada
1952	Jeanette Altwegg, Great Britain
1956	Tenley Albright, U.S.
1960	Carol Heiss, U.S.
1964	Sjoukje Dijkstra, Netherlands
1968	Peggy Fleming, U.S.
1972	Beatrix Schuba, Austria
1976	Dorothy Hamill, U.S.
1980	Anett Poetzsch, E. Germany
1984	Katarina Witt, E. Germany
1988	Katarina Witt, E. Germany
1992	Kristi Yamaguchi, U.S.
1994	Oksana Baiul, Ukraine
1998	Tara Lipinski, U.S.

	Pairs
1908#	Anna Hubler & Heinrich Burger, Germany
1920#	Ludovika & Walter Jakobsson, Finland
1924	Helene Engelman & Alfred Berger, Austria
1928	Andree Joly & Pierre Brunet, France
1932	Andree Joly & Pierre Brunet, France
1936	Maxi Herber & Ernst Baier, Germany
1948	Micheline Lannoy & Pierre Baugniet, Belgium
1952	Ria and Paul Falk, Germany
1956	Elisabeth Schwartz & Kurt Oppelt, Austria
1960	Barbara Wagner & Robert Paul, Canada
1964	Ludmila Beloussova & Oleg Protopopov, USSR
1968	Ludmila Beloussova & Oleg Protopopov, USSR
1972	Irina Rodnina & Alexei Ulanov, USSR
1976	Irina Rodnina & Aleksandr Zaitzev, USSR
1980	Irina Rodnina & Aleksandr Zaitzev, USSR
1984	Elena Valova & Oleg Vassiliev, USSR
1988	Ekaterina Gordeeva & Sergei Grinkov, USSR
1992	Natalia Mishkutienok & Artur Dimitriev, Unified Team
1994	Ekaterina Gordeeva & Sergei Grinkov, Russia
1998	Oksana Kazakova & Artur Dmitriev, Russia

	Ice Dancing
1976	Ludmila Pakhomova & Aleksandr Gorschkov, USSR
1980	Natalya Linichuk & Gennadi Karponosov, USSR
1984	Jayne Torvill & Christopher Dean, Great Britain
1988	Natalia Bestemianova & Andrei Bukin, USSR
1992	Marina Klimova & Sergei Ponomarenko, Unified Team
1994	Pasha Grishuk & Evgeny Platov, Russia
1998	Pasha Grishuk & Evgeny Platov, Russia

FREESTYLE SKIING

	Men's Moguls	Points
1992	Edgar Grospiron, France	25.81
1994	Jean-Luc Brassard, Canada	27.24
1998	Jonny Moseley, U.S.	26.93

	Men's Aerials	Points
1994	Andreas Schoenbaechler, Switzerland	234.67
1998	Eric Bergoust, U.S.	255.64

	Women's Moguls	Points
1992	Donna Weinbrecht, U.S.	23.69
1994	Stine Lise Hattestad, Norway	25.97
1998	Tae Satoya, Japan	25.06

	Women's Aerials	Points
1994	Lina Tcherjazova, Uzbekistan	166.84
1998	Nikki Stone, U.S.	193.00

ICE HOCKEY

(#) Event was held at Summer Olympics.

	Men
1920#	Canada, U.S., Czechoslovakia
1924	Canada, U.S., Great Britain
1928	Canada, Sweden, Switzerland
1932	Canada, U.S., Germany
1936	Great Britain, Canada, U.S.
1948	Canada, Czechoslovakia, Switzerland
1952	Canada, U.S., Sweden
1956	USSR, U.S., Canada
1960	U.S., Canada, USSR
1964	USSR, Sweden, Czechoslovakia
1968	USSR, Czechoslovakia, Canada
1972	USSR, U.S., Czechoslovakia
1976	USSR, Czechoslovakia, W. Germany
1980	U.S., USSR, Sweden
1984	USSR, Czechoslovakia, Sweden
1988	USSR, Finland, Sweden
1992	Unified Team, Canada, Czechoslovakia
1994	Sweden, Canada, Finland
1998	Czech Republic, Russia, Finland

	Women
1998	U.S., Canada, Finland

LUGE

	Men's Singles	Time
1964	Thomas Keohler, E. Germany	3:26.77
1968	Manfred Schmid, Austria	2:52.48
1972	Wolfgang Scheidel, E. Germany	3:27.58
1976	Detlef Guenther, E. Germany	3:27.688
1980	Bernhard Glass, E. Germany	2:54.796
1984	Paul Hildgartner, Italy	3:04.258
1988	Jens Mueller, E. Germany	3:05.548
1992	Georg Hackl, Germany	3:02.363
1994	Georg Hackl, Germany	3:21.571
1998	Georg Hackl, Germany	3:18.436

	Men's Doubles	Time
1964	Austria	1:41.62
1968	E. Germany	1:35.85
1972	Italy, E. Germany (tie)	1:28.35
1976	E. Germany	1:25.604
1980	E. Germany	1:19.331
1984	W. Germany	1:23.620
1988	E. Germany	1:31.940
1992	Germany	1:32.053
1994	Italy	1:36.720
1998	Germany	1:41.105

	Women's Singles	Time
1964	Ortun Enderlein, Germany	3:24.67
1968	Erica Lechner, Italy	2:28.66
1972	Anna M. Muller, E. Germany	2:59.18
1976	Margit Schumann, E. Germany	2:50.621
1980	Vera Zozulya, USSR	2:36.537
1984	Steffi Martin, E. Germany	2:46.570
1988	Steffi Walter, E. Germany	3:03.973
1992	Doris Neuner, Austria	3:06.696
1994	Gerda Weissensteiner, Italy	3:15.517
1998	Silke Kraushaar, Germany	3:23.779

NORDIC SKIING

Cross-Country Events

	Men's 10 Kilometers (6.2 miles)	Time
1992	Vegard Ulvang, Norway	27:36.0
1994	Bjoern Daehlie, Norway	24:20.1
1998	Bjoern Daehlie, Norway	27:24.5

	Men's 15 Kilometers (9.3 miles)	Time
1924	Thorleif Haug, Norway	1:14:31
1928	Johan Grottumsbraaten, Norway	1:37:01
1932	Sven Utterstrom, Sweden	1:23:07
1936	Erik-August Larsson, Sweden	1:14:38
1948	Martin Lundstrom, Sweden	1:13:50
1952	Hallgeir Brenden, Norway	1:01:34
1956	Hallgeir Brenden, Norway	0:49:39.0
1960	Haakon Brusveen, Norway	0:51:55.5
1964	Eero Maentyranta, Finland	0:50:54.1
1968	Harald Groenningen, Norway	0:47:54.2
1972	Sven-Ake Lundback, Sweden	0:45:28.24
1976	Nikolai Balukov, USSR	0:43:58.47
1980	Thomas Wassberg, Sweden	0:41:57.63
1984	Gunde Svan, Sweden	0:41:25.6
1988	Mikhail Deviatiarov, USSR	0:41:18.9
1992	Bjoern Daehlie, Norway	0:38:01.9
1994	Bjoern Daehlie, Norway	0:35:48.8
1998	Thomas Alsgaard, Norway	1:07:01.7

(Note: approx. 18-km course 1924-1952)

	Men's 30 Kilometers (18.6 miles)	Time
1956	Veikko Hakulinen, Finland	1:44:06.0
1956	Veikko Hakulinen, Finland	1:44:06.0
1960	Sixten Jernberg, Sweden	1:51:03.9
1964	Eero Maentyranta, Finland	1:30:50.7
1968	Franco Nones, Italy	1:35:39.2
1972	Vyacheslav Vedenine, USSR	1:36:31.15
1976	Sergei Saveliev, USSR	1:30:29.38
1980	Nikolai Zimyatov, USSR	1:27:02.80
1984	Nikolai Zimyatov, USSR	1:28:56.3
1988	Aleksei Prokourorov, USSR	1:24:26.3
1992	Vegard Ulvang, Norway	1:22:27.8
1994	Thomas Alsgaard, Norway	1:12:26.4
1998	Mika Myllylae, Finland	1:33:55.8

	Men's 50 Kilometers (31.2 miles)	Time
1924	Thorleif Haug, Norway	3:44:32.0
1928	Per Erik Hedlund, Sweden	4:52:03.0
1932	Veli Saarinen, Finland	4:28:00.0
1936	Elis Wiklund, Sweden	3:30:11.0
1948	Nils Karlsson, Sweden	3:47:48.0
1952	Veikko Hakulinen, Finland	3:33:33.0
1956	Sixten Jernberg, Sweden	2:50:27.0
1960	Kalevi Hamalainen, Finland	2:59:06.3
1964	Sixten Jernberg, Sweden	2:43:52.6
1968	Ole Ellefsaeter, Norway	2:28:45.8
1972	Paal Tyldum, Norway	2:43:14.75
1976	Ivar Formo, Norway	2:37:30.05
1980	Nikolai Zimyatov, USSR	2:27:24.60
1984	Thomas Wassberg, Sweden	2:15:55.8
1988	Gunde Svan, Sweden	2:04:30.9
1992	Bjoern Daehlie, Norway	2:03:41.5
1994	Vladimir Smirnov, Kazakhstan	2:07:20.3
1998	Bjoern Daehlie, Norway	2:05:08.2

	Men's 40-Kilometer Relay	Time
1936	Finland, Norway, Sweden	2:41:33.0
1948	Sweden, Finland, Norway	2:32:08.0
1952	Finland, Norway, Sweden	2:20:16.0
1956	USSR, Finland, Sweden	2:15:30.0
1960	Finland, Norway, USSR	2:18:45.6
1964	Sweden, Finland, USSR	2:18:34.6
1968	Norway, Sweden, Finland	2:08:33.5
1972	USSR, Norway, Switzerland	2:04:47.94
1976	Finland, Norway, USSR	2:07:59.72
1980	USSR, Norway, Finland	1:57:03.46
1984	Sweden, USSR, Finland	1:55:06.30
1988	Sweden, USSR, Czechoslovakia	1:43:58.60
1992	Norway, Italy, Finland	1:39:26.00
1994	Italy, Norway, Finland	1:41:15.00
1998	Norway, Italy, Finland	1:40:55.70

	Women's 5 Kilometers (approx. 3.1 miles)	Time
1964	Claudia Boyarskikh, USSR	17:50.5
1968	Toini Gustafsson, Sweden	16:45.2
1972	Galina Koulacova, USSR	17:00.50
1976	Helena Takalo, Finland	15:48.69
1980	Raisa Smetanina, USSR	15:06.92
1984	Marja-Liisa Haemaelainen, Finland	17:04.0
1988	Marjo Matikainen, Finland	15:04.0
1992	Marjut Lukkarinen, Finland	14:13.8
1994	Ljubov Egorova, Russia	14:08.8
1998	Larissa Lazutina, Russia	17:37.9

	Women's 10 Kilometers (6.2 miles)	Time
1952	Lydia Wideman, Finland	41:40.0
1956	Lyubov Kosyreva, USSR	38:11.0
1960	Maria Gusakova, USSR	39:46.6
1964	Claudia Boyarskikh, USSR	40:24.3
1968	Toini Gustafsson, Sweden	36:46.5
1972	Galina Koulacova, USSR	34:17.82
1976	Raisa Smetanina, USSR	30:13.41
1980	Barbara Petzold, E. Germany	30:31.54
1984	Marja-Liisa Haemaelainen, Finland	31:44.2
1988	Vida Ventsene, USSR	30:08.3
1992	Lyubov Egorova, Unified Team	25:53.7
1994	Lyubov Egorova, Russia	27:30.1
1998	Larissa Lazutina, Russia	46.06.9

	Women's 15 Kilometers (9.3 miles)	Time
1992	Lyubov Egorova, Unified Team	42:20.8
1994	Manuela Di Centa, Italy	39:44.5
1998	Olga Danilova, Russia	46:55.4

	Women's 30 Kilometers (18.6 miles)	Time
1992	Stefania Belmondo, Italy	1:22:30.1
1994	Manuela Di Centa, Italy	1:25:41.6
1998	Julija Tchepalova, Russia	1:22:01.5

	Women's 20-Kilometer Relay	Time
1956	Finland, USSR, Sweden (15 km)	1:09:01.0
1960	Sweden, USSR, Finland (15 km)	1:04:21.4
1964	USSR, Sweden, Finland (15 km)	0:59:20.2
1968	Norway, Sweden, USSR (15 km)	0:57:30.0
1972	USSR, Finland, Norway (15 km)	0:48:46.15
1976	USSR, Finland, E. Germany	1:07:49.75
1980	E. Germany, USSR, Norway	1:02:11.1
1984	Norway, Czechoslovakia, Finland	1:06:49.7
1988	USSR, Norway, Finland	0:59:51.1
1992	United Team, Norway, Italy	0:59:34.8
1994	Russia, Norway, Italy	0:57:12.5
1998	Russia, Norway, Italy	0:55:13.5

Combined Cross-Country & Jumping (Men)

Nordic Combined*

1924	Thorleif Haug, Norway
1928	Johan Grottumsbraaten, Norway
1932	Johan Grottumsbraaten, Norway
1936	Oddbjorn Hagen, Norway
1948	Heikki Hasu, Finland
1952	Simon Slattvik, Norway
1956	Sverre Stenersen, Norway
1960	Georg Thoma, W. Germany
1964	Tormod Knutsen, Norway
1968	Franz Keller, W. Germany
1972	Ulrich Wehling, E. Germany
1976	Ulrich Wehling, E. Germany
1980	Ulrich Wehling, E. Germany

Nordic Combined*

1984	Tom Sandberg, Norway	
1988	Hippolyt Kempf, Switzerland	
1992	Fabrice Guy, France	
1994	Fred Barre Lundberg, Norway	
1998	Bjarte Engen Vik, Norway	

Team Nordic Combined*

1988	W. Germany, Switzerland, Austria	
1992	Japan, Norway, Austria	
1994	Japan, Norway, Switzerland	
1998	Norway, Finland, France	

*Medals based on combination of points for jumping events and time for cross-country events.

SKI JUMPING (Men)

	Normal Hill	Points
1964	Veikko Kankkonen, Finland	229.9
1968	Jiri Raska, Czechoslovakia	216.5
1972	Yukio Kasaya, Japan	244.2
1976	Hans-Georg Aschenbach, E. Germany	252.0
1980	Toni Innauer, Austria	266.3
1984	Jens Weissflog, E. Germany	215.2
1988	Matti Nykaenen, Finland	230.5
1992	Ernst Vettori, Austria	222.8
1994	Espen Bredesen, Norway	282.0
1998	Jani Soininen, Finland	234.5

	Large Hill	Points
1924	Jacob Tullin Thams, Norway	18.960
1928	Alfred Andersen, Norway	19.208
1932	Birger Ruud, Norway	228.1
1936	Birger Ruud, Norway	232.0
1948	Petter Hugsted, Norway	228.1
1952	Arnfinn Bergmann, Norway	226.0
1956	Antti Hyvarinen, Finland	227.0
1960	Helmut Recknagel, E. Germany	227.2
1964	Toralf Engan, Norway	230.7
1968	Vladimir Beloussov, USSR	231.3
1972	Wojciech Fortuna, Poland	219.9
1976	Karl Schnabl, Austria	234.8
1980	Jouko Tormanen, Finland	271.0
1984	Matti Nykaenen, Finland	231.2
1988	Matti Nykaenen, Finland	224.0
1992	Toni Nieminen, Finland	239.5
1994	Jens Weissflog, Germany	274.5
1998	Kazuyoshi Funaki, Japan	272.3

	Team Large Hill	Points
1988	Finland, Yugoslavia, Norway	634.4
1992	Finland, Austria, Czechoslovakia	644.4
1994	Germany, Japan, Austria	970.1
1998	Japan, Germany, Austria	933.0

SNOWBOARDING

	Men's Giant Slalom	Time
1998	Ross Rebagliati, Canada	2:03.96

	Men's Halfpipe	Points
1998	Gian Simmen, Switzerland	85.2

	Women's Giant Slalom	Time
1998	Karine Ruby, France	2:17.34

	Women's Halfpipe	Points
1998	Nicola Thost, Germany	74.6

SPEED SKATING

*Better time of 2 runs. Medals based on combined times.

	Men's 500 Meters	Time*
1924	Charles Jewtraw, U.S.	0:44.0
1928	Thunberg, Finland & Evensen, Norway (tie)	0:43.4
1932	John A. Shea, U.S.	0:43.4
1936	Ivar Ballangrud, Norway	0:43.4
1948	Finn Helgesen, Norway	0:43.1
1952	Kenneth Henry, U.S.	0:43.2
1956	Evgeniy Grishin, USSR	0:40.2
1960	Evgeniy Grishin, USSR	0:40.2
1964	Terry McDermott, U.S.	0:40.1
1968	Erhard Keller, W. Germany	0:40.3
1972	Erhard Keller, W. Germany	0:39.44
1976	Evgeny Kulikov, USSR	0:39.17
1980	Eric Heiden, U.S.	0:38.03
1984	Sergei Fokichev, USSR	0:38.19
1988	Uwe-Jens Mey, E. Germany	0:36.45
1992	Uwe-Jens Mey, Germany	0:37.14
1994	Aleksandr Golubev, Russia	0:36.33
1998	Hiroyasu Shimizu, Japan	0:35.59

	Men's 1,000 Meters	Time
1976	Peter Mueller, U.S	1:19.32
1980	Eric Heiden, U.S.	1:15.18
1984	Gaetan Boucher, Canada	1:15.80
1988	Nikolai Guiliaev, USSR	1:13.03
1992	Olaf Zinke, Germany	1:14.85
1994	Dan Jansen, U.S.	1:12.43
1998	Ids Postma, Netherlands	1:10.64

	Men's 1,500 Meters	Time
1924	Clas Thunberg, Finland	2:20.8
1928	Clas Thunberg, Finland	2:21.1
1932	John A. Shea, U.S.	2:57.5
1936	Charles Mathiesen, Norway	2:19.2
1948	Sverre Farstad, Norway	2:17.6
1952	Hjalmar Andersen, Norway	2:20.4
1956	Grishin, & Mikhailov, both USSR (tie)	2:08.6
1960	Aas, Norway & Grishin, USSR (tie)	2:10.4
1964	Ants Anston, USSR	2:10.3
1968	Cornetis Verkerk, Netherlands	2:03.4
1972	Ard Schenk, Netherlands	2:02.96
1976	Jan Egil Storholt, Norway	1:59.38
1980	Eric Heiden, U.S.	1:55.44
1984	Gaetan Boucher, Canada	1:58.36
1988	Andre Hoffmann, E. Germany	1:52.06
1992	Johann Koss, Norway	1:54.81
1994	Johann Koss, Norway	1:51.29
1998	Aadne Sondral, Norway	1:47.87

	Men's 5,000 Meters	Time
1924	Clas Thunberg, Finland	8:39.0
1928	Ivar Ballangrud, Norway	8:50.5
1932	Irving Jaffee, U.S.	9:40.8
1936	Ivar Ballangrud, Norway	8:19.6
1948	Reidar Liaklev, Norway	8:29.4
1952	Hjalmar Andersen, Norway	8:10.6
1956	Boris Shilkov, USSR	7:48.7
1960	Viktor Kosichkin, USSR	7:51.3
1964	Knut Johannesen, Norway	7:38.4
1968	F. Anton Maier, Norway	7:22.4
1972	Ard Schenk, Netherlands	7:23.61
1976	Sten Stensen, Norway	7:24.48
1980	Eric Heiden, U.S.	7:02.29
1984	Sven Tomas Gustafson, Sweden	7:12.28
1988	Tomas Gustafson, Sweden	6:44.63
1992	Geir Karlstad, Norway	6:59.97
1994	Johann Koss, Norway	6:34.96
1998	Gianni Romme, Netherlands	6:22.20

	Men's 10,000 Meters	Time
1924	Julius Skutnabb, Finland	18:04.8
1928	Event not held because of thawing of ice	
1932	Irving Jaffee, U.S.	19:13.6
1936	Ivar Ballangrud, Norway	17:24.3
1948	Ake Seyffarth, Sweden	17:26.3
1952	Hjalmar Andersen, Norway	16:45.8
1956	Sigvard Ericsson, Sweden	16:35.9
1960	Knut Johannesen, Norway	15:46.6
1964	Jonny Nilsson, Sweden	15:50.1
1968	Jonny Hoeglin, Sweden	15:23.6
1972	Ard Schenk, Netherlands	15:01.35
1976	Piet Kleine, Netherlands	14:50.59
1980	Eric Heiden, U.S.	14:28.13
1984	Igor Malkov, USSR	14:39.90
1988	Tomas Gustafson, Sweden	13:48.20
1992	Bart Veldkamp, Netherlands	14:12.12
1994	Johann Koss, Norway	13:30.55
1998	Gianni Romme, Netherlands	13:15.33

	Women's 500 Meters	Time*
1960	Helga Haase, Germany	0:45.9
1964	Lydia Skoblikova, USSR	0:45.0
1968	Ludmila Titova, USSR	0:46.1
1972	Anne Henning, U.S.	0:43.33
1976	Sheila Young, U.S.	0:42.76
1980	Karin Enke, E. Germany	0:41.78
1984	Christa Rothenburger, E. Germany	0:41.02
1988	Bonnie Blair, U.S.	0:39.10
1992	Bonnie Blair, U.S.	0:40.33
1994	Bonnie Blair, U.S.	0:39.25
1998	Catriona LeMay-Doan, Canada	0:38.21

	Women's 1,000 Meters	Time
1960	Klara Guseva, USSR	1:34.1
1964	Lydia Skoblikova, USSR	1:33.2
1968	Carolina Geijssen, Netherlands	1:32.6
1972	Monika Pflug, W. Germany	1:31.40
1976	Tatiana Averina, USSR	1:28.43
1980	Natalya Petruseva, USSR	1:24.10
1984	Karin Enke, E. Germany	1:21.61
1988	Christa Rothenburger, E. Germany	1:17.65
1992	Bonnie Blair, U.S.	1:21.90
1994	Bonnie Blair, U.S.	1:18.74
1998	Marianne Timmer, Netherlands	1:16.51

	Women's 1,500 Meters	Time
1960	Lydia Skoblikova, USSR	2:52.2
1964	Lydia Skoblikova, USSR	2:22.6
1968	Kaija Mustonen, Finland	2:22.4
1972	Dianne Holum, U.S.	2:20.85
1976	Galina Stepanskaya, USSR	2:16.58
1980	Anne Borckink, Netherlands	2:10.95
1984	Karin Enke, E. Germany	2:03.42
1988	Yvonne van Gennip, Netherlands	2:00.68
1992	Jacqueline Boerner, Germany	2:05.87
1994	Emese Hunyady, Austria	2:02.19
1998	Marianne Timmer, Netherlands	1:57.58

	Women's 3,000 Meters	Time
1960	Lydia Skoblikova, USSR	5:14.3
1964	Lydia Skoblikova, USSR	5:14.9
1968	Johanna Schut, Netherlands	4:56.2
1972	Christina Baas-Kaiser, Netherlands	4:52.14
1976	Tatiana Averina, USSR	4:45.19
1980	Bjoerg Eva Jensen, Norway	4:32.13
1984	Andrea Schoene, E. Germany	4:24.79
1988	Yvonne van Gennip, Netherlands	4:11.94
1992	Gunda Niemann, Germany	4:19.90
1994	Svetlana Bazhanova, Russia	4:17.43
1998	Gunda Niemann-Stirnemann, Germany	4:07.29

	Women's 5,000 Meters	Time
1988	Yvonne van Gennip, Netherlands	7:14.13
1992	Gunda Niemann, Germany	7:31.57
1994	Claudia Pechstein, Germany	7:14.37
1998	Claudia Pechstein, Germany	6:59.61

Short-Track Speed Skating

	Men's 500 Meters	Time
1998	Takafumi Nishitani, Japan	42.862

	Men's 1,000 Meters	Time
1992	Kim Ki-Hoon, S. Korea	1:30.76
1994	Kim Ki-Hoon, S. Korea	1:34.57
1998	Dong-Sung Kim, S. Korea	1:32.375

	Men's 5,000-Meter Relay	Time
1992	S. Korea	7:14.02
1994	Italy	7:11.74
1998	Canada	7:06.075

	Women's 500 Meters	Time
1992	Cathy Turner, U.S.	47.04
1994	Cathy Turner, U.S.	45.98
1998	Annie Perreault, Canada	46.568

	Women's 1,000 Meters	Time
1998	Chun Lee-Kyung, S. Korea	1:42.776

	Women's 3,000 Meter Relay	Time
1992	Canada	4:36.62
1994	S. Korea	4:26.64
1998	S. Korea	4:16.26

Summer Olympic Games

Summer Olympic Games in 2000

Sydney, Australia, Sept. 15-Oct. 1, 2000

About 11,000 athletes from 199 countries competed in 300 events in 28 sports at the 2000 summer games. New sports introduced in Sydney included synchronized diving, trampoline, taekwondo, and triathlon. For the first time, women also competed in water polo, weight lifting, the pole vault, the hammer throw, and the modern pentathlon.

Australia's Cathy Freeman, who lit the torch in the opening ceremonies, provided one of the more memorable moments of the games on Sept. 25, when she won the women's 400m to become the 1st Aborigine to earn an individual gold medal. U.S. sprinter Marion Jones fell short of her announced goal of 5 gold medals, but not by much. She won the 100m and 200m dashes and led the U.S. 4x400 relay team to another gold. Her total of 5, which included bronzes in the long jump and 4x100m relay, is the most ever in track and field by a woman at a single Olympics. Competing in his first Olympics, American wrestler Rulon Gardner provided the biggest upset of the Games when he defeated 3-time Olympic champion Aleksandr Karelin in the Greco-Roman super heavyweight final. Karelin, a 9-time world champion, had never been beaten in international competition. The U.S. baseball team, made up largely of unknown minor league players, won its first-ever gold medal and ended Cuba's 20-year domination of international baseball. Pitcher Ben Sheets allowed only 3 hits in the 4-0 win. In swimming, hometown favorite Ian Thorpe helped set 3 relay world records and improved his own record in the 400m freestyle.

Final Medal Standings

Country	G	S	B	T
United States	40	24	33	97
Russia	32	28	28	88
China	28	16	15	59
Australia	16	25	17	58
Germany	14	17	26	57
France	13	14	11	38
Italy	13	8	13	34
Cuba	11	11	7	29
Britain	11	10	7	28
South Korea	8	10	10	28
Romania	11	6	9	26
Netherlands	12	9	4	25
Ukraine	3	10	10	23
Japan	5	8	5	18
Hungary	8	6	3	17
Belarus	3	3	11	17
Poland	6	5	3	14
Canada	3	3	8	14
Bulgaria	5	6	2	13
Greece	4	6	3	13
Sweden	4	5	3	12
Brazil	0	6	6	12
Spain	3	3	5	11
Norway	4	3	3	10
Switzerland	1	6	2	9
Ethiopia	4	1	3	8
Czech Rep.	2	3	3	8

Country	G	S	B	T
Kazakhstan	3	4	0	7
Kenya	2	3	2	7
Jamaica	0	4	3	7
Denmark	2	3	1	6
Indonesia	1	3	2	6
Mexico	1	2	3	6
Georgia	0	0	6	6
Lithuania	2	0	3	5
Slovakia	1	3	1	5
Algeria	1	1	3	5
Belgium	0	2	3	5
South Africa	0	2	3	5
Morocco	0	1	4	5
Taiwan	0	1	4	5
Turkey	3	0	2	5
Iran	3	0	1	4
Finland	2	1	1	4
Uzbekistan	1	1	2	4
New Zealand	1	0	3	4
Argentina	0	2	2	4
North Korea	0	1	3	4
Austria	2	1	0	3
Azerbaijan	2	0	1	3
Latvia	1	1	1	3
Yugoslavia	1	1	1	3
Estonia	1	0	2	3
Thailand	1	0	2	3

Country	G	S	B	T
Nigeria	0	3	0	3
Slovenia	2	0	0	2
Bahamas	1	1	0	2
Croatia	1	0	1	2
Moldova	0	1	1	2
Saudi Arabia	0	1	1	2
Trinidad & Tobago	0	1	1	2
Costa Rica	0	0	2	2
Portugal	0	0	2	2
Cameroon	1	0	0	1
Colombia	1	0	0	1
Mozambique	1	0	0	1
Ireland	0	1	0	1
Uruguay	0	1	0	1
Vietnam	0	1	0	1
Armenia	0	0	1	1
Barbados	0	0	1	1
Chile	0	0	1	1
India	0	0	1	1
Iceland	0	0	1	1
Israel	0	0	1	1
Kyrgyzstan	0	0	1	1
Kuwait	0	0	1	1
Sri Lanka	0	0	1	1
Macedonia	0	0	1	1
Qatar	0	0	1	1

Other Summer Olympics Gold Medal Winners in 2000

Archery

Men's Individual—Simon Fairweather, Australia
Men's Team—G-Korea; S-Italy; B-U.S.
Women's Individual—Mi-Jin Yun, Korea
Women's Team—G-Korea; S-Ukraine; B-Germany

Badminton

Men's Singles—Xinpeng Ji, China
Men's Doubles—Tony Gunawan & Candra Wijaya, Indonesia
Women's Singles—Zhichao Gong, China
Women's Doubles—Fei Ge & Jun Gu, China
Mixed Doubles—Ling Gao & Jun Zhang, China

Baseball

G-U.S.; S-Cuba; B-Korea

Basketball

Men—G-U.S.; S-France; B-Lithuania
Women—G-U.S.; S-Australia; B-Brazil

Canoe/Kayak

Men

Kayak Slalom—Thomas Schmidt, Germany
Kayak 500M Singles—Knut Holmann, Norway
Kayak 500M Doubles—G-Hungary; S-Australia; B-Germany
Kayak 1,000M Singles—Knut Holmann, Norway
Kayak 1,000M Doubles—G-Italy; S-Sweden; B-Hungary
Kayak 1,000M Fours—G-Hungary; S-Germany; B-Poland
Canoe Slalom Singles—Tony Estanguet, France
Canoe Slalom Doubles—G-Slovakia; S-Poland; B-Czech Rep.
Canoe 500M Singles—Gyorgy Kolonics, Hungary
Canoe 500M Doubles—G-Hungary; S-Poland; B-Romania
Canoe 1,000M Singles—Andreas Dittmer, Germany
Canoe 1,000M Doubles—G-Romania; S-Cuba; B-Germany

Women

Kayak Slalom—Stepanka Hilgertova, Czech Rep.
Kayak 500M Singles—Josefa Iden Guerrini, Italy
Kayak 500M Doubles—G-Germany; S-Hungary; B-Poland
Kayak 500M Fours—G-Germany; S-Hungary; B-Romania

Cycling

Men

Mountain Bike—Miguel Martinez, France
Individual Road Race—Jan Ullrich, Germany
Individual Time Trial—Viacheslav Ekimov, Russia
Individual Pursuit—Robert Bartko, Germany
Team Pursuit—G-Germany; S-Ukraine; B-Britain
Keirin—Florian Rousseau, France
Madison—G-Australia; S-Belgium; B-Italy
Olympic Sprint—G-France; S-Britain; B-Australia
Sprint—Marty Nothstein, U.S.
Individual Points Race—Juan Llaneras, Spain
1KM Time Trial—Jason Queally, Britain

Women

Mountain Bike—Paola Pezzo, Italy
Individual Road Race—Leontien Zijlaard, Netherlands
Individual Time Trial—Leonntien Zijlaard, Netherlands
Individual Pursuit—Leontien Zijlaard, Netherlands
Sprint—Felicia Ballanger, France
Individual Points Race—Antonella Bellutti, Italy
500M Time Trial—Felicia Ballanger, France

Diving

Men

Platform—Liang Tian, China
Springboard—Ni Xiong, China

Women

Platform—Laura Wilkinson, U.S.
Springboard—Mingxia Fu, China

Equestrian

Individual Dressage—Anky van Grunsven, Netherlands
Team Dressage—G-Germany; S-Netherlands; B-U.S.
Individual Jumping—Jeroen Dubbeldam, Netherlands
Team Jumping—G-Germany; S-Switzerland; B-Brazil
Individual Three-Day Event—David O'Connor, U.S.
Team Three-Day Event—G-Australia; S-Britain; B-U.S.

Fencing

Men

Individual Foil—Young Ho Kim, Korea
Individual Épée—Pavel Kolobkov, Russia
Team Épée—G-Italy; S-France; B-Cuba
Team Foil—G-France; S-China; B-Italy
Individual Saber—Mihai Claudiu Covaliu, Romania
Team Saber—G-Russia; S-France; B-Germany

Women

Individual Épée—Timea Nagy, Hungary
Team Épée—G-Russia; S-Switzerland; B-China
Individual Foil—Valentina Vezzali, Italy
Team Foil—G-Italy; S-Poland; B-Germany

Field Hockey

Men—G-Netherlands; S-Korea; B-Australia
Women—G-Australia; S-Argentina; B-Netherlands

Gymnastics

Men

Team—G-China; S-Ukraine; B-Russia
Individual All-Around—Alexei Nemov, Russia
Floor Exercise—Igors Vihrovs, Latvia
Vault—Gervasio Deferr, Spain
Parallel Bars—Xiaopeng Li, China
Horizontal Bar—Alexei Nemov, Russia
Pommel Horse—Marius Urzica, Romania
Rings—Szilveszter Csollany, Hungary
Trampoline—Alexandre Moskalenko, Russia

Women

Team—G-Romania; S-Russia; B-China
Individual All-Around—Simona Amanar, Romania
Floor Exercise—Elena Zamolodtchikova, Russia
Vault—Elena Zamolodtchikova, Russia
Uneven Bars—Svetlana Khorkina, Russia
Balance Beam—Xuan Liu, China

Rhythmic Gymnastics

Team—G-Russia; S-Belarus; B-Greece
Individual All-Around—Yulia Barsukova, Russia
Trampoline—Irina Karavaeva, Russia

Judo

Men

Extra Lightweight 60 kg (132 lbs)—Tadahiro Nomura, Japan
Half lightweight 66 kg (145 lbs)—Huseyelin Ozkan, Turkey
Lightweight 73 kg (161 lbs)—Giuseppe Maddaloni, Italy
Half middleweight 81 kg (178 lbs)—Makoto Takimoto, Japan
Middleweight 90 kg (198 lbs)—Mark Huizinga, Netherlands
Half heavyweight 100 kg (220 lbs)—Kosei Inoue, Japan
Heavyweight 100+ kg (**220+ lbs)**—David Douillet, France

Women

Extra lightweight 48 kg (106 lbs)—Ryoko Tamura, Japan
Half lightweight 52 kg (114 lbs)—Legna Verdecia, Cuba
Lightweight 57 kg (125 lbs)—Isabel Fernandez, Spain
Half middleweight 63 kg (139 lbs)—Severine Vandenhende, France
Middleweight 70 kg (154 lbs)—Silbelis Veranes, Cuba
Half heavyweight 78 kg (172 lbs)—Lin Tang, China
Heavyweight 78+ kg (172+ lbs)—Hua Yuan, China

Modern Pentathlon

Men—Dmitry Svatkovsky, Russia
Women—Stephanie Cook, Britain

Rowing

Men

Single Sculls—Rob Waddell, New Zealand
Double Sculls—G-Slovenia; S-Norway; B-Italy
Lightweight Double Sculls—G-Poland; S-Italy; B-France
Quadruple Sculls—G-Italy; S-Netherland; B-Germany
Coxless Pairs—G-France; S-U.S.; B-Australia
Coxless Fours—G-Britain; S-Italy; B-Australia
Lightweight Coxless Fours—G-France; S-Australia; B-Denmark
Coxed Eights—G-Britain; S-Australia; B-Croatia

Women

Single Sculls—Ekaterina Karsten, Belarus
Double Sculls—G-Germany; S-Netherlands; B-Lithuania
Lightweight Double Sculls—G-Romania; S-Germany; B-U.S.
Quadruple Sculls—G-Germany; S-Britain; B-Russia
Coxless Pairs—G-Romania; S-Australia; B-U.S.
Coxed Eights—G-Romania; S-Netherlands; B-Canada

Sailing

Men

Mistral—Christoph Sieber, Austria
Finn—Lan Percy, Britain
470—G-Australia; S-U.S.; B-Argentina

Women

Mistral—Alessandra Sensini, Italy
Europe—Shirley Robertson, Britain
470—G-Australia; S-U.S.; B-Ukraine

Open

Laser—Ben Ainslie, Britain
Tornado—G-Austria; S-Australia; B-Germany
Star—G-U.S.; S-Britain; B-Brazil
49er—G-Finland; S-Britain; B-U.S.
Soling—G-Denmark; S-Germany; B-Norway

Shooting

Men

Air Pistol—Frank Dumoulin, France
Rapid Fire Pistol—Serguei Alifirenko, Russia
Free Pistol—Tanyu Kiriakov, Bulgaria
Air Rifle—Yalin Cai, China
Three-Position Rifle—Rajmond Debevec, Slovenia
Rifle Prone—Jonas Edman, Sweden
Trap—Michael Diamond, Australia
Double Trap—Richard Faulds, Britain
Skeet—Mykola Milchev, Ukraine
Running Game Target—Ling Yang, China

Women

Air Pistol—Luna Tao, China
Sport Pistol—Maria Grozdeva, Bulgaria
Air Rifle—Nancy Johnson, U.S.
Three-Position Rifle—Renata Mauer-Rozanska, Poland
Trap—Daina Gudzineviciute, Lithuania
Double Trap—Pia Hansen, Sweden
Skeet—Zemfira Meftakhetdinova, Azerbaijan

Soccer
Men—G-Cameroon; S-Spain; B-Chile
Women—G-Norway; S-U.S.; B-Germany

Softball
G-U.S.; S-Japan; B-Australia

Synchronized Swimming
Duet—G-Russia; S-Japan; B-France
Team—G-Russia; S-Japan; B-Canada

Synchronized Diving
Men
Platform—G-Russia; S-China; B-Germany
Springboard—G-China; S-Russia; B-Australia
Women
Platform—G-China; S-Canada; B-Australia
Springboard—G-Russia; S-China; B-Ukraine

Table Tennis
Men's Singles—Linghui Kong, China
Men's Doubles—G-China; S-China; B-France
Women's Singles—Nan Wang, China
Women's Doubles—G-China; S-China; B-Korea

Taekwondo
Men
Up to 58 kg (127¾ lbs)—Michail Mouroutsis, Greece
Up to 68 kg (150 lbs)—Steven Lopez, U.S.
Up to 80 kg (176¼ lbs)—Angel Matos Fuentes, Cuba
Over 80 kg (176¼ lbs)—Kyong-Hun Kim, Korea
Women
Up to 49 kg (108 lbs)—Lauren Burns, Australia
Up to 57 kg (125½ lbs)—Jae-Eun Jung, Korea
Up to 67 kg (147¾ lbs)—Sun-Hee Lee, Korea
Over 67 kg (147¾ lbs)—Zhong Chen, China

Team Handball
Men—G-Russia; S-Sweden; B-Spain
Women—G-Denmark; S-Hungary; B-Norway

Tennis
Men's Singles—Yevgeny Kafelnikov, Russia
Men's Doubles—Sebastian Lareau & Daniel Nestor, Canada
Women's Singles—Venus Williams, U.S.
Women's Doubles—Venus Williams & Serena Williams, U.S.

Triathlon
Men—Simon Whitfield, Canada
Women—Brigitte McMahon, Switzerland

Volleyball
Men's Beach—Dan Blanton & Eric Fonoimoana, U.S.
Men's Indoor—G-Yugoslavia; S-Russia; B-Italy
Women's Beach—Natalie Cook & Kerri Ann Pottharst, Australia
Women's Indoor—G-Cuba; S-Russia; B-Italy

Water Polo
Men—G-Hungary; S-Russia; B-Yugoslavia
Women—G-Australia; S-U.S.; B-Russia

Weight Lifting
Men
Up to 56 kg (123½ lbs)—Halil Mutlu, Turkey
Up to 62 kg (136¾ lbs)—Nikolay Pechaliv, Croatia
Up to 69 kg (152 lbs)—Galabin Boevski, Bulgaria
Up to 77 kg (169¾ lbs)—Xugang Zhan, China
Up to 85 kg (187¼ lbs)—Pyrros Dimas, Greece
Up to 94 kg (207¼ lbs)—Akakios Kakiasvilis, Greece
Up to 105 kg (231½ lbs)—Hossein Tavakoli, Iran
Over 105 kg (231½ lbs)—Hossein Rezazadeh, Iran
Women
Up to 48 kg (105¾ lbs)—Tara Nott, U.S.
Up to 53 kg (116¾ lbs)—Xia Yang, China
Up to 58 kg (128 lbs)—Soraya Mendivil, Mexico
Up to 63 kg (139 lbs)—Xiaomin Chen, China
Up to 69 kg (152¼ lbs)—Weining Lin, China
Up to 75 kg (165¼ lbs)—Maria Isabel Urrutia, Columbia
Over 75 kg (165¼ lbs)—Meiyuan Ding, China

Wrestling
Freestyle
54 kg (119 lbs)—Namig Abdullayev, Azerbaijan
58 kg (127¾ lbs)—Alireza Dabir, Iran
63 kg (138¾ lbs)—Mourad Oumakhanov, Russia
69 kg (152 lbs)—Daniel Igali, Canada
76 kg (167½ lbs)—Brandon Slay, U.S.*
85 kg (187¼ lbs)—Adam Saitiev, Russia
97 kg (213¾ lbs)—Saghid Mourtasaliyev, Russia
130 kg (286 lbs)—David Moussoulbes, Russia
*Slay was awarded the gold on Nov. 15, 2001, following Alexander Leipold's disqualification for steroid use.
Greco-Roman
54 kg (119 lbs)—Kwon Ho Sim, Korea
58 kg (127¾ lbs)—Armen Nazarian, Bulgaria
63 kg (138¾ lbs)—Varteres Samourgachev, Russia
69 kg (152 lbs)—Filiberto Azcuy, Cuba
76 kg (167½ lbs)—Mourat Kardanov, Russia
85 kg (187¼ lbs)—Hamza Yerlikaya, Turkey
97 kg (213¾ lbs)—Mikael Ljungberg, Sweden
130 kg (286 lbs)—Rulon Gardner, U.S.

Summer Olympic Games Champions, 1896-2000
(*indicates Olympic record; w indicates wind-aided)

The 1980 games were boycotted by 62 nations, including the U.S. The 1984 games were boycotted by the USSR and by most Eastern bloc nations. East and West Germany competed separately, 1968-88. The 1992 Unified Team consisted of 12 former Soviet republics. The 1992 Independent Olympic Participants (I.O.P.) were athletes from Serbia, Montenegro, and Macedonia.

TRACK AND FIELD—Men
100-Meter Run

1896	Thomas Burke, United States	12.0s
1900	Francis W. Jarvis, United States	11.0s
1904	Archie Hahn, United States	11.0s
1908	Reginald Walker, South Africa	10.8s
1912	Ralph Craig, United States	10.8s
1920	Charles Paddock, United States	10.8s
1924	Harold Abrahams, Great Britain	10.6s
1928	Percy Williams, Canada	10.8s
1932	Eddie Tolan, United States	10.3s
1936	Jesse Owens, United States	10.3s
1948	Harrison Dillard, United States	10.3s
1952	Lindy Remigino, United States	10.4s
1956	Bobby Morrow, United States	10.5s
1960	Armin Hary, Germany	10.2s
1964	Bob Hayes, United States	10.0s
1968	Jim Hines, United States	9.95s
1972	Valery Borzov, USSR	10.14s
1976	Hasely Crawford, Trinidad	10.06s
1980	Allan Wells, Great Britain	10.25s
1984	Carl Lewis, United States	9.99s
1988	Carl Lewis, United States	9.92s
1992	Linford Christie, Great Britain	9.96s
1996	Donovan Bailey, Canada	9.84s*
2000	Maurice Greene, United States	9.87s

200-Meter Run

1900	Walter Tewksbury, United States	22.2s
1904	Archie Hahn, United States	21.6s
1908	Robert Kerr, Canada	22.6s
1912	Ralph Craig, United States	21.7s
1920	Allan Woodring, United States	22.0s
1924	Jackson Scholz, United States	21.6s
1928	Percy Williams, Canada	21.8s
1932	Eddie Tolan, United States	21.2s
1936	Jesse Owens, United States	20.7s
1948	Mel Patton, United States	21.1s
1952	Andrew Stanfield, United States	20.7s
1956	Bobby Morrow, United States	20.6s
1960	Livio Berruti, Italy	20.5s
1964	Henry Carr, United States	20.3s
1968	Tommie Smith, United States	19.83s
1972	Valeri Borzov, USSR	20.00s
1976	Donald Quarrie, Jamaica	20.23s
1980	Pietro Mennea, Italy	20.19s
1984	Carl Lewis, United States	19.80s
1988	Joe DeLoach, United States	19.75s
1992	Mike Marsh, United States	20.01s
1996	Michael Johnson, United States	19.32s*
2000	Konstantinos Kenteris, Greece	20.09s

400-Meter Run

1896	Thomas Burke, United States	54.2s
1900	Maxey Long, United States	49.4s
1904	Harry Hillman, United States	49.2s
1908	Wyndham Halswelle, Great Britain, walkover	50.0s
1912	Charles Reidpath, United States	48.2s
1920	Bevil Rudd, South Africa	49.6s
1924	Eric Liddell, Great Britain	47.6s
1928	Ray Barbuti, United States	47.8s

400-Meter Run

1932	William Carr, United States	46.2s
1936	Archie Williams, United States	46.5s
1948	Arthur Wint, Jamaica	46.2s
1952	George Rhoden, Jamaica	45.9s
1956	Charles Jenkins, United States	46.7s
1960	Otis Davis, United States	44.9s
1964	Michael Larrabee, United States	45.1s
1968	Lee Evans, United States	43.86s
1972	Vincent Matthews, United States	44.66s
1976	Alberto Juantorena, Cuba	44.26s
1980	Viktor Markin, USSR	44.60s
1984	Alonzo Babers, United States	44.27s
1988	Steven Lewis, United States	43.87s
1992	Quincy Watts, United States	43.50s
1996	Michael Johnson, United States	43.49s*
2000	Michael Johnson, United States	43.84s

800-Meter Run

1896	Edwin Flack, Australia	2m. 11s
1900	Alfred Tysoe, Great Britain	2m. 1.2s
1904	James Lightbody, United States	1m. 56s
1908	Mel Sheppard, United States	1m. 52.8s
1912	James Meredith, United States	1m. 51.9s
1920	Albert Hill, Great Britain	1m. 53.4s
1924	Douglas Lowe, Great Britain	1m. 52.4s
1928	Douglas Lowe, Great Britain	1m. 51.8s
1932	Thomas Hampson, Great Britain	1m. 49.8s
1936	John Woodruff, United States	1m. 52.9s
1948	Mal Whitfield, United States	1m. 49.2s
1952	Mal Whitfield, United States	1m. 49.2s
1956	Thomas Courtney, United States	1m. 47.7s
1960	Peter Snell, New Zealand	1m. 46.3s
1964	Peter Snell, New Zealand	1m. 45.1s
1968	Ralph Doubell, Australia	1m. 44.3s
1972	Dave Wottle, United States	1m. 45.9s
1976	Alberto Juantorena, Cuba	1m. 43.50s
1980	Steve Ovett, Great Britain	1m. 45.40s
1984	Joaquim Cruz, Brazil	1m. 43.00s
1988	Paul Ereng, Kenya	1m. 43.45s
1992	William Tanui, Kenya	1m. 43.66s
1996	Vebjoern Rodal, Norway	1m. 42.58s*
2000	Nils Schumann, Germany	1m. 45.08

1,500-Meter Run

1896	Edwin Flack, Australia	4m. 33.2s
1900	Charles Bennett, Great Britain	4m. 6.2s
1904	James Lightbody, United States	4m. 5.4s
1908	Mel Sheppard, United States	4m. 3.4s
1912	Arnold Jackson, Great Britain	3m. 56.8s
1920	Albert Hill, Great Britain	4m. 1.8s
1924	Paavo Nurmi, Finland	3m. 53.6s
1928	Harry Larva, Finland	3m. 53.2s
1932	Luigi Beccali, Italy	3m. 51.2s
1936	Jack Lovelock, New Zealand	3m. 47.8s
1948	Henri Eriksson, Sweden	3m. 49.8s
1952	Joseph Barthel, Luxembourg	3m. 45.2s
1956	Ron Delany, Ireland	3m. 41.2s
1960	Herb Elliott, Australia	3m. 35.6s
1964	Peter Snell, New Zealand	3m. 38.1s
1968	Kipchoge Keino, Kenya	3m. 34.9s
1972	Pekka Vasala, Finland	3m. 36.3s
1976	John Walker, New Zealand	3m. 39.17s
1980	Sebastian Coe, Great Britain	3m. 38.4s
1984	Sebastian Coe, Great Britain	3m. 32.53s
1988	Peter Rono, Kenya	3m. 35.96s
1992	Fermin Cacho Ruiz, Spain	3m. 40.12s
1996	Noureddine Morceli, Algeria	3m. 35.78s
2000	Noah Kiprono Ngenyi, Kenya	3m. 32.07s*

5,000-Meter Run

1912	Hannes Kolehmainen, Finland	14m. 36.6s
1920	Joseph Guillemot, France	14m. 55.6s
1924	Paavo Nurmi, Finlands	14m. 31.2
1928	Willie Ritola, Finland	14m. 38s
1932	Lauri Lehtinen, Finland	14m. 30s
1936	Gunnar Hockert, Finland	14m. 22.2s
1948	Gaston Reiff, Belgium	14m. 17.6s
1952	Emil Zatopek, Czechoslovakia	14m. 6.6s
1956	Vladimir Kuts, USSR	13m. 39.6s
1960	Murray Halberg, New Zealand	13m. 43.4s
1964	Bob Schul, United States	13m. 48.8s
1968	Mohamed Gammoudi, Tunisia	14m. 05.0s
1972	Lasse Viren, Finland	13m. 26.4s
1976	Lasse Viren, Finland	13m. 24.76s
1980	Miruts Yifter, Ethiopia	13m. 21.0s
1984	Said Aouita, Morocco	13m. 05.59s*
1988	John Ngugi, Kenya	13m. 11.70s
1992	Dieter Baumann, Germany	13m. 12.52s
1996	Venuste Niyongabo, Burundi	13m. 07.96s
2000	Millon Wolde, Ethiopia	13m. 35.49s

10,000-Meter Run

1912	Hannes Kolehmainen, Finland	31m. 20.8s
1920	Paavo Nurmi, Finland	31m. 45.8s
1924	Willie Ritola, Finland	30m. 23.2s
1928	Paavo Nurmi, Finland	30m. 18.8s
1932	Janusz Kusocinski, Poland	30m. 11.4s
1936	Ilmari Salminen, Finland	30m. 15.4s
1948	Emil Zatopek, Czechoslovakia	29m. 59.6s
1952	Emil Zatopek, Czechoslovakia	29m. 17.0s
1956	Vladimir Kuts, USSR	28m. 45.6s
1960	Pyotr Bolotnikov, USSR	28m. 32.2s
1964	Billy Mills, United States	28m. 24.4s
1968	Naftali Temu, Kenya	29m. 27.4s
1972	Lasse Viren, Finland	27m. 38.4s
1976	Lasse Viren, Finland	27m. 40.4s
1980	Miruts Yifter, Ethiopia	27m. 42.7s
1984	Alberto Cova, Italy	27m. 47.54s
1988	Brahim Boutaib, Morocco	27m. 21.46s
1992	Khalid Skah, Morocco	27m. 46.70s
1996	Haile Gebrselassie, Ethiopia	27m. 07.34s*
2000	Haile Gebrselassie, Ethiopia	27m. 18.20s

110-Meter Hurdles

1896	Thomas Curtis, United States	17.6s
1900	Alvin Kraenzlein, United States	15.4s
1904	Frederick Schule, United States	16.0s
1908	Forrest Smithson, United States	15.0s
1912	Frederick Kelly, United States	15.1s
1920	Earl Thomson, Canada	14.8s
1924	Daniel Kinsey, United States	15.0s
1928	Sydney Atkinson, South Africa	14.8s
1932	George Saling, United States	14.6s
1936	Forrest Towns, United States	14.2s
1948	William Porter, United States	13.9s
1952	Harrison Dillard, United States	13.7s
1956	Lee Calhoun, United States	13.5s
1960	Lee Calhoun, United States	13.8s
1964	Hayes Jones, United States	13.6s
1968	Willie Davenport, United States	13.3s
1972	Rod Milburn, United States	13.24s
1976	Guy Drut, France	13.30s
1980	Thomas Munkelt, E. Germany	13.39s
1984	Roger Kingdom, United States	13.20s
1988	Roger Kingdom, United States	12.98s
1992	Mark McCoy, Canada	13.12s
1996	Allen Johnson, United States	12.95s*
2000	Anier Garcia, Cuba	13.00s

400-Meter Hurdles

1900	J.W.B. Tewksbury, United States	57.6s
1904	Harry Hillman, United States	53.0s
1908	Charles Bacon, United States	55.0s
1920	Frank Loomis, United States	54.0s
1924	F. Morgan Taylor, United States	52.6s
1928	Lord Burghley, Great Britain	53.4s
1932	Robert Tisdall, Ireland	51.7s
1936	Glenn Hardin, United States	52.4s
1948	Roy Cochran, United States	51.1s
1952	Charles Moore, United States	50.8s
1956	Glenn Davis, United States	50.1s
1960	Glenn Davis, United States	49.3s
1964	Rex Cawley, United States	49.6s
1968	Dave Hemery, Great Britain	48.12s
1972	John Akii-Bua, Uganda	47.82s
1976	Edwin Moses, United States	47.64s
1980	Volker Beck, E. Germany	48.70s
1984	Edwin Moses, United States	47.75s
1988	Andre Phillips, United States	47.19s
1992	Kevin Young, United States	46.78s*
1996	Derrick Adkins, United States	47.54s
2000	Angelo Taylor, Atlanta	47.50s

400-Meter Relay

1912	Great Britain	42.4s
1920	United States	42.2s
1924	United States	41.0s
1928	United States	41.0s
1932	United States	40.0s
1936	United States	39.8s
1948	United States	40.6s
1952	United States	40.1s
1956	United States	39.5s
1960	Germany (U.S. disqualified)	39.5s
1964	United States	39.0s
1968	United States	38.2s
1972	United States	38.19s
1976	United States	38.33s
1980	USSR	38.26s
1984	United States	37.83s
1988	USSR (U.S. disqualified)	38.19s

400-Meter Relay

1992	United States	37.40s*
1996	Canada	37.69s
2000	United States	37.61s

1,600-Meter Relay

1908	United States	3m. 29.4s
1912	United States	3m. 16.6s
1920	Great Britain	3m. 22.2s
1924	United States	3m. 16s
1928	United States	3m. 14.2s
1932	United States	3m. 8.2s
1936	Great Britain	3m. 9s
1948	United States	3m. 10.4s
1952	Jamaica	3m. 03.9s
1956	United States	3m. 04.8s
1960	United States	3m. 02.2s
1964	United States	3m. 00.7s
1968	United States	2m. 56.16s
1972	Kenya	2m. 59.8s
1976	United States	2m. 58.65s
1980	USSR	3m. 01.1s
1984	United States	2m. 57.91s
1988	United States	2m. 56.16s
1992	United States	2m. 55.74s*
1996	United States	2m. 55.99s
2000	United States	2m. 56.35s

3,000-Meter Steeplechase

1920	Percy Hodge, Great Britain	10m. 0.4s
1924	Willie Ritola, Finland	9m. 33.6s
1928	Toivo Loukola, Finland	9m. 21.8s
1932	Volmari Iso-Hollo, Finland (About 3,450 m; extra lap by error.)	10m. 33.4s
1936	Volmari Iso-Hollo, Finland	9m. 3.8s
1948	Thore Sjoestrand, Sweden	9m. 4.6s
1952	Horace Ashenfelter, United States	8m. 45.4s
1956	Chris Brasher, Great Britain	8m. 41.2s
1960	Zdzislaw Krzyszkowiak, Poland	8m. 34.2s
1964	Gaston Roelants, Belgium	8m. 30.8s
1968	Amos Biwott, Kenya	8m. 51s
1972	Kipchoge Keino, Kenya	8m. 23.6s
1976	Anders Garderud, Sweden	8m. 08.2s
1980	Bronislaw Malinowski, Poland	8m. 09.7s
1984	Julius Korir, Kenya	8m. 11.8s
1988	Julius Kariuki, Kenya	8m. 05.51s*
1992	Matthew Birir, Kenya	8m. 08.84s
1996	Joseph Keter, Kenya	8m. 07.12s
2000	Reuben Kosgei, Kenya	8m. 21.43s

20-Kilometer Walk

1956	Leonid Spirin, USSR	1h. 31m. 27.4s
1960	Vladimir Golubnichy, USSR	1h. 33m. 7.2s
1964	Kenneth Mathews, Great Britain	1h. 29m. 34.0s
1968	Vladimir Golubnichy, USSR	1h. 33m. 58.4s
1972	Peter Frenkel, E. Germany	1h. 26m. 42.4s
1976	Daniel Bautista, Mexico	1h. 24m. 40.6s
1980	Maurizio Damilano, Italy	1h. 23m. 35.5s
1984	Ernesto Canto, Mexico	1h. 23m. 13.0s
1988	Josef Pribilinec, Czechoslovakia	1h. 19m. 57.0s
1992	Daniel Plaza Montero, Spain	1h. 21m. 45.0s
1996	Jefferson Perez, Ecuador	1h. 20m.7s
2000	Robert Korzeniowski, Poland	1h. 18m. 59.0s*

50-Kilometer Walk

1932	Thomas W. Green, Great Britain	4h. 50m. 10s
1936	Harold Whitlock, Great Britain	4h. 30m. 41.4s
1948	John Ljunggren, Sweden	4h. 41m. 52s
1952	Giuseppe Dordoni, Italy	4h. 28m. 07.8s
1956	Norman Read, New Zealand	4h. 30m. 42.8s
1960	Donald Thompson, Great Britain	4h. 25m. 30s
1964	Abdon Pamich, Italy	4h. 11m. 12.4s
1968	Christoph Hohne, E. Germany	4h. 20m. 13.6s
1972	Bern Kannenberg, W. Germany	3h. 56m. 11.6s
1980	Hartwig Gauter, E. Germany	3h. 49m. 24.0s
1984	Raul Gonzalez, Mexico	3h. 47m. 26.0s
1988	Vayachselav Ivanenko, USSR	3h. 38m. 29.0s*
1992	Andrei Perlov, Unified Team	3h. 50m. 13.0s
1996	Robert Korzeniowski, Poland	3h. 43m. 30s
2000	Robert Korzeniowski, Poland	3h. 42m. 22s

Marathon

1896	Spiridon Loues, Greece	2h. 58m. 50s
1900	Michel Théato, France	2h. 59m. 45s
1904	Thomas Hicks, United States	3h. 28m. 63s
1908	John J. Hayes, United States	2h. 55m. 18.4s
1912	Kenneth McArthur, South Africa	2h. 36m. 54.8s
1920	Hannes Kolehmainen, Finland	2h. 32m. 35.8s
1924	Albin Stenroos, Finland	2h. 41m. 22.6s
1928	A.B. El Ouafi, France	2h. 32m. 57s
1932	Juan Zabala, Argentina	2h. 31m. 36s
1936	Kijung Son, Japan (Korean)	2h. 29m. 19.2s
1948	Delfo Cabrera, Argentina	2h. 34m. 51.6s

Marathon

1952	Emil Zatopek, Czechoslovakia	2h. 23m. 03.2s
1956	Alain Mimoun, France	2h. 25m.
1960	Abebe Bikila, Ethiopia	2h. 15m. 16.2s
1964	Abebe Bikila, Ethiopia	2h. 12m. 11.2s
1968	Mamo Wolde, Ethiopia	2h. 20m. 26.4s
1972	Frank Shorter, United States	2h. 12m. 19.8s
1976	Waldemar Cierpinski, E. Germany	2h. 09m. 55s
1980	Waldemar Cierpinski, E. Germany	2h. 11m. 03s
1984	Carlos Lopes, Portugal	2h. 09m. 21s*
1988	Gelindo Bordin, Italy	2h. 10m. 32s
1992	Hwang Young-Cho, S. Korea	2h. 13m. 23s
1996	Josia Thugwane, South Africa	2h. 12m. 36s
2000	Gezahgne Abera, Ethiopia	2h. 10m. 11s

High Jump

1896	Ellery Clark, United States	1.81m. (5'11¼")
1900	Irving Baxter, United States	1.90m. (6' 2¾")
1904	Samuel Jones, United States	1.80m. (5' 11")
1908	Harry Porter, United States	1.90m. (6' 2¾")
1912	Alma Richards, United States	1.93m. (6' 4")
1920	Richmond Landon, United States	1.93m. (6' 4")
1924	Harold Osborn, United States	1.98m. (6' 6")
1928	Robert W. King, United States	1.94m. (6' 4¼")
1932	Duncan McNaughton, Canada	1.97m. (6' 5½")
1936	Cornelius Johnson, United States	2.03m. (6' 8")
1948	John L. Winter, Australia	1.98m. (6' 6")
1952	Walter Davis, United States	2.04m. (6' 8¼")
1956	Charles Dumas, United States	2.12m. (6' 11½")
1960	Robert Shavlakadze, USSR	2.16m. (7' 1")
1964	Valery Brumel, USSR	2.18m. (7' 1¾")
1968	Dick Fosbury, United States	2.24m. (7' 4¼")
1972	Jüri Tarmak, USSR	2.23m. (7' 3¾")
1976	Jacek Wszola, Poland	2.25m. (7' 4½")
1980	Gerd Wessig, E. Germany	2.36m. (7' 8¾")
1984	Dietmar Mögenburg, W. Germany	2.35m. (7' 8½")
1988	Hennady Avdeyenko, USSR	2.38m. (7' 9¾")
1992	Javier Sotomayor Sanabria, Cuba	2.34m. (7' 8")
1996	Charles Austin, United States	2.39m. (7' 10")*
2000	Sergey Kliugin, Russia	2.35m. (7' 8½")

Long Jump

1896	Ellery Clark, United States	6.35m. (20' 10")
1900	Alvin Kraenzlein, United States	7.18m. (23' 6¾")
1904	Meyer Prinstein, United States	7.34m. (24' 1")
1908	Frank Irons, United States	7.48m. (24' 6½")
1912	Albert Gutterson, United States	7.60m. (24' 11¼")
1920	William Petterssen, Sweden	7.15m. (23' 5½")
1924	William DeHart Hubbard, U.S.	7.44m. (24' 5")
1928	Edward B. Hamm, United States	7.73m. (25' 4½")
1932	Edward Gordon, United States	7.64m. (25' ¾")
1936	Jesse Owens, United States	8.06m. (26' 5½")
1948	Willie Steele, United States	7.82m. (25' 8")
1952	Jerome Biffle, United States	7.57m. (24' 10")
1956	Gregory Bell, United States	7.83m. (25' 8¼")
1960	Ralph Boston, United States	8.12m. (26' 7¾")
1964	Lynn Davies, Great Britain	8.07m. (26' 5¾")
1968	Bob Beamon, United States	8.90m. (29' 2½")*
1972	Randy Williams, United States	8.24m. (27' ½")
1976	Arnie Robinson, United States	8.35m. (27' 4¾")
1980	Lutz Dombrowski, E. Germany	8.54m. (28' ¼")
1984	Carl Lewis, United States	8.54m. (28' ¼")
1988	Carl Lewis, United States	8.72m. (28' 7½")
1992	Carl Lewis, United States	8.67m. (28' 5½")
1996	Carl Lewis, United States	8.50m. (27' 10¾")
2000	Ivan Pedroso, Cuba	8.55m. (28' ¾")

Triple Jump

1896	James Connolly, United States	13.71m.(44' 11¾")
1900	Meyer Prinstein, United States	14.47m.(47' 5¾")
1904	Meyer Prinstein, United States	14.35m.(47' 1")
1908	Timothy Ahearne, G.B.-Ireland	14.92m.(48' 11½")
1912	Gustaf Lindblom, Sweden	14.76m.(48' 5")
1920	Vilho Tuulos, Finland	14.50m.(47' 7")
1924	Anthony Winter, Australia	15.52m.(50' 11")
1928	Mikio Oda, Japan	15.21m.(49' 11")
1932	Chuhei Nambu, Japan	15.72m.(51' 7")
1936	Naoto Tajima, Japan	16.00m.(52' 6")
1948	Arne Ahman, Sweden	15.40m.(50' 6¼")
1952	Adhemar Ferreira da Silva, Brazil	16.22m.(53' 2¾")
1956	Adhemar Ferreira da Silva, Brazil	16.35m.(53' 7¾")
1960	Jozef Schmidt, Poland	16.81m.(55' 1½")
1964	Jozef Schmidt, Poland	16.85m.(55' 3½")
1968	Viktor Saneyev, USSR	17.39m.(57' ¾")
1972	Viktor Saneyev, USSR	17.35m.(56' 11¼")
1976	Viktor Saneyev, USSR	17.29m.(56' 8¾")
1980	Jaak Uudmae, USSR	17.35m.(56' 11")
1984	Al Joyner, United States	17.26m.(56' 7½")
1988	Khristo Markov, Bulgaria	17.61m.(57' 9½")
1992	Mike Conley, United States	18.17m.(59' 7½")w
1996	Kenny Harrison, United States	18.09m.(59' 4¼")*
2000	Jonathan Edwards, Britain	17.71m.(58' 1¼")

Discus Throw

1896	Robert Garrett, United States	29.15m.(95' 7")
1900	Rudolf Bauer, Hungary	36.04m.(118' 3")
1904	Martin Sheridan, United States	39.28m.(128' 10")
1908	Martin Sheridan, United States	40.89m.(134' 1")
1912	Armas Taipale, Finland	45.21m.(148' 3")
1920	Elmer Niklander, Finland	44.68m.(146' 7")
1924	Clarence Houser, United States	46.15m.(151' 4")
1928	Clarence Houser, United States	47.32m.(155' 3")
1932	John Anderson, United States	49.49m.(162' 4")
1936	Ken Carpenter, United States	50.48m.(165' 7")
1948	Adolfo Consolini, Italy	52.78m.(173' 2")
1952	Sim Iness, United States	55.03m.(180' 6")
1956	Al Oerter, United States	56.36m.(184' 11")
1960	Al Oerter, United States	59.18m.(194' 2")
1964	Al Oerter, United States	61.00m.(200' 1")
1968	Al Oerter, United States	64.78m.(212' 6")
1972	Ludvik Danek, Czechoslovakia	64.40m.(211' 3")
1976	Mac Wilkins, United States	67.50m.(221' 5")
1980	Viktor Rashchupkin, USSR	66.64m.(218' 8")
1984	Rolf Dannenberg, W. Germany	66.60m.(218' 6")
1988	Jurgen Schult, E. Germany	68.82m.(225' 9")
1992	Romas Ubartas, Lithuania	65.12m.(213' 8")
1996	Lars Riedel, Germany	69.40m.(227' 8")*
2000	Virgilijus Alekna, Lithuania	69.30m.(227' 4")

Hammer Throw

1900	John Flanagan, United States	49.73m.(163' 1")
1904	John Flanagan, United States	51.22m.(168' 0")
1908	John Flanagan, United States	51.92m.(170' 4")
1912	Matt McGrath, United States	54.74m.(179' 7")
1920	Pat Ryan, United States	52.86m.(173' 5")
1924	Fred Tootell, United States	53.28m.(174' 10")
1928	Patrick O'Callaghan, Ireland	51.38m.(168' 7")
1932	Patrick O'Callaghan, Ireland	53.92m.(176' 11")
1936	Karl Hein, Germany	56.48m.(185' 4")
1948	Imre Németh, Hungary	56.06m.(183' 11")
1952	József Csérmák, Hungary	60.34m.(197' 11")
1956	Harold Connolly, United States	63.18m.(207' 3")
1960	Vasily Rudenkov, USSR	67.10m.(202' 0")
1964	Romuald Klim, USSR	69.74m.(228' 10")
1968	Gyula Zsivótsky, Hungary	73.36m.(240' 8")
1972	Anatoly Bondarchuk, USSR	75.50m.(247' 8")
1976	Yuri Syedykh, USSR	77.52m.(254' 4")
1980	Yuri Syedykh, USSR	81.80m.(268' 4")
1984	Juha Tiainen, Finland	78.08m.(256' 2")
1988	Sergei Litvinov, USSR	84.80m.(278' 2")*
1992	Andrey Abduvaliyev, Unified Team	82.54m.(270' 9")
1996	Balázs Kiss, Hungary	81.24m.(266' 6")
2000	Szymon Ziolkowski, Poland	80.02m.(262' 6")

Javelin Throw

1908	Erik Lemming, Sweden	54.82m. (179' 10")
1912	Erik Lemming, Sweden	60.64m. (198' 11")
1920	Jonni Myyrä, Finland	64.78m. (215' 10")
1924	Jonni Myyrä, Finland	62.96m. (206' 7")
1928	Eric Lundkvist, Sweden	66.60m. (218' 6")
1932	Matti Järvinen, Finland	72.70m. (238' 6")
1936	Gerhard Stöck, Germany	71.84m. (235' 8")
1948	Kai Tapio Rautavaara, Finland	69.76m. (228' 11")
1952	Cy Young, United States	73.78m. (242' 1")
1956	Egil Danielsen, Norway	85.70m. (281' 2")
1960	Viktor Tsibulenko, USSR	84.64m. (277' 8")
1964	Pauli Nevala, Finland	82.66m. (271' 2")
1968	Janis Lusis, USSR	90.10m. (295' 7")
1972	Klaus Wolfermann, W. Germany	90.48m. (296' 10")
1976	Miklós Németh, Hungary	94.58m. (310' 4")
1980	Dainis Kula, USSR	91.20m. (299' 2")
1984	Arto Härkönen, Finland	86.76m. (284' 8")
1988	Tapio Korjus, Finland	84.28m. (276' 6")
1992	Jan Zelezny, Czechoslovakia (a)	89.66m. (294' 2")
1996	Jan Zelezny, Czech Republic	88.16m. (289' 3")
2000	Jan Zelezny, Czech Republic	90.17m. (295' 9½")*

(a) New records were kept after javelin was modified in 1986.

Pole Vault

1896	William Welles Hoyt, United States	3.30m.(10' 10")
1900	Irving Baxter, United States	3.30m.(10' 10")
1904	Charles Dvorak, United States	3.50m.(11' 6")
1908	A. C. Gilbert, United States Edward Cooke Jr., United States	3.71m (12' 2")
1912	Harry Babcock, United States	3.95m.(12' 11½")
1920	Frank Foss, United States	4.09m.(13' 5")
1924	Lee Barnes, United States	3.95m.(12' 11½")
1928	Sabin W. Carr, United States	4.20m.(13' 9¼")
1932	William Miller, United States	4.31m.(14' 1¾")
1936	Earle Meadows, United States	4.35m.(14' 3¼")
1948	Guinn Smith, United States	4.30m.(14' 1¼")
1952	Robert Richards, United States	4.55m.(14' 11¼")
1956	Robert Richards, United States	4.56m.(14' 11½")
1960	Don Bragg, United States	4.70m.(15' 5")
1964	Fred Hansen, United States	5.10m.(16' 8¾")
1968	Bob Seagren, United States	5.40m.(17' 8½")
1972	Wolfgang Nordwig, E. Germany	5.50m.(18' ½")
1976	Tadeusz Slusarski, Poland	5.50m.(18' ½")
1980	Wladyslaw Kozakiewicz, Poland	5.78m.(18' 11½")
1984	Pierre Quinon, France	5.75m.(18' 10¼")
1988	Sergei Bubka, USSR	5.90m.(19' 4¼")
1992	Maksim Tarassov, Unified Team	5.80m.(19' ¼")
1996	Jean Galfione, France	5.92m.(19' 5")*
2000	Nick Hysong, United States	5.90m.(19' 4¼")

16-lb. Shot Put

1896	Robert Garrett, United States	11.22m.(36' 9¾")
1900	Richard Sheldon, United States	14.10m.(46' 3¼")
1904	Ralph Rose, United States	14.81m.(48' 7")
1908	Ralph Rose, United States	14.21m.(46' 7½")
1912	Pat McDonald, United States	15.34m.(50' 4")
1920	Ville Pörhölä, Finland	14.81m.(48' 7¼")
1924	L. Clarence Houser, United States	14.99m.(49' 2¼")
1928	John Kuck, United States	15.87m.(52' ¾")
1932	Leo Sexton, United States	16.00m.(52' 6")
1936	Hans Woellke, Germany	16.20m.(53' 1¾")
1948	Wilbur Thompson, United States	17.12m.(56' 2")
1952	W. Parry O'Brien, United States	17.41m.(57' 1½")
1956	W. Parry O'Brien, United States	18.57m.(60' 11¼")
1960	William Nieder, United States	19.68m.(64' 6¾")
1964	Dallas Long, United States	20.33m.(66' 8½")
1968	Randy Matson, United States	20.54m.(67' 4¾")
1972	Wladyslaw Komar, Poland	21.18m.(69' 6")
1976	Udo Beyer, E. Germany	21.05m.(69' ¾")
1980	Vladimir Kyselyov, USSR	21.35m.(70' ½")
1984	Alessandro Andrei, Italy	21.26m.(69' 9")
1988	Ulf Timmermann, E. Germany	22.47m.(73' 8¾")*
1992	Michael Stulce, United States	21.70m.(71' 2½")
1996	Randy Barnes, United States	21.62m.(70' 11¼")
2000	Arsi Harju, Finland	21.29m.(69' 10¼")

Decathlon (not held 1908)

1904	Thomas Kiely, Ireland	6,036 pts.
1912	Hugo Wieslander, Sweden (a)	7,724.49 pts.
1920	Helge Lovland, Norway	6,804.35 pts.
1924	Harold Osborn, United States	7,710.77 pts.
1928	Paavo Yrjola, Finland	8,053.29 pts.
1932	James Bausch, United States	8,462.23 pts.
1936	Glenn Morris, United States	7,900 pts.
1948	Robert Mathias, United States	7,139 pts.
1952	Robert Mathias, United States	7,887 pts.
1956	Milton Campbell, United States	7,937 pts.
1960	Rafer Johnson, United States	8,392 pts.
1964	Willi Holdorf, Germany (b)	7,887 pts.
1968	Bill Toomey, United States	8,193 pts.
1972	Nikolai Avilov, USSR	8,454 pts.
1976	Bruce Jenner, United States	8,617 pts.
1980	Daley Thompson, Great Britain	8,495 pts.
1984	Daley Thompson, Great Britain (c)	8,798 pts.*
1988	Christian Schenk, E. Germany	8,488 pts.
1992	Robert Zmelik, Czechoslovakia	8,611 pts.
1996	Dan O'Brien, United States	8,824 pts.
2000	Erki Nool, Estonia	8,641 pts.

(a) Jim Thorpe of the U.S. won the 1912 Decathlon with 8,413 pts. but was disqualified and had to return his medals because he had played pro baseball prior to the Olympics. The IOC in 1982 posthumously restored his decathlon and pentathlon golds. (b) Former point systems used prior to 1964. (c) Scoring change effective Apr. 1985; Thompson's readjusted score is 8,847 pts.

TRACK AND FIELD—Women

100-Meter Run

1928	Elizabeth Robinson, United States	12.2s
1932	Stella Walsh, Poland (a)	11.9s
1936	Helen Stephens, United States	11.5s
1948	Francina Blankers-Koen, Netherlands	11.9s
1952	Marjorie Jackson, Australia	11.5s
1956	Betty Cuthbert, Australia	11.5s
1960	Wilma Rudolph, United States	11.0s
1964	Wyomia Tyus, United States	11.4s
1968	Wyomia Tyus, United States	11.0s
1972	Renate Stecher, E. Germany	11.07s
1976	Annegret Richter, W. Germany	11.08s
1980	Lyudmila Kondratyeva, USSR	11.6s
1984	Evelyn Ashford, United States	10.97s
1988	Florence Griffith-Joyner, United States	10.54s*
1992	Gail Devers, United States	10.82s
1996	Gail Devers, United States	10.94s
2000	Marion Jones, United States	10.75s

(a) A 1980 autopsy determined that Walsh was a man.

200-Meter Run

1948	Francina Blankers-Koen, Netherlands	24.4s
1952	Marjorie Jackson, Australia	23.7s
1956	Betty Cuthbert, Australia	23.4s
1960	Wilma Rudolph, United States	24.0s

200-Meter Run

1964	Edith McGuire, United States	23.0s
1968	Irena Szewinska, Poland	22.5s
1972	Renate Stecher, E. Germany	22.40s
1976	Barbel Eckert, E. Germany	22.37s
1980	Barbel Wockel, E. Germany	22.03s
1984	Valerie Brisco-Hooks, United States	21.81s
1988	Florence Griffith-Joyner, United States	21.34s*
1992	Gwen Torrence, United States	21.81s
1996	Marie-Jose Perec, France	22.12s
2000	Marion Jones, United States	21.84s

400-Meter Run

1964	Betty Cuthbert, Australia	52.0s
1968	Colette Besson, France	52.0s
1972	Monika Zehrt, E. Germany	51.08s
1976	Irena Szewinska, Poland	49.29s
1980	Marita Koch, E. Germany	48.88s
1984	Valerie Brisco-Hooks, United States	48.83s
1988	Olga Bryzgina, USSR	48.65s
1992	Marie-Jose Perec, France	48.83s
1996	Marie-Jose Perec, France	48.25s*
2000	Cathy Freeman, Australia	49.11s

800-Meter Run

1928	Lina Radke, Germany	2m. 16.8s
1960	Ludmila Shevtsova, USSR	2m. 4.3s
1964	Ann Packer, Great Britain	2m. 1.1s
1968	Madeline Manning, United States	2m. 0.9s
1972	Hildegard Falck, W. Germany	1m. 58.6s
1976	Tatyana Kazankina, USSR	1m. 54.94s
1980	Nadezhda Olizayrenko, USSR	1m. 53.5s*
1984	Doina Melinte, Romania	1m. 57.6s
1988	Sigrun Wodars, E. Germany	1m. 56.10s
1992	Ellen Van Langen, Netherlands	1m. 55.54s
1996	Svetlana Masterkova, Russia	1m. 57.73s
2000	Maria Mutola, Mozambique	1m. 56.15s

1,500-Meter Run

1972	Lyudmila Bragina, USSR	4m. 01.4s
1976	Tatyana Kazankina, USSR	4m. 05.48s
1980	Tatyana Kazankina, USSR	3m. 56.6s
1984	Gabriella Dorio, Italy	4m. 03.25s
1988	Paula Ivan, Romania	3m. 53.96s*
1992	Hassiba Boulmerka, Algeria	3m. 55.30s
1996	Svetlana Masterkova, Russia	4m. 00.83s
2000	Nouria Benida Merah, Algeria	4m. 05.10s

3,000-Meter Run

1984	Maricica Puica, Romania	8m. 35.96s
1988	Tatyana Samolenko, USSR	8m. 26.53s*
1992	Elena Romanova, Unified Team	8m. 46.04s

5,000-Meter Run

1996	Wang Junxia, China	14m. 59.88s
2000	Gabriela Szabo, Romania	14m. 40.79s*

10,000-Meter Run

1988	Olga Boldarenko, USSR	31m. 44.69s
1992	Derartu Tulu, Ethiopia	31m. 06.02s
1996	Fernanda Ribeiro, Portugal	31m. 01.63s
2000	Derartu Tulu, Ethiopia	30m. 17.49s*

100-Meter Hurdles

1972	Annelie Ehrhardt, E. Germany	12.59s
1976	Johanna Schaller, E. Germany	12.77s
1980	Vera Komisova, USSR	12.56s
1984	Benita Brown-Fitzgerald, United States	12.84s
1988	Jordanka Donkova, Bulgaria	12.38s*
1992	Paraskevi Patoulidou, Greece	12.64s
1996	Ludmila Enquist, Sweden	12.58s
2000	Olga Shishigina, Kazakhstan	12.65s

400-Meter Hurdles

1984	Nawal el Moutawakii, Morocco	54.61s
1988	Debra Flintoff-King, Australia	53.17s
1992	Sally Gunnell, Great Britain	53.23s
1996	Deon Hemmings, Jamaica	52.82s*
2000	Irina Privalova, Russia	53.02s

1,600-Meter Relay

1972	East Germany	3m. 23s
1976	East Germany	3m. 19.23s
1980	USSR	3m. 20.02s
1984	United States	3m. 18.29s
1988	USSR	3m. 15.18s*
1992	Unified Team	3m. 20.20s
1996	United States	3m. 20.91s
2000	United States	3m. 22.62s

400-Meter Relay

1928	Canada	48.4s
1932	United States	46.9s
1936	United States	46.9s
1948	Netherlands	47.5s
1952	United States	45.9s
1956	Australia	44.5s
1960	United States	44.5s
1964	Poland	43.6s
1968	United States	42.8s
1972	West Germany	42.81s
1976	East Germany	42.55s
1980	East Germany	41.60s*
1984	United States	41.65s
1988	United States	41.98s
1992	United States	42.11s
1996	United States	41.95s
2000	Bahamas	41.95s

10 Kilometer Walk

1992	Chen Yueling, China	44m. 32s
1996	Elena Nikolayeva, Russia	41m. 49s*

20 Kilometer Walk

2000	Wang Liping, China	1m. 29.05s*

Marathon

1984	Joan Benoit, United States	2h. 24m. 52s
1988	Rosa Mota, Portugal	2h. 25m. 40s
1992	Valentina Yegorova, Unified Team	2h. 32m. 41s
1996	Fatuma Roba, Ethiopia	2h. 26m. 05s
2000	Naoko Takahashi, Japan	2h. 23m. 14s*

High Jump

1928	Ethel Catherwood, Canada	1.59m.	(5' 2½")
1932	Jean Shiley, United States	1.65m.	(5' 5")
1936	Ibolya Csák, Hungary	1.60m.	(5' 3")
1948	Alice Coachman, U. S.	1.68m.	(5' 6")
1952	Esther Brand, South Africa	1.67m.	(5' 5¾")
1956	Mildred L. McDaniel, U. S.	1.76m.	(5' 9¼")
1960	Iolanda Balas, Romania	1.85m.	(6' ¾")
1964	Iolanda Balas, Romania	1.90m.	(6' 2¾")
1968	Miloslava Resková, Czech.	1.82m.	(5' 11½")
1972	Ulrike Meyfarth, W. Germany	1.92m.	(6' 3½")
1976	Rosemarie Ackermann, E. Ger.	1.93m.	(6' 4")
1980	Sara Simeoni, Italy	1.97m.	(6' 5½")
1984	Ulrike Meyfarth, W. Germany	2.02m.	(6' 7½")
1988	Louise Ritter, United States	2.03m.	(6' 8")
1992	Heike Henkel, Germany	2.02m.	(6' 7½")
1996	Stefka Kostadinova, Bulgaria	2.05m.	(6' 8¾")*
2000	Yelena Yelesina, Russia	2.01m.	(6' 7")

Long Jump

1948	Olga Gyarmati, Hungary	5.69m.	(18' 8")
1952	Yvette Williams, New Zealand	6.24m.	(20' 5¼")
1956	Elzbieta Krzeskinska, Poland	6.35m.	(20' 10")
1960	Vira Krepkina, USSR	6.37m.	(20' 10¾")
1964	Mary Rand, Great Britain	6.76m.	(22' 2¼")
1968	Viorica Viscopoleanu, Romania	6.82m.	(22' 4½")
1972	Heidemarie Rosendahl, W. Ger.	6.78m.	(22' 3")
1976	Angela Voigt, E. Germany	6.72m.	(22' ¾")
1980	Tatyana Kolpakova, USSR	7.06m.	(23' 2")
1984	Anisoara Cusmir-Stanciu, Rom.	6.96m.	(22' 10")
1988	Jackie Joyner-Kersee, United States	7.40m.	(24' 3½")*
1992	Heike Drechsler, Germany	7.14m.	(23' 5¼")
1996	Chioma Ajunwa, Nigeria	7.12m.	(23' 4½"
2000	Heike Drechsler, Germany	6.99m.	(22' 11¼")

Triple Jump

1996	Inessa Kravets, Ukraine	15.33m	(50' 3½")*
2000	Tereza Marinova, Bulgaria	15.20m	(49' 10½")

Discus Throw

1928	Halina Konopacka, Poland	39.62m.	(130' 0")
1932	Lillian Copeland, United States	40.58m.	(133' 2")
1936	Gisela Mauermayer, Germany	47.62m.	(156' 3")
1948	Micheline Ostermeyer, France	41.92m.	(137' 6")
1952	Nina Ponomareva, USSR	51.42m.	(168' 8")
1956	Olga Fikotová, Czech.	53.68m.	(176' 1")
1960	Nina Ponomareva, USSR	55.10m.	(180' 9")
1964	Tamara Press, USSR	57.26m.	(187' 10")
1968	Lia Manoliu, Romania	58.28m.	(191' 2")
1972	Faina Melnik, USSR	66.62m.	(218' 7")
1976	Evelin Jahl, E. Germany	69.00m.	(226' 4")
1980	Evelin Jahl, E. Germany	69.96m.	(229' 6")
1984	Ria Stalman, Netherlands	65.36m.	(214' 5")
1988	Martina Hellmann, E. Germany	72.30m.	(237' 2")*
1992	Maritza Martén Garcia, Cuba	70.06m.	(229' 10")
1996	Ilke Wyludda, Germany	69.66m.	(228' 6")
2000	Ellina Zvereva, Belarus	68.40m.	(224' 5")

Hammer Throw

2000	Kamila Skolimowska, Poland	71.16m.	(233' 5¾")*

Pole Vault

2000	Stacy Dragila, United States	4.60m.	(15' 1")*

Shot Put (8 lb., 13 oz.)

1948	Micheline Ostermeyer, France	13.75m.	(45' 1½")
1952	Galina Zybina, USSR	15.28m.	(50' 1½")
1956	Tamara Tyshkyevich, USSR	16.59m.	(54' 5¼")
1960	Tamara Press, USSR	17.32m.	(56' 10")
1964	Tamara Press, USSR	18.14m.	(59' 6¼")

Shot Put (8 lb., 13 oz.)

1968	Margitta Gummel, E. Germany	19.61m. (64' 4")
1972	Nadezhda Chizova, USSR	21.03m. (69' 0")
1976	Ivanka Khristova, Bulgaria	21.16m. (69' 5¼")
1980	Ilona Slupianek, E. Germany	22.41m. (73' 6¼")*
1984	Claudia Losch, W. Germany	20.49m. (67' 2¼")
1988	Natalya Lisovskaya, USSR	22.24m. (72' 11¾")
1992	Svetlana Krivelyova, Unified Team	21.06m. (69' 1¼")
1996	Astrid Kumbernuss, Germany	20.56m. (67' 5½")
2000	Yanina Karolchik, Belarus	20.56m. (67' 5½")

Javelin Throw

1932	"Babe" Didrikson, United States	43.68m. (143' 4")
1936	Tilly Fleischer, Germany	45.18m. (148' 3")
1948	Herma Bauma, Austria	45.56m. (149' 6")
1952	Dana Zátopková, Czech.	50.46m. (165' 7")
1956	Inese Jaunzeme, USSR	53.86m. (176' 8")
1960	Elvira Ozolina, USSR	55.98m. (183' 8")
1964	Mihaela Penes, Romania	60.54m. (198' 7")
1968	Angéla Németh, Hungary	60.36m. (198' 0")
1972	Ruth Fuchs, E. Germany	63.88m. (209' 7")
1976	Ruth Fuchs, E. Germany	65.94m. (216' 4")
1980	Maria Colón Ruenes, Cuba	68.40m. (224' 5")
1984	Tessa Sanderson, Great Britain	69.56m. (228' 2")
1988	Petra Felke, E. Germany	74.68m. (245' 0")
1992	Silke Renke, Germany	68.34m. (224' 2")
1996	Heli Rantanen, Finland	67.94m. (222' 11")
2000	Trine Hattestad, Norway	68.91m. (226' 1")*

Heptathlon

1984	Glynis Nunn, Australia	6,390 pts.
1988	Jackie Joyner-Kersee, U.S.	7,215 pts.*
1992	Jackie Joyner-Kersee, U.S.	7,044 pts.
1996	Ghada Shouaa, Syria	6,780 pts.
2000	Denise Lewis, Britain	6,584 pts.

SWIMMING AND DIVING—Men

50-Meter Freestyle

1988	Matt Biondi, U.S.	22.14
1992	Aleksandr Popov, Unified Team	21.91*
1996	Aleksandr Popov, Russia	22.13
2000	Anthony Ervin, U.S.	21.98
2000	Gary Hall Jr., U.S.	21.98

100-Meter Freestyle

1896	Alfred Hajos, Hungary	1:22.2
1904	Zoltan de Halmay, Hungary (100 yards)	1:02.8
1908	Charles Daniels, U.S.	1:05.6
1912	Duke P. Kahanamoku, U.S.	1:03.4
1920	Duke P. Kahanamoku, U.S.	1:01.4
1924	John Weissmuller, U.S.	59.0
1928	John Weissmuller, U.S.	58.6
1932	Yasuji Miyazaki, Japan	58.2
1936	Ferenc Csik, Hungary	57.6
1948	Wally Ris, U.S.	57.3
1952	Clark Scholes, U.S.	57.4
1956	Jon Henricks, Australia	55.4
1960	John Devitt, Australia	55.2
1964	Don Schollander, U.S.	53.4
1968	Mike Wenden, Australia	52.2
1972	Mark Spitz, U.S.	51.22
1976	Jim Montgomery, U.S.	49.99
1980	Jorg Woithe, E. Germany	50.40
1984	Rowdy Gaines, U.S.	49.80
1988	Matt Biondi, U.S.	48.63
1992	Aleksandr Popov, Unified Team	49.02
1996	Aleksandr Popov, Russia	48.74
2000	Pieter van den Hoogenband, Netherlands	48.30

200-Meter Freestyle

1968	Mike Wenden, Australia	1:55.2
1972	Mark Spitz, U.S.	1:52.78
1976	Bruce Furniss, U.S.	1:50.29
1980	Sergei Kopliakov, USSR	1:49.81
1984	Michael Gross, W. Germany	1:47.44
1988	Duncan Armstrong, Australia	1:47.25
1992	Yevgeny Sadovyi, Unified Team	1:46.70
1996	Danyon Loader, New Zealand	1:47.63
2000	Pieter van den Hoogenband, Netherlands	1:45.35*

400-Meter Freestyle

1904	C. M. Daniels, U.S. (440 yards)	6:16.2
1908	Henry Taylor, Great Britain	5:36.8
1912	George Hodgson, Canada	5:24.4
1920	Norman Ross, U.S.	5:26.8
1924	John Weissmuller, U.S.	5:04.2
1928	Albert Zorilla, Argentina	5:01.6
1932	Clarence Crabbe, U.S.	4:48.4
1936	Jack Medica, U.S.	4:44.5
1948	William Smith, U.S.	4:41.0
1952	Jean Boiteux, France	4:30.7
1956	Murray Rose, Australia	4:27.3
1960	Murray Rose, Australia	4:18.3
1964	Don Schollander, U.S.	4:12.2

400-Meter Freestyle

1968	Mike Burton, U.S.	4:09.0
1972	Brad Cooper, Australia	4:00.27
1976	Brian Goodell, U.S.	3:51.93
1980	Vladimir Salnikov, USSR	3:51.31
1984	George DiCarlo, U.S.	3:51.23
1988	Ewe Dassler, E. Germany	3:46.95
1992	Yevgeny Sadovyi, Unified Team	3:45.00
1996	Danyon Loader, New Zealand	3:47.97
2000	Ian Thorpe, Australia	3:40.59*

1,500-Meter Freestyle

1908	Henry Taylor, Great Britain	22:48.4
1912	George Hodgson, Canada	22:00.0
1920	Norman Ross, U.S.	22:23.2
1924	Andrew Charlton, Australia	20:06.6
1928	Arne Borg, Sweden	19:51.8
1932	Kusuo Kitamura, Japan	19:12.4
1936	Noboru Terada, Japan	19:13.7
1948	James McLane, U.S.	19:18.5
1952	Ford Konno, U.S.	18:30.3
1956	Murray Rose, Australia	17:58.9
1960	Jon Konrads, Australia	17:19.6
1964	Robert Windle, Australia	17:01.7
1968	Mike Burton, U.S.	16:38.9
1972	Mike Burton, U.S.	15:52.58
1976	Brian Goodell, U.S.	15:02.40
1980	Vladimir Salnikov, USSR	14:58.27
1984	Michael O'Brien, U.S.	15:05.20
1988	Vladimir Salnikov, USSR	15:00.40
1992	Kieren Perkins, Australia	14:43.48*
1996	Kieren Perkins, Australia	14:56.40
2000	Grant Hackett, Australia	14:48.33

100-Meter Backstroke

1904	Walter Brack, Germany (100 yds.)	1:16.8
1908	Arno Bieberstein, Germany	1:24.6
1912	Harry Hebner, U.S.	1:21.2
1920	Warren Kealoha, U.S.	1:15.2
1924	Warren Kealoha, U.S.	1:13.2
1928	George Kojac, U.S.	1:08.2
1932	Masaji Kiyokawa, Japan	1:08.6
1936	Adolph Kiefer, U.S.	1:05.9
1948	Allen Stack, U.S.	1:06.4
1952	Yoshi Oyakawa, U.S.	1:05.4
1956	David Thiele, Australia	1:02.2
1960	David Thiele, Australia	1:01.9
1968	Roland Matthes, E. Germany	58.7
1972	Roland Matthes, E. Germany	56.58
1976	John Naber, U.S.	55.49
1980	Bengt Baron, Sweden	56.33
1984	Rick Carey, U.S.	55.79
1988	Daichi Suzuki, Japan	55.05
1992	Mark Tewksbury, Canada	53.98
1996	Jeff Rouse, U.S.	54.10
2000	Lenny Krayzelburg, U.S.	53.72*

200-Meter Backstroke

1964	Jed Graef, U.S.	2:10.3
1968	Roland Matthes, E. Germany	2:09.6
1972	Roland Matthes, E. Germany	2:02.82
1976	John Naber, U.S.	1:59.19
1980	Sandor Wladar, Hungary	2:01.93
1984	Rick Carey, U.S.	2:00.23
1988	Igor Polianski, USSR	1:59.37
1992	Martin Lopez-Zubero, Spain	1:58.47
1996	Brad Bridgewater, U.S.	1:58.54
2000	Lenny Krayzelburg, U.S.	1:56.76*

100-Meter Breaststroke

1968	Don McKenzie, U.S.	1:07.7
1972	Nobutaka Taguchi, Japan	1:04.94
1976	John Hencken, U.S.	1:03.11
1980	Duncan Goodhew, Great Britain	1:03.44
1984	Steve Lundquist, U.S.	1:01.65
1988	Adrian Moorhouse, Great Britain	1:02.04
1992	Nelson Diebel, U.S.	1:01.50
1996	Fred Deburghgraeve, Belgium	1:00.60
2000	Domenico Fioravanti, Italy	1:00.46*

200-Meter Breaststroke

1908	Frederick Holman, Great Britain	3:09.2
1912	Walter Bathe, Germany	3:01.8
1920	Haken Malmroth, Sweden	3:04.4
1924	Robert Skelton, U.S.	2:56.6
1928	Yoshiyuki Tsuruta, Japan	2:48.8
1932	Yoshiyuki Tsuruta, Japan	2:45.4
1936	Tetsuo Hamuro, Japan	2:41.5
1948	Joseph Verdeur, U.S.	2:39.3
1952	John Davies, Australia	2:34.4
1956	Masura Furukawa, Japan	2:34.7
1960	William Mulliken, U.S.	2:37.4
1964	Ian O'Brien, Australia	2:27.8

200-Meter Breaststroke

1968	Felipe Munoz, Mexico	2:28.7
1972	John Hencken, U.S.	2:21.55
1976	David Wilkie, Great Britain	2:15.11
1980	Robertas Zhulpa, USSR	2:15.85
1984	Victor Davis, Canada	2:13.34
1988	Jozsef Szabo, Hungary	2:13.52
1992	Mike Barrowman, U.S.	2:10.16*
1996	Norbert Rozsa, Hungary	2:12.57
2000	Domenico Fioravanti, Italy	2:10.87

100-Meter Butterfly

1968	Doug Russell, U.S.	55.9
1972	Mark Spitz, U.S.	54.27
1976	Matt Vogel, U.S.	54.35
1980	Par Arvidsson, Sweden	54.92
1984	Michael Gross, W. Germany	53.08
1988	Anthony Nesty, Suriname	53.00
1992	Pablo Morales, U.S.	53.32
1996	Denis Pankratov, Russia	52.27
2000	Lars Froelander, Sweden	52.00

200-Meter Butterfly

1956	William Yorzyk, U.S.	2:19.3
1960	Michael Troy, U.S.	2:12.8
1964	Kevin J. Berry, Australia	2:06.6
1968	Carl Robie, U.S.	2:08.7
1972	Mark Spitz, U.S.	2:00.70
1976	Mike Bruner, U.S.	1:59.23
1980	Sergei Fesenko, USSR	1:59.76
1984	Jon Sieben, Australia	1:57.04
1988	Michael Gross, W. Germany	1:56.94
1992	Mel Stewart, U.S.	1:56.26
1996	Denis Pankratov, Russia	1:56.51
2000	Tom Malchow, U.S.	1:55.35*

200-Meter Individual Medley

1968	Charles Hickcox, U.S.	2:12.0
1972	Gunnar Larsson, Sweden	2:07.17
1984	Alex Baumann, Canada	2:01.42
1988	Tamas Darnyi, Hungary	2:00.17
1992	Tamas Darnyi, Hungary	2:00.76
1996	Attila Czene, Hungary	1:59.91
2000	Massimiliano Rosolino, Italy	1:58.98*

400-Meter Individual Medley

1964	Dick Roth, U.S.	4:45.4
1968	Charles Hickcox, U.S.	4:48.4
1972	Gunnar Larsson, Sweden	4:31.98
1976	Rod Strachan, U.S.	4:23.68
1980	Aleksandr Sidorenko, USSR	4:22.89
1984	Alex Baumann, Canada	4:17.41
1988	Tamas Darnyi, Hungary	4:14.75
1992	Tamas Darnyi, Hungary	4:14.23
1996	Tom Dolan, U.S.	4:14.90
2000	Tom Dolan, U.S.	4:11.76*

400-Meter Freestyle Relay

1964	United States	3:31.2
1968	United States	3:31.7
1972	United States	3:26.42
1984	United States	3:19.03
1988	United States	3:16.53
1992	United States	3:16.74
1996	United States	3:15.41
2000	Australia	3:13.67*

800-Meter Freestyle Relay

1908	Great Britain	10:55.6
1912	Australia	10:11.6
1920	United States	10:04.4
1924	United States	9:53.4
1928	United States	9:36.2
1932	Japan	8:58.4
1936	Japan	8:51.5
1948	United States	8:46.0
1952	United States	8:31.1
1956	Australia	8:23.6
1960	United States	8:10.2
1964	United States	7:52.1
1968	United States	7:52.33
1972	United States	7:35.78
1976	United States	7:23.22
1980	USSR	7:23.50
1984	United States	7:15.69
1988	United States	7:12.51
1992	Unified Team	7:11.95
1996	United States	7:14.84
2000	Australia	7:07.05*

400-Meter Medley Relay

1960	United States	4:05.4
1964	United States	3:58.4
1968	United States	3:54.9
1972	United States	3:48.16
1976	United States	3:42.22
1980	Australia	3:45.70
1984	United States	3:39.30
1988	United States	3:36.93
1992	United States	3:36.93
1996	United States	3:34.84
2000	United States	3:33.73*

	Springboard Diving	**Points**
1908	Albert Zurner, Germany	85.5
1912	Paul Guenther, Germany	79.23
1920	Louis Kuehn, U.S	675.40
1924	Albert White, U.S.	97.46
1928	Pete Desjardins, U.S.	185.04
1932	Michael Galitzen, U.S.	161.38
1936	Richard Degener, U.S.	163.57
1948	Bruce Harlan, U.S.	163.64
1952	David Browning, U.S.	205.29
1956	Robert Clotworthy, U.S.	159.56
1960	Gary Tobian, U.S.	170.00
1964	Kenneth Sitzberger, U.S.	159.90
1968	Bernie Wrightson, U.S.	170.15
1972	Vladimir Vasin, USSR	594.09
1976	Phil Boggs, U.S.	619.52
1980	Aleksandr Portnov, USSR	905.02
1984	Greg Louganis, U.S.	754.41
1988	Greg Louganis, U.S.	730.80
1992	Mark Lenzi, U.S.	676.53
1996	Xiong Ni, China	701.46
2000	Xiong Ni, China	708.72

	Platform Diving	**Points**
1904	Dr. G.E. Sheldon, U.S.	112.75
1908	Hjalmar Johansson, Sweden	183.75
1912	Erik Adlerz, Sweden	73.94
1920	Clarence Pinkston, U.S.	100.67
1924	Albert White, U.S.	97.46
1928	Pete Desjardins, U.S.	98.74
1932	Harold Smith, U.S.	124.80
1936	Marshall Wayne, U.S.	113.58
1948	Sammy Lee, U.S.	130.05
1952	Sammy Lee, U.S.	156.28
1956	Joaquin Capilla, Mexico	152.44
1960	Robert Webster, U.S.	165.56
1964	Robert Webster, U.S.	148.58
1968	Klaus Dibiasi, Italy	164.18
1972	Klaus Dibiasi, Italy	504.12
1976	Klaus Dibiasi, Italy	600.51
1980	Falk Hoffmann, E. Germany	835.65
1984	Greg Louganis, U.S.	710.91
1988	Greg Louganis, U.S.	638.61
1992	Sun Shuwei, China	677.31
1996	Dmitri Sautin, Russia	692.34
2000	Tian Liang, China	724.53

SWIMMING AND DIVING—Women

50-Meter Freestyle

1988	Kristin Otto, E. Germany	25.49
1992	Yang Wenyi, China	24.76*
1996	Amy Van Dyken, U.S.	24.87
2000	Inge de Bruijn, Netherlands	24.32

100-Meter Freestyle

1912	Fanny Durack, Australia	1:22.2
1920	Ethelda Bleibtrey, U.S.	1:13.6
1924	Ethel Lackie, U.S.	1:12.4
1928	Albina Osipowich, U.S.	1:11.0
1932	Helene Madison, U.S.	1:06.8
1936	Hendrika Mastenbroek, Holland	1:05.9
1948	Greta Andersen, Denmark	1:06.3
1952	Katalin Szoke, Hungary	1:06.8
1956	Dawn Fraser, Australia	1:02.0
1960	Dawn Fraser, Australia	1:01.2
1964	Dawn Fraser, Australia	59.5
1968	Jan Henne, U.S.	1:00.0
1972	Sandra Neilson, U.S.	58.59
1976	Kornelia Ender, E. Germany	55.65
1980	Barbara Krause, E. Germany	54.79
1984	(tie) Carrie Steinseifer, U.S.	55.92
	Nancy Hogshead, U.S.	55.92
1988	Kristin Otto, E. Germany	54.93
1992	Zhuang Yong, China	54.64
1996	Li Jingyi, China	54.50
2000	Inge de Bruijn, Netherlands	53.83

200-Meter Freestyle

1968	Debbie Meyer, U.S.	2:10.5
1972	Shane Gould, Australia	2:03.56
1976	Kornelia Ender, E. Germany	1:59.26
1980	Barbara Krause, E. Germany	1:58.33
1984	Mary Wayte, U.S.	1:59.23
1988	Heike Friedrich, E. Germany	1:57.65*

200-Meter Freestyle

1992	Nicole Haislett, U.S.	1:57.90
1996	Claudia Poll, Costa Rica	1:58.16
2000	Susie O'Neill, Australia	1:58.24

400-Meter Freestyle

1924	Martha Norelius, U.S.	6:02.2
1928	Martha Norelius, U.S.	5:42.8
1932	Helene Madison, U.S.	5:28.5
1936	Hendrika Mastenbroek, Netherlands	5:26.4
1948	Ann Curtis, U.S.	5:17.8
1952	Valerie Gyenge, Hungary	5:12.1
1956	Lorraine Crapp, Australia	4:54.6
1960	Susan Chris von Saltza, U.S.	4:50.6
1964	Virginia Duenkel, U.S.	4:43.3
1968	Debbie Meyer, U.S.	4:31.8
1972	Shane Gould, Australia	4:19.44
1976	Petra Thuemer, E. Germany	4:09.89
1980	Ines Diers, E. Germany	4:08.76
1984	Tiffany Cohen, U.S.	4:07.10
1988	Janet Evans, U.S.	4:03.85*
1992	Dagmar Hase, Germany	4:07.18
1996	Michelle Smith, Ireland	4:07.25
2000	Brooke Bennett, U.S.	4:05.80

800-Meter Freestyle

1968	Debbie Meyer, U.S.	9:24.0
1972	Keena Rothhammer, U.S.	8:53.68
1976	Petra Thuemer, E. Germany	8:37.14
1980	Michelle Ford, Australia	8:28.90
1984	Tiffany Cohen, U.S.	8:24.95
1988	Janet Evans, U.S.	8:20.20
1992	Janet Evans, U.S.	8:25.52
1996	Brooke Bennett, U.S.	8:27.89
2000	Brooke Bennett, U.S.	8:19.67*

100-Meter Backstroke

1924	Sybil Bauer, U.S.	1:23.2
1928	Marie Braun, Netherlands	1:22.0
1932	Eleanor Holm, U.S.	1:19.4
1936	Dina Senff, Netherlands	1:18.9
1948	Karen Harup, Denmark	1:14.4
1952	Joan Harrison, South Africa	1:14.3
1956	Judy Grinham, Great Britain	1:12.9
1960	Lynn Burke, U.S.	1:09.3
1964	Cathy Ferguson, U.S.	1:07.7
1968	Kaye Hall, U.S.	1:06.2
1972	Melissa Belote, U.S.	1:05.78
1976	Ulrike Richter, E. Germany	1:01.83
1980	Rica Reinisch, E. Germany	1:00.86
1984	Theresa Andrews, U.S.	1:02.55
1988	Kristin Otto, E. Germany	1:00.89
1992	Krisztina Egerszegi, Hungary	1:00.68
1996	Beth Botsford, U.S.	1:01.19
2000	Diana Mocanu, Romania	1:00.21*

200-Meter Backstroke

1968	Pokey Watson, U.S.	2:24.8
1972	Melissa Belote, U.S.	2:19.19
1976	Ulrike Richter, E. Germany	2:13.43
1980	Rica Reinisch, E. Germany	2:11.77
1984	Jolanda De Rover, Netherlands	2:12.38
1988	Krisztina Egerszegi, Hungary	2:09.29
1992	Krisztina Egerszegi, Hungary	2:07.06*
1996	Krisztina Egerszegi, Hungary	2:07.83
2000	Diana Mocanu, Romania	2:08.16

100-Meter Breaststroke

1968	Djurdjica Bjedov, Yugoslavia	1:15.8
1972	Cathy Carr, U.S.	1:13.58
1976	Hannelore Anke, E. Germany	1:11.16
1980	Ute Geweniger, E. Germany	1:10.22
1984	Petra Van Staveren, Netherlands	1:09.88
1988	Tania Dangalakova, Bulgaria	1:07.95
1992	Elena Roudkovskaia, Unified Team	1:08.00
1996	Penny Heyns, South Africa	1:07.73
2000	Megan Quann, U.S.	1:07.05

200-Meter Breaststroke

1924	Lucy Morton, Great Britain	3:33.2
1928	Hilde Schrader, Germany	3:12.6
1932	Clare Dennis, Australia	3:06.3
1936	Hideko Maehata, Japan	3:03.6
1948	Nelly Van Vliet, Netherlands	2:57.2
1952	Eva Szekely, Hungary	2:51.7
1956	Ursula Happe, Germany	2:53.1
1960	Anita Lonsbrough, Great Britain	2:49.5
1964	Galina Prozumenschikova, USSR	2:46.4
1968	Sharon Wichman, U.S.	2:44.4
1972	Beverly Whitfield, Australia	2:41.71
1976	Marina Koshevaia, USSR	2:33.35
1980	Lina Kachushite, USSR	2:29.54
1984	Anne Ottenbrite, Canada	2:30.38
1988	Silke Hoerner, E. Germany	2:26.71
1992	Kyoko Iwasaki, Japan	2:26.65
1996	Penny Heyns, South Africa	2:25.41
2000	Agnes Kovacs, Hungary	2:24.35

100-Meter Butterfly

1956	Shelley Mann, U.S.	1:11.0
1960	Carolyn Schuler, U.S.	1:09.5
1964	Sharon Stouder, U.S.	1:04.7
1968	Lynn McClements, Australia	1:05.5
1972	Mayumi Aoki, Japan	1:03.34
1976	Kornelia Ender, E. Germany	1:00.13
1980	Caren Metschuck, E. Germany	1:00.42
1984	Mary T. Meagher, U.S.	59.26
1988	Kristin Otto, E. Germany	59.00
1992	Qian Hong, China	58.62
1996	Amy Van Dyken, U.S.	59.13
2000	Inge de Bruijn, Netherlands	56:61*

200-Meter Butterfly

1968	Ada Kok, Netherlands	2:24.7
1972	Karen Moe, U.S.	2:15.57
1976	Andrea Pollack, E. Germany	2:11.41
1980	Ines Geissler, E. Germany	2:10.44
1984	Mary T. Meagher, U.S.	2:06.90
1988	Kathleen Nord, E. Germany	2:09.51
1992	Summer Sanders, U.S.	2:08.67
1996	Susan O'Neill, Australia	2:07.76
2000	Misty Hyman, U.S.	2:05.88*

200-Meter Individual Medley

1968	Claudia Kolb, U.S.	2:24.7
1972	Shane Gould, Australia	2:23.07
1984	Tracy Caulkins, U.S.	2:12.64
1988	Daniela Hunger, E. Germany	2:12.59
1992	Lin Li, China	2:11.65
1996	Michelle Smith, Ireland	2:13.93
2000	Yana Klochkova, Ukraine	2:10.68*

400-Meter Individual Medley

1964	Donna de Varona, U.S.	5:18.7
1968	Claudia Kolb, U.S.	5:08.5
1972	Gail Neall, Australia	5:02.97
1976	Ulrike Tauber, E. Germany	4:42.77
1980	Petra Schneider, E. Germany	4:36.29
1984	Tracy Caulkins, U.S.	4:39.24
1988	Janet Evans, U.S.	4:37.76
1992	Krisztina Egerszegi, Hungary	4:36.54
1996	Michelle Smith, Ireland	4:39.18
2000	Yana Klochkova, Ukraine	4:33.59*

400-Meter Freestyle Relay

1912	Great Britain	5:52.8
1920	United States	5:11.6
1924	United States	4:58.8
1928	United States	4:47.6
1932	United States	4:38.0
1936	Netherlands	4:36.0
1948	United States	4:29.2
1952	Hungary	4:24.4
1956	Australia	4:17.1
1960	United States	4:08.9
1964	United States	4:03.8
1968	United States	4:02.5
1972	United States	3:55.19
1976	United States	3:44.82
1980	East Germany	3:42.71
1984	United States	3:43.43
1988	East Germany	3:40.63
1992	United States	3:39.46
1996	United States	3:39.29
2000	United States	3:36.61*

800-Meter Freestyle Relay

1996	United States	7:59.87
2000	United States	7:57.80*

400-Meter Medley Relay

1960	United States	4:41.1
1964	United States	4:33.9
1968	United States	4:28.3
1972	United States	4:20.75
1976	East Germany	4:07.95
1980	East Germany	4:06.67
1984	United States	4:08.34
1988	East Germany	4:03.74
1992	United States	4:02.54
1996	United States	4:02.88
2000	United States	3:58.30*

Springboard Diving Points

1920	Aileen Riggin, U.S.	539.90
1924	Elizabeth Becker, U.S.	474.50
1928	Helen Meany, U.S.	78.62
1932	Georgia Coleman U.S.	87.52
1936	Marjorie Gestring, U.S.	89.27
1948	Victoria M. Draves, U.S.	108.74
1952	Patricia McCormick, U.S.	147.30
1956	Patricia McCormick, U.S.	142.36
1960	Ingrid Kramer, Germany	155.81
1964	Ingrid Engel-Kramer, Germany	145.00
1968	Sue Gossick, U.S.	150.77
1972	Micki King, U.S.	450.03
1976	Jenni Chandler, U.S.	506.19
1980	Irina Kalinina, USSR	725.91
1984	Sylvie Bernier, Canada	530.70
1988	Gao Mln, China	580.23
1992	Gao Min, China	572.40
1996	Fu Mingxia, China	547.68
2000	Fu Mingxia, China	609.42

Platform Diving Points

1912	Greta Johansson, Sweden	39.90
1920	Stefani Fryland-Clausen, Denmark	34.60
1924	Caroline Smith, U.S.	33.20
1928	Elizabeth B. Pinkston, U.S.	31.60
1932	Dorothy Poynton, U.S.	40.26
1936	Dorothy Poynton Hill, U.S.	33.93
1948	Victoria M. Draves, U.S.	8.87
1952	Patricia McCormick, U.S.	79.37
1956	Patricia McCormick, U.S.	84.85
1960	Ingrid Kramer, Germany	91.28
1964	Lesley Bush, U.S.	99.80
1968	Milena Duchkova, Czech.	109.59
1972	Ulrika Knape, Sweden	390.00
1976	Elena Vaytsekhouskaya, USSR	406.59
1980	Martina Jaschke, E. Germany	596.25
1984	Zhou Jihong, China	435.51
1988	Xu Yanmei, China	445.20
1992	Fu Mingxia, China	461.43
1996	Fu Mingxia, China	521.58
2000	Laura Wilkinson, U.S.	543.75

BOXING

Lt. Flyweight (48 kg/106 lbs)

1968 Francisco Rodriguez, Venezuela
1972 Gyorgy Gedo, Hungary
1976 Jorge Hernandez, Cuba
1980 Shamil Sabyrov, USSR
1984 Paul Gonzalez, U.S.
1988 Ivailo Hristov, Bulgaria
1992 Rogelio Marcelo, Cuba
1996 Daniel Petrov, Bulgaria
2000 Brahim Asloum, France

Flyweight (51 kg/112 lbs)

1904 George Finnegan, U.S.
1920 William Di Gennara, U.S.
1924 Fidel LaBarba, U.S.
1928 Antal Kocsis, Hungary
1932 Istvan Enekes, Hungary
1936 Willi Kaiser, Germany
1948 Pascual Perez, Argentina
1952 Nathan Brooks, U.S.
1956 Terence Spinks, Great Britain
1960 GyulaTorok, Hungary
1964 Fernando Atzori, Italy
1968 Ricardo Delgado, Mexico
1972 Georgi Kostadinov, Bulgaria
1976 Leo Randolph, U.S.
1980 Peter Lessov, Bulgaria
1984 Steve McCrory, U.S.
1988 Kim Kwang Sun, S. Korea
1992 Su Choi Choi, N. Korea
1996 Maikro Romero, Cuba
2000 Wijan Ponlid, Thailand

Bantamweight (54 kg /119 lbs)

1904 Oliver Kirk, U.S.
1908 A. Henry Thomas, Great Britain
1920 Clarence Walker, South Africa
1924 William Smith, South Africa
1928 Vittorio Tamagnini, Italy
1932 Horace Gwynne, Canada
1936 Ulderico Sergo, Italy
1948 Tibor Csik, Hungary
1952 Pentti Hamalainen, Finland
1956 Wolfgang Behrendt, E. Germany
1960 Oleg Grigoryev, USSR
1964 Takao Sakurai, Japan
1968 Valery Sokolov, USSR
1972 Orlando Martinez, Cuba
1976 Yong-Jo Gu, N. Korea
1980 Juan Hernandez, Cuba
1984 Maurizio Stecca, Italy
1988 Kennedy McKinney, U.S.
1992 Joel Casamayor, Cuba
1996 Istvan Kovacs, Hungary
2000 Guillermo Rigondeaux, Cuba

Featherweight (57 kg/125 lbs)

1904 Oliver Kirk, U.S.
1908 Richard Gunn, Great Britain
1920 Paul Fritsch, France
1924 John Fields, U.S.
1928 Lambertus van Klaveren, Netherlands
1932 Carmelo Robledo, Argentina
1936 Oscar Casanovas, Argentina
1948 Ernesto Formenti, Italy
1952 Jan Zachara, Czechoslovakia
1956 Vladimir Safronov, USSR
1960 Francesco Musso, Italy
1964 Stanislav Stephashkin, USSR
1968 Antonin Roldan, Mexico
1972 Boris Kousnetsov, USSR
1976 Angel Herrera, Cuba
1980 Rudi Fink, E. Germany
1984 Meldrick Taylor, U.S.
1988 Giovanni Parisi, Italy
1992 Andreas Tews, Germany
1996 Somluck Kamsing, Thailand
2000 Bekzat Sattarkhanov, Kazakhstan

Lightweight (60 kg/132 lbs)

1904 Harry Spanger, U.S.
1908 Frederick Grace, Great Britain
1920 Samuel Mosberg, U.S.
1924 Hans Nielsen, Denmark
1928 Carlo Orlandi, Italy
1932 Lawrence Stevens, South Africa
1936 Imre Harangi, Hungary
1948 Gerald Dreyer, South Africa
1952 Aureliano Bolognesi, Italy
1956 Richard McTaggart, Great Britain
1960 Kazimierz Pazdzior, Poland
1964 Jozef Grudzien, Poland
1968 Ronald Harris, U.S.
1972 Jan Szczepanski, Poland
1976 Howard Davis, U.S.
1980 Angel Herrera, Cuba
1984 Pernell Whitaker, U.S.
1988 Andreas Zuelow, E. Germany
1992 Oscar De La Hoya, U.S.
1996 Hocine Soltani, Algeria
2000 Mario Kindelan, Cuba

Lt. Welterweight (63.5 kg/139 lbs)

1952 Charles Adkins, U.S.
1956 Vladimir Yengibaryan, USSR
1960 Bohumil Nemecek, Czechoslovakia
1964 Jerzy Kulej, Poland
1968 Jerzy Kulej, Poland
1972 Ray Seales, U.S.
1976 Ray Leonard, U.S.
1980 Patrizio Oliva, Italy
1984 Jerry Page, U.S.
1988 Viatcheslav Janovski, USSR
1992 Hector Vinent, Cuba
1996 Hector Vinent, Cuba
2000 Mahamadkadyz Abdullaev, Uzbekistan

Welterweight (67 kg/147 lbs)

1904 Albert Young, U.S.
1920 Albert Schneider, Canada
1924 Jean Delarge, Belgium
1928 Edward Morgan, New Zealand
1932 Edward Flynn, U.S.
1936 Sten Suvio, Finland
1948 Julius Torma, Czechoslovakia
1952 Zygmunt Chychia, Poland
1956 Nicolae Linca, Romania
1960 Giovanni Benvenuti, Italy
1964 Marian Kasprzyk, Poland
1968 Manfred Wolke, E. Germany
1972 Emilio Correa, Cuba
1976 Jochen Bachfeld, E. Germany
1980 Andres Aldama, Cuba
1984 Mark Breland, U.S.
1988 Robert Wangila, Kenya
1992 Michael Carruth, Ireland
1996 Oleg Saitov, Russia
2000 Oleg Saitov, Russia

Lt. Middleweight (71 kg/156 lbs)

1952 Laszlo Papp, Hungary
1956 Laszlo Papp, Hungary
1960 Wilbert McClure, U.S.
1964 Boris Lagutin, USSR
1968 Boris Lagutin, USSR
1972 Dieter Kottysch, W. Germany
1976 Jerzy Rybicki, Poland
1980 Armando Martinez, Cuba
1984 Frank Tate, U.S.
1988 Park Si Hun, S. Korea
1992 Juan Lemus, Cuba
1996 David Reid, U.S.
2000 Yermakhan Ibraimov, Kazakhstan

Middleweight (75 kg/165 lbs)

1904 Charles Mayer, U.S.
1908 John Douglas, Great Britain
1920 Harry Mallin, Great Britain
1924 Harry Mallin, Great Britain
1928 Piero Toscani, Italy
1932 Carmen Barth, U.S.
1936 Jean Despeaux, France
1948 Laszlo Papp, Hungary
1952 Floyd Patterson, U.S.
1956 Gennady Schatkov, USSR
1960 Edward Crook, U.S.
1964 Valery Popenchenko, USSR
1968 Christopher Finnegan, Great Britain
1972 Vyacheslav Lemechev, USSR
1976 Michael Spinks, U.S.
1980 Jose Gomez, Cuba
1984 Joon-Sup Shin, S. Korea
1988 Henry Maske, E. Germany
1992 Ariel Hernandez, Cuba
1996 Ariel Hernandez, Cuba
2000 Jorge Gutierrez, Cuba

Lt. Heavyweight (81 kg/178 lbs)

1920 Edward Eagan, U.S.
1924 Harry Mitchell, Great Britain
1928 Victor Avendano, Argentina
1932 David Carstens, South Africa
1936 Roger Michelot, France
1948 George Hunter, South Africa
1952 Norvel Lee, U.S.
1956 James Boyd, U.S.
1960 Cassius Clay, U.S.
1964 Cosimo Pinto, Italy
1968 Dan Poznyak, USSR
1972 Mate Parlov, Yugoslavia
1976 Leon Spinks, U.S.
1980 Slobodan Kacar, Yugoslavia
1984 Anton Josipovic, Yugoslavia
1988 Andrew Maynard, U.S.
1992 Torsten May, Germany
1996 Vassili Jirov, Kazakhstan
2000 Alexander Lebziak, Russia

Heavyweight (91 kg/201 lbs)

1984 Henry Tillman, U.S.
1988 Ray Mercer, U.S.
1992 Felix Savon, Cuba
1996 Felix Savon, Cuba
2000 Felix Savon, Cuba

Super Heavyweight (91+ kg/201+ lbs)
(known as heavyweight, 1904-80)

1904 Samuel Berger, U.S.
1908 Albert Oldham, Great Britain
1920 Ronald Rawson, Great Britain
1924 Otto von Porat, Norway
1928 Arturo Rodriguez Jurado, Argentina
1932 Santiago Lovell, Argentina
1936 Herbert Runge, Germany
1948 Rafael Iglesias, Argentina
1952 H. Edward Sanders, U.S.
1956 T. Peter Rademacher, U.S.
1960 Franco De Piccoli, Italy
1964 Joe Frazier, U.S.
1968 George Foreman, U.S.
1972 Teofilo Stevenson, Cuba
1976 Teofilo Stevenson, Cuba
1980 Teofilo Stevenson, Cuba
1984 Tyrell Biggs, U.S.
1988 Lennox Lewis, Canada
1992 Roberto Balado, Cuba
1996 Vladimir Klitchko, Ukraine
2000 Audley Harrison, Britain

Sites of Summer Olympic Games

1896 Athens, Greece
1900 Paris, France
1904 St. Louis, U.S.
1906* Athens, Greece
1908 London, England
1912 Stockholm, Sweden
1920 Antwerp, Belgium
1924 Paris, France
1928 Amsterdam, Netherlands
1932 Los Angeles, U.S.
1936 Berlin, Germany
1948 London, England
1952 Helsinki, Finland
1956 Melbourne, Australia
1960 Rome, Italy
1964 Tokyo, Japan
1968 Mexico City, Mexico
1972 Munich, W. Germany
1976 Montreal, Canada
1980 Moscow, USSR
1984 Los Angeles, U.S.
1988 Seoul, South Korea
1992 Barcelona, Spain
1996 Atlanta, U.S.
2000 Sydney, Australia
2004 Athens, Greece
2008 Beijing, China

*Games not recognized by International Olympic Committee. Games 6 (1916), 12 (1940), and 13 (1944) were not celebrated.

Olympic Information

The modern Olympic Games, first held in Athens, Greece, in 1896, were the result of efforts by Baron Pierre de Coubertin, a French educator, to promote interest in education and culture and to foster better international understanding through love of athletics. His source of inspiration was the ancient Greek Olympic Games, most notable of the 4 Panhellenic celebrations. The games were combined patriotic, religious, and athletic festivals held every 4 years. The first such recorded festival was held in 776 BC, the date from which the Greeks began to keep their calendar by "Olympiads," or 4-year spans between the games.

Baron de Coubertin enlisted 13 nations to send athletes to the first modern Olympics in 1896; now athletes from nearly 200 nations and territories compete in the Summer Olympics. The Winter Olympic Games were started in 1924.

Symbol: Five rings or circles, linked together to represent the sporting friendship of all peoples. They also symbolize 5 geographic areas—Europe, Asia, Africa, Australia, and America. Each ring is a different color—blue, yellow, black, green, or red.

Flag: The symbol of the 5 rings on a plain white background.

Creed: "The most important thing in the Olympic Games is not to win but to take part, just as the most important thing in life is not the triumph but the struggle. The essential thing is not to have conquered but to have fought well."

Motto: "Citius, Altius, Fortius." Latin meaning "swifter, higher, stronger."

Oath: "In the name of all competitors I promise that we will take part in these Olympic Games, respecting and abiding by the rules which govern them, in the true spirit of sportsmanship for the glory of sport and the honor of our teams."

Flame: The modern version of the flame was adopted in 1936. The torch used to kindle it is first lit by the sun's rays at Olympia, Greece, then carried to the site of the Games by relays of runners. Ships and planes are used when necessary.

PARALYMPICS

The first Olympic games for the disabled were held in Rome after the 1960 Summer Olympics; use of the name "paralympic" began with the 1964 games in Tokyo. The Paralympics are held by the Olympic host country in the same year and usually same city or venue. A goal of the Paralympics is to provide elite competition to athletes with functional disabilities that prevent their involvement in the Olympics. In 1976 the first Winter Paralympic Games were held, in Ornskoldsvik, Sweden.

The XI Paralympic Summer Games, the largest ever, were held Oct. 18-Oct. 29, 2000 in Sydney, Australia. The games featured more than 4,000 athletes from 125 nations competing in 18 sports, including new additions, wheelchair rugby and sailing. Australia took home the most medals, 149 (63 gold), followed by Great Britain, 131 (41 gold), and Spain, 107 (39 gold). The VIII Paralympic Winter Games were scheduled to be held Mar. 7-Mar. 16, 2002, in Salt Lake City, Utah, and the XII Paralympic Summer Games were scheduled to be held in Athens, Greece, 2 weeks after the 2004 Summer Olympics.

SPECIAL OLYMPICS

Special Olympics is an international program of year-round sports training and athletic competition for children and adults with special needs. All 50 U.S. states, Washington, DC, and Guam have chapter offices. In addition, there are accredited Special Olympics programs in nearly 150 countries. Persons wishing to volunteer or find out more about Special Olympics can contact Special Olympics International Headquarters, 1325 G St. NW, Suite 500, Washington, DC 20005, or access the Special Olympics website at http://www.specialolympics.org

Special Olympics: 2001 World Winter Games, 2003 World Summer Games

The 7th Special Olympics World Winter Games were held Mar. 4-11, 2001, in Anchorage, AK. About 2,400 athletes from 69 countries competed for the 1,894 medals that were awarded. Also participating were more than 3,000 coaches and family members as well as some 6,000 volunteers. The competition included Alpine Skiing, Cross-Country Skiing, Floor Hockey, Figure Skating, Speed Skating, and Snowshoeing; and for the first time, Snowboarding was held as a demonstration sport.

The 11th Special Olympics World Summer Games, the first to be held outside the U.S., were scheduled for June 20-29, 2003, in Dublin, Ireland. More than 7,000 athletes, 3,000 coaches and official delegates, and 28,000 friends and family members were expected to participate. Scheduled individual competition included Athletics, Aquatics, Badminton, Bocce, Bowling, Cycling, Equestrian Sports, Golf, Gymnastics (Artistic and Rhythmic), Power lifting, Rollerskating, Table Tennis, and Tennis. Scheduled team sports were Basketball, Handball, Sailing, Soccer, and Volleyball. Kayaking and Pitch-and-Putt (a form of golf) were to be included as demonstration sports.

TRACK AND FIELD

World Record Progression for the One-Mile Run

Note: Records shown since Roger Bannister broke the 4-minute mile on May 6, 1954

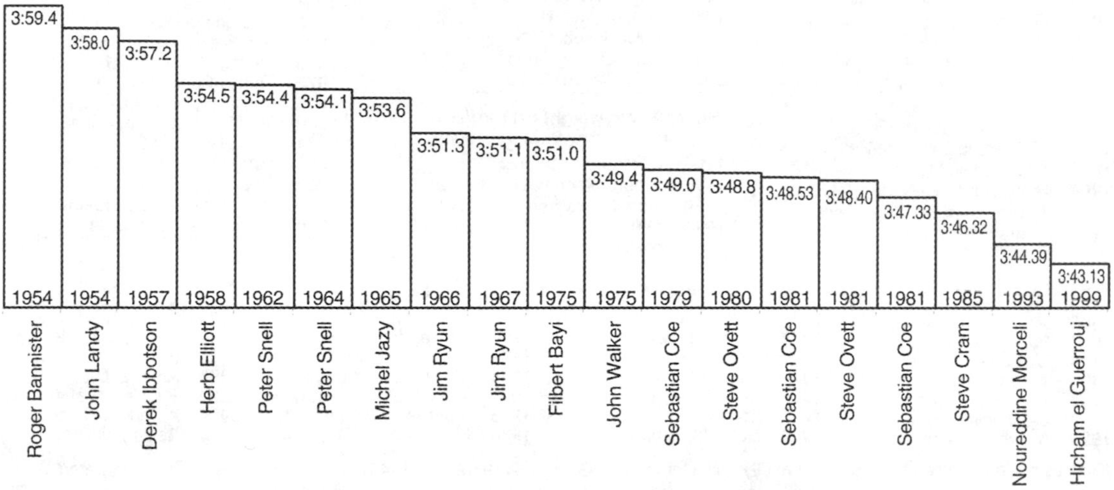

World Track and Field Outdoor Records

As of Oct. 10, 2001

The International Amateur Athletic Federation, the world body of track and field, recognizes only records in metric distances, except for the mile. *Pending ratification.**World best; marathon records not officially recognized by IAAF.

Men's Records

Running

Event	Record	Holder	Country	Date	Where made
100 meters	**9.79 s.**	Maurice Greene	U.S.	June 16, 1999	Athens, Greece
200 meters	**19.32 s.**	Michael Johnson	U.S.	Aug. 1, 1996	Atlanta, GA
400 meters	**43.18 s.**	Michael Johnson	U.S.	Aug. 26, 1999	Seville, Spain
800 meters	**1 m., 41.11 s.**	Wilson Kipketer	Denmark	Aug. 24, 1997	Cologne, Germany
1,000 meters	**2 m., 11.96 s.**	Noah Ngeny	Kenya	Sept. 5, 1999	Rieti, Italy
1,500 meters	**3 m., 26.00 s.**	Hicham El Guerrouj	Morocco	July 14, 1998	Rome, Italy
1 mile	**3 m., 43.13 s.**	Hicham El Guerrouj	Morocco	July 7, 1999	Rome, Italy
2,000 meters	**4 m., 44.79 s.**	Hicham El Guerrouj	Morocco	Sept. 7, 1999	Berlin, Germany
3,000 meters	**7 m., 20.67 s.**	Daniel Komen	Kenya	Sept. 1, 1996	Rieti, Italy
5,000 meters	**12 m., 39.36 s.**	Haile Gebrselassie	Ethiopia	June 13, 1998	Helsinki, Finland
10,000 meters	**26 m., 22.75 s.**	Haile Gebrselassie	Ethiopia	June 1,1998	Hengelo, Netherlands
20,000 meters	**56 m., 55.6 s.**	Arturo Barrios	Mexico	Mar. 30, 1991	La Fléche, France
25,000 meters	**1 hr., 13 m., 55.8 s.**	Toshihiko Seko	Japan	Mar. 22, 1981	Christchurch, NZ
3,000 meter stpl.	**7 m., 55.28 s.**	Brahim Boulami	Morocco	Aug. 8, 2001	Brussels, Belgium
Marathon**	**2 hr., 5m., 42 s.**	Khalid Khannouchi	Morocco	Oct. 24, 1999	Chicago, IL

Hurdles

110 meters	**12.91 s.**	Colin Jackson	Gr. Britain	Aug. 20, 1993	Stuttgart, Germany
400 meters	**46.78 s.**	Kevin Young	U.S.	Aug. 6, 1992	Barcelona, Spain

Relay Races

400 mtrs. (4x100)	**37.40 s.**	(Marsh, Burrell, Mitchell, Lewis)	U.S.	Aug. 8, 1992	Barcelona, Spain
		(Drummond, Cason, Mitchell, Burrell)	U.S.	Aug. 21, 1993	Stuttgart, Germany
800 mtrs. (4×200)	**1 m., 18.68 s.**	(Marsh, Burrell, Heard, Lewis)	U.S.	Apr. 17, 1994	Walnut, CA
1,600 mtrs. (4×400)	**2 m., 54.20 s.**	(Young, Pettigrew, Washington, Johnson)	U.S.	July 22, 1998	Long Island, NY
3,200 mtrs. (4×800)	**7 m., 03.89 s.**	(Elliott, Cook, Cram, Coe)	Gr. Britain	Aug. 30, 1982	London, England

Field Events

High jump	**2.45m** (8' ½")	Javier Sotomayor	Cuba	July 27, 1993	Salamanca, Spain
Long jump	**8.95m** (29' 4½")	Mike Powell	U.S.	Aug. 30, 1991	Tokyo, Japan
Triple jump	**18.29m** (60' ¼")	Jonathan Edwards	Gr. Britain	Aug. 7, 1995	Göteborg, Sweden
Pole vault	**6.14m** (20' 1¾")	Sergei Bubka	Ukraine	July 31, 1994	Sestriere, Italy
16-lb. shot put	**23.12m** (75' 10¼")	Randy Barnes	U.S.	May 20, 1990	Los Angeles, CA
Discus	**74.08m** (243' 0")	Juergen Schult	E. Germany	June 6, 1986	Neubrandenburg, Germany
Javelin	**98.48m** (323' 1")	Jan Zelezny	Czech Rep.	May 25, 1996	Jena, Germany
16-lb. hammer	**86.74m** (284' 7")	Yuri Sedykh	USSR	Aug. 30, 1986	Stuttgart, W. Germany
Decathlon	**9,026 pts.**	Roman Šebrle	Czech Rep.	May 27, 2001	Götzis, Austria

Women's Records

Running

Event	Record	Holder	Country	Date	Where made
100 meters	**10.49 s.**	Florence Griffith Joyner	U.S.	July 16, 1988	Indianapolis, IN
200 meters	**21.34 s.**	Florence Griffith Joyner	U.S.	Sept. 29, 1988	Seoul, S. Korea
400 meters	**47.60 s.**	Marita Koch	E. Germany	Oct. 6, 1985	Canberra, Australia
800 meters	**1 m., 53.28 s.**	Jarmila Kratochvilova	Czech Rep.	July 26, 1983	Munich, Germany

Event	Record	Holder	Country	Date	Where made
1,000 meters	**2 m., 28.98 s.**	Svetlana Masterkova	Russia	Aug. 23, 1996	Brussels, Belgium
1,500 meters	**3 m., 50.46 s.**	Qu Yunxia	China	Sept. 11, 1993	Beijing, China
1 mile	**4 m., 12.56 s.**	Svetlana Masterkova	Russia	Aug. 14, 1996	Zurich, Switzerland
2,000 meters	**5 m., 25.36 s.**	Sonia O'Sullivan	Ireland	July 8, 1994	Edinburgh, Scotland
3,000 meters	**8 m., 06.11 s.**	Junxia Wang	China	Sept. 13, 1993	Beijing, China
3,000 meter stpl.	**9 m., 25.31 s.**	Justyna Bak	Poland	July 9, 2001	Nice, France
5,000 meters	**14 m., 28.09 s.**	Bo Jiang	China	Oct. 23, 1997	Shanghai, China
10,000 meters	**29 m., 31.78 s.**	Junxia Wang	China	Sept. 8, 1993	Beijing, China
Marathon**	**2 h., 18 m., 47 s.**	Catherine Ndereba	Kenya	Oct. 7, 2001	Chicago, IL

Hurdles

100 meters	**12.21 s.**	Yordanka Donkova	Bulgaria	Aug. 20, 1988	Stara Zagora, Bulgaria
400 meters	**52.61 s.**	Kim Batten	U.S.	Aug. 11, 1995	Göteborg, Sweden

Relay Races

400 mtrs. (4×100)	**41.37 s.**	(Gladisch, Rieger, Auerswald, Goehr)	E. Germany	Oct. 6, 1985	Canberra, Australia
800 mtrs. (4×200)	**1 m., 28.15 s.**	(Goehr, Mueller, Woeckel, Koch)	E. Germany	Aug. 9, 1980	Jena, E. Germany
1,600 mtrs. (4×400)	**3 m., 15.17 s.**	(Ledovskaya, Nazarova, Pinigina, Bryzgina)	USSR	Oct. 1, 1988	Seoul, S. Korea
3,200 mtrs. (4×800)	**7 m., 50.17 s.**	(Olizarenko, Gurina, Borisova, Podyalovskaya)	USSR	Aug. 5, 1984	Moscow

Field Events

High jump	**2.09m** (6' 10¼")	Stefka Kostadinova	Bulgaria	Aug. 30, 1987	Rome, Italy
Long jump	**7.52m** (24' 8¼")	Galina Chistyakova	USSR	June 11, 1988	Leningrad
Triple jump	**15.50m** (50' 10¼)	Inessa Kravets	Ukraine	Aug. 10, 1995	Göteborg, Sweden
Pole vault	**4.81m** (15' 9¼")	Stacy Dragila	U.S.	June 9, 2001	Palo Alto, CA
Shot put	**22.63m** (74' 3")	Natalya Lisovskaya	USSR	June 7, 1987	Moscow, Russia
Discus	**76.80m** (252' 0")	Gabriele Reinsch	E. Germany	July 9, 1988	Neubrandenburg, Germany
Hammer	**76.07m** (249' 7")	Mihaela Melinte	Romania	Aug. 29, 1999	Rüdlingen, Switzerland
Javelin	**71.54m** (234' 8")	Osleidys Menéndez	Cuba	July 1, 2001	Réthymno, Greece
Heptathlon	**7,291 pts.**	Jackie Joyner-Kersee	U.S.	Sept. 23-24, 1988	Seoul, S. Korea

World Track and Field Indoor Records

As of Oct. 10, 2001

The International Amateur Athletic Federation began recognizing world indoor track and field records as official on Jan. 1, 1987. World indoor bests set prior to Jan. 1, 1987, are subject to approval as world records providing they meet the IAAF world records criteria, including drug testing. To be accepted as a world indoor record, a performance must meet the same criteria as a world record outdoors, except that a track performance cannot be set on an indoor track larger than 200 meters. (a)=altitude.

Men's Records

Event	Record	Holder	Country	Date	Where made
50 meters	**5.56** (a)	Donovan Bailey	Canada	Feb. 9, 1996	Reno, NV
	5.56	Maurice Greene	U.S.	Feb. 13, 1999	Los Angeles, CA
60 meters	**6.39**	Maurice Greene	U.S.	Mar. 3, 2001	Atlanta, GA
200 meters	**19.92**	Frankie Fredericks	Namibia	Feb. 18, 1996	Lievin, France
400 meters	**44.63**	Michael Johnson	U.S.	Mar. 4, 1995	Atlanta, GA
800 meters	**1:42.67**	Wilson Kipketer	Denmark	Mar. 9, 1997	Paris, France
1,000 meters	**2:14.96**	Wilson Kipketer	Denmark	Feb. 20, 2000	Birmingham, AL
1,500 meters	**3:31.18**	Hicham el-Guerrouj	Morocco	Feb. 2, 1997	Stuttgart, Germany
1 mile	**3:48.45**	Hicham el-Guerrouj	Morocco	Feb. 12, 1997	Ghent, Belgium
3,000 meters	**7:24.90**	Daniel Komen	Kenya	Feb. 6, 1998	Budapest, Hungary
5,000 meters	**12:50.38**	Haile Gebrselassie	Ethiopia	Feb. 14, 1999	Birmingham, England
50-meter hurdles	**6.25**	Mark McKoy	Canada	Mar. 5, 1986	Kobe, Japan
60-meter hurdles	**7.30**	Colin Jackson	Gr. Britain	Mar. 6, 1994	Sindelfingen, Germany
High jump	**2.43m** (7' 11½")	Javier Sotomayor	Cuba	Mar. 4, 1989	Budapest, Hungary
Pole vault	**6.15m** (20' 2")	Sergei Bubka	Ukraine	Feb. 21, 1993	Donyetsk, Ukraine
Long jump	**8.79m** (28' 10¼")	Carl Lewis	U.S.	Jan. 27, 1984	New York, NY
Triple jump	**17.83** (58' 6")	Aliecer Urrutia	Cuba	Mar. 1, 1997	Sindelfingen, Germany
Shot put	**22.66m** (74' 4¼")	Randy Barnes	U.S.	Jan. 20, 1989	Los Angeles, CA

Women's Records

Event	Record	Holder	Country	Date	Where made
50 meters	**5.96**	Irina Privalova	Russia	Feb. 9, 1995	Madrid, Spain
	6.92	Irina Privalova	Russia	Feb. 9, 1995	Madrid, Spain
60 meters		Irina Privalova	Russia	Feb. 11, 1993	Madrid, Spain
200 meters	**21.87**	Merlene Ottey	Jamaica	Feb. 13, 1993	Lievin, France
400 meters	**49.59**	Jarmila Kratochvilova	Czechoslov.	Mar. 7, 1982	Milan, Italy
800 meters	**1:56.40**	Christine Wachtel	E. Germany	Feb. 13, 1988	Vienna, Austria
1,000 meters	**2:30.94**	Maria Mutola	Mozambique	Feb. 25, 1999	Stockholm, Sweden
1,500 meters	**4:00.27**	Doina Melinte	Romania	Feb. 9, 1990	E. Rutherford, NJ
1 mile	**4:17.14**	Doina Melinte	Romania	Feb. 9, 1990	E. Rutherford, NJ
3,000 meters	**8:32.88**	Gabriela Szabo	Romania	Feb. 18, 2001	Birmingham, AL
5,000 meters	**14:47.35**	Gabriela Szabo	Romania	Feb. 13, 1999	Dortmund, Germany
50-meter hurdles	**6.58**	Cornelia Oschkenat	E. Germany	Feb. 20, 1988	Berlin, Germany
60-meter hurdles	**7.69**	Lyudmila Engquist	USSR	Feb. 4, 1990	Chelyabinsk, USSR
High jump	**2.07m** (6' 9½")	Heike Henkel	Germany	Feb. 8, 1992	Karlsruhe, Germany
Pole vault	**4.70m** (15' 5")	Stacy Dragila	U.S.	Feb. 17, 2001	Pocatello, ID
Long jump	**7.37m** (24' 2¼")	Heike Drechsler	E. Germany	Feb. 13, 1988	Vienna, Austria
Triple jump	**15.16m** (49' 9")	Ashia Hansen	Gr. Britain	Feb. 28, 1998	Valencia, Spain
Shot put	**22.50m** (73' 10")	Helena Fibingerova	Czechoslovakia	Feb. 19, 1977	Jablonec, Czech Rep.

BASEBALL

Seattle Wins 116 Games; Bonds Cranks 73 Homers; Post-Season Delayed

The Seattle Mariners tied the 1906 Cubs' Major League record by winning 116 games on their way to the AL West title. Manager Lou Piniella's Mariners—without superstar shortstop Alex Rodriguez who had signed a record $252 million, 10-year deal with Texas in the off season—were spurred by OF Ichiro Suzuki, a 7-time Japanese League batting champ. The Seattle "rookie" was the AL batting champ (.350), and led the majors in hits (242) and stolen bases (56). Giants OF Barry Bonds, who hit 73 homers to break Mark McGwire's 3-year-old home run record of 70, set a number of offensive marks (see p. 927). Padres OF Rickey Henderson broke Ty Cobb's career run record and Ruth's career walk record, finishing the season with 2,248 and 2,141, respectively. He notched the 3,000th hit of his 23-year career on the season's last day (Oct. 7). That day also saw the final game for the Padres' Tony Gwynn. The Orioles' Cal Ripken Jr. played his last game, at home, on Oct. 6. With 64, the Cubs' Sammy Sosa hit more than 60 homers a record 3d time. The individual achievements were notable in a year in which offensive numbers dipped slightly (and the strike zone rose slightly) from the previous year. Strikeouts were up 3.3% (the Brewers struck out a record 1,399 times). Home runs dropped to 5,458, from a league record 5,693 in 2000. The season ended a week later than scheduled owing to 6 days (92 games) of postponements after the Sept. 11 terror attacks.

Major League Pennant Winners, 1901–1968

National League						American League					
Year	Winner	Won	Lost	Pct	Manager	Year	Winner	Won	Lost	Pct	Manager
1901	Pittsburgh	90	49	.647	Clarke	1901	Chicago	83	53	.610	Griffith
1902	Pittsburgh	103	36	.741	Clarke	1902	Philadelphia	83	53	.610	Mack
1903	Pittsburgh	91	49	.650	Clarke	1903	Boston	91	47	.659	Collins
1904	New York	106	47	.693	McGraw	1904	Boston	95	59	.617	Collins
1905	New York	105	48	.686	McGraw	1905	Philadelphia	92	56	.622	Mack
1906	Chicago	116	36	.763	Chance	1906	Chicago	93	58	.616	Jones
1907	Chicago	107	45	.704	Chance	1907	Detroit	92	58	.613	Jennings
1908	Chicago	99	55	.643	Chance	1908	Detroit	90	63	.588	Jennings
1909	Pittsburgh	110	42	.724	Clarke	1909	Detroit	98	54	.645	Jennings
1910	Chicago	104	50	.675	Chance	1910	Philadelphia	102	48	.680	Mack
1911	New York	99	54	.647	McGraw	1911	Philadelphia	101	50	.669	Mack
1912	New York	103	48	.682	McGraw	1912	Boston	105	47	.691	Stahl
1913	New York	101	51	.664	McGraw	1913	Philadelphia	96	57	.627	Mack
1914	Boston	94	59	.614	Stallings	1914	Philadelphia	99	53	.651	Mack
1915	Philadelphia	90	62	.592	Moran	1915	Boston	101	50	.669	Carrigan
1916	Brooklyn	94	60	.610	Robinson	1916	Boston	91	63	.591	Carrigan
1917	New York	98	56	.636	McGraw	1917	Chicago	100	54	.649	Rowland
1918	Chicago	84	45	.651	Mitchell	1918	Boston	75	51	.595	Barrow
1919	Cincinnati	96	44	.686	Moran	1919	Chicago	88	52	.629	Gleason
1920	Brooklyn	93	60	.604	Robinson	1920	Cleveland	98	56	.636	Speaker
1921	New York	94	56	.614	McGraw	1921	New York	98	55	.641	Huggins
1922	New York	93	61	.604	McGraw	1922	New York	94	60	.610	Huggins
1923	New York	95	58	.621	McGraw	1923	New York	98	54	.645	Huggins
1924	New York	93	60	.608	McGraw	1924	Washington	92	62	.597	Harris
1925	Pittsburgh	95	58	.621	McKechnie	1925	Washington	96	55	.636	Harris
1926	St. Louis	89	65	.578	Hornsby	1926	New York	91	63	.591	Huggins
1927	Pittsburgh	94	60	.610	Bush	1927	New York	110	44	.714	Huggins
1928	St. Louis	95	59	.617	McKechnie	1928	New York	101	53	.656	Huggins
1929	Chicago	98	54	.645	McCarthy	1929	Philadelphia	104	46	.693	Mack
1930	St. Louis	92	62	.597	Street	1930	Philadelphia	102	52	.662	Mack
1931	St. Louis	101	53	.656	Street	1931	Philadelphia	107	45	.704	Mack
1932	Chicago	90	64	.584	Grimm	1932	New York	107	47	.695	McCarthy
1933	New York	91	61	.599	Terry	1933	Washington	99	53	.651	Cronin
1934	St. Louis	95	58	.621	Frisch	1934	Detroit	101	53	.656	Cochrane
1935	Chicago	100	54	.649	Grimm	1935	Detroit	93	58	.616	Cochrane
1936	New York	91	62	.597	Terry	1936	New York	102	51	.667	McCarthy
1937	New York	95	57	.625	Terry	1937	New York	102	52	.662	McCarthy
1938	Chicago	89	63	.586	Hartnett	1938	New York	99	53	.651	McCarthy
1939	Cincinnati	97	57	.630	McKechnie	1939	New York	106	45	.702	McCarthy
1940	Cincinnati	100	53	.654	McKechnie	1940	Detroit	90	64	.584	Baker
1941	Brooklyn	100	54	.649	Durocher	1941	New York	101	53	.656	McCarthy
1942	St. Louis	106	48	.688	Southworth	1942	New York	103	51	.669	McCarthy
1943	St. Louis	105	49	.682	Southworth	1943	New York	98	56	.636	McCarthy
1944	St. Louis	105	49	.682	Southworth	1944	St. Louis	89	65	.578	Sewell
1945	Chicago	98	56	.636	Grimm	1945	Detroit	88	65	.575	O'Neill
1946	St. Louis	98	58	.628	Dyer	1946	Boston	104	50	.675	Cronin
1947	Brooklyn	94	60	.610	Shotton	1947	New York	97	57	.630	Harris
1948	Boston	91	62	.595	Southworth	1948	Cleveland	97	58	.626	Boudreau
1949	Brooklyn	97	57	.630	Shotton	1949	New York	97	57	.630	Stengel
1950	Philadelphia	91	63	.591	Sawyer	1950	New York	98	56	.636	Stengel
1951	New York	98	59	.624	Durocher	1951	New York	98	56	.636	Stengel
1952	Brooklyn	96	57	.627	Dressen	1952	New York	95	59	.617	Stengel
1953	Brooklyn	105	49	.682	Dressen	1953	New York	99	52	.656	Stengel
1954	New York	97	57	.630	Durocher	1954	Cleveland	111	43	.721	Lopez
1955	Brooklyn	98	55	.641	Alston	1955	New York	96	58	.623	Stengel
1956	Brooklyn	93	61	.604	Alston	1956	New York	97	57	.630	Stengel
1957	Milwaukee	95	59	.617	Haney	1957	New York	98	56	.636	Stengel
1958	Milwaukee	92	62	.597	Haney	1958	New York	92	62	.597	Stengel
1959	Los Angeles	88	68	.564	Alston	1959	Chicago	94	60	.610	Lopez
1960	Pittsburgh	95	59	.617	Murtaugh	1960	New York	97	57	.630	Stengel
1961	Cincinnati	93	61	.604	Hutchinson	1961	New York	109	53	.673	Houk
1962	San Francisco	103	62	.624	Dark	1962	New York	96	66	.593	Houk
1963	Los Angeles	99	63	.611	Alston	1963	New York	104	57	.646	Houk
1964	St. Louis	93	69	.574	Keane	1964	New York	99	63	.611	Berra
1965	Los Angeles	97	65	.599	Alston	1965	Minnesota	102	60	.630	Mele
1966	Los Angeles	95	67	.586	Alston	1966	Baltimore	97	63	.606	Bauer
1967	St. Louis	101	60	.627	Schoendienst	1967	Boston	92	70	.568	Williams
1968	St. Louis	97	65	.599	Schoendienst	1968	Detroit	103	59	.636	Smith

Major League Pennant Winners, 1969-2000

National League

	East					West					Pennant
Year	Winner	W	L	Pct	Manager	Winner	W	L	Pct	Manager	Winner
1969	N.Y. Mets	100	62	.617	Hodges	Atlanta	93	69	.574	Harris	New York
1970	Pittsburgh	89	73	.549	Murtaugh	Cincinnati	102	60	.630	Anderson	Cincinnati
1971	Pittsburgh	97	65	.599	Murtaugh	San Francisco	90	72	.556	Fox	Pittsburgh
1972	Pittsburgh	96	59	.619	Virdon	Cincinnati	95	59	.617	Anderson	Cincinnati
1973	N.Y. Mets	82	79	.509	Berra	Cincinnati	99	63	.611	Anderson	New York
1974	Pittsburgh	88	74	.543	Murtaugh	Los Angeles	102	60	.630	Alston	Los Angeles
1975	Pittsburgh	92	69	.571	Murtaugh	Cincinnati	108	54	.667	Anderson	Cincinnati
1976	Philadelphia	101	61	.623	Ozark	Cincinnati	102	60	.630	Anderson	Cincinnati
1977	Philadelphia	101	61	.623	Ozark	Los Angeles	98	64	.605	Lasorda	Los Angeles
1978	Philadelphia	90	72	.556	Ozark	Los Angeles	95	67	.586	Lasorda	Los Angeles
1979	Pittsburgh	98	64	.605	Tanner	Cincinnati	90	71	.559	McNamara	Pittsburgh
1980	Philadelphia	91	71	.562	Green	Houston	93	70	.571	Virdon	Philadelphia
1981(a)	Philadelphia	34	21	.618	Green	Los Angeles	36	21	.632	Lasorda	(c)
1981(b)	Montreal	30	23	.566	Williams, Fanning	Houston	33	20	.623	Virdon	Los Angeles
1982	St. Louis	92	70	.568	Herzog	Atlanta	89	73	.549	Torre	St. Louis
1983	Philadelphia	90	72	.556	Corrales, Owens	Los Angeles	91	71	.562	Lasorda	Philadelphia
1984	Chicago	96	65	.596	Frey	San Diego	92	70	.568	Williams	San Diego
1985	St. Louis	101	61	.623	Herzog	Los Angeles	95	67	.586	Lasorda	St. Louis
1986	N.Y. Mets	108	54	.667	Johnson	Houston	96	66	.593	Lanier	New York
1987	St. Louis	95	67	.586	Herzog	San Francisco	90	72	.556	Craig	St. Louis
1988	N.Y. Mets	100	60	.625	Johnson	Los Angeles	94	67	.584	Lasorda	Los Angeles
1989	Chicago	93	69	.571	Zimmer	San Francisco	92	70	.568	Craig	San Francisco
1990	Pittsburgh	95	67	.586	Leyland	Cincinnati	91	71	.562	Piniella	Cincinnati
1991	Pittsburgh	98	64	.605	Leyland	Atlanta	94	68	.580	Cox	Atlanta
1992	Pittsburgh	96	66	.593	Leyland	Atlanta	98	64	.605	Cox	Atlanta
1993	Philadelphia	97	65	.599	Fregosi	Atlanta	104	58	.642	Cox	Philadelphia

Year	Division	Winner	W	L	Pct	Manager	Playoffs	Pennant Winner
1994(d)	East	Montreal	74	40	.649	Alou	—	—
	West	Cincinnati	66	48	.579	Johnson		
	Central	Los Angeles	58	56	.509	Lasorda		
1995	East	Atlanta	90	54	.625	Cox	Atlanta 3, Colorado* 1	Atlanta
	Central	Cincinnati	85	59	.590	Johnson	Cincinnati 3, Los Angeles 0	
	West	Los Angeles	78	66	.542	Lasorda	Atlanta 4, Cincinnati 0	
1996	East	Atlanta	96	66	.593	Cox	Atlanta 3, Los Angeles* 0	Atlanta
	Central	St. Louis	88	74	.543	La Russa	St. Louis 3, San Diego 0	
	West	San Diego	91	71	.562	Bochy	Atlanta 4, St. Louis 3	
1997	East	Atlanta	101	61	.623	Cox	Atlanta 3, Houston 0	Florida* (e)
	Central	Houston	84	78	.519	Dierker	Florida* 3, San Francisco 0	
	West	San Francisco	90	72	.556	Baker	Florida* 4, Atlanta 2	
1998	East	Atlanta	106	56	.654	Cox	Atlanta 3, Chicago* 0	San Diego
	Central	Houston	102	60	.630	Dierker	San Diego 3, Houston 1	
	West	San Diego	97	64	.602	Bochy	San Diego 4, Atlanta 2	
1999	East	Atlanta	103	59	.636	Cox	Atlanta 3, Houston 1	Atlanta
	Central	Houston	97	65	.599	Dierker	New York* 3, Arizona 1	
	West	Arizona	100	62	.617	Showalter	Atlanta 4, New York 2	
2000	East	Atlanta	95	67	.586	Cox	St. Louis 3, Atlanta 0	New York* (f)
	Central	St. Louis	95	67	.586	La Russa	New York* 3, San Francisco 1	
	West	San Francisco	97	65	.599	Baker	New York* 4, St. Louis 1	

American League

	East					West					Pennant
Year	Winner	W	L	Pct	Manager	Winner	W	L	Pct	Manager	Winner
1969	Baltimore	109	53	.673	Weaver	Minnesota	97	65	.599	Martin	Baltimore
1970	Baltimore	108	54	.667	Weaver	Minnesota	98	64	.605	Rigney	Baltimore
1971	Baltimore	101	57	.639	Weaver	Oakland	101	60	.627	Williams	Baltimore
1972	Detroit	86	70	.551	Martin	Oakland	93	62	.600	Williams	Oakland
1973	Baltimore	97	65	.599	Weaver	Oakland	94	68	.580	Williams	Oakland
1974	Baltimore	91	71	.562	Weaver	Oakland	90	72	.556	Dark	Oakland
1975	Boston	95	65	.594	Johnson	Oakland	98	64	.605	Dark	Boston
1976	New York	97	62	.610	Martin	Kansas City	90	72	.556	Herzog	New York
1977	New York	100	62	.617	Martin	Kansas City	102	60	.630	Herzog	New York
1978	New York	100	63	.613	Martin, Lemon	Kansas City	92	70	.568	Herzog	New York
1979	Baltimore	102	57	.642	Weaver	California	88	74	.543	Fregosi	Baltimore
1980	New York	103	59	.636	Howser	Kansas City	97	65	.599	Frey	Kansas City
1981(a)	New York	34	22	.607	Michael	Oakland	37	23	.617	Martin	(c)
1981(b)	Milwaukee	31	22	.585	Rodgers	Kansas City	30	23	.566	Frey, Howser	New York
1982	Milwaukee	95	67	.586	Rodgers, Kuenn	California	93	69	.574	Mauch	Milwaukee
1983	Baltimore	98	64	.605	Altobelli	Chicago	99	63	.611	La Russa	Baltimore
1984	Detroit	104	58	.642	Anderson	Kansas City	84	78	.519	Howser	Detroit
1985	Toronto	99	62	.615	Cox	Kansas City	91	71	.562	Howser	Kansas City
1986	Boston	95	66	.590	McNamara	California	92	70	.568	Mauch	Boston
1987	Detroit	98	64	.605	Anderson	Minnesota	85	77	.525	Kelly	Minnesota
1988	Boston	89	73	.549	McNamara, Morgan	Oakland	104	58	.642	La Russa	Oakland
1989	Toronto	89	73	.549	Williams, Gaston	Oakland	99	63	.611	La Russa	Oakland
1990	Boston	88	74	.543	Morgan	Oakland	103	59	.636	La Russa	Oakland
1991	Toronto	91	71	.562	Gaston	Minnesota	95	67	.586	Kelly	Minnesota
1992	Toronto	96	66	.593	Gaston	Oakland	96	66	.593	La Russa	Toronto
1993	Toronto	95	67	.586	Gaston	Chicago	94	68	.580	Lamont	Toronto

Year	Division	Winner	W	L	Pct	Manager	Playoffs	Pennant Winner
1994(d)	East	New York	70	43	.619	Showalter	—	—
	Central	Chicago	67	46	.593	Lamont		
	West	Texas	52	62	.456	Kennedy		

1995	East	Boston	86	58	.597	Kennedy	Cleveland 3, Boston 0	Cleveland
	Central	Cleveland	100	44	.694	Hargrove	Seattle 3, New York* 2	
	West	Seattle	79	66	.545	Piniella	Cleveland 4, Seattle 2	
1996	East	New York	92	70	.568	Torre	Baltimore* 3, Cleveland 1	New York
	Central	Cleveland	99	62	.615	Hargrove	New York 3, Texas 1	
	West	Texas	90	72	.556	Oates	New York 4, Baltimore* 1	
1997	East	Baltimore	98	64	.605	Johnson	Baltimore 3, Seattle 1	Cleveland
	Central	Cleveland	86	75	.534	Hargrove	Cleveland 3, New York* 2	
	West	Seattle	90	72	.556	Piniella	Cleveland 4, Baltimore 2	
1998	East	New York	114	48	.704	Torre	New York 3, Texas 0	New York
	Central	Cleveland	89	73	.549	Hargrove	Cleveland 3, Boston* 1	
	West	Texas	88	74	.543	Oates	New York 4, Cleveland 2	
1999	East	New York	98	64	.605	Torre	New York 3, Texas 0	New York
	Central	Cleveland	97	65	.599	Hargrove	Boston* 3, Cleveland 2	
	West	Texas	95	67	.586	Oates	New York 4, Boston* 1	
2000	East	New York	87	74	.540	Torre	New York 3, Oakland 2	New York
	Central	Chicago	95	67	.586	Manuel	Seattle* 3, Chicago 0	
	West	Oakland	91	70	.565	Howe	New York 4, Seattle* 21	

*Wild card team. (a) First half. (b) Second half. (c) Montreal, L.A., N.Y. Yankees, and Oakland won the divisional playoffs. (d) In Aug. 1994, a players' strike began that caused the cancellation of the remainder of the season, the playoffs, and the World Series. Teams listed as division "winners" for 1994 were leading their divisions at the time of the strike. (e) Florida manager: Jim Leyland. (f) New York manager Bobby Valentine.

The Rawlings Gold Glove Awards in 2000

National League

Greg Maddux, Atlanta, p
Mike Matheny, St. Louis, c
J. T. Snow, San Francisco, 1b
Pokey Reese, Cincinnati, 2b
Scott Rolen, Philadelphia, 3b
Neifi Perez, Colorado, ss
Jim Edmonds, St. Louis, of
Steve Finley, Arizona, of
Andruw Jones, Atlanta, of

American League

Kenny Rogers, Texas, p
Ivan Rodriguez, Texas, c
John Olerud, Seattle, 1b
Roberto Alomar, Cleveland, 2b
Travis Fryman, Cleveland, 3b
Omar Vizquel, Cleveland, ss
Jermaine Dye, Kansas City, of
Darin Erstad, Anaheim, of
Bernie Williams, N.Y. of

The following are the players at each position who have won the most Gold Gloves since the award was instituted in 1957.

Pitcher:	Jim Kaat	16
	Greg Maddux	11
Catcher:	Johnny Bench	10
	Ivan Rodriguez	9
First base:	Keith Hernandez	11
	Don Mattingly	9
Second base:	Ryne Sandberg	9
	Roberto Alomar	9
	Bill Mazeroski	8
	Frank White	8
Third base:	Brooks Robinson	16
	Mike Schmidt	10
Shortstop:	Ozzie Smith	13
	Luis Aparicio	9
Outfield:	Roberto Clemente	12
	Willie Mays	12
	Al Kaline	10
	Ken Griffey Jr.	10

Home Run Leaders

Note: Asterisk (*) indicates the all-time single-season record for each league.

National League			American League		
Year	Player, Team	HR	Year	Player, Team	HR
1901	Sam Crawford, Cincinnati	16	1901	Napoleon Lajoie, Philadelphia	13
1902	Thomas Leach, Pittsburgh	6	1902	Socks Seybold, Philadelphia	16
1903	James Sheckard, Brooklyn	9	1903	Buck Freeman, Boston	13
1904	Harry Lumley, Brooklyn	9	1904	Harry Davis, Philadelphia	10
1905	Fred Odwell, Cincinnati	9	1905	Harry Davis, Philadelphia	8
1906	Timothy Jordan, Brooklyn	12	1906	Harry Davis, Philadelphia	12
1907	David Brain, Boston	10	1907	Harry Davis, Philadelphia	8
1908	Timothy Jordan, Brooklyn	12	1908	Sam Crawford, Detroit	7
1909	Red Murray, New York	7	1909	Ty Cobb, Detroit	9
1910	Fred Beck, Boston; Frank Schulte, Chicago	10	1910	Jake Stahl, Boston	10
1911	Frank Schulte, Chicago	21	1911	J. Franklin Baker, Philadelphia	9
1912	Henry Zimmerman, Chicago	14	1912	J. Franklin Baker, Philadelphia; Tris Speaker, Boston	10
1913	Gavvy Cravath, Philadelphia	19	1913	J. Franklin Baker, Philadelphia	13
1914	Gavvy Cravath, Philadelphia	19	1914	J. Franklin Baker, Philadelphia	9
1915	Gavvy Cravath, Philadelphia	24	1915	Robert Roth, Chicago-Cleveland	7
1916	Dave Robertson, N.Y.; Fred (Cy) Williams, Chi.	12	1916	Wally Pipp, New York	12
1917	Dave Robertson, N.Y.; Gavvy Cravath, Phi.	12	1917	Wally Pipp, New York	9
1918	Gavvy Cravath, Philadelphia	8	1918	Babe Ruth, Boston; Tilly Walker, Philadelphia	11
1919	Gavvy Cravath, Philadelphia	12	1919	Babe Ruth, Boston	29
1920	Cy Williams, Philadelphia	15	1920	Babe Ruth, New York	54
1921	George Kelly, New York	23	1921	Babe Ruth, New York	59
1922	Rogers Hornsby, St. Louis	42	1922	Ken Williams, St. Louis	39
1923	Cy Williams, Philadelphia	41	1923	Babe Ruth, New York	41
1924	Jacques Fournier, Brooklyn	27	1924	Babe Ruth, New York	46
1925	Rogers Hornsby, St. Louis	39	1925	Bob Meusel, New York	33
1926	Hack Wilson, Chicago	21	1926	Babe Ruth, New York	47
1927	Hack Wilson, Chicago; Cy Williams, Philadelphia	30	1927	Babe Ruth, New York	60
1928	Hack Wilson, Chicago; Jim Bottomley, St. Louis	31	1928	Babe Ruth, New York	54
1929	Chuck Klein, Philadelphia	43	1929	Babe Ruth, New York	46
1930	Hack Wilson, Chicago	56	1930	Babe Ruth, New York	49
1931	Chuck Klein, Philadelphia	31	1931	Babe Ruth, Lou Gehrig, both New York	46
1932	Chuck Klein, Philadelphia; Mel Ott, New York	38	1932	Jimmie Foxx, Philadelphia	58
1933	Chuck Klein, Philadelphia	28	1933	Jimmie Foxx, Philadelphia	48
1934	Rip Collins, St. Louis; Mel Ott, New York	35	1934	Lou Gehrig, New York	49
1935	Walter Berger, Boston	34	1935	Jimmie Foxx, Philadelphia; Hank Greenberg, Detroit	36
1936	Mel Ott, New York	33	1936	Lou Gehrig, New York	49
1937	Mel Ott, New York; Joe Medwick, St. Louis	31	1937	Joe DiMaggio, New York	46
1938	Mel Ott, New York	36	1938	Hank Greenberg, Detroit	58
1939	John Mize, St. Louis	28	1939	Jimmie Foxx, Boston	35
1940	John Mize, St. Louis	43	1940	Hank Greenberg, Detroit	41
1941	Dolph Camilli, Brooklyn	34	1941	Ted Williams, Boston	37
1942	Mel Ott, New York	30	1942	Ted Williams, Boston	36
1943	Bill Nicholson, Chicago	29	1943	Rudy York, Detroit	34
1944	Bill Nicholson, Chicago	33	1944	Nick Etten, New York	22
1945	Tommy Holmes, Boston	28	1945	Vern Stephens, St. Louis	24
1946	Ralph Kiner, Pittsburgh	23	1946	Hank Greenberg, Detroit	44

National League			American League		
Year	Player, Team	HR	Year	Player, Team	HR
1947	Ralph Kiner, Pittsburgh; John Mize, New York	51	1947	Ted Williams, Boston	32
1948	Ralph Kiner, Pittsburgh; John Mize, New York	40	1948	Joe DiMaggio, New York	39
1949	Ralph Kiner, Pittsburgh	54	1949	Ted Williams, Boston	43
1950	Ralph Kiner, Pittsburgh	47	1950	Al Rosen, Cleveland	37
1951	Ralph Kiner, Pittsburgh	42	1951	Gus Zernial, Chicago-Philadelphia	33
1952	Ralph Kiner, Pittsburgh; Hank Sauer, Chicago	37	1952	Larry Doby, Cleveland	32
1953	Ed Mathews, Milwaukee	47	1953	Al Rosen, Cleveland	43
1954	Ted Kluszewski, Cincinnati	49	1954	Larry Doby, Cleveland	32
1955	Willie Mays, New York	51	1955	Mickey Mantle, New York	37
1956	Duke Snider, Brooklyn	43	1956	Mickey Mantle, New York	52
1957	Hank Aaron, Milwaukee	44	1957	Roy Sievers, Washington	42
1958	Ernie Banks, Chicago	47	1958	Mickey Mantle, New York	42
1959	Ed Mathews, Milwaukee	46	1959	Rocky Colavito, Cleve.; Harmon Killebrew, Wash.	42
1960	Ernie Banks, Chicago	41	1960	Mickey Mantle, New York	40
1961	Orlando Cepeda, San Francisco	46	1961	Roger Maris, New York	*61
1962	Willie Mays, San Francisco	49	1962	Harmon Killebrew, Minnesota	48
1963	Hank Aaron, Milwaukee; Willie McCovey, S.F.	44	1963	Harmon Killebrew, Minnesota	45
1964	Willie Mays, San Francisco	47	1964	Harmon Killebrew, Minnesota	49
1965	Willie Mays, San Francisco	52	1965	Tony Conigliaro, Boston	32
1966	Hank Aaron, Atlanta	44	1966	Frank Robinson, Baltimore	49
1967	Hank Aaron, Atlanta	39	1967	Carl Yastrzemski, Boston; Harmon Killebrew, Minn.	44
1968	Willie McCovey, San Francisco	36	1968	Frank Howard, Washington	44
1969	Willie McCovey, San Francisco	45	1969	Harmon Killebrew, Minnesota	49
1970	Johnny Bench, Cincinnati	45	1970	Frank Howard, Washington	44
1971	Willie Stargell, Pittsburgh	48	1971	Bill Melton, Chicago	33
1972	Johnny Bench, Cincinnati	40	1972	Dick Allen, Chicago	37
1973	Willie Stargell, Pittsburgh	44	1973	Reggie Jackson, Oakland	32
1974	Mike Schmidt, Philadelphia	36	1974	Dick Allen, Chicago	32
1975	Mike Schmidt, Philadelphia	38	1975	George Scott, Milwaukee; Reggie Jackson, Oakland	36
1976	Mike Schmidt, Philadelphia	38	1976	Graig Nettles, New York	32
1977	George Foster, Cincinnati	52	1977	Jim Rice, Boston	39
1978	George Foster, Cincinnati	40	1978	Jim Rice, Boston	46
1979	Dave Kingman, Chicago	48	1979	Gorman Thomas, Milwaukee	45
1980	Mike Schmidt, Philadelphia	48	1980	Reggie Jackson, New York; Ben Oglivie, Milwaukee	41
1981	Mike Schmidt, Philadelphia	31	1981	Bobby Grich, California; Tony Armas, Oakland; Dwight Evans, Boston; Eddie Murray, Baltimore	22
1982	Dave Kingman, New York	37	1982	Gorman Thomas, Milwaukee; Reggie Jackson, Cal.	39
1983	Mike Schmidt, Philadelphia	40	1983	Jim Rice, Boston	39
1984	Mike Schmidt, Phi.; Dale Murphy, Atlanta	36	1984	Tony Armas, Boston	43
1985	Dale Murphy, Atlanta	37	1985	Darrell Evans, Detroit	40
1986	Mike Schmidt, Philadelphia	37	1986	Jesse Barfield, Toronto	40
1987	Andre Dawson, Chicago	49	1987	Mark McGwire, Oakland	49
1988	Darryl Strawberry, New York	39	1988	Jose Canseco, Oakland	42
1989	Kevin Mitchell, San Francisco	47	1989	Fred McGriff, Toronto	36
1990	Ryne Sandberg, Chicago	40	1990	Cecil Fielder, Detroit	51
1991	Howard Johnson, New York	38	1991	Cecil Fielder, Detroit; Jose Canseco, Oakland	44
1992	Fred McGriff, San Diego	35	1992	Juan Gonzalez, Texas	43
1993	Barry Bonds, San Francisco	46	1993	Juan Gonzalez, Texas	46
1994	Matt Williams, San Francisco	43	1994	Ken Griffey Jr., Seattle	40
1995	Dante Bichette, Colorado	40	1995	Albert Belle, Cleveland	50
1996	Andres Galarraga, Colorado	47	1996	Mark McGwire, Oakland	52
1997[1]	Larry Walker, Colorado	49	1997[1]	Ken Griffey Jr., Seattle	56
1998	Mark McGwire, St. Louis	70	1998	Ken Griffey Jr., Seattle	56
1999	Mark McGwire, St. Louis	65	1999	Ken Griffey Jr., Seattle	48
2000	Sammy Sosa, Chicago	50	2000	Troy Glaus, Anaheim	47
2001	Barry Bonds, San Francisco	*73	2001	Alex Rodriguez, Texas	52

(1) In 1997, Mark McGwire hit 58 home runs; 34 with the Oakland Athletics (AL) and 24 with the St. Louis Cardinals (NL).

Runs Batted In Leaders

Note: Asterisk (*) indicates the all-time single-season record for each league since beginning of "modern" era in 1901.

National League			American League		
Year	Player, Team	RBI	Year	Player, Team	RBI
1907	Sherwood Magee, Philadelphia	85	1907	Ty Cobb, Detroit	116
1908	Honus Wagner, Pittsburgh	109	1908	Ty Cobb, Detroit	108
1909	Honus Wagner, Pittsburgh	100	1909	Ty Cobb, Detroit	107
1910	Sherwood Magee, Philadelphia	123	1910	Sam Crawford, Detroit	120
1911	Frank Schulte, Chicago	121	1911	Ty Cobb, Detroit	144
1912	Henry Zimmerman, Chicago	103	1912	J. Franklin Baker, Philadelphia	133
1913	Gavvy Cravath, Philadelphia	128	1913	J. Franklin Baker, Philadelphia	126
1914	Sherwood Magee, Philadelphia	103	1914	Sam Crawford, Detroit	104
1915	Gavvy Cravath, Philadelphia	115	1915	Sam Crawford, Detroit; Robert Veach, Detroit	112
1916	Henry Zimmerman, Chicago-NewYork	83	1916	Del Pratt, St. Louis	103
1917	Henry Zimmerman, New York	102	1917	Robert Veach, Detroit	103
1918	Sherwood Magee, Philadelphia	76	1918	Robert Veach, Detroit	78
1919	Hi Myers, Boston	73	1919	Babe Ruth, Boston	114
1920	George Kelly, N.Y.; Rogers Hornsby, St. Louis	94	1920	Babe Ruth, New York	137
1921	Rogers Hornsby, St. Louis	126	1921	Babe Ruth, New York	171
1922	Rogers Hornsby, St. Louis	152	1922	Ken Williams, St. Louis	155
1923	Emil Meusel, New York	125	1923	Babe Ruth, New York	131
1924	George Kelly, New York	136	1924	Goose Goslin, Washington	129
1925	Rogers Hornsby, St. Louis	143	1925	Bob Meusel, New York	138
1926	Jim Bottomley, St. Louis	120	1926	Babe Ruth, New York	145
1927	Paul Waner, Pittsburgh	131	1927	Lou Gehrig, New York	175
1928	Jim Bottomley, St. Louis	136	1928	Babe Ruth, New York; Lou Gehrig, New York	142
1929	Hack Wilson, Chicago	159	1929	Al Simmons, Philadelphia	157
1930	Hack Wilson, Chicago	*191	1930	Lou Gehrig, New York	174
1931	Chuck Klein, Philadelphia	121	1931	Lou Gehrig, New York	*184
1932	Don Hurst, Philadelphia	143	1932	Jimmie Foxx, Philadelphia	169

National League

Year	Player, Team	RBI
1933	Chuck Klein, Philadelphia	120
1934	Mel Ott, New York	135
1935	Walter Berger, Boston	130
1936	Joe Medwick, St. Louis	138
1937	Joe Medwick, St. Louis	154
1938	Joe Medwick, St. Louis	122
1939	Frank McCormick, Cincinnati	128
1940	John Mize, St. Louis	137
1941	Adolph Camilli, Brooklyn	120
1942	John Mize, New York	110
1943	Bill Nicholson, Chicago	128
1944	Bill Nicholson, Chicago	122
1945	Dixie Walker, Brooklyn	124
1946	Enos Slaughter, St. Louis	130
1947	John Mize, New York	138
1948	Stan Musial, St. Louis	131
1949	Ralph Kiner, Pittsburgh	127
1950	Del Ennis, Philadelphia	126
1951	Monte Irvin, New York	121
1952	Hank Sauer, Chicago	121
1953	Roy Campanella, Brooklyn	142
1954	Ted Kluszewski, Cincinnati	141
1955	Duke Snider, Brooklyn	136
1956	Stan Musial, St. Louis	109
1957	Hank Aaron, Milwaukee	132
1958	Ernie Banks, Chicago	129
1959	Ernie Banks, Chicago	143
1960	Hank Aaron, Milwaukee	126
1961	Orlando Cepeda, San Francisco	142
1962	Tommy Davis, Los Angeles	153
1963	Hank Aaron, Milwaukee	130
1964	Ken Boyer, St. Louis	119
1965	Deron Johnson, Cincinnati	130
1966	Hank Aaron, Atlanta	127
1967	Orlando Cepeda, St. Louis	111
1968	Willie McCovey, San Francisco	105
1969	Willie McCovey, San Francisco	126
1970	Johnny Bench, Cincinnati	148
1971	Joe Torre, St. Louis	137
1972	Johnny Bench, Cincinnati	125
1973	Willie Stargell, Pittsburgh	119
1974	Johnny Bench, Cincinnati	129
1975	Greg Luzinski, Philadelphia	120
1976	George Foster, Cincinnati	121
1977	George Foster, Cincinnati	149
1978	George Foster, Cincinnati	120
1979	Dave Winfield, San Diego	118
1980	Mike Schmidt, Philadelphia	121
1981	Mike Schmidt, Philadelphia	91
1982	Dale Murphy, Atlanta; Al Oliver, Montreal	109
1983	Dale Murphy, Atlanta	121
1984	Gary Carter, Montreal; Mike Schmidt, Phi.	106
1985	Dave Parker, Cincinnati	125
1986	Mike Schmidt, Philadelphia	119
1987	Andre Dawson, Chicago	137
1988	Will Clark, San Francisco	109
1989	Kevin Mitchell, San Francisco	125
1990	Matt Williams, San Francisco	122
1991	Howard Johnson, New York	117
1992	Darren Daulton, Philadelphia	109
1993	Barry Bonds, San Francisco	123
1994	Jeff Bagwell, Houston	116
1995	Dante Bichette, Colorado	128
1996	Andres Galarraga, Colorado	150
1997	Andres Galarraga, Colorado	140
1998	Sammy Sosa, Chicago	158
1999	Mark McGwire, St. Louis	147
2000	Todd Helton, Colorado	147
2001	Sammy Sosa, Chicago	160

American League

Year	Player, Team	RBI
1933	Jimmie Foxx, Philadelphia	163
1934	Lou Gehrig, New York	165
1935	Hank Greenberg, Detroit	170
1936	Hal Trosky, Cleveland	162
1937	Hank Greenberg, Detroit	183
1938	Jimmie Foxx, Boston	175
1939	Ted Williams, Boston	145
1940	Hank Greenberg, Detroit	150
1941	Joe DiMaggio, New York	125
1942	Ted Williams, Boston	137
1943	Rudy York, Detroit	118
1944	Vern Stephens, St. Louis	109
1945	Nick Etten, New York	111
1946	Hank Greenberg, Detroit	127
1947	Ted Williams, Boston	114
1948	Joe DiMaggio, New York	155
1949	Ted Williams, Bos.; Vern Stephens, Bos.	159
1950	Walt Dropo, Bos.; Vern Stephons, Bos.	144
1951	Gus Zernial, Chicago-Philadelphia	129
1952	Al Rosen, Cleveland	105
1953	Al Rosen, Cleveland	145
1954	Larry Doby, Cleveland	126
1955	Ray Boone, Detroit; Jackie Jensen, Boston	116
1956	Mickey Mantle, New York	130
1957	Roy Sievers, Washington	114
1958	Jackie Jensen, Boston	122
1959	Jackie Jensen, Boston	112
1960	Roger Maris, New York	112
1961	Roger Maris, New York	142
1962	Harmon Killebrew, Minnesota	126
1963	Dick Stuart, Boston	118
1964	Brooks Robinson, Baltimore	118
1965	Rocky Colavito, Cleveland	108
1966	Frank Robinson, Baltimore	122
1967	Carl Yastrzemski, Boston	121
1968	Ken Harrelson, Boston	109
1969	Harmon Killebrew, Minnesota	140
1970	Frank Howard, Washington	126
1971	Harmon Killebrew, Minnesota	119
1972	Dick Allen, Chicago	113
1973	Reggie Jackson, Oakland	117
1974	Jeff Burroughs, Texas	118
1975	George Scott, Milwaukee	109
1976	Lee May, Baltimore	109
1977	Larry Hisle, Minnesota	119
1978	Jim Rice, Boston	139
1979	Don Baylor, California	139
1980	Cecil Cooper, Milwaukee	122
1981	Eddie Murray, Baltimore	78
1982	Hal McRae, Kansas City	133
1983	Cecil Cooper, Milwaukee; Jim Rice, Boston	126
1984	Tony Armas, Boston	123
1985	Don Mattingly, New York	145
1986	Joe Carter, Cleveland	121
1987	George Bell, Toronto	134
1988	Jose Canseco, Oakland	124
1989	Ruben Sierra, Texas	119
1990	Cecil Fielder, Detroit	132
1991	Cecil Fielder, Detroit	133
1992	Cecil Fielder, Detroit	124
1993	Albert Belle, Cleveland	129
1994	Kirby Puckett, Minnesota	112
1995	Albert Belle, Cleveland; Mo Vaughn, Boston	126
1996	Albert Belle, Cleveland	148
1997	Ken Griffey Jr., Seattle	147
1998	Juan Gonzalez, Texas	157
1999	Manny Ramirez, Cleveland	165
2000	Edgar Martinez, Seattle	145
2001	Bret Boone, Seattle	141

Batting Champions

Note: Asterisk (*) indicates the all-time single-season record for each league since the beginning of the "modern" era in 1901.

National League

Year	Player	Team	Avg.
1901	Jesse C. Burkett	St. Louis	**.382**
1902	Clarence Beaumont	Pittsburgh	**.357**
1903	Honus Wagner	Pittsburgh	**.355**
1904	Honus Wagner	Pittsburgh	**.349**
1905	James Seymour	Cincinnati	**.377**
1906	Honus Wagner	Pittsburgh	**.339**
1907	Honus Wagner	Pittsburgh	**.350**
1908	Honus Wagner	Pittsburgh	**.354**
1909	Honus Wagner	Pittsburgh	**.339**
1910	Sherwood Magee	Philadelphia	**.331**
1911	Honus Wagner	Pittsburgh	**.334**
1912	Henry Zimmerman	Chicago	**.372**
1913	Jacob Daubert	Brooklyn	**.350**
1914	Jacob Daubert	Brooklyn	**.329**

American League

Year	Player	Team	Avg.
1901	Napoleon Lajoie	Philadelphia	***.426**
1902	Ed Delahanty	Washington	**.376**
1903	Napoleon Lajoie	Cleveland	**.355**
1904	Napoleon Lajoie	Cleveland	**.381**
1905	Elmer Flick	Cleveland	**.306**
1906	George Stone	St. Louis	**.358**
1907	Ty Cobb	Detroit	**.350**
1908	Ty Cobb	Detroit	**.324**
1909	Ty Cobb	Detroit	**.377**
1910[1]	Ty Cobb	Detroit	**.385**
1911	Ty Cobb	Detroit	**.420**
1912	Ty Cobb	Detroit	**.410**
1913	Ty Cobb	Detroit	**.390**
1914	Ty Cobb	Detroit	**.368**

National League				American League			
Year	Player	Team	Avg.	Year	Player	Team	Avg.
1915	Larry Doyle	New York	.320	1915	Ty Cobb	Detroit	.369
1916	Hal Chase	Cincinnati	.339	1916	Tris Speaker	Cleveland	.386
1917	Edd Roush	Cincinnati	.341	1917	Ty Cobb	Detroit	.383
1918	Zach Wheat	Brooklyn	.335	1918	Ty Cobb	Detroit	.382
1919	Edd Roush	Cincinnati	.321	1919	Ty Cobb	Detroit	.384
1920	Rogers Hornsby	St. Louis	.370	1920	George Sisler	St. Louis	.407
1921	Rogers Hornsby	St. Louis	.397	1921	Harry Heilmann	Detroit	.394
1922	Rogers Hornsby	St. Louis	.401	1922	George Sisler	St. Louis	.420
1923	Rogers Hornsby	St. Louis	.384	1923	Harry Heilmann	Detroit	.403
1924	Rogers Hornsby	St. Louis	*.424	1924	Babe Ruth	New York	.378
1925	Rogers Hornsby	St. Louis	.403	1925	Harry Heilmann	Detroit	.393
1926	Eugene Hargrave	Cincinnati	.353	1926	Henry Manush	Detroit	.378
1927	Paul Waner	Pittsburgh	.380	1927	Harry Heilmann	Detroit	.398
1928	Rogers Hornsby	Boston	.387	1928	Goose Goslin	Washington	.379
1929	Lefty O'Doul	Philadelphia	.398	1929	Lew Fonseca	Cleveland	.369
1930	Bill Terry	New York	.401	1930	Al Simmons	Philadelphia	.381
1931	Chick Hafey	St. Louis	.349	1931	Al Simmons	Philadelphia	.390
1932	Lefty O'Doul	Brooklyn	.368	1932	Dale Alexander	Detroit-Boston	.367
1933	Chuck Klein	Philadelphia	.368	1933	Jimmie Foxx	Philadelphia	.356
1934	Paul Waner	Pittsburgh	.362	1934	Lou Gehrig	New York	.363
1935	Arky Vaughan	Pittsburgh	.385	1935	Buddy Myer	Washington	.349
1936	Paul Waner	Pittsburgh	.373	1936	Luke Appling	Chicago	.388
1937	Joe Medwick	St. Louis	.374	1937	Charlie Gehringer	Detroit	.371
1938	Ernie Lombardi	Cincinnati	.342	1938	Jimmie Foxx	Boston	.349
1939	John Mize	St. Louis	.349	1939	Joe DiMaggio	New York	.381
1940	Debs Garms	Pittsburgh	.355	1940	Joe DiMaggio	New York	.352
1941	Pete Reiser	Brooklyn	.343	1941	Ted Williams	Boston	.406
1942	Ernie Lombardi	Boston	.330	1942	Ted Williams	Boston	.356
1943	Stan Musial	St. Louis	.357	1943	Luke Appling	Chicago	.328
1944	Dixie Walker	Brooklyn	.357	1944	Lou Boudreau	Cleveland	.327
1945	Phil Cavarretta	Chicago	.355	1945	George Stirnweiss	New York	.309
1946	Stan Musial	St. Louis	.365	1946	Mickey Vernon	Washington	.353
1947	Harry Walker	St.L.-Phi.	.363	1947	Ted Williams	Boston	.343
1948	Stan Musial	St. Louis	.376	1948	Ted Williams	Boston	.369
1949	Jackie Robinson	Brooklyn	.342	1949	George Kell	Detroit	.343
1950	Stan Musial	St. Louis	.346	1950	Billy Goodman	Boston	.354
1951	Stan Musial	St. Louis	.355	1951	Ferris Fain	Philadelphia	.344
1952	Stan Musial	St. Louis	.336	1952	Ferris Fain	Philadelphia	.327
1953	Carl Furillo	Brooklyn	.344	1953	Mickey Vernon	Washington	.337
1954	Willie Mays	New York	.345	1954	Roberto Avila	Cleveland	.341
1955	Richie Ashburn	Philadelphia	.338	1955	Al Kaline	Detroit	.340
1956	Hank Aaron	Milwaukee	.328	1956	Mickey Mantle	New York	.353
1957	Stan Musial	St. Louis	.351	1957	Ted Williams	Boston	.388
1958	Richie Ashburn	Philadelphia	.350	1958	Ted Williams	Boston	.328
1959	Hank Aaron	Milwaukee	.355	1959	Harvey Kuenn	Detroit	.353
1960	Dick Groat	Pittsburgh	.325	1960	Pete Runnels	Boston	.320
1961	Roberto Clemente	Pittsburgh	.351	1961	Norm Cash	Detroit	.361
1962	Tommy Davis	Los Angeles	.346	1962	Pete Runnels	Boston	.326
1963	Tommy Davis	Los Angeles	.326	1963	Carl Yastrzemski	Boston	.321
1964	Roberto Clemente	Pittsburgh	.339	1964	Tony Oliva	Minnesota	.323
1965	Roberto Clemente	Pittsburgh	.329	1965	Tony Oliva	Minnesota	.321
1966	Matty Alou	Pittsburgh	.342	1966	Frank Robinson	Baltimore	.316
1967	Roberto Clemente	Pittsburgh	.357	1967	Carl Yastrzemski	Boston	.326
1968	Pete Rose	Cincinnati	.335	1968	Carl Yastrzemski	Boston	.301
1969	Pete Rose	Cincinnati	.348	1969	Rod Carew	Minnesota	.332
1970	Rico Carty	Atlanta	.366	1970	Alex Johnson	California	.329
1971	Joe Torre	St. Louis	.363	1971	Tony Oliva	Minnesota	.337
1972	Billy Williams	Chicago	.333	1972	Rod Carew	Minnesota	.318
1973	Pete Rose	Cincinnati	.338	1973	Rod Carew	Minnesota	.350
1974	Ralph Garr	Atlanta	.353	1974	Rod Carew	Minnesota	.364
1975	Bill Madlock	Chicago	.354	1975	Rod Carew	Minnesota	.359
1976	Bill Madlock	Chicago	.339	1976	George Brett	Kansas City	.333
1977	Dave Parker	Pittsburgh	.338	1977	Rod Carew	Minnesota	.388
1978	Dave Parker	Pittsburgh	.334	1978	Rod Carew	Minnesota	.333
1979	Keith Hernandez	St. Louis	.344	1979	Fred Lynn	Boston	.333
1980	Bill Buckner	Chicago	.324	1980	George Brett	Kansas City	.390
1981	Bill Madlock	Pittsburgh	.341	1981	Carney Lansford	Boston	.336
1982	Al Oliver	Montreal	.331	1982	Willie Wilson	Kansas City	.332
1983	Bill Madlock	Pittsburgh	.323	1983	Wade Boggs	Boston	.361
1984	Tony Gwynn	San Diego	.351	1984	Don Mattingly	New York	.343
1985	Willie McGee	St. Louis	.353	1985	Wade Boggs	Boston	.368
1986	Tim Raines	Montreal	.334	1986	Wade Boggs	Boston	.357
1987	Tony Gwynn	San Diego	.370	1987	Wade Boggs	Boston	.363
1988	Tony Gwynn	San Diego	.313	1988	Wade Boggs	Boston	.366
1989	Tony Gwynn	San Diego	.336	1989	Kirby Puckett	Minnesota	.339
1990	Willie McGee	St. Louis	.335	1990	George Brett	Kansas City	.329
1991	Terry Pendleton	Atlanta	.319	1991	Julio Franco	Texas	.341
1992	Gary Sheffield	San Diego	.330	1992	Edgar Martinez	Seattle	.343
1993	Andres Galarraga	Colorado	.370	1993	John Olerud	Toronto	.363
1994	Tony Gwynn	San Diego	.394	1994	Paul O'Neill	New York	.359
1995	Tony Gwynn	San Diego	.368	1995	Edgar Martinez	Seattle	.356
1996	Tony Gwynn	San Diego	.353	1996	Alex Rodriguez	Seattle	.358
1997	Tony Gwynn	San Diego	.372	1997	Frank Thomas	Chicago	.347
1998	Larry Walker	Colorado	.363	1998	Bernie Williams	New York	.339
1999	Larry Walker	Colorado	.379	1999	Nomar Garciaparra	Boston	.357
2000	Todd Helton	Colorado	.372	2000	Nomar Garciaparra	Boston	.372
2001	Larry Walker	Colorado	.350	2001	Ichiro Suzuki	Seattle	.350

(1) Some baseball researchers have concluded that Ty Cobb actually hit .382 in 1910 while Napoleon Lajoie, Cleveland, hit .383.

Cy Young Award Winners

Year	Player, Team
1956	Don Newcombe, Dodgers
1957	Warren Spahn, Braves
1958	Bob Turley, Yankees
1959	Early Wynn, White Sox
1960	Vernon Law, Pirates
1961	Whitey Ford, Yankees
1962	Don Drysdale, Dodgers
1963	Sandy Koufax, Dodgers
1964	Dean Chance, Angels
1965	Sandy Koufax, Dodgers
1966	Sandy Koufax, Dodgers
1967	(NL) Mike McCormick, Giants
	(AL) Jim Lonborg, Red Sox
1968	(NL) Bob Gibson, Cardinals
	(AL) Dennis McLain, Tigers
1969	(NL) Tom Seaver, Mets
	(AL) (tie) Dennis McLain, Tigers
	Mike Cuellar, Orioles
1970	(NL) Bob Gibson, Cardinals
	(AL) Jim Perry, Twins
1971	(NL) Ferguson Jenkins, Cubs
	(AL) Vida Blue, A's
1972	(NL) Steve Carlton, Phillies
	(AL) Gaylord Perry, Indians
1973	(NL) Tom Seaver, Mets
	(AL) Jim Palmer, Orioles
1974	(NL) Mike Marshall, Dodgers
	(AL) Jim (Catfish) Hunter, A's
1975	(NL) Tom Seaver, Mets
	(AL) Jim Palmer, Orioles
1976	(NL) Randy Jones, Padres
	(AL) Jim Palmer, Orioles
1977	(NL) Steve Carlton, Phillies
	(AL) Sparky Lyle, Yankees
1978	(NL) Gaylord Perry, Padres
	(AL) Ron Guidry, Yankees
1979	(NL) Bruce Sutter, Cubs
	(AL) Mike Flanagan, Orioles
1980	(NL) Steve Carlton, Phillies
	(AL) Steve Stone, Orioles
1981	(NL) Fernando Valenzuela, Dodgers
	(AL) Rollie Fingers, Brewers
1982	(NL) Steve Carlton, Phillies
	(AL) Pete Vuckovich, Brewers
1983	(NL) John Denny, Phillies
	(AL) LaMarr Hoyt, White Sox
1984	(NL) Rick Sutcliffe, Cubs
	(AL) Willie Hernandez, Tigers
1985	(NL) Dwight Gooden, Mets
	(AL) Bret Saberhagen, Royals
1986	(NL) Mike Scott, Astros
	(AL) Roger Clemens, Red Sox
1987	(NL) Steve Bedrosian, Phillies
	(AL) Roger Clemens, Red Sox
1988	(NL) Orel Hershiser, Dodgers
	(AL) Frank Viola, Twins
1989	(NL) Mark Davis, Padres
	(AL) Bret Saberhagen, Royals
1990	(NL) Doug Drabek, Pirates
	(AL) Bob Welch, A's
1991	(NL) Tom Glavine, Braves
	(AL) Roger Clemens, Red Sox
1992	(NL) Greg Maddux, Cubs
	(AL) Dennis Eckersley, A's
1993	(NL) Greg Maddux, Braves
	(AL) Jack McDowell, White Sox
1994	(NL) Greg Maddux, Braves
	(AL) David Cone, Royals
1995	(NL) Greg Maddux, Braves
	(AL) Randy Johnson, Mariners
1996	(NL) John Smoltz, Braves
	(AL) Pat Hentgen, Blue Jays
1997	(NL) Pedro Martinez, Expos
	(AL) Roger Clemens, Blue Jays
1998	(NL) Tom Glavine, Braves
	(AL) Roger Clemens, Blue Jays
1999	(NL) Randy Johnson, Diamondbacks
	(AL) Pedro Martinez, Red Sox
2000	(NL) Randy Johnson, Diamondbacks
	(AL) Pedro Martinez, Red Sox

Most Valuable Player

(As selected by the Baseball Writers' Assoc. of America. Prior to 1931, MVP honors were named by various sources.)

National League

Year	Player, team
1931	Frank Frisch, St. Louis
1932	Chuck Klein, Philadelphia
1933	Carl Hubbell, New York
1934	Dizzy Dean, St. Louis
1935	Gabby Hartnett, Chicago
1936	Carl Hubbell, N.Y.
1937	Joe Medwick, St. Louis
1938	Ernie Lombardi, Cincinnati
1939	Bucky Walters, Cincinnati
1940	Frank McCormick, Cincinnati
1941	Dolph Camilli, Brooklyn
1942	Mort Cooper, St. Louis
1943	Stan Musial, St. Louis
1944	Martin Marion, St. Louis
1945	Phil Cavarretta, Chicago
1946	Stan Musial, St. Louis
1947	Bob Elliott, Boston
1948	Stan Musial, St. Louis
1949	Jackie Robinson, Brooklyn
1950	Jim Konstanty, Philadelphia
1951	Roy Campanella, Brooklyn
1952	Hank Sauer, Chicago
1953	Roy Campanella, Brooklyn
1954	Willie Mays, N.Y.
1955	Roy Campanella, Brooklyn
1956	Don Newcombe, Brooklyn
1957	Hank Aaron, Milwaukee
1958	Ernie Banks, Chicago
1959	Ernie Banks, Chicago
1960	Dick Groat, Pittsburgh
1961	Frank Robinson, Cincinnati
1962	Maury Wills, L.A.
1963	Sandy Koufax, L.A.
1964	Ken Boyer, St. Louis
1965	Willie Mays, San Francisco
1966	Roberto Clemente, Pittsburgh
1967	Orlando Cepeda, St. Louis
1968	Bob Gibson, St. Louis
1969	Willie McCovey, San Francisco
1970	Johnny Bench, Cincinnati
1971	Joe Torre, St. Louis
1972	Johnny Bench, Cincinnati
1973	Pete Rose, Cincinnati
1974	Steve Garvey, L.A.
1975	Joe Morgan, Cincinnati
1976	Joe Morgan, Cincinnati
1977	George Foster, Cincinnati
1978	Dave Parker, Pittsburgh
1979	Willie Stargell, Pittsburgh
(tie)	Keith Hernandez, St. Louis
1980	Mike Schmidt, Philadelphia
1981	Mike Schmidt, Philadelphia
1982	Dale Murphy, Atlanta
1983	Dale Murphy, Atlanta
1984	Ryne Sandberg, Chicago
1985	Willie McGee, St. Louis
1986	Mike Schmidt, Philadelphia
1987	Andre Dawson, Chicago
1988	Kirk Gibson, L.A.
1989	Kevin Mitchell, San Francisco
1990	Barry Bonds, Pittsburgh
1991	Terry Pendleton, Atlanta
1992	Barry Bonds, Pittsburgh
1993	Barry Bonds, San Francisco
1994	Jeff Bagwell, Houston
1995	Barry Larkin, Cincinnati
1996	Ken Caminiti, San Diego
1997	Larry Walker, Colorado
1998	Sammy Sosa, Chicago
1999	Chipper Jones, Atlanta
2000	Jeff Kent, San Francisco

American League

Year	Player, team
1931	Lefty Grove, Philadelphia
1932	Jimmie Foxx, Philadelphia
1933	Jimmie Foxx, Philadelphia
1934	Mickey Cochrane, Detroit
1935	Hank Greenberg, Detroit
1936	Lou Gehrig, N.Y.
1937	Charley Gehringer, Detroit
1938	Jimmie Foxx, Boston
1939	Joe DiMaggio, N.Y.
1940	Hank Greenberg, Detroit
1941	Joe DiMaggio, N.Y.
1942	Joe Gordon, N.Y.
1943	Spurgeon Chandler, N.Y.
1944	Hal Newhouser, Detroit
1945	Hal Newhouser, Detroit
1946	Ted Williams, Boston
1947	Joe DiMaggio, N.Y.
1948	Lou Boudreau, Cleveland
1949	Ted Williams, Boston
1950	Phil Rizzuto, N.Y.
1951	Yogi Berra, N.Y.
1952	Bobby Shantz, Philadelphia
1953	Al Rosen, Cleveland
1954	Yogi Berra, N.Y.
1955	Yogi Berra, N.Y.
1956	Mickey Mantle, N.Y.
1957	Mickey Mantle, N.Y.
1958	Jackie Jensen, Boston
1959	Nellie Fox, Chicago
1960	Roger Maris, N.Y.
1961	Roger Maris, N.Y.
1962	Mickey Mantle, N.Y.
1963	Elston Howard, N.Y.
1964	Brooks Robinson, Baltimore
1965	Zoilo Versalles, Minnesota
1966	Frank Robinson, Baltimore
1967	Carl Yastrzemski, Boston
1968	Denny McLain, Detroit
1969	Harmon Killebrew, Minnesota
1970	John (Boog) Powell, Baltimore
1971	Vida Blue, Oakland
1972	Dick Allen, Chicago
1973	Reggie Jackson, Oakland
1974	Jeff Burroughs, Texas
1975	Fred Lynn, Boston
1976	Thurman Munson, N.Y.
1977	Rod Carew, Minnesota
1978	Jim Rice, Boston
1979	Don Baylor, California
1980	George Brett, Kansas City
1981	Rollie Fingers, Milwaukee
1982	Robin Yount, Milwaukee
1983	Cal Ripken, Jr., Baltimore
1984	Willie Hernandez, Detroit
1985	Don Mattingly, N.Y.
1986	Roger Clemens, Boston
1987	George Bell, Toronto
1988	Jose Canseco, Oakland
1989	Robin Yount, Milwaukee
1990	Rickey Henderson, Oakland
1991	Cal Ripken, Jr., Baltimore
1992	Dennis Eckersley, Oakland
1993	Frank Thomas, Chicago
1994	Frank Thomas, Chicago
1995	Mo Vaughn, Boston
1996	Juan Gonzalez, Texas
1997	Ken Griffey Jr., Seattle
1998	Juan Gonzalez, Texas
1999	Ivan Rodriguez, Texas
2000	Jason Giambi, Oakland

Rookie of the Year

(As selected by the Baseball Writers' Assoc. of America)

1947—Combined selection—Jackie Robinson, Brooklyn, 1b; 1948—Combined selection—Alvin Dark, Boston, N.L., ss

National League

Year	Player, team
1949	Don Newcombe, Brooklyn, p
1950	Sam Jethroe, Boston, of
1951	Willie Mays, N.Y., of
1952	Joe Black, Brooklyn, p
1953	Jim Gilliam, Brooklyn, 2b
1954	Wally Moon, St. Louis, of
1955	Bill Virdon, St. Louis, of
1956	Frank Robinson, Cincinnati, of
1957	Jack Sanford, Philadelphia, p
1958	Orlando Cepeda, S.F., 1b
1959	Willie McCovey, S.F., 1b
1960	Frank Howard, L.A., of
1961	Billy Williams, Chicago, of
1962	Ken Hubbs, Chicago, 2b
1963	Pete Rose, Cincinnati, 2b
1964	Richie Allen, Philadelphia, 3b
1965	Jim Lefebvre, L.A., 2b
1966	Tommy Helms, Cincinnati, 2b
1967	Tom Seaver, N.Y., p
1968	Johnny Bench, Cincinnati, c
1969	Ted Sizemore, L.A., 2b
1970	Carl Morton, Montreal, p
1971	Earl Williams, Atlanta, c
1972	Jon Matlack, N.Y., p
1973	Gary Matthews, S.F., of
1974	Bake McBride, St. Louis, of
1975	John Montefusco, S.F., p
1976	Butch Metzger, San Diego, p
(tie)	Pat Zachry, Cincinnati, p
1977	Andre Dawson, Montreal, of
1978	Bob Horner, Atlanta, 3b
1979	Rick Sutcliffe, L.A., p
1980	Steve Howe, L.A., p
1981	Fernando Valenzuela, L.A., p
1982	Steve Sax, L.A., 2b
1983	Darryl Strawberry, N.Y., of
1984	Dwight Gooden, N.Y., p
1985	Vince Coleman, St. Louis, of
1986	Todd Worrell, St. Louis, p
1987	Benito Santiago, San Diego, c
1988	Chris Sabo, Cincinnati, 3b
1989	Jerome Walton, Chicago, of
1990	Dave Justice, Atlanta, 1b
1991	Jeff Bagwell, Houston, 1b
1992	Eric Karros, L.A., 1b
1993	Mike Piazza, L.A., c
1994	Raul Mondesi, L.A., of
1995	Hideo Nomo, L.A., p
1996	Todd Hollandsworth, L.A., of
1997	Scott Rolen, Philadelphia, 3b
1998	Kerry Wood, Chicago, p
1999	Scott Williamson, Cincinnati, p
2000	Rafael Furcal, Atlanta, ss

American League

Year	Player, team
1949	Roy Sievers, St. Louis, of
1950	Walt Dropo, Boston, 1b
1951	Gil McDougald, N.Y., 3b
1952	Harry Byrd, Philadelphia, p
1953	Harvey Kuenn, Detroit, ss
1954	Bob Grim, N.Y., p
1955	Herb Score, Cleveland, p
1956	Luis Aparicio, Chicago, ss
1957	Tony Kubek, N.Y., if-of
1958	Albie Pearson, Washington, of
1959	Bob Allison, Washington, of
1960	Ron Hansen, Baltimore, ss
1961	Don Schwall, Boston, p
1962	Tom Tresh, N.Y., if-of
1963	Gary Peters, Chicago, p
1964	Tony Oliva, Minnesota, of
1965	Curt Blefary, Baltimore, of
1966	Tommie Agee, Chicago, of
1967	Rod Carew, Minnesota, 2b
1968	Stan Bahnsen, N.Y., p
1969	Lou Piniella, Kansas City, of
1970	Thurman Munson, N.Y., c
1971	Chris Chambliss, Cleveland, 1b
1972	Carlton Fisk, Boston, c
1973	Al Bumbry, Baltimore, of
1974	Mike Hargrove, Texas, 1b
1975	Fred Lynn, Boston, of
1976	Mark Fidrych, Detroit, p
1977	Eddie Murray, Baltimore, dh
1978	Lou Whitaker, Detroit, 2b
1979	John Castino, Minnesota, 3b
(tie)	Alfredo Griffin, Toronto, ss
1980	Joe Charboneau, Cleveland, of
1981	Dave Righetti, N.Y., p
1982	Cal Ripken, Jr., Baltimore, ss
1983	Ron Kittle, Chicago, of
1984	Alvin Davis, Seattle, 1b
1985	Ozzie Guillen, Chicago, ss
1986	Jose Canseco, Oakland, of
1987	Mark McGwire, Oakland, 1b
1988	Walt Weiss, Oakland, ss
1989	Gregg Olson, Baltimore, p
1990	Sandy Alomar, Jr., Cleveland, c
1991	Chuck Knoblauch, Minnesota, 2b
1992	Pat Listach, Milwaukee, ss
1993	Tim Salmon, California, of
1994	Bob Hamelin, Kansas City, dh
1995	Marty Cordova, Minnesota, of
1996	Derek Jeter, N.Y., ss
1997	Nomar Garciaparra, Boston, ss
1998	Ben Grieve, Oakland, of
1999	Carlos Beltran, Kansas City, of
2000	Kazuhiro Sasaki, Seattle, p

National League Final Standings, 2001

Eastern Division

	W	L	Pct.	GB	Home	vs. East	vs. Central	vs. West	vs. AL
Atlanta	88	74	.543	—	40-41	42-34	22-14	15-17	9-9
Philadelphia	86	76	.531	2	47-34	41-35	19-17	19-13	7-11
New York	82	80	.506	6	44-37	43-33	15-21	14-18	10-8
Florida	76	86	.469	12	46-34	34-42	19-17	11-21	12-6
Montreal	68	94	.420	20	34-47	30-46	14-22	16-16	8-10

Central Division

	W	L	Pct.	GB	Home	vs. East	vs. Central	vs. West	vs. AL
Houston	93	69	.574	—	44-37	18-12	50-34	16-17	9-6
St. Louis*	93	69	.574	—	54-28	19-11	49-35	17-16	8-7
Chicago Cubs	88	74	.543	5	48-33	16-14	48-36	15-18	9-6
Milwaukee	68	94	.420	25	36-45	15-15	37-47	11-22	5-10
Cincinnati	66	96	.407	27	27-54	16-14	32-52	14-19	4-11
Pittsburgh	62	100	.383	31	38-43	7-23	36-48	11-22	8-7

Western Division

	W	L	Pct.	GB	Home	vs. East	vs. Central	vs. West	vs. AL
Arizona	92	70	.568	—	48-33	18-14	22-17	45-31	7-8
San Francisco	90	72	.556	2	49-32	18-14	21-18	41-35	10-5
Los Angeles	86	76	.531	6	44-37	17-15	23-16	40-36	6-9
San Diego	79	83	.488	13	35-46	17-15	24-15	32-44	6-9
Colorado	73	89	.451	19	41-40	15-17	24-18	32-44	2-10

*Wild card team.

IT'S A FACT: With 73 home runs in 2001, San Francisco Giants OF Barry Bonds broke the much-publicized Major League mark of 70, set by St. Louis Cardinal OF Mark McGwire in 1998. But Bonds also broke the *professional* baseball mark of 72, set by first baseman Joe Bauman playing for the Roswell (NM) Rockets in the long-defunct Class C Longhorn League in 1954.

National League Statistics, 2001

(Individual Statistics: Batting—at least 150 at-bats; Pitching—at least 70 innings or 10 saves; *changed teams within NL during season; entry includes statistics for more than 1 team; # changed teams to or from AL during season; entry includes only NL stats)

Team Batting

Team	AVG	AB	R	H	HR	RBI
Colorado	.292	5,690	923	1,663	213	874
Houston	.271	5,528	847	1,500	208	805
St. Louis	.270	5,450	814	1,469	199	768
Arizona	.267	5,595	818	1,494	208	776
San Francisco	.266	5,612	799	1,493	235	775
Florida	.264	5,542	742	1,461	166	713
Cincinnati	.262	5,583	735	1,464	176	690
Chicago	.261	5,406	777	1,409	194	748
Atlanta	.260	5,498	729	1,432	174	696
Philadelphia	.260	5,497	746	1,431	164	708
Los Angeles	.255	5,493	758	1,399	206	714
Montreal	.253	5,379	670	1,361	131	622
San Diego	.252	5,482	789	1,379	161	753
Milwaukee	.251	5,488	740	1,378	209	712
New York	.249	5,459	642	1,361	147	608
Pittsburgh	.247	5,398	657	1,333	161	618

Team Pitching

Team	ERA	IP	H	SO	BB	SV
Atlanta	3.59	1,447.1	1,363	1,133	499	41
Arizona	3.87	1,459.2	1,352	1,297	461	34
St. Louis	3.93	1,435.1	1,389	1,083	526	38
Chicago	4.03	1,437.0	1,357	1,344	550	41
New York	4.07	1,445.2	1,418	1,191	438	48
Philadelphia	4.15	1,445.1	1,417	1086	527	47
San Francisco	4.18	1,463.1	1,437	1,080	579	47
Los Angeles	4.25	1,450.2	1,387	1,212	524	46
Florida	4.32	1,438.0	1,397	1,119	617	32
Houston	4.37	1,454.2	1,453	1,228	486	48
San Diego	4.52	1,440.2	1,519	1,088	476	46
Milwaukee	4.64	1,436.1	1,452	1,057	667	28
Montreal	4.68	1,431.1	1,509	1,103	525	28
Cincinnati	4.77	1,442.2	1,572	943	515	35
Pittsburgh	5.05	1,416.1	1,493	908	549	36
Colorado	5.29	1,430.0	1,522	1,058	598	26

Arizona Diamondbacks

BATTERS	AB	R	H	HR	RBI	SO	SB	BA
Gonzalez	609	128	198	57	142	83	1	.325
Bautista	222	26	67	5	26	31	3	.302
Grace	476	66	142	15	78	36	1	.298
Dellucci	217	28	60	10	40	52	2	.276
Counsell	458	76	126	4	38	76	6	.275
Finley	495	66	136	14	73	67	11	.275
Williams	408	58	112	16	65	70	1	.275
Miller	380	45	103	13	47	80	0	.271
Durazo	175	34	47	12	38	49	0	.269
Womack	481	66	128	3	30	54	28	.266
Sanders	441	84	116	33	90	126	14	.263
Spivey	163	33	42	5	21	47	3	.258
Bell	428	59	106	13	46	79	0	.248

PITCHERS	W	L	ERA	IP	H	BB	SO	SV
Johnson	21	6	2.49	249.2	181	71	372	0
Kim	5	6	2.94	98.0	58	44	113	19
Schilling	22	6	2.98	256.2	237	39	293	0
Batista	11	8	3.36	139.1	113	60	90	0
Lopez#	4	7	4.00	81.0	74	24	69	0
Anderson	4	9	5.20	133.1	156	30	55	0
Ellis	6	5	5.77	92.0	106	34	41	0

Manager-Bob Brenly

Atlanta Braves

BATTERS	AB	R	H	HR	RBI	SO	SB	BA
C. Jones	572	113	189	38	102	82	9	.330
Jordan	560	82	165	25	97	88	3	.295
Martinez	237	33	68	2	20	44	3	.287
DeRosa	164	27	47	3	20	19	2	.287
Furcal	324	39	89	4	30	56	22	.275
Surhoff	484	68	131	10	58	48	9	.271
Lopez	438	45	117	17	66	82	1	.267
Giles	244	36	64	9	31	37	2	.262
Veras	258	39	65	3	25	52	7	.252
A. Jones	625	104	157	34	104	142	11	.251
Brogna	206	15	51	3	21	46	3	.248
Sanchez#	154	10	35	0	9	15	2	.227
Caminiti#	171	12	38	6	16	44	0	.222
Helms	216	28	48	10	36	56	1	.222
Lockhart	178	17	39	3	12	22	1	.219

PITCHERS	W	L	ERA	IP	H	BB	SO	SV
Remlinger	3	3	2.76	75.0	67	23	93	1
Burkett	12	12	3.04	219.1	187	70	187	0

PITCHERS	W	L	ERA	IP	H	BB	SO	SV
Remlinger	3	3	2.76	75.0	67	23	93	1
Maddux	17	11	3.05	233.0	220	27	173	0
Rocker#	2	2	3.09	32.0	25	16	36	19
Smoltz	3	3	3.36	59.0	53	10	57	10
Marquis	5	6	3.48	129.1	113	59	98	0
Glavine	16	7	3.57	219.1	213	97	116	0
Millwood	7	7	4.31	121.0	121	40	84	0
O. Perez	7	8	4.91	95.1	108	39	71	0

Manager-Bobby Cox

Chicago Cubs

BATTERS	AB	R	H	HR	RBI	SO	SB	BA
Sosa	577	146	189	64	160	153	0	.328
White	323	43	99	17	50	56	1	.307
Mueller	210	38	62	6	23	19	1	.295
Gutierrez	528	76	153	10	66	56	4	.290
McGriff#	170	27	48	12	41	37	0	.282
Young	603	98	168	6	42	45	31	.279
DeShields#	163	26	45	2	16	35	12	.276
Coomer	349	25	91	8	53	70	0	.261
Girardi	229	22	58	3	25	50	0	.253
Tucker*	436	62	110	12	61	102	16	.252
Stairs	340	48	85	17	61	76	2	.250
Hundley	246	23	46	12	31	89	0	.187

PITCHERS	W	L	ERA	IP	H	BB	SO	SV
Weathers*	4	5	2.41	86.0	65	34	66	4
Van Poppel	4	1	2.52	75.0	63	38	90	0
Farnsworth	4	6	2.74	82.0	65	29	107	2
Wood	12	6	3.36	174.1	127	92	217	0
Gordon	1	2	3.38	45.1	32	16	67	27
Fassero	4	4	3.42	73.2	66	23	79	12
Lieber	20	6	3.80	232.1	226	41	148	0
Bere	11	11	4.31	188.0	171	77	175	0
Tapani	9	14	4.49	168.1	186	40	149	0
Tavarez	10	9	4.52	161.1	172	69	107	0

Manager-Don Baylor

Cincinnati Reds

BATTERS	AB	R	H	HR	RBI	SO	SB	BA
Casey	533	69	165	13	89	63	3	.310
Young	540	68	163	21	69	77	8	.302
Walker*	551	93	163	17	75	82	1	.296
Boone	381	54	112	14	62	71	6	.294
Griffey	364	57	104	22	65	72	2	.286
Dunn	244	54	64	19	43	74	4	.262
Stinnett	187	27	48	9	25	61	2	.257
Larkin	156	29	40	2	17	25	3	.256
Rivera	263	37	67	10	34	83	6	.255
LaRue	364	39	86	12	43	106	3	.236
Reese	428	50	96	9	40	82	25	.224
Castro	242	27	54	3	13	50	0	.223

PITCHERS	W	L	ERA	IP	H	BB	SO	SV
Sullivan	7	1	3.31	103.1	94	36	82	0
Brower	7	10	3.97	129.1	119	60	94	1
Graves	6	5	4.15	80.1	83	18	49	32
Dessens	10	14	4.48	205.0	221	56	128	0
Davis	8	4	4.74	106.1	124	34	53	0
Reitsma	7	15	5.29	182.0	209	49	96	0
Acevedo	5	7	5.44	96.0	101	34	68	0
O. Fernandez	5	6	6.92	79.1	103	33	35	0

Manager-Bob Boone

Colorado Rockies

BATTERS	AB	R	H	HR	RBI	SO	SB	BA
L. Walker	497	107	174	38	123	103	14	.350
Helton	587	132	197	49	146	104	7	.336
Mayne#	160	15	53	0	20	24	0	.331
Pierre	617	108	202	2	55	29	46	.327
Cirillo	528	72	165	17	83	63	12	.312
Uribe	273	32	82	8	53	55	3	.300
Perez#	382	65	114	7	47	49	6	.298
Shumpert	242	37	70	4	24	44	14	.289
Ochoa	536	73	148	8	52	76	17	.276
Norton	225	30	60	13	40	65	1	.267
Gant#	171	31	44	8	22	56	3	.257
Ortiz#	204	38	52	13	35	36	3	.255
Petrick	244	41	58	11	39	67	3	.238

PITCHERS	W	L	ERA	IP	H	BB	SO	SV
Powell*	5	3	3.24	75.0	75	31	54	7
Thomson	4	5	4.04	93.2	84	25	68	0
Jimenez	6	1	4.09	55.0	56	22	37	17
Chacon	6	10	5.06	160.0	157	87	134	0
Neagle	9	8	5.38	170.2	192	60	139	0

PITCHERS	W	L	ERA	IP	H	BB	SO	SV
Hampton	14	13	5.41	203.0	236	85	122	0
Elarton*	4	10	7.06	132.2	146	59	87	0
Bohanon	5	8	7.14	97.0	127	47	47	0

Manager-Buddy Bell

Florida Marlins

BATTERS	AB	R	H	HR	RBI	SO	SB	BA
Floyd	555	123	176	31	103	101	18	.317
Millar	449	62	141	20	85	70	0	.314
Lowell	551	65	156	18	100	79	1	.283
Lee	561	83	158	21	75	126	4	.282
Wilson	468	70	128	23	71	107	20	.274
Castillo	537	76	141	2	45	90	33	.263
Johnson	451	51	117	18	75	133	0	.259
Owens	400	51	101	5	28	59	8	.252
Gonzalez	515	57	129	9	48	107	2	.250
Berg	215	26	52	4	16	39	0	.242
Mabry*	154	14	32	6	20	46	1	.208

PITCHERS	W	L	ERA	IP	H	BB	SO	SV
Nunez	4	5	2.74	92.0	79	30	64	0
Alfonseca	4	4	3.06	61.2	68	15	40	28
Looper	3	3	3.55	71.0	63	30	52	3
Penny	10	10	3.69	205.0	183	54	154	0
Burnett	11	12	4.05	173.1	145	83	128	0
Smith	5	5	4.70	88.0	89	35	71	0
Dempster	15	12	4.94	211.1	218	112	171	0
Clement	9	10	5.05	169.1	172	85	134	0

Manager-John Boles, Tony Perez

Houston Astros

BATTERS	AB	R	H	HR	RBI	SO	SB	BA
Alou	513	79	170	27	108	57	5	.331
Berkman	577	110	191	34	126	121	7	.331
Biggio	617	118	180	20	70	100	7	.292
Bagwell	600	126	173	39	130	135	11	.288
Vizcaino	256	38	71	1	14	33	3	.277
Hidalgo	512	70	141	19	80	107	3	.275
Castilla	445	62	120	23	82	86	1	.270
Lugo	513	93	135	10	37	116	12	.263
Ward	213	21	56	9	39	48	0	.263
Eusebio	154	16	39	5	14	34	0	.253
Ausmus	422	45	98	5	34	64	4	.232

PITCHERS	W	L	ERA	IP	H	BB	SO	SV
Dotel	7	5	2.66	105.0	79	47	145	2
Wagner	2	5	2.73	62.2	44	20	79	39
Oswalt	14	3	2.73	141.2	126	24	144	0
Miller	16	8	3.40	212.0	183	76	183	0
Williams*	6	4	3.80	64.0	60	35	59	22
Cruz	3	3	4.15	82.1	72	24	75	2
Reynolds	14	11	4.34	182.2	208	36	102	0
Mlicki#	7	3	5.09	86.2	85	33	49	0
Astacio*	8	14	5.09	169.2	181	54	144	0
Villone*	6	10	5.89	114.2	133	53	113	0

Manager-Larry Dierker

Los Angeles Dodgers

BATTERS	AB	R	H	HR	RBI	SO	SB	BA
Lo Duca	460	71	147	25	90	30	2	.320
Sheffield	515	98	160	36	100	67	10	.311
Green	619	121	184	49	125	107	20	.297
Grudzielnak	539	83	146	13	55	83	4	.271
Reboulet	214	35	57	3	22	48	0	.266
Beltre	475	59	126	13	60	82	13	.265
Karros	438	42	103	15	63	101	3	.235
Goodwin	286	51	66	4	22	58	22	.231
Grissom	448	56	99	21	60	107	7	.221
Cora	405	38	88	4	29	58	0	.217
Kreuter	191	21	41	6	17	52	0	.215

PITCHERS	W	L	ERA	IP	H	BB	SO	SV
Brown	10	4	2.65	115.2	94	38	104	0
Carrara	6	1	3.16	85.1	73	24	70	0
Herges	9	8	3.44	99.1	97	46	76	1
Park	15	11	3.50	234.0	183	91	218	0
Shaw	3	5	3.62	74.2	63	18	58	43
Baldwin#	3	6	4.20	79.1	82	25	53	0
Adams	12	8	4.33	166.1	172	54	141	0
Gagne	6	7	4.75	151.2	144	46	130	0
Prokopec	8	7	4.88	138.1	146	40	91	0
Dreifort	4	7	5.13	94.2	89	47	91	0

Manager-Jim Tracy

Milwaukee Brewers

BATTERS	AB	R	H	HR	RBI	SO	SB	BA
Houston	235	36	68	12	38	62	0	.289
Loretta	384	40	111	2	29	46	1	.289
White	390	52	108	14	47	95	18	.277
Sexson	598	94	162	45	125	178	2	.271
Lopez	222	22	60	4	18	44	0	.270
Jenkins	397	60	105	20	63	120	4	.264
Belliard	364	69	96	11	36	65	5	.264
Casanova	192	21	50	11	33	29	0	.260
Burnitz	562	104	141	34	100	150	0	.251
Hernandez	542	67	135	25	78	185	5	.249
Hammonds	174	20	43	6	21	42	5	.247
Blanco	314	33	66	6	31	72	3	.210

PITCHERS	W	L	ERA	IP	H	BB	SO	SV
DeJean	4	2	2.77	84.1	75	39	68	2
Leskanic	2	6	3.63	69.1	63	31	64	17
Sheets	11	10	4.76	151.1	166	48	94	0
Haynes	8	17	4.85	172.2	182	78	112	0
Wright	11	12	4.90	194.2	201	98	129	0
Rigdon	3	5	5.79	79.1	86	46	49	0
Levrault	6	10	6.06	130.2	146	59	80	0

Manager-Davey Lopes

Montreal Expos

BATTERS	AB	R	H	HR	RBI	SO	SB	BA
Vidro	486	82	155	15	59	49	4	.319
Guerrero	599	107	184	34	108	88	37	.307
Mordecai	254	28	71	3	32	53	2	.280
Cabrera	626	64	173	14	96	54	19	.276
Tatis	145	20	37	2	11	43	0	.255
Barrett	472	42	118	6	38	54	2	.250
Stevens	542	77	133	25	95	157	2	.245
Smith	194	28	47	6	18	38	0	.242
Blum	453	57	107	9	50	94	9	.236
Bradley#	220	19	49	1	19	62	7	.223
Bergeron	375	53	79	3	16	87	10	.211

PITCHERS	W	L	ERA	IP	H	BB	SO	SV
Strickland	2	6	3.21	81.1	67	41	85	9
Vazquez	16	11	3.42	223.2	197	44	208	0
Armas	9	14	4.03	196.2	180	91	176	0
Urbina#	2	1	4.24	46.2	42	21	57	15
Lloyd	9	5	4.35	70.1	74	21	44	1
Yoshii	4	7	4.78	113.0	127	26	63	0
Thurman	9	11	5.33	147.0	172	50	96	0
Reames	4	8	5.59	95.0	101	48	86	0

Manager-Felipe Alou, Jeff Torborg

New York Mets

BATTERS	AB	R	H	HR	RBI	SO	SB	BA
Relaford	301	43	91	8	36	65	13	.302
Piazza	503	81	151	36	94	87	0	.300
McEwing	283	41	80	8	30	57	8	.283
Agbayani	296	28	82	6	27	73	4	.277
Shinjo	400	46	107	10	56	70	4	.267
Zeile	531	66	141	10	62	102	1	.266
Payton	361	44	92	8	34	52	4	.255
Ordonez	461	31	114	3	44	43	3	.247
Perez	239	26	59	5	22	25	1	.247
Lawton#	183	24	45	3	13	34	10	.246
Alfonzo	457	64	111	17	49	62	5	.243
Ventura	456	70	108	21	61	101	2	.237

PITCHERS	W	L	ERA	IP	H	BB	SO	SV
Leiter	11	11	3.31	187.1	178	46	142	0
Reed#	8	6	3.48	134.2	119	17	99	0
Appier	11	10	3.57	206.2	181	64	172	0
Benitez	6	4	3.77	76.1	59	40	93	43
Trachsel	11	13	4.46	173.2	168	47	144	0
Rusch	8	12	4.63	179.0	216	43	156	0
Chen*	7	7	4.87	146.0	146	59	126	0

Manager-Bobby Valentine

Philadelphia Phillies

BATTERS	AB	R	H	HR	RBI	SO	SB	BA
Anderson	522	69	153	11	61	74	8	.293
Abreu	588	118	170	31	110	137	36	.289
Rolen	554	96	160	25	107	127	16	.289
Rollins	656	97	180	14	54	108	46	.274
Glanville	634	74	166	14	55	91	28	.262
Burrell	539	70	139	27	89	162	2	.258
Lee	555	75	143	20	90	109	3	.258
Estrada	298	26	68	8	37	32	0	.228
Pratt*	173	18	32	4	11	61	1	.185

PITCHERS	W	L	ERA	IP	H	BB	SO	SV
Mesa	3	3	2.34	69.1	65	20	59	42
Wolf	10	11	3.70	163.0	150	51	152	0
Figueroa	4	5	3.94	89.0	95	37	61	0
Coggin	6	7	4.17	95.0	99	39	62	0
Person	15	7	4.19	208.1	179	80	183	0
Daal	13	7	4.46	185.2	199	56	107	0
Telemaco	5	5	5.54	89.1	93	32	59	0

Manager-Larry Bowa

Pittsburgh Pirates

BATTERS	AB	R	H	HR	RBI	SO	SB	BA
C. Wilson	158	27	49	13	32	53	3	.310
Giles	576	116	178	37	95	67	13	.309
Ramirez	603	83	181	34	112	100	5	.300
Mackowiak	214	30	57	4	21	52	4	.266
Kendall	606	84	161	10	53	48	13	.266
Nunez	301	30	79	1	21	53	8	.262
Rios*	319	38	83	14	50	74	3	.260
Young	449	53	104	14	65	119	15	.232
Matthews*	405	63	92	14	44	100	8	.227
J. Wilson	390	44	87	3	25	70	1	.223
Meares	270	27	57	4	25	45	0	.211
Bell	156	14	27	5	13	38	0	.173

PITCHERS	W	L	ERA	IP	H	BB	SO	SV
Manzanillo	3	2	3.39	79.2	60	26	80	2
D. Williams	3	7	3.71	114.0	100	45	57	0
Ritchie	11	15	4.47	207.1	211	52	124	0
McKnight*	3	6	4.95	87.1	109	24	46	0
Arroyo	5	7	5.09	88.1	99	34	39	0
Anderson	9	17	5.10	206.1	232	83	89	0
Beimel	7	11	5.23	115.1	131	49	58	0
Olivares	6	9	6.55	110.0	123	42	69	1

Manager-Lloyd McClendon

St. Louis Cardinals

BATTERS	AB	R	H	HR	RBI	SO	SB	BA
Pujols	590	112	194	37	130	93	1	.329
Drew	375	80	121	27	73	75	13	.323
Polanco	564	87	173	3	38	43	12	.307
Edmonds	500	95	152	30	110	136	5	.304
Vina	631	95	191	9	56	35	17	.303
Cairo*	156	25	46	3	16	23	2	.295
Robinson	186	34	53	1	15	20	11	.285
Paquette	340	47	96	15	64	67	3	.282
Marrero	203	37	54	6	23	36	6	.266
Renteria	493	54	128	10	57	73	17	.260
Matheny	381	40	83	7	42	76	0	.218
Bonilla	174	17	37	5	21	53	1	.213
McGwire	299	48	56	29	64	118	0	.187

PITCHERS	W	L	ERA	IP	H	BB	SO	SV
Kline	3	3	1.80	75.0	53	29	54	9
Kile	16	11	3.09	227.1	228	65	179	0
Morris	22	8	3.16	216.1	218	54	185	0
Matthews	3	4	3.24	89.0	74	33	72	1
Veres	3	2	3.70	65.2	57	28	61	15
Smith	6	3	3.83	84.2	79	24	59	0
Stechschulte	1	5	3.86	70.0	71	30	51	6
Williams*	15	9	4.05	220.0	224	56	154	0
Timlin	4	5	4.09	72.2	78	19	47	3
Hermanson	14	13	4.45	192.1	195	73	123	0
An. Benes	7	7	7.38	107.1	122	61	78	0

Manager-Tony LaRussa

San Diego Padres

BATTERS	AB	R	H	HR	RBI	SO	SB	BA
Nevin	546	97	167	41	126	147	4	.306
Kotsay	406	67	118	10	58	58	13	.291
Klesko	538	105	154	30	113	89	23	.286
Darr	289	36	80	2	34	72	6	.277
Jimenez	308	45	85	3	33	68	2	.276
Gonzalez	160	16	44	8	27	28	2	.275
Trammell	490	66	128	25	92	78	2	.261
Lankford*	389	58	98	19	58	145	10	.252
Jackson	440	67	106	4	38	128	23	.241
Davis	448	56	107	11	57	112	4	.239
Henderson	379	70	86	8	42	84	25	.227
Crespo	153	27	32	4	12	50	6	.209

PITCHERS	W	L	ERA	IP	H	BB	SO	SV
Hoffman	3	4	3.43	60.1	48	21	63	43
Lawrence	5	5	3.45	114.2	107	34	84	0
Tollberg	10	4	4.30	117.1	133	25	71	0
Eaton	8	5	4.32	116.2	108	40	109	0
Jarvis	12	11	4.79	193.1	189	49	133	0
Jones	8	19	5.12	195.0	250	38	113	0

Manager-Bruce Bochy

San Francisco Giants

BATTERS	AB	R	H	HR	RBI	SO	SB	BA
Bonds	476	129	156	73	137	93	13	.328
Aurilia	636	114	206	37	97	83	1	.324
Kent	607	84	181	22	106	96	7	.298
Galarraga#	156	17	45	7	35	49	0	.288
Dunston	186	26	52	9	25	32	3	.280
Vander Wal*	452	58	122	14	70	122	8	.270
Benard	392	70	104	15	44	66	10	.265
Santiago	477	39	125	6	45	78	5	.262
R. Davis	167	16	43	7	17	49	1	.257
Martinez	391	48	99	5	37	52	1	.253
Snow	285	43	70	8	34	81	0	.246
Murray	326	54	80	6	25	57	8	.245
Feliz	220	23	50	7	22	50	2	.227
E. Davis	156	17	32	4	22	38	1	.205

PITCHERS	W	L	ERA	IP	H	BB	SO	SV
Rodriguez	9	1	1.68	80.1	53	27	91	0
Nen	4	5	3.01	77.2	58	22	93	45
Ortiz	17	9	3.29	218.2	187	91	169	0
Worrell	2	5	3.45	78.1	71	33	63	0
Estes	9	8	4.02	159.0	151	77	109	0
Schmidt*	13	7	4.07	150.1	138	61	142	0
Rueter	14	12	4.42	195.1	213	66	83	0
Fultz	3	1	4.56	71.0	70	21	67	1
Hernandez	13	15	5.24	226.2	266	85	138	0
Gardner	5	5	5.40	91.2	93	34	53	0

Manager-Dusty Baker

American League Final Standings, 2001

Eastern Division

	W	L	Pct.	GB	Home	vs. East	vs. Central	vs. West	vs. NL
New York	95	65	.594	—	51-28	50-24	23-13	12-20	10-8
Boston	82	79	.509	13.5	41-40	41-34	16-20	15-17	10-8
Toronto	80	82	.494	16	40-42	37-39	19-13	16-20	8-10
Baltimore	63	98	.391	32.5	30-50	31-44	16-16	10-26	6-12
Tampa Bay	62	100	.383	34	37-44	29-47	13-19	10-26	10-8

Central Division

	W	L	Pct.	GB	Home	vs. East	vs. Central	vs. West	vs. NL
Cleveland	91	71	.562	—	44-36	22-14	47-29	15-17	7-11
Minnesota	85	77	.525	6	47-34	13-19	47-29	16-20	9-9
Chicago Sox	83	79	.512	8	46-35	16-16	42-34	13-23	12-6
Detroit	66	96	.407	25	37-44	17-19	24-52	15-17	10-8
Kansas City	65	97	.401	26	35-46	13-19	30-46	14-22	8-10

Western Division

	W	L	Pct.	GB	Home	vs. East	vs. Central	vs. West	vs. NL
Seattle	116	46	.716	—	57-24	33-12	31-10	40-18	12-6
Oakland*	102	60	.630	14	53-28	31-14	27-14	32-26	12-6
Anaheim	75	87	.463	41	39-42	24-17	24-21	17-41	10-8
Texas	73	89	.451	43	41-41	21-20	17-28	27-31	8-10

*Wild card team.

IT'S A FACT: Hall of Famer Burleigh Grimes, who retired as a player in 1934, was the last pitcher to throw a legal spitball. Though the spitball was banned by Major League Baseball in 1920, 17 veteran pitchers then using the pitch were exempted for the duration of their careers.

American League Team Statistics, 2001

(Individual Statistics: Batting—at least 150 at-bats; Pitching—at least 70 innings or 10 saves; *changed teams within AL during season, entry includes statistics for more than one team; # changed teams to or from NL during season, entry includes only AL stats)

Team Batting

Team	AVG	AB	R	H	HR	RBI
Seattle	.288	5,680	927	1,637	169	881
Cleveland	.278	5,600	897	1,559	212	868
Texas	.275	5,685	890	1,566	246	844
Minnesota	.272	5,560	771	1,514	164	717
Chicago	.268	5,464	798	1,463	214	770
New York	.267	5,577	804	1,488	203	774
Boston	.266	5,605	772	1,493	198	739
Kansas City	.266	5,643	729	1,503	152	691
Oakland	.264	5,573	884	1,469	199	835
Toronto	.263	5,663	767	1,489	195	728
Anaheim	.261	5,551	691	1,447	158	662
Detroit	.260	5,537	724	1,439	139	691
Tampa Bay	.258	5,524	672	1,426	121	645
Baltimore	.248	5,472	687	1,359	136	663

Team Pitching

Team	ERA	IP	H	SO	BB	SV
Seattle	3.54	1,465.0	1,293	1,051	465	56
Oakland	3.59	1,463.1	1,384	1,117	440	44
New York	4.02	1,451.1	1,429	1,266	465	57
Boston	4.15	1,448.0	1,412	1,259	544	48
Anaheim	4.20	1,437.2	1,452	947	525	43
Toronto	4.28	1,462.2	1,553	1,041	490	41
Minnesota	4.51	1,441.1	1,494	965	445	45
Chicago	4.55	1,433.1	1,465	921	500	51
Cleveland	4.64	1,446.2	1,512	1,218	573	42
Baltimore	4.67	1,432.1	1,504	938	528	31
Kansas City	4.87	1,440.0	1,537	911	576	30
Tampa Bay	4.94	1,423.2	1,513	1,030	569	30
Detroit	5.01	1,429.1	1,624	859	553	34
Texas	5.71	1,438.1	1,670	951	596	37

Anaheim Angels

BATTERS	AB	R	H	HR	RBI	SO	SB	BA
Wooten	221	24	69	8	32	42	2	.312
Gil	260	33	77	8	39	57	3	.296
Anderson	672	83	194	28	123	100	13	.289
Eckstein	582	82	166	4	41	60	29	.285
Spiezio	457	57	124	13	54	65	5	.271
Kennedy	478	48	129	6	40	71	12	.270
B. Molina	325	31	85	6	40	51	0	.262
Erstad	631	89	163	9	63	113	24	.258
Glaus	588	100	147	41	108	158	10	.250
Palmeiro	230	29	56	2	23	24	6	.243
Salmon	475	63	108	17	49	121	9	.227

PITCHERS	W	L	ERA	IP	H	BB	SO	SV
Levine	8	10	2.38	75.2	71	28	40	2
Percival	4	2	2.65	57.2	39	18	71	39
Washburn	11	10	3.77	193.1	196	54	126	0
Pote	2	0	4.15	86.2	88	32	66	2
Ortiz	13	11	4.36	208.2	223	76	135	0
Valdes	9	13	4.45	163.2	177	50	100	0
Rapp	5	12	4.76	170.0	169	71	82	0
Schoeneweis	10	11	5.08	205.1	227	77	104	0

Manager—Mike Scioscia

Baltimore Orioles

BATTERS	AB	R	H	HR	RBI	SO	SB	BA
Conine	524	75	163	14	97	75	12	.311
Segui	292	48	88	10	46	61	1	.301
Kinkade	160	19	44	4	16	31	2	.275
Richard	483	74	128	15	61	100	11	.265
B. Roberts	273	42	69	2	17	36	12	.253
Mora	436	49	109	7	48	91	11	.250
Bordick	229	32	57	7	30	36	9	.249
Lunar	167	8	41	0	16	32	0	.246
Ripken	477	43	114	14	68	63	0	.239
Batista*	579	70	138	25	87	113	5	.238
Gibbons	225	27	53	15	36	39	0	.236
Hairston	532	63	124	8	47	73	29	.233
Fordyce	292	30	61	5	19	56	1	.209
Anderson	430	50	87	8	45	77	12	.202
DeShields#	188	29	37	3	21	42	11	.197

PITCHERS	W	L	ERA	IP	H	BB	SO	SV
Groom	1	4	3.55	66.0	64	9	54	11
Johnson	10	12	4.09	196.0	194	77	114	0
Maduro	5	6	4.23	93.2	83	36	51	0
Towers	8	10	4.49	140.1	165	16	58	0
W. Roberts	9	10	4.91	132.0	142	55	95	6
Ponson	5	10	4.94	138.1	161	37	84	0
Mercedes	8	17	5.82	184.0	219	63	123	0

Manager—Mike Hargrove

Boston Red Sox

BATTERS	AB	R	H	HR	RBI	SO	SB	BA
Ramirez	529	93	162	41	125	147	0	.306
Varitek	174	19	51	7	25	35	0	.293
Bichette	391	45	112	12	49	76	2	.286
Lewis	164	18	46	1	12	25	5	.280
Nixon	535	100	150	27	88	113	7	.280
Stynes	361	52	101	8	33	56	4	.280
Offerman	524	76	140	9	49	97	5	.267
Daubach	407	54	107	22	71	108	1	.263
Hillenbrand	468	52	123	12	49	61	3	.263
Everett	409	61	105	14	58	104	9	.257
Lansing	352	45	88	8	34	50	3	.250
Hatteberg	278	34	68	3	25	26	1	.245
O'Leary	341	50	82	13	50	73	1	.240
Mirabelli*	190	20	43	11	29	57	0	.226

PITCHERS	W	L	ERA	IP	H	BB	SO	SV
Martinez	7	3	2.39	116.2	84	25	163	0
Arrojo	5	4	3.48	103.1	88	35	78	5
Lowe	5	10	3.53	91.2	103	29	82	24
Wakefield	9	12	3.90	168.2	156	73	148	3
Beck	6	4	3.90	80.2	77	28	63	6
F. Castillo	10	9	4.21	136.2	138	35	89	0
Cone	9	7	4.31	135.2	148	57	115	0
Nomo	13	10	4.50	198.0	171	96	220	0

Manager—Jimy Williams, Joe Kerrigan

Chicago White Sox

BATTERS	AB	R	H	HR	RBI	SO	SB	BA
Ordonez	593	97	181	31	113	70	25	.305
Singleton	392	57	117	7	45	61	12	.298
Konerko	582	92	164	32	99	89	1	.282
Lee	558	75	150	24	84	85	17	.269
Durham	611	104	163	20	65	110	23	.267
Clayton	433	62	114	9	60	72	10	.263
Valentin	438	74	113	28	68	114	9	.258
Canseco	256	46	66	16	49	75	2	.258
Perry	285	38	73	7	32	55	2	.256
Liefer	254	36	65	18	39	69	0	.256
Johnson	173	21	43	5	18	31	2	.249
Alomar	220	17	54	4	21	17	1	.245

PITCHERS	W	L	ERA	IP	H	BB	SO	SV
Foulke	4	9	2.33	81.0	57	22	75	42
Buehrle	16	8	3.29	221.1	188	48	126	0
Lowe	9	4	3.61	127.0	123	32	71	3
Garland	6	7	3.69	117.0	123	55	61	1
D. Wells	5	7	4.47	100.2	120	21	59	0
Baldwin#	7	5	4.61	95.2	109	38	42	0
Howry	4	5	4.69	78.2	85	30	64	5
K. Wells	10	11	4.79	133.1	145	61	99	0
Glover	5	5	4.93	100.1	98	32	63	0
Biddle	7	8	5.39	128.2	137	52	85	0

Manager—Jerry Manuel

Cleveland Indians

BATTERS	AB	R	H	HR	RBI	SO	SB	BA
Alomar	575	113	193	20	100	71	30	.336
Gonzalez	532	97	173	35	140	94	1	.325
Cordova	409	61	123	20	69	81	0	.301
Thome	526	101	153	49	124	185	0	.291
Burks	439	83	123	28	74	85	5	.280
Diaz	437	54	121	4	56	44	1	.277
Fryman	334	34	88	3	38	63	1	.263
Cabrera	287	50	75	1	38	41	10	.261
Lofton	517	91	135	14	66	69	16	.261
Vizquel	611	84	156	2	50	72	13	.255
Cordero	268	30	67	4	21	50	0	.250
Branyan	315	48	73	20	54	132	1	.232

PITCHERS	W	L	ERA	IP	H	BB	SO	SV
Wickman	5	0	2.39	67.2	61	14	66	32
Colon	14	12	4.09	222.1	220	90	201	0
Sabathia	17	5	4.39	180.1	149	95	171	0
Woodard	3	3	5.20	97.0	129	17	52	0
Finley	8	7	5.54	113.2	131	35	96	0
Burba	10	10	6.21	150.2	188	54	118	0
Nagy	5	6	6.40	70.1	102	20	29	0

Manager—Charlie Manuel

Detroit Tigers

BATTERS	AB	R	H	HR	RBI	SO	SB	BA
Simon	256	28	78	6	37	28	0	.305
Cedeno	523	79	153	6	48	83	55	.293
T. Clark	428	67	123	16	75	108	0	.287
Halter	450	53	128	12	65	100	3	.284
Higginson	541	84	150	17	71	65	20	.277
Fick	401	62	109	19	61	62	0	.272
Macias	488	62	131	8	51	54	21	.268
Cruz	414	39	106	7	52	46	4	.256
Easley	585	77	146	11	65	90	10	.250
Encarnacion	417	52	101	12	52	93	9	.242
Palmer	216	34	48	11	40	59	4	.222
Magee	207	26	44	5	17	44	3	.213
Inge	189	13	34	0	15	41	1	.180

PITCHERS	W	L	ERA	IP	H	BB	SO	SV
Santos	2	2	3.30	76.1	62	49	52	0
Sparks	14	9	3.65	232.0	244	64	116	0
Weaver	13	16	4.08	229.1	235	68	152	0
Lima#	5	10	4.71	112.2	120	22	43	0
Anderson	3	1	4.82	56.0	56	18	52	22
Holt	7	9	5.77	151.1	197	57	80	0
Mlicki#	4	8	7.33	81.0	118	41	48	0

Manager—Phil Garner

Kansas City Royals

BATTERS	AB	R	H	HR	RBI	SO	SB	BA
Beltran	617	106	189	24	101	120	31	.306
Sweeney	559	97	170	29	99	64	10	.304
Sanchez#	390	46	118	0	28	34	9	.303
Ibanez	279	44	78	13	54	51	0	.280
Alicea	387	44	106	4	32	56	8	.274
Quinn	453	57	122	17	60	69	9	.269
Randa	581	59	147	13	83	80	3	.253
McCarty	200	26	50	7	26	45	0	.250
Ortiz	154	12	38	0	11	24	1	.247
Brown	380	39	93	7	40	81	5	.245
Perez#	199	18	48	1	12	19	3	.241
Mayne#	166	13	40	2	20	17	1	.241
Febles	292	45	69	8	25	58	5	.236

PITCHERS	W	L	ERA	IP	H	BB	SO	SV
Grimsley	1	5	3.02	80.1	71	28	61	0
Byrd#	6	6	4.05	93.1	110	22	49	0
Hernandez	5	6	4.12	67.2	69	26	46	28
Suppan	10	14	4.37	218.1	227	74	120	0
Stein	7	8	4.74	131.0	112	79	113	1
Durbin	9	16	4.93	179.0	201	58	95	0
Wilson	6	5	5.19	109.1	132	32	67	1
George	4	8	5.59	74.0	83	18	32	0
Reichert	8	8	5.63	123.0	131	67	77	0
Henry	2	2	6.07	75.2	75	45	57	0

Manager—Tony Muser

Minnesota Twins

BATTERS	AB	R	H	HR	RBI	SO	SB	BA
Mientkiewicz	543	77	166	15	74	92	2	.306
Guzman	493	80	149	10	51	78	25	.302
Lawton#	376	71	110	10	51	46	19	.293
Pierzynski	381	51	110	7	55	57	1	.289
Koskie	562	100	155	26	103	118	27	.276
J. Jones	475	57	131	14	49	92	12	.276
Buchanan	197	28	54	10	32	58	1	.274
Rivas	563	70	150	7	47	99	31	.266
Allen	175	20	46	4	20	37	1	.263
Hunter	564	82	147	27	92	125	9	.261
Hocking	327	34	82	3	25	67	6	.251
Ortiz	303	46	71	18	48	68	1	.234
Prince	196	19	43	7	23	39	3	.219

PITCHERS	W	L	ERA	IP	H	BB	SO	SV
Mays	17	13	3.16	233.2	205	64	123	0
Guardado	7	1	3.51	66.2	47	23	67	12
Radke	15	11	3.94	226.0	235	26	137	0
T. Jones*	5	5	4.24	68.0	87	29	54	13
Milton	15	7	4.32	220.2	222	61	157	0
Carrasco	4	3	4.64	73.2	77	30	70	1
Lohse	4	7	5.68	90.1	102	29	64	0
Hawkins	1	5	5.96	51.1	59	39	36	28

Manager—Tom Kelly

New York Yankees

BATTERS	AB	R	H	HR	RBI	SO	SB	BA
Jeter	614	110	191	21	74	99	27	.311
B. Williams	540	102	166	26	94	67	11	.307
Brosius	428	57	123	13	49	83	3	.287
Martinez	589	89	165	34	113	89	1	.280
Velarde*	342	50	95	9	32	86	6	.278
Posada	484	59	134	22	95	132	2	.277
Soriano	574	77	154	18	73	125	43	.268
O'Neill	510	77	136	21	70	59	22	.267
Spencer	283	40	73	10	46	58	4	.258
Knoblauch	521	66	130	9	44	73	38	.250
Justice	381	58	92	18	51	83	1	.241
G. Williams	279	42	56	4	19	55	13	.201

PITCHERS	W	L	ERA	IP	H	BB	SO	SV
M. Rivera	4	6	2.34	80.2	61	12	83	50
Stanton	9	4	2.58	80.1	80	29	78	0
Mussina	17	11	3.15	228.2	202	42	214	0
Clemens	20	3	3.51	220.1	205	72	213	0
Mendoza	8	4	3.75	100.2	89	23	70	6
Pettitte	15	10	3.99	200.2	224	41	164	0
O. Hernandez	4	7	4.85	94.2	90	42	77	0
Lilly	5	6	5.37	120.2	126	51	112	0

Manager—Joe Torre

Oakland Athletics

BATTERS	AB	R	H	HR	RBI	SO	SB	BA
Ja. Giambi	520	109	178	38	120	83	2	.342
Chavez	552	91	159	32	114	99	8	.288
Je. Giambi	371	64	105	12	57	83	0	.283
Long	629	90	178	12	85	103	9	.283
Dye*	599	91	169	26	106	112	9	.282
Tejada	622	107	166	31	113	89	11	.267
Damon	644	108	165	9	49	70	27	.256
Hernandez	453	55	115	15	60	68	1	.254
Menechino	471	82	114	12	60	97	2	.242
Myers	161	24	36	11	31	38	0	.224
Saenz	305	33	67	9	32	64	0	.220

PITCHERS	W	L	ERA	IP	H	BB	SO	SV
Isringhausen	4	3	2.65	71.1	54	23	74	34
Tam	2	4	3.01	74.2	68	29	44	3
Hudson	18	9	3.37	235.0	216	71	181	0
Mulder	21	8	3.45	229.1	214	51	153	0
Zito	17	8	3.49	214.1	184	80	205	0
Lidle	13	6	3.59	188.0	170	47	118	0
Heredia	7	8	5.58	109.2	144	29	48	0

Manager—Art Howe

Seattle Mariners

BATTERS	AB	R	H	HR	RBI	SO	SB	BA
Suzuki	692	127	242	8	69	53	56	.350
Boone	623	118	206	37	141	110	5	.331
Martinez	470	80	144	23	116	90	4	.306
Olerud	572	91	173	21	95	70	3	.302
Javier	281	44	82	4	33	47	11	.292
McLemore	409	78	117	5	57	84	39	.286
Cameron	540	99	144	25	110	155	34	.267
Wilson	377	44	100	10	42	69	3	.265
Bell	470	62	122	15	64	59	2	.260
Guillen	456	72	118	5	53	89	4	.259
Martin	283	41	68	7	42	59	9	.240
Lampkin	204	28	46	5	22	41	1	.225

PITCHERS	W	L	ERA	IP	H	BB	SO	SV
Pineiro	6	2	2.03	75.1	50	21	56	0
Garcia	18	6	3.05	238.2	199	69	163	0
Sasaki	0	4	3.24	66.2	48	11	62	45
Moyer	20	6	3.43	209.2	187	44	119	0
Franklin	5	1	3.56	78.1	76	24	60	0
Sele	15	5	3.60	215.0	216	51	114	0
Abbott	17	4	4.25	163.0	145	87	118	0
Halama	10	7	4.73	110.1	132	26	50	0

Manager—Lou Piniella

Tampa Bay Devil Rays

BATTERS	AB	R	H	HR	RBI	SO	SB	BA
McGriff#	343	40	109	19	61	69	1	.318
Gomez	189	31	57	8	36	24	3	.302
Hall	188	28	56	4	30	16	2	.298
Johnson	248	32	73	4	33	57	2	.294
Tyner	396	51	111	0	21	42	31	.280
Winn	429	54	117	6	50	81	12	.273
Abernathy	304	43	82	5	33	35	8	.270
Grieve	542	72	143	11	72	159	7	.264
Rolls	237	33	62	2	12	47	12	.262
Cox	342	37	88	12	51	75	2	.257
Huff	411	42	102	8	45	72	1	.248
Martinez	219	24	54	1	14	46	6	.247
Flaherty	248	20	59	4	29	33	1	.238
Vaughn	485	74	113	24	82	130	11	.233
Sheets	153	10	30	1	14	35	2	.196

PITCHERS	W	L	ERA	IP	H	BB	SO	SV
Yan	4	6	3.90	62.1	64	11	64	22
Sturtze	11	12	4.42	195.1	200	79	110	1
Kennedy	7	8	4.44	117.2	122	34	78	0
Wilson	8	9	4.88	151.1	165	52	119	0
Lopez#	5	12	5.34	124.2	152	51	67	0
Rekar	3	13	5.89	140.2	167	45	87	0
Rupe	5	12	6.59	143.1	161	48	123	0

Manager—Larry Rothschild, Hal McRae

Texas Rangers

BATTERS	AB	R	H	HR	RBI	SO	SB	BA
Catalanotto	463	77	153	11	54	55	15	.330
A. Rodriguez	632	133	201	52	135	131	18	.318
I. Rodriguez	442	70	136	25	65	73	10	.308
Lamb	284	42	87	4	35	27	2	.306
Sierra	344	55	100	23	67	52	2	.291
Greer	245	38	67	7	29	32	1	.273
Palmeiro	600	98	164	47	123	90	1	.273
Kapler	483	77	129	17	72	70	23	.267
Young	386	57	96	11	49	91	3	.249
Galarraga#	243	33	57	10	34	68	1	.235
Caminiti#	185	24	43	9	25	41	0	.232
Ledee	242	33	56	2	36	58	3	.231

PITCHERS	W	L	ERA	IP	H	BB	SO	SV
Zimmerman	4	4	2.40	71.1	48	16	72	28
Michalak*	8	9	4.41	136.2	157	55	67	1
Davis	11	10	4.45	186.0	220	69	115	0
Helling	12	11	5.17	215.2	256	63	154	0
Mahomes	7	6	5.70	107.1	115	55	61	0
Oliver	11	11	6.02	154.0	189	65	104	0
Rogers	5	7	6.19	120.2	150	49	74	0
Petkovsek	1	2	6.69	76.2	103	28	42	0
Myette	4	5	7.14	80.2	94	37	67	0
Bell#	5	5	7.18	105.1	130	47	64	0

Manager—Johnny Oates, Jerry Narron

Toronto Blue Jays

BATTERS	AB	R	H	HR	RBI	SO	SB	BA
Stewart	640	103	202	12	60	72	27	.316
Bush	271	32	83	3	27	50	13	.306
Delgado	574	102	160	39	102	136	3	.279
Fullmer	522	71	143	18	83	88	5	.274
Cruz	577	92	158	34	88	138	32	.274
F. Lopez	177	21	46	5	23	39	4	.260
Gonzalez	636	79	161	17	76	149	18	.253
Mondesi	572	88	144	27	84	128	30	.252
Frye	175	24	43	2	15	18	2	.246
Fletcher	416	36	94	11	56	43	0	.226

PITCHERS	W	L	ERA	IP	H	BB	SO	SV
Quantrill	11	2	3.04	83.0	86	12	58	2
Halladay	5	3	3.16	105.1	97	25	96	0
File	5	3	3.27	74.1	57	29	38	0
Escobar	6	8	3.50	126.0	93	52	121	0
Carpenter	11	11	4.09	215.2	229	75	157	0
Parris	4	6	4.60	105.2	126	41	49	0
Koch	2	5	4.80	69.1	69	33	55	36
Loaiza	11	11	5.02	190.0	239	40	110	0
Hamilton	5	8	5.89	122.1	170	38	82	0

Manager—Buck Martinez

National Baseball Hall of Fame and Museum, Cooperstown, NY[1]

#Aaron, Hank
Alexander, Grover Cleveland
Alston, Walt
Anderson, Sparky
Anson, Cap
Aparicio, Luis
Appling, Luke
Ashburn, Richie
Averill, Earl
Baker, Home Run
Bancroft, Dave
#Banks, Ernie
Barlick, Al
Barrow, Edward G.
Beckley, Jake
Bell, Cool Papa
#Bench, Johnny
Bender, Chief
Berra, Yogi
Bottomley, Jim
Boudreau, Lou
Bresnahan, Roger
#Brett, George
#Brock, Lou
Brouthers, Dan
Brown, Mordecai (Three Finger)
Bulkeley, Morgan C.
Bunning, Jim
Burkett, Jesse C.
Campanella, Roy
#Carew, Rod
Carey, Max
#Carlton, Steve
Cartwright, Alexander
Cepeda, Orlando
Chadwick, Henry
Chance, Frank
Chandler, Happy
Charleston, Oscar
Chesbro, John
Chylak, Nestor
Clarke, Fred
Clarkson, John
Clemente, Roberto
Cobb, Ty[2]
Cochrane, Mickey
Collins, Eddie
Collins, James
Combs, Earle
Comiskey, Charles A.
Conlan, Jocko
Connolly, Thomas H.
Connor, Roger
Coveleski, Stan
Crawford, Sam
Cronin, Joe
Cummings, Candy
Cuyler, Kiki
Dandridge, Ray
Davis, George "Gorgeous"
Day, Leon
Dean, Dizzy
Delahanty, Ed
Dickey, Bill
DiHigo, Martin
DiMaggio, Joe
Doby, Larry
Doerr, Bobby
Drysdale, Don
Duffy, Hugh
Durocher, Leo
Evans, Billy
Evers, John
Ewing, Buck
Faber, Urban
#Feller, Bob
Ferrell, Rick
Fingers, Rollie
Fisk, Carlton
Flick, Elmer H.
Ford, Whitey
Foster, Andrew (Rube)
Foster, Bill
Fox, Nellie
Foxx, Jimmie
Frick, Ford
Frisch, Frank
Galvin, Pud
Gehrig, Lou
Gehringer, Charles
#Gibson, Bob
Gibson, Josh
Giles, Warren
Gomez, Lefty
Goslin, Goose
Greenberg, Hank
Griffith, Clark
Grimes, Burleigh
Grove, Lefty
Hafey, Chick
Haines, Jesee
Hamilton, Bill
Hanlon, Ned
Harridge, Will
Harris, Bucky
Hartnett, Gabby
Heilmann, Harry
Herman, Billy
Hooper, Harry
Hornsby, Rogers
Hoyt, Waite
Hubbard, Cal
Hubbell, Carl
Huggins, Miller
Hulbert, William
Hunter, Catfish
Irvin, Monte
#Jackson, Reggie
Jackson, Travis
Jenkins, Ferguson
Jennings, Hugh
Johnson, Byron
Johnson, William (Judy)
Johnson, Walter[2]
Joss, Addie
#Kaline, Al
Keefe, Timothy
Keeler, William
Kell, George
Kelley, Joe
Kelly, George
Kelly, King
Killebrew, Harmon
Kiner, Ralph
Klein, Chuck
Klem, Bill
#Koufax, Sandy
Lajoie, Napoleon
Landis, Kenesaw M.
Lasorda, Tom
Lazzeri, Tony
Lemon, Bob
Leonard, Buck
Lindstrom, Fred
Lloyd, Pop
Lombardi, Ernie
Lopez, Al
Lyons, Ted
Mack, Connie
MacPhail, Larry
MacPhail, Lee
#Mantle, Mickey
Manush, Henry
Maranville, Rabbit
Marichal, Juan
Marquard, Rube
Mathews, Eddie
Mathewson, Christy[2]
#Mays, Willie
*Mazeroski, Bill
McCarthy, Joe
McCarthy, Thomas
#McCovey, Willie
McGinnity, Joe
McGowan, Bill
McGraw, John
McKechnie, Bill
McPhee, John "Bid"
Medwick, Joe
Mize, Johnny
#Morgan, Joe
#Musial, Stan
Newhouser, Hal
Nichols, Kid
Niekro, Phil
O'Rourke, James
Ott, Mel
Paige, Satchel
#Palmer, Jim
Pennock, Herb
Perez, Tony
Perry, Gaylord
Plank, Ed
*#Puckett, Kirby
Radbourn, Charlie
Reese, Pee Wee
Rice, Sam
Rickey, Branch
Rixey, Eppa
Rizzuto, Phil (Scooter)
Roberts, Robin
#Robinson, Brooks
#Robinson, Frank
#Robinson, Jackie
Robinson, Wilbert
Rogan, Joe "Bullet"
Roush, Edd
Ruffing, Red
Rusie, Amos
Ruth, Babe[2]
#Ryan, Nolan
Schalk, Ray
#Schmidt, Mike
Schoendienst, Red
#Seaver, Tom
Selee, Frank
Sewell, Joe
Simmons, Al
Sisler, George
Slaughter, Enos
*Smith, Hilton
Snider, Duke
#Spahn, Warren
Spalding, Albert
Speaker, Tris
#Stargell, Willie
Stearnes, Norman "Turkey"
Stengel, Casey
Sutton, Don
Terry, Bill
Thompson, Sam
Tinker, Joe
Traynor, Pie
Vance, Dazzy
Vaughan, Arky
Veeck, Bill
Waddell, Rube
Wagner, Honus[2]
Wallace, Roderick
Walsh, Ed
Waner, Lloyd
Waner, Paul
Ward, John
Weaver, Earl
Weiss, George
Welch, Mickey
Wells, Willie
Wheat, Zach
Wilhelm, Hoyt
Williams, Billy
Williams, Smokey Joe
#Williams, Ted
Williams, Vic
Wilson, Hack
*#Winfield, Dave
Wright, George
Wright, Harry
Wynn, Early
#Yastrzemski, Carl
Yawkey, Tom
Young, Cy
Youngs, Ross
#Yount, Robin

(1) Player must generally be retired for five complete seasons before being eligible for induction. (2) Players inducted in 1936 (the year the Hall of Fame began). # Denotes players chosen in first year of Hall of Fame eligibility. *Denotes 2000 inductees. **NOTE:** Four players, Babe Ruth (1936), Lou Gehrig (1939), Joe DiMaggio (1955), and Roberto Clemente (1973), were inducted less than five years after retirement or, in Clemente's case, death.

All-Star Baseball Games, 1933-2001

Year	Winner, Score	Host team
1933*	American, 4-2	Chicago (AL)
1934*	American, 9-7	New York (NL)
1935*	American, 4-1	Cleveland
1936*	National, 4-3	Boston (NL)
1937*	American, 8-3	Washington
1938*	National, 4-1	Cincinnati
1939*	American, 3-1	New York (AL)
1940*	National, 4-0	St. Louis (NL)
1941*	American, 7-5	Detroit
1942	American, 3-1	New York (NL)
1943	American, 5-3	Philadelphia (AL)
1944	National, 7-1	Pittsburgh
1945	(Not played)	
1946*	American, 12-0	Boston (AL)
1947*	American, 2-1	Chicago (NL)
1948*	American, 5-2	St. Louis (AL)
1949*	American, 11-7	Brooklyn
1950*	National, 4-3[1]	Chicago (AL)
1951*	National, 8-3	Detroit
1952*	National, 3-2	Philadelphia (NL)
1953*	National, 5-1	Cincinnati
1954*	American, 11-9	Cleveland
1955*	National, 6-5[2]	Milwaukee
1956*	National, 7-3	Washington
1957*	American, 6-5	St. Louis
1958*	American, 4-3	Baltimore
1959*	National, 5-4	Pittsburgh
1959*	American, 5-3	Los Angeles
1960*	National, 5-3	Kansas City
1960*	National, 6-0	New York (AL)
1961*	National, 5-4[3]	San Francisco
1961*	Called–rain, 1-1	Boston
1962*	National, 3-1[3]	Washington
1962*	American, 9-4	Chicago (NL)
1963*	National, 5-3	Cleveland
1964*	National, 7-4	New York (NL)
1965*	National, 6-5	Minnesota
1966*	National, 2-1[3]	St. Louis
1967*	National, 2-1[4]	California
1968	National, 1-0	Houston
1969*	National, 9-3	Washington
1970	National, 5-4[2]	Cincinnati
1971	American, 6-4	Detroit
1972	National, 4-3[3]	Atlanta
1973	National, 7-1	Kansas City
1974	National, 7-2	Pittsburgh
1975	National, 6-3	Milwaukee
1976	National, 7-1	Philadelphia
1977	National, 7-5	New York (AL)
1978	National, 7-3	San Diego
1979	National, 7-6	Seattle
1980	National, 4-2	Los Angeles
1981	National, 5-4	Cleveland
1982	National, 4-1	Montreal
1983	American, 13-3	Chicago (AL)
1984	National, 3-1	San Francisco
1985	National, 6-1	Minnesota
1986	American, 3-2	Houston
1987	National, 2-0[5]	Oakland
1988	American, 2-1	Cincinnati
1989	American, 5-3	California
1990	American, 2-0	Chicago (NL)
1991	American, 4-2	Toronto
1992	American, 13-6	San Diego
1993	American, 9-3	Baltimore
1994	National, 8-7[3]	Pittsburgh
1995	National, 3-2	Texas
1996	National, 6-0	Philadelphia
1997	American, 3-1	Cleveland
1998	American, 13-8	Colorado
1999	American, 4-1	Boston
2000	American, 6-3	Atlanta
2001	American, 4-1	Seattle

*Denotes day game. (1) 14 innings. (2) 12 innings. (3) 10 innings. (4) 15 innings. (5) 13 innings.

Major League Leaders in 2001

American League

Batting

I. Suzuki, Seattle, .350; Ja. Giambi, Oakland, .342; R. Alomar, Cleveland, .336; B. Boone, Seattle, .331; F. Catalanotto, Texas, .330.

Runs

A. Rodriguez, Texas, 133; I. Suzuki, Seattle, 127; B. Boone, Seattle, 118; R. Alomar, Cleveland, 113; D. Jeter, N.Y., 110.

Runs Batted In

B. Boone, Seattle, 141; J. Gonzalez, Cleveland, 140; A. Rodriguez, Texas, 135; M. Ramirez, Boston, 125; J. Thome, Cleveland, 124.

Hits

I. Suzuki, Seattle, 242; B. Boone, Seattle, 206; S. Stewart, Toronto, 202; A. Rodriguez, Texas, 201; G. Anderson, Anaheim, 194.

Doubles

Ja. Giambi, Oakland, 47; M. Sweeney, Kansas City, 46; S. Stewart, Toronto, 44; E. Chavez, Oakland, 43; R. Durham, Chicago, 42.

Triples

C. Guzman, Minnesota, 14; R. Alomar, Cleveland, 12; C. Beltran, Kansas City, 12; R. Cedeno, Detroit, 11; R. Durham, Chicago, 10.

Home Runs

A. Rodriguez, Texas, 52; J. Thome, Cleveland, 49; R. Palmeiro, Texas, 47; T. Glaus, Anaheim, 41; M. Ramirez, Boston, 41.

Stolen Bases

I. Suzuki, Seattle, 56; R. Cedeno, Detroit, 55; A. Soriano, N.Y., 43; M. McLemore, Seattle, 39; C. Knoblauch, N.Y., 38.

Pitching
(Most wins: W-L, ERA, Pct.)

M. Mulder, Oakland, 21-8, 3.45, .724; R. Clemens, New York Yankees, 20-3, 3.51, .870; J. Moyer, Seattle, 20-6, 3.43, .769; F. Garcia, Seattle, 18-6, 3.05, .750; T. Hudson, Oakland, 18-9, 3.37, .667.

Strikeouts

H. Nomo, Boston, 220; M. Mussina, NY, 214; R. Clemens, N.Y., 188; B. Zito, Oakland, 205; B. Colon, Cleveland, 201.

Saves

M. Rivera, N.Y., 50; K. Sasaki, Seattle, 45; K. Foulke, Chicago, 42; T. Percival, Anaheim, 39; B. Koch, Toronto, 36.

National League

Batting

L. Walker, Colorado, .350; T. Helton, Colorado, .336; M. Alou, Houston, .331; L. Berkman, Houston, .331; C. Jones, Atlanta, .330.

Runs

S. Sosa, Chicago, 146; T. Helton, Colorado, 132; B. Bonds, San Francisco, 129; L. Gonzalez, Arizona, 128; J. Bagwell, Houston, 126.

Runs Batted In

S. Sosa, Chicago, 160; T. Helton, Colorado, 146; L. Gonzales, Arizona, 142; B. Bonds, SF, 137; J. Bagwell, Houston, 132; A. Pujols, St. Louis, 132.

Hits

R. Aurilia, San Francisco, 206; J. Pierre, Colorado, 202; L. Gonzalez, Arizona, 198; T. Helton, Colorado, 197; A. Pujols, St. Louis, 194.

Doubles

L. Berkman, Houston, 55; T. Helton, Colorado, 54; J. Kent, San Francisco, 49; B. Abreu, Philadelphia, 48; A. Pujols, St. Louis, 47.

Triples

J. Rollins, Philadelphia, 12; J. Pierre, Colorado, 11, J. Uribe, Colorado, 11; L. Castillo, Florida, 10; N. Perez, Colorado, 8; M. Tucker, Chicago Cubs, 8; F. Vina, St. Louis, 8.

Home Runs

B. Bonds, San Francisco, 73; S. Sosa, Chicago, 64; L. Gonzalez, Arizona, 57; S. Green, Los Angeles, 49; T. Helton, Colorado, 49.

Stolen Bases

J. Pierre, Colorado, 46; J. Rollins, Philadelphia, 46; V. Guerro, Montreal, 37; B. Abreu, Philadelphia, 36; L. Castillo, Florida, 33.

Pitching
(Most wins: W-L, ERA, Pct.)

M. Morris, St. Louis, 22-8, 3.16, .733; C. Schilling, Arizona, 22-6, 2.98, .786; R. Johnson, Arizona, 21-6, 2.49, .778; J. Lieber, Chicago, 20-6, 3.80, .769; G. Maddux, Atlanta, 17-11, 3.05, .607; R. Ortiz, San Francisco, 17-9, 3.29, .654.

Strikeouts

R. Johnson, Arizona, 372; C. Schilling, Arizona, 293; C. Park, L.A., 218; K. Wood, Chicago, 217; J. Vazquez, Montreal, 208.

Saves

R. Nen, San Francisco, 45; A. Benitez, N.Y., 43; T. Hoffman, San Diego, 43; J. Shaw, Los Angeles, 43; J. Mesa, Philadelphia, 42.

Barry Bonds Hits 73 Home Runs

San Francisco Giants' OF Barry Bonds (right) reacts to his record-setting 73d home run of the 2001 season on Oct. 7, in Pacific Bell Park. Bonds tied Mark McGwire's 1998 single-season mark of 70 homers on Oct. 4, at Houston's Enron Field. At home against the Los Angeles Dodgers the next night, Bonds broke the record with his 1st swing in the 1st inning—a 442-foot shot into right-center off Chan Ho Park. In the 3d, Bonds tagged Park for number 72. His 73d came on the season's last day against L.A.'s Dennis Springer.

Bonds also broke Babe Ruth's 1923 single-season walk record of 170, with 177, and his .863 slugging percentage bettered Ruth's 1920 record of .847. Bonds was the 1st player to have an on-base percentage over .500 (.515) since Mickey Mantle and Ted Williams in 1957. An astounding 61.5% of Bonds' hits were home runs. In 2001, Bonds moved from 17th to 6th on the all-time career home run list, finishing with 567.

The Sluggers: Month by Month

Player, year	Mar.	Apr.	May	June	July	Aug.	Sept.	Oct.	Tot.	GP	AB	AB/HR
Barry Bonds, 2001 . . .	0	11	17	11	6	12	12	4	73	153	476	6.52
Mark McGwire, 1998 .	1	10	16	10	8	10	15	0	70	155	509	7.27
Sammy Sosa, 1998 . .	0	6	7	20	9	13	11	0	66	159	643	9.74
Roger Maris, 1961 . . .	0	1	11	15	13	11	9	1	61	161	590	9.67
Babe Ruth, 1927	0	4	12	9	9	9	17	0	60	151	540	9.00

GP = games played. AB = at bats. AB/HR = at bats per home run.

50 Home Run Club

Only Mark McGwire and Barry Bonds have ever hit 70 or more home runs in a season. Five players—including Babe Ruth and Roger Maris—have hit 60 or more, a feat Sammy Sosa accomplished for the 3d time in 2001. These 5 are at the pinnacle of a select group of players to have hit 50 or more homers in a season. The following list shows each time a player achieved this mark.

HR	Player, team	Year
73	Barry Bonds, San Francisco Giants	2001
70	Mark McGwire, St. Louis Cardinals	1998
66	Sammy Sosa, Chicago Cubs	1998
65	Mark McGwire, St. Louis Cardinals	1999
64	Sammy Sosa, Chicago Cubs	2001
63	Sammy Sosa, Chicago Cubs	1999
61	Roger Maris, N.Y. Yankees	1961
60	Babe Ruth, N.Y. Yankees	1927
59	Babe Ruth, N.Y. Yankees	1921
58	Jimmie Foxx, Philadelphia Athletics	1932
58	Hank Greenberg, Detroit Tigers	1938
58	Mark McGwire, Oakland A's/St. Louis Cardinals	1997
57	Luis Gonzalez, Arizona Diamondbacks	2001
56	Hack Wilson, Chicago Cubs	1930
56	Ken Griffey Jr., Seattle Mariners	1997
56	Ken Griffey Jr., Seattle Mariners	1998
54	Babe Ruth, N.Y. Yankees	1920
54	Babe Ruth, N.Y. Yankees	1928
54	Ralph Kiner, Pittsburgh Pirates	1949
54	Mickey Mantle, N.Y. Yankees	1961
52	Mickey Mantle, N.Y. Yankees	1956
52	Willie Mays, San Francisco Giants	1965
52	George Foster, Cincinnati Reds	1977
52	Mark McGwire, Oakland A's	1996
52	Alex Rodriguez, Texas Rangers	2001
51	Ralph Kiner, Pittsburgh Pirates	1947
51	Johnny Mize, N.Y. Giants	1947
51	Willie Mays, N.Y. Giants	1955
51	Cecil Fielder, Detroit Tigers	1990
50	Jimmie Foxx, Boston Red Sox	1938
50	Albert Belle, Cleveland Indians	1995
50	Brady Anderson, Baltimore Orioles	1996
50	Greg Vaughn, San Diego Padres	1998
50	Sammy Sosa, Chicago Cubs	2000

Earned Run Average Leaders

Year	National League				Year	American League			
	Player, team	G	IP	ERA		Player, team	G	IP	ERA
1977	John Candelaria, Pittsburgh	33	231	**2.34**	1977	Frank Tanana, California	31	241	**2.54**
1978	Craig Swan, New York	29	207	**2.43**	1978	Ron Guidry, New York	35	274	**1.74**
1979	J. R. Richard, Houston	38	292	**2.71**	1979	Ron Guidry, New York	33	236	**2.78**
1980	Don Sutton, Los Angeles	32	212	**2.21**	1980	Rudy May, New York	41	175	**2.47**
1981	Nolan Ryan, Houston	21	149	**1.69**	1981	Steve McCatty, Oakland	22	186	**2.32**
1982	Steve Rogers, Montreal	35	277	**2.40**	1982	Rick Sutcliffe, Cleveland	34	216	**2.96**
1983	Atlee Hammaker, San Francisco	23	172	**2.25**	1983	Rick Honeycutt, Texas	25	174	**2.42**
1984	Alejandro Pena, Los Angeles	28	199	**2.48**	1984	Mike Boddicker, Baltimore	34	261	**2.79**
1985	Dwight Gooden, New York	35	276	**1.53**	1985	Dave Stieb, Toronto	36	265	**2.48**
1986	Mike Scott, Houston	37	275	**2.22**	1986	Roger Clemens, Boston	33	254	**2.48**
1987	Nolan Ryan, Houston	34	211	**2.76**	1987	Jimmy Key, Toronto	36	261	**2.76**
1988	Joe Magrane, St. Louis	24	165	**2.18**	1988	Allan Anderson, Minnesota	30	202	**2.45**
1989	Scott Garrelts, San Francisco	30	193	**2.28**	1989	Bret Saberhagen, Kansas City	36	262	**2.16**
1990	Danny Darwin, Houston	48	162	**2.21**	1990	Roger Clemens, Boston	31	228	**1.93**
1991	Dennis Martinez, Montreal	31	222	**2.39**	1991	Roger Clemens, Boston	35	271	**2.62**
1992	Bill Swift, San Francisco	30	164	**2.08**	1992	Roger Clemens, Boston	32	246	**2.41**
1993	Greg Maddux, Atlanta	36	267	**2.36**	1993	Kevin Appier, Kansas City	34	238	**2.56**
1994	Greg Maddux, Atlanta	25	202	**1.56**	1994	Steve Ontiveros, Oakland	27	115	**2.65**
1995	Greg Maddux, Atlanta	28	209	**1.63**	1995	Randy Johnson, Seattle	30	214	**2.48**
1996	Kevin Brown, Florida	32	233	**1.89**	1996	Juan Guzman, Toronto	27	187	**2.93**
1997	Pedro Martinez, Montrea	31	241	**1.90**	1997	Roger Clemens, Toronto	34	264	**2.05**
1998	Greg Maddux, Atlanta	34	251	**2.22**	1998	Roger Clemens, Toronto	33	234	**2.65**
1999	Randy Johnson, Arizona	35	271	**2.48**	1999	Pedro Martinez, Boston	31	213	**2.07**
2000	Kevin K. Brown, Los Angeles	33	230	**2.58**	2000	Pedro Martinez, Boston	29	217	**1.74**
2001	Randy Johnson, Arizona	35	249	**2.49**	2001	Freddy Garcia, Seattle	34	238	**3.05**

ERA is computed by multiplying earned runs allowed by 9, then dividing by innings pitched.

Strikeout Leaders

Note: Asterisk (*) indicates the all-time single-season record for each league.

National League			American League		
Year	Pitcher, Team	SO	Year	Pitcher, Team	SO
1901	Noodles Hahn, Cincinnati	239	1901	Cy Young, Boston	158
1902	Vic Willis, Boston	225	1902	Rube Waddell, Philadelphia	210
1903	Christy Mathewson, New York	267	1903	Rube Waddell, Philadelphia	302
1904	Christy Mathewson, New York	212	1904	Rube Waddell, Philadelphia	349

National League

Year	Pitcher, Team	SO
1905	Christy Mathewson, New York	206
1906	Fred Beebe, Chicago-St. Louis	171
1907	Christy Mathewson, New York	178
1908	Christy Mathewson, New York	259
1909	Orval Overall, Chicago	205
1910	Earl Moore, Philadelphia	185
1911	Rube Marquard, New York	237
1912	Grover Alexander, Philadelphia	195
1913	Tom Seaton, Philadelphia	168
1914	Grover Alexander, Philadelphia	214
1915	Grover Alexander, Philadelphia	241
1916	Grover Alexander, Philadelphia	167
1917	Grover Alexander, Philadelphia	201
1918	Hippo Vaughn, Chicago	148
1919	Hippo Vaughn, Chicago	141
1920	Grover Alexander, Chicago	173
1921	Burleigh Grimes, Brooklyn	136
1922	Dazzy Vance, Brooklyn	134
1923	Dazzy Vance, Brooklyn	197
1924	Dazzy Vance, Brooklyn	262
1925	Dazzy Vance, Brooklyn	221
1926	Dazzy Vance, Brooklyn	140
1927	Dazzy Vance, Brooklyn	184
1928	Dazzy Vance, Brooklyn	200
1929	Pat Malone, Chicago	166
1930	Bill Hallahan, St. Louis	177
1931	Bill Hallahan, St. Louis	159
1932	Dizzy Dean, St. Louis	191
1933	Dizzy Dean, St. Louis	199
1934	Dizzy Dean, St. Louis	195
1935	Dizzy Dean, St. Louis	190
1936	Van Lingle Mungo, Brooklyn	238
1937	Carl Hubbell, New York	159
1938	Clay Bryant, Chicago	135
1939	Claude Passeau, Philadelphia-Chicago Bucky Walters, Cincinnati	137
1940	Kirby Higbe, Philadelphia	137
1941	John Vander Meer, Cincinnati	202
1942	John Vander Meer, Cincinnati	186
1943	John Vander Meer, Cincinnati	174
1944	Bill Voiselle, New York	161
1945	Preacher Roe, Pittsburgh	148
1946	Johnny Schmitz, Cincinnati	135
1947	Ewell Blackwell, Cincinnati	193
1948	Harry Brecheen, St. Louis	149
1949	Warren Spahn, Boston	151
1950	Warren Spahn, Boston	191
1951	Warren Spahn, Boston Don Newcombe, Brooklyn	164
1952	Warren Spahn, Boston	183
1953	Robin Roberts, Philadelphia	198
1954	Robin Roberts, Philadelphia	185
1955	Sam Jones, Chicago	198
1956	Sam Jones, Chicago	176
1957	Jack Sanford, Philadelphia	188
1958	Sam Jones, St. Louis	225
1959	Don Drysdale, Los Angeles	242
1960	Don Drysdale, Los Angeles	246
1961	Sandy Koufax, Los Angeles	269
1962	Don Drysdale, Los Angeles	232
1963	Sandy Koufax, Los Angeles	306
1964	Bob Veale, Pittsburgh	250
1965	Sandy Koufax, Los Angeles	*382
1966	Sandy Koufax, Los Angeles	317
1967	Jim Bunning, Philadelphia	253
1968	Bob Gibson, St. Louis	268
1969	Ferguson Jenkins, Chicago	273
1970	Tom Seaver, New York	283
1971	Tom Seaver, New York	289
1972	Steve Carlton, Philadelphia	310
1973	Tom Seaver, New York	251
1974	Steve Carlton, Philadelphia	240
1975	Tom Seaver, New York	243
1976	Tom Seaver, New York	235
1977	Phil Niekro, Atlanta	262
1978	J.R. Richard, Houston	303
1979	J.R. Richard, Houston	313
1980	Steve Carlton, Philadelphia	286
1981	Fernando Valenzuela, Los Angeles	180
1982	Steve Carlton, Philadelphia	286
1983	Steve Carlton, Philadelphia	275
1984	Dwight Gooden, New York	276
1985	Dwight Gooden, New York	268
1986	Mike Scott, Houston	306
1987	Nolan Ryan, Houston	270
1988	Nolan Ryan, Houston	228
1989	Jose DeLeon, St. Louis	201

American League

Year	Pitcher, Team	SO
1905	Rube Waddell, Philadelphia	287
1906	Rube Waddell, Philadelphia	196
1907	Rube Waddell, Philadelphia	232
1908	Ed Walsh, Chicago	269
1909	Frank Smith, Chicago	177
1910	Walter Johnson, Washington	313
1911	Ed Walsh, Chicago	255
1912	Walter Johnson, Washington	303
1913	Walter Johnson, Washington	243
1914	Walter Johnson, Washington	225
1915	Walter Johnson, Washington	203
1916	Walter Johnson, Washington	228
1917	Walter Johnson, Washington	188
1918	Walter Johnson, Washington	162
1919	Walter Johnson, Washington	147
1920	Stan Coveleski, Cleveland	133
1921	Walter Johnson, Washington	143
1922	Urban Shocker, St. Louis	149
1923	Walter Johnson, Washington	130
1924	Walter Johnson, Washington	158
1925	Lefty Grove, Philadelphia	116
1926	Lefty Grove, Philadelphia	194
1927	Lefty Grove, Philadelphia	174
1928	Lefty Grove, Philadelphia	183
1929	Lefty Grove, Philadelphia	170
1930	Lefty Grove, Philadelphia	209
1931	Lefty Grove, Philadelphia	175
1932	Red Ruffing, New York	190
1933	Lefty Gomez, New York	163
1934	Lefty Gomez, New York	158
1935	Tommy Bridges, Detroit	163
1936	Tommy Bridges, Detroit	175
1937	Lefty Gomez, New York	194
1938	Bob Feller, Cleveland	240
1939	Bob Feller, Cleveland	246
1940	Bob Feller, Cleveland	261
1941	Bob Feller, Cleveland	260
1942	Tex Hughson, Boston Bobo Newsom, Washington	113
1943	Allie Reynolds, Cleveland	151
1944	Hal Newhouser, Detroit	187
1945	Hal Newhouser, Detroit	212
1946	Bob Feller, Cleveland	348
1947	Bob Feller, Cleveland	196
1948	Bob Feller, Cleveland	164
1949	Virgil Trucks, Detroit	153
1950	Bob Lemon, Cleveland	170
1951	Vic Raschi, New York	164
1952	Allie Reynolds, New York	160
1953	Billy Pierce, Chicago	186
1954	Bob Turley, Baltimore	185
1955	Herb Score, Cleveland	245
1956	Herb Score, Cleveland	263
1957	Early Wynn, Cleveland	184
1958	Early Wynn, Chicago	179
1959	Jim Bunning, Detroit	201
1960	Jim Bunning, Detroit	201
1961	Camilo Pacual, Minnesota	221
1962	Camilo Pacual, Minnesota	206
1963	Camilo Pacual, Minnesota	202
1964	Al Downing, New York	217
1965	Sam McDowell, Cleveland	325
1966	Sam McDowell, Cleveland	225
1967	Jim Lonborg, Boston	246
1968	Sam McDowell, Cleveland	283
1969	Sam McDowell, Cleveland	279
1970	Sam McDowell, Cleveland	304
1971	Mickey Lolich, Detroit	308
1972	Nolan Ryan, California	329
1973	Nolan Ryan, California	*383
1974	Nolan Ryan, California	367
1975	Frank Tanana, California	269
1976	Nolan Ryan, California	327
1977	Nolan Ryan, California	341
1978	Nolan Ryan, California	260
1979	Nolan Ryan, California	223
1980	Len Barker, Cleveland	187
1981	Len Barker, Cleveland	127
1982	Floyd Bannister, Seattle	209
1983	Jack Morris, Detroit	232
1984	Mark Langston, Seattle	204
1985	Bert Blyleven, Cleveland-Minnesota	206
1986	Mark Langston, Seattle	245
1987	Mark Langston, Seattle	262
1988	Roger Clemens, Boston	291
1989	Nolan Ryan, Texas	301

National League

Year	Pitcher, Team	SO
1990	David Cone, New York	233
1991	David Cone, New York	241
1992	John Smotlz, Atlanta	215
1993	Jose Rijo, Cincinnati	227
1994	Andy Benes, San Diego	189
1995	Hideo Nomo, Los Angeles	236
1996	John Smotlz, Atlanta	276
1997	Curt Schilling, Philadelphia	319
1998	Curt Schilling, Philadelphia	300
1999	Randy Johnson, Arizona	364
2000	Randy Johnson, Arizona	347
2001	Randy Johnson, Arizona	372

American League

Year	Pitcher, Team	SO
1990	Nolan Ryan, Texas	232
1991	Roger Clemens, Boston	241
1992	Randy Johnson, Seattle	241
1993	Randy Johnson, Seattle	308
1994	Randy Johnson, Seattle	204
1995	Randy Johnson, Seattle	294
1996	Roger Clemens, Boston	257
1997	Roger Clemens, Toronto	292
1998	Roger Clemens, Toronto	271
1999	Pedro Martinez, Boston	313
2000	Pedro Martinez, Boston	284
2001	Hideo Nomo, Boston	220

Victory Leaders

Note: Asterisk (*) indicates the all-time single-season record for each league in the "modern" era beginning in 1901.

National League

Year	Pitcher, Team	Wins
1901	Bill Donavan, Brooklyn	25
1902	Jack Chesbro, Pittsburgh	28
1903	Joe McGinnity, New York	31
1904	Joe McGinnity, New York	35
1905	Christy Mathewson, New York	31
1906	Joe McGinnity, New York	27
1907	Christy Mathewson, New York	24
1908	Christy Mathewson, New York	*37
1909	Mordecai Brown, Chicago	27
1910	Christy Mathewson, New York	27
1911	Grover Alexander, Chicago	28
1912	Rube Marquard, New York	26
1913	Tom Seaton, Philadelphia	27
1914	Grover Alexander, Philadelphia	27
1915	Grover Alexander, Philadelphia	31
1916	Grover Alexander, Philadelphia	33
1917	Grover Alexander, Philadelphia	30
1918	Hippo Vaughn, Chicago	22
1919	Jesse Barnes, New York	25
1920	Grover Alexander, Philadelphia	27
1921	Burleigh Grimes, Brooklyn	22
1922	Eppa Rixey, Cincinnati	25
1923	Dolf Luque, Cincinnati	27
1924	Dazzy Vance, Brooklyn	28
1925	Dazzy Vance, Brooklyn	22
1926	Flint Rhem, St. Louis	20
1927	Charlie Root, Chicago	26
1928	Burleigh Grimes, Pittsburgh	25
1929	Pat Malone, Chicago	22
1930	Pat Malone, Chicago	20
1931	Heine Meine, Pittsburgh	19
1932	Lon Warneke, Chicago	22
1933	Carl Hubbell, New York	23
1934	Dizzy Dean, St. Louis	30
1935	Dizzy Dean, St. Louis	28
1936	Carl Hubbell, New York	26
1937	Carl Hubbell, New York	22
1938	Bill Lee, Chicago	22
1939	Bucky Walters, Cincinnati	27
1940	Bucky Walters, Cincinnati	22
1941	Whit Wyatt, Brooklyn	22
1942	Mort Cooper, St. Louis	22
1943	Rip Sewell, Pittsburgh	21
1944	Bucky Walters, Cincinnati	23
1945	Red Barrett, Boston-St. Louis	23
1946	Howie Pollet, St. Louis	21
1947	Ewell Blackwell, Cincinnati	22
1948	Johnny Sain, Boston	24
1949	Warren Spahn, Boston	21
1950	Warren Spahn, Boston	21
1951	Sal Maglie, New York	23
1952	Robin Roberts, Philadelphia	28
1953	Warren Spahn, Milwaukee	23
1954	Robin Roberts, Philadelphia	23
1955	Robin Roberts, Philadelphia	23
1956	Don Newcombe, Brooklyn	27
1957	Warren Spahn, Milwaukee	21
1958	Warren Spahn, Milwaukee	22
1959	Warren Spahn, Milwaukee	21
1960	Warren Spahn, Milwaukee	21
1961	Warren Spahn, Milwaukee	21
1962	Don Drysdale, Los Angeles	25
1963	Juan Marichal, San Francisco	25
1964	Larry Jackson, Chicago	24
1965	Sandy Koufax, Los Angeles	26
1966	Sandy Koufax, Los Angeles	27
1967	Mike McCormick, San Francisco	22
1968	Juan Marichal, San Francisco	26
1969	Tom Seaver, New York	25
1970	Gaylord Perry, San Francisco	23

American League

Year	Pitcher, Team	Wins
1901	Cy Young, Boston	33
1902	Cy Young, Boston	32
1903	Cy Young, Boston	28
1904	Jack Chesbro, New York	*41
1905	Rube Waddell, Philadelphia	27
1906	Al Orth, New York	27
1907	Doc White, Chicago	27
1908	Ed Walsh, Chicago	40
1909	George Mullin, Detroit	29
1910	Jack Coombs, Philadelphia	31
1911	Jack Coombs, Philadelphia	28
1912	Joe Wood, Boston	34
1913	Walter Johnson, Washington	36
1914	Walter Johnson, Washington	28
1915	Walter Johnson, Washington	27
1916	Walter Johnson, Washington	25
1917	Eddie Cicotte, Chicago	28
1918	Walter Johnson, Washington	23
1919	Eddie Cicotte, Chicago	29
1920	Jim Bagby, Cleveland	31
1921	Urban Shocker, St. Louis	27
1922	Eddie Rommel, Philadelphia	27
1923	George Uhle, Cleveland	26
1924	Walter Johnson, Washington	23
1925	Eddie Rommel, Philadelphia	21
1926	George Uhle, Cleveland	27
1927	Ted Lyons, Chicago	22
1928	George Pipgras, New York	24
1929	George Earnshaw, Philadelphia	24
1930	Lefty Grove, Philadelphia	28
1931	Lefty Grove, Philadelphia	31
1932	Alvin Crowder, Washington	26
1933	Lefty Grove, Philadelphia	24
1934	Lefty Gomez, New York	26
1935	Wes Ferrell, Boston	25
1936	Tommy Bridges, Detroit	23
1937	Lefty Gomez, New York	21
1938	Red Ruffing, New York	21
1939	Bob Feller, Cleveland	24
1940	Bob Feller, Cleveland	27
1941	Bob Feller, Cleveland	25
1942	Tex Hughson, Boston	22
1943	Dizzy Trout, Detroit	20
1944	Hal Newhouser, Detroit	29
1945	Hal Newhouser, Detroit	25
1946	Hal Newhouser, Detroit	26
1947	Bob Feller, Cleveland	20
1948	Hal Newhouser, Detroit	21
1949	Mel Parnell, Boston	25
1950	Bob Lemon, Cleveland	23
1951	Bob Feller, Cleveland	22
1952	Bobby Shantz, Philadelphia	24
1953	Bob Porterfield, Washington	22
1954	Early Wynn, Cleveland	23
1955	Frank Sullivan, Boston	18
1956	Frank Lary, Detroit	21
1957	Billy Pierce, Chicago	20
1958	Bob Turley, New York	21
1959	Early Wynn, Chicago	22
1960	Jim Perry, Cleveland	18
1961	Whitey Ford, New York	25
1962	Ralph Terry, New York	23
1963	Whitey Ford, New York	24
1964	Gary Peters, Chicago	20
1965	Mudcat (Jim) Grant, Minnesota	21
1966	Jim Kaat, Minnesota	25
1967	Earl Wilson, Detroit	22
1968	Denny McLain, Detroit	31
1969	Denny McLain, Detroit	24
1970	Jim Perry, Minnesota	24

National League

Year	Pitcher, Team	Wins
1971	Fergie Jenkins, Chicago	24
1972	Steve Carlton, Philadelphia	27
1973	Ron Bryant, San Francisco	24
1974	Phil Niekro, Atlanta	20
1975	Tom Seaver, New York	22
1976	Randy Jones, San Diego	22
1977	Steve Carlton, Philadelphia	23
1978	Gaylord Perry, San Diego	21
1979	Phil Niekro, Atlanta	21
1980	Steve Carlton, Philadelphia	24
1981	Tom Seaver, Cincinnati	14
1982	Steve Carlton, Philadelphia	23
1983	John Denny, Philadelphia	19
1984	Joaquin Andujar, St. Louis	20
1985	Dwight Gooden, New York	24
1986	Fernando Valezuela, Los Angeles	21
1987	Rick Sutcliffe, Chicago	18
1988	Danny Jackson, Cincinnati	23
1989	Mike Scott, Houston	20
1990	Doug Drabek, Pittsburgh	22
1991	John Smiley, Pittsburgh	20
1992	Greg Maddux, Chicago	20
1993	Tom Glavine, Atlanta	22
1994	Greg Maddux, Atlanta	16
1995	Greg Maddux, Atlanta	19
1996	John Smotz, Atlanta	24
1997	Denny Neagle, Atlanta	20
1998	Tom Glavine, Atlanta	20
1999	Mike Hampton, Houston	22
2000	Tom Glavine, Atlanta	21
2001	Matt Morris, St. Louis; Curt Schilling, Arizona	22

American League

Year	Pitcher, Team	Wins
1971	Mickey Lolich, Detroit	25
1972	Wilbur Wood, Chicago	24
1973	Wilbur Wood, Chicago	24
1974	Fergie Jenkins, Texas	25
1975	Jim Palmer, Baltimore	23
1976	Jim Palmer, Baltimore	22
1977	Jim Palmer, Baltimore	20
1978	Ron Guidry, New York	25
1979	Mike Flanagan, Baltimore	23
1980	Steve Stone, Baltimore	25
1981	Pete Vuckovich, Milwaukee	14
1982	La Marr Hoyt, Chicago	19
1983	La Marr Hoyt, Chicago	24
1984	Mike Boddicker, Baltimore	20
1985	Ron Guidry, New York	22
1986	Roger Clemens, Boston	24
1987	Dave Stewart, Oakland	20
1988	Frank Viola, Minnesota	24
1989	Bret Saberhagen, Kansas City	23
1990	Bob Welch, Oakland	27
1991	Bill Gullickson, Detroit	20
1992	Jack Morris, Toronto	21
1993	Jack McDowell, Chicago	22
1994	Jimmy Key, New York	17
1995	Mike Mussina, Baltimore	19
1996	Andy Pettitte, New York	21
1997	Roger Clemens, Toronto	21
1998	Rick Helling, Texas; Roger Clemens, Toronto	20
1999	Pedro Martinez, Boston	23
2000	David Wells, Toronto	20
2001	Mark Mulder, Oakland	21

World Series Results, 1903-2000

1903	Boston AL 5, Pittsburgh NL 3
1904	No series
1905	New York NL 4, Philadelphia AL 1
1906	Chicago AL 4, Chicago NL 2
1907	Chicago NL 4, Detroit AL 0, 1 tie
1908	Chicago NL 4, Detroit AL 1
1909	Pittsburgh NL 4, Detroit AL 3
1910	Philadelphia AL 4, Chicago NL 1
1911	Philadelphia AL 4, New York NL 2
1912	Boston AL 4, NewYork NL 3, 1 tie
1913	Philadelphia AL 4, New York NL 1
1914	Boston NL 4, Philadelphia AL 0
1915	Boston AL 4, Philadelphia NL 1
1916	Boston AL 4, Brooklyn NL 1
1917	Chicago AL 4, New York NL 2
1918	Boston AL 4, Chicago NL 2
1919	Cincinnati NL 5, Chicago AL 3
1920	Cleveland AL 5, Brooklyn NL 2
1921	New York NL 5, New York AL 3
1922	New York NL 4, NewYork AL 0, 1 tie
1923	New York AL 4, New York NL 2
1924	Washington AL 4, New York NL 3
1925	Pittsburgh NL 4, Washington AL 3
1926	St. Louis NL 4, New York AL 3
1927	New York AL 4, Pittsburgh NL 0
1928	New York AL 4, St. Louis NL 0
1929	Philadelphia AL 4, Chicago NL 1
1930	Philadelphia AL 4, St. Louis NL 2
1931	St. Louis NL 4, Philadelphia AL 3
1932	New York AL 4, Chicago NL 0
1933	New York NL 4, Washington AL 1
1934	St. Louis NL 4, Detroit AL 3
1935	Detroit AL 4, Chicago NL 2
1936	New York AL 4, New York NL 2
1937	New York AL 4, New York NL 1
1938	New York AL 4, Chicago NL 0
1939	New York AL 4, Cincinnati NL 0
1940	Cincinnati NL 4, Detroit AL 3
1941	New York AL 4, Brooklyn NL 1
1942	St. Louis NL 4, New York AL 1
1943	New York AL 4, St. Louis NL 1
1944	St. Louis NL 4, St. Louis AL 2
1945	Detroit AL 4, Chicago NL 3
1946	St. Louis NL 4, Boston AL 3
1947	New York AL 4, Brooklyn NL 3
1948	Cleveland AL 4, Boston NL 2
1949	New York AL 4, Brooklyn NL 1
1950	New York AL 4, Philadelphia NL 0
1951	New York AL 4, New York NL 2
1952	New York AL 4, Brooklyn NL 3
1953	New York AL 4, Brooklyn NL 2
1954	New York NL 4, Cleveland AL 0
1955	Brooklyn NL 4, New York AL 3
1956	New York AL 4, Brooklyn NL 3
1957	Milwaukee NL 4, New York AL 3
1958	New York AL 4, Milwaukee NL 3
1959	Los Angeles NL 4, Chicago AL 2
1960	Pittsburgh NL 4, New York AL 3
1961	New York AL 4, Cincinnati NL 1
1962	New York AL 4, San Francisco NL 3
1963	Los Angeles NL 4, New York AL 0
1964	St. Louis NL 4, New York AL 3
1965	Los Angeles NL 4, Minnesota AL 3
1966	Baltimore AL 4, Los Angeles NL 0
1967	St. Louis NL 4, Boston AL 3
1968	Detroit AL 4, St. Louis NL 3
1969	New York NL 4, Baltimore AL 1
1970	Baltimore AL 4, Cincinnati NL 1
1971	Pittsburgh NL 4, Baltimore AL 3
1972	Oakland AL 4, Cincinnati NL 3
1973	Oakland AL 4, New York NL 3
1974	Oakland AL 4, Los Angeles NL 1
1975	Cincinnati NL 4, Boston AL 3
1976	Cincinnati NL 4, New York AL 0
1977	New York AL 4, Los Angeles NL 2
1978	New York AL 4, Los Angeles NL 2
1979	Pittsburgh NL 4, Baltimore AL 3
1980	Philadelphia NL 4, Kansas City AL 2
1981	Los Angeles NL 4, New York AL 2
1982	St. Louis NL 4, Milwaukee AL 3
1983	Baltimore AL 4, Philadelphia NL 1
1984	Detroit AL 4, San Diego NL 1
1985	Kansas City AL 4, St. Louis NL 3
1986	New York NL 4, Boston AL 3
1987	Minnesota AL 4, St. Louis NL 3
1988	Los Angeles NL 4, Oakland AL 1
1989	Oakland AL 4, San Francisco NL 0
1990	Cincinnati NL 4, Oakland AL 0
1991	Minnesota AL 4, Atlanta NL 3
1992	Toronto AL 4, Atlanta NL 2
1993	Toronto AL 4, Philadelphia NL 2
1994	No series
1995	Atlanta NL 4, Cleveland AL 2
1996	New York AL 4, Atlanta NL 2
1997	Florida NL 4, Cleveland AL 3
1998	New York AL 4, San Diego NL 0
1999	New York AL 4, Atlanta NL 0
2000	New York AL 4, New York NL 1

World Series MVP

Year	Player, Position, Team
1955	Johnny Podres, p, Brooklyn
1956	Don Larsen, p, New York, AL
1957	Lew Burdette, p, Milwaukee, NL
1958	Bob Turley, p, NY AL
1959	Larry Sherry, p, LA
1960[1]	Bobby Richardson, 2b, NY, AL
1961	Whitey Ford, p, NY, AL
1962	Ralph Terry, p, NY, AL
1963	Sandy Koufax, p, Los Angeles, NL
1964	Bob Gibson, p, St. Louis
1965	Sandy Koufax, p, Los Angeles, NL
1966	Frank Robinson, of, Baltimore
1967	Bob Gibson, p, St. Louis
1968	Mickey Lolich, p, Detroit
1969	Donn Clendenon, 1b, NY, NL
1970	Brooks Robinson, 3b, Baltimore

Year	Player, Position, Team
1971	Roberto Clemente, of, Pittsburgh
1972	Gene Tenace, c, Oakland
1973	Reggie Jackson, of, Oakland
1974	Rollie Fingers, p, Oakland
1975	Pete Rose, 3b, Cincinnati
1976	Johnny Bench, c, Cincinnati
1977	Reggie Jackson, of, NY, AL
1978	Bucky Dent, ss, NY, AL
1979	Willie Stargell, 1b, Pittsburgh
1980	Mike Schmidt, 3b, Philadelphia
1981	Ron Cey, 3b, LA
	Pedro Guerrero, of, LA
	Steve Yeager, c, LA
1982	Darrell Porter, c, St. Louis
1983	Rick Dempsey, c, Baltimore
1984	Alan Trammell, ss, Detroit

Year	Player, Position, Team
1985	Bret Saberhagen, p, Kansas City
1986	Ray Knight, 3b, NY, NL
1987	Frank Viola, p, Minnesota
1988	Orel Hershiser, p, LA
1989	Dave Stewart, p, Oakland
1990	Jose Rijo, p, Cincinnati
1991	Jack Morris, p, Minnesota
1992	Pat Borders, c, Toronto
1993	Paul Molitor, dh, Toronto
1994	no series
1995	Tom Glavine, p, Atlanta
1996	John Wetteland, p, NY, AL
1997	Livan Hernandez, p, Florida
1998	Scott Brosius, 3b, NY, AL
1999	Mariano Rivera, p, NY, AL
2000	Derek Jeter, ss, NY, AL

(1) Bobby Richardson won the MVP although Pittsburgh beat New York.

World Series Won-Lost Records, by Franchise[1]

Team	Wins	Losses
New York Yankees	26	11
Philadelphia/Kansas City/Oakland A's	9	5
St. Louis Cardinals	9	6
Brooklyn/Los Angeles Dodgers	6	12
Pittsburgh Pirates	5	2
Boston Red Sox	5	4
Cincinnati Reds	5	4
New York/San Francisco Giants	5	11
Detroit Tigers	4	5
Washington Senators/Minnesota Twins	3	3
St. Louis Browns/Baltimore Orioles	3	4

(1) Through 2000.

Team	Wins	Losses
Boston/Milwaukee/Atlanta Braves	3	6
Toronto Blue Jays	2	0
New York Mets	2	2
Chicago White Sox	2	2
Cleveland Indians	2	3
Chicago Cubs	2	8
Florida Marlins	1	0
Kansas City Royals	1	1
Philadelphia Phillies	1	4
Seattle Pilots/Milwaukee Brewers	0	1
San Diego Padres	0	2

All-Time Major League Leaders

(*player in 2001 season)

Games	
Pete Rose	3,562
Carl Yastrzemski	3,308
Hank Aaron	3,298
Ty Cobb	3,035
Eddie Murray	3,026
Stan Musial	3,026
Cal Ripken Jr.*	3,001
Willie Mays	2,992
Rickey Henderson*	2,979
Dave Winfield	2,973

At Bats	
Pete Rose	14,053
Hank Aaron	12,364
Carl Yastrzemski	11,988
Cal Ripken Jr.*	11,551
Ty Cobb	11,434
Eddie Murray	11,336
Robin Yount	11,008
Dave Winfield	11,003
Stan Musial	10,972
Willie Mays	10,881

Runs Batted In	
Hank Aaron	2,297
Babe Ruth	2,213
Lou Gehrig	1,995
Stan Musial	1,951
Ty Cobb	1,937
Jimmie Foxx	1,922
Eddie Murray	1,917
Willie Mays	1,903
Cap Anson	1,879
Mel Ott	1,860

Runs	
Ricky Henderson*	2,248
Ty Cobb	2,246
Hank Aaron	2,174
Babe Ruth	2,174
Pete Rose	2,165
Willie Mays	2,062
Stan Musial	1,949
Lou Gehrig	1,888
Tris Speaker	1,882
Mel Ott	1,859

Stolen Bases	
Rickey Henderson*	1,395
Lou Brock	938
Billy Hamilton	912
Ty Cobb	892
Tim Raines*	808
Vince Coleman	752
Eddie Collins	744
Arlie Latham	739
Max Carey	738
Honus Wagner	722

Triples	
Sam Crawford	309
Ty Cobb	295
Honus Wagner	252
Jake Beckley	243
Roger Connor	233
Tris Speaker	222
Fred Clarke	220
Dan Brouthers	205
Joe Kelley	194
Paul Waner	191

Doubles	
Tris Speaker	792
Pete Rose	746
Stan Musial	725
Ty Cobb	724
George Brett	665
Nap Lajoie	657
Carl Yastrzemski	646
Honus Wagner	640
Hank Aaron	624
Paul Moliter	605
Paul Waner	605

Walks	
Ricky Henderson*	2,141
Babe Ruth	2,062
Ted Williams	2,019
Joe Morgan	1,865
Carl Yastrzemski	1,845
Mickey Mantle	1,733
Barry Bonds*	1,724
Mel Ott	1,708
Eddie Yost	1,614
Darrell Evans	1,605

Strikeouts	
Nolan Ryan	5,714
Steve Carlton	4,136
Roger Clemens*	3,717
Bert Blyleven	3,701
Tom Seaver	3,640
Don Sutton	3,574
Gaylord Perry	3,534
Walter Johnson	3,509
Randy Johnson*	3,412
Phil Niekro	3,342

Saves	
Lee Smith	478
John Franco*	422
Dennis Eckersley	390
Jeff Reardon	367
Randy Myers	347
Rollie Fingers	341
John Wetteland	330
Rick Aguilera*	318
Trevor Hoffman*	314
Tom Henke	311

Shutouts	
Walter Johnson	110
Grover Alexander	90
Christy Mathewson	79
Cy Young	76
Eddie Plank	69
Warren Spahn	63
Nolan Ryan	61
Tom Seaver	61
Bert Blyleven	60
Don Sutton	58

Losses	
Cy Young	316
Jim Galvin	308
Nolan Ryan	292
Walter Johnson	279
Phil Niekro	274
Gaylord Perry	265
Don Sutton	256
Jack Powell	254
Eppa Rixey	251
Bert Blyleven	250

All-Time Home Run Leaders

Player	HR
Hank Aaron	755
Babe Ruth	714
Willie Mays	660
Frank Robinson	586
Mark McGwire*	583
Harmon Killebrew	573
Barry Bonds*	567
Reggie Jackson	563
Mike Schmidt	548
Mickey Mantle	536
Jimmie Foxx	534
Willie McCovey	521
Ted Williams	521

Player	HR
Ernie Banks	512
Ed Mathews	512
Mel Ott	511
Eddie Murray	504
Lou Gehrig	493
Stan Musial	475
Willie Stargell	475
Dave Winfield	465
Jose Canseco*	462
Ken Griffey Jr.*	460
Carl Yastrzemski	452
Sammy Sosa*	450
Fred McGriff*	448

Player	HR
Rafael Palmeiro*	447
Dave Kingman	442
Andre Dawson	438
Cal Ripken Jr.*	431
Billy Williams	426
Darrell Evans	414
Duke Snider	407
Al Kaline	399
Dale Murphy	398
Juan Gonzalez*	397
Joe Carter	396
Graig Nettles	390

Player	HR
Johnny Bench	389
Dwight Evans	385
Harold Baines*	384
Frank Howard	382
Jim Rice	382
Albert Belle	381
Orlando Cepeda	379
Tony Perez	379
Norm Cash	377
Andres Galarraga*	377
Carlton Fisk	376
Rocky Colavito	374

Players With 3,000 Major League Hits

Player	Hits
Pete Rose	4,256
Ty Cobb	4,189
Hank Aaron	3,771
Stan Musial	3,630
Tris Speaker	3,514
Carl Yastrzemski	3,419

Player	Hits
Honus Wagner	3,415
Paul Molitor	3,319
Eddie Collins	3,315
Willie Mays	3,283
Eddie Murray	3,255
Nap Lajoie	3,242

Player	Hits
Cal Ripken Jr.*	3,184
George Brett	3,154
Paul Waner	3,152
Robin Yount	3,142
Tony Gwynn*	3,141
Dave Winfield	3,110

Player	Hits
Rod Carew	3,053
Lou Brock	3,023
Wade Boggs	3,010
Al Kaline	3,007
Roberto Clemente	3,000
Rickey Henderson*	3,000

Pitchers With 300 Major League Wins

Cy Young	511	Kid Nichols	361	Eddie Plank	326	Tom Seaver	311
Walter Johnson	417	Pud Galvin	360	Nolan Ryan	324	Charley Radbourn	309
Grover Alexander	373	Tim Keefe	342	Don Sutton	324	Mickey Welch	307
Christy Mathewson	373	Steve Carlton	329	Phil Niekro	318	Lefty Grove	300
Warren Spahn	363	John Clarkson	328	Gaylord Perry	314	Early Wynn	300

All-Time Major League Single-Season Leaders

(*player active in 2001 season; records for "modern" era beginning 1901)

Home Runs

Barry Bonds* (2001)	73
Mark McGwire*(1998)	70
Sammy Sosa* (1998)	66
Mark McGwire* (1999)	65
Sammy Sosa* (2001)	64

Batting Average

Nap Lajoie (1901)	.426
Rogers Hornsby (1924)	.424
George Sisler (1922)	.420
Ty Cobb (1912)	.420
Ty Cobb (1912)	.409

Earned Run Average

Dutch Leonard (1914)	0.96
Mordecai Brown (1906)	1.04
Bob Gibson (1968)	1.12
Walter Johnson (1913)	1.14
Christy Mathewson (1909)	1.14

Runs

Babe Ruth (1921)	177
Lou Gehrig (1936)	167
Lou Gehrig (1931)	163
Babe Ruth (1928)	163
Chuck Klein (1930)	158
Babe Ruth (1920, 1927)	158

Stolen Bases

Rickey Henderson* (1982)	130
Lou Brock (1974)	118
Vince Coleman (1985)	110
Vince Coleman (1987)	109
Rickey Henderson* (1983)	108

Wins

Jack Chesbro (1904)	41
Ed Walsh (1908)	40
Christy Mathewson (1908)	37
Walter Johnson (1913)	36
Joe McGinnity (1904)	35

Hits

George Sisler (1920)	257
Bill Terry (1930)	254
Lefty O'Doul (1929)	254
Al Simmons (1925)	253
Chuck Klein (1930)	250
Rogers Hornsby (1922)	250

Walks (Batter)

Barry Bonds* (2001)	177
Babe Ruth (1923)	170
Mark McGwire* (1998)	162
Ted Williams (1949)	162
Ted Williams (1947)	162

Strikeouts

Nolan Ryan (1973)	383
Sandy Koufax (1965)	382
Randy Johnson* (2001)	372
Nolan Ryan (1974)	367
Randy Johnson* (1999)	364

Runs Batted In

Hack Wilson (1930)	191
Lou Gehrig (1931)	184
Hank Greenberg (1937)	183
Jimmie Foxx (1938)	175
Lou Gehrig (1927)	175

Strikeouts (Batter)

Bobby Bonds (1970)	189
Preston Wilson (2000)	187
Bobby Bonds (1969)	187
Rob Deer (1987)	186
Jim Thome* (2001)	185
Pete Incaviglia (1986)	185

Saves

Bobby Thigpen (1990)	57
Trevor Hoffman* (1998)	53
Randy Myers (1993)	53
Rod Beck* (1998)	51
Dennis Eckersley (1992)	51

The All-Century Dream Team

Eighteen of the greatest baseball players of all time gathered at Atlanta's Turner Field Oct. 24, 1999, before Game 2 of the World Series; they were the living members selected as part of a 30-player All-Century Team. Announcer Vin Scully read the 30 names to a cheering crowd. Twenty-five were chosen by fans in nationwide balloting; the final 5 were picked by a panel of baseball executives and experts. The list below shows the members of this "dream team" in their positions on the field in order of votes received for each position (outfielders not listed by field position).

Pitchers:
- **Nolan Ryan**
- **Sandy Koufax**
- Cy Young
- Roger Clemens
- Bob Gibson
- Walter Johnson
- Warren Spahn*
- Christy Mathewson*
- Lefty Grove*

Catchers:
- **Johnny Bench**
- Yogi Berra

First Base:
- **Lou Gehrig**
- Mark McGwire

Second Base:
- **Jackie Robinson**
- Rogers Hornsby

Shortstop:
- **Cal Ripken, Jr.**
- Ernie Banks
- Honus Wagner*

Third Base:
- **Mike Schmidt**
- Brooks Robinson

Outfielders:
- **Babe Ruth**
- **Ted Williams**
- **Willie Mays**
- Hank Aaron
- Joe DiMaggio
- Mickey Mantle
- Pete Rose
- Ty Cobb
- Ken Griffey, Jr.
- Stan Musial*

(*) Denotes player selected by panel of baseball executives and experts. **Boldface = starting players.**

Major League Franchise Shifts and Additions

1953—Boston Braves (NL) became Milwaukee Braves.
1954—St. Louis Browns (AL) became Baltimore Orioles.
1955—Philadelphia Athletics (AL) became Kansas City Athletics.
1958—New York Giants (NL) became San Francisco Giants.
1958—Brooklyn Dodgers (NL) became L.A. Dodgers.
1961—Washington Senators (AL) became Minnesota Twins.
1961—L.A. Angels (renamed California Angels in 1965 and Anaheim Angels in 1997) enfranchised by the American League.
1961—Washington Senators enfranchised by the American League (a new team, replacing the former Washington club, whose franchise was moved to Minneapolis-St. Paul).
1962—Houston Colt .45's (renamed the Houston Astros in 1965) enfranchised by the National League.
1962—New York Mets enfranchised by the National League.
1966—Milwaukee Braves (NL) became Atlanta Braves.
1968—Kansas City Athletics (AL) became Oakland Athletics.
1969—Kansas City Royals and Seattle Pilots enfranchised by the American League; Montreal Expos and San Diego Padres enfranchised by the National League.
1970—Seattle Pilots became Milwaukee Brewers.
1971—Washington Senators became Texas Rangers (Dallas-Fort Worth area).
1977—Toronto Blue Jays and Seattle Mariners enfranchised by the American League.
1993—Colorado Rockies (Denver) and Florida Marlins (Miami) enfranchised by the National League.
1998—Tampa Bay Devil Rays began play in the American League; Arizona Diamondbacks (Phoenix) began play in the National League (both teams enfranchised in 1995). Milwaukee Brewers moved from the AL to the NL.

Baseball Stadiums[1]

National League

Team	Stadium (year opened)	Surface	Home run distances (ft.) LF	Center	RF	Seating capacity
Arizona Diamondbacks	Bank One Ballpark (1998)	Grass	330	407	334	48,500
Atlanta Braves	Turner Field (1997)	Grass	335	401	330	50,062
Chicago Cubs	Wrigley Field (1914)	Grass	355	400	353	38,902
Cincinnati Reds	Cinergy Field (1970)	Artificial	330	404	330	52,953
Colorado Rockies	Coors Field (1995)	Grass	347	415	350	50,381
Florida Marlins	Pro Player Stadium (1987)	Grass	325	410	345	42,531
Houston Astros	Enron Field (2000)	Grass	315	435	326	42,000
Los Angeles Dodgers	Dodger Stadium (1962)	Grass	330	395	330	56,000
Milwaukee Brewers	Miller Park (2001)	Grass	342	400	356	43,000
Montreal Expos	Olympic Stadium (1976)	Artificial	325	404	325	46,500
New York Mets	Shea Stadium (1964)	Grass	338	410	338	55,775
Philadelphia Phillies	Veterans Stadium (1971)	Artificial	330	408	330	62,409
Pittsburgh Pirates	PNC Park (2001)	Grass	325	399	320	38,127
St. Louis Cardinals	Busch Stadium (1966)	Grass	330	402	330	49,625
San Diego Padres	Qualcomm Stadium (1967)	Grass	327	405	330	56.133
San Francisco Giants	Pacific Bell Park (2000)	Grass	335	404	307	40,800

American League

Team	Stadium (year opened)	Surface	LF	Center	RF	Seating capacity
Anaheim Angels	Edison Intl. Field of Anaheim (1966)	Grass	333	408	333	45,050
Baltimore Orioles	Oriole Park at Camden Yards (1992)	Grass	333	400	318	48,876
Boston Red Sox	Fenway Park (1912)	Grass	310	420	302	33,871
Chicago White Sox	Comiskey Park (1991)	Grass	347	400	347	44,321
Cleveland Indians	Jacobs Field (1994)	Grass	325	405	325	43,368
Detroit Tigers	Comerica Park (2000)	Grass	345	402	330	40,000
Kansas City Royals	Kauffman Stadium (1973)	Grass	330	400	330	40,625
Minnesota Twins	Hubert H. Humphrey Metrodome (1982)	Artificial	343	408	327	48,678
New York Yankees	Yankee Stadium (1923)	Grass	318	408	314	55,070
Oakland A's	Network Associates Coliseum (1968)	Grass	330	400	330	43,662
Seattle Mariners	Safeco Field (1999)	Grass	331	405	327	47,000
Tampa Bay Devil Rays	Tropicana Field (1990)	Artificial	315	407	322	45,200
Texas Rangers	The Ballpark in Arlington (1994)	Grass	332	400	325	49,166
Toronto Blue Jays	SkyDome (1989)	Artificial	328	400	328	50,516

(1) As of 2001 season.

Little League World Series

The Little League World Series is played annually in Williamsport, PA. The team from Tokyo, Japan, won the 2001 series by defeating the team from Apopka, FL, 2-1, on Aug. 26. It was the 2d championship in 3 years for Japan, which last won in 1999, and their 5th overall. In 2001, the series expanded from 8 teams to a 16-team format.

Year	Winning / Losing Team	Score
1947	Williamsport, PA; Lock Haven, PA	16-7
1948	Lock Haven, PA; St. Petersburg, FL	6-5
1949	Hammonton, NJ; Pensacola, FL	5-0
1950	Houston, TX; Bridgeport, CT	2-1
1951	Stamford, CT; Austin, TX	3-0
1952	Norwalk, CT; Monongahela, PA	4-3
1953	Birmingham, AL; Schenectady, NY	1-0
1954	Schenectady, NY; Colton, CA	7-5
1955	Morrisville, PA; Merchantville, NJ	4-3
1956	Roswell, NM; Delaware, NJ	3-1
1957	Mexico; La Mesa, CA	4-0
1958	Mexico; Kankakee, IL	10-1
1959	Hamtramck, MI; Auburn, CA	12-0
1960	Levittown, PA; Ft. Worth, TX	5-0
1961	El Cajon, CA; El Campo, TX	4-2
1962	San Jose, CA; Kankakee, IL	3-0
1963	Granada Hills, CA; Stratford, CT	2-1
1964	Staten Island, NY; Mexico	4-0
1965	Windsor Locks, CT; Ontario, Canada	3-1
1966	Houston, TX; W. New York, NJ	8-2
1967	Tokyo, Japan; Chicago, IL	4-1
1968	Osaka, Japan; Richmond, VA	1-0
1969	Taiwan; Santa Clara, CA	5-0
1970	Wayne, NJ; Campbell, CA	2-0
1971	Taiwan; Gary, IN	12-3
1972	Taiwan; Hammond, IN	6-0
1973	Taiwan; Tucson, AZ	12-0
1974	Taiwan; Red Bluff, CA	12-1
1975	Lakewood, NJ; Tampa, FL	4-3
1976	Tokyo, Japan; Campbell, CA	10-3
1977	Taiwan; El Cajon, CA	7-2
1978	Taiwan; Danville, CA	11-1
1979	Taiwan; Campbell, CA	2-1
1980	Taiwan; Tampa, FL	4-3
1981	Taiwan; Tampa, FL	4-2
1982	Kirkland, WA; Taiwan	6-0
1983	Marietta, GA; Dominican Rep.	3-1
1984	South Korea; Altamonte Springs, FL	6-2
1985	South Korea; Mexico	7-1
1986	Taiwan; Tucson, AZ	12-0
1987	Chinese Taipei; Irvine, CA	21-1
1988	Chinese Taipei; Pearl City, HI	10-0
1989	Trumbull, CT; Chinese Taipei	5-2
1990	Chinese Taipei; Shippensburg, PA	9-0
1991	Chinese Taipei; Danville, CA	11-0
1992	Long Beach, CA; Philippines*	6-0
1993	Long Beach, CA; Panama	3-2
1994	Venezuela; Northridge, CA	4-3
1995	Taiwan; Spring, TX	17-3
1996	Taiwan; Cranston, RI	13-3
1997	Mexico; Mission Viejo, CA	5-4
1998	Toms River, NJ; Japan	12-9
1999	Japan; Phenix City, AL	5-0
2000	Venezuela; Bellaire, TX	3-2
2001	Japan; Apopka, FL	2-1

*Philippines won 15-4, but was disqualified for using ineligible players. Long Beach was awarded title by forfeit 6-0 (1 run per inning).

NCAA Baseball Champions

1960	Minnesota
1961	USC
1962	Michigan
1963	USC
1964	Minnesota
1965	Arizona St.
1966	Ohio St.
1967	Arizona St.
1968	USC
1969	Arizona St.
1970	USC
1971	USC
1972	USC
1973	USC
1974	USC
1975	Texas
1976	Arizona
1977	Arizona St.
1978	USC
1979	Cal. St.-Fullerton
1980	Arizona
1981	Arizona St.
1982	Miami (FL)
1983	Texas
1984	Cal. St.-Fullerton
1985	Miami (FL)
1986	Arizona
1987	Stanford
1988	Stanford
1989	Wichita St.
1990	Georgia
1991	LSU
1992	Pepperdine
1993	LSU
1994	Oklahoma
1995	Cal. St.-Fullerton
1996	LSU
1997	LSU
1998	USC
1999	Miami (FL)
2000	LSU
2001	Miami (FL)

NATIONAL BASKETBALL ASSOCIATION

2000-2001 Season: Lakers Repeat, Award-winning Sixers, New Rules, Air Jordan is Back!

The L.A. Lakers capped a near-perfect run through the playoffs to earn their 2d consecutive NBA title and 13th in franchise history. Only Boston, with 16 NBA titles, has more. L.A. went a record 15-1 in post-season play, sweeping Portland (3-0), Sacramento (4-0), and San Antonio (4-0) before dropping the opening game of the NBA Finals to the Philadelphia 76ers, 101-107, in overtime. Runner-up Philadelphia nearly swept the NBA's post-season awards, taking 4 of 6 individual honors, including MVP (Allen Iverson, the scoring leader, with a 31.1 average), Coach of the Year (Larry Brown, who led the Sixers to 56 wins and their 1st Finals appearance in 18 years), Sixth Man Award (Aaron McKie), and Defensive Player of the Year (Dikembe Motumbo).

In Apr. 2001, NBA owners approved rule changes for 2001-2002, including elimination of a long-standing ban on zone defenses and establishment of a defensive 3-sec. rule, limiting a player's time in the lane when not closely guarding an opponent.

Michael Jordan, who retired for the 2d time in Jan. 1999, announced his return as a player on Sept. 25, ending months of rumor and speculation. Jordan, 38, signed with the Washington Wizards and, under NBA rules, must sell his stake in the team and resign as president of basketball operations.

Final Standings, 2000-2001 Season

(playoff seedings in parentheses; in each conference the 2 division winners automatically get the number 1 and 2 seeds)

Eastern Conference

Atlantic Division	W	L	Pct	GB
Philadelphia (1)	56	26	.683	—
Miami (3)	50	32	.610	6
New York (4)	48	34	.585	8
Orlando (7)	43	39	.524	13
Boston	36	46	.439	20
New Jersey	26	56	.317	30
Washington	19	63	.232	37
Central Division				
Milwaukee (2)	52	30	.634	—
Toronto (5)	47	35	.573	5
Charlotte (6)	46	36	.561	6
Indiana (8)	41	41	.500	11
Detroit	32	50	.390	20
Cleveland	30	52	.366	22
Atlanta	25	57	.305	27
Chicago	15	67	.183	37

Western Conference

Midwest Division	W	L	Pct	GB
San Antonio (1)	58	24	.707	—
Utah (4)	53	29	.646	5
Dallas (5)	53	29	.646	5
Minnesota (8)	47	35	.573	11
Houston	45	37	.549	13
Denver	40	42	.488	18
Vancouver	23	59	.280	35
Pacific Division				
L.A. Lakers (2)	56	26	.683	—
Sacramento (3)	55	27	.671	1
Phoenix (6)	51	31	.622	5
Portland (7)	50	32	.610	6
Seattle	44	38	.537	12
L.A. Clippers	31	51	.378	25
Golden State	17	65	.207	39

NBA Regular Season Individual Highs in 2000-2001

Most minutes played, game — 63: Vince Carter, Toronto v. Sacramento, Feb. 23 (3 OT).
Most points, game — 57: Jerry Stackhouse, Detroit at Chicago, Apr. 3.
Most field goals made, game — 24: Chris Webber, Sacramento v. Indiana, Jan. 5.
Most field goal attempts, game — 47: Chris Webber, Sacramento v. Indiana, Jan. 5.
Most 3-pt. field goals made, game — 9: Antoine Walker, Boston at Sacramento, Jan. 17.
Most 3-pt. field goal attempts, game — 15: Chris Whitney, Washington v. Milwaukee, Jan. 15.
Most free throws made, game — 23: Kobe Bryant, L.A. Lakers at Cleveland, Jan. 30.
Most free throw attempts, game — 27: Vince Carter, Toronto at Phoenix, Dec. 30.
Most rebounds, game — 29: Dikembe Mutombo, Atlanta v. Toronto, Jan. 31.
Most assists, game — 220: George McCloud, Denver at Chicago, Mar. 26.
Most steals, game — 10: Michael Finley, Dallas v. Philadelphia, Jan. 23.
Most blocked shots, game — 12: Keon Clark, Toronto v. Atlanta, Mar. 23.
Most minutes played, season — 3,443: Michael Finley, Dallas.
Most offensive rebounds, season — 307: Dikembe Mutombo, Atlanta-Philadelphia.
Most defensive rebounds, season — 749: Ben Wallace, Detroit.

2001 NBA Playoff Results

Eastern Conference

Philadelphia defeated Indiana 3 games to 1
Milwaukee defeated Orlando 3 games to 1
Charlotte defeated Miami 3 games to 0
Toronto defeated New York 3 games to 2
Philadelphia defeated Toronto 4 games to 3
Milwaukee defeated Charlotte 4 games to 3
Philadelphia defeated Milwaukee 4 games to 3

Western Conference

San Antonio defeated Minnesota 3 games to 1
L.A. Lakers defeated Portland 3 games to 0
Sacramento defeated Phoenix 3 games to 1
Dallas defeated Utah 3 games to 2
San Antonio defeated Dallas 4 games to 1
L.A. Lakers defeated Sacramento 4 games to 0
L.A. Lakers defeated San Antonio 4 games to 0

Championship

Los Angeles defeated Philadelphia 4 games to 1 [101-107 OT, 98-89, 96-91, 100-86, 108-96].

Lakers Repeat in 2001

The L.A. Lakers defeated the Philadelphia 76ers, 108-96, in Game 5 of the NBA Finals on June 15, in Philadelphia. After losing Game 1, the Lakers swept the next 4 to repeat as NBA champs. L.A. center Shaquille O'Neal also repeated as MVP of the Finals, only the 3d player to do so. Philadelphia's Allen Iverson led all scorers in the series, with 35.6 points per game. O'Neal averaged 33, and teammate Kobe Bryant added 24.6 per game. The title was L.A. coach Phil Jackson's 8th in 10 years (he won 6 with Chicago).

NBA Finals Composite Box Scores

L.A. Lakers	FG M-A	FT M-A	Reb O-T	Ast	Avg
Shaquille O'Neal	63-110	39-76	31-79	24	33.0
Kobe Bryant	44-106	32-38	5-39	29	24.6
Derek Fisher	17-39	5-6	1-6	10	9.8
Rick Fox	15-34	12-13	3-23	19	9.8
Robert Horry	14-25	6-6	10-25	6	8.4
Horace Grant	10-34	6-8	12-28	3	5.2
Ron Harper	5-8	2-3	0-5	3	4.3
Tyronn Lue	7-12	0-0	1-4	7	3.6
Brian Shaw	6-20	3-5	3-16	14	3.6
Mark Madsen	0-1	0-0	1-1	0	0.0

Philadelphia 76ers	FG M-A	FT M-A	Reb O-T	Ast	Avg
Allen Iverson	66-162	35-48	5-28	19	35.6
Dikembe Mutombo	33-55	18-26	20-61	2	16.8
Eric Snow	22-54	19-26	11-22	30	12.6
Aaron McKie	15-48	6-9	7-27	30	8.0
Tyrone Hill	13-33	7-9	7-33	2	6.6
Matt Geiger	12-18	2-2	2-5	2	5.2
Raja Bell	4-13	5-10	2-9	4	2.6
Todd MacCulloch	5-12	3-4	3-7	0	2.6
Jumaine Jones	4-10	0-0	2-10	1	2.0
Kevin Ollie	1-3	3-3	1-1	1	1.0
George Lynch	1-3	0-0	2-5	1	1.0
Rodney Buford	1-6	0-0	2-6	0	0.7

NBA Finals MVP

1969 Jerry West, Los Angeles
1970 Willis Reed, New York
1971 Lew Alcindor (Kareem Abdul-Jabbar), Milwaukee
1972 Wilt Chamberlain, Los Angeles
1973 Willis Reed, New York
1974 John Havlicek, Boston
1975 Rick Barry, Golden State
1976 Jo Jo White, Boston
1977 Bill Walton, Portland
1978 Wes Unseld, Washington
1979 Dennis Johnson, Seattle
1980 Magic Johnson, Los Angeles
1981 Cedric Maxwell, Boston
1982 Magic Johnson, Los Angeles
1983 Moses Malone, Philadelphia
1984 Larry Bird, Boston
1985 Kareem Abdul-Jabbar, L.A. Lakers
1986 Larry Bird, Boston
1987 Magic Johnson, L.A. Lakers
1988 James Worthy, L.A. Lakers
1989 Joe Dumars, Detroit
1990 Isiah Thomas, Detroit
1991 Michael Jordan, Chicago
1992 Michael Jordan, Chicago
1993 Michael Jordan, Chicago
1994 Hakeem Olajuwon, Houston
1995 Hakeem Olajuwon, Houston
1996 Michael Jordan, Chicago
1997 Michael Jordan, Chicago
1998 Michael Jordan, Chicago
1999 Tim Duncan, San Antonio
2000 Shaquille O'Neal, L.A. Lakers
2001 Shaquille O'Neal, L.A. Lakers

NBA Finals All-Time Statistical Leaders

(at the end of the 2001 NBA season finals; *denotes active in 2000-2001)

Scoring Average (Minimum 10 games)

	G	Pts.	Avg
Rick Barry	10	363	36.3
*Shaquille O'Neal	15	505	33.7
Michael Jordan	35	1,176	33.6
Jerry West	55	1,679	30.5
Bob Pettit	25	709	28.4
*Hakeem Olajuwon	17	467	27.5
Elgin Baylor	44	1,161	26.4
Julius Erving	22	561	25.5
Joe Fulks	11	272	24.7
Clyde Drexler	15	367	24.5

Games Played

Bill Russell	70
Sam Jones	64
Kareem Abdul-Jabbar	56
Jerry West	55
Tom Heinsohn	52

Rebounds

Bill Russell	1,718
Wilt Chamberlain	862
Elgin Baylor	593
Kareem Abdul-Jabbar	507
Tom Heinsohn	473

Assists

Magic Johnson	584
Bob Cousy	400
Bill Russell	315
Jerry West	306
Dennis Johnson	228

NBA Scoring Leaders

Year	Scoring champion	Pts	Avg
1947	Joe Fulks, Philadelphia	1,389	23.2
1948	Max Zaslofsky, Chicago	1,007	21.0
1949	George Mikan, Minneapolis	1,698	28.3
1950	George Mikan, Minneapolis	1,865	27.4
1951	George Mikan, Minneapolis	1,932	28.4
1952	Paul Arizin, Philadelphia	1,674	25.4
1953	Neil Johnston, Philadelphia	1,564	22.3
1954	Neil Johnston, Philadelphia	1,759	24.4
1955	Neil Johnston, Philadelphia	1,631	22.7
1956	Bob Pettit, St. Louis	1,849	25.7
1957	Paul Arizin, Philadelphia	1,817	25.6
1958	George Yardley, Detroit	2,001	27.8
1959	Bob Pettit, St. Louis	2,105	29.2
1960	Wilt Chamberlain, Philadelphia	2,707	37.9
1961	Wilt Chamberlain, Philadelphia	3,033	38.4
1962	Wilt Chamberlain, Philadelphia	4,029	50.4
1963	Wilt Chamberlain, San Francisco	3,586	44.8
1964	Wilt Chamberlain, San Francisco	2,948	36.5
1965	Wilt Chamberlain, San Francisco, Philadelphia	2,534	34.7
1966	Wilt Chamberlain, Philadelphia	2,649	33.5
1967	Rick Barry, San Francisco	2,775	35.6
1968	Dave Bing, Detroit	2,142	27.1
1969	Elvin Hayes, San Diego	2,327	28.4
1970	Jerry West, Los Angeles	2,309	31.2
1971	Lew Alcindor (Kareem Abdul-Jabbar), Milwaukee	2,596	31.7
1972	Kareem Abdul-Jabbar, Milwaukee	2,822	34.8
1973	Nate Archibald, Kans. City-Omaha	2,719	34.0
1974	Bob McAdoo, Buffalo	2,261	30.6
1975	Bob McAdoo, Buffalo	2,831	34.5
1976	Bob McAdoo, Buffalo	2,427	31.1
1977	Pete Maravich, New Orleans	2,273	31.1
1978	George Gervin, San Antonio	2,232	27.2
1979	George Gervin, San Antonio	2,365	29.6
1980	George Gervin, San Antonio	2,585	33.1
1981	Adrian Dantley, Utah	2,452	30.7
1982	George Gervin, San Antonio	2,551	32.3
1983	Alex English, Denver	2,326	28.4
1984	Adrian Dantley, Utah	2,418	30.6
1985	Bernard King, New York	1,809	32.9
1986	Dominique Wilkins, Atlanta	2,366	30.3
1987	Michael Jordan, Chicago	3,041	37.1
1988	Michael Jordan, Chicago	2,868	35.0
1989	Michael Jordan, Chicago	2,633	32.5
1990	Michael Jordan, Chicago	2,753	33.6
1991	Michael Jordan, Chicago	2,580	31.5
1992	Michael Jordan, Chicago	2,404	30.1
1993	Michael Jordan, Chicago	2,541	32.6
1994	David Robinson, San Antonio	2,383	29.8
1995	Shaquille O'Neal, Orlando	2,315	29.3
1996	Michael Jordan, Chicago	2,465	30.4
1997	Michael Jordan, Chicago	2,431	29.6
1998	Michael Jordan, Chicago	2,357	28.7
1999	Allen Iverson, Philadelphia	1,284	26.8
2000	Shaquille O'Neal, L.A. Lakers	2,344	29.7
2001	Allen Iverson, Philadelphia	2,207	31.1

NBA Most Valuable Player

1956 Bob Pettit, St. Louis
1957 Bob Cousy, Boston
1958 Bill Russell, Boston
1959 Bob Pettit, St. Louis
1960 Wilt Chamberlain, Philadelphia
1961 Bill Russell, Boston
1962 Bill Russell, Boston
1963 Bill Russell, Boston
1964 Oscar Robertson, Cincinnati
1965 Bill Russell, Boston
1966 Wilt Chamberlain, Philadelphia
1967 Wilt Chamberlain, Philadelphia
1968 Wilt Chamberlain, Philadelphia
1969 Wes Unseld, Baltimore
1970 Willis Reed, New York
1971 Lew Alcindor (Kareem Abdul-Jabbar), Milwaukee
1972 Kareem Abdul-Jabbar, Milwaukee
1973 Dave Cowens, Boston
1974 Kareem Abdul-Jabbar, Milwaukee
1975 Bob McAdoo, Buffalo
1976 Kareem Abdul-Jabbar, Los Angeles
1977 Kareem Abdul-Jabbar, Los Angeles
1978 Bill Walton, Portland
1979 Moses Malone, Houston
1980 Kareem Abdul-Jabbar, Los Angeles
1981 Julius Erving, Philadelphia
1982 Moses Malone, Houston
1983 Moses Malone, Philadelphia
1984 Larry Bird, Boston
1985 Larry Bird, Boston
1986 Larry Bird, Boston
1987 Magic Johnson, L.A. Lakers
1988 Michael Jordan, Chicago
1989 Magic Johnson, L.A. Lakers
1990 Magic Johnson, L.A. Lakers
1991 Michael Jordan, Chicago
1992 Michael Jordan, Chicago
1993 Charles Barkley, Phoenix
1994 Hakeem Olajuwon, Houston
1995 David Robinson, San Antonio
1996 Michael Jordan, Chicago
1997 Karl Malone, Utah
1998 Michael Jordan, Chicago
1999 Karl Malone, Utah
2000 Shaquille O'Neal, L.A. Lakers
2001 Allen Iverson, Philadelphia

NBA Champions, 1947-2001

	Regular season		Playoffs		
Year	Eastern Conference	Western Conference	Winner	Coach	Runner-up
1947	Washington Capitols	Chicago Stags	Philadelphia	Ed Gottlieb	Chicago
1948	Philadelphia Warriors	St. Louis Bombers	Baltimore	Buddy Jeannette	Philadelphia
1949	Washington Capitols	Rochester	Minneapolis	John Kundla	Washington
1950	Syracuse	Minneapolis	Minneapolis	John Kundla	Syracuse
1951	Philadelphia Warriors	Minneapolis	Rochester	Lester Harrison	New York
1952	Syracuse	Rochester	Minneapolis	John Kundla	New York
1953	New York	Minneapolis	Minneapolis	John Kundla	New York
1954	New York	Minneapolis	Minneapolis	John Kundla	Syracuse
1955	Syracuse	Ft. Wayne	Syracuse	Al Cervi	Ft. Wayne
1956	Philadelphia Warriors	Ft. Wayne	Philadelphia	George Senesky	Ft. Wayne
1957	Boston	St. Louis	Boston	Red Auerbach	St. Louis
1958	Boston	St. Louis	St. Louis	Alex Hannum	Boston
1959	Boston	St. Louis	Boston	Red Auerbach	Minneapolis
1960	Boston	St. Louis	Boston	Red Auerbach	St. Louis
1961	Boston	St. Louis	Boston	Red Auerbach	St. Louis
1962	Boston	Los Angeles	Boston	Red Auerbach	Los Angeles
1963	Boston	Los Angeles	Boston	Red Auerbach	Los Angeles
1964	Boston	San Francisco	Boston	Red Auerbach	San Francisco
1965	Boston	Los Angeles	Boston	Red Auerbach	Los Angeles
1966	Philadelphia	Los Angeles	Boston	Red Auerbach	Los Angeles
1967	Philadelphia	San Francisco	Philadelphia	Alex Hannum	San Francisco
1968	Philadelphia	St. Louis	Boston	Bill Russell	Los Angeles
1969	Baltimore	Los Angeles	Boston	Bill Russell	Los Angeles
1970	New York	Atlanta	New York	Red Holzman	Los Angeles

Year	Atlantic	Central	Midwest	Pacific	Winner	Coach	Runner-up
1971	New York	Baltimore	Milwaukee	Los Angeles	Milwaukee	Larry Costello	Baltimore
1972	Boston	Baltimore	Milwaukee	Los Angeles	Los Angeles	Bill Sharman	New York
1973	Boston	Baltimore	Milwaukee	Los Angeles	New York	Red Holzman	Los Angeles
1974	Boston	Capital	Milwaukee	Los Angeles	Boston	Tom Heinsohn	Milwaukee
1975	Boston	Washington	Chicago	Golden State	Golden State	Al Attles	Washington
1976	Boston	Cleveland	Milwaukee	Golden State	Boston	Tom Heinsohn	Phoenix
1977	Philadelphia	Houston	Denver	Los Angeles	Portland	Jack Ramsay	Philadelphia
1978	Philadelphia	San Antonio	Denver	Portland	Washington	Dick Motta	Seattle
1979	Washington	San Antonio	Kansas City	Seattle	Seattle	Len Wilkens	Washington
1980	Boston	Atlanta	Milwaukee	Los Angeles	Los Angeles	Paul Westhead	Philadelphia
1981	Boston	Milwaukee	San Antonio	Phoenix	Boston	Bill Fitch	Houston
1982	Boston	Milwaukee	San Antonio	Los Angeles	Los Angeles	Pat Riley	Philadelphia
1983	Philadelphia	Milwaukee	San Antonio	Los Angeles	Philadelphia	Billy Cunningham	Los Angeles
1984	Boston	Milwaukee	Utah	Los Angeles	Boston	K.C. Jones	Los Angeles
1985	Boston	Milwaukee	Denver	L.A. Lakers	L.A. Lakers	Pat Riley	Boston
1986	Boston	Milwaukee	Houston	L.A. Lakers	Boston	K.C. Jones	Houston
1987	Boston	Atlanta	Dallas	L.A. Lakers	L.A. Lakers	Pat Riley	Boston
1988	Boston	Detroit	Denver	L.A. Lakers	L.A. Lakers	Pat Riley	Detroit
1989	New York	Detroit	Utah	L.A. Lakers	Detroit	Chuck Daly	L.A. Lakers
1990	Philadelphia	Detroit	San Antonio	L.A. Lakers	Detroit	Chuck Daly	Portland
1991	Boston	Chicago	San Antonio	Portland	Chicago	Phil Jackson	L.A. Lakers
1992	Boston	Chicago	Utah	Portland	Chicago	Phil Jackson	Portland
1993	New York	Chicago	Houston	Phoenix	Chicago	Phil Jackson	Phoenix
1994	New York	Atlanta	Houston	Seattle	Houston	Rudy Tomjanovich	New York
1995	Orlando	Indiana	San Antonio	Phoenix	Houston	Rudy Tomjanovich	Orlando
1996	Orlando	Chicago	San Antonio	Seattle	Chicago	Phil Jackson	Seattle
1997	Miami	Chicago	Utah	Seattle	Chicago	Phil Jackson	Utah
1998	Miami	Chicago	Utah	L.A. Lakers	Chicago	Phil Jackson	Utah
1999	Miami	Indiana	San Antonio	Portland	San Antonio	Gregg Popovich	New York
2000	Miami	Indiana	Utah	L.A. Lakers	L.A. Lakers	Phil Jackson	Indiana
2001	Philadelphia	Milwaukee	San Antonio	L.A. Lakers	L.A. Lakers	Phil Jackson	Philadelphia

NBA Coach of the Year, 1963-2001

1963 Harry Gallatin, St. Louis Hawks
1964 Alex Hannum, San Francisco Warriors
1965 Red Auerbach, Boston Celtics
1966 Dolph Schayes, Philadelphia 76ers
1967 Johnny Kerr, Chicago Bulls
1968 Richie Guerin, St. Louis Hawks
1969 Gene Shue, Baltimore Bullets
1970 Red Holzman, New York Knicks
1971 Dick Motta, Chicago Bulls
1972 Bill Sharman, Los Angeles Lakers
1973 Tom Heinsohn, Boston Celtics
1974 Ray Scott, Detroit Pistons
1975 Phil Johnson, Kansas City-Omaha Kings
1976 Bill Fitch, Cleveland Cavaliers
1977 Tom Nissalke, Houston Rockets
1978 Hubie Brown, Atlanta Hawks
1979 Cotton Fitzsimmons, Kansas City Kings
1980 Bill Fitch, Boston Celtics
1981 Jack McKinney, Indiana Pacers
1982 Gene Shue, Washington Bullets
1983 Don Nelson, Milwaukee Bucks
1984 Frank Layden, Utah Jazz
1985 Don Nelson, Milwaukee Bucks
1986 Mike Fratello, Atlanta Hawks
1987 Mike Schuler, Portland Trail Blazers
1988 Doug Moe, Denver Nuggets
1989 Cotton Fitzsimmons, Phoenix Suns
1990 Pat Riley, Los Angeles Lakers
1991 Don Chaney, Houston Rockets
1992 Don Nelson, Golden State Warriors
1993 Pat Riley, New York Knicks
1994 Lenny Wilkens, Atlanta Hawks
1995 Del Harris, Los Angeles Lakers
1996 Phil Jackson, Chicago Bulls
1997 Pat Riley, Miami Heat
1998 Larry Bird, Indiana Pacers
1999 Mike Dunleavy, Portland Trail Blazers
2000 Glenn "Doc" Rivers, Orlando Magic
2001 Larry Brown, Philadelphia 76ers

NBA All-League and All-Defensive Teams, 2000-2001

All-League Team			All-Defensive Team	
First team	Second team	Position	First team	Second team
Tim Duncan, San Antonio	Kevin Garnett, Minnesota	Forward	Tim Duncan, San Antonio	Bruce Bowen, Miami
Chris Webber, Sacramento	Vince Carter, Toronto	Forward	Kevin Garnett, Minnesota	P. J. Brown, Charlotte
Shaquille O'Neal, L.A. Lakers	Dikembe Mutombo, Atl.-Phil.	Center	Dikembe Mutombo, Atl.-Phil.	Shaquille O'Neal, L.A. Lakers
Allen Iverson, Philadelphia	Kobe Bryant, L.A. Lakers	Guard	Gary Payton, Seattle	Kobe Bryant, L.A. Lakers
Jason Kidd, Phoenix	Tracy McGrady, Orlando	Guard	Jason Kidd, Phoenix	Doug Christie, Sacramento

NBA Statistical Leaders, 2000-2001

Scoring Average
(Minimum 70 games or 1,400 pts)

	G	FG	FT	Pts	Avg
Iverson, Philadelphia	71	762	585	2,207	31.1
Stackhouse, Detroit	80	774	666	2,380	29.8
O'Neal, L.A. Lakers	74	813	499	2,125	28.7
Bryant, L.A. Lakers	68	701	475	1,938	28.5
Carter, Toronto	75	762	384	2,027	27.6
Webber, Sacramento	70	786	324	1,898	27.1
McGrady, Orlando	77	788	430	2,065	26.8
Pierce, Boston	82	687	550	2,071	25.3
Jamison, Golden State	82	800	382	2,044	24.9
Marbury, New Jersey	67	563	362	1,598	23.9

Rebounds per Game
(Minimum 70 games or 800 rebounds)

	G	Off	Def	Tot	Avg
Mutombo, Atlanta-Philadelphia	75	307	708	1,015	13.5
Wallace, Detroit	80	303	749	1,052	13.2
O'Neal, L.A. Lakers	74	291	649	940	12.7
Duncan, San Antonio	82	259	738	997	12.2
McDyess, Denver	70	240	605	845	12.1
Garnett, Minnesota	81	219	702	921	11.4
Webber, Sacramento	70	179	598	777	11.1
Marion, Phoenix	79	220	628	848	10.7
Davis, Toronto	78	274	513	787	10.1
Brand, Chicago	74	285	461	746	10.1

Field Goal Percentage
(Minimum 300 field goals made)

	FGM	FGA	Pct
O'Neal, L.A. Lakers	813	1,422	.572
Wells, Portland	387	726	.533
Camby, New York	304	580	.524
Thomas, New York	314	614	.511
Szczerbiak, Minnesota	469	920	.510
Miles, L.A. Clippers	318	630	.505
Stockton, Utah	328	651	.504
Marshall, Utah	427	849	.503
Williamson, Detroit	325	647	.502
Weatherspoon, Cleveland	347	692	.501
Wallace, Portland	590	1,178	.501

Free Throw Percentage
(Minimum 125 free throws made)

	FTM	FTA	Pct
Miller, Indiana	323	348	.928
Houston, New York	279	307	.909
Christie, Sacramento	280	312	.897
Nash, Dallas	231	258	.895
Richmond, Washington	143	160	.894
Smith, Portland	309	347	.890
Allen, Milwaukee	348	392	.888
Armstrong, Orlando	220	249	.884
Piatkowski, L.A. Clippers	158	181	.873
Brandon, Minnesota	195	224	.871

3-Point Field Goal Percentage
(Minimum 55 3-point field goals made)

	FG	FGA	Pct
Barry, Seattle	109	229	.476
Stockton, Utah	61	132	.462
Williams, Seattle	61	133	.459
Davis, Washington	78	171	.456
Ferry, San Antonio	70	156	.449
Kukoc, Atlanta	70	157	.446
Garrity, Orlando	97	224	.433
Allen, Milwaukee	202	467	.433
Lewis, Seattle	123	285	.432
Curry, Toronto	62	145	.428

Assists per Game
(Minimum 70 games or 400 assists)

	G	No	Avg
Kidd, Phoenix	77	753	9.8
Stockton, Utah	82	713	8.7
Van Exel, Denver	71	600	8.5
Bibby, Vancouver	82	685	8.4
Payton, Seattle	79	642	8.1
Jackson, New York	83	661	8.0
Miller, Cleveland	82	657	8.0
Cassell, Milwaukee	76	580	7.6
Marbury, New Jersey	67	506	7.6
Brandon, Minnesota	78	583	7.5

Steals per Game
(Minimum 70 games or 125 steals)

	G	No	Avg
Iverson, Philadelphia	71	178	2.51
Blaylock, Golden State	69	163	2.36
Christie, Sacramento	81	183	2.26
Kidd, Phoenix	77	166	2.16
Davis, Charlotte	82	170	2.07
Brandon, Minnesota	78	161	2.06
Artest, Chicago	76	152	2.00
Armstrong, Orlando	75	135	1.80
Francis, Houston	80	141	1.76
Walker, Boston	81	138	1.70

Blocked Shots per Game
(Minimum 70 games or 100 blocked shots)

	G	Blk	Avg
Ratliff, Philadelphia-Atlanta	50	187	3.74
O'Neal, Indiana	81	228	2.81
Bradley, Dallas	82	228	2.78
O'Neal, L.A. Lakers	74	204	2.76
Mutombo, Atlanta-Philadelphia	75	203	2.71
Foyle, Golden State	58	156	2.69
LaFrentz, Denver	78	206	2.64
Robinson, San Antonio	80	197	2.46
Duncan, San Antonio	82	192	2.34
Wallace, Detroit	80	186	2.33

NBA Rookie of the Year

Year	Player
1953	Don Meineke, Ft. Wayne
1954	Ray Felix, Baltimore
1955	Bob Pettit, Milwaukee
1956	Maurice Stokes, Rochester
1957	Tom Heinsohn, Boston
1958	Woody Sauldsberry, Philadelphia
1959	Elgin Baylor, Minneapolis
1960	Wilt Chamberlain, Philadelphia
1961	Oscar Robertson, Cincinnati
1962	Walt Bellamy, Chicago
1963	Terry Dischinger, Chicago
1964	Jerry Lucas, Cincinnati
1965	Willis Reed, New York
1966	Rick Barry, San Francisco
1967	Dave Bing, Detroit
1968	Earl Monroe, Baltimore
1969	Wes Unseld, Baltimore
1970	Lew Alcindor, Milwaukee
1971	Dave Cowens, Boston; Geoff Petrie, Portland (tie)
1972	Sidney Wicks, Portland
1973	Bob McAdoo, Buffalo
1974	Ernie DiGregorio, Buffalo
1975	Keith Wilkes, Golden State
1976	Alvan Adams, Phoenix
1977	Adrian Dantley, Buffalo
1978	Walter Davis, Phoenix
1979	Phil Ford, Kansas City
1980	Larry Bird, Boston
1981	Darrell Griffith, Utah
1982	Buck Williams, New Jersey
1983	Terry Cummings, San Diego
1984	Ralph Sampson, Houston
1985	Michael Jordan, Chicago
1986	Patrick Ewing, New York
1987	Chuck Person, Indiana
1988	Mark Jackson, New York
1989	Mitch Richmond, Golden State
1990	David Robinson, San Antonio
1991	Derrick Coleman, New Jersey
1992	Larry Johnson, Charlotte
1993	Shaquille O'Neal, Orlando
1994	Chris Webber, Golden State
1995	Grant Hill, Detroit; Jason Kidd, Dallas (tie)
1996	Damon Stoudamire, Toronto
1997	Allen Iverson, Philadelphia
1998	Tim Duncan, San Antonio
1999	Vince Carter, Toronto
2000	Elton Brand, Chicago; Steve Francis, Houston (tie)
2001	Mike Miller, Orlando

NBA Individual Statistics, 2000-2001

(more than 600 minutes played; players ranked by scoring average)

Atlanta Hawks

	Min	FG%	FT%	Reb	Ast	Pts	Avg
Jason Terry	3,089	.436	.846	269	403	1,617	19.7
Theo Ratliff	1,800	.499	.760	413	58	621	12.4
Lorenzen Wright	1,988	.448	.718	535	87	881	12.4
Toni Kukoc	1,597	.473	.631	259	199	721	11.1
Alan Henderson	1,810	.444	.638	406	50	769	10.5
Nazr Mohammed	912	.477	.706	307	19	441	7.6
Chris Crawford	901	.452	.819	110	37	318	6.8
Matt Maloney	1,403	.420	.765	117	154	369	6.7
Brevin Knight	1,457	.375	.818	168	311	333	6.3
Dion Glover	929	.421	.681	131	69	340	6.0
Larry Robinson	632	.364	.875	87	36	199	5.9
Dermarr Johnson	1,307	.376	.736	178	64	397	5.0
Hanno Mottola	986	.444	.811	172	25	317	4.3

Coach-Lon Kruger

Boston Celtics

	Min	FG%	FT%	Reb	Ast	Pts	Avg
Paul Pierce	3,120	.454	.745	522	253	2,071	25.3
Antoine Walker	3,396	.414	.716	719	445	1,892	23.4
Bryant Stith	2,506	.401	.845	284	168	756	9.7
Vitaly Potapenko	1,901	.476	.728	495	64	611	7.5
Kenny Anderson	849	.388	.831	73	134	246	7.5
Eric Williams	1,746	.362	.714	207	112	535	6.6
Tony Battie	846	.537	.638	233	16	260	6.5
Milt Palacio	1,126	.470	.848	102	150	340	6.0
Randy Brown	1,239	.422	.575	99	154	223	4.1
Mark Blount	1,098	.505	.697	231	32	248	3.9

Coach-Rick Pitino, Jim O'Brien

Charlotte Hornets

	Min	FG%	FT%	Reb	Ast	Pts	Avg
Jamal Mashburn	2,989	.413	.766	576	411	1,528	20.1
David Wesley	3,106	.422	.799	224	361	1,414	17.2
Baron Davis	3,192	.427	.677	408	598	1,131	13.8
Elden Campbell	2,337	.440	.709	608	104	1,022	13.1
PJ Brown	2,811	.444	.852	742	127	676	8.4
Derrick Coleman	683	.380	.685	184	39	277	8.2
Eddie Robinson	1,201	.531	.727	198	59	498	7.4
Jamaal Magloire	1,095	.450	.655	295	27	339	4.6
Hersey Hawkins	681	.409	.857	80	72	183	3.1
Otis Thorpe	647	.450	.833	145	29	138	2.8

Coach-Paul Silas

Chicago Bulls

	Min	FG%	FT%	Reb	Ast	Pts	Avg
Elton Brand	2,906	.476	.708	746	240	1,490	20.1
Ron Mercer	2,535	.446	.825	236	201	1,202	19.7
Ron Artest	2,363	.401	.750	294	228	907	11.9
Marcus Fizer	1,580	.430	.727	313	76	683	9.5
Fred Hoiberg	2,247	.438	.866	308	263	673	9.1
Brad Miller	1,434	.435	.743	419	107	505	8.9
Bryce Drew	1,305	.379	.737	69	185	302	6.3
Khalid El Amin	936	.370	.778	81	145	314	6.3
AJ Guyton	630	.406	.833	36	64	198	5.8
Corey Benjamin	857	.381	.675	100	69	307	4.7
Jamal Crawford	1,050	.352	.794	89	141	282	4.6
Michael Ruffin	879	.444	.506	262	39	119	2.6

Coach-Tim Floyd

Cleveland Cavaliers

	Min	FG%	FT%	Reb	Ast	Pts	Avg
Andre Miller	2,848	.452	.833	360	657	1,296	15.8
Lamond Murray	2,225	.423	.735	340	124	998	12.8
Zydrunas Ilgauskas	616	.487	.679	160	18	281	11.7
Jimmy Jackson	1,577	.384	.829	214	153	619	11.7
Chris Gatling	1,670	.449	.684	391	61	842	11.4
Clarence Weatherspoon	2,774	.501	.790	796	103	92	11.3
Matt Harpring	1,615	.454	.812	242	102	623	11.1
Chris Mihm	1,166	.442	.794	280	16	446	7.6
Wesley Person	958	.438	.800	130	64	314	7.1
Trajan Langdon	1,116	.430	.895	89	81	389	6.0
Robert Traylor	1,212	.497	.567	300	63	402	5.7
Bimbo Coles	804	.382	.857	48	138	232	4.9
Cedric Henderson	961	.389	.652	90	79	235	4.3

Coach-Randy Wittman

Dallas Mavericks

	Min	FG%	FT%	Reb	Ast	Pts	Avg
Dirk Nowitzki	3,125	.474	.838	754	173	1,784	21.8
Michael Finley	3,443	.458	.775	425	360	1,765	21.5
Juwan Howard	2,974	.479	.773	573	223	1,464	18.1
Steve Nash	2,387	.487	.895	223	509	1,092	15.6
Howard Eisley	2,426	.393	.825	197	295	741	9.0
Shawn Bradley	2,001	.490	.787	608	38	579	7.1
Greg Buckner	820	.438	.728	157	49	229	6.2
Calvin Booth	933	.476	.679	246	42	293	5.3
Vernon Maxwell	660	.327	.656	66	49	201	4.7

Coach-Don Nelson

Denver Nuggets

	Min	FG%	FT%	Reb	Ast	Pts	Avg
Antonio McDyess	2,555	.495	.700	845	146	1,458	20.8
Nick Van Exel	2,688	.414	.819	241	600	1,259	17.7
Raef LaFrentz	2,457	.477	.698	607	107	1,008	12.9
Voshon Lenard	2,331	.397	.797	231	190	972	12.2
George McCloud	2,007	.382	.840	224	279	729	9.6
Kevin Willis	1,830	.441	.769	532	50	722	9.3
James Posey	2,255	.412	.816	431	163	666	8.1
Robert Pack	1,260	.425	.766	137	293	479	6.5
Ryan Bowen	696	.556	.614	113	30	191	3.4

Coach-Dan Issel

Detroit Pistons

	Min	FG%	FT%	Reb	Ast	Pts	Avg
Jerry Stackhouse	3,215	.402	.822	315	410	2,380	29.8
Joe Smith	1,941	.403	.805	491	79	847	12.3
Chucky Atkins	2,363	.399	.692	173	330	971	12.0
Corliss Williamson	1,686	.502	.637	321	61	801	11.6
Dana Barros	1,079	.444	.850	94	110	478	8.0
Ben Wallace	2,760	.491	.336	1,052	123	513	6.4
Mateen Cleaves	1,268	.400	.708	132	207	422	5.4
Michael Curry	1,485	.455	.849	121	132	356	5.2
Mikki Moore	1,154	.493	.731	316	33	359	4.4
Billy Owens	793	.383	.475	205	55	198	4.4
Jud Buechler	737	.460	.750	94	39	191	3.4

Coach-George Irvine

Golden State Warriors

	Min	FG%	FT%	Reb	Ast	Pts	Avg
Antawn Jamison	3,394	.443	.713	715	164	2,011	24.9
Larry Hughes	1,846	.383	.766	276	223	823	16.5
Marc Jackson	1,410	.467	.802	361	59	633	13.2
Bobby Sura	1,684	.390	.714	226	242	586	11.1
Mookie Blaylock	2,352	.396	.697	272	462	760	11.0
Chris Porter	1,147	.394	.681	189	61	426	8.6
Erick Dampier	1,038	.350	.693	250	59	480	7.4
Vonteego Cummings	1,495	.401	.524	137	227	307	7.3
Adonal Foyle	1,457	.416	.446	405	48	333	5.9
Corie Blount	1,305	.447	.614	400	59	307	4.6
Adam Keefe	836	.401	.619	209	36	166	2.5

Coach-Dave Cowens

Houston Rockets

	Min	FG%	FT%	Reb	Ast	Pts	Avg
Steve Francis	3,194	.451	.817	553	517	1,591	19.9
Cuttino Mobley	3,002	.434	.831	397	195	1,538	19.5
Maurice Taylor	1,972	.489	.735	378	104	899	13.0
Hakeem Olajuwon	1,545	.498	.621	431	72	689	11.9
Shandon Anderson	2,396	.446	.734	333	189	710	8.7
Walt Williams	1,583	.394	.770	245	97	599	8.3
Kenny Thomas	1,820	.443	.722	417	77	528	7.1
Moochie Norris	1,654	.446	.778	198	283	544	6.6
Matt Bullard	1,000	.423	.714	130	42	354	5.8
Kelvin Cato	624	.577	.649	141	11	165	4.7

Coach-Rudy Tomjanovich

Indiana Pacers

	Min	FG%	FT%	Reb	Ast	Pts	Avg
Jalen Rose	2,943	.457	.828	359	435	1,478	20.5
Reggie Miller	3,181	.440	.928	284	260	1,527	18.9
Jermaine O'Neal	2,641	.465	.601	795	98	1,041	12.9
Travis Best	2,457	.440	.827	222	473	918	11.9
Austin Croshere	1,872	.394	.866	387	92	822	10.2
Al Harrington	1,892	.444	.656	380	130	586	7.5
Zan Tabak	777	.527	.426	213	33	216	3.9
Sam Perkins	1,001	.381	.842	168	41	242	3.8
Jeff Foster	1,152	.469	.516	389	33	249	3.5
Derrick Mckey	987	.441	.778	176	74	145	2.2

Coach-Isiah Thomas

Los Angeles Clippers

	Min	FG%	FT%	Reb	Ast	Pts	Avg
Lamar Odom	2,835	.460	.679	592	392	1,304	17.2
Jeff McInnis	2,831	.463	.807	220	447	1,046	12.9
Eric Piatkowski	2,153	.433	.873	241	96	860	10.6
Corey Maggette	1,359	.462	.774	291	82	690	10.0
Darius Miles	2,133	.504	.521	477	99	761	9.4
Michael Olowokandi	2,126	.435	.545	525	46	701	8.6
Quentin Richardson	1,357	.442	.627	257	62	613	8.1
Keyon Dooling	1,233	.410	.698	89	177	449	5.9
Sean Rooks	1,552	.428	.748	303	77	446	5.4
Cherokee Parks	1,054	.489	.705	229	46	299	4.6

Coach-Alvin Gentry

Los Angeles Lakers

	Min	FG%	FT%	Reb	Ast	Pts	Avg
Shaquille O'Neal	2,924	.572	.513	940	277	2,125	28.7
Kobe Bryant	2,783	.464	.853	399	338	1,938	28.5
Derek Fisher	709	.412	.806	59	87	229	11.5
Rick Fox	2,291	.444	.779	325	262	787	9.6
Horace Grant	2,390	.462	.775	545	121	657	8.5
Isaiah Rider	1,206	.426	.855	156	111	507	7.6
Ron Harper	1,139	.469	.708	166	113	307	6.5
Brian Shaw	1,834	.399	.797	304	258	421	5.3
Robert Horry	1,587	.387	.711	296	128	407	5.2
Mike Penberthy	850	.414	.903	63	71	267	5.0
Mark Madsen	641	.487	.703	152	24	137	2.0

Coach-Phil Jackson

Miami Heat

	Min	FG%	FT%	Reb	Ast	Pts	Avg
Eddie Jones	2,282	.445	.844	292	171	1,094	17.4
Anthony Mason	3,254	.482	.781	770	248	1,290	16.1
Brian Grant	2,771	.479	.797	718	101	1,250	15.2
Tim Hardaway	2,613	.392	.801	204	483	1,150	14.9
Bruce Bowen	2,685	.363	.609	245	132	623	7.6
Anthony Carter	1,630	.406	.631	180	268	461	6.4
Dan Majerle	1,306	.336	.818	166	88	267	5.0
A.C. Green	1,411	.444	.712	313	39	367	4.5

Coach-Pat Riley

Milwaukee Bucks

	Min	FG%	FT%	Reb	Ast	Pts	Avg
Glenn Robinson	2,813	.468	.820	526	252	1,674	22.0
Ray Allen	3,129	.480	.888	428	374	1,806	22.0
Sam Cassell	2,709	.474	.858	290	580	1,381	18.2
Tim Thomas	2,086	.430	.771	313	138	954	12.6
Lindsey Hunter	2,002	.381	.802	170	222	825	10.1
Jason Caffey	1,460	.488	.673	353	53	500	7.1
Scott Williams	1,272	.474	.857	364	35	403	6.1
Ervin Johnson	1,981	.545	.538	613	40	266	3.2
Mark Pope	942	.437	.629	147	38	151	2.4

Coach-George Karl

Minnesota Timberwolves

	Min	FG%	FT%	Reb	Ast	Pts	Avg
Kevin Garnett	3,205	.477	.764	921	401	1,784	22.0
Terrell Brandon	2,822	.451	.871	298	583	1,250	16.0
Wally Szczerbiak	2,858	.510	.870	447	260	1,145	14.0
Anthony Peeler	2,128	.421	.862	192	192	791	10.6
Laphonso Ellis	1,948	.464	.790	494	93	772	9.4
Chauncey Billups	1,790	.422	.842	158	259	713	9.3
Felipe Lopez	1,565	.441	.699	234	107	550	7.9
Reggie Slater	681	.514	.673	186	26	254	4.6
Radoslav Nesterovic	1,233	.461	.523	286	45	328	4.5
Sam Mitchell	980	.408	.727	123	57	285	3.5
Dean Garrett	833	.481	.692	217	24	177	2.5

Coach-Flip Saunders

New Jersey Nets

	Min	FG%	FT%	Reb	Ast	Pts	Avg
Stephon Marbury	2,550	.441	.790	205	506	1,598	23.9
Keith Van Horn	1,738	.435	.806	347	82	831	17.0
Kenyon Martin	2,274	.445	.630	502	131	814	12.0
Johnny Newman	2,051	.419	.855	176	115	895	10.9
Aaron Williams	2,333	.456	.787	590	88	836	10.2
Lucious Harris	2,073	.425	.770	288	135	683	9.4
Kendall Gill	892	.333	.722	131	87	285	9.2
Stephen Jackson	1,657	.426	.719	208	140	635	8.3
Sherman Douglas	1,095	.403	.748	74	144	338	5.7
Mark Strickland	719	.431	.643	161	24	252	4.6
Evan Eschmeyer	1,331	.460	.657	366	40	251	3.4

Coach-Byron Scott

New York Knickerbockers

	Min	FG%	FT%	Reb	Ast	Pts	Avg
Allan Houston	2,858	.449	.909	283	173	1,459	18.7
Latrell Sprewell	3,017	.430	.783	347	269	1,364	17.7
Marcus Camby	2,127	.524	.667	723	52	759	12.1
Glen Rice	2,212	.440	.852	307	89	899	12.0
Kurt Thomas	2,125	.511	.814	515	63	800	10.4
Larry Johnson	2,105	.411	.797	363	127	645	9.9
Othella Harrington	1,815	.487	.763	388	56	665	9.0
Mark Jackson	2,588	.419	.785	305	661	631	7.6
Charlie Ward	1,492	.416	.800	159	273	433	7.1

Coach-Jeff Van Gundy

Orlando Magic

	Min	FG%	FT%	Reb	Ast	Pts	Avg
Tracy McGrady	3,091	.457	.733	580	352	2,065	26.8
Darrell Armstrong	2,773	.412	.884	343	524	1,189	15.9
Mike Miller	2,389	.436	.711	327	140	975	11.9
Pat Garrity	1,570	.387	.867	210	51	628	8.3
John Amaechi	1,722	.400	.631	268	74	650	7.9
Bo Outlaw	2,534	.614	.573	619	225	582	7.3
Michael Doleac	1,397	.417	.847	273	65	490	6.4
Monty Williams	1,206	.445	.639	243	79	410	5.0
Troy Hudson	1,002	.336	.817	105	162	357	4.8
Andrew DeClercq	903	.554	.573	236	32	261	3.9
Don Reid	764	.566	.613	242	21	210	3.2

Coach-Doc Rivers

Philadelphia 76ers

	Min	FG%	FT%	Reb	Ast	Pts	Avg
Allen Iverson	2,979	.420	.814	274	325	2,207	31.1
Aaron McKie	2,394	.473	.768	311	377	878	11.6
Dikembe Mutombo	2,591	.484	.725	1014	76	749	10.0
Eric Snow	1,740	.418	.792	166	369	491	9.8
Roshown McLeod	922	.437	.882	120	58	337	9.6
Tyrone Hill	2,363	.474	.630	688	48	728	9.6
George Lynch	2,652	.446	.719	590	139	686	8.4
Jumaine Jones	866	.444	.755	189	32	304	4.7
Kevin Ollie	925	.396	.708	95	146	216	3.1

Coach-Larry Brown

Phoenix Suns

	Min	FG%	FT%	Reb	Ast	Pts	Avg
Shawn Marion	2,857	.480	.810	847	160	1,369	17.3
Jason Kidd	3,065	.411	.814	494	753	1,299	16.9
Cliff Robinson	2,751	.422	.709	334	237	1,345	16.4
Tony Delk	2,288	.415	.787	261	160	1,005	12.3
Rodney Rogers	2,182	.430	.761	359	180	998	12.2
Tom Gugliotta	1,158	.392	.792	255	55	362	6.4
Iakovos Tsakalidis	947	.470	.593	242	19	256	4.5
Mario Elie	1,506	.423	.797	155	131	299	4.4
Vinny Del Negro	922	.453	.932	82	126	254	3.9
Chris Dudley	613	.397	.389	183	18	72	1.4

Coach-Scott Skiles

Portland Trail Blazers

	Min	FG%	FT%	Reb	Ast	Pts	Avg
Rasheed Wallace	2,940	.501	.766	602	212	1,477	19.2
Steve Smith	2,542	.456	.890	272	213	1,105	13.6
Damon Stoudamire	2,655	.434	.831	303	468	1,066	13.0
Bonzi Wells	1,995	.533	.660	367	208	948	12.6
Scottie Pippen	2,133	.451	.739	333	294	721	11.3
Arvydas Sabonis	1,299	.479	.776	331	91	616	10.1
Rod Strickland	1,371	.424	.751	140	304	498	9.2
Dale Davis	2,162	.498	.632	605	103	580	7.2
Shawn Kemp	1,083	.407	.771	259	65	441	6.5
Greg Anthony	856	.383	.676	61	82	284	4.9
Stacey Augmon	1,182	.477	.655	159	98	311	4.7

Coach-Mike Dunleavy

Sacramento Kings

	Min	FG%	FT%	Reb	Ast	Pts	Avg
Chris Webber	2,836	.481	.703	776	294	1,898	27.1
Predrag Stojakovic	2,905	.470	.856	434	165	1,529	20.4
Doug Christie	2,939	.395	.897	355	289	996	12.3
Vlade Divac	2,420	.482	.691	674	231	974	12.0
Jason Williams	2,290	.407	.789	185	416	720	9.4
Bobby Jackson	1,648	.439	.739	246	161	566	7.2
Scot Pollard	1,658	.468	.749	465	47	498	6.5
Hidayet Turkoglu	1,245	.412	.777	210	69	391	5.3
Jon Barry	1,010	.404	.877	94	130	316	5.1
Lawrence Funderburke	698	.496	.623	196	17	288	4.9

Coach-Rick Adelman

San Antonio Spurs

	Min	FG%	FT%	Reb	Ast	Pts	Avg
Tim Duncan	3,174	.499	.618	997	245	1,820	22.2
Derek Anderson	2,859	.416	.851	363	301	1,269	15.5
David Robinson	2,371	.486	.747	691	116	1,151	14.4
Antonio Daniels	2,060	.468	.776	163	304	745	9.4
Sean Elliott	1,229	.434	.714	170	81	409	7.9
Malik Rose	1,219	.435	.713	308	48	437	7.7
Terry Porter	1,678	.448	.793	201	251	573	7.2
Avery Johnson	1,290	.447	.683	85	237	310	5.6
Danny Ferry	1,688	.475	.733	223	71	448	5.6
Samaki Walker	963	.480	.629	243	29	321	5.3
Steve Kerr	650	.421	.933	35	57	181	3.3

Coach-Gregg Popovich

Seattle SuperSonics

	Min	FG%	FT%	Reb	Ast	Pts	Avg
Gary Payton	3,244	.456	.766	361	642	1,823	23.1
Rashard Lewis	2,720	.480	.826	541	125	1,151	14.8
Ruben Patterson	2,059	.494	.681	382	161	988	13.0
Vin Baker	2,129	.422	.723	430	90	927	12.2
Patrick Ewing	2,107	.430	.685	585	92	760	9.6
Brent Barry	1,778	.494	.816	211	225	589	8.8
Shammond Williams	1,238	.438	.875	132	190	467	6.8
Desmond Mason	1,522	.431	.736	249	63	463	5.9
Emanuel Davis	1,290	.418	.818	154	137	361	5.8
Jelani Mccoy	1,143	.523	.441	252	57	317	4.5

Coach-Paul Westphal, Nate McMillan

Toronto Raptors

	Min	FG%	FT%	Reb	Ast	Pts	Avg
Vince Carter	2,979	.460	.765	416	291	2,070	27.6
Antonio Davis	2,729	.433	.754	787	106	1,069	13.7
Alvin Williams	2,394	.430	.752	212	407	802	9.8
Charles Oakley	2,767	.388	.836	741	264	748	9.6
Morris Peterson	1,809	.431	.717	259	105	747	9.3
Keon Clark	1,720	.480	.592	434	72	640	7.9
Jerome Williams	1,182	.463	.741	382	45	372	6.3
Dell Curry	956	.424	.843	85	75	429	6.0
Chris Childs	1,859	.403	.845	202	355	362	4.7
Eric Montross	649	.406	.258	173	19	120	2.2

Coach-Lenny Wilkins

Utah Jazz

	Min	FG%	FT%	Reb	Ast	Pts	Avg
Karl Malone	2,895	.498	.793	669	361	1,878	23.2
Donyell Marshall	2,326	.503	.751	566	133	1,100	13.6
Bryon Russell	2,473	.440	.779	330	160	933	12.0
John Stockton	2,397	.504	.817	227	713	944	11.5
John Starks	2,122	.398	.802	154	178	699	9.3
Danny Manning	1,305	.494	.729	214	92	603	7.4
Jacque Vaughn	1,620	.433	.780	150	323	498	6.1
Olden Polynice	1,619	.496	.262	378	31	429	5.3
Greg Ostertag	1,491	.495	.556	415	22	363	4.5

Coach-Jerry Sloan

Vancouver Grizzlies

	Min	FG%	FT%	Reb	Ast	Pts	Avg
Shareef Abdur-Rahim	3,241	.472	.834	735	250	1,663	20.5
Michael Dickerson	2,618	.417	.763	229	233	1,142	16.3
Mike Bibby	3,190	.454	.761	304	685	1,301	15.9
Bryant Reeves	1,832	.460	.796	452	80	622	8.3
Damon Jones	1,415	.409	.712	124	224	461	6.5
Grant Long	1,507	.439	.713	274	83	396	6.0
Erick Strickland	830	.303	.861	128	95	260	5.2
Stromile Swift	1,312	.451	.603	284	28	391	4.9
Tony Massenburg	823	.462	.700	210	9	233	4.5
Isaac Austin	845	.356	.700	222	58	226	4.4
Kevin Edwards	634	.329	.811	82	52	160	3.5

Coach-Sidney Lowe

Washington Wizards

	Min	FG%	FT%	Reb	Ast	Pts	Avg
Richard Hamilton	2,519	.438	.868	238	224	1,411	18.1
Mitch Richmond	1,216	.407	.894	109	111	598	16.2
Courtney Alexander	1,382	.417	.820	143	62	618	9.5
Chris Whitney	1,532	.387	.894	106	248	558	9.5
Christian Laettner	1,663	.503	.833	365	124	728	9.3
Jahidi White	1,609	.498	.567	520	20	583	8.6
Tyrone Nesby	1,556	.357	.802	173	76	512	8.3
Hubert Davis	1,692	.453	.871	139	110	524	7.9
Gerard King	706	.509	.800	129	31	214	4.8
Laron Profit	605	.394	.733	64	89	152	4.3
Michael Smith	1,610	.486	.578	562	101	301	3.8
Popeye Jones	638	.392	.745	220	31	162	3.6

Coach-Leonard Hamilton

2001 NBA Player Draft, First-Round Picks

(held June 27, 2001)

	Team	Player, College/Team
1.	Washington	Kwame Brown, F, Glynn Academy (GA)
2.	L.A. Clippers	Tyson Chandler[1], F, Dominguez HS (CA)
3.	Atlanta	Pau Gasol, F, Barcelona (Spain)
4.	Chicago	Eddy Curry, F/C, Thornwood HS (IL)
5.	Golden State	Jason Richardson, G/F, Michigan State
6.	Vancouver	Shane Battier, F, Duke
7.	New Jersey	Eddie Griffin[2], F, Seton Hall
8.	Cleveland	DeSagana Diop, C, Oak Hill Academy (VA)
9.	Detroit	Rodney White, F, Charlotte
10.	Boston	Joe Johnson, G/F, Arkansas
11.	Boston[3]	Kedrick Brown, F, Okaloosa-Walton CC (FL)
12.	Seattle	Vladimir Radmanovic, F/C, FMP Zeleznik (Yugoslavia)
13.	Houston	Richard Jefferson[4], F, Arizona
14.	Golden State[5]	Troy Murpy, F, Notre Dame
15.	Orlando	Steven Hunter, C, DePaul
16.	Charlotte	Kirk Haston, F, Indiana
17.	Toronto	Michael Bradley, F, Villanova
18.	Houston[6]	Jason Collins[7], C, Stanford
19.	Portland	Zach Randolph, F, Michigan State
20.	Cleveland[8]	Brendan Haywood[9], C, North Carolina
21.	Boston[10]	Joseph Forte, G, North Carolina
22.	Orlando[11]	Jeryl Sasser, G, Southern Methodist
23.	Houston[12]	Brandon Armstrong[13], G, Pepperdine
24.	Utah	Raul Lopez, G, Real Madrid (Spain)
25.	Sacramento	Gerald Wallace, F, Alabama
26.	Philadelphia	Samuel Dalembert, C, Seton Hall
27.	Vancouver[14]	Jamaal Tinsley, G, Iowa State
28.	San Antonio	Tony Parker, G, Paris Basket Racing (France)
29.	Minnesota	Forfeited*

(1) Traded to Chicago. (2) Traded to Houston. (3) From Denver. (4) Traded to New Jersey. (5) From Indiana. (6) From New York through Phoenix and Orlando. (7) Traded to New Jersey. (8) From Miami. (9) Traded to Orlando. (10) From Phoenix. (11) From Milwaukee through Houston. (12) From Dallas through Orlando. (13) Traded to New Jersey. (14) From Utah. *NOTE: As part of the penalties from the Joe Smith Salary Cap circumvention case, Minnesota's 2001 and 3 future first-round picks were forfeited.

Number-One First-Round NBA Draft Picks, 1966-2001

Year	Team	Player, college
1966	New York	Cazzie Russell, Michigan
1967	Detroit	Jimmy Walker, Providence
1968	Houston	Elvin Hayes, Houston
1969	Milwaukee	Lew Alcindor[1], UCLA
1970	Detroit	Bob Lanier, St. Bonaventure
1971	Cleveland	Austin Carr, Notre Dame
1972	Portland	LaRue Martin, Loyola-Chicago
1973	Philadelphia	Doug Collins, Illinois St.
1974	Portland	Bill Walton, UCLA
1975	Atlanta	David Thompson[2], N.C. State
1976	Houston	John Lucas, Maryland
1977	Milwaukee	Kent Benson, Indiana
1978	Portland	Mychal Thompson, Minnesota
1979	L.A. Lakers	Magic Johnson, Michigan St.
1980	Golden State	Joe Barry Carroll, Purdue
1981	Dallas	Mark Aguirre, DePaul
1982	L.A. Lakers	James Worthy, North Carolina
1983	Houston	Ralph Sampson, Virginia
1984	Houston	Akeem Olajuwon, Houston
1985	New York	Patrick Ewing, Georgetown
1986	Cleveland	Brad Daugherty, North Carolina
1987	San Antonio	David Robinson, Navy
1988	L.A. Clippers	Danny Manning, Kansas
1989	Sacramento	Pervis Ellison, Louisville
1990	New Jersey	Derrick Coleman, Syracuse
1991	Charlotte	Larry Johnson, UNLV
1992	Orlando	Shaquille O'Neal, LSU
1993	Orlando	Chris Webber[3], Michigan
1994	Milwaukee	Glenn Robinson, Purdue
1995	Golden State	Joe Smith, Maryland
1996	Philadelphia	Allen Iverson, Georgetown
1997	San Antonio	Tim Duncan, Wake Forest
1998	L.A. Clippers	Michael Olowokandi, Pacific
1999	Chicago Bulls	Elton Brand, Duke
2000	New Jersey	Kenyon Martin, Cincinnati
2001	Washington	Kwame Brown, Glynn Academy (HS)

(1) Later Kareem Abdul-Jabbar. (2) Signed with Denver of the ABA. (3) Traded to Golden State.

All-Time NBA Statistical Leaders

(At the end of the 2000-2001 season. *Player active in 2000-2001 season.)

Scoring Average
(Minimum 400 games or 10,000 points)

Scoring Average
(Minimum 400 goals or 10,000 points)

	G	Pts.	Avg
Michael Jordan	930	29,277	31.5
Wilt Chamberlain	1,045	31,419	30.1
*Shaquille O'Neal	608	16,812	27.7
Elgin Baylor	846	23,149	27.4
Jerry West	932	25,192	27.0
Bob Pettit	792	20,880	26.4
George Gervin	791	20,708	26.2
*Karl Malone	1,273	32,919	25.9
Oscar Robertson	1,040	26,710	25.7
Dominique Wilkins	1,074	26,668	24.8

Field Goal Percentage
(Minimum 2,000 field goals made)

	FGA	FGM	Pct.
Artis Gilmore	9,570	5,732	.599
Mark West	4,356	2,528	.580
*Shaquille O'Neal	11,632	6,709	.577
Steve Johnson	4,965	2,841	.572
Darryl Dawkins	6,079	3,477	.572
James Donaldson	5,442	3,105	.571
Jeff Ruland	3,734	2,105	.564
Kareem Abdul-Jabbar	28,307	15,837	.559
Kevin McHale	12,334	6,830	.554
Bobby Jones	6,199	3,412	.550

Free Throw Percentage
(Minimum 1,200 free throws made)

	FTA	FTM	Pct.
Mark Price	2,362	2,135	.904
Rick Barry	4,243	3,818	.900
Calvin Murphy	3,864	3,445	.892
Scott Skiles	1,741	1,548	.889
Larry Bird	4,471	3,960	.886
*Reggie Miller	6,038	5,338	.884
Bill Sharman	3,559	3,143	.883
Jeff Hornacek	3,390	2,973	.877
Ray Allen	1,625	1,424	.876
Ricky Pierce	3,871	3,389	.875

3-Point Field Goal Percentage
(Minimum 250 3-point field goals made)

	3-FGA	3-FGM	Pct.
*Steve Kerr	1,409	651	.462
*Hubert Davis	1,488	659	.443
Drazen Petrovic	583	255	.437
Tim Legler	603	260	.431
B.J. Armstrong	1,026	436	.425
*Dana Barros	2,581	1,066	.413
*Wesley Person	1,974	811	.411
Trent Tucker	1,410	575	.408
*Dell Curry	2,967	1,200	.404
*Glen Rice	3,550	1,435	.404

Games Played

Robert Parish	1,611
Kareem Abdul-Jabbar	1,560
*John Stockton	1,340
Moses Malone	1,329
Buck Williams	1,307
Elvin Hayes	1,303
*Sam Perkins	1,286
*A.C. Green	1,278
*Karl Malone	1,273
John Havlicek	1,270

Points

Kareem Abdul-Jabbar	38,387
*Karl Malone	32,919
Wilt Chamberlain	31,419
Michael Jordan	29,277
Moses Malone	27,409
Elvin Hayes	27,313
Oscar Robertson	26,710
Dominique Wilkins	26,668
*Hakeem Olajuwon	26,511
John Havlicek	26,395

Rebounds

Wilt Chamberlain	23,924
Bill Russell	21,620
Kareem Abdul-Jabbar	17,440
Elvin Hayes	16,279
Moses Malone	16,212
Robert Parish	14,715
Nate Thurmond	14,464
Walt Bellamy	14,241
Wes Unseld	13,769
*Hakeem Olajuwon	13,382

Minutes Played

Kareem Abdul-Jabbar	57,446
Elvin Hayes	50,000
Wilt Chamberlain	47,859
*Karl Malone	47,503
John Havlicek	46,471
Robert Parish	45,704
Moses Malone	45,071
Oscar Robertson	43,886
*John Stockton	42,923
*Hakeem Olajuwon	42,844

Field Goals Attempted

Kareem Abdul-Jabbar	28,307
Elvin Hayes	24,272
John Havlicek	23,930
Wilt Chamberlain	23,497
*Karl Malone	23,122
Michael Jordan	21,686
Dominique Wilkins	21,589
Alex English	21,036
*Hakeem Olajuwon	20,573
Elgin Baylor	20,171

3- Point Field Goals Attempted

*Reggie Miller	5,093
*Dale Ellis	4,266
*Tim Hardaway	3,960
*Vernon Maxwell	3,931
*Mookie Blaylock	3,774
*Dan Majerle	3,563
*Glen Rice	3,550
*John Starks	3,496
Chuck Person	3,370
*Mitch Richmond	3,355

Assists

*John Stockton	14,503
Magic Johnson	10,141
Oscar Robertson	9,887
*Mark Jackson	9,235
Isiah Thomas	9,061
Maurice Cheeks	7,392
Lenny Wilkens	7,211
*Rod Strickland	7,026
Bob Cousy	6,955
*Terry Porter	6,955
Guy Rodgers	6,917

Field Goals Made

Kareem Abdul-Jabbar	15,837
Wilt Chamberlain	12,681
*Karl Malone	12,105
Elvin Hayes	10,976
Michael Jordan	10,962
Alex English	10,659
*Hakeem Olajuwon	10,555
John Havlicek	10,513
Dominique Wilkins	9,963
Robert Parish	9,614

3- Point Field Goals Made

*Reggie Miller	2,037
*Dale Ellis	1,719
*Glen Rice	1,435
*Tim Hardaway	1,408
*Mitch Richmond	1,308
*Dan Majerle	1,281
*Mookie Blaylock	1,268
*Vernon Maxwell	1,256
*Terry Porter	1,238
*Dennis Scott	1,214
*Hersey Hawkins	1,226

Blocked Shots

*Hakeem Olajuwon	3,740
Kareem Abdul-Jabbar	3,189
Mark Eaton	3,064
*Patrick Ewing	2,849
*David Robinson	2,703
*Dikembe Mutombo	2,646
Tree Rollins	2,542
Robert Parish	2,361
Manute Bol	2,086
George T. Johnson	2,082

Steals

*John Stockton	2,976
Maurice Cheeks	2,310
Michael Jordan	2,306
Clyde Drexler	2,207
Alvin Robertson	2,112
*Hakeem Olajuwon	2,088
*Scottie Pippen	2,080
*Mookie Blaylock	2,051
Derek Harper	1,957
*Gary Payton	1,883

Personal Fouls

Kareem Abdul-Jabbar	4,657
Robert Parish	4,443
Buck Williams	4,267
*Hakeem Olajuwon	4,236
Elvin Hayes	4,193
*Charles Oakley	4,148
*Otis Thorpe	4,146
James Edwards	4,042
Jack Sikma	3,879
Hal Greer	3,855

Basketball Hall of Fame, Springfield, MA

(2001 inductees have an asterisk*)

PLAYERS

Abdul-Jabbar, Kareem
Archibald, Nate
Arizin, Paul
Barlow, Thomas
Barry, Rick
Baylor, Elgin
Beckman, John
Bellamy, Walt
Belov, Sergei
Bing, Dave
Bird, Larry
Blazejowski, Carol
Borgmann, Bennie
Bradley, Bill
Brennan, Joseph
Cervi, Al
Chamberlain, Wilt
Cooper, Charles
Cosic, Kresimir
Cousy, Bob
Cowens, Dave
Crawford, Joan
Cunningham, Billy
Curry, Denise
Davies, Bob
DeBernardi, Forrest
DeBusschere, Dave
Denhart, Dutch
Donovan, Anne
Endacott, Paul
English, Alex
Erving, Julius (Dr. J)
Foster, Bud
Frazier, Walt
Friedman, Max
Fulks, Joe
Gale, Lauren
Gallatin, Harry
Gates, Pop
Gervin, George
Gola, Tom
Goodrich, Gail
Greer, Hal
Gruenig, Ace
Hagan, Cliff
Hanson, Victor
Harris-Stewart, Luisa
Havlicek, John
Hawkins, Connie
Hayes, Elvin
Haynes, Marques
Heinsohn, Tom
Holman, Nat
Houbregs, Bob
Howell, Bailey
Hyatt, Chuck
Issel, Dan
Jeannette, Buddy
Johnson, William
Johnston, Neil
Jones, K.C.
Jones, Sam
Krause, Moose
Kurland, Bob
Lanier, Bob
Lapchick, Joe
Lieberman-Cline, Nancy
Lovellette, Clyde
Lucas, Jerry
Luisetti, Hank
Macauley, Ed
*Malone, Moses
Maravich, Pete
Martin, Slater
McAdoo, Bob
McCracken, Branch
McCracken, Jack
McDermott, Bobby
McGuire, Dick
McHale, Kevin
Meyers, Ann
Mikan, George
Mikkelsen, Vern
Miller, Cheryl
Monroe, Earl
Murphy, Calvin
Murphy, Stretch
Page, Pat
Pettit, Bob
Phillip, Andy
Pollard, Jim
Ramsey, Frank
Reed, Willis
Risen, Arnie
Robertson, Oscar
Roosma, John S.
Russell, Bill
Russell, Honey
Schayes, Adolph
Schmidt, Ernest
Schommer, John
Sedran, Barney
Semjonova, Uljana
Sharman, Bill
Steinmetz, Christian
Thomas, Isiah
Thompson, Cat
Thompson, David
Thurmond, Nate
Twyman, Jack
Unseld, Wes
Vandivier, Fuzzy
Wachter, Edward
Walton, Bill
Wanzer, Bobby
West, Jerry
White, Nera
Wilkens, Lenny
Wooden, John
Yardley, George

COACHES

Allen, Forrest (Phog)
Anderson, Harold
Auerbach, Red
Barry, Sam
Blood, Ernest
Cann, Howard
Carlson, Dr. H. C.
Carnesecca, Lou
Carnevale, Ben
Carril, Pete
Case, Everett
*Chaney, John
Conradt, Jody
Crum, Denny
Daly, Chuck
Dean, Everett
Diaz-Miguel, Antonio
Diddle, Edgar
Drake, Bruce
Gaines, Clarence
Gardner, Jack
Gill, Slats
Gomelsky, Aleksandr
Hannum, Alex
Harshman, Marv
Haskins, Don
Hickey, Edgar
Hobson, Howard
Holzman, Red
Iba, Hank
Julian, Alvin
Keaney, Frank
Keogan, George
Knight, Bob
*Krzyzewski, Mike
Kundla, John
Lambert, Ward
Litwack, Harry
Loeffler, Kenneth
Lonborg, Dutch
McCutchan, Arad
McGuire, Al
McGuire, Frank
McLendon, John
Meanwell, Dr. W. E.
Meyer, Ray
Miller, Ralph
Moore, Billie
Newell, Pete
Nikolic, Aleksandar
Ramsay, Jack
Rubini, Cesare
Rupp, Adolph
Sachs, Leonard
Shelton, Everett
Smith, Dean
Summitt, Pat
Taylor, Fred
Thompson, John
Wade, Margaret
Watts, Stan
Wilkens, Lenny
Wooden, John
Woolpert, Phil
Wootten, Morgan

REFEREES

Enright, James
Hepbron, George
Hoyt, George
Kennedy, Matthew
Leith, Lloyd
Mihalik, Red
Nucatola, John
Quigley, Ernest
Shirley, J. Dallas
Strom, Earl
Tobey, David
Walsh, David

CONTRIBUTORS

Abbott, Senda B.
Bee, Clair
Biasone, Danny
Brown, Walter
Bunn, John
Douglas, Bob
Duer, Al O.
Embry, Wayne
Fagan, Cliff
Fisher, Harry
Fleisher, Larry
Gottlieb, Edward
Gulick, Dr. L. H.
Harrison, Lester
Hepp, Dr. Ferenc
Hickox, Edward
Hinkle, Tony
Irish, Ned
Jones, R. W.
Kennedy, Walter
Liston, Emil
Mokray, Bill
Morgan, Ralph
Morgenweck, Frank
Naismith, Dr. James
Newton, C. M.
O'Brien, John
O'Brien, Larry
Olsen, Harold
Podoloff, Maurice
Porter, H. V.
Reid, William
Ripley, Elmer
St. John, Lynn
Saperstein, Abe
Schabinger, Arthur
Stagg, Amos Alonzo
Stankovich, Boris
Steitz, Edward
Taylor, Chuck
Teague, Bertha
Tower, Oswald
Trester, Arthur
Wells, Clifford
Wilke, Lou
Zollner, Fred

TEAMS

First Team
Original Celtics
Buffalo Germans
NY Renaissance

All-Time NBA Coaching Victories

(At the end of the 2000-2001 season. *Active through 2000-2001 season.)

Coach	W-L	Pct.
Lenny Wilkens*	1,226-1,016	.547
Pat Riley*	1,049-466	.692
Don Nelson*	979-781	.556
Bill Fitch	944-1,106	.460
Red Auerbach	938-479	.662
Dick Motta	935-1,017	.479
Jack Ramsay	864-783	.525
Cotton Fitzsimmons	832-775	.518
Larry Brown*	788-612	.563
Gene Shue	784-861	.477
Jerry Sloan*	784-448	.636
John MacLeod	707-657	.518
Red Holzman	696-604	.535
Phil Jackson*	668-234	.741
Chuck Daly	638-437	.593
Doug Moe	628-529	.543
George Karl*	625-418	.599
Mike Fratello	572-465	.552
Alvin Attles	557-518	.518
Del Harris	556-457	.549

NBA Home Courts

Team	Name (built)	Capacity
Atlanta	Philips Arena (1999)	20,000
Boston	FleetCenter (1995)	18,624
Charlotte	Charlotte Coliseum (1988)	23,799
Chicago	United Center (1994)	21,500
Cleveland	Gund Arena (1994)	20,562
Dallas	American Airlines Center (2001)	19,200
Denver	Pepsi Center (1999)	19,099
Detroit	The Palace of Auburn Hills (1988)	22,076
Golden State	Arena in Oakland[1] (1966)	19,596
Houston	Compaq Center[2] (1975)	16,285
Indiana	Conseco Fieldhouse (1999)	18,345
L.A. Clippers	Staples Center (1999)	19,282
L.A. Lakers	Staples Center (1999)	18,964
Miami	American Airlines Arena (1999)	19,600
Milwaukee	Bradley Center (1988)	18,600
Minnesota	Target Center (1990)	19,006
New Jersey	Continental Airlines Arena[3] (1981)	20,049
New York	Madison Square Garden (1968)	19,763
Orlando	TD Waterhouse Centre[4] (1989)	17,248
Philadelphia	First Union Center[5] (1996)	20,444
Phoenix	America West Arena (1992)	19,023
Portland	The Rose Garden (1995)	19,980
Sacramento	ARCO Arena (1988)	17,317
San Antonio	Alamodome (1993)	20,557[6]
Seattle	KeyArena at Seattle Center[7] (1962)	17,072
Toronto	Air Canada Centre (1999)	19,800
Utah	Delta Center (1991)	19,911
Vancouver	GM Place (1995)	19,193
Washington	MCI Center (1997)	20,674

(1) Oakland Coliseum Arena, 1966-96; renovated and renamed in 1997. (2) The Summit, 1975-97. (3) Brendan Byrne/Meadowlands Arena, 1981-96. (4) Orlando Arena, 1989-2000. (5) CoreStates Center, 1996-98. (6) Normal capacity; can seat up to 35,000. (7) Seattle Center Coliseum, 1962-94; renovated, expanded, and renamed in 1995.

WOMEN'S PROFESSIONAL BASKETBALL

WNBA 2001: L.A. Sparks Take First Title, Three MVPs for Lisa Leslie

For the 1st time in the 5-year history of the WNBA, the Houston Comets didn't win the championship. The L.A. Sparks, who finished with a 39-5 record (19-1 at home), swept the Charlotte Sting in the finals (after sweeping Houston, 2 games to 0, in the opening round), taking the deciding 2d game in the best-of-3 WNBA finals, 82-54, at the Staples Center in Los Angeles, Sept. 1. The victory margin was the biggest in championship series history and gave the city its 2d pro basketball championship in 2001 (the L.A. Lakers won the NBA title in June). L.A.'s Lisa Leslie had 24 points, 13 rebounds, 6 assists, and a title-game-record 7 blocked shots in the final game. Leslie, who was also named MVP of the regular season and the All-Star game, is the 1st WNBA player to take home all 3 MVP trophies. L.A. coach Michael Cooper became the 1st person to win championships in both the NBA (5, playing for the Lakers in the 1980s) and the WNBA.

WNBA Final Standings, 2001 Season

x-clinched playoff berth; y-clinched top seed

Eastern Conference

	W	L	Pct	GB
y-Cleveland Rockers	22	10	.688	—
x-New York Liberty	21	11	.656	1
x-Miami Sol	20	12	.625	2
x-Charlotte Sting	18	14	.563	4
Orlando Miracle	13	19	.406	9
Indiana Fever	10	22	.313	12
Detroit Shock	10	22	.313	12
Washington Mystics	10	22	.313	12

Western Conference

	W	L	Pct	GB
y-Los Angeles Sparks	28	4	.875	—
x-Sacramento Monarchs	20	12	.625	8
x-Utah Starzz	19	13	.594	9
x-Houston Comets	19	13	.594	9
Phoenix Mercury	13	19	.406	15
Minnesota Lynx	12	20	.375	16
Portland Fire	11	21	.344	17
Seattle Storm	10	22	.313	18

2001 WNBA Playoffs

(Playoff seeding in parentheses; Conference winner automatically gets top seed)

Eastern Conference

Charlotte (4) defeated Cleveland (1) 2 games to 1
New York (2) defeated Miami (3) 2 games to 1
Charlotte defeated New York 2 games to 1

Western Conference

Los Angeles (1) defeated Houston (4) 2 games to 0
Sacramento (2) defeated Utah (3) 2 games to 0
Los Angeles defeated Sacramento 2 games to 1

WNBA Championship (Best of 3)

Los Angeles defeated Charlotte 2 games to 0 [75-66, 82-54].

2001 All-WNBA Teams

First Team	Position	Second Team
Katie Smith, Minnesota	Forward	Tina Thompson, Houston
Natalie Williams, Utah	Forward	Chamique Holdsclaw, Washington
Lisa Leslie, Los Angeles	Center	Yolanda Griffith, Sacramento
Janeth Arcain, Houston	Guard	Ticha Penicheiro, Sacramento
Merlakia Jones, Cleveland	Guard	Tamecka Dixon, Los Angeles

WNBA Statistical Leaders and Awards in 2001

Minutes played — 1,234: Katie Smith, Minnesota.
Total points — 739: Katie Smith, Minnesota.
Points per game — 23.1: Katie Smith, Minnesota.
Highest field goal % — .602: Latasha Byears, Los Angeles.
Highest 3-pt. field goal % — .514: Jennifer Azzi, Utah.
Highest free throw % — .930: Elena Baranova, Miami.
Total rebounds — 357: Yolanda Griffith, Sacramento.
Rebounds per game — 11.2: Yolanda Griffith, Sacramento.
Total assists — 203: Teresa Weatherspoon, New York.
Assists per game — 7.5: Ticha Penicheiro, Sacramento.
Total steals — 82: Debbie Black, Miami.
Steals per game — 2.56: Debbie Black, Miami.
Total blocked shots — 113: Margo Dydek, Utah.
Coach of the year — Dan Hughes, Cleveland.
Defensive player of the year — Debbie Black, Miami.
Most improved player of the year — Janeth Arcain, Houston.

WNBA Champions

	Regular season		Playoffs		
Year	Eastern Conference	Western Conference	Winner	Coach	Runner-up
1997	Phoenix Mercury	Houston Comets	Houston	Van Chancellor	New York
1998	Cleveland Rockers	Houston Comets	Houston	Van Chancellor	Phoenix
1999	New York Liberty	Houston Comets	Houston	Van Chancellor	New York
2000	New York Liberty	Los Angeles Sparks	Houston	Van Chancellor	New York
2001	Cleveland Rockers	Los Angeles Sparks	Los Angeles	Michael Cooper	Charlotte

WNBA Scoring Leaders

Year	Scoring champion	Pts	Avg
1997	Cynthia Cooper, Houston	621	22.2
1998	Cynthia Cooper, Houston	680	22.7
1999	Cynthia Cooper, Houston	686	22.1
2000	Sheryl Swoopes, Houston	643	20.7
2001	Katie Smith, Minnesota	739	23.1

WNBA Finals MVP

1997 Cynthia Cooper, Houston
1998 Cynthia Cooper, Houston
1999 Cynthia Cooper, Houston
2000 Cynthia Cooper, Houston
2001 Lisa Leslie, Los Angeles

WNBA Most Valuable Player

1997 Cynthia Cooper, Houston
1998 Cynthia Cooper, Houston
1999 Yolanda Griffith, Sacramento
2000 Sheryl Swoopes, Houston
2001 Lisa Leslie, Los Angeles

WNBA Rookie of theYear

1997 no award
1998 Tracy Reid, Charlotte
1999 Chamique Holdsclaw, Washington
2000 Betty Lennox, Minnesota
2001 Jackie Stiles, Portland

IT'S A FACT: In 2001, Portland Fire guard Jackie Stiles was named WNBA Rookie of the Year, finishing 9th overall in scoring, with 14.9 points per game and 6th in 3-point field goal percentage (43.1%). In March, playing for Southwest Missouri State, she set the women's NCAA Division I all-time scoring record, with 3,393.

COLLEGE BASKETBALL

Final NCAA Division I Conference Standings, 2000-2001

(*conference tournament champion)

America East

Team	Conference W	Conference L	All Games W	All Games L
Hofstra*	16	2	26	5
Delaware	14	4	20	10
Drexel	12	6	15	12
Maine	10	8	18	11
Boston U.	9	9	14	14
Northeastern	8	10	10	19
Vermont	7	11	12	17
Towson	7	11	12	17
New Hampshire	6	12	7	21
Hartford	1	17	4	24

Atlantic Coast

Team	Conference W	Conference L	All Games W	All Games L
North Carolina	13	3	26	7
Duke*	13	3	35	4
Maryland	10	6	25	11
Virginia	9	7	20	9
Georgia Tech	8	8	17	13
Wake Forest	8	8	19	11
North Carolina St.	5	11	13	16
Florida St.	4	12	9	21
Clemson	2	14	12	19

Atlantic 10

Team	Conference W	Conference L	All Games W	All Games L
St. Joseph's (PA)	14	2	26	7
Xavier (OH)	12	4	21	8
Temple*	12	4	24	13
Massachusetts	11	5	15	15
St. Bonaventure	9	7	18	12
Dayton	9	7	21	13
George Washington	6	10	14	18
La Salle	5	11	12	17
Fordham	4	12	12	17
Duquesne	3	13	9	21
Rhode Island	3	13	7	23

Big East

Team	Conference W	Conference L	All Games W	All Games L
East Division				
Boston College*	13	3	27	5
Providence	11	5	21	10
Connecticut	8	8	20	12
Villanova	8	8	18	13
St. John's (NY)	8	8	14	15
Miami (FL)	8	8	16	13
Virginia Tech	2	14	8	19
West Division				
Notre Dame	11	5	20	10
Georgetown	10	6	25	7
Syracuse	10	6	25	9
West Virginia	8	8	17	12
Pittsburgh	7	9	19	13
Seton Hall	5	11	16	15
Rutgers	3	13	11	16

Big Sky

Team	Conference W	Conference L	All Games W	All Games L
Cal. St. Northridge*	13	3	22	10
Eastern Wash.	11	5	17	11
Idaho St.	10	6	14	14
Montana St.	8	8	16	14
Weber St.	8	8	15	14
Northern Arizona	8	8	15	14
Montana	6	10	11	16
Portland St.	6	10	9	18
Sacramento St.	2	14	5	22

Big South

Team	Conference W	Conference L	All Games W	All Games L
Radford	12	2	19	10
Winthrop*	11	3	18	3
UNC Asheville	9	5	15	13
Charleston Southern	6	8	10	19
Coastal Carolina	6	8	8	20
Liberty	5	9	13	15
Elon	4	10	9	20
High Point	3	11	7	21

Big Ten

Team	Conference W	Conference L	All Games W	All Games L
Illinois	13	3	27	8
Michigan St.	13	3	28	4
Ohio St.	12	4	20	11
Indiana*	10	6	21	13
Wisconsin	9	7	18	11
Penn St.	7	9	21	12
Iowa	7	9	23	12
Purdue	6	10	17	15
Minnesota	5	11	10	18
Michigan	4	12	12	16
Northwestern	3	13	11	19

Big 12

Team	Conference W	Conference L	All Games W	All Games L
Iowa St.*	13	3	25	6
Kansas	12	4	26	6
Oklahoma	12	4	26	7
Texas	12	4	25	9
Oklahoma St.	10	6	20	10
Missouri	9	7	20	13
Nebraska	7	9	14	16
Baylor	6	10	19	12
Colorado	5	11	15	15
Kansas St.	4	12	11	18
Texas A&M	3	13	10	20
Texas Tech	3	13	9	19

Big West

Team	Conference W	Conference L	All Games W	All Games L
UC Irvine	15	1	25	5
Utah St.*	13	3	28	6
Long Beach St.	10	6	18	3
UC Santa Barbara	9	7	13	5
Pacific (CA)	8	8	18	12
Boise St.	8	8	17	14
Cal. St. Fullerton	3	13	5	23
Cal. Poly	3	13	9	19
Idaho	3	13	6	21

Colonial Athletic Association

Team	Conference W	Conference L	All Games W	All Games L
Richmond	12	4	22	7
UNC Wilmington	11	5	19	11
George Mason*	11	5	18	12
Va. Commonwealth	9	7	16	14
William & Mary	7	9	11	17
Old Dominion	7	9	13	18
James Madison	6	10	12	17
East Carolina	6	10	14	14
American U.	3	13	7	20

Conference USA

Team	Conference W	Conference L	All Games W	All Games L
American Division				
Cincinnati	11	5	25	10
UNC Charlotte*	10	6	22	11
Marquette	9	7	15	14
St. Louis	8	8	17	14
Louisville	8	8	12	19
DePaul	4	12	12	18
National Division				
So. Mississippi	11	5	22	9
Memphis	10	6	21	15
South Florida	9	7	18	13
Alabama-Birmingham	8	8	17	14
Houston	6	10	9	20
Tulane	2	14	9	21

Ivy Group[1]

Team	Conference W	Conference L	All Games W	All Games L
Princeton	11	3	16	11
Pennsylvania	9	5	12	17
Brown	9	5	15	12
Harvard	7	7	14	12
Columbia	7	7	12	15
Yale	7	7	10	17
Dartmouth	3	11	8	19
Cornell	3	11	7	20

Metro Atlantic Athletic

Team	Conference W	Conference L	All Games W	All Games L
Iona*	12	6	22	11
Niagara	12	6	15	13
Siena	12	6	20	11
Marist	11	7	17	13
Rider	11	7	16	12
Manhattan	11	7	14	15
Canisius	9	9	20	11
Fairfield	8	10	12	16
Loyola (MD)	2	16	6	23
St. Peter's	2	16	4	24

Mid-American

Team	Conference W	Conference L	All Games W	All Games L
East Division				
Kent St.*	13	5	24	10
Marshall	12	6	18	9
Ohio U.	12	6	19	11
Miami (OH)	10	8	17	16
Bowling Green	10	8	15	14
Akron	9	9	12	16
Buffalo	2	16	4	24
West Division				
Central Michigan	14	4	20	8
Toledo	12	6	22	10
Ball St.	12	6	18	12
Western Michigan	7	11	7	21
Northern Illinois	4	14	5	23
Eastern Michigan	1	17	3	25

Mid-Continent

Team	Conference W	Conference L	All Games W	All Games L
Valparaiso	13	3	24	8
Southern Utah*	13	3	25	6
Youngstown St.	11	5	19	11
Missouri-K.C.	9	7	14	16
Oakland	8	8	12	16
Indiana/Purdue-Indianapolis	6	10	11	18
Oral Roberts	5	11	10	19
Western Illinois	5	11	5	23
Chicago St.	2	14	5	23

Mid-Eastern Athletic

Team	Conference W	Conference L	All Games W	All Games L
Hampton*	14	4	25	7
South Carolina St.	14	4	19	13
Delaware St.	11	7	13	15
Norfolk St.	11	7	12	17
Coppin St.	11	8	13	15
MD-Eastern Shore	10	8	12	16
North Carolina A&T	8	10	13	17
Howard	8	10	10	18
Bethune-Cookman	5	13	10	19
Florida A&M	4	14	6	22
Morgan St.	4	15	6	23

Midwestern Collegiate

Team	Conference W	Conference L	All Games W	All Games L
Butler*	11	3	24	8
Detroit	10	4	25	12
Cleveland St.	9	5	19	13
Wright St.	8	6	18	11
Wisconsin-Milwaukee	7	7	15	13
Illinois-Chicago	5	9	11	17
Wisconsin-Green Bay	4	10	11	17
Loyola (IL)	2	12	7	21

Missouri Valley

Team	Conference W	Conference L	All Games W	All Games L
Creighton	14	4	24	8
Illinois St.	12	6	21	9
Bradley	12	6	19	12
Indiana St.*	10	8	22	12
Southern Illinois	10	8	16	14
Evansville	9	9	14	16
Drake	8	10	12	16
SW Missouri St.	8	10	13	16
Wichita St.	4	14	9	19
Northern Iowa	3	15	7	24

Mountain West

Team	Conference W	Conference L	All Games W	All Games L
Utah	10	4	19	12
BYU*	10	4	24	9
Wyoming	10	4	20	10
UNLV	7	7	16	13
New Mexico	6	8	20	12
Colorado St.	6	8	15	13
San Diego St.	4	10	14	14
Air Force	3	11	8	21

Northeast

Team	Conference W	Conference L	All Games W	All Games L
St. Francis (NY)	16	4	18	11
Monmouth (NJ)*	15	5	21	10
MD- Baltimore County	13	7	18	11
LIU-Brooklyn	12	8	12	16
Wagner	11	9	16	13
Cent. Connecticut St.	11	9	14	14
Fairleigh Dickinson	10	10	13	15
St. Francis (PA)	9	11	9	18
Mt. St. Mary's (MD)	7	13	7	21
Robert Morris	7	13	7	22
Sacred Heart	6	14	7	21
Quinnipiac	3	17	6	21

Ohio Valley

Team	Conference W	Conference L	All Games W	All Games L
Tennessee Tech	13	3	20	9
Eastern Illinois*	11	5	21	10
Murray St.	11	5	17	12
Austin Peay	10	6	22	10
SE Missouri St.	8	8	18	12
Tennessee St.	7	9	10	19
Morehead St.	6	10	12	16
Tennessee-Martin	5	11	10	18
Eastern Kentucky	1	15	7	19

Pacific-10[1]

Team	Conference W	Conference L	All Games W	All Games L
Stanford	16	2	31	3
Arizona	15	3	28	8
UCLA	14	4	23	9
USC	11	7	24	10
California	11	7	20	11
Oregon	5	13	14	14
Arizona St.	5	13	13	16
Washington St.	5	13	12	16
Oregon St.	4	14	10	20
Washington	4	14	10	20

Patriot League

Team	Conference W	Conference L	All Games W	All Games L
Holy Cross*	10	2	22	8
Navy	9	3	19	12
Colgate	6	6	13	15
Lehigh	6	6	13	16
Lafayette	4	8	12	16
Bucknell	4	8	14	15
Army	3	9	9	19

Southeastern

Team	Conference W	Conference L	All Games W	All Games L
Eastern Division				
Kentucky*	12	4	24	10
Florida	12	4	24	7
Georgia	9	7	16	15
Tennessee	8	8	22	11
South Carolina	6	10	15	15
Vanderbilt	4	12	15	15
Western Division				
Mississippi	11	5	27	8
Arkansas	10	6	20	11
Alabama	8	8	25	11
Mississippi St.	7	9	18	13
Auburn	7	9	18	14
LSU	7	9	13	16

(1) Conference does not hold a tournament.

Southern

Team	Conference W	Conference L	All Games W	All Games L
North Division				
East Tennessee St.	13	3	18	10
UNC-Greensboro*	10	6	19	12
Appalachian St.	7	9	11	20
Davidson	7	9	15	17
VMI	5	11	9	19
Western Carolina	3	13	6	25
South Division				
Col. of Charleston	12	4	22	7
Tenn.-Chattanooga	9	7	17	13
The Citadel	9	7	16	12
Georgia Southern	9	7	15	15
Wofford	7	9	12	16
Furman	5	11	10	16

Southland

Team	Conference W	Conference L	All Games W	All Games L
McNeese St.	17	3	22	9
Nicholls St.	12	8	14	14
Texas-San Antonio	12	8	14	15
Northwestern St.*	11	9	19	13
Sam Houston St.	11	9	16	13
Texas-Arlington	11	9	13	15
SW Texas St.	10	10	13	15
Louisiana-Monroe	8	12	11	17
Lamar	7	13	9	18
Stephen F. Austin	6	14	9	17
SE Louisiana	5	15	8	21

Southwestern Athletic

Team	Conference W	Conference L	All Games W	All Games L
Alabama St.*	15	13	22	9
Mississippi Valley St.	14	4	18	9
Alabama A&M	13	5	17	11
Alcorn St.	13	5	15	15
Southern U.	8	10	11	16
Grambling	8	10	8	18
Jackson St.	7	11	7	23
Prairie View A&M	5	13	6	22
Texas Southern	5	13	7	22
Arkansas-Pine Bluff	2	16	2	25

Sun Belt

Team	Conference W	Conference L	All Games W	All Games L
East Division				
Western Kentucky*	14	2	24	7
Arkansas St.	10	6	17	13
Louisiana Tech	10	6	17	12
Arkansas-Little Rock	9	7	18	11
Florida International	5	11	8	21
Middle Tennessee	1	15	5	22
West Division				
South Alabama	11	5	22	11
U. Louisiana-Lafayette	10	6	16	13
New Mexico St.	10	6	14	14
New Orleans	10	6	17	12
Denver	5	11	10	18
North Texas	1	15	4	24

Trans America Athletic

Team	Conference W	Conference L	All Games W	All Games L
Georgia St.*	16	2	29	5
Troy St.	12	6	19	12
Stetson	11	7	17	12
Jacksonville	11	7	18	10
Samford	11	7	15	14
Mercer	10	8	13	15
Jacksonville St.	6	12	9	19
Campbell	5	13	7	21
Florida Atlantic	5	13	7	24
Central Florida	3	15	8	23

West Coast

Team	Conference W	Conference L	All Games W	All Games L
Gonzaga*	13	1	26	6
Pepperdine	12	2	22	8
Santa Clara	10	4	20	12
San Diego	7	7	16	13
San Francisco	5	9	12	17
Loyola Marymount	5	9	9	19
Portland	4	10	11	17
St. Mary's (CA)	0	14	2	26

Western Athletic

Team	Conference W	Conference L	All Games W	All Games L
Fresno St.	13	3	26	7
Tulsa	10	6	26	11
Texas-El Paso	10	6	23	9
Texas Christian	9	7	20	11
Hawaii*	8	8	17	14
Southern Methodist	8	8	18	12
San Jose St.	6	10	14	14
Rice	5	11	14	16
Nevada	3	13	10	18

All-Time Winningest Division I College Teams by Percentage

(through 2000-2001 season)

School	Years	Won	Lost	Pct.
Kentucky	98	1,795	558	.763
North Carolina	91	1,781	630	.739
UNLV	43	886	341	.723
Kansas	103	1,738	740	.702
UCLA	82	1,489	642	.699
St. John's (NY)	94	1,621	738	.688
Duke	96	1,649	764	.684
Syracuse	100	1,549	719	.683
Western Kentucky	82	1,414	710	.666
Arkansas	78	1,354	758	.657
Utah	93	1,446	708	.657
Indiana	101	1,494	800	.652
Temple	105	1,571	843	.651
Louisville	87	1,387	758	.647
Notre Dame	96	1,483	817	.645
Purdue	103	1,445	796	.645
DePaul	78	1,233	684	.644
Illinois	96	1,407	782	.643
Weber State	39	712	396	.643
Arizona	96	1,386	775	.642
Pennsylvania	101	1,508	863	.637
Villanova	81	1,327	759	.637

Major College Basketball Tournaments

The National Invitation Tournament (NIT), first played in 1938, is the nation's oldest basketball tournament. The first National Collegiate Athletic Association (NCAA) national championship tournament was played one year later. Selections for both tournaments are made in Mar., with the NCAA selecting first from among the top Division I teams.

National Invitation Tournament Champions

Year	Champion
1938	Temple
1939	Long Island Univ.
1940	Colorado
1941	Long Island Univ.
1942	West Virginia
1943	St. John's
1944	St. John's
1945	De Paul
1946	Kentucky
1947	Utah
1948	St. Louis
1949	San Francisco
1950	CCNY
1951	Brigham Young
1952	LaSalle
1953	Seton Hall
1954	Holy Cross
1955	Duquesne
1956	Louisville
1957	Bradley
1958	Xavier (Ohio)
1959	St. John's
1960	Bradley
1961	Providence
1962	Dayton
1963	Providence
1964	Bradley
1965	St. John's
1966	Brigham Young
1967	Southern Illinois
1968	Dayton
1969	Temple
1970	Marquette
1971	North Carolina
1972	Maryland
1973	Virginia Tech
1974	Purdue
1975	Princeton
1976	Kentucky
1977	St. Bonaventure
1978	Texas
1979	Indiana
1980	Virginia
1981	Tulsa
1982	Bradley
1983	Fresno State
1984	Michigan
1985	UCLA
1986	Ohio State
1987	Southern Mississippi
1988	Connecticut
1989	St. John's
1990	Vanderbilt
1991	Stanford
1992	Virginia
1993	Minnesota
1994	Villanova
1995	Virginia Tech
1996	Nebraska
1997	Michigan
1998	Minnesota
1999	California
2000	Wake Forest
2001	Tulsa

2001 MEN'S NCAA BASKETBALL TOURNAMENT

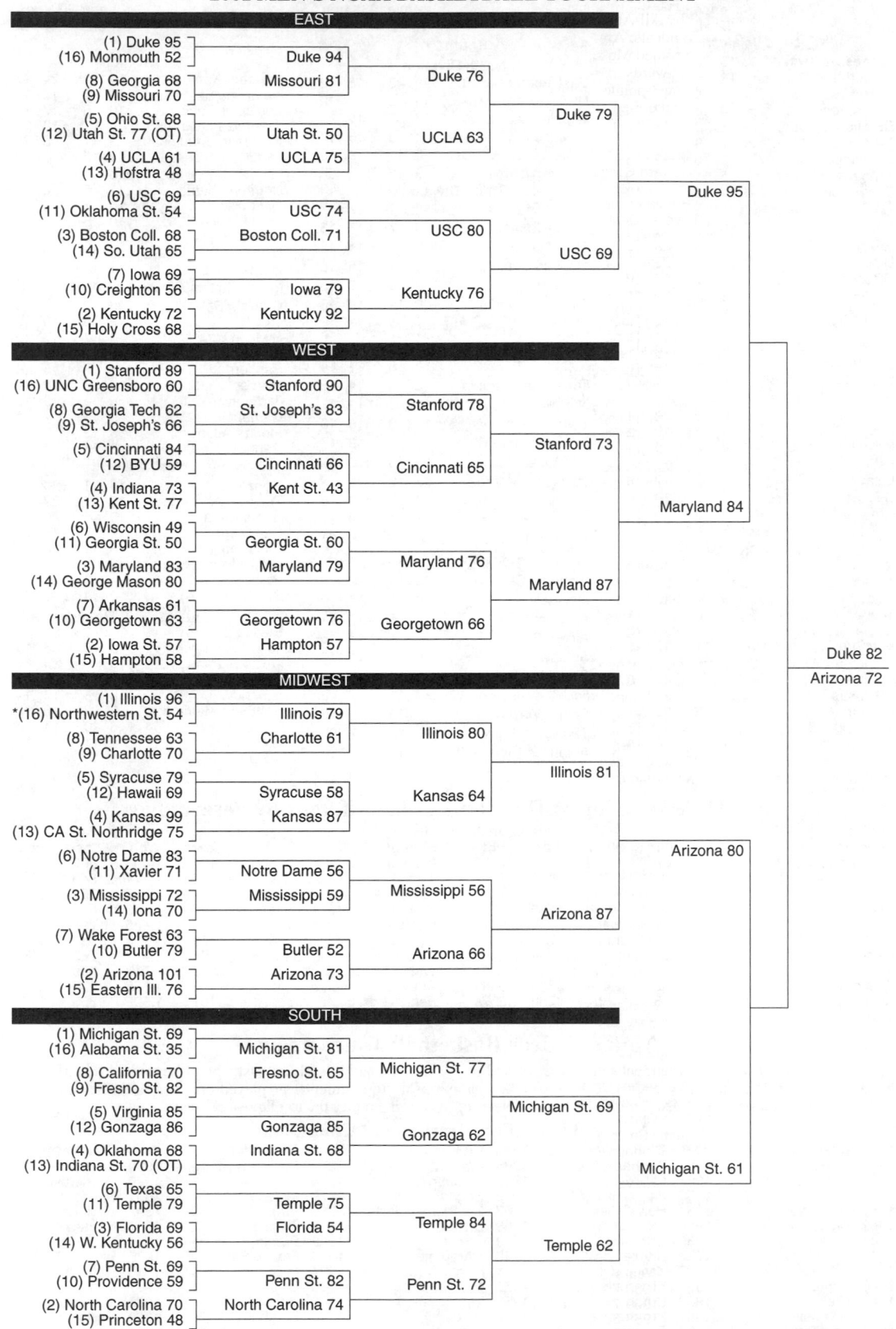

EAST

(1) Duke 95
(16) Monmouth 52
(8) Georgia 68
(9) Missouri 70
(5) Ohio St. 68
(12) Utah St. 77 (OT)
(4) UCLA 61
(13) Hofstra 48
(6) USC 69
(11) Oklahoma St. 54
(3) Boston Coll. 68
(14) So. Utah 65
(7) Iowa 69
(10) Creighton 56
(2) Kentucky 72
(15) Holy Cross 68

Duke 94
Missouri 81
Utah St. 50
UCLA 75
USC 74
Boston Coll. 71
Iowa 79
Kentucky 92

Duke 76
UCLA 63
USC 80
Kentucky 76

Duke 79
USC 69

Duke 95

WEST

(1) Stanford 89
(16) UNC Greensboro 60
(8) Georgia Tech 62
(9) St. Joseph's 66
(5) Cincinnati 84
(12) BYU 59
(4) Indiana 73
(13) Kent St. 77
(6) Wisconsin 49
(11) Georgia St. 50
(3) Maryland 83
(14) George Mason 80
(7) Arkansas 61
(10) Georgetown 63
(2) Iowa St. 57
(15) Hampton 58

Stanford 90
St. Joseph's 83
Cincinnati 66
Kent St. 43
Georgia St. 60
Maryland 79
Georgetown 76
Hampton 57

Stanford 78
Cincinnati 65
Maryland 76
Georgetown 66

Stanford 73
Maryland 87

Maryland 84

MIDWEST

(1) Illinois 96
*(16) Northwestern St. 54
(8) Tennessee 63
(9) Charlotte 70
(5) Syracuse 79
(12) Hawaii 69
(4) Kansas 99
(13) CA St. Northridge 75
(6) Notre Dame 83
(11) Xavier 71
(3) Mississippi 72
(14) Iona 70
(7) Wake Forest 63
(10) Butler 79
(2) Arizona 101
(15) Eastern Ill. 76

Illinois 79
Charlotte 61
Syracuse 58
Kansas 87
Notre Dame 56
Mississippi 59
Butler 52
Arizona 73

Illinois 80
Kansas 64
Mississippi 56
Arizona 66

Illinois 81
Arizona 87

Arizona 80

SOUTH

(1) Michigan St. 69
(16) Alabama St. 35
(8) California 70
(9) Fresno St. 82
(5) Virginia 85
(12) Gonzaga 86
(4) Oklahoma 68
(13) Indiana St. 70 (OT)
(6) Texas 65
(11) Temple 79
(3) Florida 69
(14) W. Kentucky 56
(7) Penn St. 69
(10) Providence 59
(2) North Carolina 70
(15) Princeton 48

Michigan St. 81
Fresno St. 65
Gonzaga 85
Indiana St. 68
Temple 75
Florida 54
Penn St. 82
North Carolina 74

Michigan St. 77
Gonzaga 62
Temple 84
Penn St. 72

Michigan St. 69
Temple 62

Michigan St. 61

Final

Duke 82
Arizona 72

*Northwestern St. defeated Winthrop, 71-67, in a special play-in game on Mar. 13 to earn the 16th seed in the Midwest.

2001 Men's NCAA Tournament: Duke Downs Arizona for Third Championship

The Duke Blue Devils, led by All-American senior Shane Battier, earned their 3d NCAA Championship in 11 years with an 82-72 win over the Arizona Wildcats, Apr. 2, at the Metrodome in Minneapolis, MN. Battier, who had 18 points, 11 rebounds, and 6 assists in the game, was named Most Outstanding Player of the Final Four. He also won the 2001 AP Player of the Year, and the Naismith and Wooden awards. Arizona's Loren Woods led all scorers, with 22 points, followed by Duke's Mike Dunleavy, who had 21, including five 3-pointers. Coach Mike Krzyewski, who led Duke to all 3 of its national championships (1991-92, 2001), tied Bobby Knight and trailed only legends John Wooden (10) and Adolph Rupp (4) in career NCAA titles.

NCAA Division I Champions

Year	Champion	Coach	Final opponent	Score	Outstanding player	Site
1939	Oregon	Howard Hobson	Ohio St.	46-33	None	Evanston, IL
1940	Indiana	Branch McCracken	Kansas	60-42	Marvin Huffman, Indiana	Kansas City, MO
1941	Wisconsin	Harold Foster	Washington St.	39-34	John Kotz, Wisconsin	Kansas City, MO
1942	Stanford	Everett Dean	Dartmouth	53-38	Howard Dallmar, Stanford	Kansas City, MO
1943	Wyoming	Everett Shelton	Georgetown	46-34	Ken Sailors, Wyoming	New York, NY
1944	Utah	Vadal Peterson	Dartmouth	42-40[1]	Arnold Ferrin, Utah	New York, NY
1945	Oklahoma St.[2]	Henry Iba	NYU	49-45	Bob Kurland, Oklahoma St.	New York, NY
1946	Oklahoma St.[2]	Henry Iba	North Carolina	43-40	Bob Kurland, Oklahoma St.	New York, NY
1947	Holy Cross	Alvin Julian	Oklahoma	58-47	George Kaftan, Holy Cross	New York, NY
1948	Kentucky	Adolph Rupp	Baylor	58-42	Alex Groza, Kentucky	New York, NY
1949	Kentucky	Adolph Rupp	Oklahoma St.	46-36	Alex Groza, Kentucky	Seattle, WA
1950	CCNY	Nat Holman	Bradley	71-68	Irwin Dambrot, CCNY	New York, NY
1951	Kentucky	Adolph Rupp	Kansas St.	68-58	None	Minneapolis, MN
1952	Kansas	Forrest Allen	St. John's	80-63	Clyde Lovellette, Kansas	Seattle, WA
1953	Indiana	Branch McCracken	Kansas	69-68	B.H. Born, Kansas	Kansas City, MO
1954	La Salle	Kenneth Loeffler	Bradley	92-76	Tom Gola, La Salle	Kansas City, MO
1955	San Francisco	Phil Woolpert	LaSalle	77-63	Bill Russell, San Francisco	Kansas City, MO
1956	San Francisco	Phil Woolpert	Iowa	83-71	Hal Lear, Temple	Evanston, IL
1957	North Carolina	Frank McGuire	Kansas	54-53[1]	Wilt Chamberlain, Kansas	Kansas City, MO
1958	Kentucky	Adolph Rupp	Seattle	84-72	Elgin Baylor, Seattle	Louisville, KY
1959	California	Pete Newell	West Virginia	71-70	Jerry West, West Virginia	Louisville, KY
1960	Ohio St.	Fred Taylor	California	75-55	Jerry Lucas, Ohio St.	San Francisco, CA
1961	Cincinnati	Edwin Jucker	Ohio St.	70-65[1]	Jerry Lucas, Ohio St.	Kansas City, MO
1962	Cincinnati	Edwin Jucker	Ohio St.	71-59	Paul Hogue, Cincinnati	Louisville, KY
1963	Loyola (IL)	George Ireland	Cincinnati	60-58[1]	Art Heyman, Duke	Louisville, KY
1964	UCLA	John Wooden	Duke	98-83	Walt Hazzard, UCLA	Kansas City, MO
1965	UCLA	John Wooden	Michigan	91-80	Bill Bradley, Princeton	Portland, OR
1966	Texas-El Paso[3]	Don Haskins	Kentucky	72-65	Jerry Chambers, Utah	College Park, MD
1967	UCLA	John Wooden	Dayton	79-64	Lew Alcindor, UCLA	Louisville, KY
1968	UCLA	John Wooden	North Carolina	78-55	Lew Alcindor, UCLA	Los Angeles, CA
1969	UCLA	John Wooden	Purdue	92-72	Lew Alcindor, UCLA	Louisville, KY
1970	UCLA	John Wooden	Jacksonville	80-69	Sidney Wicks, UCLA	College Park, MD
1971	UCLA	John Wooden	Villanova*	68-62	Howard Porter, Villanova*	Houston, TX
1972	UCLA	John Wooden	Florida St.	81-76	Bill Walton, UCLA	Los Angeles, CA
1973	UCLA	John Wooden	Memphis St.	87-66	Bill Walton, UCLA	St. Louis, MO
1974	North Carolina St.	Norm Sloan	Marquette	76-64	David Thompson, N.C. St.	Greensboro, NC
1975	UCLA	John Wooden	Kentucky	92-85	Richard Washington, UCLA	San Diego, CA
1976	Indiana	Bob Knight	Michigan	86-68	Kent Benson, Indiana	Philadelphia, PA
1977	Marquette	Al McGuire	North Carolina	67-59	Butch Lee, Marquette	Atlanta, GA
1978	Kentucky	Joe Hall	Duke	94-88	Jack Givens, Kentucky	St. Louis, MO
1979	Michigan St.	Jud Heathcote	Indiana St.	75-64	Magic Johnson, Michigan St.	Salt Lake City, UT
1980	Louisville	Denny Crum	UCLA*	59-54	Darrell Griffith, Louisville	Indianapolis, IN
1981	Indiana	Bob Knight	North Carolina	63-50	Isiah Thomas, Indiana	Philadelphia, PA
1982	North Carolina	Dean Smith	Georgetown	63-62	James Worthy, N. Carolina	New Orleans, LA
1983	North Carolina St.	Jim Valvano	Houston	54-52	Hakeem Olajuwon, Houston	Albuquerque, NM
1984	Georgetown	John Thompson	Houston	84-75	Patrick Ewing, Georgetown	Seattle, WA
1985	Villanova	Rollie Massimino	Georgetown	66-64	Ed Pinckney, Villanova	Lexington, KY
1986	Louisville	Denny Crum	Duke	72-69	Pervis Ellison, Louisville	Dallas, TX
1987	Indiana	Bob Knight	Syracuse	74-73	Keith Smart, Indiana	New Orleans, LA
1988	Kansas	Larry Brown	Oklahoma	83-79	Danny Manning, Kansas	Kansas City, MO
1989	Michigan	Steve Fisher	Seton Hall	80-79[1]	Glen Rice, Michigan	Seattle, WA
1990	UNLV	Jerry Tarkanian	Duke	103-73	Anderson Hunt, UNLV	Denver, CO
1991	Duke	Mike Krzyzewski	Kansas	72-65	Christian Laettner, Duke	Indianapolis, IN
1992	Duke	Mike Krzyzewski	Michigan	71-51	Bobby Hurley, Duke	Minneapolis, MN
1993	North Carolina	Dean Smith	Michigan	77-71	Donald Williams, N. Carolina	New Orleans, LA
1994	Arkansas	Nolan Richardson	Duke	76-72	Corliss Williamson, Arkansas	Charlotte, NC
1995	UCLA	Jim Harrick	Arkansas	89-78	Ed O'Bannon, UCLA	Seattle, WA
1996	Kentucky	Rick Pitino	Syracuse	76-67	Tony Delk, Kentucky	E. Rutherford, NJ
1997	Arizona	Lute Olson	Kentucky	84-79[1]	Miles Simon, Arizona	Indianapolis, IN
1998	Kentucky	Tubby Smith	Utah	78-69	Jeff Sheppard, Kentucky	San Antonio, TX
1999	Connecticut	Jim Calhoun	Duke	77-74	Richard Hamilton, Connecticut	St. Petersburg, FL
2000	Michigan St.	Tom Izzo	Florida	89-76	Mateen Cleaves, Michigan St.	Indianapolis, IN
2001	Duke	Mike Krzyzewski	Arizona	82-72	Shane Battier, Duke	Minneapolis, MN

*Declared ineligible after the tournament. (1) Overtime. (2) Then known as Oklahoma A&M. (3) Then known asTexas Western.

Top Division I Career Scorers

(minimum 1,500 points; ranked by average)

Player, school	Years	Points	Avg.
Pete Maravich, LSU	1968-70	3,667	44.2
Austin Carr, Notre Dame	1969-71	2,560	34.6
Oscar Robertson, Cincinnati	1958-60	2,973	33.8
Calvin Murphy, Niagara	1968-70	2,548	33.1
Dwight Lamar, SW Louisiana	1972-73	1,862	32.7
Frank Selvy, Furman	1952-54	2,538	32.5
Rick Mount, Purdue	1968-70	2,323	32.3
Darrell Floyd, Furman	1954-56	2,281	32.1
Nick Werkman, Seton Hall	1962-64	2,273	32.0
Willie Humes, Idaho State	1970-71	1,510	31.5

John R. Wooden Award

Awarded to the nation's outstanding college basketball player by the Los Angeles Athletic Club.

1977 Marques Johnson, UCLA
1978 Phil Ford, North Carolina
1979 Larry Bird, Indiana State
1980 Darrell Griffith, Louisville
1981 Danny Ainge, Brigham Young
1982 Ralph Sampson, Virginia
1983 Ralph Sampson, Virginia
1984 Michael Jordan, North Carolina
1985 Chris Mullin, St. John's
1986 Walter Berry, St. John's
1987 David Robinson, Navy
1988 Danny Manning, Kansas
1989 Sean Elliott, Arizona
1990 Lionel Simmons, La Salle
1991 Larry Johnson, UNLV
1992 Christian Laettner, Duke
1993 Calbert Cheaney, Indiana
1994 Glenn Robinson, Purdue
1995 Ed O'Bannon, UCLA
1996 Marcus Camby, Massachusetts
1997 Tim Duncan, Wake Forest
1998 Antawn Jamison, North Carolina
1999 Elton Brand, Duke
2000 Kenyon Martin, Cincinnati
2001 Shane Battier, Duke

Most Coaching Victories in the NCAA Tournament Through 2001

(Coaches active in 2000-2001 season in bold)

Coach, School(s), First/Last appearance	Wins	Tourns.
Dean Smith, North Carolina, 1967/1997	65	27
Mike Krzyzewski, Duke, 1984/2001	56	17
John Wooden, UCLA, 1950/1975	47	16
Denny Crum, Louisville, 1972/2000	42	23
Bob Knight, Indiana, 1973/2000	42	24
Lute Olson, Iowa, Arizona, 1979/2001	37	22
John Thompson, Georgetown, 1975/1997	34	20
Jim Boeheim, Syracuse, 1977/2001	32	21
Eddie Sutton, Creighton, Arkansas, Kentucky, Oklahoma St., 1974/2001	32	22
Jerry Tarkanian, Long Beach St., UNLV, Fresno St.,1970/2001	32*	15

*Does not include 6 wins in the 1971-73 tournaments which were later vacated for NCAA rule violations.

Women's College Basketball

2001 Women's NCAA Tournament: Notre Dame Holds off Purdue for First Title

The Notre Dame Fighting Irish edged out the Purdue Boilermakers, 68-66, to win their first NCAA women's title, Apr. 1, at the Savvis Center in St. Louis. Ruth Riley hit 2 free throws with 5.8 seconds left in a tie game to give Notre Dame the victory. The 6'5" Riley led all players in scoring (28), rebounds (13), and blocked shots (7), and was named Most Outstanding Player of the Final Four. She also won the Naismith Award and was named AP Player of the Year. Notre Dame coach Muffet McGraw was also honored with the AP and the Naismith Award as Coach of the Year.

NCAA Division I Women's Champions

Year	Champion	Coach	Final opponent	Score	Outstanding player	Site
1982	Louisiana Tech	Sonja Hogg	Cheyney	76-62	Janice Lawrence, La. Tech	Norfolk, VA
1983	USC	Linda Sharp	Louisiana Tech	69-67	Cheryl Miller, USC	Norfolk, VA
1984	USC	Linda Sharp	Tennessee	72-61	Cheryl Miller, USC	Los Angeles, CA
1985	Old Dominion	Marianne Stanley	Georgia	70-65	Tracy Claxton, Old Dominion	Austin, TX
1986	Texas	Jody Conradt	USC	97-81	Clarissa Davis, Texas	Lexington, KY
1987	Tennessee	Pat Summitt	Louisiana Tech	67-44	Tonya Edwards, Tennessee	Austin, TX
1988	Louisiana Tech	Leon Barmore	Auburn	56-54	Erica Westbrooks, La. Tech	Tacoma, WA
1989	Tennessee	Pat Summitt	Auburn	76-60	Bridgette Gordon, Tennessee	Tacoma, WA
1990	Stanford	Tara VanDerveer	Auburn	88-81	Jennifer Azzi, Stanford	Knoxville, TN
1991	Tennessee	Pat Summitt	Virginia	70-67*	Dawn Staley, Virginia	New Orleans, LA
1992	Stanford	Tara VanDerveer	W. Kentucky	78-62	Molly Goodenbour, Stanford	Los Angeles, CA
1993	Texas Tech	Marsha Sharp	Ohio St.	84-82	Sheryl Swoopes, Texas Tech	Atlanta, GA
1994	North Carolina	Sylvia Hatchell	Louisiana Tech	60-59	Charlotte Smith, North Carolina	Richmond, VA
1995	Connecticut	Geno Auriemma	Tennessee	70-64	Rebecca Lobo, Connecticut	Minneapolis, MN
1996	Tennessee	Pat Summitt	Georgia	83-65	Michelle Marciniak, Tennessee	Charlotte, NC
1997	Tennessee	Pat Summitt	Old Dominion	68-59	Chamique Holdsclaw, Tennessee	Cincinnati, OH
1998	Tennessee	Pat Summitt	Louisiana Tech	93-75	Chamique Holdsclaw, Tennessee	Kansas City, MO
1999	Purdue	Carolyn Peck	Duke	62-45	Ukari Figgs, Purdue	San Jose, CA
2000	Connecticut	Geno Auriemma	Tennessee	71-52	Shea Ralph, Connecticut	Philadelphia, PA
2001	Notre Dame	Muffet McGraw	Purdue	68-66	Ruth Riley, Notre Dame	St. Louis, MO

* Overtime.

Wade Trophy

Awarded by National Assn. for Girls and Women in Sport for academics, community service, and player performance.

Year	Player, school
1978	Carol Blazejowski, Montclair St.
1979	Nancy Lieberman, Old Dominion
1980	Nancy Lieberman, Old Dominion
1981	Lynette Woodard, Kansas
1982	Pam Kelly, Louisiana Tech
1983	LaTaunya Pollard, Long Beach St.
1984	Janice Lawrence, Louisiana Tech
1985	Cheryl Miller, USC
1986	Kamie Ethridge, Texas
1987	Shelly Pennefeather, Villanova
1988	Teresa Weatherspoon, Louisiana Tech
1989	Clarissa Davis, Texas
1990	Jennifer Azzi, Stanford
1991	Daedra Charles, Tennessee
1992	Susan Robinson, Penn St.
1993	Karen Jennings, Nebraska
1994	Carol Ann Shudlick, Minnesota
1995	Rebecca Lobo, Connecticut
1996	Jennifer Rizzotti, Connecticut
1997	DeLisha Milton, Florida
1998	Chamique Holdsclaw, Tennessee
1999	Stephanie White-McCarty, Purdue
2000	Edwina Brown, Texas
2001	Jackie Stiles, SW Missouri St.

Top Division I Women's Career Scorers

(Minimum 1,500 points; ranked by average)

Player, school	Years	Points	Avg.
Patricia Hoskins, Miss. Valley St.	1985-89	3,122	28.4
Sandra Hodge, New Orleans	1981-84	2,860	26.7
Jackie Stiles, SW Missouri St.	1997-2001	3,393	26.3
Lorri Bauman, Drake	1981-84	3,115	26.0
Andrea Congreaves, Mercer	1989-93	2,796	25.9
Cindy Blodgett, Maine	1994-98	3,005	25.5
Valorie Whiteside, Appalachian St.	1984-88	2,944	25.4
Joyce Walker, LSU	1981-84	2,906	24.8
Tarcha Hollis, Grambling	1988-91	2,058	24.2
Korie Hlede, Duquesne	1994-98	2,631	24.1
Erma Jones, Bethune-Cookman	1982-84	2,095	24.1
Karen Pelphrey, Marshall	1983-86	2,746	24.1

2001 WOMEN'S NCAA BASKETBALL TOURNAMENT

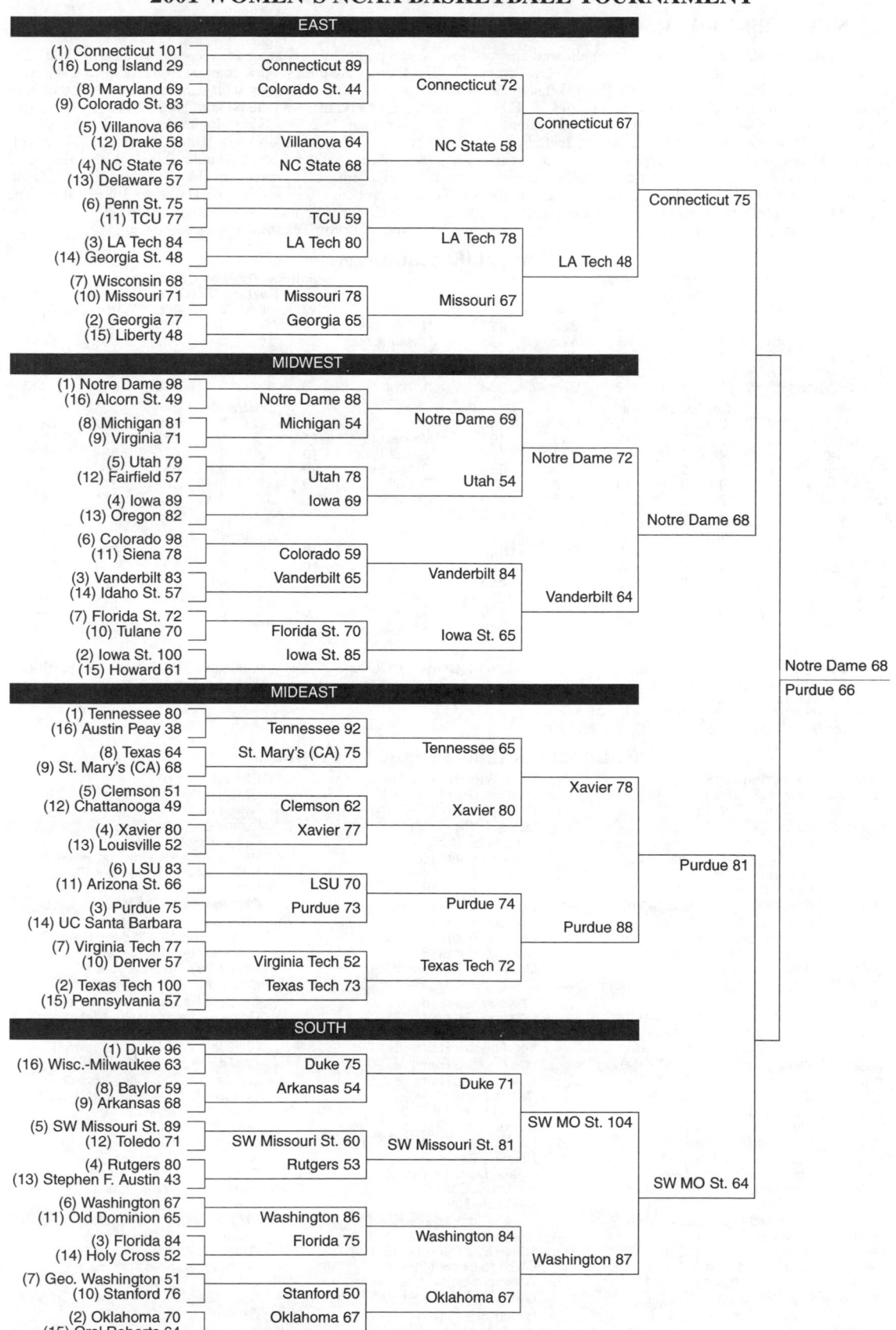

EAST

First Round	Second Round	Regional Semifinals	Regional Final	National Semifinals
(1) Connecticut 101	Connecticut 89	Connecticut 72	Connecticut 67	Connecticut 75
(16) Long Island 29	Colorado St. 44	NC State 58	LA Tech 48	Notre Dame 68
(8) Maryland 69	Villanova 64	LA Tech 78		
(9) Colorado St. 83	NC State 68	Missouri 67		
(5) Villanova 66	TCU 59			
(12) Drake 58	LA Tech 80			
(4) NC State 76	Missouri 78			
(13) Delaware 57	Georgia 65			
(6) Penn St. 75				
(11) TCU 77				
(3) LA Tech 84				
(14) Georgia St. 48				
(7) Wisconsin 68				
(10) Missouri 71				
(2) Georgia 77				
(15) Liberty 48				

MIDWEST

First Round	Second Round	Regional Semifinals	Regional Final
(1) Notre Dame 98	Notre Dame 88	Notre Dame 69	Notre Dame 72
(16) Alcorn St. 49	Michigan 54	Utah 54	Vanderbilt 64
(8) Michigan 81	Utah 78	Vanderbilt 84	
(9) Virginia 71	Iowa 69	Iowa St. 65	
(5) Utah 79	Colorado 59		
(12) Fairfield 57	Vanderbilt 65		
(4) Iowa 89	Florida St. 70		
(13) Oregon 82	Iowa St. 85		
(6) Colorado 98			
(11) Siena 78			
(3) Vanderbilt 83			
(14) Idaho St. 57			
(7) Florida St. 72			
(10) Tulane 70			
(2) Iowa St. 100			
(15) Howard 61			

Championship

Notre Dame 68
Purdue 66

MIDEAST

First Round	Second Round	Regional Semifinals	Regional Final	National Semifinals
(1) Tennessee 80	Tennessee 92	Tennessee 65	Xavier 78	Purdue 81
(16) Austin Peay 38	St. Mary's (CA) 75	Xavier 80	Purdue 88	SW MO St. 64
(8) Texas 64	Clemson 62	Purdue 74		
(9) St. Mary's (CA) 68	Xavier 77	Texas Tech 72		
(5) Clemson 51	LSU 70			
(12) Chattanooga 49	Purdue 73			
(4) Xavier 80	Virginia Tech 52			
(13) Louisville 52	Texas Tech 73			
(6) LSU 83				
(11) Arizona St. 66				
(3) Purdue 75				
(14) UC Santa Barbara				
(7) Virginia Tech 77				
(10) Denver 57				
(2) Texas Tech 100				
(15) Pennsylvania 57				

SOUTH

First Round	Second Round	Regional Semifinals	Regional Final
(1) Duke 96	Duke 75	Duke 71	SW MO St. 104
(16) Wisc.-Milwaukee 63	Arkansas 54	SW Missouri St. 81	Washington 87
(8) Baylor 59	SW Missouri St. 60	Washington 84	
(9) Arkansas 68	Rutgers 53	Oklahoma 67	
(5) SW Missouri St. 89	Washington 86		
(12) Toledo 71	Florida 75		
(4) Rutgers 80	Stanford 50		
(13) Stephen F. Austin 43	Oklahoma 67		
(6) Washington 67			
(11) Old Dominion 65			
(3) Florida 84			
(14) Holy Cross 52			
(7) Geo. Washington 51			
(10) Stanford 76			
(2) Oklahoma 70			
(15) Oral Roberts 64			

NATIONAL FOOTBALL LEAGUE

NFL 2000-2001: Ravens Rule, Faulk Honored, Aikmen Retires, NFL Realigns

The Baltimore Ravens—whose defense allowed a record low of 165 points in a 16-game season—dominated the N.Y. Giants, 34-7, in Super Bowl XXXV, Jan. 28, 2001, in Tampa, FL. The Ravens are the 1st NFL champs from Baltimore since the Colts won Super Bowl V in 1971, and join Oakland (1981) and Denver (1998) as the only wild-card winners. St. Louis running back Marshall Faulk, who scored a record 26 TDs (18 rushing, 8 receiving), was the NFL's 2000 MVP, AP Offensive Player of the Year, and *The Sporting News* and Miller Lite Player of the Year. In 2000, 23 running backs gained 1,000 yds. or more, and 11 games had a 200-yd. rusher, including a 278-yd. game by Cincinnati's Corey Dillon on Oct. 22—all NFL records. Minnesota's Gary Anderson, with 2,059 career pts. at season's end, passed George Blanda (2,002) with a field goal on Oct. 19 to become the NFL's all-time leading scorer. In Apr. 2001, quarterback Troy Aikman, 34, who had suffered his 9th concussion in 2000, retired after 12 seasons. He was the top NFL draft pick in 1989 and helped Dallas to 3 NFL titles. The NFL, May 22, approved a plan for realignment into 8 four-team divisions (East, South, North, West) for the 2002 season. Seattle will move from the AFC to the NFC to accommodate the new AFC Houston Texans.

Final 2000 Standings

American Football Conference

Eastern Division

	W	L	T	Pct.	Pts.	Opp.
Miami	11	5	0	.688	323	226
Indianapolis*	10	6	0	.625	429	326
N.Y. Jets	9	7	0	.562	321	321
Buffalo	8	8	0	.500	315	350
New England	5	11	0	.312	276	338

Central Division

	W	L	T	Pct.	Pts.	Opp.
Tennessee	13	3	0	.812	346	191
Baltimore*	12	4	0	.750	333	165
Pittsburgh	9	7	0	.562	321	255
Jacksonville	7	9	0	.438	367	327
Cincinnati	4	12	0	.250	185	359
Cleveland	3	13	0	.188	161	419

Western Division

	W	L	T	Pct.	Pts.	Opp.
Oakland	12	4	0	.750	479	299
Denver*	11	5	0	.688	485	369
Kansas City	7	9	0	.438	355	354
Seattle	6	10	0	.375	320	405
San Diego	1	15	0	.062	269	440

National Football Conference

Eastern Division

	W	L	T	Pct.	Pts.	Opp.
N.Y. Giants	12	4	0	.750	328	246
Philadelphia*	11	5	0	.688	351	245
Washington	8	8	0	.500	281	269
Dallas	5	11	0	.312	294	361
Arizona	3	13	0	.188	210	443

Central Division

	W	L	T	Pct.	Pts.	Opp.
Minnesota	11	5	0	.688	397	371
Tampa Bay*	10	6	0	.625	388	269
Green Bay	9	7	0	.562	353	323
Detroit	9	7	0	.562	307	307
Chicago	5	11	0	.312	216	355

Western Division

	W	L	T	Pct.	Pts.	Opp.
New Orleans	10	6	0	.625	354	305
St. Louis*	10	6	0	.625	540	471
Carolina	7	9	0	.438	310	310
San Francisco	6	10	0	.375	388	422
Atlanta	4	12	0	.250	252	413

* Wild card team.

AFC Playoffs—Miami 23, Indianapolis 17 (OT); Baltimore 21, Denver 3; Oakland 27, Miami 0; Baltimore 24, Tennessee 10; **Championship:** Baltimore 16, Oakland 3.

NFC Playoffs—New Orleans 31, St. Louis 28; Philadelphia 21, Tampa Bay 3; Minnesota 34, New Orleans 16; N.Y. Giants 20, Philadelphia 10; **Championship:** N.Y. Giants 41, Minnesota 0.

Super Bowl—Baltimore 34, N.Y. Giants 7.

National Football League Champions

Year	East Winner (W-L-T)	West Winner (W-L-T)	Playoff
1933	New York Giants (11-3-0)	Chicago Bears (10-2-1)	Chicago Bears 23, New York 21
1934	New York Giants (8-5-0)	Chicago Bears (13-0-0)	New York 30, Chicago Bears 13
1935	New York Giants (9-3-0)	Detroit Lions (7-3-2)	Detroit 26, New York 7
1936	Boston Redskins (7-5-0)	Green Bay Packers (10-1-1)	Green Bay 21, Boston 6
1937	Washington Redskins (8-3-0)	Chicago Bears (9-1-1)	Washington 28, Chicago Bears 21
1938	New York Giants (8-2-1)	Green Bay Packers (8-3-0)	New York 23, Green Bay 17
1939	New York Giants (9-1-1)	Green Bay Packers (9-2-0)	Green Bay 27, New York 0
1940	Washington Redskins (9-2-0)	Chicago Bears (8-3-0)	Chicago Bears 73, Washington 0
1941	New York Giants (8-3-0)	Chicago Bears (10-1-1)(a)	Chicago Bears 37, New York 9
1942	Washington Redskins (10-1-1)	Chicago Bears (11-0-0)	Washington 14, Chicago Bears 6
1943	Washington Redskins (6-3-1)(a)	Chicago Bears (8-1-1)	Chicago Bears, 41, Washington 21
1944	New York Giants (8-1-1)	Green Bay Packers (8-2-0)	Green Bay 14, New York 7
1945	Washington Redskins (8-2-0)	Cleveland Rams (9-1-0)	Cleveland 15, Washington 14
1946	New York Giants (7-3-1)	Chicago Bears (8-2-1)	Chicago Bears 24, New York 14
1947	Philadelphia Eagles (8-4-0)(a)	Chicago Cardinals (9-3-0)	Chicago Cardinals 28, Philadelphia 21
1948	Philadelphia Eagles (9-2-1)	Chicago Cardinals (11-1-0)	Philadelphia 7, Chicago Cardinals 0
1949	Philadelphia Eagles (11-1-0)	Los Angeles Rams (8-2-2)	Philadelphia 14, Los Angeles 0
1950	Cleveland Browns (10-2-0)(a)	Los Angeles Rams (9-3-0)(a)	Cleveland 30, Los Angeles 28
1951	Cleveland Browns (11-1-0)	Los Angeles Rams (8-4-0)	Los Angeles 24, Cleveland 17
1952	Cleveland Browns (8-4-0)	Detroit Lions (9-3-0)(a)	Detroit 17, Cleveland 7
1953	Cleveland Browns (11-1-0)	Detroit Lions (10-2-0)	Detroit 17, Cleveland 16
1954	Cleveland Browns (9-3-0)	Detroit Lions (9-2-1)	Cleveland 56, Detroit 10
1955	Cleveland Browns (9-2-1)	Los Angeles Rams (8-3-1)	Cleveland 38, Los Angeles 14
1956	New York Giants (8-3-1)	Chicago Bears (9-2-1)	New York 47, Chicago Bears 7
1957	Cleveland Browns (9-2-1)	Detroit Lions (8-4-0)(a)	Detroit 59, Cleveland 14
1958	New York Giants (9-3-0)(a)	Baltimore Colts (9-3-0)	Baltimore 23, New York 17(b)
1959	New York Giants (10-2-0)	Baltimore Colts (9-3-0)	Baltimore 31, New York 16
1960	Philadelphia Eagles (10-2-0)	Green Bay Packers (8-4-0)	Philadelphia 17, Green Bay 13
1961	New York Giants (10-3-1)	Green Bay Packers (11-3-0)	Green Bay 37, New York 0
1962	New York Giants (12-2-0)	Green Bay Packers (13-1-0)	Green Bay 16, New York 7
1963	New York Giants (11-3-0)	Chicago Bears (11-1-2)	Chicago 14, New York 10
1964	Cleveland Browns (10-3-1)	Baltimore Colts (12-2-0)	Cleveland 27, Baltimore 0
1965	Cleveland Browns (11-3-0)	Green Bay Packers (10-3-1)(a)	Green Bay 23, Cleveland 12
1966	Dallas Cowboys (10-3-1)	Green Bay Packers (12-2-0)	Green Bay 34, Dallas 27

(a) Won divisional playoff. (b) Won at 8:15 of sudden death overtime period.

Year	Conference	Division	Winner (W-L-T)	Playoffs(c)	Year
1967	East	Century	Cleveland Browns (9-5-0)	Dallas 52, Cleveland 14	1967
		Capitol	Dallas Cowboys (9-5-0)		
	West	Central	Green Bay Packers (9-4-1)	Green Bay 28, Los Angeles 7	
		Coastal	Los Angeles Rams (11-1-2)(a)	Green Bay 21, Dallas 17	
1968	East	Century	Cleveland Browns (10-4-0)	Cleveland 31, Dallas 20	1968
		Capitol	Dallas Cowboys (12-2-0)		
	West	Central	Minnesota Vikings (8-6-0)	Baltimore 24, Minnesota 14	
		Coastal	Baltimore Colts (13-1-0)	Baltimore 34, Cleveland 0	
1969	East	Century	Cleveland Browns (10-3-1)	Cleveland 38, Dallas 14	1969
		Capitol	Dallas Cowboys (11-2-1)		
	West	Central	Minnesota Vikings (12-2-0)	Minnesota 23, Los Angeles 20	
		Coastal	Los Angeles Rams (11-3-0)	Minnesota 27, Cleveland 7	
1970	American	Eastern	Baltimore Colts (11-2-1)	Baltimore 17, Cincinnati 0	1970
		Central	Cincinnati Bengals (8-6-0)	Oakland 21, Miami* 14	
		Western	Oakland Raiders (8-4-2)	Baltimore 27, Oakland 17	
	National	Eastern	Dallas Cowboys (10-4-0)	Dallas 5, Detroit* 0	
		Central	Minnesota Vikings (12-2-0)	San Francisco 17, Minnesota 14	
		Western	San Francisco 49ers (10-3-1)	Dallas 17, San Francisco 10	
1971	American	Eastern	Miami Dolphins (10-3-1)	Miami 27, Kansas City* 24	1971
		Central	Cleveland Browns (9-5-0)	Baltimore 20, Cleveland 3	
		Western	Kansas City Chiefs(10-3-1)	Miami 21, Baltimore 0	
	National	Eastern	Dallas Cowboys (11-3-0)	Dallas 20, Minnesota 12	
		Central	Minnesota Vikings (11-3-0)	San Francisco 24, Washington* 20	
		Western	San Francisco 49ers (9-5-0)	Dallas 14, San Francisco 3	
1972	American	Eastern	Miami Dolphins (14-0-0)	Miami 20, Cleveland* 14	1972
		Central	Pittsburgh Steelers (11-3-0)	Pittsburgh 13, Oakland 7	
		Western	Oakland Raiders (10-3-1)	Miami 21, Pittsburgh 17	
	National	Eastern	Washington Redskins (11-3-0)	Washington 16, Green Bay 3	
		Central	Green Bay Packers (10-4-0)	Dallas* 30, San Francisco 28	
		Western	San Francisco 49ers (8-5-1)	Washington 26, Dallas* 3	
1973	American	Eastern	Miami Dolphins (12-2-0)	Miami 34, Cincinnati 16	1973
		Central	Cincinnati Bengals (10-4-0)	Oakland 33, Pittsburgh* 14	
		Western	Oakland Raiders (9-4-1)	Miami 27, Oakland 10	
	National	Eastern	Dallas Cowboys (10-4-0)	Dallas 27, Los Angeles 16	
		Central	Minnesota Vikings (12-2-0)	Minnesota 27, Washington* 20	
		Western	Los Angeles Rams (12-2-0)	Minnesota 27, Dallas 10	
1974	American	Eastern	Miami Dolphins (11-3-0)	Oakland 28, Miami 26	1974
		Central	Pittsburgh Steelers (10-3-1)	Pittsburgh 32, Buffalo* 14	
		Western	Oakland Raiders (12-2-0)	Pittsburgh 24, Oakland 13	
	National	Eastern	St. Louis Cardinals (10-4-0)	Minnesota 30, St. Louis 14	
		Central	Minnesota Vikings (10-4-0)	Los Angeles 19, Washington* 10	
		Western	Los Angeles Rams (10-4-0)	Minnesota 14, Los Angeles 10	
1975	American	Eastern	Baltimore Colts (10-4-0)	Pittsburgh 28, Baltimore 10	1975
		Central	Pittsburgh Steelers (12-2-0)	Oakland 31, Cincinnati* 28	
		Western	Oakland Raiders (11-3-0)	Pittsburgh 16, Oakland 10	
	National	Eastern	St. Louis Cardinals (11-3-0)	Dallas* 17, Minnesota 14	
		Central	Minnesota Vikings (12-2-0)	Los Angeles 35, St. Louis 23	
		Western	Los Angeles Rams (12-2-0)	Dallas* 37, Los Angeles 7	
1976	American	Eastern	Baltimore Colts (11-3-0)	Pittsburgh 40, Baltimore 14	1976
		Central	Pittsburgh Steelers (10-4-0)	Oakland 24, New England* 21	
		Western	Oakland Raiders (13-1-0)	Oakland 24, Pittsburgh 7	
	National	Eastern	Dallas Cowboys (11-3-0)	Minnesota 35, Washington* 20	
		Central	Minnesota Vikings (11-2-1)	Los Angeles 14, Dallas 12	
		Western	Los Angeles Rams (10-3-1)	Minnesota 24, Los Angeles 13	
1977	American	Eastern	Baltimore Colts (10-4-0)	Oakland* 37, Baltimore 31	1977
		Central	Pittsburgh Steelers (9-5-0)	Denver 34, Pittsburgh 21	
		Western	Denver Broncos (12-2-0)	Denver 20, Oakland* 17	
	National	Eastern	Dallas Cowboys (12-2-0)	Dallas 37, Chicago* 7	
		Central	Minnesota Vikings (9-5-0)	Minnesota 14, Los Angeles 7	
		Western	Los Angeles Rams (10-4-0)	Dallas 23, Minnesota 6	
1978	American	Eastern	New England Patriots (11-5-0)	Pittsburgh 33, Denver 10	1978
		Central	Pittsburgh Steelers (14-2-0)	Houston* 31, New England 14	
		Western	Denver Broncos (10-6-0)	Pittsburgh 34, Houston* 5	
	National	Eastern	Dallas Cowboys (12-4-0)	Dallas 27, Atlanta* 20	
		Central	Minnesota Vikings (8-7-1)	Los Angeles 34, Minnesota 10	
		Western	Los Angeles Rams (12-4-0)	Dallas 28, Los Angeles 0	
1979	American	Eastern	Miami Dolphins (10-6-0)	Houston* 17, San Diego 14	1979
		Central	Pittsburgh Steelers (12-4-0)	Pittsburgh 34, Miami 14	
		Western	San Diego Chargers (12-4-0)	Pittsburgh 27, Houston* 13	
	National	Eastern	Dallas Cowboys (11-5-0)	Tampa Bay 24, Philadelphia* 17	
		Central	Tampa Bay Buccaneers (10-6-0)	Los Angeles 21, Dallas 19	
		Western	Los Angeles Rams (9-7-0)	Los Angeles 9, Tampa Bay 0	
1980	American	Eastern	Buffalo Bills (11-5-0)	San Diego 20, Buffalo 14	1980
		Central	Cleveland Browns (11-5-0)	Oakland* 14, Cleveland 12	
		Western	San Diego Chargers (11-5-0)	Oakland* 34, San Diego 27	
	National	Eastern	Philadelphia Eagles (12-4-0)	Philadelphia 31, Minnesota 16	
		Central	Minnesota Vikings (9-7-0)	Dallas* 30, Atlanta 27	
		Western	Atlanta Falcons (12-4-0)	Philadelphia 20, Dallas* 7	
1981	American	Eastern	Miami Dolphins (11-4-1)	San Diego 41, Miami 38	1981
		Central	Cincinnati Bengals (12-4-0)	Cincinnati 28, Buffalo* 21	
		Western	San Diego Chargers (10-6-0)	Cincinnati 27, San Diego 7	
	National	Eastern	Dallas Cowboys (12-4-0)	Dallas 38, Tampa Bay 0	
		Central	Tampa Bay Buccaneers (9-7-0)	San Francisco 38, N.Y. Giants* 24	
		Western	San Francisco 49ers (13-3-0)	San Francisco 28, Dallas 27	
1982 (d)	American		Los Angeles Raiders (8-1-0)	Strike-shortened season (see	1982 (d)
	National		Washington Redskins (8-1-0)	playoff results after footnote)	
1983	American	Eastern	Miami Dolphins (12-4-0)	Seattle* 27, Miami 20	1983
		Central	Pittsburgh Steelers (10-6-0)	L.A. Raiders 38, Pittsburgh 10	
		Western	Los Angeles Raiders (12-4-0)	L.A. Raiders 30, Seattle* 14	
	National	Eastern	Washington Redskins (14-2-0)	Washington 51, L.A. Rams* 7	
		Central	Detroit Lions (9-7-0)	San Francisco 24, Detroit 23	
		Western	San Francisico 49ers (10-6-0)	Washington 24, San Francisco 21	

Year	Conference	Division	Winner (W-L-T)	Playoffs(c)	Year
1984	American	Eastern	Miami Dolphins (14-2-0)	Miami 31, Seattle* 10	1984
		Central	Pittsburgh Steelers (9-7-0)	Pittsburgh 24, Denver 17	
		Western	Denver Broncos (13-3-0)	Miami 45, Pittsburgh 28	
	National	Eastern	Washington Redskins (11-5-0)	Chicago 23, Washington 19	
		Central	Chicago Bears (10-6-0)	San Francisco 21, N.Y. Giants* 10	
		Western	San Francisco 49ers (15-1-0)	San Francisco 23, Chicago 0	
1985	American	Eastern	Miami Dolphins (12-4-0)	New England* 27, L.A. Raiders 20	1985
		Central	Cleveland Browns (8-8-0)	Miami 24, Cleveland 21	
		Western	Los Angeles Raiders (12-4-0)	New England* 31, Miami 14	
	National	Eastern	Dallas Cowboys (10-6-0)	Chicago 21, N.Y. Giants* 0	
		Central	Chicago Bears (15-1-0)	L.A. Rams 20, Dallas 0	
		Western	Los Angeles Rams (11-5-0)	Chicago 24, L.A. Rams 0	
1986	American	Eastern	New England Patriots (11-5-0)	Denver 22, New England 17	1986
		Central	Cleveland Browns (12-4-0)	Cleveland 23, N.Y. Jets* 20	
		Western	Denver Broncos (11-5-0)	Denver 23, Cleveland 20	
	National	Eastern	New York Giants (14-2-0)	N.Y. Giants 49, San Francisco 3	
		Central	Chicago Bears (14-2-0)	Washington* 27, Chicago 13	
		Western	San Francisco 49ers (10-5-1)	N.Y. Giants 17, Washington* 0	
1987	American	Eastern	Indianapolis Colts (9-6-0)	Cleveland 38, Indianapolis 21	1987
		Central	Cleveland Browns (10-5-0)	Denver 34, Houston* 10	
		Western	Denver Broncos (10-4-1)	Denver 38, Cleveland 33	
	National	Eastern	Washington Redskins (11-4-0)	Washington 21, Chicago 17	
		Central	Chicago Bears (11-4-0)	Minnesota* 36, San Francisco 24	
		Western	San Francisco 49ers (13-2-0)	Washington 17, Minnesota* 10	
1988	American	Eastern	Buffalo Bills (12-4-0)	Buffalo 17, Houston* 10	1988
		Central	Cincinnati Bengals (12-4-0)	Cincinnati 21, Seattle 13	
		Western	Seattle Seahawks (9-7-0)	Cincinnati 21, Buffalo 10	
	National	Eastern	Philadelphia Eagles (10-6-0)	Chicago 20, Philadelphia 12	
		Central	Chicago Bears (12-4-0)	San Francisco 34, Minnesota* 9	
		Western	San Francisco 49ers (10-6-0)	San Francisco 28, Chicago 3	
1989	American	Eastern	Buffalo Bills (9-7-0)	Cleveland 34, Buffalo 30	1989
		Central	Cleveland Browns (9-6-1)	Denver 24, Pittsburgh* 23	
		Western	Denver Broncos (11-5-0)	Denver 37, Cleveland 21	
	National	Eastern	New York Giants (12-4-0)	San Francisco 41, Minnesota 13	
		Central	Minnesota Vikings (10-6-0)	L.A. Rams* 19, N.Y. Giants 13	
		Western	San Francisco 49ers (14-2-0)	San Francisco 30, L.A. Rams* 3	
1990	American	Eastern	Buffalo Bills (13-3-0)	L.A. Raiders 20, Cincinnati 10	1990
		Central	Cincinnati Bengals (9-7-0)	Buffalo 44, Miami* 34	
		Western	Los Angeles Raiders (12-4-0)	Buffalo 51, L.A. Raiders 3	
	National	Eastern	New York Giants (13-3-0)	San Francisco 28, Washington* 10	
		Central	Chicago Bears (11-5-0)	N.Y. Giants 31, Chicago 3	
		Western	San Francisco 49ers (14-2-0)	N.Y. Giants 15, San Francisco 13	
1991	American	Eastern	Buffalo Bills (13-3-0)	Denver 26, Houston 24	1991
		Central	Houston Oilers (11-5-0)	Buffalo 37, Kansas City* 14	
		Western	Denver Broncos (12-4-0)	Buffalo 10, Denver 7	
	National	Eastern	Washington Redskins (14-2-0)	Washington 24, Atlanta* 7	
		Central	Detroit Lions (12-4-0)	Detroit 38, Dallas* 6	
		Western	New Orleans Saints (11-5-0)	Washington 41, Detroit 10	
1992	American	Eastern	Miami Dolphins (11-5-0)	Miami 31, San Diego 0	1992
		Central	Pittsburgh Steelers (11-5-0)	Buffalo* 24, Pittsburgh 3	
		Western	San Diego Chargers (11-5-0)	Buffalo* 29, Miami 10	
	National	Eastern	Dallas Cowboys (13-3-0)	Dallas 34, Philadelphia* 10	
		Central	Minnesota Vikings (11-5-0)	San Francisco 20, Washington* 13	
		Western	San Francisco 49ers (14-2-0)	Dallas 30, San Francisco 20	
1993	American	Eastern	Buffalo Bills (12-4-0)	Buffalo 29, L.A. Raiders* 23	1993
		Central	Houston Oilers (12-4-0)	Kansas City 28, Houston 20	
		Western	Kansas City Chiefs (11-5-0)	Buffalo 30, Kansas City 13	
	National	Eastern	Dallas Cowboys (12-4-0)	Dallas 27, Green Bay* 17	
		Central	Detroit Lions (10-6-0)	San Francisco 44, N.Y. Giants* 3	
		Western	San Francisco 49ers (10-6-0)	Dallas 38, San Francisco 21	
1994	American	Eastern	Miami Dolphins (10-6-0)	Pittsburgh 29, Cleveland* 9	1994
		Central	Pittsburgh Steelers (12-4-0)	San Diego 22, Miami 21	
		Western	San Diego Chargers (11-5-0)	San Diego 17, Pittsburgh 13	
	National	Eastern	Dallas Cowboys (12-4-0)	San Francisco 44, Chicago* 15	
		Central	Minnesota Vikings (10-6-0)	Dallas 35, Green Bay* 9	
		Western	San Francisco 49ers (13-3-0)	San Francisco 38, Dallas 28	
1995	American	Eastern	Buffalo Bills (10-6-0)	Indianapolis* 10, Kansas City 7	1995
		Central	Pittsburgh Steelers (11-5-0)	Pittsburgh 40, Buffalo 21	
		Western	Kansas City Chiefs (13-3-0)	Pittsburgh 20, Indianapolis* 16	
	National	Eastern	Dallas Cowboys (12-4-0)	Dallas 30, Philadelphia* 11	
		Central	Green Bay Packers (11-5-0)	Green Bay 27, San Francisco 17	
		Western	San Francisco 49ers (11-5-0)	Dallas 38, Green Bay 27	
1996	American	Eastern	New England Patriots (11-5-0)	Jacksonville* 30, Denver 27	1996
		Central	Pittsburgh Steelers (10-6-0)	New England 28, Pittsburgh 3	
		Western	Denver Broncos (13-3-0)	New England 20, Jacksonville* 6	
	National	Eastern	Dallas Cowboys (10-6-0)	Green Bay 35, San Francisco* 14	
		Central	Green Bay Packers (13-3-0)	Carolina 26, Dallas 17	
		Western	Carolina Panthers (12-4-0)	Green Bay 30, Carolina 13	
1997	American	Eastern	New England Patriots (10-6-0)	Pittsburgh 7, New England 6	1997
		Central	Pittsburgh Steelers (11-5-0)	Denver* 14, Kansas City 10	
		Western	Kansas City Chiefs (13-3-0)	Denver* 24, Pittsburgh 21	
	National	Eastern	New York Giants (10-5-1)	San Francisco 38, Minnesota* 22	
		Central	Green Bay Packers (13-3-0)	Green Bay 21, Tampa Bay* 7	
		Western	San Francisco 49ers (13-3-0)	Green Bay 23, San Francisco 10	
1998	American	Eastern	N.Y. Jets (12-4-0)	Denver 38, Miami* 3	1998
		Central	Jacksonville Jaguars (11-5-0)	N.Y. Jets 34, Jacksonville 24	
		Western	Denver Broncos (14-2-0)	Denver 23, N.Y. Jets 10	
	National	Eastern	Dallas Cowboys (10-6-0)	Atlanta 20, San Francisco* 18	
		Central	Minnesota Vikings (15-1-0)	Minnesota 41, Arizona* 21	
		Western	Atlanta Falcons (14-2-0)	Atlanta 30, Minnesota 27 (OT)	

Year	Conference	Division	Winner (W-L-T)	Playoffs(c)	Year
1999	American	Eastern	Indianapolis Colts (13-3-0)	Jacksonville 62, Miami* 7	1999
		Central	Jacksonville Jaguars (14-2-0)	Tennessee* 19, Indianapolis 16	
		Western	Seattle Seahawks (9-7-0)	Tennessee* 33, Jacksonville 14	
	National	Eastern	Washington Redskins (10-6-0)	Tampa Bay 14, Washington 13	
		Central	Tampa Bay Buccaneers (11-5-0)	St. Louis 49, Minnesota* 37	
		Western	St. Louis Rams (13-3-0)	St. Louis 11, Tampa Bay 6	
2000	American	Eastern	Miami Dolphins (11-5-0)	Oakland 27, Miami 0	2000
		Central	Tennessee Titans (13-3-0)	Baltimore* 24, Tennessee 10	
		Western	Oakland (12-4-0)	Baltimore* 16, Oakland 3	
	National	Eastern	New York Giants (12-4-0)	Minnesota 34, New Orleans 16	
		Central	Minnesota Vikings (11-5-0)	N.Y. Giants 20, Philadelphia* 10	
		Western	New Orleans Saints (10-6-0)	N.Y. Giants 41, Minnesota 0	

*Wild card team. (c) From 1978 on, only the final 2 conference playoff rounds are shown. (d) A strike shortened the 1982 season from 16 to 9 games. The top 8 teams in each conference played in a tournament to determine the conference champion. See below. **AFC playoffs**—Miami 28, New England 13; L.A. Raiders 27, Cleveland 10; N.Y. Jets 44, Cincinnati 17; San Diego 31, Pittsburgh 28; N.Y. Jets 17, L.A. Raiders 14; Miami 34, San Diego 13; Miami 14, N.Y. Jets 0. **NFC playoffs**—Washington 31, Detroit 7; Green Bay 41, St. Louis 16; Dallas 30, Tampa Bay 17; Minnesota 30, Atlanta 24; Washington 21, Minnesota 7; Dallas 37, Green Bay 26; Washington 31, Dallas 17. **AFC Champion**—Miami Dolphins. **NFC Champion**—Washington Redskins.

Baltimore Ravens Stifle NY Giants, 34-7, in Super Bowl XXXV

On Jan. 28, 2001, the Baltimore Ravens dominated the NY Giants, 34-7, in Super Bowl XXXV in Tampa, FL. The Baltimore defense, led by Super Bowl MVP linebacker Ray Lewis, never allowed the Giants' offense inside the Ravens' 29-yard-line and tied a Super Bowl record by intercepting Giant quarterback Kerry Collins 4 times. Both offenses struggled, together setting Super Bowl records for fewest combined total yards (393) and most punts (21). The game's most exciting moment—3 touchdowns on 3 plays in 36 seconds—came late in the 3d quarter with the Ravens leading, 10-0. Baltimore's Duane Starks intercepted a Collins pass and returned it 49 yards for a score. Ron Dixon returned the ensuing kickoff 97 yards for New York's only touchdown. On the following kickoff, Jermaine Lewis countered for Baltimore with an 84-yard touchdown return.

Score by Quarters

Baltimore	7	3	14	10—34
New York Giants	0	0	7	0— 7

Scoring

Baltimore—Stokley 38 yd. pass from Dilfer (Stover kick)
Baltimore—Stover 47 yd. field goal
Baltimore—Starks 49 yd. interception return (Stover kick)
NY Giants—Dixon 97 yd. kickoff return (Daluiso kick)
Baltimore—Je. Lewis 84 yd. kickoff return (Stover kick)
Baltimore—Ja. Lewis 3 yd. run (Stover kick)
Baltimore—Stover 34 yd. field goal

Individual Statistics

Rushing — Baltimore, Ja. Lewis 27-102, Holmes 4-8, Je. Lewis 1-1, Dilfer 1-0. NY Giants, Barber 11-49, Collins 3-12, Montgomery 2-5.

Passing — Baltimore, Dilfer 12-25-1-153. NY Giants, Collins 15-39-0-112.

Receiving — Baltimore, Stokley 3-52, Coates 3-30, Ismail 1-44, Johnson 1-8, Je. Lewis 1-6, Sharpe 1-5, Ja. Lewis 1-4, Holmes 1-4. NY Giants, Barber 6-26, Hilliard 3-30, Toomer 2-24, Dixon 1-16, Cross 1-7, Mitchell 1-7, Comella 1-2.

Team Statistics	BAL	NYG
First downs	13	11
Total net yards	244	152
Rushes-yards	33-111	16-66
Passing yards, net	133	86
Punt returns-yards	3-34	5-46
Kickoff returns-yards	2-111	7-171
Interception returns-yards	4-59	0-0
Att.-comp.-int.	26-12-0	39-15-4
Field goals made-attempts	2-3	0-0
Sacked-yards lost	1-6	0-0
Punts-average	10-43	11-38.4
Fumbles-lost	2-0	2-1
Penalties-yards	9-70	6-27
Time of possession	34:06	25:54

Attendance—71,921. **Time**—3:23.

Super Bowl Single-Game Statistical Leaders

Passing Yards

	Year	Att/Comp	Yds	TDs
Kurt Warner, Rams	2000	45/24	414	2
Joe Montana, 49ers	1989	36/23	357	2
Doug Williams, Redskins	1988	29/18	340	4

Receiving Yards

	Year	Recept.	Yds	TDs
Jerry Rice, 49ers	1989	11	215	1
Ricky Sanders, Redskins	1988	9	193	2
Lynn Swann, Steelers	1976	4	161	1

Rushing Yards

	Year	Attempts	Yds	TDs
Timmy Smith, Redskins	1988	22	204	2
Marcus Allen, Raiders	1984	20	191	2
John Riggins, Redskins	1983	38	166	1

Passing Touchdowns

	Year	Att/Comp	Yds	TDs
Steve Young, 49ers	1995	36/24	325	6
Joe Montana, 49ers	1990	29/22	297	5
Troy Aikman, Cowboys	1993	30/22	273	4
Doug Williams, Redskins	1988	29/18	340	4
Terry Bradshaw, Steelers	1979	30/17	318	4

Scoring

	Year	Points	
Terrell Davis, Broncos	1998	18	3 TDs
Jerry Rice, 49ers	1995	18	3 TDs
Ricky Watters, 49ers	1995	18	3 TDs
Jerry Rice, 49ers	1990	18	3 TDs
Roger Craig, 49ers	1985	18	3 TDs
Don Chandler, Packers	1968	15	4 FG, 3 PATs

Super Bowl Results

	Year	Winner	Loser	Winning coach	Site
I	1967	Green Bay Packers, 35	Kansas City Chiefs, 10	Vince Lombardi	Los Angeles Coliseum, CA
II	1968	Green Bay Packers, 33	Oakland Raiders, 14	Vince Lombardi	Orange Bowl, Miami, FL
III	1969	New York Jets, 16	Baltimore Colts, 7	Weeb Ewbank	Orange Bowl, Miami, FL
IV	1970	Kansas City Chiefs, 23	Minnesota Vikings, 7	Hank Stram	Tulane Stadium, New Orleans, LA
V	1971	Baltimore Colts, 16	Dallas Cowboys, 13	Don McCafferty	Orange Bowl, Miami, FL
VI	1972	Dallas Cowboys, 24	Miami Dolphins, 3	Tom Landry	Tulane Stadium, New Orleans, LA
VII	1973	Miami Dolphins, 14	Washington Redskins, 7	Don Shula	Los Angeles Coliseum, CA
VIII	1974	Miami Dolphins, 24	Minnesota Vikings, 7	Don Shula	Rice Stadium, Houston, TX
IX	1975	Pittsburgh Steelers, 16	Minnesota Vikings, 6	Chuck Noll	Tulane Stadium, New Orleans, LA
X	1976	Pittsburgh Steelers, 21	Dallas Cowboys, 17	Chuck Noll	Orange Bowl, Miami, FL
XI	1977	Oakland Raiders, 32	Minnesota Vikings, 14	John Madden	Rose Bowl, Pasadena, CA
XII	1978	Dallas Cowboys, 27	Denver Broncos, 10	Tom Landry	Superdome, New Orleans, LA
XIII	1979	Pittsburgh Steelers, 35	Dallas Cowboys, 31	Chuck Noll	Orange Bowl, Miami, FL
XIV	1980	Pittsburgh Steelers, 31	Los Angeles Rams, 19	Chuck Noll	Rose Bowl, Pasadena, CA
XV	1981	Oakland Raiders, 27	Philadelphia Eagles, 10	Tom Flores	Superdome, New Orleans, LA
XVI	1982	San Francisco 49ers, 26	Cincinnati Bengals, 21	Bill Walsh	Silverdome, Pontiac, MI
XVII	1983	Washington Redskins, 27	Miami Dolphins, 17	Joe Gibbs	Rose Bowl, Pasadena, CA
XVIII	1984	Los Angeles Raiders, 38	Washington Redskins, 9	Tom Flores	Tampa Stadium, FL
XIX	1985	San Francisco 49ers, 38	Miami Dolphins, 16	Bill Walsh	Stanford Stadium, Palo Alto, CA
XX	1986	Chicago Bears, 46	New England Patriots, 10	Mike Ditka	Superdome, New Orleans, LA

	Year	Winner	Loser	Winning coach	Site
XXI	1987	New York Giants, 39	Denver Broncos, 20	Bill Parcells	Rose Bowl, Pasadena, CA
XXII	1988	Washington Redskins, 42	Denver Broncos, 10	Joe Gibbs	San Diego Stadium, CA
XXIII	1989	San Francisco 49ers, 20	Cincinnati Bengals, 16	Bill Walsh	Joe Robbie Stadium, Miami, FL
XXIV	1990	San Francisco 49ers, 55	Denver Broncos, 10	George Seifert	Superdome, New Orleans, LA
XXV	1991	New York Giants, 20	Buffalo Bills, 19	Bill Parcells	Tampa Stadium, FL
XXVI	1992	Washington Redskins, 37	Buffalo Bills, 24	Joe Gibbs	Metrodome, Minneapolis, MN
XXVII	1993	Dallas Cowboys, 52	Buffalo Bills, 17	Jimmy Johnson	Rose Bowl, Pasadena, CA
XXVIII	1994	Dallas Cowboys, 30	Buffalo Bills, 13	Jimmy Johnson	Georgia Dome, Atlanta, GA
XXIX	1995	San Francisco 49ers, 49	San Diego Chargers, 26	George Seifert	Joe Robbie Stadium, Miami, FL
XXX	1996	Dallas Cowboys, 27	Pittsburgh Steelers, 17	Barry Switzer	Sun Devil Stadium, Tempe, AZ
XXXI	1997	Green Bay Packers, 35	New England Patriots, 21	Mike Holmgren	Superdome, New Orleans, LA
XXXII	1998	Denver Broncos, 31	Green Bay Packers, 24	Mike Shanahan	Qualcomm Stadium, San Diego, CA
XXXIII	1999	Denver Broncos, 34	Atlanta Falcons, 19	Mike Shanahan	Pro Player Stadium, Miami, FL
XXXIV	2000	St. Louis Rams, 23	Tennessee Titans, 16	Dick Vermeil	Georgia Dome, Atlanta, GA
XXXV	2001	Baltimore Ravens, 34	New York Giants, 7	Brian Billick	Raymond James Stad., Tampa, FL

Super Bowl MVPs

1967 Bart Starr, Green Bay
1968 Bart Starr, Green Bay
1969 Joe Namath, N.Y. Jets
1970 Len Dawson, Kansas City
1971 Chuck Howley, Dallas
1972 Roger Staubach, Dallas
1973 Jake Scott, Miami
1974 Larry Csonka, Miami
1975 Franco Harris, Pittsburgh
1976 Lynn Swann, Pittsburgh
1977 Fred Biletnikoff, Oakland
1978 Randy White, Harvey Martin, Dallas
1979 Terry Bradshaw, Pittsburgh
1980 Terry Bradshaw, Pittsburgh
1981 Jim Plunkett, Oakland
1982 Joe Montana, San Francisco
1983 John Riggins, Washington
1984 Marcus Allen, L.A. Raiders
1985 Joe Montana, San Francisco
1986 Richard Dent, Chicago
1987 Phil Simms, N.Y. Giants
1988 Doug Williams, Washington
1989 Jerry Rice, San Francisco
1990 Joe Montana, San Francisco
1991 Ottis Anderson, N.Y. Giants
1992 Mark Rypien, Washington
1993 Troy Aikman, Dallas
1994 Emmitt Smith, Dallas
1995 Steve Young, San Francisco
1996 Larry Brown, Dallas
1997 Desmond Howard, Green Bay
1998 Terrell Davis, Denver
1999 John Elway, Denver
2000 Kurt Warner, St. Louis
2001 Ray Lewis, Baltimore

American Football Conference Leaders

(American Football League, 1960-69)

	Passing[1]					Receiving			
Player, team	Att	Com	YG	TD	Year	Player, team	Rec.	YG	TD
Jack Kemp, L.A. Chargers	406	211	3,018	20	**1960**	Lionel Taylor, Denver	92	1,235	12
George Blanda, Houston	362	187	3,330	36	**1961**	Lionel Taylor, Denver	100	1,176	4
Len Dawson, Dallas Texans	310	189	2,759	29	**1962**	Lionel Taylor, Denver	77	908	4
Tobin Rote, San Diego	286	170	2,510	20	**1963**	Lionel Taylor, Denver	78	1,101	10
Len Dawson, Kansas City	354	199	2,879	30	**1964**	Charley Hennigan, Houston	101	1,546	8
John Hadl, San Diego	348	174	2,798	20	**1965**	Lionel Taylor, Denver	85	1,131	6
Len Dawson, Kansas City	284	159	2,527	26	**1966**	Lance Alworth, San Diego	73	1,383	13
Daryle Lamonica, Oakland	425	220	3,228	30	**1967**	George Sauer, N.Y. Jets	75	1,189	6
Len Dawson, Kansas City	224	131	2,109	17	**1968**	Lance Alworth, San Diego	68	1,312	10
Greg Cook, Cincinnati	197	106	1,854	15	**1969**	Lance Alworth, San Diego	64	1,003	4
Daryle Lamonica, Oakland	356	179	2,516	22	**1970**	Marlin Briscoe, Buffalo	57	1,036	8
Bob Griese, Miami	263	145	2,089	19	**1971**	Fred Biletnikoff, Oakland	61	929	9
Earl Morrall, Miami	150	83	1,360	11	**1972**	Fred Biletnikoff, Oakland	58	802	7
Ken Stabler, Oakland	260	163	1,997	14	**1973**	Fred Willis, Houston	57	371	1
Ken Anderson, Cincinnati	328	213	2,667	18	**1974**	Lydell Mitchell, Baltimore Colts	72	544	2
Ken Anderson, Cincinnati	377	228	3,169	21	**1975**	Reggie Rucker, Cleveland	60	770	3
						Lydell Mitchell, Baltimore Colts	60	554	4
Ken Stabler, Oakland	291	194	2,737	27	**1976**	MacArthur Lane, Kansas City	66	686	1
Bob Griese, Miami	307	180	2,252	22	**1977**	Lydell Mitchell, Baltimore Colts	71	620	4
Terry Bradshaw, Pittsburgh	368	207	2,915	28	**1978**	Steve Largent, Seattle	71	1,168	8
Dan Fouts, San Diego	530	332	4,082	24	**1979**	Joe Washington, Baltimore Colts	82	750	3
Brian Sipe, Cleveland	554	337	4,132	30	**1980**	Kellen Winslow, San Diego	89	1,290	9
Ken Anderson, Cincinnati	479	300	3,754	29	**1981**	Kellen Winslow, San Diego	88	1,075	10
Ken Anderson, Cincinnati	309	218	2,495	12	**1982**	Kellen Winslow, San Diego	54	721	6
Dan Marino, Miami	296	173	2,210	20	**1983**	Todd Christensen, L.A. Raiders	92	1,247	12
Dan Marino, Miami	564	362	5,084	48	**1984**	Ozzie Newsome, Cleveland	89	1,001	5
Ken O'Brien, N.Y. Jets	488	297	3,888	25	**1985**	Lionel James, San Diego	86	1,027	6
Dan Marino, Miami	623	378	4,746	44	**1986**	Todd Christensen, L.A. Raiders	95	1,153	8
Bernie Kosar, Cleveland	389	241	3,033	22	**1987**	Al Toon, N.Y. Jets	68	976	5
Boomer Esiason, Cincinnati	388	223	3,572	28	**1988**	Al Toon, N.Y. Jets	93	1,067	5
Boomer Esiason, Cincinnati	455	258	3,525	28	**1989**	Andre Reed, Buffalo	88	1,312	9
Jim Kelly, Buffalo	346	219	2,829	24	**1990**	Haywood Jeffires, Houston	74	1,048	8
						Drew Hill, Houston	74	1,019	5
Jim Kelly, Buffalo	474	304	3,844	33	**1991**	Haywood Jeffires, Houston	100	1,181	7
Warren Moon, Houston	346	224	2,521	18	**1992**	Haywood Jeffires, Houston	90	913	9
John Elway, Denver	551	348	4,030	25	**1993**	Reggie Langhorne, Indianapolis	85	1,038	3
Dan Marino, Miami	615	385	4,453	30	**1994**	Ben Coates, New England	96	1,174	7
Jim Harbaugh, Indianapolis	314	200	2,575	17	**1995**	Carl Pickens, Cincinnati	99	1,234	17
John Elway, Denver	466	287	3,328	26	**1996**	Carl Pickens, Cincinnati	100	1,180	12
Mark Brunell, Jacksonville	435	264	3,281	18	**1997**	Tim Brown, Oakland	104	1,408	5
Vinny Testaverde, N.Y. Jets	421	259	3,256	29	**1998**	O.J. McDuffie, Miami	90	1,050	7
Peyton Manning, Indianapolis	533	331	4,135	26	**1999**	Jimmy Smith, Jacksonville	116	1,636	6
Brian Griese, Denver	336	216	2,688	19	**2000**	Marvin Harrison, Indianapolis	102	1,413	14

	Scoring					Rushing			
Player, team	TD	PAT	FG	Pts	Year	Player, team	Yds	Att	TD
Gene Mingo, Denver	6	33	18	123	**1960**	Abner Haynes, Dallas Texans	875	156	9
Gino Cappelletti, Boston	8	48	17	147	**1961**	Billy Cannon, Houston	948	200	6
Gene Mingo, Denver	4	32	27	137	**1962**	Cookie Gilchrest, Buffalo	1,096	214	13
Gino Cappelletti, Boston	2	35	22	113	**1963**	Clem Daniels, Oakland	1,099	215	3
Gino Cappelletti, Boston	7	36	25	155	**1964**	Cookie Gilchrest, Buffalo	981	230	6
Gino Cappelletti, Boston	9	27	17	132	**1965**	Paul Lowe, San Diego	1,121	222	7
Gino Cappelletti, Boston	6	35	16	119	**1966**	Jim Nance, Boston	1,458	299	11

Player, team	Scoring TD	PAT	FG	Pts	Year	Rushing Player, team	Yds	Att	TD
George Blanda, Oakland	0	56	20	116	**1967**	Jim Nance, Boston	1,216	269	7
Jim Turner, N.Y. Jets	0	43	34	145	**1968**	Paul Robinson, Cincinnati	1,023	238	8
Jim Turner, N.Y. Jets	0	33	32	129	**1969**	Dick Post, San Diego	873	182	6
Jan Stenerud, Kansas City	0	26	30	116	**1970**	Floyd Little, Denver	901	209	3
Garo Yepremian, Miami	0	33	28	117	**1971**	Floyd Little, Denver	1,133	284	6
Bobby Howfield, N.Y. Jets	0	40	27	121	**1972**	O.J. Simpson, Buffalo	1,251	292	6
Roy Gerela, Pittsburgh	0	36	29	123	**1973**	O.J. Simpson, Buffalo	2,003	332	12
Roy Gerela, Pittsburgh	0	33	20	93	**1974**	Otis Armstrong, Denver	1,407	263	9
O.J. Simpson, Buffalo	23	0	0	138	**1975**	O.J. Simpson, Buffalo	1,817	329	16
Toni Linhart, Baltimore Colts	0	49	20	109	**1976**	O.J. Simpson, Buffalo	1,503	290	8
Errol Mann, Oakland	0	39	20	99	**1977**	Mark van Eeghen, Oakland	1,273	324	7
Pat Leahy, N.Y. Jets	0	41	22	107	**1978**	Earl Campbell, Houston	1,450	302	13
John Smith, New England	0	46	23	115	**1979**	Earl Campbell, Houston	1,697	368	19
John Smith, New England	0	51	26	129	**1980**	Earl Campbell, Houston	1,934	373	13
Jim Breech, Cincinnati	0	49	22	115	**1981**	Earl Campbell, Houston	1,376	361	10
Nick Lowery, Kansas City	0	37	26	115					
Marcus Allen, L.A. Raiders	14	0	0	84	**1982**	Freeman McNeil, N.Y. Jets	786	151	6
Gary Anderson, Pittsburgh	0	38	27	119	**1983**	Curt Warner, Seattle	1,446	335	13
Gary Anderson, Pittsburgh	0	45	24	117	**1984**	Earnest Jackson, San Diego	1,179	296	8
Gary Anderson, Pittsburgh	0	40	33	139	**1985**	Marcus Allen, L.A. Raiders	1,759	380	11
Tony Franklin, New England	0	44	32	140	**1986**	Curt Warner, Seattle	1,481	319	13
Jim Breech, Cincinnati	0	25	24	97	**1987**	Eric Dickerson, L.A. Rams-Ind.	1,288*	283	6
Scott Norwood, Buffalo	0	33	32	129	**1988**	Eric Dickerson, Indianapolis	1,659	388	14
David Treadwell, Denver	0	39	27	120	**1989**	Christian Okoye, Kansas City	1,480	370	12
Nick Lowery, Kansas City	0	37	34	139	**1990**	Thurman Thomas, Buffalo	1,297	271	11
Pete Stoyanovich, Miami	0	28	31	121	**1991**	Thurman Thomas, Buffalo	1,407	288	7
Pete Stoyanovich, Miami	0	34	30	124	**1992**	Barry Foster, Pittsburgh	1,690	390	11
Jeff Jaeger, L.A. Raiders	0	27	35	132	**1993**	Thurman Thomas, Buffalo	1,315	355	6
John Carney, San Diego	0	33	34	135	**1994**	Chris Warren, Seattle	1,545	333	9
Norm Johnson, Pittsburgh	0	39	34	141	**1995**	Curtis Martin, New England	1,487	368	14
Cary Blanchard, Indianapolis	0	27	36	135	**1996**	Terrell Davis, Denver	1,538	345	13
Mike Hollis, Jacksonville	0	41	31	134	**1997**	Terrell Davis, Denver	1,750	369	15
Steve Christie, Buffalo	0	41	33	140	**1998**	Terrell Davis, Denver	2,008	392	21
Mike Vanderjagt, Indianapolis	0	43	34	145	**1999**	Edgerrin James, Indianapolis	1,553	369	13
Matt Stover, Baltimore	0	30	35	135	**2000**	Edgerrin James, Indianapolis	1,709	387	13

*Includes 277 yards after being traded to NFC; 1,011 yards led AFC. (1) Based on quarterback ranking points.

National Football Conference Leaders

(National Football League, 1960-69)

Player, team	Passing[1] Att	Com	YG	TD	Year	Receiving Player, team	Rec.	YG	TD
Milt Plum, Cleveland	250	151	2,297	21	**1960**	Raymond Berry, Baltimore Colts	74	1,298	10
Milt Plum, Cleveland	302	177	2,416	18	**1961**	Jim Phillips, L.A. Rams	78	1,092	5
Bart Starr, Green Bay	285	178	2,438	12	**1962**	Bobby Mitchell, Washington	72	1,384	11
Y.A. Tittle, N.Y. Giants	367	221	3,145	36	**1963**	Bobby Joe Conrad, St. Louis Cardinals	73	967	10
Bart Starr, Green Bay	272	163	2,144	15	**1964**	Johnny Morris, Chicago	93	1,200	10
Rudy Bukich, Chicago	312	176	2,641	20	**1965**	Dave Parks, San Francisco	80	1,344	12
Bart Starr, Green Bay	251	156	2,257	14	**1966**	Charley Taylor, Washington	72	1,119	12
Sonny Jurgensen, Washington	508	288	3,747	31	**1967**	Charley Taylor, Washington	70	990	9
Earl Morrall, Baltimore Colts	317	182	2,909	26	**1968**	Clifton McNeil, San Francisco	71	994	7
Sonny Jurgensen, Washington	442	274	3,102	22	**1969**	Dan Abramowicz, New Orleans	73	1,015	7
John Brodie, San Francisco	378	223	2,941	24	**1970**	Dick Gordon, Chicago	71	1,026	13
Roger Staubach, Dallas	211	126	1,882	15	**1971**	Bob Tucker, N.Y. Giants	59	791	4
Norm Snead, N.Y. Giants	325	196	2,307	17	**1972**	Harold Jackson, Philadelphia	62	1,048	4
Roger Staubach, Dallas	286	179	2,428	23	**1973**	Harold Carmichael, Philadelphia	67	1,116	9
Sonny Jurgensen, Washington	167	107	1,185	11	**1974**	Charles Young, Philadelphia	63	696	3
Fran Tarkenton, Minnesota	425	273	2,994	25	**1975**	Chuck Foreman, Minnesota	73	691	9
James Harris, L.A. Rams	158	91	1,460	8	**1976**	Drew Pearson, Dallas	58	806	6
Roger Staubach, Dallas	361	210	2,620	18	**1977**	Ahmad Rashad, Minnesota	51	681	2
Roger Staubach, Dallas	413	231	3,190	25	**1978**	Rickey Young, Minnesota	88	704	5
Roger Staubach, Dallas	461	267	3,586	27	**1979**	Ahmad Rashad, Minnesota	80	1,156	9
Ron Jaworski, Philadelphia	451	257	3,529	27	**1980**	Earl Cooper, San Francisco	83	567	4
Joe Montana, San Francisco	488	311	3,565	19	**1981**	Dwight Clark, San Francisco	85	1,105	4
Joe Thiesmann, Washington	252	161	2,033	13	**1982**	Dwight Clark, San Francisco	60	913	5
Steve Bartkowski, Atlanta	432	274	3,167	22	**1983**	Roy Green, St. Louis Cardinals	78	1,227	14
						Charlie Brown, Washington	78	1,225	8
						Earnest Gray, N.Y. Giants	78	1,139	5
Joe Montana, San Francisco	432	279	3,630	28	**1984**	Art Monk, Washington	106	1,372	7
Joe Montana, San Francisco	494	303	3,653	27	**1985**	Roger Craig, San Francisco	92	1,016	6
Tommy Kramer, Minnesota	372	208	3,000	24	**1986**	Jerry Rice, San Francisco	86	1,570	15
Joe Montana, San Francisco	398	266	3,054	31	**1987**	J.T. Smith, St. Louis Cardinals	91	1,117	8
Wade Wilson, Minnesota	332	204	2,746	15	**1988**	Henry Ellard, L.A. Rams	86	1,414	10
Joe Montana, San Francisco	386	271	3,521	26	**1989**	Sterling Sharpe, Green Bay	90	1,423	12
Phil Simms, N.Y. Giants	311	184	2,284	15	**1990**	Jerry Rice, San Francisco	100	1,502	13
Steve Young, San Francisco	279	180	2,517	17	**1991**	Michael Irvin, Dallas	93	1,523	8
Steve Young, San Francisco	402	268	3,465	25	**1992**	Sterling Sharpe, Green Bay	108	1,461	13
Steve Young, San Francisco	462	314	4,023	29	**1993**	Sterling Sharpe, Green Bay	112	1,274	11
Steve Young, San Francisco	461	324	3,969	35	**1994**	Cris Carter, Minnesota	122	1,256	7
Brett Favre, Green Bay	570	359	4,413	38	**1995**	Herman Moore, Detroit	123	1,686	14
Steve Young, San Francisco	316	214	2,410	14	**1996**	Jerry Rice, San Francisco	108	1,254	8
Steve Young, San Francisco	356	241	3,029	19	**1997**	Herman Moore, Detroit	104	1,293	8
Randall Cunningham, Minnesota	425	259	3,704	34	**1998**	Frank Sanders, Arizona	89	1,145	3
Kurt Warner, St. Louis	499	325	4,353	41	**1999**	Muhsin Muhammad, Carolina	96	1,253	8
Trent Green, St. Louis	240	145	2,063	16	**2000**	Muhsin Muhammad, Carolina	102	1,183	6

Scoring: Player, team	TD	PAT	FG	Pts	Year	Rushing: Player, team	Yds	Att	TD
Paul Hornung, Green Bay	15	41	15	176	**1960**	Jim Brown, Cleveland	1,257	215	9
Paul Hornung, Green Bay	10	41	15	146	**1961**	Jim Brown, Cleveland	1,408	305	8
Jim Taylor, Green Bay	19	0	0	114	**1962**	Jim Taylor, Green Bay	1,474	272	19
Don Chandler, N.Y. Giants	0	52	18	106	**1963**	Jim Brown, Cleveland	1,863	291	12
Lenny Moore, Baltimore Colts	20	0	0	120	**1964**	Jim Brown, Cleveland	1,446	280	7
Gale Sayers, Chicago	22	0	0	132	**1965**	Jim Brown, Cleveland	1,544	289	17
Bruce Gossett, L.A. Rams	0	29	28	113	**1966**	Gale Sayers, Chicago	1,231	229	8
Jim Bakken, St. Louis Cardinals	0	36	27	117	**1967**	Leroy Kelly, Cleveland	1,205	235	11
Leroy Kelly, Cleveland	20	0	0	120	**1968**	Leroy Kelly, Cleveland	1,239	248	16
Fred Cox, Minnesota	0	43	26	121	**1969**	Gale Sayers, Chicago	1,032	236	8
Fred Cox, Minnesota	0	35	30	125	**1970**	Larry Brown, Washington	1,125	237	5
Curt Knight, Washington	0	27	29	114	**1971**	John Brockington, Green Bay	1,105	216	4
Chester Marcol, Green Bay	0	29	33	128	**1972**	Larry Brown, Washington	1,216	285	8
David Ray, L.A. Rams	0	40	30	130	**1973**	John Brockington, Green Bay	1,144	265	3
Chester Marcol, Green Bay	0	19	25	94	**1974**	Lawrence McCutcheon, L.A. Rams	1,109	236	3
Chuck Foreman, Minnesota	22	0	0	132	**1975**	Jim Otis, St. Louis Cardinals	1,076	269	5
Mark Moseley, Washington	0	31	22	97	**1976**	Walter Payton, Chicago	1,390	311	13
Walter Payton, Chicago	16	0	0	96	**1977**	Walter Payton, Chicago	1,852	339	14
Frank Corral, L.A. Rams	0	31	29	118	**1978**	Walter Payton, Chicago	1,395	333	11
Mark Moseley, Washington	0	39	25	114	**1979**	Walter Payton, Chicago	1,610	369	14
Ed Murray, Detroit	0	35	27	116	**1980**	Walter Payton, Chicago	1,460	317	6
Ed Murray, Detroit	0	46	25	121	**1981**	George Rogers, New Orleans	1,674	378	13
Rafael Septien, Dallas	0	40	27	121					
Wendell Tyler, L.A. Rams	13	0	0	78	**1982**	Tony Dorsett, Dallas	745	177	5
Mark Moseley, Washington	0	62	33	161	**1983**	Eric Dickerson, L.A. Rams	1,808	390	18
Ray Wersching, San Francisco	0	56	25	131	**1984**	Eric Dickerson, L.A. Rams	2,105	379	14
Kevin Butler, Chicago	0	51	31	144	**1985**	Gerald Riggs, Atlanta	1,719	397	10
Kevin Butler, Chicago	0	36	28	120	**1986**	Eric Dickerson, L.A. Rams	1,821	404	11
Jerry Rice, San Francisco	23	0	0	138	**1987**	Charles White, L.A. Rams	1,374	324	11
Mike Cofer, San Francisco	0	40	27	121	**1988**	Herschel Walker, Dallas	1,514	361	5
Mike Cofer, San Francisco	0	49	29	136	**1989**	Barry Sanders, Detroit	1,470	280	14
Chip Lohmiller, Washington	0	41	30	131	**1990**	Barry Sanders, Detroit	1,304	255	13
Chip Lohmiller, Washington	0	56	31	149	**1991**	Emmitt Smith, Dallas	1,563	365	12
Morten Andersen, New Orleans	0	33	29	120	**1992**	Emmitt Smith, Dallas	1,713	373	18
Chip Lohmiller, Washington	0	30	30	120					
Jason Hanson, Detroit	0	28	34	130	**1993**	Emmitt Smith, Dallas	1,486	283	9
Fuad Reveiz, Minnesota	0	30	34	132	**1994**	Barry Sanders, Detroit	1,883	331	7
Emmitt Smith, Dallas	22	0	0	132					
Emmitt Smith, Dallas	25	0	0	150	**1995**	Emmitt Smith, Dallas	1,773	377	25
John Kasay, Carolina	0	34	37	145	**1996**	Barry Sanders, Detroit	1,553	307	11
Richie Cunningham, Dallas	0	24	34	126	**1997**	Barry Sanders, Detroit	2,053	335	11
Gary Anderson, Minnesota	0	59	35	164	**1998**	Jamal Anderson, Atlanta	1,846	410	14
Jeff Wilkins, St. Louis	0	64	20	124	**1999**	Stephen Davis, Washington	1,405	290	17
Marshall Faulk, St. Louis	26	0	0	156	**2000**	Robert Smith, Minnesota	1,521	295	7

(1) Based on quarterback ranking points.

2000 NFL Individual Leaders

American Football Conference

PASSING	Att	Comp	Pct comp	Yds	Avg gain	Long	TD	Pct TD	Int	Rating points
Brian Griese, Denver	336	216	64.3	2,688	8.0	61	19	5.7	4	102.9
Peyton Manning, Indianapolis	571	357	62.5	4,413	7.7	78td	33	5.8	15	94.7
Rich Gannon, Oakland	473	284	60.0	3,430	7.3	84td	28	5.9	11	92.4
Elvis Grbac, Kansas City	547	326	59.6	4,169	7.6	81td	28	5.1	14	89.9
Doug Flutie, Buffalo	231	132	57.1	1,700	7.4	52	8	3.5	3	86.5
Mark Brunell, Jacksonville	512	311	60.7	3,640	7.1	67td	20	3.9	14	84.0
Steve McNair, Tennessee	396	248	62.6	2,847	7.2	56td	15	3.8	13	83.2
Rob Johnson, Buffalo	306	175	57.2	2,125	6.9	74td	12	3.9	7	82.2
Gus Frerotte, Denver	232	138	59.5	1,776	7.7	44	9	3.9	8	82.1
Drew Bledsoe, New England	531	312	58.8	3,291	6.2	59	17	3.2	13	77.3

RUSHING	Att	Yds	Avg	Long	TD
Edgerrin James, Indianapolis	387	1,709	4.4	30	13
Eddie George, Tennessee	403	1,509	3.7	35td	14
Mike Anderson, Denver	297	1,500	5.1	80td	15
Corey Dillon, Cincinnati	315	1,435	4.6	80td	7
Fred Taylor, Jacksonville	292	1,399	4.8	71	12
Jamal Lewis, Baltimore	309	1,364	4.4	45	6
Jerome Bettis, Pittsburgh	355	1,341	3.8	30	8
Ricky Watters, Seattle	278	1,242	4.5	55	7
Curtis Martin, N.Y. Jets	316	1,204	3.8	55	9
Lamar Smith, Miami	309	1,139	3.7	68td	14

RECEIVING	Rec.	Yds	Avg	Long	TD
Marvin Harrison, Indianapolis	102	1,413	13.9	78td	14
Ed McCaffrey, Denver	101	1,317	13.0	61	9
Rod Smith, Denver	100	1,602	16.0	49	8
Eric Moulds, Buffalo	94	1,326	14.1	52	5
Keenan McCardell, Jacksonville	94	1,207	12.8	67td	5
Tony Gonzalez, Kansas City	93	1,203	12.9	39	9
Jimmy Smith, Jacksonville	91	1,213	13.3	65td	8
Richie Anderson, N.Y. Jets	88	853	9.7	41	2
Troy Brown, New England	83	944	11.4	44td	4
Terry Glenn, New England	79	963	12.2	39td	6

SCORING—KICKERS	PAT	FG	Long	Pts
Matt Stover, Baltimore	30/30	35/39	51	135
Mike Vanderjagt, Indianapolis	46/46	25/27	48	121
Al Del Greco, Tennessee	37/38	27/33	50	118
Olindo Mare, Miami	33/34	28/31	49	117
Sebastian Janikowski, Oakland	46/46	22/32	54	112

SCORING—NON-KICKERS	TD	Rush	Pass	2 Pt	Pts
Edgerrin James, Indianapolis	18	13	5	0	108
Eddie George, Tennessee	16	14	2	0	96
Lamar Smith, Miami	16	14	2	0	96
Mike Anderson, Denver	15	15	0	0	90
Marvin Harrison, Indianapolis	14	0	14	0	84
Fred Taylor, Jacksonville	14	12	2	0	84

INTERCEPTIONS	No.	Yds	Avg	Long	TD
Samari Rolle, Tennessee	7	140	20.0	81td	1
Brian Walker, Miami	7	110	15.7	31	0
Eric Allen, Oakland	6	145	24.2	50td	3
Victor Green, N.Y. Jets	6	144	24.0	43	1
Duane Starks, Baltimore	6	125	20.8	64	0
Terrell Buckley, Denver	6	110	18.3	33	1
Rodney Harrison, San Diego	6	97	16.2	63td	1
William Thomas, Oakland	6	68	11.3	46td	1

KICKOFF RETURNS	No.	Yds	Avg	Long	TD
Derrick Mason, Tennessee	42	1,132	27.0	66	0
Kevin Williams, N.Y. Jets-Miami	21	551	26.2	97td	1
Autry Denson, Miami	20	495	24.8	56	0
Charlie Rogers, Seattle	66	1,629	24.7	81td	1
David Dunn, Oakland	44	1,073	24.4	88td	1

PUNT RETURNS	No.	Yds	Avg	Long	TD
Jermaine Lewis, Baltimore	36	578	16.1	89td	2
Charlie Rogers, Seattle	26	363	14.0	43	0
Hank Poteat, Pittsburgh	36	467	13.0	54	1
Derrick Mason, Tennessee	51	662	13.0	69td	1
Troy Brown, New England	39	504	12.9	66td	1

PUNTING	No.	Yds	Long	Avg
Darren Bennett, San Diego	92	4,248	66	46.2
Shane Lechler, Oakland	65	2,984	69	45.9
Chris Gardocki, Cleveland	108	4,919	67	45.5
Hunter Smith, Indianapolis	65	2,906	65	44.7
Tom Tupa, N.Y Jets	83	3,714	70	44.7

SACKS	No.
Trace Armstrong, Miami	16.5
Jason Taylor, Miami	14.5
Eric Hicks, Kansas City	14.0
Jason Gildon, Pittsburgh	13.5
Trevor Pryce, Denver	12.0

National Football Conference

PASSING	Att	Comp	Pct comp	Yds	Avg gain	Long	TD	Pct TD	Int	Rating points
Trent Green, St. Louis	240	145	60.4	2,063	8.6	64	16	6.7	5	101.8
Kurt Warner, St. Louis	347	235	67.7	3,429	9.9	85td	21	6.1	18	98.3
Daunte Culpepper, Minnesota	474	297	62.7	3,937	8.3	78td	33	7.0	16	98.0
Jeff Garcia, San Francisco	561	355	63.3	4,278	7.6	69td	31	5.5	10	97.6
Kerry Collins, N.Y. Giants	529	311	58.8	3,610	6.8	59	22	4.2	13	83.1
Jeff Blake, New Orleans	302	184	60.9	2,025	6.7	49td	13	4.3	9	82.7
Steve Beuerlein, Carolina	533	324	60.8	3,730	7.0	54	19	3.6	18	79.7
Brett Favre, Green Bay	580	338	58.3	3,812	6.6	67td	20	3.4	16	78.0
Donavan McNabb, Philadelphia	569	330	58.0	3,365	5.9	70td	21	3.7	13	77.8
Shaun King, Tampa Bay	428	233	54.4	2,769	6.5	75	18	4.2	13	75.8

RUSHING	Att	Yds	Avg	Long	TD
Robert Smith, Minnesota	295	1,521	5.2	72td	7
Marshall Faulk, St. Louis	253	1,359	5.4	36	18
Stephen Davis, Washington	332	1,318	4.0	50td	11
Emmitt Smith, Dallas	294	1,203	4.1	52	9
James Stewart, Detroit	339	1,184	3.5	34	10
Ahman Green, Green Bay	263	1,175	4.5	39td	10
Charlie Garner, San Francisco	258	1,142	4.4	42	7
Warrick Dunn, Tampa Bay	248	1,133	4.6	70td	8
James Allen, Chicago	290	1,120	3.9	29	2
Jamal Anderson, Atlanta	282	1,024	3.6	42	6

RECEIVING	Rec.	Yds	Avg	Long	TD
Muhsin Muhammad, Carolina	102	1,183	11.6	36	6
Terrell Owens, San Francisco	97	1,451	15.0	69td	13
Cris Carter, Minnesota	96	1,274	13.3	53	9
Joe Horn, New Orleans	94	1,340	14.3	52	8
Isaac Bruce, St. Louis	87	1,471	16.9	78td	9
Torry Holt, St. Louis	82	1,635	19.9	85td	6
Marshall Faulk, St. Louis	81	830	10.2	72td	8
Larry Centers, Washington	80	600	7.5	26	3
Amani Toomer, N.Y. Giants	78	1,094	14.0	54td	7
Randy Moss, Minnesota	77	1,437	18.7	78td	15

SCORING—KICKERS	PAT	FG	Long	Pts
Ryan Longwell, Green Bay	32/32	33/38	52	131
Martin Gramatica, Tampa Bay	42/42	28/34	55	126
Joe Nedney, Carolina	24/24	34/38	52	126
David Akers, Philadelphia	34/36	29/33	51	121
Gary Anderson, Minnesota	45/45	22/23	49	111

SCORING—NON-KICKERS	TD	Rush	Pass	2 Pt	Pts
Marshall Faulk, St. Louis	26	18	8	0	156
Randy Moss, Minnesota	15	0	15	0	90
Ahman Green, Green Bay	13	10	3	0	78
Terrell Owens, San Francisco	13	0	13	0	78
Stephen Davis, Washington	11	11	0	0	66
James Stewart, Detroit	11	10	1	0	66

INTERCEPTIONS	No.	Yds	Avg	Long	TD
Darren Sharper, Green Bay	9	109	12.1	47	0
Dexter McCleon, St. Louis	8	28	3.5	23	0
Donnie Abraham, Tampa Bay	7	82	11.7	23	0
Kurt Shultz, Detroit	7	53	7.6	19	0
Bryant Westbrook, Detroit	6	126	21	101td	1
Ray Buchanan, Atlanta	6	114	19	60	0
Emmanuel McDaniel, N.Y. Giants	6	30	5	17	0
Damien Robinson, N.Y. Jets	6	1	0.2	1	0

KICKOFF RETURNS	No.	Yds	Avg	Long	TD
Darrick Vaughn, Atlanta	39	1,082	27.7	100td	3
MarTay Jenkins, Arizona	82	2,186	26.7	98td	1
Allen Rossum, Green Bay	50	1,288	25.8	92td	1
Desmond Howard, Detroit	57	1,401	24.6	70	0
Tony Horne, St. Louis	57	1,379	24.2	103td	1

PUNT RETURNS	No.	Yds	Avg	Long	TD
Az-Zahir Hakim, St. Louis	32	489	15.3	86td	1
Desmond Howard, Detroit	31	457	14.7	95td	1
Wayne McGarity, Dallas	30	353	11.8	64td	2
Brian Mitchell, Philadelphia	32	335	10.5	72td	1
Tim Dwight, Atlanta	33	309	9.4	70td	1

PUNTING	No.	Yds	Long	Avg
Mitch Berger, Minnesota	62	2,773	60	44.7
Scott Player, Arizona	65	2,871	55	44.2
John Jett, Detroit	93	4,044	59	43.5
Micah Knorr, Dallas	58	2,485	60	42.8
Sean Landeta, Philadelphia	86	3,635	60	42.3

SACKS	No.
La'Roi Glover, New Orleans	17.0
Warren Sapp, Tampa Bay	16.5
Hugh Douglas, Philadelphia	15.0
Marcus Jones, Tampa Bay	13.0
Marco Coleman, Washington	12.0
Joe Johnson, New Orleans	12.0

First-Round Selections in the 2001 NFL Draft

Team	Player	Pos	College
1. Atlanta[1]	Michael Vick	QB	Virginia Tech
2. Arizona	Leonard Davis	OL	Texas
3. Cleveland	Gerard Warren	DL	Florida
4. Cincinnati	Justin Smith	DL	Missouri
5. San Diego[2]	LaDainian Tomlinson	RB	Texas Christian
6. New England	Richard Seymour	DL	Georgia
7. San Francisco[3]	Andre Carter	DL	California
8. Chicago	David Terrell	WR	Michigan
9. Seattle[4]	Koren Robinson	WR	N. Carolina St.
10. Green Bay[5]	Jamal Reynolds	DL	Florida St.
11. Carolina	Dan Morgan	LB	Miami (FL)
12. St. Louis[6]	Damione Lewis	DL	Miami (FL)
13. Jacksonville	Marcus Stroud	DL	Georgia
14. Tampa Bay[7]	Kenyatta Walker	OL	Florida
15. Washington	Rod Gardner	WR	Clemson
16. N.Y. Jets[8]	Santana Moss	WR	Miami (FL)
17. Seattle[9]	Steve Hutchinson	OL	Michigan
18. Detroit	Jeff Backus	OL	Michigan
19. Pittsburgh[10]	Casey Hampton	DL	Texas
20. St. Louis	Adam Archuleta	S	Arizona St.
21. Buffalo[11]	Nate Clements	DB	Ohio St.
22. N.Y. Giants[12]	Will Allen	DB	Syracuse
23. New Orleans	Deuce McAllister	RB	Mississippi
24. Denver	Willie Middlebrooks	DB	Minnesota
25. Philadelphia	Freddie Mitchell	WR	UCLA
26. Miami	Jamar Fletcher	DB	Wisconsin
27. Minnesota	Michael Bennett	RB	Wisconsin
28. Oakland	Derrick Gibson	S	Florida St.
29. St. Louis[13]	Ryan Pickett	DL	Ohio St.
30. Indianapolis[14]	Reggie Wayne	WR	Miami (FL)
31. Baltimore	Todd Heap	TE	Arizona St.

(1) From San Diego. (2) From Atlanta. (3) From Dallas through Seattle. (4) From San Francisco. (5) From Seattle. (6) From Kansas City. (7) From Buffalo. (8) From Pittsburgh. (9) From Green Bay. (10) From N.Y. Jets. (11) From Tampa Bay. (12) From Tampa Bay. (13) From Tennessee. (14) From N.Y. Giants.

Number One NFL Draft Choices, 1936-2001

Year	Team	Player, Pos., College	Year	Team	Player, Pos., College
1936	Philadelphia	Jay Berwanger, HB, Chicago	1969	Buffalo	O.J. Simpson, RB, USC
1937	Philadelphia	Sam Francis, FB, Nebraska	1970	Pittsburgh	Terry Bradshaw, QB, La.Tech
1938	Cleveland Rams	Corbett Davis, FB, Indiana	1971	New England	Jim Plunkett, QB, Stanford
1939	Chicago Cards	Ki Aldrich, C, TCU	1972	Buffalo	Walt Patulski, DE, Notre Dame
1940	Chicago Cards	George Cafego, HB, Tennessee	1973	Houston	John Matuszak, DE, Tampa
1941	Chicago Bears	Tom Harmon, HB, Michigan	1974	Dallas	Ed "Too Tall" Jones, DE, Tenn. St.
1942	Pittsburgh	Bill Dudley, HB, Virginia	1975	Atlanta	Steve Bartkowski, QB, Cal.
1943	Detroit	Frank Sinkwich, HB, Georgia	1976	Tampa Bay	Lee Roy Selmon, DE, Oklahoma
1944	Boston Yanks	Angelo Bertelli, QB, Notre Dame	1977	Tampa Bay	Ricky Bell, RB, USC
1945	Chicago Cards	Charley Trippi, HB, Georgia	1978	Houston	Earl Campbell, RB, Texas
1946	Boston Yanks	Frank Dancewicz, QB, Notre Dame	1979	Buffalo	Tom Cousineau, LB, Ohio St.
1947	Chicago Bears	Bob Fenimore, HB, Okla. A&M	1980	Detroit	Billy Sims, RB, Oklahoma
1948	Washington	Harry Gilmer, QB, Alabama	1981	New Orleans	George Rogers, RB, S.Carolina
1949	Philadelphia	Chuck Bednarik, C, Penn	1982	New England	Kenneth Sims, DT, Texas
1950	Detroit	Leon Hart, E, Notre Dame	1983	Baltimore Colts	John Elway, QB, Stanford
1951	N.Y. Giants	Kyle Rote, HB, SMU	1984	New England	Irving Fryar, WR, Nebraska
1952	L.A. Rams	Bill Wade, QB, Vanderbilt	1985	Buffalo	Bruce Smith, DE, Va.Tech
1953	San Francisco	Harry Babcock, E, Georgia	1986	Tampa Bay	Bo Jackson, RB, Auburn
1954	Cleveland	Bobby Garrett, QB, Stanford	1987	Tampa Bay	Vinny Testaverde, QB, Miami (FL)
1955	Baltimore Colts	George Shaw, QB, Oregon	1988	Atlanta	Aundray Bruce, LB, Auburn
1956	Pittsburgh	Gary Glick, DB, Col. A&M	1989	Dallas	Troy Aikman, QB, UCLA
1957	Green Bay	Paul Hornung, QB, Notre Dame	1990	Indianapolis	Jeff George, QB, Illinois
1958	Chicago Cards	King Hill, QB, Rice	1991	Dallas	Russell Maryland, DL, Miami (FL)
1959	Green Bay	Randy Duncan, QB, Iowa	1992	Indianapolis	Steve Emtman, DL, Washington
1960	L.A. Rams	Billy Cannon, HB, LSU	1993	New England	Drew Bledsoe, QB, Washington St.
1961	Minnesota	Tommy Mason, HB, Tulane	1994	Cincinnati	Dan Wilkinson, DT, Ohio St.
1962	Washington	Ernie Davis, HB, Syracuse	1995	Cincinnati	Ki-Jana Carter, RB, Penn State
1963	L.A. Rams	Terry Baker, QB, Oregon St.	1996	N.Y. Jets	Keyshawn Johnson, WR, USC
1964	San Francisco	Dave Parks, E, Texas Tech	1997	St. Louis	Orlando Pace, T, Ohio St.
1965	N.Y. Giants	Tucker Frederickson, HB, Auburn	1998	Indianapolis	Peyton Manning, QB, Tennessee
1966	Atlanta	Tommy Nobis, LB, Texas	1999	Cleveland	Tim Couch, QB, Kentucky
1967	Baltimore Colts	Bubba Smith, DT, Michigan St.	2000	Cleveland	Courtney Brown, DE, Penn State
1968	Minnesota	Ron Yary, T, USC	2001	Atlanta	Michael Vick, QB, Virginia Tech

NFL MVP, Defensive Player of the Year, and Rookie of the Year

The Most Valuable Player and Defensive Player of the Year are two of many awards given out annually by the Associated Press. Rookie of the Year is one of many awards given out annually by *The Sporting News*. Many other organizations give out annual awards honoring the NFL's best players.

Most Valuable Player

1957	Jim Brown, Cleveland
1958	Gino Marchetti, Baltimore Colts
1959	Charley Conerly, N.Y. Giants
1960	Norm Van Brocklin, Philadelphia; Joe Schmidt, Detroit
1961	Paul Hornung, Green Bay
1962	Jim Taylor, Green Bay
1963	Y.A. Tittle, N.Y. Giants
1964	John Unitas, Baltimore Colts
1965	Jim Brown, Cleveland
1966	Bart Starr, Green Bay
1967	John Unitas, Baltimore Colts
1968	Earl Morrall, Baltimore Colts
1969	Roman Gabriel, L.A. Rams
1970	John Brodie, San Francisco
1971	Alan Page, Minnesota
1972	Larry Brown, Washington
1973	O.J. Simpson, Buffalo
1974	Ken Stabler, Oakland
1975	Fran Tarkenton, Minnesota
1976	Bert Jones, Baltimore
1977	Walter Payton, Chicago
1978	Terry Bradshaw, Pittsburgh
1979	Earl Campbell, Houston
1980	Brian Sipe, Cleveland
1981	Ken Anderson, Cincinnati
1982	Mark Moseley, Washington
1983	Joe Theismann, Washington
1984	Dan Marino, Miami
1985	Marcus Allen, L.A. Raiders
1986	Lawrence Taylor, N.Y. Giants
1987	John Elway, Denver
1988	Boomer Esiason, Cincinnati
1989	Joe Montana, San Francisco
1990	Joe Montana, San Francisco
1991	Thurman Thomas, Buffalo
1992	Steve Young, San Francisco
1993	Emmitt Smith, Dallas
1994	Steve Young, San Francisco
1995	Brett Favre, Green Bay
1996	Brett Favre, Green Bay
1997	(tie) Brett Favre, Green Bay; Barry Sanders, Detroit
1998	Terrell Davis, Denver
1999	Kurt Warner, St. Louis
2000	Marshall Faulk, St. Louis

Defensive Player of the Year

1966	Larry Wilson, St. Louis
1967	Deacon Jones, Los Angeles
1968	Deacon Jones, Los Angeles
1969	Dick Butkus, Chicago
1970	Dick Butkus, Chicago
1971	Carl Eller, Minnesota
1972	Joe Greene, Pittsburgh
1973	Alan Page, Minnesota
1974	Joe Greene, Pittsburgh
1975	Curley Culp, Houston
1976	Jerry Sherk, Cleveland
1977	Harvey Martin, Dallas
1978	Randy Gradishar, Denver
1979	Lee Roy Selmon, Tampa Bay
1980	Lester Hayes, Oakland
1981	Joe Klecko, N.Y. Jets
1982	Mark Gastineau, N.Y. Jets
1983	Jack Lambert, Pittsburgh
1984	Mike Haynes, L.A. Raiders
1985	Howie Long, L.A. Raiders; Andre Tippett, New England
1986	Lawrence Taylor, N.Y. Giants
1987	Reggie White, Philadelphia
1988	Mike Singletary, Chicago
1989	Tim Harris, Green Bay
1990	Bruce Smith, Buffalo
1991	Pat Swilling, New Orleans
1992	Junior Seau, San Diego
1993	Bruce Smith, Buffalo
1994	Deion Sanders, San Francisco
1995	Bryce Paup, Buffalo
1996	Bruce Smith, Buffalo
1997	Dana Stubblefield, San Francisco
1998	Reggie White, Green Bay
1999	Warren Sapp, Tampa Bay
2000	Ray Lewis, Baltimore

Rookie of the Year

1964	Charley Taylor, Washington
1965	Gale Sayers, Chicago
1966	Tommy Nobis, Atlanta
1967	Mel Farr, Detroit
1968	Earl McCullouch, Detroit
1969	Calvin Hill, Dallas
1970	NFC: Bruce Taylor, San Francisco; AFC: Dennis Shaw, Buffalo
1971	NFC: John Brockington, Green Bay; AFC: Jim Plunkett, New England
1972	NFC: Chester Marcol, Green Bay; AFC: Franco Harris, Pittsburgh
1973	NFC: Chuck Foreman, Minnesota; AFC: Boobie Clark, Cincinnati
1974	NFC: Wilbur Jackson, San Francisco; AFC: Don Woods, San Diego
1975	NFC: Steve Bartkowski, Atlanta; AFC: Robert Brazile, Houston
1976	NFC: Sammy White, Minnesota; AFC: Mike Haynes, New England
1977	NFC: Tony Dorsett, Dallas; AFC: A. J. Duhe, Miami
1978	NFC: Al Baker, Detroit; AFC: Earl Campbell, Houston
1979	NFC: Ottis Anderson, St. Louis; AFC: Jerry Butler, Buffalo
1980	Billy Sims, Detroit
1981	George Rogers, New Orleans
1982	Marcus Allen, L.A. Raiders
1983	Dan Marino, Miami
1984	Louis Lipps, Pittsburgh
1985	Eddie Brown, Cincinnati
1986	Rueben Mayes, New Orleans
1987	Robert Awalt, St. Louis
1988	Keith Jackson, Philadelphia
1989	Barry Sanders, Detroit
1990	Richmond Webb, Miami
1991	Mike Croel, Denver
1992	Santana Dotson, Tampa Bay
1993	Jerome Bettis, L.A. Rams
1994	Marshall Faulk, Indianapolis
1995	Curtis Martin, New England
1996	Eddie George, Houston
1997	Warrick Dunn, Tampa Bay
1998	Randy Moss, Minnesota
1999	Edgerrin James, Indianapolis
2000	Brian Urlacher, Chicago

The Sporting News 2000 NFL All-Pro Team

Offense—Quarterback: Rich Gannon, Oakland. Running backs: Marshall Faulk, St. Louis; Edgerrin James, Indianapolis. Wide receivers: Marvin Harrison, Indianapolis; Randy Moss, Minnesota. Tight end: Tony Gonzalez, Kansas City. Tackles: Jonathan Ogden, Baltimore; Orlando Pace, St. Louis. Guards: Larry Allen, Dallas; Bruce Matthews, Tennessee. Center: Tom Nalen, Denver.

Defense—Linebackers: Ray Lewis, Baltimore; Derrick Brooks, Tampa Bay; Junior Seau, San Diego. Defensive ends: Hugh Douglas, Philadelphia; Jason Taylor, Miami. Defensive tackles: La'Roi Glover, New Orleans; Warren Sapp, Tampa Bay. Cornerbacks: Sam Madison, Miami; Samari Rolle, Tennessee. Safeties: John Lynch, Tampa Bay; Darren Sharper, Green Bay.

Special Teams—Kicker: Matt Stover, Baltimore. Punter: Shane Lechler, Oakland. Punt returner: Az-Zahir Hakim, St. Louis. Kick returner: Derrick Mason, Tennessee.

All-Time NFL Coaching Victories

(at end of 2000 season; ranked by career wins; *active in 2000)

			Regular Season				Career			
Coach	Years	Teams	W	L	T	Pct	W	L	T	Pct
Don Shula	33	Colts, Dolphins	328	156	6	.676	347	173	6	.665
George Halas	40	Bears	318	148	31	.671	324	151	31	.671
Tom Landry	29	Cowboys	250	162	6	.605	270	178	6	.601
Curly Lambeau	33	Packers, Cardinals, Redskins	226	132	22	.624	229	134	22	.623
Chuck Noll	23	Steelers	193	148	1	.566	209	156	1	.572
Chuck Knox	22	Rams, Bills, Seahawks	186	147	1	.558	193	158	1	.550
Dan Reeves*	20	Broncos, Giants, Falcons	171	140	1	.550	181	148	1	.550
Paul Brown	21	Browns, Bengals	166	100	6	.621	170	109	6	.607
Bud Grant	18	Vikings	158	96	5	.620	168	108	5	.607
Steve Owen	23	Giants	153	100	17	.598	155	108	17	.584
Marv Levy	17	Chiefs, Bills	143	112	0	.561	154	120	0	.562
M. Schottenheimer	15	Browns, Chiefs	145	85	1	.630	150	96	1	.609
Bill Parcells	15	Giants, Patriots, Jets	138	100	1	.579	149	106	1	.584
Joe Gibbs	12	Redskins	124	60	0	.674	140	65	0	.683
Hank Stram	17	Chiefs, Saints	131	97	10	.571	136	100	10	.573
Weeb Ewbank	20	Colts, Jets	130	129	7	.502	134	130	7	.507
Mike Ditka	14	Bears, Saints	121	95	0	.560	127	101	0	.557
Sid Gillman	18	Rams, Chargers, Oilers	122	99	7	.550	123	104	7	.541
George Seifert*	10	49ers, Panthers	113	47	0	.706	123	52	0	.703
George Allen	12	Rams, Redskins	116	47	5	.705	120	54	5	.684

All-Time Professional (NFL and AFL) Football Records

(at end of 2000 season; *active in 2000; (a) includes AFL statistics)

Leading Lifetime Scorers

Player	Yrs	TD	PAT	FG	Total
Gary Anderson*	19	0	676	461	2,059
George Blanda (a)	26	9	943	335	2,002
Morten Andersen*	19	0	615	441	1,938
Norm Johnson	18	0	638	366	1,736
Nick Lowery	18	0	562	383	1,711
Jan Stenerud (a)	19	0	580	373	1,699
Eddie Murray*	19	0	538	352	1,594
Al Del Greco*	17	0	543	347	1,584
Pat Leahy	18	0	558	304	1,470
Jim Turner (a)	16	1	521	304	1,439
Matt Bahr	17	0	522	300	1,422
Mark Moseley	16	0	482	300	1,382
Jim Bakken	17	0	534	282	1,380
Fred Cox	15	0	519	282	1,365
Lou Groza	17	1	641	234	1,349
Jim Breech	14	0	517	243	1,246
Pete Stoyanovich*	12	0	420	272	1,236
Chris Bahr	14	0	490	241	1,213
Kevin Butler	13	0	413	265	1,208
Steve Christie*	11	0	358	272	1,174

Leading Lifetime Touchdown Scorers

Player	Yrs	Rush	Rec	Ret	Total
Jerry Rice*	16	10	176	1	187
Emmit Smith*	11	145	11	0	156
Marcus Allen	16	123	21	1	145
Jim Brown	9	106	20	0	126
Walter Payton	13	110	15	0	125
Cris Carter*	14	0	123	1	124
John Riggins	14	104	12	0	116
Lenny Moore	12	63	48	2	113
Barry Sanders	10	99	10	0	109
Don Hutson	11	3	99	3	105
Steve Largent	14	1	100	0	101
Franco Harris	13	91	9	0	100
Eric Dickerson	11	90	6	0	96
Jim Taylor	10	83	10	0	93
Tony Dorsett	12	77	13	1	91
Bobby Mitchell	11	18	65	8	91
Tim Brown*	13	1	86	3	90
Leroy Kelly	10	74	13	3	90
Charley Taylor	13	11	79	0	90
Ricky Watters*	9	77	13	0	90

Most Points, Season — 176, Paul Hornung, Green Bay Packers, 1960 (15 TDs, 41 PATs, 15 FGs).
Most Points, Game — 40, Ernie Nevers, Chicago Cardinals vs. Chicago Bears, Nov. 28, 1929 (6 TDs, 4 PATs).
Most Touchdowns, Season — 26, Marshall Faulk, St. Louis Rams, 2000 (18 rushing, 8 receiving).
Most Touchdowns, Game — 6, Ernie Nevers, Chicago Cardinals vs. Chicago Bears, Nov. 28, 1929 (6 rushing); Dub Jones, Cleveland Browns vs. Chicago Bears, Nov. 25, 1951 (4 rushing, 2 pass receptions); Gale Sayers, Chicago Bears vs. San Francisco 49ers, Dec. 12, 1965 (4 rushing, 1 pass reception, 1 punt return).
Most Points After TD, Season — 66, Uwe von Schamann, Miami Dolphins, 1984.
Most Consecutive Points After TD — 301, Norm Johnson, Atlanta Falcons-Pittsburgh Steelers-Philadelphia Eagles, 1991-99.
Most Field Goals, Season — 39, Olindo Mare, Miami Dolphins, 1999.
Most Field Goals, Game — 7, Jim Bakken, St. Louis Cardinals vs. Pittsburgh Steelers, Sept. 24, 1967; Rich Karlis, Minnesota vs. L.A. Rams, Nov. 5, 1989 (OT); Chris Boniol, Dallas vs. Green Bay, Nov. 18, 1996.
Most Field Goals Career — 461, Gary Anderson, Pitts. Steelers-Phil. Eagles-SF 49ers-Minn. Vikings, 1982-2000.
Longest Field Goal — 63 yds., Tom Dempsey, New Orleans Saints vs. Detroit Lions, Nov. 8, 1970; Jason Elam, Denver Broncos vs. Jacksonville Jaguars, Oct. 25, 1998.

Defensive Records

Most Interceptions, Career — 81, Paul Krause, Washington Redskins-Minnesota Vikings, 1964-79.

Most Interceptions, Season — 14, Dick "Night Train" Lane, L. A. Rams, 1952.

Most Touchdowns, Career — 9, Ken Houston, Houston Oilers-Washington Redskins, 1967-80; Rod Woodson, Pittsburgh Steelers-San Francisco 49ers-Baltimore Ravens, 1987-1999.

Most Touchdowns, Season — 4, Ken Houston, Houston Oilers, 1971; Jim Kearney, Kansas City Chiefs, 1972; Eric Allen, Philadelphia Eagles, 1993.

Most Sacks, Career (Since 1982) — 198, Reggie White, Philadelphia Eagles-Green Bay Packers, 1985-2000.

Most Sacks, Season (Since 1982) — 22, Mark Gastineau, N.Y. Jets, 1984.

Most Sacks, Game (Since 1982) — 7, Derrick Thomas, Kansas City Chiefs vs. Seattle Seahawks, Nov. 11, 1990.

Leading Lifetime Rushers

(ranked by rushing yards)

Player	Yrs	Att	Yards	Avg	Long	TD
Walter Payton	13	3,838	16,726	4.4	76	110
Barry Sanders	10	3,062	15,269	5.0	85	99
Emmitt Smith*	11	3,537	15,166	4.3	75	145
Eric Dickerson	11	2,996	13,259	4.4	85	90
Tony Dorsett	12	2,936	12,739	4.3	99	77
Jim Brown	9	2,359	12,312	5.2	80	106
Marcus Allen	16	3,022	12,243	4.1	61	123
Franco Harris	13	2,949	12,120	4.1	75	91
Thurman Thomas*	13	2,877	12,074	4.2	80	65
John Riggins	14	2,916	11,352	3.9	66	104
O.J. Simpson (a)	11	2,404	11,236	4.7	94	61
Ricky Watters*	9	2,550	10,325	4.0	57	77
Ottis Anderson	14	2,562	10,273	4.0	76	81
Jerome Bettis*	8	2,461	9,804	4.0	71	49
Earl Campbell	8	2,187	9,407	4.3	81	74
Jim Taylor	10	1,941	8,597	4.4	84	83
Joe Perry	14	1,737	8,378	4.8	78	53
Earnest Byner	15	2,095	8,261	3.9	54	56
Herschel Walker	12	1,954	8,225	4.2	91	61
Roger Craig	11	1,991	8,189	4.1	71	56

Most Yards Gained, Season — 2,105, Eric Dickerson, L.A. Rams, 1984.
Most Yards Gained, Game — 278, Corey Dillon, Cincinnati Bengals vs. Denver Broncos, Oct. 22, 2000.
Most Touchdowns Rushing, Career — 145, Emmitt Smith, Dallas Cowboys, 1990-2000.
Most Touchdowns Rushing, Season — 25, Emmitt Smith, Dallas Cowboys, 1995.
Most Touchdowns Rushing, Game — 6, Ernie Nevers, Chicago Cardinals vs. Chicago Bears, Nov. 28, 1929.
Most Rushing Attempts, Game — 45, Jamie Morris, Washington Redskins vs. Cincinnati Bengals, Dec. 17, 1988 (overtime).
Longest Run From Scrimmage — 99 yds., Tony Dorsett, Dallas Cowboys vs. Minnesota Vikings, Jan. 3, 1983 (touchdown).

Leading Lifetime Receivers

(ranked by number of receptions)

Player	Yrs	No.	Yards	Avg	Long	TD
Jerry Rice*	16	1,281	19,247	15.0	96	176
Cris Carter*	14	1,020	12,962	12.7	80	123
Andre Reed*	16	951	13,198	13.9	83	87
Art Monk	16	940	12,721	13.5	79	68
Irving Fryar*	17	851	12,785	15.0	80	84
Tim Brown*	13	846	12,072	14.3	80	86
Steve Largent	14	819	13,089	16.0	74	100
Henry Ellard	16	814	13,777	16.9	81	65
James Lofton	16	764	14,004	18.3	80	75
Michael Irvin	12	750	11,904	15.9	87	65
Charlie Joiner (a)	18	750	12,146	16.2	87	65
Andre Rison*	12	741	10,173	13.7	80	84
Gary Clark	11	699	10,856	15.5	84	65
Larry Centers*	11	685	5,683	8.3	54	25
Herman Moore*	10	666	9,098	13.7	93	62
Ozzie Newsome	13	662	7,980	12.1	74	47
Charley Taylor	13	649	9,110	14.0	88	79
Drew Hill	15	634	9,831	15.5	81	60
Don Maynard (a)	15	633	11,834	18.7	87	88
Raymond Berry	13	631	9,275	14.7	70	68

Most Yards Gained, Career — 19,247, Jerry Rice, San Francisco 49ers, 1985-2000.
Most Yards Gained, Season — 1,848, Jerry Rice, San Francisco 49ers, 1995.
Most Yards Gained, Game — 336, Willie "Flipper" Anderson, L. A. Rams vs. New Orleans, Nov. 26, 1989 (overtime).
Most Pass Receptions, Season — 123, Herman Moore, Detroit Lions, 1995.
Most Pass Receptions, Game — 20, Terrell Owens, San Francisco 49ers vs. Chicago Bears, Dec. 17, 2000 (283 yards).
Most Touchdown Receptions, Career — 176, Jerry Rice, San Francisco 49ers, 1985-2000.
Most Touchdown Receptions, Season — 22, Jerry Rice, San Francisco 49ers, 1987.
Most Touchdown Receptions, Game — 5, Bob Shaw, Chicago Cardinals vs. Baltimore Colts, Oct. 2, 1950; Kellen Winslow, San Diego Chargers vs. Oakland Raiders, Nov. 22, 1981; Jerry Rice, San Francisco 49ers vs. Atlanta Falcons, Oct. 14, 1990.

Leading Lifetime Passers

(minimum 1,500 attempts; ranked by quarterback rating points)

Player	Yrs	Att	Comp	Yds	TD	Int	Pts[1]
Steve Young	15	4,149	2,667	33,124	232	107	96.8
Joe Montana	15	5,391	3,409	40,551	273	139	92.3
Dan Marino	17	8,358	4,967	61,361	420	252	86.4
Brett Favre*	10	4,932	2,997	34,706	255	157	86.0
Peyton Manning*	3	1,679	1,014	12,287	85	58	85.4
Mark Brunell*	8	2,672	1,608	19,212	106	66	85.1
Brad Johnson*	7	1,820	1,125	12,973	79	57	84.7
Jim Kelly	11	4,779	2,874	35,467	237	175	84.4
Roger Staubach	11	2,958	1,685	22,700	153	109	83.4
Neil Lomax	8	3,153	1,817	22,771	136	90	82.7
S. Jurgensen	18	4,262	2,433	32,224	255	189	82.63
Len Dawson (a)	19	3,741	2,136	28,711	239	183	82.56
Neil O'Donnell*	11	3,121	1,802	20,938	116	65	81.86
Ken Anderson	16	4,475	2,654	32,838	197	160	81.85
Bernie Kosar	12	3,365	1,994	23,301	124	87	81.82
Danny White	13	2,950	1,761	21,959	155	132	81.71
Elvis Grbac*	7	1,978	1,181	13,741	84	63	81.66
Troy Aikman*	12	4,715	2,898	32,942	165	141	81.61
Dave Krieg	19	5,311	3,105	38,147	261	199	81.49
R. Cunningham*	15	4,200	2,375	29,406	204	132	81.47

(1) Rating points based on performances in the following categories: Percentage of completions, percentage of touchdown passes, percentage of interceptions, and average gain per pass attempt.

Most Yards Gained, Career — 61,361, Dan Marino, Miami Dolphins, 1983-99.
Most Yards Gained, Season — 5,084, Dan Marino, Miami Dolphins, 1984.
Most Yards Gained, Game — 554, Norm Van Brocklin, L. A. Rams vs. N.Y. Yanks, Sept. 18, 1951 (27 completions in 41 attempts).
Most Touchdowns Passing, Career — 420, Dan Marino, Miami Dolphins, 1983-99.
Most Touchdowns Passing, Season — 48, Dan Marino, Miami Dolphins, 1984.
Most Touchdowns Passing, Game — 7, Sid Luckman, Chicago Bears vs. N.Y. Giants, Nov. 14, 1943; Adrian Burk, Phil. Eagles vs. Washington Redskins, Oct. 17, 1954; George Blanda, Houston Oilers vs. N.Y. Titans, Nov. 19, 1961; Y.A. Tittle, N.Y. Giants vs. Washington Redskins, Oct. 28, 1962; Joe Kapp, Minnesota Vikings vs. Baltimore Colts, Sept. 28, 1969.
Most Passes Completed, Career — 4,967, Dan Marino, Miami Dolphins, 1983-99.
Most Passes Completed, Season — 404, Warren Moon, Houston Oilers, 1991.
Most Passes Completed, Game — 45, Drew Bledsoe, New England Patriots vs. Minnesota Vikings, Nov. 13, 1994 (overtime).

The NFL All-Time Team

(Selected in 2000 by the Pro Football Hall of Fame voters.)

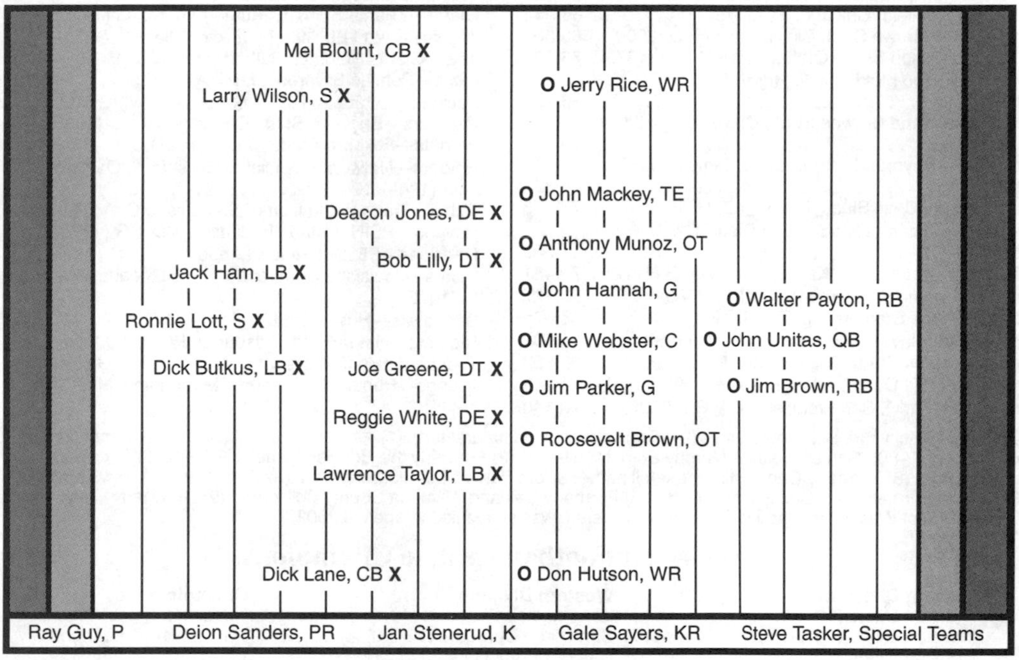

Pro Football Hall of Fame, Canton, Ohio

(Asterisks indicate 2001 inductees.)

Herb Adderley
Lance Alworth
Doug Atkins
Morris "Red" Badgro
Lem Barney
Cliff Battles
Sammy Baugh
Chuck Bednarik
Bert Bell
Bobby Bell
Raymond Berry
Charles Bidwill
Fred Biletnikoff
George Blanda
Mel Blount
Terry Bradshaw
Jim Brown
Paul Brown
Roosevelt Brown
Willie Brown
Buck Buchanan
*Nick Buoniconti
Dick Butkus
Earl Campbell
Tony Canadeo
Joe Carr
Guy Chamberlin
Jack Christiansen
Earl "Dutch" Clark
George Connor
Jim Conzelman
Lou Creekmur
Larry Csonka
Al Davis
Willie Davis
Len Dawson
Eric Dickerson
Dan Dierdorf
Mike Ditka
Art Donovan
Tony Dorsett
John "Paddy" Driscoll
Bill Dudley
Glen "Turk" Edwards
Weeb Ewbank
Tom Fears
Jim Finks
Ray Flaherty
Len Ford
Dr. Daniel Fortmann
Dan Fouts
Frank Gatski
Bill George
Joe Gibbs
Frank Gifford
Sid Gillman
Otto Graham
Red Grange
Bud Grant
Joe Greene
Forrest Gregg
Bob Griese
Lou Groza
Joe Guyon
George Halas
Jack Ham
John Hannah
Franco Harris
Mike Haynes
Ed Healey
Mel Hein
Ted Hendricks
Wilbur "Pete" Henry
Arnold Herber
Bill Hewitt
Clarke Hinkle
Elroy "Crazylegs" Hirsch
Paul Hornung
Ken Houston
Cal Hubbard
Sam Huff
Lamar Hunt
Don Hutson
Jimmy Johnson
John Henry Johnson
Charlie Joiner
David "Deacon" Jones
Stan Jones
Henry Jordan
Sonny Jurgensen
Leroy Kelly
Walt Kiesling
Frank "Bruiser" Kinard
Paul Krause
Earl "Curly" Lambeau
Jack Lambert
Tom Landry
Dick "Night Train" Lane
Jim Langer
Willie Lanier
Steve Largent
Yale Lary
Dante Lavelli
Bobby Layne
Alphonse "Tuffy" Leemans
*Marv Levy
Bob Lilly
Larry Little
Vince Lombardi
Howie Long
Ronnie Lott
Sid Luckman
Roy "Link" Lyman
Tom Mack
John Mackey
Tim Mara
Wellington Mara
Gino Marchetti
George Preston Marshall
Ollie Matson
Don Maynard
George McAfee
Mike McCormack
Tommy McDonald
Hugh McElhenny
Johnny "Blood" McNally
Mike Michalske
Wayne Millner
Bobby Mitchell
Ron Mix
Joe Montana
Lenny Moore
Marion Motley
*Mike Munchak
Anthony Munoz
George Musso
Bronko Nagurski
Joe Namath
Earle "Greasy" Neale
Ernie Nevers
Ozzie Newsome
Ray Nitschke
Chuck Noll
Leo Nomellini
Merlin Olsen
Jim Otto
Steve Owen
Alan Page
Clarence "Ace" Parker
Jim Parker
Walter Payton
Joe Perry
Pete Pihos
Hugh "Shorty" Ray
Dan Reeves
Mel Renfro
John Riggins
Jim Ringo
Andy Robustelli
Art Rooney
Dan Rooney
Pete Rozelle
Bob St. Clair
Gale Sayers
Joe Schmidt
Tex Schramm
Lee Roy Selmon
Billy Shaw
Art Shell
Don Shula
O.J. Simpson
Mike Singletary
*Jackie Slater
Jackie Smith
Bart Starr
Roger Staubach
Ernie Stautner
Jan Stenerud
Dwight Stephenson
Ken Strong
Joe Stydahar
*Lynn Swann
Fran Tarkenton
Charley Taylor
Jim Taylor
Lawrence "LT" Taylor
Jim Thorpe
Y.A. Tittle
George Trafton
Charley Trippi
Emlen Tunnell
Clyde "Bulldog" Turner
Johnny Unitas
Gene Upshaw
Norm Van Brocklin
Steve Van Buren
Doak Walker
Bill Walsh
Paul Warfield
Bob Waterfield
Mike Webster
Arnie Weinmeister
Randy White
Dave Wilcox
Bill Willis
Larry Wilson
Kellen Winslow
Alex Wojciechowicz
Willie Wood
*Ron Yary
*Jack Young

NFL Stadiums[1]

Team—Stadium, Location, Turf (Year Built)	Capacity
Bears—Soldier Field, Chicago, IL, G (1924)	66,944
Bengals—Paul Brown Stad, Cincinnati, OH, G (2000)	65,600
Bills—Ralph Wilson Stad., Orchard Park, NY, A (1973)	73,800
Broncos—Invesco Field at Mile High[1], Denver, CO, G (2001)	76,125
Browns—Cleveland Browns Stad., Cleveland, OH, G (1999)	73,000
Buccaneers—Raymond James Stad., Tampa, FL, G (1998)	66,321
Cardinals—Sun Devil Stad., Tempe, AZ, G (1958)	73,273
Chargers—Qualcomm Stad.[2], San Diego, CA, G (1967)	71,000
Chiefs—Arrowhead Stad., Kansas City, MO, G (1972)	79,451
Colts—RCA Dome [3], Indianapolis, IN, A (1983)	56,127
Cowboys—Texas Stad., Irving, TX, A (1971)	65,675
Dolphins—Pro Player Stad.[4], Miami, FL, G (1987)	75,192
Eagles—Veterans Stad., Philadelphia, PA, A (1971)	65,352
Falcons—Georgia Dome, Atlanta, GA, A (1992)	71,228
49ers—3Com Park[5], San Francisco, CA, G (1960)	70,140
Giants—Giants Stad., E. Rutherford, NJ, G (1976)	79,466
Jaguars—ALLTEL Stad.[6], Jacksonville, FL, G (1946)	73,000
Jets—Giants Stad., E. Rutherford, NJ, G (1976)	79,466
Lions—Pontiac Silverdome, MI, A (1975)	80,311
Packers—Lambeau Field[7], Green Bay, WI, G (1957)	60,890
Panthers—Ericsson Stad., Charlotte, NC, G (1996)	73,250
Patriots—Foxboro Stad., MA, G (1971)	60,292
Raiders—Network Associates Coliseum[8], Oakland, CA, G (1966)	63,132
Rams—Trans World Dome, St. Louis, MO, A (1995)	66,000
Ravens—PSINet Stad., Baltimore, MD, SG (1998)	69,354
Redskins—FEDEX Field[9], Landover, MD, G (1997)	80,116
Saints—Louisiana Superdome, New Orleans, A (1975)	70,200
Seahawks—Husky Stad.[10], Seattle, WA, G (1920)	72,500
Steelers—Heinz Field, Pittsburgh, PA, A (2001)	65,000 est.
Titans—Adelphia Coliseum, Nashville, TN, G (1999)	67,000
Vikings—Hubert H. Humphrey Metrodome, Minn., MN, A (1982)	64,121

G=Grass. A=Artificial turf. SG=Sport Grass (hybrid of artificial and natural turf). (1) As of the start of the 2001 season. (2) Formerly San Diego Stad. (1967-80), San Diego Jack Murphy Stad. (1981-97). (3) Formerly the Hoosier Dome (1983-94). (4) Formerly Joe Robbie Stad. (1987-96). (5) Formerly Candlestick Park; full name: 3Com Park at Candlestick Point. (6) Formerly Jacksonville Municipal Stad. (1946-97). (7) Formerly City Stadium (1957-65). (8) Formerly Oakland/Alameda County Coliseum. (9) Formerly Jack Kent Cooke Stad. (10) Univ. of Washington facility. Seahawks' new stad. was scheduled to open in 2002.

American Football League Champions

Year	Eastern Division	Western Division	Championship
1960	Houston Oilers (10-4-0)	Los Angeles Chargers (10-4-0)	Houston 24, Los Angeles 16
1961	Houston Oilers (10-3-1)	San Diego Chargers (12-2-0)	Houston 10, San Diego 3
1962	Houston Oilers (11-3-0)	Dallas Texans (11-3-0)	Dallas 20, Houston 17 (2 overtimes)
1963	Boston Patriots (7-6-1)(a)	San Diego Chargers (11-3-0)	San Diego 51, Boston 10
1964	Buffalo Bills (12-2-0)	San Diego Chargers (8-5-1)	Buffalo 20, San Diego 7
1965	Buffalo Bills (10-3-1)	San Diego Chargers (9-2-3)	Buffalo 23, San Diego 0
1966	Buffalo Bills (9-4-1)	Kansas City Chiefs (11-2-1)	Kansas City 31, Buffalo 7
1967	Houston Oilers (9-4-1)	Oakland Raiders (13-1-0)	Oakland 40, Houston 7
1968	New York Jets (11-3-0)	Oakland Raiders (12-2-0)(b)	New York 27, Oakland 23
1969	New York Jets (10-4-0)	Oakland Raiders (12-1-1)	Kansas City 17, Oakland 7(c)

(a) Defeated Buffalo Bills in divisional playoff. (b) Defeated Kansas City Chiefs in divisional playoff. (c) Kansas City Chiefs defeated N.Y. Jets and Oakland Raiders defeated Houston Oilers in divisional playoffs.

Future Sites of the Super Bowl

(Information subject to change.)

No.	Site	Date
XXXVI	Louisiana Superdome, New Orleans, LA	Feb. 3, 2002*
XXXVII	Qualcomm Stadium, San Diego, CA	Jan. 26, 2003
XXXVIII	Reliant Stadium, Houston, TX	Feb. 1, 2004
XXXIX	ALLTEL Stadium, Jacksonville, FL	Feb. 6, 2005
XL	Ford Field, Detroit, MI	Feb. 5, 2006

*Originally set for Jan. 27, the game was rescheduled due to a 1-week suspension of NFL play following the terrorist attacks on Sept. 11, 2001.

CANADIAN FOOTBALL LEAGUE

Grey Cup Championship Game, 1954-2000

1954	Edmonton Eskimos 26, Montreal Alouettes 25
1955	Edmonton Eskimos 34, Montreal Alouettes 19
1956	Edmonton Eskimos 50, Montreal Alouettes 27
1957	Hamilton Tiger-Cats 32, Winnipeg Blue Bombers 7
1958	Winnipeg Blue Bombers 35, Hamilton Tiger-Cats 28
1959	Winnipeg Blue Bombers 21, Hamilton Tiger-Cats 7
1960	Ottawa Rough Riders 16, Edmonton Eskimos 6
1961	Winnipeg Blue Bombers 21, Hamilton Tiger-Cats 14
1962	Winnipeg Blue Bombers 28, Hamilton Tiger-Cats 27
1963	Hamilton Tiger-Cats 21, British Columbia Lions 10
1964	British Columbia Lions 34, Hamilton Tiger-Cats 24
1965	Hamilton Tiger-Cats 22, Winnipeg Blue Bombers 16
1966	Saskatchewan Roughriders 29, Ottawa Rough Riders 14
1967	Hamilton Tiger-Cats 24, Saskatchewan Roughriders 1
1968	Ottawa Rough Riders 24, Calgary Stampeders 21
1969	Ottawa Rough Riders 29, Saskatchewan Roughriders 11
1970	Montreal Alouettes 23, Calgary Stampeders 10
1971	Calgary Stampeders 14, Toronto Argonauts 11
1972	Hamilton Tiger-Cats 13, Saskatchewan Roughriders 10
1973	Ottawa Rough Riders 22, Edmonton Eskimos 18
1974	Montreal Alouettes 20, Edmonton Eskimos 7
1975	Edmonton Eskimos 9, Montreal Alouettes 8
1976	Ottawa Rough Riders 23, Saskatchewan Roughriders 20
1977	Montreal Alouettes 41, Edmonton Eskimos 6
1978	Edmonton Eskimos 20, Montreal Alouettes 13
1979	Edmonton Eskimos 17, Montreal Alouettes 9
1980	Edmonton Eskimos 48, Hamilton Tiger-Cats 10
1981	Edmonton Eskimos 26, Ottawa Rough Riders 23
1982	Edmonton Eskimos 32, Toronto Argonauts 16
1983	Toronto Argonauts 18, British Columbia Lions 17
1984	Winnipeg Blue Bombers 47, Hamilton Tiger-Cats 17
1985	British Columbia Lions 37, Hamilton Tiger-Cats 24
1986	Hamilton Tiger-Cats 39, Edmonton Eskimos 15
1987	Edmonton Eskimos 38, Toronto Argonauts 36
1988	Winnipeg Blue Bombers 22, British Columbia Lions 21
1989	Saskatchewan Roughriders 43, Hamilton Tiger-Cats 40
1990	Winnipeg Blue Bombers 50, Edmonton Eskimos 11
1991	Toronto Argonauts 36, Calgary Stampeders 21
1992	Calgary Stampeders 24, Winnipeg Blue Bombers 10
1993	Edmonton Eskimos 33, Winnipeg Blue Bombers 23
1994	British Columbia Lions 26, Baltimore Football Club* 23
1995	Baltimore Stallions 37, Calgary Stampeders 20
1996	Toronto Argonauts 43, Edmonton Eskimos 37
1997	Toronto Argonauts 47, Saskatchewan Roughriders 23
1998	Calgary Stampeders 26, Hamilton Tiger-Cats 24
1999	Hamilton Tiger-Cats 32, Calgary Stampeders 21
2000	British Columbia Lions 28, Montreal Alouettes 26

*Later Baltimore Stallions.

COLLEGE FOOTBALL

Undefeated Oklahoma Wins Orange Bowl, 2000 National Title

The top-ranked and undefeated Oklahoma Sooners beat the defending champion Florida State Seminoles, 13-2, in the Orange Bowl in Miami, Jan. 3, 2001, to clinch their 7th national title. The Oklahoma victory kept the Bowl Championship Series'(BCS) string of undisputed national champions intact. An FSU win might have meant a shared title with the Sugar Bowl champion Miami (FL) Hurricanes. The anticipated shootout between the nation's top 2 quarterbacks (FSU's Heisman Trophy winner, Chris Weinke, and OU's Heisman runner-up and AP Player of the Year, Josh Heupel) never materialized. The Sooner defense, led by Orange Bowl MVP linebacker Torrance Marshall, dominated the heavily favored Seminoles. Florida St., a team that averaged 42 points per game during the season, was held to a safety, scored with 55 seconds remaining.

National College Football Champions, 1936-2000

The unofficial champion as selected by the AP poll of writers and USA Today/ESPN (until 1991, UPI; 1991-1996 USA Today/CNN) poll of coaches. In years the polls disagreed, both teams are listed (AP winner first). The AP poll started in 1936; the UPI poll in 1950.

1936 Minnesota
1937 Pittsburgh
1938 Texas Christian
1939 Texas A&M
1940 Minnesota
1941 Minnesota
1942 Ohio St.
1943 Notre Dame
1944 Army
1945 Army
1946 Notre Dame
1947 Notre Dame
1948 Michigan
1949 Notre Dame
1950 Oklahoma
1951 Tennessee
1952 Michigan St.
1953 Maryland
1954 Ohio St., UCLA
1955 Oklahoma
1956 Oklahoma
1957 Auburn, Ohio St.
1958 Louisiana St.
1959 Syracuse
1960 Minnesota
1961 Alabama
1962 USC
1963 Texas
1964 Alabama
1965 Alabama, Mich. St.
1966 Notre Dame
1967 USC
1968 Ohio St.
1969 Texas
1970 Nebraska, Texas
1971 Nebraska
1972 USC
1973 Notre Dame, Alabama
1974 Oklahoma, USC
1975 Oklahoma
1976 Pittsburgh
1977 Notre Dame
1978 Alabama, USC
1979 Alabama
1980 Georgia
1981 Clemson
1982 Penn St.
1983 Miami (FL)
1984 Brigham Young
1985 Oklahoma
1986 Penn St.
1987 Miami (FL)
1988 Notre Dame
1989 Miami (FL)
1990 Colorado, Georgia Tech
1991 Miami (FL), Washington
1992 Alabama
1993 Florida St.
1994 Nebraska
1995 Nebraska
1996 Florida
1997 Michigan, Nebraska
1998 Tennessee
1999 Florida St.
2000 Oklahoma

2000 Final Associated Press and USA Today/ESPN NCAA Football Polls

Associated Press Rankings

1. Oklahoma (13-0)
2. Miami (FL) (11-1)
3. Washington (11-1)
4. Oregon St. (11-1)
5. Florida St. (11-2)
6. Virginia Tech (11-1)
7. Oregon (10-2)
8. Nebraska (10-2)
9. Kansas St. (11-3)
10. Florida (10-3)
11. Michigan (9-3)
12. Texas (9-3)
13. Purdue (8-4))
14. Colorado St. (10-2)
15. Notre Dame (9-3)
16. Clemson (9-3)
17. Georgia Tech (9-3)
18. Auburn (9-4)
19. S. Carolina (8-4)
20. Georgia (8-4)
21. TCU (10-2)
22. LSU (8-4)
23. Wisconsin (9-4)
24. Mississippi St. (8-4)
25. Iowa St. (9-3)

USA Today/ESPN Rankings

1. Oklahoma
2. Miami (FL)
3. Washington
4. Florida St.
5. Oregon St.
6. Virginia Tech
7. Nebraska
8. Kansas St.
9. Oregon
10. Michigan
11. Florida
12. Texas
13. Purdue
14. Clemson
15. Colorado St.
16. Notre Dame
17. Georgia
18. TCU
19. Georgia Tech
20. Auburn
21. South Carolina
22. Mississippi St.
23. Iowa St.
24. Wisconsin
25. Tennessee

Note: Team records include bowl games. Won-loss records for team in USA Today/ESPN poll are in AP poll, except Tennessee (8-4).

Annual Results of Major Bowl Games

(Dates indicate year the game was played; bowl games are generally played in late December or early January.)

Rose Bowl, Pasadena, CA

1902 (Jan.) Michigan 49, Stanford 0
1916 Washington St. 14, Brown 0
1917 Oregon 14, Pennsylvania 0
1918-19 Service teams
1920 Harvard 7, Oregon 6
1921 California 28, Ohio St. 0
1922 Wash. & Jeff. 0, California 0
1923 USC 14, Penn St. 3
1924 Navy 14, Washington 14
1925 Notre Dame 27, Stanford 10
1926 Alabama 20, Washington 19
1927 Alabama 7, Stanford 7
1928 Stanford 7, Pittsburgh 6
1929 Georgia Tech 8, California 7
1930 USC 47, Pittsburgh 14
1931 Alabama 24, Wash. St. 0
1932 USC 21, Tulane 12
1933 USC 35, Pittsburgh 0
1934 Columbia 7, Stanford 0
1935 Alabama 29, Stanford 13
1936 Stanford 7, SMU 0
1937 Pittsburgh 21, Washington 0
1938 California 13, Alabama 0
1939 USC 7, Duke 3
1940 USC 14, Tennessee 0
1941 Stanford 21, Nebraska 13
1942* Oregon St. 20, Duke 16
1943 Georgia 9, UCLA 0
1944 USC 29, Washington 0
1945 USC 25, Tennessee 0
1946 Alabama 34, USC 14
1947 Illinois 45, UCLA 14
1948 Michigan 49, USC 0
1949 Northwestern 20, California 14
1950 Ohio St. 17, California 14
1951 Michigan 14, California 6
1952 Illinois 40, Stanford 7
1953 USC 7, Wisconsin 0
1954 Mich. St. 28, UCLA 20
1955 Ohio St. 20, USC 7
1956 Mich. St. 17, UCLA 14
1957 Iowa 35, Oregon St. 19
1958 Ohio St. 10, Oregon 7
1959 Iowa 38, California 12
1960 Washington 44, Wisconsin 8
1961 Washington 17, Minnesota 7
1962 Minnesota 21, UCLA 3
1963 USC 42, Wisconsin 37
1964 Illinois 17, Washington 7
1965 Michigan 34, Oregon St. 7
1966 UCLA 14, Mich. St. 12
1967 Purdue 14, USC 13
1968 USC 14, Indiana 3
1969 Ohio St. 27, USC 16
1970 USC 10, Michigan 3
1971 Stanford 27, Ohio St. 17
1972 Stanford 13, Michigan 12
1973 USC 42, Ohio St. 17
1974 Ohio St. 42, USC 21
1975 USC 18, Ohio St. 17
1976 UCLA 23, Ohio St. 10
1977 USC 14, Michigan 6
1978 Washington 27, Michigan 20
1979 USC 17, Michigan 10
1980 USC 17, Ohio St. 16
1981 Michigan 23, Washington 6
1982 Washington 28, Iowa 0
1983 UCLA 24, Michigan 14
1984 UCLA 45, Illinois 9
1985 USC 20, Ohio St. 17
1986 UCLA 45, Iowa 28
1987 Arizona St. 22, Michigan 15
1988 Mich. St. 20, USC 17
1989 Michigan 22, USC 14
1990 USC 17, Michigan 10
1991 Washington 46, Iowa 34
1992 Washington 34, Michigan 14
1993 Michigan 38, Washington 31
1994 Wisconsin 21, UCLA 16
1995 Penn St. 38, Oregon 20
1996 USC 41, Northwestern 32
1997 Ohio St. 20, Arizona St. 17
1998 Michigan 21, Wash. St. 16
1999 Wisconsin 38, UCLA 31
2000 Wisconsin 17, Stanford 9
2001 Washington 34, Purdue 24

*Played at Durham, NC.

Orange Bowl, Miami, FL

1935 (Jan.) Bucknell 26, Miami (FL) 0
1936 Catholic U. 20, Mississippi 19
1937 Duquesne 13, Mississippi St. 12
1938 Auburn 6, Michigan St. 0
1939 Tennessee 17, Oklahoma 0
1940 Georgia Tech 21, Missouri 7
1941 Mississippi St. 14, Georgetown 7
1942 Georgia 40, TCU 26
1943 Alabama 37, Boston Coll. 21
1944 LSU 19, Texas A&M 14
1945 Tulsa 26, Georgia Tech 12
1946 Miami (FL) 13, Holy Cross 6
1947 Rice 8, Tennessee 0
1948 Georgia Tech 20, Kansas 14
1949 Texas 41, Georgia 28
1950 Santa Clara 21, Kentucky 13
1951 Clemson 15, Miami (FL) 14
1952 Georgia Tech 17, Baylor 14
1953 Alabama 61, Syracuse 6
1954 Oklahoma 7, Maryland 0
1955 Duke 34, Nebraska 7
1956 Oklahoma 20, Maryland 6
1957 Colorado 27, Clemson 21
1958 Oklahoma 48, Duke 21
1959 Oklahoma 21, Syracuse 6
1960 Georgia 14, Missouri 0
1961 Missouri 21, Navy 14
1962 LSU 25, Colorado 7
1963 Alabama 17, Oklahoma 0
1964 Nebraska 13, Auburn 7
1965 Texas 21, Alabama 17
1966 Alabama 39, Nebraska 28
1967 Florida 27, Georgia Tech 12
1968 Oklahoma 26, Tennessee 24
1969 Penn St. 15, Kansas 14
1970 Penn St. 10, Missouri 3
1971 Nebraska 17, LSU 12
1972 Nebraska 38, Alabama 6
1973 Nebraska 40, Notre Dame 6
1974 Penn St. 16, LSU 9
1975 Notre Dame 13, Alabama 11
1976 Oklahoma 14, Michigan 6
1977 Ohio St. 27, Colorado 10
1978 Arkansas 31, Oklahoma 6
1979 Oklahoma 31, Nebraska 24
1980 Oklahoma 24, Florida St. 7
1981 Oklahoma 18, Florida St. 17
1982 Clemson 22, Nebraska 15
1983 Nebraska 21, LSU 20
1984 Miami (FL) 31, Nebraska 30
1985 Washington 28, Oklahoma 17
1986 Oklahoma 25, Penn St. 10
1987 Oklahoma 42, Arkansas 8
1988 Miami (FL) 20, Oklahoma 14
1989 Miami (FL) 23, Nebraska 3
1990 Notre Dame 21, Colorado 6
1991 Colorado 10, Notre Dame 9
1992 Miami (FL) 22, Nebraska 0
1993 Florida St. 27, Nebraska 14
1994 Florida St. 18, Nebraska 16
1995 Nebraska 24, Miami (FL) 17
1996 Florida St. 31, Notre Dame 26
1996 (Dec.) Nebraska 41, Virginia Tech 21
1998 (Jan.) Nebraska 42, Tennessee 17
1999 Florida 31, Syracuse 10
2000 Michigan 35, Alabama 34 (OT)
2001 Oklahoma 13, Florida St. 2

Sugar Bowl, New Orleans, LA

1935 (Jan.) Tulane 20, Temple 14
1936 TCU 3, LSU 2
1937 Santa Clara 21, LSU 14
1938 Santa Clara 6, LSU 0
1939 TCU 15, Carnegie Tech 7
1940 Texas A&M 14, Tulane 13
1941 Boston Col. 19, Tennessee 13
1942 Fordham 2, Missouri 0
1943 Tennessee 14, Tulsa 7
1944 Georgia Tech 20, Tulsa 18
1945 Duke 29, Alabama 26
1946 Oklahoma A&M 33, St. Mary's 13
1947 Georgia 20, N. Carolina 10
1948 Texas 27, Alabama 7
1949 Oklahoma 14, N. Carolina 6
1950 Oklahoma 35, LSU 0
1951 Kentucky 13, Oklahoma 7
1952 Maryland 28, Tennessee 13
1953 Georgia Tech 24, Mississippi 7
1954 Georgia Tech 42, West Virginia 19
1955 Navy 21, Mississippi 0
1956 Georgia Tech 7, Pittsburgh 0
1957 Baylor 13, Tennessee 7
1958 Mississippi 39, Texas 7
1959 LSU 7, Clemson 0
1960 Mississippi 21, LSU 0
1961 Mississippi 14, Rice 6
1962 Alabama 10, Arkansas 3
1963 Mississippi 17, Arkansas 13
1964 Alabama 12, Mississippi 7
1965 LSU 13, Syracuse 10
1966 Missouri 20, Florida 18
1967 Alabama 34, Nebraska 7
1968 LSU 20, Wyoming 13
1969 Arkansas 16, Georgia 2
1970 Mississippi 27, Arkansas 22
1971 Tennessee 34, Air Force 13
1972 Oklahoma 40, Auburn 22
1972* (Dec.) Oklahoma 14, Penn St. 0
1973 Notre Dame 24, Alabama 23
1974 Nebraska 13, Florida 10
1975 Alabama 13, Penn St. 6
1977 (Jan.) Pittsburgh 27, Georgia 3
1978 Alabama 35, Ohio St. 6
1979 Alabama 14, Penn St. 7
1980 Alabama 24, Arkansas 9
1981 Georgia 17, Notre Dame 10
1982 Pittsburgh 24, Georgia 20
1983 Penn St. 27, Georgia 23
1984 Auburn 9, Michigan 7
1985 Nebraska 28, LSU 10
1986 Tennessee 35, Miami (FL) 7
1987 Nebraska 30, LSU 15
1988 Syracuse 16, Auburn 16
1989 Florida St. 13, Auburn 7
1990 Miami (FL) 33, Alabama 25
1991 Tennessee 23, Virginia 22
1992 Notre Dame 39, Florida 28
1993 Alabama 34, Miami (FL) 13
1994 Florida 41, West Virginia 7
1995 Florida St. 23, Florida 17
1995 (Dec.) Virginia Tech 28, Texas 10
1997 (Jan.) Florida 52, Florida St. 20
1998 Florida St. 31, Ohio St. 14
1999 Ohio St. 24, Texas A&M 14
2000 Florida St. 46, Virginia Tech 29
2001 Miami (FL) 37, Florida 20

* Penn St. awarded game by forfeit.

Fiesta Bowl, Tempe, AZ

1971 (Dec.) Arizona St. 45, Florida St. 38
1972 Arizona St. 49, Missouri 35
1973 Arizona St. 28, Pittsburgh 7
1974 Okla. St. 16, Brigham Young 6
1975 Arizona St. 17, Nebraska 14
1976 Oklahoma 41, Wyoming 7
1977 Penn St. 42, Arizona St. 30
1978 UCLA 10, Arkansas 10
1979 Pittsburgh 16, Arizona 10
1980 Penn St. 31, Ohio St. 19
1982 (Jan.) Penn St. 26, USC 10
1983 Arizona St. 32, Oklahoma 21
1984 Ohio St. 28, Pittsburgh 23
1985 UCLA 39, Miami (FL) 37
1986 Michigan 27, Nebraska 23
1987 Penn St. 14, Miami (FL) 10
1988 Florida St. 31, Nebraska 28
1989 Notre Dame 34, W. Virginia 21
1990 Florida St. 41, Nebraska 17
1991 Louisville 34, Alabama 7
1992 Penn St. 42, Tennessee 17
1993 Syracuse 26, Colorado 22
1994 Arizona 29, Miami (FL) 0
1995 Colorado 41, Notre Dame 24
1996 Nebraska 62, Florida 24
1997 Penn St. 38, Texas 15
1997 (Dec.) Kansas St. 35, Syracuse 18
1999 (Jan.) Tennessee 23, Florida St. 16
2000 Nebraska 31, Tennessee 21
2001 Oregon St. 41, Notre Dame 9

Cotton Bowl, Dallas, TX

1937 (Jan.) TCU 16, Marquette 6
1938 Rice 28, Colorado 14
1939 St. Mary's 20, Texas Tech 13
1940 Clemson 6, Boston Coll. 3
1941 Texas A&M 13, Fordham 12
1942 Alabama 29, Texas A&M 21
1943 Texas 14, Georgia Tech 7
1944 Randolph Field 7, Texas 7
1945 Oklahoma A&M 34, TCU 0
1946 Texas 40, Missouri 27
1947 Arkansas 0, LSU 0
1948 SMU 13, Penn St. 13
1949 SMU 21, Oregon 13
1950 Rice 27, North Carolina 13
1951 Tennessee 20, Texas 14
1952 Kentucky 20, TCU 7
1953 Texas 16, Tennessee 0
1954 Rice 28, Alabama 6
1955 Georgia Tech 14, Arkansas 6
1956 Mississippi 14, TCU 13
1957 TCU 28, Syracuse 27
1958 Navy 20, Rice 7
1959 TCU 0, Air Force 0
1960 Syracuse 23, Texas 14
1961 Duke 7, Arkansas 6
1962 Texas 12, Mississippi 7
1963 LSU 13, Texas 0
1964 Texas 28, Navy 6
1965 Arkansas 10, Nebraska 7
1966 LSU 14, Arkansas 7
1966 (Dec.) Georgia 24, SMU 9
1968 (Jan.) Texas A&M 20, Alabama 16
1969 Texas 36, Tennessee 13
1970 Texas 21, Notre Dame 17
1971 Notre Dame 24, Texas 11
1972 Penn St. 30, Texas 6
1973 Texas 17, Alabama 13
1974 Nebraska 19, Texas 3
1975 Penn St. 41, Baylor 20
1976 Arkansas 31, Georgia 10
1977 Houston 30, Maryland 21
1978 Notre Dame 38, Texas 10
1979 Notre Dame 35, Houston 34
1980 Houston 17, Nebraska 14
1981 Alabama 30, Baylor 2
1982 Texas 14, Alabama 12
1983 SMU 7, Pittsburgh 3
1984 Georgia 10, Texas 9
1985 Boston Coll. 45, Houston 28
1986 Texas A&M 36, Auburn 16
1987 Ohio St. 28, Texas A&M 12
1988 Texas A&M 35, Notre Dame 10
1989 UCLA 17, Arkansas 3
1990 Tennessee 31, Arkansas 27
1991 Miami (FL) 46, Texas 3
1992 Florida St. 10, Texas A&M 2
1993 Notre Dame 28, Texas A&M 3
1994 Notre Dame 24, Texas A&M 21
1995 USC. 55, Texas Tech 14
1996 Colorado 38, Oregon 6
1997 Brigham Young 19, Kansas St. 15
1998 UCLA 29, Texas A&M 23
1999 Texas 38, Mississippi St. 11
2000 Arkansas 27, Texas 6
2001 Kansas St. 35, Tennessee 21

Sun Bowl, El Paso, TX (John Hancock Bowl, 1989-93)

1936 (Jan.) Hardin-Simmons 14, New Mexico St. 14
1937 Hardin-Simmons 34, Texas Mines 6
1938 West Virginia 7, Texas Tech 6
1939 Utah 26, New Mexico 0
1940 Catholic U. 0, Arizona St. 0
1941 Western Reserve 26, Arizona St. 13
1942 Tulsa 6, Texas Tech 0
1943 2d Air Force 13, Hardin-Simmons 7
1944 Southwestern (TX) 7, New Mexico 0
1945 Southwestern (TX) 35, Univ. of Mexico 0
1946 New Mexico 34, Denver 24
1947 Cincinnati 18, Virginia Tech 6
1948 Miami (OH) 13, Texas Tech 12
1949 West Virginia 21, Texas Mines 12
1950 Texas Western 33, Georgetown 20
1951 West Texas St. 14, Cincinnati 13
1952 Texas Tech 25, Pacific (CA) 14
1953 Pacific (CA) 26, S. Mississippi 7
1954 Texas Western 37, S. Miss. 14
1955 Texas Western 47, Florida St. 20
1956 Wyoming 21, Texas Tech 14
1957 Geo. Washington 13, TX Western 0
1958 Louisville 34, Drake 20
1958 (Dec.) Wyoming 14, Hardin-Simmons 6
1959 New Mexico St. 28, N. Texas St. 8
1960 New Mexico St. 20, Utah St. 13
1961 Villanova 17, Wichita 9
1962 West Texas St. 15, Ohio U. 14
1963 Oregon 21, SMU 14
1964 Georgia 7, Texas Tech 0
1965 Texas Western 13, TCU 12
1966 Wyoming 28, Florida St. 20
1967 UTEP 14, Mississippi 7
1968 Auburn 34, Arizona 10
1969 Nebraska 45, Georgia 6
1970 Georgia Tech. 17, Texas Tech 9
1971 LSU 33, Iowa St. 15
1972 North Carolina 32, Texas Tech 28
1973 Missouri 34, Auburn 17
1974 Mississippi St. 26, North Carolina 24
1975 Pittsburgh 33, Kansas 19
1977 (Jan.) Texas A&M 37, Florida 14
1977 (Dec.) Stanford 24, LSU 14
1978 Texas 42, Maryland 0
1979 Washington 14, Texas 7
1980 Nebraska 31, Mississippi St. 17
1981 Oklahoma 40, Houston 14
1982 North Carolina 26, Texas 10
1983 Alabama 28, SMU 7
1984 Maryland 28, Tennessee 27
1985 Georgia 13, Arizona 13
1986 Alabama 28, Washington 6
1987 Oklahoma St. 35, West Virginia 33
1988 Alabama 29, Army 28
1989 Pittsburgh 31, Texas A&M 28
1990 Michigan St. 17, USC 16
1991 UCLA 6, Illinois 3
1992 Baylor 20, Arizona 15
1993 Oklahoma 41, Texas Tech 10
1994 Texas 35, North Carolina 31
1995 Iowa 38, Washington 18
1996 Stanford 38, Michigan St. 0
1997 Arizona St. 17, Iowa 7
1998 TCU 28, USC 19
1999 Oregon 24, Minnesota 20
2000 Wisconsin 21, UCLA 20

Gator Bowl, Jacksonville, FL

1946 (Jan.) Wake Forest 26, S. Carolina 14
1947 Oklahoma 34, N. Carolina St. 13
1948 Maryland 20, Georgia 20
1949 Clemson 24, Missouri 23
1950 Maryland 20, Missouri 7
1951 Wyoming 20, Washington & Lee 7
1952 Miami (FL) 14, Clemson 0
1953 Florida 14, Tulsa 13
1954 Texas Tech 35, Auburn 13
1954 (Dec.) Auburn 33, Baylor 13
1955 Vanderbilt 25, Auburn 13
1956 Georgia Tech 21, Pittsburgh 14
1957 Tennessee 3, Texas A&M 0
1958 Mississippi 7, Florida 3
1960 (Jan.) Arkansas 14, Georgia Tech 7
1960 (Dec.) Florida 13, Baylor 12
1961 Penn St. 30, Georgia Tech 15
1962 Florida 17, Penn St. 7
1963 N. Carolina 35, Air Force 0
1965 (Jan.) Florida St. 36, Okla.19
1965 (Dec.) GA Tech 31, Texas Tech 21
1966 Tennessee 18, Syracuse 12
1967 Penn St. 17, Florida St. 17
1968 Missouri 35, Alabama 10
1969 Florida 14, Tennessee 13
1971 (Jan.) Auburn 35, Mississippi 28
1971 (Dec.) Georgia 7, N. Carolina 3
1972 Auburn 24, Colorado 3
1973 Texas Tech 28, Tennessee 19
1974 Auburn 27, Texas 3
1975 Maryland 13, Florida 0
1976 Notre Dame 20, Penn St. 9
1977 Pittsburgh 34, Clemson 3
1978 Clemson 17, Ohio St. 15
1979 N. Carolina 17, Michigan 15
1980 Pittsburgh 37, S. Carolina 9
1981 N. Carolina 31, Arkansas 27
1982 Florida St. 31, West Virginia 12
1983 Florida 14, Iowa 6
1984 Oklahoma St. 21, S. Carolina 14
1985 Florida St. 34, Oklahoma St. 23
1986 Clemson 27, Stanford 21
1987 LSU 30, S. Carolina 13
1989 (Jan.) Georgia 34, Michigan St. 27
1989 (Dec.) Clemson 27, W. Virginia 7
1991 (Jan.) Michigan 35, Mississippi 3
1991 (Dec.) Oklahoma 48, Virginia 14
1992 Florida 27, N. Carolina St. 10
1993 Alabama 24, N. Carolina 10
1994 Tennessee 45, Virginia Tech 23
1996 (Jan.) Syracuse 41, Clemson 0
1997 N. Carolina 20, W. Virginia 13
1998 N. Carolina 42, Virginia Tech 3
1999 Georgia Tech 35, Notre Dame 28
2000 Miami (FL) 28, Georgia Tech 13
2001 Virginia Tech 41, Clemson 20

Liberty Bowl, Memphis, TN

1959 (Dec.) Penn St. 7, Alabama 0
1960 Penn St. 41, Oregon 12
1961 Syracuse 15, Miami (FL) 14
1962 Oregon St. 6, Villanova 0
1963 Mississippi St. 16, N. Carolina St. 12
1964 Utah 32, West Virginia 6
1965 Mississippi 13, Auburn 7
1966 Miami (FL) 14, Virginia Tech 7
1967 N. Carolina St. 14, Georgia 7
1968 Mississippi 34, Virginia Tech 17
1969 Colorado 47, Alabama 33
1970 Tulane 17, Colorado 3
1971 Tennessee 14, Arkansas 13
1972 Georgia Tech 31, Iowa St. 30
1973 N. Carolina St. 31, Kansas 18
1974 Tennessee 7, Maryland 3
1975 USC 20, Texas A&M 0
1976 Alabama 36, UCLA 6
1977 Nebraska 21, N. Carolina 17
1978 Missouri 20, LSU 15
1979 Penn St. 9, Tulane 6
1980 Purdue 28, Missouri 25
1981 Ohio St. 31, Navy 28
1982 Alabama 21, Illinois 15
1983 Notre Dame 19, Boston Coll. 18
1984 Auburn 21, Arkansas 15
1985 Baylor 21, LSU 7
1986 Tennessee 21, Minnesota 14
1987 Georgia 20, Arkansas 17
1988 Indiana 34, S. Carolina 10
1989 Mississippi 42, Air Force 29
1990 Air Force 23, Ohio St. 11
1991 Air Force 38, Mississippi St. 15
1992 Mississippi 13, Air Force 0
1993 Louisville 18, Michigan St. 7
1994 Illinois 30, East Carolina 0
1995 East Carolina 19, Stanford 13
1996 Syracuse 30, Houston 17
1997 So. Mississippi 41, Pittsburgh 7
1998 Tulane 41, Brigham Young 27
1999 So. Mississippi 23, Colorado St. 17
2000 Colorado St. 22, Louisville 17

Florida Citrus Bowl, Orlando, FL (Tangerine Bowl Until 1983)

1947 (Jan.) Catawba 31, Maryville 6
1948 Catawba 7, Marshall 0
1949 Murray St. 21, Sul Ross St. 21
1950 St. Vincent 7, Emory & Henry 6
1951 Morris Harvey 35, Emory & Henry 14
1952 Stetson 35, Arkansas St. 20
1953 East Texas St. 33, Tenn. Tech 0
1954 East Texas St. 7, Arkansas St. 7
1955 Neb.-Omaha 7, E. Kentucky 6
1956 Juniata 6, Missouri Valley 6
1957 West Texas St. 20, So. Miss. 13
1958 East Texas St. 10, So. Miss. 9
1958 (Dec.) East Texas St. 26, Missouri Valley 7
1960 (Jan.) Middle Tennessee 21, Presbyterian 12
1960 (Dec.) Citadel 27, Tenn. Tech 0
1961 Lamar 21, Middle Tennessee 14
1962 Houston 49, Miami (OH) 21
1963 Western Ky. 27, Coast Guard 0
1964 E. Carolina 14, Massachusetts 13
1965 E. Carolina 31, Maine 0
1966 Morgan St. 14, West Chester 6
1967 Tenn.-Martin 25, West Chester 8
1968 Richmond 49, Ohio U. 42
1969 Toledo 56, Davidson 33
1970 Toledo 40, William & Mary 12
1971 Toledo 28, Richmond 3
1972 Tampa 21, Kent St. 18
1973 Miami (OH) 16, Florida 7
1974 Miami (OH) 21, Georgia 10
1975 Miami (OH) 20, S. Carolina 7
1976 Okla. St. 49, Brigham Young 21
1977 Florida St. 40, Texas Tech 17
1978 N. Carolina St. 30, Pittsburgh 17
1979 LSU 34, Wake Forest 10
1980 Florida 35, Maryland 20
1981 Missouri 19, So. Mississippi 17
1982 Auburn 33, Boston College 26
1983 Tennessee 30, Maryland 23
1984 Georgia 17, Florida St. 17
1985 Ohio St. 10, Brigham Young 7
1987 (Jan.) Auburn 16, USC 7
1988 Clemson 35, Penn St. 10
1989 Clemson 13, Oklahoma 6
1990 Illinois 31, Virginia 21
1991 Georgia Tech 45, Nebraska 21
1992 California 37, Clemson 13
1993 Georgia 21, Ohio St. 14
1994 Penn St. 31, Tennessee 13
1995 Alabama 24, Ohio St. 17
1996 Tennessee 20, Ohio St. 14
1997 Tennessee 48, Northwestern 28
1998 Florida 21, Penn St. 6
1999 Michigan 45, Arkansas 31
2000 Michigan St. 37, Florida 34
2001 Michigan 31, Auburn 28

Peach Bowl, Atlanta, GA

1968 (Dec.) LSU 31, Florida St. 27
1969 W. Virginia 14, S. Carolina 3
1970 Arizona St. 48, N. Carolina 26
1971 Mississippi 41, Georgia Tech 18
1972 N. Carolina St. 49, W. Virginia 13
1973 Georgia 17, Maryland 16
1974 Vanderbilt 6, Texas Tech 6
1975 W. Virginia 13, N. Carolina St. 10
1976 Kentucky 21, N. Carolina 0
1977 N. Carolina St. 24, Iowa St. 14
1978 Purdue 41, Georgia Tech. 21
1979 Baylor 24, Clemson 18
1981 (Jan.) Miami (FL) 20, Virginia Tech 10
1981 (Dec.) W. Virginia 26, Florida 6
1982 Iowa 28, Tennessee 22
1983 Florida St. 28, N. Carolina 3
1984 Virginia 27, Purdue 22
1985 Army 31, Illinois 29
1986 Va. Tech 25, N. Carolina St. 24
1988 (Jan.) Tennessee 28, Indiana 22
1988 (Dec.) N. Carolina St. 28, Iowa 23
1989 Syracuse 19, Georgia 18
1990 Auburn 27, Indiana 23
1992 (Jan.) E. Carolina 37, NC St. 34
1993 N. Carolina 21, Mississippi St. 17
1993 (Dec.) Clemson 14, Kentucky 13
1995 (Jan.) N. Carolina St. 28, Miss. St. 24
1995 (Dec.) Virginia 34, Georgia 27
1996 LSU 10, Clemson 7
1998 (Jan.) Auburn 21, Clemson 17
1998 (Dec.) Georgia 35, Virginia 33
1999 Mississippi St. 27, Clemson 7
2000 LSU 28, Georgia Tech 14

Holiday Bowl, San Diego, CA

1978 (Dec.) Navy 23, Brigham Young 16
1979 Indiana 38, BrighamYoung 37
1980 Brigham Young 46, SMU 45
1981 Brigham Young 38, Washington St. 36
1982 Ohio St. 47, Brigham Young 17
1983 Brigham Young 21, Missouri 17
1984 Brigham Young 24, Michigan 17
1985 Arkansas 18, Arizona St. 17
1986 Iowa 39, San Diego St. 38
1987 Iowa 20, Wyoming 19
1988 Oklahoma St. 62, Wyoming 14
1989 Penn St. 50, Brigham Young 39
1990 Texas A&M 65, Brigham Young 14
1991 Iowa 13, Brigham Young 13
1992 Hawaii 27, Illinois 17
1993 Ohio St. 28, Brigham Young 21
1994 Michigan 24, Colorado St. 14
1995 Kansas St. 54, Colorado St. 21
1996 Colorado 33, Washington 21
1997 Colorado St. 35, Missouri 24
1998 Arizona 23, Nebraska 20
1999 Kansas St. 24, Washington 20
2000 Oregon 35, Texas 30

Aloha Bowl, Honolulu, HI

1982 (Dec.) Washington 21, Md. 20
1983 Penn St. 13, Washington 10
1984 SMU 27, Notre Dame 20
1985 Alabama 24, USC 3
1986 Arizona 30, North Carolina 21
1987 UCLA 20, Florida 16
1988 Washington St. 24, Houston 22
1989 Michigan St. 33, Hawaii 13
1990 Syracuse 28, Arizona 0
1991 Georgia Tech 18, Stanford 17
1992 Kansas 23, Brigham Young 20
1993 Colorado 41, Fresno St. 30
1994 Boston Coll. 12, Kansas St. 7
1995 Kansas 51, UCLA 30
1996 Navy 42, California 38
1997 Washington 51, Michigan St. 23
1998 Colorado 51, Oregon 23
1999 Wake Forest 23, Arizona St. 3
2000 Boston Coll. 31, Arizona St. 17

Oahu Bowl, Honolulu, HI

1998 Air Force 45, Washington 25
1999 Hawaii 23, Oregon St. 17
2000 Georgia 37, Virginia 14

Other Bowl Results in 2000-2001

Alamo Bowl, San Antonio, TX: Nebraska 66, Northwestern 17
Galleryfurniture.com Bowl, Houston, TX: E. Carolina 40, Texas Tech 27
Humanitarian Bowl, Boise, ID: Boise St. 38, UTEP 23
Independence Bowl, Shreveport, LA: Miss. St. 43, Texas A&M 41 (OT)
Insight.com Bowl, Tucson, AZ: Iowa St. 37, Pittsburgh 29
Las Vegas Bowl, Las Vegas, NV: UNLV 31, Arkansas 14
MicronPC.com Bowl, Miami, FL: N. Carolina St. 38, Minnesota 30
Mobile Alabama Bowl, Mobile, AL: So. Mississippi 28, TCU 21
Motor City Bowl, Pontiac, MI: Marshall 25, Cincinnati 14
Music City Bowl, Nashville, TN: W. Virginia 49, Mississippi 38
Outback Bowl, Tampa, FL: S. Carolina 24, Ohio St. 7
Silicon Valley Bowl, San Jose, CA: Air Force 37, Fresno St. 34

All-Time NCAA Division I-A Statistical Leaders

(at end of 2000 season)

Career Rushing Yards

Player, team	Yrs	Carries	Yds	Avg
Ron Dayne, Wisconsin	1996-99	1,115	6,397	5.74
Ricky Williams, Texas	1995-98	1,011	6,279	6.21
Tony Dorsett, Pittsburgh	1973-76	1,074	6,082	5.66
Charles White, USC	1976-79	1,023	5,598	5.47
Travis Prentice, Miami (OH)	1996-99	1,138	5,596	4.92

Career Rushing Yard/Game (min. 2,500 yds.)

Player, team	Yrs	Carries	Yds	Avg/ Game
Ed Marinaro, Cornell	1969-71	918	4,715	174.6
O.J. Simpson, USC	1967-68	621	3,124	164.4
Herschel Walker, Georgia	1980-82	994	5,259	159.4
LeShon Johnson, N. Illinois	1992-93	592	3,314	150.6
Ron Dayne, Wisconsin	1996-99	1,115	6,397	148.8

Career Passing Yards

Player, team	Yrs	Comp/Att	Yds
Ty Detmer, BYU	1988-91	958/1,530	15,031
Tim Rattay, Louisiana Tech	1997-99	1,015/1,552	12,746
Chris Redman, Louisville	1996-99	1,031/1,679	12,541
Todd Santos, San Diego St.	1984-87	910/1,484	11,425
Tim Lester, Western Mich.	1996-99	875/1,507	11,299

Career Receiving Yards

Player, team	Yrs	Rec	Yds	Avg
Trevor Insley, Nevada	1996-99	298	5,005	16.8
Marcus Harris, Wyoming	1993-96	259	4,518	17.4
Ryan Yarborough, Wyoming	1990-93	229	4,357	19.0
Troy Edwards, Louisiana Tech	1996-98	280	4,352	15.5
Aaron Turner, Pacific (CA)	1989-92	266	4,345	16.3

Selected College Division I Football Teams in 2000

(2000 record does not include bowl games or Division I-AA playoff games; coaches at the start of 2001 season)

Team	Nickname	Team colors	Conference	Coach	2000 record (W-L)
Air Force	Falcons	Blue & silver	Mountain West	Fisher DeBerry	9-3
Akron	Zips	Blue & gold	Mid-American	Lee Owens	6-5
Alabama	Crimson Tide	Crimson & white	Southeastern	Dennis Franchione	3-8
Arizona	Wildcats	Cardinal & navy	Pacific Ten	John Mackovic	5-6
Arizona State	Sun Devils	Maroon & gold	Pacific Ten	Dirk Koetter	6-6
Arkansas	Razorbacks	Cardinal & white	Southeastern	Houston Nutt	6-6
Arkansas State	Indians	Scarlet & black	Big West	Joe Hollis	1-10
Army	Cadets, Black Knights	Black, gold, gray	Conference USA	Todd B. Terry	1-10
Auburn	Tigers	Burnt orange & navy	Southeastern	Tommy Tuberville	9-4
Ball State	Cardinals	Cardinal & white	Mid-American	Bill Lynch	5-6
Baylor	Bears	Green & gold	Big Twelve	Kevin Steele	2-9
Boston College	Eagles	Maroon & gold	Big East	Tom O'Brien	7-5
Bowling Green	Falcons	Orange & brown	Mid-American	Urban Meyer	2-9
Brigham Young (BYU)	Cougars	Royal blue & white	Mountain West	Gary Crowton	6-6
Brown	Bears	Brown, cardinal, white	Ivy League	Phil Estes	7-3

Team	Nickname	Team colors	Conference	Coach	2000 record (W-L)
California	Golden Bears	Blue & gold	Pacific Ten	Tom Holmoe	3-8
Central Michigan	Chippewas	Maroon & gold	Mid-American	Mike DeBord	2-9
Cincinnati	Bearcats	Red & black	Conference USA	Rick Minter	7-5
Citadel	Bulldogs	Blue & white	Southern	Ellis Johnson	2-9
Clemson	Tigers	Purple & orange	Atlantic Coast	Tommy Bowden	9-3
Colgate	Red Raiders	Maroon, gray, & white	Patriot League	Dick Biddle	7-4
Colorado	Golden Buffaloes	Silver, gold, & black	Big Twelve	Gary Barnett	3-8
Colorado State	Rams	Green & gold	Mountain West	Sonny Lubick	10-2
Columbia	Lions	Columbia blue & white	Ivy League	Ray Tellier	3-7
Connecticut	Huskies	Blue & white	Independent	Randy Edsall	3-8
Cornell	Big Red	Carnelian & white	Ivy League	Steve Miller	3-7
Dartmouth	Big Green	Dartmouth green & white	Ivy League	John Lyons	2-8
Delaware	Fightin' Blue Hens	Blue & gold	Atlantic Ten	Harold Raymond	12-2
Delaware State	Hornets	Red & blue	Mid-Eastern Athletic	Ben Blacknall	7-4
Duke	Blue Devils	Royal blue & white	Atlantic Coast	Carl Franks	0-11
East Carolina	Pirates	Purple & gold	Conference USA	Steve Logan	8-4
East Tennessee State	Buccaneers	Blue & gold	Southern	Paul Hamilton	6-5
Eastern Illinois	Panthers	Blue & gray	Ohio Valley	Bob Spoo	8-4
Eastern Kentucky	Colonels	Maroon & white	Ohio Valley	Roy Kidd	6-5
Eastern Michigan	Eagles	Dark green & white	Mid-American	Jeff Woodruff	3-8
Eastern Washington	Eagles	Red & white	Big Sky	Paul Wulff	6-5
Florida	Gators	Orange & blue	Southeastern	Steve Spurrier	10-3
Florida A&M	Rattlers	Orange & green	Mid-Eastern Athletic	Billy Joe	9-3
Florida State	Seminoles	Garnet & gold	Atlantic Coast	Bobby Bowden	11-2
Fresno State	Bulldogs	Cardinal & blue	Western Athletic	Pat Hill	7-5
Furman	Paladins	Purple & white	Southern	Bobby Johnson	9-3
Georgia	Bulldogs	Red & black	Southeastern	Mark Richt	8-4
Georgia Southern	Eagles	Blue & white	Southern	Paul Johnson	13-2
Georgia Tech	Yellow Jackets	Old gold & white	Atlantic Coast	George O'Leary	9-3
Grambling State	Tigers	Black & gold	Southwestern	Doug Williams	10-2
Harvard	Crimson	Crimson, black, white	Ivy League	Tim Murphy	5-5
Holy Cross	Crusaders	Royal purple	Patriot League	Dan Allen	7-4
Houston	Cougars	Scarlet & white	Conference USA	Dana Dimel	3-8
Howard	Bison	Blue, white & red	Mid-Eastern Athletic	Steve Wilson	3-8
Idaho	Vandals	Silver & gold	Big West	Tom Cable	5-6
Idaho State	Bengals	Orange & black	Big Sky	Larry Lewis	6-5
Illinois	Fighting Illini	Orange & blue	Big Ten	Ron Turner	5-6
Illinois State	Redbirds	Red & white	Gateway	Denver Johnson	7-4
Indiana	Hoosiers	Cream & crimson	Big Ten	Cam Cameron	3-8
Indiana State	Sycamores	Blue & white	Gateway	Tim McGuire	1-10
Iowa	Hawkeyes	Old gold & black	Big Ten	Kirk Ferentz	3-9
Iowa State	Cyclones	Cardinal & gold	Big Twelve	Dan McCarney	9-3
Jackson State	Tigers	Blue & white	Southwestern	Robert Hughes	7-4
James Madison	Dukes	Purple & gold	Atlantic Ten	Mickey Matthews	6-5
Kansas	Jayhawks	Crimson & blue	Big Twelve	Terry Allen	4-7
Kansas State	Wildcats	Purple & white	Big Twelve	Bill Snyder	11-3
Kent State	Golden Flashes	Navy blue & gold	Mid-American	Dean Pees	1-10
Kentucky	Wildcats	Blue & white	Southeastern	Guy Morriss	2-9
Lafayette	Leopards	Maroon & white	Patriot League	Frank Tavani	2-9
Lehigh	Mountain Hawks	Brown & white	Patriot League	Pete Lembo	12-1
Liberty	Flames	Red, white, blue	Independent	Ken Karcher	3-8
Louisiana-Lafayette	Ragin' Cajuns	Vermilion & white	Independent	Jerry Baldwin	1-10
Louisiana-Monroe	Indians	Maroon & gold	Independent	Bobby Keasler	1-10
Louisiana State (LSU)	Fighting Tigers	Purple & gold	Southeastern	Nick Saban	8-4
Louisiana Tech	Bulldogs	Red & blue	Independent	Jack Bicknell, III	3-9
Louisville	Cardinals	Red, black, white	Conference USA	John L. Smith	9-3
Maine	Black Bears	Blue & white	Atlantic Ten	Jack Cosgrove	5-6
Marshall	Thundering Herd	Green & white	Mid-American	Bob Pruett	8-5
Maryland	Terrapins	Red, white, black, gold	Atlantic Coast	Ralph Friedgen	5-6
Massachusetts	Minutemen	Maroon & white	Atlantic Ten	Mark Whipple	7-4
McNeese State	Cowboys	Blue & gold	Southland	Tommy Tate	8-4
Memphis	Tigers	Blue & gray	Conference USA	Tommy West	4-7
Miami (Florida)	Hurricanes	Orange, green, white	Big East	Larry Coker	11-1
Miami (Ohio)	RedHawks	Red & white	Mid-American	Terry Hoeppner	6-5
Michigan	Wolverines	Maize & blue	Big Ten	Lloyd Carr	9-3
Michigan State	Spartans	Green & white	Big Ten	Bobby Williams	5-6
Middle Tennessee St.	Blue Raiders	Blue & white	Independent	Andy McCollum	6-5
Minnesota	Golden Gophers	Maroon & gold	Big Ten	Glen Mason	6-6
Mississippi	Rebels	Cardinal red & navy	Southeastern	David Cutcliffe	7-5
Mississippi State	Bulldogs	Maroon & white	Southeastern	Jackie Sherrill	8-4
Mississippi Valley	Delta Devils	Green & white	Southwestern	LaTraia Jones	2-9
Missouri	Tigers	Old gold & black	Big Twelve	Gary Pinkel	3-8
Montana	Grizzlies	Copper, silver, gold	Big Sky	Joe Glenn	13-2
Montana State	Bobcats	Blue & gold	Big Sky	Mike Kramer	0-11
Morehead State	Eagles	Blue & gold	Independent	Matt Ballard	6-3
Morgan State	Bears	Blue & orange	Mid-Eastern Athletic	Stanley Mitchell	1-10
Murray State	Racers	Blue & gold	Ohio Valley	Joe Pannunzio	6-5
Navy	Midshipmen	Navy blue & gold	Independent	Charlie Weatherbie	1-10
Nebraska	Cornhuskers	Scarlet & cream	Big Twelve	Frank Solich	10-2
Nevada	Wolf Pack	Silver & blue	Western Athletic	Chris Tormey	2-10
Nev.-Las Vegas (UNLV)	Rebels	Scarlet & gray	Mountain West	John Robinson	8-5
New Hampshire	Wildcats	Blue & white	Atlantic Ten	Sean McDonnell	6-5
New Mexico	Lobos	Cherry & silver	Mountain West	Rocky Long	5-7
New Mexico State	Aggies	Crimson & white	Big West	Tony Samuel	3-8
Nicholls St.	Colonels	Red & gray	Southland	Daryl Daye	1-10
North Carolina	Tar Heels	Carolina blue & white	Atlantic Coast	John Bunting	6-5
North Carolina A & T	Aggies	Blue & gold	Mid-Eastern Athletic	Bill Hayes	8-3
North Carolina State	Wolfpack	Red & white	Atlantic Coast	Chuck Amato	8-4

Team	Nickname	Team colors	Conference	Coach	2000 record (W-L)
North Texas	Mean Green Eagles	Green & white	Big West	Darrell Dickey	3-8
Northeastern	Huskies	Red & black	Atlantic Ten	Don Brown	4-7
Northern Arizona	Lumberjacks	Blue & gold	Big Sky	Jerome Souers	3-8
Northern Illinois	Huskies	Cardinal & black	Mid-American	Joe Novak	6-5
Northern Iowa	Panthers	Purple & old gold	Gateway	Mark Farley	7-4
Northwestern	Wildcats	Purple & white	Big Ten	Randy Walker	8-4
Northwestern State	Demons	Purple, white, & burnt orange	Southland	Steve Roberts	6-5
Notre Dame	Fighting Irish	Gold & blue	Independent	Bob Davie	9-3
Ohio	Bobcats	Ohio green & white	Mid-American	Brian Knorr	7-4
Ohio State	Buckeyes	Scarlet & gray	Big Ten	Jim Tressel	8-4
Oklahoma	Sooners	Crimson & cream	Big Twelve	Bob Stoops	13-0
Oklahoma State	Cowboys	Orange & black	Big Twelve	Les Miles	3-8
Oregon	Ducks	Green & yellow	Pacific Ten	Mike Bellotti	10-2
Oregon State	Beavers	Orange & black	Pacific Ten	Dennis Erickson	11-1
Penn State	Nittany Lions	Blue & white	Big Ten	Joe Paterno	5-7
Pennsylvania	Quakers	Red & blue	Ivy League	Al Bagnoli	7-3
Pittsburgh	Panthers	Blue & gold	Big East	Walt Harris	7-5
Princeton	Tigers	Orange & black	Ivy League	Roger Hughes	3-7
Purdue	Boilermakers	Old gold & black	Big Ten	Joe Tiller	8-4
Rhode Island	Rams	Light & dark blue, white	Atlantic Ten	Tim Stowers	3-8
Rice	Owls	Blue & gray	Western Athletic	Ken Hatfield	3-8
Richmond	Spiders	Red & blue	Atlantic Ten	Jim Reid	10-3
Rutgers	Scarlet Knights	Scarlet	Big East	Greg Schiano	3-8
Sam Houston State	Bearkats	Orange & white	Southland	Ron Randleman	7-4
Samford	Bulldogs	Crimson & blue	Independent	Pete Hurt	4-7
San Diego State	Aztecs	Scarlet & black	Mountain West	Ted Tollner	3-8
San Jose State	Spartans	Gold, white, blue	Western Athletic	Fritz Hill	7-5
South Carolina	Fighting Gamecocks	Garnet & black	Southeastern	Lou Holtz	8-4
South Carolina State	Bulldogs	Garnet & blue	Mid-Eastern Athletic	Willie E. Jeffries	3-8
SE Missouri State	Indians	Red & black	Ohio Valley	Tim Billings	3-8
Southern California (USC)	Trojans	Cardinal & gold	Pacific Ten	Pete Carroll	5-7
Southern Illinois	Salukis	Maroon & white	Gateway	Jerry Kill	3-8
Southern Methodist (SMU)	Mustangs	Red & blue	Western Athletic	Mike Cavan	3-9
Southern Mississippi	Golden Eagles	Black & gold	Conference USA	Jeff Bower	8-4
SW Missouri State	Bears	Maroon & white	Gateway	Randy Ball	5-6
SW Texas State	Bobcats	Maroon & gold	Southland	Bob DeBesse	7-4
Stanford	Cardinal	Cardinal & white	Pacific Ten	Tyrone Willingham	5-6
Stephen F. Austin	Lumberjacks	Purple & white	Southland	Mike Santiago	6-5
Syracuse	Orangemen	Orange	Big East	Paul Pasqualoni	6-5
Temple	Owls	Cherry & white	Big East	Bobby Wallace	4-7
Tennessee	Volunteers	Orange & white	Southeastern	Phillip Fulmer	8-4
Tennessee-Chattanooga	Mocs	Navy blue & gold	Southern	Donnie Kirkpatrick	5-6
Tennessee-Martin	Skyhawks	Orange, white, blue	Ohio Valley	Sam McCorkle	2-9
Tennessee State	Tigers	Royal blue & white	Ohio Valley	James Reese	3-8
Tennessee Tech	Golden Eagles	Purple & gold	Ohio Valley	Mike Hennigan	8-3
Texas	Longhorns	Burnt orange & white	Big Twelve	Mack Brown	9-3
Texas A & M	Aggies	Maroon & white	Big Twelve	R. C. Slocum	7-5
Texas Christian (TCU)	Horned Frogs	Purple & white	Western Athletic	Gary Patterson	10-2
Texas Southern	Tigers	Maroon & gray	Southwestern	Bill Thomas	8-3
Texas Tech	Red Raiders	Scarlet & black	Big Twelve	Mike Leach	7-6
Toledo	Rockets	Blue & gold	Mid-American	Tom Amstutz	10-1
Troy State	Trojans	Cardinal, gray, black	Southland	Larry Blakeney	9-3
Tulane	Green Wave	Olive green & sky blue	Conference USA	Chris Scelfo	6-5
Tulsa	Golden Hurricane	Blue & gold	Western Athletic	Keith Burns	5-7
UCLA	Bruins	Blue & gold	Pacific Ten	Bob Toledo	6-6
Utah	Utes	Crimson & white	Mountain West	Ron McBride	4-7
Utah State	Aggies	Navy blue & white	Big West	Mike Dennehy	5-6
UTEP (Texas-El Paso)	Miners	Orange, blue, white	Western Athletic	Gary Nord	8-4
Vanderbilt	Commodores	Black & gold	Southeastern	Woody Widenhofer	3-8
Villanova	Wildcats	Blue & white	Atlantic Ten	Andy Talley	5-6
Virginia	Cavaliers	Orange & blue	Atlantic Coast	Al Groh	6-6
Virginia Military Inst. (VMI)	Keydets	Red, white & yellow	Southern	Cal McCombs	2-9
Virginia Tech	Gobblers, Hokies	Orange & maroon	Big East	Frank Beamer	11-1
Wake Forest	Demon Deacons	Old gold & black	Atlantic Coast	Jim Grobe	2-9
Washington	Huskies	Purple & gold	Pacific Ten	Rick Neuheisel	11-1
Washington State	Cougars	Crimson & gray	Pacific Ten	Mike Price	4-7
Weber State	Wildcats	Royal purple & white	Big Sky	Jerry Graybeal	7-4
West Virginia	Mountaineers	Old gold & blue	Big East	Rich Rodriquez	7-5
Western Carolina	Catamounts	Purple & gold	Southern	Bill Bleil	4-7
Western Illinois	Leathernecks	Purple & gold	Gateway	Don Patterson	9-3
Western Kentucky	Hilltoppers	Red & white	Ohio Valley	Jack Harbaugh	11-2
Western Michigan	Broncos	Brown & gold	Mid-American	Gary Darnell	9-3
William & Mary	Tribe	Green, gold, silver	Atlantic Ten	Jimmye Laycock	5-6
Wisconsin	Badgers	Cardinal & white	Big Ten	Barry Alvarez	9-4
Wyoming	Cowboys	Brown & yellow	Mountain West	Vic Koenning	1-10
Yale	Bulldogs, Elis	Yale blue & white	Ivy League	Jack Siedlecki	7-3
Youngstown State	Penguins	Red & white	Gateway	John Heacock	9-3

Heisman Trophy Winners

Awarded annually to the nation's outstanding college football player by the Downtown Athletic Club.

1935 Jay Berwanger, Chicago, HB
1936 Larry Kelley, Yale, E
1937 Clinton Frank, Yale, HB
1938 David O'Brien, Texas Christian, QB
1939 Nile Kinnick, Iowa, HB
1940 Tom Harmon, Michigan, HB
1941 Bruce Smith, Minnesota, HB
1942 Frank Sinkwich, Georgia, HB
1943 Angelo Bertelli, Notre Dame, QB
1944 Leslie Horvath, Ohio St., QB
1945 Felix Blanchard, Army, FB
1946 Glenn Davis, Army, HB
1947 John Lujack, Notre Dame, QB
1948 Doak Walker, SMU, HB
1949 Leon Hart, Notre Dame, E
1950 Vic Janowicz, Ohio St., HB
1951 Richard Kazmaier, Princeton, HB
1952 Billy Vessels, Oklahoma, HB
1953 John Lattner, Notre Dame, HB
1954 Alan Ameche, Wisconsin, FB
1955 Howard Cassady, Ohio St., HB

1956 Paul Hornung, Notre Dame, QB
1957 John Crow, Texas A & M, HB
1958 Pete Dawkins, Army, HB
1959 Billy Cannon, LSU, HB
1960 Joe Bellino, Navy, HB
1961 Ernest Davis, Syracuse, HB
1962 Terry Baker, Oregon St., QB
1963 Roger Staubach, Navy, QB
1964 John Huarte, Notre Dame, QB
1965 Mike Garrett, USC, HB
1966 Steve Spurrier, Florida, QB
1967 Gary Beban, UCLA, QB
1968 O. J. Simpson, USC, RB
1969 Steve Owens, Oklahoma, RB
1970 Jim Plunkett, Stanford, QB
1971 Pat Sullivan, Auburn, QB
1972 Johnny Rodgers, Nebraska, RB-WR
1973 John Cappelletti, Penn St., RB
1974 Archie Griffin, Ohio St., RB
1975 Archie Griffin, Ohio St., RB
1976 Tony Dorsett, Pittsburgh, RB
1977 Earl Campbell, Texas, RB
1978 Billy Sims, Oklahoma, RB
1979 Charles White, USC, RB
1980 George Rogers, S. Carolina, RB
1981 Marcus Allen, USC, RB
1982 Herschel Walker, Georgia, RB
1983 Mike Rozier, Nebraska, RB
1984 Doug Flutie, Boston College, QB
1985 Bo Jackson, Auburn, RB
1986 Vinny Testaverde, Miami, QB
1987 Tim Brown, Notre Dame, WR
1988 Barry Sanders, Oklahoma St., RB
1989 Andre Ware, Houston, QB
1990 Ty Detmer, BYU, QB
1991 Desmond Howard, Michigan, WR
1992 Gino Torretta, Miami, QB
1993 Charlie Ward, Florida St., QB
1994 Rashaan Salaam, Colorado, RB
1995 Eddie George, Ohio St., RB
1996 Danny Wuerffel, Florida, QB
1997 Charles Woodson, Michigan, CB
1998 Ricky Williams, Texas, RB
1999 Ron Dayne, Wisconsin, RB
2000 Chris Weinke, Florida St., QB

Outland Award Winners

Honoring the outstanding interior lineman selected by the Football Writers Association of America.

1946 George Connor, Notre Dame, T
1947 Joe Steffy, Army, G
1948 Bill Fischer, Notre Dame, G
1949 Ed Bagdon, Michigan St., G
1950 Bob Gain, Kentucky, T
1951 Jim Weatherall, Oklahoma, T
1952 Dick Modzelewski, Maryland, T
1953 J. D. Roberts, Oklahoma, G
1954 Bill Brooks, Arkansas, G
1955 Calvin Jones, Iowa, G
1956 Jim Parker, Ohio St., G
1957 Alex Karras, Iowa, T
1958 Zeke Smith, Auburn, G
1959 Mike McGee, Duke, T
1960 Tom Brown, Minnesota, G
1961 Merlin Olsen, Utah St., T
1962 Bobby Bell, Minnesota, T
1963 Scott Appleton, Texas, T
1964 Steve Delong, Tennessee, T
1965 Tommy Nobis, Texas, G
1966 Loyd Phillips, Arkansas, T
1967 Ron Yary, Southern Cal, T
1968 Bill Stanfill, Georgia, T
1969 Mike Reid, Penn St., DT
1970 Jim Stillwagon, Ohio St., MG
1971 Larry Jacobson, Nebraska, DT
1972 Rich Glover, Nebraska, MG
1973 John Hicks, Ohio St., OT
1974 Randy White, Maryland, DE
1975 Lee Roy Selmon, Oklahoma, DT
1976 Ross Browner, Notre Dame, DE
1977 Brad Shearer, Texas, DT
1978 Greg Roberts, Oklahoma, G
1979 Jim Ritcher, North Carolina St., C
1980 Mark May, Pittsburgh, OT
1981 Dave Rimington, Nebraska, C
1982 Dave Rimington, Nebraska, C
1983 Dean Steinkuhler, Nebraska, G
1984 Bruce Smith, Virginia Tech, DT
1985 Mike Ruth, Boston College, NG
1986 Jason Buck, BYU, DT
1987 Chad Hennings, Air Force, DT
1988 Tracy Rocker, Auburn, DT
1989 Mohammed Elewonibi, BYU, G
1990 Russell Maryland, Miami (FL), DT
1991 Steve Emtman, Washington, DT
1992 Will Shields, Nebraska, G
1993 Rob Waldrop, Arizona, NG
1994 Zach Wiegert, Nebraska, OT
1995 Jonathan Ogden, UCLA, OT
1996 Orlando Pace, Ohio St., OT
1997 Aaron Taylor, Nebraska, OT
1998 Kris Farris, UCLA, OT
1999 Chris Samuels, Alabama, OT
2000 John Henderson, Tennessee, DT

All-Time Division I-A Percentage Leaders

(Classified as Division I-A for the last 10 years; record includes bowl games; ties computed as half won and half lost)

	Years	Won	Lost	T	Pct.	Bowl Games** W	L	T
Notre Dame	112	776	241	42	.753	13	11	0
Michigan	121	805	262	36	.746	17	15	0
Alabama*	106	737	276	43	.718	28	19	3
Nebraska	111	753	299	40	.708	20	19	0
Ohio St.	111	724	287	53	.705	14	18	0
Oklahoma	106	702	278	53	.705	21	12	1
Texas	108	744	302	33	.705	18	20	2
Tennessee*	104	707	292	52	.697	22	19	0
Penn St.	114	739	312	41	.696	23	11	2
USC	108	678	288	54	.691	25	14	0
Florida St.*	54	392	185	17	.674	17	9	2
Washington*	111	617	337	50	.639	14	12	1
Miami (OH) *	112	597	332	44	.636	5	2	0
Georgia	107	641	362	54	.632	19	14	3
LSU	107	618	360	47	.626	15	16	1
Arizona St.	88	490	290	24	.624	10	8	1
Central Michigan	100	512	301	36	.624	0	2	0
Miami (FL)	74	472	282	19	.623	14	11	0
Auburn*	108	610	365	47	.620	14	11	2
Army	111	618	374	51	.617	2	2	0
Colorado	111	611	376	36	.615	11	12	0
Florida	94	564	347	40	.614	13	15	0
Texas A&M	106	609	386	48	.607	12	14	0
Syracuse	111	638	411	49	.603	11	8	1
UCLA	82	484	314	37	.602	11	11	1

*Includes games that were forfeited or changed by action of NCAA Council and/or Committee on Infractions. **Includes major bowl games only; that is, those where team's opponent was classified as a major college team that season or at the time of the bowl game.

College Football Coach of the Year

The Division I-A Coach of the Year has been selected by the American Football Coaches Assn. since 1935 and selected by the Football Writers Assn. of America since 1957. When polls disagree, both winners are indicated.

1935 Lynn Waldorf, Northwestern
1936 Dick Harlow, Harvard
1937 Edward Mylin, Lafayette
1938 Bill Kern, Carnegie Tech
1939 Eddie Anderson, Iowa
1940 Clark Shaughnessy, Stanford
1941 Frank Leahy, Notre Dame
1942 Bill Alexander, Georgia Tech
1943 Amos Alonzo Stagg, Pacific
1944 Carroll Widdoes, Ohio St.
1945 Bo McMillin, Indiana
1946 Earl "Red" Blaik, Army
1947 Fritz Crisler, Michigan
1948 Bennie Oosterbaan, Michigan
1949 Bud Wilkinson, Oklahoma
1950 Charlie Caldwell, Princeton
1951 Chuck Taylor, Stanford
1952 Biggie Munn, Michigan St.
1953 Jim Tatum, Maryland
1954 Henry "Red" Sanders, UCLA
1955 Duffy Daugherty, Michigan St.
1956 Bowden Wyatt, Tennessee
1957 Woody Hayes, Ohio St.
1958 Paul Dietzel, LSU
1959 Ben Schwartzwalder, Syracuse
1960 Murray Warmath, Minnesota
1961 Paul "Bear" Bryant, Ala. (AFCA)
Darrell Royal, Texas (FWAA)
1962 John McKay, USC
1963 Darrell Royal, Texas
1964 Ara Parseghian, Notre Dame, & Frank Broyles, Arkansas (AFCA); Ara Parseghian (FWAA)
1965 Tommy Prothro, UCLA (AFCA); Duffy Daugherty, Mich. St. (FWAA)
1966 Tom Cahill, Army
1967 John Pont, Indiana
1968 Joe Paterno, Penn St. (AFCA); Woody Hayes, Ohio St. (FWAA)
1969 Bo Schembechler, Michigan
1970 Charles McClendon, LSU, & Darrell Royal, Texas (AFCA); Alex Agase, Northwestern (FWAA)
1971 Paul "Bear" Bryant, Alabama (AFCA); Bob Devaney, Nebraska (FWAA)
1972 John McKay, USC
1973 Paul "Bear" Bryant, Alabama (AFCA); Johnny Majors, Pittsburgh (FWAA)
1974 Grant Teaff, Baylor
1975 Frank Kush, Arizona St. (AFCA); Woody Hayes, Ohio St. (FWAA)
1976 Johnny Majors, Pittsburgh
1977 Don James, Washington (AFCA); Lou Holtz, Arkansas (FWAA)
1978 Joe Paterno, Penn St.
1979 Earle Bruce, Ohio St.
1980 Vince Dooley, Georgia
1981 Danny Ford, Clemson
1982 Joe Paterno, Penn St.
1983 Ken Hatfield, Air Force (AFCA); Howard Schnellenberger, Miami (FL) (FWAA)
1984 LaVell Edwards, Brigham Young
1985 Fisher De Berry, Air Force
1986 Joe Paterno, Penn St.
1987 Dick MacPherson, Syracuse
1988 Don Nehlen, W. Virginia (AFCA); Lou Holtz, Notre Dame (FWAA)
1989 Bill McCartney, Colorado
1990 Bobby Ross, Georgia Tech
1991 Don James, Washington
1992 Gene Stallings, Alabama
1993 Barry Alvarez, Wisconsin (AFCA); Terry Bowden, Auburn (FWAA)
1994 Tom Osborne, Nebraska (AFCA) Rich Brooks, Oregon (FWAA)
1995 Gary Barnett, Northwestern
1996 Bruce Snyder, Arizona St.
1997 Mike Price, Washington St.
1998 Phillip Fulmer, Tennessee
1999 Frank Beamer, Virginia Tech
2000 Bob Stoops, Oklahoma

All-Time Division I-A Coaching Victories (Including Bowl Games)

Paul "Bear" Bryant	323	Hayden Fry	232	Dan McGugin	197	Gil Dobie	180
*Joe Paterno	322	*Lou Holtz	224	Fielding Yost	196	Carl Snavely	180
Glenn "Pop" Warner	319	Jess Neely	207	Howard Jones	194	Jerry Claiborne	179
*Bobby Bowden	315	Warren Woodson	203	*John Cooper	192	Ben Schwartzwalder	178
Amos Alonzo Stagg	314	*Don Nehlen	202	John Vaught	190	Frank Kush	176
*LaVell Edwards	257	Eddie Anderson	201	*George Welsh	189	Don James	176
Tom Osborne	255	Vince Dooley	201	John Heisman	185	Ralph Jordan	176
Woody Hayes	238	Jim Sweeney	200	Johnny Majors	185		
Bo Schembechler	234	Dana X. Bible	198	Darrell Royal	184		

Coaches active in 2000 are denoted by an asterisk (*). Eddie Robinson of Grambling State Univ. (Div. I-AA) , who retired after the 1997 season, holds the record for most college football victories, with 408.

Selected College Football Conference Champions (1980-2000)

Atlantic Coast

1980 North Carolina
1981 Clemson
1982 Clemson
1983 Maryland
1984 Maryland
1985 Maryland
1986 Clemson
1987 Clemson
1988 Clemson
1989 Virginia, Duke
1990 Georgia Tech
1991 Clemson
1992 Florida St.
1993 Florida St.
1994 Florida St.
1995 Virginia, Florida St.
1996 Florida St.
1997 Florida St.
1998 Florida St., Georgia Tech
1999 Florida St.
2000 Florida St.

Ivy Group

1980 Yale
1981 Yale, Dartmouth
1982 Harvard, Dartmouth, Penn
1983 Harvard, Penn
1984 Penn
1985 Penn
1986 Penn
1987 Harvard
1988 Penn, Cornell
1989 Yale, Princeton
1990 Cornell, Dartmouth
1991 Dartmouth
1992 Dartmouth, Princeton
1993 Penn
1994 Penn
1995 Princeton
1996 Dartmouth
1997 Harvard
1998 Penn
1999 Brown, Yale
2000 Penn

Big Eight*

1980 Oklahoma
1981 Nebraska
1982 Nebraska
1983 Nebraska
1984 Nebraska, Oklahoma
1985 Oklahoma
1986 Oklahoma
1987 Oklahoma
1988 Nebraska
1989 Colorado
1990 Colorado
1991 Nebraska, Colorado
1992 Nebraska
1993 Nebraska
1994 Nebraska
1995 Nebraska

Big Ten

1980 Michigan
1981 Iowa, Ohio St.
1982 Michigan
1983 Illinois
1984 Ohio St.
1985 Iowa
1986 Michigan, Ohio St.
1987 Michigan St.
1988 Michigan
1989 Michigan
1990 Iowa, Ill., Mich., Mich. St.
1991 Michigan
1992 Michigan
1993 Ohio St., Wisconsin
1994 Penn St.
1995 Northwestern
1996 Ohio St., Northwestern
1997 Michigan
1998 Ohio St., Wisconsin, Michigan
1999 Wisconsin
2000 Michigan, Northwestern, Purdue

Mid-American Athletic

1980 Central Michigan
1981 Toledo
1982 Bowling Green
1983 Northern Illinois
1984 Toledo
1985 Bowling Green
1986 Miami (OH)
1987 E. Michigan
1988 W. Michigan
1989 Ball St.
1990 Central Michigan
1991 Bowling Green
1992 Bowling Green
1993 Ball St.
1994 Central Michigan
1995 Toledo
1996 Ball St.
1997 Marshall
1998 Marshall
1999 Marshall
2000 Marshall

Southern

1980 Furman
1981 Furman
1982 Furman
1983 Furman
1984 Tenn.-Chattanooga
1985 Furman
1986 Appalachian St.
1987 Appalachian St.
1988 Marshall, Furman
1989 Furman
1990 Furman
1991 Appalachian St.
1992 Citadel
1993 Georgia Southern
1994 Marshall
1995 Appalachian St.
1996 Marshall
1997 Georgia Southern
1998 Georgia Southern
1999 Appalachian St., GA Southern, Furman
2000 Georgia Southern

Southeastern

1980 Georgia
1981 Georgia, Alabama
1982 Georgia
1983 Auburn
1984 Florida (title vacated)
1985 Tennessee
1986 LSU
1987 Auburn
1988 Auburn, LSU
1989 Ala., Tenn., Auburn
1990 Tennessee
1991 Florida
1992 Alabama
1993 Florida
1994 Florida
1995 Florida
1996 Florida
1997 Tennessee
1998 Tennessee
1999 Florida, Alabama
2000 Florida

Southwest*

1980 Baylor
1981 Texas
1982 SMU
1983 Texas
1984 SMU, Houston
1985 Texas A&M
1986 Texas A&M
1987 Texas A&M
1988 Arkansas
1989 Arkansas
1990 Texas
1991 Texas A&M
1992 Texas A&M
1993 Texas A&M
1994 Baylor, Rice, Texas, TCU, Texas Tech
1995 Texas

Pacific Ten

1980 Washington
1981 Washington
1982 UCLA
1983 UCLA
1984 USC
1985 UCLA
1986 Arizona St.
1987 UCLA, USC
1988 USC
1989 USC
1990 Washington
1991 Washington
1992 Washington, Stanford
1993 UCLA, Arizona, USC
1994 Oregon
1995 USC, Washington
1996 Arizona St.
1997 Washington St., UCLA
1998 UCLA
1999 Stanford
2000 Washington, Oregon St., Oregon

Big East

1991 Miami (FL), Syracuse
1992 Miami (FL)
1993 West Virginia
1994 Miami (FL)
1995 Virginia Tech, Miami (FL)
1996 Virginia Tech, Miami (FL), Syracuse
1997 Syracuse
1998 Syracuse
1999 Virginia Tech
2000 Miami (FL)

Western Athletic

1980 Brigham Young (BYU)
1981 Brigham Young
1982 Brigham Young
1983 Brigham Young
1984 Brigham Young
1985 BYU, Air Force
1986 San Diego St.
1987 Wyoming
1988 Wyoming
1989 Brigham Young
1990 Brigham Young
1991 Brigham Young
1992 Hawaii, BYU, Fresno St.
1993 Wyoming, Fresno St., BYU
1994 Colorado St.
1995 Colorado St., Air Force, Utah, BYU
1996 Brigham Young
1997 Colorado St.
1998 Air Force
1999 Fresno St., Hawaii, TCU
2000 Texas Christian, UTEP

Big 12*

1996 Texas
1997 Nebraska
1998 Texas A&M
1999 Nebraska
2000 Oklahoma

Big West

1980 Long Beach St.
1981 San Jose St.
1982 Fresno St.
1983 Cal St.-Fullerton
1984 Cal St.-Fullerton
1985 Fresno St.
1986 San Jose St.
1987 San Jose St.
1988 Fresno St.
1989 Fresno St.
1990 San Jose St.
1991 San Jose St., Fresno St.
1992 Nevada
1993 SW Louisiana, Utah St.
1994 Nevada, SW Louisiana, UNLV
1995 Nevada
1996 Nevada, Utah St.
1997 Nevada, Utah St.
1998 Idaho
1999 Boise St.
2000 Boise St.

Conference USA

1996 So. Mississippi, Houston
1997 So. Mississippi
1998 Tulane
1999 So. Mississippi
2000 Louisville

Mountain West**

1999 BYU, Colorado St., Utah
2000 Colorado St.

(*) After the 1995 season, the Big Eight and Southwest conferences disbanded. In 1996 all former Big Eight Conference teams joined with 4 of the 8 Southwest Conference teams to form the Big 12 Conference. (**) After the 1998 season, 8 members of the 16-team Western Athletic Conference split off to form the Mountain West Conference.

NATIONAL HOCKEY LEAGUE

2000-2001 Season: Colorado Wins, Bourque Goes Out on Top, Lemieux Returns

The Colorado Avalanche beat the defending champion New Jersey Devils in Stanley Cup finals, 4 games to 3, taking Game 7, 3-1, at Denver's Pepsi Center, June 9, 2001. Colorado goalie Patrick Roy, who allowed 1 goal in the final 2 games, won the Conn Smythe Trophy (playoff MVP) a record 3d time. At an emotional post-game presentation Joe Sakic, Colorado's team captain, gave up his traditional right to the first skate with the Cup and passed it to Ray Bourque. After 1,826 games over 22 seasons (20½ in Boston) and a record 19 consecutive All-Star teams, the highest-scoring defenseman in NHL history finally held the Stanley Cup aloft at age 40. He retired 17 days later, on June 26. Mario Lemieux, retired and voted to the Hall of Fame in 1997, became the 1st owner-player in modern sports when he skated for his slumping Pittsburgh Penguins, Dec. 27, 2000. The 6-time NHL scoring champ had a goal and 2 assists in the win over Toronto. Lemieux, 35, was a 2001 All-Star and led the Penguins to the Eastern Conf. finals, where they lost to New Jersey.

Final Standings 2000-2001

(playoff seeding in parentheses; in each conference the three division winners automatically get the number 1, 2, and 3 seeds)

Eastern Conference

Atlantic Division

	W	L	T	OTL	GF	GA	Pts
New Jersey (1)	48	19	12	3	295	195	111
Philadelphia (4)	43	25	11	3	240	207	100
Pittsburgh (6)	42	28	9	3	281	256	96
N.Y. Rangers	33	43	5	1	250	290	72
N.Y. Islanders	21	51	7	3	185	268	52

Northeast Division

	W	L	T	OTL	GF	GA	Pts
Ottawa (2)	48	21	9	4	274	205	109
Buffalo (5)	46	30	5	1	218	184	98
Toronto (7)	37	29	11	5	232	207	90
Boston	36	30	8	8	227	249	88
Montreal	28	40	8	6	206	232	70

Southeast Division

	W	L	T	OTL	GF	GA	Pts
Washington (3)	41	27	10	4	233	211	96
Carolina (8)	38	32	9	3	212	225	88
Florida	22	38	13	9	200	246	66
Atlanta	23	45	12	2	211	289	60
Tampa Bay	24	47	6	5	201	280	59

Western Conference

Central Division

	W	L	T	OTL	GF	GA	Pts
Detroit (2)	49	20	9	4	253	202	111
St. Louis (4)	43	22	12	5	249	195	103
Nashville	34	36	9	3	186	200	80
Chicago	29	40	8	5	210	246	71
Columbus	28	39	9	6	190	233	71

Northwest Division

	W	L	T	OTL	GF	GA	Pts
Colorado (1)	52	16	10	4	270	192	118
Edmonton (6)	39	28	12	3	243	222	93
Vancouver (8)	36	28	11	7	239	238	90
Calgary	27	36	15	4	197	236	73
Minnesota	25	39	13	5	168	210	68

Pacific Division

	W	L	T	OTL	GF	GA	Pts
Dallas (3)	48	24	8	2	241	187	106
San Jose (5)	40	27	12	3	217	192	95
Los Angeles (7)	38	28	13	3	252	228	92
Phoenix	35	27	17	3	214	212	90
Anaheim	25	41	11	5	188	245	66

2001 Stanley Cup Playoff Results

Eastern Conference

New Jersey defeated Carolina 4 games to 2
Toronto defeated Ottawa 4 games to 0
Pittsburgh defeated Washington 4 games to 2
Buffalo defeated Philadelphia 4 games to 2
New Jersey defeated Toronto 4 games to 3
Pittsburgh defeated Buffalo 4 games to 3
New Jersey defeated Pittsburgh 4 games to 1

Western Conference

Colorado defeated Vancouver 4 games to 0
Los Angeles defeated Detroit 4 games to 2
Dallas defeated Edmonton 4 games to 2
St. Louis defeated San Jose 4 games to 2
Colorado defeated Los Angeles 4 games to 3
St. Louis defeated Dallas 4 games to 0
Colorado defeated St. Louis 4 games to 1

Finals

Colorado defeated New Jersey 4 games to 3 [5-0, 1-2, 3-1, 2-3, 1-4, 4-0, 3-1].

Stanley Cup Champions Since 1927

Year	Champion	Coach	Final opponent
1927	Ottawa	Dave Gill	Boston
1928	N.Y. Rangers	Lester Patrick	Montreal
1929	Boston	Cy Denneny	N.Y. Rangers
1930	Montreal	Cecil Hart	Boston
1931	Montreal	Cecil Hart	Chicago
1932	Toronto	Dick Irvin	N.Y. Rangers
1933	N.Y. Rangers	Lester Patrick	Toronto
1934	Chicago	Tommy Gorman	Detroit
1935	Montreal Maroons	Tommy Gorman	Toronto
1936	Detroit	Jack Adams	Toronto
1937	Detroit	Jack Adams	N.Y. Rangers
1938	Chicago	Bill Stewart	Toronto
1939	Boston	Art Ross	Toronto
1940	N.Y. Rangers	Frank Boucher	Toronto
1941	Boston	Cooney Weiland	Detroit
1942	Toronto	Hap Day	Detroit
1943	Detroit	Jack Adams	Boston
1944	Montreal	Dick Irvin	Chicago
1945	Toronto	Hap Day	Detroit
1946	Montreal	Dick Irvin	Boston
1947	Toronto	Hap Day	Montreal
1948	Toronto	Hap Day	Detroit
1949	Toronto	Hap Day	Detroit
1950	Detroit	Tommy Ivan	N.Y. Rangers
1951	Toronto	Joe Primeau	Montreal
1952	Detroit	Tommy Ivan	Montreal
1953	Montreal	Dick Irvin	Boston
1954	Detroit	Tommy Ivan	Montreal
1955	Detroit	Jimmy Skinner	Montreal
1956	Montreal	Toe Blake	Detroit
1957	Montreal	Toe Blake	Boston
1958	Montreal	Toe Blake	Boston
1959	Montreal	Toe Blake	Toronto
1960	Montreal	Toe Blake	Toronto
1961	Chicago	Rudy Pilous	Detroit
1962	Toronto	Punch Imlach	Chicago
1963	Toronto	Punch Imlach	Detroit
1964	Toronto	Punch Imlach	Detroit
1965	Montreal	Toe Blake	Chicago
1966	Montreal	Toe Blake	Detroit
1967	Toronto	Punch Imlach	Montreal
1968	Montreal	Toe Blake	St. Louis
1969	Montreal	Claude Ruel	St. Louis
1970	Boston	Harry Sinden	St. Louis
1971	Montreal	Al MacNeil	Chicago
1972	Boston	Tom Johnson	N.Y. Rangers
1973	Montreal	Scotty Bowman	Chicago
1974	Philadelphia	Fred Shero	Boston
1975	Philadelphia	Fred Shero	Buffalo
1976	Montreal	Scotty Bowman	Philadelphia
1977	Montreal	Scotty Bowman	Boston
1978	Montreal	Scotty Bowman	Boston
1979	Montreal	Scotty Bowman	N.Y. Rangers
1980	N.Y. Islanders	Al Arbour	Philadelphia
1981	N.Y. Islanders	Al Arbour	Minnesota
1982	N.Y. Islanders	Al Arbour	Vancouver
1983	N.Y. Islanders	Al Arbour	Edmonton
1984	Edmonton	Glen Sather	N.Y. Islanders
1985	Edmonton	Glen Sather	Philadelphia
1986	Montreal	Jean Perron	Calgary
1987	Edmonton	Glen Sather	Philadelphia
1988	Edmonton	Glen Sather	Boston
1989	Calgary	Terry Crisp	Montreal
1990	Edmonton	John Muckler	Boston
1991	Pittsburgh	Bob Johnson	Minnesota
1992	Pittsburgh	Scotty Bowman	Chicago
1993	Montreal	Jacques Demers	Los Angeles
1994	N.Y. Rangers	Mike Keenan	Vancouver
1995	New Jersey	Jacques Lemaire	Detroit
1996	Colorado	Marc Crawford	Florida
1997	Detroit	Scotty Bowman	Philadelphia
1998	Detroit	Scotty Bowman	Washington
1999	Dallas	Ken Hitchcock	Buffalo
2000	New Jersey	Larry Robinson	Dallas
2001	Colorado	Bob Hartley	New Jersey

Individual Leaders, 2000-2001

Points

Jaromir Jagr, Pittsburgh, 121; Joe Sakic, Colorado, 118; Patrik Elias, New Jersey, 96; Jason Allison, Boston, 95; Alexei Kovalev, Pittsburgh, 95; Martin Straka, Pittsburgh, 95.

Goals

Pavel Bure, Florida, 59; Joe Sakic, Colorado, 54; Jaromir Jagr, Pittsburgh, 52; Peter Bondra, Washington, 45; Alexei Kovalev, Pittsburgh, 44.

Assists

Jaromir Jagr, Pittsburgh, 69; Adam Oates, Washington, 69; Martin Straka, Pittsburgh, 68; Doug Weight, Edmonton, 65; Joe Sakic, Colorado, 64.

Power-play goals

Peter Bondra, Washington, 22; Joe Thornton, Boston, 19; Joe Sakic, Colorado, 19; Pavel Bure, Florida, 19; Paul Kariya, Anaheim, 18; Markus Naslund, Vancouver, 18.

Shorthanded goals

Steve Sullivan, Chicago, 8; Wes Walz, Minnesota, 7; Theoren Fleury, N.Y. Rangers, 5; 5 players tied with 4.

Shooting percentage
(minimum 82 shots)

Gary Roberts, Toronto, 21.0; Keith Primeau, Chi.-Buf., 20.6; Mario Lemieux, Pittsburgh, 20.5; Joe Thornton, Boston, 20.4; Alex Tanguay, Colorado, 20.0.

Plus/Minus

Joe Sakic, Colorado, 45; Patrik Elias, New Jersey, 45; Scott Stevens, New Jersey, 40; Petr Sykora, New Jersey, 36; Brian Rafalski, New Jersey, 36.

Penalty minutes

Matthew Barnaby, Tampa Bay, 265; Peter Worrell, Florida, 248; Stu Grimson, Los Angeles, 235; Andrei Nazarov, Boston, 229; Jeff Odgers, Atlanta, 226.

Goaltending Leaders
(minimum 25 games)

Goals against average

Marty Turco, Dallas, 1.90; Roman Cechmanek, Philadelphia, 2.01; Manny Legace, Detroit, 2.05; Dominik Hasek, Buffalo, 2.11; Brent Johnson, St. Louis, 2.17.

Wins

Martin Brodeur, New Jersey, 42; Patrick Roy, Colorado, 40; Dominik Hasek, Buffalo, 37; Olaf Kolzig, Washington, 37; Arturs Irbe, Carolina, 37.

Save percentage

Marty Turco, Dallas, .925; Mike Dunham, Nashville, .923; Sean Burke, Phoenix, .922; Dominik Hasek, Buffalo, .921; Roman Cechmanek, Philadelphia, .921.

Shutouts

Dominik Hasek, Buffalo, 11; Roman Cechmanek, Philadelphia, 10; Martin Brodeur, New Jersey, 9; Ed Belfour, Dallas, 8; Tommy Salo, Edmonton, 8.

All-Time Leading Scorers

Player	Goals	Assists	Points	Player	Goals	Assists	Points
Wayne Gretzky	894	1,963	2,857	Stan Mikita	541	926	1,467
Gordie Howe	801	1,049	1,850	Bryan Trottier	524	901	1,425
Marcel Dionne	731	1,040	1,771	Dale Hawerchuk	518	891	1,409
Mark Messier	627	1,087	1,714	Jari Kurri	601	797	1,398
Ron Francis*	487	1,137	1,624	John Bucyk	556	813	1,369
Steve Yzerman*	645	969	1,614	Guy Lafleur	560	793	1,353
Phil Esposito	717	873	1,590	Doug Gilmour*	429	914	1,343
Ray Bourque*	410	1,169	1,579	Denis Savard	473	865	1,338
Mario Lemieux*	648	922	1,570	Mike Gartner	708	627	1,335
Paul Coffey*	396	1,135	1,531	Gilbert Perreault	512	814	1,326

Note: Through end of 2000-2001 season. *Active in the 2000-01 season.

Most NHL Goals in a Season

Player	Team	Season	Goals	Player	Team	Season	Goals
Wayne Gretzky	Edmonton	1981-82	92	Jari Kurri	Edmonton	1984-85	71
Wayne Gretzky	Edmonton	1983-84	87	Brett Hull	St. Louis	1991-92	70
Brett Hull	St. Louis	1990-91	86	Mario Lemieux	Pittsburgh	1987-88	70
Mario Lemieux	Pittsburgh	1988-89	85	Bernie Nicholls	Los Angeles	1988-89	70
Phil Esposito	Boston	1971-72	76	Mike Bossy	N.Y. Islanders	1978-79	69
Alexander Mogilny	Buffalo	1992-93	76	Mario Lemieux	Pittsburgh	1992-93	69
Teemu Selanne	Winnipeg	1992-93	76	Mario Lemieux	Pittsburgh	1995-96	69
Wayne Gretzky	Edmonton	1984-85	73	Mike Bossy	N.Y. Islanders	1980-81	68
Brett Hull	St. Louis	1989-90	72	Phil Esposito	Boston	1973-74	68
Wayne Gretzky	Edmonton	1982-83	71	Jari Kurri	Edmonton	1985-86	68

Art Ross Trophy (Leading Points Scorer)

1927 Bill Cook, N.Y. Rangers
1928 Howie Morenz, Montreal
1929 Ace Bailey, Toronto
1930 Cooney Weiland, Boston
1931 Howie Morenz, Montreal
1932 Harvey Jackson, Toronto
1933 Bill Cook, N.Y. Rangers
1934 Charlie Conacher, Toronto
1935 Charlie Conacher, Toronto
1936 Dave Schriner, N.Y. Americans
1937 Dave Schriner, N.Y. Americans
1938 Gordie Drillon, Toronto
1939 Toe Blake, Montreal
1940 Milt Schmidt, Boston
1941 Bill Cowley, Boston
1942 Bryan Hextall, N.Y. Rangers
1943 Doug Bentley, Chicago
1944 Herbie Cain, Boston
1945 Elmer Lach, Montreal
1946 Max Bentley, Chicago
1947 Max Bentley, Chicago
1948 Elmer Lach, Montreal
1949 Roy Conacher, Chicago
1950 Ted Lindsay, Detroit
1951 Gordie Howe, Detroit
1952 Gordie Howe, Detroit
1953 Gordie Howe, Detroit
1954 Gordie Howe, Detroit
1955 Bernie Geoffrion, Montreal
1956 Jean Beliveau, Montreal
1957 Gordie Howe, Detroit
1958 Dickie Moore, Montreal
1959 Dickie Moore, Montreal
1960 Bobby Hull, Chicago
1961 Bernie Geoffrion, Montreal
1962 Bobby Hull, Chicago
1963 Gordie Howe, Detroit
1964 Stan Mikita, Chicago
1965 Stan Mikita, Chicago
1966 Bobby Hull, Chicago
1967 Stan Mikita, Chicago
1968 Stan Mikita, Chicago
1969 Phil Esposito, Boston
1970 Bobby Orr, Boston
1971 Phil Esposito, Boston
1972 Phil Esposito, Boston
1973 Phil Esposito, Boston
1974 Phil Esposito, Boston
1975 Bobby Orr, Boston
1976 Guy Lafleur, Montreal
1977 Guy Lafleur, Montreal
1978 Guy Lafleur, Montreal
1979 Bryan Trottier, N.Y. Islanders
1980 Marcel Dionne, Los Angeles
1981 Wayne Gretzky, Edmonton
1982 Wayne Gretzky, Edmonton
1983 Wayne Gretzky, Edmonton
1984 Wayne Gretzky, Edmonton
1985 Wayne Gretzky, Edmonton
1986 Wayne Gretzky, Edmonton
1987 Wayne Gretzky, Edmonton
1988 Mario Lemieux, Pittsburgh
1989 Mario Lemieux, Pittsburgh
1990 Wayne Gretzky, Los Angeles
1991 Wayne Gretzky, Los Angeles
1992 Mario Lemieux, Pittsburgh
1993 Mario Lemieux, Pittsburgh
1994 Wayne Gretzky, Los Angeles
1995 Jaromir Jagr, Pittsburgh
1996 Mario Lemieux, Pittsburgh
1997 Mario Lemieux, Pittsburgh
1998 Jaromir Jagr, Pittsburgh
1999 Jaromir Jagr, Pittsburgh
2000 Jaromir Jagr, Pittsburgh
2001 Jaromir Jagr, Pittsburgh

Maurice "Rocket" Richard Trophy (Most Goals)

1999 Teemu Selanne, Anaheim
2000 Pavel Bure, Florida
2001 Pavel Bure, Florida

James Norris Memorial Trophy (Outstanding Defenseman)

1954 Red Kelly, Detroit
1955 Doug Harvey, Montreal
1956 Doug Harvey, Montreal
1957 Doug Harvey, Montreal
1958 Doug Harvey, Montreal
1959 Tom Johnson, Montreal
1960 Doug Harvey, Montreal
1961 Doug Harvey, Montreal
1962 Doug Harvey, N.Y. Rangers
1963 Pierre Pilote, Chicago
1964 Pierre Pilote, Chicago
1965 Pierre Pilote, Chicago
1966 Jacques Laperriere, Montreal
1967 Harry Howell, N.Y. Rangers
1968 Bobby Orr, Boston
1969 Bobby Orr, Boston
1970 Bobby Orr, Boston
1971 Bobby Orr, Boston
1972 Bobby Orr, Boston
1973 Bobby Orr, Boston
1974 Bobby Orr, Boston
1975 Bobby Orr, Boston
1976 Denis Potvin, N.Y. Islanders
1977 Larry Robinson, Montreal
1978 Denis Potvin, N.Y. Islanders
1979 Denis Potvin, N.Y. Islanders
1980 Larry Robinson, Montreal
1981 Randy Carlyle, Pittsburgh
1982 Doug Wilson, Chicago
1983 Rod Langway, Washington
1984 Rod Langway, Washington
1985 Paul Coffey, Edmonton
1986 Paul Coffey, Edmonton
1987 Ray Bourque, Boston
1988 Ray Bourque, Boston
1989 Chris Chelios, Montreal
1990 Ray Bourque, Boston
1991 Ray Bourque, Boston
1992 Brian Leetch, N.Y. Rangers
1993 Chris Chelios, Chicago
1994 Ray Bourque, Boston
1995 Paul Coffey, Detroit
1996 Chris Chelios, Chicago
1997 Brian Leetch, N.Y. Rangers
1998 Rob Blake, Los Angeles
1999 Al MacInnis, St. Louis
2000 Chris Pronger, St. Louis
2001 Nicklas Lidstrom, Detroit

Vezina Trophy (Outstanding Goalie)*

1927 George Hainsworth, Montreal
1928 George Hainsworth, Montreal
1929 George Hainsworth, Montreal
1930 Tiny Thompson, Boston
1931 Roy Worters, N.Y. Americans
1932 Charlie Gardiner, Chicago
1933 Tiny Thompson, Boston
1934 Charlie Gardiner, Chicago
1935 Lorne Chabot, Chicago
1936 Tiny Thompson, Boston
1937 Normie Smith, Detroit
1938 Tiny Thompson, Boston
1939 Frank Brimsek, Boston
1940 Dave Kerr, N.Y. Rangers
1941 Turk Broda, Toronto
1942 Frank Brimsek, Boston
1943 Johnny Mowers, Detroit
1944 Bill Durnan, Montreal
1945 Bill Durnan, Montreal
1946 Bill Durnan, Montreal
1947 Bill Durnan, Montreal
1948 Turk Broda, Toronto
1949 Bill Durnan, Montreal
1950 Bill Durnan, Montreal
1951 Al Rollins, Toronto
1952 Terry Sawchuk, Detroit
1953 Terry Sawchuk, Detroit
1954 Harry Lumley, Toronto
1955 Terry Sawchuk, Detroit
1956 Jacques Plante, Montreal
1957 Jacques Plante, Montreal
1958 Jacques Plante, Montreal
1959 Jacques Plante, Montreal
1960 Jacques Plante, Montreal
1961 John Bower, Toronto
1962 Jacques Plante, Montreal
1963 Glenn Hall, Chicago
1964 Charlie Hodge, Montreal
1965 Sawchuk, Bower, Toronto
1966 Worsley, Hodge, Montreal
1967 Hall, DeJordy, Chicago
1968 Worsley, Vachon, Montreal
1969 Hall, Plante, St. Louis
1970 Tony Esposito, Chicago
1971 Giacomin, Villemure, N.Y. Rangers
1972 Esposito, Smith, Chicago
1973 Ken Dryden, Montreal
1974 Bernie Parent, Philadelphia; Tony Esposito, Chicago
1975 Bernie Parent, Philadelphia
1976 Ken Dryden, Montreal
1977 Dryden, Larocque, Montreal
1978 Dryden, Larocque, Montreal
1979 Dryden, Larocque, Montreal
1980 Sauve, Edwards, Buffalo
1981 Sevigny, Larocque, Herron, Montreal
1982 Bill Smith, N.Y. Islanders
1983 Pete Peeters, Boston
1984 Tom Barrasso, Buffalo
1985 Pelle Lindbergh, Philadelphia
1986 John Vanbiesbrouck, N.Y. Rangers
1987 Ron Hextall, Philadelphia
1988 Grant Fuhr, Edmonton
1989 Patrick Roy, Montreal
1990 Patrick Roy, Montreal
1991 Ed Belfour, Chicago
1992 Patrick Roy, Montreal
1993 Ed Belfour, Chicago
1994 Dominik Hasek, Buffalo
1995 Dominik Hasek, Buffalo
1996 Jim Carey, Washington
1997 Dominik Hasek, Buffalo
1998 Dominik Hasek, Buffalo
1999 Dominik Hasek, Buffalo
2000 Olaf Kolzig, Washington
2001 Dominik Hasek, Buffalo

*Before 1982, awarded to the goalie or goalies who played a minimum of 25 games for the team that allowed the fewest goals; sinc e 1982, awarded to the outstanding goalie, as determined by a vote of NHL general managers.

Frank J. Selke Trophy (Best Defensive Forward)

1978 Bob Gainey, Montreal
1979 Bob Gainey, Montreal
1980 Bob Gainey, Montreal
1981 Bob Gainey, Montreal
1982 Steve Kasper, Boston
1983 Bobby Clarke, Philadelphia
1984 Doug Jarvis, Washington
1985 Craig Ramsay, Buffalo
1986 Troy Murray, Chicago
1987 Dave Poulin, Philadelphia
1988 Guy Carbonneau, Montreal
1989 Guy Carbonneau, Montreal
1990 Rick Meagher, St. Louis
1991 Dirk Graham, Chicago
1992 Guy Carbonneau, Montreal
1993 Doug Gilmour, Toronto
1994 Sergei Fedorov, Detroit
1995 Ron Francis, Pittsburgh
1996 Sergei Federov, Detroit
1997 Michael Peca, Buffalo
1998 Jere Lehtinen, Dallas
1999 Jere Lehtinen, Dallas
2000 Steve Yzerman, Detroit
2001 John Madden, New Jersey

Calder Memorial Trophy (Rookie of the Year)

1933 Carl Voss, Detroit
1934 Russ Blinco, Montreal Maroons
1935 Dave Schriner, N.Y. Americans
1936 Mike Karakas, Chicago
1937 Syl Apps, Toronto
1938 Cully Dahlstrom, Chicago
1939 Frank Brimsek, Boston
1940 Kilby Macdonald, N.Y. Rangers
1941 John Quilty, Montreal
1942 Grant Warwick, N.Y. Rangers
1943 Gaye Stewart, Toronto
1944 Gus Bodnar, Toronto
1945 Frank McCool, Toronto
1946 Edgar Laprade, N.Y. Rangers
1947 Howie Meeker, Toronto
1948 Jim McFadden, Detroit
1949 Pentti Lund, N.Y. Rangers
1950 Jack Gelineau, Boston
1951 Terry Sawchuk, Detroit
1952 Bernie Geoffrion, Montreal
1953 Gump Worsley, N.Y. Rangers
1954 Camille Henry, N.Y. Rangers
1955 Ed Litzenberger, Chicago
1956 Glenn Hall, Detroit
1957 Larry Regan, Boston
1958 Frank Mahovlich, Toronto
1959 Ralph Backstrom, Montreal
1960 Bill Hay, Chicago
1961 Dave Keon, Toronto
1962 Bobby Rousseau, Montreal
1963 Kent Douglas, Toronto
1964 Jacques Laperriere, Montreal
1965 Roger Crozier, Detroit
1966 Brit Selby, Toronto
1967 Bobby Orr, Boston
1968 Derek Sanderson, Boston
1969 Danny Grant, Minnesota
1970 Tony Esposito, Chicago
1971 Gilbert Perreault, Buffalo
1972 Ken Dryden, Montreal
1973 Steve Vickers, N.Y. Rangers
1974 Denis Potvin, N.Y. Islanders
1975 Eric Vail, Atlanta
1976 Bryan Trottier, N.Y. Islanders
1977 Willi Plett, Atlanta
1978 Mike Bossy, N.Y. Islanders
1979 Bobby Smith, Minnesota
1980 Ray Bourque, Boston
1981 Peter Stastny, Quebec
1982 Dale Hawerchuk, Winnipeg
1983 Steve Larmer, Chicago
1984 Tom Barrasso, Buffalo
1985 Mario Lemieux, Pittsburgh
1986 Gary Suter, Calgary
1987 Luc Robitaille, Los Angeles
1988 Joe Nieuwendyk, Calgary
1989 Brian Leetch, N.Y. Rangers
1990 Sergei Makarov, Calgary
1991 Ed Belfour, Chicago
1992 Pavel Bure, Vancouver
1993 Teemu Selanne, Winnipeg
1994 Martin Brodeur, New Jersey
1995 Peter Forsberg, Quebec
1996 Daniel Alfredsson, Ottawa
1997 Bryan Berard, N.Y. Islanders
1998 Sergei Samsonov, Boston
1999 Chris Drury, Colorado
2000 Scott Gomez, New Jersey
2001 Evgeni Nabokov, San Jose

Lady Byng Memorial Trophy (Most Gentlemanly Player)

1925	Frank Nighbor, Ottawa
1926	Frank Nighbor, Ottawa
1927	Billy Burch, N.Y. Americans
1928	Frank Boucher, N.Y. Rangers
1929	Frank Boucher, N.Y. Rangers
1930	Frank Boucher, N.Y. Rangers
1931	Frank Boucher, N.Y. Rangers
1932	Joe Primeau, Toronto
1933	Frank Boucher, N.Y. Rangers
1934	Frank Boucher, N.Y. Rangers
1935	Frank Boucher, N.Y. Rangers
1936	Doc Romnes, Chicago
1937	Marty Barry, Detroit
1938	Gordie Drillon, Toronto
1939	Clint Smith, N.Y. Rangers
1940	Bobby Bauer, Boston
1941	Bobby Bauer, Boston
1942	Syl Apps, Toronto
1943	Max Bentley, Chicago
1944	Clint Smith, Chicago
1945	Bill Mosienko, Chicago
1946	Toe Blake, Montreal
1947	Bobby Bauer, Boston
1948	Buddy O'Connor, N.Y. Rangers
1949	Bill Quackenbush, Detroit
1950	Edgar Laprade, N.Y. Rangers
1951	Red Kelly, Detroit
1952	Sid Smith, Toronto
1953	Red Kelly, Detroit
1954	Red Kelly, Detroit
1955	Sid Smith, Toronto
1956	Earl Reibel, Detroit
1957	Andy Hebenton, N.Y. Rangers
1958	Camille Henry, N.Y. Rangers
1959	Alex Delvecchio, Detroit
1960	Don McKenney, Boston
1961	Red Kelly, Toronto
1962	Dave Keon, Toronto
1963	Dave Keon, Toronto
1964	Ken Wharram, Chicago
1965	Bobby Hull, Chicago
1966	Alex Delvecchio, Detroit
1967	Stan Mikita, Chicago
1968	Stan Mikita, Chicago
1969	Alex Delvecchio, Detroit
1970	Phil Goyette, St. Louis
1971	John Bucyk, Boston
1972	Jean Ratelle, N.Y. Rangers
1973	Gil Perreault, Buffalo
1974	John Bucyk, Boston
1975	Marcel Dionne, Detroit
1976	Jean Ratelle, N.Y.R.-Boston
1977	Marcel Dionne, Los Angeles
1978	Butch Goring, Los Angeles
1979	Bob MacMillan, Atlanta
1980	Wayne Gretzky, Edmonton
1981	Rick Kehoe, Pittsburgh
1982	Rick Middleton, Boston
1983	Mike Bossy, N.Y. Islanders
1984	Mike Bossy, N.Y. Islanders
1985	Jari Kurri, Edmonton
1986	Mike Bossy, N.Y. Islanders
1987	Joe Mullen, Calgary
1988	Mats Naslund, Montreal
1989	Joe Mullen, Calgary
1990	Brett Hull, St. Louis
1991	Wayne Gretzky, Los Angeles
1992	Wayne Gretzky, Los Angeles
1993	Pierre Turgeon, N.Y. Islanders
1994	Wayne Gretzky, Los Angeles
1995	Ron Francis, Pittsburgh
1996	Paul Kariya, Anaheim
1997	Paul Kariya, Anaheim
1998	Ron Francis, Pittsburgh
1999	Wayne Gretzky, N.Y. Rangers
2000	Pavol Demitra, St. Louis
2001	Joe Sakic, Colorado

Hart Memorial Trophy (MVP)

1927	Herb Gardiner, Montreal
1928	Howie Morenz, Montreal
1929	Roy Worters, N.Y. Americans
1930	Nels Stewart, Montreal Maroons
1931	Howie Morenz, Montreal
1932	Howie Morenz, Montreal
1933	Eddie Shore, Boston
1934	Aurel Joliat, Montreal
1935	Eddie Shore, Boston
1936	Eddie Shore, Boston
1937	Babe Siebert, Montreal
1938	Eddie Shore, Boston
1939	Toe Blake, Montreal
1940	Ebbie Goodfellow, Detroit
1941	Bill Cowley, Boston
1942	Tom Anderson, N.Y. Americans
1943	Bill Cowley, Boston
1944	Babe Pratt, Toronto
1945	Elmer Lach, Montreal
1946	Max Bentley, Chicago
1947	Maurice Richard, Montreal
1948	Buddy O'Connor, N.Y. Rangers
1949	Sid Abel, Detroit
1950	Chuck Rayner, N.Y. Rangers
1951	Milt Schmidt, Boston
1952	Gordie Howe, Detroit
1953	Gordie Howe, Detroit
1954	Al Rollins, Chicago
1955	Ted Kennedy, Toronto
1956	Jean Beliveau, Montreal
1957	Gordie Howe, Detroit
1958	Gordie Howe, Detroit
1959	Andy Bathgate, N.Y. Rangers
1960	Gordie Howe, Detroit
1961	Bernie Geoffrion, Montreal
1962	Jacques Plante, Montreal
1963	Gordie Howe, Detroit
1964	Jean Beliveau, Montreal
1965	Bobby Hull, Chicago
1966	Bobby Hull, Chicago
1967	Stan Mikita, Chicago
1968	Stan Mikita, Chicago
1969	Phil Esposito, Boston
1970	Bobby Orr, Boston
1971	Bobby Orr, Boston
1972	Bobby Orr, Boston
1973	Bobby Clarke, Philadelphia
1974	Phil Esposito, Boston
1975	Bobby Clarke, Philadelphia
1976	Bobby Clarke, Philadelphia
1977	Guy Lafleur, Montreal
1978	Guy Lafleur, Montreal
1979	Bryan Trottier, N.Y. Islanders
1980	Wayne Gretzky, Edmonton
1981	Wayne Gretzky, Edmonton
1982	Wayne Gretzky, Edmonton
1983	Wayne Gretzky, Edmonton
1984	Wayne Gretzky, Edmonton
1985	Wayne Gretzky, Edmonton
1986	Wayne Gretzky, Edmonton
1987	Wayne Gretzky, Edmonton
1988	Mario Lemieux, Pittsburgh
1989	Wayne Gretzky, Los Angeles
1990	Mark Messier, Edmonton
1991	Brett Hull, St. Louis
1992	Mark Messier, N.Y. Rangers
1993	Mario Lemieux, Pittsburgh
1994	Sergei Fedorov, Detroit
1995	Eric Lindros, Philadelphia
1996	Mario Lemieux, Pittsburgh
1997	Dominik Hasek, Buffalo
1998	Dominik Hasek, Buffalo
1999	Jaromir Jagr, Pittsburgh
2000	Chris Pronger, St. Louis
2001	Joe Sakic, Colorado

Conn Smythe Trophy (MVP in Playoffs)

1965	Jean Beliveau, Montreal
1966	Roger Crozier, Detroit
1967	Dave Keon, Toronto
1968	Glenn Hall, St. Louis
1969	Serge Savard, Montreal
1970	Bobby Orr, Boston
1971	Ken Dryden, Montreal
1972	Bobby Orr, Boston
1973	Yvan Cournoyer, Montreal
1974	Bernie Parent, Philadelphia
1975	Bernie Parent, Philadelphia
1976	Reg Leach, Philadelphia
1977	Guy Lafleur, Montreal
1978	Larry Robinson, Montreal
1979	Bob Gainey, Montreal
1980	Bryan Trottier, N.Y. Islanders
1981	Butch Goring, N.Y. Islanders
1982	Mike Bossy, N.Y. Islanders
1983	Billy Smith, N.Y. Islanders
1984	Mark Messier, Edmonton
1985	Wayne Gretzky, Edmonton
1986	Patrick Roy, Montreal
1987	Ron Hextall, Philadelphia
1988	Wayne Gretzky, Edmonton
1989	Al MacInnis, Calgary
1990	Bill Ranford, Edmonton
1991	Mario Lemieux, Pittsburgh
1992	Mario Lemieux, Pittsburgh
1993	Patrick Roy, Montreal
1994	Brian Leetch, N.Y. Rangers
1995	Claude Lemieux, New Jersey
1996	Joe Sakic, Colorado
1997	Mike Vernon, Detroit
1998	Steve Yzerman, Detroit
1999	Joe Nieuwendyk, Dallas
2000	Scott Stevens, New Jersey
2001	Patrick Roy, Colorado

National Hockey Hall of Fame, Toronto, Ontario

(2001 inductees have an asterisk*)

PLAYERS

Abel, Sid
Adams, Jack
Apps, Syl
Armstrong, George
Bailey, Ace
Bain, Dan
Baker, Hobey
Barber, Bill
Barry, Marty
Bathgate, Andy
Bauer, Bobby
Beliveau, Jean
Benedict, Clint
Bentley, Doug
Bentley, Max
Blake, Toe
Boivin, Leo
Boon, Dickie
Bossy, Mike
Bouchard, Butch
Boucher, Frank
Boucher, George
Bower, Johnny
Bowie, Dubbie
Brimsek, Frank
Broadbent, Punch
Broda, Turk
Bucyk, John
Burch, Billy
Cameron, Harry
Cheevers, Gerry
Clancy, King
Clapper, Dit
Clarke, Bobby
Cleghorn, Sprague
Colville, Neil
Conacher, Charlie
Conacher, Lionel
Conacher, Roy
Connell, Alex
Cook, Bill
Cook, Bun
Coulter, Art
Cournoyer, Yvan
Cowley, Bill
Crawford, Rusty
Darragh, Jack
Davidson, Scotty
Day, Hap
Delvecchio, Alex
Denneny, Cy
Dionne, Marcel
Drillon, Gordie
Drinkwater, Graham
Dryden, Ken
Dumart, Woody
Dunderdale, Tommy
Durnan, Bill
Dutton, Red
Dye, Babe
Esposito, Phil
Esposito, Tony
Farrel, Arthur
*Fetisov, Viacheslav
Flaman, Fernie
Foyston, Frank
Fredrickson, Frank
Gadsby, Bill
Gainey, Bob
Gardiner, Chuck
Gardiner, Herb
Gardiner, Jimmy
*Gartner, Mike

Geoffrion, Bernie
Gerard, Eddie
Giacomin, Eddie
Gilbert, Rod
Gilmour, Billy
Goheen, Moose
Goodfellow, Ebbie
Goulet, Michel
Grant, Mike
Green, Shorty
Gretzky, Wayne
Griffis, Si
Hainsworth, George
Hall, Glenn
Hall, Joe
Harvey, Doug
*Hawerchuk, Dale
Hay, George
Hern, Riley
Hextall, Bryan
Holmes, Hap
Hooper, Tom
Horner, Red
Horton, Tim
Howe, Gordie
Howe, Syd
Howell, Harry
Hull, Bobby
Hutton, Bouse
Hyland, Harry
Irvin, Dick
Jackson, Busher
Johnson, Ching
Johnson, Ernie
Johnson, Tom
Joliat, Aurel
Keats, Duke
Kelly, Red
Kennedy, Ted
Keon, Dave
*Kurri, Jari
Lach, Elmer
Lafleur, Guy
Lalonde, Newsy
Laperriere, Jacques
Lapointe, Guy
Laprade, Edgar
Laviolette, Jack
LeSueur, Percy
Lehman, Hughie
Lemaire, Jacques
Lemieux, Mario
Lewis, Herbie
Lindsay, Ted
Lumley, Harry
MacKay, Mickey
Mahovlich, Frank
Malone, Joe
Mantha, Sylvio
Marshall, Jack
Maxwell, Fred
McDonald, Lanny
McGee, Frank
McGimsie, Billy
McNamara, George
Mikita, Stan
Moore, Dickie
Moran, Paddy
Morenz, Howie
Mosienko, Bill
Mullen, Joe
Nighbor, Frank
Noble, Reg
O'Connor, Buddy
Oliver, Harry
Olmstead, Bert
Orr, Bobby
Parent, Bernie
Park, Brad
Patrick, Lester
Patrick, Lynn
Perreault, Gilbert
Phillips, Tom
Pilote, Pierre
Pitre, Didier
Plante, Jacques
Potvin, Denis
Pratt, Babe
Primeau, Joe
Pronovost, Marcel
Pulford, Bob
Pulford, Harvey
Quackenbush, Bill
Rankin, Frank
Ratelle, Jean
Rayner, Chuck
Reardon, Kenny
Richard, Henri
Richard, Maurice
Richardson, George
Roberts, Gordie
Robinson, Larry
Ross, Art
Russel, Blair
Russell, Ernie
Ruttan, Jack
Salming, Borje
Savard, Denis
Savard, Serge
Sawchuk, Terry
Scanlan, Fred
Schmidt, Milt
Schriner, Sweeney
Seibert, Earl
Seibert, Oliver
Shore, Eddie
Shutt, Steve
Siebert, Babe
Simpson, Joe
Sittler, Darryl
Smith, Alf
Smith, Billy
Smith, Clint
Smith, Hooley
Smith, Tommy
Stanley, Allan
Stanley, Barney
Stastny, Peter
Stewart, Jack
Stewart, Nels
Stuart, Bruce
Stuart, Hod
Taylor, Cyclone
Thompson, Tiny
Tretiak, Vladislav
Trihey, Harry
Trottier, Bryan
Ullman, Norm
Vezina, Georges
Walker, Jack
Walsh, Marty
Watson, Harry (Moose)
Watson, Harry Percival
Weiland, Cooney
Westwick, Harry
Whitcroft, Fred
Wilson, Phat
Worsley, Gump
Worters, Roy

BUILDERS

Adams, Charles
Adams, Weston
Ahearn, Bunny
Ahearn, Frank
Allan, Sir Montagu
Allen, Keith
Arbour, Al
Ballard, Harold
Bauer, Father David
Bickell, J.P.
Bowman, Scotty
Brown, George
Brown, Walter
Buckland, Frank
Bush, Walter, Jr.
Butterfield, Jack
Calder, Frank
Campbell, Angus
Campbell, Clarence
Cattarinich, Joseph
Dandurand, Leo
Dilio, Frank
Dudley, George
Dunn, James
Francis, Emile
Gibson, Jack
Gorman, Tommy
Griffiths, Frank
Hanley, Bill
Hay, Charles
Hendy, Jim
Hewitt, Foster
Hewitt, William
Hume, Fred
Imlach, Punch
Ivan, Tommy
Jennings, William
Johnson, Bob
Juckes, Gordon
Kilpatrick, John
Knox, Seymour
LeBel, Robert
Leader, Al
Lockhart, Thomas
Loicq, Paul
Mariucci, John
Mathers, Frank
McLaughlin, Frederic
Milford, Jake
Molson, Sen. Hartland
Morrison, Ian "Scotty"
Murray, Pere Athol
Nelson, Francis
Norris, Bruce
Norris, James
Norris, James Sr.
Northey, William
O'Brien, J. Ambrose
O'Neill, Brian Francis
Page, Frederick
*Patrick, Craig
Patrick, Frank
Pickard, Allan
Pilous, Rudy
Poile, Bud
Pollock, Sam
Raymond, Sen. Donat
Robertson, John Ross
Robinson, Claude
Ross, Phillip
Sabetzki, Gunther
Sather, Glen
Selke, Frank
Sinden, Harry
Smith, Frank
Smythe, Conn
Snider, Ed
Stanley, Lord (of Preston)
Sutherland, Capt. James T.
Tarasov, Anatoli
Torrey, Bill
Turner, Lloyd
Tutt, William
Voss, Carl
Waghorne, Fred
Wirtz, Arthur
Wirtz, Bill
Ziegler, John A., Jr.

REFEREES AND LINESMEN

Armstrong, Neil
Ashley, John
Chadwick, Bill
D'Amico, John
Elliott, Chaucer
Hayes, George
Hewiston, Bobby
Ion, Mickey
Pavelich, Matt
Rodden, Mike
Smeaton, Cooper
Storey, Red
Udvari, Frank
Van Hellemond, Andy

NHL Home Ice

Team	Name (built)	Capacity
Anaheim	The Arrowhead Pond of Anaheim (1993)	17,174
Atlanta	Philips Arena (1999)	18,545
Boston	FleetCenter (1995)	17,565
Buffalo	HSBC Arena[1] (1996)	18,690
Calgary	Pengrowth Saddledome (1983)	17,139
Carolina	Entertainment & Sports Arena (1999)	18,730
Chicago	United Center (1994)	20,500
Colorado	Pepsi Center (1999)	22,076
Columbus	Nationwide Arena (2000)	18,524
Dallas	American Airlines Center (1980)	18,000
Detroit	Joe Louis Arena (1979)	19,983
Edmonton	Skyreach Centre[2] (1974)	17,100
Florida	National Car Rental Center (1999)	19,250
Los Angeles	Staples Center (1999)	18,118
Minnesota	Xcel Energy Arena (2000)	18,600
Montreal	Le Centre Molson (1996)	21,273
Nashville	Gaylord Entertainment Center[3] (1996)	17,500
New Jersey	Continental Airlines Arena[4] (1981)	19,040
N.Y. Islanders	Nassau Veterans Memorial Col. (1972)	16,297
N.Y. Rangers	Madison Square Garden (1968)	18,200
Ottawa	Corel Centre (1996)	18,500
Philadelphia	First Union Center (1996)	19,519
Phoenix	America West Arena (1992)	16,210
Pittsburgh	Mellon Arena[5] (1961)	16,958
St. Louis	Savvis Center[6] (1994)	21,000
San Jose	Compaq Center[7] (1993)	17,496
Tampa Bay	Ice Palace (1996)	19,758
Toronto	Air Canada Centre (1999)	18,800
Vancouver	GM Place (1995)	18,422
Washington	MCI Center (1997)	18,672

(1) Marine Midland Arena, 1996-2000. (2) Northlands Col., 1974-79; Edmonton Col., 1979-98. (3) Nashville Arena, 1997-1999. (4) Brendan Byrne/Meadowlands Arena, 1981-96. (5) Civic Arena, 1961-99. (6) Kiel Center, 1994-2000. (7) San Jose Arena, 1993-2000.

NCAA HOCKEY CHAMPIONS

1948	Michigan
1949	Boston College
1950	Colorado College
1951	Michigan
1952	Michigan
1953	Michigan
1954	RPI
1955	Michigan
1956	Michigan
1957	Colorado College
1958	Denver
1959	North Dakota
1960	Denver
1961	Denver
1962	Michigan Tech
1963	North Dakota
1964	Michigan
1965	Michigan Tech
1966	Michigan State
1967	Cornell
1968	Denver
1969	Denver
1970	Cornell
1971	Boston Univ.
1972	Boston Univ.
1973	Wisconsin
1974	Minnesota
1975	Michigan Tech
1976	Minnesota
1977	Wisconsin
1978	Boston Univ.
1979	Minnesota
1980	North Dakota
1981	Wisconsin
1982	North Dakota
1983	Wisconsin
1984	Bowling Green
1985	RPI
1986	Michigan State
1987	North Dakota
1988	Lake Superior St.
1989	Harvard
1990	Wisconsin
1991	N. Michigan
1992	Lake Superior St.
1993	Maine
1994	Lake Superior St.
1995	Boston Univ.
1996	Michigan
1997	North Dakota
1998	Michigan
1999	Maine
2000	North Dakota
2001	Boston College

GOLF

Men's All-Time Major Professional Championship Leaders

(Through the 2001 season; *active PGA player; (a)=amateur.)

Player	Masters	U.S. Open	British Open	PGA	Total
Jack Nicklaus*	1963, '65-66, '72, '75, '86	1962, '67, '72, '80	1966, '70, '78	1963, '71, '73, '75, '80	18
Walter Hagen	—	1914, '19	1922, '24, '28-29	1921, '24-27	11
Ben Hogan	1951, '53	1948, '50-51, '53	1953	1946, '48	9
Gary Player	1961, '74, '78	1965	1959, '68, '74	1962, '72	9
Tom Watson*	1977, '81	1982	1975, '77, '80, '82-83	—	8
Bobby Jones[a]	—	1923, '26, '29-30	1926-27, '30	—	7
Arnold Palmer	1958, '60, '62, '64	1960	1961-62	—	7
Gene Sarazen	1935	1922, '32	1932	1922-23, '33	7
Sam Snead	1949, '52, '54	—	1946	1942, '49, '51	7
Harry Vardon	—	1900	1896, '98-99, 1903, '11, '14	—	7
Nick Faldo*	1989-90, '96	—	1987, '90, '92	—	6
Lee Trevino	—	1968, '71	1971-72	1974, '84	6
Tiger Woods*	1997, 2001	2000	2000	1999, 2000	6

IT'S A FACT: Tiger Woods's Masters win at Augusta National on Apr. 8, 2001, gave him an unprecedented 4th consecutive major championship. In 2000, he had won the U.S. Open, the British Open, and the PGA Championship.

Professional Golfers' Association Leading Money Winners

Year	Player	Earnings	Year	Player	Earnings	Year	Player	Earnings
1946	Ben Hogan	$42,556	1965	Jack Nicklaus	$140,752	1983	Hal Sutton	$426,668
1947	Jimmy Demaret	27,936	1966	Billy Casper	121,944	1984	Tom Watson	476,260
1948	Ben Hogan	36,812	1967	Jack Nicklaus	188,988	1985	Curtis Strange	542,321
1949	Sam Snead	31,593	1968	Billy Casper	205,168	1986	Greg Norman	653,296
1950	Sam Snead	35,758	1969	Frank Beard	175,223	1987	Curtis Strange	925,941
1951	Lloyd Mangrum	26,088	1970	Lee Trevino	157,037	1988	Curtis Strange	1,147,644
1952	Julius Boros	37,032	1971	Jack Nicklaus	244,490	1989	Tom Kite	1,395,278
1953	Lew Worsham	34,002	1972	Jack Nicklaus	320,542	1990	Greg Norman	1,165,477
1954	Bob Toski	65,819	1973	Jack Nicklaus	308,362	1991	Corey Pavin	979,430
1955	Julius Boros	65,121	1974	Johnny Miller	353,201	1992	Fred Couples	1,344,188
1956	Ted Kroll	72,835	1975	Jack Nicklaus	323,149	1993	Nick Price	1,478,557
1957	Dick Mayer	65,835	1976	Jack Nicklaus	266,438	1994	Nick Price	1,499,927
1958	Arnold Palmer	42,407	1977	Tom Watson	310,653	1995	Greg Norman	1,654,959
1959	Art Wall, Jr.	53,167	1978	Tom Watson	362,429	1996	Tom Lehman	1,780,159
1960	Arnold Palmer	75,262	1979	Tom Watson	462,636	1997	Tiger Woods	2,066,833
1961	Gary Player	64,540	1980	Tom Watson	530,808	1998	David Duval	2,591,031
1962	Arnold Palmer	81,448	1981	Tom Kite	375,699	1999	Tiger Woods	6,616,585
1963	Arnold Palmer	128,230	1982	Craig Stadler	446,462	2000	Tiger Woods	9,188,321
1964	Jack Nicklaus	113,284						

Masters Golf Tournament Winners

Year	Winner	Year	Winner	Year	Winner	Year	Winner
1934	Horton Smith	1953	Ben Hogan	1970	Billy Casper	1986	Jack Nicklaus
1935	Gene Sarazen	1954	Sam Snead	1971	Charles Coody	1987	Larry Mize
1936	Horton Smith	1955	Cary Middlecoff	1972	Jack Nicklaus	1988	Sandy Lyle
1937	Byron Nelson	1956	Jack Burke	1973	Tommy Aaron	1989	Nick Faldo
1938	Henry Picard	1957	Doug Ford	1974	Gary Player	1990	Nick Faldo
1939	Ralph Guldahl	1958	Arnold Palmer	1975	Jack Nicklaus	1991	Ian Woosnam
1940	Jimmy Demaret	1959	Art Wall Jr.	1976	Ray Floyd	1992	Fred Couples
1941	Craig Wood	1960	Arnold Palmer	1977	Tom Watson	1993	Bernhard Langer
1942	Byron Nelson	1961	Gary Player	1978	Gary Player	1994	Jose Maria Olazabal
1943-45	Not Played	1962	Arnold Palmer	1979	Fuzzy Zoeller	1995	Ben Crenshaw
1946	Herman Keiser	1963	Jack Nicklaus	1980	Seve Ballesteros	1996	Nick Faldo
1947	Jimmy Demaret	1964	Arnold Palmer	1981	Tom Watson	1997	Tiger Woods
1948	Claude Harmon	1965	Jack Nicklaus	1982	Craig Stadler	1998	Mark O'Meara
1949	Sam Snead	1966	Jack Nicklaus	1983	Seve Ballesteros	1999	Jose Maria Olazabal
1950	Jimmy Demaret	1967	Gay Brewer, Jr.	1984	Ben Crenshaw	2000	Vijay Singh
1951	Ben Hogan	1968	Bob Goalby	1985	Bernhard Langer	2001	Tiger Woods
1952	Sam Snead	1969	George Archer				

United States Open Winners

(First contested in 1895)

Year	Winner	Year	Winner	Year	Winner	Year	Winner
1934	Olin Dutra	1954	Ed Furgol	1970	Tony Jacklin	1986	Ray Floyd
1935	Sam Parks, Jr.	1955	Jack Fleck	1971	Lee Trevino	1987	Scott Simpson
1936	Tony Manero	1956	Cary Middlecoff	1972	Jack Nicklaus	1988	Curtis Strange
1937	Ralph Guldahl	1957	Dick Mayer	1973	Johnny Miller	1989	Curtis Strange
1938	Ralph Guldahl	1958	Tommy Bolt	1974	Hale Irwin	1990	Hale Irwin
1939	Byron Nelson	1959	Billy Casper	1975	Lou Graham	1991	Payne Stewart
1940	Lawson Little	1960	Arnold Palmer	1976	Jerry Pate	1992	Tom Kite
1941	Craig Wood	1961	Gene Littler	1977	Hubert Green	1993	Lee Janzen
1942-45	Not Played	1962	Jack Nicklaus	1978	Andy North	1994	Ernie Els
1946	Lloyd Mangrum	1963	Julius Boros	1979	Hale Irwin	1995	Corey Pavin
1947	L. Worsham	1964	Ken Venturi	1980	Jack Nicklaus	1996	Steve Jones
1948	Ben Hogan	1965	Gary Player	1981	David Graham	1997	Ernie Els
1949	Cary Middlecoff	1966	Billy Casper	1982	Tom Watson	1998	Lee Janzen
1950	Ben Hogan	1967	Jack Nicklaus	1983	Larry Nelson	1999	Payne Stewart
1951	Ben Hogan	1968	Lee Trevino	1984	Fuzzy Zoeller	2000	Tiger Woods
1952	Julius Boros	1969	Orville Moody	1985	Andy North	2001	Retief Goosen
1953	Ben Hogan						

British Open Winners

(First contested in 1860)

Year	Winner	Year	Winner	Year	Winner	Year	Winner
1934	Henry Cotton	1955	Peter Thomson	1971	Lee Trevino	1987	Nick Faldo
1935	Alf Perry	1956	Peter Thomson	1972	Lee Trevino	1988	Seve Ballesteros
1936	Alf Padgham	1957	Bobby Locke	1973	Tom Weiskopf	1989	Mark Calcavecchia
1937	T.H. Cotton	1958	Peter Thomson	1974	Gary Player	1990	Nick Faldo
1938	R.A. Whitcombe	1959	Gary Player	1975	Tom Watson	1991	Ian Baker-Finch
1939	Richard Burton	1960	Kel Nagle	1976	Johnny Miller	1992	Nick Faldo
1940-45	Not Played	1961	Arnold Palmer	1977	Tom Watson	1993	Greg Norman
1946	Sam Snead	1962	Arnold Palmer	1978	Jack Nicklaus	1994	Nick Price
1947	Fred Daly	1963	Bob Charles	1979	Seve Ballesteros	1995	John Daly
1948	Henry Cotton	1964	Tony Lema	1980	Tom Watson	1996	Tom Lehman
1949	Bobby Locke	1965	Peter Thomson	1981	Bill Rogers	1997	Justin Leonard
1950	Bobby Locke	1966	Jack Nicklaus	1982	Tom Watson	1998	Mark O'Meara
1951	Max Faulkner	1967	Roberto de Vicenzo	1983	Tom Watson	1999	Paul Lawrie
1952	Bobby Locke	1968	Gary Player	1984	Seve Ballesteros	2000	Tiger Woods
1953	Ben Hogan	1969	Tony Jacklin	1985	Sandy Lyle	2001	David Duval
1954	Peter Thomson	1970	Jack Nicklaus	1986	Greg Norman		

PGA Championship Winners

(First contested in 1916)

Year	Winner	Year	Winner	Year	Winner	Year	Winner
1934	Paul Runyan	1951	Sam Snead	1968	Julius Boros	1985	Hubert Green
1935	Johnny Revolta	1952	James Turnesa	1969	Ray Floyd	1986	Bob Tway
1936	Denny Shute	1953	Walter Burkemo	1970	Dave Stockton	1987	Larry Nelson
1937	Denny Shute	1954	Melvin Harbert	1971	Jack Nicklaus	1988	Jeff Sluman
1938	Paul Runyan	1955	Doug Ford	1972	Gary Player	1989	Payne Stewart
1939	Henry Picard	1956	Jack Burke	1973	Jack Nicklaus	1990	Wayne Grady
1940	Byron Nelson	1957	Lionel Hebert	1974	Lee Trevino	1991	John Daly
1941	Victor Ghezzi	1958	Dow Finsterwald	1975	Jack Nicklaus	1992	Nick Price
1942	Sam Snead	1959	Bob Rosburg	1976	Dave Stockton	1993	Paul Azinger
1943	Not Played	1960	Jay Hebert	1977	Lanny Wadkins	1994	Nick Price
1944	Bob Hamilton	1961	Jerry Barber	1978	John Mahaffey	1995	Steve Elkington
1945	Byron Nelson	1962	Gary Player	1979	David Graham	1996	Mark Brooks
1946	Ben Hogan	1963	Jack Nicklaus	1980	Jack Nicklaus	1997	Davis Love III
1947	Jim Ferrier	1964	Bob Nichols	1981	Larry Nelson	1998	Vijay Singh
1948	Ben Hogan	1965	Dave Marr	1982	Ray Floyd	1999	Tiger Woods
1949	Sam Snead	1966	Al Geiberger	1983	Hal Sutton	2000	Tiger Woods
1950	Chandler Harper	1967	Don January	1984	Lee Trevino	2001	David Toms

Women's All-Time Major Professional Championship Leaders

(Through the 2001 season; *active LPGA player.)

Player	Nabisco[1]	LPGA	U.S. Open[2]	du Maurier[3]	Titleholders[4]	Western Open[5]	Total
Patty Berg	—	—	1946	—	1937-39, '48, '53, '55, '57	1941, '43, '48, '51, '55, '57-58	15
Mickey Wright	—	1958, '60-61, '63	1958-59, '61, '64	—	1961-62	1962-63, '66	13
Louise Suggs	—	1957	1949, '52	—	1946, '54, '56, '59	1946-47, '49, '53	11
Babe Zaharias	—	—	1948, '50, '54	—	1947, '50, '52	1940, '44-45, '50	10
Betsy Rawls	—	1959, '69	1951, '53, '57, '60	—	—	1952, '59	8
Pat Bradley*	1986	1986	1981	1980, '85-86	—	—	6
Juli Inkster*	1984, '89	1999-2000	1999	1984	—	—	6
Betsy King*	1987, '90, '97	1992	1989-90	—	—	—	6
Patty Sheehan*	1996	1983-84, '93	1992, '94	—	—	—	6
Kathy Whitworth	—	1967, '71, '75	—	—	1965-66	1967	6

Tournaments: (1) Nabisco Championship, formerly Nabisco Dinah Shore (1982-1999), designated major in 1983. (2) U.S. Women's Open (3) du Maurier Classic, formerly Peter Jackson Classic (1974-1982); designated major in 1979. (4) Titleholders Championship; major from 1930 to 1972. (5) Western Open; major from 1937 to 1967.

Ladies Professional Golf Association Leading Money Winners

Year	Player	Earnings	Year	Player	Earnings	Year	Player	Earnings
1954	Patty Berg	$16,011	1970	Kathy Whitworth	$30,235	1986	Pat Bradley	$492,021
1955	Patty Berg	16,492	1971	Kathy Whitworth	41,181	1987	Ayako Okamoto	466,034
1956	Marlene Hagge	20,235	1972	Kathy Whitworth	65,063	1988	Sherri Turner	347,255
1957	Patty Berg	16,272	1973	Kathy Whitworth	82,854	1989	Betsy King	654,132
1958	Beverly Hanson	12,629	1974	JoAnne Carner	87,094	1990	Beth Daniel	863,578
1959	Betsy Rawls	26,774	1975	Sandra Palmer	94,805	1991	Pat Bradley	763,118
1960	Louise Suggs	16,892	1976	Judy Rankin	150,734	1992	Dottie Mochrie	693,335
1961	Mickey Wright	22,236	1977	Judy Rankin	122,890	1993	Betsy King	595,992
1962	Mickey Wright	21,641	1978	Nancy Lopez	189,813	1994	Laura Davies	687,201
1963	Mickey Wright	31,269	1979	Nancy Lopez	215,987	1995	Annika Sorenstam	666,533
1964	Mickey Wright	29,800	1980	Beth Daniel	231,000	1996	Karrie Webb	1,002,000
1965	Kathy Whitworth	28,658	1981	Beth Daniel	206,977	1997	Annika Sorenstam	1,236,789
1966	Kathy Whitworth	33,517	1982	JoAnne Carner	310,399	1998	Annika Sorenstam	1,092,748
1967	Kathy Whitworth	32,937	1983	JoAnne Carner	291,404	1999	Karrie Webb	1,591,959
1968	Kathy Whitworth	48,379	1984	Betsy King	266,771	2000	Karrie Webb	1,876,853
1969	Carol Mann	49,152	1985	Nancy Lopez	416,472			

Nabisco Championship Winners[1]

Year	Winner	Year	Winner	Year	Winner	Year	Winner
1983	Amy Alcott	1988	Amy Alcott	1993	Helen Alfredsson	1998	Pat Hurst
1984	Juli Inkster	1989	Juli Inkster	1994	Donna Andrews	1999	Dottie Pepper
1985	Alice Miller	1990	Betsy King	1995	Nanci Bowen	2000	Karrie Webb
1986	Pat Bradley	1991	Amy Alcott	1996	Patty Sheehan	2001	Annika Sorenstam
1987	Betsy King	1992	Dottie Pepper	1997	Betsy King		

(1) Formerly the Colgate Dinah Shore (1972-81), the Nabisco Dinah Shore (1982-99). Designated as a major championship in 1983.

LPGA Championship Winners

Year	Winner	Year	Winner	Year	Winner	Year	Winner
1955	Beverly Hanson	1967	Kathy Whitworth	1979	Donna Caponi	1991	Meg Mallon
1956	Marlene Hagge	1968	Sandra Post	1980	Sally Little	1992	Betsy King
1957	Louise Suggs	1969	Betsy Rawls	1981	Donna Caponi	1993	Patty Sheehan
1958	Mickey Wright	1970	Shirley Englehorn	1982	Jan Stephenson	1994	Laura Davies
1959	Betsy Rawls	1971	Kathy Whitworth	1983	Patty Sheehan	1995	Kelly Robbins
1960	Mickey Wright	1972	Kathy Ahern	1984	Patty Sheehan	1996	Laura Davies
1961	Mickey Wright	1973	Mary Mills	1985	Nancy Lopez	1997	Chris Johnson
1962	Judy Kimball	1974	Sandra Haynie	1986	Pat Bradley	1998	Se Ri Pak
1963	Mickey Wright	1975	Kathy Whitworth	1987	Jane Geddes	1999	Juli Inkster
1964	Mary Mills	1976	Betty Burfeindt	1988	Sherri Turner	2000	Juli Inkster
1965	Sandra Haynie	1977	Chako Higuchi	1989	Nancy Lopez	2001	Karrie Webb
1966	Gloria Ehret	1978	Nancy Lopez	1990	Beth Daniel		

U.S. Women's Open Winners

Year	Winner	Year	Winner	Year	Winner	Year	Winner
1946	Patty Berg	1961	Mickey Wright	1974	Sandra Haynie	1988	Liselotte Neumann
1947	Betty Jameson	1962	Murle Lindstrom	1975	Sandra Palmer	1989	Betsy King
1948	"Babe" Zaharias	1963	Mary Mills	1976	JoAnne Carner	1990	Betsy King
1949	Louise Suggs	1964	Mickey Wright	1977	Hollis Stacy	1991	Meg Mallon
1950	"Babe" Zaharias	1965	Carol Mann	1978	Hollis Stacy	1992	Patty Sheehan
1951	Betsy Rawls	1966	Sandra Spuzich	1979	Jerilyn Britz	1993	Lauri Merten
1952	Louise Suggs	1967	Catherine Lacoste	1980	Amy Alcott	1994	Patty Sheehan
1953	Betsy Rawls		(amateur)	1981	Pat Bradley	1995	Annika Sorenstam
1954	"Babe" Zaharias	1968	Susie Maxwell Berning	1982	Janet Alex	1996	Annika Sorenstam
1955	Fay Crocker	1969	Donna Caponi	1983	Jan Stephenson	1997	Alison Nicholas
1956	Mrs. K. Cornelius	1970	Donna Caponi	1984	Hollis Stacy	1998	Se Ri Pak
1957	Betsy Rawls	1971	JoAnne Carner	1985	Kathy Baker	1999	Juli Inkster
1958	Mickey Wright	1972	Susie Maxwell Berning	1986	Jane Geddes	2000	Karrie Webb
1959	Mickey Wright	1973	Susie Maxwell Berning	1987	Laura Davies	2001	Karrie Webb
1960	Betsy Rawls						

du Maurier Classic Winners[1]

Year	Winner	Year	Winner	Year	Winner	Year	Winner
1979	Amy Alcott	1985	Pat Bradley	1991	Nancy Scranton	1996	Laura Davies
1980	Pat Bradley	1986	Pat Bradley	1992	Sherri Steinhauer	1997	Colleen Walker
1981	Jan Stephenson	1987	Jody Rosenthal	1993	Brandie Burton	1998	Brandie Burton
1982	Sandra Haynie	1988	Sally Little	1994	Martha Nause	1999	Karrie Webb
1983	Hollis Stacy	1989	Tammie Green	1995	Jenny Lidback	2000	Meg Mallon
1984	Juli Inkster	1990	Cathy Johnston				

(1) Formerly La Canadienne (1973), the Peter Jackson Classic (1974-82). Designated a major championship from 1979-2000. In 2001, the Women's British Open became the LPGA's 4th major championship and was won by South Korea's Se Ri Pak.

International Golf

Ryder Cup

Began as a biennial team competition between pro golfers from the U.S. and Great Britain. The British team was expanded in 1973 to include players from Ireland and in 1979 to include players from the rest of Europe. Scheduled for Sept. 2001, at The Belfry in Sutton Coldfield, England, the 34th Ryder Cup was postponed a year due to the terrorist attacks of Sept. 11, 2001, and will move permanently to even-numbered years.

Year	Winner	Year	Winner	Year	Winner	Year	Winner
1927	U.S., 9½-2½	1951	U.S., 9½-2½	1967	U.S., 23½-8½	1985	Europe, 16½-11½
1929	Britain-Ireland, 7-5	1953	U.S., 6½-5½	1969	Draw, 16-16	1987	Europe, 15-13
1931	U.S., 9-3	1955	U.S., 8-4	1971	U.S., 18½-13½	1989	Draw, 14-14
1933	Britain, 6½-5½	1957	Britain-Ireland,	1973	U.S., 19-13	1991	U.S., 14½-13½
1935	U.S., 9-3		7½-4½	1975	U.S., 21-11	1993	U.S., 15-13
1937	U.S., 8-4	1959	U.S., 8½-3½	1977	U.S., 12½-7½	1995	Europe, 14½-13½
1939-45	Not played	1961	U.S., 14½-9½	1979	U.S., 17-11	1997	Europe, 14½-13½
1947	U.S., 11-1	1963	U.S., 23-9	1981	U.S., 18½-9½	1999	U.S., 14½-13½
1949	U.S., 7-5	1965	U.S., 19½-12½	1983	U.S., 14½-13½		

Solheim Cup

The Solheim Cup began in 1990 as a biennial team competition between women professional golfers from Europe and the U.S. The 2002 competition was scheduled to be held Sept. 20-22, at Interlachen Country Club in Edina, MN.

Year	Winner	Year	Winner	Year	Winner
1990	U.S., 11½-4½	1994	U.S., 13-7	1998	U.S., 16-12
1992	Europe, 11½-6½	1996	U.S., 17-11	2000	Europe, 14½-11½

TENNIS

All-Time Grand Slam Singles Titles Leaders

Men	Australian Open	French Open[1]	Wimbledon	U.S. Open	Total
Pete Sampras*	1994, '97	—	1993-95, '97-2000	1990, '93, '95-96	13
Roy Emerson	1961, '63-67	1963, '67	1964-65	1961, '64	12
Bjorn Borg	—	1974-75, 1978-81	1976-80	—	11
Rod Laver	1960, '62, '69	1962, '69	1961-62, '68-69	1962, '69	11
Bill Tilden	—	—	1920-21, '30	1920-25, '29	10
Jimmy Connors	1974	—	1974, '82	1974, '76, '78, '82-83	8
Ivan Lendl	1989-90	1984, '86-87	—	1985-87	8
Fred Perry	1934	1935	1934-36	1933-34, '36	8
Ken Rosewall	1953, '55, '71-72	1953, '68	—	1956, '70	8
Women					
Margaret Smith Court	1960-66, '69-71, '73	1962, '64, '69-70, '73	1963, '65, '70	1962, '65, '69-70, '73	24
Steffi Graf	1988-90, '94	1987-88, '93, '95-96, '99	1988-89, '91-93, '95-96	1988-89, '93, '95-96	22
Helen Wills Moody	—	1928-30, '32	1927-30, '32-33, '35, '38	1923-25, '27-29, '31	19
Chris Evert Lloyd	1982, '84	1974-75, '79-80, '83, '85-86	1974, '76, '81	1975-78, '80, '82	18
Martina Navratilova	1981, '83, '85	1982, '84	1978-79, '82-87, '90	1983-84, '86-87	18
Billie Jean King	1968	1972	1966-68, '72-73, '75	1967, '71-72, '74	12
Suzanne Lenglen	—	1920-23, '25-26	1919-23, '25	—	12
Maureen Connolly	1953	1953-54	1952-54	1951-53	9
Monica Seles*	1991-93, '96	1990-92	—	1991-92	9

*active player (1) Prior to 1925, French Open entry was limited to members of French clubs.

Australian Open Singles Champions, 1969-2001

(First contested 1905 for men, 1922 for women)

*Two tournaments held in 1977 (Jan. & Dec.). **Tournament moved forward to Jan. 1987, so no championship was decided in 1986.

Men's Singles

Year	Champion	Final Opponent	Year	Champion	Final Opponent
1969	Rod Laver	Andres Gimeno	1985**	Stefan Edberg	Mats Wilander
1970	Arthur Ashe	Dick Crealy	1987	Stefan Edberg	Pat Cash
1971	Ken Rosewall	Arthur Ashe	1988	Mats Wilander	Pat Cash
1972	Ken Rosewall	Mal Anderson	1989	Ivan Lendl	Miloslav Mecir
1973	John Newcombe	Onny Parun	1990	Ivan Lendl	Stefan Edberg
1974	Jimmy Connors	Phil Dent	1991	Boris Becker	Ivan Lendl
1975	John Newcombe	Jimmy Connors	1992	Jim Courier	Stefan Edberg
1976	Mark Edmondson	John Newcombe	1993	Jim Courier	Stefan Edberg
1977*	Roscoe Tanner	Guillermo Vilas	1994	Pete Sampras	Todd Martin
	Vitas Gerulaitis	John Lloyd	1995	Andre Agassi	Pete Sampras
1978	Guillermo Vilas	John Marks	1996	Boris Becker	Michael Chang
1979	Guillermo Vilas	John Sadri	1997	Pete Sampras	Carlos Moya
1980	Brian Teacher	Kim Warwick	1998	Petr Korda	Marcelo Rios
1981	Johan Kriek	Steve Denton	1999	Yevgeny Kafelnikov	Thomas Enqvist
1982	Johan Kriek	Steve Denton	2000	Andre Agassi	Yevgeny Kafelnikov
1983	Mats Wilander	Ivan Lendl	2001	Andre Agassi	Arnaud Clement
1984	Mats Wilander	Kevin Curren			

Women's Singles

Year	Champion	Final Opponent	Year	Champion	Final Opponent
1969	Margaret Smith Court	Billie Jean King	1985**	Martina Navratilova	Chris Evert Lloyd
1970	Margaret Smith Court	Kerry Melville Reid	1987	Hana Mandlikova	Martina Navratilova
1971	Margaret Smith Court	Evonne Goolagong	1988	Steffi Graf	Chris Evert
1972	Virginia Wade	Evonne Goolagong	1989	Steffi Graf	Helena Sukova
1973	Margaret Smith Court	Evonne Goolagong	1990	Steffi Graf	Mary Joe Fernandez
1974	Evonne Goolagong	Chris Evert	1991	Monica Seles	Jana Novotna
1975	Evonne Goolagong	Martina Navratilova	1992	Monica Seles	Mary Joe Fernandez
1976	Evonne Goolagong	Renata Tomanova	1993	Monica Seles	Steffi Graf
1977*	Kerry Reid	Dianne Balestrat	1994	Steffi Graf	Arantxa Sánchez Vicario
	Evonne Goolagong	Helen Gourlay	1995	Mary Pierce	Arantxa Sánchez Vicario
1978	Chris O'Neill	Betsy Nagelsen	1996	Monica Seles	Anke Huber
1979	Barbara Jordan	Sharon Walsh	1997	Martina Hingis	Mary Pierce
1980	Hana Mandlikova	Wendy Turnbull	1998	Martina Hingis	Conchita Martínez
1981	Martina Navratilova	Chris Evert Lloyd	1999	Martina Hingis	Amelie Mauresmo
1982	Chris Evert Lloyd	Martina Navratilova	2000	Lindsay Davenport	Martina Hingis
1983	Martina Navratilova	Kathy Jordan	2001	Jennifer Capriati	Martina Hingis
1984	Chris Evert Lloyd	Helena Sukova			

French Open Singles Champions, 1968-2001

(First contested 1925)

Men's Singles

Year	Champion	Final Opponent	Year	Champion	Final Opponent
1968	Ken Rosewall	Rod Laver	1985	Mats Wilander	Ivan Lendl
1969	Rod Laver	Ken Rosewall	1986	Ivan Lendl	Mikael Pernfors
1970	Jan Kodes	Zeljko Franulovic	1987	Ivan Lendl	Mats Wilander
1971	Jan Kodes	Ilie Nastase	1988	Mats Wilander	Henri Leconte
1972	Andres Gimeno	Patrick Proisy	1989	Michael Chang	Stefan Edberg
1973	Ilie Nastase	Nikki Pilic	1990	Andres Gomez	Andre Agassi
1974	Bjorn Borg	Manuel Orantes	1991	Jim Courier	Andre Agassi
1975	Bjorn Borg	Guillermo Vilas	1992	Jim Courier	Petr Korda
1976	Adriano Panatta	Harold Solomon	1993	Sergi Bruguera	Jim Courier
1977	Guillermo Vilas	Brian Gottfried	1994	Sergi Bruguera	Alberto Berasategui
1978	Bjorn Borg	Guillermo Vilas	1995	Thomas Muster	Michael Chang
1979	Bjorn Borg	Victor Pecci	1996	Yevgeny Kafelnikov	Michael Stich
1980	Bjorn Borg	Vitas Gerulaitis	1997	Gustavo Kuerten	Sergei Bruguera
1981	Bjorn Borg	Ivan Lendl	1998	Carlos Moya	Alex Corretja
1982	Mats Wilander	Guillermo Vilas	1999	Andre Agassi	Andrei Medvedev
1983	Yannick Noah	Mats Wilander	2000	Gustavo Kuerten	Magnus Norman
1984	Ivan Lendl	John McEnroe	2001	Gustavo Kuerten	Alex Corretja

Women's Singles

Year	Champion	Final Opponent
1968	Nancy Richey	Ann Jones
1969	Margaret Smith Court	Ann Jones
1970	Margaret Smith Court	Helga Niessen
1971	Evonne Goolagong	Helen Gourlay
1972	Billie Jean King	Evonne Goolagong
1973	Margaret Smith Court	Chris Evert
1974	Chris Evert	Olga Morozova
1975	Chris Evert	Martina Navratilova
1976	Sue Barker	Renata Tomanova
1977	Mima Jausovec	Florenza Mihai
1978	Virginia Ruzici	Mima Jausovec
1979	Chris Evert Lloyd	Wendy Turnbull
1980	Chris Evert Lloyd	Virginia Ruzici
1981	Hana Mandlikova	Sylvia Hanika
1982	Martina Navratilova	Andrea Jaeger
1983	Chris Evert Lloyd	Mima Jausovec
1984	Martina Navratilova	Chris Evert Lloyd
1985	Chris Evert Lloyd	Martina Navratilova
1986	Chris Evert Lloyd	Martina Navratilova
1987	Steffi Graf	Martina Navratilova
1988	Steffi Graf	Natalia Zvereva
1989	Arantxa Sánchez Vicario	Steffi Graf
1990	Monica Seles	Steffi Graf
1991	Monica Seles	Arantxa Sánchez Vicario
1992	Monica Seles	Steffi Graf
1993	Steffi Graf	Mary Joe Fernandez
1994	Arantxa Sánchez Vicario	Mary Pierce
1995	Steffi Graf	Arantxa Sánchez Vicario
1996	Steffi Graf	Arantxa Sánchez Vicario
1997	Iva Majoli	Martina Hingis
1998	Arantxa Sánchez Vicario	Monica Seles
1999	Steffi Graf	Martina Hingis
2000	Mary Pierce	Conchita Martinez
2001	Jennifer Capriati	Kim Clijsters

All-England Champions, Wimbledon, 1925-2001

Men's Singles

(First contested 1877)

Year	Champion	Final Opponent
1925	Rene Lacoste	Jean Borotra
1926	Jean Borotra	Howard Kinsey
1927	Henri Cochet	Jean Borotra
1928	Rene Lacoste	Henri Cochet
1929	Henri Cochet	Jean Borotra
1930	Bill Tilden	Wilmer Allison
1931	Sidney B. Wood	Francis X. Shields
1932	Ellsworth Vines	Henry Austin
1933	Jack Crawford	Ellsworth Vines
1934	Fred Perry	Jack Crawford
1935	Fred Perry	Gottfried von Cramm
1936	Fred Perry	Gottfried von Cramm
1937	Donald Budge	Gottfried von Cramm
1938	Donald Budge	Henry Austin
1939	Bobby Riggs	Elwood Cooke
1940-45	Not held	Not held
1946	Yvon Petra	Geoff E. Brown
1947	Jack Kramer	Tom P. Brown
1948	Bob Falkenburg	John Bromwich
1949	Ted Schroeder	Jaroslav Drobny
1950	Budge Patty	Frank Sedgman
1951	Dick Savitt	Ken McGregor
1952	Frank Sedgman	Jaroslav Drobny
1953	Vic Seixas	Kurt Nielsen
1954	Jaroslav Drobny	Ken Rosewall
1955	Tony Trabert	Kurt Nielsen
1956	Lew Hoad	Ken Rosewall
1957	Lew Hoad	Ashley Cooper
1958	Ashley Cooper	Neale Fraser
1959	Alex Olmedo	Rod Laver
1960	Neale Fraser	Rod Laver
1961	Rod Laver	Chuck McKinley
1962	Rod Laver	Martin Mulligan
1963	Chuck McKinley	Fred Stolle
1964	Roy Emerson	Fred Stolle
1965	Roy Emerson	Fred Stolle
1966	Manuel Santana	Dennis Ralston
1967	John Newcombe	Wilhelm Bungert
1968	Rod Laver	Tony Roche
1969	Rod Laver	John Newcombe
1970	John Newcombe	Ken Rosewall
1971	John Newcombe	Stan Smith
1972	Stan Smith	Ilie Nastase
1973	Jan Kodes	Alex Metreveli
1974	Jimmy Connors	Ken Rosewall
1975	Arthur Ashe	Jimmy Connors
1976	Bjorn Borg	Ilie Nastase
1977	Bjorn Borg	Jimmy Connors
1978	Bjorn Borg	Jimmy Connors
1979	Bjorn Borg	Roscoe Tanner
1980	Bjorn Borg	John McEnroe
1981	John McEnroe	Bjorn Borg
1982	Jimmy Connors	John McEnroe
1983	John McEnroe	Chris Lewis
1984	John McEnroe	Jimmy Connors
1985	Boris Becker	Kevin Curren
1986	Boris Becker	Ivan Lendl
1987	Pat Cash	Ivan Lendl
1988	Stefan Edberg	Boris Becker
1989	Boris Becker	Stefan Edberg
1990	Stefan Edberg	Boris Becker
1991	Michael Stich	Boris Becker
1992	Andre Agassi	Goran Ivanisevic
1993	Pete Sampras	Jim Courier
1994	Pete Sampras	Goran Ivanisevic
1995	Pete Sampras	Boris Becker
1996	Richard Krajicek	MaliVai Washington
1997	Pete Sampras	Cedric Pioline
1998	Pete Sampras	Goran Ivanisevic
1999	Pete Sampras	Andre Agassi
2000	Pete Sampras	Patrick Rafter
2001	Goran Ivanisevic	Patrick Rafter

Women's Singles

(First contested 1884)

Year	Champion	Final Opponent
1925	Suzanne Lenglen	Joan Fry
1926	Kathleen McKane Godfree	Lili de Alvarez
1927	Helen Wills	Lili de Alvarez
1928	Helen Wills	Lili de Alvarez
1929	Helen Wills	Helen Jacobs
1930	Helen Wills Moody	Elizabeth Ryan
1931	Cilly Aussem	Hilde Kranwinkel
1932	Helen Wills Moody	Helen Jacobs
1933	Helen Wills Moody	Dorothy Round
1934	Dorothy Round	Helen Jacobs
1935	Helen Wills Moody	Helen Jacobs
1936	Helen Jacobs	Hilde Kranwinkel Sperling
1937	Dorothy Round	Jadwiga Jedrzejowska
1938	Helen Wills Moody	Helen Jacobs
1939	Alice Marble	Kay Stammers
1940-45	Not held	Not held
1946	Pauline Betz	Louise Brough
1947	Margaret Osborne	Doris Hart
1948	Louise Brough	Doris Hart
1949	Louise Brough	Margaret Osborne duPont
1950	Louise Brough	Margaret Osborne duPont
1951	Doris Hart	Shirley Fry
1952	Maureen Connolly	Louise Brough
1953	Maureen Connolly	Doris Hart
1954	Maureen Connolly	Louise Brough
1955	Louise Brough	Beverly Fleitz
1956	Shirley Fry	Angela Buxton
1957	Althea Gibson	Darlene Hard
1958	Althea Gibson	Angela Mortimer
1959	Maria Bueno	Darlene Hard
1960	Maria Bueno	Sandra Reynolds
1961	Angela Mortimer	Christine Truman
1962	Karen Hantze-Susman	Vera Sukova
1963	Margaret Smith	Billie Jean Moffitt
1964	Maria Bueno	Margaret Smith
1965	Margaret Smith	Maria Bueno
1966	Billie Jean King	Maria Bueno
1967	Billie Jean King	Ann Haydon Jones
1968	Billie Jean King	Judy Tegart
1969	Ann Haydon-Jones	Billie Jean King
1970	Margaret Smith Court	Billie Jean King
1971	Evonne Goolagong	Margaret Smith Court
1972	Billie Jean King	Evonne Goolagong
1973	Billie Jean King	Chris Evert
1974	Chris Evert	Olga Morozova
1975	Billie Jean King	Evonne Goolagong Cawley

Year	Champion	Final Opponent
1976	Chris Evert	Evonne Goolagong Cawley
1977	Virginia Wade	Betty Stove
1978	Martina Navratilova	Chris Evert
1979	Martina Navratilova	Chris Evert Lloyd
1980	Evonne Goolagong	Chris Evert Lloyd
1981	Chris Evert Lloyd	Hana Mandlikova
1982	Martina Navratilova	Chris Evert Lloyd
1983	Martina Navratilova	Andrea Jaeger
1984	Martina Navratilova	Chris Evert Lloyd
1985	Martina Navratilova	Chris Evert Lloyd
1986	Martina Navratilova	Hana Mandlikova
1987	Martina Navratilova	Steffi Graf
1988	Steffi Graf	Martina Navratilova
1989	Steffi Graf	Martina Navratilova
1990	Martina Navratilova	Zina Garrison
1991	Steffi Graf	Gabriela Sabatini
1992	Steffi Graf	Monica Seles
1993	Steffi Graf	Jana Novotna
1994	Conchita Martinez	Martina Navratilova
1995	Steffi Graf	Arantxa Sánchez Vicario
1996	Steffi Graf	Arantxa Sánchez Vicario
1997	Martina Hingis	Jana Novotna
1998	Jana Novotna	Nathalie Tauziat
1999	Lindsay Davenport	Steffi Graf
2000	Venus Williams	Lindsay Davenport
2001	Venus Williams	Justine Henin

U.S. Open Champions, 1925-2001

Men's Singles

(First contested 1881)

Year	Champion	Final Opponent
1925	Bill Tilden	William Johnston
1926	Rene Lacoste	Jean Borotra
1927	Rene Lacoste	Bill Tilden
1928	Henri Cochet	Francis Hunter
1929	Bill Tilden	Francis Hunter
1930	John Doeg	Francis Shields
1931	H. Ellsworth Vines	George Lott
1932	H. Ellsworth Vines	Henri Cochet
1933	Fred Perry	John Crawford
1934	Fred Perry	Wilmer Allison
1935	Wilmer Allison	Sidney Wood
1936	Fred Perry	Don Budge
1937	Don Budge	Baron G. von Cramm
1938	Don Budge	C. Gene Mako
1939	Robert Riggs	S. Welby Van Horn
1940	Don McNeill	Robert Riggs
1941	Robert Riggs	F. L. Kovacs
1942	F. R. Schroeder Jr.	Frank Parker
1943	Joseph Hunt	Jack Kramer
1944	Frank Parker	William Talbert
1945	Frank Parker	William Talbert
1946	Jack Kramer	Thomas Brown Jr.
1947	Jack Kramer	Frank Parker
1948	Pancho Gonzales	Eric Sturgess
1949	Pancho Gonzales	F. R. Schroeder Jr.
1950	Arthur Larsen	Herbert Flam
1951	Frank Sedgman	E. Victor Seixas Jr.
1952	Frank Sedgman	Gardnar Mulloy
1953	Tony Trabert	E. Victor Seixas Jr.
1954	E. Victor Seixas Jr.	Rex Hartwig
1955	Tony Trabert	Ken Rosewall
1956	Ken Rosewall	Lewis Hoad
1957	Malcolm Anderson	Ashley Cooper
1958	Ashley Cooper	Malcolm Anderson
1959	Neale A. Fraser	Alejandro Olmedo
1960	Neale A. Fraser	Rod Laver
1961	Roy Emerson	Rod Laver
1962	Rod Laver	Roy Emerson
1963	Rafael Osuna	F. A. Froehling 3d
1964	Roy Emerson	Fred Stolle
1965	Manuel Santana	Cliff Drysdale
1966	Fred Stolle	John Newcombe
1967	John Newcombe	Clark Graebner
1968	Arthur Ashe	Tom Okker
1969	Rod Laver	Tony Roche
1970	Ken Rosewall	Tony Roche
1971	Stan Smith	Jan Kodes
1972	Ilie Nastase	Arthur Ashe
1973	John Newcombe	Jan Kodes
1974	Jimmy Connors	Ken Rosewall
1975	Manuel Orantes	Jimmy Connors
1976	Jimmy Connors	Bjorn Borg
1977	Guillermo Vilas	Jimmy Connors
1978	Jimmy Connors	Bjorn Borg
1979	John McEnroe	Vitas Gerulaitis
1980	John McEnroe	Bjorn Borg
1981	John McEnroe	Bjorn Borg
1982	Jimmy Connors	Ivan Lendl
1983	Jimmy Connors	Ivan Lendl
1984	John McEnroe	Ivan Lendl
1985	Ivan Lendl	John McEnroe
1986	Ivan Lendl	Miloslav Mecir
1987	Ivan Lendl	Mats Wilander
1988	Mats Wilander	Ivan Lendl
1989	Boris Becker	Ivan Lendl
1990	Pete Sampras	Andre Agassi
1991	Stefan Edberg	Jim Courier
1992	Stefan Edberg	Pete Sampras
1993	Pete Sampras	Cedric Pioline
1994	Andre Agassi	Michael Stich
1995	Pete Sampras	Andre Agassi
1996	Pete Sampras	Michael Chang
1997	Patrick Rafter	Greg Rusedski
1998	Patrick Rafter	Mark Philippoussis
1999	Andre Agassi	Todd Martin
2000	Marat Safin	Pete Sampras
2001	Lleyton Hewitt	Pete Sampras

Women's Singles

(First contested 1887)

Year	Champion	Final Opponent
1925	Helen Willis	Kathleen McKane
1926	Molla B. Mallory	Elizabeth Ryan
1927	Helen Wills	Betty Nuthall
1928	Helen Wills	Helen Jacobs
1929	Helen Wills	M. Watson
1930	Betty Nuthall	L. A. Harper
1931	Helen Wills Moody	E. B. Whittingstall
1932	Helen Jacobs	Carolin A. Babcock
1933	Helen Jacobs	Helen Wills Moody
1934	Helen Jacobs	Sarah H. Palfrey
1935	Helen Jacobs	Sarah Palfrey Fabyan
1936	Alice Marble	Helen Jacobs
1937	Anita Lizana	Jadwiga Jedrzejowska
1938	Alice Marble	Nancye Wynne
1939	Alice Marble	Helen Jacobs
1940	Alice Marble	Helen Jacobs
1941	Sarah Palfrey Cooke	Pauline Betz
1942	Pauline Betz	Louise Brough
1943	Pauline Betz	Louise Brough
1944	Pauline Betz	Margaret Osborne
1945	Sarah Palfrey Cooke	Pauline Betz
1946	Pauline Betz	Doris Hart
1947	Louise Brough	Margaret Osborne
1948	Margaret Osborne duPont	Louise Brough
1949	Margaret Osborne duPont	Doris Hart
1950	Margaret Osborne duPont	Doris Hart
1951	Maureen Connolly	Shirley Fry
1952	Maureen Connolly	Doris Hart
1953	Maureen Connolly	Doris Hart
1954	Doris Hart	Louise Brough
1955	Doris Hart	Patricia Ward
1956	Shirley Fry	Althea Gibson
1957	Althea Gibson	Louise Brough
1958	Althea Gibson	Darlene Hard
1959	Maria Bueno	Christine Truman
1960	Darlene Hard	Maria Bueno
1961	Darlene Hard	Ann Haydon
1962	Margaret Smith	Darlene Hard
1963	Maria Bueno	Margaret Smith
1964	Maria Bueno	Carole Graebner
1965	Margaret Smith	Billie Jean Moffitt
1966	Maria Bueno	Nancy Richey
1967	Billie Jean King	Ann Haydon Jones
1968	Virginia Wade	Billie Jean King
1969	Margaret Smith Court	Nancy Richey
1970	Margaret Smith Court	Rosemary Casals
1971	Billie Jean King	Rosemary Casals
1972	Billie Jean King	Kerry Melville
1973	Margaret Smith Court	Evonne Goolagong
1974	Billie Jean King	Evonne Goolagong

Year	Champion	Final Opponent
1975	Chris Evert	Evonne Goolagong
1976	Chris Evert	Evonne Goolagong
1977	Chris Evert	Wendy Turnbull
1978	Chris Evert	Pam Shriver
1979	Tracy Austin	Chris Evert Lloyd
1980	Chris Evert Lloyd	Hana Mandlikova
1981	Tracy Austin	Martina Navratilova
1982	Chris Evert Lloyd	Hana Mandlikova
1983	Martina Navratilova	Chris Evert Lloyd
1984	Martina Navratilova	Chris Evert Lloyd
1985	Hana Mandlikova	Martina Navratilova
1986	Martina Navratilova	Helena Sukova
1987	Martina Navratilova	Steffi Graf
1988	Steffi Graf	Gabriela Sabatini
1989	Steffi Graf	Martina Navratilova
1990	Gabriela Sabatini	Steffi Graf
1991	Monica Seles	Martina Navratilova
1992	Monica Seles	Arantxa Sanchez Vicario
1993	Steffi Graf	Helena Sukova
1994	Arantxa Sanchez Vicario	Steffi Graf
1995	Steffi Graf	Monica Seles
1996	Steffi Graf	Monica Seles
1997	Martina Hingis	Venus Williams
1998	Lindsay Davenport	Martina Hingis
1999	Serena Williams	Martina Hingis
2000	Venus Williams	Lindsay Davenport
2001	Venus Williams	Serena Williams

Davis Cup Challenge Round, 1900-2000

Year	Result
1900	United States 3, British Isles 0
1901	Not held
1902	United States 3, British Isles 2
1903	British Isles 4, United States 1
1904	British Isles 5, Belgium 0
1905	British Isles 5, United States 0
1906	British Isles 5, United States 0
1907	Australia 3, British Isles 2
1908	Australasia 3, United States 2
1909	Australasia 5, United States 0
1910	Not held
1911	Australasia 5, United States 0
1912	British Isles 3, Australasia 2
1913	United States 3, British Isles 2
1914	Australasia 3, United States 2
1915-18	Not held
1919	Australasia 4, British Isles 1
1920	United States 5, Australasia 0
1921	United States 5, Japan 0
1922	United States 4, Australasia 1
1923	United States 4, Australasia 1
1924	United States 5, Australasia 0
1925	United States 5, France 0
1926	United States 4, France 1
1927	France 3, United States 2
1928	France 4, United States 1
1929	France 3, United States 2
1930	France 4, United States 1
1931	France 3, Great Britain 2
1932	France 3, United States 2
1933	Great Britain 3, France 2
1934	Great Britain 4, United States 1
1935	Great Britain 5, United States 0
1936	Great Britain 3, Australia 2
1937	United States 4, Great Britain 1
1938	United States 3, Australia 2
1939	Australia 3, United States 2
1940-45	Not held
1946	United States 5, Australia 0
1947	United States 4, Australia 1
1948	United States 5, Australia 0
1949	United States 4, Australia 1
1950	Australia 4, United States 1
1951	Australia 3, United States 2
1952	Australia 4, United States 1
1953	Australia 3, United States 2
1954	United States 3, Australia 2
1955	Australia 5, United States 0
1956	Australia 5, United States 0
1957	Australia 3, United States 2
1958	United States 3, Australia 2
1959	Australia 3, United States 2
1960	Australia 4, Italy 1
1961	Australia 5, Italy 0
1962	Australia 5, Mexico 0
1963	United States 3, Australia 2
1964	Australia 3, United States 2
1965	Australia 4, Spain 1
1966	Australia 4, India 1
1967	Australia 4, Spain 1
1968	United States 4, Australia
1969	United States 5, Romania 0
1970	United States 5, W. Germany 0
1971	United States 3, Romania 2
1972	United States 3, Romania 2
1973	Australia 5, United States 0
1974	South Africa (default by India)
1975	Sweden 3, Czechoslovakia 2
1976	Italy 4, Chile 1
1977	Australia 3, Italy 1
1978	United States 4, Great Britain 1
1979	United States 5, Italy 0
1980	Czechoslovakia 4, Italy 1
1981	United States 3, Argentina 1
1982	United States 4, France, 1
1983	Australia 3, Sweden 2
1984	Sweden 4, United States 1
1985	Sweden 3, W. Germany 2
1986	Australia 3, Sweden 2
1987	Sweden 5, India 0
1988	W. Germany 4, Sweden 1
1989	W. Germany 3, Sweden 2
1990	United States 3, Australia 2
1991	France 3, United States 1
1992	United States 3, Switzerland 1
1993	Germany 4, Australia 1
1994	Sweden 4, Russia 1
1995	United States 3, Russia 2
1996	France 3, Sweden 2
1997	Sweden 5, United States 0
1998	Sweden 4, Italy 1
1999	Australia 3, France 2
2000	Spain 3, Australia 1

RIFLE AND PISTOL INDIVIDUAL CHAMPIONSHIPS

Source: National Rifle Association

National Outdoor Rifle and Pistol Championships in 2001

Pistol—MSG Brian Zins, USMC, Quantico, VA, 2654-126X
Civilian Pistol—Dr. Darius R. Young, Malo, WA, 2653-148X
Woman Pistol—Kimberly Hobart, New Philadelphia, OH, 2545-68X
Smallbore Rifle Prone—MAJ Steve Goff, USA, Columbus, GA, 6378-453X
Civilian Smallbore Rifle Prone—Carolyn D. Millard-Sparks, Dunwoody, GA, 6375-466X
Woman Smallbore Rifle Prone—Carolyn D. Millard-Sparks, Dunwoody, GA, 6375-466X
Smallbore Rifle NRA 3-Position—Troy A. Basham, Phenix City, AL, 2267-81X
Civilian Smallbore Rifle NRA 3-Position—Troy A. Bassham, Phenix City, AL 2267-81X
Woman Smallbore Rifle NRA 3-Position—Jamie L. Beyerle, Lebanon, PA, 2241-64X
High Power Rifle—G. David Tubb, Canadian, TX, 2380-112X
Civilian High Power Rifle—G. David Tubb, Canadian, TX, 2380-112X
Woman High Power Rifle—Alonda J. Roy, Houston, TX, 2342-82X
High Power Rifle Long Range—Michelle M. Gallagher, Prescott, AZ, 1436-63X
Woman High Power Rifle Long Range—Michelle M. Gallagher, Prescott, AZ, 1436-63X

National Indoor Rifle and Pistol Championships in 2001

Smallbore Rifle 4-Position—Glen Dubis, Columbus, GA, 800-77x
Woman Smallbore Rifle 4-Position—Jamie Beyerle, Lebanon, PA, 798-73X
Smallbore Rifle NRA 3-Position—Glen Dubis, Columbus, GA, 1192-95X
Woman Smallbore Rifle NRA 3-Position—Jamie Beyerle, Lebanon, PA, 1178-80X
International Smallbore Rifle—Glen Dubis, Columbus, GA, 1189-97X
Woman International Smallbore Rifle—Melissa Mulloy, Fairbanks, AK, 1176-76X
Air Rifle—Matthew Emmons, Browns Mill, NJ, 597
Woman Air Rifle—Melissa Mulloy, Fairbanks, AK, 587
Conventional Pistol—John Vespa, Hibbing, MN, 887-37X
Woman Conventional Pistol—Judy Tant, East Lansing, MI, 870-28X
International Free Pistol—Robert Fleming, Aurora, CO, 545
Woman International Free Pistol—Laura Tyler, Craig, CO, 477
International Standard Pistol—John Bickar, Colorado Springs, CO, 580
Woman International Standard Pistol—Kathy Chatterton, Glen Rock, NJ, 549
Air Pistol—Robert Patton, Charleston, SC, 576
Woman Air Pistol—Kathy Chatterton, Glen Rock, NJ, 533

NRA Bianchi Cup National Action Pistol Championships in 2001

Action Pistol—Doug Koeing, Albertus, PA, 1920.184
Woman Action Pistol—Vera Koo, Menlo Park, CA, 1910.137
Junior Action Pistol—Mitch Conrad, Tulsa, OK, 1912.145

AUTO RACING

Indianapolis 500 Winners

Year	Winner, Car (Chassis-Engine)	MPH[1]
1911	Ray Harroun, Marmon	74.602
1912	Joe Dawson, National	78.719
1913	Jules Goux, Peugeot	75.933
1914	Rene Thomas, Delage	82.474
1915	Ralph DePalma, Mercedes	89.840
1916	Dario Resta, Peugeot	84.001
1917-18	—Not held	
1919	Howdy Wilcox, Peugeot	88.050
1920	Gaston Chevrolet, Frontenac	88.618
1921	Tommy Milton, Frontenac	89.621
1922	Jimmy Murphy, Duesenberg-Miller	94.484
1923	Tommy Milton, Miller	90.954
1924	L.L. Corum-Joe Boyer, Duesenberg	98.234
1925	Peter DePaolo, Duesenberg	101.127
1926	Frank Lockhart, Miller	95.904
1927	George Souders, Duesenberg	97.545
1928	Louie Meyer, Miller	99.482
1929	Ray Keech, Miller	97.585
1930	Billy Arnold, Summers-Miller	100.448
1931	Louis Schneider, Stevens-Miller	96.629
1932	Fred Frame, Wetteroth-Miller	104.144
1933	Louie Meyer, Miller	104.162
1934	Bill Cummings, Miller	104.863
1935	Kelly Petillo, Wetteroth-Offy	106.240
1936	Louie Meyer, Stevens-Miller	109.069
1937	Wilbur Shaw, Shaw-Offy	113.580
1938	Floyd Roberts, Wetteroth-Miller	117.200
1939	Wilbur Shaw, Maserati	115.035
1940	Wilbur Shaw, Maserati	114.277
1941	Floyd Davis-Mauri Rose, Wetteroth-Offy	115.117
1942-45	—Not held	
1946	George Robson, Adams-Sparks	114.820
1947	Mauri Rose, Deidt-Offy	116.338
1948	Mauri Rose, Deidt-Offy	119.814
1949	Bill Holland, Deidt-Offy	121.327
1950	Johnnie Parsons, Kurtis-Offy	124.002
1951	Lee Wallard, Kurtis-Offy	126.244
1952	Troy Ruttman, Kuzma-Offy	128.922
1953	Bill Vukovich, KK500A-Offy	128.740
1954	Bill Vukovich, KK500A-Offy	130.840
1955	Bob Sweikert, KK500C-Offy	128.213
1956	Pat Flaherty, Watson-Offy	128.490
1957	Sam Hanks, Salih-Offy	135.601
1958	Jimmy Bryan, Salih-Offy	133.791
1959	Rodger Ward, Watson-Offy	135.857
1960	Jim Rathmann, Watson-Offy	138.767
1961	A.J. Foyt Jr., Trevis-Offy	139.130
1962	Rodger Ward, Watson-Offy	140.293
1963	Parnelli Jones, Watson-Offy	143.137
1964	A.J. Foyt Jr., Watson-Offy	147.350
1965	Jim Clark, Lotus-Ford	150.686
1966	Graham Hill, Lola-Ford	144.317
1967	A.J. Foyt Jr., Coyote-Ford	151.207
1968	Bobby Unser, Eagle-Offy	152.882
1969	Mario Andretti, Hawk-Ford	156.867
1970	Al Unser, P.J. Colt-Ford	155.749
1971	Al Unser, P.J. Colt-Ford	157.735
1972	Mark Donohue, McLaren-Offy	162.962
1973	Gordon Johncock, Eagle-Offy	159.036
1974	Johnny Rutherford, McLaren-Offy	158.589
1975	Bobby Unser, Eagle-Offy	149.213
1976	Johnny Rutherford, McLaren-Offy	148.725
1977	A.J. Foyt Jr., Coyote-Foyt	161.331
1978	Al Unser, Lola-Cosworth	161.363
1979	Rick Mears, Penske-Cosworth	158.899
1980	Johnny Rutherford, Chaparral-Cosworth	142.862
1981	Bobby Unser, Penske-Cosworth	139.084
1982	Gordon Johncock, Wildcat-Cosworth	162.029
1983	Tom Sneva, March-Cosworth	162.117
1984	Rick Mears, March-Cosworth	163.612
1985	Danny Sullivan, March-Cosworth	152.982
1986	Bobby Rahal, March-Cosworth	170.722
1987	Al Unser, March-Cosworth	162.175
1988	Rick Mears, Penske-Chevy Indy V8	144.809
1989	Emerson Fittipaldi, Penske-Chevy Indy V8	167.581
1990	Arie Luyendyk, Lola-Chevy Indy V8	185.981*
1991	Rick Mears, Penske-Chevy Indy V8	176.457
1992	Al Unser Jr., Galmer-Chevy Indy V8A	134.477
1993	Emerson Fittipaldi, Penske-Chevy Indy V8C	157.207
1994	Al Unser Jr., Penske-Mercedes Benz	160.872
1995	Jacques Villeneuve, Reynard-Ford Cosworth XB	153.616
1996	Buddy Lazier, Reynard-Ford Cosworth	147.956
1997	Arie Luyendyk, G Force-Aurora	145.827
1998	Eddie Cheever, Dallara-Aurora	145.155
1999	Kenny Brack, Dallara-Aurora	153.176
2000	Juan Montoya, G Force-Aurora	167.607
2001	Helio Castroneves, Reynard-Honda	131.294

*Race record. **Note:** The race was less than 500 mi in the following years: 1916 (300 mi), 1926 (400 mi), 1950 (345 mi), 1973 (332.5 mi), 1975 (435 mi), 1976 (255 mi). (1) Average speed.

FedEx Championship Series PPG Cup Winners

(U.S. Auto Club Champions prior to 1979; Championship Auto Racing Teams [CART] Champions, 1979-2000)

Year	Driver	Year	Driver	Year	Driver	Year	Driver
1959	Roger Ward	1970	Al Unser	1981	Rick Mears	1991	Michael Andretti
1960	A. J. Foyt	1971	Joe Leonard	1982	Rick Mears	1992	Bobby Rahal
1961	A. J. Foyt	1972	Joe Leonard	1983	Al Unser	1993	Nigel Mansell
1962	Rodger Ward	1973	Roger McCluskey	1984	Mario Andretti	1994	Al Unser Jr.
1963	A. J. Foyt	1974	Bobby Unser	1985	Al Unser	1995	Jacques Villeneuve
1964	A. J. Foyt	1975	A. J. Foyt	1986	Bobby Rahal	1996	Jimmy Vasser
1965	Mario Andretti	1976	Gordon Johncock	1987	Bobby Rahal	1997	Alex Zanardi
1966	Mario Andretti	1977	Tom Sneva	1988	Danny Sullivan	1998	Alex Zanardi
1967	A. J. Foyt	1978	Tom Sneva	1989	Emerson Fittipaldi	1999	Juan Montoya
1968	Bobby Unser	1979	Rick Mears	1990	Al Unser Jr.	2000	Gil de Ferran
1969	Mario Andretti	1980	Johnny Rutherford				

Notable One-Mile Land Speed Records

Andy Green, a Royal Air Force pilot, broke the sound barrier and set the first supersonic world speed record on land, Oct. 15, 1997, in Black Rock Desert, NV. Green, driving a car built by Richard Noble, had 2 runs at an average speed of 763.035 mph, as calculated according to the rules of the Federation Internationale Automobiliste (FIA). This record and speed exceeded the speed of sound, calculated at 751.251 mph for that place and time. On Sept. 25, Green had set a new world mark at 714.144 mph, which eclipsed the old record of 633.468 mph. Both 1997 records were recorded by the U.S. Auto Club and recognized by the FIA.

Date	Driver	Car	MPH
1/26/06	Marriott	Stanley (Steam)	127.659
3/16/10	Oldfield	Benz	131.724
4/23/11	Burman	Benz	141.732
2/12/19	DePalma	Packard	149.875
4/27/20	Milton	Dusenberg	155.046
4/28/26	Parry-Thomas	Thomas Spl.	170.624
3/29/27	Seagrave	Sunbeam	203.790
4/22/28	Keech	White Triplex	207.552
3/11/29	Seagrave	Irving-Napier	231.446
2/05/31	Campbell	Napier-Campbell	246.086
2/24/32	Campbell	Napier-Campbell	253.96
2/22/33	Campbell	Napier-Campbell	272.109
9/03/35	Campbell	Bluebird Special	301.13
11/19/37	Eyston	Thunderbolt 1	311.42
9/16/38	Eyston	Thunderbolt 1	357.5
8/23/39	Cobb	Railton	368.9
9/16/47	Cobb	Railton-Mobil	394.2
8/05/63	Breedlove	Spirit of America	407.45
10/27/64	Arfons	Green Monster	536.71
11/15/65	Breedlove	Spirit of America	600.601
10/23/70	Gabelich	Blue Flame	622.407
10/09/79	Barrett	Budweiser Rocket	638.637*
10/04/83	Noble	Thrust 2	633.468
9/25/97	Green	Thrust SSC	714.144
10/15/97	Green	Thrust SSC	763.035

*Not recognized as official by sanctioning bodies.

2001 Le Mans 24 Hours Race

Frank Biela (Germany), Tom Kristensen (Denmark), and Emmanuele Pirro (Italy) drove their Audi R8 to a 2d-consecutive victory in the "24 Hours of Le Mans" race, held June 17, 2001. In rainy conditions, they completed 321 laps of the 8.456-mile (13.608-km) circuit at an average speed of 113.074 mph (181.975 km/hour). The Audi team, which swept the top 3 places in 2000, also took 2d. Bentley Motors, competing for the 1st time since 1930, finished 15 laps back in 3d. Only 21 out of 48 starters finished the race.

World Formula One Grand Prix Champions, 1950-2001

Year	Driver	Year	Driver	Year	Driver
1950	Nino Farini, Italy	1968	Graham Hill, England	1985	Alain Prost, France
1951	Juan Fangio, Argentina	1969	Jackie Stewart, Scotland	1986	Alain Prost, France
1952	Alberto Ascari, Italy	1970	Jochen Rindt, Austria	1987	Nelson Piquet, Brazil
1953	Alberto Ascari, Italy	1971	Jackie Stewart, Scotland	1988	Ayrton Senna, Brazil
1954	Juan Fangio, Argentina	1972	Emerson Fittipaldi, Brazil	1989	Alain Prost, France
1955	Juan Fangio, Argentina	1973	Jackie Stewart, Scotland	1990	Ayrton Senna, Brazil
1956	Juan Fangio, Argentina	1974	Emerson Fittipaldi, Brazil	1991	Ayrton Senna, Brazil
1957	Juan Fangio, Argentina	1975	Niki Lauda, Austria	1992	Nigel Mansell, Britain
1958	Mike Hawthorne, England	1976	James Hunt, England	1993	Alain Prost, France
1959	Jack Brabham, Australia	1977	Niki Lauda, Austria	1994	Michael Schumacher, Germany
1960	Jack Brabham, Australia	1978	Mario Andretti, United States	1995	Michael Schumacher, Germany
1961	Phil Hill, United States	1979	Jody Scheckter, South Africa	1996	Damon Hill, England
1962	Graham Hill, England	1980	Alan Jones, Australia	1997	Jacques Villeneuve, Canada
1963	Jim Clark, Scotland	1981	Nelson Piquet, Brazil	1998	Mika Hakkinen, Finland
1964	John Surtees, England	1982	Keke Rosberg, Finland	1999	Mika Hakkinen, Finland
1965	Jim Clark, Scotland	1983	Nelson Piquet, Brazil	2000	Michael Schumacher, Germany
1966	Jack Brabham, Australia	1984	Niki Lauda, Austria	2001	Michael Schumacher, Germany
1967	Denis Hulme, New Zealand				

NASCAR Racing

Winston Cup Champions, 1949-2000

Year	Driver	Year	Driver	Year	Driver	Year	Driver
1949	Red Byron	1962	Joe Weatherly	1975	Richard Petty	1988	Bill Elliott
1950	Bill Rexford	1963	Joe Weatherly	1976	Cale Yarborough	1989	Rusty Wallace
1951	Herb Thomas	1964	Richard Petty	1977	Cale Yarborough	1990	Dale Earnhardt
1952	Tim Flock	1965	Ned Jarrett	1978	Cale Yarborough	1991	Dale Earnhardt
1953	Herb Thomas	1966	David Pearson	1979	Richard Petty	1992	Alan Kulwicki
1954	Lee Petty	1967	Richard Petty	1980	Dale Earnhardt	1993	Dale Earnhardt
1955	Tim Flock	1968	David Pearson	1981	Darrell Waltrip	1994	Dale Earnhardt
1956	Buck Baker	1969	David Pearson	1982	Darrell Waltrip	1995	Jeff Gordon
1957	Buck Baker	1970	Bobby Isaac	1983	Bobby Allison	1996	Terry Labonte
1958	Lee Petty	1971	Richard Petty	1984	Terry Labonte	1997	Jeff Gordon
1959	Lee Petty	1972	Richard Petty	1985	Darrell Waltrip	1998	Jeff Gordon
1960	Rex White	1973	Benny Parsons	1986	Dale Earnhardt	1999	Dale Jarrett
1961	Ned Jarrett	1974	Richard Petty	1987	Dale Earnhardt	2000	Bobby Labonte

NASCAR Rookie of the Year, 1958-2000

Year	Driver	Year	Driver	Year	Driver	Year	Driver
1958	Shorty Rollins	1969	Dick Brooks	1980	Jody Riley	1991	Bobby Hamilton
1959	Richard Petty	1970	Bill Dennis	1981	Ron Bouchard	1992	Jimmy Hensley
1960	David Pearson	1971	Walter Ballard	1982	Geoff Bodine	1993	Jeff Gordon
1961	Woodie Wilson	1972	Larry Smith	1983	Sterling Martin	1994	Jeff Burton
1962	Tom Cox	1973	Lennie Pond	1984	Rusty Wallace	1995	Ricky Craven
1963	Billy Wade	1974	Earl Ross	1985	Ken Schrader	1996	Johnny Benson
1964	Doug Cooper	1975	Bruce Hill	1986	Alan Kulwicki	1997	Mike Skinner
1965	Sam McQuagg	1976	Skip Manning	1987	Davey Allison	1998	Kenny Irwin
1966	James Hylton	1977	Ricky Rudd	1988	Ken Bouchard	1999	Tony Stewart
1967	Donnie Allison	1978	Ronnie Thomas	1989	Dick Trickle	2000	Matt Kenseth
1968	Pete Hamilton	1979	Dale Earnhardt	1990	Rob Moroso		

IT'S A FACT: Known as the "King" of stock car racing, Richard Petty retired in 1992 as the all-time leader in many Winston Cup series statistics, including wins (200), races started (1,184), top-5 finishes (555), top-10 finishes (712), laps completed (307,836), and consecutive wins (10). As of 2001, these records all stood.

Daytona 500 Winners, 1959-2001

Year	Driver, car	Avg. MPH	Year	Driver, car	Avg. MPH
1959	Lee Petty, Oldsmobile	135.521	1981	Richard Petty, Buick	169.651
1960	Junior Johnson, Chevrolet	124.740	1982	Bobby Allison, Buick	153.991
1961	Marvin Panch, Pontiac	149.601	1983	Cale Yarborough, Pontiac	155.979
1962	Fireball Roberts, Pontiac	152.529	1984	Cale Yarborough, Chevrolet	150.994
1963	Tiny Lund, Ford	151.566	1985	Bill Elliott, Ford	172.265
1964	Richard Petty, Plymouth	154.334	1986	Geoff Bodine, Chevrolet	148.124
1965	Fred Lorenzen, Ford (a)	141.539	1987	Bill Elliott, Ford	176.263
1966	Richard Petty, Plymouth (b)	160.627	1988	Bobby Allison, Buick	137.531
1967	Mario Andretti, Ford	146.926	1989	Darrell Waltrip, Chevrolet	148.466
1968	Cale Yarborough, Mercury	143.251	1990	Derrike Cope, Chevrolet	165.761
1969	Lee Roy Yarborough, Ford	160.875	1991	Ernie Irvan, Chevrolet	148.148
1970	Pete Hamilton, Plymouth	149.601	1992	Davey Allison, Ford	160.256
1971	Richard Petty, Plymouth	144.456	1993	Dale Jarrett, Chevrolet	154.972
1972	A. J. Foyt, Mercury	161.550	1994	Sterling Marlin, Chevrolet	156.931
1973	Richard Petty, Dodge	157.205	1995	Sterling Marlin, Chevrolet	141.710
1974	Richard Petty, Dodge (c)	140.894	1996	Dale Jarrett, Ford	154.308
1975	Benny Parsons, Chevrolet	153.649	1997	Jeff Gordon, Chevrolet	148.295
1976	David Pearson, Mercury	152.181	1998	Dale Earnhardt, Chevrolet	172.712
1977	Cale Yarborough, Chevrolet	153.218	1999	Jeff Gordon, Chevrolet	161.551
1978	Bobby Allison, Ford	159.730	2000	Dale Jarrett, Ford	155.669
1979	Richard Petty, Oldsmobile	143.977	2001	Michael Waltrip, Chevrolet	161.794
1980	Buddy Baker, Oldsmobile	177.602			

(a) 322.5 mi. (b) 495 mi. (c) 450 mi.

BOXING

Champions by Classes

There are many governing bodies in boxing, including the World Boxing Council, World Boxing Assn., International Boxing Federation, World Boxing Org., U.S. Boxing Assn., North American Boxing Federation, and European Boxing Union. Others are recognized by TV networks and the print media. All the governing bodies have their own champions and assorted boxing divisions. The following are the recognized champions—as of Oct. 10, 2001—in the principal divisions of the WBA, WBC, and IBF.

Class, Weight limit	WBA	WBC	IBF
Heavyweight	John Ruiz, U.S.	Hasim Rahman, U.S.	Hasim Rahman, U.S.
Cruiserweight (190 lb)	Virgil Hill, U.S	Juan Carlos Gomez, Cuba/Germany	Vassiliy Jirov, U.S./Kazakhstan
Light Heavyweight (175 lb)	Vacant (a)	Roy Jones Jr., U.S.	Roy Jones Jr., U.S.
Super Middleweight (168 lb)	Byron Mitchell, U.S	Eric Lucas, Canada	Sven Ottke, Germany
Middleweight (160 lb)	Bernard Hopkins, U.S.	Bernard Hopkins, U.S.	Bernard Hopkins, U.S.
Jr. Middleweight (154 lb)	Fernando Vargas, U.S.	Oscar de la Hoya, U.S.	Vacant
Welterweight (147 lb)	Andrew Lewis, Guyana	Shane Mosley, U.S.	Vernon Forrest, U.S.
Jr. Welterweight (140 lb)	Vacant (b)	Kostya Tszyu, Australia	Zab Judah, U.S.
Lightweight (135 lb)	Raul Balbi, Argentina	Jose Luis Castillo, Mexico	Paul Spadafora, U.S.
Jr. Lightweight (130 lb)	Joel Casamayor, U.S.	Floyd Mayweather Jr., U.S.	Steve Forbes, U.S.
Featherweight (126 lb)	Derrick Gainer, U.S.	Erik Morales, Mexico	Frankie Toledo, U.S.
Jr. Featherweight (122 lb)	Vacant	Willie Jorrin, U.S.	Manny Pacquiao, Philippines
Bantamweight (118 lb)	Paulie Ayala, U.S.	Veeraphol Sahaprom, Thailand	Tim Austin, U.S.
Jr. Bantamweight (115 lb)	Setsuo Kobayashi, Japan	Masanori Tokuyama, Japan	Felix Machado, Venezuela
Flyweight (112 lb)	Eric Morel, U.S./P.R.	P.S. Wonjongkam, Thailand	Irene Pacheco, Colombia
Jr. Flyweight (108 lb)	Rosendo Alvarez, Nicaragua	Yo-Sam Choi, S. Korea	Ricardo Lopez, Mexico
Strawweight (105 lb)	Yutaka Niida, Japan	Jose Antonio Aguirre, Mexico	Robert Leyva, Mexico

(a) The WBA calls Roy Jones Jr. and Kostya Tszyu "Super World Champions" in their respective classes and designates the WBA title in those categories as vacant.

Ring Champions by Years

(*abandoned the title or was stripped of it; IBF champions listed only for heavyweight division)

Heavyweights

1882-1892	John L. Sullivan (a)
1892-1897	James J. Corbett (b)
1897-1899	Robert Fitzsimmons
1899-1905	James J. Jeffries* (c)
1905-1906	Marvin Hart
1906-1908	Tommy Burns
1908-1915	Jack Johnson
1915-1919	Jess Willard
1919-1926	Jack Dempsey
1926-1928	Gene Tunney*
1928-1930	Vacant
1930-1932	Max Schmeling
1932-1933	Jack Sharkey
1933-1934	Primo Carnera
1934-1935	Max Baer
1935-1937	James J. Braddock
1937-1949	Joe Louis*
1949-1951	Ezzard Charles
1951-1952	Joe Walcott
1952-1956	Rocky Marciano*
1956-1959	Floyd Patterson
1959-1960	Ingemar Johansson
1960-1962	Floyd Patterson
1962-1964	Sonny Liston
1964-1967	Cassius Clay* (Muhammad Ali) (d)
1970-1973	Joe Frazier
1973-1974	George Foreman
1974-1978	Muhammad Ali
1978-1979	Muhammad Ali* (WBA)
1978	Leon Spinks (WBC*, WBA) (e); Ken Norton (WBC)
1978-1983	Larry Holmes* (WBC) (f)
1979-1980	John Tate (WBA)
1980-1982	Mike Weaver (WBA)
1982-1983	Michael Dokes (WBA)
1983-1984	Gerrie Coetzee (WBA)
1983-1985	Larry Holmes (IBF) (f)
1984	Tim Witherspoon (WBC)
1984-1985	Greg Page (WBA)
1984-1986	Pinklon Thomas (WBC)
1985-1986	Tony Tubbs (WBA)
1985-1987	Michael Spinks* (IBF)
1986	Tim Witherspoon (WBA)
	Trevor Berbick (WBC)
1986-1987	Mike Tyson (WBC); James "Bonecrusher" Smith (WBA)
1987	Tony Tucker (IBF)
1987-1990	Mike Tyson (WBC, WBA, IBF)
1990	"Buster" Douglas (WBA, WBC, IBF)
1990-1992	Evander Holyfield (WBA, WBC, IBF)
1992-1993	Riddick Bowe (WBA, IBF, WBC*)
1992-1994	Lennox Lewis (WBC)
1993-1994	Evander Holyfield (WBA, IBF)
1994	Michael Moorer (WBA, IBF)
1994-1995	Oliver McCall (WBC)
	George Foreman (WBA*, IBF*)
1995	Frans Botha* (IBF)
1995-1996	Bruce Seldon (WBA)
	Frank Bruno (WBC)
1996	Mike Tyson (WBC*, WBA)
1996-1997	Michael Moorer (IBF)
1996-1999	Evander Holyfield (WBA, IBF)
1997-2001	Lennox Lewis (WBC)
1999-2001	Lennox Lewis (WBA*, WBC, IBF)
2000-2001	Evander Holyfield (WBA)
2001	John Ruiz (WBA); Hasim Rahman (WBC, IBF)

(a) London Prize Ring (bare knuckle champion). (b) First Marquis of Queensberry champion. (c) Jeffries vacated title (1905) and designated Marvin Hart and Jack Root as logical contenders. Hart def. Root in 12 rounds (1905) and in turn was def. by Tommy Burns (1906), who claimed the title. Jack Johnson def. Burns (1908) and was recognized as champ. Johnson won the title by defeating Jeffries in an attempted comeback (1910). (d) Title declared vacant by the WBA and others in 1967 after Ali's refusal to fulfill his military obligation. Joe Frazier was recognized as champ by 6 states, Mexico, and South America. Jimmy Ellis was declared champ by the WBA. Frazier KOd Ellis, Feb. 16, 1970. (e) After Spinks defeated Ali, the WBC recognized Ken Norton as champ. Ali def.Spinks in 1978 rematch for WBA title, retired in 1979. (f) Holmes relinquished WBC title in Dec. 1983, to fight as champ of the new IBF.

Light Heavyweights

1903	Jack Root, George Gardner
1903-1905	Bob Fitzsimmons
1905-1912	Philadelphia Jack O'Brien*
1912-1916	Jack Dillon
1916-1920	Battling Levinsky
1920-1922	George Carpentier
1922-1923	Battling Siki
1923-1925	Mike McTigue
1925-1926	Paul Berlenbach
1926-1927	Jack Delaney*
1927-1929	Tommy Loughran*
1930-1934	Maxey Rosenbloom
1934-1935	Bob Olin
1935-1939	John Henry Lewis*
1939	Melio Bettina
1939-1941	Billy Conn*
1941	Anton Christoforidis (won NBA title)
1941-1948	Gus Lesnevich, Freddie Mills
1948-1950	Freddie Mills
1950-1952	Joey Maxim
1952-1962	Archie Moore
1962-1963	Harold Johnson
1963-1965	Willie Pastrano
1965-1966	Jose Torres
1966-1968	Dick Tiger
1968-1974	Bob Foster*
1974-1977	John Conteh (WBC)
1974-1978	Victor Galindez (WBA)
1977-1978	Miguel Cuello (WBC)
1978	Mate Parlov (WBC)
1978-1979	Mike Rossman (WBA); Marvin Johnson (WBC)
1979-1981	Matthew Saad Muhammad (WBC)
1979-1980	Marvin Johnson (WBA)
1980-1981	Eddie Mustafa Muhammad (WBA)
1981-1983	Michael Spinks (WBA); Dwight Braxton (WBC)
1983-1985	Michael Spinks*
1985-1986	J. B. Williamson (WBC)
1986-1987	Marvin Johnson (WBA); Dennis Andries (WBC)
1987	Leslie Stewart (WBA)
1987-1991	Virgil Hill (WBA)
1987	Thomas Hearns* (WBC)
1987-1988	Don Lalonde (WBC)
1988	Sugar Ray Leonard* (WBC)
1989	Dennis Andries (WBC)
1989-1990	Jeff Harding (WBC)
1990-1991	Dennis Andries (WBC)
1991-1994	Jeff Harding (WBC)
1991-1992	Thomas Hearns (WBA)
1992	Iran Barkley* (WBA)
1992-1997	Virgil Hill (WBA)
1994-1995	Mike McCallum (WBC)
1995-1996	Fabrice Tiozzo* (WBC)
1996-1997	Roy Jones Jr. (WBC)
1997	Montell Griffin (WBC); Roy Jones Jr. (WBC); Darius Michalczewski*(WBA)
1997-1998	Lou Del Valle (WBA)
1998	Roy Jones Jr. (WBA,WBC)

Middleweights

1884-1891	Jack "Nonpareil" Dempsey
1891-1897	Bob Fitzsimmons*
1897-1907	Tommy Ryan*
1907-1908	Stanley Ketchel, Billy Papke
1908-1910	Stanley Ketchel
1911-1913	vacant
1913	Frank Klaus; George Chip
1914-1917	Al McCoy
1917-1920	Mike O'Dowd
1920-1923	Johnny Wilson
1923-1926	Harry Greb
1926-1931	Tiger Flowers;
	Mickey Walker
1931-1932	Gorilla Jones (NBA)
1932-1937	Marcel Thil
1938	Al Hostak (NBA);
	Solly Krieger (NBA)
1939-1940	Al Hostak (NBA)
1941-1947	Tony Zale
1947-1948	Rocky Graziano
1948	Tony Zale; Marcel Cerdan
1949-1951	Jake LaMotta
1951	Ray Robinson; Randy Turpin;
	Ray Robinson*
1953-1955	Carl (Bobo) Olson
1955-1957	Ray Robinson
1957	Gene Fullmer; Ray Robinson
1957-1958	Carmen Basilio
1958	Ray Robinson
1959	Gene Fullmer (NBA);
	Ray Robinson (NY)
1960	Gene Fullmer (NBA);
	Paul Pender (NY and MA)
1961	Gene Fullmer (NBA);
	Terry Downes (NY, MA, Europe)
1962	Gene Fullmer;
	Dick Tiger (NBA);
	Paul Pender (NY and MA)*
1963	Dick Tiger (universal)
1963-1965	Joey Giardello
1965-1966	Dick Tiger
1966-1967	Emile Griffith
1967	Nino Benvenuti
1967-1968	Emile Griffith
1968-1970	Nino Benvenuti
1970-1977	Carlos Monzon*
1977-1978	Rodrigo Valdez
1978-1979	Hugo Corro
1979-1980	Vito Antuofermo
1980	Alan Minter
1980-1987	Marvin Hagler
1987	Sugar Ray Leonard* (WBC)
1987-1989	Sumbu Kalambay (WBA)
1987-1988	Thomas Hearns (WBC)
1988-1989	Iran Barkley (WBC)
1989-1990	Roberto Duran* (WBC)
1989-1991	Mike McCallum (WBA)
1990-1993	Julian Jackson (WBC)
1992-1993	Reggie Johnson (WBA)
1993-1995	Gerald McClellan* (WBC)
1993-1994	John David Jackson (WBA)
1994-1997	Jorge Castro (WBA)
1995	Julian Jackson (WBC)
1995-1996	Quincy Taylor (WBC);
	Shinji Takehara (WBA)
1996-1998	Keith Holmes (WBC)
1996-1997	William Joppy (WBA)
1997	Julio Cesar Green (WBA)
1998-2001	William Joppy (WBA)
1998-1999	Hassine Cherifi (WBC)
1999-2001	Keith Holmes (WBC)
2001	Bernard Hopkins (WBC);
	Felix Trinidad (WBA)

Welterweights

1892-1894	Mysterious Billy Smith
1894-1896	Tommy Ryan
1896	Kid McCoy*
1900	Rube Ferns; Matty Matthews
1901	Rube Ferns
1901-1904	Joe Walcott
1904-1906	Dixie Kid; Joe Walcott;
	Honey Mellody
1907-1911	Mike Sullivan
1911-1915	Vacant
1915-1919	Ted Lewis
1919-1922	Jack Britton
1922-1926	Mickey Walker
1926	Pete Latzo
1927-1929	Joe Dundee
1929	Jackie Fields
1930	Jack Thompson;
	Tommy Freeman
1931	Tommy Freeman; Jack
	Thompson; Lou Brouillard
1932	Jackie Fields
1933	Young Corbett;
	Jimmy McLarnin
1934	Barney Ross; Jimmy McLarnin
1935-1938	Barney Ross
1938-1940	Henry Armstrong
1940-1941	Fritzie Zivic
1941-1946	Fred Cochrane
1946	Marty Servo*
1946-1951	Ray Robinson* (a)
1951	Johnny Bratton (NBA)
1951-1954	Kid Gavilan
1954-1955	Johnny Saxton
1955	Tony De Marco
1955-1956	Carmen Basilio
1956	Johnny Saxton
1956-1957	Carmen Basilio*
1958	Virgil Akins
1958-1960	Don Jordan
1960-1961	Benny Paret
1961	Emile Griffith
1961-1962	Benny Paret
1962-1963	Emile Griffith
1963	Luis Rodriguez
1963-1966	Emile Griffith*
1966-1969	Curtis Cokes
1969-1970	Jose Napoles
1970-1971	Billy Backus
1971-1975	Jose Napoles
1975-1976	John Stracey (WBC);
	Angel Espada (WBA)
1976-1979	Carlos Palomino (WBC)
1976-1980	Jose Cuevas (WBA)
1979	Wilfredo Benitez (WBC)
1979-1980	Sugar Ray Leonard (WBC)
1980	Roberto Duran (WBC)
1980-1981	Thomas Hearns (WBA)
1980-1982	Sugar Ray Leonard*
1983-1985	Donald Curry (WBA);
	Milton McCrory (WBC)
1985-1986	Donald Curry
1986-1987	Lloyd Honeyghan (WBC)
1987	Mark Breland (WBA)
1987-1988	Marlon Starling (WBA);
	Jorge Vaca (WBC)
1988-1989	Tomas Molinares (WBA);
	Lloyd Honeyghan (WBC)
1989-1990	Marlon Starling (WBC);
	Mark Breland (WBA)
1990-1991	Maurice Blocker (WBC);
	Aaron Davis (WBA)
1991	Simon Brown (WBC)
1991-1992	Meldrick Taylor (WBA)
1991-1993	Buddy McGirt (WBC)
1992-1994	Crisanto Espana (WBA)
1993-1997	Pernell Whitaker (WBC)
1994-1998	Ike Quartey* (WBA)
1997-1999	Oscar De La Hoya* (WBC)
1998	James Page (WBA*)
1999-2000	Felix Trinidad* (WBC)
2000	Oscar De La Hoya (WBC)(b);
	Shane Mosley (WBC)
2001	Andrew Lewis (WBA)

(a) Robinson gained the title by defeating Tommy Bell in an elimination agreed to by the New York Commission and the National Boxing Association. Both claimed Robinson waived his title when he won the middleweight crown from LaMotta in 1951. (b) Trinidad was stripped of the WBC title when he moved up to super welterweight; De La Hoya def. Derrell Coley for the title in Feb. 2000.

Lightweights

1896-1899	Kid Lavigne
1899-1902	Frank Erne
1902-1908	Joe Gans
1908-1910	Battling Nelson
1910-1912	Ad Wolgast
1912-1914	Willie Ritchie
1914-1917	Freddie Welsh
1917-1925	Benny Leonard*
1925	Jimmy Goodrich;
	Rocky Kansas
1926-1930	Sammy Mandell
1930	Al Singer; Tony Canzoneri
1930-1933	Tony Canzoneri
1933-1935	Barney Ross*
1935-1936	Tony Canzoneri
1936-1938	Lou Ambers
1938	Henry Armstrong
1939	Lou Ambers
1940	Lew Jenkins
1941-1943	Sammy Angott
1944	S. Angott (NBA);
	J. Zurita (NBA)
1945-1951	Ike Williams (NBA: later
	universal)
1951-1952	James Carter
1952	Lauro Salas; James Carter
1953-1954	James Carter
1954	Paddy De Marco; James Carter
1955	James Carter; Bud Smith
1956	Bud Smith; Joe Brown
1956-1962	Joe Brown
1962-1965	Carlos Ortiz
1965	Ismael Laguna
1965-1968	Carlos Ortiz
1968-1969	Teo Cruz
1969-1970	Mando Ramos
1970	Ismael Laguna
1970-1972	Ken Buchanan (WBA)
1971-1972	Pedro Carrasco (WBC)
1972-1979	Roberto Duran* (WBA)
1972	Mando Ramos (WBC);
	Chango Carmona (WBC)
1972-1974	Rodolfo Gonzalez (WBC)
1974-1976	Ishimatsu Suzuki (WBC)
1976-1978	Esteban De Jesus (WBC)
1979-1981	Jim Watt (WBC)
1979-1980	Ernesto Espana (WBA)
1980-1981	Hilmer Kenty (WBA)
1981	Sean O'Grady (WBA);
	Claude Noel (WBA)
1981-1983	Alexis Arguello* (WBC)
1981-1982	Arturo Frias (WBA)
1982-1984	Ray Mancini (WBA)
1983-1984	Edwin Rosario (WBC)
1984-1986	Livingstone Bramble (WBA)
1984-1985	Jose Luis Ramirez (WBC)
1985-1986	Hector (Macho) Camacho
	(WBC)
1986-1987	Edwin Rosario (WBA)
1987-1988	Julio Cesar Chavez (WBA);
	Jose Luis Ramirez (WBC)
1988-1989	Julio Cesar Chavez (WBA,
	WBC)
1989-1990	Edwin Rosario (WBA);
	Pernell Whitaker (WBC)
1990	Juan Nazario (WBA)
1990-1992	Pernell Whitaker*
1992	Joey Gamache (WBA)
1992-1996	Miguel Angel Gonzalez* (WBC)
1992-1993	Tony Lopez (WBA)
1993	Dingaan Thobela (WBA)
1993-1998	Orzubek Nazarov (WBA)
1996-1997	Jean-Baptiste Mendy (WBC)
1997-1998	Steve Johnston (WBC)
1998-1999	Jean-Baptiste Mendy (WBA);
	Cesar Bazan (WBC)
1999	Julian Lorcy (WBA);
	Stefano Zoff (WBA)
1999-2000	Gilberto Serrano (WBA);
	Steve Johnston (WBC)
2000-2001	Takanori Hatakeyama (WBA);
2000	Jose Luis Castillo (WBC)
2001	Julien Lorcy (WBA);
	Raul Balbi (WBA)

Featherweights

1892-1900	George Dixon (disputed)
1900-1901	Terry McGovern; Young Corbett*
1901-1912	Abe Attell
1912-1923	Johnny Kilbane
1923	Eugene Criqui; Johnny Dundee
1923-1925	Johnny Dundee*
1925-1927	Kid Kaplan*
1927-1928	Benny Bass; Tony Canzoneri
1928-1929	Andre Routis
1929-1932	Battling Battalino*
1932-1934	Tommy Paul (NBA)
1933-1936	Freddie Miller
1936-1937	Petey Sarron
1937-1938	Henry Armstrong*
1938-1940	Joey Archibald (a)
1940-1941	Harry Jeffra
1942-1948	Willie Pep
1948-1949	Sandy Saddler
1949-1950	Willie Pep
1950-1957	Sandy Saddler*
1957-1959	Hogan (Kid) Bassey
1959-1963	Davey Moore
1963-1964	Sugar Ramos
1964-1967	Vicente Saldivar*
1968	Paul Rojas (WBA)
1968-1969	Jose Legra (WBC)
1968-1971	Shozo Saijyo (WBA)
1969-1970	Johnny Famechon (WBC)
1970	Vicente Salvidar (WBC)
1970-1972	Kuniaki Shibata (WBC)
1971-1972	Antonio Gomez (WBA)
1972	Clemente Sanchez* (WBC)
1972-1974	Ernesto Marcel* (WBA)
1972-1973	Jose Legra (WBC)
1973-1974	Eder Jofre* (WBC)
1974	Ruben Olivares (WBA)
1974-1975	Bobby Chacon (WBC)
1974-1976	Alexis Arguello* (WBA)
1975	Ruben Olivares (WBC)
1975-1976	David Kotey (WBC)
1976-1980	Danny Lopez (WBC)
1977	Rafael Ortega (WBA)
1977-1978	Cecilio Lastra (WBA)
1978-1985	Eusebio Pedrosa (WBA)
1980-1982	Salvador Sanchez (WBC)
1982-1984	Juan LaPorte (WBC)
1984	Wilfredo Gomez (WBC)
1984-1988	Azumah Nelson (WBC)
1985-1986	Barry McGuigan (WBA)
1986-1987	Steve Cruz (WBA)
1987-1991	Antonio Esparragoza (WBA)
1988-1990	Jeff Fenech* (WBC)
1990-1991	Marcos Villasana (WBC)
1991-1993	Park Yung Kyun (WBA); Paul Hodkinson (WBC)
1993	Goyo Vargas (WBC)
1993-1995	Kevin Kelley (WBC)
1993-1996	Eloy Rojas (WBA)
1995	Alejandro Gonzalez (WBC)
1995-1996	Manuel Medina (WBC)
1995-1999	Luisito Espinosa (WBC)
1996-1997	Wilfredo Vasquez* (WBA)
1998	Freddie Norwood (WBA)
1998-1999	Antonio Ceremeno (WBA)
1999	Cesar Soto (WBC); Naseem Hamed* (WBC); Freddie Norwood (WBA)
2000-2001	Guty Espadas (WBC); Derrick Gaines (WBA)
2001	Erik Morales (WBC)

(a) After Petey Scalzo knocked out Archibald in an overweight match and was refused a title bout, the NBA named Scalzo champion. NBA title succession: Scalzo, 1938-1941; Richard Lemos, 1941; Jackie Wilson, 1941-1943; Jackie Callura, 1943; Phil Terranova, 1943-1944; Sal Bartolo, 1944-1946.

History of Heavyweight Championship Bouts

(bouts in which title changed hands)

1889—July 8—John L. Sullivan def. Jake Kilrain, 75, Richburg, MS. (Last championship bare knuckles bout.)
1892—Sept. 7—James J. Corbett def. John L. Sullivan, 21, New Orleans. (Big gloves used for first time.)
1897—Bob Fitzsimmons def. James J. Corbett, 14, Carson City, NV.
1899—June 9, James J. Jeffries def. Bob Fitzsimmons, 11, Coney Island, NY. (Jeffries retired as champion in 1905.)
1905—July 3, Marvin Hart KOd Jack Root, 12, Reno, NV. (Jeffries refereed and gave the title to Hart. Jack O'Brien also claimed the title.)
1906—Feb. 23, Tommy Burns def. Marvin Hart, 20, Los Angeles.
1908—Dec. 26, Jack Johnson KOd Tommy Burns, 14, Sydney, Australia. (Police halted contest.)
1915—April 5, Jess Willard KOd Jack Johnson, 26, Havana, Cuba.
1919—July 4, Jack Dempsey KOd Jess Willard, Toledo, OH. (Willard failed to answer bell for 4th round.)
1926—Sept. 23, Gene Tunney def. Jack Dempsey, 10, Philadelphia. (Tunney retired as champion in 1928.)
1930—June 12, Max Schmeling def. Jack Sharkey, 4, NY. (Sharkey fouled Schmeling in a bout generally considered to have resulted in the election of a successor to Tunney.)
1932—June 21, Jack Sharkey def. Max Schmeling, 15, NY.
1933—June 29, Primo Carnera KOd Jack Sharkey, 6, NY.
1934—June 14, Max Baer KOd Primo Carnera, 11, NY.
1935—June 13, James J. Braddock def. Max Baer, 15, NY.
1937—June 22, Joe Louis KOd James J. Braddock, 8, Chicago. (Louis retired as champion in 1949.)
1949—June 22, Ezzard Charles def. Joe Walcott, 15, Chicago; NBA recognition only.
1951—July 18, Joe Walcott KOd Ezzard Charles, 7, Pittsburgh.
1952—Sept. 23, Rocky Marciano KOd Joe Walcott, 13, Philadelphia. (Marciano retired as champion in 1956.)
1956—Nov. 30, Floyd Patterson KOd Archie Moore, 5, Chicago.
1959—June 26, Ingemar Johansson KOd Floyd Patterson, 3, NY.
1960—June 20, Floyd Patterson KOd Ingemar Johansson, 5, NY. (Patterson was 1st heavyweight to regain title.)
1962—Sept. 25, Sonny Liston KOd Floyd Patterson, 1, Chicago.
1964—Feb. 25, Cassius Clay (Muhammad Ali) KOd Sonny Liston, 7, Miami Beach, FL. (In 1967, Ali was stripped of his title by the WBA and others for refusing military service.)
1970—Feb. 16, Joe Frazier KOd Jimmy Ellis, 5, NY. (Frazier def. Ali in 15 rounds, Mar. 8, 1971, in NY.)
1973—Jan. 22, George Foreman KOd Joe Frazier, 2, Kingston, Jamica
1974—Oct. 30, Muhammad Ali KOd George Foreman, 8, Kinshasa, Zaire.
1978—Feb. 15, Leon Spinks def. Muhammad Ali, 15, Las Vegas. (WBC recognized Ken Norton as champion after Spinks refused to fight him before his rematch with Ali.)
1978—June 9, (WBC) Larry Holmes def. Ken Norton, 15, Las Vegas. (Holmes gave up title in Dec. 1983.)
1978—Sept. 15, (WBA) Muhammad Ali def. Leon Spinks, 15, New Orleans. (Ali retired as champion in 1979.)
1979—Oct. 20, (WBA) John Tate def. Gerrie Coetzee, 15, Pretoria, South Africa.
1980—Mar. 31, (WBA) Mike Weaver KOd John Tate, 15, Knoxville.
1982—Dec. 10, (WBA) Michael Dokes KOd Mike Weaver, 1, Las Vegas.
1983—Sept. 23, (WBA) Gerrie Coetzee KOd Michael Dokes, 10, Richfield, OH.
1983—In Dec., Larry Holmes relinquished the WBC title and was named champion of the newly formed IBF.
1984—Mar. 9, (WBC) Tim Witherspoon def. Greg Page, 12, Las Vegas.
1984—Aug. 31, (WBC) Pinklon Thomas def. Tim Witherspoon, 12, Las Vegas.
1984—Dec. 2, WBA) Greg Page KOd Gerrie Coetzee, 8, Sun City, Bophuthatswana.
1985—Apr. 29, (WBA) Tony Tubbs def. Greg Page, 15, Buffalo, NY.
1985—Sept. 21, (IBF) Michael Spinks def. Larry Holmes, 15, Las Vegas. (Spinks relinquished title in Feb. 1987.)
1986—Jan. 17, (WBA) Tim Witherspoon def. Tony Tubbs, 15, Atlanta, GA.
1986—Mar. 23, (WBC) Trevor Berbick def. Pinklon Thomas, 12, Miami.
1986—Nov. 22, (WBC) Mike Tyson KOd Trevor Berbick, 2, Las Vegas.
1986—Dec. 12, (WBA) James "Bonecrusher" Smith KOd Tim Witherspoon, 1, NY.
1987—Mar. 7, (WBA, WBC) Mike Tyson def. James "Bonecrusher" Smith, 12, Las Vegas.
1987—May 30, (IBF) Tony Tucker KO'd James "Buster" Douglas, 10, Las Vegas.
1987—Aug. 1, (WBA, WBC, IBF) Mike Tyson def. Tony Tucker, 12, Las Vegas. (Tyson became undisputed champion.)
1990—Feb. 11, (WBA, WBC, IBF) James "Buster" Douglas KOd Mike Tyson, 10, Tokyo.
1990—Oct. 25, (WBA, WBC, IBF) Evander Holyfield KOd James "Buster" Douglas, 3, Las Vegas.
1992—Nov. 13, (WBA, WBC, IBF) Riddick Bowe def. Evander Holyfield, 12, Las Vegas. (Lennox Lewis was later named WBC champion when Bowe refused to fight him.)
1993—Nov. 6, (WBA, IBF) Evander Holyfield def. Riddick Bowe, 12, Las Vegas.
1994—Apr. 22, (WBA, IBF) Michael Moorer def. Evander Holyfield, 12, Las Vegas.
1994—Sept. 24, (WBC) Oliver McCall KOd Lennox Lewis, 2, London.
1994—Nov. 5, (WBA, IBF) George Foreman KOd Michael Moorer, 10, Las Vegas. (In Mar. 1995, Foreman was stripped of the WBA title. In June, Foreman relinquished the IBF title.)
1995—Sept. 2, (WBC) Frank Bruno def. Oliver McCall, 12, London.
1995—Dec. 9, (IBF) Frans Botha def. Axel Schulz, 12, Las Vegas. (Botha was subsequently stripped of title.)
1996—Mar. 16, (WBC) Mike Tyson KOd Frank Bruno, 3, Las Vegas.
1996—June 22, (IBF) Michael Moorer def. Axel Schulz, 12, Dortmund, Germany.
1996—Sept. 7, (WBA, WBC) Mike Tyson KOd Bruce Seldon, 1, Las Vegas. (Tyson was subsequently stripped of WBC title.)
1996—Nov. 9, (WBA) Evander Holyfield KOd Mike Tyson, 11, Las Vegas.
1997—Feb. 7, (WBC) Lennox Lewis KOd Oliver McCall, 5, Las Vegas.
1997—Nov. 8, (IBF) Evander Holyfield def. Michael Moorer, 8, Las Vegas.
1999—Nov. 13, (WBA, WBC, IBF) Lennox Lewis def. Evander Holyfield, 12, Las Vegas. (Lewis became undisputed champion. In April 2000, Lewis was stripped of his WBA title.)
2000—Aug. 12, (WBA) Evander Holyfield def. John Ruiz, 12, Las Vegas.
2001—Mar. 3, (WBA) John Ruiz def. Evander Holyfield, 12, Las Vegas.
2001—Apr. 21, (WBC, IBF) Hasim Rahman KOd Lennox Lewis, 5, Brakpan, South Africa.

THOROUGHBRED RACING

Triple Crown Winners

Since 1920, colts have carried 126 lb. in triple crown events; fillies, 121 lb.
(Kentucky Derby, Preakness, and Belmont Stakes)

Year	Horse	Jockey	Trainer
1919	Sir Barton	J. Loftus	H. G. Bedwell
1930	Gallant Fox	E. Sande	J. Fitzsimmons
1935	Omaha	W. Sanders	J. Fitzsimmons
1937	War Admiral	C. Kurtsinger	G. Conway
1941	Whirlaway	E. Arcaro	B. A. Jones
1943	Count Fleet	J. Longden	G. D. Cameron
1946	Assault	W. Mehrtens	M. Hirsch
1948	Citation	E. Arcaro	H. A. Jones
1973	Secretariat	R. Turcotte	L. Laurin
1977	Seattle Slew	J. Cruguet	W. H. Turner Jr.
1978	Affirmed	S. Cauthen	L. S. Barrera

Kentucky Derby

Churchill Downs, Louisville, KY; inaug. 1875; distance 1-1/4 mi; 1-1/2 mi until 1896. 3-year-olds.
Best time: 1:59 2/5, by Secretariat, 1973; 2001 time:1:59.97.

Year	Winner	Jockey
1875	Aristides	O. Lewis
1876	Vagrant	R. Swim
1877	Baden Baden	W. Walker
1878	Day Star	Carter
1879	Lord Murphy	C. Schauer
1880	Fonso	G. Lewis
1881	Hindoo	J. McLaughlin
1882	Apollo	B. Hurd
1883	Leonatus	W. Donohue
1884	Buchanan	I. Murphy
1885	Joe Cotton	E. Henderson
1886	Ben Ali	P. Duffy
1887	Montrose	I. Lewis
1888	Macbeth II.	G. Covington
1889	Spokane	T. Kiley
1890	Riley	I. Murphy
1891	Kingman	I. Murphy
1892	Azra	A. Clayton
1893	Lookout	E. Kunze
1894	Chant	F. Goodale
1895	Halma	J. Perkins
1896	Ben Brush	W. Simms
1897	Typhoon II.	F. Garner
1898	Plaudit	W. Simms
1899	Manuel	F. Taral
1900	Lieut. Gibson	J. Boland
1901	His Eminence	J. Winkfield
1902	Alan-a-Dale	J. Winkfield
1903	Judge Himes	H. Booker
1904	Elwood	F. Prior
1905	Agile	J. Martin
1906	Sir Huon	R. Troxler
1907	Pink Star	A. Minder
1908	Stone Street	A. Pickens
1909	Wintergreen	V. Powers
1910	Donau	F. Herbert
1911	Meridian	G. Archibald
1912	Worth	C.H. Shilling
1913	Donerail	R. Goose
1914	Old Rosebud	J. McCabe
1915	Regret*	J. Notter
1916	George Smith	J. Loftus
1917	Omar Khayyam	C. Borel
1918	Exterminator	W. Knapp
1919	Sir Barton	J. Loftus
1920	Paul Jones	T. Rice
1921	Behave Yourself	C. Thompson
1922	Morvich	A. Johnson
1923	Zev	E. Sande
1924	Black Gold	J. D. Mooney
1925	Flying Ebony	E. Sande
1926	Bubbling Over	A. Johnson
1927	Whiskery	L. McAtee
1928	Reigh Count	C. Lang
1929	Clyde Van Dusen	L. McAtee
1930	Gallant Fox	E. Sande
1931	Twenty Grand	C. Kurtsinger
1932	Burgoo King	E. James
1933	Brokers Tip	D. Meade
1934	Cavalcade	M. Garner
1935	Omaha	W. Saunders
1936	Bold Venture	I. Hanford
1937	War Admiral	C. Kurtsinger
1938	Lawrin	E. Arcaro
1939	Johnstown	J. Stout
1940	Gallahadion	C. Bierman
1941	Whirlaway	E. Arcaro
1942	Shut Out	W. D. Wright
1943	Count Fleet	J. Longden
1944	Pensive	C. McCreary
1945	Hoop, Jr.	E. Arcaro
1946	Assault	W. Mehrtens
1947	Jet Pilot	E. Guerin
1948	Citation	E. Arcaro
1949	Ponder	S. Brooks
1950	Middleground	W. Boland
1951	Count Turf	C. McCreary
1952	Hill Gail	E. Arcaro
1953	Dark Star	H. Moreno
1954	Determine	R. York
1955	Swaps	W. Shoemaker
1956	Needles	D. Erb
1957	Iron Liege	W. Hartack
1958	Tim Tam	I. Valenzuela
1959	Tomy Lee	W. Shoemaker
1960	Venetian Way	W. Hartack
1961	Carry Back	J. Sellers
1962	Decidedly	W. Hartack
1963	Chateaugay	B. Baeza
1964	Northern Dancer	W. Hartack
1965	Lucky Debonair	W. Shoemaker
1966	Kauai King	D. Brumfield
1967	Proud Clarion	R. Ussery
1968	Dancer's Image#	R. Ussery
1969	Majestic Prince	W. Hartack
1970	Dust Commander	M. Manganello
1971	Canonero II	G. Avila
1972	Riva Ridge	R. Turcotte
1973	Secretariat	R. Turcotte
1974	Cannonade	A. Cordero
1975	Foolish Pleasure	J. Vasquez
1976	Bold Forbes	A. Cordero
1977	Seattle Slew	J. Cruguet
1978	Affirmed	S. Cauthen
1979	Spectacular Bid	R. Franklin
1980	Genuine Risk*	J. Vasquez
1981	Pleasant Colony	J. Velasquez
1982	Gato del Sol	E. Delahoussaye
1983	Sunny's Halo	E. Delahoussaye
1984	Swale	L. Pincay
1985	Spend a Buck	A. Cordero
1986	Ferdinand	W. Shoemaker
1987	Alysheba	C. McCarron
1988	Winning Colors*	G. Stevens
1989	Sunday Silence	P. Valenzuela
1990	Unbridled	C. Perret
1991	Strike the Gold	C. Antley
1992	Lil E. Tee	P. Day
1993	Sea Hero	J. Bailey
1994	Go for Gin	C. McCarron
1995	Thunder Gulch	G. Stevens
1996	Grindstone	J. Bailey
1997	Silver Charm	G. Stevens
1998	Real Quiet	K. Desormeaux
1999	Charismatic	C. Antley
2000	Fusaichi Pegasus	K. Desormeaux
2001	Monarchos	J. Chavez

*Regret, Genuine Risk, and Winning Colors are the only fillies to have won the Derby. # Dancer's Image was disqualified from purse money after tests disclosed that he had run with a pain-killing drug, phenylbutazone, in his system. All wagers were paid on Dancer's Image. Forward Pass was awarded first place money.The Kentucky Derby has been won 5 times by 2 jockeys: Eddie Arcaro, 1938, 1941, 1945, 1948, and 1952; and Bill Hartack, 1957, 1960, 1962, 1964, and 1969. It was won 4 times by Willie Shoemaker, 1955, 1959, 1965, and 1986; and 3 times by each of 4 jockeys: Isaac Murphy, 1884, 1890, and 1891; Earle Sande, 1923, 1925, and 1930; Angel Cordero, 1974, 1976, and 1985; and Gary Stevens, 1988, 1995, and 1997.

Top 10 Fastest Winning Times for the Kentucky Derby

(Official Kentucky Derby times measured in fifths of a second.)

Time	Horse	Jockey	Year
1m. 59 2/5 s.	Secretariat	Ron Turcotte	1973
1m. 59 4/5 s.	Monarchos	Jorge Chavez	2001
2m.	Northern Dancer	Bill Hartack	1964
2m. 1/5 s.	Spend a Buck	Angel Cordero Jr.	1985
2m. 2/5 s.	Decidedly	Bill Hartack	1962
2m. 3/5 s.	Proud Clarion	Bill Shoemaker	1967
2m. 1 s.	Fusaichi Pegasus	Kent Desormeaux	2000
	Grindstone	Jerry Bailey	1996
2m. 1 1/5 s.	Thunder Gulch	Gary Stevens	1995
	Affirmed	Steve Cauthen	1978
	Lucky Debonair	Bill Shoemaker	1965
2m. 1 2/5 s.	Whirlaway	Eddie Arcaro	1941
2m. 1 3/5 s.	Bold Forbes	Angel Cordero Jr.	1976
	Hill Gail	Eddie Arcaro	1952
	Middleground	William Boland	1950

Preakness Stakes

Pimlico Race Course, Baltimore, MD; inaug. 1873; distance 1-3/16 mi. 3-year-olds.
Best time: 1:53 2/5, by Tank's Prospect (1985) and Louis Quatorze (1996); 2001 time: 1:55.51.

Year	Winner	Jockey
1873	Survivor	G. Barbee
1874	Culpepper	M. Donohue
1875	Tom Ochiltree	L. Hughes
1876	Shirley	G. Barbee
1877	Cloverbrook	C. Holloway
1878	Duke of Magenta	C. Holloway
1879	Harold	L. Hughes
1880	Grenada	L. Hughes
1881	Saunterer	W. Costello
1882	Vanguard	W. Costello
1883	Jacobus	G. Barbee
1884	Knight of Ellerslie	S. H. Fisher

Year	Winner	Jockey
1885	Tecumseh	J. McLaughlin
1886	The Bard	S. H. Fisher
1887	Dunboyne	W. Donohue
1888	Refund	F. Littlefield
1889	Buddhist	G. Anderson
1890	Montague	W. Martin
1894	Assignee	F. Taral
1895	Belmar	F. Taral
1896	Margrave	H. Griffin
1897	Paul Kauvar	C. Thorpe
1898	Sly Fox	W. Simms
1899	Half Time	R. Clawson
1900	Hindus	H. Spencer
1901	The Parader	F. Landry
1902	Old England	L. Jackson
1903	Flocarline	W. Gannon
1904	Bryn Mawr	E. Hildebrand
1905	Cairngorm	W. Davis
1906	Whimsical	W. Miller
1907	Don Enrique	G. Mountain
1908	Royal Tourist	E. Dugan
1909	Effendi	W. Doyle
1910	Layminster	R. Estep
1911	Watervale	E. Dugan
1912	Colonel Holloway	C. Turner
1913	Buskin	J. Butwell
1914	Holiday	A. Schuttinger
1915	Rhine Maiden	D. Hoffman
1916	Damrosch	L. McAtee
1917	Kalitan	E. Haynes
1918	War Cloud	J. Loftus
	Jack Hare	Jr. C. Peak
1919	Sir Barton	J. Loftus
1920	Man o' War	C. Kummer
1921	Broomspun	F. Coltiletti
1922	Pillory	L. Morris
1923	Vigil	B. Marinelli
1924	Nellie Morse	J. Merimee
1925	Coventry	C. Kummer
1926	Display	J. Malben
1927	Bostonian	A. Abel
1928	Victorian	R. Workman
1929	Dr. Freeland	L. Schaefer
1930	Gallant Fox	E. Sande
1931	Mate	G. Ellis
1932	Burgoo King	E. James
1933	Head Play	C. Kurtsinger
1934	High Quest	R. Jones
1935	Omaha	W. Saunders
1936	Bold Venture	G. Woolf
1937	War Admiral	C. Kurtsinger
1938	Dauber	M. Peters
1939	Challedon	G. Seabo
1940	Bimelech	F.A. Smith
1941	Whirlaway	E. Arcaro
1942	Alsab	B. James
1943	Count Fleet	J. Longden
1944	Pensive	C. McCreary
1945	Polynesian	W.D. Wright
1946	Assault	W. Mehrtens
1947	Faultless	D. Dodson
1948	Citation	E. Arcaro
1949	Capot	T. Atkinson
1950	Hill Prince	E. Arcaro
1951	Bold	E. Arcaro
1952	Blue Man	C. McCreary
1953	Native Dancer	E. Guerin
1954	Hasty Road	J. Adams
1955	Nashua	E. Arcaro
1956	Fabius	W. Hartack
1957	Bold Ruler	E. Arcaro
1958	Tim Tam	I. Valenzuela
1959	Royal Orbit	W. Harmatz
1960	Bally Ache	R. Ussery
1961	Carry Back	J. Sellers
1962	Greek Money	J.L. Rotz
1963	Candy Spots	W. Shoemaker
1964	Northern Dancer	W. Hartack
1965	Tom Rolfe	R. Turcotte
1966	Kauai King	D. Brumfield
1967	Damascus	W. Shoemaker
1968	Forward Pass	I. Valenzuela
1969	Majestic Prince	W. Hartack
1970	Personality	E. Belmonte
1971	Canonero II	G. Avila
1972	Bee Bee Bee	E. Nelson
1973	Secretariat	R. Turcotte
1974	Little Current	M. Rivera
1975	Master Derby	D. McHargue
1976	Elocutionist	J. Lively
1977	Seattle Slew	J. Cruguet
1978	Affirmed	S. Cauthen
1979	Spectacular Bid	R. Franklin
1980	Codex	A. Cordero
1981	Pleasant Colony	J. Velasquez
1982	Aloma's Ruler	J. Kaenel
1983	Deputed Testamony	D. Miller
1984	Gate Dancer	A. Cordero
1985	Tank's Prospect	P. Day
1986	Snow Chief	A. Solis
1987	Alysheba	C. McCarron
1988	Risen Star	E. Delahoussaye
1989	Sunday Silence	P. Valenzuela
1990	Summer Squall	P. Day
1991	Hansel	J. Bailey
1992	Pine Bluff	C. McCarron
1993	Prairie Bayou	M. Smith
1994	Tabasco Cat	P. Day
1995	Timber Country	P. Day
1996	Louis Quatorze	P. Day
1997	Silver Charm	G. Stevens
1998	Real Quiet	K. Desormeaux
1999	Charismatic	C. Antley
2000	Red Bullet	J. Bailey
2001	Point Given	G. Stevens

Belmont Stakes

Belmont Park, Elmont, NY; inaug. 1867; distance 1-1/2 mi. 3-year-olds. Best time: 2:24, Secretariat, 1973; 2001 time: 2:26.56.

Year	Winner	Jockey
1867	Ruthless	J. Gilpatrick
1868	General Duke	R. Swim
1869	Fenian	C. Miller
1870	Kingfisher	W. Dick
1871	Harry Bassett	W. Miller
1872	Joe Daniels	J. Rowe
1873	Springbok	J. Rowe
1874	Saxon	G. Barbee
1875	Calvin	R. Swim
1876	Algerine	W. Donohue
1877	Cloverbrook	C. Holloway
1878	Duke of Magenta	L. Hughes
1879	Spendthrift	S. Evans
1880	Grenada	L. Hughes
1881	Saunterer	T. Costello
1882	Forester	J. McLaughlin
1883	George Kinney	J. McLaughlin
1884	Panique	J. McLaughlin
1885	Tyrant	P. Duffy
1886	Inspector	B.J. McLaughlin
1887	Hanover	J. McLaughlin
1888	Sir Dixon	J. McLaughlin
1889	Eric	W. Hayward
1890	Burlington	S. Barnes
1891	Foxford	E. Garrison
1892	Patron	W. Hayward
1893	Comanche	W. Simms
1894	Henry of Navarre	W. Simms
1895	Belmar	F. Taral
1896	Hastings	H. Griffin
1897	Scottish Chieftain	J. Scherrer
1898	Bowling Brook	F. Littlefield
1899	Jean Bereaud	R. R. Clawson
1900	Ildrim	N. Turner
1901	Commando	H. Spencer
1902	Masterman	J. Bullman
1903	Africander	J. Bullman
1904	Delhi	G. Odom
1905	Tanya	E. Hildebrand
1906	Burgomaster	L. Lyne
1907	Peter Pan	G. Mountain
1908	Colin	J. Notter
1909	Joe Madden	E. Dugan
1910	Sweep	J. Butwell
1913	Prince Eugene	R. Troxler
1914	Luke McLuke	M. Buxton
1915	The Finn	G. Byrne
1916	Friar Rock	E. Haynes
1917	Hourless	J. Butwell
1918	Johren	F. Robinson
1919	Sir Barton	J. Loftus
1920	Man o' War	C. Kummer
1921	Grey Lag	E. Sande
1922	Pillory	C. H. Miller
1923	Zev	E. Sande
1924	Mad Play	E. Sande
1925	American Flag	A. Johnson
1926	Crusader	A. Johnson
1927	Chance Shot	E. Sande
1928	Vito	C. Kummer
1929	Blue Larkspur	M. Garner
1930	Gallant Fox	E. Sande
1931	Twenty Grand	C. Kurtsinger
1932	Faireno	T. Malley
1933	Hurryoff	M. Garner
1934	Peace Chance	W. D. Wright
1935	Omaha	W. Saunders
1936	Granville	J. Stout
1937	War Admiral	C. Kurtsinger
1938	Pasteurized	J. Stout
1939	Johnstown	J. Stout
1940	Bimelech	F. A. Smith
1941	Whirlaway	E. Arcaro
1942	Shut Out	E. Arcaro
1943	Count Fleet	J. Longden
1944	Bounding Home	G. L. Smith
1945	Pavot	E. Arcaro
1946	Assault	W. Mehrtens
1947	Phalanx	R. Donoso
1948	Citation	E. Arcaro
1949	Capot	T. Atkinson
1950	Middleground	W. Boland
1951	Counterpoint	D. Gorman
1952	One Count	E. Arcaro
1953	Native Dancer	E. Guerin
1954	High Gun	E. Guerin
1955	Nashua	E. Arcaro
1956	Needles	D. Erb
1957	Gallant Man	W. Shoemaker
1958	Cavan	P. Anderson
1959	Sword Dancer	W. Shoemaker
1960	Celtic Ash	W. Hartack
1961	Sherluck	B. Baeza
1962	Jaipur	W. Shoemaker
1963	Chateaugay	B. Baeza
1964	Quadrangle	M. Ycaza
1965	Hail to All	J. Sellers
1966	Amberoid	W. Boland
1967	Damascus	W. Shoemaker
1968	Stage Door Johnny	H. Gustines
1969	Arts and Letters	B. Baeza
1970	High Echelon	J. L. Rotz
1971	Pass Catcher	W. Blum
1972	Riva Ridge	R. Turcotte
1973	Secretariat	R. Turcotte
1974	Little Current	M. Rivera
1975	Avatar	W. Shoemaker
1976	Bold Forbes	A. Cordero
1977	Seattle Slew	J. Cruguet
1978	Affirmed	S. Cauthen
1979	Coastal	R. Hernandez
1980	Temperence Hill	E. Maple
1981	Summing	G. Martens
1982	Conquistador Cielo	L. Pincay
1983	Caveat	L. Pincay
1984	Swale	L. Pincay
1985	Creme Fraiche	E. Maple
1986	Danzig Connection	C. McCarron
1987	Bet Twice	C. Perret
1988	Risen Star	E. Delahoussaye
1989	Easy Goer	P. Day
1990	Go and Go	M. Kinane
1991	Hansel	J. Bailey
1992	A.P. Indy	E. Delahoussaye
1993	Colonial Affair	J. Krone
1994	Tabasco Cat	P. Day
1995	Thunder Gulch	G. Stevens
1996	Editor's Note	R. Douglas
1997	Touch Gold	C. McCarron
1998	Victory Gallop	G. Stevens
1999	Lemon Drop Kid	J. Santos
2000	Commendable	P. Day
2001	Point Given	G. Stevens

Annual Leading Jockey — Money Won[1]

Year	Jockey	Earnings	Year	Jockey	Earnings	Year	Jockey	Earnings
1957	Bill Hartack	$3,060,501	1972	Laffit Pincay, Jr.	$3,225,827	1987	Jose Santos	$12,375,433
1958	Willie Shoemaker	2,961,693	1973	Laffit Pincay, Jr.	4,093,492	1988	Jose Santos	14,877,298
1959	Willie Shoemaker	2,843,133	1974	Laffit Pincay, Jr.	4,251,060	1989	Jose Santos	13,838,389
1960	Willie Shoemaker	2,123,961	1975	Braulio Baeza	3,695,198	1990	Gary Stevens	13,881,198
1961	Willie Shoemaker	2,690,819	1976	Angel Cordero, Jr.	4,709,500	1991	Chris McCarron	14,441,083
1962	Willie Shoemaker	2,916,844	1977	Steve Cauthen	6,151,750	1992	Kent Desormeaux	14,193,006
1963	Willie Shoemaker	2,526,925	1978	Darrel McHargue	6,029,885	1993	Mike Smith	14,024,815
1964	Willie Shoemaker	2,649,553	1979	Laffit Pincay, Jr.	8,193,535	1994	Mike Smith	15,979,820
1965	Braulio Baeza	2,582,702	1980	Chris McCarron	7,663,300	1995	Jerry Bailey	16,311,876
1966	Braulio Baeza	2,951,022	1981	Chris McCarron	8,397,604	1996	Jerry Bailey	19,465,376
1967	Braulio Baeza	3,088,888	1982	Angel Cordero, Jr.	9,483,590	1997	Jerry Bailey	18,320,743
1968	Braulio Baeza	2,835,108	1983	Angel Cordero, Jr.	10,116,697	1998	Gary Stevens	19,622,855
1969	Jorge Velasquez	2,542,315	1984	Chris McCarron	12,045,813	1999	Pat Day	18,092,845
1970	Laffit Pincay, Jr.	2,626,526	1985	Laffit Pincay, Jr.	13,353,299	2000	Pat Day	17,479,838
1971	Laffit Pincay, Jr.	3,784,377	1986	Jose Santos	11,329,297			

(1) Total earnings for all horses that jockey raced in year listed; does not reflect jockey's earnings.

Breeders' Cup

The Breeders' Cup was inaugurated in 1984 and consists of 7 races at one track on one day late in the year to determine Thoroughbred racing's champion contenders. It has been held at the following locations:

1984	Hollywood Park, CA	1990	Belmont Park, NY	1996	Woodbine Racetrack, Ontario
1985	Aqueduct Racetrack, NY	1991	Churchill Downs, KY	1997	Hollywood Park, CA
1986	Santa Anita Park, CA	1992	Gulfstream Park, FL	1998	Churchill Downs, KY
1987	Hollywood Park, CA	1993	Santa Anita Park, CA	1999	Gulfstream Park, FL
1988	Churchill Downs, KY	1994	Churchill Downs, KY	2000	Churchill Downs, KY
1989	Gulfstream Park, FL	1995	Belmont Park, NY		

Juvenile

Distances: 1 mi 1984-85, 1987; 1-1/16 mi 1986 and since 1988

Year		Jockey	Year		Jockey	Year		Jockey
1984	Chief's Crown	D. MacBeth	1990	Fly So Free	J. Santos	1996	Boston Harbor	J. Bailey
1985	Tasso	L. Pincay, Jr.	1991	Arazi	P. Valenzuela	1997	Favorite Trick	P. Day
1986	Capote	L. Pincay, Jr.	1992	Gilded Time	C. McCarron	1998	Answer Lively	J. Bailey
1987	Success Express	J. Santos	1993	Brocco	G. Stevens	1999	Anees	G. Stevens
1988	Is It True	L. Pincay, Jr.	1994	Timber Country	P. Day	2000	Macho Uno	J. Bailey
1989	Rhythm	C. Perret	1995	Unbridled's Song	M. Smith			

Juvenile Fillies

Distances: 1 mi 1984-85, 1987; 1-1/16 mi 1986 and since 1988

Year		Jockey	Year		Jockey	Year		Jockey
1984	*Outstandingly	W. Guerra	1990	Meadow Star	J. Santos	1996	Storm Song	C. Perret
1985	Twilight Ridge	J. Velasquez	1991	Pleasant Stage	E. Delahoussaye	1997	Countess Diana	S. Sellers
1986	Brave Raj	P. Valenzuela	1992	Eliza	P. Valenzuela	1998	Silverbulletday	G. Stevens
1987	Epitome	P. Day	1993	Phone Chatter	L. Pincay, Jr.	1999	Cash Run	J. Bailey
1988	Open Mind	A. Cordero, Jr.	1994	Flanders	P. Day	2000	Caressing	J. Velazquez
1989	Go for Wand	R. Romero	1995	My Flag	J. Bailey			

*By disqualification.

Filly & Mare Turf

Distance: 1-3/8 mi

Year		Jockey	Year		Jockey
1999	Soaring Softly	J. Bailey	2000	Perfect Sting	J. Bailey

Sprint

Distance: 6 furlongs

Year		Jockey	Year		Jockey	Year		Jockey
1984	Eillo	C. Perret	1990	Safely Kept	C. Perret	1996	Lit De Justice	C. Nakatani
1985	Precisionist	C. McCarron	1991	Sheikh Albadou	P. Eddery	1997	Elmhurst	C. Nakatani
1986	Smile	J. Vasquez	1992	Thirty Slews	E. Delahoussaye	1998	Reraise	C. Nakatani
1987	Very Subtle	P. Valenzuela	1993	Cardmania	E. Delahoussaye	1999	Artax	J. Chaves
1988	Gulch	A. Cordero, Jr.	1994	Cherokee Run	M. Smith	2000	Kona Gold	A. Solis
1989	Dancing Spree	A. Cordero, Jr.	1995	Desert Stormer	K. Desormeaux			

Mile

Year		Jockey	Year		Jockey	Year		Jockey
1984	Royal Heroine	F. Toro	1990	Royal Academy	L. Piggott	1996	Da Hoss	G. Stevens
1985	Cozzene	W. Guerra	1991	Opening Verse	P. Valenzuela	1997	Spinning World	C. Asmussan
1986	Last Tycoon	Y. St.-Martin	1992	Lure	M. Smith	1998	Da Hoss	J. Velazquez
1987	Miesque	F. Head	1993	Lure	M. Smith	1999	Silic	C. Nakatani
1988	Miesque	F. Head	1994	Barathea	L. Dettori	2000	War Chant	G. Stevens
1989	Steinlen	J. Santos	1995	Ridgewood Pearl	J. Murtagh			

Distaff

Distances: 1-1/4 mi 1984-87; 1-1/8 mi since 1988

Year		Jockey
1984	Princess Rooney	E. Delahoussaye
1985	Life's Magic	A. Cordero, Jr.
1986	Lady's Secret	P. Day
1987	Sacahuista	R. Romero
1988	Personal Ensign	R. Romero
1989	Bayakoa	L. Pincay, Jr.
1990	Bayakoa	L. Pincay, Jr.
1991	Dance Smartly	P. Day
1992	Paseana	C. McCarron
1993	Hollywood Wildcat	E. Delahoussaye
1994	One Dreamer	G. Stevens
1995	Inside Information	M. Smith
1996	Jewel Princess	C. Nakatani
1997	Ajina	M. Smith
1998	Escena	G. Stevens
1999	Beautiful Pleasure	J. Chaves
2000	Spain	V. Espinoza

Turf

Distance: 1-1/2 mi

Year		Jockey
1984	Lashkari	Y. St.-Martin
1985	Pebbles	P. Eddery
1986	Manila	J. Santos
1987	Theatrical	P. Day
1988	Great Communicator	R. Sibille
1989	Prized	E. Delahoussaye
1990	In The Wings	G. Stevens
1991	Miss Alleged	E. Legrix
1992	Fraise	P. Valenzuela
1993	Kotashaan	K. Desormeaux
1994	Tikkanen	M. Smith
1995	Northern Spur	C. McCarron
1996	Pilsudski	W. Swinburn
1997	Chief Bearhart	J. Santos
1998	Buck's Boy	S. Sellers
1999	Daylami	L. Dettori
2000	Kalanisi	J. Murtagh

Classic

Distance: 1-1/4 mi

Year		Jockey
1984	Wild Again	P. Day
1985	Proud Truth	J. Velasquez
1986	Skywalker	L. Pincay, Jr.
1987	Ferdinand	W. Shoemaker
1988	Alysheba	C. McCarron
1989	Sunday Silence	C. McCarron
1990	Unbridled	P. Day
1991	Black Tie Affair	J. Bailey
1992	A.P. Indy	E. Delahoussaye
1993	Arcangues	J. Bailey
1994	Concern	J. Bailey
1995	Cigar	J. Bailey
1996	Alphabet Soup	C. McCarron
1997	Skip Away	M. Smith
1998	Awesome Again	P. Day
1999	Cat Thief	P. Day
2000	Tiznow	C. McCarron

Eclipse Awards

The Eclipse Awards, honoring the Horse of the Year and other champions of the sport, began in 1971 and are sponsored by the *Daily Racing Form,* the Thoroughbred Racing Associations, and the National Turf Writers Assn. Prior to 1971, the *DRF* (1936-70) and the TRA (1950-70) issued separate selections for Horse of the Year.

Eclipse Awards for 2000

Horse of the Year—Tiznow
2-year-old colt or gelding—Macho Uno
2-year-old filly—Caressing
3-year-old colt or gelding—Tiznow
3-year-old filly—Surfside
Older male (4-year-olds & up)—Lemon Drop Kid
Older female (4-year-olds & up)—Riboletta
Male turf horse—Kalanisi
Turf filly or mare—Perfect Sting
Sprinter—Kona Gold
Steeplechase horse—All Gong
Trainer—Robert Frankel
Jockey—Jerry Bailey
Apprentice jockey—Tyler Baze
Breeder—Frank Stronach
Owner—Frank Stronach

Horse of the Year

Year	Horse
1936	Granville
1937	War Admiral
1938	Seabiscuit
1939	Challedon
1940	Challedon
1941	Whirlaway
1942	Whirlaway
1943	Count Fleet
1944	Twilight Tear
1945	Busher
1946	Assault
1947	Armed
1948	Citation
1949	Capot
1950	Hill Prince
1951	Counterpoint
1952	One Count (DRF) Native Dancer (TRA)
1953	Tom Fool
1954	Native Dancer
1955	Nashua
1956	Swaps
1957	Bold Ruler (DRF) Dedicate (TRA)
1958	Round Table
1959	Sword Dancer
1960	Kelso
1961	Kelso
1962	Kelso
1963	Kelso
1964	Kelso
1965	Roman Brother (DRF) Moccasin (TRA)
1966	Buckpasser
1967	Damascus
1968	Dr. Fager
1969	Arts and Letters
1970	Fort Marcy (DRF) Personality (TRA)
1971	Ack Ack
1972	Secretariat
1973	Secretariat
1974	Forego
1975	Forego
1976	Forego
1977	Seattle Slew
1978	Affirmed
1979	Affirmed
1980	Spectacular Bid
1981	John Henry
1982	Conquistador Cielo
1983	All Along
1984	John Henry
1985	Spend A Buck
1986	Lady's Secret
1987	Ferdinand
1988	Alysheba
1989	Sunday Silence
1990	Criminal Type
1991	Black Tie Affair
1992	A.P. Indy
1993	Kotashaan
1994	Holy Bull
1995	Cigar
1996	Cigar
1997	Favorite Trick
1998	Skip Away
1999	Charismatic
2000	Tiznow

HARNESS RACING

Harness Horse of the Year

(Chosen by the U.S. Trotting Assn. and the U.S. Harness Writers Assn.)

Year	Horse
1947	Victory Song
1948	Rodney
1949	Good Time
1950	Proximity
1951	Pronto Don
1952	Good Time
1953	Hi Lo's Forbes
1954	Stenographer
1955	Scott Frost
1956	Scott Frost
1957	Torpid
1958	Emily's Pride
1959	Bye Bye Byrd
1960	Adios Butler
1961	Adios Butler
1962	Su Mac Lad
1963	Speedy Scot
1964	Bret Hanover
1965	Bret Hanover
1966	Bret Hanover
1967	Nevele Pride
1968	Nevele Pride
1969	Nevele Pride
1970	Fresh Yankee
1971	Albatross
1972	Albatross
1973	Sir Dalrae
1974	Delmonica Hanover
1975	Savoir
1976	Keystone Ore
1977	Green Speed
1978	Abercrombie
1979	Niatross
1980	Niatross
1981	Fan Hanover
1982	Cam Fella
1983	Cam Fella
1984	Fancy Crown
1985	Nihilator
1986	Forrest Skipper
1987	Mack Lobell
1988	Mack Lobell
1989	Matt's Scooter
1990	Beach Towel
1991	Precious Bunny
1992	Artsplace
1993	Staying Together
1994	Cam's Card Shark
1995	CR Kay Suzie
1996	Continentalvictory
1997	Malabar Man
1998	Moni Maker
1999	Moni Maker
2000	Gallo Blue Chip

SOCCER

World Cup

1999 Women's World Cup

The U.S. team won the women's soccer World Cup by defeating China, 5-4, on penalty kicks, July 10, 1999, in the Rose Bowl in Pasadena, CA, before 90,185 fans, the largest crowd ever at a U.S. women's sporting event. It was the 2d World Cup victory for the U.S.; the U.S. had won the inaugural event, held in China in 1991, by defeating Norway, 2-1.

The next Women's World Cup will be held in China in 2003.

Women's World Cup, 1991-99

Year	Winner	Final Opponent	Score	Site	Third Place
1991	U.S.	Norway	2-1	China	Germany
1995	Norway	Germany	2-0	Sweden	U.S.
1999	U.S.	China	0-0*	U.S.	Brazil

* U.S. 5-4, penalty kicks

1998 Men's World Cup

In 1998, France became the first host country since 1978 to win the men's soccer World Cup, defeating Brazil, 3-0, on July 12. It was the 2d time the event was held in France; the first time was in 1938. The 2002 World Cup was scheduled to be held jointly in Japan and South Korea in late May and June.

Men's World Cup, 1930-98

Year	Winner	Final opponent	Site
1930	Uruguay	Argentina	Uruguay
1934	Italy	Czechoslovakia	Italy
1938	Italy	Hungary	France
1950	Uruguay	Brazil	Brazil
1954	W. Germany	Hungary	Switzerland
1958	Brazil	Sweden	Sweden
1962	Brazil	Czechoslovakia	Chile
1966	England	W. Germany	England
1970	Brazil	Italy	Mexico
1974	W. Germany	Netherlands	W. Germany
1978	Argentina	Netherlands	Argentina
1982	Italy	W. Germany	Spain
1986	Argentina	W. Germany	Mexico
1990	W. Germany	Argentina	Italy
1994	Brazil	Italy	U.S.
1998	France	Brazil	France

Major League Soccer

2001 Final Standings

Eastern Division	W	L	T	GF	GA	Pts
Miami Fusion	16	5	5	57	36	53
NY/NJ MetroStars*	13	10	3	38	35	42
New England Revolution	7	14	6	35	52	27
Washington, DC United	8	16	2	42	50	16

Central Division	W	L	T	GF	GA	Pts
Chicago Fire	16	6	5	50	30	53
Columbus Crew*	13	7	6	49	36	45
Dallas Burn*	10	11	5	48	47	35
Tampa Bay Mutiny	4	21	2	32	68	14

Western Division	W	L	T	GF	GA	Pts
Los Angeles Galaxy	14	7	5	52	36	47
San Jose Earthquakes*	13	7	6	47	29	45
Kansas City Wizards*	11	13	3	33	53	36
Colorado Rapids	5	13	8	36	47	23

*Clinched playoff berth. **Note:** 3 points for a win, 1 point for a tie.

2001 MLS Statistical Leaders

Leading Scorers (2 points for a goal, 1 point for an assist)

Name	Team	GP	G	A	Pts
1. Alex Pineda Chacon	Miami	25	19	9	47
2. Diego Serna	Miami	22	15	15	45
3. John Spencer	Colorado	23	14	7	35
4. Jeff Cunningham	Columbus	22	10	13	33
5. John Wilmar Perez	Columbus	25	8	15	31
6. Preki	Miami	24	8	14	30
Ariel Graziani	Dallas	25	11	8	30
8. Abdul Thompson Conteh	DC	25	14	1	29
9. Ronald Cerritos	San Jose	25	11	6	28
10. Eric Wynalda	Chicago*	21	10	5	25

* Played for more than 1 team, most recent shown.

Goalkeeping Leaders

Goalkeeping Leaders (minimum 1,000 minutes)

Name	Team	Games	Minutes	Shots[1]	Saves	GA	GAA	W	L	T
1. Zach Thornton	Chicago	27	2,496	145	111	30	1.08	16	6	5
2. Joe Cannon	San Jose	25	2,306	134	101	28	1.09	13	6	6
3. Nick Rimando	Miami	25	2,300	155	116	33	1.29	14	5	5
4. Tim Howard	MetroStars	26	2,370	190	146	35	1.33	13	10	3
5. Tom Presthus	Columbus	25	2,309	178	136	35	1.36	12	7	6

Note: GA = goals against; GAA = goals against average. (1) Not shots on goal; includes shots over the goal or just past the post.

MLS Cup Champions, 1996-2000

Year	Winner	Final opponent	Score	Site	MVP
1996	Washington, DC	Los Angeles	3–2 (OT)	Foxboro, MA	Marco Etcheverry
1997	Washington, DC	Colorado	2–1	Washington, DC.	Jaime Moreno
1998	Chicago	Washington, DC	2–0	Pasadena, CA	Peter Nowak
1999	Washington, DC	Los Angeles	2–0	Foxboro, MA	Ben Olsen
2000	Kansas City	Chicago	1-0	Washington, DC.	Tony Meola

2001 Women's United Soccer Association

The 1st-ever international professional women's soccer league, the Women's United Soccer Association (WUSA), kicked off its inaugural season in 2001 with 8 franchises: the Atlanta Beat, Bay Area CyberRays, Boston Breakers, Carolina Courage, New York Power, Philadelphia Charge, San Diego Spirit, and Washington Freedom. In its 1st year the league featured 20 founding players from the U.S. national team, including such World Cup stars as Brandi Chastain, Julie Foudy, Mia Hamm, and Briana Scurry, as well as international players selected in its Dec. 2000 draft. The league made its debut on Apr. 14, 2001 when the CyberRays took on the Freedom at RFK stadium in Washington, DC. The 1st championship game took place on Aug. 25, 2001. The CyberRays defeated the Beat on penalty kicks (4-2) after a 3-3 double overtime tie at Foxboro (MA) Stadium before a crowd of 21,078. The CyberRays' Julie Murray was named MVP of the championship game. New York's Tiffany Millbrett was the WUSA MVP and Offensive Player of the Year.

NCAA Soccer Champions, 1982-2000

Year[1]	Men	Women
1982	Indiana	North Carolina
1983	Indiana	North Carolina
1984	Clemson	North Carolina
1985	UCLA	George Mason
1986	Duke	North Carolina
1987	Clemson	North Carolina
1988	Indiana	North Carolina
1989	Santa Clara (tie, 2 ot) Virginia	North Carolina
1990	UCLA	North Carolina
1991	Virginia	North Carolina
1992	Virginia	North Carolina
1993	Virginia	North Carolina
1994	Virginia	North Carolina
1995	Wisconsin	Notre Dame
1996	St. John's (NY)	North Carolina
1997	UCLA	North Carolina
1998	Indiana	Florida
1999	Indiana	North Carolina
2000	Connecticut	North Carolina

(1) NCAA Championships began in 1959 for men, in 1982 for women.

NCAA WRESTLING CHAMPIONS

Year	Champion
1964	Oklahoma State
1965	Iowa State
1966	Oklahoma State
1967	Michigan State
1968	Oklahoma State
1969	Iowa State
1970	Iowa State
1971	Oklahoma State
1972	Iowa State
1973	Iowa State
1974	Oklahoma
1975	Iowa
1976	Iowa
1977	Iowa State
1978	Iowa
1979	Iowa
1980	Iowa
1981	Iowa
1982	Iowa
1983	Iowa
1984	Iowa
1985	Iowa
1986	Iowa
1987	Iowa State
1988	Arizona State
1989	Oklahoma State
1990	Oklahoma State
1991	Iowa
1992	Iowa
1993	Iowa
1994	Oklahoma State
1995	Iowa
1996	Iowa
1997	Iowa
1998	Iowa
1999	Iowa
2000	Iowa
2001	Minnesota

CHESS

World Chess Champions

Source: U.S. Chess Federation

Official world champions since the title was first used are as follows:

1866-1894	Wilhelm Steinitz, Austria
1894-1921	Emanuel Lasker, Germany
1921-1927	Jose R. Capablanca, Cuba
1927-1935	Alexander A. Alekhine, France
1935-1937	Max Euwe, Netherlands
1937-1946	Alexander A. Alekhine, France (a)
1948-1957	Mikhail Botvinnik, USSR
1957-1958	Vassily Smyslov, USSR
1958-1959	Mikhail Botvinnik, USSR
1960-1961	Mikhail Tal, USSR
1961-1963	Mikhail Botvinnik, USSR
1963-1969	Tigran Petrosian, USSR
1969-1972	Boris Spassky, USSR
1972-1975	Bobby Fischer, U.S. (b)
1975-1985	Anatoly Karpov, USSR
1985-1993	Garry Kasparov, USSR/Russia (c)
1993-1995	Garry Kasparov, Russia (PCA) (d)
1993-1999	Anatoly Karpov, Russia (FIDE)
1999	Aleksandr Khalifman, Russia (FIDE)
2000	Viswanathan Anand, India (FIDE)

(a) After Alekhine died in 1946, the title was vacant until 1948, when Botvinnik won the 1st championship match sanctioned by the International Chess Federation (FIDE). (b) Defaulted championship after refusal to accept FIDE rules for a championship match, Apr. 1975. (c) Kasparov broke with FIDE, Feb. 26, 1993. FIDE stripped Kasparov of his title Mar. 23. Kasparov defeated Nigel Short of Great Britain in a world championship match played Sept.-Oct. 1993 under the auspices of a new organization the two had founded, the Professional Chess Association (PCA). FIDE held a championship match between Anatoly Karpov (Russia) and Jan Timman (the Netherlands), which Karpov won in Nov. 1993. (d) The PCA folded in 1995.

Recent matches: In Feb. 1996, Kasparov defeated Deep Blue (3 wins, 1 loss, 2 draws), a computer designed by IBM, in the 1st multigame regulation match between a world chess champion and a computer. In a May 1997 rematch, however, Kasparov was defeated by the computer; he scored 1 win, 2 losses, 3 draws. Karpov successfully defended the FIDE title in June-July 1996 against 1991 U.S. chess champion Gata Kamsky of New York City, 10 to 7. In Aug. 1999, after Karpov had refused to play under the controversial format, Aleksandr Khalifman (Russia) earned the FIDE title by defeating Vladmir Akopian (Armenia), 3 to 2, in Las Vegas. In Nov. 2000, Vladimir Kramnik (Russia) defeated Garry Kasparov (Russia), widely recognized as the unofficial world champion, 8½-6½, at the Braingames World Chess Championships in London. In Dec. 2000, Viswanathan Anand (India) def. Alexi Shirov (Spain), 3½-½, in Tehran, Iran, for the FIDE title. The next FIDE World Chess Championship was scheduled for Nov. 25 through Dec. 12, 2001, in Moscow, Russia, with the final match scheduled for Jan. 15-25, 2002.

Further information: More information on chess and chess champions may be accessed on the U.S. Chess Federation's Internet site: http://www.uschess.org

BOWLING

Professional Bowlers Association

Hall of Fame

(Asterisks indicate 2000 inductees. 2001 induction delayed by PBA following Sept. 11 attacks.)

PERFORMANCE

Bill Allen
Glenn Allison
Earl Anthony
Barry Asher
Mike Aulby
Tom Baker
*Parker Bohn III
Roy Buckley
Nelson Burton, Jr.
Don Carter
Pat Colwell
Steve Cook
Dave Davis
Gary Dickinson
Mike Durbin
Buzz Fazio
Dave Ferraro
Skee Foremsky
Jim Godman
Johnny Guenther
Billy Hardwick
Tommy Hudson
Dave Husted
Don Johnson
Joe Joseph
Larry Laub
Mike Limongello
Don McCune
Mike McGrath
Amleto Monacelli
David Ozio
George Pappas
Johnny Petraglia
Dick Ritger
Mark Roth
Jim St. John
Carmen Salvino
Ernie Schlegel
Teata Semiz
Bob Strampe
Harry Smith
Dave Soutar
Jim Stefanich
Brian Voss
Wayne Webb
Dick Weber
Pete Weber
Billy Welu
Walter Ray Williams, Jr.
Wayne Zahn

MERITORIOUS SERVICE

Joe Antenora
John Archibald
Chuck Clemens
Eddie Elias
Frank Esposito
Dick Evans
Raymond Firestone
E. A. "Bud" Fisher
*Jim Fitzgerald
Lou Frantz
Harry Golden
Ted Hoffman, Jr.
John Jowdy
Joe Kelley
Larry Lichstein
Steve Nagy
Keijiro Nakano
Chuck Pezzano
Jack Reichert
Joe Richards
Chris Schenkel
Lorraine Stilzlein
Al Thompson
Roger Zeller
Chuck Pezzano

Tournament of Champions

Year	Winner	Year	Winner	Year	Winner	Year	Winner
1965	Billy Hardwick	1974	Earl Anthony	1983	Joe Berardi	1992	Marc McDowell
1966	Wayne Zahn	1975	Dave Davis	1984	Mike Durbin	1993	George Branham, 3d
1967	Jim Stefanich	1976	Marshall Holman	1985	Mark Williams	1994	Norm Duke
1968	Dave Davis	1977	Mike Berlin	1986	Marshall Holman	1996	Dave D'Entremont
1969	Jim Godman	1978	Earl Anthony	1987	Pete Weber	1997	John Gant
1970	Don Johnson	1979	George Pappas	1988	Mark Williams	1998	Bryan Goebel
1971	Johnny Petraglia	1980	Wayne Webb	1989	Del Ballard, Jr.	1999	Jason Couch
1972	Mike Durbin	1981	Steve Cook	1990	Dave Ferraro	2000	Jason Couch
1973	Jim Godman	1982	Mike Durbin	1991	David Ozio		

PBA Leading Money Winners

Total winnings are from PBA, ABC Masters, and BPAA All-Star tournaments only and do not include numerous other tournaments or earnings from special television shows and matches.

Year	Bowler	Amount	Year	Bowler	Amount	Year	Bowler	Amount
1962	Don Carter	$49,972	1975	Earl Anthony	$107,585	1988	Brian Voss	$225,485
1963	Dick Weber	46,333	1976	Earl Anthony	110,833	1989	Mike Aulby	298,237
1964	Bob Strampe	33,592	1977	Mark Roth	105,583	1990	Amleto Monacelli	204,775
1965	Dick Weber	47,674	1978	Mark Roth	134,500	1991	David Ozio	225,585
1966	Wayne Zahn	54,720	1979	Mark Roth	124,517	1992	Marc McDowell	174,215
1967	Dave Davis	54,165	1980	Wayne Webb	116,700	1993	Walter Ray Williams, Jr.	296,370
1968	Jim Stefanich	67,377	1981	Earl Anthony	164,735	1994	Norm Duke	273,753
1969	Billy Hardwick	64,160	1982	Earl Anthony	134,760	1995	Mike Aulby	219,792
1970	Mike McGrath	52,049	1983	Earl Anthony	135,605	1996	Walter Ray Williams, Jr.	241,330
1971	Johnny Petraglia	85,065	1984	Mark Roth	158,712	1997	Walter Ray Williams, Jr.	240,544
1972	Don Johnson	56,648	1985	Mike Aulby	201,200	1998	Walter Ray Williams, Jr.	238,225
1973	Don McCune	69,000	1986	Walter Ray Williams, Jr	145,550	1999	Parker Bohn III	240,912
1974	Earl Anthony	99,585	1987	Pete Weber	175,491	2000	Norm Duke	143,325

Leading PBA Averages by Year

Year	Bowler	Average	Year	Bowler	Average	Year	Bowler	Average
1962	Don Carter	212.844	1975	Earl Anthony	219.060	1988	Mark Roth	218.036
1963	Billy Hardwick	210.346	1976	Mark Roth	215.970	1989	Pete Weber	215.432
1964	Ray Bluth	210.512	1977	Mark Roth	218.174	1990	Amleto Monacelli	218.158
1965	Dick Weber	211.895	1978	Mark Roth	219.834	1991	Norm Duke	218.208
1966	Wayne Zahn	208.663	1979	Mark Roth	221.662	1992	Dave Ferraro	219.702
1967	Wayne Zahn	212.342	1980	Earl Anthony	218.535	1993	Walter Ray Williams, Jr.	222.980
1968	Jim Stefanich	211.895	1981	Mark Roth	216.699	1994	Norm Duke	222.830
1969	Bill Hardwick	212.957	1982	Marshall Holman.	212.844	1995	Mike Aulby	225.490
1970	Nelson Burton, Jr.	214.908	1983	Earl Anthony	216.645	1996	Walter Ray Williams, Jr.	225.370
1971	Don Johnson	213.977	1984	Marshall Holman	213.911	1997	Walter Ray Williams, Jr.	222.008
1972	Don Johnson	215.290	1985	Mark Baker	213.718	1998	Walter Ray Williams, Jr.	226.130
1973	Earl Anthony	215.799	1986	John Gant	214.378	1999	Parker Bohn III	228.040
1974	Earl Anthony	219.394	1987	Marshall Holman	216.801	2000	Chris Barnes	220.930

American Bowling Congress

ABC Masters Tournament Champions

Year	Winner	Year	Winner	Year	Winner
1980	Neil Burton, St. Louis, MO	1988	Del Ballard, Jr., Richardson, TX	1995	Mike Aulby, Indianapolis, IN
1981	Randy Lightfoot, St. Charles, MO	1989	Mike Aulby, Indianapolis, IN	1996	Ernie Schlegel, Vancouver, WA
1982	Joe Berardi, Brooklyn, NY	1990	Chris Warren, Dallas, TX	1997	Jason Queen, Decatur, IL
1983	Mike Lastowski, Havre de Grace, MD	1991	Doug Kent, Canandaigua, NY	1998	Mike Aulby, Indianapolis, IN
1984	Earl Anthony, Dublin, CA	1992	Ken Johnson, N. Richmond Hills, TX	1999	Brian Boghosian, Middletown, CT
1985	Steve Wunderlich, St. Louis, MO	1993	Norm Duke, Oklahoma City, OK	2000	MIka Koivuniemi, Finland
1986	Mark Fahy, Chicago, IL	1994	Steve Fehr, Cincinnati, OH	2001	Parker Bohn III, Jackson, NJ
1987	Rick Steelsmith, Wichita, KS				

Champions in 2001

Regular Singles	Nicholas J. Hoagland, Bloomington, IN	**Classified Singles**	Kevin F. Noble, Elmwood, IL
Regular Doubles	Gregg Zicha, Glen Ellyn, IL & Bob Udseth, Rolling Meadows, IL	**Classified Doubles**	Christina Singh, Sacramento CA & Darrell D. Moore, Pinole, CA
Regular All Events	D.J. Archer, Amarillo, TX	**Classified All Events**	Charles Perry, Wauwatosa, WI
Regular Team	Joliet Town/Country Lanes #2, Joliet, IL	**Classified Team**	Five T's, Wichita Falls, TX

Most Sanctioned 300 Games

Jeff Carter, Springfield, IL	70	Randy Choat, Granite City, IL	53	Jason Hurd, Tulare, CA	44
Joe Jimenez, Saginaw, MI	62	Bob Buckery, McAdoo, PA	52	Dave Frascatore Jr., Amsterdam, NY	44
Robert Faragon, Albany, NY	59	Jerry Kessler, Dayton, OH	50	John Delp III, West Lawn, PA	43
Bob Learn Jr., EriePA	57	Ralph Burley Jr., Dayton, OH.	49	John Chacko Jr., Larksville, PA	43
Jeff Jensen, Wichita, KS	55	Ken Hall, Schenectady, NY	47	Bob J. Johnson, Dayton, OH.	42
Jim Johnson Jr., Tampa, FL	54	John Wilcox Jr., Lewisburg, PA	46	Randy Lightfoot, St. Charles, MO	42
Dean Wolf, Reading, PA.	54	Ron Krippelcz, St. Louis, MO	46	Steve Gehringer, Reading, PA.	42
Mike Whalin, Cincinnati, OH	53				

Women's International Bowling Congress

Champions in 2001

Queens Tournament: Carolyn Dorin-Ballard, Richland Hills, TX
Classic Singles: Lisa Wagner, Palmetto, FL
Classic Doubles: Nancy Fehr, Cincinnati, OH, & Lisa Wagner, Palmetto, FL
Classic All Events: Jonquay Armon, Lake in the Hills, IL
Classic Team: The Replacements, Fort Worth, TX
Div. I Singles: Sissy Neugent, Waxahachie, TX
Div. I Doubles: Amy Ciammaichella & Edith Meerdo, Imperial, PA
Div. I All Events: Kathy Street, Union City, IN
Div. I Team: Together Again, Waldorf, MD

Most Sanctioned 300 Games

Tish Johnson, Panorama City, CA 28
Aleta Sill, Dearborn, MI 25
Jodi Musto, Schenectady, NY........ 24
Jeanne Naccarato, Tacoma, WA 23
Leanne Barrette, Yukon, OK.......... 23
Dede Davidson, Woodland Hills, CA .. 23
Vicki Fischel, Wheat Ridge, CO...... 21
Debbie McMullen, Denver, CO....... 21
Anne-Marie Duggan, Edmond, OK ... 20
Jodi Hughes, Greenville, SC 19
Cheryl Daniels, Detroit, MI.......... 18
Shannon Duplantis, New Orleans, LA 18
Carolyn Dorin-Ballard, N. Richland Hills, TX......................,18
Mandy Wilson, Dayton, OH 17
Kim Terrell, San Francisco, CA 17
Marianne DiRupo, Succasunna, NJ .. 15
Cindy Coburn-Carroll, Tonawanda, NY.......................... 14
Donna Adamek, Apple Valley, CA.... 12
Jackie Mitskavich, DuBois, PA 12
Stacy Rider, LaHabra, CA........... 12
Charita Williams, Indianapolis, IN 12

FIGURE SKATING

U.S. and World Individual Champions, 1952-2001

U.S. Champions			World Champions	
MEN	**WOMEN**	**YEAR**	**MEN**	**WOMEN**
Dick Button	Tenley Albright	**1952**	Dick Button, U.S.	Jacqueline du Bief, France
Hayes Jenkins	Tenley Albright	**1953**	Hayes Jenkins, U.S.	Tenley Albright, U.S.
Hayes Jenkins	Tenley Albright	**1954**	Hayes Jenkins, U.S.	Gundi Busch, W. Germany
Hayes Jenkins	Tenley Albright	**1955**	Hayes Jenkins, U.S.	Tenley Albright, U.S.
Hayes Jenkins	Tenley Albright	**1956**	Hayes Jenkins, U.S.	Carol Heiss, U.S.
Dave Jenkins	Carol Heiss	**1957**	Dave Jenkins, U.S.	Carol Heiss, U.S.
Dave Jenkins	Carol Heiss	**1958**	Dave Jenkins, U.S.	Carol Heiss, U.S.
Dave Jenkins	Carol Heiss	**1959**	Dave Jenkins, U.S.	Carol Heiss, U.S.
Dave Jenkins	Carol Heiss	**1960**	Alain Giletti, France	Carol Heiss, U.S.
Bradley Lord	Laurence Owen	**1961**	none	none
Monty Hoyt	Barbara Roles Pursley	**1962**	Don Jackson, Canada	Sjoukje Dijkstra, Netherlands
Tommy Litz	Lorraine Hanlon	**1963**	Don McPherson, Canada	Sjoukje Dijkstra, Netherlands
Scott Allen	Peggy Fleming	**1964**	Manfred Schnelldorfer, W. Germany	Sjoukje Dijkstra, Netherlands
Gary Visconti	Peggy Fleming	**1965**	Alain Calmat, France	Petra Burka, Canada
Scott Allen	Peggy Fleming	**1966**	Emmerich Danzer, Austria	Peggy Fleming, U.S.
Gary Visconti	Peggy Fleming	**1967**	Emmerich Danzer, Austria	Peggy Fleming, U.S.
Tim Wood	Peggy Fleming	**1968**	Emmerich Danzer, Austria	Peggy Fleming, U.S.
Tim Wood	Janet Lynn	**1969**	Tim Wood, U.S.	Gabriele Seyfert, E. Germany
Tim Wood	Janet Lynn	**1970**	Tim Wood, U.S.	Gabriele Seyfert, E. Germany
John Misha Petkevich	Janet Lynn	**1971**	Ondrej Nepela, Czechoslovakia	Beatrix Schuba, Austria
Ken Shelley	Janet Lynn	**1972**	Ondrej Nepela, Czechoslovakia	Beatrix Schuba, Austria
Gordon McKellen, Jr.	Janet Lynn	**1973**	Ondrej Nepela, Czechoslovakia	Karen Magnussen, Canada
Gordon McKellen, Jr.	Dorothy Hamill	**1974**	Jan Hoffmann, E. Germany	Christine Errath, E. Germany
Gordon McKellen, Jr.	Dorothy Hamill	**1975**	Sergei Volkov, USSR	Dianne de Leeuw, Neth.-U.S.
Terry Kubicka	Dorothy Hamill	**1976**	John Curry, Gr. Britain	Dorothy Hamill, U.S.
Charles Tickner	Linda Fratianne	**1977**	Vladimir Kovalev, USSR	Linda Fratianne, U.S.
Charles Tickner	Linda Fratianne	**1978**	Charles Tickner, U.S.	Anett Poetzsch, E. Germany
Charles Tickner	Linda Fratianne	**1979**	Vladimir Kovalev, USSR	Linda Fratianne, U.S.
Charles Tickner	Linda Fratianne	**1980**	Jan Hoffmann, E. Germany	Anett Poetzsch, E. Germany
Scott Hamilton	Elaine Zayak	**1981**	Scott Hamilton, U.S.	Denise Biellmann, Switzerland
Scott Hamilton	Rosalynn Sumners	**1982**	Scott Hamilton, U.S.	Elaine Zayak, U.S.
Scott Hamilton	Rosalynn Sumners	**1983**	Scott Hamilton, U.S.	Rosalynn Sumners, U.S.
Scott Hamilton	Rosalynn Sumners	**1984**	Scott Hamilton, U.S.	Katarina Witt, E. Germany
Brian Boitano	Tiffany Chin	**1985**	Aleksandr Fadeev, USSR	Katarina Witt, E. Germany
Brian Boitano	Debi Thomas	**1986**	Brian Boitano, U.S.	Debi Thomas, U.S.
Brian Boitano	Jill Trenary	**1987**	Brian Orser, Canada	Katarina Witt, E. Germany
Brian Boitano	Debi Thomas	**1988**	Brian Boitano, U.S.	Katarina Witt, E. Germany
Christopher Bowman	Jill Trenary	**1989**	Kurt Browning, Canada	Midori Ito, Japan
Todd Eldredge	Jill Trenary	**1990**	Kurt Browning, Canada	Jill Trenary, U.S.
Todd Eldredge	Tonya Harding	**1991**	Kurt Browning, Canada	Kristi Yamaguchi, U.S.
Christopher Bowman	Kristi Yamaguchi	**1992**	Viktor Petrenko, Ukraine	Kristi Yamaguchi, U.S.
Scott Davis	Nancy Kerrigan	**1993**	Kurt Browning, Canada	Oksana Baiul, Ukraine
Scott Davis	vacant[1]	**1994**	Elvis Stojko, Canada	Yuka Sato, Japan
Todd Eldredge	Nicole Bobek	**1995**	Elvis Stojko, Canada	Chen Lu, China
Rudy Galindo	Michelle Kwan	**1996**	Todd Eldredge, U.S.	Michelle Kwan, U.S.
Todd Eldredge	Tara Lipinski	**1997**	Elvis Stojko, Canada	Tara Lipinski, U.S.
Todd Eldredge	Michelle Kwan	**1998**	Alexei Yagudin, Russia	Michelle Kwan, U.S.
Michael Weiss	Michelle Kwan	**1999**	Alexei Yagudin, Russia	Maria Butyrskaya, Russia
Michael Weiss	Michelle Kwan	**2000**	Alexei Yagudin, Russia	Michelle Kwan, U.S.
Timothy Goebel	Michelle Kwan	**2001**	Yevgeny Plushchenko, Russia	Michelle Kwan, U.S.

(1) Tonya Harding was stripped of title.

SKIING

World Cup Alpine Champions, 1967-2001

Men

1967 Jean Claude Killy, France
1968 Jean Claude Killy, France
1969 Karl Schranz, Austria
1970 Karl Schranz, Austria
1971 Gustavo Thoeni, Italy
1972 Gustavo Thoeni, Italy
1973 Gustavo Thoeni, Italy
1974 Piero Gros, Italy
1975 Gustavo Thoeni, Italy
1976 Ingemar Stenmark, Sweden
1977 Ingemar Stenmark, Sweden
1978 Ingemar Stenmark, Sweden
1979 Peter Luescher, Switzerland
1980 Andreas Wenzel, Liechtenstein
1981 Phil Mahre, U.S.
1982 Phil Mahre, U.S.
1983 Phil Mahre, U.S.
1984 Pirmin Zurbriggen, Switzerland
1985 Marc Girardelli, Luxembourg
1986 Marc Girardelli, Luxembourg
1987 Pirmin Zurbriggen, Switzerland
1988 Pirmin Zurbriggen, Switzerland
1989 Marc Girardelli, Luxembourg
1990 Pirmin Zurbriggen, Switzerland
1991 Marc Girardelli, Luxembourg
1992 Paul Accola, Switzerland
1993 Marc Girardelli, Luxembourg
1994 Kjetil Andre Aamodt, Norway
1995 Alberto Tomba, Italy
1996 Lasse Kjus, Norway
1997 Luc Alphand, France
1998 Hermann Maier, Austria
1999 Lasse Kjus, Norway
2000 Hermann Maier, Austria
2001 Hermann Maier, Austria

Women

1967 Nancy Greene, Canada
1968 Nancy Greene, Canada
1969 Gertrud Gabl, Austria
1970 Michele Jacot, France
1971 Annemarie Proell, Austria
1972 Annemarie Proell, Austria
1973 Annemarie Proell, Austria
1974 Annemarie Proell, Austria
1975 Annemarie Proell, Austria
1976 Rose Mittermaier, W. Germany
1977 Lise-Marie Morerod, Switzerland
1978 Hanni Wenzel, Liechtenstein
1979 Annemarie Proell Moser, Austria
1980 Hanni Wenzel, Liechtenstein
1981 Marie-Theres Nadig, Switzerland
1982 Erika Hess, Switzerland
1983 Tamara McKinney, U.S.
1984 Erika Hess, Switzerland
1985 Michela Figini, Switzerland
1986 Maria Walliser, Switzerland
1987 Maria Walliser, Switzerland
1988 Michela Figini, Switzerland
1989 Vreni Schneider, Switzerland
1990 Petra Kronberger, Austria
1991 Petra Kronberger, Austria
1992 Petra Kronberger, Austria
1993 Anita Wachter, Austria
1994 Vreni Schneider, Switzerland
1995 Vreni Schneider, Switzerland
1996 Katja Seizinger, Germany
1997 Pernilla Wiberg, Sweden
1998 Katja Seizinger, Germany
1999 Alexandra Meissnitzer, Austria
2000 Renate Goetschl, Austria
2001 Janica Kostelic, Croatia

LACROSSE

Lacrosse Champions in 2001

U.S. Club Lacrosse Association Championship—Baltimore, MD, Jun. 10: Long Island 12, NY Athletic Club, 8.

National Lacrosse League Championship—Toronto, Ontario, Canada, Apr. 27: Philadelphia def. Toronto 9-8.

NCAA Men's Division I Championship—Piscataway, NJ, May 28: Princeton 10, Syracuse 9.

NCAA Women's Division I Championship—Baltimore, MD, May 20: Maryland 14, Georgetown 13.

2001 Men's NCAA Division I All-AmericaTeam

Attack: Mike Powell, Syracuse; Conor Gill, Virginia; Tom Glatzel, Notre Dame.

Midfield: Doug Shanahan, Hofstra; Josh Coffman, Syracuse; Steve Dusseau, Georgetown; Gavin Prout, Loyola.

Defense: John Glatzel, Syracuse; Ryan Mollett, Princeton; Mark Koontz, Virginia.

Goal: Pat McGinnis, Maryland; Trevor Tierney, Princeton.

Coach of the Year: Tony Seaman, Towson.

2001 Women's NCAA Division I All-America Team

Attack: Allison Comito, Maryland; Erin Elbe, Georgetown; Stacy Morlang, Loyola; Sheehan Stanwick, Georgetown.

Midfield: Jen Adams, Maryland; Quinn Carney, Maryland; Suzy Gibbons, Dartmouth; Kate Kaiser, Duke; Christine McPike, North Carolina; Julie Shaner, Princeton.

Defense: Rachael Becker, Princeton; Stacey Brown, Syracuse; Courtney Martinez, Maryland; Caitland McLean, Georgetown.

Goal: Bowen Holden, Georgetown; Kristen Foster, Duk.e

Coach of the Year: Kim Simons, Georgetown.

NCAA Division I Lacrosse Champions 1982-2001

Year[1]	Men	Women
1982	North Carolina	Massachusetts
1983	Syracuse	Delaware
1984	Johns Hopkins	Temple
1985	Johns Hopkins	New Hampshire
1986	North Carolina	Maryland
1987	Johns Hopkins	Penn St.
1988	Syracuse	Temple
1989	Syracuse	Penn St.
1990	vacated	Harvard
1991	North Carolina	Virginia
1992	Princeton	Maryland
1993	Syracuse	Virginia
1994	Princeton	Princeton
1995	Syracuse	Maryland
1996	Princeton	Maryland
1997	Princeton	Maryland
1998	Princeton	Maryland
1999	Virginia	Maryland
2000	Syracuse	Maryland
2001	Princeton	Maryland

(1) NCAA Championships began in 1971 for men, in 1982 for women.

SWIMMING

World Swimming Records

(Long course, as of Oct. 1, 2001.)

Men's Records

Freestyle

Distance	Time	Holder	Country	Where made	Date
50 meters	**0:21.64**	Alexander Popov	Russia	Moscow, Russia	June 16, 2000
100 meters	**0:47.84**	Pieter van den Hoogenband	Netherlands	Sydney, Australia	Sept. 19, 2000
200 meters	**1:44.06**	Ian Thorpe	Australia	Fukuoka, Japan	July 25, 2001
400 meters	**3:40.17**	Ian Thorpe	Australia	Fukuoka, Japan	July 22, 2001
800 meters	**7:39.16**	Ian Thorpe	Australia	Fukuoka, Japan	July 24, 2001
1,500 meters	**14:34:56**	Grant Hackett	Australia	Fukuoka, Japan	July 29, 2001

Breaststroke

Distance	Time	Holder	Country	Where made	Date
50 meters	**0:27.39**	Ed Moses	U.S.	Austin, TX	Mar. 31, 2001
100 meters	**0:59.94**	Roman Sloudnov	Russia	Fukuoka, Japan	July 23, 2001
200 meters	**2:10.16**	Mike Barrowman	U.S.	Barcelona, Spain	July 29, 1992

Butterfly

50 meters	**0:23.44**	Geoffrey Huegill	Australia	Fukuoka, Japan	July 27, 2001
100 meters	**0:51.81**	Michael Klim	Australia	Canberra, Australia	Dec. 12, 1999
200 meters	**1:54.58**	Michael Phelps	U.S.	Fukuoka, Japan	July 24, 2001

Backstroke

50 meters	**0:24.99**	Lenny Krayzelburg	U.S.	Sydney, Australia	Aug. 28, 1999
100 meters	**0:53.60**	Lenny Krayzelburg	U.S.	Sydney, Australia	Aug. 24, 1999
200 meters	**1:55.87**	Lenny Krayzelburg	U.S.	Sydney, Australia	Aug. 27, 1999

Individual Medley

200 meters	**1:58.16**	Jani Sievinen	Finland	Rome, Italy	Sept. 11, 1994
400 meters	**4:11.76**	Tom Dolan	U.S.	Sydney, Australia	Sept. 17, 2000

Medley Relay

400 m. (4×100)	**3:33.73**	(Krayzelburg, Moses, Crocker, Hall)	U.S.	Sydney, Australia	Sept. 23, 2000

Freestyle Relays

400 m. (4×100)	**3:13.67**	(Klim, Fydler, Callus, Thorpe)	Australia	Sydney, Australia	Sept. 16, 2000
800 m. (4×200)	**7:04.66**	(Hackett, Klim, Kirby, Thorpe)	Australia	Fukuoka, Japan	July 27, 2001

Women's Records

Freestyle

Distance	Time	Holder	Country	Where made	Date
50 meters	**0:24.13**	Inge de Bruijn	Netherlands	Sydney, Australia	Sept. 22, 2000
100 meters	**0:53.77**	Inge de Bruijn	Netherlands	Sydney, Australia	Sept. 20, 2000
200 meters	**1:56.78**	Franziska Van Almsick	Germany	Rome, Italy	Sept. 6, 1994
400 meters	**4:03.85**	Janet Evans	U.S.	Seoul, South Korea	Sept. 22, 1988
800 meters	**8:16.22**	Janet Evans	U.S.	Tokyo, Japan	Aug. 20, 1989
1,500 meters	**15:52.10**	Janet Evans	U.S.	Orlando, FL.	Mar. 26, 1988

Breaststroke

50 meters	**0:30.83**	Penny Heyns	South Africa	Canberra, Australia	Aug. 28, 1999
100 meters	**1:06.52**	Penny Heyns	South Africa	Canberra, Australia	Aug. 23, 1999
200 meters	**2:22.99**	Hui Qi	China	Hangzhou, China	Apr. 13, 2001

Butterfly

50 meters	**0:25.64**	Inge de Bruijn	Netherlands	Sheffield, England	May 26, 2000
100 meters	**0:56.61**	Inge de Bruijn	Netherlands	Sydney, Australia	Sept. 17, 2000
200 meters	**2:05.81**	Susann O'Neill	Australia	Sydney, Australia	May 17, 2000

Backstroke

50 meters	**0:28.25**	Sandra Voelker	Germany	Berlin, Germany	June 17, 2000
100 meters	**1:00.16**	Cihong He	China	Rome, Italy	Sept. 10, 1994
200 meters	**2:06.62**	Krisztina Egerszegi	Hungary	Athens, Greece	Aug. 25, 1991

Individual Medley

200 meters	**2:09.72**	Yanyan Wu	China	Shanghai, China	Oct. 17, 1997
400 meters	**4:33.59**	Yana Klochkova	Ukraine	Sydney, Australia	Sept. 16, 2000

Freestyle Relays

400 m. (4×100)	**3:36.61**	(Van Dyken, Torres, Shealy, Thompson)	U.S.	Sydney, Australia	Sept. 16, 2000
800 m. (4×200)	**7:55.47**	(Stellmach, Strauss, Mohring, Friedrich)	E. Germany	Strasbourg, France	Aug. 18, 1987

Medley Relay

400 m. (4×100)	**3:58.30**	(Bedford, Quann, Thompson, Torres)	U.S.	Sydney, Australia	Sept. 23, 2000

YACHTING

The America's Cup in 2000

In the 30th America's Cup, held in the Hauraki Gulf off the coast of Auckland, New Zealand, Team New Zealand defeated Italy's Prada Challenge to become the first non-American syndicate to successfully defend the oldest trophy in sports. On Mar. 2, 2000, the Kiwis' *New Zealand* sailed to a 48-sec. victory over Luna Rossa, to complete a 5-0 sweep in the best-of-nine series. Team New Zealand, which also swept the U.S. yacht Young America in 1995, set a record of 10 straight wins in Cup finals. In an surprising move, veteran skipper Russell Coutts stepped aside after his record-tying 9th consecutive Cup race victory to allow 26-year-old Dean Barker to take the helm in the final race. The 31st America's Cup was scheduled to be held in New Zealand in 2003.

Competition for the America's Cup grew out of the first contest to establish a world yachting championship, one of the carnival features of the London Exposition of 1851. The race covered a 60-mile course around the Isle of Wight; the prize was a cup worth about $500, donated by the Royal Yacht Squadron of England, known as the "America's Cup" because it was first won by the U.S. yacht *America*.

Winners of the America's Cup

1851 America
1870 Magic defeated Cambria, England, (1-0)
1871 Columbia (first three races) and Sappho (last two races) defeated Livonia, England, (4-1)
1876 Madeline defeated Countess of Dufferin, Canada, (2-0)
1881 Mischief defeated Atalanta, Canada, (2-0)
1885 Puritan defeated Genesta, England, (2-0)
1886 Mayflower defeated Galatea, England, (2-0)
1887 Volunteer defeated Thistle, Scotland, (2-0)
1893 Vigilant defeated Valkyrie II, England, (3-0)
1895 Defender defeated Valkyrie III, England, (3-0)
1899 Columbia defeated Shamrock, England, (3-0)
1901 Columbia defeated Shamrock II, England, (3-0)
1903 Reliance defeated Shamrock III, England, (3-0)
1920 Resolute defeated Shamrock IV, England, (3-2)
1930 Enterprise defeated Shamrock V, England, (4-0)
1934 Rainbow defeated Endeavour, England, (4-2)
1937 Ranger defeated Endeavour II, England, (4-0)
1958 Columbia defeated Sceptre, England, (4-0)
1962 Weatherly defeated Gretel, Australia, (4-1)
1964 Constellation defeated Sovereign, England, (4-0)
1967 Intrepid defeated Dame Pattie, Australia, (4-0)
1970 Intrepid defeated Gretel II, Australia, (4-1)
1974 Courageous defeated Southern Cross, Australia, (4-0)
1977 Courageous defeated Australia, Australia, (4-0)
1980 Freedom defeated Australia, Australia, (4-1)
1983 Australia II, Australia, defeated Liberty, (4-3)
1987 Stars & Stripes defeated Kookaburra III, Australia, (4-0)
1988 Stars & Stripes defeated New Zealand, New Zealand, (2-0)
1992 America^3 defeated Il Moro di Venezia, Italy, (4-1)
1995 Black Magic 1, New Zealand, defeated Young America, (5-0)
2000 New Zealand, NZ, defeated Luna Rossa, Italy, (5-0)

POWER BOATING

American Power Boat Assn. Gold Cup Champions, 1978-2001

Year	Boat	Driver	Year	Boat	Driver
1978	Atlas Van Lines	Bill Muncey	1990	Miss Budweiser	Tom D'Eath
1979	Atlas Van Lines	Bill Muncey	1991	Winston Eagle	Mark Tate
1980	Miss Budweiser	Dean Chenoweth	1992	Miss Budweiser	Chip Hanauer
1981	Miss Budweiser	Dean Chenoweth	1993	Miss Budweiser	Chip Hanauer
1982	Atlas Van Lines	Chip Hanauer	1994	Smokin' Joe's	Mark Tate
1983	Atlas Van Lines	Chip Hanauer	1995	Miss Budweiser	Chip Hanauer
1984	Atlas Van Lines	Chip Hanauer	1996	Pico American Dream	Dave Villwock
1985	Miller American	Chip Hanauer	1997	Miss Budweiser	Dave Villwock
1986	Miller American	Chip Hanauer	1998	Miss Budweiser	Dave Villwock
1987	Miller American	Chip Hanauer	1999	Miss PICO	Chip Hanauer
1988	Circus Circus	Chip Hanauer	2000	Miss Budweiser	Dave Villwock
1989	Miss Budweiser	Tom D'Eath	2001	Miss Tubby's Subs	Mike Hanson

RODEO

Pro Rodeo Cowboy All-Around Champions, 1977-2000

Year	Winner	Money won	Year	Winner	Money won
1977	Tom Ferguson, Miami, OK	$76,730	1989	Ty Murray, Odessa, TX	$134,806
1978	Tom Ferguson, Miami, OK	103,734	1990	Ty Murray, Stephenville, TX	213,772
1979	Tom Ferguson, Miami, OK	96,272	1991	Ty Murray, Stephenville, TX	244,230
1980	Paul Tierney, Rapid City, SD	105,568	1992	Ty Murray, Stephenville, TX	225,992
1981	Jimmie Cooper, Monument, NM	105,862	1993	Ty Murray, Stephenville, TX	297,896
1982	Chris Lybbert, Coyote, CA.	123,709	1994	Ty Murray, Stephenville, TX	246,170
1983	Roy Cooper, Durant, OK	153,391	1995	Joe Beaver, Huntsville, TX	141,753
1984	Dee Pickett, Caldwell, ID.	122,618	1996	Joe Beaver, Huntsville, TX	166,103
1985	Lewis Feild, Elk Ridge, UT	130,347	1997	Dan Mortensen, Manhattan, MT	184,559
1986	Lewis Feild, Elk Ridge, UT	166,042	1998	Ty Murray, Stephenville, TX	264,673
1987	Lewis Feild, Elk Ridge, UT	144,335	1999	Fred Whitfield, Hockley, TX	217,819
1988	Dave Appleton, Arlington, TX	121,546	2000	Joe Beaver, Huntsville, TX	225,396

DOGS

Westminster Kennel Club, 1989-2001

Year	Best-in-show	Breed	Owner(s)
1989	Ch. Royal Tudor's Wild As The Wind	Doberman	Sue & Art Kemp, Richard & Carolyn Vida, Beth Wilhite
1990	Ch. Wendessa Crown Prince	Pekingese	Ed Jenner
1991	Ch. Whisperwind on a Carousel	Poodle	Joan & Frederick Hartsock
1992	Ch. Registry's Lonesome Dove	Fox Terrier	Marion & Sam Lawrence
1993	Ch. Salilyn's Condor	English Springer Spaniel	Donna & Roger Herzig
1994	Ch. Chidley Willum	Norwich Terrier	Ruth Cooper & Patricia Lussier
1995	Ch. Gaelforce Post Script	Scottish Terrier	Dr. Vandra Huber & Dr. Joe Kinnarney
1996	Ch. Clussexx Country Sunrise	Clumber Spaniel	Judith & Richard Zaleski
1997	Ch. Parsifal Di Casa Netzer	Standard Schnauzer	Rita Holloway & Gabrio Del Torre
1998	Ch. Fairewood Frolic	Norwich Terrier	Sandina Kennels
1999	Ch. Loteki Supernatural Being	Papillon	John Oulton
2000	Ch. Salilyn 'N Erin's Shameless	English Springer Spaniel	Carl Blain, Fran Sunseri, & Julia Gasow
2001	Ch. Special Times Just Right	Bichons Frises	Cecilia Ruggles, E. McDonald, & F. Werneck

2001 Iditarod Trail Sled Dog Race

Doug Swingley of Lincoln, MT, won his 3d consecutive (4th overall, 1995) Iditarod Trail Sled Dog Race, Mar. 14, 2001, completing the 1,100-mile course from Anchorage to Nome, AK, in 9 days, 19 hours, 55 minutes, 50 seconds. He is still the oldest (47) and only non-Alaskan winner ever. Hindered by adverse weather and trail conditions, Swingley took about 19 hours longer than in 2000. Linwood Fiedler, of Willow, AK, finished 2d (10 d, 3 h, 58 m, 57s). First prize was $63,000 and a pickup truck valued at $39,000. Swingley ties Susan Butcher (1986-88, 1990) on the all-time list, behind 5-time winner Rick Swenson (1977, 1979, 1981-82, 1991).

MARATHONS

Boston Marathon

In the 105th Boston Marathon, Apr. 16, 2001, Lee Bong-ju of South Korea won in 2:09:43, ending the Kenyan men's 10-year winning streak. Ecuadorian Silvio Guerra finished second, 24 seconds back. The Kenyan runners finished third, fourth, fifth, and tenth. Rod DeHaven (Wisconsin) was the first U.S. finisher, sixth overall, in 2:12:41. In the women's race, Kenya fared better as Catherine Ndereba won her second consecutive Boston title with a time of 2:23:53. The top American woman was Jill Gaitenby (Rhode Island) who finished 14th in 2:36.45. All times in hour:minute:second format. *Course records.

Men's Winner	Time	Year	Women's Winner	Time
O. Suomalainen, Finland	2:15:39	**1972**	N. Kuscsik, U.S.	3:10:26
J. Anderson, U.S.	2:16:03	**1973**	J. Hansen, U.S.	3:05:59
N. Cusack, Ir.	2:13:39	**1974**	M. Gorman, U.S.	2:47:11
B. Rogers, U.S.	2:09:55	**1975**	L. Winter, West Ger.	2:42:24
J. Fultz, U.S.	2:20:19	**1976**	K. Merritt, U.S.	2:47:10
J. Drayton, Can.	2:14:46	**1977**	M. Gorman, U.S.	2:48:33
B. Rogers, U.S.	2:10:13	**1978**	G.S. Barron, U.S.	2:44:52
B. Rogers, U.S.	2:09:27	**1979**	J. Benoit, U.S.	2:35:15
B. Rogers, U.S.	2:12:11	**1980**	J. Gareau, Can.	2:34:28
T. Seko, Japan	2:09:26	**1981**	A. Roe, N. Zealand	2:26:46
A. Salazar, U.S.	2:08:52	**1982**	C. Teske, West Ger.	2:29:33
G. Myer, U.S.	2:09:00	**1983**	J. Benoit, U.S.	2:22:43
G. Smith, G.B.	2:10:34	**1984**	L. Moller, N. Zealand	2:29:28
G. Smith, G.B.	2:14:05	**1985**	L. Weidenbach, U.S.	2:34:06
R. de Castella, Australia	2:07:51	**1986**	I. Kristiansen, Nor.	2:24:55

Men's Winner	Time	Year	Women's Winner	Time	Men's Winner	Time	Year	Women's Winner	Time
T. Seko, Japan	2:11:50	**1987**	R. Mota, Portugal	2:25:21	C. Ndeti, Kenya	2:07:15*	**1994**	U. Pippig, Germany	2:21:45*
I. Hussein, Ken.	2:08:43	**1988**	R. Mota, Portugal	2:24:30	C. Ndeti, Kenya	2:09:22	**1995**	U. Pippig, Germany	2:25:11
A. Mekonnen, Eth.	2:09:06	**1989**	I. Kristiansen, Nor.	2:24:33	M. Tanui, Kenya	2:09:15	**1996**	U. Pippig, Germany	2:27:12
G. Bordin, Italy	2:08:19	**1990**	R. Mota, Portugal	2:25:24	L. Aguta, Kenya	2:10:34	**1997**	F. Roba, Ethiopia	2:26:23
I. Hussein, Kenya	2:11:06	**1991**	W. Panfil, Poland	2:24:18	M. Tanui, Kenya	2:07:34	**1998**	F. Roba, Ethiopia	2:23:21
I. Hussein, Kenya	2:08:14	**1992**	O. Markova, CIS	2:23:43	J. Chebet, Kenya	2:09:52	**1999**	F. Roba, Ethiopia	2:23:25
C. Ndeti, Kenya	2:09:33	**1993**	O. Markova, CIS	2:25:27	E. Lagat, Kenya	2:09:47	**2000**	C. Ndereba, Kenya	2:26:11

Note: The first Boston Marathon was held in 1897. Women were officially accepted into the race in 1972.

New York City Marathon

All times in hour:minute:second format. *Course records.

Men's Winner	Time	Year	Women's Winner	Time	Men's Winner	Time	Year	Women's Winner	Time
G. Muhrcke, U.S.	2:31:38	**1970**	no finisher	—	S. Jones, G.B.	2:08:20	**1988**	G.Waitz, Norway	2:28:07
N. Higgins, U.S.	2:22:54	**1971**	B. Bonner, U.S.	2:55:22	J. Ikangaa, Tanz.	2:08:01*	**1989**	I. Kristiansen, Norway	2:25:30
S. Karlin, U.S.	2:27:52	**1972**	N. Kuscsik, U.S.	3:08:41	D. Wakiihuri, Ken.	2:12:39	**1990**	W. Panfil, Poland	2:30:45
T. Fleming, U.S.	2:19:25	**1973**	N. Kuscsik, U.S.	2:57:07	S. Garcia, Mexico	2:09:28	**1991**	L. McColgan, G.B.	2:27:32
N. Sander, U.S.	2:26:30	**1974**	K. Switzer, U.S.	3:07:29	W. Mtolo, S. Afr.	2:09:29	**1992**	L. Ondieki, Australia	2:24:40*
T. Fleming, U.S.	2:19:27	**1975**	K. Merritt, U.S.	2:46:14	A. Espinosa, Mex.	2:10:04	**1993**	U. Pippig, Germany	2:26:24
B. Rogers, U.S.	2:10:10	**1976**	M. Gorman, U.S.	2:39:11	G. Silva, Mexico	2:11:21	**1994**	T. Loroupe, Kenya	2:27:37
B. Rogers, U.S.	2:11:28	**1977**	M. Gorman, U.S.	2:43:10	G. Silva, Mexico	2:11:00	**1995**	T. Loroupe, Kenya	2:28:06
B. Rogers, U.S.	2:12:12	**1978**	G. Waitz, Norway	2:32:30	G. Leone, Italy	2:09:54	**1996**	A. Catuna, Romania	2:28:43
B. Rogers, U.S.	2:11:42	**1979**	G. Waitz, Norway	2:27:33	J. Kagwe, Kenya	2:08:12	**1997**	F. Rochat-Moser, Switzerland	2:28:43
A. Salazar, U.S.	2:09:41	**1980**	G. Waitz, Norway	2:25:42	J. Kagwe, Kenya	2:08:45	**1998**	F. Fiacconi, Italy	2:25:17
A. Salazar, U.S.	2:08:13	**1981**	A. Roe, N. Zealand	2:25:29	J. Chebet, Kenya	2:09:14	**1999**	A. Fernandez, Mex.	2:25:06
A. Salazar, U.S.	2:09:29	**1982**	G. Waitz, Norway	2:27:14	A. Mouaziz, Morocco	2:10:09	**2000**	L. Petrova, Russia	2:25:45
R. Dixon, N.Z.	2:08:59	**1983**	G. Waitz, Norway	2:27:00					
O. Pizzolato, Italy	2:14:53	**1984**	G. Waitz, Norway	2:29:30					
O. Pizzolato, Italy	2:11:34	**1985**	G. Waitz, Norway	2:28:34					
G.Poli, Italy	2:11:06	**1986**	G. Waitz, Norway	2:28:06					
I. Hussein, Kenya	2:11:01	**1987**	P. Welch, G.B.	2:30:17					

Other Marathon Results in 2001

Los Angeles Marathon—Mar. 4. Men: Steven Ndungu, Kenya, 2:13:13. Women: Elana Paramonova, Russia, 2:35:58.

Paris Marathon—April 8. Men: Simon Biwott, Kenya, 2:09:40. Women: Ruth Kutol, Kenya, 2:27:53.

Fortis Rotterdam Marathon—April 22. Men: Josephat Kiprono, Kenya, 2:06:50. Women: Susan Chepkemei, Kenya, 2:25:45.

London Marathon—April 22. Men: Abdelkhader El Mouaziz, Morocco, 2:07:11. Women: Derartu Tulu, Ethiopia, 2:23:56.

Berlin Marathon—Sept. 30. Men: Joseph Ngolepus, Kenya, 2:08:47. Women: Naoko Takahashi, Japan, 2:19:46.

Chicago LaSalle Bank Marathon—Oct. 7. Men: Ben Kimondiu, Kenya, 2:08:52. Women: Catherine Ndereba, Kenya, 2:18:47 (world best).

Ironman Triathlon World Championships

The Ironman Triathlon World Championships—a 2.4-mile ocean swim, 112-mile bike ride and 26.2-mile run—are held annually at Kailua-Kona, Hawaii. On Oct. 6, 2001, the men's race was won by American Tim DeBoom in 8:31:17. New Zealand's Cameron Brown won the women's race in 8:47:40. All times in hour:minute:second format. *Course records.

Men's Winner	Time	Year	Women's Winner	Time	Men's Winner	Time	Year	Women's Winner	Time
G. Haller, U.S.	11:46:58	**1978**	no finisher	—	M. Allen, U.S.	8:18:32	**1991**	P. Newby-Fraser, Zimbabwe	9:07:52
T. Warren, U.S.	11:15:56	**1979**	L. Lemaire, U.S.	12:55:00	M. Allen, U.S.	8:09:08	**1992**	P. Newby-Fraser, Zimbabwe	8:55:28*
D. Scott, U.S.	9:24:33	**1980**	R. Beck, U.S.	11:21:24	M. Allen, U.S.	8:07:45	**1993**	P. Newby-Fraser, Zimbabwe	8:58:23
J. Howard, U.S.	9:38:29	**1981**	L. Sweeney, U.S.	12:00:32	G. Welch, Australia	8:20:27	**1994**	P. Newby-Fraser, Zimbabwe	9:20:14
D. Scott, U.S.	9:08:23	**1982**	J. Leach, U.S.	10:54:08	M. Allen, U.S.	8:20:34	**1995**	K. Smyers, U.S.	9:16:46
D. Scott, U.S.	9:05:57	**1983**	S. Puntous, Canada	10:43:36	L. Van Lierde, Belgium	8:04:08*	**1996**	P. Newby-Fraser, Zimbabwe	9:06:49
D. Scott, U.S	8:54:20	**1984**	S. Puntous, Canada	10:25:13	T. Hellriegel, Germany	8:33:01	**1997**	H. Fuhr, Canada	9:31:43
S.Tinley, U.S.	8:50:54	**1985**	J. Ernst, U.S.	10:25:22	P. Reid, Canada	8:24:20	**1998**	N. Badmann, Switz.	9:24:16
D. Scott, U.S.	8:28:37	**1986**	P. Newby-Fraser, Zimbabwe	9:49:14	L. Van Lierde, Belgium	8:17:17	**1999**	Lori Bowden, U.S.	9:13:02
D. Scott, U.S.	8:34:13	**1987**	E. Baker, New Zealand	9:35:25	P. Reid, Canada	8:21:01	**2000**	N. Badmann, Switz.	9:26:16
S. Molina, U.S.	8:31:00	**1988**	P. Newby-Fraser, Zimbabwe	9:01:01					
M. Allen, U.S.	8:09:15	**1989**	P. Newby-Fraser, Zimbabwe	9:00:56					
M. Allen, U.S.	8:28:17	**1990**	E. Baker, New Zealand	9:13:42					

CYCLING

2001 Tour de France

Lance Armstrong won his 3d straight Tour de France on July 29, 2001 becoming the 1st from the U.S., and 5th rider in history, to win the race 3 consecutive times. The win ties him with Greg LeMond (1986, 1989-90) for most wins by an American. Miguel Indurain, of Spain, is the only cyclist to win 5 consecutive Tours (1991-1995). Armstrong finished the 20-stage, 2,150-mile (3,460-km) race in 86 hr., 17 min., 28 sec. Jan Ullrich (Germany), the 1997 winner, trailed Armstrong by 6 min., 44 sec., finishing behind the Texan for the 2d year in row. Spanish rider Joseba Beloki finished 3d, 9 min., 5 sec., behind. The 3-time champion U.S. Postal Service team remained the only American team ever to win the race. Armstrong had battled back from a near-fatal bout with cancer in 1996 before beginning his record run by winning his 1st Tour in 1999.

FISHING

Selected IGFA Saltwater & Freshwater All-Tackle World Records

Source: International Game Fish Association; records confirmed to Oct. 1, 2001

Saltwater Fish Records

Species	Weight	Where caught	Date	Angler
Albacore	88 lbs. 2 oz.	Canary Islands, Spain	Nov. 19, 1977	Siegfried Dickemann
Amberjack, greater	155 lbs. 12 oz.	Bermuda	Aug. 16, 1992	Larry Trott
Barracuda, great	85 lbs.	Christmas Island, Kiribati	Apr. 11, 1992	John W. Helfrich
Barracuda, Mexican	21 lbs.	Phantom Isle, Costa Rica	Mar. 27, 1987	E. Greg Kent
Barracuda, Pacific	26 lbs. 8 oz.	Playa Matapalo, Costa Rica	Jan. 3, 1999	Doug Hettinger
Bass, barred sand	13 lbs. 3 oz.	Huntington Beach, CA	Aug. 29, 1988	Robert Halal
Bass, black sea	10 lbs. 4 oz.	Virginia Beach, VA	Jan. 1, 2000	Allan P. Paschall
Bass, giant sea	563 lbs. 8 oz.	Anacapa Island, CA	Aug. 20, 1968	James D. McAdam Jr.
Bass, striped	78 lbs. 8 oz.	Atlantic City, NJ	Sept. 21, 1982	Albert R. McReynolds
Bluefish	31 lbs. 12 oz.	Hatteras Inlet, NC	Jan. 30, 1972	James M. Hussey
Bonefish	19 lbs.	Zululand, South Africa	May 26, 1962	Brian W. Batchelor
Bonito, Atlantic	18 lbs. 4 oz.	Faial Island, Azores	July 8, 1953	D. Gama Higgs
Bonito, Pacific	21 lbs. 3 oz.	Malibu, CA	July 30, 1978	Gino M. Picciolo
Cabezon	23 lbs.	Juan De Fuca Strait, WA	Aug. 4, 1990	Wesley S. Hunter
Cobia	135 lbs. 9 oz.	Shark Bay, Australia	July 9, 1985	Peter W. Goulding
Cod, Atlantic	98 lbs. 12 oz.	Isle of Shoals, NH	June 8, 1969	Alphonse J. Bielevich
Cod, Pacific	35 lbs.	Unalaska Bay, AK	June 16, 1999	Jim Johnson
Conger	133 lbs. 4 oz.	Berry Head, S. Devon, England	June 5, 1995	Vic Evans
Dolphin	88 lbs.	Exuma, Bahamas	May 5, 1998	Richard D. Evans
Drum, black	113 lbs. 1 oz.	Lewes, DE	Sept. 15, 1975	Gerald M. Townsend
Drum, red	94 lbs. 2 oz.	Avon, NC	Nov. 7, 1984	David G. Deuel
Eel, American	9 lbs. 4 oz.	Cape May, NJ	Nov. 9, 1995	Jeff Pennick
Eel, marbled	36 lbs. 1 oz.	Hazelmere Dam, South Africa	June 10, 1984	Ferdie Van Nooten
Flounder, southern	20 lbs. 9 oz.	Nassau Sound, FL	Dec. 23, 1983	Larenza W. Mungin
Flounder, summer	22 lbs. 7 oz.	Montauk, NY	Sept. 15, 1975	Charles Nappi
Grouper, Warsaw	436 lbs. 12 oz.	Gulf of Mexico, Destin, FL	Dec. 22, 1985	Steve Haeusler
Halibut, Atlantic	355 lbs. 6 oz.	Valevag, Norway	Oct. 20, 1997	Odd Arve Gunderstad
Halibut, California	58 lbs. 9 oz.	Santa Rosa Island, CA	June 26, 1999	Roger W. Borrell
Halibut, Pacific	459 lbs.	Dutch Harbor, AK	June 11, 1996	Jack Tragis
Jack, crevalle	58 lbs. 6 oz.	Barra do Kwanza, Angola	Dec. 10, 2000	Nuno Abohbot Po da Silva
Jack, horse-eye	29 lbs. 8 oz.	Ascencion Island, South Atlantic	May 28, 1993	Mike Hanson
Jack, Pacific crevalle	39 lbs.	Playa Zancudo, Costa Rica	Mar. 3, 1997	Ingrid Callaghan
Jewfish	680 lbs.	Fernandina Beach, FL	May 20, 1961	Lynn Joyner
Kawakawa	29 lbs.	Clarion Island, Mexico	Dec. 17, 1986	Ronald Nakamura
Lingcod	69 lbs. 3 oz.	Waterfall Resort, AK	Aug. 18, 1999	Rizwan Sheikh
Mackerel, cero	17 lbs. 2 oz.	Islamorada, FL	Apr. 5, 1986	G. Michael Mills
Mackerel, king	93 lbs.	San Juan, PR	Apr. 18, 1999	Steve Perez Graulau
Mackerel, Spanish	13 lbs.	Ocracoke Inlet, NC	Nov. 4, 1987	Robert Cranton
Marlin, Atlantic blue	1,402 lbs. 2 oz.	Vitoria, Brazil	Feb. 29, 1992	Paulo Roberto A. Amorim
Marlin, black	1,560 lbs.	Cabo Blanco, Peru	Aug. 4, 1953	Alfred C. Glassell Jr.
Marlin, Pacific blue	1,376 lbs.	Kaaiwi Pt., Kona, HI	May 31, 1982	Jay W. deBeaubien
Marlin, striped	494 lbs.	Tutukaka, New Zealand	Jan. 16, 1986	Bill Boniface
Marlin, white	181 lbs. 14 oz.	Vitoria, Brazil	Dec. 8, 1979	Evandro Luiz Coser
Permit	56 lbs. 2 oz.	Ft. Lauderdale, FL	June 30, 1997	Thomas Sebestyen
Pollack, European	27 lbs. 6 oz.	Salcombe, Devon, England	Jan. 16, 1986	Robert Samuel Milkins
Pollock	50 lbs.	Salstraumen, Norway	Nov. 30, 1995	Thor-Magnus Lekang
Pompano, African	50 lbs. 8 oz.	Daytona Beach, FL	Apr. 21, 1990	Tom Sargent
Roosterfish	114 lbs.	La Paz, Baja Cal., Mexico	June 1, 1960	Abe Sackheim
Runner, blue	11 lbs. 2 oz.	Dauphin Isl., AL	June 28, 1997	Stacey Michelle Moiren
Runner, rainbow	37 lbs. 9 oz.	Clarion Island, Mexico	Nov. 21, 1991	Tom Pfleger
Sailfish, Atlantic	141 lbs. 1 oz.	Luanda, Angola	Feb. 19, 1994	Alfredo de Sousa Neves
Sailfish, Pacific	221 lbs.	Santa Cruz Island, Ecuador	Feb. 12, 1947	C. W. Stewart
Seabass, white	83 lbs. 12 oz.	San Felipe, Mexico	Mar. 31, 1953	L. C. Baumgardner
Seatrout, spotted	17 lbs. 7 oz.	Ft. Pierce, FL	May 11, 1995	Craig F. Carson
Shark, bigeye thresher	802 lbs.	Tutukaka, New Zealand	Feb. 8, 1981	Dianne North
Shark, bignose	369 lbs. 14 oz.	Markham R., Papua New Guinea	Oct. 23, 1993	Lester J. Rohrlach
Shark, blue	454 lbs.	Martha's Vineyard, MA	July 19, 1996	Pete Bergin
Shark, great hammerhead	991 lbs.	Sarasota, FL	May 30, 1982	Allen Ogle
Shark, Greenland	1,708 lbs. 9 oz.	Trondheimsfjord, Norway	Oct. 18, 1987	Terje Nordtvedt
Shark, porbeagle	507 lbs.	Caithness, Scotland	Mar. 9, 1993	Christopher Bennett
Shark, shortfin mako	1,115 lbs.	Black River, Mauritius	Nov. 16, 1988	Patrick Guillanton
Shark, tiger	1,780 lbs.	Cherry Grove, SC	June 14, 1964	Walter Maxwell
Shark, white	2,664 lbs.	Ceduna, S.A., Australia	Apr. 21, 1959	Alfred Dean
Sheepshead	21 lbs. 4 oz.	New Orleans, LA	Apr. 16, 1982	Wayne Desselle
Skipjack, black	26 lbs.	Thetis Bank, Baja Cal., Mexico	Oct. 23, 1991	Clifford Hamaishi
Snapper, cubera	121 lbs. 8 oz.	Cameron, LA	July 5, 1982	Mike Hebert
Snapper, red	50 lbs. 4 oz.	Gulf of Mexico, LA	June 23, 1996	Capt. Doc Kennedy
Snook, common	53 lbs. 10 oz.	Parismina Ranch, Costa Rica	Oct. 18, 1978	Gilbert Ponzi
Spearfish, Mediterranean	90 lbs. 13 oz.	Madeira Island, Portugal	June 2, 1980	Joseph Larkin
Swordfish	1,182 lbs.	Iquique, Chile	May 7, 1953	L. B. Marron
Tarpon	283 lbs. 4 oz.	Sherbro Island, Sierra Leone	Apr. 16, 1991	Yvon Sebag
Tautog	25 lbs.	Ocean City, NJ	Jan. 20, 1998	Anthony R. Monica
Trevally, bigeye	31 lbs. 8 oz.	Poivre Isl., Seychelles	Apr. 23, 1997	Les Sampson
Trevally, giant	145 lbs. 8 oz.	Makena, Maui, HI	Mar. 28, 1991	Russell Mori
Tuna, Atlantic bigeye	392 lbs. 6 oz.	Canary Islands, Spain	July 15, 1996	Dieter Vogel
Tuna, blackfin	45 lbs. 8 oz.	Key West, FL	May 4, 1996	Sam J. Burnett
Tuna, bluefin	1,496 lbs.	Aulds Cove, Nova Scotia	Oct. 26, 1979	Ken Fraser
Tuna, longtail	79 lbs. 2 oz.	Montague Isl., N.S.W., Australia	Apr. 12, 1982	Tim Simpson
Tuna, Pacific bigeye	435 lbs.	Cabo Blanco, Peru	Apr. 17, 1957	Dr. Russel V. A. Lee

Species	Weight	Where caught	Date	Angler
Tuna, skipjack	45 lbs. 4 oz.	Flathead Bank, Baja Cal., Mexico	Nov. 16, 1996	Brian Evans
Tuna, southern bluefin	348 lbs. 5 oz.	Whakatane, New Zealand	Jan. 16, 1981	Rex Wood
Tuna, yellowfin	388 lbs. 12 oz.	San Benedicto Island, Mexico	Apr. 1, 1977	Curt Wiesenhutter
Tunny, little	35 lbs. 2 oz.	Cap de Garde, Algeria	Dec. 14, 1988	Jean Yves Chatard
Wahoo	158 lbs. 8 oz.	Loreto, Baja Cal., Mexico	June 10, 1996	Keith Winter
Weakfish	19 lbs. 2 oz.	Jones Beach Inlet, NY	Oct. 11, 1984	Dennis Roger Rooney
		Delaware Bay, DE	May 20, 1989	William E. Thomas
Yellowtail, California	80 lbs. 11 oz.	Alijos Rocks, Baja Cal., Mexico	Nov. 12, 1998	Brian Buddell
Yellowtail, southern	114 lbs. 10 oz.	Tauranga, New Zealand	Feb. 5, 1984	Mike Godfrey
		White Island, New Zealand	Jan. 9, 1987	David Lugton

Freshwater Fish Records

Species	Weight	Where caught	Date	Angler
Barramundi	83 lbs. 7 oz.	Lake Tinaroo, N. Queensland, Australia	Sept. 23, 1999	David Powell
Bass, largemouth	22 lbs. 4 oz.	Montgomery Lake, GA	June 2, 1932	George W. Perry
Bass, rock	3 lbs.	York River, Ontario	Aug. 1, 1974	Peter Gulgin
	3 lbs.	Lake Erie, PA	June 18, 1998	Herbert G. Ratner, Jr.
Bass, shoal	8 lbs. 12 oz.	Apalachicola River, FL	Jan. 28, 1995	Carl W. Davis
Bass, smallmouth	10 lbs. 14 oz.	Dale Hollow Lake, TN	Apr. 24, 1969	John T. Gorman
Bass, white	6 lbs. 13 oz.	Lake Orange, VA	July 31, 1989	Ronald L. Sprouse
Bass, whiterock	27 lbs. 5 oz.	Greers Ferry Lake, AR	April 24, 1997	Jerald C. Shaum
Bass, yellow	2 lbs. 9 oz.	Waverly, TN	Feb. 27, 1998	John T. Chappell
Bluegill	4 lbs. 12 oz.	Ketona Lake, AL	Apr. 9, 1950	T. S. Hudson
Bowfin	21 lbs. 8 oz.	Florence, SC	Jan. 29, 1980	Robert L. Harmon
Buffalo, bigmouth	70 lbs. 5 oz.	Bastrop, LA	Apr. 21, 1980	Delbert Sisk
Buffalo, black	63 lbs. 6 oz.	Mississippi River, IA	Aug. 14, 1999	Jim Winters
Buffalo, smallmouth	82 lbs. 3 oz.	Athens Lake, AR	June 6, 1993	Randy Collins
Bullhead, brown	6 lbs. 1 oz.	Waterford, NY	Apr. 26, 1998	Bobby Triplett
Bullhead, yellow	4 lbs. 4 oz.	Mormon Lake, AZ	May 11, 1984	Emily Williams
Burbot	18 lbs. 11 oz.	Angenmanalren, Sweden	Oct. 22, 1996	Margit Agren
Carp, common	75 lbs. 11 oz.	Lac de St. Cassien, France	May 21, 1987	Leo van der Gugten
Catfish, blue	111 lbs.	Wheeler Reservoir, Tenn. R.	July 5, 1996	William P. McKinley
Catfish, channel	58 lbs.	Santee-Cooper Res., SC	July 7, 1964	W. B. Whaley
Catfish, flathead	123 lbs. 9 oz.	Independence, KS	May 14, 1998	Ken Paulie
Catfish, white	21 lbs. 8 oz.	Gorton Pond, CT	Apr. 22, 2001	Thomas Urguhart
Char, Arctic	32 lbs. 9 oz.	Tree River, Canada	July 30, 1981	Jeffrey L. Ward
Crappie, white	5 lbs. 3 oz.	Enid Dam, MS	July 31, 1957	Fred L. Bright
Dolly Varden	19 lbs. 4 oz.	Unnamed river, AK	Sept. 4, 1998	Gary D. Ordway
DoradoYW	51 lbs. 5 oz.	Toledo (Corrientes), Argentina	Sept. 27, 1984	Armando Giudice
Drum, freshwater	54 lbs. 8 oz.	Nickajack Lake, TN	Apr. 20, 1972	Benny E. Hull
Gar, alligator	279 lbs.	Rio Grande, TX	Dec. 2, 1951	Bill Valverde
Gar, Florida	9 lbs. 7 oz.	Lake Lawne, FL	Mar. 25, 2001	Patric A. McDaniel
Gar, longnose	50 lbs. 5 oz.	Trinity River, TX	July 30, 1954	Townsend Miller
Gar, shortnose	5 lbs. 12 oz.	Ren Lake, IL	July 16, 1995	Donna K. Willmert
Gar, spotted	9 lbs. 12 oz.	Lake Mexia, TX	Apr. 7, 1994	Rick Rivard
Grayling, Arctic	5 lbs. 15 oz.	Katseyedie River, N.W.T.	Aug. 16, 1967	Jeanne P. Branson
Inconnu	53 lbs.	Pah River, AK	Aug. 20, 1985	Lawrence E. Hudnall
Kokanee	9 lbs. 6 oz.	Okanagan Lake, Vernon, B.C.	June 18, 1988	Norm Kuhn
Muskellunge	67 lbs. 8 oz.	Lake Court Oreilles, WI	July 24, 1949	Cal Johnson
Muskellunge, tiger	51 lbs. 3 oz.	Lac Vieux-Desert, MI	July 16, 1919	John Knobla
Perch, Nile	213 lbs.	Lake Nasser, Egypt	Dec. 18, 1997	Adrian Brayshaw
Perch, white	4 lbs. 12 oz.	Messalonskee Lake, ME	June 4, 1949	Earl Small
Perch, yellow	4 lbs. 3 oz.	Bordentown, NJ	May, 1865	Dr. C. C. Abbot
Pickerel, chain	9 lbs. 6 oz.	Homerville, GA	Feb. 17, 1961	Baxley McQuaig Jr.
Pike, northern	55 lbs. 1 oz.	Lake of Grefeern, W. Germany	Oct. 16, 1986	Lothar Louis
Redhorse, greater	9 lbs. 3 oz.	Salmon River, Pulaski, NY	May 11, 1985	Jason Wilson
Redhorse, silver	11 lbs. 7 oz.	Plum Creek, WI	May 29, 1985	Neal Long
Salmon, Atlantic	79 lbs. 2 oz.	Tana River, Norway	1928	Henrik Henriksen
Salmon, chinook	97 lbs. 4 oz.	Kenai River, AK	May 17, 1985	Les Anderson
Salmon, chum	35 lbs.	Edye Pass, BC	July 11, 1995	Todd A. Johansson
Salmon, coho	33 lbs. 4 oz.	Salmon River, Pulaski, NY	Sept. 27, 1989	Jerry Lifton
Salmon, pink	13 lbs. 1 oz.	St. Mary's River, Ontario	Sept. 23, 1992	Ray Higaki
Salmon, sockeye	15 lbs. 3 oz.	Kenai River, AK	Aug. 9, 1987	Stan Roach
Sauger	8 lbs. 12 oz.	Lake Sakakawea, ND	Oct. 6, 1971	Mike Fischer
Shad, American	11 lbs. 4 oz.	Connecticut River, MA	May 19, 1986	Bob Thibodo
Sturgeon, beluga	224 lbs. 13 oz.	Guryev, Kazakhstan	May 3, 1993	Merete Lehne
Sturgeon, white	468 lbs.	Benicia, CA	July 9, 1983	Joey Pallotta 3d
Sunfish, green	2 lbs. 2 oz.	Stockton Lake, MO	June 18, 1971	Paul M. Dilley
Sunfish, redbreast	1 lb. 12 oz.	Suwannee River, FL	May 29, 1984	Alvin Buchanan
Sunfish, redear	5 lbs. 7oz.	Diverson Canal, GA	Nov. 6, 1998	Amos M. Gay
Tigerfish, giant	97 lbs.	Zaire River, Kinshasa, Zaire	July 9, 1988	Raymond Houtmans
Tilapia, Nile	9 lbs. 8 oz.	Antelope Isl., Karibe, Zimbabwe	Apr. 22, 2001	David Barnard
Trout, Apache	5 lb. 3 oz.	Apache Res., AZ	May 29, 1991	John Baldwin
Trout, brook	14 lbs. 8 oz.	Nipigon River, Ontario	July, 1916	Dr. W. J. Cook
Trout, bull	32 lbs.	Lake Pend Oreille, ID	Oct. 27, 1949	N. L. Higgins
Trout, cutthroat	41 lbs.	Pyramid Lake, NV	Dec., 1925	John Skimmerhorn
Trout, golden	11 lbs.	Cooks Lake, WY	Aug. 5, 1948	Charles S. Reed
Trout, lake	72 lbs.	Great Bear Lake, N.W.T.	Aug. 9, 1995	Lloyd E. Bull
Trout, rainbow	42 lbs. 2 oz.	Bell Island, AK	June 22, 1970	David Robert White
Trout, tiger	20 lbs. 13 oz.	Lake Michigan, WI	Aug. 12, 1978	Pete M. Friedland
Walleye	25 lbs.	Old Hickory Lake, TN	Aug. 2, 1960	Mabry Harper
Warmouth	2 lbs. 7 oz.	Yellow River, Holt, FL	Oct. 19, 1985	Tony D. Dempsey
Whitefish, lake	14 lbs. 6 oz.	Meaford, Ontario	May 21, 1984	Dennis M. Laycock
Whitefish, mountain	5 lbs. 8 oz.	Elbow River, Calgary, AB	Aug. 1, 1995	Randy G. Woo
Whitefish, round	6 lbs. 7 oz.	Shasta Lakes, CA	June 13, 2001	James W. Schmidt
Zander	25 lbs. 2 oz.	Trosa, Sweden	June 12, 1986	Harry Lee Tennison

SULLIVAN AWARD

James E. Sullivan Memorial Trophy Winners

The James E. Sullivan Memorial Trophy, named after the former president of the Amateur Athletic Union (AAU) and inaugurated in 1930, is awarded annually by the AAU to the athlete who "by his or her performance, example and influence as an amateur, has done the most during the year to advance the cause of sportsmanship."

Year	Winner	Sport
1930	Bobby Jones	Golf
1931	Barney Berlinger	Track
1932	Jim Bausch	Track
1933	Glenn Cunningham	Track
1934	Bill Bonthron	Track
1935	Lawson Little	Golf
1936	Glenn Morris	Track
1937	Don Budge	Tennis
1938	Don Lash	Track
1939	Joe Burk	Rowing
1940	Greg Rice	Track
1941	Leslie MacMitchell	Track
1942	Cornelius Warmerdam	Track
1943	Gilbert Dodds	Track
1944	Ann Curtis	Swimming
1945	Doc Blanchard	Football
1946	Arnold Tucker	Football
1947	John Kelly, Jr.	Rowing
1948	Robert Mathias	Track
1949	Dick Button	Skating
1950	Fred Wilt	Track
1951	Rev. Robert Richards	Track
1952	Horace Ashenfelter	Track
1953	Dr. Sammy Lee	Diving
1954	Mal Whitfield	Track
1955	Harrison Dillard	Track
1956	Patricia McCormick	Diving
1957	Bobby Joe Morrow	Track
1958	Glenn Davis	Track
1959	Parry O'Brien	Track
1960	Rafer Johnson	Track
1961	Wilma Rudolph Ward	Track
1962	James Beatty	Track
1963	John Pennel	Track
1964	Don Schollander	Swimming
1965	Bill Bradley	Basketball
1966	Jim Ryun	Track
1967	Randy Matson	Track
1968	Debbie Meyer	Swimming
1969	Bill Toomey	Track
1970	John Kinsella	Swimming
1971	Mark Spitz	Swimming
1972	Frank Shorter	Track
1973	Bill Walton	Basketball
1974	Rick Wohlhutter	Track
1975	Tim Shaw	Swimming
1976	Bruce Jenner	Track
1977	John Naber	Swimming
1978	Tracy Caulkins	Swimming
1979	Kurt Thomas	Gymnastics
1980	Eric Heiden	Speed Skating
1981	Carl Lewis	Track
1982	Mary Decker	Track
1983	Edwin Moses	Track
1984	Greg Louganis	Diving
1985	Joan Benoit Samuelson	Marathon
1986	Jackie Joyner-Kersee	Track
1987	Jim Abbott	Baseball
1988	Florence Griffith Joyner	Track
1989	Janet Evans	Swimming
1990	John Smith	Wrestling
1991	Mike Powell	Track
1992	Bonnie Blair	Speed Skating
1993	Charlie Ward	Football, Basketball
1994	Dan Jansen	Speed Skating
1995	Bruce Baumgartner	Wrestling
1996	Michael Johnson	Track
1997	Peyton Manning	Football
1998	Chamique Holdsclaw	Basketball
1999	Kelly Miller and Coco Miller	Basketball
2000	Rulon Gardner	Wrestling

DIRECTORY OF SPORTS ORGANIZATIONS

Major League Baseball

Website: http://www.majorleaguebaseball.com
Note: Teams and leagues as of 2001 season.

Commissioner's Office
245 Park Ave., 31st fl.
New York, NY 10167

American League

Anaheim Angels
2000 Gene Autry Way
Anaheim, CA 92806

Baltimore Orioles
333 W. Camden St.
Baltimore, MD 21201

Boston Red Sox
4 Yawkey Way
Boston, MA 02215

Chicago White Sox
333 W. 35th St.
Chicago, IL 60616

Cleveland Indians
2401 Ontario St.
Cleveland, OH 44115

Detroit Tigers
2100 Woodward Ave.
Detroit, MI 48201

Kansas City Royals
1 Royal Way
Kansas City, MO 64141

Minnesota Twins
34 Kirby Puckett Place
Minneapolis, MN 55415

New York Yankees
161st St. and River Ave.
Bronx, NY 10451

Oakland Athletics
7677 Oakport, Suite 200
Oakland, CA 94621

Seattle Mariners
PO Box 4100
Seattle, WA 98104

Tampa Bay Devil Rays
One Tropicana Dr.
St. Petersburg, FL 33705

Texas Rangers
1000 Ballpark Way
Arlington, TX 76011

Toronto Blue Jays
1 Blue Jays Way, Suite 3200
Toronto, Ont. M5V 1J1

National League

Arizona Diamondbacks
401 E. Jefferson St.
Phoenix, AZ 85001

Atlanta Braves
755 Hank Aaron Drive
Atlanta, GA 30315

Chicago Cubs
1060 W. Addison
Chicago, IL 60613

Cincinnati Reds
100 Cinergy Field
Cincinnati, OH 45202

Colorado Rockies
2001 Blake St.
Denver, CO 80205

Florida Marlins
2269 Dan Marino Blvd.
Miami, FL 33056

Houston Astros
501 Crawford St.
Houston, TX 77002

Los Angeles Dodgers
1000 Elysian Park Ave.
Los Angeles, CA 90012

Milwaukee Brewers
One Brewers Way
Milwaukee, WI 53214

Montreal Expos
4549 Ave. Pierre de Coubertin
Montreal, Que. H1V 3N7

New York Mets
123-01 Roosevelt Ave.
Flushing, NY 11368

Philadelphia Phillies
3501 S. Broad St.
Philadelphia, PA 19148

Pittsburgh Pirates
115 Federal St.
Pittsburgh, PA 15212

St. Louis Cardinals
250 Stadium Plaza
St. Louis, MO 63102

San Diego Padres
8880 Rio San Diego Dr.
San Diego, CA 92112

San Francisco Giants
24 Willie Mays Plaza
San Francisco, CA 94107

National Basketball Association

Website: http://www.nba.com

League Office
Olympic Tower 645 5th Ave.
New York, NY 10022

Atlanta Hawks
One CNN Center, Ste. 405, South Tower
Atlanta, GA 30303

Boston Celtics
151 Merrimac St.
Boston, MA 02114

Charlotte Hornets
100 Hive Dr.
Charlotte, NC 28217

Chicago Bulls
1901 W. Madison St.
Chicago, IL 60612

Cleveland Cavaliers
1 Center Court
Cleveland, OH 44115

Dallas Mavericks
2909 Taylor St.
Dallas, TX 75226

Denver Nuggets
1000 Chopper Pl.
Denver, CO 80204

Detroit Pistons
Two Championship Dr.
Auburn Hills, MI 48326

Golden State Warriors
1011 Broadway
Oakland, CA 94607

Houston Rockets
Two Greenway Plaza, Ste. 400
Houston, TX 77046

Indiana Pacers
125 S. Pennsylvania St.
Indianapolis, IN 46204

Los Angeles Clippers
1111 S. Figueroa St., Ste. 1100
Los Angeles, CA 90037

Los Angeles Lakers
555 Nash St.
El Segundo, CA 90245

Memphis Grizzlies
P.O. Box 3463
Memphis, TN 38173

Miami Heat
601 Biscayne Blvd.
Miami, FL 33132

Milwaukee Bucks
1001 N. 4th St.
Milwaukee, WI 53203

Minnesota Timberwolves
600 1st Ave. North
Minneapolis, MN 55403

New Jersey Nets
390 Murray Hill Parkway
E. Rutherford, NJ 07073

New York Knickerbockers
Two Pennsylvania Plaza
New York, NY 10121

Orlando Magic
Two Magic Place
8701 Maitland Summit Blvd.
Orlando, FL 32801

Philadelphia 76ers
3601 S. Broad St.
Philadelphia, PA 19148

Phoenix Suns
201 E. Jefferson
Phoenix, AZ 85004

Portland Trail Blazers
7325 SW Childs Rd.
Portland, OR 97224

Sacramento Kings
One Sports Parkway
Sacramento, CA 95834

San Antonio Spurs
100 Montana St.
San Antonio, TX 78203

Seattle SuperSonics
351 Elliott Ave., West
Suite 500
Seattle, WA 98119

Toronto Raptors
40 Bay St., Ste. 400
Toronto, Ont. M5J 2X2

Utah Jazz
301 W. South Temple
Salt Lake City, UT 84101

Washington Wizards
601 F St., NW
Washington, DC 20004

National Hockey League

Website: http://www.nhl.com

League Headquarters
1251 Ave. of the Americas
New York, NY 10020

Mighty Ducks of Anaheim
2695 E. Katella Ave.
Anaheim, CA 92803

Atlanta Thrashers
1CNN Ctr., 12th Fl., S. Tower
Atlanta, GA 30348

Boston Bruins
One FleetCenter, Ste. 250
Boston, MA 02114

Buffalo Sabres
HSBC Arena
One Seymour H. Knox III Plaza
Buffalo, NY 14203

Calgary Flames
PO Box 1540, Station M
Calgary, Alta. T2P 3B9

Carolina Hurricanes
1400 Edwards Mill Rd.
Raleigh, NC 27607

Chicago Blackhawks
1901 W. Madison St.
Chicago, IL 60612

Colorado Avalanche
1000 Chopper Cr.
Denver, CO 80204

Columbus Blue Jackets
200 W. Nationwide Blvd.
Columbus, OH 43215

Dallas Stars
211 Cowboys Parkway
Irving, TX 75063

Detroit Red Wings
600 Civic Center Dr.
Detroit, MI 48226

Edmonton Oilers
11230 110 St.
Edmonton, Alta. T5G 3H7

Florida Panthers
One Panther Parkway
Sunrise, FL 33323

Los Angeles Kings
1111 S. Figueroa St.
Los Angeles, CA 90015

Minnesota Wild
317 Washington St.
St. Paul, MN 55102

Montreal Canadiens
1260 rue de La Gauchetiere Ouest
Montreal, Que. H3B 5E8

Nashville Predators
501 Broadway
Nashville, TN 37203

New Jersey Devils
50 Rte. 120 N. PO Box 504
E. Rutherford, NJ 07073

New York Islanders
Nassau Veterans Memorial Coliseum
Uniondale, NY 11553

New York Rangers
Two Pennsylvania Plaza
New York, NY 10121

Ottawa Senators
1000 Palladium Dr.
Kanata, Ont. K2V 1A5

Philadelphia Flyers
First Union Center
3601 South Broad St.
Philadelphia, PA 19148

Phoenix Coyotes
9375 E. Bell Rd.
Scottsdale, AZ 85260

Pittsburgh Penguins
66 Mario Lemieux Place
Pittsburgh, PA 15219

St. Louis Blues
1401 Clark Ave.
St. Louis, MO 63103

San Jose Sharks
525 W. Santa Clara St.
San Jose, CA 95113

Tampa Bay Lightning
401 Channelside Dr.
Tampa, FL 33602

Toronto Maple Leafs
40 Bay St.
Toronto, Ont. M5J 2X2

Vancouver Canucks
800 Griffiths Way
Vancouver, B.C. V6B 6G1

Washington Capitals
601 F St. NW
Washington, DC 20004

National Football League

Website: http://www.nfl.com

League Office
280 Park Avenue
New York, NY 10017

Arizona Cardinals
PO Box 888
Phoenix, AZ 85001

Atlanta Falcons
4400 Falcon Parkway
Flowery Branch, GA 30542

Baltimore Ravens
11001 Owings Mills Blvd.
Owings Mills, MD 2117

Buffalo Bills
One Bills Drive
Orchard Park, NY 14127

Carolina Panthers
800 S. Mint St.
Charlotte, NC 28202

Chicago Bears
1000 Football Dr.
Lake Forest, IL 60045

Cincinnati Bengals
One Paul Brown Stadium
Cincinnati, OH 45202

Cleveland Browns
76 Lou Groza Blvd.
Berea, OH 44017

Dallas Cowboys
One Cowboys Parkway
Irving, TX 75063

Denver Broncos
13655 Broncos Parkway
Englewood, CO 80112

Detroit Lions
1200 Featherstone Rd.
Pontiac, MI 48342

Green Bay Packers
1265 Lombardi Ave.
Green Bay, WI 54304

Indianapolis Colts
PO Box 535000
Indianapolis, IN 46253

Jacksonville Jaguars
One ALLTELL Stadium Place
Jacksonville, FL 32202

Kansas City Chiefs
One Arrowhead Drive
Kansas City, MO 64129

Miami Dolphins
7500 SW 30th St.
Davie, FL 33314

Minnesota Vikings
9520 Viking Dr.
Eden Prairie, MN 55344

New England Patriots
60 Washington St.
Foxboro, MA 02035

New Orleans Saints
5800 Airline Drive
Metairie, LA 70003

New York Giants
Giants Stadium
E. Rutherford, NJ 07073

New York Jets
1000 Fulton Ave.
Hempstead, NY 11550

Oakland Raiders
1220 Harbor Bay Parkway
Alameda, CA 94502

Philadelphia Eagles
One Novacare Way
Philadelphia, PA 19145

Pittsburgh Steelers
3400 S. Water St.
Pittsburgh, PA 15203

St. Louis Rams
One Rams Way
St. Louis, MO 63045

San Diego Chargers
PO Box 609609
San Diego, CA 92160

San Francisco 49ers
4949 Centennial Blvd.
Santa Clara, CA 95054

Seattle Seahawks
11220 NE 53d St.
Kirkland, WA 98033

Tampa Bay Buccaneers
One Buccaneer Place
Tampa, FL 33607

Tennessee Titans
460 Great Circle Rd.
Nashville, TN 37228

Washington Redskins
21300 Redskin Park Dr.
Ashburn, VA 20147

Other Sports Organizations

Amateur Athletic Union
PO Box 10000
Lake Buena Vista, FL 32830
http://www.aausports.org

Amateur Softball Assn.
2801 NE 50th St.
Oklahoma City, OK 73111
http://www.softball.org

American Kennel Club
260 Madison Ave., 4th Fl.
New York, NY 10016
http://www.akc.org

Canadian Football League
110 Eglinton Ave. W, 5th Fl.
Toronto, Ont. M4R 1A3
http://www.cfl.ca

CART
755 W. Big Beaver Rd.
Troy, MI 48084
http://www.cart.com

Intl. Game Fish Assn.
300 Gulfstream Way
Dania Beach, FL 33004
http://www.igfa.org

LPGA
100 International Golf Dr.
Daytona Beach, FL 32124
http://www.lpga.com

Little League Baseball
PO Box 3485
Williamsport, PA 17701
http://www.littleleague.org

Major League Soccer
110 E. 42d St., 10th Fl.
New York, NY 10017
http://www.mlsnet.com

NASCAR
1801 W. Intl. Speedway Blvd.
Daytona Beach, FL 32120
http://www.nascar.com

NCAA
700 W. Washington St.
PO Box 6222
Indianapolis, IN 46206
http://www.ncaa.org

National Rifle Assn.
11250 Waples Mill Rd.
Fairfax, VA 22030
http://www.nra.org

Pro Bowlers Assn.
999 3d Ave. , Ste. 2810
Seattle, WA 98104
http://www.pbatour.com

PGA
100 Ave. of the Champions
Box 109601
Palm Beach Gardens, FL 33410
http://www.pga.com

Pro Rodeo Cowboys Assn.
101 Pro Rodeo Dr.
Colorado Springs, CO 80919
http://www.prorodeo.com

Special Olympics
1325 G St., NW, Ste. 500
Washington, DC 20005
http://www.special olympics.org

Thoroughbred Racing Assn.
420 Fair Hill DR.
Elkton, MD 21921
http://www.tra-online.com

USA Equestrian
4047 Iron Works Pkwy.
Lexington, KY 40511
http://www.equestrian.org

USA Swimming
One Olympic Plaza
Colorado Springs, CO 80909
http://www.usa-swimming.org

USA Track & Field
1 RCA Dome, Ste. 140
Indianapolis, IN 46225
http://www.usatf.org

U.S. Auto Club
4910 W. 16th St.
Speedway, IN 46224

U.S. Figure Skating Assn.
20 First St.
Colorado Springs, CO 80906
http://www.usfsa.org

U.S. Olympic Committee
One Olympic Plaza
Colorado Springs, CO 80909
http://www.usoc.org

U.S. Skiing Assn.
1500 Kearns Blvd.
PO Box 100
Park City, UT 84060
http://www.usskiteam.com

U.S. Soccer Federation
1801-1811 S. Prairie Ave.
Chicago, IL 60616
http://www.us-soccer.com

U.S. Tennis Assn.
PO Box 5046
White Plains, NY 10602
http://www.usta.com

U.S. Trotting Assn.
750 Michigan Ave.
Columbus, OH 43215
http://www.ustrotting.com

WNBA
645 5th Ave., 10th Fl.
New York, NY 10022
http://www.wnba.com

NOTABLE SPORTS PERSONALITIES

Henry (Hank) Aaron, b. 1934: Milwaukee-Atlanta outfielder; hit record 755 home runs, led NL 4 times; record 2,297 RBIs.
Kareem Abdul-Jabbar, b. 1947: Milwaukee, L.A. Lakers center; MVP 6 times; all-time leading NBA scorer, 38,387 points.
Andre Agassi, b. 1970: won career Grand Slam: Wimbledon, '92; U.S. Open, '94; Australian Open, '95; French Open, '99.
Troy Aikman, b. 1966: quarterback, led Dallas Cowboys to Super Bowl wins in 1993-94, 1996; Super Bowl MVP, 1993.
Amy Alcott, b. 1956: golfer, 29 career wins (5 majors), inducted into World Golf Hall of Fame in 1999.
Grover Cleveland "Pete" Alexander (1887-1950): pitcher; won 373 NL games; pitched 16 shutouts, 1916.
Muhammad Ali, b. 1942: 3-time heavyweight champion.
Gary Anderson, b. 1959: kicker, NFL's career points leader, with 2,059 through the end of the 2000 seaon.
Sparky Anderson, b. 1934: only manager to win World Series in the NL (Cincinnati, 1975-76) and the AL (Detroit, 1984).
Mario Andretti, b. 1940: won Daytona 500 (1967), Indy 500 (1969); Formula 1 world title (1978).
Earl Anthony (1938-2001): bowler, won record 6 PBA Championships (1973-75, 1981-83), 41 career PBA tournaments.
Eddie Arcaro, (1916-97): only jockey to win racing's Triple Crown twice, 1941,1948; rode 4,779 winners in his career.
Henry Armstrong (1912-88): boxer, held feather-, welter-, lightweight titles simultaneously, 1937-38.
Lance Armstrong, b. 1971: cyclist, 3-time winner of the Tour de France (1999-2001).
Arthur Ashe (1943-93): tennis, won U.S. Open (1968); Wimbledon (1975); died of AIDS.
Evelyn Ashford, b.1957: sprinter, won 100m gold (1984) and silver (1988); member of 5 U.S. Olympic teams (1976-1992).
Red Auerbach, b. 1917: coached Boston to 9 NBA titles.
Tracy Austin, b. 1962: youngest player to win U.S. Open tennis title (age 16 in 1979), 2-time AP Female Athlete of the Year.
Ernie Banks, b. 1931: Chicago Cubs slugger; hit 512 NL homers; twice MVP; never played in World Series.
Roger Bannister, b. 1929: British physician, ran first sub 4-minute mile, May 6, 1954 (3 min. 59.4 sec.).
Charles Barkely, b. 1963: NBA MVP, 1993; 4th player ever to surpass 20,000 pts, 10,000 rebounds, and 4,000 assists.
Rick Barry, b. 1944: NBA scoring leader, 1967; ABA, 1969.
Sammy Baugh, b. 1914: Washington Redskins quarterback; held numerous records upon retirement after 16 pro seasons.
Elgin Baylor, b. 1934: L.A. Lakers forward; 10-time all-star.
Bob Beamon, b. 1946: Olympic long jump gold medalist in 1968; world record jump of 29' 2½" stood until 1991.
Boris Becker, b. 1967: German tennis star; won U.S. Open 1989; Wimbledon champ 3 times.
Jean Beliveau, b. 1931: Montreal Canadiens center; scored 507 goals; twice MVP.
Johnny Bench, b. 1947: Cincinnati Reds catcher; MVP twice; led league in home runs twice, RBIs 3 times.
Patty Berg, b. 1918: won more than 80 golf tournaments; AP Woman Athlete-of-the-Year 3 times.
Yogi Berra, b. 1925: Yankee catcher (1946-63); 3-time MVP.
Abebe Bikila (1932-73): Ethiopian runner, won consecutive Olympic marathon gold medals in 1960 (barefoot), 1964.
Matt Biondi, b. 1965: swimmer, won 5 golds, 1988 Olympics.
Larry Bird, b. 1956: Boston Celtics forward; chosen MVP 1984-86; 1998 coach of the year with Indiana Pacers.
Bonnie Blair, b. 1964: speed skater won 5 individual gold medals in 3 Olympics (1988, '92, '94).
George Blanda, b. 1927: quarterback, kicker; 26 years as active player, scored 2,002 career points.
Fanny Blankers-Koen, b. 1918: track, won 4 golds in 1948.
Wade Boggs, b. 1958: AL batting champ, 1983, 1985-88; reached 3,000 career hits, 1999 (3,010).
Barry Bonds, b. 1964: outfielder, hit record 73 homers and walked record 177 times in 2001; NL MVP 1990, 1992-93; only player in 400 home runs/400 stolen bases club (467/484); 7th all-time in home runs (567).
Bjorn Borg, b. 1956: led Sweden to first Davis Cup, 1975; Wimbledon champion 5 times.
Mike Bossy, b. 1957: N.Y. Islanders right wing scored more than 50 goals 8 times.
Ray Bourque, b. 1960: Boston defenseman,1979-2000, 5-time Norris Trophy winner; won Stanley Cup with Colorado, 2001.
Bill Bradley, b. 1943: basketball All-America at Princeton, won Sullivan Award, Player of the Year, Final 4 Most Outstanding Player in 1965; led NY Knicks to 2 NBA titles (1970, '73).
Terry Bradshaw, b. 1948: quarterback, led Pittsburgh to 4 Super Bowl wins (1975-76, 1979-80); NFL MVP, 1978.
George Brett, b. 1953: Kansas City Royals infielder, led AL in batting, 1976, 1980, 1990; MVP, 1980.
Lou Brock, b. 1939: St. Louis Cardinals outfielder, stole NL record 118 bases, 1974; led NL 8 times.
Jim Brown, b. 1936: Clev. fullback, 12,312 yds.; 3-time MVP.
Paul Brown (1908-91): football owner, coach; led eponymous Cleveland Browns to 3 NFL championships.
Paul "Bear" Bryant (1913-83): college football coach with 323 wins; led Alabama to 5 national titles (1961, '64, '65, '78, '79).
Sergei Bubka, b. 1963: Ukrainian pole vaulter; first to clear 20 feet; gold medal, 1988 Olympics.
Don Budge, (1915-2000): won numerous amateur and pro tennis titles; "grand slam," 1938.
Maria Bueno, b. 1939: tennis, 4 U.S. titles, 3 Wimbledon.
Dick Butkus, b. 1942: Chicago Bears linebacker, twice chosen best NFL defensive player.
Dick Button, b. 1929: figure skater; won 1948, 1952 Olympic gold medals; world titlist, 1948-52.
Walter Camp, (1859-1925): Yale football player, coach, athletic director; established many rules.
Roy Campanella (1921-93): Brooklyn catcher (1948-57); 3-time MVP.
Earl Campbell, b. 1955: NFL running back; MVP 1978-79.
Rod Carew, b. 1945: AL infielder; 7 batting titles, 1977 MVP.
Steve Carlton, b. 1944: NL pitcher; won 20 games 5 times, Cy Young award 4 times.
Billy Casper, b. 1931: PGA Player-of-the-Year 3 times; U.S. Open champ twice.
Tracy Caulkins, b. 1963: swimmer, won 3 Olympic golds, 1984; set 63 U.S. and 5 world records; won 48 individual U.S. titles.
Wilt Chamberlain (1936-99): center; was NBA leading scorer 7 times, MVP 4 times; scored 100 pts. in a game, 1962.
Bobby Clarke, b. 1949: Philadelphia Flyers center; led team to 2 Stanley Cup championships; MVP 3 times.
Roger Clemens, b. 1962: pitcher, 1986 AL MVP; only 5-time Cy Young winner (1986-87, '91, 1997-98); twice struck out record 20 batters in a game; all-time AL strikeout leader (3,717).
Roberto Clemente (1934-72): Pittsburgh Pirates outfielder; won 4 batting titles; MVP, 1966; killed in plane crash.
Ty Cobb (1886-1961): Detroit Tigers outfielder; had record .367 lifetime batting average, 12 batting titles.
Sebastian Coe, b. 1956: British runner, won Olympic 1,500m gold medal and 800m silver medal in 1980 and 1984.
Nadia Comaneci, b. 1961: Romanian gymnast, won 3 gold medals, achieved 7 perfect scores, 1976 Olympics.
Maureen Connolly (1934-69): won tennis "grand slam," 1953; AP Woman-Athlete-of-the-Year 3 times.
Jimmy Connors, b. 1952: tennis, 5 U.S. titles, 2 Wimbledon.
Cynthia Cooper, b. 1963: basketball, 4-time MVP of the WNBA finals and 2-time league MVP for the Houston Comets.
James J. Corbett (1866-1933): heavyweight champion, 1892-97; credited with being the first "scientific" boxer.
Angel Cordero, b. 1942: leading money winner, 1976, 1982-83; rode 3 Kentucky Derby winners.
Howard Cosell (1920-95): commentator for ABC's *Monday Night Football* and *Wide World of Sports.*
Margaret Smith Court, b. 1942: Australian tennis great, won 24 grand slam events.
Bob Cousy, b. 1928: Boston guard; 6 NBA titles; 1957 MVP.
Bjoern Daehlie, b. 1967: Norwegian cross-country skier; won record 8 Winter Olympic gold medals.
Lindsay Davenport, b. 1976: tennis, won Olympic gold (1996), U.S. Open (1998), Wimbledon (1999), Austral. Open (2000).
Dizzy Dean (1911-74): colorful pitcher for St. Louis Cardinals "Gashouse Gang" in the 30s; MVP, 1934.
Mary Decker Slaney, b. 1958: runner, has held 7 separate American records from the 800m to 10,000m.
Oscar De La Hoya, b. 1972: won lightweight title, 1995; super lightweight title, 1996; welterweight title, 1997.
Donna de Varona, b. 1947: 2 Olympic swimming golds,1964; 1st female sportscaster at a major network (ABC), 1965.
Jack Dempsey (1895-1983): heavyweight champ, 1919-26.
Gail Devers, b. 1966: Olympic 100m gold medalist, 1992, '96.
Eric Dickerson, b. 1960: NFL record 2,105 rushing yds.,1984.
Joe DiMaggio (1914-99): N.Y. Yankees outfielder; hit safely in record 56 consecutive games, 1941; AL MVP 3 times.
Tony Dorsett, b. 1954: Heisman winner who led the Dallas Cowboys to an NFL title in his rookie year (1977); 5th all-time in career rushing yards (12,739).
Roberto Duran, b. 1951: Panamanian boxer, held titles at 3 weights; lost 1980 "no mas" fight to Sugar Ray Leonard.
Leo Durocher (1906-91): manager, won 3 NL pennants (Brooklyn-1941, NY Giants-1951, '54) and 1954 World Series.
Dale Earnhardt (1951-2001): 7-time NASCAR Winston Cup champ; died in a last-lap crash at 2001 Daytona 500.
Stefan Edberg, b. 1966: U.S. singles champ, 1991, 1992; Wimbledon champ, 1988, 1990.
Gertrude Ederle, b. 1906: first woman to swim English Channel, broke existing men's record, 1926.
Teresa Edwards, b. 1964: basketball, 5-time Olympian; gold medalist in 1984, '88, '96, 2000 and bronze medal in 1992.
Hicham El Guerrouj, b. 1974: Moroccan miler, set world record of 3:43.13 in Rome on July 7, 1999.

John Elway, b. 1960: quarterback; led Denver Broncos to 2 Super Bowl wins, 1998, 1999; regular-season MVP, 1987.
Julius Erving, b. 1950: 3-time ABA MVP, 1981 NBA MVP.
Phil Esposito, b. 1942: NHL scoring leader 5 times.
Janet Evans, b. 1971: 4 Olympic swimming golds, 1988-92.
Lee Evans, b. 1947: Olympic 400m gold medalist in 1968 with a 43.86 sec. world record not broken until 1988.
Chris Evert, b. 1954: U.S. Open tennis champ 6 times, Wimbledon champ 3 times.
Ray Ewry (1873-1937): track-and-field star, won 8 gold medals, 1900, 1904, and 1908 Olympics.
Nick Faldo, b. 1957: won Masters, British Open 3 times each.
Juan Fangio (1911-95): Argentinian, 5-time World Grand Prix driving champ (1951, 1954-57).
Brett Favre, b. 1969: led Green Bay Packers to Super Bowl win, 1997; NFL regular-season MVP, 1995, 1996; co-MVP, 1997.
Bob Feller, b. 1918: Cleveland Indians pitcher; won 266 games; pitched 3 no-hitters, 12 one-hitters.
Rollie Fingers, b. 1946: pitcher, 341 career saves; AL MVP, Cy Young Award, 1982; World Series MVP, 1974.
Peggy Fleming, b. 1948: world figure skating champion, 1966-68; gold medalist, 1968 Olympics.
Whitey Ford, b. 1928: N.Y. Yankees pitcher, won record 10 World Series games.
George Foreman, b. 1949: heavyweight champion, 1973-74, 1994-95; at 45, the oldest to win a heavyweight title.
Dick Fosbury, b. 1947: high jumper; won 1968 Olympic gold medal; developed the "Fosbury Flop."
Dan Fouts, b. 1951: quarterback (San Diego), 5th in career passing yards (43,040); TV analyst on *Monday Night Football.*
Jimmie Foxx (1907-67): Red Sox, Athletics slugger; MVP 3 times; triple crown, 1933.
A. J. Foyt, b. 1935: won Indy 500 4 times; U.S. Auto Club champ 7 times.
Joe Frazier, b. 1944: heavyweight champion, 1970-73.
Walt Frazier, b. 1945: Hall of Fame guard for N.Y. Knicks NBA championship teams (1970, '73); NBA radio-TV commentator.
Haile Gebrselassie, b. 1973: Ethiopian, world record holder in 5,000m and 10,000m; 10,000m gold medalist in 1996, 2000.
Lou Gehrig (1903-41): N.Y. Yankees 1st baseman; MVP, 1927, 1936; triple crown, 1934; AL record 184 RBIs, 1931.
George Gervin, b. 1952: top NBA scorer, 1978-80, 1982.
Althea Gibson, b. 1927: 2-time U.S. and Wimbledon champ.
Bob Gibson, b. 1935: St. Louis Cardinals pitcher; won Cy Young award twice; struck out 3,117 batters.
Josh Gibson (1911-47): Hall of Fame catcher, known as "Babe Ruth of the Negro Leagues"; credited with as many as 84 homers in 1 season and about 800 in his career.
Marc Girardelli, b. 1963: skier (Lux.), won 5 World Cup titles.
Jeff Gordon, b. 1971: race car driver, youngest to win NASCAR Winston Cup 3 times (1995, 1997-98).
Steffi Graf, b. 1969: German; won tennis "grand slam," 1988; U.S. champ 5 times; Wimbledon champ 7 times.
Otto Graham, b. 1921: Cleveland quarterback, 4-time all-pro.
Red Grange (1903-91): All-America at Univ. of Illinois, 1923-25; played for Chicago Bears, 1925-35.
Joe Greene, b. 1946: Pittsburgh Steelers lineman; twice NFL outstanding defensive player.
Maurice Greene, b. 1974: 3-time 100m world champ ('97, '99, 2001) and Olympic champ (2000); world record holder.
Wayne Gretzky, b. 1961: top scorer in NHL history with record 894 goals, 1,963 assists, 2,857 points; MVP, 1980-87, 1989.
Bob Griese, b. 1945: All-Pro quarterback led Miami Dolphins to 17-0 season (1972) and 2 Super Bowl titles (1973-74).
Ken Griffey Jr., b. 1969: outfielder, led AL in homers 1994, 1997-1999; 1997 AL MVP; 10 gold gloves.
Archie Griffin, b. 1954: Ohio State running back is the only 2-time winner of the Heisman Trophy (1974-75).
Florence Griffith Joyner, (1959-98): sprinter; won 3 gold medals at 1988 Olympics; Olympic record for 100m.
Lefty Grove (1900-75): pitcher; won 300 AL games.
Janet Guthrie, b. 1938: 1st woman driver in Indy 500 (1977).
Tony Gwynn, b. 1960: 8-time NL batting champ, 1984, 1987-89, 1994-97; 3,141 career hits; retired in 2001.
Walter Hagen (1892-1969): golf; 5 PGA , 4 British Open titles.
George Halas (1895-1983): founder-coach of Chicago Bears; won 5 NFL championships.
Dorothy Hamill, b. 1956: figure skater, gold medalist at the Olympics and World championships in 1976.
Scott Hamilton, b. 1958: U.S. and world figure skating champion, 1981-84; Olympic gold medalist, 1984.
Mia Hamm, b. 1972: led U.S. to World Cup (1995, '99) and Olympic ('96) titles; most career goals in women's soccer.
Franco Harris, b. 1950: running back, led Steelers to 4 Super Bowls (1975-76, 1979-80); 1,000+ yds. in a season 8 times.
Bill Hartack, b. 1932: jockey, rode 5 Kentucky Derby winners.
Dominik Hasek, b. 1965: Buffalo Sabres goalie; won Vezina Trophy, 1994-95, 1997-99, 2001; NHL MVP, 1997-98.
John Havlicek, b. 1940: Boston Celtics forward scored more than 26,000 NBA points.
Eric Heiden, b. 1958: speed skater, won 5 Olympic golds, 1980.
Rickey Henderson, b. 1958; outfielder, 1990 AL MVP; record 130 stolen bases, 1982; all-time leader in steals (1,395), runs (2,248), and walks (2,141).
Sonja Henie (1912-69): world champion figure skater, 1927-36; Olympic gold medalist, 1928, 1932, 1936.
Martina Hingis, b. 1980: won Australian Open, Wimbledon, and U.S. Open; youngest No. 1 player (16 yrs., 6 m.) in 1997.
Ben Hogan (1912-97): golfer, won 4 U.S. Open championships, 2 PGA, 2 Masters.
Chamique Holdsclaw, b. 1977: basketball, 2-time national player of the year, led Tennessee to 3 NCAA titles (1996-98).
Evander Holyfield, b. 1962: 4-time heavyweight champion.
Rogers Hornsby (1896-1963): NL 2d baseman; batted record .424 in 1924; twice won triple crown; batting leader, 1920-25.
Paul Hornung, b. 1935: Green Bay Packers runner-placekicker, scored record 176 points, 1960.
Gordie Howe, b. 1928: hockey forward; NHL MVP 6 times; scored 801 goals in 26 NHL seasons.
Carl Hubbell (1903-88): N.Y. Giants pitcher; 20-game winner 5 consecutive years, 1933-37.
Bobby Hull, b. 1939: NHL all-star 10 times; MVP, 1965-66.
Brett Hull, b. 1964: St. Louis Blues forward; led NHL in goals, 1990-92; MVP, 1991.
Catfish Hunter (1946-99); pitched perfect game, 1968; 20-game winner 5 times.
Don Hutson (1913-97): Packers receiver, caught 99 TD passes.
Julie Inkster, b. 1960: Hall of Fame golfer, 2d to win all 4 of LPGA's modern majors; won 8 career major titles.
Phil Jackson, b. 1945: won 8 NBA titles as coach with Bulls (1991-93, 1996-98) and Lakers (2000-2001); NBA championship with Knicks as a player, 1973.
Reggie Jackson, b. 1946: slugger; led AL in home runs 4 times; MVP, 1973; hit 5 World Series home runs, 1977.
"Shoeless" Joe Jackson (1889-1951): outfielder, 3rd highest career batting average (.356); one of the "Black Sox" banned for allegedly throwing 1919 World Series.
Jaromir Jagr, b. 1972: Czech hockey player, NHL MVP in 1999; won Art Ross Trophy (leading scorer) 1995, 1998-2001.
Bruce Jenner, b. 1949: Olympic decathlon gold medalist, 1976.
Lynn Jennings, b. 1960: runner, 3-time World and 9-time U.S. cross country champ; bronze at 1992 Olympics (10,000m).
Earvin (Magic) Johnson, b. 1959: NBA MVP, 1987, 1989, 1990; Playoff MVP, 1980, 1982, 1987; 2d in career assists.
Jack Johnson (1878-1946): heavyweight champion, 1908-15.
Michael Johnson, b. 1967: 5-time Olympic gold medalist (1996, 2000); world and Olympic record holder, 200m and 400m.
Randy Johnson, b. 1963: 3-time Cy Young winner; strikeout leader: 1993-94, 1998-2001; 3,412 career strikeouts (9th).
Walter Johnson (1887-1946): Washington Senators pitcher; won 416 games; record 110 shutouts.
Bobby Jones (1902-71): won "grand slam of golf" 1930; U.S. Amateur champ 5 times, U.S. Open champ 4 times.
David "Deacon" Jones, b.1938: Hall of Fame defensive end; 5-time All-Pro with LA Rams (1965-69); quarterback "sack" specialist credited with inventing the term.
Marion Jones, b. 1975: 2000 Olympic 100m, 200m, 1,600m relay gold medalist, bronze in long jump and 400m relay; Most track and field medals won by a woman at 1 Olympics.
Roy Jones Jr., b. 1969: undisputed Light heavyweight champ.
Michael Jordan, b. 1963: leading NBA scorer, 1987-93, 1996-98; MVP, 1988, 1991-92, '96, '98; Playoff MVP, 1991-93, 1996-98; ESPN Athlete of the Century.
Dorothy Kamenshek, b. 1925: 1st baseman who led Rockford (IL) Peaches to 4 All-American Girls Baseball League championships in the 1940s.
Jackie Joyner-Kersee, b. 1962: Olympic gold medalist in heptathlon (1988 ,'92) and long jump (1988).
Harmon Killebrew, b. 1936: Minnesota Twins slugger; led AL in home runs 6 times; 573 lifetime.
Jean Claude Killy, b. 1943: French skier; 3 1968 Olympic golds.
Ralph Kiner, b. 1922: Pittsburgh Pirates slugger, led NL in home runs 7 consecutive years, 1946-52.
Billie Jean King, b. 1943: U.S. singles champ 4 times; Wimbledon champ 6 times; beat Bobby Riggs, 1973.
Bob Knight, b. 1940: basketball coach, led Indiana U. to NCAA title in 1976, '81, '87.
Olga Korbut, b. 1955: Soviet gymnast; 3 1972 Olympic golds.
Sandy Koufax, b. 1935: Dodgers pitcher; won Cy Young award 3 times; lowest ERA in NL, 1962-66; pitched 4 no-hitters, one a perfect game.
Ingrid Kristiansen, b. 1956: Norwegian, only runner ever to hold world records in 5,000m, 10,000m, and marathon.
Julie Krone, b. 1963: winningest female jockey, only woman to ride a winner in a Triple Crown race (Belmont, 1993); 1st woman elected to horse racing Hall of Fame (2000).
Michelle Kwan, b.1980: figure skater, U.S. Champion (1996, 1998-2001) and World Champion (1996, '98, 2000, 2001); silver medalist at 1998 Olympics.

Guy Lafleur, b. 1951: 3-time NHL scoring leader; 1977-78 MVP.
Kennesaw Mountain Landis (1866-1944): 1st commissioner of baseball (1920-44); banned the 8 "Black Sox" involved in the fixing of the 1919 World Series.
Tom Landry (1924-2000): Dallas Cowboys head coach, 1960-88; won 2 Super Bowls (1972, '78); 3d in career wins (270).
Dick "Night train" Lane, b. 1928: Hall of Fame defensive back, intercepted an NFL single-season record 14 passes (1952).
Don Larsen, b. 1929: As NY Yankee, pitched only World Series perfect game, Oct. 8, 1956—a 2-0 win over Brooklyn.
Rod Laver, b. 1938: Australian; won tennis "grand slam" twice, 1962, 1969; Wimbledon champ 4 times.
Mario Lemieux, b. 1965: 6-time NHL leading scorer; MVP, 1988, 1993, 1996; Playoff MVP, 1991-92.
Greg Lemond, b. 1961: 3-time Tour de France winner (1986, '89-90); first American to win the event.
Ivan Lendl, b. 1960: U.S. Open tennis champ, 1985-87.
Sugar Ray Leonard, b. 1956: boxer, held titles in 5 different weight classes.
Carl Lewis, b. 1961: track-and-field star, won 9 Olympic gold medals in sprinting and the long jump.
Lennox Lewis, b. 1965: Brit. boxer, heavyweight champion, 1997-2001.
Tara Lipinski, b. 1982: youngest figure skater to win U.S. and world championships, 1997, and Winter Olympic gold, 1998.
Vince Lombardi (1913-70): Green Bay Packers coach, led team to 5 NFL championships and 2 Super Bowl victories.
Nancy Lopez, b. 1957: Hall of Fame golfer, 4-time LPGA Player of the Year, 3-time winner of the LPGA Championship.
Greg Louganis, b. 1960: won Olympic gold medals in both springboard and platform diving, 1984, 1988.
Joe Louis (1914-81): heavyweight champion, 1937-49.
Sid Luckman (1916-98): Chicago Bears quarterback; led team to 4 NFL championships; MVP, 1943.
Connie Mack (1862-1956): Philadelphia Athletics manager, 1901-50; won 9 pennants, 5 championships.
John Madden, b. 1936: won Super Bowl as coach of the Oakland Raiders (1977); NFL TV analyst since 1982.
Greg Maddux, b. 1966: NL pitcher, won 4 consecutive Cy Young awards, 1992-95.
Karl Malone, b. 1963: Utah Jazz forward; was MVP, 1997, 1999; 11-time All-Star; 32,919 career points (2d all-time).
Moses Malone, b. 1955: NBA center, MVP, 1979, 1982-83.
Mickey Mantle (1931-95): N.Y. Yankees outfielder; triple crown, 1956; 18 World Series home runs; MVP 3 times.
Pete Maravich (1948-88): guard, scored NCAA record 44.2 ppg during collegiate career; led NBA in scoring, 1977.
Rocky Marciano (1923-69): heavyweight champion, 1952-56; retired undefeated.
Dan Marino, b. 1961: Miami Dolphins quarterback; passed for NFL record 5,084 yds and 48 touchdowns, 1984; career records for touchdowns, yds passing, completions.
Roger Maris (1934-85): N.Y. Yankees outfielder; hit AL record 61 home runs, 1961; MVP, 1960 and 1961.
Eddie Mathews, b. 1931: Milwaukee-Atlanta 3d baseman, hit 512 career home runs.
Christy Mathewson (1880-1925): N.Y. Giants pitcher, 373 wins.
Bob Mathias, b. 1930: decathlon gold medalist, 1948, 1952.
Willie Mays, b. 1931: N.Y.-S.F. Giants center fielder; hit 660 home runs, led NL 4 times; had 3,283 hits; twice MVP.
Willie McCovey, b. 1938: S.F. Giants slugger; hit 521 home runs; led NL 3 times; MVP, 1969.
John McEnroe, b. 1959: U.S. Open tennis champ, 1979-81, 1984; Wimbledon champ, 1981, 1983-84.
John McGraw (1873-1934): N.Y. Giants manager, led team to 10 pennants, 3 championships.
Mark McGwire, b. 1963: hit then-record 70 home runs in 1998; 583 career home runs (5th).
Tamara McKinney, b. 1962: 1st U.S. skier to win overall Alpine World Cup championship (1983).
Mary T. Meagher, b. 1964: swimmer, "Madame Butterfly" won 3 Olympic gold medals in 1984.
Mark Messier, b. 1961: center, chosen NHL MVP, 1990, 1992; Conn Smythe Trophy, 1984.
Debbie Meyer, b. 1952: won 3 swimming gold medals, 1968.
George Mikan, b. 1924: Minn. Lakers center, considered the best basketball player of the first half of the century.
Stan Mikita, b. 1940: Chicago Black Hawks center, led NHL in scoring 4 times; MVP twice.
Billy Mills, b. 1938: runner, upset winner of the 1964 Olympic 10,000m; only American man ever to win the event.
Joe Montana, b. 1956: S.F. 49ers quarterback; Super Bowl MVP, 1982, 1985, 1990.
Archie Moore (1913-98): light-heavyweight champ, 1952-62.
Howie Morenz (1902-37): Montreal Canadiens forward, considered best hockey player of first half of the century.
Edwin Moses, b. 1955: undefeated in 122 consecutive 400m hurdles races, 1977-87; Olympic gold medalist, 1976, '84.
Shirley Muldowney, b. 1940: 1st woman to race National Hot Rod Assoc. Top Fuel dragsters; 3-time NHRA points champ.
Eddie Murray, b. 1956: durable slugger; 3d player to combine 3,000+ hits with 500+ home runs.
Stan Musial, b. 1920: St. Louis Cardinals star; won 7 NL batting titles; MVP 3 times.
Bronko Nagurski (1908-90): Chicago Bears fullback and tackle; gained more than 4,000 yds. rushing.
Joe Namath, b. 1943: Jets quarterback, 1969 Super Bowl MVP.
Martina Navratilova, b. 1956: Wimbledon champ 9 times, U.S. Open champ 1983-84, 1986-87.
Byron Nelson, b. 1912: won 11 consecutive golf tournaments in 1945; twice Masters and PGA titlist.
Ernie Nevers (1903-76): Stanford star, selected as best college fullback to play between 1919-69.
Paula Newby-Fraser, b. 1972: 8-time winner of the Ironman Triathlon World Championships in Hawaii; holds course record.
John Newcombe, b. 1943: Australian; twice U.S. Open tennis champ; Wimbledon titlist 3 times.
Jack Nicklaus, b. 1940: PGA Player-of-the-Year, 1967, 1972; leading money winner 8 times; won 18 majors (6 Masters).
Chuck Noll, b. 1931: Pittsburgh coach; won 4 Super Bowls.
Paavo Nurmi (1897-1973): Finnish distance runner, won 6 Olympic gold medals, 1920, 1924, 1928.
Al Oerter, b. 1936: discus thrower, won gold medal at 4 consecutive Olympics, 1956-68.
Hakeem Olajuwon, b. 1963: Houston center; NBA MVP, 1994, Playoffs MVP, 1994-95; career leader in blocked shots.
Barney Oldfield, (1878-1946): pioneer auto racer was first to drive a car 60 mph (1903).
Shaquille O'Neal, b. 1972: center, led L.A. Lakers to NBA titles in 2000, 2001; 2000, 2001 Finals MVP; 2000 NBA MVP.
Bobby Orr, b. 1948: Boston Bruins defenseman; Norris Trophy 8 times; led NHL in scoring twice, assists 5 times.
Mel Ott (1909-1958): N.Y. Giants outfielder hit 511 home runs; led NL 6 times.
Jesse Owens (1913-80): track and field star, won 4 1936 Olympic gold medals.
Satchel Paige (1906-82): pitcher, starred in Negro leagues, 1924-48; entered major leagues at age 42.
Se Ri Pak, b. 1977: youngest (at 20) LPGA, and U.S. Women's Open Champ, 1998; won British Open, 2001.
Arnold Palmer, b. 1929: golf's first $1 million winner; won 4 Masters, 2 British Opens.
Jim Palmer, b. 1945: Baltimore Orioles pitcher; Cy Young award 3 times; 20-game winner 8 times.
Joe Paterno, b. 1926: winningest active NCAA football coach; led Penn St. to 2 national titles, 1982, 1986.
Floyd Patterson, b. 1935: 2-time heavyweight champion.
Walter Payton (1954-1999): Chicago Bears running back; most rushing yards in NFL history; top NFC rusher, 1976-80.
Pelé, b. 1940: soccer star, led Brazil to 3 World Cup titles (1958, '62, '70); scored 1,281 goals in 22-year career.
Bob Pettit, b. 1932: first NBA player to score 20,000 points; twice NBA scoring leader.
Richard Petty, b. 1937: NASCAR national champ 7 times; 7-time Daytona 500 winner.
Laffit Pincay Jr., b. 1946: jockey, leading money-winner, 1970-74, 1979, 1985.
Jacques Plante (1929-86): goalie; 7 Vezina trophies; first goalie to wear a mask in a game.
Gary Player, b. 1936: South African golfer, won 3 Masters, 3 British Opens, 2 PGA Championships, and the U.S. Open.
Steve Prefontaine, (1951-75): runner, 1st to win 4 NCAA titles in same event (5,000m, 1970-73); died in auto accident.
Kirby Puckett, b. 1961: Minnesota Twins outfielder; won AL batting title, 1989; led AL in hits, 1987-89, 1992; RBIs, 1994.
Willis Reed, b. 1942: N.Y. Knicks center; MVP, 1970; Playoff MVP, 1970, 1973.
Mary Lou Retton, b. 1968: 1st American woman to win a gold medal in gymnastics (1984).
Jerry Rice, b. 1962: receiver, 1989 Super Bowl MVP; NFL record for career touchdowns, receptions.
Maurice Richard, (1921-2000): Montreal Canadiens forward scored 544 regular season goals, 82 playoff goals.
Branch Rickey (1881-1965): executive; helped break baseball's color barrier, 1947; initiated farm system, 1919.
Cal Ripken Jr., b. 1960: Baltimore shortstop; AL MVP 1983, 1991; most consecutive games played, 2,632; retired 2001.
Oscar Robertson, b. 1938: guard; averaged career 25.7 points per game; 3d most career assists; MVP, 1964.
Brooks Robinson, b. 1937: Baltimore Orioles 3d baseman; played in 4 World Series; MVP, 1964; 16 gold gloves.
Frank Robinson, b. 1935: MVP in both NL and AL; triple crown, 1966; 586 career home runs; first black manager in majors.
Jackie Robinson (1919-72): broke baseball's color barrier with Brooklyn Dodgers, 1947; MVP, 1949.
Sugar Ray Robinson (1920-89): middleweight champion 5 times, welterweight champion.
Knute Rockne (1888-1931): Notre Dame football coach, 1918-31; revolutionized game by stressing forward pass.

Dennis Rodman, b. 1961: eccentric forward, led NBA in rebounding 1991-98.
Bill Rogers, b. 1947: runner, won Boston and New York City marathons 4 time each, 1975-80.
Pete Rose, b. 1941: won 3 NL batting titles; hit safely in 44 consecutive games, 1978; has most career hits, 4,256; banned from baseball for alleged gambling, 1989.
Ken Rosewall, b. 1934: Australian; 2-time U.S. Open champ, 8 grand slam singles titles.
Patrick Roy, b. 1965: Montreal-Colorado goalie; only 3-time NHL Playoffs MVP (Conn Smythe Trophy), 1986, '93, 2001.
Wilma Rudolph (1940-94): sprinter, won 3 1960 Olympic golds.
Adolph Rupp (1901-77): NCAA basketball coach; led Kentucky to 4 national titles, 1948-49, 1951, 1958.
Bill Russell, b. 1934: Boston Celtics center, led team to 11 NBA titles; MVP 5 times; first black coach of major pro sports team.
Babe Ruth (1895-1948): N.Y. Yankees outfielder; hit 60 home runs, 1927; 714 lifetime; led AL 12 times.
Johnny Rutherford, b. 1938: auto racer, won 3 Indy 500s.
Nolan Ryan, b. 1947: pitcher; holds season (383), career (5,714) strikeout records; won 324 games (7 no-hitters).
Pete Sampras, b. 1971: tennis star; 1st man in Open era to win 7 Wimbledons; most career Grand Slam wins (13).
Joan Benoit Samuelson, b. 1968: won 1st Olympic women's marathon (1984), Boston Marathon (1979, '83).
Barry Sanders, b. 1968: rushed for 2,053 yards in 1997; led NFL in rushing, 1990, 1994, 1996, 1997.
Gene Sarazen, (1902-99): won PGA championship 3 times, U.S. Open twice; developed the sand wedge.
Gale Sayers, b. 1943: Chicago back, twice led NFL in rushing.
Mike Schmidt, b. 1949: Phillies 3d baseman; led NL in home runs 8 times; 548 lifetime; NL MVP, 1980, 1981, 1986.
Tom Seaver, b. 1944: pitcher; won NL Cy Young award 3 times; won 311 major league games.
Monica Seles, b. 1973: won U.S. ('91-92), Australian ('91-93, '96), French ('90-92) Opens; stabbed on court by fan, 1993.
Patty Sheehan, b. 1956: Hall of Fame golfer, 3 LPGA Championships (1983-84, '93).
Willie Shoemaker, b. 1931: jockey; rode 4 Kentucky Derby and 5 Belmont Stakes winners; leading career money winner.
Eddie Shore (1902-85): Boston Bruins defenseman; MVP 4 times, first-team all-star 7 times.
Frank Shorter, b. 1947: runner, only American to win men's Olympic marathon (1972) since 1908; silver medalist in 1976.
Don Shula, b. 1930: all-time winningest NFL coach.
Al Simmons (1902-56): AL outfielder batted .334 lifetime.
O. J. Simpson, b. 1947: running back; rushed for 2,003 yds., 1973; AFC leading rusher 4 times; acquitted of murder, 1995.
George Sisler (1893-1973): St. Louis Browns 1st baseman; had record 257 hits, 1920; batted .340 lifetime.
Dean Smith, b. 1931: basketball coach, most career Division I wins (879); led North Carolina to 2 NCAA titles (1982, '93).
Emmitt Smith, b. 1969: Dallas running back; NFL and Super Bowl MVP, 1993; record 25 rushing touchdowns, 1995.
Lee Smith, b. 1957: relief pitcher, all-time saves leader, 478.
Conn Smythe, (1895-1980): won 7 Stanley Cups as Toronto GM (1929-1961); playoff MVP award named in his honor.
Sam Snead, b. 1912: PGA and Masters champ 3 times each.
Annika Sorenstam, b. 1970: golfer, set LPGA 18-hole record of 59 (–13), and 72-hole record of 27-under-par in Mar. 2001.
Sammy Sosa, b. 1968: Cubs outfielder; 66 homers, NL MVP, 1998; 1st to hit 60+ homers 3 times (63 in 1999, 64 in 2001).
Warren Spahn, b. 1921: pitcher; won 363 NL games; 20-game winner 13 times; Cy Young award, 1957.
Tris Speaker (1885-1958): AL outfielder; batted .345 over 22 seasons; hit record 792 career doubles.
Mark Spitz, b. 1950: swimmer, won 7 golds at 1972 Olympics.
Amos Alonzo Stagg (1862-1965): Univ. of Chicago football coach for 41 years, including 5 undefeated seasons; introduced huddle, man-in-motion, and end-around play.
Bart Starr, b. 1934: Green Bay Packers quarterback, led team to 5 NFL titles and 2 Super Bowl victories.
Roger Staubach, b. 1942: Dallas Cowboys quarterback; leading NFC passer 5 times.
Casey Stengel (1890-1975): managed Yankees to 10 pennants, 7 championships, 1949-60.
Jackie Stewart, b. 1939: Scot auto racer, 27 Grand Prix wins.
Payne Stewart (1957-99): golfer; won 2 U.S. Opens (1991, 1999), 1 PGA Championship (1989); died in plane crash.
John Stockton b. 1962: Utah Jazz guard; NBA career leader in assists, steals; NBA assists leader, 1988-96.
Picabo Street, b. 1971: skier, 2-time World Cup downhill champion (1995-96); Olympic super G gold medalist in 1998.
Louise Suggs, b. 1923: U.S. Women's Open champ., 1949, '52; 11 major victories, ranks 3d all-time.
John L. Sullivan (1858-1918): last bareknuckle heavyweight champion, 1882-1892.
Pat Summit, b. 1952: basketball coach, led Tennessee Lady Vols to 6 NCAA titles (1987, '89, '91, '96-98).
Fran Tarkenton, b. 1940: quarterback, Minnesota, N.Y. Giants, 2d in career touchdown passes; 1975 Player of the Year.
Lawrence Taylor, b. 1959; linebacker; led N.Y. Giants to 2 Super Bowl titles; played in 10 Pro Bowls.
Frank Thomas, b. 1968: Chicago White Sox 1st baseman; was AL MVP, 1993-94; won AL batting title, 1997.
Jenny Thompson, b. 1973: swimmer, most decorated U.S. woman with 10 Olympic medals (8 gold) in 1992, 1996, 2000.
Daley Thompson, b. 1958: British decathlete, Olympic gold medalist in 1980, '84.
Jim Thorpe (1888-1953): football All-America, 1911, 1912; won pentathlon and decathlon, 1912 Olympics.
Bill Tilden (1893-1953): U.S. singles champ 7 times; played on 11 Davis Cup teams.
Y. A. Tittle, b. 1926: N.Y. Giants quarterback; MVP, 1961, 1963.
Alberto Tomba, b. 1966: Italian skier, all-time Olympic alpine medalist (3 golds, 2 silver).
Lee Trevino, b. 1939: golfer, won U.S., British Open twice.
Bryan Trottier, b. 1956: center for 6 Stanley Cup champs.
Gene Tunney, (1897-1978): heavyweight champion, 1926-28.
Mike Tyson, b. 1966: Undisputed heavyweight champ, 1987-1990; at 19, youngest to win a heavyweight title (WBC, 1986).
Wyomia Tyus, b. 1945: Olympic 100m gold medalist, 1964, '68.
Johnny Unitas, b. 1933: Baltimore Colts quarterback; passed for more than 40,000 yds; MVP, 1957, 1967.
Al Unser, b. 1939: Indy 500 winner 4 times.
Bobby Unser, b. 1934: Indy 500 winner 3 times.
Norm Van Brocklin (1926-83): quarterback; passed for game record 554 yds., 1951; MVP, 1960.
Amy Van Dyken, b. 1973: swimmer, first American woman to win 4 gold medals in one Olympics (1996).
Lasse Viren, b. 1949: Finnish runner; Olympic 5,000m and 10,000m gold medalist in 1972 and 1976.
Honus Wagner (1874-1955): Pittsburgh Pirates shortstop, won 8 NL batting titles.
Grete Waitz, b. 1953: Norwegian, 9-time winner of the New York City Marathon (1978-80, 1982-86, '88).
"Jersey" Joe Walcott, (1914-94): boxer, became heavyweight champion at age 37, 1951-52.
Bill Walton, b. 1952: center led Portland Trail Blazers to 1977 NBA title; MVP, 1978; NBA TV commentator.
Kurt Warner, b. 1971: quarterback, NFL MVP, Super Bowl MVP; record 414 passing yds. in Super Bowl XXXIV (2000).
Tom Watson, b. 1949: 6-time PGA Player of the Year, won 5 British Opens, 2 Masters, U.S. Open.
Karrie Webb, b. 1974: Australian golfer; youngest (26 yrs. 6 mos.) to win career Grand Slam: du Maurier ('99), Nabisco Championship, U.S. Women's Open (2000), LPGA (2001).
Johnny Weissmuller (1903-84): swimmer; won 52 national championships, 5 Olympic gold medals; set 67 world records.
Jerry West, b. 1938: L.A. Lakers guard; had career average 27 points per game; first team all-star 10 times.
Byron "Whizzer" White, b. 1917: running back, led NCAA in scoring and rushing at Colorado (1937), led NFL in rushing twice (1938, '40); Supreme Court justice, 1962-93.
Reggie White, b. 1961: defensive end, all-time NFL sack leader.
Kathy Whitworth, b. 1939: 7-time LPGA Player of the Year (1966-69, 1971-73); 88 tour wins most on LPGA or PGA tour.
Lenny Wilkens, b. 1937: winningest coach in NBA history; in Hall of Fame as player and coach.
Serena Williams, b. 1981: U.S. Open singles champ (1999); won doubles titles with sister, Venus, at U.S. and French Opens (1999), Wimbledon and Sydney Olympics (2000).
Ted Williams, b. 1918: Boston Red Sox outfielder; won 6 batting titles, two triple crowns; hit .406 in 1941.
Venus Williams, b. 1980: Wimbledon, U.S. Open, Olympic champ (2000); Wimbledon, U.S. Open champ (2001).
Helen Willis Moody (1905-98): tennis star; won U.S. Open 7 times, Wimbledon 8 times.
Katarina Witt, b. 1965: German figure skater; won Olympic gold medal, 1984, 1988; world champ, 1984-84, 1987-88.
John Wooden, b. 1910: coached UCLA basketball team to 10 national championships.
Tiger Woods, b. 1975: golfer, youngest ever to win the sport's career Grand Slam (all 4 majors), at age 24, when he won the British Open, 2000.
Mickey Wright, b. 1935: won LPGA championship 4 times, Vare Trophy 5 times; twice AP Woman-Athlete-of-the-Year.
Kristi Yamaguchi, b. 1971: figure skater, won national, world, and Olympic titles in 1992.
Carl Yastrzemski, b. 1939: Boston Red Sox slugger; won 3 batting titles; triple crown, 1967.
Cy Young (1867-1955): pitcher, won record 511 games.
Steve Young, b. 1961: 49ers quarterback; led NFL in passing, 1991-94, 1996, 1997; Super Bowl MVP, 1995.
Babe Didrikson Zaharias (1914-56): track star; won 2 1932 Olympic gold medals; won numerous golf tournaments.
Emil Zátopek, (1922): Czech runner won 3 gold medals at 1952 Olympics (5,000m,10,000m, and marathon).

SPORTS QUICK REFERENCE INDEX

For complete Index, see pages 4-32.

WORLD ALMANAC EDITORS' PICKS

The *World Almanac* staff ranked the following as their favorite chapters of the book:

1. Noted Personalities (284-332)
2. Notable Quotes (63)
3. Nations of the World (767-808, 817-876)
4. Year in Pictures (193-200, 809-816)
5. Arts and Media (267-283)

HERKIMER COUNTY COMMUNITY COLLEGE
LIBRARY
HERKIMER, NY 13350-1598